Y0-BEE-995

American Men & Women of Science

1992-93 • 18th Edition

The 18th edition of *AMERICAN MEN & WOMEN OF SCIENCE* was prepared by the R.R. Bowker Database Publishing Group.

Stephen L. Torpie, Managing Editor
Judy Redel, Managing Editor, Research
Richard D. Lanam, Senior Editor
Tanya Hurst, Research Manager
Karen Hallard, Beth Tanis, Associate Editors

Peter Simon, Vice President, Database Publishing Group
Dean Hollister, Director, Database Planning
Edgar Adcock, Jr., Editorial Director, Directories

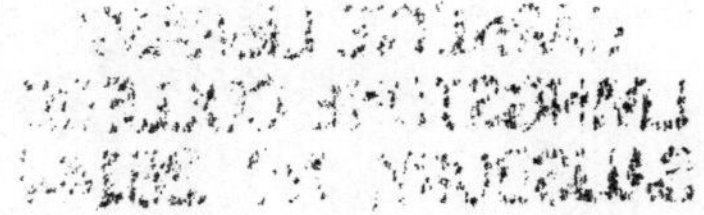

American Men & Women of Science

1992-93 • 18th Edition

A Biographical Directory of Today's Leaders in Physical, Biological and Related Sciences.

Volume 7 • T-Z

R. R. BOWKER
New Providence, New Jersey

Published by R.R. Bowker, a division of Reed Publishing, (USA) Inc.

International Standard Book Number

Set:	0-8352-3074-0
Volume I:	0-8352-3075-9
Volume II:	0-8352-3076-7
Volume III:	0-8352-3077-5
Volume IV:	0-8352-3078-3
Volume V:	0-8352-3079-1
Volume VI:	0-8352-3080-5
Volume VII:	0-8352-3081-3
Volume VIII:	0-8352-3082-1

International Standard Serial Number: 0192-8570
Library of Congress Catalog Card Number: 6-7326
Printed and bound in the United States of America.

8 Volume Set

ISBN 0-8352-3074-0

9 780835 230742

Contents

Advisory Committee

Dr. Robert F. Barnes
Executive Vice President
American Society of Agronomy

Dr. John Kistler Crum
Executive Director
American Chemical Society

Dr. Charles Henderson Dickens
Section Head, Survey & Analysis Section
Division of Science Resource Studies
National Science Foundation

Mr. Alan Edward Fechter
Executive Director
Office of Scientific & Engineering Personnel
National Academy of Science

Dr. Oscar Nicolas Garcia
Prof Electrical Engineering
Electrical Engineering & Computer Science Department
George Washington University

Dr. Charles George Groat
Executive Director
American Geological Institute

Dr. Richard E. Hallgren
Executive Director
American Meteorological Society

Dr. Michael J. Jackson
Executive Director
Federation of American Societies for Experimental Biology

Dr. William Howard Jaco
Executive Director
American Mathematical Society

Dr. Shirley Mahaley Malcom
Head, Directorate for Education and Human Resources Programs
American Association for the Advancement of Science

Mr. Daniel Melnick
Sr Advisor Research Methodologies
Sciences Resources Directorate
National Science Foundation

Ms. Beverly Fearn Porter
Division Manager
Education & Employment Statistics Division
American Institute of Physics

Dr. Terrence R. Russell
Manager
Office of Professional Services
American Chemical Society

Dr. Irwin Walter Sandberg
Holder, Cockrell Family Regent Chair
Department of Electrical & Computer Engineering
University of Texas

Dr. William Eldon Splinter
Interim Vice Chancellor for Research,
Dean, Graduate Studies
University of Nebraska

Ms. Betty M. Vetter
Executive Director, Science Manpower Comission
Commission on Professionals in Science & Technology

Dr. Dael Lee Wolfe
Professor Emeritus
Graduate School of Public Affairs
University of Washington

Preface

American Men and Women Of Science remains without peer as a chronicle of North American scientific endeavor and achievement. The present work is the eighteenth edition since it was first compiled as *American Men of Science* by J. Mckeen Cattell in 1906. In its eighty-six year history *American Men & Women of Science* has profiled the careers of over 300,000 scientists and engineers. Since the first edition, the number of American scientists and the fields they pursue have grown immensely. This edition alone lists full biographies for 122,817 engineers and scientists, 7021 of which are listed for the first time. Although the book has grown, our stated purpose is the same as when Dr. Cattell first undertook the task of producing a biographical directory of active American scientists. It was his intention to record educational, personal and career data which would make "a contribution to the organization of science in America" and "make men [and women] of science acquainted with one another and with one another's work." It is our hope that this edition will fulfill these goals.

The biographies of engineers and scientists constitute seven of the eight volumes and provide birthdates, birthplaces, field of specialty, education, honorary degrees, professional and concurrent experience, awards, memberships, research information and adresses for each entrant when applicable. The eighth volume, the discipline index, organizes biographees by field of activity. This index, adapted from the National Science Foundation's Taxonomy of Degree and Employment Specialties, classifies entrants by 171 subject specialties listed in the table of contents of Volume 8. For the first time, the index classifies scientists and engineers by state within each subject specialty, allowing the user to more easily locate a scientist in a given area. Also new to this edition is the inclusion of statistical information and recipients of theNobel Prizes, the Craaford Prize, the Charles Stark Draper Prize, and the National Medals of Science and Technology received since the last edition.

While the scientific fields covered by *American Men and Women Of Science* are comprehensive, no attempt has been made to include all American scientists. Entrants are meant to be limited to those who have made significant contributions in their field. The names of new entrants were submitted for consideration at the editors' request by current entrants and by leaders of academic, government and private research programs and associations. Those included met the following criteria:

1. Distinguished achievement, by reason of experience, training or accomplishment, including contributions to the literature, coupled with continuing activity in scientific work;

   or

2. Research activity of high quality in science as evidenced by publication in reputable scientific journals; or for those whose work cannot be published due to governmental or industrial security, research activity of high quality in science as evidenced by the judgement of the individual's peers;

   or

3. Attainment of a position of substantial responsibility requiring scientific training and experience.

This edition profiles living scientists in the physical and biological fields, as well as public health scientists, engineers, mathematicians, statisticians, and computer scientists. The information is collected by means of direct communication whenever possible. All entrants receive forms for corroboration and updating. New entrants receive questionaires and verification proofs before publication. The information submitted by entrants is included as completely as possible within the boundaries of editorial and space restrictions. If an entrant does not return the form and his or her current location can be verified in secondary sources, the full entry is repeated. References to the previous edition are given for those who do not return forms and cannot be located, but who are presumed to be still active in science or engineering. Entrants known to be deceased are noted as such and a reference to the previous edition is given. Scientists and engineers who are not citizens of the United States or Canada are included if a significant portion of their work was performed in North America.

The information in AMWS is also available on CD-ROM as part of *SciTech Reference Plus*. In adition to the convenience of searching scientists and engineers, *SciTech Reference Plus* also includes *The Directory of American Research & Technology*, *Corporate Technology Directory*, sci-tech and medical books and serials from *Books in Print* and *Bowker International Series*. *American Men and Women Of Science* is available for online searching through the subscription services of DIALOG Information Services, Inc. (3460 Hillview Ave, Palo Alto, CA 94304) and ORBIT Search Service (800 Westpark Dr, McLean, VA 22102). Both CD-Rom and the on-line subscription services allow all elements of an entry, including field of interest, experience, and location, to be accessed by key word. Tapes and mailing lists are also available through the Cahners Direct Mail (John Panza, List Manager, Bowker Files 245 W 17th St, New York, NY, 10011, Tel: 800-537-7930).

A project as large as publishing *American Men and Women Of Science* involves the efforts of a great many people. The editors take this opportunity to thank the eighteenth edition advisory committee for their guidance, encouragement and support. Appreciation is also expressed to the many scientific societies who provided their membership lists for the purpose of locating former entrants whose addresses had changed, and to the tens of thousands of scientists across the country who took time to provide us with biographical information. We also wish to thank Bruce Glaunert, Bonnie Walton, Val Lowman, Debbie Wilson, Mervaine Ricks and all those whose care and devotion to accurate research and editing assured successful production of this edition.

Comments, suggestions and nominations for the nineteenth edition are encouraged and should be directed to The Editors, *American Men and Women Of Science*, R.R. Bowker, 121 Chanlon Road, New Providence, New Jersey, 07974.

Edgar H. Adcock, Jr.
Editorial Director

Major Honors & Awards

Nobel Prizes

Nobel Foundation

The Nobel Prizes were established in 1900 (and first awarded in 1901) to recognize those people who "have conferred the greatest benefit on mankind."

1990 Recipients

Chemistry:

Elias James Corey

Awarded for his work in retrosynthetic analysis, the synthesizing of complex substances patterned after the molecular structures of natural compounds.

Physics:

Jerome Isaac Friedman

Henry Way Kendall

Richard Edward Taylor

Awarded for their breakthroughs in the understanding of matter.

Physiology or Medicine:

Joseph E. Murray

Edward Donnall Thomas

Awarded to Murray for his kidney transplantation achievements and to Thomas for bone marrow transplantation advances.

1991 Recipients

Chemistry:

Richard R. Ernst

Awarded for refinements in nuclear magnetic resonance spectroscopy.

Physics:

Pierre-Gilles de Gennes*

Awarded for his research on liquid crystals.

Physiology or Medicine:

Erwin Neher

Bert Sakmann*

Awarded for their discoveries in basic cell function and particularly for the development of the patch clamp technique.

Crafoord Prize

Royal Swedish Academy of Sciences
(Kungl. Vetenskapsakademien)

The Crafoord Prize was introduced in 1982 to award scientists in disciplines not covered by the Nobel Prize, namely mathematics, astronomy, geosciences and biosciences.

1990 Recipients

Paul Ralph Ehrlich

Edward Osborne Wilson

Awarded for their fundamental contributions to population biology and the conservation of biological diversity.

1991 Recipient

Allan Rex Sandage

Awarded for his fundamental contributions to extragalactic astronomy, including observational cosmology.

Charles Stark Draper Prize

National Academy of Engineering

The Draper Prize was introduced in 1989 to recognize engineering achievement. It is awarded biennially.

1991 Recipients

Hans Joachim Von Ohain

Frank Whittle

Awarded for their invention and development of the jet aircraft engine.

National Medal of Science

National Science Foundation

The National Medals of Science have been awarded by the President of the United States since 1962 to leading scientists in all fields.

1990 Recipients:

Baruj Benacerraf
Elkan Rogers Blout
Herbert Wayne Boyer
George Francis Carrier
Allan MacLeod Cormack
Mildred S. Dresselhaus
Karl August Folkers
Nick Holonyak Jr.
Leonid Hurwicz
Stephen Cole Kleene
Daniel Edward Koshland Jr.
Edward B. Lewis
John McCarthy
Edwin Mattison McMillan**
David G. Nathan
Robert Vivian Pound
Roger Randall Dougan Revelle**
John D. Roberts
Patrick Suppes
Edward Donnall Thomas

1991 Recipients

Mary Ellen Avery
Ronald Breslow
Alberto Pedro Calderon
Gertrude Belle Elion
George Harry Heilmeier
Dudley Robert Herschbach
George Evelyn Hutchinson**
Elvin Abraham Kabat
Robert Kates
Luna Bergere Leopold
Salvador Edward Luria**
Paul A. Marks
George Armitage Miller
Arthur Leonard Schawlow
Glenn Theodore Seaborg
Folke Skoog
H. Guyford Stever
Edward Carroll Stone Jr
Steven Weinberg
Paul Charles Zamecnik

National Medal of Technology

U.S. Department of Commerce,
Technology Administration

The National Medals of Technology, first awarded in 1985, are bestowed by the President of the United States to recognize individuals and companies for their development or commercialization of technology or for their contributions to the establishment of a technologically-trained workforce.

1990 Recipients

John Vincent Atanasoff
Marvin Camras
The du Pont Company
Donald Nelson Frey
Frederick W. Garry
Wilson Greatbatch
Jack St. Clair Kilby
John S. Mayo
Gordon Earle Moore
David B. Pall
Chauncey Starr

1991 Recipients

Stephen D. Bechtel Jr
C. Gordon Bell
Geoffrey Boothroyd
John Cocke
Peter Dewhurst
Carl Djerassi
James Duderstadt
Antonio L. Elias
Robert W. Galvin
David S. Hollingsworth
Grace Murray Hopper
F. Kenneth Iverson
Frederick M. Jones**
Robert Roland Lovell
Joseph A. Numero**
Charles Eli Reed
John Paul Stapp
David Walker Thompson

*These scientists' biographies do not appear in *American Men & Women of Science* because their work has been conducted exclusively outside the US and Canada.

**Deceased [Note that Frederick Jones died in 1961 and Joseph Numero in May 1991. Neither was ever listed in *American Men and Women of Science*.]

Statistics

Statistical distribution of entrants in *American Men & Women of Science* is illustrated on the following five pages. The regional scheme for geographical analysis is diagrammed in the map below. A table enumerating the geographic distribution can be found on page xvi, following the charts. The statistics are compiled by tallying all occurrences of a major index subject. Each scientist may choose to be indexed under as many as four categories; thus, the total number of subject references is greater than the number of entrants in *AMWS*.

All Disciplines

	Number	Percent
Northeast	58,325	34.99
Southeast	39,769	23.86
North Central	19,846	11.91
South Central	12,156	7.29
Mountain	11,029	6.62
Pacific	25,550	15.33
TOTAL	**166,675**	**100.00**

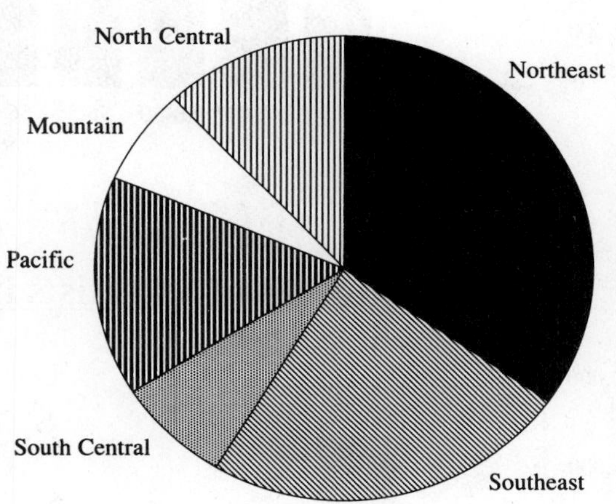

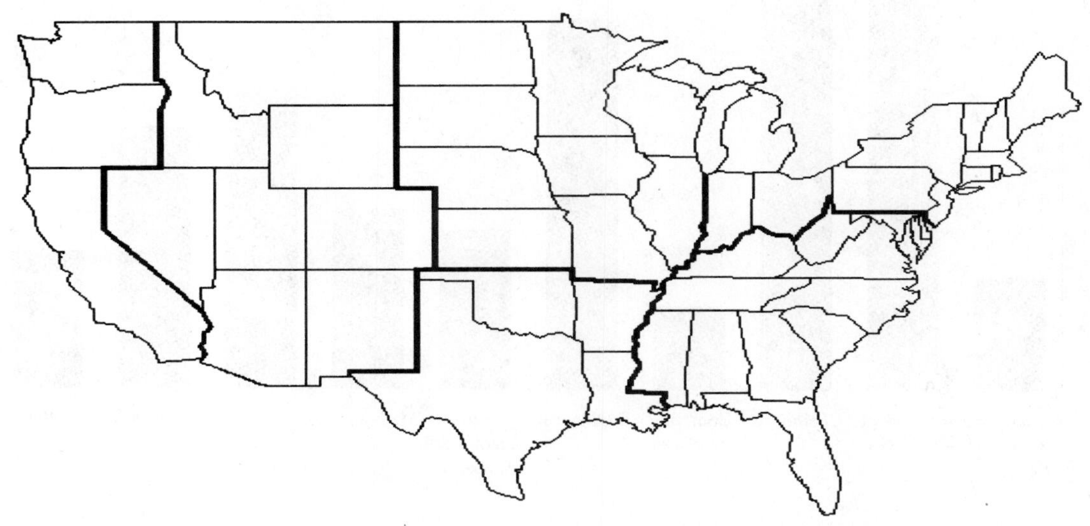

Age Distribution of American Men & Women of Science

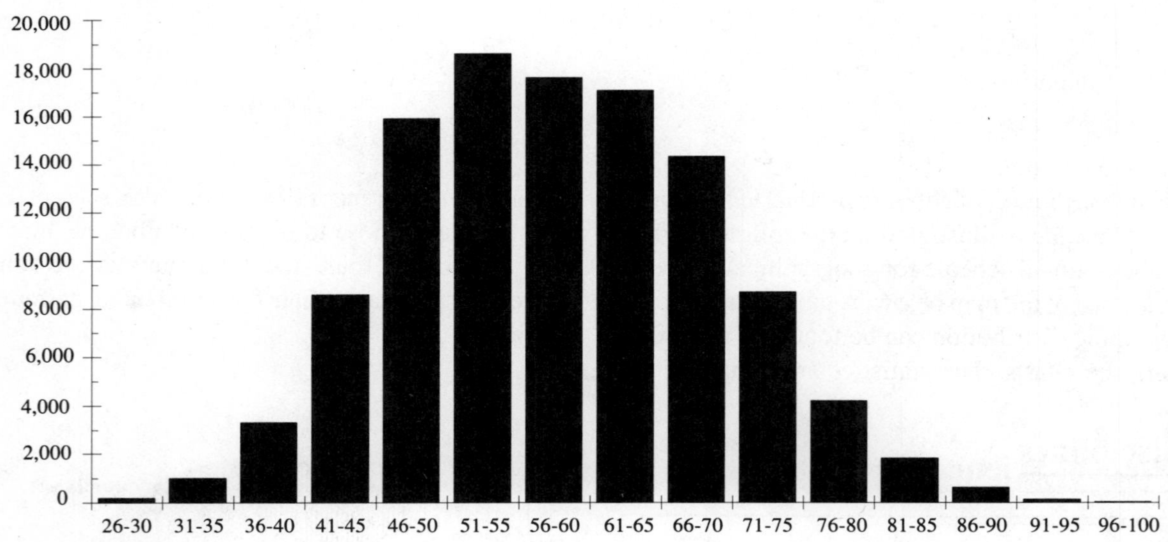

Number of Scientists in Each Discipline of Study

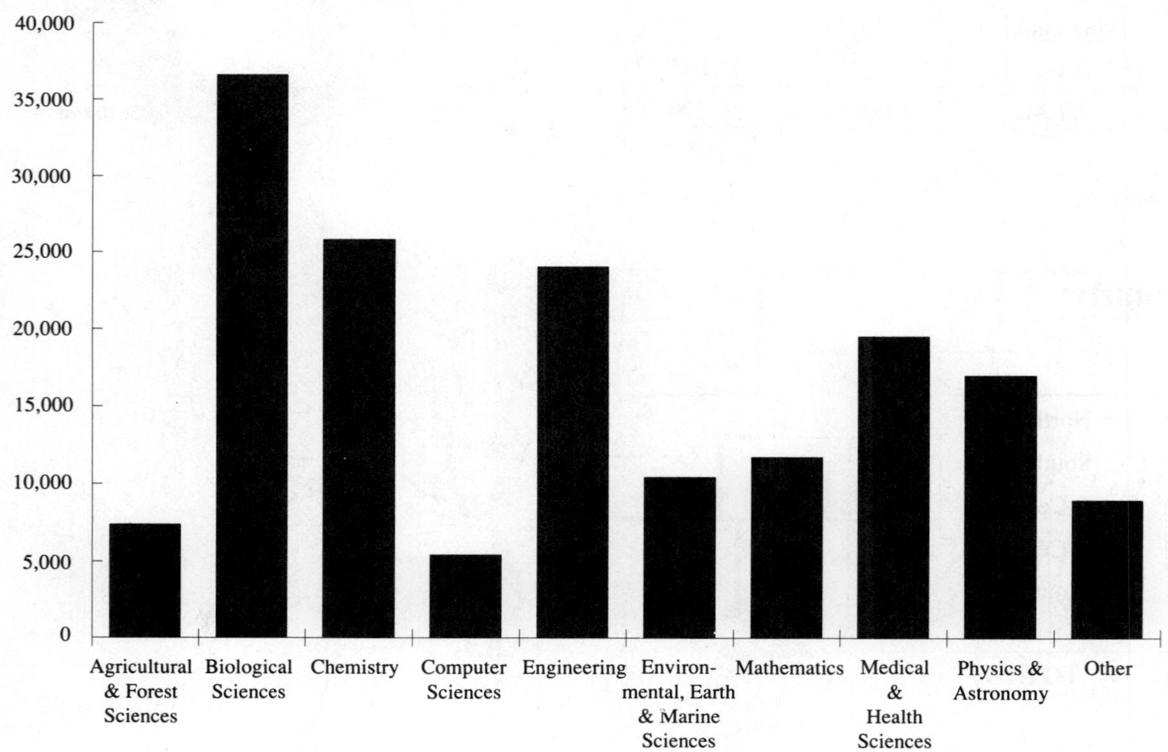

Agricultural & Forest Sciences

	Number	Percent
Northeast	1,574	21.39
Southeast	1,991	27.05
North Central	1,170	15.90
South Central	609	8.27
Mountain	719	9.77
Pacific	1,297	17.62
TOTAL	**7,360**	**100.00**

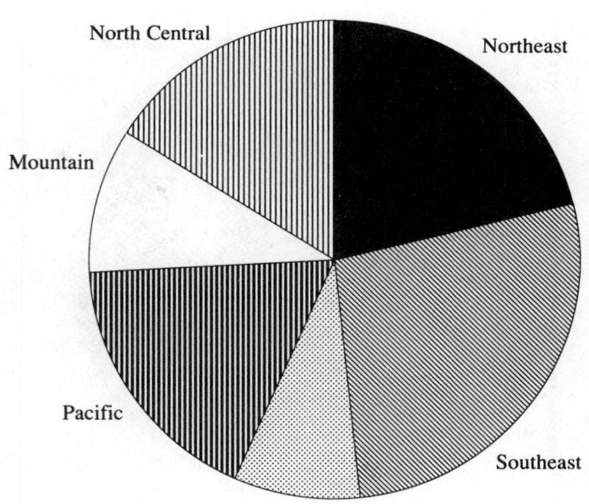

Biological Sciences

	Number	Percent
Northeast	12,162	33.23
Southeast	9,054	24.74
North Central	5,095	13.92
South Central	2,806	7.67
Mountain	2,038	5.57
Pacific	5,449	14.89
TOTAL	**36,604**	**100.00**

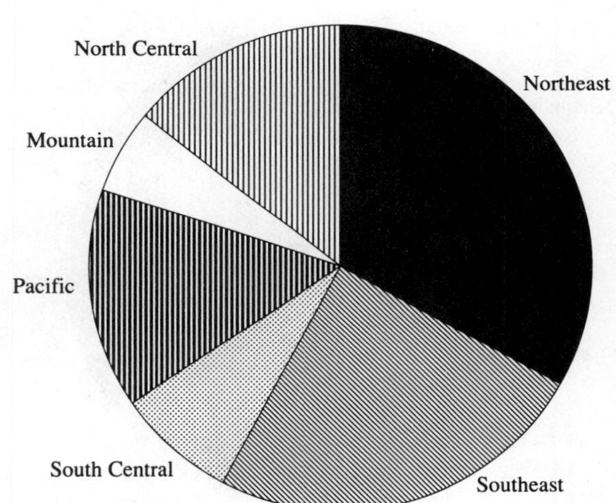

Chemistry

	Number	Percent
Northeast	10,343	40.15
Southeast	6,124	23.77
North Central	3,022	11.73
South Central	1,738	6.75
Mountain	1,300	5.05
Pacific	3,233	12.55
TOTAL	**25,760**	**100.00**

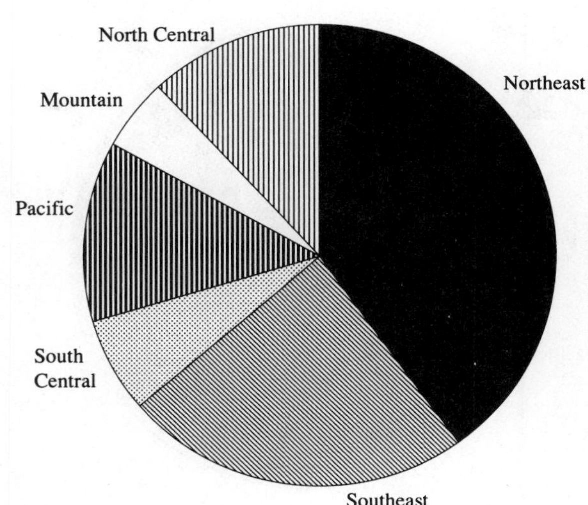

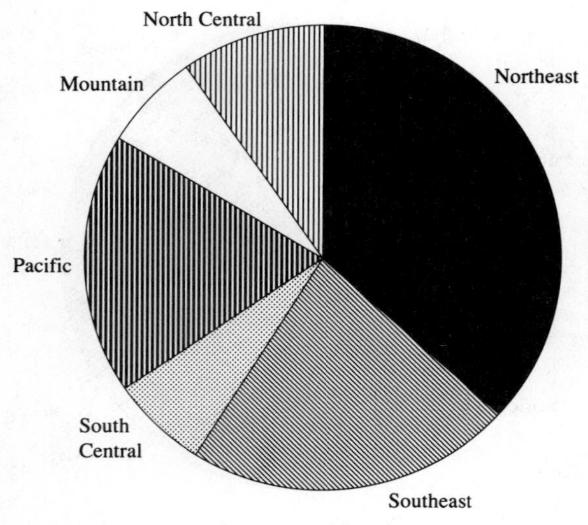

Computer Sciences

	Number	Percent
Northeast	1,987	36.76
Southeast	1,200	22.20
North Central	511	9.45
South Central	360	6.66
Mountain	372	6.88
Pacific	976	18.05
TOTAL	**5,406**	**100.00**

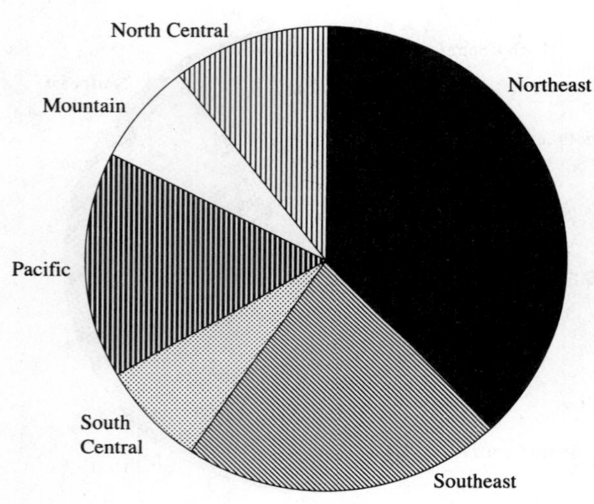

Engineering

	Number	Percent
Northeast	9,122	38.01
Southeast	5,202	21.68
North Central	2,510	10.46
South Central	1,710	7.13
Mountain	1,646	6.86
Pacific	3,807	15.86
TOTAL	**23,997**	**100.00**

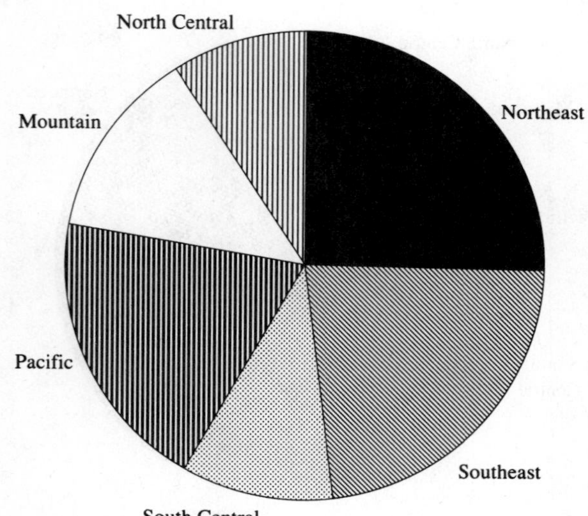

Environmental, Earth & Marine Sciences

	Number	Percent
Northeast	2,657	25.48
Southeast	2,361	22.64
North Central	953	9.14
South Central	1,075	10.31
Mountain	1,359	13.03
Pacific	2,022	19.39
TOTAL	**10,427**	**100.00**

Mathematics

	Number	Percent
Northeast	4,211	35.92
Southeast	2,609	22.26
North Central	1,511	12.89
South Central	884	7.54
Mountain	718	6.13
Pacific	1,789	15.26
TOTAL	**11,722**	**100.00**

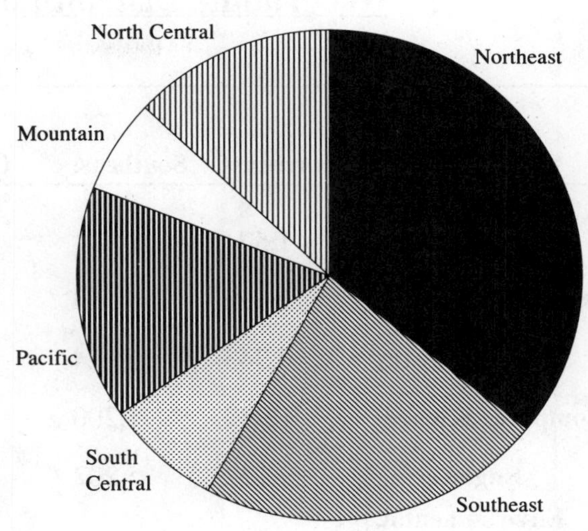

Medical & Health Sciences

	Number	Percent
Northeast	7,115	36.53
Southeast	5,004	25.69
North Central	2,577	13.23
South Central	1,516	7.78
Mountain	755	3.88
Pacific	2,509	12.88
TOTAL	**19,476**	**100.00**

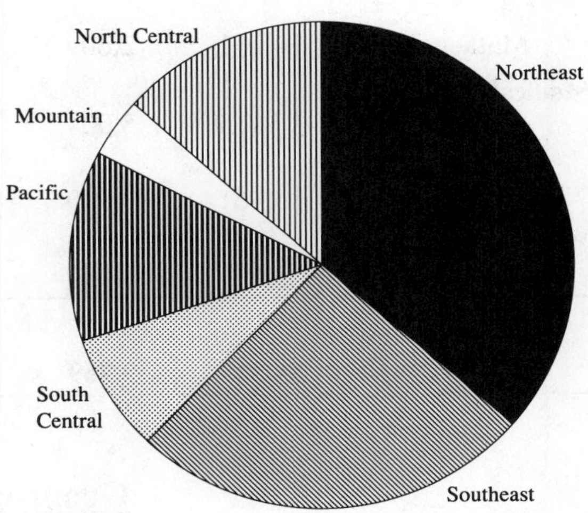

Physics & Astronomy

	Number	Percent
Northeast	5,961	35.12
Southeast	3,670	21.62
North Central	1,579	9.30
South Central	918	5.41
Mountain	1,607	9.47
Pacific	3,238	19.08
TOTAL	**16,973**	**100.00**

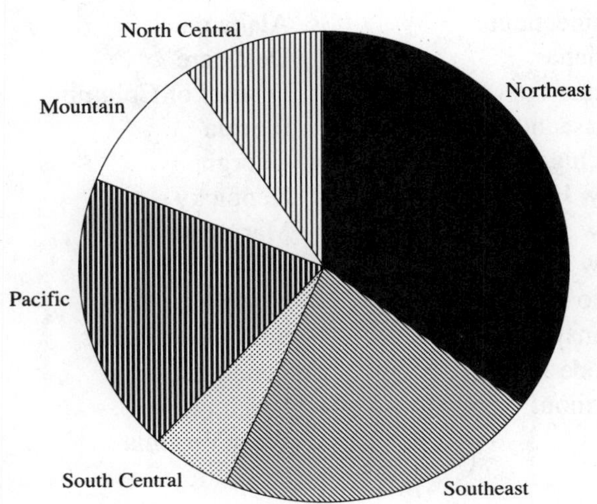

Geographic Distribution of Scientists by Discipline

	Northeast	Southeast	North Central	South Central	Mountain	Pacific	TOTAL
Agricultural & Forest Sciences	1,574	1,991	1,170	609	719	1,297	**7,360**
Biological Sciences	12,162	9,054	5,095	2,806	2,038	5,449	**36,604**
Chemistry	10,343	6,124	3,022	1,738	1,300	3,233	**25,760**
Computer Sciences	1,987	1,200	511	360	372	976	**5,406**
Engineering	9,122	5,202	2,510	1,710	1,646	3,807	**23,997**
Environmental, Earth & Marine Sciences	2,657	2,361	953	1,075	1,359	2,022	**10,427**
Mathematics	4,211	2,609	1,511	884	718	1,789	**11,722**
Medical & Health Sciences	7,115	5,004	2,577	1,516	755	2,509	**19,476**
Physics & Astronomy	5,961	3,670	1,579	918	1,607	3,238	**16,973**
Other Professional Fields	3,193	2,554	918	540	515	1,230	**8,950**
TOTAL	**58,325**	**39,769**	**19,846**	**12,156**	**11,029**	**25,550**	**166,675**

Geographic Definitions

Northeast
Connecticut
Indiana
Maine
Massachusetts
Michigan
New Hampshire
New Jersey
New York
Ohio
Pennsylvania
Rhode Island
Vermont

Southeast
Alabama
Delaware
District of Columbia
Florida
Georgia
Kentucky
Maryland
Mississippi
North Carolina
South Carolina
Tennessee
Virginia
West Virginia

North Central
Illinois
Iowa
Kansas
Minnesota
Missouri
Nebraska
North Dakota
South Dakota
Wisconsin

South Central
Arkansas
Louisiana
Texas
Oklahoma

Mountain
Arizona
Colorado
Idaho
Montana
Nevada
New Mexico
Utah
Wyoming

Pacific
Alaska
California
Hawaii
Oregon
Washington

Sample Entry

American Men & Women of Science (AMWS) is an extremely useful reference tool. The book is most often used in one of two ways: to find more information about a particular scientist or to locate a scientist in a specific field.

To locate information about an individual, the biographical section is most helpful. It encompasses the first seven volumes and lists scientists and engineers alphabetically by last name. The fictitious biographical listing shown below illustrates every type of information an entry may include.

The Discipline Index, volume 8, can be used to easily find a scientist in a specific subject specialty. This index is first classified by area of study, and within each specialty entrants are divided further by state of residence.

CARLETON, PHYLLIS B(ARBARA), b Glenham, SDak, April 1, 30. m 53, 69; c 2. ORGANIC CHEMISTRY. *Educ:* Univ Notre Dame, BSc, 52, MSc, 54, Vanderbilt Univ, PhD(chem), 57. *Hon Degrees:* DSc, Howard Univ, 79. *Prof Exp:* Res chemist, Acme Chem Corp, 54-59, sr res chemist, 59-60; from asst to assoc prof chem 60-63, prof chem, Kansas State Univ, 63-72; prof chem, Yale Univ, 73-89; CONSULT, CARLETON & ASSOCS, 89-. *Concurrent Pos:* Adj prof, Kansas State Univ 58-60; vis lect, Oxford Univ, 77, consult, Union Carbide, 74-80. *Honors & Awards:* Gold Medal, Am Chem Society, 81; *Mem:* AAAS, fel Am Chem Soc, Sigma Chi. *Res:* Organic synthesis, chemistry of natural products, water treatment and analysis. *Mailing Address:* Carleton & Assocs 21 E 34th St Boston MA 02108

Abbreviations

AAAS—American Association for the Advancement of Science
abnorm—abnormal
abstr—abstract
acad—academic, academy
acct—Account, accountant, accounting
acoust—acoustic(s), acoustical
ACTH—adrenocorticotrophic hormone
actg—acting
activ—activities, activity
addn—addition(s), additional
Add—Address
adj—adjunct, adjutant
adjust—adjustment
Adm—Admiral
admin—administration, administrative
adminr—administrator(s)
admis—admission(s)
adv—adviser(s), advisory
advan—advance(d), advancement
advert—advertisement, advertising
AEC—Atomic Energy Commission
aerodyn—aerodynamic
aeronaut—aeronautic(s), aeronautical
aerophys—aerophsical, aerophysics
aesthet—aesthetic
AFB—Air Force Base
affil—affiliate(s), affiliation
agr—agricultural, agriculture
agron—agronomic, agronomical, agronomy
agrost—agrostologic, agrostological, agrostology
agt—agent
AID—Agency for International Development
Ala—Alabama
allergol—allergological, allergology
alt—alternate
Alta—Alberta
Am—America, American
AMA—American Medical Association
anal—analysis, analytic, analytical
analog—analogue
anat—anatomic, anatomical, anatomy
anesthesiol—anesthesiology
angiol—angiology
Ann—Annal(s)
ann—annual
anthrop—anthropological, anthropology
anthropom—anthropometric, anthropometrical, anthropometry
antiq—antiquary, antiquities, antiquity
antiqn—antiquarian
apicult—apicultural, apiculture
APO—Army Post Office
app—appoint, appointed
appl—applied
appln—application
approx—approximate(ly)
Apr—April
apt—apartment(s)
aquacult—aquaculture
arbit—arbitration
arch—archives
archaeol—archaeological, archaeology
archit—architectural, architecture
Arg—Argentina, Argentine
Ariz—Arizona
Ark—Arkansas
artil—artillery
asn—association
assoc(s)—associate(s), associated
asst(s)—assistant(s), assistantship(s)
assyriol—Assyriology
astrodyn—astrodynamics
astron—astronomical, astronomy
astronaut—astonautical, astronautics
astronr—astronomer
astrophys—astrophysical, astrophysics
attend—attendant, attending
atty—attorney
audiol—audiology
Aug—August
auth—author
AV—audiovisual
Ave—Avenue
avicult—avicultural, aviculture

b—born
bact—bacterial, bacteriologic, bacteriological, bacteriology
BC—British Colombia
bd—board
behav—behavior(al)
Belg—Belgian, Belgium
Bibl—biblical
bibliog—bibliographic, bibliographical, bibliography
bibliogr—bibliographer
biochem—biochemical, biochemistry
biog—biographical, biography
biol—biological, biology
biomed—biomedical, biomedicine
biomet—biometric(s), biometrical, biometry
biophys—biophysical, biophysics
bk(s)—book(s)
bldg-building
Blvd—Boulevard
Bor—Borough
bot—botanical, botany
br—branch(es)
Brig—Brigadier
Brit—Britain, British
Bro(s)—Brother(s)
byrol—byrology
bull—Bulletin
bur—bureau
bus—business
BWI—British West Indies

c—children
Calif—California
Can—Canada, Canadian
cand—candidate
Capt—Captain
cardiol-cardiology
cardiovasc—cardiovascular
cartog—cartographic, cartographical, cartography
cartogr—cartographer
Cath—Catholic
CEngr—Corp of Engineers
cent—central
Cent Am—Central American
cert—certificate(s), certification, certified
chap—chapter
chem—chemical(s), chemistry
chemother—chemotherapy
chg—change
chmn—chairman
citricult—citriculture
class—classical
climat—climatological, climatology
clin(s)—clinic(s), clinical
cmndg—commanding
Co—County
co—Companies, Company
co-auth—coauthor
co-dir—co-director
co-ed—co-editor
co-educ—coeducation, coeducational
col(s)—college(s), collegiate, colonel
collab—collaboration, collaborative
collabr—collaborator
Colo—Colorado
com—commerce, commercial
Comdr—Commander

commun—communicable, communication(s)
comn(s)—commission(s), commissioned
comndg—commanding
comnr—commissioner
comp—comparitive
compos—composition
comput—computation, computer(s), computing
comt(s)—committee(s)
conchol—conchology
conf—conference
cong—congress, congressional
Conn—Connecticut
conserv—conservation, conservatory
consol—consolidated, consolidation
const—constitution, constitutional
construct—construction, constructive
consult(s)—consult, consultant(s), consultantship(s), consultation, consulting
contemp—contemporary
contrib—contribute, contributing, contribution(s)
contribr—contributor
conv—convention
coop—cooperating, cooperation, cooperative
coord—coordinate(d), coordinating, coordination
coordr—coordinator
corp—corporate, corporation(s)
corresp—correspondence, correspondent, corresponding
coun—council, counsel, counseling
counr—councilor, counselor
criminol—criminological, criminology
cryog—cryogenic(s)
crystallog—crystallographic, crystallographical, crystallography
crystallogr—crystallographer
Ct—Court
Ctr—Center
cult—cultural, culture
cur—curator
curric—curriculum
cybernet—cybernetic(s)
cytol—cytological, cytology
Czech—Czechoslovakia

DC—District of Columbia
Dec—December
Del—Delaware
deleg—delegate, delegation
delinq—delinquency, delinquent
dem—democrat(s), democratic
demog—demographic, demography
demogr—demographer
demonstr—demontrator
dendrol—dendrologic, dendrological, dendrology
dent—dental, dentistry
dep—deputy
dept—department
dermat—dermatologic, dermatological, dermatology
develop—developed, developing, development, developmental
diag—diagnosis, diagnostic
dialectol-dialectological, dialectology
dict—dictionaries, dictionary
Dig—Digest
dipl—diploma, diplomate
dir(s)—director(s), directories, directory
dis—disease(s), disorders
Diss Abst—Dissertation Abstracts
dist—district
distrib—distributed, distribution, distributive
distribr—distributor(s)
div—division, divisional, divorced
DNA—deoxyribonucleic acid
doc—document(s), documentary, documentation
Dom—Dominion
Dr—Drive
E—east
ecol—ecological, ecology
econ(s)—economic(s), economical, economy
economet—econometric(s)
ECT—electroconvulsive or electroshock therapy
ed—edition(s), editor(s), editorial
ed bd—editorial board
educ—education, educational
educr—educator(s)
EEG—electroencephalogram, electroencephalographic, electroencephalography
Egyptol—Egyptology
EKG—electrocardiogram
elec—elecvtric, electrical, electricity
electrochem-electrochemical, electrochemistry
electroph—electrophysical, electrophysics
elem—elementary
embryol—embryologic, embryological, embryology
emer—emeriti, emeritus
employ—employment
encour—encouragement
encycl—encyclopedia
endocrinol—endocrinologic, endocrinology
eng—engineering
Eng—England, English
engr(s)—engineer(s)
enol—enology
Ens—Ensign
entom—entomological, entomology
environ-environment(s), environmental
enzym—enzymology
epidemiol—epideiologic, epidemiological, epidemiology
equip—equipment
ERDA—Energy Research & Development Administration
ESEA—Elementary & Secondary Education Act
espec—especially
estab—established, establishment(s)
ethnog—ethnographic, ethnographical, ethnography
ethnogr—ethnographer
ethnol—ethnologic, ethnological, ethnology
Europ—European
eval—evaluation
Evangel—evangelical
eve—evening
exam—examination(s), examining
examr—examiner
except—exceptional
exec(s)—executive(s)
exeg—exegeses, exegesis, exegetic, exegetical
exhib(s)—exhibition(s), exhibit(s)
exp—experiment, experimental
exped(s)—expedition(s)
explor—exploration(s), exploratory
expos—exposition
exten—extension

fac—faculty
facil—facilities, facility
Feb—February
fed—federal
fedn—federation
fel(s)—fellow(s), fellowship(s)
fermentol—fermentology
fertil—fertility, fertilization
Fla—Florida
floricult—floricultural, floriculture
found—foundation
FPO—Fleet Post Office
Fr—French
Ft—Fort

Ga—Georgia
gastroenterol—gastroenterological, gastroenterology
gen—general
geneal—genealogical, genealogy
geod—geodesy, geodetic
geog—geographic, geographical, geography
geogr—geographer
geol—geologic, geological, geology
geom—geometric, geometrical, geometry
geomorphol—geomorphologic, geomorphology
geophys—geophysical, geophysics
Ger—German, Germanic, Germany
geriat—geriatric
geront—gerontological, gerontology
GES—Gesellschaft
glaciol—glaciology
gov—governing, governor(s)
govt—government, governmental
grad—graduate(d)
Gt Brit—Great Britain
guid—guidance
gym—gymnasium
gynec—gynecologic, gynecological, gynecology

handbk(s)—handbook(s)
helminth—helminthology
hemat—hematologic, hematological, hematology
herpet—herpetologic, herpetological, herpetology
HEW—Department of Health, Education & Welfare
Hisp—Hispanic, Hispania
hist—historic, historical, history
histol—histological, histology
HM—Her Majesty
hochsch—hochschule
homeop—homeopathic, homeopathy
hon(s)—honor(s), honorable, honorary
hort—horticultural, horticulture
hosp(s)—hospital(s), hospitalization
hq—headquarters

ABBREVIATIONS

HumRRO—Human Resources Research Office
husb—husbandry
Hwy—Highway
hydraul—hydraulic(s)
hydrodyn—hydrodynamic(s)
hydrol—hydrologic, hydrological, hydrologics
hyg—hygiene, hygienic(s)
hypn—hypnosis

ichthyol—ichthyological, ichthyology
Ill—Illinois
illum—illuminating, illumination
illus—illustrate, illustrated, illustration
illusr—illustrator
immunol—immunologic, immunological, immunology
Imp—Imperial
improv—improvement
Inc—Incorporated
in-chg—in charge
incl—include(s), including
Ind—Indiana
indust(s)—industrial, industries, industry
Inf—infantry
info—information
inorg—inorganic
ins—insurance
inst(s)—institute(s), institution(s)
instnl—institutional(ized)
instr(s)—instruct, instruction, instructor(s)
instrnl—instructional
int—international
intel—intellligence
introd—introduction
invert—invertebrate
invest(s)—investigation(s)
investr—investigator
irrig—irrigation
Ital—Italian

J—Journal
Jan—January
Jct—Junction
jour—journal, journalism
jr—junior
jurisp—jurisprudence
juv—juvenile

Kans—Kansas
Ky—Kentucky

La—Louisiana
lab(s)—laboratories, laboratory
lang—language(s)
laryngol—larygological, laryngology
lect—lecture(s)
lectr—lecturer(s)
legis—legislation, legislative, legislature
lett—letter(s)
lib—liberal
libr—libraries, library
librn—librarian
lic—license(d)
limnol—limnological, limnology
ling—linguistic(s), linguistical
lit—literary, literature
lithol—lithologic, lithological, lithology

Lt—Lieutenant
Ltd—Limited

m—married
mach—machine(s), machinery
mag—magazine(s)
maj—major
malacol—malacology
mammal—mammalogy
Man—Manitoba
Mar—March
Mariol—Mariology
Mass—Massechusetts
mat—material(s)
mat med—materia medica
math—mathematic(s), mathematical
Md—Maryland
mech—mechanic(s), mechanical
med—medical, medicinal, medicine
Mediter—Mediterranean
Mem—Memorial
mem—member(s), membership(s)
ment—mental(ly)
metab—metabolic, metabolism
metall—metallurgic, metallurgical, metallurgy
metallog—metallographic, metallography
metallogr—metallographer
metaphys—metaphysical, metaphysics
meteorol—meteorological, meteorology
metrol—metrological, metrology
metrop—metropolitan
Mex—Mexican, Mexico
mfg—manufacturing
mfr—manufacturer
mgr—manager
mgt—management
Mich—Michigan
microbiol—microbiological, microbiology
micros—microscopic, microscopical, microscopy
mid—middle
mil—military
mineral—mineralogical, mineralogy
Minn—Minnesota
Miss—Mississippi
mkt—market, marketing
Mo—Missouri
mod—modern
monogr—monograph
Mont—Montana
morphol—morphological, morphology
Mt—Mount
mult—multiple
munic—municipal, municipalities
mus—museum(s)
musicol—musicological, musicology
mycol—mycologic, mycology

N—north
NASA—National Aeronautics & Space Administration
nat—national, naturalized
NATO—North Atlantic Treaty Organization
navig—navigation(al)
NB—New Brunswick
NC—North Carolina
NDak—North Dakota
NDEA—National Defense Education Act
Nebr—Nebraska

nematol—nematological, nematology
nerv—nervous
Neth—Netherlands
neurol—neurological, neurology
neuropath—neuropathological, neuropathology
neuropsychiat—neuropsychiatric, neuropsychiatry
neurosurg—neurosurgical, neurosurgery
Nev—Nevada
New Eng—New England
New York—New York City
Nfld—Newfoundland
NH—New Hampshire
NIH—National Institute of Health
NIMH—National Institute of Mental Health
NJ—New Jersey
NMex—New Mexico
No—Number
nonres—nonresident
norm—normal
Norweg—Norwegian
Nov—November
NS—Nova Scotia
NSF—National Science Foundation
NSW—New South Wales
numis—numismatic(s)
nutrit—nutrition, nutritional
NY—New York State
NZ—New Zealand

observ—observatories, observatory
obstet—obstetric(s), obstetrical
occas—occasional(ly)
occup—occupation, occupational
oceanog—oceanographic, oceanographical, oceanography
oceanogr—oceanographer
Oct—October
odontol—odontology
OEEC—Organization for European Economic Cooperation
off—office, official
Okla—Oklahoma
olericult—olericulture
oncol—oncologic, oncology
Ont—Ontario
oper(s)—operation(s), operational, operative
ophthal—ophthalmologic, ophthalmological, ophthalmology
optom—optometric, optometrical, optometry
ord—ordnance
Ore—Oregon
org—organic
orgn—organization(s), organizational
orient—oriental
ornith—ornithological, ornithology
orthod—orthodontia, orthodontic(s)
orthop—orthopedic(s)
osteop—osteopathic, osteopathy
otol—otological, otology
otolaryngol—otolaryngological, otolaryngology
otorhinol—otorhinologic, otorhinology

Pa—Pennsylvania
Pac—Pacific
paleobot—paleobotanical, paleontology
paleont—paleontology

Pan-Am—Pan-American
parisitol—parasitology
partic—participant, participating
path—pathologic, pathological, pathology
pedag—pedagogic(s), pedagogical, pedagogy
pediat—pediatric(s)
PEI—Prince Edward Islands
penol—penological, penology
periodont—periodontal, periodontic(s)
petrog—petrographic, petrographical, petrography
petrogr—petrographer
petrol—petroleum, petrologic, petrological, petrology
pharm—pharmacy
pharmaceut—pharmaceutic(s), pharmaceutical(s)
pharmacog—pharmacognosy
pharamacol—pharmacologic, pharmacological, pharmacology
phenomenol—phenomenologic(al), phenomenology
philol—philological, philology
philos—philosophic, philosophical, philosophy
photog—photographic, photography
photogeog—photogeographic, photogeography
photogr—photographer(s)
photogram—photogrammetric, photogrammetry
photom—photometric, photometrical, photometry
phycol—phycology
phys—physical
physiog—physiographic, physiographical, physiography
physiol—physiological, phsysiology
Pkwy—Parkway
Pl—Place
polit—political, politics
polytech—polytechnic(s)
pomol—pomological, pomology
pontif—pontifical
pop—population
Port—Portugal, Portuguese
Pos:—Position
postgrad—postgraduate
PQ—Province of Quebec
PR—Puerto Rico
pract—practice
practr—practitioner
prehist—prehistoric, prehistory
prep—preparation, preparative, preparatory
pres—president
Presby—Presbyterian
preserv—preservation
prev—prevention, preventive
prin—principal
prob(s)—problem(s)
proc—proceedings
proctol—proctologic, proctological, proctology
prod—product(s), production, productive
prof—professional, professor, professorial
Prof Exp—Professional Experience
prog(s)—program(s), programmed, programming
proj—project(s), projection(al), projective
prom—promotion
protozool—protozoology
Prov—Province, Provincial
psychiat—psychiatric, psychiatry
psychoanal—psychoanalysis, psychoanalytic, psychoanalytical
psychol—psychological, psychology
psychomet—psychometric(s)
psychopath—psychopathologic, psycho pathology
psychophys—psychophysical, psychophysics
psychophysiol—psychophysiological, psychophysiology
psychosom—psychosomtic(s)
psychother—psychoterapeutic(s), psychotherapy
Pt—Point
pub—public
publ—publication(s), publish(ed), publisher, publishing
pvt—private

Qm—Quartermaster
Qm Gen—Quartermaster General
qual—qualitative, quality
quant—quantitative
quart—quarterly
Que—Quebec

radiol—radiological, radiology
RAF—Royal Air Force
RAFVR—Royal Air Force Volunteer Reserve
RAMC—Royal Army Medical Corps
RAMCR—Royal Army Medical Corps Reserve
RAOC—Royal Army Ornance Corps
RASC—Royal Army Service Corps
RASCR—Royal Army Service Corps Reserve
RCAF—Royal Canadian Air Force
RCAFR—Royal Canadian Air Force Reserve
RCAFVR—Royal Canadian Air Force Volunteer Reserve
RCAMC—Royal Canadian Army Medical Corps
RCAMCR—Royal Canadian Army Medical Corps Reserve
RCASC—Royal Canadian Army Service Corps
RCASCR—Royal Canadian Army Service Corps Reserve
RCEME—Royal Canadian Electrical & Mechanical Engineers
RCN—Royal Canadian Navy
RCNR—Royal Canadian Naval Reserve
RCNVR—Royal Canadian Naval Volunteer Reserve
Rd—Road
RD—Rural Delivery
rec—record(s), recording
redevelop—redevelopment
ref—reference(s)
refrig—refrigeration
regist—register(ed), registration
registr—registrar
regt—regiment(al)
rehab—rehabilitation
rel(s)—relation(s), relative
relig—religion, religious
REME—Royal Electrical & Mechanical Engineers
rep—represent, representative
Repub—Republic
req—requirements
res—research, reserve
rev—review, revised, revision
RFD—Rural Free Delivery
rhet-rhetoric, rhetorical
RI—Rhode Island
Rm—Room
RM—Royal Marines
RN—Royal Navy
RNA—ribonucleic acid
RNR—Royal Naval Reserve
RNVR—Royal Naval Volunteer Reserve
roentgenol—roentgenologic, roentgenological, roentgenology
RR—Railroad, Rural Route
Rte—Route
Russ—Russian
rwy—railway

S—south
SAfrica—South Africa
SAm—South America, South American
sanit—sanitary, sanitation
Sask—Saskatchewan
SC—South Carolina
Scand—Scandinavia(n)
sch(s)—school(s)
scholar—scholarship
sci—science(s), scientific
SDak—South Dakota
SEATO—Southeast Asia Treaty Organization
sec—secondary
sect—section
secy—secretary
seismog—seismograph, seismographic, seismography
seismogr—seismographer
seismol—seismological, seismology
sem—seminar, seminary
Sen—Senator, Senatorial
Sept—September
ser—serial, series
serol—serologic, serological, serology
serv—service(s), serving
silvicult—silvicultural, silviculture
soc(s)—societies, society
soc sci—social science
sociol—sociologic, sociological, sociology
Span—Spanish
spec—special
specif—specification(s)
spectrog—spectrograph, spectrographic, spectrography
spectrogr—spectrographer
spectrophotom—spectrophotometer, spectrophotometric, spectrophotometry
spectros—spectroscopic, spectroscopy
speleol—speleological, speleology
Sq—Square
sr—senior
St—Saint, Street(s)
sta(s)—station(s)
stand—standard(s), standardization
statist—statistical, statistics
Ste—Sainte
steril—sterility

stomatol—stomatology
stratig—stratigraphic, stratigraphy
stratigr—stratigrapher
struct—structural, structure(s)
stud—student(ship)
subcomt—subcommittee
subj—subject
subsid—subsidiary
substa—substation
super—superior
suppl—supplement(s), supplemental, supplementary
supt—superintendent
supv—supervising, supervision
supvr—supervisor
supvry—supervisory
surg—surgery, surgical
surv—survey, surveying
survr—surveyor
Swed—Swedish
Switz—Switzerland
symp—symposia, symposium(s)
syphil—syphilology
syst(s)—system(s), systematic(s), systematical

taxon—taxonomic, taxonomy
tech—technical, technique(s)
technol—technologic(al), technology
tel—telegraph(y), telephone
temp—temporary
Tenn—Tennessee
Terr—Terrace
Tex—Texas
textbk(s)—textbook(s)
text ed—text edition
theol—theological, theology
theoret—theoretic(al)
ther—therapy
therapeut—therapeutic(s)
thermodyn—thermodynamic(s)
topog—topographic, topographical, topography
topogr—topographer
toxicol—toxicologic, toxicological, toxicology
trans—transactions
transl—translated, translation(s)
translr—translator(s)
transp—transport, transportation
treas—treasurer, treasury
treat—treatment
trop—tropical
tuberc—tuberculosis
TV—television
Twp—Township

UAR—United Arab Republic
UK—United Kingdom
UN—United Nations
undergrad—undergraduate
unemploy—unemployment
UNESCO—United Nations Educational Scientific & Cultural Organization
UNICEF—United Nations International Childrens Fund
univ(s)—universities, university
UNRRA—United Nations Relief & Rehabilitation Administration
UNRWA—United Nations Relief & Works Agency
urol—urologic, urological, urology
US—United States
USAAF—US Army Air Force
USAAFR—US Army Air Force Reserve
USAF—US Air Force
USAFR—US Air Force Reserve
USAID—US Agency for International Development
USAR—US Army Reserve
USCG—US Coast Guard
USCGR—US Coast Guard Reserve
USDA—US Department of Agriculture
USMC—US Marine Corps
USMCR—US Marine Corps Reserve
USN—US Navy
USNAF—US Naval Air Force
USNAFR—US Naval Air Force Reserve
USNR—US Naval Reserve
USPHS—US Public Health Service
USPHSR—US Public Health Service Reserve
USSR—Union of Soviet Socialist Republics

Va—Virginia
var—various
veg—vegetable(s), vegetation
vent—ventilating, ventilation
vert—vertebrate
Vet—Veteran(s)
vet—veterinarian, veterinary
VI—Virgin Islands
vinicult—viniculture
virol—virological, virology
vis—visiting
voc—vocational
vocab—vocabulary
vol(s)—voluntary, volunteer(s), volume(s)
vpres—vice president
vs—versus
Vt—Vermont

W—west
Wash—Washington
WHO—World Health Organization
WI—West Indies
wid—widow, widowed, widower
Wis—Wisconsin
WVa—West Virginia
Wyo—Wyoming

Yearbk(s)—Yearbook(s)
YMCA—Young Men's Christian Association
YMHA—Young Men's Hebrew Association
Yr(s)—Year(s)
YT—Yukon Territory
YWCA—Young Women's Christian Association
YWHA—Young Women's Hebrew Association

zool—zoological, zoology

American Men & Women of Science

T

TAAGEPERA, MARE, b Narva, Estonia, May 16, 38; US citizen; m 61; c 3. PHYSICAL ORGANIC CHEMISTRY. *Educ:* Univ Del, BS, 60, MS, 63; Univ Pa, PhD(chem), 70. *Prof Exp:* Chemist, E I Du Pont de Nemours & Co, Inc, 62-64; FEL CHEM, UNIV CALIF, IRVINE, 71- *Mem:* Am Chem Soc; Sigma Xi; Asn Advan Baltic Studies. *Res:* Mechanisms of organic reactions; rate and equilibrium studies; mass and ion cyclotron resonance spectroscopy. *Mailing Add:* 19 Mayapple Way Irvine CA 92715-2714

TAAM, RONALD EVERETT, b New York, NY, 48; m 74; c 2. HYDRODYNAMICS, STELLAR STRUCTURE & EVOLUTION. *Educ:* Polytech Inst Brooklyn BS, 69; Columbia Univ, MA, 71, PhD(astron), 73. *Prof Exp:* Fel, Univ Calif, Santa Cruz, 73-76, vis prof astron, Berkeley, 76-78; from asst prof to assoc prof, 79-86, PROF PHYSICS & ASTRON, NORTHWESTERN UNIV, 86- *Mem:* Am Astron Soc; Royal Astron Soc; Am Phys Soc; Int Astron Union. *Res:* Application of fluid mechanics, radiation transfer, and nuclear physics to astrophysical problems; structure and evolution of close binary stars; high energy astrophysics. *Mailing Add:* Dept Physics & Astron Northwestern Univ Evanston IL 60208

TABACHNICK, IRVING I A, b New York, NY, July 20, 24; m 51; c 2. PHARMACOLOGY. *Educ:* Harvard Univ, AB, 48; Yale Univ, PhD(pharmacol), 53. *Prof Exp:* Pharmacologist & statistician, Baxter Labs, 53-55; pharmacologist, 55-58, sr pharmacologist, 58-60, sect head, 60-61, head dept biochem pharm, 61-64, head dept physiol & biochem, 64-68, assoc dir biol res div, 68-70, dir, 70-72, sr dir biol res & develop, Schering Corp, 72-77, VPRES, DRUG SAFETY & METAB, SCHERING-PLOUGH CORP, 77- *Mem:* AAAS; Am Soc Pharmacol & Exp Therapeut; Soc Exp Biol & Med; NY Acad Sci; Am Acad Allergy; Sigma Xi. *Res:* Histamine; diabetes; insulin; catecholamines; adrenergic receptors; anti-hormones; toxicology; drug metabolism. *Mailing Add:* Schering Plough Corp Biol Res & Develop 86 Orange St Bloomfield NJ 07003

TABACHNICK, JOSEPH, b New York, NY, May 14, 19; m 61; c 2. RADIOIMMUNOASSAY. *Educ:* Univ Calif, Berkeley, BS, 42, MS, 47, PhD(comp biochem), 50. *Prof Exp:* Res asst food technol, Univ Calif, Berkeley, 46-51, lab instr, 47-49; res chemist, Turner Hall Corp, NJ, 51-52; res biochemist, 52-59, sr res biochemist & assoc mem, Div of Labs & head, Lab Exp Dermat, 59-78, HEAD, HORMONE RECEPTOR LAB, ALBERT EINSTEIN MED CTR, 78- *Concurrent Pos:* Adj assoc prof, NY Med Col, 73- *Mem:* Sigma Xi; Clin Ligand Assay Soc; Am Asn Path; Radiation Res Soc. *Res:* Estrogen and progesterone receptors in human breast cancer; radiation biology and biochemistry of the skin; cytokinetics of repair in beta-irradiated skin; formation and metabolism of pyroglutamic acid and free amino acid in skin; suppression of radiation fibrosis. *Mailing Add:* Hormone Receptor Lab Dept Surg A312 Korman Albert Einstein Med Ctr Philadelphia PA 19141

TABACHNICK, MILTON, b New York, NY, June 25, 22; m 52. BIOCHEMISTRY. *Educ:* Univ Calif, AB, 47, MA, 49, PhD(biochem), 53. *Prof Exp:* Asst, Univ Calif, 49-52; Am Cancer Soc res fel biochem, Sch Med, Duke Univ, 52-53; instr, Inst Indust Med, Post-Grad Med Sch, NY Univ, 53-55; vis investr, Div Nutrit & Physiol, Pub Health Res Inst of City New York, 55-57; res assoc chem, Mt Sinai Hosp, 57-59; res assoc, NY State Psychiat Inst, 59-60; from asst prof to assoc prof, 61-69, dean Grad Sch Basic Med Sci, 71-80, PROF BIOCHEM, NEW YORK MED COL, 69- *Mem:* Am Soc Biol Chem; Am Chem Soc; Endocrine Soc; Am Thyroid Asn; AAAS. *Res:* Protein chemistry; transport and mechanism of action of thyroid hormone; purification and characterization of thyroxine-binding globulin. *Mailing Add:* Dept Biochem New York Med Col Valhalla NY 10595

TABACHNICK, WALTER J, b Brooklyn, NY, June 14, 47. POPULATION GENETICS, EVOLUTIONARY BIOLOGY. *Educ:* Brooklyn Col, City Univ New York, BS, 68; Rutgers Univ, MS, 71, PhD(zool), 74. *Prof Exp:* Asst prof genetics, Univ Wis, Parkside, 73-75; NIH fel, Yale Univ, 75-78; lectr biol & genetics, Univ Calif, Los Angeles, 78-79; res assoc genetics, Yale Univ, 79-82; prof biol, Loyola Univ, 82-85; RES SCIENTIST, USDA-ARS, 87- *Concurrent Pos:* Nat Defense Educ Act Title IV fel, Rutgers Univ, 70-73; prin investr, Yale Univ, US Army, NIH res grants, 75-87. *Mem:* Entom Soc Am; Sigma Xi; Am Soc Study Evolution; Genetics Soc Am; Am Soc Trop Med & Hyg. *Res:* Studying the genetics and evolution of insect disease vectors with the goal of using this information to understand and control insect borne animal disease. *Mailing Add:* Anthropod-borne Animal Dis Res Lab USDA-ARS PO Box 3965 Univ Sta Laramie WY 82071

TABAK, DANIEL, b Wilna, Poland, June 16, 34; US citizen; m 60; c 2. AUTOMATIC CONTROL SYSTEMS. *Educ:* Israel Inst Technol, BS, 59, MS, 63; Univ Ill, Urbana, PhD(elec eng), 67. *Prof Exp:* Asst nuclear sci, Israel Inst Technol, 61-63; elec engr, Univ Ill, Urbana, 63-66; guid & control systs engr, Missile & Space Div, Gen Elec Co, Pa, 66-68; sr consult, Wolf Res & Develop Corp, Md, 68-70; assoc prof systs eng, Rensselaer Polytech Inst, 70-72; assoc prof elec eng, Ben-Gurion Univ of the Negev, Israel, 72-76, chmn, Indust Eng Dept, 73-74, chmn, Elec Eng Dept, 76-77, prof elec eng, 76-; PROF ELEC & COMPUTER ENG, GEORGE MASON UNIV. *Concurrent Pos:* Consult, Missile & Space Div, Gen Elec Co, 66, Wolf Res & Develop Corp, 70-72; assoc ed, J Automatica, Int Fedn Automatic Control, 72-; on leave, NRC sr res assoc, NASA Langley Res Ctr, 77-78 & vis prof elec eng, Univ Tex, Austin, 78-79. *Mem:* Sr mem Inst Elec & Electronics Engrs; Opers Res Soc Am. *Res:* Computational methods of optimal control; digital systems; computer architecture; computer engineering. *Mailing Add:* Elec & Computer Eng Dept George Mason Univ Fairfax VA 22030

TABAK, MARK DAVID, b Philadelphia, Pa, Dec 2, 37; m 61; c 2. SOLID STATE PHYSICS, ELECTRICAL ENGINEERING. *Educ:* Univ Pa, BS, 59; Princeton Univ, MA, 62, PhD(elec eng), 65. *Prof Exp:* Instr, Princeton Univ, 62-63; scientist xerographic sci, Res Labs Div, Xerox Res, 65-70, area mgr, 70-72, sect mgr, 72-73, mgr, Imaging Sci Lab, 73-77, mgr, Advan Marking Prog, Corp Res & Develop Group, 77-78, vpres & mgr, Webster Res Ctr, 78-84, vpres & mgr, Electronic Pub Bur, 84-85, vpres technol transfer, 85-87; RETIRED. *Concurrent Pos:* Mem, Advan Mgt Prog, Harvard Bus Sch, 82. *Mem:* Am Phys Soc; Inst Elec & Electronics Engrs; Soc Photog Engrs & Scientists. *Res:* Electrophotographic materials and systems; photoconductivity; electronic transport and photogeneration; amorphous semiconductors. *Mailing Add:* 16244 Avenida Florencia Poway CA 92064

TABAKIN, BURTON SAMUEL, b Philadelphia, Pa, July 6, 21; m 72; c 5. MEDICINE. *Educ:* Univ Pa, AB, 43, MD, 47; Am Bd Internal Med, dipl, 55. *Prof Exp:* Fel physiol, Univ Vt & Trudeau Found, 51-52; from instr to assoc prof, 54-67, dir Cardiopulmonary Lab, 54-80, dir Cardiol Univ, 72-80, actg chmn dept, 74-76, PROF MED, COL MED, UNIV VT, 67- *Concurrent Pos:* Attend physician, Mary Fletcher Hosp & DeGoesbriand Mem Hosp, Burlington, Vt, 60-66; fel coun clin cardiol, Am Heart Asn. *Mem:* Fel Am Col Physicians; Am Heart Asn; Am Col Chest Physicians; Am Fedn Clin Res; fel Am Col Cardiol. *Res:* Clinical cardiopulmonary and exercise physiology; echocardiography. *Mailing Add:* Med Ctr Hosp Vt Burlington VT 05401

TABAKIN, FRANK, b Newark, NJ, Sept 20, 35; m 63; c 2. THEORETICAL PHYSICS. *Educ:* Queens Col, NY, BS, 56; Mass Inst Technol, PhD(physics), 63. *Prof Exp:* Res assoc physics, Columbia Univ, 63-65; from asst prof to assoc prof, 65-74, PROF PHYSICS, UNIV PITTSBURGH, 74- *Concurrent Pos:* Res visitor, Oxford Univ, 70. *Mem:* Am Phys Soc. *Res:* Nuclear forces and matter; three-body problems; properties of nuclei; meson physics; meson photoproduction; antiproton reactions. *Mailing Add:* 623 S Linden Ave Pittsburgh PA 15208

TABAKOFF, BORIS, b Tien-Tsin, China, Sept 27, 42. NEUROPHARMACOLOGY, ALCOHOLISM. *Educ:* Univ Colo, Boulder, BA & BPh, 66, PhD(pharmacol), 70. *Prof Exp:* Asst prof biochem, Chicago Med Sch, 71-73, assoc prof, 73-75; from assoc prof to prof, dept physiol, Med Ctr, Univ Ill, 75-84; DIR, IRP, NAT INST ALCOHOL ABUSE & ALCOHOLISM/ALCOHOL, DRUG ABUSE & MENTAL HEALTH ADMIN/DEPT HEALTH & HUMAN SERV, 84- *Concurrent Pos:* Bd trustees res award, Chicago Med Sch, 73; Hoffman LaRoche Found award & NIH-Swiss NSF award, 75; res award, Interstate Postgrad Med Asn NAm, 76; mem, NIH/Alcohol, Drug Abuse & Ment Health Admin Study Sect, Nat Inst Alcohol Abuse & Alcoholism, 78-81; res scientist, West Side Vet Admin Med Ctr, Chicago, 79-84; dir, Alcohol & Drug Abuse Res & Training Prog,

Univ Ill Med Ctr, 80-84. *Honors & Awards:* Vet Admin Award, 79. *Mem:* Am Soc Pharmacol & Exp Therapeut; Am Soc Neurochem; Soc Neurosci; Res Soc Alcoholism; Int Soc Biomed Res Alcoholism; Am Col Neuropsychopharmacol. *Res:* Neurochemical and behavioral effects of alcohol and their contribution to development of tolerance and physical dependence; development of therapies to ameliorate tolerance and physical dependence. *Mailing Add:* DICBR/Nat Inst Alcohol Abuse & Alcoholism NIH Clin Ctr 10/3C103 Bethesda MD 20892

TABAKOFF, WIDEN, b Stakevzi, Bulgaria, Dec 14, 19; US citizen; m 43; c 2. PROPULSION, GAS DYNAMICS. *Educ:* Prague Tech Univ, MS, 42; Univ Berlin, MS, 46, PhD(sci), 45. *Prof Exp:* Asst technician, Univ Berlin, 42-46; designer, Aerotallers Argentinos, 48-50; tech dir prod, Helamet, Arg, 50-56; supvr res, Knapsack Grisheim, Ger, 56-58; from instr to assoc prof, 58-66, PROF AEROSPACE ENG & APPL MECH, UNIV CINCINNATI, 67- *Concurrent Pos:* Consult indust & govt, 54- *Mem:* Fel Am Inst Aeronaut & Astronaut; Am Soc Eng Educ; Am Soc Mech Engrs; Am Asn Univ Prof; Am Soc Testing & Mat. *Res:* Turbomachinery components, inlets, compressors, combustion chambers, turbine nozzles and exhaust engine nozzles; ramjet components; aerodynamic heating for hypersonic vehicles; internal gas dynamics flows; erosion problems in propulsion systems. *Mailing Add:* 421 Deanview Dr Cincinnati OH 45224

TABARROK, B(EHROUZ), b Tehran, Iran, June 16, 38; m 63; c 3. MECHANICAL ENGINEERING. *Educ:* Wolverhampton & Staffordshire Col Technol, BSc, 62; Oxford Univ, DPhil(structural dynamics), 65. *Prof Exp:* Ford Found fel mech eng, 65-66, from asst prof to prof mech eng, Univ Toronto, 66-87; PROF MECH ENG, UNIV TORONTO, 87- *Concurrent Pos:* Alexander von Humboldt fel, Hannover Tech Univ, 69; sr res fel, Sci Res Coun, UK, 74. *Honors & Awards:* Robert W Angus Medal, Can Soc Mech Eng, 85. *Mem:* Fel Can Soc Mech Engr; fel Eng Inst Can. *Res:* Applied mechanics, particularly in dynamics of structures; computational mechanics, variational principles in mechanics. *Mailing Add:* Dept Mech Eng Univ Victoria Victoria BC V8W 2Y2 Can

TABATA, SUSUMU, b Steveston, BC, Dec 9, 25; m 59; c 3. PHYSICAL OCEANOGRAPHY. *Educ:* Univ BC, BA, 50, MA, 54; Univ Tokyo, DSc, 65. *Prof Exp:* Phys oceanogr, Pac Oceanog Group, Fisheries Res Bd Can, 52-70; res phys oceanogr, 71-84, RES SCIENTIST, OCEAN PHYSICS DIV, INST OCEAN SCI, 84- *Concurrent Pos:* Asst scientist, Pac Oceanog Group, Fisheries Res Bd Can, 52-58, assoc scientist, 59-65, sr scientist, 66-70. *Mem:* Am Soc Limnol & Oceanog; Am Meteorol Soc; Am Geophys Union; Oceanog Soc Japan. *Res:* Circulation of inshore and offshore waters; processes affecting water properties; variability in the oceans; large-scale air-sea interactions; interpretation of satellite imagery; ocean climatology. *Mailing Add:* Inst Ocean Sci Patricia Bay PO Box 6000 Sidney BC V8L 4B2 Can

TABATABAI, LOUISA BRAAL, b Oostvoorne, Neth, Dec 18, 39; US citizen; m 62; c 3. BIOCHEMISTRY. *Educ:* Univ Calif, Berkeley, BA, 62; Iowa State Univ, MS, 66, PhD(biochem), 76. *Prof Exp:* Res assoc food technol, 64-70, res assoc muscle biochem, 71-74, assoc biochem, 76-77, ASST PROF BIOCHEM , IOWA STATE UNIV, 85-; RES CHEMIST BIOCHEM, NAT ANIMAL DIS CTR, 77- *Concurrent Pos:* Nat Res Coun fel, Nat Animal Dis Ctr, 77-78. *Mem:* Am Chem Soc; Am Soc Microbiol; AAAS; Protein Soc. *Res:* Virulence and pathogenicity of Brucella organisms; biochemical immunological properties of Brucella cell surface proteins, mechanism of action on host tissues; structure/function of Brucella proteins. *Mailing Add:* Nat Animal Dis Ctr PO Box 70 Ames IA 50010-0070

TABATABAI, M ALI, b Karbala, Iraq, Feb 25, 34; US citizen; m 62; c 3. SOIL CHEMISTRY, SOIL BIOCHEMISTRY. *Educ:* Univ Baghdad, BS, 58; Okla State Univ, MS, 60; Iowa State Univ, PhD(soil chem), 65. *Prof Exp:* Res assoc soil biochem, 66-72, from asst prof to assoc prof, 72-78, PROF SOIL CHEM & BIOCHEM, IOWA STATE UNIV, 78- *Concurrent Pos:* Consult, Elec Power Res Inst, Palo Alto, Calif, 78-83. *Mem:* Am Soc Agron; Am Chem Soc; Am Soc Microbiol; AAAS; Int Soc Soil Sci. *Res:* Soil enzymology and chemistry of sulfur, nitrogen and phosphorus in soils; nutrient cycling in the environment. *Mailing Add:* Dept Agron Iowa State Univ Ames IA 50011

TABB, DAVID LEO, b Louisville, Ky, Feb 8, 46; m 66; c 2. POLYMER SCIENCE. *Educ:* Univ Louisville, BChE, 69; Case Western Reserve Univ, MS, 72, PhD(polymer sci), 74. *Prof Exp:* Res eng chem eng, 69-70, SR RES ENGR, POLYMER PRODS DEPT, SPECIALTY POLYMERS DIV, E I DU PONT DE NEMOURS & CO, INC, 74- *Res:* Characterization of polymer structure and correlation to physical properties; thermoset and thermoplastic elastomers; polymer blends; polymer processing technology; Fourier transform infrared spectroscopy. *Mailing Add:* Polymers Prods Dept/ Specialty Polymers Div Chestnut Run Plaza Bldg 711 Wilmington DE 19808

TABBERT, ROBERT L, b Ripon, Wis, Sept 6, 28; m 52; c 2. GEOLOGY. *Educ:* Univ Wis, BS, 52, MSc, 54. *Prof Exp:* Micropaleontologist, Magnolia Petrol Co, 54-59; micropaleontologist, Socony Mobil's Field Res Lab, 59-62; palynologist, Atlantic Richfield Co, 62-67, supvr palynology group, 67-73, dir struct & stratig res, 73-75, sr res assoc, Geol Sci Group, 75-76, explor geologist, 76-86; RETIRED. *Mem:* Soc Econ Paleont & Mineral; Am Asn Petrol Geologists; Am Asn Stratig Palynologists; Geol Soc Am; Asn Prof Geol Scientists. *Res:* Biostratigraphy of Cretaceous-Tertiary sediments of Alaska; regional correlations in Mesozoic and Tertiary sediments of Arctic; structural geology, stratigraphy and petroleum exploration of Arctic; exploration and geothermal energy of Western United States. *Mailing Add:* 211 Ursline Lafayette LA 70506

TABBUTT, FREDERICK DEAN, b Philadelphia, Pa, Dec 28, 31; m 56; c 4. COMPUTER SIMULATIONS, CATASTROPHE THEORY. *Educ:* Haverford Col, BS, 53; Harvard Univ, MA, 55, PhD(chem), 58. *Prof Exp:* Instr, Reed Col, 57-59, asst prof chem, 59-63, assoc prof, 63-67, prof, 67-70; MEM FAC CHEM, EVERGREEN STATE COL, 70- *Concurrent Pos:* NSF fac fel, Southhampton Univ, UK, 63-64, Univ Warwick, 74-75. *Mem:* Am Chem Soc. *Res:* Oscillating chemical reactions, particularly the Belousov-Zhabotinsky reaction; applications of castastrophe theory to oscillating systems; gas phase kinetics of unimolecular reactions; laser flash photolysis. *Mailing Add:* Evergreen State Col Olympia WA 98501

TABER, CHARLES ALEC, b Texanna, Okla, Dec 10, 37; m 70. ICHTHYOLOGY. *Educ:* Northeastern State Col, BS, 61; Univ Okla, PhD(zool), 69. *Prof Exp:* Asst prof biol, 69-77, assoc prof life sci, 77-82, PROF BIOL, SOUTHWEST MO STATE UNIV, 82- *Mem:* Am Soc Ichthyologists & Herpetologists; Am Fisheries Soc. *Res:* Natural history and ecology of fishes. *Mailing Add:* Dept Biol Southwest Mo State Univ 901 S National Springfield MO 65804

TABER, DAVID, b New York, NY, July 14, 22; m 48; c 3. ORGANIC CHEMISTRY. *Educ:* NY Univ, AB, 48; Polytech Inst Brooklyn, PhD(chem), 53. *Prof Exp:* Jr chemist, Air Reduction Co, Inc, 48-49; res assoc indust med, Postgrad Med Sch, NY Univ, 53-55; chemist, Gen Aniline & Film Corp, 55-58; sr chemist, Koppers Co, Inc, 58-63; sect head, Org Chem, Armour Grocery Prod Co, 63-65, res mgr new chem, Household Prod Res & Develop Dept, 65-68, mgr biol & med progs, Armour-Dial, Inc, 68-69, asst res dir, 69-75; vpres & dir tech serv, Hollister, Inc, 75-77; dir regulatory affairs, Wesley-Jessen Inc, 77-84; vpres & dir regulatory affairs, Am Bioactive Colloids, Inc, 84-87; CONSULT, DAVID TABER & ASSOC, 78-; DIR, REGULATORY AFFAIRS, WATSON LABS, INC 87- *Mem:* Am Chem Soc; Soc Cosmetic Chemists; Sigma Xi. *Res:* Organic synthesis, germicides, microbiology; product safety; regulatory affairs; clinical testing; medical and hospital devices; quality assurance. *Mailing Add:* 7328 N Oakley Ave Chicago IL 60645-1252

TABER, DOUGLASS FLEMING, b Berkeley, Calif, Nov 11, 48; m 69; c 6. ENANTIOSELECTIVE SYNTHESIS. *Educ:* Stanford Univ, BS, 70; Columbia Univ, PhD(org chem), 74. *Prof Exp:* Res fel org chem, Univ Wis-Madison, 74-75; res instr, dept pharmacol, Vanderbilt Univ, 75-77, asst prof, 77-82; asst prof, 82-84, ASSOC PROF ORG CHEM, UNIV DEL, 84- *Concurrent Pos:* Consult, div clin pharmacol, Sch Med, Vanderbilt Univ, 82-; Alfred P Sloan Found fel, 83. *Mem:* Am Chem Soc. *Res:* Development of new methods in synthetic organic chemistry; natural product synthesis; organometallics in organic synthesis. *Mailing Add:* Dept Chem & Biochem Univ Del Newark DE 19716

TABER, ELSIE, b Columbia, SC, May 3, 15. EMBRYOLOGY, REPRODUCTIVE PHYSIOLOGY. *Educ:* Univ SC, BS, 35; Stanford Univ, MA, 36; Univ Chicago, PhD(zool), 47. *Prof Exp:* Teacher biol, Greenwood High Sch, SC, 36-38; instr, Lander Col, 38-41; instr, Univ Chicago, 44-48, asst dean students, Div Biol Sci, 47-48; from asst prof to assoc prof, 48-65, PROF ANAT, MED UNIV SC, 65- *Mem:* AAAS; Am Soc Zool; Am Asn Anatomists; Am Inst Biol Sci; Soc Study Reproduction; Sigma Xi. *Res:* Developmental biology; endocrinology of reproductive systems. *Mailing Add:* 216 Molasses Lane Mt Pleasant SC 29464

TABER, HARRY WARREN, b Longview, Wash, Oct 30, 35; c 2. MICROBIOLOGY, MOLECULAR BIOLOGY. *Educ:* Reed Col, BA, 57; Univ Rochester, PhD(biochem), 63. *Prof Exp:* AEC fel, Univ Rochester, 62-64; USPHS fel, Nat Inst Neurol Dis & Stroke, Bethesda, Md, 64-66; USPHS spec fel, Photosynthesis Lab, Gif-sur-Yvette, France, 66-67; asst prof microbiol, Sch Med, Univ Rochester, 68-73, assoc prof, 73-79; PROF MICROBIOL, DEPT MICROBIOL & IMMUNOL, ALBANY MED COL UNION UNIV, 80- *Concurrent Pos:* NIH res career develop award, 74-79, mem, study sect, 79-88. *Mem:* Am Soc Microbiol; Soc Gen Microbiol; AAAS; Am Asn Univ Prof. *Res:* Genetic regulation of respiratory chain components in bacteria; antibiotic transport by bacteria; biosynthesis of vitamin K. *Mailing Add:* Dept Microbiol & Immunol Albany Med Col Union Univ Albany NY 12208

TABER, JOSEPH JOHN, b Adena, Ohio, Feb 6, 20; m 47; c 4. PETROLEUM ENGINEERING, OIL RECOVERY RESEARCH. *Educ:* Muskingum Col, BS, 42; Univ Pittsburgh, PhD, 55. *Prof Exp:* Asst prof naval sci, Ohio State Univ, 46; instr chem, Washington & Jefferson Col, 46-50; sr proj chemist, Gulf Res & Develop Co, 54-64; from asst prof to assoc prof chem, 64-72, adj assoc prof petrol eng, 67-72, prof chem, grad fac, 72-76, officer, Admin Dept, 64-76, prof petrol eng, 72-76, dir NMex Petrol Recovery Res Ctr, 76-87, EMER DIR & ADJ PROF PETROL ENG, NMEX PETROL RECOVERY RES CTR, NMEX INST MINING & TECHNOL, 87- *Concurrent Pos:* Co prin investr, res on CO2 for enhanced oil recovery, 78-87; lectr, Int Energy Agency Workshops & Symp, enhanced oil recovery, 84; consult, Argonne Nat Lab, 84-, Enhanced Recovery Comt, Interstate Oil Compoct Comn, 80-; tech auditor, Norway Oil Recovery Res, 86-; mem adv bd, Sch Earth Sci, Stanford Univ, 89-91. *Honors & Awards:* Enhanced Oil Recovery Pioneer Award, Soc Petrol Engrs, 90. *Mem:* Am Chem Soc; Am Inst Mining, Metall & Petrol Engrs; AAAS; Sigma Xi. *Res:* Surface chemistry; liquid-liquid and liquid-solid interfaces; effect of interfacial energies on capillarity and fluid flow in porous media; new methods of petroleum recovery; effect of phase behavior on solvent displacement of oil from rocks; use of CO2 and other gases for enhanced oil recovery; environmental problems of oil recovery. *Mailing Add:* 700 Leroy Pl Socorro NM 87801

TABER, RICHARD DOUGLAS, b San Francisco, Calif, Nov 22, 20; m 46; c 3. WILDLIFE ECOLOGY. *Educ:* Univ Calif, AB, 42, PhD, 51; Univ Wis, MS, 49. *Prof Exp:* Wildlife researcher & asst specialist, Univ Calif, 51-55, actg asst prof zool, 55-56; asst prof forestry, Univ Mont, 56-57, from assoc prof to prof, 58-68, assoc dir, Mont Forest & Conserv Exp Sta, 64-68; prof, 68-85, EMER PROF FOREST ZOOL, UNIV WASH, 85- *Concurrent Pos:* US specialist forest-wildlife rels, Germany & Czech, 60 & Poland, 60 & 64; Fulbright res scholar, West Pakistan, 63-64; Guggenheim fel, 64; mem comn threatened deer, Int Union Conserv Nature, 73- *Mem:* Wildlife Soc; AAAS; Am Soc Mammal; Am Inst Biol Sci. *Res:* Biology and conservation of free-living birds and mammals; wildlife and human culture; ungulate biology and effects on ecosystem. *Mailing Add:* Col Forest Resources 625 Continental Way Missoula MT 59803

TABER, RICHARD LAWRENCE, b Pontiac, Mich, Nov 9, 35; m 57; c 2. BIO-ORGANIC CHEMISTRY. *Educ:* Colo State Col, AB, 58; Univ NMex, PhD(org chem), 63. *Prof Exp:* From asst prof to assoc prof, 63-75, PROF CHEM, COLO COL, 75- *Honors & Awards:* Meritorious Serv Award, Am Chem Soc, 69. *Mem:* Am Chem Soc; Sigma Xi. *Res:* Enzymology of dihydroorotate dehydrogenase; organic liquid scintillation solutes. *Mailing Add:* Chem Dept Colo Col Colorado Springs CO 80903

TABER, ROBERT IRVING, b Perth Amboy, NJ, June 28, 36; m 60; c 3. PSYCHOPHARMACOLOGY, NEUROPHARMACOLOGY. *Educ:* Rutgers Univ, BS, 58; Med Col Va, PhD(pharmacol), 63. *Prof Exp:* Pharmacologist, Schering Corp, 62-66, sr pharmacologist, 66-71, mgr pharmacol, 71-74, dir biol res, 74-77, sr dir biol res, 77-; BIOMED PROD DEPT, E I DU PONT DE NEMOURS & CO INC. *Mem:* AAAS; assoc Am Col Neuropsychopharmacol; Am Soc Pharmacol & Exp Therapeut; Acad Pharmaceut Sci; Am Pharmaceut Asn. *Res:* Analgesics; drug effects on learning and memory. *Mailing Add:* Biomed Prod Dept E I Du Pont de Nemours & Co Inc E400-2418A Wilmington DE 19898

TABER, STEPHEN, III, b Columbia, SC, Apr 17, 24; m 45; c 8. APICULTURE. *Educ:* Univ Wis, BS, 49. *Prof Exp:* Apiculturist, USDA, La, 50-65, apiculturist, Ariz, 65-79; RES DIR, HONEY BEE GENETICS, 79- *Mem:* AAAS; Entom Soc Am; Am Genetic Asn; Bee Res Asn. *Res:* Bee behavior and genetics. *Mailing Add:* PO Box 1672 Vacaville CA 95688

TABER, WILLARD ALLEN, b Marshalltown, Iowa, Feb 18, 25; m 50; c 3. MICROBIOLOGY, HISTORY & PHILOSOPHY OF SCIENCE. *Educ:* Univ Iowa, AB, 49, MS, 51; Rutgers Univ, PhD(microbiol), 54. *Prof Exp:* Asst, Univ Iowa, 49-51; fel, Rutgers Univ, 54-55; from asst res officer to sr res officer, Prarie Regional Lab, Nat Res Coun Can, 55-63; prof biol, Tex A&M Univ, 63-88; RETIRED. *Concurrent Pos:* NIH consult, 67-70; consult to chem indust. *Mem:* Soc Indust Microbiol; Sigma Xi; Am Chem Soc; AAAS; Am Forestry Asn; Am Phy to Path Soc. *Res:* Antifungal antibiotics; ecology of soil fungi; metabolism of sugar carbon sources; mycology; secondary metabolism; transport in fungi, mycorrhizae, streptomyces; biodegradation. *Mailing Add:* 1100 Village Dr College Station TX 77840

TABI, RIFAT, b Simferopol, Crimea, June 5, 38; US citizen; m 67; c 1. MECHANICAL ENGINEERING. *Educ:* Aachen Tech Univ, BSME, 61, MSME, 62; Vienna Tech Univ, Dr Eng Sci (mech eng), 65; State Univ NY, PE, 72. *Prof Exp:* Sci res asst, Testing & Res Inst Mat, Vienna Tech Univ, 62-65; proj engr, Tech Ctr, Gen Motors Corp, Mich, 65-66; asst prof mech eng technol, 66-67, assoc prof, 67-70, prof, 70-72, PROF MECH, INDUST & AEROSPACE ENG & CHMN DEPT, NY INST TECHNOL, 72-, ASSOC DIR, CTR TECHNOL, 80- *Concurrent Pos:* Consult, private practice, 67-; qualified fallout shelter analyst & instr, Univ Hawaii, 68. *Res:* Materials-mechanics; fracture mechanics of materials and failure analysis. *Mailing Add:* Dept Mech Indust & Aerospace Eng NY Inst Technol PO Box 170 Old Westbury NY 11568

TABIBIAN, RICHARD, b Detroit, Mich, June 1, 29; m 55; c 3. POLYMER CHEMISTRY. *Educ:* Wayne State Univ, BS, 51, MS, 54, PhD(chem), 56. *Prof Exp:* CHEMIST, E I DU PONT DE NEMOURS & CO, INC, 55- *Mem:* Am Chem Soc. *Res:* Colloid chemistry; elastomers. *Mailing Add:* Polymer Prods Dept Chestnut Run Plaza 711 E I Du Pont de Nemours & Co Inc Wilmington DE 19880-0711

TABISZ, GEORGE CONRAD, b New York, NY, Aug 28, 39; Can citizen; m 66; c 2. MOLECULAR PHYSICS. *Educ:* Univ Toronto, BASc, 61, MA, 63, PhD(physics), 68. *Prof Exp:* Nat Res Coun Can fel, High Pressure Lab, Nat Ctr Sci Res, France, 68-70; from asst prof to assoc prof, 70-80, PROF PHYSICS, UNIV MANITOBA, 80- *Concurrent Pos:* Vis scientist, Univ Cambridge, 76-77, Nat Res Coun Can, Ottawa, 83-84, Joint Inst Lab Astrophys, Univ Colo, Boulder, 90-91. *Mem:* Optical Soc Am; Can Asn Physicists; Am Phys Soc. *Res:* Molecular interactions in gases and liquids; visible and infrared absorption spectroscopy; laser raman scattering; theory of spectral lineshapes. *Mailing Add:* Dept Physics Univ Man Winnipeg MB R3T 2N2 Can

TABITA, F ROBERT, b Bronx, NY, Oct 1, 43. MICROBIAL BIOCHEMISTRY, PHOTOSYNTHETIC METABOLISM. *Educ:* St Johns Univ, BS, 65, MS, 67; Syracuse Univ, PhD(microbiol), 71. *Prof Exp:* NIH fel biochem, Wash State Univ, 71-73; from asst prof to assoc prof, 73-85, PROF MICROBIOL, UNIV TEX, AUSTIN, 85- *Concurrent Pos:* Panel mem, Biochem Study Sect, NIH, 76-77 & Microbiol, Physiol & Genetics Sect, 84-88; mem, Competitive Res Grants Panel, USDA, 81 & 84; asst dir, Ctr Appl Microbiol, Univ Tex Austin, 81- *Mem:* Am Soc Biol Chemists; Am Chem Soc; Am Soc Microbiol; AAAS; Am Soc Plant Physiologists; Int Soc Plant Microbiol. *Res:* Molecular basis and biochemistry of carbon dioxide fixation; regulation of biological nitrogen fixation; protection of nitrogenase from oxygen inactivation in vivo. *Mailing Add:* Dept Microbiol Ohio State Univ 484 W 12th St Columbus OH 43210-1214

TABLER, RONALD DWIGHT, b Denver, Colo, May 18, 37; m 64; c 2. HYDROLOGY. *Educ:* Colo State Univ, BS, 59, PhD(watershed mgt), 65. *Prof Exp:* Res hydrologist, Rocky Mountain Forest & Range Exp Sta, US Forest Serv, 59-85; CONSULT SNOW & WIND ENG, TABLER & ASSOCS, 86- *Concurrent Pos:* Adj prof mech eng, Univ Wyo, 86-, adj prof range mgt, 86-; res affil, Univ Alaska, Fairbanks, 86-; vis scientist, Civil Eng Res Inst, Japan, 80-81. *Honors & Awards:* USDA Super Serv Award, 76; D Grant Mickle Award, Nat Acad Sci, 79. *Mem:* AAAS; Am Rwy Eng Asn; Am Geophys Union; Soil Conserv Soc; Am Water Resources Asn; Int Glaciol Soc. *Res:* Physics of snow transport by wind; design of snow fence systems for highway protection; management of snow in windswept areas to increase water yields; watershed management. *Mailing Add:* 1416 Tabler Assocs PO Box 483 Niwot CO 80544-0483

TABOADA, JOHN, b Tampico, Mex, Sept 8, 43; US citizen; m 68; c 1. PHYSICAL OPTICS, APPLIED PHYSICS. *Educ:* Trinity Univ, BA, 66; Tex A&M Univ, MS, 68, PhD(physics), 73. *Prof Exp:* Asst physics, Tex A&M Univ, 66-68; res physicist radiation sci div, US Air Force Sch Aerospace Med, 68-80, res physicist data sci div, 80-84, RES PHYSICIST, CLIN SCI DIV, 84- *Concurrent Pos:* Mem comt laser measurements, Am Nat Standards Inst, 76- *Honors & Awards:* Outstanding Tech Achievement Award, USAF Systs Command, 71, Outstanding Sci Achievement Award, 75. *Mem:* Am Phys Soc; Optical Soc Am. *Res:* Biophysics of ultrashort pulsed lasers; laser spectroscopy; nonlinear optics; applied optics; robotics. *Mailing Add:* 12530 Elm Country San Antonio TX 78230

TABOR, CELIA WHITE, b Boston, Mass, Nov 15, 18; m 46; c 4. BIOCHEMISTRY. *Educ:* Radcliffe Col, BA, 40; Columbia Univ, MD, 43. *Prof Exp:* Intern med, Mass Gen Hosp, Boston, 44-45; asst resident, Univ Hosp, Vanderbilt Univ, 45-46; res assoc pharmacol, George Washington Univ, 47-49; MEM STAFF, LAB BIOCHEM PHARMACOL, NAT INST ARTHRITIS, METAB & DIGESTIVE DIS & USPHS, 52- *Mem:* AAAS; Am Soc Biol Chemists; Am Soc Pharmacol & Exp Therapeut. *Res:* Biochemistry; enzymatic and metabolic studies of the polyamines. *Mailing Add:* Four North Dr Bethesda MD 20814

TABOR, CHRISTOPHER ALAN, b Frederick, Md, Jan 1, 37; div; c 1. PLANT PHYSIOLOGY. *Educ:* NC State Univ, BS, 59, MS, 65; Univ Md, PhD, 82. *Prof Exp:* Mem staff bioprod res & develop, Dow Chem Co, 64-65; res forester, 67-77, PLANT PHYSIOLOGIST, FOREST SERV, USDA, 77- *Mem:* Am Soc Plant Physiologists. *Res:* Physiology of the effects of environmental stress on tree growth and development. *Mailing Add:* NE Forest Exp Sta PO Box 968 Burlington VT 05402

TABOR, EDWARD, b Washington, DC, Apr 30, 47; m 70; c 4. VIROLOGY, CANCER ETIOLOGY. *Educ:* Harvard Univ, BA, 69; Columbia Univ, MD, 73. *Prof Exp:* Intern pediat, Columbia-Presby Med Ctr, New York, 73-74, resident, 74-75; res investr, Food & Drug Admin, 75-81, sr med investr, Bur Biologics, 81-83, dir, Div Anti-Infective Drug Prods, 83-88; ASSOC DIR, BIOL CARCINOGENESIS, NAT CANCER INST, 88- *Concurrent Pos:* Clin asst prof pediat, Georgetown Univ, 80; asst clin prof child health & develop, George Washington Univ, 88- *Honors & Awards:* US PHS Commendation Medal, 79; Unit Commendation, USPHS, 85; Outstanding Serv Medal, USPHS, 86; Outstanding Unit Commendation, USPHS, 87; Spec Citation, US Food & Drug Admin, 88. *Mem:* Fel Infectious Dis Soc Am. *Res:* Non-A and non-B hepatitis; hepatocellular carcinoma; hepatitis A and B; safety and effectiveness of vaccines and antiviral drugs. *Mailing Add:* Nat Cancer Inst Nat Inst Health Bldg 41 Rm A100 Bethesda MD 20892

TABOR, HERBERT, b New York, NY, Nov 28, 18; m 46; c 4. PHARMACOLOGY, BIOCHEMISTRY. *Educ:* Harvard Univ, AB, 37, MD, 41. *Prof Exp:* Biochem researcher, Harvard Med Sch, 41-42; intern med, New Haven Hosp, Conn, 42; CHIEF LAB BIOCHEM PHARMACOL, 43, NAT INST DIABETES, DIGESTIVE & KIDNEY DIS, NIH, 61- *Concurrent Pos:* Field ed, J Pharmacol & Exp Therapeut, 60-68; assoc ed, J Biol Chem, 68-71, ed-in-chief, 71- *Mem:* Nat Acad Sci; Am Soc Pharmacol & Exp Therapeut; Am Soc Biol Chem; Am Chem Soc; Am Acad Arts & Sci. *Res:* Biochemistry of amino acids and amines. *Mailing Add:* Nat Inst Diabetes Digestive & Kidney Dis Bldg 8 Rm 223 Bethesda MD 20892

TABOR, JOHN MALCOLM, b Harrisburg, Pa, May 30, 52. BIOTECHNOLOGY, DRUG DEVELOPMENT OF BIOTHERAPEUTICS. *Educ:* Elizabethtown Col, BS, 74; Kans State Univ, PhD(molecular biol), 78. *Prof Exp:* Postdoctoral, Roche Inst Molecular Biol, 79-80; postdoctoral, Mass Inst Technol, 80-81, vis scientist, 81-82, instr exp biol, 82; sr scientist, Bristol-Myers Squibb Co, 82-84, dept head, 84-87, mgr, 87, asst dir, 87-90, ASSOC DIR, BRISTOL-MYERS SQUIBB CO, 90- *Concurrent Pos:* Adj res asst prof, Biol Dept, Syracuse Univ, 82-83. *Mem:* Sigma Xi. *Res:* Planning genetic engineering support strategies for research discovery areas and development activities for the manufacture of recumbinant DNA and munoclonal antibody therapeutic products. *Mailing Add:* Bristol-Myers Squibb Co PO Box 4755 Syracuse NY 13221-4755

TABOR, MARVIN WILSON, b Bluefield, WVa, Nov 24, 44; c 2. ENVIRONMENTAL ANALYTICAL CHEMISTRY, MICROBIAL DEGRADATION OF HAZARDOUS WASTES. *Educ:* Emory & Henry Col, BS, 66; Marshall Univ, MS, 68; Univ Cincinnati, PhD(biol chem), 74. *Prof Exp:* Res assoc biochem, Dept Obstet & Gynec, Univ Cincinnati Med Ctr, 76-78; asst prof toxicol, 78-84, ASSOC PROF TOXICOL & ENVIRON CHEM, DEPT ENVIRON HEALTH, UNIV CINCINNATI, 84- *Concurrent Pos:* Course dir biochem, Dept Chem, Univ Cincinnati, 73-, dir mass spectrometry, Mass Spectrometry Facil, 81-, dir environ analytical chem, Inst Environ Health, 85-; sci adv/mentor, Int Vis Scientist Prog, World Health Orgn, 85-; vis scientist, Dept Environ Chem, Delft Tech Univ, Netherlands, 87; mem, Test & Eval Facil Tech Adv Coun, US Environ Protection Agency, 87-89; chmn bd trustees, Ohio River Basin Consortium for Res & Educ, 87-89; vis prof, Instituto Investigaciones Biomedicas, Universidad Nacional Autonoma Mex, 88; sci adv, Secretaria Desarrollo Urbano Y Ecolugia, Gov Mex, 88-, Secretaria Salud, 88-; adv & mem, Interagency Groundwater & Environ Adv Coun, State Ohio Gov Cabinet Cluster, 88- *Honors & Awards:* Miguel Aleman Mem Award, Miguel Aleman Found, 90. *Mem:* Am Chem Soc; Sigma Xi; Am Soc Microbiol; Int Soc Study Xenobiotics; Am Soc Mass Spectrometry; Am Waterworks Asn. *Res:* Identification and assessment of human exposure to environmental anthropogenic pollutants; development and assessment of microbial methods for the biodegradation, metabolism and detoxification of hazardous wastes. *Mailing Add:* Inst Environ Health Univ Cincinnati Med Ctr Cincinnati OH 45267-0056

TABOR, ROWLAND WHITNEY, b Denver, Colo, June 10, 32; m 57, 70; c 2. GEOLOGY. *Educ:* Stanford Univ, BS, 54; Univ Wash, MS, 58, PhD(geol), 61. *Prof Exp:* Teaching asst geol, Univ Wash, 57-61; GEOLOGIST, US GEOL SURV, 61- *Mem:* Fel Geol Soc Am. *Res:* Field mapping northwestern Washington, stratigraphy, igneous and metamorphic petrology and structure of Olympic and North Cascade Mountains; emphasis on geologic history and hazards; KAr geochronology; popular science writing. *Mailing Add:* US Geol Surv MS 975 345 Middlefield Rd Menlo Park CA 94025

TABOR, SAMUEL LYNN, b Tylertown, Miss, June 10, 45; m. NUCLEAR PHYSICS. *Educ:* Tulane Univ, BS, 67; Stanford Univ, MS, 68, PhD(physics), 72. *Prof Exp:* Res assoc nuclear physics, Univ Pa, 72-75 & Argonne Nat Lab, 75-77; sr res assoc nuclear physics, Univ Md, 77-79; from asst prof to assoc prof, 79-88, PROF PHYSICS, FLA STATE UNIV, 88- *Mem:* Am Phys Soc; Sigma Xi. *Res:* Nuclear structure and reactions. *Mailing Add:* Dept Physics Fla State Univ Tallahassee FL 32306

TABOR, THEODORE EMMETT, b Great Falls, Mont, Dec 28, 40; m 59; c 3. ORGANIC CHEMISTRY. *Educ:* Univ Mont, BA, 62; Kans State Univ, PhD(org chem), 67. *Prof Exp:* Res chemist, Spec Assignment Prog, Dow Chem Co, 67, res chemist, Prod Dept Lab, 67-72; res specialist, Halogens Res Lab, 72-75, mgr acad educ, 75-78, res leader org chem dept, 78-81; MGR COOP RES, DOW CHEM CO, 81-, PROG MGR, DOW CHEM CO FOUND, 89- *Concurrent Pos:* Adj prof, Cent Mich Univ, Mt Pleasant, 79-83. *Mem:* Am Chem Soc; Soc Res Adminr. *Res:* Mechanisms of epoxide rearrangements; solvent processing of textiles; fire retardant chemicals for textiles and plastics; specialty organic chemicals product development. *Mailing Add:* Coop Res Dow Chem Co 1801 Bldg Midland MI 48674

TABORSKY, GEORGE, b Budapest, Hungary, Feb 12, 28; nat US; m 53; c 2. PROTEIN CHEMISTRY. *Educ:* Brown Univ, BS, 51; Yale Univ, PhD(biochem), 56. *Prof Exp:* Am Cancer Soc fel, Carlsberg Lab, Denmark, 56-57; from instr to assoc prof biochem, Yale Univ, 57-70, dir grad studies in biochem, 64-67; assoc prof biochem, 70-72, chmn dept, 73-78, PROF BIOCHEM, UNIV CALIF, SANTA BARBARA, 72- *Concurrent Pos:* Prin investr, NIH & NSF res grants, 63- *Mem:* Am Chem Soc; Biophys Soc; Am Soc Biochem & Molecular Biol; Am Inst Chemists; NY Acad Sci; Protein Soc. *Res:* Biochemistry of phosphoproteins; protein chemistry; biochemistry of egg yolk; phosphoprotein-metal interactions. *Mailing Add:* Dept Biol Sci Univ Calif Santa Barbara CA 93106

TABORSKY, GERALD J, JR, b Madison, Wis, Oct 26, 48; m; c 2. MEDICINE. *Educ:* Marquette Univ, BS, 71; Univ Southern Calif, PhD(biomed eng), 76. *Prof Exp:* Res assoc prof, 76-90, RES PROF, DEPT MED, UNIV WASH, 90- *Mem:* Am Diabetes Asn; Am Physiol Soc; Endocrine Soc. *Res:* Pancreatic neuropeptides; neurocontrol of pancreatic islets. *Mailing Add:* Div Endocrinol & Metab 151 Vet Admin Med Ctr 1660 S Columbian Way Seattle WA 98108

TACHIBANA, DORA K, b Cupertino, Calif, Dec 19, 34. IMMUNOLOGY, MEDICAL MICROBIOLOGY. *Educ:* San Jose State Col, BA, 56; Stanford Univ, MA, 62, PhD(immunol), 64. *Prof Exp:* Pub health microbiologist trainee, Calif State Dept Pub Health, 56-57; clin lab technician trainee, San Jose Hosp, Calif, 57; pub health microbiologist, Santa Clara County Pub Health, 57-59; NIH fel immunol, Czech Acad Sci, 64-65; lab instr & res asst, Stanford Univ, 65-66; asst prof microbiol & immunol, San Francisco State Col, 66-69; res assoc med, Sch Med, Stanford Univ, 70; assoc prof, 70-75, PROF IMMUNOL & MICROBIOL, CALIF COL PODIATRIC MED, 75- *Concurrent Pos:* Fac res award, San Francisco State Col, 68-69; clin lab dir, Calif Podiatry Hosp, San Francisco, 70-79; consult, Bur Med Devices-Immunol Div, Food & Drug Admin, 80-88. *Mem:* Am Soc Microbiol; Am Asn Immunol; Sigma Xi; Am Asn Univ Prof; AAAS. *Res:* Microbial flora of the foot. *Mailing Add:* Dept Basic Sci Calif Col Podiat Med 1770 Eddy St San Francisco CA 94132

TACHIBANA, HIDEO, b Los Altos, Calif, June 30, 25; m 64, 84; c 2. PLANT PATHOLOGY. *Educ:* Univ Calif, Davis, BS, 57; Wash State Univ, PhD(plant path), 63. *Prof Exp:* RES PLANT PATHOLOGIST, USDA, 63- *Mem:* Am Phytopath Soc; Int Soc Plant Path; Crop Sci Soc Am; AAAS; Am Soc Agr; Sigma Xi. *Res:* Soybean diseases; brown stem rot; breeding and screening for disease resistance; gene and pest management. *Mailing Add:* Plant Path Iowa State Univ 415 Bessey Ames IA 50011

TACHIBANA, TAKEHIKO, b Toyana, Toyana, Apr 29, 28; m 55; c 2. TUMOR IMMUNOLOGY, PSYCHONEUROIMMUNOLOGY. *Educ:* Kanazawa Univ, BM, 52, DMS, 58. *Prof Exp:* Res assoc embryol, 2nd Dept Path, Med Sch, Kanazawa Univ, 53-59, instr tumor immunol 59-63; asst prof tumor immunol, Dept Cancer Prev, Res Inst Infect Dis, Tokyo Univ, 63-64; sect head tumor immunol, Virol Div, Nat Cancer Ctr Res Inst, Tokyo, 64-74; PROF TUMOR IMMUNOL, RES INST TUBERCULOSIS & CANCER, TOKYO UNIV, SENDAI, 74- *Concurrent Pos:* Lectr, Fac Med, Tohoku Univ, 50- & Fac Pharmaceut, Kanazawa Univ, 78-81; res assoc immunochem, Dept Zool, Univ Mich, Ann Arbor, 57-59; vis investr tumor immunol, Dept Tumor Biol, Karolinska Inst, Stockholm, 68-69; lectr tumor immunol, Res Inst Med Sci, Tokyo Univ, 70-71, from assoc prof to prof, 71-76. *Mem:* Japan Cancer Asn; Japan Soc Allergol; Japan Immunol Asn; Japan Soc Cell Biol; Japan Soc Virol; Japan Soc Reticuloendothelial Syst. *Res:* Relationship between immunological responses in local tumor tissue and lymph-node metastasis of tumor cells; cellular and immunological properties of tumor cells metastasizing to the lymph node. *Mailing Add:* 3-21-206 Hachiman 1-Chome Aoba-Ku Sendai 980 Japan

TACHMINDJI, ALEXANDER JOHN, b Athens, Greece, Feb 16, 28; US citizen; m 65. HYDRODYNAMICS. *Educ:* Durham Univ, BSc, 49, BSc(hons), 50; Mass Inst Technol, SM, 51. *Prof Exp:* Eng, Swann, Hunter & W Richardson, 45-46; res assoc naval archit, Mass Inst Technol, 50-51; mem res staff, Ship Div, David W Taylor Model Basin, US Navy, 51-54, head, Res & Propeller Br, 54-59; head tactical warfare group, Weapons Syst Eval Div, Inst Defense Anal, 59-64, asst dir res & eng support div, 64-67, dep dir, Sci Technol Div, 67-69, dir, Systs Eval Div, 69-72; dir, Tactical Technol Off, Defense Advan Res Proj Agency, Dept Defense, 72-73, dep dir, Defense Advan Res Proj Agency, 73-75; chief scientist, Mitre Corp, 75-76, vpres, 76-84, gen mgr, 79-84, sr vpres, 84-88; RETIRED. *Concurrent Pos:* Consult, Am Bur Shipping, 56-59, Anti-Submarine Warfare Comt, Defense Sci Bd, 65-70 & Naval Surface Warfare Panel, 75; ed, J Defense Res, 69-; mem publ comt, Am Inst Aeronaut & Astronaut. *Mem:* Fel AAAS; Soc Naval Architects & Marine Engrs; assoc fel Am Inst Aeronaut & Astronaut; Opers Res Soc Am; fel Royal Inst Naval Architects. *Res:* Cavitation; super cavitation; potential theory; ship vibration and noise; systems analyses; hydroelasticity. *Mailing Add:* 5314 Falmouth Rd Bethesda MD 20816

TACHOVSKY, THOMAS GREGORY, b Los Angeles, Calif, Feb 1, 47; m 68; c 2. IMMUNOLOGY, VIROLOGY. *Educ:* Gonzaga Univ, BS, 68; Univ Rochester, PhD(microbiol), 74. *Prof Exp:* Lab technician microbiol, Sch Med, Univ Rochester, 71-74; fel immunol, 74-76, res asst, 76-77, asst prof immunol, Wistar Inst, 77-; MEM TECH STAFF, CAMBRIDGE RES LAB. *Mem:* Sigma Xi; Am Soc Microbiol; AAAS. *Res:* Neuroimmunology; multiple sclerosis; immunopathology of central nervous system diseases; neurovirology. *Mailing Add:* Surg Assocs Bethlehem Bethlehem PA 18015

TACK, PETER ISAAC, b Marion, NY, Apr 15, 11; m 37; c 2. FISH BIOLOGY. *Educ:* Cornell Univ, BS, 34, PhD(agr), 39. *Prof Exp:* Asst biologist, State Conserv Dept, NY, 39-40, biologist, 40; instr zool, 40-43, asst prof & res asst, 43-46, assoc prof & res assoc, 46-50, prof fisheries & wildlife, 50-70, chmn dept, 50-69, prof fisheries, wildlife & zool, 70-78, EMER PROF FISHERIES & WILDLIFE, MICH STATE UNIV, 77- *Concurrent Pos:* Mem, Governor's Botulism Control Comn, 64-65. *Mem:* Am Soc Limnol & Oceanog; Am Fisheries Soc; Am Soc Ichthyologists & Herpetologists; fel AAAS. *Res:* Pond fish culture; population fluctuations of whitefish in Northern Lake Michigan and effects of pumped storage generating facility on Lake Michigan ecology. *Mailing Add:* 314 Brittany Dr Lansing MI 48906-1613

TACKER, MARTHA MCCLELLAND, b Mineral Wells, Tex, Jan 16, 43; m 67; c 3. BIOCHEMISTRY, SCIENCE WRITING. *Educ:* Concordia Col, MN(Chem), 64; Baylor Col Med, PhD(biochem), 69. *Prof Exp:* Res assoc biochem, Baylor Col Med, 69-70; res asst gastroenterol, Mayo Grad Sch Med, Univ Minn & Mayo Found, 70-71; instr physiol, Baylor Col Med, 71-74; BIOMED COMMUN CONSULT, 74- *Concurrent Pos:* Sci writer and ed, 74- *Mem:* AAAS; Sigma Xi; fel Am Med Writers Asn; Coun Biol Ed; Soc Tech Commun; Europ Asn Sci Ed. *Mailing Add:* 300 Forest Hill Dr West Lafayette IN 47906

TACKER, WILLIS ARNOLD, JR, b Tyler, Tex, May 24, 42; m 67; c 3. CARDIOVASCULAR PHYSIOLOGY, MEDICAL EDUCATION. *Educ:* Baylor Univ, BS, 64, MD & PhD(physiol), 70. *Prof Exp:* Intern med, Mayo Clin, 70-71; mem fac physiol, Baylor Col Med, 71-74; MEM FAC BIOMED ENG, PURDUE UNIV, WEST LAFAYETTE, 74- *Mem:* Am Physiol Soc; Asn Advan Med Instrumentation. *Res:* Life-threatening arrhythmia therapy; new teaching techniques and devices; diagnostic and therapeutic devices development. *Mailing Add:* Biomed Eng Ctr Purdue Univ West Lafayette IN 47906

TACKETT, JAMES EDWIN, JR, b Los Angeles, Calif, Oct 8, 37; m 62; c 3. ANALYTICAL CHEMISTRY. *Educ:* Occidental Col, BA, 60; Univ Calif, Riverside, PhD(chem), 64. *Prof Exp:* Chemist, Union Carbide Corp, WVa, 64-65; sr res chemist, 65-77, mgr, Analysis Dept, 77-83, SR RES CHEMIST, DENVER RES CTR, MARATHON OIL CO, 83- *Mem:* Am Chem Soc; Sigma Xi. *Res:* Applied spectroscopy; thermal analysis; analytical separations; application of nuclear magnetic resonance and Fourier transform infrared to the study of aqueous polymer systems. *Mailing Add:* Denver Res Ctr Marathon Oil Co PO Box 269 Littleton CO 80160-0269

TACKETT, JESSE LEE, b Dublin, Tex, Sept 20, 35; m 55; c 3. SOIL PHYSICS. *Educ:* Tex A&M Univ, BS, 57; Auburn Univ, MS, 61, PhD(soil physics), 63. *Prof Exp:* Instr agron, Tarleton State Col, 57-59; asst, Tex A&M Univ, 59-60; res soil scientist, Soil & Water Conserv Res Div, Agr Res Serv, USDA, 63-65; assoc prof agr, 65-70, PROF AGR & DEAN SCH AGR & BUS, TARLETON STATE UNIV, 70- *Mem:* Am Soc Agron; Soil Sci Soc Am. *Res:* Soil aeration and strength; plant growth relations. *Mailing Add:* Dean Agr & Bus Tarleton State Univ Tarleton Sta Stephenville TX 76402

TACKETT, STANFORD L, b Virgie, Ky, Sept 5, 30; m 51; c 4. ANALYTICAL CHEMISTRY. *Educ:* Ohio State Univ, BS, 57, PhD(chem), 62. *Prof Exp:* Instr chem, Ohio State Univ, 61-62; asst prof, Ariz State Univ, 62-66; assoc prof, 66-69, chmn dept, 74-83, PROF CHEM, INDIANA UNIV PA, 69- *Concurrent Pos:* Consult, Off Res Analysis, Holloman AFB, 63; NIH res grant, 64-66. *Mem:* Am Chem Soc; Meteoritical Soc. *Res:* Analytical chemistry and electro-analytical techniques; chemistry of meteorites; chemistry of cyanocobalamin; environmental chemistry. *Mailing Add:* Weyandt 143 Indiana Univ Pa Indiana PA 15705

TADE, WILLIAM HOWARD, pathology, physiology; deceased, see previous edition for last biography

TADEPALLI, ANJANEYULU S, CARDIOVASCULAR PHARMACOLOGY, PHYSIOLOGY. *Educ:* Univ Pittsburgh, PhD(pharmacol), 72. *Prof Exp:* SR RES SCIENTIST, BURROUGHS WELLCOME RES LABS, 80- *Mem:* Am Soc Pharmacol; Soc Neurosci; Int Soc Hypertension. *Mailing Add:* Dept Pharmacol Burroughs Wellcome Co Res Labs 3030 Cornwallis Rd Research Triangle Park NC 22709

TADJERAN, HAMID, b Teheran, Nov 30, 52. NUMERICAL ANALYSIS. *Educ:* Am Univ Beirut, BS, 74; Univ Colo Boulder, MS, 78, PhD(appl math), 80. *Prof Exp:* NSF res asst, Univ Colo, 78-80, US Geol Surv res assoc, 80; lectr math & computer sci, Univ Calif, Davis, 80-81; asst prof computer sci & math,

Univ NC, Wilmington, 81-; DEPT MATH SCI, UNIV TULSA. *Mem:* Soc Indust & Appl Math; Am Math Soc. *Res:* Numerical solution of differential equations; step size control. *Mailing Add:* 18920 Barnhart Ave Cupertino CA 95014

TAEGTMEYER, HEINRICH, b Forst, Ger, Feb 14, 41; m 66; c 3. CARDIOLOGY, MUSCLE PHYSIOLOGY. *Educ:* Univ Freiburg, MD, 68, Univ Oxford, PhD(biochem), 81. *Prof Exp:* Clin fel med, Harvard Med Sch, 71-73, res fel, 73-76, instr, 76-79; asst prof, 81-87, ASSOC PROF MED, UNIV TEX MED SCH, HOUSTON, 87- *Honors & Awards:* LB Johnson Award, Am Heart Asn, 83. *Mem:* Am Heart Asn; Am Col Cardiol; Am Col Physicians; Am Fedn Clin Res; Int Soc Heart Res. *Res:* Metabolic regulation and control of energy metabolism in heart and skeletal muscle. *Mailing Add:* Div Cardiol Univ Tex Med Sch Houston 6431 Fannin Houston TX 77030

TAEUSCH, H WILLIAM, PULMONARY DEVELOPMENT, PULMONARY SURFACTANT. *Educ:* West Reserve Med Sch, MD, 65. *Prof Exp:* From assoc prof to prof pediat, Sch Med, Harvard Univ, 74-87; scientist, Dept Neonatology, Brigham Women's Hosp, 80-87; DIR, DIV NEONATOLOGY, KING/DREW MED CTR, 87- *Concurrent Pos:* Vis scientist, Cardiovasc Res Inst, Univ Calif, San Francisco, 82-83. *Res:* Co-author of one book. *Mailing Add:* King/Drew Med Ctr 12021 S Wilmington Ave Los Angeles CA 90059

TAFF, LAURENCE GORDON, astrometry, celestial mechanics, for more information see previous edition

TAFFE, WILLIAM JOHN, b Albany, NY, Feb 3, 43; m 65; c 2. COMPUTER SCIENCE, EDUCATION. *Educ:* Le Moyne Col, NY, BS, 64; Univ Chicago, SM, 67, PhD(geophys), 68. *Prof Exp:* Res physicist, Air Force Cambridge Res Labs, 68-69; asst prof physics, Colby Col, 69-71; from asst prof to prof natural sci, 71-81, PROF COMPUTER SCI & CHMN DEPT, PLYMOUTH STATE COL, 81- *Concurrent Pos:* Vis prof, Univ NH, Durham, 79-; mem, NH Post Secondary Educ Accreditation Teams, 82-88. *Mem:* Asn Comput Mach; Am Asn Univ Prof. *Res:* Computer science education. *Mailing Add:* Dept Computer Sci Plymouth State Col Plymouth NH 03264

TAFLOVE, ALLEN, b Chicago, Ill, June 14, 49; m 77; c 2. ELECTROMAGNETIC FIELDS & WAVES, NUMERICAL MODELING. *Educ:* Northwestern Univ, BS, 71, MS, 72, PhD(elec eng), 75. *Prof Exp:* Assoc engr, IIT Res Inst, Chicago, 75-77, res engr, 77-80, sr engr & group leader electronics, 80-84; assoc prof, 84-88, PROF ELEC ENG, NORTHWESTERN UNIV, 88- *Concurrent Pos:* Prin investr, US Air Force Rome Air Develop Ctr, 77-81 & 82-84, Global Anal, 81-82, Elec Power Res Inst, 81-83 & 84, Sci Appln Inc, 82-84, Lawrence Livermore Lab, 85-87, NASA, 85-87 & 90, NSF, 88-, Lockheed Corp, 85-87, Off Naval Res, 88-, Gen Dynamics, 88- & Northrop, 90-; consult, Lawrence Livermore Nat Lab, 85-87, Lockheed Corp, 85-87, Naval Res Lab, 87-, MRJ, Inc, 87-90; guest ed, Wave Motion, 88; nat lectr, Antennas & Propagation Soc, Inst Elec & Electronics Engrs, 90- *Mem:* Sigma Xi; AAAS; fel Inst Elec & Electronics Engrs; NY Acad Sci; Int Union Radio Sci; Electromagnetics Acad. *Res:* Development and application of supercomputing computational electromagnetics models; numerical modeling of electromagnetic wave interactions with complex structures; first-principles computational modeling of ultrahigh-speed electronic and optical switching directly from Maxwell's equations. *Mailing Add:* Dept Elec Eng & Computer Sci McCormick Sch Eng Northwestern Univ Evanston IL 60208-3118

TAFT, BRUCE A, b San Francisco, Calif, Jan 29, 30; m 61; c 1. PHYSICAL OCEANOGRAPHY. *Educ:* Stanford Univ, BS, 51; Univ Calif, San Diego, MS, 61, PhD(oceanog), 65. *Prof Exp:* Statistician, US Fish & Wildlife Serv, Calif, 55-58; grad res oceanogr, Scripps Inst, Univ Calif, 58-59; res asst, Johns Hopkins Univ, 59-60; grad res oceanogr, Scripps Inst Oceanog, Univ Calif, 60-65, asst res oceanogr, 65-68, asst prof oceanog, 68-73; from res assoc prof to res prof oceanog, Univ Wash, 73-83; supvry oceanogr, Pac Marine Environ Lab, Nat Oceanic & Atmospheric Admin, 81-89, WOCE INT PROJ OFF, IOS DEACON LAB, UK, 89- *Concurrent Pos:* NSF grant, US-Japan Coop Sci Prog, 66-67; affil prof oceanog, Univ Wash, 83-; chmn, CCCO Trop Pac Panel, sr fel, Joint Inst Study Atmosphere & Ocean, Wash & Joint Inst Marine Atmospheric Res, Hawaii; assoc ed, J Marine Res, 75-90. *Mem:* AAAS; Am Geophys Union; Oceanog Soc Japan. *Res:* Description of large scale oceanic circulation; velocity structure and distribution of properties in ocean currents. *Mailing Add:* Pac Marine Environ Lab/NOAA 7600 Sand Point Way NE Seattle WA 98115

TAFT, CHARLES KIRKLAND, b Cleveland, Ohio, July 24, 28; m 51; c 3. CONTROL ENGINEERING. *Educ:* Amherst Col, BA, 51; Mass Inst Technol, BS, 53; Case Inst Technol, MS, 56, PhD(feedback control systs), 60. *Prof Exp:* Spec apprentice to res engr, Warner & Swasey, 53-58, chief servo engr, 60-61; from asst prof to assoc prof eng, Case Inst Technol, 61-67; PROF MECH ENG, UNIV NH, 67-, CHMN, DEPT MECH ENG, 85- *Honors & Awards:* Charles Strosacker Award, Case Inst Technol, 66; Achievement Award, Nat Fluid Power Asn, 66; Rail Transp Award, Am Soc Mech Engrs, 80. *Mem:* Am Soc Mech Engrs; Inst Elec & Electronics Engrs. *Res:* Electromechanical systems; control system synthesis; digital and discontinuous control and fluidic systems. *Mailing Add:* Dept Mech Eng Univ NH Durham NH 03824

TAFT, DAVID DAKIN, b Cleveland, Ohio, Mar 27, 38; m 61; c 3. POLYMER CHEMISTRY, TECHNICAL MANAGEMENT. *Educ:* Kenyon Col, AB, 60; Mich State Univ, PhD(org chem), 63. *Prof Exp:* Sr res chemist, Archer Daniels Midland Co, 64-67; group leader, Resins, Ashland Chem Co, 67-70, mgr polymer chem, Ashland Oil, Inc, 70-72; vpres & gen mgr, Oxyplast, Inc, 74; asst to pres, 72-74, dir res & develop, 74-76, from vpres to exec vpres, Consumer & Spec Chem, Henkel Corp, 77-82; Group Mgr, Telecom Group, 83-86, VPRES MFG, RAYCHEM CORP, 86- *Concurrent Pos:* dir com develop, Gen Mills Chem Inc, 72-73. *Mem:* Am Chem Soc; Indust Res Inst; Fedn Socs Paint Technol; Com Develop Asn; Asn Mfg Excellence. *Res:* Acrylic, polyester, epoxy, urethane coating and adhesive systems; water soluble coating and adhesive polymers; functional monomers; powder coatings; non-yellowing isocyanates; nylon polymers; hydrophilic polymers; surface active agents; specialty chemicals for cosmetics; consumer adhesive products; compounding of polymers; 24 US patents. *Mailing Add:* 45 Melanie Lane Atherton CA 94027

TAFT, EARL J, b New York, NY, Aug 27, 31; m 59; c 2. ALGEBRA. *Educ:* Amherst Col, BA, 52; Yale Univ, MA, 53, PhD(math), 56. *Prof Exp:* Instr math, Columbia Univ, 56-59; from asst prof to assoc prof, 59-66, PROF MATH, RUTGERS UNIV, 66- *Concurrent Pos:* NSF res grants, 63-67, 70-71, 73, 76 & 87; exec ed, Commun in Algebra, 74-, Exec Ed Monographs & Lectr Notes Math, Marcel Dekker Inc, New York. *Mem:* Am Math Soc; Math Asn Am. *Res:* Nonassociative algebras; Hopf algebras; rings; groups. *Mailing Add:* Dept Math Rutgers Univ New Brunswick NJ 08903

TAFT, EDGAR BRECK, b New Haven, Conn, Nov 16, 16; m 43. PATHOLOGY. *Educ:* Yale Univ, MD, 42; Univ Kans, MS, 50. *Prof Exp:* Nat Res Coun fel, Univ Kans, 48-49; Runyon clin res fel, Univ Kans & Univ Stockholm, 49-50; from asst pathologist to assoc pathologist, Mass Gen Hosp, 52-60, pathologist, 61-88; assoc clin prof path, Harvard Med Sch, 63-88; RETIRED. *Mem:* AAAS; Histochem Soc; Soc Develop Biol; AMA; Am Asn Pathologists; Sigma Xi. *Res:* Liver disease; methods of histochemistry and cytochemistry. *Mailing Add:* PO Box 635 Stockbridge MA 01262

TAFT, JAY LESLIE, b Rockville Centre, NY, Mar 19, 44. BIOLOGICAL OCEANOGRAPHY. *Educ:* Lafayette Col, BA, 67; Johns Hopkins Univ, MA, 73, PhD(biol oceanog), 74. *Prof Exp:* Res assoc, 74, ASSOC RES SCIENTIST BIOL OCEANOG, CHESAPEAKE BAY INST, 74-; DIR ADMIN, DEPT ORGANISMIC & EVOLUTIONARY BIOL, HARVARD UNIV. *Mem:* AAAS; Am Soc Limnol & Oceanog. *Res:* Production and utilization of dissolved organic matter in estuaries and coastal ocean; nutrient cycling in estuaries. *Mailing Add:* 26 Oxford St Harvard Univ Cambridge MA 02138

TAFT, KINGSLEY ARTER, JR, b Cleveland, Ohio, Nov 17, 30; m 55; c 3. FORESTRY, GENETICS. *Educ:* Amherst Col, AB, 53; Univ Mich, BS, 57; NC State Univ, MS, 62, PhD(forestry, genetics), 66. *Prof Exp:* Forester, Nebo Oil Co, La, 57-59; res asst forestry, NC State, 59-63; forest geneticist, Div Forestry, Tenn Valley Authority, 63- 74, chief, Forest & Wildlife Resources Br, Fisheries & Wildlife Develop, 74-79, spec projs coordr & asst to dir, Div Land & Forest Resources, 79-88; RETIRED. *Mem:* Soc Am Foresters. *Res:* Hardwood tree improvement and genetics. *Mailing Add:* 325 Trossachs Ln NW Knoxville TN 37922

TAFT, ROBERT WHEATON, JR, b Lawrence, Kans, Dec 17, 22; m 44; c 3. PHYSICAL CHEMISTRY, ORGANIC CHEMISTRY. *Educ:* Univ Kans, BS, 44, MS, 46; Ohio State Univ, PhD(chem), 49. *Prof Exp:* Lab asst chem, Univ Kans, 44-46; asst, Ohio State Univ, 46-49; res assoc, Columbia Univ, 49-50; from asst prof to prof, Pa State Univ, 50-65; PROF CHEM, UNIV CALIF, IRVINE, 65- *Concurrent Pos:* Sloan fel, 55-57; Guggenheim fel, Harvard Univ, 58; consult, Sun Oil Co, 58- *Mem:* Am Chem Soc. *Res:* Kinetics; effect of molecular structure on reactivity; mechanisms of organic reactions; rate and equilibrium studies; fluorine nuclear magnetic resonance shielding. *Mailing Add:* Dept Chem Univ Calif Irvine CA 92717

TAFURI, JOHN FRANCIS, b St Barbara, Italy, Aug 4, 24; nat US; m 58; c 3. ENTOMOLOGY. *Educ:* Fordham Univ, BS, 44, MS, 48, PhD, 51. *Prof Exp:* Lab instr comp anat, Fordham Univ, 47-48, lab instr entom, 48, instr comp anat, Sch Educ, 50-51; from instr to assoc prof biol, 51-68, PROF BIOL, XAVIER UNIV, OHIO, 68- *Concurrent Pos:* Asst, Dept Animal Behav, Am Mus Natural Hist, NY, 48-51. *Mem:* AAAS; Sigma Xi. *Res:* Electrical changes in tissues; social behavior in insects; aging in cells and tissues. *Mailing Add:* Albers Biol Labs 3800 Victory Pkwy Cincinnati OH 45207

TAG, PAUL MARK, b Meyersdale, Pa, Sept 8, 45; m 84. METEOROLOGY, ATMOSPHERIC NUMERICAL MODELING. *Educ:* Pa State Univ, BS, 66, MS, 68, PhD(meteorol), 77. *Prof Exp:* Res meteorologist, Navy Weather Res Facil, 68-71; res meteorologist, Naval Environ Prediction Res Facil, 71-89; CONSULT METEOROLOGIST, 89- *Concurrent Pos:* Weather forecaster, Air Nat Guard, 70-76. *Mem:* Am Meteorol Soc; Sigma Xi. *Res:* Numerical weather prediction; cloud physics; atmospheric turbulence; boundary layer; atmospheric numerical modeling; artificial intelligence; satellite imagery interpretation. *Mailing Add:* Naval Oceanog & Atmospheric Res Lab Monterey CA 93943-5006

TAGER, IRA BRUCE, US citizen. CLINICAL EPIDEMIOLOGY, INFECTIOUS DISEASES. *Educ:* Colgate Univ, AB, 65; Univ Rochester, MD, 69; Harvard Med Sch, MPH, 73. *Prof Exp:* Res fel infectious dis, Harvard Med Sch, 72-73, instr med, 73-76, asst prof, 76-; CHIEF INFECTIOUS DIS, VET ADMIN MED CTR, SAN FRANCISCO. *Concurrent Pos:* Edward Elliot Trudeau fel, Am Lung Asn, 77-81; consult, Dept Health, Div Epidemiol, State of RI, 78- *Mem:* Am Thoracic Soc; Am Soc Microbiol; fel Infectious Dis Soc Am; fel Am Col Epidemiol. *Res:* Epidemiology chronic lung disease and identification of determinants; determinants interventions for hospital acquired infections. *Mailing Add:* Chief Infectious Dis Med Ctr 4150 Clement St San Francisco CA 94121

TAGGART, G(EORGE) BRUCE, b Philadelphia, Pa, Apr 8, 42; div. CONDENSED MATTER THEORY, TECHNICAL MANAGEMENT. *Educ:* Col William & Mary, BS, 64; Temple Univ, PhD(physics), 71. *Prof Exp:* Instr physics, Drexel Univ, 70; vis asst prof, Temple Univ, 70-71; asst prof, Va Commonwealth Univ, 71-76, assoc prof physics, 76-82, prof, 82-83; mgr Mat Sci Technol, Advan Technol Div, BDM Corp, 83-90; PROG DIR, MAT THEORY, DIV MAT RES, NSF, 89- *Concurrent Pos:* Consult, Temple Univ, 74, res assoc, 75; res assoc, Oak Ridge Nat Lab, 74; vis prof, Dept Theoret Physics, Oxford Univ, 78; vis assoc prof physics, Univ Ill, Urbana-Champaign, 78-79; guestworker, Thermophysics Div, Nat Bur Standards, Gaithersburg,

Md, 78-88; vis prof physics, Fed Univ Pernambuco, Brazil, 80. *Mem:* AAAS; Am Phys Soc; Fedn Am Scientists; Mat Res Soc. *Res:* Materials theory. *Mailing Add:* Div Materials Research NSF, 1800 G St NW Rm 408 Washington DC 20550

TAGGART, JOHN VICTOR, b Brigham, Utah, Aug 29, 16; m 59. PHYSIOLOGY, MEDICINE. *Educ:* Univ Southern Calif, MD, 41. *Prof Exp:* Intern, Los Angeles County Hosp, Calif, 40-41; asst resident med, Clins, Univ Chicago, 41-42; res resident, Goldwater Mem Hosp, New York, 42-43; from instr to prof med, Col Physicians & Surgeons, Columbia Univ, 46-82, Dalton prof physiol & chmn dept, 62-82; MEM, RUSSELL SAGE INST PATH, 55- *Concurrent Pos:* Nat Res Coun Welch fel internal med, 46-52; consult, Comn on Growth, Nat Res Coun, 53-54; Nathanson mem lectr, Univ Southern Calif, 55; consult, USPHS, 55-60; mem bd dirs, Russell Sage Inst Path, 55-, pres, 59-72; career investr, Am Heart Asn, 58-62; mem exec comt, Health Res Coun, City of New York, 63-70, vchmn, 67-70; mem res career award comt, NIH, 66-69, chmn, 67-69; mem bd sci counselors, Nat Heart & Lung Inst, 69-72. *Mem:* Am Physiol Soc; Am Soc Clin Invest; Harvey Soc (pres, 69); Asn Am Physicians. *Res:* Metabolic aspects of renal transport mechanisms; biochemistry; enzymology. *Mailing Add:* Greenwich Rd Bedford NY 10506

TAGGART, KEITH ANTHONY, b Cleveland, Ohio, July 10, 44; m 68; c 3. PHYSICS, COMPUTATIONAL PHYSICS. *Educ:* Case Inst Technol, BS, 66, MS, 68; Case Western Reserve Univ, PhD(physics), 70. *Prof Exp:* Staff scientist plasma physics, Air Force Weapons Lab, 70-72; fel, Plasma Physics Lab, Princeton Univ, 73; staff mem, Laser Fusion, Los Almos Nat Lab, 73-79, staff mem, Cumputational Physics, 79-83, prog mgr, Strategic Defense Res, 83-85; dir countermeasures, Strategic Defense Initiative Orgn, 85-88; DIR, STRATEGIC DEFENSE TECHNOL, SCI APPLN INT CORP, 88- *Mem:* Am Phys Soc. *Res:* Laser fusion target simulation on digital computers; computational simulation of complex systems; computational fluid dynamics. *Mailing Add:* 4004 Patricia St Annandale VA 22003

TAGGART, R(OBERT) THOMAS, b Long Beach, Calif, July 14, 51. HUMAN GENETICS, MEDICAL GENETICS. *Educ:* Univ Denver, BS, 73; Ind Univ, PhD(med genetics & molecular biol), 78. *Prof Exp:* Fel human genetics, Sch Med, Yale Univ, 78-80; asst prof med, Univ Calif, Los Angeles, 80-87; chief, human genetics, Vet Admin Med Ctr, Sepulveda, Calif, 81-87; ASSOC PROF, DEPT MOLECULAR BIOL & GENETICS, WAYNE STATE UNIV SCH MED, 87- *Concurrent Pos:* Investr, Ctr Ulcer Res & Educ, Univ Calif, Los Angeles, 81-; mem, Jonsson Comprehensive Cancer Ctr, 81- & Inflammatory Bowel Dis Ctr, 84- *Mem:* AAAS; Am Soc Biol Chemists; Am Soc Human Genetics; Genetics Soc Am; Sigma Xi. *Res:* Molecular genetics analysis of gastrointestinal proteases, growth factors and hormones; chromosomal localization and isolation of the genes encoding specific proteins. *Mailing Add:* Dept Molecular Biol & Genetics Wayne State Sch Med 3216 Scott Hall 540 E Canfield Detroit MI 48201

TAGGART, RAYMOND, b Bradford, Eng, July 9, 22; m 49. MECHANICAL ENGINEERING. *Educ:* Univ London, BSc, 48; Queen's Univ, Belfast, PhD, 56. *Prof Exp:* Eng draftsman, Harland & Wolffe, Ltd, N Ireland, 38-43; instr mech eng, Col Technol, Belfast, 43-48, sr lectr, 48-56; lectr, Queen's Univ, Belfast, 56-57; Nat Res Coun Can fel metall, Univ Alta, 57-58, res assoc, 58-59; from asst prof to assoc prof, 59-68, PROF MECH ENG, UNIV WASH, 68- *Concurrent Pos:* Blair fel, London County Coun, London, 57-58; consult, failure anal, accident reconstruction, aircraft, helicopters, fracture & fatigue anal, design anal (mechanical), mat anal & failure. *Mem:* Fel Am Soc Metals; Brit Inst Mech Engrs; Sigma Xi. *Res:* Fatigue of metals with reference to crack propagation; effects of microstructural changes and precipitate morphology on properties of materials, especially crack propagation; relation of constitution to mechanical properties in binary alloys; superconducting properties of zirconium and titanium base alloys. *Mailing Add:* Dept Mech Eng FU10 Univ Wash Seattle WA 98195

TAHA, HAMDY ABDELAZIZ, b Egypt, Apr 19, 37; US citizen; m 65; c 3. OPERATIONS RESEARCH. *Educ:* Univ Alexandria, BS, 58; Stanford Univ, MS, 62; Ariz State Univ, PhD(indust eng), 64. *Prof Exp:* Instr elec eng, Univ Assiut, 58-59; mgr planning, Suez Oil Co, Egypt, 64-67; asst prof indust eng, Univ Okla, 67-69; assoc prof, 69-75, PROF INDUST ENG, UNIV ARK, FAYETTEVILLE, 75- *Concurrent Pos:* Lectr, Cairo Univ, 64-67; consult, Tenneco Oil Co, Okla, 67, Sun Oil Co, Tex, 70, Ford Found, 72, Petromin, Saudi Arabia, 75 & Hylsa, SA, Puebla, Mex, 77-78; vis prof, Univ Americas, Puebla, Mex, 77-78. *Mem:* Am Inst Indust Engrs; Inst Mgt Sci; Opers Res Soc Am. *Res:* Mathematical programming with emphasis on integer programming. *Mailing Add:* Dept Indust Eng Univ Ark Main Campus Fayetteville AR 72701

TAHAN, THEODORE WAHBA, b Alexandria, Egypt, Jan 22, 36; m 62; c 1. RADIATION ONCOLOGY. *Educ:* Alexandria Univ, MBBCh, 61, DMR&E, 64, MD & PhD(radiother), 69. *Prof Exp:* Resident radiother, Alexandria Univ Hosps, 62-64, clin demonstr, Fac Med, 64-69, lectr, 69-71; from asst prof to assoc prof radiation oncol, 72-85, chmn fac med, 79-85, CHMN DEPT NUCLEAR MED & RADIOBIOL, SHERBROOKE UNIV, 85- *Concurrent Pos:* Consult radiother, Hotel Dieu Hosp, St Vincent Hosp & Sherbrooke Hosp, Sherbrooke, 73; coordr postgrad med teaching, Fac Med, Sherbrooke Univ, 84- *Mem:* Can Radiol Asn; Can Oncol Soc; Egyptian Cancer Soc; Egyptian Radiol Soc; Radiol Soc NAm. *Res:* Elemental diet in irradiated patients; bronchial carcinoma, carcinoma of the bilharzial bladder, tongue and nasopharynx; misonidasole, carcinoma of bladder and protate. *Mailing Add:* Dept Nuclear Med Univ Sherbrooke Sherbrooke PQ J1H 5N4 Can

TAHILIANI, VASU H, b Baroda, India, Oct 26, 42; US citizen; m 70; c 3. INSULATION COORDINATION, DIELECTRICS. *Educ:* MS Univ Baroda, BE, 64; WVa Univ, MSEE, 70. *Prof Exp:* Dep engr, Gujarat Elec Bd, India, 64-65; design engr, Jyoti Elec Ltd, India, 65; engr, Power Syst Div, McGraw Edison, 66-71; sr engr, ITE Imp Corp, 71-74, proj mgr, 74-76; proj mgr, 77-84, prog mgr, 84-90, SR PROG MGR, ELEC POWER RES INST, 91- *Concurrent Pos:* Prin investr, High Temp Gas Insulated Cables, 74-76, 765 Kv Gas Insulated Substas, 75-77; proj mgr & chmn, ad hoc comt Gas Insulated Substas Technol, 78-80. *Mem:* Sr mem Inst Elec & Electronics Engrs. *Res:* Gas insulated substations technology; gas insulated cables; insulation coordination for electrical power systems; metal oxide varistor technology; distribution engineering; cable materials technology. *Mailing Add:* 3412 Hillview Ave Palo Alto CA 94304

TAHIR-KHELI, RAZA ALI, b Hazara, West Pakistan, May 1, 36; m 62; c 1. THEORETICAL MAGNETISM. *Educ:* Oxford Univ, BA, 58, DPhil(physics), 62. *Prof Exp:* Res assoc physics, Univ Pa, 62-64; sr scientific officer, Pakistan Atomic Energy Comn, 64-66; from asst prof to assoc prof, 66-71, PROF PHYSICS, TEMPLE UNIV, 71- *Concurrent Pos:* Assoc, Exp Sta, E I Du Pont de Nemours & Co, Del, Inc, 67-70. *Mem:* Am Phys Soc; Sigma Xi. *Res:* Solid state physics; many body physics; magnetism; order-disorder phenomena; random magnetism; atomic diffusion; non-equilibrium thermodynamics. *Mailing Add:* Dept Physics Temple Univ Col Lib Arts Philadelphia PA 19122

TAI, CHEN-TO, b Soochow, China, Dec 30, 15; US citizen; m 41; c 5. ELECTRICAL ENGINEERING. *Educ:* Tsinghua Univ, China, BS, 37; Harvard Univ, DSc(commun), 47. *Prof Exp:* Res fel elec eng, Harvard Univ, 47-49; sr res engr, Stanford Res Inst, 49-54; assoc prof elec eng, Ohio State Univ, 54-56; prof electronics, Tech Inst Aeronaut, Brazil, 56-60; prof elec eng, Ohio State Univ, 60-64; PROF ELEC ENG, UNIV MICH, ANN ARBOR, 64- *Concurrent Pos:* Mem, Comn B, Int Union Radio Sci, 62- *Honors & Awards:* Centennial Award, Inst Elec & Electronics Engrs. *Mem:* Nat Acad Engrs; fel Inst Elec & Electronics Engrs. *Res:* Electromagnetic and antenna theories. *Mailing Add:* Dept EECS Univ Mich Ann Arbor MI 48109

TAI, DOUGLAS L, b Hong Kong, Nov 6, 40; m 70; c 2. RADIOLOGICAL PHYSICS. *Educ:* Chinese Univ Hong Kong, BSc, 64; Cornell Univ, PhD(chem), 69. *Prof Exp:* Res assoc phys chem, Cornell Univ, 69-71; fac res assoc solid state chem, Ariz State Univ, 71-72, asst prof, 72-73; instr, Dept Chem, Univ Ky, 73-74 & Dept Physics & Astron, 75-77; instr & radiol physicist, Dept Radiol, Univ Miss Med Ctr, 77-79, asst prof & radiol physicist, 79-81; ASST PROF & RADIOL PHYSICIST, DEPT RADIOL & ONCOL, UNIV TENN, MEMPHIS, 81- *Concurrent Pos:* Res fel, Dept Radiation Med, Albert B. Chandler Med Ctr, Univ Ky, 75- *Mem:* Am Asn Physicians Med; Health Physics Soc; Soc Photo-Optical Instrumentation Engrs; Am Col Med Physicians. *Res:* Teletherapy and intracavitary dosimetry. *Mailing Add:* Dept Radiation & Oncol Univ Tenn Ctr Health Sci Memphis TN 38163

TAI, HAN, b Yang Chow, China, Mar 20, 24; m 61; c 2. ANALYTICAL CHEMISTRY. *Educ:* Nanking Univ, BS, 49; Emory Univ, MS, 55, PhD(chem), 58. *Prof Exp:* Asst geol, Nat Taiwan Univ, Formosa, 51-54; res chemist, A E Staley Mfg Co, 58-66, head instrumental analytical lab, 66-67, group leader analytical labs, 67-70; supvr pesticides monitoring lab, 70-80, CHEMIST, ENVIRON CHEM LAB, ENVIRON PROTECTION AGENCY, 80- *Mem:* Am Chem Soc; Soc Appl Spectros. *Res:* Chemical analysis of rocks and minerals, carbohydrates, polymers, pesticides; infrared spectroscopy; chromatography; environmental monitoring on inorganic and organic pollutants and pesticide residues. *Mailing Add:* Environ Chem Lab Environ Protection Agency Bay St Louis MS 39529

TAI, JULIA CHOW, b Shanghai, China, 1935; US citizen; m 60; c 3. CHEMISTRY. *Educ:* Nat Taiwan Univ, BS, 57; Univ Okla, MS, 59; Univ Ill, Urbana, PhD(chem), 63. *Prof Exp:* Fels, Wayne State Univ, 63-68; from asst prof to assoc prof, 69-79, PROF CHEM, UNIV MICH, DEARBORN, 79- *Concurrent Pos:* vis assoc prof, Nat Taiwan Univ, Taipei, Taiwan, 68-69. *Mem:* Am Chem Soc; Quantum Chem Prog Exchange. *Res:* Quantum mechanical calculations of electronic spectra of molecules; structure of molecules. *Mailing Add:* Dept Natural Sci Univ Mich-Dearborn Dearborn MI 48128

TAI, PETER YAO-PO, b Chutung, Taiwan, July 6, 37; US citizen; m 64; c 2. PLANT BREEDING. *Educ:* Nat Taiwan Univ, BS, 61; Tex A&M Univ, MS, 66; Okla State Univ, PhD(crop sci), 72. *Prof Exp:* Res assoc, 72-75, instr plant breeding, Agr Exp Sta, Univ Ga, 75-77; RES GENETICIST PLANT, SCI & EDUC ADMIN-AGR RES, USDA, 77- *Mem:* Am Soc Agron; Crop Sci Soc Am; Am Soc Sugar Cane Technol; AAAS. *Res:* Development of new breeding lines of sugarcane (Saccharum sp), improvement of cold tolerance in sugarcane and selection methodology. *Mailing Add:* Sugarcane Field Station Star Rte Box 8 USDA-RAF Canal Point FL 33438

TAI, TSZE CHENG, b Shaoxiang, China, Apr 29, 33; US citizen; m 65; c 3. FLUID MECHANICS, COMPUTATIONAL AERODYNAMICS. *Educ:* Chung Cheng Inst of Technol, Taiwan, dipl, 57; Clemson Univ, MS, 65; Va Polytech Inst, PhD(aerospace eng), 69. *Prof Exp:* Aircraft eng, Taoyuan Airbase, Taiwan, 58-63; res asst fluid mechanics, Clemson Univ, 63-65; teaching asst aerodyn, Va Polytech Inst, 65-67, instr aerospace eng, 67-68; res scientist, 68-84, SR RES SCIENTIST, FLUID MECHANICS, DAVID TAYLOR RES CTR, 84- *Concurrent Pos:* Chmn, Air Inlet & Diffusers Panel, Navy Aeroballistics Comt, 79-81; invited lectr, von Karman Inst Fluid Dynamics, Belg, 80; sci officer, Off Naval Res, 85-86. *Honors & Awards:* E N Brooks Award, David Taylor Res Ctr, 79. *Mem:* Assoc fel Am Inst Aeronaut & Astronaut; Sigma Xi. *Res:* Transonic aerodynamics; computational fluid dynamics; numerical optimization; three-dimensional flow separation. *Mailing Add:* David Taylor Res Ctr Code 1205 Bethesda MD 20084-5000

TAI, WILLIAM, b Yangzhow, China, Mar 9, 34; m 63; c 4. CYTOGENETICS, CROP BREEDING. *Educ:* Nat Chung Hsing Univ, BSc, 56; Utah State Univ, MSc, 64; Univ Utah, PhD(genetics), 67. *Prof Exp:* NIH fel biol sci, Stanford Univ, 67-69; from asst prof to prof bot & plant path, Mich State Univ, 69-82; PROF PLANT SCI, UNIV MAN, 82- *Concurrent Pos:* Vis prof, Nat Chung-Hsing Univ, 75-76, Genetic Inst, Acad Sinica, Beijing,

China; hon prof, Acad Sci China, Jiangsu Agr Col, China. *Mem:* Bot Soc Am; Am Inst Biol Sci; Genetic Soc Can; Am Soc Agron; Crop Sci Soc Am. *Res:* Plant cytogenetics; cytotaxonomy; plant breeding. *Mailing Add:* 313 Carmel Woods St Louis MO 63021

TAIBLESON, MITCHELL H, b Oak Park, Ill, Dec 31, 29; m 49; c 3. MATHEMATICS. *Educ:* Univ Chicago, SM, 60, PhD(math), 62. *Prof Exp:* From asst prof to assoc prof math, 62-69, chmn dept, 70-73, res prof, Dept Psychiat, Med Sch, 73-75, PROF MATH, WASHINGTON UNIV, 69- *Concurrent Pos:* Mem, Inst Advan Study, 66-67; vis prof, Higher Normal Sch, Pisa, Italy, 80-81 & Nanjing Univ, People's Repub China, 85. *Mem:* Am Math Soc; Math Asn Am. *Res:* Several dimensional harmonic analysis on real and local fields; Lipschitz and potential spaces; special functions; Hardy spaces. *Mailing Add:* Dept Math Washington Univ PO Box 1146 St Louis MO 63130

TAICHMAN, NORTON STANLEY, b Toronto, Ont, May 27, 36; m 58; c 5. PATHOLOGY, IMMUNOLOGY. *Educ:* Univ Toronto, DDS, 61, PhD(immunopath), 67; Harvard Univ, dipl periodont, 64. *Prof Exp:* Assoc dent, Fac Dent, Univ Toronto, 65-68, from lectr to asst prof path, 67-71; PROF PATH & CHMN DEPT, SCH DENT MED, UNIV PA, 72- *Concurrent Pos:* Assoc prof dent, Fac Dent, Univ Toronto, 68-72, assoc prof path, 71-72, mem, Inst Immunol, 70-72. *Honors & Awards:* Lindback Found Award, 77. *Mem:* Am Soc Exp Path; Am Soc Microbiol; Reticuloendothel Soc; Int Asn Dent Res. *Res:* Inflammation, immunopathology, periodontal disease; immune deficiency syndromes; polymorphonuclear-leukocytes. *Mailing Add:* Dept Path Univ Pa Sch Dent Med 4010 Locust St Philadelphia PA 19104-6002

TAIGANIDES, E PAUL, b Polymylos, Greece, Oct 6, 34; m 61; c 3. SANITARY ENGINEERING, ENVIRONMENTAL ENGINEERING. *Educ:* Univ Maine, BS, 57; Iowa State Univ, MS, 60, PhD(environ eng), 63. *Prof Exp:* Asst prof agr eng, Iowa State Univ, 63-65; proj mgr & tech adv, Food & Agr Org, United Nations, 75-87; assoc prof, 65-69, PROF AGR ENG, OHIO STATE UNIV, 69-; EPT CONSULT, 87- *Concurrent Pos:* Consult various indust & govt; proj mgr, US Feed Grains Coun, Tokyo, 73 & Food & Agr Orgn, UN, Singapore, 75- *Honors & Awards:* Young Educator Award, Am Soc Agr Engrs, 74. *Mem:* Fel Am Soc Agr Engrs; Am Soc Eng Educ; Am Acad Environ Engrs. *Res:* Environmental engineering for food industry wastewater control. *Mailing Add:* 1800 Willow Forge Dr Columbus OH 43220

TAIGEN, THEODORE LEE, b Seattle, Wash, Nov 22, 52; m 73; c 2. PHYSIOLOGICAL ECOLOGY. *Educ:* Colo State Univ, BS, 76, MS, 78; Cornell Univ, PhD(ecol), 81. *Prof Exp:* ASST PROF ECOL, UNIV CONN, 81- *Mem:* Am Soc Ichthyologists & Herpetologists; Am Ornithologists Union; Am Soc Zoologists; Ecol Soc Am; Sigma Xi. *Res:* Ecological and behavioral correlates of metabolic characteristics of terrestrial vertebrates; physiological variables, including resting and activity metabolism, are analyzed for their association with foraging strategies, habitat selection, predator avoidance mechanisms and modes of locomotion. *Mailing Add:* Ecol Evolution Biol Dept Univ Conn Storrs CT 06268

TAIMUTY, SAMUEL ISAAC, b West Newton, Pa, Dec 20, 17; m 53; c 2. PHYSICS. *Educ:* Carnegie Inst Technol, BS, 40; Univ Southern Calif, PhD(physics), 51. *Prof Exp:* Asst res physicist, Am Soc Heat & Ventilating Eng Lab, 40-42; physicist, Philadelphia Naval Shipyard, 42-44 & Long Beach Naval Shipyard, 44-46; sr physicist, US Naval Radiol Defense Lab, 50-52 & Stanford Res Inst, 52-72; sr physicist, Lockheed Missiles & Space Co, 72-89; RETIRED. *Concurrent Pos:* Consult, 89- *Mem:* Am Phys Soc; Sigma Xi. *Res:* Heat transmission; magnetism; nuclear and radiation physics; radiation effects in solids; radiation dosimetry; industrial applications of radiation, ferroelectricity, thin films and organic dielectrics. *Mailing Add:* 3346 Kenneth Dr Palo Alto CA 94303-4217

TAIT, JAMES SIMPSON, b Charlottetown, PEI, Feb 25, 30; m 58; c 2. FISH BIOLOGY. *Educ:* Dalhousie Univ, BSc, 50, MSc, 52; Univ Toronto, PhD(zool), 59. *Prof Exp:* Res scientist, Res Br, Ont Dept Lands & Forests, 58-64; asst prof biol, 64-70, ASSOC PROF BIOL, YORK UNIV, 70- *Concurrent Pos:* Consult, Res Br, Ont Dept Lands & Forests, 64-70. *Mem:* Am Fisheries Soc; Can Soc Zool. *Res:* Environmental physiology of fish particularly buoyancy regulation, behavioral thermoregulation, tolerance and avoidance of low oxygen and effects of insecticides on neuroendocrine systems; studies on stressed populations of salmonids. *Mailing Add:* Dept Biol York Univ Downsview ON M3J 1P3 Can

TAIT, JOHN CHARLES, b Vancouver, BC, Sept 23, 45; m. NUCLEAR WASTE MANAGEMENT. *Educ:* Univ BC, BSc, 67, PhD(chem), 74. *Prof Exp:* Fel phys chem, Nat Res Coun Can, 74-77; res assoc, Univ BC, 77-78; RES CHEMIST PHYS CHEM, ATOMIC ENERGY CAN, LTD, 78- *Mem:* Chem Inst Can. *Res:* Kinetics and dissolution properties of used fuel; surface adsorption; surface analysis by secondary ion mass spectrometry, electron spectroscopy for chemical analysis scanning electron microscopy, radiation chemistry. *Mailing Add:* AECL Res Whiteshell Labs Pinawa MB R0E 1L0 Can

TAIT, KEVIN S, b New York, NY, Nov 24, 33; m 59; c 4. APPLIED MATHEMATICS. *Educ:* Princeton Univ, AB, 55; Harvard Univ, PhD(appl math), 65. *Prof Exp:* Res staff mem, Sperry Rand Res Ctr, 64-67; assoc prof math, Boston Univ, 67-74; mem staff, Aerodyne Res, 74-75; STAFF ANALYST, THE ANALYTICAL SCI CORP, 75- *Mem:* Soc Indust & Appl Math. *Res:* Optimal control and estimation; mathematical programming; numerical analysis. *Mailing Add:* The Analytical Sci Corp 55 Walkers Brook Dr Reading MA 01867

TAIT, ROBERT JAMES, b Glasgow, Scotland, Aug 28, 37. CONTINUM MATHEMATICS, WAVE MOTION. *Educ:* Univ Glasgow, PhD(math), 62. *Prof Exp:* PROF MATH, UNIV ALTA, 82- *Res:* Incompressible elastic materials and shock capturing techniques from a numerical point of view. *Mailing Add:* Dept Math Univ Alta Edmonton AB T6G 2G1 Can

TAIT, WILLIAM CHARLES, b Waterloo, Iowa, Feb 9, 32; m 54; c 4. THEORETICAL SOLID STATE PHYSICS. *Educ:* Wabash Col, BA, 54; Cornell Univ, MA, 58; Purdue Univ, PhD, 62. *Prof Exp:* Teaching asst, Cornell Univ, 54-58; physicist, Res Lab, Bendix Corp, 58; instr physics, Wabash Col, 58-61; sr physicist, Cent Res, 62-66, res specialist, 67-68, supvr, 69-70, mgr cent res, 71-72, MGR DUPLICATING PROD, MINN MINING & MFG CO, 72- *Mem:* Am Phys Soc; Sigma Xi. *Res:* Solid state and quantum field theory; semiconductor lasers; photoconductors; electrophotography. *Mailing Add:* 14452 57 St N Stillwater MN 55082

TAIZ, LINCOLN, b Philadelphia, Pa, Nov 5, 42; m 63; c 1. PLANT PHYSIOLOGY. *Educ:* Univ Utah, BS, 67; Univ Calif, Berkeley, PhD(bot), 71. *Prof Exp:* Actg asst prof bot, Univ Calif, Berkeley, 72-73; from asst prof to assoc prof, 73-83, PROF BIOL, UNIV CALIF, SANTA CRUZ, 83- *Concurrent Pos:* NSF fel, Univ Calif, Berkeley, 71-72. *Mem:* Am Soc Plant Physiologists; Bot Soc Am; AAAS. *Res:* The structure, function and evolution of the plant vacuolar ATPase; hormonal control of plant development. *Mailing Add:* Dept Biol Univ Calif Santa Cruz CA 95064

TAJIMA, TOSHIKI, b Ngoya, Japan, Jan 18, 48; c 2. PLASMAS PHYSICS. *Educ:* Univ Tokyo, BS, 71, MS, 73; Univ Calif, Irvine, PhD(physics), 75. *Prof Exp:* Asst res physicist, Univ Calif, Los Angeles, 76-80, assoc res physicist, 80; from asst prof to assoc prof, 80-89, PROF PHYSICS, UNIV TEX, AUSTIN, 90- *Concurrent Pos:* Consult, TRW, Inc, 78-81, Jaycor, 78-79, Toshiba, 79-80, Western Res Inc, 78-, FM Technol, 88-, SSC Lab, 89- & INEL, 91-; prin investr, NSF, 81-, Dept Energy, 83- & NASA, 85-; mem, Int Sci Radio Union/Nat Res Coun; vis scientist, Los Alamos, 83-85; co-ed, J Particle Accelerator, 85; auth. *Mem:* Fel Am Phys Soc; Am Geophys Union; fel Japan Soc Prom Sci; Am Astron Soc; Phys Soc Japan; Acad Japan. *Res:* Physics of nuclear fusion, plasma physics, astrophysics, particle accelerators and computational physics; muon calalyzed fusion-fission hybrid reactor; direct energy conversion of x-ray to electricity; fusion reactor concept for muon fusion with magnetic-inertial confinement; author of one book. *Mailing Add:* Physics Dept Univ Tex Austin TX 78712

TAKACS, GERALD ALAN, b Edmonton, Alta, Sept 10, 43; c 2. ATMOSPHERIC CHEMISTRY, PLASMA CHEMISTRY. *Educ:* Univ Alta, Edmonton, BS, 65; Univ Wis-Madison, PhD(phys chem), 71. *Prof Exp:* Teaching asst chem, Univ Wis-Madison, 65-67, res asst phys chem, 67-71; postdoctoral fel phys chem, Rice Univ, 71-72; teaching intern & res assoc phys chem, Univ Ala, Huntsville, 72-73; vis asst prof phys chem, Rochester Inst Technol, 73-75, from asst prof to assoc prof, 75-82, PROF PHYS CHEM, ROCHESTER INST TECHNOL, 82- *Concurrent Pos:* Sr res assoc, Nat Oceanic & Atmospheric Admin, Nat Res Coun, 82-83; prin investr, IBM, Endicott, NY, 84-; head, Dept Chem, Rochester Inst Technol, 85- *Mem:* Am Chem Soc; Am Vacuum Soc; Coun Undergrad Res. *Res:* Atmospheric chemistry; chem kinetics; photochemistry; physical chemistry; plasma chemistry. *Mailing Add:* Dept Chem Rochester Inst Technol Rochester NY 14623

TAKAGI, SHOZO, b Nishinomiya, Japan, Apr 2, 43; m 67; c 2. CRYSTALLOGRAPHY. *Educ:* Kwansei Gakuin Univ, BSc, 66; Univ Pittsburgh, PhD(crystallog), 71. *Prof Exp:* Res assoc chem, Vanderbilt Univ, 71-75; res assoc, Dept Chem, Brookhaven Nat Lab, 75-78; RES ASSOC, AM DENT ASN HEALTH FOUND RES UNIT, PAFFENBARGER RES CTR, NAT INST STANDARDS & TECHNOL, 78- *Mem:* Int Asn Dent Res; Am Crystallog Asn. *Res:* To determine structures of components of importance to dental health by x-ray and neutron diffraction techniques; conduct research on topical fluoridation of tooth and calcium phosphate cement. *Mailing Add:* Am Dent Asn Health Found Paffenbarger Res Ctr Nat Inst Standards & Technol Gaithersburg MD 20899

TAKAHASHI, AKIO, b Andong, Korea, July 15, 32; m 64; c 4. POLYMER CHEMISTRY. *Educ:* Tokyo Col Sci, BS, 57; Tokyo Inst Technol, MS, 60, PhD(polymer chem), 63. *Prof Exp:* Sr chemist, Mitsui Chem Indust, Inc, 63-65; sr chemist, Gaylord Assocs, Inc, NJ, 65-67, sect mgr, 67-70; res assoc polymer chem, Hooker Chem & Plastics Co, 70-73, sr res assoc, 73-75, scientist & discipline/prog leader, 75-78; res mgr, Am Can Co, 78-83, Air Prods & Chem, 83-85 & Henkel Corp, 85-88; VPRES TECHNOL, SUNKYONG AM, 88- *Mem:* Am Chem Soc; Japanese Soc Polymer Sci. *Res:* Polymer-organic chemistry; polymerization by free radical, Ziegler-Natta and ionic catalysts; polymerization through charge-transfer complexes; polybutadiene; polyvinyl chloride; polyethers; graft copolymers. *Mailing Add:* Prides Crossing Flanders NJ 07836

TAKAHASHI, ELLEN SHIZUKO, b Berkeley, Calif. PHYSIOLOGICAL OPTICS, OPTOMETRY. *Educ:* Univ Calif, Berkeley, BS, 52, MOpt, 53, PhD(physiol optics), 68. *Prof Exp:* Optometrist in pvt pract, 53-56; optometrist orthoptics, Stanford Univ Hosps, 56-62; clin instr & actg asst prof optom, Univ Calif, 62-67; NIH fel visual neurophysiol, Australian Nat Univ, 68-70; asst prof, 70-72, assoc prof, 72-81 dir grad studies, Sch Optom, 74-76, PROF PHYSIOL OPTICS, UNIV ALA, 81- *Concurrent Pos:* Mem, Nat Adv Coun Health Prof Educ, 72-75, & Visual Sci B Study Sect, NIH, 79-82; mem, Papers Prog, Am Acad Optom, 81- *Mem:* Am Acad Optom; Asn Res Vision & Ophthal; Soc Neurosci. *Res:* Visual neurophysiology and anatomy; binocular vision. *Mailing Add:* Sch Optom Med Ctr Univ Ala Birmingham AL 35294

TAKAHASHI, FRANCOIS IWAO, microbial genetics, for more information see previous edition

TAKAHASHI, HIRONORI, b Tokyo, Japan, June 5, 42. NUCLEAR FUSION. *Educ:* Keio Univ, Japan, BEng, 65; Mass Inst Technol, MS, 67, DSc(aeronaut & astronaut), 70. *Prof Exp:* Fel plasma physics, Univ Stuttgart, 71-72, guest lectr, 72-73; res assoc, Plasma Physics Lab, Princeton Univ, 73-75, res staff mem, 75-79; group leader & co-prin investr, Nat Magnetic Lab, Mass Inst Technol, 79-80; res staff mem, 79-85, RES PHYSICIST, PLASMA PHYSICS LAB, PRINCETON UNIV, 85- *Mem:* Am Phys Soc. *Res:* Plasma physics in connection with controlled thermonuclear fusion research; magnetic diagnostics and operation of tokamaks. *Mailing Add:* Plasma Physics Lab Princeton Univ PO Box 451 Princeton NJ 08543-0451

TAKAHASHI, JOSEPH S, b Dec 16, 51; m 85; c 1. CIRCADIAN RHYTHMS, PHOTORECEPTION. *Educ:* Swarthmore Col, BA, 74; Univ Ore, Eugene, PhD(neurosci), 81. *Prof Exp:* Res assoc pharmacol, NIMH, 81-83; from asst prof to prof neurobiol, 83-91, ACTG ASSOC DIR, INST NEUROSCI, NORTHWESTERN UNIV, 88- *Concurrent Pos:* Alfred P Sloan Found award, 83; mem, adv bd, J Biol Rhythms, 84-, adv comt, Soc Res Biol Rhythms, 86-, NIMH Psychobiol & Behav Rev Comt, 88-92; Searle Scholars award, Chicago Community Trust, 85; Presidential young investr award, NSF, 85. *Honors & Awards:* Honma Prize in Biol Rhythms, Honma Found, 86. *Mem:* AAAS; Soc Neurosci; Asn Res Vision & Ophthal; Soc Res Biol Rhythms. *Res:* Cell and molecular biology of circadian rhythms; molecular neurobiology; signal transduction; photoreceptor gene expression and regulation. *Mailing Add:* Dept Neurobiol & Physiol Northwestern Univ 2153 Sheridan Rd Evanston IL 60208

TAKAHASHI, KOZO, b Kagawa, Japan, Jan 17, 48; m 72; c 2. OCEANOGRAPHY, EARTH SCIENCES. *Educ:* Univ Wash, BSc, 75, MSc, 77; Mass Inst Technol, PhD(oceanog), 81. *Prof Exp:* Asst scientist oceanog, Scripps Inst Oceanog, Univ Calif, San Diego, 82-84; ASSOC SCIENTIST OCEANOG, WOODS HOLE OCEANOG INST, 84- *Mem:* Am Geophys Union. *Res:* Examining marine biogenic particle fluxes to the deep sea and trying to link global climate changes with fossil record. *Mailing Add:* Woods Hole Oceanog Inst Woods Hole MA 02543

TAKAHASHI, LOREY K, b Honolulu, Hawaii, July 25, 53; m 82; c 1. DEVELOPMENTAL PSYCHOBIOLOGY, BEHAVIORAL NEUROENDOCRINOLOGY. *Educ:* Univ Hawaii, Manoa, BA, 75, MA, 78; Rutgers Univ, NB, PhD(psychol), 82. *Prof Exp:* Undergrad teaching asst, Dept Psychol, Rutgers Univ, 78-82; NIMH postdoctoral trainee, Dept Biol, Princeton Univ, 82-83, NIMH postdoctoral fel, 83-85, res assoc, 85-86; asst scientist, 86-90, ASST PROF, DEPT PSYCHIAT, MED SCH, UNIV WIS, 90- *Concurrent Pos:* Clin asst prof, Dept Psychiat, Med Sch, Univ Wis, 89-90. *Mem:* Sigma Xi; AAAS; Am Psychol Asn; Am Psychol Soc; Am Soc Zoologists; Animal Behav Soc; Soc Neurosci. *Res:* Neurobiology of emotional expression using animal models; combining neuroendocrine, neurochemical, neuroanatomical and ethological methods to study how brain processes influence the development of social and sexual patterns of behavior. *Mailing Add:* Dept Psychiat Univ Wis Med Sch Madison WI 53792

TAKAHASHI, MARK T, b Holtville, Calif, Feb 26, 36; m 61; c 2. PHYSICAL BIOCHEMISTRY. *Educ:* Oberlin Col, BA, 58; Univ Wis-Madison, PhD(phys chem), 63. *Prof Exp:* Res biochemist, Battelle Mem Inst, 64-67; asst prof, 70-77, ASSOC PROF PHYSIOL, RUTGERS MED SCH, COL MED & DENT NJ, 77- *Concurrent Pos:* NIH fel, Univ Mass, Amherst, 67-70. *Mem:* AAAS; Biophys Soc; Am Soc Biol Chemists; Am Chem Soc. *Res:* Physical enzymology; membranes. *Mailing Add:* Dept Physiol Rob't Wood Johnson Med Sch Piscataway NJ 08854-5635

TAKAHASHI, PATRICK KENJI, b Honolulu, Hawaii, Sept 6, 40; m 62. CHEMICAL ENGINEERING, ENERGY. *Educ:* Stanford Univ, BS, 62; La State Univ, Baton Rouge, MS, 69, PhD(chem eng), 71. *Prof Exp:* Sugar processing engr, C Brewer, 62-67; proj engr computerized optimization, Hawaiian Sugar Planters Asn, 67-68; asst prof eng, 71-75, ASSOC PROF CIVIL ENG, UNIV HAWAII, 75-, ACTG ASSOC DEAN ENG, 76-, ACAD ASST CHANCELLOR, 78-, PROF ENG, 82- *Concurrent Pos:* Consult, Lawrence Livermore Lab, Gen Tel & Electronics Corp, 77-; spec asst energy to Sen Spark Matsunaga, 79-82; dir, Hawaii Nat Energy Inst, Univ Hawaii, 84, vpres develop, 88. *Mem:* Am Chem Soc; Am Inst Chem Engrs. *Res:* Energy (biomass, geothermal, wind, solar); laser applications. *Mailing Add:* Hawaii Natural Energy Inst Univ Hawaii at Monoa 540 Dole St Honolulu HI 96822

TAKAHASHI, TARO, b Tokyo, Japan, Nov 15, 30; US citizen; m 66; c 2. GEOCHEMISTRY, GEOPHYSICS. *Educ:* Univ Tokyo, BEng, 53; Columbia Univ, PhD(geol), 57. *Prof Exp:* Res scientist, Lamont Geol Observ, 57-59, res assoc, 59-77, ADJ PROF, DEPT GEOL SCI & SR SCIENTIST, LAMONT-DOHERTY GEOL OBERV, COLUMBIA UNIV, 77-, ASSOC DIR, 81- *Concurrent Pos:* Lectr, Queens Col, NY, 58; res chemist, Scripps Inst Oceanog, Univ Calif, 59; asst prof, State Univ NY Col Ceramics, Alfred, 59-62, vis prof, 63; assoc prof, Univ Rochester, 62-70, prof, 70; vis asst prof, Columbia Univ, 66; vis assoc prof, Calif Inst Technol, 71; distinguished prof, Queens Col, NY, 71-77. *Mem:* Am Geophys Union; Geol Soc Am; Geochem Soc (secy, 81-84); Sigma Xi. *Res:* Thermodynamic and physical properties of metal oxides under extremely high pressures and temperatures; geochemistry of carbon dioxide in the ocean atmosphere system; thermodynamics of ore forming minerals. *Mailing Add:* 350 Hennessy St Haworth NJ 07641

TAKAHASHI, YASUNDO, b Nagoya-shi, Japan, June 12, 12; m 39; c 1. MECHANICAL ENGINEERING. *Educ:* Tokyo Imp Univ, BSc, 35; Univ Tokyo, DrEng, 46. *Hon Degrees:* Dr, Nat Polytech Inst, France, 78. *Prof Exp:* Asst engr, Govt Rwys, Japan, 35-37; asst prof heat power systs, Yokohama Tech Col, 37-40 & Nagoya Imp Univ, 40-44; from asst prof to prof automatic control, Univ Tokyo, 44-58; vis prof, Univ Calif, Berkeley, 55-56, prof automatic control, 58-79; emer prof, Toyohashi Univ Technol, Japan, 79-82; SR TECH CONSULT, MIKUNI BERKELEY R & D, 82- *Concurrent Pos:* Fulbright res scholar & vis fel, Mass Inst Technol, 54-55; consult, Mitsubishi Heavy Industs, Japan, 56-57, Ajinomoto Co, 56-57 & Nucleonics Div, Aerojet Gen Corp, 59-60; vis prof, Polytech Inst Grenoble, France, 65 & 70, Tokyo Inst Technol, 65 & 70, Nat Polytech Inst, Mex, 72, 75 & 78, Univ LaPlata, Arg, 73 & Keio Univ, Japan, 77. *Honors & Awards:* Am Soc Mech Eng Medal, 78. *Mem:* Am Soc Mech Engrs; Instrument Soc Am; Japanese Soc Mech Engrs; Japanese Instrument Soc. *Res:* Theory of automatic control; dynamical systems design; digital control algorithms. *Mailing Add:* 135 York Ave Berkeley CA 94708

TAKAHASHI, YASUSHI, b Osaka, Japan, Dec 12, 24; m 59; c 2. THEORETICAL PHYSICS. *Educ:* Nagoya Univ, BSc, 51, DSc(physics), 54. *Prof Exp:* Nat Res Coun Can fel, 54-55; res assoc physics, Iowa State Univ, 55-57; res scholar, Dublin Inst Advan Studies, 57, from asst prof to prof, 57-68; dir, Theoret Physics Inst, 69-85, PROF PHYSICS, UNIV ALTA, 68- *Mem:* Fel Am Phys Soc; Royal Irish Acad; fel Royal Soc Can. *Res:* Quantization of relativistic fields; quantum electrodynamics and field theory; many-body theory. *Mailing Add:* Dept Physics Univ Alta Edmonton AB T6G 2S1 Can

TAKAI, YASUYUKI, b Osaka, Japan, Aug 23, 51; m 82; c 2. CELLULAR IMMUNOLOGY, INTERNAL MEDICINE. *Educ:* Osaka Univ, MD, 77, PhD(med), 84. *Prof Exp:* Postdoctoral fel immunol, Div Pediat Oncol, Dana-Farber Cancer Inst, 84-87; res fel immunol, Inst Cancer Res, 79-84, asst prof, Biomed Res Ctr, 87-90, INTERNAL MED, SECOND DEPT MED, OSAKA UNIV, 79-; PHYSICIAN INTERNAL MED, IKEDA CITY HOSP, 90- *Mem:* Am Asn Immunologists; Fedn Am Socs Exp Biol. *Res:* Cellular immunology; T cell differentiation cytokine regulation in T cell activation. *Mailing Add:* 2-7-8 Toneyama Toyonaka-city Osaka 560 Japan

TAKANO, MASAHARU, b Tainan, Taiwan, Jan 20, 35; nat US; m 65; c 3. CHEMICAL ENGINEERING, BIOTECHNOLOGY. *Educ:* Hokkaido Univ, BSc, 57; Univ Tokyo, MSc, 59, DrSc(rheol), 63. *Prof Exp:* Nat Res Coun fel, McGill Univ, 63-65, fel, 65-67; res specialist, Corp Res Dept, 67-75, SR RES SPECIALIST & TECH TRANSLATOR, ANIMAL SCI DIV, MONSANTO AGR CO, 75- *Mem:* Soc Rheology; Am Chem Soc; AAAS; Am Phys Soc; Am Inst Chemists; NY Acad Sci; Sigma Xi; Fine Particle Soc. *Res:* Rheology and physical chemistry of polymers and disperse systems; polymer and composite technologies; industrial process technologies; crystallization solvent extraction, reaction kinetics and purification formulation, bovine and procine growth hormones from genetic engineered E Coli. *Mailing Add:* 13146 Roundstone Ct St Louis MO 63146

TAKARO, TIMOTHY, b Budapest, Hungary, Aug 30, 20; US citizen; m 49; c 4. MEDICINE. *Educ:* Dartmouth Col, BA, 41; NY Univ, MD, 43; Univ Minn, MS, 50. *Prof Exp:* Dir surg, Wanless Tuberc Sanatorium, India, Presby Bd Foreign Missions, 54-57; dir res & resident educ, 57-60, chief cardiovasc surg sect, 60-62, CHIEF SURG SERV, VET ADMIN MED CTR, ASHEVILLE, 62-, SR PHYSICIAN, 71-, CHIEF OF STAFF, 83- *Concurrent Pos:* Exchange scientist, US-USSR Exchange Agreement, Moscow, 62; co-chmn, Vet Admin Coop Study of Surg for Coronary Arterial Occlusive Dis, 68-; assoc clin prof surg, Duke Univ Med Ctr, 68-80, clin prof, 80-; chmn, Cardiac Surg Adv Group, Vet Admin Cent Off, 70-; assoc ed, Annals Thoracic Surg, 70- *Mem:* Fel Am Col Surgeons; Am Asn Thoracic Surg; Soc Thoracic Surgeons; Soc Vascular Surg. *Res:* Pulmonary emphysema; coronary arterial occlusive disease. *Mailing Add:* VA Hosp Asheville NC 28805

TAKASHIMA, SHIRO, b Japan, May 12, 23; m 53; c 2. PHYSICAL BIOCHEMISTRY, MEMBRANE BIOPHYSICS. *Educ:* Univ Tokyo, BS, 47, PhD(biochem), 55. *Prof Exp:* Res fel phys chem, Univ Minn, 55-57; res assoc biomed eng, Univ Pa, 57-59; from assoc prof to prof protein res, Osaka Univ, 59-62; vis scientist, Walter Reed Med Ctr, 62-63; res assoc biomed eng, 63-64, asst prof, 64-70, assoc prof, 70-76, PROF BIOMED ENG, UNIV PA, 76- *Mem:* Biophys Soc; Inst Elec & Electronics Engrs; NY Acad Sci; AAAS. *Res:* Dielectric relaxation of desoxyribonucleic acid; synthetic polynucleotides; polyamino acids and proteins; electrical properties of excitable and passive membranes. *Mailing Add:* Dept Bioeng Univ Pa Philadelphia PA 19104-6392

TAKASUGI, MITSUO, b Tacoma, Wash, Jan 28, 28; m 54; c 4. CANCER, IMMUNOLOGY. *Educ:* Univ Calif, Los Angeles, BA, 52, PhD(immunogenetics), 68; Univ Ore, MS, 62. *Prof Exp:* Sci teacher, Los Angeles City Schs, 53-64; fel, 53-64; Dept Tumor Biol, Karolinska Inst, Stockholm, Sweden, 68-69; RESEARCHER CANCER, DEPT SURG, UNIV CALIF, LOS ANGELES, 69- *Mem:* AAAS; Am Asn Cancer Res; Am Asn Immunologists; Transplantation Soc. *Res:* Investigations into the cellular and humoral immune response to cancer. *Mailing Add:* Tissue Typing Lab Univ Calif Rehab Ctr 950 Veteran Ave Los Angeles CA 90024

TAKATS, STEPHEN TIBOR, b West Englewood, NJ, May 24, 30; m 60; c 3. CYTOLOGY, GENETICS. *Educ:* Cornell Univ, BS, 52; Univ Wis, MS, 54, PhD(genetics), 58. *Prof Exp:* Asst genetics, Univ Wis, 52-55; res collabr & USPHS res fel, Biol Dept, Brookhaven Nat Lab, 57-60; USPHS res fel biochem, Univ Glasgow, 60-61, Med Res Coun grant, 61; from asst prof to assoc prof biol, 61-69, chmn dept, 69-75, PROF BIOL, TEMPLE UNIV, 69- *Mem:* Bot Soc Am; Am Soc Plant Physiol; Genetics Soc Am; Am Soc Cell Biol. *Res:* Biochemical cytology; control of DNA synthesis in plant development. *Mailing Add:* Dept Biol Temple Univ Meetinghouse Rd Ambler PA 19112

TAKAYAMA, KUNI, b Wapato, Wash, Feb 28, 32; m 59; c 2. BIOCHEMISTRY, MICROBIOLOGY. *Educ:* Ore State Univ, BS, 56; Univ Idaho, MS, 61, PhD(biochem), 64. *Prof Exp:* NIH grant, Inst Enzyme Res, Univ Wis-Madison, 64-65; proj assoc biochem of mycobact, Vet Admin Hosp, Madison, Wis, 65-67, res chemist, Tuberc Res Lab, 67-71, actg chief chemist, 68-69, chief chemist, 71-74, CHIEF RES CHEMIST, TUBERC RES LAB, VET ADMIN HOSP, MADISON, WIS, 74-; ASSOC PROF BACT, UNIV WIS-MADISON, 86- *Concurrent Pos:* Proj assoc, Inst Enzyme Res, Univ Wis-Madison, 65-67, asst prof, 67-85; NSF grants, Vet Admin Hosp & Univ Wis-Madison, 69-72, NIH grant, 73- *Mem:* AAAS; Am Soc Microbiol; Am Soc Biol Chem; Am Soc Cell Biol; Am Inst Biol Sci. *Res:* Biochemistry of mycobacteria; biosynthesis of mannophospholipids; synthesis of lipids; mode of action of isoniazid; structure of lipopolysaccharides and lipid A. *Mailing Add:* Mycobact Lab Vet Admin Hosp 2500 Overlook Terrace Madison WI 53705

TAKEDA, YASUHIKO, b Nagano, Japan, Mar 16, 27; m 57; c 4. PHYSIOLOGY, CLINICAL PATHOLOGY. *Educ:* Shinshu Univ, 46-48; Chiba Univ, MD, 52; Am Bd Path, dipl, 70. *Prof Exp:* From instr to prof, Med Ctr, Univ Colo, Denver, 63-88; RETIRED. *Concurrent Pos:* Res fel, Div Lab Med, Med Ctr, Univ Colo, 58-60; Nat Res Coun Can res fel, McGill Univ, 60-63; Colo Heart Asn sr res fel med, Med Ctr, Univ Colo, Denver, 63-64, Am Heart Asn advan res fel, 64-66; NIH career develop award, 67-72; mem coun thrombosis, Am Heart Asn. *Mem:* AAAS; Am Physiol Soc; Am Soc Clin Path; Int Soc Thrombosis & Hemorrhagic Dis; Col Am Pathologists. *Res:* Regulation of plasma protein metabolism in health and disease; dynamics of thrombus formation and dissolution. *Mailing Add:* 635 Dexter St Denver CO 80220

TAKEMORI, AKIRA EDDIE, b Stockton, Calif, Dec 9, 29; m 58; c 2. PHARMACOLOGY. *Educ:* Univ Calif, AB, 51, MS, 53; Univ Wis, PhD(pharmacol), 58. *Prof Exp:* Res asst pharmacol, Univ Calif, 51-53; res asst, Univ Wis, 55-57, Am Cancer Soc fel, Enzyme Inst, 58-59; instr pharmacol, State Univ NY Upstate Med Ctr, 59-61, asst prof, 61-63; from asst prof to assoc prof, 61-69, PROF PHARMACOL, HEALTH SCI CTR, UNIV MINN, MINNEAPOLIS, 69-, DIR GRAD STUDIES, 87- *Concurrent Pos:* Mem pharmacol A study sect, NIH, 70-74; mem, rev panel on new drug regulation, Dept Health, Educ & Welfare, 76, bd sci counr, Addiction Res Ctr, Nat Inst on Drug Abuse, 82-86 & Comt on Probs Drug Dependence, 83-87. *Honors & Awards:* Vis Scientist Award, Japan Soc Promotion Sci, 71; Alan Gregg Fellow Med Educ, China Med Bd NY, 71. *Mem:* AAAS; Am Soc Pharmacol & Exp Therapeut; Soc Exp Biol & Med; Am Asn Univ Prof; Am Chem Soc; Am Col Neuropsychopharmacol. *Res:* Mechanism of action of opioid analgesics and opioid antagonists. *Mailing Add:* Dept Pharmacol 3-249 Millard Hall Univ Minn 435 Delaware St SE Minneapolis MN 55455

TAKEMOTO, DOLORES JEAN, b Indianapolis, Ind, May 5, 49; m 73; c 2. CYCLIC NUCLEOTIDE METABOLISM. *Educ:* Ball State Univ, BS, 71; Colo State Univ, MS, 73; Univ Southern Calif, PhD(molecular biol), 79. *Prof Exp:* Sr res technician physiol, Colo State Univ, 73-75; asst molecular biol, Univ Southern Calif, 75-79; res assoc, 79-81, vis asst prof, 81-84, ASST PROF BIOCHEM, KANS STATE UNIV, 84- *Concurrent Pos:* Prin investr, Nat Cancer Inst grant, 80-83; mem, Kans State Ctr Basic Cancer Res, 80-; prin investr, Nat Eye Inst grant, 85- *Mem:* Asn Res Vision & Ophthal; Am Soc Biol Chem; Am Asn Cancer Res. *Res:* Cyclic nucleotide metabolism in leukemic versus normal human lymphocytes; retinal biochemistry of dystrophies. *Mailing Add:* Dept Biochem Willard Hall Kans State Univ Manhattan KS 66506

TAKEMOTO, JON YUTAKA, b Chicago, Ill, Sept 13, 44; m 73; c 2. MICROBIAL PHYSIOLOGY, BIOCHEMISTRY. *Educ:* Univ Calif, Los Angeles, BS, 67, PhD(microbiol), 73. *Prof Exp:* Fel biol & Maria Moor Cabot fel, Harvard Univ, 73-74; from asst prof to assoc prof, 75-85, PROF BIOL, UTAH STATE UNIV, 85-, DIR, MOLECULAR BIOL PROG, 89- *Concurrent Pos:* NIH fel, Harvard Univ, 74; staff mem, Utah Agr Exp Sta, Logan, 75-; prin investr, NSF grants, Utah State Univ, 76-; vis prof biol, Freiburg Univ, WGermany, 81-82, Alexander von Humboldt Found fel, 81-82; vis scientist, Hiroshima Univ, Japan, 88-89; Japan Soc Prom Sci fel, 88-89. *Mem:* Am Soc Microbiol; Sigma Xi; AAAS; Am Soc Plant Physiol. *Res:* Mechanisms of action of phytopathogenic bacterial toxins on plant cells; plant-microbe interactions. *Mailing Add:* Dept Biol Utah State Univ Logan UT 84322

TAKEMURA, KAZ H(ORACE), b San Juan Bautista, Calif, Nov 2, 21; m 59; c 7. ORGANIC CHEMISTRY. *Educ:* Univ Calif, Los Angeles, BS, 47; Univ Ill, MS, 48, PhD(chem), 50. *Prof Exp:* Res fel chem, Ohio State Univ, 50-52; res chemist, Univ Calif, Berkeley, 52-53; chemist southern regional res lab, USDA, La, 53-56; asst prof chem, Loyola Univ, La, 56-58 & Univ Tulsa, 58-59; chemist, Sahyun Labs, 59-60; from asst prof to assoc prof, 60-66, chmn dept, 70-73, PROF CHEM, CREIGHTON UNIV, 66- *Concurrent Pos:* Vis prof, Univ Calif, Berkeley, 79. *Mem:* Royal Soc Chem; Am Chem Soc. *Res:* Organic synthesis and mechanisms. *Mailing Add:* Dept Chem Creighton Univ Omaha NE 68178

TAKESAKI, MASAMICHI, b Sendai, Japan, July 18, 33; m 59; c 1. MATHEMATICS. *Educ:* Tohoku Univ, Japan, MS, 58, DSc(math), 65. *Prof Exp:* Res asst math, Tokyo Inst Technol, 58-63; assoc prof, Tohoku Univ, Japan, 63-70; PROF MATH, UNIV CALIF, LOS ANGELES, 70- *Concurrent Pos:* Fel, Sakkokai Found, 65-68; vis assoc prof, Univ Pa, 68-69 & Univ Calif, Los Angeles, 69-70; vis prof, Univ Aix-Marseille, 73-74 & Univ Bielefeld, 75-76. *Mem:* Am Math Soc; Math Soc Japan; Math Soc France. *Res:* Functional analysis; operator algebras; mathematical physics. *Mailing Add:* Dept Math 6356 Math Sci Univ Calif 405 Hilgard Ave Los Angeles CA 90024

TAKESHITA, TSUNEICHI, b Tokyo, Japan, Sept 13, 26; m 56; c 3. PHYSICAL ORGANIC CHEMISTRY. *Educ:* Waseda Univ, Japan, BSEng, 50; Univ Del, PhD(org chem), 62. *Prof Exp:* Chemist, Cent Res Inst, Japan Monopoly Corp, Tokyo, 50-64; assoc prof catalysis, Res Inst Catalysis, Hokkaido Univ, 64-66; chemist, Elastomer Chem Dept, E I Du Pont de Nemours & Co, Inc, Wilmington, 66-80. *Mem:* Am Chem Soc; NY Acad Sci; Chem Soc Japan; Sigma Xi. *Res:* Catalytic studies in organic chemistry; polymer chemistry. *Mailing Add:* Five Hillock Lane Chadds Ford PA 19317-9704

TAKETA, FUMITO, biochemistry; deceased, see previous edition for last biography

TAKETOMO, YASUHIKO, b Tokyo, Japan; US citizen; c 2. PSYCHIATRY. *Educ:* First Col, Tokyo, BA, 42; Osaka Univ, MD, 45, DMedSc, 49; Columbia Univ, cert psychoanal med, 59; Univ State New York, MD, 81. *Prof Exp:* Spec res fel, Ministry Educ, Japanese Govt, Med Sch, Osaka Univ, 45-47, asst biochem & neuropsychiat, 47-50; asst resident neuropsychiat, Albany Med Col, 50-51; resident/asst res psychiatrist, Worcester State Hosp, 51-52; res psychiatrist/sr res scientist, Res Facil, Rockland State Hosp, 52-64; res assoc neurol & psychiat, St Vincent's Hosp & Med Ctr, 64-67; asst prof psychiat, New York Med Col, 68-69; asst prof, 70-73, assoc prof psychiat, 73-77, ASSOC CLIN PROF, ALBERT EINSTEIN COL MED, 77-; SR PSYCHIATRIST & ASSOC DIR, ALBERT EINSTEIN PSYCHIAT RESIDENCY PROG, BRONX PSYCHIAT CTR, 79- *Concurrent Pos:* Garioa scholar, US State Dept, Albany Med Col & Worcester State Hosp, 50-52; from res asst to res assoc, Col Physicians & Surgeons, Columbia Univ, 52-64; lectr, Med Sch, Osaka Univ, 54; vis scientist, NIH, 56; assoc, New York Sch Psychiat, 62-67; New York City Health Res Coun career scientist award, St Vincent's Hosp & New York Med Col, 64-69; mem acad fac, State Conn Dept Ment Health, 65-67; symp assoc, Ctr Res Math, Morphol & Psychol, 67-73; consult, Asn for Help of Retarded Children, 73-76; consult, Sch Med, Uniformed Serv Univ, 78. *Mem:* Am Psychiat Asn; fel Am Acad Psychoanalysis; Asn Psychoanalysis Med; Am Psychoanalysis Asn. *Res:* Psychopathology and treatment of psychiatric disorders of mental retardation; communicational behavior; cognitive and semiotic psychiatry; phenomenology of adaptational and existential crises; methodology of psychiatric research, especially the issue of temporality; transcultural psychiatry. *Mailing Add:* 1198 Post Rd Scarsdale NY 10583

TAKEUCHI, ESTHER SANS, b Kansas City, Mo, Sept 8, 53; m 82. LITHIUM BATTERY DEVELOPMENT, IMPLANTABLE MEDICAL COMPONENT. *Educ:* Univ Pa, BA, 75; Ohio State Univ, PhD(org chem), 81. *Prof Exp:* Sr chemist, Union Carbide Corp, 81-82; Postdoctoral work,Univ NC, 82-83,State Univ NY, Buffalo, 83-84; ASSOC DIR RES & DEVELOP, WILSON GREATBATCH LTD, 84- *Mem:* Electrochem Soc; Am Chem Soc. *Res:* Development of high technology batteries with a specialty in lithium implantable batteries for medical use; characterization and utilization of solid state materials for energy storage. *Mailing Add:* Wilson Greatbatch Ltd 10000 Wehrle Dr Clarence NY 14031

TAKEUCHI, KENJI, b Tokyo, Japan, Dec 11, 34; m 64; c 3. ENGINEERING SCIENCE, PHYSICS. *Educ:* Tokyo Inst Technol, BS, 58; Univ Mich, PhD(nulcear eng), 67. *Prof Exp:* Res assoc, Argonne Nat Lab, 66-70; res fel, Univ Manchester, 70-73; ADV ENGR REACTOR SAFETY, WESTINGHOUSE ELEC CORP, 73- *Mem:* Am Phys Soc; Am Nuclear Soc. *Res:* Analysis methods development of fluid structural interactions for the purpose of nuclear reactor safety evaluation; theoretical nuclear physics; development of artificial intelligence expert systsms; assessment of thermo-hydraulic code; mb-2 tests and analyses for steam generator transients. *Mailing Add:* Westinghouse Elec Corp MNC 326A Box 355 Pittsburgh PA 15230

TAKEUCHI, KIYOSHI HIRO, b Furubira, Japan, Oct 20, 48; m 79; c 2. BIOCHEMISTRY, CELL BIOLOGY. *Educ:* Hokkaido Univ, Japan, BS, 74; Tokyo Metrop Univ, Japan, PhD(biochem), 79. *Prof Exp:* Res assoc, Dept Biochem, Molecular & Cell Biol, Northwestern Univ, Evanston, 79-83 & Dept Anat & Cell Biol, Northwestern Univ, Chicago, 83-84; cancer res scientist, Roswell Park Mem Inst, Buffalo, 84-88; asst investr, Dept Biomed Res, St Elizabeth's Hosp, Boston, 88-89; ASST BIOCHEMIST, RALPH LOWELL LABS, MCLEAN HOSP & ASST PROF, DEPT PSYCHIAT, HARVARD MED SCH, 89- *Concurrent Pos:* Investr, Marine Biol Lab, Woods Hole, 82-83. *Honors & Awards:* Deguchi Award, Sapporo, Japan, 90. *Mem:* Am Soc Biochem & Molecular Biol; AAAS; Am Soc Cell Biol; NY Acad Sci; Int Soc Exp Hemat; Am Soc Neurochem. *Res:* Role of the proteolylysis in neurodegenerative diseases and aging, includng Alzheimer's disease; proteinase inhibitors in Chediak-Higashi (beige) neutrophils; embryo stem cells; developmental biology. *Mailing Add:* Ralph Lowell Labs McLean Hosp Harvard Med Sch Belmont MA 02178

TAKEUTI, GAISI, b Isikawa, Japan, Jan 25, 26; m 47; c 2. MATHEMATICAL LOGIC. *Educ:* Univ Tokyo, PhD(math logic), 56. *Prof Exp:* Instr math, Univ Tokyo, 49-50; from asst prof to prof, Tokyo Univ Educ, 50-66; PROF MATH, UNIV ILL, URBANA, 66- *Concurrent Pos:* Mem, Inst Advan Study, 59-60 & 66-68. *Mem:* Am Math Soc; Asn Symbolic Logic; Math Soc Japan; Japan Asn Philos Sci. *Res:* Proof-theory; set-theory. *Mailing Add:* Dept Math 375 Altgeld-MC-382 Univ Ill Urbana IL 61801

TAKMAN, BERTIL HERBERT, b Stockholm, Sweden, Aug 15, 21; m 43; c 5. CHEMISTRY. *Educ:* Univ Stockholm, PhD(org chem), 63. *Prof Exp:* Head chem sect, Astra Pharmaceut Prod, Astra Res Labs, 58- 77, internal consult, 77-91; RETIRED. *Concurrent Pos:* Guest prof med chem, Northeastern Univ, 73-79; asst pharmacologist, Brigham Womens Hosp, 82-85; clin instr anesthesia pharmacol, Harvard Med Sch, 83-84. *Mem:* Am Chem Soc; NY Acad Sci. Medicinal and organic chemistry. *Res:* Chemistry and pharmacology of local anesthetics and local anesthesia; chemical interactions. *Mailing Add:* 167 Richmond Ave Worcester MA 01602

TAKVORIAN, KENNETH BEDROSE, b Philadelphia, Pa, Aug 24, 43; m 66; c 2. POLYMER CHEMISTRY, TEXTILE CHEMISTRY. *Educ:* Philadelphia Col Textiles & Sci, BS, 65; Clemson Univ, MS, 67, PhD(chem), 69. *Prof Exp:* RES CHEMIST, TEXTILE RES LAB, E I DU PONT DE NEMOURS & CO, INC, 69- *Mem:* Am Chem Soc; Am Asn Textile Chemists & Colorists. *Res:* All areas of textile chemistry; dyeing, finishing, and textile technology; polymer synthesis and morphology; metal chelates and organic synthesis and mechanisms. *Mailing Add:* 732 Westcliff Wilmington DE 19803

TALAAT, MOSTAFA E(ZZAT), b Cairo, Egypt, May 16, 24; nat US; m 51; c 5. ELECTRICAL ENGINEERING. *Educ:* Cairo Univ, BSc, 46, MSc, 47; Univ Pa, PhD(elec eng), 51. *Prof Exp:* Asst, Mass Inst Technol, 47-49; instr, Univ Pa, 51; proj engr, Westinghouse Elec Corp, 51-52; spec mach designer, Star Kimble Motor Div, Miehle Press, 53; sr res engr, Elliott Co Div, Carrier Corp, 53-59; mgr energy conversion & asst dir eng res, Nuclear Div, Martin-Marietta Corp, 59-64; PROF MECH ENG, COL ENG, UNIV MD, COLLEGE PARK, 64- *Concurrent Pos:* Consult. *Honors & Awards:* Inst Elec & Electronics Engrs Awards, 56-58; Awards, Martin-Marietta Corp,

60-63. *Mem:* AAAS; Inst Elec & Electronics Engrs; Am Phys Soc; Am Inst Aeronaut & Astronaut. *Res:* Energy conversion, including magnetoplasmadynamics, thermionic, thermoelectric and fuel cell energy conversion as well as biological power sources; solar-thermal energy storage and conversion. *Mailing Add:* Dept Mech Eng Univ Md College Park MD 20742

TALALAY, PAUL, b Berlin, Ger, Mar 31, 23; nat US; m 53; c 4. MOLECULAR PHARMACOLOGY. *Educ:* Mass Inst Technol, SB, 44; Yale Univ, MD, 48. *Hon Degrees:* DSc, Acadia Univ, 74. *Prof Exp:* Intern & asst resident, Surg Serv, Mass Gen Hosp, 48-50; asst prof, Ben May Lab Cancer Res, Univ Chicago, 50-57, from assoc prof to prof, Lab & Dept Biochem & Med, Univ, 57-63; John Jacob Abel prof pharmacol & exp therapeut & dir dept pharmacol, 63-75, JOHN JACOB ABEL DISTINGUISHED SERV PROF PHARMACOL & EXP THERAPEUT, SCH MED, JOHNS HOPKINS UNIV, 75- *Concurrent Pos:* Am Cancer Soc scholar, 54-58; Am Cancer Soc res prof, 58-63; mem pharm B study sect, NIH, 63-67; mem, Nat Adv Cancer Coun, 67-71; ed-in-chief, Molecular Pharm, 68-71; mem bd sci consults, Sloan-Kettering Inst, 71-81; mem bd sci adv, Jane Coffin Childs Mem Fund for Med Res, 71-80; Guggenheim Mem fel, 73-74; Am Cancer Soc prof, 77-; bd overseers, Comt Visitation, Div Med Sci, Harvard Univ. *Honors & Awards:* Theobold Smith Award, AAAS, 57; Premio Int La Madonnina, Milan, 78. *Mem:* Nat Acad Sci; Am Philos Soc; Am Soc Biol Chemists; Am Soc Pharmacol & Exp Therapeut; Am Soc Clin Invest; Am Cancer Soc; fel Am Acad Arts & Sci; fel AAAS. *Res:* Molecular pharmacology; biochemistry; chemoprotection against cancer; metabolism and mechanism of action of steroid hormones. *Mailing Add:* Dept Pharmacol Sch Med Johns Hopkins Univ Baltimore MD 21205

TALAMANTES, FRANK, b July 8, 43. ENDOCRINOLOGY. *Educ:* Univ St Thomas, Tex, BA, 66; Sam Houston State Univ, MA, 70; Univ Calif, Berkeley, PhD(endocrinol), 74. *Prof Exp:* PROF ENDOCRINOL, DIV NATURAL SCI, THIMANN LABS, UNIV CALIF, SANTA CRUZ, 74- *Mem:* Sigma Xi; NY Acad Sci; Am Asn Anatomists; AAAS; Am Physiol Soc; Soc Exp Biol & Med. *Mailing Add:* Div Natural Sci Univ Calif Sinsheimer Labs Santa Cruz CA 95064

TALAMO, BARBARA LISANN, b Washington, DC, May 30, 39; m 58, 85; c 3. NEUROBIOLOGY, BIOCHEMISTRY. *Educ:* Radcliffe Col, AB, 60; Harvard Univ, PhD(biochem), 72. *Prof Exp:* Tutor biochem sci, Harvard Col, 71-74; asst prof neurol & physiol chem, Med Sch, Johns Hopkins Univ, 74-80; asst prof, 80-83, DIR, GRAD PROG NEUROSCI, ASSOC PROF NEUROL & PHYSIOL, MED SCH, TUFTS UNIV, 83- *Concurrent Pos:* NSF fel neurobiol, Harvard Med Sch, 72-74; mem study sects, 79-83, prin investr, NIH grants, 76-, ad hoc mem, NSF study sect, NSF, 87- *Mem:* Soc Neurosci; Am Soc Neurochem; Int Soc Neurochem; Asn Chemoreception Sci. *Res:* regulation and mechanism of neurotransmitter sensitivity; developmental regulation of mechanisms of secretion and second messenger systems; olfactory system Alzheimer's disease. *Mailing Add:* Neurosci Labs Tufts Med Sch 136 Harrison Ave Boston MA 02111

TALAPATRA, DIPAK CHANDRA, b Bangladesh, Jan 20, 42; US citizen; m 73; c 2. FINITE ELEMENT ANALYSIS, STRUCTURAL DYNAMICS. *Educ:* Indian Inst Technol, BTech, 63; McGill Univ, MEng, 68; Univ BC, PhD(mech eng),72. *Prof Exp:* Sr scientist, Ensco Inc, 72-77; sr res engr, Gen Tire & Rubber Co, 77-80; mech engr, Dept Navy-Naval Ord Sta, 80; aerospace engr, 80-83, mgr flight support syst, 83-85, mgr dynamic analysis, Goddard Space Flight Ctr, 85-87, TECH PROG MGT, SPACE STA PROG OFF, NASA HQ, 87- *Honors & Awards:* Merit Performance Award, NASA Goddard Space Flight Ctr, 84, 85, 86. *Mem:* Am Soc Mech Engrs. *Res:* Structural mechanics investigation; fracture mechanics; finite element analysis; vibration testing; analysis of composites; modal analysis and testing; space systems engineering. *Mailing Add:* 7213 Adrienne Glen Ave Springfield VA 22152

TALATY, ERACH R, b Nagpur, India, Oct 20, 26; nat US; m 60. ORGANIC CHEMISTRY, ELECTROCHEMISTRY. *Educ:* Univ Nagpur, BSc(honors), 48, MSc, 49, PhD(electrochem), 54; Ohio State Univ, PhD(org chem), 57. *Prof Exp:* Lectr chem, Col Sci, Univ Nagpur, 48-54; asst, Ohio State Univ, 56-57; sr res chemist, Columbia-Southern Chem Corp, 57-61; sr res chemist, Bridesburg Labs, Rohm and Haas Co, Pa, 61; fel, Harvard Univ, 61-62; assoc prof, Univ SDak, 62-64; res assoc, Iowa State Univ, 64-66; asst prof chem, La State Univ, New Orleans, 66-69; RES PROF CHEM, WICHITA STATE UNIV, 69- *Mem:* Am Chem Soc; Royal Soc Chem; Indian Chem Soc. *Res:* Isomerization of azobenzenes; electrodeposition of metals; phosgene chemistry; reactions of chloroformates and carbonates; addition and condensation polymers; acid chlorides; electron spin resonance; steroids; natural products; small-ring compounds; theoretical studies. *Mailing Add:* Dept Chem Wichita State Univ Wichita KS 67208-1587

TALBERT, GEORGE BRAYTON, b Ripon, Wis, July 21, 21; m 47; c 2. AGING, ANATOMY. *Educ:* Univ NDak, BS, 41; Univ Wis, MA, 42, PhD(zool), 50. *Prof Exp:* From res assoc to assoc prof, State Univ NY Downstate Med Ctr, 50-68, actg chmn, 82-85, prof anat & cell biol, Col Med, 68-88; RETIRED. *Concurrent Pos:* USPHS spec fel anat, Univ Birmingham, 63-64; USPHS res grant, Nat Inst Child Health & Human Develop, 66-76. *Mem:* Endocrine Soc; Am Asn Anatomists; Brit Soc Fertil; Geront Soc; Soc Study Reproduction. *Res:* Pituitary-gonadal relationship; sexual maturation; longevity; aging of reproductive system. *Mailing Add:* PO Box 33 Quinby VA 23423

TALBERT, JAMES LEWIS, b Cassville, Mo, Sept 26, 31; m 58; c 2. PEDIATRIC SURGERY, THORACIC SURGERY. *Educ:* Vanderbilt Univ, BA, 53, MD, 57. *Prof Exp:* Intern surg, Johns Hopkins Hosp, 56-57, resident, 59-60 & 62-64, resident pediat surg, 64-65, instr surg, Sch Med, Johns Hopkins Univ, 65-66, asst prof surg & pediat surg, 66-67; assoc prof, 67-70, PROF SURG & PEDIAT, COL MED, UNIV FLA, 70- *Concurrent Pos:* Sr asst surgeon, Nat Heart Inst, 60-62; Garrett scholar pediat surg, Sch Med, Johns Hopkins Univ, 65-66; consult surg, Univ Hosp Jacksonville & Vet Admin Hosp, Gainesville, 72- *Mem:* Am Surg Asn; Soc Pediat Res; Soc Univ Surgeons; Am Pediat Surg Asn; Am Pediat Soc. *Res:* Congenital anomalies; cancer in childhood; metabolic responses to surgical stress in infants and children. *Mailing Add:* Dept Surg Univ Fla Col Med J Hillis Miller Health Ctr Box J286 Gainesville FL 32610

TALBERT, LUTHER M, b Abingdon, Va, Dec 30, 26; m 49; c 3. OBSTETRICS & GYNECOLOGY. *Educ:* Emory & Henry Col, BA, 49; Univ Va, MD, 53; Am Bd Obstet & Gynec, dipl. *Prof Exp:* From instr to assoc prof, 58-69, PROF OBSTET & GYNEC, SCH MED, UNIV NC, CHAPEL HILL, 69- *Mem:* Endocrine Soc; Soc Gynec Invest; Am Gynec Soc; Am Asn Obstet & Gynec; Am Fertil Soc; Am Col Obstet & Gynec. *Res:* Reproductive endocrinology and infertility. *Mailing Add:* Dept Obstet & Gynec-CB7570 Univ NC Med Sch Chapel Hill NC 27599

TALBERT, NORWOOD K(EITH), b Felixville, La, July 29, 21; m 51; c 3. CHEMICAL ENGINEERING, PHYSICAL CHEMISTRY. *Educ:* Tex A&M Univ, BS, 49. *Prof Exp:* Chem engr, Exp Sta, E I du Pont de Nemours & Co, Del, 49-55; group leader & staff specialist, Chem Res Dept, Spencer Chem Co, 55-60, mgr new prod develop, Chemetron Corp, 60-63; mgr com develop, Texas City Refinery, 63-66, mgr nitrogen ctr, 66-67, mgr nitrogen div, 67-71, DIR ENVIRON QUAL, AGWAY INC, 71- *Mem:* Am Chem Soc; Am Inst Chem Engrs. *Res:* Chemical process studies and engineering development; economic evaluation and market appraisal; product and commercial development; management. *Mailing Add:* 14306 N Alamo Canyon Dr Tucson AZ 85737-9170

TALBERT, PRESTON TIDBALL, b Washington, DC, Feb 17, 25; m 56. ORGANIC CHEMISTRY. *Educ:* Howard Univ, BS, 50, MS, 52; Washington Univ, PhD(chem), 55. *Prof Exp:* Asst, Washington Univ, 51-52; res assoc, Univ Wash, 55-56, res instr, 56-59, NIH fel, 57-59; from asst prof to assoc prof 59-70, assoc chmn dept, 66-87, PROF BIO-ORG CHEM, HOWARD UNIV, 70- *Concurrent Pos:* Mem postdoc fel panel, Nat Res Coun, 80 & Ford Found, 83. *Mem:* Fel AAAS; fel Am Inst Chemists; Am Chem Soc; NY Acad Sci. *Res:* Mechanism of function, synthesis and degradation of biologically important compounds, especially nucleic, nucleosides, proteins and vitamins; enzyme function and mechanism of action. *Mailing Add:* Dept Chem Howard Univ Washington DC 20059

TALBERT, RONALD EDWARD, b Toulon, Ill, May 20, 36; m 55; c 3. AGRONOMY, WEED SCIENCE. *Educ:* Univ Mo, BS, 58, MS, 60, PhD(field crops), 63. *Prof Exp:* Instr field crops, Univ Mo, 60-63; from asst prof to assoc prof agron, 63-73, prof, 73-90, UNIV PROF AGRON, UNIV ARK, FAYETTEVILLE, 90- *Mem:* AAAS; fel Weed Sci Soc Am; Am Soc Agron; Coun Agr Sci & Technol. *Res:* Use of herbicides in crops; behavior of herbicides in soils; physiological selectivity and action of herbicides. *Mailing Add:* Altheimer Lab Univ Ark 276 Altheimer Dr Fayetteville AR 72703

TALBERT, WILLARD LINDLEY, JR, b Casper, Wyo, Mar 8, 32; m 52; c 4. NUCLEAR PHYSICS. *Educ:* Univ Colo, BA, 54; Iowa State Univ, PhD(physics), 60. *Prof Exp:* Res physicist, Ohio Oil Co, 59-62; from asst prof to prof physics, Iowa State Univ, 62-76; group leader, 77-79 & 81-82, STAFF MEM, LOS ALAMOS NAT LAB, 76- *Concurrent Pos:* Prog dir nuclear sci, Ames Lab, US Energy Res & Develop Admin, 74-76. *Mem:* Fel Am Phys Soc. *Res:* Experimental nuclear spectroscopy, especially shortlived isotopes using on-line isotope separator. *Mailing Add:* Two La Flora Ct Los Alamos NM 87544

TALBOT, BERNARD, b New York, NY, Oct 6, 37; m 63; c 2. MEDICAL SCIENCES. *Educ:* Columbia Col, BA, 58; Columbia Univ, MD, 62; Mass Inst Technol, PhD(biol), 67. *Prof Exp:* Fel, Mass Inst Technol, 67-69 & Univ Rome, 69-70; grants assoc, NIH, 70-71, med officer, Nat Cancer Inst, 71-75, special asst intramural affairs, 75-78, special asst to dir, 78-81, dep dir, Nat Inst Allergy & Infectious Dis, 81-87, MED OFFICER, DRR, NIH, 87- *Mailing Add:* NIH Bldg WB Rm 10A03 Bethesda MD 20892

TALBOT, DONALD R(OY), b Bridgeport, Conn, Jan 23, 31; m 53; c 5. NUCLEAR ENGINEERING. *Educ:* State Univ NY, BS, 52. *Prof Exp:* Test engr, Gen Elec Co, 52, engr, Knolls Atomic Power Lab, 53-56, proj engr, Atomic Power Equip Dept, 56-57, shift supvr, Vallecitos Atomic Lab, 57-58; mgr eng labs, Nuclear Div, Martin Co, 58-62, proj dir, Floating Nuclear Power Plant, 62-67, dir water resources progs, Chem Div, Martin Marietta Corp, NY, 67-69, dir spec studies, Corp Hq, 68-70, dir environ progs, Corp Res Lab, Md, 70-74, dir, Environ Technol Ctr, 74-84, gen mgr environ systs, Martin Marietta Corp, 84-87; VPRES & GEN MGR, VERSAR, INC, 87- *Concurrent Pos:* Mem natural resources comt, US Chamber Com, 68-80; chmn environ steering comt, Ctr Int Mgt Studies, 73-82; mem bus & indust adv comt, Orgn Econ Coop & Develop, 73-; mem, Com Tech Adv Bd, Panel Proj Independence, Dept Com, 74- 75, Panel Energy Policy, 76-77, mem directorate, man & biosphere, UNESCO, 77-79; consult, Nat Comn Air Qual, 79-80; mem, Environ Protection Agency Task Force, Pres Pvt Sector Surv Cost Control, 83-84. *Honors & Awards:* Dept Defense Antarctica Serv Medal. *Mem:* Water Pollution Control Fedn; Air Pollution Control Asn. *Mailing Add:* Virsar Inc ESM Opers 9200 Rumsey Rd Columbia MD 21045-1934

TALBOT, EUGENE L(EROY), b Ogden, Utah, Jan 18, 21; m 42; c 4. METALLURGY, MATERIALS ENGINEERING. *Educ:* La State Univ, BS, 44; Univ Utah, BS, 47, MS, 48, PhD(metall), 50. *Prof Exp:* Chemist, Tri State Oil & Refining Co, 41-42; explosives, Ogden Ord Depot, 42; inorg applns leader, Minn Mining & Mfg Co, 50-63; supt mat develop, 63-75, MAT SPECIALIST, HERCULES INC, 75- *Concurrent Pos:* Lectr, Brigham Young Univ, 65 & Univ Utah, 67; adj prof, Univ Utah, 69- *Mem:* Soc Advan Mat & Process Eng (nat vpres, 73-75, nat pres, 75-76); Am Soc Metals; Soc Logistics Engrs. *Res:* Materials research; high temperature insulation; graphite; metals; stress corrosion; fractography; protective coatings; surface chemistry; adhesives; sealers; composite structures; ceramics; rocket motor systems. *Mailing Add:* 8670 Russell Park Rd Salt Lake City UT 84121

TALBOT, JAMES LAWRENCE, b Epsom, Eng, Sept 6, 32; div; c 3. GEOLOGY, STRUCTURAL. *Educ:* Cambridge Univ, BA, 54; Univ Calif, Berkeley, MA, 57; Univ Adelaide, PhD(geol), 63. *Prof Exp:* Lectr struct geol, Univ Adelaide, 58-63, sr lectr, 63-67; assoc prof geol, Lakehead Univ, 67-70; prof geol & chmn dept, 70-75, actg acad vpres, Univ Mont, 75-76; vpres acad affairs, 76-84, PROF GEOL, WESTERN WASH UNIV, 84- *Concurrent Pos:* Alexander von Humboldt Found fel, Univ Bonn, 63-64. *Mem:* Geol Soc Am; Sigma Xi. *Res:* Structural analysis of basement cover complexes; analysis of strain in metamorphic rocks; studies on rock cleavage and mylonites. *Mailing Add:* Geol Dept Western Wash Univ Bellingham WA 98225

TALBOT, JOHN MAYO, b Sebastopol, Calif, May 8, 13; m 46; c 3. AEROSPACE MEDICINE, RADIOBIOLOGY. *Educ:* Univ Ore, AB, 35, MD, 38. *Prof Exp:* Med officer, US Army & US Air Force, 39-73; dir sci commun div, George Wash Univ Med Ctr, 73-74; CONSULT BIOMED RES, LIFE SCI RES OFF, FEDN AM SOCS EXP BIOL, 74- *Concurrent Pos:* Consult aerospace med, NASA, 73-76; med consult, Environ Protection Agency, 73-75; pvt consult occup med, 78- *Honors & Awards:* Theodore C Lyster Award, Aerospace Med Asn, 67. *Mem:* Aerospace Med Asn (pres, 68-69); Int Acad Astronaut; Am Col Prev Med; Pan Am Med Asn; Air Force Asn. *Res:* General medical sciences and nutrition; investigations, reviews and reports in aerospace medicine; nutrition safety, nutritional epidemiology and environmental toxicology. *Mailing Add:* 2509 Carrollton Rd Annapolis MD 21403

TALBOT, LAWRENCE, b Brooklyn, NY, Dec 30, 25; m 59; c 2. MECHANICAL ENGINEERING. *Educ:* Univ Mich, PhD(eng mech), 52. *Prof Exp:* Asst prof mech eng, 52-58, assoc prof aeronaut sci, 58-63, vchmn dept, 70-74, PROF MECH ENG, UNIV CALIF, BERKELEY, 63- *Concurrent Pos:* Lectr & consult, Adv Group Aeronaut Res & Develop, Europe, 56-57 & 67-68; Guggenheim fel & vis res fel, All Souls Col, Oxford Univ, 67-68. *Mem:* Am Phys Soc; Am Inst Aeronaut & Astronaut; Sigma Xi. *Res:* Fluid mechanics; rarefied gas dynamics; physiological fluid mechanics. *Mailing Add:* 6117 Etcheverry Hall Univ Calif Col Eng Berkeley CA 94720

TALBOT, LEE MERRIAM, b New Beford, Mass, Aug 2, 30; m 59; c 2. ECOLOGICAL BASIS FOR SUSTAINABLE DEVELOPMENT, ENVIRONMENTAL CONSIDERATIONS IN INTERNATIONAL DEVELOPMENT. *Educ:* Univ Calif, AB, 53, MA & PhD(geog range & vert ecol), 63. *Prof Exp:* Field biologist, Arctic Res Lab, Alaska, 51; staff ecologist, UNESCO-Int Union Conserv, Belg, 54-56; ecologist & dir, EAfrican wildlife & Wild Land Res Proj, Nat Acad Sci-Nat Res Coun, Rockefeller Found, NY Zool Soc & Govt Kenya, 59-63; dir SE Asia proj, Int Union Conserv, 64-65; field rep int affairs ecol & conserv & res ecologist, Smithsonian Inst, 66-70, sci coordr conserv sect, Int Biol Prog, 66-70; sr scientist, President's Coun on Environ Qual, 70-75, asst to chmn & dir int & sci affairs, 75-78; dir conserv & spec sci adv, World Wildlife Fund Int, 78-80; sr sci adv conserv & natural resources, Int Council Sci Unions, 78-83; dir-gen, Int Union Conserv Nature & Natural Resources, 80-83; res fel, Environ & Policy Inst, East-West Ctr, Honolulu, 83-89; vis fel, World Resources Inst, Washington, DC, 84-89; SR ENVIRON ADV, WORLD BANK, 84- *Concurrent Pos:* UNESCO lectr, Southeast Asia, 55, consult, 65-; leader, African wildlife exp, Wildlife Mgt Inst & Am Comt Int Wildlife Protection, 56; Taussig traveling fel, Univ Calif, 58-59; consult, various African, Am, Asian, Australian & Europ govts, 59-90; wildlife adv, UN Spec Fund & EAfrican Agr & Forestry Res Orgn, 63-64; assoc ecol, US Nat Zool Park, 66-; spec adv, Mus Natural Hist, Smithsonian Inst, 67-, res assoc, 74-; overseas consult, Fauna Preservation Soc, 67-; consult, UN Spec Fund, 62-63, Pac Sci Bd, Nat Acad Sci-Nat Res Coun, 64-65, Peace Corps, 66, Int Comn Nat Parks, 66-, Nat Park Serv, 68, AID, 69, WHO, 72, World Bank, 84-, Inter-Am Develop Bank, 87-90, UN Develop Prog, 88-, Asian Develop Bank, 90-; chmn, Am Comt Int Conserv, 74-78; vpres, Int Union Conserv Nature & Natural Resources, 75-78; Regent's lectr, Univ Calif, Santa Barbara, 85-86; adj prof ecology, Union Experimenting Col & Univs, 85-88; vpres, Fauna Soc, London, 86-; adj prof biol, George Mason Univ, 87-90. *Honors & Awards:* Wildlife Soc Award, 63; Albert Schweitzer Medal, Animal Welfare Inst, 75; Distinguished Serv Award, Am Inst Biol Sci, 79. *Mem:* Fel AAAS; Am Soc Mammal; Ecol Soc Am; Int Asn Ecol; Wildlife Soc; Sigma Xi; fel Pop Ref Bur. *Res:* International conservation; wildlife, especially ecology and management; tropical land use and savannah ecology; conservation of renewable natural resources; methodology of ecological research and survey; environmental impact analysis; endangered species; incorporation environmental considerations into economic development. *Mailing Add:* 6656 Chilton Ct McLean VA 22101

TALBOT, NATHAN BILL, b Boston, Mass, Nov 25, 09; m 34; c 2. PEDIATRICS. *Educ:* Harvard Univ, AB, 32, MD, 36. *Prof Exp:* Intern, Children's Hosp, 36-38; Charles Wilder prof pediat, Mass Gen Hosp, head dept & chief Children's Serv, 60-76; from asst to assoc prof, 39-62, EMER PROF, HARVARD MED SCH, 76- *Concurrent Pos:* Asst resident, Children's Hosp, 39-40, asst physician, 41-42; res fel pediat, Harvard Med Sch, 40-41; from asst physician to physician, Mass Gen Hosp, 42-62, hon pediat; consult, Children's Hosp Med Ctr, Boston Lying-In Hosp & Cambridge City Hosp. *Honors & Awards:* Mead Johnson Award & Borden Award, Am Acad Pediat. *Mem:* Fel Am Acad Arts & Sci; fel Am Soc Clin Invest; fel Am Pediat Soc; fel Soc Pediat Res; fel Endocrine Soc. *Res:* Interplay of physical, biologic, social, psychologic and behavioral factors on human development, health and disease. *Mailing Add:* 176 Warren St Brookline MA 02146

TALBOT, PIERRE J, b Quebec City, Que, July 11, 56; m 77; c 3. VIROLOGY, IMMUNOLOGY. *Educ:* Univ Laval, Ste-Foy, BSc, 77; Univ BC, Vancouver, PhD(biochem), 81. *Prof Exp:* Teaching asst biochem, Univ BC, Vancouver, 78-79; res assoc virol, Scripps Clin & Res Found, La Jolla, 81-84; asst prof, 84-89, ASSOC PROF VIROL, INST ARMAND-FRAPPIER, UNIV QUE, 89- *Concurrent Pos:* Prog dir, MSc & PhD prog, Inst Armand-Frappier, Univ Que, 88- *Honors & Awards:* Fisher Sci Award, Can Soc Microbiologists. *Mem:* Am Soc Microbiol; Am Soc Virol; Can Soc Microbiologists; French-Can Asn Advan Sci. *Res:* Molecular studies on viruses, especially coronaviruses, and their relationship to the immune system and pathogenesis; research on viral proteins with the help of modern techniques of cellular fusion and recombinant DNA. *Mailing Add:* Virol Res Ctr Inst Armand-Frappier 531 boul des Prairies Laval PQ H7N 4Z3 Can

TALBOT, PRUDENCE, b Mass, June 9, 44; m 68. REPRODUCTIVE PHYSIOLOGY, CELL BIOLOGY. *Educ:* Wilson Col, BA, 66; Wellesley Col, MA, 68; Univ Houston, PhD(cell biol), 72. *Prof Exp:* Res assoc mammalian fertil, Univ Houston, 72-77; from asst prof to assoc prof, 81-86, PROF BIOL, UNIV CALIF, RIVERSIDE, 86- *Mem:* Am Soc Cell Biol; Am Soc Zoologists; Soc Study Reproduction; Sigma Xi. *Res:* Morphology, physiology and biochemistry of mammalian fertilization and the mechanism of mammalian ovulation; lobster reproductive biology. *Mailing Add:* Dept Biol Univ Calif Riverside CA 92521

TALBOT, RAYMOND JAMES, JR, b Portsmouth, Va, Sept 17, 41; m 68. IMAGE PROCESSING, ASTROPHYSICS. *Educ:* Mass Inst Technol, SB, 63, PhD(physics), 69. *Prof Exp:* Res assoc, Space Sci Physics & Astron, Rice Univ, 69-71, asst prof, 71-76, assoc prof, 76-81; MEM TECH STAFF, AEROSPACE CORP, 81- *Mem:* Am Astron Soc; Sigma Xi. *Res:* Evolution of stars and galaxies; nucleosynthesis; image processing. *Mailing Add:* 1927 Curtis Ave Redondo Beach CA 90278

TALBOT, RICHARD BURRITT, b Waterville, Kans, Jan 4, 33; m 53; c 2. TOXICOLOGY. *Educ:* Kans State Univ, BS, 54, DVM, 58; Iowa State Univ, PhD(physiol), 63. *Prof Exp:* From instr to assoc prof physiol, Iowa State Univ, 58-65; prof physiol & chmn dept, Univ Ga, 65-68, prof toxicol & dean, Col Vet Med, 68-75; dean, 75-85, PROF TOXICOL & MED ETHICS, VA POLYTECH INST & STATE UNIV, 75- *Concurrent Pos:* Prin investr, NIH, Dept Energy, USDA & Food & Drug Admin grants, 59-; fel, Baylor Col Med, 59; consult, NIH, 70-77 & Food & Drug Admin, 84-; ed, J Vet Med Educ, 79-; mem, Comt Vet Manpower, Nat Acad Sci, 81-83; vis scientist, Food & Drug Admin, 84-85. *Mem:* Am Physiol Soc; Am Acad Vet & Comp Toxicol; Am Acad Vet Pharmacol & Therapeut; Am Vet Med Asn; Conf Res Workers Animal Dis; Am Soc Vet Physiologists & Pharmacologists. *Res:* Manpower prediction models and models for management of large data bases of drug information; toxicology of stable rare earth compounds. *Mailing Add:* Col Vet Med Va Tech Campus Blacksburg VA 24061-0442

TALBOT, T(HOMAS) F, b Birmingham, Ala, July 31, 30; m 57; c 3. MECHANICAL ENGINEERING, METALLURGY. *Educ:* Auburn Univ, BME, 52; Calif Inst Technol, MSME, 53; Ga Inst Technol, PhD(mech eng), 64. *Prof Exp:* Sr practice man, Tenn Coal & Iron Div, US Steel Corp, 56-58; asst prof mech eng, Ga Inst Technol, 58-65; assoc prof mat sci & mech design, Vanderbilt Univ, 65-67; from assoc prof to prof eng, Univ Ala, Birmingham, 67-70, dir continuing eng educ, 72-79, prof & chmn mech eng, 83-89; RETIRED. *Concurrent Pos:* Consult, Humble Oil & Refining Co, 59, Chicago Bridge & Iron Co, 65 & Rust Eng Co, 66; chmn bd dirs, Am Alloy Prods, 77-; mem, Bd Regist Prof Engrs & Land Survs, Ala, 85-; B/G USAF Res; consult engr, 90- *Mem:* Am Soc Metals; Am Soc Mech Engrs; Am Soc Eng Educ; Am Welding Soc; Soc Automotive Engrs; Nat Coun Eng Examrs. *Res:* Machine design; materials processing; design aspects of automobile safety and accident reconstruction; design aspects of product liability. *Mailing Add:* 3837 Brook Hollow Lane Birmingham AL 35243

TALBOT, TIMOTHY RALPH, JR, research administration; deceased, see previous edition for last biography

TALBOTT, EDWIN M, b Baltimore, Md, Oct 3, 23; m 47; c 3. ELECTRONICS ENGINEERING, ENGINEERING MECHANICS. *Educ:* Johns Hopkins Univ, BE, 43; Univ Baltimore, JD, 50. *Prof Exp:* Engr, US Indust Chem Div, Airco, 46-50; engr & mgr, Bendix Corp, 50-60; prog mgr, Martin Marietta Corp, 60-64; pres, Chesapeake Systs Corp, 64-68; pres, Varigas Industs, Inc, 68-76, PRES, VARIGAS RES, INC, 77- *Concurrent Pos:* Pres, Etalon Corp, 80-; mem, bd dirs, Sci Assocs, Inc, 81-85, Pneumatics, Inc, 84- *Mem:* Am Defense Preparedness Asn. *Res:* Energy economics; energy conservation; compressed air systems; powered hand tools; fluorescent lighting control. *Mailing Add:* 1314 Doves Cove Rd Timonium MD 21204

TALBOTT, RICHARD LLOYD, b Chicago, Ill, July 15, 35; m 58; c 3. ORGANIC CHEMISTRY. *Educ:* DePauw Univ, BA, 57; Univ Ill, PhD(org chem), 60. *Prof Exp:* NSF fel org chem, Mass Inst Technol, 60-61; sr chemist, Cent Res Dept, 61-66, sr chemist, Indust Tape Div, 66-68, supvr adhesives res, 68-74, res mgr, Packing Systs Div, 74-84, CORP SCIENTIST, 3M CO, 84- *Mem:* Am Chem Soc; Tech Asn Pulp & Paper Indust. *Res:* Pressure-sensitive adhesives chemistry; chemistry of fluorinated oxidants and fluorinated peroxides; oriented plastic films. *Mailing Add:* 4398 Fisher Ln White Bear Lake MN 55110

TALBOTT, TED DELWYN, b Sudan, Tex, Oct 18, 29; m 55; c 2. REGISTRATIONS CHEMISTRY. *Educ:* NTex State Col, BS, 51, MS, 55. *Prof Exp:* Indust chemist, E I Du Pont de Nemours & Co, Inc 55-57; anal chemist, Res Ctr, US Rubber Co, 57-64; SR RES CHEMIST, ANALYTICAL RES SECT, AGR DIV, MOBAY CHEM CORP, 64- *Mem:* Am Chem Soc; Soc Appl Spectros. *Res:* Preparation and assembly of information on pesticides for product chemistry brochures for registrations of technicals and formulations in the United States and Canada; the development of a computer data-base for this information. *Mailing Add:* 6714 N Bales Kansas City MO 64119-1331

TALENT, DAVID LEROY, b Springfield, Mo, Apr 1, 52; m 79; c 4. ORBITAL DEBRIS STUDIES, ORBITAL DEBRIS MODELING. *Educ:* Southwest Mo State Univ, BS, 74; Rice Univ, MS, 79, PhD(space physics), 81. *Prof Exp:* William Gaertner fel astron, Yerkes Observ, Univ Chicago, 80-82; asst prof physics, Abilene Christian Univ, 82-86; PRIN SCIENTIST, SOLAR SYST EXPLOR DEPT, LOCKHEED ENG & SCI CO, 86- *Concurrent Pos:* Sci reporter, KRBC-TV, Abilene, Tex, 84-85; mem, peer rev panel Int Ultraviolet

Explorer Satellite, NASA, 85; chief scientist, Debris Collision Warning Sensor Proj, NASA, Lockheed, 86- *Mem:* Am Astron Soc; Astron Soc Pac; AAAS; Am Inst Aeronaut & Astronaut. *Res:* Observation and modeling of the earth orbital debris environment; design of a space shuttle payload known as the Debris Collision Warning Sensor (DCWS) experiment; author of numerous publications for science journals and one astronomy book. *Mailing Add:* Solar Syst Explor Dept Lockheed Eng & Sci Co Houston TX 77058

TALENT, LARRY GENE, b Cerrogordo, Ark, May 4, 46; m 65; c 2. ECOLOGY, ZOOLOGY. *Educ:* Calif State Univ, Fresno, BA, 70, MA, 73; Ore State Univ, PhD(wildlife), 80. *Prof Exp:* Instr biol, Hartnell Community Col, 73-74; asst prof, 80-85, ASSOC PROF ZOOL, OKLA STATE UNIV, 85- *Mem:* Wildlife Soc; Wilson Ornith Soc; Am Ornithologists Union; Cooper Ornith Soc; Herpetologists League. *Res:* Vertebrate ecology; habitat use and competitive interactions of sympatric species. *Mailing Add:* Dept Zool Okla State Univ Stillwater OK 74078

TALESNIK, JAIME, b Santiago, Chile, May 18, 16; m 41; c 2. PHARMACOLOGY, PHYSIOLOGY. *Educ:* Univ Chile, MD, 41. *Prof Exp:* Second chief instr physiol, Sch Med, Univ Chile, 41-49, from asst prof physiopath & dir dept exp med, 61-63; vis prof, 67-69, PROF PHARMACOL, FAC MED, UNIV TORONTO, 69-, EMER PROF, 82. *Concurrent Pos:* Rockefeller Found grant, Banting & Best Dept Med Res, Univ Toronto, 46-47; Brit Coun grant, Nat Inst Med Res, London, Eng, 52. *Mem:* Chilean Biol Soc; Brit Physiol Soc; Am Soc Pharmacol & Exp Therapeut; Pharmacol Soc Can; NY Acad Sci. *Res:* Physiopharmacology of the coronary circulation; central modulation of cardiovascular reflexes. *Mailing Add:* Dept Pharmacol Univ Toronto Med Sci Bldg Rm 4318 Toronto ON M5S 1A8 Can

TALHAM, ROBERT J, b Cohoes, NY, May 27, 29; m 56; c 4. APPLIED MATHEMATICS, ACOUSTICS. *Educ:* State Univ NY Albany, BA, 55, MS, 56; Rensselaer Polytech Inst, PhD(appl math), 60. *Prof Exp:* Nat Acad Sci-Nat Res Coun res fel, Naval Res Lab, DC, 60-61; mem tech staff, Bell Tel Labs, NJ, 61-64; mgr undersea defense systs eng, Heavy Mil Electronics Dept, Gen Elec Co, 64-77, mgr oper planning, 77-89; RETIRED. *Mem:* Am Math Soc; Soc Indust & Appl Math; Acoust Soc Am. *Res:* Sonar systems; underwater acoustics; sound propagation in non-homogeneous medium; acoustic array design and development; signal processing. *Mailing Add:* 107 Old Powder Mill Rd Fayetteville NY 13066

TALHOUK, RABIH SHAKIB, b Lebanon, Feb 20, 59; US citizen; m 83. EXTRACELLULAR MATRIX, EPITHELIAL. *Educ:* Am Univ Beirut, BSc, 81, MSc, 83; Ohio State Univ, PhD(dairy sci), 88. *Prof Exp:* POSTDOCTORAL FEL CELL & MOLECULAR BIOL, LAWRENCE BERKELEY LAB, UNIV CALIF, BERKELEY & SAN FRANCISCO, 88- *Honors & Awards:* Upjohn Ag/Vet Res Award, 83. *Mem:* Am Soc Cell Biol; Am Dairy Sci Asn. *Res:* Effect of extracellular matrix and extracellular matrix degrading proteinases on epithelial cell function; epithelial tissue growth and development. *Mailing Add:* Div Cell & Molecular Biol Lawrence Berkeley Lab Bldg 83 One Cyclotron Rd Berkeley CA 94720

TALIAFERRO, CHARLES M, b Leon, Okla, Mar 1, 40; m 60; c 3. PLANT BREEDING, PLANT GENETICS. *Educ:* Okla State Univ, BS, 62; Tex A&M Univ, MS, 65, PhD(plant breeding & genetics), 66. *Prof Exp:* Res agronomist, Agr Res Serv, USDA, 65-68; from asst prof to assoc prof, 68-76, PROF FORAGE BREEDING & GENETICS, OKLA STATE UNIV, 76 - *Mem:* Am Soc Agron; Am Genetic Asn. *Res:* Basic genetic and breeding studies involving forage crops. *Mailing Add:* Dept Agron Okla State Univ Stillwater OK 74078

TALIAFERRO, STEVEN DOUGLAS, b Honolulu, Hawaii, Apr 4, 49. MATHEMATICS. *Educ:* San Diego State Univ, BS, 71; Stanford Univ, PhD(math), 76. *Prof Exp:* Teaching asst math, Stanford Univ, 72-76; ASST PROF MATH, TEX A&M UNIV, 76- *Mem:* Am Math Soc. *Res:* Asymptotic behavior and stability of solutions of ordinary differential equations. *Mailing Add:* Dept Math Tex A&M Univ College Station TX 77843

TALL, FRANKLIN DAVID, b New York, NY, Apr 21, 44. SET THEORY, SET THEORETIC TOPOLOGY. *Educ:* Harvard Col, AB, 64; Univ Wis Madison, PHD(math), 69. *Prof Exp:* From asst prof to assoc prof, 69-80, PROF MATH, UNIV TORONTO, 80- *Concurrent Pos:* Vis res prof, Dartmouth Col, 82-83. *Mem:* Am Math Soc; Can Math Soc; Asn Symbolic Logic. *Res:* Topology with emphasis on establishing that topological propositions are not decided by usual axioms for set theory. *Mailing Add:* Univ Toronto Toronto ON M5S 1A1 Can

TALLAL, PAULA, b Austin, Tex, May 12, 47; m 72. EXPERIMENTAL PSYCHOLOGY, SPEECH SCIENCE. *Educ:* NY Univ, BA, 69; Cambridge Univ, PhD(exp psychol), 73. *Prof Exp:* Instr pediat, Sch Med, Johns Hopkins Univ, 74-75, asst prof neurol, 75-79; asst prof, Univ Calif, San Diego, 79-80, assoc prof psychiat, Sch Med, 80-86, prof, 86-88; PROF & CO-DIR, CTR MOLECULAR & BEHAV NEUROSCI, RUTGERS UNIV, 88- *Concurrent Pos:* Prin investr, Nat Inst Neurol Dis & Stroke, NIH, 76-85, Nat Inst Neurol Dis, Commun Dis & Stroke/NIMH, 85- *Honors & Awards:* Distinguished Young Scientist of the Year, Md Acad Sci, 76; President's Award, Notre Dame Col, 77. *Mem:* Int Neuropsychol Soc; Acoust Soc Am; Am Asn Phonetic Sci; Acad Aphasia; Am Speech-Lang-Hearing Asn; Soc Neurosci; Am Psychiat Asn. *Res:* Sensory, perceptual and cognitive function in normal and delayed language development; neural processing of nonverbal and verbal information in various sensory modalities. *Mailing Add:* Ctr Molecular & Behav Neurosci Rutgers Univ 195 Univ Ave Newark NJ 07102

TALLAN, HARRIS H, b New York, NY, July 9, 24; c 3. BIOCHEMISTRY. *Educ:* NY Univ, BA, 47; Yale Univ, PhD(biochem), 50. *Prof Exp:* Asst, Rockefeller Inst, 50-53, asst prof biochem, 53-59; biochemist, Res Labs, Geigy Chem Corp, 59-68; RES SCIENTIST V, DEPT DEVELOP BIOCHEM, NY STATE INST BASIC RES IN DEVELOP DISABILITIES, 68- *Concurrent Pos:* NIH spec fel, Oxford Univ, 57-58. *Mem:* Am Soc Biol Chemists; NY Acad Sci. *Res:* Enzymology; amino acid metabolism; inborn errors of metabolism; neurochemistry. *Mailing Add:* NY State Inst Basic Res Develop Disabilities 1050 Forest Hill Rd Staten Island NY 10314

TALLAN, IRWIN, b New York, NY, June 26, 27; m 59. GENETICS. *Educ:* Rutgers Univ, BA, 49; Ind Univ, PhD(genetics), 57. *Prof Exp:* Technician to H J Muller, Ind Univ, 50, asst to T M Sonneborn, 50-53; lectr zool, 56-58, from asst prof to assoc prof, 58-74, PROF ZOOL, UNIV TORONTO, 74- *Mem:* Am Soc Zool; Brit Soc Gen Microbiol. *Res:* Genetics; nucleo-cytoplasmic interactions in protozoans; infectivity of kappa and other plasmids. *Mailing Add:* Bot Univ Toronto St George Campus 150 College St Toronto ON M5S 1A1 Can

TALLAN, NORMAN M, b Newark, NJ, Sept 24, 32; m 58; c 4. MATERIALS SCIENCE, CERAMICS. *Educ:* Rutgers Univ, BSc, 54; Ohio State Univ, MS, 55; Alfred Univ, PhD(solid state ceramics), 59. *Prof Exp:* Res physicist, Aerospace Res Labs, 59-61, supvr ceramic res, 61-70, dir, Metals & Ceramics Res Lab, 70-76, chief, Processing & High Temperture Mat Br, Air Force Mat lab, 76-77, actg chief, Metals & Ceramics Div, 77-78, chief scientist, Mat Lab, 78-83, CHIEF, METALS & CERAMICS DIV, MAT LAB, WRIGHT-PATTERSON AFB, 83- *Concurrent Pos:* Res asst, Israel Inst Technol, 66-67. *Honors & Awards:* Ross Coffin Purdy Award, 69. *Mem:* Fel Am Ceramic Soc; Am Phys Soc. *Res:* Equilibrium point defect structure and the transport of charge and mass in metal oxides at high temperatures; physical and mechanical properties of ceramic materials. *Mailing Add:* Mat Div WL/MLL Wright-Patterson AFB Dayton OH 45433

TALLARIDA, RONALD JOSEPH, b Philadelphia, Pa, May 26, 37; m 58; c 3. BIOMATHEMATICS, PHARMACOLOGY. *Educ:* Drexel Inst, BS, 59, MS, 63; Temple Univ, PhD(pharmacol), 67. *Prof Exp:* Coop student, Philco Corp-Drexel Inst, 55-59; jr engr, Philco Corp, 59-60; from instr to asst prof math, Drexel Inst, 60-67; asst prof, 67-71, assoc prof pharmacol, 71-78, mem fac, 78-79, PROF PHARMACOL, TEMPLE UNIV, 79- *Concurrent Pos:* Lectr, Philadelphia Col Pharm, 60, PMC Col, 61-62 & cardiovascular training grant prog, Med Sch, Temple Univ, 63-64; consult, Drexel Inst, 67- *Mem:* AAAS; Math Asn Am; Am Soc Pharmacol & Exp Therapeut. *Res:* Mathematical models for application to biology and medicine; drug receptor theory; pharmacology of vascular smooth muscle; pharmacology of morphine; drug induced disease. *Mailing Add:* Dept Pharmacol Temple Univ Health Sci Campus Bd & Ont Philadelphia PA 19140

TALLEDO, OSCAR EDUARDO, b Sullana, Peru, Aug 1, 29; US citizen; m 59; c 3. OBSTETRICS & GYNECOLOGY. *Educ:* San Marcos Univ, Lima, BS, 48, MD, 55; Am Bd Obstet & Gynec, spec cert div gynec oncol, 75. *Prof Exp:* Intern, San Marcos Univ, Lima, 54-55; intern, Crawford W Long Hosp, Emory Univ, 56-57, resident obstet & gynec, 57-58; resident, 58-60, fel, 60-61, from instr to assoc prof, 61-71, actg chmn, 81-82, PROF OBSTET & GYNEC, MED COL GA, 71-, CHIEF GYNEC SERV, 74- *Concurrent Pos:* Nat Heart Inst grant obstet & gynec, Med Col Ga, 65, NIH grant, 68, dir obstet/gynec residency training prog, 74-; consult, Cent State Hosp, Macon City Hosp, Greenville Mem Hosp & Mem Med Ctr, Univ Hosp & Humana Hosp, 69-, Vet Admin Med-Surg Ctr, 74- *Mem:* Fel Am Col Obstet & Gynec; Soc Gynec Invest; AMA; Am Fertil Soc; Soc Gynec Oncol; Gynec-Urol Soc; Gynec Laser Soc. *Res:* Physiology of pregnancy; vascular reactivity in pregnancy; fetal electrocardiography; uterine contractility studies; amniotic fluid; laser surgery and evaluation of chemotherapy in gynecological malignancies; vaginal flora in patients with abnormal Pap smears; urodynamics and cancer of the cervix. *Mailing Add:* 4700 Waters Ave Savannah GA 31404-0996

TALLENT, WILLIAM HUGH, b Akron, Ohio, May 28, 28; m 52; c 3. BIOCHEMISTRY, ORGANIC CHEMISTRY. *Educ:* Univ Tenn, BS, 49, MS, 50; Univ Ill, PhD(biochem), 53. *Prof Exp:* Asst, Univ Tenn, 49-50 & Univ Ill, 50-53; asst scientist, Nat Heart Inst, 53-57; res chemist, G D Searle & Co, 57-64; invests leader, Northern Regional Res Ctr, 64-69, chief indust crops res, 69-75, dir Northern Regional Res Ctr, 75-83, dep adminr, Northeastern Region, 83-84, ASST ADMINR, AGR RES SERV, BELTSVILLE, MD, 84- *Concurrent Pos:* Assoc ed, J Am Oil Chemists Soc, 70-83. *Mem:* Am Chem Soc; Am Oil Chemists Soc; Soc Econ Bot; AAAS; Inst Food Technologists. *Res:* Application of chromatographic and spectroscopic methods to analysis, biochemistry of fats and oils, isolation and structure determination of terpenes, plant lipids and natural insecticides; plant enzymes; useful derivatives and synthetic modifications of natural products; research management. *Mailing Add:* USDA-Agr Res Serv Rm 358 A Admin Bldg Washington DC 20250

TALLERICO, PAUL JOSEPH, b New York, NY, Nov 30, 38; m 62; c 2. ELECTRICAL ENGINEERING. *Educ:* Mass Inst Technol, BS & MS, 61; Univ Mich, PhD(elec eng), 68. *Prof Exp:* Staff mem, Int Bus Mach Res Labs, 62-63; ASSOC GROUP LEADER, LOS ALAMOS NAT LAB, 68- *Concurrent Pos:* Consult, various orgns, 79, 80, 88-90. *Mem:* Sigma Xi; Inst Elec & Electronics Engrs. *Res:* Microwave generation and amplification especially as applied to accelerator power sources. *Mailing Add:* 238 Loma del Escolar Los Alamos NM 87544

TALLEY, CHARLES PETER, b New York, NY, Aug 15, 41; m 68; c 2. ORGANIC CHEMISTRY, POLYMER CHEMISTRY. *Educ:* St Peter's Col, NJ, BS, 63; Polytech Inst New York, PhD(phys org chem), 74. *Prof Exp:* Sr res chemist, Merck Sharp & Dohme Res Labs, Div Merck & Co, 68-72; res chemist, Anal Chem Div, Nat Bur Standards, 73; sr group leader anal res, Calgon Corp, Subsid Merck & Co, Inc, 73-78; dir res & develop, 79-84, GEN MGR INT, GAF CORP, 85- *Mem:* Am Chem Soc; AAAS. *Res:* Gas and high pressure liquid chromatography, especially as applied to the analysis of trace organics in biological, environmental and polymeric matrices. *Mailing Add:* 157 Pleasant Plains Ave Staten Island NY 10309-3699

TALLEY, EUGENE ALTON, b Glenn Allen, Va, June 5, 11; m 44; c 2. AGRICULTURAL CHEMISTRY. *Educ:* Col William & Mary, BCh, 36; Univ Richmond, MS, 38; Ohio State Univ, PhD(org chem), 42. *Prof Exp:* Asst chemist carbohydrate div, Eastern Regional Res Lab, Bur Agr & Indust Chem, USDA, 42-44, chemist, 44-53, sr chemist, 53-80, emer protein chemist, Plant Prod Lab, Eastern Regional Res Lab, Agr Res Serv, 80-86; RETIRED. *Mem:* Am Chem Soc; Am Potato Asn. *Res:* Synthesis of oligosaccharides; preparation of starch esters and ethers; nitrogen compounds in plants; glycoalkaloids. *Mailing Add:* 3100 Quarry Lane Lafayette Hill PA 19444-2006

TALLEY, JOHN HERBERT, b Wilmington, Del, Jan 16, 44; m 67; c 2. GROUND-WATER GEOLOGY. *Educ:* Univ Del, BA, 69; Franklin & Marshall Col, MS, 74. *Prof Exp:* Eng geologist, Geo-Del, Ltd, 71-72; proj geologist, 72-74, SCIENTIST & HYDROGEOLOGIST, DEL GEOL SURV, 74- *Concurrent Pos:* Comnr, Bd Reg Geologists, Del. *Mem:* Asn Eng Geologists; Asn Ground Water Scientist & Engrs. *Res:* Geologic mapping; subsurface stratigraphy and structural interpretation; hydrologic mapping; ground water exploration; ground water-surface water relationships; borehole geophysics. *Mailing Add:* Del Geol Surv Univ Del Newark DE 19716

TALLEY, ROBERT BOYD, b Scottsbluff, Nebr, 1931. INTERNAL MEDICINE. *Educ:* Colo Univ, MD, 56; Am Bd Internal Med, dipl. *Prof Exp:* Intern, Wayne County Gen Hosp, 56-57; resident,Iowa Univ, 59-62, instr med, 62-63; PHYSICIAN INTERNAL MED, PVT PRACT, STOCKTON, CALIF, 63- *Concurrent Pos:* NIH fel, Div Gastroenterol, Univ Iowa, 62-63; consult gastroenterol, San Joaquin County Hosp, 63-70; clin instr med, Med Ctr, Univ Calif, San Francisco, 65-70; chief staff, St Joseph's Hosp, Stockton, Calif, 71, chief med, 75-77, mem bd trustees, 84-86; mem, Community Serv Comt, Am Col Physicians, 71-81, Adv Comt Nat health Ins, House Ways & Means Comt, US Cong, 78-80 & Tech Adv Comt, Group Health Asn Am, 82-85; vpres, United Found Med Care, 82-84, pres, 84-86; med dir & vpres, Concurrent Rev Technol, 87- *Mem:* Inst Med-Nat Acad Sci; fel Am Col Physicians. *Res:* Gastroenterology. *Mailing Add:* 1617 N California 2E Stockton CA 95204

TALLEY, ROBERT MORRELL, b Erwin, Tenn, Mar 13, 24; m 48; c 2. INFRARED DETECTORS. *Educ:* Univ SC, BS, 45; Univ Tenn, MS, 48, PhD(physics), 50. *Prof Exp:* Chief infrared br, US Naval Ord Lab, 51-57, chief solid state div, 57-58; vpres & mgr labs, 58-76, PRES, SANTA BARBARA RES CTR, 76- *Mem:* Am Phys Soc; Optical Soc Am; Sigma Xi. *Res:* Infrared spectroscopy; intermetallic semiconductors; energy bands in solids; photodetectors; military infrared systems. *Mailing Add:* 4511 Carriage Hill Dr Santa Barbara CA 93110

TALLEY, SPURGEON MORRIS, b Atkins, Ark, May 6, 18; m 48; c 1. ANIMAL NUTRITION. *Educ:* Agr, Mech & Norm Col, Ark, BSA, 47; Kans State Univ, MS, 53, PhD(nutrit), 66. *Prof Exp:* Asst prof poultry sci & prod mgr, 54-66, assoc prof animal nutrit, 66-77, PROF ANIMAL SCI, LINCOLN UNIV, MO, 77- *Mem:* Poultry Sci Asn; Am Soc Animal Sci. *Res:* Monogastric animals; nutrition of poultry and swine; plant proteins as sources of protein for the avian species; level of dietary protein and phase feeding on esophagoulcerogenesis of market swine; metabolizable energy requirements of market-type swine. *Mailing Add:* RR 1 Hartsburg MO 64050

TALLEY, THURMAN LAMAR, b Portales, NMex, July 26, 37; m 62; c 2. PHYSICS. *Educ:* Eastern NMex Univ, BS, 59, MS, 60; Fla State Univ, PhD(physics), 68. *Prof Exp:* Instr eng sci, Fla State Univ, 64-65; MEM STAFF, LOS ALAMOS NAT LAB, UNIV CALIF, 66- *Concurrent Pos:* Chmn, Joint AEC-Dept Defense Working Group Safeguard Sprint Nuclear Vulnerability & Effects, 68-71. *Mem:* Am Phys Soc; Sigma Xi. *Res:* Nuclear reaction theory; nuclear weapons design; nuclear weapons effects; computer simulation of complex physical phenomena. *Mailing Add:* 606 Rim Rd Los Alamos NM 87544

TALLEY, WILSON K(INTER), b St Louis, Mo, Jan 27, 35; m 81; c 3. NUCLEAR ENGINEERING, APPLIED MATHEMATICS. *Educ:* Univ Calif, Berkeley, BS, 56, PhD(nuclear eng), 63; Univ Chicago, SM, 57. *Prof Exp:* Physicist, Lawrence Radiation Lab, 59; from asst prof to assoc prof appl sci, 63-72, vchmn dept appl sci, 66-67, actg chmn, 68-69, PROF APPL SCI, UNIV CALIF, DAVIS, 72- *Concurrent Pos:* Consult, Lawrence Livermore Lab, 63-, leader, Theoret Physics Div, 71 & Governor's Select Comt, NY, 66-68; mem, Stanford Res Inst, 67-70; White House fel, Dept Health, Educ & Welfare, 69-70; consult, Hazardous Mat Adv Comt, Environ Protection Agency, 71-74; asst vpres acad planning, Univ Calif, 71-74; pres, Fannie & John Hertz Found, 72-; study dir, Comn Critical Choices Am, 74; asst admin res & develop, US Environ Protection Agency, Washington, DC, 74-77; mem, US Army Sci Bd, 78-86, chmn, 83-86; dir, Helionetics, Inc, 81-86, chmn, 84-86; mem, US Army Med Res Adv Bd, 88- *Honors & Awards:* Except Civilian Serv Medal, US Army, 86. *Mem:* Am Phys Soc. *Res:* Linear transport theory; peaceful uses of nuclear explosives; applications of radioisotopes; energy and environmental policy. *Mailing Add:* 3167 Paseo Granada Pleasanton CA 94566

TALLIAN, TIBOR E(UGENE), b Budapest, Hungary, Oct 18, 20; nat US; m 50; c 2. MECHANICAL ENGINEERING, MECHANICS. *Educ:* Budapest Tech Univ, ME, 43. *Prof Exp:* Supvr eng & res, Ball Bearing Factories, Hungary, 52-56; supvr metrology, 57-58, mgr res lab, 58-68, vpres res, 68-73, vpres technol serv, SKF Industs Inc, 73-85; TRIBOLOGY CONSULT, 85- *Honors & Awards:* Nat Award, Am Soc Lubrication Engrs, 75. *Mem:* Fel Soc Tribology & Lubrication Engrs; fel Am Soc Mech Engrs. *Res:* Tribology, fatigue of metals; applied mechanics of bearings; surface geometry and vibrations of rolling systems; mathematical statistics; random processes; electromechanical instrumentation; failure diagnosis; expert systems. *Mailing Add:* 36 Dunminning Rd Newtown Square PA 19073

TALLITSCH, ROBERT BOYDE, b Oak Park, Ill, June 3, 50; m 71. PHYSIOLOGY. *Educ:* NCent Col, BA, 71; Univ Wis-Madison, MS, 72, PhD(physiol), 75. *Prof Exp:* Res fel, Wis Heart Asn, 74-75; asst prof, 75-83, PROF, BIOL DEPT, AUGUSTANA COL, 83- *Concurrent Pos:* NIH res fel, 81 vis scientist, Geront Res Ctr, Nat Inst Aging, NIH, 81; mem hypertension coun, Am Heart Asn. *Mem:* Assoc Am Physiol Soc; Am Heart Asn; Int Soc Heart Res NAm Chap. *Res:* Ion transport in skeletal and cardiac muscle cells. *Mailing Add:* Dept Biol Augustana Col Biol Bldg Rm 102 Rock Island IL 61201

TALLMADGE, J(OHN) A(LLEN), JR, b Allentown, Pa, Feb 19, 28; wid; c 3. CHEMICAL ENGINEERING, ENVIRONMENTAL SCIENCE. *Educ:* Lehigh Univ, BS, 48; Carnegie-Mellon Univ, MS, 50, PhD(chem eng), 54. *Prof Exp:* Res engr process develop, E I Du Pont de Nemours & Co, Inc 53-56; from asst prof to assoc prof chem eng, Yale Univ, 56-65,; actg head dept, 67 & 73-74, PROF CHEM ENG, DREXEL UNIV, 66- *Concurrent Pos:* Fulbright prof, Univ NSW, Australia, 74; vis prof, Imperial Col London, 65, Univ Calif, Berkeley, 82 & Univ Wash, 83. *Mem:* Am Chem Soc; Am Soc Eng Educ; Am Inst Chem Engrs; Sigma Xi. *Res:* Fluid dynamics; heat transfer, powder metallurgical atomization; industrial water treatment; mass transfer; packed beds; ion exchange; coating processes. *Mailing Add:* Dept Chem Eng Drexel Univ Philadelphia PA 19104

TALLMAN, DENNIS EARL, b Bellefontaine, Ohio, Apr 23, 42; c 3. ELECTROCHEMISTRY, LABORATORY COMPUTERS. *Educ:* Ohio State Univ, BSc, 64, PhD(anal chem), 68. *Prof Exp:* NIH fel chem, Cornell Univ, 68-70; asst prof anal chem, 70-73, assoc prof, 73-78, chmn dept, 77-79, PROF ANAL CHEM, NDAK STATE UNIV, 78- *Concurrent Pos:* Res Corp grant, NDak State Univ, 71-73; Off Water Resources res grants, 71-91; NIH res grant, 74-76; EPA res grant, 75-81; NSF res grants, 87-91. *Mem:* Am Chem Soc; Soc Electroanalytical Chemists. *Res:* Electroanalytical chemistry; high-performance liquid chromatography; environmental chemistry; laboratory applications of small computers; solution dynamics. *Mailing Add:* Dept Chem NDak State Univ Fargo ND 58105-5516

TALLMAN, J(OHN) C(ORNWELL), b Auburn, NY, June 12, 18; m 41; c 4. CHEMICAL ENGINEERING. *Educ:* Cornell Univ, BCh, 39, ChemE, 40. *Prof Exp:* Jr Res engr, Ammonia Dept, Exp Sta, E I du Pont de Nemours & Co, Del, 40-41, WVa, 41-46, asst tech supt, 46-47, res engr, 47-52, econ studies supvr, Textile Fibers Dept, 52-56, indust mkt analyst, Develop Dept, 56-62, mgr mkt res, Latin Am Div, Int Dept, 62-69, mgr develop, Du Pont Do Brasil, 69-71, tech investr, Tech Div, Int Dept, 71-76, sr bus analyst, Finance Div, 77-81; RETIRED. *Concurrent Pos:* Abstractor, Chem Abstr, 47-75, sect ed, 59-62. *Mem:* Am Chem Soc; Chem Mkt Res Asn; Am Inst Chem Engrs. *Res:* Plastics; synthetic fibers; economic studies; market research; foreign exchange. *Mailing Add:* 119 Marcella Rd Wilmington DE 19803

TALLMAN, JOHN FRANCIS, b New York, NY, Jan 24, 47; m 71. PHARMACOLOGY, PSYCHIATRY. *Educ:* Georgetown Univ, BS, 68, PhD(biochem), 72. *Prof Exp:* Staff fel, Nat Inst Neurol & Commun Dis & Stroke, NIH, 72-74, sect chief, NIMH, Biol Psychiat Br, 74-83; assoc prof pharmacol & psychiat, Yale Univ Sch Med, 83-89; SCI DIR & EXEC VPRES, NEUROGEN CORP, 89- *Concurrent Pos:* Adj assoc prof pharmacol & psychiat, Yale Univ Sch Med, 89- *Mem:* Soc Neurosci; Am Soc Biochem & Molecular Biol; Am Soc Pharmacol & Exp Therapeut; Am Col Neuropsychopharmacol; Int Col Neuropsychopharmacol. *Res:* Gaba receptors; molecular biology and pharmacology of benzodiazepines. *Mailing Add:* Neurogen Corp 35 NE Industrial Rd Branford CT 06405

TALLMAN, JOHN GARY, b Sistersville, WVa, Mar 20, 50. GENETICS, BIOCHEMISTRY. *Educ:* West Liberty State Col, AB, 71; WVa Univ, PhD(genetics), 76. *Prof Exp:* Res assoc biochem, Kans State Univ, 76-78; from asst prof to assoc prof biol & genetics, 78-86, PROF BIOL & GENETICS, PEPPERDINE UNIV, 87- *Concurrent Pos:* Found fel genetics, WVa Univ, 71-75, Gulf Oil Found fel, 75-76; Vis scholar, Stanford Univ, 86. *Mem:* Am Genetic Asn; Am Soc Plant Physiol; AAAS. *Res:* Heterosis; chromatin proteins; transcription; DNA tumor viruses; plant protoplasting and tissue culture; stomatal isolation and biochemistry; plant gas exchange; senescence of plant leaves. *Mailing Add:* Div Nat Sci Pepperdine Univ Malibu CA 90263

TALLMAN, RALPH COLTON, organic chemistry, for more information see previous edition

TALLMAN, RICHARD DALE (JUNIOR), CARDIOPULMONARY PHYSIOLOGY, VENTILATORY CONTROL. *Educ:* Ohio State Univ, PhD(physiol), 79. *Prof Exp:* ASST PROF CARDIOPULMONARY RENOPHYSIOL, SCH MED, OHIO STATE UNIV, 79- *Res:* Extracortoreal oxygenation. *Mailing Add:* Dept Allied Med Prof Ohio State Univ Col Med 1583 Perry St Columbus OH 43210

TALLMAN, RICHARD LOUIS, b Wheeling, WVa, Apr 24, 31; m 56; c 3. CORROSION, STRUCTURAL CHEMISTRY. *Educ:* Kenyon Col, AB, 53; Univ Wis, PhD(phys chem), 60. *Prof Exp:* Sr chemist, Res Labs, Westinghouse Elec Corp, 59-73; sr res chemist, Gen Motors Res Labs, 73-76; assoc scientist, 77-80, scientist, EG&G Idaho Inc, 80-86. *Concurrent Pos:* Supvry res mat scientist, Mont Technol Cos, Inc, 89. *Honors & Awards:* Res & Develop 100 Award, 88. *Mem:* Am Chem Soc; Am Nuclear Soc. *Res:* Corrosion; metal oxidation; combustion; crystallography; microscopy; gravimetry; radwaste leaching; ceramic joining. *Mailing Add:* 1653 Halsey Idaho Falls ID 83401

TALMAGE, DAVID WILSON, b Kwangju, Korea, Sept 15, 19; US citizen; m 44; c 5. TRANSPLANTATION, CANCER. *Educ:* Davidson Col, BS, 41; Washington Univ, MD, 44. *Hon Degrees:* DSc, Buena Vista Col, 69, Colo State Univ, 80. *Prof Exp:* USPHS res fel, Washington Univ, 50-51; asst res prof path, Sch Med, Univ Pittsburgh, 51-52; from asst prof to assoc prof med,

Sch Med, Univ Chicago, 52-59; chmn, Dept Microbiol, 63-66, assoc dean, 66-68, actg dean, 68-69, dean, 69-71, dir, Webb-Waring Lung Inst, 73-83, assoc dean, 83-86, prof med, 59-87, prof microbiol, 60-87, DISTINGUISHED PROF MED & MICROBIOLOGY, SCH MED, UNIV COLO, 87- *Concurrent Pos:* Markle scholar med sci, 55-60; consult, Vet Admin Hosp, 59-71; ed, J Allergy, 63-67. *Mem:* Nat Acad Sci; AAAS; Am Soc Clin Invest; Am Asn Immunologists (pres, 78); Am Acad Allergy (pres, 65). *Res:* Effect of oxygen during culture on survival of mouse thyroid allografts; immunological tolerance in animals bearing cultured allografts. *Mailing Add:* Box C 321 Univ Colo 4200 E Ninth Ave Denver CO 80262

TALMAGE, ROY VAN NESTE, b Moppo, Korea, Feb 9, 17; US citizen; m 42; c 3. PHYSIOLOGY. *Educ:* Maryville Col, AB, 38; Univ Richmond, MA, 40; Harvard Univ, PhD(endocrinol), 47. *Prof Exp:* Instr biol, Univ Richmond, 40-41; asst, Harvard Univ, 41-42 & 46-47; from instr to prof, Rice Univ, 47-70, chmn dept, 56-64, master, Wiess Col, 57-70; dir orthop res & prof surg & pharmacol, 70-84, EMER PROF SURG & PHARMACOL, SCH MED, UNIV NC, CHAPEL HILL, 84- *Concurrent Pos:* NIH res fel, State Univ Leiden, 64; mem, NIH Study Sects, 64-68 & 70-; gen chmn & co-chmn parathyroid confs, Houston, 60, Leiden, 64, Montreal, 67, Chapel Hill, 71, Oxford, Eng, 74, Vancouver, 77 & Denver, 80; staff biochemist, AEC, 69-70, mem nat adv dent res coun, Nat Inst Dent Res, 74-77; pres, Int Conf Calcium Regulatory Hormones, 79-81. *Honors & Awards:* William Neuman Award, Am Soc Bone & Mineral Res, 84. *Mem:* AAAS; Orthop Res Soc; Am Soc Zoologists; Soc Exp Biol & Med; Am Physiol Soc; Endocrine Soc. *Res:* Calcium regulating hormones and ion transport processes in bone; osteoporosis and bone density in women. *Mailing Add:* Orthop Res Labs Univ NC Sch Med Clin Sci Bldg 229-H Chapel Hill NC 27514

TALMAN, JAMES DAVIS, b Toronto, Ont, July 24, 31; m 57; c 4. THEORETICAL PHYSICS. *Educ:* Univ Western Ont, BA, 53, MSc, 54; Princeton Univ, PhD, 59. *Prof Exp:* Instr physics, Princeton Univ, 57-59; asst prof, Am Univ Beirut, 59-60; from asst prof to prof math, 60-67, PROF APPL MATH, UNIV WESTERN ONT, 67- *Concurrent Pos:* Res asst, Univ Calif, Davis, 63-64; vis, Niels Bohr Inst, Copenhagen, Denmark, 69-70; vis prof physics & aeronomy, Univ Fla, Gainesville, 78-79. *Mem:* Am Phys Soc; Can Asn Physicists. *Res:* Quantal many-body problem; atomic structure theory; numerical methods. *Mailing Add:* Dept Appl Math Univ Western Ont Engr & Math Sci Bldg London ON N6A 5B9 Can

TALMAN, RICHARD MICHAEL, b Toronto, Ont, Sept 24, 34; m 57; c 4. PHYSICS. *Educ:* Univ Western Ont, BA, 56, MA, 57; Calif Inst Technol, PhD(physics), 63. *Prof Exp:* From asst prof to assoc prof, 62-71, PROF PHYSICS, CORNELL UNIV, 71- *Mem:* Am Phys Soc. *Res:* Elementary and experimental particle physics; accelerator physics. *Mailing Add:* Lab Nuclear Study Cornell Univ Ithaca NY 14853

TALNER, NORMAN STANLEY, b Mt Vernon, NY, Sept 28, 25; m 50; c 3. PEDIATRICS, CARDIOLOGY. *Educ:* Univ Mich, Ann Arbor, BS, 45; Yale Univ, MD, 49. *Hon Degrees:* MA, Yale Univ, 69. *Prof Exp:* Intern & resident pediat, Kings County Hosp, State Univ NY, 49-51; resident, Univ Hosp, Univ Mich, 51-52, instr, Med Sch, 54-56, Mich Heart Asn fel pediat cardiol, Hosp, 56-58, asst prof pediat, Univ, 58-60; from asst prof to assoc prof, 60-69, PROF PEDIAT, SCH MED, YALE UNIV, 69- *Concurrent Pos:* Attend physician, Yale-New Haven Hosp, 60-; USPHS career develop award, 62-72; examr, Sub-Bd Pediat Cardiol, Am Bd Pediat, 69-74; prog chmn, Am Heart Asn, 69-72; consult, Vet Admin, 72-; Exam ed, Sub-Bd Pediat Cardiol. *Mem:* Soc Pediat Res (mem secy, 69-72); Am Pediat Soc; Am Col Cardiol (asst secy, 72-74); cor resp mem Asn Europ Pediat Cardiol. *Res:* Cardiopulmonary physiology in infants and children. *Mailing Add:* Childrens Clin Res Ctr 333 Cedar St New Haven CT 06510

TALVACCHIO, JOHN, b Cleveland, Ohio, Aug 11, 55. SUPERCONDUCTIVITY, DEVICE PHYSICS. *Educ:* Case Western Reserve Univ, BS, 77; Stanford Univ, PhD(appl physics), 82. *Prof Exp:* Res asst, Condensed Matter Physics Group, Case Western Reserve Univ, 75-77; res asst, Hansen Lab, Stanford Univ, 77-82; SR SCIENTIST, CRYOGENIC ELECTRONICS & TECHNOL DEPT, WESTINGHOUSE RES & DEVELOP CTR, 82- *Concurrent Pos:* Vis scientist, NTT Ibaraki Lab, 87; mem Prog Comt Appl Superconductivity Conf; fel, NSF, 78-80; co-prin investr, Air Force Off Sci Res, 88-90. *Mem:* Am Phys Soc; Mat Res Soc; Metall Soc Am Inst Mech Engrs. *Res:* Materials for applied superconductivity; author of over 40 publications in thin film growth and characterization, surface and interface science, electronic device development, and fundamental physics of superconductors. *Mailing Add:* Westinghouse R & D Ctr 1310 Beulah Rd Pittsburgh PA 15235-5098

TALWANI, MANIK, b India, Aug 22, 33; m 58; c 3. GEOPHYSICS. *Educ:* Univ Delhi, BSc, 51, MSc, 53; Columbia Univ, PhD(geol), 59. *Hon Degrees:* PhD, Univ Oslo, Norway, 81. *Prof Exp:* Mem staff, Lamont-Doherty Geol Observ, Columbia Univ, 57-81, prof geol, Univ, 70-81, dir, Observ, 73-81; dir, Ctr Crustal Studies, Gulf Res & Develop Co, 81-83, chief scientist, Co, 83-85; SCHLUMBERGER PROF GEOPHYSICS, RICE UNIV, 85-; DIR, GEOTECHNOL RES INST, HOUSTON ADVAN RES CTR, 85- *Concurrent Pos:* Mem ocean affairs bd & exec comt, Joint Oceanog Inst Deep Earth Sampling; mem, Ocean Policy Comt, Nat Acad Sci & Bd Gov Joint Oceanog Inst, Inc; Fulbright-Hays Fel, 73; Guggenheim fel, 74. *Honors & Awards:* Indian Geophys Union First Krishnan Medal, 65; James B Macelwane Award, Am Geophys Union, 67; NASA Exceptional Sci Achievement Award, 73; Maurice Ewing Award, Am Geophys Union, 81; George Woollard Award, Geol Soc Am, 83. *Mem:* Am Soc Explor Geophys; fel Am Geophys Union; Seismol Soc Am; fel Geol Soc Am; fel Royal Astron Soc; Sigma Xi. *Res:* Marine geophysics; oceanography, geodesy. *Mailing Add:* 1111 Hermann Dr Apt 10D Houston TX 77004

TAM, ANDREW CHING, b Canton, China, Oct 13, 44; m 70; c 2. ATOMIC PHYSICS, MOLECULAR PHYSICS. *Educ:* Univ Hong Kong, BSc, 68, MSc, 70; Columbia Univ, PhD(physics), 75. *Prof Exp:* Fac fel physics, Columbia Radiation Lab, Columbia Univ, 70-72, preceptor, 71-72, res asst physics, 72-74, asst prof, 75-77; mem tech staff, Bell Labs, 78-79; res staff mem, 79-84, MGR, IBM RES DIV, 85- *Mem:* Fel Optical Soc Am; sr mem Inst Elec & Electronics Engrs; fel Am Phys Soc; Acoust Soc Am; Int Soc Optical Eng. *Res:* Atomic and molecular spectroscopy; optical pumping; excimer lasers applications; laser interaction with matter & surfaces; photoacoustics; photothermal sensors. *Mailing Add:* Dept K66-803 IBM Almaden Res Ctr San Jose CA 95120-6099

TAM, CHICK F, b Toishan, China, Jan 17, 46. AGING, IMMUNOLOGY. *Educ:* Univ Calif, Los Angeles, PhD(pub health), 74. *Prof Exp:* ASSOC PROF NUTRIT, CALIF STATE UNIV, LOS ANGELES, 76- *Concurrent Pos:* Cert home economist. *Mem:* Am Inst Nutrit; Sigma Xi; NY Acad Sci; Inst Food Technol; Am Dietetics Asn; Am Home Econ Asn. *Mailing Add:* Dept Family Studies & Consumer Sci Rm FA255 Calif State Univ Los Angeles CA 90032

TAM, CHRISTOPHER K W, US citizen; m; c 2. FLUIDS, NOISE. *Educ:* McGill Univ, BEng, 62; Calif Inst Technol, MSc, 63, PhD(appl mech), 66. *Prof Exp:* Res fel, Calif Inst Technol, 66-67; asst prof, Mass Inst Technol, 67-71; assoc prof, 71-76, PROF MATH, FLA STATE UNIV, 76- *Honors & Awards:* Aero Acoust Award, Am Inst Aeronaut & Astronaut, 87. *Mem:* Acoust Soc Am; Soc Indust & Appl Math; Am Inst Aeronaut & Astronaut; Am Phys Soc; Am Geophys Union. *Res:* Physics of noise generation and propagation in aeroacoustics, including jet noise; turbulence and hydrodynamic stability theory; applied mathematics, computational mathematics and computational fluid dynamics. *Mailing Add:* Dept of Math Fla State Univ Tallahassee FL 32306-3027

TAM, JAMES PINGKWAN, b Hong Kong, Mar 25, 47; m 72; c 2. MEDICINAL CHEMISTRY, BIOCHEMISTRY. *Educ:* Univ Wis, BS, 71, PhD(pharm), 76. *Prof Exp:* fel, 76-77, res assoc, 77-79, asst prof, 80-82, ASSOC PROF BIOCHEM, ROCKEFELLER UNIV, 82- *Mem:* Am Chem Soc; NY Acad Sci; Am Soc Biol Chem; AAAS. *Res:* New synthetic methods; solution and solid phase peptide syntheses; hormonal, immunological peptides and enzyme inhibitors; design and synthesis of vaccine; biologic functions of tumor growth factors. *Mailing Add:* Rockefeller Univ 1230 York Ave Box 294 New York NY 10021

TAM, KWOK KUEN, b Hong Kong, Oct 30, 38; m 64; c 2. APPLIED MATHEMATICS. *Educ:* Univ Toronto, BASc, 62, MA, 63, PhD(appl math), 65. *Prof Exp:* Asst prof, 65-69, assoc prof, 70-79, PROF APPL MATH, MCGILL UNIV, 80- *Concurrent Pos:* Res fel, Harvard Univ, 71-72. *Mem:* Can Math Cong. *Res:* Fluid mechanics; construction of approximate solutions to some nonlinear boundary value problems. *Mailing Add:* Dept Math & Sci McGill Univ 853 Sherbrooke St W Montreal PQ H3A 2M5 Can

TAM, KWOK-WAI, b Hong Kong, Mar 16, 38; US citizen; m 68; c 3. MATHEMATICAL ANALYSIS, OPERATIONS RESEARCH. *Educ:* Univ Wash, BS, 60, PhD(math), 67. *Prof Exp:* Teaching asst math, Univ Wash, 61-66; asst prof, 66-75, ASSOC PROF MATH, PORTLAND STATE UNIV, 75- *Mem:* Am Math Soc. *Res:* Mathematical programming. *Mailing Add:* Dept Math Portland State Univ Box 751 Portland OR 97207

TAM, PATRICK YUI-CHIU, b Canton, China, June 7, 48; US citizen; m 76; c 3. BIOMEDICAL ENGINEERING FOR THIRD WORLD, TECHNOLOGY TRANSFER. *Educ:* Mass Inst Technol, BS, 71, MS, 72; Univ Calif, Berkeley, PhD(mech eng), 78. *Prof Exp:* NIH fel bioeng, Univ Wash, 78-79, affil asst prof, 84-87; prog dir, Prog Appropriate Technol Health, 79-84; pres, Wash Res Found, 84-87 & Chemfet Int, 87-88; exec vpres, US Tech, 88-89; PRIN, DANUBE INT, 89- *Concurrent Pos:* Consult, WHO, 80-82; expert witness, US Senate Comt Sci & Technol, 86. *Mem:* NY Acad Sci; Biophys Soc; AAAS. *Res:* Physical properties of biological materials; technology transfer policy issues. *Mailing Add:* 6150 NE 192nd Seattle WA 98155

TAM, SANG WILLIAM, US citizen. ANTIPSYCHOTICS, COGNITIVE ENHANCERS. *Educ:* Univ Wis-Oshkosh, BS, 74; Univ Nebr, MS, 76; State Univ NY, PhD(biochem), 79. *Prof Exp:* Teaching asst biochem, Univ Nebr-Lincoln, 74-76; res asst, Downstate Med Ctr, State Univ NY, 76-79; postdoctoral fel, Sch Med, Yale Univ, 79-81; res pharmacologist, E I du Pont de Nemours & Co, Inc, 81-86, sr res pharmacologist, 86-88, group leader, 88-89; SR GROUP LEADER, DUPONT MERCK PHARMACEUT CO, 90- *Concurrent Pos:* Vis prof, Shanghai Med Univ, Shanghai, China, 87-88. *Mem:* Am Soc Pharmacol & Exp Therapeut; Soc Neurosci; Mid-Atlantic Pharmacol Soc; Soc Chinese Bioscientists Am. *Res:* Discovery of novel therapeutic agents for the treatment of central nervous system diseases; antipsychic, analgesic and cognitive enhancer research: biochemical pharmacology, behavior and molecular biology. *Mailing Add:* CNS Dis Res DuPont Merck Pharmaceut Co PO Box 80400 Wilmington DE 19880-0400

TAM, WING YIM, b Hong Kong, Dec 12, 53; m 87; c 1. CHAOS, FRACTAL. *Educ:* Univ Calif, Santa Barbara, PhD(physics), 85. *Prof Exp:* Res fel res, Univ Tex, Austin, 85-88; ASST PROF TEACHING & RES, UNIV ARIZ, TUCSON, 88- *Concurrent Pos:* Postdoctoral fel, IBM, 85-86; Sloan fel, Alfred P Sloan Found, 90-92. *Mem:* Am Phys Soc. *Res:* Nonlinear dynamics; chaos; fractal formation. *Mailing Add:* Physics Dept Univ Ariz Tucson AZ 85721

TAM, WING-GAY, Can citizen. PHYSICS. *Educ:* Hong Kong Univ, BSc, 60; Univ BC, MSc, 64, PhD(physics), 67. *Prof Exp:* Nat Res Coun overseas fel theoret physics, Univ Nijmegan, Neth, 67-69; res assoc molecular physics, Laval Univ, Can, 69-72, asst prof, 72-74; DEFENSE SCIENTIST OPTICAL PHYSICS, DEFENSE RES ESTAB, VALCARTIER, QUE, 74- *Mem:* Optical Soc Am. *Res:* Atmospheric propagation of electromagnetic waves; energy transfer in molecular systems; atmospheric aerosols. *Mailing Add:* Five McClure Crescent Kanata ON K2L 2H1 Can

TAMANO, TERUO, b Tokyo, Japan, Feb 13, 37; m 64; c 2. NUCLEAR FUSION, PLASMA PHYSICS. *Educ:* Univ Tokyo, BS, 61, MS, 63, PhD(physics), 66. *Prof Exp:* Res assoc physics, Univ Tokyo, 66-72; fel plasma physics, Plasma Physics Lab, Princeton Univ, 69-70; assoc scientist plasma physics, Gen Atomic Co, Gulf Oil Corp, 70, sr scientist, 70-74, dept mgr, 74-78, PROG MGR PLASMA PHYSICS, GEN ATOMICS, 78- *Mem:* Fel Am Phys Soc; Phys Soc Japan. *Res:* Development of confinement concepts in nuclear fusion; plasma physics and manage experimental programs; investigation of the toroidal pinch concept, called OHTE. *Mailing Add:* Gen Atomics PO Box 85608 San Diego CA 92138

TAMAOKI, TAIKI, b Miki, Hyogo-Ken, Japan, Dec 3, 28; m 61; c 4. BIOCHEMISTRY, PLANT PATHOLOGY. *Educ:* Univ Tokyo, BSc, 51; Purdue Univ, MS, 58; Univ Wis, PhD(plant path), 60. *Prof Exp:* Fel oncol, McArdle Lab, Univ Wis, 61-64; asst prof biochem, Cancer Res Unit, Univ Alta, 68-80; at Southern Alta Cancer Ctr, 80-; AT ONCOL RES GROUP, FAC MED, UNIV CALGARY. *Mem:* AAAS; Am Chem Soc; Am Soc Biol Chem; Am Asn Cancer Res; Can Biochem Soc. *Res:* Regulation of protein and RNA synthesis in mammalian cells. *Mailing Add:* Dept Biochem Univ Calgary FAC Med 3330 Hospital Dr NW Calgary AB T2N 4N1 Can

TAMAR, HENRY, b Vienna, Austria, Sept 15, 29; nat US; m 55; c 3. PROTOZOOLOGY, PHYSIOLOGY. *Educ:* NY Univ, AB, 49, MS, 51; Fla State Univ, PhD(physiol), 57. *Prof Exp:* Researcher, Lebanon Hosp, New York, 51; asst physiol, Fla State Univ, 51-55; asst prof biol, Am Int Col, 55-57; prof & head div, Pembroke State Col, 57-62; assoc prof, 62-77, PROF ZOOL, IND STATE UNIV, TERRE HAUTE, 77- *Concurrent Pos:* Vis prof, Stephen F Austin State Col, 59, NC State Col, 61-62 & Marine Biol Labs, Woods Hole, 65, 72 & 74, Marine Sci Ctr, Santa Catalina Island, 81. *Mem:* Am Micro Soc; Soc Protozool. *Res:* Principles of sensory physiology; locomotion, responses and structure of ciliates; jumping ciliates and their forward avoidance reactions. *Mailing Add:* Dept Life Sci Ind State Univ Terre Haute IN 47809

TAMARELLI, ALAN WAYNE, b Wilkinsburg, Pa, Aug 13, 41; m 63; c 2. BATCH SYNTHESIS OF POLYMERS, CATALYTIC POLLUTION CONTROL. *Educ:* Carnegie-Mellon Univ, BS, 63, MS, 65, PhD (chem eng), 66; NY Univ, MBA, 72. *Prof Exp:* Asst prof chem eng, Carnegie-Mellon Univ, 65-66; res engr, Exxon Corp, 66-70; sr vpres, Engelhard Corp, 70-83; CHMN, DOCK RESINS CORP, 89- *Concurrent Pos:* Pres, Linden Indust Asn, 90-; chmn, Chem Indust Coun NJ, 90-; vchmn, Synthetic Org Chem Mfrs Asn, 91- *Mem:* Am Chem Soc; Am Inst Chem Engrs; Am Soc Safety Engrs; Fedn Socs Coatings Technol. *Res:* Chemical kinetics; catalysis; reactor engineering; polymerization; organic coatings applications. *Mailing Add:* 49 Wexford Way Basking Ridge NJ 07920

TAMARI, DOV, b Fulda, Ger, Apr 29, 11; m 48; c 3. MATHEMATICS. *Educ:* Hebrew Univ, Israel, MSc, 39; Univ Paris, Dr es Sc(math), 51. *Prof Exp:* Res fel math, Nat Ctr Sci Res, Paris, France, 49-53; sr lectr, Israel Inst Technol, 53-55, assoc prof, 55-59; prof, Univ Rochester, 59-60; mem, Inst Advan Study, 60-61 & 67-68; Orgn Am States vis prof, Univ Brazil, 61-62; res assoc, Univ Utrecht, 62; prof, Univ Caen, 62-63; chmn dept, 64-67, prof, 63-81, EMER PROF MATH, STATE UNIV NY BUFFALO, 81- *Concurrent Pos:* Vis prof, Hebrew Univ, Israel, 53-59; mem, Inst Advan Study, 67-68; vis prof, Israel Inst Technol, 73. *Mem:* Am Math Soc; Math Asn Am; Asn Symbolic Logic. *Res:* Algebra; semi-group, group and ring theory; embedding and word problems; topological semi-groups, groups and fields; mathematical logic; binary relations; partial algebras; combinatorial analysis; associativity theory and the four-color-map problem; theory of lists and standard polyhedra with Whitney cycles. *Mailing Add:* 175 W 76 St Apt 1E New York NY 10023

TAMARIN, ARNOLD, b Chicago, Ill, Mar 27, 23; m 45; c 2. ORAL BIOLOGY, HISTOLOGY. *Educ:* Univ Ill, BS, 49, DDS, 51; Univ Wash, MSD, 60. *Prof Exp:* Assoc prof, 66-69, PROF ORAL BIOL, UNIV WASH, 69-, ADJ PROF, DEPT BIOL STRUCT, 74- *Concurrent Pos:* Hon res fel, Dept Anat & Embryol, Univ Col London, 82-; vis res prof, Middlesex Hosp Med Sch, 82-83. *Mem:* Fel Royal Micros Soc; Am Asn Anatomists; fel Zool Soc London; Am Soc Cell Biol; Pan-Am Anat Asn. *Res:* Comparative odontology; cell kinematics in the exocrine secretory process; exocrine collagen secretion in mytilus; ultrastructural morphology; embryological morphogenesis. *Mailing Add:* Dept Oral Biol SB-22 Univ Wash Seattle WA 98195

TAMARIN, ROBERT HARVEY, b Brooklyn, NY, Dec 14, 42; m 68; c 2. GENETICS, ECOLOGY. *Educ:* Brooklyn Col, BS, 63; Ind Univ, PhD(zool), 68. *Prof Exp:* Comt Instnl Coop traveling scholar, Univ Wis, 67-68; USPHS fel genetics, Univ Hawaii, 68-70; Ford Found fel, Princeton Univ, 70-71; from asst prof to assoc prof, 71-83, PROF BIOL, BOSTON UNIV, 83- *Concurrent Pos:* NIH & NSF res grants. *Mem:* AAAS; Sigma Xi; Am Soc Mammal; Am Soc Naturalists; Ecol Soc Am. *Res:* Population biology, including genetics, demography, reproductive physiology, behavior and general ecology of insular and mainland field mice to understand population regulation; tropical and radiation studies of field mice. *Mailing Add:* Dept Biol Boston Univ Boston MA 02215

TAMASHIRO, MINORU, b Hilo, Hawaii, Sept 16, 24; m 52. INSECT PATHOLOGY. *Educ:* Univ Hawaii, BS, 51, MS, 54; Univ Calif, PhD, 60. *Prof Exp:* Asst entom, 51-54, jr entomologist, 54-55, from asst prof entom & asst entomologist to assoc prof entom & assoc entomologist, 57-73, PROF ENTOM & ENTOMOLOGIST, UNIV HAWAII, 73- *Concurrent Pos:* WHO consult, 63; NIH fel, 64-65. *Mem:* Entom Soc Am; Sigma Xi. *Res:* Microbial control; effect of pathogens and insecticides on biological control; termites, biology, ecology, control. *Mailing Add:* Dept Entom GIL 601 Univ Hawaii at Manoa 2500 Campus Rd Honolulu HI 96822

TAMBASCO, DANIEL JOSEPH, b Amsterdam, NY, Mar 10, 36; m 66; c 2. THEORETICAL PHYSICS. *Educ:* Union Col, BS, 58; Univ Iowa, PhD(physics), 65. *Prof Exp:* Asst prof, 65-69, ASSOC PROF PHYSICS, MERRIMACK COL, 69- *Mem:* Am Phys Soc. *Res:* Field theory theory; statistical mechanics. *Mailing Add:* Dept Physics Merrimack Col North Andover MA 01845

TAMBORLANE, WILLIAM VALENTINE, b New York, NY, Aug 25, 46; m 69; c 3. PEDIATRICS, DIABETOLOGY. *Educ:* Georgetown Univ, BS, 68, MD, 72. *Prof Exp:* Resident pediat, Georgetown Univ; fel pediat endocrinol, 75-76, endocrinol & metabolism, 76-77, asst prof pediat to assoc prof pediat, 77-86, PROF PEDIAT, YALE UNIV SCH MED, 86- *Concurrent Pos:* Attending physician, Yale-New Haven Hosp, 77-; sect chief, Yale Univ Sch Med, 85; dir, Yale Childrens clin Res Ctr, 86- *Honors & Awards:* Mary Jane Kugel Award, Juv Diabetes Found Int, 85; Peter May Award, Am Diabetes Assoc, 85. *Mem:* Soc Pediat Res; Am Fedn Clin Res; Am Bd Pediat; Lawson Wilkens Pediat Endocrine Soc; Am Acad Pediat; Endocrine Soc. *Res:* Disorders of metabolism with special emphasis on diabetes mellitus in children; effect of treatment on diabetes control and diabetic complications currently under investigation. *Mailing Add:* Childrens Clin Res Ctr Yale Univ 333 Cedar St New Haven CT 06510

TAMBORSKI, CHRIST, b Buffalo, NY, Nov 12, 26; m 49; c 7. ORGANIC CHEMISTRY, FLUORINE CHEMISTRY. *Educ:* Univ Buffalo, BA, 49, PhD(org chem), 53. *Prof Exp:* Fel, Univ Buffalo, 53; sr scientist, Air Force Mat Lab, Wright-Patterson AFB, 55-86; CONSULT, FLUIDICS INC, 86- *Concurrent Pos:* Consult, Childrens Hosp Res Found, 74-, Sun Oil Co, 80-; chmn, Fluorine Div, Am Chem Soc, 72. *Honors & Awards:* Jacobowitz Award, 52; US Dept Com Inventors Award. *Mem:* Am Chem Soc; Sigma Xi. *Res:* High temperature stable fluids and elastomers for advanced aerospace applications; synthesis of organometallic compounds, heterocyclic compounds, organoaliphatic and aromatic fluorine compounds, anti-oxidants. *Mailing Add:* Fluidics Inc PO Box 291886 Dayton OH 45429

TAMBURIN, HENRY JOHN, b Passaic, NJ, July 24, 44; m 68; c 2. INDUSTRIAL ORGANIC CHEMISTRY. *Educ:* Seton Hall Univ, BS, 66; Univ Md, College Park, PhD(org chem), 71. *Prof Exp:* From teaching asst to instr org chem, Univ Md, College Park, 66-72; res & develop chemist, Toms River Chem Corp, 72-75, sr develop chemist, 75-77, actg group leader, 78-79, sr prod chemist, 79-81, GROUP LEADER, CIBA-GEIGY CORP, 81- *Mem:* Am Chem Soc. *Res:* Modern chemical and processing technology in the dyestuff manufacturing process. *Mailing Add:* 5770 Kingston Court Mobile AL 36609-3313

TAMBURINO, LOUIS A, b Pittsburgh, Pa, May 9, 36; m 58; c 1. THEORETICAL PHYSICS. *Educ:* Carnegie Inst Technol, BS, 57; Univ Pittsburgh, PhD(physics), 62. *Prof Exp:* Res assoc, Syracuse Univ, 63-64; res physicist, Aerospace Res Labs, 64-72, MATH PHYSICIST, AIR FORCE AVIONICS LAB, USAF, 72- *Mem:* Asn Comput Mach; Inst Elec & Electronics Engrs; Sigma Xi; Am Phys Soc. *Res:* General relativity; airborne electronic terrain map and display systems; inertial navigation; pattern recognition; optics; image processing; neural networks and learning systems. *Mailing Add:* 2930 E Stroop Rd Kettering OH 45440

TAMBURRO, CARLO HORACE, b Caserta, Italy, Jan 20, 36; US citizen; m 71; c 4. INTERNAL MEDICINE, HEPATOLOGY. *Educ:* Georgetown Univ, BS, 58; Seton Hall Univ, MD, 62; Columbia Univ, MPH, 85. *Prof Exp:* Intern med, Jersey City Med Ctr, NJ, 62-63, resident, 63-64; asst, Sch Med, Tufts Univ, 64-65; instr med, NJ Med Sch, 67-68; asst prof, Col Med & Dent NJ, Newark, 69-74; assoc prof, NJ Med Sch, 74; assoc prof med, Sch Med, Univ Louisville, 74-77, assoc oncol, Cancer Ctr, chief, Div Digestive Dis & Nutrit, & dir, Vinyl Chloride Proj, 74-80, prof commun health, 81-89, PROF MED, UNIV LOUISVILLE, 74-, PROF PHARMACOL & TOXICOL, 90- *Concurrent Pos:* Resident med, New Eng Ctr Hosp, Boston, 64-65; NIH fel hepatic dis, 65-68; assoc pediat, Sch Med, Univ Louisville, 77-, dir, Liver Res Ctr, 80-, chief, Div Occup Toxicol, 81-, assoc pharmacol & toxicol, 82-84. *Mem:* Int Asn Study Liver; Am Asn Study Liver Dis; Am Soc Human Genetics; Am Col Toxicol; Am Fedn Clin Res; Soc Toxicol; fel Am Col Physicians; fel Am Col Nutrit; Int Soc Environ Epidemiol; Am Col Epidemiol; AAAS; Am Col Occup Med; Am Pub Health Asn; Am Soc Parenteral & Enteral Nutrit. *Res:* Hepatic cancer; industrial chemical carcinogenesis, vinyl monomer; clinical toxicology; alcoholism, drug addiction, and withdrawal syndromes; viral hepatitis; liver disease and nutrition; vitamin metabolism and deficiency; liver regeneration and metabolism; hepatic collagen formation; immunology and liver injury. *Mailing Add:* Dir Liver Res Ctr/ch Div Occup Univ Louisville HSC Bldg Rm 119 Louisville KY 40292

TAMBURRO, KATHLEEN O'CONNELL, b New York, NY, Oct 30, 42; m 71; c 4. PROTOZOOLOGY. *Educ:* Marymount Manhattan Col, BA, 64; Fordham Univ, MS, 65, PhD(biol, protozool), 68. *Prof Exp:* From res asst to res assoc biochem & physiol of protozoa, Haskins Labs, 65-74; admin assoc & grant coordr-med/ed, Div Digestive Dis & Nutrit, 76-80, SR MED ED & ADMIN ASSOC, DIV OCCUP DIS, SCH MED, UNIV LOUISVILLE, 80- *Mem:* AAAS; Sigma Xi; Soc Protozool; Am Soc Trop Med & Hyg; Am Soc Microbiologists. *Res:* Protozoa as pharmacological tools; chemotherapy of trypanosomatid parasites; nutrition; biochemistry and physiology of Trypanosomatidae. *Mailing Add:* 512 Brandon Rd Louisville KY 40207

TAMERIUS, JOHN, b Bremertown, Wash, June 27, 45. MICROBIOLOGY, IMMUNOLOGY. *Educ:* Univ Wash, PhD(microbiol & immunol), 76. *Prof Exp:* Vpres & dir res & develop, Cytotech, Inc, San Diego, 82-89; VPRES REGULATORY AFFAIRS QUAL ASSURANCE, QUIDEL CORP, 89- *Mailing Add:* Quidel Corp 10165 McKellar Ct San Diego CA 92121

TAMHANE, AJIT CHINTAMAN, b Bhiwandi, India, Nov 12, 46; m 75; c 2. APPLIED STATISTICS, MATHEMATICAL STATISTICS. *Educ:* Indian Inst Technol, Bombay, BTech(Hon), 68; Cornell Univ, MS, 73, PhD(statist), 75. *Prof Exp:* Jr engr design, Larsen & Toubro Ltd, Bombay, 68-70; from asst prof to assoc prof, 75-87, PROF INDUST ENG & MGT SCI & STATIST, 87- *Concurrent Pos:* Statist consult. *Honors & Awards:* WJ Youdan Prize, 85. *Mem:* Inst Math Statist; Am Statist Asn; Biometrics Soc; Am Soc Qual Control. *Res:* Ranking and selection procedures; multiple comparisons; design of experiments; biostatistics; engineering statistics. *Mailing Add:* Dept Indust Eng McCormick Sch Eng & Appl Sci Northwestern Univ Evanston IL 60208

TAMIMI, YUSUF NIMR, b Nablus, Jordan, Nov 15, 31; m 63; c 3. SOIL CHEMISTRY. *Educ:* Purdue Univ, BS, 57; NMex State Univ, MS, 60; Univ Hawaii, PhD(soil chem), 64. *Prof Exp:* Asst agronomist, Univ, 63-70, assoc soil scientist, 70-75, PROF SOIL SCI, AGR EXP STA, UNIV HAWAII, 75- *Concurrent Pos:* Vis prof, Purdue Univ, 70-71; prof & dept head, Dept Soil Sci & Irrig, Univ Jordan; Sr Soil Scientist, Wash State Univ, Jordan, 85-87; consult. *Mem:* Am Soc Agron; Int Soc Soil Sci; Am Soil Sci Soc; Sigma Xi. *Res:* Chemistry of soil phosphorous; field crops, tropical pasture fertilization and forest tree nutrition; forest soils. *Mailing Add:* Agr Exp Sta Univ Hawaii 461 W Lanikaula Hilo HI 96720

TAMIR, HADASSAH, b Haifa, Israel, Oct 5, 30; US citizen; m 49; c 2. NEUROSCIENCE, BIOCHEMISTRY. *Educ:* Hebrew Univ, Jerusalem, MSc, 55; Israel Inst Technol, DSc(chem), 59. *Prof Exp:* Mem res staff, Princeton Univ, 65-67; res assoc, Med Sch, Columbia Univ, 67-71; SR RES SCIENTIST, DIV NEUROSCI, PSYCHIAT INST STATE NY, 71- *Concurrent Pos:* Res fel biochem, Pub Health Res Inst, City of New York, 59-63; res fel biochem & bact, Med Sch, NY Univ, 63-65. *Res:* Effects of drugs on release and uptake of biogenic amines in nerve endings and cell-free systems; storage and release of Serotinon in neurons and paraneurons (parafollicular cells of the thyroid); comparison with storage of the amines with non-neuronal cells such as platelets, mast cells and the enterochromaffin cells of the gut mucosa. *Mailing Add:* Psychiat Inst State NY 722 W 168th St New York NY 10032

TAMIR, THEODOR, b Bucharest, Roumania, Sept 17, 27; m 49; c 2. ELECTROPHYSICS, OPTICS. *Educ:* Israel Inst Technol, BS, 53, Dipl Ing, 54, MS, 58; Polytech Inst Brooklyn, PhD(electrophys), 62. *Prof Exp:* Res engr, Sci Dept, Ministry Defense, Israel, 53-56; instr elec eng, Israel Inst Technol, 56-58; res assoc, 58-62, from asst prof to assoc prof, 62-69, PROF ELECTROPHYS, POLYTECH UNIV, 69- *Concurrent Pos:* Consult indust & govt labs; co-ed, Springer Series in Optical Sciences, 79-; adv ed, Optics Commun, 75-84; head, Dept Elec Eng, Polytech Inst NY, 74-79; NSF res grants. *Honors & Awards:* Inst Prem, Inst Elec Engrs, UK, 65; Spec Recognition, Antennas & Propagation Soc, Inst Elec & Electronics Engrs, 68; Citation Distinguished Res, Polytech Chap, Sigma Xi, 78. *Mem:* Fel Inst Elec & Electronics Engrs; Int Union Radio Sci; fel Optical Soc Am. *Res:* Electromagnetic wave propagation in non-uniform media and periodic structures; radiation and diffraction phenomena; properties of configurations supporting surface, leaky, lateral and other wave types; elastic and optical waves; integrated optics. *Mailing Add:* Dept Elec Eng Polytech Univ Brooklyn NY 11201

TAMM, IGOR, b Tapa, Estonia, Apr 27, 22; nat US; m 53; c 3. VIROLOGY, MEDICINE. *Educ:* Yale Univ, MD, 47. *Prof Exp:* Intern med, Grace-New Haven Community Hosp Univ Serv, 47-48, asst resident, 48-49; from asst prof to assoc prof, 49-64, univ prof, 64-86, ABBY ROCKEFELLER MAUZE PROF, ROCKEFELLER UNIV HOSP, 86-, SR PHYSICIAN, 64- *Concurrent Pos:* Asst, Sch Med, Yale Univ, 47-49; from asst physician to physician, Rockefeller Univ Hosp, 49-64; assoc mem comn acute respiratory dis, Armed Forces Epidemiol Bd, 61-73; mem virol & rickettsiology study sect, NIH, 64-68; mem bd sci consults, Sloan-Kettering Inst Cancer Res, 66-75, vchmn, 71-72, chmn, 72-73; centennial lectr, Univ Ill, 68; mem study panel allergy & infectious dis, Health Res Coun City of New York, 68-75; mem adv comt, Am Cancer Soc, 69-72; gen chmn, task force virol, Nat Insts Allergy & Infectious Dis, NIH, 76-78. *Honors & Awards:* Alfred Benzon Prize, 67; Sarah L Poiley Award, NY Acad Sci, 77. *Mem:* Nat Acad Sci; Am Soc Cell Biol; Asn Am Physicians; Am Soc Microbiol; Am Asn Immunol; Am Soc Clin Invest; Sigma Xi; fel AAAS; fel NY Acad Sci; corresp mem Deutsche Ges Hyg Mikrobiol. *Res:* Cell adhesion; cell cycle and motility; effects of cytokines and growth factors on cells; gene regulation. *Mailing Add:* Dept Cell Physiol & Virol Rockefeller Univ New York NY 10021

TAMMINGA, CAROL ANN, b Grand Rapids, Mich, Jan 26, 46; c 2. SCHIZOPHRENIA, PSYCHOPHARMACOLOGY. *Educ:* Calvin Col, BS, 66; Vanderbilt Univ Sch Med, MD, 71. *Prof Exp:* Chief res psychiat, Univ Chicago, 75-79; instr psychiat, Univ Chicago, 75-78, asst prof, 78-79; res fel schizophrenia, NIMH, 78-79; chief, clin biochem, Exp Therapeut Br, Nat Inst Neurol & Commun Disorders & Stroke, 79-85; assoc prof, 79- 85, PROF PSYCHIAT, SCH MED, UNIV MD, 85-; CHIEF SCHIZOPHRENIA, INPATIENT RES PROG, 79- *Concurrent Pos:* Villian Allen fel, Med Sch, Vanderbilt Univ, 68; mem, TDA rev comt, NIMH, 81-85; consult, psychopharmacol adv comt, Food & Drug Admin, 82-85, Orphan Prod Develop, 83- & dept psychiat, Va Med Ctr, Baltimore, 85- *Honors & Awards:* Sandoz Award Psychiat Res, Univ Chicago, 75; McAlpin Award, Nat Asn Ment Health, 79. *Mem:* Am Psychiat Asn; Am Col Neuropsychopharmacol; AAAS; Soc Neurosci. *Res:* Pathophysiology and treatment for schizophrenia and hyperkinetic motor disorders. *Mailing Add:* Dept Psychiat Univ Md MPRC Box 21247 Baltimore MD 21228

TAMOR, STEPHEN, b New York, NY, Nov 29, 25; m 49; c 4. THEORETICAL PHYSICS. *Educ:* City Col New York, BS, 44; Univ Rochester, PhD, 50. *Prof Exp:* Physicist, Oak Ridge Nat Lab, 50-52; physicist, Radiation Lab, Univ Calif, 52-55; physicist, Res Lab, Gen Elec Co, NY, 55-66, Space Sci Lab, Pa, 66-71; physicist, Sci Appln Inc, 71-86; CONSULT, 87- *Concurrent Pos:* Guggenheim fel, 63-64. *Mem:* Am Phys Soc; AAAS. *Res:* Meson theory; nuclear and plasma physics; reactor theory. *Mailing Add:* 2165 Via Don Benito La Jolla CA 92037

TAMORRIA, CHRISTOPHER RICHARD, b Washington, DC, June 20, 32; m 61; c 1. CHEMISTRY. *Educ:* Georgetown Univ, BS, 54, MS, 58; Univ Md, PhD(med chem), 61. *Prof Exp:* Asst org chem, Georgetown Univ, 54-55; chemist, Food & Drug Admin, 55-56; asst org chem, Georgetown Univ, 56-57; asst inorg chem, Univ Md, 57-58; org chemist, Pharmaceut Prod Develop Sect, Lederle Labs, Am Cyanamid Co, 60-68, mgr regulatory agencies & info processing, Med Res Div, Cyanamid Int, 68-70; sr tech assoc, US Pharmacopeia, Md, 70-73; dir sci commun, Purdue Frederick Co & Affil, 73-77; mgr tech info, Toxicol Sect, Lederle Labs, 77; sr regulatory assoc, Drug Regulatory Affairs, Ayerst Labs, 78-80; dep dir, Drug Regulatory Affairs, Sterling Drug Inc, 80-84; ASSOC DIR DRUG REGULATORY AFFAIRS, BOEHRINGER INGELHEIM PHARMACEUT, 84- *Honors & Awards:* Gold Medal, Am Inst Chemists, 54. *Mem:* Am Chem Soc; Am Pharmaceut Asn; fel Am Found Pharmaceut Educ; Am Inst Chemists; Regulatory Affairs Prof Soc; Drug Info Asn. *Res:* Partial synthesis of steroids; correlation of structure and biological activity, especially in the synthesis of new steroid homologs and tetracycline antibiotics. *Mailing Add:* 27 Maymont Lane Trumbull CT 06611

TAMPAS, JOHN PETER, b Burlington, Vt, May 18, 29; m 62; c 4. RADIOLOGY. *Educ:* Univ Vt, BS, 51, MD, 54. *Prof Exp:* Teaching fel pediat radiol, Children's Hosp of Los Angeles, Univ Southern Calif, 60-61; NIH res fel cardiovasc radiol, Nat Heart Inst, 61-62; from asst prof to assoc prof, 62-69, PROF RADIOL & CHMN DEPT, UNIV VT, 70- *Concurrent Pos:* James Picker Found scholar radiol res, Univ Vt, 62-65; from asst attend radiologist to attend radiologist, Mary Fletcher Hosp & DeGoesbriand Mem Hosp, 62-; physician-in-residence, Vet Admin Hosp, 72- *Mem:* AMA; fel Am Col Radiol; Soc Pediat Radiol; Am Roentgen Ray Soc; Radiol Soc NAm. *Res:* Basic and clinical problems in radiology; pediatric and cardiovascular radiology. *Mailing Add:* Dept Radiol Mary Fletcher Unit Med Ctr Hosp Vt Burlington VT 05401

TAMPICO, JOSEPH, b Baltimore, Md, Apr 28, 16; div; c 3. ELECTRICAL ENGINEERING. *Educ:* Johns Hopkins Univ, BE, 37, Dr Eng, 41. *Prof Exp:* Mem staff, Lab Appl Physics, Johns Hopkins Univ, 45-54 & Hycon Mfg Co, 54-55; vpres, Assoc Missile Prod Corp, Am Mach & Foundry Co, 55-58; mgr independent res & develop, Marquardt Corp, 59-65; mgr bus planning, Spacecraft Dept, 65-72, mgr oper planning, Locomotive Dept, 72-76, strategic planning analyst, Locomotive Opers, 77-81, MGR MKT ANALYSIS, LOCOMOTIVE MKT, MKT DIV, GEN ELEC CO, 81- *Mem:* Assoc fel Am Inst Aeronaut & Astronaut; sr mem Inst Elec & Electronics Engrs. *Res:* Properties of dielectrics; resistance welding; jet propulsion engines; interfacial contact resistance; electronic test equipment; research administration; aerospace business planning. *Mailing Add:* 10850 Green Mountain Circle Apt 315 Columbia MD 21044

TAMPLIN, MARK LEWIS, b Rantoul, Ill, Feb 4, 55; m 79; c 2. ENVIRONMENTAL MICROBIOLOGY, MEDICAL MICROBIOLOGY. *Educ:* Univ SFla, Tampa, BA, 78, MA, 81, PhD(med sci), 85. *Prof Exp:* Res assoc, Ctr Marine Biotechnol, Univ Md, 85-87, asst res scientist, 87; res microbiologist, Fishery Res Br, Food & Drug Admin, Dauphin Island, Ala, 87-90; ASSOC PROF FOOD SAFETY, UNIV FLA, 90- *Concurrent Pos:* Consult, Jamaican Oyster Cult Prog, 82, Int Ctr Diarrheal Dis Res, 87 & Peruvian Govt Ministry Health, 91; comt chmn, Environ Panel, Food & Drug Admin Vibrio Vulnificus Workshop, 8; mem, Microbiol Comt, Interstate Shellfish Sanit Conf, 88-89, Depuration Comt, 88 & Vibrio Vulnificus Work Group, 89. *Mem:* Am Soc Microbiol; Nat Shellfisheries Asn. *Res:* Environmental influence on pathogenicity and epidemiology of water borne infections. *Mailing Add:* McCarty Hall, Rm 3031 Univ Fla Gainesville FL 32611-0130

TAMRES, MILTON, b Warsaw, Poland, Mar 12, 22; US citizen; m 60; c 2. PHYSICAL INORGANIC CHEMISTRY. *Educ:* Brooklyn Col, BA, 43; Northwestern Univ, PhD(phys chem), 49. *Prof Exp:* Anal chemist, Celanese Corp Am, Md, 43-44; asst, Northwestern Univ, 44-47; from instr to asst prof chem, Univ Ill, 48-53; from asst prof to assoc prof, 53-63, prof, 63-87, EMER PROF CHEM, UNIV MICH, ANN ARBOR, 87- *Concurrent Pos:* Guggenheim fel, 59-60; mem, Adv Coun Col Chem, 62-66; Am Chem Soc-Petrol Res Fund int fel, 66-67; vis scholar, Univ Tokyo, 74. *Mem:* Fel AAAS; Am Chem Soc; fel Am Inst Chemists. *Res:* Electron donor-acceptor interactions; basicities of cyclic compounds. *Mailing Add:* Dept Chem Univ Mich Ann Arbor MI 48109

TAMSITT, JAMES RAY, b Big Spring, Tex, Nov 22, 28. ZOOLOGY, MAMMALIAN SYSTEMATICS. *Educ:* Univ Tex, BA, 51, MA, 53, PhD(vert ecol), 58. *Prof Exp:* Lectr zool, Univ Man, 57-58; instr biol, ETex State Univ, 58-59; prof zool, Univ of the Andes, Colombia, 59-65; NIH fel med zool, Sch Med, Univ PR, San Juan, 65-67; assoc cur, cur, Dept Mammal, Royal Ont Mus, 67-73, 73-85; assoc prof zool, Univ Toronto, 69-85; lectr biol, Univ Tex, Austin, 86-88; INSTR BIOL, AUSTIN COMMUNITY COL, 88- *Concurrent Pos:* Vis prof biol, Pontificia Univ Javeriana, Colombia, 75, 82 & 85; sr Fulbright-Hays fel, 75 & 82; vis prof, Nat Univ Colombia, 82, 84 & 85, Univ Indust Santander, Colombia, 82; assoc ed, J Mammal, 68-70; ed, Royal Ont Mus Life Sci Publs, 71-73 & 81-84; resolutions comt, Am Soc Mammal, 70-72. *Mem:* Fel AAAS; Am Soc Mammalogists. *Res:* Ecology, natural history, taxonomy and ectoparasites of Neotropical mammals. *Mailing Add:* 2903 Cedarview Dr Austin TX 78704-4608

TAMSKY, MORGAN JEROME, b St Louis, Mo, July 26, 42; m 66; c 2. LABORATORY TECHNICAL DIRECTOR. *Educ:* Washington Univ, BA, 64; Univ Kans, PhD(chem), 70. *Prof Exp:* Sr chemist, Cent Res Labs, 3M Co, 69-74, res specialist polymer physics, 74-77, suprv, 77-80, res mgr, 80-82, tech mgr, Com Tape Div, 82-84, lab mgr, Com Off Supply Div, 84-85, tech dir, Health Care Specialties Div, 87-89, TECH DIR, DISPOSABLE PROD DIV, 3M CO, 89- *Mem:* Am Chem Soc; Adhesion Soc. *Res:* Surface phenomenon; adhesion; polymer physics; medical devices; pharmacentrical aerosol formulation filling; medical tapes. *Mailing Add:* 1920 Bayard Ave St Paul MN 55116

TAMURA, TSUNEO, b Hawaii, Nov 15, 25; div; c 3. SOILS, WASTE MANAGEMENT. *Educ:* Univ Hawaii, BS, 48; Univ Wis, MS, 51, PhD(soils), 52. *Prof Exp:* From asst soil scientist to assoc soil scientist, Conn Agr Exp Sta, 52-57; chemist, Westinghouse Elec Corp, 57; sr res staff mem, Oak Ridge Nat Lab, 57-77, earth sci sect head, 77-82, sr res adv, 82-90; SR DEVELOP SCI, MARTIN MARIETTA ENERGY SYSTS, 90- *Mem:* Fel AAAS; fel Am Soc Agron; Am Chem Soc; Sigma Xi; fel Soil Sci Soc Am; fel Am Inst Chem; Clay Minerals Soc. *Res:* Soil chemistry and genesis; soil clay mineralogy; radioactive waste disposal; health physics; toxic metals in environment. *Mailing Add:* Martin Marietta Energy Systs PO Box 2003 Oak Ridge TN 37831-7606

TAN, AGNES W H, biochemistry, for more information see previous edition

TAN, AH-TI CHU, b Amoy, China, Sept 24, 35; Can citizen. CHEMISTRY, BIOCHEMISTRY. *Educ:* Mapua Inst Technol, BSChem, 57; Adamson Univ, Manila, BSChE, 58; McGill Univ, MSc, 62, PhD(chem kinetics), 66. *Prof Exp:* Lectr phys chem, Adamson Univ, Manila, 58-60; res chemist, Bathurst Paper Co, 65-66; assoc biochem, Col Med, Univ Vt, 66-68, vis asst prof, 68-69; asst prof ophthal, Fac Med, McGill Univ, 69-72, asst prof anesthesia, 72-78; assoc prof, Fac Med, Univ Montreal, 78-79; SR CONSULT, TAN & ASSOCS, 82- *Concurrent Pos:* NIH grant biochem, Col Med, Univ Vt, 66-68; Que Med Res Coun grant ophthal, Fac Med, McGill Univ, 69-72; prof assoc anesthesia res, McGill Univ, 72-78. *Mem:* Am Chem Soc; Chem Inst Can; Can Biochem Soc; Soc Neurosci; AAAS; Sigma Xi. *Res:* Physicochemical studies of proteins; brain cell membranes; neurotransmitters; molecular mechanism of synaptic transmission; neuroendocrinology; molecular mechanism of depression. *Mailing Add:* 1951 de Maisonneuve E Apt 1108 Montreal PQ H2K 2C9 Can

TAN, ARJUN, b Santiniketan, W Bengal, India, Aug 6, 43. IONOSPHERIC MODELLING, ORBITAL DEBRIS. *Educ:* Univ Calcutta, MSc, 65; Univ Fla, MS, 74; Univ Ala, Huntsville, PhD(physics), 79. *Prof Exp:* Instr phys sci, Visva-Bharati Univ, India, 66-67; lectr physics, Krishnagar Women's Col, India, 67-71; res assoc physics, Univ Ala, Huntsville, 79; post-doctoral assoc physics, Univ Fla, 79-80; instr math-physics, Newberry Col, 80-81; from asst prof to assoc prof, 81-88, PROF PHYSICS, ALA A&M UNIV, 88- *Concurrent Pos:* Summer fac, Jadavpur Univ, India, 68, NASA Marshall Space Flight Ctr, 81, Arnold Air Force Sta, US Air Force, 86, NASA Johnson Space Ctr, 87, 88, US Army, Redstone Arsenal, 89; guest worker, NOAA Space Environ Lab, 77-79; lectr, Univ Ala, Huntsville, 81; physicist, US Army TMDE Lab, 82, 83 & 84; physicist, Lawrence Livermore Nat Lab, 85; prin investr, NSF grant, 83-85, 86-88, NASA grant, 87-91. *Mem:* Sigma Xi; Am Geophys Union; Am Asn Physics Teachers; Math Asn Am; Nat Coun Teachers Math; Am Phys Soc. *Res:* Planetary and space science; author of 80 journal articles which appeared in various journals. *Mailing Add:* Box 447 Normal AL 35762

TAN, BARRIE, b Ipoh, Malaysia, Oct 7, 53; US citizen. CHEMICAL TOXICOLOGY, PHARMACOLOGY. *Educ:* Univ Otago, NZ, BS, 76, PhD(anal chem), 79. *Prof Exp:* Res assoc fel environ toxicol, Auburn Univ, Ala, 79-81; asst prof Anal Chem, 82-90, ADJ PROF FOOD SCI, UNIV MASS, AMHERST, 90- *Mem:* Am Chem Soc; Am Oil Chem Soc; Soc Environ Toxicol Chem; Inst Food Technol. *Res:* Molecular spectroscopic and luminescence detectors in high pressure liquid chromatography; bioanalytical techniques in cancer research (chemical carcinogenesis, biochemical toxicology); effects of acid precipitation; carotenoid biochemistry; analyses and pharmacokinetics of polynuclear aromatic hydrocarbons and polychlorinated biphenyls; analyses of lipid-soluble carotenoids, vitamins A and E from foods. *Mailing Add:* Dept Food Sci Univ Mass Amherst MA 01003

TAN, BOEN HIE, b Pandangan, Indonesia, Dec 14, 26; US citizen. BIOMEDICAL SCIENCE, MOLECULAR BIOLOGY. *Educ:* State Univ Leiden, Holland, BSc, 52, MSc, 55, DSc(pharmacol, toxicol, pharmaco & anal biochem), 62. *Prof Exp:* Asst prof anal pharmacol chem, Univ Leiden, Holland, 53-55 & 62-64; res fel anal chem, Univ Minn, 55-61, res assoc anal biochem, 64-68, res specialist anal chem, 72-73; res fel phys biochem, Max Planck Inst, Ger, 61-62; res assoc anal biochem, New York Hosp-Cornell Med Ctr, 68-72; res assoc pharmacol, Univ Groningen, Holland, 73-81; RES ASSOC BIOCHEM, UNIV SALA, MOBILE, 82- *Mem:* Am Asn Clin Chem; Am Chem Soc; AAAS; NY Acad Sci. *Res:* Purification, analysis, pharmacokinetic, pharmacological activities of anti-arrhythmic, new drugs; alpha-1-antitrypsin, plasma proteins, leucocyte-enzymes and liver functions; fibrin formation, inhibition and lysis; heart perfusions and type I, type II-diabetes; sulfhydryl, disulfides and protein denaturation-renaturations; vanadate-sulfhydryl complexes and PDE-activities; DNA damage and repair; clinical chemistry, nuclear medicine in diagnosis and the cure of diseases. *Mailing Add:* Dept Biochem MSB2158 Univ SAla Mobile AL 36688

TAN, CHARLOTTE, b Kiang-Si, China, Apr 19, 23; US citizen; m 59; c 1. CANCER. *Educ:* Hsiang-Ya Med Col, China, MD, 47; Am Bd Pediat, dipl, 54. *Prof Exp:* Resident & rotating internal, Nanking Cent Hosp, China, 47-48; rotating intern & gen resident, St Barnabas Hosp, Newark, NJ, 48-50; res resident hemat & pediat resident, Children's Hosp Philadelphia, 50-51; pediat resident, Philadelphia Gen Hosp, 52; res fel chemother, Sloan-Kettering Inst, 52-54, res assoc, 55-57, asst, 57-60, assoc mem, 60-84; assoc prof, 70-78, PROF PEDIAT MED SCH, CORNELL UNIV, 78-, MEM, DEPT PEDIAT, SLOAN KETTERING CANCER CTR, 84- *Concurrent Pos:* Spec fel med, Memorial Ctr, New York, 52-55, spec fel pediat, 55-57, clin asst, Pediat Serv, 57-58; instr med, Sloan-Kettering Div, Grad Sch Med Sci, Med Col, Cornell Univ, 54-55, instr med, Med Col, 55-57, instr pediat, 58-62; clin asst pediatrician, James Ewing Hosp, 57-58, from asst vis pediatrician, to assoc vis pediatrician, 58-68; from asst attend pediatrician to assoc attend pediatrician, Memorial Hosp, 58-70, attend pediatrician, 70-, assoc chmn chemother, 74-; vis prof, Nat Taiwan Univ Med Col, 66-67; assoc & attend pediatrician, NY Hosp, 78- *Mem:* Am Acad Pediat; Am Asn Cancer Res; Am Fedn Clin Res; AMA; Am Soc Clin Oncol; NY Acad Sci; Am Soc Hemat; Soc Surg Oncol; Int Soc Pediat Oncol. *Res:* Cancer chemotherapy. *Mailing Add:* Mem-Sloan-Kettering Res Ctr 1275 York Ave New York NY 10021

TAN, CHIN SHENG, b Taiwan, Mar 15, 47; Can citizen; m 73; c 2. AGROMETEOROLOGY, IRRIGATION. *Educ:* Nat Chung-Hsing Univ, Taiwan, BSc, 69; Univ NH, MSc, 72; Univ BC, PhD(agrometeorol), 77. *Prof Exp:* Vis scientist, Commonwealth Sci & Indust Res Orgn, Div Water Resources, Griffith, NSW Australia, 87-88; RES SCIENTIST AGROMETEOROL & IRRIG, HARROW RES STA, CAN DEPT AGR, 78- *Honors & Awards:* Carroll R Miller Award, Am Soc Hort Sci-Nat Peach Coun, 82 & 85. *Mem:* Am Soc Agron; Can Soc Hort Sci; Int Soc Hort Sci. *Res:* Soil and plant water relations; water requirements of crops in relation to production; interrelationship of irrigation, climate, soil and crop management factors; evapotranspiration models for various crops; climatic model for scheduling irrigation; soil moisture effects on root growth. *Mailing Add:* Res Sta Agr Can Harrow ON N0R 1G0 Can

TAN, CHOR-WENG, b Canton, China, Apr 20, 36; US citizen; m 63; c 2. MECHANICAL ENGINEERING. *Educ:* Evansville Col, BS, 59; Univ Ill, MS, 61, PhD(mech eng), 63. *Prof Exp:* Prof mech eng, Cooper Union, 63-90, dean, Sch Eng, 76-87; CONSULT, 90- *Concurrent Pos:* Prog dir, NSF; exec dir, Cooper Union Res Found; dir bd, Tround Int, Inc, Nomura Pac Basin Fund, Japan OTC Fund, Jakarta Growth Fund, UHT Corp, APT, Inc, Appl Biomed Asn. *Mem:* Am Soc Mech Engrs; Am Soc Eng Educ. *Res:* Thermodynamic and transport properties of partially ionized gases; magnetohydrodynamics; electrogasdynamics; environmental engineering. *Mailing Add:* Am Soc Mech Engrs 345 E 47th St New York NY 10017

TAN, ENG M, b Malaysia, Aug 26, 26; US citizen; m 62; c 2. IMMUNOLOGY. *Educ:* Johns Hopkins Univ, AB, 52, MD, 56. *Prof Exp:* Res fel, Rockfeller Univ, 62-65; asst prof med, Wash Univ, 65-67; assoc mem, dept Exp Path, Scripps Clin & Res Found, 67-70, head, Div Allergy & Immunol, 70-77; head, Div Rheumatic Dis, Univ Colo Med Ctr, 77-82; DIR, W M KECK AUTOIMMUNE DIS CTR, SCRIPPS CLIN & RES FOUND, 82- *Concurrent Pos:* Nesbitt vis prof, Minneapolis, 79; chmn, Allergy, Immunol & Transplantation Res Comt, NIH, 81; Macy Found fac scholar, 81. *Honors & Awards:* Dunlop-Dottridge Lectr, Can Rheumatism Asn, Ottawa, 80; Mclaughlin Lectr, Galveston, 81; Alexander von Humboldt Sr Sci Scientist Award, 86. *Mem:* Asn Am Physicians; Am Soc Clin Invest; Am Asn Immunologists; Am Asn Pathologists; Am Rheumatism Asn (pres, 84-85). *Res:* Autoimmune diseases; immunological aspects of rheumatic diseases; antinuclear and other autoantibodies in systemic lupus erythematosus, rheumatoid arthritis, Sjogren's syndrome, scleroderma, dermatomyositis and polymyositis. *Mailing Add:* Univ Colo Sch Med 4200 E Ninth Ave Denver CO 80262

TAN, FRANCIS C, b Manila, Philippines, Sept 21, 39; m 71. CHEMICAL OCEANOGRAPHY. *Educ:* Cheng Kung Univ, Taiwan, BSc, 61; McGill Univ, MSc, 65; Pa State Univ, PhD(geochem), 69. *Prof Exp:* NSF fel, Pa State Univ, 69-70; geochemist, Minn Geol Surv, Univ Minn, 70-72; RES SCIENTIST, DEPT FISHERIES & OCEANS, BEDFORD INST OCEANOG, CAN DEPT FISHERIES & OCEANS, 72- *Concurrent Pos:* Hon res fel, Third Inst Oceanog, State Oceanic Admin, Xiamen, People's Repub China; hon res assoc, Dept Oceanog, Dalhousie Univ, Halifax, Can, 79-90. *Mem:* Am Geophys Union. *Res:* Stable isotope oceanography, marine geochemistry. *Mailing Add:* Bedford Inst Oceanog Dartmouth NS B2Y 4A2 Can

TAN, HENRY HARRY, b Sukabumi, Indonesia, Dec 15, 24; US citizen; m 59; c 5. ORGANIC CHEMISTRY. *Educ:* Hope Col, AB, 55; Univ Mich, Ann Arbor, MS, 58, PhD(chem), 62. *Prof Exp:* Res chemist, E I Du Pont de Nemours & Co, Inc, 62-64, tech field rep, Appl Chem & Mkt, 64-68, res chemist, 68-73, mem tech serv & develop staff, 73-81, develop spec geotextiles, 81-86; BUS MGR, REEMAY INC, 86- *Mem:* Am Chem Soc; Am Soc Testing & Mat. *Res:* Organic syntheses; applied chemistry; political, social and economic affairs of Asia. *Mailing Add:* 103 Colorado Ave Shipley Heights Wilmington DE 19803

TAN, HENRY S I, b Bandung, Indonesia, Mar 26, 32; US citizen; m. SYNTHETIC ORGANIC & NATURAL PRODUCTS CHEMISTRY. *Educ:* Univ Indonesia, BSPharm, 54, MSPharm, 56; Univ Ky, PhD(pharmaceut sci), 71. *Prof Exp:* Instr & assoc prof, Bandung Inst Technol, 57-66; from asst prof to assoc prof pharm, 71-82, PROF PHARM CHEM, UNIV CINCINNATI, 82- *Concurrent Pos:* Prin & co-investr funded grants; elected mem, USP, Comt Rev. *Honors & Awards:* Ten-Year Serv Award Chem Abstract Serv, Asn Clin Scientists. *Mem:* Am Asn Pharmaceut Scientists. *Res:* Development of analysis procedures for drugs in dosage forms, biological fluids, animal feed and cosmetic prepns. *Mailing Add:* Univ Cincinnati Cincinnati OH 45267-0004

TAN, JAMES CHIEN-HUA, b Nanchang, China, Oct 8, 35; m 65; c 2. GENETICS, STATISTICS. *Educ:* Chung-Shing Univ, Taiwan, BS, 57; Mont State Univ, MS, 61; NC State Univ, PhD(genetics), 68. *Prof Exp:* Asst prof biol, Slippery Rock State Col, 65-66; from asst prof to assoc prof, 66-78, PROF BIOL, VALPARAISO UNIV, 78-, UNIV RES PROF, 80- *Concurrent Pos:* Res fel, Roswell Park Mem Inst,77; fel, O P Kretzemann Mem Wheat Ridge Found. *Mem:* AAAS; Am Genetics Soc; Environ Mutagen Soc; Genetics Soc Can. *Res:* Cytogenetics and statistical biology; mutagenicity and carcinogenicity testing. *Mailing Add:* Dept Biol Valparaiso Univ Valparaiso IN 46383

TAN, JULIA S, b Taipei, Taiwan; US citizen. POLYMER SCIENCE, PHYSICAL CHEMISTRY. *Educ:* Nat Taiwan Univ, BA, 61; Wesleyan Univ, MA, 63; Yale Univ, PhD(chem), 66. *Prof Exp:* Asst prof chem, Wesleyan Univ, 66-69; res assoc biophys, Univ Rochester, 69-70; res chemist, 70-77, RES ASSOC, RES LABS, EASTMAN KODAK CO, 77- *Mem:* Am Chem Soc. *Res:* Solution properties of polyelectrolytes. *Mailing Add:* Res Labs B-81 Eastman Kodak Rochester NY 14650

TAN, KIM H, b Djakarta, Indonesia, Mar 24, 26; US citizen; m 57; c 1. TROPICAL AGRICULTURE, SOIL CHEMISTRY. *Educ:* Univ Indonesia, MSc, 55, PhD(soil sci), 58. *Prof Exp:* Assoc prof soil sci fac agr, Univ Indonesia, 58-64, prof fac agr & agr acad, 64-67, head dept soil sci, 65-67; technician soil anal nitrogen lab, Agr Res Serv, USDA, Colo, 67-68; asst prof soil sci, Dept Agron, 68-73, assoc prof agron, 73-77, PROF AGRON, UNIV GA, 77- *Concurrent Pos:* Rockefeller Found grant/fel, NC State Univ, 60-61 & Cornell Univ, 61; mem comt VIII, Southern Regional Coop Soil Surv, Soil Conserv Serv, USDA, 72- *Mem:* Clay Mineral Soc Am; fel Am Soc Agron; fel Soil Sci Soc Am; Int Soc Soil Sci. *Res:* Pedology; genesis and characterization of soils and organic matter in soils; effect of organic matter on soil properties and plant growth; chemistry and mineralogy of soils. *Mailing Add:* Dept Agron Univ Ga Athens GA 30602

TAN, KOK-KEONG, b Shanghai, China, June 1, 43; m 69; c 4. MATHEMATICS. *Educ:* Nanyang Univ, BSc, 66; Univ BC, PhD(Math), 70. *Prof Exp:* Teacher high sch, Malaysia, 66; from asst prof to assoc prof, 70-83, PROF MATH, DALHOUSIE UNIV, 83- *Concurrent Pos:* Vis res prof, Nat Tsing Hua Univ, Taiwan, 80-81; vis res expert, Acad Sinica, Taiwan, 81 & Nat Central Univ, Taiwan, 84; vis res prof, Nat Central Univ, Taiwan, 87. *Mem:* Am Math Soc; SE Asian Math Soc; Can Math Soc. *Res:* Functional analysis and topology, in particular, fixed point theorems; convex analysis, in particular, minimax inequalities, variational inequalities. *Mailing Add:* Dept Math Statist & Computer Sci Dalhousie Univ Halifax NS B3H 3J5 Can

TAN, LIAT, b Semarang, Java, Apr 1, 29; div; c 2. BIOCHEMISTRY, ORGANIC CHEMISTRY. *Educ:* Univ Amsterdam, BSc, 53; Univ Münster, MSc, 55; Univ Freiburg, Dr rer nat, 58. *Prof Exp:* Res chemist, Steroid Res Lab, Leo Pharmaceut Prod, Denmark, 59-60; sr res chemist, Union Chimique Belge SA, 60-62; res fel org chem, Laval Univ, 62-63; examr steroid chem, Can Patent Off, 63-66; Welch res fel & instr biochem & nutrit, Med Br, Univ Tex, 66-67; from asst prof to assoc prof, 67-80, PROF BIOCHEM, UNIV SHERBROOKE, 80- *Concurrent Pos:* Vis scientist, Hormone Res Lab, Univ Calif, 76-77. *Mem:* AAAS; Am Chem Soc; NY Acad Sci; Can Biochem Soc. *Res:* Synthesis and biochemistry of steroids; steroid oxygenases, specificity and mechanism of action; physiologically active natural products; mechanism of biological oxidations at inactive sites in steroids; cytochrome P-450 aromatase; androgen synthetase; breast cancer. *Mailing Add:* Dept Biochem Univ Sherbrooke Sherbrooke PQ J1H 5N4 Can

TAN, MENG HEE, b Kuala Pilah, Malaysia, Mar 30, 42; Can citizen; m 70; c 3. INTERNAL MEDICINE, MEDICAL SCIENCES. *Educ:* Dalhousie Univ, BSc, 65, MD, 69; FRCP(C), 75; FACP, 78. *Prof Exp:* Lectr, 74-75, from asst prof to assoc prof, 75-83, HEAD, DIV ENDOCRINOL, DALHOUSIE UNIV, 81-, PROF MED, 83- *Concurrent Pos:* Res fel, Harvard Med Sch, 71-73; res fel, Med Coun Can, 71-74, centennial fel, 74-75; res fel, Cardiovasc Res Inst, San Francisco, 73-75. *Mem:* Am Col Physicians; Am Diabetes Asn; Am Fedn Clin Res; Royal Col Physicians & Surgeons Can. *Res:* Lipoprotein metabolism in secondary hyperlipidemia; triglyceride metabolism in diabetes mellitus; insulin receptors. *Mailing Add:* 5303 Morris St Suite 813 Halifax NS B3J 1B6 Can

TAN, OWEN T, b Indramaju, Indonesia, Aug 30, 31; US citizen; m 56; c 3. ELECTRICAL ENGINEERING. *Educ:* Bandung Technol Faculty, MSc, 55; Eindhoven Technol Univ, PhD(elec eng), 61. *Prof Exp:* Res & develop engr, Willem Smit & Co, Neth, 56-62; lectr elec eng, Bandung Technol Inst, 62-64, sr lectr, 64-66; from asst prof to assoc prof, 67-72, PROF ELEC ENG, LA STATE UNIV, BATON ROUGE, 78- *Concurrent Pos:* Res fel, Siemens Schuckert, WGer, 62, Eindhoven Univ Technol, 77 & Delft Univ Technol, 91. *Mem:* Inst Elec & Electronics Engrs; Neth Royal Inst Eng. *Res:* Energy conversion; power systems; artificial neural networks. *Mailing Add:* Dept Elec & Computer Eng La State Univ Baton Rouge LA 70803

TAN, PETER CHING-YAO, mathematical statistics, civil engineering, for more information see previous edition

TAN, VICTOR, b Manila, Philippines, Aug 8, 44; m 76. POLYMER ENGINEERING, MATERIAL SCIENCE. *Educ:* Adamson Univ, BS, 67; Univ Pittsburgh, MS, 70; Stevens Inst Technol, PhD(chem eng), 75. *Prof Exp:* Fel, 74-76, RES ASSOC CHEM ENG, MCGILL UNIV, 78-; SR ENGR POLYMER PROCESSING INST, STEVENS INST TECHNOL. *Mem:* Soc Plastics Engrs. *Res:* Polymer characterization and polymer processing; computer simulation of polymer processing; polymer physics. *Mailing Add:* Dept Chem Eng Stevens Inst Technol Castle Pt Hoboken NJ 07030

TAN, WAI-YUAN, b China, Aug 14, 34; m 64; c 2. STATISTICS. *Educ:* Taiwan Prov Col Agr, BA, 55; Nat Taiwan Univ, MS, 59; Univ Wis, MS(math) & MS(statist), 63, PhD(statist), 64. *Prof Exp:* Asst res fel biostatist, Inst Bot, Acad Sinica, Taiwan, 59-61, assoc res fel, 64-67 & res fel, 67-68; assoc prof statist, Nat Taiwan Univ, 65-67; from assoc prof to prof biostatist, Biol Res Ctr, Taiwan, 65-68; asst prof statist, Univ Wis-Madison, 68-72; assoc prof math, Wash State Univ, 73-75; Cancer Expert, Nat Cancer Inst, Bethesda, Md, 84-85; RES PROF MATH, MEMPHIS STATE UNIV, 75- *Concurrent Pos:* Vis assoc prof, Tsing Hua Univ, Taiwan, 65-67; statist adv, Joint Inst Indust Res, Taiwan, 65-67; sr res statistician, Oak Ridge Nat Lab, Tenn, 78, 79; vis prof & consult, Dept Genetics, Univ Hawaii, Honolulu, 73, 74, 76; consult, Pig Res Inst, Taiwan, 75 & Fox Chase Cancer Ctr, Philadelphia, Pa, 83; distinguished res award, Memphis State Univ, 83, 89 & SPUR res award, 84, 88 & 90; vis prof, Emory Univ, 90; math statistician, CDC, Atlanta, Ga, 90. *Mem:* Am Statist Asn; Biomet Soc; Inst Math Statist; Royal Statist Soc; AAAS; Am Asn Cancer Res. *Res:* Statistical inferences; multivariate analysis; mathematical genetics and quantitative genetics; robust statistics; biostatistics; statistical methods for mutagenicity and carcinogenesis; Cancer stochastic models; applied stochastic processes; robust procedures; AIDS stochastic models. *Mailing Add:* Dept Math Memphis State Univ Memphis TN 38152

TAN, Y H, b Singapore, Sept 2, 41; c 2. CELL BIOLOGY, MOLECULAR BIOLOGY. *Educ:* Univ Singapore, BSc Hons, 65; Univ Man, PhD(biochem), 69. *Prof Exp:* Med Res Coun Can fel biochem & virol, Univ Pittsburgh, 69-71; Med Res Coun Can fel somatic cell genetics, Yale Univ, 72-74; asst prof pediat, Johns Hopkins Univ, 74-75; assoc prof, 75-83, prof, Dept Med Biochem, Univ Calgary, 83-87; DIR, INST MOLECULAR & CELL BIOL, NAT UNIV SINGAPORE, 87- *Concurrent Pos:* Sect chief molecular genetics, Lab Cellular Comp Physiol, Nat Inst Aging, 74-75; Nat Cancer Inst scholar, 76-80. *Mem:* Am Soc Microbiol. *Res:* Regulation of host cell-virus interaction; human interferon receptors; membrane transduction of the interferon signal. *Mailing Add:* Inst Molecular & Cell Biol Nat Univ Singapore Ten Kent Ridge Crescent Singapore Singapore

TAN, YEN T, b Hong Kong, Feb 12, 40. SOLID STATE PHYSICS, SURFACE PHYSICS. *Educ:* Columbia Univ, BS, 62; Yale Univ, PhD(chem), 66. *Prof Exp:* RES ASSOC, RES LABS, EASTMAN KODAK CO, 66- *Concurrent Pos:* Adj prof, Rochester Inst Technol, 74-75. *Mem:* Am Chem Soc; Am Inst Mining, Metall & Petrol Engrs; Am Vacuum Soc; Sigma Xi. *Res:* Surface properties of solids; transport phenomena; thermodynamics. *Mailing Add:* 437 True Hickory Dr Rochester NY 14615

TANABE, MASATO, b Stockton, Calif, Jan 18, 25; m 55; c 3. ORGANIC CHEMISTRY. *Educ:* Univ Calif, BS, 47, PhD(chem), 51. *Prof Exp:* Chemist, US Naval Radiation Defense Lab, 47-48; res chemist, Riker Labs, Inc, 51-57; sr org chemist, 57-72, DIR, DEPT BIO-ORG CHEM, SRI INT, MENLO PARK, 72- *Concurrent Pos:* Fulbright res fel, Japan, 54-55. *Mem:* Am Chem Soc. *Res:* Medicinal chemistry; steroids; alkaloids; natural products; biosynthesis. *Mailing Add:* 972 Moreno St Palo Alto CA 94303-3733

TANABE, MICHAEL JOHN, b Keaau, Hawaii, Sept 15, 47. HORTICULTURE, PLANT PHYSIOLOGY. *Educ:* Univ Hawaii, Manoa, BS, 69, MS, 72, PhD(hort), 76. *Prof Exp:* PROF HORT, COL AGR, UNIV HAWAII, HILO, 75- *Mem:* Am Soc Hort Sci; Sigma Xi. *Res:* Phenylalanine ammonia lyase activity as it affects anthocyanin production; determining regenerative capabilities of several tropical plant species by plant tissue culture; plant organ culture. *Mailing Add:* Col Agr Hilo HI 96720-4051

TANABE, TSUNEO Y, physiology, for more information see previous edition

TANADA, TAKUMA, b Honolulu, Hawaii, Oct 30, 19; m 47; c 2. PLANT PHYSIOLOGY. *Educ:* Univ Hawaii, BS, 42, MS, 44; Univ Ill, PhD(bot), 50. *Prof Exp:* Asst soil chemist, Agr Exp Sta, Univ Hawaii, 42-44; sci consult natural resources sect, Supreme Comdr Allied Powers, US Army, Tokyo, 46-47; plant physiologist, Agr Res Serv, USDA, 50-57; agron res adv, Int Coop Admin, Ceylon, 57-60; RES PLANT PHYSIOLOGIST, SCI & EDUC ADMIN-AGR RES, USDA, 60- *Mem:* Am Soc Plant Physiol; Am Inst Biol Sci. *Res:* Photobiology; photosynthesis; mineral nutrition. *Mailing Add:* 19 Skycrest Way Napa CA 94558

TANADA, YOSHINORI, b Puuloa, Oahu, Hawaii, June 8, 17; m 49; c 2. INSECT PATHOLOGY, INSECT VIROLOGY. *Educ:* Univ Hawaii, BS, 40, MS, 45; Univ Calif, PhD(entom), 53. *Prof Exp:* Asst zool, Univ Hawaii, 43-45, jr entomologist exp sta, 45-53, asst entomologist, 53-56, asst prof zool & entom, 54-56, assoc prof & assoc entomologist, 56; asst insect pathologist, Lab Insect Path, Univ Calif, Berkeley, 56-59, assoc insect pathologist, 59-64, lectr, 61-65, chmn, Div Invert Path, 64-65, insect pathologist, Exp Sta, 64-87, prof entom, 65-87; RETIRED. *Concurrent Pos:* Consult, US Army, Okinawa, 50, SPac Comn, Pac Sci Bd, Nat Res Coun, 59, UN Develop Prog, Western Samoa, 71 & Food & Agr Orgn, UN, Thailand, 71; Fulbright res scholar, Japan, 62-63; spec vis prof, Univ Tokyo, 80. *Honors & Awards:* Founder's lect, Soc Invert Pathol, 84. *Mem:* Fel AAAS; Entom Soc Am; Soc Protozool; Soc Invert Path; Am Inst Biol Scientists; Am Soc Virol. *Res:* Insect virology; general insect pathology; epizootiology of diseases of insects. *Mailing Add:* Ten Truitt Lane Oakland CA 94618

TANAKA, FRED SHIGERU, b Shoshone, Idaho, Aug 1, 37; m 66; c 3. PESTICIDE CHEMISTRY. *Educ:* Ore State Univ, BS, 59, PhD(org chem), 66. *Prof Exp:* Chief chemist, Wash State Dept Health, 66-67; RES CHEMIST, USDA, 67- *Concurrent Pos:* Foreign specialists, Japanese Govt Res Award, 77; tech expert, Int Atomic Energy Agency, 83 & 88. *Mem:* Am Chem Soc. *Res:* Organic microsynthesis of radiochemically labeled compounds; photochemical degradation of pesticides; identification of biological metabolites; organic and biological reaction mechanisms. *Mailing Add:* Biosci Res Lab State Univ Sta PO Box 5674 Fargo ND 58105

TANAKA, JOHN, b San Diego, Calif, June 18, 24; m 59; c 2. INORGANIC CHEMISTRY. *Educ:* Univ Calif, Los Angeles, BA, 51; Iowa State Univ, PhD, 56. *Hon Degrees:* Dr, Univ Paul Sapatier, France. *Prof Exp:* From asst prof to assoc prof chem, SDak State Univ, 56-63; NASA fel, Univ Pittsburgh, 63-65; from asst prof to assoc prof, 65-75, PROF CHEM, UNIV CONN, 75-, DIR HONORS PROG, 71- *Mem:* Am Chem Soc; Royal Soc Chem; fel Inst Elec & Electronics Engrs. *Res:* Synthesis and properties of ternary hydrides; materials for electrical insulation; reactions of boron hydrides; vacuum line syntheses. *Mailing Add:* Dept Chem Univ Conn Storrs CT 06268

TANAKA, KATSUMI, b San Francisco, Calif, Mar 1, 25; m 53; c 1. PHYSICS. *Educ:* Univ Calif, AB, 49, PhD(physics), 52. *Prof Exp:* Assoc physicist, Argonne Nat Lab, 52-64; PROF PHYSICS, OHIO STATE UNIV, 64- *Concurrent Pos:* Vis prof, Univ Naples, 60-61. *Mem:* Fel Am Phys Soc. *Res:* Elementary particle physics. *Mailing Add:* Dept Physics Ohio State Univ Columbus OH 43210

TANAKA, KAY, b Osaka, Japan, Mar 2, 29; m; c 2. GENETICS, BIOCHEMISTRY. *Educ:* Univ Tokyo, Japan, MD, 56. *Hon Degrees:* MA, Yale Univ, 82. *Prof Exp:* PROF HUMAN GENETICS, YALE UNIV, 82- *Honors & Awards:* Merit Award, NIH. *Mem:* Am Soc Biol Chem; Am Soc Human Genetics. *Res:* Genetic metabolic disorders; molecular biology. *Mailing Add:* Dept Human Genetics 393 NSB SHM Yale Univ New Haven CT 06510

TANAKA, KOUICHI ROBERT, b Fresno, Calif, Dec 15, 26; m 65; c 3. MEDICINE, HEMATOLOGY. *Educ:* Wayne State Univ, BS, 49, MD, 52. *Prof Exp:* Intern, Los Angeles County Gen Hosp, 52-53; resident path, Detroit Receiving Hosp, Mich, 53-54, resident med, 54-57; instr med & jr res hematologist, 57-59, asst prof med, Sch Med & asst res hematologist, Med Ctr, 59-61, assoc prof med, Sch Med, 61-68, PROF MED, SCH MED, UNIV CALIF, LOS ANGELES, 68-, ATTEND PHYSICIAN, MED CTR, 58-; CHIEF, DIV HEMAT, HARBOR-UCLA MED CTR, 61-, ASSOC CHMN DEPT MED, 71- *Concurrent Pos:* Consult, US Naval Hosp, Long Beach, St Mary Med Ctr. *Mem:* AAAS; fel Am Col Physicians; Asn Am Physicians; Sigma Xi; Am Soc Clin Invest; Am Fedn Clin Res. *Res:* Internal medicine; red cell metabolism. *Mailing Add:* Dept Med Bin 414 Harbor-UCLA Med Ctr 1000 W Carson St Torrance CA 90509

TANAKA, NOBUYUKI, b Tokyo, Japan, May 12, 37; US citizen. NUCLEAR PHYSICS, NUCLEAR ENGINEERING. *Educ:* Harvard Univ, AB, 62; Tufts Univ, PhD(physics), 69. *Prof Exp:* Res asst physics, Tufts Univ, 65-69; STAFF MEM PHYSICS, LOS ALAMOS SCI LAB, UNIV CALIF, 69- *Res:* Intermediate-energy nuclear research using high resolution proton and pion spectrometers; improving the existing spectrometers and associated equipment; high energy physics. *Mailing Add:* 940 Old Bridge Ct Santa Fe NM 87505

TANAKA, TOYOICHI, b Nagaoka, Japan, Jan 4, 46; m 70; c 2. BIOPHYSICS. *Educ:* Univ Tokyo, BS, 68, MA, 70, DSc(physics), 72. *Prof Exp:* Fel biophys, 72-75, from asst prof to assoc prof, 75-82, PROF PHYSICS, MASS INST TECHNOL, 82- *Concurrent Pos:* Res assoc med physics, Retina Found, 73-75; vis prof, Univ Louis Pasteur, 80-81. *Honors & Awards:* Nishina Mem Prize, 85; Award Polymer Soc Japan, 86. *Mem:* Am Phys Soc; Phys Soc Japan; Biophys Soc Japan. *Res:* Laser scattering spectroscopy; critical phenomena of macromolecular solutions with applications to cataract disease; critical phenomena and phase transition in polymers and gels; physics of microemulsion. *Mailing Add:* Mass Inst Technol Rm 13-2153 77 Massachusetts Ave Cambridge MA 02139

TANAKA, YASUOMI, b Tokyo, Japan, Dec 5, 39; m 74; c 2. FOREST ECOLOGY. *Educ:* Tokyo Univ Educ, BS, 62; Duke Univ, MF, 67, PhD(forest ecol), 70. *Prof Exp:* Silviculturist, Agr Farm Monte D'Este, Brazil, 62-63; res fel forestry, Univ Sao Paulo, Brazil, 64-65; instr, Univ Parana, Brazil, 65-66; res assoc ecol, Ecol Sci Div, Oak Ridge Nat Lab, 70-71; FOREST NURSERY ECOLOGIST, WEYERHAEUSER FORESTRY RES CTR, 71- *Mem:* Japanese Forestry Soc; Soc Am Foresters. *Res:* Physiological and ecological aspect of seedling production; seedling dormancy, growth, nutrition in the greenhouse and openbed nursery; seed technology. *Mailing Add:* 1111 N Washington Centralia WA 98310

TANANBAUM, HARVEY DALE, b Buffalo, NY, July 17, 42; m 64; c 2. ASTROPHYSICS. *Educ:* Yale Univ, BA, 64; Mass Inst Technol, PhD(physics), 68. *Prof Exp:* Staff scientist, Am Sci & Eng, Inc, 68-73; ASTROPHYSICIST, SMITHSONIAN INST ASTROPHYS OBSERV, 73- *Concurrent Pos:* Assoc astron, Harvard Col Observ, Harvard Univ, 73-, lectr, dept astron, 80-; assoc dir, Harvard-Smithsonian Ctr Astrophys, 81-; mem, comt data mgt & comput, Space Sci Bd, 78-82, comt space astron & astrophys, 81-84, astron surv comt, Extragalactic Astron Working Group, 78-80, Res Briefing Panel Astron & Astrophys, Advan X-Ray Astrophys Facil Sci Working Group, 78-83, Astrophys Subcomt, NASA, 85-, Asn Univ Res Astron, Comt Future Directions, NOAO, 86-87, Space Sci & Applications Adv Comt, 88-, Automatic Planetary Sta Priorities Comt, 88, Astron & Astrophys Surv Comt, High Energy Panel, NRC, 89- *Honors & Awards:* Except Sci Achievement Medal, NASA, 80, Pub Serv Medal, 88. *Mem:* Fel AAAS; Am Astron Soc. *Res:* X-ray astronomy, especially discrete cosmic x-ray sources with satellite payloads. *Mailing Add:* Ctr Astrophys 60 Garden St Cambridge MA 02138

TANCRELL, ROGER HENRY, b Whitinsville, Mass, Feb 17, 35. PHYSICS, ELECTRICAL ENGINEERING. *Educ:* Worcester Polytech Inst, BS, 56; Mass Inst Technol, MS, 58; Harvard Univ, PhD(appl physics), 68. *Prof Exp:* Mem staff digital electronics, Lincoln Labs, Mass Inst Technol, 56-60; PRIN SCIENTIST MED ELECTRONICS, RAYTHEON RES, 68- *Mem:* Inst Elec & Electronics Engrs; Am Inst Ultrasound Med; Am Asn Physicists Med. *Res:* Medical imaging systems including ultrasound, x-ray and nuclear; surface acoustic wave devices for filters and analog signal processing; sonar (naval) transducers. *Mailing Add:* Raytheon Co Res 131 Spring St Lexington MA 02173

TANCZOS, FRANK I, b Northampton, Pa, Jan 2, 21. CHEMICAL PHYSICS. *Educ:* Moravian Col, BS, 39; Cath Univ, PhD(physics), 56. *Prof Exp:* Res chemist cent labs, Lehigh Portland Cement Co, Pa, 39-43; phys chemist, Bur Ord, 46-59, tech dir supporting res, Bur Naval Weapons, 59-66, tech dir res & technol, Air Systs Command, Dept Navy, 66-80; CONSULT, 80- *Concurrent Pos:* Mem, NASA res adv comts, Chem Energy Processes, 59-60, mem chem energy systs, 60-61 & air-breathing propulsion systs, 62-; lectr grad sch eng, Cath Univ, 60-61. *Mem:* Am Chem Soc; Am Phys Soc; assoc fel Am Inst Aeronaut & Astronaut. *Res:* New cement composition chemistry; molecular vibrational relaxation theory; energy conversion; propellant chemistry and thermodynamics; rocket and air-breathing jet propulsion principles; hypersonic air-breathing propulsion principles. *Mailing Add:* 1500 Massachusetts Ave NW Apt 249 Washington DC 20005

TANDBERG-HANSSEN, EINAR ANDREAS, b Bergen, Norway, Aug 6, 21; US citizen; m 51; c 2. ASTRONOMY, PHYSICS. *Educ:* Univ Oslo, PhD(astron), 60. *Prof Exp:* Res assoc astrophys, Univ Oslo, 50-57; mem sr res staff solar physics, High Altitude Observ, Boulder, Colo, 57-74; sr res scientist solar physics, 74-83, dep lab dir, 83-87, LAB DIR, NASA MARSHALL SPACE FLIGHT CTR, 87- *Concurrent Pos:* Adj prof, Physics Dept, Univ Ala, Huntsville, 82- *Mem:* Int Astron Union; Am Astron Soc; Norweg Geophys Soc; Norweg Acad Sci. *Res:* Solar physics, particularly flare and prominence research; solar corona and interplanetary space. *Mailing Add:* ESOI NASA Marshall Space Flight Ctr Huntsville AL 35812

TANDLER, BERNARD, b Brooklyn, NY, Feb 18, 33; wid; c 2. CYTOLOGY, ELECTRON MICROSCOPY. *Educ:* Brooklyn Col, BS, 55; Columbia Univ, AM, 57; Cornell Univ, PhD(cytol), 61. *Prof Exp:* Res fel, Sloan-Kettering Inst, 61-62; instr anat, Sch Med, NY Univ, 62-63; assoc biol, Sloan-Kettering Inst Cancer Res, Cornell Univ, 63-67, asst prof, Grad Sch, Sloan-Kettering Div, 66-67; assoc prof oral biol & med, 67-72, assoc prof anat, 67-79, PROF ORAL BIOL & MED, SCH DENT, CASE WESTERN RESERVE UNIV, 72-, PROF ANAT, SCH MED, 79- *Concurrent Pos:* Lectr, Brooklyn Col, 61-63; vis prof anat, Univ Copenhagen, 73; vis assoc prof, Sch Med, Stanford Univ, 75; vis prof, Univ Cagliari, 83 & Col Med, Northeastern Ohio Univ, 81; actg chmn oral biol, Sch Dent, Case Western Reserve Univ, 87-88. *Mem:* Am Soc Cell Biol; Am Asn Anatomists; Electron Micros Soc Am; Int Asn Dent Res. *Res:* Mitochondrial biogenesis; ultrastructure of normal and neoplastic salivary glands; pulmonary ultrastructure. *Mailing Add:* Dept Oral Biol Case Western Reserve Univ 2123 Adelbert Rd Cleveland OH 44106

TANEJA, VIDYA SAGAR, b India, Sept 7, 31; m 62; c 2. MATHEMATICAL STATISTICS. *Educ:* Panjab Univ India, BA, 50, MA, 52; Univ Minn, MA, 63; Univ Conn, PhD(statist), 66. *Prof Exp:* Lectr math, Doaba Col, 53-59; instr, Univ Minn, Morris, 64-65; asst prof math statist, NMex State Univ, 66-70; assoc prof, 70-74, PROF MATH STATIST, WESTERN ILL UNIV, 74- *Concurrent Pos:* Vis prof, Ohio State Univ, 77-78. *Mem:* Inst Math Statist; Am Statist Asn. *Res:* Statistical methodology, statistical analysis, time series, ranking and selection; heteroscedastic methods; probability distributions. *Mailing Add:* Dept Math Western Ill Univ Macomb IL 61455

TANENBAUM, BASIL SAMUEL, b Providence, RI, Dec 1, 34; m 56; c 3. IONOSPHERIC PHYSICS. *Educ:* Brown Univ, BS, 56; Yale Univ, MS, 57, PhD(physics), 60. *Prof Exp:* Staff physicist res div, Raytheon Co, 60-63; prof eng, Case Western Reserve Univ, 63-75; DEAN FAC, HARVEY MUDD COL, 75- *Concurrent Pos:* Vis scientist, Arecibo Observ, 67-68; Sigma Xi Res Award, 69; sci adv comt, Nat Astron & Ionospheric Ctr, 72-77; vis scientist, Southern Calif Edison Co, 90. *Mem:* Am Phys Soc; AAAS; Am Soc Eng Educ; Inst Elec & Electronics Engrs; Am Asn Univ Prof; Sigma Xi. *Res:* Sound propagation; theory of turbulence; ionospheric physics; waves in plasmas; radiation in a plasma; kinetic theory of gas mixtures and plasmas; shock wave theory; energy conversion; electromagnetic fields. *Mailing Add:* 611 Delaware Dr Claremont CA 91711

TANENBAUM, MORRIS, b Huntington, WVa, Nov 10, 28; m 50; c 2. CHEMICAL PHYSICS, METALLURGY. *Educ:* Johns Hopkins Univ, AB, 49; Princeton Univ, Am, 51, PhD(phys chem), 52. *Prof Exp:* Asst, Princeton Univ, 49-50, instr, 50-51; mem tech staff, Bell Tel Labs, 52-56, subdept head, 56-60, asst metall dir, 60-62, dir solid state devices lab, 62-64; dir res & develop, Western Elec Co, Inc, 64-68, gen mgr eng, 68-71, vpres eng, 71-72, vpres, Transmission Equip Div, 72-75; exec vpres systems eng & develop, Bell Labs, 75-76; vpres eng & network serv, Am Tel & Tel Co, 76-78; pres, NJ Bell Tel Co, 78-80; exec vpres, AT&T, 80-86, vchmn & chief financial officer, 86-91; RETIRED. *Concurrent Pos:* Mem mat adv bd, Nat Res Coun-Nat Acad Sci; consult, Dept Defense, NASA & Nat Bur Standards; mem vis comts, Mass Inst Technol, Princeton Univ, Carnegie Inst Technol, Univ Pa & Lehigh Univ. *Mem:* Nat Acad Eng; Am Chem Soc; fel Am Phys Soc; Am Inst Mining Metall & Petrol Engrs; Inst Elec & Electronics Engrs. *Res:* Chemistry and physics of solids; solid-state device physics; engineering research in manufacturing processes. *Mailing Add:* AT&T 295 N Maple Ave Rm 4449I3 Basking Ridge NJ 07920

TANENBAUM, STUART WILLIAM, b New York, NY, July 15, 24; m 62; c 2. BIOCHEMISTRY, MICROBIOLOGY. *Educ:* City Col New York, BS, 44; Columbia Univ, PhD, 51. *Prof Exp:* Instr chem, City Col New York, 48; Am Cancer Soc fel, Stanford Univ, 51-52; lectr bact, Univ Calif, 52; res assoc biol, Stanford Univ, 52-53; from res assoc to prof microbiol, Col Physicians & Surgeons, Columbia Univ, 53-73; dean Sch Biol Chem & Ecol, 73-85, PROF BIOTECHNOL, STATE UNIV NY COL ENVIRON SCI & FORESTRY, 85-; ADJ PROF MICROBIOL, STATE UNIV NY UPSTATE MED CTR, 73- *Concurrent Pos:* Mem panel molecular biol, NSF, 62-63 & 72-73, resident prog dir molecular biol sect, 71-72; State Univ NY fac exchange scholar, 74-77; trustee, Forestry Found, Syracuse, NY, 74-; prog dir, Biomed Res Support Prog, State Univ NY Col Environ Seci & Forestry, 77-84; mem competitive res, NSF, 80-; chmn, oversight panel, molecular & genetic biosci, NSF, 82; vis prof, istituto superiore di sanita, Rome, 63-64. *Honors & Awards:* Sigma Xi Fac Res Award, 86. *Mem:* Am Soc Biol Chemists; Am Soc Microbiol; Am Chem Soc; Soc Indust Microbiol. *Res:* Fungal metabolism; antibiotic biosynthesis; bacterial physiology; immunochemistry; cell and molecular biology; biotechnology. *Mailing Add:* Chem Dept State Univ NY Col Environ Sci & Forestry Syracuse NY 13210

TANFORD, CHARLES, b Halle, Ger, Dec 29, 21; nat US; div; c 3. MEMBRANES & TRANSPORT. *Educ:* NY Univ, BA, 43; Princeton Univ, MA, 44, PhD(phys chem), 47. *Prof Exp:* Asst, Princeton Univ, 43-44; chemist, Tenn Eastman Corp, 44-45; asst, Princeton Univ, 45-46; Lalor fel phys chem, Harvard Med Sch, 47-49; from asst prof to prof, Univ Iowa, 49-60; prof, 60-71, JAMES B DUKE PROF, DUKE UNIV, MED CTR, 71- *Concurrent Pos:* Guggenheim fel, Yale Univ, 56-57; consult, USPHS, 59-63; USPHS res career award, 62; vis prof, Harvard Univ, 66; mem, Whitehead Med Res Inst, 77-81; George Eastman vis prof, Univ Oxford, 77-78; Walker-Ames prof, Univ Washington, 79. *Honors & Awards:* Reilly Lectr, Univ Notre Dame, 79; Alexander von Humboldt Prize, 84. *Mem:* Nat Acad Sci; Am Acad Arts & Sci; Am Chem Soc; Biophys Soc (pres, 79-80). *Res:* Physical chemistry of proteins, especially transport proteins and related substances; structure and function of membranes. *Mailing Add:* Dept Cell Biol Duke Univ Med Ctr Durham NC 27710

TANG, ALFRED SHO-YU, b Shanghai, China, Sept 9, 34; US citizen; m 69; c 2. ALGEBRA. *Educ:* Univ Hong Kong, BSc, 56; Univ SC, MS, 60; Univ Calif, Berkeley, PhD(math), 69. *Prof Exp:* Assoc prof, 66-80, PROF MATH, SAN FRANCISCO STATE UNIV, 80- *Mem:* AAAS; Am Math Soc; Math Asn Am. *Mailing Add:* Dept Math San Francisco State Univ 1600 Holloway Ave San Francisco CA 94132

TANG, ANDREW H, b Canton, China, Feb 10, 36; US citizen; m 64; c 2. PHARMACOLOGY. *Educ:* Howard Col, BS, 60; Purdue Univ, MS, 62, PhD(pharmacol), 64. *Prof Exp:* Res assoc, 64-70, SR RES SCIENTIST PHARMACOL, UPJOHN CO, 70- *Mem:* AAAS; Am Soc Pharmacol & Exp Therapeut. *Res:* Pharmacology of the central nervous system; spinal cord physiology; behavioral pharmacology. *Mailing Add:* Res Dept Upjohn Co 301 Henrietta St Kalamazoo MI 49001

TANG, CHUNG LIANG, b Shanghai, China, May 14, 34; US citizen; m 58; c 3. PHYSICS. *Educ:* Univ Wash, BS, 55; Calif Inst Technol, MS, 56; Harvard Univ, PhD(appl physics), 60. *Prof Exp:* Res staff mem, Raytheon Co, 60-61, sr res scientist, 61-63, prin res scientist, 63-64; assoc prof elec eng, 64-68, SPENCER T OLIN PROF ENG, CORNELL UNIV, 68-; PRES, ITHACA RES CORP. *Concurrent Pos:* Consult res div, Raytheon Co, 64-72; assoc ed,

J Quantum Electronics, Inst Elec & Electronics Eng, 69- *Mem:* Nat Acad Eng; fel Inst Elec & Electronics Engrs; fel Am Phys Soc; fel Optical Soc Am. *Res:* Quantum electronics; electromagnetic theory. *Mailing Add:* Dept Elec Eng Sch Eng Cornell Univ 418 Phillips Hall Ithaca NY 14853

TANG, CHUNG-MUH, b Tungkang, Taiwan, Oct 20, 36; m 65; c 2. ATMOSPHERIC DYNAMICS. *Educ:* Nat Taiwan Univ, BS, 59; Univ Calif, Los Angeles, MA, 65, CPhil, 69, PhD(meteorol), 70. *Prof Exp:* Res asst, Taiwan Rain Stimulation Res Inst, 61-62; NSF grant & asst res meteorologist, Univ Calif, Los Angeles, 70; Defense Dept grant & res staff meteorologist, Yale Univ, 70-75; asst prof physics & atmospheric sci, Drexel Univ, 75-80; res scientist, 80-82, CONSULT, UNIV SPACE RES ASN, 82-88; fac, 89, ASST PROF PHYSICS, VILLANOVA UNIV, 90- *Mem:* Am Meteorol Soc. *Res:* Theoretical studies of large scale atmospheric motions; theoretical studies of shear-flow instability. *Mailing Add:* 241 Marple Rd Haverford PA 19041

TANG, CHUNG-SHIH, b China, Jan 8, 38; m 65; c 3. PLANT CHEMISTRY. *Educ:* Taiwan Univ, BS, 60, MS, 62; Univ Calif, Davis, PhD(agr chem), 67. *Prof Exp:* Res chemist, Univ Calif, Davis, 67-68; from asst prof to assoc prof, 68-79, PROF AGR BIOCHEM, UNIV HAWAII, 79-, CHMN, DEPT AGR BIOCHEM, 89- *Concurrent Pos:* Hon res prof, SChina Inst Bot, Acad Sinica, Guangzhou, China, 87-, Kunming Inst Bot, Acad Sinica, Kunming, China, 88- *Mem:* AAAS; Am Chem Soc; Am Soc Plant Physiol. *Res:* Bioactive plant secondary metabolite; allelopathy; rhizospheric chemistry. *Mailing Add:* Dept Agr Henke Hall 329 Univ Hawaii 1800 E W Hwy Honolulu HI 96822

TANG, DENNY DUAN-LEE, b China; US citizen; m 71; c 3. MAGNETICS. *Educ:* Univ Mich, Ann Arbor, PhD(elec eng), 75. *Prof Exp:* Mem res staff, Int Bus Mach, 75-82, mgr, 82-89, mem tech planning staff, 89-90, MEM RES STAFF, IBM ALMADEN RES CTR, 90- *Mem:* Fel Inst Elec & Electronics Engrs; Inst Elec & Electronics Engrs Electron Devices Soc. *Res:* Silicon bipolar technology. *Mailing Add:* IBM Almaden Res Ctr K66/803 650 Harry Rd San Jose CA 95120

TANG, DONALD T(AO-NAN), b China, May 9, 32; m 62; c 2. ELECTRICAL ENGINEERING. *Educ:* Nat Taiwan Univ, BS, 53; Univ Ill, PhD(elec eng), 60. *Prof Exp:* Instr elec eng, Univ Ill, 55-60; MEM RES STAFF, IBM CORP, 60- *Mem:* Inst Elec & Electronics Engrs; Sigma Xi. *Res:* Network theory, communication theory and information theory with applications to filter design, pattern recognition, error-correcting codes and data compaction. *Mailing Add:* 49 Fox Den Rd Mt Kisco NY 10549

TANG, HOMER H(O), b Kwangtung, China, Apr 4, 34; m 60; c 3. PHYSICAL SCIENCE, AERONAUTICAL ENGINEERING. *Educ:* Nat Cheng-Kung Univ, BS, 56; Okla State Univ, MS, 61, PhD(aeronaut eng), 64. *Prof Exp:* Instr mech eng, Okla State Univ, 63; res engr, Douglas Aircraft Co, McDonnell Douglas Astronautics Co, 63-64, sr scientist, 64-68, sect chief, Gas Dynamics, 68-73, PRIN ENGR, MCDONNELL DOUGLAS SPACE SYSTS CO, 74- *Concurrent Pos:* Pres, Scott Refrig & Tang Int. *Mem:* Am Inst Aeronaut & Astronaut; Am Soc Mech Engrs; Am Soc Heating, Refrigerating & Air Conditioning Engr. *Res:* Jet mixing and separated flow; blastwave interaction, gasdynamics and nuclear weapon effects; missile space and reentry vehicles technology. *Mailing Add:* PO Box 1217 Huntington Beach CA 92647

TANG, HWA-TSANG, b Shanghai, China, Nov 29, 37; US citizen; m 69. LINEAR ALGEBRA, SET THEORY. *Educ:* Univ Chicago, SM, 58, PhD(math), 65. *Prof Exp:* Asst prof math, Calif State Univ, Northridge, 59-61; Asst prof math, Temple Univ, Philadelphia, 64-69; ASSOC PROF MATH CALIF STATE UNIV, HAYWARD, 69- *Mem:* Am Math Soc. *Res:* Linear algebra; set theory; group theory; partial differential equations. *Mailing Add:* Dept Math Calif State Univ Hayward CA 94542

TANG, IGNATIUS NING-BANG, b Nanking, China, July 7, 33; m 62; c 2. AEROSOL SCIENCE, ATMOSPHERIC CHEMISTRY. *Educ:* Nat Taiwan Univ, BS, 55; Univ NDak, MS, 60; State Univ NY Stony Brook, MS, 75, PhD, 82. *Prof Exp:* Asst engr, Taiwan Fertilizer Corp, Taiwan, Repub of China, 56-58; CHEM ENGR, BROOKHAVEN NAT LAB, 64- *Mem:* Am Asn Aerosol Res; Am Chem Soc. *Res:* Thermodynamics of concentrated electrolyte solutions; mechanisms of gas-to-particle conversion and atmospheric nucleation phenomena; condensational growth and light scattering of atmospheric aerosols; gas phase reaction kinetics; mass transfer at gas-liquid interface. *Mailing Add:* Brookhaven Nat Lab Bldg 815 Upton NY 11973

TANG, JAMES JUH-LING, b Tientsin, China, Mar 8, 37; US citizen; m 65; c 2. APPLIED MECHANICS, HEAT TRANSFER. *Educ:* Nat Taiwan Univ, BS, 59; Univ Mo-Rolla, MS, 63; Yale Univ, PhD(appl mech), 70. *Prof Exp:* Engr, Weiskopf & Pickworth, 63-65; sr res engr, Princeton, NJ, 70-75, res assoc, 75-77, supvr machine design, 77-78, supvr develop eng, 78-82, mgr eng & prod develop, Am Can Co, Barrington, Ill, 82-87; MGR, CAN DEVELOP, CONTINENTAL CAN CO, OAK BROOK, IL, 87- *Mem:* Am Soc Mech Engrs; Soc Rheology; Am Inst Physics. *Res:* Elastic-plastic material behavior; heat transfer of industrial processes; computerized process control; finite element analysis; computer-aided engineering. *Mailing Add:* 833 S Elm St Palatine IL 60067

TANG, JORDAN J N, b Foochow, China, Mar 23, 31; m 58; c 2. BIOCHEMISTRY. *Educ:* Taiwan Prov Col, BS, 53; Okla State Univ, MS, 57; Univ Okla, PhD(biochem), 61. *Prof Exp:* Res asst biochem, Okla State Univ, 55-57; res asst, 57-58, biochemist, 61-63, assoc biochem, 63-65, assoc prof, 65-69, head, neurosci sect & actg head, 70-71, HEAD, LAB PROTEIN STUDIES, OKLA MED RES FOUND, 71-; PROF BIOCHEM, SCH MED, UNIV OKLA, 71- *Concurrent Pos:* Res assoc biochem, Sch Med, Univ Okla, 62-63, asst prof & asst head dept, 63-67, assoc prof, 67-70, vis scientist, Lab Molecular Biol, Cambridge, Eng, 65-66; Guggenheim Fel, 65; J G Puterbaugh chair med res, 85- *Mem:* AAAS; Am Chem Soc; Am Soc Biol Chemists. *Res:* Structure and function of proteins; structure of gastric, lysosomal and retroviral proteolytic enzymes. *Mailing Add:* Okla Med Res Found 825 NE 13th St Oklahoma City OK 73104

TANG, KWONG-TIN, b Feb 24, 36; US citizen. PHYSICS, PHYSICAL CHEMISTRY. *Educ:* Univ Wash, BS, 58, MA, 59; Columbia Univ, PhD(physics), 65. *Prof Exp:* Res assoc chem, Columbia Univ, 65-66; physicist, Collins Radio Co, 66-67; from asst prof to assoc prof, 67-72, chmn dept, 72-77, PROF PHYSICS, PAC LUTHERAN UNIV, 72- *Concurrent Pos:* Consult, Boeing Co, 70 & 72; vis lectr, Univ Wash, 71; Res Corp grant, Pac Lutheran Univ, 71-74. *Mem:* Am Phys Soc; Am Asn Physics Teachers. *Res:* Atomic and molecular collision; scattering theory; reaction rates; intermolecular forces; optical dispersion; lattice vibration. *Mailing Add:* Dept Physics Pac Lutheran Univ S 121st & Park Ave Tacoma WA 98447

TANG, PEI CHIN, b Hupei, China, Sept 14, 14; nat US; m; c 6. NEUROPHYSIOLOGY, NEUROANATOMY. *Educ:* Tsing Hua Univ, China, BS, 42; Univ Wash, PhD(physiol), 53. *Prof Exp:* Instr physiol, Med Sch, Peking Univ, 45-48; instr pharmacol, Univ Wash, 52-55; asst prof anat, Univ Tex Southwestern Med Sch Dallas, 55-56; neurophysiologist, Air Force Sch Aviation Med, 56-60; chief neurophysiol br, Civil Aeromed Res Inst, Fed Aviation Agency, Okla, 60-66; chief physiol sci div, Res Dept, Naval Aerospace Med Inst, 66-70; assoc prof, 70-77, PROF PHYSIOL, CHICAGO MED SCH, 77- *Mem:* Am Physiol Soc; Am Asn Anat; Aerospace Med Asn; Soc Exp Biol & Med. *Res:* Central nervous control of respiration, micturition and vasomotor activity. *Mailing Add:* Dept Physiol & Biophys Chicago Med Sch Univ Health Sci 3333 Green Bay Rd North Chicago IL 60064

TANG, RUEN CHIU, b Kiangsu, China, Oct 31, 34; m 60; c 3. FOREST PRODUCTS. *Educ:* Nat Chung-Hsin Univ, Taiwan, BS, 57; NC State Univ, PhD(wood sci), 68. *Prof Exp:* Teacher, Kung Hua Sch Elec Technol, 56-57; wood technologist, Taiwan Forest Bur, 59-63; res asst, US Naval Res, NC State Univ, 63-66; teaching asst wood mech, Univ Wash, 66-67; State of Ky & USAF res assoc, Inst Theoret & Appl Mech, Univ Ky, 68, State of Ky res assoc wood mech, Univ, 69, from asst prof to assoc prof wood sci, 74-77; PROF WOOD SCI, AUBURN UNIV, 78- *Mem:* Soc Wood Sci & Technol; Soc Exp Stress Analysis; Forest Prod Res Soc; Soc Am Foresters; Am Soc Testing & Mat; Sigma Xi. *Res:* Anisotropic elasticity; composite materials; fiber mechanics; noise control; reliability in structural design; math modeling. *Mailing Add:* Dept Forestry Auburn Univ Auburn AL 36830

TANG, STEPHEN SHIEN-PU, b Changsha, China, Nov 13, 35; m 72; c 2. SYSTEMS ENGINEERING. *Educ:* Nat Taiwan Univ, BS, 59; Univ Cailf, Berkeley, MS, 64; Princeton Univ, PhD(aerosci), 69. *Prof Exp:* Res assoc molecular beams, Dept Appl Sci, Yale Univ, 68-71; vis assoc prof fluid mech, Nat Univ Taiwan, 71-72, vis assoc prof molecular beams, Tech Univ Hannover, Ger, 72-74; res scientist chem laser, Defense & Space Syst Group, TRW, Inc, 74-80; PROJ ENGR SPACE TECHNOL PLANNING, AEROSPACE CORP, LOS ANGELES, 80- *Mem:* Am Inst Aeronaut & Astronaut; Am Phys Soc; Sigma Xi. *Res:* Molecular beams; gasdynamics and fluid mechanics; gas phase kinetics; gas-solid interactions; space technology planning; high power lasers; high beam energy particle beams; satellite systems and technologies. *Mailing Add:* 1611 Toscanini Dr San Pedro CA 90732

TANG, TING-WEI, b Taiwan, China, May 27, 34; m 63; c 2. ELECTRICAL ENGINEERING, PLASMA PHYSICS. *Educ:* Taiwan Univ, BSEE, 57; Brown Univ, MS, 61, PhD(eng), 64. *Prof Exp:* Teaching asst elec eng, Taiwan Univ, 57-59; res asst eng, Brown Univ, 59-63; instr elec eng, Univ Conn, 63-64, asst prof aerospace eng, 64-68; assoc prof elec eng, 68-74, PROF ELEC & COMPUTER ENG, UNIV MASS, AMHERST, 74- *Concurrent Pos:* NSF res initiation grant, 65-66, res grants, 66-68, 70-72, 74-76 & 76-78. *Mem:* Inst Elec & Electronics Engrs; Am Phys Soc. *Res:* Electromagnetic theory; nonlinear wave interactions in plasmas; antenna design; solid-state device modeling. *Mailing Add:* Dept Elec & Computer Eng Univ Mass Amherst MA 01003

TANG, VICTOR KUANG-TAO, b Peiping, China, Mar 13, 29; US citizen; m 63. STATISTICS, MATHEMATICS. *Educ:* Nat Taiwan Univ, BA, 56; Univ Wash, MA, 63; Iowa State Univ, PhD(statist), 71. *Prof Exp:* From assoc prof to prof math, Humboldt State Univ, 72-85; ANALYST, CTR NAVAL ANALYSIS, 85- *Mem:* Inst Math Statist; Int Asn Survey Statisticians; Royal Statist Soc; Am Statist Asn. *Res:* Applied and mathematical statistics. *Mailing Add:* Ctr Naval Analysis 4401 Ford Ave Alexandria VA 22302-0268

TANG, WALTER KWEI-YUAN, chemical engineering, polymer chemistry; deceased, see previous edition for last biography

TANG, WILSON H, b Hong Kong, Aug 16, 43; US citizen; m 69; c 2. RISK & RELIABILITY ANALYSIS, ENGINEERING EDUCATION. *Educ:* Mass Inst Technol, BS, 66, MS, 67; Stanford Univ, PhD(civil eng), 69. *Prof Exp:* From asst prof to assoc prof, 69-80, assoc head, 89-91, PROF CIVIL ENG, UNIV ILL, URBANA-CHAMPAIGN, 80- *Concurrent Pos:* Guggenheim fel, John S Guggenheim Found, 76; vis prof, Norweg Geotech Inst, Oslo, 76-77, Imp Col, London, 77 & Nat Univ Singapore, 83; prin investr, NSF & Am Petrol Inst, 79-; chmn, Offshore Reliability Comt, Am Soc Civil Engrs, 85-88; mem, Geotech Bd, Nat Res Coun. *Honors & Awards:* State-of-the-Art Award, Am Soc Civil Engrs, 90. *Mem:* Am Soc Civil Engrs; Am Soc Eng Educ; Int Soc Soil Mech & Found Eng; Int Geostatist Asn; Int Asn Struct Safety & Reliability; Int Asn Civil Eng Reliability & Risk Analysis. *Res:* Risk and reliability modeling and assessment in geotechnical, structural, hydraulic and offshore engineering. *Mailing Add:* 310 Willard Urbana IL 61801

TANG, Y(U) S(UN), b Nanking, China, Oct 24, 22; US citizen; m 50; c 3. HEAT TRANSFER, NUCLEAR WASTE MANAGEMENT. *Educ:* Nat Cent Univ, China, BSME, 44; Univ Wis, MS, 48; Univ Fla, PhD(chem eng), 52. *Prof Exp:* Sr process engr, Gen Chem Div, Allied Chem & Dye Corp, 52-54, sr proj engr, 54-56; sr develop engr, Steam Div, Westinghouse Elec Corp, 56-59; group leader heat transfer, Allison Div, Gen Motors Corp, 59-64, prin scientist, 64-66; adv engr, Astronuclear Lab, Westinghouse Elec Corp, 66-71, adv engr, Advan Reactors Div, 71-84; res assoc prof energy res, Univ

Pittsburgh, 84-86, res prof chem eng, 87-89, res prof mech eng, 89-90, CONSULT ENG, UNIV PITTSBURGH, 91- *Concurrent Pos:* Lectr, Univ Pittsburgh, 67-68; vis specialist, Nat Cent Univ, Repub China, 84; vis prof, Nanyang Technol Inst, Singapore, 85, Nat Tsing Hua Univ, Repub China, 86. *Mem:* Fel Am Inst Chem Engrs; Am Nuclear Soc; Am Soc Mech Engrs. *Res:* Heat transfer and fluid flow; liquid metal boiling and two-phase flow; space power generation; liquid-metal fast breed reactor thermal analysis; nuclear waste management. *Mailing Add:* 1552 Holly Hill Dr Bethel Park PA 15102-3508

TANG, YAU-CHIEN, b China, Aug 7, 28; nat US; m 60; c 2. PHYSICS. *Educ:* Univ Ill, PhD(physics), 58. *Prof Exp:* Res assoc, Fla State Univ, 58-62 & Brookhaven Nat Lab, 62-64; assoc prof, 64-70, PROF PHYSICS, UNIV MINN, MINNEAPOLIS, 70- *Honors & Awards:* Alexander von Humboldt Sr Scientist Award. *Mem:* Am Phys Soc. *Res:* Low energy nuclear physics. *Mailing Add:* Dept Physics Univ Minn 148 Physics Minneapolis MN 55455

TANG, YI-NOO, b Hunan, China, Feb 28, 38; m 64; c 2. PHYSICAL CHEMISTRY, RADIOCHEMISTRY. *Educ:* Chung Chi Col, Hong Kong, BA, 59; Univ Kans, PhD(chem), 64. *Prof Exp:* Fel, Univ Kans, 64-65; fel, Univ Calif, Irvine, 65-66, instr chem, 66-67; from asst prof to assoc prof chem, 67-77, PROF CHEM, TEX A&M UNIV, 77- *Mem:* Am Chem Soc; Am Inst Physics. *Res:* Hot atom chemistry; unimolecular reactions; photochemistry; carbene chemistry; gas chromatography; silicon chemistry. *Mailing Add:* Dept Chem Tex A&M Univ College Station TX 77843

TANGEL, O(SCAR) F(RANK), b Philadelphia, Pa, Jan 11, 10; m 36; c 3. METALLURGY. *Educ:* Lafayette Col, BS, 32; Mont Sch Mines, MS, 34. *Prof Exp:* Mill supt & metall engr, Mont Coal & Iron Co, 35; metall engr, New Bonanza Mine, Nev, 36; mill supt & metall engr, Ambassador Gold Mines, Ltd, Nev, 36 & Goldfields of Am, Ltd, 36-37; metall engr, Fresnillo Co, Mex, 37-40, asst mill supt, 40-41; res engr, Battelle Mem Inst, 41-42, asst supvr, 45-53, div chief, 53-65, tech adv, 65-66; consult, Newmont Mining Corp, 66-67, chief metall engr, 68-71, vpres res & develop, 72-77; PVT CONSULT, 78- *Concurrent Pos:* Plant shift boss, Lakeview Gold Mines, Mont, 35; assayer, King Solomon Gold Mines, Calif, 35-36; metall engr, Pan-Am Eng Co, Calif, 37; dir, Atlantic Cement Co, 67-85, Idarado Mining Co, 69-78, Newmont Explor Ltd, 69-, Foote Mineral Co, 74- & Magma Copper Co, 75-78; vpres, Newmont Explor Ltd, 69-78. *Mem:* Am Inst Mining, Metall & Petrol Engrs; Can Inst Mining & Metall. *Res:* Beneficiation of metallic and non-metallic ores. *Mailing Add:* 25 Thunder Mountain Rd Greenwich CT 06831

TANGHERLINI, FRANK R, b Boston, Mass, Mar 14, 24; div; c 4. PHYSICS. *Educ:* Harvard Univ, BS, 48; Univ Chicago, MS, 52; Stanford Univ, PhD(physics), 59. *Prof Exp:* NSF fel physics, Niels Bohr Inst, Copenhagen, Denmark, 58-59 & Sch Theoret & Nuclear Physics, Naples, Italy, 59-60; res assoc, Univ NC, Chapel Hill, 60-61; asst prof, Duke Univ, 61-64; assoc prof, George Washington Univ, 64-66; sci assoc space res, Ion Lab, Tech Univ Denmark & Danish Space Res Inst, 66-67; ASSOC PROF PHYSICS, COL HOLY CROSS, 67- *Concurrent Pos:* Sr res engr, Gen Dynamics/Convair, 52-55; lectr, Univ NC, Chapel Hill, 61; vis scientist, Int Ctr Theoret Physics, 73-74; vis scholar, Harvard Univ, 88-89. *Mem:* Am Phys Soc; Sigma Xi; Am Asn Univ Prof. *Res:* Mathematical biology; cybernetics; high energy physics; foundations of special relativity; general relativity and gravitation; dimensionality of space; classical electron theory; ionosphere and space physics; theoretical physics; elementary particles; cosmology; relativity and quantum optics. *Mailing Add:* Dept Physics Col Holy Cross Worcester MA 01610

TANGNEY, JOHN FRANCIS, b Evanston, Ill, Aug 4, 49; m 85. HUMAN PATTERN RECOGNITION, VISUAL PSYCHOPHYSICS. *Educ:* Loyola Univ Chicago, BS, 71; State Univ NY Buffalo, PhD(psychol), 78. *Prof Exp:* Res asst, Parmly Hearing Inst, 73-74; res asst, Psychol Dept, State Univ NY Buffalo, 74-78; fac res asst, Univ MD, Col Park, 79-80, fac res assoc, 81-85; PROG MGR, AIR FORCE SCI RES, 85; DIV DIR, NAT RES COUN, 90- *Concurrent Pos:* Co-prin investr, NIH grants, 80- *Mem:* Asn Res Vision & Ophthal; Acoust Soc Am; Optical Soc Am; Soc Neurosci. *Res:* Visual psychophysical experiments--to discriminate between models of human pattern recognition and to describe the dynamics and the interaction of neural structures that underlie vision. *Mailing Add:* Div Dir Div Human Behav & Performance Nat Res Coun 2101 Constitution Ave NW Washington DC 20418

TANGONAN, GREGORY LIGOT, b Springfield, Mass, Oct 26, 47. APPLIED PHYSICS, PHYSICS. *Educ:* Ateneo Manila Univ, BS, 69; Calif Inst Technol, MS, 72, PhD(appl physics), 75. *Prof Exp:* Mem tech staff, 71-78, HEAD, INTEGRATED OPTICS SECT, HUGHES RES LABS, HUGHES AIRCRAFT CO, 78- *Mem:* Sigma Xi; Optical Soc Am. *Res:* Development of optical circuits for use in high speed optical data processing with emphasis on development of new fabrication techniques for optical waveguide devices. *Mailing Add:* 141 Santa Rosa Ave Oxnard CA 93035-4475

TANGORA, MARTIN CHARLES, b New York, NY, June 21, 36; m 73; c 2. TOPOLOGY. *Educ:* Calif Inst Technol, BS, 57; Northwestern Univ, MS, 58, PhD(math), 66. *Prof Exp:* Instr math, Northwestern Univ, 66-67 & Univ Chicago, 67-69; temp lectr, Univ Manchester, 69-70; asst prof, 70-72, ASSOC PROF MATH, UNIV ILL, CHICAGO, 72- *Concurrent Pos:* Sr vis fel, Univ Oxford, 73-74. *Mem:* Am Math Soc; Math Asn Am. *Res:* Algebraic topology; homotopy theory; cohomological methods. *Mailing Add:* Dept Math m/c 249 Univ Ill Box 4348 Chicago IL 60680-4348

TANGUAY, A(RMAND) R(ENE), b Can, Feb 1, 24; nat US; m 48; c 3. ELECTRICAL ENGINEERING. *Educ:* Univ Mass, BS, 50; Mass Inst Technol, MS, 51. *Prof Exp:* Asst comput lab, Mass Inst Technol, 50-51; sect head systs anal, Cornell Aeronaut Lab, Inc, 51-57; dept head systs res, Res Div, Radiation, Inc, Fla, 57-60; dir advan systs, Ryan Aerolab, Aerolab Develop Co, 60-61; assoc mgr advan electronics & info systs div, Electro-Optical Systs, Inc, 61-63, mgr, 63, mgr energy conversion div, 64, mgr info systs div, 65-67; div mgr med diag opers, Xerox Corp, 67-70, mgr, Micrographics Progs Res & Eng Div, 70-71, mgr strategic tech planning info technol group, 72-80, tech specialist & prog mgr, 80-83; CONSULT, 83- *Mem:* Sr mem Inst Elec & Electronics Engrs. *Res:* Systems and technology development and engineering in information sciences; automation; control systems; technical planning and program management; technology and systems planning and development management in digital and graphical information processing and management with specialties in micrographics and office information automation. *Mailing Add:* 5740 Barnes Rd Canandaigua NY 14424

TANGUAY, ROBERT M, b Sherbrooke, Que, Dec 17, 44; m 67; c 2. CELLULAR & DEVELOPMENTAL BIOLOGY. *Educ:* Sherbrooke Sem, BA, 63; Sherbrooke Univ, BSc, 66; Laval Univ, DSc, 71. *Prof Exp:* Fel molecular biol, Karolinska Inst, Stockholm, 71-73; from asst prof to assoc prof genetics, 71-83, dir, Dept Ontogenesis & Molecular Genetics, 81-86, PROF GENETICS, DEPT MED, LAVAL UNIV, 83- *Concurrent Pos:* Vis prof, Europ Molecular Biol Lab, Heidelberg, 80-81 & Inst Genetics, Munich, 88-89; pres, State Ministry of Cult & Sci Develop, Can, 82-83; mem, Comt Terminology in Genetic Eng, Med Res Coun, Can, 84-85, Human Genome Adv Comt, Can Network Centres Excellence; asst ed, Biochem & Cell Biol, 84- *Mem:* Can Soc Cell Biol (dir, 77-79, treas, 86-89); Genetics Soc Can (dir, 83-85); Can Biochem Soc; NY Acad Sci. *Res:* Cellular and molecular biology of the heat shock response in Drosophila; biochemistry and function of the various heat shock proteins; regulation of gene activity during stress; study of the molecular basis of the hereditary disease, tyrosinemia (type 1); molecular entomology of black flies. *Mailing Add:* Ontogenesis & Molecular Genetics Res Ctr CHUL Ste-Foy PQ G1V 4G2 Can

TANI, SMIO, b Tokyo, Japan, Feb 24, 25; US citizen; m 58; c 3. THEORETICAL PHYSICS. *Educ:* Univ Tokyo, BS, 46, ScD, 55. *Prof Exp:* Res assoc, Kyoto Univ, 51-54, Tokyo Univ Educ, 54-57, Case Western Reserve Univ, 57-59 & Wash Univ, 59-60; from res scientist to sr res scientist physics, NY Univ, 60-65; assoc prof, 65-68, PROF PHYSICS, MARQUETTE UNIV, 68- *Mem:* Am Phys Soc; Am Asn Physics Teachers. *Res:* Theory of scattering; canonical transformation in classical and quantum mechanics; atomic physics; elementary particle physics. *Mailing Add:* Dept Physics Marquette Univ 1515 W Wisconsin Ave Milwaukee WI 53233

TANIGAKI, NOBUYUKI, b Tokyo, Japan, Oct 22, 29. CANCER. *Educ:* Tokyo Univ, MD, 56. *Prof Exp:* Res fel path, Sch Med, Hokkaido Univ, 57-61, asst, 61-63, asst, Inst Cancer Immunopath, 63-66, instr, 66-67; asst prof, Cancer Res Inst, Kanazawa Univ, 67-68; sr cancer res scientist, Dept Biochem Res, 69-72, assoc cancer res scientist, 72-75, PRIN CANCER RES SCIENTIST, ROSWELL PARK MEM INST, 75- *Concurrent Pos:* Sloan-Kettering Inst Cancer Res fel, 63-64; cancer res scientist, Dept Biochem Res, Roswell Park Mem Inst, 64-66. *Mem:* Japanese Cancer Asn; Japanese Path Soc; Am Asn Immunologists. *Mailing Add:* Dept Molecular Immunol Roswell Park Mem Inst Buffalo NY 14263

TANIKELLA, MURTY SUNDARA SITARAMA, b Amalapuram, India, Dec 5, 38; m 67; c 1. PHYSICAL CHEMISTRY, POLYMER CHEMISTRY. *Educ:* Osmania Univ, India, BSc, 57, MSc, 59; Princeton Univ, MA, 64; Univ Pittsburgh, PhD(phys, physico-org chem), 67. *Prof Exp:* Lectr chem, Osmania Univ, India, 59-62; fel with Prof K S Pitzer, Rice Univ, 67-68; fel, Nat Res Coun Can, 68-69; Nat Res Coun Can fel, Univ Calgary, 69-70; res chemist, Carothers Res Lab, 70-74, sr res chemist, 74-76, sr res chemist, Chattanooga Nylon Tech, 77-80, mem res staff, feedstocks div, cent res & develop dept, exp sta, 80-82, res assoc, Nomex indust appl res, 82-85, res assoc composites, Chestnut Run, Wilmington, Del, E I Du Pont de Nemours & Co, Inc, 85-90. *Concurrent Pos:* Fulbright travel grant, 62. *Mem:* Am Chem Soc; Soc Advan Mat & Process Eng. *Res:* Thermodynamics of hydrogen bonding; spectroscopy; structure of water; problems in textile fibers chemistry; renewable resources; nomex and aromatic polyamides; composites, polyimides. *Mailing Add:* Bldg 702 Chestnut Run E I du Pont de Nemours & Co Inc Wilmington DE 19898

TANIMOTO, TAFFEE TADASHI, b Kobe, Japan, Dec 15, 17; nat US; m 46; c 4. MATHEMATICS, GEOMETRY. *Educ:* Univ Calif, Los Angeles, AB, 42; Univ Chicago, MS, 46; Univ Pittsburgh, PhD(math), 50. *Prof Exp:* Instr math, Ill Inst Technol, 46-49; instr math, Allegheny Col, 49-51, asst prof, 51-54; mathematician, Int Bus Mach Corp, 54-61; head pattern recognition lab, Melpar, Inc, 61-63; staff mathematician, Honeywell, Inc, 63-65; prof math & chmn dept, Univ Mass, Boston, 65-90, dir grad progs, 75-76; RETIRED. *Mem:* Am Math Soc; Math Asn Am. *Res:* Geometry and analysis. *Mailing Add:* 156 Bellevue St Newton MA 02158

TANINO, KAREN KIKUMI, b Toronto, Ont, Feb 15, 58. PHYSIOLOGY, ENVIRONMENTAL STRESS. *Educ:* Univ Guelph, BSc, 81, MSc, 83; Ore State Univ, PhD(hort), 90. *Prof Exp:* ASST PROF ENVIRON STRESS PHYSIOL, UNIV SASK, 89- *Mem:* AAAS; Am Soc Plant Physiologists; Am Soc Hort Sci; Int Soc Hort Sci; Sigma Xi. *Res:* Low temperature stress physiology; using whole plant and tissue culture systems to understand the role of abscisic acid in freezing resistance; examining the relationship between dormancy and stress resistance. *Mailing Add:* Dept Hort Univ Sask Saskatoon SK S7N 0W0 Can

TANIS, ELLIOT ALAN, b Grand Rapids, Mich, Apr 23, 34; m 59; c 3. COMPUTER ART, COMPUTERS IN EDUCATION. *Educ:* Cent Col, Iowa, BA, 56; Univ Iowa, MS, 60, PhD(math), 63. *Prof Exp:* Lectr math, Univ Iowa, 63; asst prof math statist, Univ Nebr, 63-65; assoc prof, 65-71, PROF MATH, HOPE COL, 71- *Mem:* Am Math Soc; Sigma Xi; Math Asn Am; Am Statist Asn. *Res:* Writing educational materials in statistics-textbooks and computer based laboratory materials; developing computer programs for drawing artistic designs and repeating patterns; use of the computer in statistics. *Mailing Add:* Dept Math Hope Col Holland MI 49423-3698

TANIS, JAMES IRAN, b Zeeland, Mich, Oct 8, 34; m 70; c 1. EXPLORATION GEOPHYSICS. *Educ:* Mich Technol Univ, BS, 57, MS, 58; Univ Utah, PhD(geophys), 63. *Prof Exp:* Geophysicist, Shell Oil Co, 62-70; GEOPHYSICIST, CONOCO, INC 70- *Mem:* Soc Explor Geophysicists; Europ Asn Explor Geophysicists. *Res:* Crustal studies. *Mailing Add:* Conoco El Hamma Ltd Three rue Enee Tunis Belvedere 1002 Tunisia

TANIS, STEVEN PAUL, b Newport, RI, Sept 1, 52; m 79. SYNTHETIC ORGANIC CHEMISTRY. *Educ:* Rutgers Univ, BA, 74; Columbia Univ, MA, 77, MPhil, 78, PhD(chem), 80. *Prof Exp:* NIH fel, Calif Inst Technol, 79-80; asst prof chem, Mich State Univ, 80-86; res scientist, 86-90, ASSOC DIR, MED CHEM RES, UPJOHN CO, 91- *Concurrent Pos:* Camille & Henry Dreyfuss award young fac chem, 80-84. *Mem:* Am Chem Soc. *Res:* Isolation; structure determination; synthesis of biologically active natural products; medicinal chemistry. *Mailing Add:* Med Chem Res Upjohn Co 7246-209-6 Kalamazoo MI 49001

TANIUCHI, HIROSHI, b Japan, April 13, 30; US citizen; m 57; c 2. PROTEIN CHEMISTRY, CHEMICAL BIOLOGY. *Educ:* Kyoto Univ, MD, 62, PhD(biochem), 62. *Prof Exp:* Res assoc, dept med chem, Kyoto Univ Fac Med, 61-63; vis scientist lab chem biol, Nat Inst Arthritis & Metab Dis, 63-69, RES CHEMIST LAB CHEM BIOL, NAT INST ARTHRITIS, DIABETES & DIGESTIVE & KIDNEY DIS, NIH, 69-, CHIEF PROTEIN CHEM, 85- *Concurrent Pos:* Res fel dept biol chem, Harvard Med Sch, 63; chief sect protein conformation, lab chem biol, Nat Inst Arthritis, Diabetes & Digestive & Kidney Dis, NIH, 72-85. *Mem:* Am Soc Biochem & Molecular Biol. *Res:* Protein folding mechanism; antigen recognition mechanism; mechanism of long range interaction in proteins. *Mailing Add:* NIH Bldg 10 Rm 9N313 Bethesda MD 20892

TANIYAMA, TADAYOSHI, b Aichi, Japan, Dec 21, 47; m 76; c 2. DIFFERENTIATION, GENE EXPRESSION. *Educ:* Osaka Univ, PhD(med sci), 76. *Prof Exp:* Instr immunol, Saga Med Sch, 80-82; res scientist, 82-84, sr investr, 84-90, CHIEF IMMUNOL, NIH, 90- *Concurrent Pos:* Vis asst prof, Sch Educ, Waseda Univ, 87-, Sch Med, Osaka Univ, 90- & Sch Med, Kobe Univ, 90-; vis prof, Sch Med, McGill Univ, 89. *Mem:* Am Asn Immunologists; NY Acad Sci; Japanese Soc Immunol; Molecular Biol Soc Japan. *Res:* Regulation of gene expression of human myeloid cells during differentiation. *Mailing Add:* Immunobiol Div Dept Cellular Immunol NIH Shinagawa-Ku Tokyo 141 Japan

TANK, PATRICK WAYNE, b Charlotte, Mich, Jan 9, 50; m 73. PATTERN FORMATION, REGENERATION. *Educ:* Western Mich Univ, BS, 72; Univ Mich, Ann Arbor, MS, 73, PhD(anat), 76. *Prof Exp:* Fel develop biol, Univ Calif, Irvine, 76-78; asst prof, 78-83, assoc prof anat, 83-89, PROF ANAT, UNIV ARK MED SCI, 89- *Mem:* Am Asn Anatomists; Soc Develop Biol; Sigma Xi; Am Soc Zoologists. *Res:* Morphogenesis during development and regeneration in vertebrates; pattern formation during regeneration of the limbs of urodele amphibians; larval and embryonic systems. *Mailing Add:* Dept Anat 510 Univ Ark Med Sci 4301 W Markham Little Rock AR 72205

TANKARD, ANTHONY JAMES, b Pietermaritzburg, SAfrica, June 11, 42; Can citizen; m 69; c 2. HYDROCARBON EXPLORATION. *Educ:* Univ Natal, SAfrica, BSc, 66, Hons, 69; Rhodes Univ, SAfrica, PhD(geol), 74. *Prof Exp:* Chemist, Lever Bros SAfrica, 67; metallurgist, Alcan SAfrica, 68-69; geologist, SAfrican Mus, 70-78; res assoc, Univ SC, 78-79; asst prof sedimentology-coal, Univ Tenn, 79-81; sr geologist, 81-83, chief geologist, 83-85, explor mgr, 85-86, CHIEF SURGEON, PETRO-CANADA RESOURCES, 86- *Concurrent Pos:* Mem, Can Nat Comt Ocean Drilling Proj, 85-; mem, Geosci Coun Can, 86- *Honors & Awards:* Tracks Award, Can Soc Petrol Geologists, 85. *Mem:* Am Asn Petrol Geologists; Can Soc Petrol Geologists; Soc Econ Paleontologists & Mineralogists; Geol Asn Can; Int Asn Sedimentologists. *Res:* Basin analysis. *Mailing Add:* Petro Canada Inc Calgary AB T2P 3E3 Can

TANKERSLEY, DONALD, b Stockton, Calif, Nov 29, 39. IMMUNOCHEMISTRY. *Educ:* Calif State Univ, BS, 60; Univ Nev, MS, 62. *Prof Exp:* CHIEF, LAB PLASMA DERIVATIVES, CTR BIOL EVAL & RES, NIH, 84- *Mem:* Am Chem Soc; AAAS. *Mailing Add:* Lab Plasma Derivatives Ctr Biol Eval & Res NIH Bldg 29 Rm 303 8800 Rockville Pike Bethesda MD 20892

TANKERSLEY, ROBERT WALKER, JR, b Watsonville, Calif, June 18, 27; m 51; c 3. VIROLOGY, BACTERIOLOGY. *Educ:* Stanford Univ, AB, 52, MA, 54, PhD(med microbiol), 56. *Prof Exp:* Instr virol, Med Sch, Univ Minn, 56-58, instr bact & virol, 58-60; from asst prof to assoc prof microbiol, Med Col Va, 60-68; dir microbiol res, 68-80, DIR MOLECULAR BIOL, A H ROBINS PHARMACEUT CO, 80- *Concurrent Pos:* USPHS fel, Med Sch, Univ Minn, 56-58. *Mem:* Am Soc Microbiol. *Res:* Antiviral chemotherapy; cell-virus relationships. *Mailing Add:* 1703-B Choate Pl Richmond VA 23233

TANKIN, RICHARD S, b Baltimore, Md, July 14, 24; m 56; c 3. FLUID MECHANICS. *Educ:* Johns Hopkins Univ, AB, 48, BS, 50; Mass Inst Technol, MS, 53; Harvard Univ, PhD(mech eng), 60. *Prof Exp:* Asst prof civil eng, Univ Del, 60-61; from asst prof to assoc prof mech eng, 61-69, chmn dept, 72-77, PROF MECH ENG, NORTHWESTERN UNIV, EVANSTON, 70- *Mem:* Am Soc Mech Engrs. *Res:* Hydrodynamic stability; plasma properties; combustion; two phase flow. *Mailing Add:* Dept Mech Eng Northwestern Univ 633 Clark St Evanston IL 60208

TANKINS, EDWIN S, b Midland, Pa, Sept 12, 27; m 55; c 3. PHYSICAL METALLURGY. *Educ:* Univ Wis, BS, 54; Univ Pa, MS, 57. *Prof Exp:* Jr engr air mat lab, Naval Air Eng Ctr, 54-55, res asst, 56-59, res metallurgist, 59, mat engr, 59-61, res metallurgist, 61-67, asst to chief scientist, 66-67, RES PHYS METALLURGIST, MAT LAB, AIRCRAFT & CREW SYSTS TECHNOL DIRECTORATE, NAVAL AIR DEVELOP CTR, 67- *Concurrent Pos:* Adj prof, Eng Mgt Prog, Drexel Univ, 87-, Widener Univ, 91- *Mem:* AAAS; Am Soc Metals; Am Inst Mining, Metall & Petrol Engrs; Am Chem Soc; Sigma Xi. *Res:* Chemical metallurgy; neutron activation studies; yield and fracture stress of refractory metals as a function of temperature, grain, size and strain rate; equilibria of hydrogen and oxygen in iron group metals; binary and ternary alloys related to thermodynamic studies; gas analysis in metals; applied mathematics. *Mailing Add:* 1575 Revere Rd Yardley PA 19067

TANNAHILL, MARY MARGARET, b Weatherford, Tex, Apr 30, 44. POLYMER CHEMISTRY. *Educ:* Tex Tech Univ, BS, 66; Mich State Univ, PhD(phys chem), 73. *Prof Exp:* Asst chem, Mich State Univ, 66-72; trainee physiol, Univ Tex Med Br, Galveston, 72-73; res chemist, Union Carbide Corp, 74-75; dir mats control, High Density Polyethylene, Gulf Oil Chem Co, 75-79; sr res chemist res & develop, 79, PROD & QUAL SUPVR, MOBIL CHEM CO, 80- *Mem:* Am Chem Soc; Soc Plastics Engrs. *Res:* Product development for polypropylene; catalyst preparation and testing; high density polyethylene. *Mailing Add:* Vista Chem Co PO Box 200135 Austin TX 78720-0135

TANNEHILL, JOHN CHARLES, b Salem, Ill, Oct 14, 43; m 67; c 2. COMPUTATIONAL FLUID DYNAMICS, AERODYNAMICS. *Educ:* Iowa State Univ, BS, 65, MS, 67, PhD(aerospace eng & mech eng), 69. *Prof Exp:* Aerospace engr, Flight Res Ctr, NASA, 65; mem tech staff, Aerospace Corp, 68; from asst prof to assoc prof, 69-79, PROF AEROSPACE ENG, IOWA STATE UNIV, 79-, MGR, COMPUT FLUID DYNAMICS CTR, 85-; PRES, ENG ANAL, INC, 84- *Concurrent Pos:* NASA-ASEE fac fel, Ames Res Ctr, NASA, 70 & 71; consult, Eng Anal, Inc, 75-84. *Mem:* Assoc fel Am Inst Aeronaut & Astronaut; Am Soc Eng Educ. *Res:* Computational fluid dynamics; computation of a variety of flow fields using navier-stokes equations; computation of the flow around the space shuttle orbiter. *Mailing Add:* Iowa State Univ 402 Town Eng Bldg Ames IA 50011

TANNEN, RICHARD L, b New York, NY, Aug 31, 37; m 65; c 3. INTERNAL MEDICINE, NEPHROLOGY. *Educ:* Univ Tenn, MD, 60. *Prof Exp:* Asst Med, Harvard Med, Sch, Boston, 65-66; asst prof & co-dir nephrology, Univ Vt, Burlington, 69-73, assoc prof & dir, 73-78, actg assoc chmn, Dept Med, 75-76; vis scientist clin biochem, Radcliffe Infirmary, Oxford, Eng, 76-77; PROF & DIR NEPHROLOGYV MICH, 78- *Concurrent Pos:* Estab investr, Am Heart Asn; actg chief nephrology, Vet Admin Med Ctr, Mich, 78-82; sci adv bd, Nat Kidney Found, 81-; exec comt, Coun Kidney Dis, Am Heart Asn, 82- *Mem:* Fel Am Col Physicians; Am Fedn Clin Res; Am Soc Nephrology; Int Soc Nephrology; Am Soc Clin Invest; Asn Am Physicians. *Res:* Acid-base physiology; renal ammonia metabolism; potassium regulation. *Mailing Add:* Div Nephrology Med Sch Univ 3914 Taubman Ctr Ann Arbor MI 48109-0364

TANNENBAUM, CARL MARTIN, b New York, NY, Apr 1, 40. BIOCHEMISTRY. *Educ:* City Col New York, BS, 60; Univ Ariz, MS, 68; Univ Nebr, PhD(chem), 74. *Prof Exp:* Chemist, Ciba Pharmaceut Co, 61-64; fel physiol, Med Sch, Yale Univ, 72-75; asst prof biochem, Swiss Fed Inst Technol, 75-78; RES BIOLOGIST, PROCTER & GAMBLE CO, 78- *Mem:* Am Chem Soc. *Res:* Epithelial membrane transport of sugars and amino acids; characterization of the factors involved in transport. *Mailing Add:* 1941 S Staunton Dr Fairfield OH 45014-3629

TANNENBAUM, GLORIA SHAFFER, b Montreal, Que, July 9, 38; m 59; c 4. NEUROENDOCRINOLOGY, GROWTH HORMONE. *Educ:* McGill Univ, BSc, 59, MSc, 73, PhD(neuroendocrinol), 76. *Prof Exp:* Res fel neuroendocrinol, Montreal Gen Hosp, 76-78; from asst prof to assoc prof, 78-89, PROF PEDIAT, NEUROL & NEUROSURG, MCGILL UNIV, 89- *Concurrent Pos:* Prin investr, McGill Univ-Montreal Children's Hosp Res Inst, 78-; mem, Assessment Panel, Can Coun Animal Care, 85- *Mem:* Am Endocrine Soc; Am Physiol Soc; Soc Neurosci; Can Soc Clin Invest; Can Soc Endocrinol & Metab. *Res:* Elucidation of the neuroendocrine control mechanisms governing the rhythmic secretion of growth hormone. *Mailing Add:* Dept Neurol & Pediat Montreal PQ H3G 1Y6 Can

TANNENBAUM, HAROLD E, b New York, NY, Dec 31, 14; m 37; c 2. SCIENCE EDUCATION. *Educ:* Columbia Univ, MA, 37, EdD, 50. *Prof Exp:* Teacher, Park Sch, Ohio, 37-42; head sci dept, Elisabeth Irwin High Sch, NY, 44-52; prof sci educ, State Univ NY Teachers Col, New Paltz, 52-61; chmn div curric & instruct, Grad Sch Educ, Yeshiva Univ, 61-64; prof, 64-78, chmn dept curric & teaching, 68-72, EMER PROF SCI EDUC, HUNTER COL, 78- *Concurrent Pos:* Sci Manpower Comt fel, Columbia Univ, 58-59; consult, State Educ Depts, NH, Va, NDak & NY. *Mem:* AAAS; Nat Sci Teachers Asn. *Mailing Add:* Box 295 Phoenicia NY 12464

TANNENBAUM, HARVEY, b New York, NY, June 26, 23; m 46; c 3. PHYSICAL CHEMISTRY. *Educ:* NY Univ, BS, 48. *Prof Exp:* Mem staff, CB Detection & Alarms Div, Chem Systs Lab, Edgewood Arsenal, 49-65, chief remote sensing, 65-79; prin staff engr, Chem Defense Ctr, Honeywell, Inc, Clearwater, Fla, 79-83; consult, Chem-Biol Defense Technol, 84-85; sr prog dir, SRI Int, Edgewood, Md, 87-88; CONSULT, 88- *Concurrent Pos:* Partic, NATO Experts Panel Laser Remote Sensing of Atmosphere, 75-; mem, Joint Army Navy NASA Air Force Comt Propulsion Hazards, 75- *Mem:* Optical Soc Am; Sigma Xi. *Res:* Infrared physics; trace gas detection; pollution monitoring instrumentation; remote sensing instrumentation; electro-optical systems; spectroscopy. *Mailing Add:* 12611 Mt Laurel Ctr Reisterstown MD 21136

TANNENBAUM, IRVING ROBERT, b Spring Lake, NJ, Feb 24, 26; m 51; c 4. PHYSICAL CHEMISTRY. *Educ:* Va Polytech Inst, BS, 46, MS, 48; Univ Ill, PhD(phys chem), 51. *Prof Exp:* Instr math, Va Polytech Inst, 46-47; asst phys chem, Univ Ill, 47-50; mem staff chem res, Los Alamos Sci Lab, Univ Calif, 51-56; sr phys chemist, Atomics Int Div, NAm Aviation, Inc, 56-61; scientist, Electro-Optical Systs, Inc, 61-63; sr scientist, Heliodyne Corp, 63-65; PRES & TECH DIR, CHEMATICS RES CORP, RESEDA, 65-; PROF CHEM, W LOS ANGELES COL, 79- *Concurrent Pos:* Adj prof, Univ Calif, Los Angeles, 58- *Mem:* Am Chem Soc; Sigma Xi. *Res:* Theory of liquid mixtures; inorganic and plutonium chemistry; hydrides; x-ray crystallography; chemical kinetics; re-entry physics. *Mailing Add:* 8354 Etiwanda Ave Northridge CA 91325

TANNENBAUM, JANET, CELL BIOLOGY, CYTOSKELETON PROTEINS. *Educ:* Columbia Univ, PhD(microbiol), 75. *Prof Exp:* ASST PROF PATH, COL PHYSICIANS & SURGEONS, COLUMBIA UNIV, 80- *Mailing Add:* Dept Editorial/Annals Assoc Editor NY Acad Sci Two E 63rd St New York NY 10021

TANNENBAUM, MICHAEL GLEN, b New York, NY, July 19, 53; m 81; c 1. BIOLOGY. *Educ:* Cornell Univ, BS, 75; Clemson Univ, PhD(zool), 85. *Prof Exp:* Asst prof biol sci, Marshall Univ, 84-88; ASST PROF BIOL, NE MO STATE UNIV, 88- *Concurrent Pos:* Postdoctoral res assoc, Dept Cellular & Struct Biol, Univ Tex Health Sci Ctr, San Antonio, 86 & 87. *Mem:* AAAS; Am Soc Zoologists; Am Soc Mammalogists; Sigma Xi. *Res:* Endocrine changes underlying seasonal changes in function, behavior, and morphology in small wild rodents associated with overwinter survival. *Mailing Add:* Div Sci NE Mo State Univ Kirksville MO 63501

TANNENBAUM, MICHAEL J, b Bronx, NY, Mar 10, 39; m 73; c 2. HIGH ENERGY PHYSICS, RELATIVISTIC HEAVY ION PHYSICS. *Educ:* Columbia Univ, AB, 59, MA, 60, PhD(physics), 65. *Prof Exp:* Vis scientist, Europ Orgn Nuclear Res, 65-66; from asst prof to assoc prof physics, Harvard Univ, 66-71; assoc prof physics, Rockefeller Univ, 71-80; head planning & anal, Isabelle Magnet Div, 80-81, head superconductor procurement & magnet lamination physics, 81-82, physicist, Physics Dept, 82-87, SR PHYSICIST, BROOKHAVEN NAT LAB, 87- *Concurrent Pos:* Ernest Kempton Adams traveling fel from Columbia Univ, 65-66; NSF fel, 66; Alfred P Sloan Found fel, 67-69; mem prog adv comt, Fermi Nat Lab, 72-75; attache sci, Europ Orgn Nuclear Res, 73-80. *Mem:* Fel Am Phys Soc; fel AAAS; Sigma Xi. *Res:* Muon elastic and inelastic scattering; muon g-2; muon tridents; photoproduction with a tagged beam; single leptons, lepton pairs and high transverse momentum phenomena in proton-proton interactions; first direct measurement of the constituent scattering angular distribution; superconductive magnetics; relativistic heavy ion physics. *Mailing Add:* Bldg 510A Physics Brookhaven Nat Lab Upton NY 11973

TANNENBAUM, STANLEY, b New York, NY, Mar 1, 25; m 47; c 4. INORGANIC CHEMISTRY. *Educ:* City Col New York, BS, 46; Ohio State Univ, PhD(chem), 49. *Prof Exp:* Res chemist, Nat Adv Comt Aeronaut, 50-53; res chemist reaction motors div, Thiokol Chem Corp, 53-59, sect head, 59-66, prod mgr, 66-69; mgr tech serv, Ronson Metals Corp, 69-76; SR SCIENTIST, HAZARD RES CORP, 76- *Mem:* Am Chem Soc. *Res:* Synthesis of silicon and boron containing chemicals; physical and thermochemical properties of materials; alteration of properties of rocket propellants; determination of fire and explosive hazards of chemicals and chemical processes. *Mailing Add:* 18 A Celtis Plaza Cranbury NJ 08512-3199

TANNENBAUM, STEVEN ROBERT, b New York, NY, Feb 23, 37; m 59; c 2. TOXICOLOGY. *Educ:* Mass Inst Technol, BS, 58, PhD(food sci), 62. *Prof Exp:* From asst prof to assoc prof nutrit & food sci, Mass Inst Technol, 64-74, prof food chem, 74-88, prof toxicol, 81-88, regist & admis officer, dept appl biol sci, 82-88; PROF CHEM & TOXICOL & REGIST & ADMIS OFFICER, PROG TOXICOL, WHITAKER COL, 88- *Concurrent Pos:* Vis prof, Hebrew Univ Jerusalem Fac Agr, 73-74; consult, Inst Nutrit Cent Am & Panama, 68, Protein Adv Group, UN Develop Prog, 70-74, FDA, 71-73, Am Cancer Soc, 77-81, Nat Cancer Inst-NIH, 78-82; mem, Comt Food Standards & Fortification Policy, Nat Res Coun, Nat Acad Sci, 70-73, Safe Drinking Water Comt, Comt Amines, Inst Food Technologists Expert Panel Food Safety & Nutrit, 71-73, co-chmn, 76-77, chmn, 77-78; mem adv comt, Biochem & Chem Carcinogenesis, Am Cancer Soc, 77-81, Cancer Spec Prog Adv Comt, 79-82, Peer Rev Comt, Nat Toxicol Prog, 83-85, Frederick Cancer Fac adv comt, Nat Cancer Inst, 89- *Honors & Awards:* Samuel Cate Prescott Res Award, Inst Food Technologists, 70, Babcock Hart Award, 80. *Mem:* AAAS; Am Chem Soc; Inst Food Technologists; Am Cancer Soc; Am Col Toxicol; Am Asn Cancer Res; Sigma Xi. *Res:* Chemistry of nitrates, nitrites and nitrosamines; molecular dosimetry of carcinogens; analysis of cancer risk in human populations. *Mailing Add:* Dept Chem Div Toxicol Rm 56-311 Mass Inst Technol Cambridge MA 02139

TANNENWALD, PETER ERNEST, b Kiel, Ger, Mar 30, 26; nat US; div; c 1. PHYSICS. *Educ:* Univ Calif, AB, 47, PhD(physics), 52. *Prof Exp:* Asst physics, Univ Calif, 47-51, Radiation Lab, 50-52; physicist, Mass Inst Technol, 52-59, asst group leader, 59-62, group leader, 62-63, asst div head, 63-65, assoc div head, 65-74, sr scientist, 74-86, CONSULT, SOLID STATE DIV, LINCOLN LAB, MASS INST TECHNOL, 86- *Mem:* Fel Am Phys Soc. *Res:* Solid state physics and quantum electronics; microwave resonance in ferrites; spin wave resonance in magnetic films; masers; millimeter waves; microwave ultrasonics; lasers; Raman and Brillouin spectroscopy; far infrared quantum electronics and submillimeter wave technology. *Mailing Add:* Solid State Div Lincoln Lab Mass Inst Technol Lexington MA 02173

TANNER, ALAN ROGER, b Port Lavaca, Tex, Jan 2, 41; m 69; c 2. INDUSTRIAL ORGANIC CHEMISTRY, ELECTROCHEMISTRY. *Educ:* Univ Tex, Austin, BS, 64, PhD(org chem), 69. *Prof Exp:* Appln res chemist, Jefferson Chem Co, 69-71; RES & DEVELOP CHEMIST, SOUTHWESTERN ANALYTICAL CHEM, INC, 71 - *Concurrent Pos:* Instr org chem, St Edward's Univ, 74-75. *Mem:* Am Chem Soc; Electrochem Soc. *Res:* Conventional and electrochemical synthetic approaches to new and existing marketable products; electrochemical cell and plant design. *Mailing Add:* 1415 Fairfield Dr Austin TX 78758-7243

TANNER, ALLAN BAIN, b New York, NY, May 27, 30; m 80; c 3. GEOPHYSICS, GEOCHEMISTRY. *Educ:* Mass Inst Technol, SB, 52. *Prof Exp:* Geophysicist, Theoret Geophys Br, US Geol Surv, DC, 54-69, geophysicist, Isotope Geol Br, 69-89, Sedimentary Processes Br, 89-90, EMER SCIENTIST, US GEOL SURV, 90- *Concurrent Pos:* Comt mem, Nat Coun Radiation Protection & Measurements, 73-74 & 85-; policy coordr, Off Sci & Technol Policy, 84-90. *Mem:* Soc Explor Geophys. *Res:* Behavior of radon isotopes in natural environments; nuclear geophysics and geochemistry; x-ray fluorescence; isotope geology; gamma-ray spectrometry; in situ neutron activation analysis; health physics. *Mailing Add:* US Geol Surv 990 Nat Ctr Reston VA 22092

TANNER, CHAMP BEAN, microclimatology, soil physics; deceased, see previous edition for last biography

TANNER, CHARLES E, b Preston, Cuba, Sept 10, 32; Can citizen; m 57; c 2. IMMUNOLOGY, PARASITOLOGY. *Educ:* Purdue Univ, BS, 53; McGill Univ, MS, 56, PhD, 57. *Prof Exp:* Teaching fel bact & immunol, McGill Univ, 55-57; Nat Res Coun Can overseas fel, 57-58; from asst prof to assoc prof parasitol, 58-71, PROF PARASITOL, INST PARASITOL, MACDONALD COL, McGILL UNIV, 71-, ASSOC MEM MICROBIOL, FAC AGR, UNIV, 73- *Concurrent Pos:* Mem, Int Comn Trichinellosis, 73- *Mem:* AAAS; Am Soc Parasitologists; Can Soc Microbiol; Can Soc Immunol; Can Soc Zool; Sigma Xi. *Res:* Immunology of host-parasite relations; immunochemistry of parasite antigens. *Mailing Add:* Inst Parasitol Macdonald Col McGill Univ 21111 Lakeshore Rd Ste Anne de Bellevue PQ H9X 1C0 Can

TANNER, DANIEL, b New York, NY, Sept 22, 26; m 48. HISTORY & PHILOSOPHY OF SCIENCE, SCIENCE EDUCATION. *Educ:* Mich State Univ, BS, 49, MS, 51; Ohio State Univ, PhD(educ), 55. *Prof Exp:* Instr biol & dept chair, Galesburg Community Schs, Mich, 49-51; teaching asst sci educ, Mich State Univ, 51-52; instr biol sci & dept chair, State Univ NY Agr & Tech Inst, Morrisville, 52-53; asst prof sec educ, San Francisco State Univ, 55-60; assoc prof educ & coordr, Midwest Prog Airborne TV Instr, Purdue Univ, 60-62; assoc prof educ & assoc dir, Master Arts Teaching Prog, Northwestern Univ, 62-64, assoc dir, Int Prog Educ Leaders, 62-63; assoc prof educ res & dir Col Discovery & Develop Prog, City Univ New York, 64-66; prof educ & dir, Ctr Urban Educ, Univ Wis-Milwaukee, 66-67; PROF EDUC & DIR, GRAD PROG CURRIC THEORY & DEVELOP, RUTGERS UNIV, 67- *Concurrent Pos:* Vis lectr, Emory Univ, 58, Teachers Col, Columbia Univ, 66, State Univ NY, Binghamton, 68; consult, US Off Educ, 65 & AT&T, 68-71; vis prof scholar, Univ London, 74-75; fel, Rutgers Res Coun, 74-75; distinguished lectr, State Univ NY, Buffalo, 83; mem bd dirs, John Dewey Soc, 85-87, archivist, 89- *Honors & Awards:* John Dewey Mem Lect, 84; Raths Mem Lect, 84. *Mem:* Fel AAAS; fel John Dewey Soc; Am Educ Res Asn; Soc Study Curric Hist. *Res:* Science in intellectual history; science and society; curriculum design and development for interdisciplinary science in school and college; philosophy of science. *Mailing Add:* Highwood Rd Somerset NJ 08873

TANNER, DAVID, b Brooklyn, NY, Aug 7, 28; m 53; c 4. POLYMER CHEMISTRY. *Educ:* NY Univ, BA, 50; Brooklyn Polytech Inst, PhD(polymer chem), 54. *Prof Exp:* Res asst polymer chem, Univ Ill, 53-54; res chemist Dupont exp sta, Nylon Res Div, 54-59, res assoc, 59-62, supvr res, 62-63, sr supvr technol, Nylon Tech Div, 63-65, res mgr, Orlon-Lycra Res Div, 65-68, lab dir, Benger Lab, 68-69, tech mgr, Dacron Div, 70-72, tech mgr, Orlon-Acetate-Lycra Div, 72-76, res dir, Textile Fibers Dept, 76-80, mgr, Strategic Planning Div, 80-82, TECH DIR, INDUST FIBERS DIV, E I DU PONT DE NEMOURS & CO, INC, 82- *Mem:* Am Chem Soc; Asn Res Dirs; NAm Planning Soc; Asn Mgmt Innovation. *Res:* Organic polymer synthesis; radiation chemistry; fiber technology including polyesters, polyamides, acrylics, elastomers, aramids; creativity and innovation management. *Mailing Add:* 712 Hertford Rd Wilmington DE 19803-1618

TANNER, DENNIS DAVID, b Montreal, Que, Mar 6, 30; US & Can citizen; m 60; c 2. ORGANIC CHEMISTRY. *Educ:* Univ Calif, Los Angeles, BSc, 53; Stanford Univ, MSc, 57; Univ Colo, PhD(chem), 61. *Prof Exp:* Asst, Stanford Univ, 56-57 & Univ Colo, 57-60; res fel chem, Columbia Univ, 61-63; annual asst prof, 63-65, from asst prof to assoc prof, 65-75, PROF CHEM, UNIV ALTA, 75- *Mem:* Am Chem Soc; Chem Inst Can. *Res:* Mechanisms of free radical reactions; free radical and ionic rearrangement mechanisms. *Mailing Add:* Dept Chem Univ Alta Edmonton AB T5H 0Z8 Can

TANNER, GEORGE ALBERT, b Vienna, Austria, Aug 2, 38; US citizen; m 62; c 2. RENAL FUNCTION. *Educ:* Cornell Univ, AB, 59; Harvard Univ, PhD(physiol), 64. *Prof Exp:* Nat Heart Inst res trainee physiol, Med Col, Cornell Univ, 64-67; from asst prof to assoc prof, 67-78, PROF PHYSIOL, SCH MED, IND UNIV, INDIANAPOLIS, 78- *Concurrent Pos:* Vis prof, Heidelberg Univ, 74-75; sr postdoctoral fel, Yale Univ, 86-87. *Mem:* Am Heart Asn; Soc Exp Biol & Med; Int Soc Nephrology; Am Soc Nephrology; Am Physiol Soc. *Res:* Renal function. *Mailing Add:* Dept Physiol & Biophys MS Rm 339 635 Barnhill Dr Indianapolis IN 46202-5120

TANNER, JAMES MERVIL, b Jesup, Ga, Dec 29, 34; m 56; c 3. THEORETICAL PHYSICS. *Educ:* Ga Inst Technol, BS, 56, MS, 61, PhD(physics), 64. *Prof Exp:* Assoc scientist, Westinghouse Elec Corp, 56-58; instr physics, Ga Inst Technol, 58-62, asst prof, 64-65; assoc prof, Univ NC, Charlotte, 65-67; ASSOC PROF PHYSICS, GA INST TECHNOL, 67- *Mem:* Am Phys Soc; Am Asn Physics Teachers. *Res:* Mathematical physics. *Mailing Add:* Dept Physics Ga Inst Technol 225 North Ave NW Atlanta GA 30332

TANNER, JAMES TAYLOR, animal ecology; deceased, see previous edition for last biography

TANNER, JAMES THOMAS, b Franklin, Ky, Apr 23, 39; m 64. RADIOCHEMISTRY. *Educ:* Eastern Ky State Col, BS, 61; Univ Ky, PhD(radiochem), 66. *Prof Exp:* Res chemist, Carnegie-Mellon Univ, 66-68, lectr chem, 68-69; res chemist, 69-79, CHIEF, NUTRIENT SURVEILLANCE BR, FOOD & DRUG ADMIN, 79- *Mem:* Fel AAAS; fel Meteoritical Soc; Am Chem Soc; Sigma Xi; fel Am Nuclear Soc. *Res:* Neutron activation analysis; trace elements in meteorites; trace element distribution in foods, drugs and consumer products; nutrient content of foods; standard reference materials. *Mailing Add:* Div Nutrit HFF-266 Food & Drug Admin Washington DC 20204

TANNER, JOHN EYER, JR, b Cleveland, Ohio, Apr 30, 30; m 66; c 3. NUCLEAR CRITICALITY SAFETY. *Educ:* Oberlin Col, AB, 51; Ind Univ, MS, 54; Univ Wis, PhD(phys chem), 66. *Prof Exp:* Res asst phys chem, Am Found Biol Res, Wis, 59-62; res fel, Pa State Univ, 66-68; res asst, Max Planck Inst Med Res, Ger, 68-69; fel, Sci Res Staff, Ford Motor Co, Mich, 69-71; res chemist, Naval Weapons Support Ctr, 71-79; sr res engr, Exxon Nuclear Idaho Co, 79-82; SR ENGR, WESTINGHOUSE IDAHO NUCLEAR CO, 82- *Mem:* Am Chem Soc; Am Nuclear Soc; Am Phys Soc. *Res:* Nuclear magnetic resonance; emission and absorption spectroscopy; thermodynamics; nuclear waste management; co-developer of the pulsed-field-gradient, nuclear-magnetic- resonance method of diffusion measurement; pioneered the application to viscous liquids, polymers, emulsions, plastic crystals, and biological cells. *Mailing Add:* Safety Dept Box 4000 Westinghouse Idaho Nuclear Co Idaho Falls ID 83403-5222

TANNER, LEE ELLIOT, b Brooklyn, NY, May 28, 31; m 56; c 2. PHYSICAL METALLURGY. *Educ:* NY Univ, BS, 53; Univ Pa, MS, 58. *Prof Exp:* Res asst phys metall, Armour Res Found, 53, assoc metallurgist, 56-59; sr metallurgist, ManLabs, Inc, 59-63; sr res metallurgist, Ledgemont Lab, Kennecott Copper Corp, 63-73; staff metallurgist, Mat Res Ctr, Allied Chem Corp, 73-78, sr metallurgist, ManLabs, Inc, 78-80, PRIN INVESTR, LAWRENCE LIVERMORE NAT LAB, 80- *Honors & Awards:* Outstanding Sci Accomplishment in Metall & Ceramics Res, US Dept Energy, 88. *Mem:* AAAS; Am Soc Metals; Am Inst Mining, Metall & Petrol Engrs; Am Inst Physics; Electron Micros Soc Am. *Res:* Precipitation, ordering, martensitic transitions; relationships microstructure to physical properties; crystalline and amorphous; phase transformations, precipitation, ordering; relationships of microstructure to physical properties. *Mailing Add:* Lawrence Livermore Nat Lab PO Box 808/L-355 Livermore CA 94550

TANNER, LLOYD GEORGE, b Cozad, Nebr, Oct 3, 18; c 4. GEOLOGY. *Educ:* Univ Nebr, BS, 51, MS, 56. *Prof Exp:* Field asst & party leader, Univ Nebr State Mus, 38-39 & 51-75, field supvr, Works Progress Admin, 39-40, asst cur vert paleontol, 51-56, assoc cur, 56-77, asst prof geol, Univ Nebr, 71-73, res assoc vert paleontol, 73-78, COORDR SYST COLLECTIONS, UNIV NEBR STATE MUS, 70- & CUR VERT PALEONT, 77-, ASSOC PROF MUS & GEOL, UNIV NEBR, 78- *Concurrent Pos:* Co-prin investr fauna & stratig sequence, Big Bone Lick, Ky, 62-66; mem, Yale Peabody Mus Egyptian Expeds, 62-66; fac fel, Univ Nebr, 71-; mem, Duke Univ Egyptian Expeds, 77-81. *Res:* Vertebrate paleontology; stratigraphy; Pleistocene geology. *Mailing Add:* RR 2 Box 469 North Platte NE 69101

TANNER, MARTIN ABBA, b Highland Park, Ill, Oct 19, 57; m 84; c 1. SOFTWARE DEVELOPMENT. *Educ:* Univ Chicago, BA, 78, MS, 81, PhD(statist), 82. *Prof Exp:* Asst prof, Math Res Ctr, Univ Wis-Madison, 82-83, asst prof statist, 82-87, asst prof human oncol, 83-87, assoc prof statist, human oncol, 87-90, dir lab statist, 84-90; PROF & DIR, DIV BIOSTATIST, UNIV ROCHESTER MED CTR, 90- *Concurrent Pos:* Consult, Kirkland & Ellis, 80-82, Vet Admin, dept surg & med, 82- & Bur Justice Statist, 83-; prin investr, NSF, 83-85 & tech reviewer, 84-; prin investr, reviewer, NIH, 86-; assoc ed, J Am Stat Asn, 88- *Honors & Awards:* Young Investr Award, NIH, 84. *Mem:* AAAS; Am Statist Asn; Sigma Xi; fel Royal Statist Soc. *Res:* Design and analysis of clinical trials and laboratory experiments; development of statistical methodology for the analysis of missing data problems; development of statistical methodology for the analysis of agreement among raters. *Mailing Add:* Biostatist Box 630 601 Elmwood Ave Rochester NY 14618

TANNER, NOALL STEVAN, b Ogden, Utah, Sept 21, 34; m 55; c 2. PHARMACOLOGY, PHARMACOGNOSY. *Educ:* Univ Utah, BS, 56, PhD(pharmacog), 66. *Prof Exp:* Asst prof pharmacog, Col Pharm, Butler Univ, 63-66; ASSOC PROF PHARMACOL & DIR PROF RELATIONS, COL PHARM, NDAK STATE UNIV, 67- *Mem:* Am Pharmaceut Asn; Acad Pharmaceut Sci; Soc Hosp Pharmacists. *Res:* Biopharmaceutical sciences; environmental study of Veratrum californicum; relation of neurotransmitter to behavior in rats; study and screening of natural products. *Mailing Add:* Dept Pharmacol NDak State Univ Fargo ND 58105

TANNER, RAYMOND LEWIS, b Memphis, Tenn, Dec 11, 31; m 68; c 3. RADIOLOGICAL PHYSICS, HEALTH PHYSICS. *Educ:* Memphis State Univ, BS, 53; Vanderbilt Univ, MS, 55; Univ Calif, Los Angeles, PhD(med physics), 67. *Prof Exp:* Asst prof physics, Memphis State Univ, 55-62; vis physicist, Harbor Gen Hosp, Torrance, Calif, 63-64; assoc prof, 67-70, asst to chancellor for facil planning, 77-82, asst dean Grad Sch, 82-85, PROF MED PHYSICS, UNIV TENN, MEMPHIS, 70-,. *Concurrent Pos:* Consult self-radiation protection, 55-88; chmn, Comn Physics, Am Col Radiol, 82- *Mem:* Am Asn Physicists in Med (pres, 73-74); Health Physics Soc; Am Col Radiol; Radiol Soc NAm (vpres, 80-81); Sigma Xi. *Res:* Radiation dosimetry; x-ray quality control. *Mailing Add:* Univ Tenn 800 Madison Ave Memphis TN 38163

TANNER, ROBERT DENNIS, b Detroit, Mich, Jan 17, 39; m 63; c 2. CHEMICAL ENGINEERING, BIOCHEMICAL ENGINEERING. *Educ:* Univ Mich, BSE, 61 & 62, MSE, 63; Case Western Reserve Univ, PhD(chem eng), 67. *Prof Exp:* Engr, Diamond Shamrock Corp, 63; eng assoc res, Merck & Co, Inc, 67-72; from asst prof to assoc prof, 72-84, PROF CHEM ENG, VANDERBILT UNIV, 84- *Concurrent Pos:* Chmn, Div Microbiol & Biochem Technol, Sigma Xi, 79-80; vis prof, Eidgenossiche Technische Hochschule, Zurich, 81-82. *Mem:* Am Inst Chem Engrs; Am Chem Soc; Sigma Xi. *Res:* Fermentation modeling; enzyme kinetics; yeast technology; on-line recovery of protein from yeast; solid and semi-solid state fermentation processes in air fluidized beds. *Mailing Add:* Dept Chem Eng Vanderbilt Univ Nashville TN 37235

TANNER, ROBERT H, b London, Eng, July 22, 15; m; c 4. NOISE CONTROL. *Educ:* Univ London, BSc, 36, MSc, 61. *Hon Degrees:* LLD, Concordia Univ, 89. *Prof Exp:* TV engr & res engr, Brit Broadcasting Corp, 36-47; from engr to mgr, Northern Elec Co, 47-70; dir info, Bell-Northern Res, 70-75; ACOUST CONSULT, 75- *Honors & Awards:* McNaughton Medal, Inst Elec & Electronics Engrs, 74, Haraden Pratt Award, 81. *Mem:* Fel Inst Elec & Electronics Engrs (pres, 72); fel Acoust Soc Am; fel Inst Elect Engrs; foreign fel Eng Inst Can; Inst Noise Control Engrs. *Mailing Add:* PO Box 655 Naples FL 33939-0655

TANNER, ROBERT MICHAEL, b Spanish Fork, Utah, Mar 22, 46; m 65; c 2. INFORMATION SCIENCES. *Educ:* Stanford Univ, BS, 66, MS, 67, PhD(elec eng), 71. *Prof Exp:* Asst prof elec eng, Tenn State Univ, 70-71; from asst prof to assoc prof info sci, 71-84, actg dean, Nat Sci, 88-89, PROF COMP & INFO SCI, UNIV CALIF, SANTA CRUZ, 84-, ACAD VCHANCELLOR, 89- *Concurrent Pos:* Consult, Technol Commun Int, 71-79, Ford Aerospace, 85-89, Optimem, 82; vis assoc prof, Stanford, 81; vis scientist, IBM, 84. *Mem:* Inst Elec & Electronics Engrs. *Res:* Information theory, coding and complexity. *Mailing Add:* Univ Calif Santa Cruz CA 95064

TANNER, ROGER IAN, b Wells, Eng, July 25, 33; m 57; c 5. ENGINEERING. *Educ:* Bristol Univ, BSc, 56; Univ Calif, Berkeley, MS, 58; Manchester Univ, PhD(eng), 61. *Prof Exp:* Tech asst engr, Bristol-Siddeley Aero Engines Ltd, UK, 57-58; asst lectr mech eng, Univ Manchester, 58-59, lectr, 59-61; sr lectr, Sydney Univ, 61-64, reader, 64-66; from assoc prof to prof eng, Brown Univ, 66-75; P N RUSSELL PROF, SYDNEY UNIV, 75- *Honors & Awards:* David Medal, Royal Soc NSW, 67. *Mem:* Australian Acad Sci; Am Soc Mech Engrs; Australian Acad Tech Sci & Eng. *Res:* Flow of non-newtonian fluids; application of computer methods to polymer processing, including extrusion. *Mailing Add:* Dept Mech Eng Univ Sydney Sydney NSW 2006 Australia

TANNER, ROGER LEE, b Union City, Pa, Sept 17, 43; c 4. ANALYTICAL & ATMOSPHERIC CHEMISTRY. *Educ:* Pa State Univ, BS, 64; Univ Ill, Urbana, PhD(analytical chem), 69. *Prof Exp:* Temp asst prof chem, Portland State Univ, 68-69; asst prof chem, Univ Okla, 69-71; res assoc chem, Univ Ill, Urbana, 72-73; asst chemist, 73-75, assoc chemist, 75-77, CHEMIST, ENVIRON CHEM DIV, BROOKHAVEN NAT LAB, 77-, HEAD, ANALYTICAL CHEM GROUP, 86- *Concurrent Pos:* Consult, Sci Adv Bd, Environ Protection Agency, 77-81, sci rev panel, 87-; environ comt, NY Sect, Am Chem Soc, 82-86. *Mem:* Am Chem Soc; AAAS. *Res:* Trace analytical chemistry of environmental samples; chemistry of wet deposition processes; atmospheric chemistry of sulfur, nitrogen and carbon compounds; chemical speciation and source allocation of aerosols; real-time air monitoring instrumentation. *Mailing Add:* Energy Environ Eng Ctr Desert Res Inst PO Box 60220 Reno NV 89506-0220

TANNER, WARD DEAN, JR, b Jacksonville, Fla, Dec 6, 18. WILDLIFE MANAGEMENT, ECOLOGY. *Educ:* Univ Minn, BS, 41; Pa State Univ, MS, 48; Iowa State Univ, PhD(zool), 53. *Prof Exp:* Asst wildlife mgt, Pa State Univ, 41, 46-48; refuge mgr, Bombay Hook Nat Wildlife Refuge, US Fish & Wildlife Serv, 48-50; from instr to assoc prof biol, 53-74, PROF BIOL, GUSTAVUS ADOLPHUS COL, 74- *Mem:* Wildlife Soc; Sigma Xi. *Res:* Ecology of wildlife and ruffed grouse; botany; forestry; fisheries management. *Mailing Add:* 1407 18th Ave NE Minneapolis MN 55418

TANNER, WILLIAM FRANCIS, JR, b Milledgeville, Ga, Feb 4, 17; m 38; c 3. SEDIMENT TRANSPORT, PALEOGEOGRAPHY. *Educ:* Baylor Univ, BA, 37; Tex Tech Col, MA, 39; Univ Okla, PhD(geol), 53. *Prof Exp:* Asst geol, Baylor Univ, 35-37; oil ed, Amarillo Times, 39-41 & 45-46; asst prof geol & journalism, Okla Baptist Univ, 46-51; spec instr, Univ Okla, 51-54; vis lectr geol, 54-56, from assoc prof to prof, 56-74, REGENTS PROF GEOL, FLA STATE UNIV, 74- *Concurrent Pos:* Geologist, Shell Oil Co, 54; ed, Coastal Res, 62-; NSF vis scientist, 65-66. *Mem:* Fel AAAS; fel Geol Soc Am; Soc Econ Paleontologists & Mineralogists; Int Asn Sedimentology; Am Geophys Union. *Res:* Sedimentology; stratigraphy; geomorphology; hydrodynamics; beach and shore processes; structural, areal, field and subsurface geology; rheology and deformation of materials; circular statistics; paleogeography; paleoclimatology; petroleum exploration and resources; other energy sources; suite statistics. *Mailing Add:* Dept Geol Fla State Univ Tallahassee FL 32306-3026

TANNOCK, IAN FREDERICK, b Hatfield, Eng, Nov 22, 43; Can citizen; m 67; c 3. MEDICAL ONCOLOGY, EXPERIMENTAL CANCER RESEARCH. *Educ:* Cambridge Univ, BA, 65; Inst Cancer Res, PhD(biophys), 68; Univ Pa, MD, 74. *Prof Exp:* SR SCIENTIST & STAFF PHYSICIAN, ONT CANCER INST & PRINCESS MARGARET HOSP, 78-, CHIEF MED, 90-; PROF MED & MED BIOPHYS, UNIV TORONTO, 89- *Concurrent Pos:* Mem, Med Res Coun Panel, 86-88, chmn, 88-; assoc ed, Cancer Res, 83-87. *Mem:* Am Soc Clin Oncol; Am Asn Cancer Res; Cell Kinetics Soc; Radiation Res Soc. *Res:* Laboratory-based research in biology and therapy of solid tumors; clinical trials methodology; clinical trials in genitouriary, head and neck, and breast cancer. *Mailing Add:* Ont Cancer Inst 500 Sherbourne St Toronto ON M4X 1K9 Can

TANNY, GERALD BRIAN, b Montreal, Que, Dec 26, 45; m 68; c 3. POLYMER CHEMISTRY, MEMBRANE SCIENCE. *Educ:* McGill, BS, 66; PhD(polymer chem), 70. *Prof Exp:* fel, 70-72, scientist, 72-74, sr scientist, 74-77; assoc res dir, Gelman Sci Inc, 77-86, vpres, 86-90; CHIEF EXEC OFFICER, OSMOTEK, 91- *Concurrent Pos:* Managing dir, Gelman Sci Technol, 82-88. *Honors & Awards:* H Dudley Wright Award for Membrane Achievement. *Mem:* Am Chem Soc; Israel Chem Soc; Europ Membrane Soc. *Res:* Membrane science and technology; application of microporous membranes in crossflow microfiltration; UV curing for manufacturing microporous membranes; thermodynamics of membrane formations. *Mailing Add:* Osmotek PO Box 550 Rehovot Israel

TANPHAICHITR, VICHAI, b Bangkok, Thailand, Oct 5, 40; m 70; c 2. HUMAN NUTRITION, ENTERAL & PARENTERAL NUTRITION. *Educ:* Univ Med Sci, MD, 64, MSc, 68; Vanderbilt Univ, PhD(biochem), 73. *Prof Exp:* From asst prof to assoc prof, 75-80, PROF INTERNAL MED,

FAC MED RAMATHIBODI HOSP, 81-, DIR RES CTR, MED RES, 88- *Concurrent Pos:* Lectr internal med, Fac Med, Ramathibodi Hosp, 66-74; mem, Int Glutamate Tech Comt, 83-; ed-in-chief, J Internal Med, Royal Col Physicians of Thailand, 85-; prin invesr, Bristol-Myers Squibb Unrestricted Nutrit Res Grant, 89-93; mem, Comt Functional Consequences of Vitamin Deficiency, 89-93. *Honors & Awards:* First Ajinomoto Award, Nat Res Coun Thailand, 85; First Govt Pharmaceut Orgn Thailand Award, Thailand, 91. *Mem:* Am Soc Parenteral & Enteral Nutrit; Am Inst Nutrit; Am Col Physicians; Am Col Nutrit. *Res:* Nutrient requirements in tube feeding and parenteral nutrition; therapeutic nutrition in various diseases; thiamin, riboflavin, pyridoxine, vitamin A, carnitine, iron, and essential fatty acid metabolism in health and diseases; hyperlipoprotienemia. *Mailing Add:* Dept Med & Res Ctr Fac Med Ramathibodi Hosp Mahidol Univ Bangkok 10400 Thailand

TANQUARY, ALBERT CHARLES, b Columbus, Kans, Mar 9, 29; m 49, 80; c 5. POLYMER SCIENCE. *Educ:* Kans State Col, Pittsburg, BS, 50; Okla State Univ, MS, 52, PhD(chem), 54. *Prof Exp:* Sr chemist cent res dept, 3M Co, 54-55, group supvr, 55-58, res mgr, fibers dept, 59-61, mgr, 62-63; group leader res & develop div, Union Camp Corp, 63-65; assoc dir, Southern Res Inst, 66-80; dir, Gulf South Res Inst, 80-81; vpres, Res & Develop, Gulf Aviation Corp, 86-88; sr proj dir, Family Health Int, 88-90; PRES, TANQUARY ASSOCS, 81-; PRES, TELESIS CORP, 90- *Mem:* Am Chem Soc; Controlled Release Soc; Licensing Exec Soc; Sigma Xi. *Res:* Polymer chemistry; characterization of polymers; mechanical properties of adhesives, fibers and films; fiber spinning; biomedical materials; microcapsules; membrane processes; controlled release systems. *Mailing Add:* PO Box 12885 Research Triangle Park NC 27709-2885

TANSEY, MICHAEL RICHARD, b Oakland, Calif, Mar 27, 43; m 63, 88; c 2. MYCOLOGY. *Educ:* Univ Calif, Berkeley, BA, 65, PhD(bot), 70. *Prof Exp:* NSF fel & res assoc microbiol, 70-71, asst prof, 71-77, ASSOC PROF BIOL, IND UNIV, BLOOMINGTON, 78- *Mem:* AAAS; Mycol Soc Am; Brit Mycol Soc; Int Soc Human Animal Mycol; Am Inst Biol Sci. *Res:* Biology of thermophilic fungi; heated habitats; medical mycology; anaerobic biology of Candida albicans. *Mailing Add:* Dept Biol Ind Univ Bloomington IN 47405

TANSEY, ROBERT PAUL, SR, b Newark, NJ, Apr 27, 14; m 41; c 4. PHARMACY. *Educ:* Rutgers Univ, BS, 38, MS, 50. *Prof Exp:* Control pharmacist, Res & Develop Labs, Burroughs Wellcome Co, Inc, 40-43; asst dept head prod & control, E R Squibb & Sons, 43-45; head, Pharmaceut Res & Develop Lab, Maltbie Labs, 45-50; assoc dept head, Merck & Co, 50-53; sect leader, Schering Co, Inc, 53-58; res coordr, Strong Cobb Arner Co, Inc, 58-62; regist pharmacist, Saywell Pharm, Ohio, 62-63; tech dir, Vet Labs, Inc, 63-88; CONSULT, 88- *Mem:* Animal Health Inst; Am Pharmaceut Asn. *Res:* Pharmaceutical development; formulation and methods analysis; production processing techniques and control methods; pharmaceutical plant and equipment design; development of special techniques for control and sustained release medicinal forms; plant management. *Mailing Add:* 11141 Glen Arbor Rd Kansas City MO 64114

TANSY, MARTIN F, b Wilkes Barre, Pa, Mar 8, 37; m 64; c 3. PHYSIOLOGY. *Educ:* Wilkes Col, BA, 59; Jefferson Med Col, MS, 61, PhD(physiol), 64. *Prof Exp:* Res asst physiol, Jefferson Med Col, 59-61, fel, 62-64; from asst prof to assoc prof, 64-72, PROF PHYSIOL, SCHS DENT, PHARM & ALLIED HEALTH PROF, TEMPLE UNIV, 72-, CHMN DEPT, 64-, BASIC SCI COORDR, 79- *Mem:* Fel Am Col Nutrit; Soc Exp Biol & Med; Am Physiol Soc; Am Pharmaceut Asn; Am Fedn Clin Res. *Res:* Gastrointestinal and radiation physiology. *Mailing Add:* 412 St Davids Rd St Davids PA 19087

TANTRAPORN, WIROJANA, b Chonburi, Thailand, Apr 17, 31; m 53; c 1. SOLID STATE MATERIALS & DEVICES. *Educ:* Univ Denver, BS, 52; Univ Mich, MS, 53, PhD(physics), 58. *Prof Exp:* Lectr physics, Univ Mich, 58-59; physicist electronics lab, 59-62, PHYSICIST, RES & DEVELOP CTR, GEN ELEC CO, 62-; DEP DIR & TECH ADV, SCI & TECHNOL DEVELOP BD, THAILAND. *Res:* Thin films; semiconductors; insulators; solid state microwave devices; electron devices; computer simulations; infrared detection materials; high voltage power devices. *Mailing Add:* Jaran Ins Sixth Floor 401 Rachadaphisek Rd Bangkok 10310 Thailand

TANTTILA, WALTER H, b Sax, Minn, Nov 21, 22; m 51; c 6. PHYSICS. *Educ:* Univ Minn, BChE, 48, MA, 50; Univ Wash, PhD(physics), 55. *Prof Exp:* Instr physics, Minn State Teachers Col, Winona, 50-51; res physicist, Minneapolis-Honeywell Regulator Co, 51-52; asst physics, Univ Wash, 52-55; asst prof, Mich State Univ, 55-58; from asst prof to prof, 58-87, EMER PROF & VIS PROF PHYSICS, UNIV COLO, BOULDER, 87- *Mem:* Am Phys Soc. *Res:* Glass, solid and liquid properties; low temperature phenomena. *Mailing Add:* Dept Physics Univ Colo Boulder CO 80309

TAN-WILSON, ANNA L(I), b Manila, Philippines, Mar 16, 46. BIOCHEMISTRY, PLANT PHYSIOLOGY. *Educ:* Univ Philippines, BS, 66; Univ Col London, MSc, 67; State Univ NY, Buffalo, PhD(biochem), 73. *Prof Exp:* Instr chem, Univ Philippines, 67-69; res assoc biochem & immunol, State Univ NY, Buffalo, 73-75; res assoc biochem, Purdue Univ, 75-76; asst prof, 76-82, ASSOC PROF BIOL, STATE UNIV NY, BINGHAMTON, 82- *Concurrent Pos:* Res Found grant, State Univ NY, 77-80, chmn, Dept Biol Sci, Binghamton, 89-91; co-prin investr, NFS grant, 81-; co-chmn-elect, Binghamton Sect, Sigma Xi. *Mem:* Am Soc Plant Physiologists; AAAS; Am Chem Soc; Sigma Xi. *Res:* Immunochemical and biochemical analysis of soybean proteinase inhibitor and storage protein metabolism. *Mailing Add:* Dept Biol Sci State Univ NY PO Box 6000 Binghamton NY 13902-6000

TANZ, RALPH, b New York, NY, Oct 10, 25; m 52; c 3. PHARMACOLOGY, PHYSIOLOGY. *Educ:* Univ Rochester, BA, 48; Univ Colo, PhD(pharmacol), 58. *Prof Exp:* Asst pharmacol, Sch Med, Univ Colo, 54-57; instr, Med Units, Univ Tenn, 57-59; sr instr, Sch Med, Western Reserve Univ, 59-62; asst prof pharmacol, New York Med Col, 62-63; head cardiovasc sect, Dept Pharmacol, Geigy Res Labs, NY, 63-69; from assoc prof to prof, 69-90, EMER PROF PHARMACOL, SCH MED, ORE HEALTH SCI UNIV, 90- *Concurrent Pos:* NIH career develop award, 61; chmn sect, Gordon Res Conf, 66; Fogarty sr int fel, Heart Res Labs, Dept Med, Univ Cape Town Med Sch, SAfrica, 76-77, Univ Melbourne & Univ Auckland, 84-85. *Mem:* AAAS; Am Soc Pharmacol & Exp Therapeut; Cardiac Muscle Soc (pres, 64); Am Heart Asn; Int Heart Res Soc. *Res:* Isolated cardiac tissue; effect of cardiac glycosides and catecholamines on cardiac muscle; antihypertensives; physiological and biochemical correlates of arrhythmogenesis and antiarrhythinic drugs. *Mailing Add:* Dept Pharmacol Sch Med Ore Health Sci Univ 3181 SW Sam Jackson Portland OR 97201

TANZER, CHARLES, b New York, NY, Dec 4, 12; m 58. SCIENCE EDUCATION, MEDICAL MICROBIOLOGY. *Educ:* Long Island Univ, BS, 33; NY Univ, MS, 36, PhD(bact), 41. *Hon Degrees:* LHD, Long Island Univ, 80. *Prof Exp:* Asst bact, Col Med, NY Univ, 34-38; instr, Bronx High Sch Sci, 38-43 & Dewitt Clinton High Sch, 45-49; chmn sci dept, Seward Park High Sch, 49-57; prin jr high schs, New York, 57-65; prof & coordr sci, Teacher Educ Prog, 65-73, EMER PROF, TEACHER EDUC PROG, HUNTER COL, 73- *Concurrent Pos:* Adj assoc prof, Long Island Univ, 51-65; Ford Found fel, 57. *Honors & Awards:* Meritorious Serv Award, Am Cancer Soc, 73. *Mem:* AAAS; Sigma Xi; Am Cancer Soc; Nat Asn Biol Teachers. *Res:* Serology; parasitology; intestinal parasites. *Mailing Add:* 600 W 218th St Apt 2H New York NY 10034

TANZER, MARVIN LAWRENCE, b New York, NY, Jan 26, 35; m 54; c 4. BIOCHEMISTRY. *Educ:* Mass Inst Technol, SB, 55; NY Univ, MD, 59. *Prof Exp:* Intern med, Johns Hopkins Hosp, Baltimore, 59-60, asst resident, 60-61; res fel, Mass Gen Hosp-Harvard Med Sch, 61-62 & 64-65, asst biologist, 65-68, assoc, 67-68; from asst prof to assoc prof biochem, 68-75, prof orthop surg & dir orthop labs, 78-88, PROF BIOCHEM, HEALTH CTR, UNIV CONN, 75-, PROF BIOSTRUCT & FUNCTION, 86- *Concurrent Pos:* Arthritis Found fel, Mass Gen Hosp, Boston, 61-62, Am Heart Asn fel, 64-66; Am Heart Asn estab investr, Mass Gen Hosp, Boston, 66-68 & Med Sch, Univ Conn, 68-71; investr, Marine Biol Lab, Woods Hole, 66-71; tutor biochem sci, Harvard Univ, 67-68; Josiah Macy, Jr Found fac scholar award, 74-75; vis prof dermat, Univ Liege, Belg, 74-75; NIH Study Sect, Biophys & Biophys Chem B, 76-80; vis prof, Univ Calif, Los Angeles, Bone Res Lab, 76; vis prof, Jap Soc Promotion Sci, Tokyo, 77, Univ Lund, Sweden, 81 & Univ Claude Bernard, Lyon, France, 85; mem NIH study sect, pathobiochem, 86-, chmn, 88-90; prof & head dept biostructure & function, Univ Conn, 86-; mem, Breast Cancer Task Force comt, NIH, 82-86; assoc ed, J Cellular Biochem, 90- *Mem:* Am Soc Bone & Mineral Res; Orthop Res Soc; Am Soc Cell Biol; Am Chem Soc; Am Soc Biol Chemists; NY Acad Sci. *Res:* Properties, function and synthesis of connective tissue components. *Mailing Add:* Dept Biostruct & Function Univ Conn Health Ctr Farmington CT 06030

TAO, FRANK F, b Chang Sha, China; US citizen; c 2. SYNTHETIC FUEL, LUBRICATION. *Educ:* Univ Mo, MS, 57; Univ Mo-Rolla, PhD(chem eng), 64. *Prof Exp:* Res engr, Dorr-Oliver Inc, 57-62; instr, Univ Mo-Rolla, 62-64; sr res engr, Exxon Res & Eng Co, 64-71, eng assoc, 72-81, sr eng assoc, 82-87; ADJ PROF, LAMAR UNIV, 87- *Honors & Awards:* Arch T Colwell Award, Soc Automobile Engrs, 75. *Mem:* Am Inst Chem Engrs; Sigma Xi. *Res:* Fluidification; lubrication; synthetic fuel (coal liquefaction). *Mailing Add:* 5110 Ashwood Dr Baytown TX 77521

TAO, L(UH) C(HENG), b Wusih, China, Feb 6, 22; m 50; c 4. CHEMICAL ENGINEERING. *Educ:* Univ Nanking, China, BS, 46; Univ Wis, MS, 49, PhD, 52. *Prof Exp:* Asst, Univ Nanking, 46-47; chem process engr, Singh Co, Ill, 52-55; res engr, Titanium Metals Corp Am, Nev, 55-59; from assoc prof to prof, 59-70, HOWARD S WILSON PROF CHEM ENG, UNIV NEBR-LINCOLN, 70-, CHMN DEPT, 78- *Mem:* AAAS; Am Chem Soc; Am Inst Chem Engrs; Am Soc Eng Educ; Sigma Xi. *Res:* Heat and mass transfer; phase equilibrium. *Mailing Add:* Dept Chem Eng Avery Lab Rm 228 Univ Nebr Lincoln NE 68588-0126

TAO, LIANG NENG, b China, June 27, 27; nat US; m 57; c 3. ENGINEERING. *Educ:* Chiao Tung Univ, BS, 49; Univ Ill, MS, 50, PhD(mech eng), 53. *Prof Exp:* Res engr, Worthington Corp, 53-55; from asst prof to assoc prof, 55-61, PROF MECH, ILL INST TECHNOL, 61- *Concurrent Pos:* Consult, Armour Res Found Am. *Mem:* Soc Eng Educ; Am Soc Mech Engrs; Am Inst Aeronaut & Astronaut; Am Asn Univ Prof; Sigma Xi. *Res:* Engineering sciences; applied mathematics; fluid mechanics; heat transfer; thermodynamics; lubrication; magnetohydrodynamics. *Mailing Add:* Dept Mech & Math Ill Inst Tech 3300 S Federal St Chicago IL 60616

TAO, MARIANO, b Davao City, Philippines, Mar 3, 38; US citizen; m 67; c 2. ENZYMOLOGY, MEMBRANE BIOCHEMISTRY. *Educ:* Cheng Kung Univ, Taiwan, BS, 62; Univ Washington, Seattle, PhD(biochem), 67. *Prof Exp:* Sr fel biochem, Univ Wash, Seattle, 67-68; guest investr biochem, Rockefeller Univ, NY, 68-70; actg head, 79-80, PROF BIOCHEM, UNIV ILL CHICAGO, 70- *Concurrent Pos:* Vis prof, Nat Taiwan Univ, 81; mem, Biochem Study Sect, NIH, 85-89. *Mem:* Am Chem Soc; AAAS; Am Soc Biochem & Molecular Biol. *Res:* Enzymology and control mechanisms; regulation of erythrocyte membrane cytoskeletal protein interactions and assembly by phosphorylation-dephosphorylation; structure-function relationships of protein kinases and the Fru-2,6-P2-regulated PPi-dependent phosphofructokinase. *Mailing Add:* Dept Biochem M/C 536 Univ Ill Chicago 1853 W Polk St Chicago IL 60612

TAO, RONGJIA, b Shanghai, China, Jan 28, 47; m 77; c 2. MATHEMATICAL PHYSICS, COMPUTATIONAL PHYSICS. *Educ:* Univ Sci & Technol China, BS, 70; Columbia Univ, MA, 80, MPhil & PhD(physics), 82. *Prof Exp:* Res assoc physics, Univ Wash, 82-84; asst prof physics, Univ Southern Calif, 84-85; ASST PROF PHYSICS, NORTHEASTERN UNIV, 85- *Concurrent Pos:* Vis scientist, Univ Cambridge, Eng, 83-84 & IBM Bergen Sci Ctr, 87. *Mem:* Am Phys Soc. *Res:* Properties of electron systems under strong magnetic fields; functional integration method in statistical mechanics; application of computer in physics; diffusion in disordered media; exact solution for physics models. *Mailing Add:* Dept Physics Southern Ill Univ Carbondale IL 62901

TAO, SHU-JEN, b Soochow, China, Oct 7, 28; m 58; c 3. NUCLEAR SCIENCE, PHYSICAL CHEMISTRY. *Educ:* Amoy Univ, BSc, 49; Univ NSW, MEngSc, 61, PhD(nuclear chem), 64. *Prof Exp:* Scientist, Taiwan Rain Stimulation Res Inst, 51-59; res fel nuclear & radiation chem, Australian AEC, 60-61; SR STAFF SCIENTIST, NEW ENG INST MED RES, 65- *Mem:* AAAS; Am Phys Soc; Am Chem Soc; NY Acad Sci; Sigma Xi. *Res:* Positron physics; positronium chemistry; fast timing electronic instruments; applied statistics. *Mailing Add:* 12 Woodchuck Lane Wilton CT 06897

TAOKA, GEORGE TAKASHI, b Honolulu, Hawaii, Feb 19, 35. ENGINEERING MECHANICS, CIVIL ENGINEERING. *Educ:* Ore State Univ, BS, 58; Univ Ill, MS, 60, PhD(mech), 64. *Prof Exp:* Struct engr, NAm Aviation, 58-59; instr mech, Univ Ill, 60-64; tech staff mech, Sandia Corp, 64-65; PROF CIVIL ENG, UNIV HAWAII, 65- *Concurrent Pos:* NASA fel, Jet Propulsion Lab, 68; vis fel, Princeton Univ, 72-73; vis prof, Tokyo Inst Technol, 74. *Mem:* Am Acad Mech; Am Soc Civil Engrs; Soc Exp Stress Analysis; Inst Transp Engrs. *Res:* Traffic accident analysis; structural dynamics. *Mailing Add:* Dept Civil Eng HC1 339 Univ Hawaii at Manoa 2540 Campus Rd Honolulu HI 96822

TAPAROWSKY, ELIZABETH JANE, b Worcester, Mass, Apr 12, 54. BIOLOGY. *Educ:* Emmanuel Col, BA; 76; Brown Univ, PhD(biol), 82. *Prof Exp:* Postdoctoral fel, Cold Spring Harbor Lab, NY, 82-84, Univ Va Sch Med, 84-86; ASST PROF BIOL SCI, PURDUE UNIV, 86- *Mem:* AAAS; Am Soc Cell Biol; Am Soc Microbiol. *Res:* Structure and function of genes involved in mammalian cell transformation; role of the ras oncoprotein in eukaryotic signal transduction; analysis of myc oncoprotein function. *Mailing Add:* Dept Biol Sci Purdue Univ West Lafayette IN 47907

TAPE, GERALD FREDERICK, b Ann Arbor, Mich, May 29, 15; m 39; c 3. SCIENCE ADMINISTRATION. *Educ:* Eastern Mich Univ, AB, 35; Univ Mich, MS, 36, PhD(physics), 40. *Hon Degrees:* DSc, Eastern Mich Univ, 64. *Prof Exp:* Asst physics, Eastern Mich Univ, 33-35 & Univ Mich, 36-39; instr, Cornell Univ, 39-42; staff mem radiation lab, Mass Inst Technol, 42-46; from asst prof to assoc prof physics, Univ Ill, 46-50; asst to dir, Brookhaven Nat Lab, 50-51, dep dir, 51-62; comnr, US AEC, 63-69; from vpres to pres, Assoc Univs, Inc, 62-63, pres, 69-80, spec asst pres, 80-82, CONSULT, ASSOC UNIVS, INC, 82- *Concurrent Pos:* Mem, President's Sci Adv Comn, 69-73; mem high energy adv panel, US AEC, 69-74, sr tech adv Geneva IV, 71; mem, Defense Sci Bd, 70-74, chmn, 70-73; mem bd dirs, Atomic Indust Forum, 70-73; bd trustees, Sci Serv, 70-; mem sci adv comt, Int Atomic Energy Agency, 72-74, US rep, 73-77; mem gen adv comt, US Energy Res & Develop Admin, 75-77; mem adv coun, Elec Power Res Inst, 78-85; mem, Univ Chicago Bd Gov, Argonne Nat Lab, 82-85. *Honors & Awards:* Henry DeWolf Smyth Nuclear Statement Award, 78; Enrico Fermi Award, Dept Energy, 87. *Mem:* Nat Acad Eng; fel Am Phys Soc; fel Am Nuclear Soc; Am Astron Soc; fel AAAS. *Res:* Nuclear physics; particle physics; accelerator development; reactor development; applications of atomic energy; radioastronomy. *Mailing Add:* Assoc Univs Inc 1400 16th St NW Suite 730 Washington DC 20036

TAPER, LILLIAN JANETTE, TRACE MINERAL REQUIREMENTS. *Educ:* Va Polytech Inst & State Univ, PhD(human nutrit), 76. *Prof Exp:* ASSOC PROF HUMAN NUTRIT, COL HUMAN RESOURCES, 76- *Mailing Add:* Human Nutrit Dept Va Tech Rm 206 Wallace Hall Blacksburg VA 24061-0430

TAPIA, FERNANDO, b Panama, Apr 8, 22; m 47; c 3. PSYCHIATRY. *Educ:* Univ Iowa, BA, 44, MD, 47; Am Bd Psychiat & Neurol, dipl psychiat, 60, dipl child psychiat, 66. *Prof Exp:* Intern, Santo Thomas Hosp, Panama, 48; dir, Boquette Sanit Unit-Panama, 48-54; resident psychiat, Barnes Hosp, Wash Univ, 54-57; asst dir out-patient clin, Malcolm Bliss Ment Health Ctr, 57-58; chief psychiatrist, Child Guid Clin, St Louis County Health Ctr, 58-59, dir ment health div, 59-61; from asst prof to prof psychiat, Sch Med, Univ Mo-Columbia, 61-72, chief sect child psychiat, 61-72; PROF PSYCHIAT & BEHAV SCI, COL MED, UNIV OKLA, 72- CHIEF, MENT HEALTH SERV, UNIV HOSP & CLINS, 72- *Concurrent Pos:* Instr, Wash Univ, 57-61; consult, St Louis County Juv Court, 57-61 & Convent of the Good Shepherd, 57-61; dir children's serv, Mid-Mo Ment Health Ctr, 66-72. *Mem:* Fel Am Psychiat Asn; Sigma Xi. *Mailing Add:* 23029 Fernando Scottsdale AZ 85255

TAPIA, M(OIEZ) A(HMEDALE), b Surat, India, Nov 17, 35; m 72. ELECTRICAL ENGINEERING. *Educ:* Univ Poona, BE, 60; Univ Ill, Urbana, MS, 62; Univ Notre Dame, PhD(elec eng), 66. *Prof Exp:* Asst lectr elec eng, Polytech Inst, India, 60-61; asst prof, Ga Inst Technol, 66-67 & Univ Miami, 67-68; asst prof, Ga Inst Technol, 68-76; ASST PROF ELEC ENG, UNIV MIAMI, 76- *Concurrent Pos:* Jr engr, Koyna Elec, India, 60-61; Am Soc Eng Educ-Ford Found resident fel, NASA-Langley Res Ctr, 72-73; NASA grant fel prog comput sci, 72- *Mem:* Inst Elec & Electronics Engrs; Sigma Xi. *Res:* Computer engineering and science; network topology; communications; linear systems. *Mailing Add:* Elec Eng Dept Univ Miami Box 248294 Coral Gables FL 33124

TAPIA, RICHARD, b Santa Monica, Calif, Mar 25, 39; m 60; c 3. NUMERICAL ANALYSIS, OPTIMIZATION. *Educ:* Univ Calif, Los Angeles, BA, 61, MA, 66, PhD(math), 67. *Prof Exp:* Mathematician, Todd Shipyards, Calif, 61-63 & Int Bus Mach Corp, 63-65; actg asst prof math, Univ Calif, Los Angeles, 67-68; vis asst prof, US Army Math Res Ctr, Univ Wis-Madison, 68-70; from asst prof to assoc prof, 70-76, PROF MATH SCI, RICE UNIV, 76- *Mem:* Soc Indust & Appl Math; Inst Math Statist; Soc Advan Chicano & Native Am Scientists; Am Math Soc. *Mailing Add:* Dept Math Sci Rice Univ PO Box 1892 Houston TX 77251-1892

TAPIA, SANTIAGO, b Santiago, Chile, Nov 13, 39. ASTROPHYSICS. *Educ:* Univ Chile, Lic, 65; Univ Ariz, PhD(astron), 75. *Prof Exp:* Lab asst crystallog, Univ Chile, 61-62, teaching asst physics, 63-65, res asst astron, 66-69; res assoc astron, Univ Ariz, 75-79, asst astronr, 80-; MEM STAFF, LINCOLN LABS. *Concurrent Pos:* Lectr astron, Tech Sch Aeronaut, Chilean Air Force, 69. *Mem:* Am Astron Soc. *Res:* Observational study of optical properties of BL Lacertae objects and their relation to quasi-stellar objects; observational study of optical polarization in AM Herculis type objects and x-ray binary sources. *Mailing Add:* Lincoln Labs Rm 91 244 Wood St Lexington MA 02173

TAPLEY, BYRON D(EAN), b Charleston, Miss, Jan 16, 33; m 59; c 2. AEROSPACE ENGINEERING, ASTRODYNAMICS. *Educ:* Univ Tex, BS, 56, MS, 58, PhD(eng mech), 60. *Prof Exp:* Asst prof aerospace eng & eng mech, Univ Tex, 60-64, assoc prof, 64-66, assoc prof aerospace eng, 66-68, chmn dept, 66-77, prof, 68-74, W R Woolrich prof aerospace eng & eng mech, 74-84, DIR, CTR SPACE RES, UNIV TEX, 82-, CLARE COCKRELL WILLIAMS CENTENNIAL CHAIR ENG, 84-, DIR, TEX SPACE GRANT CONSORTIUM, 89- *Concurrent Pos:* Ford Found fel, Univ Tex, 61-62; mem, Adv Comt Guidance, Control & Navig, NASA, 65-66, Eos Sci Adv group, 86-88, Space & Earth Sci Adv Comt, 87-88; chmn, Comt Geodesy, Nat Res Coun, 78-82, mem, Aeronaut & Space Eng Bd, 81-85, mem ad hoc comt, Space Station Eng & Technol Develop, 84-85, chmn, Comt Earth Sci, 87-88, mem, Space Sic Bd, 87-88, Comt Earth Studies, 89-, Space Studies Bd, 89-, chmn, Geophys Study Comt, 90-; assoc ed, Geophys Res Letters, Bull Geodesique, Celestial Mech J & J Guidance & Control; mem, Tech Panel on Dynamics of Artificial Satellites & Space Probes, Int Comt Space Res, 79- *Honors & Awards:* Except Sci Achievement Medal, NASA, 83; Mech & Control of Flight Award, Am Inst Aeronaut & Astronaut, 87. *Mem:* Nat Acad Sci; Am Astronaut Soc; fel Am Inst Aeronaut & Astronaut; Soc Eng Sci; fel AAAS; Am Soc Mech Engrs; fel Am Geophys Union; Am Acad Mech; Sigma Xi; Am Astron Soc. *Res:* Theory of the guidance and navigation of continuous powered space vehicles including the development and application of methods in numerical optimization theory of stochastic processes and orbit determination theory. *Mailing Add:* Univ Tex Ctr Space Res Bldg WRW Rm 402 Austin TX 78712

TAPLEY, DONALD FRASER, b Woodstock, NB, May 19, 27; nat US; m 57; c 3. INTERNAL MEDICINE. *Educ:* Acadia Univ, BSc, 48; Univ Chicago, MD, 52. *Prof Exp:* Intern & asst resident, Presby Hosp, New York, 52-54; fel physiol chem, Johns Hopkins Univ, 54-56; from asst prof to assoc prof, 56-72, actg dean fac med, 73-74, dean fac med, 74-84, PROF MED, COL PHYSICIANS & SURGEONS, COLUMBIA UNIV, 72-, ASSOC DEAN FAC AFFAIRS, 70-, ALUMNI PROF & SR DEP VPRES, 84- *Concurrent Pos:* Fel physiol chem, Oxford Univ, 56-57; from asst attend physician to attend, Presby Hosp, 57- *Mem:* Am Soc Clin Invest; Am Thyroid Asn; Endocrine Soc; Harvey Soc. *Res:* Intermediary metabolism; thyroid physiology. *Mailing Add:* Dept Med Columbia Univ 630 W 168th St New York NY 10032

TAPLIN, DAVID MICHAEL ROBERT, b Chesterfield, Eng, July 19, 39; c 5. PHYSICAL METALLURGY, MECHANICAL ENGINEERING. *Educ:* Univ Aston, BSc, 61, DSc, 79; Oxford Univ, DPhil(metall), 64. *Prof Exp:* Metallurgist, Imp Chem Industs, Birmingham, Eng, 57-61; lectr phys metall, Univ Melbourne, 64-68; res fel metall, Banaras Hindu Univ, 67-68; prof mat, Univ Calif, 87-88; assoc prof, 68-72, PROF MECH ENG & PHYSICS, UNIV WATERLOO, 72- *Concurrent Pos:* Mem coun, Int Cong Fracture, 69-77, dir & vpres, 77-81, pres, 81-; pres, Can Fracture Corp, 76-; dir, Pergamon Press of Can, 78-; ed-in-chief, Int Conf on Fracture, 77. *Honors & Awards:* Kamani Gold Medal, Indian Inst Metals, 71. *Mem:* Fel Brit Inst Metall; fel Royal Soc Arts; Am Inst Mining, Metall & Petrol Engrs; hon fel Int Cong Fracture; fel Brit Inst Mech Eng; fel Inst Engrs Ireland. *Res:* Creep and creep fracture in metals; fatigue at elevated temperatures; superplasticity; life prediction and drainage mechanics; thermal cycling; ductile-brittle transition; grain size effects on mechanical behavior; fracture maps; metal matrix composites; ceramic matrix composites; manufacturing of composites. *Mailing Add:* Dept Mech Eng & Mfg Eng Trinity Col Dublin 2 Ireland

TAPLIN, LAEL BRENT, b Blackwell, Okla, Jan 5, 27; m 51; c 4. MECHANICAL ENGINEERING, ELECTRICAL ENGINEERING. *Educ:* Ore State Col, BS, 48; Univ Ill, MS, 51. *Prof Exp:* Test engr, Gen Elec Co, 48-49; res asst theoret & appl mech, Univ, Ill, 49-50, res assoc, 50-51; sr engr, Vickers Inc, Sperry Rand Corp, 51-55, proj engr, 55-58; proj engr, Res Labs Div, Bendix Corp, 58-61, dept head, Lab Flight Controls, 61-64, mgr energy conversion & dynamic controls, 64-66, dir mech sci & controls lab, 66-74, consult scientist, Res Labs, 74-80; ENG-ADVAN TECH & SYST, SPERRY VICKERS, 80- *Concurrent Pos:* Instr, Wayne State Univ, 57-58. *Mem:* Inst Elec & Electronics Engrs; Sigma Xi. *Res:* Dynamics of fluid power servovalves, motors and control systems; hot gas servos; pneumatic controls; gas generators and controls; pneumatic flight controls; fluidic elements, sensors and systems; fluidic circuit analysis; fuel management; electronic fuel injection; emissions control. *Mailing Add:* 8396 Golfside Dr Union Lake MI 48387

TAPP, CHARLES MILLARD, b Memphis, Tenn, Nov 9, 36; m 55, 78; c 3. ENGINEERING. *Educ:* Union Univ, BA, 58; Memphis State Univ, BS, 60; Univ Va, MS, 62, PhD(physics), 64. *Prof Exp:* Staff mem radiation calibration, Nat Bur Standards, 60, 62; tech staff mem radiation damage, 64-66, div supvr vacuum tube physics & technol, 66-69, dept mgr vacuum tube devices, 69-71, dept mgr microelectronic components, 71-77, mgr info systs dept, 77-80, MGR, ELECTRONIC INSTRUMENTATION DEPT, SANDIA NAT LABS, 80- *Concurrent Pos:* Ed, Transactions Components, Hybrids & Mfg Technol & gen chmn, Electronic Components Conf, Inst Elec & Electronics Engrs, 74- *Mem:* Sr mem Inst Elec & Electronics Engrs; Am Phys Soc. *Res:* Solid state electronics and circuits; microelectronic thin and thick film processes; vacuum tube design, development; radiation effects in devices; neutron sources. *Mailing Add:* 9201 Lona Lane Albuquerque NM 87111

TAPP, WILLIAM JOUETTE, b Quincy, Ill, July 26, 18; m 46; c 1. ORGANIC CHEMISTRY. *Educ:* Univ Ill, BS, 39; Cornell Univ, PhD(org chem), 43. *Prof Exp:* Asst chem, Cornell Univ, 40-41 & 41-43, Nat Defense Res Comt fel, 40-41; res & develop chemist, Union Carbide Corp, 43-46, proj leader, 46-54, staff asst, 55-56, patent adminr, 57-66, asst dir pharmaceut tech, 66-67, mgr admin, 67-81; CONSULT, 81- *Mem:* Am Chem Soc. *Res:* Organic nitrogen and sulfur compounds; synthetic lubricants; industrial organic synthesis. *Mailing Add:* 1031 W Sterlington Pl Apex NC 27502

TAPPAN, DONALD VESTER, biochemistry; deceased, see previous edition for last biography

TAPPEINER, JOHN CUMMINGS, II, b Los Angeles, Calif, Dec 15, 34; m 65; c 2. FOREST ECOLOGY, SILVICULTURE. *Educ:* Univ Calif, Berkeley, BS, 57, MS, 61, PhD(forestry), 66. *Prof Exp:* Forester, US Forest Serv, 59-63; res asst forest ecol, Univ Calif, Berkeley, 63-66; Ford Found teaching fel forestry, Agr Univ Minas Gerais, 66-67; from asst prof to assoc prof, Forest Res Ctr, Univ Minn, St Paul, 68-73; regional silviculturist, US Forest Serv, 73-81; PROF FORESTRY, SILVICULTURE & FOREST ECOL, ORE STATE UNIV, 81- *Mem:* Soc Am Foresters; Ecol Soc Am. *Res:* Natural regeneration of Sierra Nevada Douglas fir and ponderosa pine; ecology of hazel and understory vegetation; biomass and nutrient content of shrubs and herbs; effect of mechanized harvesting of forest soils; ecology of shrubs and hardwood in forests of Oregon. *Mailing Add:* Dept Forest Mgt Ore State Univ Corvallis OR 97331

TAPPEL, ALOYS LOUIS, b St Louis, Mo, Nov 21, 26; m 51; c 6. BIOCHEMISTRY. *Educ:* Iowa State Univ, BS, 48; Univ Minn, PhD(biochem), 51. *Prof Exp:* From instr to assoc prof, 51-61, PROF FOOD SCI & BIOCHEMIST, UNIV CALIF DAVIS, 61- *Concurrent Pos:* Guggenheim fel, 65-66. *Honors & Awards:* Borden Award, Am Inst Nutrit, 73, Osborne & Mendel Award, 87. *Mem:* Am Chem Soc; Am Oil Chem Soc; Am Soc Biol Chem; Am Inst Nutrit. *Res:* Oxidant molecular damage and biological protection systems. *Mailing Add:* Dept Food Sci & Technol Univ Calif Davis CA 95616

TAPPEN, NEIL CAMPBELL, b Jacksonville, Fla, Feb 26, 20; m 52; c 2. PHYSICAL ANTHROPOLOGY, PRIMATOLOGY. *Educ:* Univ Fla, AB, 41; Univ Chicago, MA, 49, PhD(anthrop), 52. *Prof Exp:* Res assoc human biol, Univ Mich, 51-52; from assoc anthrop to instr, Univ Pa, 52-54; from instr anat to asst prof, Emory Univ, 54-59; assoc prof phys anthrop, Tulane Univ, 59-65; prof, 65-69, Earnest A Hooten prof, 69-90, EMER PROF ANTHROP, UNIV WIS, MILWAUKEE, 90- *Concurrent Pos:* NIH grants, Emory Univ, 55-59, Tulane Univ, 60-65 & Univ Wis, Milwaukee, 65-71; Fulbright sr res scholar, Makerere Col, Uganda, 56-57; NSF grant, Univ Wis, Milwaukee, 71-76. *Mem:* AAAS; Am Anthrop Asn; Am Asn Phys Anthropologists; Am Anat Asn; Int Primatological Soc. *Res:* Organization of bone; non-human primates and their relationship to human evolution; structure of bone in fossil hominids; problems of human evolution. *Mailing Add:* Dept Anthrop Univ Wis Milwaukee WI 53201

TAPPER, DANIEL NAPHTALI, b Philadelphia, Pa, Dec 5, 29; m 59. NEUROPHYSIOLOGY. *Educ:* Rutgers Univ, BS, 51; Univ Pa, VMD, 55; Cornell Univ, PhD(physiol), 59. *Prof Exp:* Asst physiol, 55-59, res assoc radiation biol, 59-61, from asst prof to assoc prof, 61-69, PROF, CORNELL UNIV, 69- *Concurrent Pos:* NIH spec fel, Stockholm, 65-66; NIH spec fel, 72-73, adj prof, Rockefeller Univ, 73-74. *Mem:* AAAS; Soc Neurosci; Am Physiol Soc. *Res:* Behavior; receptor physiology; neurophysiology of skin sensibility. *Mailing Add:* NY State Vet Coll Cornell Univ Ithaca NY 14850

TAPPERT, FREDERICK DRACH, b Philadelphia, Pa, Apr 21, 40; m 79; c 2. PHYSICS. *Educ:* Pa State Univ, BS, 62; Princeton Univ, PhD(physics), 67. *Prof Exp:* Mem tech staff, Bell Labs, 67-73; sr res scientist, Courant Inst Math Sci, NY Univ, 74-78; PROF APPL MARINE PHYSICS, 78- *Concurrent Pos:* Vis staff mem, Los Alamos Sci Lab, 74-; consult, Sci Applns, Inc, 74-; consult, Nat Oceanog & Atmospheric Admin, US Govt, 80- , Daubin Systs Corp, 82- , Naval Underwater Syst Cen, 83-84 & 89-90. *Mem:* Am Phys Soc; AAAS; Soc Indust & Appl Math; Am Geophys Union; fel Acoust Soc Am; Sigma Xi. *Res:* Theory and numerical simulation of wave propagation effects in plasmas, gases, liquids and solids. *Mailing Add:* 907 Jeronimo Dr Coral Gables FL 33146

TAPPHORN, RALPH M, b Grinnell, Kans, July 26, 44; m 69; c 2. NUCLEAR PHYSICS. *Educ:* Ft Hays Kans State Univ, BS, 66; Kans State Univ, PhD(physics), 70. *Prof Exp:* Res assoc nuclear physics, Ballistics Res Lab, Aberdeen Proving Ground, Md, 70-72; sr scientist nuclear physics, Schlumberger-Doll Res Ctr, 72-77; sr mem tech staff, Ball Aerospace Systs Div, 77-80, prog mgr, 80-82; bus mgr, Tapphorn Conserv, Ltd, 82-87; prin scientist, 87-90, SCI SUPVR, LOCKHEED ENG & SCI CO, 90- *Honors & Awards:* Eagle Manned Mission Success Award, Nat Space Club, 90. *Mem:* Am Phys Soc; Sigma Xi; Instrument Soc Am; Am Soc Nondestructive Testing. *Res:* Aerospace instrumentation of gamma-ray spectrometers for astrophysical studies; geophysical exploration with gamma-ray spectroscopy; nuclear detectors and instrumentation; radiation damage investigations; infrared fiber-optic sensors; flame combustion; Fourier transfer infrared combustion analysis spectroscopy; Laser induced combustion spectroscopy; flash X-ray radiography; neutron radiography; ultrasonic imaging; aerospace instrumentation for materials testing. *Mailing Add:* NASA White Sands Test Facil PO Drawer MM Las Cruces NM 88004

TAPPMEYER, WILBUR PAUL, b Owensville, Mo, May 19, 22; m 47; c 5. INORGANIC CHEMISTRY. *Educ:* Southeast Mo State Col, AB, 45; Univ Mo, BS, 47, PhD(inorg chem), 61. *Prof Exp:* Teacher high sch, Mo, 44-45; asst chem, Mo Sch Mines, 45-46; prof, Southwest Baptist Col, 47-60; from asst prof to assoc prof, 60-66, PROF CHEM, UNIV MO, ROLLA, 66- *Mem:* Am Chem Soc. *Res:* Solid phase extraction of metal chelates; dimeric and polymeric properties of certain metal acetates and other alkonates. *Mailing Add:* 705 E Fifth St Rolla MO 65401-3417

TAPSCOTT, ROBERT EDWIN, b Terre Haute, Ind, June 10, 38; m 67; c 1. COMBUSTION CHEMISTRY, ENVIRONMENTAL CHEMISTRY. *Educ:* Univ Colo, BS, 64; Univ Ill, Urbana, PhD, 68. *Prof Exp:* From asst prof to assoc prof chem, 68-84, SR SCIENTIST, UNIV NMEX, 84- *Concurrent Pos:* Vis prof, Univ NC, 78. *Mem:* Sigma Xi; AAAS; Royal Soc Chem; Am Chem Soc; Am Inst Chem Engrs; Air Pollution Control Asn. *Res:* Chemistry of combustion and extinguishment; environmental chemistry; hazardous materials; waste disposal technology; emissions from large fires. *Mailing Add:* NMERI Box 25 Univ NMex Albuquerque NM 87131

TARAGIN, MORTON FRANK, b Washington, DC, Feb 1, 44; m 68; c 5. NUCLEAR EXPERIMENTAL PHYSICS. *Educ:* George Washington Univ, BS, 65, MPh, 69, PhD(physics), 70. *Prof Exp:* ASSOC PROF PHYSICS, GEORGE WASHINGTON UNIV, 70- *Concurrent Pos:* Consult comput systs; vis prof, Weizmann Inst, 79-81. *Mem:* Am Phys Soc. *Res:* Pion-nucleon interactions; computer systems and simulations. *Mailing Add:* 1512 Red Oak Dr Silver Spring MD 20910

TARAMAN, KHALIL SHOWKY, b Cairo, Egypt, July 9, 39; US citizen; m 69; c 3. MANUFACTURING PRODUCTIVITY, MANUFACTURING SYSTEMS. *Educ:* Ain Shams Univ, Cairo, BSc, 64, MSc, 67; Univ Wis-Madison, MS, 69; Tax Tech Univ, Lubbock, PhD(indust eng), 71. *Prof Exp:* Instr prod eng, Ain Shams Univ, Cairo, 64-67; from asst prof to prof mfg eng, Univ Detroit, 70-86, dir, Mfg Eng Inst, 75-86, chmn dept, 77-86; assoc dean eng & DIT endowed chmn, 86-89, DEAN ENG, LAWRENCE TECHNOL UNIV, 89- *Concurrent Pos:* Sr tech consult, Ford, Gen Elec, Bendix & other co, 70-; chmn, Mat Removal Coun, Soc Mfg Engrs, 79-84, int dir, 82-87, chmn, Ref Publ Comt, 89-90, Publ & Ref Mgt Coun, 89-; dir, Mfg Div, Am Soc Eng Educ, 88-90; mem, Mfg Studies Bd, Nat Acad Engrs & Nat Res Coun, 89-90. *Mem:* Sr mem Soc Mfg Engrs; sr mem Am Inst Indust Engrs; Am Soc Mech Engrs; Am Soc Eng Educ. *Res:* Manufacturing engineering; author of numerous publications and two books. *Mailing Add:* Lawrence Technol Univ 21000 W Ten Mile Southfield MI 48075-1058

TARANIK, JAMES VLADIMIR, b Los Angeles, Calif, Apr 23, 40; m 71; c 2. EXPLORATION GEOLOGY, PHOTOGEOLOGY. *Educ:* Stanford Univ, BS, 64; Colo Sch Mines, PhD(geol), 74. *Prof Exp:* Chief remote sensing, Iowa Geol Surv, 71-74; prin remote sensing scientist, Earth Resources Observ Syst Data Ctr, US Geol Surv, 75-78; chief Non-Renewable Resources Br, NASA HQ, Washington, DC, 79-82; dean, Mackay Sch Mines, Univ Nev, Reno, 82-87; PROF GEOL & GEOPHYS, DEPT GEOL SCI, UNIV NEV, RENO, 82-; DIR, INST AEROSPACE SCI & TERRESTRIAL APPLNS, DESERT RES INST, UNIV NEV SYSTS, RENO, 86-, PRES, DESERT RES INST, 87- *Concurrent Pos:* Adj prof geol, Univ Iowa, 71-; vis prof civil eng, Iowa State Univ, 72-74; consult, Earth Resources Technol Satellite Follow on Eval Panel Geol, Goddard Space Flight Ctr, NASA, Synchronous Observ Satellite Prog Eval, Geol Applns, Active microwave Syst Eval Workshop Earth-Land Panel, Geol Landuse Water, Johnson Space Ctr, 74; adj prof Earth Sci, Univ SDak, 76-79; chmn working group on instrumentation for remote sensor data processing & anal, Int Soc Photogrammetry, 77-80; chmn, working group non-renewable resources, Int Soc Photogram & Remote Sensing; scientist, various space shuttle progs; chmn, Dept Com Working Group Commercialization and & Weather Satellites, 82-84; chmn, NASA Space Applns Adv Comt, Subcomt Remote Sensing Earth, 86-88; Fel, Explorers Club; mem, NASA Space Sci Ctr & Appln adv comt, 88-90, chmn subcomt Remote Sensing, NASA space applns adv comt, 86-88; comt Strategic Reloctable Targets, Air Force Studies Bd, NAS, 89-; prog dir, Space Grant Prog, Univ Nev Syst Space Grant Consortium, 91- *Honors & Awards:* Except Sci Achievement Medal, NASA, 82, Group Achievement Award, OSTA-1, STS-2, 82; Johnson Space Ctr Group Achievement Award, STS-41G, 84; NASA Group Achievement Award, Shuttle Imaging Radar-B, STS-41G, 90. *Mem:* Fel Geol Soc Am; Am Asn Petrol Geologists; Am Inst Aeronaut & Astronaut; Am Inst Prof Geologists; Soc Explor Geophysicists; fel AAAS; Int Acad Astronaut; sr mem, Am Aeronaut & Astronaut; Am Astron Soc. *Res:* Development of applications of remote sensing technology to mineral and mineral fuel exploration; assessment of environmental and engineering geologic aspects of mineral resource development; engineering geology and geohydrology of civil works site selection. *Mailing Add:* Desert Res Inst Univ Nev Syst PO Box 60220 Reno NV 89506

TARANTINE, FRANK J(AMES), b Youngstown, Ohio, May 27, 35; m 57; c 4. MECHANICAL ENGINEERING. *Educ:* Youngstown Univ, BE, 57; Univ Akron, MS, 61; Carnegie Inst Technol, PhD(fluid dynamics), 65. *Prof Exp:* From instr to assoc prof, 57-70, PROF MECH ENG, YOUNGSTOWN STATE UNIV, 70- *Concurrent Pos:* engr, Com Sharing Inc, 53-56 & 59-63, Automatic Sprinkler Corp Am, 56-59; NSF sci fac fel, 64-65. *Mem:* Am Soc Mech Engrs; Am Soc Eng Educ; Sigma Xi. *Res:* Vibrations and experimental stress analysis; mechanical design; acoustics. *Mailing Add:* 1221 Cherokee Dr Youngstown OH 44511

TARANTINO, LAURA M(ARY), b Exeter, Pa, Feb 6, 47. BIOCHEMISTRY. *Educ:* Col Misericordia, BS, 68; Cornell Univ, PhD(biochem), 75. *Prof Exp:* Assoc res scientist med, Col Physicians & Surgeons, Columbia Univ & Roosevelt Hosp, 75-78; asst prof biochem, Eastern Va Med Sch, 79-87; CONSUMER SAFETY OFFICER, FOOD & DRUG ADMIN, 87- *Mem:* AAAS; Sigma Xi; NY Acad Sci; Am Chem Soc. *Res:* Enzyme regulation in the central nervous system; role and control of hexosemonophosphate shunt in cells; metabolic effects of insulin; diabetes. *Mailing Add:* Food & Drug Admin-HFF330 200 C St SW Washington DC 20204

TARANTO, MICHAEL VINCENT, food science, biochemistry, for more information see previous edition

TARAPCHAK, STEPHEN J, b Staten Island, NY, Mar 20, 42; m 65. LIMNOLOGY, PHYCOLOGY. *Educ:* Clarion State Col, BS, 64; Ohio Univ, MS, 66; Univ Minn, PhD(ecol), 73. *Prof Exp:* Res fel, Limnol Res Ctr, Univ Minn, Minneapolis, 73-74; res scientist biol oceanog, Great Lakes Environ Res Lab, Nat Oceanic & Atmospheric Admin, 74-81; ECOLOGIST, 81- *Concurrent Pos:* Consult, Environ Statements Syst Div, Argonne Nat Lab, 73-74. *Mem:* Am Phycol Soc; Am Soc Limnol & Oceanog; Int Limnol Soc; Int Asn Great Lakes Res. *Res:* Phytoplankton ecology; ecology. *Mailing Add:* 1803 Crestland Dr Ann Arbor MI 48104

TARAS, MICHAEL ANDREW, b Olyphant, Pa, Sept 6, 21; m 48; c 4. FOREST PRODUCTS, WOOD TECHNOLOGY. *Educ:* Pa State Univ, BS, 42, MF, 48; NC State Univ, PhD(wood technol), 65. *Prof Exp:* Forest prod technologist, Forest Prod Lab, US Forest Serv, 48-54 & Southeastern Forest

Exp Sta, Forestry Sci Lab, 54-79; prof forestry, 79-82, DEPT HEAD, CLEMSON UNIV, 83- *Mem:* Forest Prod Res Soc; Soc Wood Sci & Technol; Int Asn Wood Anat. *Res:* Forestry; wood anatomy related to wood identification, quality, seasoning and moisture movement through wood; in log and tree classification systems and wood weight-volume relationships; forest tree biomass prediction and evaluation. *Mailing Add:* Forestry Dept Col Forest & Recreation Clemson Univ Clemson SC 29634-1003

TARAS, PAUL, b Tunisia, May 12, 41; Can citizen; m 63; c 2. HIGH SPIN PHYSICS, RELATIVISTIC HEAVY ION PHYSICS. *Educ:* Univ Toronto, BScEng Phys, 62, MSc, 63, PhD(nuclear physics), 65. *Prof Exp:* Asst prof, 65-70, assoc prof, 70-76, PROF PHYSICS, UNIV MONTREAL, 76- *Concurrent Pos:* Attached staff, Chalk River Nuclear Labs, Atomic Energy Can Ltd, 78-; vis prof, Univ Heidelberg, 71-72. *Mem:* Am Phys Soc; Can Asn Physicists. *Res:* High spin nuclear spectroscopy and reaction mechanisms via heavy ion induced reactions; elementary particle physics. *Mailing Add:* Dept Physics Univ Montreal Montreal PQ H3C 3J7 Can

TARASUK, JOHN DAVID, b St Walburg, Sask, Dec 24, 36; m 61; c 3. THERMODYNAMICS, MECHANICAL ENGINEERING. *Educ:* Univ Toronto, BASc, 59, MASc, 61; Univ Sask, PhD(mech eng), 69. *Prof Exp:* Demonstr thermodyn, Univ Toronto, 59-61; res & develop engr, John Inglis, Toronto, 61-62; asst prof heat transfer, NS Tech Col, 62-65; lectr thermodyn & heat transfer, Univ Sask, 65-68; MEM FAC MECH ENG, UNIV WESTERN ONT, 68- *Concurrent Pos:* Consult, G Graner & Assoc, Toronto, 58-64; NSF grants, Okla State Univ, 63 & Univ Calif, Los Angeles, 64. *Res:* Natural convection; natural and forced convection; thermodynamic properties of engineering fluids. *Mailing Add:* Dept Mech Engrs Engr & Math Sci Bldg Univ Western Ont London ON N6A 5B9 Can

TARASZKA, ANTHONY JOHN, b Wallington, NJ, Feb 19, 35; m 60. ANALYTICAL CHEMISTRY. *Educ:* Rutgers Univ, BS, 56, MS, 58; Univ Wis, PhD(pharmaceut chem), 62. *Prof Exp:* Res assoc analytical res & develop, 62-63, head dept, 63-66, mgr control res & develop, 66-70, dir control, 70-74, vpres control, 74-80, CORP VPRES CONTROL, UPJOHN CO, 80- *Mem:* Am Chem Soc; Am Pharmaceut Asn. *Res:* Trace component analysis; separation techniques; reaction mechanisms; pharmaceutical chemistry. *Mailing Add:* The Upjohn Co 7171 Portage Rd Kalamazoo MI 49001-0101

TARBELL, DEAN STANLEY, b Hancock, NH, Oct 19, 13; m 42; c 3. ORGANIC CHEMISTRY. *Educ:* Harvard Univ, AB, 34, MA, 35, PhD(org chem), 37. *Prof Exp:* Asst, Radcliffe Col, 36-37; fel, Univ Ill, 37-38; from instr to prof org chem, Univ Rochester, 38-60, Houghton prof chem, 60-66, chmn dept, 64-66; Harvie Branscom distinguished prof, 75-76, distinguished prof chem, 67-81, EMER PROF CHEM, VANDERBILT UNIV, 81- *Concurrent Pos:* Guggenheim fels, Oxford Univ, 46-47 & Stanford Univ, 61-62; Fuson lectr, Univ Nev, 72; mem, NIH study sects; consult, Walter Reed Army Inst Res, 72- *Honors & Awards:* C H Herty Medal, Am Chem Soc, 73, Dexter Award, Hist Chem Div, 89. *Mem:* Nat Acad Sci; Am Chem Soc; Am Acad Arts & Sci; The Chem Soc; Hist Sci Soc. *Res:* Reaction phenolic ethers; organic sulfur compounds; structure and synthesis of natural products; structural and theoretical organic chemistry; structure of antibiotics; history of organic chemistry in the United States. *Mailing Add:* Dept Chem Vanderbilt Univ Nashville TN 37235

TARBELL, THEODORE DEAN, b Rochester, NY, Nov 11, 50; m 73; c 1. SOLAR PHYSICS, SPACE INSTRUMENT DESIGN. *Educ:* Harvard Univ, AB, 71, Calif Inst Technol, PhD(astrophys), 76. *Prof Exp:* SR STAFF SCIENTIST ASTROPHYS, LOCKHEED PALO ALTO RES LAB, 76- *Concurrent Pos:* Co-investr on spacelab 2 mission, NASA Solar Optical Universal Polarimeter, 76-85, prin investr, Exp Sunlab Shuttle Mission, 85-, mem, Data Systems Users Working Group, 78-83, co-investr, Coord Instrument Package Orbiting Solar Lab, 80-, mem, Solar & Heliospheric Physics, 86-, mem, NSF Global Oscillations Network Group, 87-, mem, NAS Comt Solar & Space Physics, 87-, co-investr, Solar & Heliospheric Observ, 87-; prin investr, Air Force Geophys Lab Study Spectral Imaging, 80-85; mem, Max Steering Comt for Flare Res, 87- *Mem:* Am Astron Soc; Am Phys Soc; Astron Soc Pac; Optical Soc Am. *Res:* Concentration on the structure of the sun and stars, first through theoretical modeling and more recently through observations of magnetohydrodynamic processes in the solar atmosphere; using spacecraft and mountain-top observation. *Mailing Add:* Lockheed Res Lab 3251 Hanover St Dept 91-30 Bldg 256 Palo Alto CA 94304

TARBY, STEPHEN KENNETH, b Braddock, Pa, July 15, 34; m 58; c 2. METALLURGICAL ENGINEERING. *Educ:* Carnegie Inst Technol, BS, 56, MS, 59, PhD(metall eng), 62. *Prof Exp:* Asst prof metall eng, 61-62, from asst prof to assoc prof, 63-76, PROF METALL & MAT SCI, LEHIGH UNIV, 76- *Mem:* Am Inst Mining, Metall & Petrol Engrs; Metall Soc; Sigma Xi. *Res:* Kinetics of slag-metal reactions; thermodynamics of liquid metals and alloys; computer applications to metallurgical processes. *Mailing Add:* 2694 E Blvd Bethlehem PA 18017

TARBY, THEODORE JOHN, b Auburn, NY, May 9, 41; m 64; c 1. ANATOMY, NEUROPHYSIOLOGY. *Educ:* Calif Inst Technol, BS, 64; Univ Calif, Los Angeles, PhD(anat), 68. *Prof Exp:* Asst prof anat, Med Ctr, Univ Colo, Denver, 68-80; BARROW NEUROL INST, 80- *Concurrent Pos:* Milhelm Found Cancer Res grant, Med Ctr, Univ Colo, Denver, 71- *Mem:* AAAS; Am Asn Anat; Soc Neurosci. *Res:* Cerebral tissue impedance and extracellular space; blood-brain barrier; olfactory function in normal salmon; glial physiology. *Mailing Add:* Barrow Neurol Inst 350 W Thomas Rd Phoenix AZ 85013

TARDIF, HENRI PIERRE, MATERIALS TECHNOLOGY. *Educ:* Univ Laval, BASc; Carnegie Inst Technol, MSc; Univ Birmingham, Eng, PhD, 53. *Prof Exp:* Head, Mat Lab, Defense Res Bd Can, Valcartier, Que, 56-65, dep chief Can Defense Res Staff, London, Eng, 66-68, dir, Armaments Div, Valcartier, Que, 69-72, asst chief, 73-77, dep chief, 74-88, chief, 84-90; RETIRED. *Concurrent Pos:* Nat Defense Col, Kingston, Ont, 72-73; mem, Can Coun Non-Destructive Testing Comt Aeronaut Structures & Mat, Nat Res Coun. *Mem:* Fel Inst Metals Eng; fel Am Soc Metals; Can Inst Mining & Metall. *Res:* Application of materials technology to armaments; behavior of materials subjected to ballistic explosive and dynamic stresses; explosive forming of metals; dynamic properties of materials; determination of sabotage by explosives in airplane crashes; development of new alloys and their use in antiarmour munitions; more than 45 technical documents. *Mailing Add:* 1257 Jean Dequen Ave Ste-Foy PQ G1W 3H5 Can

TARDIFF, ROBERT G, b Lowell, Mass, Feb 1, 42; m 70; c 3. TOXICOLOGY, ENVIRONMENTAL HEALTH. *Educ:* Merrimack Col, BA, 64; Univ Chicago, PhD(toxicol & pharmacol), 68. *Prof Exp:* Res toxicologist org contaminants, USPHS, 68-70; br chief toxicol assessments, Environ Protection Agency, 70-77; EXEC DIR TOXICOL & ENVIRON HEALTH, NAT ACAD SCI, 77-; DIR, RISK FOCUS DIV, BURSAR INC. *Concurrent Pos:* USPHS fel, 64-68; assoc prof, Med Col Va, 79-; assoc prof, Georgetown Univ Sch Med, 87-90. *Mem:* Am Col Toxicol; Soc Toxicol; Environ Mutagen Soc; NY Acad Sci; Soc Risk Analysis. *Res:* Hazard assessment; interactions; metabolism; organic compounds in drinking water; mutagens; carcinogens; toxicology; pharmacology. *Mailing Add:* Risk Focus 6850 Versar Ctr Springfield VA 22151

TAREN, JAMES A, b Toledo, Ohio, Nov 10, 24. NEUROSURGERY. *Educ:* Univ Toledo, BS, 48; Univ Mich, MD, 52; Am Bd Neurol Surg, dipl, 60. *Prof Exp:* Intern surg, Univ Hosp, Univ Mich, 52-53, resident, 53-54; teaching fel neurosurg, Harvard Med Sch, 54-55; resident, 55-57, from instr to assoc prof, 57-69, dir Neurobehav Sci Prog, 75-78, assoc dean educ & student affairs, 78-87, PROF NEUROSURG, MED SCH, UNIV MICH, ANN ARBOR, 69-; DIR, MED CTR INFO TECHNOL INTEGRATION, 88- *Concurrent Pos:* Res fel, Boston Children's Hosp, 54-55; asst surg, Peter Bent Brigham Hosp, Boston, 54-55; actg chief neurosurg, Vet Admin Hosp, Ann Arbor, 58-73; chief neurosurg, Wayne County Gen Hosp, 58-72; vis fel, Karolinska Inst, Stockholm, 81; vis prof Hosp Foch, Paris, 66, 73 & 80, Gumma Univ, Japan, 89, 90. *Mem:* AAAS; Asn Am Med Cols; Cong Neurol Surg; NY Acad Sci; Soc Neurosci. *Res:* Central nervous system; stereotaxic neurosurgery. *Mailing Add:* Neurosurg Box 03038 1500 E Med Ctr Dr Ann Arbor MI 48109

TARESKI, VAL GERARD, b Bottineau, NDak, Dec 20, 41; m 66; c 3. COMPUTER ENGINEERING, COMPUTER NETWORKS. *Educ:* NDak State Univ, BS, 63, MS, 69; Univ Ill, Urbana, PhD(computer sci), 73. *Prof Exp:* Engr, Collins Radio, 62; tech writer, AC Electronics Div, Gen Motors Corp, 63; asst elec eng, NDak State Univ, 63-64, instr, 64-67; res asst computer sci, Univ Ill, Urbana, 67-69 & 70-71; asst prog dir, NSF, 71-72, assoc prog dir, 72-74; asst prof, 71-83, ASSOC PROF, NDAK STATE UNIV, 83- *Concurrent Pos:* Instr, Moorhead State Col, 67 & NDak State Univ, 67-71; assoc prof, Univ Nebr, 72. *Mem:* Inst Elec & Electronics Engrs; Asn Computer Mach; Sigma Xi; Soc Indust & Appl Math. *Res:* Computer architecture; microcomputer systems; computer communications and networks. *Mailing Add:* Elec Eng Dept NDak State Univ Fargo ND 58105

TARG, RUSSELL, b Chicago, Ill, Apr 11, 34; m 58; c 3. PHYSICS, PARAPSYCHOLOGY. *Educ:* Queens Col, NY, BS, 54. *Prof Exp:* Res asst physics, Columbia Univ, 54-56; engr, Sperry Gyroscope Co, 56-59; res assoc plasmas, Polytech Inst Brooklyn, 59; physicist, TRG, Inc, 59-62; eng specialist, Sylvania Elec Co, 62-72; sr res physicist, radio physics lab, Stanford Res Inst, 72-; PHYSICIST, DELPHI ASSOC. *Mem:* Am Phys Soc; Parapsychology Asn; Inst Elec & Electronics Engrs. *Res:* Extra sensory perception, electron beam-plasma interactions; gas laser research; laser detection; modulation and frequency control; research in extra sensory perception. *Mailing Add:* Lockheed Res & Dev Ctr Dept 9701 3251 Hanover St Palo Alto CA 94304

TARGETT, NANCY MCKEEVER, b Pittsburgh, Pa, Dec 23, 50; m 75; c 1. CHEMICAL ECOLOGY. *Educ:* Univ Pittsburgh, BS, 72; Univ Miami, MS, 75; Univ Maine, PhD(oceanog), 79. *Prof Exp:* Res assoc, 80-82, asst res prof, Skidaway Inst Oceanog, 82-84; asst prof, 84-88, ASSOC PROF, UNIV DEL, 88- *Concurrent Pos:* Vis prof, Friday Harbor Lab, Univ Washington, 88. *Mem:* Am Chem Soc; Am Soc Limnol & Oceanog; Asn Women Sci; Int Soc Chem Ecol (secy, 87-90). *Res:* Chemical ecology, specifically the role of secondary metabolites in chemical-biological interactions among marine organisms; marine biofouling. *Mailing Add:* Univ Del Col Marine Studies Lewes DE 19958

TARGETT, TIMOTHY ERWIN, b Farmington, Maine, Aug 1, 50; m 75; c 1. FISH ECOLOGY, TROPHIC ENERGETICS. *Educ:* Univ Maine, BS, 72, PhD(zool), 79; Univ Miami, MS, 75. *Prof Exp:* asst res prof, Skidaway Inst Oceanog, 80-84; assoc scientist, 84-86, asst prof, 86-90, ASSOC PROF, COL MARINE STUDIES, UNIV DEL, 91- *Concurrent Pos:* Adj asst prof, Dept Zool, Univ Ga, 81-84; US rep fish ecol, Biomass Working Party, 84; comnr, Atlantic States Marine Fisheries Comn, 87- *Honors & Awards:* Antarctica Serv Medal, NSF, 80. *Mem:* Am Soc Ichthyologists & Herpetologists; Am Fisheries Soc; Estuarine Res Fedn. *Res:* Ecology of estuarine and coastal marine fishes; trophic biology of fishes (physiological ecology of feeding, digestion, assimilation and growth); energetics; food webs. *Mailing Add:* Col Marine Studies Univ Del Lewes DE 19958

TARGOWSKI, STANISLAW P, immunology, veterinary microbiology; deceased, see previous edition for last biography

TARJAN, ARMEN CHARLES, b Cambridge, Mass, Dec 10, 20; m 45; c 2. TERRESTRIAL & MARINE NEMATOLOGY. *Educ:* Rutgers Univ, BS, 47; Univ Md, MS, 49, PhD(plant path), 51. *Prof Exp:* Asst nematologist, USDA, Md, 50-51; asst res prof plant path, Univ RI, 51-55; PROF NEMATOL, AGR RES & EDUC CTR, UNIV FLA, 55- & DEPT ENTOM-NEMATOL, 78- *Mem:* Soc Nematol; Europ Soc Nematol; Orgn Trop Am Nematol; Int Asn Meiobenthologists. *Res:* Biological control of plant nematodes; marine nematology; taxonomy and systematics. *Mailing Add:* Dept Entom Univ Fla Gainesville FL 32611

TARJAN, GEORGE, b Zsolna, Hungary, June 18, 12; US citizen; m 41; c 2. PSYCHIATRY. *Educ:* Pazmany Peter Univ, Hungary, MD, 35. *Prof Exp:* Resident physician, Mercy Hosp, Janesville, Wis, 40-41; asst physician, Utah State Hosp, Provo, 41-43; dir clin serv, Peoria State Hosp, Ill, 46-47; dir clin serv, Pac State Hosp, Pomona, Calif, 47-49, supt & med dir, 49-65; from asst clin prof to clin prof, Sch Pub Health, 53-65, prof psychiat, 65-76, prog dir ment retardation & child psychiat, Neuropsychiat Inst, 65-82, prof psychiat, Sch Med & Pub Health, 65-79, EMER PROF PSYCHIAT, SCH MED, UNIV CALIF, LOS ANGELES, 79- *Concurrent Pos:* Mem, President's Panel Ment Retardation, 61-62 & President's Comt, 66-71; consult, President's Comt, 71-83, President's Comn Ment Health, 77-78. *Honors & Awards:* Seymour Vestermark Award & Agnes Purcell McGavin Award, Am Psychiat Asn, 78. *Mem:* Am Psychiat Asn (pres, 83-84); Am Asn Ment Deficiency (pres, 58-59); AMA; Am Acad Child Psychiat (pres, 77-79); Group Advan Psychiat (pres, 71-73). *Res:* Child psychiatry; mental retardation. *Mailing Add:* Neuropsychiat Inst 760 Westwood Plaza Los Angeles CA 90024

TARJAN, ROBERT ENDRE, b Pomona, Calif, Apr 30, 48; m 78; c 3. COMPUTER SCIENCE. *Educ:* Calif Inst Technol, BS, 69; Stanford Univ, MS, 71, PhD(computer sci), 72. *Prof Exp:* Asst prof, Cornell Univ, 72-74; asst prof, 74-77, assoc prof computer sci, Stanford Univ, 77-80; mem tech staff, AT&T Bell Labs, 80-90; JAMES S MCDONNELL DISTINGUISHED UNIV PROF, DEPT COMPUTER SCI, PRINCETON UNIV, 85- *Concurrent Pos:* Miller fel, Univ Calif, Berkeley, 73-75; Guggenheim fel, Stanford Univ, 78-79; ed, numerous jours, 85-; mem comt, Math Sci: Status & Future Directions, Nat Res Coun, 89-90; mem, Computer Sci & Eng Peer Comt, Nat Acad Eng, 89-92; corresp, Math Intelligencer, 91- *Honors & Awards:* Nevanlinna Prize Info Sci, 83; Nat Acad Sci Award, Initiatives in Res, 84; A M Turing Award, 86. *Mem:* Nat Acad Sci; Nat Acad Eng; Soc Indust & Appl Math; Asn Comput Mach; Math Asn Am; Sigma Xi; Am Acad Arts & Sci; fel AAAS; Am Philos Soc. *Res:* Analysis of algorithms; computational complexity; combinatorics; data structures. *Mailing Add:* Dept Computer Sci Princeton Univ 35 Olden St Princeton NJ 08544-2087

TARKOY, PETER J, b Budapest, Hungary, Nov 13, 41; m 83. GEOLOGY. *Educ:* City Col NY, BS, 64; Univ Tenn, Knoxville, MS, 67; Univ, Urbana-Champaign, PhD(civil eng), 75. *Prof Exp:* GEOTECH & CONSTRUCTION CONSULT, 71- *Concurrent Pos:* Eng geologist, Ill Geol Surv, Champaign, 70-71, teaching & res asst, Univ Ill, Urbana-Champaign, 71-75; geotech consult, Perini Corp, Framingham, Mass, 75-78; vis assoc prof, Boston Univ, 79, Cornell Univ, 78; lectr, Univ Wis Exten, Milwaukee, 75- *Mem:* Am Soc Civil Engrs; Geol Soc Am; Int Asn Geologist; Asn Eng Geologists; Brit Tunnelling Soc; Tunnel Asn Can; Am Inst Mining Engrs. *Res:* Author of over 35 articles. *Mailing Add:* 102 N Main St Sherborn MA 01770

TARLE, GREGORY, b New York, NY, June 13, 51; m 72. PARTICLE ASTROPHYSICS, NON-ACCELERATOR HIGH ENERGY PHYSICS. *Educ:* Calif Inst Technol, BS, 72; Univ Calif, Berkeley, PhD(physics), 78. *Prof Exp:* asst prof, 83-87, ASSOC PROF PHYSICS, UNIV MICH, 88-; ASST RES PHYSICIST, SPACE SCI LAB, UNIV CALIF, BERKELEY, 78- *Mem:* Am Phys Soc. *Res:* Composition of cosmic radiation; underground astrophysics. *Mailing Add:* Randall Lab Univ Mich Ann Arbor MI 48109

TARLETON, GADSON JACK, JR, b Sumter, SC, Apr 29, 20; m 49; c 2. RADIOLOGY. *Educ:* Morris Col, AB, 39; Meharry Med Col, MD, 44; Am Bd Radiol, dipl, 49. *Prof Exp:* Resident radiol & orthop, Hubbard Hosp & Meharry Med Col, 44, resident radiol, 45-48; assoc prof, Meharry Med Col, 49-52, chmn dept & dir tumor clin, 49-78, prof radiol, 52-78; asst chief, dept radiol, Vet Admin Med Ctr, Nashville, 79-83; RADIOL, YORK VET ADMIN MED CTR, MURFREESBORO, TENN, 83- *Concurrent Pos:* Fel radiother, Bellevue Hosp, NY, 48-49; vis scholar, Columbia Presby Hosp, NY, 49; guest examr, Am Bd Radiol, 71-76 & 78-87. *Mem:* Radiol Soc NAm; fel Am Col Radiol. *Mailing Add:* 1714 Windover Dr Nashville TN 37208

TARLOV, ALVIN RICHARD, b South Norwalk, Conn, July 11, 29; m 56; c 5. INTERNAL MEDICINE, BIOCHEMISTRY. *Educ:* Dartmouth Col, BA, 51; Univ Chicago, MD, 56. *Prof Exp:* Intern, Philadelphia Gen Hosp, 56-57; resident med, Univ Chicago, 57-58, res assoc, 58-62; res assoc biochem, Harvard Med Sch, 62-64; from asst prof to prof med, Univ Chicago, 64- 85, chmn dept, 70-81, head, gen internal med, 81-83; pres, Henry J Kaiser Family Found, 84-90; SR SCIENTIST, HEALTH INST, NEW ENG MED CTR, 90-; PROF MED, TUFTS UNIV, 90- *Concurrent Pos:* USPHS res career develop award, 62-69; Markle scholar, 66-71; counr, Fed Coun Internal Med, 75-78, dir, Nat Study Internal Med Manpower Needs, 76-80, chmn, 78-79; chmn, Task Force Manpower Needs Internal Med, Asn Professors Med, 76-80, Grad Med Educ Nat Adv Comt, Health Resources Admin, US Dept Health & Human Serv, 77-81 & Comt Nat Agenda Prev Dis, Inst Med-Nat Acad Sci, 89-; mem, Comn Human Resources, Nat Res Coun, Nat Acad Sci, 80-82, Comt Plan Study Med Educ, Inst Med, 82-84, Comt Health Policy Agenda Am People, AMA, 83-86 & Res & Educ Adv Panel, US Gen Acct Off, 87-; prof health prom, Sch Pub Health, Harvard Univ, 90- *Honors & Awards:* Alan Gregg Lectr, Asn Am Med Cols, 87. *Mem:* Inst Med-Nat Acad Sci; master Am Col Physicians; Am Fedn Clin Res; Am Soc Hemat; Sigma Xi; Cent Soc Clin Res; Asn Am Physicians; Asn Prof Med (secy-treas, 75-78, pres-elect, 78-80, pres, 79-80). *Res:* Metabolism of red blood cells; inherited disorders of red cell metabolism; biochemistry of red cell membranes; health manpower, supply, productivity, needs. *Mailing Add:* Health Inst New Eng Med Ctr Box 345 750 Washington St Boston MA 02111

TARN, TZYH-JONG, b Szechwan, China, Nov 16, 37; m 67; c 2. MATHEMATICAL SYSTEMS THEORY. *Educ:* Cheng Kung Univ, Taiwan, BSc, 59; Stevens Inst Technol, MEng, 65; Washington Univ, DSc(control systs), 68. *Prof Exp:* Res assoc control theory, 68-69, asst prof chem eng, 69-72, assoc prof control systs, 72-77, PROF CONTROL SYSTS, WASHINGTON UNIV, 77-, DIR, CTR ROBOTICS & AUTOMATION, 87- *Concurrent Pos:* NSF grant, 70-85; assoc ed, Trans on Automatic Control, Inst Elec & Electronics Engrs, 86-89, tech ed, J Robotics & Automation, 88-; vpres, Inst Elec & Electronics Engrs Robotics & Automation Soc, 89-90. *Mem:* Soc Indust & Appl Math; Inst Elec & Electronics Engrs. *Res:* Control theory; stochastic systems; process optimization; robot arm control. *Mailing Add:* Campus Box 1040 Washington Univ St Louis MO 63130-4899

TARNEY, ROBERT EDWARD, b Hammond, Ind, Jan 8, 31; m 66; c 3. ORGANIC CHEMISTRY. *Educ:* Purdue Univ, BS, 52; Univ Wis, PhD, 58. *Prof Exp:* RES CHEMIST, E I DU PONT DE NEMOURS & CO, INC, 57- *Res:* Synthesis of monomers for and polymers of elastomeric materials. *Mailing Add:* 505 Summit Dr Hockessin DE 19707

TARONE, ROBERT ERNEST, b Modesto, Calif, Sept 11, 46; m 76; c 2. SURVIVAL ANALYSIS, CATEGORICAL DATA ANALYSIS. *Educ:* Univ Calif, Davis, BS, 68, MS, 69, PhD(math), 74. *Prof Exp:* MATH STATISTICIAN, NAT CANCER INST, 74- *Concurrent Pos:* Assoc ed, Am Statist Asn, 78-83. *Mem:* Fel Am Statist Asn; Biomet Soc. *Res:* Analysis methods for censored survival data; categorical data analysis; empirical bayes methods for frequency data; multiple comparisons problems; methods for analyzing epidemiologic studies. *Mailing Add:* Nat Cancer Inst EPN Bldg Rm 4070 Bethesda MD 20205

TARPLEY, ANDERSON RAY, JR, b New Orleans, La, Sept 19, 44; m 69. PHYSICAL CHEMISTRY, SPECTROSCOPY. *Educ:* Ga Inst Technol, BS, 66; Emory Univ, MS, 70, PhD(phys chem), 71; Ga State Univ, MBA, 72. *Prof Exp:* Res chemist, Eastman Kodak Co, NY, 66-67; instr res, Emory Univ, 71-72; SR CHEMIST ANALYTICAL SERV, TENN EASTMAN CO, EASTMAN KODAK CO, 72- *Concurrent Pos:* NIH fel med chem, Emory Univ, 71. *Mem:* AAAS; Am Chem Soc; Am Inst Chem; Am Mgt Asn. *Res:* Nuclear magnetic resonance spectroscopy; technical management; uses of computers in science; molecular orbital calculations; mass spectrometry. *Mailing Add:* 4520 Preston Dr Kingsport TN 37664-2136

TARPLEY, JERALD DAN, b Lubbock, Tex, July 13, 42; m 65; c 2. ATMOSPHERIC PHYSICS, REMOTE SENSING. *Educ:* Tex Technol Col, BS, 64; Univ Colo, PhD(astrogeophys), 69. *Prof Exp:* Advan Study Prog fel, Nat Ctr Atmospheric Res, 69-70, physicist, Environ Res Labs, 70-73, PHYSICIST, SATELLITE RES LAB, NAT ENVIRON SATELLITE DATA & INFO SERV, NAT OCEANIC & ATMOSPHERIC ADMIN, 73- *Mem:* Am Geophys Union; Am Meteorol Soc. *Res:* Remote sensing of the environment; atmospheric radiation. *Mailing Add:* Nat Environ Satellite Data & Info Serv Nat Oceanic & Atmospheric Admin Washington DC 20233

TARPLEY, WALLACE ARMELL, b Norwood, Ga, Feb 13, 34; m 59; c 2. INSECT ECOLOGY. *Educ:* Univ Ga, BSEd, 54, PhD(zool), 67; Clemson Univ, MS, 56. *Prof Exp:* Asst prof entom, Va Polytech Inst & State Univ, 60-64; ASSOC PROF BIOL, E TENN STATE UNIV, 64- *Mem:* AAAS; Entom Soc Am; Ecol Soc Am. *Res:* Ecological terminology, specifically the preparation of an ecological glossary; history of ecological terms; ecology of fresh water insects. *Mailing Add:* Dept Biol S427 E Tenn State Johnson City TN 37614

TARQUIN, ANTHONY JOSEPH, b Follansbee, WVa, July 10, 41; m 65; c 2. SANITARY ENGINEERING, ENVIRONMENTAL ENGINEERING. *Educ:* WVa Univ, BSIE, 64, MSE, 65, PhD(environ eng), 69. *Prof Exp:* Asst prof environ eng, Univ Tex, 69-73, asst dean eng, 76-79 & 87-89, assoc prof, 73-91, PROF CIVIL ENG, UNIV TEX, EL PASO, 91- *Concurrent Pos:* NSF & Environ Protection Agency grants. *Mem:* Water Pollution Control Fedn; Am Soc Civil Engrs. *Res:* Combined disposal of liquid and solid wastes; land disposal of wastewater and sludges; water treatment. *Mailing Add:* Dept Civil Eng Univ Tex El Paso TX 79968-0516

TARR, CHARLES EDWIN, b Johnstown, Pa, Jan 14, 40; m 77. PHYSICS. *Educ:* Univ NC, Chapel Hill, BS, 61, PhD(physics), 66. *Prof Exp:* Asst physics, Univ NC, Chapel Hill, 62-66, res assoc, summer 66 & Univ Pittsburgh, 66-68; from asst prof to assoc prof physics, 68-78, assoc dean, Col Arts & Sci, 79-81, actg dean, grad sch, 81-87, actg vpres res, 84-87, PROF PHYSICS, UNIV MAINE, ORONO, 78-, DEAN GRAD SCH, 87- *Mem:* Am Phys Soc; Inst Elec & Electronics Engrs. *Res:* Nuclear magnetic resonance; electron paramagnetic resonance; electronic instrumentation. *Mailing Add:* Off Dean Grad Sch Two Winslow Hall Univ Maine Orono ME 04469

TARR, DONALD ARTHUR, b Norfolk, Nebr, Aug 1, 32; m 57; c 2. INORGANIC CHEMISTRY. *Educ:* Doane Col, AB, 54; Yale Univ, MS, 56, PhD(chem), 59. *Prof Exp:* From instr to asst prof chem, Col Wooster, 58-65; asst prof, 65-67, ASSOC PROF CHEM, ST OLAF COL, 67- *Concurrent Pos:* Danforth teaching fel, 58-; res assoc, Univ Colo, 64-65; vis res fel, Univ Kent, 72-73. *Mem:* Fed Am Scientists; Am Asn Univ Prof; AAAS; Am Chem Soc. *Res:* Structure and stability of metal complexes; kinetics and mechanism of reaction of coordination compounds. *Mailing Add:* Dept Chem St Olaf Col Northfield MN 55057

TARR, JOEL ARTHUR, b Jersey City, NJ, May 8, 34. HISTORY OF TECHNOLOGY. *Educ:* Rutgers Univ, BS, 56, MA, 57; Northwestern Univ, PhD(hist), 63. *Prof Exp:* Asst prof hist, Calif State Univ, Long Beach, 61-66; vis asst prof, Univ Calif, Santa Barbara, 66-67; prof hist technol, 67-90, RICHARD S CALIGUIRI PROF URBAN STUDIES, CARNEGIE-MELLON UNIV, 90- *Concurrent Pos:* Consult pollution hist; consult, Infrastructure hist. *Honors & Awards:* Abel Wolman Award, 89. *Mem:* Soc Hist Technol; Sigma Xi; Orgn Am Historians. *Res:* The city and technology, interaction of the processes of urbanization and technological innovation; history of environmental pollution; wastewater systems; communication systems; transportation systems; energy systems; retrospective technology assessment. *Mailing Add:* Dept Hist Schenley Park Pittsburgh PA 15213

TARR, MELINDA JEAN, b San Francisco, Calif, Aug 18, 48; m 85. IMMUNOPHARMACOLOGY, COMPARATIVE PATHOLOGY. *Educ:* Univ Calif, Davis, BS, 71, DVM, 73; Ohio State Univ, MS, 76, PhD(vet pathobiol), 79; Am Col Vet Path, dipl, 83. *Prof Exp:* Grad res assoc equine med, Ohio State Univ, 74-76, fel vet path, 76-79, res assoc, 79-81, asst prof, 81-86, assoc prof vet path, 86-91, EMER PROF VET PATH, OHIO STATE UNIV, 91- *Concurrent Pos:* Prin investr, Air Force Off Sci Res, Dept Defense, 86 - *Mem:* Am Col Vet Path; Int Acad Pathologists; Int Asn Immunopharmacol; Inter-Am Soc Chemother; Am Vet Med Asn; Am Asn Vet Immunologists. *Res:* Mechanisms of immunotoxicology and immunomodulation; pathogenesis of familial amyloidosis; reproductive and hepatic diseases of captive cheetah; retrovirology. *Mailing Add:* Dept Vet Pathobiol Ohio State Univ 1925 Coffey Rd Columbus OH 43210

TARRANT, PAUL, b Birmingham, Ala, Nov 1, 14; m 37, 72; c 3. CHEMISTRY. *Educ:* Howard Col, BS, 36; Purdue Univ, MS, 38; Duke Univ, PhD(chem), 44. *Prof Exp:* Instr, Ala Bd Educ, 38-40; chemist, Shell Develop Co, Calif, 40-41; asst, Duke Univ, 41-44; res chemist, Am Cyanamid Co, Conn, 44-46; from instr to assoc prof, 46-57, PROF CHEM, UNIV FLA, 57- *Concurrent Pos:* Chief investr, Off Naval Res Proj, 47-50 & 53-56; dir, Off Qm Gen Res Proj, 51-67, US Air Force Res Proj, 54-66 & 72-74, NSF, 68-71, NASA, 69-71 & Mass Inst Technol, 69-71; dir res, PCR, 53-66, vpres, 66-68; consult, Redstone Arsenal, US Dept Army & Naval Ord Lab, Calif; adv, Cotton Chem Lab, USDA; ed, Fluorine Chem Rev. *Mem:* Am Chem Soc. *Res:* Preparations and reactions of fluorine containing organic compounds and inert polymers. *Mailing Add:* Dept Chem Univ Fla Gainesville FL 32601

TARRANTS, WILLIAM EUGENE, b Liberty, Mo, Dec 9, 27; div; c 2. SAFETY ENGINEERING, HUMAN FACTORS ENGINEERING. *Educ:* Ohio State Univ, BIE, 51, MSc, 59; NY Univ, PhD(indust safety), 63. *Prof Exp:* Chief, Ground Safety Div, US Air Force, 51-57; instr indust eng, Ohio State Univ, 58-59; instr & res assoc indust safety, NY Univ, 59-63, asst prof, 63-64; chief div accident res, Bur Labor Statist, US Dept Labor, 65-67; chief manpower develop div, 67-80, chief scientist, Off Prog & Demonstration Eval, 80-84, prog analyst, off occupant protection, hwy safety, US Dept Transp, 84-87, PROG ANALYST, EVAL STAFF, HWY SAFETY, US DEPT TRANSP, 87- *Concurrent Pos:* Consult evaluation res, Indust Comn Ohio, 59; mem safety standards bd, Am Nat Standards Inst, 67-69; mem res proj comt, Nat Safety Coun, 73-76; chmn sci & tech info adv bd, Nat Hwy Traffic Safety Admin, 73-; mem comt planning & admin transp safety, Nat Acad Sci, 74-; ed-in-chief, Traffic Safety Eval Res Rev, 83-; instr, Occup Safety & Health Mgt, Johns Hopkins Univ, Baltimore, Md, 84- *Mem:* AAAS; fel Am Soc Safety Engrs (pres, 77-78); sr mem Am Inst Indust Engrs; Human Factors Soc; Eval Res Soc; Soc Risk Analysis. *Res:* Measurement of safety performance; accident causation; psychological factors in accidents; risk acceptance; highway traffic safety; evaluation of safety programs. *Mailing Add:* 606 Woodsmans Way Crownsville MD 21032

TARSHIS, IRVIN BARRY, b Portland, Ore, May 12, 14; m 44; c 2. PARASITOLOGY, ZOOLOGY. *Educ:* Ore State Col, BA, 38; Univ Calif, Berkeley, PhD(parasitol), 53. *Prof Exp:* Res scientist, State Dept Fish & Game, Calif, 48-51; agr res scientist, USDA, Ga, 53-54; sect dep chief & med entomologist, Ft Detrick, Md, 54-57; asst prof entom, Univ Calif, Los Angeles, 57-63; res parasitologist, Patuxent Wildlife Res Ctr, 64-84, res zoologist, 76-84; RETIRED. *Concurrent Pos:* Scripps res fel, Zool Soc San Diego, 60-61; consult, Israel, 60-61; del, Int Cong Entom, Montreal, Can, 56, Vienna, Austria, 60, Canberra, Australia, 72; del Int Cong Trop Med & Malaria, Rio de Janeiro, Brazil, 63; production ceramicist, 77-; sci adv panel, World Health Asn, Volta River Basin Area, West Africa, 65-70. *Honors & Awards:* Gorgas Mem Award, 34; Animal Care Panel Res Award, 63. *Mem:* Fel AAAS; Sigma Xi; Entom Soc Am. *Res:* Biology and epidemiology of arthropods affecting man and animal; host-parasite relationships; transmission of arthropod-borne diseases to wildlife; biology of Simuliidae; membrane feeding haematophagous arthropods; uptake and depuration of petroleum hydrocarbons by aquatic organisms; accumulation and elimination of heavy metals by terrestrial arthropods; utilization of organic chemicals in the production of high fire ceramics. *Mailing Add:* 17219 Emerson Dr Silver Spring MD 20905

TARTAGLIA, PAUL EDWARD, b New York, NY, Sept 30, 44; div; c 2. ENGINEERING. *Educ:* Univ Detroit, BME, 67, DEng, 70; Northwestern Univ, Evanston, MSME, 68. *Prof Exp:* Engr, Space Div, Chrysler Corp, La, 64-65; test engr, Elec Boat Div, Gen Dynamics Corp, Conn, 65; proj engr, Eng Off, Chrysler Corp, Mich, 69-70; asst prof mech eng, Univ Mass, Amherst, 70-75; proj engr, Rodney Hunt Co, 75-76; chief engr, Computerized Biomech Anal, Inc, 76-77; assoc prof & head mech eng dept, 77-86, HEAD, ENG & TECHNOL DIV, NORWICH UNIV, 86- *Concurrent Pos:* NSF, Norton Co & Kollmorgen, Inc grants, Univ Mass, Amherst, 72-73. *Mem:* Am Soc Mech Engrs; Am Soc Eng Educ. *Res:* Mechanical engineering; transportation; engineering systems design; automatic control systems; sports equipment design and analysis; biomechanics of sports. *Mailing Add:* Div Eng & Technol Norwich Univ Northfield VT 05663

TARTAR, VANCE, b Corvallis, Ore, Sept 15, 11; m 50; c 3. EXPERIMENTAL MORPHOLOGY. *Educ:* Univ Wash, Seattle, BS, 33, MS, 34; Yale Univ, PhD(zool), 38. *Prof Exp:* Seessel fel, Yale Univ, 39; instr zool, Univ Vt, 39-42 & Yale Univ, 42; asst aquatic biologist, US Fish & Wildlife Serv, 42; biologist, Wash State Dept Fisheries, 43-50; res assoc prof, 51-61, RES PROF ZOOL, UNIV WASH, 61- *Mem:* Am Soc Zool; hon mem Soc Protozool; Soc Develop Biol. *Res:* Oyster and cell biology; experimental morphogenesis in ciliates; protozoology; cell biology; the biology of Stentor. *Mailing Add:* Univ Wash Field Lab Ocean Park WA 98640

TARTER, CURTIS BRUCE, b Louisville, Ky, Sept 26, 39; m 87; c 1. ASTROPHYSICS, THEORETICAL PHYSICS. *Educ:* Mass Inst Technol, SB, 61; Cornell Univ, PhD(astrophys), 67. *Prof Exp:* Sr scientist, Aeronutronic Div, Philco-Ford Corp, 67; physicist, 67-69, group leader, 69-73, dep div leader, 73-78, div leader theoret physics, 78-84, dep assoc dir, 84-88, ASSOC DIR PHYSICS, LAWRENCE LIVERMORE NAT LAB, 88- *Concurrent Pos:* Lectr, Dept appl sci, Univ Calif, Davis, 71- & Army Sci Bd, 89- *Mem:* Int Astron Union; Am Phys Soc; Am Astron Soc. *Res:* Theoretical description of the properties of matter at high temperatures and densities; theoretical astrophysics, particularly quasars, X-ray sources and stellar evolution. *Mailing Add:* Lawrence Livermore Lab PO Box 808 L-295 Livermore CA 94550

TARTER, DONALD CAIN, b Somerset, Ky, July 22, 36; m 60; c 3. ZOOLOGY. *Educ:* Georgetown Col, Ky, BS, 58; Miami Univ, Ohio, MAT, 62; Univ Louisville, PhD(zool), 68. *Prof Exp:* Teacher chem biol, Bradford High Sch, Ohio, 58-60 & Tipp City High Sch, Ohio, 60-64; teacher biol, Ky Southern Col, 64-68; instr zool, 68-81, ASSOC PROF BIOL, MARSHALL UNIV, 81- *Mem:* Am Fisheries Soc; Am Entom Soc; Sigma Xi; Am Soc Ichthyologists & Herpetologists. *Res:* Taxonomy and ecology of fishes and aquatic insects. *Mailing Add:* Dept Biol Sci Marshall Univ Huntington WV 25701

TARTER, MICHAEL E, b New York, NY, Dec 20, 38; m 61; c 2. BIOSTATISTICS. *Educ:* Univ Calif, Los Angeles, AB, 59, MA, 61, PhD(biostatist), 64. *Prof Exp:* From asst prof to assoc prof biostatist, Univ Mich, Ann Arbor, 64-67; assoc prof, 70-76, PROF, UNIV CALIF, BERKELEY, 77- *Concurrent Pos:* Consult, Upjohn Drug Co, Med Diag Corp, Regional Med Asn, Pac Med Ctr & Presch & Adolescent Proj; Calif State Health Dept, NIH Cancer & Heart, Lung & Blood Insts; FDA Comm on Neuropsyciatic Drugs; assoc prof, 68-78, prof biostat, Depth Med & Math, Univ Calif, Irvine, 78- *Mem:* Fel Am Statist Asn; Asn Comput Mach; Int Statist Inst. *Res:* Graphical biometry; computational aspects of statistical procedures; biostatistical consultation training; nonparametric density estimation; programmed and computer assisted instruction; sorting theory. *Mailing Add:* Dept Biostatist Univ Calif Berkeley CA 94720

TARTOF, DAVID, b Detroit, Mich, Sept 15, 45. HUMAN CELLULAR IMMUNOLOGY. *Educ:* Univ Mich, MD, 70; Univ Chicago, PhD(med), 78. *Prof Exp:* ASST PROF MED, UNIV CHICAGO, 80- *Concurrent Pos:* Attend physician, Michael Reese Med Ctr. *Mem:* Sigma Xi; Am Asn Immunol. *Mailing Add:* 5511 S Kimbark Chicago IL 60637-1618

TARTOF, KENNETH D, b Detroit, Mich, Dec 30, 41; m 67. GENETICS. *Educ:* Univ Mich, BS, 63, PhD(genetics), 68. *Prof Exp:* NIH res fel, 68-70, res assoc, 70-71, asst mem, 71-76, MEM, INST CANCER RES, 76- *Mem:* AAAS; Genetics Soc Am. *Res:* Structure and function of genes; genetic control of gene redundancy; regulation of DNA and RNA metabolism. *Mailing Add:* Inst Cancer Res Rm 251 7701 Burholme Ave Philadelphia PA 19111

TARTT, THOMAS EDWARD, b Martin, TN, July 3, 40; m 23; c 2. RESEARCH ADMINISTRATION. *Educ:* Auburn Univ, BS, 64. *Prof Exp:* Indust engr, 64-71, staff accountant, 71-73, asst controller, 73-77, DIR INDUST ENGR, ECUSTA DIV, TECH DEPT, PH GLATFELTER CO, 77- *Concurrent Pos:* Dir , Inst Indust engrs, 74-76. *Mem:* Inst Indust Engrs. *Mailing Add:* Tech Dept P H Glatfelter Co PO Box 200 Pisgah Forest NC 28768

TARUI, YASUO, b Tokyo, Japan, June 4, 29; m 57; c 2. SEMICONDUCTOR DEVICES, SEMICONDUCTOR MATERIAL. *Educ:* Waseda Univ, BE, 51. *Hon Degrees:* Dr, Univ Tokyo, 65. *Prof Exp:* Res staff mem, Electrotech Lab, 51-65, chief sect, 65-76; dir, VLSI Coop Labs, 76-80; PROF SEMICONDUCTOR DEVICE, TOKYO UNIV AGR & TECHNOL, 81- *Concurrent Pos:* Res assoc, Stanford Univ, 62-63. *Mem:* Fel Inst Elec & Electronics Engrs. *Res:* Schottky TTL; diffusion self-aligned MOS transistor; electrically erasable non-volatile stacked-gate memory. *Mailing Add:* Tokyo Univ Agr & Technol 2-24-16 Nakamachi Koganei Tokyo 184 Japan

TARVER, FRED RUSSELL, JR, b Knoxville, Tenn, Mar 7, 25; m 50; c 3. FOOD TECHNOLOGY, BACTERIOLOGY. *Educ:* Univ Tenn, BSA, 50, MA, 54; Univ Ga, PhD(food technol, bact), 63. *Prof Exp:* Mem staff, Security Mills, Inc, Tenn, 50-54; instr poultry, Univ Tenn, 54-56; asst prof, Univ Fla, 56-60, 62-63; exten assoc prof, 63-75, EXTEN PROF FOOD SCI, NC STATE UNIV, 75. *Mem:* Inst Food Technol; Poultry Sci Asn; World Poultry Sci Asn; fel Inst Food Technologists. *Res:* Product development and marketing of new poultry and egg products; sanitation in processing plants, with emphasis on ecology; 4-H club activities connected with poultry and egg products. *Mailing Add:* 5501 Parkwood Dr Brookhaven Raleigh NC 27612

TARVER, HAROLD, b Wigan, Eng, June 7, 08; nat US; c 2. BIOCHEMISTRY, MEDICAL & HEALTH SCIENCES. *Educ:* Univ Alta, BS, 32, MSc, 35; Univ Calif, PhD(biochem), 39. *Prof Exp:* Asst, Univ Alta, 32-35; asst, Sch Med, Univ Calif, Berkeley, 36-39, fel, 39-41; from instr to prof biochem, 41-75, EMER PROF BIOCHEM, SCH MED, UNIV CALIF, SAN FRANCISCO, 75- *Mem:* AAAS; Am Soc Biol Chem. *Res:* Metabolism of sulfur and protein-isotopic studies. *Mailing Add:* 1715 Wawona St San Francisco CA 94116

TARVER, JAMES H, JR, clinical drugs, for more information see previous edition

TARVER, MAE-GOODWIN, b Selma, Ala, Aug 9, 16. STATISTICS. *Educ:* Univ Ala, BS, 39, MS, 40. *Prof Exp:* Res engr, Continental Can Co, Inc, 41-48, proj engr statist, 48-54, res statistician & internal qual control consult res & develop dept, 54-77; PRIN CONSULT, QUEST ASSOCS, 78- *Concurrent Pos:* Instr, Ill Inst Technol, 57-62, adj assoc prof, 63- *Honors & Awards:* Lisy Award, Am Soc Qual Control, 61; Edward J Oakley Award, Am Soc Qual Control, 75, E L Grant Award, 83. *Mem:* Am Statist Asn; fel Am Soc Qual Control; sr mem Soc Women Engrs; Inst Food Technol. *Res:* Multivariate statistical analysis of industrial research data; statistical and information theory appraisal of sensory tests; quality engineering applied to metal fabrication, processing and packaging of food products; pulp, paper, glass and electronic assembly. *Mailing Add:* Quest Assocs 130 26th St Park Forest IL 60466

TARVIN, ROBERT FLOYD, b Montezuma, Iowa, Feb 20, 42; m 61; c 3. POLYMER APPLICATION, RESEARCH MANAGEMENT. *Educ:* Univ Northern Iowa, BA, 65; Univ Iowa, PhD(org chem), 72. *Prof Exp:* Res asst drug interaction, Univ Iowa, 65-67; NAS vis prof liquid crystalline polymers, Univ Rio de Janeiro, 72-74; proj leader water treating & mining chem, Am Cyanamid Co, 74-80, mgr paper chem, chem res div, 80-84; mgr res, 84-85, DIR RES, KIMBERLY-CLARK CORP, 85- *Mem:* Am Chem Soc; Sigma Xi. *Res:* Organic polymer synthesis; water treating chemicals; slow release drug delivery; liquid crystalline polymers; paper chemicals, nonwovens, adhesives. *Mailing Add:* Kimberly-Clark Corp 518 E Water St Troy OH 45373

TARWATER, JAN DALTON, b Ft Worth, Tex, Sept 30, 37; m 58; c 3. ALGEBRA. *Educ:* Tex Tech Col, BS, 59; Univ NMex, MA, 61, PhD(math), 65. *Prof Exp:* Asst prof math, Western Mich Univ, 65-67 & NTex State Univ, 67-68; from asst prof to assoc prof, 68-73, assoc chmn dept, 72-73, chmn dept, 73-78, PROF MATH, TEX TECH UNIV, 73- *Mem:* Math Asn Am; Soc Indust & Appl Math. *Res:* Algebra, especially Abelian groups and homological algebra; graph theory; history of mathematics. *Mailing Add:* Dept Math Tex Tech Univ Lubbock TX 79409

TARWATER, OLIVER REED, b Chattanooga, Tenn, Mar 12, 44; m 66; c 2. AGRICHEMICAL FORMULATION SCIENCE. *Educ:* Maryville Col, BS, 66; Purdue Univ, Lafayette, MS, 69, PhD(med chem), 70. *Prof Exp:* Res chemist, Personal Care Div, Gillette Co, Chicago, 70-74; sr chemist, Southern Res Inst, Birmingham, 74-76; sr chemist, Lilly Res Labs, Greenfield, Ind, 76-80, res scientist, 81-87; MGR, PHARMACEUT PROJS, LILLY CORD CTR, INDIANAPOLIS, IND, 87- *Mem:* Am Chem Soc; Sigma Xi. *Mailing Add:* 331 Hickory Dr RR 7 Greenfield IN 46140

TARZWELL, CLARENCE MATTHEW, b Deckerville, Mich, Sept 29, 07; m 38; c 3. WATER POLLUTION. *Educ:* Univ Mich, AB, 30, MS, 32, PhD(aquatic biol, fisheries, Trout, Stream improvement) 36. *Hon Degrees:* ScD, Baldwin-Wallace Col, 67. *Prof Exp:* Stream improv supvr, Mich State Dept Conserv, 33-34; asst aquatic biologist, US Bur Fisheries, 34; asst range examr, US Forest Serv, NMex, 35, 36-38; from asst aquatic biologist to assoc aquatic biologist, Tenn Valley Authority, Ala, 38-43; asst & sr sanitarian, USPHS, Ga, 43-46, sr biologist, Tech Develop Div, 46-48, chief aquatic biol sect, Environ Health Ctr, Cincinnati, Ohio, 48-53, sci dir, Robert A Taft Sanit Eng Ctr, 53-65; dir, Nat Marine Water Qual Lab, Environ Protection Agency, RI, 65-72; sr res advr, Environ Protection Agency Nat Environ Res Ctr, Corvallis Ore, 72-75; ENVIRON CONSULT, 75- *Concurrent Pos:* Am deleg first & second int meetings on sci res in water pollution, Orgn Econ Coop & Develop, Paris France, 61-62, chmn sub-group III, 62-65; mem expert adv panel on environ health, WHO, Geneva, Switz, 62-65; men, Food & Agr Orgn Panel Fishery Experts UN, Rome, Italy; mem, Nat Acad Eng, Nat Acad Sci, Adv Comt Power Plant Siting, 71-72; mem, Nat Temperature Comt, Environ Protection Agency, 71-73. *Honors & Awards:* Aldo Leopold Medal, Wildlife Soc, 63; Bronze Medal, USPHS, 74. *Mem:* Sigma Xi; Am Soc Limnol & Oceanog; Soc Int Limnol; hon mem Am Fisheries Soc; Am Soc Ichthyologists & Herpetologists; Am Inst Biol Scientists; AAAS; Nat Malaria Soc; Am Soc Tropical Med; Water Pollution Control Fedn. *Res:* Water pollution control; determination of water quality criteria and standards for aquatic life; standard bioassay methods; environmental improvement and management; toxicity of wastes to aquatic life; aquatic biology and ecology; author of more than 120 scientific publications. *Mailing Add:* 380 Post Rd No B Wakefield RI 02879

TASAKI, ICHIJI, b Fukushim-Ken, Japan, Oct 21, 10; nat US; c 2. NEUROPHYSIOLOGY. *Educ:* Keio Univ, Japan, MD, 38. *Hon Degrees:* DS, Uppsala Univ, 72. *Prof Exp:* Privat-docent physiol, Keio Univ, 38-42, privat-docent physics & prof physiol, Med Col, 42-51; prof physiol, Nihon Univ, Tokyo, 51; res assoc, Cent Inst Deaf, St Louis, Mo, 51-53; chief spec senses sect, Lab Neurophysiol, Nat Inst Neurol Dis & Blindness, 53-61; mem staff, 61-66, CHIEF, LAB NEUROBIOL, NIMH, 66- *Concurrent Pos:* Mem, Marine Biol Lab, Woods Hole, Mass. *Mem:* Am Acad Neurol; Physiol Soc Japan. *Res:* Nerve and sense organs; electrophysiology. *Mailing Add:* NIMH Bldg 36 2B-16 Bethesda MD 20892

TASCH, AL FELIX, JR, b Corpus Christi, Tex, May 12, 41; m 63; c 2. ELECTRICAL ENGINEERING, SOLID STATE PHYSICS. *Educ:* Univ Tex, Austin, BS, 63; Univ Ill, Urbana, MS, 65, PhD(physics), 69. *Prof Exp:* RES SCIENTIST & ENGR, CENT RES LABS, TEX INSTRUMENTS INC, 69-; DIR RES & DEVELOP, MOTOROLA INC. *Concurrent Pos:* Prof elec eng, Univ Tex Austin. *Honors & Awards:* J J Ebers Award, Inst Elec & Electronics Engrs, 88. *Mem:* Electrochem Soc; Inst Elec & Electronics Engrs; Mat Res Soc. *Res:* Solid state device physics and silicon processing technology; process and device modeling; metal-oxide-silicon field effect transistors; charge-coupled devices. *Mailing Add:* Dept Elec Eng Univ Texas Austin 433 Ens Austin TX 78712

TASCH, PAUL, b New York, NY, Nov 28, 10. PALEOBIOLOGY, GEOLOGY. *Educ:* City Col New York, BS, 48; Pa State Col, MS, 50; Iowa State Univ, PhD(geol), 52. *Prof Exp:* Instr geol, Pa State Col, 48-49; instr, Univ Conn, 52-53; asst prof, NDak Agr Col, 53-54; assoc prof, Moorhead State Univ, 54-55; PROF GEOL, WICHITA STATE UNIV, 55-, DISTINGUISHED PROF NAT SCI, 77- *Concurrent Pos:* Chief investr earth sci grant, NSF, 56-73 & off polar progs, 66-76 & 77- *Honors & Awards:* Antarctic Serv Medal, US Cong, 70. *Mem:* Paleont Soc; fel Geol Soc Am; fel AAAS; fel Geol Soc India; fel Geol Soc London; Sigma Xi. *Res:* Non-marine fossil biotas of Antarctica and other Gondwana continents, especially conchostracans palynomorphs and spoor; geomicrobiology; evaporites; branchiopoda, fossil and living; history of science. *Mailing Add:* 1346 Parkwood Lane Wichita KS 67208

TASCHDJIAN, CLAIRE LOUISE, medicine, mycology, for more information see previous edition

TASCHEK, RICHARD FERDINAND, b Chicago, Ill, June 5, 15; m 42; c 4. PHYSICS. *Educ:* Univ Wis, BA, 36, PhD(physics), 41; Univ Fla, MS, 38. *Prof Exp:* Teaching asst, Univ Wis, 38-40; res physicist, Oldbury Electro-Chem Co, NY, 41-42; Nat Defense Res Comt physicist, Princeton Univ, 42-43; physicist, 43-62, div leader exp physics, 62-70, asst dir res, 71-72, ASSOC DIR RES, LOS ALAMOS NAT LAB, 72- *Concurrent Pos:* Mem nuclear cross sect adv group, AEC, 48-57, chmn, 53-57, mem tripartite nuclear cross sect comt, 56-61, chmn, 61; mem Euro-Am nuclear data comt, 57-72, chmn, 60-62; mem int nuclear data working group, 63-72; adv, Int Nuclear Data Comt, 63-73, ex-officio mem, 72-74; mem standing comt controlled thermonuclear reactions & ad hoc adv comt on Los Alamos meson proj; co-chmn, Vis Comt Lab Nuclear Sci, Mass Inst Technol, 67; mem adv comt neutron physics, Oak Ridge Nat Lab; mem Nuclear Physics Panel, Physics Surv Comt, Div Phys Sci, Nat Res Coun; mem Off Stand Ref Data Eval Panel, Nat Bur Stand, & Inst Basic Stand Eval Panel, 71-75; mem, Ctr Radiation Res, Nat Bur Stand Panel, 71-76, chmn, 72-; mem adv comt, Univ Alaska Geophys Inst, 73-76; mem, Numerical Data Adv Bd, US Nat Comt for CODATA, 73-, chmn, 74- *Mem:* Fel AAAS; fel Am Phys Soc; fel Am Nuclear Soc; Am Geophys Union; Sigma Xi. *Res:* Nuclear reactions and scattering; nuclear properties; neutron physics; accelerators and detectors; space physics; Vela satellite program; controlled thermonuclear reactions. *Mailing Add:* 2035 47th St Los Alamos NM 87544

TASCHNER, MICHAEL J, b Milwaukee, Wis, Sept 5, 53; m 77; c 2. ORGANIC CHEMISTRY. *Educ:* Univ Wis-Eau Claire, BSc, 76; Iowa State Univ, PhD(org chem), 80. *Prof Exp:* NIH fel chem, Univ Calif, Berkeley, 80-82; ASST PROF CHEM, UNIV AKRON, 82- *Mem:* Am Chem Soc. *Res:* The total synthesis of natural products which possess biological activity; development of new synthetic methodology. *Mailing Add:* Dept Chem Univ Akron Akron OH 44325-3601

TASHIAN, RICHARD EARL, b Cranston, RI, Oct 7, 22; m 68; c 1. MOLECULAR GENETICS. *Educ:* Univ RI, BS, 47; Purdue Univ, MS, 49, PhD(zool), 51. *Prof Exp:* Asst, Purdue Univ, 48-51; asst prof biol, Long Island Univ, 51-54, actg chmn dept, 54; sci assoc, Dept Trop Res, NY Zool Soc, 54-55; res assoc, Inst Study Human Variation, Columbia Univ, 56-57; from res assoc to assoc prof, 57-70, PROF HUMAN GENETICS, MED SCH, UNIV MICH, ANN ARBOR, 70- *Concurrent Pos:* Vis scientist, dept chem, Carlsberg Lab, Copenhagen, Denmark, 68-69, 79. *Mem:* Fel AAAS; Am Soc Human Genetics; Am Soc Biochem & Molecular Biol; Brit Biochem Soc; NY Acad Sci. *Res:* Structure-function relationships of carbonic anhydrase isozymes; organization, structure, expression, and evolution of carbonic anhydrase genes. *Mailing Add:* Dept Human Genetics 4708 MS II Box 0618 Univ Mich Med Sch Ann Arbor MI 48109-0618

TASHIRO, HARUO, b Selma, Calif, Mar 24, 17; m 42; c 3. ENTOMOLOGY. *Educ:* Wheaton Col, BS, 45; Cornell Univ, MS, 46, PhD(entom), 50. *Prof Exp:* Asst entom, Cornell Univ, 47-50; entomologist, USDA, 50-67; prof entom, Agr Exp Sta, Cornell Univ, 67-83; RETIRED. *Concurrent Pos:* Assoc prof, Cornell Univ, 58-63; res assoc, Univ Calif, Riverside, 63-67. *Mem:* Entom Soc Am. *Res:* Biology; biological and chemical control of turf insects; insects, ornamental plants, permanent plantings; insect pathology. *Mailing Add:* 31 Denton Ave Geneva NY 14456

TASHJIAN, ARMEN H, JR, b Cleveland, Ohio, May 2, 32; m 55; c 3. ENDOCRINOLOGY, CELL BIOLOGY. *Educ:* Yale Univ, 50-53; Harvard Univ, MD, 57. *Prof Exp:* Intern med, Harvard Serv, Boston City Hosp, 57-58, asst resident, 58-59; clin res assoc, Metab Dis Br, Nat Inst Arthritis & Metab Dis, 59-61; Nat Found res fel, 61-63, from instr to assoc prof, 63-70, prof pharmacol, Sch Dent Med & Sch Med, 70-78, PROF TOXICOL, SCH PUB HEALTH & PROF PHARMACOL, SCH MED, HARVARD UNIV, 78- *Concurrent Pos:* Assoc ed, Metab, 70-80 & Cancer Res, 80-84; mem cell biol study sect, div res grants, NIH, 73-77; mem res comt, Med Found Inc, 77-80; mem bd dirs, int confs calcium regulating hormones, 83-; Guggenheim fel, 84-85; vis prof biol chem, Hebrew Univ, Jerusalem, 84-85; vis fel commoner, Trinity Col, Cambridge Univ, 85. *Honors & Awards:* H B Van Dyke Mem Award & Lectr, Columbia Univ, 77; Edwin B Astwood Lectr, Endocrine Soc, 77. *Mem:* Fel AAAS; Am Soc Pharmacol & Exp Therapeut; Endocrine Soc; Am Soc Cell Biol; Tissue Cult Asn; Am Fedn Clin Res; fel NY Acad Sci; Int Soc Biochem Pharmacol; Am Soc Bone & Mineral Res; Soc Toxicol. *Res:* Molecular mechanisms of action of protein and peptide hormones; hormone receptor transduction mechanisms and regulation of cell calcium; cell biology; establishment, control of function and growth of differentiated, clonal strains of animal and human cells in culture; biochemical mechanisms of action of tumor promoters; toxicology. *Mailing Add:* Dept Physiol Harvard Sch Pub Health 665 Huntington Ave Boston MA 02115

TASHJIAN, ROBERT JOHN, b Worcester, Mass, Feb 4, 30. VETERINARY MEDICINE. *Educ:* Clark Univ, AB, 51; Univ Pa, VMD, 56. *Hon Degrees:* DSc, Hartwick Col, 68. *Prof Exp:* Staff vet, Animal Med Ctr, New York, 56-58, head med, 58-62, chief of staff, 62-63, dir, 63-74; PRES, NEW ENG INST COMP MED, W BOYLSTON, MASS, 74- *Concurrent Pos:* Res assoc, Vet Admin Hosp, Bronx, NY, 60-64; instr, Sch Vet Med, Univ Pa, 61-63; prin investr, Nat Heart Inst grant, 62-66, sponsor grant, 66-69; lectr animal sci & mem grad fac, Univ Maine, Orono, 67-, vis prof, 74-; affil prof, Colo State Univ, 67-; adj prof, Univ Mass, Amherst, 69-70; dir, Duke Farms, Somerville, NJ, 71-74; consult to bd trustees, Animal Med Ctr, New York, 74-; mem coun clin cardiol, Am Heart Asn. *Mem:* AAAS; Am Asn Lab Animal Sci; Am Asn Vet Clinicians; Am Vet Med Asn; Am Vet Radiol Soc. *Res:* Internal medicine, especially comparative cardiovascular disease. *Mailing Add:* 65 S Webster Haverhill MA 01830

TASI, JAMES, b New York, NY, Dec 6, 33; m 60; c 3. MECHANICS. *Educ:* NY Univ, BCE, 55; Univ Ill, MS, 56; Columbia Univ, PhD(mech), 62. *Prof Exp:* Engr, Martin Co, Md, 57-58, assoc res scientist, Colo, 61-65; fel mech, Johns Hopkins Univ, 65-66; assoc prof, 66-72, PROF MECH, STATE UNIV NY STONY BROOK, 72- *Mem:* Am Soc Mech Engrs; Sigma Xi. *Res:* Acoustic vibrations of structures; thermoelastic dissipation in crystalline solids; wave propagation; stability; mechanical properties of solids; shock response of crystal lattices. *Mailing Add:* Dept Mech Eng State Univ NY Stony Brook NY 11794-2300

TASKER, CLINTON WALDORF, b Syracuse, NY, Sept 14, 18; m 41; c 3. CHEMISTRY. *Educ:* Syracuse Univ, BSc, 41, MSc, 44; McGill Univ, PhD(cellulose chem), 47. *Prof Exp:* Res chemist, Sylvania Indust Corp, Va, 41-43; lab asst pulp & paper technol, State Univ NY Col Forestry, Syracuse Univ, 43-44; lab demonstr org & inorg chem, McGill Univ, 44-46; sr res chemist, Sylvania Div, Am Viscose Corp, 47-53, tech supt, 53-59, mgr supt, 59-61; dir tech res & develop, Tenneco Inc, 61-65, vpres res & develop, 65-68, vpres & gen mgr, Filer Mill, Paperboard Div, 68-73, vpres corp res & develop, Packaging Corp Am, 73-86; RETIRED. *Mem:* Am Inst Chemists; NY Acad Sci; Am Chem Soc; Tech Asn Pulp & Paper Indust. *Res:* Alkaline chemical pulping processes; chemistry and structure of cellulose ethers; synthesis of plasticizers for cellulose; tosyl and iodo derivatives of some hydroxyethyl ethers. *Mailing Add:* 3075 Baker Park Dr SE Grand Rapids MI 49508

TASKER, JOHN B, b Concord, NH, Aug 28, 33; m 61; c 3. VETERINARY MEDICINE, CLINICAL PATHOLOGY. *Educ:* Cornell Univ, DVM, 57, PhD(vet path), 63; Am Col Vet Path, Dipl, 72. *Prof Exp:* Am Vet Med Asn fel, Cornell Univ, 61-63; from asst prof to assoc prof vet med, Colo State Univ, 63-67; assoc prof vet path, NY State Vet Col, Cornell Univ, 67-69, prof clin path, 69-78; prof clin path & assoc dean, Sch Vet Med, La State Univ, 78-84; PROF PATH & DEAN, COL VET MED, MICH STATE UNIV, 84- *Mem:* Am Vet Med Asn; Am Soc Vet Clin Path. *Res:* Veterinary clinical pathology. *Mailing Add:* Col Vet Med A-133 E Fee Hall Mich State Univ East Lansing MI 48824-1316

TASKER, RONALD REGINALD, b Toronto, Ont, Dec 18, 27; m 55; c 4. NEUROSURGERY. *Educ:* Univ Toronto, BA, 48, MD, 52, MA, 54; FRCS(C), 59. *Prof Exp:* Asst physiol, Banting & Best Dept Med Res, Univ Toronto, resident fel neurosurg, Fac Med, 58-59, mem clin & res staff, 61-66, assoc prof surg, 66-78, PROF SURG, FAC MED, UNIV TORONTO, 78- *Concurrent Pos:* McLaughlin traveling fel, Mass Gen Hosp, 59 & Univ Wis, 60; Markle fel, 61-66; mem clin & res staff, Toronto Gen Hosp, 61-66, asst prof neurosurg, 66-73, assoc prof, 73-78; mem coun, Int Asn Study Pain, 84-90. *Mem:* Am Soc Stereotactic & Functional Neurosurg (pres, 80-81); Int Asn Study Pain; World Soc Stereotactic & Functional Neurosurg (pres, 85-89); Am Acad Neurol Surg. *Res:* Hyperkinetic disorders; dyskinesias and stereotatic surgery; sensory physiology and pain. *Mailing Add:* Toronto Western Hosp 399 Bathurst St MP 15-312 Toronto ON M5T 2S8 Can

TASLITZ, NORMAN, b New York, NY, Feb 12, 29; m 56; c 3. NEUROANATOMY. *Educ:* NY Univ, BS, 51; Univ Pa, cert phys ther, 52; Stanford Univ, PhD(anat), 63. *Prof Exp:* Physical therapist, Univ Wis Hosps, Madison, 52-54 & Wis Neurol Found, Madison, 54-58; asst prof anat, Sch Phys Ther, Case Western Reserve Univ, 63-77, from instr to sr instr, Dept Anat, 63-67, asst prof, 68-76; assoc prof anat & prog dir human anat, 77-82, CHMN, DEPT ANAT & ASSOC DEAN ACAD AFFAIRS, NORTHEASTERN OHIO UNIVS COL MED, 82- *Concurrent Pos:* Adj prof, Sch Law, Case Western Res Univ, 77- *Mem:* Am Asn Anatomists; Sigma Xi; AAAS; Am Asn Univ Prof; Asn Advan Med Educ. *Res:* Development of biological models to demonstrate the hemodynamic, metabolic and electrophysiologic performance of intact central nervous system tissue under control conditions, at various subnormal temperature levels and following trauma or periods of circulatory arrest. *Mailing Add:* Dept Anat Northeastern Ohio Univs Col Med Rootstown OH 44272

TASMAN, WILLIAM S, US citizen; m 62; c 3. OPHTHALMOLOGY. *Educ:* Haverford Col, BA, 51; Temple Univ, MD, 55. *Prof Exp:* ASSOC RETINA SERV, RETINOVITREOUS ASSOC, 62-; OPHTHALMOLOGIST-IN-CHIEF, WILLS EYE HOSP, 85- *Concurrent Pos:* Heed fel, 61-62; Retina Found fel, 62. *Mem:* Am Col Physicians; Am Col Surg; Am Ophthal Soc; Retina Soc; Am Bd Ophthal. *Res:* Retinal diseases in children; retinal detachment surgery; retinopathy of prematurity. *Mailing Add:* 910 E Willow Grove Ave Philadelphia PA 19118

TASSAVA, ROY A, b Ironwood, Mich, July 5, 37; m 61; c 3. REGENERATION, DEVELOPMENTAL BIOLOGY. *Educ:* Northern Mich Univ, BS, 59; Brown Univ, MAT, 65; Mich State Univ, PhD(zool), 68. *Prof Exp:* Pub sch teacher, 59-64; NIH res fel zool, Mich State Univ, 68-69; asst prof, 69-73, assoc prof, 73-76, PROF ZOOL, OHIO STATE UNIV, 76- *Concurrent Pos:* Sigma Xi res award, 68. *Mem:* Am Inst Biol Sci; Am Soc Zoologists. *Res:* Role of nerves, wound epithelium and hormones in amphibian limb regeneration; pituitary, thyroid and adrenal hormone physiology in amphibians. *Mailing Add:* 963 Biol Sci Bldg Ohio State Univ Main Campus Columbus OH 43210

TASSINARI, SILVIO JOHN, b New York, NY, June 2, 22; m 52; c 2. NUCLEAR MEDICINE, RADIOCHEMISTRY. *Educ:* St Michael's Col, Vt, BS, 42, MS, 47; Int Univ, PhD(chem, nuclear med), 77. *Prof Exp:* Supv radiochemist, Brookhaven Nat Lab, 51-71; supv radiochemist, Vet Admin Med Ctr, Brooklyn, 71-72, supv radiochemist nuclear med, Northport, 72-84; PRES & OWNER, CERT HAZARDOUS MAT MGT, LONG ISLAND LABS INC, 80-, CERT HAZARD MAT CONTROL MGR, 84- *Concurrent Pos:* Consult, Cath Med Ctr Brooklyn & Queens, 73-83. *Mem:* Fel Am Inst Chemists; Radiation Res Soc; Health Physics Soc; Soc Nuclear Med; Am Soc Radiologic Technologists; fel NY Acad Sci. *Res:* Diagnostic nuclear medicine using new and innovative radioisotopes and radiopharmaceuticals; investigation of new procedures in organ and system dianosis; utilizing computer assistance such as computed-assisted diagnosis and computed-aided monitoring. *Mailing Add:* 47 Moriches Rd Nissequogue St James NY 11780

TASSOUL, JEAN-LOUIS, b Brussels, Belg, Nov 1, 38; m 66. STELLAR HYDRODYNAMICS, ROTATING STARS. *Educ:* Free Univ Brussels, LSc, 61, DSc, 64. *Prof Exp:* Res fel, Nat Found Sci Res, Belg, 65-66; res assoc, Univ Chicago, 66-67 & Princeton Univ, 67-68; from asst prof to assoc prof, 68-75, PROF PHYSICS, UNIV MONTREAL, 75- *Mem:* Int Astron Union. *Res:* Stellar structure. *Mailing Add:* Dept Physics Univ Montreal PO Box 6128 Montreal PQ H3C 3J7 Can

TASSOUL, MONIQUE, b Brussels, Belg, Sept 23, 42; m 66. ASTROPHYSICS. *Educ:* Univ Brussels, Belg, LSc, 63; Univ Montreal, PhD(physics), 74. *Prof Exp:* RESEARCHER ASTROPHYS, UNIV MONTREAL, 74- *Mem:* Int Astron Union. *Res:* Stellar structure and oscillations. *Mailing Add:* Dept of Physics PO Box 6128 Montreal PQ H3C 3J7 Can

TASWELL, HOWARD FILMORE, b Paterson, NJ, July 21, 28; m 52; c 6. IMMUNOHEMATOLOGY, LABORATORY MEDICINE. *Educ:* Harvard Col, AB, 49; New York Univ, MD, 53; Univ Minn, MS, 61; Am Bd Path, cert, anat & clin path, 61 & blood banking & immunohemat, 73. *Prof Exp:* Asst prof path, Hahnemann Med Sch, 61-63; from instr to asst prof clin path, 63-73, assoc prof lab med, 73-77, PROF LAB MED, MAYO MED SCH, UNIV MINN, 77- *Concurrent Pos:* Assoc pathologist, Harrisburg Hosp, Pa, 61-63; NIH prin & co-investr, Nat Inst Arthritis & Metab Dis, 63-67, 66-71, 71-74 & 75-; pres, Minn Asn Blood Banks, 69-70 & Minn Soc Clin Pathologists, 72-73; consult, Food & Drug Admin, Bur Biologics, 74-75; dir blood bank, Mayo Clin, Rochester, Minn, 63- *Mem:* Am Asn Blood Banks; Am Soc Clin Pathologists; Am Blood Comn; Sigma Xi. *Res:* Clinical aspects of immunohematology, blood banking, blood transfusion and transfusion reactions; post-transfusion hepatitis; blood resource management and quality control; histocompatibility testing for tissue transplantation. *Mailing Add:* 3202 Mayowood Hills Dr SW Rochester MN 55902

TATA, XERXES RAMYAR, b Bombay, India, Apr 27, 54; m 86. ELEMENTARY PARTICLE PHYSICS. *Educ:* Bombay Univ, BSc, 74; Indian Inst Technol, MSc, 76; Univ Tex Austin, PhD(physics), 81. *Prof Exp:* Res assoc physics, Univ Tex Austin, 81-83 & Univ Ore, Eugene, 83-84 & 85-86; sci assoc physics, Cern, Geneva, Switz, 84-85; asst scientist physics, Univ Wis, Madison, 86-88; ASSOC PROF PHYSICS, UNIV HAWAII, 88- *Concurrent Pos:* Vis scientist, Kek, Nat Lab High Energy Physics, Japan, 86-87. *Mem:* Am Phys Soc. *Res:* Fundamental and phenomenological studies in elementary particle physics; electroweak theory; supersymmetry; new particle searches at high energy colliders; unification of fundamental forces; model-building; search for physics beyond the standard model. *Mailing Add:* Physics Dept Univ Hawaii Manoa Honolulu HI 96822

TATARCZUK, JOSEPH RICHARD, b Portland, Maine, June 15, 36; m 64; c 4. NUCLEAR PHYSICS, COMPUTER ENGINEERING. *Educ:* Col of the Holy Cross, BS, 58; Rensselaer Polytech Inst, MS, 61, PhD(physics), 65. *Prof Exp:* Res asst nuclear physics, Rensselaer Polytech Inst, 59-65; res assoc, Nuclear Physics Div, Max Planck Inst Chem, 65-66; res assoc neutron physics, 66-70, ASST TO DIR LINAC OPERS & SUPPORT SERV, LINEAR ACCELERATOR LAB, RENSSELAER POLYTECH INST, 70-; ASSOC PROF, ALBANY MED COL, 74- *Concurrent Pos:* Chief physicist, Nuclear Med Serv, Albany Vet Admin Hosp, 72-; adj prof nuclear eng & sci, Rensselaer Polytech Inst, 76-79, adj prof biomed eng, 80- *Mem:* Am Phys Soc. *Res:* Neutron physics; gamma ray spectroscopy; nuclear, accelerator and computer instrumentation. *Mailing Add:* Three Todd Dr Troy NY 12180

TATE, BRYCE EUGENE, b Girard, Ill, Apr 15, 20; m 59; c 2. ORGANIC CHEMISTRY. *Educ:* Univ Ill, BS, 42; Univ Wis, MS, 44, PhD(chem), 50. *Prof Exp:* Res fel chem, Harvard Univ, 53-55; chemist, Chas Pfizer & Co, 55-59, proj leader, 59-61, group supvr, 61-68, mgr chem prod res, 68-77, ASST DIR INDUST SPECIALTY, CHEM DEPT, PFIZER INC, 77- *Mem:* Am Chem Soc; Soc Petrol Engrs. *Res:* Synthetic organic chemistry; mechanism of organic reactions; polymer chemistry; oil recovery chemicals. *Mailing Add:* 11 Laurel Hill Dr Niantic CT 06357

TATE, CHARLOTTE ANNE, b Mt Clemens, Mich, Sept 15, 44. EXERCISE PHYSIOLOGY, CARDIOVASCULAR SCIENCES. *Educ:* Tex Woman's Univ, BS, 69; Southwest Tex State Univ, MA, 72; Univ Tex, Austin, PhD(phys educ & exercise physiol), 76. *Prof Exp:* Teacher adaptive phys educ, Northeast ISD, 69-72; teaching asst, Univ Tex, Austin, 72-76; res physiologist biochem, Inst Environ Stress, 76-77; fel biochem res muscle, 77-79, instr, 79-80, ASST PROF, BAYLOR COL MED, 80- *Concurrent Pos:* Res asst, Univ Res Coun, Univ Tex, Austin, 74-75; NIH trainee, Inst Environ Stress, 76-77 & NIH fel, Baylor Col Med, 77-79. *Mem:* Am Heart Asn; Am Col Sports Med; Am Physiol Soc; AAAS; NY Acad Sci; Biophys Soc. *Res:* Calcium fluxes in subcellular organelles of cardiac and skeletal muscle in normal and diseased states; biochemical adaptations to exercise and fatigue in muscle. *Mailing Add:* Dept Pharmacol Houston Col Pharm Houston TX 77204-5515

TATE, DAVID PAUL, b Chicago, Ill, Dec 10, 31; m 53; c 3. ORGANIC & POLYMER CHEMISTRY. *Educ:* Hamline Univ, BS, 53; Purdue Univ, MS, 55, PhD, 58. *Prof Exp:* Sr chemist, Standard Oil Co, Ohio, 57-63; mgr polymerization, Firestone Tire & Rubber Co, 63-71, asst dir res, 71-80, res assoc, 80-86; EXEC OFFICER, CTR FOR ADHESIVES, SEALANTS & COATINGS, CASE WESTERN RESERVE UNIV, 87- *Mem:* Am Chem Soc. *Res:* Lithium amine reductions; phosphorous compounds; organometallic chemistry; elastomer synthesis; inorganic polymers, phosphazene polymers. *Mailing Add:* 3801 Deer Run Richfield OH 44286

TATE, JOHN T, b Minneapolis, Minn, Mar 13, 25; m 56; c 3. MATHEMATICS. *Educ:* Harvard Univ, BA, 46; Princeton Univ, PhD(math), 50. *Prof Exp:* Higgins res assoc & instr, Princeton Univ, 50-53; vis asst prof math, Columbia Univ, 53-54; from asst prof to assoc prof, 54-59, PROF MATH, HARVARD UNIV, 59-; PROF & SID W RICHARDSON CHAIR MATH, UNIV TEX, 90- *Concurrent Pos:* Sloan fel, 59-61; vis prof, Univ Calif, Berkeley, 63-, IHES, France, 68-69, Orsay, France, 80-81 & Univ Tex, 89-90; Guggenheim fel, IHES, France, 65-66. *Honors & Awards:* Cole Prize in Number Theory, Am Math Soc, 56. *Mem:* Nat Acad Sci; Am Math Soc; Math Asn Am; fel AAAS. *Res:* Algebra; algebraic number theory; diophantine algebraic geometry. *Mailing Add:* Dept Math RLM Hall 10-166 Univ Tex Austin TX 78712

TATE, LAURENCE GRAY, b Cambridge, Eng, Feb 10, 45; US citizen. INSECT TOXICOLOGY, INSECT PHYSIOLOGY. *Educ:* Limestone Col, BS, 66; Univ SC, MS, 68, PhD(biol), 71. *Prof Exp:* USPHS res assoc insect toxicol, NC State Univ, 71-74; asst prof, 74-77, ASSOC PROF BIOL, UNIV SALA, 77- *Mem:* Sigma Xi. *Res:* Metabolism of xenobiotics in tissues of marine organisms; carbohydrate metabolism in insects. *Mailing Add:* Dept Biol Sci Univ SAla 307 University Blvd Mobile AL 36688

TATE, R(OGER) W(ALLACE), b Chicago, Ill, Jan 31, 25; m 58; c 3. CHEMICAL ENGINEERING. *Educ:* Ill Inst Technol, BS, 48; Univ Wis, MS, 48, PhD(chem eng), 50. *Prof Exp:* Chem engr, Standard Oil Develop Co, NJ, 50-54; staff engr, Kearney & Trecker Corp, Wis, 54-57; DIR RES, DELAVAN INC, 57- *Concurrent Pos:* Adj prof, Newark Col Eng, 51-53. *Mem:* Am Inst Chem Engrs; Am Soc Testing & Mat. *Res:* Atomization and spray analysis; development of fuel injectors and spray nozzles. *Mailing Add:* Delavan Inc 811 Fourth St West Des Moines IA 50265-0100

TATE, ROBERT FLEMMING, b Oakland, Calif, Dec 15, 21. MATHEMATICAL STATISTICS. *Educ:* Univ Calif, AB, 44, PhD(math statist), 52; Univ NC, MS, 49. *Prof Exp:* Lectr, Univ Calif, 51-53; from instr to assoc prof math, Univ Wash, Seattle, 53-65; assoc prof, 65-67, PROF MATH, UNIV ORE, 67- *Concurrent Pos:* Mem, Math Inst & Inst Statist, Univ Vienna, 64. *Mem:* Sigma Xi; fel Inst Math Statist (assoc secy, 63-73); Am Statist Asn. *Res:* Theory of correlation and of estimation; multivariate analysis. *Mailing Add:* Dept Math Univ Ore Eugene OR 97403-1222

TATE, ROBERT LEE, III, b Victoria, Tex, Dec 1, 44; m 71; c 2. SOIL MICROBIOLOGY, ENVIRONMENTAL MICROBIOLOGY. *Educ:* Univ Ariz, BS, 66, MS, 67; Univ Wis-Madison, PhD(bact), 70. *Prof Exp:* Scholar bact, Univ Calif, Los Angeles, 70-72, res assoc, Dept Agron, Cornell Univ, 72-75; asst prof microbiol, Agr Res & Educ Ctr, Univ Fla, 75-80, assoc prof, 80-81; asst prof, 81-84, ASSOC PROF SOIL MICROBIOL, DEPT SOILS & CROPS, RUTGERS UNIV, 84- *Concurrent Pos:* Consult, Brookhaven Nat Lab, 82-89; assoc ed, Soil Sci, 84-, Soil Sci Soc Am J, 85-90. *Mem:* Am Soc Microbiol; Soil Sci Soc Am; Am Soc Agron. *Res:* Microbial interactions with soil organic matter; biogeochemical cycles in soils, nitrogen cycle, carbon cycle; behavior of xenobiotics in soil; denitrification; soil enzymes. *Mailing Add:* Dept Environ Sci Rutgers Univ New Brunswick NJ 08903-0231

TATE, SURESH S, b India, Dec 2, 36. PROTEIN CHEMISTRY, CELL BIOLOGY. *Educ:* London Univ, PhD(biochem), 63. *Prof Exp:* ASSOC PROF BIOCHEM, MED COL, CORNELL UNIV, 78- *Mem:* Am Soc Biol Chemists; Harvey Soc; NY Acad Sci. *Mailing Add:* Dept Biochem Med Col Cornell Univ 1300 York Ave New York NY 10021

TATELMAN, MAURICE, b Omaha, Nebr, Dec 6, 17; m 61; c 2. MEDICINE. *Educ:* Univ Nebr, AB, 40, MD, 42. *Prof Exp:* From asst prof to assoc prof, Col Med, Wayne State Univ, 50-61, prof radiol, 61-86; RETIRED. *Concurrent Pos:* Chmn, Dept Radiol, Sinai Hosp Detroit, 68-83; consult, Detroit Mem Hosp. *Mem:* Fel Am Col Radiol; Am Roentgen Ray Soc; Radiol Soc NAm; sr mem Am Soc Neuroradiol; Am Soc Head & Neck Radiol; Sigma Xi. *Res:* Clinical diagnostic radiology; neuroradiology. *Mailing Add:* 7525 E Gainey Ranch Rd No 150 Scottsdale AZ 85258-1607

TATINA, ROBERT EDWARD, b Chicago, Ill, May 18, 42; m 78; c 2. PRAIRIE ECOLOGY. *Educ:* Northern Ill Univ, BS, 65; Southern Ill Univ, MA, 71, PhD(bot), 81. *Prof Exp:* Teacher sci & math, Trewyn Jr High Sch, 65-66, biol, Evergreen Park Community High Sch, 66-69; PROF BIOL, DAKOTA WESLEYAN UNIV, 76- *Mem:* Sigma Xi; Bot Soc Am; Am Biol Teachers Asn. *Res:* Flora of South Dakota; gradient analysis of mixed grass prairie. *Mailing Add:* Dept Biol Dakota Wesleyan Univ Mitchell SD 57301-4398

TATINI, SITA RAMAYYA, b Mortha, India, Oct 6, 35; m 54; c 3. MICROBIOLOGY, FOOD SCIENCE. *Educ:* Univ Madras, BVSc, 57; Univ Minn, MS, 66, PhD(food sci & indust), 69. *Prof Exp:* Vet asst surgeon, Andhra Animal Husb Dept, State of Andhra Pradesh, India, 58-61; res fel, 69, NIH-Food & Drug Admin res grant, 71-74, asst prof food microbiol, 69-77, prof food sci & nutrit, Univ Minn, St Paul, 77-; DEPT FOOD SCI & NUTRIT, UNIV MINN, MINNEAPOLIS. *Mem:* AAAS; Inst Food Technol; Am Soc Microbiol; Am Dairy Sci Asn; Int Asn Milk, Food & Environ Sanit. *Res:* Growth, survival and production of enterotoxins by staphylococci in food products; developing rapid methods for assessment of psychrophilic bacteria in milk. *Mailing Add:* Dept Food Sci & Nutrit Univ Minn St Paul MN 55108

TATLOW, RICHARD H(ENRY), III, b Denver, Colo, May 27, 06; m 32; c 2. CIVIL ENGINEERING. *Educ:* Univ Colo, BS, 27, CE, 32. *Prof Exp:* Jr hwy engr, US Bur Pub Rds, DC, 27-29; engr, Harrington & Cortelyou, Mo, 29-34, partner, DC, 34-41; chmn, Abbott, Merkt & Co, 46-84, chmn bd, Abbott, Merkt Int Inc, 72-84; RETIRED. *Concurrent Pos:* Mem comt supersonic transport-sonic boom, Nat Acad Sci, 64; mem, Mayor's Sci & Tech Adv Comt, New York, 66-; dir, NUS Corp. *Mem:* Nat Acad Eng; Am Soc Civil Engrs (pres, 68); fel Am Soc Mech Engrs; fel Am Inst Consult Engrs. *Res:* Structural engineering; design of industrial buildings and department stores. *Mailing Add:* 11 Rutland Rd Scarsdale NY 10583

TATOMER, HARRY NICHOLAS, b Jersey City, NJ, Feb 13, 13; m 40; c 4. CHEMICAL ENGINERRING. *Educ:* Univ Ill, BS, 37. *Prof Exp:* Develop engr, Olin Mathieson Chem Corp, 37-40, asst proj supvr, 40-43, proj supvr, 43-46, mgr pilot opers, 46-52, process develop, 52-57, tech asst prod, Energy Div, 57-60; tech asst prod, Chem Div, Union Carbide Corp, 60-64, admin assoc to dir rocket propulsion & staff coordr res & develop admin, South Charleston Tech Ctr, 64-77, chem eng, 77-78; RETIRED. *Mem:* Am Chem Soc; Am Inst Chem Engrs. *Res:* Pilot plant development of sodium chlorite; chlorine dioxide generation; sodium amalgam processes; hydrazine and derivatives; boron hydrides; high energy fuels; rocket propellants; government contract administration; research administration. *Mailing Add:* 2018 Weberwood Dr Charleston WV 25303-3029

TATOR, CHARLES HASKELL, b Toronto, Ont, Aug 24, 36; m 60; c 3. NEUROSURGERY. *Educ:* Univ Toronto, MD, 61, MA, 63, PhD(neuropath), 65; FRCS(c); FACS. *Prof Exp:* From assoc to asst prof, 69-74, assoc prof, 74-80, PROF SURG, UNIV TORONTO, 80-; SURGEON, DIV NEUROSURG, TORONTO WESTERN HOSP, 69- *Concurrent Pos:* Co-dir, Playfair Neurosci Unit, Univ Toronto. *Res:* Brain tumor research and spinal cord injury research. *Mailing Add:* Dept Surg Univ Toronto Toronto ON M5S 1A8 Can

TATRO, CLEMENT A(USTIN), b Kingman Co, Kans, May 16, 24; m 46; c 2. MECHANICS. *Educ:* Friends Univ, BA, 49; Purdue Univ, MS, 51, PhD(physics), 56. *Prof Exp:* Res asst, Purdue Univ, 49-56; asst prof appl mech, Mich State Univ, 56-60, assoc prof metall, mech & mat sci, 60-62; prof mech eng, Tulane Univ, 62-66; head mat eng sect, Dept Mech Eng, 66-77, RES & ADVAN DEVELOP ENGR, LAWRENCE LIVERMORE LAB, UNIV CALIF, 77- *Mem:* Am Soc Testing & Mat; Am Phys Soc; Soc Exp Stress Analysis; Am Soc Nondestructive Testing; Inst Environ Sci (pres, 80-81); Sigma Xi. *Res:* Acoustic emission; wave propagation in solids; experimental stress analysis; material dynamics. *Mailing Add:* Lawrence Livermore Lab 384 Martin Ave Livermore CA 94550

TATRO, PETER RICHARD, b Winthrop, Mass, Jan 20, 36; m 57; c 3. PHYSICAL OCEANOGRAPHY, ACOUSTICS. *Educ:* Ga Inst Technol, BME, 57; Mass Inst Technol, PhD(oceanog), 66. *Prof Exp:* Res oceanogr, Fleet Numerical Weather Ctr, Calif, 66-69; spec asst for ocean sci, Off Naval Res, Washington, DC, 69-72, head, Acoust Environ Support Detachment, Arlington, 72-75, spec asst oceanogr of the Navy, 75-76; mgr, Ocean Sci Div, Sci Appln Int Corp, 77-80, corp vpres, Ocean Sci Dept, McLean, Va, 80-91; CONSULT, 91- *Concurrent Pos:* Consult, Int Decade Ocean Explor, NSF, 72-75. *Honors & Awards:* Navy Achievement Medal; Navy Expeditionary Medal. *Mem:* Acoust Soc Am; Marine Technol Soc. *Res:* Application of advanced digital technology to the problem of predicting the acoustic characteristics of the oceans. *Mailing Add:* 1965 Lakeport Way Reston VA 22091

TATSUMOTO, MITSUNOBU, b Junsho, Japan, Mar 19, 23; m 48; c 2. GEOCHEMISTRY, GEOCHRONOLOGY. *Educ:* Tokyo Bunrika Univ, BS, 48, DSc(inorg chem), 57. *Hon Degrees:* Dr, Univ Paris VII, 75. *Prof Exp:* Res asst chem, Tokyo Kyoiku Univ, 48-57, lectr, 57; res fel oceanog, Scripps Inst, Calif, 57-58; Welch Found fel, 59; res fel oceanog, Tex A&M Univ, 59; res fel geochem, Calif Inst Technol, 59-62; CHEMIST, US GEOL SURV, 62- *Concurrent Pos:* Vis prof, Inst Geophys, Univ Paris VI, 71 & 74; adj prof dept geol sci, Univ Rochester, NY, 78- *Mem:* Geochem Soc; fel Am Geophys Union; Meteoritical Soc. *Res:* Isotope geochemistry; natural radioactivity; meteorite and lunar chronology; marine geochemistry. *Mailing Add:* Isotope Geol Br US Geol Surv MS 963 Box 25046 Denver CO 80225

TATTAR, TERRY ALAN, b Port Chester, NY, May 9, 43; m 69; c 2. FOREST PATHOLOGY, SHADE TREE PATHOLOGY. *Educ:* Northeastern Univ, BA, 67; Univ New Hampshire, PhD(bot), 71. *Prof Exp:* Plant pathologist forest path, USDA Forest Serv, 71-73; from asst prof to assoc prof, 73-84, PROF SHADE TREE PATH, UNIV MASS, 85- *Concurrent Pos:* Consult, Tree Health, 75- *Mem:* Am Phytopath Soc; Sigma Xi; Am Soc Consult Arborists. *Res:* Determining the effects of vascular wilt pathogens on trees; developing diagnostic techniques for early detection of diseases of trees; detection of hazard trees. *Mailing Add:* Dept Plant Path Univ Mass Amherst MA 01003

TATTER, DOROTHY, b Chicago, Ill, Apr 11, 22; m 49; c 3. MEDICINE. *Educ:* Rosary Col, BS, 43; Univ Ill, MD, 47. *Prof Exp:* From instr assoc prof, 49-83, EMER PROF PATH, SCH MED, UNIV SOUTHERN CALIF, 83- *Concurrent Pos:* Resident path, Los Angeles County Gen Hosp, 49-52, head physician, Autopsy Dept Labs, 52-83. *Mem:* Sigma Xi. *Res:* Pathology. *Mailing Add:* 555 Madeline Dr Pasadena CA 91105

TATTERSALL, IAN MICHAEL, b Paignton, Devon, Eng, May 10, 45; m. PHYSICAL ANTHROPOLOGY, PRIMATOLOGY. *Educ:* Cambridge Univ, BA, 67, MA, 70; Yale Univ, MPhil, 70, PhD(geol), 71. *Prof Exp:* Asst cur, 71-76, assoc cur, 76-81, CUR PHYS ANTHROP, AM MUS NATURAL HIST, 81- *Concurrent Pos:* Vis lectr, Grad Fac, New Sch Social Res, 71-72; adj asst prof, Lehman Col, City Univ New York, 71-74; adj assoc prof, Columbia Univ, 78-79. *Mem:* Am Asn Phys Anthropologists; Soc Vert Paleont; Am Soc Primatology; Int Primatology Soc; AAAS; Sigma Xi. *Res:* Evolution, functional anatomy, ecology and behavior of the primates, particularly of the Malagasy lemurs; human evolution; primate systematics; evolutionary theory in relation to phylogenetic reconstruction and other systematic applications. *Mailing Add:* Dept Anthrop Am Mus Natural Hist Cent Park W at 79th St New York NY 10024

TATUM, CHARLES MARIS, b Philadelphia, Pa, Oct 10, 47; m 70; c 2. BIOCHEMISTRY, ORGANIC CHEMISTRY. *Educ:* Amherst Col, BA, 69; Pa State Univ, PhD(org chem), 76. *Prof Exp:* Biochemist clin chem, Gen Rose Mem Hosp, 70-72; asst prof chem, Middlebury Col, 76-79; sr scientist, Explor Agr Res, Rohm & Haas Co, 79-82, res sect mgr, 82-84, res dept mgr, 84-90, VPRES, CORP DIR RES, ROHM & HAAS CO, 90- *Mem:* Am Chem Soc. *Res:* Strategic planning, agricultural applications of genetic engineering and pesticide biochemistry; pesticide synthesis; pesticide biology. *Mailing Add:* 475 W Prospect Ave North Wales PA 19454-2629

TATUM, FREEMAN A(RTHUR), physical electronics, mathematics, for more information see previous edition

TATUM, JAMES PATRICK, b Dallas, Tex, July 6, 38. PHYSICAL CHEMISTRY. *Educ:* Rice Univ, BA, 61; Fla State Univ, PhD(phys chem), 66. *Prof Exp:* Res assoc chem, Univ Ill, 66-68; from asst prof to prof chem, Ind State Univ, Terre Haute, 68-82; CONSULT, 78- *Concurrent Pos:* Vis prof quantum theory proj, Univ Fla, Gainesville, 78. *Mem:* Am Phys Soc. *Res:* Theoretical chemistry. *Mailing Add:* Dept Chem Ind State Univ Terre Haute IN 47809

TATUM, WILLIAM EARL, b Ft Payne, Ala, Sept 13, 33; m 52; c 4. ORGANIC CHEMISTRY. *Educ:* Chattanooga Univ, BS, 55; Univ Tenn, PhD(chem), 58. *Prof Exp:* Res chemist, Exp Sta, 58-61, tech rep, Venture Develop Sect, 61-63, staff scientist, Yerkes Res Lab, 63-64, res supvr, Circleville Res Lab, Ohio, 64-67, develop supvr, Circleville Plant, 67-68, tech supt, Florence Plant, SC, 68-70, cellophane prod mgr, Del, 70-71, dir prod & tech div, 71-73, venture mgr, 73-74, dir specialty mkt div, 75-76, dir packaging films div, 76-78, dir, Fluoropolymers Div, Polymer Prod Dept, 78-80, dir safety, health & environ affairs, 80-81, gen mgr, Energy & Mat Dept, 81-82, vpres, Mat & Logistics, Dept, 82-84, vpres Int Dept, 84-85, SR VPRES, MAT & LOGISTICS DEPT, E I DUPONT DE NEMOURS & CO, INC, DEL, 85- *Concurrent Pos:* Mem steering comt, Bus Roundtable, Off Technol Assessment, US Cong, Conserv Found & Nat Environ Develop Asn; mem, Environ Assessment Coun. *Mem:* Am Chem Soc. *Res:* Polymer chemistry and engineering; synthetic organic chemistry. *Mailing Add:* Mat Logistics & Serv E I du Pont de Nemours & Co Inc Wilmington DE 19898

TATYREK, ALFRED FRANK, b Hillside, NJ, Jan 23, 30. POLYMER MATERIALS IDENTIFICATION. *Educ:* Seton Hall Univ, BS, 54. *Prof Exp:* Res chemist polymer synthesis, Bakelite Div, Union Carbide Corp, 53-58; res chemist polymer res & appln, US Radium Corp, 59-62; anal chemist chem items, US Army, NY, 62-64, res chemist chemiluminescent & pyrotechnics, Armament Res & Develop Ctr, 64-73, chem engr environ, 73-84, MATS ENG & CHEM, US ARMAMENT RES, DEVELOP & ENG CTR, 84- *Concurrent Pos:* Pres Picatinny Chap, Sigma Xi, 74-75, 79-80, 85-86, lectr, 81- *Mem:* Sigma Xi. *Res:* Experimental exploratory research on new polymers and their synthesis, analysis and applications as structural materials, adhesives and coatings; developing electroluminescent light emiting cells based upon highly polar polymer systems; chemical reaction methods of generating colored smokes; kinetics and mechanism of thermal decomposition of explosives; applications and chemical reactions useful for increasing the light yield of chemiluminescent compounds; development and engineering on new industrial systems for the control and disposal of hazardous wastes generated at Army ammunition plants; analytical investigations into the composition of organic materials using FTIR infrared spectrometry, thermal analysis and mechanical analysis; design of an innovative vacuum crankcase system for internal combustion engines for control of air pollution and engine oil contamination; 6 US patents. *Mailing Add:* 27 Orchard Rd Maplewood NJ 07040

TAUB, AARON M, b Jersey City, NJ, Dec 21, 35; m 67; c 2. PHARMACY. *Educ:* Wagner Col, BS, 60; State Univ NY Buffalo, PhD(biol), 65. *Prof Exp:* Instr zool, Penn State Univ, 64-65, asst prof, 65-68; asst prof anat, Ont Vet Col, Univ Guelph, 68-69; mgr med serv, Fisons (Can) Ltd, 69-72; mgr & dir qual control, Fisons Corp, 72-81, dir regulatory affairs, 81-83, dir proj mgt, 83-88, DIR NEW PROD COORD, FISONS CORP, 89- *Mem:* Sigma Xi; Soc Study Amphibians & Reptiles; Am Asn Aerosol Res. *Res:* Pharmaceutical development; new product acquisition. *Mailing Add:* Fisons Corp 755 Jefferson Rd Rochester NY 14623

TAUB, ABRAHAM, b New York, NY, Sept 21, 01. PHARMACEUTICAL CHEMISTRY. *Educ:* Columbia Univ, BS, 22, AM, 27. *Hon Degrees:* ScD, Columbia Univ, 76. *Prof Exp:* From instr to prof chem, 22-65, distinguished serv prof, 65-69, EMER DISTINGUISHED SERV PROF CHEM, COL PHARMACEUT SCI, COLUMBIA UNIV, 69- *Concurrent Pos:* Asst, Revision Comt, US Pharmacopeia, 20-30; consult chemist, 22-; asst, Nat Formulary, 37; consult, Nat Asn Pharmaceut Mfrs, 70-86. *Honors & Awards:* Rusby Award, 62; Man of Year Award, Nat Asn Pharmaceut Mfrs, 72. *Mem:* AAAS; Am Chem Soc; Am Pharmaceut Asn; fel Am Inst Chem; NY Acad Sci. *Res:* Quantitative color standards; deterioration of medicinals; development of analytical methods; stability of parenteral solutions; chromatography; radioisotope tracer techniques. *Mailing Add:* 1080 Fifth Ave New York NY 10128

TAUB, ABRAHAM HASKEL, b Chicago, Ill, Feb 1, 11; m 33; c 3. MATHEMATICS. *Educ:* Univ Chicago, BS, 31; Princeton Univ, PhD(math physics), 35. *Prof Exp:* Asst, Princeton Univ, 34-35 & Inst Adv Study, 35-36 & 40-41; from instr to prof math, Univ Wash, Seattle, 36-48; res prof appl math, Univ Ill, Urbana, 48-64, head digital comput lab, 61-64; dir comput ctr, 64-68, prof, 64-78, EMER PROF MATH, UNIV CALIF, BERKELEY, 78- *Concurrent Pos:* Theoret physicist, Div 2, Nat Defense Res Comt, Princeton Univ, 42-45; mem, Guggenheim Post-Serv, 47-48 & Guggenheim fel, 53 & 58; mem, Appl Math Adv Coun, Nat Bur Stand, 49-54, chmn adv panel, Appl Math Div, 51-60; mem comt on training & res in math, Nat Res Coun, 52-54; mem rev comt, Appl Math Div, Argonne Nat Lab, 60-62. *Mem:* Fel AAAS; Am Math Soc; fel Am Phys Soc; Math Asn Am; fel Am Acad Arts & Sci. *Res:* Relativity; interaction of shock waves; digital computers; numerical analysis. *Mailing Add:* Dept Math Univ Calif Berkeley CA 94720

TAUB, ARTHUR, b New York, NY, Jan 4, 32; m 63; c 1. NEUROLOGY, NEUROPHYSIOLOGY. *Educ:* Yeshiva Univ, BA, 52; Mass Inst Technol, SM, 53, PhD(neurophysiol), 64; Yale Univ, MD, 57; Am Bd Psychiat Neurol, dipl, 75. *Prof Exp:* NIH fel neurophysiol, Mass Inst Technol, 64-66, res assoc, 66-68; dir,neurosurg res lab, 69-75, dir pain clin, 71-75, asst prof, 68-72, assoc prof neurophysiol & neurol, 72-75, prof clin anesthesiol & dir sect study & treatment of pain, 75-76, CLIN PROF ANESTHESIOL, DEPT ANESTHESIOL, SCH MED, YALE UNIV, 76- *Concurrent Pos:* NIH res career develop award, Mass Inst Technol & Yale Univ, 66-73; resident neurol, Sch Med, Yale Univ, 69-72; assoc neurologist, Yale New Haven Hosp, 71-75, attend neurologist, 75-; Royal Soc Med Found traveling fel, UK, 72. *Mem:* Am Asn Study Headache; Am Acad Neurol; Am Soc Function Stereot Neurosurg; Soc Neurosci; Int Asn Study Pain; Am Pain Soc; Am Soc Regional Anesthesia. *Res:* Pain; application of basic scientific approach to modalities for control; neuropharmacology; computer applications medicolegal studies; clinical control by medical, surgical and anesthesiological means. *Mailing Add:* 46-48 Prince St New Haven CT 06510

TAUB, DAVID, b New York, NY, Nov 13, 19; m 44; c 2. PHARMACEUTICAL CHEMISTRY. *Educ:* City Col NY, BS, 40; Harvard Univ, AM, 46, PhD, 50. *Prof Exp:* Chemist, Manhattan Proj, Kellex Corp, 42-46; USPHS fel, Harvard Univ, 49-51; sr chemist, 51-65, sect head process res, 65-68, sr res fel, 68-77, SR INVESTR, MERCK SHARP & DOHME RES LABS, 77- *Mem:* Am Chem Soc. *Res:* Organic synthesis; natural products; synthetic medicinals. *Mailing Add:* 54 Wistar Ave Metuchen NJ 08840

TAUB, EDWARD, b Brooklyn, NY, Oct 22, 31; m 59. PHYSIOLOGICAL PSYCHOLOGY. *Educ:* Brooklyn Col, BA, 53; Columbia Univ, MA, 59; NY Univ, PhD(psychol), 70. *Prof Exp:* Res asst psychol, Columbia Univ, 56; res asst, Dept Exp Neurol, Jewish Chronic Dis Hosp, 57-60, res assoc, 60-68; chief, Behav Biol Ctr, 68-83, dir, Feedback Res Ctr, 83-87; PROF PSYCHOL, UNIV ALA, BIRMINGHAM, 86- *Concurrent Pos:* Asst prof, Sch Med, Johns Hopkins Univ, 70-82; Guggenheim fel, 83-84; mem Psychol Exp Comt Div 6, Steering Comt Sect Nervous Syst, Am Physiol Soc. *Honors & Awards:* Distinguished Res Award, Biofeedback Soc Am, 88; Pioneering Res Contrib Award, Asn Appl Psychophysiol & Biofeedback, 89. *Mem:* Fel Am Psychol Asn; Soc Neurosci; Biofeedback Soc Am (pres, 79); Am Physiol Soc; fel Behav Med Soc; fel AAAS; fel Am Psychol Soc. *Res:* The role of somatosensory feedback and spinal reflexes in movement and learning; biofeedback and self-regulation of human hand temperature; rehabilitation of movement; methods to reduce aerophagia. *Mailing Add:* Dept Psychol Univ Ala 201 Campbell Hall Birmingham AL 35294

TAUB, FRIEDA B, b Newark, NJ, Oct 11, 34; m 54; c 3. ECOLOGY, POLLUTION. *Educ:* Rutgers Univ, BA, 55, MA, 57, PhD(zool), 59. *Prof Exp:* Fisheries biologist, 59-61, biologist, Inst Food Sci, 61-62, from res asst prof to res assoc prof, 62-71, PROF FISHERIES, SCH FISHERIES, UNIV WASH, 71- *Concurrent Pos:* Res grants, 61-91; sr assoc, Nat Res Coun, 90-91. *Mem:* Fel AAAS; Ecol Soc Am; Am Soc Limnol & Oceanog; Sigma Xi; Am Fisheries Soc; Am Soc Testing & Mat; Soc Environ Toxicol & Chem; Am Soc Microbiol; Am Inst Fisheries Res Biologists; Soc Int Limnol. *Res:* Aquatic food chains; ecosystems; closed ecological systems; environmental risk; genetically engineered organisms. *Mailing Add:* Sch Fisheries HF 15 Univ Wash Seattle WA 98195

TAUB, HASKELL JOSEPH, b Princeton, NJ, June 8, 45; m 72; c 1. SURFACE PHYSICS. *Educ:* Stanford Univ, BS, 66. *Prof Exp:* Assoc res scientist, dept physics, New York, Univ, 71-73; asst physicist, dept physics, Brookhaven Nat Lab, 73-75; from asst prof to assoc prof, 75-84, PROF PHYSICS, DEPT PHYSICS & ASTRON, UNIV MO, COLUMBIA, 84- *Concurrent Pos:* Vis prof, French Nat Comt Physics, Univ d'Aix-Marseille II, 81-82; res assoc, Ames Lab, 82-84; vis assoc prof, Sch Appl & Eng Physics, Cornell Univ, 82. *Mem:* Am Phys Soc; AAAS. *Res:* Experimental surface physics; studies of the structure and dynamics of adsorbed films by neutron scattering, low-energy electron diffraction and x-ray scattering techniques; structural and magnetic phase transitions in bulk condensed matter and films. *Mailing Add:* Dept Physics & Astron Univ Mo 223 Physics Bldg Columbia MO 65211

TAUB, HERBERT, b New York, NY, Dec 23, 18; m 43; c 3. ELECTRICAL ENGINEERING. *Educ:* City Col New York, BS, 40; Columbia Univ, MA, 43, PhD, 49. *Prof Exp:* Tutor, Dept Physics, City Col New York, 40-43; elec engr, Nat Defense Res Proj, Princeton Univ, 43-44; tutor, 44-46, from instr to assoc prof, 47-59, PROF ELEC ENG, CITY COL NEW YORK, 60- *Concurrent Pos:* Elec engr, Allen B Du Mont, Inc, NJ, 41; consult, Gen Tel & Electronics, Inc, NY, 69-70, Bell Labs, 72-83 & Paine-Webber Inc, 84-86. *Mem:* Sr mem Inst Elec & Electronics Engrs; Sigma Xi. *Res:* Determination of nuclear magnetic moments by molecular beam methods; electronics, especially pulse circuitry. *Mailing Add:* Dept Elec Eng City Col NY New York NY 10031

TAUB, IRWIN A(LLEN), b Brooklyn, NY, July 18, 34; m 59; c 2. FOOD PRESERVATION, FOOD IRRADIATION. *Educ:* Queens Col, NY, BS, 55; Univ Minn, PhD(inorg chem), 61. *Prof Exp:* Resident res assoc, Argonne Nat Lab, 61-63; fel, Carnegie-Mellon Univ, 63-69; supvry chemist & chief, Cobalt Br, Radiation Sources Div, 69-80, chief plant prod group, develop labs, 80-84, CHIEF, TECHNOL ACQUISITION & DEVELOP BR, FOOD TECHNOL DIV, US ARMY NATICK RES, DEVELOP & ENG CTR, 84- *Concurrent Pos:* Radiation consult, 85- *Mem:* Am Chem Soc; Sigma Xi; AAAS; Inst Food Technologists; Int Inst Refrig. *Res:* Radiation chemistry; kinetics and spectra of free radicals; food irradiation; inorganic fast reactions; protein and lipid radicals; myoglobin radiolysis; autoxidation; vitamin, protein and lipid stability; food dehydration, thermoprocessing and extrusion; protein-carbohydrate crosslinking. *Mailing Add:* 29 William J Heights Framingham MA 01701

TAUB, JAMES M, b Cleveland, Ohio, July 26, 18. NUCLEAR ENGINEERING. *Educ:* Case Western Reserve Univ, BS, 40, MS, 43. *Prof Exp:* Group leader, Los Alamos Nat Lab, 45-75; RETIRED. *Honors & Awards:* E O Lawrence Award, AEC, 65. *Mem:* Fel Am Soc Metals Int; Am Nuclear Soc. *Mailing Add:* 5686 N Camino de la Noche Tucson AZ 85718

TAUB, JESSE J, b New York, NY, Apr 27, 27; m 55; c 3. ELECTRICAL ENGINEERING, PHYSICS. *Educ:* City Col New York, BEE, 48; Polytech Inst Brooklyn, MEE, 49. *Prof Exp:* Engr, US Naval Mat Lab, 49-51, group leader, 51-55; engr, AIL Div, Eaton Corp, 55-58, group leader, 58, sect head, 58-60, dept consult, 60-66, div consult, 66-75; CHIEF SCIENTIST, AIL SYSTS INC, 75- *Concurrent Pos:* Mem staff, Grad Sch, City Col New York, 59-61. *Honors & Awards:* Centennial Medal, IEEE, 84. *Mem:* Fel Inst Elec & Electronics Engrs; Sigma Xi; Asn Old Crows. *Res:* Microwave device development; millimeter and submillimeter techniques; mixers and mixer diodes; multimode power measurements; microwave network synthesis; gallium arsenide integrated circuits. *Mailing Add:* AIL Systs Inc Walt Whitman Rd Melville NY 11746

TAUB, JOHN MARCUS, b Chicago, Ill, July 26, 47. PSYCHOPHYSIOLOGY. *Educ:* Univ Calif, Santa Cruz, AB, 69, MS & PhD(biopsychol), 72. *Prof Exp:* Res biopsychologist, Univ Calif, Santa Cruz, 72-73; fel neurosci, Brain Res Inst, Univ Calif, Los Angeles, 73-75; dir, Sleep & Dream Lab & asst prof psychiat & psychol, Sch Med, Univ Va, 75-78; assoc res prof & dir, Sleep & Performance Lab, St Louis Univ, 78-83. *Concurrent Pos:* NIMH fel, Univ Calif, Los Angeles Brain Res Inst & Dept Psychiat,73-75. *Mem:* AAAS; Sleep Res Soc; Soc Neurosci. *Res:* Investigations on the behavioral and psychophysiological effects of acute and chronic variations in the length and time of sleep in young adults; biological rhythms; sleep and human performance. *Mailing Add:* 7323 Hoover Ave Apt B3 Richmond Heights MO 63117

TAUB, MARY L, b Chicago, Ill, Sept 26, 48. ANIMAL CELL CULTURE, ENDOCRINOLOGY. *Educ:* Univ Calif, San Diego, BA, 71, Santa Barbara, MA, 75, PhD(biol), 76. *Prof Exp:* Teaching fel biochem & cell biol, dept biol, Univ Calif, San Diego, 76-79; asst prof, 79-85, ASSOC PROF BIOCHEM, SCH MED, STATE UNIV NY BUFFALO, 85- *Mem:* Am Soc Cell Biol; AAAS; Am Women Sci. *Res:* Mechanisms by which hormones and other regulatory factors control animal cell growth and expression of differentiated function; kidney cell culture, serum free medium; somatic genetics; primary cell culture. *Mailing Add:* Dept Biochem 102 Cary Hall State Univ NY Health Sci Ctr 3435 Main St Buffalo NY 14214

TAUB, ROBERT NORMAN, b Brooklyn, NY, Apr 21, 36; m 68; c 1. ONCOLOGY, HEMATOLOGY. *Educ:* Yeshiva Univ, AB, 57; Yale Univ, MD, 61; Univ London, PhD(biol), 69. *Prof Exp:* Intern med, New Eng Med Ctr Hosps, Boston, 61-62; intern path, Sch Med, Yale Univ, 62-63; clin & res fel hemat, New Eng Med Ctr Hosps, 63-65, asst resident med, 65-66; NIH res fel immunol, Nat Inst Med Res, London, 66-68; assoc hemat, Mt Sinai Sch Med, 68-69, asst prof med, 69-72, assoc prof med, 72-76, head transplantation immunol lab, Mt Sinai Hosp, 70-76; prof & chmn, Med Col Va, Richmond, 76-81, assoc dir, Va Commonwealth Univ Cancer Ctr, 77-81; co-dir, Comprehensive Cancer Ctr, 81-85, PROF CLIN MED, COL PHYSICIANS & SURGEONS, COLUMBIA UNIV, 81- *Concurrent Pos:* Res fel path, Sch Med, Yale Univ, 62-63; Leukemia Soc Am scholar award, 68-73; attend physician, Mt Sinai Hosp, 69-70; USPHS res career develop award, 75-; Am Cancer Soc prof clin oncol, 77-; co-dir, Columbia Comprehensive Cancer Ctr, 81- *Honors & Awards:* Emil Conason Mem Res Award, Mt Sinai Sch Med, 71. *Mem:* Transplantation Soc; Am Soc Clin Invest; Am Soc Hemat; Am Asn Immunologists; Am Soc Exp Path; Am Soc Clin Oncol; Am Asn Cancer Res. *Res:* Cancer cell biology; immunology of cancer and leukemia. *Mailing Add:* Dept Med Columbia Univ 630 W 168th St New York NY 10032

TAUB, STEPHAN ROBERT, b Jamaica, NY, Nov 30, 33; m 63; c 2. COMPUTER, APPLICATIONS IN BIOLOGY. *Educ:* Rochester Univ, AB, 55; Univ Ind, PhD(zool), 60. *Prof Exp:* Instr biol, Harvard Univ, 60-63; asst prof, Princeton Univ, 63-69; from asst prof to assoc prof, Richmond Col, City Univ New York, 69-74; chmn dept, 74-77, PROF BIOL, GEORGE MASON UNIV, FAIRFAX, 74- *Mem:* AAAS; Genetics Soc Am; Am Soc Cell Biol. *Res:* Extra-chromosomal and developmental genetics; population genetics, trends in amair populations. *Mailing Add:* Dept Biol George Mason Univ Fairfax VA 22030

TAUBE, HENRY, b Neudorf, Sask, Nov 30, 15; nat US; m 40, 52; c 4. INORGANIC CHEMISTRY. *Educ:* Univ Sask, BS, 35, MS, 37; Univ Calif, PhD(chem), 40. *Hon Degrees:* DSc, Univ Chicago, 83, Polytech Inst NY, 84 & State Univ NY, 85, Univ Guelph, 87. *Prof Exp:* Instr chem, Univ Calif, 40-41; from instr to asst prof, Cornell Univ, 41-46; res assoc, Nat Defense Res Comt, 44-45; from asst prof to prof chem, Univ Chicago, 46-61, chmn dept, 55-59; prof chem, Stanford Univ, 62-76, chmn dept, 72-74 & 78-79, Marguerite Blake Wilbur Endowed prof, 76-88, EMER MARGUERITE BLAKE WILBUR PROF CHEM, STANFORD UNIV, 88- *Concurrent Pos:* Guggenheim fel, 49 & 55; corresp mem, Acad Arts & Sci PR, 85; consult, Catalytica Assocs, Inc, Mountain View, Calif. *Honors & Awards:* Nobel Prize in Chem, 83; Award, Am Chem Soc, 55, Howe Award, 60; Chandler Award, Columbia Univ, 64; Kirkwood Award, Yale Univ & Am Chem Soc, 66, Monsanto Co, 81; Nat Medal Sci, 77; Chem Sci Award, Nat Acad Sci, 83; Baylor Medal, Univ Ill, 83; Robert A Welch Found Award, 83; Priestly Medal, Am Chem Soc, 85; hon mem, Hungarian Acad Sci; Distinguished Achievement Award, Int Precious Metals Inst, 86. *Mem:* Nat Acad Sci; Am Chem Soc; Am Acad Arts & Sci; Am Philos Soc; Royal Phys Soc London; hon mem Can Chem Soc; Sigma Xi; Royal Danish Acad Sci & Lett; foreign mem Finnish Acad Sci & Lett; corresp mem Brazilian Acad Sci. *Res:* Chemistry of complex ions; new aquo ions, nitrogen as a ligand; mechanisms of "atom" and electron transfer reactions; mixed valence molecules; charge transfer as affecting properties including the reactivity of ligands; basic chemistry of osmium and nuthenium; over 300 scientific articles and a book published. *Mailing Add:* Dept Chem Stanford Univ Stanford CA 94305

TAUBE, SHEILA EFRON, b New York, NY, Aug 5, 41; m 65; c 2. CANCER DIAGNOSIS, DEVELOPMENTAL RESEARCH. *Educ:* Brandeis Univ, BA, 63; Univ Pittsburg, PhD(microbiol), 70. *Prof Exp:* Postdoctoral assoc human genetics, Yale Univ Sch Med, 71-73; from instr to asst prof microbiol, Univ Conn Med Sch, 73-81; prog dir genetics, NSF, 81-82; grants assoc, NIH, 82-83; PROG DIR BIOCHEM & GENETICS, NAT CANCER INST, NIH, 83-, BR CHIEF CANCER DIAG, 86- *Concurrent Pos:* Prin investr, NIH, 76-79 & Univ Conn res found grant, 79- 80. *Honors & Awards:* Dir's Award, NIH. *Mem:* Am Asn Cancer Res; Am Soc Human Genetics; AAAS; Asn Women Sci. *Res:* Direct program to develop improved approaches to cancer diagnosis and prognosis. *Mailing Add:* Nat Cancer Inst Exec Plaza S Rm 638 Bethesda MD 20892-9904

TAUBENECK, WILLIAM HARRIS, b Marshall, Ill, Aug 27, 23. PETROLOGY. *Educ:* Ore State Col, BS, 49, MS, 50; Columbia Univ, PhD(geol), 55. *Prof Exp:* From instr to prof, 51-84, EMER PROF GEOL, ORE STATE UNIV, 84- *Concurrent Pos:* NSF fel, 59-61; Guggenheim fel, 63-64. *Mem:* AAAS; Geol Soc Am; Mineral Soc Am. *Res:* Layering in igneous rocks; petrogenesis of granite rocks; Columbia River basalt; thermal metamorphism; rock forming minerals; general geology of northeastern Oregon and western Idaho; evolution of the Pacific Northwest. *Mailing Add:* Dept Geosci Ore State Univ Corvallis OR 97331-5506

TAUBER, ALFRED IMRE, b Washington, DC, June 24, 47; m 66; c 4. NEUTROPHIL PHYSIOLOGY. *Educ:* Tufts Univ, BS, 69, MD, 73. *Prof Exp:* From instr to asst prof med, Med Sch, Harvard Univ, 78-82; jr assoc, Peter Bent Brigham Hosp-Brigham & Women's Hosp, 79-82; assoc res prof biochem, 82-86, assoc prof med, 82-86, assoc prof path, 85-86, PROF MED, SCH MED, BOSTON UNIV, 86-, PROF PATH, 87- *Concurrent Pos:* Vis scientist, Mass Inst Technol, 81-82; Consult, US Public Serv Hosp, Brighton, MA, 80-82, Boston Vet Admin Hosp, 83-; assoc staff, Univ Hosp, 83-; dir, Boston Univ Sch Med-Hebrew Univ Exchange Prog, 85-87; assoc physician, 82-, dir labs, Boston City Hosp, 84- *Mem:* Am Fedn Clin Res; Am Soc Hemat; Am Col Physicians; Soc Free Radical Res; Am Soc Biol Chemists; Am Soc Clin Invest; Am Soc Cell Biol; Am Asn Immunol; Asn Am Physicians. *Res:* Elucidation of the mechanisms of oxidative microbial killing by human neutrophil; cell activation pathways; oxidative enzymology; chemistry of the oxygen-derived reactive species. *Mailing Add:* Boston Univ Sch Med 80 E Concord St Boston MA 02118

TAUBER, ARTHUR, b New York, NY, June 2, 28; m 56; c 3. CHEMISTRY, MAGNETISM. *Educ:* NY Univ, BA, 51, MA, 52; Polytech Inst Brooklyn, MS, 59, PhD, 72. *Prof Exp:* Phys chemist, Sig Corps Eng Labs, 52-56, res phys scientist, Electronics Res & Develop Labs, 56-63; electronics command, US Army, 63-75, electronic res & develop command, ET&D Lab, 75-80, res phys scientist supvr, electronic res develop command, ET&D, Lab, 80-85, res phys scientist supvr, Labcom, ET&D Lab, 85-87; CONSULT, 87- *Concurrent Pos:* Res & Develop Award, US Army, 63 & 83. *Honors & Awards:* Meritorious Civilian Serv Medal, US Dept Army, 63. *Mem:* Am Chem Soc; Inst Elec & Electronics Engrs; Am Crystallog Asn; Am Asn Crystal Growth. *Res:* Synthesize and characterize polycrystalline and single crystal microwave/ millimeter wave ferrites; samarium cobalt permanent magnets; amorphous magnetics; intermetallic hydrogen absorbers; high Tc superconductors. *Mailing Add:* 927 Woodgate Ave Long Branch NJ 07740

TAUBER, CATHERINE A, b San Francisco, Calif. BIOSYSTEMATICS, INSECT SEASONALITY. *Educ:* Univ Calif, Berkeley, BS, 62, MS, 64, PhD(emtom), 67. *Prof Exp:* NIH res fel, 67-69, res assoc, 69-76, SR RES ASSOC, CORNELL UNIV, 66- *Mem:* Entom Soc Am; Soc Study Evolution; Soc Syst Zoologists; Sigma Xi. *Res:* Seasonal adaptations, evolutionary diversification and speciation of insects with special emphasis on the Neuroptora. *Mailing Add:* Dept Entom Cornell Univ Ithaca NY 14853

TAUBER, GERALD ERICH, theoretical physics; deceased, see previous edition for last biography

TAUBER, MAURICE JESSE, b Cracow, Can, Oct 21, 37; m 66; c 3. ENTOMOLOGY. *Educ:* Univ Man, BS, 58, MS, 59; Univ Calif, Berkeley, PhD, 66. *Prof Exp:* chmn dept, 81-86, PROF ENTOM, CORNELL UNIV, 66- *Mem:* Fel AAAS; Am Soc Zool; Animal Behav Soc; Ecol Soc Am; Am Inst Biol Sci; Sigma Xi; fel Entom Soc Can. *Res:* Insect behavior; biological control; insect phenology. *Mailing Add:* Dept Entomol Comstock Hall Cornell Univ Ithaca NY 14853-0999

TAUBER, RICHARD NORMAN, b New York, NY, Apr 7, 40; m 63; c 2. SEMICONDUCTOR DEVICE TECHNOLOGY, VERY LARGE SCALE & ULTRA LARGE SCALE INTEGRATION DEVICE PROCESSING. *Educ:* NY Univ, BA & BMet E, 61, MS, 64, PhD(mat sci), 66. *Prof Exp:* Asst prof mat sci, Lehigh Univ, 66-69; mem tech staff, Bell Tel Labs, 69-74; mgr, Gould Inc, 74-75, Xerox Corp, 75-77 & Hughes, 77-78; chief scientist, TRW, 78-89; PRES, MICROELECTRONICS EDUC ASN, 89- *Mem:* Inst Elec & Electronics Engrs. *Res:* Microelectronics industry; ultra large scale integration technology and manufacturing; technological issues; process equipment; manufacturing issues; marketing directions and seminar development; author of one book. *Mailing Add:* 6923 Starstone Dr Rancho Palos Verdes CA 90274

TAUBER, SELMO, b Shanghai, China, Aug 24, 20; nat US; m 50; c 1. APPLIED MATHEMATICS. *Educ:* Beirut Sch Eng, Lebanon, dipl, 43; Univ Lyons, France, Licès Sc, 47; Univ Vienna, Austria, DrPhil, 50. *Prof Exp:* Head sci dept, High Sch, Lebanon, 51-55; design engr, 56-57; assoc prof, Portland State Univ, 57-63, prof math, 63-90; RETIRED. *Concurrent Pos:* From instr to asst prof, Univ Kans, 57-59. *Res:* Engineering structural problems; finite differences; differential equations; combinatorial analysis; system analysis; mathematical models in air pollution; mathematical m dels in physiology. *Mailing Add:* 2839 SW Fairmont Blvd Portland OR 97201

TAUBERT, KATHRYN ANNE, b Lufkin, Tex, Jan 3, 45. CARDIOVASCULAR PHYSIOLOGY. *Educ:* Stephen F Austin State Univ, BS, 65, MS, 66; Univ Tex Southwestern Med Sch, PhD(physiol), 75. *Prof Exp:* Res asst cardiol, Univ Tex Southwestern Med Sch, 66-68; physiologist, Dallas Vet Admin Hosp, 68-75; instr, Univ Tex Southwestern Med Sch, 75-77; asst prof physiol, Univ Calif, Riverside, 77-80; assoc prof physiol & pharmacol, Sch Pharm, Univ Pac, 80-85; SCI CONSULT, NAT OFF, AM HEART ASN, 85 - *Concurrent Pos:* Fel, Univ Tex Health Sci Ctr, 75; NIH investr, Univ Calif, Riverside, 77-80; prin investr, Calif Heart Asn, 78-81 & 81-85. *Mem:* Am Heart Asn; Am Fedn Clin Res; Am Physiol Soc; Int Soc Heart Res. *Res:* Actions of cardiac glycosides on the heart, especially in circumstances similar to those seen in patients with ischemic heart disease or myocardial infarctions; effects of hypertension on cardiac muscle mechanics. *Mailing Add:* Am Heart Asn 7320 Greenville Ave Dallas TX 75231

TAUBLER, JAMES H, b Cokeville, Pa, Mar 30, 35; m 59, 80; c 3. MICROBIOLOGY. *Educ:* St Vincent Col, BA, 57; Cath Univ, MS, 59, PhD(microbiol), 65. *Prof Exp:* Microbiologist, Philadelphia Gen Hosp, 59-69; asst prof, 69-71, PROF BIOL, ST VINCENT COL, 71-, CHMN DEPT, 85-

Concurrent Pos: Lectr, Holy Family Col, Pa, 66-; dir, Delmont Labs, Swarthmore, 66-; Environ Protection Agency grant, 75-77; mem, Pa State Task Force, health care needs AIDS patients. *Mem:* NY Acad Sci. *Res:* Staphylococcal alpha toxin and its relationship to pathogenesis; delayed hypersensitivity and its effects in staphylococcal infections. *Mailing Add:* Dept Biol St Vincent Col Latrobe PA 15650

TAUBMAN, MARTIN ARNOLD, b New York, NY, July 10, 40; m 65; c 2. IMMUNOLOGY, ORAL BIOLOGY. *Educ:* Brooklyn Col, BS, 61; Columbia Univ, DDS, 65; State Univ NY Buffalo, PhD(oral biol), 70. *Prof Exp:* Sr staff mem & head immunol dept, Forsyth Dent Ctr, 70-76; asst clin prof, 76-80, ASSOC CLIN PROF, DEPT ORAL BIOL & PATHOPHYSIOL, HARVARD UNIV SCH DENT MED, BOSTON, 80- & HEAD, IMMUNOL DEPT, FORSYTH DENT CTR, 70- *Concurrent Pos:* Nat Inst Dent Res grant, Forsyth Dent Ctr, 72-, NIH career develop award, 72-77; mem, Oral Biol & Med Study Sect, 80-84; pres, Am Asn Dent Res, 87-91, vpres, 91- *Mem:* Am Asn Immunologists; Int Asn Dent Res. *Res:* Secretory immunoglobulins; effect of secretory antibodies on oral microorganisms. *Mailing Add:* Forsyth Dent Ctr 140 Fenway Boston MA 02115

TAUBMAN, ROBERT EDWARD, b New York, NY, Jan 12, 21; m 43; c 3. PSYCHIATRY, PSYCHOLOGY. *Educ:* City Col New York, BA, 41; Columbia Univ, MS, 42, PhD(psychol), 48; Univ Nebr, MD, 60. *Prof Exp:* Chief psychol serv, Hastings State Hosp, Nebr, 52-56; resident physician psychiat, Ore State Hosp, Salem, 61-64; assoc prof, 64-70, PROF PSYCHIAT, MED SCH, UNIV ORE, 70- *Concurrent Pos:* NIMH fel ment retardation, Letchworth Village, Thiells, NY, 62; consult, Ore Fairview Home, Div Voc Rehab, Marion County Juv Dept, 62-64; pvt pract, 64-; attend physician, Vet Admin Hosp, Portland, Ore, 64-; psychiat dir, Physicians Inst, Ore Acad Family Physicians, 65- *Mem:* Am Pyschiat Asn; AMA; Soc Teachers Family Med. *Res:* Aging; dying; death; self growth among physicians and other professionals; emotional components of life-threatening diseases; psychiatry for non-psychiatric physicians. *Mailing Add:* 5242 SW Humphrey Blvd Portland OR 97221

TAUBMAN, SHELDON BAILEY, b Cleveland, Ohio, Oct 8, 36; m 60; c 2. IMMUNOLOGY, PATHOLOGY. *Educ:* Northwestern Univ, Evanston, BA, 58; Case Western Reserve Univ, PhD(biochem), 64, MD, 66. *Prof Exp:* Asst prof path, 68-77, asst prof lab med, Health Ctr Univ Conn, 77-; CHIEF, DIV CLIN PATH, MT SINAI HOSP, CLEVELAND. *Concurrent Pos:* NIH fel, 64-66, fel path, NY Univ, 66-68 & grant, Health Ctr, Univ Conn. *Mem:* AAAS; Harvey Soc; Am Asn Path & Bact. *Res:* Biochemical mechanisms of inflammation. *Mailing Add:* 52 Pleasant Ridge Dr Poughkeepsie NY 12603

TAUC, JAN, b Pardubice, Czech, Apr 15, 22; m 47; c 2. AMORPHOUS SEMICONDUCTORS. *Educ:* Czech Tech Univ, IngDr, 49; Czech Acad Sci, DrSc(physics), 56; Charles Univ, Prague, RNDr, 56. *Prof Exp:* Scientist microwave res, Sci & Tech Res Inst, Tanvald & Prague, Czech, 49-52; dept head inst solid state physics, Czech Acad Sci, 53-69; prof exp physics, Charles Univ, Prague, 64-69, dir inst physics, 68-69; mem tech staff, Bell Labs, 69-70; dir, Mat Res Lab, 83-88, L HERBERT BALLOU PROF ENG & PHYSICS, BROWN UNIV, 70- *Concurrent Pos:* UNESCO res fel, Harvard Univ, 61-62; vis prof, Univ Paris, 69; consult, Bell Labs, 70-78; vis prof, Stanford Univ, 77; vis scientist, Max Planck Inst Solid State Res, Stuttgart, WGermany, 82. *Honors & Awards:* Nat Prize for Sci, Czech Govt, 55 & 69; Alexander von Humboldt US Sr Scientist Award, 81; Isakson Prize, Am Phys Soc, 82; D Adler Lect Award, Am Phys Soc, 88. *Mem:* Fel Am Phys Soc; Europ Phys Soc; fel AAAS. *Res:* Optical properties and electronic states of crystalline and amorphous solids; picosecond spectroscopy of amorphous semiconductors; real time studies of ultra-high frequency phonons. *Mailing Add:* Div Eng Brown Univ Providence RI 02912

TAUCHERT, THEODORE R, b New York, NY, Sept 3, 35; m 58; c 5. SOLID MECHANICS. *Educ:* Princeton Univ, BSE, 57; Yale Univ, MEng, 60, DEng(solid mech), 64. *Prof Exp:* Struct engr, Sikorsky Aircraft Div, United Aircraft Corp, 57-61; res assoc & lectr solid mech, Princeton Univ, 64-65, asst prof, 65-70; assoc prof, 70-76, PROF SOLID MECH, UNIV KY, 76-, CHMN, DEPT ENG MECH, 80-84, 89- *Mem:* Am Soc Eng Educ; Sigma Xi; Am Soc Civil Engrs; Am Soc Mech Engrs; Soc Eng Sci. *Res:* Composite materials; pressure vessels; thermal stresses. *Mailing Add:* Dept Eng Mech Univ Ky Lexington KY 40506-0046

TAUER, KENNETH J, b Minn, Apr 5, 23; m 44; c 4. PHYSICAL CHEMISTRY, CHEMICAL PHYSICS. *Educ:* Univ Minn, PhD, 51. *Prof Exp:* Researcher, Gen Elec Co, Wash, 51-53; asst prof chem, Boston Col, 53-56; MEM RES STAFF, MAT RES AGENCY LAB, US ARMY MAT & MECH RES CTR, 57- *Mem:* Am Chem Soc; Am Phys Soc. *Res:* Electronics structure of transition metals and alloys; magnetism; electronic transport; magnetoresistance; Hall effect. *Mailing Add:* 23 Old Tavern Rd Cochituate MA 01778

TAULBEE, CARL D(ONALD), b Detroit, Mich, Oct 18, 28; div; c 2. MECHANICAL ENGINEERING, NUCLEAR ENGINEERING. *Educ:* Wayne State Univ, BSME, 53, MSME, 59. *Prof Exp:* Engr, Studebaker-Packard Corp, 53-56; supvry engr, Res Lab Div, Bendix Corp, 56-73; spec projs mgr, Door-Man Mfg Co, 73-79; PRES, AM INDUST DOORS, 79- *Mem:* Am Welding Soc; Am Nuclear Soc; Am Soc Mech Engrs; Am Soc Qual Control. *Res:* Nuclear power plant products; earth resources; radiation effects; nuclear power and propulsion and gas turbine technologies. *Mailing Add:* 26 Oakland Park Blvd Pleasant Ridge MI 48069

TAULBEE, DALE B(RUCE), b Detroit, Mich, Nov 17, 36; m 58; c 2. AEROSPACE & ENGINEERING SCIENCE. *Educ:* Mich State Univ, BS, 58, MS, 60; Univ Ill, PhD(appl mech), 64. *Prof Exp:* Asst prof aerospace eng, State Univ NY, Buffalo, 63-70, assoc prof eng sci, 70-79; DEPT MECH, AEROSPACE ENG, 79- *Concurrent Pos:* Tech consult, Bell Aerosysts Co, 66- *Mem:* Am Inst Aeronaut & Astronaut; Am Soc Mech Engrs; Sigma Xi. *Res:* Fluid mechanics, laminar and turbulent boundary layers; gas dynamics, high speed viscous flows. *Mailing Add:* 30 Harbour Pl Buffalo NY 14202-4305

TAUNTON-RIGBY, ALISON, b Barnsley, Eng, Apr 23, 44; m 66; c 4. BIOTECHNOLOGY. *Educ:* Bristol Univ, BSc, 65, PhD(chem), 68. *Prof Exp:* Vpres res & develop, Collab Res Inc, 69-83; vpres bus develop, Biogen Inc, 83-84; vpres & gen mgr, Vivotech Inc, 84-86; health industs mgr, Arthur D Little Co, 86-87; SR VPRES THERAPEUT, GENZYME CORP, 87- *Concurrent Pos:* Dir, Asn Biotechnol Co, 84-88, Centaur Inc, 85-87, New Eng Brit Bus Asn, 86-90 & Mass Biotech Coun, 90- *Mem:* AAAS; Am Chem Soc. *Res:* Molecular biology; recombinant DNA; immunology; microbiology; cell biology; glycoproteins; enzymology; biopolymers; carbohydrate chemistry. *Mailing Add:* Farrar Rd Lincoln MA 01773

TAUROG, ALVIN, b St Louis, Mo, Dec 5, 15; m 40; c 2. BIOCHEMISTRY, PHYSIOLOGY. *Educ:* Univ Calif, Los Angeles, BA, 37, MA, 39; Univ Calif, Berkeley, PhD(physiol), 43. *Prof Exp:* Asst chem, Univ Calif, Los Angeles, 37-38; asst physiol, Univ Calif, Berkeley, 42-43, chemist, Radiation Lab, 44-45, res assoc physiol, 46-59; from assoc prof to prof, 59-87, ADJ PROF PHARMACOL, UNIV TEX SOUTHWESTERN MED CTR, DALLAS, 87- *Concurrent Pos:* Distinguished lectr, Am Thyroid Asn, 80. *Mem:* AAAS; Am Chem Soc; Am Soc Biol Chem; Endocrine Soc; Am Thyroid Asn; Sigma Xi. *Res:* Thyroid physiology and biochemistry; iodine metabolism. *Mailing Add:* Dept Pharmacol Univ Tex Southwestern Med Ctr 5323 Harry Hines Blvd Dallas TX 75235-9041

TAUSSIG, ANDREW, b Budapest, Hungary, Dec 6, 29; Can citizen; m 59; c 3. MICROBIOLOGY, BIOCHEMISTRY. *Educ:* McGill Univ, BSc, 52, PhD(biochem), 55. *Prof Exp:* Res fel biochem, McGill Univ, 55-61; from asst bacteriologist to bacteriologist, Jewish Gen Hosp, 61-67; vis scientist, Dept Biochem, McGill Univ, 67-70; BACTERIOLOGIST, MICROBIOL LAB, 69- *Concurrent Pos:* Bacteriologist, Bellechasse Hosp, Montreal, 73-84, Jewish Convalescent Hosp, Chomedey & Mt Sinai Hosp, St Agathe, Que; dir labs, Mt Sinai Hosp, St Agathe, Que, 74- *Mem:* Am Soc Microbiol; Can Soc Microbiol; Can Col Microbiolists. *Res:* Nucleic acids of bacteria and bacteriophage; induced enzyme synthesis in bacteria; diagnostic bacteriology. *Mailing Add:* Microbiol Lab 5845 Cote de Neiges Rd Montreal PQ H3S 1Z4 Can

TAUSSIG, STEVEN J, b Timisoara, Rumania, June 2, 14; US citizen; m 65; c 1. INDUSTRIAL MICROBIOLOGY, ENZYMOLOGY. *Educ:* Univ Prague, Czech, Chem Eng, 37; Bucharest Polytech Inst, PhD(biochem), 58. *Prof Exp:* Chemist, Solventul SA, Timisoara, 37-48; asst prof, Polytech Inst Timisoara, 48-57; res fel, Agr Res Inst, Timisoara, 57-59; mgr plant & equip sales, Int Chem Corp, NY, 60-61; tech dir, Pac Labs, Inc, Honolulu, 61-62 & Pac Enzyme Prod, 62-63; res biochemist, Dole Co, Inc, 63-64, dir lab serv, 64-73; PRES, CHEM CONSULTS INT INC, HONOLULU, 73 - *Concurrent Pos:* Tech consult, Pac Labs, Inc & Pac Biochem Co, 63- & Monsanto Co, Mo, 64-65. *Mem:* AAAS; Am Chem Soc; Soc Indust Microbiol; Inst Food Technologists. *Res:* Fermentations; bacterial and other hydrolytic enzymes; chemical equilibria in esterification reactions; cancer research; study of effects of Bromelain on cancer. *Mailing Add:* 469 Ena Rd Apt 3212 Honolulu HI 96815

TAUSSKY, OLGA, b Olomouc, Czech Socialist Repub, 1906; US citizen; m 38. ALGEBRAIC NUMBER THEORY, INTEGRAL MATRICES. *Educ:* Univ Vienna, DPhil(math), 30. *Hon Degrees:* MA, Univ Cambridge, Eng, 37; DSc, Univ Southern Calif, 88. *Prof Exp:* Asst math, Univ Göttingen, 31-32 & Univ Vienna, 32-34; fel, Girton Col, Cambridge Univ, 34-40; asst lectr, Univ London, 37-43; sci officer, Ministry Aircraft Prod, 43-46; mathematician, US Nat Bur Standards, 47-57; res assoc, 57-71, PROF MATH, CALIF INST TECHNOL, 71- *Concurrent Pos:* Lectr, Queen's Univ, Belfast, Northern Ireland, 39-40; res grant, Dept Sci & Indust Res, London, 46-47; mem, Inst Advan Study, Princeton, 48; adj prof, Courant Inst, New York Univ, 55; Fulbright res prof, Univ Vienna, 65. *Honors & Awards:* Ford Prize, Math Asn Am, 71; Gold Cross Honor, Austrian Govt, 78. *Mem:* Corresp mem Austrian Acad Sci; corresp mem Bavarian Acad Sci; Am Math Soc (vpres, 86-88); fel AAAS. *Res:* Algebraic number theory; integral matrices; matrices in algebra and analysis; translating number theory into integral matrices; topological algebra. *Mailing Add:* Dept Math 253-37 Calif Inst Technol Pasadena CA 91125

TAUTVYDAS, KESTUTIS JONAS, b Telsiai, Lithuania, Jan 1, 40; US citizen; m 62; c 4. BIOPROCESSING. *Educ:* Univ Md, BS, 63; Cornell Univ, MS, 65; Yale Univ, PhD(cell & molecular develop biol), 69. *Prof Exp:* Res assoc cell biol, Univ Colo, 69-71; asst prof plant physiol, Marquette Univ, 71-79; supvr biol screening & biochem, 79-85, BIOTECHNOL SUPVR, 3M CO, 85- *Mem:* Am Soc Microbiol. *Res:* Biotransformations. *Mailing Add:* 8182 Hidden Bay Trail St Paul MN 55042

TAUXE, WELBY NEWLON, b Knoxville, Tenn, May 24, 24; c 4. NUCLEAR MEDICINE, CLINICAL PATHOLOGY. *Educ:* Univ Tenn, Knoxville, BS, 44, MD, 50; Univ Minn, Minneapolis, MS, 58. *Prof Exp:* Mayo Found fel, Mayo Clin, 54-58, chief nuclear med dept, 58-72; prof nuclear med & clin path, 72-77, PROF DIAG RADIOL & PATH, COL MED, UNIV ALA, BIRMINGHAM, 77- *Concurrent Pos:* Consult, AEC, 62-70; treas, Am Bd Nuclear Med, 71; consult, Am Nat Standards Inst, 72. *Mem:* Am Soc Clin Path; Soc Nuclear Med. *Res:* Copper kinetics in Wilson's disease; use of radioactive materials in diagnosis. *Mailing Add:* Dept Nuclear Med Univ Pittsburgh Presby Hosp Pittsburgh PA 15213

TAVANO, DONALD C, b Newark, NY, Aug 26, 36; m 58; c 2. PUBLIC HEALTH EDUCATION, MEDICAL EDUCATION. *Educ:* State Univ NY Col Cortland, BS, 60; Univ Ill, MS, 61; Univ Mich, MPH, 63; Mich State Univ, PhD(educ), 71. *Prof Exp:* Instr health educ, State Univ NY Col Cortland, 61-62; dir health educ, Saginaw City & County Health Depts, 63-65; consult health educ, Mich Dept Pub Health, 65-66; from asst prof to assoc prof community med, 66-83, PROF, MED EDUC RES & DEVELOP, MICH STATE UNIV, 83- *Concurrent Pos:* Vis lectr, Sch Pub Health, Univ Mich, 72; consult, Mich Dept Educ, 66-, Gov Off Health & Med Affairs & Nat Bd Examrs Osteop Physicians & Surgeons Inc, 75. *Mem:* Soc Pub Health

Educ; Am Pub Health Asn; Asn Behav Sci Med Educ; Asn Teachers Prev Med. *Res:* Patient education; the impact of patient education on health care cost containment, patient compliance and treatment outcomes. *Mailing Add:* Med Educ Res A214c E Ree Hall Mich State Univ East Lansing MI 48824

TAVARES, DONALD FRANCIS, organic chemistry; deceased, see previous edition for last biography

TAVARES, ISABELLE IRENE, b Merced, Calif, Oct 6, 21. MYCOLOGY, LICHENOLOGY. *Educ:* Univ Calif, PhD(bot), 59. *Prof Exp:* Sr lab technician protozool, 49-52, from herbarium botanist to sr herbarium botanist, 52-68, assoc specialist, 68-84, SPECIALIST BOT, UNIV CALIF, BERKELEY, 84- *Mem:* Bot Soc Am; Mycol Soc Am; Am Bryological & Lichenological Soc; Brit Lichenology Soc; Int Asn Lichenology. *Res:* Laboulbeniales; Usnea. *Mailing Add:* 6701 San Pablo Ave Marchant Bldg Oakland CA 94608

TAVARES, STAFFORD EMANUEL, b Kingston, Jamaica, WI, May 11, 40; Can citizen. ELECTRICAL ENGINEERING. *Educ:* McGill Univ, BEng, 62, PhD(elec eng), 68; Calif Inst Technol, MS, 64. *Prof Exp:* Jr res off elec eng, Nat Res Coun Can Labs, Ottawa, 62-65, asst res off, 65-70; asst prof, 70-74, assoc prof, 74-80, PROF ELEC ENG, QUEEN'S UNIV, ONT, 80- *Concurrent Pos:* Lectr, Carleton Univ, 68-70; Natural Sci & Eng Res Coun Can res grants, Queen's Univ, Ont, 71-; vis assoc prof, Stanford Univ, 77-78; prin invstr, Proj Secure Wireless Commun; sponsor, Telecommunication Res Inst Ont. *Mem:* Inst Elec & Electronics Engrs. *Res:* Computer communications; digital communications; error-correcting codes; data encryption and security in computer communications networks. *Mailing Add:* Dept Elec Eng Queen's Univ Kingston ON K7L 3N6 Can

TAVASSOLI, MEHDI, b Tehran, Iran, Mar 30, 33; US citizen; m 66; c 3. HEMATOLOGY. *Educ:* Tehran Univ Sch Med, MD, 61. *Prof Exp:* Intern, Cambridge City Hosp, Mass, 61-62; resident, Cook County Hosp, Chicago, 63-64 & Carney Hosp, Boston, 64-66; hematologist, Tufts-New Eng Med Ctr, 66-72 & from instr to asst prof med, Sch Med, Tufts Univ, 69-72; hematologist, Scripps Clin & Res Found, 72-81; PROF MED, SCH MED, UNIV MISS, 81-; CHIEF HEMAT-ONCOL & DIR, CELL BIOL LAB, VET ADMIN HOSP, JACKSON, MISS, 81- *Concurrent Pos:* Charlton fel, Tufts Univ, 68; res fel, The Med Found, Boston, 70; vis investr anat, Med Sch, Johns Hopkins Univ, 70-71; consult hemat, Vet Admin Hosp & Univ Calif, San Diego, 73-; res career develop award, NIH, 74-79. *Honors & Awards:* John Larkin Award, The Guild of St Luke, 66. *Mem:* Am Soc Hemat; Am Col Physicians; Int Soc Exp Hemat; Am Asn Cancer Res; Am Soc Clin Oncol; Asn Am Physicians. *Res:* Structural basis of hemopoiesis and the function of microenvironmental factor in immunohemopoiesis; membrane structure and function. *Mailing Add:* Dept Res Vets Admin Med Ctr 1500 E Woodrow Wilson Dr Jackson MS 39216

TAVE, DOUGLAS, b Oxford, Eng, Dec 13, 49; US citizen; m 78; c 1. AQUACULTURE, FISH BREEDING. *Educ:* Coe Col, BA, 71; Univ Ill, MS, 73; Auburn Univ, PhD(aquacult), 79, MEd, 85. *Prof Exp:* Fel fish breeding, Dept Fisheries & Allied Aquacultures, Auburn Univ, 74-80, asst prof, 80-81; asst prof aquacult, dept entom, fisheries & wildlife, Univ Minn, 81-84; vis scientist, Dept Fisheries & Allied Aquacult, Auburn Univ, 84-89; ASSOC PROF AQUACULT, DEPT AQUACULT, UNIV ARK, PINE BLUFF, 89- *Concurrent Pos:* Peace Corps, 75-76. *Mem:* Sigma Xi; World Maricult Soc. *Res:* Quantitative and mendelian genetic research on tropical and temperate food fish: tilapia, catfishes and baitfishes. *Mailing Add:* Dept Agr Univ Ark Pine Bluff Pine Bluff AR 71601

TAVEL, MORTON, b Brooklyn, NY, June 14, 39; m 69; c 1. THEORETICAL PHYSICS. *Educ:* City Col New York, BS, 60; Stevens Inst Technol, MS, 62; Yeshiva Univ, PhD(physics), 64. *Prof Exp:* Res assoc, Brookhaven Nat Lab, 64-66, asst scientist, 66-67; asst prof, 67-70, assoc prof, 70-74, PROF PHYSICS, VASSAR COL, 74- *Mem:* AAAS; Am Phys Soc. *Res:* Quantum field theory; plasma physics; transport theory. *Mailing Add:* Dept Physics Vassar Col Poughkeepsie NY 12601

TAVERAS, JUAN M, b Dominican Repub, Sept 27, 19; nat US; m 47; c 3. RADIOLOGY. *Educ:* Norm Sch Santiago, Dominican Repub, BS, 37; Univ Santo Domingo, MD, 43; Univ Pa, MD, 49. *Hon Degrees:* MA, Harvard Univ, 71; DSc, Univ Pedro Henriquez, 87. *Prof Exp:* Instr radiol, Col Physicians & Surgeons, Columbia Univ, 50-52, from asst prof to prof, 52-65; prof radiol & chmn dept, Sch Med, Wash Univ, radiologist-in-chief, Barnes & Allied Hosps, Univ Med Ctr & dir, Mallinckrodt Inst Radiol, 65-71; RADIOLOGIST-IN-CHIEF, MASS GEN HOSP, BOSTON, 71-; PROF RADIOL, HARVARD MED SCH, 71- *Concurrent Pos:* Dir radiol, Neurol Inst, New York, 52-65; from asst to attend radiologist, Presby Hosp, New York, 50-65; mem neurol study sect, Nat Inst Neurol Dis & Stroke, 64-68; consult, US Marine Hosp, Staten Island, NY, Bronx Vet Admin Hosp, St Barnabas Hosp Chronic Dis, New York, Morristown Mem Hosp, NJ, St Louis City Hosp & Jewish Hosp, St Louis, 65-71; hon prof med, Univ Santo Domingo & Univ Chile. *Honors & Awards:* Gold Medal, Am Col Radiol; Gold Medal, Am Roentgen Ray Soc; Gold Medal, Radiol Soc NAm, 80; Gold Medal, Asn Univ Radiol. *Mem:* AMA; Am Neurol Asn; fel Am Col Radiol; Am Roentgen Ray Soc; Am Soc Neuroradiol (pres, 62-64); Radiol Soc NAm; Asn Univ Radiol. *Res:* Radiologic aspect of neurological science, especially cerebral angiography and cerebral vascular disease. *Mailing Add:* Dept Radiol Mass Gen Hosp Fruit St Boston MA 02114

TAVES, MILTON ARTHUR, b Aberdeen, Idaho, Aug 14, 25; m 45; c 2. ORGANIC CHEMISTRY. *Educ:* Univ Utah, BS, 45; Mass Inst Technol, PhD(org chem), 48. *Prof Exp:* Lab asst chem, Univ Utah, 43-45; asst, Mass Inst Technol, 45-46; res chemist, Res Ctr, Hercules, Inc, 48-54, res supvr, 54-60 & 62-64, tech asst to dir res, 60-62, res mgr, Synthetic Res Div, 64-77, new technol coordr, 77-81; CONSULT, 81- *Mem:* Am Chem Soc; Sigma Xi. *Res:* Heterocyclics; hydrogen peroxide; hydroperoxide chemistry; auto-oxidation and catalytic oxidation of organic compounds; terpenes; condensation polymers; resins; plasticizers; chemicals via fermentation; immobilized enzyme technology; process development; agricultural chemicals. *Mailing Add:* 210 N Spring Valley Rd Wilmington DE 19807

TAVILL, ANTHONY SYDNEY, b Manchester, Eng, July 15, 36; m 59; c 3. GASTROENTEROLOGY, HEPATOLOGY. *Educ:* Univ Manchester, MB & ChB, 60, MD, 70; Royal Col Physicians London, MRCP, 63, FRCP, 78. *Prof Exp:* Med Res Coun travelling fel med, Albert Einstein Col Med, 66-68; lectr, Royal Free Hosp, Sch Med, Univ London, 68-71; sr clin scientist, Div Clin Invest, Med Res Coun Clin Res Ctr, Eng, 71-72, consult gastroenterol & liver dis, 72-75; HEAD GASTROENTEROL & PROF MED, CASE WESTERN RESERVE UNIV, CLEVELAND METROP GEN HOSP, 75- *Mem:* Med Res Soc Gt Brit; Brit Soc Gastroenterol; Europ Soc Clin Invest; Am Asn Study Liver Dis; Int Asn Study Liver; Cent Soc Clin Res. *Res:* Control mechanisms in hepatic protein metabolism in gastrointestinal and renal disease, homeostasis of plasma transport proteins and metabolic interrelationships between iron transport and storage proteins of the liver; hepatotoxicity. *Mailing Add:* Dept Gastroenterol Cleveland Metrop Gen Hosp Cleveland OH 44109

TAVLARIDES, LAWRENCE LASKY, b Wilkinsburg, Pa, Jan 8, 42; m 65; c 2. SEPARATIONS, KINETICS. *Educ:* Univ Pittsburgh, BSChE, 63, MS, 64, PhD(chem eng), 68. *Prof Exp:* Engr, Mobay Chem Co, 62; chem engr, Gulf Res & Develop Co, 64-66 & 68; fel photochem reactions, Delft Univ Technol, 68-69; from asst prof to prof, Ill Inst Technol, 69-81; chmn, dept chem engr & mat sci, 81-85, PROF CHEM ENGR, SYRACUSE UNIV, 85- *Concurrent Pos:* Consult, CPC Int, Inc, 70-78, Res Inst, Ill Inst Technol, 73-77 & Exxon, 84-; NSF res grants, 71-72, 75, 77, 78, 82, 83, 85, 86-87, 87-88 & 91-93; Dept Energy grants, 79-86; Dept Defense, Naval Surface Weapons Ctr, 85-87; NY State Ctr Hazardous Waste Mgt, 88-91; NY State Energy & Res Develop Authority, 89-91; Nat Inst Environ Health & Sci, 90-91; ed, Solvent Extraction & Ion Exchange. *Mem:* AAAS; fel Am Inst Chem Engrs; Am Chem Soc. *Res:* Metal extraction in liquid dispersions; droplet rate processes in liquid dispersions; mixing in liquid dispersions; kinetics of metal extraction; Fischer-Tropsch kinetics; chemical reaction engineering; plasma reactors and reactions; supercritical extraction and supercritical wet oxidation for soils remediation; metal ion separation from dilute waste streams by supported liquid membranes, separations, control of extractors. *Mailing Add:* Dept Chem Eng & Mat Sci 320 Hinds Hall Syracuse Univ Syracuse NY 13244

TAVOULARIS, STAVROS, b Athens, Greece, June 16, 50; Can citizen; m 78; c 2. FLUID MECHANICS, EXPERIMENTAL TECHNIQUES. *Educ:* Nat Tech Univ Greece, dipl eng, 73; Va Polytech Inst & State Univ, MSc, 74; Johns Hopkins Univ, PhD(fluid mech), 78. *Prof Exp:* Assoc res scientist, Dept Chem Eng, Johns Hopkins Univ, 79-80; from asst prof to assoc prof, 80-83, PROF MECH ENG, UNIV OTTAWA, 88- *Concurrent Pos:* Chmn, Dept Mech Eng, Univ Ottawa, 87-90; dir, Ottawa-Carleton Inst Mech & Aeronaut Eng, 90-93. *Mem:* Am Phys Soc; Sigma Xi. *Res:* Fluid mechanics, especially on turbulent flows; structure of turbulent shear flows; turbulent diffusion and mixing; jet cavitation; flow in nuclear reactor rod bundles; hemodynamics of cardiac assist devices; low Reynolds number flows. *Mailing Add:* Dept Mech Eng Univ Ottawa 770 King Edward Ave Ottawa ON K1N 6N5 Can

TAX, ANNE, b New York, NY, May 7, 44; m 68; c 2. IMMUNOLOGY. *Educ:* Rutgers Univ, BA, 66; Cornell Univ, PhD(microbiol), 71. *Prof Exp:* Head radioimmunoassay lab, Meloy Labs, 71-73; ASST PROF TUMOR IMMUNOL, WISTAR INST, 73- *Mem:* AAAS; Am Asn Immunologists; Am Asn Cancer Res. *Res:* Immunology to produce and utilize monoclonal antibodies to detect cell surface antigens; virology; tumor viruses; molecular biology. *Mailing Add:* 9733 Bustleton Ave Suite 2N Philadelphia PA 19115

TAYA, MINORU, b Yokosuka, Japan, Sept 19, 44; m 76; c 3. APPLIED MECHANICS, MATERIALS SCIENCE. *Educ:* Univ Tokyo, BS, 68; Northwestern Univ, MS, 73, PhD(ductile fracture), 77. *Prof Exp:* Design engr stress anal, Sumitomo Heavy Industs Co Ltd, Japan, 68-71 & 73-76; fel mech property of porous media, Northwestern Univ, 77-78; from asst prof to assoc prof, Univ Del, 83-86; assoc prof, 86-89, PROF MECH ENG, UNIV WASH, 90- *Mem:* Am Acad Mech; Am Soc Mech Engrs; Am Inst Mineral Eng; Am Ceramic Soc. *Res:* Solid mech; thermo-mechanical properties of composite materials; manufacturing process of composite materials; impact physics; electrical packaging materials. *Mailing Add:* Dept Mech Eng Univ Wash Seattle WA 98195

TAYAMA, HARRY K, b Los Angeles, Calif, May 26, 35; m 61; c 2. HORTICULTURE, FLORICULTURE. *Educ:* Univ Ill, BS, 58, MS, 59; Ohio State Univ, PhD(hort), 63. *Prof Exp:* Asst prof hort, Pa State Univ, 63-64; from asst prof to assoc prof, 64-67, PROF HORT, OHIO AGR RES & DEVELOP CTR, OHIO STATE UNIV, 70- *Concurrent Pos:* Exec dir & ed, Ohio Florists' Asn Bull, 77- *Honors & Awards:* Kenneth Post Award, 66. *Mem:* Am Soc Hort Sci. *Res:* Ecological factors affecting growth and flowering of florist crops; plant growth regulators. *Mailing Add:* Dept Hort Ohio State Univ 2001 Fyffe Ct Columbus OH 43210-1096

TAYBACK, MATTHEW, b Tarrytown, NY, June 30, 19; m 45; c 3. PUBLIC HEALTH. *Educ:* Harvard Univ, AB, 39; Columbia Univ, AM, 40; Johns Hopkins Univ, ScD(biostatist), 53. *Prof Exp:* Res assoc, NY State Psychiat Inst, 40-42; res statistician, NY State Health Dept, 46-48, dir bur biostatist, Baltimore, 48-53, dir statist sect, 53, asst comnr health, 57-63, dep comnr health, 63-69, asst secy health & ment hyg & sci affairs, 69-73; from asst prof to assoc prof, 52-65, PROF HYG & PUB HEALTH, SCH MED, UNIV MD, BALTIMORE CITY, 65-; ADJ PROF, JOHNS HOPKINS UNIV, 83- *Concurrent Pos:* Lectr, Johns Hopkins Univ, 51-; vis prof, Univ Philippines, 57; consult, US Army, 57-, WHO, 60- & USAID, 61-; vchmn, State Comn on Aging, Md, 61-75; chmn, State Adv Bd Price Comn, 72-74; chmn, State Emp Ret Rev Bd, 74-; hon prof community med, Pahlavi Univ, Iran, 78; state dir on aging, 74-83. *Mem:* Am Pub Health Asn. *Res:* Epidemiology; demography; health services administration. *Mailing Add:* Dept Med Johns Hopkins Univ 4940 Eastern Ave Baltimore MD 21224

TAYLOR, A, b Manchester, Eng, Aug 20, 11. METALLURGY, PHYSICS. *Educ:* Manchester Univ, BSc, 33, MSc, 34, PhD(physics), 36. *Hon Degrees:* DSc, Manchester Univ, 67. *Prof Exp:* Sr physicist, Mond Nickel Co, 47-52; sr fel engr, Westinghouse Res Lab, 54-73; PRES, A & R TAYLOR TRANSL & SCI CONSULT SERV, 73- *Mem:* Am Phys Soc; Inst Phys Soc Eng. *Mailing Add:* 2415 Beechwood Blvd Pittsburgh PA 15217

TAYLOR, ALAN D, b Melrose, Mass, Oct 27, 47. SET THEORIST, MATHEMATICAL POLITICAL SCIENCE. *Educ:* Dartmouth Col, PhD(math), 75. *Prof Exp:* From asst prof to assoc prof math, 75-82, chmn, 85-88, PROF MATH, UNION COL, 82- *Mailing Add:* Dept Math Union Col Schenectady NY 12308

TAYLOR, ALAN NEIL, b Franklin, NY, Sept 10, 34; m 55; c 2. ANATOMY, PHYSIOLOGY. *Educ:* Ohio State Univ, BS, 57; Cornell Univ, MS, 60, PhD(phys biol), 69. *Prof Exp:* Res assoc mineral metab, NY State Vet Col, Cornell Univ, 60-66, NIH traineeship, 66-69, sr res assoc membrane transport, 69-75; from assoc prof to prof, 77-85, PROF & CHMN, DEPT ANAT, BAYLOR COL DENT & GRAD SCH, BAYLOR UNIV, 85- *Mem:* AAAS; Am Inst Nutrit; Am Asn Anatomists; Soc Exp Biol & Med; Int Asn Dent Res. *Res:* Mineral metabolism; membrane transport; mechanisms of vitamin D action; biochemistry. *Mailing Add:* Dept Anat Baylor Col Dent 3302 Gaston Ave Dallas TX 75246

TAYLOR, ALBERT CECIL, b Hopei, China, Aug 15, 05; US citizen; m 35; c 2. BIOLOGY. *Educ:* Taylor Univ, BA, 30; Univ Ky, MA, 34; Univ Chicago, PhD(zool), 42. *Prof Exp:* Teacher biol acad, Chicago Evangel Inst, 30-32; teacher sci, Asheland Jr High Sch, Louisville, 34-35; teacher zool, NC State Teachers Col, Asheville, 35-39; asst zool, Univ Chicago, 42-43; res assoc, 43-45; asst prof anat, Col Dent, NY Univ, 45-49, assoc prof, Col Dent & Grad Sch, 49-55; assoc, Rockefeller Inst, 55-58, assoc prof develop biol, 58-65; prof anat, Univ Tex Dent Br Houston & Grad Sch Biomed Sci & mem dent sci inst, Houston, 65-74; RETIRED. *Mem:* Soc Exp Biol & Med; Am Asn Anatomists; Am Soc Cell Biol; Soc Develop Biol; Soc Cryobiol. *Res:* Development nervous system; freezing and freeze-drying of tissues; cell interactions in development; cell contact and adhesion; lysis of tissue collagen. *Mailing Add:* 3202 Blue Bonnet Houston TX 77025

TAYLOR, ALLEN, b New York, NY, Jan 11, 46; m 77; c 1. CATARACT REMEDIATION, ENZYMOLOGY. *Educ:* City Col New York, BS, 67; Rutgers Univ, PhD(chem & biochem), 73. *Prof Exp:* Fel biochem, Univ Calif, Berkeley, 73-75, lectr, 76; asst prof chem & nutrit, Williams Col, 77-81, fel, 81-83; PROF BIOCHEM & NUTRIT, TUFTS UNIV, 83-, DIR, LAB NUTRIT & VISION RES, USDA HUMAN NUTRIT RES CTR ON AGING, 83- *Concurrent Pos:* NIH fel, 74, res grant, 78-94; vis scientist, Harvard Univ, 81-83; consult, Biogen USA Consult Capacities Group, 83-; Guggenheim fel, 86-88. *Mem:* Am Chem Soc; Sigma Xi; NY Acad Sci; Asn Res Vision Ophthal; Am Soc Biochem & Molecular Biol. *Res:* Use of nutrition to delay the onset of cataracts and other visual disorders; molecular and structural studies on leucine aminopeptidase; relationships between aging, antioxidant function and proteolytic capability; investigation of the ubiquitin and energy requirement for proteolytic activity in the aging eye lens; development of alternatives to surgery for cataract victims. *Mailing Add:* USDA Human Nutrit Res Ctr on Aging Tufts Univ 711 Washington St Boston MA 02111

TAYLOR, ANDREW RONALD ARGO, b Ottawa, Ont, July 6, 21; m 47; c 4. BOTANY. *Educ:* Univ Toronto, BA, 43, PhD(bot), 55. *Prof Exp:* Asst bot, Univ Toronto, 43-46; from asst prof to assoc prof biol, 46-62, actg dean sci, 74-75, prof, 62-87, EMER PROF BOT, UNIV NB, FREDERICTON, 87- *Concurrent Pos:* Asst biologist, Fisheries Res Bd Can, 48-60; hon lectr, Univ St Andrews, 60-61, 76; vis prof, Univ Adelaide, 75-76; hon res assoc, Univ Western Australia, 82-83. *Mem:* Int Phycol Soc (treas, 77-82); Phycol Soc Am; Can Bot Asn; Brit Phycol Soc. *Res:* Developmental morphology, ecology and taxonomy of marine algae and sea grasses. *Mailing Add:* Dept Biol Univ NB Fredericton NB E3B 5A3 Can

TAYLOR, ANGUS ELLIS, b Craig, Colo, Oct 13, 11; m 36; c 3. MATHEMATICS. *Educ:* Harvard Univ, SB, 33; Calif Inst Technol, PhD(math), 36. *Prof Exp:* Instr math, Calif Inst Technol, 36-37; Nat Res Coun fel, Princeton Univ, 37-38; from instr to prof math, Univ Calif, Los Angeles, 38-66; vpres acad affairs, Univ Calif Systemwide Admin, 65-75, Univ provost, 75-77; univ provost & chancellor, Univ Calif, Santa Cruz, 76-77, EMER PROF, UNIV CALIF, LOS ANGELES & BERKELEY, EMER UNIV PROVOST & EMER CHANCELLOR, SANTA CRUZ, 77- *Concurrent Pos:* Fulbright res fel, Univ Mainz, 55. *Mem:* Am Math Soc; Math Asn Am. *Res:* Theory of functions; linear operators and spectral theory; history of mathematics. *Mailing Add:* 82 Norwood Ave Kensington CA 94707-1150

TAYLOR, ANNA NEWMAN, b Vienna, Austria, Oct 28, 33; US citizen; m 59. PHYSIOLOGY, ANATOMY. *Educ:* Western Reserve Univ, AB, 55, PhD(physiol), 61. *Prof Exp:* Am Heart Asn res fel physiol, Western Reserve Univ, 61-63, instr, 62-63; Am Heart Asn adv res fel, Lab Neurophysiol, Henri-Rousselle Hosp, Paris, France, 63-64; asst prof physiol, Western Reserve Univ, 64-65; asst prof anat, Dept Anat & Psychiat, Col Med, Baylor Univ, 65-67; asst res anatomist, 67-68, from asst prof to assoc prof, 68-79, PROF ANAT, UNIV CALIF, LOS ANGELES, 79-; CHIEF, ALCOHOL RES LAB, US VET ADMIN, BRENTWOOD, LOS ANGELES, 79- *Concurrent Pos:* USPHS fel, 61; res specialist, Houston State Psychiat Inst, 65-67; NIMH res scientist develop award, 72-77; mem biomed panel, Nat Insts Drug Abuse Res Review Comt, 77-81; US Vet Admin Alcohol res award, 79-81. *Mem:* Fel AAAS; Am Physiol Soc; Endocrine Soc; Am Asn Anat; Soc Neurosci. *Res:* Neuroendocrinology; central nervous system regulation of pituitary-adrenal function; central actions of hormones; developmental and long-term effects of perinatal exposure to hormones, drugs of abuse and alcohol. *Mailing Add:* Dept Anat 73-235 Univ Calif 405 Hilgard Ave Los Angeles CA 90024

TAYLOR, ARCHER S, b Longmont, Colo, Feb 14, 16; m 44; c 4. ELECTRONICS ENGINEERING. *Educ:* Anitoch Col, BS, 38. *Prof Exp:* Physicist & engr, Nat Bur Standards, 38-43; engr, Paul F Godley Co, 44-47, Missoula, Mont, 47-64; ENGR & SR VPRES ENG, MALARKEY-TAYLOR ASSOC, 65- *Concurrent Pos:* Technician & instr, Mont State Univ, Missola, 50-62. *Honors & Awards:* Matti S Siukola Award, Inst Elec & Electronics Engrs. *Mem:* Fel Inst Elec & Electronics Engrs; Soc Motion Picture & TV Eng. *Mailing Add:* Malarkey-Taylor Assoc Inc 1130 Connecticut Ave NW Washington DC 20036

TAYLOR, ARDELL NICHOLS, b Terral, Okla, Jan 19, 17; m 43; c 2. PHYSIOLOGY. *Educ:* Tex Tech Col, BS, 39; Univ Tex, MA, 41, PhD(zool sci), 43. *Hon Degrees:* DSc, Lincoln Col, 71. *Prof Exp:* Tutor zool, Univ Tex, 39-42, from instr to asst prof physiol, Sch Med, 43-46; from asst prof to assoc prof, Sch Med, Univ Okla, 46-51, prof, chmn dept & assoc dean, 51-60; assoc secy, Coun Med Educ & dir, Dept Allied Med Prof & Serv, AMA, 60-67; dean sch related health sci, Chicago Med Sch, 67-69, pres, 69-76, EMER PRES, UNIV HEALTH SCI-CHICAGO MED SCH, 76- *Concurrent Pos:* Ord Episcopal priest, 70, assoc rector, Christ Church, 70- *Honors & Awards:* Distinguished Serv Award. *Mem:* Soc Exp Biol & Med; Am Physiol Soc; Am Math Asn; Am Asn Med Rec Librn. *Res:* Nucleic acid metabolism in ova; nerve conduction and facilitation; hypertension; experimental vascular physiology; dynamics of circulation; medical education. *Mailing Add:* 503 Hawthorn Winnetka IL 60093

TAYLOR, AUBREY ELMO, b El Paso, Tex, June 4, 33; m 54; c 3. PHYSIOLOGY, BIOPHYSICS. *Educ:* Tex Christian Univ, AB, 60; Univ Miss, PhD(physiol, biophys), 64. *Prof Exp:* Res asst learning theory, Bell Helicopter Co, Tex, 59-60; prof math, Exten Ctr, Univ Miss, 60-65, from asst prof physiol & biophys to prof, Med Ctr, 71-77; PROF PHYSIOL & CHMN DEPT, UNIV S ALA, 77- *Concurrent Pos:* Fel, Harvard Med Sch Biophys Lab, 64-67; assoc ed, J Appl Physiol; ed, J Critical Care; bd dirs, Fed Am Soc Exp Biol, NAS & Int Physiol Unit; mem comt, Am Heart Asn. *Honors & Awards:* Landis Award, Microcirculatory Soc; Wiggers Award, Am Physiol Soc, 87; C Drinker Award NAm Soc Lymphol, 88; Dickinson Richards Award, Am Heart Asn, 88. *Mem:* Am Physiol Soc (counr & pres, 88-89); Microcirculatory Soc (pres & counr); Am Heart Asn; Biophys Soc; AAAS; NY Acad Sci; Europ Microcirculatory Soc; NAm Soc Lymphol; Int Lymphol Soc. *Res:* Irreversible thermodynamics and membrane biophysics to mammalian physiology, especially in fields of cardio-pulmonary, intestinal dynamics and capillary exchange; author or coauthor of over 500 publications and 4 textbooks. *Mailing Add:* Dept Physiol Univ SAla Mobile AL 36688

TAYLOR, AUSTIN LAURENCE, b Vancouver, BC, Jan 23, 32; US citizen; m 56; c 2. BACTERIAL GENETICS, BACTERIOPHAGE GENETICS. *Educ:* Western Md Col, BS, 54; Univ Calif, Berkeley, PhD(bact), 61. *Prof Exp:* Res assoc bact genetics, Brookhaven Nat Lab, 61-62; res microbiologist, NIH, 62-65; asst prof, 65-70, ASSOC PROF MICROBIOL, SCH MED, UNIV COLO, DENVER, 70- *Concurrent Pos:* Prin investr, USPHS res grant, 66- *Mem:* AAAS; Genetics Soc Am; Am Soc Microbiol. *Res:* Molecular genetic studies on the mechanism of bacteriophage Mu DNA replication and transposition. *Mailing Add:* Dept Microbiol & Immunol Med Ctr Univ Colo 4200 E Ninth Ave Denver CO 80262

TAYLOR, B GRAY, b Booneville, Miss, Aug 11, 24; m 49; c 2. ONCOLOGY, SURGERY. *Educ:* Harvard Med Sch, MD, 48. *Prof Exp:* From intern to resident surg, Grady Mem Hosp, Atlanta, Ga, 48-50; resident, Hosp, Emory Univ, 50-51; resident, Mem Ctr Cancer & Allied Dis, New York, 53-57; instr surg, 57-61, assoc prof clin surg, 64-72, assoc prof surg, 72-74, prof surg, La State Univ Med Ctr, New Orleans, 74-86; ASST CONSULT PROF SURGERY, DUKE UNIV MED CTR, 86-; CHIEF OF STAFF VA MED CTR, FAYETTEVILLE, NC, 86- *Concurrent Pos:* Spec fel head & neck surg with Dr Hayes Martin, Mem Ctr Cancer & Allied Dis, New York, 57; chief surg, Vet Admin Hosp, New Orleans, 64-, sr physician, 72-; consult, USPHS Hosp, New Orleans, 64-81; active staff, Touro Infirmary, New Orleans, 64-86; sr vis surgeon, Charity Hosp, New Orleans, 64-86; assoc staff, St Charles Hosp, New Orleans, 72-86. *Mem:* James Ewing Soc; Soc Head & Neck Surg; fel Am Col Surg. *Res:* Oncologic and head and neck surgery; physical and chemical properties of human cadaver blood; clincial studies of ameloblastoma of the mandible and carotid body tumors; clinical studies on carcinoma of the male breast, the parotid salivary gland and the larynx. *Mailing Add:* 2300 Ramsey St Fayetteville NC 28301

TAYLOR, BARNEY EDSEL, b Elizabethton, Tenn, Dec 10, 51; m 71. EXPERIMENTAL SOLID STATE PHYSICS. *Educ:* ETenn State Univ, BS, 73; Clemson Univ, PhD(physics), 78. *Prof Exp:* Asst prof physics, Denison Univ, 78-79; asst prof physics, Jackson State Univ, 79-; AT DEPT PHYS, WRIGHT STATE UNIV. *Mem:* Am Phys Soc; Am Asn Physics Teachers. *Res:* Transport studies in ionic solids by means of electrical conductivity; ionic thermocurrents and radiotracer diffusion. *Mailing Add:* Dept Physics Wright State Univ Colonel PO Box 48168 Dayton OH 45435

TAYLOR, BARRIE FREDERICK, b Nottingham, Eng, Nov 21, 39; m 65; c 2. MICROBIOLOGY, BIOCHEMISTRY. *Educ:* Univ Leeds, BSc, 62, PhD(biochem), 65. *Prof Exp:* Postdoctoral, Rutgers Univ, 65-67; res assoc microbiol, Univ Tex, Austin, 67-69; asst prof marine sci, Biol & Living Resources Div, Sch Marine & Atmospheric Sci, Univ Miami, 70-74, assoc prof, 74-78; prof biol & living resources, Univ Miami, Coral Gables, 78-80; PROF MARINE & ATMOSPHERIC CHEM DIV, ROSENSTIEL SCH MARINE & ATMOSPHERIC SCI, UNIV MIAMI, 80- *Concurrent Pos:* NSF res grants, 70-72 & 73-81, 86-89 & 90-93; NIH grant, 74-77 & 81-84. *Mem:* AAAS; Am Soc Microbiol; Am Soc Limnol & Oceanog; Am Chem Soc; Soc Gen Microbiol. *Res:* Microbial biochemistry; autotrophic and lithotrophic micro-organisms; aromatic degradation by microbes. *Mailing Add:* Mac-RSMAS-UM 4600 Rickenbacker Causeway Miami FL 33149

TAYLOR, BARRY EDWARD, b Potsdam, NY, July 7, 47; m 68; c 1. SOLID STATE CHEMISTRY. *Educ:* State Univ NY Col Fredonia, BS, 69; Brown Univ, PhD(chem), 74. *Prof Exp:* Teaching asst chem, State Univ NY Col Fredonia, 66-67, res asst, 68-69; teaching assoc chem, Brown Univ, 69-73; chemist, 73-74, res chemist, exp sta, 74-78, sr res chemist, 78-81, RES ASSOC, PHOTO EMD, E I DU PONT DE NEMOURS & CO, INC, NIAGARA FALLS, NY, 81- *Mem:* Am Chem Soc; AAAS; Int Soc Hybrid Microelectronics; Am Ceramic Soc. *Res:* Preparative solid state chemistry; crystallographic and physical properties of oxides, halides and sulfides; chemistry of alkali metal compounds; study of ionic conductivity in solids; thick film technology of resistors, conductors, dielectrics and solder compositions. *Mailing Add:* Du Pont Japan Ltd 4997 Shin-Yoshida-cho Kohoku-ku Yokohama-shi Kanagawa 223 Japan

TAYLOR, BARRY L, b Sydney, Australia, May 7, 37; c 3. MICROBIOLOGY, BACTERIOLOGY. *Educ:* Avondale Col, Australia, BA, 59; Univ NSW, Australia, BSc, 66; Case Western Reserve Univ, PhD(biochem), 73. *Prof Exp:* From asst prof to assoc prof, 76-83, PROF BIOCHEM, LOMA LINDA UNIV, 83-, PROF & CHMN MICROBIOL, 88- *Concurrent Pos:* Res assoc, Univ Calif, Berkeley, 73-75; vis postdoctoral fel, Australian Nat Univ, 75-76; vis scientist biochem, Univ Utah, 82-83; interim dir, Ctr Molecular Biol, Loma Linda Univ, 89- *Mem:* Am Soc Microbiol; Am Soc Biochem & Molecular Biol; AAAS. *Res:* Bacterial chemotaxis by pathways that are independent of receptor methylation; mechanism by which oxygen chemoreceptors detect changes in oxygen concentration; oxygen receptores in bacteria are used as a model system. *Mailing Add:* Dept Microbiol Loma Linda Univ Sch Med Loma Linda CA 92350

TAYLOR, BARRY NORMAN, b Philadelphia, Pa, Mar 27, 36; m 58; c 3. FUNDAMENTAL CONSTANTS, PRECISION MEASUREMENTS. *Educ:* Temple Univ, AB, 57; Univ Pa, MS, 60, PhD(physics), 63. *Prof Exp:* From instr to asst prof physics, Univ Pa, 63-66; physicist, RCA Labs, NJ, 66-70; chief, Absolute Elec Measurements Sect, Nat Bur Standards, 70-74, chief elec div, 74-88, MGR, FUNDAMENTAL CONSTANTS DATA CTR, NAT INST STANDARDS & TECHNOL, 88- *Concurrent Pos:* Consult, Philco Corp, 64-65; instr math, Rider Col, 69-70; mem, Nat Acad Sci-Nat Res Coun-Nat Acad Eng Adv Comt Fundamental Constants, and Standards, 69-, mem adv panel to Elec Div of Inst Basic Standards, Nat Bur Standards, 69, mem Comt Data Sci & Technol Task Group on Fundamental Constants, 76-; Nat Bur Standards deleg, 14th session Comt Consult Elec, Comt Int Poids & Measures, Paris, France, 75, deleg 15th Session, 78, 16th Session, 83, 17th Session, 86 & 18th Session, 88, 10th Session, Comt Int Poids & Measures, Nat Inst Standards & Technol, 90-; ed, Metrologia, 76-84 & 88-; mem, NSF Interagency Atomic & Molecular Physics Group, 76-83; charter mem, Sr Exec Serv, US Govt, 79-; comptroller, Conf Precision Electro Magnetic Measurements, 82-; vchmn, Topical Group on Fundamental Constants and Precise Tests of Phys Laws, Am Phys Soc, 88-90, chmn, 90-92; adminr, Precision Measurement Grants Prog, Nat Inst Standards & Technol, 75-, rep, TC 25, Working Group 1, 89-, US Tech Adv Group, Int Orgn Standardization TC 12, 89-, ISO TAG 4/WG3, Meterol, 89-, ISO TAG 4/WG1 Int Vocab Meterol, 89-, Int Adv Panel, TC 12, 90-, Subcomt Standards & Metric Practices Metrication Operating Comt, Interagency Comt Metric Policy, 90-, Comt E-43 SI Pract, Am Soc Testing & Mat, 90-, authorized off interpretation, Int Syst Units US, 90-; mem, Nat Conf Standards Labs, 87-91; tech adv, US Nat Comt, Int Electrotech Comn, 88-; mem, Standards Coord Comt 14, Inst Elec & Electronics Engrs, 88-; chief ed, J Res Nat Inst Standards & Technol, 89-; mem adv comt, Particle Data Group, Lawrence Berkeley Lab, 89- *Honors & Awards:* RCA Outstanding Achievement Award in Sci, 69; John Price Wetherill Medal, Franklin Inst, Wash, Philadelphia & Silver Medal Award, Dept Com, 75, Gold Medal Award, 89. *Mem:* Fel Am Phys Soc; fel Inst Elec & Electronics Engrs; Sigma Xi; Sci Res Soc. *Res:* Precision measurement and fundamental constants; data analysis; absolute electrical measurements; superconductivity; electron tunneling; Josephson effects. *Mailing Add:* Nat Inst Standards & Technol Bldg 221 Rm 160 Gaithersburg MD 20899

TAYLOR, BENJAMIN JOSEPH, b Pasadena, Calif, July 5, 42; m 70; c 6. PHYSICS, ASTRONOMY. *Educ:* Pasadena City Col, AA, 62; Univ Calif, Berkeley, BA, 64, PhD(astron), 69. *Prof Exp:* Res assoc, Princeton Univ, 69-71; res assoc, Univ Wash, 71-73, instr astron, 74; Nat Res Coun assoc, Ames Res Ctr, NASA, 74-76, assoc, 77-80; ASST PROF PHYSICS & ASTRON, BRIGHAM YOUNG UNIV, 80- *Concurrent Pos:* Instr physics & astron, San Jose State Univ, 77-80. *Mem:* Astron Soc Pac. *Res:* Spectrophotometry of secondary standards and solar analogs; differential broad-band photometry of clusters for blanketing analyses; spectrophotometry of K giants for abundance analyses. *Mailing Add:* Dept Physics & Astron 296 ESC Brigham Young Univ Provo UT 84602

TAYLOR, BERNARD FRANKLIN, b Charles Town, WVa, Mar 21, 30; m 57; c 2. VIROLOGY, SERODIAGNOSIS. *Educ:* Storer Col, Harpers Ferry, WVa, BS, 52; Mich State Univ, MS, 58; Rutgers Univ, New Brunswick, NJ, PhD(microbiol), 72; Rider Col, Lawrenceville, NJ, MA, 80. *Prof Exp:* Bacteriologist I virol, Mich State Dept Health, 52-54, virologist II, 56-59; instr sci, Elizabeth City State Col, NC, 59-60; from virologist to chief virologist, 60-79, DIR, PUB HEALTH LAB SERV DIAG MICROBIOL, NJ STATE DEPT HEALTH, 79- *Concurrent Pos:* Med technician, Helene Fuld Hosp, Trenton, NJ, 61-64; adj prof biol, Trenton State Col, NJ, 72-; Nat Defense exec reservist, Fed Emergency Mgt Agency Reg III, Philadelphia, Pa, 84-87 & 87-90; mem, Med Technician Adv Comn, Mercer County Community Col, Trenton, NJ, 84-87. *Honors & Awards:* Ella Stewart Biol Award. *Mem:* Theobald Smith Soc; Nat Soc Biol Teachers; Conf Pub Health Admin; NY Acad Sci; Sigma Xi. *Res:* Development of cost-effective serodiagnostic techniques which are specific, sensitive and yield rapid results. *Mailing Add:* 438 Walnut Ave Trenton NJ 08609

TAYLOR, BEVERLEY ANN PRICE, b Kingsport, Tenn, Nov 24, 51; m 71. PHYSICS, QUANTUM FIELD THEORY. *Educ:* ETenn State Univ, BS, 73; Clemson Univ, PhD(physics), 78. *Prof Exp:* asst prof physics, Denison Univ, 78-79; asst prof physics, Jackson State Univ, 79-84; MIAMI UNIV, HAMILTON, OH, 84- *Mem:* Am Phys Soc; Am Asn Phys Teachers; Sigma Xi. *Res:* Intrinsically nonlinear quantum field theories; elementary school science curriculum and instruction. *Mailing Add:* Miami Univ 1601 Peck Blvd Hamilton OH 45011

TAYLOR, BRUCE CAHILL, b Cleveland, Ohio, June 5, 42; div; c 4. BIOMEDICAL ENGINEERING, INSTRUMENTATION. *Educ:* Hiram Col, BA, 64; Kent State Univ, MA, 66, PhD(physiol), 71. *Prof Exp:* Assoc dir res, Vascular Res Lab, Akron City Hosp, 71-75; sr res scientist, Abcor, Inc, 75-76; dir dept med eng, Akron City Hosp, 78-84; RES ASSOC PROF BIOENG, UNIV AKRON, 81-, DIR, DEPT BIOMED ENG RES, 85- *Concurrent Pos:* Biomed consult, 75-; prog dir, Proj Hope, People's Repub China, 84-85; developer, Biomed Eng Support Serv. *Mem:* Am Soc Artifical Internal Organs; Asn Advan Med Instrumentation; Inst Elec & Electronics Engrs; Eng Med & Biol Soc. *Res:* Design and development of new types of medical instrumentation; blood pressure monitoring; comuter modeling and simulation. *Mailing Add:* Dept Biomed Eng Univ Akron Akron OH 44325

TAYLOR, C(HARLES) E(DWIN), b West Lafayette, Ind, Mar 24, 24; m 46; c 2. OPTIC METHODS. *Educ:* Purdue Univ, BSME, 46, MS, 48; Univ Ill, PhD(theoret & appl mech), 53. *Prof Exp:* Instr eng mech, Purdue Univ, 46-48; from instr to asst prof theoret & appl mech, Univ Ill, 48-52, from asst prof to prof, Univ Ill, Urbana, 54-80; struct res engr, David Taylor Model Basin, DC, 52-54; PROF ENG SCI, UNIV FLA, 81- *Concurrent Pos:* Vis prof, India, 66, 69 & Univ Calif, Berkeley, 68. *Honors & Awards:* M M Frocht Award, Soc Exp Stress Analysis, 69, M Hetenyi Award, 70, & F G Notatnall Award, 83. *Mem:* Nat Acad Eng; Am Soc Mech Engrs; Am Soc Eng Educ; Am Soc Eng Sci; hon mem Soc Exp Stress Analysis. *Res:* Three-dimensional photoelasticity; applications of lasers to experimental mechanics; shell theory; holography. *Mailing Add:* 8322 SW Fifth Pl Gainesville FL 32607

TAYLOR, C P(ATRICK) S(TIRLING), b Toronto, Ont, May 11, 30; m 55; c 4. BIOPHYSICS. *Educ:* Univ BC, BA, 52; Oxford Univ, BA, 54, MA, 57; Univ Pa, PhD(biophys), 60. *Prof Exp:* Childs Mem Fund fel biophys, Cambridge Univ, 60-61; asst prof physics, Univ BC, 61-67; assoc prof, 68-74, PROF BIOPHYS, UNIV WESTERN ONT, 74- *Mem:* Sigma Xi; Biophys Soc Can. *Res:* Mathematical modelling; surface energy effects in biology. *Mailing Add:* Dept Med Biophys Univ Western Ont London ON N6A 5C1 Can

TAYLOR, CARL ERNEST, b Mussoorie, India, July 26, 16; m 43; c 3. PREVENTIVE MEDICINE, EPIDEMIOLOGY. *Educ:* Muskingum Col, BS, 37; Harvard Univ, MD, 41, MPH, 51, DrPH, 53; FRCP(C), 45. *Hon Degrees:* DSc, Muskingum Col, 62; DHL, Towson State Univ, Baltimore, 78. *Prof Exp:* Med officer, Gorgas Hosp, 41-44; chief med serv, USPHS Marine Hosp, Pittsburgh, 44-46; hosp supt, Fatehgarh, India, 47-50; instr, Sch Pub Health, Harvard Univ, 51-53; prof prev med, Christian Med Col, India, 53-56; assoc prof prev med & pub health & dir prog for teachers, Sch Pub Health, Harvard Univ, 56-61; prof & chair, Dept Int Health, Sch Hyg & Pub Health, Johns Hopkins Univ, 61-84; US rep, Unicer, China, 84-87; EMER PROF INT HEALTH, JOHNS HOPKINS UNIV, 87- *Honors & Awards:* Ryan Prize; NCIH Award. *Mem:* Inst Med-Nat Acad Sci; Asn Teachers Prev Med; Am Pub Health Asn; Royal Soc Trop Med & Hyg; Am Soc Trop Med & Hyg. *Res:* International health; health planning in developing countries; population dynamics; medical education; epidemiology of leprosy and nutrition and infections; integration of health and family planning. *Mailing Add:* Johns Hopkins Inst Int Progress 103 E Mt Royal Ave Baltimore MD 21202

TAYLOR, CARSON WILLIAM, b Superior, Wis, May 24, 42; m 66; c 1. ELECTRICAL POWER ENGINEERING, CONTROL ENGINEERING. *Educ:* Univ Wis, BS, 65; Rensselaer Polytech Inst, MS, 69. *Prof Exp:* Elec engr, US Bur Reclamation, 67-68; elec engr, 69-89, PRIN ENGR, BONNEVILLE POWER ADMIN, 89- *Concurrent Pos:* Adj prof, Univ Portland, 82-86; working group chmn, Inst Elec & Electronic Engrs, 82-, mem, Int Conf Large High Voltage Elec Systs, 82-; prin, Carson Taylor Seminars, 86- *Mem:* Fel Inst Elec & Electronics Engrs. *Res:* Methods to enhance performance of large interconnected electric power systems. *Mailing Add:* Bonneville Power Admin PO Box 3621 MS EOH Portland OR 97208

TAYLOR, CHARLES ARTHUR, JR, plant taxonomy, for more information see previous edition

TAYLOR, CHARLES BRUCE, b Hecla, SDak, Feb 6, 15; m 38; c 2. PATHOLOGY. *Educ:* Univ Minn, BS, 38, MB, 40, MD, 41. *Prof Exp:* Intern, Univ Hosp, Univ Minn, 40-41, fel physiol, 41-42; res assoc & asst prof path, Col Med, Univ Ill, 46-51, assoc prof, 54-59; assoc prof, Univ NC, 51-54; prof, Med Sch, Northwestern Univ, 69-70; prof, Med Ctr, Univ Ala, Birmingham, 70-72; PROF PATH, ALBANY MED COL, 72-; prof, 80-82, EMER PROF PATH, MED COL UNION UNIV, 82- *Concurrent Pos:* Res assoc, Presby Hosp, Chicago, 46-51, attend pathologist, 54-59; chmn dept path, Evanston Hosp, Ill, 59-70; mem coun arteriosclerosis, Am Heart Asn; dir res, Vet Admin Hosp, Albany, 72-80. *Mem:* Am Physiol Soc; Am Soc Exp Path; Am Heart Asn; Am Asn Path & Bact; AMA. *Res:* Arteriosclerosis and its pathogenesis; human metabolism of cholesterol; degenerative pulmonary diseases; pathogenesis of gall stones. *Mailing Add:* 23 Sunset Dr Delmar NY 12054

TAYLOR, CHARLES ELLETT, b Chicago, Ill, Sept 9, 45; m 69; c 1. POPULATION GENETICS. *Educ:* Univ Calif, Berkeley, AB, 68; State Univ NY, Stony Brook, PhD(ecol & evolution), 73. *Prof Exp:* Asst prof biol, Univ Calif, Riverside, 74-79; assoc prof biol, 80-87, PROF BIOL, UNIV CALIF, LOS ANGELES, 87- *Concurrent Pos:* Co-dir, cognitive sci res prog, Univ Calif, Los Angeles. *Mem:* Genetics Soc Am; Soc Study Evolution; Am Soc Naturalists. *Res:* Population genetics and ecology; artificial life. *Mailing Add:* Dept Biol Univ Calif Los Angeles CA 90024

TAYLOR, CHARLES EMERY, b White Plains, NY, Mar 2, 40; m 70; c 1. NUCLEAR MAGNETIC RESONANCE, ZERO GRAVITY MATERIALS SCIENCE. *Educ:* Williams Col, BA, 61, MA, 63; Mich State Univ, PhD(physics), 67. *Prof Exp:* From asst prof to assoc prof, 67-80, PROF PHYSICS, ANTIOCH COL, 80- *Concurrent Pos:* Dir, Sci Inst, 82-86. *Mem:* Am Phys Soc; Int Solar Energy Asn; Am Asn Physics Teachers. *Res:* Nuclear spin lattice relaxation in antiferromagnetics; holography; holaesthetics; solar energy and alternative energy sources; zero gravity experiments, for future shuttle flight, on solidification. *Mailing Add:* Dept Physics Inst Sci Antioch Col 795 Livermore St Yellow Springs OH 45387

TAYLOR, CHARLES JOEL, b Portland, Ore, Apr 12, 19; m 52; c 3. PHYSICS. *Educ:* Univ Ill, BS, 40, MS, 48, PhD(physics), 51. *Prof Exp:* Asst nuclear physics, Univ Ill, 50-51; physicist, NAm Aviation, Inc, 51-52; physicist, Lawrence Livermore Nat Lab, Univ Calif, 52-85, tech mgt systs anal nuclear weapons, 75-81, asst assoc dir, 81-85; RETIRED. *Concurrent Pos:* Consult, US Deleg, conf disarmament, Geneva, Switz, 79, 80. *Mem:* Am Phys Soc. *Res:* Scintillation counters; neutron physics; nuclear weapons design; systems analysis. *Mailing Add:* 4275 Cornell Way Livermore CA 94550-4906

TAYLOR, CHARLES RICHARD, b Phoenix, Ariz, Sept 8, 39. COMPARATIVE PHYSIOLOGY, ENVIRONMENTAL PHYSIOLOGY. *Educ:* Occidental Col, BA, 60; Harvard Univ, MA, 62, PhD(biol), 63. *Prof Exp:* Res fel mammal, Mus Comp Zool, Harvard Univ, 64-67; res assoc zool, Duke Univ, 68-70; assoc prof, 70-73, PROF BIOL, DEPT ORGANISMIC & EVOLUTIONARY BIOL, HARVARD UNIV, 73-, ALEXANDER AGASSIZ PROF ZOOL, MUS COMP ZOOL, 73- & DIR, CONCORD FIELD STA, 70- *Concurrent Pos:* Hon lectr, Univ Col, Nairobi, Kenya & attached res officer, EAfrican Vet Res Orgn, 64-67; res assoc, Mus Comp Zool, Harvard Univ, 68-; vis prof, Univ Nairobi, 77; Guggenheim fel, 77; mem, Interunion Comn Comp Physiol, 78-, chmn, 81-; mem, Int Physiol Comt, 90-92. *Mem:* Nat Acad Sci; fel AAAS; Am Soc Mammal; Am Physiol Soc; Am Soc Zoologists; NY Acad Sci; Sigma Xi. *Res:* Water metabolism; temperature regulation; respiratory and exercise physiology; physiology of wild and domestic ruminants. *Mailing Add:* Concord Field Sta Harvard Univ Old Causeway Rd Bedford MA 01730

TAYLOR, CHARLES WILLIAM, b Duluth, Minn, Sept 26, 30; m 53; c 4. ORGANIC CHEMISTRY. *Educ:* Univ Minn, BA, 52; Univ Wis, PhD(chem), 57. *Prof Exp:* Sr chemist, 57-72, res specialist, 72-81, sr res specialist, Med Prod Div, Cent Res Dept, 81-87, DIV SCIENTIST, 3M CO, 87- *Mem:* Am Chem Soc; Sigma Xi; Adhesion Soc. *Res:* Fluorocarbon chemistry; biomedical materials; thermosetting acrylics; pressure sensitive adhesives. *Mailing Add:* 4677 Birchbark Trail N Lake Elmo MN 55042

TAYLOR, CHRISTOPHER E, IMMUNOLOGY, NUTRITION. *Educ:* Johns Hopkins Univ, ScD, 81. *Prof Exp:* ASST PROF IMMUNOL, MED COL PA, 83- *Mailing Add:* Sr Staff Fellow N1-AID Twinbrook II Res Fac Rockville MD 20852

TAYLOR, CLAYBORNE D, b Kokomo, Miss, July 15, 38; m 63; c 3. ELECTROMAGNETICS. *Educ:* Miss State Univ, BS, 61; NMex State Univ, MS, 64, PhD(physics), 65. *Prof Exp:* Staff mem, Sandia Corp, 65-67; from asst prof to assoc prof physics, Miss State Univ, 67-71; prof elec eng, Univ Miss, 71-72; prof elec eng, 72-74, PROF ELEC ENG & PHYSICS, MISS STATE UNIV, 74-86 & 88- *Concurrent Pos:* Various consulting activities and short course instr; Stocker vis prof & chair, Ohio Univ, 86-88; consult, USAF; teaching emp, R&B Enterprises, Praxis Int; Herrin Hess prof, Miss State Univ, 90-91. *Mem:* Nat Soc Prof Engrs; Inst Elec & Electronics Engrs; Int Radio Union Radio Sci. *Res:* Field and antenna theories; electromagnetic boundary value problems. *Mailing Add:* Dept Elec Eng/Box EE Miss State Univ PO Box 5328 Mississippi State MS 39762

TAYLOR, CLIVE ROY, b Cambridge, Eng, July 24, 44; m 67; c 4. PATHOLOGY. *Educ:* Cambridge Univ, MBBChir, 69, MD, 80; Oxford Univ, PhD(immunol), 74. *Prof Exp:* Lectr path, Univ Oxford, 70-75; fel cancer res, UK Res Coun, 75-76; chmn path, 83, PROF & CHMN PATH, UNIV SOUTHERN CALIF, LOS ANGELES, 76- *Res:* Immunopathology, immunohistology and cancer diagnosis, with particular reference to leukemia & lymphoma. *Mailing Add:* Dept Path Univ Southern Calif 2011 Zonal Ave Los Angeles CA 90033

TAYLOR, CONSTANCE ELAINE SOUTHERN, b Washington, DC, Nov 14, 37; m 59; c 3. ECOLOGY, SYSTEMATIC BOTANY. *Educ:* Univ Okla, BS, 59, MS, 61, PhD(plant ecol & syst bot), 75. *Prof Exp:* Teacher pub schs, Okla, 63-64; from instr to assoc prof, 70-85, PROF BIOL, SOUTHEASTERN OKLA STATE UNIV, 85- *Mem:* Am Soc Plant Taxonomists; Nat Wildlife Fedn. *Res:* The genus Solidago and Euthamia, goldenrods, Oklahoma vascular plants, endangered and rare species. *Mailing Add:* Dept Biol Southeastern Okla State Univ Durant OK 74701

TAYLOR, D DAX, b Chicago, Ill, Oct 10, 37; m 61; c 3. ANATOMIC PATHOLOGY, CLINICAL PATHOLOGY. *Educ:* Amherst Col, AB, 59; Univ Mo-Columbia, MD, 63; Am Bd Path, dipl & cert anat path & clin path, 68. *Prof Exp:* Resident path, Sch Med, Univ Mo-Columbia, 63-68, instr, 68-69, asst prof path & asst dean sch med, 69-72; assoc prof & assoc dean med educ, Southern Ill Univ, Springfield, 72-76, prof path & assoc dean acad affairs, Sch Med, 76-79, exec assoc dean, 79-80; vpres Evaluation Prog, Nat Bd Med Examiners, 80-87; MED DIR, METPATH, 87- *Mem:* Asn Am Med Cols; Am Soc Clin Path; Col Am Physicians. *Res:* Medical education, student and curriculum evaluation; platelet patho-physiology. *Mailing Add:* MetPath Inc 1355 Mittel Blvd Wood Dale IL 60191

TAYLOR, D(OROTHY) JANE, b Waco, Tex. BIOLOGY. *Educ:* Rice Univ, BA, 43; Iowa State Univ, MS, 47; George Washington Univ, PhD(biol), 57. *Prof Exp:* Tech asst, Synthetic Rubber Lab, Humble Oil Co, 43-45; lab instr zool, biol & physiol, Iowa State Univ, 45-47; parasitologist, Lab Trop Dis, NIH, 47-58; head endocrine-related tumor syst sect, Cancer Chemother Nat Serv Ctr, 58-69, head, Gen Lab & Clin, 64-73; head exp biol proj sect, Breast Cancer Prog Coord Br, Nat Cancer Inst, NIH 73-82, chief, Breast Cancer Prog Coord Br, Div Cancer Biol & Diag & Exec Secy, Breast Cancer Task Force Comt, 75-82; sci adminr, Stehlin Found Cancer Res, Houston, Tex, 82-88; CONSULT, BREAST CANCER RES, ROSE JOAN GORDON CTR, HOUSTON, TEX, 90- *Concurrent Pos:* Mem bd, Nat Alliance Breast Cancer Orgn, New York, NY, 87-; mem adv bd, Cancer Fighter's of Houston, 89-, Rose Mammography Ctr, Houston, Tex, 90. *Honors & Awards:* Super Serv Award, Dept Health & Human Serv, Pub Health Serv, 81. *Mem:* Fel AAAS; Am Asn Cancer Res; Sigma Xi. *Res:* Experimental biology of breast cancer; malaria and amebiasis; in vitro cultivation; experimental chemotherapy; nutritional aspects; endocrine tumors; host-tumor biology and therapy; chemotherapy of human tumors in athymic mice; immune competence of women with stage one breast cancer. *Mailing Add:* 5001 Woodway Dr Houston TX 77056

TAYLOR, D LANSING, b Baltimore, Md, Dec 26, 46; m 69; c 3. CELL BIOLOGY. *Educ:* Univ Md, BS, 68; State Univ NY, Albany, PhD(biol), 73. *Prof Exp:* Fel biophysics, Marine Biol Labs, 73-74; asst prof biol, Harvard Univ, 74-78, assoc prof, 78-82; PROF BIOL, CARNEGIE-MELLON UNIV, 82- *Concurrent Pos:* Ed, J Cell Biol & J Cell Motility, 81- *Mem:* Am Soc Cell Biol; Biophys Soc; NY Acad Sci. *Res:* Molecular basis of amoeboid movements, utilizing biochemical, cell biological and biophysical approaches and fluorescence spectroscopy. *Mailing Add:* Dept Biol Sci Ctr Fluorescence Res Carnegie-Mellon Univ 5000 Forbes Ave Pittsburgh PA 15213

TAYLOR, DALE FREDERICK, b Evansville, Ind, June 16, 44; m 67; c 2. ELECTROCHEMISTRY. *Educ:* Univ Toronto, BSc, 66, MSc, 68, PhD(chem), 70. *Prof Exp:* Staff scientist battery res, 70-73, mgr personnel admin, 74, STAFF SCIENTIST, MAT LABS, GEN ELEC CORP RES & DEVELOP, 75- *Mem:* Electrochem Soc. *Res:* Corrosion of boiling water reactor structural materials. *Mailing Add:* Gen Elec Res & Develop Ctr PO Box 8 Schenectady NY 12301

TAYLOR, DAVID COBB, b Portland, Maine, June 7, 39; m 71. ELECTROCHEM, COMPUTER TECHNIQUES. *Educ:* Bowdoin Col, AB, 61; Wesleyan Univ, MA, 63; Univ Conn, PhD(chem), 70. *Prof Exp:* From asst prof to assoc prof, 68-83, PROF CHEM, SLIPPERY ROCK UNIV, 83- *Concurrent Pos:* Consult anal methods & environ systs eng. *Mem:* AAAS; Am Chem Soc; fel Am Inst Chemists. *Res:* Electroanalytical chemistry and multicomponent systems in the realm of industrial methods development. *Mailing Add:* Dept Chem Slippery Rock Univ Slippery Rock PA 16057-9989

TAYLOR, DAVID JAMES, b Ottawa, Ont, May 12, 51. FAULT TOLERANCE, DISTRIBUTED SYSTEMS. *Educ:* Univ Sask, BSc, 72; Univ Waterloo, MSc, 74, PhD(computer sci), 77. *Prof Exp:* Asst prof, 77-87, ASSOC PROF COMPUTER SCI, UNIV WATERLOO, 87- *Concurrent Pos:* SERC vis fel, Comput Lab, Univ Newcastle upon Tyne, 83-84. *Mem:* Asn Comput Mach; Inst Elec & Electronics Engrs. *Res:* Software fault tolerance particularly robust storage structures; software structure for distributed computer systems. *Mailing Add:* Dept Computer Sci Univ Waterloo Waterloo ON N2L 3G1 Can

TAYLOR, DAVID NEELY, b Ann Arbor, Mich, July 31, 48; m 80. INFECTIOUS DISEASES. *Educ:* Kenyon Col, BA, 70; Dartmouth Med Sch, DMS, 72; Harvard Med Sch, MD, 74; London Sch Hygiene & Trop Med, MSc, 78. *Prof Exp:* Med resident, State Univ NY, Buffalo, 74-77; res fel geographic med, Sch Med, Johns Hopkins Univ, 78-80; EPIDEMIOLOGIST ENTERIC DIS, CTR DIS CONTROL, WALTER REED ARMY INST RES, 80- *Concurrent Pos:* Consult, Gorga's Hosp, Panama, 78-80 & Chilalongkorn Hosp, Thailand, 83- *Mem:* Am Soc Microbiol; Am Soc Trop Med & Hyg. *Res:* Epidemiologic studies in infectious causes of diarrheal disease, including studies in salmonella, typhoid, campylobacter and intestinal parasites; development of rapid diagnostic methods to detect enteric disease agents. *Mailing Add:* Ctr Vaccine Develop Ten Pine St Baltimore MD 21201

TAYLOR, DAVID WARD, b Chesterfield, Eng, Aug 18, 38; m 65; c 2. THEORETICAL PHYSICS, SOLID STATE PHYSICS. *Educ:* Oxford Univ, BA, 61, MA, 65, PhD(physics), 65. *Prof Exp:* Mem tech staff, Bell Tel Labs, NJ, 64-67; from asst prof to assoc prof, 67-77, assoc chmn, Physics Dept, 80-84, PROF THEORET SOLID STATE PHYSICS, MCMASTER UNIV, 77- *Mem:* Can Phys Soc; Am Phys Soc. *Res:* Dynamics of disordered crystals; calculations of phonons and phonon dependent properties of metals and alloys. *Mailing Add:* Dept Physics McMaster Univ Hamilton ON L8S 4M1 Can

TAYLOR, DERMOT BROWNRIGG, b Ireland, Mar 30, 15; US citizen; m 45, 65; c 1. PHARMACOLOGY. *Educ:* Trinity Col, Dublin, MD, 37, MB, BCh & BAO, 38. *Prof Exp:* Asst physiol, Trinity Col, Dublin, 38-39; lectr, King's Col, Univ London, 39-45, lectr pharmacol, 45-50; assoc prof, Univ Calif, San Francisco, 50-53; chmn dept, 53-68, PROF PHARMACOL, UNIV CALIF, LOS ANGELES, 53- *Concurrent Pos:* Univ London traveling fel, Yale Univ, 48. *Mem:* Am Soc Pharmacol & Exp Therapeut; Brit Physiol Soc; Royal Soc Chem; Brit Biochem Soc; Brit Pharmacol Soc. *Res:* Mode of action of neuromuscular blocking agents. *Mailing Add:* Dept Pharmacol - Inst Environ Stress Univ Calif Santa Barbara CA 93106

TAYLOR, DIANE WALLACE, b Covina, Calif; m. TROPICAL MEDICINE, HYBRIDOMA TECHNOLOGY. *Educ:* Univ Hawaii, BA, 68, MS, 70, PhD(zool), 75. *Prof Exp:* Instr biol, Sch Med, Univ Hawaii, 70-73, fel trop med, 75-78; fel immunol, Lab Microbiol Immunol, NIH, 78-72; ASSOC PROF, DEPT BIOL, GEORGETOWN UNIV, WASHINGTON, DC, 82- *Mem:* Am Soc Trop Med Hyg; Sigma Xi; Am Soc Microbiologists. *Res:* Immune studies of parasitic infections with special emphasis on malaria. *Mailing Add:* Dept Biol Georgetown Univ 37th & 0 Sts NW Washington DC 20057

TAYLOR, DONALD CURTIS, b London, Ky, June 16, 39; m 67; c 1. MATHEMATICS. *Educ:* Univ Ky, BS, 61, MS, 64, PhD(math), 67. *Prof Exp:* Elec eng, Westinghouse Elec Corp, 61-62; assoc prof math, Univ Mo-Columbia, 67-73; assoc prof, 73-77, PROF MATH, MONT STATE UNIV, 77- *Concurrent Pos:* Fel, La State Univ, Baton Rouge, 69-70. *Mem:* Am Math Soc. *Res:* Functional analysis. *Mailing Add:* Dept Math Mont State Univ Bozeman MT 59717

TAYLOR, DONALD JAMES, b Dayton, Ohio, Mar 6, 33; m 55, 85; c 2. ASTRONOMY. *Educ:* Calif Inst Technol, BS, 55, MS, 58; Univ Wis, PhD(astron), 63. *Prof Exp:* Proj assoc space astron lab, Univ Wis, 63-65; asst prof astron, Univ Ariz, 65-71; ASSOC PROF PHYSICS, UNIV NEBR, LINCOLN, 71- *Mem:* Int Astron Union; Am Astron Soc; Astron Soc Pac. *Res:* Planetary astronomy; astronomical instrumentation; pulsars; clusters; nebulae. *Mailing Add:* Behlen Lab Physics Univ Nebr Lincoln NE 68588

TAYLOR, DOUGLAS HIRAM, b Doddsville, Miss, Dec 15, 39; m 61; c 2. ETHOLOGY, ECOLOGY. *Educ:* Univ Dayton, BS, 66, MS, 68; Miss State Univ, PhD(zool), 70. *Prof Exp:* NSF fel, Univ Notre Dame, 70-71; from asst prof to assoc prof, 71-78, PROF ZOOL, MIAMI UNIV, 78- *Concurrent Pos:* Res biologist, US Environ Protection Agency. *Mem:* AAAS; Am Soc Zoologists; Animal Behav Soc; Ecol Soc Am; Sigma Xi. *Res:* Animal behavior, ecology and orientation; agonistic behavior; behavioral aspects of ecology of vertebrates; behavioral toxicology. *Mailing Add:* Dept Zool Miami Univ Oxford OH 45056

TAYLOR, DUANE FRANCIS, b Iowa City, Iowa, Sept 30, 25; m 50; c 7. DENTAL MATERIALS. *Educ:* Univ Mich, BSE, 49, MSE, 50; Georgetown Univ, PhD(biochem), 61. *Prof Exp:* Head dept dent mat, Sch Dent, Wash Univ, St Louis, 50-54; phys metallurgist, Dent Sect, Nat Bur Stand, 54-61; dir mat res, CMP Industs, 61-63; prof dent, Sch Dent, 63-74, PROF OPER DENT-DENT SCI, DENT RES CTR, UNIV NC, CHAPEL HILL, 74- *Concurrent Pos:* Consult, US Army, 67- & NIH, 69- *Mem:* AAAS; Am Soc Metals; Int Asn Dent Res. *Res:* Dental amalgams; cobalt-chromium alloys; denture base materials; polymerization mechanisms; properties of multiphase solids; materials for implant prosthesis. *Mailing Add:* Box 1328 Chapel Hill NC 27514

TAYLOR, DUNCAN PAUL, b Bremerton, Wash, Feb 4, 49; m 72; c 1. NEUROPHARMACOLOGY. *Educ:* Calif Inst Technol, BS, 71; Ore State Univ, PhD(biochem), 77. *Prof Exp:* Technician, Analytical Serv, Carnation Co Res Labs, 67-70; volunteer, US Peace Corps, 71-73; res asst, Dept Biochem & Biophys, Ore State Univ, 73-77; res assoc, Sect Biochem & Pharmacol, NIMH, 77-79; scientist & neuropharmacologist, Mead Johnson & Co, 79-80, res assoc & neuropharmacologist, 80, sr scientist & neuropharmacologist biol res, Pharmaceut Div, 80-82; sr scientist, 82-83, sr res scientist, 83-85, res fel, 85-89, SR RES FEL, CENTRAL NERVOUS SYST NEUROPHARMACOL DEPT, PHARMACEUT RES INST, BRISTOL-MYERS SQUIBB CO, 89- *Concurrent Pos:* Res fel, Comt Advan Sci Training, NSF, 65 & 70; teaching asst, Dept Biochem & Biophysics, Ore State Univ, 74; partic, Advan Study Inst Cyclic Nucleotides, NATO, 77; Nat Res Serv Award, Nat Inst Drug Abuse, 77-79. *Mem:* Am Chem Soc; AAAS; Am Soc Neurochem; Soc Exp Biol & Med; Am Soc Pharmacol Exp Ther; Soc Neurosci. *Res:* Receptors in nervous tissue membranes; receptor coupling to second messengers; linkage of changes in receptors to pathology behavior; author or coauthor of over 50 publications. *Mailing Add:* 49 Black Walnut Dr Middletown CT 06457-6131

TAYLOR, DWIGHT WILLARD, b Pasadena, Calif, Jan 18, 32. INVERTEBRATE ZOOLOGY. *Educ:* Univ Mich, BS, 53; Univ Calif, MA, 54, PhD, 57. *Prof Exp:* Geologist, US Geol Surv, DC, 55-67; assoc prof zool, Ariz State Univ, 67-69; res assoc malacol, San Diego Mus Natural Hist, 69-74; prof zool, Pac Marine Sta, Univ Pac, 74-77; PROF ZOOL, TIBURON CTR ENVIRON STUDIES, SAN FRANCISCO STATE UNIV, 77- *Mem:* AAAS; Soc Syst Zool; Am Malacol Union. *Res:* Freshwater mollusks, ecology, taxonomy and distribution; biogeography. *Mailing Add:* 98 Main St PO Box 308 Tiburon CA 94920

TAYLOR, EDWARD CURTIS, b Springfield, Mass, Aug 3, 23; m 46; c 2. ORGANIC CHEMISTRY. *Educ:* Cornell Univ, AB, 46, PhD(org chem), 49. *Hon Degrees:* DSc, Hamilton Col, 69. *Prof Exp:* Merck fel, Zurich Tech Univ, 49-50; Du Pont fel, Univ Ill, 50-51, instr chem, 51-53, asst prof org chem, 53-54; from asst prof to assoc prof, 54-64, chmn dept chem, 74-79, PROF CHEM, PRINCETON UNIV, 64-, A BARTON HEPBURN PROF ORG CHEM, 66- *Concurrent Pos:* Consult, Procter & Gamble Co, 52-, Eastman Kodak Co, 65-82, Eli Lilly & Co, 70-, Tenn Eastman Co, 71-82, Burroughs-Wellcome Co, 83-88 & du Pont, 85-; NSF sr fac fel, Harvard Univ, 59; Fulbright scholar, 60; vis prof inst org chem, Stuttgart Tech Univ, 60; vis lectr, Weizmann Inst, 60; mem, Chem Adv Comt, Air Force Off Sci Res, 62-70; distinguished vis prof, Univ Buffalo, 69; ed org chem, Wiley Intersci, Inc, 69-; vis prof, Univ E Anglia, 69 & 72; ed, Advances in Org Chem; co-ed, Gen Heterocyclic Chem & Chem of Heterocyclic Compounds; Guggenheim Mem fel, 79-80; Alexander von Humboldt sr US scientist award, 84-85. *Honors & Awards:* Res Awards, Smith Kline & French Found, 55, Hoffmann-La Roche Found, 64 & 65, S B Penick Found, 69, 70, 71 & 72 & Ciba Pharmaceut Co, 71; Creative Work Award, Am Chem Soc, 74; H J Backer Lectr, Univ Groningen, 71; 5th Int Award in Heterocyclic Chem, 89. *Mem:* AAAS; Am Chem Soc; fel NY Acad Sci; Royal Soc Chem; Ger Chem Soc. *Res:* Organic synthesis; heterocyclic chemistry, particularly pyrimidines, purines and pteridines; organothallium chemistry; natural products; photochemistry; medicinal chemistry. *Mailing Add:* Frick Chem Lab Princeton Univ Princeton NJ 08540

TAYLOR, EDWARD DONALD, b Clifton, Tex, Sept 30, 40; m 65; c 1. PHYSICS & CHEMISTRY. *Educ:* Univ Tex, Austin, BS, 63; Tex Tech Univ, PhD(chem), 67. *Prof Exp:* Asst prof & fel chem, Fla State Univ, 67-68; PROF CHEM, ODESSA COL, 68- *Concurrent Pos:* Fel, Tex A&M Univ, 71. *Honors & Awards:* Eastman Fel, Eastman Kodak Co, 68. *Mem:* Am Chem Soc. *Res:* Investigations of thermoluminescence via electron paramagnetic resonance to understand the mechanism of the process. *Mailing Add:* Odessa Col 201 W Univ Odessa TX 79764

TAYLOR, EDWARD MORGAN, b Rapid City, SDak, Dec 27, 33; m 58; c 2. GEOLOGY, PETROLOGY. *Educ:* Ore State Univ, BS, 57, MS, 60; Wash State Univ, PhD(geol), 67. *Prof Exp:* Instr geol, Ore State Univ, 62-63 & Wash State Univ, 64-65; asst prof, 66-71, ASSOC PROF GEOL, ORE STATE UNIV, 71- *Mem:* Geol Soc Am; Mineral Soc Am; Am Geophys Union. *Res:* Volcanic petrology of Cascade Range of California, Oregon and Washington. *Mailing Add:* Dept Geol Ore State Univ Corvallis OR 97331

TAYLOR, EDWARD S(TORY), engineering; deceased, see previous edition for last biography

TAYLOR, EDWARD STEWART, b Hecla, SDak, Aug 20, 11; m 40; c 3. OBSTETRICS & GYNECOLOGY. *Educ:* Univ Iowa, BA, 33, MD, 36; Am Bd Obstet & Gynec, dipl, 46. *Prof Exp:* Prof obstet & gynec & head dept, 47-76, dir, Am Bd Obstet & Gynec, 60-69, clin prof, 77-80, EMER PROF OBSTET & GYNEC, MED CTR, UNIV COLO, DENVER, 80- *Concurrent Pos:* Consult Surgeon Gen, US Dept Air Force; coun mem, Nat Inst Child Health & Human Develop; ed, Obstet & Gynec Surv, 67- *Honors & Awards:* Distinguished Serv Award, Am Col Obstet & Gynec, 84. *Mem:* Am Gynec Soc; Am Asn Obstet & Gynec; Am Col Surg; Am Col Obstet & Gynec; Am Gynec & Obstet Soc. *Res:* Cancer of the cervix; physiology of pregnancy. *Mailing Add:* Dept Obstet & Gynec Univ Colo Med Ctr Denver CO 80220

TAYLOR, EDWIN FLORIMAN, b Oberlin, Ohio, June 22, 31; m 55; c 3. PHYSICS. *Educ:* Oberlin Col, AB, 53; Harvard Univ, MA, 54, PhD(physics), 58. *Prof Exp:* Asst prof physics, Wesleyan Univ, 56-64; vis assoc prof, Educ Res Ctr, Mass Inst Technol, 64-66, sr res scientist, 66-73; ed, Am J Physics, 73-78; DIR, EDUC VIDEO RESOURCES, MASS INST TECHNOL, 79- *Mem:* Am Phys Soc; Am Asn Physics Teachers. *Res:* Solid state physics; educational writing and research in mechanics, special relativity and quantum physics; computer-assisted learning. *Mailing Add:* Mass Inst Technol Bldg 126 Rm 147 Cambridge MA 02139

TAYLOR, EDWIN WILLIAM, b Toronto, Ont, June 8, 29; m 56; c 3. BIOPHYSICS. *Educ:* Univ Toronto, BA, 52; McMaster Univ, MSc, 55; Univ Chicago, PhD(biophys), 57. *Prof Exp:* Instr physics, Ont Agr Col, 52-53; res assoc biol, Mass Inst Technol, 57-59; from instr to prof biophys, 59-74, PROF BIOPHYS & THEORET BIOL, UNIV CHICAGO, 74-, MASTER BIOL SCI, COL DIV & ASSOC DEAN, 76- *Mem:* Biophys Soc. *Res:* Mechanochemical systems; muscle; flagella; protoplasmic streaming; mechanism of cell division; physical protein chemistry. *Mailing Add:* Molecular Genetics-Cell Biol/CLSC 339 Univ Chicago 920 E 58th St Chicago IL 60637

TAYLOR, ELIZABETH BEAMAN HESCH, b Sumter, SC, Oct 27, 21; m 48, 77; c 2. MATHEMATICS. *Educ:* Winthrop Col, BA, 43; Duke Univ, MA, 46; Columbia Univ, PhD, 55. *Prof Exp:* Instr math, Winthrop Col, 43; teacher pub schs, SC, 43-45; asst prof, Radford Col, 46-52 & Westhampton Col, 52-70; from assoc prof to prof math, Univ Richmond, 70-85; RETIRED. *Mem:* Math Asn Am. *Res:* Nature of mathematical evidence and its significance for the teaching of secondary school mathematics. *Mailing Add:* 2431 Swathmore Rd Richmond VA 23235

TAYLOR, ELLISON HALL, b Kalamazoo, Mich, Sept 6, 13; m 38; c 2. PHYSICAL CHEMISTRY. *Educ:* Cornell Univ, BChem, 35; Princeton Univ, MA, 37, PhD(phys chem), 38. *Prof Exp:* Instr chem, Univ Utah, 38-40; instr chem eng, Cornell Univ, 40-42; res chemist, Div War Res, Columbia Univ, 42-45; res chemist, Clinton Labs, Tenn, 45-48, asst dir chem div, 46-48; actg dir, Chem Div, Oak Ridge Nat Lab, 48, prob leader, 48-49, from assoc dir to dir, Chem Div, 49-51, asst dir res, 51-54, dir, Chem Div, 54-75, sr res staff mem, 75-84; RETIRED. *Concurrent Pos:* Vis prof, Cornell Univ, 65; consult, 84- *Honors & Awards:* S C Lind Lectr. *Mem:* AAAS; Am Chem Soc; Am Phys Soc. *Res:* Heterogeneous catalysis; chemical problems related to isotope separations; radiation chemistry; chemical kinetics; molecular beams in chemistry; trace element analysis by resonance ionization. *Mailing Add:* 143 Orchard Lane Oak Ridge TN 37830

TAYLOR, ERIC ROBERT, b Quincy, Mass, Oct 31, 47; m 74; c 1. BIOPHYSICS. *Educ:* Ohio State Univ, BS, 72; Rutgers Univ, PhD(biochem), 81. *Prof Exp:* Postdoctoral res asst, Dept Chem, Renesselaer Polytech Inst, 81-84; ASST PROF BIOCHEM, UNIV SOUTHWESTERN LA, 84- *Mem:* NY Acad Sci; Sigma Xi; AAAS. *Res:* Computer modeling of DNA drug interactions. *Mailing Add:* Dept Chem Univ Southwestern La Lafayette LA 70504-4370

TAYLOR, EUGENE M, b Cheyenne, Wyo, Dec 25, 32; m 78; c 5. PHYSIOLOGY, PHARMACOLOGY. *Educ:* Idaho State Col, BS, 58; Univ Wash, MS, 59, PhD(physiol & psychol), 64. *Prof Exp:* res asst prof rehab med, Sch Med, Univ Wash, 72-74; investr physiol, Va Mason Res Ctr, 74-77, assoc mem sci staff physiol, 77-81; lectr psychol, Univ Wash,81-82; res scientist, Dept Hemat/Oncol, Children's Hosp & Med Ctr, Seattle, 82-91; RETIRED. *Concurrent Pos:* Consult, Vet Admin Hosp, Phoenix, Ariz, 68-72; fel, Dept Bioeng, Univ Wash, 71-72, affil assoc prof, Dept Otolaryngol, 76-82, res affil, Regional Primate Res Ctr, 77-80, res assoc, Dept Pediat, 85-; affil investr, Virginia Mason Res Ctr, 71-74. *Res:* Evaluation of pediatric oncology, chemotherapy and radiation therapy, especially long-term physiologic changes associated with cancer therapy; neuropathology. *Mailing Add:* 2640 Perkins Lane Seattle WA 98199

TAYLOR, FLETCHER BRANDON, JR, b Aug 24, 29; m 54; c 4. INTERNAL MEDICINE. *Educ:* Stanford Univ, BS, 52; Univ Calif, San Francisco, MD, 56. *Hon Degrees:* MS, Univ Pa, 71. *Prof Exp:* Intern, Southern Pac Hosp, San Francisco, 56-57, resident surg, 57-58; res assoc protein chem, London Hosp, Eng, 58-59; mem res staff thrombosis, Cardiovasc Res Inst, San Francisco, 59-65; from asst prof to assoc prof med, Hosp Univ Pa, Philadelphia, 65-74, co-chmn div allergy & immunol, 68-74; prof path & med & dir div exp path & med, Health Sci Ctr, Univ Okla, 74-; OKLA MED RES FOUND. *Concurrent Pos:* Resident med, Univ Calif Hosp, San Francisco, 59-62; consult hemat, NASA Manned Space Flight Ctr, 68-; mem thrombosis coun, Am Heart Asn, 71-; head, Am Acad Allergy Post-Graduate Educ Comt; clin prof res med, Okla Univ Health Sci Ctr, prof res biochem, head sect exp path & med; dir clin hemat serv, Univ Hosp, Oklahoma City; dir, Oklahoma City Children's Mem Hosp Coagulation Lab. *Honors & Awards:* Cochems Prize Cardiovasc Res, 68; Louis Pasteur Lectr Award, Univ Paris, 69. *Mem:* Int Soc Thrombosis & Haemostasis; Am Asn Immunol; Am Soc Clin Invest; Am Physiol Soc; Am Fedn Clin Res; Sigma Xi. *Res:* Thrombosis and protein chemistry. *Mailing Add:* Hemat-Thrombosis Res Okla Med Res Found 825 NE 13th St Oklahoma City OK 73104

TAYLOR, FLOYD HECKMAN, b North Versailles, Pa, May 6, 26; m 55; c 2. MATHEMATICS. *Educ:* Bucknell Univ, BS, 49; Univ Pittsburgh, MS, 53, ScD, 63. *Prof Exp:* Asst proj engr math, Sperry Gyroscope Co, NY, 53-54; instr, Hood Col, 54-56; mathematician, Atlantic Div, Aerojet Gen Corp Div, Gen Tire & Rubber Co, 56-57 & US Army Biol Labs, 57-64; asst prof biostatist, Grad Sch Pub Health & asst prof prev med, Sch Med, 64-70, assoc prof community med, 70-75, RES PROF COMMUNITY MED, SCH MED, UNIV PITTSBURGH, 75- *Concurrent Pos:* Instr, Frederick Community Col, 57-61. *Mem:* Sigma Xi; Am Statist Soc; Biomet Soc. *Res:* Numerical analysis; applications of mathematics to digital computers; mathematical models as applied to biology; biostatistics. *Mailing Add:* 5023 Frew St Pittsburgh PA 15213

TAYLOR, FRANCIS B, b New York, NY, June 15, 25. MATHEMATICS. *Educ:* Manhattan Col, AB, 44; Columbia Univ, AM, 47, PhD(math educ), 59. *Prof Exp:* From instr to assoc prof, 47-65, head dept, 64-72, PROF MATH, MANHATTAN COL, 65- *Concurrent Pos:* Lectr, Col Mt St Vincent, 53-56; NSF fac fel, 57-58; lectr, Sch Gen Studies, Hunter Col, 59-60. *Mem:* Math Asn Am; Am Statist Asn; Inst Math Statist; Nat Coun Teachers Math; Sigma Xi. *Res:* Mathematical statistics. *Mailing Add:* Dept Math & Computer Sci Manhattan Col Bronx NY 10471

TAYLOR, FRANK EUGENE, b Richmond, Va, Apr 1, 42; m 68; c 3. EXPERIMENTAL HIGH ENERGY PHYSICS. *Educ:* Mich State Univ, BS, 64; Cornell Univ, PhD(exp high energy physics), 70. *Prof Exp:* Res assoc exp high energy physics, Lab Nuclear Studies, Cornell Univ, 70-71, Deutsches Elektronen Synchrotron-Hamburg, Ger, 71-72; from asst prof to assoc prof exp high energy physics, 77-83, prof physics, Northern Ill Univ, 83-87; SR RES SCIENTIST, MASS INST TECHNOL, 83- *Concurrent Pos:* Prin physicist, Dept Energy grant, 78-83. *Mem:* Am Phys Soc. *Res:* Experiment and phenomenology of inclusive reactions in strong interactions; experimental high energy neutrino physics; electron-positron collisions at ZO pole. *Mailing Add:* Bin 96 SLAC Box 4349 Stanford CA 94305

TAYLOR, FRANK JOHN RUPERT (MAX), b Cairo, Egypt, July 17, 39; m 63; c 3. MARINE BIOLOGY, PROTISTOLOGY. *Educ:* Univ Cape Town, BSc, 59, Hons, 60, PhD(marine bot), 65. *Prof Exp:* Res asst marine phytoplankton, Inst Oceanog, Univ Cape Town, 60-64; from asst prof to assoc prof, 65-75, PROF BIOL & OCEANOG, DEPT OCEANOG, UNIV BC, 75- *Concurrent Pos:* Can-France Exchange fel, France, 72; vis scientist, Phuket Marine Biol Ctr, Thailand, 73; res assoc, Int Develop Res Ctr, Bellairs Res Inst, Barbados, 79-80, Plant Sci, Univ Oxford, 86-87; Christensen fel, St Catherine's Col, Oxford, 86, hon mem, 89-; Killam sr fel, 86-87. *Mem:* Soc Evolutionary Protistology (pres, 79-81). *Res:* Taxonomy and distributional ecology of unicellular marine organisms, principally diatoms and dinoflagellates; undergraduate biology; marine phytoplankton ecology; red tides; intracellular symbiosis; evolution, especially origin of eukaryotes and protist evolution. *Mailing Add:* Dept Oceanog Univ BC Vancouver BC V6T 1W5 Can

TAYLOR, FRED M, b Chanute, Kans, Aug 21, 19; m 42; c 5. PEDIATRICS. *Educ:* Stanford Univ, AB, 41, MD, 44. *Prof Exp:* From instr to prof pediat, Baylor Col Med, 48-69; prof, Univ Tex Med Sch San Antonio, 69-72, from asst dean to assoc dean acad develop, 69-72; dir, Off Continuing Educ, 72-78, spec asst to exec vpres & dean, 78-80, prof pediat, 72-85, exec asst to pres, 80-85, EMER PROF & ADV TO PRES, BAYLOR COL MED, 85- *Mem:* AAAS; Am Acad Pediat; Soc Res Child Develop. *Res:* Infant nutrition and feeding; developmental behavior of infants, children and adolescents. *Mailing Add:* 810 Alhambra Sugar Land TX 77478

TAYLOR, FRED WILLIAM, b Springcreek, WVa, Jan 17, 32; m 54; c 4. WOOD SCIENCE & TECHNOLOGY. *Educ:* Va Polytech Inst, BS, 53; NC State Col, MWT, 54, PhD(wood sci & technol), 65. *Prof Exp:* Asst prof forestry, Univ Vt, 54-55; asst timber buyer, J B Belcher Lumber Co, 55-56; wood technologist, Pulaski Veneer & Furniture Co, 56-59; res asst, NC State Univ, 59-62; wood utilization exten specialist, Univ Mo, 63-65; PROF WOOD SCI & TECHNOL & ASST DIR, MISS FORESTRY PROD UTILIZATION LAB, MISS STATE UNIV, 65- *Concurrent Pos:* Coun Sci & Indust Res grant, SAfrica, 71-72. *Mem:* Forest Prod Res Soc; Soc Wood Sci & Technol; Int Asn Wood Anatomists; Inst Wood Sci. *Res:* Natural variation in the anatomical structure of angiosperm xylem, especially genetic implications of property variations. *Mailing Add:* Dept Wood Sci/Box FP Miss State Univ PO Box 5328 Mississippi State MS 39762

TAYLOR, FREDRIC WILLIAM, b Amble, Eng, Sept 24, 44; m 69. ATMOSPHERIC PHYSICS. *Educ:* Univ Liverpool, BSc, 66; Oxford Univ, DPhil(atmospheric physics), 70. *Prof Exp:* Res scientist tech staff atmos physics, Jet Propulsion Lab, Calif Inst Technol, 71-79; HEAD DEPT ATMOSPHERIC PHYSICS, OXFORD UNIV, 79- *Honors & Awards:* Exceptional Sci Achievement Award Rank Prize, NASA, 89. *Mem:* Fel Royal Meteorol Soc; Am Meteorol Soc; Optical Soc Am; assoc Am Astron Asn; fel Royal Soc Arts. *Res:* Physics of the atmospheres of the Earth and planets, specializing in atmospheric radiation, remote sensing techniques and infrared observational methods. *Mailing Add:* Dept Physics Clarendon Lab Oxford Univ Oxford 0X1 3PU England

TAYLOR, FREDRICK JAMES, b Wisconsin Rapids, Wis, Apr 28, 40; m 68; c 3. ELECTRICAL ENGINEERING, COMPUTER SCIENCE. *Educ:* Milwaukee Sch Eng, BS, 65; Univ Colo, MS, 66, PhD(elec eng), 69. *Prof Exp:* Researcher elec eng, Tex Instruments, Dallas, 69-70; assoc prof, Univ Tex, El Paso, 70-75; PROF ELEC ENG, UNIV CINCINNATI, 75- *Concurrent Pos:* Prin investr, US Army Atmospheric Sci Lab grant, 73-75, Eng Found grant, 77-78 & NSF grant, 78- *Mem:* Inst Elec & Electronics Engrs. *Res:* Digital systems; digital signal processing; medical signal analysis; ultrasound; finite mathematics. *Mailing Add:* Dept Elec Eng Univ Fla Gainsville FL 32611

TAYLOR, G JEFFREY, b Port Jefferson, NY, June 27, 44; m 65; c 5. PLANETARY SCIENCE. *Educ:* Colgate Univ, AB, 66; Rice Univ, MA, 68, PhD(geol), 70. *Prof Exp:* Smithsonian Res Found res fel, lunar mineral & petrol, Smithsonian Astrophys Observ, 70-72, res assoc, 72-73; asst prof, Wash Univ, 73-76; SR RES ASSOC, INST METEORITICS, UNIV NMEX, 76- *Concurrent Pos:* Assoc, Harvard Univ, 70-; vis scientist, Lunar Sci Inst, 74-76. *Honors & Awards:* Nininger Meteorite Prize, Ctr Meteorite Studies, Ariz State Univ, 69. *Mem:* Am Geophys Union; Geochem Soc; Meteoritical Soc; AAAS. *Res:* Petrologic and chemical nature of the moon, meteorites and earth, with emphasis on their origins and thermal histories. *Mailing Add:* 1920 Vassar Dr No 11 Albuquerque NM 87106

TAYLOR, GARY, b London, Eng, Sept 23, 52; US citizen; m 83. PLASMA PHYSICS, MICROWAVE & FAR-INFRARED MEASUREMENTS. *Educ:* Univ Manchester, BSc, 74; Oxford Univ, MSc, 75, DPhil(plasma physics), 77. *Prof Exp:* Res assoc, 77-79, staff res physicist, Plasma Physics Lab, 79-85, RES PHYSICIST, PRINCETON UNIV, 85- *Mem:* Am Phys Soc. *Res:* Application of microwave and far-infrared techniques to high temperature plasma confinement in Tokamak magnetic fusion machines. *Mailing Add:* Princeton Plasma Physics Lab Princeton Univ PO Box 451 Princeton NJ 08543

TAYLOR, GARY N, b Plainfield, NJ, Oct 19, 42; m 64; c 3. ORGANIC CHEMISTRY. *Educ:* Princeton Univ, BA, 64; Yale Univ, MS, 66, PhD(org chem), 68. *Prof Exp:* Air Force Off Sci Res fel, Calif Inst Technol, 68-69; MEM TECH STAFF, BELL LABS, 69-, SUPVR, 79- *Mem:* Am Chem Soc; Electrochem Soc. *Res:* Synthetic organic chemistry; photochemistry; polymer chemistry; radiation chemistry; lithography; plasma chemistry. *Mailing Add:* Bell Labs 1D247 600 Mountain Ave Murray Hill NJ 07974-2008

TAYLOR, GENE WARREN, b Abilene, Tex, Nov 9, 36; m 64; c 2. PHYSICAL CHEMISTRY, MATERIALS SCIENCE. *Educ:* Tex Western Col, BS, 63; NMex State Univ, PhD(phys chem), 69. *Prof Exp:* Mem staff chem, El Paso Natural Gas Prod, 63-65; teacher math, Ysleta Independent Schs, 64-65; mem fac chem, Kans State Univ, 69-72; STAFF MEM CHEM, LOS ALAMOS NAT LAB, 72- *Concurrent Pos:* Mem staff, Schlesinger Res Found, 64-65; fel, Kans State Univ, 69-72. *Honors & Awards:* Withens Award for Weapons Res. *Mem:* Am Chem Soc. *Res:* Rare gas, metastable atom flowing afterglow reactions for chemical analysis; gas phase chemical kinetics, and solid phase thermal decomposition reaction kinetics; composite materials. *Mailing Add:* 760 42nd St Los Alamos NM 87545

TAYLOR, GEOFF W, DEVICE PHYSICS, MATERIAL PHYSICS & PHYSICS OF OPTICAL DEVICES. *Educ:* Queens Univ, Kingston, Ont, BASc, 66; Univ Toronto, MASc, 68 & PhD(elec eng), 72. *Prof Exp:* Mem tech staff, Honeywell, 74-76; mem tech staff, AT&T Bell Labs, Murray Hill, 76-86, MEM TECH STAFF, AT&T BELL LABS, HOLMDEL, 86-t7. *Concurrent Pos:* Comt mem, Device Res Conf, 80-85, tech prog chmn, 85 & CONF CHMN, 86- *Mem:* Sr mem Inst Elec & Electronics Engrs; Optical Soc Am; Soc Photo-Optical Instrumentation Engrs. *Res:* Integrated opto-electronics using transistors, lasers, detectors, and integrated optics components. *Mailing Add:* Rm 4D-423 AT&T Bell Labs Crawfords Corner Rd Holmdel NJ 07733-1988

TAYLOR, GEORGE EVANS, JR, b Richmond, Va, June 3, 49; m 72; c 2. EVOLUTIONARY BIOLOGY. *Educ:* Randolph-Macon Col, BS, 71; Emory Univ, PhD(biol), 76. *Prof Exp:* Instr biol, Agnes Scott Col, 76-77; res assoc bot, Nat Acad Sci, Nat Res Coun, 77-80; mem staff, Oak Ridge Nat Lab, 80-90; MEM STAFF, BIOL SCI CTR, DESERT RES INST, 90- *Mem:* Am Soc Plant Physiologist; Bot Soc Am; Am Chem Soc; AAAS; Sigma Xi. *Res:* Plant stress physiology; evolutionary biology of plant populations including genetics, physiological ecology and population biology; rapid microevolutionary events in response to man generated stesses. *Mailing Add:* Desert Res Inst Biol Sci Ctr PO Box 60220 Reno NV 89506-0220

TAYLOR, GEORGE RUSSELL, physical chemistry, for more information see previous edition

TAYLOR, GEORGE STANLEY, b Jackson, NC, Nov 29, 20; m 47; c 3. SOIL PHYSICS. *Educ:* NC State Col, BS, 43, MS, 48; Iowa State Univ, PhD(soil physics), 50. *Prof Exp:* Asst, NC State Col, 47-48 & Iowa State Univ, 49-50; from asst prof to assoc prof, 51-61, prof agron, 61-85, EMER PROF AGRON, OHIO STATE UNIV, 85- *Concurrent Pos:* Vis assoc prof, Univ Calif, 58-59; consult, US Agency Int Develop, Punjab, India, 68 & 70; UN Univ, Nigeria, 85. *Mem:* Am Soc Agron; Am Soc Agr Engrs; Am Geophys Union; Soil Sci Soc Am. *Res:* Water flow in porous media; land drainage; two-dimensional modeling of heat and water flow in porous media by numerical techniques; disposal wastes in soil. *Mailing Add:* Dept Agron Ohio State Univ 2021 Coffey Rd Columbus OH 43210

TAYLOR, GEORGE THOMAS, b Asheboro, NC, July 18, 35. DEVELOPMENTAL BIOLOGY, ZOOLOGY. *Educ:* Guilford Col, AB, 57; Univ NC, Chapel Hill, MA, 64; Univ Mass, Amherst, PhD(zool), 70. *Prof Exp:* Instr zool, Atlantic Christian Col, 62-65; asst prof human anat, Col Osteop Med & Surg, 69-73; asst prof physiol, Southern Ill Univ, 73-77; asst prof dept biol sci & dir electron micros lab, 77-82, assoc prof anat, Am Univ, Caribbean, 84-85, RES ASSOC, DEPT BIOL SCI, FLA INT UNIV, 88- *Concurrent Pos:* Educ consult & dir, Cameo Health Sci Inc, 85- *Mem:* AAAS; Am Soc Cell Biol; Am Soc Zoologists; Soc Develop Biol; Am Ornithologists Union. *Res:* Cytochemical and cytological aspects of oocyte differentiation and early development in marine invertebrates; special interest in changes in subcellular morphology and function during embryonic cytodifferentiation. *Mailing Add:* 19510 SW 87th Ave Miami FL 33157

TAYLOR, GEORGE WILLIAM, b Perth, Australia, June 16, 34; US citizen; m 57; c 4. SOLID STATE PHYSICS, ELECTRICAL ENGINEERING. *Educ:* Western Australia Univ, BE, 57, DEng, 81; London Univ, PhD(ferroelec), 61. *Hon Degrees:* DSc, Don Pedag Inst, USSR, 89. *Prof Exp:*

Group engr, telecommun, Australian Post Off, 61; lectr elec eng, Sydney Univ, 61-62; mem tech staff comput res, RCA Corp Labs, 62-70; vpres res & eng, Princeton Mat Sci Inc, 71-75, PRES, PRINCETON RESOURCES INC, 75- & PRINCETON RES ASSOCS, 77- *Concurrent Pos:* Ed, Int J Ferroelectrics, 70-, Int J Ferroelectrics Letters, 82- emer ed, Electronic Display World; prof elec eng, Rutgers Univ, 76-77; travel fel, US NSF, 76. *Honors & Awards:* Fisk Prize, IRE, 56; Achievement Award, RCA Labs, 66. *Mem:* Sr mem Inst Elec & Electronics Engrs; fel Australian Inst Eng; Am Phys Soc; fel Brit Inst Elec Eng. *Res:* Properties, synthesis, applications of ferroelectric materials; computer memories; displays and electro-optics; piezoelectric and pyroelectric materials & devices; author of two books. *Mailing Add:* 305 Dodds Lane Princeton NJ 08540

TAYLOR, GERALD C, b Oregon, Mo, Sept 10, 19; m 50; c 4. BACTERIOLOGY, VIROLOGY. *Educ:* Univ Kans, BA, 49, MA, 51, PhD(bact), 55. *Prof Exp:* Lab instr bact & virol, Univ Kans, 50-52, asst rickettsiae, 52-55; asst, Lab Br, Commun Dis Ctr, USPHS, 55-57, lab supvr & unit chief, Tissue Cult Unit, 57-71, LAB SUPVR & UNIT CHIEF, LAB BR & CHIEF, CELL CULT & MEDIA SECT, SCI SERV DIV, CTR DIS CONTROL, USPHS, 71- *Mem:* AAAS; Am Soc Microbiol; Sigma Xi. *Res:* Human and animal virology; development of tissue culture in the field of virology. *Mailing Add:* Tissue Culture-Medi Sect CDC 6-418 1600 Clifton Rd Atlanta GA 30333

TAYLOR, GERALD REED, JR, b Bloxom, Va, Apr 17, 37; m 60; c 2. PHYSICS. *Educ:* Va Polytech Inst & State Univ, BS, 59, MS, 61; Univ Va, PhD(physics), 67. *Prof Exp:* Assoc physicist, Texaco Exp Inc, 60-62; res physicist solid state physics, Linde Div, Union Carbide Corp, 67-69; assoc prof physics, Madison Col, 69-80, PROF PHYSICS, JAMES MADISON UNIV, HARRISONBURG, VA, 80- *Concurrent Pos:* Dir, Vis Scientists Prog, Va Acad Sci, 77-79. *Honors & Awards:* J Shelton Horsley Res Award, Va Acad Sci, 62. *Mem:* Am Phys Soc; Am Asn Physics Teachers; Sigma Xi. *Res:* Low Temperature solid state physics; ferromagnetism; magneto-thermal conductivity; resistivity; physics education and instruction; plasma physics and plasma diagnostics. *Mailing Add:* Dept Physics Madison Col Harrisonburg VA 22807

TAYLOR, GLADYS GILLMAN, b Bloomfield, NJ, Dec 6, 26. FOUNDATION & PHILOSOPHY OF MATHEMATICS. *Educ:* Radcliffe Univ, MA, 52; Ind Univ, PhD(hist & philos sci), 80. *Prof Exp:* From assoc prof to prof math, Ind State Univ, 80-89; RETIRED. *Mem:* Philos Sci Asn; Am Math Asn. *Res:* Logic & set theory. *Mailing Add:* 2711 Wilson Dr Terre Haute IN 47803

TAYLOR, GORDON STEVENS, b Danbury, Conn, Nov 12, 21; m 46; c 1. PLANT PATHOLOGY. *Educ:* Univ Conn, BS, 47; Iowa State Col, MS, 49, PhD(corn stalk rot), 52. *Prof Exp:* Asst corn leaf blight & stalk rot, Iowa State Col, 48-52; asst agr scientist, 52-53, assoc agr scientist, Valley Lab, 53-55, agr scientist, 55-60, CHIEF AGR SCIENTIST, VALLEY LAB, CONN AGR EXP STA, 60- *Mem:* AAAS; Am Phytopath Soc; Am Inst Biol Sci; Air Pollution Control Asn. *Res:* Helminthosporium leaf blight of corn; stalk rot of corn in relation to the corn borer; verticillium wilt of potatoes; tobacco diseases; canker disease on maples; nematodes; effects of air pollution on plants. *Mailing Add:* 812 Matianuck Ave Windsor CT 06095

TAYLOR, HAROLD ALLISON, JR, b Richmond, Va, Oct 18, 42; m 65; c 2. HUMAN GENETICS, BIOCHEMISTRY. *Educ:* Univ Tenn, BS, 65, MS, 67, PhD(zool), 71. *Prof Exp:* Fel pediat, Sch Med, Johns Hopkins Univ, 71-73, instr, 73-75; asst dir, Genetics Lab, John F Kennedy Inst, Baltimore, 73-75; DIR LABS, GREENWOOD GENETIC CTR, 75- *Mem:* AAAS; Am Soc Human Genetics. *Res:* Lysosomal storage diseases; inborn errors of metabolism; genetics of mental retardation. *Mailing Add:* Greenwood Genetics Ctr Genetics Lab 1 Gregor Mendel Circle Greenwood SC 29646

TAYLOR, HAROLD EVANS, b Philadelphia, Pa, Sept 13, 39; m 64; c 4. PLASMA PHYSICS, SPACE PHYSICS. *Educ:* Haverford Col, BA, 61; Mass Inst Technol, MS, 62; Univ Iowa, PhD(physics), 66. *Prof Exp:* Res assoc physics, Univ Iowa, 66; Nat Acad Sci assoc, Goddard Spaceflight Ctr, NASA, 66-68; res assoc plasma physics, Princeton Univ, 68-71; from asst prof to prof, 71-88, PROF ASTROPHYS, STOCKTON STATE COL, 88- *Concurrent Pos:* Consult, Los Alamos Sci Labs, vis prof mech eng, Univ Pa, 78; vis prof, Princeton Univ, 86-87, vis res sci, Princeton Univ, 86-87. *Mem:* Am Assoc Phys Technol; Fedn Am Sci; Int Solar Energy Soc; Am Phys Soc; Inst Elec & Electronics Engrs. *Res:* Energy conservation in buildings; space plasma physics, including physics of the magnetosphere; physics of the interplanetary medium; cosmic rays; solar and wind energy devices; astronomical photometry; personal computer hardware. *Mailing Add:* Fac Natural Sci & Math Stockton State Col Pomona NJ 08240

TAYLOR, HAROLD LELAND, b Cambridge, Kans, May 3, 20; m 43; c 1. BIOCHEMISTRY. *Educ:* Southwestern Col, Kans, AB, 42; Univ Kans, PhD(biochem), 55. *Hon Degrees:* DSc, Southwestern Col, Kans, 73. *Prof Exp:* Res assoc, Harvard Univ, 42-46; biochemist, State Health Dept, Mich, 46-51 & 53-56; mgr immunochem dept, Pitman-Moore Div, Dow Chem Co, 56-63, tech asst to dir biol labs, 63-64, res chemist, Dow Res Labs, Zionsville, 64-72, clin monitor, Med Dept, 72-74, CLIN INVESTR, MERRELL DOW PHARMACEUT, 74- *Mem:* Am Chem Soc; Sigma Xi; NY Acad Sci; Am Soc Clin Pharmacol & Therapeut. *Res:* Fractionation of human blood plasma; physiological effects of antithyroid drugs in rats; isolation of immune fraction from hyperimmune canine plasma; control of biologicals production; drug metabolism; serum cholesterol lowering drugs. *Mailing Add:* 719 W 750 S Hebron IN 46341

TAYLOR, HAROLD LEROY, mechanical engineering, for more information see previous edition

TAYLOR, HAROLD MELLON, b Lucama, NC, May 14, 29; m 53; c 3. ORGANIC CHEMISTRY. *Educ:* Univ NC, AB, 51, PhD(org chem), 59. *Prof Exp:* Researcher, 59-80, RES ADV, ELI LILLY & CO, 80- *Concurrent Pos:* Sr adv, Dow Elanco, 91- *Res:* Discovery and development of new agricultural pesticides. *Mailing Add:* PO Box 708 Greenfield IN 46140

TAYLOR, HAROLD NATHANIEL, b Baltimore, Md, May 18, 21; m 42; c 3. CHEMISTRY. *Educ:* Johns Hopkins Univ, BE, 42, MS, 45; Cornell Univ, PhD(phys chem), 49. *Prof Exp:* Jr chem engr, Tenn Valley Authority, Wilson Dam, Ala, 42-43; jr instr chem eng, Johns Hopkins Univ, 43-44, asst, Off Rubber Reserve, 44 & Manhattan Proj, SAM Labs, Columbia Univ, 44-45; res chemist, Manhattan Proj, Carbide & Carbon Chem Co, 45-46 & Ammonia Dept, Exp Sta, E I du Pont de Nemours & Co, 49-53; PRES, HAGERSTOWN LEATHER GOODS CO, 53- *Concurrent Pos:* Dir, Cent Chem Co, 70- & Antietam Bank Co, 74- *Mem:* AAAS; Am Chem Soc. *Res:* Polymer chemistry; polymerization; plastic applications. *Mailing Add:* 1877 Fountain Head Rd Hagerstown MD 21740

TAYLOR, HARRY ELMER, b Easton, Pa, July 1, 31; m 53; c 3. ELECTRICAL ENGINEERING. *Educ:* Okla State Univ, BS, 57, MS, 58. *Prof Exp:* Mem tech staff, 58-68, SUPVR, INTEGRATED CIRCUIT DESIGN GROUP, BELL TEL LABS, 68- *Res:* Transistor applications in Bell System use; consult and advise circuit designers desiring custom integrated circuits for Bell System use. *Mailing Add:* Bell Tel Labs 555 Union Blvd Allentown PA 18103

TAYLOR, HARRY WILLIAM, b Saskatchewan, Can, Sept 28, 25; m 49; c 2. NUCLEAR PHYSICS. *Educ:* Univ Man, BSc, 51, MSc, 52, PhD(physics), 54. *Prof Exp:* Lectr physics, Univ Man, 52-53; fel, Cosmic Ray Sect, Nat Res Coun Can, 54-55; asst prof physics, Queen's Univ, Ont, 55-61; assoc prof, Univ Alta, 61-65; assoc prof, 65-69, PROF PHYSICS, UNIV TORONTO, 69- *Mem:* Fel AAAS; fel Am Phys Soc; fel Brit Inst Physics; fel Inst Nuclear Eng. *Res:* Gamma-ray spectroscopy; angular correlations of successive gamma-rays; distribution of fallout in Arctic plants; environmental radioactivity. *Mailing Add:* Dept Physics Erindale Col Univ Toronto Mississauga ON M5S 1A1 Can

TAYLOR, HERBERT LYNDON, b Van Alstyne, Tex, Aug 11, 31; m 56; c 2. PHYSICS, ELECTRICAL ENGINEERING. *Educ:* Austin Col, BA, 51; Rice Inst, MA, 52, PhD(physics), 55. *Prof Exp:* Mem tech staff res & develop, Tex Instruments Inc, 55-63; vis lectr, Univ Tex, Austin, 63-65, assoc prof, 65-80; failure analyst, Mostek, 80-81; consult engr, 81-83; COMPONENT ENGR, ROCKWELL INT, 83- *Mem:* Am Phys Soc; Inst Elec & Electronics Engrs; Am Vacuum Soc; Electrochem Soc. *Res:* Semiconductors; solid state devices; solid surfaces and interfaces; telecommunications defect analysis; optical systems; fiber optics. *Mailing Add:* 7014 Mason Dells Dr Dallas TX 75230

TAYLOR, HOWARD EDWARD, b Ft Worth, Tex, Feb 1, 22; m 44; c 3. MATHEMATICS. *Educ:* Rice Univ, BA, 42, MA, 48, PhD(math), 50; Calif Inst Technol, MS, 43. *Prof Exp:* Instr meteorol, Calif Inst Technol, 43-44; asst math, Rice Inst, 46-50; from instr to assoc prof, Fla State Univ, 50-69; callaway prof, 69-87, EMER PROF MATH, WGA COL, 87- *Concurrent Pos:* Vis assoc prof, Univ Chicago, 57-58; assoc chmn dept math, Fla State Univ, 64-69; mem comt, Math Achievement Test, Col Entrance Exam Bd, 70-80, chmn, 75-80, mem math adv comt, 76-80. *Mem:* Am Meteorol Soc; Math Asn Am. *Res:* Functions of a complex variable; analysis. *Mailing Add:* 145 E Greenwood Dr Carrollton GA 30117

TAYLOR, HOWARD LAWRENCE, b Kansas City, Mo, May 23, 38; m 60; c 2. APPLIED MATHEMATICS, RESEARCH ADMINISTRATION. *Educ:* Austin Col, AB, 59; Univ Kans, MA, 62, PhD(math), 68. *Prof Exp:* Lab asst physics, Austin Col, 57-59; asst math, Univ Kans, 59-67; mathematician, Sun Oil Co, 60-70, mgr res, 70-77, mgr geophysics, 77-78, consult, 78-85; GEOPHYS CONSULT, ARAMCO, 85- *Concurrent Pos:* Lectr, Univ Tex, Arlington & Univ Dallas. *Mem:* Math Asn Am; Soc Indust & Appl Math; Soc Petrol Engrs; Soc Explor Geophysicists; Europ Asn Explor Geophysicists; Asn Comput Mach. *Res:* Development of mathematical methods to analyze geophysical data and solve reservoir engineering problems; simulation models and inverse problems; deconvolution with the li norem. *Mailing Add:* Aramco Box 1094 Badge 195055 Dhahran 31311 Saudi Arabia

TAYLOR, HOWARD MELVIN, b Pride, Tex, Jan 20, 24; m 48; c 2. SOIL SCIENCE. *Educ:* Tex Tech Col, BS, 49; Univ Calif, PhD, 57. *Prof Exp:* Soil scientist, Soil Conserv Serv, USDA, 49-54; res asst irrig, Univ Calif, 55-57, instr, 57; soil scientist, Agr Res Serv, USDA, 57-80; prof agron, Iowa State Univ, 80-82; ROCKWELL PROF, TEX TECH UNIV, 82- *Concurrent Pos:* FAO Andre Mayer fel, Australia, 65-66. *Mem:* Fel AAAS; fel Am Soc Agron; fel Soil Sci Soc Am; Am Soc Agr Engrs; Soil Conserv Soc Am. *Res:* Soil physics, particularly soil-plant water relations and effects of soil compaction on plant growth. *Mailing Add:* Dept Plant & Soil Sci Tex Tech Univ Lubbock TX 79409

TAYLOR, HOWARD MILTON, III, b Baltimore, Md, May 9, 37; div. APPLIED PROBABILITY, STATISTICS. *Educ:* Cornell Univ, BME, 60, MIndustEng, 61; Stanford Univ, PhD(math, statist), 65. *Prof Exp:* Res assoc & lectr appl probability, Stanford Univ, 64-65; from asst prof to assoc prof, 65-80, PROF OPERS RES, CORNELL UNIV, 80- *Concurrent Pos:* NSF fel & vis asst prof, Univ Calif, Berkeley, 68-69; on leave at Math Inst, Oxford Univ, 72-73; grad fac rep, Cornell Univ, 74-77. *Mem:* Fel Inst Math Statist; Am Math Soc; Math Asn Am. *Res:* Applied probability; Markov sequential decision processes; optimal stopping problems; quality control. *Mailing Add:* Dept Math Sci Univ Del Newark DE 19716

TAYLOR, HOWARD S, b New York, NY, Sept 17, 35; m 59; c 3. PHYSICAL MATHEMATICS, MECHANICS. *Educ:* Columbia Univ, BA, 56; Univ Calif, Berkeley, PhD(chem physics), 59. *Prof Exp:* NSF fel chem, Free Univ Brussels, 59-61; from asst prof to prof, 61-74, Humboldt prof, 74-75, PROF CHEM & PHYSICS, UNIV SOUTHERN CALIF, 75- *Concurrent Pos:* Consult, Jet Propulsion Lab, Calif Inst Technol, 60-, NAm Aviation Inc, 65-66, Lawrence Livermore Nat Lab & Los Alamos Sci Lab; guest prof, Univ Freiburg, 67; staff scientist, Los Alamos Nat Lab, 73; vis prof , Univ Amsterdam, Univ Paris-Sud & Freiberg Univ. *Mem:* Am Chem Soc; fel Am Phys Soc. *Res:* Atomic and molecular physics; lasers; dynamics; spectroscopy; chaotic phenomena. *Mailing Add:* Dept Chem Univ Southern Calif Univ Park Los Angeles CA 90089-0482

TAYLOR, HUGH P, JR, b Holbrook, Ariz, Dec 27, 32; m 59. ORE DEPOSITS, ISOTOPE GEOCHEMISTRY. *Educ:* Calif Inst Technol, BS, 54, PhD(geochem), 59; Harvard Univ, AM, 55. *Prof Exp:* Asst prof geol, Calif Inst Technol, 59-61; asst prof geochem, Pa State Univ, 61-62; from asst prof to prof, 62-82, ROBERT P SHARP PROF GEOL, CALIF INST TECHNOL, 82- *Mem:* Nat Acad Sci; fel AAAS; fel Geol Soc Am; fel Am Geophys Union; fel Mineral Soc Am; Am Acad Arts & Sci. *Res:* Oxygen, hydrogen, carbon and silicon isotopic compositions of igneous and metamorphic minerals and rocks, meteorites and the moon; ore deposits and hydrothermal alteration; ultramafic rocks of southeast Alaska. *Mailing Add:* Div Geol & Planetary Sci Calif Inst Technol Pasadena CA 91125

TAYLOR, IAIN EDGAR PARK, b Chester, Eng, Aug 18, 38; m 67; c 1. PLANT PHYSIOLOGY, PLANT BIOCHEMISTRY & BIOPHYSICS. *Educ:* Univ Liverpool, BSc, 61, PhD(bot), 64. *Prof Exp:* Teacher, Blundell's Sch, Tiverton, Eng, 64-66; res assoc bot, Univ Tex, Austin, 67-68, vis asst prof, 68; asst prof, 68-71, assoc prof, 71-86, PROF BOT, UNIV BC, 86- *Concurrent Pos:* Chair, Can Nat Comt, Int Union Biol Sci, 87, exec comt mem, 89-; ed, Can J Bot, 89- *Honors & Awards:* Mary Elliott Award, Can Bot Asn, 89. *Mem:* Am Soc Plant Physiol; Can Bot Asn (pres 85-86); Can Soc Plant Physiol; Brit Biochem Soc; Brit Inst Biol. *Res:* Biophysics and biochemistry of plant cell wall structure and growth; tissue culture and micropropagation of conifers. *Mailing Add:* Dept Bot Univ BC Vancouver BC V6T 2B1 Can

TAYLOR, ISAAC MONTROSE, b Morganton, NC, June 15, 21; m 46, 79; c 6. MEDICINE. *Educ:* Univ NC, AB, 42; Harvard Univ, MD, 45. *Prof Exp:* Intern, Mass Gen Hosp, 45-46, resident physician, 47; asst med adv, Harvard Med Sch, 48; chief med res, Mass Gen Hosp, 51; from asst prof to assoc prof, Sch Med, Univ NC, Chapel Hill, 52-64, dean sch med, 64-71, prof med, 64-78, res prof dept community med, 78-80; DEP DIR ADMIN, HUBERT H HUMPHREY CANCER RES CTR, BOSTON UNIV, 80-, ADJ PROF MED, SCH MED, 81- *Concurrent Pos:* Nat Res Coun fel, Harvard Med Sch, 48-50; Markle scholar, 54-61; manpower consult, Tristate Regional Med Prog, 71-72, assoc dir manpower, 72-74. *Mem:* AMA; Am Fedn Clin Res. *Res:* Metabolism of electrolytes. *Mailing Add:* Nine Charles River Sq Boston MA 02114

TAYLOR, J(AMES) HERBERT, b Corsicana, Tex, Jan 14, 16; m 46; c 3. BIOLOGY, GENETICS. *Educ:* Southeastern State Col, BS, 39; Univ Okla, MS, 41; Univ Va, PhD(biol), 44. *Prof Exp:* Teacher high sch, Okla, 39-40; asst prof plant sci, Univ Okla, 46-47; assoc prof bot, Univ Tenn, 47-51; from asst prof to assoc prof, Columbia Univ, 51-58, prof cell biol, 58-64; assoc dir, Fla State Univ, 72-80, prof biol sci, 64-83, dir, Inst Molecular Biophys, 80-85, Robert O Lawton distinguished prof, 83-90, EMER PROF, FLA STATE UNIV, 90- *Concurrent Pos:* Guggenheim Found fel, Calif Inst Technol, Pasadena, 58-59. *Mem:* Nat Acad Sci; Am Soc Cell Biol (pres, 70-71); Genetics Soc Am; Biophys Soc. *Res:* Autoradiographic studies of nucleic acid and protein synthesis at the cellular level; chromosome duplication and structure; mechanisms of DNA replication in chromosomes; molecular organization of chromosomes; cloning origins of replication and modifications of DNA (methylation). *Mailing Add:* Inst Molecular Biophys Fla State Univ Tallahassee FL 32306

TAYLOR, JACK ELDON, b Emporia, Kans, Jan 16, 26; m 48; c 2. OPTICS. *Educ:* Univ Wis, PhB, 46, MS, 49, PhD(physics), 51. *Prof Exp:* Asst physics, Univ Wis, 47-51; res assoc, Gen Elec Co, 51-61; sr res staff mem, Gen Dynamics/Electronics, 61-71; prin engr, Stromberg Carlson Corp, 71-81; prin engr, Eastman Kodak Apparatus Div, 81-86, PRIN ENGR, EASTMAN KODAK RES LAB, 86- *Concurrent Pos:* Consult, semiconductor mfg, 79-82; adj prof, Rochester Inst Technol, 80-82. *Mem:* Am Phys Soc; Inst Elec & Electronics Engrs; sr mem Optical Soc Am; Sigma Xi. *Res:* Ultra high vacuum techniques; gas lasers; optical communication systems; atmospheric optical transmission; solid state devices; LSI logic arrays; switching matrices; material and manufacturing problems. *Mailing Add:* 31 Old Pond Rd Rochester NY 14625

TAYLOR, JACK HOWARD, b Memphis, Tenn, July 7, 22; m 44; c 4. PHYSICS. *Educ:* Southwestern at Memphis, BS, 44; Johns Hopkins Univ, PhD(physics), 52. *Prof Exp:* Physicist, US Naval Res Lab, 44-46; instr physics, Southwestern at Memphis, 46-47; asst, Radiation Lab, Johns Hopkins Univ, 48-50; physicist, Exp Sta, E I du Pont de Nemours & Co, 52-53; asst prof physics, Univ of the South, 53-54; consult infrared, US Naval Res Lab, 54-56; assoc prof, 56-60, PROF PHYSICS, SOUTHWESTERN AT MEMPHIS, 60-, PRIN INVESTR INFRARED, AIR FORCE CAMBRIDGE RES CTR CONTRACT, 59-, DIR LAB ATMOSPHERIC & OPTICAL PHYSICS, 64- *Concurrent Pos:* Consult, Electro-Optics Group, Pan Am World Airways, Inc & Patrick Air Force Base, Fla, 63- *Mem:* Fel AAAS; Am Phys Soc; fel Optical Soc Am; Am Asn Physics Teachers; Sigma Xi. *Res:* Physics and military applications of infrared; atmospheric physics and transmission in infrared; time-dependent infrared phenomena and infrared techniques. *Mailing Add:* 671 East Dr Memphis TN 38112

TAYLOR, JACKSON JOHNSON, b Winnabow, NC, Nov 20, 18; m 51; c 3. PHYSICS. *Educ:* Univ Richmond, BS, 42; Cornell Univ, MS, 48. *Prof Exp:* From instr to assoc prof, 48-69, chmn div sci, 57-72, chmn dept physics, 51-54, 55-58 & 63-69, PROF PHYSICS, UNIV RICHMOND, 69- *Concurrent Pos:* Assoc prof, Sch Pharm, Med Col Va, 51-61. *Mem:* Am Phys Soc; Am Asn Physics Teachers. *Res:* Evaporation of chlorine atoms from silver chloride crystals; teaching undergraduate physics, especially lecture demonstrations and curriculum development; computer-assisted instruction in introductory physics. *Mailing Add:* 2431 Swathmore Rd Richmond VA 23235

TAYLOR, JAMES A, b Woonsocket, RI, 1939; m 60; c 3. BIOCHEMISTRY, PHARMACOLOGY. *Educ:* Providence Col, BS, 60; Purdue Univ, MS, 63, PhD(biochem), 66. *Prof Exp:* Resident res assoc, Div Biol & Med, Argonne Nat Lab, 65-67; sr res scientist, Med Res Labs, Pfizer Inc, 67-75, liaison officer, Food & Drug Admin, Pfizer Cent Res, 75-79; mgr, New Drug Affairs, ICI Americas, 79-80; dir drug regulatory affairs dept, ICI Americas Inc, 80-83; vpres regulatory affairs, Carter-Wallace Inc, 83-87; VPRES & CHIEF REGULATORY OFFICER, IMMOGEN INC, 87- *Mem:* Am Soc Pharmacol & Exp Therapeut; Sigma Xi; Regulatory Affairs Prof Soc; Drug Info Asn. *Res:* Drug metabolism and pharmacokinetics. *Mailing Add:* ImmunoGen 148 Sidney St Cambridge MA 02139

TAYLOR, JAMES EARL, b Beverly, Ohio, Sept 7, 16; m 45; c 3. RESEARCH ADMINISTRATION. *Educ:* Western Reserve Univ, AB, 38, MS, 40; Univ Pa, PhD(physics), 43. *Prof Exp:* Asst physics, Western Reserve Univ, 38-40 & Univ Pa, 40-42; physicist, Norden Labs Corp, 43-44, head res & develop sect, 44-50; from supvr electronics group & dir exp lab to lab dir, M Ten Bosch, Inc, 50-63; mem tech planning dept, Xerox Corp, 63-66, mgr res, Tech Planning Off, 66-69 & commun & educ, 69-70, tech staff specialist, 70-74, mgr res tech staff, 74-83; CONSULT, 83- *Mem:* Am Phys Soc; Am Radio Relay League. *Res:* Ultrasonic studies; mass spectroscopy; isotope separation; ordnance research and development; research and development management; education. *Mailing Add:* 1257 Wildflower Dr Webster NY 14580

TAYLOR, JAMES H(OBERT), b Tishomingo, Miss, Jan 28, 29; m 51; c 2. SOIL COMPACTION BY MACHINERY. *Educ:* Miss State Univ, BS, 51; Auburn Univ, PhD(agr eng), 64. *Prof Exp:* Jr engr, Int Harvester Co, 51-52, asst zone mgr farm equip, 54-59; asst agr eng, Auburn Univ, 59-62; res agr engr, 62-72, res leader, traction res, Nat Tillage Mach Lab, Agr Res, 72-85, NAT TECH ADV, TRACTION & CONTROLLED TRAFFIC, USDA, 82- *Concurrent Pos:* Res lectr, Grad Fac, Auburn Univ, 68- *Mem:* Am Soc Agr Engrs; Int Soc Terrain-Vehicle Systs; Soc Automotive Engrs; Am Soc Testing & Mat. *Res:* Terrain vehicle systems; off-road-locomotion; soil-machine systems; mobility; force-deformation relationships; systems analysis; soil dynamics; traffic-induced soil compaction. *Mailing Add:* Nat Soil Dynamics Lab US Dept Agr Box 792 Auburn AL 36831-0792

TAYLOR, JAMES HUGH, b San Jose, Calif, June 21, 40; m 63; c 4. NONLINEAR SYSTEMS THEORY, EXPERT SYSTEMS. *Educ:* Univ Rochester, BSc, 63, MSc, 64; Yale Univ, MPh, 68, PhD(eng & appl sci), 69. *Prof Exp:* Vis asst prof elec eng, Indian Inst Sci, Bangalore, India, 69-72; sr engr & proj leader systs eng, The Anal Sci Corp, Reading, Mass, 73-78; assoc prof mech & aerospace eng, Okla State Univ, 78-81; SYSTS ENGR, GEN ELEC CORP RES & DEVELOP, 81- *Concurrent Pos:* Consult, Syst Eng Tech Assoc Corp, Stillwater, Okla, 79-81; short-course summer instr, Mass Inst Tech, 84-87. *Mem:* Am Soc Mech Engrs; Inst Elec & Electronics Engrs; Am Asn Artificial Intel; Sigma Xi. *Res:* Nonlinear systems, especially analysis and design using describing function methods and stability criteria; stochastic systems, especially nonlinear estimation and control; modeling; simulation; expert systems for computer-aided design and real-time control. *Mailing Add:* Gen Elec Corp Res & Develop Bldg KW Rm D209A PO Box 8 Schenectady NY 12301

TAYLOR, JAMES KENNETH, b Fall River, Mass, July 28, 29; m 53; c 2. BIOLOGY. *Educ:* State Teachers Col Bridgewater, BScEd, 51; Columbia Univ, MA, 55. *Prof Exp:* Teacher pub schs, Mass, 51-56; from instr to assoc prof, 56-85, chmn dept, 74-89, PROF BIOL, WESTFIELD STATE COL, 85- *Concurrent Pos:* Chmn, Westfield Conserv Comn, 62-; consult, Rand-McNally Publ Co, 72-83 & D C Health Publ Co, 83- *Honors & Awards:* Distinguished Serv Conserv Award, Trout Unlimited, 86. *Mem:* AAAS; Nat Asn Biol Teachers; Ecol Soc Am; Nat Sci Teachers Asn. *Res:* Ecological basis of conservation; dissemination of ecological principles and their applications to solution of local problems; acid rain monitoring. *Mailing Add:* Dept Biol Westfield State Col Westfield MA 01085

TAYLOR, JAMES LEE, b Berkey, Ohio, Jan 6, 31; m 58; c 2. HORTICULTURE. *Educ:* Ohio State Univ, BS, 53; Mich State Univ, MS, 57; Univ Ill, PhD(bot), 60. *Prof Exp:* Prof hort & exten specialist, 60-89, EMER PROF HORT, MICH STATE UNIV, 89- *Concurrent Pos:* Treas, Nat Jr Hort Found, Inc. *Mem:* Am Hort Soc; Am Soc Hort Sci; Nat Jr Hort Asn. *Res:* Nut tree culture and physiology. *Mailing Add:* 6132 Shoeman Rd Haslett MI 48840

TAYLOR, JAMES VANDIGRIFF, b New York, NY, Dec 27, 31; m 55; c 1. FORENSIC ANTHROPOLOGY. *Educ:* Columbia Univ, BS, 57, PhD(anthrop), 68. *Prof Exp:* Lectr anthrop, Hunter Col, City Univ New York, 64-65; lectr anthrop, New York Univ, 66-67; asst prof, 67-71, ASSOC PROF ANTHROP, LEHMAN COL, CITY UNIV NEW YORK, 71-, DEPT ANTHROP. *Concurrent Pos:* Prin investr res award, City Univ New York, 75-76, co-prin investr, 80-81, prin investr, 79-; dir Metrop Forensic Anthrop Team, Lehman Col, City Univ New York, 79- *Res:* Identification of human skeletal and dental remains in a forensic context. *Mailing Add:* Bedford Park Blvd W Herbert H Lehman Col/CUNY Bronx NY 10468

TAYLOR, JAMES WELCH, b Newton, Miss, Sept 17, 35; m 57; c 2. ANALYTICAL CHEMISTRY. *Educ:* Vanderbilt Univ, BA, 56; Ga Inst Technol, MS, 58; Univ Ill, Urbana, PhD(chem), 64. *Prof Exp:* Develop chemist, Mobil Oil Co, 58-61; asst prof chem, Tulane Univ, La, 64-66; from asst prof to assoc prof, 66-70, PROF CHEM, UNIV WIS-MADISON, 70- *Mem:* Am Chem Soc; The Chem Soc; Sigma Xi. *Res:* Photoionization mass spectrometry; photoelectron spectroscopy; isotope kinetics; rates and mechanisms of reactions; analytical instrumentation. *Mailing Add:* Dept Chem Univ Wis 1101 University Ave Madison WI 53706

TAYLOR, JAVIN MORSE, b Lancaster, Wis, Jan 20, 36; m 59; c 4. ELECTRICAL ENGINEERING, COMPUTER ENGINEERING. *Educ:* Univ Ill, Urbana, BS, 57; Univ Southern Calif, MS, 62; Univ Wyo, PhD(elec eng), 70. *Prof Exp:* Field engr, Hughes Aircraft Co, 57-59; engr, Ramo-Wooldridge Div, TRW, Inc, 59-61, TRW Comput Div, 61-62; eng specialist, Guid & Control Systs Div, Litton Industs, 62-66; instr & res engr, Natural Resources Res Inst, Univ Wyo, 66-69; res engr, Autonetics Div, NAm Rockwell, Inc, 69-70; from asst prof to assoc prof elec eng, Univ Mo-Rolla,

70-76; dept head, Elec & Computer Eng, 87-89, PROF ELEC & COMPUTER ENG, NMEX STATE UNIV, 76- *Concurrent Pos:* Consult, TRW Comput Div, TRW, Inc, 63, Guid & Control Systs Div, Litton Industs, 66-68 & John S Bereman & Co Eng Consult, 68-69; lectr, Calif State Univ, Los Angeles, 70; consult, Instrumentation Directorate, White Sands Missile Range, 77-80; elec & computer eng prog evaluator, Accrediting Bd Eng & Tech. *Mem:* Inst Elec & Electronics Engrs; Inst Elec & Electronics Engrs Computer Soc; Inst Elec & Electronicer Engrs Educ Soc; Am Soc Eng Educ. *Res:* Computer architecture and design; digital communication networks; computer networks; digital signal processing. *Mailing Add:* Dept Elec & Computer Eng NMex State Univ Las Cruces NM 88003-0001

TAYLOR, JAY EUGENE, b Stayton, Ore, Feb 2, 18; m 48; c 4. CHEMISTRY. *Educ:* Ore State Col, BA, 40; Univ Wis, MS, 43; Purdue Univ, PhD, 47. *Prof Exp:* Instr chem, Univ Wis, 46-48; asst prof chem, Miami Univ, 48-52; fel phys chem, Ohio State Univ, 52-54; asst prof, Univ Nebr, 54-60; from assoc prof to prof 60-84, RES PROF CHEM, KENT STATE UNIV, 84- *Mem:* Am Chem Soc; Sigma Xi. *Res:* Mechanisms of oxidation and decomposition reactions; flow techniques applied to kinetic studies of both liquid and gaseous systems; homogeneous versus heterogeneous gas-phase pyrolysis studies using wall-less and homogeneous front reactors. *Mailing Add:* Dept Chem Kent State Univ Kent OH 44242

TAYLOR, JEAN ELLEN, b San Mateo, Calif, Sept 17, 44; m 73; c 3. GEOMETRIC MEASURE THEORY, SOME ASPECTS OF THEORETICAL METALLURGY. *Educ:* Mt Holyoke Col, BA, 66; Univ Calif, Berkeley, MSc, 68 & Univ Warwick, 71; Princeton Univ, PhD(math), 73. *Prof Exp:* Instr math, Mass Inst Technol, 72-73; from asst prof to assoc prof, 73-82, prof math, 82-87, PROF II MATH, RUTGERS UNIV, 87- *Concurrent Pos:* Prin investr, NSF grants, 73-, Air Force Off Sci Res grants, 88-; mem, Inst Adv Study, 74-75, 77-78 & 85; Alfred P Sloan Found fel, 76-78; mem, Nominating Comt, Am Math Soc, 77-78, coun, 84-88, exec comt, 85-88 & chair, Nat Prog Comt; mem comt appl math training, Nat Acad Sci/Nat Res Coun, 77-78; vis, Princeton Univ, 80-81, Stanford Univ, 89; consult, Nat Inst Standards & Technol, 90-91. *Mem:* Am Math Soc; Math Asn Am; Asn Women Math; fel AAAS; Mat Res Soc; Soc Indust & Appl Math. *Res:* Equilibrium and growth shapes of crystalline materials. *Mailing Add:* Dept Math Rutgers Univ New Brunswick NJ 08903

TAYLOR, JEAN MARIE, b Protection, NY, Nov 21, 32. TOXICOLOGY, PHARMACOLOGY. *Educ:* Keuka Col, BA, 54; Univ Rochester, MS, 56, PhD(pharm), 59. *Prof Exp:* Res assoc pharm, Atomic Energy Proj, Univ Rochester, 54-59; pharmacologist, US Food & Drug Admin, 59-71, actg chief, Chronic Toxicol Br, Div Toxicol, 71-89; RETIRED. *Mem:* Soc Toxicol; Sigma Xi. *Res:* Toxicity of flavoring agents; hepatotoxins. *Mailing Add:* 941 Broadview Dr Harrisonburg VA 22801

TAYLOR, JERRY DUNCAN, b Plumerville, Ark, June 5, 38; m 64; c 3. MATHEMATICS. *Educ:* State Col Ark, BA, 60; Univ Ark, MS, 64; Fla State Univ, PhD(math educ), 69. *Prof Exp:* PROF MATH, CAMPBELL UNIV, NC, 62- *Concurrent Pos:* Co-auth, Prentice-Hall Encycl Math, 83. *Mem:* Math Asn Am. *Res:* New methods of extending the field of rational numbers to the field of real numbers. *Mailing Add:* Dept Math Campbell Univ Buies Creek NC 27506

TAYLOR, JERRY LYNN, b Washington, Mo, Jan 12, 47; m 71. INTERFERON, VIROLOGY. *Educ:* Univ Mo, BA, 69, MA, 71; Southern Ill Univ, Carbondale, PhD(microbiol), 76. *Prof Exp:* Asst prof virol, Calif State Univ, Long Beach, 76-77; postdoctoral fel virol, Med Col Wis, Milwaukee, 77-79, from instr to asst prof microbiol, 79-84, asst prof ophthal, 81-84, assoc prof microbiol & ophthal, 84-90, PROF MICROBIOL & OPHTHAL, MED COL WIS, MILWAUKEE, 90- *Mem:* Am Soc Microbiol; Am Soc Virol; Int Soc Antiviral Res; Asn Res Vision & Ophthal; AAAS; Sigma Xi. *Res:* Actions of antiviral agents including nucleoside analogues, interferons and interferon inducers on viral diseases especially herpes simplex virus keratitis; effects of interferons on cellular differentiation. *Mailing Add:* Dept Microbiol Med Col Wis 8701 Watertown Plank Rd Milwaukee WI 53226

TAYLOR, JOCELYN MARY, b Portland, Ore, May 30, 31; m 72. ZOOLOGY. *Educ:* Smith Col, BA, 52; Univ Calif, Berkeley, MA, 53, PhD(zool), 59. *Prof Exp:* Asst zool, Conn Col, 53-54; Fulbright grantee, Australia, 54-55; assoc zool, Univ Calif, 59; from instr to asst prof, Wellesley Col, 59-65; from assoc prof to prof zool, Univ BC, 65-82; dir, Cowan Vertebrate Mus, 65-82; collab scientist, Ore Regional Primate Res Ctr, 83-87; PROF, ORE STATE UNIV, 84-; ADJ PROF BIOL, CASE WESTERN RESERVE UNIV, 87-; DIR, CLEVELAND MUS NATURAL HIST, 87- *Concurrent Pos:* Sigma Xi grant, 61-62; Lalor Found grant, 62-63; NSF grants, 63-71 & Australia, 63-64 & 65; Nat Res Coun Can res grants, 66-84, travel grant, Div Animal Physiol, Commonwealth Sci & Indust Res Orgn, Australia, 71-72; Killiam Sr fel, 78-79; assoc ed, J Mammalog, 81-82; mem rodent specialist group, endangered species, Int Union Conserv Nature & Natural resources, 80- *Mem:* Sigma Xi; Am Soc Mammalogists (vpres, 78-82, pres, 82-84, bd dirs, 84-); Cooper Ornithol Soc; Australian Mammal Soc; Soc Women Geogrs. *Res:* Reproductive biology of mammals; evolution of Australasian murid rodents; marsupial placentation. *Mailing Add:* 2718 SW Old Orchard Rd Portland OR 97201

TAYLOR, JOHN BRYAN, crystal chemistry, thermodynamics, for more information see previous edition

TAYLOR, JOHN CHRISTOPHER, b Chelmsford, Eng, Jan 17, 36; Can citizen; m 69; c 3. PROBABILITY. *Educ:* Acadia Univ, BSc, 55; Queen's Univ, Ont, MA, 57; McMaster Univ, PhD(math), 60. *Prof Exp:* J F Ritt instr math, Columbia Univ, 60-63; from asst prof to assoc prof, 63-74, PROF MATH, MCGILL UNIV, 74- *Mem:* Am Math Soc; Can Math Cong. *Res:* Analysis; potential theory; probability theory. *Mailing Add:* Dept Math McGill Univ 805 Shetbrooke St W Montreal PQ H3A 2K6 Can

TAYLOR, JOHN DIRK, b Mecca, Calif, Mar 31, 39; m 63; c 2. CANCER BIOLOGY. *Educ:* Univ Ariz, BS, 62, PhD(biol), 67. *Prof Exp:* NSF, US-Japan Coop Sci Prog fel biol, Keio Univ, Japan, 67-68; from asst prof to assoc prof, Wayne State Univ, 68-75, assoc prof comp med, 72-75, chmn dept biol, 74-87, dep dean, Lib Arts, 83-84, PROF BIOL, WAYNE STATE UNIV, 75-, PROF RADIATION ONCOL, SCH MED, 80- *Concurrent Pos:* Asian Found guest lectr, Univs in Seoul, Korea, 68. *Mem:* Fel AAAS; Am Soc Cell Biol; Am Soc Zool; Soc Develop Biol. *Res:* Biochemical and ultrastructural aspects of developmental processes with particular emphasis on intracellular movements and metastasis. *Mailing Add:* Dept Biol Sci Wayne State Univ Detroit MI 48202

TAYLOR, JOHN EDGAR, b Cheyenne, Wyo, Oct 17, 31; wid; c 3. RANGE SCIENCE, PLANT ECOLOGY. *Educ:* Idaho State Univ, BS, 58 & Wash State Univ, 60; Mont State Univ, MS, 67; NDak State Univ, PhD(bot & plant ecol), 76. *Prof Exp:* From instr to asst prof, 63-76, ASSOC PROF RANGE SCI, MONT STATE UNIV, 76- *Mem:* Soc Range Mgt. *Res:* Rangeland analysis and measurements; remote sensing of natural resources data. *Mailing Add:* Dept Range Mgt Mont State Univ Bozeman MT 59715

TAYLOR, JOHN FULLER, b Jamestown, NY, June 10, 12; m 35; c 2. BIOCHEMISTRY. *Educ:* Cornell Univ, AB, 33; Johns Hopkins Univ, PhD(physiol chem), 37; Am Bd Clin Chem, dipl. *Prof Exp:* Nat Res Coun fel med sci, Harvard Med Sch, 37-39, teaching fel biol chem, 39-41; biochemist, Lederle Labs, Inc, 41-43; asst prof biol chem, Sch Med, Washington Univ, 43-52; chmn dept, 52-72, prof, 52-78, EMER PROF BIOCHEM, SCH MED, UNIV LOUISVILLE, 78- *Concurrent Pos:* Vis prof, Univ Oslo, 55; Commonwealth Fund fel, Cambridge, Eng & Rome, Italy, 61-62. *Mem:* Am Chem Soc; Am Soc Biol Chemists; Soc Exp Biol & Med; Biophys Soc; Brit Biochem Soc. *Res:* Hemoglobin; enzyme proteins; quinones; hemochromogens; oxidation-reduction potentials of biological systems; sulfhydryl groups of proteins; physical chemistry of proteins and polysaccharides; ultracentrifugation. *Mailing Add:* Dept Biochem Health Sci Ctr Univ Louisville Louisville KY 40292

TAYLOR, JOHN GARDINER VEITCH, b Toronto, Ont, Sept 22, 26; m 55. PHYSICS. *Educ:* McMaster Univ, BSc, 50; Univ Sask, MSc, 52. *Prof Exp:* Res physicist photonuclear reactions, Univ Sask, 53; res physicist atomic mass measurements, McMaster Univ, 54-55; res physicist radioactivity standards, 56-76, head, Counter Develop Sect, 76-82, HEAD, NUCLEAR DETECTORS & METROL SECT, ATOMIC ENERGY CAN, LTD, 82- *Concurrent Pos:* Ed, Int J Appl Radiation Isotopes, 68-88; chmn, Sect II (Measure of Radionuclides), Consult Comt Standards of Measure Ionizing Radiation, Int Bur Weights & Measures, 85- *Mem:* Am Phys Soc; Can Asn Physicists; AAAS; Inst Elec & Electronics Engrs; Int Comt Radionuclide Metrol. *Res:* Radioactivity; radiation detectors. *Mailing Add:* Physics Div AECL Res Chalk River ON K0J 1J0 Can

TAYLOR, JOHN HALL, vision; deceased, see previous edition for last biography

TAYLOR, JOHN JACOB, b Dayton, Ohio, June 26, 28. MICROBIOLOGY. *Educ:* Heidelberg Col, BS, 50; Univ Ohio, MS, 52; Ohio State Univ, PhD, 57. *Prof Exp:* Instr bot, Univ Ohio, 50-52; res specialist toxicol, Nat Cash Register Co, 52-54; asst, Res Found, Ohio State Univ, 56-57; from asst prof to assoc prof, 57-69, PROF MICROBIOL, UNIV MONT, 69- *Mem:* Mycol Soc Am; Med Mycol Soc Americas; Can Soc Microbiol. *Res:* Morphology and physiology of fungi; host parasite relationships; antibiotic resistance in pathogenic fungi. *Mailing Add:* Dept Animal Sci Mont State Univ Bozeman MT 59717

TAYLOR, JOHN JOSEPH, b Hackensack, NJ, Feb 27, 22; m 43; c 3. MATHEMATICS. *Educ:* St John's Univ, NY, BS, 42; Univ Notre Dame, MS, 46. *Hon Degrees:* DSc, St John's Univ, NY, 74. *Prof Exp:* Appl mathematician, Bendix Aviation Corp, 46-47; scientist, Physics Dept, Kellex Corp, NY, 47-50; sr scientist, Bettis Atomic Power Lab, Westinghouse Elec Corp, 50-S2, mgr shielding physics, 52-55, mgr surface ship proj, Physics Dept, 55-58, mgr reator develop, 58-65, mgr mat develop, 65-67, mgr eng, Atomic Power Div, 67, eng mgr, Power Plant Div, 67-70, gen mgr, Breeder Reactor Div, 70-74, vpres, Advan Nuclear Systs, 74-76, vpres & gen mgr, Water Reactor Bus Unit, 76-81; VPRES, NUCLEAR POWER DIV, ELEC POWER RES INST, 81- *Concurrent Pos:* Mem bd dirs, Am Nuclear Soc, 71-74, Advan Reactor Corp, Inst Nuclear Power Opers Adv Coun, Vis Comt Dept Nuclear Eng, Brookhaven Nat Lab; consult, US Govt Acct Off, 74-; mem adv comt, Argonne Nat Lab, Oak Ridge Nat Lab & Lawrence Livermore Nat Lab; mem bd trustees, Argonne Univs Asn. *Honors & Awards:* Westinghouse Award of Merit, 57; George Westinghouse Gold Medal Award, 90. *Mem:* Nat Acad Eng; fel AAAS; Am Phys Soc; fel Am Nuclear Soc. *Res:* Nuclear reactors; shielding and reactors for atomic submarines; digital computer techniques in reactor design; nuclear fuel development. *Mailing Add:* PO Box 10412 Elec Power Res Inst Palo Alto CA 94303

TAYLOR, JOHN JOSEPH, b Westhampton Beach, NY, May 15, 25; wid; c 2. ANATOMY, ELECTRON MICROSCOPY. *Educ:* Hofstra Col, BA, 53; Cornell Univ, MS, 56; Univ Buffalo, PhD(anat), 59. *Prof Exp:* Asst to Dr G N Papanicolaou, Med Col, Cornell Univ, 50-53, asst anat, 54-56; from asst to instr, Sch Med, Univ Buffalo, 56-60; from asst prof to assoc prof, Sch Med, Univ NDak. 60-66; assoc prof anat, 66-85, ASSOC PROF NEUROANAT IN NEUROL, SCH MED, ST LOUIS UNIV, 67-, PROF ANAT, 85- *Concurrent Pos:* Lectr, Drexel Inst, 62-64. *Mem:* Soc Neurosci; Am Asn Anatomists; Electron Micros Soc Am; Am Soc Cell Biol; Am Asn Hist Med. *Res:* Electron microscopy and histochemistry of connective tissue, central nervous system and conduction system of heart. *Mailing Add:* Dept Anat St Louis Univ Sch Med St Louis MO 63104

TAYLOR, JOHN KEENAN, b Mt Rainier, Md, Aug 14, 12; m 38; c 2. PHYSICAL CHEMISTRY, ANALYTICAL CHEMISTRY. *Educ:* George Washington Univ, BS, 34; Univ Md, MS, 38, PhD(chem), 41. *Prof Exp:* Sci aide, Nat Bur Standards, 29-36, chemist, 36-40, asst chemist, 40-42, assoc chemist, 42-44, chemist, 44-61, chief microchem analytical sect, 61-73, chief, Air & Water Pollution Analysis Sect, 73-77, chief, Gas & Particulate Sci Div, 77-80, qual assurance & vol standards coord, 80-87; CONSULT, 87- *Concurrent Pos:* Adj prof, Am Univ. *Honors & Awards:* Rosa Award Stand Activ, Nat Bur Standards, 74. *Mem:* Am Chem Soc; Electrochem Soc; Air Pollution Control Soc. *Res:* Preparation of pure platinum metals; electrochemistry of solutions; standard electrode potentials; separation of isotopes; polarography; coulometry; microchemical and environmental analysis; development of analytical methods and standard reference materials for air and water pollution analysis; quality assurance procedures for chemical analysis. *Mailing Add:* 12816 Tern Dr Gaithersburg MD 20878

TAYLOR, JOHN LANGDON, JR, b Brooklyn, NY, Nov 3, 28; m 61. MEDICAL EDUCATION. *Educ:* Univ Chicago, PhB, 50; Univ Calif, Los Angeles, PhD(physiol), 57. *Prof Exp:* Grad res anatomist, Univ Calif, Los Angeles, 56-57, grad res physiologist, 57, asst res physiologist, 57-59; from instr to asst prof physiol, Col Med, Univ Ill, 59-65; assoc prof biol, Wayne State Univ, 65-68; assoc prof physiol & asst to dean, Col Osteop Med, Mich State Univ, 69-71, assoc prof med educ, & asst dean student affairs, Col Osteop Med, 71-72, assoc prof acad affairs, Col Osteop Med, 72-75; assoc dean student affairs, 75-79, PROF OSTEOP MED, COL OSTEOP MED, OHIO UNIV, 75- *Mem:* AAAS; Am Educ Res Asn; Nat Coun Measurement in Educ; Soc Gen Physiol; Biophys Soc; Sigma Xi. *Res:* Neuro-regulatory mechanism; biological rhythms; test scoring methods; career choices in the health professions; computer assisted instruction. *Mailing Add:* Nine Fort St Athens OH 45701

TAYLOR, JOHN MARSTON, b Melbourne, Australia, Sept 10, 41; m 60; c 2. VIROLOGY. *Educ:* Univ Melbourne, BSc, 62, MSc, 64; Univ Toronto, PhD(cell biol), 68. *Prof Exp:* Mem staff, Inst Cancer Res, 74-; CHIEF CHEMIST, GLADSTONE FEDN LABS, SAN FRANCISCO. *Res:* Virus replication; cell biology; carcinogenesis. *Mailing Add:* Gladstone Fedn Labs PO Box 40608 San Francisco CA 94140-0608

TAYLOR, JOHN ROBERT, b London, Eng, Feb 2, 39; m 62; c 2. THEORETICAL PHYSICS. *Educ:* Cambridge Univ, BA, 60; Univ Calif, Berkeley, PhD(physics), 63. *Prof Exp:* NATO fel physics, Cambridge Univ, 62-64; instr, Princeton Univ, 64-66; from asst prof to assoc prof, 66-72, PROF PHYSICS, UNIV COLO, BOULDER, 72- *Mem:* Am Phys Soc. *Res:* Quantum theory; quantum theory of scattering. *Mailing Add:* Dept Physics/ DP6t F-1035 Univ Colo Boulder CO 80309-0390

TAYLOR, JOHN WILLIAM, b Austin, Tex, Dec 14, 46; m 68; c 2. ANALYTICAL & PHARMACEUTICAL CHEMISTRY. *Educ:* Univ Cincinnati, BS, 69; Duke Univ, PhD(analytical chem), 78. *Prof Exp:* REVIEWING CHEMIST RX DRUGS, DIV GENERIC DRUG MONOGRAPHS, BUR DRUGS, FOOD & DRUG ADMIN, HEW, 74- *Mem:* Sigma Xi; Asn Off Analytical Chemists; Am Chem Soc. *Res:* Prediction of ion exchange selectivities in mixed solvents; isotope separations; polymorphism and bioavailability. *Mailing Add:* White Meadow Rd Rockaway NJ 07866-1106

TAYLOR, JOSEPH HOOTON, JR, b Philadelphia, Pa, Mar 29, 41; m 63; c 2. RADIO ASTRONOMY, PULSARS. *Educ:* Haverford Col, BA, 63; Harvard Univ, PhD(astron), 68. *Hon Degrees:* DSc, Univ Chicago, 85. *Prof Exp:* Lectr astron & res assoc, Harvard Col Observ, Harvard Univ, 68-69; from asst prof to prof astron, Univ Mass, 69-81; prof, 81-86, JAMES MCDONNELL DISTINGUISHED PROF PHYSICS, PRINCETON UNIV, 86- *Concurrent Pos:* Grant, Res Corp, Univ Mass, 68-69, NSF, 69-82 & NASA, 71-72; consult, Mass Gen Hosp, 71-73; Tomalla Found Prize Gravitation & Cosmology, Tomalla Found, 87. *Honors & Awards:* Dannie Heineman Prize, Am Astron Sci, 80; Henry Draper Medal, Nat Acad Sci, 85; Magellanic Premium, Am Philos Soc, 90; John J Carty Award Advan Sci, US Nat Acad Sci, 91; Einstein Prize Laureate, Albert Einstein Soc, 91. *Mem:* Nat Acad Sci; Am Astron Soc; Int Union Radio Sci; Int Astron Union; fel Am Phys Soc; fel Am Acad Arts & Sci. *Res:* Radio astronomy; pulsars; design and development of radio telescopes and information processing systems. *Mailing Add:* Dept Physics Princeton Univ Princeton NJ 08544

TAYLOR, JOSEPH LAWRENCE, b Apr 7, 41; US citizen; m 59; c 3. MATHEMATICAL ANALYSIS. *Educ:* La State Univ, BS, 63, PhD(math), 64. *Prof Exp:* Instr math, Harvard Univ, 64-65; from asst prof to assoc prof, Univ Utah, 65-71, chmn dept, 79-82, dean sci, 85-87, acad vpres, 87-90, PROF MATH, UNIV UTAH, 71- *Concurrent Pos:* Sloan fel. *Honors & Awards:* Steele Prize, Am Math Soc, 75. *Mailing Add:* Dept Math Univ Utah Salt Lake City UT 84112

TAYLOR, JULIUS DAVID, b Erie, Pa, Dec 18, 13; m 38; c 1. BIOCHEMISTRY, TOXICOLOGY. *Educ:* Univ Pittsburgh, BS, 36; Univ Rochester, PhD(biochem), 40. *Prof Exp:* Instr biochem, Sch Med & Dent, Univ Rochester, 40-41; res org chemist, Distillation Prod, Inc, NY, 41-43; res biochemist, Eaton Labs, Inc, 43-47; chem pharmacologist, Abbott Labs, 48-60, from asst dept head to dept head pharmacol, 60-64, drug eval specialist, Dept Drug Regist, 65-69 & Div Regulatory Affairs, 69-70, data specialist, Dept Exp Biomet, 70-77, control record coordr, 78-79; RETIRED. *Concurrent Pos:* Lectr, Med Sch, Northwestern Univ, Chicago, 59-62; lectr, Sch Med, Univ Chicago, 62-65. *Mem:* AAAS; Am Soc Pharmacol & Exp Therapeut; Soc Toxicol. *Res:* Drug enzymology, metabolism and kinetics; pharmacology; toxicology; bionics; simulation. *Mailing Add:* Apt A8 905 Baldwin Ave Waukegan IL 60085

TAYLOR, KATHLEEN C, b Cambridge, Mass, Mar 16, 42. PHYSICAL CHEMISTRY. *Educ:* Rutgers Univ, New Brunswick, AB, 64; Northwestern Univ, Evanston, PhD(phys chem), 68. *Prof Exp:* Fel, Univ Edinburgh, 68-70; assoc sr res chemist, 70-74, sr res chemist, 74-75, asst dept head, 75-83, DEPT HEAD, RES LABS, GEN MOTORS CORP, 83- *Honors & Awards:* Garvan Medal, Am Chem Soc. *Mem:* Am Chem Soc; Catalysis Soc; Royal Chem Soc; Mat Res Soc; Soc Automotive Engrs. *Res:* Surface chemistry; heterogeneous catalysis; catalytic control of automobile exhaust emissions. *Mailing Add:* Phys Chem Dept Gen Motors Res Labs 12 Mile & Mound Rds Warren MI 48090

TAYLOR, KEITH EDWARD, b Toronto, Ont, Dec 21, 46; m 68; c 2. BIOCHEMISTRY. *Educ:* Univ Toronto, BSc, 69; PhD(bioorg chem), 74. *Prof Exp:* Fel enzymatic stereochem, Lab Org Chem, Swiss Fed Inst Technol, 73-76; fel protein chem, Harvard Univ, 76; asst prof, 76-82, ASSOC PROF BIOCHEM, UNIV WINSOR, 82- *Concurrent Pos:* Sabbaticant, Lab Biochem, Switz Fed Inst Technol, 84-85. *Mem:* Am Chem Soc; Chem Inst Can; Can Biochem Soc. *Res:* Protein chemistry enzymology; glycoproteins; immobilized enzymes; enzyme-based reactors. *Mailing Add:* Dept Chem & Biochem Univ Windsor Windsor ON N9B 3P4 Can

TAYLOR, KENNETH BOIVIN, b Columbus, Ohio, Aug 7, 35; m 58; c 2. BIOCHEMISTRY, ENZYMOLOGY. *Educ:* Oberlin Col, AB, 57; Case Western Reserve Univ, MD, 61; Mass Inst Technol, PhD, 67. *Prof Exp:* Res assoc, Mass Inst Technol, 64-67, asst prof biol, 67-70; ASSOC PROF BIOCHEM, UNIV ALA, BIRMINGHAM, 70- *Concurrent Pos:* Mgr, Fermentation Facil, Univ Ala, Birmingham, 83- *Mem:* Am Soc Biochem & Molecular Biol; Am Chem Soc; AAAS. *Res:* Enzyme mechanisms & kinetics; genetic engineering in plants; biotechnology-fermentation, purification. *Mailing Add:* Dept Biochem Univ Ala Birmingham AL 35294

TAYLOR, KENNETH DOYLE, b Hartford, Conn, Nov 5, 49; m 72; c 1. BIOENGINEERING & BIOMEDICAL ENGINEERING, ELECTRICAL ENGINEERING. *Educ:* Univ Conn, BS, 71, MS, 74, PhD(biol eng), 81; Rensselaer Polytechnic Inst, MBA, 88. *Prof Exp:* Coordr, Res Lab, St Francis Hosp & Med Ctr, 71-73, mgr, 74-79; design engr, Nuclear & Ultrasound Div, Picker Corp, 73-74; sr proj engr, United Technologies Res Ctr, 79-90; ASST DIR TECHNOL ASSESSMENT, PFIZER HOSP PROD GROUP, 90- *Concurrent Pos:* Lectr, Univ Conn, 77-82; adj prof, Trinity Col, 88-90; adj asst prof, Hartford Grad Ctr, 80-; spec asst to dir, Div Heart & Vascular Dis, Nat Heart, Lung & Blood Inst, 85-86. *Mem:* Inst Elec & Electronics Engrs; Sigma Xi. *Res:* Development of medical devices using electronic and/or electro-optic technology. *Mailing Add:* Pfizer Hosp Prod Group 235 E 42nd St New York NY 10017

TAYLOR, KENNETH GRANT, b Paterson, NJ, May 12, 36; m 61; c 3. ORGANIC CHEMISTRY. *Educ:* Calvin Col, AB, 57; Wayne State Univ, PhD(org chem), 63. *Prof Exp:* Res assoc, Mass Inst Technol, 63-64; sr res assoc chem, Wayne State Univ, 64-66; from asst prof to assoc prof, 66-73, chmn dept, 78-87, PROF CHEM, UNIV LOUISVILLE, 73- *Concurrent Pos:* Assoc prof, Univ Nancy I, France, 74-75, 82-83; vis prof, Univ Lund, Sweden, 91. *Mem:* Am Chem Soc; AAAS; NY Acad Sci. *Res:* Synthesis of strained carbocyclic and heterocyclic compounds; carbohydrate chemistry; chemical carcinogenesis; cycloaddition reactions. *Mailing Add:* Dept Chem Univ Louisville Louisville KY 40292-0001

TAYLOR, KENNETH J W, b Essex, Gt Brit, Mar 8, 39; m 75; c 2. RADIOLOGY. *Educ:* London Univ, BSc, 61, MB & BS, 64, MD, 75, PhD(biophys), 73; Yale Univ, MA, 79. *Prof Exp:* Lectr anat, London Univ, 67-72; sr fel nuclear med, Royal Hosp, 73-75; assoc prof radiol, 75-79, PROF DIAG IMAGING, YALE UNIV, 79- *Concurrent Pos:* Chmn Bd Clin Diag US, 76-; vis prof, Bristol Univ, 82-83; co-dir, Ctr for Sonics & Ultrasonics, Yale Univ, 83-; prin investr, NIH, 88--1. *Mem:* Am Inst Ultrasound Med (bd govs, 79-82); Am Roentgen Ray Soc; Radiol Soc Am; Royal Col Surgeons. *Res:* Applications of ultrasound in medicine; imaging techniques; doppler investigation of flow 7 physiologic function. *Mailing Add:* 333 Cedar St New Haven CT 06510

TAYLOR, KENNETH LAPHAM, b Los Angeles, Calif, May 16, 41; m 69; c 3. HISTORY OF MODERN SCIENCE, HISTORY OF GEOLOGY. *Educ:* Harvard Univ, AB, 62, AM, 65, PhD(hist of sci), 68. *Prof Exp:* From asst prof to assoc prof hist of sci, 67-72, CHMN DEPT HIST SCI, UNIV OKLA, 79-, PROF HIST SCI, 85- *Concurrent Pos:* Fel, Alexandre Koyre Ctr, Ecole Pratique des Hautes Etudes, 73-74; corresp mem, Int Comn Hist Geol Sci, 84-; chmn, US Nat Comt Hist Geol, 90- *Mem:* Hist Sci Soc; Hist Earth Sci Soc (treas, 85-91); Soc Hist Technol; Brit Soc Hist Sci; Comt Francais Hist Geol. *Res:* History of geology in 18th & early 19th centuries; history of science in 18th century. *Mailing Add:* Dept Hist Sci Univ Okla Phys Sci No 622 Norman OK 73019-0315

TAYLOR, KIRMAN, b Yorkshire, Eng, Sept 30, 20; nat US; m 41; c 2. ANALYTICAL CHEMISTRY, CLINICAL CHEMISTRY. *Educ:* Queen's Col, NY, BS, 41; Polytech Inst Brooklyn, MS, 43. *Prof Exp:* Instr, Polytech Inst Brooklyn, 46; res chemist, Celotex Corp, 47-48; res chemist & group leader, Westvaco Chem Div, Food Machinery & Chem Corp, 48-52; res assoc, George Washington Univ, 52-54; group leader, Diamond Alkali Co, 54-65, proj mgr, 65-67, assoc dir res dept, 68-75, dir res admin, res ctr, Diamond Shamrock Corp, 76-79; PRES, KIRMAN ASSOCS, TECH & MKT CONSULTS, 80- *Concurrent Pos:* Lectr, Wagner Col, 49-52 & Lake Erie Col, 60-67. *Mem:* Am Chem Soc; NY Acad Sci; AAAS. *Res:* Building materials technology; inorganic phosphates; coordination compounds; inorganic polymers; plastic and metal coatings; nuclear fuels; water chemistry; electroless and electrolytic plating; research administration; hydrometallurgy; management of analytical environmental chemical facilities. *Mailing Add:* 512 Whisperwood Dr Greenville TN 37743-6646

TAYLOR, LARRY THOMAS, b Woodruff, SC, Dec 31, 39; m 60; c 2. POLYMER CHEMISTRY, ANALYTICAL CHEMISTRY. *Educ:* Clemson Univ, BS, 62, PhD(chem), 65. *Prof Exp:* Res assoc chem, Ohio State Univ, 65-67; from asst prof to assoc prof, 67-78, PROF CHEM, VA POLYTECH INST & STATE UNIV, 78- *Mem:* Am Chem Soc; Sigma Xi. *Res:* Supercritical fluid chromatography; fourier transform infrared spectrometry; liquid chromatography; surface analysis; modification of polymers by metal ion addition. *Mailing Add:* 2101 Walnut Dr Blacksburg VA 24060-1812

TAYLOR, LAURISTON SALE, b Brooklyn, NY, June 1, 02; m 25, 73; c 6. MEDICAL & RADIOLOGICAL PHYSICS, BIOPHYSICS. *Educ:* Cornell Univ, AB, 26. *Hon Degrees:* DSc, Univ Pa, 60 & St Procopius Col, 65. *Prof Exp:* Asst, Heckscher Found, Cornell Univ, 24-27; from asst physicist to assoc physicist, Nat Bur Standards, 27-35, sr physicist & chief, X-ray Sect, 35-41, chief proving ground group, 40-43, chief x-ray sect, 42-43, 46-49, asst chief, Atomic Physics Div, 47-51, chief, Radiol Physics Lab, 49-51, Atomic & Radiation Physics Div, 51-60, Radiation Physics Div, 60-62, assoc dir, 62-64; spec asst to pres, Nat Acad Sci, 65-69, exec dir adv comt emergency planning, 65-71; pres, 64-77, HON PRES, NAT COUN RADIATION PROTECTION & MEASUREMENTS, 77-; RADIATION PHYSICS CONSULT, 77- *Concurrent Pos:* Mem, Int Comn Radiol Protection, 28-69, secy, 37-50, emer mem, 69-; mem, Int Comn Radiol Units & Measurements 28-69, secy, 34-50, chmn, 53-69, hon chmn & emer mem, 69-; chmn, Nat Coun Radiation Protection & Measurements, 29-64, chmn subcomt regulation of radiation exposure, 53-57, subcomt permissible exposure under emergency conditions, 55-59; dir Pan-Am Cancer Union, 39; mem sect E, div A, Nat Defense Res Comt, Off Sci Res & Develop, 42; nuclear comt Z-54, safety code, indust use of radiation, Am Nat Stand Comt, 42-46, nuclear stand bd, 56-; chief opers res, Eigth Fighter Command, 43 & Ninth Air Force, 43-45, consult, Air Force Opers Anal Div, Dept Defense, 47-52, mem, guided missiles countermeasures panel, Res & Develop Bd, 47-52, adv comt radiol defense, 48-51, comt weapons systs eval with Nat Acad Sci, 53-55, consult, Weapons Systs Eval Group, Joint Chiefs of Staff, 54-65, mem interagency comt biomed weapons effects tests, 57, consult, Inst Defense Anal, 57-67, mem ad hoc sci adv comt, Armed Forces Radiobiol Res Facil & chmn panel radiol instruments, 62-65, mem nuclear weapons effects res med adv group, 65-71, chmn rev comt, Armed Forces Radiobiol Res Inst, 66; chief biophys br, AEC, 48-49; mem subcomt radiobiol, Comt Nuclear Sci, Nat Acad Sci-Nat Res Coun, 49-54, consult comt med & surg, 54, comt radiol, 54, mem adv comt civil defense, 54-65, chmn, 57-65; sr consult, Civil Serv Bd Expert Exam, Civil Serv Comn, 52-58, chmn, 62-65. *Honors & Awards:* Bronze Star, USAF, 45; Sylvanus Thompson Medal, Brit Inst Radiol, 50; Lester Medal Lectr, Soc Nondestructive Testing, 54; Janeway Medal, Am Radium Soc, 56; Gold Medal, Am Col Radiol; Edward Bennett Rosa Award, Nat Bur Standards; Failla Mem Lectr, Health Physics Soc NY; Gold Medal, Royal Swedish Acad Sci & 13th Int Cong Radiol, Madrid, 73; Landauer Award, Am Asn Physicists Med & Health Physics Soc, 79; Sievert Award, Int Radiol Protection Agency, 79; Antoine Beclere Prize & Gold Medal, Int Soc Radiol, 81; Launiston S Taylor Lectr, Nat Coun Radiation Protection & Measurement, 77. *Mem:* Am Asn Physicists Med; fel Am Phys Soc; Am Roentgen Ray Soc; fel Am Col Radiol; hon mem Ger Roentgen Soc; Health Physics Soc (pres, 58-59); Radiation Res Soc; Radiol Soc NAm; Sigma Xi; hon mem Nippon Soc Radiol; hon fel Am Col Dentists. *Mailing Add:* 10450 Lottsford Rd Unit 3011 Mitchellville MD 20721-2734

TAYLOR, LAWRENCE AUGUST, b Paterson, NJ, Sept 14, 38; m 60; c 2. GEOCHEMISTRY, PETROLOGY. *Educ:* Ind Univ, Bloomington, BS, 61, MS, 63; Lehigh Univ, PhD(geochem), 68. *Prof Exp:* Instr geol, Univ Del, 63-64; fel geochem, Geophys Lab, Carnegie Inst Wash, 68-70; Fulbright fel, Max Planck Inst Nuclear Physics, Heidelberg, 70-71; asst prof, Purdue Univ, West Lafayette, 71-73; mem fac, dept geol sci, 73, PROF GEOL SCI, UNIV TENN, KNOXVILLE, 73- *Concurrent Pos:* Res grants, Geochem Sect, NSF; prin investr, Lunar Sample Prog, NASA, 72-; co-investr, Max Planck Inst Nuclear Physics; prin investr, NASA Planetary Mat Prog, NASA Meteorite Prog, mem, Lunar Sample Rev Panel, 74-75, Lunar & Planetary Rev Panel, 75-76, Lunar Sample Anal Prog Team, 76-78, Lunar & Planetary Sample Team, 78-81, NASA meteorite steering comt, 81-82, mem, Lunar & Planetary Sci Conf Prog Comt, 79-87, chmn, Lunar & Planetary Sample Team, 82-86, Lunar & Planetary Rev Panel, 82-84, Geosci Rev Panel, 84-85, Planetary Geosci Working Group, 84-86, Planetary Meetings Steering Comt, 84-86, Lunar Base Comt, 86-,Lunar Base Site Selection Comt, NASA Space Indignous Mat Res Utilization, 89-; discipline scientist & prog mgr, Solar Syst Explor, NASA, Washington, DC, 81-82; assoc ed, Proc Lunar & Planetary Sci Conf, 75, 77-79 & 81, J Geophys Res, 82-86. *Honors & Awards:* Sci Achievement Award, NASA, 78 & 83; Chancellor's Scholar, Univ Tenn, 81. *Mem:* Fel Mineral Soc Am; Mineral Asn Can; Am Geophys Union; fel Meteoritical Soc; Int Mineral Asn; Geochem Soc. *Res:* Experimental geochemistry and petrology into the stability relations of sulfide, oxide and silicate compounds at low and high pressures and application of these data to natural minerals and rocks; mineralogy, petrology and geochemistry of lunar rocks, meteorites, Kimberlites and mantle xenoliths; kinetics of melt crystallization and cooling rates of rock formations. *Mailing Add:* Dept Geol Sci Univ Tenn Knoxville TN 37996

TAYLOR, LAWRENCE DOW, b Boston, Mass, Oct 6, 32; m 55; c 2. GLACIOLOGY, GEOMORPHOLOGY. *Educ:* Dartmouth Col, BA, 54, MA, 58; Ohio State Univ, PhD(geol), 62. *Prof Exp:* Geologist, US Geol Surv, Greenland, 54-55; res assoc, Northwest Greenland Glaciol, Dartmouth Col & Air Force Cambridge Res Labs, 57-58; res assoc, Southeast Alaska Glaciol, Inst Polar Studies, Ohio State Univ, 59-62; chief glaciologist, SPole Traverse, US Antarctic Res Prog, NSF, 62-63; asst prof geol, Col Wooster, 63-64; from asst prof to assoc prof geol, 64-76, chmn dept, 64-85, PROF GEOL, ALBION COL, 77- *Concurrent Pos:* NSF res grants, 60-61 & 62-63, field inst grant, Can Rockies, 68; Kellogg Found res & teaching grant, 71-72; Hewlett-Mellow Found res grant, 81, Pew Sci Prog grant, 91; pres, E Cent Sect, Nat Asn Geol Teachers, 84-85. *Mem:* AAAS; fel Geol Soc Am; Int Glaciological Soc; Nat Asn Geol Teachers; Arctic Inst NAm; Sigma Xi. *Res:* Glacial geology of Michigan; structure of lake ice, Greenland; structure and flow of Alaskan glaciers; snow stratigraphy, Antarctica; microparticles in Antarctic ice. *Mailing Add:* Dept Geol Sci Albion Col Albion MI 49224

TAYLOR, LEIGHTON ROBERT, JR, b Glendale, Calif, Nov 17, 40; m 63; c 2. ICHTHYOLOGY. *Educ:* Occidental Col, BA, 62; Univ Hawaii, MS, 65; Scripps Inst Oceanog, Univ Calif, San Diego, PhD(marine biol), 72, Chaminade Univ, Honolulu, MBA, 85. *Prof Exp:* Res asst, Scripps Inst Oceanog, 69-71, mus cur, Scripps Aquarium-Mus, La Jolla, Calif, 71-72; asst leader res, Hawaii Coop Fishery Res Unit, 72-75; DIR, WAIKIKI AQUARIUM, UNIV HAWAII, 75- *Concurrent Pos:* Mem grad fac, Dept Zool, Univ Hawaii, 72-; res assoc, Calif Acad Sci & B P Bishop Mus; pres, Phoenix Sci Ctr. *Mem:* Am Soc Ichthyologists & Herpetologists; Am Fisheries Soc; Am Soc Zoologists; Sigma Xi. *Res:* Taxonomy and ecology of tropical marine fishes and sharks. *Mailing Add:* PO Box 2791 Sausalito CA 94966-2791

TAYLOR, LEONARD S, b New York, NY, Dec 28, 28; m 54; c 2. ELECTRICAL ENGINEERING, BIOELECTROMAGNETICS. *Educ:* Harvard Univ, AB, 51; NMex State Univ, MS, 56, PhD(physics), 60. *Prof Exp:* Electronics engr, Raytheon Mfg Co, 51-54; electronics scientist, White Sands Missile Range, 54-59; theoret physicist, Gen Elec Co, 60-64; assoc prof eng, Case Western Reserve Univ, 64-67; PROF ELEC ENG & RADIOL ONCOL, UNIV MD, COLLEGE PARK, 67- *Concurrent Pos:* Sr Fulbright lectr, Univ Madrid, 62-63; consult, Ford Found & vis prof, Ford Found prog, Ctr Adv Res & Studies, Nat Polytech Inst, Mex, 64-65; mem, Comn VI, US Nat Comt/Int Union Radio Sci, 71- & US Nat Comn, Int Sci Radio Union, 78-82; assoc ed, Bioelectromagnetics J, 83-85, Inst Elec & Electronics Engrs-Biomed Electronics, 86-87. *Mem:* Fel Inst Elec & Electronics Engrs; fel Am Soc Laser Med & Surg; Am Phys Soc; Bioelectromagnetics Soc; Optical Soc Am. *Res:* Biological effects of microwaves; microwave surgery and hyperthermia; optical and radio communication systems; remote sensing; microwave engineering. *Mailing Add:* Dept Elec Eng Univ Md College Park MD 02181

TAYLOR, LINCOLN HOMER, b Wolsey, SDak, Oct 26, 20; m 46; c 5. AGRONOMY. *Educ:* SDak State Col, BS, 42; Iowa State Univ, MS, 49, PhD, 51. *Prof Exp:* From asst prof to assoc prof agron, Univ Maine, 51-55; prof agron, Va Polytech Inst & State Univ, 55-86; RETIRED. *Concurrent Pos:* Grass breeding consult. *Mem:* Crop Sci Soc Am; Am Soc Agron. *Res:* Forage crop and turfgrass breeding; genetics. *Mailing Add:* Dept Agron Va Polytech Inst Blacksburg VA 24061

TAYLOR, LLOYD DAVID, b Boston, Mass, Jan 11, 33; m 57; c 3. OTHER CHEMISTRY. *Educ:* Boston Col, BS, 54; Mass Inst Technol, PhD(org chem), 58. *Prof Exp:* Res assoc, Mass Inst Technol, 57-58; scientist, 58-65, res group leader, 65-68, mgr polymer res lab, 68-78, tech dir polymer sci, 78-80, dir chem res, 80-83, SR RES FEL & CORP OFFICER, POLAROID CORP, CAMBRIDGE, 80- *Concurrent Pos:* Adj prof chem, Boston Col, 83- *Mem:* AAAS; Am Chem Soc; Am Inst Chem; Soc Photog Sci & Eng; NY Acad Sci. *Res:* Polymer chemistry; photographic chemistry; syntheses of novel monomers and polymers; plastics; solubility and diffusional phenomena; critical temperature behavior; chemistry of molecular switches activated by heat or light; molecular recognition. *Mailing Add:* One Maureen Rd Lexington MA 02173-2103

TAYLOR, LYLE HERMAN, b Paton, Iowa, Oct 23, 36; m 59; c 5. LASERS. *Educ:* Iowa State Univ, BS, 58; NMex State Univ, MS, 61; Univ Kans, PhD(physics), 68. *Prof Exp:* Asst physicist, White Sands Missile Range, 58-61; assoc physicist, Midwest Res Inst, 61-64, tech consult, 64-67; sr scientist, 67-84, FEL SCIENTIST, WESTINGHOUSE RES LABS, 84- *Mem:* Am Phys Soc. *Res:* Laser pumps; holographic strain analysis; gas laser computer simulation; laser radar; inertial confinement fusion studies; molecular spectroscopy; acoustic-optic devices. *Mailing Add:* 3317 Benden Dr Murrysville PA 15668

TAYLOR, LYNN JOHNSTON, organic chemistry, polymer chemistry; deceased, see previous edition for last biography

TAYLOR, MALCOLM HERBERT, b Annapolis, Md, Apr 7, 42; m 65; c 4. ENVIRONMENTAL PHYSIOLOGY, ENDOCRINOLOGY. *Educ:* Franklin & Marshall Col, BA, 64; Johns Hopkins Univ, PhD(physiol), 69. *Prof Exp:* NIH fel physiol, Med Sch, Univ Pittsburgh, 69-71; res assoc marine sci, Col Marine Studies, 71-73, asst prof biol sci, 73-79, ASSOC PROF LIFE & HEALTH SCI, UNIV DEL, 79- *Concurrent Pos:* Joint appt, Col Marine Studies, Univ Del, 77- *Mem:* AAAS; Am Soc Zoologists; Am Fisheries Soc. *Res:* Environmental control of reproduction in fish. *Mailing Add:* Sch Life & Health Sci Univ Del Newark DE 19716

TAYLOR, MARTHA LOEB, b Birdsboro, Pa, May 9, 49; m 72; c 1. NUTRITION. *Educ:* Univ Del, Newark, BS, 71; Univ Md, College Park, MS, 72, PhD(nutrit sci), 77. *Prof Exp:* Proj coordr & res asst, Ohio State Univ, Columbus, 77-81; asst prof human nutrit, Drexel Univ, 81-85; ASST PROF HUMAN NUTRIT, UNIV MD, COLLEGE PARK, 85- *Mem:* AAAS; Sigma Xi; Am Dietetic Asn; Am Col Nutrit; Am Inst Nutrit; Soc Nutrit Educ. *Res:* Growth, development and nutritional status of children with developmental disabilities; nutritional status of individuals with chronic health problems. *Mailing Add:* Dept Food Nutrit & Food Systs Univ Md College Park MD 20742

TAYLOR, MARY LOWELL BRANSON, b Coeur d'Alene, Idaho, Nov 24, 32. MICROBIAL PHYSIOLOGY. *Educ:* Univ Idaho, BS, 54; Univ Ill, PhD(bact), 59. *Prof Exp:* Asst, Univ Ill, 54-57; res assoc, Emory Univ, 57-59; USPHS fel, Oak Ridge Nat Lab, 59-61; res assoc biol, 61-77, ASSOC PROF ENVIRON SCI & RESOURCES, PORTLAND STATE UNIV, 77- *Mem:* Am Soc Microbiol; Am Soc Plant Physiol. *Res:* Bacterial physiology; carbohydrate synthesis and dissimilation; alcohol oxidation; natural products oxidation; enzyme biosynthesis; role of metals and metaloids in microbial metabolism biological oxidation. *Mailing Add:* Dept Environ Sci & Biol Portland State Univ PO Box 751 Portland OR 97207

TAYLOR, MARY MARSHALL, microbiology, for more information see previous edition

TAYLOR, MERLIN GENE, b Zanesville, Ohio, May 11, 36; m 63; c 2. PHYSICS. *Educ:* Muskingum Col, BS, 58; Brown Univ, MSc, 65, PhD(physics), 67. *Prof Exp:* Res asst, High Energy Physics Lab, Brown Univ, 58-66; asst prof physics, Wilkes Col, 66-67 & Am Univ Cairo, 67-69; asst prof, 69-71, assoc prof, 71-81, PROF PHYSICS, BLOOMSBURG STATE COL, 81- *Mem:* Am Phys Soc; Am Asn Physics Teachers. *Res:* Use of computers in physics teaching; nuclear physics; activation analysis. *Mailing Add:* Dept Physics Bloomsburg Univ Bloomsburg PA 17815

TAYLOR, MICHAEL ALAN, b New York, NY, Mar 6, 40; m 63; c 2. NEUROPSYCHIATRY. *Educ:* Cornell Univ, BA, 61; New York Med Col, MD, 65. *Prof Exp:* Residency psychiat, New York Med Col, 69, asst prof & chief acute treatment univ, 71-73; assoc prof & dir residency training, State Univ NY, Stony Brook, 73-76; dir residency training, Chicago Med Sch, 76-89, chmn, Dept Psychiat & Behav Sci, 77-80, actg chmn, 86-87, PROF PSYCHOL, CHICAGO MED SCH, 77-, PROF PSYCHIAT & BEHAV SCI, 76- *Concurrent Pos:* Consult, Pilgrim Psychiat Ctr, West Brentwood, NY, 73-76, Kings Park Psychiat Ctr, Kings Park, NY, 74-76, South Oaks Hosp, Amityville, NY, 74-76 & Psychiat Serv, North Chicago Vet Admin Hosp, North Chicago, Ill, 76-; mem res subcomt, Prov Ment Health Adv Coun, Alberta, Can, 76-80; mem psychiat adv bd, Ill Dept Ment Health & Deviation Dis, Springfield, Ill, 77-; actg chmn, Dept Psychol, Sch Grad & Postdoctoral Studies, Chicago Med Sch, Univ Health Sci, 77-80. *Honors & Awards:* A E Bennett Clin Res Award, Soc Biol Psychiat, 69; First Prize Clin Res, NY Acad Med, 69; Morris L Parker Award, Chicago Med Sch, Univ Health Sci, 78. *Mem:* Am Psychiat Asn; Am Psychopath Asn; Psychiat Res Soc; AAAS; Int Neuropsychol Soc. *Res:* Validity of diagnoses of schizophrenia and manic depressive illness by relating clinical phenomenology of these groups to demographic, family illness, cerebral latoralization of cortical dysfunction, neuropsychological and treatment response variables; functional relationships between neuronal groups and behavior utilizing neuropsychological techniques; genetics of major psyches. *Mailing Add:* Dept Psychiat & Behav Sci Univ Health Sci Chicago Med Sch 3333 Greenbay Rd North Chicago IL 60064

TAYLOR, MICHAEL DEE, b New York, NY, Dec 17, 40; m 70. MATHEMATICS. *Educ:* Univ Fla, BA, 63; Fla State Univ, MS, 65, PhD(math), 69. *Prof Exp:* Asst prof, 68-72, ASSOC PROF MATH, UNIV CENT FLA, 72- *Mem:* Math Asn Am; Am Math Soc. *Res:* Nondeterministic analysis and cellular automata. *Mailing Add:* Dept Math Fla Tech Univ PO Box 25000 Orlando FL 32816

TAYLOR, MICHAEL E, b Salt Lake City, Utah, Aug 28, 39. GEOLOGICAL RESEARCH. *Educ:* Utah State Univ, BS, 62 & MS, 64; Univ Calif, Berkeley, PhD(palentol), 71. *Prof Exp:* RES GEOLOGIST, US GEOL SURV, 69- *Concurrent Pos:* Guest scientist, Acad Sinica, 84 & 86, Peoples Rep China & USSR Acad Sci, 85, 87 & 90. *Honors & Awards:* G K Gilbert Award, US Geol Surv. *Mem:* Geol Soc Am; Soc Econ Paleontologists & Mineralogists; AAAS; Paleont Soc; Am Asn Petrol Geologists; Inst Cambrian Studies. *Res:* Lower Paleozoic stratigraphy and paleontology. *Mailing Add:* US Geol Surv Mail Stop 919 Box 25046 Fed Ctr Denver CO 80225

TAYLOR, MICHAEL LEE, b Rockville, Ind, May 27, 41; m 65; c 2. ANALYTICAL & ENVIRONMENTAL CHEMISTRY, TECHNOLOGY. *Educ:* Purdue Univ, Lafayette, BS, 63, MS, 65, PhD(med chem), 67. *Prof Exp:* Nat Res Coun resident res assoc, Chem Lab, Aerospace Res Labs, Wright-Patterson AFB, Ohio, 70-71, res scientist anal mass spectros, 71-75; res assoc prof chem, Wright State Univ, 75-87, assoc prof pharmacol & assoc dir, Brehm Lab, 78-87; DIR RES & DEVELOP & SR TECH ASSOC, IT CORP, CINCINNATI, 87- *Mem:* Am Soc Mass Spectrometry; Am Chem Soc; Soc Toxicol. *Res:* Use of ultrasensitive mass spectral techniques to assess relationships between molecular structure and elicited toxicological and pharmacological response and to determine environmental distribution and the fate of toxic chemicals; develop bench and pilot scale processes for detoxifying hazardous chemicals and chemically contaminated soil and debris. *Mailing Add:* 10561 Cranwood Ct Cincinnati OH 45240-3421

TAYLOR, MILTON WILLIAM, b Glasgow, Scotland, Dec 10, 31; US citizen; m 57; c 2. GENETICS, MOLECULAR BIOLOGY. *Educ:* Cornell Univ, BS, 61; Stanford Univ, PhD(biol), 66. *Prof Exp:* NIH fel virol, Univ Calif, Irvine, 66-67; from asst prof to assoc prof, 67-75, PROF MICROBIOL & GENETICS, IND UNIV, BLOOMINGTON, 75- *Concurrent Pos:* Vis prof, Univ Rome, Italy, 83-84; consult, AMGEN Corp, 89- *Honors & Awards:* Fogarty Int Fel, NIH, 83-84. *Mem:* AAAS; Am Soc Microbiol; Am Soc Biol Chemists; Int Soc Interferon Res. *Res:* Cancer research; recombinant DNA and gene cloning; somatic cell genetics; purine metabolism; microbiology; interferon. *Mailing Add:* Dept Biol Ind Univ Bloomington IN 47401

TAYLOR, MITCHELL KERRY, life history analysis, for more information see previous edition

TAYLOR, MORRIS CHAPMAN, b Fulton, Ky, May 28, 39; m 60; c 2. INSTRUMENTATION. *Educ:* Univ Tenn, BS, 62; Univ Calif, Los Angeles, MS, 64; Rice Univ, MA, 66, PhD(physics), 68. *Prof Exp:* Lab technician, Oak Ridge Nat Lab, 61-62; mem tech staff, Hughes Aircraft Co, 62-64; asst prof physics, St Louis Univ, 68-69; mem staff, Columbia Sci Res Inst, Houston, 69-71; AEC & State of Tex joint sr fel, Rice Univ & Univ Tex M D Anderson Hosp & Tumor Inst, 71; chief scientist, Columbia Sci Industs, 72-76, dir eng, 76-82; PRES & CHIEF EXEC OFFICER, NAT BUS CONTROL SYSTS, 82- *Mem:* Am Soc Testing & Mat; Air Pollution Control Asn; Instrument Soc Am; Am Phys Soc. *Res:* Experimental nuclear physics; radiological physics; air quality monitoring; nondestructive elemental analysis. *Mailing Add:* 11102 Aerie Cove Austin TX 78759

TAYLOR, MORRIS D, b Mitchell, Ind, Apr 14, 34; m 56; c 3. SCIENCE EDUCATION, PHYSICAL CHEMISTRY. *Educ:* Purdue Univ, BS, 56, PhD(sci educ), 66. *Prof Exp:* Instr physics & educ, Purdue Univ, 60-63; from asst prof to prof chem & educ, 63-70, PROF CHEM, EASTERN KY UNIV, 70- *Mem:* AAAS; Nat Sci Teachers Asn. *Mailing Add:* Dept Chem Eastern Ky Univ Richmond KY 40475

TAYLOR, MURRAY EAST, b Casselton, NDak, Apr 14, 15; m 50; c 1. ANALYTICAL CHEMISTRY. *Educ:* Univ Wash, BS, 48, MS, 52, PhD(chem), 60. *Prof Exp:* Microchemist, Univ Wash, 48-52, res instr chemother, 52-58; res chemist, Boeing Airplane Co, 58-63; assoc staff scientist, Missile Div, Chrysler Corp, 63-64; prin engr, 64-80, SR PRIN ENGR, BOEING CO, SEATTLE, 80- *Mem:* AAAS; Am Chem Soc; NY Acad Sci. *Res:* Instrumental analysis; microchemistry; thermochemistry; radiation effects on materials; polymer chemistry. *Mailing Add:* 4401 138th Ave SE Bellevue WA 98006

TAYLOR, NORMAN FLETCHER, b Newcastle Upon Tyne, Eng, Mar 4, 28; m 48; c 2. BIOCHEMISTRY. *Educ:* Univ Oxford, BA, 53, MA, 56, DPhil(biochem), 56; FRSC. *Prof Exp:* Exchange vis scientist, Dept Pharmacol, Univ Calif, Los Angeles, 57-59; Sci Res Coun fel, Dept Biochem, Univ Oxford, 59-60; sr lectr chem, Bristol Col Sci & Technol, Eng, 62-65; reader & head biochem, Univ Bath, Eng, 65-73; PROF BIOCHEM, UNIV WINDSOR, 73- *Concurrent Pos:* Vis prof, Dept Chem, Temple Univ, 67, Dept Biochem, Univ Cambridge, UK & Weizmann Ins Sci, Israel, 83; fel, Canterbury Col, Ont, 74- *Mem:* Am Soc Biochem & Molecular Biol; Can Biochem Soc; Brit Biochem Soc; Brit Chem Soc; Am Chem Soc; Am Soc Microbiol. *Res:* Microbial and mammalian metabolism of synthetic fluorinated carbohydrates and related compounds; mechanism of transport across biological membranes; insect biochemistry. *Mailing Add:* Dept Chem & Biochem Univ Windsor Windsor ON N9B 3P4 Can

TAYLOR, NORMAN LINN, b Augusta, Ky, July 18, 26; m 51; c 5. AGRONOMY. *Educ:* Univ Ky, BS, 49, MS, 51; Cornell Univ, PhD(plant breeding), 53. *Prof Exp:* Asst agronomist, 53-56, assoc prof, 56-66, ASSOC AGRONOMIST, UNIV KY, 56-, PROF AGRON, 66- *Mem:* AAAS; fel Crop Sci Soc Am; fel Am Soc Agron; Am Genetic Asn; Genetics Soc Can. *Res:* Forage crops genetics and breeding; interspecific hybridization in the genus Trifolium. *Mailing Add:* Dept Agron Univ Ky Lexington KY 40506

TAYLOR, OLIVER CLIFTON, b Hallett, Okla, Nov 29, 18; m 40; c 3. HORTICULTURE. *Educ:* Okla Agr & Mech Col, MS, 51; Mich State Col, PhD, 53. *Prof Exp:* Asst county agent, Okla Exten Serv, 47-49; instr agr, Northeast Okla Agr & Mech Col, 50-51; horticulturist, 53, lectr plant sci, 74-76, HORTICULTURIST, STATEWIDE AIR POLLUTION RES CTR, UNIV CALIF, RIVERSIDE, 53-, PROF PLANT SCI, 76- *Mem:* AAAS; Air Pollution Control Asn; Am Soc Hort Sci. *Res:* Citriculture; biological effects of air pollutants and plant physiological responses to air pollutants. *Mailing Add:* 4762 Windsor Rd Riverside CA 92507

TAYLOR, PALMER WILLIAM, b Stevens Point, Wis, Oct 3, 38; m 65; c 3. PHARMACOLOGY. *Educ:* Univ Wis-Madison, BS, 60, PhD(pharm), 64. *Prof Exp:* Res assoc pharmacol, NIH, Bethesda, Md, 64-68; NIH vis fel, Molecular Pharmacol Res Unit, Cambridge Univ, 68-70; assoc prof, 74-78, prof & head, Div Pharmacol, 78-86, PROF & CHMN, DEPT PHARMACOL, SCH MED, UNIV CALIF, SAN DIEGO, 80- *Mem:* Am Soc Biol Chemists; Am Soc Pharmacol & Exp Therapeut. *Res:* Cholinergic neurotransmission; cholinergic receptors; acetylcholinesterass. *Mailing Add:* Med/Sch Med Univ Calif San Diego Box 109 La Jolla CA 92093

TAYLOR, PATRICK TIMOTHY, b Mt Vernon, NY, Mar 20, 38; m 73; c 2. GEOMAGNETICS. *Educ:* Mich State Univ, BS, 60; Pa State Univ, MS, 62; Stanford Univ, PhD(geophys), 65. *Prof Exp:* Res asst marine geophys, Lamont-Doherty Geol Observ, Columbia Univ, 65-66; geophysicist, US Naval Oceanog Off, 66-76 & Naval Ocean Res & Develop Activity, 76-78, GEOPHYSICIST, GEOPHYS BR, GODDARD SPACE FLIGHT CTR, NASA, 78- *Concurrent Pos:* Assoc prof lectr, George Washington Univ. *Honors & Awards:* Kaminiski Award, Sci Res Soc Am, 70. *Mem:* Am Geophys Union; Geol Soc Am; Soc Explor Geophys. *Res:* Interpretation of satellite derived geophysical data, such as gravity and magnetics with supporting geophysical and geological information, for example petrology, remotely sensed photographs, seismicity and heat flow. *Mailing Add:* Code 922 NASA Greenbelt MD 20771

TAYLOR, PAUL DUANE, b Warren, Ohio, July 8, 40; m 65; c 4. CHEMISTRY. *Educ:* Ind Inst Technol, BS, 62; Long Island Univ, MS, 64; Univ Cincinnati, PhD(inorg chem), 69. *Prof Exp:* Teaching asst, Long Island Univ, 62-64; res chemist, Res & Eng Develop Dept, M W Kellogg Co, 64-66; teaching asst, Univ Cincinnati, 66-68; from res chemist to sr res chemist, Celanese Res Co, 69-74, res supvr, 74-77, sect leader, Celanese Chem Co, Tex, 77-78; res mgr, Oxirane Int, 78-81; mgr catalysis develop, Arco Chem Co, Pa, 81-82; mgr planning, Lummus, 82-83; mgr com develop, PQ Corp, 83-85; DIR CHEM & PROCESS RES, GAF CORP, WAYNE, NJ, 85- *Concurrent Pos:* Chmn, NY Catalysis Soc, 76-77. *Mem:* Com Develop Asn; Am Chem Soc; Chem Mkt Res Asn. *Res:* Research and management of catalysis; speciality chemicals and speciality polymers. *Mailing Add:* GAF Corp 1361 Alps Rd Wayne NJ 07470

TAYLOR, PAUL JOHN, b Chicago, Ill, Jan 30, 39; m 60; c 1. ANALYTICAL CHEMISTRY. *Educ:* Northern Ill Univ, BS, 64, PhD(analytical chem), 71. *Prof Exp:* US AEC fel, Purdue Univ, Lafayette, 70-72; from asst prof to assoc prof chem, Wright State Univ, 72-78; lectr, 78-81, PROF, UNIV WIS-LACROSSE, 81- *Mem:* Am Chem Soc. *Res:* Coordination compounds and their analytical applications, gas chromatography; computers and their applications to analytical chemistry; liquid crystals and mass spectroscopy. *Mailing Add:* Dept Chem Univ Wis Lacrosse WI 54601

TAYLOR, PAUL M, b Baltimore, Md, June 26, 27; m 55; c 4. PEDIATRICS, PHYSIOLOGY. *Educ:* Johns Hopkins Univ, AB, 47, MD, 51. *Prof Exp:* Res fel, 54-56, from instr to assoc prof, 56-71, PROF PEDIAT, OBSTET & GYNEC, SCH MED, UNIV PITTSBURGH, 71- *Concurrent Pos:* USPHS fel, Nuffield Inst Med Res, Oxford Univ, 59-60; dir div neonatology & chief dept pediat, Magee-Women's Hosp, 65-; vis prof, Inst Path, Univ Geneva, 71-72. *Mem:* Soc Pediat Res; Am Physiol Soc. *Res:* Physiology of the newborn infant; development of parent-infant attachment. *Mailing Add:* Magee-Women's Hosp Forbes & Halket Sts Pittsburgh PA 15213

TAYLOR, PAUL PEAK, b Childress, Tex, May 11, 21; m 45; c 2. PEDODONTICS. *Educ:* Baylor Univ, DDS, 44; Univ Mich, MS, 51. *Prof Exp:* Prof Grad Pedodontics & Chmn Dept, Col Dent, Baylor Univ, 60-86; Dir Training & Chief Dent Serv, Children's Med Ctr, 60-87, Dir Dent, 66-87; RETIRED. *Concurrent Pos:* Staff, Tex Scottish Rite Hosp for Crippled Children, Dallas, Tex, 60-87 & Denton State Sch, 64-87; proctor, Am Bd Pedodont, 66; exam mem, Am Bd Pedodont, 77-84. *Mem:* Fel Am Col Dent; Am Acad Pedodont; Am Dent Asn; Am Soc Dent Children. *Res:* Physiological responses of pulp tissues and the testing of patient responses to dental stimuli. *Mailing Add:* 2615 Briarcove Plano TX 75074

TAYLOR, PETER, b Warkworth, Eng, Sept 12, 49; m 73; c 1. INORGANIC CHEMISTRY, CRYSTALLOGRAPHY. *Educ:* Univ Birmingham, BSc, 69, PhD(inorg chem), 72. *Prof Exp:* Fel struct inorg chem, Univ NB, 72-75; fel, 75-77, RES OFFICER STRUCT INORG CHEM, RES CHEM BR, WHITESHELL NUCLEAR RES ESTAB, ATOMIC ENERGY CAN LTD, 77- *Mem:* Chem Inst Can. *Res:* Structural chemistry, phase relations, agueous and surface chemistry of inorganic oxide systems; radioactive waste management. *Mailing Add:* Res Chem Br Lab Sta 39 Whiteshell Atomic Energy Can Ltd Pinawa MB R0E 1L0 Can

TAYLOR, PETER ANTHONY, b Liverpool, Eng, June 9, 32; m 68; c 2. PHYSICAL CHEMISTRY, TEXTILE ENGINEERING. *Educ:* Liverpool Col Technol, Eng, ARIC, 56; Univ Manchester, PhD(chem), 63. *Prof Exp:* Jr asst analyst, Liverpool City Pub Health Dept, 49-52; analyst, Distillers Co (Biochem), Ltd, 52-56, res chemist, 58-60 & Fibers Div, Allied Chem Corp, 63-66; PROJ MGR, PHILLIPS FIBERS CORP, 66- *Mem:* Royal Inst Chem; Am Chem Soc. *Res:* Polymerization kinetics, free radical and condensation; man-made fiber rheology, processing and dyeing; polymer pigmentation; textile and fiber finishing; spin finish development. *Mailin- Add:* 105 Jamestown Dr Greenville SC 29615-3213

TAYLOR, PETER BERKLEY, b Yonkers, NY, Dec 1, 33; m 58; c 3. MARINE ECOLOGY. *Educ:* Cornell Univ, BS, 55; Univ Calif, Los Angeles, MS, 60; Univ Calif, San Diego, PhD(marine biol), 64. *Prof Exp:* NSF res fel, 63-64; asst prof oceanog, Univ Wash, 64-71; MEM FAC, EVERGREEN STATE COL, 71- *Res:* Coastal and estuarine benthic ecology; ecology of marine fishes; venomous marine animals. *Mailing Add:* Dept Oceanog Lab 1 Evergreen State Col Olympia WA 98505

TAYLOR, PETER D, b Vancouver, BC, Dec 7, 42. POPULATION MODELING, GENETIC MODELS. *Educ:* Queens Univ, BSc, 64; Harvard Univ, PhD(math), 69. *Prof Exp:* From asst prof to assoc prof math, 69-75, PROF MATH & BIOL, QUEENS UNIV, 81- *Mem:* AAAS; Can Math Soc; Math Asn Am. *Res:* Mathematics education. *Mailing Add:* Dept Math & Statist Queens Univ Kingston ON K7L 3N6 Can

TAYLOR, PHILIP CRAIG, b Paterson, NJ, March 17, 42; m 69; c 2. CRYSTALLINE & AMORPHOUS SEMICONDUCTORS. *Educ:* Carleton Col, BA, 64; Brown Univ, PhD(physics), 69. *Prof Exp:* Nat Acad Sci res assoc, Naval Res Lab, 69-71, res physicist, 71-80, supvr res physicist, 80-82; PROF PHYSICS, UNIV UTAH, 82-, CHMN, 89- *Concurrent Pos:* Vis prof, Heriot-Watt Univ, Edinburgh & Cambridge Univ, 76-77; adj prof, mat sci eng dept, Univ Utah, 85-, assoc dir, Lazer Inst, 87- *Mem:* Fel Am Phys Soc; AAAS; Mat Res Soc. *Res:* Optical, electronic and structural properties of crystalline and amorphous semiconductors including localized electronic states in amorphoys semiconductors, metastabilities in disordered solids and electronic properties of very thin layered structures. *Mailing Add:* Dept Physics Univ Utah Salt Lake City UT 84112

TAYLOR, PHILIP LIDDON, b London, Eng, Oct 17, 37; m 66; c 2. PHYSICS OF POLYMERS. *Educ:* Univ London, BSc, 59; Cambridge Univ, PhD(physics), 62. *Prof Exp:* Magnavox res fel mat sci, Case Western Reserve Univ, 62-64, from asst prof to prof physics, 64-88, PROF MACROMOLECULAR SCI, CASE WESTERN RESERVE UNIV, 77-, PERKINS PROF PHYSICS, 88- *Concurrent Pos:* Mem, comt recommendations for US Army basic sci res, Nat Res Coun Assembly Math & Phys Sci. *Mem:* Fel Am Phys Soc; fel AAAS. *Res:* Theoretical solid state physics. *Mailing Add:* Dept Physics Case Western Res Univ Cleveland OH 44106

TAYLOR, PHILLIP R, b Mason City, Iowa, Feb 20, 48; m 74; c 3. CANCER RESEARCH. *Educ:* Iowa State Univ, BS, 69; Univ Iowa, MD, 73; Harvard Sch Pub Health, SM, 82, ScD, 88; Am Bd Intern Med, cert, 76. *Prof Exp:* Intern, Vanderbilt Univ Med Ctr, Nashville, 73-74, resident, 74-76; med epidemiologist, Field Serv Div, Bur Epidemiol, Ctr Dis Control, USPH, Dept HEW, Acute Commun Dis Control, Los Angelas, Calif, Calif, 76-78, Spec Studies Br, Chronic Dis Div, Bur Epidemiol, Div Epidemiol, NY State Dept Health, Albany, 78-80, Indust Wide Studies Br, Div Surveillance, Hazard Eval & Field Studies, Nat Inst Occup Safety & Health, 80-81; grad student, Dept Epidemiol, Harvard Sch Pub Health, Boston, 81-83; actg dep br chief, 83-86, actg br chief, 86-87, SR INVESTR, CANCER PREV STUDIES BR, CANCER PREV RES PROG, DIV CANCER PREV & CONTROL, NAT CANCER INST, USPHS, DEPT HEALTH & HUMAN SERV, NIH, 83-, BR CHIEF, 87- *Concurrent Pos:* Co-prin investr, Nutrit Intervention Studies, Esophageal Cancer, Linxian, China, co-investr, Isotretinoin Basal Cell Carcinoma Prev Trial, Alpha-Tocopherol Beta Carotene Lung Prev Trial & Lung Cancer Intervention Feasibility Study Among Yunnan Tin Miners. *Mem:* Am Pub Health Asn; Soc Epidemiol Res; AAAS; Am Col Epidemiol; Am Soc Prev Oncol. *Res:* Nutrition and cancer; nutritional intervention studies; clinical nutrition/metabolic studies; environmental epidemiology. *Mailing Add:* NIH Nat Cancer Inst Cancer Prev Studies Br Exec Plaza N Rm 211 9000 Rockville Pike Rockville MD 20892

TAYLOR, R(AYMOND) JOHN, b Ada, Okla, Jan 20, 30; m 59; c 3. PLANT TAXONOMY. *Educ:* ECent State Col, BSEd, 54; Univ Okla, MNS, 61, PhD(plant ecol), 67. *Prof Exp:* Teacher high sch, Okla, 54-55 & Okla City Pub Sch Syst, 55-61; asst prof biol, Southeastern State Col, 61-63; asst bot, Univ Okla, 63-65; from assoc prof to prof biol, Southeastern State Univ, 74-91; RETIRED. *Concurrent Pos:* Grants, Southeastern State Col Res Found, 68 & NIH, 72. *Mem:* Am Soc Plant Taxonomists. *Res:* Dauphine Island, Oklahoma, Alaska, Costa Rica and Alabama plants; aquatic macrophytes; endangered plant species. *Mailing Add:* Rte 1 Box 157 Durant OK 74701

TAYLOR, RALPH DALE, b Boonville, Mo, Dec 24, 45; div; c 2. HIGH ENERGY PULSE MEASUREMENTS, ADMINISTRATION OF RESEARCH & DEVELOPMENT. *Educ:* Univ Mo, Rolla, BS, 68; Univ Mo, Columbia, MS, 70; Univ Mo, Kansas City, MBA, 78. *Prof Exp:* Sr engr, Bendix, 68-74; chief engr, 74-91, DIR ENG, DIT-MCO, 91- *Concurrent Pos:* Adj prof logic, Univ Mo, Columbia, 74-78. *Mem:* Inst Elec & Electronics Engrs. *Res:* Basic design and development of large computer directed test equipment. *Mailing Add:* 12708 E 62nd Ct Kansas City MO 64133

TAYLOR, RALPH E, geology, for more information see previous edition

TAYLOR, RALPH WILSON, b Whitesburg, Ky, June 1, 37. FIELD BIOLOGY, MALACOLOGY. *Educ:* Murray State Univ, BS, 60; Univ Louisville, MS, 68, PhD(herpet), 72. *Prof Exp:* Pub sch teacher biol, Durrett High Sch, Louisville, Ky, 60-66; instr, Spalding Col, 69-70; from asst prof to assoc prof, 72-84, PROF BIOL, MARSHALL UNIV, 84- *Concurrent Pos:* Pres, WVa Acad Sci, 86-87. *Mem:* Am Malacological Union; Sigma Xi. *Res:* Ecology, taxonomy, and distribution of freshwater and terrestrial mollusks of West Virginia and surrounding states. *Mailing Add:* Dept Biol Sci Marshall Univ Huntington WV 25701

TAYLOR, RAYMOND DEAN, b Okemah, Okla, Aug 18, 28; m 61; c 2. LOW TEMPERATURE PHYSICS, THERMOMETRY. *Educ:* Kans State Col, Pittsburg, BS, 50; Rice Univ, PhD(phys chem), 54. *Prof Exp:* Asst, Rice Inst, 51-52; assoc group leader, 73-86, mem staff, 54-90, CONSULT, LOS ALAMOS NAT LAB, 90- *Concurrent Pos:* Humble Res fel, 52-54. *Mem:* Am Chem Soc; Am Phys Soc. *Res:* Low temperature calorimetry; cryogenics; transport and state properties of liquid helium; Mössbauer effect; superconductivity; magnetism; high pressures. *Mailing Add:* Los Alamos Nat Lab P-10 MS-K764 Los Alamos NM 87545

TAYLOR, RAYMOND ELLORY, b Ames, Iowa, Oct 19, 29; m 52; c 2. THERMOPHYSICAL PROPERTIES, HEAT TRANSFER. *Educ:* Iowa State Univ, BS, 51; Univ Idaho, MS, 56; Pa State Univ, PhD(solid state technol), 67. *Prof Exp:* Chemist, Gen Elec, Richland, Wash, 51-56; sr res engr, Atomic Int, 57-64; res fel, NAm Rockwell, 64-67; assoc sr researcher & assoc prof, Thermophys Properties Res Ctr, 67-75, SR RESEARCHER & DIR, THERMOPHYS PROPERTIES RES LAB, SCH MECH, PURDUE UNIV, 75- *Concurrent Pos:* Consult numerous industs, 74-91; mem, exec bd, E-37 Thermal Anal, Am Soc Testing & Mat, 75-91; chmn, Int Thermal Conductivity Conf, 86-90, mem, Head Placement Comt & By-Laws Comt; ed, Rev Sci Instruments, 86-89. *Honors & Awards:* Thermal Conductivity Award, Int Thermal Conductivity Conf, 77; Cert Recognition, NASA, 82 & 84; Europ Thermophys Award, Europ Conf Thermophys, 90. *Mem:* Am Soc Testing & Mat; Int Thermal Conductivity Conf (secy, 90-). *Res:* Thermophysical properties of solids, especially at elevated temperatures; thermal conductivity; thermal diffusivity; specific heat capacity; thermal expansion; emissivity. *Mailing Add:* Thermophys Properties Res Lab Purdue Univ 2595 Yeager Rd West Lafayette IN 47906

TAYLOR, RAYMOND L, b Providence, RI, July 3, 30; m 55; c 5. LASERS, CHEMICAL PHYSICS. *Educ:* Brown Univ, ScB, 55; Calif Inst Technol, PhD(chem), 60. *Prof Exp:* Asst, Calif Inst Technol, 55-59; prin res scientist, Avco Everett Res Lab, Inc, 59-73; prin scientist & mem bd dirs, Phys Sci Inc, 73-78, mgr laser devices, 78-80; mem staff, Res & Laser Tech Inc, 80-82; VPRES RES & ENG, CVD INC, 82- *Concurrent Pos:* Chmn Atomic Physics Res Comt, Avco Everett Res Lab, Inc, 70-73; mem, Comt Stratospheric Chem, Dept Transp, 73-75. *Honors & Awards:* Silver Combustion Medal, Combustion Inst, 68. *Mem:* Am Chem Soc; Am Phys Soc; Combustion Inst; Sigma Xi. *Res:* Radiation and energy transfer processes in gases; optical experiments and instrumentation; molecular gas laser device research and development; laser applications. *Mailing Add:* 1413 Sheffield Way Saugus MA 01906

TAYLOR, RICHARD EDWARD, b Medicine Hat, Alta, Nov 2, 29; m 50; c 1. HIGH ENERGY PHYSICS. *Educ:* Univ Alta, BSc, 50, MSc, 52; Stanford Univ, PhD(physics), 62. *Hon Degrees:* Dr, Univ Paris, 80. *Prof Exp:* Physicist, Lawrence Radiation Lab, Univ Calif, 61-62; staff mem, 62-68, assoc prof, 68-70, assoc dir Stanford Linear Accelerator Ctr Res, 82-86, PROF PHYSICS, STANFORD UNIV, 70- *Concurrent Pos:* Guggenheim fel, 71. *Honors & Awards:* Nobel Prize Physics, 90; Alexander von Humboldt Award, 82; W K H Panofsky Award, Am Phys Soc, 89. *Mem:* Am Phys Soc; Can Asn Physicists; Royal Soc Can. *Res:* High energy physics, interactions of electrons and photons with matter; high energy electron scattering. *Mailing Add:* Stanford Linear Accelerator Ctr Stanford Univ PO Box 4349 Stanford CA 94309

TAYLOR, RICHARD G, b Rochester, Minn, Nov 9, 52; div; c 1. CELL BIOLOGY, ELECTRON MICROSCOPY. *Educ:* Mont State Univ, BS, 75; Wake Forest Univ, MS, 84, PhD(path), 87. *Prof Exp:* Instr path, Bowman Gray Sch Med, NC, 88-90; ELECTRON MICROSCOPIST, UNIV ARIZ, TUCSON, 90- *Mem:* Am Soc Cell Biol; Fedn Am Socs Exp Biol; Am Heart Asn; Electron Micros Soc Am. *Res:* Cell biology of smooth muscle, skeletal muscle, and blood platelets; electron microscopy and activities involving ultrastructional examination, enzyme cytochemistry and immunolocalization. *Mailing Add:* 632 Shantz Bldg Univ Ariz Tucson AZ 85721

TAYLOR, RICHARD L, b South Bend, Ind, Aug 11, 39; m 61; c 2. CIVIL ENGINEERING, SANITARY & ENVIRONMENTAL ENGINEERING. *Educ:* Purdue Univ, BS, 63, MS, 65. *Prof Exp:* Jr engr sanit eng, Clark-Dietz & Assoc, 65-67; design engr sanit eng, Boyd E Phelps & Assoc, 67-68; from asst prof to assoc prof civil eng technol, NCent Campus, 68-86, actg head, 84-86, HEAD, TECH ENG DEPT, PURDUE UNIV, 86- *Concurrent Pos:* Consult, 67-80. *Res:* Anaerobic sludge digestion; land surveying relocation problems. *Mailing Add:* N Cent Campus Dept Eng & Technol Purdue Univ Westville IN 46391

TAYLOR, RICHARD MELVIN, b Salt Lake City, Utah, Aug 19, 29; m 55; c 4. AGRONOMY, PLANT PHYSIOLOGY. *Educ:* Utah State Univ, BS, 58, MS, 59; Iowa State Univ, PhD(agron, plant physiol), 64. *Prof Exp:* Instr agron, Iowa State Univ, 59-64; ASSOC PROF AGRON, TEX A&M UNIV, 64- *Mem:* Am Soc Agron; Crop Sci Soc Am; Am Soc Hort. *Res:* Effects of soluble salts and temperature upon the germination and emergence of seeds and fruiting patterns of cotton; production and management; root physiology, native plant domestication and vegetable production. *Mailing Add:* Tex Agr Res Ctr 1380 A&M Circle El Paso TX 79927

TAYLOR, RICHARD N, b Denver, Colo, Dec 11, 52. COMPUTER SCIENCE. *Educ:* Univ Colo, Denver, BS, 74; Univ Colo, Boulder, MS, 76, PhD(computer sci), 80. *Prof Exp:* Anal & programmer, US Bur Reclamation, Denver, Colo, 74-76; teaching asst, Dept Computer Sci, Univ BC, 76-77; sr software engr, Boeing Computer Serv, Seattle, 77-79; res asst, Univ Colo, Boulder, 79-80; asst prof, Dept Computer Sci, Univ Victoria, 81-83; asst prof, 82-85, ASSOC PROF, DEPT INFO & COMPUTER SCI, UNIV CALIF, IRVINE, 85- *Concurrent Pos:* Consult, Boeing Computer Serv Co, Seattle, Wash, 79-82, Joint Syst Develop Corp, Tokyo, Japan, 83, Res Triangle Inst, Res Triangle Park, NC, 83, ITT Adv Technol Ctr, Stratford, Conn, 84, Aerospace Corp, El Segundo, Calif, 83-88, GSI-TECSI Indus, Paris, France, 87-88, Inst Defense Anal, Arlington, Va, 88, TRW, Redondo Beach, Calif, 88, Telesoft, San Diego, Calif, 88-89, Nimble Computer Corp, Encino, Calif, 89-90; grants, Pres Nat Sci & Eng Res Coun, Res Prog, 81-82, operating, 82-83, 83-84, Microelectronics Innovation & Computer Sci Res Prog, Univ Calif, 85-86 & 86-87, Alcoa Found, 85-87; exec comt, Inst Elec & Electronics Engrs Computer Soc Tech Comt, Software Eng, 85-86; conf prog comt, Sixth Int Conf Distrib Comput Systs, 86, Computer Languager, Inst Elec & Electronics Engrs, 86; vchair, Asn Comput Mach SIGSOFT, 89-89, chair, 89-91, prog chair, 90; comt, Asn Comput Mach Software Syst Award, 90-95. *Res:* Computer languages; practical software development; author of several book chapters and articles. *Mailing Add:* Dept Info & Computer Sci Univ Calif Irvine CA 92717

TAYLOR, RICHARD TIMOTHY, b Coatesville, Pa, June 12, 50. ORGANIC CHEMISTRY. *Educ:* Univ Del, BS, 72; Ohio State Univ, PhD(chem), 77. *Prof Exp:* Asst chem, Ohio State Univ, 73-75, fel, 75-77; NIH fel org chem, Cornell Univ, 77-78; ASST PROF ORG CHEM, MIAMI UNIV, 78- *Mem:* Am Chem Soc; Sigma Xi. *Res:* Synthetic aspects of organosilicon chemistry; synthesis of strained ring compounds. *Mailing Add:* Chem Dept Miami Univ Oxford OH 45056

TAYLOR, ROBERT BURNS, JR, b Downingtown, Pa, Oct 13, 20; m 45, 71; c 1. ORGANIC CHEMISTRY, INFORMATION SCIENCE. *Educ:* Swarthmore Col, AB, 41; Ohio State Univ, MSc, 42; Pa State Col, PhD(org chem), 45. *Prof Exp:* Instr chem, Pa State Col, 45-46; res chemist, E I Dupont de Nemours & Co, Inc, 46-53, res supvr, 53-56, sr patent chemist, 56-60, from supvr to sr supvr textile fibers patent div, 60-71, mgr cent patent index, 71-76, asst to div mgr, info syst dept, 76-82; RETIRED. *Mem:* Sigma Xi. *Res:* Synthetic antimalarials; synthetic fibers; patents. *Mailing Add:* 1306 Grayson Rd Wilmington DE 19803

TAYLOR, ROBERT CLEMENT, b Mankato, Minn, Dec 2, 35; m 57; c 6. PHYSIOLOGY, ZOOLOGY. *Educ:* Mankato State Col, BS, 57; Univ SDak, MS, 61; Univ Ariz, PhD(physiol), 66. *Prof Exp:* Asst prof, 66-76, ASSOC PROF ZOOL, UNIV GA, 77- *Mem:* AAAS; Am Soc Zoologists. *Res:* Comparative physiology; neurophysiology. *Mailing Add:* Dept Zool & Biol Sci Univ Ga Athens GA 30602

TAYLOR, ROBERT COOPER, b Colorado Springs, Colo, May 5, 17; m 42; c 2. PHYSICAL CHEMISTRY, MOLECULAR SPECTROSCOPY. *Educ:* Kalamazoo Col, AB, 41; Brown Univ, PhD(phys chem), 47. *Prof Exp:* Asst chem, Brown Univ, 41-42, res chemist, Manhattan Proj, 42-46, instr phys chem, 47-49; from instr to assoc prof, 49-62, actg chmn, 66, prof, 62-87, assoc chmn, 67-87, EMER PROF CHEM, UNIV MICH, ANN ARBOR, 87- *Concurrent Pos:* Nat Res Coun fel, 46-47; consult, W J Barrow Res Lab, 74-77; vis staff mem, Los Alamos Nat Lab, 75-85. *Mem:* AAAS; Am Chem Soc; Am Phys Soc; Am Crystallog Asn. *Res:* Molecular spectroscopy and structure; boron hydride derivatives; Lewis complexes; hydrogen bonded substances; uranium chemistry. *Mailing Add:* Dept Chem Univ Mich Ann Arbor MI 48109-1055

TAYLOR, ROBERT CRAIG, b Franklin, Pa, Jan 26, 39; m 66; c 1. INORGANIC CHEMISTRY. *Educ:* Col Wooster, BA, 60; Princeton Univ, MA, 62, PhD(chem), 64. *Prof Exp:* NATO fel, Imp Col, Univ London, 64-65; asst prof chem, Univ Ga, 65-72; ASSOC PROF CHEM, OAKLAND UNIV, 72- *Mem:* Am Chem Soc; Royal Soc Chem; Sigma Xi; Am Asn Univ Prof; NY Acad Sci. *Res:* Transition metal chemistry; organophosphorous chemistry; lanthanide shift reagents; transition metal catalyzed stereospecific polymerizations of diolefins; transition metal complexes as antitumor agents; role of molybdenum in enzymes; Nuclear magnetic resonance and electrochemistry of metal clusters. *Mailing Add:* Dept Chem Oakland Univ Rochester MI 48309

TAYLOR, ROBERT DALTON, b Greenville, Ala, June 29, 50; m 83. MICROBIOLOGY OF CLOSED SYSTEMS. *Educ:* Southeastern La Univ, BS, 72, MS, 73; La State Univ, PhD(microbiol), 79. *Prof Exp:* Postdoctoral fel, Nat Cancer Inst, 79-82; asst prof microbiol, Dept Microbiol, Univ Southern Miss, 82-83, from asst prof to assoc prof, Dept Biol Sci, 83-87; environ microbiologist, Microbiol Lab, Johnson Space Ctr, NASA, 87-89, group mgr, Biomed Opers & Res Group, 89-91; DIR, ENVIRON SERV, KRUG LIFE SCI, 91- *Concurrent Pos:* NIH fel, 79-82; consult, US Army, Camp Shelby, Miss, 84-86 & Nat Marine Fisheries Serv, 85-88. *Mem:* Am Soc Microbiol; AAAS; Inst Food Technol; NY Acad Sci; Proj Mgt Inst; Nat Mgt Asn. *Res:* Microbial ecology of closed systems, particularly space craft environments; problems arising from recycling potable and hygiene water in long duration spaceflight. *Mailing Add:* 15203 Woodhorn Dr Houston TX 77062

TAYLOR, ROBERT E, b Havelock, Nebr, July 24, 20; div; c 3. PHYSIOLOGY, BIOPHYSICS. *Educ:* Univ Ill, BS, 42; Univ Rochester, PhD(physiol), 50. *Prof Exp:* Mem staff, Radiation Lab, Off Sci Res & Develop, Mass Inst Technol, 42-45; Merck fel physiol, Univ Chicago, 50-51, Nat Heart Inst fel, 51-52; asst prof neurophysiol, Col Med, Univ Ill, 52-53; NSF fel, Physiol Lab, Cambridge Univ, 53-54, Nat Inst Neurol Dis & Blindness fel, 54-55 & Univ Col, Univ London, 55-56; physiologist, lab biophys, nat inst neurol & commun disorders & stroke, 56-88; RETIRED. *Mem:* Am Physiol Soc; Biophys Soc; NY Acad Sci; Soc Gen Physiol; hon mem Chilean Biol Soc. *Res:* Properties of natural excitable membranes; muscle contraction activation. *Mailing Add:* 20 Harbor Hill Rd Woods Hole MA 02543-1215

TAYLOR, ROBERT EMERALD, JR, b Polk City, Fla, Aug 21, 30; m 51, 77; c 4. ENDOCRINOLOGY. *Educ:* Mercer Univ Pharm, BS, 58; Univ Fla, MS, 61, PhD(physiol), 63. *Prof Exp:* NIH fel pharmacol, Univ Vt, 63-65; from asst prof to prof physiol & biophys, Sch Med, Univ Ala, Birmingham, 65-73, coord correlated basic med sci educ, 71-73, assoc dir off undergrad med educ, 72-73; prof physiol & biophys & assoc dean, 73-84, PROF PHYSIOL & BIOPHYS, GRAD SCH MED, UNIV TENN, 86- *Concurrent Pos:* Prin investr & dir, Spec Prog for Minority access to Health Careers. *Mem:* Am Physiol Soc. *Res:* Experimental endocrinology; control of water and electrolyte balance; active ion transport; developmental physiology. *Mailing Add:* Univ Tenn 62 S Dunlap Memphis TN 38163

TAYLOR, ROBERT GAY, b Cleveland, Ohio, July 8, 40. MICROBIOLOGY, ENVIRONMENTAL SCIENCES. *Educ:* Wittenberg Univ, BS, 63; John Carroll Univ, MS, 66; Tex A&M Univ, PhD(environ studies), 69. *Prof Exp:* Sr bacteriologist, Cleveland Dept Pub Health, 63-65; asst prof, 69-73, ASSOC PROF MICROBIOL, EASTERN NMEX UNIV, 73-, DIR, SCH NATURAL SCI, 77- *Concurrent Pos:* Dept Interior, Water Resources Res Inst grant, Eastern NMex Univ, 71-72; vis scientist, Lawrence Berkeley Lab, Univ Calif, 72; consult, Dept Civil Eng, Univ Tex, El Paso, 72- *Honors & Awards:* Outstanding Res Award, Nat Air Pollution Control Asn, 67. *Mem:* AAAS; Am Chem Soc. *Res:* Biomedical biochemical mechanisms; environmental microbiology with respect to water treatment and contamination. *Mailing Add:* PO Box 2296 Portales NM 88130

TAYLOR, ROBERT JOE, b Pomona, Calif, May 1, 45; m 67; c 2. POPULATION ECOLOGY. *Educ:* Stanford Univ, AB, 67; Univ Calif, Santa Barbara, MS, 70, PhD(biol), 72. *Prof Exp:* Res assoc ecol, Princeton Univ, 71-72; asst prof ecol, Univ Minn, St Paul, 72-78; assoc prof zool, Clemson Univ, 78-85; ASSOC PROF FISHERIES & WILDLIFE, UTAH STATE UNIV, 85- *Mem:* Ecol Soc Am; Brit Ecol Soc; Soc Pop Ecol. *Res:* The influence of predatory behavior upon population in space and time; artificial intelligence models in ecology. *Mailing Add:* Dept Fish & Wildlife Utah State Univ Logan UT 84322-5210

TAYLOR, ROBERT JOSEPH, b Salt Lake City, Utah, Dec 10, 41; m 67; c 6. PHYSICS, ELECTROMAGNETISM. *Educ:* Univ Utah, BA, 67; Cornell Univ, PhD(appl physics), 71. *Prof Exp:* Scientist metall, Res Ctr, Kennecott Copper Co, 67; scientist acoust, Interand Corp, 71-72; PHYSICIST, JOHNS HOPKINS UNIV APPL PHYSICS LAB, 72- *Concurrent Pos:* Teaching & consult. *Mem:* Am Geophys Union. *Res:* Electromagnetic wave propagation through the ionosphere for global dissemination of submicrosecond time from satellites; tropospheric propagation and RF ducting; acoustics; impact of ultrasound on colonial hydroids; infrared propagation. *Mailing Add:* Appl Physics Lab Johns Hopkins Univ Johns Hopkins Rd Laurel MD 20723-6099

TAYLOR, ROBERT LEE, b Palmer, Nebr, July 17, 25; m 45. MEDICAL MYCOLOGY. *Educ:* Nebr Wesleyan Univ, AB, 47; Univ Nebr, MA, 50; Duke Univ, PhD(microbiol), 54; Am Bd Microbiol, dipl. *Prof Exp:* Asst chief bact & chief mycol sect, Lab Serv, Fitzsimons Army Hosp, Med Serv Corp, US Army, Colo, 50-55, chief bact sect, Med Res & Develop Unit, 55-58 & mycol sect, Middle Am Res Unit, Ancon, CZ, 58-61, mycologist, Walter Reed Army Inst Res, 61-63, chief mycol sect, 63-65, chief dept bact & mycol, SEATO Med Res Lab, Bangkok, Thailand, 65-67; dep chmn dept, 72-78, PROF MICROBIOL, UNIV TEX HEALTH SCI CTR, SAN ANTONIO, 68-,. *Mem:* Fel Am Acad Microbiol; Am Soc Microbiol; Mycol Soc Am; Am Thoracic Soc; Med Mycol Soc Am; Sigma Xi. *Mailing Add:* Dept Microbiol Univ Tex Health Sci Ctr San Antonio TX 78285

TAYLOR, ROBERT LEE, b Tenn, July 23, 43; m 68. MATHEMATICS, STATISTICS. *Educ:* Univ Tenn, Knoxville, BS, 66; Fla State Univ, MS, 69, PhD(statist), 70. *Prof Exp:* Teaching asst math, Univ Tenn, Knoxville, 66-67; asst prof, 71-74, assoc prof math, 74-80, PROF MATH & STATIST, UNIV SC, 80- *Mem:* Math Asn Am; Am Statist Asn; Inst Math Statist. *Res:* Probability; probabilistic functional analysis and statistical applications. *Mailing Add:* Dept Math & Statist Univ Ga Athens GA 30602

TAYLOR, ROBERT LEROY, b Riverside, Calif, July 14, 34; m 63; c 5. CIVIL ENGINEERING. *Educ:* Univ Calif, Berkeley, BS, 56, MS, 58, PhD(civil eng), 63. *Hon Degrees:* Hon fel, Univ Wales, UK, 88. *Prof Exp:* From asst prof to assoc prof, 62-72, PROF CIVIL ENG, UNIV CALIF, BERKELEY, 72- *Concurrent Pos:* Founder & dir solid mech, Centric Eng Systs, Inc, 90- *Mem:* Nat Acad Eng; Am Soc Civil Engrs; Int Asn Computational Mech; Soc Computational Mech. *Res:* Development of finite methods and software for problems in solid and structural mechanics; non-linear applications. *Mailing Add:* Dept Civil Eng Univ Calif 714 Davis Hall Berkeley CA 94720

TAYLOR, ROBERT MORGAN, b Orange, NJ, May 13, 41; m 65; c 2. ANALYTICAL CHEMISTRY, ELECTROCHEMISTRY. *Educ:* Williams Col, BA, 63; Pa State Univ, PhD(chem), 68; Drexel Univ, MBA, 74. *Prof Exp:* From scientist to corp scientist, 68-84, DIR RES, TECH CTR, LEEDS & NORTHRUP CO, 85- *Mem:* Am Chem Soc; Electrochem Soc; Inst Elec & Electronics Engrs; Instrument Soc Am. *Res:* Potentiometric and voltammetric analysis; high temperature electrochemistry; molten and solid electrolytes; thermometric analysis; chemical instrumentation. *Mailing Add:* Leeds & Northrup Co Sumneytown Pike North Wales PA 19454

TAYLOR, ROBERT THOMAS, b Harrison, Ark, Sept 14, 36; m 58; c 2. BIOCHEMISTRY. *Educ:* Univ Calif, Los Angeles, BA, 59; Univ Calif, Berkeley, PhD(biochem), 64. *Prof Exp:* USPHS res fel, Nat Heart Inst, 64-66, staff res fel, 66-68; RES BIOCHEMIST, BIOMED SCI DIV, LAWRENCE LIVERMORE NAT LAB, 68- *Mem:* Inter-Am Photochem Soc; Soc Environ Geochem & Health; Am Soc Biol Chem; AAAS; Environ Mutagen Soc; Am Chem Soc; Inst Food Technologists. *Res:* Enzymatic methylation, folate one carbon cell mutants and metabolism; mammalian cell mutagenesis, especially auxotrophic reversion mutagenesis; methanol fuel related genotoxicity; mechanisms of mutagenic heterocyclic amine formation during cooking of meat; metal alkylation and mutagenesis; microbiol sulfur and selenium metabolism. *Mailing Add:* 174 Elvira St Livermore CA 94550-3907

TAYLOR, ROBERT TIECHE, b San Diego, Calif, June 29, 32; m 64; c 2. MEDICAL ENTOMOLOGY. *Educ:* Okla State Univ, BS, 54, MS, 57, PhD(entom), 60. *Prof Exp:* Asst prev med officer, US Army, Ft Stewart, Ga, 54-56; malaria specialist, Pan Am Health Orgn, 60-61; spec asst wood preserv & entom, Are Pub Works Off, Chesapeake, US Navy, 61-63; malaria specialist, Malaria Eradication Prog, Port-au-Prince, Haiti, 63-65; sr scientist, 65-72, SCIENTIST DIR ENTOM, PARASITIC DIS DIV, CTR INFECTIOUS DIS, USPHS, 72- *Concurrent Pos:* Consult, WHO, Pan Am Health Orgn & USAID. *Mem:* Am Soc Trop Med & Hyg; Royal Soc Trop Med & Hyg; Entom Soc Am; Am Mosquito Control Asn. *Res:* Conducting and supervising investigations on chemical control of mosquitoes, triatomidae and simuliidae. *Mailing Add:* 635 Grecken Green Peachtree City GA 30269

TAYLOR, ROBERT WILLIAM, b Dallas, Tex, Feb 10, 32; m 55; c 3. COMPUTER SCIENCE, RESEARCH MANAGEMENT. *Educ:* Univ Tex, Austin, BA, 57, MA, 64. *Prof Exp:* Res scientist psychoacoust, Defense Res Lab, Univ Tex, Austin, 55-59; teacher math, Howey Acad, Fla, 59-60; systs engr systs design, Martin Co, Fla, 60-61; sr res scientist man-mach systs res, ACF Electronics, Md, 61-62; res mgr electronics & control, Off Advan Res & Technol, NASA Hq, Washington, DC, 62-65; res dir computer sci, Advan Res Projs Agency, Off Secy Defense, 65-69; res dir, Info Res Lab, Univ Utah, 69-70; prin scientist & assoc mgr, 70-80, mgr, Computer Sci Lab, Xerox Palo Alto Res Ctr, 80-; DIR, SYSTS RES CTR, DIGITAL EQUIP CORP. *Concurrent Pos:* Mem comts vision & bioacoust, Nat Res Coun, 62-65; mem computer sci & eng bd, Nat Acad Sci, 67-69; mem electronic data processing adv bd, Dept Defense, 68-69; mem computer sci adv bd, Stanford Univ, 71-81 & chmn, 78-79, lectr, 75- & mem computer sci adv comt, NSF, 78-81 & Univ Calif, Berkeley, 80- *Honors & Awards:* Cert Appreciation, Advan Res Projs Agency, Off Secy Defense, 69. *Mem:* Nat Acad Eng; Inst Elec & Electronics Engrs; Asn Comput Mach. *Res:* Interactive information processing and communications systems; central nervous system; computer graphics; artificial intelligence; research and development management. *Mailing Add:* Digital Equip Corp Syst Res Ctr 130 Lytton Ave Palo Alto CA 94301

TAYLOR, ROGER, solid state physics, theoetical physics, for more information see previous edition

TAYLOR, RONALD, b Victor, Idaho, Oct 16, 32; m 55; c 2. BOTANY, GENETICS. *Educ:* Idaho State Univ, BS, 56; Univ Wyo, MS, 60; Wash State Univ, PhD(bot, genetics), 64. *Prof Exp:* Asst prof bot, 64-68, assoc prof biol, 68-72, PROF BIOL, WESTERN WASH UNIV, 72- *Concurrent Pos:* Environ consult. *Mem:* AAAS; Bot Soc Am; Genetics Soc Am; Am Soc Plant Taxon; Sigma Xi; Torrey Bot Club. *Res:* Chemotaxonomy and evolution of selected higher plant taxa; biosystematics of Taraxacum; biosystematics of North American/Mexican Picea. *Mailing Add:* Dept Bot Western Wash Univ Bellingham WA 98225

TAYLOR, RONALD D, b Baltimore, Md, Dec 16, 50. SCIENCE POLICY. *Educ:* Johns Hopkins Univ, BA, 72; Col William & Mary, MS, 74, PhD(physics), 79. *Prof Exp:* Instr physics, Embry-Riddle Aeronaut Univ, 75-77; res assoc, Dept Chem, Univ Toronto, 79-83; asst prof physics, Villanova Univ, 83-84; staff scientist, Berkeley Res Assocs, 84-90; prog officer, 90-91, SR PROG OFFICER, BD PHYSICS & ASTRON, NAT ACAD SCI, 91- *Mem:* Am Phys Soc. *Res:* Atomic and molecular collision theory; atomic processes in plasmas; science policy. *Mailing Add:* Nat Acad Sci 2101 Constitution Ave NW Washington DC 20418

TAYLOR, RONALD PAUL, b Oct 18, 45; m 82; c 3. BIOPHYSICAL CHEMISTRY, IMMUNE COMPLEX CHEMISTRY. *Educ:* City Col NY, BS, 66; Princeton Univ, PhD(chem), 70. *Prof Exp:* PROF BIOCHEM, SCH MED, UNIV VA, 73- *Concurrent Pos:* Consult, NIH, NSF, Vet Admin & Toxicol Div, Environ Protection Agency; consult & ad hoc reviewer, Can Arthritis Asn; fel rev comt, Arthritis Found; Fogert fel, US-USSR Health Scientist Exchange. *Mem:* Am Rheumatism Asn; Am Chem Soc; Biophys Soc; Union Concerned Scientists; Am Soc Biol Chemists; Fedn Am Scientists. *Res:* Lupus pathogenesis; complement and complement receptors; immunochemistry. *Mailing Add:* Dept Biochem Univ Va Charlottesville VA 22908

TAYLOR, ROSCOE L, agronomy; deceased, see previous edition for last biography

TAYLOR, ROSS, b Welwyn Garden City, Eng, Oct 25, 54; m 79; c 3. SEPARATION PROCESS SIMULATION, MULTICOMPONENT MASS TRANSFER. *Educ:* Univ Manchester Inst Sci & Technol, BSc, 76, MSc, 78, PhD(chem eng), 80. *Prof Exp:* From asst prof to assoc prof, 80-89, PROF CHEM ENG, CLARKSON UNIV, 89- *Mem:* Am Inst Chem Engrs; Am Soc Eng Educ. *Res:* Mathematical modeling of multicomponent mass transfer rate governed processes like distillation, absorption and condensation; development of algorithms for solving process models equations. *Mailing Add:* Dept Chem Eng Clarkson Univ Potsdam NY 13699

TAYLOR, ROY JASPER, b Salvisa, Ky, Nov 1, 18; m 46; c 3. CHEMICAL ENGINEERING, CHEMISTRY. *Educ:* Purdue Univ, Lafayette, BSChE, 42; Polytech Inst Brooklyn, PhD(chem), 70. *Prof Exp:* Res engr, Com Solvents Corp, 42-44; chem engr, Allegany Ballistics Lab, 44-45; chem engr, 45-47, supvr process develop, 47-52, mgr, 52-61, dir chem prod & develop, 61-64, MGR PROD PROCESS DEVELOP, CHEM DIV, PFIZER, INC, 64- *Mem:* Am Chem Soc; NY Acad Sci; Sigma Xi. *Res:* Processes for separation and purification; chemical reaction kinetics; reverse osmosis; ultrafiltration; mass transfer kinetics. *Mailing Add:* Pfizer Inc Groton CT 06340-5196

TAYLOR, ROY LEWIS, b Olds, Alta, Apr 12, 32; m 79. PLANT TAXONOMY. *Educ:* Sir George Williams Univ, BSc, 57; Univ Calif, Berkeley, PhD(bot), 62. *Prof Exp:* Teaching bot, Univ Calif, Berkeley, 58-60, assoc, 61-62; res officer, Plant Res Inst, Can Dept Agr, 62-65, chief taxon sect, 65-68; prof plant sci & dir Bot Garden, Univ BC, 68-85; DIR, CHICAGO BOT GARDEN, 85-; PRES & CHIEF ELECTED OFF, CHICAGO AGR SOC, 85- *Concurrent Pos:* Mem exec comt & coun, Pac Sci Asn, 79-83; mem gov bd, Biol Coun Can, 66-69, secy, 69-72, from vpres to pres, 72-74; mem accreditation comn, Am Asn Mus, 80-, chmn, 85- *Honors & Awards:* Queen's Silver Jubilee Medal, 77; Mary E Elliott Serv Award, Can Bot Asn, 83. *Mem:* Int Orgn Biosyst; Am Soc Plant Taxon; fel Linnean Soc London; Can Bot Asn (secy, 65-66, vpres, 66-67, pres, 67-68); Am Asn Bot Gardens & Arboretums (vpres, 73-75, pres, 75-77). *Res:* Systematic botany of western North American vascular plants; systematics and cytotaxonomy of the vascular plants of British Columbia. *Mailing Add:* Chicago Bot Garden PO Box 400 Glencoe IL 60022

TAYLOR, RUSSELL JAMES, JR, b Rockville, Conn, Mar 8, 35. CLINICAL PHARMACOLOGY. *Educ:* Bates Col, BS, 57; Ohio State Univ, MSc, 62, PhD(biochem), 64. *Prof Exp:* Res chemist, Parke, Davis & Co, Mich, 57-59; asst biochem, Ohio State Univ, 59-64; res biochemist, Lederle Labs, Am Cyanamid Co, NY, 64-69; sr res biochemist, McNeil Labs, Inc, 69-70, group leader biochem, 70-79, asst dir med res, 79-80; sr med assoc, 80-85, asst dir clin res, 85-87, ASST DIR PROF SERV, MILES PHARMACEUT, 87- *Mem:* Am Soc Clin Pharmacol & Therapeut; Am Chem Soc; NY Acad Sci; Am Med Writers Asn; AAAS; Am Soc Pharmacol & Exp Therapeut. *Res:* Clinical pharmacology; cardiovascular drugs. *Mailing Add:* Miles Inc Pharmaceut Div 400 Morgan Lane West Haven CT 06516

TAYLOR, SAMUEL EDWIN, b Tuskegee, Ala, Oct 19, 41; m 61; c 2. PHARMACOLOGY. *Educ:* Univ Ala, Tuscaloosa, BS, 63; Univ Ala, Birmingham, PhD(pharmacol), 71. *Prof Exp:* NIH trainee pharmacol, Univ Tenn Med Units, 71-72; asst prof, Sch Dent, Univ Ore Health Sci Ctr, 72-77; asst prof, 77-78, ASSOC PROF, PHARMACOL DEPT, BAYLOR COL DENT, BAYLOR UNIV, 78-, GRAD FAC, 80- *Mem:* Am Soc Pharmacol & Exp Therapeut; Am Asn Dent Res; Soc Exp Biol & Med; Sigma Xi. *Res:* Autonomic/cardiovascular pharmacology; effects of autonomic agents on salivary gland and smooth muscle; cardiovascular shock. *Mailing Add:* Pharmacol Dept Baylor Col Dent 3302 Gaston Ave Dallas TX 75246

TAYLOR, SAMUEL G, III, b Elmhurst, Ill, Sept 2, 04; m 38, 80; c 3. ONCOLOGY. *Educ:* Yale Univ, BA, 27; Univ Chicago, MD, 32. *Prof Exp:* Dir steroid tumor clin, Univ Ill, 47-71; dir cancer ctr planning, Rush Presby-St Luke's Med Ctr, 72-75, assoc dir, Rush Cancer Ctr, 75-78, prof, 71-78, EMER PROF MED, RUSH UNIV, 78- *Concurrent Pos:* Assoc attend physician, Presby Hosp, 48-60; from asst prof to prof, Univ Ill Col Med, 48-72; rep dept med, Tumor Coun, Univ Ill, 48-72; mem consult staff, Lake Forest Hosp, 54-61; from attend physician to sr attend physician, Presby-St Luke's Hosp, Chicago, 61-78, head sect oncol, 61-71, consult, Sect Oncol, 71-; consult, Cancer Control Prog, USPHS, 59-63; dir, Ill Cancer Coun Comprehensive Cancer Ctr, 73-78. *Mem:* Endocrine Soc; Am Asn Cancer Res; Am Col Physicians; Am Radium Soc; Soc Surg Oncol. *Res:* Systematic therapy for cancer. *Mailing Add:* Rush Univ 1725 W Harrison Chicago IL 60612

TAYLOR, SNOWDEN, b New York, NY, June 25, 24; m 49; c 6. PARTICLE PHYSICS, TECHNICAL EDUCATION FOR DISADVANTAGED STUDENTS. *Educ:* Stevens Inst Technol, ME, 50; Columbia Univ, AM, 57, PhD(physics), 59. *Prof Exp:* Fac, Oak Ridge Sch Reactor Technol, 51-52; asst & lectr physics, Columbia Univ, 52-58; from instr to assoc prof, 58-66, prof, 66-90, EMER PROF PHYSICS, STEVENS INST TECHNOL, 91- *Honors & Awards:* Ottens Res Award, 63; STEP Award, 75 & 82; Arthur Schomberg Award, Asn Equality & Excellence Educ, 87. *Mem:* Am Phys Soc; Sigma Xi; Am Asn Univ Prof; Fed Am Scientists; Union Concerned Scientists. *Res:* Horology; particle physics; technical education for minority and disadvantaged students. *Mailing Add:* Dept Physics Stevens Inst Technol Hoboken NJ 07030

TAYLOR, STEPHEN KEITH, b Los Angeles, Calif, Mar 28, 44; m 69; c 2. INDUSTRIAL CHEMISTRY, ORGANIC CHEMISTRY. *Educ:* Pasadena Col, BA, 69; Univ Nev, Reno, PhD(org chem), 74. *Prof Exp:* Res chemist, E I Du Pont de Nemours & Co, Inc, 73-78, sr res chemist, 78; from asst prof to prof chem, Olivet Nazarene Col, 78-85; ASSOC PROF CHEM, HOPE COL, 85- *Mem:* Am Chem Soc; Sigma Xi; Soc Photog Scientist & Engrs. *Res:* Influence of metal ions in organic reactions; reactions of epoxides; stereoselective reactions. *Mailing Add:* Dept Chem Hope College Holland MI 49423

TAYLOR, STEVE L, b Portland, Ore, July 19, 46; m 73; c 2. FOOD SCIENCE, TOXICOLOGY. *Educ:* Ore State Univ, BS, 68, MS, 69; Univ Calif, Davis, PhD, 73. *Prof Exp:* Res assoc, Dept Food Sci, Univ Calif, Davis, 73-74, fel toxicol, 74-76; res chemist food toxicol, Dept Nutrit, Letterman Army Inst Res, 76-78; from asst prof to assoc prof, Food Res Inst, 83-87, DIR FOOD PROCESSING CTR & HEAD FOOD SCI TECH, UNIV NEBR, 87- *Concurrent Pos:* Fel, Nat Inst Environ Health Sci, 74-76. *Mem:* Inst Food Technologists; Am Chem Soc; Am Acad Allergy Immunol. *Res:* Toxicological evaluation of food chemicals and products including improved analytical methods, evaluation of relative hazard, analysis of toxin levels in foods; food allergy studies. *Mailing Add:* Food Processing Ctr Univ Nebr Lincoln NE 68583-0919

TAYLOR, STUART ROBERT, b Brooklyn, NY, July 15, 37; m 63; c 3. PHYSIOLOGY, BIOPHYSICS. *Educ:* Cornell Univ, BA, 58; Columbia Univ, MA, 61; NY Univ, PhD(physiol), 66. *Prof Exp:* Lab asst zool, Columbia Univ, 59-60, lectr, 60-61; res asst physiol, Inst Muscle Dis, Inc, 62-67; Dept Health, Educ & Welfare rehab res fel, Univ Col, Univ London, 67-69; from instr to asst prof pharmacol, State Univ NY Downstate Med Ctr, 69-71; assoc prof, 75-78, PROF PHYSIOL & BIOPHYS, PHARMACOL, MAYO MED SCH, UNIV MINN, 78-, CONSULT, 71- *Concurrent Pos:* Mem res allocations comt, Minn Heart Asn, 72-74, mem bd dirs, 73-75; mem physiol study sect, Div Res Grants, NIH, 73-77; estab investr, Am Heart Asn, 74-79; adv panel gen prof educ of the physician, Asn Am Med Col, 81-84; guest prof, State of Baden-Wurttemberg, Univ Ulm, 83-85; adv panel cellular physiol prog, 85-88; bd scientific counr, NIH & Nat Inst Neurol Commun Disorders & Stroke, 87-91. *Mem:* Am Physiol Soc; Biophys Soc; Brit Biophys Soc; assoc Brit Physiol Soc; Soc Gen Physiol; Am Soc Pharmacol Exp Therapeut; NY Acad Sci; Soc Neurosci. *Res:* Physiology of stimulus-response coupling in contractile cells; computerized image analysis of rapid cellular events. *Mailing Add:* Mayo Grad Sch Med Mayo Found 711 Guggenheim Rochester MN 55905

TAYLOR, SUSAN SEROTA, b Racine, Wis, June 20, 42; m 65; c 3. BIOCHEMISTRY. *Educ:* Univ Wis-Madison, BA, 64; Johns Hopkins Univ, PhD(biochem), 68. *Prof Exp:* NIH fel, Med Res Coun Lab Molecular Biol, Cambridge Univ, 68-70; NIH fel, 71-72, asst prof, 72-79, assoc prof, 79-85, PROF CHEM, UNIV CALIF, SAN DIEGO, 85- *Concurrent Pos:* NIH career develop award & res grant, Univ Calif, San Diego, 72-77; NIH biochem study sect, 78-82; Fogarty fel, Cambridge Univ, 80-81; Am Cancer Soc Rev Group, 83- *Mem:* Am Soc Biol Chemists; Am Chem Soc. *Res:* Protein chemistry; cAMP-dep protein kinase; LDH; amino acid sequencing; structure-function relationships; regulation of kinase genes. *Mailing Add:* Dept Chem/D-006 Univ Calif San Diego M-054 La Jolla CA 92093-0654

TAYLOR, THEODORE BREWSTER, b Mexico City, Mex, July 11, 25; nat US; m 48; c 5. APPLIED PHYSICS. *Educ:* Calif Inst Technol, BS, 45; Cornell Univ, PhD(theoret physics), 54. *Prof Exp:* Physicist, Radiation Lab, Univ Calif, 46-49, theoret physicist, Los Alamos Sci Lab, 49-56; nuclear physicist, High Energy Fluid Dynamics Dept & chmn, Gen Atomic Div, Gen Dynamics Corp, 56-64; dep dir, Defense Atomic Support Agency, 64-66; chmn bd, Int Res & Tech Corp, 67-76; prof mech & aerospace eng, Princeton Univ, 76-80; PRES, NOVA, INC, WEST CLARKSVILLE, NY, 80- *Concurrent Pos:* Consult, Govt & Indust, 56-; lectr, San Diego State Col, 57, Univ Calif, Santa Cruz, 88- *Honors & Awards:* Lawrence Mem Award, Atomic Energy Comn, 65; Boris Pregel Award, NY Acad Sci, 81. *Mem:* AAAS; Am Phys Soc; Solar Energy Soc. *Res:* Renewable energy systems; controlled environment agriculture; international control and development of nuclear energy; nuclear explosives and effects of nuclear explosions; space propulsion; pollution control; technology assessment; energy conservation. *Mailing Add:* PO Box 39 West Clarksville NY 14786

TAYLOR, THOMAS NEWTON, b Cedar Rapids, Iowa, June 21, 44. SURFACE PHYSICS. *Educ:* Iowa State Univ, BS, 66; Brown Univ, MS & PhD(physics), 73. *Prof Exp:* Fel surface physics, Lawrence Berkeley Lab & Lawrence Livermore Lab, 73-75; STAFF PHYSICIST, LOS ALAMOS NAT LAB, 75- *Concurrent Pos:* Guest scientist, FOM-Inst Atomic & Molecular Physics, Amsterdam, Holland, 88. *Mem:* Am Chem Soc; Am Vacuum Soc. *Res:* Surface properties of metallic and ceramic advanced materials; emphasis on interface growth and corrosion; low energy electron diffraction; ion scattering; Auger electron spectroscopy and allied techniques. *Mailing Add:* Los Alamos Nat Lab MS-G738 CLS-2 Los Alamos NM 87545

TAYLOR, THOMAS NORWOOD, b Lakewood, Ohio, June 14, 37; m 59; c 5. PALEOBOTANY. *Educ:* Miami Univ, AB, 60; Univ Ill, Urbana, PhD(bot), 64. *Prof Exp:* NSF res fel, Yale Univ, 64-65; from asst prof to prof biol sci, Univ Ill, Chicago Circle, 65-72, dir scanning electron microscope lab, 67-72; prof bot, Ohio Univ, 72-74; chmn dept, 74-78, PROF BOT, OHIO STATE UNIV, 74-, PROF, DEPT GEOL & MINERAL, 79-, RES SCIENTIST, BYRD POLAR RES CTR, 89- *Concurrent Pos:* NSF res grants, 64-; NSF res grants paleobot, 65-78; res assoc, Geol Dept, Field Mus Natural Hist, Chicago, 67-; Ill Acad Sci res grants, 70-72; assoc ed, Rev Palaeobot Palynol, Paleobiol, Grana, Modern Geol, Symbiosis. *Honors & Awards:* Merit Award, Bot Soc. *Mem:* Bot Soc Am; Brit Paleont Asn; Int Orgn Paleobot; fel Linnean Soc London; Am Asn Stratig Palynologists. *Res:* Structure and evolution of Paleozoic vascular plants; electron microscopy of fossil pollen and spores; morphology of extant vascular plants; antarctica fossil plants; fossil fungi. *Mailing Add:* Bot/168 B&Z The Ohio State Univ Columbus OH 43210

TAYLOR, THOMAS TALLOTT, SR, b Montpelier, Ind, Apr 18, 21; m 58; c 2. PHYSICS. *Educ:* Purdue Univ, BS, 42; Calif Inst Technol, MS, 53, PhD(physics), 58. *Prof Exp:* Engr, Gen Elec Co, 42-46; res physicist, Hughes Aircraft Co, 46-54; asst prof physics, Univ Calif, Riverside, 58-63; from asst prof to prof physics, 63-80, chmn dept, 67-77, EMER PROF PHYSICS, LOYOLA MARYMOUNT UNIV, LOS ANGELES, 80- *Concurrent Pos:* Instr, Engr Exten, Univ Calif, Los Angeles, 47-50 & 57-58. *Mem:* Am Phys Soc; fel Inst Elec & Electronics Engrs; Am Asn Physics Teachers. *Res:* Electromagnetic theory; antennas; lattice sums in crystals. *Mailing Add:* 6622 W 87th St Los Angeles CA 90045

TAYLOR, TIMOTHY H, b Sawyer, Ky, July 4, 18; m 45; c 2. AGRONOMY. *Educ:* Univ Ky, BS, 48, MS, 50; Pa State Univ, PhD(agron), 55. *Prof Exp:* Asst agronomist, Va Agr Exp Sta, 49-55; assoc agronomist, Agr Exp Sta, 55-60, from assoc prof to prof 60-84, EMER PROF AGRON, 84- *Concurrent Pos:* Consult, 59-; vis scientist, Am Soc Agron; vis biologist, Am Inst Biol Sci. *Mem:* Fel Am Soc Agron; fel Crop Sci Am; Am Forage & Grassland Coun; Sigma Xi. *Res:* Ecology of humid temperate grasslands; forage crop ecology and physiology; ecology of cultivated grasslands. *Mailing Add:* Gen Delivery Sawyer KY 42643-9999

TAYLOR, WALTER FULLER, b Boston, Mass, Dec 5, 40; m 67. MATHEMATICS. *Educ:* Swarthmore Col, AB, 62; Harvard Univ, MA, 63, PhD(math), 68. *Prof Exp:* From asst prof to assoc prof, 67-77, NSF grants, 69-74 & 76-79, PROF MATH, UNIV COLO, 77- *Concurrent Pos:* Fulbright Found sr res fel, Univ New SWales, Australia, 75; vis prof, Univ Hawaii, 77-78; ed, Algebra Universalis, 77- *Mem:* Am Math Soc; Sigma Xi. *Res:* Model theory, universal algebra and topology. *Mailing Add:* Dept Math Univ Colo Campus Box 426 Boulder CO 80309-0426

TAYLOR, WALTER HERMAN, JR, b Laurens, SC, July 5, 31; m 55; c 2. MICROBIAL PHYSIOLOGY. *Educ:* Duke Univ, MA, 54; Univ Ill, PhD(bact), 59. *Prof Exp:* Asst bact, Univ Ill, 54-57 & Emory Univ, 57-59; res assoc biol div, Oak Ridge Nat Lab, 59, USPHS fel, 59-61; from asst prof to assoc prof, 61-69, PROF BIOL & HEAD DEPT, PORTLAND STATE UNIV, 69- *Mem:* AAAS; Am Soc Microbiol; Brit Soc Gen Microbiol; Am Chem Soc. *Res:* Bacterial physiology; carbohydrate metabolism; protein biosynthesis; pyrimidine metabolism in microorganisms. *Mailing Add:* Dept Biol Portland State Univ Box 751 Portland OR 97207

TAYLOR, WALTER KINGSLEY, b Calhoun, Ky, Nov 12, 39; m 68; c 1. ORNITHOLOGY, VERTEBRATE ECOLOGY. *Educ:* Murray State Univ, BS, 62; La Tech Inst, MS, 64; Ariz State Univ, PhD(zool), 67. *Prof Exp:* Asst, La Tech Univ, 62-64 & Ariz State Univ, 64-67; from asst prof to assoc prof, 69-82, PROF BIOL, UNIV CENT FLA, 82- *Mem:* Am Ornith Union; Wilson Ornith Soc; Cooper Ornith Soc; Sigma Xi. *Res:* Breeding biology; migratory biology; population ecology of birds; vocalizations of birds. *Mailing Add:* Dept Biol Univ Cent Fla PO Box 25000 Orlando FL 32816

TAYLOR, WALTER ROWLAND, b Baltimore, Md, Dec 31, 18; m 44; c 2. OCEANOGRAPHY. *Educ:* Wash Col, Md, BS, 40; Univ Wis, MS, 47, PhD(biochem), 49. *Prof Exp:* Asst chem, Ga Inst Technol, 40-41; asst biochem, Univ Wis, 46-49; res assoc chem, Univ Ill, 49-51; instr physiol chem, Sch Med, 51-56, res assoc, Chesapeake Bay Inst, 56-74, from asst prof to assoc prof oceanog, 58-68, assoc prof earth & planetary sci, 68-74, asst dir res, 75-78, actg dir, 78-81, PRIN RES SCIENTIST, CHESAPEAKE BAY INST, JOHNS HOPKINS UNIV, 74-, DIR, 81- *Concurrent Pos:* Instr marine ecol, Marine Biol Lab, Woods Hole, Mass, 60-70, instr-in-chg, 63-67. *Mem:* AAAS; Am Soc Limnol & Oceanog; Phycol Soc Am; Ecol Soc Am; Int Phycol Soc; Sigma Xi. *Res:* Primary production and nutrient recycling in marine and estuarine environments; physiology and ecology of marine organisms. *Mailing Add:* 1725 Ceder Park Rd Annapolis MD 21401

TAYLOR, WARREN EGBERT, b Colorado Springs, Colo, Nov 15, 20; m 47; c 3. METROLOGY. *Educ:* Kalamazoo Col, BA, 47; Ohio State Univ, PhD(physics), 52. *Prof Exp:* Mem tech staff, Sandia Nat Labs, 52-70, proj leader, Vacuum Metrol Group, 70-86; RETIRED. *Mem:* Am Phys Soc; Am Vacuum Soc; Sigma Xi. *Res:* Vacuum technology; nuclear radiation measurements; health physics. *Mailing Add:* 5123 Royene Ave NE Albuquerque NM 87110

TAYLOR, WELTON IVAN, b Birmingham, Ala, Nov 12, 19; m 45; c 2. MICROBIOLOGY. *Educ:* Univ Ill, Urbana, AB, 41, MS, 47, PhD(bact), 48; Am Bd Med Microbiol, dipl, 68. *Prof Exp:* From instr to asst prof bact, Univ Ill Col Med, 48-54; res bacteriologist, Swift & Co, 54-59; supvr clin microbiol, Children's Mem Hosp, 59-64; bacteriologist in chief, West Suburban Hosp, Oak Park, Ill, 64-69; assoc prof microbiol, Med Ctr, Univ Ill, 61-87; RETIRED. *Concurrent Pos:* Consult microbiol, Northwest Community Hosp, Arlington Heights, Ill, 63-70, Jackson Park, 64-81, Englewood Hosps, Chicago, 64-88, Armour & Co, 66-68, Resurrection Hosp, Chicago, 67-88, Grant Hosp, Chicago, 69-75, St Mary Nazareth Hosp Ctr, Chicago, 73-80 & Swed Covenant Hosp, Chicago, 74-88; Nat Inst Allergy & Infectious Dis spec res fel, Inst Pasteur, France & Cent Pub Health Lab, Eng, 61-62; bd sci adv, Am Asn Bioanalysts, 70-82, bd dirs, 73-82; Dept Army res contract, 71-73; pres & owner, Micro-Palettes, Inc, 77-88. *Mem:* Fel Am Soc Microbiol; fel Am Acad Microbiol. *Res:* Detection of Vibrio parahemolyticus in routine stool analysis; methods for detection of Salmonella and Shigella with minimal laboratory facilities; rapid indentification procedures for non-enteric pathogens, enteric pathogens, anaerobes, and sexually-transmitted diseases. *Mailing Add:* 7621 S Prairie Ave Chicago IL 60619

TAYLOR, WESLEY GORDON, b Melfort, Sask, Mar 29, 47; m 69. MEDICINAL CHEMISTRY, PESTICIDE CHEMISTRY. *Educ:* Univ Sask, BSP, 69; PhD(pharm), 73. *Prof Exp:* Fel pharm, Dept Med Chem & Pharmacog, Sch Pharm & Pharm Sci, Purdue Univ, 73-75; RES SCIENTIST PESTICIDE CHEM, RES BR, RES STA, AGR CAN, 76- *Mem:* Am Chem Soc; Chem Inst Can. *Res:* Biological alkylating agents; organic synthesis; drug metabolism; pesticide toxicology. *Mailing Add:* Animal Parasitol Sec Agr CA Res Sta PO Box 3000 Main Lethbridge AB T1J 4B1 Can

TAYLOR, WILLIAM CLYNE, b Aberdeen, Scotland, Mar 26, 24; Can citizen; m 54; c 3. PEDIATRICS. *Educ:* Aberdeen Univ, MB & ChB, 45; Univ London, DCH, 50; FRCP(C), 58. *Prof Exp:* PROF PEDIAT, UNIV ALTA, 57- *Concurrent Pos:* Mead Johnson res fel, Univ Alta, 58-59; Schering traveling fel, Africa, Australia & NZ, 66-67; Brit Commonwealth grant, UK, 66. *Mem:* Can Soc Clin Invest; Am Acad Pediat; Can Pediat Soc. *Res:* Evaluation of undergraduate and postgraduate medical education; delivery of health care in the Northwest Territories of Canada. *Mailing Add:* Dept Pediat Univ Alta Edmonton AB T6G 2E2 Can

TAYLOR, WILLIAM DANIEL, b Cardiff, Gt Brit, May 25, 34. DNA REPAIR, MUTAGENESIS. *Educ:* Univ Manchester, BSc, 56, PhD(phys chem), 59. *Prof Exp:* Fel physics, Pa State Univ, 59-61; res fel chem, Univ Manchester, 61-63; asst prof biophys, 63-68, assoc prof, 68-71, head dept, 71-75, PROF BIOPHYS, PA STATE UNIV, UNIVERSITY PARK, 71-, CHMN, MOLECULAR BIOL PROG, 84- *Mem:* AAAS; Radiol Res Soc; Biophys Soc; Fedn Am Sci; The Chem Soc; Am Soc Photobiol. *Res:* Biophysical chemistry; chemical carcinogenesis; nucleic acids; radiation biology. *Mailing Add:* Dept Biophys Pa State Univ University Park PA 16802

TAYLOR, WILLIAM F, b Cincinnati, Ohio, Oct 14, 21; m 43, 81; c 3. BIOSTATISTICS. *Educ:* Univ Calif, PhD(math statist), 51. *Prof Exp:* Instr biostatist, Univ Calif, 50-51; chief dept biometrics, Sch Aviation Med, US Air Force, 51-58; assoc prof biostatist, Sch Pub Health, Univ Calif, Berkeley, 58-63, prof, 63-67; head med res statist sect, 67-76, MEM FAC, DEPT MED STATIST & EPIDEMIOL, MAYO CLIN, 76- *Mem:* Biomet Soc; fel Am Pub Health Asn; fel Am Statist Asn; Sigma Xi. *Res:* Design of clinical trials; sequential analysis in medicine; reference value problems. *Mailing Add:* 1524 Wilshire Dr NE Rochester MN 55906

TAYLOR, WILLIAM FRANCIS, b Washington, DC, Apr 20, 31; m 61; c 3. FUEL TECHNOLOGY, PETROLEUM ENGINEERING. *Educ:* Cath Univ Am, BChE, 53; Ohio State Univ, MS, 57; Rutgers Univ, MS, 62; Stevens Inst Technol, ScD(chem eng), 67. *Prof Exp:* Chem engr, Goodyear Tire & Rubber Co, Ohio, 53-54; res engr, 57-71, sr res engr, 71-75, eng assoc, 75-80, SR ENGR ASSOC, EXXON RES & ENG CO, 80- *Concurrent Pos:* mem, coord res coun, Am Soc Testing & Mat. *Mem:* Am Inst Chem Engrs; Am Chem Soc; Sigma Xi. *Res:* Aviation fuels product quality; jet fuel thermal stability; synthetic fuels; heterogeneous kinetics and catalysis. *Mailing Add:* 1598 Brookside Rd Mountainside NJ 07092

TAYLOR, WILLIAM GEORGE, CELL BIOLOGY. *Educ:* Univ Ill, PhD(microbiol), 70. *Prof Exp:* RES BIOLOGIST, NAT CANCER INST, NIH, 70- *Mailing Add:* NCI NIH 9000 Rockville Pk NIH Bldg 37 Rm 1C09 Bethesda MD 20892

TAYLOR, WILLIAM H, II, b Philadelphia, Pa, Dec 17, 38; m 57; c 2. SOLID STATE PHYSICS. *Educ:* Johns Hopkins Univ, BES, 60; Princeton Univ, MSE, 61, AM, 61, PhD(solid state sci), 64. *Prof Exp:* Staff mem, Redstone Arsenal Res Div, Rohm and Haas Co, Ala, 63-64; chief, Solid State Br, Explosives Lab, Picatinny Arsenal, 66-68; vpres, Data Sci Ventures, Inc, 68-72, White, Weld & Co, Inc, 72-73 & Crocker Capital Corp, 73-79; GEN PARTNER, TAYLOR & TURNER, VENTURE CAPITAL, 79- *Mem:* Am Phys Soc. *Res:* Investment banking, venture capital financing of technology companies; point defects in solids; properties of solid explosives; soft contact lens materials. *Mailing Add:* 2452 Francisco St San Francisco CA 94123

TAYLOR, WILLIAM IRVING, b NZ, July 23, 23; m 52; c 3. ORGANIC CHEMISTRY. *Educ:* Univ Auckland, PhD(chem), 48, DSc, 68. *Prof Exp:* Nat Res Coun scholar, Switz, 48-49; Nat Res Coun Can fel, 50; Imp Chem Industs fel, Cambridge Univ, 51-52; assoc prof chem, Univ NB, 52-55; chemist, Ciba Pharmaceut Co, 55-62, dir natural prod, 63-67, dir biochem res, 67-68; dir fragrance res, Int Flavors & Fragrances, Inc, 68-71, vpres res & develop, 71-88, dir, chem synthesis & develop, 72-88, vpres, 78-88; RETIRED. *Concurrent Pos:* Guest investr, Rockefeller Univ, 66-67. *Honors & Awards:* Res Achievement Award Natural Prod, Am Pharmaceut Asn, 68. *Mem:* Am Chem Soc; NY Acad Sci. *Res:* Synthesis and structural elucidation of natural products, especially in field of flavor and aroma chemicals. *Mailing Add:* Four Hickory Hill Radford VA 24141

TAYLOR, WILLIAM JAPE, b Booneville, Miss, Sept 5, 24; m 48; c 4. INTERNAL MEDICINE, CARDIOLOGY. *Educ:* Yale Univ, BS, 44; Harvard Med Sch, MD, 47. *Prof Exp:* Intern, Second Harvard Med Serv, Boston City Hosp, 47-48; asst resident med, Duke Univ, 48-50, instr, Sch Med & Hosp, 54-55; instr, Sch Med, Univ Pittsburgh, 55-58; chief, Div Cardiol, 58-74, from asst prof to assoc prof, 58-64, prof med, 64-74, DISTINGUISHED SERV PROF MED, COL MED, UNIV FLA, 74- *Concurrent Pos:* Fel coun clin cardiol, Am Heart Asn; Am Col Physicians res fel med, Duke Univ, 50-51, univ res fel, Sch Med & Hosp, 51-52, USPHS res fel, 54-55; vis prof med, Fac Health Sci, Univ Ife, Ile-Ife, Nigeria, 74-75. *Mem:* Am Fedn Clin Res; Asn Univ Cardiol; fel Am Col Cardiol; fel Am Col Physicians. *Mailing Add:* Dept Med Univ Fla Col Med Gainesville FL 32610

TAYLOR, WILLIAM JOHNSON, b Chengdu, China, Dec 3, 16; US citizen; m 49; c 4. THEORETICAL CHEMISTRY, PHYSICAL CHEMISTRY. *Educ:* Denison Univ, AB, 37; Ohio State Univ, PhD(phys chem), 42. *Prof Exp:* Instr chem, Univ Calif, Berkeley, 41-42; sr res assoc, Thermochem Sect, Nat Bur Standards, 42-47; spectros & low-temperature res, Cryogenic Lab, 47-50, from asst prof to prof chem, 50-85, EMER PROF CHEM, OHIO STATE UNIV, 85- *Concurrent Pos:* Vis prof, Lab Molecular Structure & Spectra, Univ Chicago, 63. *Mem:* Am Chem Soc; Am Phys Soc; AAAS; Am Asn Univ Prof. *Res:* Statistical and quantum mechanics; statistics of long-chain molecules; molecular vibrations; dipole moments; localized molecular orbitals; configuration interaction method and correlation energy; quantum integrals; applications of group theory. *Mailing Add:* Dept Chem Ohio State Univ 120 W 18th Ave Columbus OH 43210

TAYLOR, WILLIAM L, b Corsicana, Tex, Oct 17, 26; m 50; c 3. METEOROLOGY, ELECTROMAGNETISM. *Educ:* Okla State Univ, BS, 50. *Prof Exp:* Physicist, Nat Bur Standards, Washington, DC, 50-51, Alaska, 51-52 & Colo, 52-65; physicist, Inst Telecommun Sci & Aeronomy, Environ Sci Serv Admin, US Dept Com, 65-70, physicist, Environ Res Labs, Nat Oceanic & Atmospheric Admin, 70-84; PHYSICIST, CIMMS, UNIV OKLA, 84- *Mem:* AAAS; Am Geophys Union; Sigma Xi; Int Union Radio Sci; Am Meteorol Soc. *Res:* Radio propagation between the earth and ionosphere in the lower frequency bands, using lightning discharges as the source; measurement of radio noise from severe thunderstorms and tornadoes; lightning discharge characteristics at all radio frequencies; relationships between thunderstorm precipitation, windfields, turbulence and lightning. *Mailing Add:* 4300 Brookline Pl Norman OK 73069

TAYLOR, WILLIAM RANDOLPH, botany; deceased, see previous edition for last biography

TAYLOR, WILLIAM ROBERT, b Borger, Tex, Oct 25, 39; m 59; c 2. ENGINEERING, APPLIED STATISTICS. *Educ:* Okla State Univ, BS, 63; Univ Ark, Fayetteville, MS, 67, PhD(eng), 69. *Prof Exp:* Indust engr, Southwestern Bell Tel Co, Okla, 63-65; teaching asst, Univ Ark, Fayetteville, 68-69; asst prof, 69-77, ASSOC PROF INDUST & MGT ENG, MONT STATE UNIV, 77-, US FOREST SERV GRANT, 71- *Concurrent Pos:* Consult, Morrison-Knudsen Co, Inc, 72- *Mem:* Am Inst Indust Engrs; Am Soc Eng Educ. *Res:* Operations research applications for harvesting timber; application of management science principles to hospital systems; application of engineering principles in designing disease diagnostic equipment. *Mailing Add:* Dept Indust Eng Mont State Univ Bozeman MT 59717

TAYLOR, WILLIAM WALLER, b Rochester, NY, Nov 20, 50; c 1. LIMNOLOGY, FISHERIES BIOLOGY. *Educ:* Hartwick Col, BA, 72; WVa Univ, MS, 75; Ariz State Univ, PhD(zool), 78. *Prof Exp:* Asst prof fisheries, Univ Mo-Columbia, 78-80; ASST PROF, DEPT FISHERIES & WILDLIFE, MICH STATE UNIV, 80- *Concurrent Pos:* Vis scientist, Hydrobiol Inst, Lake Ohrid, Yugoslavia, Smithsonian Inst, 76- *Mem:* AAAS; Am Fisheries Soc; Am Soc Limnol & Oceanog; Int Asn Theoret & Appl Limnol; Sigma Xi. *Res:* Biological limnology; population dynamics and production of heterotrophs; ecosystem structure and function. *Mailing Add:* Dept Fish & Wildlife Mich State Univ East Lansing MI 48824

TAYLOR, WILLIAM WEST, b Northampton Co, NC, Dec 4, 23; m 53; c 4. PHARMACY. *Educ:* Univ NC, BS, 47, PhD(pharm), 62. *Prof Exp:* Intern hosp pharm, Duke Hosp, Durham, NC, 46-47, staff pharmacist, 48; chief pharmacist, Strong Mem Hosp, Rochester, NY, 48; instr, Univ NC, Chapel Hill, 52-62, asst prof hosp pharm, Div Pharmaceut, 62-83; RETIRED. *Concurrent Pos:* Chief pharmacist, NC Mem Hosp, 52-68, assoc dir, Div Pharm Serv, 68-74, special formulations pharmacist, 75- *Mem:* Am Pharmaceut Asn; Am Soc Hosp Pharmacists; Am Asn Cols Pharm. *Res:* Hospital pharmacy; drug control; special compounding and dosage preparation; purification and formulation of dyes for clinical purposes. *Mailing Add:* Whitfield Rd Chapel Hill NC 27514

TAYLOR-CADE, RUTH ANN, b Yazoo City, Miss, Nov 17, 37; m 74; c 1. ENGINEERING, COMPUTER SCIENCE. *Educ:* Miss State Univ, BS, 63; Univ Ala, MA, 68, PhD(eng), 69. *Prof Exp:* Chem engr, Southern Res Inst, 63-64; instr eng mech, Univ Ala, 67-68; asst prof math, 68-75, assoc prof computer sci, 75-83, PROF CONSTRUCT & ARCHIT ENG TECH, UNIV SOUTHERN MISS, 83-, DIR, SCH ENG TECHNOL, 89- *Mem:* Am Soc Eng Educ. *Res:* Women in science and engineering. *Mailing Add:* Dept Archit Eng Tech Univ Southern Miss Box 9257 Hattiesburg MS 39406-9257

TAYLOR-MAYER, RHODA E, b Hartford City, Ind, Feb 20, 36; div; c 2. GENETIC TOXICOLOGY, MARINE BIOLOGY. *Educ:* Asbury Col, BA, 57; Purdue Univ, MS, 63, PhD(physiol), 65. *Prof Exp:* Asst prof biol, Ind Univ, Kokomo, 65-67; assoc prof, 67-80, dept chmn, 84-89, PROF BIOL, SLIPPERY ROCK UNIV, 80- *Concurrent Pos:* Cong affairs specialist, Intergovt Personnel Agreement, Appt Nat Oceanic & Atmospheric Admin, 80-83; NSF pre-col teacher develop sci prog grants, 79 & 80. *Res:* Mutagenesis and aneuploidy induction; reproductive physiology and behavior; maternal behavior. *Mailing Add:* Dept Biol Slippery Rock Univ Slippery Rock PA 16057

TAYSOM, ELVIN DAVID, b Rockland, Idaho, Aug 5, 17; m 39; c 3. ANIMAL HUSBANDRY. *Educ:* Univ Idaho, BS, 40; Utah State Univ, MS, 50; Wash State Univ, PhD, 61. *Prof Exp:* Asst animal sci, Utah State Univ, 49-50; asst, Wash State Univ, 50-53, sheep specialist, 53; from asst prof to prof, 53-82, EMER PROF ANIMAL SCI, ARIZ STATE UNIV, 82- *Concurrent Pos:* Asst dir, Int Stockmen's Sch, Ariz & Tex, 63-81. *Mem:* Am Soc Animal Sci. *Res:* Mineral metabolism and its relation to the formation to urinary calculi and hormone functions in the body. *Mailing Add:* 2028 S College Ave Tempe AZ 85282

TAZUMA, JAMES JUNKICHI, b Seattle, Wash, July 17, 24; m 54; c 2. CHEMISTRY. *Educ:* Univ Wash, BSc, 48, PhD(org chem), 53. *Prof Exp:* Parke Davis Co fel, Wayne State Univ, 52-53; res chemist, Henry Ford Hosp, 53-54 & Food Mach & Chem Corp, NJ, 55-58; sr res chemist, Goodyear Tire & Rubber Co, 58-65, sect head spec assignment, 65-79, res scientist, 79-90; RETIRED. *Mem:* Am Chem Soc. *Res:* Petroleum chemistry; catalysis; reaction mechanism; process development; polymer additives; monomers; rubber chemicals development. *Mailing Add:* 15800 Village Green Dr Unit 5 Mill Creek Bothell WA 98012-1299

TCHAO, RUY, b China. CELL BIOLOGY. *Educ:* Univ Nottingham, BSc, 60; Univ Manchester, PhD(biochem), 64. *Prof Exp:* Res fel, Inst Cancer Res, Chester Beatty Inst, London, 64-66, from mem res staff to sr biochemist, 66-72; asst prof path, 72-76, ASSOC PROF PATH, MED COL PA, 76- *Concurrent Pos:* Vis assoc prof, Med Col, Nat Taiwan Univ, 69; vis prof, Institut fur Zellforschungszentrum, Heidelberg, Ger, 77; res fel, Int Agency Res Cancer, Lyon, France, 80-81. *Mem:* Int Soc Differentiation; Europ Tissue Culture Asn; NY Acad Sci; Am Asn Cancer Res. *Res:* Tumor cell differentiation, invasion. *Mailing Add:* Dept Pharmacol & Toxicol Philadelphia Col Pharm & Sci 43rd & Woodland Ave Philadelphia PA 19104

TCHEN, TCHE TSING, b Peiping, China, Oct 1, 24; nat US; m 60. BIOCHEMISTRY. *Educ:* Aurora Univ, China, ChemE, 48; Univ Chicago, PhD(biochem), 54. *Prof Exp:* Asst, Univ Chicago, 51-54, res assoc, 54-55; res fel, Harvard Univ, 55-58; assoc prof biochem, 58-61, PROF BIOCHEM, WAYNE STATE UNIV, 61-, ADJ PROF BIOL, 80- *Concurrent Pos:* Am Cancer Soc scholar, 56-58; mem, Physiol Chem Study Sect, USPHS, 60-64; mem res comt, Am Heart Asn, 66-71; ed, Arch Biochem & Biophys, 68-72, J Biol Chem, Am Soc Biol Chem, 68-72 & Endocrine Res Commun, 74- *Mem:* Am Chem Soc; Am Soc Biol Chem. *Res:* Cellular differentiation; mechanism of cell death; action of ACTH and MSH; pigment cells. *Mailing Add:* Chem/435 Chem Bldg Wayne State Univ 5950 Cass Ave Detroit MI 48202

TCHERTKOFF, VICTOR, b Lausanne, Switz, Aug 7, 19; US citizen; m 42; c 2. PATHOLOGY. *Educ:* City Col New York, BS, 40; New York Med Col, MD, 43; Am Bd Path, dipl anat, 61. *Prof Exp:* Asst pathologist, Metrop Hosp, New York, 57-60; Assoc prof, 61-67, PROF PATH, DEPT PATH, NY MED COL, 67-, ACTG CHMN, 88-; DIR PATH LABS, METROP HOSP, 67- *Concurrent Pos:* Pathologist-in-charge, Metrop Hosp, 61-66. *Mem:* Col Am Path; Am Soc Clin Path; Int Acad Path. *Mailing Add:* Off Dir Labs Metrop Hosp 1901 First Ave New York NY 10029

TCHEUREKDJIAN, NOUBAR, b Beirut, Lebanon, Jan 4, 37; m 71; c 2. PHYSICAL CHEMISTRY. *Educ:* Ill Inst Technol, BS, 58; Lehigh Univ, MS, 60, PhD(phys chem), 63. *Prof Exp:* Sr res chemist, S C Johnson & Son, Inc, 63-80, res assoc, 80-82, sect mgr, 82, sr group leader, 82-89, MGR, S C JOHNSON & SON, INC, 89- *Mem:* Am Chem Soc; Soc Rheology; Sigma Xi; Int Asn Colloid & Surface Scientists. *Res:* Colloid and surface chemistry; personal care; chemical specialties; aerosol technology. *Mailing Add:* Phys Res S C Johnson & Son Inc Racine WI 53403-5011

TCHOBANOGLOUS, GEORGE, b Patterson, Calif, May 24, 35; m 57; c 3. ENVIRONMENTAL ENGINEERING. *Educ:* Univ of the Pac, BS, 58; Univ Calif, Berkeley, MS, 60; Stanford Univ, PhD(sanit eng), 69. *Prof Exp:* Res engr, Univ Calif, Berkeley, 60-62 & Water Resources Engrs, Inc, 62-63; actg asst prof sanit eng, Stanford Univ, 66-70; assoc prof, 70-76, PROF CIVIL ENG, UNIV CALIF, DAVIS, 76- *Concurrent Pos:* Consult, George S Nolte & Assoc, 81- *Honors & Awards:* Gordon Masken Fair Medal, Water Pollution Control Fedn. *Mem:* AAAS; Water Pollution Control Fedn; Am Soc Civil Engrs; Am Geophys Union; Am Water Works Asn. *Res:* Physical processes in water and wastewater treatment; operation of small treatment systems; solid waste management; aquatic treatment systems. *Mailing Add:* Dept Civil Eng Univ Calif Davis CA 95616

TCHOLAKIAN, ROBERT KEVORK, b Apr 26, 38; US citizen; c 3. BIOCHEMISTRY, REPRODUCTIVE BIOLOGY. *Educ:* Berea Col, BS, 58; Fla State Univ, MS, 63; Med Col Ga, PhD(physiol, biochem), 67. *Prof Exp:* NIH fels, Steroid Biochem Inst, Univ Utah, 67-68 & Univ Southern Calif, 68-69; instr reproduction, Univ Southern Calif, 69-70; asst prof endocrinol, M D Anderson Hosp & Tumor Inst, 70-71; asst prof reproductive med & biol, 71-81, ASSOC PROF OBSTET, GYNEC & REPRODUCTIVE SCI, MED SCH, UNIV TEX, HOUSTON, 71-, ASSOC PROF REPRODUCTIVE MED & BIOL, 81- *Concurrent Pos:* Assoc mem, Grad Sch Biomed Sci, Univ Tex, Houston. *Mem:* AAAS; Am Soc Zoologists; Soc Study Reproduction; Am Soc Andrology; Endocrine Soc. *Res:* Reproductive biology and steroid biochemistry as related to action of hormones; endocrine and paracrine regulation of testicular testosterone synthesis. *Mailing Add:* Med Sch Dept Obstet Gynec & Reprod Sci Univ Tex Health Sci Ctr 6431 Fannin Suite 3204 Houston TX 77030

TEABEAUT, JAMES ROBERT, II, b Fayetteville, NC, Aug 27, 24. PATHOLOGY. *Educ:* Duke Univ, MD, 47; Am Bd Path, dipl, 53. *Prof Exp:* Intern path, Duke Hosp, Durham, NC, 48, intern internal med, 49; Rockefeller res fel legal med, Harvard Med Sch, 49-51; chief div forensic path, Armed Forces Inst Path, 51-54; asst prof path, Sch Med, Univ Tenn, 54-59; coroner, Shelby County, Tenn, 55-59; assoc prof, 59-69, PROF PATH, MED COL GA, 69- *Concurrent Pos:* Lederle med fac award, 54-55; med examr, State of Ga, 59-; consult, Vet Admin Hosp, Augusta, Ga, 59- & US Army Hosp, Ft Gordon, Ga, 60-, lectr, Mil Police Sch, 63-, hon mem staff, 66- *Mem:* Col Am Path; Am Soc Clin Path; AMA; Int Acad Path; Int Acad Forensic Sci. *Res:* Forensic pathology; pathology of human cardiovascular disease. *Mailing Add:* Dept Path Med Col Ga 1120 15th St Augusta GA 30912

TEACH, EUGENE GORDON, b Hayward, Calif, Oct 27, 26; m 54; c 1. ORGANIC CHEMISTRY. *Educ:* St Mary's Col, Calif, BS, 51; Univ Notre Dame, PhD(chem), 53. *Prof Exp:* Res assoc, Univ Calif, Los Angeles, 53-54, res chemist, USDA, 54-57; res assoc, ICI Americas, 87-91; RES ASSOC, STAUFFER CHEM CO, 57-, SUPVR, 74- *Mem:* Am Chem Soc. *Res:* Actylene-allene chemistry; high temperature polymers; fluorine chemistry; agricultural chemistry. *Mailing Add:* 1929 Downey Pl El Cerrito CA 94530

TEAF, CHRISTOPHER MORRIS, b Philadelphia, Pa, May 5, 53; m 81; c 1. TOXICOLOGY, ENVIRONMENTAL RISK ASSESSMENT. *Educ:* Pa State Univ, BS, 75; Fla State Univ, MS, 80; Univ Ark, PhD(toxicol), 85. *Prof Exp:* Res staff, Hazardous Waste Mgt Prog, 79-81, res assoc, Ctr Biomed & Toxicol Res, 81-83, res asst, Nat Ctr Toxicol Res, 82-85, ASSOC DIR, CTR BIOMED & TOXICOL RES, FLA STATE UNIV, 85- *Mem:* Soc Toxicol; Soc Risk Analysis; AAAS; Am Soc Ichthyologists & Herpetologists; Sigma Xi; Nat Asn Underwater Instrs. *Res:* Modulation of mutagenic processes; reproductive toxicology; establishment of acceptable concentrations of toxic contaminants in water, soil and air. *Mailing Add:* Ctr Biomed & Toxicol Res Fla State Univ 361 Bellamy Bldg Tallahassee FL 32306

TEAFORD, MARGARET ELAINE, b Union Star, Mo, Feb 2, 28. BIOCHEMISTRY, CLINICAL CHEMISTRY. *Educ:* Northwest Mo State Col, BS, 50; Univ Mo-Columbia, MS, 59, PhD(biochem), 64; Am Bd Clin Chemists, dipl. *Prof Exp:* Med technologist, Methodist Hosp, St Joseph, Mo, 51-52, chief med technol, 52-56; NIH traineeship, Univ Wash, 64-66; tech dir lab med, Allen Med Labs, Ltd, 66-78, assoc dir, 78-83; CONSULT, 83- *Mem:* AAAS; Am Asn Clin Chemists; assoc Am Soc Clin Path; hon mem Sigma Xi. *Res:* Methodology in clinical chemistry and establishing normal values for the various biochemical parameters in the human. *Mailing Add:* 1964 Dougherty Ferry Kirkwood MO 63122

TEAGER, HERBERT MARTIN, b Canton, Ohio, Mar 20, 30; m 53; c 2. ELECTRICAL ENGINEERING, BIOMEDICAL ENGINEERING. *Educ:* Mass Inst Technol, SB, 52, ScD(control eng), 55. *Prof Exp:* From asst prof to assoc prof, 59-66, RES PROF MED & CHIEF BIOMED ENG, SCH MED, BOSTON UNIV, 66- *Concurrent Pos:* Consult, President's Sci Adv, Sprague Elec Co, Elec Boat Div, Gen Dynamics Corp, Am Optical Co & Compagnie Europeene D'Automatisme Electronique; lectr elec eng, Mass Inst Technol, 66- *Mem:* AAAS; sr mem Inst Elec & Electronics Engrs; Asn Comput Mach; NY Acad Sci; Am Soc Acoust. *Res:* Application of information processing techniques to the collection and analysis of diagnostic information from machine perceived sonic, tactile and visual information; instrumentation; computer design and man-machine interaction; physiology and pathology of speech production and hearing; diagnostic uses of sound (passive) and vibration; speech recognition. *Mailing Add:* Dept Med Boston Univ Sch Med 80 E Concord Boston MA 02118

TEAGUE, ABNER F, b Gainesville, Tex, May 25, 19; m 71; c 3. CHEMICAL ENGINEERING. *Educ:* Tex Tech Col, BS, 43; Univ Southern Calif, MS, 69. *Prof Exp:* Chemist, Naval Ord Test Sta, 46-48, res chemist, 48-51; unit head, Bur Ord, US Navy, Washington, DC, 51-56; proj engr, Astrodyn, McGregor, Tex, 56-58; engr, TRW Space Tech Labs, 59-63, proj mgr, TRW Systs, Inc, 63-65, proj engr, 65-67, mgr propulsion subproj, TRW, Inc, 67-71; engr, Naval Weapons Eng Support Activ Navy Space Projs-Fleet Commun Satellite Proj, Naval Electronic Systs Command, 71-77, head, Mech Systs Br & dir missile develop, Joint Cruise Missile Projs Off Cruise Missile, Navy Mat Command, Washington, DC, 77-84; RETIRED. *Mem:* NY Acad Sci; Am Inst Chem Engrs; Nat Geog Soc. *Res:* Developing hydrazine propulsion systems for satellite attitude control and station keeping; development and processing of solid propellant rockets; cruise missile mechanical systems including booster and pyrotechnic systems. *Mailing Add:* 950 E Henry St Hamilton TX 76531

TEAGUE, CLAUDE EDWARD, JR, b Sanford, NC, Sept 9, 24; div; c 3. CHEMISTRY. *Educ:* Univ NC, AB, 47, PhD(chem), 50. *Prof Exp:* Res chemist, Am Viscose Corp, 50-51; res chemist, R J Reynolds Tobacco Co, 52-60, mgr chem res, 60-70, asst dir res, 70-75, planning mgr, 76-77, dir corp res, R J Reynolds Indust, 78-81, dir, Res & Develop Admin, 81-87; RETIRED. *Mem:* Am Chem Soc. *Res:* Synthetic organic chemistry; research planning; polymers and synthetic fibers; tobacco chemistry. *Mailing Add:* 716 Archer Rd Winston-Salem NC 27106

TEAGUE, DAVID BOYCE, b Franklin, NC, May 17, 37; m 64, 85; c 2. APPLIED MATHEMATICS. *Educ:* NC State Col, BSEE, 59, MS, 61; NC State Univ, PhD(appl math), 65. *Prof Exp:* Instr math, NC State Univ, 64-65; asst prof, Univ NC, Charlotte, 65-68; ASSOC PROF MATH, WESTERN CAROLINA UNIV, 68- *Concurrent Pos:* Vis assoc prof computer sci, UTK 82. *Mem:* Asn Comput Mach; Sigma Xi. *Res:* Elasticity; mathematical theory of elasticity; mixed boundary value problems in elasticity; computer science; operating systems. *Mailing Add:* Dept Math Western Carolina Univ Cullowhee NC 28723

TEAGUE, HAROLD JUNIOR, b Fayetteville, NC, Nov 5, 41; m 70. ORGANIC CHEMISTRY. *Educ:* Methodist Col, NC, BS, 64; NC State Univ, MS, 67, PhD(org chem), 70. *Prof Exp:* Assoc prof, 70-77, PROF CHEM, PEMBROKE STATE UNIV, 77- *Mem:* Am Chem Soc. *Res:* Mechanistic rearrangement studies of sulfur containing ring compounds; biochemical mechanisms. *Mailing Add:* 109 Elmhurst Dr Lumberton NC 28358-7731

TEAGUE, HOWARD STANLEY, b Rockville, Ind, Jan 16, 22; m 43; c 4. NUTRITION, REPRODUCTIVE PHYSIOLOGY. *Educ:* Univ Nebr, BS, 48, MS, 49; Univ Minn, PhD, 52. *Prof Exp:* From asst prof to prof animal sci, Ohio Agr Res & Develop Ctr, 52-72; mem staff, Animal Sci Res Div, US Meat Animal Res Ctr, 72-76; MEM STAFF, COOP STATE RES SERV, SCI EDUC, USDA, 76- *Mem:* Am Soc Animal Sci; Am Inst Nutrit; Soc Study Reproduction. *Res:* Nutrition and physiology relative to growth and reproductive processes in domestic animals. *Mailing Add:* 9207 Leamington Ct Fairfax VA 22031

TEAGUE, KEFTON HARDING, b Siler City, NC, Sept 30, 20; m 43; c 2. ECONOMIC GEOLOGY. *Educ:* NC State Col, BS, 41. *Prof Exp:* Jr assoc geologist, Tenn Valley Authority, 42-51; geologist, US Geol Surv, 51-55; geologist, 55-62, chief div geologist, 62-73, SR GEOLOGIST GROUP, INT MINERALS & CHEM CORP, 73- *Mem:* Geol Soc Am; Soc Econ Geol. *Res:* Areal and structural geology of Georgia; mica pegmatites; sillimanite; kyanite; talc; feldspar; olivine; barite; bentonite; quartz. *Mailing Add:* 1830 Broadway Dr Graham NC 27253

TEAGUE, MARION WARFIELD, b Arkadelphia, Ark, July 6, 41; m 62; c 3. PHYSICAL INORGANIC CHEMISTRY. *Educ:* Ouachita Baptist Col, BS, 63; Purdue Univ, MS, 68, PhD(chem), 71. *Prof Exp:* Res chemist, Aberdeen Res & Develop Ctr, 68-70; ASSOC PROF CHEM, HENDRIX COL, 70- *Mem:* AAAS; Am Chem Soc. *Res:* Non-coplanar aromatic systems; electron transfer mechanism. *Mailing Add:* Dept Chem Hendrix Col Conway AR 72032-3099

TEAGUE, PERRY OWEN, b Marshall, Tex, July 13, 36; m 64; c 2. IMMUNOLOGY. *Educ:* NTex State Univ, BA, 58, MA, 61; Univ Okla, PhD(immunol), 66. *Prof Exp:* Res assoc immunol, Univ Southern Calif, 66; res fel, Univ Minn, 66-68; asst prof, 68-73, ASSOC PROF PATH, MED SCH, UNIV FLA, 73- *Concurrent Pos:* NIH fel pediat, Univ Minn, 67-68. *Mem:* Am Asn Immunol; Soc Exp Biol & Med; assoc Am Soc Clin Path. *Res:* Age-associated and early decline of thymus dependent lymphocyte function; diagnostic clinical immunology. *Mailing Add:* North West Labs 4444 N Classem Oklahoma City OK 73118

TEAGUE, PEYTON CLARK, b Montgomery, Ala, June 26, 15; m 37; c 1. ORGANIC CHEMISTRY. *Educ:* Auburn Univ, BS, 36; Pa State Univ, MS, 37; Univ Tex, PhD(org chem, biochem), 42. *Prof Exp:* Res chemist, Am Agr Chem Co, NJ, 37-39; instr chem, Auburn Univ, 41-42; res chemist, US Naval Res Lab, DC, 42-45; asst prof chem, Univ Ga, 45-48 & Univ Ky, 48-50; from assoc prof to prof, 50-82, assoc dean grad sch, 66-68, DISTINGUISHED EMER PROF CHEM, UNIV SC, 82- *Concurrent Pos:* Vis prof, Univ Col, Dublin, 63-64 & 77. *Mem:* Am Chem Soc; Phytochem Soc NAm (pres, 69-70). *Res:* Chemistry and stereochemistry of flavonoids. *Mailing Add:* Dept Chem Univ SC Columbia SC 29208

TEAGUE, TOMMY KAY, b Crossett, Ark, July 11, 43; m 65; c 1. DATA COMPRESSION, GROUP THEORY. *Educ:* Hendrix Col, AB, 65; Univ Kans, MA, 67; Mich State Univ, PhD(math), 71. *Prof Exp:* Asst prof math, Gustavus Adolphus Col, 71; asst prof math, Hendrix Col, 71-76, coordr comput syst & serv, 75-76; systs rep, Burroughs, 76-77, sr syst rep, 77-79, systs specialist, 79-80, sr systs specialist, 80-84; vpres, Fed Home Loan Bank Dallas, 84-85; mgr operating systs, 85-87, CONSULT PROGRAMMER, UNISYS, 87- *Mem:* Am Math Soc; Math Asn Am. *Res:* Varieties of groups; embeddings of groups; operating systems; data compression. *Mailing Add:* 25839 Lochmoor Dr Valencia CA 91355-2406

TEAL, GORDON KIDD, b Dallas, Tex, Jan 10, 07; m 31; c 3. PHYSICAL-INORGANIC CHEMISTRY. *Educ:* Baylor Univ, AB, 27; Brown Univ, ScM, 28, PhD(phys-inorg chem), 31. *Hon Degrees:* LLD, Baylor Univ, 69; ScD, Brown Univ, 69. *Prof Exp:* Chem solid state physicist, Bell Tel Labs Inc, 30-53; asst vpres & dir mat & components res, Tex Instruments, Dallas, 53-55, asst vpres & dir Cent Res Labs, 55-61, asst vpres res & eng, 61-63, asst vpres & int tech dir, London, Paris, Rome, 63-65; first dir, Inst Mat Res, Nat Bur Standards, Washington, DC 65-67; asst vpres tech develop, Equip Group, Tex Instruments, Dallas, 67-68, vpres & chief scientist, 68-72; consult indust & govt, 72-78; RETIRED. *Concurrent Pos:* Res assoc, Columbia Univ, 32-35; consult, Dept Defense, 56-64, 70-72, NASA, 70-72, Nat Bur Standards, 72-73, Tex Instruments, 72-77; mem panel selenium, Nat Acad Sci-Nat Res Coun, 56, mem panel semiconductors, 57, mem mat adv bd, 60-64, mem ad hoc comt mat & processes for electronic devices, 70-71; mem mat panel, Adv Group Electronic Parts, Off Asst Secy Defense, 56-59; dir at large, Inst Radio Eng, 59 & 62; mem adv panels, Nat Acad Sci, Nat Acad Eng & Nat Res Res Coun to Inst Appl Technol, Nat Bur Standards, 69-75, consult, comt electronic technol issues study, 72-73, chmn panel evaluate electronic technol div, 72-75; mem, US-India Nat Acad Sci Workshop Indust Res Mgt, 70; chmn US Nat Acad Sci deleg to Ceylon, Indust Res Mgt Workshop, 70; mem aeronaut & space eng bd, Nat Acad Eng, 70-72; mem, Nat Acad Eng Comn Int Activ, 70-71; trustee, Brown Univ, 69-74, emer trustee, 74-, chmn corp comt comput educ, 71-75, mem, 76-81; trustee, Baylor Univ, 70-79, med ctr, Dallas, 70-79; mem adv coun, Col Arts & Sci, Univ Tex, Austin, 72-78, Col Educ, 77-, Col Nat Sci, 79-; contribr, Comt Surv Mat Sci & Eng, Nat Acad Sci, 72-75; mem US Nat Acad Sci deleg to Joint Repub China-US Workshop Indust Innovation & Prod Develop, Taiwan, 75; chmn panel, Nat Acad Sci-Nat Res Coun, res facil & sci opportunities in use of low energy neutrons, 77-78. *Honors & Awards:* Inventor of the Year Award, Patent, Trademark & Copyright Res Inst, George Washington Univ, 66; Golden Plate Award, Am Acad Achievement, 67; Cert Appreciation & Honor Scroll, Nat Bur Standards & US Dept Com, 67; Medal of Honor, Inst Elec & Electronics Engrs, 68; Creative Invention Award, Am Chem Soc, 70; Inst Elec & Electronics Engrs Centennial Medal, 84; Semmy Award, Semiconductor Mat & Equip Inst, 84. *Mem:* Nat Acad Eng; fel AAAS; fel Inst Elec & Electronics Engrs; fel Am Inst Chem; Am Chem Soc; Am Phys Soc; Electrochem Soc; emer mem Indust Res Inst; Sigma Xi. *Res:* Raman spectra deuterium isotopic effects; photoelectric and secondary emission phenomena; pyrolytically deposited hard or semiconducting films; microwave attenuator materials; silicon carbide varistors; germanium and silicon single crystals; transistors; junction transistor; recipient of 64 patents from US and abroad. *Mailing Add:* 5222 Park Ln Dallas TX 75220

TEAL, JOHN MOLINE, b Omaha, Nebr, Nov 9, 29; m 50, 79; c 2. COASTAL & WETLANDS ECOLOGY. *Educ:* Harvard Univ, AB, 51, MA, 52, PhD, 55. *Prof Exp:* Asst prof zool, Marine Inst, Univ Ga, 55-59; asst zool & oceanog, Inst Oceanog, Dalhousie Univ, 59-61; assoc scientist, 61-71, SR SCIENTIST, WOODS HOLE OCEANOG INST, 71- *Mem:* AAAS; Am Soc Limnol & Oceanog; Ecol Soc Am. *Res:* Ecology, chemical cycling, waste treatment productivity of coastal wetlands; hydrocarbon biogeochemistry; coastal pollution. *Mailing Add:* Woods Hole Oceanog Inst Woods Hole MA 02543

TEANEY, DALE T, b Monrovia, Calif, May 19, 33; div; c 3. SOLID STATE PHYSICS, ACOUSTICS. *Educ:* Pomona Col, BA, 55; Univ Calif, Berkeley, PhD(physics), 60. *Prof Exp:* Res assoc physics, Atomic Energy Res Estab, Eng, 60-62; res staff mem, Watson Res Ctr, IBM Corp, 62-83; prof elec eng, NJ Inst Tech, 83-90; PRES, SYNCHROVOICE INC, 90- *Concurrent Pos:* Staff scientist, Nat Acad Sci Phys Surv Comt, 70-71; hon res fel, Univ Col, London, 80-81; Voice Found Fel, 80-81. *Mem:* AAAS; fel Am Phys Soc; NY Acad Sci. *Res:* Magnetic resonance; calorimetry; liquid crystals; lipid bilayers; acoustic spectroscopy; voice science. *Mailing Add:* Synchrovoice Inc 400 Harrison Ave Harrison NJ 07029

TEARE, FREDERICK WILSON, b Lacombe, Alta, June 9, 25; m 50; c 4. PHARMACEUTICAL CHEMISTRY, RADIOPHARMACY. *Educ:* Univ Alta, BSc, 49, MSc, 51; Univ NC, PhD(pharmaceut chem), 55. *Prof Exp:* Res pharmacist, Chas Pfizer Co, NY, 55-57; from asst prof to assoc prof pharmaceut chem, Univ Toronto, 57-77, prof, 77-; RETIRED. *Concurrent Pos:* Drugs Metab Labs, Ciba Ltd, Switz, 69-70. *Mem:* Can Pharmaceut Asn; Soc Nuclear Med; Can Inst Chem; Asn Fac Pharm Can. *Res:* Toxicology; radiopharmaceutics; quantitative analysis of drugs and metabolites in products and in biological media; drug metabolism; electrochemical and chemical radiocodination of various drugs (or substrates) used as potential diagnostic agents or to study and/or optimize parameters for their synthesis; separation and purification; characteriztion and stability of new compounds. *Mailing Add:* 17 Clansman Blvd Willodale ON M2H 1X5 Can

TEARE, IWAN DALE, b Moscow, Idaho, July 24, 31; m 52; c 4. INTEGRATED PEST MANAGEMENT. *Educ:* Univ Idaho, BS, 53; Wash State Univ, MS, 59; Purdue Univ, PhD(crop physiol, ecol), 63. *Prof Exp:* County agent, Idaho, 56-57; instr agron, Purdue Univ, 61-63; asst prof, Wash State Univ, 63-69; assoc prof agron, Kans State Univ, 69-77, prof, 77-79; prof agr & dir, Agr Res & Educ Ctr, Quincy-Marianna, 79-82, RES SCHOLAR & SCIENTIST, NFLA RES & EDUC CTR, UNIV FLA, QUINCY, 82- *Concurrent Pos:* Vis prof, Duke Univ, 78-79; assoc ed, Agron J, 79-85, tech ed, 85-; bio-space technol training prog, Univ Va, 69. *Mem:* Crop Sci Soc Am; fel Soc Agron; Sigma Xi; Entom Soc Am. *Res:* Modeling crop responses to insect and disease pests, the microclimate and the soil-plant-air continuum; developing hardware and software for conducting integrated pest and crop management research. *Mailing Add:* NFla Res & Educ Ctr Univ Fla Rte 3 Box 4370 Quincy FL 32351-9500

TEARNEY, RUSSELL JAMES, b Aug 10, 38; m; c 3. HYPERTENSION. *Educ:* Howard Univ, PhD(physiol), 73. *Prof Exp:* DIR PHYSIOL LABS, MED SCH, HOWARD UNIV, 80-; ADJ ASSOC PROF, UNIV DIST COLUMBIA. *Mem:* Porter fel Am Physiol Soc; Am Col Sports Med; Am Hypertension Soc. *Res:* Vascular dispensibility in hypertensive subjects. *Mailing Add:* Dept Physiol & Biophys Howard Univ 520 W St NW Washington DC 20059

TEAS, DONALD CLAUDE, physiological psychology, for more information see previous edition

TEAS, HOWARD JONES, b Rolla, Mo, Sept 4, 20; m 42; c 4. GENETICS. *Educ:* La State Univ, AB, 42; Stanford Univ, MA, 46; Calif Inst Technol, PhD(genetics), 47. *Prof Exp:* Asst genetics, Carnegie Inst, 42-43; biologist, Oak Ridge Nat Lab, 47-48; res fel, Calif Inst Technol, 48-49, sr res fel, 50-53; plant physiologist, USDA, 53-56; assoc prof biochem, Univ Fla, 56-60; head agr bio-sci div, Nuclear Ctr, Univ PR, 60-62; prog dir, NSF, 62-64; chmn div biol sci, Univ Ga, 64-67; PROF BIOL, UNIV MIAMI, 67- *Concurrent Pos:* Mem bd dirs, Orgn Trop Studies, 67-72. *Mem:* AAAS; Ecol Soc Am; Am Soc Plant Physiol; Am Soc Biol Chem; Radiation Res Soc. *Res:* Plant physiology; tropical biology; physiological ecology. *Mailing Add:* Dept Biol 215 Cox Sci Bldg Univ Miami Coral Gables FL 33139

TEASDALE, JOHN G, b Utah, June 11, 13; m 42; c 3. PHYSICS. *Educ:* Univ Calif, Los Angeles, AB, 36, PhD(physics), 50. *Prof Exp:* Physicist, US Navy Radio & Sound Lab, 41 & Manhattan Proj, Radiation Lab, Univ Calif, 42-45; res fel, Calif Inst Technol, 50-52, sr res fel, 52-56; from asst prof to assoc prof physics, 56-62, PROF PHYSICS, SAN DIEGO STATE UNIV, 62- *Concurrent Pos:* Consult, Convair Div, Gen Dynamics Corp, 60. *Mem:* Am Phys Soc; Am Inst Physics; Am Asn Physics Teachers; Sigma Xi. *Res:* Nuclear physics. *Mailing Add:* 23732 Villena Mission Viejo CA 92692

TEASDALE, WILLIAM BROOKS, b Brownsville, Pa, July 19, 39; m 63; c 3. ORGANIC CHEMISTRY. *Educ:* Geneva Col, BS, 61. *Prof Exp:* Prod chemist, 61-62, develop chemist, 62-73, sr develop chemist, 73-76 & Dept Tech Staff, 76-77, SR DEVELOP CHEMIST, DIV TECH STAFF, EASTMAN KODAK CO, 77- *Mem:* Am Chem Soc; Soc Photog Scientists & Engrs. *Res:* Organic chemical processes to be used in the production of photographic chemicals. *Mailing Add:* 104 Paddy Hill Dr Rochester NY 14616-1138

TEASDALL, ROBERT DOUGLAS, neurology; deceased, see previous edition for last biography

TEATE, JAMES LAMAR, b Moultrie, Ga, Mar 4, 32; m 53; c 3. FOREST ECOLOGY. *Educ:* Univ Ga, BS, 54, MF, 56; NC State Univ, PhD(forestry, ecol), 67. *Prof Exp:* Info & educ forester, Fla Forest Serv, 56-58; instr forestry & res asst, Auburn Univ, 58-60; instr & asst forester, Miss State Univ, 60-62; asst prof forestry, Wis State Univ-Stevens Point, 65-67; assoc prof forest recreation, Okla State Univ, 67-76; res specialist, Okla Agr Exp Sta, 67-76; DIR & PROF, SCH FORESTRY, LA TECH UNIV, 76- *Concurrent Pos:* Proj consult statewide comprehensive outdoor recreation plan, Okla Indust Develop & Parks Dept, 69-70; chmn, Southern Regional Educ Comt, mem, Nat Educ Comt, Nat Asn Prof Forestry Schs & Cols, 88-89. *Mem:* Fel Soc Am Foresters; Forest Farmers Asn; Sigma Xi; Am Forestry Asn; Int Soc Trop Foresters. *Mailing Add:* Sch Forestry La Tech Univ Ruston LA 71272

TEATER, ROBERT WOODSON, b Ky, Feb 27, 27; m 52; c 4. AGRONOMY. *Educ:* Univ Ky, BS, 51; Ohio State Univ, MS, 55, PhD(agron), 57. *Prof Exp:* Asst prof agron, Ohio State Univ & Agr Exp Sta, 57; exec asst to dir, Ohio Dept Natural Resources, 61-63, asst dir, 63-69; dir, Ohio Dept Natural Resources, 75-83; dir, Sch Natural Resources, 71-74, chmn & prof, Dept Natural Resources & prof agron, 73-74, ASSOC DEAN, COL AGR & HOME ECON, OHIO STATE UNIV, 69- *Concurrent Pos:* Pres, Robert W Teater & Assocs, 83- *Mem:* Am Soc Agron; Soil Conserv Soc Am. *Res:* Soil fertility; plant nutrition; conservation; natural resources; nature conservancy; environmental science. *Mailing Add:* 286 W Wisheimer Rd Columbus OH 43214

TEBBE, DENNIS LEE, b St Louis, Mo, Oct 21, 42; m 64; c 2. ELECTRICAL ENGINEERING, STATISTICS. *Educ:* Univ Mo-Columbia, BS, 64, MS, 65, PhD(elec eng, statist), 68. *Prof Exp:* Asst prof elec eng, Univ Mo-Columbia, 68-74; scientist, Geometric Data, Div Smithkline Corp, 74-82; SR PRIN ENGR, GCS DIV, HARRIS CORP, 82- *Mem:* Inst Elec & Electronics Engrs; AAAS. *Res:* Pattern recognition; information theory and multivariate data analysis; medical instrumentation. *Mailing Add:* Harris Corp GCS Div PO Box 91000 MS 1-5361 Melbourne FL 32902

TE BEEST, DAVID ORIEN, b Baldwin, Wis, Nov 9, 46; m 72; c 2. PLANT PATHOLOGY. *Educ:* Univ Wis-Stevens Point, BS, 68, Univ Wis-Madison, MS, 71, PhD(plant path), 74. *Prof Exp:* Res asst plant path, Univ Wis-Madison, 68-74; res assoc, 75-78, from asst prof to assoc prof, 78-85, PROF, UNIV ARK, FAYETEVILLE, 85- *Honors & Awards:* Super Serv in Res, USDA, 90. *Mem:* Am Phytopath Soc; Int Soc Plant Path; Sigma Xi; Am Soc Microbiol. *Res:* Biological control of weeds with plant pathogens; ecological epidemiology; physiology of plant disease. *Mailing Add:* Dept Plant Path Univ Ark Fayetteville AR 72701

TEBO, HEYL GREMMER, b Atlanta, Ga, Oct 17, 16; m 40. ANATOMY. *Educ:* Oglethorpe Univ, AB, 37, MA, 39; Emory Univ, DDS, 47. *Prof Exp:* Instr anat, Oglethorpe Univ, 38-39; teaching fel oral surg & anat, Univ Tex Dent Br Houston, 47-48, instr anat & surg, 48-50; asst prof diag & radiol, Sch

Dent, Univ Ala, Birmingham, 50-52; asst chief dent serv, Vet Admin Hosp, Houston, Tex, 52-61; chmn gross anat, 78-84, prof anat, Univ Tex Dent Br-Houston, 62-, prof anat, Med Sch, 74-, adj prof anat, Dent Br & Med Sch, 84-; RETIRED. *Concurrent Pos:* Clin assoc prof, Univ Tex Dent Br Houston, 52-61; consult, Vet Admin Hosp, Houston, Tex, 62- *Mem:* Fel AAAS; Am Asn Anat; Int Asn Dent Res; Am Acad Dent Radiol; Am Asn Phys Anthrop. *Res:* Osteology of head; radiographic anatomy; personality characteristics of patients; oral pathology related to radiography of head. *Mailing Add:* 5822 Queensloca Dr Houston TX 77092

TECHO, ROBERT, b New York, NY, Jan 1, 31; m 55; c 3. INFORMATION SCIENCE, DATA COMMUNICATIONS. *Educ:* Ga Inst Technol, BChE, 53, MS, 58, PhD(chem eng), 61. *Prof Exp:* Sr res engr, Eng Exp Sta, Ga Inst Technol 59-65; assoc prof, 69-74, PROF INFO SYSTS, GA STATE UNIV, 74- *Concurrent Pos:* Consult chem eng & computer anal, 65- *Mem:* Am Chem Soc; Am Inst Chem Eng; Asn Comput Mach. *Res:* Computer applications of engineering problems, including systems analysis for pipeline operations and hydraulic transients; computer science, including teleprocessing information systems, data communications design and computer communication networks. *Mailing Add:* PO Box 11802 University Plaza Atlanta GA 30305

TECKLENBURG, HARRY, b Seattle, Wash, Nov 3, 27; m 51; c 2. CHEMICAL ENGINEERING. *Educ:* Mass Inst Technol, BS, 50; Univ Wash, MS, 52. *Prof Exp:* Chem engr, Procter & Gamble Co, 52-54, group leader soap process develop, 54-57, sect leader res & develop, 57-60, dir paper prod develop, 60-61, mgr paper mfg & prod develop, 61-67, dir res & develop, 67-69, mgr, 69-70, vpres, 70-76, sr vpres res & develop, 76-84, SR VPRES & GEN MGR, NORWICH DIV, PROCTER & GAMBLE CO, 84- *Mem:* AAAS; Am Inst Chem Engrs; Am Chem Soc; Tech Asn Pulp & Paper Indust. *Res:* Research administration. *Mailing Add:* 78 S Broad St Norwich NY 13815-1736

TECOTZKY, MELVIN, b Chicago, Ill, Feb 17, 24; m 56; c 2. INORGANIC CHEMISTRY. *Educ:* Univ Ill, BS, 48, PhD(chem), 53. *Prof Exp:* Res asst inorg chem, Univ Ill, 51-53, fel, 53-54; res chemist, Diversey Corp, 54-56; sr chemist, W R Grace & Co, 56-59; res chemist & proj leader, FMC Corp, 59-61; staff scientist, Missiles & Space Div, Lockheed Aircraft Corp, Calif, 61-68; dir res, Chem Prod Div, Radium Corp, 68-80; vpres, Optonix Inc, Hackettstown, NJ, 80-84; vpres, Digirad Corp, Palo Alto, Calif, 84-88, consult, AGFA, 88-91; PVT CONSULT, 91- *Mem:* Electrochem Soc; Soc Info Display. *Res:* Rare earths; solid state chemistry; luminescence; chelates; thorium; uranium; sulfides; hydrazine; phosphates; transition elements; electronic materials; magnetic materials; non-aqueous solvents. *Mailing Add:* 27 N Linden Lane Mendham NJ 07945

TEDESCHI, DAVID HENRY, b Newark, NJ, Feb 20, 30. PHARMACOLOGY. *Educ:* Rutgers Univ, BSc, 52; Univ Utah, PhD(pharmacol), 55. *Prof Exp:* Asst pharmacol, Univ Utah, 52-54; assoc dir pharmacol, Smith Kline & French Labs, Pa, 55-68; dir pharmacol, Geigy Pharmaceut, NY, 68-70, dir pharmacol & dep dir biol res, Ciba-Geigy Corp, 70-72; dir cent nerv syst dis ther, Res Sect, Lederle Labs Div, Am Cyanamid Co, 72-77; dir res, Biobasics, 77-78; dir biosci res, 78-88, STAFF SCIENTIST BIOSCI, 3M CO, 89- *Honors & Awards:* Philemon Hommell Prize Pharmacol, 52; Am Pharmaceut Asn Found Award in Pharmacodynamics, 69. *Mem:* Am Col Neuropsychopharmacol; Am Soc Pharmacol & Exp Therapeut; Soc Exp Biol & Med; Int Col Neuropsychopharmacol; Int Soc Biochem Pharmacol. *Res:* Neuropsychopharmacology; site and mechanism of action of drugs on central nervous system. *Mailing Add:* Basic Lab Life Sci Sector 3M Co 3M Ctr Bldg 270-2A-08 St Paul MN 55144

TEDESCHI, HENRY, b Novara, Italy, Feb 3, 30; nat US; m 57; c 3. PHYSIOLOGY. *Educ:* Univ Pittsburgh, BS, 50; Univ Chicago, PhD(physiol), 55. *Prof Exp:* Res assoc & instr, Univ Chicago, 55-57, asst prof, 57-60; from asst prof to assoc prof physiol, Univ Ill Col Med, 60-65; chairperson dept biol sci, 82-85, PROF BIOL, STATE UNIV NY, ALBANY, 65- *Concurrent Pos:* NIH sr res fel, Oxford Univ, 71-72. *Mem:* Biophys Soc; Am Soc Cell Biol; Am Physiol Soc; Soc Gen Physiol; Am Soc Biol Chem. *Res:* Cell physiology; structural and functional organization of the cell; intracellular membranes. *Mailing Add:* Dept Biol Sci State Univ NY 1400 Washington Ave Albany NY 12222

TEDESCHI, RALPH EARL, b Newark, NJ, Nov 20, 27; m 51; c 2. PHARMACOLOGY. *Educ:* Rutgers Univ, BS, 51; Med Col Va, PhD(pharmacol), 54. *Prof Exp:* Resident pharmacol, Oxford Univ, 54-55; res assoc, Div Metab Res, Jefferson Med Col, 55-56; assoc dir pharmacol, Smith Kline & French Labs, 56-69; head, Dept Pharmacol, Wm S Merrell Co, Ohio, 69-71; head dept, Dow Chem Co, Zionsville, 71-72, dir clin pharmacol, 72-73, dir develop, 73-74, tech asst to dir pharmaceut res & develop, 74-75, tech asst to med dir, Human Health Res & Develop Labs, 75-, GROUP DIR CLIN RES, MARION-MERRELL DOW. *Mem:* Am Soc Pharmacol & Exp Therapeut; Acad Pharmaceut Sci; Am Pharmaceut Asn; Soc Exp Biol & Med. *Res:* Neuropharmacology; pharmacology of the autonomic nervous system; cardiovascular pharmacology. *Mailing Add:* Merrell Dow Rest Inst 10123 Alliance Rd Suite 105 Cincinnati OH 45242-9553

TEDESCHI, ROBERT JAMES, b Woodside, NY, July 25, 21; m 52; c 3. ORGANIC CHEMISTRY. *Educ:* Cornell Univ, AB, 44, MS, 45, PhD(org chem), 47. *Prof Exp:* Microanal chemist, Wyeth Inst, Pa, 46; res chemist, Calco Chem Div, Am Cyanamid Co, NJ, 47-53; sect head, Cent Res Labs, Air Reduction Co, 53-59, proj leader, Air Prod & Chem Co, 59-64, supvr chem div, Airco Chem & Plastics Div, 64-69, supvr org chem, 69-71, dir res & develop, Acetylenic Chem Div, 71-74, assoc dir res, Acetylenic Chem Div, Air Prod & Chemicals, Inc, Middlesex, 74-79; PRES, TEDESCHI & ASSOCS, 80- *Concurrent Pos:* Invited lectr, Acetylene Chem, USSR Acad Sci, Irbutok, Siberia, 87. *Mem:* Am Chem Soc; fel Am Inst Chem; NY Acad Sci. *Res:* Acetylene chemistry; high pressure synthesis; catalytic reactions; organic chemicals development; industrial applications for acetylenic chemicals; organo metallics; chelates and complexes; corrosion inhibitors; perfumery intermediates; agricultural chemicals; acetylenic surfactants; specialty monomers and polymers. *Mailing Add:* 21 Court St Winslow ME 04901

TEDESCO, FRANCIS J, b Derby, Conn, March 8, 44; m 70; c 1. GASTROENTEROLOGY. *Educ:* Fairfield Univ, BS, 65; St Louis Univ Sch Med, MD, 69. *Prof Exp:* Asst instr, Hosp Univ Pa, 71-72; asst prof, Wash Univ Sch Med, 74-75; from asst prof to assoc prof, Univ Miami, 75-78, co-dir clin res, 76-78; assoc prof, 78-81, PROF, MED COL GA, 81- *Concurrent Pos:* Chief gastroenterol, Med Col Ga, 78-88, actg vpres clin activ, 84, vpres clin activ, 84-88, interim dean, 86-88, pres, 88; consult, Dwight D Eisenhower Army Med Ctr, Vet Admin Med Ctr, Augusta, Walter Reed Army Med Ctr; chmn Nat Digestive Dis Adv Bd, 87-88. *Honors & Awards:* Eddie Palmer Award Gastrointestinal Endoscopy, 83. *Mem:* Am Soc Gastrointestinal Endoscopy (treas, 81-84, pres, 85-86); Am Col Physicians; Am Col Gastroenterol. *Res:* Gastroenterology. *Mailing Add:* Pres Off Med Col Ga Augusta GA 30912

TEDESCO, THOMAS ALBERT, b York, Pa, Dec 5, 35. HUMAN GENETICS, REPRODUCTION. *Educ:* Franklin & Marshall Col, BS, 60; Univ Pa, PhD(biol, genetics), 69. *Prof Exp:* Res technician pediat, Hosp, 60-61, res asst, 61-62, res assoc genetics, 62-65, from instr to asst prof pediat, Sch Med, 65-72, asst prof pediat & med genetics, 72-74; asst prof, Col Med, Univ SFla, 74-81, assoc prof, 81-90, assoc prof with tenure, 79, PROF PEDIAT, COL MED, UNIV SFLA, 90-, DIR, PEDIAT LABS, 85-; DIR, EMBRYOL & ANDROLOGY LABS, IVF/GIFT PROG, UNIV SFLA/HUMANA WOMENS HOSP, TAMPA, 85- *Concurrent Pos:* Prin investr, NIH grant; prog dir, Nat Found-March Dimes Med Serv grant; co-dir, Regional Genetics Prog, Children's Med Serv, Dept Health & Rehab Serv, State Fla. *Mem:* NY Acad Sci; AAAS; Am Soc Human Genetics; Am Chem Soc; Am Genetics Asn; Am Fertility Soc. *Res:* Inborn errors in metabolism; early embryonic development; prenatal and preimplantation genetic diagnosis. *Mailing Add:* Dept Pediat Univ SFla Med Ctr 12901 Bruce B Downs Blvd Tampa FL 33612

TEDESKO, ANTON, b Gruenberg, Ger, May 25, 03; nat US; m 38; c 2. STRUCTURAL ENGINEERING. *Educ:* Inst Technol, Vienna, Austria, CE, 26, DSc, 51; Univ Berlin, dipl eng, 30. *Hon Degrees:* DEng, Lehigh Univ, 66; DTechSc, Univ Vienna, 78. *Prof Exp:* Construct engr, Vienna, 26; mem staff, Fairbanks-Morse Co, Chicago, 27 & Miss Valley Struct Steel Co, Melrose Park, Ill, 27-28; asst prof, Inst Technol, Vienna, 29; designer dams, bridges, shells and indust struct, Dyckerhoff & Widmann, Wiesbaden, Ger, 30-32; mem staff, Roberts & Schaefer Co, Engrs, Chicago, 32-43, eng mgr, Washington, DC, 43-44, struct mgr, Chicago, 44-54, mgr, New York, 55, vpres, 56-67; CONSULT ENGR, 67- *Concurrent Pos:* Consult, Hq USAF, 55-70; dir, Thompson-Starrett Co, 60-61; struct eng manned lunar landing prog assembly & launch facil, Kennedy Space Ctr, 62-66; mem & arbitrator, Res Coun Performance Struct. *Honors & Awards:* Alfred Lindau Award, 61; Arthur J Boase Award, Reinforced Concrete Res Coun, 78; Int Award Merit Struct Eng, 78; Henry C Turner Medal, 88. *Mem:* Nat Acad Eng; hon mem Am Soc Civil Engrs; hon mem Int Asn Shell & Spatial Struct; Int Asn Bridge & Struct Eng; hon mem Am Concrete Inst. *Res:* Design and construction of arenas, air terminals, bridges, toll roads, industrial structures, ballistic missile and space rocket launching facilities, wide-span hangars and shell structures; concrete structures of long span; prestressed concrete structures. *Mailing Add:* 26 Brookside Circle Bronxville NY 10708

TEDFORD, RICHARD HALL, b Los Angeles, Calif, Apr 25, 29; m 54. VERTEBRATE PALEONTOLOGY, STRATIGRAPHY. *Educ:* Univ Calif, Los Angeles, BS, 51; Univ Calif, Berkeley, PhD(paleont), 60. *Prof Exp:* Instr geol, Univ Calif, Riverside, 59-60, lectr, 60-61, from asst prof to assoc prof, 61-66; assoc cur vert paleont, 66-69, CUR VERT PALEONT, AM MUS NATURAL HIST, 69-, CHMN DEPT, 77- *Mem:* Australian Mammal Soc; Soc Vert Paleont; Paleont Soc; Am Soc Mammal; Sigma Xi. *Res:* Phylogeny, geographic distribution and paleoecology of Carnivora, Marsupials and other mammals; stratigraphy and chronology of Cenozoic rocks. *Mailing Add:* Dept Vert Paleont Am Mus Natural Hist New York NY 10024

TEDMON, CRAIG SEWARD, JR, metallurgy, electrochemistry, for more information see previous edition

TEDROW, JOHN CHARLES FREMONT, b Rockwood, Pa, Apr 21, 17; m 43; c 2. SOILS. *Educ:* Pa State Univ, BS, 39; Mich State Univ, MS, 40; Rutgers Univ, PhD, 50. *Prof Exp:* Jr soil surveyor, Soil Conserv Serv, USDA, 41, soil scientist, 46-47; from instr to prof, 48-83, EMER PROF SOILS, RUTGERS UNIV, NEW BRUNSWICK, 83- *Concurrent Pos:* Sr pedologist, Arctic Soil Invests, 53; consult indust & govt; prin investr, Arctic Inst NAm, 55-67; Antarctic pedologic investr, NSF, 61-63; ed in chief, Soil Sci, 69-79; Lindback res award, Rutgers Univ. *Mem:* Fel Soil Sci Soc Am; Am Arbit Asn; Am Geophys Union; fel Am Soc Agron; Sigma Xi; fel Arctic Inst NAm. *Res:* Soil morphology; genesis and survey; soils of the Arctic and Alpine regions. *Mailing Add:* Dept Environ Resources Rutgers Univ New Brunswick NJ 08903

TEEBOR, GEORGE WILLIAM, b Vienna, Austria, July 22, 35; US citizen; m 58; c 3. CARCINOGENESIS, DNA REPAIR MECHANISMS. *Educ:* Yale Univ, BS, 56; Yeshiva Univ, MD, 61. *Prof Exp:* PROF PATH, NY UNIV MED CTR, 65- *Concurrent Pos:* Merit Award, NCI-NIH, 88. *Mem:* Am Asn Cancer Res; Am Asn Pathologists; Am Chem Soc; Radiation Res Soc. *Res:* Mammalian DNA repair enzymology; characterization of DNA damage. *Mailing Add:* Dept Path Med Sci Bldg Rm 605 NY Univ Med Ctr 550 First Ave New York NY 10016-6451

TEEGARDEN, BONNARD JOHN, b Elizabeth, NJ, Aug 23, 40; m 62; c 2. ASTROPHYSICS. *Educ:* Mass Inst Technol, BS, 62; Univ Md, PhD(physics), 67. *Prof Exp:* PHYSICIST ASTROPHYS, NASA GODDARD SPACE FLIGHT CTR, 63- *Mem:* Am Phys Soc; Am Astron Soc. *Res:* Gamma-ray astronomy. *Mailing Add:* NASA Goddard Space Flight Ctr Code 661 Greenbelt MD 20771

TEEGARDEN, DAVID MORRISON, b Dayton, Ohio, Jan 10, 41; m 66; c 3. MONOMER & POLYMER SYNTHESIS. *Educ:* Ohio Wesleyan Univ, AB, 63; Univ Mich, MS, 65, PhD(org chem), 72. *Prof Exp:* Asst prof chem, Univ Wis-Platteville, 69-73; from asst prof to prof chem, John Fisher Col, 73-86; mem res staff, Xerox Webster Res Ctr, 80-82; RES CHEMIST, EASTMAN KODAK CO, 86- *Concurrent Pos:* Fel, Xerox Webster Res Ctr, 79-80. *Mem:* Am Chem Soc. *Res:* Synthesis and characterization of functional monomers and polymers; polymer blends, photoresists. *Mailing Add:* 159 Village Lane Rochester NY 14610

TEEGARDEN, KENNETH JAMES, b Chicago, Ill, May 13, 28; m 59. PHYSICS. *Educ:* Univ Chicago, AB, 47, BS, 50; Univ Ill, MS, 51, PhD, 54. *Prof Exp:* Res assoc, 54-58, asst prof, 58-59, sr res assoc, 60-61, assoc prof, 61-66, PROF OPTICS, INST OPTICS, UNIV ROCHESTER, 66-, DIR, INST OPTICS, 81-, DIR, CTR ADV OPTICAL TECHNOL, 83- *Concurrent Pos:* Alfred P Sloan Found fel, Univ Rochester, 59-63. *Mem:* Fel Am Phys Soc; fel Optical Soc Am. *Res:* Electronic properties of ionic solids and the solid rare gases. *Mailing Add:* 82 Westland Ave Rochester NY 14618

TEEGUARDEN, DENNIS EARL, b Gary, Ind, Aug 21, 31; m 54; c 3. FORESTRY ECONOMICS. *Educ:* Mich Tech Univ, BS, 53; Univ Calif, Berkeley, MF, 58, PhD(agr econ), 64. *Prof Exp:* Res asst, Pac Southwest Forest & Range Exp Sta, US Forest Serv, Berkeley, Calif, 57; asst specialist, Agr Exp Sta, 58-63, actg asst prof, Sch Forestry, 63, from asst prof to assoc prof, 64-73, chmn, dept forestry & resource mgt, 78-86, actg dir, Forest Prod Lab, 87-88, PROF FORESTRY, SCH FORESTRY, UNIV CALIF, BERKELEY, 73-, ASSOC DEAN ACAD AFFAIRS, COL NAT RESOURCES, 90- *Mem:* Fel Soc Am Foresters. *Res:* Application of operations research techniques to problems of resource allocation in public and private forestry enterprises. *Mailing Add:* Dept Forestry & Resource Mgt Univ Calif Berkeley CA 94720

TEEKELL, ROGER ALTON, b Elmer, La, Mar 3, 30; m 53; c 4. METABOLISM, BIOCHEMISTRY. *Educ:* La State Univ, BS, 51, MS, 55, PhD(nutrit, biochem), 58. *Prof Exp:* Asst, La State Univ, 54-58; res scientist, Agr Res Lab, Univ Tenn-AEC, 58-61; res chemist, Dow Chem Co, Tex, 61-63; from assoc prof to prof physiol, La State Univ, Baton Rouge, 63-67, prof poultry sci, 67-76; head, Dept Poultry Sci, Va Polytech Inst & State Univ, Blacksburg, 76-79, assoc dean, Grad Sch, 79-82, dean, 82-90, PROF POULTRY SCI, VA POLYTECH INST & STATE UNIV, BLACKSBURG, 90- *Mem:* Am Inst Biol Sci; Sigma Xi; Poultry Sci Asn; World Poultry Sci Asn. *Res:* Intermediary metabolism of amino acids and lipids using labeled compounds. *Mailing Add:* Poultry Sci Dept Va Polytech Inst & State Univ 3310 Litton Reaves Hall Blacksburg VA 24061-0332

TEERI, ARTHUR EINO, b Dover, NH, July 29, 16; m 41; c 2. BIOCHEMISTRY. *Educ:* Univ NH, BS, 37, MS, 40; Rutgers Univ, PhD(biochem), 43. *Prof Exp:* Asst physiol chem, Univ NH, 38-40; res fel biochem, Res Labs, US Dept Interior, 40-41; asst prof, 43-53, PROF BIOCHEM, UNIV NH, 53- *Mem:* Fel AAAS; Am Chem Soc; Am Inst Nutrit; NY Acad Sci. *Res:* Physiological chemistry; animal and human nutrition; vitamins; clinical methods. *Mailing Add:* Prof Biochem Univ NH PO Box 146 Durham NH 03824

TEERI, JAMES ARTHUR, b Exeter, NH, Feb 28, 44; m 67. ECOLOGY, POLAR BIOLOGY. *Educ:* Univ NH, BS, MS, 68; Duke Univ, PhD(bot), 72. *Prof Exp:* Asst prof biol, Univ Chicago, 72-88; ASST VPRES PERSONNEL, OFF VPRES & CHIEF FINANCIAL OFFICER, UNIV MICH, ANN ARBOR, 88- *Concurrent Pos:* Assoc ed, Paleobiol, 74-; mem, Comn Optical Radiation Measurement, 74; actg chmn, Comt Evol Biol, 78. *Mem:* Fel AAAS; Sigma Xi; Ecol Soc Am; Am Inst Biol Sci; Arctic Inst NAm. *Res:* Evolution of plant growth responses to environmental fluctuation; evolutionary biology. *Mailing Add:* Univ Mich 6048 Fleming Admin BLdg Postal Code 1340 Ann Arbor MI 48109-1340

TEETER, JAMES WALLIS, b Hamilton, Ont, Mar 14, 37; m 60; c 3. MICROPALEONTOLOGY, PALEOECOLOGY. *Educ:* McMaster Univ, BSc, 60, MSc, 62; Rice Univ, PhD(paleont), 66. *Prof Exp:* From asst prof to assoc prof, 65-76, PROF GEOL, UNIV AKRON, 76- *Concurrent Pos:* Faculty res grants, Univ Akron, 69, 73, 74, 78 & 83. *Mem:* Paleont Soc; Sigma Xi; Geol Soc Am; Int Oceanog Found; Soc Econ Paleont & Mineral. *Res:* Post Pleistocene depositional history, San Salvador Island, Bahamas; Key Largo limestone facies; Ordovician Nautiloid touchmarks; Ostracoda and environments of Caloosahatchee Formation; living Pelecypod behavior; marine Ostracoda dispersal. *Mailing Add:* Dept Geol Univ Akron Akron OH 44304

TEETER, MARTHA MARY, b Boston, Mass, Oct 15, 44; m; c 2. PROTEIN CRYSTALLOGRAPHY. *Educ:* Wellesley Col, BA, 66; Pa State Univ, PhD(inorg chem), 73. *Prof Exp:* Res scientist, Rohm & Haas, 73-74; Nat Cancer Inst fel, Dept Biol, Mass Inst Technol, 74-76, Naval Res Lab, 76-77; vis asst prof phys chem, Dept Chem, Boston Univ, 77-78, instr life sci chem, 78-80, res asst prof, 80-86, ASSOC PROF, DEPT CHEM, BOSTON COL, 86- *Concurrent Pos:* Vis scientist, Dept Biol, Mass Inst Technol, 77- *Mem:* Am Chem Soc; Am Crystallog Asn; Biophys Soc; AAAS. *Res:* High resolution protein crystal structure determination (x-ray and neutron) as well as molecular dynamics; protein structure; water structure around proteins; molecular modeling. *Mailing Add:* Dept Chem Boston Col Chestnut Hill MA 02167

TEETER, RICHARD MALCOLM, b Berkeley, Calif, Feb 24, 26; div; c 2. MASS SPECTROMETRY. *Educ:* Univ Calif, BS, 49; Univ Wash, PhD(chem), 54. *Prof Exp:* Res chemist, Chevron Res Co, 54-64, sr res chemist, 64-68, sr res assoc, 68-86; RETIRED. *Concurrent Pos:* Pres, PCMASPEC. *Mem:* Am Chem Soc; Am Soc Mass Spectrometry. *Res:* Analytical mass spectrometry; preparation of derivatives to aid analysis; reaction mechanisms in mass spectrometry; application of computers to mass spectrometry. *Mailing Add:* 203 Del Valle Cir El Sobrante CA 94803-3373

TEFFT, MELVIN, b Dec 15, 32; US citizen. RADIOTHERAPY. *Educ:* Harvard Univ, AB, 54; Boston Univ, MD, 58; Am Bd Radiol, dipl, 63. *Prof Exp:* Intern med, Boston City Hosp, 58-59; resident radiol, Mass Mem Hosp, 59-62; asst, Harvard Med Sch, 62-65, instr, 66-67, clin assoc, 67-69, asst prof, 69-70, from asst prof to assoc prof radiation ther, 70-73; prof radiol, Cornell Univ, 73-75; PROF RADIATION MED, BROWN UNIV, 75-; RADIOTHERAPIST, DEPT RADIATION ONCOL & ASSOC MEM, DEPT PEDIAT, RI HOSP, 75- *Concurrent Pos:* Asst radiol, Children's Hosp Med Ctr, 62-64, radiotherapist, Med Ctr & Tumor Ther Div, 64-70, chief, Div Radiother & Nuclear Med, 67-69, radiotherapist-in-chief, Dept Radiation Ther, 69-70; consult radiol, Lemuel Shattuck Hosp, 62-68, Mass Gen Hosp, 64-71 & Boston Lying-In Hosp, 66-74; consult radiother, Lemuel Shattuck Hosp, 66-70; consult, Dept Radiother, Tufts Med Sch at Lemuel Shattuck Hosp, 69-70; mem, Children's Cancer Chemother Group A, 69-; consult, Dept Radiation Ther, Mem Hosp, New York, 72-74; asst radiol, Sch Med, Boston Univ, 62-68; NIH clin fel radiation ther, Mass Gen Hosp, 63-64; assoc radiol, Peter Bent Brigham Hosp, 66-70; assoc prof therapeut radiol, Med Sch, Tufts Univ, 69-70; assoc attend radiotherapist, Mem Hosp, 70-71, attend radiotherapist, 73-75, dir med educ, Dept Radiation Ther, 73-75; assoc mem, Sloan-Kettering Inst, 70-71; assoc radiotherapist, Mass Gen Hosp, 71-73; attend radiologist, New York Hosp, 73-75. *Mem:* Fel Am Col Radiol; Am Soc Therapeut Radiologists; Am Radium Soc; Int Soc Pediat Oncol; AMA. *Res:* Pediatric oncology, combined radiation therapy and chemotherapy to enhance local control and disease-free survival; evaluation of normal tissue sensitivity by combined modalities of treatment. *Mailing Add:* Dept Radiation Oncol RI Hosp 593 Eddy St Providence RI 02902

TEGGE, B(RUCE) R(OBERT), b Haddon Heights, NJ, May 31, 17; m 40; c 4. CHEMICAL ENGINEERING. *Educ:* Pa State Univ, BS, 38, MS, 40, PhD(chem eng), 42. *Prof Exp:* Asst petrol ref, Pa State Col, 40; sr eng assoc, Exxon Res & Eng Co, 42-86; RETIRED. *Mem:* Am Textile Inst; Am Inst Chem Engrs; Am Chem Soc; Soc Plastics Engrs. *Res:* Equilibrium relationships between sulfur dioxide and pure binary hydrocarbon mixtures; performance of solvent extraction equipment; butyl rubber; paraxylene; petroleum resins; synthetic polymer plants; polypropylene; synthetic textile fibers; synthetic fertilizers; organic fungicides; polyesters; mastics and polyisobutylenes; detergents. *Mailing Add:* 68 Prospect St Madison NJ 07940

TEGGINS, JOHN E, b Wallasey, Eng, Jan 6, 37; m 60; c 3. INORGANIC CHEMISTRY. *Educ:* Univ Sheffield, BSc, 58; Boston Univ, AM, 60, PhD(chem), 65. *Prof Exp:* Res chemist, Courtaulds Can, Ltd, 60-62; res assoc radiochem, Iowa State Univ, 65-66; from asst prof to assoc prof, 66-75, PROF CHEM, AUBURN UNIV, MONTGOMERY, 75-, HEAD DEPT, 81- *Mem:* Am Chem Soc; sr mem Chem Inst Can. *Res:* Determination of thermodynamic and kinetic data for coordination complexes in aqueous solution. *Mailing Add:* Dept Physical Sci Auburn Univ Montgomery AL 36193

TEGNER, MIA JEAN, b Santa Monica, Calif, July 7, 47; m 80; c 1. ZOOLOGY. *Educ:* Univ Calif, San Diego, BA, 69, PhD(marine biol), 74. *Prof Exp:* Res asst develop biol, 71-74, res biologist marine biol, 74-88, ASSOC RES BIOLOGIST MARINE BIOL, SCRIPPS INST OCEANOG, UNIV CALIF, SAN DIEGO, 84- *Concurrent Pos:* Researcher, Sea Grant, 74-, Develop Multispecies Mgt Kelp Bed Resources, 77-81 & Exp Abalone Enhancement Prog, 77-87; chairperson, Joint Univ Calif Sea Grant Col & Calif Dept Fish & Game Exp Abalone Enhancement Prog, 77-81. *Mem:* AAAS; Am Soc Limnol & Oceanog; Am Soc Zoologists. *Res:* Kelp forest community ecology; living marine resources of the nearshore environment; how man's activities have affected the structure and dynamics of this community. *Mailing Add:* 8840 Villa La Jolla Dr No 219 La Jolla CA 92037

TEGTMEYER, CHARLES JOHN, b Hamilton, NY, July 25, 39; m 63. ANGIOGRAPHY & INTERVENTIONAL RADIOLOGY. *Educ:* Colgate Univ, BA, 61; George Washington Univ Sch Med, MD, 65. *Prof Exp:* Intern, George Washington Univ Hosp, 65-66, residency surg, 66-68, diagnostic radiol, 68-71; fel, Peter Bent Brigham Hosp, Harvard, 71-72; from asst prof to assoc prof radiol, 72-78, from asst prof to assoc prof anat, 73-87, dir angiography & spec procedures, 74-87, PROF RADIOL, UNIV VA MED CTR, 78-, PROF ANAT & CELL DIV ANGIOGRAPY, INTERVENTIONAL RADIOL & SPEC PROCEDURES, 87- *Concurrent Pos:* Founder & coordr, Sch Special Procedure, Univ Va Med Ctr, 72, dir, radiol med students, 72-81, bus mgr, Radiol Dept, 80-; mem adv coun radiol, Minn Mining & Manufacturing Co, 78; examr, Am Bd Radiol, 79, 81, 84-86; course dir, New Dimensions Tech Space Processes, 80 & 81; vis prof, dept radiol,WVa Univ, 80, Southwestern Med Sch, 82, Dalhousie Med Sch, 82, Univ Nebr, 83, Univ Wis, 84, Thomas Jefferson Univ, 84, Col Physicians & Surgeons, Columbia Univ,85, Univ Conn, 87; vis staff, Martha Jefferson Hosp, 81-; ed pro-tem, Soc Cardiovasc Radiol, 81-; cardiovasc ed, Radiographics, 82-; prof anat & cell biol. *Mem:* Soc Cardiovasc & Interventional Radiol (secy-treas, 83, pres, 85); fel Am Col Radiol; Am Roentgen Ray Soc; Radiol Soc NAm; Sigma Xi; fel Soc Cardiovasc & Intradiol. *Res:* Angiography, arthrography, lymphography and other specialized invasive radiographic procedures including interventional radiology. *Mailing Add:* 2040 Earlysville Rd Earlysville VA 22936

TEH, HUNG-SIA, b Telok Anson, Malaysia, Oct 2, 45; Can citizen; m 69; c 2. IMMUNOLOGY. *Educ:* Univ Alta, BSc, 69, PhD(biochem), 75. *Prof Exp:* Res fel immunol, Ont Cancer Inst, Toronto, 75-77; from asst prof to assoc prof, 77-90, PROF MICROBIOL, UNIV BC, VANCOUVER, 90- *Concurrent Pos:* Mem, Basel Inst Immunol, 81-82; counr, Can Soc Immunol, 85-88; vis scientist, Basel Inst Immunol, 88. *Mem:* Can Soc Immunol; Am Asn Immunologists. *Res:* Cellular immunology; transgenic mice; T cell development; T cel tolerance. *Mailing Add:* Dept Microbiol Univ BC Vancouver BC V6T 1W5 Can

TEHON, STEPHEN WHITTIER, b Shenandoah, Iowa, Oct 20, 20; m 42; c 5. PIEZOELECTRICITY. *Educ:* Univ Ill, BS, 42, MS, 47, PhD(elec eng), 58. *Prof Exp:* Sr engr, Curtiss-Wright Corp, 47; instr elec eng, Univ Ill, 47-52; engr, Electronics Lab, Gen Elec Co, 52-60, consult engr, 60-66; res elec engr, Res Lab, Tecumseh Prod Co, 66-67; consult scientist, 67-80, prin staff scientist, Electronics Lab, Gen Elec Co, 80-87; STAFF SCIENTIST, GMK CONSULT SERV, 87- *Concurrent Pos:* Adj prof, Univ Mich, 66-67 & Syracuse Univ, 76-; vis prof, Clarkson Col Technol, Potsdam, NY, 79. *Mem:* Fel Inst Elec & Electronics Engrs; Acoust Soc Am. *Res:* Ultrasonic transducers; nondestructive testing; ultrasonic medical imaging; digital sensors. *Mailing Add:* 6056 Pine Grove Rd Cicero NY 13039

TEICH, MALVIN CARL, b New York, NY, May 4, 39. QUANTUM ELECTRONICS. *Educ:* Mass Inst Technol, SB, 61; Stanford Univ, MS, 62; Cornell Univ, PhD(quantum electronics), 66. *Prof Exp:* Res scientist, Lincoln Lab, Mass Inst Technol, 66-67; chmn, Dept Elec Eng & Comput Sci, 78-80, PROF ENG SCI, COLUMBIA UNIV, 67- *Concurrent Pos:* Dep ed, Quantum Optics. *Honors & Awards:* Browder J Thompson Mem Prize Award, Inst Elec & Electronics Engrs, 69; John Simon Guggenheim Award, 73; Just Sci Info Citation Classic Award, 81. *Mem:* Fel Am Phys Soc; fel Inst Elec & Electronics Engrs; fel Optical Soc Am; NY Acad Sci; Acoust Soc Am; fel AAAS; Asn Res Otolaryngol; Sigma Xi. *Res:* Optical and infrared detection; lightwave communications; sensory perception; quantum optics; co-author of book on photonics. *Mailing Add:* Dept Elec Eng Columbia Univ New York NY 10027

TEICHER, HARRY, b Middle Village, NY, Jan 11, 27; m 51; c 3. FOOD SCIENCE & TECHNOLOGY. *Educ:* Queens Col, NY, BS, 48; Syracuse Univ, MS, 50, PhD(chem), 53. *Prof Exp:* Asst chem, Syracuse Univ, 49-53; res chemist, Monsanto Co, 53-56, res group leader, 56-81, res mgr, res & develop dept, 81-85, tech serv mgr, 86-88, MKT TECH SERV PRIN, MONSANTO CO, 88- *Mem:* AAAS; Am Chem Soc; Sigma Xi; Am Asn Cereal Chemists; Inst Food Technol. *Res:* Silica; food technology. *Mailing Add:* Monsanto Co Technol Dept Detergents & Phosphates Div St Louis MO 63167

TEICHER, HENRY, b Jersey City, NJ, July 9, 22. MATHEMATICAL STATISTICS. *Educ:* Univ Iowa, BA, 46; Columbia Univ, MA, 47, PhD(math statist), 50. *Prof Exp:* Asst prof math, Univ Del, 50-51; from asst prof to prof math statist, Purdue Univ, 51-67; vis prof, Columbia Univ, 67-68; PROF MATH STATIST, RUTGERS UNIV, NEW BRUNSWICK, 68- *Concurrent Pos:* Vis asst prof, Stanford Univ, 55-56; vis assoc prof & mem inst math sci, NY Univ, 60-61. *Mem:* Am Math Soc; fel Inst Math Statist. *Res:* Probability and mathematical statistics, especially stopping rules, limit distributions and mixtures of distributions; law of the iterated logarithms. *Mailing Add:* Dept Statist Rutgers State Univ New Brunswick NJ 08903

TEICHER, JOSEPH D, b New York, NY, Aug 1, 17; m 42; c 2. PSYCHIATRY. *Educ:* City Col New York, BS, 33; Columbia Univ, MA, 34; NY Univ, MD, 40; NY Psychoanal Inst, cert, 51. *Prof Exp:* Consult, Pub Schs, Bronxville, NY, 46-52; assoc clin prof psychiat, 53-60, PROF PSYCHIAT & BEHAV SCI, SCH MED, UNIV SOUTHERN CALIF, 60- *Concurrent Pos:* Dir child guid clin, St Luke's Hosp, New York, 47-52; dir, Child Guid Clin Los Angeles, 52-; chief psychiat, Children's Hosp, 53-58; mem fac, Southern Calif Psychoanal Inst, 54- *Mem:* Fel Am Psychiat Asn; Am Psychoanal Asn. *Res:* Normal and atypical child development; communication with the elderly; studies in attempted suicide; alcoholism and drug abuse. *Mailing Add:* 152 S Lasky Dr Beverly Hills CA 90212

TEICHERT, CURT, b Koenigsberg, Prussia, May 8, 05; m 28. GEOLOGY, PALEONTOLOGY. *Educ:* Univ Koenigsberg, PhD(geol), 28; Univ Western Australia, DSc, 44. *Prof Exp:* Asst, Freiburg Univ, 27-29; Rockefeller fel geol, 30; res fel, Univ Copenhagen, 33-37; res lectr, Univ Western Australia, 37-46; asst chief govt geologist, Victoria Mines Dept, 46-47; sr lectr, Univ Melbourne, 47-53; prof geol, NMex Inst Mining & Technol, 53-54; geologist, US Geol Surv, 54-64, chief petrol geol lab, 54-58, staff geologist, 58-61, geol adv, US AID, Pakistan, 61-64; regents distinguished prof geol, Univ Kans, 64-75; ADJ PROF GEOL SCI, UNIV ROCHESTER, 77- *Concurrent Pos:* Fulbright traveling scholar, 51-52; guest prof, Univs Göttingen, Bonn & Freiburg, 58, Univ Tex, 61 & Free Univ Berlin, 74; consult, Caltex, 40-41, Stand Vacuum, 48-50, Australian Bur Mineral Resources, 48-51 & Shell Oil Co, 53-54; mem, Danish Exped, Greenland, 31-32; US coordr, Cento Treaty Orgn Stratig Working Group, 63-76. *Honors & Awards:* David Syme Prize, Univ Melbourne, 49; Raymond Cecil Moore Medal, Soc Econ Paleontologists & Mineralogists, 82; Paleont Soc Medal, 84. *Mem:* Fel Geol Soc Am; fel AAAS; Paleont Soc (pres, 71-72); hon corresp Australian Geol Soc (secy, 51-53); Am Asn Petrol Geol; Int Paleont Asn (pres, 76-80); Sigma Xi; hon fel Geol Soc London; hon mem Geol Soc Belg; hon mem Royal Soc Western Australia; Soc Econ Paleontologists & Mineralogists; foreign assoc Geol Soc France; foreign mem Danish Acad Sci; hon mem Ger Paleont Soc; foreign corresp Senckenberg Natural Sci Soc; foreign corresp Paleont Soc India. *Res:* Paleozoic stratigraphy and paleontology; stratigraphy of southwestern Asia; ancient & modern coral reefs; fossil cephalopods; sedimentation; paleoecology. *Mailing Add:* Dept Geol Sci Univ Rochester Rochester NY 14627

TEICHLER-ZALLEN, DORIS, b Brooklyn, NY, Mar 7, 41; m 64; c 2. HUMAN GENETICS, BIOETHICS. *Educ:* Brooklyn Col, BS, 61; Harvard Univ, AM, 63, PhD(biol), 66. *Prof Exp:* NIH fel biol, Univ Rochester, 66-69, asst prof, 69-70; asst prof, Nazareth Col, Rochester, 77-83; assoc prof humanities, 83-90, ASSOC PROF SCI STUDIES & HUMANITIES, VA POLYTECH INST & STATE UNIV, 90- *Concurrent Pos:* Asst prof biol, Univ Rochester, 74, res fel pediat, Genetics Div, Sch Med, 77-83; interim assoc dir, Ctr Progs in Humanities, Va Polytech Inst & State Univ, 86-87, dir, Choices & Challenges Proj, 84-, adj biochem & nutrit, 88-; mem, subcomt Human Gene Therapy, NIH, 89-; acad vis, Imp Col Sci, Technol & Med, London, 91; hon res fel, Wellcome Inst Hist Med, London, 91. *Honors & Awards:* Prog Execellence Award, Choices & Challenges Proj, Nat Univ Continuing Educ Asn, 88. *Mem:* AAAS; Am Soc Human Genetics; Sigma Xi; Hist Sci Soc. *Res:* Development of new methods of prenatal detection of genetic disorders; genetic and biomedical basis of chloroplast development; social and ethical issues arising in genetic and reproductive technologies; history of genetics. *Mailing Add:* 351 Lane Hall Va Polytech Inst & State Univ Blacksburg VA 24061-0227

TEICHMANN, THEODOR, b Koenigsberg, Ger, Sept 16, 23; nat US; div; c 2. APPLIED PHYSICS. *Educ:* Univ Cape Town, BSc, 43, MSc, 45; Princeton Univ, Am, 47, PhD(physics), 49. *Prof Exp:* Jr lectr elec eng, Univ Cape Town, 44-46; res assoc physics, Princeton, 50-52; res physicist, Res & Develop Labs, Hughes Aircraft Co, 52-55; mgr systs anal & simulation, Missiles & Space Div, Lockheed Aircraft Corp, 55-56, mgr nuclear physics, 56-57, sci asst to dir res, 57-60; consult scientist satellite systs, 60; mem res & develop staff, Spec Nuclear Effects Lab, Gen Atomic Div, Gen Dynamics Corp, 60-68; prin scientist, KMS Technol Ctr, Calif, 68-72, prin scientist, KMS Fusion Inc, Ann Arbor, 72-75; consult, Phys Dynamics, Inc, 75-77; SCIENTIST, BROOKHAVEN NAT LAB, 78- *Res:* Operations research; systems analysis; nuclear energy probabilistic risk; assessment and safety analysis; nuclear materials safeguards. *Mailing Add:* PO Box 424 Upton NY 11973-0424

TEICHROEW, DANIEL, b Can, Jan 5, 25; nat US; m 50; c 1. MATHEMATICS. *Educ:* Univ Toronto, BA, 48, MA, 49; Univ NC, PhD(statist), 53. *Prof Exp:* Res assoc, Univ NC, 51-52; mathematician, Nat Bur Stand, DC & Inst Numerical Anal, Univ Calif, Los Angeles, 52-55; sr electronics appln specialist, Nat Cash Register Co, 55, spec rep prod develop, 55-56, head bus systs anal, 56-57; from assoc prof to prof mgt, Grad Sch Bus, Stanford Univ, 57-64; prof orgn sci & head div, Case Western Reserve Univ, 64-68; chmn dept, 68-73, PROF INDUST ENG, UNIV MICH, ANN ARBOR, 68-; PRES, ISDOS, INC, 83- *Concurrent Pos:* Lectr, Sch Bus Admin, Univ Southern Calif, 56-57; ed sci & bus appln sect, Communications, Asn Comput Mach, 63- *Mem:* Asn Comput Mach; Inst Mgt Sci (vpres, 67-); Opers Res Soc Am; Inst Math Statist; Am Math Soc; Soc Mgt Info Systs (pres, 76). *Res:* Development and application of scientific techniques to organizational problems, particularly operations research, management science and computer techniques; computer-aided tools and techniques for system development and software engineering; problem statement languages and analyzers. *Mailing Add:* Dept Indust Eng Univ Mich Ann Arbor MI 48109-2117

TEIGER, MARTIN, b New York, Dec 30, 36; m 64; c 1. PHYSICS, ASTRONOMY. *Educ:* Columbia Univ, AB, 58, MA, 60, PhD(physics), 65. *Prof Exp:* Lectr physics, City Col New York, 61-65, instr, 65-66; from asst prof to assoc prof, 66-75, chmn dept, 69-79, PROF PHYSICS, LONG ISLAND UNIV, 75- *Mem:* AAAS; Am Phys Soc; Am Pub Health Asn; NY Acad Sci. *Res:* Planetary atmospheres and surface environments; radiative transfer theory; numerical methods for computers; environmental management. *Mailing Add:* 125 Ann St Valley Stream NY 11580

TEIPEL, JOHN WILLIAM, b Covington, Ky, Feb 17, 43; m 66; c 2. BIOCHEMISTRY, IMMUNOASSAYS. *Educ:* Rockhurst Col, BA, 64; Duke Univ, PhD(biochem), 68. *Prof Exp:* Am Cancer Soc fel biochem, Univ Calif, Berkeley, 68-70; asst prof chem & biochem, Univ Ill, Urbana, 70-72; scientist, 72-74, sr scientist, 74-75, prin scientist, 75-76, RES DIR, ORTHO DIAG SYSTS, INC, 76- *Mem:* Am Chem Soc. *Res:* Physical biochemistry of proteins and enzymes; labelled immunoassays; immunochemistry. *Mailing Add:* Ortho Diag US Rte 202 Raritan NJ 08869

TEITEL, ROBERT J(ERRELL), b Indianapolis, Ind, Aug 4, 22. METALLURGY. *Educ:* Purdue Univ, BS, 44; Mass Inst Technol, ScD(metall), 48. *Prof Exp:* Assoc metallurgist, Brookhaven Nat Labs, 48-53, metallurgist, 53-55; group leader, Nuclear & Basic Res Lab, Dow Chem Co, Mich, 55-60; sr tech specialist, Rocketdyne Div, NAm Aviation, Inc, 60-61; prin scientist, Douglas Aircraft Co, 61-72; head mat eng dept, KMS Fusion, Inc, Ann Arbor, Mich, 72-76; PRES, ROBERT J TEITEL ASSOCS, 76- *Concurrent Pos:* Civilian with AEC, 44. *Mem:* AAAS; Am Soc Metals; fel Am Nuclear Soc. *Res:* Material sciences; nuclear fission and fusion reactor materials and design; nuclear fuel cycles; missile materials; space nuclear power plants; hydrogen production and storage systems; fossil fuel systems; metallurgical thermodynamics; liquid metal corrosion; alloy phase diagrams. *Mailing Add:* 5025 Santorini Way Oceanside CA 92056-5858

TEITELBAUM, CHARLES LEONARD, b Brooklyn, NY, June 14, 25; m 50. ANALYTICAL CHEMISTRY. *Educ:* Brooklyn Col, BA, 45; Purdue Univ, MS, 48, PhD(org chem), 51. *Prof Exp:* Res chemist, Heyden Chem Corp, 50-53; prin chemist, Battelle Mem Inst, 53-58; chemist, Coty, Inc Div, Chas Pfizer & Co, 58-65 & Florasynth, Inc, 65-66; res specialist, Gen Foods Corp, 66-86; ADJ ASST PROF, NY CITY UNIV, 88- *Mem:* Am Chem Soc. *Res:* Analysis of natural products relating to odor and flavor. *Mailing Add:* 85-46 Midland Pkwy Jamaica NY 11432

TEITELBAUM, PHILIP, b Brooklyn, NY, Oct 9, 28; c 5. NEUROSCIENCE, PHYSIOLOGICAL PSYCHOLOGY. *Educ:* City Col New York, BS, 50; Johns Hopkins Univ, MA, 52, PhD, 54. *Prof Exp:* Instr, 54-56, asst prof physiol psychol, Harvard Univ, 56-59; assoc prof, Univ Pa, 59-63, prof, 63-73; prof psychol, Univ Ill, Champaign, 73-85, prof, Ctr Advan Study, 80-85; GRAD RES PROF PSYCHOL, UNIV FLA, 85- *Concurrent Pos:* Fel behav sci, Ctr Advan Study, Stanford Univ, 75-76; Fulbright fel, Dept Zool, Tel Aviv Univ, 78-79; Guggenheim fel, 84-85. *Honors & Awards:* Distinguished Sci Contrib Award, Am Psychol Asn. *Mem:* Nat Acad Sci; Am Physiol Soc; Am Psychol Asn (pres, 77); AAAS; Soc Exp Psychologists. *Res:* Recovery of movement and of eating and drinking. *Mailing Add:* Dept Psychol Univ Fla Gainesville FL 32611

TEITLER, SIDNEY, b New York, NY, July 1, 30. THEORETICAL PHYSICS. *Educ:* Long Island Univ, BS, 51; Univ Ill, MS, 53; Syracuse Univ, PhD(physics), 57. *Prof Exp:* PHYSICIST, ELEC SCI & TECH DIV, US NAVAL RES LAB, 57- *Mem:* Sigma Xi; fel Am Phys Soc. *Mailing Add:* Assoc Supt Elec Sci & Tech Div Code 6801 US Naval Res Lab Washington DC 20375

TEIXEIRA, ARTHUR ALVES, b Fall River, Mass, Jan 30, 44; m 86; c 3. FOOD ENGINEERING, AGRICULTURAL ENGINEERING. *Educ:* Univ Mass, BS, 66, MS, 68, PhD, 71. *Prof Exp:* Proj leader res, Ross Div, Abbott Labs, 71-73, group leader mgt, 73-77; sr consult, Arthur D Little Inc, 77-82; assoc prof, 82-89, PROF FOOD ENG, UNIV FLA, 89- *Concurrent Pos:* Bd dir, Food Eng Div, Sigma Xi, 88; NATO Sr Guest lectr, 88 & 89; Am Fulbright Scholar, Portugal, 90. *Mem:* Inst Food Technologists; Am Soc Agr Engrs; Sigma Xi. *Res:* Thermal processing of canned foods; food product and process development; applications of energy-saving technologies in food processing; computer simulation and control of food processing operations. *Mailing Add:* Agr Eng Dept Frazier-Rogers Hall Univ Fla Gainesville FL 32611

TEJA, AMYN SADRUDDIN, b Zanzibar, Tanzania, May 11, 46; UK citizen; m 71; c 2. PHASE EQUILIBRIA, CRITICAL PROPERTIES. *Educ:* Imp Col London, BSc, & ACGI, 68, PhD(chem eng) & DIC, 72. *Prof Exp:* Vis assoc prof chem eng, Ohio State Univ, 80; res fel, Loughborough Univ, Eng, 71-74, lectr, 74-80; vis assoc prof, Univ Del, 78-79; assoc prof, 80-84, prof chem eng, 84-90, REGENTS PROF, GA INST TECHNOL, 90- *Concurrent Pos:* Assoc ed, Chem Eng J, 73-; tech dir, Fluid Properties Res, Inc, 85- *Honors & Awards:* Hinchley Medal, 68; Sustained Res Award, Sigma Xi, 87. *Mem:* Am Inst Chem Engrs; Am Soc Eng Educ; Sigma Xi. *Res:* Measurement, correlation and prediction of the thermodynamic and transport properties of mixtures with emphasis on the critical region and on mixtures of technological interest; bioseparations using supercritical fluids. *Mailing Add:* Sch Chem Eng Ga Inst Technol Atlanta GA 30332-0100

TEJA, EDWARD RAY, electronics, economics, for more information see previous edition

TEJWANI, GOPI ASSUDOMAL, b Dadu, India, March 1, 46; m 73; c 1. OPIOID RECEPTORS & PEPTIDES, ANESTHETICS. *Educ:* Nagpur Univ, India, BS, 66, MS, 68; All-India Inst Med Sci, PhD(biochem), 73. *Prof Exp:* Fel enzyme, Sch Med, St Louis Univ, 73-74 & Roche Inst Molecular Biol, 74-76; clin asst prof, 76-78, asst prof, 78-88, ASSOC PROF PHARMACOL, COL MED, OHIO STATE UNIV, 88- *Concurrent Pos:* Vis prof, Univ Sao Paulo, Brazil, 78 & Moscow State Univ, 81; Lectr, Univ Chile, 78, Univ Wroclaw, Poland, 81 & Univ Ioannina, Greece, 81; consult, Immunobiol Res Inst, NJ, 80; prin investr, numerous res grants biochem pharmacol, 78- *Mem:* AAAS; Am Soc Biochem & Molecular Biol; Soc Neurosci; Am Soc Pharmacol & Exp Therpeut. *Res:* Regulation of key enzymes involved in glycolysis and gluconegenesis; role of endorphins in obesity and cardiovascular diseases; role of stress in facilitation of mammary tumorigenesis; modulation of opioid receptors by anesthetics; pharmacologic effects of nicotine and neurotoxins. *Mailing Add:* Dept Pharmacol Ohio State Univ 333 W Tenth Ave Columbus OH 43210-1239

TEKEL, RALPH, b New York, NY, May 27, 20; m 60; c 2. ORGANIC CHEMISTRY. *Educ:* Polytech Inst NY, BS, 41; Purdue Univ, MS, 47, PhD(chem), 49. *Prof Exp:* Asst tech dir, Vitamins, Inc, Ill, 48-49; res assoc, Carter Prod, NJ, 49-51; mgr pilot plant, Nat Drug Co, 51-60; asst to mgr chem div, Wyeth Labs, 60-63; dir org res, Betz Lab, 63-65; from asst prof to assoc prof org chem, 65-74, ASSOC PROF CHEM, LA SALLE COL, 74- *Concurrent Pos:* Lectr, Holy Family Col, Pa, 66-67; consult, Am Electronic Labs, 66- *Mem:* AAAS; fel Am Inst Chem; Am Chem Soc. *Res:* Medicinals; biochemicals; halogen chemicals; pilot plant development; continuous thin layer chromatography. *Mailing Add:* 21 Linden Dr Elkins Park PA 19117-1333

TEKELI, SAIT, b Samsun, Turkey, June 14, 32; m 59; c 1. VETERINARY MEDICINE, PATHOLOGY. *Educ:* Univ Ankara, DVM, 54, DSc(path), 58; Univ Wis-Madison, MS, 62, PhD(avian leukosis), 64. *Prof Exp:* Dist vet, Dept Agr, Samsun, Turkey, 54-55; res asst animal path, Univ Ankara, 55-60; res asst vet sci, Univ Wis-Madison, 60-64, res asst bovine leukosis, 66-67; res pathologist, Norwich Pharmacal Corp, 67-69; SR RES PATHOLOGIST, ABBOTT LABS, 69- *Mem:* Soc Toxicol; Int Acad Path; Soc Toxicol Pathologists; Sigma Xi. *Res:* Drug toxicity; drug-induced lesions as well as chemical carcinogens. *Mailing Add:* Dept Path Dept 469 Abbott Labs Abbott Park IL 60064

TELANG, VASANT G, b Kumta, India, July 18, 35. MEDICINAL CHEMISTRY. *Educ:* Univ Bombay, BS, 56, MS, 64; Univ RI, PhD(pharmaceut chem), 68. *Prof Exp:* Lab instr pharmaceut chem, Univ Bombay, 59-62, asst prof pharm, 62-63; teaching asst pharmaceut chem, Col Pharm, Univ RI, 63-68; NIH res specialist, Col Pharm, Univ Minn, Minneapolis, 68-74; mem staff, Col Pharm, 74-80, ASSOC PROF BIOMED CHEM, HOWARD UNIV, 80-, ASSOC DEAN, SCH PHARM, 80- *Concurrent Pos:* Consult, Suneeta Labs, India, 70- *Mem:* Indian Pharmaceut Asn; Am Chem Soc. *Res:* Mechanism of action of narcotic analgesics and their antagonists. *Mailing Add:* Pharm Col Howard Univ Washington DC 20059

TELEB, ZAKARIA AHMED, b Siuz, Egypt, Sept 6, 49; m; c 2. TOXICOLOGY. *Educ:* Ain-Shams Univ Egypt, BSc, 68, MSc, 78, PhD(biochem), 82. *Prof Exp:* Researcher molecular drug eval, 69-88, ASST PROF BIOCHEM, NAT ORG DRUG CONTROL & RES, 88- *Concurrent Pos:* Postdoctoral position, Dept Biochem, Ga Univ, Athens, 86-87. *Mem:* Am Soc Cell Biol. *Res:* Isolation and purification of subcellular particles as, Lysosomal enzymes, Mitochondrial compartment and specific nuclear proteins; side effect of some pharmacologically active agents viz: steroida hormones, contraceptives, non-steroidal analgesic anti-inflammator drugs, molluscicide, besticide, anti-biharzial drugs; determination and assessment of the potentiation of the cytotoxic effect of drugs on cellular & subcellular. *Mailing Add:* Six Abou Hazem St Pyramids Ave PO Box 29 Cairo Egypt

TELEGDI, VALENTINE LOUIS, b Budapest, Hungary, Jan 11, 22; nat US; m 50. PARTICLE PHYSICS, WEAK INTERACTION. *Educ:* Univ Lausanne, MSc, 46; Swiss Fed Inst Technol, PhD(physics), 50. *Prof Exp:* Asst physics, Swiss Fed Inst Technol, 47-50; from instr to prof physics, Univ Chicago, 50-71, Enrico Fermi distinguished serv prof, 71-76; PROF PHYSICS, SWISS FED INST TECHNOL, 78- *Concurrent Pos:* Vis mem, H H Wills Lab, Univ Bristol, 48; lectr, Northwestern Univ, 53-54; vis res fel, Calif Inst Technol, 53; Ford fel & NSF vis scientist, Europ Orgn Nuclear Res, Geneva, 59; Loeb vis prof, Harvard Univ, 66; univ lectr, NY Univ, 67; vis prof, Calif Inst Technol, 78- *Mem:* Nat Acad Sci; Am Acad Arts & Sci; Padova Acad Sci; Torino Acad Sci. *Res:* Nuclear emulsion technique; experiment and theory of interaction of nuclei with photons; Compton effect of proton; symmetry properties of weak interactions; muon decay and absorption; decay of free neutron; magnetic properties of the muon; hypernuclei; long-lived strange particles. *Mailing Add:* Dept Physics ETH-Zentron Swiss Fed Inst Technol Zurich 8092 Switzerland

TELETZKE, GERALD H(OWARD), sanitary engineering, for more information see previous edition

TELFAIR, RAYMOND CLARK, II, b Ennis, Tex, Mar 5, 41. FISH & WILDLIFE SCIENCES. *Educ:* NTex State Univ, BA, 65, MA, 67; Tex A&M Univ, PhD(wildlife & fisheries sci), 79. *Prof Exp:* Inst, NTex State Univ, 65-68; instr wildlife & fisheries sci biol, Tex A&M Univ, 72-84; ENVIRON ASSESSMENT BIOLOGIST, RESOURCE PROTECTION DIV, TEX PARKS & WILDLIFE DEPT, 86- *Mem:* Am Ornithologists' Union; Am Soc Zoologists; Sigma Xi; Wildlife Soc. *Res:* Biology and ecology of colonial waterbirds with emphasis on the Cattle Egret, Bubulcus ibis; evaluate various project impacts on fish-wildlife; review environmental documents; provide guidelines on construction, reclamation and mitigation; develop data bases-reports; serve as expert witness. *Mailing Add:* 11780 S Hill Creek Rd Whitehouse TX 75791-9601

TELFAIR, WILLIAM BOYS, b Richmond, Ind, Apr 8, 47; m 69; c 2. PHYSICS, ENGINEERING. *Educ:* Pa State Univ, MS, 72, PhD(physics), 74. *Prof Exp:* Physicist, Wilks Sci Corp, 74-77; physicist indust res instrument, 77-80, mgr, Foxboro/Wilks, Inc, 80-82; engr mgr, Nanometrics, 82-84; sr prog mgr, Perkin-Elmer Corp, 84-86; VPRES R&D, TAUNTON TECHNOL, 86- *Honors & Awards:* Indust Res 100 Award, 77. *Mem:* Am Asn Physics Teachers; Optical Soc Am; Soc Appl Spectros; Coblentz Soc. *Res:* Developing laser surgery system for eyes; studying biological response of eye to surgery; developing exciser and UV laser diagnostic equipment; developing ultraviolet laser delivery systems. *Mailing Add:* Pocono Rd No R7 Newtown CT 06470

TELFER, NANCY, b San Francisco, Calif, Apr 15, 30. MEDICINE, NUCLEAR MEDICINE. *Educ:* Stanford Univ, AB, 51; Med Col Pa, MD, 56; Am Bd Internal Med, dipl, 63 & 77; Am Bd Nuclear Med, dipl, 72. *Prof Exp:* Intern, Los Angeles County-Univ Southern Calif Med Ctr, 56-57, resident internal med, 57-60; instr med, Ctr Health Sci, Univ Calif, Los Angeles, 60-61, asst prof, 62-67; from asst prof to assoc prof radiol & med, Los Angeles County-Univ Southern Calif Med Ctr, 67-84. *Concurrent Pos:* Los Angeles County Heart Asn res fel, Isotope Lab Med, Cantonal Hosp, Geneva, Switz, 61-62; Kate Meade Hurd fel, Woman's Med Col Pa, 61-62; vis assoc prof nuclear med, Beth Israel Hosp-Harvard Med Sch, 81-82. *Mem:* Am Fedn Clin Res; fel Am Col Physicians; Soc Nuclear Med; fel Am Col Nuclear Physicians. *Res:* Body electrolyte composition using radioactive tracers and the dilution principle; computer analysis of radionuclide cardiac and pulmonary function studies; soft tissue deposition of 99m technetium diphorphonate; red blood cell 86rubidium uptake. *Mailing Add:* PO Box 3142 Warrenton VA 22186

TELFER, WILLIAM HARRISON, b Seattle, Wash, June 21, 24; m 50; c 2. REPRODUCTIVE BIOLOGY, DEVELOPMENTAL BIOLOGY. *Educ:* Reed Col, BA, 48; Harvard Univ, MS, 49, PhD(biol), 52. *Prof Exp:* Jr fel, Harvard Soc Fels, 52-54; from asst prof to prof biol, 54-73, chmn grad group, 60-70, prof zool & chmn dept biol, 73-77, PROF BIOL, UNIV PA, 78- *Concurrent Pos:* Guggenheim fel, Stanford Univ, 60-61; NSF sr fel, Univ Miami, 68-69; staff mem, NIH training prog in fertilization & gamete physiology, Marine Biol Lab, Woods Hole, Mass, 71-; res assoc biochem, Univ Ariz, 81-82. *Mem:* Am Soc Zool; Soc Develop Biol. *Res:* Physiology and developmental aspects of egg formation and blood proteins in insects. *Mailing Add:* Dept Biol/203 LI/G7 Univ Pa Philadelphia PA 19104

TELFORD, IRA ROCKWOOD, b Idaho Falls, Idaho, May 6, 07; m 33; c 4. ANATOMY. *Educ:* Univ Utah, AB, 31, AM, 33; George Washington Univ, PhD(anat), 42. *Prof Exp:* Sch teacher, Idaho, 33-37; from instr to assoc prof anat, Sch Med, George Washington Univ, 41-47, prof anat, Schs Med, 53-72; prof & chmn dept, Sch Dent, Univ Tex, 47-53; prof anat, Sch Med & Dent, Georgetown Univ, 72-78; VIS PROF ANAT, UNIFORMED SERV UNIV, 78- *Mem:* Soc Exp Biol & Med; Am Asn Anat; Am Acad Neurol. *Res:* Histology, muscle and nerve studies in vitamin E deficiency; vitamins; cancer in dietary deficiencies; muscular dystrophy. *Mailing Add:* 3424 Garrison St NW Washington DC 20008

TELFORD, JAMES WARDROP, b Merbein, Australia, Aug 16, 27; m 54; c 3. ATMOSPHERIC PHYSICS, COMPUTER SCIENCE. *Educ:* Univ Melbourne, BSc, 50, DSc(atmospheric convection), 70; Univ Sydney, dipl numerical anal automatic comput, 62. *Prof Exp:* Sr res scientist, Radiophys Div, Commonwealth Sci & Indust Res Orgn, 50-65; vis scientist, Dept Cloud Physics, Imp Col, Univ London, 65-66; sr res scientist, Radiophys Div, Commonwealth Sci & Indust Res Orgn, 66-67; dep dir lab atmospheric physics & res prof, 67-78, DIR AIR MOTION LAB & RES PROF, DESERT RES INST, UNIV NEV SYST, RENO, 78- *Concurrent Pos:* Dept Defense res contract, Atmospheric Sci Ctr, Desert Res Inst, Reno, 67-88, NSF res grants, 70-88, lectr, Univ Nev, Reno, 78; NASA res contracts, 75-88. *Mem:* Fel Am Meteorol Soc; fel Royal Meteorol Soc; Sigma Xi; fel AAAS. *Res:* Experimental and theoretical work on stochastic coalescence mechanisms in warm clouds; marine boundary layer, airborne observations and theory; observations and theory of clear air and cloudy air convection; airborne air motion measuring system; thunderstorm theory. *Mailing Add:* 1975 Fallen Leaf Ct Reno NV 89509

TELFORD, SAM ROUNTREE, JR, b Winter Haven, Fla, Aug 25, 32; m 57; c 3. EPIZOOTIOLOGY. *Educ:* Univ Va, BA, 55; Univ Fla, MS, 61; Univ Calif, Los Angeles, PhD(zool), 64. *Prof Exp:* Lectr zool, Univ Calif, Los Angeles, 64-65; Nat Inst Allergy & Infectious Dis res fel parasitol, Inst Infectious Dis, Univ Tokyo, 65-67; mem staff, Gorgas Mem Lab, CZ, 67-70; int assoc cur, 70, asst cur, 70-73, asst prof biol sci & zool, 70-73, FIELD RES ASSOC & ADJ CUR, FLA STATE MUS, UNIV FLA, 73- *Concurrent Pos:* Med Zoologist, WHO, Geneva, Switz, 73, Chagas Dis Vector Res Unit, Acarigua, Venezuela, 73-75 & Vertebrate Pest Control Ctr, Karachi, Pakistan, 75-77; WHO Spec Prog for Res & Training in Trop Dis-Div Malaria, Geneva, Switz, 77-78; Rodent Control Demonstration Unit, WHO, Rangoon, Burma, 78-80; proj leader, Denmark-Tanzania Rodent Control, Morogoro, 81-85; WHO consult med zool, Govt Zaire, 81; res assoc entomol & nematol, 85-87 & Dept Infectious Dis, Col Vet Med, Univ Fla, 89-; res assoc, Fla Mus Natural Hist, 87-88; consult publ health vector-borne & parasitic dis, Govt Ecuador Inter-Am Develop Bank Trop Res & Develop, Gainesville, Fla, 88; consult upland ecol & endangered species, Wetlands Mgt, Inc, Jensen Beach, Fla, 88- *Mem:* Am Soc Ichthyologists & Herpetologists; Am Soc Parasitol; Soc Protozool; Soc Study Amphibians & Reptiles. *Res:* Herpetology; parasitology; ecology; population dynamics of reptilian host-parasite associations; lower vertebrate parasitology; ecology and systematics of reptiles and amphibians; saurian malaria; zoonotic disease; rodent control. *Mailing Add:* 1712 NW 49th Terr Gainesville FL 32609

TELFORD, SAM ROUNTREE, III, b Gainesville, Fla, Aug 29, 61. BIOLOGY OF LYME DISEASE SPIROCHAETES, ECOLOGY OF PARASITES. *Educ:* Johns Hopkins Univ, BA, 83; Harvard Univ, MS, 87, DSc(parasitol), 90. *Prof Exp:* Curatorial asst, Mus Comp Zool, 84-86; grad res asst, 84-90, RES FEL, HARVARD UNIV, 90- *Concurrent Pos:* Lectr & lab instr, Dept Trop Pub Health, Harvard Univ, 85-; consult, Trop Dis Diag, Becton Dickinson Co, 88- *Mem:* Soc Study Evolution; Soc Study Amphibians & Reptiles; Herpetologists League; Am Soc Trop Med & Hyg; Am Soc Parasitologists; Am Soc Mammalogists. *Res:* Medical zoology; parasite ecology; evolutionary theory; role that parasites may or may not play in the ecology and evolution of their hosts and use principles from such work towards bettering the public health. *Mailing Add:* Dept Trop Pub Health Harvard Univ 665 Huntington Ave Boston MA 02115

TELIONIS, DEMETRI PYRROS, b Athens, Greece, Mar 17, 41; m 67. AERODYNAMICS, APPLIED MATHEMATICS. *Educ:* Nat Tech Univ Athens, dipl, 64; Cornell Univ, MS, 69, PhD(aerospace eng), 70. *Prof Exp:* Mech engr, Royal Greek Navy Shipyards, 64-67; asst prof aerospace eng, 70-74, assoc prof eng mech, 74-78, PROF ENG MECH, VA POLYTECH INST & STATE UNIV, 78- *Concurrent Pos:* Consult, Commun Orgn of Greece, 67 & Hellenic Air Force, 77-; Soc Eng Res Award, Am Soc Eng Educ, 87. *Mem:* Assoc fel Am Inst Aeronaut & Astronaut; Sigma Xi; Tech Chamber Greece; Am Soc Mech Engrs. *Res:* Incompressible and compressible aerodynamics; viscous flows; boundary-layer theory; acoustics; applied mechanics; turbulent flows; unsteady aerodynamics; experimental fluid mechanics. *Mailing Add:* 3138 Indian Meadow Dr Blacksburg VA 24060

TELKES, MARIA, b Budapest, Hungary, Dec 12, 00; nat US. SOLAR ENERGY CONVERSION, THERMAL ENERGY STORAGE. *Educ:* Univ Budapest, BA, 20, PhD(phys chem), 24. *Hon Degrees:* DSc, St Joseph Col, Conn, 57. *Prof Exp:* Instr physics, Univ Budapest, 23-24; biophysicist, Cleveland Clin Found, 26-37; engr, Res Dept, Westinghouse Elec & Mfg Co, 37-39; res assoc, Mass Inst Technol, 39-53; proj dir solar energy prog, Res Div, NY Univ, 53-58; res dir, Solar Energy Lab, Curtiss-Wright Corp, 58-60; dir res, Cryo-Therm, Inc, Pa, 60-64; mgr thermodyn lab, Melpar Inc, Westinghouse Air Brake Co, 64-69; sr res specialist, Nat Ctr Energy Mgt & Power, Univ Pa, 69-72; sr scientist, Inst Energy Conversion, Univ Del, 72-77; dir solar thermal storage develop, Am Technol Univ, 77-80; CONSULT, 80- *Honors & Awards:* CG Abbot Award, Am Solar Energy Soc, 77. *Mem:* Am Chem Soc; Solar Energy Soc; Soc Women Engrs. *Res:* Solar-thermal storage materials used in solar heated and cooled buildings; thermoelectric generators; semiconductors; phase-change thermal control of terrestrial and space applications. *Mailing Add:* 1475 NE 125th Terr Suite 414 North Miami FL 33161

TELL, BENJAMIN, b Philadelphia, Pa, Dec 11, 36; m 66; c 2. PHYSICS. *Educ:* Columbia Univ, BA, 58; Univ Mich, Ann Arbor, MS, 60, PhD(physics), 63. *Prof Exp:* MEM TECH STAFF, BELL TEL LABS, 63- *Concurrent Pos:* Vis mem staff, Philips Industs N V Philips Res Lab, Eindhoven Holland, 76. *Mem:* Am Phys Soc. *Res:* Optical and electrical properties of semiconductors; ion implantation and III-V devices. *Mailing Add:* AT&T Bell Labs 4C-403 Crawford Corner Rd Holmdel NJ 07733

TELLE, JOHN MARTIN, b Akron, Ohio, Nov 3, 47; m 68; c 3. LASERS. *Educ:* Univ Colo, BS, 69; Cornell Univ, MS, 72, PhD(physics), 75. *Prof Exp:* RES SCIENTIST LASER PHYSICS, LOS ALAMOS NAT LAB, 75- *Mem:* Am Phys Soc; Optical Soc Am. *Res:* Optically pumped infrared gas lasers. *Mailing Add:* 210 Garver Lane Los Alamos NM 87544

TELLEP, DANIEL M, b Forest City, Pa, Nov 20, 31. MECHANICAL ENGINEERING, FLUIDS. *Educ:* Univ Calif, Berkeley, BS, 54, MS, 55. *Prof Exp:* Sr scientist, Lockheed Missiles & Space Co, 55-57, head thermal res, 58-61, mgr missile thermodyn, 63-66, chief Poseidon reentry systs eng, asst chief engr develop & chief engr, Missile Systs Div, 66-75, vpres & asst gen mgr, Advan Systs Div, 75-83, exec vpres, 83-84, pres, 84-87, group pres, Lockheed Missiles & Space Systs, 86-87, pres, Lockheed Corp, 88-89, CHMN BD & CHIEF EXEC OFFICER, LOCKHEED CORP, 89- *Concurrent Pos:* Mem bd dirs, Lockheed Missiles & Space Systs, 87- *Honors & Awards:* Lawrence B Sperry Award, Am Inst Aeronaut & Astronaut, 64 & Missile Syst Award, 86; Syst Award, Am Inst Aeronaut & Astronaut, 86. *Mem:* Nat Acad Eng; Sigma Xi; fel Am Inst Aeronaut & Astronaut; fel Am Astronaut Soc. *Mailing Add:* Lockheed Corp 4500 Park Granada Blvd Calabasas CA 91399

TELLER, AARON JOSEPH, b Brooklyn, NY, June 30, 21; m 46; c 1. ENVIRONMENTAL & CHEMICAL ENGINEERING. *Educ:* Cooper Union, BChE, 43; Polytech Inst Brooklyn, MChE, 49; Case Western Reserve Univ, PhD(chem), 51. *Prof Exp:* Res engr chem eng, Columbia Univ-Manhattan Proj, 42-44 & Martin Labs, 44-45; develop engr, City Chem Corp, 45-47; chmn, dept chem eng, Cleveland State Univ, 47-56; res prof chem eng & chmn dept, Univ Fla, 56-61; vpres eng, Colonial Iron-Patterson Industs, 61-63; dean, Col Eng & Sci, Cooper Union, 63-70; pres, Teller Environ Systs Inc, 70-86; consult, 86-89, SR TECH ADV & VPRES TECHNOL, RES COTTRELL CO, 89- *Concurrent Pos:* Consult, Davy Power Gas, Bechtel, C F Braun, Borden Co, Am Cyanamide, Exxon & Tenn Valley Auth, 56-70; mem, Nat Adv Comt Air Pollution Technol, Environ Protection Agency, 68-71. *Honors & Awards:* Ann Lectr Award, Am Inst Chem Engrs, 72; Valeur Award, 75. *Mem:* Am Inst Chem Engrs; Am Chem Soc; Tech Asn Pulp & Paper Indust. *Res:* Environmental-chemical engineering; diffusional operations, packing; nucleation; chromatographic absorption; applications of chemical engineering to environmental solutions. *Mailing Add:* 3140 S Ocean Blvd Palm Beach FL 33480

TELLER, CECIL MARTIN, II, b Galveston, Tex, Oct 25, 39; m 66; c 2. NONDESTRUCTIVE EVALUATION, FATIGUE & FRACTURE. *Educ:* Univ Tex Austin, BS, 64, MS, 66 & PhD(mat sci & eng), 71. *Prof Exp:* Mgr, Tracor Inc, 72-74; br chief, US Govt, 74-77; mgr, Southwest Res Inst, 77-83; tech dir, Tex Res Inst Inc, 83-85, vpres, 85-88; pres, 88-89, CHIEF ENGR, TEX RES INT/APPL RES & TECHNOL INC, 89-; CORP SCI OFFICER, TEX RES INT, INC, 89- *Concurrent Pos:* Chief engr, Tex Res Inst Inc, 89- *Mem:* Am Soc Mech Engrs; Am Soc Metals Int; Am Soc Nondestructive Testing. *Res:* Research and development of new nondestructive evaluation methods for metals and composites; bonded structures and thick composites. *Mailing Add:* 1801 Congressional Circle Austin TX 78746

TELLER, DANIEL MYRON, b Nashville, Tenn, Feb 10, 30; m 58; c 2. ORGANIC CHEMISTRY. *Educ:* Northwestern Univ, BS, 52; Loyola Univ, Ill, MS, 54; Mich State Univ, PhD(org chem), 59. *Prof Exp:* Res chemist, Fabrics & Finishes Dept, E I du Pont de Nemours & Co, 59-60; RES SCIENTIST, WYETH LABS, INC, PHILADELPHIA, 60- *Mem:* Am Chem Soc. *Res:* Thiophene derivatives; steroid synthesis; medicinal chemistry. *Mailing Add:* 824 Devon State Rd Devon PA 19333-1059

TELLER, DAVID CHAMBERS, b Wilkes-Barre, Pa, July 25, 38; m 60; c 1. PHYSICAL BIOCHEMISTRY. *Educ:* Swarthmore Col, BA, 60; Univ Calif, Berkeley, PhD(biochem), 65. *Prof Exp:* Asst prof, 65-70, ASSOC PROF BIOCHEM, UNIV WASH, 70- *Concurrent Pos:* Consult, Spinco Div, Beckman Instruments, 66. *Mem:* Am Soc Biol Chem; Sigma Xi. *Res:* Physical chemistry and equilibria of proteins; non-covalent association. *Mailing Add:* Dept Biochem Univ Wash Seattle WA 98195

TELLER, DAVID NORTON, b New York, NY, Oct 1, 36; m 59. NEUROCHEMISTRY, PSYCHOPHARMACOLOGY. *Educ:* Brooklyn Col, BS, 57; NY Univ, MS, 60, PhD(cytochem), 64. *Prof Exp:* Biologist, Fine Organics, Inc, 56-57; res asst hemat & nutrit, New York Med Col, 57-59; sr res scientist, NY State Ment Hyg, Manhattan State Hosp, NY State Res Inst, 59-66, assoc res scientist, 66-76; assoc prof, 76-79, PROF, DEPT PSYCHIAT & BEHAV SCI, MED SCH, UNIV LOUISVILLE, 79- *Concurrent Pos:* Lectr, Dept Psychiat, New York Med Col, 66-67 & Grad Div, Fairleigh Dickinson Univ, 68-71. *Mem:* Am Chem Soc; Am Soc Neurochem; Am Soc Pharmacol & Exp Therapeut; Am Soc Testing & Mat; Int Col Neuropsychopharmacol; Biochem Soc Brit. *Res:* Drug binding and transport; subcellular particle preparation; molecular pharmacology; evaluation of medical education. *Mailing Add:* Univ Louisville Med Sch Louisville KY 40292

TELLER, DAVIDA YOUNG, b Yonkers, NY, July 25, 38; m 90; c 2. VISION, VISUAL DEVELOPMENT. *Educ:* Swarthmore Col, BA, 60; Univ Calif, Berkeley, PhD(psychol), 65. *Prof Exp:* Res asst prof psychol, 65-67, actg asst prof, 67-68, from asst prof to assoc prof psychol & physiol, 68-74, PROF PSYCHOL PHYSIOL & BIOPHYS, UNIV WASH, 74- *Concurrent Pos:* Nat Inst Neurol Dis & Blindness res grant, 68-71; Nat Eye Inst res grants, 71-; mem comt vision, Nat Res Coun, 71-80; mem vision res & training comt, Nat Eye Inst, 72-76; affil, Regional Primate Res Ctr, 73- & Child Develop & Ment Retardation Ctr, 75-; NSF res grant, 75-85; mem, Vision B Study Sect, NIH, 81-85, chair, 83-85. *Honors & Awards:* Glenn Fry Award, Am Acad Optom, 82. *Mem:* Fel AAAS; fel Optical Soc Am; Asn Res Vision & Ophthal; Asn Women Sci; Sigma Xi; fel Coun Am Asn Advan Sci. *Res:* Development of vision in human infants; philosophical aspects of visual sciences; visual psychophysics. *Mailing Add:* Dept Psychol NI-25 Univ Wash Seattle WA 98195

TELLER, EDWARD, b Budapest, Hungary, Jan 15, 08; nat US; m 34; c 2. PHYSICS. *Educ:* Univ Leipzig, PhD, 30. *Hon Degrees:* Many from various cols & univ in US, 54-64. *Prof Exp:* Res assoc, Univ Leipzig, 29-31 & Univ Göttingen, 31-33; Rockefeller fel, Copenhagen, 34; lectr, Univ London, 34-35; prof physics, George Washington Univ, 35-41 & Columbia Univ, 41-42; physicist, Manhattan Eng Dist, Univ Chicago, 42-43 & Los Alamos Sci Lab, 43-46; prof physics, Univ Chicago, 46-52; prof, 53-60, univ prof, 60-75, EMER PROF PHYSICS, UNIV CALIF, BERKELEY, 75-; SR RES FEL, HOOVER INST WAR, PEACE & REVOLUTION, STANFORD UNIV, 75- *Concurrent Pos:* Asst dir, Los Alamos Sci Lab, 49-52; consult, Lawrence Radiation Lab, Livermore, 52-53, dir, 58-60, assoc dir, Lawrence Livermore Lab, Univ Calif, 54-58 & 72-; mem sci adv bd, US Air Force; gen adv comt, USAEC; dir, Thermo Electron Corp; mem, President's Foreign Intel Adv Bd; consult, Comn Critical Choices of Americans, 74-; mem, White House Sci Coun, adv bd, Technol in Soc, bd dirs, Hertz Found & bd gov, Am Acad Achievement & Am Friends Tel Aviv Univ; emer bd mem, Defense Intel Sch; hon trustee, Asn Unmanned Space Vehicle Systs; consult, Defense Sci Bd, Undersecy Defense; mem adv bd, Fed Emergency Mgt Agency. *Honors & Awards:* Priestley Mem Award, Dickinson Col, 57; Einstein Award, 59; Gen Donovan Mem Award, 59; Award, Midwest Res Inst, 60;

Living Hist Award, Res Inst Am, 60; Golden Plate Award, 61; White & Fermi Awards, 62; Robins Award Am, 63; Leslie R Groves Gold Medal Award, 74; Harvey Prize, Technion Inst, Israel, 75; Albert Einstein Award, Inst Israel, 77; Gold Medal, Am Col Nuclear Med, 80 & Am Acad Achievement, 82; A C Eringen Award, Soc Eng Sci, Inc, 80; Nat Medal Sci for 1982, Pres Ronald Reagan, 83; Sylvanus Thayer Award, 86. *Mem:* Nat Acad Sci; fel Am Nuclear Soc; fel Am Phys Soc; Am Ord Asn; fel Am Acad Arts & Sci; fel AAAS; Am Defense Preparedness Asn; Am Geophys Union; Int Platform Asn; Scientists & Engrs for Secure Energy. *Res:* Chemical, molecular and nuclear physics; quantum theory. *Mailing Add:* Hoover Inst Stanford CA 94305-6010

TELLER, J C, engineering, for more information see previous edition

TELLER, JAMES TOBIAS, b Evanston, Ill, Aug 1, 40; m 63; c 2. GEOLOGY. *Educ:* Univ Cincinnati, BS, 62, PhD(geol), 70; Ohio State Univ, MS, 64. *Prof Exp:* Field geologist, Inst Polar Studies, 64-65; petrol geologist, Atlantic Richfield Co, 65-67; from asst prof to assoc prof, 70-81, PROF GEOL, UNIV MAN, 81- *Concurrent Pos:* Nat Sci & Eng Res Coun Can grants, Univ Man, 70-75 & 77-89, Geol Surv Can grants, 71-72 & 81-83; geol consult sand, gravel & petrol, 75-; vis assoc prof geol, Univ Cincinnati, 76; res fel, Australian Nat Univ, 77; assoc ed, Geosci Can, 79-; vis scientist, Univ Cape Town, 83; mem, NAm Comn Stratig Nomenclature, 84-87, Can Nat Comm INQUA, 88-92. *Honors & Awards:* Stillwell Medal, Geol Soc Australia, 87. *Mem:* Am Asn Quaternary Res; Soc Econ Paleont & Mineral; Can Quaternary Asn; Int Asn Sedimentologists; Hist Earth Sci Soc. *Mailing Add:* Dept Geol Sci Univ Man Winnipeg MB R3T 2N2 Can

TELLER, JOHN ROGER, b Cincinnati, Ohio, June 30, 32; m 60; c 2. ALGEBRA. *Educ:* Univ Cincinnati, BS, 55, MA, 59; Tulane Univ, PhD(math), 64. *Prof Exp:* Asst prof math, Univ NH, 64-65; ASSOC PROF MATH, GEORGETOWN UNIV, 65- *Mem:* Am Math Soc; Math Asn Am. *Res:* Mathematical research in partially ordered groups. *Mailing Add:* Dept Math 246 Reiss Georgetown Univ 37th & O Sts NW Washington DC 20057

TELLINGHUISEN, JOEL BARTON, b Cedar Falls, Iowa, May 27, 43; m 72; c 2. MOLECULAR SPECTROSCOPY, MOLECULAR PHYSICS. *Educ:* Cornell Univ, AB, 65; Univ Calif, Berkeley, PhD(chem), 69. *Prof Exp:* Res assoc chem, Univ Canterbury, 69-71; res assoc physics, Univ Chicago, 71-73; Nat Res Coun res assoc, Nat Oceanic & Atmospheric Admin, Boulder, Co, 73-75; from asst prof to assoc prof, 75-83, PROF CHEM, VANDERBILT UNIV, 83- *Mem:* Am Chem Soc; Am Phys Soc. *Res:* Molecular and atomic physics; optical spectroscopy; lasers. *Mailing Add:* Dept Chem Vanderbilt Univ Nashville TN 37235

TELSCHOW, KENNETH LOUIS, b St Paul, Minn, Jan 4, 47; m 74; c 1. PHYSICS, MATERIALS SCIENCE. *Educ:* Univ Calif, Los Angeles, BS, 69, PhD(physics), 73. *Prof Exp:* Adj asst prof physics, Univ Calif, Los Angeles, 74-75; lectr physics, Univ Mass, Amherst, 75-77; ASST PROF PHYSICS, SOUTHERN ILL UNIV, 77- *Concurrent Pos:* Teaching fel physics, Univ Calif, Los Angeles, 69-73; fel physics, Univ Mass, Amherst, 75-77; prin investr, Res Corp res grant, 77-79; prin investr, NSF res grant, 79-81. *Mem:* Am Phys Soc; Acoust Soc Am. *Res:* Low temperature physics; liquid helium; acoustics; quantum fluids and solids. *Mailing Add:* Idaho Nat Eng Lab EG&G Idaho Inc NDE PO Box 1625 Idaho Falls ID 83415

TELSER, ALVIN GILBERT, b Chicago, Ill, May 11, 39; m 67; c 2. BIOCHEMISTRY, CELL BIOLOGY. *Educ:* Univ Chicago, BS, 61, PhD(biochem), 68. *Prof Exp:* Helen Hay Whitney Found fel develop biol, Brandeis Univ, 68-70; Helen Hay Whitney Found fel cell biol, Yale Univ, 70-71; asst prof, 71-77, ASSOC PROF ANAT & CELL BIOL, MED SCH, NORTHWESTERN UNIV, CHICAGO, 77- *Mem:* AAAS; Soc Develop Biol; Am Soc Cell Biol; NY Acad Sci; Sigma Xi. *Res:* Biochemical aspects of differentiation and development in eukaryotic systems with emphasis on understanding regulatory mechanisms at a molecular level. *Mailing Add:* Dept Cell Biol & Anat Northwestern Univ Med Sch 303 E Chicago Ave Chicago IL 60611

TEMES, CLIFFORD LAWRENCE, b Jersey City, NJ, Feb 4, 30; m 63; c 3. ELECTRICAL ENGINEERING. *Educ:* Cooper Union, BEE, 51; Case Inst Technol, MS, 54; Columbia Univ, EE, 60; Polytech Inst Brooklyn, PhD(elec eng), 65. *Prof Exp:* Electronic scientist instrumentation, Nat Adv Comt Aeronaut, 51-54; lab supvr radar systs, Electronics Res Lab, Columbia Univ, 56-60; sr proj engr elec eng, Fed Sci Corp, 60-65; mem tech staff, Gen Res Corp, 65-74; mem dept staff, Mitre Corp, 74-77; HEAD SEARCH RADAR BR, NAVAL RES LAB, 77- *Concurrent Pos:* Consult, Electronics Res Lab, Columbia Univ, 60-65; reviewer, Inst Elec & Electronics Engrs, 60-75, Prentice-Hall, 68-69. *Honors & Awards:* Qual Award, Naval Res Lab, 78. *Mem:* Sigma Xi; sr mem Inst Elec & Electronics Engrs. *Res:* Radar systems and technology; signal processing and wave form design; surveillance and tracking radar; clutter rejection; pulse compression and high resolution systems. *Mailing Add:* Naval Res Lab 4555 Overlook Ave SW Washington DC 20375

TEMES, GABOR CHARLES, b Budapest, Hungary, Oct 14, 29; Can citizen; m 54; c 2. ELECTRICAL ENGINEERING, COMPUTER SCIENCES. *Educ:* Budapest Tech Univ, Dipl Ing, 52; Eotvos Lorand Univ, Budapest, dipl phys, 54; Univ Ottawa, PhD(elec eng), 61. *Prof Exp:* Asst prof elec eng, Budapest Tech Univ, 52-56; proj engr, Measurement Eng Ltd, 57-59; dept head networks, Northern Elec Co Ltd, 59-64; group leader light electronics, Stanford Linear Accelerator Ctr, 64-66; corp consult networks, Ampex Corp, Calif, 66-69; chmn dept elec sci & eng, 75-80, PROF ELEC ENG, UNIV CALIF, LOS ANGELES, 69- *Concurrent Pos:* Ed, Trans on Circuit Theory, 69-71; consult, TRW, Rockwell Int, 75-, Am Microsysts, Inc, 78- & Xerox Corp, 80- *Honors & Awards:* Darlington Award, Inst Elec & Electronics Engrs, Circuits & Systs Soc, 69, 81 & Centennial Medal, 84. *Mem:* Fel Inst Elec & Electronics Engrs. *Res:* Integrated circuit design, filters and digital signal processing. *Mailing Add:* Rm 7731 Boelter Hall Univ Calif Los Angeles CA 90024

TEMEYER, KEVIN BRUCE, b Independence, Iowa, July 15, 51. BIOLOGICAL CONTROL OF INSECTS, PLASMID BIOLOGY. *Educ:* Iowa State Univ, BS, 73; Univ Mo, MA, 77, PhD(biol sci), 82. *Prof Exp:* Res asst, Vet Med Res Inst, Iowa State Univ, Ames, Iowa, 73; teaching asst gen biol, Univ Mo, Columbia, Mo, 75-78 & 80-82, res asst, 78-82; res microbiologist, 82-89, RES MOLECULAR BIOLOGIST, AGR RES SERV, USDA, KERRVILLE, TEX, 90-91. *Honors & Awards:* Young Investr Award, Am Soc Microbiol, 84. *Mem:* Soc Invert Path; Am Soc Microbiol; Soc Indust Microbiol; Southwestern Asn Parasitol. *Res:* Molecular genetics, physiology, plasmid biology and toxicology of entomopathogenic bacteria, particularly Bacillus thuringiensis; nutritional physiology and microecology of horn fly larvae; molecular phylogeny of ectoparasitic Dipteran insects. *Mailing Add:* US Livestock Insects Lab USDA Agr Res Serv PO Box 232 Kerrville TX 78028

TEMIN, HOWARD MARTIN, b Philadelphia, Pa, Dec 10, 34; m 62; c 2. ONCOLOGY, VIROLOGY. *Educ:* Swarthmore Col, BA, 55; Calif Inst Technol, PhD(biol), 59. *Hon Degrees:* Dr, Swarthmore Col, 72, New York Med Col, 72, Univ Pa, 76, Hahnemann Med col, 76, Lawrence Univ, 76, Temple Univ, 76 & Med Col Wis, 81, DSc, Colo State Univ, 87; Dr, Univ Pierre et Marie Curie (Paris), 88. *Prof Exp:* Asst prof, 60-64, assoc prof, 64-69, PROF ONCOL, UNIV WIS- MADISON, 69-, WIS ALUMNI RES FOUND PROF CANCER RES, 71-, AM CANCER SOC PROF VIRAL ONCOL & CELL BIOL, 74-, RUSCH PROF CANCER RES, 80-, STEENBOCK PROF BIOL SCI, 82- *Concurrent Pos:* Mem virol & rickettsiol study sect, NIH, 71-74, consult, working Group Human Gene Ther, 84-89; assoc ed, J Cell Physiol, 66-77 & Cancer Res, 71-74; lectr, Japanese Found Promotion Cancer Res, 85; vis prof, First Wilmot, Univ Rochester, distinguished vis prof, Univ Tenn, 87, Zickler vis prof, State Univ NY, Stony Brook, 89; mem, working group, Nat Cancer Inst, 72-73, Int Comt Virus Nomenclature Study Group, 73-75, Fundamental Res Panel, Nat Conf Health Res Prin, 78 & adv comt to dir, NIH, 79-83; consult, Off Technol Assessment Panel Saccharin, 77; mem, adv bd, Ctr Hist Microbiol, Am Soc Microbiol, 86-, Comt Nat Strategy AID, Inst Med, Nat Acad Sci, 86-88, Oversight Comt AIDS Activities, 88-90; mem, bd dirs, Found Advan Cancer Studies, 88-, Nat Cancer Adv Bd, 87-, chmn, subcomt AIDS; chmn, Adv Comt Genetic Variation Immunodeficiency Viruses, Vaccine Br, Div AIDS, Nat Inst Allergies & Infectious Dis, NIH, 88-; mem, Steering Comt Biomed Res Global Prog AIDS, 88-90, sci adv bd, Coord Coun Cancer Res, 89- *Honors & Awards:* Nobel Prize in Physiol & Med, 75 (shared); Warren Triennial Prize, Mass Gen Hosp, 71 (shared); Pap Award, Pap Inst Miami, 72; Bertner Award, Univ Tex M D Anderson Hosp & Tumor Inst Houston, 72; US Steel Award, Nat Acad Sci, 72; Enzyme Chem Award, Am Chem Soc, 73; Griffuel Prize, Asn Develop Res Cancer, Villejuif, France, 73; Dyer Lectr Award, NIH, 74; Clowes Lectr Award, Am Asn Cancer Res, 74; Int Award, Gaindner Found, Toronto, 74 (shared); Albert Lasker Award Basic Med Sci, 74; Lucy Wortham James Award, Soc Surg Oncologists, 76; Lila Gruber Award, Am Acad Dermat, 81; Bitterman Mem Lectr & Cetus Lectr, Univ Calif, Berkeley, 84. *Mem:* Nat Acad Sci; Inst Med; fel Am Acad Arts & Sci; Am Soc Microbiol; Am Asn Cancer Res; fel AAAS; hon mem, Tissue Cult Asn; Am Philos Soc; foreign mem Royal Soc London. *Res:* Replication of and mechanism of neoplastic transformation by RNA tumor viruses; RNA-directed DNA synthesis and protoviruses; retrovirus vectors, retrovirus variation and evolution. *Mailing Add:* McArdle Lab Univ Wis 1400 University Ave Madison WI 53706

TEMIN, RAYLA GREENBERG, b New York, NY, May 4, 36; m 62; c 2. GENETICS. *Educ:* Brooklyn Col, BS, 56; Univ Wis, MS, 58, PhD(genetics), 63. *Prof Exp:* Proj assoc, 63-72, asst scientist med genetics & genetics, 72-82, adj assoc prof, 82-88, ADJ PROF, UNIV WIS-MADISON, 88- *Concurrent Pos:* Fulbright scholar, Inst Animal Genetics, Edinburgh, 59-60. *Mem:* Genetics Soc Am. *Res:* Studies of the enhancer of segregation distorter gene, a meiotic drive gene in Drosophila melanogaster; teaching of undergraduate genetics. *Mailing Add:* Dept Genetics Univ Wis Madison WI 53706

TEMIN, SAMUEL CANTOR, b Washington, DC, Nov 4, 19; m 47; c 2. POLYMER CHEMISTRY. *Educ:* Wilson Teachers Col, BS, 39; Univ Md, MS, 43, PhD(org chem), 49. *Prof Exp:* Res chemist, Army Chem Ctr, Md, 42-44; asst gen org chem & biochem, Univ Md, 46-48; res chemist, Indust Rayon Corp, Ohio, 49-53, res supvr, 53-58; mgr polymer chem group, Explor Sect, Koppers Co, Inc, 58-65; asst dir, Fabric Res Labs, Inc, Mass, 65-72; sect head polymer chem, Lexington Lab, Kendall Co, Colgate-Palmolive Co, Lexington, 72-83, sr scientist, 83-85; CONSULT, 85- *Concurrent Pos:* Lectr, Pa State Univ, 63-65, Northeastern Univ, 66-77, Tufts Univ, 78 & Fla Atlantic Univ, 88, 90. *Mem:* Am Chem Soc; Fiber Soc; Int Asn Dent Res; Soc Plastics Engrs. *Res:* Polymer synthesis and structural relationships, adhesives, dental materials, monomer synthesis. *Mailing Add:* 20 Rainbow Pond Dr Walpole MA 02081

TEMKIN, AARON, b Morristown, NJ, Aug 15, 29; m 58; c 2. ATOMIC PHYSICS. *Educ:* Rutgers Univ, BS, 51; Mass Inst Technol, PhD(physics), 56. *Prof Exp:* Fulbright fel, Ger, 56-57; physicist, US Naval Res Lab, 57-58; physicist, Nat Bur Standards, 58-60; SR SCIENTIST, LAB ASTRON & SOLAR PHYSICS, GODDARD SPACE FLIGHT CTR, NASA, 60- *Honors & Awards:* Goddard Except Performance Award, NASA, 71. *Mem:* Fel Am Phys Soc. *Res:* Scattering of electrons from atoms, polarized orbitals, nonadiabatic theory; oxygen, hydrogen; threshold law for electron-atom impact ionization; symmetric Euler angle decomposition of three body problem; calculation of autoionization states; resonance projection operators; fixed-nuclei, adiabatic-nuclei and hybrid theories of electron-molecule scattering; non-iterative numerical technique for solution of elliptic partial differential equations. *Mailing Add:* Code 680 NASA-Goddard Space Flight Ctr Greenbelt MD 20771

TEMKIN, OWSEI, b Minsk, Russia, Oct 6, 02; US citizen; m 32; c 2. HISTORY OF MEDICINE & SCIENCE. *Educ:* Univ Leipzig, MD, 27. *Hon Degrees:* LLD, Johns Hopkins Univ, 73; DSc, Med Col Ohio, 75. *Prof Exp:* Intern, St Jacob Hosp, Leipzig, Ger, 27-28; asst hist of med, Univ Leipzig, 28-

32, pvt dozent, 31-33; assoc, Johns Hopkins Univ, 32-35, from assoc prof to prof, 35-58, William H Welch prof & dir dept, 58-68, EMER WILLIAM H WELCH PROF HIST MED, JOHNS HOPKINS UNIV, 68- *Concurrent Pos:* Actg ed & ed, Bull Hist Med, 48-68. *Honors & Awards:* William H Welch Medal, Am Asn Hist Med, 52; Sarton Medal, Hist of Sci Soc, 60; Hideyo Noguchi lectr, Johns Hopkins Univ, 69; Messenger lectr, Cornell Univ, 70. *Mem:* Nat Acad Sci; Am Philos Soc; Am Acad Arts & Sci; Am Asn Hist of Med (pres, 58-60); Hist of Sci Soc. *Res:* History of irritability; life and work of Hippocrates and Galen. *Mailing Add:* 830 W 40th St Baltimore MD 21211

TEMKIN, RICHARD JOEL, b Boston, Mass, Jan 8, 45; m 72; c 3. LASERS, PLASMA PHYSICS. *Educ:* Harvard Col, BA, 66; Mass Inst Technol, PhD(physics), 71. *Prof Exp:* Res fel physics, Harvard Univ, 71-74; staff mem physics, Francis Bitter Nat Magnet Lab, 74-79, asst group leader, 79, group leader, 80-85, DIV HEAD, PLASMA FUSION CTR, 86-, SR SCIENTIST, PHYS DEPT, MASS INST TECHNOL, 86- *Concurrent Pos:* IBM Corp fel, 72-74; assoc ed, Int J Infrared & Millimeter Waves, Trans Electron Devices, Inst Elec & Electronics Engrs; prog comt, Conf Infrared & Millimeter Waves, Inst Elec & Electronics Engrs. *Mem:* Am Phys Soc; Fusion Power Asn; Inst Elec & Electronics Engrs. *Res:* Submillimeter lasers, both theory and experiment; laser breakdown and heating of gases; optical and submillimeter diagnostics of plasmas; plasma heating; cyclotron resonance masers and gyrotrons, free electron lasers. *Mailing Add:* Plasma Fusion Ctr Mass Inst Technol Cambridge MA 02139

TEMKIN, SAMUEL, b Mexico City, Mex, Jan 10, 36; m 65; c 2. FLUID DYNAMICS, ACOUSTICS. *Educ:* Univ Nuevo Leon, ME, 60; Brown Univ, ScM, 64, PhD(eng), 66. *Prof Exp:* Sr scientist acoust, Bolt Beranek & Newman, Inc, 66-67; from asst prof to assoc prof eng, Rutgers Univ, New Brunswick, 67-73, grad prog dir mech & aerospace engr, 76-89, chmn, Dept Mech & Aerospace Eng, 80-89, PROF MECH ENG, RUTGERS UNIV, NEW BRUNSWICK, 73- *Concurrent Pos:* Consult, US Army Ballistic Res Labs, 69-73; Lady Davis Trust Fund fel, 74-75; consult, Naval Res Lab, Washington, DC, 86-; vis prof, Royal Inst Technol, Stockholm, Sweden, 89 & Univ Twente, Enschede, Neth, 90. *Honors & Awards:* Victor & Erna Hasselblad Found Award, Sweden, 86. *Mem:* AAAS; Am Phys Soc; Acoust Soc Am; Am Soc Mech Engrs. *Res:* Acoustic wave propagation; fluid dynamics of aerosols; droplet dynamics; bubble dynamics. *Mailing Add:* Dept Mech & Aero Eng Rutgers Univ Piscataway NJ 08855-0909

TEMME, DONALD H(ENRY), b Winside, Nebr, Jan 12, 28; m 55; c 2. ELECTRICAL ENGINEERING. *Educ:* Univ Nebr, BS, 49; Mass Inst Technol, MS, 55. *Prof Exp:* Asst physics, Univ Nebr, 49-51; assoc group leader, 58-76, GROUP LEADER, LINCOLN LAB, MASS INST TECHNOL, 76- *Mem:* Inst Elec & Electronics Engrs. *Res:* Phased array radar components. *Mailing Add:* 126 Silver Hill Rd Concord MA 01742

TEMMER, GEORGES MAXIME, b Vienna, Austria, Apr 10, 22; nat US; m 43, 79. NUCLEAR PHYSICS. *Educ:* Queens Col, NY, BS, 43; Univ Calif, MA, 44, PhD(physics), 49. *Prof Exp:* Asst physics, Univ Calif, 43-44 & 46-49; res assoc, Univ Rochester, 49-51; physicist, Nat Bur Standards, 51-53; mem staff terrestrial magnetism, Carnegie Inst, 53-63; dir, Nuclear Physics Lab, 63-85, prof, 63-85, UNIV PROF PHYSICS, RUTGERS UNIV, NEW BRUNSWICK, 85- *Concurrent Pos:* Guest investr, Cryogenic Sect, Nat Bur Standards, 53-55; Guggenheim Mem fel, Paris & Copenhagen, 56-57; vis prof, Univ Md, 59; prof, Fla State Univ, 60-63; Rutgers Res Coun fac fel, 68-69 & 75; Nat Acad Sci sr exchange scholar, People's Repub China, 80. *Honors & Awards:* Lindback Award for Excellence in Res, Rutgers Univ, 73; A v Humboldt Prize, 84. *Mem:* Fel Am Phys Soc. *Res:* Nuclear reaction mechanisms; very short lifetimes; scattering; angular correlation; low temperature nuclear alignment; gamma-ray spectroscopy; Coulomb excitation; polarized particle sources; electron channeling radiation; beam-foil spectroscopy; nuclear arms race concerns. *Mailing Add:* Dept Physics Rutgers Univ New Brunswick NJ 08903

TEMPEL, GEORGE EDWARD, b Feb 14, 44; m 68; c 2. CARDIOVASCULAR, AUTONOMIC. *Educ:* Ind Univ, PhD(physiol), 72. *Prof Exp:* ASSOC PROF PHYSIOL, MED UNIV SC, 78- *Mem:* Am Physiol Soc; Sigma Xi. *Mailing Add:* Med Univ SC 171 Ashley Ave Charleston SC 29403

TEMPELIS, CONSTANTINE H, b Superior, Wis, Aug 27, 27; m 55; c 2. IMMUNOLOGY. *Educ:* Univ Wis, Superior, BS, 50; Univ Wis-Madison, MS, 53, PhD(med microbiol), 55. *Prof Exp:* Proj assoc immunol, Univ Wis, 55-57; instr microbiol, Sch Med, Univ WVa, 57-58; from asst res immunologist to assoc res immunologist, 58-66, lectr, 60-66, assoc prof-in-residence immunol, 67-70, assoc prof, 70-72, PROF IMMUNOL, SCH PUB HEALTH, UNIV CALIF, BERKELEY, 72- *Concurrent Pos:* NIH career develop award, 65-70; vis scientist, Wellcome Res Labs, Eng, 77-78, gen & exp path, Univ Innsbruck, Austria, 85, 90; Fogarty Sr Int Fel, 77-78. *Mem:* AAAS; Am Asn Immunol; NY Acad Sci; Sigma Xi; Fedn Am Soc Exp Biol. *Res:* Studies of immune regulation in the chicken. *Mailing Add:* Sch Pub Health Univ Calif Berkeley CA 94720

TEMPERLEY, JUDITH KANTACK, b Meriden, Conn, Feb 12, 36; m 56. WEAPONS SYSTEMS ANALYSIS. *Educ:* Univ Rochester, BS, 57; Univ Ore, MS, 59, PhD(physics), 65. *Prof Exp:* Res physicist, 65-84, chief, air defense systs br, 84-86, CHIEF, BALLISTIC WEAPONS SYSTS ENG BR, US ARMY NUCLEAR DEFENSE LAB & US ARMY BALLISTIC RES LAB, 86- *Mem:* Am Phys Soc; Sigma Xi. *Res:* Engineering analysis and effectiveness studies of new concepts in weapon systems; computer simulation methodology. *Mailing Add:* US Army Ballistic Res Lab SLCBR-SE-B Aberdeen Proving Ground MD 21005-5066

TEMPEST, BRUCE DEAN, b Catasauqua, Pa, Nov 3, 35; m 59; c 3. INFECTIOUS DISEASES. *Educ:* Lafayette Col, AB, 57; Univ Pa, MD, 61; Am Bd Internal Med, cert, 68 & 74. *Prof Exp:* Resident med, Philadelphia Gen Hosp, 61-65; fel allergy & immunol, Univ Pa Hosp, 65-67; chief internal med, Dept Health, Educ & Welfare, USPHS, Tuba City, 67-70 & Gallup Indian Med Ctr, 70-71; CLIN DIR SURVEILLANCE PROJ, GALLUP INDIAN MED CTR, 71-, CHIEF INTERNAL MED, 82- *Concurrent Pos:* Clin asst prof, Dept Med, Sch Med, Univ NMex, 73-78, clin assoc prof, 78-; clin dir, Pneumococcal Surveillance Proj, 71-76, dep chief internal med, 75-82. *Mem:* Fel Am Col Physicians. *Res:* Epidemiology of pneumonia, especially pneumococcal pneumonia and the study of clinical manifestations; efficacy of vaccines in pneumonia prevention. *Mailing Add:* 1603 Monterey Dr Gallup NM 87301

TEMPLE, AUSTIN LIMIEL, (JR), b Leesville, La, Nov 3, 40; m 62; c 2. APPLIED MATHEMATICS. *Educ:* Centenary Col La, BS, 62; La State Univ, Baton Rouge, MA, 64; George Peabody Col, PhD(math), 71. *Prof Exp:* Instr math, Northwestern State Univ, 64-65, asst prof, 67-69; instr, Vanderbilt Univ, 69-70; asst prof, 70-75, ASSOC PROF MATH, NORTHWESTERN STATE UNIV, 75- *Mem:* Math Asn Am. *Res:* Mathematics education, in-service curriculum for elementary school teachers; develop and standardize tests for college freshmen math courses; develop computer statistical library. *Mailing Add:* Dept Math Northwestern State Univ Natchitoches LA 71457

TEMPLE, CARROLL GLENN, b Hickory, NC, Mar 7, 32:; m 56; c 2. ORGANIC CHEMISTRY, MEDICINAL CHEMISTRY. *Educ:* Lenoir-Rhyne Col, BS, 54; Birmingham-Southern Col, MS, 58; Univ NC, PhD(org chem), 62. *Prof Exp:* Assoc chemist, Southern Res Inst, 55-59, res chemist, 60-64, sr chemist, 64-80, HEAD, PHARMACEUT CHEM DIV, SOUTHERN RES INST, 81-, DIR, ORG CHEM RES INST, 91- *Mem:* Am Chem Soc. *Res:* Synthesis of potential antimalarian and anticancer drugs. *Mailing Add:* Southern Res Inst PO Box 55305 Birmingham AL 35255-5305

TEMPLE, DAVIS LITTLETON, JR, b Tupelo, Miss, June 10, 43; m 66. MEDICINAL CHEMISTRY. *Educ:* Univ Miss, BS, 66, PhD(med chem), 69. *Prof Exp:* Res assoc, La State Univ, New Orleans, 69-70; sr investr, 70-80, DIR CHEM RES, MEAD JOHNSON & CO, DIV BRISTOL-MYERS CO, 80- *Concurrent Pos:* Mem, Bio-Org & Nat Prod Study Sect, NIH, 81; adv bd, Advan Develop Chem Series, Am Chem Soc. *Mem:* AAAS; Am Chem Soc; Soc Neurosci; hon mem British Brain Asn; NY Acad Sci. *Res:* Central nervous system, cardiovascular and respiratory drugs; chemistry and biology. *Mailing Add:* Bristol Myers Co Five Research Pkwy PO Box 5100 Wallingford CT 06492-1927

TEMPLE, KENNETH LOREN, b St Paul, Minn, Mar 22, 18; m 43; c 3. GEOENVIRONMENTAL SCIENCE. *Educ:* Middlebury Col, AB, 40; Univ Wis, MS, 42; Rutgers Univ, PhD(microbiol), 48. *Prof Exp:* Chemist, US Naval Res Lab, DC, 42-45; instr bact, Univ RI, 48; assoc res specialist, Eng Exp Sta, WVa Univ, 48-53; microbiologist, Tex Co, 53-55; assoc prof microbiol, Agr Exp Sta, Mont State Univ, 55-61; sr res specialist, Commonwealth Sci & Indust Res Orgn, Australia, 61-63; prof microbiol, Mont State Univ, 64-83; RETIRED. *Res:* Autotrophic bacteria; microbiology of thermal waters; coal mines; geomicrobiology. *Mailing Add:* 6950 Tepee Ridge Rd Bozeman MT 59715-8631

TEMPLE, PETER LAWRENCE, b Springfield, Mass, Dec 13, 46; m 78; c 1. SOLAR ENERGY. *Educ:* Dartmouth Col, AB, 69. *Prof Exp:* Instr physics, Mount Holyoke Col, 70-73; instr physics eng, Holyoke Community Col, 76-77; sr res engr, res & develop solar energy, Total Environ Action, Inc, 77-82; CONSULT, PETER TEMPLE & ASSOC, 82- *Concurrent Pos:* Tech ed, Solar Age, Solar Vision Inc, 77- *Mem:* Int Solar Energy Soc. *Res:* Solar energy system design; building energy analysis; solar materials science; appropriate technology for developing countries; innovative solar product development; photovoltaics; passive solar systems. *Mailing Add:* The Island Box 65 Harrisville NH 03450

TEMPLE, ROBERT DWIGHT, b Des Moines, Iowa, July 1, 41; div; c 4. ORGANIC CHEMISTRY. *Educ:* Clemson Col, BS, 62; Fla State Univ, PhD(chem), 66. *Prof Exp:* Chemist, E I du Pont de Nemours & Co, 62; RES CHEMIST, PROCTER & GAMBLE CO, 66-, SECT HEAD, 72- *Mem:* Am Chem Soc. *Res:* Physical organic chemistry. *Mailing Add:* 1421 Hillcrest Rd Cincinnati OH 45224-2608

TEMPLE, STANLEY, b New York, NY, Aug 17, 30; div; c 2. ORGANIC CHEMISTRY. *Educ:* NY Univ, AB, 52, PhD(org chem), 58. *Prof Exp:* Res fel cancer steroids, Med Sch, Univ Va, 58-60; res chemist, Plant Tech Sect, Org Chem Dept, 60, res chemist, Process Dept, 60-61, res chemist, Res & Develop Div, Org Chem Dept, 61-74, SR RES CHEMIST, JACKSON LABS, CHAMBERS WORKS, E I DU PONT DE NEMOURS & CO, INC, 74- *Mem:* Sigma Xi; Lepidopterists' Soc; Royal Soc Chem; NY Acad Sci; Am Soc Testing & Mat. *Res:* Fluorochemicals; fluorinated polymers; surface active agents; bio-organic chemistry; cosmetics; environmental chemistry; pollution control; water treatment; food processing. *Mailing Add:* Jackson Lab E I du Pont de Nemours & Co Inc Wilmington DE 19898

TEMPLE, STANLEY A, b Cleveland, Ohio, Sept 26, 46; m 80. ECOLOGY, CONSERVATION. *Educ:* Cornell Univ, BSc, 68, MSc, 70, PhD(ecol), 73. *Prof Exp:* Teaching asst ecol, Cornell Univ, 68-73; res biologist, World Wildlife Fund, 73-75; res assoc ornith, Cornell Univ, 75-76; from asst prof to assoc prof, 76-84, PROF WILDLIFE ECOL, UNIV WIS-MADISON, 84-, BEERS-BASCOM PROF CONSERV, 80- *Concurrent Pos:* Dir, Int Coun Bird Preserv, 76-78, secy, 78-, dir, 80-; dir, Raptor Res Found, 77- *Mem:* Am Ornithologists Union; Ecol Soc Am; Wildlife Soc; Wilson Ornith Soc; AAAS. *Res:* Ornithology; wildlife management; endangered species; physiology; behavior. *Mailing Add:* Dept Wildlife Ecol/226 Russell Lab Univ Wis 1630 Linden Dr Madison WI 53706

TEMPLE, VICTOR ALBERT KEITH, b Winnipeg, Man, Apr 3, 44. SOLID STATE PHYSICS. *Educ:* Univ Man, BSc, 67; MacMaster Univ, MEng, 68, PhD(physics), 72. *Prof Exp:* PHYSICIST, RES & DEVELOP CTR, GEN ELEC, 74- *Concurrent Pos:* Nat Adv Coun fel, MacMaster Univ, 72-74. *Mem:* Inst Elec & Electronics Engrs. *Res:* Physics of power semiconductor carriers and the design; fabrication and development of new and improved power semiconductor devices. *Mailing Add:* Gen Elec Res & Develop One River Rd Schenectady NY 12301

TEMPLEMAN, GARETH J, b Little Falls, NY, Apr 21, 37; m 70; c 1. PHYSICAL CHEMISTRY, ANALYTICAL CHEMISTRY. *Educ:* Ohio Wesleyan Univ, BA, 60; State Univ NY Buffalo, PhD(chem), 70. *Prof Exp:* Res fel phys org chem, State Univ NY Buffalo, 69-70; scientist, Pillsbury Co, 70-71, group leader instrumentation, Corp Res, 71-78; mgr appl res, Pepsico Inc, 78-80; dir anal serv, Standard Brands, Inc, 80-81; group dir, Res Serv, Nabisco Brands, Inc, 81-84, group dir sci res, 85-86; group dir, Sci Serv, 86-88, GROUP DIR, SCI SERV, NABISCO BISCUIT CO, 89- *Mem:* Am Chem Soc; Inst Food Technologists; Am Asn Cereal Chemists. *Res:* Nuclear magnetic resonance; mass spectrometry; analytical robotics; flavors; perception; food chemistry; microwave heating; fats and oils; food analysis; beverage technology; emulsions; chemical kinetics; food chemistry; food microbiology; cereal chemistry; polymer chemistry; sensory analysis; information science and systems. *Mailing Add:* Nabisco Biscuit Co 200 De Forrest Ave PO Box 1944 East Hanover NJ 07936-1944

TEMPLEMAN, WILFRED, marine biology, fisheries; deceased, see previous edition for last biography

TEMPLER, DAVID ALLEN, b Chicago, Ill, July 23, 42; m 65; c 1. POLYMER CHEMISTRY. *Educ:* Northwestern Univ, Evanston, BS, 64; Ind Univ, Bloomington, PhD(org chem), 68. *Prof Exp:* Sr chemist, 68-75, lab head, Latin Am oper, 75-78, RES & DEVELOP MGR, INDUST CHEMS, LATIN AM REGION, TECH SERV LAB, ROHM & HAAS, 78- *Mem:* AAAS; Am Chem Soc; Royal Soc Chem. *Res:* Polymer synthesis and characterization with regard to applications in the area of organic coatings; medicinal chemistry; leather chemistry; textile chemistry; ion exchange. *Mailing Add:* Rohm & Haas Latin Am Oper 12906 SW 89th Ct Miami FL 33176

TEMPLETON, ARCH W, b Madison, Wis, Mar 30, 32; m 55; c 4. MEDICINE, RADIOLOGY. *Educ:* Univ Omaha, BA, 54; Univ Nebr, MD, 57. *Prof Exp:* Asst prof med, Wash Univ, 63-64; assoc prof, Univ Mo, 64-68; PROF RADIOL & CHMN DEPT, MED CTR, UNIV KANS, 68- *Mem:* Radiol Soc NAm; AMA; Asn Univ Radiol. *Res:* Computer research in medicine; vascular radiology. *Mailing Add:* 4821 Adams Shawnee Mission KS 66205

TEMPLETON, CHARLES CLARK, b Houston, Tex, Oct 4, 21; m 44; c 4. PHYSICAL CHEMISTRY. *Educ:* La Polytech Inst, BS, 42; Univ Wis, MS, 47, PhD(chem), 48. *Prof Exp:* Jr res chemist, Shell Oil Co, 42; res chemist, Univ Wis, 46-48; instr, Univ Mich, 48-50; staff res chemist, Bellaire Res Ctr, Shell Develop Co, 50-80; RETIRED. *Mem:* Am Chem Soc; Soc Petrol Engrs; Am Inst Mech Engrs. *Res:* Solvent extraction; phase equilibria; electrochemistry; multiphase fluid flow; petroleum production. *Mailing Add:* 6119 Reamer St Houston TX 77074

TEMPLETON, DAVID HENRY, b Houston, Tex, Mar 2, 20; m 48; c 2. PHYSICAL CHEMISTRY. *Educ:* La Polytech Inst, BS, 41; Univ Tex, MA, 43; Univ Calif, PhD(chem), 47. *Hon Degrees:* Fil Dr, Univ Uppsala, 77. *Prof Exp:* Instr chem, Univ Tex, 42-44; res chemist, Metall Lab, Univ Chicago, 44-46; res chemist, Radiation Lab, 46-47, from instr to assoc prof, Univ, 47-58, dean, Col Chem, 70-75, PROF CHEM, UNIV CALIF, BERKELEY, 58- *Concurrent Pos:* Guggenheim Mem fel, 53; lectr, Univ Lausanne, 82. *Honors & Awards:* Patterson Award, Am Crystallog Asn, 87. *Mem:* AAAS; Am Chem Soc; Am Crystallog Asn (pres, 85). *Res:* Properties of radioactive isotopes; nuclear reactions; structures of crystals; anomalous scattering of x-rays. *Mailing Add:* Dept Chem Univ of Calif Berkeley CA 94720

TEMPLETON, FREDERIC EASTLAND, b Portland, Ore, May 11, 05; m 36, 75; c 2. RADIOLOGY. *Educ:* Washington Univ, BS, 27; Univ Ore, MD, 31. *Prof Exp:* Univ Chicago fel radiol, Univ Stockholm, 33-34; from instr to assoc prof roentgenol, Univ Chicago, 35-43; head dept radiol, Cleveland Clin, 43-45; prof & exec officer, 47-53, clin prof, 53-68, prof, 68-75, EMER PROF RADIOL, MED SCH, UNIV WASH, 75- *Mem:* Am Gastroenterol Asn; Am Roentgen Ray Soc (1st vpres, 54); fel AMA; Radiol Soc NAm; hon mem Mex Soc Radiol & Phys Ther; Sigma Xi. *Res:* Radiologic gastroenterology. *Mailing Add:* PO Box 103 Medina WA 98039

TEMPLETON, GEORGE EARL, b Little Rock, Ark, June 27, 31; m 58; c 4. PLANT PATHOLOGY. *Educ:* Univ Ark, BS, 53, MS, 54; Univ Wis, PhD(plant path), 58. *Prof Exp:* Asst, Univ Wis, 56-58; from asst prof to prof, 58-85, UNIV PROF PLANT PATH, UNIV ARK, FAYETTEVILLE, 85- *Concurrent Pos:* Joint planning & eval staff, US Dept Agr, Sci & Educ Admin, 80; Underwood fel UK Agr Res Coun, 83; vis scholar, Univ Guelph, Ont, Can, 85. *Honors & Awards:* John White Award Excellence Agr Res, 79; Super Serv Award, USDA, 90. *Mem:* Fel Am Phytopath Soc; Mycol Soc Am; Sigma Xi. *Res:* Physiology of parasitism; diseases of rice; biological control of weeds with plant pathogen. *Mailing Add:* Dept Plant Path Univ Ark Fayetteville AR 72701

TEMPLETON, GORDON HUFFINE, b Edowah, Tenn, July 17, 40; m 64; c 1. PHYSIOLOGY. *Educ:* Univ Tenn, Knoxville, BS, 63; Southern Methodist Univ, MS, 68; Univ Tex Southwestern Med Sch Dallas, PhD(biophys), 70. *Prof Exp:* Instrumentation engr, Gen Dynamics Corp, Tex, 63-66; instr, 70-71, asst prof, 71-77, ASSOC PROF PHYSIOL, UNIV TEX SOUTHWESTERN MED SCH DALLAS, 78- *Mem:* Am Heart Asn; Inst Elec & Electronics Engrs; Am Fedn Clin Res; Am Physiol Soc. *Res:* Muscle mechanics; detection of ventricular asynergy by three-dimensional imaging. *Mailing Add:* Dept Physiol Univ Tex Southwestern Med Ctr Dallas TX 75235-9040

TEMPLETON, IAN M, b Rugby, Eng, July 31, 29; m 56; c 2. METAL PHYSICS, LOW TEMPERATURE PHYSICS. *Educ:* Oxford Univ, MA, 50, DPhil(physics), 53. *Prof Exp:* Fel physics, Nat Res Coun Can, 53-54; mem staff, Res Lab, Assoc Elec Industs, Rugby, 55-57; from asst res officer to sr res officer physics div, 57-71, PRIN RES OFFICER PHYSICS DIV, NAT RES COUN CAN, 71- *Mem:* Fel Brit Inst Physics; fel Royal Soc Can; Am Phys Soc; Can Asn Physicists. *Res:* Noise in semiconductors; superconductive devices; thermoelectricity; Fermi surfaces; focused ion beams. *Mailing Add:* Nat Res Coun Inst Microstruct Sci Ottawa ON K1A 0R6 Can

TEMPLETON, JOE WAYNE, b Loraine, Tex, July 18, 41; m 62; c 1. IMMUNOLOGY, GENETICS. *Educ:* Abilene Christian Col, BS, 64; Ore State Univ, PhD(genetics), 68. *Prof Exp:* Res fel genetics, Ore State Univ, 65-68; asst prof med genetics, Med Sch, Univ Ore, 68-75; asst prof microbiol & immunol, Baylor Col Med, Houston, 75-78; assoc prof vet med & surg, 75-80, genetics & vet pathobiol, 80-87, PROF GENETICS & VET PATHOBIOL, COL VET MED, TEX A&M UNIV, 87- *Mem:* Genetics Soc Am; Am Genetic Asn. *Res:* Immunogenetics of organ transplantation, especially in the dog; general canine genetics. *Mailing Add:* PO Box 9796 College Station TX 77842-0796

TEMPLETON, JOHN CHARLES, b Buffalo, NY, June 7, 43; m 67; c 2. INORGANIC CHEMISTRY. *Educ:* Col Wooster, BA, 65; Wesleyan Univ, MA, 67; Univ Colo, Boulder, PhD(inorg chem), 70. *Prof Exp:* Asst prof, 70-76, ASSOC PROF CHEM, WHITMAN COL, 76- *Concurrent Pos:* NSF grant, Whitman Col, 71-73. *Mem:* Am Chem Soc; Sigma Xi (vpres, 72-73, pres, 73-74). *Res:* Studies of transition-metal complex ions in concentrated acid solutions, correlations with acidity functions; reaction kinetics and mechanisms; ultraviolet-visible spectroscopy. *Mailing Add:* Dept Chem Whitman Col Walla Walla WA 99362

TEMPLETON, JOHN Y, III, b Portsmouth, Va, July 1, 17; m 43; c 4. SURGERY. *Educ:* Davidson Col, BS, 37; Jefferson Med Col, MD, 41. *Prof Exp:* Clin prof surg, Jefferson Med Col, 57-64; prof, Univ Pa, 65-67; Samuel D Gross prof & head dept, 67-70, PROF SURG, JEFFERSON MED COL, 70- *Concurrent Pos:* Am Cancer Soc clin fel, Jefferson Hosp, 50-51, Runyon fel, 51-52. *Mem:* Am Surg Asn; Soc Thoracic Surg; Am Col Surg; Soc Vascular Surg; Int Soc Surg; Sigma Xi. *Res:* General, cardiac and gastrointestinal surgery. *Mailing Add:* 311 Airdale Rd Rosemont PA 19010

TEMPLETON, JOSEPH LESLIE, b Knoxville, Iowa, Nov 3, 48; m 71; c 2. EARLY TRANSITION METALS, METAL CLUSTERS. *Educ:* Calif Inst Technol, BS, 71; Iowa State Univ, PhD(inorg), 75. *Prof Exp:* NATO fel, Imperial Col Sci & Technol, 75-76; asst prof, 76-81, assoc prof inorg, 81-86, PROF, UNIV NC, 86- *Mem:* Am Chem Soc. *Res:* Early transition metal organometallic chemistry; ligand pi donation in electron deficient complexes; metal cluster chemistry; organometallic reaction mechanisms. *Mailing Add:* Dept Chem Univ NC Chapel Hill NC 27514

TEMPLETON, MCCORMICK, b Cincinnati, Ohio, May 12, 23; m 54; c 2. ANATOMY. *Educ:* Columbia Univ, AB, 48; Univ Kans, PhD(anat), 58. *Prof Exp:* Asst zool, Columbia Univ, 48-49, asst oncol & path, Med Ctr, Univ Kans, 49-51, instr anat, 51-54, asst neuroembryol, 54-57; instr anat, Med Sch, Northwestern Univ, 57-65; asst prof, 65-69, actg chmn dept, 71-76, chmn dept, 76-80, ASSOC PROF ANAT, SCH DENT, UNIV SOUTHERN CALIF, 69-, CO-DIR TEMPOROMANDIBULAR JOINT & OROFACIAL PAIN CLIN, 78- *Mem:* AAAS; Am Asn Anat; Biol Stain Comn; Am Soc Cell Biol. *Res:* Medical anatomy; histochemistry and microchemistry of nervous system of embryo and adult vertebrates; electrophoresis of esterases in blood and digestive tract. *Mailing Add:* 4529 Valdina Pl Los Angeles CA 90043

TEMPLETON, WILLIAM LEES, b London, Eng, Apr 15, 26; m 52; c 3. RADIOLOGICAL ASSESSMENT, MARINE DUMPING. *Educ:* Univ St Andrews, BSc, 50, Hons, 51. *Prof Exp:* Sr biologist, UK Atomic Energy Authority, Windscale, Eng, 51-65; sr res scientist radioecol, 65-68, mgr aquatic ecol, 68-69, assoc dept mgr ecosysts, 69-85, sr staff scientist, Earth & Environ Sci dept, 85-86, MGR NEPA & RADIOL PROTECTION, PAC NORTHWEST LABS, OFF HANFORD ENVIRON, 86- *Concurrent Pos:* Consult, Int Atomic Energy Agency, 60-; mem panel radioactivity in marine environ, Nat Acad Sci-Nat Res Coun, 68-72, mem panel energy & environ, 75-78; chmn, Exec Comt Coord Res & Environ Surveillance Prog, Orgn Econ Coop & Develop/Nuclear Energy Agency, 80-85; mem coun, Nat Coun Radiation Protection & Measurements, 85- *Mem:* Fel AAAS; Marine Biol Asn UK; Am Soc Limnol & Oceanog; UK Freshwater Biol Asn. *Res:* Waste management practice as related to radioecology and limnology of fresh and marine waters; effects of low level chronic pollution; radiological assessment. *Mailing Add:* Pac Northwest Labs Off Hanford Environ K1-30 Battelle Mem Inst PO Box 999 Richland WA 99352

TEMS, ROBIN DOUGLAS, b London, UK, Nov 28, 52; m 77; c 2. CORROSION ENGINEERING. *Educ:* Univ Newcastle, UK, BSc, 74, PhD(stress corrosion), 78. *Prof Exp:* Jr res assoc, Univ Newcastle upon Tyne, 74-77, res assoc, 77-78; res engr, Pedco Environ, Inc, 78-79; sr staff metallurgist, 80-84, sr staff corrosion engr, Mobil Explor, Norway, 84-86, SR RES METALLURGIST, MOBIL RES & DEVELOP CORP, 86- *Mem:* Inst Metals; Nat Asn Corrosion Engrs; Norweg Corrosion Soc. *Res:* Corrosion and stress corrosion in oil and gas; gas transmission; pollution control. *Mailing Add:* Mobil Res & Develop Corp Dallas Res Lab 13777 Midway Rd Dallas TX 75244

TENAZA, RICHARD REUBEN, b San Mateo, Calif, Mar 22, 39. ETHOLOGY, ECOLOGY. *Educ:* San Francisco State Univ, BA, 64; Univ Calif, Davis, PhD(zool), 74. *Prof Exp:* Asst prof biol, Univ of the Pac, 75-76; res scientist wildlife biol, Sci Applns, Inc, 76-77; ASST PROF BIOL, UNIV OF THE PAC, 77- *Concurrent Pos:* Consult wildlife biol, Sci Applns, Inc, 77- *Mem:* AAAS. *Res:* Primatology; behavioral ecology; animal communication; marine birds and mammals; environmental impacts of oil development. *Mailing Add:* Dept Biol/Sci Univ of the Pac 3601 Pacific Ave Stockton CA 95211

TEN BRINK, NORMAN WAYNE, b Shelby, Mich, May 17, 43; m 67; c 2. QUATERNARY GEOLOGY. *Educ:* Univ Mich, Ann Arbor, BS, 66; Franklin & Marshall Col, MS, 68; Univ Wash, PhD(geol), 71. *Prof Exp:* Contract geologist, Geol Surv Greenland, 69-71; res fel geol, Inst Polar Studies & Dept Geol, Ohio State Univ, 71-72, asst dir, Inst Polar Studies, 72-73; chmn dept, 83-85, PROF GEOL, GRAND VALLEY STATE COL, 73- *Concurrent Pos:* Field leader NSF grant, Univ Wash, 70-71; prin investr NSF grants, Ohio State Univ, 72-74 & 73-75; consult geohydrologist, Environ Protection Agency, 76-77; prin investr, Nat Park Serv & Nat geog grants, Grant Valley State Col & Univ Alaska, 77-83 & Alaska Geol Surv grants, 82-83; vis prof, S Ill Univ, 79-80; consult geologist, Woodward-Clyde consults, 80-81. *Mem:* Am Quaternary Asn; Arctic Inst NAm; Geol Soc Am. *Res:* Glacial geology and geomorphology of Arctic, Antarctic and alpine areas. *Mailing Add:* Dept Geol Grand Valley State Cols Allendale MI 49401

TENBROEK, BERNARD JOHN, b Grand Rapids, Mich, Mar 29, 24; m 48; c 4. ZOOLOGY. *Educ:* Calvin Col, BA, 49; Univ Colo, MA, 55, PhD(zool), 60. *Prof Exp:* From instr to assoc prof, Calvin Col, 55-66, chmn dept, 61-73, prof biol, 66-86; RETIRED. *Mem:* AAAS; Am Soc Zool; Sigma Xi. *Res:* Studies on thyroid and pituitary function in neotenic forms of Ambystoma tigrinum. *Mailing Add:* 2307 Edgewood St SE Grand Rapids MI 49546

TENCA, JOSEPH IGNATIUS, b Bay Shore, NY, Mar 6, 29; m 55; c 3. ENDODONTICS. *Educ:* Holy Cross Col, AB, 50; Georgetown Univ, DDS, 54; George Washington Univ, MA, 74; Am Bd Endodontics, dipl. *Prof Exp:* Captain dent, US Navy, 53-75; PROF & CHMN ENDODONTICS DEPT, SCH DENT MED, 75-, DIR, ADVAN EDUC, TUFTS UNIV, 83- *Concurrent Pos:* Dir, secy-treas & pres, Am Bd Endodontics, 79-85; chmn & dir grad educ, Nat Naval Dent Ctr, Md, 71-75; prof & lectr oral biol, Grad Sch George Washington Univ, 71-75; consult, Nat Naval Dent Ctr, 76- & Comn Dent Accreditation, Am Dent Asn, 80-; vis lectr, Dept Restorative Dent, Sch Clin Dent, Univ Sheffield, UK, 85-86. *Mem:* Int Asn Dent Res; Am Asn Endodontists; Am Dent Asn; Am Asn Dent Schs; fel Int Col Dentists. *Res:* Clinical endodontics, more specifically in radiographic interpretation and reliability of various endodontic instruments, filling materials; restoration of endodontically treated teeth. *Mailing Add:* Endodontics Dept Sch Dent Med Tufts Univ One Kneeland St Boston MA 02111

TEN CATE, ARNOLD RICHARD, b Accrington, Eng, Oct 21, 33; m 56; c 3. ANATOMY, DENTISTRY. *Educ:* Univ London, BDS, 60, BSc, 55, PhD(anat), 58. *Hon Degrees:* DSc, McGill, 89, Univ Western Ont, 89. *Prof Exp:* Sr lectr dent sci, Royal Col Surgeons Eng, 61-63; sr lectr anat in dent, Guy's Hosp Med Sch, Univ London, 63-68; prof anat & dent, Univ Toronto, 68-71, chmn div, 71-77, prof biol sci & dean fac dent, 77-89, VPROVOST, HEALTH SCI, UNIV TORONTO, 89- *Honors & Awards:* Colyer Prize, Royal Soc Med, 62; Milo Hellman Award, Am Asn Orthod, 75; Isaac Schour Mem Award, Int Asn Dent Res, 78. *Mem:* Int Asn Dent Res (vpres, 82, pres-elect, 83, pres, 84). *Res:* Dental histology; development of periodontium and connective tissue remodeling. *Mailing Add:* Univ Toronto 27 Kings College Circle Toronto ON M5S 1A1 Can

TENCZA, THOMAS MICHAEL, b Wallington, NJ, July 8, 32; m 59. ORGANIC CHEMISTRY. *Educ:* Columbia Univ, AB, 54; Seton Hall Univ, MS, 64, PhD(chem), 66; Fairleigh Dickinson Univ, MBA, 71. *Prof Exp:* Res chemist, S B Penick Co, 57-60; DIR PROD DEVELOP, BRISTOL-MYERS CO, HILLSIDE, 60- *Mem:* Am Chem Soc; Am Pharmaceut Asn; Am Mgt Asn; Soc Cosmetic Chemists. *Res:* Research and development management; product development; rearrangement reactions of small ring compounds; botanical drugs. *Mailing Add:* 31 Wagner Ave Wallington NJ 17057-1638

TENDAM, DONALD JAN, b Hamilton, Ohio, May 28, 16; m 39; c 3. NUCLEAR PHYSICS. *Educ:* Miami Univ, AB, 40; Purdue Univ, MS, 42, PhD(physics), 49. *Prof Exp:* Asst, 40-42, from instr to assoc prof, 42-60, PROF PHYSICS, PURDUE UNIV, WEST LAFAYETTE, 60-, ASSOC HEAD DEPT, 66- *Mem:* AAAS; Am Phys Soc; Am Asn Physics Teachers. *Res:* Particle accelerators; radioactive tracers; nuclear reactions; deuteron-bombarded semiconductors. *Mailing Add:* 332 Meridian St West Lafayette IN 47906

TENDLER, MOSES DAVID, b New York, NY, Aug 7, 26; m 48; c 8. MICROBIOLOGY. *Educ:* NY Univ, BA, 47, MA, 51; Columbia Univ, PhD, 57. *Prof Exp:* From instr to assoc prof, 52-63, asst dean, 56-59, PROF BIOL, YESHIVA UNIV, 63-, PROF TALMUD, 65- & PROF MED ETHICS, 86- *Concurrent Pos:* Consult, Eli Lilly & Co, 59-61 & Hoffmann-La Roche Inc, 63-; res dir, Thermobios Pharmaceut Corp, 63-; mem res adv coun, NY Cancer Res Inst, 65-71; vchmn, Kashruth Adv Bd, NY State Dept Agr, 68- *Honors & Awards:* Maimonides Award, 87. *Mem:* AAAS; Am Soc Microbiol; Asn Orthodox Jewish Scientists; NY Acad Sci. *Res:* Nutrition of thermophilic actinomycetes; antibiotic and antitumor agents proliferated by thermophilic organisms; physiological problems of thermophily; discoverer of Anthramycin, an antitumor antibiotic. *Mailing Add:* Dept Biol Yeshiva Univ 500 W 185th St New York NY 10033

TEN EICK, ROBERT EDWIN, b Portchester, NY, Oct 14, 37; m 62; c 2. CELLULAR ELECTROPHYSIOLOGY, PHARMACOLOGY. *Educ:* Columbia Univ, BS, 63, PhD(pharmacol), 68. *Prof Exp:* Guest investr cardiac electrophysiol, Rockefeller Univ, 68; from asst prof to assoc prof, 68-81, PROF PHARMACOL, MED SCH, NORTHWESTERN UNIV, CHICAGO, 81- *Concurrent Pos:* NIH trainee cardiac electrophysiol, Rockefeller Univ, 68; vis prof II, Physiol Inst, Univ Saarland, WGer, 74-75; NIH res career develop award, 75-80; consult, Heart, Lung & Blood Inst & Physiol & Cardiovasc Study Sect, Pharmacol Study Sect, NIH, 88-92; Warren McDonald int scholar, Australian Heart Coun, 90. *Mem:* Am Physiol Soc; Am Heart Asn; Am Soc Pharmacol & Exp Therapeut; Cardiac Muscle Soc; Int Soc Heart Res. *Res:* Cellular electrophysiological basis of cardiac dysrhythmias associated with heart disease; cellular electrophysiology of the heart; myocardial membrane currents and their relation to cardiac electrical activity during cardiac hypertrophy. *Mailing Add:* Dept Pharmacol Northwestern Univ Med Sch Chicago IL 60611

TENENBAUM, JOEL, b Brooklyn, NY, Dec 17, 40; m 67; c 3. DYNAMIC METEOROLOGY, COMPUTER GRAPHICS. *Educ:* Calif Inst Technol, BS, 62; Harvard Univ, AM, 63, PhD(physics), 69. *Prof Exp:* Res assoc physics, Stanford Linear Accelerator Ctr, 68-71; asst prof, State Univ NY Col Purchase, 71-76, actg dean, Natural Sci, 78 & 79, assoc prof physics, 76-89, PROF PHYSICS & SCI COMPUT, STATE UNIV NY COL PURCHASE, 89- *Concurrent Pos:* Consult, Inst Space Studies, NASA Goddard Space Flight Ctr, 72-75; vis assoc prof meteorol, Mass Inst Technol, 80-82; vis scholar meteorol, Harvard, 80-82. *Mem:* AAAS; Am Phys Soc; Am Asn Physics Teachers; Am Meteorol Soc. *Res:* Modeling of large scale processes in dynamic meteorology; computer graphics representations of atmospheric phenomena. *Mailing Add:* Div Natural Sci State Univ NY Col Purchase NY 10577-0556

TENENBAUM, MICHAEL, b St Paul, Minn, July 23, 13; m 41; c 2. BASIC OXYGEN FURNACE STEELMAKING TECHNOLOGY. *Educ:* Univ Minn, BS, 36, MS, 37, PhD(metall, phys chem), 40. *Hon Degrees:* DSc, Northwestern Univ, 74. *Prof Exp:* Raw mat res TC&I RR, 39-40; metallurgist, Metall Dept, Inland Steel Co, 41-50, asst supt qual control, 50-56, supt metall dept, 56-59, asst gen mgr tech serv, 59-61, gen mgr res & qual control, 61-66, vpres res, 66-68, vpres, Steel Mfg, 68-71, pres, 71-78, dir, 71-85; RETIRED. *Concurrent Pos:* Dir, Blast Furnace Res, Inc, 64-69; US rep, Int Iron & Steel Inst; dir, Paxall Ind, Cont Ill Bank, 72-80. *Honors & Awards:* Nat Open Hearth Comt Award, Am Inst Mining, Metall & Petrol Engrs, 47 & 48, Raymond Award, 49, Hunt Award, 50 & Fairless Award, 75; Bessemgr Medal, Brit Metal Soc, 80. *Mem:* Nat Acad Eng; distinguished mem Am Inst Mining, Metall & Petrol Engra; Metall Soc (pres, 68); distinguished mem Am Soc Metals; Am Iron & Steel Inst; Asn Iron & Steel Engrs; Brit Inst Metals. *Res:* Metallurgy and chemistry of iron and steel manufacture. *Mailing Add:* 4049 220 Pl SE Issaquah WA 98027

TENENBAUM, SAUL, b New York, NY, Nov 3, 17; m 41; c 3. MICROBIOLOGY. *Educ:* Wash State Univ, BS, 43; Long Island Univ, MS, 64. *Prof Exp:* Jr seafood inspector, US Food & Drug Admin, 42-44; asst chemist, US Maritime Comn, 44-45; sr biochemist, Standard Brands, 45-46; bacteriologist, Atlantic Yeast Co, 46-48; chief bacteriologist, Premo Pharmaceut Lab, 48-57; group leader, Revlon Res Ctr, 57-65, mgr, 65-67, asst dir microbiol, 67-79, dir res microbiol, 79-84; RETIRED. *Concurrent Pos:* Lectr, Fairleigh-Dickinson Univ, 51-56; adj assoc prof pharmaceut sci, Sch Pharm, St John's Univ, 73-85; course dir, Ctr Continuing Educ, 77-84. *Mem:* Soc Indust Microbiol; Am Soc Microbiol; Soc Cosmetic Chemistry; NY Acad Sci. *Res:* Development and evaluation of biostatic and biocidal agents; pseudomonads; preservation; microbial content; skin microbiology; immunology; hypersensitive state and agents; mutagenicity, topical and ocular infection; clinical evaluation of irritants and allergens; sterilization, disinfection and antisepsis; phototoxicity. *Mailing Add:* 2143 Sunhaven Cir Fairfield CA 94533

TENENHOUSE, ALAN M, b Montreal, Que, Aug 8, 35; m 61; c 2. BIOCHEMISTRY, ENDOCRINOLOGY. *Educ:* McGill Univ, BSc, 55, PhD(biochem), 59, MD & CM, 62. *Prof Exp:* Asst prof, 68-77, ASSOC PROF PHARMACOL & THERAPEUT, McGILL UNIV, 78- *Concurrent Pos:* Fel biochem, Univ Wis, 63-65; NIH fel, 64-66; fel biochem, Univ Pa, 65-68; Univ Pa plan scholar, 66-68. *Mem:* Endocrine Soc; Can Biochem Soc; Can Pharmacol Soc. *Res:* Mechanism of hormone action with emphasis on role of 3' 5' AMP and calcium in hormone action; biochemistry and physiology of parathyroid hormone and calcitonin. *Mailing Add:* Dept Pharmacol McGill Univ Sherbrooke St W Montreal PQ H3A 2M5 Can

TENENHOUSE, HARRIET SUSIE, b Montreal, Que, Apr 15, 40; m 61; c 2. GENETICS. *Educ:* McGill Univ, BSc, 61, MSc, 63, PhD(biochem), 72. *Prof Exp:* Lectr pediat, 77-81, asst prof pediat, 81-87, ASSOC PROF PEDIAT, MCGILL CTR HUMAN GENETICS, MCGILL UNIV, 87- *Mem:* Am Soc Biol Chemists; Am Soc Bone & Mineral Res; Can Biochem Soc; Sigma Xi. *Res:* The regulation of renal phosphate, calcium and vitamin D metabolism; the nature of the primary mutation in a mouse model (Hyp) of X-linked hypophosphatemic rickets in man. *Mailing Add:* MRC Genetics Group Montreal Children's Hosp 2300 Tupper St Montreal PQ H3H 1P3 Can

TENER, GORDON MALCOLM, b Vancouver, BC, Nov 24, 27. BIOCHEMISTRY. *Educ:* Univ BC, BA, 49; Univ Wis, MS, 51, PhD(biochem), 53. *Prof Exp:* Rockefeller fel biochem, Inst Phys-Chem Biol, Paris, France, 53-54; res scientist, BC Res Coun, Vancouver, 54-60; from asst prof to assoc prof, 60-67, PROF BIOCHEM, UNIV BC, 67- *Honors & Awards:* Merck Sharp & Dohme Award, Chem Inst Can, 64. *Mem:* Am Chem Soc; Am Soc Biol Chem; Can Biochem Soc; Royal Soc Chem. *Res:* Purification and properties of transfer ribonucleic acids; gene localization in Drosophila; gene structure; recombinant DNA; molecular biology of aging. *Mailing Add:* Dept Biochem Univ BC 2146 Health Sci Mall Vancouver BC V6T 1W5 Can

TEN EYCK, EDWARD H(ANLON), JR, b Pearl River, NY, Sept 6, 23; m 52; c 2. CHEMICAL ENGINEERING, INDUSTRIAL & MANUFACTURING ENGINEERING. *Educ:* Syracuse Univ, BChE, 43; Polytech Inst Brooklyn, MChE, 48, DChE, 50. *Prof Exp:* Chem engr, Johns Manville Corp, 44; from asst tech supt to lab adminr, E I DuPont de Nemours Co, Inc, Wilmington, 49-71, prog mgr, 71-76, mfg mgr, plastics develop, 76-85; INDUST CONSULT, 85- *Mem:* Am Chem Soc; Am Inst Chem Engrs; Sigma Xi. *Res:* Polymerization and plasticizers; phase relations of petroleum hydrocarbons; process development; organic chemicals; energy economics. *Mailing Add:* PO Box 3656 Greenville DE 19807-0656

TENFORDE, THOMAS SEBASTIAN, b Middletown, Ohio, Dec 15, 40; m 79; c 2. BIOPHYSICS, SCIENCE ADMINISTRATION. *Educ:* Harvard Univ, AB, 62; Univ Calif, Berkeley, PhD(biophys), 69. *Prof Exp:* Fel, Univ Calif, Berkeley, 69-73; biophysicist, Lawrence Berkeley Lab, Univ Calif, 73-87; sr scientist, 82-87, group leader, Radiation Biophys Group, 81-82, dept dir Donner Lab, Biol & Med Div, 82-83; physiol group leader, 83-87, CHIEF

SCIENTIST, LIFE SCI CTR, BATTELLE PAC NORTHWEST LAB, 88- *Concurrent Pos:* Mem, Comt SC-67, Nat Coun Radiation Protection & Measurements, 81-; mem, tech panel magnetic fusion energy, Energy Res Adv Bd, US Dept Energy, 83-84; mem, Physiol Working Group Comt 95.4, Am Nat Standards Inst, 83-90; mem Fla Sci Adv comn elec & magnetic fields, 84-85; mem, comt biol & human health effects extremely low frequency fields, Am Inst Biol Sci, 84-85; mem adv comt, biol effects elec & magnetic fields, Elec Power Res Inst, 85-; mem energy eng bd, comt energy conserv Res, Nat Res Coun, Nat Acad Sci, 85-86; comt man & radiation, Inst Elec & Electronics Engrs, 88-; coun mem, Nat Coun Radiation Protect & Measurements, 88-; mem, bd radiation effects res, Nat Res Coun, 89-, phys agents comt, Am Conf Govt Indust Hygienists, 89-, bd dir, Nat Coun Radiation Protected Measurements 91-; chmn, comt health effects ground wave commun syst, Nat Res Coun, 90- *Mem:* AAAS; Bioelectromagnetics Soc (pres 87-88); Biophys Soc; NY Acad Sci; Radiation Res Soc. *Res:* Radiation biology; biological effects of electromagnetic fields; surface chemistry of normal and cancer cells. *Mailing Add:* Life Sci Ctr (K1-50) Battelle Pac Northwest Lab PO Box 999 Richland WA 99352

TENG, CHING SUNG, b Amoy, Fukien, Nov 20, 37; US citizen; m 64; c 2. REPRODUCTIVE BIOLOGY, BIOCHEMISTRY. *Educ:* Tunghai Univ, Taiwan, BS, 60; Univ Tex, Austin, MS, 64, PhD(biochem), 67. *Prof Exp:* Res assoc biochem, Univ Tex, Austin, 67-69; guest investr, Rockefeller Univ, 69-71; asst prof develop biol, State Univ NY Stony Brook, 71-73; assoc prof cell biol, Baylor Col Med, 73-80; PROF ANAT & PHYSIOL, NC STATE UNIV, 81- *Concurrent Pos:* NIH res fel, Cancer Inst, 69-70, NIH spec res fel, 70-71, NIH grant award, 73-; NSF grant award, 82-; Rockefeller Found grant award, 83- *Mem:* Am Soc Cell Biol; Endocrine Soc; Sigma Xi. *Res:* Steroid hormone-controlled sex organ differentiation. *Mailing Add:* Dept Anat Physiol Sci & Radiol NC State Univ 4700 Hillsborough Rd Raleigh NC 27606

TENG, CHOJAN, b Taipei, Taiwan, Aug 31, 47; m 71, 83; c 3. NONCOOPERATIVE TARGET RECOGNITION & RADAR COMMUNICATION, SYSTEMS DESIGN & ANALYSTS. *Educ:* Nat Taiwan Univ, BS, 69; Wash State Univ, MS, 72; Univ Wis-Madison, MS, 76, PhD(elec eng), 78. *Prof Exp:* MEM TECH STAFF RADAR SYST, DIGITAL SIGNAL PROCESSING & WAVE PROPAGATION, MITRE CORP, 78- *Concurrent Pos:* Mem staff, Mass Inst Technol, Lincoln Lab, 80-83; adj prof, Univ Lowell, 85- *Mem:* Inst Elec & Electronics Engrs; Sigma Xi; Appl Computational Electromagnetic Soc. *Res:* Noncooperative target recognition techniques including radar signal modulation, neural network, inverse SAR, high range resolution, lasar radar detection of vibration signatures; radar system design and analysis; digital signal processing; numerical analysis; applied mathematics. *Mailing Add:* MITRE Corp MS D205 Burlington Rd Bedford MA 01730-0208

TENG, CHRISTINA WEI-TIEN TU, b Kuming, Yunnan, China, July 23, 42; m 64; c 2. GENE REGULATION. *Educ:* Tunghai Univ, Taiwan, BS, 63; Univ Tex, Austin, PhD(biol), 69. *Prof Exp:* Guest investr cell biol, Rockefeller Univ, 69-71; sr res assoc med, Brookhaven Nat Lab, 71-73; asst prof cell biol, Baylor Col Med, 73-81; vis asst prof, 81-83, ADJ ASSOC PROF, NC STATE UNIV, 83-; EXPERT, NAT INST ENVIRON HEALTH SCI, NIH, 83- *Mem:* Am Soc Cell Biol; Sigma Xi. *Res:* Hormone controlled sex organ differentiation; study of the regulatory mechanism for gene activation in mammalian system. *Mailing Add:* Nat Inst Environ Health Sci NIH PO Box 12233 MD1301 Research Triangle Park NC 27709

TENG, EVELYN LEE, b Chungking, China, Feb 8, 38; m 63; c 2. PSYCHOBIOLOGY, NEUROPSYCHOLOGY. *Educ:* Taiwan Univ, BS, 59; Stanford Univ, MA, 60, PhD(psychol), 63. *Prof Exp:* Res fel psychobiol, Calif Inst Technol, 63-69, sr res fel, 69-72; asst prof neurol, 72-75, ASSOC PROF NEUROL, SCH MED, UNIV SOUTHERN CALIF, 75- *Mem:* Am Psychol Asn; Int Neuropsychol Soc. *Res:* Functional relationship between brain and behavior, especially higher cognitive functions in man. *Mailing Add:* 1474 Rose Villa St Pasadena CA 91106

TENG, JAMES, b Hong Kong, Dec 4, 29; US citizen; m 57; c 2. ENGINEERING. *Educ:* Tri-State Col, BS, S2; Case Western Reserve Univ, MS, 61, PhD(org chem), 67. *Prof Exp:* Chem engr, Radio Receptor Co, 52-53; process engr, Nylonge Corp, 53-56, res supvr, 56-61, tech supvr, 61-66; fel, Purdue Univ, 66-67; res proj mgr, 68-75, res mgr advan prod, 75-78, mgr, Process Optimization Ctr, 78-83, DIR, PROCESS DEVELOP & OPTIMIZATION CTR, ANHEUSER BUSCH CO, 83-, SR DIR, TECHNOL PLANNING. *Mem:* AAAS; Am Inst Chem; NY Acad Sci; Am Chem Soc; Master Brewers Asn Am. *Res:* Carbohydrate chemistry; regenerated cellulose; cellulose derivatives; starch derivatives; carbohydrates in brewing; brewing process. *Mailing Add:* Corp Res Anheuser Busch Co St Louis MO 63118

TENG, JON IE, b Kienow, China, Oct 19, 30; m 58; c 2. STEROID CHEMISTRY, NATURAL PRODUCTS CHEMISTRY. *Educ:* Nat Taiwan Univ, BS, 55; SDak State Univ, MS, 62; Univ Fla, PhD(agr biochem), 65. *Prof Exp:* Agr scientist, Taiwan Sugar Corp, Inc, 56-60; res assoc nitrogen metab in hort plants, Univ Ill, Urbana, 65-66; res chemist, Am Crystal Sugar Co, 66-68; NIH fel steroid biochem, Med Sch, Univ Minn, 68-69, res assoc pharmacol, 69-70; res biochemist, 70-75, RES SCIENTIST, UNIV TEX MED BR GALVESTON, 75- *Mem:* Am Chem Soc; Inst Am Chemists; Sigma Xi. *Res:* Steroid biosynthesis and metabolism; drug metabolism; plant nutrition and biochemistry; steroid biochemistry; cholesterol metabolism in mammalian liver, kidney, brain and aortal tissues. *Mailing Add:* Dept Biochem Hendrix Bldg Univ Tex Med Br Galveston TX 77550

TENG, LEE CHANG-LI, b Peiping, China, Sept 5, 26; nat US; m 61; c 1. THEORETICAL PHYSICS. *Educ:* Fu Jen Univ, China, BS, 46; Univ Chicago, MS, 48, PhD(physics), 51. *Prof Exp:* Cyclotron asst, Univ Chicago, 49-51; lectr physics, Univ Minn, 51-52, asst prof, 52-53; assoc prof, Univ Wichita, 53-55; assoc physicist, Particle Accelerator Div, Argonne Nat Lab, 55-61, head theory group, 56-62, sr physicist, 61-67, dir, 62-67; head accelerator theory sect, Fermi Nat Accelerator Lab, 67-72, assoc head, Accelerator Div, 72-75, head adv proj sect, 75-83, dir Synchrotron Radiation Res Ctr, 83-85, dirs off spec assignment, 87-89, BD DIRS SYNCHROTRON RADIATION RES CTR, FERMI NAT ACCELERATOR LAB, 85-; HEAD ACCELERATOR PHYSICS, ADVAN PHOTON SOURCE PROJ, ARGONNE NAT LAB, 89- *Concurrent Pos:* Argonne fel, Argonne Nat Lab, 84-89. *Mem:* Fel Am Phys Soc. *Res:* High energy accelerators and instrumentation; high energy and nuclear physics. *Mailing Add:* Argonne Nat Lab 9700 S Cass Ave Argonne IL 60439

TENG, LINA CHEN, b Fukien, China, Dec 8, 39; US citizen; m 65. DRUG METABOLISM. *Educ:* Nat Taiwan Univ, BS, 63; Utah State Univ, PhD(org chem), 67. *Prof Exp:* Res chemist, Philip Morris Inc, 67-71; sr res chemist, 73-78, res assoc, 78-84, SR RES ASSOC, A H ROBINS CO, 84- *Mem:* Am Chem Soc; Sigma Xi; AAAS. *Res:* Studies of the metabolism, mainly isolation and identification of the metabolites, of the existing or research drugs in animals and human beings; synthesis of radiolabelled compounds for drug research. *Mailing Add:* A H Robins Co 1211 Sherwood Ave Richmond VA 23220

TENG, TA-LIANG, b China, July 3, 37; m 63; c 2. GEOPHYSICS. *Educ:* Univ Taiwan, BS, 59; Calif Inst Technol, PhD(geophys, appl math), 66. *Prof Exp:* Res fel geophys, Seismol Lab, Calif Inst Technol, 66-67; from asst prof to assoc prof geophys, 67-74, assoc prof, 74-77, PROF GEOL SCI, UNIV SOUTHERN CALIF, 78- *Mem:* AAAS; Am Geophys Union; Seismol Soc Am. *Res:* Elastic wave propagations; observational and theoretical seismology; elastic and anelastic properties of the earth and planetary interiors. *Mailing Add:* Dept Geol Sci Univ Southern Calif Los Angeles CA 90089

TENGERDY, ROBERT PAUL, b Budapest, Hungary, Dec 17, 30; US citizen; m 53; c 2. ADJUVANT VACCINES, MICROBIOLOGY. *Educ:* Budapest Tech Univ, Dipl Chem Eng, 53; St John's Univ, NY, PhD(microbial biochem), 61. *Prof Exp:* Asst prof biochem eng, Budapest Tech Univ, 53-56; res biochemist, Chas Pfizer & Co, NY, 57-61; asst prof chem & microbiol, 61-64, assoc prof biochem & microbiol, 64-71, PROF MICROBIOL, COLO STATE UNIV, 71- *Concurrent Pos:* Europ Molecular Biol Orgn fel, Pasteur Inst Paris, 68; Humboldt fel, Max Planck Inst, Univ G-ettingen, 68; Fulbright fel, Peru, 85. *Mem:* AAAS; Soc Indust Microbiol; Am Soc Microbiol. *Res:* Agricultural biotechnology; nutritional aspects of immunology; applied microbiology; waste conversion by microbes; vaccine development; bioconversion of agricultural residues into animal feed or fuel using combined enzymatic hydrolysis and ensiling, solid substrate fermentation by fungi; fluidized bed technology development of vitamin E adjuvant vaccines for veterinary use. *Mailing Add:* Dept Microbiol Colo State Univ Ft Collins CO 80523

TENHOVER, MICHAEL ALAN, b Cincinnati, Ohio, Nov 9, 53. CATALYSIS. *Educ:* Univ Cincinnati, BS, 76; Calif Inst Technol, MS, 78, PhD(appl physics), 81. *Prof Exp:* SR RES PHYSICIST, DEPT RES, STANDARD OIL CO, 81- *Mem:* Am Phys Soc. *Res:* Atomic and electronic structure of amorphous metals and semiconductors; high fluid superconductivity; nuclear spectroscopy; oxidation catalysis. *Mailing Add:* Standard Oil Co 4440 Warrensville Center Rd Warrensville Heights OH 44128

TENN, JOSEPH S, b Los Angeles, Calif, May 11, 40; m 67; c 2. MODERN HISTORY ASTRONOMY. *Educ:* Stanford Univ, BS, 62; Univ Wash, Seattle, MS, 66, PhD(physics), 70. *Prof Exp:* Teacher, physics, math, US Peace Corps, Ethiopia, 62-64; teaching asst, physics, Univ Wash, 64-66, assoc res, 66-70; res scientist, physics, NASA Ames Res Ctr, 65; from asst prof to assoc prof, 70-80, PROF, PHYSICS & ASTRON, SONOMA STATE UNIV, 80- *Concurrent Pos:* Postgrad res Astronomer, Lick Obser, 75-76; adj prof, physics & astron, Univ Mass, Amherst, 84-85. *Mem:* Am Phys Soc; Am Asn Physics Teachers; Am Astron Soc; Astron Soc Pac. *Res:* Astronomy, history of astronomy. *Mailing Add:* Dept Phys & Astron Sonoma State Univ Rohnert Park CA 94928

TENNANKORE, KANNAN NAGARAJAN, b Madras, India, Oct 30, 46; m 71; c 1. CHEMICAL ENGINEERING. *Educ:* Univ Madras, BSc, 65; Indian Inst Technol, Madras, BTech, 68, MTech, 70; Univ NB, PhD(chem eng), 75. *Prof Exp:* Asst eng design & develop distillation columns, Engrs India Ltd, India, 70-71; fel combustion, Univ NB, 75-77, res assoc dispersion of aerial sprays, 77-78; ENG RES ANALYST FLOW MASS & HEAT TRANSFER ENCLOSURES, WHITESHELL NUCLEAR RES ESTAB, ATOMIC ENERGY CAN LTD, 78- *Res:* Experimental study and modeling of flow, heat and mass transfer in enclosures. *Mailing Add:* Seven Cauchon Rd Pinawa MB R0E 1L0 Can

TENNANT, BUD C, b Burbank, Calif, Nov 10, 33; m 63; c 3. VETERINARY MEDICINE, GASTROENTEROLOGY. *Educ:* Univ Calif, BS , 57, DVM, 59. *Prof Exp:* Intern, Sch of Vet Med, Univ Calif, 59, from asst prof to assoc prof, 62-72; res assoc, Dept Surg, Albert Einstein Col Med, 62; res fel, Gastrointestinal Unit, Mass Gen Hosp, 6869; PROF COMP GASTROENTEROL, NY STATE COL VET MED, CORNELL UNIV, 72- *Mem:* Am Col Vet Internal Med; Am Gastroenterol Asn; Am Inst Nutrit; Am Vet Med Asn; Soc Exp Biol Med. *Res:* Diseases of the gastrointestinal tract and liver of domestic animals; pathogenesis of viral hepatitis; mechanisms of hepatic injury. *Mailing Add:* Four Sunny Knoll Ithaca NY 14850

TENNANT, DONALD L, b Mt Gilead, Ohio, Jan 27, 27; m 56; c 3. FISH BIOLOGY, LIMNOLOGY. *Educ:* Ohio State Univ, BS, 52. *Prof Exp:* Fish mgt supvr, Ohio Div Wildlife, 52-57; fishery res biologist, 57-59, fishery biologist, 59-67, fish & wildlife biologist, US Fish & Wildlife Serv, 67-; RETIRED. *Honors & Awards:* Fisheries Scientist Award, Am Fisheries Soc, 68. *Mem:* Am Fisheries Soc; fel Am Inst Fishery Res Biologists. *Res:* In stream flow regimens for fish, wildlife, recreation and related environmental resources; reservoir, lake and pond limnology and management; artificial propagation of muskellunge and fish hybridization. *Mailing Add:* 1809 Darlene Lane Billings MT 59102

TENNANT, RAYMOND WALLACE, b West Frankfort, Ill, Sept 19, 37; m 60; c 4. GENETIC TOXICOLOGY, CARCINOGENESIS. *Educ:* Univ Notre Dame, MS, 61; Georgetown Univ, PhD(microbiol), 63. *Prof Exp:* Virologist, Dept Virus Res, Microbiol Assocs, Inc, Md, 61-65; USPHS fel replication of DNA viruses, Albert Einstein Med Ctr, 65-66; sr staff scientist, Biol Div, Oak Ridge Nat Lab, 66-80; CHIEF, EXP CARCINOGENESIS & MUTAGENESIS BR, NAT TOXICOL PROG, NAT INST ENVIRON HEALTH SCI, RESEARCH TRIANGLE PARK, NC, 80- *Concurrent Pos:* mem adv comt, Am Cancer Soc. *Mem:* Environ Mutagen Soc; Am Asn Cancer Res. *Res:* Cancer biology; RNA tumor virus cell biology and genetics; cellular transformation; genetic toxicology; chemical carcinogenesis; transgenic mice. *Mailing Add:* Nat Inst Environ Health Sci Research Triangle Park NC 27709

TENNANT, WILLIAM EMERSON, b Washington, DC, Oct 8, 45; m 68; c 3. SOLID STATE PHYSICS, MATERIALS SCIENCE ENGINEERING. *Educ:* Harvard Univ, AB, 67; Univ Calif, Berkeley, PhD(solid state physics), 74. *Prof Exp:* Mem tech staff physics, 73-79, mgr, Infrared Detector Mat Group, 79-86, prin scientist, Electronic Imaging Function, 86-88, DIR, IMAGING, ROCKWELL INT SCI CTR, 88- *Mem:* Am Phys Soc; Sigma Xi; sr mem Inst Elec & Electronics Engrs. *Res:* Infrared detectors and imagers; semiconductor devices; collective excitations in solids; optical nondestructive evaluation methods; radiation damage; crystal alloys and defects; solar energy collection. *Mailing Add:* Rockwell Int Sci Ctr 1049 Camino Dos Rios Thousand Oaks CA 91360

TENNEBAUM, JAMES I, allergy; deceased, see previous edition for last biography

TENNENT, DAVID MADDUX, b Bryn Mawr, Pa, Oct 2, 14; m 45; c 4. BIOCHEMISTRY, SCIENCE ADMINISTRATION. *Educ:* Yale Univ, AB, 36, PhD(org chem), 40. *Prof Exp:* Asst appl physiol, Yale Univ, 40-42; biochemist, Merck Inst Therapeut Res, 42-60; asst dir res, Hess & Clark Div, Richardson-Merrell Inc, 60-63, dir res & develop, 63-69, vpres & dir res & develop, 69-75; consult, Vet Affairs, Rhodia Inc, 75-79; RETIRED. *Concurrent Pos:* Fel, Coun Arteriosclerosis, Am Heart Asn. *Mem:* Fel AAAS; Am Chem Soc; Am Soc Biol Chem; Soc Exp Biol & Med. *Res:* Pharmacology of drugs; bacterial pyrogens; cholesterol metabolism and experimental atherosclerosis; medications to improve performance and health of production animals; FDA applications. *Mailing Add:* 981 Forest Lane Ashland OH 44805

TENNENT, HOWARD GORDON, b Quebec, Que, Feb 29, 16; US citizen; m 48; c 4. ORGANOMETALLIC CHEMISTRY. *Educ:* Rensselaer Polytech Inst, BS, 37, MS, 39; Univ Wis, PhD(phys chem), 42. *Prof Exp:* Res chemist, Hercules, Inc, 42-47, mgr res div cellulose prod, 47-51, exp sta, 52, cent res div, 53-57, res assoc, 58-66, sr res assoc, 66-81; SR SCIENTIST, HYPERION CATALYSIS INT, LEXINGTON, MASS, 85- *Mem:* Am Chem Soc; Sci Res Soc Am. *Res:* Synthesis and applications of carbon fibrils. *Mailing Add:* 301 Chandler Mill Rd Kennett Square PA 19348

TENNER, THOMAS EDWARD, JR, b Pittsburgh, Pa, June 2, 49; m 72; c 3. CARDIOVASCULAR PHARMACOLOGY. *Educ:* Univ Dallas, BA, 71; Univ Tex Health Sci Ctr, PhD(pharmacol), 76. *Prof Exp:* Fel, Fac Pharmaceut Sci, Univ BC, 76-78; from asst prof to assoc prof, 78-90, PROF DEPT PHARMACOL, TEX TECH UNIV HEALTH SCI CTR, 90- *Mem:* Am Soc Pharm & Exp Therapeut; Am Heart Asn. *Res:* Drug-induced modulation of sensitivity and responsiveness of the cardiovascular system in particular reserpine and propranolol withdrawal induced supersensitivity phenomena. *Mailing Add:* Tex Tech Univ Health Sci Ctr 3601 Fourth St Lubbock TX 79430

TENNESSEN, KENNETH JOSEPH, b Ladysmith, Wis, June 10, 46; m 67; c 2. FRESH WATER ECOLOGY, ODONATA SYSTEMATICS. *Educ:* Univ Wis, BS, 68; Univ Fla, MS, 73, PhD(entom), 75. *Prof Exp:* BIOLOGIST ECOL EFFECTS AQUATIC INSECTS & MOSQUITOES, TENN VALLEY AUTHORITY, 75- *Mem:* Am Mosquito Control Asn; NAm Benthological Soc; Int Soc Odontol. *Res:* Investigation of effects of heated discharge and toxic substances from electric generating plants on the survival, growth, reproductive capacity and distribution of aquatic insects in the Tennessee River Valley; biology and ecology of mosquitoes in relation to control operations; identification of chironomidae. *Mailing Add:* Aquatic Biol Dept Off Serv Annex 1 B Tenn Valley Authority Muscle Shoals AL 35660

TENNEY, AGNES, b Boston, Mass. THEORETICAL PHYSICAL CHEMISTRY. *Educ:* Regis Col, AB, 68; Ind Univ, PhD(chem), 75. *Prof Exp:* Assoc instr chem, Ind Univ, 68-73, syst analyst comput sci, 73-76, vis asst prof, 76-77; ASST PROF CHEM, UNIV PORTLAND, 77- *Concurrent Pos:* Fac res grant, Univ Portland, 78-79. *Mem:* Am Chem Soc; Am Phys Soc; Int Asn Hydrogen. *Res:* Theory, particularly, electron atom molecule scattering in the intermediate energy range; renewable energy storage via hydrogen; computer controlled experiments. *Mailing Add:* Dept Chem Univ Portland 5000 N Williamette Portland OR 97203-5750

TENNEY, ALBERT SEWARD, III, b Lakewood, NJ, Mar 31, 43; m 66; c 2. PHYSICAL CHEMISTRY. *Educ:* Rutgers Univ, AB, 65, PhD(phys chem), 71. *Prof Exp:* Phys chemist mat sci, Gen Elec Co, 69-74; group leader solar cells, SES, Inc, 74-76; prin scientist temp measurement, 76-80, PRIN SCIENTIST SENSOR DEVELOP, LEEDS & NORTHRUP CO, 80- *Mem:* Am Chem Soc; Instrument Soc Am. *Res:* Materials research on molten salts, thin films, glasses and semiconductors; temperature measurement; development of light-emitting diodes, solar cells, temperature measuring instruments and analytical sensors. *Mailing Add:* 801 E Walnut St Gwynedd Greene Apts North Wales PA 19454-2823

TENNEY, GEROLD H, engineering; deceased, see previous edition for last biography

TENNEY, MARK W, b Chicago, Ill, Dec 10, 36; m 74; c 2. ENVIRONMENTAL HEALTH ENGINEERING, CIVIL ENGINEERING. *Educ:* Mass Inst Technol, SB, 58, SM, 59, ScD(civil & sanit eng), 65; Environ Eng Intersoc, dipl. *Prof Exp:* Design engr, Greeley & Hansen Eng, 59-61; from asst prof to assoc prof civil eng, Univ Notre Dame, 65-73; CHIEF EXEC OFFICER, TENECH ENG, INC, 69- *Honors & Awards:* Harrison Prescott Eddy Medal, Water Pollution Control Fedn, 73. *Mem:* Fel Am Soc Civil Engrs; Am Water Works Asn; Water Pollution Control Fedn; Nat Soc Prof Engrs; Am Acad Environ Engrs; Am Consult Eng Coun. *Res:* Sanitary engineering. *Mailing Add:* TenEch Eng Inc 744 W Washington Ave South Bend IN 46537

TENNEY, STEPHEN MARSH, b Bloomington, Ill, Oct 22, 22; m 47; c 3. PHYSIOLOGY. *Educ:* Dartmouth Col, AB, 43; Cornell Univ, MD, 46. *Hon Degrees:* Dsc, Univ Rochester, 84. *Prof Exp:* Asst prof physiol, Dartmouth Med Sch, 51-54; from asst prof to assoc prof physiol & med, Sch Med & Dent, Univ Rochester, 52-56; prof physiol, 56-74, chmn dept, 56-77, NATHAN SMITH PROF PHYSIOL, DARTMOUTH MED SCH, 74- *Concurrent Pos:* Markle scholar, 54-59. *Mem:* Am Physiol Soc; Am Acad Arts & Sci; Nat Inst Med; Am Soc Clin Invest. *Res:* Physiology of circulation and respiration. *Mailing Add:* Dept Physiol Dartmouth Med Sch Hanover NH 03756

TENNEY, WILTON R, b Buckhannon, WVa, July 2, 28; m 51. PLANT PATHOLOGY. *Educ:* WVa Wesleyan Col, BS, 50; Univ WVa, MS, 52, PhD(plant path), 55. *Prof Exp:* Plant scientist, Chem Res & Develop Labs, Army Chem Ctr, Md, 57; from asst prof to assoc prof, 57-71, PROF BIOL, UNIV RICHMOND, 71- *Mem:* Sigma Xi. *Res:* Physiology of fungi; host-parasite relations in fungus diseases of plants; toxicity of freshwater bryozoans. *Mailing Add:* Dept Biol Univ Richmond Richmond VA 23173

TENNILLE, AUBREY W, b Baker Co, Ga, Feb 4, 29; m 53; c 2. SOIL FERTILITY, SOIL MICROBIOLOGY. *Educ:* Univ Ga, BSA, 50; Okla State Univ, MSA, 55; Univ Fla, PhD(soils), 59. *Prof Exp:* Lab asst soil microbiol, Univ Fla, 58-59; asst county agent, Coop Exten, Univ Ga, 59-60, exten specialist, 60-62; from asst prof to prof agron, Ark State Univ, 62-88; RETIRED. *Mem:* Am Soc Agron; Soil Sci Soc Am; Soil Conserv Soc Am. *Res:* Fertility research on zinc and manganese of rice soils of Arkansas and soil salt problems of eastern Arkansas. *Mailing Add:* 206 Karen Ct Jonesboro AR 72401

TENNILLE, NEWTON BRIDGEWATER, veterinary radiology; deceased, see previous edition for last biography

TENNISON, ROBERT L, mathematics; deceased, see previous edition for last biography

TENNYSON, RICHARD HARVEY, b Minneapolis, Minn, Oct 11, 21; m 45; c 9. ORGANIC CHEMISTRY. *Educ:* St Mary's Col, Minn, BS, 46; Univ Ill, PhD(org chem), 52. *Prof Exp:* Res chemist, Corn Prod Co, 50-55, anal res supvr, 56-61; sect leader, Qual Control Dept, Mead Johnson & Co, 61-68, dir nutrit qual control, 68-76, vpres qual assurance, 76-86; RETIRED. *Mem:* Am Chem Soc; Inst Food Technol; Am Soc Qual Control; Sigma Xi. *Res:* Quality assurance adminstration; analytical methods development; physical, chemical, biological and microbiological control; nutritionals and pharmaceuticals. *Mailing Add:* 1365 Mesker Park Dr Evansville IN 47712

TENORE, KENNETH ROBERT, b Boston, Mass, Mar 22, 43. BIOLOGICAL OCEANOGRAPHY. *Educ:* St Anselm Col, AB, 65; NC State Univ, MS, 67, PhD(zool), 70. *Prof Exp:* Investr biol oceanog, Woods Hole Oceanog Inst, 70-72, asst scientist, 72-75; asst prof biol oceanog, Skidaway Inst Oceanog, 75-77; adj prof, Grad Sch Oceanog, Univ RI, 78-80; mem fac, Skidaway Inst Oceanog, Univ Ga, 80-83; PROF & HEAD CHESAPEAKE BIOL LAB, CTR ENVIRON & ESTUARINE STUDIES, UNIV MD, SOLOMONS, 84- *Mem:* Am Soc Limnol & Oceanog; Estuarine Res Fedn; Ecol Soc Am; AAAS. *Res:* Bioenergetics of detrital food chains in marine benthic communities. *Mailing Add:* Chesapeake Biol Lab PO Box 38 Solomons MD 20688-0038

TENOSO, HAROLD JOHN, US citizen. IMMUNOLOGY, MEDICAL MICROBIOLOGY. *Educ:* Univ Calif, Los Angeles, BA, 60, PhD, 66. *Prof Exp:* From sr biologist to mgr microbiol, Aerojet Med & Biol Syst, Aerojet-Gen Corp, 66-72, mgr biol oper, 72-74; dir biol oper, Organon Inc, 74-76, dir res & develop, Organon Diag, 76-80, gen mgr, diag div, 80-85; exec vpres, 85-86, pres, 86-88, CEO, UNIMED INC, 89- *Concurrent Pos:* Consult, 88- *Mem:* Am Asn Clin Chem; Am Soc Microbiol; AAAS. *Res:* Clinical chemistry; immunology and infectious disease, especially as it relates to immunoassays and early disease detection; pharmacology; oncology. *Mailing Add:* 203 Woodland Ave Westfield NJ 07091

TENSMEYER, LOWELL GEORGE, b Pocatello, Idaho, Feb 21, 28; m 54; c 4. PHYSICAL CHEMISTRY. *Educ:* Univ Utah, BA, 52, PhD(combustion), 57. *Prof Exp:* Asst prof chem, Ohio Univ, 56-57 & Utah State Univ, 57-59; Petrol Inst fel ceramics, Pa State Univ, 59; res scientist, Linde Div, Union Carbide Corp, 60-63; sr phys chemist, 63-72, RES SCIENTIST, ELI LILLY & CO, 72- *Mem:* Am Chem Soc; Am Phys Soc; Coblentz Soc. *Res:* Molecular spectroscopy; adsorption; crystal growth and purification; lasers; photochemistry and photobiology. *Mailing Add:* 35 W 59th St Indianapolis IN 46208

TENZER, RUDOLF KURT, b Jena, Ger, Oct 9, 20; nat US; m 47; c 4. MAGNETIC MATERIALS, X-RAY ANALYSIS. *Educ:* Univ Frankfurt, Dipl & Dr rer nat, 50. *Prof Exp:* Scientist radiation temperature measurements, Hartmann & Braun Co, Ger, 48-53; scientist magnetics; Ind Gen Corp, 53-65, mgr res, 65-69; mgr res, Electronic Memories & Magnetics Corp, 69-74, mgr res & mfg eng, 74-76, tech dir, IGC Div, 76-84; OWNER & CONSULT, TENZER ASSOCS, 84- *Mem:* Inst Elec & Electronics Engrs; Am Phys Soc; Am Ceramic Soc. *Res:* Permanent magnets; magnetization process; domain theory; ferrites; high temperature properties; temperature measurements by radiation; color pyrometers; magnetic bubble memories bias field assemblies. *Mailing Add:* 1643 Brookdale Dr Martinsville NJ 08836-9733

TEPAS, DONALD IRVING, b Buffalo, NY, Apr 7, 33; m; c 2. PHYSIOLOGICAL PSYCHOLOGY, NEUROSCIENCE. *Educ:* Univ Buffalo, BA, 55; State Univ NY Buffalo, PhD(psychol), 63. *Prof Exp:* Instr psychol, Univ Buffalo, 58-59; res scientist neuropsychiat, Walter Reed Army Inst Res, 59-62; sr res scientist human factors, Honeywell, Inc, 62-66; prof psychol, St Louis Univ, 66-78; prof, Ill Inst Technol, 78-85, chmn dept, 78-81; PROF PSYCHOL & DIR, DIV INDUST & ORGN PSYCHOL, UNIV CONN, 85- *Concurrent Pos:* Asst prof ophthal res, Univ Minn, 63-66; prin investr, USAF, NSF, NIMH, Nat Inst Occup Safety & Health grants, exchange scientist, Nat Acad Sci, Czech, 67; mem, Sci Comt on Night & Shift Work, Int Comn Occup Health, 79-, Comt Outer Continental Shelf Safety Info & Anal, Nat Res Coun, 82-84, exec comt, Sleep Res Soc, 84-87; US ed, Shiftwork Int Newslet, 85-, NAm ed, Work & Stress, 88- *Mem:* Fel Am Psychol Asn; fel AAAS; Psychonomic Soc; Soc Neurosci; Soc Comput Psychol (pres, 72-73); Sleep Res Soc; Human Factors Soc; Int Comn Occup Health; fel Am Psychol Soc. *Res:* Shiftwork; ergonomics; occupational safety and health; human electrophysiology; human factors aspects of computer hardware and software USC. *Mailing Add:* Dept Psychol U 20 Univ Conn 406 Babbidge Rd Storrs CT 06269-1020

TE PASKE, EVERETT RUSSELL, b Sheldon, Iowa, Sept 15, 30; m 51; c 4. BIOLOGY, ANIMAL BEHAVIOR. *Educ:* Westmar Col, AB, 51; State Col Iowa, MA, 57; Okla State Univ, PhD(zool), 63. *Prof Exp:* Teacher pub schs, Iowa, 52-61; from asst prof to prof biol, Univ Northern Iowa, 63-89; RETIRED. *Mem:* AAAS; Animal Behav Soc; Nat Asn Biol Teachers; Mammal Soc. *Res:* Morphology and taxonomy of Chiroptera; breeding behavior in the Japanese quail; social behavior in chickens. *Mailing Add:* PO Box 732 Cedar Falls IA 50613

TEPFER, SANFORD SAMUEL, b Brooklyn, NY, Mar 24, 18; m 42; c 4. PLANT MORPHOLOGY. *Educ:* City Col New York, BS, 38; Cornell Univ, MS, 39; Univ Calif, PhD(bot), 50. *Prof Exp:* Asst bot, Univ Calif, 47-50; instr, Univ Ariz, 50-53, res assoc agr, 53-54; instr biol, Ore Col Educ, 54-55; from asst prof to prof, 55-83, co-chmn dept, 68-71, head dept, 72-78, EMER PROF BIOL, UNIV ORE, 84- *Concurrent Pos:* NSF sci fac fel, 65; Fulbright lectr, Univ Paris, 71-72, vis prof, 71-72 & 78-79. *Mem:* AAAS; Bot Soc Am. *Res:* Developmental studies of shoot apex and flowers; culture of floral buds; floral morphogenesis. *Mailing Add:* Dept Biol Univ Ore Eugene OR 97403

TEPHLY, THOMAS R, b Norwich, Conn, Feb 1, 36; m 60; c 2. PHARMACOLOGY. *Educ:* Univ Conn, BS, 57; Univ Wis, PhD(pharmacol), 62; Univ Minn, MD, 65. *Prof Exp:* Instr pharmacol, Univ Wis, 62; from asst prof to assoc prof, Univ Mich, Ann Arbor, 65-71; PROF PHARMACOL, UNIV IOWA, 71- *Concurrent Pos:* Am Cancer Soc res scholar, 62-65. *Honors & Awards:* John J Abel Award, Am Soc Pharmacol & Exp Therapeut, 71. *Mem:* Am Soc Pharmacol & Exp Therapeut; Am Soc Biol Chem; Soc Toxicol. *Res:* Biochemical pharmacology and toxicology; drug metabolism; methanol and ethanol metabolism; heme biosynthesis. *Mailing Add:* Dept Pharmacol Univ Iowa Iowa City IA 52240

TEPLEY, NORMAN, b Denver, Colo, Dec 14, 35; m 68; c 3. MEDICAL PHYSICS, SOLID STATE PHYSICS. *Educ:* Mass Inst Technol, SB, 57, PhD(physics), 63. *Prof Exp:* Asst prof physics, Wayne State Univ, 63-69; assoc prof, 69-77, PROF PHYSICS, OAKLAND UNIV, 77-, CHMN DEPT, 83- *Concurrent Pos:* Vis prof, Dept Physics, Univ Lancaster, 70; sci dir, Neuromagnetism Lab, Henry Ford Hosp, 88- *Mem:* AAAS; Am Phys Soc. *Res:* Magnetic fields arising from living systems; neuromagnetism; physics of metals; ultrasonic studies of Fermi surfaces; electronic structures of metals; properties of superconductors. *Mailing Add:* Dept Physics Oakland Univ Rochester MI 48309

TEPLICK, JOSEPH GEORGE, b Philadelphia, Pa, Sept 29, 11; m 37; c 3. MEDICINE, RADIOLOGY. *Educ:* Univ Pa, AB, 31, MS, 32, MD, 36, MSc, 42. *Prof Exp:* Assoc radiol, Jefferson Med Col, 43-48; chief & dir radiol, Kensington Hosp, 49-63; clin assoc prof radiol, Hahnemann Med Col, 63-69, clin assoc prof diag radiol, 69-71, prof radiol, 71-90, dir div gen diag, 74-90, EMER PROF RADIOL, HAHNEMANN MED COL, 90- *Concurrent Pos:* Dir, Curtis X-ray Dept, Jefferson Med Col, 45-48; chief radiol, Albert Einstein Med Ctr, 50-53; vis radiologist, Philadelphia Gen Hosp, 60-; assoc, Sch Med, Univ Pa, 60-; mem staff, Hahnemann Hosp, 63-; expert in medico legal litigation. *Mem:* Fel Am Col Radiol; Radiol Soc NAm; Roentgen Ray Soc; Am Thoracic Soc; NY Acad Sci. *Res:* Hapato-splenography; intravenous and parenteral radiopaque emulsions; computed tomography of the spine. *Mailing Add:* Dept Diag Radiol Hahnemann Med Col 85230 N Broad St Philadelphia PA 19102

TEPLITZ, VIGDOR LOUIS, b Cambridge, Mass, Feb 5, 37; m 61; c 2. ASTROPHYSICS. *Educ:* Mass Inst Technol, SB, 58; Univ Md, PhD(physics), 62. *Prof Exp:* Physicist, Lawrence Radiation Lab, Univ Calif, Berkeley, 62-64; NATO fel physics, Europ Orgn Nuclear Res, 64-65; from asst prof to assoc prof, Mass Inst Technol, 65-73; head dept, Va Polytech Inst & State Univ, 73-77, prof physics, 73-78; phys sci officer, US Arms Control & Disarmament Agency, 78-80, dep div chief, 81-90; CHAIR PHYSICS DEPT, SOUTHERN METHODIST UNIV, 90- *Concurrent Pos:* Coun mem, Fedn Am Scientists, 72-76; coun mem forum on physics & soc, Am Phys Soc, 77-79; mem, US ASAT deleg, 78-79, START deleg, 90; prog dir theoret physics, NSF, 87-88. *Mem:* Fel Am Phys Soc; Fedn Am Sci; Am Astron Soc. *Res:* Elementary particle theory; phenomenology and data analysis; applications of particle theory to cosmology and astrophysics. *Mailing Add:* Physics Dept Southern Methodist Univ Dallas TX 75275-0175

TEPLY, LESTER JOSEPH, b Muscoda, Wis, Apr 22, 20; m 50; c 3. BIOCHEMISTRY. *Educ:* Univ Wis, BA, 40, MS, 44, PhD(biochem), 45. *Prof Exp:* Asst biochem, Univ Wis, 40-45; tech secy, Food Composition Comt, Nutrit Biochemist Coord, Nat Res Coun, 45; biochemist, USPHS, 45-46; res biochemist, Columbia Univ, 46-48; res biochemist, Enzyme Inst, Univ Wis, 48-51, asst dir labs, Wis Alumni Res Found, 51-55, dir lab projs, 55-60; sr nutritionist, UNICEF, NY, 60-85; ADJ PROF, NY MED COL, 85- *Concurrent Pos:* Chmn, Int Vitamin A Consult Group. *Mem:* Am Chem Soc; Inst Food Tech; Am Pub Health Asn; Am Inst Nutrit. *Res:* Nutrition; vitamins; enzymes; animal nutrition; microbiological nutrition and metabolism; B-complex vitamins; food technology. *Mailing Add:* 32 Colonial Ave Larchmont NY 10538

TEPLY, MARK LAWRENCE, b Lincoln, Nebr, Jan 11, 42; m 68, 83; c 4. ALGEBRA. *Educ:* Univ Nebr, BA, 63, MA, 65, PhD(math), 68. *Prof Exp:* From asst prof to assoc prof, 68-81, prof math, Univ Fla, 81-85; PROF MATH, UNIV WIS-MILWAUKEE, 85- *Concurrent Pos:* Ed, Commun in Algebra, 81-; investr NSF grants, 73, 77, 78; vis assoc prof, Fla State Univ, 76. *Mem:* Am Math Soc; Math Asn Am; Sigma Xi. *Res:* Noncommutative rings and their modules; torsion theories; filters of ideals; direct sum decompositions of modules; idealizer subrings; semigroup rings. *Mailing Add:* Univ Wis Milwaukee WI 53201

TEPPER, BYRON SEYMOUR, b New Bedford, Mass, Apr 12, 30; m 55; c 2. MICROBIOLOGY. *Educ:* Northeastern Univ, BS, 51; Univ Mass, MS, 53; Univ Wis, PhD(microbiol), 57. *Prof Exp:* Res assoc biochem, Univ Ill Col Med, 57-59; asst prof, 60-68, assoc prof pathobiol, Sch Hyg, 68-77, ASSOC PROF ENVIRON HEALTH, JOHNS HOPKINS UNIV, 78-, EXEC SECY, COMT USE INFECTIOUS AGENTS & BIOHAZARDOUS MATS, 78- *Concurrent Pos:* Assoc biochemist, Leonard Wood Mem Leprosy Res Lab, Baltimore, 59-65, microbiologist, 65-74; biohazards safety officer, Johns Hopkins Med Insts, 74-77. *Mem:* AAAS; Am Soc Microbiol; Soc Gen Microbiol; Int Leprosy Asn; Am Acad Microbiol. *Res:* Host dependent microorganisms; human and murine leprosy; mycobacterial physiology; photobiology. *Mailing Add:* Dept Environ Health 2021 E Monument St Baltimore MD 21205

TEPPER, FREDERICK, b Brooklyn, NY, Apr 9, 34; m 54; c 2. PHYSICAL CHEMISTRY, METALLURGY. *Educ:* NY Univ, BA, 54. *Prof Exp:* Chemist, Turner-Hall Corp, 54-55; phys chemist, Radiation Res Corp, 55-56; metallurgist, Atomic Energy Div, Sylvania-Elec Corp, 56-57; phys chemist, Mine Safety Appliances Res Corp, 57-60, sect head mat res, 60-69, dir res, 69-70, GEN MGR, INSTRUMENT DIV, MINE SAFETY APPLIANCE CO, 70-, VPRES, 71- *Mem:* Electrochem Soc; Instrument Soc Am. *Res:* Alkali metals; physical, thermodynamic and chemical properties; phase diagrams; production and purification techniques; corrosive effects on structural materials; gas sorption phenomena by activated carbon, metal oxides and ion exchange resins; molten salt electrochemistry; batteries. *Mailing Add:* Gateway Towers Pittsburgh PA 15222

TEPPER, HERBERT BERNARD, b Brooklyn, NY, Dec 25, 31; m 59; c 2. PLANT ANATOMY. *Educ:* State Univ NY Col Forestry, Syracuse Univ, BS, 53, MS, 58; Univ Calif, Davis, PhD(bot), 62. *Prof Exp:* Res asst forest bot, State Univ NY Col Forestry, Syracuse Univ, 56-58; res forester, US Forest Serv, 58-59; res asst, Univ Calif, Davis, 59-62; from instr to assoc prof, 62-67, PROF FOREST BOT, STATE UNIV NY COL ENVIRON SCI & FORESTRY, SYRACUSE, 67- *Mem:* AAAS; Int Soc Plant Morphol; Sigma Xi; Tissue Cult Asn. *Res:* Morphogenesis in the shoot apex; seed germination; bud and cambial reactivation. *Mailing Add:* Dept Environ & Forest Biol Environ Sci & Forestry Syracuse NY 13210

TEPPER, LLOYD BARTON, b Los Angeles, Calif, Dec 21, 31; m 57; c 2. TOXICOLOGY, OCCUPATIONAL HEALTH. *Educ:* Dartmouth Col, AB, 54; Harvard Univ, MD, 57, MIH, 60, ScD(occup med), 62; Am Bd Prev Med, dipl, 64. *Prof Exp:* Fel, Mass Gen Hosp, 58-60; fel, Mass Inst Technol, 59-61; physician, Eastman Kodak Co, 61-62 & AEC, 62-65; assoc dir occup med & inst environ health, Kettering Lab, Univ Cincinnati, 65-72, assoc prof environ health, Univ, 65-71, prof, 71-72, assoc prof med, 69-72; assoc comnr sci, Food & Drug Admin, 72-76; CORP MED DIR, AIR PROD & CHEM, INC, 76-; ADJ PROF ENVIRON MED, UNIV PA, 77- *Concurrent Pos:* Ed, J Occup Med, 79- *Mem:* Am Acad Occup Med (pres, 80-81); Am Occup Med Asn. *Res:* Industrial and environmental toxicology, especially as related to toxicology of beryllium, lead and other industrial metals; environmental and medical standards. *Mailing Add:* Air Prod & Chem Inc Allentown PA 18195-1501

TEPPER, MORRIS, b Palestine, Mar 1, 16; nat US; c 2. METEOROLOGY, SCIENCE ADMINISTRATION. *Educ:* Brooklyn Col, BA, 36, MA, 38; Johns Hopkins Univ, PhD(fluid mech), 52. *Prof Exp:* Qualifications analyst & chief, Phys Sci Unit, US Civil Serv Comn, 39-43; chief, Severe Local Storms Res Unit, US Weather Bur, 46-59; dir meteorol prog, NASA, 59-65, dep dir space applications progs & dir meteorol, 66-69, dep dir earth observs progs & dir meteorol, 69-78, head, Spec Proj Off, Goddard Space Flight Ctr, 78-79; prof math physics, Capitol Col, Md, 79-90; CONSULT, DEPT ENERGY, 87- *Concurrent Pos:* Mem staff, USDA Grad Sch, 52-79; mem, US Nat Comt Int Hydrol Decade & chmn work group remote sensing in hydrol, Nat Acad Sci, 71-75, liaison rep, US Comt Global Atmospheric Res Prog, mem, Comt Int Environ Progs, US Interagency Comts; chmn, Working Group 6, Comt Space Res, Int Coun Sci Unions; mem, Int Comn Space Sci Bd, 65-79; consult, Systs Gen Corp, 83- *Honors & Awards:* Meissinger Award, Am Meteorol Soc, 50; Except Serv Medal, NASA, 66; Gold Medal, Nat Ctr Space Studies, France, 72; Am Meteorol Soc Spec Award, 78; Nordberg Mem Award, Comt Space Res, 79. *Mem:* Fel Am Meteorol Soc; AAAS. *Res:* Satellite meteorology; mesometeorology; severe local storms; space applications; earth observation satellites; remote sensing; global atmospheric research; climate; education. *Mailing Add:* 107 Bluff Terr Silver Spring MD 20902

TEPPERMAN, BARRY LORNE, b Toronto, Ont, Jan 29, 47; m 72; c 2. MEDICAL SCIENCES. *Educ:* Univ Toronto, BSc, 69, MSc, 72; Univ Calgary, PhD(physiol), 75. *Prof Exp:* Fel physiol, Univ Tex, Houston, 75-77; asst prof, 77-82, ASSOC PROF PHYSIOL, UNIV WESTERN ONT, 82- *Mem:* Am Gastroenterol Asn; Can Physiol Soc. *Res:* Factors regulating the integrity of gastrointestinal mucosa, specifically prostaglandins and prostaglandin receptors and salivary gland factors; role of gastrointestinal peptides of central origin in the regulation of gastrointestinal function. *Mailing Add:* Dept Physiol Health Sci Ctr Univ Western Ont London ON N6A 5C1 Can

TEPPERMAN, HELEN MURPHY, b Hartford, Conn, Jan 9, 17; m 43; c 3. PHYSIOLOGICAL CHEMISTRY. *Educ:* Mt Holyoke Col, BA, 38; Yale Univ, PhD(physiol chem), 42. *Hon Degrees:* DSc, SUNY, 87. *Prof Exp:* Asst, Mem Hosp, NY, 42-43 & Yale Univ, 43-44; pharmacologist, Med Res Lab, Edgewood Arsenal, Md, 44-45; from instr to prof, 46-85, EMER PROF PHARMACOL, STATE UNIV NY UPSTATE MED CTR, 85- *Mem:* Am Physiol Soc; Endocrine Soc. *Res:* Endocrinology and metabolism. *Mailing Add:* Dept Pharmacol 766 Irving Ave State Univ NY Health Sci Ctr Syracuse NY 13210

TEPPERMAN, JAY, b Newark, NJ, Mar 23, 14; m 43; c 3. MEDICINE. *Educ:* Univ Pa, AB, 33; Columbia Univ, MD, 38. *Hon Degrees:* DSc, 87. *Prof Exp:* Intern, Bassett Hosp, NY, 38-40; hon fel, Sch Med, Yale Univ, 40-41, Coxe fel, 41-42, asst, Aeromed Unit, Dept Physiol, 42-44; assoc prof pharmacol, 46-53, prof exp med, Col Med, State Univ NY Upstate Med Ctr, 53-85; RETIRED. *Concurrent Pos:* Mem metab study sect, NIH & physiol comt, Nat Bd Med Examr, 63-67, mem pharmacol comt, 72-75; consult, Vet Admin, DC, 64-67 & Food & Drug Admin, 69-; fac exchange scholar, State Univ NY. *Honors & Awards:* Purkinje Medalist, Czech Med Soc. *Mem:* Soc Exp Biol & Med; Am Physiol Soc; Endocrine Soc; Am Soc Pharmacol & Exp Therapeut; Am Soc Biol Chem. *Res:* Endocrinology and metabolism. *Mailing Add:* 7684 Stonehedge Lane Manlius NY 13104

TEPPERMAN, KATHERINE GAIL, b Syracuse, NY, May 31, 47; m 79; c 2. CELLULAR PHYSIOLOGY, BIOINORGANIC CHEMISTRY. *Educ:* Middlebury Col, BA, 68; Univ Conn, PhD(develop biol), 73. *Prof Exp:* From instr to asst prof biol, Middlebury Col, Vt, 72-74; fel, Dept Embryol, Carnegie Inst, 74-76; asst prof, 76-83, ASSOC PROF BIOL, UNIV CINCINNATI, 83- *Mem:* Am Soc Cell Biol. *Res:* Metal based drugs and interactions of metal ions in biological systems; gold based antiarthritis drugs, effects of silver drugs and electrochemically generated silver on tissue culture cells and renal damage induced by platinum anti-cancer drugs. *Mailing Add:* Dept Biol Sci ML No 6 Univ Cincinnati Cincinnati OH 45221-0006

TEPPERT, WILLIAM ALLAN, SR, b Oshkosh, Nebr, Oct 10, 15; m 39; c 2. PHARMACOLOGY. *Educ:* Univ Ill, BS, 43 & 48, MS, 47; Univ Iowa, PhD(zool), 58. *Prof Exp:* Asst mammalian physiol, Univ Ill, 42-43 & 46-48; from instr to asst prof biol, 48-57, from asst prof to assoc prof pharmacol, 49-64, PROF PHARMACOL, COL PHARM, DRAKE UNIV, 64- *Mem:* Sigma Xi. *Res:* Cell physiology and pharmacology; cellular neuropharmacology; convulsive disorders; geriatric medication. *Mailing Add:* 1533 48th St Des Moines IA 50311-2456

TEPPING, BENJAMIN JOSEPH, b Philadelphia, Pa, Jan 29, 13; m 40; c 2. MATHEMATICAL STATISTICS, APPLIED STATISTICS. *Educ:* Ohio State Univ, BA, 33, MA, 35, PhD(math), 39. *Prof Exp:* Math statistician, US Bur Census, 40-55; mathematician, Nat Analysts, Inc, 55-60; chief statist adv group, Surv & Res Corp, Korea, 60-63; chief, Res Ctr Measurement Methods, US Bur Census, 63-73; STATIST CONSULT, 73- *Concurrent Pos:* Lectr, USDA Grad Sch, 41-52, Am Univ, 43 & Univ Mich, 48-53; adj assoc prof, Univ Pa, 56-60; assoc ed, J Am Statist Asn, 64-66; mem vis lectr prog, Inst Math Statist, 69-72. *Honors & Awards:* Cult Medal, Repub Korea, 63. *Mem:* Fel AAAS; fel Am Statist Asn; Am Math Soc; Inst Math Statist; Int Statist Inst. *Res:* Sampling theory and methods; measurement problems in censuses and surveys; problems of matching lists. *Mailing Add:* 401 Apple Grove Rd Silver Spring MD 20904

TERADA, KAZUJI, b Honolulu, Hawaii, Jan 4, 27. INORGANIC CHEMISTRY. *Educ:* Univ Hawaii, BA, 52, MS, 54; Univ Utah, PhD, 61. *Prof Exp:* chemist, Res & Develop, Rocky Flats Div, Rockwell Int, Golden, 60-88; RETIRED. *Mem:* Am Chem Soc. *Mailing Add:* 1443 W Beach Rd Oak Harbor WA 98277

TERAMURA, ALAN HIROSHI, b Los Angeles, Calif, Dec 26, 48; m 74. PHYSIOLOGICAL ECOLOGY, ENVIRONMENTAL STRESS PHYSIOLOGY. *Educ:* Calif State Univ, Fullerton, BA, 71, MA, 73; Duke Univ, PhD(physiol ecol), 78. *Prof Exp:* Fel photobiol, Univ Fla, 77-78; from asst prof to assoc prof ecol, 79-88, PROF & CHAIR, UNIV MD, 88- *Concurrent Pos:* Vis prof ecol, Utah State Univ, 79-; guest prof photobiol, Univ Karlsruhe, Fed Repub Ger, 82-83; consult, Environ Protection Agency, 80-, Nat Acad Sci, 84 & 88; vis prof, Univ Hawaii, 87 & 88; chmn, sci adv bd, Ctr Global Change, Univ Md, 89- *Mem:* Ecol Soc Am; Am Soc Plant Physiologists; Scand Soc Plant Physiol; Bot Soc Am. *Res:* The effects of environmental stress on plant growth, physiology, and ecology; adaptive strategies of widespread, weedy plants; the effects of global climate change on plant productivity. *Mailing Add:* Dept Bot Univ Md College Park MD 20742

TERANGO, LARRY, b Clarksburg, WVa, Nov 30, 25; m 51; c 2. SPEECH PATHOLOGY, AUDIOLOGY. *Educ:* Kent State Univ, BA, 50, MA, 54; Case Western Reserve Univ, PhD(speech path & audiol), 66. *Prof Exp:* Clinician, Painesville City Schs, 52-59; instr speech path & audiol, Kent State Univ, 61-62; asst prof, San Jose State Col, 62-63; instr speech path & audiol, Kent State Univ, 63-64; instr speech & dir speech & hearing clin, Ohio State Univ, 64-66; assoc prof speech & dir speech & hearing clin, Univ Wyo, 66-68; prof, chmn dept spec educ & dir speech & hearing clin, ETenn State Univ, 68-74; prof health sci & dir speech & hearing ctr, Western Carolina Univ, 74-78; PROF SPEC EDUC & COORDR COMMUN PROG, EASTERN KY UNIV, 78- *Concurrent Pos:* Vpres, Wyo Cleft Palate Eval Team, 66-68; audiologist & hearing aid consult, 78- *Mem:* Am Speech & Hearing Asn; Coun Except Children; Nat Educ Asn; Am Cleft Palate Asn. *Res:* Vocal characteristics in the male voice; language dysfunction; multidisciplinary approach to study of neurological disturbances. *Mailing Add:* 352 High St Richmond KY 40475

TERANISHI, ROY, b Stockton, Calif, Aug 1, 22; m 44. ORGANIC CHEMISTRY. *Educ:* Univ Calif, BS, 50; Ore State Col, PhD, 54. *Hon Degrees:* Dr Agr Sci, Univ Gent, Belg. *Prof Exp:* Instr chem, Portland State Col, 53-54; RES CHEMIST, USDA, 54- *Mem:* Am Chem Soc; hon fel Japan Soc Prom Sci. *Res:* Gas chromatography; flavor chemistry. *Mailing Add:* 89 Kingston Rd Kensington CA 94707-1321

TERASAKI, PAUL ICHIRO, b Los Angeles, Calif, Sept 10, 29; m 56; c 4. IMMUNOLOGY. *Educ:* Univ Calif, Los Angeles, BA, 50, MA, 52, PhD(zool), 56. *Prof Exp:* Res asst zool, 52-54, res asst, Atomic Energy Proj, 54-55, res zoologist, Dept Surg, 55-56, jr res zoologist, 56-57, asst res zoologist, 58-61, assoc res zoologist, 61-62, assoc prof surg, 62-66, PROF SURG, CTR HEALTH SCI, UNIV CALIF, LOS ANGELES, 66- *Concurrent Pos:* Res fel zool with Prof P B Medawar, Univ Col, Univ London, 57-58; USPHS career develop award, 63-; mem transplantation & immunol adv comt, NIH, 67-70; mem nomenclature comt leukocyte antigens, WHO. *Honors & Awards:* Modern Med Award Distinguished Achievement. *Mem:* Am Soc Cell Biol; Am Asn Immunol; Soc Cryobiol; Am Soc Immunol; Int Transplantation Soc. *Res:* Transplantation immunology; homotransplantation; leucocyte typing. *Mailing Add:* Dept Surg Univ Calif Med Sch Rehab Ctr 1000 Veterans Ave Los Angeles CA 90024

TERASAWA, EI, b Ihda City, Japan, Apr 8, 38; m 75; c 1. NEUROENDOCRINOLOGY, REPRODUCTIVE PHYSIOLOGY. *Educ:* Univ Tokyo, BS, 61; Yokohama City Univ, PhD(physiol), 66. *Prof Exp:* Res physiologist anat, Dept Physiol, Univ Calif, Berkeley, 66-67; res fel physiol, Dept Anat, Univ Calif, Los Angeles, 67-68; instr, Dept Physiol, Med Sch, Yokohama City Univ, 68-70, asst prof, 70-73; assoc scientist, 73-80, SR SCIENTIST, PHYSIOL, PRIMATE RES CTR, UNIV WIS-MADISON, 80- *Concurrent Pos:* NIHDRG mem Reproductive Biol Study Sect, 87- *Mem:* Soc Study Reproduction; AAAS; Soc Neurosci; Am Physiol Soc; Endocrine Soc. *Res:* Integral function of the hypothalamus in control of the pituitary-gonadal system; control mechamisms of the Lutenizing-hormone releasing hormone neuronal system, which may serve as a model of peptidergic neurons in the mammalian hypothalamus. *Mailing Add:* Regional Primate Res Ctr Univ Wis 1223 Capitol Ct Madison WI 53715-1299

TERASMAE, JAAN, b Estonia, May 28, 26; nat Can; m 54. GEOLOGY, PALYNOLOGY. *Educ:* Univ Uppsala, Fil Kand, 51; McMaster Univ, PhD, 55. *Prof Exp:* Asst, Palynological Lab Stockholm, 50-51; geologist, Geol Surv Can, 55-68; chmn dept geol sci, 69-75, PROF GEOL, BROCK UNIV, 68- *Honors & Awards:* WA Johnston Medal, 90. *Mem:* AAAS; Int Glaciol Soc; Int Limnol Soc; Royal Soc Can; Geol Soc Am; Am Asn Stratig Palynologist. *Res:* Pleistocene chronology, geology and stratigraphy; paleobotany of Pleistocene deposits. *Mailing Add:* Dept Geol Sci Brock Univ St Catherines ON L2S 3A1 Can

TERBORGH, JOHN J, b Washington, DC, Apr 16, 36. POPULATION BIOLOGY, PLANT PHYSIOLOGY. *Educ:* Harvard Univ, AB, 58, AM, 60, PhD(biol), 63. *Prof Exp:* Staff scientist, Tyco Labs, Inc, 63-65; asst prof bot, Univ Md, 65-71; from assoc prof to prof biol, Princeton Univ, 71-89; RUTH F DEVARNEY PROF, DUKE UNIV, 89-, DIR, CTR TROP CONSERV, 91- *Concurrent Pos:* Res grants, Am Philos Soc, Am Mus Natural Hist & Nat Geog Soc, 64-67; NSF res grants, 68- *Mem:* Fel AAAS; Am Soc Naturalists; Soc Study Evolution; Ecol Soc Am. *Res:* Tropical ecology; population biology of birds. *Mailing Add:* Ctr Trop Conserv Duke Univ 3705-C Erwin Rd Durham NC 27705

TERDIMAN, JOSEPH FRANKLIN, b New York, NY, Feb 14, 40; m 65; c 2. BIOMEDICAL ENGINEERING. *Educ:* Cornell Univ, BEngPhys, 61; NY Univ, MD, 65; Univ Ill Med Ctr, PhD(physiol), 72. *Prof Exp:* Res scientist pop exposure studies, Nat Ctr Radiol Health, 67-69; med info scientist, Kaiser Found Res Inst, 69-79; ASST TO DIR, MED METHODS RES, KAISER-PERMANENTE, 80- *Concurrent Pos:* Lectr, Univ Ill, Chicago Circle, 65-67 & Sch Optom, Univ Calif, Berkeley, 69-78; NIH special fel, 67-69; clin asst prof, Sch Optom, Univ Calif, Berkeley, 79- *Mem:* AAAS; Soc Advan Med Systs; Biomed Eng Soc. *Res:* Development of medical information systems for hospital automation; patient monitoring, diagnosis and therapy; integration of computers and engineering methods with classical neurophysiological techniques; neurophysiology. *Mailing Add:* 5001 Branciforte Dr Santa Cruz CA 95065

TERENZI, JOSEPH F, b Marlboro, NY, Aug 21, 32; m 54; c 3. PLASTICS ENGINEERING, CHEMICAL & ENVIRONMENTAL ENGINEERING. *Educ:* Rensselaer Polytech Inst, BS, 53; Princeton Univ, MS, 55, PhD, 58. *Prof Exp:* Mem staff res & develop, Am Cyanamid Co, 58-65, tech mgr plastics div, 65-71, mgr mfg, 71-75, dir mfg technol, 75-80, dir environ toxicol, 80-90; VPRES, SAFETY HEALTH & ENVIRON, 91- *Concurrent Pos:* Mem fac, Dept Chem Eng, Univ Conn, 64; lectr, Princeton Univ, Columbia Univ, Pa State Univ & Lowell Tech Inst. *Mailing Add:* 688 King Rd Franklin Lakes NJ 07417

TERENZIO, JOSEPH V, b New Haven, Ct, Feb 4, 18; m 45; c 3. PUBLIC HEALTH. *Educ:* Yale Univ, BA, 39; Fordham Univ Law Sch, JD, 47; Columbia Univ Sch Pub Health, MS, 52. *Prof Exp:* Exec vpres admin, Brooklyn Cumberland Med Ctr, NYC, 60-66; commr hosp, Dept Hosps, City of New York, 66-70; exec vpres, Evanston Hosp, Ill, 70-72; pres, United Hosp fund at New York, 75-83; CHMN BD, META HEALTH TECHNOL INC, 83- *Concurrent Pos:* Spec adv to the pres, Columbia Univ, 72-80; adj prof, Pub Health Admin, Columbia Univ, 72- *Mem:* Nat Acad Sci; fel Am Pub Health Asn; hon fel Am Health Care Execs. *Mailing Add:* Five Bote Ct Greenwich CT 06830

TERESA, GEORGE WASHINGTON, b Osceola, Ark, Nov 23, 23; m 54. BACTERIOLOGY. *Educ:* Ark Agr & Mech Col, BS, 52; Univ Ark, MS, 55; Kans State Univ, PhD(microbiol), 59. *Prof Exp:* Asst prof bact, Auburn Univ, 59-61 & Univ RI, 61-62; assoc prof biol, Wichita State Univ, 62-68; from asst prof to assoc prof bact & biochem, 68-81, PROF BACT, UNIV IDAHO, 81- *Mem:* AAAS; Am Soc Microbiol; Sigma Xi. *Res:* Immunology and pathogenic bacteriology; immunology and pathogenic mechanism of Gram anaerobes. *Mailing Add:* PO Box 8374 Moscow ID 83843

TERESHKOVICH, GEORGE, b New York, NY, Mar 18, 30; m 55; c 1. HORTICULTURE. *Educ:* La Tech Univ, BS, 52; Univ Ga, MS, 57; La State Univ, PhD(hort, agron), 63. *Prof Exp:* Res asst, Univ Ga, 52-54, asst prof hort, 56-60; res assoc, La State Univ, 60-63; asst prof, Univ Ga, 63-68; assoc prof hort, 68-74, PROF & ASSOC CHAIRPERSON, DEPT AGR-HORT-ENTOM, TEX TECH UNIV, 75- *Mem:* Am Soc Hort Sci. *Res:* Cultural and adaptability studies with vegetable and ornamental crops. *Mailing Add:* Dept Agr-Hort-Entom Tex Tech Univ Lubbock TX 79409

TERESI, JOSEPH DOMINIC, b San Jose, Calif, Aug 18, 15; m 47; c 6. BIOCHEMISTRY, HEALTH PHYSICS. *Educ:* San Jose State Col, AB, 38; Univ Wis, PhD(biochem), 43. *Prof Exp:* Res assoc, Manhattan Proj, Chicago, 43, Clinton Labs, Tenn, 44, & Univ Chicago, 45; res assoc, Stanford Univ, 47-51, actg instr, 48-50, actg asst prof, 50-51; chemist, US Naval Radiol Defense Lab, 51-69; sr biochemist, Stanford Res Inst, 69-71; sr scientist, San Francisco Bay Marine Res Ctr, 71-73; res assoc div nuclear med, Stanford Univ, 73-74; sr eng, Advan Reactor Systs Dept, Gen Elec Co, 74-80; RETIRED. *Mem:* Fel AAAS; Am Chem Soc; Health Physics Soc; Am Nuclear Soc. *Res:* Analysis of radionuclides in biological materials; radiation effects; aerospace nuclear safety; radiation protection; radiation ecology and internal emitters; bay sediment analysis; radiological assessment; nuclear reactor safety analysis. *Mailing Add:* 1395 Villa Dr Los Altos CA 94024-5338

TERHAAR, CLARENCE JAMES, b Cottonwood, Idaho, Apr 29, 26; m 57; c 4. TOXICOLOGY. *Educ:* Univ Idaho, BS, 53, MS, 54; Kans State Col, PhD(parasitol), 57. *Prof Exp:* Asst prof entom, Kans State Col, 57-58; toxicologist, Eastman Kodak Co, 58-86; DIR, DRUG SAFETY EVAL PENNWALT PHARMACEUT, 86- *Mem:* Entom Soc Am; Am Soc Pharmacol & Exp Therapeut; Am Micros Soc; Soc Toxicol; Am Indust Hyg Asn; NY Acad Sci. *Res:* Invertebrate and mammalian toxicology. *Mailing Add:* Pennwalt Pharmaceut 755 Jefferson Rd Rochester NY 14623

TER HAAR, GARY L, b Zeeland, Mich, May 2, 36; m 56; c 2. INORGANIC CHEMISTRY. *Educ:* Hope Col, BA, 58; Univ Mich, MS, 60, PhD(chem), 62. *Prof Exp:* Res assoc, 62-76, DIR TOXICOL & INDUST HYG, ETHYL CORP, 76-, VPRES, HEALTH & ENVIRON, 85- *Mem:* Am Chem Soc; Soc Toxicol. *Res:* Environmental research. *Mailing Add:* Ethyl Corp 451 Florida St Baton Rouge LA 70801

TERHUNE, ROBERT WILLIAM, b Detroit, Mich, Feb 7, 26; m 47; c 2. QUANTUM ELECTRONICS. *Educ:* Univ Mich, BS, 47 & PhD(hysics), 57; Dartmouth Col, Ma, 48. *Prof Exp:* Supvr, Digital Comput & Logic Des Sect, Willow Run Labs, Univ Mich, 51-54, res physicist, 54-59, mgr, Solid State Physics Lab, 59-60; res physicist, Sci Lab, Ford Motor Co, Dearborn, Mich, 60-65, mgr, Physics Electronics Dept, 65-75; vis scholar, Stanford Univ, 75-76; sr staff scientist, Eng & Res Staff, Ford Motor Co, 76-87; SR MEM TECH STAFF, JPL-CALIF TECH, 88- *Concurrent Pos:* Ed, Optics Let J, 77-83, J Ont Soc Am, Optical Soc Am, 84-87. *Honors & Awards:* Sci & Eng Award, Drexel Inst Technol, 64. *Mem:* Optical Soc Am; Am Phys Soc; Inst Elec & Electronic Engrs. *Res:* Quantum electronics; nonlinear optics; optical properties of solids and surfaces; molecular spectroscopy; advanced instrumentation. *Mailing Add:* 1460 Peyfair Estates Dr Pasadena CA 91103

TERMAN, CHARLES RICHARD, b Mansfield, Ohio, Sept 8, 29; m 51; c 2. ANIMAL ECOLOGY, ANIMAL BEHAVIOR. *Educ:* Albion Col, AB, 52; Mich State Univ, MS, 54, PhD(behav pop dynamics), 59. *Prof Exp:* Assoc prof biol, Taylor Univ, 61-63, actg dir res, 62-63; from asst prof to assoc prof, 63-69, PROF BIOL, COL WILLIAM & MARY, 69- *Concurrent Pos:* Nat Inst Ment Health fel, Sch Hyg & Pub Health, Johns Hopkins Univ, 59 & Penrose Res Lab, 59-61; NATO sr sci fel, Eng & exchange scientist, US & Polish Nat Acads Sci, 74; NIH career develop award, 70-74. *Mem:* Fel AAAS; Animal Behav Soc; Am Sci Affil; Am Soc Mammal; Am Soc Nat. *Res:* Population dynamics; socio-biological factors influencing the growth and physiology of populations; reproductive physiology; behavioral ecology. *Mailing Add:* Dept Biol Col William & Mary Williamsburg VA 23185

TERMAN, LEWIS MADISON, b San Francisco, Calif, Aug 26, 35; m 58. ELECTRONICS. *Educ:* Stanford Univ, BS, 56, MS, 58, PhD(elec eng), 61. *Prof Exp:* Res staff mem, IBM Corp, 61-63, mgr read only storage res, 63-65, mgr integrated memory circuit res, 65-70, mgr semiconductor storage-circuits & systs, 70-75, res staff mem, 75- 79, mem, Dir Res Tech Planning Staff, 79-80, mgr, Very-Large-Scale Integration Circuits, 80-82, MGR VERY-LARGE SCALE INTEGRATION LOGIC & MEMORY, IBM CORP, 82- *Concurrent Pos:* Ed, J Solid State Circuits, 74-77; mem, Circuits & Systs Admin Comt, Inst Elec & Electronics Engrs, 81-83 & Electron Devices Soc Admin Comt, 84-90; chmn, Int Solid-State Circuits Conf, 83; chmn, Symp Very Large Scale Integration Technol, 85-86 & Symp Very Large Scale Integration Circuits, 88-89; vpres, Inst Elec & Electronic Engrs Electron Devices Soc, 88-89, pres, 90-91; treas, Inst Elec & Electronics Engrs Solid-State Circuits Coun, 88-89. *Mem:* Fel AAAS; fel Inst Elec & Electronic Engrs. *Res:* Integrated circuits, memory systems and semiconductor device research and development; logic design. *Mailing Add:* IBM Res Ctr PO Box 218 Yorktown Heights NY 10598

TERMAN, MAX R, b Mansfield, Ohio, Apr 15, 45; m 68; c 2. ECOLOGY, ETHOLOGY. *Educ:* Spring Arbor Col, BA, 67; Mich State Univ, MS, 69, PhD(zool), 73. *Prof Exp:* PROF BIOL, TABOR COL, 69- *Concurrent Pos:* Consult, Environ Land Use; Ausable Inst, 81- *Mem:* Ecol Soc Am; Am Soc Mammalogists; Animal Behav Soc; Am Inst Biol Sci; Sigma Xi. *Res:* Interspecific competition between rodents; ecosystem ecology; rodent population dynamics; behavior of rodents; earth-sheltered housing. *Mailing Add:* RR 2 Box 78B Hillsboro KS 67063

TERMINE, JOHN DAVID, b Brooklyn, NY, Sept 25, 38; m 61; c 4. BIOCHEMISTRY, MOLECULAR BIOLOGY. *Educ:* St John's Univ, NY, BS, 60; Univ Md, MS, 63; Cornell Univ, PhD(biochem), 66. *Prof Exp:* Teaching asst chem, Univ Md, 60-63; from asst res scientist to assoc res scientist, Hosp Spec Surg, New York, 63-66; from instr to asst prof biochem, Med Col, Cornell Univ, 66-70; spec res fel, NIH, 70-73, res biochemist, Molecular Struct Sect, 73-80, chief, Skeletal Matrix Biochem Sect, Lab Biol Struct, 80-83, Bone Res Br, Nat Inst Dent Res, 83-91; EXEC DIR, OSTEOPOROSIS/OSTEOARTHRITIS RES, LILLY RES LABS, ELI LILLY & CO, INDIANAPOLIS, IND, 91- *Concurrent Pos:* Counr, Am Soc Bone & Mineral Res, 84-87; Assoc Ed, Calcified Tissue Int, 86-; Board dr, Int Conf Calcium Regulating Hormones & Bone Metab, 87- *Honors & Awards:* NIH Dir Award, 83; Biol Mineralization Res Award, Int Asn Dent Res, 88. *Mem:* AAAS; Int Asn Dent Res; Am Chem Soc; Biophys Soc; Am Soc Biochem & Molecular Biol; Am Soc Bone & Mineral Res; Orthop Res Soc; Am Soc Cell Biol. *Res:* Molecular and cellular biology of bone and tooth; molecular biology; protein biochemistry; osteoporosis and osteoarthritis research. *Mailing Add:* Lilly Res Labs Lilly Corp Ctr Eli Lilly & Co Indianapolis IN 46285

TERMINIELLO, LOUIS, enzymology, for more information see previous edition

TERNAY, ANDREW LOUIS, JR, b New York, NY, Aug 29, 39; m 61; c 2. ORGANIC CHEMISTRY, MEDICINAL CHEMISTRY. *Educ:* City Col New York, BS, 59; NY Univ, MS, 62, PhD(chem), 63. *Prof Exp:* NSF fel & res assoc chem, Univ Ill, 63-64; instr, Case Western Reserve Univ, 64-65, asst prof, 65-69; assoc prof, 70-76, PROF CHEM, UNIV TEX, ARLINGTON, 77-, DIR, MED CHEM RES, 88- *Concurrent Pos:* Grants, Nat Cancer Inst, Dept Chem, Case Western Reserve Univ, 66-69 & Welch Found, Univ Tex, 72-; consult, Arbrook, Inc, 71-; adj prof med chem, Univ Houston, 81; US Army Med Res & Develop Command, 84- *Mem:* AAAS; Am Chem Soc; Royal Soc Chem; Sigma Xi. *Res:* Drug design; molecular spectroscopy; application of stereochemistry to synthesis of new drugs, especially psychoactive materials; radioprotective drugs. *Mailing Add:* Dept Chem Univ Tex Arlington TX 76019

TERNBERG, JESSIE L, b Corning, Calif, May 28, 24. MEDICAL EDUCATION, PEDIATRIC SURGERY. *Educ:* Grinnell Col, AB, 46; Univ Tex, PhD(biochem), 50; Washington Univ, MD, 53. *Hon Degrees:* DSc, Grinnell Col, 72 & Univ Mo-St Louis, 81. *Prof Exp:* Intern med, Boston City Hosp, 53-54; asst resident surg, Barnes Hosp, St Louis, 54-57, chief & admin resident surg, 58-59; res fel surg, dept surg, 57-58, from instr to assoc prof, 59-71, PROF SURG, SCH MED, WASHINGTON UNIV, 71-, PROF PEDIAT SURG, 75-, CHIEF, DIV PEDIAT SURG, DEPT SURG, 72- *Concurrent Pos:* Pediat surgeon-in-chief, St Louis Children's Hosp, 74-; chmn surg sect, pediat oncol group, US Food & Drug Admin. *Honors & Awards:* Horatio Alger Award. *Mem:* Fel Am Col Surgeons; Am Pediat Surg Asn; Soc Surg Alimentary Tract; Soc Pelvic Surgeons; Brit Asn Pediat Surg. *Res:* Formation of neo-mucosa of the intestine; nutritional aspects of short gut. *Mailing Add:* Washington Univ Sch Med Box 8109 4960 Audubon Ave St Louis MO 63110

TERNER, CHARLES, b Lublin, Poland, Apr 30, 16; nat US; m 45; c 3. BIOCHEMISTRY. *Educ:* Univ London, BSc, 44, DSc(biochem), 69; Univ Sheffield, PhD(biochem), 49. *Prof Exp:* Mem staff, Med Res Coun Unit for Res Cell Metab, Dept Biochem, Univ Sheffield, 47-50; sr sci officer, Dept Physiol, Nat Inst Res in Dairying, Eng, 50-55; mem staff, Worcester Found Exp Biol, 55-59; prof, 59-87, EMER PROF BIOL, BOSTON UNIV, 87- *Concurrent Pos:* Vis scientist, Pop Coun, Rockefeller Univ, 72. *Mem:* AAAS; Am Soc Biol Chem; Biochem Soc (Gt Brit); Soc Exp Biol & Med; Soc Study Reproduction. *Res:* Biochemistry of male reproductive tissues; control of male fertility; embryonic development of fish. *Mailing Add:* Biol Sci Ctr Boston Univ Five Cummington St Boston MA 02215

TERNER, JAMES, b Reading, Eng, Mar 27, 51; US citizen; m 79; c 2. RESONANCE RAMAN SPECTROSCOPY, HEME ENZYMES. *Educ:* Brandeis Univ, BA, 73; Univ Calif Los Angeles, PhD(chem), 79. *Prof Exp:* Asst prof, 81-86, ASSOC PROF CHEM, VA COMMONWEALTH UNIV, 86- *Concurrent Pos:* Alfred P Sloan Found fel, 85. *Mem:* Am Chem Soc; Biophys Soc; AAAS. *Res:* Resonance Raman spectroscopy of transient species; intermediates and excited states of photochemical and biological significance. *Mailing Add:* Chem Dept Va Commonwealth Univ 1001 Main St Richmond VA 23284-2006

TER-POGOSSIAN, MICHEL MATHEW, b Berlin, Ger, Apr 21, 25; nat US. MEDICAL PHYSICS. *Educ:* Univ Paris, BA, 42; Washington Univ, MS, 48, PhD(nuclear physics), 50. *Prof Exp:* From instr to prof radiation physics, 50-73, PROF BIOPHYS PHYSIOL, SCH MED, WASH UNIV, 64-, PROF RADIATION SCI, 73-; DIR, DIV RADIATION SCI, MALLINCKRODT INST RADIOL, 61-, PROF RADIOL, 71- *Concurrent Pos:* Mem, Diag Radiol & Nuclear Med Study Sect, 79-81 & var comts, NIH & Dept Energy; ed, Inst Elec & Electronics Engrs Trans Med Imaging, 82-83; Amy Bowles Lawrence distinguished scientist res med award, Lawrence Berkeley Lab, 89. *Honors & Awards:* Wendell Scott Lectr, Wash Univ, 73; Benedict Cassen Lectr, 76; Paul C Aebersold Award, Soc Nuclear Med, 76; David Gould Lectr, Johns Hopkins Univ, 77; R S Landauer Mem Lectr, 81; Hans Hecht Lectr, Univ Chicago, 81; Herrman L Blumgart Pioneer Award, 84; Georg Charles de Hevesy Nuclear Med Pioneer Award, Soc Nuclear Med, 85. *Mem:* Inst Med-Nat Acad Sci; fel Am Phys Soc; Am Nuclear Soc; Am Radium Soc; Radiation Res Soc; hon fel Am Col Radiol; Soc Nuclear Med. *Res:* Medical applications of short-lived isotopes; gamma ray spectroscopy; scintillation counters; radiation dosimetry; radiobiology; lasers in biology; reconstructive tomography in radiologic imaging; positron emission imaging. *Mailing Add:* Mallinckrodt Inst 510 S Kingshighway St Louis MO 63110

TERRAGLIO, FRANK PETER, b Portland, Ore, May 19, 28; m 64. ENVIRONMENTAL SCIENCE, CHEMISTRY. *Educ:* Univ Portland, BS, 49; Rutgers Univ, MS, 62, PhD(environ sci), 64. *Prof Exp:* Chemist, Ore State Bd Health, 49-50 & Ore Air Pollution Authority, 52-56; asst instr air pollution, Rutgers Univ, 56-63; res chemist, Calif Dept Pub Health, 63-64; asst prof civil eng, Ore State Univ, 64-66; from asst prof to assoc prof, 66-72, PROF APPL SCI, PORTLAND STATE UNIV, 72- *Mem:* Am Chem Soc;

Air Pollution Control Asn; Am Water Works Asn; Water Pollution Control Fedn; Am Indust Hyg Asn; Sigma Xi. *Res:* Reactions of atmospheric sulfur dioxide. *Mailing Add:* Dept Appl Sci & Eng Portland State Univ PO Box 751 Portland OR 97207

TERRANOVA, ANDREW CHARLES, b Cleveland, Ohio, Aug 29, 35. ENTOMOLOGY, TOXICOLOGY. *Educ:* Ohio State Univ, BSc, 60, MSc, 61, PhD(entom), 65. *Prof Exp:* Insect physiologist, Metab & Radiation Res Lab, USDA, 65-76, res entomologist, Sci & Educ Admin-Agr Res, 76-90; RETIRED. *Mem:* AAAS; Entom Soc Am; Am Chem Soc. *Res:* Metabolism and mode of action of insect chemosterilants; biochemistry and physiology of insect reproduction; population genetics. *Mailing Add:* 911 Oak Ave Sanford FL 32771

TERRAS, AUDREY ANNE, b Wash, DC, Sept 10, 42; div. NUMBER THEORY, ANALYSIS. *Educ:* Univ Md, College Park, BS, 64; Yale Univ, MA, 66, PhD(math), 70. *Prof Exp:* Instr math, Univ Ill, Urbana, 68-70; asst prof Univ PR, Mayaguez, 70-71 & Brooklyn Col, 71-72; from asst prof to assoc prof, 72-83, PROF MATH, UNIV CALIF, SAN DIEGO, 83- *Concurrent Pos:* NSF grant, 74-88; vis assoc prof, Mass Inst Technol, 83; mem, Inst Advan Study, Princeton, NJ, 84; mem coun, Am Math Soc, 85-87; mem comts, Am Math Soc, 87-88. *Mem:* Am Math Soc; Math Asn Am; Am Women in Math; Soc Indust & Appl Math; fel AAAS; Asn Women Sci. *Res:* Zeta functions; automorphic forms of matrix argument; harmonic analysis on homogeneous spaces; fundamental domains of discrete transformation groups. *Mailing Add:* Dept Math C-012 Univ Calif San Diego La Jolla CA 92093

TERREAULT, BERNARD J E J, b Montreal, Que, Mar 29, 40; m 68; c 2. SURFACE PHYSICS, THERMONUCLEAR FUSION TECHNOLOGY. *Educ:* Univ Montreal, BSc, 60, MSc, 62; Univ Ill, Urbana-Champaign, PhD(physics), 68. *Prof Exp:* Fel, Lab High Energy Physics, Polytech Sch, Paris, 69-70; res assoc high energy physics, Ohio Univ, 70-71, asst prof, 71-72; assoc prof, 72-77, PROF NUCLEAR SCI, ENERGY CTR, NAT INST SCI RES, UNIV QUE, 77- *Concurrent Pos:* Foreign scientist, French Atomic Energy Comn, 82-83. *Mem:* Am Phys Soc; Can Asn Physicists; Am Vacuum Soc; Mat Res Soc. *Res:* Surface diagnostics in controlled thermonuclear fusion devices; surface radiation effects; fuel recycling and impurity production by plasma-surface interactions in fusion devices. *Mailing Add:* Energy Ctr Nat Inst Sci Res Univ Que CP 1020 Varennes PQ J0L 2P0 Can

TERREL, RONALD LEE, b Klamath Falls, Ore, Sept 2, 36; m 59, 81; c 3. CIVIL ENGINEERING, CONSTRUCTION MATERIALS. *Educ:* Purdue Univ, Lafayette, BSCE, 60, MSCE, 61; Univ Calif, Berkeley, PhD(civil eng), 67. *Prof Exp:* Estimator, J H Pomeroy & Co, Inc, 55-56; mat engr, US Bur Reclamation, Denver, 61-64; proj engr, J H Pomeroy & Co, Inc, 64-65; res asst, Univ Calif, Berkeley, 65-67; pres, Pavements Systs, Inc, 72-80; dir, Wash State Transp Ctr, 81-83; head transp, Seattle Eng Int, Inc, 67-75, vpres, 80-82; vpres, Pavement, Technol, Inc, 85-86; from asst prof to prof, 67-75, EMER PROF CIVIL ENG, UNIV WASH, 85-; OWNER, TERREL RES, 86- *Concurrent Pos:* Chmn, Triaxial Inst Struct Pavement Design, 71-73; mem trans res bd, Nat Acad Sci; consult eng res & develop, gov & private indust, States, Fed Hwy Admin, UN; vpres, Seattle Eng Int, Inc, 80-82; prof civil eng, Oregon State Univ, 89-; vpres, Hydrogenesis, Inc, 91- *Honors & Awards:* Walter Emmons Award, Asn Asphalt Paving Technologists, 83. *Mem:* Asn Asphalt Paving Technologists; Am Soc Civil Engrs; Am Soc Testing & Mat. *Res:* Pavement and construction materials technology including asphalt, polymer modified asphalt, concrete, aggregates, waste materials such as sulphur, lignin, ash, and recycled pavement materials for highway and airports; construction engineering and management; systems for environmental storage of solid waste. *Mailing Add:* 9703 241st Pl SW Edmonds WA 98020-6512

TERRELL, C(HARLES) W(ILLIAM), b Louisville, Ky, May 10, 27; m 52; c 2. NUCLEAR ENGINEERING, ENGINEERING EDUCATION. *Educ:* Purdue Univ, BSEE, 52, PhD(nuclear eng), 70; NC State Univ, Raleigh, BS, 54, MS, 55. *Prof Exp:* Res engr, Bendix Res Labs, Mich, 52-54; mem res & teaching staff physics, NC State Univ, Raleigh, 54-57; supvr reactor opers, Armour Res Found, 57-59, mgr reactor res, 59-61, mgr nuclear res, 61-63; asst dir physics res, IIT Res Inst, 63-65, dir physics res div, 65-67; assoc prof nuclear eng & supvr comput opers, Purdue Univ, 69-70; dist mgr appl sci & mkt support, Comput Sci Corp, Ill, 70-72; pres, Ind Inst Technol, 72-77; prof nuclear eng, Univ Okla, 77-87; SR TECHNOLOGIST, PHILLIPS LAB, KIRKLIN AFB, 87- *Mem:* Am Nuclear Soc; Am Soc Eng Educ; sr mem Inst Elec & Electronics Engrs. *Res:* Higher education administration; nuclear engineering education; research management. *Mailing Add:* Phillips Lab Kirtland AFB Albuquerque NM 87117-6008

TERRELL, CHARLES R, MARINE BIOLOGY. *Educ:* Boston Univ, BA, 65; Northeastern Univ, MS, 68. *Prof Exp:* Asst prof & lab dir biol sci, Salem State Col, 67-74; dir, Coastal Rev Ctr, Coastal Zone Mgt, 75-76; assoc, Conserv Found, 76-776-77; actg sect chief, Environ Protection Agency, 77-80; NAT WATER QUAL SPECIALIST, SOIL CONSERV SERV, WASHINGTON, DC, 80- *Concurrent Pos:* Mem Shellfish Adv Comn, 69-74; consult ecologist, Edwards & Kelcey Inc, 71-72; legis fel US Senate, 85 & 86. *Mem:* Fed Water Qual Asn (secy, 89-91); Water Pollution Control Fed; Am Inst Biol Sci; Am Water Resources Asn. *Res:* Establish national technical policies on abating agricultural pollution; water conservation; environmental problems; numerous publications and audio-visuals. *Mailing Add:* Soil Conserv Serv USDA PO Box 2980 Washington DC 20013

TERRELL, EDWARD EVERETT, b Wilmington, Ohio, Oct 6, 23; m 50; c 3. PLANT TAXONOMY. *Educ:* Wilmington Col, AB, 47; Cornell Univ, MS, 49; Univ Wis, PhD, 52. *Prof Exp:* Muellhaupt scholar bot, Ohio State Univ, 52-53; assoc prof biol & head dept sci, Pembroke State Col, 54-56; assoc prof biol, Guilford Col, 56-60; botanist, Agr Res Serv, USDA, 60-85; RETIRED. *Concurrent Pos:* Vis prof bot, Univ Md, 86; res assoc bot, Univ Md, 86-91. *Mem:* Bot Soc Am; Am Soc Plant Taxon; Torrey Bot Club; Int Asn Plant Taxon; Soc Econ Bot. *Res:* Plant taxonomy and ecology; taxonomy of Houstonia; taxonomy of grasses. *Mailing Add:* 14001 Wildwood Dr Silver Spring MD 20905

TERRELL, GLEN EDWARD, b Humble, Tex, Nov 17, 39; m 59; c 2. NUCLEAR PHYSICS. *Educ:* Univ Tex, Austin, BS, 62, MA, 64, PhD(physics), 66. *Prof Exp:* Asst prof, 66-74, ASSOC PROF PHYSICS, UNIV TEX, ARLINGTON, 74- *Mem:* Am Phys Soc. *Res:* Computer-assisted instruction; low energy nuclear physics; polarization of protons elastically scattered by several nuclei; gamma ray directional correlation. *Mailing Add:* Dept Physics Univ Tex Box 19088 Uta Station Arlington TX 76019

TERRELL, MARVIN PALMER, b Pine Bluff, Ark, May 19, 34; m 59; c 2. INDUSTRIAL ENGINEERING, OPERATIONS RESEARCH. *Educ:* Univ Ark, Fayetteville, BSIE, 57, MSIE, 60; Univ Tex, Austin, PhD(opers res), 66. *Prof Exp:* Instr indust eng, Univ Ark, Fayetteville, 57-60; mfg engr, Gen Elec Co, 60-63; indust engr, Tex Instruments Co, 63-64; asst eng & Ford Found & Alcoa fels, Univ Tex, Austin, 64-66, Nat Tau Beta Pi-Ford fel, 65-66; from asst prof to assoc prof, 66-77, PROF INDUST ENG, OKLA STATE UNIV, 77- *Concurrent Pos:* Consult, Continental Oil Co, 67, NAm Rockwell Corp, 68-69 & Phillips Petrol Co, 72 & Bray Truck Lines, 79-82. *Mem:* Am Inst Indust Engrs; Opers Res Soc Am; Inst Mgt Sci; Am Soc Eng Educ. *Res:* Mathematical programming; optimization theory; combinatorics; operations modeling and analysis; management science; quality control and reliability. *Mailing Add:* Dept Indust Eng Okla State Univ Stillwater OK 74078

TERRELL, N(ELSON) JAMES, JR, b Houston, Tex, Aug 15, 23; m 45; c 3. PHYSICS, ASTROPHYSICS. *Educ:* Rice Univ, BA, 44, MA, 47, PhD(physics), 50. *Prof Exp:* Res asst physics, Rice Univ, 50; asst prof, Case Western Reserve Univ, 50-51; mem staff, 51-89, ASSOC, LOS ALAMOS NAT LAB, UNIV CALIF, 89- *Concurrent Pos:* USAEC fel, 48-50; vis prof, Highlands Univ, Las Vegas, 59; vis scientist, Univ Calif, Lawrence Berkeley Lab, 63. *Mem:* Fel AAAS; fel Am Phys Soc; Am Astron Soc; Sigma Xi; Int Astron Union. *Res:* Astrophysics; relativity; fission; diffraction; Fourier analysis; x-ray astronomy. *Mailing Add:* Los Alamos Nat Lab Mail Stop D436 Group SST-9 Los Alamos NM 87545

TERRELL, ROSS CLARK, organic chemistry, for more information see previous edition

TERRELL, TERRY LEE TICKHILL, b Wheeling, WVa, Nov 23, 49. AQUATIC ECOLOGY. *Educ:* Ohio State Univ, BS, 71; Univ Ga, PhD(zool), 76. *Prof Exp:* Instr bot, Univ Wyo, 75-76, guest lectr bot, 76, asst prof, 76; aquatic ecologist, 77-85, RES ADMINR, US FISH & WILDLIFE SERV, 85- *Concurrent Pos:* Aquatic ecologist, Res Eval Tech Prog, US Forest Serv, 79-81; leader Aquatic Classification Tech Working Group, Five-Way Interagency Agreement Comt Nat Assessments, 79-81. *Mem:* Phycol Soc Am; Am Soc Limnol & Oceanog; Int Soc Limnol; AAAS. *Res:* Classification, inventory and analysis of aquatic ecosystems in relation to natural resource management; most emphasis on such management perturbations as mining and hydroelectric development; classical limnological studies and their application to natural resources management; lake trophic dynamics; acid precipitation. *Mailing Add:* US Fish & Wildlife Serv Nat Environ Res Ctr 4512 McMurray Ave Ft Collins CO 80525

TERRES, GERONIMO, b Santa Barbara, Calif, July 1, 25; m 47; c 5. IMMUNOBIOLOGY. *Educ:* Univ Calif, BA, 50; Stanford Univ, MA, 51; Calif Inst Technol, PhD(biol), 56. *Prof Exp:* Asst, Calif Inst Technol, 52-55; assoc scientist microbiol, Brookhaven Nat Lab, 55-60; asst prof human physiol, Sch Med, Stanford Univ, 60-69; assoc prof to prof, 69-91, EMER PROF PHYSIOL, SCH MED, TUFTS UNIV, 91- *Concurrent Pos:* Vis asst prof, Harvard Med Sch, 67-68; vis scientist, Mass Inst Technol, 75-76, Swiss Inst Allergy & Asthma Res, Davos, 89; res collabr, Swiss Inst Exp Cancer Res, Lausanne, Switz, 79; vis assoc, Calif Inst Technol, 81; mem, Basel Inst Immunol, Basel, Switz, 82-83; NIH Sr Res Fel, 60-62; USPHS Res Career Develop Award, 62-69. *Honors & Awards:* Yamagiwa-Yoshida Fel Award, 79. *Mem:* AAAS; Am Asn Immunol; Soc Exp Biol & Med; Am Physiol Soc; Radiation Res Soc; Sigma Xi. *Res:* Immune degradation; tumor (leukemias) cell rejection; T cell hybridization; acquired immune tolerance in mice; initiation and control of the immune response. *Mailing Add:* 335 Iris Way Palo Alto CA 94303

TERRIERE, ROBERT T, b Seattle, Wash, July 17, 26; m 58; c 2. GEOLOGY. *Educ:* Calif Inst Technol, BS, 49; Pa State Col, MS, 51; Univ Tex, PhD(geol), 60. *Prof Exp:* Geologist, US Geol Surv, 51-58; res geologist, Cities Serv Oil Co, 58-70, res assoc, 70-78, sr geol assoc, 78-80, region explor geologist, Cities Serv Co-Petrol Explor, 80-83, sr geol assoc, 83-85; RETIRED. *Mem:* Fel Geol Soc Am; Am Asn Petrol Geol. *Res:* Physical stratigraphy; sedimentary petrography; petroleum geology. *Mailing Add:* 2618 S Allison St Lakewood CO 80227

TERRILE, RICHARD JOHN, b New York, NY, Mar 22, 51; m 81. PLANETARY SCIENCE. *Educ:* State Univ NY, Stony Brook, BS, 72; Calif Inst Technol, MS, 73, PhD(planetary sci), 78. *Prof Exp:* Res asst, Calif Inst Technol, 72-78, assoc scientist, 78; physicist, Trend Western Tech Corp, 78-80, sr scientist, 80-81, MEM TECH STAFF, JET PROPULSION LAB, 81- *Concurrent Pos:* Mem, Voyager Target Selection Working Group, 78-; planetary astron prin investr, NASA grant, 78-, planetary atmospheres prin investr, 79-; guest investr, Voyager Imaging Sci Team, 78-80. *Honors & Awards:* Except Sci Achievement Medal, NASA, 88. *Mem:* Am Astron Soc; Am Geophys Union; Int Astron Union; Sigma Xi. *Res:* Ground-based planetary astromony; planetary atmospheres; comparative geology; photographic interpretation of spacecraft data and the study of planetary ring systems. *Mailing Add:* 2121 Woodlyn Rd Pasadena CA 91104

TERRILL, CLAIR ELMAN, b Rippey, Iowa, Oct 27, 10; m 32; c 2. ANIMAL BREEDING. *Educ:* Iowa State Col, BS, 32; Univ Mo, PhD, 36. *Prof Exp:* Asst animal husb, Univ Mo, 32-36; asst animal husbandman, Exp Sta, Univ Ga, 36; asst animal husbandman, Sheep Exp Sta, Bur Animal Indust, Idaho, 36-37, assoc animal husbandman, Western Sheep Breeding Lab & Sheep Exp

Sta, Agr Res Serv, 53-55, chief sheep & fur animal res br, Animal Husb Res Div, 55-72, nat prog staff scientist for sheep & other animals, Agr Res Serv, USDA, 72-80; COLLABR, 81- *Concurrent Pos:* Dir, Am Forage & Grass Land Coun, 63-65; mem, World Asn Animal Prod Coun, 63- *Honors & Awards:* Achievement Award, Ital Exp Inst & Ital Soc Advan Zootech; Distinguished Achievement Award, Sheep Indust Develop; Distinguished Serv Award, Am Soc Animal Sci. *Mem:* Fel AAAS; Genetics Soc Am; hon fel Am Soc Animal Sci (secy-treas, 60-62, vpres, 63, pres, 64); Am Meat Sci Asn; Am Genetic Asn (vpres, 69, pres, 70). *Res:* Animal genetics; reproductive physiology; sheep, goat and fur animal breeding and production. *Mailing Add:* 318 Apple Grove Rd Silver Springs MD 20904

TERRIS, JAMES MURRAY, b May 6, 41; m; c 3. HYPERTENSION, COMPARATIVE KIDNEY PHYSIOLOGY. *Educ:* Mich State Univ, PhD(physiol), 74. *Prof Exp:* ASSOC PROF KIDNEY PHYSIOL, UNIFORMED SERV, UNIV HEALTH SCI, F EDWARD HERBERT SCH MED, 78- *Mem:* Am Physiol Soc; Am Chem Soc; AAAS. *Res:* Hypertension; renal physiology. *Mailing Add:* Dept Physiol Uniformed Serv Univ Health Sci 4301 Jones Bridge Rd Bethesda MD 20814-4779

TERRIS, MILTON, b New York, NY, Apr 22, 15; m 41, 71; c 2. EPIDEMIOLOGY. *Educ:* Columbia Col, AB, 35; NY Univ, MD, 39; Johns Hopkins Univ, MPH, 44. *Prof Exp:* Intern, Harlem Hosp, New York, 39-41 & Bellevue Psychiat Hosp, New York, 41-42; apprentice epidemiologist, State Dept Health, NY, 42-43, asst dist health officer, 44-46; med assoc subcomt on med care, Am Pub Health Asn, 46-48, staff dir, 48-51; assoc prev med & pub health, Sch Med, Univ Buffalo, 52-54, from asst prof to assoc prof, 54-58; prof epidemiol, Sch Med, Tulane Univ, 58-60; head chronic dis unit, Div Epidemiol, Pub Health Res Inst New York, 60-64; prof prev med, New York Med Col, 64-80, chmn dept community & prev med, 68-80; RETIRED. *Concurrent Pos:* Asst dean post-grad educ, Univ Buffalo, 51-58. *Mem:* Am Pub Health Asn (pres, 66-67); Soc Epidemiol Res (pres, 67-69); Am Epidemiol Soc; Asn Teachers Prev Med (pres, 61-62); Int Epidemiol Asn. *Res:* Epidemiology of cancer; cirrhosis of liver; prematurity; heart disease. *Mailing Add:* 208 Meadowood Dr South Burlington VT 05403

TERRY, DAVID LEE, b Burkley, Ky, Mar 22, 36; div; c 3. SOIL SCIENCE. *Educ:* Univ Ky, BS, 58, MS, 61; NC State Univ, PhD(soil sci), 68. *Prof Exp:* Agronomist, Univ Ky, 59-60; from instr to asst prof soil sci, NC State Univ, 71-74; COORDR FERTILIZER REGULATORY PROG, UNIV KY, 74- *Mem:* Asn Am Plant Food Control Officials; Am Soc Agron. *Res:* Soil fertility; fertilizer control in state of Kentucky. *Mailing Add:* 3521 Coltneck Lane Lexington KY 40502

TERRY, FRED HERBERT, b Bedford, Ind, July 29, 40; m 63; c 2. ELECTRICAL ENGINEERING, BIOENGINEERING. *Educ:* Rose-Hulman Inst Technol, BS, 62, MS, 64; Case Western Reserve Univ, PhD(eng), 67. *Prof Exp:* Asst prof, 67-69, head elec eng, 69-72, assoc prof, 70-80, PROF ELEC ENG, CHRISTIAN BROS COL, 80-, CHMN, DEPT ENG, 72- *Concurrent Pos:* Instr med units, Univ Tenn, 67-; gen res eng, Memphis Vet Admin Hosp, 75-78. *Mem:* Inst Elec & Electronics Engrs; Am Soc Eng Educ; Nat Soc Prof Engrs; Sigma Xi. *Res:* Computer acquisition and analysis of electrophysiological data; linear electrical properties of canine purkinje tissue; modeling of cardiac conduction abnormalities; electroacoustic analysis of heart sounds. *Mailing Add:* Dept Eng Christian Bros Col 650 E Pkwy S Memphis TN 38104

TERRY, HERBERT, b New York, NY, Jan 30, 22; m 57; c 2. OPERATIONS RESEARCH. *Educ:* City Col New York, BS, 42; Polytech Inst Brooklyn, BChE, 46. *Prof Exp:* Group leader, Foster D Snell, Inc, 42-46; tech serv mgr, Shawinigan Resins Corp, 47-63; tech mgr coatings div, Hooker Chem Corp, 63-67; res dir polymer appln, Foster D Snell, Inc, Subsidiary Booz, Allen & Hamilton, Inc, 67-69, res dir, 69-71, vpres, 71-74; PRES, DECISIONEX, INC, 74- *Mem:* Am Chem Soc; Opers Res Soc Am; Inst Mgt Sci. *Res:* Applications of decision theory and operations research to computer programming for financial analysis particularly in real estate, facility planning and site location. *Mailing Add:* Decisionex Inc 1200 Post Rd E Westport CT 06880

TERRY, JAMES LAYTON, b Peoria, Ill, Apr 8, 51. ATOMIC & HIGH TEMPERATURE PHYSICS. *Educ:* Denison Univ, BS, 73; Johns Hopkins Univ, MA, 75, PhD(physics), 78. *Prof Exp:* RES STAFF PHYSICS, MASS INST TECHNOL, 78- *Mem:* Am Phys Soc. *Res:* Diagnostics of high temperature plasmas. *Mailing Add:* Mass Inst Technol NW16-282 Cambridge MA 02139

TERRY, LEON CASS, b Northville, Mich, Dec 22, 40; m 64; c 2. NEUROENDOCRINOLOGY, NEUROSCIENCE. *Educ:* Univ Mich, Dr Pharm, 64; Marquette Univ, MD, 69; McGill Univ, PhD(neuroendocrinol), 82. *Prof Exp:* Med intern, Univ Rochester, NY, 69-70; staff assoc, Nat Inst Arthritis, Diabetes & Digestive & Kidney Dis, NIH, 70-72; resident neurol, McGill Univ, 72-75, fel neuroendocrinol, 75-78; assoc prof neurol, Ctr Health Sci, Univ Tenn, 78-81; staff phys, Vet Admin Med Ctr, Memphis, 80-81, clin investr, 81, asst chief, Ann Arbor, 82-85; assoc prof, 81-84, PROF NEUROL, UNIV MICH, 84-, ASSOC PROF PHYSIOL, 83-; ACTG CHIEF NEUROL, VET ADMIN MED CTR, ANN ARBOR, 85- *Concurrent Pos:* Consult, Baptist Hosp, Memphis, Tenn, 79-81 & Methodist Hosp, 80-81; prin investr, Vet Admin Merit Ref Res grants, 80-, NIH res grants, 81-; co-investr, NIH res grants, 80- *Mem:* Am Soc Clin Invest; Am Neurol Asn; Endocrine Soc; Am Acad Neurol. *Res:* Neural regulation of anterior pituitary hormone secretion focussing on hypothalamic regulation of endocrine rhythms; neurotransmitter and neuropeptide regulation of growth hormone secretion. *Mailing Add:* Dept Neurol Med Col Wis 9200 W Wisconsin Ave Milwaukee WI 53226

TERRY, LUCY IRENE, b Atlanta, Ga, Jan 18, 50. INSECT ECOLOGY. *Educ:* Fla Southern Col, BS, 72; Univ Fla, MS, 76; NC State Univ, PhD(entom), 83. *Prof Exp:* Res technician plant virol, Univ Fla, 72-74, res asst entom & morphol, Grad Sch, 74-76; entom experimentalist, Ciga-Geigy Corp, 76-79; res asst, entom dept, Grad Sch, NC State Univ, 79-83; ASST PROF & RES SCIENTIST, INSECT ECOL & HOST PLANT RESISTANCE, UNIV ARIZ, 83- *Mem:* AAAS; Entom Soc Am; Sigma Xi. *Res:* Entomological research on cotton pests, predator prey interactions and population dynamics; control of insect pests; sampling and economic thresholds; effects of different irrigation systems on insect populations. *Mailing Add:* Dept Entom Univ Ariz Tucson AZ 85721

TERRY, MAURICE ERNEST, plant physiology, for more information see previous edition

TERRY, NORMAN, b Maidstone, Eng, Sept 5, 39; m 68; c 4. PLANT PHYSIOLOGY. *Educ:* Southampton Univ, BSc, 61; Nottingham Univ, MSc, 63, PhD(plant physiol), 66. *Prof Exp:* Res fel plant physiol, Div Biosci, Nat Res Coun Can, 66-68; asst specialist plant physiol, Dept Soils & Plant Nutrit, 68-72, from asst prof to assoc prof, 72-84, PROF PLANT BIOL, UNIV CALIF, BERKELEY, 84- *Mem:* Am Soc Plant Physiologists; AAAS; Crop Sci Soc Am; Brit Soc Exp Biol. *Res:* Environmental and internal factors involved in the regulation of photosynthesis; chloroplast development; mineral nutrition and salinity effects on plant function. *Mailing Add:* Dept Plant Biol Univ Calif 111 GPBB Berkeley CA 94720

TERRY, PAUL H, b Fall River, Mass, June 22, 28. ORGANIC & ANALYTICAL CHEMISTRY. *Educ:* Southeastern Mass Univ, BS, 51; Univ Mass, MS, 59, PhD(org chem), 63. *Prof Exp:* Chemist, Dept Geront, Wash Univ, 52-53; res chemist, Insect Chemosterilants Lab, Agr Environ Qual Inst, USDA, Beltsville, Md, 63-79, res chemist, Plant Hormone Lab, Plant Physiol Inst, Agr Res Serv, 79-88; RETIRED. *Mem:* Am Chem Soc; Am Inst Chem. *Res:* Synthesis of compounds to sexually sterilize insects; analysis of plant hormones, especially abscisic acid and indole-3-acetic acid, in plants; comparison of hormone levels with plant morphology; physiological responses of plants to various stress factors. *Mailing Add:* 3102 Craiglawn Rd Beltsville MD 20705-3437

TERRY, RAYMOND DOUGLAS, b Southampton, NY, Apr 19, 45; m 81. MATHEMATICS. *Educ:* State Univ NY, Stony Brook, BS, 66; Mich State Univ, MS, 68, PhD(math), 72. *Prof Exp:* Teaching asst math, Mich State Univ, 66-72; instr math, Ga Inst Technol, 72-74; asst prof, 74-78, ASSOC PROF MATH, CALIF POLYTECH STATE UNIV, SAN LUIS OBISPO, 78- *Concurrent Pos:* Vis assoc prof, Mich State Univ, 81-82. *Mem:* Am Math Soc; Math Asn Am; Soc Indust & Appl Math. *Res:* Higher-order delay and functional differential equations. *Mailing Add:* Dept Math Calif Polytech State Univ San Luis Obispo CA 93407

TERRY, RICHARD D, b Salt Lake City, Utah, Jan 29, 24; m 48. OCEANOGRAPHY, SPACE SCIENCES. *Educ:* Univ Southern Calif, AB, 50, MS, 56, PhD, 65. *Prof Exp:* Res asst oceanog, Allan Hancock Found, Univ Southern Calif, 50-55, res assoc, 55-60; res specialist, Autonetics Div, NAm Aviation Inc, 60-64, dir ocean eng, Gen Off, 64-65, spec asst oceanology, Ocean Systs Oper & Mgr, Earth Resources Group, 66-68; pres, Seaonics Int Inc, 68-70; mem tech staff, Interstate Electronics, 80-82, Rockwell Int, Space Systs Div, 83-90; PRES, RICHARD TERRY & ASSOCS/ENVIRON SCI & SERV, 70-79, 91- *Concurrent Pos:* Consult oceanog, marine geology, 55-, Dep Asst Secy Defense, 64 & US Dept Navy, 64-68; mem, Community Regional & Natural Resources Develop Group Comt, US Chamber Commerce, 66-67; prog dir, Nat Oceanog Govt & Indust Report, 64-65; chmn bd, Red Sea Enterprises Co, Ltd, 70- *Honors & Awards:* Seamount named in honor. *Mem:* Am Geophys Union; fel Geol Soc Am. *Res:* Marine geology; submarine topography; sediment; physical and chemical oceanography; ocean engineering; deep submergence technology; oceanography from space science; environmental sciences; basic and applied research in environmental sciences; strategic defense. *Mailing Add:* 3903 Calle Abril San Clemente CA 92672

TERRY, RICHARD ELLIS, b Rigby, Idaho, Feb 8, 49; m 72; c 3. SOIL BIOCHEMISTRY, SOIL MICROBIOLOGY. *Educ:* Brigham Young Univ, BS, 72; Purdue Univ, MS, 74, PhD(soil Sci), 76. *Prof Exp:* Asst soil sci, Purdue Univ, 74-75, res asst, 72-76; asst prof soil bioche, Univ Fla, 77-80; assoc prof, 80-86, PROF AGRON, BRIGHAM YOUNG UNIV, 86- *Mem:* Am Soc Agron; Soil Sci Soc Am; Int Humic Substances Soc. *Res:* Microbial oxidation of soil organic matter; subsidence of organic soils; nitrogen transformations in soils and sediments; denitrification in terrestrial and aquatic systems; soil iron nutrition. *Mailing Add:* Dept Agron Brigham Young Univ 263 WIDB Provo UT 84602

TERRY, ROBERT DAVIS, b Hartford, Conn, Jan 13, 24; m 52; c 1. NEUROPATHOLOGY. *Educ:* Williams Col, BA, 46; Union Univ, MD, 50. *Hon Degrees:* DSc, Williams Col, 91. *Prof Exp:* Asst pathologist, Montefiore Hosp, 55-59; from assoc prof to prof path, Einstein Col Med, 59-84, chmn dept, 69-84; PROF NEUROSCI & PATH, UNIV CALIF, SAN DIEGO, 84- *Concurrent Pos:* Res fel cancer, Inst Cancer Res, 65-66; mem, Med & Sci Adv Bd, Alzheimer Dis & Related Dis, 78-88. *Honors & Awards:* Potamkin Prize, Am Acad Neurol, 88; Distinguished Serv Award, Am Asn Neuropathologists, 89; Metrop Life Found Award, 91. *Mem:* Fel AAAS; Am Neurol Asn; Am Asn Neuropathologists (pres, 69-70); Int Soc Neuropath (vpres, 82-84). *Res:* Morphological and quantitative studies of the brain in Alzheimer disease. *Mailing Add:* Dept Neurosci Univ Calif La Jolla CA 92093-0064

TERRY, ROBERT JAMES, zoology, developmental biology; deceased, see previous edition for last biography

TERRY, ROBERT LEE, cell physiology; deceased, see previous edition for last biography

TERRY, ROGER, b Waterville, NY, May 8, 17; m 42; c 2. SURGICAL PATHOLOGY. *Educ:* Colgate Univ, AB, 39; Univ Rochester, MD, 44. *Prof Exp:* Intern path, Med Ctr, Univ Rochester, 44-45, instr, Univ, 45-51, from asst prof to prof, 51-69; head surg pathologist, Los Angeles County, 69-82, prof, 69-82, EMER PROF PATH, MED CTR, UNIV SOUTHERN CALIF, 82- *Concurrent Pos:* Actg pathologist, Park Ave Hosp, Rochester, NY, 45-47; resident, Med Ctr, Univ Rochester, 45-51; actg pathologist, Genesee Hosp, Rochester, 47-50 & Highland Hosp, 49-51; co-exec dir, Calif Tumor Tissue Registry, Los Angeles, 69-82; pathologist, San Gabriel Valley Med Ctr, 82- *Mem:* Am Soc Exp Path; Am Soc Clin Path; Am Asn Pathologists; Col Am Pathologists; Int Acad Path; AMA; Am Soc Cytol; Int Soc Dermatopath. *Res:* Metabolic bone diseases. *Mailing Add:* 2841 Shakespeare Dr San Marino CA 91108

TERRY, SAMUEL MATTHEW, b Nashville, Tenn, Jan 23, 15; m 43; c 2. CHEMISTRY. *Educ:* Vanderbilt Univ, BA, 43. *Prof Exp:* Jr chemist, Winthrop Chem Co, NY, 43-44; chemist, Publicker Industs, Inc, 44-46; res engr, Battelle Mem Inst, 46-51; res fel, Mellon Inst, 51-54; res chemist, Pittsburgh Plate Glass Co, 54-56; mgr prod develop, Reynolds Chem Prods Co, 56-66; vpres & tech dir, Atlantis Chem Corp, 66-67; tech dir, Hoover Chem Prod Div, Hoover Ball & Bearing Co, 67-81; RETIRED. *Mem:* Am Chem Soc. *Res:* Physical, organic and polymer chemistry. *Mailing Add:* 1560 Marian Ave Ann Arbor MI 48103-5730

TERRY, STUART LEE, b Chicago, Ill, Apr 8, 42; m 80; c 4. COMMERCIAL DEVELOPMENT, APPLICATIONS TECHNOLOGY. *Educ:* Cornell Univ, BChemE, 65, PhD(chem eng), 69; Rensselaer Polytech Inst, MS, 74. *Prof Exp:* NSF trainee polymer chem, Cornell Univ, 64-68 & asst, 65-68; sr res chemist, Monsanto Co, 68-75, group leader polymer synthesis, 75-78, prod develop, 78-82, mgr, commercial develop, 82-84, bus develop, 85, technol planning, 86-87, mgr technol acquisition, 87-89; DIR TECHNOL, SONOCO, 89- *Mem:* Soc Plastic Engrs; Tech Trans Soc. *Res:* Identification of product opportunities and preparation of polymeric materials with morphologies required for industrial and consumer end use performance. *Mailing Add:* Sonoco Second St Hartsville SC 29550

TERRY, THOMAS MILTON, b Knoxville, Tenn, Apr 2, 39; m 76; c 2. BIOPHYSICS, MICROBIOLOGY. *Educ:* Yale Univ, BA, 61, MS, 63, PhD(molecular biophys), 67. *Prof Exp:* USPHS fel biophys, Univ Geneva, 67; asst prof microbiol, Albert Einstein Col Med, 68-69; asst prof microbiol & biol, 77-81, ASSOC PROF MOLECULAR & CELL BIOL, UNIV CONN, 81- *Mem:* AAAS; Am Soc Microbiol. *Res:* Biological membrane structure and function; molecular biology of mycoplasma; ultrastructure of bacteria. *Mailing Add:* Dept Molecular & Cell Biol Univ Conn U-125 75 N Eaglevil Storrs CT 06268

TERRY, WILLIAM DAVID, b New York, NY, Oct 22, 33; m 66; c 4. IMMUNOLOGY. *Educ:* Cornell Univ, BA, 54; State Univ NY Downstate Med Ctr, MD, 58. *Prof Exp:* Intern, Jewish Hosp Brooklyn, NY, 58-59, asst resident, 59-61; NIH trainee, Sch Med, Univ Calif, San Francisco, 61-62; res assoc, Immunol Sect, Gen Labs & Clins, Nat Cancer Inst, 62-64, sr investr, 64-71, br chief, 71-87, assoc dir immunol, Div Cancer Biol & Diag, 73-87; SR VPRES & GEN MGR ADMIN, DAMON BIOTECH INC, 88- *Concurrent Pos:* Adminr, Cancer Ctrs Prog, Nat Cancer Inst, 78-81, Cancer Control Prog, 79-80, ctrs & community activ, Div Resources, 80-81. *Mem:* Am Asn Immunol; Am Fedn Clin Res; Am Soc Clin Invest; Am Asn Cancer Res. *Res:* Nature of the immune response, particularly as it relates to the recognition of and reaction against tumors by the tumor bearing host. *Mailing Add:* Abbott Biotech Inc 119 Fourth Ave Needham Heights MA 02194

TERSHAKOVEC, GEORGE ANDREW, biochemistry; deceased, see previous edition for last biography

TERSOFF, JERRY DAVID, b Wash DC, June 12, 55. THEORETICAL SOLID STATE PHYSICS, SURFACES & INTERFACES. *Educ:* Swarthmore Col, BA, 77; Univ Calif, Berkeley, PhD(physics), 82. *Prof Exp:* Postdoctoral, Bell Labs, 82-84; RES STAFF MEM, IBM THOMAS J WATSON RES CTR, 84- *Honors & Awards:* Peter Mark Mem Award, Am Vacuum Soc, 88. *Mem:* Am Phys Soc. *Res:* Electronic structures of surfaces and interfaces; theory of scanning tunnelling microscopy; semiconductor heterojunction band lineups; Schottky barries. *Mailing Add:* IBM Thomas J Watson Res Ctr PO Box 218 Yorktown Heights NY 10598

TERSS, ROBERT H, b East St Louis, Ill, Sept 13, 25; m 55. ORGANIC CHEMISTRY. *Educ:* Wash Univ, AB, 49; Univ Kans, PhD(chem), 53. *Prof Exp:* Asst instr chem, Univ Kans, 49-51; res chemist, Dept Org Chem, E I du Pont de Nemours & Co, Inc, 53-57, res supvr dyes, 57-58, head div dyes, 59-63, patents & intel, 63-65, photochem, 65-70, supt dyes & chem qual control, 70-71, supt dye mfg, 71-73, div head dyes process, 73-75, div head patents, 76-77, mem, Res & Develop Staff Serv, 76-79, mem, Personnel Develop, Employee Relations Dept, E I du Pont de Nemours & Co, Inc, 80-86; MGT CONSULT, 86- *Mem:* Am Chem Soc; Am Soc Training & Develop. *Res:* Dyes; heterocycles; photochemistry; information handling systems; photographic materials. *Mailing Add:* 708 Ambleside Dr Wilmington DE 19808-1503

TERWEDOW, HENRY ALBERT, JR, b Hoboken, NJ, July 22, 46; m 67; c 3. MEDICAL ENTOMOLOGY, AGRICULTURAL ENTOMOLOGY. *Educ:* Univ Notre Dame, BS, 68, PhD(biol), 74; Montclair State Col, MA, 69. *Prof Exp:* Res assoc entom, Sch Med, Univ Md, 73-75; fel entom, Univ Calif, Berkeley, 75-77; res entomologist, Celanese Res Co, 77-79; VELSICOL CHEM CORP, 79- *Mem:* Entom Soc Am; Am Soc Trop Med & Hyg; Am Soc Parasitologists; Am Phytopath Soc; Am Mosquito Control Asn; Weed Sci Soc Am. *Res:* Vectorial capacity and genetic control of mosquito species involved in the transmission of filariae and arboviruses; entomological aspects of insecticidal improvement. *Mailing Add:* Sandoz Crop Protection Co 975 California Ave Palo Alto CA 94304

TERWILLIGER, DON WILLIAM, b Klamath Falls, Ore, Mar 27, 42; m 70. COMPUTER SCIENCE. *Educ:* Calif Inst Technol, BS, 64; Univ Ore, MA, 66, PhD(physics), 70. *Prof Exp:* Asst prof physics, Middlebury Col, 70-75; MGR COMPUTER RES, TEKTRONIX LABS, 76- *Mem:* Inst Elec & Electronics Engrs; Comput Soc; Comput Applns & Instrumentation. *Res:* Low temperature physics; electron scattering from imperfections in metals; Fermi surface studies. *Mailing Add:* 14120 SW Barlow Rd Beaverton OR 97005

TERWILLIGER, JAMES PAUL, b Miami Beach, Fla, Feb 25, 43; c 2. CHEMICAL ENGINEERING. *Educ:* Rensselaer Polytech Inst, BChE, 65, MChE, 66, PhD(chem eng), 69. *Prof Exp:* Sr res chemist, 69-74, res assoc, 74-75, res lab head, 75-80, asst div dir, 80-86, DIV DIR, EASTMAN KODAK CO, 86- *Mem:* Am Inst Chem Engrs; Instrument Soc Am; Soc Photog Scientists & Engrs. *Res:* Heat and mass transfer; reaction kinetics; process control; crystallization. *Mailing Add:* 71 Woodstone Lane Rochester NY 14626

TERWILLIGER, KENT MELVILLE, high energy physics; deceased, see previous edition for last biography

TERWILLIGER, NORA BARCLAY, b Hartford, Conn, Oct 9, 41; wid; c 2. RESPIRATORY PROTEINS, COMPARATIVE BIOCHEMISTRY. *Educ:* Univ Vt, BS, 63; Univ Wis, MS, 65; Univ Ore, PhD(biol), 81. *Prof Exp:* Res asst, Boston Univ, 67-69, lectr embryol, 68; res asst, Inst Marine Biol, Univ Ore, 71-78; instr biol, Southwestern Ore Community Col, Coos Bay, 80-81; res assoc, 81-89, ASSOC PROF, DEPT BIOL, INST MARINE BIOL, UNIV ORE, 89- *Concurrent Pos:* Instr marine biol, Div Continuing Educ, State of Ore, 74-75; consult, Coos Bay Sch Dist, Ore, 75-76; vis scientist, Marine Biol Asn Lab, Plymouth, UK, 83-84. *Mem:* AAAS; Am Soc Zoologists; Crustacean Soc. *Res:* Comparative biochemistry and physiology of respiratory proteins; structure and function of invertebrate hemoglobins, hemocyanins and hemerythrins. *Mailing Add:* Ore Inst Marine Biol Univ Ore Charleston OR 97420

TERWILLIGER, PAUL M, b Ann Arbor, Mich, June 24, 55; m 82. COMBINATORICS & FINITE MATHEMATICS, ALGEBRA. *Educ:* Univ Mich, BS, 77; Univ Ill, PhD(math), 82. *Prof Exp:* Instr math, Ohio State Univ, 82-85; asst prof, 85-89, ASSOC PROF MATH, UNIV WIS, 89- *Mem:* Am Math Soc; Math Asn Am. *Res:* Combinatorics and graph theory. *Mailing Add:* Dept Math van Vleck Hall Univ Wis 480 Lincoln Dr Madison WI 53706

TERZAGHI, MARGARET, b Boston, Mass, May 7, 41. CANCER. *Educ:* Boston Univ, AB, 64, MS, 69; Harvard Univ, MS, 70, DSc(radiation biol, physiol), 74. *Prof Exp:* Res asst cancer res, Harvard Univ, 70-75, res assoc, 75; RES ASSOC CANCER RES, OAK RIDGE NAT LAB, 75- *Mem:* Sigma Xi. *Res:* An examination of possible in vitro models of in vivo carcinogenesis induced by chemical carcinogens and/or radiation. *Mailing Add:* 107 Orkney Rd Oak Ridge TN 37830

TERZAGHI, RUTH DOGGETT, b Chicago, Ill, Oct 14, 03. EARTH SCIENCES, GEOLOGY. *Educ:* Univ Chicago, BS, 24, MS, 25; Harvard Univ, PhD(geol), 30. *Prof Exp:* Instr geol, Goucher Col, 25-26, Wellesley Col, 26-28; consult, 40-70; lect eng geol, Harvard Univ, 57-61; RETIRED. *Honors & Awards:* Clemence Hersch Prize, Boston Soc Civil Engrs, 50. *Mem:* Hon mem Asn Eng Geologists; Geol Soc Am. *Res:* Getrology; concrete deterioration; compaction of sedimentary rocks. *Mailing Add:* Unit 48 171 Swanton St Winchester MA 01890

TERZAKIS, JOHN A, b Bridgeport, Conn, Sept 13, 35; m 61; c 3. DERMATOPATHOLOGY, ELECTRON MICROSCOPY. *Educ:* New York Univ, MD, 61. *Prof Exp:* CHIEF SURG PATH, LENOX HILL HOSP, NEW YORK,86-, ATTEND PATHOLOGIST, 76- *Mem:* Am Soc Cell Biol; NY Acad Sci. *Res:* Morphologic ultrastructure of human tissues, normal and diseased; x-ray microanalysis of diseased human tissue. *Mailing Add:* Dept Path Lenox Hill Hosp 100 E 77th St New York NY 10021

TERZIAN, YERVANT, b Feb 9, 39; m 66; c 2. ASTRONOMY. *Educ:* Am Univ Cairo, BSc, 60; Univ Ind, MA, 63, PhD(astron), 65. *Hon Degrees:* DSc, Ind Univ, 89. *Prof Exp:* Res assoc radio astron, Cornell Univ Ctr Radiophysics & Space Res & Arecibo Ionospheric Observ, 65-67, from asst prof to assoc prof, 67-77, asst dir, Ctr Radiophysics & Space Res, 68-74, grad fac rep, 74-79, PROF ASTRON, CORNELL UNIV, 77-, CHMN, DEPT ASTRON, 79-, DIR, PEW SCI EDUC NY PROG, 88-, JAMES A WEEKS PROF PHYS SCI, 90- *Concurrent Pos:* Vis prof, Univs Montreal, Thessaloniki & Florence; assoc ed, Astrophys J. *Mem:* Int Union Radio Sci; Int Astron Union; Am Astron Soc; foreign mem Armenian Acad Sci. *Res:* Radio astronomical studies of interstellar matter; radio properties of galaxies and other radio sources; radio emission from planetary nebulae; pulsars; author of numerous scientific publications and editor of 4 books. *Mailing Add:* Dept Astron Space Sci Bldg Cornell Univ Ithaca NY 14853

TERZUOLI, ANDREW JOSEPH, b Brooklyn, NY, Oct 5, 14; m 42; c 4. MATHEMATICS. *Educ:* Brooklyn Col, BA, 36; NY Univ, MS, 48. *Prof Exp:* prof, 46-86, PROF EMER MATH, POLYTECH INST BROOKLYN, 86- *Concurrent Pos:* Consult statist. *Mem:* AAAS; Am Math Soc; Am Meteorol Soc; Math Asn Am; Inst Math Statist. *Res:* Probability; mathematical statistics. *Mailing Add:* 2481 Stuart St Brooklyn NY 11229

TERZUOLO, CARLO A, b Acqui, Italy, Sept 2, 25; nat US; m 54; c 1. NEUROSCIENCE. *Educ:* Univ Torino, MD, 49. *Prof Exp:* Asst, Univ Torino, 48-49; Ital Res Coun fel, 50-51; asst prof, Free Univ Brussels, 51-53; asst, Univ Calif, Los Angeles, 54-56, res assoc, 57-59; PROF PHYSIOL, UNIV MINN, MINNEAPOLIS, 59- *Concurrent Pos:* Multiple Sclerosis Soc fel, 56-57; Fulbright res fel, Univ Pisa, 66-67. *Mem:* Am Physiol Soc; Soc Neurosci. *Res:* Nerve cell and receptor physiology; dynamic characteristics of neuronal systems controlling movements; vestibular, cerebellar and segmental reflex mechanisms; motor control. *Mailing Add:* Dept Physiol Sch Med Univ Minn Minneapolis MN 55455

TESAR, DELBERT, b Beaver Crossing, Nebr, Sept 2, 35; m 57; c 4. MECHANICAL ENGINEERING. *Educ:* Univ Nebr, BSc, 58, MSc, 60; Ga Inst Technol, PhD(mech eng), 64. *Prof Exp:* Instr eng mech, Univ Nebr, 57-59 & appl mech, Kans State Univ, 59-61; lectr mech eng, Ga Inst Technol, 61-64; assoc prof, 64-69, PROF MECH ENG & ENG SCI, UNIV FLA, 69- *Concurrent Pos:* Chief investr, US Army Res Off res grant, 61; NSF res grants, 64, 69-71 & 72-74; NSF-NATO fel, Vienna Tech Univ, 64-65; consult, Procter & Gamble Co, Ohio, 66-; vis prof, Wash State Univ, 67; consult, Wayne S Colony Co, Fla, 71- & Deering Milliken Res Corp, SC, 72-; vis res prof, Liverpool Polytech, Eng, 71-72; mem, Sci Adv Bd Air Force, 82-; dir & founder, Ctr Intelligent Machines & Robotics. *Mem:* Am Soc Eng Educ; Am Soc Mech Engrs. *Res:* Kinematic synthesis; dynamic analysis and synthesis; vibrations; machine design; lubrication; experimental stress analysis; strength of materials; continuum mechanics. *Mailing Add:* Dept Mech Engrs Univ Tex Austin TX 78712

TESAR, MILO B, b Nebr, Apr 7, 20; m 44; c 4. PLANT BREEDING & GENETICS. *Educ:* Univ Nebr, BS, 41; Univ Wis, MS, 47, PhD(agron), 49. *Hon Degrees:* DSc, Univ Nebr, 89. *Prof Exp:* From asst prof to assoc prof, Mich state Univ, 49-58, actg chmn dept, 64-66, prof crop sci, 58-88, EMER PROF, MICH STATE UNIV, 88- *Concurrent Pos:* NATO fel, Grassland Res Inst, Eng, 59-60; consult, Univ Ryukus, 67 & Univ Federale Rio Grande do Sul, Brazil, 76; mem, Int Grassland Cong, USA, 52, 81, Australia, 70, Russia, 74, Japan, 85, Europ Grassland Fedn, 77; mem, USSR Forage Surv, 74 & AID, Somalia Mission, 78; lectr, Inner Mongolia, China, 85. *Honors & Awards:* Agron Achievement Award, Am Soc Agron, 84; Career Award, Crop Sci Soc Am, 88. *Mem:* Crop Sci Soc Am; fel Am Soc Agron; Sigma Xi. *Res:* Forage physiology, management, and digestibility; maximum yield of alfalfa; legume seeding establishment; no-till pasture renovation; alfalfa breeding; autotoxicity/allelopathy in alfalfa; alfalfa to reduce effluent damage to environment. *Mailing Add:* 2379 Emerald Forest Circle East Lansing MI 48823

TESCHAN, PAUL E, b Milwaukee, Wis, Dec 15, 23; m 48; c 2. MEDICINE, NEPHROLOGY. *Educ:* Univ Minn, BS, 46, MD & MS, 48; Am Bd Internal Med, dipl, 55. *Prof Exp:* Intern, Res & Educ Hosp, Univ Ill, 48-49, resident internal med, Presby Hosp, Chicago, 49-50, ward officer, Metab Ward, Dept Hepatic & Metab Dis, Walter Reed Army Inst Res, 50-53, resident internal med, Barnes Hosp, St Louis, Mo, 53-54, chief renal br, Surg Res Unit, Brooke Army Med Ctr, Ft Sam Houston, Tex, 54-60, asst commandant, Walter Reed Army Inst Res, Walter Reed Army Med Ctr, 60-63, dep dir div basic surg res, 64-65, dep dir div surg, 65-66, chief dept metab & dir div med, 66-69; PROF MED & ASSOC PROF UROL & BIOMED ENG, VANDERBILT UNIV, 69- *Concurrent Pos:* Med Corps, US Army, 48-69; fel cardiorenal dis, Peter Bent Brigham Hosp, Boston, 50; chief renal insufficiency ctr, Korea, 52-53; consult, Surgeon Gen, US Army, 60-69; chief dept surg physiol, Walter Reed Army Inst Res, Walter Reed Army Med Ctr, 61-66, dep dir div basic surg res, 62-63, chief renal-metab serv, Walter Reed Gen Hosp, 66-69; chief US Army med res team, Vietnam, 63-64; dir, Tenn Mid-South Regional Med Prog, 69-72. *Mem:* Am Fedn Clin Res; fel Am Col Physicians; Soc Artificial Internal Organs; Am Physiol Soc; Int Soc Nephrology. *Res:* Pathogenesis and prevention of acute renal failure; prophylactic dialysis; uremia; prevention of progression in chronic renal failure. *Mailing Add:* Dept Med & Urol/Biomed Vanderbilt Univ Sch Med Nashville TN 37232

TESH, ROBERT BRADFIELD, b Wilmington, Del, Jan 22, 36; m 60; c 2. MICROBIOLOGY. *Educ:* Franklin & Marshall Col, BS, 57; Jefferson Med Col, MD, 61; Tulane Univ, MS, 67. *Prof Exp:* Intern, San Francisco Gen Hosp, Calif, 61-62; resident pediat, Gorgas Hosp, 62-63; physician, USPHS, Peace Corps, Recife, Brazil, 63-65; NIH fel infectious dis, Depts Pediat & Epidemiol, Sch Med, Tulane Univ, 65-67; epidemiologist, Mid Am Res Unit, NIH, 67-72, Pac Res Sect, 72-80; PROF, DEPT EPIDEMIOL & PUB HEALTH, YALE ARBOVIRUS RES UNIT, SCH MED, YALE UNIV, 80- *Concurrent Pos:* Counr, Am Soc Trop Med Hyg, Exec Comt, Am Comt Arboviruses. *Mem:* AAAS; Am Soc Trop Med & Hyg; Am Soc Microbiol; Royal Soc Trop Med & Hyg; Am Soc Virol; Am Col Epidemiol. *Res:* Entomology; microbiology; public health; virology. *Mailing Add:* Sch Med Yale Univ PO Box 3333 New Haven CT 06510

TESK, JOHN A, b Chicago, Ill, Oct 19, 34; m. DENTAL RESEARCH. *Educ:* Northwestern Univ, BS, 57, MS, 60, PhD(mat sci), 63. *Prof Exp:* Asst prof metall, Univ Ill, Chicago, 64-68; asst metallurgist, Metall Div, Argonne Nat Lab, 68-70; asst mgr res & develop, How-Medica, Inc, Chicago, 70-71, dir res & develop, Dent Div, 71-77; dir educ serv, Inst Gas Technol, Chicago, 77-78; gen phys scientist, Nat Bureau Standards, 78-83, GROUP LEADER, DENT & MED MATS, NAT INST STANDARDS & TECHNOL, 83- *Concurrent Pos:* Consult, Argonne Nat Lab, 64-67; dent res, 77-; pres, Wash Chap, Int Asn Dent Res, 84-85, treas, DMG, 87- *Honors & Awards:* Grainger Award, 64; Bronze Medal, Nat Bur Standard, 87. *Mem:* Int Asn Dent Res; Am Inst Mining & Metall Engrs; Am Soc Metals; Biomat Soc; Sigma Xi; fel Dent Mats Acad. *Res:* Radiation damage in metals at low temperature; point defects in metals; dental and medical materials and devices; biomaterials. *Mailing Add:* A 143 Polymers Nat Inst Standards & Technol Washington DC 20235

TESKA, WILLIAM REINHOLD, b Chicago, Ill, Oct 11, 50. ECOLOGY, MAMMALOGY. *Educ:* Univ Idaho, BS, 72; Mich State Univ, MSc, 74, PhD(zool), 78. *Prof Exp:* ASSOC PROF BIOL, FURMAN UNIV, 77- *Mem:* Am Soc Mammalogists; Ecol Soc Am; Wildlife Soc; Sigma Xi. *Res:* Mammalian ecology; population ecology; vertebrate ecology. *Mailing Add:* Dept Biol Furman Univ Greenville SC 29613

TESKE, RICHARD GLENN, b Cleveland, Ohio, Aug 16, 30; m 75. ASTRONOMY. *Educ:* Bowling Green State Univ, BS, 52; Ohio State Univ, MA, 56; Harvard Univ, PhD, 61. *Prof Exp:* From instr to assoc prof, McMath-Hulbert Observ, 61-77, PROF ASTRON, UNIV MICH, ANN ARBOR, 77- *Concurrent Pos:* Dir, Mich Dartmouth MIT Observ, 88- *Mem:* Int Astron Union; Am Astron Soc. *Res:* Supernova remnants; solar physics; x-rays from supernova remnants. *Mailing Add:* Dept Astron Univ Mich Ann Arbor MI 48109

TESKE, RICHARD H, b Christiansburg, Va, July 22, 39; m 61; c 2. VETERINARY MEDICINE, TOXICOLOGY. *Educ:* Va Polytech Inst, BA, 62; Univ Ga, DVM, 65; Univ Fla, MS, 66; Am Bd Vet Toxicol, dipl. *Prof Exp:* Asst prof vet sci, Univ Fla, 67; dir toxicol, Hill Top Res, Inc, Ohio, 67-70; VET MED OFFICER, CTR VET MED, FOOD & DRUG ADMIN, 70- *Mem:* Fel Am Acad Vet & Comp Toxicol; fel Am Acad Vet Pharm & Therapeut. *Res:* Comparative pharmacology and toxicology. *Mailing Add:* Ctr Vet Med Food & Drug Admin 5600 Fishers Lane Rockville MD 20857

TESKEY, HERBERT JOSEPH, b Grande Prairie, Alta, June 9, 28; m 53; c 2. ENTOMOLOGY, SYSTEMATICS. *Educ:* Univ Alta, BSc, 51; Univ Toronto, MSA, 56; Cornell Univ, PhD(entom), 67. *Prof Exp:* Res scientist entom, Can Dept Agr, Guelph, 51-67; RES SCIENTIST ENTOM, BIOSYST RES INST, AGR CAN, 67- *Mem:* Entom Soc Can. *Res:* Systematics of the Diptera, especially of the lower Brachycera and of the immature stages of Diptera. *Mailing Add:* 569 Brierwood Ottawa ON K2A 2H6 Can

TESMER, IRVING HOWARD, b Buffalo, NY, May 31, 26; m 64; c 2. STRATIGRAPHY, PALEONTOLOGY. *Educ:* Univ Buffalo, BA, 46, MA, 48; Syracuse Univ, PhD(geol), 54. *Prof Exp:* From instr to asst prof geol, Univ NH, 50-55; instr, Rutgers Univ, 55-57; from asst prof to assoc prof, 57-63, chmn dept, 66-69, PROF GEOL, STATE UNIV NY COL BUFFALO, 63- *Concurrent Pos:* Mem, Paleont Res Inst; res assoc geol, Buffalo Mus Sci, 70- *Mem:* Fel AAAS; fel Geol Soc Am; Am Asn Petrol Geol; Paleont Soc. *Res:* Devonian stratigraphy and paleontology; geology of western New York; history of geology of Western New York. *Mailing Add:* 127 Fayette Ave Buffalo NY 14223

TESMER, JOSEPH RANSDELL, b Lafayette, Ind, Sept 9, 39; m 62; c 2. EXPERIMENTAL NUCLEAR PHYSICS. *Educ:* Purdue Univ, BS, 62; Univ Wash, PhD(nuclear physics), 71. *Prof Exp:* Engr, Boeing Co, 62-64; res assoc nuclear physics, Purdue Univ, 71-73; asst scientist physics, Univ Wis, 73-75; MEM STAFF NUCLEAR PHYSICS, LOS ALAMOS NAT LAB, 75- *Mem:* Am Phys Soc; AAAS. *Res:* Stripping of high energy negative hydrogen beams, and negative hydrogen beam production; accelerator development; accelerator based mass spectrometry. *Mailing Add:* 408 Rover Blvd White Rock NM 87544

TESORIERO, JOHN VINCENT, b Brooklyn, NY, Feb 10, 41; m 64; c 2. CELL BIOLOGY, BIOLOGY OF AGING. *Educ:* Fairfield Univ, BS, 63; Adelphi Univ, MS, 68; State Univ NY, Downstate Med Sch, PhD(anat), 76. *Prof Exp:* ASST PROF BASIC SCI & EDUC NE MED SCH, UNIV MED & DENT, NJ, 75- *Mem:* Am Asn Anatomists; Am Soc Cell Biol. *Res:* Cell biology of mammalian oogenesis with particular emphasis on age related changes in maturing oocytes; age related changes of the tubulin-microtubular system in mammalian oocytes. *Mailing Add:* Basic Sci & Educ Dent NJ 100 Bergen St Newark NJ 07103

TESORO, GIULIANA C, b Venice, Italy, June 1, 21; nat US; m 43; c 2. ORGANIC POLYMER CHEMISTRY. *Educ:* Yale Univ, PhD(org chem), 43. *Prof Exp:* Res chemist, Calco Chem Co, NJ, 34-44; res chemist, Onyx Oil & Chem Co, 44-46, head org synthesis dept, 46-55, asst dir res, 55-57, assoc dir, 57-58; asst dir org res, Cent Res Lab, J P Stevens & Co, Inc, 58-68; sr scientist, Textile Res Inst, NJ, 68-69; sr scientist, Burlington Industs, Inc, 69-71, dir chem res, 71-72; RES PROF, POLYTECH UNIV, 82- *Concurrent Pos:* Vis prof, Mass Inst Technol, 72-76, adj prof & sr res scientist, 76-82; mem comt on fire safety aspects of polymeric materials, Nat Acad Sci, 72-78, mem nat mat adv bd, 76-78; mem comt military personnel supplies, Nat Res Coun, 79-81, comt toxicol combustion prod, 84-89. *Honors & Awards:* Olney Medal, Am Asn Textile Chem & Colorists. *Mem:* Am Chem Soc; Am Asn Textile Chem & Colorists; Fiber Soc (pres, 74); Am Inst Chemists; AAAS. *Res:* Synthesis of pharmaceuticals; textile chemicals; germicides; polymers; chemical modification of fibers; synthesis and rearrangement of glycols in the hydrogenated naphthalene series; polymer flammability and flame retardants; polymers for electronics; thermally stable polymers and coupling agents in composites. *Mailing Add:* 278 Clinton Ave Dobbs Ferry NY 10522

TESS, ROY WILLIAM HENRY, b Chicago, Ill, Apr 25, 15; m 44; c 2. COATINGS, SOLVENTS. *Educ:* Univ Ill, BS, 39; Univ Minn, PhD(org chem), 44. *Prof Exp:* Lab asst, Underwriters Labs, Inc, Ill, 36-37; asst chem, Univ Minn, 39-44; chemist, Shell Develop Co, 44-59, res supvr, Shell Chem Co, 59-62 & 64-67, Royal Dutch Shell Plastics Lab, Holland, 62-63, tech planning supvr, Coatings, Shell Chem NY, 67-70, Tex, 70-73, tech supvr solvents bus ctr, 73-78, consult solvents & resin prod, 78-79; INDEPENDENT CONSULT, 79- *Concurrent Pos:* Med air qual comt, Nat Paint & Coatings Asn, 68-79; trustee, Paint Res Inst, 71-79, pres, 73-76; ed, Solvents Theory & Pract, 73, Appl Polymer Sci, 75 & Appl Polymer Sci Second Ed, 85; dir, Fedn Soc Coatings Technol, 73-76; chmn, div org coating & plastic chem, Am Chem Soc, 78, exec comt, div polymeric mats, 77- *Honors & Awards:* Roon Award, Fedn Soc Coatings Technol, 56, Heckel Award, 78. *Mem:* Fedn Soc Coatings Technol; fel Am Inst Chem; Am Chem Soc; AAAS. *Res:* Epoxy, alkyd, polyester and hydrocarbon resins; surface coatings; varnishes and drying oils; polyols; polar and hydrocarbon solvents; high polymer latexes; atmospheric chemistry; polymer chemistry; writer on applied polymer science. *Mailing Add:* PO Box 577 Fallbrook CA 92028

TESSEL, RICHARD EARL, b Cincinnati, Ohio, June 9, 44. NEUROPHARMACOLOGY, PSYCHOPHARMACOLOGY. *Educ:* Univ Calif, Los Angeles, BA, 66; Univ Ill, Chicago Circle, MA, 69; Univ Mich, PhD(pharmacol), 74. *Prof Exp:* Nat Inst Drug Abuse fel pharmacol, Sch Med, Univ Colo, 74-75; ASST PROF PHARMACOL, SCH PHARM, UNIV KANS, 75-, ASSOC TOXICOL, 81- *Res:* Role of biogenic amine disposition in brain in modulating the reinforcing, locomotor-stimulant, stereotypic and operant-schedule effects of amphetamines and its congeners. *Mailing Add:* Dept Pharmacol Univ Kans 5040 Mal Lawrence KS 66045-2505

TESSER, HERBERT, b Jersey City, NJ, Mar 25, 39; m 61; c 2. PHYSICS. *Educ:* Polytech Inst Brooklyn, BS, 60; Stevens Inst Technol, MS, 63, PhD(physics), 68. *Prof Exp:* Metrol engr, Kearfott Corp, 60-62; res asst physics, Stevens Inst Technol, 64-67; assoc prof, 67-81, PROF PHYSICS, PRATT INST, 81- *Concurrent Pos:* Res consult, NRA, Inc, 71-72, Procedyne, 73-74 & Stevens Inst, 75-78. *Mem:* Am Phys Soc. *Res:* Electrodynamics; statistical mechanics; relativity; plasma physics. *Mailing Add:* 120 Park St Woodmere NY 11598

TESSIER, CLAIRE ADRIENNE, b Staten Island, NY, Oct 24, 53; m 78; c 1. SYNTHETIC INORGANIC & ORGANOMETALLIC CHEMISTRY. *Educ:* Univ Vt, BS, 75; State Univ NY-Buffalo, PhD(chem), 82. *Prof Exp:* Postdoctoral res assoc, Northwestern Univ, 80-82; vis asst prof gen chem, Univ RI, 82-83; vis asst prof org chem, Case Western Reserve Univ, 83-85, sr instr, 85-86, sr res assoc, 86-90; ASST PROF INORG CHEM, UNIV AKRON, 90- *Concurrent Pos:* Vis prof, Chem Dept, Univ Wis-Madison, 91; adj prof, Chem Dept, Case Western Reserve Univ, 91- *Mem:* Am Chem Soc; AAAS. *Mailing Add:* Dept Chem Univ Akron Akron OH 44325-3601

TESSIERI, JOHN EDWARD, b Vineland, NJ, Sept 3, 20; m 43; c 3. RESEARCH ADMINISTRATION. *Educ:* Pa State Univ, BS, 42, MS, 47; Stanford Univ, PhD(org chem), 50. *Prof Exp:* Chemist & asst to asst dir res, Texaco, Inc, NY, 49, group leader, 55, asst supvr lubricants res, 55, Tex, 55-56, supvr chem res, 56-57, asst dir res, 57-60, dir fuels & chem res, 60-62, vpres, Texaco Exp Inc, Va, 62-63, exec vpres, 63-65, pres, 65-66, mgr sci planning, Texaco Inc, 67-68, asst to pres, 68-69, staff coordr strategic planning group exec off, 69-70, gen mgr strategic planning, 70-71, vpres Res Environ & Safety Dept, Texaco, Inc, 71-81; RETIRED. *Mem:* Am Chem Soc; Indust Res Inst; AAAS; Sci Res Soc Am; Dirs Indust Res. *Res:* Product and process development, including petrochemicals, fuels and lubricants; exploration and production research; coal beneficiation, gasification and liquefaction. *Mailing Add:* 27 Lincoln Dr Poughkeepsie NY 12601

TESSLER, ARTHUR NED, b New York, NY, Feb 21, 27; m 53; c 4. UROLOGY. *Educ:* NY Univ, AB, 48, MD, 52; Am Bd Urol, dipl. *Prof Exp:* Investr, USPHS grant, 59-62, PROF CLIN UROL, SCH MED, NY UNIV, 72- *Concurrent Pos:* Consult, Vet Admin Hosp, NY, 70- *Honors & Awards:* Carl Hartman Award, Am Fertil Soc, 63. *Mem:* AMA; Am Fertil Soc; Am Col Surg. *Mailing Add:* Dept Urol NY Univ Sch Med 530 First Ave New York NY 10016

TESSLER, GEORGE, b Brooklyn, NY, Mar 7, 36; m 69; c 2. PHYSICS. *Educ:* Brooklyn Col, BS, 57; Univ Pa, MS, 59, PhD(physics), 64. *Prof Exp:* FEL SCIENTIST, BETTIS ATOMIC POWER LAB, WESTINGHOUSE ELEC CORP, 63- *Mem:* Am Phys Soc; AAAS; NY Acad Sci; Am Nuclear Soc. *Res:* Acquisition of neutron cross section data for use in reactor design; nuclear reactor design. *Mailing Add:* Bettis Atomic Power Lab Westinghouse Elec Corp PO Box 79 West Mifflin PA 15122

TESSLER, MARTIN MELVYN, b Brooklyn, NY, Sept 12, 37; m 62; c 2. ORGANIC CHEMISTRY, POLYSACCHARIDE CHEMISTRY. *Educ:* Brooklyn Col, BS, 58; Univ Kans, PhD(chem), 62. *Prof Exp:* Chemist, Enjay Chem Intermediates Lab, Esso Res & Eng Co, 65-68; proj supvr, 68-72, from res assoc to sr res assoc, 72-81, sr res scientist, 81-83, assoc dir, 83-85, DIR NATURAL POLYMER RES, NAT STARCH & CHEM CORP, BRIDGEWATER, 85- *Mem:* AAAS; Am Chem Soc; Am Asn Cereal Chemists. *Res:* Starch chemistry. *Mailing Add:* 507 Darwin Blvd Edison NJ 08820

TESSMAN, IRWIN, b New York, NY, Nov 24, 29; m 49; c 1. MICROBIOLOGY, MOLECULAR BIOLOGY. *Educ:* Cornell Univ, AB, 50; Yale Univ, MS, 51, PhD(physics), 54. *Prof Exp:* NSF fel, Cornell Univ, 54-55, Am Cancer Soc fel, 55-57; fel, Mass Inst Technol, 57-58, res assoc biol, 58-59; assoc prof biophys, 59-62, prof biochem & molecular biol, Univ Calif, Irvine, 70-72; PROF BIOL, PURDUE UNIV, 62- *Concurrent Pos:* NSF sr fel, Harvard Med Sch, 67; prof molecular biol, Univ Calif, 69-72. *Honors & Awards:* Gravity Res Found Prize, 53; Sigma Xi Res Award, Purdue Univ, 66. *Mem:* Am Soc Biochem & Molecular Biol; AAAS; Genetics Soc Am; Am Soc Microbiol. *Res:* Radiocarbon dating; ionization by charged particles; molecular genetics; reproduction of bacterial viruses; molecular studies of repair, mutation, recombination and function of genetic material. *Mailing Add:* Dept Biol Sci Purdue Univ West Lafayette IN 47907

TESSMAN, JACK ROBERT, physics, for more information see previous edition

TESSMER, CARL FREDERICK, b North Braddock, Pa, May 28, 12; m 39; c 2. PATHOLOGY. *Educ:* Univ Pittsburgh, BS, 33, MD, 35; Am Bd Path, dipl, 41. *Prof Exp:* Resident path, Presby Hosp, Pittsburgh, 36-37; fel, Mayo Clin, 37-38; resident pathologist, Queen's Hosp, Honolulu, Hawaii, 39-40; chief lab, Tripler Gen Hosp, Honolulu, Med Corps, US Army, 42-45, chief radiologic safety, Bikini, 46, pathologist, US Naval Med Res Inst, 46-48, dir atomic bomb casualty comn, Nat Res Coun, 48-51, commanding officer, Army Med Res Lab, Ft Knox, Ky, 51-54, chief basic sci div & radiation path br, Armed Forces Inst Path, Walter Reed Army Med Ctr, DC, 54-60, commanding officer, 406th Med Gen Lab, Japan, 60-62, pathologist, Walter Reed Army Inst Res, 62-63; chief exp path sect, Univ Tex M D Anderson Hosp & Tumor Inst Houston, 63-71, prof path, 63-73; chief lab serv, Vet Admin Ctr, Temple, Tex, 73-85; CONSULT, WALTER REED ARMY MED CTR, 64- *Concurrent Pos:* Hektoen Lect, 60; Armed Forces Inst Path Centennial lect, 62; mem grad fac, Univ Tex Grad Sch Biomed Sci, 63-73, path coordr, Univ Tex Med Sch Houston, 71-73; consult, Walter Reed Army Med Ctr, 64-; mem, USPHS Adv Comt, Collab Radiol Health Animal Res Lab, 65-70; mem, subcomt 34, Nat Comt Radiation Protection, 70-76; consult, Radiation Bioeffects & Epidemiol Adv Comn, Food & Drug Admin, Dept Health, Educ & Welfare, 72-75. *Honors & Awards:* Hektoen lectr, 60. *Mem:* Fel Am Soc Clin Path; Radiation Res Soc; Am Asn Path & Bact; fel Col Am Path; Int Acad Path. *Res:* Morphologic and experimental radiation pathology; trace elements; copper metabolism. *Mailing Add:* Rte Five Box 5291 Belton TX 76513

TEST, CHARLES EDWARD, b Indianapolis, Ind, Jan 10, 16; m 38; c 4. MEDICINE. *Educ:* Princeton Univ, AB, 37; Univ Chicago, MD, 41. *Prof Exp:* Instr med, Univ Chicago, 49-51; from asst prof to prof, Sch Med, Ind Univ, Indianapolis, 53-89; RETIRED. *Mem:* Endocrine Soc; Am Diabetes Asn; fel Am Col Physicians. *Res:* Endocrinology; metabolism. *Mailing Add:* 4430 N Meridian St Indianapolis IN 46208

TEST, FREDERICK L(AURENT), b Philadelphia, Pa, June 15, 25; m 47; c 3. MECHANICAL ENGINEERING. *Educ:* Mass Inst Technol, SB, 45, SM, 47; Pa State Univ, PhD(mech eng), 56. *Prof Exp:* Res engr, Mass Inst Technol, 47-48; instr mech eng, Northeastern Univ, 48-49; from instr to assoc prof, 49-72, chmn dept, 72-76, PROF MECH ENG, UNIV RI, 72- *Concurrent Pos:* NSF fac fel, 59-60; consult, US Navy, 60-66; Fulbright res fel, Neth, 66-67. *Mem:* Am Soc Mech Engrs; Am Inst Aeronaut & Astronaut; Sigma Xi. *Mailing Add:* 258 Blackberry Hill Dr Wakefield RI 02879

TESTA, ANTHONY CARMINE, b New York, NY, Nov 19, 33; m 62. PHYSICAL CHEMISTRY, PHOTOCHEMISTRY. *Educ:* City Col New York, BS, 55; Columbia Univ, MA, 58, PhD(chem), 61. *Prof Exp:* Res chemist, Cent Res Div, Lever Bros Co, 61-63; from asst prof to assoc prof, 63-71, PROF CHEM, ST JOHN'S UNIV, NY, 71- *Concurrent Pos:* Res leave, Max-Planck Inst Spectros, Goettingen, 70-71. *Mem:* NY Acad Sci; Am Chem Soc. *Res:* Photochemistry and flash photolysis of molecules in solution; luminescence spectroscopy; fluorescence and phosphorescence. *Mailing Add:* Dept Chem St John's Univ Jamaica NY 11439

TESTA, RAYMOND THOMAS, b New Haven, Conn, Dec 21, 37; m 62; c 3. MICROBIOLOGY, BIOCHEMISTRY. *Educ:* Providence Col, BS, 59; Syracuse Univ, MS, 64, PhD(microbiol), 66. *Prof Exp:* Res asst microbiol, Syracuse Univ, 64-66; scientist, Schering Corp, 66-68, sr scientist, 68-72, prin microbiologist, 72-74, mgr, antibiotics screening & fermentation dept, 74-77; MEM STAFF, MED RES DIV, AM CYANAMID CO, 77- *Concurrent Pos:* Counr, NY Acad Sci, 88-; chair-elect, Div A, Am Soc Microbiol, 91-92. *Mem:* Am Soc Microbiol; Soc Indust Microbiol (secy, 76-79, pres elec, 79-80, pres, 80-81); NY Acad Sci. *Res:* Clinical microbiology; antibiotics; resistance mechanisms; factors affecting the production and biosynthesis of antibiotics; spore formation. *Mailing Add:* Antimicrobiol Chemother Dept Am Cyanamid Co Lederle Lab Pearl River NY 10965

TESTA, RENE B(IAGGIO), b Montreal, Can, May 30, 37; m 59; c 4. CIVIL ENGINEERING, ENGINEERING MECHANICS. *Educ:* McGill Univ, BEng, 59; Columbia Univ, MS, 60, Eng ScD(eng mech), 63. *Prof Exp:* From asst prof to assoc prof, 63-75, PROF CIVIL ENG & ENG MECH, COLUMBIA UNIV, 75-, DIR, ROBERT A W CARLETON MAT LAB, 65- *Concurrent Pos:* Consult var govt & indust orgns. *Mem:* Am Soc Mech Engrs; Am Soc Civil Engrs; Am Soc Testing & Mat; Struct Stability Res Coun. *Res:* Solid mechanics; experimental mechanics of materials and structures; mechanical properties of coated fabrics. *Mailing Add:* Dept Civil Eng Columbia Univ Broadway & W 116th New York NY 10027

TESTA, STEPHEN M, b Fitchburg, Mass, July 17, 51; m 86; c 1. GEOLOGY, HYDROGEOLOGY. *Educ:* Calif State Univ, Northridge, BS, 76, MS, 78. *Prof Exp:* Engr geologist, R T Franklian & Assocs, 76-78; geologist & chief petrologist, Bechtel Inc, 78-80; proj geologist, Converse Consults, Wash, 80-82; chief hydrologist ecol & environ, Wash, 82-83; proj mgr & hydrogeologist, Dames & Moore, 83-86; vpres west coast opers, Eng Enterprises, Inc, 86-90; PRES, APPL ENVIRON SERV, 90- *Concurrent Pos:* Adv bd trustee, Geol Soc Am; co-chmn, Am Asn Engr Geologists, Wash State Sect, 86; mem, Nat Comt Hydrogeol & Waste Mgt, 89-90; instr geol, mineral, geochem & hazardous waste mgt, Calif State Univ, Fullerton; mem, Nat Comt Prof Develop, Nat Comt Continuing Educ & Nat Comt Ann Meeting, Am Inst Prof Geologists. *Honors & Awards:* Presidential Cert of Merit, Am Inst Prof Geologists, 87. *Mem:* AAAS; Am Asn Petrol Geologists; Am Inst Prof Geologists; Mineral Soc Can; Mineral Soc Am; Nat Water Well Asn; Asn Eng Geologists; Geol Soc Am; Hazardous Mat Control Res Inst; Sigma Xi. *Res:* Environmental science; fate and transport of contaminants in geologic and hydrogeologic systems; remediation design and methods; public policy issues; environmental law; igneous and metamorphic petrology; historical development of applied geosciences. *Mailing Add:* 23113 Plaza Pointe Dr No 100 Laguna Hills CA 92653-1425

TESTARDI, LOUIS RICHARD, b Philadelphia, Pa, Sept 23, 30; m 57; c 4. SOLID STATE PHYSICS. *Educ:* Univ Calif, Berkeley, AB, 56; Univ Pa, MS, 60, PhD(physics), 63. *Prof Exp:* Res physicist, Elec Storage Battery Co, 57-58 & Franklin Inst Labs, 58-62; res asst, Univ Pa, 63; res physicist, Bell Tel Labs, 63-80; dir mat processing in space, NASA Hq, 80-82; CHIEF, METALL DIV, NAT BUR STANDARDS, GAITHERSBURG, MD, 82- *Mem:* Fel Am Phys Soc. *Res:* Transport, optical, magnetic and ultrasonic properties of solids; low temperature physics; superconductivity. *Mailing Add:* 2731 Blairstone Rd 175 Tallahassee FL 32301

TESTER, CECIL FRED, b Boone, NC, May 23, 38; m 67; c 3. BIOCHEMISTRY, BIOTECHNOLOGY. *Educ:* Appalachian State Univ, BS, 60; Univ Ga, PhD(biochem), 67. *Prof Exp:* Teacher city schs, NC, 60-61; res asst biochem, Univ Ga, 64-65, teaching asst, 65-66; AEC fel, Purdue Univ, Lafayette, 67-68; res chemist & USDA grant, NC State Univ, 68-75; RES CHEMIST, BELTSVILLE AGR RES CTR, AGR RES SERV, USDA, 75- *Concurrent Pos:* Prof leader org matter transformations & mineral nutrit. *Honors & Awards:* Super Serv Award, USDA, 77. *Mem:* Am Soc Agron; Soil Sci Soc; Crop Sci Soc. *Res:* Biochemistry of rhizosphere plant-microbial interactions; characterization and use of monodonial antibodies for genetic determinants. *Mailing Add:* 11402 Westview Ct Beltsville MD 20705

TESTER, JEFFERSON WILLIAM, b New York, NY, Mar 27, 45; m 67; c 1. CHEMICAL ENGINEERING, PHYSICAL CHEMISTRY. *Educ:* Cornell Univ, BS, 66, MS, 67; Mass Inst Technol, PhD(chem eng), 71. *Prof Exp:* Fel, Los Alamos Sci Lab, Univ Calif, 71-72; asst prof chem eng & dir, Sch Chem Eng Pract at Oak Ridge Nat Lab, Mass Inst Technol, 72-74; group leader-geothermal technol, Los Alamos Sci Lab, 74-80; assoc prof, 80-88, DIR, SCH CHEM ENG PRACT, MASS INST TECHNOL, 80-, PROF CHEM ENG, 88-, DIR, ENERGY LAB, 89- *Concurrent Pos:* Vis staff mem, Los Alamos Sci Lab, Univ Calif, 72-74; adj prof chem eng, Univ NMex, 75-80; consult, Los Alamos Nat Lab & Sandia Nat Labs, 88- *Mem:* Am Inst Chem Engrs; Am Chem Soc; Soc Petrol Engrs. *Res:* Applied thermodynamics; surface chemistry and physics; chemical kinetics; heat and mass transfer; environmental control technology; electrochemistry; geothermal energy, reservoir engineering. *Mailing Add:* 19 Liberty Rd Hingham MA 02043

TESTER, JOHN ROBERT, b New Ulm, Minn, Nov 18, 29; m 60; c 2. ECOLOGY. *Educ:* Univ Minn, BS, 51; Colo State Univ, MS, 53; Univ Minn, PhD(wildlife ecol), 60. *Prof Exp:* Res asst wildlife biol, Colo Game & Fish Dept, 51-53; game biologist, Minn Div Game & Fish, 54-56; asst scientist ecol, Mus Natural Hist, 56-60, from instr to assoc prof, 60-70, head dept, 73-76, PROF ECOL & BEHAV BIOL, UNIV MINN, MINNEAPOLIS, 70- *Concurrent Pos:* NSF fel, Aschoff Div, Max Planck Inst Physiol of Behav & Aberdeen Univ, 69-70; mem behav sci training comt, NIH; Nat Acad Sci exchange scientist, 82; hon fel, Aberdeen Univ, Scotland; dir, Cedar Creek Nat Hist Area, 84- *Mem:* Fel AAAS; Ecol Soc Am; Am Soc Mammal; Wildlife Soc. *Res:* Population ecology; biotelemetry; wildlife management; behavioral ecology; fire ecology. *Mailing Add:* Dept Ecol & Behav Biol Univ Minn Minneapolis MN 55455

TESTERMAN, JACK DUANE, b Marietta, Okla, Dec 13, 33; m 53; c 3. STATISTICS, COMPUTER SCIENCE. *Educ:* Okla State Univ, BS, 55, MS, 57; Univ Tex, PhD, 72. *Prof Exp:* Res statistician, Jersey Prod Res Co, 57-63 & Phillips Petrol Co, 63; assoc prof, Univ Southwestern La, 63-69, registr, 65-70, prof math & statist, 69-84, dir, instnl res, 70-84, vpres univ relations, 73-84; consult, 84-87; PROF MGT, SOUTHEASTERN OKLA STATE UNIV, 86- *Concurrent Pos:* Chmn inst studies & opers analytical comt, Am Asn Col Registrars & Admis Officers, 74-; chmn ad hoc comt ways & means, Am Statist Asn, 75- *Mem:* Am Statist Asn; Asn Comput Mach; Am Mgt Asn. *Res:* Application of statistics; data analysis and use of computers. *Mailing Add:* PO Box 5609 Ardmore OK 73403

TESTERMAN, JOHN KENDRICK, comparative physiology, for more information see previous edition

TETENBAUM, MARVIN, b Brooklyn, NY, June 27, 21; m 54; c 3. PHYSICAL CHEMISTRY. *Educ:* NY Univ, BChE, 42; Polytech Inst Brooklyn, MChE, 47, PhD(chem), 54. *Prof Exp:* Res asst, Metall Lab, Univ Chicago, 42-43; res engr sam labs, Columbia Univ, 43-44; chem engr, Ballistics Res Lab, Ord Dept, US Dept Army, 47-48; radio chemist, US Naval Res Lab, 48-56; sr engr, Aircraft Nuclear Propulsion Dept, Gen Elec Co, Ohio, 56-59; CHEMIST, ARGONNE NAT LAB, 59- *Concurrent Pos:* Mem staff, Atomic Energy Res Estab, Harwell, Eng, 66-67; vis scientist, Europ Inst Transuranium Elements, Karlsruhe, WGer, 83-84. *Honors & Awards:* Fel, Am Ceramic Soc. *Mem:* Am Nuclear Soc; Am Ceramic Soc; AAAS; Am Chem Soc; Sigma Xi. *Res:* High temperature chemistry; Superconductivity. *Mailing Add:* Chem Technol Div Argonne Nat Lab 9700 S Cass Ave Argonne IL 60439

TETERIS, NICHOLAS JOHN, b Martins Ferry, Ohio, Jan 14, 29; m 61; c 2. OBSTETRICS & GYNECOLOGY. *Educ:* Washington & Jefferson Col, BA, 50; Ohio State Univ, MD, 54, MSc, 61; Am Bd Obstet & Gynec, dipl, 65. *Prof Exp:* Asst prof, 65-67, assoc prof & asst dean col med, 67-70, PROF OBSTET & GYNEC, COL MED, OHIO STATE UNIV, 70- *Concurrent Pos:* Cancer fel obstet & gynec, Col Med, Ohio State Univ, 62-64; consult, USAF Hosps, Wright-Patterson & Lockborne AFB, 62; asst dir, Ohio State Univ Hosps, 62- *Mem:* AMA; Am Col Obstet & Gynec; Am Col Surg. *Res:* Fetology; gynecologic cancer; obstetrical emergencies. *Mailing Add:* Obstet-Gyn Ohio State Univ Coll Med 1654 Upham Dr Columbus OH 43210

TETLOW, NORMAN JAY, b Downs, Kans, Dec 9, 34; m 57; c 2. CHEMICAL ENGINEERING. *Educ:* Kans State Univ, BS, 57, MS, 59; Tex A&M Univ, PhD(chem eng), 66. *Prof Exp:* Chem engr, Mason & Hanger, Silas Mason Co, 57-58 & 59-60; res & develop engr, Dow Chem Co, 60-63 & Tex Instruments, Inc, 66-68; process syst specialist, 68-79, PROCESS ENG MGR, TEX DIV, DOW CHEM CO, 79- *Mem:* Am Inst Chem Engrs; Am Chem Soc. *Res:* Application of computers and numerical methods to chemical process engineering and process control. *Mailing Add:* 120 Clover Lake Jackson TX 77566-4606

TETRAULT, ROBERT CLOSE, b Walhalla, NDak, Nov 25, 33; m 58; c 4. ENTOMOLOGY. *Educ:* NDak State Univ, BS, 58, MS, 63; Univ Wis, PhD(entom), 67. *Prof Exp:* From asst prof to assoc prof entom, Pa State Univ, University Park, 71-86; RETIRED. *Mem:* Entom Soc Am. *Res:* Taxonomy of Coleoptera, especially the family Helodidae. *Mailing Add:* PO Box 63 Walhalla ND 58282

TETREAULT, FLORENCE G, statistics, for more information see previous edition

TETTENHORST, RODNEY TAMPA, b St Louis, Mo, Feb 1, 34; m 60; c 3. MINERALOGY. *Educ:* Wash Univ, BS, 55, MA, 57; Univ Ill, Urbana, PhD(mineral), 60. *Prof Exp:* From asst prof to assoc prof, 60-75, PROF MINERAL, OHIO STATE UNIV, 75- *Mem:* Clay Minerals Soc; Mineral Soc Am; Mineral Soc Gt Brit & Ireland. *Res:* Clay mineralogy; x-ray diffraction and physical properties of small crystalline particles. *Mailing Add:* Dept Geol 107 Mendenhall Lab Ohio State Univ Columbus OH 43210

TEUBER, LARRY ROSS, b Prescott, Ariz, July 8, 51; m 83. PLANT BREEDING, POLLINATION BIOLOGY. *Educ:* NMex State Univ, BS, 73, MS, 74; Univ Minn, PhD(plant breeding), 78. *Prof Exp:* asst prof, 77-84, ASSOC PROF PLANT BREEDING, UNIV CALIF, DAVIS, 84- *Concurrent Pos:* Assoc agronomist, Calif Agr Exp Sta, 77-; mem, NAm Alfalfa Improv Conf. *Mem:* Crop Sci Soc Am; Am Soc Agron; Int Comn Bee Bot; Sigma Xi. *Res:* Alfalfa breeding and genetics; pollinator activity and seed production; forage yield and quality; nitrogen fixation and disease resistance including: Colletotrichum, Stagnospora, and Stemphylium. *Mailing Add:* Dept Agron & Range Sci Univ Calif Davis CA 95616

TEUKOLSKY, SAUL ARNO, b Johannesburg, SAfrica, Aug 2, 47; m 71; c 2. THEORETICAL ASTROPHYSICS, RELATIVITY. *Educ:* Univ Witwatersrand, BSc Hons(physics) & BSc Hons(appl math), 70; Calif Inst Technol, PhD(physics), 73. *Prof Exp:* Res assoc physics, Calif Inst Technol, 73-74; from asst prof to assoc prof physics, 74-83, PROF PHYSICS & ASTRON, CORNELL UNIV, 83- *Concurrent Pos:* Alfred P Sloan Found res fel, 75-77; John Simon Guggenheim Mem fel, 81-82. *Mem:* Am Phys Soc; Am Astron Soc. *Res:* General relativity and relativistic astrophysics. *Mailing Add:* Lab Nuclear Studies Cornell Univ Ithaca NY 14853

TEUSCHER, GEORGE WILLIAM, b Chicago, Ill, Jan 11, 08; m 34; c 2. DENTISTRY. *Educ:* Northwestern Univ, DDS, 29, MSD, 36, MA, 40, PhD(educ), 42. *Hon Degrees:* ScD, NY Univ, 65. *Prof Exp:* From instr to assoc prof, 31-45, dean, Dent Sch, 53-72, prof, 45-72, EMER PROF PEDODONTICS, DENT SCH, NORTHWESTERN UNIV, 72- *Concurrent Pos:* Regent, Nat Libr Med; ed, J Am Soc Dent Children, 68- *Mem:* Am Soc Dent Children; Am Dent Asn; Am Acad Pedodontics; fel Am Col Dent; Int Asn Dent Res. *Res:* Dental caries; reactions of dental pulp in children; sodium fluoride; prevention in clinical dentistry for children; principles of dental education. *Mailing Add:* Am Soc Dent Children 211 E Chicago Ave Chicago IL 60611-2616

TEVAULT, DAVID EARL, b Evansville, Ind, July 23, 48; m 73; c 2. INORGANIC CHEMISTRY. *Educ:* Univ Evansville, BA, 70; Univ Va, PhD(chem), 74. *Prof Exp:* Fel chem, Marquette Univ, 74-76; fel chem, Naval Res Lab, 76-78, staff scientist, 78-87; US ARMY CHEM RES, DEVELOP & ENG CTR, 87- *Mem:* Am Chem Soc; Sigma Xi. *Res:* Mechanisms and kinetics of atmospheric, combustion related, heterogeneous catalysis, and infrared laser promoted chemical reactions by cryogenic spectroscopic methods; air purification. *Mailing Add:* 802 Hayden Ct E Bel Air MD 21014

TEVEBAUGH, ARTHUR DAVID, b Knox Co, Ind, Nov 25, 17; m 43; c 2. PHYSICAL CHEMISTRY, INORGANIC CHEMISTRY. *Educ:* Purdue Univ, BS, 40; Iowa State Univ, PhD(phys chem), 47. *Prof Exp:* Asst chem, Iowa State Univ, 40-42; res chemist, Manhattan Proj, 42-47; res chemist, Knolls Atomic Power Lab, Gen Elec Co, 47-55; supvr reactor chem unit, 50-54, actg mgr chem & chem eng sect, 54-55, sr res chemist, Res & Develop Lab, 55-63; sr chemist & sect mgr chem eng div, Argonne Nat Lab, 63-69, assoc dir div, 69-72, dir lab prog planning off, 72-73, dir coal progs, 73-77, assoc dir chem div, 77-81; RETIRED. *Mem:* Am Chem Soc. *Res:* Analytical, radio and soil chemistry; production and handling of fluorine; chemical problems related to development and operation of nuclear reactors and reactor fuel reprocessing; polymer research; thermoelectric materials; fuel cells and batteries; electrochemistry; nuclear reactor safety. *Mailing Add:* Five Park Rd Hatfield PA 19440

TEVETHIA, MARY JUDITH (ROBINSON), b Ft Wayne, Ind, Feb 25, 39; m 65; c 2. MOLECULAR BIOLOGY, GENETICS. *Educ:* Mich State Univ, BS, 60, MS, 62, PhD(microbiol), 64. *Prof Exp:* Fel microbiol, Emory Univ, 64-65; fel biol, Univ Tex, M D Anderson Hosp & Tumor Inst, Houston, 65-72, asst biologist & asst prof biol, 72-73; asst prof path, Sch Med, Tufts Univ, 73-; STAFF MEM, DEPT MICROBIOL, PA STATE UNIV. *Concurrent Pos:* NIH fels, 64-65, 65- *Mem:* Am Soc Microbiol; Am Soc Virol. *Res:* Genetic studies on simian papova virus SV4O; human cytomegalovirus. *Mailing Add:* Dept Microbiol Pa State Univ 500 University Dr Hershey PA 17033

TEVETHIA, SATVIR S, b Buland Shahr, India, Aug 5, 36; m 65; c 1. VIROLOGY, IMMUNOLOGY. *Educ:* Agra Univ, BSc, 54, BVSc, 58; Mich State Univ, MS, 62, PhD(microbiol), 64. *Prof Exp:* Vet asst surg, Indian Govt, 58-59; res asst microbiol, Mich State Univ, 60-64; from asst prof to assoc prof virol, Baylor Col Med, 71-73; assoc prof path, Sch Med, Tufts Univ, 73-77, prof, 77-78; PROF MICRO BIOL, COL MED, PA STATE UNIV, 78- *Concurrent Pos:* Fel virol, Baylor Col Med, 64-66; Nat Cancer Inst res career develop award, 67-71. *Mem:* AAAS; Am Soc Microbiol; Am Asn Cancer Res. *Res:* Tumor viruses and immunology. *Mailing Add:* Milton S Hershey Med Ctr Pa State Univ PO Box 850 Hershey PA 17033

TEVIOTDALE, BETH LUISE, b Long Beach, Calif, July 17, 40; div. PHYTOPATHOLOGY. *Educ:* Pomona Col, BA, 62; Univ Calif, Davis, MS, 70, PhD(plant path), 74. *Prof Exp:* Lab technician bot, Calif Inst Technol, 62-63; lab technician immunol, Univ Calif, Los Angeles, 65-68; EXTEN SPECIALIST PLANT PATH, SAN JOAQUIN VALLEY RES & EXTEN CTR, UNIV CALIF, 75- *Mem:* Sigma Xi. *Res:* Vegetable and tree crops, with major emphasis on disease-free potatoes for seed and deep bark canker of walnuts. *Mailing Add:* San Joaquin Valley Res & Exten Ctr Univ Calif 9240 S Riverbend Rd Parlier CA 93648

TEW, JOHN GARN, b Mapleton, Utah, Oct 26, 40; m 65; c 6. IMMUNOLOGY, MICROBIOLOGY. *Educ:* Brigham Young Univ, BS, 66, MS, 67, PhD(microbiol), 70. *Prof Exp:* NIH fel, Case Western Reserve Univ, 70-72; from asst prof to assoc prof, 72-82, PROF MICROBIOL, VA COMMONWEALTH UNIV, 82- *Concurrent Pos:* Vis prof, Walter & Elizabeth Hall Inst Med Res, Melbourne, Australia, 77-78. *Mem:* Am Soc Microbiol; Am Asn Immunologists; Reticuloendothelial Soc; Sigma Xi. *Res:* Role of persisting antigen and follicular dendritic cells in the induction maintenance and regulation of the humoral immune response; role of beta-lysin in innate immunity; immunobiology of periodontal disease. *Mailing Add:* Dept Microbiol Med Col Va PO Box 678 MCV Sta Richmond VA 23298

TEW, KENNETH DAVID, b Dumbarton, Scotland, Apr 20, 52; US citizen; m 87. CANCER CHEMOTHERAPY, DRUG RESISTANCE. *Educ:* Univ London, PhD(biochem pharmacol), 76. *Prof Exp:* CHMN PHARMACOL, FOX CHASE CANCER CTR, PHILADELPHIA, 85- *Concurrent Pos:* Adj assoc prof pharmacol, Univ Pa, 85-90, adj prof, 90-; mem, Study Sect ET 1, NIH, 88-, chmn, 90-92; mem, Sci Adv Bd, 87-; chmn, Gordon Conf Chemother Cancer, 89; mem, Subcomt Space Sta Pharmacodynamics, NASA/ACS, 88-; vis prof, Nat Cancer Inst, Tokyo, 88-89. *Mem:* Am Soc Pharmacol & Exp Therapeut; Am Assoc Cancer Res; Am Soc Cell Biol; AAAS. *Res:* Focused upon understanding the mechanisms of action of alkylating agent classes of anticancer drugs, with special emphasis on determinants of tumor cell resistance to these agents. *Mailing Add:* Chmn Dept Pharmacol Fox Chase Cancer Ctr 7701 Burholme Rd Philadelphia PA 19111

TEWARI, SUJATA, b Murshidabad, India, July 1, 38; m 64; c 2. MOLECULAR BIOLOGY, NEUROCHEMISTRY. *Educ:* Agra Univ, BSc, 55; Univ Lucknow, MS, 57; McGill Univ, PhD(biochem), 62. *Prof Exp:* Neurochemist co-investr brain protein synthesis & neurotransmitters, Vet Admin Hosp, City of Hope, Sepulveda, Calif, 66-70; asst researcher neurochemist step III ethanol & cent nervous syst protein synthesis, 70-72, asst researcher neurochemist step IV, 72, asst prof res III, 72-76, asst prof res IV, 76-78, ASSOC PROF RES ETHANOL & CENT NERVOUS SYST PROTEIN SYNTHESIS, DEPT PSYCYHIAT & HUMAN BEHAV, UNIV CALIF, IRVINE, 78- *Concurrent Pos:* Fel biochem, Lucknow Univ, India, 63-64; NSF res proj grant, 75; prin investr, Nat Inst Alcohol Abuse & Alcoholism, 76-, NIMH grant, 78-; mem, Pub Comt Alcoholism, Clin & Exp Res, 77- & biomed panel Calif state supported ctr, 78-; sponsor, Nat Inst Alcohol Abuse & Alcoholism, Ronald L Alkana & Eugene Fleming fel, 78-; sci dir, Nat Inst Alcohol Abuse & Alcoholism, Alcohol Res Ctr grant, 78- *Mem:* AAAS; Biochem Soc; Int Soc Neurochem; Res Soc Alcoholism; Int Soc Biomed Res on Alcoholism; Am Soc Neurochem. *Res:* Neurobiology combining the disciplines of molecular biology, biochemistry, neurochemistry and neuropharmacology; biomedical research in alcoholism and psychoactive drugs. *Mailing Add:* Dept Psychiat & Human Behav Univ Calif Sch Med Irvine CA 92717-0001

TEWARSON, REGINALD P, b Pauri, India, Nov 17, 30; div; c 2. BIOMATHEMATICS. *Educ:* Univ Lucknow, BSc, 50; Agra Univ, MSc, 52; Boston Univ, PhD(appl math), 61. *Prof Exp:* Lectr physics, Messmore Col, India, 50-51; lectr math, Univ Lucknow, 52-57; sr mathematician, Honeywell Inc, 60-64; from asst prof to prof, 64-89, actg chmn, 83-84, LEADING PROF APPL MATH, STATE UNIV NY STONY BROOK, 89- *Concurrent Pos:* Vis prof, Oxford Univ, 70-71. *Mem:* Am Math Soc; Soc Indust & Appl Math; Soc Math Biol. *Res:* Sparse matrices; linear algebra; numerical analysis; mathematical modelling in biology. *Mailing Add:* Dept Appl Math & Statist State Univ NY Stony Brook NY 11794-3600

TEWELL, JOSEPH ROBERT, b Albany, NY, May 19, 34; m 60; c 3. ADVANCED PROGRAMS, PROJECT DEVELOPMENT. *Educ:* Rensselaer Polytech Inst, BEE, 55, MEE, 60. *Prof Exp:* Instr elec eng, Rensselaer Polytech Inst, 55-64; sr res scientist, Martin Marietta Corp, 65-72, mgr, Space Teleopers, 73-75, Space Lab, 76-77, Space Vehicles, 78-79, Advan Progs, 79-86 & Launch Systs, 87-90, MGR COMPUTER AIDED PROD, MARTIN MARIETTA CORP, 90- *Concurrent Pos:* Res engr, NAm Aviation, Inc, 55; assoc res engr, Lockheed Aircraft Corp, 56; consult, Redford Corp, 61-62. *Honors & Awards:* Achievement Award, NASA, 74, New Technol Award, 76. *Mem:* Am Inst Aeronaut & Astronaut; Unmanned Vehicle Syst Asn; fel Explorers Club; Smithsonian Inst; Air & Space Mus. *Res:* Astronaut space maneuvering units; spacecraft docking and retrieval mechanisms; on-orbit teleoperators; launch vehicle design; stereo visual systems for space and undersea systems; atmospheric re-entry systems; solar power satellites; launch system technology requirements; author of over 35 publications. *Mailing Add:* 619 Legendre Dr Slidell LA 70460

TEWES, HOWARD ALLAN, b Los Angeles, Calif, May 1, 24; m 53; c 3. PHYSICAL CHEMISTRY. *Educ:* Univ Calif, Los Angeles, BS, 48, MS, 50, PhD(chem), 52. *Prof Exp:* Asst chem, Univ Calif, Los Angeles, 49-52; chemist, Calif Res & Develop Co, 52-53; chemist, Lawrence Livermore Nat Lab, Univ Calif, 53-89; RETIRED. *Mem:* AAAS; Am Chem Soc; Am Phys Soc. *Res:* Proton induced reactions of thorium; spallation reactions; high energy neutron induced nuclear reactions; industrial applications of nuclear explosives; analysis of environmental impacts of advanced energy resource recovery technologies; nuclear waste isolation technology. *Mailing Add:* 4151 Cologate Way Livermore CA 94550

TEWHEY, JOHN DAVID, b Lewiston, Maine, Feb 14, 43; m 65; c 3. HYDROGEOCHEMISTRY, HYDROGEOLOGY. *Educ:* Colby Col, BA, 65; Univ SC, MS, 68; Brown Univ, PhD(geol), 75. *Prof Exp:* Geologist, Lawrence Livermore Lab, Univ, Calif, 74-80; mgr, E C Jordan, Co, 81-87; PRES, TEWHEY ASSOCS, 87- *Concurrent Pos:* Mem geosci fac, Chabot Col, Livermore, Calif, 73-80, Univ Southern Maine, Gorham, 87-89. *Mem:* Geol Soc Am; Am Geophys Union; Am Chem Soc; Asn Groundwater Scientists & Engrs; Am Inst Hydrol; Asn Eng Geologists. *Res:* Evaluation of contamination in ground water; evaluation of the geochemical controls of soil-water interaction; radionuclide migration; geological applications to engineering. *Mailing Add:* John D Tewhey Assoc 500 Southborough Dr South Portland ME 04106

TEWINKEL, G CARPER, b Lamona, Wash, Jan 20, 09. PHOTOGRAMMETRY. *Educ:* Wash State Univ, BSc, 32; Syracuse Univ, MCE, 40. *Prof Exp:* Draftsman cartog, US Soil Conserv Serv, 35-40; gen engr photogram, Nat Ocean Surv, US Nat Oceanic, Atmospheric Admin, Rockville, Md, 40-72; RETIRED. *Concurrent Pos:* Instr Math, George Washington Univ, 45-48; ed in chief, Am Soc Photogram, 65-75. *Honors & Awards:* Fairchild Award, Am Soc Photogram, 71; Schwidefsky Award, Int Soc Photogram, 88. *Mem:* Am Soc Photogram (pres, 60); Am Cong Surv & Mapping; Am Soc Civil Engrs; Inst Soc Photogram (vpres, 68-72). *Res:* Development of a computational system and a computer program called Analytic Aerotriangulation which greatly improved the accuracy of aerial mapping and enabled a reduction of costs for applications to nautical and aeronautical charting. *Mailing Add:* 62012 Starr Lane La Grande OR 97850

TEWKSBURY, CHARLES ISAAC, b Portsmouth, NH, Feb 26, 25; m 49; c 3. RUBBER CHEMISTRY. *Educ:* Univ NH, BS, 48, MS, 49; Princeton Univ, PhD(chem), 53. *Prof Exp:* Lab asst, Univ NH, 48-49; asst, Princeton Univ, 49-53; phys chemist, Nat Res Corp, 53-54, proj mgr, 54-57, sr chemist, 57-59; res chemist, Monsanto Chem Co, 59; group leader, Cabot Corp, 59-61; asst assoc dir new prod res, 61-63; RES DIR, ODELL CO, 63-; TREAS, FAY SPECIALTIES, INC, 71-, PRES, 74- *Mem:* Am Chem Soc. *Res:* Gas kinetics; hydrocarbon oxidation; heterogeneous catalysis; metals; high temperature phenomena; radiation chemistry; polymerization; polymer characterization; elastomers; adhesives. *Mailing Add:* 17 Adams Way Wayland MA 01778

TEWKSBURY, DUANE ALLAN, b Osceola, Wis, Oct 4, 36; div; c 2. BIOCHEMISTRY. *Educ:* St Olaf Col, BA, 58; Univ Wis, MS, 60, PhD(biochem), 64. *Prof Exp:* Res biochemist, 64-80, SR RES BIOCHEMIST, MARSHFIELD MED FOUND, 80- *Concurrent Pos:* Mem, Coun on High Blood Pressure Res, Am Heart Asn. *Mem:* Am Chem Soc; NY Acad Sci; Am Soc Biol Chemists; Inter-Am Soc Hypertension; Am Heart Asn. *Res:* Biochemical studies of peptide hormone systems; isolation and characterization of the protein components of the renin-angiotensin system. *Mailing Add:* Marshfield Med Found 510 N St Joseph Ave Marshfield WI 54449

TEWKSBURY, L BLAINE, b Boston, Mass, Sept 1, 17; m 49; c 8. INDUSTRIAL RESEARCH, EDUCATION. *Educ:* Yale Univ, BS, 38, PhD(org chem), 41. *Prof Exp:* Chief chemist, Tom's of Maine Inc, 72-90; RETIRED. *Mem:* Sigma Xi; Am Chem Soc. *Mailing Add:* Box 783 Kennebunk ME 04043

TEWKSBURY, STUART K, b Manchester, NH, Apr 22, 42; m 65; c 2. SYSTEMS DESIGN & SYSTEMS SCIENCE. *Educ:* Univ Rochester, NY, BS, 64, PhD(physics), 69. *Prof Exp:* Mem res staff, AT&T Bell Labs, 69-90; PROF COMPUTER ENG, WVA UNIV, 90- *Mem:* Sr mem Inst Elec & Electronics Engrs; Am Inst Physics; Int Soc Optical Eng. *Res:* Integrated electronic systems with emphasis on limits facing present device/system technologies, new emerging technologies and enabled new system architectures/organizations. *Mailing Add:* Dept Elec & Computer Eng WVa Univ PO Box 6101 Morgantown WV 26506-6101

TEWS, JEAN KRING, b Ogdensburg, NY, May 21, 28; m 56; c 1. BIOCHEMISTRY. *Educ:* St Lawrence Univ, BS, 49; Univ Wis, MS, 52, PhD(biochem), 54. *Prof Exp:* Asst biochem, Univ Wis, 50-54; biochemist, Galesburg State Res Hosp, 54-55; proj assoc physiol, 55-62, res assoc, 63-66, res assoc biochem, 67-77, ASSOC SCIENTIST BIOCHEM, UNIV WIS-MADISON, 77- *Concurrent Pos:* Acad vis, Inst Psychiat, London, 85. *Mem:* Int Soc Neurochem; Am Inst Nutrit; Am Soc Neurochem. *Res:* Enzyme activities and nutrition; neurochemistry; factors influencing chemical components of brain; amino acids; amino acid transport. *Mailing Add:* 5445 Lake Mendora Dr Madison WI 53705

TEWS, LEONARD L, b Rush Lake, Wis, May 28, 34; m 60; c 4. BOTANY, MYCOLOGY. *Educ:* Wis State Univ, Oshkosh, BS, 56; Ind Univ, MA, 58; Univ Wis, PhD(bot, mycol), 65. *Prof Exp:* Teaching asst bot, Ind Univ, 56-58; teacher high sch, 58-61; res asst mycol, Univ Wis, 61-63; from instr to assoc prof, 64-75, PROF BIOL, UNIV WIS-OSHKOSH, 75- *Concurrent Pos:* Water Resources res grant, Water Resources Ctr, 69-70; vis prof, Univ RI, 83. *Mem:* Mycol Soc Am; Brit Mycol Soc. *Res:* Vesicular-arbuscular endomycorrhizal fungi. *Mailing Add:* Dept Biol Univ Wis Oshkosh 800 Algoma Blvd Oshkosh WI 54901

TEXON, MEYER, b New York, NY, Apr 23, 09; m 41; c 2. MEDICINE. *Educ:* Harvard Univ, AB, 30; NY Univ, MD, 34; Am Bd Internal Med & Am Bd Cardiovasc Dis, dipl, 44. *Prof Exp:* Asst prof, 57-73, ASSOC PROF FORENSIC MED, NY UNIV, SCH MED, 73- *Concurrent Pos:* Asst med examr, City of New York, 57-; sr clin asst med, Mt Sinai Hosp; consult cardiovasc dis, Bur Hearings & Appeals, Social Security Agency, Dept Health, Educ & Welfare; fel coun clin cardiol & coun atherosclerosis, Am Heart Asn; attend physician, Doctors Hosp; pres, NY County Med Soc, 82-83. *Honors & Awards:* Hektoen Silver Medal, AMA, 58. *Mem:* AMA; Am Heart Asn; Am Col Physicians; fel Am Col Cardiol; NY Acad Med. *Res:* Cardiovascular disease; internal medicine; atherosclerosis; hemodynamics; heart disease and industry; role of vascular dynamics in the development of atherosclerosis; forensic med - can the cardiac stand trial!. *Mailing Add:* Three E 68th St New York NY 10021

TEXTER, E CLINTON, JR, b Detroit, Mich, June 12, 23; m 49; c 3. MEDICAL EDUCATION, GASTROENTEROLOGY. *Educ:* Mich State Univ, BA, 43; Wayne State Univ, MD, 46. *Prof Exp:* Asst resident, Goldwater Mem Hosp, NY Univ, 50-51; instr, Sch Med, Duke Univ, 51-53; assoc, Med Sch, Northwestern Univ, Chicago, 53-56, asst chief gastroenterol clins, 53-55, dir training prog in gastroenterol, 55-63, from asst prof to assoc prof med, 58-68; chmn dept clin physiol, Olen B Culbertson Res Ctr, Scott & White Clin, Temple, Tex, 68-72; asst dean, Col Health Related Prof, Med Ctr, 72-73, assoc dean, 73-75, dir, Div Gastroenterol, 73-85, PROF MED PHYSIOL & BIOPHYS, COL MED, UNIV ARK, LITTLE ROCK, 72-, JEROME S LEVY PROF MED, 85- *Concurrent Pos:* Res fel med, Col Med, Cornell Univ, 48-50; attend physician, Passavant Mem Hosp, 53-68; co-founder, Gastroenterol Res Group, 55, chmn, 59-60; ed, Am J Digestive Dis, 56-68; chief gastroenterol sect, Vet Admin Res Hosp, 59-63; consult, US Naval Hosp, Great Lakes, 63-68; attend physician, Cook County Hosp, Ill, 66-68; mem sr staff, Scott & White Mem Hosp, Temple Univ, 68-72; consult, Santa Fe Hosp, Vet Admin Ctr, Temple, 68-72; William Beaumont Hosp, Dept Army, El Paso, 68-, consult to Surgeon Gen, US Army, 70; lectr, Univ Fla, 69; adj prof physiol, Univ Tex Southwestern Med Sch Dallas, 69-72; fac coordr allied health training prog, Temple Jr Col, 69-72; attend physician, Univ Hosp, Univ Ark, Little Rock, 72-75; assoc chief of staff for educ & actg chief gastroenterol, Vet Admin Hosp, 72-79. *Mem:* Am Physiol Soc; Am Gastroenterol Asn; fel Am Col Physicians; Am Fedn Clin Res; Am Med Writers' Asn (pres, 73-74); Am Col Gastroenterol. *Res:* Gastrointestinal physiology and pathophysiology; health care delivery systems. *Mailing Add:* Univ Ark Med Sci 4301 W Markham No 567 Little Rock AR 72205

TEXTER, JOHN, b Lancaster, Pa, Aug 9, 49; m 84; c 1. PHYSICAL CHEMISTRY, BIOPHYSICS. *Educ:* Lehigh Univ, BSEE, 71, MS, 73, MS, 76, PhD(chem), 76. *Prof Exp:* Instr, chem, Lafeyette Col, 73-74; assoc physiologist, Biophys Spectros Lab, Univ Calif, Irvine, 76-77; res assoc, Chem Dept, State Univ NY, Binghamton, 77-78; res chemist, Emulsion Phys Chem Lab, 78-80, sr res scientist, Dispersion Tech Lab, 80-90, SR RES SCIENTIST, COLOR PAPER MAT LAB, EASTMAN KODAK CO, 90- *Concurrent Pos:* Consult, Strider Res Corp, 87. *Mem:* Am Phys Soc; Sigma Xi; Inst Elec & Electronics Engrs; Am Chem Soc; Int Asn Colloid Interface Scientists; Am Inst Chem Engrs; Soc Photo Sci Engrs. *Res:* Heterogeneous chemistry of image dye formation in photographic coupler dispersions; reaction-diffusion kinetics in colloidal systems; oil/water interfacial structure and trans-interfacial transport; microemulsion polymerization; stabilization of pigment dispersions. *Mailing Add:* Res Lab Eastman Kodak Co Rochester NY 14650-2109

TEXTOR, ROBIN EDWARD, b Detroit, Mich, Mar 19, 43; m 65; c 2. MATHEMATICS. *Educ:* Tenn Polytech Inst, BS, 64; Univ Tenn, Knoxville, MS, 68; Drexel Univ, PhD(math), 72. *Prof Exp:* Comput appln programmer, Comput Technol Ctr, Union Carbide Corp, Tenn, 64-69; asst prof math, Univ SC, 71-72; COMPUT APPLN ANALYST, OAK RIDGE NAT LAB, UNION CARBIDE CORP, 72-, SECT HEAD, 73- *Mem:* Math Asn Am; Am Math Soc; Soc Indust & Appl Math. *Res:* Singular hyperbolic partial differential equations; numerical solution of fluid flow problems and partial differential equations. *Mailing Add:* Martin Marietta Energy Syst Bldg 9983-62MS-816 PO Box 2009 Oak Ridge TN 37831-8163

TEXTORIS, DANIEL ANDREW, b Cleveland, Ohio, Jan 19, 36; m 59; c 3. GEOLOGY. *Educ:* Case Western Reserve Univ, BA, 58; Ohio State Univ, MS, 60; Univ Ill, PhD(geol), 63. *Prof Exp:* Asst prof geol, Univ Ill, 63-65; from asst prof to assoc prof, 65-73, asst dean res admin, 68-74, assoc dean res admin, 74-83, PROF GEOL, UNIV NC, CHAPEL HILL, 73- *Concurrent Pos:* Geologist, Diamond Alkali Co, 57-60; consult, Southern Ill-Pa Coal, 61-65; coord NSF sci develop prog, 67-74, Carbonate dredging, 68-70 & Chevron, Texaco & Exxon, 82-; sr assoc, Dept of Energy, 78-, consult, 79-81. *Mem:* AAAS; Geol Soc Am; Soc Econ Paleont & Mineral; Am Asn Petrol Geol. *Res:* Sedimentary geology; carbonate petrography; diagenesis of sediments; petrology and geochemistry of volcanic tuff; interpretation of ancient carbonate environments; paleoecology; geology of fossil fuels. *Mailing Add:* Dept Geol Univ NC Chapel Hill NC 27599-3315

TEYLER, TIMOTHY JAMES, b Portland, Ore, Nov 25, 42; m 66; c 1. NEUROSCIENCES. *Educ:* Ore State Univ, BS, 64; Univ Ore, MS, 68, PhD(psychol), 69. *Prof Exp:* Asst prof psychol, Univ Southern Calif, 68-69; lectr psychobiol, Univ Calif, Irvine, 69-72, assoc res psychobiologist, 73-74; lectr psychol, Harvard Univ, 74-75, assoc prof, 75-77; assoc prof, 77-81, PROF NEUROBIOL, COL MED, NORTHEASTERN OHIO UNIV, 81- *Concurrent Pos:* NIMH fel, Univ Calif, Irvine, 70-72; NSF sr fel & Fulbright scholar, Inst Neurophysiol, Univ Oslo, 73-74. *Mem:* Soc Neurosci; Psychonomic Soc. *Res:* Neurobiological correlates of behavioral plasticity; neurolinguistics; magnetoencephalography; neuroendocrinology. *Mailing Add:* Neurobiol/Physiol Col Med Northeastern Ohio Univ 4209 St Rte 44 Box 95 Rootstown OH 44272

THACH, ROBERT EDWARDS, b Oklahoma City, Okla, Feb 2, 39; m 68; c 3. PROTEIN BIOSYNTHESIS, VIROLOGY. *Educ:* Princeton Univ, BA, 61; Harvard Univ, PhD(biochem), 64. *Prof Exp:* Res fel biochem and molecular biol, Harvard Univ, 64-66, from asst prof to assoc prof, 66-70; assoc prof biol chem, 70-72, dir Ctr Basic Cancer Res, 72-77, dir grad prog molecular biol, 74-77, chmn dept biol, 77-81, PROF BIOL CHEM, WASHINGTON UNIV, 72-, PROF BIOL, 77- *Concurrent Pos:* Fel, Guggenheim Mem Found, 69-; mem, res grant rev comt, United Cancer Coun, Inc, 77-79; inst biosafety Comt, Monsanto Co, 80-83; biochem rev panel, NSF, 80-81. *Mem:* Am Soc Biol Chem; Am Soc Virol; NY Acad Sci; AAAS; Sigma Xi. *Res:* Mech and regulation of protein biosynthesis; viral replication and effects on host cells. *Mailing Add:* Dept Biol Wash Univ Lindell & Skinker Blvds St Louis MO 63130

THACH, WILLIAM THOMAS, b Oklahoma City, Okla, Jan 3, 37; m 63; c 3. NEUROPHYSIOLOGY, NEUROLOGY. *Educ:* Princeton Univ, AB, 59; Harvard Med Sch, MD, 64. *Prof Exp:* Intern & resident med & neurol, Mass Gen Hosp, 64-66, 69-71; staff assoc neurophysiol, NIMH, 66-69; from asst prof to assoc prof physiol & neurol, Med Sch, Yale Univ, 71-75; assoc prof, 75-80, PROF NEUROBIOL & NEUROL, SCH MED, WASHINGTON UNIV, 80-, DIR, IWJ REHAB INST, 91- *Concurrent Pos:* Prin investr, Nat Inst Neurol Dis & Stroke res grant, 71-; neurologist, Barnes Hosp, Jewish Hosp & St Louis Regional Hosp, 75- *Mem:* Soc Neurosci; Am Physiol Soc; Am Acad Neurol. *Res:* Physiology of behavior; cerebellar control of posture and movement; mechanisms of neurorehabilitation. *Mailing Add:* Dept Anat & Neurobiol Wash Univ Sch Med 660 Euclid Ave St Louis MO 63110

THACHER, HENRY CLARKE, JR, b New York, NY, Aug 8, 18; m 42; c 5. COMPUTER SCIENCE, NUMERICAL ANALYSIS. *Educ:* Yale Univ, AB, 40; Harvard Univ, MA, 42; Yale Univ, PhD(phys chem), 49. *Prof Exp:* Instr chem, Yale Univ, 46-48; asst prof, Ind Univ, 49-54; task scientist, Aeronaut Res Lab, Wright Air Develop Ctr, USAF, Ohio, 54-58; assoc chemist, Argonne Nat Lab, 58-66; prof computer sci, Univ Notre Dame, 66-71; prof, 71-84, EMER PROF COMPUTER SCI, UNIV KY, 84- *Concurrent Pos:* Consult, Argonne Nat Lab, 66-77. *Mem:* AAAS; NY Acad Sci. *Res:* Numerical approximation and computer programming; computation and approximation of special functions. *Mailing Add:* 211 Quarry Lane Santa Cruz CA 95060

THACHER, PHILIP DURYEA, b Palo Alto, Calif, Jan 13, 37; m 63; c 2. SOLID STATE PHYSICS. *Educ:* Calif Inst Technol, BS, 58; Cornell Univ, PhD(physics), 65. *Prof Exp:* STAFF MEM, SANDIA LABS, 65- *Mem:* AAAS; Am Phys Soc; Inst Elec & Electronics Engrs. *Res:* Optical effects in ferroelectric ceramics and crystals; laser energy; radiometry; 14 MeV neutron detection. *Mailing Add:* 524 Camino Del Bosque NW Albuquerque NM 87114

THACKER, HARRY B, b Pittsburgh, Pa, May 4, 47; m 76; c 1. HIGH ENERGY PHYSICS, PARTICLE PHYSICS. *Educ:* Calif Inst Technol, BS, 68; Univ Calif, Los Angeles, MS, 71, PhD(particle physics theory), 73. *Prof Exp:* Res assoc, State Univ NY Stony Brook, 73-76; THEORET PHYSICIST PARTICLE PHYSICS, FERMI NAT ACCELERATOR LAB, 76- *Res:* Quantum field theory; gauge theories of quarks and leptons; connections between statistical mechanics and quantum field theory. *Mailing Add:* 1848 Fendall Ave Charlottesville VA 22903

THACKER, JOHN CHARLES, b Clinton, Okla, Oct 29, 43; m 68; c 2. STATISTICAL ANALYSIS. *Educ:* Cornell Univ, BS, 66; Brown Univ, PhD(appl math), 74. *Prof Exp:* Mem tech staff statist, 74-91, PRIN DIR SYSTS ANALYSIS SUB SYST, ENG DIV ENG GROUP, AEROSPACE CORP, 91- *Mem:* Am Math Soc; Inst Math Statist; Am Statist Asn; Soc Indust & Appl Math. *Res:* Statistical inference on stochastic processes; time series analysis; stochastic point processes; application of statistical techniques to air and water pollution problems. *Mailing Add:* PO Box 92957 M4954 Los Angeles CA 90009

THACKER, RAYMOND, b Ashton-U-Lyne, UK, May 9, 32. ELECTROCHEMISTRY, PHYSICAL CHEMISTRY. *Educ:* Univ Manchester, BSc, 52, MSc, 53, PhD(phys chem), 55. *Prof Exp:* Sci officer, UK Atomic Energy Auth Indust Group, 55-58; res assoc electrode kinetics, Univ Pa, 58-60; STAFF RES SCIENTIST, RES LABS, GEN MOTORS CORP, 60- *Mem:* Electrochem Soc; Royal Soc Chem. *Res:* Thermodynamic properties of nonelectrolyte solutions; electrochemistry of surfaces; fuel cell electrode processes; batteries. *Mailing Add:* PO Box 75 12 Mile & Mound Rds Franklin MI 48025

THACKER, STEPHEN BRADY, b Independence, Mo, Dec 30, 47; m 76; c 2. TECHNOLOGY ASSESSMENT, PUBLIC HEALTH SURVEILLANCE. *Educ:* Princeton Univ, NJ, AB, 69; Mt Sinai Sch Med, NY, MD, 73; London Sch Hyg & Trop Med, MSc, 84. *Prof Exp:* Chief, consolidated surveillance & commun activ, Epidemiol Prog Off, Ctr Dis Control, 78-83, dir, Div Surveillance & Epidemiol Studies, 83-86, asst dir sci, Ctr Environ Health & Injury Control, 86-89, DIR, EPIDEMIOL PROG OFF, CTR DIS CONTROL, 89- *Concurrent Pos:* Mem steering comt, Asn Behav Sci Med Educ, 71-74; Robert Wood Johnson Found Clin scholar, 74-75; assoc, Dept Community Med, Med Ctr, Duke Univ, Durham, NC, 75-76; lectr, Dept Community Med, Mt Sinai Sch Med, NY, 78- & Sch Med, Emory Univ, Atlanta, Ga, 85-86; consult epidemiol, Arab Rep Egypt, 79-91; clin asst prof community health, Sch Med, Emory Univ, 86-; ed, Am J Epidemiol, Baltimore, Md, 90- *Honors & Awards:* Mosby Bood Award for Excellence, 73; Saul Horowitz Jr Mem Award, 90; Supvry Award for Contrib to the Advan of Women, 91. *Mem:* Am Epidemiol Soc; Am Pub Health Asn; Am Col Epidemiol. *Res:* Published broad range of fields in epidemiology including public health surveillance; infectious disease; environmental health; alcohol abuse; health care delivery; meta-analysis; technology assessment. *Mailing Add:* Dis Control Centers MS C08 Atlanta GA 30333

THACKRAY, ARNOLD, b Eng, July 30, 39; c 3. HISTORY & PHILOSOPHY OF SCIENCE. *Educ:* Bristol Univ, Eng, BSc, 60; Cambridge Univ, Eng, MA, 65, PhD(hist sci), 66. *Prof Exp:* Res fel, Churchill Col, Cambridge Univ, 65-68; prof hist & social sci, Univ Pa, 68-85, chmn, 70-77; DEAN GRAD STUDIES & RES, UNIV MD, 85- *Concurrent Pos:* Vis lectr, Harvard Univ, 67-68, Nat lectr, Sigma Xi, 76; dir, Ctr Hist Chem, 82-; ed, Isis, 78-85, Osiris, 85-; mem bd dir & treas, Am Coun Learned Soc, 85- *Honors & Awards:* Dexre Award, Am Chem Soc, 83; George Savon Mem Lectr, AAAS, 84. *Mem:* Fel AAAS; fel Royal Hist Soc; fel Royal Soc Chem; Soc Social Studies Sci (pres, 82-84); Hist Sci Soc. *Res:* History of chemistry and chemical technology; the social sciences in modern America; reseach administration. *Mailing Add:* Dept Hist Sci Univ Pa Philadelphia PA 19104

THACKSTON, EDWARD LEE, b Nashville, Tenn, Apr 29, 37; m 61; c 2. ENVIRONMENTAL ENGINEERING, WATER RESOURCES ENGINEERING. *Educ:* Vanderbilt Univ, BE, 61, PhD(environ & water resources eng), 66; Univ Ill, Urbana, MS, 63. *Prof Exp:* City engr, Lebanon, Tenn, 58-59; design engr, City of Nashville, 61-62; from asst prof to assoc prof sanit & water resources eng, 66-76, prof environ & water resources eng, 76-80, PROF CIVIL & ENVIRON ENG & CHMN DEPT, VANDERBILT UNIV, 80- *Concurrent Pos:* Consult engr, Vanderbilt Univ, 66-; on leave as staff asst environ affairs to Gov State of Tenn, 72-73. *Honors & Awards:* Z Cartter Patten Award, 83. *Mem:* Am Soc Civil Engrs; Am Water Works Asn; Water Pollution Control Fedn; Asn Environ Eng Prof. *Res:* Mixing and reaeration in streams; effects of impoundments on water quality; industrial waste treatment; environmental policy; dredged material disposal. *Mailing Add:* 2010 Priest Rd Nashville TN 37215

THACORE, HARSHAD RAI, b Ahmedabad, India, Dec 1, 39; US citizen; m 65; c 1. VIROLOGY. *Educ:* Univ Lucknow, BSc, 58, MSc, 60; Duke Univ, PhD(microbiol), 65. *Prof Exp:* Fel virol, Ohio State Univ, 65-67; res assoc, Univ Pittsburgh, 67-69, instr, 69-71, asst res prof, 71-74; asst prof, 74-79, ASSOC PROF MICROBIOL, STATE UNIV NY BUFFALO, 79- *Mem:* Am Soc Microbiol; Am Soc Virol; Int Soc Interferon Res. *Res:* Interferon, especially mechanism and induction of; rescue of interferon sensitive virus by poxviruses; persistent viral infections, especially initiation and maintenance of. *Mailing Add:* Dept Microbiol Rm 218 Sherman Hl Health Sci Ctr State Univ NY 2435 Main St Buffalo NY 14214

THADDEUS, PATRICK, b June 6, 32; US citizen; m 63; c 2. PHYSICS, ASTROPHYSICS. *Educ:* Univ Del, BSc, 53; Oxford Univ, MA, 55; Columbia Univ, PhD(physics), 60. *Prof Exp:* Res physicist, Radiation Lab, Columbia Univ, 60-61; Nat Acad Sci fel astrophys, 61-64, res physicist, Goddard Inst, Space Studies, 64-86; PROF ASTRON & APPL PHYSICS, HARVARD UNIV, 86- *Concurrent Pos:* Fulbright fel, 53-55; adj asst prof, Columbia Univ, 64-66, adj prof, 71-; adj assoc prof, NY Univ, 63-; vis assoc prof, State Univ NY, Stony Brook, 66-; mem, Space Sci Adv Comt, NASA, 79, Astron Surv Comt, 78-80, 88-90; sr vis fel, Univ Cambridge, 83. *Honors & Awards:* Alexander von Humboldt Award, WGer, 73; Lindsey Mem Award, 76; Medal for Except Sci Achievement, NASA, 70 & 85. *Mem:* Nat Acad Sci; Am Phys Soc; Am Astron Soc; Sigma Xi; Int Astron Union; Am Acad Arts & Sci. *Res:* Radio astronomy; interstellar molecules; microwave spectroscopy. *Mailing Add:* 58 Garfield St Cambridge MA 02138

THAELER, CHARLES SCHROPP, JR, b Philadelphia, Pa, Jan 9, 32; m 57; c 3. ZOOLOGY. *Educ:* Earlham Col, AB, 54; Univ Calif, Berkeley, MA, 60, PhD(zool), 64. *Prof Exp:* Actg instr zool, Univ Calif, Berkeley, 63-64, actg asst cur, Mus Vert Zool, 63-64; asst prof zool, South Bend Campus, Ind Univ, 64-66; from asst prof to assoc prof, 66-74, PROF BIOL, NMEX STATE UNIV, 74- *Concurrent Pos:* NSF res grants, 68-70 & 72-74. *Mem:* Fel AAAS; Am Soc Mammal; Soc Study Evolution; Ecol Soc Am; Soc Syst Zool; Sigma Xi. *Res:* Mammalian systematics and cytogenetics; evolution and ecology, especially taxonomy; evolution, cytogenetics and distribution of geomyids; taxonomy of microtine rodents. *Mailing Add:* Dept Biol NMex State Univ Box 10003 Las Cruces NM 88003

THAEMERT, JONA CARL, anatomy, for more information see previous edition

THAKAR, JAY H, b Bombay, India, Dec 4, 40; Can citizen; m 73; c 2. NEUROPHARMACOLOGY. *Educ:* Univ Bombay, BSc Hons, 62, MSc, 64; Univ Manitoba, PhD(biochem), 73. *Prof Exp:* Sci officer, Bhabha Atomic Res Ctr, Bombay, India, 64-68; fel animal sci, Univ Cailf, Davis, 74-76; prof asst biochem, Univ Western Ont, 76-78; clin chemist, Civic Hosp, Ottawa, 78-80; chief, Neuropharmacol Lab, Royal Ottawa Hosp, 80-84; CHIEF, NEUROPHARMACOL LAB, OTTAWA CIVIC HOSP, 85- *Res:* Neuropsychopharmacology of psychiatric disorders; normal and abnormal functions of muscle organelle; animal models of muscular dystrophy; Parkinson's Disease and movement disorders. *Mailing Add:* Childrens Res Hosp 332 N Lauderdale PO Box 318 Memphis TN 38101

THAKKAR, AJIT JAMNADAS, b Poona, India, Oct 23, 50; Can citizen; m 81; c 2. INTERMOLECULAR FORCES, SCATTERING THEORY. *Educ:* Queen's Univ, Can, BSc, 73, PhD(chem), 76. *Prof Exp:* Fel chem, Univ Waterloo, 76-78; res assoc, Queen's Univ, Can, 78-79, asst prof, 79-80; asst prof, Univ Waterloo, 80-84; assoc prof, 84-88, PROF CHEM, UNIV NB, 88- *Concurrent Pos:* Hon adj prof, Dalhouse Univ, 89. *Honors & Awards:* Noranda Lectr, Can Soc Chem, 91. *Mem:* Chem Inst Can; Can Soc Chem; Can Asn Theoret Chemists; Am Asn Physics Teachers. *Res:* Theoretical studies of polarizabilities; dispersion coefficients; intermolecular forces; electronic momentum densities; x-ray and high energy electron scattering; generalized oscillator strengths; electron pair densities; correlation holes, density matrices and density functionals. *Mailing Add:* Dept Chem Univ NB Fredericton NB E3B 6E2 Can

THAKKAR, ARVIND LAVJI, b Karachi, Pakistan, Apr 19, 39; c 3. PHYSICAL PHARMACY. *Educ:* Univ Bombay, BPharm, 61; Columbia Univ, MS, 64; Univ Wash, PhD(phys pharm), 67. *Prof Exp:* Teaching asst col pharm, Columbia Univ, 61-63; col pharm, Univ Wash, 64-66; sr pharmaceut chemist, 67-75, RES SCIENTIST, RES LABS, ELI LILLY & CO, 75- *Mem:* Am Pharmaceut Asn; Acad Pharmaceut Sci; NY Acad Sci. *Res:* Surface activity of drugs; micellar solubilization; drug absorption; controlled release of drugs; solubility. *Mailing Add:* Lilly Res Labs Indianapolis IN 46285

THAKKER, DHIREN R, b Broach, India, Jan 25, 49. DRUG METABOLISM, CHEMICAL CARCINOGENESIS. *Educ:* Univ Kans, PhD(biochem), 76. *Prof Exp:* SR INVESTR, NIH, 83- *Mem:* Am Chem Soc; Am Asn Cancer Res; Am Soc Pharmaceut & Exp Therapeut. *Mailing Add:* Glaxo Res Labs Five Moore Dr Research Triangle Park NC 27709

THAKOR, NITISH VYOMESH, b Bombay, India, Feb 19, 52; m; c 2. BIOMEDICAL COMPUTING, MEDICAL INSTRUMENTATION. *Educ:* Indian Inst Technol, BTech, 74; Univ Wis, Madison, MS, 78, PhD(elec eng), 81. *Prof Exp:* Electronics engr instrument design & mkt, Philips India, Ltd, 74-76; res asst biomed eng, Univ Wis, Madison, 77-81; asst prof elec & comput eng, Northwestern Univ, 81-83; ASSOC PROF BIOMED ENG, JOHNS HOPKINS SCH MED, 84- *Concurrent Pos:* Assoc ed, Inst Elec & Electronics Engrs Trans Biomed Eng, J Ambulatory Monitoring, Med Design & Mat. *Honors & Awards:* Presidential Young Investr Award, NSF; Res Career Develop Award, NIH. *Mem:* Inst Elec & Electronics Engrs Comput Soc; Inst Elec & Electronics Engrs Biomed Soc; Sigma Xi. *Res:* Medical instrumentation and computer applications in medical care, including microprocessors, very large scale integrations, signal processing and pattern recognition in medicine. *Mailing Add:* Dept Biomed Eng 720 Rutland Ave Baltimore MD 21205

THALACKER, VICTOR PAUL, b Muscatine, Iowa, Apr 21, 41; m 70; c 2. POLYMER SCIENCE & ENGINEERING, TECHNICAL MANAGEMENT. *Educ:* Wis State Univ-Stevens Point, BS, 63; Univ Ariz, PhD(org chem), 68. *Prof Exp:* Sr chemist org chem, 67-77, res supvr, 77-80, tech mgr, 80-84, LAB MGR, 3M, 84- *Concurrent Pos:* mem, Polymer Div, Am Chem Soc & Rad Tech Intl. *Mem:* Am Chem Soc. *Res:* Radiation curing; radiation curable coatings and adhesives; ultraviolet curable inks and polymers; laser processes; polymer processing; extrusion; plant culture; abrasives; polymer morphology. *Mailing Add:* 3M Co Bldg 208-1 3M Ctr St Paul MN 55144

THALE, THOMAS RICHARD, b Indianapolis, Ind, June 4, 15; m 41; c 5. PSYCHIATRY. *Educ:* Loyola Univ Chicago, BSM, 38, MD, 39. *Prof Exp:* Resident psychiat, Manteno State Hosp, Ill, 39 & Norwich State Hosp, Conn, 39-43; instr psychol, Univ Conn, 42-45; instr psychiat & pub health, Yale Univ, 43-45; instr psychiat, Wash Univ, 45-50; from instr to assoc prof, 50-69, PROF PSYCHIAT, MED SCH & ASSOC PROF SOCIAL WORK, SCH SOCIAL SERV, ST LOUIS UNIV, 70-, CLIN PROF PSYCHIAT, 80- *Concurrent Pos:* Med adv, Bur Hearings & Appeals, Social Security Admin, 67-; consult, Family & Childrens Soc Greater St Louis & Vet Admin Asn Retarded Children. *Mem:* AMA; fel Am Psychiat Asn; Am Group Psychother Asn. *Res:* Psychological evaluation chemotherapy. *Mailing Add:* 351 Meadowbrook Dr Ballwin MO 63011

THALER, ALVIN ISAAC, b New York, NY, Aug 26, 38; m 59; c 2. ALGEBRA, NUMBER THEORY. *Educ:* Columbia Univ, AB, 59; Johns Hopkins Univ, PhD(math), 66. *Prof Exp:* Assoc math appl physics lab, Johns Hopkins Univ, 62-64; instr, Col Notre Dame, Md, 64-66; asst prof, Univ Md, 66-71; prog dir algebra & number theory, 71-81, PROG DIR SPEC PROJS MATH SCI, NSF, 81- *Mem:* Am Math Soc; Math Asn Am. *Res:* Algebraic number theory; algebraic geometry. *Mailing Add:* NSF Math Sci Sect Washington DC 20550

THALER, BARRY JAY, b Brooklyn, NY, June 10, 50; m 82; c 2. ADVANCED ELECTRONIC PACKAGING DEVELOPMENT, CERAMICS DEVELOPMENT. *Educ:* State Univ NY, Stony Brook, BS, 72; Mich State Univ, MS, 74, PhD(physics), 77. *Prof Exp:* Fel, Northwestern Univ, 77-79; MEM TECH STAFF, RCA LAB, 79- *Concurrent Pos:* Vis scientist, Argonne Nat Lab, 78-79. *Mem:* Am Phys Soc; Inst Elec & Electronics Engrs. *Res:* Insulating properties of organic and inorganic; electronic packaging applications. *Mailing Add:* SRI Int David Sarnoff Res Ctr CN 5300 Princeton NJ 08543-5300

THALER, ERIC RONALD, b Denver, Colo, May 19, 60; m 83; c 2. SYNOPTIC METEOROLOGY. *Educ:* Univ Utah, BS, 82; Colo Sch Mines, MS, 89. *Prof Exp:* Meteorologist & forecaster, 82-90, SCI & OPERS OFFICER, NAT WEATHER SERV, 90- *Concurrent Pos:* Mem, Atmospheric Technol Comt, Nat Weather Asn, 90- *Mem:* Am Meteorol Soc; Nat Weather Asn. *Res:* Finding ways to use new observational and theoretical advances to improve weather forecasts; numerical modelling using advanced datasets. *Mailing Add:* Nat Weather Serv 10230 Smith Rd Denver CO 80239

THALER, G(EORGE) J(ULIUS), b Baltimore, Md, Mar 15, 18; m 44; c 4. ELECTRICAL ENGINEERING. *Educ:* Johns Hopkins Univ, BE, 40, DrEng, 47. *Prof Exp:* Instr & asst elec eng, Johns Hopkins Univ, 42-47; asst prof, Univ Notre Dame, 47-51; from asst prof to prof, 51-76, DISTINGUISHED PROF ELEC ENG, NAVAL POSTGRAD SCH, 76- *Mem:* Am Soc Eng Educ; Inst Elec & Electronics Engrs. *Res:* Theory of automatic control, particularly discontinuous and nonlinear systems. *Mailing Add:* Dept Elec Eng US Naval Postgrad Educ Monterey CA 93940

THALER, JON JACOB, b Richland, Wash, Feb 3, 47; m 68; c 1. EXPERIMENTAL HIGH ENERGY PHYSICS. *Educ:* Columbia Univ, BA, 67, MA, 69, PhD(physics), 72. *Prof Exp:* From instr to asst prof physics, Princeton Univ, 71-77; ASST PROF PHYSICS, UNIV ILL, URBANA, 77- *Mem:* Am Phys Soc. *Res:* Investigation of high energy processes as tests of the Quark-Parton model, scaling, and the possible existence of new quantum numbers. *Mailing Add:* Dept Physics Univ Ill 1110 W Green St Urbana IL 61801

THALER, M MICHAEL, Can citizen; m 66; c 2. PEDIATRICS, DEVELOPMENTAL BIOLOGY. *Educ:* Univ Toronto, MD, 58. *Prof Exp:* Intern, Mt Zion Hosp, San Francisco, 58-59; jr pediat resident, Children's Hosp, Detroit, 59-60; sr resident, Boston City Hosp, Mass, 60-61; asst resident path, Hosp Sick Children, Toronto, 61-62; res fel pediat path, Univ Toronto, 62; vis resident, Hosp St Antoine, Paris, 63; res fel path chem, Hosp Sick Children, Toronto, 63-65; res fel develop biochem, Harvard Med Sch, 65-67; from instr to assoc prof,67-76, PROF PEDIAT, SCH MED, UNIV CALIF, SAN FRANCISCO, 77- *Concurrent Pos:* Vis scientist, Wash Univ, 64 & Weizmann Inst Sci, Israel; Josiah Macy Jr Found fac scholar, 74-75. *Mem:* Soc Pediat Res; Am Soc Clin Invest; Am Soc Biol Chemists; Am Asn Study Liver Dis; Int Asn Study Liver. *Res:* Liver disease of newborn infants; perinatal and developmental aspects of hepatic metabolism; bilirubin metabolism; pediatric gastrointestinal function and pathology. *Mailing Add:* Dept Pediat M680 Box 0136 Univ Calif Med Sch San Francisco CA 94143

THALER, OTTO FELIX, b Vienna, Austria, June 17, 23; US citizen; m 47; c 3. PSYCHIATRY, PSYCHOANALYSIS. *Educ:* Univ Rochester, MD, 49; State Univ NY Downstate Med Ctr, cert psychoanalysis, 66. *Prof Exp:* PROF PSYCHIAT, SCH MED, UNIV ROCHESTER, 55- *Mem:* Am Psychiat Asn. *Mailing Add:* 300 Crittenden Blvd Rochester NY 14642

THALER, RAPHAEL MORTON, b Brooklyn, NY, May 19, 25; m 52; c 4. THEORETICAL PHYSICS. *Educ:* NY Univ, AB, 47; Brown Univ, ScM, 49, PhD(physics), 50. *Prof Exp:* Res assoc theoret physics, Yale Univ, 50-52; mem staff, Los Alamos Sci Lab, 53-60; assoc prof, Case Western Reserve Univ, 60-64, prof physics, 64-, vchmn dept, 67-; RETIRED. *Concurrent Pos:* Res assoc, Mass Inst Technol, 57-58; consult, Argonne Nat Lab & Lewis Res Lab, NASA. *Mem:* Fel Am Phys Soc; Ital Phys Soc. *Res:* Nuclear and elementary particle physics. *Mailing Add:* MSH 850 Lamps St Los Alamos NM 87545

THALER, WARREN ALAN, b New York, NY, Jan 7, 34; m 56; c 1. ORGANIC CHEMISTRY. *Educ:* City Col New York, BS, 56; Columbia Univ, MA, 58, PhD(chem), 61. *Prof Exp:* Res assoc, 60-77, SR CORP RES LAB, EXXON RES & ENG CO, 77- *Mem:* Am Chem Soc. *Res:* Free radical chemistry; sulfur chemistry; additive substitution reactions; stereochemistry of free radical reactions; cationic polymerization; elastomer chemistry. *Mailing Add:* Five Pleasant View Way Flemington NJ 08822-9245

THALER, WILLIAM JOHN, b Baltimore, Md, Dec 4, 25; m 51; c 2. PHYSICS. *Educ:* Loyola Col, BS, 47; Cath Univ, MS, 49, PhD(physics), 51. *Prof Exp:* Physicist, Baird Assocs, Inc, 47; instr physics, Cath Univ, 47-48, asst, 48-51; PHYSICIST, US OFF NAVAL RES, 51-; PROF PHYSICS, GEORGETOWN UNIV, 60- *Concurrent Pos:* Chief scientist, Off Telecommun Policy, Exec Off President, 76-77, dir, 77-78. *Mem:* Am Phys Soc; Acoust Soc Am. *Res:* Ultrasonic studies of relaxation phenomena in gases; propagation of shock waves in liquids, solids and gases; effects of ultrasonics on biological media; laser research. *Mailing Add:* 5532 Summit St Centreville VA 22020

THALL, PETER FRANCIS, b Stillwater, Okla, Aug 5, 49. MATHEMATICAL STATISTICS, PROBABILITY. *Educ:* Mich State Univ, BS, 71; Fla State Univ, MS, 73, PhD(statist & probability), 75. *Prof Exp:* Asst Prof, Prog Math Sci, Univ Tex, 75-80; sci corres, Hazardous Waste & Pollution Adv Letter, Bur Bus Pract, Prentice Hall Publ Co, 82-83; statistician, Nat Coop Gallstone Study, 82, Diabetes Control & Complications Trial, Biostatistic Ctr, Dept Statistics, NIH, Nat Inst Arthritis, Diabetes & Digestive Kidney Diseases, 83 & Dept Radiation Oncol, 85; asst prof, 80-84, ASSOC PROF, DEPT STATIST, GEORGE WASHINGTON UNIV, 84- *Mem:* Am Statist Asn; Inst Math Statist; Soc Clin Trials. *Res:* Biostatistics; clinical trials; stochastic point processes and random measures; reliability theory. *Mailing Add:* Dept Statist George Washington Univ 2121 Eye St NW Washington DC 20052

THALMANN, ROBERT H, b San Antonio, Tex, Nov 14, 39. NEUROSCIENCE. *Educ:* Univ Tex, BA, 61; Univ Mich, MA, 64, PhD(psychol), 67. *Prof Exp:* USPHS fel, Emory Univ, 67-69; STAFF MEM, BAYLOR COL MED, 69-, ASSOC PROF, 84- *Concurrent Pos:* Vis asst prof, Univ Calif, Irvine, 78; res vis, AT&T Bell Labs, 85-86. *Mem:* Soc Neurosci; AAAS. *Res:* Neurophysiology of neurotransmitters. *Mailing Add:* Dept Cell Biol Baylor Col Med Houston TX 77030

THAM, MIN KWAN, b Rangoon, Burma, July 29, 39; US citizen; m 70; c 2. CHEMICAL ENGINEERING, PHYSICAL CHEMISTRY. *Educ:* Rangoon Univ, BSc, 62; Univ Fla, MSE, 68, PhD(chem eng), 70. *Prof Exp:* Asst lectr chem eng, Rangoon Inst Technol, 63-66; fel, Univ Fla, 70-72, res assoc, 72-74, asst engr, 74-76; chem engr, Bartlesville Energy Technol Ctr, US Dept Energy, 76-83; tech adv, 83-88, SR SCI ADV, NAT INST PETROL & ENERGY RES, 88- *Concurrent Pos:* Hon tech adv, Sci Res Inst Petrol Explor & Develop, People's Repub of China. *Mem:* Am Inst Chem Engrs; Soc Petrol Engrs. *Res:* Statistical thermodynamics; fuel cells and batteries research; physical and chemical aspects of tertiary oil recovery; interfacial phenomena, adsorptions, mass transfer. *Mailing Add:* 837 SE Belmont Rd Bartlesville OK 74006

THAMER, B(URTON) J(OHN), b Kitchener, Ont, June 22, 21; nat US. NUCLEAR ENGINEERING, CHEMISTRY. *Educ:* Univ Calif, BS, 43; Iowa State Col, PhD(chem), 50; Univ Ariz, MS, 73. *Prof Exp:* Asst chemist, Manhattan Proj Calif, 43-44, Clinton Labs, 44 & Hanford Eng Works, 44-45; asst chem, Inst Atomic Res, Iowa State Univ, 46-50; mem staff, Los Alamos Sci Lab, NMex, 50-71; metallurgist, Magma Copper Co, Ariz, 73-77; nuclear engr, Ford, Bacon & Davis, Utah, 77-84; chemist, USA Dugway Proving Ground, Utah, 85-86; CONSULT, 86- *Mem:* Am Chem Soc; Am Nuclear Soc. *Res:* Chemistry of nuclear reactors, management of radioactive waste, chemical equilibria; diffusion of radon. *Mailing Add:* PO Box 8337 Salt Lake City UT 84108

THAMES, HOWARD DAVIS, JR, b Monroe, La, Aug 3, 41; m 66; c 2. BIOMATHEMATICS. *Educ:* Rice Univ, BA, 63, PhD(chem), 70. *Prof Exp:* Proj investr, 70-71, from asst prof to assoc prof, 71-83, PROF BIOMATH, UNIV TEX-M D ANDERSON CANCER CTR, HOUSTON, 84- *Mem:* Radiation Res Soc; Brit Inst Radiol. *Res:* Experimental design analysis in radiology. *Mailing Add:* Dept Biomath Univ Tex M D Anderson Cancer Ctr 1515 Holcomb Houston TX 77030

THAMES, JOHN LONG, b Richmond, Va, Sept 29, 24; m 48; c 4. WATERSHED MANAGEMENT. *Educ:* Univ Fla, BSF, 50; Univ Miss, MS, 59; Univ Ariz, PhD, 66. *Prof Exp:* Res forester, US Forest Serv, 50-67; assoc prof watershed hydrol, 67-69, PROF WATERSHED HYDROL, UNIV ARIZ, 69-, PROG CHMN WATERSHED HYDROL, 75- *Concurrent Pos:* Coordr, Int Biol Prog; consult, Argonne Nat Lab, Am Smelting & Refining Co & Shelly Loy. *Mem:* Soil Sci Soc Am; Soc Am Foresters; Sigma Xi; Am Geophys Union. *Res:* Plant-soil-water relations; hydrologic modeling; decision analyses; hydrology of surface mined lands. *Mailing Add:* 5470 Cumberland Dr Univ Ariz Tucson AZ 85704

THAMES, MARC, b Houston, Tex, Sept 25, 44. CARDIOLOGY. *Educ:* Va Commonwealth Univ, MD, 70. *Prof Exp:* PROF & CHIEF INTERNAL MED, MED COL VA, 82-; CHIEF CARDIOL, VET ADMIN MED CTR, RICHMOND, VA, 82- *Mailing Add:* Cardiol Sect Vet Admin Med Ctr Richmond VA 23249

THAMES, SHELBY FRELAND, b Hattiesburg, Miss, Aug 10, 36; m 54; c 3. POLYMER CHEMISTRY, ORGANIC CHEMISTRY. *Educ:* Univ Southern Miss, BS, 59, MS, 61; Univ Tenn, PhD(org chem), 64. *Prof Exp:* From instr to assoc prof chem, 60-70, dean col sci, 71-74, dean col sci & technol, 74-77, PROF POLYMER SCI, UNIV SOUTHERN MISS, 70-, VPRES ADMIN & REGIONAL CAMPUSES, 77-, DISTINGUISHED UNIV RES PROF POLYMER SCI, 86-; EXEC VPRES, SOUTHERN INST SURFACE COATINGS, 80- *Concurrent Pos:* Res grants, Walter Reed Inst Res, 64-68, Diamond-Shamrock Corp, 66-68, Inst Copper Res Asn, NASA, Masonite Corp, Stand Paint & Varnish Co & Dept Defense, USDA; Petrol Res Fund fel award, 68-69; dir, Southern Inst Surface Coatings, 68-70. *Honors & Awards:* Mattiello Lectr, Fedn Socs Coatings Technol. *Mem:* Am Chem Soc; Am Inst Chem; Fedn Socs Coatings Technol; Sigma Xi. *Res:* Organosilicon, synthetic organic and organometallic chemistry; polymeric coatings and water soluble polymers synthesis, characterization and properties. *Mailing Add:* Dept Polymer Sci Box 10076 Southern Sta Hattiesburg MS 39406

THAMES, WALTER HENDRIX, JR, b Richmond, Va, July 29, 18; m 43; c 2. NEMATOLOGY. *Educ:* Univ Fla, BSA, 47, MS, 48, PhD, 59. *Prof Exp:* Asst entomologist, Everglades Exp Sta, Univ Fla, 48-55, asst soil microbiol, 57-59; assoc prof, 59-68, prof, 68-80, EMER PROF PLANT NEMATOL, TEX A&M UNIV, 80- *Concurrent Pos:* Prin instr, Nematode Identification Short Course, Clemson Univ, 81-88. *Mem:* Soc Nematol. *Res:* Biology and control of nematodes; plant nematology. *Mailing Add:* 705 Pershing St College Station TX 77840

THAMM, RICHARD C, JR, organic chemistry, for more information see previous edition

THANASSI, JOHN WALTER, b St Louis, Mo, Oct 2, 37; m 64; c 3. BIOCHEMISTRY. *Educ:* Lafayette Col, BA, 59; Yale Univ, PhD(biochem), 63. *Prof Exp:* Fel chem, Cornell Univ, 63-64; fel, Univ Calif, Santa Barbara, 64-65; staff fel, Lab Chem, NIH, 65-67; from asst prof to assoc prof biochem, 67-78, PROF BIOCHEM, COL MED, UNIV VT, 78- *Concurrent Pos:* USPHS res grants, 68-71, 72-76 & 78-, NASA res grant, 74-75. *Mem:* AAAS; Am Soc Biol Chem; Am Inst Nutrit; Am Asn Univ Prof; Am Asn Cancer Res. *Res:* Vitamin B6 metabolism. *Mailing Add:* 24 Victoria Dr South Burlington VT 05401

THANGARAJ, SANDY, b Madras, India, Dec 3, 34; m; c 2. TECHNICAL ACTIVITIES, NEW PRODUCT RESEARCH. *Educ:* Univ Madras, India, BS, 54, MS, 56, dipl rubber technol, 57, PhD(chem eng), 60. *Prof Exp:* Plant mgr & chief tech officer, Plastics & Rubber Industs, Nairobi, Kenya, EAfrica, 65-71; mgr res & develop, Globe Superior Prod, 71-72; tech dir & chief chemist, Avon Sole Co & Subsid, 72-75; mgr res & develop rubber, Teknon Apex Co, 75-78; dir tech & mfg, Industs Modernon LTDA, Colombia, SAm, 79-80; mgr res & develop, Int Shoe Co, Tex, 80-82; mgr res & develop prod, O'Sullivan-Vulcan Corp, Tenn, 82-87; TECH & LAB MGR, RUBBER INDUSTS INC, SHAKOPEE, MINN, 88- *Mem:* Am Chem Soc; Am Inst Chemists; Am Chem Soc Rubber Div. *Res:* Setting standards for products especially rubber and plastics. *Mailing Add:* Rubber Industs Inc 200 Cavanaugh Dr Shakopee MN 55379

THAPAR, MANGAT RAI, b Khanna, India, Apr 10, 39; m 68; c 2. SEISMOLOGY, GEOPHYSICS. *Educ:* Indian Sch Mines, Dhanbad, BSc, 61, MSc & AISM, 62; Univ Western Ont, PhD(geophys), 68. *Prof Exp:* Sr sci officer seismol, Coun Sci & Indust Res, New Delhi, 62-64; teaching fel seismol & geophys, Univ Western Ont, 68-71; NSF res fel seismol, Univ Pittsburgh, 71-72; chief res engr, Seismograph Serv Corp, 72; analyst seismol, Phillips Petrol Co, 72-74; geophys assoc, Cities Serv Co, 74-80, sr geophys assoc, 80-85; PRES, INT GEOPHYS CO, 85- *Concurrent Pos:* Lectr, Univ Pittsburgh, 71-72. *Mem:* Am Geophys Union; Soc Exp Geophys; Indian Soc Earthquake Technol; Europ Asn Exp Geophysicists. *Res:* Applied geophysics; experimental, theoretical, model earthquake and lunar seismology; digital data processing techniques; geophysical exploration techniques; seismic interpretation research; seismic wave propagation; 2-dimensional digital seismic modeling; color processing of seismic data. *Mailing Add:* 4325 E 51st St Suite 116 Tulsa OK 74135

THAPAR, NIRWAN T, b Raikot, India, Jan 26, 38; US citizen; m 67; c 2. VETERINARY PATHOLOGY, VETERINARY MEDICINE. *Educ:* Punjab Univ FSc, 57; Punjab State Vet Col, India, BVScAH, 61; Univ SDak, Brookings, MS, 64. *Prof Exp:* Instr, Punjab State Vet Col, 61-62; grad res fel, SDak State Univ, 62-64; res asst, Univ Wis-Madison, 64-66; veterinarian, Norwich Pharmaceut Co, 66-68 & Toxicol Path Sect, Wis Alumni Res Found, 68-69; instr path & neuropath, dept path, Univ Wis-Madison, 69-72; vet pathologist, San Diego County, Calif, 72-77; dir, State Animal Health Diag Lab, Centreville, Md, 77-87; VET PATHOLOGIST, ANIMAL HEALTH LAB, MD DEPOT AGR, 87- *Concurrent Pos:* Rockefeller travel fel award, 62-64; res fel, Univ SDak, 62-64 & Univ Wis, 64-66, teaching fel, 69-72; monitor, study site review prog vet path, CL Davis Found Southern Calif, 72; veterinarian-in-charge, Meat Inspection Plant, Wis State Dept Agr, 71. *Mem:* Sigma Xi; Am Vet Med Asn; Soc Pharmacol & Environ Pathologists; Asn Indian Vet Am (pres, 90-). *Res:* Experimental pathology; diagnostic and toxicologic pathology. *Mailing Add:* Md Dept Agr Animal Health Lab 3740 Metzerott Rd Col Park MD 20740

THARIN, JAMES COTTER, b West Palm Beach, Fla, Mar 22, 31; m 55; c 2. GEOLOGY. *Educ:* St Joseph's Col, Ind, BS, 54; Univ Ill, Urbana, MS, 58, PhD(geol), 60. *Prof Exp:* Instr phys sci, Univ Ill, Urbana, 59-61; petrol geologist, Chevron Oil Co, 61-63; asst prof geol, Wesleyan Univ, 63-67; assoc prof, 67-74, PROF GEOL, HOPE COL, 74-, CHMN DEPT, 67-80 & 88- *Mem:* AAAS; Geol Soc Am; Soc Econ Geol; Soc Econ Paleont & Mineral; Sigma Xi. *Res:* Sedimentation; Pleistocene geology; glacial geology and clay mineralogy; textural studies of glacial drift in northwestern Pennsylvania, Calgary area, Alberta and Connecticut Valley; petroleum exploration in Mississippi Delta. *Mailing Add:* Dept Geol Hope Col Holland MI 49423

THARP, A G, b Franklinton, Ky, Jan 6, 27. INORGANIC CHEMISTRY. *Educ:* Univ Ky, BS, 51; Purdue Univ, PhD, 57. *Prof Exp:* Asst, Purdue Univ, 51-54; asst res engr, Univ Calif, 54-55, res engr, 59; res chemist, Oak Ridge Nat Lab, 56-59; from asst prof to prof inorg chem, Calif State Univ, Long Beach, 59-74, prof chem, 74-; RETIRED. *Mem:* Am Chem Soc. *Res:* Structural investigations of high melting silicides, germanides and carbides; thermodynamic properties of high melting inorganic compounds. *Mailing Add:* 2510 Steeple Chase Jeffersontown KY 40299

THARP, GERALD D, b Wahoo, Nebr, Aug 9, 32; m 57; c 4. VERTEBRATE PHYSIOLOGY. *Educ:* Univ Nebr, BS, 58, MS, 61; Univ Iowa, PhD(physiol), 65. *Prof Exp:* Instr physiol, Univ Iowa, 64-65; asst prof, Wis State Univ, Oshkosh, 65-67; from asst prof to assoc prof, 67-76, PROF PHYSIOL, UNIV NEBR, LINCOLN, 76- *Mem:* Am Physiol Soc; Am Col Sports Med. *Res:* Exercise and stress physiology especially the effects of training. *Mailing Add:* Sch Biol Sci Univ Nebr Lincoln NE 68588

THARP, VERNON LANCE, b Hemlock, Ohio, Mar 13, 17; m 40; c 6. VETERINARY MEDICINE. *Educ:* Ohio State Univ, DVM, 40. *Prof Exp:* Field vet, US Bur Animal Indust, 40-42; instr vet surg & clin, Ohio State Univ, 42-47, from asst prof to prof vet med, 47-83, dir vet clin, 47-71, chmn dept vet clin sci, 60-71, dir equine res ctr, 65-71, dir food animal res ctr, 70-71, assoc dean, 72-83, EMER PROF VET MED, COL VET MED, OHIO STATE UNIV, 83- *Concurrent Pos:* Harness Tracks Am, NY Racing Asn & New York Jockey Club grants equine res col vet med, Ohio State Univ, 65-72; consult, Am Holstein-Friesian Asn, 82-; comnr, Ohio Racing Comn, 83- *Mem:* World Vet Asn (vpres, 79-); Am Asn Bovine Practitioners (pres, 76); Am Vet Med Asn (pres, 78-79); Am Asn Vet Clinicians (secy-treas, 62-72); World Equine Vet Asn (secy-treas, 85-). *Res:* Funding, administration and performance of research in environmental health, nutrition, reproduction and diseases of domestic and laboratory animals. *Mailing Add:* Col Vet Med Ohio State Univ 1900 Coffey Rd Columbus OH 43210

THATCHER, C(HARLES) M(ANSON), b Milwaukee, Wis, Apr 4, 22; m 46; c 2. CHEMICAL ENGINEERING. *Educ:* Univ Mich, BSE, 43, MSE, 50, PhD(chem eng), 55. *Prof Exp:* From instr to asst prof chem eng & metall eng, Univ Mich, 47-58; prof chem eng & chmn dept, Pratt Inst, 58-63, dean, Sch Eng & Sci, 63-70; DISTINGUISHED PROF CHEM ENG, UNIV ARK, FAYETTEVILLE, 70- *Honors & Awards:* Western Elec Award, 67; Burlington Northern Award, 87. *Mem:* Am Chem Soc; Am Inst Chem Engrs; Nat Soc Prof Engrs. *Res:* Computer systems programming and data processing; mathematical modelling and computer simulation; educational methods. *Mailing Add:* Dept Chem Eng Univ Ark Fayetteville AR 72701

THATCHER, EVERETT WHITING, b Jefferson, Ohio, Jan 24, 04; m 28; c 2. PHYSICS. *Educ:* Oberlin Col, AB, 26, AM, 27; Univ Mich, PhD(physics), 31. *Prof Exp:* Asst physics, Purdue Univ, 26-27; instr, Univ Nebr, 27-29; instr, Univ Mich, 29-31; from asst prof to assoc prof, Union Univ, NY, 31-46, instr aerodyn, meteorol & radio civilian pilot training prog, 39-43, coordr, 40-43, instr electronics eng defense training, 40-41; head res dept, US Naval Electronics Lab, 46-53, head spec res div, 53-65; CHMN SPACE SCI & MEM EXEC COUN & ADV COMT, COMN EDUC RESOURCES, 65- *Concurrent Pos:* Res assoc, Princeton Univ, 43; sci liaison officer, OSRD, Washington, London & Europe, 44-45; dep tech dir joint task force I, Washington, DC & Bikini, 46. *Mem:* AAAS; fel Am Phys Soc; Inst Elec & Electronics Engrs; Am Geophys Union. *Res:* Propagation of radio waves; statistical fluctuations in electron currents under space charge; thermal agitation of electric charge in conductors; multiple space-charge; properties of monomolecular films. *Mailing Add:* 3803 Liggett Dr San Diego CA 92106

THATCHER, ROBERT CLIFFORD, b Boonville, NY, Jan 11, 29; m 49; c 4. FOREST ENTOMOLOGY. *Educ:* State Univ NY, BS, 53, MS, 54; Auburn Univ, PhD, 71. *Prof Exp:* Biol aide, Southern Forest Exp Sta, USDA, La, 54, entomologist, 54-68, asst br chief forest insects, 68-73, proj leader, 73-74, prog mgr, Off Secy, Southern Pine Beetle Prog, 74-80, prog mgr, Integrated Pest Mgt, Res Develop & Appln Prog, Southern Forest Exp Sta, 81-85, asst sta dir, Southeast Forest Exp Sta, US Forest Serv, 85-89; RETIRED. *Concurrent Pos:* Instr, Stephen F Austin State Univ, 57 & 63. *Mem:* Soc Am Foresters; Entom Soc Am. *Res:* Forest ecology; interdisciplinary research and development activities relating to southern pine beetle pest management. *Mailing Add:* 12 Bevlyn Dr Asheville NC 28803-3331

THATCHER, WALTER EUGENE, b Evanston, Ill, Jan 22, 27; m 49; c 11. ANALYTICAL CHEMISTRY. *Educ:* Northwestern Univ, BS, 50; Univ Ill, PhD(chem), 55. *Prof Exp:* RES SPECIALIST, CENT RES LABS, 3M CO, 54- *Mem:* Am Chem Soc; Am Crystallog Asn. *Res:* Analytical chemistry in x-ray diffraction and fluorescence. *Mailing Add:* Cent Res Labs 3M Ctr 201 BW-09 PO Box 33221 St Paul MN 55133

THATCHER, WAYNE RAYMOND, b Montreal, Que, May 23, 42; m 78; c 1. SEISMOLOGY. *Educ:* McGill Univ, BSc, 64; Calif Inst Technol, MS, 67, PhD(geophys), 72. *Prof Exp:* chief, Br Tectonophysics, US Geol Surv, 84-89, GEOPHYSICIST, OFF EARTHQUAKE STUDIES, 71-84, 89- *Concurrent Pos:* Vis prof, Stanford Univ, 72-76; mem, Nat Acad Sci comt on geodesy, 76-78; assoc ed, Geophys Res Lett, 77-79 & J Geophys Res, 81-83; pres-elect, Tectonophysics Sect, Am Geophys Union, 88-90, pres, 90- *Honors & Awards:* Bradley Lectr, Geol Div Western Region, US Geol Surv, 84. *Mem:* Am Geophys Union; Seismol Soc Am. *Res:* Aspects of earthquake source mechanism; crustal deformation and earthquake prediction; seismic surface wave propagation; crustal structure; microearthquakes. *Mailing Add:* Off Earthquake Studies MS 977 US Geol Surv 345 Middlefield Rd Menlo Park CA 94025

THATCHER, WILLIAM WATTERS, b Baltimore, Md, Jan 12, 42; m 62; c 3. REPRODUCTIVE PHYSIOLOGY, REPRODUCTION ENDOCRINOLOGY. *Educ:* Univ Md, BS, 63, MS, 65; Mich State Univ, PhD(dairy sci), 68. *Prof Exp:* NIH fel, Mich State Univ, 68-69; asst prof, 69-74, assoc prof, 74-78, PROF REPRODUCTIVE PHYSIOL & ENDOCRINOL, UNIV FLA, 78- *Concurrent Pos:* Sabbaticals, France, 77, 85. *Honors & Awards:* Physiol Award, Am Dairy Sci Asn, 81; Phycol & Endocrin Award, Am Soc Animal Sci, 85. *Mem:* Am Dairy Sci Asn; Soc Study Reproduction; Am Soc Animal Sci. *Res:* Embryos; uterus; corpus luteum; pregnancy; post partum. *Mailing Add:* Dept Dairy Sci Univ Fla Gainesville FL 32611

THAU, FREDERICK E, b Bronx, NY, Dec 2, 38; m 62; c 2. ELECTRICAL ENGINEERING. *Educ:* NY Univ, BEE, 59, MEE, 61, DEngSc(modern control theory), 64. *Prof Exp:* Mem tech staff digital systs, Bell Labs, NY, 59-65; sr staff scientist modern control theory, Kearfott Div, Singer Co, NJ, 65-69; asst prof, City Col New York, 69-76; assoc prof, 76-78, PROF, DEPT ELEC ENG, CITY COL NEW YORK, 78- *Concurrent Pos:* City Univ New York fac res grant, 70-71. *Mem:* Inst Elec & Electronics Engrs. *Res:* Modern control theory and applications to process control; biological control problems; power systems; large space structures; adaptive signal processing. *Mailing Add:* Dept Elec Eng City Col New York Convent at 138th St New York NY 10031

THAU, ROSEMARIE B ZISCHKA, b Vienna, Austria, Mar 15, 36; m 70. IMMUNOLOGY, ENDOCRINOLOGY. *Educ:* Univ Vienna, BS, 54, PhD(chem), 63. *Prof Exp:* Res assoc exp surg med ctr, Duke Univ, 63-65; instr pediat, State Univ NY Downstate Med Ctr, 67-70, asst prof, 67-72; scientist, 72-87, SR SCIENTIST, DIR CONTRACEPTIVE DEVELOP, POP COUN, ROCKEFELLER UNIV, 87- *Mem:* Sigma Xi; Endocrine Soc; Am Soc Immunol Reproduction. *Res:* Contraceptive development, endocrinology of reproduction, antifertility vaccines. *Mailing Add:* Pop Coun Rockefeller Univ York Ave & 66th St New York NY 10021

THAW, RICHARD FRANKLIN, b Denver, Colo, Nov 13, 20; m 42; c 1. BOTANY, NATURAL HISTORY. *Educ:* Ore State Col, BS, 43, MEd, 47, MS, 53, EdD(gen sci, sci educ), 58. *Prof Exp:* Teacher high sch, Ore, 47-58; sci coordr, San Diego County Schs, Calif, 58-59; from assoc prof to prof biol & natural sci, 59-83, EMER PROF BIOL & NATURAL SCI, SAN JOSE STATE UNIV, 83- *Concurrent Pos:* Crown-Zellerbach fel, 54. *Mem:* Nat Sci Teachers Asn; Nat Audubon Soc. *Res:* Plant taxonomy; science teaching methods. *Mailing Add:* 1255 Olive Branch Lane San Jose State Univ San Jose CA 95192

THAWLEY, DAVID GORDEN, b Hastings, NZ, Oct 4, 46; m 71; c 2. EPIDEMIOLOGY, VETERINARY MEDICINE. *Educ:* Massey Univ, NZ, BVSc, 70; Univ Guelph, PhD(vet prevent med), 75; Am Col Vet Prev Med, dipl, 78. *Prof Exp:* Vet pvt pract, Huntly Vet Club, 69-70, Morrinsville Club, 70-71; asst vet med, Univ Guelph, 71-75; clin teaching fel, Massey Univ, 75; asst prof, 76-79, ASSOC PROF VET MED, UNIV MO, 80- *Concurrent Pos:* Coop investr epidemiologist, US Dept Agr, 78-; prin investr, USPHS-NIH, 78-80, Mo Pork Producers & Nat Pork Producers, 78- *Mem:* Am Vet Med Asn; US Animal Health Asn; Am Acad Vet Prev Med; Asn Teachers Vet Pub Health & Prev Med. *Res:* Investigation of pseudorabies epidemiology and strategies for control; investigation of interactions among heavy metals and the resulting toxicities produced; general infectious disease epidemiology. *Mailing Add:* 32 E Oaks Rd St Paul MN 55127

THAXTON, GEORGE DONALD, b Richmond, Va, Feb 28, 31; m 54; c 4. NUCLEAR PHYSICS, SOLID STATE PHYSICS. *Educ:* Richmond Univ, BS, 59; Univ NC, PhD(physics), 65. *Prof Exp:* Res assoc physics, Fla State Univ, 64-66; asst prof, 66-77, ASSOC PROF PHYSICS, AUBURN UNIV, 77- *Mem:* Am Phys Soc; Am Asn Physics Teachers. *Res:* Theory of direct nuclear reactions; theory of band structure of solids. *Mailing Add:* Dept Physics Auburn Univ Auburn AL 36830

THAXTON, JAMES PAUL, b Longview, Miss, Sept 6, 41; m 65; c 3. AVIAN PHYSIOLOGY. *Educ:* Miss State Univ, BS, 64, MS, 66; Univ Ga, PhD(physiol), 71. *Prof Exp:* Instr biol, Northeast La Univ, 66-67; from asst prof to prof poultry sci, NC State Univ, 71-85; DIR RES & DEVELOP, EMBREX, INC, RALEIGH, NC, 85- *Concurrent Pos:* Consult, Nat Inst Environ Health Sci, 74-84. *Honors & Awards:* Poultry Sci Assoc Res Award, Poultry Sci Asn, 74; Res Award, Sigma Xi. *Mem:* AAAS; Poultry Sci Asn; World Poultry Sci Asn; Sigma Xi; Soc Develop & Comp Immunol. *Res:* Effects of environmental parameters, such as temperatures, toxins and heavy metals, on the immunological responsiveness of the avian species. *Mailing Add:* Dept Poultry Sci NC State Univ PO Box 5307 Raleigh NC 27650

THAYER, CHARLES WALTER, b Springfield, Vt, May 18, 44; m 67; c 2. INVERTEBRATE PALEONTOLOGY, PALEOECOLOGY. *Educ:* Dartmouth Col, BA, 66; Yale Univ, MPhil, 69, PhD(geol), 72. *Prof Exp:* Asst prof, 71-78, ASSOC PROF GEOL, UNIV PA, 78- *Concurrent Pos:* Res assoc, Acad Natural Sci, Philadelphia, 78- *Mem:* AAAS; Paleont Soc; Sigma Xi. *Res:* Pattern and process of evolution; marine ecology; sedimentary environments. *Mailing Add:* Dept Geol Univ Pa Philadelphia PA 19104

THAYER, CHESTER ARTHUR, b Stillwater, Okla, July 30, 48; m 69; c 2. MOLECULAR PHOTOPHYSICS, LASER ISOTOPE SEPARATION. *Educ:* Okla State Univ, BS, 70; Univ Ill, PhD(phys chem), 74. *Prof Exp:* Res chemist, Victoria, Tex, 74-78, asst div supt, 78-80, res supvr, Savannah River Lab, 80-83, RES ASSOC, E I DU PONT EXP STA, E I DU PONT DE NEMOURS & CO, INC, 83- *Concurrent Pos:* Instr, Victoria Col, 77-80. *Mem:* Am Chem Soc; Sigma Xi. *Res:* Development of flexible printed circuit materials for the electronics industry; photophysics of laser excited molecules; industrial synthesis of nylon intermediates. *Mailing Add:* 123 Parrish Lane E I du Pont de Nemours & Co Inc Wilmington DE 19810-3457

THAYER, DONALD WAYNE, b Kansas City, Mo, Jan 15, 37; m 69. MICROBIOLOGY, BIOCHEMISTRY. *Educ:* Kans State Univ, BS, 62, MS, 63; Colo State Univ, PhD(microbiol), 66. *Prof Exp:* Nat Acad Sci, NRC resident res assoc, Naval Med Res Inst, 66-67, res chemist, 66-69; from asst prof to prof biol, Tex Tech Univ, 69-79; prog mgr appl biol, NSF, 78-81; RES LEADER, FOOD SAFETY RES, ARS, USDA, 81- *Concurrent Pos:* Lectr, Soc Sigma Xi-Sci Res Soc Am regional lect exchange prog, 71; indust consult, 77-79; rep Interdept Man & Biosphere Task Force, NSF, 79 & Interagency Integrated Pest Mgt Task Force for Pub Health, 80-81; mem Adv Panel Biotechnol, NSF, 83; mem Subj-Matter Expert Rating Panel for Life Sci, ARS, 83-85; US deleg, FAO working group, health impact & control methods of irradiated foods, Neuhrberg, 86; adv bd, Food Res Inst, Univ Wis, 88-; US deleg, harmonization regulations for food irradiation, Inter-Am meeting, Int Atomic Energy Agency, 89; US deleg, Int Consult Group, food irradiation-irradiation as quarantine treatment, 90. *Mem:* Am Soc Microbiol; Am Chem Soc; Soc Indust Microbiol; Sigma Xi; Inst Food Technol; fel Am Acad Microbiol; Asn Milk Food Environ Sanitarians. *Res:* Single-cell protein and carbohydrate metabolism; food borne microorganisms; microbiological, chemical and nutritional safety of foods treated with ionizing radiation. *Mailing Add:* Eastern Regional Res Ctr USDA ARS NAA 600 E Mermaid Lane Philadelphia PA 19118

THAYER, DUANE M, b Kingsford, Mich, June 15, 34; m 57; c 5. METALLURGICAL ENGINEERING. *Educ:* Mich Tech Univ, BS, 59, MS, 62. *Prof Exp:* From instr to assoc prof, 59-70, PROF METALL ENG, MICH TECHNOL UNIV, 70-, ADMIN ASST, 66- *Mem:* Am Inst Mining Metall & Petrol Engrs. *Res:* Mineral processing; flotation and agglomeration of iron oxides; fine particle flotation; reclamation of industrial wastes. *Mailing Add:* Dept Metall Eng Mich Technol Univ Houghton MI 49931

THAYER, GORDON WALLACE, b Weymouth, Mass, Feb 28, 40; m 63; c 2. WETLAND ECOLOGY, WETLAND RESTORATION ECOLOGY. *Educ:* Gettysburg Col, BA, 62; Oberlin Col, MA, 64; NC State Univ, PhD(zool), 69. *Prof Exp:* Fishery biologist, Southeast Fisheries Ctr, 68-77, BR LEADER, BEAUFORT LAB, NAT MARINE FISHERIES SERV, NAT OCEANIC & ATMOSPHERIC ADMIN, 77- *Concurrent Pos:* Habitat coordr, Nat Marine Fisheries Serv. *Mem:* Estuarine Res Fedn; Sigma Xi. *Res:* Ecology of seagrass; wetland habitat use; dynamics of zooplankton and estuarine fishery populations; influence of detritus in invertebrate and vertebrate food webs; restoration ecology. *Mailing Add:* Beaufort Lab Nat Oceanic-Atmospheric Admin Beaufort NC 28516

THAYER, JOHN STEARNS, b Glen Ridge, NJ, Apr 1, 38. INORGANIC CHEMISTRY, ORGANOMETALLIC CHEMISTRY. *Educ:* Cornell Univ, BA, 60; Univ Wis, PhD(chem), 64. *Prof Exp:* Asst prof chem, Ill Inst Technol, 64-66; from asst prof to assoc prof, 66-88, PROF CHEM, UNIV CINCINNATI, 89- *Concurrent Pos:* Frederick Gardner Cottrell grant, 67-73; vis res assoc prof, Chesapeake Biol Lab, 80-84; guest worker, Nat Bur Standards, 83-; vis res prof, Leicester Polytechnic, UK, 87. *Mem:* AAAS; Am Chem Soc. *Res:* Transalkylation of metals in aqueous media; biological aspects of organometallic chemistry. *Mailing Add:* Dept Chem Univ Cincinnati Cincinnati OH 45221

THAYER, KEITH EVANS, b Lime Springs, Iowa, Feb 5, 28; m 53; c 4. DENTISTRY. *Educ:* Cornell Col, BA, 51; Univ Iowa, DDS, 55, MS, 56. *Prof Exp:* From instr to assoc prof, 56-63, head dept, 60-80, PROF FIXED PROSTHODONTICS, COL DENT, UNIV IOWA, 63- *Concurrent Pos:* Attend dentist, Vet Admin Hosp, Iowa City; vis Fulbright prof, Univ Singapore, 68-69; consult, Am Dent Asn Vietnam Educ Proj, 72; consult, Vet Admin Hosp. *Mem:* Int Dent Fedn; Int Asn Dent Res; Am Dent Asn; Sigma Xi. *Res:* Rubber base and silicone impression materials; gingival retraction agents and their effect on oral tissues; occlusion. *Mailing Add:* Fixed Pros Col Univ Iowa Iowa City IA 52242

THAYER, PAUL ARTHUR, b New York, NY, Apr 30, 40; m 66; c 1. MARINE GEOLOGY, SEDIMENTOLOGY. *Educ:* Rutgers Univ, BA, 61; Univ NC, PhD(geol), 67. *Prof Exp:* Develop geologist, Calif Co Div, Chevron Oil Co, 67-68; asst prof geol, Tex A&I Univ, 68-70; asst prof, 70-76, ASSOC PROF MARINE SCI RES, UNIV NC, WILMINGTON, 76- *Concurrent Pos:* Tex A&I Univ fac res grants, 68-69 & 69-70; Soc Sigma Xi grant, 69; petrol geologist, BP Alaska Explor Inc, 75-; instrnl sci equipment prog award, NSF, 72-74 & 75-77; res grant, AEC, 74. *Mem:* AAAS; Geol Soc Am; Int Asn Sedimentol; Soc Econ Paleont & Mineral; Am Asn Petrol Geol. *Res:* Petrology of clastic sedimentary rocks; reconstruction of depositional environments within ancient sedimentary rocks; Triassic nonmarine stratigraphy; provenance, dispersal and origin of modern and ancient sediments; sedimentology of modern carbonate sediments. *Mailing Add:* Dept Marine Sci Res 601 S College Rd Wilmington NC 28403-3297

THAYER, PAUL LOYD, b Centralia, WVa, Feb 25, 28; m 53; c 3. PLANT PATHOLOGY. *Educ:* Marietta Col, BS, 52; Ohio State Univ, MS, 55, PhD, 58. *Prof Exp:* Asst plant pathologist, Everglades Exp Sta, Univ Fla, 58-65; plant pathologist, Eli Lilly & Co, Greenfield, 65-72, northeastern regional plant sci res mgr, 72-80, head crop protection res, 80-87; RETIRED. *Mem:* Am Phytopath Soc. *Res:* Bacterial and fungus diseases of vegetable crops; fungicide and nematocide evaluation. *Mailing Add:* 4145 NW 67th Terr Gainesville FL 32606

THAYER, PHILIP STANDISH, b Pelham, Mass, Oct 1, 23; div; c 3. MICROBIOLOGY. *Educ:* Amherst Col, BA, 48, MA, 49; Calif Inst Technol, PhD(biochem), 52. *Prof Exp:* Merck fel bact, Univ Calif, 51-53; instr chem, Univ Calif, Los Angeles, 53-55; mem staff biol, Arthur D Little, Inc, 55-81, vpres, 72-81; CONSULT, 81- *Mem:* Sigma Xi. *Res:* Tissue culture; chemotherapy, radiation biology, toxicology and carcinogenesis. *Mailing Add:* 14 Lynn Ave Hull MA 02045

THAYER, ROLLIN HAROLD, b St Francis Mission, SDak, Dec 30, 16; m 44; c 1. POULTRY NUTRITION. *Educ:* Okla State Univ, BS, 40; Univ Nebr, MS, 42; Wash State Univ, PhD, 55. *Prof Exp:* From asst prof to prof poultry sci, 43-55, prof poultry sci & nutrit, 55-80, EMER PROF POULTRY SCI & NUTRIT, OKLA STATE UNIV, 80- *Mem:* AAAS; Am Poultry Sci Asn; Am Inst Nutrit; World Poultry Sci Asn; Sigma Xi. *Res:* Estrogens in poultry fattening; layer breeder hen requirements; nutritive requirements for breeder turkeys. *Mailing Add:* 105 N Stallard Stillwater OK 74075

THAYER, WALTER RAYMOND, JR, b Providence, RI, Apr 16, 29; m 55; c 3. MEDICINE, GASTROENTEROLOGY. *Educ:* Providence Col, BS, 50; Tufts Univ, MD, 54. *Hon Degrees:* MA, Brown Univ, 66. *Prof Exp:* Resident gastroenterol, Sch Med, Yale Univ, 61-62, instr med, 60-62, asst prof, 62-66; assoc prof, 66-70, PROF MED, BROWN UNIV, 70-; CHIEF GASTROENTEROL, RI HOSP, 66- *Concurrent Pos:* Fel clin gastroenterol, Sch Med, Yale Univ, 59-60; NSF fel, Wenner-Gren Inst, Stockholm, 71-72. *Mem:* Am Soc Clin Invest; Am Gastroenterol Asn. *Res:* Immunology in gastrointestinal diseases; gastric secretion. *Mailing Add:* Dept Med Brown Univ Brown Sta Providence RI 02902

THAYER, WILLIAM, b Plymouth, In, Sept 23, 48; m 78. BIOENERGETICS, BIOCHEMICAL PHARMACOLOGY. *Educ:* Ind Univ, BS, 70; Cornell Univ, PhD(biochem), 75. *Prof Exp:* Asst prof, 77-83, ASSOC PROF BIOCHEM, HAHNEMANN UNIV, 83- *Honors & Awards:* Res Scientist Develop Award, Nat Inst Alcohol Abuse & Alcoholism, 87. *Mem:* Am Chem Soc; Am Soc Biochem & Molecular Biol; Am Asn Pharmaceut Scientists; Oxygen Soc; AAAS; NY Acad Sci. *Res:* Bioenergetic aspects of alcohol consumption; role of oxygen radicals and peroxides in cardiomyopathy; biochemical toxicology; membrane biochemistry and biophysics. *Mailing Add:* Dept Biol Chem MS 411 Hahnemann Univ Broad & Vine Sts Philadelphia PA 19102

THAYNE, WILLIAM V, b Binghamton, NY, July 23, 41; m 63; c 3. EXPERIMENTAL DESIGN. *Educ:* Cornell Univ, BS, 63; Univ Ill, MS, 67, PhD(dairy sci), 71. *Prof Exp:* Res asst dairy sci, Univ Ill, 63-67; instr animal sci, 67-70, asst prof statist & comput sci, 73-75, from asst prof animal sci to assoc prof, 70-85, PROF STATIST & COMPUTER SCI, WVA UNIV, 85- *Mem:* Biomet Soc; Am Genetic Asn. *Res:* Statistical applications and agriculture. *Mailing Add:* 1016 AS WVa Univ Morgantown WV 26506-6108

THEDFORD, ROOSEVELT, b Greene Co, Ala, Apr 16, 37; m 60; c 5. ORGANIC BIOCHEMISTRY, MOLECULAR BIOLOGY. *Educ:* Clark Col, BS, 59; Univ Buffalo, MA, 62; State Univ NY, Buffalo, PhD(biochem), 73. *Prof Exp:* Cancer res scientist, Roswell Park Mem Inst, 61-69 & 72-74; assoc prof, Clark Col, 74-84, PROF CHEM, CLARK ATLANTA UNIV, 84- *Mem:* Am Chem Soc; Sigma Xi. *Res:* Study of the physicochemical and biological properties of alkylated synthetic homopoly ribonucleotides and determination of the functions of strategically located modified nucleosides as found in transfer RNA. *Mailing Add:* 2916 Edna Lane Decatur GA 30032

THEEUWES, FELIX, b Duffel, Belgium, May 25, 37; m 62; c 3. PHYSICAL CHEMISTRY, PHARMACY. *Educ:* Cath Univ Louvain, Licentiaat physics, 61, DrSc(physics), 66. *Prof Exp:* Res assoc chem, Univ Kans, 66-68, asst prof, 68-70; res scientist pharm chem, 70-74, PRIN SCIENTIST, ALZA CORP, 74-, VPRES PROD, RES & DEVELOP, 80- *Concurrent Pos:* High sch teacher, St Vincent Sch, Westerlo, Belgium, 61-64. *Honors & Awards:* Louis Busse Lectr, Dept Pharmacol, Univ Wis, 81. *Mem:* Acad Pharmaceut Sci; NY Acad Sci; Am Chem Soc. *Res:* Osmosis, diffusion; solid state physics; cryogenics; high pressure; thermodynamics; pharmacology; pharmacokinetics; calorimetry. *Mailing Add:* 950 Page Mill Rd Palo Alto CA 94304-1012

THEIL, ELIZABETH, b Jamaica, NY, Mar 29, 36; div; c 2. BIOCHEMISTRY, MOLECULAR & DEVELOPMENTAL BIOLOGY. *Educ:* Cornell Univ, BS, 57; Columbia Univ, PhD(biochem), 62. *Prof Exp:* Res assoc chem, Fla State Univ, 64-66; res assoc, 67-69, from instr to prof, 69-88, UNIV PROF BIOCHEM, NC STATE UNIV, 88- *Concurrent Pos:* NIH merit grant, 87. *Honors & Awards:* O Max Gardener Award, 88. *Mem:* Sigma Xi; Am Chem Soc; Soc Develop Biol; Am Soc Biol Chemists. *Res:* Molecular biology of ferritin; cell-specificity of gene expression; developmental changes in gene expression; memory RNA structure/function (translational control of protein synthesis); biophysics of iron proteins (x-ray absorption spectroscopy fine structure). *Mailing Add:* Dept Biochem 339 Polk Hall NC State Univ Raleigh NC 27695

THEIL, MICHAEL HERBERT, b Brooklyn, NY, Nov 2, 33; div; c 2. POLYMER CHEMISTRY. *Educ:* Cornell Univ, AB, 54; Polytech Inst Brooklyn, PhD(chem), 63. *Prof Exp:* Sr res chemist, Res Ctr Tex, US Chem Co, 62-64; res assoc chem, Fla State Univ, 64-66; from asst prof to assoc prof, 66-80, PROF TEXTILE CHEM, NC STATE UNIV, 80- *Mem:* Am Chem Soc; Am Phys Soc; Sigma Xi; Fiber Soc. *Res:* Phase transitions of polymers; polymerization mechanisms; copolymer statistics. *Mailing Add:* Dept Textile Chem NC State Univ Raleigh NC 27695-8302

THEILEN, ERNEST OTTO, b Columbus, Nebr, June 4, 23. INTERNAL MEDICINE. *Educ:* Univ Nebr, BA, 45, MD, 47; Am Bd Internal Med, dipl, 55; Am Bd Cardiovasc Dis, dipl, 65. *Prof Exp:* Instr med, 51-52, assoc, 52, from asst prof to assoc prof, 55-63, PROF INTERNAL MED, COL MED, UNIV IOWA, 63- *Concurrent Pos:* Fel coun clin cardiol & coun circulation, Am Heart Asn. *Mem:* AMA; Am Fedn Clin Res; Am Col Cardiol; Am Soc Int Med; Am Col Physicians. *Res:* Cardiovascular physiology; electrocardiography; clinical cardiology. *Mailing Add:* Dept Int Med Univ Iowa Iowa City IA 52242

THEILEN, GORDON H, b Montevideo, Minn, May 29, 28; m 53; c 3. VETERINARY MEDICINE. *Educ:* Univ Calif, BS, 53, DVM, 55. *Prof Exp:* Pvt pract, Ore, 55-56; lectr vet med, 56-57, from instr to asst prof, 57-62, from asst prof to prof clin sci, 62-74, PROF SURG, SCH VET MED, UNIV CALIF, DAVIS, 74- *Concurrent Pos:* Spec fel, Leukemia Prog, Nat Cancer Inst, 64-65; NY Cancer Res Inst fel tumor immunol, with Chester Beatty, Univ London, 72-73; mem sci & rev comt & bd dirs, Leukemia Soc Am, 71-76; World Comt mem, Int Asn Comp Res Leukemia & Related Dis, 90-94. *Honors & Awards:* Alexander von Humboldt Sr Scientist Award, WGer Govt, 79-80; Small Animal Res Award, Ralston Purina, 82. *Mem:* Am Vet Med Asn; Am Asn Cancer Res; Am Asn Vet Clin; Int Asn Comp Res Leukemia & Related Dis. *Res:* Leukemia-sarcoma and myeloproliferative disease complex and subjects dealing with clinical oncology, particularly tumor biology. *Mailing Add:* Dept Surg Sch Vet Med Univ Calif Davis CA 95616

THEILHEIMER, FEODOR, b Gunzenhausen, Ger, June 18, 09; nat US; m 48; c 1. MATHEMATICS. *Educ:* Berlin Univ, PhD(math), 36. *Prof Exp:* From instr to asst prof math, Trinity Col, 42-48; mathematician naval ord lab, 48-53, mathematician, Naval Ship Res & Develop Ctr USN, Bethesda, Md, 53-78; RETIRED. *Mem:* Am Math Soc; Math Asn Am; Soc Indust & Appl Math; Asn Comput Mach. *Res:* Numerical analysis; fluid dynamics. *Mailing Add:* 2608 Spencer Rd Chevy Chase MD 20815

THEILHEIMER, WILLIAM, b Augsburg, Ger, Oct 11, 14; nat US; wid; c 1. ORGANIC CHEMISTRY. *Educ:* Basel Univ, PhD(org chem), 40. *Prof Exp:* Asst, Basel Univ, 40-47; consult, 48-59, lit chemist sci info dept, 59-63, RESIDENT CONSULT, HOFFMANN-LA ROCHE, INC, 64- *Concurrent Pos:* Ed, Synthetic Methods Org Chem, 44-81. *Honors & Awards:* Herman Skolnik Award, 87. *Mem:* Am Chem Soc. *Res:* Synthetic methods. *Mailing Add:* 318 Hillside Ave Nutley NJ 07110

THEIMER, EDGAR E, b Newark, NJ, June 29, 15; m 58; c 1. ANALYTICAL CHEMISTRY, PHARMACEUTICAL CHEMISTRY. *Educ:* Polytech Inst Brooklyn, BChE, 36; NY Univ, MS, 39. *Prof Exp:* Chief chemist, Metrop Labs, Inc, 40-58; chief chemist, Pharmich Div, Mich Chem Corp, 58-61; staff chemist, Int Flavors & Fragrances, Inc, 61-64; head anal chem sect res & develop, Smith, Miller & Patch, Inc, 64-72; head anal chem sect prod develop, Cooper Labs, Inc, 72-78; sr sci assoc, US Pharmacopeial Conv, Inc, Rockville, Md, 78-87; CONSULT, 87- *Mem:* Am Chem Soc. *Res:* Analytical methods development. *Mailing Add:* 3919 Brooke Meadow Lane Olney MD 20832-1303

THEIMER, OTTO, physics; deceased, see previous edition for last biography

THEINE, ALICE, b Menomonee, Wis, Feb 23, 38. ORGANIC CHEMISTRY. *Educ:* Alverno Col, BA, 59; Marquette Univ, MS, 65; La State Univ, PhD(chem), 72. *Prof Exp:* Teacher sci & math, St Boniface High Sch, Iowa, 59-63 & St Joseph High Sch, Wis, 63-66; from instr to assoc prof chem, Alverno Col, 66-85; CONSULT, 85- *Mem:* Am Chem Soc; Sigma Xi. *Res:* Development of improved methods of competence-based instruction in undergraduate chemistry courses; investigation of cationic intermediates in oxidative decarboxylation reactions by product analysis studies. *Mailing Add:* 1208 N Chicago Ave South Milwaukee WI 53172

THEINER, MICHA, biochemistry; deceased, see previous edition for last biography

THEIS, GAIL GOODMAN, immunology, for more information see previous edition

THEIS, JEROLD HOWARD, b Richmond, Calif, July 29, 38; m 67; c 2. MEDICAL MICROBIOLOGY, PARASITOLOGY. *Educ:* Univ Calif, Berkeley, AB, 60; Univ Calif, Davis, DVM, 64, PhD(comp path), 72. *Prof Exp:* Asst res vet, George Williams Hooper Found, Univ Calif, San Francisco Med Ctr & Repub Singapore, 64-67; asst prof, 70-77, assoc prof, 77-86, PROF MED MICROBIOL, SCH MED, UNIV CALIF, DAVIS, 86- *Concurrent Pos:* USPHS fel, Univ Calif, Davis, 67-69; consult, Sacramento Med Ctr, Calif, 71- & Primate Res Ctr, Davis, 72- *Honors & Awards:* Grand Prize, Int Med Film Festival, Brussels, Belg, 72. *Mem:* Am Vet Med Asn; Am Soc Laser Med & Surg. *Res:* Mechanisms of transmission of arthropod borne disease agents at the host-arthropod interface; laser treatment of Atherosclerosis. *Mailing Add:* Dept Med Microbiol Univ Calif Sch Med Davis CA 95616

THEIS, RICHARD JAMES, b Cincinnati, Ohio, Nov 30, 37; m 61; c 3. ORGANIC POLYMER CHEMISTRY & POLYESTER & POLYIMIDE FILMS IN ELECTRICAL END USES. *Educ:* Xavier Univ, BS, 60, MS, 62; Univ Cincinnati, PhD(chem), 66. *Prof Exp:* Res Chemist, E I du Pont de Nemours & Co, Inc, 66-72, staff scientist, Electronics Dept, 73-87; TECH CONSULT, 87- *Mem:* Am Chem Soc; Am Soc Testing & Mat. *Res:* Barrier coatings for films; films for packaging uses; adherable films; filled films; polyester films for capacitors; polyester films for printed circuits and membrane switches; polyimide films for magnet wire insulation. *Mailing Add:* E I du Pont de Nemours & Co Inc PO Box 89 Circleville OH 43113

THEISEN, CHARLES THOMAS, anatomy, embryology, for more information see previous edition

THEISEN, CYNTHIA THERES, b Dearborn, Mich. ORGANIC CHEMISTRY. *Educ:* Siena Heights Col, BS, 60; Purdue Univ, Lafayette, MS, 63; St John's Univ, PhD(org chem), 67. *Prof Exp:* Assoc ed org chem, 67-75, sr assoc ed, 75-81, SR ED ORG CHEM, CHEM ABSTR SERV, 81- *Mem:* Am Chem Soc. *Mailing Add:* 1182 Stanhope Dr Columbus OH 43221

THEISEN, WILFRED ROBERT, b Sept 5, 29; US citizen. HISTORY OF SCIENCE, PHYSICS. *Educ:* St John's Univ, BA, 52; Univ Colo, MS, 63; Univ Wis, PhD(hist of sci), 72. *Prof Exp:* PROF PHYSICS & HIST OF SCI, ST JOHN'S UNIV, MINN, 55- *Res:* Medieval optical manuscripts of Euclid; alchemy. *Mailing Add:* St John's Univ Collegeville MN 56321

THEISS, JEFFREY CHARLES, b Stamford, Conn, Aug 29, 46; m 68; c 2. GENETIC TOXICOLOGY. *Educ:* Univ RI, BS, 68; Brown Univ, PhD(med sci), 73. *Prof Exp:* Res assoc biochem pharmacol, Roger Williams Gen Hosp & Brown Univ, 71-73; res fel biochem pharmacol, Univ Calif, San Diego & L C Strong Res Found, 73-75; asst res sci pharmacol & toxicol, Univ Calif, San Diego, 75-77, asst prof res community med, 77-80; assoc prof environ health, Sch Pub Health, Univ Tex, 80-85; dir genetic toxicol, 85-90, DIR MOLECULAR TOXICOL, WARNER LAMBERT CO, 90- *Honors & Awards:* Nat Res Award, Am Soc Hosp Pharm Res & Ed Found. *Mem:* Sigma Xi; AAAS; Am Asn Cancer Res; Soc Toxicol; Environ Mutagen Soc. *Res:* Mechanistic studies in chemical carinogenesis and cocarcinogenesis; screening of chemicals for carcinogenic potency by the A mouse lung tumor bioassy; testing of chemicals and body fluids for mutagenic activity. *Mailing Add:* Warner Lambert Co 2800 Plymouth Rd Ann Arbor MI 48105

THEKDI, ARVIND C, b Ahmedabad, India, Aug 5, 41; US citizen. COMBUSTION ENGINEERING, HEAT TRANSFER. *Educ:* Gujarat Univ, India, BS, 63; Indian Inst Sci, Bangalore, MS, 65; Pa State Univ, PhD(fuel sci), 70. *Prof Exp:* Res engr, Surface Combustion, Midland-Ross Corp, 70-73, mgr thermal systs, 73-77, mgr thermal & mech eng, 78-80, asst dir develop, 80-84, dir res & develop, Tech Ctr, 84-87; VPRES, INDUGAS, 87- *Concurrent Pos:* Res asst, Pa State Univ, 67-70; instr, Toledo Univ, 72-74. *Mem:* Am Soc Mech Engrs; Combustion Inst; Air Pollution Control Asn. *Res:* Energy conservation and conversion; combustion heat transfer; heat recovery; process development in carbon and graphite industry. *Mailing Add:* Indugas 5924 American Rd E Toledo OH 43612

THELEN, CHARLES JOHN, b Cedar Rapids, Nebr, Mar 2, 21; m 47; c 1. ORGANIC CHEMISTRY, AEROSPACE TECHNOLOGY. *Educ:* Univ Iowa, BS, 42, PhD(org chem), 49. *Prof Exp:* Chemist, B F Goodrich Co, 42; asst, Univ Iowa, 47-49; fel, Univ Calif, Los Angeles, 49-50; res chemist, US Naval Ord Test Sta, Naval Weapons Ctr, 50-58, head explosives & pyrotech div, 58-60, head propellants div, 60-68, missile propulsion technol adminr, 69-80, air weaponry technol adminr, 80-82, weapons cookoff prog adminr, 82-85; TECH ADV, PROPULSION DIV, ATLANTIC RES CORP, 85- *Concurrent Pos:* Michelson Lab fel mgt, Naval Weapons Ctr, 72; mem bd dirs, Ridgecrest Community Hosp, 76-85 & 88- *Honors & Awards:* L T E Thompsin Award, 80. *Mem:* Am Chem Soc; Sigma Xi. *Res:* Missile propulsion technology and high polymers. *Mailing Add:* 344 E Monte Vista Ave Ridgecrest CA 93555

THELEN, THOMAS HARVEY, b Albany, Minn, Aug 11, 41; m 64; c 2. GENETICS. *Educ:* St John's Univ, Minn, BS, 64; Univ Minn, PhD(genetics), 69. *Prof Exp:* Cytogeneticist, Minn Dept Health, 69-70; from asst prof to assoc prof, 70-82, PROF BIOL, CENT WASH UNIV, 82- *Res:* Behavioral genetics, population modeling. *Mailing Add:* Dept Biol Cent Wash Univ Ellensburg WA 98926

THELIN, JACK HORSTMANN, b Kearny, NJ, Aug 15, 12; m 39; c 2. CHEMISTRY. *Educ:* Maryville Col, BA, 38; Univ Tenn, MS, 39; Rutgers Univ, PhD(phys chem), 43. *Prof Exp:* Res chemist, Calco Chem Co Div, Am Cyanamid Co, 36-41, develop chemist, 41-45, asst chief develop chemist dye intermediates div, 45-46, chief develop chemist, 46-53, res chemist, Res Div, 53-77; Horst Instrument Design Co Inc, 77-90; RETIRED. *Mem:* Am Soc Testing & Mat; Am Chem Soc; Sigma Xi. *Res:* Sulfonation; nitration and alkylation; system ammonium sulfamate; sodium nitrate; ammonium nitrate; polymer physical testing equipment; polymer tribology. *Mailing Add:* 126 E Spring St Somerville NJ 08876

THELIN, LOWELL CHARLES, b Plainfield, NJ, July 31, 46; m 71; c 2. TECHNICAL MANAGEMENT. *Educ:* Muhlenberg Col, AB, 68; Rutgers Univ, MS, 73; Fairleigh- Dickinson Univ, BSEE, 90. *Prof Exp:* Radiation safety officer, Cambridge Nuclear Radiopharmaceut Corp, NL Industs, 73-74, Indust Reactor Labs, Inc, 74-75 & Sterling Forest Res Ctr, Union Carbide Corp, 75-87; STAFF HEALTH PHYSICIST, CINTICHEM INC SUBSID OF HOFFMAN-LAROCHE INC, 87- *Concurrent Pos:* Appointee, Comprehensive Cert Panel Examr, Am Bd Health Physics, 85-88. *Mem:* Health Physics Soc. *Res:* Technical analysis for decommissioning plan of major hot laboratory and nuclear research reactor facility; noble gas effluent dosimetry; beta dosimetry; fission product internal dose assessment; nuclear worker radiation protection program. *Mailing Add:* 126 Greenlawn Ave Clifton NJ 07013

THELLMANN, EDWARD L, b Cleveland, Ohio. METALLURGY. *Educ:* Cleveland State Univ, BS. *Prof Exp:* Mgr, Appl Mat Technol, Gould, Inc, 59-87; RETIRED. *Honors & Awards:* Am Soc Metals Award; John C Valler Award; IR-100 Award. *Mem:* Fel Am Soc Metals. *Res:* Ten patents; milestone advances in titanium powder metallurgy; fuel cells. *Mailing Add:* Village Walton Hills Munic Bldg 7595 Walton Rd Walton Hills OH 44146

THELMAN, JOHN PATRICK, b Richmond Hill, NY, Dec 25, 42; m 65; c 3. PAPER CHEMISTRY. *Educ:* State Univ NY, Stony Brook, BS, 64; State Univ NY, Buffalo, PhD(org chem), 69. *Prof Exp:* Res chemist, ITT Rayonnier Inc, Whippany, NJ, 68-71, res group leader acetate sect, 71-74, res supvr acetate sect, Eastern Res Div, 74-77; chief res proj chemist, 77-81, CHIEF DEVELOP ASSOC, SCOTT PAPER CO, PHILADELPHIA, 81- *Mem:* Tech Asn Pulp & Paper Indust. *Res:* Papermaking; product development. *Mailing Add:* 855 Old Horseshoe Pike Downingtown PA 19335

THEMELIS, NICKOLAS JOHN, b Athens, Greece, Apr 25, 33; US citizen; c 3. CHEMICAL ENGINEERING. *Educ:* McGill Univ, BEng, 56, PhD(chem eng), 61. *Prof Exp:* Engr, Pulp & Paper Res Inst Can, 56-57; res consult, Strategic Mat Corp, 60-62; head, dept chem eng, Noranda Res Ctr, Can, 62-66, mgr eng div, 67-72; vpres res & eng, metal mining div, Kennecott Copper Corp, 72-79, vpres technol, 79-80; chmn, Henry Krumb Sch Mines, 85-87, PROF MINERAL ENG, COLUMBIA UNIV, 80- *Honors & Awards:* ERCO Award, 71; McConnell Award, 85, Kohnstamm Award, 87. *Mem:* Nat Acad Eng; Can Inst Mining & Metall; Can Soc Chem Engrs; Am Inst Mining Metall & Petrol Engrs; fel Metall Soc. *Res:* Process metallurgy. *Mailing Add:* Henry Krumb Sch Mines Columbia Univ New York NY 10027

THENEN, SHIRLEY WARNOCK, b San Mateo, Calif; m 62; c 2. VITAMINS. *Educ:* Univ Calif, Berkeley, AB, 57, PhD(nutrit), 70. *Prof Exp:* Res fel hemat, Harvard Med Sch, 70-72; from asst prof to assoc prof nutrit, Dept Nutrit, Harvard Univ, 72-84; PROF NUTRIT, DEPT FOOD SCI & NUTRIT, UNIV MINN, 84- *Concurrent Pos:* NIH postdoc trainee, 70-72 & res career develop award, 75-80; NSF scholar, 68-69. *Mem:* Am Inst Nutrit; Soc Exp Biol & Med; Geront Soc Am; NY Acad Sci; Soc Nutrit Ed. *Res:* Folate and vitamin B-12 coenzyme function and bioavailability; nutritional, biochemical and endocrine abnormalities in obesity and diabetes; vitamin C interactions with vitamin B-12. *Mailing Add:* Dept Food Sci & Nutrit Univ Minn 1334 Eckles Ave St Paul MN 55108

THEOBALD, CHARLES EDWIN, JR, b Hackensack, NJ, Aug 14, 27; m 52; c 3. ELECTRICAL ENGINEERING, OPERATIONS RESEARCH. *Educ:* Columbia Univ, AB, 47, AM, 48; Mass Inst Technol, SM, 59. *Prof Exp:* Aerodynamicist, Curtiss-Wright Corp, NJ, 49-56; sr aerodynamicist, Kaman Aircraft Corp, Conn, 56-58; teaching asst instrumentation & control, Mass Inst Technol, 58-59; sr scientist, Syst Develop Corp, 59-67; prin engr, Systs Electronics Lab, Raytheon Co, 67-74; ENG CONSULT, 74- *Concurrent Pos:* Engr, H H Aerospace Design Co, 77- *Res:* Adaptive learning and optimal control; statistical decision theory; queueing theory; reliability theory; stability and control of fixed and rotary wing aircraft; public transit operations; railroad track dynamics and geometry. *Mailing Add:* 37 Old Billerica Rd Bedford MA 01730

THEOBALD, J KARL, b Prescott, Ariz, Feb 18, 21; m 42, 67. PHYSICS. *Educ:* Univ Calif, PhD(physics), 52. *Prof Exp:* Physicist, Los Alamos Sci Lab, 52-76; physicist, EG&G, Inc, 76-86; RETIRED. *Mem:* Sigma Xi. *Res:* Fiber optics; radiation effects. *Mailing Add:* 1030 Granite Dr Granite Shoals TX 78654

THEOBALD, JOHN J(ACOB), engineering; deceased, see previous edition for last biography

THEOBALD, WILLIAM L, b New York, NY, Feb 12, 36. SYSTEMATIC BOTANY, HORTICULTURE. *Educ:* Rutgers Univ, BS, 58, MS, 59; Univ Calif, Los Angeles, PhD(bot), 63. *Prof Exp:* Lectr biol, Univ Calif, Santa Barbara, 63-65; NSF fel, Jodrell Lab, Royal Bot Gardens, Kew, Eng, 65-66; fel, Evolutionary Biol, Gray Herbarium, Harvard Univ, 66-67; asst prof biol, Occidental Col, 67-71; assoc prof bot, Univ Hawaii, 71-75; DIR, NAT TROP BOT GARDEN, LAWAI, HAWAII, 75- *Concurrent Pos:* Res assoc bot, Univ Hawaii & Bishop Mus, 75- *Mem:* Bot Soc Am; Am Soc Plant Taxon; fel Linnean Soc London; Int Asn Plant Taxon; Am Asn Bot Gardens & Arboreta. *Res:* Flora of Hawaii, Pacific Islands and Ceylon; systematics of Gesneriaceae, Acanthaceae, Araliaceae, Bignoniaceae, Umbelliferae; trichome anatomy and classification in angiosperms; comparative anatomical studies. *Mailing Add:* Nat Trop Bot Garden PO Box 340 Lawai Kauai HI 96765-0340

THEODORE, JOSEPH M, JR, b Fall River, Mass, Apr 29, 31; m 55; c 5. PHARMACY. *Educ:* New Eng Col Pharm, BS, 55; Univ Wis, MS, 58; Mass Col Pharm, PhD(pharm), 65. *Prof Exp:* Instr pharm, New Eng Col Pharm, 58-61; from instr to asst prof, Northeastern Univ, 62-66; assoc prof, 66-74, PROF PHARM, OHIO NORTHERN UNIV, 74- *Mem:* Am Pharmaceut Asn; Am Col Apothecaries; Am Soc Hosp Pharmacists. *Res:* Spectrofluorometric analysis of drugs; solid state reactions occurring in certain tablet formulations. *Mailing Add:* Dept Pharm Ohio Northern Univ 500 S Main Ada OH 45810

THEODORE, TED GEORGE, b Los Angeles, Calif, Aug 19, 37; m 61; c 2. ECONOMIC GEOLOGY. *Educ:* Univ Calif, Los Angeles, AB, 61, PhD(geol), 67. *Prof Exp:* GEOLOGIST, US GEOL SURV, 67- *Mem:* AAAS; Geol Soc Am; Soc Econ Geol. *Res:* Genesis of porphyrytype, molybdenum and copper deposits; geochemistry of base-metal ore deposits; fabrics of metamorphic terranes. *Mailing Add:* 2015 Camino De Los Robles Menlo Park CA 93110

THEODORE, THEODORE SPIROS, b Braddock, Pa, Nov 6, 33; m 57; c 3. MICROBIAL PHYSIOLOGY. *Educ:* Univ Pitteburgh, BSc, 55, MSc, 57, PhD(bact), 62. *Prof Exp:* Asst bact, Univ Pittsburgh, 57-62; staff fel, NIH, 62-65, res microbiologist, 65-76; MEM STAFF, LAB STEPOCOCCAL DIS, NAT INST ALLERGY & INFECTIOUS DIS, NIH, 76- *Mem:* AAAS; Am Soc Microbiol. *Res:* Bacterial physiology and nutrition; intermediary and mineral metabolism; biochemical genetics. *Mailing Add:* Inst Allergy & Infectious Dis NIH Bldg 4 Rm 316 Bethesda MD 20205

THEODORIDES, VASSILIOS JOHN, b Konstantia, Greece, Feb 20, 31; US citizen; m 58; c 3. DRUG TOXICOLOGY. *Educ:* Univ Thessaloniki, DVM, 56; Boston Univ, MA, 60, PhD(parasitol), 63. *Prof Exp:* Asst to prof clins vet sch, Univ Thessaloniki, 56-57; teaching fel microbiol, Boston Univ, 58-62; lectr micros anat, 62-63; res parasitologist, Charles Pfizer & Co, Inc, 63-65; sr microbiologist parasitologist, SmithKline & French Labs, 65-66, group leader microbiol, 66-68, group leader, Animal Health Dept, 68, assoc dir res chemother, 68-73, mgr parasitol, SmithKline Corp, 73-82, DIR TOXICOL, SMITHKLINE & BEECHAM AH, 82- *Concurrent Pos:* Adj prof, Sch Vet Med, Univ Pa, 72- *Mem:* Am Soc Microbiol; Am Vet Med Asn; Am Soc Parasitol; Am Soc Trop Med & Hyg; NY Acad Sci. *Res:* Morphology, physiology and electron microscopy of Trichomonas; development of chemotherapeutic agents for the control of gastrointestinal nematodes of domestic animals; toxicologic evaluvation of potential animal drugs. *Mailing Add:* SmithKline & Beecham AH 1600 Paoli Pike West Chester PA 19380

THEODORIDIS, GEORGE CONSTANTIN, b Braila, Romania, Dec 3, 35; US citizen; m 75; c 1. BIOMEDICAL ENGINEERING. *Educ:* Nat Tech Univ Athens, dipl elec eng, 59. *Hon Degrees:* ScD, Mass Inst Technol, 64. *Prof Exp:* Res assoc biol, Mass Inst Technol, 64; sr scientist space res, Am Sci Eng, Mass, 64-68; assoc prof physiol optics, Univ Calif, Berkeley, 68-70; ASSOC PROF BIOMED ENG, UNIV VA, 70- *Concurrent Pos:* Prof, Univ Patras, Greece, 76-83; consult, Food & Drug Admin, 77-78 & Appl Physics Lab, Johns Hopkins Univ, 78. *Mem:* Biomed Eng Soc; Am Phys Soc; Inst Elec & Electronics Engrs; Am Geophys Union; NY Acad Sci. *Res:* Speech perception; evolution; space physics. *Mailing Add:* Dept Biomed Eng Univ Va Charlottesville VA 22908

THEODOROU, DOROS NICOLAS, b Athens, Greece, Oct 29, 57. COMPUTER MODELING OF MATTER, POLYMER SCIENCE & ENGINEERING. *Educ:* Nat Tech Univ Athens, Greece, dipl chem eng, 82; Mass Inst Technol, MS, 83, PhD(chem eng), 85. *Prof Exp:* Res asst chem eng, Mass Inst Technol, 81-85, teaching asst, 85; instr computer prog, Hellenic Naval Acad, Piraeus, Greece, 85-86; asst prof, 86-90, ASSOC PROF CHEM ENG, UNIV CALIF, BERKELEY, 90-, PROJ LEADER, CTR ADVAN MAT, LAWRENCE BERKELEY LAB, 86- *Concurrent Pos:* NSF presidential young investr award, 88; consult, Polymer Prod Dept, E I du Pont de Nemours & Co, Inc, 88-, W R Grace & Co, B F Goodrich Co & Biosym Technologies, 89- *Mem:* Am Inst Chem Engrs; Am Chem Soc; Am Phys Soc. *Res:* Using the principles of statistical mechanics and molecular simulations, methods are developed for predicting the macroscopic (thermodynamic, mechanical, transport, interfacial) properties of materials from chemical constitution; polymers in the amorphous glassy and melt states; zeolites. *Mailing Add:* Dept Chem Eng 218 Gilman Hall Univ Calif Berkeley CA 94720

THEOFANOUS, THEOFANIS GEORGE, b Athens, Greece, May 21, 42; m 69; c 2. CHEMICAL ENGINEERING. *Educ:* Nat Tech Univ Athens, dipl, 65; Univ Minn, Minneapolis, PhD(chem eng), 69. *Prof Exp:* Instr chem eng, Univ Minn, Minneapolis, 68-69; from asst prof to prof nuclear eng, Purdue Univ, 69-85; PROF CHEM & NUCLEAR ENG, UNIV CALIF, SANTA BARBARA, 85- *Concurrent Pos:* Consult, Adv Comt on Reactor Safeguards, 71-89; vpres, FGH&T Ltd, 77-81; pres, Theofanus & Co, Inc, 81- *Mem:* Fel Am Nuclear Soc; Am Inst Chem Engrs. *Res:* Transport phenomena in turbulent and multiphase systems with particular emphasis on nuclear and chemical reactor safety applications. *Mailing Add:* Dept Chem & Nuclear Eng Univ Calif Santa Barbara CA 93106

THEOFILOPOULOS, ARGYRIOS N, IMMUNOLOGY. *Educ:* Univ Athens, Greece, MD, 70. *Prof Exp:* MEM, DEPT IMMUNOL, SCRIPPS CLIN & RES FOUND, 73- *Mailing Add:* Dept Immunol IMM3 Scripps Clin & Res Found 10666 N Torrey Pines Rd La Jolla CA 92037

THEOKRITOFF, GEORGE, b Eng, Apr 7, 24; US citizen; m. PALEONTOLOGY. *Educ:* Univ London, BSc, 45, MSc, 48, PhD(geol), 61. *Prof Exp:* Instr geol, Mt Holyoke Col, 54-56 & Bucknell Univ, 56-60; asst prof geol, Univ NH, 60-61; assoc prof geol, St Lawrence Univ, 64-67; assoc prof, 67-85, PROF GEOL, RUTGERS UNIV, NEWARK, 85- *Mem:* Geol Soc Am; Geol Soc London; Paleont Soc. *Res:* Cambrian paleontology and stratigraphy, including morphology, taxonomy and evolution of Cambrian trilobites; ecology of Cambrian organisms; biogeography and biostratigraphy of North Atlantic region. *Mailing Add:* Dept Geol Rutgers Univ Newark NJ 07102

THEOLOGIDES, ATHANASIOS, b Ptolemais, Greece, Feb 5, 31; US citizen; m 65; c 2. INTERNAL MEDICINE, ONCOLOGY. *Educ:* Aristoteles Univ, MD, 55; Univ Minn, Minneapolis, PhD(med & biochem), 67. *Prof Exp:* From instr to assoc prof, 65-74, Nat Cancer Inst grant, Med Ctr, 69-75, PROF MED, MED SCH, UNIV MINN, MINNEAPOLIS, 74- *Mem:* Am Asn Cancer Res; Am Fedn Clin Res; Am Soc Clin Oncol; Soc Exp Biol & Med; Am Soc Hemat. *Res:* Medical oncology; tumor-host metabolic interrelationships; hematology. *Mailing Add:* Dept Med Univ Minn Med Sch Hennepin County Med Ctr 701 Park Ave Minneapolis MN 55415

THEON, JOHN SPERIDON, b Wash, DC, Dec 12, 34; m 65; c 2. MICROWAVE RADIOMETRY. *Educ:* Univ Md, BS, 57; Pa State Univ, BS, 59, MS, 62; Univ Tenn, PhD(eng sci & mech), 85. *Prof Exp:* Aeronaut eng, Douglas Aircraft Co, 57-58; Weather officer, USAF, 58-60; mech engr, US Naval Ord Lab, 60; res meteorologist, Goddard Space Flight Ctr, NASA, 62-74, head, Meteorol Br, 74-77, asst chief, Lab Atmospheric Sci, 77-78, mgr, Global Weather Res, Earth Sci Div, NASA Hq, 78-82, chief, Atmosphere Dynamics & Radiation, 82-91, CHIEF, ATMOSPHERE DYNAMICS, RADIATION & HYDROL, EARTH SCI DIV, NASA HQ, 91- *Concurrent Pos:* Proj scientist, Nimbus 5 & Nimbus 6 projs, Goddard Space Flight Ctr, 72-78; prog scientist, Spacelab 3 Prog, NASA Hq, 78-86 & Trop Rainfall Measuring Mission Prog, 88-; mem, Laser in-Space Tech Prog, 82-; chmn, Atmospheric Environ Tech Comn, Am Inst Aeronaut & Astronaut, 86-89; instrument scientist, EOS Prog, 88- *Honors & Awards:* Losey Medal, Am Inst Aeronaut & Astronaut, 86. *Mem:* Am Inst Aeronaut & Astronaut; Am Meteorol Soc; Am Geophys Union; Sigma Xi; AAAS. *Res:* Explored the meteorology of the mesosphere using sounding rockets; investigation of circulation, noctilucent clouds, gravity waves and turbulence; developed techniques to observe precipitation from space using passive microwave radiometry. *Mailing Add:* 6801 Lupine Lane McLean VA 22101

THEOPHANIDES, THEOPHILE, b Platamon-Kavala, Greece, Nov 12, 32; Can citizen; m 59; c 2. BIOCHEMISTRY, PHARMACY. *Educ:* Univ Bologna, BSc, 57; Univ Toronto, MA, 61, PhD(chem), 63. *Prof Exp:* Asst prof chem, Univ NB, 65-66; from asst prof to assoc prof, 66-76, PROF CHEM, UNIV MONTREAL, 77- *Concurrent Pos:* Nat Res Coun dir phys chem, Nat Cent Sci Res, Greece, 76-78. *Honors & Awards:* Silver Award, Nat Med Acad France, 84. *Mem:* Fel Can Chem Inst; Am Chem Soc; Sigma Xi. *Res:* Structure & properties of metal complexes; fourier transfer infrared & ATR spectroscopy of metal complexes with biological molecules such as nucleic acids. *Mailing Add:* CP 535 Morin Heights PQ J0R 1H0 Can

THEOPOLD, KLAUS HELLMUT, b Berlin, Ger, Apr 18, 54; m 82; c 3. MATERIALS SCIENCE. *Educ:* Univ Hamburg, vordiplom, 77; Univ Calif, Berkeley, PhD(chem), 82. *Prof Exp:* Assoc fel, Mass Inst Technol, 82-83; asst prof inorg chem, Cornell Univ, 83-90; ASSOC PROF, DEPT CHEM & BIOCHEM, UNIV DEL, 90- *Mem:* Am Chem Soc; Ger Chem Soc. *Res:* Structure and reactivity of organometallic and inorganic compounds; models for catalysis; semiconductor clusters. *Mailing Add:* Dept Chem & Biochem Univ Del Newark DE 19716

THÉRIAULT, GILLES P, b Montreal, Que. OCCUPATIONAL CANCER EPIDEMIOLOGY, OCCUPATIONAL MEDICINE. *Educ:* Col St-Joseph, Trois-Riviëres, BA, 62; Univ Laval, MD, 66; Harvard Univ, MIH, 71, DrPH, 73. *Prof Exp:* Teacher human biol, Centre des Hautes études collégiales de Trois-Rivi06res, 68-70; teaching fel health & environ, Harvard Univ, 72; course supvr occup health, Univ Laval, 73-77, from asst prof to assoc prof, 77-82; assoc prof occup health & epidemiol, 82-89, DIR, SCH OCCUP HEALTH, MCGILL UNIV, 83-, PROF OCCUP HEALTH & EPIDEMIOL, 89- *Concurrent Pos:* Consult specialist, Gen Hosp Montreal, 82- & Hosp Sacré-Coeur, Montreal, 88- *Res:* Occupational cancer epidemiology; epidemiology of occupational diseases; exposure to electromagnetic fields and cancer; poly aromatic hydrocarbons and cancer. *Mailing Add:* Sch Occup Health McGill Univ 1130 Pine Ave W Montreal PQ H3A 1A3 Can,

THERIOT, EDWARD DENNIS, JR, b Baton Rouge, La, Mar 19, 38; m 60; c 2. PHYSICS. *Educ:* Duke Univ, BS, 60; Yale Univ, MS, 61, PhD(physics), 67. *Prof Exp:* NATO vis scientist fel physics, Europ Orgn Nuclear Res, Geneva, Switz, 67-68; res fel, Los Alamos Sci Lab, 68-69; head neutrino dept,

76-78, dept head collider Detector, 81-88, PHYSICIST, FERMI NAT ACCELERATOR LAB, 69-, ASSOC DIR, 89- *Mem:* Am Phys Soc. *Res:* Elementary particle physics, particularly relating to weak and electromagnetic interactions; neutrino interactions; hyperon decays; particle production; positronium; proton-antiproton colliding beams. *Mailing Add:* Fermi Nat Accelerator Lab PO Box 500 Batavia IL 60510

THERIOT, KEVIN JUDE, b Houma, La, June 16, 58; m 91. POLYALPHAOLEFIN. *Educ:* NE La Univ, BSc, 80; La State Univ, PhD(chem), 89. *Prof Exp:* SR RES & DEVELOP CHEMIST, ETHYL CORP, 89- *Mem:* Am Chem Soc. *Res:* Polyalphaolefins; synthetic lubricants; organopalladium compounds; organoplatinum compounds. *Mailing Add:* 8000 GSRI Ave Baton Rouge LA 70820

THERIOT, LEROY JAMES, b Port Arthur, Tex, Apr 11, 35; m 58; c 3. INORGANIC CHEMISTRY. *Educ:* Southwestern La Univ, BS, 57; Tulane Univ, PhD(chem), 62. *Prof Exp:* Chemist, Ethyl Corp, 62-63; fel, Harvard Univ, 63-64 & Univ Tex, 64-65; asst prof, 65-70, assoc prof, 70-80, PROF CHEM, NTEX STATE UNIV, 80- *Mem:* Am Chem Soc. *Res:* Preparation and electronic structure of metal complexes. *Mailing Add:* Dept Chem NTex State Univ Denton TX 76203

THERN, ROYAL EDWARD, b Winona, Minn, May 11, 42; m 71; c 3. ACCELERATOR PHYSICS. *Educ:* St Olaf Col, BA, 64; Mass Inst Technol, PhD(physics), 72. *Prof Exp:* Res specialist, Univ Pa, 72-79; SR PHYSICS ASSOC, BROOKHAVEN NAT LAB, 79- *Mem:* Am Phys Soc. *Res:* Accelerator physics; beam lines; instrumentation. *Mailing Add:* Little Bay Rd Wading River NY 11792

THERRIAULT, DONALD G, biochemistry, for more information see previous edition

THERRIEN, CHESTER DALE, b Coos Bay, Ore, June 18, 36; m 83; c 4. MYCOLOGY, CYTOCHEMISTRY. *Educ:* St Ambrose Col, BA, 62; Univ Tex, PhD(bot), 66. *Prof Exp:* From asst prof to assoc prof biol, 65-87, ASSOC PROF BIOL & PLANT PATH, PA STATE UNIV, 87- *Mem:* Bot Soc Am; Genetics Soc Am. *Res:* Cytology and genetics of phytophthora infestans. *Mailing Add:* Dept Biol Pa State Univ University Park PA 16802

THET, LYN AUNG, BIOCHEMISTRY, PATHOLOGY. *Educ:* Inst Med, Burma, MD, 71. *Prof Exp:* Asst prof med, Sch Med, Duke Univ, 80-90; ASSOC PROF MED, UNIV WIS, 90- *Mailing Add:* Dept Med Univ Wis Med Ser Vet Admin Med Ctr Madison WI 53705

THEUER, PAUL JOHN, b Hoboken, NJ, Jan 9, 36; m 60; c 2. TECHNOLOGY TRANSFER-MANAGEMENT. *Educ:* St Peter's Col, BS, 57; Iowa State Univ, BS, 66; Pa State Univ, MEng, 73; Army War Col, dipl, 80. *Prof Exp:* Comput porgrammer, Nat Coun Compensation Ins, 57; reliability engr missiles, Douglas Aricraft Co, 57-58; comd & staff assignments, US Army, 58-68, comdr construct mgt, 808th Eng Battalion, 68-69, chief opers log mgt, US Army Support Command, Cam Ranh Bay, SVietnam, 69-70, commandant of cadets, Mil Sci, Pa State Univ, 70-73, div chief construct mgt, US Army Eng Command, Europe, 73-74, exec chief staff, Off Eng, HQ US Army, Europe, 74-76, rep hq staff HQ, US Army, Europe, 76-78, asst dir construct mgt, HQ US Army CEngrs, 78-83, comdr & dir, US Army Construct Eng Res Lab, 83-86; VPRES, STANLEY CONSULTS, INC, WASHINGTON, DC, 86- *Concurrent Pos:* Consult, HP Gauff Engrs, Inc, Frankfurt, WGer, 80-83. *Mem:* Am Soc Civil Engrs; Sigma Xi; Am Consult Engrs Coun; Soc Am Mil Engrs. *Res:* Construction engineering and facilities management; solid waste management and disposal; solid-liquid fuel boiler combustion and conversion; industrial sanitary-environmental facilities design; technology transfer; systems maintenance management. *Mailing Add:* Two Chester Mill Ct Silver Spring MD 20906

THEUER, RICHARD C, b Hoboken, NJ, June 15, 39; m 62; c 3. NUTRITION. *Educ:* St Peter's Col, NJ, BS, 60; Univ Wis, Madison, MS, 62, PhD(biochem), 65; Ind State Univ, MBA, 73. *Prof Exp:* Res asst biochem, Univ Wis, 60-65; sr scientist, Mead Johnson Res Ctr, 65-67, group leader, 67-68, sect leader, 68-70, dir dept nutrit res, 70-75; dir nutrit res, Int Div, Bristol-Myers Co, 76-80; asst vpres nutrit serv, Nestec, 80-83, vpres res & develop, 83-86, PRES BEECHNUT NUTRIT CORP, 86- *Mem:* AAAS; Am Chem Soc; Am Inst Nutrit; Sigma Xi. *Res:* Infant nutrition; research and development of nutritional specialty products. *Mailing Add:* Beechnut Nutrit Corp/Ralston Purina Checkerboard Sq St Louis MO 63164

THEUER, WILLIAM JOHN, b New York, NY, Nov 27, 35; m 61; c 3. ORGANIC CHEMISTRY. *Educ:* Queens Col, BS, 57; Univ Del, PhD(chem), 65. *Prof Exp:* Sr scientist, Sandoz Pharmaceut, 65-67; res chemist, Celanese Res Co, 67-68, sr res chemist, Celanese Fibers Co, 68-71, group leader, Celanese Fibers Mkt Co, 71-74, tech mgr, 74-76, prod mgr, 76-77, tech dir resins, 78, TECH DIR EMULSIONS, CELANESE POLYMER-SPECIALTIES CO, 79-, BUS MGR, BASF, 85- *Mem:* Am Chem Soc; Am Asn Textile Chemists & Colorists; Int Disposable & Nonwoven Asn; Sigma Xi. *Res:* Heterocyclic chemistry; photochemistry; cellulose and fiber chemistry; spinning research; pharmaceutical chemistry; industrial uses of fibers; textile polymer chemistry. *Mailing Add:* One Pine Knoll Dr Lake Wylie SC 29710-9245

THEURER, CLARK BRENT, b Logan, Utah, Oct 17, 34; m 56; c 4. ANIMAL NUTRITION. *Educ:* Utah State Univ, BS, 56; Iowa State Univ, MS, 60, PhD(animal nutrit), 62. *Prof Exp:* Asst prof animal sci, Va Polytech Inst, 62-64; assoc prof, 64-71, PROF ANIMAL SCI, UNIV ARIZ, 71-, ANIMAL SCIENTIST, AGR EXP STA, 74- *Concurrent Pos:* Dept head animal sci, 81-88. *Mem:* Am Soc Animal Sci (pres, pres-elect, secy-treas, 83-88); Am Dairy Sci Asn; Am Inst Nutrit; Soc Range Mgt; Sigma Xi. *Res:* regulation of voluntary feed intake in ruminants; Starch and protein metabolism in ruminants. *Mailing Add:* Dept Animal Sci Univ Ariz Tucson AZ 85721

THEURER, JESSOP CLAIR, b Logan, Utah, Sept 4, 28; m 53; c 4. PLANT GENETICS, PLANT BREEDING. *Educ:* Utah State Univ, BS, 53, MS, 57; Univ Minn, PhD(plant genetics), 62. *Prof Exp:* Res fel oats radiation, Univ Minn, 61-62; res agronomist, 62-63, GENETICIST, CROPS RES DIV, AGR RES SERV, USDA, 63- *Concurrent Pos:* Int farm youth exchange student, Lebanon & Syria, 52; assoc ed, Crop Sci Soc Am, 70-72. *Mem:* Am Soc Agron; Am Soc Sugar Beet Technol. *Res:* Agronomy; cytology; plant pathology; breeding, genetics, male sterility, and disease resistance of sugar beets. *Mailing Add:* Crop & Soil Sci Mich State Univ East Lansing MI 48824

THEUSCH, COLLEEN JOAN, b Milwaukee, Wis, Dec 18, 32. NUMBER THEORY, OPERATIONS RESEARCH. *Educ:* Col Racine, Wis, BEd, 61; Univ Detroit, MA, 66; Mich State Univ, PhD(math), 71. *Prof Exp:* Instr math, Col Racine, 70-71; mathematician, Res & Develop Dept, Richman Bros Co, 71-84, sr res assoc, 84-89; LASER CAM/SR APPLICATIONS ANALYST, AM GREETINGS, 89- *Concurrent Pos:* Instr, Cleveland State Univ & Cuyahoga Community Col, 71-72. *Mem:* Am Math Soc. *Res:* Development of grading and marker making system and numerically controlled cutting via laser cutters in men's clothing manufacturing; software for laser cutters; CAD systems. *Mailing Add:* 22962 Maple Ridge Rd N Olmsted OH 44070-1471

THEWS, ROBERT L(EROY), b Fairmont, Minn, June 13, 39; c 2. THEORETICAL PHYSICS. *Educ:* Mass Inst Technol, SB, 62, PhD(physics), 66. *Prof Exp:* Physicist, Lawrence Berkeley Lab, Univ Calif, 66-68; res assoc & asst prof physics, Univ Rochester, 68-70; asst prof, 70-75, assoc prof, 75-80, PROF PHYSICS, UNIV ARIZ, 80- *Mem:* Am Phys Soc. *Res:* Theoretical high energy elementary particle physics. *Mailing Add:* Dept Physics Univ Ariz Tucson AZ 85721

THIBAULT, ROGER EDWARD, b Salem, Mass, June 28, 47; m 70; c 1. ECOLOGY. *Educ:* Univ Wis, BS, 69; Univ Conn, PhD(ecol), 74. *Prof Exp:* NDEA fel ecol, Univ Conn, 71-74; instr zool, Iowa State Univ, 74-75; ASST PROF BIOL, BOWLING GREEN STATE UNIV, 75- *Mem:* Am Soc Ichthyol & Herpet; Ecol Soc Am; Soc Study Evolution; Sigma Xi; AAAS. *Res:* Aquatic ecology; aquatic entomology and ichthyology; evolution of unisexual vertebrates. *Mailing Add:* Dept Biol Sci Bowling Green State Univ Bowling Green OH 43403

THIBAULT, THOMAS DELOR, b Claremont, NH, Aug 14, 42; m 69; c 4. ORGANIC CHEMISTRY. *Educ:* Providence Col, BS, 64; Mass Inst Technol, PhD(org chem), 69. *Prof Exp:* Sr org chemist, 69-74, RES SCIENTIST, ELI LILLY & CO, 74- *Mem:* Am Chem Soc. *Res:* Synthesis of biologically active structures; reaction mechanisms; new synthetic reactions. *Mailing Add:* DowElanco Box 708 Greenfield IN 46140

THIBEAULT, JACK CLAUDE, b Lowell, Mass, June 23, 46; div; c 1. COLLOID CHEMISTRY, COATINGS TECHNOLOGY. *Educ:* Lowell Technol Inst, BS, 67; Calif Inst Technol, PhD(inorg chem), 72. *Prof Exp:* Fel theoret chem, Cornell Univ, 72-74; RES CHEMIST, ROHM & HAAS CO, 74- *Mem:* Am Chem Soc. *Res:* Physical chemistry of polymer colloids and solutions. *Mailing Add:* Res Div Rohm & Haas Co 727 Norristown Rd Spring House PA 19477

THIBERT, ROGER JOSEPH, B Tecumseh, Ont, Aug 29, 29; m 54; c 2. CHEMISTRY. *Educ:* Univ Western Ont, BA, 51; Univ Detroit, MS, 54; Wayne State Univ, PhD(biochem), 58. *Prof Exp:* Lectr chem, Univ Windsor, 53-56, from asst prof to assoc prof, 57-67, assoc dean arts & sci, 64-70, PROF CHEM & BIOCHEM, UNIV WINDSOR, 67-, DIR CLIN CHEM, 72-; DIV HEAD, CLIN CHEM LAB, DETROIT RECEIVING HOSP, UNIV HEALTH CTR, 71- *Concurrent Pos:* Instr sch nursing, Grace Hosp, 54-73; res assoc sch med, Wayne State Univ, 71-72, prof path, 72- *Honors & Awards:* Union Carbide Award, Chem Inst Can, 78; SmithKline Clin Labs Award, 80; Ames Award, Can Soc Clin Chemists, 88. *Mem:* Fel AAAS; fel Nat Acad Clin Biochem; fel Chem Inst Can; fel Can Acad Clin Biochem; Am Asn Clin Chem; Can Soc Clin Chemists; Am Soc Biochem & Molecular Biol; Am Chem Soc; Can Biochem Soc; Sigma Xi. *Res:* Clinical biochemistry; development of new reagents or methods for the measurement of bilirubin, phospholipids, pyrophosphate, adenosine triphosphate, trace metals, enzymes, hydrogen peroxide and glycated proteins. *Mailing Add:* Dept Chem & Biochem Univ Windsor Windsor ON N9B 3P4 Can

THIBODEAU, GARY A, b Sioux City, Iowa, Sept 26, 38; m 64; c 2. PHYSIOLOGY, PHARMACOLOGY. *Educ:* Creighton Univ, BS, 62; SDak State Univ, MS, 67, MS, 70, PhD(physiol), 71. *Prof Exp:* Mem prof serv staff, Baxter Labs, Inc, 63-65; from asst prof to prof entom-zool, 65-80, vpres, 80-85, PROF BIOL, SDAK STATE UNIV, 80-; CHANCELLOR, UNIV WIS, RIVER FALLS, 85- *Mem:* AAAS; Am Inst Biol Sci; Am Pub Health Asn. *Res:* Animal physiology; pharmacology of hypolipedemic agents; pathological dyslipemias; thyroid physiology; vascular morphology. *Mailing Add:* 439 River Hills Rd N River Falls WI 54022-2936

THIBODEAUX, LOUIS J, b Church Point, La, Nov 14, 39; m 59; c 2. ENGINEERING. *Educ:* La State Univ, BS, 62, MS, 66, PhD(ionic diffusion), 68. *Prof Exp:* Engr, E I du Pont de Nemours & Co, 62-64, Uniroyal, Inc, 64 & Nat Coun Air Stream Improv, 64-68; from asst prof to prof chem eng, Univ Ark, Fayetteville, 68-84; PROF CHEM ENG, LA STATE UNIV, BATON ROUGE, 84- *Concurrent Pos:* Consult to numerous chem companies; consult, Ga Kraft Co, 69 & Int Paper Co, 70- *Mem:* AAAS; Am Inst Chem Engrs; Am Chem Soc; Sigma Xi (pres, 77-78). *Res:* Movement of chemicals in the environment, environmental chemistry and interphase mass transfer; fate, life-time, transport rates, direction of movement, chemodynamics of trace chemicals in the natural environment. *Mailing Add:* 3449 Tezcucco Dr Baton Rouge LA 70808

THICH, JOHN ADONG, b London, Eng, Nov 16, 48; US citizen; m 74. AEROSPACE LUBRICANTS. *Educ:* Temple Univ, Pa, BA, 70; Rutgers Univ, NJ, PhD(inorg chem), 75. *Prof Exp:* Res fel chem, Calif Inst Technol, 75-77; scientist, Rohm and Haas Co, Pa, 77-79; LAB DIR, ROYAL LUBRICANTS CO, INC, NJ, 79- *Mem:* Am Chem Soc; Am Soc Lubrication Engrs; Am Soc Testing & Mat; Soc Automotive Engrs. *Res:* Development of high performance fluids and lubricants for the aerospace industry and the United States military. *Mailing Add:* Royal Lubricant Co Inc PO Box 518 River Rd East Hanover NJ 07936

THICKSTUN, WILLIAM RUSSELL, JR, b Washington, DC, Oct 14, 22; m 54. MATHEMATICS. *Educ:* Univ Md, BSc, 47, MA, 49, PhD, 52. *Prof Exp:* Mathematician, US Naval Ord Lab, Md, 53-68; assoc prof math, Clarkson Col Technol, 68-76 & Univ Petrol & Mineral, Saudi Arabia, 76-78; ASSOC PROF MATH, ST LAWRENCE UNIV, 76- *Mem:* Am Math Soc; Soc Indust & Appl Math; Am Inst Aeronaut & Astronaut. *Res:* Applied mathematics; fluid dynamics. *Mailing Add:* 143 Henlopen Ave Rehoboth Beach DE 19971

THIEBAUX, HELEN JEAN, b Washington, DC, Aug 17, 35; m 62; c 5. APPLIED STATISTICS. *Educ:* Reed Col, BA, 57; Univ Ore, MA, 60; Stanford Univ, PhD(statist), 64. *Prof Exp:* Res asst econ & indust, Ivan Block & Assoc, Econ & Indust Consult, 58; statist analyst, Med Sch, Univ Ore, 60; asst pub health analyst, Calif Dept Pub Health, 61; res assoc, Hanson Physics Labs, Stanford Univ, 63-64; asst prof statist, Univ Conn, 64-65; asst prof, Univ Mass, Amherst, 65-66 & 67-71; lectr, Univ Colo, Boulder, 72-73; consult & vis scientist, Nat Ctr Atmospheric Res, 72-74; from assoc prof to prof, Dalhousie Univ, 75-87; RES MATH STATIST, NAT OCEANIC & ATMOSPHERIC ADMIN/NAT WEATHER SERV/NAT METEOROL CTR, 88- *Concurrent Pos:* NSF vis prof, Pa State Univ, 90-91. *Mem:* Fel Royal Meteorol Soc; Am Statist Asn. *Res:* Statistical modelling and estimation for spatially coherent systems; applications to meteorology and oceanography. *Mailing Add:* 9708 Old Allentown Rd Ft Washington MD 20744

THIEBERGER, PETER, b Vienna, Austria, Sept 19, 35; m 63; c 1. EXPERIMENTAL NUCLEAR PHYSICS. *Educ:* Balseiro Inst Physics, Argentina, MS, 59; Univ Stockholm, Fil lic, 61, Fil Dr(physics), 62. *Prof Exp:* Physicist, Bariloche Atomic Ctr, Argentine AEC, 62-65; res asst, 65-67, from asst physicist to physicist, 67-74, head Tandem Van De Graaff Facil Opers, 71-75, GROUP LEADER, TANDEM VAN DE GRAAFF FACIL OPERS & DEVELOP GROUP, 75-, HEAD, TANDEM VAN DE GRAAFF RES, 85-, SR PHYSICIST, BROOKHAVEN NAT LAB, 74- *Concurrent Pos:* Asst prof, Balseiro Inst Physics, 62-63, prof, 63-65; vis scientist, Res Inst Physics, Stockholm, Sweden, 68-69; consult, Tennelec Instrument Co, 67-71, Arg AEC, 77- & Techint SACI, Arg, 78-80; NSF grant for coop res with Latin Am, 76-85; dep mgr, heavy ion transfer line proj prod relativistic heavy ions, 84-; Yale Univ A W Wright Nuclear Struct Lab, 85-87. *Mem:* Fel Am Phys Soc. *Res:* Nuclear structure; measurements of half-lives and g-factors of nuclear states; development of nuclear instruments and methods; nuclear reactions with heavy ions; molecular ion structure measurements; accelerator development and operation; applied uses of accelerators and nuclear instrumentation. *Mailing Add:* Physics Dept Brookhaven Nat Lab Upton NY 11973

THIEDE, EDWIN CARL, b Richland Co, Wis, Nov 11, 37; m 64. MMW-IR SENSORS, SIGNAL PROCESSING. *Educ:* Univ Wis-Madison, BS, 59; Univ Calif, Los Angeles, MS, 64; Stanford Univ, PhD(elec eng), 68. *Prof Exp:* Dynamics engr, Gen Dynamics/Astronaut, 59-62; tech staff engr, Hughes Aircraft Co, 63-65; res asst radar astron, Stanford Electronics Lab, 65-68; asst prof elec eng, Univ Minn, Minneapolis, 68-73; sr prin scientist, Honeywell-SRC, 73-84, sr staff engr, Honeywell-PWO, Minneapolis, 84-89; SR ENGR FEL, ALLIANT TECHSYSTEMS, 90- *Concurrent Pos:* Consult, Univac Defense Systs Div, 72-74; instr, Hennepin County Vo-Tech, 74-80; lectr, Technol Training Corp, 89- *Mem:* Inst Elec & Electronics Engrs. *Res:* Multisensor fusion, target detection and classification, infrared and radar signal processing in arionics and guided missiles, digital filtering, optimal filtering, statistical communication theory. *Mailing Add:* Alliant Techsystems MN 38-2100 10400 Yellow Circle Dr Minnetonka MN 55343

THIEDE, HENRY A, b Rochester, NY, Oct 2, 26; m 51; c 2. OBSTETRICS & GYNECOLOGY. *Educ:* Univ Buffalo, MD, 49. *Prof Exp:* Intern surg, Buffalo Gen Hosp, 49-50; asst resident obstet & gynec, Genesee & Strong Mem Hosps, Rochester, 52-54; resident, Genesee Hosp, 54-56; from instr to assoc prof, Sch Med, Univ Rochester, 57-66; prof obstet & gynec & chmn dept, Sch Med, Univ Miss, 67-77, asst dean, Sch Med, 70-73, assoc dean, 73-77; PROF & CHMN DEPT OBSTET & GYNEC, OBSTETRICIAN & GYNECOLOGIST-IN-CHIEF, UNIV ROCHESTER MED SCH, 77- *Mem:* Am Col Obstet & Gynec; Soc Gynec Invest; Am Gynec & Obstet Soc; Am Gynec Club; Am Fertil Soc. *Res:* Biology of reproduction and reproduction wastage. *Mailing Add:* 601 Elmwood Ave Rochester NY 14642

THIEL, FRANK L(OUIS), b Buffalo, NY, July 22, 42; m 63; c 2. FIBER OPTICS, OPTICAL WAVEGUIDES. *Educ:* Rensselaer Polytech Inst, BEE, 64, MEE, 65, PhD(electrophys), 69. *Prof Exp:* Instr elec eng, Rensselaer Polytech Inst, 65-69; sr physicist, Corning Glass Works, 69-71, sr res physicist, 71-75, mgr, 75-80; chief engr & tech mgr optical fibers, Deeside, Clwyd, Wales, UK, 80-84; mgr, Corning Glass Works, 84-87; CONSULT, 87- *Concurrent Pos:* Lectr optical waveguide courses, Univ Colo, 74-76 & George Washington Univ, 74-75; mem working group, Brit Standards Inst, 83-84. *Mem:* Inst Elec & Electronics Engrs; Am Phys Soc. *Res:* Electronic materials; physical electronics; opto-electronics; optical waveguides and associated components and processes. *Mailing Add:* Corning Glass 12117 Ohbuchi Osuka-Cho Ogasa-Gun Shizuoka 437-13 Japan

THIEL, PATRICIA ANN, b Adrian, Minn, Feb 20, 53. SURFACE CHEMISTRY. *Educ:* MacAlester Col, BA, 75; Calif Inst Technol, PhD(chem), 81. *Prof Exp:* Assoc scientist, Control Data Corp, 75-76; scientist, Sandia Nat Lab, Livermore, 81-83; from asst prof to assoc prof, 83-91, PROF CHEM, AMES LAB, IOWA STATE UNIV, 91- *Concurrent Pos:* Sloan award, Alfred P Sloan Found, 84; NSF presidential young investr, 85; Dreyfus teacher scholar, Camille & Henry Dreyfus Found, 86; prog dir mat chem, Ames Lab, 88-; NSF fac awardee women in sci & eng, 91. *Mem:* Am Vacuum Soc; Am Chem Soc; Mat Res Soc; AAAS. *Res:* Surface chemistry; gas-solid interactions; metal film growth; metal oxidation; surface reactions of fluorocarbons; electrocatalysis at fuel cells; water-solid interactions; surface polymerization. *Mailing Add:* Ames Lab Iowa State Univ Ames IA 50011

THIEL, THOMAS J, b Upper Sandusky, Ohio, Dec 31, 28; m 55, 84; c 4. APPLIED COMPUTER SCIENCE, SOIL PHYSICS. *Educ:* Ohio State Univ, BS, 56, MS, 59. *Prof Exp:* Soil scientist, 57-63, soil scientist & asst adminr, 63-72, prog analyst, prog planning & rev, 72-84, PROJ MGR, ADVAN SYSTS DEVELOP, USDA & DEPT DEFENSE, 84- *Concurrent Pos:* Staff mem, Ohio State Univ, 60-63, Univ Minn, 63-72; consult, laser-optic systs; coun rep, Soil Conserv Soc, 81-84. *Mem:* Am Soc Agron; Soil Sci Soc Am; Soil Conserv Soc; Asn Comput Mach; Assoc Asn Info & Image Mgt; Soc Appl Learning Technol; fel Soil Conserv Soc. *Res:* Analog, numerical, model and field studies on the drainage of agricultural lands, especially lake-bed, sloping fragipan and peat soils; computer and systems analysis and information retrieval; program and administrative management especially in the area of soil and water conservation; computer systems analysis and development; laser-optic information systems applications; compact disk-read only memory, optical data disk and videodisc systems; laser-optic memories. *Mailing Add:* 4541 Saucon Valley Ct Alexandria VA 22312

THIELE, ELIZABETH HENRIETTE, b Portland, Ore, Apr 26, 20. BIOCHEMISTRY. *Educ:* Univ Wash, BS, 42; Univ Pa, MS, 43, PhD(bact), 51. *Prof Exp:* Technician, Merck Sharp & Dohme, 43-44, res assoc, Merck Inst Therapeut Res, 44-78; RETIRED. *Mem:* Am Soc Microbiol; Reticuloendothelial Soc. *Res:* Enzymology; immunology. *Mailing Add:* 17389 Plaza Dolores San Diego CA 92128

THIELE, ERNEST, b Chicago, Ill, Dec 8, 1895. CHEMICAL ENGINEERING. *Educ:* Loyola Univ, Chicago, AB, 16; Univ Ill, Urbana, BS, 19; Mass Inst Technol, MS, 23, ScD, 25. *Hon Degrees:* LLD, Univ Notre Dame, 71. *Prof Exp:* Vis prof chem eng, Univ Notre Dame, 60-70; RETIRED. *Mem:* Nat Acad Eng; Sigma Xi; Am Chem Soc; Am Inst Chem Engrs. *Mailing Add:* 1555 Oak Ave Evanston IL 60201

THIELE, GARY ALLEN, b Cleveland, Ohio, May 5, 38; m 60; c 3. ELECTRICAL ENGINEERING, ELECTROMAGNETISM. *Educ:* Purdue Univ, BSEE, 60; Ohio State Univ, MSc, 64, PhD(elec eng), 68. *Prof Exp:* Elec engr, Gen Elec Co, 60-61; res assoc, Ohio State Univ, 63-68, from asst prof to assoc prof elec eng, 68-80; ASSOC DEAN ENG GRAD STUDIES, UNIV DAYTON, 80- *Concurrent Pos:* Consult, Lockheed Missiles & Space Co, 71, Jet Propulsion Lab, 80-83 & Army Sci Bd, 89-91. *Honors & Awards:* NATO Res Award, 77. *Mem:* Int Union Radio Sci; fel Inst Elec & Electronics Engrs; Sigma Xi; Am Inst Aeronaut & Astronaut. *Res:* Electromagnetics; radiation and scattering problems via numerical methods; microwave techniques; propagation of electromagnetic waves; antennas; radar cross section. *Mailing Add:* Univ Dayton 262 KL 300 College Park Ave Dayton OH 45469-0227

THIELEN, LAWRENCE EUGENE, b Chicago, Ill, Sept 2, 21; m 50; c 2. ORGANIC CHEMISTRY. *Educ:* Loyola Univ, Ill, BS, 42. *Prof Exp:* Chemist, Kankakee Ord Works, E I du Pont de Nemours & Co, 42; res chemist, Pure Oil Co, 43 & 46, G D Searle & Co, 46-56, Nalco Chem Co, 56-58 & Abbott Labs, 58-62; group leader polymer & org chem, R R Donnelley & Sons Co, 62-76; safety & health dir, Printing Ink Technol, Inmont Corp, Chicago, 76-87; CONSULT, THIELEN INC, 87- *Concurrent Pos:* mem bd dirs, Chicago Sect, Am Chem Soc. *Mem:* Am Chem Soc. *Res:* Pharmaceuticals; synthesis of plastics and pharmaceuticals; inks; coatings; printing process technology. *Mailing Add:* 110 E Madison St Villa Park IL 60181

THIELGES, BART A, b Chicago, Ill, June 16, 38; m 60; c 3. FOREST GENETICS. *Educ:* Southern Ill Univ, BS, 63; Yale Univ, MF, 64, MPhil, 67, PhD(forest genetics), 67. *Prof Exp:* Res asst plant anat, Southern Ill Univ, 62-63; res asst forest genetics, Yale Univ, 63-67; asst prof, Ohio Agr Res & Develop Ctr, 67-71; assoc prof, Sch Forestry & Wildlife Mgt, La State Univ, Baton Rouge, 71-76; proj leader, US Forest Serv, Starkville, Miss, 76-77; prof & chmn dept forestry, Univ Ky, 77-90; ASSOC DEAN RES, COL FORESTRY, ORE STATE UNIV, 90- *Concurrent Pos:* Vis prof, Univ Oxford, Eng, 83-84. *Mem:* Soc Am Foresters; Forest Prod Res Soc; AAAS. *Res:* Breeding of forest trees; natural variation studies; genetics of disease resistance; management of research. *Mailing Add:* Sch Forestry Ore State Univ Corvallis OR 97331-5704

THIELMANN, VERNON JAMES, b Hastings, Minn, June 4, 37; m 63; c 2. ANALYTICAL CHEMISTRY. *Educ:* Northern State Col, BS, 63; Univ SDak, MNS, 68; Baylor Univ, PhD(chem), 74. *Prof Exp:* Teacher pub schs, SDak & Iowa, 63-69; asst prof chem, Morningside Col, 69-72; from asst prof to assoc prof, 74-82, head dept, 84-90, PROF CHEM, SOUTHWEST MO STATE UNIV, 82- *Mem:* Am Chem Soc; Sigma Xi. *Res:* Preparation of complex cation exchanged montmorillonite clays and subsequent study of their structure, thermal stability, surface area and effectiveness for use as gas chromatographic packing materials; clay and clay minerals. *Mailing Add:* Dept Chem Southwest Mo State Univ Springfield MO 65804-0089

THIELSCH, HELMUT, b Berlin, Nov 16, 22; nat US; m 52; c 4. MECHANICAL ENGINEERING. *Educ:* Auburn Univ, BS, 43. *Prof Exp:* Res engr, Allis Chalmers Co, Milwaukee, 45-46; metall engr, Black, Sivalls & Bryson, Inc, Kansas City, Mo, 46-47; res engr, Lukens Steel Co, Coatsville, Pa, 48-49; engr, Welding Res Coun, NY, 49-52; dir res, Eutectic Welding Alloys Co, NY, 52-53; vpres & dir res, develop & eng, ITT Grinnell Corp, Providence, 54-84; PRES, THIELSCH ENG ASSOCS, INC, PROVIDENCE, 84- *Concurrent Pos:* Consult on failure anal to indust, pub utilities, equip builders, 54-; lectr, confs on failures & failure prev, var univs; mem, Component Tech Comt, Argonne Nat Lab, Ill; mem, Comt Am Soc Testing & Mat, E-7 Comt Nondestructive Testing & Tech Coun; mem, Fabrication, Inspection & Testing Comt, Am Soc Mech Engrs; chmn, Metall Eng Comt, Pipe Fabrication Inst; consult, Adv Comt, Reactor Safeguards Nuclear Regulatory Comn. *Honors & Awards:* Comfort A Adams Lectr Award, Am Welding Soc, 82. *Mem:* Fel Am Soc Mech Engrs; Am Soc Testing & Mat; fel Am Soc Nondestructive Testing; Nat Asn Corrosion Engrs; Tech Asn Pulp & Paper Indust; fel Am Soc Metals; Am Welding Soc; Am Chem Soc; Am Nuclear Soc; Am Soc Qual Control. *Res:* Failure analysis; materials evaluation; quality assurance; author of various publications; granted one patent. *Mailing Add:* 195 Frances Ave Cranston RI 02910

THIEME, CORNELIS LEO HANS, b Arnhem, Neth, Dec 7, 48. SUPERCONDUCTORS, COMPOSITE MATERIALS. *Educ:* Twente Univ Technol, Neth, BEngSc, 73, MEngSc, 75. *Prof Exp:* Res staff, Twente Univ Technol, Neth, 75-80; RES STAFF MAT SCI, FRANCES BITTER MAGNET LAB & PLASMA FUSION CTR, MASS INST TECHNOL, 81- *Mem:* Netherlands Ceramic Soc. *Res:* Development of multifilamentary ductile superconductor materials. *Mailing Add:* NW 14 3115 Mass Inst Technol Cambridge MA 02139

THIEME, MELVIN T, b Decatur, Ind, Oct 27, 25; m 49; c 3. SHOCK HYDRODYNAMICS, COMPUTER SCIENCE. *Educ:* Purdue Univ, BS, 49, MS, 51, PhD(nuclear physics), 55. *Prof Exp:* Staff mem, Lawrence Livermore Lab, 63-64; staff mem, Los Alamos Nat Lab, 55-63, 64-72 & 76-86, asst group leader, 72-76; RETIRED. *Concurrent Pos:* Teacher, Pima Community Col, Tucson, Az & Northern NMex Community Col, Esponola, NMex. *Mem:* Am Phys Soc. *Res:* Beta-ray spectroscopy; nuclear weapon design and development with specialization in shock hydrodynamics. *Mailing Add:* 35 Timber Ridge Rd Los Alamos NM 87544-2317

THIEN, LEONARD B, b Breese, Ill, Nov 11, 38; m 67; c 2. EVOLUTION, SYSTEMATICS. *Educ:* Southern Ill Univ, BS, 59; Washington Univ, MS, 61; Univ Calif, Los Angeles, PhD(bot), 68. *Prof Exp:* Asst prof cytol & systs, Dept Bot, Univ Wis-Madison, 67-71; from asst prof to assoc prof ecol, 71-81, PROF BIOL, TULANE UNIV, 81- *Concurrent Pos:* Hormel & NSF fel, Univ Minn, 75. *Mem:* Bot Soc Am; AAAS; Int Asn Plant Taxonomists; Ecol Soc Am. *Res:* Pollination biology of primitive angiosperms; cytology and evolution of gayophytum; productivity of marshes. *Mailing Add:* Dept Biol Tulane Univ New Orleans LA 70118

THIEN, STEPHEN JOHN, b Clarence, Iowa, Apr 11, 44; m 66; c 2. AGRONOMY. *Educ:* Iowa State Univ, BS, 66; Purdue Univ, Lafayette, MS, 68, PhD(agron), 71. *Prof Exp:* Asst agron, Purdue Univ, 66-70; from asst prof to assoc prof, 70-81, PROF AGRON, KANS STATE UNIV, 81- *Concurrent Pos:* Danforth assoc. *Mem:* Fel Am Soc Agron; Soil Sci Soc Am. *Res:* Soil fertility; plant nutrition and physiology; soil management; soil biochemistry. *Mailing Add:* Dept Agron Throck Morton Hall Kans State Univ Manhattan KS 66506

THIENE, PAUL G(EORGE), b Pasadena, Calif, Dec 10, 19; m 46; c 3. PHYSICAL ELECTRONICS. *Educ:* Calif Inst Technol, BS, 43, PhD(phys electronics), 52. *Prof Exp:* Asst physicist, Calif Inst Technol, 43-45, res engr, Jet Propulsion Lab, 46-47; dep chief & res adminr, Western Div, Off Sci Res, USAF, 52-55, consult, Air Tech Intel Ctr, 55-57; sr physicist, Res Lab, Giannini Plasmadyne Corp, 58-62; sr staff scientist, MHD Res Inc, 62-66; supvr, Optical Physics Sect, Res Lab, Aeronutronic Div, Philco-Ford Corp, 66-72; CONSULT, 72- *Concurrent Pos:* Consult, Litech, Inc, P W Webster Co, Surg Mech Res, Inc & Curt Deckert Assocs, 73-88. *Mem:* Am Phys Soc. *Res:* Electromedical technology, laser technology; semiconductor devices; superconductivity; plasma physics; magneto gas dynamics; electro-optics; microwave spectroscopy and radiometry. *Mailing Add:* 24833 Outlook Ct Carmel CA 93923

THIER, SAMUEL OSIAH, b Brooklyn, NY, June 23, 37; m 58; c 3. HEALTH SCIENCE POLICY. *Educ:* State Univ NY, Syracuse, MD, 60; Am Bd Internal Med, dipl, 67, recert, 74. *Hon Degrees:* MA, Univ Pa, 71 & Yale Univ, 75; DSc, State Univ NY, 87, Tufts Univ & George Washington Univ, 88, City Univ NY & Hahnemann Univ, 89; LHD, Rush Univ, 89. *Prof Exp:* Instr med, Harvard Med Sch, 67, assoc, 67-68, asst prof, 69; from assoc prof to prof, Sch Med, Univ Pa, 69-74; prof internal med & chmn dept, Sch Med, Yale Univ, 75-85, chief med, Yale-New Haven Hosp, 75-85; pres, Inst Med, Nat Acad Sci, 86-91; PRES, BRANDEIS UNIV, 91- *Concurrent Pos:* Asst med & chief, Renal Unit, Mass Gen Hosp, 67-69; assoc dir med serv, Univ Pa Hosp, 69-74; vchmn med, Sch Med, Univ Pa, 71-74; mem bd dirs, Hospice Inc, 76-82 & Yale-New Haven Hosp, 78-85; vis prof & lectr, numerous univs, 76-90; mem, Comt Study Airline Pilot Age, Health & Performance, Inst Med, Nat Acad Sci, 82, chmn, Bd Health Sci Policy, 83, mem, Comt Study Orgn Struct NIH, 85; chmn, Am Bd Internal Med, 84. *Honors & Awards:* Helen & Payne Whitney Lectr, Cornell Univ, 76; E Stanley Emery Jr Mem Lectr, Brigham & Women's Hosp, 81; Gunnar Lectr, Univ Ill, 81; Blankenhorn Lectr, Univ Cincinnati, 85; Michael & Irene Karl Master Sci Ser Lectr, Wash Univ, 86; John E Franklin Lectr, Cornell Univ, 87; Moshe Prywes Lectr, Ben-Gurion Univ, 87; Heller Lectr, Mt Sinai Hosp, 87; Martin E Rehfuss Lectr, Thomas Jefferson Univ, 87; Thomas M Durant Mem Lectr, Temple Univ, 87; Donald P Shiley Lectr, Scripps Clin & Res Found, 88; Jacob Yules Lectr, Tufts Univ, 88; Ralph Major Lectr, Univ Kans, 88; Florence Mahoney Lectr, NIH, 88; Ernest J & Elena B Bruno Mem Lectr, Lenox Hill Hosp, 88. *Mem:* Inst Med-Nat Acad Sci; fel Am Col Physicians; Am Fedn Clin Res (pres, 76); Asn Am Med Cols; Am Physiol Soc; Am Soc Nephrology. *Res:* Amino acid transport in the kidney. *Mailing Add:* 415 South St Waltham MA 02554-9110

THIERET, JOHN WILLIAM, b Chicago, Ill, Aug 1, 26; m 50; c 5. BOTANY. *Educ:* Utah State Univ, BS, 50, MS, 51; Univ Chicago, PhD(bot), 53. *Prof Exp:* Cur econ bot, Field Mus Natural Hist, 53-62; from assoc prof to prof biol, Univ Southwestern La, 62-73, Edwin Lewis Stephens prof sci, 72-73; chmn, Dept Biol Sci, 73-80, PROF BOT, NORTHERN KY UNIV, 73- *Concurrent Pos:* Adv, Encycl Britannica, 59-; ed, Econ Bot, 87-90. *Mem:* Soc Econ Bot. *Res:* Flora of central and southeastern United States; taxonomy of Scrophulariaceae and Gramineae; seed and fruit classification and morphology; economic botany. *Mailing Add:* Dept Biol Sci Northern Ky Univ Highland Heights KY 41076-1448

THIERMANN, ALEJANDRO BORIES, b Valparaiso, Chile, July 15, 47; US citizen; m 73; c 2. VETERINARY MEDICINE, EPIDEMIOLOGY. *Educ:* Univ Chile, Santiago, DVM, 71; Wayne State Univ, PhD(microbiol, immunol), 79. *Prof Exp:* Vet small animals, pvt pract, 71-72; med ecologist pub health, Lockheed Elec Co, NASA, Houston, 72-73; instr, Lab Animal Med, Sch Med, Wayne State Univ, 73-79; res leader leptospirosis & mycobacteriosis, Nat Animal Dis Ctr, 79-87; NAT PROG LEADER ANIMAL HEALTH, USDA ARS, 87- *Concurrent Pos:* Mem, Am Leptospirosis Res Conf, 74-, chmn epidemiol sect, 77-, pres, 84, secy-treas, 85-; mem biohazards comt, Wayne State Univ, mem recombinant DNA subcomt, 77-79; res leader, Leptospirosis & Mycobacteriosis Labs & Nat Leptospirosis Reference Ctr, 85-87. *Mem:* Am Soc Microbiologists; Sigma Xi; Am Vet Med Asn; US Animal Health Asn; Am Asn Vet Lab Diagnosticians; Wildlife Dis Asn. *Res:* Leptospirosis, its epidemiology, pathogenesis and molecular biology; paratuberculosis, its pathogenesis and molecular biology. *Mailing Add:* Nat Animal Dis Lab PO Box 70 Ames IA 50010

THIERRIN, GABRIEL, b Surpierre, Switz, Dec 22, 21; m 51; c 2. MATHEMATICS, COMPUTER SCIENCE. *Educ:* Univ Fribourg, DSc, 51; Univ Paris, DSc, 54. *Prof Exp:* Sci assoc, Nat Ctr Sci Res, France, 52-54 & NFS Res, Switz, 54-57; prof math, Inst Higher Studies, Tunisia, 57-58 & Univ Montreal, 58-70; PROF MATH, UNIV WESTERN ONT, 70- *Mem:* Asn Comput Mach; Am Math Soc; Can Math Soc. *Res:* Algebra; theories of rings and semi-groups; systems theory; theory of automata and languages. *Mailing Add:* Dept Math Univ Western Ont London ON N6A 5B7 Can

THIERS, EUGENE ANDRES, b Santiago, Chile, Aug 25, 41; US citizen; m 83; c 3. PRICE FORMATION MECHANISMS, ROBUSTNESS ANALYSIS. *Educ:* Univ Chile, Santiago, BS, 59; Columbia Univ, MS, 65, DESc(mineral eng), 70. *Prof Exp:* Mgr, Minbanco Corp, 66-70; sr engr, Battelle Mem Inst, 70-72, dir, 72-75; dir, Metals Ctr, 79-83, SR CONSULT, SRI INT, 75- *Concurrent Pos:* Consult prof, Stanford Univ, 82-; mgr inorganics, SRI Int. *Mem:* Fel AAAS; Am Inst Mining Metall & Petrol Engrs. *Res:* Mineral economics, particularly price formation mechanisms in industrial minerals and metals; competitive analysis of resource industries. *Mailing Add:* 426 27th Ave San Mateo CA 94403

THIERS, HARRY DELBERT, b Ft McKavett, Tex, Jan 22, 19; m 53; c 1. MYCOLOGY. *Educ:* Schreiner Inst, AB, 38; Univ Tex, BA, 41, MA, 47; Univ Mich, PhD, 55. *Prof Exp:* Asst, Univ Tex, 39-41, tutor bot, 45-47, instr bot & mycol, Tex A&M Univ, 47-50, asst biol, 50-55, assoc prof, 55-59; assoc prof, 59-63, PROF BIOL, SAN FRANCISCO STATE UNIV, 63- *Concurrent Pos:* Assoc plant pathologist, USDA, 48. *Mem:* Bot Soc Am; Mycol Soc Am; Am Bryol & Lichenological Soc. *Res:* Taxonomy of fleshy fungi of California and West Coast of North America. *Mailing Add:* Dept Biol San Francisco State Univ 1600 Holloway Ave San Francisco CA 94132

THIERSTEIN, GERALD E, b Whitewater, Kans, Apr 30, 31; m 76. CROP PROTECTION, CONSERVATION TILLAGE. *Educ:* Kans State Univ, BS, 57, MS, 63. *Prof Exp:* Asst prof agr eng, WVa Univ, 62-67; res assoc, McGill Univ, 67-71; sr lectr, Makerere Univ, Kampala, 71-73; vis assoc prof, Agr Mechanization, Fac Agr Sci, Am Univ-Beirut, 73-76; prin scientist agr eng, Int Crops Res Inst Semi-Arid Tropics, India, 76-84; ASSOC PROF & RES ENGR, DEPT AGR ENG, KANS STATE UNIV, MANHATTAN, KS, 84- *Concurrent Pos:* Adj prof, Agr Eng Dept, Univ Fla, Gainesville, 84. *Honors & Awards:* Jyoti Award, Indian Soc Agr Engrs. *Mem:* Sigma Xi; Am Soc Agr Engrs; Soil & Water Conserv Soc. *Res:* Conservation tillage; crop protection; surface irrigation. *Mailing Add:* 2115 R St Belleville KS 66935

THIERSTEIN, HANS RUDOLF, b Zürich, Switz, May 27, 44; m 69; c 2. GEOLOGY, MICROPALEONTOLOGY. *Educ:* Univ Zürich, dipl geol, 69, DrPhil, 72. *Prof Exp:* Teaching asst geol, Eidgenoessische Tech Hochschule, Z rich, Switz, 69-71, res asst, 71-73; fel, Swiss NFS, 73-76; asst prof, 76-80, ASSOC PROF GEOL, SCRIPPS INST OCEANOG, UNIV CALIF, SAN DIEGO, 80- *Mem:* Geol Soc Am; Swiss Geol Soc; Am Geophys Union. *Res:* Paleoceanography, calcareous nannoplankton, biostratigraphy, sedimentology, geochemistry. *Mailing Add:* Oceanog A-015 Univ Calif San Diego Box 109 La Jolla CA 92093

THIERY, JEAN PAUL, b Remiremont, France, Apr 25, 47. MOLECULAR EMBRYOLOGY. *Educ:* Nancy Univ, Baccalaureate, 64; Paris Univ, Docteur d'Etat(biochem), 76. *Prof Exp:* Sr researcher, 68-82, DIR RES, NAT CTR SCI RES, PARIS, 82- *Concurrent Pos:* Dir, Lab Physiopath Develop, Nat Ctr Sci Res & Ecole Normale Super, 87- *Honors & Awards:* Karger Prize, 90. *Mem:* Am Soc Cell Biol; Int Soc Develop Biol; Europ Molecular Biol Orgn. *Res:* Molecular embryology, cell recognition, cell migration and pattern formation; cancer invasion and metastasis; molecular immunology, hematopoietic precursors; adhesion molecules; growth factors and receptors. *Mailing Add:* Dept Biol CNRS 46 rue d'Ulm 8th Fl Cedex 05 Paris 75230 France

THIES, RICHARD WILLIAM, b Detroit, Mich, Sept 16, 41; m 76; c 3. ORGANIC CHEMISTRY. *Educ:* Univ Mich, BS, 63; Univ Wis, PhD(org chem), 67. *Prof Exp:* NIH fel org chem, Univ Calif, Los Angeles, 67-68; from asst prof to assoc prof, 68-87, PROF ORG CHEM, ORE STATE UNIV, 88- *Concurrent Pos:* NATO sr scientists fel, 79-80; prog officer synthetic, org & natural prod chem, NSF, 83-84, asst dean sci, 87-89, assoc dean, 89- *Mem:* Am Chem Soc; Royal Soc Chem. *Res:* Medium sized ring chemistry; carbonium ion chemistry; thermal rearrangements; synthesis of hormone analogs; polymers. *Mailing Add:* Dept Chem Gilbert Hall 153 Ore State Univ Corvallis OR 97331-4003

THIES, ROGER E, b Bronxville, NY, June 30, 33; m 85; c 3. PHYSIOLOGY. *Educ:* Bates Col, BS, 55; Harvard Univ, AM, 57; Rockefeller Univ, PhD(physiol), 61; Univ Okla, MA, 75. *Prof Exp:* Guest investr neurophysiol & NIH fel, Rockefeller Univ, 60-61; from instr to asst prof physiol, Sch Med, Wash Univ, 61-65; lectr, Makerere Univ Col, Uganda, 65-67; ASSOC PROF PHYSIOL & BIOPHYS, MED CTR, UNIV OKLA, 67-, ASSOC PROF BIOL PSYCHOL, HEALTH SCI CTR, 72- *Concurrent Pos:* Hon res asst & NIH spec fel, Univ Col, Univ London, 64-65. *Mem:* Am Physiol Soc; Soc Neurosci. *Res:* Etiology of myopia; effect of vagal activity on relay of cardiac pain to the brain; effective education and innovative teaching, especially problem solving for health professional and minority students. *Mailing Add:* Dept Physiol & Biophys Univ Okla Health Sci Ctr PO Box 26901 Oklahoma City OK 73190

THIESFELD, VIRGIL ARTHUR, b Glencoe, Minn, Oct 26, 37; m 59; c 2. BOTANY, PLANT PHYSIOLOGY. *Educ:* Luther Col, Iowa, BA, 59; Univ SDak, Vermillion, MA, 63; Univ Okla, PhD(bot), 65. *Prof Exp:* Teacher high sch, Iowa, 59-60 & Minn, 60-62; res asst bot, Univ Okla, 63-65; asst prof, 65-68, assoc prof biol, 68-76, CHMN DEPT, UNIV WIS-STEVENS POINT, 68-, PROF BIOL, 76- *Concurrent Pos:* Res grant, Univ Wis-Stevens Point, 68-69. *Mem:* Am Soc Plant Physiol; Sigma Xi. *Res:* Plant growth regulators. *Mailing Add:* 1003 Fifth St Stevens Point WI 54481

THIESSEN, HENRY ARCHER, b Teaneck, NJ, Nov 8, 40; m 62; c 1. ELEMENTARY PARTICLE PHYSICS, ACCELERATOR PHYSICS. *Educ:* Calif Inst Technol, BS, 61, MS, 62, PhD(physics), 67. *Prof Exp:* STAFF MEM MEDIUM ENERGY PHYSICS, LOS ALAMOS NAT LAB, 66- *Mem:* Fel Am Phys Soc. *Res:* Design of accelerators for medium energy physics; nuclear physics with kaon, pion and high energy proton beams; elementary particle physics. *Mailing Add:* 531 Rover Blvd Los Alamos NM 87544

THIESSEN, JACOB WILLEM, b Rotterdam, Neth, Apr 22, 28; US citizen; c 3. RESEARCH ADMINISTRATION. *Educ:* Univ Utrecht Med Sch, MD, 53; Am Bd Prev Med, dipl, 75, Am Bd Health Physics, dipl, 77. *Prof Exp:* Adv human & environ radioactivity, Neth Health Control, 63-70; vpres med affairs, Radiation Mgt Corp, 70-75; dir occup & environ health, US Army Environ Hyg Agency, 75-80; dep assoc dir health & environ res, Off Energy Res, US Dept Energy, 80-87; VCHMN, RADIATION EFFECTS RES FOUND, 87- *Concurrent Pos:* Adj assoc prof, Uniformed Serv Univ Health Sci, Bethesda, 80-; mem Nat Coun Radiation Protection & Measurements, 86- *Mem:* Fel Am Col Prev Med; fel Health Physics Soc; Am Acad Health Physics. *Res:* Metabolism and internal dosimetry of radioactive materials; health protection against ionizing radiation; medical management of radiation accidents. *Mailing Add:* Radiation Effects Res Found 5-2 Hijiyama Park Minami-Ku Hiroshima 732 Japan

THIESSEN, REINHARDT, JR, b Kiel, Wis, Oct 20, 13; m 38; c 2. BIOCHEMISTRY. *Educ:* Univ Pittsburgh, BS, 34. *Prof Exp:* Asst, Univ Pittsburgh, 35-37; head chemist res & control lab, Repub Yeast Corp, NJ, 37-41; proj leader cent labs, Gen Foods Corp, 42-46, head biol sect, 46-48, head nutrit sect, 48-55, from asst lab dir to lab dir, 55-62, sr res specialist, Tech Ctr, 62-69, res assoc & area mgr nutrit sci, 69-72, corp res mgr, 72-81, prin scientist nutrit sci, Tech Ctr, 81-82; RETIRED. *Mem:* AAAS; Am Pub Health Asn; Am Chem Soc; Inst Food Technol; Am Inst Nutrit. *Res:* Nutrition; toxicology; bacteriology; protein nutrition; carbohydrate nutrition; cacao chemistry; tracers in nutrition and toxicology; food chemistry; dental caries. *Mailing Add:* 586 Gilbert Ave Pearl River NY 10965-3320

THIESSEN, WILLIAM ERNEST, b Kans City, Mo, Sept 17, 34; m 60, 85; c 1. ORGANIC CHEMISTRY, CALORIMETRY. *Educ:* Univ Calif, Berkeley, BS, 56, PhD(chem), 60. *Prof Exp:* Instr chem, Univ Wash, 60-62; asst prof, Univ Calif, Davis, 62-68; NIH spec fel, 68-70, res chemist, Chem Div, Oak Ridge Nat Lab, 70-90; CONSULT, 90- *Mem:* Am Chem Soc; Am Crystallog Asn. *Res:* Structure elucidation of complex natural products and accurate molecular geometry of organic compounds by x-ray and neutron diffraction; development of expert systems for planning complex organic synthesis; enthalpies of dilution of electrolyte solutions at elevated temperature and pressure. *Mailing Add:* 108 Olney Lane Oak Ridge TN 37830-3913

THIGPEN, J(OSEPH) J(ACKSON), b Ruston, La, Feb 4, 17; m 41; c 2. MECHANICAL ENGINEERING. *Educ:* La Polytech Inst, BS, 36; US Mil Acad, BS, 41; Univ Tex, MS, 51, PhD(mech eng), 59. *Prof Exp:* From asst prof to prof mech eng, La Tech Univ, 47-75, head dept, 53-75, head dept, 53-75, assoc dean, 76, dean col eng, 76-82; RETIRED. *Concurrent Pos:* Consult, Opers Res Off, Johns Hopkins Univ, 48-50. *Mem:* Am Soc Mech Engrs; Am Soc Eng Educ; Nat Soc Prof Engrs; Sigma Xi. *Res:* Thermodynamics; heat transfer. *Mailing Add:* 1116 Carey Ave Ruston LA 71270

THILENIUS, OTTO G, b Bad Soden, Ger, July 7, 29; US citizen; m 56; c 3. PEDIATRIC CARDIOLOGY, PHYSIOLOGY. *Educ:* Univ Frankfurt, MD, 53; Univ Chicago, PhD(physiol), 62. *Prof Exp:* Resident, 57-59, instr, 61-62, asst prof pediat & physiol, 64-69, assoc prof pediat, 69-72, PROF PEDIAT, UNIV CHICAGO, 72- *Concurrent Pos:* USPHS fel physiol, Univ Chicago, 59-61; fel cardiol, Harvard Univ, 62-64. *Mem:* Am Acad Pediat; Am Col Cardiol. *Mailing Add:* Dept Pediat Christ Hosp & Med Ctr 4440 W 95th St Oak Lawn IL 60453

THILL, DONALD CECIL, b Colfax, Wash, Aug 30, 50; m 71; c 5. WEED SCIENCE, PLANT PROTECTION. *Educ:* Wash State Univ, BS, 72, MS, 76; Ore State Univ, PhD(crop sci), 79. *Prof Exp:* Plant physiologist, Agr Res Serv, USDA, 75-79; biochem field specialist, PPG Industs, Inc, 79-80; ASSOC PROF WEED SCI, UNIV IDAHO, 80- *Concurrent Pos:* Prin investr, Weed Control Systs, Univ Idaho, 81- *Mem:* Weed Sci Soc Am; Sigma Xi. *Res:* Develop cultural and herbicidal weed control tactics for use in small grains; herbicide mode and mechanism of action experiments; weed biology; herbicide resistance. *Mailing Add:* Dept Plant Soil & Entom Sci Univ Idaho Moscow ID 83843

THILL, RONALD E, b Tonopah, Nev, Oct 22, 44; m 66; c 2. RANGE-WILDLIFE INTERACTIONS. *Educ:* Humboldt State Col, Calif, BS, 67; SDak State Univ, MS, 69; Univ Ariz, PhD(range mgt), 81. *Prof Exp:* Range Scientist, Forest Ser, USDA, 77-86, RES WILDLIFE BIOLOGIST, S FOREST EXP STA, 86- *Mem:* Soc Range Mgt; Wildlife Soc. *Res:* Deer and cattle interactions; riparian zone wildlife; effects of forest management practices on wildlife; forest management effects (spacing and site preparation) on cattle and deer forage resources; radiotelemetry, deer behavior. *Mailing Add:* Box 7600 SFA Sta Nacogdoches TX 75962

THIMANN, KENNETH VIVIAN, b Ashford, Eng, Aug 5, 04; nat US; m 29; c 3. PLANT PHYSIOLOGY, PLANT HORMONES. *Educ:* Univ London, BSc, 24, PhD(biochem), 28. *Hon Degrees:* AM, Harvard Univ, 38; PhD, Univ Basel, 60; Dr, Univ Clermont-Ferrand, 61; DS, Brown Univ, 86. *Prof Exp:* Instr bact, Univ London, 26-28, Beit fel, 29-30; instr biochem, Calif Inst Technol, 30-35; lectr bot, 35-36, asst prof plant physiol, 36-39, tutor, 36-42, assoc prof, 39-46, dir biol labs, 46-50, prof, 46-62, Higgins prof biol, 62-65, EMER HIGGINS PROF BIOL, HARVARD UNIV, 65-; EMER PROF BIOL, UNIV CALIF, SANTA CRUZ, 78- *Concurrent Pos:* Guggenheim fel, 50-51 & 58; exchange prof, France, 54-55; provost, Crown Col, Univ Calif, Santa Cruz, 65-72; pres, Int Bot Cong, Seattle, 69 & Int Plant Growth Substance Cong, Tokyo, 73; mem exec comt, Assembly Life Sci, Nat Res Coun, 73-77; vis prof, Univ Mass, Amherst, 74, Univ Tex, Austin, 76; prof biol, Univ Calif, Santa Cruz, 65-78. *Honors & Awards:* Hales Prize, Am Soc Plant Physiol, 36; Balzan Prize, 82; Int Medallist, Plant Growth Subst Asn, 73. *Mem:* Fel Nat Acad Sci; Am Soc Naturalists (pres, 54); Bot Soc Am (pres, 60); Am Soc Plant Physiol (pres, 50); Soc Gen Physiol (pres, 49); Am Inst Biol Sci (pres); Am Philos Soc; Assoc Royal Col Sci; foreign mem Royal Soc London. *Res:* Physiology of bacteria, protozoa and fungi; co-discoverer of auxins, the first known growth-controlling hormones in plants; interactions of auxins with other plant hormones; senescence in leaves and its control by hormones and other factors. *Mailing Add:* The Quadrangle 3300 Darby Rd Apt 3314 Haverford PA 19043

THIND, GURDARSHAN S, b Lyallpore, India, Oct 17, 40; c 2. HYPERTENSION, INTERNAL MEDICINE. *Educ:* Punjab Univ, MD, 62; Univ Pa, MS, 66. *Prof Exp:* DIR, HYPERTENSION SECT, UNIV LOUISVILLE, 76-, PROF MED, 85- *Concurrent Pos:* Head, Chair Heart Res, Univ Louisville, 77-82. *Mem:* Fel Am Col Physicians; fel Am Col Cardiol; Am Physiol Soc; NY Acad Sci; Am Heart Asn; AAAS. *Res:* Hypertension and hypertensive cardio-renal diseases. *Mailing Add:* Dept Med Univ Louisville HSC ACB Rm A3G12 Louisville KY 40292

THIO, ALAN POO-AN, b Jatinegara, Indonesia, Jan 17, 31; US citizen; m 57; c 3. MEDICINAL CHEMISTRY, PESTICIDE CHEMISTRY. *Educ:* Univ Indonesia, BS, 54, MS, 57; Univ Ky, PhD(org chem), 61. *Prof Exp:* Assoc prof orgc chem, Bandung Inst Technol, 60-67; assoc med chem, Col Pharm, 67-70, REGULATORY SPECIALIST, DIV REGULATORY SERV, UNIV KY, 70- *Concurrent Pos:* Abstractor, Chem Abstr Serv, 66- *Mem:* Am Chem Soc. *Res:* Development of analytical procedures for the quantitative determination of a pesticide residues in feeds, fertilizers and soils; amino acids in feeds. *Mailing Add:* 242 Aberdeen Dr Lexington KY 40517

THIRGOOD, JACK VINCENT, b Northumberland, Eng, Apr 5, 24; m 49; c 3. FORESTRY. *Educ:* Univ Wales, BSc, 50; Univ BC, MF, 61; Ore State Univ, MF, 65; State Univ NY Col Forestry, Syracuse, PhD, 71. *Prof Exp:* With, UK Forestry Comn, 40-44, silviculture asst, Res Br, 50-54; silviculturist, Cyprus Forest Serv, 54-56; adv & dir Nat Forest Res Inst Iraq, Food & Agr Orgn, UN, 56-57; forestry consult, UK, 57-58; forester, BC Forest Serv, 58; res forester, Celgar Ltd, BC, 59; asst prof silviculture, State Univ NY Col Forestry, Syracuse, 60-62; prof silviculture & forest bot & head dept, Univ Liberia, Found Mutual Asst in Africa & UK Ministry Overseas Develop, 62-64; forestry consult & area dir, Tilhill Forestry & Adv Ltd, Eng, 64-67; assoc prof, 68-80, PROF FOREST POLICY/HIST & INT FORESTRY, UNIV BC, 80- *Concurrent Pos:* Consult indust land reclamation. *Honors & Awards:* Noranda Land Reclamation Award, Can Land Reclamation Asn. *Mem:* Can Land Reclamation Asn; Sigma Xi; Can Inst Forestry; Forest Hist Soc; Commonwealth Forestry Asn. *Res:* Silviculture, forest management, forest history and policy; temperate, tropical and arid zone forestry; industrial land reclamation. *Mailing Add:* Fac Forestry Univ BC Westbrook Pl Vancouver BC V6T 1W5

THIRION, JEAN PAUL JOSEPH, b Metz, France, July 30, 39; Can citizen; m 67; c 2. SOMATIC CELL GENETICS, RECOMBINANT DNA. *Educ:* Univ Nancy, France, BS, 60; Univ Wis-Madison, PhD(biochem), 66; Pasteur's Inst, France, Doct d'etat, 69. *Prof Exp:* Attache genetics, Nat Ctr Sci Res, Pasteur's Inst, 67-68, charge, 68-72; From asst prof to assoc prof, 72-80, PROF MICROBIOL, UNIV SHERBROOKE, 80- *Honors & Awards:* Fulbright Fel, 63-67; Scholar Med Res Coun Can, 72-77. *Mem:* Am Genetics Soc. *Res:* Genetic analysis of gene regulation. *Mailing Add:* Dept Microbiol Chu-Univ Sherbrooke Sherbrooke PQ J1H 5N4 Can

THIRKILL, JOHN D, b Soda Springs, Idaho, Apr 26, 29; m 50; c 3. CHEMICAL ENGINEERING. *Educ:* State Col Wash, BS, 53. *Prof Exp:* Chem engr, E I du Pont de Nemours & Co, Inc, 53; proj engr, Thiokol Chem Corp, Md, 55-58, dept head rocket eng, Wasatch Div, 58-60, preliminary design & anal, 60-63, div mgr, 63-64, proj eng, 64-67, div mgr, 68-71, dep dir space shuttle prog, 71-78, dir eng, 78-89; RETIRED. *Concurrent Pos:* Adv, NASA Internal Comt Design Criteria for Chem Propulsion, 66-68. *Mem:* Am Ord Asn; Am Inst Aeronaut & Astronaut. *Res:* Rocket propulsion, design and development of solid propellant propulsion systems. *Mailing Add:* 5295 S 1300 E South Ogden UT 84403

THIRUMALAI, DEVARAJAN, b Madras, India, June 6, 56; m 82. STATISTICAL MECHANICS, THEORY OF LIQUIDS GLASSES & POLYMERS. *Educ:* Indian Inst Technol, Kanpur, India, MSc, 77; Univ Minn, PhD(theoret chem), 82. *Prof Exp:* Res assoc statist mech, Columbia Univ, 82-85; asst prof, 85-89, ASSOC PROF STATIST MECH, INST PHYS

SCI & TECHNOL, UNIV MD, COLLEGE PARK, 89- PARK, 85- *Honors & Awards:* Presidential Young Investr Award. *Mem:* Am Phys Soc. *Res:* Equilibrium and non-equilibrium statistical mechanics; theory of disordered systems. *Mailing Add:* Inst Phys Sci & Technol Univ Md College Park MD 20742

THIRUVATHUKAL, JOHN VARKEY, b Shertallay, India, Aug 4, 39; m 71; c 2. GEOPHYSICS, OCEANOGRAPHY. *Educ:* St Louis Univ, BS, 61; Mich State Univ, MS, 63; Ore State Univ, PhD(geophys), 68. *Prof Exp:* Res asst geophys, Ore State Univ, 63-67; from instr to asst prof geol, DePauw Univ, 67-70; from asst prof to assoc prof, 70-88, GEOSCI COORDR, MONTCLAIR STATE COL, 85-, PROF GEOL, 88- *Concurrent Pos:* Consult, Bd Earth Sci, Nat Acad Sci, 69-87; adj prof, Fairleigh Dickinson Univ, 71-75. *Mem:* Am Geophys Union; Soc Explor Geophys. *Mailing Add:* Dept Phys-Geosci Montclair State Col Upper Montclair NJ 07043

THIRUVATHUKAL, KRIS V, b Shertallay, Kerala, India, May 1, 25; div; c 3. ZOOLOGY, MORPHOLOGY. *Educ:* Univ Kerala, BSc, 53; Boston Col, MS, 56; St Louis Univ, PhD(biol, histol), 60. *Prof Exp:* Instr zool, anat, histol & animal tech, Duquesne Univ, 59-60; asst prof zool & histol, Aquinas Col, 60-62; asst prof biol, Gonzaga Univ, 62-65 & Canisius Col, 65-68; chmn dept, 68-71, PROF BIOL, LEWIS UNIV, 68- *Mem:* Am Micros Soc; Soc Syst Zool; Indian Soc Animal Morphol & Physiol; Am Soc Zoologists; Am Asn Univ Profs (secy, 75-). *Res:* Histology and morphology of reptiles; vertebrate zoology; coelacanth morphology; research in reptilia and coelacanth. *Mailing Add:* 2213 Arden Pl Joliet IL 60435

THIRUVENGADAM, ALAGU PILLAI, b Madurai, India, Aug 16, 35; US citizen; m 61; c 3. MECHANICAL ENGINEERING, SYSTEMS DESIGN. *Educ:* Univ Madras, BE, 57; Indian Inst Sci, MSc, 59, PhD(eng), 61. *Prof Exp:* Govt India sr res scholar, 58-60; Univ Grants Comn sr res fel, 60-62; res scientist, Hydronautics Inc, 62-65, sr res scientist, 65-67, prin res scientist, 67-69; prof mech eng, Cath Univ Am, 69-75; CHMN BD, DAEDALEAN INC, 72-, PRES, 80- *Concurrent Pos:* Lectr, Cath Univ Am, 65-69, prof, 69-78; fel, Fac Arts & Sci, Johns Hopkins Univ, 67-69; consult, Hydronautics Inc, 69-, Brookhaven Nat Lab, 69-71, Waukesha Motors, Wis, 71- & Naval Ship Res & Develop Ctr, 71- *Honors & Awards:* Hess Award, Am Soc Mech Engrs, 63; Outstanding engr, Eng Joint Coun & Archit Socs, Nat Acad Sci, 64. *Mem:* Am Soc Mech Engrs; Am Soc Testing & Mat. *Res:* Dynamic response of materials; cavitation erosion, wet steam erosion, rain erosion, stress corrosion; corrosion fatigue; modeling erosion; open channel flow; fluid mechanics of turbomachinery; fluid dynamics; material engineering; power systems; materials science. *Mailing Add:* Daedalean Inc 15110 Frederick Rd Woodbine MD 21797

THISSEN, WIL A, Neth citizen. SYSTEMS ANALYSIS & MODELING, POLICY ANALYSIS. *Educ:* Eindhoven Univ Technol, MEng, 73, PhD(systs & control eng), 78. *Prof Exp:* Asst prof systs & control eng, Eindhoven Univ Technol, 73-78; asst prof & sr scientist, Univ Va, 78-80; head, Policy Anal Div, Rykswaterstaat, Dutch Pub Works Dept, 80-85; ASSOC PROF SYSTS & POLICY ANALYSIS, DELFT UNIV TECHNOL, 86- *Mem:* Inst Elec & Electronics Engrs, Systs Man & Cybernet Soc; Asn Pub Policy Anal & Mgt; Inst Mgt Sci; Int Asn Impact Assessment. *Res:* Systems engineering and policy analysis methodology, specifically applied to integrated water resource management, transportation and infrastructures; environmental problems. *Mailing Add:* Delft Univ Technol PO Box 5048 Delft 2600 GA Netherlands

THISTED, RONALD AARON, b Los Angeles, Calif, Mar 2, 51; m 73; c 1. DATA ANALYSIS, COMPUTATIONAL STATISTICS. *Educ:* Pomona Col, BA, 72; Stanford Univ, MS, 73, PhD(statist), 77. *Prof Exp:* Asst prof, 76-82, ASSOC PROF STATIST, UNIV CHICAGO, 82-, ASSOC PROF ANESTHESIA & CRITICAL CARE, 89- *Concurrent Pos:* Vis lectr, Soc Indust & Appl Math, 79-80; assoc ed, J Am Statist Asn, 79-85; prin investr, NSF grant, 84- *Mem:* Fel Am Statist Asn; Inst Math Statist; Asn Comput Mach; Am Asn Artificial Intel; AAAS. *Res:* Statistical computation; development of algorithms; statistical graphics; biostatistics; clinical trials. *Mailing Add:* Dept Statist Univ Chicago 5734 Univ Ave Chicago IL 60637

THODE, E(DWARD) F(REDERICK), b New York, NY, May 31, 21; m 44; c 3. INDUSTRIAL CHEMISTRY, MATERIALS. *Educ:* Mass Inst Technol, SB, 42, SM, 43, ScD(chem eng), 47. *Prof Exp:* Chem engr, Boston Woven Hose & Rubber Co, Mass, 43-45; from asst prof to assoc prof chem eng, Univ Maine, 47-54; chem engr, Cent Res Dept, Minn Mining & Mfg Co, 54-55; res assoc phys chem, Inst Paper Chem, 55-57; chem eng, 57-59, chief pulping & papermaking sect, 59-60, adminr eng & tech sect, 60-61, coordr info processing, 61-63; prof chem eng & head dept, 63-74, assoc dir eng exp sta, 65-66, actg dir, Ctr Bus Res & Serv, 77, prof mgt, 74-86, EMER PROF, CHEM ENG & MGT, NMEX STATE UNIV, 86- *Concurrent Pos:* Affil, Los Alamos Nat Lab, 65-90; consult, Gen Elec Co, 64-66 & Bell Tel Labs, 66-70. *Mem:* Sigma Xi; Am Inst Chem Engrs; Am Soc Eng Educ. *Res:* Technico-economic studies of energy alternatives; environmental control technology; management science; production/operations management. *Mailing Add:* 905 Conway 45 Las Cruces NM 88005-3775

THODE, LESTER ELSTER, b Alameda, Calif, Apr 8, 43; m 67; c 3. PLASMA PHYSICS, CHARGED PARTICLE BEAM PHYSICS. *Educ:* Univ Calif, Berkeley, BS, 69; Cornell Univ, PhD(appl physics), 74. *Prof Exp:* Staff mem, 73-78, alt group leader, 78-79, GROUP LEADER INTENSE PARTICLE BEAM PHYSICS, LOS ALAMOS NAT LAB, 79- *Mem:* Am Phys Soc. *Res:* Collective behavior of particle beams. *Mailing Add:* 292 Aragon Ave Los Alamos NM 87544

THODOS, GEORGE, b Chicago, Ill, Sept 15, 15. CHEMICAL ENGINEERING. *Educ:* Armour Inst Technol, BS, 38, MS, 39; Univ Wis, PhD(chem eng), 43. *Prof Exp:* Jr chem engr, Standard Oil Co Ind, 39-40 & Phillips Petrol Co, Okla, 40-41, chem engr, 43-46; chem engr, Pure Oil Co, Ill, 46-47; from asst prof to prof chem eng, 47-77, WALTER P MURPHY PROF, TECH INST NORTHWESTERN UNIV, EVANSTON, 77- *Mem:* Am Chem Soc; Am Inst Chem Engrs. *Res:* Petroleum processing; mass transfer studies; vapor pressures and critical constants of hydrocarbons; transport properties of substances; vapor-liquid equilibrium studies. *Mailing Add:* Dept Chem Eng Tech Bldg Northwestern Univ 633 Clark St Evanston IL 60201

THOE, ROBERT STEVEN, b Pensacola, Fla, Aug 19, 45; m 68; c 1. ATOMIC PHYSICS. *Educ:* Baylor Univ, BS, 68; Univ Conn, MS, 70, PhD(physics), 73. *Prof Exp:* Asst res prof physics, Univ Tenn, 73-76, asst prof, 76-80; CONSULT, 75- *Mem:* Am Phys Soc. *Res:* The study of atomic collision phenomena, primarily by the detection and measurement of the non-characteristic radiations emitted during violent ion-atom collisions. *Mailing Add:* Lawrence Livermore Lab L43 Livermore CA 94550

THOENE, JESS GILBERT, b Bakersfield, Calif, Aug 4, 42; m 65. CLINICAL BIOCHEMICAL GENETICS. *Educ:* Stanford Univ, BS, 64; Johns Hopkins Univ, MD, 68. *Prof Exp:* Asst clin prof pediat, Univ Calif, San Diego, 75-77; from asst prof to assoc prof pediat, 77-86, asst prof biochem, 84-87, ASSOC PROF BIOCHEM, UNIV MICH, 87-, PROF PEDIAT, 86- *Concurrent Pos:* Dir, Mich Dept Ment Health Genetic Screening Lab, 77- & Pediat Clin Res Ctr, 86-; Welcome Found Travel Grant, 84; vchmn, Nat Comn Orphan Dis, 86-89, chair, 89; Kennedy Found Fel, 88-89. *Honors & Awards:* Pub Health Serv Award Outstanding Achievement Orphan Prod Develop, 86. *Mem:* Soc Pediat Res; Soc Inherited Metab Disorders; Am Soc Human Genetics; Am Chem Soc; Nat Orgn Rare Dis (pres, 85-); Am Pediat Soc; Am Soc Biochem & Molecular Biol. *Res:* Inborn errors of aminoacid, organic acid and vitamin metabolism; cellular proteolysis; lysosomol physiology. *Mailing Add:* Dept Pediat 2612 SPHI Univ Mich Ann Arbor MI 48109

THOLEN, ALBERT DAVID, b Philadelphia, Pa, Aug 23, 27; m 56; c 4. CIVIL ENGINEERING, METROLOGY. *Educ:* Drexel Univ, BSCE, 49. *Prof Exp:* Vpres opers res, Gen Res Corp, 62-77; dep dir standards anal div, 77, CHIEF, OFF WEIGHTS & MEASURES, NAT BUR STANDARDS, US DEPT COM, 77- *Concurrent Pos:* Adv comt mem, Int Legal Metrol, US Dept Com, 78- *Mem:* Fel AAAS; Opers Res Soc Am. *Res:* Measurement science; electronics; microprocessing. *Mailing Add:* 7121 Thomas Branch Dr Bethesda MD 20817

THOM, RONALD MARK, b Long Beach, Calif, June 6, 48; m 71; c 1. MARINE BENTHIC ALGAE, MARINE POLLUTION. *Educ:* Calif State Col, Dominguez Hills, BA, 71; Calif State Univ, Long Beach, MA, 76; Univ Wash, PhD(phycol), 78. *Prof Exp:* Biologist, Co Sanitation Dist, Los Angeles, 71-74 & US Army Corps Engrs, 79-82; res scientist, Univ Wash, 82-90; INSTR OCEANOG, CHAPMAN COL, 85-; RES SCIENTIST, BATTELLE, 90- *Concurrent Pos:* Consult, City of Seattle, 82-; adv, Off Puget Sound, US Environ Protection Agency, 84- *Mem:* AAAS; Ecol Soc Am; Phycol Soc Am; Am Soc Limnol & Oceanog; Estuarine Res Fedn. *Res:* Marine algal ecology; nearshore marine systems ecology; primary productivity of seaweed dominated systems; pollution effects on nearshore marine systems; wetland construction. *Mailing Add:* 358 Dunlap Ave N Sequim WA 98382

THOMA, GEORGE EDWARD, b Dayton, Ohio, Aug 9, 22; m 49; c 8. INTERNAL MEDICINE, NUCLEAR MEDICINE. *Educ:* Univ Dayton, BS, 43; St Louis Univ, MD, 47. *Prof Exp:* From instr to assoc prof internal med, 51-76, head, Sect Nuclear Med, 59-73, asst to vpres, 62-67, asst vpres, Med Ctr, 67-73, PROF INTERNAL MED, ST LOUIS UNIV, 76-, VPRES, MED CTR, 73- *Concurrent Pos:* Dir radioisotopes lab, Med Ctr, St Louis Univ, 49-51 & 54-73; consult, Health Physics Div, Oak Ridge Nat Lab, 54-, Lockheed Aircraft Corp, 54-, US Army, 56-, Med Div, Oak Ridge Inst Nuclear Studies, 58-, Div Radiol Health, USPHS, 60-, Div Compliance, US Nuclear Res Coun, 62- & Am Pub Health Asn, 62-; ed, J Nuclear Med, 59-70. *Mem:* Soc Nuclear Med; Radiation Res Soc; Health Physics Soc; Am Thyroid Asn; Am Soc Internal Med. *Res:* Radiobiology; thyroid function; clinical application of radioisotopes. *Mailing Add:* Dept Internal Med Sch Med St Louis Univ 1402 S Grand Blvd St Louis MO 63104-1083

THOMA, GEORGE RANJAN, b India, Mar 1, 44; US citizen. ELECTRICAL ENGINEERING, COMMUNICATIONS ENGINEERING. *Educ:* Swarthmore Col, BS, 65; Univ Pa, MS, 67, PhD(elec eng), 71. *Prof Exp:* Ford Found fel, 65-67; res assoc, Moore Sch Elec Eng, Univ Pa, 68-71; systs engr, AII Systs, Moorestown, NJ, 71-73, Gen Elec Co, 73-74; SR ELECTRONICS ENGR & CHIEF COMMUN ENG, NAT LIBR MED, 74- *Concurrent Pos:* Consult, NSF & var industs. *Honors & Awards:* Quality Serv Award, 83. *Mem:* Inst Elec & Electronics Engrs; Am Soc Info Sci; Soc Photo-Optical & Instrumentation Engrs. *Res:* Telecommunications; signal processing; satellite aided video and voice communications; satellite aided radio navigation; digital image processing; video disc and optical disk technologies; electronic document storage and retrieval; information science and systems. *Mailing Add:* Nat Libr Med 8600 Rockville Pike Bethesda MD 20894

THOMA, JOHN ANTHONY, b Springfield, Ill, Dec 6, 32; m 58; c 4. BIOCHEMISTRY. *Educ:* Bradley Univ, AB, 54; Iowa State Univ, PhD(biochem), 58. *Prof Exp:* Asst prof, Ind Univ, 60-66; PROF CHEM, UNIV ARK, 70- *Concurrent Pos:* Vis fel, Univ Sydney, 72; vis scholar, Univ Calif, Los Angeles, 77-78. *Mem:* Am Chem Soc; Sigma Xi; Am Asn Biol Chemists. *Res:* Anti-viral substances; enzymology; theory and practice of chromatography. *Mailing Add:* Dept Chem Univ Ark Fayetteville AR 72701-1202

THOMA, RICHARD WILLIAM, b Milwaukee, Wis, Dec 7, 21; m 52; c 4. ENVIRONMENTAL SCIENCES. *Educ:* Univ Wis, BSc, 47, MSc, 49, PhD(biochem), 51. *Prof Exp:* Res assoc sect microbiol, Squibb Inst Med Res, 51-61, res supvr microbiol develop, 62-68, asst dir biol process develop, 68-79, sr res fel, biol process develop, E R Squibb & Sons, 79-82; dir, process develop, New Brunswick Sci Co, Inc, 82-84; PRES, THOMA CONSULT, INC, 84-; SAFETY CONSULT, HARBOR BR OCEANOG INST, INC, 88-

Concurrent Pos: Consult, Chem Health & Safety, Hazardous Mat Mgt, Regulatory Compliance, 88- *Mem:* AAAS; Am Inst Biol Sci; Soc Indust Microbiol; Brit Soc Gen Microbiol; Am Chem Soc; NY Acad Sci; Am Soc Safety Engrs; Sigma Xi. *Res:* Fermentation process research and development. *Mailing Add:* 3772 Outrigger Ct Ft Pierce FL 34946

THOMA, ROY E, b San Antonio, Tex, May 12, 22; m 53; c 2. INORGANIC CHEMISTRY, ENVIRONMENTAL ASSESSMENTS. *Educ:* Univ Tex, MA, 48. *Prof Exp:* Assoc prof, Sam Houston State Col, 48-51; asst prof, Tex Tech Col, 51-52; chemist, 52-60, proj chemist molten salt reactor progs, 67-71, task group dir, 71-78, tech asst to assoc dir admin, 78-80, environ assessments analyst, 80-85, PRIN INVESTR, ENVIRON HEALTH & SAFETY, SYNFUELS PROJ, OAK RIDGE NAT LAB, 85- *Concurrent Pos:* Mem bd dirs, Environ Systs Corp, 73-77; mem, environ qual adv bd, City of Oak Ridge, 85- *Mem:* Am Chem Soc; Sigma Xi; fel Am Ceramic Soc; Am Nuclear Soc; fel AAAS. *Res:* Physical chemistry of inorganic fused salts, particularly determinations of phase equilibria and interrelationships of crystal structures in condensed systems of these materials; environmental effects of nuclear and synfuels facilities. *Mailing Add:* 119 Underwood Rd Oak Ridge TN 37830

THOMAN, CHARLES JAMES, b Wilkes-Barre, Pa, Nov 4, 28; m 82; c 2. MODEL MEMBRANE TRANSPORT, SYNTHESIS OF ANTIVIRALS. *Educ:* Spring Hill Col, BS, 53; Fordham Univ, MS, 56; Woodstock Col, STB, 59, STM, 60; Univ Mass, Amherst, PhD(org chem), 66. *Prof Exp:* Instr chem, Univ Scranton, 53-55, from asst prof to prof, 66-82, chmn chem, 78-82; res assoc toxicol, Sch Med, La State Univ, Shreveport, 82-83; vis prof chem, Univ Ala, Tuscaloosa, 83-84; prof chem, Univ Scranton, 84-87; prof & chair chem, Stephen F Austin State Univ, 87-89; PROF & CHAIR CHEM, PHILADELPHIA COL PHARM & SCI, 89- *Mem:* Am Chem Soc; Sigma Xi. *Res:* Chemistry and biochemistry of polysorbate 80; chemistry of n-nitrosoketimines; sydnone chemistry. *Mailing Add:* Chem Dept Philadelphia Col Pharm & Sci Philadelphia PA 19104

THOMAN, MARILYN LOUISE, CELLULAR IMMUNOLOGY. *Educ:* Univ Calif, Berkeley, PhD(molecular biol), 78. *Prof Exp:* ASST MEM IMMUNOL DEPT, SCRIPPS CLIN & RES FOUND, 84- *Mailing Add:* Dept Immunol Scripps Clin & Res Found 10666 N Torrey Pines Rd La Jolla CA 92037

THOMANN, GARY C, b Burlington, Iowa, July 23, 42. ENERGY RESEARCH. *Educ:* Univ Kans, BS, 65, MS, 67, PhD(elec eng), 70. *Prof Exp:* Res engr, Univ Kans, 67-70; prin investr, NASA, 70-75; ASSOC PROF ELEC ENG, WICHITA STATE UNIV, 75- *Mem:* Inst Elec & Electronics Engrs; Sigma Xi; Am Soc Eng Educ. *Res:* Use of wind energy in Kansas; electric utility research, particularly in energy storage, residential heating and cooling systems, and air source, well water and earth coil heat pumps. *Mailing Add:* Power Technol Inc PO Box 1058 1482 Erie Blvd Schenectady NY 12301

THOMANN, ROBERT V, b New York, NY, Sept 1, 34; m 57; c 7. CIVIL ENGINEERING, OCEANOGRAPHY. *Educ:* Manhattan Col, BCE, 56; NY Univ, MCE, 60, PhD(oceanog), 63. *Prof Exp:* Engr, Delaware River & Bay Study, USPHS, 56-59, engr in charge Narragansett Bay Study, 59-60; tech dir estuary water qual mgt, Fed Water Pollution Control Admin, 62-66; PROF CIVIL ENG, MANHATTAN COL, 66- *Concurrent Pos:* Consult. *Mem:* Am Chem Soc; Am Soc Limnol & Oceanog; Am Soc Civil Engrs; Water Pollution Control Fedn; Sigma Xi. *Res:* Interrelationships of environment on waste water discharge, water quality and water use. *Mailing Add:* 326 N Maple Ave Ridgewood NJ 07450

THOMAS, ADRIAN WESLEY, b Edgefield, SC, June 23, 39; m 64; c 2. RISK STRATEGIES, CONSERVATION PRODUCTION SYSTEMS. *Educ:* Clemson Univ, BSAE, 62, MSAE, 65; Colo State Univ, PhD(agr eng), 72. *Prof Exp:* Res scientist, Agr Res Serv, USDA, Tifton, Ga, 65-69, Ft Collins, Colo, 69-72, Watkinsville, Ga, 72-78, res leader, Watkinsville, Ga, 78-89, RES LEADER, AGR RES SERV, USDA, TIFTON, GA, 89- *Concurrent Pos:* Acad fac, Colo State Univ, Ft Collins, 69-72; adj prof, Univ Ga, Athens, 73 -; consult, USDA-Agr Res Serv & US State Dept, 81. *Mem:* Am Soc Agr Engrs; Am Soc Agron; Soil & Water Conserv Soc; Soil Sci Soc Am; Sigma Xi. *Res:* Development of technologies that will integrate soil, water, and climate resources for optimization of conservation and production strategies in agricultural systems for the Southeast. *Mailing Add:* USDA-Agr Res Serv PO Box 946 Tifton GA 31793

THOMAS, ALBERT LEE, JR, b Auburn, Ala, Sept 4, 23; m 49; c 5. ELECTRONICS. *Educ:* Cornell Univ, BEE, 49. *Prof Exp:* Res assoc, Ala Polytech Inst, 49; res engr, 49-53, head, instrument develop sect, 53-62, head eng physics div, Southern Res Inst, 62-83, Project Develop Lab, 83-88; SR RES ADV, SOUTHERN RES TECHNOL INC, 88- *Mem:* Inst Elec & Electronics Engrs; SPIE. *Res:* Electro-optical instrumentation. *Mailing Add:* PO Box 827 3555 N College St Auburn AL 36831-0827

THOMAS, ALEXANDER, b New York, NY, Jan 11, 14; m 38; c 4. PSYCHIATRY. *Educ:* City Col New York, BS, 32; NY Univ, MD, 36; Am Bd Psychiat & Neurol, dipl, 48. *Prof Exp:* From instr to assoc prof, 48-66, PROF PSYCHIAT, SCH MED, NY UNIV, 66- *Concurrent Pos:* Assoc attend psychiatrist, Bellevue & Univ Hosps, 58-68, attend psychiatrist, 68-; dir psychiat div, Bellevue Hosp, New York, 68-78. *Honors & Awards:* Ittleson Award, Am Psychiat Asn; Baum Award, NY Ment Health Asn. *Mem:* Fel Am Psychiat Asn; Sigma Xi. *Res:* Longitudinal study of child development; psychosomatic medicine. *Mailing Add:* Dept Psychiat Sch Med NY Univ 550 First Ave New York NY 10016

THOMAS, ALEXANDER EDWARD, III, b Chicago, Ill, May 3, 30; m 56; c 3. ORGANIC CHEMISTRY, ANALYTICAL CHEMISTRY. *Educ:* Univ Ill, BS, 55; DePaul Univ, MS, 61. *Prof Exp:* Res chemist, Cent Org Res Lab, Glidden Co, 55-58, sect head anal chem, Durkee Foods Group, 58-66, mgr chem res dept, 66-71, mgr chem res, Dwight P Joyce Res Ctr, 71-78, mgr applns res, Glidden-Durkee Div, 71-78, assoc dir appl sci, Dwight P Joyce Res Ctr, Durkee Div, SCM Corp, 78-85, tech consult, 86-90; RETIRED. *Mem:* Am Chem Soc; Am Oil Chem Soc. *Res:* Analytical chemistry of glycerides, surfactants and protective coatings; chromatographic and instrumental methods; synthesis of organic azides. *Mailing Add:* 16335 Ramona Dr Middleburg Heights OH 44130

THOMAS, ALFORD MITCHELL, b Bunnlevel, NC, July 24, 42; m 70; c 1. MEDICINAL CHEMISTRY, PEPTIDE CHEMISTRY. *Educ:* Campbell Col, NC, BA, 64; Univ NC, Chapel Hill, PhD(org chem), 69. *Prof Exp:* NIH fel, Univ Va, 69-71, res assoc org chem, 71-72; sr chemist, 72-77, RES INVESTR, ABBOTT LABS, 78- *Mem:* Am Chem Soc; AAAS; Sigma Xi. *Res:* Synthesis of biological peptides. *Mailing Add:* 38660 Shagbark Lane Wadsworth IL 60083

THOMAS, ALVIN DAVID, JR, b Gary, Ind, Nov 3, 28; m 52; c 3. PHYSICAL METALLURGY. *Educ:* Case Inst Technol, BS, 40; Purdue Univ, MS, 59, PhD(metall eng), 61. *Prof Exp:* Res engr, Repub Steel Res Ctr, 60-63; asst prof mech eng, Univ Tex, 63-67; head mat res sect, Tracor, Inc, 67-72, dep dir, Environ & Phys Sci Div, 72-76; DIR, MAT SCI DIV, RADIAN CORP, 76- *Mem:* Am Inst Mining, Metall & Petrol Engrs; Am Soc Metals; Am Soc Testing & Mat. *Res:* Mechanisms in precipitation hardening; strengthening mechanisms in steels; properties of Invar and Elinvar type alloys; resistance welding; composite materials; failure analysis and accident prevention. *Mailing Add:* RFD 2 W Hill Rd Maple Dale Farm Winsted CT 06098

THOMAS, ANTHONY, b May 3, 31; US citizen; c 2. MECHANICAL ENGINEERING, ENGINEERING PHYSICS. *Educ:* Univ Mich, BS, 54, MS, 63. *Prof Exp:* Res proj engr aerial photog, Chicago Aerial Industs, 57-61; group leader, Bubble Chamber, 61-78, PROG MGR OCEAN THERMAL ENERGY CONVERSION, ARGONNE NAT LAB, 78-, SECT MGR EXP SYSTS ENG, 80-, MGR, THERMAL TECHNOL PROG, 87- *Concurrent Pos:* Mem, US Bubble Chamber Working Group, 77-78; chmn, Deep Ocean Environ, Workshop on Conserv related heat Transfer Res, 87. *Res:* Design and operation of cryogenic devices; devices for energy conservation, conversion and efficiency enchancement. *Mailing Add:* Argonne Nat Lab 9700 S Cass Ave Argonne IL 60439

THOMAS, ARTHUR L, b New York, NY, July 24, 28; m 77. CHEMISTRY. *Educ:* Columbia Col, AB, 51; Princeton Univ, PhD, 56. *Prof Exp:* Engr photo prod, E I du Pont de Nemours & Co, Inc, 55-58, res supvr, 58-59; chem engr, Standard Ultramarine & Color Co, 60-65 & MHD, Inc & Plasmachem Inc, 65-68; from instr to asst prof chem, Calif State Polytech Col, San Luis Obispo, 69-72; vis asst prof immunochem, Columbia Univ, 73; sci ed, Ronald Press Co, 74-77; ed, Chem Mkt Res, Hull & Co, 78-87; ABSTRACTOR, AM PETROL INST, NEW YORK, 88- *Mem:* AAAS. *Res:* Pigments; high temperature reactions. *Mailing Add:* Two Putnam Park Greenwich CT 06830-5747

THOMAS, ARTHUR NORMAN, b Los Angeles, Calif, Jan 27, 31; m 50; c 5. SURGERY, THORACIC SURGERY. *Educ:* Stanford Univ, BS, 53; Univ BC, MD, 57. *Prof Exp:* Intern surg, San Francisco Gen Hosp, 57-58, resident, 58-59; resident, Univ Hosp, 60-62, clin instr, Sch Med, 63-68, from asst prof to assoc prof, 68-83, PROF SURG, SCH MED, UNIV CALIF, SAN FRANCISCO, 83-; CHIEF THORACIC SURG, SAN FRANCISCO GEN HOSP, 70- *Concurrent Pos:* NIH res fel, Univ Hosp, Univ Calif, San Francisco, 59-60, NIH fel, Cardiovasc Res Inst, 63-65; asst chief surg, Vet Admin Hosp, San Francisco, 66-70, attend physician thoracic surg, 70- *Mem:* Soc Thoracic Surg; Samson Thoracic Surg Soc; Am Asn Thoracic Surg. *Res:* Thoracic surgery; cardiopulmonary research. *Mailing Add:* 1001 Potrero Ave San Francisco CA 94110

THOMAS, AUBREY STEPHEN, JR, b Wolfeboro, NH, Nov 4, 33; m 56. PLANT PHYSIOLOGY & ECOLOGY, PLANT CELL CULTURE. *Educ:* Keene State Col, BEd, 62; Univ NH, MS, 64, PhD(bot, plant physiol), 67. *Prof Exp:* From asst prof to assoc prof, Merrimack Col, 67-87, chmn dept, 75-81, actg dean, Div Sci & Eng, 84-86, PROF BIOL, MERRIMACK COL, 87- *Mem:* AAAS; Am Soc Plant Physiol; Bot Soc Am; Am Soc Hort Sci; Nat Asn Biol Teachers; Am Inst Biol Sci. *Res:* Plant physiological ecology; effect of light, radiation, electrical fields and other environmental factors upon plant growth and development; plant allelopathy; plant cell & tissue culture. *Mailing Add:* Dept Biol Merrimack Col North Andover MA 01845

THOMAS, BARRY, b Eng, Dec 31, 41; US citizen. ZOOLOGY, ENVIRONMENTAL EDUCATION. *Educ:* Calif State Univ, Fullerton, BA, 67, MA, 68; Univ BC, PhD(zool), 71. *Prof Exp:* Lectr ecol, Univ Calgary, 71-72; asst prof, 72-75, PROF ENVIRON EDUC, CALIF STATE UNIV, FULLERTON, 75- *Concurrent Pos:* Pres, BioReCon, 71-; dir, Tucker Wildlife Sanctuary, 72- *Mem:* Am Soc Mammal; Audubon Soc. *Res:* Karyotaxonomy of island rodents; urbanization effects on wild animal populations; biological impact statements. *Mailing Add:* Dept Sci Educ Calif State Univ Fullerton CA 92634

THOMAS, BARRY HOLLAND, b Lancaster, Eng, June 1, 39; m 66; c 2. BIOCHEMISTRY, PHARMACOLOGY. *Educ:* Univ Liverpool, BSc, 62, PhD(pharmacol), 65. *Prof Exp:* Asst lectr pharmacol, Univ Liverpool, 65-67, lectr, 67-69; RES SCIENTIST, HEALTH PROTECTION BR, DEPT NAT HEALTH & WELFARE CAN, 69- *Concurrent Pos:* Assoc ed, Can J Physiol Pharmacol, 80-86. *Mem:* Soc Toxicol Can (pres, 86-87); Pharmacol Soc Can; Soc Toxicol USA. *Res:* Role of drug metabolism in the toxicity of drugs; toxicity of drug interactions. *Mailing Add:* PO Box 51 Kars ON K0A 2E0 Can

THOMAS, BERT O, b Lead, SDak, May 15, 26; m 49; c 2. ZOOLOGY. *Educ:* Colo State Univ, BS, 50, MS, 52; Univ Minn, PhD, 59. *Prof Exp:* Biologist, Hydrol Surv, US Fish & Wildlife Serv, 50; asst zool, Univ Minn, 52-57; biologist, State Dept Health, Minn, 57-59; from asst prof to prof zool & chmn dept biol sci, 59-74, prof biol, 74-77, PROF ZOOL, UNIV NORTHERN COLO, 77- *Concurrent Pos:* NSF lectr, Univ Colo, 59-61; tech adv, Continental Mach, Inc, 55-57; consult, Wilkie Found, 57-59. *Mem:* Fel AAAS; Am Soc Limnol & Oceanog. *Res:* Plankton communities; ichthyology; radio ecology; aquatic biology. *Mailing Add:* Dept Sci Educ Calif State Univ Fullerton CA 92634

THOMAS, BERWYN BRAINERD, b Iowa City, Iowa, Apr 6, 19; m 50; c 3. CHEMISTRY. *Educ:* Univ Ariz, BS, 40; Lawrence Univ, PhD(paper chem), 44. *Prof Exp:* Res chemist, Olympic Res Div, ITT Rayonier Inc, 47-60, group leader, 60-72, tech info specialist, 72-80; RETIRED. *Mem:* Am Chem Soc; Tech Asn Pulp & Paper Indust. *Res:* Preparation and use of cellulose in chemical conversion processes and paper manufacture; methods of evaluation; information retrieval. *Mailing Add:* E 2370 Highway 3 Shelton WA 98584

THOMAS, BILLY SEAY, b Tenn, Dec 31, 26; m 53; c 1. THEORETICAL PHYSICS. *Educ:* Wayne State Univ, BS, 53; Vanderbilt Univ, MS, 55, PhD(physics), 59. *Prof Exp:* Nat Res Coun fel, Argonne Nat Lab, 58-59; asst prof physics, Vanderbilt Univ, 59-60; asst prof, 60-77, ASSOC PROF PHYSICS, UNIV FLA, 77- *Mem:* Am Phys Soc. *Res:* Atomic and molecular scattering; elementary particle physics. *Mailing Add:* Dept Physics Univ Fla Gainesville FL 32611

THOMAS, BRUCE ROBERT, b Guthrie Center, Iowa, Jan 1, 38; m 60; c 3. EXPERIMENTAL PHYSICS, ELECTRONICS. *Educ:* Grinnell Col, BA, 60; Cornell Univ, PhD(theoret physics), 65. *Prof Exp:* Asst prof physics, Grinnell Col, 65-67; asst prof, 67-70, assoc prof, 70-77, PROF PHYSICS, CARLETON COL, 77- *Concurrent Pos:* Guest asst prof, Univ Heidelberg, Germany, 74-75. *Mem:* Am Phys Soc; Am Asn Physics Teachers. *Res:* Experimental atomic physics. *Mailing Add:* Dept Physics & Astron Carleton Col Northfield MN 55057-4025

THOMAS, BYRON HENRY, b Oakland, Calif, Oct 9, 97; m 22; c 3. NUTRITION, BIOCHEMISTRY. *Educ:* Univ Calif, BS, 22; Univ Wis, MS, 24, PhD(animal nutrit), 29. *Prof Exp:* Instr animal husb, Univ Calif, 22-23; asst, Univ Wis, 24-25; dir nutrit, Walker-Gordon Lab Co, NJ, 29-31; prof animal husb & head animal nutrit & chem, Exp Sta, 31-49, PROF BIOCHEM, IOWA STATE UNIV, 49- *Mem:* Am Chem Soc; Am Soc Animal Sci; Soc Exp Biol & Med; Am Dairy Sci Asn; Am Inst Nutrit. *Res:* Irregularities in nutrition in the production of congenital abnormalities in mammals, and their effect on the nutrition, embryology, and histology of affected fetuses. *Mailing Add:* Dept Biochem & Biophys Iowa State Univ 360 Gilman Hall Ames IA 50011

THOMAS, CARL H(ENRY), agricultural engineering; deceased, see previous edition for last biography

THOMAS, CAROLYN EYSTER, b Toledo, Ohio, July 14, 28; m 53. NEUROANATOMY, HISTOLOGY. *Educ:* Univ Toledo, BS, 48; Northwestern Univ, Chicago, MS, 51, PhD(anat), 53. *Prof Exp:* Instr anat, Med Sch, Northwestern Univ, Chicago, 53-58, asst prof, 58-66; asst prof, 66-68, ASSOC PROF ANAT, CHICAGO MED SCH-UNIV HEALTH SCI, 68- *Mem:* Am Asn Anat; Am Soc Zool; Sigma Xi. *Res:* Electrophysiological studies of cat cerebellum; spinal cord structure; muscular architecture of the heart; electron microscope studies of the spleen. *Mailing Add:* Dept Anat Chicago Med Sch-Univ Health Sci 3333 Green Bay Rd North Chicago IL 60064

THOMAS, CECIL OWEN, JR, b East Cleveland, Ohio, Sept 6, 42; m 68; c 1. NUCLEAR ENGINEERING. *Educ:* Univ Tenn, Knoxville, BS, 64, MS, 66, PhD(nuclear eng), 71. *Prof Exp:* Res technician biomed res, Mem Res Ctr & Hosp, Univ Tenn, Knoxville, 63-64, asst, Univ, 67-68; Oak Ridge fel, Savannah River Lab, E I Du Pont de Nemours & Co, Inc, 68-71; nuclear engr, Nuclear Eng Br, Tenn Valley Authority, 71-75; MEM STAFF, NUCLEAR REGULATORY COMN, 75- *Mem:* Am Nuclear Soc. *Res:* Nuclear power plant safety; nuclear instrumentation and control systems. *Mailing Add:* US Nuclear Regulatory Comn Washington DC 20555

THOMAS, CECIL WAYNE, b Dry Ridge, Ky, May 17, 41; m 63; c 2. BIOMEDICAL ENGINEERING, VISUAL PERCEPTION. *Educ:* Univ Ky, BS, 63, MS, 65; Univ Tex, PhD(bioeng), 74. *Prof Exp:* Res engr electronics, Martin Marietta Corp, 65-69; PROF BIOENG, CASE WESTERN RESERVE UNIV, 73- *Concurrent Pos:* Consult, Proj Hope, 82; vis prof, Cairo Univ, Egypt, 82 & Australian Nat Univ, 90. *Mem:* Inst Elec & Electronics Engrs; Psychomet Soc. *Res:* Imaging; visual perception; biomedical signals; computer pattern recognition; intelligent systems; biological applications. *Mailing Add:* Dept Biomed Eng Case Western Reserve Univ Cleveland OH 44106

THOMAS, CHARLES ALLEN, JR, b Dayton, Ohio, July 7, 27; m 51; c 2. BIOPHYSICAL CHEMISTRY. *Educ:* Princeton Univ, AB, 50; Harvard Univ, PhD(phys chem), 54. *Prof Exp:* Res scientist, Eli Lilly & Co, Ind, 54-55; Nat Res Coun fel physics, Univ Mich, 55-56 , instr, 56-57; from asst prof to prof biophys, Johns Hopkins Univ, 57-67; prof biol chem, Harvard Med Sch, 67-77; mem & chmn dept cellular biol, Scripps Clin & Res Found, 77-81; FOUNDER, DIR, HELICON FOUND, 81- *Concurrent Pos:* NSF sr fel, Weizmann Inst, 65; ed bd, Plasmid, Mechanisms Aging & Develop. *Mem:* Am Acad Arts & Sci; Genetics Soc Am; Am Soc Biol Chemists. *Res:* Molecular anatomy of viral and bacterial chromosomes; genetic recombination, organization and function of higher chromosomes; role of free radicals in cellular aging. *Mailing Add:* Helicon Found 4622 Santa Fe St San Diego CA 92109

THOMAS, CHARLES CARLISLE, JR, b Rochester, NY, Aug 18, 25; m 45; c 4. NUCLEAR SAFETY, NUCLEONICS. *Educ:* Univ Iowa, BS, 47; Univ Rochester, MS, 50. *Prof Exp:* Nuclear chemist, US Bur Mines, Okla, 50-51; sr engr, Bausch & Lomb Optical Co, NY, 51-52; tech engr, Aircraft Nuclear Propulsion Dept, Gen Elec Co, Ohio, 52-53; prin chemist, Battelle Mem Inst, 53-55; fel engr, Westinghouse Elec Corp, Pa, 55-60; proj leader radiation chem, Quantum, Inc, 60-62; res mgr, Western NY Nuclear Res Ctr, Inc, 62-72; actg dir, Nuclear Sci & Technol Fac, State Univ NY, Buffalo, 72-74, dir, 74-78; mem staff, Los Alamos Nat Lab, Univ Calif, 78-90; INSTR, NORTHERN NMEX COMMUNITY COL, 85- *Concurrent Pos:* Int Atomic Energy Agency vis prof, Tsing Hua, China, 64-65; adj assoc prof eng sci, aerospace & nuclear eng, State Univ NY Buffalo, 73-78; consult, 90- *Honors & Awards:* Fel, Am Nuclear Soc, 87. *Mem:* Am Chem Soc; Am Nuclear Soc; Sigma Xi; Health Physics Soc. *Res:* Nuclear materials production, safety; neutron activation analysis; reactor technology; environmental analysis; radiation effects on materials. *Mailing Add:* 3373 La Avenida de San Marcos Santa Fe NM 87505

THOMAS, CHARLES HILL, b Dexter, Ga, Jan 31, 22; m 45; c 1. POULTRY SCIENCE, GENETICS. *Educ:* Univ Ga, BSA, 52, MSA, 53; NC State Univ, PhD(genetics), 56. *Prof Exp:* Asst prof poultry husb, Miss State Univ & asst poultry husbandman, Agr Exp Sta, 56-58, assoc prof & assoc poultry husbandman, 58-66, PROF POULTRY SCI, MISS STATE UNIV & POULTRY GENETICIST, AGR EXP STA, 66-, PROF SCI BASIC TO MED, COL VET MED, 77- *Mem:* Am Poultry Sci Asn; Am Genetic Asn. *Res:* Inheritance of resistance to insecticides in Drosophila melanogaster. *Mailing Add:* Dept Poultry Sci Miss State Univ Box 298 Mississippi State MS 39762-0298

THOMAS, CHARLES L(AMAR), b Hendersonville, NC, Oct 13, 05; m 30; c 2. PETROLEUM, CHEMISTRY. *Educ:* Univ NC, BS, 28, MS, 29; Northwestern Univ, PhD, 31. *Prof Exp:* Fel, Northwestern Univ, 29-31; res chemist & assoc dir res, Universal Oil Prods Co, 31-45; dir, res & develop, Great Lakes Carbon Corp, 45-51; staff asst, Sun Oil Co, 51-52, mgr, chem res lab, 52-53, assoc dir, dept res & develop, 53-57, dir, 57-59, sci adv, dept res eng, 59-68; RETIRED. *Mem:* Am Chem Soc; Soc Automotive Eng; fel Inst Chem; Inst Chem Eng. *Res:* Thermal and catalytic reactions of hydrocarbons; catalysts; heterogeneous catalysis; chemistry and physics of solid carbons. *Mailing Add:* 2625 E Southern Ave C-290 Tempe AZ 85282-7601

THOMAS, CLAUDE EARLE, b Spartanburg, SC, Dec 4, 40; m 60; c 3. SCIENCE ADMINISTRATION, PHYTOPATHOLOGY. *Educ:* Wofford Col, AB, 62; Clemson Col, MS, 64; Clemson Univ, PhD(plant path), 66. *Prof Exp:* Res plant pathologist, Subtrop Res Lab, 66-82, Veg Lab, 82-90, LAB DIR, VEG LAB, AGR RES SERV, USDA, CHARLESTON, SC, 90- *Concurrent Pos:* Assoc ed, Plant Dis, 86-88. *Mem:* Am Phytopath Soc. *Res:* Identification, characterization and genetics of resistance to foliar fungal diseases in vegetable crops. *Mailing Add:* Veg Lab Agr Res Serv USDA 2875 Savannah Hwy Charleston SC 29414

THOMAS, CLAUDEWELL SIDNEY, b New York, NY, Oct 5, 32; m 68; c 3. PSYCHIATRY, PUBLIC HEALTH. *Educ:* Columbia Univ, BA, 52; State Univ NY Downstate Med Ctr, MD, 56; Am Bd Psychiat, dipl, 62; Yale Univ, MPH, 64. *Prof Exp:* Chief emergency treatment serv, Ment Health Ctr, New Haven, Conn, 65-67; educ dir psychiat emergency serv, Yale Univ, 67-68, dir social & community psychiat training, 68-73; dir div ment health prog, NIMH, 73-74; chmn dept psychiat, Col Med & Dent, NJ Med Sch, 76-83; CHMN, DEPT PSYCHIAT & HUMAN BEHAV, CHAS DREW MED SCH, 83-; PROF & VCHMN, DEPT PSYCHIAT & BIOBEHAV SCI, SCH MED, UNIV CALIF, LOS ANGELES, 83- *Concurrent Pos:* Consult, Compass Club, New Haven, 63-65; consult psychiatrist, Div Alcoholism, State of Conn, 63-65; vol consult, Caribbean Fed Ment Health, New York, 64; dir Hill-West Haven div & chief unit III, Conn Ment Health Ctr, 67-68; mem ad hoc comt minority admin, Yale Univ, 68-70; soc sci mem, Nat Ctr Health Res & Develop, 69-71; mem assembly behav & soc sci, Nat Acad Sci; consult, Wash Sch Psychiat, A K Rice Inst Wash, 70-77; vis prof sociol, Rutgers Univ, 73-80; attend psychiatrist, Harrison Martland Hosp, Newark; consult, Carrier Clin Belle Mead, St Joseph's Hosp, Patterson, 73-83; chief serv psychiat, Martin Luther King Hosp, Los Angeles, Calif; dep dir mental health, Los Angeles County Dept Mental Health. *Mem:* Fel Am Psychiat Asn; fel Am Pub Health Asn; fel Royal Soc Health; fel NY Acad Med; fel NY Acad Sci. *Res:* Application of theory and concepts in the areas of social and community psychiatry to further the understanding of mental health needs of individuals and groups. *Mailing Add:* Dept Psychiat & Human Behav Charles Drew Postgrad Med School 1720 E 120 St Los Angeles CA 90059

THOMAS, CLAYTON LAY, b Metropolis, Ill, Dec 23, 21; m 50; c 4. MEDICINE. *Educ:* Univ Ky, BS, 44; Med Col Va, MD, 46; Harvard Univ, MPH, 58. *Prof Exp:* Intern med, Montreal Gen Hosp, 46-47; instr, US Naval Sch Aviation Med, 53-54; instr, Col Med, Univ Utah, 54-56; med dir, 58-70, vpres, 69-87, EMER DIR MED AFFAIRS, TAMPAX INC, 87- *Concurrent Pos:* Clin asst, Harvard Med Serv, Boston City Hosp, 56-57; res fel path, Mallory Inst Path, 56-57; consult, Flight Safety Found, 57-58 & Parachutes, Inc, Mass, 60; consult, Dept Pop Sci, Sch Pub Health, Harvard Univ, fel epidemiol, 57-58; mem med & training serv comt, US Olympic Comt, 66-73; pres, Balloon Sch Mass, Inc, 70-; Fed Aviation Admin pilot exam-lighter-than-air-free balloon, 72-; pres, Pop Res Found, 74-; mem med dept vis comt, Mass Inst Technol, 75-78; mem med & training serv comt, US Olympic Comt Sports Med, 78-80. *Mem:* Am Col Physicians; Aerospace Med Asn. *Res:* Medical ecology; physiology of menstruation; medical aspects of sport parachuting and hot air ballooning; aerospace medicine; epidemiology; medical lexicography; sports medicine; health and sex education; physiology of reproduction. *Mailing Add:* Festiniog Farm Dingley Dell 69 Sutcliffe Rd Palmer MA 01069-2122

THOMAS, COLIN GORDON, JR, b Iowa City, Iowa, July 25, 18; m 46; c 4. SURGERY. *Educ:* Univ Chicago, BS, 40, MD, 43. *Prof Exp:* Assoc surg, Col Med, Univ Iowa, 50-51, asst prof, 51-52; from asst prof to prof, Sch Med, Univ NC, Chapel Hill, 52-84, chmn dept, 66-84, chief gen surg, 84-89, PROF SURG, SCH MED, UNIV NC, CHAPEL HILL, 89- *Mem:* Am Col Surgeons; Am Asn Cancer Res; Am Thyroid Asn; Am Surg Asn; Soc Surg Alimentary Tract; Am Asn Endocrine Surgeons; Int Asn Endocrine

Surgeons. *Res:* Thyroid cancer; gastrointestinal disorders; biological characerístics of thyroid neoplasms related to the influence of thyroid-stimulating hormone on their genesis, growth, function and management. *Mailing Add:* Dept Surg 136 Clin Sci Bldg Chapel Hill NC 27514

THOMAS, CRAIG EUGENE, b Clearfield, Pa, Apr 30, 58; m 85; c 2. ROLE OF OXIDATIVE DAMAGE IN ATHEROSCLEROSIS, BIOCHEMICAL MECHANISMS OF TOXICITY. *Educ:* Pa State Univ, BS, 80, MS, 82; Mich State Univ, PhD(biochem & environ toxicol), 86. *Prof Exp:* Nat Inst Environ Health Sci postdoctoral fel biochem/environ health sci, Ore State Univ, 86-88; sr biochem toxicol, Rohm & Haas Co, 88-89; SR RES BIOCHEMIST, MARION MERRELL DOW RES INST, 89- *Mem:* Fedn Am Socs Exp Biol; Am Soc Biochem & Molecular Biol; Oxygen Soc. *Res:* Study of cell death resulting from oxidative damage to biomolecules; role of low density lipoprotein oxidation in atherosclerosis; contribution of oxygen radicals to cerebral ischemic/reperfusion injury. *Mailing Add:* Marion Merrell Dow Res Inst 2110 E Galbraith Rd Cincinnati OH 45215-6300

THOMAS, DAN ANDERSON, b Ooltewah, Tenn, Oct 1, 22; m 44; c 2. PHYSICS. *Educ:* Univ Chattanooga, BS, 45; Vanderbilt Univ, PhD(physics), 52. *Prof Exp:* Asst prof physics, Univ of the South, 49-51; res physicist, US Naval Ord Lab, 51-52; from assoc prof to prof physics, Rollins Col, 52-63; prof physics & dean faculties, 63-80, vpres, 67-80, TRUSTEE PROF PHYSICS, JACKSONVILLE UNIV, 80- *Concurrent Pos:* Consult, US Naval Underwater Sound Reference Lab, 53-63. *Mem:* Fel AAAS; Am Phys Soc; Acoust Soc Am; Am Asn Physics Teachers. *Res:* Beta ray spectroscopy; underwater acoustics; wave motion in solids; vibration of plates. *Mailing Add:* 1990 River Bluff Rd N Jacksonville FL 32211

THOMAS, DAVID ALDEN, b Baltimore, Md, Sept 8, 30; m 56; c 3. MATERIALS SCIENCE. *Educ:* Cornell Univ, BMetE, 53; Mass Inst Technol, ScD(metall), 58. *Prof Exp:* From instr to assoc prof metall, 53-63; chief mat sci res, Ingersoll-Rand Co, 63-68; assoc prof metall & mat sci, 68-70, PROF METALL & MAT SCI, LEHIGH UNIV, 70-, ASSOC DIR, MAT RES CTR, 68- *Mem:* AAAS; Am Soc Metals; Am Inst Mining, Metall & Petrol Engrs; Sigma Xi. *Res:* Structure and mechanical properties of metals, polymers and composites; materials applications; electro microscopy. *Mailing Add:* 1629 Pleasant Dr Bethlehem PA 18015

THOMAS, DAVID BARTLETT, b Butte, Mont, Sept 8, 37; m 61; c 2. MEDICINE, EPIDEMIOLOGY. *Educ:* Univ Washington, MD, 63; Johns Hopkins Univ, DrPH(epidemiol), 72. *Prof Exp:* From asst prof to assoc prof epidemiol, Johns Hopkins Univ, 71-75; assoc prof, 75-81, PROF EPIDEMIOL, UNIV WASH, 81- *Concurrent Pos:* Mem epidemiol, Fred Hutchinson Cancer Res Ctr, 75-; mem breast cancer task force, Nat Cancer Inst, 76-78; consult epidemiol, WHO, 78-79. *Mem:* Soc Epidemiol Res; Am Pub Health Asn; AAAS; Am Epidemiol Soc; fel Am Col Epidemiol. *Res:* Epidemiologic studies of the etiology of gynecologic and breast cancers; studies of the potential carcinogenic effects of steroid contraceptives; nutritional factors and cancer. *Mailing Add:* Fred Hutchinson Cancer Res Ctr 1124 Columbia St Seattle WA 98104

THOMAS, DAVID DALE, b Lansing, Mich, Sept 18, 49; m 75. MOLECULAR BIOPHYSICS. *Educ:* Stanford Univ, BS, 71, PhD(biophys), 76. *Prof Exp:* Res asst physics, High Energy Physics Lab, Stanford Univ, 70; res asst, Dept Genetics, Stanford Univ, 71; fel biophys, Dept Muscle Res, Boston Biomed Res Inst, 76-77; mem fac, Dept Biophys, Stanford Univ, 77-80; ASST PROF BIOCHEM, DEPT BIOL, SCH MED, UNIV MINN, 80- *Concurrent Pos:* NSF fel, 71; fel, Muscular Dystrophy Asns Am, Inc, 76-77. *Mem:* AAAS; Fedn Am Scientists; Biophys Soc. *Res:* Molecular dynamics in muscle contraction as studied by spectroscopic probe methods; electron paramagnetic resonance studies of spin-labeled muscle proteins, such as actin and myosin. *Mailing Add:* Dept Biochem 4-225 Millard Univ Minn Med Sch 435 Delaware St SE Minneapolis MN 55455

THOMAS, DAVID GILBERT, b London, Eng, Aug 4, 28; US citizen; m 57; c 2. TELECOMMUNICATION TRANSMISSION. *Educ:* Oxford Univ, BA, 49, MA, 50, DPhil(chem), 52. *Prof Exp:* Head semiconductor electronics res lab, 62-68, dir electron device process & battery lab, 68-69, exec dir electronic device, Process Mat Div, Murray Hills, NJ, 69-76, EXEC DIR TRANSMISSION SYSTS, BELL LABS, HOLMDEL, NJ, 76- *Honors & Awards:* Oliver E Buckley Solid State Physics Award, Am Phys Soc, 69. *Mem:* Fel Am Phys Soc; fel Inst Elec & Electronics Engrs; AAAS. *Res:* Optical and electrical properties of semiconductors. *Mailing Add:* Bell Labs 3B 631 Crawfords Corner Rd Holmdel NJ 07733

THOMAS, DAVID GLEN, b St Clairsville, Ohio, July 21, 26; m 59. HYDRODYNAMICS, CHEMICAL ENGINEERING. *Educ:* Ohio State Univ, BChE, 47, MSc, 48, PhD(chem eng), 53. *Prof Exp:* Res engr, Battelle Mem Inst, 48-50; res engr, 52-77, PROG MGR, OAK RIDGE NAT LAB, 77-, FORD FOUND PROF FLUID MECH, UNIV TENN, KNOXVILLE, 64- *Concurrent Pos:* NSF sr fel, Cambridge Univ, 60-61; consult, Aerojet-Gen Nucleonics Div, Gen Tire & Rubber Co, 62-63. *Mem:* AAAS; fel Am Inst Chem Engrs; Am Chem Soc; Sigma Xi; Soc Rheology. *Res:* Transport characteristics and non-Newtonian characteristics of suspensions; heat and mass transfer enhancement; vortex interactions; natural convection; turbulence promotion; reverse osmosis; reactor safety thermal hydraulics; hypervelocity projectile shields. *Mailing Add:* 113 Morningside Dr Oak Ridge TN 37830

THOMAS, DAVID LEE, b Dodgeville, Wis, May 22, 49; m 71; c 3. ANIMAL BREEDING, ANIMAL PRODUCTION. *Educ:* Univ Wis-Madison, BS, 71; Okla State Univ, Stillwater, MS, 75, PhD(animal breeding), 77. *Prof Exp:* Asst prof, Ore State Univ, 77-81; asst prof, 81-84, ASSOC PROF ANIMAL SCI, UNIV ILL, 84- *Concurrent Pos:* US Peace Corps, Kenya, 71-73. *Mem:* Am Soc Animal Sci; Am Genetic Asn; Sigma Xi. *Res:* Animal breeding research with sheep and sheep production with primary emphasis on improvement of reproductive efficiency. *Mailing Add:* Animal Sci Univ Ill 156 Animal Sci Lab Urbana IL 61801

THOMAS, DAVID TIPTON, b Barnesville, Ohio, Dec 13, 37; m 66; c 2. ELECTRICAL ENGINEERING. *Educ:* Carnegie Inst Technol, BS, 59, MS, 60; Ohio State Univ, PhD(elec eng), 62. *Prof Exp:* Res assoc elec eng, Antenna Labs, Ohio State Univ, 61-62, assoc supvr & asst prof, 62-63; asst prof elec eng, Carnegie-Mellon Univ, 63-68; mem tech staff, Radio Transmission Lab, Bell Tel Labs, 68-73; prin engr, 73-78, MGR ANTENNA & MICROWAVE ENG, ELECTROMAGNETIC SYSTS DIV, RAYTHEON CORP, 78- *Concurrent Pos:* NSF grant, Carnegie-Mellon Univ, 66-68. *Mem:* Inst Elec & Electronics Engrs. *Res:* Electromagnetic field theory; computer solutions in electromagnetics; scattering, antennas and radiation. *Mailing Add:* Raytheon Co Electromagnetics Systs 6380 Hollister Ave Goleta CA 93017

THOMAS, DAVID WARREN, ANTIGEN PRESENTATION, T-LYMPHOCYTE ACTIVATION. *Educ:* Univ Colo, PhD(microbiol & immunol), 75. *Prof Exp:* Asst prof immunol, Wash Univ, 77-85; PROF IMMUNOL, UNIV MICH, 85- *Mailing Add:* Dept Cell Biol & Immunol Res Biogen Inc 14 Cambridge Ctr Cambridge MI 02142

THOMAS, DON WYLIE, b Spanish Fork, Utah, Aug 8, 23; m 52; c 6. VETERINARY SCIENCE, ANIMAL SCIENCE. *Educ:* Utah State Univ, BS, 49; Iowa State Univ, DVM, 53. *Prof Exp:* Vet pathologist, Calif State Dept Agr, 53-54; prof vet & animal sci & exten vet, Utah State Univ, 54-86; RETIRED. *Concurrent Pos:* State del nat plans conf, USDA, 56, 58 & 62; vchmn, Utah Herd Health & Mastitis Comt, 58- *Mem:* Am Asn Exten Vet; Am Asn Vet Nutritionists. *Res:* Prevention of diseases in cattle, sheep, horses and poultry. *Mailing Add:* 234 E Center Hyde Park PO Box 98 Hyde Park UT 84318

THOMAS, DONALD CHARLES, b Cincinnati, Ohio, Sept 26, 36; m 57; c 3. MICROBIOLOGY, PATHOLOGY. *Educ:* Xavier Univ, BS, 57; Univ Cincinnati, MS, 59; St Louis Univ, PhD(microbiol & virol), 68. *Prof Exp:* Asst dir, Dept Surg, Surg Bact Labs, Univ Cincinnati, 59-61; instr, Dept Biol, Villa Madonna Col, 61-63; grad student, Dept Microbiol & Molecular Virol Inst, St Louis Univ, 63-68; instr, Dept Med Microbiol & Pediat, Col Med, Ohio State Univ, 68-69, asst prof, 69-72, assoc dir, Prog Develop Asst Div, 72-77; dir contracts & grants mgt, Wright State Univ, 77-78, asst dean res, Sch Grad Studies, 79-80, actg dean & assoc dean res & dir, Univ Res Serv, 80-83, dean & assoc vpres for res, 83-90, ASSOC PROF, DEPT PATH & ASSOC PROF, DEPT MICROBIOL & IMMUNOL, SCH MED & COL SCI & MATH, WRIGHT STATE UNIV, 78- *Concurrent Pos:* Adj asst prof, Dept Med Microbiol & Pediat, Col Med, Ohio State Univ, 72-77; admin adv, Nat Reyes Syndrome Found, Ohio, 75-; lab consult, Vet Admin Ctr, Ohio, 78-; fed liaison rep, Am Coun Educ, Wright State Univ, 79-; Reagents Adv Comt Grad Studies, 83-90, chmn, 88-89; Governor's Technol Task Force, 82. *Mem:* Am Soc Microbiol; AAAS; Fedn Am Scientists; Licensing Execs Soc. *Res:* diagnostic virology; molecular aspects of virus replication, pathogenesis of virus diseases and developments in tumor viruses. *Mailing Add:* Dept Microbiol & Immunol Wright State Univ Dayton OH 45435

THOMAS, DONALD E(ARL), b Pittsburgh, Pa, Oct 27, 18; m 42; c 3. MATERIALS SCIENCE ENGINEERING. *Educ:* Carnegie Inst Technol, BS, 42, MS, 49; DSc, 50. *Prof Exp:* Asst phys metall, Carnegie Inst Technol, 46-47; supvry scientist corrosion & phys metall, Atomic Power Div, Westinghouse Elec Corp, 49-58, mgr naval reactor metall, 58-59, mgr mat dept, Astronuclear Lab, 59-66, eng mgr systs & technol, 66-71, consult scientist, Res labs, 71-82; RETIRED. *Concurrent Pos:* Instr, Carnegie Inst Technol, 51-52. *Honors & Awards:* Eng Mat Achievement Award, Am Soc Metals, 72; Outstanding Achievement Award, Am Nuclear Soc, 85. *Mem:* Fel Am Nuclear Soc; Am Inst Mining, Metall & Petrol Engrs; fel Am Soc Metals. *Res:* Materials research and development. *Mailing Add:* 317 Old Farm Rd Pittsburgh PA 15228-2543

THOMAS, DONALD H(ARVEY), b Phoenixville, Pa, Dec 1, 33; div; c 2. COMPUTER AIDED INSTRUCTION. *Educ:* Drexel Inst, BSME, 56, MSME, 59; Case Inst Technol, PhD(design eng), 65. *Prof Exp:* From instr to asst prof, Drexel Univ, 56-61, assoc prof mech eng, 65-88, chmn mech eng, 70-73, assoc dir, Ctr Teaching Innovation, 74-84, assoc dean eng, 80-88, PROF MECH ENG, DREXEL UNIV, 88- *Concurrent Pos:* Consult, 59- *Honors & Awards:* Fred Merryfield Design Award, Am Soc Eng Educ, 86. *Mem:* Sigma Xi; Am Soc Mech Engrs; Am Soc Eng Educ; Soc Automotive Engrs. *Res:* Medical engineering; innovative teaching in engineering; biomechanics; product design liability; system dynamics and control; vehicle dynamics; computer-aided design engineering. *Mailing Add:* Mech Eng Drexel Univ Philadelphia PA 19104

THOMAS, DUDLEY WATSON, biochemistry; deceased, see previous edition for last biography

THOMAS, EDWARD DONNALL, b Mart, Tex, Mar 15, 20; m 42; c 3. INTERNAL MEDICINE, ONCOLOGY. *Educ:* Univ Tex, BA, 41, MA, 43; Harvard Med Sch, MD, 46; Am Bd Internal Med, dipl, 53. *Prof Exp:* From intern to sr asst resident med, Peter Bent Brigham Hosp, Boston, 46-52, chief med res, 52-53, hematologist, 53; res assoc, Cancer Res Found, Children's Med Ctr, 53-55; hematologist & asst physician, Mary Imogene Bassett Hosp, 55-56; assoc clin prof med, Col Physicians & Surgeons, Columbia Univ, 56-63; PROF MED, SCH MED, UNIV WASH, 63- *Concurrent Pos:* Fel med, Mass Inst Technol, 50-51; instr, Harvard Med Sch, 53; physician-in-chief, Mary Imogene Bassett Hosp, 56-63; clin prof, Albany Med Col, 58-63; mem, Fred Hutchinson Cancer Res Ctr, Seattle, 63- *Honors & Awards:* Nobel Prize Med, 90; Nat Award Basic Sci, Am Cancer Soc, 80; Kettering Prize, Gen Motors Cancer Res Found, 81; Karl Lansteiner Mem Award, 87; Nat Medal Sci, 90. *Mem:* Nat Acad Sci; Am Soc Clin Invest; Am Asn Cancer Res; Am Soc Hemat (pres, 88); Exp Hemat Soc; Transplantation Soc. *Res:* Marrow biochemistry and transplantation; irradiation effects; hematology. *Mailing Add:* Fred Hutchinson Cancer Res Ctr 1124 Columbia St Seattle WA 98104

THOMAS, EDWARD SANDUSKY, JR, b Kansas City, Mo, Jan 11, 38; m 60. MATHEMATICS. *Educ:* Whittier Col, BA, 59; Univ Wash, MS, 61; Univ Calif, Riverside, PhD(math), 64. *Prof Exp:* Asst prof math, Univ Mich, Ann Arbor, 65-69; from assoc prof to prof, 69-85, chmn dept math & statist, DISTINGUISHED TEACHING PROF MATH, STATE UNIV NY-ALBANY, 85- *Mem:* Am Math Soc. *Res:* Topology, dynamical systems; differential equations. *Mailing Add:* Dept Math & Statist State Univ NY ES Bldg Albany NY 12222

THOMAS, EDWARD WILFRID, b Croydon, Eng, May 9, 40. PHYSICS. *Educ:* Univ London, BSc, 61, PhD(physics), 64. *Prof Exp:* Asst res physicist, 64-65, from asst prof to assoc prof, 65-73, PROF PHYSICS, GA INST TECHNOL, 73-, DIR, SCH PHYSICS, 82- *Concurrent Pos:* Consult, Oak Ridge Nat Lab, 65- *Mem:* Am Phys Soc; Optical Soc Am; Brit Inst Physics. *Res:* Collisions between atomic, ionic and molecular systems, particularly on the formation of excited states; development of photon and particle detection techniques. *Mailing Add:* Sch Physics Ga Inst Technol Atlanta GA 30332-0430

THOMAS, EDWIN LEE, b Sandusky, Ohio, Nov 30, 43. BIOCHEMISTRY. *Educ:* Miami Univ, BA, 65; Univ Mich, PhD(biochem), 70. *Prof Exp:* Fel biochem, Roche Inst Molecular Biol, 70-72; ASSOC MEM STAFF BIOCHEM, ST JUDE CHILDREN'S HOSP & ASST PROF, UNIV TENN, 72- *Concurrent Pos:* Prin investr, Dent Res Inst, NIH, 76- *Mem:* Am Soc Microbiol; AAAS; Am Soc Biol Chemists. *Res:* Biological membrane structure and function; mechanisms of resistance to infection. *Mailing Add:* Dent Res Ctr Univ Tenn 210 Nash Bldg Memphis TN 38163

THOMAS, EDWIN LORIMER, b Attleboro, Mass, June 14, 47; m 70; c 3. POLYMER SCIENCE, MATERIALS ENGINEERING. *Educ:* Univ Mass, BS, 69; Cornell Univ, PhD(mat sci), 73. *Prof Exp:* Asst prof chem engr, Univ Minn, 73-77; assoc prof, Univ Mass, 77-82, prof, 82-88, head dept, 85-88; PROF MAT SCI & ENGR, MASS INST TECH, 89- *Concurrent Pos:* Humboldt fel, Univ Freiburg, Ger, 81; vis prof, Univ Minn, 83 & Cambridge Univ, 84. *Honors & Awards:* Young Polymer Chemist Award, Am Chem Soc, 85; Fiber Soc Distinguished Award, 86; Ford Prize Polymer Physics, Am Phys Soc, 91. *Mem:* Am Chem Soc; Electron Micros Soc Am; Mat Res Soc; Royal Micros Soc; fel Am Phys Soc. *Res:* Polymer physics and engineering; application of electron microscopy; electron and x-ray diffraction to materials characterization; structure property relationships. *Mailing Add:* Dept Mat Sci & Engr Mass Inst Technol Cambridge MA 02139

THOMAS, ELIZABETH WADSWORTH, b Washington, DC, May 23, 44; m 70; c 1. ANALYTICAL CHEMISTRY, BIOPHARMACEUTICS. *Educ:* ECarolina Univ, BS, 66; Univ Va, PhD(phys & org chem), 70. *Prof Exp:* Res assoc pharmacol, Med Sch, Univ Va, 70-72; lectr, Univ Wis, Parkside, 73-74; analytical chemist, Abbott Labs, 75-79, pharmacologist, 80-86, group leader & sr scientist, 87-88, MGR, ABBOTT LABS, 89- *Concurrent Pos:* Pres, Abbott Chap, Sigma Xi, 82-83; Chair, Abbott's Tech Adv Bd, 89-90. *Mem:* Am Chem Soc; Am Asn Pharmaceut Sci; Sigma Xi. *Res:* Created and validated stability assays for bulk chemicals and formulated products; directed animal studies for evaluation of new drug formulations; determined pharmacokinetics from animal/human data; develop analytical methods for new drug delivery systems; established department that offered unique drug delivery capabilities. *Mailing Add:* 38660 Shagbark Lane Wadsworth IL 60083

THOMAS, ELLIDEE DOTSON, b Huntsville, Ark, July 20, 26; m 60; c 1. NEUROLOGY. *Educ:* Univ Ark, BA, 47, BSM, 58, MD, 58; Univ Tex Med Br, cert, 48. *Prof Exp:* Asst prof child neurol, Sch Med, 65-68, assoc prof, 70-75, PROF CHILD DEVELOP NEUROL, UNIV OKLA HEALTH SCI CTR, 75-, DIR, CHILD STUDY CTR, 69-, CHIEF, SECT DEVELOP PEDIAT, 80- *Mem:* Child Neurol Soc; Am Acad Neurol; Am Acad Pediat; Am Acad Cerebral Palsy & Develop Med. *Res:* Normal and abnormal development in early life and the impact of a handicapped child on parents individually, family relationships and impact on siblings. *Mailing Add:* Child Study Ctr 1100 NE 13th St Oklahoma City OK 73117

THOMAS, ELMER LAWRENCE, b Springfield, Ohio, Jan 18, 16; wid; c 3. FOOD SCIENCE. *Educ:* Ohio State Univ, BSc, 41; Univ Minn, MSc, 43, PhD, 50. *Prof Exp:* Res asst, Univ Minn, St Paul, 41-45, instr dairy prod, 46-50, asst prof dairy technol, 51-54, from assoc prof to prof dairy indust, 54-81, emer prof food sci, Univ Minn, St Paul, 81-; RETIRED. *Concurrent Pos:* Consult food indust, 50- *Mem:* Am Dairy Sci Asn; Inst Food Technologists. *Res:* Physical, chemical and engineering aspects of dairy products processing; sensory testing of foods. *Mailing Add:* 1987 Aldine St Roseville MN 55113

THOMAS, ELVIN ELBERT, b Osceola, Iowa, Nov 23, 44; m 69; c 3. RUMINANT NUTRITION, ANIMAL NUTRITION. *Educ:* Iowa State Univ, BS, 68, MS, 74, PhD(animal nutrit), 77. *Prof Exp:* Voc agr instr, Newell Providence Community Schs, 68-71; instr animal sci, Iowa State Univ, 73-76; ASST PROF ANIMAL SCI, AUBURN UNIV, 77- *Res:* Improvement of the efficiency of feedstuff utilization by ruminants with emphasis on protein utilization and metabolism. *Mailing Add:* Box 708 Greenfield IN 46140

THOMAS, ESTES CENTENNIAL, III, b Plaquemine, La, Dec 13, 40; m 64; c 3. PETROPHYSICS, CORE ANALYSIS. *Educ:* La State Univ, Baton Rouge, BS, 62; Stanford Univ, PhD(phys chem), 66. *Prof Exp:* Fel, Princeton Univ, 66-67; chemist, 67-72, sr engr, 75-82, sr staff engr, 82-86, PETRO PHYS ADV, SHELL OIL CO, 86- *Mem:* Soc Prof Well Log Analysts; Soc Petrol Engrs; Sigma Xi; Soc Care Analysts. *Res:* Physical characteristics and responses to transport of energy through interstices of subterranean earthen formations, particularly those containing hydrocarbons; analysis of these data to deduce volume and type of hydrocarbons with subterranean strata and their propensity to expel hydrocarbons. *Mailing Add:* Shell Oil Co PO Box 2463 Houston TX 77252

THOMAS, EVERETT DAKE, b Stamford, Conn, Jan 24, 43; m 66; c 3. FARM MANAGEMENT. *Educ:* Univ Conn, BS, 65; Cornell Univ, MS, 67. *Prof Exp:* Agr Exten Agent, Coop Extension Asn Clinton County, 66-69; Exten Agronomist, Cornell Univ, 69-81, AGRONOMIST, W H MINER AGRICULT RES INST, 81- *Concurrent Pos:* Consult, US Holstein Asn, 84-; Adj fac, Univ Vermont, 82-, State Univ NY Col Plattsburgh, 82- *Mem:* Am Soc Agron; Soil Sci Soc; Forage & Grassland Coun. *Res:* Soil fertility and crop management; author of articles writes for regional and national magazines on a variety of agricultural crop subjects and edits a small newsletter. *Mailing Add:* Miner Inst Rte 191 Chazy NY 12921

THOMAS, FLOYD W, JR, b Columbia, SC, Apr 7, 38; m. MECHANICAL ENGINEERING. *Educ:* Univ SC, BS, 61; NC State Univ, MS, 63, PhD(mech eng), 67. *Prof Exp:* Mem tech staff, Bell Tel Labs, 61-64; TRW Systs, 67-68; sr eng scientist, McDonnell Douglas Astronautics, 68-69; assoc prof eng & chmn mech aerospace eng, 69-76, assoc dean, 83-88, PROF ENG, CALIF STATE UNIV, FULLERTON, 76- *Mem:* Am Soc Mech Engrs; Am Soc Eng Educ. *Mailing Add:* Mech Engr Dept Calif State Univ Fullerton CA 92634

THOMAS, FORREST DEAN, II, b Provo, Utah, Dec 27, 30; m 52; c 2. INORGANIC CHEMISTRY. *Educ:* Brigham Young Univ, BS, 55; Pa State Univ, PhD(chem), 59. *Prof Exp:* From asst prof to assoc prof, 59-73, PROF CHEM, UNIV MONT, 73- *Mem:* Am Chem Soc. *Res:* Organic and inorganic synthesis; preparation of coordination compounds. *Mailing Add:* Dept Chem Univ Mont Missoula MT 59812

THOMAS, FRANCIS T, b Minneapolis, Minn, June 24, 39; m 68; c 3. THORACIC SURGERY. *Educ:* Univ Minn, BS, 62 & MD, 64. *Prof Exp:* Fel thoracic surg, Case Western Reserve Univ, 69-71; instr surg, Med Col Va, 71-72, from asst prof to assoc prof, 71-79; CHIEF TRANSPLANTATION SURG, DIR RESIDENT RES TRAINING & SURG RES LAB & PROF SURG, ECAROLINA UNIV SCH MED, 79- *Concurrent Pos:* Wangensteen fel foreign study, Birmingham, Eng, 64; mem comt issues, Assoc Acad Surg, 76-79, Human Biol Coun; chmn, Va State Comt Brain Death, Am Soc Transport Surgeon, 78-79, mem prog & publ comt, 88; adv bd, Dialysis & Transplantation; bd dirs, Kidney Found NC, planning & info comt, NC Kidney Coun, Carolina Organ Procurement Agency; med adv bd, Nat Kidney Found NC, 81-84; ad hoc sci consult, Nat Heart, Lung & Blood Inst, Nat Inst Allergy & Infectious Dis & NIH res ctr grants; bd dirs, sci proj & publ comt, Southeastern Organ Procurement Found; mem, Coun Transplantation, Pan-Am Med Asn; reviewer, Am J Kidney Dis, 83-84; ad hoc site reviewer, Transplant Training Prog, Tumor Registrar Asn NC; comt publ, Coun Nominating Comt, Soc Univ Surgeons; guest lectr, Rutgers Med Sch & Univ Calif Los Angeles, 75, Univ Fla, Gainesville & Univ Wash, 76, Cornell Med Sch & Bowman-Gray Sch Med, 77, Med Col Pa & Naval Med Res Ctr, 78, Va Med Soc & Marshfield Clin, 79, Int Conf Immunol Monitoring, Netherlands & Int Transplant Soc, Mass, 80, Barnes Hosp & Univ Minn, Minneapolis, 81, 9th Int Cong Transplantation Soc, Brighton, UK & Am Col Surgeons, Chicago, 82, Int Cong Kidney Transplantation, Ger & Am Col Surgeons, Atlanta, 83, Lenoir-Green Med Soc, NC & 10th Int Cong Transplantation Soc, 84 & other various univ & conf, 85-88; adv comt cardiac transplant, Pitt County Mem Hosp, med dir ECU Transplantation & Procurement, human organ transplantation comt, nursing liaison comt, renal care comt & rep End Stage Res Dis Coun SE; dir resident res activ & surg res lab, dept surg exec, chmn liver transplantation comt & ad hoc comt fac tenure, E Carolina Sch Med; physician's comt comput-based hosp info syst. *Mem:* AAAS; Am Asn Immunologists; Am Asn Tissue Banks; Am Asn Univ Prog; fel Am Cancer Soc; fel Am Col Surgeons; Am Diabetes Asn; Am Med Asn; Am Soc Microbiol; Am Soc Nephrology. *Res:* Cinical and experimental organ transplantation, immunology, surgery, kidney, pancreas and heart transplantation. *Mailing Add:* Dept Surg ECarolina Univ Sch Med Brody 45-10 Moye Blvd Greenville NC 27834

THOMAS, FRANK BANCROFT, b Camden, Del, June 14, 22; m 60; c 2. HORTICULTURE, FOOD TECHNOLOGY. *Educ:* Univ Del, BS, 48; Pa State Univ, MS, 49, PhD(hort), 55. *Prof Exp:* From instr to asst prof hort, Pa State Univ, 49-58; food processing exten specialist, 58-61, exten assoc prof, 61-66, EXTEN PROF FOOD SCI, NC STATE UNIV, 66- *Concurrent Pos:* Vis fel food technol, Mass Inst Technol, 54-55; consult cryogenic foods, 63-64; mem NC Gov Comn Com Fisheries, 64-65, mem Com & Sport Fisheries Adv Comt, 75-76; sabbatical, Dept Food Sci & Technol, Univ Hawaii, 68; partic, Food & Agr Orgn Conf Fish Qual & Inspection, Halifax, Can, 69 & Conf Fishery Prod Technol, Japan, 73; prog leader Food Sci Seafood Adv Serv, NC Sea Grant Prog, 70- *Honors & Awards:* Earl P McFee Award. *Mem:* Inst Food Technologists. *Res:* Post-harvest physiology of fruits and vegetables; food processing; chemical and microbiological changes in seafoods; flavor and color evaluation; extension and applied research on seafood utilization. *Mailing Add:* 2704 Lakeview Dr Raleigh NC 27609-7636

THOMAS, FRANK HARRY, b Alamo, Ga, Oct 16, 32; m 52; c 3. SOILS, INORGANIC CHEMISTRY. *Educ:* Univ Ga, BSA, 54, MS, 56, PhD(soil phosphorus), 59. *Prof Exp:* Asst chemist, Everglades Exp Sta, 59-66; assoc chemist, 66-68, prof chem & chmn div sci-math, 68-75, asst acad dean, 73-75, ACAD DEAN, ABRAHAM BALDWIN COL, 75- *Mem:* Am Chem Soc; AAAS; Am Soc Allied Health Professions; Sigma Xi. *Mailing Add:* 1017 Clark Ave Albany GA 31794

THOMAS, FRANK J(OSEPH), b Pocatello, Idaho, Apr 15, 30; m 49; c 4. ENGINEERING, NUCLEAR PHYSICS. *Educ:* Univ Idaho, BS, 52; Univ Calif, Berkeley, MS, 57. *Prof Exp:* Staff mem advan studies, Sandia Corp, 52-56; prog mgr mobile reactors, Aerojet-Gen Nucleonics Div, 57-61, mgr eng, 61-63, dep mgr appl sci div, 63-64; staff specialist res & eng, Off Secy Defense, 64-65, asst dir res & eng, Nuclear Progs, 65-67; phys scientist, Rand Corp, Calif, 67-71; PRES, PAC-SIERRA RES CORP, LOS ANGELES, 71- *Concurrent Pos:* Lectr, Exten Div, Univ Calif, Berkeley, 57-58; adv, Sci Adv Comt, Defense Intel Agency, 66-73; consult, US Arms Control & Disarmament Agency, 72-76. *Honors & Awards:* Master Design Award, Prod Eng Mag, 63; Meritorious Civilian Serv Award, Secy Defense, 67. *Mem:* Am Inst Aeronaut & Astronaut; AAAS. *Res:* National security and energy issues. *Mailing Add:* 21442 Paseo Portola Malibu CA 90265

THOMAS, GAIL B, b Newton, Mass, Mar 22, 44. CLINICAL MICROBIOLOGY, CHLAMYDIA. *Educ:* Elmira Col, BA, 66; Iowa State Univ, Ames, MS, 70, PhD(vet microbiol), 75. *Prof Exp:* Res assoc vet microbiol, Iowa State Univ, 71-72, res asst 72-75; assoc microbiologist, ITT Res Inst, Chicago, 75-76, res microbiologist, 76-79; res assoc, 79-81, MICROBIOLOGIST, RES & CLIN LAB, ST MARGARET'S HOSP FOR WOMEN, BOSTON, 81-; ASST PROF OBSTET & GYNEC, SCH MED, TUFTS UNIV, 82- *Concurrent Pos:* Prin investr, Nat Inst Environ Health Sci, 78-81; instr obstet & gynec, Sch Med, Tufts Univ, 79-82. *Mem:* Am Soc Microbiol; Int Orgn Mycoplasmol. *Res:* Chlamydial and genital mycoplasma infections in pregnancy; antimicrobial inhibitory activity of amniotic fluids against chlamydia and genital mycoplasmas. *Mailing Add:* 24 Fresh Pond Circle Plymouth MA 02360

THOMAS, GARETH, b Maesteg, Gt Brit, Aug 9, 32; m 60; c 1. PHYSICAL METALLURGY, MATERIALS SCIENCE. *Educ:* Univ Wales, BSc, 52; Cambridge Univ, PhD(metall), 55, ScD, 69. *Prof Exp:* Imp Chem Industs Fel metall, Cambridge Univ, 56-59; from asst prof to assoc prof, 60-66, Miller res prof, 64, assoc dean, Grad Div, 68-69, asst chancellor & act gvchancellor acad affairs, 69-72, chmn fac Col Eng, 77-78, PROF METALL, UNIV CALIF, BERKELEY, SCI DIR, NAT CTR ELECTRON MICROS, LAWRENCE BERKELEY LAB, 81- *Concurrent Pos:* Consult, Exxon; vis scientist, Alcoa Res Labs, 59; Guggenheim fel, Cambridge Univ, 71-72; Alexander von Humboldt sr scientist award, Max Planck Inst, Germany, 81. *Honors & Awards:* Curtis-McGraw Res Award, Am Soc Eng Educ, 66; Nat Award Phys Sci, Electron Micros Soc, 65, 75-76 & 80-81; Electron Micros Soc Am Prize, 65; Rosenhain Medal, Brit Metals Soc, 77; Ernest O Lawrence Award, US Dept Energy, 78. *Mem:* Nat Acad Sci; Nat Acad Eng; Electron Micros Soc Am(pres,75); Am Phys Soc; Am Inst Mining, Metall & Petrol Engrs; Brit Inst Metals; fel Am Soc Metals; fel Metall Soc; fel Royal Micros Soc. *Res:* Electron microscopy; investigations of relation of structure to properties; alloy design. *Mailing Add:* Dept Mat Sci & Mineral Eng Univ Calif 284 Hearst Mining Bldg Berkeley CA 94720

THOMAS, GARLAND LEON, b Topeka, Kans, Aug 29, 20; m 48; c 3. NUCLEAR ENGINEERING. *Educ:* Drury Col, BS, 42; Univ Mo, AM, 48, PhD(physics), 54. *Prof Exp:* Fel engr, atomic power dept, Westinghouse Elec Corp, 53-59; assoc prof physics, Drury Col, 59-66 & Fla Inst Technol, 66-71; mem staff, Fla Planning Dept, Brevard County, 71-79; prin engr, Planning Res Corp, 79-85, AEROSPACE ENGR, NASA SAFETY ENG OFF, KENNEDY SPACE CTR, FLA, 85- *Mem:* Biophys Soc; Am Phys Soc; Am Asn Physics Teachers; Am Nuclear Soc. *Res:* Remote sensing; nuclear reactor physics; ultrasonic cavitation. *Mailing Add:* 1208 E River Dr No 102 Melbourne FL 32901

THOMAS, GARTH JOHNSON, b Pittsburg, Kans, Sept 8, 16; m 45; c 2. BEHAVIORAL NEUROBIOLOGY. *Educ:* Kans State Teachers Col Pittsburg, AB, 38; Univ Kans, MA, 40; Harvard Univ, AM, 43, PhD(exp psychol), 48. *Prof Exp:* Asst instr, Univ Kans, 38-41; tutor, Harvard Univ, 41-43 & 47-48; from instr to asst prof psychol, Univ Chicago, 48-54; res assoc & assoc prof, Neuropsychiat Inst, Col Med, Univ Ill, 54-57, res prof, Biophys Res Lab, Dept Elec Eng & Dept Physiol & Biophys, 57-66; prof, 66-82, dir, 70-77, EMER PROF, CTR BRAIN RES, UNIV ROCHESTER, 82- *Concurrent Pos:* Dept Defense NIMH & NSF res grants, Univ Chicago, Univ Ill & Univ Rochester, 51-77; mem psychol sci fel rev panel, NIMH, 64-69 & psychobiol rev panel NSF, 68-71; consult ed, behav neurosci; assoc ed, J Comp & Physiol Psychol, 69-74, ed, 75-81. *Mem:* AAAS; Am Physiol Soc; Psychonomic Soc; Soc Exp Psychol; Soc Neurosci; Am Psychol Asn; Animal Behav Soc. *Res:* Brain function; studies of behavioral effects of central nervous system lesions; spatial behavior; animal memory. *Mailing Add:* Dept Neurobiol & Anat Box 603 Med Ctr Univ Rochester Rochester NY 14642

THOMAS, GARY E, b Lookout, WVa, Oct 25, 34; m 61; c 2. ATMOSPHERIC PHYSICS. *Educ:* NMex State Univ, BS, 57; Univ Pittsburgh, PhD(physics), 63. *Prof Exp:* Res assoc, Aeronomy Serv, Nat Ctr Sci Res, France, 62-63; mem tech staff space physics lab, Aerospace Corp, 65-67; assoc prof astro-geophys, 67-74, PROF, DEPT ATMOSPHERIC & PLANETARY SCI. *Mem:* Am Geophys Union. *Res:* Theoretical study of upper atmosphere; radiative transfer; application of spectroscopic remote sensing data to study of atmospheric structure; study of radiative and photochemical processes in the stratosphere and mesosphere; noctilucent clouds. *Mailing Add:* Dept Atmospheric & Planetary Sci Univ Colo Boulder CO 80309

THOMAS, GARY LEE, b El Paso, Tex, Feb 27, 47; m 70; c 4. FISHERIES ACOUSTICS, AQUATIC ECOLOGY. *Educ:* Calif Western Univ, BA, 70; San Diego State Univ, MS, 73; Univ Wash, PhD(fisheries), 78. *Prof Exp:* Res assoc, Scripps Inst, Univ Calif, 71-73; res asst, Col Fisheries, Univ Wash, 73-78, biologist, 78-79, res scientist, Appl Physics Lab, 80-85, asst leader, Wash Coop Fish Res Unit, 83-89, RES FAC, COL FISHERIES, UNIV WASH, 79-, DIR, SCI CTR, 90- *Concurrent Pos:* Prin investr, Fisheries Res Inst, 79- *Honors & Awards:* Outstanding Serv Award, US Fish & Wildlife Serv, 90. *Mem:* Am Fisheries Soc; Pac Fisheries Biologists; Am Fedn Inst Res Biologists. *Res:* Hydroacoustic measurement technology and its application to fish population dynamics and aquatic ecology. *Mailing Add:* Prince William Sound Sci Ctr PO Box 705 Cordova AK 99547

THOMAS, GARY LEE, b Willows, Calif, May 12, 37; m 77; c 3. ELECTRICAL ENGINEEERING, SOLID STATE PHYSICS. *Educ:* Univ Calif, Berkeley, BSc, 60, MA, 62, PhD(elec eng), 67. *Prof Exp:* Ed officer elec sci, Accra Polytech Inst, 62-64; instr elec eng, Univ Calif, Berkeley, 67; from asst prof to prof elec eng, State Univ NY, Stony Brook, 67-79, chmn dept, 75-79; VPRES ACAD AFFAIRS, NJ INST TECHNOL, 80- *Concurrent Pos:* NSF grants, 72-; AAAS cong fel, 74-75; mem mat prog, Off Technol Assessment, 74-75. *Mem:* Am Phys Soc; Inst Elec & Electronics Engrs; AAAS; Am Asn Univ Prof. *Res:* Solid state electronics; laser annealing; magnetoelastic surface wave. *Mailing Add:* NJ Inst Technol 323 Martin Luther King Blvd Newark NJ 07012

THOMAS, GEORGE B, b Madras, India, Feb 26, 41; US citizen; m 70; c 3. RADIATION CURING. *Educ:* Univ Kerala, India, BSc, 61, MSc, 63; Indian Inst Technol, Kharagpur, dipl high polymer & rubber technol, 64; Polytech Inst NY, MS, 66, PhD(polymer chem), 71. *Prof Exp:* Res chemist, Polymer Res Corp Am, 71-74; group leader, Floor Prod Res Lab, GAF Corp, 74-81; sr res chemist, Res Div, Gencorp, 81-87; sr res chemist, 87-90, RES SCIENTIST, GEN TIRE CO, 90- *Concurrent Pos:* Mem, Publ Comt, Tire Sci & Technol, Akron Rubber Group & Polymer Lect Group. *Mem:* Am Chem Soc. *Res:* Ultraviolet curable coatings; EB cure of rubber; vulcanization chemistry; adhesive systems; structure-property relationships of polymers. *Mailing Add:* Tire Res Gen Tire Co One General St Akron OH 44329-0026

THOMAS, GEORGE BRINTON, JR, b Boise, Idaho, Jan 11, 14; m 36, 75, 80; c 3. MATHEMATICS. *Educ:* State Col Wash, AB, 34, AM, 36; Cornell Univ, PhD(math), 40. *Prof Exp:* Instr math, Cornell Univ, 37-40; from instr to assoc prof, 44-60, asst elec eng, 43-45, exec officer dept math, 50-59, prof math, Mass Inst Technol, 60-78; RETIRED. *Mem:* Am Math Soc; Math Asn Am (1st vpres, 58-59). *Res:* Calculus and analytic geometry. *Mailing Add:* 500 E Marylyn Ave No G-106 State College PA 16801-6270

THOMAS, GEORGE HOWARD, b Minerva, Ohio, Apr 27, 36; m 60; c 3. PEDIATRICS. *Educ:* Western Md Col, AB, 59; Univ Md, PhD, 63. *Prof Exp:* Instr biochem, Sch Med, Univ Md, 63; asst pediat, Sch Med, 65-67, instr, 67-68, asst prof pediat & med, Sch Med & dir genetics lab, John F Kennedy Inst, 68-76, ASSOC PROF PEDIAT, JOHNS HOPKINS HOSP, JOHNS HOPKINS UNIV, 76- *Mem:* AAAS; Am Soc Human Genetics; Soc Inherited Metab Dis; Soc Pediat Res. *Res:* Human genetics. *Mailing Add:* Dept Pediat & Med Johns Hopkins Univ 720 Rutland Ave Baltimore MD 21205-2179

THOMAS, GEORGE JOSEPH, JR, b New Bedford, Mass, Dec 24, 41; m 66; c 3. BIOPHYSICAL CHEMISTRY. *Educ:* Boston Col, BS, 63; Mass Inst Technol, PhD(phys chem), 67. *Prof Exp:* From asst prof to prof chem, 68-87, head dept, Southeastern Mass Univ, 74-80; CURATORS PROF & HEAD, DIV CELL BIOL & BIOPHYS, SCH LIFE SCI, UNIV MO, KANSAS CITY, 87- *Concurrent Pos:* Prin investr, NIH & NSF grants, 70-; sr res fel, US-Japan Coop Sci Prog, Inst Protein Res, Osaka Univ, Japan, 75-76; mem adv comt, NIH, 79-83; vis scientist, Dept Biol, Mass Inst Technol, Cambridge, 83-84. *Honors & Awards:* Coblentz Award, Coblentz Soc & Soc Appl Spectros. *Mem:* Biophys Soc; AAAS; Sigma Xi; Am Chem Soc; Am Soc Biochem & Molecular Biol. *Res:* Raman and infrared spectroscopy; structure and function of biological molecules; nucleic acid and protein interactions and structures; virus structure and assembly. *Mailing Add:* Div Cell Biol & Biophys Sch Life Sci Univ Mo-Kansas City Kansas City MO 64110-2499

THOMAS, GEORGE RICHARD, b Bethlehem, Pa, Feb 1, 20; m 55; c 4. ORGANIC CHEMISTRY. *Educ:* Bowdoin Col, BS, 41; Northwestern Univ, PhD(chem), 48; Harvard Univ, adv mgt prog, 62. *Prof Exp:* Res fel, Univ Ill, 48-49 & Harvard Univ, 49-50; proj dir, Boston Univ, 50-54; dyestuffs res sect, Qm Res & Eng Command, Natick Labs, US Army, 54-56, asst chief res & develop, Chem & Plastics Div, 56-58, chief, 58-62, assoc dir & dir res, Clothing & Org Mat Lab, 62-68, chief, Mat Res Labs, 68-72, Org Mat Lab, 72-86, chief scientist, Army Mat & Mech Res Ctr, 86-89; RETIRED. *Concurrent Pos:* Pres, Thomason Chem, Inc, 54-60; vis prof, Boston Univ, 63-65; sr exec fel, Kennedy Sch Gov, Harvard Univ, 80; consult, 89- *Mem:* Fel Am Inst Chemists; fel AAAS; Am Chem Soc; Sigma Xi. *Res:* Military applications of polymers as fibers, films, foams, elastomers and rigid and reinforced plastics; materials research; polymer research; operations and management research; business administration; automation in manufacturing. *Mailing Add:* 258 Brayton Point Rd Westport MA 02790-5117

THOMAS, GERALD ANDREW, b Birmingham, Ala, Oct 8, 11; m 40; c 4. RADIOCHEMISTRY. *Educ:* Birmingham-Southern Col, BS & MS, 32; Univ Fla, PhD(chem), 52. *Prof Exp:* Teacher pub schs, Ala, 32-39; asst, Johns Hopkins Univ, 39-40; res chemist, Niagara Alkali Co, 40-45; from instr to assoc prof chem, Univ Fla, 46-57; chmn div sci, math & eng, San Francisco State Univ, 57-63, prof chem, 57-77; RETIRED. *Concurrent Pos:* NSF fel, 54-56; lectr & scientist, US AEC Latin-Am Prog, Atoms in Action, 65-68; consult, Oak Ridge Assoc Univs, 63-80. *Mem:* Am Chem Soc; Am Nuclear Soc. *Res:* Physical properties of organic compounds; nucleonics; radioisotopes. *Mailing Add:* 410 Hillcrest Blvd Millbrae CA 94030-2346

THOMAS, GERALD H, b Salt Lake City, Utah, Sept 3, 42; m; c 1. SOFTWARE SYSTEMS, THEORETICAL PHYSICS. *Educ:* Calif Inst Technol, BS, 64; Univ Calif, Los Angeles, MS, 66, PhD(physics), 69. *Prof Exp:* NATO fel physics, Europ Orgn Nuclear Res, 69-70; res assoc, Univ Helsinki, 70-71; postdoctoral physicist, Argonne Nat Lab, 71-73, asst physicist, 73-75, physicist, 75-81; mem tech staff, 81-84, SUPVR, AT&T BELL LABS, 84- *Mem:* Am Phys Soc; Inst Elec & Electronics Engrs Computer Soc; Asn Computer Mach; NY Acad Sci; AAAS; Math Asn Am. *Res:* Planning and development of large software systems, particularly switching systems. *Mailing Add:* AT&T Bell Labs Rm 1A-131 200 Park Plaza Naperville IL 60566-7050

THOMAS, GORDON ALBERT, b Kingston, Pa, June 8, 43; m 66; c 1. EXPERIMENTAL SOLID STATE PHYSICS. *Educ:* Brown Univ, ScB, 65; Univ Rochester, PhD(physics), 71. *Prof Exp:* Fel physics, Univ Rochester, 71-72; MEM TECH STAFF PHYSICS, BELL LABS, 72- *Concurrent Pos:* Vis prof, Univ Tokyo, 81; vis scholar, Harvard Univ, 85-86. *Mem:* Fel Am Phys Soc. *Res:* Studies of optical properties near the metal-insulator transition and in superconductors. *Mailing Add:* AT&T Bell Labs 1D238 Murray Hill NJ 07974

THOMAS, GRAHAM HAVENS, b Summit, NJ, Mar 26, 51; m 80; c 2. COMPUTERIZED SIGNAL PROCESSING & IMAGE ANALYSIS. *Educ:* Drexel Univ, BS, 74, MS, 75, PhD(mech eng), 79. *Prof Exp:* Sr mem tech staff, Sandia Nat Labs, 79-90; GROUP LEADER, LAWRENCE LIVERMORE NAT LAB, 90- *Mem:* Am Soc Nondestructive Testing. *Res:* Application of signal processing and pattern recognition techniques to ultrasonic nondestructive evaluation; weld flaw detection and sizing; bond strength determination in adhesive and solid state bonds; material characterizations. *Mailing Add:* 1679 Quail Ct Livermore CA 94550

THOMAS, GRANT WORTHINGTON, b Washington, DC, Feb 23, 31; m 51; c 5. SOIL CHEMISTRY. *Educ:* Brigham Young Univ, BS, 53; NC State Univ, MS, 56, PhD(soils), 58. *Prof Exp:* Asst prof agron, Va Polytech Inst, 58-60, assoc prof, 60-64; prof soils, Tex A&M Univ, 64-68; PROF AGRON, UNIV KY, 68- *Concurrent Pos:* Vis prof, Univ Calif, Riverside, 62-63; vis fel, St Cross Col, Univ Oxford, 76. *Mem:* Fel Am Soc Agron; fel Soil Sci Soc Am. *Res:* Reactions and movement of solutes in soils and no-tillage cropping. *Mailing Add:* Dept Agron Univ Ky Lexington KY 40506

THOMAS, H RONALD, b Auburn, Ind, June 9, 42. PHOTOELECTRON SPECTROSCOPY, SURFACE CHEMISTRY. *Educ:* Univ Durham, MSc, 75, PhD(surface chem), 77. *Prof Exp:* Chemist, Xerox Webster Res Ctr, 67-74, assoc scientist, 77-79, scientist, 79-82; SR RES SCIENTIST, MINERALS, PIGMENTS & METALS DIV, PFIZER, INC, 82- *Concurrent Pos:* Instr, Intensive Short Course Photoelectron Spectros, 79-; adj prof chem eng, Univ Wash, 82- *Mem:* Am Chem Soc; Am Phys Soc; Am Vacuum Soc. *Res:* Surface studies on organic polymeric and metallic materials using photoelectron spectroscopy; radio frequency plasma chemistry; laser assisted chemical vapor deposition; interface studies; conducting and semiconducting polymers. *Mailing Add:* 532 Main St Ranshaw PA 17866

THOMAS, HAROLD A(LLEN), JR, b Terre Haute, Ind, Aug 14, 13; m 35; c 3. CIVIL ENGINEERING. *Educ:* Carnegie Inst Technol, BS, 35; Harvard Univ, MS, 37, ScD(sanit eng), 38. *Prof Exp:* From instr to assoc prof sanit eng, Harvard Univ, 38-56, Gordon McKay prof civil & sanit eng, 56-80; RETIRED. *Concurrent Pos:* Consult, Nat Acad Sci-Nat Res Coun, 43-, Dept Health, Educ & Welfare, 49- & US Dept Defense, 52-58. *Honors & Awards:* R E Horton Medal, Am Geophys Union, 78. *Mem:* Nat Acad Eng; fel Am Acad Arts & Sci; Am Geophys Union; Am Soc Civil Engrs. *Res:* Fluid mechanics; hydrology; mathematical statistics; water supply and treatment; systems analysis for water resource development; sanitary and environmental engineering. *Mailing Add:* 61 Cotuit Rd Sandwich MA 02563

THOMAS, HAROLD LEE, b Westphalia, Kans, July 2, 34; m 56; c 3. MATHEMATICS, STATISTICS. *Educ:* Kans State Col Pittsburg, BS, 58, MS, 59; Okla State Univ, PhD(statist), 64. *Prof Exp:* Instr math, Kans State Col Pittsburg, 59-60; asst, Okla State Univ, 60-64; from asst prof to assoc prof, 64-69, PROF MATH, PITTSBURG STATE UNIV, 69- *Mem:* Math Asn Am; Am Statist Asn. *Res:* Factorial experiments in experimental designs. *Mailing Add:* Dept Math Pittsburg State Univ Pittsburg KS 66762

THOMAS, HAROLD TODD, b Seattle, Wash, Feb 3, 42; c 1. PHOTOCHEMISTRY, OPTICAL DISC MATERIALS. *Educ:* Calif Inst Technol, BS, 64; Wesleyan Univ, MA, 66; Princeton Univ, PhD(chem), 70. *Prof Exp:* Res chem, Bell Tel Labs, 69-70; sr res chemist, 70-79, res assoc, 79-84, TECH ASSOC, EASTMAN KODAK CO, 84- *Mem:* Am Chem Soc. *Res:* Photochemistry of ordered systems; spectroscopy and photochemistry of adsorbed molecules; photoresists; optical disc materials. *Mailing Add:* 60 Wintergreen Way Rochester NY 14618

THOMAS, HAZEL JEANETTE, b Birmingham, Ala, Jan 19, 32. ORGANIC CHEMISTRY, MEDICINAL CHEMISTRY. *Educ:* Birmingham Southern Col, BS, 53, MS, 58. *Prof Exp:* SR CHEMIST, RES SOUTHERN RES INST, 53- *Mem:* Am Chem Soc (secy, 76-77, chmn, 80-81). *Res:* Synthesis of potential anticancer agents; nucleosides and nucleotides. *Mailing Add:* Southern Res Inst 2000 Ninth Ave PO Box 55305 South Birmingham AL 35255-5305

THOMAS, HENRY COFFMAN, b Sacramento, Ky, Dec 29, 18; m 44; c 1. PHYSICS. *Educ:* Western Ky State Col, BS, 43; Vanderbilt Univ, MS & PhD, 49. *Prof Exp:* From asst to assoc prof physics, Miss State Col, 49-55; assoc prof & head dept, Bradley Univ, 55-58; chmn dept, 58-74, assoc dean arts & sci, 74-76, PROF PHYSICS, TEX TECH UNIV, 58- *Mem:* Am Phys Soc. *Res:* Low energy nuclear physics. *Mailing Add:* 304 Venable Hall Chapel Hill NC 27515

THOMAS, HERIBERTO VICTOR, b Panama City, Repub Panama, Mar 17, 17; US citizen; m 50; c 2. PREVENTIVE MEDICINE, PUBLIC HEALTH & EPIDEMIOLOGY. *Educ:* Univ Southern Calif, AB, 50, MS, 60; Univ Calif, Los Angeles, MPH, 64; Univ Calif, Berkeley, PhD, 68. *Prof Exp:* Res fel pharmacol & biochem, Univ Southern Calif, 53-56, res assoc, Sch Med, 57-61; res assoc physiol & biochem, St Joseph Hosp, Burbank, 61-64; res chemist, Calif State Dept Health, 64-66, res specialist, 66-72, coordr, Sickle Cell Anemia Prog, 72-74, chief genetic dis, 74-80; sr lectr, health & med sci, Univ Calif, Berkeley, 73-81; CONSULT, 80- *Concurrent Pos:* Reviewer & consult sci, AAAS & NSF, 72-80 & Nat Heart, Lung & Blood Inst, NIH, Rev Br, 75-81; mem, Med Qual Rev Comt No 5, State Med Bd Calif, 80- *Honors & Awards:* Macgee Award, Am Oil Chem Soc, 63. *Mem:* AAAS; Am Chem Soc; Am Soc Human Genetics; Sigma Xi. *Res:* Physiological chemistry of air pollutants and their health effects; lipid metabolism; structural changes in lung tissue as a consequence of adverse ambient conditions; prenatal diagnosis of disabling genetic diseases; genetic counseling as a method of prevention. *Mailing Add:* PO Box 9062 Berkeley CA 94709

THOMAS, HERMAN HOIT, b Raleigh, NC, Dec 26, 31; m 81; c 3. ROCK MAGNETISM, TRACE ELEMENT. *Educ:* Lincoln Univ, Pa, BA, 58; Univ Pa, PhD(geochem), 73. *Prof Exp:* Chemist, US Geol Surv, 58-64, Fairchild-Hiller Corp, 64-65 & Melpar Corp, 65-66; GEOPHYSICIST GEOCHEM, GODDARD SPACE FLIGHT CTR, NASA, 66- *Mem:* Am Geophys Union. *Res:* Composition of the Earth; geophysical analysis of Earth's lithosphere; composition and age of the solar system. *Mailing Add:* Goddard Space Flight Ctr NASA Code 922 Greenbelt MD 20771

THOMAS, HOWARD MAJOR, b Elmwood, Nebr, Feb 14, 18; m 43; c 6. PHYSICAL CHEMISTRY. *Educ:* Nebr State Teachers Col, 42; Univ Iowa, PhD(chem), 49. *Prof Exp:* Asst prof chem & head dept, St Ambrose Col, 49-52; from assoc prof to prof, Univ SDak, 53-58; prof chem, Univ Wis, Superior, 58-82, chmn dept, 58-79; RETIRED. *Concurrent Pos:* Vis prof, Univ Wis, 64-65; Fulbright lectr phys chem, Univ Col Cape Coast, Ghana, 68-69. *Mem:* AAAS; Am Chem Soc; Sigma Xi. *Res:* Chemical kinetics; azeotropic solution; audio-visual aids for chemistry teaching. *Mailing Add:* RFD 9 Box 9240 Hayward WI 54843

THOMAS, HUBERT JON, b Philadelphia, Pa, June 14, 40; m 81. DIGITAL SYSTEM SIMULATION, CONTROL SYSTEM ANALYSIS. *Educ:* Pa State Univ, BS, 62, MS, 64, PhD(elec eng), 69. *Prof Exp:* Asst prof elec eng, Rochester Inst Technol, 69-71; asst prof eng, Calif State Univ, Los Angeles, 71-73; res engr, Zenith Radio Corp, 73-77; SR ENG SPECIALIST, LITTON GUID & CONTROL, 77- *Mem:* Inst Elec & Electronics Engrs. *Res:* Analog and hybrid computation; logical design and switching theory; system simulation and identification; stochastic optimal controls. *Mailing Add:* Litton Guid & Control 5500 Canoga Ave Woodland Hills CA 91364

THOMAS, JACK WARD, b Ft Worth, Tex, Sept 7, 34; m 57; c 2. WILDLIFE BIOLOGY. *Educ:* Tex A&M Univ, BS, 57; WVa Univ, MS, 69; Univ Mass, PhD, 73. *Prof Exp:* Biologist, Tex Game & Fish Comn, 57-62; res biologist, Tex Parks & Wildlife Dept, 62-67; wildlife res biologist, Forestry Sci Lab, Northeastern Forest Exp Sta, US Forest Serv, 67-71, proj dir environ forestry res, Pinchot Inst Environ Forestry, 71-73, PROJ LEADER, RANGE & WILDLIFE HABITAT RES, PAC NORTHWEST FOREST EXP STA, US FOREST SERV, 73-, CHIEF RES BIOLOGIST, 80- *Concurrent Pos:* Adj prof, Wash State Univ, Ore State Univ, Univ Idaho, Eastern Ore State Col & WVa Univ. *Honors & Awards:* Ore Chapter Award, Wildlife Soc, 79; Einarsen Award, Northwest Sect, Wildlife Soc, 81; Gulf Conserv Award, 84; Earle A Chiles Award, 85; Nat Wildlife Fedn Award, 91. *Mem:* Sigma Xi; Wildlife Soc; Wilson Ornith Soc; Am Ornith Union; Am Soc Mammal; fel Soc Am Foresters. *Res:* Mobility and home range management of deer and turkeys; population dynamics of deer; disease impact on deer and antelope populations; wildlife habitat research; sociobioeconomic implications of game habitat manipulation; habitat requirements for wildlife in urbanizing areas; relationships of wild and domestic ungulates to forested ranges; non-consumptive utilization of wildlife; forestry-wildlife relationships. *Mailing Add:* USDA Forest Serv 1401 Gekeler Lane La Grande OR 97850

THOMAS, JAMES, b Spokane, Wash, 1946. COMPUTER SCIENCE, COMPUTER GRAPHICS. *Educ:* Eastern Wash Univ, BS, 68; Wash State Univ, MS, 71. *Prof Exp:* Chief scientist, Environ & Molecular Sci Lab, 87-91, COMPUTER SCI TECHNOL LAB, BATTELLE PAC NORTHWEST LABS, 91- *Concurrent Pos:* Chair, Spec Interest Group Computer Graphics, Asn Comput Mach, 89- *Mem:* Inst Elec & Electronics Engrs; Asn Comput Mach; Human Factors Soc. *Mailing Add:* Battelle Pac Northwest Labs PO Box 999 MS K7-32 Richland WA 99352

THOMAS, JAMES ARTHUR, b International Falls, Minn, Apr 22, 38; m 61; c 2. OXIDATIVE STRESS, ENZYMES. *Educ:* St Olaf Col, BA, 60; Univ Wis, MS, 63, PhD(biochem), 66. *Prof Exp:* USPHS fel biochem, Univ Minn, 67-69; from asst prof to assoc prof, 69-87, PROF BIOCHEM, IOWA STATE UNIV, 88- *Mem:* Am Soc Biochem & Molecular Biol; Am Chem Soc; Am Soc Biol Chemists; Sigma Xi; Oxygen Soc. *Res:* Protein S-thiolation; oxidative stress and proteins in intact cells. *Mailing Add:* Biochem Iowa State Univ 375 C Gilman Ames IA 50011

THOMAS, JAMES E, b Marshall, Mo, May 17, 26; m 52; c 2. PHYSICS. *Educ:* Mo Valley Col, BS, 50; Univ Mo-Rolla, MS, 55; Univ Mo-Columbia, PhD(physics), 63. *Prof Exp:* Teacher high sch, Mo, 50-52; instr math, Univ Mo-Rolla, 52-55; assoc prof physics, Mo Valley Col, 55-61; assoc prof, 63-65, PROF PHYSICS, PITTSBURG STATE UNIV, 65-, ACTG CHAIR, 90- *Mem:* Am Phys Soc; Am Asn Physics Teachers. *Res:* Small angle x-ray diffraction; particle size and structure determination, solid state; radiation damage in perfect crystals; thin films growth and characterization. *Mailing Add:* Dept Physics Pittsburg State Univ Pittsburg KS 66762

THOMAS, JAMES H, b Cardston, Alta, Jan 9, 36; m 59; c 4. PLANT BREEDING, GENETICS. *Educ:* Utah State Univ, BSc, 61, MS, 63; Univ Alta, PhD(genetics), 66. *Prof Exp:* Res asst agron, Utah State Univ, 60-63; teacher sci, Taber Sch Div, Alta, 63-64; res asst genetics, Univ Alta, 64-66; res officer, Can Dept Agr, 66-67 & Rudy Patrick Seed Co, 67-69; seed specialist & adv, Utah State Univ-AID, Bolivia, 69-72, assoc prof plant sci & int progs, Utah State Univ, 72-74, dryland farming adv, Tehran, Iran, 74-76, int progs coordr, 76-80, chief party & dir res, Lapaz, Bolivia, 80-82, forage res adv, Egypt, 82-84, dir int progs, 84-90; DIR, USDA REGIONAL RES, INDIA, 90- *Mem:* Am Soc Agron; Crop Sci Soc Am; Am Inst Biol Sci. *Res:* Administrator international programs. *Mailing Add:* Dept Plant Sci Utah State Univ Logan UT 84322

THOMAS, JAMES H, b Waltham, Mass, Mar 4, 55; m 81; c 2. HEAVY ION PHYSICS, WEAK INTERACTIONS. *Educ:* Wash State Univ, BS, 76; Yale Univ, MS, 77, MPhil, 79, PhD(physics), 82. *Prof Exp:* Millikan fel, Calif Inst Technol, 83-85, sr res fel, 86-88; PHYSICIST, LAWRENCE LIVERMORE NAT LAB, 89- *Concurrent Pos:* Mem exec comt, Brookhaven Nat Lab, 91. *Mem:* Am Phys Soc; Sigma Xi. *Res:* Relativistic heavy ion research; weak interaction physics; beta decay; nuclear physics; gravity. *Mailing Add:* Lawrence Livermore Nat Lab L-397 Livermore CA 94550

THOMAS, JAMES WARD, RHEUMATOLOGY, DIABETES. *Educ:* Univ Tenn, Memphis, MD, 73. *Prof Exp:* ASST PROF MED, MICROBIOL & IMMUNOL, BAYLOR COL MED, 81- *Mailing Add:* Dept Med & Immunol Baylor Col Med One Baylor Plaza Houston TX 77030

THOMAS, JAMES WILLIAM, b Ironwood, Mich, Nov 11, 41; m 67; c 4. MATHEMATICS. *Educ:* Mich Technol Univ, BS, 63; Univ Ariz, MS, 65, PhD(math), 67. *Prof Exp:* Asst prof math, Univ Wyo, 67-72; assoc prof, 72-77, PROF MATH, COLO STATE UNIV, 77- *Concurrent Pos:* NSF sci develop grant, Univ Ariz, 70-71. *Mem:* Am Math Soc; Soc Indust & Appl Math. *Res:* Applied mathematics; hydrodynamics; nonlinear functional analysis and application; numerical solution of partial differential equations. *Mailing Add:* Dept Math Colo State Univ Ft Collins CO 80521

THOMAS, JEROME FRANCIS, b Chicago, Ill, Jan 8, 22; c 9. AIR & WATER POLLUTION, COMBUSTION. *Educ:* DePaul Univ, BS, 43; Univ Calif, PhD(chem), 50. *Prof Exp:* Res chemist, 50-55, assoc prof, 55-72, PROF SANIT ENG, UNIV CALIF, BERKELEY, 72-, CHMN, DIV HYDRAUL & SANIT ENG, 73-, EMER PROF, 87- *Concurrent Pos:* Consult, Energy Res Develop Agency 69- & Nat Acad Sci, 73-; adj prof, Environ Protection Agency, 70- *Mem:* Am Chem Soc; Am Soc Eng Educ; Soc Appl Spectros; Am Water Works Asn. *Res:* Sanitary chemistry; chemical aspects applied to air, water pollution and quality control; applied organic chemistry. *Mailing Add:* Davis Hall Univ Calif Berkeley CA 94720

THOMAS, JOAB LANGSTON, b Holt, Ala, Feb 14, 33; m 54; c 4. BOTANY. *Educ:* Harvard Univ, AB, 55, AM, 57, PhD, 59. *Hon Degrees:* DSc, Univ Ala, 81; LLD, Stillman Col, 87. *Prof Exp:* Cytotaxonomist, Arnold Arboretum Harvard Univ, 59-61; from asst prof to prof biol, Univ Ala, 61-76, asst dean, Col Arts & Sci, 65, dean students & vpres students affairs, 69-76; chancellor, NC State Univ, Raleigh, 76-81; PRES, UNIV ALA, 81- *Concurrent Pos:* Dir, Herbarium, Univ Ala, 61-76, dir arboretum, 64-76. *Honors & Awards:* Carleton K Butler Award, 75. *Mem:* Am Soc Plant Taxon; Bot Soc Am; Int Asn Plant Taxon; Sigma Xi. *Res:* Systematics and cytogenetics of higher plants. *Mailing Add:* Univ Ala PO Box 1927 Tuscaloosa AL 35487-1498

THOMAS, JOE ED, b Moran, Tex, Oct 3, 37; m 58; c 3. ELECTRICAL ENGINEERING. *Educ:* Univ Wyo, BS, 60; Univ Idaho, MS, 62; Univ Denver, PhD(elec eng), 70. *Prof Exp:* Res engr seismic propagation, Standard Oil Co Calif, 60-61 & Denver Res Inst, 62-66; from asst prof to prof elec eng Univ Idaho, 66-89, chmn dept, 77-89; CHMN & PROF ELEC ENG, UNIV COLO, DENVER, 89- *Concurrent Pos:* NSF grant, Univ Idaho, 67-, Nat Oceanic & Atmospheric Admin grant, 71- *Mem:* Am Geophys Soc; Inst Elec & Electronics Engrs. *Res:* Acoustics, particularly acoustic-gravity wave propagation and acoustic sources. *Mailing Add:* Dept Elec Eng Univ Colo Campus Box 110 PO Box 173364 Denver CO 80217-3364

THOMAS, JOHN A, b La Crosse, Wis, Apr 6, 33; m 57; c 2. PHARMACOLOGY, TOXICOLOGY. *Educ:* Univ Wis-La Crosse, BS, 56; Univ Iowa, MA, 58, PhD(physiol), 61. *Prof Exp:* Instr physiol, Univ Iowa, 60-61; asst prof pharmacol, Sch Med, Univ Va, 61-64; assoc prof, Sch Med, Creighton Univ, 64-67; prof pharmacol, Sch Med, WVa Univ, 67-82, asst dean, 73-75, assoc dean, 75-82; vpres corp res, Baxter Int, 82-88; VPRES ACAD SERV, UNIV TEX HEALTH SCI CTR, 88- *Concurrent Pos:* Adj prof, Northwestern Univ, Univ Ill, Rush-Presby, Chicago Med, Univ Chicago, 83-88 & Med Col Wis, 83- *Honors & Awards:* Sci Recognition, Environ Protection Agency; DuBois Award Toxicol. *Mem:* Am Soc Pharmacol & Exp Therapeut; Endocrine Soc; Soc Toxicol; Pharmacol Soc Can; Am Col Toxicol; Am Acad Vet Pharmacol & Therapeut; Teratology Soc. *Res:* Endocrine pharmacology, mechanism of action of androgens; prostate gland neoplasms; pesticides and reproduction; reproductive toxicology; phthalate acid esters; genetic engineering and biotechnology. *Mailing Add:* 219 Wood Shadow San Antonio TX 78216

THOMAS, JOHN ALVA, b Berwyn, Ill, May 9, 40; m 65; c 2. BIOCHEMISTRY. *Educ:* DePauw Univ, AB, 62; Univ Ill, Urbana, PhD(biochem), 68. *Prof Exp:* NIH fel, Univ Pa, 68-70; from asst prof to assoc prof, 70-84, PROF BIOCHEM, SCH MED, UNIV SDAK, 85- *Concurrent Pos:* Vis prof, Cornell Univ, 77-78. *Mem:* AAAS; Biophys Soc; Am Soc Biochem & Molecular Biol; Sigma Xi; Soc Gen Physiologists. *Res:* Bioenergetics and oxidative phosphorylation; enzyme kinetics; intracellular pH. *Mailing Add:* Dept Biochem Univ SDak Sch Med Vermillion SD 57069

THOMAS, JOHN B(OWMAN), b New Kensington, Pa, July 14, 25; m 44; c 6. ELECTRICAL ENGINEERING. *Educ:* Gettysburg Col, AB, 44; Johns Hopkins Univ, BS, 52; Stanford Univ, MS, 53, PhD(elec eng), 55. *Prof Exp:* Elec engr, Koppers Co, Inc, 46-51, asst chief engr, 51-52; from asst prof to assoc prof, 55-62, PROF ELEC ENG, PRINCETON UNIV, 62- *Concurrent Pos:* NSF sr fel, 67-68. *Mem:* Fel Inst Elec & Electronics Engrs. *Res:* Communication and information theory; random processes; corona discharge; high voltage rectification. *Mailing Add:* Dept Elec Eng Princeton Univ Princeton NJ 08544

THOMAS, JOHN HOWARD, b Chicago, Ill, Apr 9, 41; m 62; c 2. SOLAR PHYSICS, FLUID DYNAMICS. *Educ:* Purdue Univ, BS, 62, MS, 64, PhD(eng sci), 66. *Prof Exp:* NATO fel appl math, Cambridge Univ, 66-67; from asst prof to assoc prof mech & aerospace sci, Univ Rochester, 67-81, assoc dean grad studies, 81-83, assoc, C E K Mees Observ, 72-85; PROF MECH & AEROSPACE SCI, UNIV ROCHESTER, 81-, DEAN GRAD STUDIES, 83- *Concurrent Pos:* Vis scientist, Max Planck-Inst Physics & Astrophysics, Munich, 73-74, High Altitude Observ, Boulder, Colo, 85, Nat Solar Observ, Sunspot, NMex; vis prof theoret physics, Univ Oxford, 87-88, vis fel, Worcester Col, Oxford, 87-88; prof astron, Univ Rochester, 85- *Mem:* Am Phys Soc; Am Astron Soc; Int Astron Union; Am Geophys Union; Am Soc Mech Engrs; AAAS; Sigma Xi. *Res:* Astrophysical fluid dynamics; magnetohydrodynamics; solar physics; applied mathematics. *Mailing Add:* 436 Lattimore Hall Univ Rochester River Campus Rochester NY 14627

THOMAS, JOHN HUNTER, b Beuthen, Ger, Mar 26, 28; US citizen. PLANT SYSTEMATICS. *Educ:* Calif Inst Technol, BS, 49; Stanford Univ, AM, 49, PhD, 59. *Prof Exp:* from asst cur to cur, 58-72, assoc prof, 69-77, DIR DUDLEY HERBARIUM, STANFORD UNIV, 72-, PROF BIOL SCI, 77- *Concurrent Pos:* Cur, Dept Bot, Calif Acad Sci, 69- *Mem:* Am Soc Plant Taxon; Bot Soc Am; Soc Study Evolution; Am Fern Soc. *Res:* Flora of central and lower California and Alaska; management of systematic collections; information storage and retrieval in systematic collections; botanical history. *Mailing Add:* Dept Biol Sci Stanford Univ Stanford CA 94305-5020

THOMAS, JOHN JENKS, b Boston, Mass, Dec 15, 36; m 65; c 2. GEOLOGY. *Educ:* Williams Col, BA, 61; Northwestern Univ, MS, 65; Univ Kans, PhD(geol), 68. *Prof Exp:* from asst prof to assoc prof, 68-82, PROF GEOL, SKIDMORE COL, 82- *Mem:* AAAS; Geol Soc Am. *Res:* Structural geology; tectonics; metamorphic petrology. *Mailing Add:* Dept Geol Skidmore Col Saratoga Springs NY 12866

THOMAS, JOHN KERRY, b Llanelly, South Wales, Gt Brit, May 16, 34; m 59; c 3. PHYSICAL CHEMISTRY. *Educ:* Univ Manchester, BSc, 54, PhD(chem), 57, DSc, 69. *Hon Degrees:* DSc, Univ Manchester, Eng, 69. *Prof Exp:* Nat Res Coun Can fel chem, 57-58; sci off, Atomic Energy Res Estab, Eng, 58-60; res assoc, Argonne Nat Lab, 60-62, assoc chemist, 62-70; prof chem, 70-83, JULIUS NIEUWLAND PROF CHEM, UNIV NOTRE DAME, 83- *Honors & Awards:* Res Award, Radiation Res Soc, 72. *Mem:* Radiation Res Soc; Am Chem Soc; fel Royal Soc Chem; Photobiol Soc. *Res:* Photo-induced reactions in organized media, including micellas, microemulsions, P stymes and colloidal semiconductors; techniques used in rapid laser spectroscopy; colloid chemistry; polymer radiation chemistry; photochemistry. *Mailing Add:* Dept Chem Univ Notre Dame Notre Dame IN 46556

THOMAS, JOHN M, b Wilmar, Calif, Sept 17, 36; m 57; c 3. BIOMETRICS, ECOLOGY. *Educ:* Calif State Polytech Col, BS, 58; Wash State Univ, MS, 60; Univ Ariz, PhD(biochem, nutrit), 65. *Prof Exp:* Instr avian physiol, Calif State Polytech Col, 60-61; res assoc biochem, Univ Ariz, 61-65; sr res scientist, Pac Northwest Labs, Battelle Mem Inst, 61-71; gen ecologist, Div Biol & Med, Energy Res & Develop Admin, 71-72; res assoc, 72-76, MEM STAFF, ENVIRON SCI DEPT, PAC NORTHWEST DIV, BATTELLE MEM INST, 76- *Concurrent Pos:* Mem task group, Int Comt Radiation Protection, 74-; adj prof & coordr biol prog, Wash State Univ, 77- *Mem:* Soc Environ Toxicol & Chem; Am Statist Asn; Biomet Soc. *Res:* Biology and ecology; methods development for field surveys and the prediction of effects of various insults on humans based on laboratory and field animal data; ecological effect; environmental impact; bioassessment of hazardous chemical waste sites. *Mailing Add:* 226 Wright Ave Richland WA 99301

THOMAS, JOHN MARTIN, b Omaha, Nebr, Oct 17, 10; m 36; c 4. PEDIATRICS. *Educ:* Grinnell Col, AB, 32; Yale Univ, MD, 37. *Prof Exp:* From instr to asst prof pediat, Col Med, Creighton Univ, 40-49; from asst prof to prof, 49-83, asst prof rehab, 66-83, EMER PROF PEDIAT, UNIV NEBR, MED CTR, OMAHA, 83-; STAFF PHYSICIAN, GLENWOOD STATE HOSP-SCH, 83- *Mem:* AMA; Am Acad Pediat. *Mailing Add:* 1002 N 63rd St Omaha NE 68132

THOMAS, JOHN OWEN, b Los Angeles, Calif, Nov 1, 46; c 2. BIOPHYSICAL CHEMISTRY, ELECTRON MICROSCOPY. *Educ:* San Diego State Univ, BS, 68; Cornell Univ, PhD(biochem), 72. *Prof Exp:* Fel, Biochem Dept, Stanford Univ, 72-75; asst prof biochem, 75-82, RES ASSOC PROF, MED SCH, NY UNIV, 82- *Concurrent Pos:* Fel, Damon Runyon Mem Fund, 72-74; Am Cancer Soc Sr Fel, 74-75. *Mem:* NY Acad Sci. *Res:* Nucleic acid-protein interactions; structures of large nucleoprotein complexes; electron microscopy of macromolecules; RNA metabolism. *Mailing Add:* Dept Biochem Med Sch New York Univ 550 First Ave New York NY 10016

THOMAS, JOHN PAUL, physical inorganic chemistry; deceased, see previous edition for last biography

THOMAS, JOHN PELHAM, b Ashby, Ala, Apr 18, 22; m 45; c 6. MATHEMATICS. *Educ:* Auburn Univ, BS, 46; Univ Va, MAT, 61; Univ SC, PhD(math), 65. *Prof Exp:* Asst county agt, Agr Exten Serv, Auburn, 46-53; farmer, 53-54; jr & high sch teacher, Ala, 55-60; asst prof math, Univ NC, 64-67; head dept, 67-75, PROF MATH, WESTERN CAROLINA UNIV, 67- *Mem:* Math Asn Am. *Res:* Maximal topological spaces; separation axioms; properties preserved under strengthening and weakening of topologies. *Mailing Add:* Rte 67 Box 358 Cullowhee NC 28723

THOMAS, JOHN RICHARD, b Anchorage, Ky, Aug 26, 21; m 44; c 2. PHYSICAL CHEMISTRY. *Educ:* Univ Calif, BS, 43, PhD(phys chem), 47. *Prof Exp:* Asst Nat Defense Res Comt, Univ Calif, 43-44, Manhattan Dist, 44-47; res assoc, US AEC Contract, Gen Elec Co, 47-48; res chemist, Calif Res Corp, 48-49; asst chief chem br, US AEC, 49-51; sr res scientist, Calif Res Corp, 51-67, mgr res & develop, Ortho Div, Chevron Chem Co, 67-68, asst secy, Stand Oil Co Calif, 68-70, pres, Chevron Res Co, 70-83, vpres, Chevron Corp, 83-86; RETIRED. *Honors & Awards:* Earl B Barnes Award, Am Chem Soc, 90. *Res:* Free radicals; oxidation kinetics; electron spin resonance; petroleum technology; synthetic fuels. *Mailing Add:* 847 Mcellon Way Lafayette CA 94549

THOMAS, JOHN WILLIAM, b Spanish Fork, Utah, Mar 25, 18; m 45; c 4. ANIMAL NUTRITION, DAIRY SCIENCE. *Educ:* Utah State Univ, BS, 40; Cornell Univ, PhD(nutrit), 46. *Prof Exp:* Res assoc, Nat Defense Res Comt, Northwestern Univ, 42-45 & Carnegie Inst Technol, 45; nutritionist & biochemist, Bur Dairy Indust, USDA, 46-53 & dairy husb res br, 53-60; prof, 60-87, EMER PROF DAIRY SCI, MICH STATE UNIV, 87- *Concurrent Pos:* Exten specialist dairy, 79- *Honors & Awards:* Am Feed Mfrs Asn Award, 53; Borden Award, Am Dairy Sci Asn, 74. *Mem:* Fel AAAS; fel Am Soc Animal Sci; Am Dairy Sci Asn; Am Inst Nutrit; Sigma Xi. *Res:* Mineral and vitamin requirements and functions in feeding of dairy cattle; forage evaluation and preservation; thyroid active stimulants for cattle; rumen functions; calf nutrition. *Mailing Add:* Dept Animal Sci Mich State Univ East Lansing MI 48824

THOMAS, JOHN XENIA, JR, b Birmingham, Ala, July 25, 50; m 74; c 1. PHYSIOLOGY. *Educ:* Birmingham-Southern Col, BS, 72; Univ Miss Med Ctr, PhD(physiol), 76. *Prof Exp:* Res assoc, Dept Physiol & Biophys, Univ Miss, 72-76; res assoc, 76-78, asst prof, 78-84, ASSOC PROF PHYSIOL, STRICTCH SCH MED, LOYOLA UNIV CHICAGO, 84- *Concurrent Pos:* Res fel, Chicago Heart Asn, 77-78; instr nursing physiol, Hinds Jr Col, 75-76; Schweppe Found career develop award, 78-82; chmn, Cardiovasc Inst, 85- *Mem:* Am Heart Asn; Am Physiol Soc; AAAS. *Res:* Cardiac metabolism; cardiac dynamics; coronary circulation; nervous control of circulation. *Mailing Add:* Dept Physiol Strictch Sch Med Loyola Univ Chicago 2160 S First Ave Maywood IL 60153

THOMAS, JOSEPH CALVIN, b Churubusco, Ind, May 2, 33; m 55; c 1. SCIENCE EDUCATION, ORGANIC CHEMISTRY. *Educ:* Asbury Col, AB, 54; Univ Ky, MA, 55, EdD(sci educ, chem), 61. *Prof Exp:* Instr chem, Asbury Col, 54-59 & Jessamine County High Sch, 59-60; asst prof sci, 61-65, chmn chem dept, 63-74, assoc dean Sch Arts & Sci, 79-81, dean, 81-87, dean, fac & instr, 87-90, PROF CHEM, UNIV NALA, 65-, VPRES ACAD AFFAIRS & PROVOST, 90- *Concurrent Pos:* Sci consult local pub sch systs, 63- *Mem:* AAAS; Am Chem Soc; Nat Sci Teachers Asn. *Res:* Preparation and continued education of secondary school science teachers, particularly their preparation in the physical sciences. *Mailing Add:* Box 5150 Univ NAla Florence AL 35632-0001

THOMAS, JOSEPH CHARLES, b Mt Union, Pa, Oct 16, 45; m 65; c 2. MATHEMATICS. *Educ:* Shippensburg State Col, BS, 66; Pa State Univ, MA, 68; Kent State Univ, PhD(math), 75. *Prof Exp:* Instr, 68-70, ASST PROF MATH, LOCK HAVEN STATE COL, 72- *Res:* Torsion theories generated by ideals; global dimension of associative rings with identity. *Mailing Add:* Dept Math Kutztown Univ Kutztown PA 19530

THOMAS, JOSEPH ERUMAPPETTICAL, b Piravom, Kerala State, India, Feb 11, 37; m 64; c 3. APPLIED PSYCHOPHYSIOLOGY, HYPNOTHERAPY. *Educ:* Univ Kerala, India, BA, 57, MA, 60, PhD(psychol), 69. *Prof Exp:* Res asst, Univ Kerala, 63-66, lectr psychol, 66-70; intern clin psychol, Northwestern Univ Med Sch, 71-72; psychologist drug abuse res, Univ Chicago, 72-74; psychologist, Inst Psychiat, Northwestern Univ, 74-76; psychologist chronic pain mgt, Rehab Inst Chicago, 76-80; DIR CLIN PSYCHOL, INST BEHAV HEALTH, 80- *Concurrent Pos:* Asst prof, Dept Psychiat & Behav Sci, Northwestern Univ Med Sch, 77-; consult, Michael Reese Hosp, Chicago, Ill, 80-85, NIH, 84-; dir, Psychiat Unit, Mental Health Ctr, La Salle, Ill, 74; pres, Biofeedback Soc Ill, 85. *Mem:* Am Psychol Asn; Am Soc Clin Hypnosis; Biofeedback Soc Am. *Mailing Add:* 16 W 731 89th Pl Hinsdale IL 60521

THOMAS, JOSEPH FRANCIS, JR, b Chicago, Ill, Feb 29, 40; m 67; c 3. MATERIALS ENGINEERING, MATERIALS PROCESSING. *Educ:* Cornell Univ, BEP, 63; Univ Ill, Urbana, MS, 65, PhD(physics), 68. *Prof Exp:* Asst prof physics, Univ Va, 67-72; from asst prof to assoc prof eng & physics, Wright State Univ, 72-83, prog dir, Mat Sci & Eng, 76-84, dept chair Mech & Mat Eng, 84-90, PROF MAT SCI & ENG, WRIGHT STATE UNIV, 83-, DEAN, SCH GRAD STUDIES & ASSOC VPRES RES, 90- *Concurrent Pos:* Prog Dir Metall, NSF, 83-84. *Mem:* Am Soc Metals Int; Am Phys Soc; Minerals, Metals & Mat Soc; Am Soc Mech Engrs; Am Soc Eng Educators. *Res:* Plastic deformation; applications to metal forming; material constitutive equations; physics of metals and alloys. *Mailing Add:* Grad Studies & Res Wright State Univ Colonel Glenn Hwy Dayton OH 45435-0001

THOMAS, JOSEPH JAMES, b Columbia, Pa, Sept 10, 09; m 32, 51; c 7. BIOCHEMISTRY. *Educ:* Pa State Univ, BS, 30, MS, 32, PhD(biochem), 35. *Prof Exp:* Asst res, NY Exp Sta, Geneva, 30; instr agr biochem, Pa State Univ, 31-36; biochemist, Rohm and Haas, 36-41; from asst dir to dir res, S D Warren Co, 42-68, vpres res, 68-72; tech consult, Edward C Jordan Co, Inc, 73-76; TECH PULP, PAPER & CHEM CONSULT, 73- *Concurrent Pos:* Tech adv, Int Exec Serv Corps, 80. *Mem:* Am Chem Soc; Tech Asn Pulp & Paper Indust. *Res:* Synthetic resins; functional uses of pulp and paper. *Mailing Add:* 16234 N 111th Ave Sun City AZ 85351

THOMAS, JUDITH M, Lynn, Mass, Jan 4, 44; m; c 3. TRANSPLANTATION IMMUNOLOGY. *Educ:* NY Univ, PhD(biol immunol), 72. *Prof Exp:* DIR TRANSPLANT IMMUNOL, ECAROLINA UNIV, 80-, PROF SURG, 84- & DIR HLA LAB, SCH MED. *Concurrent Pos:* Adj prof microbiol & biol, Sch Med, ECarolina Univ, 84- *Mem:* Am Asn Immunol; Int Transplantation Soc; Am Soc Histocompatibility & Immunogenetics; Sigma Xi; Nat Soc Med Res. *Mailing Add:* Dept Surg Div Transplantation ECarolina Sch Med Greenville NC 27834

THOMAS, JULIAN EDWARD, SR, b Yazoo City, Miss, Aug 1, 37; m 56; c 3. MICROBIAL PHYSIOLOGY. *Educ:* Fisk Univ, AB, 59; Atlanta Univ, MS, 67, PhD(biol), 71; Southern Univ, MST, 68. *Prof Exp:* Teacher pub schs, Ga, 59-65; instr biol & chem, SC State Col, 67-69; fel microgenetics, Argonne Nat Lab, 71-73; assoc prof, 73-77, PROF BIOL, TUSKEGEE INST, 77-, HEAD, BIOL DEPT, 79-, ASSOC DIR, CARVER RES FOUND, 79- *Concurrent Pos:* Consult, Argonne Ctr Educ Affairs, 75-76. *Mem:* Sigma Xi; Fedn Am Scientists; AAAS; Am Soc Microbiol. *Res:* Involvement of transfer RNA in the regulation of enzyme synthesis by repression, and derepression, control. *Mailing Add:* Kresege Ctr Tuskegee Univ Tuskegee AL 36088

THOMAS, KENNETH ALFRED, JR, b Oklahoma City, Okla, Nov 28, 46; m 73; c 3. PROTEIN CHEMISTRY & CRYSTALLOGRAPHY, GROWTH FACTORS. *Educ:* Univ Del, BS, 69; Duke Univ, PhD(biochem), 74. *Prof Exp:* Res fels, Duke Univ, 74-75, NIH, 75-77 & Wash Univ, 77-79; DIR GROWTH FACTOR RES, MERCK SHARP & DOHME RES LABS, MERCK INST, 79- *Mem:* Protein Chem Soc; AAAS; Am Soc Biol Chemists. *Res:* Protein growth factors; blood vessel growth control; tumor metastasis; the structure and function of proteins. *Mailing Add:* Dept Biochem Rm 80W-243 Merck Inst PO Box 2000 Rahway NJ 07065-0900

THOMAS, KENNETH EUGENE, III, b Hammond, La, Jan 31, 54; m 79. INORGANIC CHEMICAL SEPARATIONS. *Educ:* Southeastern La Univ, BS, 75; Univ Calif, Berkeley, PhD(chem), 80. *Prof Exp:* Resident, 79-80, MEM STAFF, LOS ALAMOS NAT LAB, 80- *Mem:* Am Chem Soc. *Res:* Production and isolation of large quantities of various radionuclides for use in the fields of medicine, chemistry, and physics; development of chemical separation processes applicable to hot cell handling of highly radioactive target materials. *Mailing Add:* Dept Math Andrews Univ Berrien Springs MI 49104

THOMAS, KIMBERLY W, b Albany, NY, July 3, 52; m 79. NUCLEAR CHEMISTRY, RADIOBIOLOGY. *Educ:* Middlebury Col, AB, 73; Univ Calif, Berkeley, MBioradiol, 78, PhD(nuclear chem), 78. *Prof Exp:* STAFF SCIENTIST CHEM, LOS ALAMOS NAT LAB, 78- *Mem:* Am Chem Soc; Asn Women Sci. *Res:* Radiochemistry; radioactive waste management; isotope synthesis and isolation; applications of radiochemistry to biology and medicine. *Mailing Add:* 376 Catherine Los Alamos NM 87544-3565

THOMAS, LARRY EMERSON, b Indianapolis, Ind, Dec 27, 43. APPLIED MATHEMATICS. *Educ:* Rose Polytech Inst, BS, 66; Rensselaer Polytech Inst, MS, 68, PhD(math), 70. *Prof Exp:* Asst prof, 70-72, ASSOC PROF MATH, ST PETER'S COL, NJ, 72-, COORDR PRE-ENG PROG, 76- *Mem:* Math Asn Am; Soc Indust & Appl Math. *Res:* Differential equations. *Mailing Add:* Dept Math St Peter's Col 2641 Kennedy Blvd Jersey City NJ 07306

THOMAS, LAWRENCE E, b Columbus, Ohio, Mar 15, 42. PROBABILITY, REAL ANALYSIS. *Educ:* Yale Univ, BS, 64, PhD(physics), 70. *Prof Exp:* Res asst, Fed Inst Technol, Zurich, Switz, 70-72; res asst, Univ Geneva, 72-74; from asst prof to assoc prof, 74-81, PROF MATH, UNIV VA, 81- *Mem:* Am Phys Soc; Am Math Soc; Int Asn Math Physics. *Res:* Work & theory Shrodinger operators, statistical mechanics, & stocastic processes. *Mailing Add:* Math/Astron Bldg Cabell Dr Univ Va Charlottesville VA 22903

THOMAS, LAZARUS DANIEL, b Toledo, Ohio, Oct 21, 25; m 50; c 5. PHYSICAL CHEMISTRY. *Educ:* Univ Mich, BS, 48, MS, 49. *Prof Exp:* Teaching fel, Univ Mich, 49-50; res supvr, Libbey-Owens-Ford Co, 51-88; RETIRED. *Mem:* Electrochem Soc; Am Electroplaters Soc; Am Chem Soc; Am Ceramic Soc. *Res:* Films on glass; semiconductors; surface chemistry of glass; electrochemistry. *Mailing Add:* Five Lindsay Ct Bowling Green OH 43402

THOMAS, LEE W(ILSON), b Boswell, Pa, Oct 31, 26; m 50; c 2. CHEMICAL ENGINEERING. *Educ:* Univ Pittsburgh, BS, 49. *Prof Exp:* Chem analyst, Jones & Laughlin Steel Co, 49-50; process engr, Bethlehem Steel Co, 50-52; field engr, Eng Dept, 52-63, res engr, Pigments Dept, 63-67, sr res engr, Newport, 67-72, tech serv rep, Pigments Dept, E I du Pont de Nemours & Co, Inc, 72-85; PRIN CONSULT, CECON GROUP INC, 85- *Mem:* Am Inst Chem Engrs; Am Chem Soc. *Res:* Development work in particle processes including extreme temperature ranges. *Mailing Add:* 230 Steeplechase Circle Wilmington DE 19808

THOMAS, LEO ALVON, b Gifford, Idaho, Mar 19, 22; c 2. PARASITOLOGY, MEDICAL MICROBIOLOGY. *Educ:* Univ Idaho, BS, 49; Univ Mich, MS, 50; Tulane Univ, PhD(parasitol, med microbiol), 55. *Prof Exp:* With virus labs, Rockefeller Found, NY, 55-57; med bacteriologist, Rocky Mountain Lab, USPHS, 57-80, res microbiologist, 80-85; RETIRED. *Res:* Ecology and classification of arthropod-borne viruses; biological and chemical characterization of Coxiella burnetii antigens. *Mailing Add:* Rocky Mountain Lab USPHS Hamilton MT 59840

THOMAS, LEO JOHN, b Grand Rapids, Minn, Oct 30, 36; m 58; c 4. CHEMICAL ENGINEERING. *Educ:* Univ Minn, BS, 58; Univ Ill, MS, 60, PhD(chem eng), 62. *Hon Degrees:* Dr, Worcester Polytech Inst, 88. *Prof Exp:* Chmn, Sterling Drug Inc, 88-89; res chemist, Color Photog Div, Res Labs, Eastman Kodak Inc, 61-67, head, Color Physics & Eng Lab, 67-70, asst div head, 70-72, tech asst to dir, 72-75, asst dir, Res Labs, 75-77, vpres, 77-78, sr vpres & dir, 78-84, gen mgr, Life Sci Div, 84-88, GROUP VPRES & GEN MGR, HEALTH GROUP, EASTMAN KODAK CO, 89- *Concurrent Pos:* Bd chem sci & technol, Nat Res Coun, 84; mem, Resource Develop Comt, Dept Chem Eng, Univ Ill; mem bd dirs, Rochester Tel Corp & John Wiley & Sons, Inc; mem, Bd Chem Sci & Tech, Nat Res Coun. *Mem:* Nat Acad Eng; Am Chem Soc; Am Acad Arts & Sci; Am Inst Chem Engrs; AAAS. *Res:* Photographic science; chemical engineering kinetics. *Mailing Add:* Eastman Kodak Co 343 State St Rochester NY 14650-1152

THOMAS, LEONARD WILLIAM, SR, b Birmingham, Ala, May 11, 09; m 34; c 3. ELECTROMAGNETIC COMPATIBILITY. *Educ:* Alabama Polytech Inst, BS, 31. *Prof Exp:* Radio eng, radiostation WAPI, Birmingham, 32-39 & Columbia Broadcasting Syst, Washington, DC, 39-42; ELECTRONICS ENGR, THOMAS ENG CO, 70- *Honors & Awards:* LG Cumming Award, Electromagnetic Compatibility Soc, Inst Elec & Electronics Engrs, 79. *Mem:* Inst Elec & Electronics Engrs; Soc Automotive Engrs; Nat Soc Prof Engrs; Am Soc Naval Engrs; Am Nat Standards Comt. *Res:* Development of measurement instruments; instrument specifications and standards; radiated and conducted measurement techniques. *Mailing Add:* 1604 Buchanan St NE Washington DC 20017-3121

THOMAS, LEWIS, b Flushing, NY, Nov 25, 13; m 43; c 3. INTERNAL MEDICINE, PATHOLOGY. *Educ:* Princeton Univ, BS, 33; Harvard Univ, MD, 37. *Hon Degrees:* MA, Yale Univ, 69; ScD, Rochester Univ, 74, Princeton Univ, 76, Med Col Ohio, 76, Columbia Univ, 78, Mem Univ Nfld, 78, Univ NC, 78, Williams Col, 82, Conn Col, 83, Univ Wales, 83, Univ Ariz, 85, Harvard Univ, 86, Long Island Univ, 87, Univ Ill, 89, Rockefeller Univ, 89, Univ Minn, 89; LLD, Johns Hopkins Univ, 76, Trinity Col, 80, Ursinus Col, 81; LHD, Duke Univ, 76, Reed Col, 77, NY Univ, 83, Drew Univ, 83, Mt Sinai Sch Med, 90; LittD, Dickinson Col, 80, State Univ NY, 83; PhD, Weizmann Inst, 84. *Prof Exp:* Intern, Boston City Hosp, 37-39; intern, Neurol Inst, New York, 39-41; Tilney Mem fel, Thorndike Lab, Boston City Hosp, 41-42; vis investr, Rockefeller Inst, 42-46; asst prof pediat, Sch Med, Johns Hopkins Univ, 46-48; from assoc prof to prof med, Sch Med, Tulane Univ, 48-50; prof pediat & med & dir pediat res labs, Heart Hosp, Univ Minn, 50-54; prof path & chmn dept, Sch Med, NY Univ, 54-58, prof med & chmn dept, 58-66, dean, 66-69; prof path & chmn dept, Sch Med Yale Univ, 69-72, dean sch med, 72-73; pres & chief exec officer, 73-80, chancellor, 80-83, EMER PRES, MEM SLOAN-KETTERING CANCER CTR, 83-, EMER MEM, 83-; PROF MED & PATH, MED SCH, CORNELL UNIV, 73-, CO-DIR,

GRAD SCH MED SCI, 74- *Concurrent Pos:* Mem, Comn Streptococcal & Staphylococcal Dis, Dept Defense, Armed Forces Epidemiol Bd, 50-62, Bd Health, City New York, 55-69, Pathol Study Sect, USPHS, 55-59, Pres Sci Adv Comt, 67-70; consult, Surgeon Gen, US Army, 52-66, Manhattan Vet Admin Hosp, 54-59, Comt Res, Pres Comt Heart Dis, Cancer & Stroke, 64-65; dir, III & IV Med Divs, Bellevue Hosp, 58-66, pres med bd, 63-65; dir med, Univ Hosp, 59-60; mem, Nat Adv Health Coun, NIH, 60-64, Nat Adv Child Health & Human Develop Coun, 64-68, Nat Adv Coun Aging, 82-83, Adv Comt to Dir, 83-84; mem bd dirs, Pub Health Res Inst, New York, 60-69, Squibb Corp, 69-86, Josiah Macy Jr Found, 75-84, Burke Rehab Ctr, 80-85, Richard Lounsbery Found, 82-, Am Friends Cambridge Univ, 84-; chmn narcotics adv comt, New York City Health Res Coun, 61-63; chief path, Yale-New Haven Hosp, 69-73; mem sci adv comt, Mass Gen Hosp, 69-72, Fox Chase Inst Cancer Res, 76-84, Sidney Farber Cancer Inst, 78-82; chmn, Comt Rev Nat Cancer Plan, Nat Acad Sci, 72; mem sci adv bd, Scripps Clin & Res Found, 73-78, C V Whitney Lab Exp & Marine Biol, Univ Fla, 76-80, Kennedy Inst Ethics, Georgetown Univ, 82-; mem bd trustees, C S Draper Lab, 74-79, Rockefeller Univ, 75-88, Guggenheim Found, 75-85, Hellenic Anticancer Inst, Athens, 77-, Mt Sinai Med Ctr, 79-84, Menninger Found, 81-, Col of the Atlantic, 82-85; chmn, Cluster Overview, Pres Biomed Res Panel, 75-76; mem, Comt Planetary Biol & Chem Evolution, Space Sci Bd, 77-79; mem sci coun, Int Inst Cellular & Molecular Path, Brussels, 77-82; spec med adv group, Vet Admin, 77-80; adv, Nat Hospice Orgn, 78-; mem adv coun, Cornell Univ Grad Sch Bus & Pub Admin, 78-84, Prog in Hist of Sci, Princeton Univ, 82-; chmn bd, Scientists' Inst Pub Info, 82-88, Monell Chem Senses Ctr, 82-; vis prof neuropath, Yale Univ Sch Med; adj prof, NY Univ Sch Med & Rockefeller Univ; scholar-in-residence, Cornell Univ Med Col; vis physician, Rockefeller Univ Hosp. *Honors & Awards:* Harvey Lectr, 68; Papanicolaou Award, 79; Med Educ Award, AMA, 79; Kober Medal, Asn Am Physicians, 83; Milton Helpern Mem Award, 86; Lewis Thomas Award for Commun, Am Col Physicians, 86; William B Coley Award, Cancer Res Inst, 87; Gold Headed Cane Award, Am Asn Pathologists, 88; Pub Serv Award, Fedn Am Socs Exp Biol, 88; Albert Lasker Pub Serv Award, 89; Loren Eiseley Award, 90; John Stearns Award for Lifetime Achievement, NY Acad Sci, 91. *Mem:* Nat Acad Sci; Inst Med-Nat Acad Sci; fel NY Acad Sci (pres, 89); Practitioners Soc; Am Asn Immunologists; Am Philos Soc; Am Acad Arts & Sci; Asn Am Physicians; Am Pediat Soc; Soc Exp Biol & Med. *Res:* Infectious disease; hypersensitivity; pathogenicity of mycoplasmas. *Mailing Add:* Cornell Univ Med Col 1300 York Ave New York NY 10021

THOMAS, LEWIS EDWARD, b Lima, Ohio, May 18, 13; m 40; c 4. ORGANIC CHEMISTRY. *Educ:* Ohio Northern Univ, BS, 35; Purdue Univ, MS, 37. *Prof Exp:* Asst chem, Purdue Univ, 35-39; from instr to asst prof, Va Mil Inst, 40-45; develop engr, 45-50, tech serv & lab supvr, 50-70, asst mgr lab, 70-73, mgr lab, Sun Oil Co, 74-78; RETIRED. *Concurrent Pos:* Vis scientist, Ohio Acad Sci, NSF vchmn bd trustees, Univ Toledo & Toledo-Lucas County Libr Syst. *Mem:* Nat Soc Prof Engrs; Am Chem Soc; Am Inst Chem Engrs. *Res:* Chlorination of aliphatic hydrocarbons; selective solvents for olefin and diolefin purification; pyrolysis of chlorinated aliphatic hydrocarbons; esterification of alcohol ethers. *Mailing Add:* 4148 Deepwood Lane Toledo OH 43614

THOMAS, LEWIS JONES, JR, b Philadelphia, Pa, Dec 13, 30; m 55; c 2. BIOMEDICAL COMPUTING. *Educ:* Haverford Col, BS, 53; Wash Univ, MD, 57; Am Bd Anesthesiol, dipl, 63. *Prof Exp:* Intern med, Bronx Munic Hosp, NY, 57-58; USPHS res fel, Sch Med, Wash Univ, 58-60; resident anesthesiol, Barnes Hosp, St Louis, Mo, 60-62; staff anesthesiologist, Clin Ctr, NIH, 62-64; asst prof anesthesiol, Sch Med, Wash Univ, 64-74, physiol & biophys, 70-74, biomed eng, 72-74, assoc dir, Biomed Comput Lab, 72-75, assoc prof, Inst Biomed Comput, 85-89, ASSOC PROF ANESTHESIOL, PHYSIOL/BIOPHYS & BIOMED ENG, SCH MED, WASH UNIV, 74-, ELEC ENG, 78-, PROF, INST BIOMED COMPUT, 89- *Concurrent Pos:* Alexander Berg prize undergrad res microbiol, Wash Univ, 55, dir, Biomed Comput Lab, 75- & assoc dir, Inst Biomed Comput, Wash Univ, 83-; USPHS res career develop award, 66. *Honors & Awards:* Borden Award. *Mem:* AAAS; Am Physiol Soc; AMA; NY Acad Sci. *Res:* Respiratory physiology; biomedical computer applications; anesthesia. *Mailing Add:* Biomed Comput Lab Sch Med Washington Univ 700 S Euclid St Louis MO 63110-1085

THOMAS, LLEWELLYN HILLETH, b London, Eng, Oct 21, 03; nat US; m 33; c 3. PHYSICS, MATHEMATICS. *Educ:* Cambridge Univ, BA, 24, PhD(theoret physics), 27, MA, 28, DSc, 65. *Prof Exp:* From asst prof to prof physics, Ohio State Univ, 29-43; physicist & ballistician, Ballistic Res Lab, Aberdeen Proving Ground, Md, 43-45; prof physics, Ohio State Univ, 45-46; mem sr staff, Watson Sci Comput Lab, 46-68, prof physics, 46-68, EMER PROF PHYSICS, COLUMBIA UNIV, 68- *Concurrent Pos:* Prof physics, NC State Univ, 68-76, emer prof, 76- *Honors & Awards:* Davisson-Germer Prize, 82. *Mem:* Nat Acad Sci; AAAS; fel Am Phys Soc; Royal Astron Soc. *Res:* Theoretical astrophysics; atomic physics; relativity theory; nuclear, atomic and molecular structure; field theory; computational methods. *Mailing Add:* 3012 Wycliff Rd Raleigh NC 27607

THOMAS, LLYWELLYN MURRAY, b Detroit, Mich, Sept 23, 22; m 47; c 6. NEUROSURGERY. *Educ:* Wayne State Univ, BA, 49, MD, 52. *Prof Exp:* Assoc prof, 65-68, asst chmn dept, 65-70, assoc dean hosp affairs, 72-81, PROF NEUROSURG & CHMN DEPT, SCH MED, WAYNE STATE UNIV, 70- *Concurrent Pos:* Sr attend, Detroit Gen Hosp, 65- & Grace Hosp, Detroit, 70-; consult, Harper Hosp, 71- & Children's Hosp Mich, 71- *Mem:* Am Asn Neurol Surg; Am Col Surg; AMA; Cong Neurol Surg. *Res:* Head injury. *Mailing Add:* Neuro Surg Assoc PC 22250 Providence Dr Suite 601 Southfield MI 48075

THOMAS, LOUIS BARTON, b Medicine Lodge, Kans, June 8, 19; m 44; c 3. PATHOLOGY. *Educ:* Col Idaho, AB, 40; Univ Chicago, MD, 45; Am Bd Path, dipl, 52. *Prof Exp:* Resident path, Univ Minn, 48-51; spec fel neuropath, Mayo Clin, 51-52; resident, Mem Ctr Cancer & Allied Dis, New York, 52-53; head surg path & post-mortem serv, Clin Ctr, NIH, 53-69; chief lab path, Nat Cancer Inst, 69-; RETIRED. *Concurrent Pos:* Clin prof, Schs Med & Dent, Georgetown Univ, 72- *Mem:* Am Asn Path & Bact; fel Col Am Path; Am Asn Cancer Res; Am Soc Exp Path; Int Acad Path. *Res:* Diagnostic and research pathology, particularly cancer; leukemia and malignant lympomas. *Mailing Add:* 2107 Essex Ct Ft Collins CO 80526

THOMAS, LUCIUS PONDER, b Easley, SC, June 30, 25; m 52; c 2. ELECTRONICS ENGINEERING. *Educ:* Clemson Univ, BS, 47. *Prof Exp:* Engr, RCA Corp, 47-61, leader eng TV, 61-69, mgr advan prod develop, 69-71, mgr black & white TV, 71-78, mgr eng prod safety, 78-87; RETIRED. *Mem:* Sr mem Inst Elec & Electronics Engrs. *Res:* Design, development and supervision in television receiver development. *Mailing Add:* 7311 N Lesley Ave Indianapolis IN 46250

THOMAS, LYELL JAY, JR, b Madison, Wis, Apr 17, 25; m 48; c 2. PHARMACOLOGY. *Educ:* Oberlin Col, AB, 48; Univ Pa, PhD(zool), 53. *Prof Exp:* Instr pharmacol, Woman's Med Col, Pa, 52-55; asst prof biol, 55-60, assoc prof, 60-62, ASSOC PROF PHARMACOL, UNIV SOUTHERN CALIF, 62- *Mem:* Am Physiol Soc; Soc Gen Physiol; Cardiac Muscle Soc; Sigma Xi. *Res:* Cellular physiology and pharmacology of heart muscle; excitation contraction coupling in heart muscle; mechanism of insulin secretion. *Mailing Add:* Dept Pharmacol & Nutrit Univ Southern Calif 2025 Zonal Ave Los Angeles CA 90033

THOMAS, MCCALIP JOSEPH, b Yazoo City, Miss, Jan 1, 14; m 59; c 2. CHEMISTRY. *Educ:* Miss State Col, BS, 36; Vanderbilt Univ, MS, 38, PhD(org chem), 41. *Prof Exp:* Asst chemist, Exp Sta, Miss State Col, 36-37; Glidden-Upjohn-Abbott fel, Northwestern Univ, 41-42; from res chemist to sr res chemist, A E Staley Mfg Co, 42-48, asst to mgr, Mkt Develop Dept, 48-52, asst mgr, 52-57, mgr, Chem Mkt dev, 57-60; mem staff, Applns Res Dept, Nat Cash Register Co, 61-67, sci liaison & mem staff tech support, 67-68; exec vpres, Hill Top Res, Inc, 68-73, DIR MKT, HILL TOP TESTING SERV, INC DIV, AM BIOMED CORP, 73-, VPRES MKT, 76-, MKT CONSULT, 85- *Concurrent Pos:* Mem indust adv comt soup & gravy bases, Qm Food & Container Inst, 54 & task group, Res & Develop, 55; res fel, Northwestern Univ, 41-42. *Mem:* Am Chem Soc; Sigma Xi; Am Pharmaceut Asn; Inst Food Technol; fel Am Inst Chemists; Soc Cosmetic Chemists. *Res:* Research, development and testing in the biological, toxicological, chemical, medical and microbiological fields. *Mailing Add:* 678 Hyde Park Dr Dayton OH 45429

THOMAS, MARTHA JANE BERGIN, b Boston, Mass, Mar 13, 26; m 55; c 4. ANALYTICAL CHEMISTRY, PHYSICAL CHEMISTRY. *Educ:* Radcliffe Col, AB, 45; Boston Univ, AM, 50, PhD(chem), 52; Northeastern Univ, MBA, 81. *Prof Exp:* Sr engr in chg chem lab, 45-59, group leader lamp mat eng labs, Lighting Prod Div, 59-66, sect head chem & phosphor lab, Sylvania Lighting Ctr, 66-72, mgr tech asst labs, GTE sylvania lighting prod group, Sylvania Elec Prod, Inc, 72-81, tech dir, Tech Serv Labs, 81-83, DIR, TECH QUAL CONTROL, GEN TEL & ELECTRONICS CORP, DANVERS, MA, 83- *Concurrent Pos:* Instr eve div, Boston Univ, 52-70; adj prof chem, Univ RI, 74- *Honors & Awards:* Nat Achievement Award, Soc Women Engrs, 65; Golden Plate, Am Acad Achievement, 66; Centennial Alumni Award, Boston Latin Acad, 78. *Mem:* Am Chem Soc; Electrochem Soc; fel Am Inst Chemists; Soc Women Engrs. *Res:* Phosphors; photoconductors; ion exchange membranes; complex ions; instrumental analysis. *Mailing Add:* 18 Cabot St Winchester MA 01890-3502

THOMAS, MARTIN LEWIS HALL, b Feb 9, 35; Can citizen; m 56; c 3. BIOLOGY. *Educ:* Univ Durham, BSc, 56; Univ Toronto, MSA, 62; Dalhousie Univ, PhD, 70. *Prof Exp:* Assoc scientist, Biol Sta, Fisheries Res Bd Can, Ont, 56-62, scientist, Biol Sub-Sta, 62-70; asst prof, 70-74, assoc prof, 70-79, PROF BIOL, UNIV NB, ST JOHN, 79- *Concurrent Pos:* Mem, Int Oceanog Found, 55-; bd dir, Huntsman Marine Lab, 77-80. *Mem:* Hon mem NY Acad Sci; Nat Shellfisheries Asn; Marine Biol Asn UK; Brit Ecol Soc. *Res:* Ecology of larval lampreys; estuarine ecology; marine benthic ecology; marine intertidal ecology; mangrove biology. *Mailing Add:* Dept Biol Univ NB Tucker Park St John NB E2L 4L5 Can

THOMAS, MARY BETH, b Sewanee, Tenn, Mar 2, 41. CELL BIOLOGY, INVERTEBRATE CYTOLOGY. *Educ:* Agnes Scott Col, BA, 63; Univ NC, Chapel Hill, MA, 70, PhD(zool), 71. *Prof Exp:* Vis asst prof biol, Wake Forest Univ, 71-72, asst prof, 72-76, assoc prof, 76-79; PROF BIOL, UNIV NC, CHARLOTTE, 80- *Mem:* Am Soc Zoologists; Am Micros Soc. *Res:* Invertebrate embryology and cytology; cnidarian ultrastructure; turbellarian ultrastructure and phylogeny. *Mailing Add:* Dept Biol Univ NC Charlotte NC 28223

THOMAS, MICHAEL DAVID, b Merthyr Tydfil, Wales, Jan 2, 42; m; c 2. GEOPHYSICS. *Educ:* Univ Wales, BS, 64, PhD(geol), 68. *Prof Exp:* Geol interpretation magnetic data, Geol Surv Can, 68-69; geophysicist, Survair Ltd, 69-71; res scientist gravity interpretation, Earth Physics Br, Dept Energy, Mines & Resources, Can, 72-86; RES SCIENTIST, GEOL INTERPRETATION MAGNETIC DATA, GEOL SURV CAN, 86- *Mem:* Geol Asn Can; Can Geophys Union. *Res:* Geological interpretation of gravity and magnetic anomalies over the Canadian landmass and territorial waters with a particular interest in the plate tectonic evolution of the coast since early Precambian time. *Mailing Add:* Continental Geosci Div Geol Surv Can Ottawa ON K1A 0Y3 Can

THOMAS, MICHAEL E(DWARD), b Monahans, Tex, May 10, 37; m 59; c 3. OPERATIONS RESEARCH. *Educ:* Univ Tex, BS, 60, MS, 63; Johns Hopkins Univ, PhD(opers res), 65. *Prof Exp:* Prod engr, Union Carbide Corp, 60-61; res asst optimization, Univ Tex, 61-62; jr instr chem eng, Johns Hopkins Univ, 62-63, res asst opers res, 63-64, instr, 64-65; from asst prof to prof indust & systs eng, Univ Fla, 65-78, chairperson dept, 73-78; PROF & DIR, SCH INDUST & SYSTS ENG, GA INST TECHNOL, 78- *Concurrent Pos:* Consult, US Army Corps Engrs, 70-77, MAPS, Inc, 74-78 & Hewlett Packard, Inc, 80-81; opers res analyst, Nat Bur Sci, 71-72. *Mem:* Opers Res Soc Am (secy, 80-83, pres, 84-85); Inst Mgt Sci; fel Am Inst Indust Engrs; Sigma Xi. *Res:* Optimization techniques, including decomposition techniques for nonlinear programming problems and optimal control theory. *Mailing Add:* Exec VPres Off Pres Ga Tech Atlanta GA 30332-0325

THOMAS, MIRIAM MASON HIGGINS, b Chicago, Ill, June 22, 20; m 47; c 1. NUTRITION. *Educ:* Bennett Col, NC, BS, 40; Univ Chicago, MS, 42. *Prof Exp:* Res assoc food chem, Div Biol Sci, Univ Chicago, 42-45; res chemist nutrit, Sci & Adv Tech Lab, Biol Sci Div, US Army Natick Res & Develop Labs, 45-85; RETIRED. *Concurrent Pos:* Vis fac lectr, Dept Nutrit & Food Sci, Mass Inst Technol, 74-83; Dept Defense Sec Army fel, 75. *Mem:* AAAS; Soc Nutrit Educ; Asn Vitamin Chemists; Inst Food Technol; Sigma Xi. *Res:* Chemical aspects of protein and amino acid metabolism; bioavailability of nutrients; effects of processing and storage on the nutritive quality of military rations and vitamin fortification of ration components. *Mailing Add:* 57 Eaton Rd Framingham MA 01701

THOMAS, MITCHELL, b Terre Haute, Ind, Nov 25, 36; m 64; c 3. PHYSICS, ENGINEERING. *Educ:* Harvard Univ, AB, 58; Univ Ill, Urbana, MS, 59; Calif Inst Technol, PhD(radiative transfer), 64. *Prof Exp:* Engr, McDonnell Douglas Corp, 59-61, eng consult, 62, sect chief appl res, 64-68, br chief, Advan Systs & Technol, 68-75; dir res & develop, 75-76, PRES, L'GARDE, INC, 76- *Mem:* Am Inst Aeronaut & Astronaut; AAAS. *Res:* Ablation, reentry and midcourse physics, especially radiative transfer through gases; calculation of transport properties of high-temperature gases. *Mailing Add:* L'Garde Inc 15181 Woodlawn Ave Tustin CA 92680-6487

THOMAS, MONTCALM TOM, b Brooklyn, Conn, Feb 5, 36; m 62; c 1. PHYSICS. *Educ:* Univ Conn, BA, 57, MS, 59; Brown Univ, PhD(physics), 66. *Prof Exp:* Mem tech staff, Bell Tel Labs, NJ, 65-68; asst prof physics, Wash State Univ, 68-74; MEM STAFF, BATTELLE PAC NORTHWEST LABS, 74- *Mem:* Am Phys Soc; Am Vacuum Soc. *Res:* Solid state, atomic and molecular physics; surface structure and kinetics of solids; thin films in solid state physics; photoelectric phenomena; low energy electron diffraction; high vacuum techniques. *Mailing Add:* 1708 Hunt Ave Richland WA 99352

THOMAS, MORLEY KEITH, b Middlesex Co, Ont, Aug 19, 18; m 42; c 2. METEOROLOGY, CLIMATOLOGY. *Educ:* Univ Western Ont, BA, 41; Univ Toronto, MA, 49. *Prof Exp:* Meteorologist, Atmospheric Environ Serv, Can, 41-51 & div bldg res, Nat Res Coun Can, 51-53; supt climat opers, 53-72, dir meteorol applns br, 72-75, dir gen, Cent Servs, 76-80, dir gen, Can Climate Ctr, 80-83, CONSULT & HIST METEOROL PROJ, ATMOSPHERIC ENVIRON SERV, 83- *Concurrent Pos:* Assoc comt snow & ice mech, Nat Res Coun Can, 59-65 & subcomt meteorol & atmospheric sci, 67-70; mem working group on climatic atlases, World Meteorol Orgn, 60-65, chmn, 65-69; mem, Nat Adv Comt Geog Res, 65-70; pres Comn Climatol & Appln Meteorol, World Meteorol Orgn, 78-82. *Honors & Awards:* Patterson Distinguished Service Medal For Can Meterol, 80; Thomas Award Vol Weather Observers Estab, 83; Massey Medal Outstanding Achievement in Climat, 85. *Mem:* Fel Am Meteorol Soc; Royal Meteorol Soc (treas, Can Br, 50-51, secy, 64-66, vpres, 66-67); Can Meteorol Soc (vpres, 67-68, pres, 68-70); Can Asn Geog. *Res:* Atlases; urban climates; climatic change; climatological services; meteorological applications; history of meteorology and climatology. *Mailing Add:* Atmospheric Environ Serv 4905 Dufferin St Downsview ON M3H 5T4 Can

THOMAS, NORMAN RANDALL, b Caerphilly, Wales, Dec 22, 32; m 54; c 5. DENTISTRY, PHYSIOLOGY. *Educ:* Bristol Univ, BDS, 57, BSc, 60, PhD(dent), 65, Am Bd Oral Path, cert, FRCD(C), 86. *Prof Exp:* Med Res Coun sci asst path res, Royal Col Surgeons, Eng, 60-62; lectr dent med, Bristol Univ, 62-66, lectr anat, 66-68; prof dent & hons prof med, 68-89, EMER PROF, UNIV ALTA, 89- *Mem:* Can Dent Asn; Can Asn Anat; Int Asn Dent Res; fel Int Col Craniomandibolar Orthop; Am Asn Oral Path. *Res:* Collagen formation and maturation in tooth eruption; neurophysiology of orofacial complex; TMJ dysfunction. *Mailing Add:* 5412 142 St Edmonton AB T6H 4B8 Can

THOMAS, OSCAR OTTO, animal nutrition; deceased, see previous edition for last biography

THOMAS, OWEN PESTELL, b Middelburg, SAfrica, Apr 28, 33; m 67; c 1. POULTRY NUTRITION. *Educ:* Univ Natal, BSc, 54, MSc, 62; Univ Md, College Park, PhD(poultry nutrit), 66. *Prof Exp:* Nutritionist, United Oil & Cake Mills Ltd, 55-61; asst, Univ Md, College Park, 64-66, res assoc, 66-68; nutritionist, United Oil & Cake Mills Ltd, 68-70; from asst prof to assoc prof, 70-77, PROF POULTRY SCI, UNIV MD, COLLEGE PARK, 77- CHMN DEPT, 71- *Mem:* Poultry Sci Asn; Sigma Xi. *Res:* Protein and amino acid requirements of broilers; body composition and pigmentation of broilers. *Mailing Add:* Dept Poultry Sci Univ Md College Park MD 20742

THOMAS, P(AUL) D(ANIEL), b Vaughnsville, Ohio, Jan 13, 05; m 35; c 3. METALLURGY. *Educ:* Ohio Wesleyan Univ, AB, 27. *Prof Exp:* Metallurgist, Jones & Laughlin Steel Corp, Pa, 27-33, foreman, 33-35, supvr tube invest & develop div, 35-49; mat engr, Asiatic Petrol Corp, 49-67; staff mat eng, Shell Oil Co, NY, 67-70; independent consult metallurgist, 70-74; staff metallurgist, Alyeska Pipeline Serv Co, 74-78; INDEPENDENT CONSULT METALLURGIST, 78- *Concurrent Pos:* Instr exten courses, Pa State Univ, 34-42. *Mem:* Am Petrol Inst; Soc Metals; Welding Soc; Asn Corrosion Engrs; NY Acad Sci. *Res:* Metallurgical and inspection phases in production of seamless and welded steel pipe and tubes; leak resistance of threaded joints; corrosion fatigue; ordnance tubing; high strength threaded joints and steels for pipe; welding metallurgy. *Mailing Add:* 9847 Warwana Rd Houston TX 77080

THOMAS, PAUL A V, b Guernsey, Channel Islands, Europ, Oct 6, 25; m 53; c 3. ELECTRICAL ENGINEERING. *Educ:* London Univ, BSc, 50; Glasgow Univ, PhD(elec eng), 61. *Prof Exp:* Res asst mech, Royal Col Sci & Technol, Scotland, 50-53; lectr, Glasgow Univ, 53-62; from assoc prof to prof elec eng, Univ Windsor, 63-75, head dept, 64-68; dept chmn, 75-78 & 80-85, PROF COMPUTER SCI & INFO PROCESSING, BROCK UNIV, 75- *Mem:* Sr mem Inst Elec & Electronics Engrs; Brit Computer Soc; Brit Inst Elec Eng; Asn Computer Mach; Can Info Processing Soc. *Res:* Electronic computers and graphics systems. *Mailing Add:* Dept Computer Sci & Info Processing Brock Univ St Catharines ON L2S 3A1 Can

THOMAS, PAUL CLARENCE, b Watsonville, Calif, Nov 26, 28; m 56; c 1. PLANT BREEDING, VEGETABLE CROPS. *Educ:* Col Agr, Univ Calif, BS, 50. *Prof Exp:* Veg breeder, W Atlee Burpee Co, 50-58; DIR RES, PETOSEED CO, INC, 58-, SR VPRES, 78- *Concurrent Pos:* Mem bd dirs, All Am Selections; comnr, Calif Pepper Comn. *Mem:* Am Soc Hort Sci. *Res:* Management of quality assurance and stock seed production programs of vegetable hybrid parents in northern and southern hemispheres. *Mailing Add:* Four Juniper Ct Woodland CA 95695

THOMAS, PAUL DAVID, b Bellwood, Pa, Mar 8, 26; m 51; c 4. ORGANIC CHEMISTRY. *Educ:* Rutgers Univ, BA, 49; Univ Ill, PhD(org chem), 54. *Prof Exp:* Chemist, Fries Bros Chem Mfg Co, 49-51; res chemist, Tidewater Assoc Oil Co, 51; res chemist, Pfizer Inc, 54-71, res supvr org chem res & develop, 71-73, registered patent agent, 78-87; RETIRED. *Concurrent Pos:* Pvt consult, 87- *Mem:* Am Chem Soc; Inst Food Technol. *Res:* Food additives and flavors. *Mailing Add:* 271 Plant St Groton CT 06340

THOMAS, PAUL EMERY, b Phoenix, Ariz, Feb 15, 27; m 58; c 2. NUMBER THEORY. *Educ:* Oberlin Col, BA, 50; Oxford Univ, BA, 52; Princeton Univ, PhD(math), 55. *Prof Exp:* Res instr, Columbia Univ, 55-56; from asst prof to assoc prof, 56-63, chmn, 72-73, PROF MATH, UNIV CALIF, BERKELEY, 63- *Concurrent Pos:* NSF fel, 58-59; Guggenheim Mem Found fel, 61; prof, Miller Inst, 66-67; ed, Proc, Am Math Soc, 68-71; mem div math sci, Nat Res Coun, 70-72; trustee, Am Math Soc, 80-84; dep dir, Math Sci Res Inst, 87-90; exec dir, Miller Inst Basic Res Sci, 87-89. *Mem:* Am Math Soc. *Res:* Number theory. *Mailing Add:* Dept Math Univ Calif Berkeley CA 94720

THOMAS, PAUL MILTON, b Sligo, Pa, Dec 1, 29; m 51; c 1. IMMUNOBIOLOGY, ICHTHYOLOGY. *Educ:* Allegheny Col, BS, 58; Univ Mich, MA, 59, MS, 62, PhD(sci admin, ichthyol), 64; Drew Univ, DMin, 80. *Prof Exp:* Instr biol, Houghton Col, 59-62; teacher high sch, Mich, 62-64; from asst prof to assoc prof biol, Pasadena Col, 64-67; res fel, Calif Inst Technol, 67-68; chmn dept, 69-85, PROF BIOL, EDINBORO UNIV PA, 68- *Concurrent Pos:* Res fel radiation biol, Cornell Univ, 66-67. *Mem:* Am Fisheries Soc. *Res:* Fish anesthetics; effects of industrial pollution of fish; radiation effects on elasmobranch antibody response; sexual dimorphism in fish; artificial fish shelters; administrative effects on science education; gamma globulin synthesis in fish; Lake Erie fishery; death education; acid rain and aquatic habitats. *Mailing Add:* Dept Biol Edinboro Univ Edinboro PA 16412

THOMAS, PETER, b Bridgend, UK, Apr 25, 46; m 80; c 2. GLYCOPROTEIN METABOLISM, CANCER MARKERS. *Educ:* Univ Wales, BSc, 67, PhD(biochem), 71. *Prof Exp:* A K fel chem, Inst Cancer Res, London, 71-79; sr res assoc, Mallory Gastrointestinal Res Lab, Mallory Inst Path, Boston City Hosp, 79-86; from assoc med to prin assoc med, 79-85, asst prof surg biochem, 85-90, ASSOC PROF SURG BIOCHEM, HARVARD MED SCH, 90-; ASSOC MEM, CANCER RES INST, NEW ENG DEACONESS HOSP, 85- *Mem:* Biochem Soc; Am Soc Biol Chem; Am Gastroentrol Asn; Am Asn Study Liver Dis; Protein Soc; Am Asn Cancer Res. *Res:* The metabolism of glycoproteins especially carcinoembryonic antigen; interactions between the Kupffer cell and hepatocyte in glycoprotein handling; mechanism of transfer of proteins from blood to bile; tumor cell surfaces' relationship to development of metasteses especially from colorectal cancer. *Mailing Add:* Lab Cancer Biol Shields Warren Radiation Lab New Eng Deaconess Hosp 50 Binney St Boston MA 02115

THOMAS, QUENTIN VIVIAN, b Glendale, Calif, Apr 13, 49; m 69; c 2. ANALYTICAL CHEMISTRY, MASS SPECTROMETRY. *Educ:* Ore State Univ, BS, 71; Purdue Univ, MS, 74, PhD(chem), 76. *Prof Exp:* Appln chemist mass spectrometry, Finnigan Instrument Div, 76-78, sr instr anal chem, 78-80, MGR TRAINING, FINNIGAN INST, 80- *Mem:* Am Chem Soc; Am Soc Mass Spectrometry. *Res:* Negative ion mass spectrometry; application of computers in the chemical laboratory. *Mailing Add:* 1574 Christine Dr Hamilton OH 45014-3502

THOMAS, R E, b Austin, Tex, Apr 12, 30; m 51; c 3. ELECTRICAL ENGINEERING. *Educ:* NMex State Univ, BS, 51 & 52; Stanford Univ, MS, 53; Univ Ill, PhD(elec eng), 59. *Prof Exp:* Consult engr, Wright Air Develop Ctr, US Air Force, 53-57, from instr to prof astronaut, US Air Force Acad, 59-65, prof elec eng & head dept, 66-79; sr scientist, Kaman Sci Corp, Colorado Springs, 79-82 & Mission Res Corp, 82-84; prin engr, Motorola Inc, 84-85; PRES, THOMAS CONSULT SERV, 85- *Mem:* Inst Elec & Electronics Engrs. *Res:* Control system analysis and synthesis; active network synthesis; linear systems synthesis; servomechanism analysis and synthesis. *Mailing Add:* 10685 E Ironwood Dr Scottsdale AZ 85258

THOMAS, R NOEL, b Caernarfon, NWales, Dec 25, 36. ENGINEERING. *Educ:* Univ Col NWales, UK, Bsc, 58; Univ Cambridge, Eng, PhD(physics), 61. *Prof Exp:* MGR, RES & DEVELOP CTR, WESTINGHOUSE ELEC CORP, PITTSBURGH, PA, 62- *Mem:* Fel Inst Elec & Electronics Engrs; Am Phys Soc. *Res:* Melt growth of large denominator of element of compound semi-conductors. *Mailing Add:* R & D Center Westinghouse Elec Corp 1310 Beulah Rd Pittsburgh PA 15235

THOMAS, RALPH HAROLD, b Reading, Eng, Nov 27, 32; m 58; c 3. HEALTH PHYSICS. *Educ:* Univ Col, London, BSc, 55, PhD(nuclear physics), 59 ,DSc, 79; Am Bd Health Physics, cert, 69; Univ Calif Sch Pub Health, MPH, 82. *Hon Degrees:* MA, Oxon, 85. *Prof Exp:* Res physicist, Assoc Elec Industs, UK, 58-59; prin sci officer, Rutherford High Energy Lab, Sci Res Coun, UK, 59-68; sr health physicist, Stanford Univ, 68-70; various app, Lawrence Berkeley Nat Lab, 70-87, div head, Occup Health Div, 88-90; HEAD, HAZARDS CONTROL DEPT, LAWRENCE LIVERMORE NAT LAB, 90-; ADJ PROF, SCH PUB HEALTH, UNIV CALIF, BERKELEY, 90- *Concurrent Pos:* Lectr, Reading Col Technol, 55-63; vis scientist, Lawrence Berkeley Lab, Univ Calif, 63-65, Europ Orgn Nuclear Res, Geneva, Switz, 66, Brookhaven Nat Lab, 70 & KEK Nat Lab High-Energy Physics, Oho-Machi, Japan, 77; mem working group, Int Comn Radiol Protection,

66-70; chmn adv panel accelerator radiation safety, US Atomic Energy Comn, 69-72; mem, Comt High Energy & Space Dosimetry, Int Comn Radiation Units, 73-78; mem comt 3, Int Comn Radiol Protection, 78-85, mem comt 2, 85-93; mem, comt int syst units, Nat Coun Radiation Protection & Measurement, 90-96; mem int comn, Radiation Units & Measurements Comt, Dose Equivalent Determination, 80-; fel, Keble Col, Oxford Univ, 85-86; vis scholar, Radcliffe Sci Libr, Univ Oxford. *Mem:* Fel Health Physics Soc (treas, 77-79); Radiol Res Soc; fel Brit Inst Physics; fel Royal Soc Health. *Res:* Accelerator radiation problems; high energy dosimetry; radiological protection standards. *Mailing Add:* Hazards Control Dept Lawrence Livermore Lab Univ Calif Livermore CA 94550

THOMAS, RALPH HENRY, SR, b Brooklyn, NY, July 7, 31; m 44; c 4. PACKAGE ENGINEERING, INTERNATIONAL PACKAGE RESEARCH. *Educ:* State Univ NY, BS, 79. *Prof Exp:* Dept head, Packaging Res, E R Squibb & Sons, 53-55; proj leader, Packaging Res, Gen Foods Corp, 54-55 & Colgate Palmolive Co, 55-57; packaging res engr, Bristol-Myers Co, 49-53, dir, 57-81; PRES, THOMAS PACKAGING CONSULTS, 81- *Concurrent Pos:* Instr packaging technol, Columbia Univ, Upsala Col, Ctr Prof Educ, Felician Col, Packaging Inst & Am Mgt Asn. *Mem:* Am Soc Mech Engrs; Am Mgt Asn; Soc Plastic Engrs; Packaging Inst. *Res:* Package production and manufacturing; author of 78 publications and six books; awarded 38 patents. *Mailing Add:* 2204 Morris Ave Union NJ 07083

THOMAS, RAYE EDWARD, b Cross Creek, NB, June 5, 38; m 63; c 2. SOLID STATE ELECTRONICS, ELECTRICAL ENGINEERING. *Educ:* Univ NB, BScEE, 61; Imp Col, Univ London, PhD(elec eng), 66. *Prof Exp:* Mem sci staff solid state devices, Res & Develop Labs, Northern Elec Co Ltd, 66-69, mgr physics devices, 69; from asst prof to prof, 69-84, ADJ PROF ENG, CARLETON UNIV, 84-; PRES, TPK SOLAR SYSTS INC, 79- *Concurrent Pos:* Consult, Microsysts Int Ltd, 69-70, Bell-Northern Res, 73-84 & Mitel Corp, 79-63. *Mem:* Sr mem Inst Elec & Electronics Engrs; Solar Energy Soc Can; Int Solar Energy Soc. *Res:* Solid state device physics; discrete device and integrated circuit design, fabrication and characterization; device modeling; solar energy conversion-photovoltaics. *Mailing Add:* Astro Power Canada Ltd Six Antares Dr Unit 12 Nepean ON K2E 8A9 Can

THOMAS, (JOHN) (PAUL) RICHARD, b Jacksonville, Fla, May 2, 38. VERTEBRATE SYSTEMATICS. *Educ:* Univ SFla, BA, 69; La State Univ, PhD(zool), 76. *Prof Exp:* ASSOC PROF BIOL, UNIV PR, 76- *Concurrent Pos:* Prin investr, NSF, 77-79. *Mem:* Am Soc Ichthyologists & Herpetologists; Herpetologists League; Soc Study Amphibians & Reptiles; Soc Study Evolution; Soc Syst Zoologists; AAAS; Am Soc Zool; Europ Herpet Soc; Asn Trop Biol. *Res:* Systematics, biogeography and ecology of neotropical amphibians and reptiles, especially those of the Antillean region. *Mailing Add:* Dept Biol Univ PR Rio Piedras PR 00931

THOMAS, RICHARD ALAN, b Smithville, Mo, Mar 14, 48. LOW TEMPERATURE PHYSICS. *Educ:* William Jewell Col, BA, 70; Stanford Univ, PhD(appl phys), 77. *Prof Exp:* Asst physicist, 77-80, PHYSICS ASSOC I, BROOKHAVEN NAT LAB, 77- *Mem:* Am Phys Soc; Inst Elec & Electronics Engrs; Asn Comput Mach. *Res:* Superconducting power transmission, polymeric insulation, dielectric properties of helium, automatic monitoring and control of large-scale low-temperature systems, quantum mechanical tunneling in dielectrics; particle acclerators. *Mailing Add:* Bldg 911A Brookhaven Nat Lab Upton NY 11973-5000

THOMAS, RICHARD CHARLES, b Syracuse, NY, July 22, 49; m 71. ORGANIC CHEMISTRY, MEDICINAL CHEMISTRY. *Educ:* Univ Rochester, BS, 71; Univ Calif, Los Angeles, PhD(chem), 76. *Prof Exp:* Fel chem, Mass Inst Technol, 76-77; SR RES SCIENTIST CHEM, UPJOHN CO, 77- *Mem:* Am Chem Soc; Royal Soc Chem. *Res:* Chemical synthesis and modification of antibiotics. *Mailing Add:* Upjohn Co 7246-25-6 301 Henrietta St Kalamazoo MI 49001-0199

THOMAS, RICHARD DEAN, b Payson, Utah, Feb 14, 47; m 70; c 5. TOXICOLOGY, PATHOLOGY. *Educ:* Utah State Univ, BS, 71; Colo State Univ, PhD(med chem), 74. *Prof Exp:* Sr metab chemist agr chem, Biochem Dept, Agr Div, Ciba-Geigy Corp, 74-76; toxicologist & criteria doc mgr, Ctr Occup & Environ Safety & Health, Stanford Res Inst, 76-78; sr environ systs scientist toxicol, Dept Environ Chem & Biol, Metrek Div, The Mitre Corp, 78-80; at Borriston Labs Inc, 80-82; DIR HUMAN TOXICOL & RISK ASSESSMENT, NAT ACAD SCI, NAT RES COUN, 82- *Concurrent Pos:* Adv & consult, Environ Protection Agency, Food & Drug Admin, Dept Energy, WHO, Int Agency Res Cancer, Justice Dept, CDC & UN. *Mem:* Am Chem Soc; AAAS; Am Inst Chemists; Am Soc Appl Spectros; Am Col Toxicol; Am Pub Health Asn; Environ Mutagen Soc; Environ Health Inst; Genetic Tox Asn; NY Acad Sci. *Res:* Investigation into the toxicology, metabolism and environmental impact of chemicals; physiological impairment and potential for cancer and disease production related to chemical exposure; setting tolerances as they relate to health and regulation of chemicals; risk assessment. *Mailing Add:* Nat Acad Sci Nat Res Coun HA354 2101 Constitution Ave NW Washington DC 20418

THOMAS, RICHARD EUGENE, b Logan, Ohio, Dec 29, 25; m 50; c 3. AEROSPACE ENGINEERING. *Educ:* Ohio State Univ, BAeroE, 51, BA, 53, MS, 56, PhD(aerodyn), 64. *Prof Exp:* Aeronaut engr, Air Tech Intel Ctr, Wright-Patterson AFB, Ohio, 51-52; res assoc aerodyn, Ohio State Univ, 52-56, res assoc & instr, 56-61, asst supvr & instr, 61-64; assoc prof, Tex A&M Univ, 64-66, prof, 66-69; head dept aerospace eng, Univ Md, College Park, 69-71; prof eng, 71-77, actg dean, 77-79, DIR, CTR STRATEGIC TECHNOL, TEX A&M UNIV, 79- *Concurrent Pos:* Consult, Dept Aviation, Ohio State Univ, 63-64, NAm Aviation, Inc, 64, NSF, 76-80 & Computer Aided Mfg Comt, USAF, 77-81, USAF Ballistic Missile Off, 85-87, TRW, 87-90, Nat Res Coun, 88-90, USAF Mfg Tech Prog, 91-; vis distinguished prof, Am Univ, Cairo, 78-81. *Mem:* Am Inst Aeronaut & Astronaut. *Res:* Aerodynamics; gas dynamics; aerothermochemistry; flight dynamics. *Mailing Add:* Ctr Strategic Technol Tex A&M Univ College Station TX 77843-3572

THOMAS, RICHARD GARLAND, b Houston, Tex, June 23, 23; m 71. PHYSICS. *Educ:* Hampton Inst, BS, 43; Columbia Univ, MA, 50; Univ Calif, Berkeley, PhD(physics), 59. *Prof Exp:* Sr scientist physics, Gen Elec Co, 59-63 & Lawrence Livermore Lab, 63-68; PROF PHYSICS & HEAD DEPT, PRAIRIE VIEW AGR & MECH COL, 68- *Mem:* AAAS; Am Phys Soc; Am Asn Physics Teachers; Sigma Xi. *Res:* Low energy nuclear physics; x-ray spectroscopy. *Mailing Add:* Box 698 Prairie View Agr & Mech Col Prairie View TX 77446

THOMAS, RICHARD JOSEPH, b Wilkes-Barre, Pa, Nov 29, 28; m 51; c 2. WOOD TECHNOLOGY. *Educ:* Pa State Univ, BS, 54; NC State Univ, MWT, 55; Duke Univ, DF, 67. *Prof Exp:* Tech rep, Nat Casein NJ, 55-57, sales mgr, 57; from asst prof to assoc prof wood & paper sci, 57-71, PROF WOOD & PAPER SCI, SCH FOREST RESOURCES, NC STATE UNIV, 71-, HEAD DEPT, 78- *Concurrent Pos:* Sci fac fel, NSF. *Mem:* Forest Prods Res Soc (vpres 83, pres elect 84, pres 85); Int Asn Wood Anat; Soc Wood Sci & Technol (pres elect, 81, pres, 82). *Res:* Study of wood ultrastructure, particularly relationships of ultrastructure to physical properties and function within the plant; investigations of differentiation of cell wall and cell wall markings; distribution of major chemical constituents throughout cell wall. *Mailing Add:* Dept Wood & Paper Sci NC State Univ Box 8005 Raleigh NC 27695

THOMAS, RICHARD NELSON, b Omaha, Nebr, Mar 3, 21; m 45; c 1. STELLAR ATMOSPHERES, NONEQUILIBRIUM THERMODYNAMICS. *Educ:* Harvard Univ, BS, 42, PhD(astron), 48. *Prof Exp:* Ballistician, Ballistic Res Lab, Aberdeen Proving Ground, Md, 42-45; Jewett fel, Inst Advan Study, 48-49; assoc prof astron, Univ Utah, 48-53; vis lectr, Harvard Univ, 52-53, lectr observ, 53-57; consult astrophys to dir, Boulder Labs, Nat Bur Stand, 57-62; fel, Joint Inst Lab Astrophys, 62-74, vis prof, Col de France, Paris, 73-75 & Univ Paris, 61, 75-76, res scientist, CNRS-France, 77-86; CONSULT, 86-, RADIOPHYSICS, INC, BOULDER, CO, 87- *Concurrent Pos:* Mem bd dirs, Annual Rev, Inc; adj prof Astrophys, Univ Colo, Boulder, 62-75. *Honors & Awards:* NBS Gold Medal, 63. *Mem:* Int Astron Union; Am Astron Soc; Am Phys Soc. *Res:* Stellar atmospheres; solar physics; astroballistics; non-equilibrium thermodynamics. *Mailing Add:* Pine Brook Hills 1155 Timberlane Boulder CO 80304

THOMAS, RICHARD SANBORN, b Madison, Wis, June 14, 27; m 57; c 2. SOIL MICROBIOLOGY, MICROSCOPY. *Educ:* Oberlin Col, BA, 49; Univ Calif, Berkeley, PhD(biophys), 55. *Prof Exp:* Am Cancer Soc fel cancer res & NSF fel cytochem, Carlsberg Lab, Denmark, 55-57; asst res biophysicist, Virus Lab, Univ Calif, Berkeley, 58-60; res physicist, 60-90, COLLABR, WESTERN REGIONAL RES CTR, USDA, 90- *Concurrent Pos:* USPHS spec fel, Dept Gen Botany, Swiss Fed Inst Technol, 67-68; consult microscopy & biosci appln plasma chem, Tegal Corp, Richmond, Calif, 72-79. *Mem:* AAAS; Electron Micros Soc Am; Soil Sci Soc Am; Biophys Soc; Microbeam Analytical Soc; Am Soc Agron. *Res:* Biological ultrastructure and fine cytochemistry; development of techniques for electron microscopic cytochemistry and electron probe microanalysis, especially by plasma etching; intracellular mineral deposits, bacterial spores, keratin, microfibrillar proteins, virus particles; plant tissues; cereal products; effects of endomycorrhizal fungi on host plants and soils; microscopy of microorganisms in soil. *Mailing Add:* Western Regional Res Ctr USDA Albany CA 94710

THOMAS, ROBERT, b Atlanta, Ga, Aug 27, 34; m 69. CRYSTALLOGRAPHY, PHYSICAL CHEMISTRY. *Educ:* Boston Univ, AB, 55, PhD(phys chem), 65. *Prof Exp:* NIH res assoc chem, Univ Colo, 64-66; AEC res assoc, 66-68, assoc scientist chem, 68-77, SCIENTIST CHEM, BROOKHAVEN NAT LAB, 77- *Mem:* Sigma Xi; Am Crystallog Asn; Am Chem Soc. *Res:* Crystal structure determination by x-ray and neutron diffraction; study of phase transitions, ferroelectrics and critical phenomena. *Mailing Add:* Off Educ Progs Bldg 490 Brookhaven Nat Lab Upton NY 11973

THOMAS, ROBERT E, b Salineville, Ohio, Feb 17, 36; m 62; c 2. PHYSIOLOGY, BIOCHEMISTRY. *Educ:* Kent State Univ, BS, 61, MA, 63, PhD(biol sci), 66. *Prof Exp:* Instr biol sci, Kent State Univ, 63-64; from asst prof to assoc prof, 66-74, PROF BIOL SCI, CALIF STATE UNIV, CHICO, 74- *Concurrent Pos:* Res physiologist, Nat Marine Fisheries Serv, 73- *Honors & Awards:* Charles Y Conkle Publ Award, Nat Oceanic & Atmospheric Admin, US Dept Com, 84. *Mem:* AAAS; Am Inst Biol Sci; Sigma Xi. *Res:* Sublethal effects of pollutants on marine life. *Mailing Add:* Dept Biol Sci Calif State Univ Chico CA 95929

THOMAS, ROBERT EUGENE, b Iowa, Oct 15, 19; m 43; c 4. PSYCHIATRY. *Educ:* Univ Southern Calif, AB, 42, MD, 51; Johns Hopkins Univ, MPH, 55. *Prof Exp:* Intern, Santa Fe Coastlines Hosp, Los Angeles, Calif, 50-51; psychiat resident, Vet Admin Hosp, Perry Point & Baltimore, Md, 51-53; chief div ment health, Wash County Dept Health, Hagerstown, 53; from instr to asst prof pub health admin, Sch Hyg, Johns Hopkins Univ, 54-58; regional ment health adminr, Calif Dept Ment Hyg, 58-68, regional ment health dir, div local progs, 68-69; dir, Hemet Valley Community Ment Health Ctr, 69-85; RETIRED. *Concurrent Pos:* Chief div ment health, State Dept Health, Md, 54-60; mem, Gov Comn Ment Health, 54; lectr, Sch Pub Health & asst clin prof, Dept Psychiat, Sch Med, Univ Calif, Los Angeles, 60-70; third year psychiat resident ment health admin in pub health, Sch Hyg, Johns Hopkins Univ; assoc clin prof, Dept Psychiat, Loma Linda Sch Med. *Mem:* Fel Am Psychiat Asn; AMA; fel Am Pub Health Asn; fel Am Orthopsychiat Asn. *Res:* Mental health administration in public health; administration of alcoholism programs. *Mailing Add:* PO Box 97 Little Creek CA 92358

THOMAS, ROBERT GLENN, b Watertown, NY, Oct 9, 26; m 49; c 3. RADIOBIOLOGY, BIOPHYSICS. *Educ:* St Lawrence Univ, BS, 49; Univ Rochester, PhD(radiation biol), 55. *Prof Exp:* From instr to asst prof radiation biol, Univ Rochester, 55-61; from sect head to dept head radiobiol, Lovelace

Found Med Educ & Res, 61-74; group leader mammalian biol, 74-79, HEALTH DIV OFF, LOS ALAMOS NAT LAB, 79- *Concurrent Pos:* Mem task group, Biol Effects Radiation on Lung Comt 1, Int Comn Radiol Protection, 68-; ed, Health Physics J, 81- *Mem:* Am Radiation Res Soc; Health Physics Soc; Reticuloendothelial Soc; Am Indust Hyg Asn; AAAS. *Res:* Toxicity of inhaled radioactive materials; application of experimental results to practical hazards evaluation in nuclear industry; toxicity of inhaled fossil fuel products. *Mailing Add:* 20706 Highland Hall Dr Gaithersburg MD 20879

THOMAS, ROBERT HAYNE, ceramics engineering, physical chemistry, for more information see previous edition

THOMAS, ROBERT JAMES, b Flint, Mich, July 5, 49. BOTANY, DEVELOPMENTAL PHYSIOLOGY. *Educ:* Univ Mich, Flint, AB, 71; Univ Calif, Santa Cruz, PhD(biol), 75. *Prof Exp:* From asst prof to assoc prof, 75-88, CHMN DEPT, BATES COL, 82-, PROF BIOL, 88- *Concurrent Pos:* Prin investr, NSF grant, 90-92. *Mem:* Bot Soc Am; Am Soc Plant Physiologists; Am Bryol & Lichenol Soc; Brit Bryol Soc. *Res:* Plant growth and development; physiology, biochemistry and development of bryophytes. *Mailing Add:* Dept Biol Bates Col Lewiston ME 04240

THOMAS, ROBERT JAY, b Harvey, Ill, Mar 30, 30; m 53; c 2. COMPUTER SCIENCES. *Educ:* Oberlin Col, BA, 52; Ind Univ, MS, 54; Univ Ill, MS, 58, PhD(math), 64. *Prof Exp:* Dir recreational ther, Cent State Ment Hosp, Indianapolis, Ind, 54-55; adv, 3-2 combined eng prog, DePauw Univ, 62-72, dir, Comput Ctr, 63-66, from instr to assoc prof math, 58-71, PROF MATH, DEPAUW UNIV, 71-, PROF COMPUTER SCI, 76- *Concurrent Pos:* Pace res appointment, Argonne Nat Lab, 67, comput consult, 67-71; bd dirs, Sex Info & Educ Coun US, 83. *Mem:* AAAS; Asn Comput Mach; Am Asn Sex Educrs & Counrs; Sex Info & Educ Coun US. *Res:* Pattern recognition; determination of cell motility by computer; human sexuality. *Mailing Add:* Dept Math & Computer Sci DePauw Univ Locust St Greencastle IN 46135

THOMAS, ROBERT JOSEPH, b Lowell, Mass, July 13, 12; m 42; c 4. CHEMISTRY. *Educ:* Lowell Textile Inst, BTC, 34; Univ Notre Dame, MS, 37, PhD(org chem), 39. *Prof Exp:* Textile chemist, Apponaug Co, RI, 34-36; res chemist, Tech Lab, E I du Pont de Nemours & Co, Inc, 39-42, Jackson Lab, 42-43, Manhattan Proj, Chambers Works, 43-44 & Tech Lab, 44-50, supvr dyeing develop div, 50-65, interdept liaison, Tech Lab, 65-77; consult textile chem & dyeing, 77-80; RETIRED. *Concurrent Pos:* Adj prof textiles, Clemson Univ, 77-80. *Mem:* Am Chem Soc; Am Asn Textile Chem & Colorists. *Res:* Dye application to textile fibers; textile chemistry. *Mailing Add:* 12 Sack Ave Penns Grove NJ 08069

THOMAS, ROBERT L, b Dover-Foxcroft, Maine, Oct 10, 38; m 62; c 1. SOLID STATE PHYSICS. *Educ:* Bowdoin Col, AB, 60; Brown Univ, PhD(physics), 65. *Prof Exp:* Res asst physics, Brown Univ, 60-65; res assoc physics, Wayne State Univ, 65-66, from asst prof to assoc prof, 66-76, asst chmn, 81-86, PROF PHYSICS, WAYNE STATE UNIV, 76-, DIR, INST MFG RES, 86- *Concurrent Pos:* Sr vis fel, Bedford Col, Univ London, 73-74. *Mem:* Fel Am Phys Soc; Sigma Xi; Inst Elec & Electronics Engrs. *Res:* Ultrasonics; thermal wave imaging. *Mailing Add:* Dept Physics 281 Physics Wayne State Univ 5950 Cass Ave Detroit MI 48202

THOMAS, ROBERT SPENCER DAVID, b Toronto, Ont, July 29, 41; m 65; c 2. MATHEMATICS, APPLICATIONS. *Educ:* Univ Toronto, BSc, 64; Univ Waterloo, MA, 65; Univ Southampton, PhD(math), 68. *Prof Exp:* Lectr math, Univ Waterloo, 65-66 & Univ Zambia, 68-70; from asst prof to assoc prof comput sci, 70-78, assoc prof, 78-85, PROF APPL MATH, UNIV MAN, 85- *Concurrent Pos:* Managing ed, Utilitas Math, 71-82. *Mem:* Can App Math Soc; Inst Math & Appln; Can Soc Hist & Philos Math; Can Math Soc. *Res:* Application of mathematics. *Mailing Add:* Dept Appl Math Univ Man Winnipeg MB R3T 2N2 Can

THOMAS, ROGER DAVID KEEN, b Maidstone, Kent, Eng, Oct 5, 42; m 70; c 2. EVOLUTIONARY BIOLOGY, PALEOECOLOGY. *Educ:* Imp Col Univ London, BSc, 63, ARCS, 63; Harvard Univ, AM, 65, PhD(geol), 70. *Prof Exp:* Asst prof geol, Harvard Univ, 70-75; asst prof, 75-80, ASSOC PROF GEOL & ASSOC DEAN ACAD AFFAIRS, FRANKLIN & MARSHALL COL, 80- *Concurrent Pos:* Allston Burr sr tutor, Quincy House, Harvard Univ, 70-75, asst cur invert paleont, Mus Comp Zool, Harvard Univ, 70-75; Wissenschaftlich Angestellte Palokologie Universitat Tubingen, 73-74; prin investr, NSF grants, 77-79, 81-82, 83 & 85; vis scientist, Field Mus Natural Hist, Chicago, 82; vis prof, Univ Tubingen, 84. *Mem:* Paleont Soc; Int Palaeont Asn; Geol Soc Am; Geol Soc London; AAAS; Soc Econ Paleontologists & Mineralogists. *Res:* Paleobiology of fossil bivalves; interaction of mechanical function, growth patterns and evolutionary history in the determination of organic form; paleoecology, functional morphology and the evolution of diversity. *Mailing Add:* Dept Geol Franklin & Marshall Col Lancaster PA 17604

THOMAS, ROGER JERRY, b Detroit, Mich, July 3, 42; m 66; c 1. SOLAR PHYSICS, ASTROPHYSICS. *Educ:* Univ Mich, Ann Arbor, BS, 64, MS, 66, PhD(astron), 70. *Prof Exp:* Nat Acad Sci-Nat Res Coun resident res assoc solar physics, 70-71, ASTROPHYSICIST, GODDARD SPACE FLIGHT CTR, NASA, 71- *Concurrent Pos:* Proj scientist orbiting solar observ satellite prog, Goddard Space Flight Ctr, NASA, 76-83, actg dep chief, Off Solar & Heliospheric Physics, 83-84. *Mem:* Int Astron Union; Am Astron Soc. *Res:* Solar x-ray and extreme ultraviolet astronomy; solar activity; solar flares; solar corona; x-ray and EUV optics. *Mailing Add:* Code 682-1 Goddard Space Flight Ctr NASA Greenbelt MD 20771

THOMAS, RONALD EMERSON, b Ont, Can, Apr 19, 30; m 62; c 3. MATHEMATICAL STATISTICS, OPERATIONS RESEARCH. *Educ:* Queen's Univ, Ont, BA, 52, MA, 58; Univ NC, PhD(math statist), 62. *Prof Exp:* Actuarial asst, Excelsior Life Inst Co, 52-57; mem tech staff, Bell Labs, 62-68, supvr, Appl Probability Group, 68-75, opers res methods, 75-80, supvr field performance studies, 80-84 & data networks bus planning, 84-86, SUPVR & SYSTS ENGR, AT&T INFO PRODS, 86- *Concurrent Pos:* Adj prof, Fairleigh Dickinson Univ, 64-65; assoc ed, Networks J, 79-87. *Mem:* Opers Res Soc Am; Am Statist Asn. *Res:* Mathematical studies of probability; statistical methodology; graph theory and network design. *Mailing Add:* 17 Daac Lane Little Silver NJ 07739

THOMAS, RONALD LESLIE, b Edmonton, Alta, Can, June 29, 35; m 59; c 3. AGRONOMY. *Educ:* Univ Alta, BSc, 57, MSc, 59; Ohio State Univ, PhD(soils), 63. *Prof Exp:* From asst prof to assoc prof, 63-71, PROF SOIL SCI, UNIV GUELPH, 71- *Mem:* Am Soc Agron; Agr Inst Can; Can Soc Soil Sci. *Res:* Soil organic matter chemistry, the reactions, nature and importance of organic matter and its decomposition. *Mailing Add:* Dept Land Resource Sci Univ Guelph Guelph ON N1G 2W1 Can

THOMAS, ROY DALE, b Sevier Co, Tenn, Nov 12, 36; m 59; c 3. PLANT TAXONOMY. *Educ:* Carson-Newman Col, BS, 58; Southeastern Baptist Theol Sem, BD, 62; Univ Tenn, PhD(plant taxon), 66. *Prof Exp:* Assoc prof, 66-76, CUR HERBARIUM, NORTHEAST LA UNIV, 74-, PROF BIOL, 76- *Mem:* Am Soc Plant Taxon; Bot Soc Am; Int Soc Plant Taxon; Soc Econ Bot; Am Fern Soc; Soc Wetland Scientists. *Res:* Vegetation and flora of Chilhowee Mountain in east Tennessee; flora of Louisiana; Ophioglossaceae of the Gulf South. *Mailing Add:* Dept Biol Northeast La Univ Monroe LA 71209-0502

THOMAS, ROY ORLANDO, b Oneida, Tenn, Dec 15, 21; m 45; c 1. ANIMAL NUTRITION, DAIRY HUSBANDRY. *Educ:* Berea Col, BS, 46; Univ Tenn, MS, 52; Mich State Univ, PhD(animal nutrit), 64. *Prof Exp:* Teacher, Lewis County, Ky Bd Educ, 46 & Scott County, Tenn Bd Educ, 46-51; cow tester exten serv, Univ Tenn, 52; fieldman, Nashville Milk Producers, Inc, 52-53; asst dairy husbandman, Univ Tenn, 53-61; res asst animal nutrit, Mich State Univ, 61-64; from asst prof to assoc prof dairy sci, WVa Univ, 64-90, dairy scientist, 72-90; RETIRED. *Mem:* AAAS; Am Dairy Sci Asn; Am Soc Animal Sci. *Res:* Evaluation of feed materials, methods of feeding and the effect of these materials and methods on production and well-being of animals; design of free stalls and sequential exposure of animals to stalls and feed. *Mailing Add:* Dept Dairy Sci WVa Univ Morgantown WV 26506

THOMAS, RUTH BEATRICE, b Ringgold, La. BOTANY. *Educ:* Northwestern State Col, La, BS, 40; George Peabody Col, MA, 44; Vanderbilt Univ, PhD(biol), 51. *Prof Exp:* Pub sch teacher, La, 40-44; instr, Sullins Col, 44-46; instr, George Peabody Col, 46-48; asst prof biol, Millikin Univ, 51-54; prof, Eastern NMex Univ, 54-63; assoc prof, 64-70, PROF BIOL, SAM HOUSTON STATE UNIV, 70- *Mem:* Fel AAAS; Bot Soc Am; Nat Asn Biol Teachers. *Res:* Gymnosperm gametophyte development; descriptive morphology. *Mailing Add:* Dept Biol Sam Houston State Univ Huntsville TX 77341

THOMAS, SAMUEL GABRIEL, b Youngstown, Ohio, Dec 20, 46; m 68; c 3. RESEARCH MANGEMENT. *Educ:* Youngstown State Univ, BS, 68; Univ Cincinnati, PhD(chem), 73; Mich State Univ, MBA, 82. *Prof Exp:* Teaching fel phys chem, Drexel Univ, 73-74; instrument chemist, anal chem, Amerada Hess Corp, 74-76; anal chemist, 76-80, sr anal chemist, 80-81, actg supvr, 81-84, supvr anal chem, 84-89, mgr res & develop, 89-90, STRUCT & ACTIV TESTING DIR, ETHYL COPR, 90- *Mem:* Am Chem Soc; Soc Tribologist & Lubrication Engrs. *Res:* Separation and indentification of complex hydrocarbon mixtures; separation by high performance liquid chromatography and identification of gas chromatography-mass spectrometry; applications research for materials, surfactants and lubricants. *Mailing Add:* Ethyl Corp PO Box 14799 Baton Rouge LA 70898

THOMAS, SARAH NELL, b Gainesville, Ga. PHYSIOLOGY, RADIATION BIOLOGY. *Educ:* Brenau Col, BA, 48; Univ Denver, MS, 57; Tex Woman's Univ, PhD(radiation biol), 70. *Prof Exp:* Teacher & head dept sci, Pub Schs, Ga, 48-60; assoc prof biol & chmn dept, Brenau Col, 60-67; instr, Tex Woman's Univ, 70; assoc prof, 70-75, PROF BIOL & CHMN DEPT NATURAL SCI, LANGSTON UNIV, 75- *Concurrent Pos:* NSF traineeship, 67-69. *Mem:* AAAS; Am Inst Biol Sci; Nat Asn Biol Teachers. *Res:* Gonad development in male rats irradiated the first day of postnatal life. *Mailing Add:* Dept Biol Langston Univ Langston OK 73050

THOMAS, SETH RICHARD, b Torrington, Conn, May 5, 41; m 63; c 4. METALLURGY. *Educ:* Lehigh Univ, BA, 63, BS, 64, MS, 66. *Prof Exp:* Process engr, Tex Instruments, 65-68; tech dir, Teledyne Rodney Metals, 68-77; DIR METALL SERV, THOMAS STEEL STRIP CORP, 77- *Mem:* Fel Am Soc Metals Int; Sigma Xi; Am Soc Testing & Mat; Am Soc Qual Control. *Res:* Crystallographic orientation of unidirectionally-solidified lamellar eutectics and reverse martensitic transformations in stainless steel. *Mailing Add:* 11234 Vista Dr Hiram OH 44234

THOMAS, STANISLAUS S(TEPHEN), b Barberton, Ohio, Nov 1, 19; m 46; c 5. MECHANICAL ENGINEERING, INDUSTRIAL ENGINEERING. *Educ:* Univ Akron, BME, 50; Cornell Univ, MS, 55; Purdue Univ, PhD, 67. *Prof Exp:* Chief resident inspector, Pittsburgh Chem Warfare Procurement Dist, Pa, 40-45; sr draftsman, Goodyear Tire & Rubber Co, Ohio, 46-51; engr, Army Chem Ctr, Md, 51-52; instr mach design, Sibley Sch Mech Eng, Cornell Univ, 52-55; asst prof mech eng, Notre Dame Univ, 55-60; mech engr, Midwestern Univs Res Asn, 60-61; instr, Sch Civil Eng, Purdue Univ, 61-66; eng mgr, Midwest Appl Sci Corp, Ind, 67-71; asst chmn dept indust & mgt eng, NJ Inst Technol, 71-83, assoc dean eng technol, 83-85; CONSULT, 86- *Concurrent Pos:* Assoc fac, Grad Sch Mgt, Rutgers Univ; vis prof, Purdue Univ, 85-86. *Mem:* Am Soc Mech Engrs; Am Soc Eng Educ; Am Soc Metals; Am Inst Indust Engrs; Int Mat Handling Soc; Soc Mfg Engrs. *Res:* Design of mechanisms and machines; quality control in manufacturing processes; reliability of engineering systems. *Mailing Add:* Dept Mech & Indust Eng NJ Inst Technol Newark NJ 07102

THOMAS, STEVEN P, b Lancaster, Pa, July 30, 43. BAT FLIGHT PHYSIOLOGY, EXERCISE PHYSIOLOGY. *Educ:* Ind Univ, Bloomington, PhD(physiol), 71. *Prof Exp:* PROF BIOL, DUSQUESNE UNIV, 82- *Mailing Add:* Dept Biol Dusquesne Univ Pittsburgh PA 15282

THOMAS, TELFER LAWSON, b Montreal, Que, June 1, 32; m 56; c 3. ORGANIC CHEMISTRY, SOFTWARE SYSTEMS. *Educ:* McGill Univ, BS, 53, PhD(org chem), 57. *Prof Exp:* Res chemist, Imp Oil Ltd, 57-59 & Gen Aniline & Film Co, 59-62; PRIN INVESTR, PHARMACEUT DIV, PENNWALT CORP, 62- *Mem:* Am Chem Soc. *Res:* Synthesis of new compounds for discovery of useful drugs; micro/mini computer operating systems; database management systems. *Mailing Add:* 755 Jefferson Rd Fisons Pharmaceut Rochester NY 14623

THOMAS, TERENCE MICHAEL, b Ft Dix, NJ, Apr 13, 52; m 75; c 4. SURFACE-INTERFACE CHEMISTRY, INORGANIC BONDING THEORY. *Educ:* Benedictine Col, BA, 74; Colo State Univ, MS, 76; Univ Tenn, Knoxville, PhD(inorg chem), 80. *Prof Exp:* Tech asst, phys chem lab, Colo State Univ, 74-76; surface chemist, Oak Ridge Nat Lab, 76-80, electro chemist, 78; staff chemist surface-interface mat res, Solar Energy Res Inst, 80-87; SR CHEMIST, BRUSH WELLMAN INC, 87- *Concurrent Pos:* Tech asst, gen chem lab, Univ Tenn, Knoxville, 76-80. *Mem:* Am Chem Soc; Am Vacuum Soc. *Res:* Perform XPS, ISS, SIMS and Auger spectroscopies on advanced materials; identify the physical and chemical processes which degrade the performance of advanced materials or increase the lifetime of manufactured components. *Mailing Add:* Brush Wellman Inc 17876 St Clair Ave Cleveland OH 44110

THOMAS, THOMAS DARRAH, b Glen Ridge, NJ, Apr 8, 32; m 56; c 4. PHYSICAL CHEMISTRY. *Educ:* Haverford Col, BS, 54; Univ Calif, PhD(chem), 57. *Prof Exp:* From instr to asst prof chem, Univ Calif, 57-59; vis assoc chemist, Brookhaven Nat Lab, 59-60, assoc chemist, 60-61; from asst prof to assoc prof chem, Princeton Univ, 61-71; chmn, Ore State Univ, 81-85, dir, Ctr Advan Mat Res, 86-91, PROF CHEM, ORE STATE UNIV, 71-, DISTINGUISHED PROF CHEM, 89- *Concurrent Pos:* Consult, Los Alamos Sci Lab, 65; Guggenheim fel, Univ Calif, Berkeley, 69; fel, Univ Liverpool, 84-85. *Mem:* Fel AAAS; Am Chem Soc; fel Am Phys Soc. *Res:* electron spectroscopy. *Mailing Add:* Dept Chem Gilbert Hall 153 Ore State Univ Corvallis OR 97331-4003

THOMAS, TIMOTHY FARRAGUT, b Cleveland, Ohio, June 15, 38; m 63; c 3. PHYSICAL CHEMISTRY, CHEMICAL DYNAMICS. *Educ:* Oberlin Col, AB, 60; Univ Ore, PhD(chem kinetics), 64. *Prof Exp:* Res assoc chem, Brandeis Univ, 64-66; asst prof, 66-73, ASSOC PROF CHEM, UNIV MO-KANSAS CITY, 73- *Concurrent Pos:* Nat Res Coun sr res assoc, Air Force Cambridge Res Lab, 75-76; univ resident res prog vis prof, Air Force Geophys Lab, 81- 83. *Mem:* Am Chem Soc; Am Phys Soc; Am Soc Mass Spectrometry; Royal Soc Chem; Inter-Am Photochem Soc; Sigma Xi. *Res:* Unimolecular reaction kinetics; photochemistry of gases; fluorescence lifetimes and quantum yields; mass spectrometric appearance potentials; photodissociation spectra of gaseous ions. *Mailing Add:* Dept Chem Univ Mo Kansas City MO 64110

THOMAS, TUDOR LLOYD, b Utica, NY, Apr 23, 21; m 44; c 4. PHYSICAL CHEMISTRY. *Educ:* Univ Mich, BS, 43, MS, 46, PhD(chem), 49. *Prof Exp:* Res chemist, 49-56, develop supvr, 56-58, from asst mgr to mgr molecular sieve develop, 58-63, mgr molecular sieve prod, 63-67, vpres, 77-83, GEN MGR MOLECULAR SIEVE DEPT, LINDE DIV, UNION CARBIDE CORP, 67-; MGR ZEOLITE RES PROG, GA INST TECHNOL, 83- *Mem:* AAAS; Am Chem Soc; Am Mgt Asn. *Res:* Adsorption; catalysis. *Mailing Add:* 8905 Willowbrae Lane Roswell GA 30076-3572

THOMAS, VERA, b Prague, Czech, May 2, 28; US citizen; m 67. CHEMISTRY, TOXICOLOGY. *Educ:* Charles Univ, Prague, MS, 52; Czech Acad Sci, PhD(chem), 62. *Prof Exp:* Res assoc indust toxicol, Inst Indust Hyg & Occup Dis, Prague, Czech, 49-67; asst prof, 67-77, assoc prof, 77-81, PROF DRUG METAB, SCH MED, UNIV MIAMI, 81- *Concurrent Pos:* Secy comt maximum allowable concentration toxic compounds, Czech Ministry Health, 62-67; consult, WHO, Chile & Venezuela, 67. *Mem:* NY Acad Sci; Am Conf Govt Indust Hygienists. *Res:* Uptake, distribution, metabolism and excretion of drugs, especially of volatile compounds; pesticides distribution. *Mailing Add:* Dept Anesthesiol Sch Med Univ Miami Miami FL 33101

THOMAS, VIRGINIA LYNN, b Graham, Tex, Mar 29, 43. MEDICAL MICROBIOLOGY. *Educ:* Baylor Univ, BSc, 65, MSc, 67; Univ Tex Med Sch, San Antonio, PhD(microbiol), 73. *Prof Exp:* Teaching fel biol, Baylor Univ, 65-66, res asst & technician microbiol, Col Dent & Med Ctr, Dallas, 66-68; sr res & teaching asst microbiol,; instr, 73-75, asst prof, 75-79, ASSOC PROF MICROBIOL, UNIV TEX HEALTH SCI CTR, SAN ANTONIO, 79- *Mem:* Am Soc Microbiol; Sigma Xi; AAAS. *Res:* Host-parasite relationships in bacterial diseases; immunologic aspects of urinary tract infection; opsonophagocytosis and infection susceptibility. *Mailing Add:* Dept Microbiol Univ Tex Health Sci Ctr 7703 Floyd Curl Dr San Antonio TX 78284

THOMAS, WALTER DILL, JR, b St Louis, Mo, July 3, 18; m 39; c 2. FORESTRY, HORTICULTURE. *Educ:* Colo State Univ, BS, 39; Univ Minn, MS, 43, PhD(phytopath), 47. *Prof Exp:* Instr bot, Colo State Univ, 39-41; asst phytopath, Univ Minn, 41-44, 46; from asst prof plant path & asst plant pathologist to prof & plant pathologist, Agr Exp Sta, 46-54; dir res, Arboriculture Serv & Supply Co, Colo, 54-55; lead res biologist, Ortho Div, Chevron Chem Co, 55-66, tech asst to mgr res & develop, 66-67, forestry specialist, 67-70; pres, Forest & Environ Protection Serv, 70-71; vpres, Natural Resouces Mgt Corp & mem bd dir, Environ Home & Garden Serv, 72-74; pres, Forest-Agr Corp, 74-86; CONSULT DENDROPATHOLOGIST, 86- *Mem:* Am Phytopath Soc; fel AAAS; Soc Am Foresters; Asn Consult Foresters; Am Soc Consult Arborists; Int Soc Arboriculture; Am Forestry Asn. *Res:* Disease of potatoes, beans, onions and ornamental plants; forest diseases; agricultural pesticides; mycorrhizae; air pollution damage to plants; remote sensing of forest diseases and insects. *Mailing Add:* 2435 Heatherleaf Lane Martinez CA 94553

THOMAS, WALTER E, b West Lafayette, Ind, Dec 19, 22; m 43; c 4. MANUFACTURING ENGINEERING, ENGINEERING TECHNOLOGY. *Educ:* Purdue Univ, Lafayette, BSME, 48, MSIE, 53. *Prof Exp:* Instr, Sch Eng, Purdue Univ, Lafayette, 48-53; design supvr, Chrysler Grad Inst, Mich, 53-54; asst prof, Sch Eng, Univ Mich, Ann Arbor, 54-57; asst chief draftsman, Fla Res & Develop Ctr, Pratt & Whitney Aircraft, 57-62; mgr eng serv, Atlantic Res Corp, Va, 62-63; sr design engr, Fla Res & Develop Ctr, Pratt & Whitney Aircraft, 63-64; prof mech eng & head dept, Mfg Technol, Purdue Univ, Lafayette, 64-73; prof & assoc dean, Sch Technol, Fla Int Univ, 73-76; prof & dean, Sch Technol & Appl Sci, Western Carolina Univ, 76-90; RETIRED. *Concurrent Pos:* Design engr, HydroPower Inc, Ohio, 50-52; consult, Altamil Corp, Ind, 70-; chmn eng technol comt, Engrs Coun Prof Develop, 77-78. *Mem:* Soc Mfg Engrs; Am Soc Eng Educ. *Res:* Updating of manufacturing processes in the machine tool area and in the foundry area. *Mailing Add:* Dean Sch Engr Technol Univ Ark 33 & Univ Ave Little Rock AR 72204

THOMAS, WALTER IVAN, b Elwood, Nebr, Mar 27, 19; m 41; c 1. AGRONOMY, GENETICS. *Educ:* Iowa State Univ, BS, 49, MS, 53, PhD(plant breeding genetics), 55. *Prof Exp:* Res asst agron, Iowa State Univ, 49-50, 53-55, asst prof, 55-59; from assoc prof to prof, Pa State Univ, University Park, 59-79, head dept, 64-69, assoc dean res, Col Agr & assoc dir, Agr Exp Sta, 69-79; dep dir, Sci & Educ Admin-Coop Res, 79-81, ADMINR, COOP STATE RES SERV, USDA, 81- *Mem:* Fel AAAS; fel Am Soc Agron; Sigma Xi; fel Soil Sci Soc. *Res:* Plant breeding, genetics and pathology. *Mailing Add:* Off Adminr Coop State Res Serv USDA Washington DC 20250

THOMAS, WARREN H(AFFORD), b Portsmouth, Ohio, July 15, 33; m 57; c 2. INDUSTRIAL ENGINEERING, OPERATIONS RESEARCH. *Educ:* Case Inst Technol, BSME, 55; Purdue Univ, MSIE, 61, PhD(indust eng), 64. *Prof Exp:* Instr indust eng, Purdue Univ, 62-63; from asst prof to assoc prof, 63-77, distinguished prof, 77- , chmn dept, 69-87, ASSOC DEAN INDUST ENG, STATE UNIV NY, BUFFALO, 87- *Concurrent Pos:* Vis sr res fel, Dept Oper Res, Univ Lancaster, 70-71; vis prof, Univ Nottingham, 81. *Mem:* Inst Indust Engrs; Am Soc Eng Educ. *Res:* Manufacturing systems; computer simulation; design of production control systems. *Mailing Add:* Dept Indust Eng State Univ NY 342 Bell Hall Buffalo NY 14260

THOMAS, WILBUR ADDISON, b Louisville, Miss, June 26, 22; m 49; c 2. PATHOLOGY. *Educ:* Univ Miss, BA, 41; Univ Tenn, MD, 46; Am Bd Path, dipl. *Prof Exp:* Intern, Baptist Hosp, Memphis, Tenn, 46-47, asst resident path, 49-50; asst resident & resident, Mass Gen Hosp, Boston, 50-52; instr, Harvard Med Sch, 52-53; from instr to assoc prof, Sch Med, Wash Univ, 53-59; CYRUS STRONG MERRILL PROF PATH & CHMN DEPT, ALBANY MED COL, 59- *Mem:* Am Soc Exp Path; Am Asn Path & Bact; AMA; Col Am Path. *Res:* Arteriosclerosis. *Mailing Add:* Eight Locust Lane Loudonville NY 12211

THOMAS, WILLIAM ALBERT, b Washington, DC, Apr 25, 50; m 76; c 1. CELL-CELL ADHESION, NEURONAL GUIDANCE. *Educ:* Hamilton Col, BA, 72; Princeton Univ, PhD(develop biol), 79. *Prof Exp:* ASST PROF BIOCHEM & DEVELOP BIOL, DEPT BIOL, WAKE FOREST UNIV, 83- *Mem:* Am Soc Cell Biol; AAAS. *Res:* Use of immunological and cell biological techniques to investigate the possible role of specific cell surface adhesion molecules in guiding cell rearrangements characteristic of early avian development. *Mailing Add:* Dept Biol Wake Forest Univ PO Box 7325 Reynolda Sta Winston-Salem NC 27109

THOMAS, WILLIAM ANDREW, b Berea, Ky, July 23, 36; m 57; c 2. GEOLOGY. *Educ:* Univ Ky, BS, 56, MS, 57; Va Polytech Inst, PhD(geol), 60. *Prof Exp:* Geologist, Calif Co, 59-63; from assoc prof to prof geol, Birmingham-Southern Col, 63-70, chmn dept, 67-70; assoc prof, Queens Col, NY, 70-72, chmn dept, 71-72; prof geol & chmn dept, Ga State Univ, 72-79; prof geol, Univ Ala, 79-90; PROF GEOL & CHMN DEPT, UNIV KY, 91- *Concurrent Pos:* Ed, Geol Soc Am Bull, 82-88; vis scientist, Cornell Univ, 87. *Mem:* Geol Soc Am; Am Asn Petrol Geol; Soc Econ Paleontologists & Mineralogists. *Res:* Tectonics and tectonic framework of sedimentation; stratigraphic and structural continuity of Appalachian and Ouachita Mountains; stratigraphy and structure of Gulf Coastal Plain; Appalachian structure and stratigraphy; Mississippian stratigraphy. *Mailing Add:* Dept Geol Sci Univ Ky Lexington KY 40506-0059

THOMAS, WILLIAM CLARK, JR, b Bartow, Ga, Apr 7, 19; m 46. INTERNAL MEDICINE, ENDOCRINOLOGY. *Educ:* Univ Fla, BS, 40; Cornell Univ, MD, 43. *Prof Exp:* Intern med, New York Hosp, 44, asst resident med, 46-49; pvt pract, 49-54; NIH fels, Johns Hopkins Univ, 54-57; from asst prof to assoc prof, 57-63, chief, Div Postgrad Educ, 57-60 & Endocrine Div, 57-70, dir, Clin Res Ctr, 62-68, PROF MED, COL MED, UNIV FLA, 63- *Concurrent Pos:* Chief med serv, Vet Admin Hosp, Gainesville, Fla, 68-73, assoc chief of staff for res, 73-85; hon res fel, Univ Manchester, 69-70, dir, Geriatric Res Ed Clin Ctr & Vet Admin Med Ctr, Gainesville, Fl, 85- *Mem:* AAAS; Am Clin & Climat Asn; Am Diabetes Asn; Endocrine Soc; Am Fedn Clin Res. *Res:* Mineral metabolism; clinical research; factors affecting renal calculus formation. *Mailing Add:* Med Serv Vet Admin Hosp Gainesville FL 32601

THOMAS, WILLIAM ERIC, b Nashville, Tenn, Aug, 2, 51; m 84; c 5. NEUROBIOLOGY & NEUROPHYSIOLOGY. *Educ:* Tenn State Univ, BS, 73, MS, 75; Meharry Med Col, PhD(biochem), 80. *Prof Exp:* Res assoc neurochem, Dept Physiol & Biophys, Univ Ill Med Ctr, 80; fel neurobiol, Harvard Med Col, 80-82; asst prof neurophysiol, Meharry Med Col, 82-87; ASSOC PROF NEUROBIOL, DEPT ORAL BIOL, OHIO STATE UNIV, 88- *Concurrent Pos:* Consult, Minority Biochem Res Support & Nat Sci Found, 86-; Klingenstein fel in the neuroscience. *Mem:* AAAS; Am Aging Asn; Soc Neurosci; Am Soc Cell Biol; NY Acad Sci; Am Asn Dent Res. *Res:* Determination of functional properties of various putative transmitters in the cerebral cortex; contribution of glial cells to the maintenance of cerebral cortical function. *Mailing Add:* Ohio State Univ Col Dent 305 W 12th Ave Columbus OH 43210-1241

THOMAS, WILLIAM GRADY, b Charlotte, NC, Mar 21, 34; m 55; c 3. AUDIOLOGY. *Educ:* Appalachian State Univ, BS, 57; Wash Univ, MA, 61; Univ Fla, PhD(auditory physiol), 68. *Prof Exp:* From instr to asst prof, 61-70, ASSOC PROF, DEPT SURG, MED SCH, UNIV NC, CHAPEL HILL, 70-, DIR HEARING & SPEECH, 61-, DIR, AUDITORY RES LAB, 68-, ASSOC PROF, DIV SPEECH & HEARING SCI, 70- *Concurrent Pos:* Fac res grant, Univ NC, Chapel Hill, 68-69, Off Naval Res grants, 69-79, NIH grant, 72-75; consult, Exp Diving Unit, Dept Navy, 69- & Nat Inst Environ Health Sci, 72-; Rockefeller Found grant, 75-80; NIH grant, 75-80; res scientist, Child Develop Inst, 76-; adj prof, Dept Commun & Theater, Univ NC, Greensboro, 81-86; pvt pract, 86- *Honors & Awards:* Cert of Recognition, Am Speech & Hearing Asn, 69; Cert of Appreciation, Am Speech & Hearing Soc. *Mem:* Fel Am Speech & Hearing Asn; Acoust Soc Am. *Res:* Auditory physiology and psychoacoustics, particularly the effects of drugs and environmental conditions on the ear; basic electrophysiology of the auditory system. *Mailing Add:* 4002 Barrett Dr Suite 101 Raleigh NC 27609

THOMAS, WILLIAM HEWITT, b Riverside, Calif, Dec 25, 26; m 56; c 2. BIOLOGICAL OCEANOGRAPHY, LIMNOLOGY. *Educ:* Pomona Col, BA, 49; Univ Md, MS, 52, PhD, 54. *Prof Exp:* Lab asst, Regional Salinity Lab, USDA, 46-47, lab asst plant physiol, 48; lab asst plant physiol, Citrus Exp Sta, Univ Calif, 49-50 & Univ Md, 51-54; jr res biologist, 54-56, asst res biologist, 56-64, assoc res biologist, 64-75, res biologist, 75-88, EMER RES BIOLOGIST, SCRIPPS INST OCEANOG, UNIV CALIF, SAN DIEGO, 88- *Mem:* AAAS; Am Soc Limnol & Oceanog; Phycol Soc Am; Int Phycol Soc. *Res:* Mineral nutrition and nitrogen metabolism of algae; primary production in the ocean; cultural requirements of marine phytoplankton; marine and freshwater pollution; desert and mountain algae. *Mailing Add:* Scripps Inst Oceanog Univ Calif San Diego La Jolla CA 92093

THOMAS, WILLIAM J, b Sharon, Pa, Oct 11, 24; m 49; c 2. TISSUE CULTURE, FREEZE-DRYING. *Educ:* Westminster Col, Pa, BS, 48; Univ Md, MS, 51; Univ Pa, PhD(med micro), 59. *Prof Exp:* Sect leader, Wistar Inst, Ft Detrick, Md, 50-55, res assoc, 55-58; group leader, Merrell Nat Labs, 58-78; DIR PROD, GOV SERV DIV, SALK INST, 78- *Concurrent Pos:* Pres, Mt Pocono Munic Authority, 70-72, Burnley Workshop, Pa, 78-84. *Mem:* Tissue Cult Asn; Soc Cryobiol. *Res:* Developer of certified tissue culture cell systems for vaccine propagation; studies on stabilizers for freeze drying attenuated vaccines such as Junin and Rift Valley Fever; preparation of diagnostic reagents. *Mailing Add:* Gov Serv Div Salk Inst PO Box 250 Swiftwater PA 18370

THOMAS, WILLIAM ROBB, b Toronto, Kans, Dec 17, 26; m 54; c 1. FOOD SCIENCE, DAIRY BACTERIOLOGY. *Educ:* Okla State Univ, BSc, 50; Ohio State Univ, MSc, 52; Iowa State Univ, PhD(dairy bact), 61. *Prof Exp:* Asst dairy tech, Ohio State Univ, 50-52; mem sanit stand staff, Evaporated Milk Asn, Ill, 52-53; asst prof dairy mfg, Univ Wyo, 53-59; res asst, Iowa State Univ, 59-61; assoc prof dairy mfg, Univ Wyo, 61-65; exten food technologist, Univ Calif, Davis, 65-68; assoc dean & dir resident instr, Col Agr Sci, Colo State Univ, 69-89, interim dean, 84-85; RETIRED. *Concurrent Pos:* Appointee, Int Sci & Educ Coun, 74-77. *Mem:* Nat Asn Col & Teachers Agr (pres, 75-76); Am Dairy Sci Asn; Inst Food Technol; Int Asn Milk, Food & Environ Sanit. *Res:* Dairy technology; thermoduric bacteria; lipolytic enzymes of milk; consumer and market analysis of dairy products; dairy plant operation analysis. *Mailing Add:* 2017 Stover St Ft Collins CO 80525

THOMAS, WINFRED, b Geneva, Tex, June 6, 20; m 43; c 1. AGRONOMY. *Educ:* Prairie View State Col, BS, 43; Cornell Univ, MS, 47; Ohio State Univ, PhD(agron), 54. *Prof Exp:* Instr agron, Ala Agr & Mech Col, 47-51; asst, Ohio State Univ, 51-53; agronomist, 53-81, PROF AGRON, ALA A&M UNIV, 81-, DEAN, SCH AGR & ENVIRON SCI, 73- *Mem:* Am Soc Agron; Soil Sci Soc Am. *Res:* Effect of foliar applied fertilizers on growth and composition of corn; plant population-nitrogen relationships of corn; nitrogen-sulfur-protein relationships of corn. *Mailing Add:* 2041 Winchester Rd NE Normal AL 35811

THOMASIAN, ARAM JOHN, b Boston, Mass, Aug 12, 24; m 53; c 3. ELECTRICAL ENGINEERING, STATISTICS. *Educ:* Brown Univ, BSc, 49; Harvard Univ, MA, 51; Univ Calif, PhD(math statist), 56. *Prof Exp:* PROF ELEC ENG & STATIST, UNIV CALIF, BERKELEY, 56- *Mem:* Inst Elec & Electronics Engrs. *Res:* Information theory; probability; electroencephatography. *Mailing Add:* Dept Elec Eng Univ Calif 2120 Oxford St Berkeley CA 94720

THOMASON, BERENICE MILLER, b Birmingham, Ala, Mar 10, 24; m 44; c 1. MICROBIOLOGY. *Educ:* Ga State Col, BS, 60. *Prof Exp:* Med technologist, Sta Hosp, Ft Benning, Ga, 43-45 & Thayer Gen Hosp, Nashville, Tenn, 45; pub health technologist, Muscogee County Health Dept, Columbus, Ga, 48-51; bacteriologist, Commun Dis Ctr, Atlanta, Ga, 51-53 & Third Army Labs, Ft McPherson, Ga, 53-54; bacteriologist, 54-63, RES MICROBIOLOGIST, CTR DIS CONTROL, USPHS, 63- *Mem:* Am Soc Microbiol; Sigma Xi. *Res:* Development and application of fluorescent antibody technic for rapid detection of pathogenic bacteria. *Mailing Add:* CDC L Div Mic Br 4-112 1600 Clifton Rd NE Atlanta GA 30333

THOMASON, DAVID MORTON, b Martinsville, Va, May 28, 47; m 77. POULTRY SCIENCE. *Educ:* Va Polytech Inst, BS, 69, MS, 71, PhD(genetics), 74. *Prof Exp:* Res asst, Va Polytech Inst, 69-73; res assoc, Duck Res Lab, Cornell Univ, 74-76; exten poultry scientist, Coop Exten Serv, Univ Ga, 76-79; tech training dir, Mathtech, Inc, 79-81; dir tech serv, 81-85, tech dir, 85-87, regional dir, 87-89, CONSULT, AM SOYBEAN ASN, SINGAPORE, 89- *Concurrent Pos:* Adj assoc prof genetics, Southampton Col Long Island Univ, 75; consult, Delight Menues, Baltimore Md, 78, Pinecrest Duck Farm, 78, Govt Egypt thru Mathtec, 79, Int Develop Assoc, 81 & Poultry & Egg Facs, 81. *Mem:* Sigma Xi; Poultry Sci Asn; World Poultry Sci. *Res:* Practical poultry processing technology; economical and physiological evaluation of the reproductive performance of turkeys under different environmental conditions; soybean meal use in animal nutrition. *Mailing Add:* PO Box 340023 Boca Raton FL 33434

THOMASON, DONALD BRENT, b Richland, Wash, Oct 15, 57. MUSCLE PHYSIOLOGY, CARDIOVASCULAR PHYSIOLOGY. *Educ:* Univ Va, BA, 80; Univ Calif, Irvine, PhD(physiol), 86. *Prof Exp:* Res assoc, 86-89, res asst prof, 89-90, ASST PROF PHYSIOL, UNIV TENN HEALTH SCI CTR, 90- *Mem:* AAAS; Am Chem Soc; Am Motility Soc; Protein Soc. *Res:* Mechanisms of gene expression in muscle in response to functional demand; effects of exercise and gravitation. *Mailing Add:* Dept Physiol & Biophys Univ Tenn Health Sci Ctr 894 Union Ave Memphis TN 38163

THOMASON, IVAN J, b Burney, Calif, June 27, 25; m 50; c 5. PLANT NEMATOLOGY, PLANT PATHOLOGY. *Educ:* Univ Calif, BS, 50; Univ Wis, MS, 52, PhD(plant path), 54. *Prof Exp:* Res asst plant path, Univ Wis, 50-54; jr nematologist, Citrus Res Ctr, Agr Exp Sta, 54-56, from asst nematologist to nematologist, 54-89, chmn, dept nematol, 63-70, prof nematol, 67-89, prof plant path, 73-89, EMER PROF NEMATOL & PLANT PATH, UNIV CALIF, RIVERSIDE, 89- *Concurrent Pos:* Mem, subcomt nematodes, Agr Bd, Nat Acad Sci-Nat Res Coun, 64-67; mem, Univ Calif-AID Pest Mgt Study Team, Southeast Asia, 71; mem, agr pest control adv comt, Calif State Dept Agr, 72- & pest control advisors comt, Dir, adv comt on APCA, Calif Dept Food & Agr, 74-76; asst dir, Pest & Dis Mgt Prog, Coop Exten, Univ Calif, 76-81, asst dir, Agr Exp Sta & dir, Statewide Pest Mgt Proj, 78-81. *Mem:* Fel Am Phytopath Soc; fel Soc Nematologists (pres, 75-76); Soc Europ Nematol. *Res:* Biology and control of nematodes attacking field and vegetable crops; efficacy of nematicides; nematode resistance in crop plants. *Mailing Add:* 4686 Holyoke Pl Riverside CA 92507

THOMASON, ROBERT WAYNE, b Tulsa, Okla, Nov 5, 52. ALGEBRAIC GEOMETRY AND TOPOLOGY. *Educ:* Princeton Univ, PhD(math), 77. *Prof Exp:* Instr math, Mass Inst Tech, 77-79; asst prof math, Univ Chicago, 79-80; lectr math, Mass Inst Tech, 80-82; mem, Inst Advan Study, 82-83; from asst prof to prof math, Johns Hopkins Univ, 83-90; DIR RES, NAT CTR SCI RES, 90- *Concurrent Pos:* Prin investr, Nat Sci Found, 81-90; vis prof, Rutgers Univ, 87, Univ Paris VII, 90; invited speaker, Int Cong Mathematicians. *Mem:* Am Math Soc; Math Soc France. *Res:* Algebraic K-theory to relate algebraic geometry of varieties to their topology currently; foundation of algebraic topology and homological algebra; intersection theory. *Mailing Add:* UFR Math Univ Paris VII Two Pl Jussieu Paris Cederos 75251 France

THOMASON, STEVEN KARL, b Salem, Ore, June 2, 40; m 60; c 2. MATHEMATICAL LOGIC. *Educ:* Univ Ore, BS, 62; Cornell Univ, PhD, 66. *Prof Exp:* From asst prof to assoc prof, 66-78, PROF MATH, SIMON FRASER UNIV, 78- *Concurrent Pos:* Vis asst prof, Univ Calif, Berkeley, 68-69. *Mem:* Am Math Soc; Can Math Cong; Asn Symbolic Logic. *Res:* Nonclassical logic, especially modal logic. *Mailing Add:* Dept Math Simon Fraser Univ Burnaby BC V5A 1S6 Can

THOMASON, WILLIAM HUGH, b Hampton, Ark, Apr 4, 45; m 69; c 1. PHYSICAL CHEMISTRY, CORROSION. *Educ:* Hendrix Col, BA, 67; La State Univ, Baton Rouge, PhD(phys chem), 75. *Prof Exp:* res scientist, 75-80, RES GROUP LEADER PHYS CHEM & CORROSION RES & DEVELOP, CONOCO, INC, 80- *Mem:* Am Chem Soc; Soc Petrol Engrs; Nat Asn Corrosion Engrs. *Res:* Corrosion problems in oil production, particularly those caused by hydrogen sulfide; water treating and scale problems. *Mailing Add:* 2505 Barola Lane Ponca City OK 73541

THOMASSEN, DAVID GEORGE, b Ft Collins, Colo, Aug 12, 52; m 73; c 2. CARCINOGENESIS. *Educ:* Wash State Univ, BS, 74, MS, 75; Univ Wis-Madison, PhD(genetics), 80. *Prof Exp:* Teaching asst genetics/biol, Wash State Univ, 73-75; teaching asst genetics, Univ Wis-Madison, 77-78; postdoctoral fel, Nat Inst Environ Health Sci, 81-84; sr staff fel, Nat Cancer Inst, 84-86; CELL BIOLOGIST, INHALATION TOXICOL RES INST, 86-, MGR, RADIATION MECHANISMS PROG, 89- *Concurrent Pos:* Fac affiliate, Dept Path, Colo State Univ, 87-, Col Pharm, Univ NMex, 91-; ad hoc reviewer, Nat Cancer Inst & Nat Inst Environ Health Sci, 90. *Mem:* Am Asn Cancer Res. *Res:* Mechanism of multistage progression to neoplasia in respiratory epithelium; alterations in regulation of proliferation during carcinogenesis, in vitro carcinogenesis, radiation-induced cancer and identification of transforming genes. *Mailing Add:* Inhalation Toxicol Res Inst PO Box 5890 Albuquerque NM 87185

THOMASSEN, KEITH I, b Harvey, Ill, Nov 22, 36; m 57; c 2. PLASMA PHYSICS. *Educ:* Chico State Col, BS, 58; Stanford Univ, MS, 60, PhD(elec eng), 63. *Prof Exp:* Res assoc plasma physics, Stanford Univ, 62-63, NATO fel, 63-64, res assoc, 64-65, lectr, 64-68, res physicist, 65-68; from asst prof to assoc prof elec eng, Mass Inst Technol, 68-73; asst ctr div leader, Los Alamos Sci Lab, 73-74, assoc ctr div leader, 74-77; MFTF Prog Leader, Univ Calif, Livermore, 77-83, asst assoc dir, 83-87, MTX Prog leader, 87, DEP ASSOC DIR, UNIV CALIF, LIVERMORE, 88-; PROF IN RESIDENCE, NUCLEAR ENG, UNIV CALIF, BERKELEY, 89- *Concurrent Pos:* Consult, Lincoln Labs, Lexington, Mass, 68-74. *Mem:* Fel Am Phys Soc; Am Nuclear Soc. *Res:* Fusion reactor design; component development; plasma research, energy storage and transfer. *Mailing Add:* 9030 Doubletree Lane Livermore CA 94550

THOMASSON, CLAUDE LARRY, b Blue Grass, Va, Mar 6, 32; m 57; c 3. PHARMACY. *Educ:* Univ Cincinnati, BS, 54; Univ Fla, PhD(pharm), 57. *Prof Exp:* Assoc prof pharm, Southern Col Pharm, Mercer, 57-61, prof & chmn dept, 61-64; assoc prof, WVa Univ, 64-66; ASSOC PROF PHARM, AUBURN UNIV, 66- *Mem:* Am Pharmaceut Asn; Acad Pharmaceut Sci; AAAS. *Res:* Dispensing and clinical pharmacy. *Mailing Add:* Dept Clin Pharmacol Auburn Univ Auburn University AL 36830

THOMASSON, JOSEPH R, b Hayden, Colo, June 6, 46; m 67; c 3. PALEOBOTANY, PLANT SYSTEMATICS. *Educ:* Ft Hays State Univ, BS, 68;Iowa State Univ, PhD(bot), 76. *Prof Exp:* Assoc prof bot, Black Hill State Col, 76-81; prof bot, Ft Hays State Univ, 82-87; DISTINGUISHED VIS PROF BOT, USAF ACAD, 88- *Concurrent Pos:* Prin investr, NSF Grants, 78-85 & Nat Geog Soc Grants, 78-86. *Res:* Investigations of fossil and living grasses in order to elucidate the origins and evolution of grasses. *Mailing Add:* Dept Biol Ft Hays State Univ Hays KS 67601

THOMASSON, MAURICE RAY, b Columbia, Mo, Sept 3, 30; m 56, 90; c 3. GEOLOGY. *Educ:* Univ Mo, BA, 53, MA, 54; Univ Wis, PhD(geol), 59. *Prof Exp:* Geologist, Shell Oil Co, La, 59-68, mgr geol dept, Shell Develop Co, Tex, 68-70, div explor mgr, Shell Oil Co, La, 70-72, mgr forecasting, planning & econ, 72-74, mem staff, Shell Int Petrol Co, Ltd, London, 74-76, chief geologist, Shell Oil Co, Houston, 76-77; vpres explor, McCormick Oil & Gas Corp, 77-80; pres, Spectrum Oil & Gas, 80-82; pres, Pend Orielle Oil & Gas, 83-85; PRES, THOMASSON PARTNER ASSOC, INC, 85- *Mem:* Geol Soc Am; Soc Econ Paleont & Mineral; Am Asn Petrol Geol. *Res:* Paleogeography and sedimentation of western North America; paleocurrent and basin studies; general stratigraphy and stratigraphic paleontology. *Mailing Add:* 1580 Lincoln No 1220 Denver CO 80203

THOMBORSON, CLARK D, b Minn, Jan 11, 54; m 83; c 2. VERY LARGE SCALE INTEGRATED CIRCUITS THEORY, COMPUTER-AIDED DESIGN. *Educ:* Stanford Univ, BS & MS, 75; Carnegie Mellon Univ, PhD(computer sci), 80. *Prof Exp:* Asst prof computer sci, Elec Eng Computer Sci Dept, Univ Calif, Berkeley, 79-86; PROF COMPUTER SCI, UNIV MINN, DULUTH, 86- *Concurrent Pos:* Prin investr, NSF, 81-, AT&T & Mentor Graphics Inc, 85-86. *Mem:* Asn Comput Mach; Inst Elec & Electronics Engrs, Computer Soc. *Res:* Algorithms for computer-aided design of very large scale integrated circuits; area-time complexity; special-purpose chip design; data compression; federal research policy. *Mailing Add:* Computer Sci Dept 320 HH Univ Minn Duluth MN 55812

THOMEIER, SIEGFRIED, b Aussig, Czech, Dec 19, 37; Can citizen; m 61; c 2. TOPOLOGY, ALGEBRAIC TOPOLOGY. *Educ:* Univ Frankfurt, Dipl Math, 62, Dr Phil Nat(math), 65. *Prof Exp:* Sci asst math, Frankfurt Univ, 63-65; assoc prof, Math Inst, Aarhus, Denmark, 65-68; PROF MATH, MEM UNIV NFLD, 68- *Concurrent Pos:* Nat Res Coun Can res grant, Mem Univ Nfld, 69-; vis prof math, Univ Konstanz, Ger, 75-76. *Mem:* Can Math Soc; Math Assoc Am; London Math Soc; Ger Math Soc; NY Acad Sci; Can Soc Hist & Philos Math. *Res:* Homotopy theory; homotopy groups of special topological spaces; homology theory; homological and categorical algebra; topological fixed point theory; applications of topology and error-correcting codes. *Mailing Add:* Dept Math Mem Univ Nfld St John's NF A1B 3X7 Can

THOMERSON, JAMIE E, b Ft McKavett, Tex, May 7, 35; m 57; c 2. ICHTHYOLOGY, ZOOLOGY. *Educ:* Univ Tex, BS, 57; Tex Tech Col, MS, 61; Tulane Univ, PhD(zool), 65. *Prof Exp:* High sch teacher, 58-59; from asst prof to assoc prof, 65-77, PROF ZOOL, SOUTHERN ILL UNIV, EDWARDSVILLE, 77- *Mem:* Am Soc Ichthyologists & Herpetologists; Am Fisheries Soc; Asn Trop Biol; Soc Syst Zool. *Res:* Fish systematics, ecology, behavior and genetics. *Mailing Add:* Dept Biol Sci Southern Ill Univ Edwardsville IL 62026-1001

THOMFORDE, C(LIFFORD) J(OHN), b Crookston, Minn, Dec 15, 17; m 41; c 3. ELECTRICAL ENGINEERING. *Educ:* Univ NDak, BS, 41; Iowa State Univ, MSEE, 51. *Prof Exp:* Engr, Fed Commun Comn, 41-42 & Collins Radio Co, 42-47; asst prof elec eng, 47-55, chmn dept, 55-75, PROF ELEC ENG, UNIV NDAK, 55-; CONSULT ENG. *Mem:* Am Soc Eng Educ; Inst Elec & Electronics Engrs. *Res:* Radio communications, particularly in the field of radio and television broadcasting. *Mailing Add:* 2615 Fourth Ave N Grand Forks ND 58201

THOMMES, ROBERT CHARLES, b Chicago, Ill, Aug 31, 28. ZOOLOGY. *Educ:* De Paul Univ, BS, 50, MS, 52; Northwestern Univ, PhD, 56. *Prof Exp:* From instr to assoc prof, 56-67, chmn dept, 68-70, PROF BIOL, DE PAUL UNIV, 67- *Mem:* AAAS; Soc Zool; Soc Develop Biol; Soc Exp Biol & Med; Endocrine Soc; Poultry Sci Soc. *Res:* Developmental endocrinology. *Mailing Add:* 2621 Tanglewood Dr Sarasota FL 34239

THOMOPOULOS, NICK TED, b Chicago, Ill, Aug 21, 30; m 64; c 1. OPERATIONS RESEARCH, STATISTICS. *Educ:* Univ Ill, BS, 53, MA, 58; Ill Inst Technol, PhD(indust eng), 66. *Prof Exp:* Supvr opers res, Int Harvester Co, 58-66; sr scientist, IIT Res Inst, 66-68; ASSOC PROF INDUST ENG, ILL INST TECHNOL, 68- *Mem:* Inst Mgt Sci; Opers Res Soc Am. *Res:* Manufacturing and assembly methods; uncertainty in mathematical models; statistical analysis; production and inventory control. *Mailing Add:* Dept Mgt Ill Inst Technol 3300 S Federal St Chicago IL 60616-3793

THOMPKINS, LEON, b Augusta, Ga, Nov 4, 36; m 62; c 2. PHARMACEUTICAL CHEMISTRY. *Educ:* Morehouse Col, BS, 61; Univ Calif, San Francisco, PhD(pharmaceut chem), 68. *Prof Exp:* Chemist, Hyman Labs, Fundamental Res Co, Inc, 62-63; sr pharmaceut chemist, 68-76, RES SCIENTIST, ELI LILLY & CO, 76- *Mem:* Am Pharmaceut Asn; Am Chem Soc; Sigma Xi; AAAS; NY Acad Sci. *Res:* Development and bioavailability of human drug dosage forms. *Mailing Add:* 1018 Loughery Lane Indianapolis IN 46208

THOMPSON, A(LEXANDER) RALPH, b Toronto, Ont, Sept 11, 14; nat US; wid; c 2. CHEMICAL ENGINEERING. *Educ:* Univ Toronto, BASc, 36; Univ Pa, PhD(chem eng), 45. *Prof Exp:* Chem engr, Can Industs, Ltd, Ont, 36-40; asst instr chem eng, Univ Pa, 40-41, instr, 42-45, from asst prof to assoc prof, 45-52; prof, 52-80, chmn dept, 52-72, dir, Water Resources Ctr, 66-80, EMER PROF CHEM ENG, UNIV RI, 80- *Concurrent Pos:* Res adv, Gen Refractories Co, Pa, 41-43; instr supvr eng, Sci & Mgt War Training, Radio Corp Am, NJ, 45; develop engr, Sun Oil Co, Pa, 45-53; consult, Monsanto Chem Co, Mass, 55-57; mem, Univs Coun Water Resources; chmn, Nat Asn Water Inst Directors, 80-81. *Mem:* Am Soc Eng Educ; Am Chem Soc; fel Am Inst Chem Engrs; fel AAAS. *Res:* Water resources; desalination; refraction, dispersion and densities of binary solutions; crystallization; distillation; phase equilibria; biochemical engineering. *Mailing Add:* Dept Chem Eng-Crawford Univ RI Kingston RI 02881

THOMPSON, ALAN MORLEY, physiology; deceased, see previous edition for last biography

THOMPSON, ALLAN LLOYD, b Leeds, Que, Mar 4, 20; m 43; c 3. PHYSICAL CHEMISTRY, ENGINEERING. *Educ:* Bishop's Univ, Can, BA, 40; McGill Univ, PhD(chem), 43. *Prof Exp:* Res chemist, Nat Res Coun Can, 43-46; RES ASSOC CHEM, RADIATION LAB, MCGILL UNIV, 46-, ASSOC PROF MECH ENG, 58- *Concurrent Pos:* From sci consult to assoc dir, Gas Dynamics Lab, McGill Univ, 53-58. *Mem:* Sigma Xi. *Res:* Engineering science; chemical kinetics; analytical, nuclear and high temperature chemistry; cyclotron chemical problems; ignition and corrosion studies; vacuum engineering; combustion; road safety; collision investigations; pollution control of vehicles. *Mailing Add:* Dept Mech Eng McGill Univ 817 Sherbrooke St W Montreal PQ H3A 2K6 Can

THOMPSON, ALLAN M, b Ithaca, NY, May 22, 40; m 66. PETROLOGY, SEDIMENTOLOGY. *Educ:* Carleton Col, BA, 62; Brown Univ, ScM, 64, PhD(stratig), 68. *Prof Exp:* Asst prof, 67-72, ASSOC PROF GEOL & PETROL, UNIV DEL, 72- *Concurrent Pos:* Assoc geologist, Del Geol Surv, 72- *Mem:* Geol Soc Am; Am Asn Petrol Geologists; Soc Econ Paleont & Mineral; Sigma Xi. *Res:* Sedimentology, plutonic petrology and structure; stratigraphy, structure and petrology of Appalachian orogen. *Mailing Add:* Dept Geol Univ Del Newark DE 19711

THOMPSON, ALONZO CRAWFORD, b Tifton, Ga, June 4, 28; m 55; c 3. ORGANIC CHEMISTRY, PHARMACEUTICAL CHEMISTRY. *Educ:* Berry Col, AB, 53; Univ Miss, MS, 55, PhD(chem), 62. *Prof Exp:* Prof chem, Miss Delta Jr Col, 55-56 & Southern State Col, 56-58; RES CHEMIST, BOLL WEEVIL RES LAB, USDA, 62- *Concurrent Pos:* Adj asst prof biochem, Miss State Univ, 74- *Mem:* Am Chem Soc; Am Inst Chemists. *Res:* Pharmaceutical synthesis; natural products; insects and plant stimulants. *Mailing Add:* Rte 4 Box 29 Columbus MS 39730-9406

THOMPSON, ALVIN JEROME, b Washington, DC, Apr 5, 24; m 50; c 5. GASTROENTEROLOGY. *Educ:* Howard Univ, MD, 46; Am Bd Internal Med, dipl, 53. *Prof Exp:* Intern, St Louis City Hosp, Mo, 46-47, resident, 47-51; physician & gastroenterologist, Vet Admin Hosp, Seattle, 53-57; founder & dir, Gastroenterol Lab, Providence Hosp, Seattle, 63-77, chief med, 72-74; clin prof med, Sch Med, Univ Wash, 72-74; PVT MED PRACT INT MED & GASTROENTEROL, 57- *Concurrent Pos:* med advisor, Draft Bd, 67-72; co-chmn phys health task force, King County Comprehensive Health Planning Coun, 68, vpres, 73-; mem adv bd, King County Med Serv Bur, 69-70; consult, Div Educ & Res Facilities, NIH, 70-72. *Mem:* Inst Med-Nat Acad Sci; Nat Med Asn; AMA; fel Am Col Physicians; Am Gastroenterol Asn; Am Soc Gastrointestinal Endoscopy; Am Soc Internal Med. *Mailing Add:* 1600 E Jefferson SW 620 Seattle WA 98195

THOMPSON, ANSEL FREDERICK, JR, b Birmingham, Ala, Oct 19, 41; m 63; c 3. ENVIRONMENTAL ENGINEERING, ENGINEERING MANAGEMENT. *Educ:* Pa State Univ, BS, 63; Calif Inst Technol, MS, 65, PhD(environ health eng), 68. *Prof Exp:* Proj engr & scientist, Roy F Weston Inc, 67-70, prin engr, Weston Europe SpA, 70-73, proj mgr, 73-75, vpres, Eng Design, 75-80, Quality Assurance/Finance, 80-87, exec vpres, 87-89, VCHMN, ROY F WESTON INC, 89- *Concurrent Pos:* dipl, Am Acad Environ Engrs. *Mem:* Am Water Works Asn; Water Pollution Control Fedn; Am Soc Civil Engrs; Am Acad Environ Engrs; Prof Serv Mgt Asn. *Res:* Thermodynamics and ultrafiltration of salt-polyelectrolyte solution; automatic control of biological treatment processes; Monte Carlo methods in analysis of treatment systems; systems analysis. *Mailing Add:* 656 St Mattews Rd Chester Springs PA 19425

THOMPSON, ANSON ELLIS, b Eugene, Ore, Apr 9, 24; m 45; c 4. PLANT BREEDING, PLANT GENETICS. *Educ:* Ore State Univ, BS, 48; Cornell Univ, PhD(plant breeding), 52. *Prof Exp:* Lab instr bot, Ore State Univ, 47-48; asst plant breeding, Cornell Univ, 48-51; from instr to prof plant genetics, Univ Ill, Urbana, 51-71, asst dir agr exp sta, 67-69; prof hort & head dept hort & landscape archit, Univ Ariz, 71-75, prof hort & horticulturist, Dept Plant Sci, 75-76; field staff plant scientist, Rockefeller Found, Indonesia, 76-79; prog coordr hort crops, Prog Develop & Coord Staff, Sci & Educ Admin, 79-80, nat res prog leader, 80-83, RES PLANT GENETICIST, AGR RES SERV, USDA, 83- *Concurrent Pos:* Univ Ky-Int Coop Admin vis prof, Univ Indonesia, 58-60; party chief & adv, Univ Ill-AID Col Contract Team, Uttar Pradesh Agr Univ, India, 67-69; vis prof, Gadjah Mada Univ, Indonesia, 76-79. *Honors & Awards:* Woodbury Award, Am Soc Hort Sci, 65, Asgrow Award, 66. *Mem:* Fel AAAS; fel Am Soc Hort Sci; Soc Econ Bot; Int Soc Hort Sci; Sigma Xi; Asn Adv Indust Crops. *Res:* Vegetable breeding and genetics; inheritance and breeding methods for disease resistance and quality constituents; breeding new industrial crops for arid lands. *Mailing Add:* US Water Conserv Lab 4331 E Broadway Phoenix AZ 85040

THOMPSON, ANTHONY RICHARD, b Hull, Eng, Apr 7, 31; m 63; c 1. RADIO ASTRONOMY INSTRUMENTATION, ELECTROMAGNETIC COMPATIBILITY. *Educ:* Univ Manchester, BSc, 52, PhD, 56. *Prof Exp:* Electronic engr, Elec & Musical Instruments Ltd, Eng, 55-57; res assoc, Harvard Col Observ, 57-61, res fel, 61-62; radio astronr, Stanford Univ, 62-70, sr res assoc, 70-72; VLA proj engr, Nat Radio Astron Observ, dep mgr VLA proj, 75-81, systs engr & frequency coordr VLA proj, 81-84, HEAD, ELECTRONICS DIV VLBA PROJ, 84-, DEP MGR, VLBA PROJ, 87- *Concurrent Pos:* Vis sr res fel, Calif Inst Technol, 66-71, vis assoc, 71-72; mem, US Study Group 2, Int Radio Consult Comt, 79-; chmn radio astron subcomt, Comt on Radio Frequencies, Nat Acad Sci, 80-86; mem, Comt on Radio Frequencies, 80-; sec, Interunion Comn Allocation Frequencies Radioastronomy & Space Sci, 82-88. *Mem:* Int Astron Union; Int Union Radio Sci; fel Inst Elec & Electronics Engrs; Am Astron Soc. *Res:* Design of radio telescopes; theory and practice of radio interferometry; electromagnetic compatibility and frequency protection of radio astronomy instruments; structure of cosmic radio sources; co-author of 1 publication. *Mailing Add:* Nat Radio Astron Observ Edgemont Rd Charlottesville VA 22903

THOMPSON, ANTHONY W, b Burbank, Calif, Mar 6, 40; m 63; c 2. METALLURGY, MATERIALS SCIENCE. *Educ:* Stanford Univ, BS, 62; Univ Wash, MS, 65; Mass Inst Technol, PhD(metall, mat sci), 70. *Prof Exp:* Res engr, Jet Propulsion Lab, 62-63; mem tech staff, Sandia Labs, 70-73; mem tech staff, Rockwell Int Sci Ctr, 73-77; assoc prof metall, 77-79, dept head, Metall Eng & Mat Sci Dept, 87-90, PROF METALL, CARNEGIE MELLON UNIV, 80- *Concurrent Pos:* Overseas fel, Churchill Col, Cambridge, UK, 82-83; ed, Metall Trans, 83-88. *Mem:* AAAS; fel Am Soc Metals; Am Inst Mech Engrs; Sigma Xi. *Res:* Relation between microstructure of materials and mechanical behavior, particularly strength and fracture; including environmental effects and grain size effects on polycrystal behavior; fatigue and fracture toughness of engineering materials. *Mailing Add:* Dept Metall Eng & Mat Sci Carnegie Mellon Univ 5000 Forbes Ave Pittsburgh PA 15213

THOMPSON, ARTHUR HOWARD, b Salt Lake City, Utah, Mar 3, 42; m 65; c 1. SOLID STATE PHYSICS. *Educ:* Ohio State Univ, BSc & MSc, 66; Stanford Univ, PhD(physics), 70. *Prof Exp:* Fel, Stanford Univ, 70; group leader physics, Syva Co, Calif, 70-71; sr res physicist, Exxon Res & Eng Co, Linden, 71-76, group head, Linden, NJ, 76-81, GROUP HEAD & RES ADV, EXXON PROD RES CO, HOUSTON, 81- *Mem:* Fel Am Phys Soc; AAAS; Am Geophys Union; Soc Explor Geophys. *Res:* Super conductivity; semiconductors; transport and magnetic properties; geophysics. *Mailing Add:* PO Box 2189 Exxon Prod Res Co Houston TX 77252-2189

THOMPSON, ARTHUR ROBERT, b Washington, DC, May 8, 59; m 86. NUCLEAR MAGNETIC RESONANCE SPECTROSCOPY. *Educ:* Valparaiso Univ, BS, 81; Univ Ill, PhD(phys chem), 87. *Prof Exp:* Postdoctoral, Argonne Nat Lab, 86-88; CHEMIST, NCAUR, AGR RES SERV, USDA, 88- *Mem:* Am Chem Soc. *Res:* Multinuclear nuclear magnetic resonance to study materials which do not lend themselves to easy interpretation by solution nuclear magnetic resonance; solid state nuclear magnetic resonance of plant polymers. *Mailing Add:* NCAUR Agr Res Serv USDA 1815 N University St Peoria IL 61604

THOMPSON, AYLMER HENRY, b Ill, Sept 11, 22; m 41; c 3. METEOROLOGY. *Educ:* Univ Calif, Los Angeles, MA, 48, PhD(meteorol), 60. *Prof Exp:* Lectr meteorol, Univ Calif, Los Angeles, 48-52; asst prof, Univ Utah, 52-60; prof, 60-88, EMER PROF METEOROL, TEX A&M UNIV, 88- *Concurrent Pos:* Sci adv, Found Glacier & Environ Res, Wash; sci consult, 55-; vis prof, Geophys Inst, Univ Alaska, 67-68 & 71-72. *Mem:* Am Meteorol Soc; Am Geophys Union; Royal Meteorol Soc; Int Glaciol Soc; Sigma Xi. *Res:* Synoptic meteorology of sub-tropics; satellite meteorology; inversions; meteorology of glaciated regions. *Mailing Add:* Dept Meteorol Tex A&M Univ College Station TX 77843-3146

THOMPSON, BOBBY BLACKBURN, b Lumber City, Ga, May 15, 33; m 59; c 1. MEDICINAL CHEMISTRY, ORGANIC CHEMISTRY. *Educ:* Berry Col, BA, 55; Univ Miss, MS, 56, PhD(org med chem), 63. *Prof Exp:* Asst prof, 60-70, ASSOC PROF MED CHEM, SCH PHARM, UNIV GA, 70- *Mem:* Am Chem Soc; Am Pharmaceut Asn. *Res:* Synthesis of organic and heterocyclic compounds of potential medicinal value; structural elucidation by chemical and instrumental means. *Mailing Add:* 375 Pinewood Circle Athens GA 30606-3812

THOMPSON, BONNIE CECIL, b Baird, Tex, Dec 18, 35; m 59; c 2. SOLID STATE PHYSICS. *Educ:* NTex State Univ, BA, 57, MA, 58; Univ Tex, PhD(physics), 65. *Prof Exp:* Instr physics, NTex State Univ, 58-60; teaching asst, Univ Tex, Austin 60-61, res scientist, 61-65; asst prof, 65-74, ASSOC PROF, 74-, UNIV TEX, ARLINGTON, ASSOC CHMN PHYSICS, 87- *Mem:* Am Phys Soc. *Res:* Nuclear spin-lattice relaxation processes in solids; nuclear magnetic resonance of biological molecules. *Mailing Add:* Dept Physics Univ Tex Arlington TX 76019

THOMPSON, BRIAN J, b Glossop, Eng, June 10, 32; m 56; c 2. OPTICS. *Educ:* Univ Manchester, BScTech, 55, PhD(appl physics), 59. *Prof Exp:* Demonstr physics fac tech, Univ Manchester, 55-56, asst lectr, 56-59; lectr, Leeds Univ, 59-62; sr physicist, Tech Opers, Inc, 63-65, mgr phys optics dept, 65-66, dir optics dept, 66-67, mgr tech opers west & tech dir, Beckman & Whitley Div, 67-68; dir, Inst Optics, 68-75, dean col eng & appl sci, 75-84, PROF OPTICS, UNIV ROCHESTER, 68-, PROVOST, 84- *Concurrent Pos:* Adj prof, Northeastern Univ, 66-67. *Honors & Awards:* Presidents Award, 67; Kingslake Medal & Pezuto Award, Soc Photo-Opitcal Instrumentation Engrs, 78, Gold Medal, 86. *Mem:* Am Phys Soc; fel Optical Soc Am; fel Brit Inst Physics & Phys Soc; fel Soc Photo-Optical Instrumentation Engrs (pres, 74-75, 75-76). *Res:* Diffraction; interference; partial coherence; holography; application to particle sizing; optical data processing. *Mailing Add:* 200 Admin Bldg Univ Rochester Rochester NY 14627

THOMPSON, BUFORD DALE, b Lake Wales, Fla, Oct 22, 22; m 44; c 3. HORTICULTURE, VEGETABLE CROPS. *Educ:* Univ Fla, BSA, 48, MSA, 49, PhD(hort), 54, JD, 76. *Prof Exp:* Asst prof veg crops & asst horticulturist, Univ Fla, 49-60, assoc prof & assoc horticulturist, 60-66, prof veg crops & horticulturist, 66-80; ATTY AGR LAW, AGR LEGAL CONSULT, US ARMY, 80- *Honors & Awards:* Vaughan Award, Am Soc Hort Sci, 62. *Mem:* Am Soc Plant Physiol. *Res:* Biological and chemical changes involved in the post harvest handling, transportation and storage of horticultural crops; agricultural law. *Mailing Add:* 3920 NW 29th Lane Gainesville FL 32606

THOMPSON, BYRD THOMAS, JR, b Chicago, Ill, Nov 6, 24; m 47; c 2. MECHANICAL ENGINEERING & FLUID MECHANICS. *Educ:* Univ Ala, BS, 49; Univ Ala, Huntsville, MSE, 68. *Prof Exp:* Mech engr, Tenn Valley Authority, 49-51, Monsanto Chem Corp, 51-52 & Patchen & Zimmerman Consult Engrs, 52-54; mech develop engr, Masonite Corp, 54-55; res & develop engr, Chemstrand Co Div, Monsanto Co, 55-70, sr process develop engr, 70-72; sr textile res engr, Textiles Div, 72-75, gen engr, Central Eng Dept, 75-77; mech engr, Exp Eng Dept, Oak Ridge Nat Lab, 77-81; mech engr, US Army Corps Engrs, 81-88; RETIRED. *Mem:* Am Soc Mech Engrs; Am Soc Heating, Vent & Refrig Engrs; Soc Am Mil Engrs. *Res:* Mechanical engineering and engineering mechanics as applied to design, development and research of synthetic fibers-processes and machinery. *Mailing Add:* 2010 Birch St SE Decatur AL 35601

THOMPSON, CARL EUGENE, b Lucinda, Pa, June 14, 41; m 64; c 4. ANIMAL BREEDING, ANIMAL GENETICS. *Educ:* Pa State Univ, University Park, BS, 63, MS, 68; Va Polytech Inst & State Univ, PhD(animal breeding & genetics), 71. *Prof Exp:* Asst county agr agent, Agr Exten Serv, Pa State Univ, 63-66; asst prof animal sci, Ft Hays State Univ, 71-73; asst prof, 74-77, assoc prof, 77-81, PROF BEEF CATTLE BREEDING, CLEMSON UNIV, 81- *Mem:* Am Soc Animal Sci; Am Genetics Asn; Sigma Xi. *Res:* Beef cattle breeding and genetics; cross-breeding; genetic-environmental interaction; reproductive physiology. *Mailing Add:* Rte 3 Box 231 Central SC 29630

THOMPSON, CHARLES, b Tachikawa, Japan, Jan 15, 54; US citizen; m 75; c 3. NONLINEAR & CHAOTIC SYSTEMS, FLUID MECHANICS. *Educ:* NY Univ, BS, 76; Polytech Univ, MS, 78; Mass Inst Technol, PhD(acoust), 82. *Prof Exp:* Asst prof appl mech, Va Polytech Inst & State Univ, 82-87; ASSOC PROF ELEC ENG & DIR LAB ADVAN COMPUTATION, UNIV MASS, LOWELL, 87- *Concurrent Pos:* Analog Devices prof, Analog Devices, 87. *Mem:* Acoust Soc Am; Am Phys Soc. *Res:* Theoretical and computational modeling of fluid and electronic systems; control theory; applied mathematics. *Mailing Add:* Dept Elec Eng Lab Advan Comput Univ Mass Lowell MA 01854

THOMPSON, CHARLES CALVIN, b Los Angeles, Calif, May 4, 35; m 60; c 2. ORAL PATHOLOGY. *Educ:* St Martin's Col, BA, 57; Univ Ore, DMD, 62; Emory Univ, MSD, 68; Am Bd Oral Path, dipl; Nat Bd Forensic Dent, dipl. *Prof Exp:* Pvt dent pract, Ore, 64-66; resident path, Dent Sch, Emory Univ, 66-68; asst prof oral diag & med, Dent Sch, Univ Calif, Los Angeles, 68-69; ASSOC PROF PATH, DENT SCH & ORE HEALTH SCI UNIV, 69- *Concurrent Pos:* Nat Inst Dent Res fel, Emory Univ, 66-68; attend consult, Wadsworth Vet Admin Hosp, Los Angeles, 68-69; oral pathologist & clin consult, 71-; instr, Mt Hood Community Col, 72-73. *Mem:* Fel Am Acad Oral Path; Int Soc Forensic Odontol-Stomatol. *Res:* Oncology, chemical, physical carcinogenesis; teratology, induction of developmental defects and explanation; forensic odontology; bone, especially effects of dimethyl sulfoxide on bone. *Mailing Add:* Dept Path Sch Dent Ore Sci Univ 611 SW Campus Dr Portland OR 97201

THOMPSON, CHARLES DENISON, b Niagara Falls, NY, May 4, 40. CORROSION. *Educ:* Oberlin Col, BA, 62; Am Univ, MS, 68, PhD(chem), 71. *Prof Exp:* Chemist, US Army Environ Hyg Agency, 63-65; chemist biochem, Aldridge Assocs, 69; Welch fel, Rice Univ, 71-72; CORROSION ENGR, KNOLLS ATOMIC POWER LAB, GEN ELEC CO, 73- *Mem:* Electrochem Soc. *Res:* High temperature materials corrosion and electrochemistry. *Mailing Add:* 50 Hill St Alplaus NY 12008

THOMPSON, CHARLES FREDERICK, b Dayton, Ohio, Oct 1, 43; m 67. POPULATION ECOLOGY, ORNITHOLOGY. *Educ:* Ind Univ, BA, 67, MA, 70, PhD(zool), 71. *Prof Exp:* Res fel ecol, Univ Ga, 71-72, asst prof zool, 72-73; teaching fel zool, Miami Univ, 73-75; asst prof biol, State Univ NY Col Geneseo, 75-78; asst prof, 78-82, assoc prof, 82-88, PROF ECOL, ILL STATE UNIV, 88- *Concurrent Pos:* Vis asst prof zool, Ind Univ, 74; vis scientist, DSIR Ecol Div, NZ, 84, Edward Grey Inst, Oxford Univ, 85, 91. *Mem:* Ecol Soc Am; Brit Ecol Soc; Am Ornithologists Union; Brit Ornithologists Union; Neth Ornithologists Union. *Res:* Regulation and dynamics of bird populations; avian breeding adaptations; structure and evolution of avian social organization; evolution of avian life-history traits. *Mailing Add:* Dept Biol Sci Ill State Univ Normal IL 61761-6901

THOMPSON, CHARLES WILLIAM NELSON, b Bethlehem, Pa; m 48; c 3. INDUSTRIAL ENGINEERING, MANAGEMENT SCIENCE. *Educ:* Kutztown Univ, BS, 43; Harvard Law Sch, LLB, 49; Ohio State Univ MBA, 56; Northwestern Univ, PhD(indust eng), 69. *Prof Exp:* Chief electronic reconnaissance sect, Wright Air Develop Ctr, USAF, 52-58; dir eng serv, Govt Electronics Div, Admiral Corp, 58-64; assoc prof, 68-77, PROF INDUST ENG & MGT SCI, TECHNOL INST, NORTHWESTERN UNIV, 77- *Concurrent Pos:* Consult exp technol prog, Nat Bur Standards, 75-80. *Mem:* Inst Elec & Electronics Engrs; Inst Mgt Sci; Am Inst Indust Engrs; Sigma Xi. *Res:* Theory and methodology of unstructured problems in organizations and systems, with particular emphasis on field research, including administrative experiments. *Mailing Add:* 240 Randolph St Glencoe IL 60022

THOMPSON, CHESTER RAY, b Storrs, Utah, May 27, 15; m 40; c 4. BIOCHEMISTRY. *Educ:* Utah State Univ, BS, 38; Univ Wis, MS, 41, PhD(biochem), 43. *Prof Exp:* Chemist, Rocky Mountain Packing Corp, Utah, 36-39; biochemist, Univ Wis, 39-43, Forest Prods Lab, US Forest Serv, 43 & Purdue Univ, 44-45; plant biochemist, Univ Chicago, 45-49; head forage invest, Field Crops Lab, Western Utilization Res & Develop Div, USDA, 49-60; proj leader, 60-67, RES BIOCHEMIST, STATEWIDE AIR POLLUTION RES CTR, AGR AIR RES PROG, UNIV CALIF, RIVERSIDE, 67- *Mem:* Fel AAAS; Am Soc Plant Physiol; Am Chem Soc; Air Pollution Control Asn. *Res:* Chemical stabilization of carotene in alfalfa; occurrence of anti-oxidants in natural products; carotenoids in green plants; saponins and estrogens in forages; effects of air pollution on plants and human beings. *Mailing Add:* Statewide Air Pollution Res Ctr Univ Calif Riverside CA 92521

THOMPSON, CLARENCE HENRY, JR, b Perry, Kans, May 4, 18; m 42; c 3. VETERINARY MEDICINE. *Educ:* Kans State Col, DVM, 41, MS, 47. *Prof Exp:* Asst vet private vet hosp, 41; jr vet tuberc eradication, Bur Animal Indust, Agr Res Serv, USDA, 41-46, vet pathologist path div, 47-54, adminstr state exp sta div, Agr Res Serv, 54-63, asst to dir animal disease & parasite res div, 63-71, staff asst to adminr, 71-74; RETIRED. *Mem:* Am Vet Med Asn. *Res:* Animal diseases, especially virus diseases of poultry, in fields of epizoology, bacteriology, pathology and immunology. *Mailing Add:* 6203 87th Ave Hyattsville MD 20784

THOMPSON, CLIFTON C, b Franklin, Tenn, Aug 16, 39; m 59, 78; c 2. PHYSICAL CHEMISTRY. *Educ:* Middle Tenn State Univ, BS, 61; Univ Miss, PhD(phys & inorg chem), 64. *Prof Exp:* Res assoc, Univ Tex, 64-65; asst prof spectrochem, Rutgers Univ, 65; asst prof chem, Marshall Univ, 65-66; assoc prof, Mid Tenn State Univ, 66-68; from asst prof to assoc prof, Memphis State Univ, 68-74; dean, Sch Sci & Technol, 74-84, dean, Col Sci & Math, 84-87, PROF CHEM, SOUTHWEST MO STATE UNIV, 74-, dean, DIR, CTR SCI RES, 87- *Concurrent Pos:* Lectr, Kanawha Valley grad ctr, WVa Univ, 66; NSF acad year exten grant, 66-68; mem, Med Technol Rev Comt, 74-80. *Mem:* Am Chem Soc; Royal Soc Chem; AAAS; Sigma Xi. *Res:* Spectral, thermodynamic and kinetic studies of molecular complexes; quantum chemistry; computer applications to physical systems. *Mailing Add:* 2831 E Crestview Springfield MO 65804

THOMPSON, CRAYTON BEVILLE, b Paris, Tex, Dec 28, 20; m 55; c 1. ORGANIC CHEMISTRY. *Educ:* Univ Tex, BS, 42; Univ Ill, MS, 47, PhD(chem), 49. *Prof Exp:* Chem engr, Freeport Sulphur Co, 42-46; lab asst chem, Univ Ill, 47-49; RES & DEVELOP CHEMIST, EASTMAN KODAK CO, 49- *Mem:* Am Chem Soc. *Res:* Synthesis of amino acids; antihalation backings for photographic films; abrasion resistant and anitstatic applications for plastics; adhesion. *Mailing Add:* 505 Wanda Ridge Durham NC 27712-2751

THOMPSON, D(ONALD) W(ILLIAM), b Gosport, Eng, Mar 16, 33; m 57; c 2. CHEMICAL ENGINEERING. *Educ:* Univ Birmingham, BSc, 54, PhD(chem eng), 58. *Prof Exp:* Res fel chem eng, Univ BC, 58-60; engr, Shell Develop Co, Calif, 60-67; assoc prof, 67-77, PROF CHEM ENG, UNIV BC, 77- *Honors & Awards:* Jr Moulton Medal, Brit Inst Chem Eng, 61. *Mem:* Am Inst Chem Engrs; Chem Inst Can; Brit Inst Chem Eng. *Res:* Adsorption and chromatographic processes; cyclic separation processes; membrane separations; flow visualization; optimization methods. *Mailing Add:* Dept Chem Eng Univ BC 2075 Wesbrook Mall Vancouver BC V6T 2W5 Can

THOMPSON, DANIEL JAMES, b Terre Haute, Ind, Feb 10, 42; m 65; c 2. TERATOLOGY, REPRODUCTIVE TOXICOLOGY. *Educ:* Ind State Univ, BS, 64, MA, 66; Am Bd Toxicol, dipl, 81. *Prof Exp:* Res specialist, 66-80, res leader, Dow Chem Co. 80-82, SR RES TOXICOLOGIST, MARION MERRELL DOW INC, 82- *Concurrent Pos:* Adj assoc prof, Purdue Univ. *Mem:* Teratology Soc; Environ Mutagen Soc; Soc Toxicol. *Res:* Reproductive physiology; perinatal toxicology. *Mailing Add:* Dept Toxicol Marion Merrell Dow Inc PO Box 68470 Indianapolis IN 46268-0470

THOMPSON, DANIEL QUALE, b Madison, Wis, Oct 3, 18; m 53; c 4. WILDLIFE ECOLOGY, CONSERVATION. *Educ:* Univ Wis-Madison. BS, 42, MS, 50; Univ Mo, PhD(field zool), 55. *Prof Exp:* Instr field zool, Univ Mo, 50-54; from asst prof to assoc prof biol, Ripon Col, 55-62; wildlife res biologist, US Fish & Wildlife Serv & leader, NY Coop Wildlife Res Unit, Cornell Univ, 62-75; wildlife ed, Colo State Univ, US Fish & Wildlife Serv, 75-85; CONSULT/WRITER FISH & WILDLIFE SCI, 86- *Concurrent Pos:* Mem, Grad Fel Panel, NSF, 71-72; ed, J Wildlife Mgt, 73-74. *Mem:* Wilderness Soc; Wildlife Soc. *Res:* Wildlife conservation; ecology of terrestrial vertebrates. *Mailing Add:* 623 Del Norte Pl Ft Collins CO 80521

THOMPSON, DARRELL ROBERT, b Hickory, NC, July 13, 37; m 66; c 2. BIOMATERIALS, PHYSICAL CHEMISTRY. *Educ:* Lenoir Rhyne Col, BS, 59; Univ NC, PhD(phys chem), 64. *Prof Exp:* Res chemist pigment dispersion, E I du Pont de Nemours & Co, Inc, 64-68; sr res chemist, Celanese Fibers Co, 68-70; GROUP LEADER SURG PRODS, JOHNSON & JOHNSON CORP, 70- *Mem:* Am Chem Soc; Sigma Xi; Fiber Soc. *Res:* Surgical products; surface and colloid chemistry; coatings; pigment dispersion. *Mailing Add:* 172 Windy Willow Way Somerville NJ 08876

THOMPSON, DAVID A(LFRED), b Chicago, Ill, Sept 9, 29; wid; c 5. ERGONOMICS. *Educ:* Univ Va, BME, 51; Univ Fla, BIE, 55, MSE, 56; Stanford Univ, PhD(indust eng), 61. *Prof Exp:* Asst eng & indust, Exp Sta, Univ Fla, 55-56; actg instr indust eng, 56-58, actg asst prof, 58-61, res assoc rehab med, 58-64, from asst prof to assoc prof, 61-72, prof indust eng & assoc chmn dept, 72-83, EMER PROF INDUST ENG, STANFORD UNIV, 83-; PRES, PORTOLA ASSOCS, PALO ALTO, CALIF. *Concurrent Pos:* Consult. *Mem:* Am Inst Indust Engrs; Human Factors Soc; Inst Elec & Electronics Engrs; Am Soc Eng Educ; Sigma Xi. *Res:* Analysis and design of man-machine systems, especially the physiological, neurological and psychological information processing in man. *Mailing Add:* Dept Indust Eng Stanford Univ Stanford CA 94305

THOMPSON, DAVID A, b Devils Lake, NDak, Dec 17, 40. INFORMATION STORAGE TECHNOLOGY, APPLIED MAGNETISM. *Educ:* Carnegie Inst Technol, BS, 62; MS, 63 & PhD(elec eng), 66. *Prof Exp:* Asst prof elec eng, Carnegie Inst Technol, 65-68; res staff mem, IBM T J Watson Res Ctr, 68-80; FEL, IBM RES, IBM ALMADEN RES CTR, 80-, DIR, COMPACT STORAGE LAB, IBM T J WATSON RES CTR, 87-, DIR, MAGNETIC REC INST, 87- *Concurrent Pos:* Admin comt, Inst Elec & Electronics Engrs Magnetics Soc, 76-; adv comt, Ctr Magnetic Rec Res, Univ Calif San Diego & Magnetic Tech Ctr, Carnegie Mellon Inst. *Mem:* Nat Acad Eng; fel Inst Elec & Electronics Engrs. *Res:* Information storage technology; applied magnetism; magnetic recording systems; sensors and measurement systems; servo mechansims. *Mailing Add:* IBM Almaden Res Ctr 650 Harry Rd K01 8O2 San Jose CA 95120

THOMPSON, DAVID ALLAN, b Oxford, Eng, Apr 28, 42; Can citizen; m 64; c 3. ENGINEERING PHYSICS, MATERIALS SCIENCE ENGINEERING. *Educ:* Reading Univ, UK, BSc, 63, PhD(physics), 67. *Prof Exp:* Mgr, process res & develop, Westinghouse Can, 67-72; from asst prof to assoc prof, 73-81, chmn dept, 81-87, PROF ENG PHYSICS, 81-, DIR, CENTRE ELECTRO PHOTONIC MAT & DEVICES, MCMASTER UNIV, 87- *Concurrent Pos:* Consult, various organizations; dir,67-87. *Mem:* Mat Res Soc; Can Asn Physics; Boehmische Phys Soc; Inst Physics UK. *Res:* Atomic collisions in solids; ion beam processing of solids; ion beam analysis; device technology; molecular boam epitaxy. *Mailing Add:* Dept Eng Physics McMaster Univ Hamilton ON L8S 4M1 Can

THOMPSON, DAVID ALLEN, b Gallipolis, Ohio, July 10, 50; m 72; c 2. INORGANIC CHEMISTRY, GLASS CHEMISTRY. *Educ:* Ohio State Univ, BS, 72; Univ Mich, MS, 73, PhD(chem), 77. *Prof Exp:* Res chemist, Corning Glass Works, 76-80, supvr explor res, 80-84, mgr, optical waveguide res, 84-86, optical component res, 86-90, MGR COMPONENT TECHNOL, CORNING GLASS WORKS, 91- *Mem:* Am Chem Soc; Am Ceramic Soc; Optical Soc Am. *Res:* Glass chemistry research including composition, durability, diffusion and melting of glass. *Mailing Add:* Corning Inc MP-50-0-8 Corning NY 14831

THOMPSON, DAVID DUVALL, b Ithaca, NY, June 1, 22; m 45; c 4. INTERNAL MEDICINE. *Educ:* Cornell Univ, BA, 43, MD, 46. *Prof Exp:* Intern & resident med, New York Hosp, 46-50; res fel physiol, Med Col, Cornell Univ, 50-51; instr physiol, Med Col, Cornell Univ, 51-53; resident physician, NIH, 53-55; asst prof physiol, 55-57, assoc prof med, 57-64, dir, 67-87, PROF MED, MED COL, CORNELL UNIV, 64-; ATTEND PHYSICIAN, DEPT MED, NEW YORK HOSP, 64-, COORDR CLIN EDUC, 87-, CONSULT, 87- *Concurrent Pos:* Clin instr, Sch Med, George Washington Univ, 54-55; resident physician, NIH, 55-56; Lederle award, 55-57; from asst attend to actg physician-in-chief, Dept Med, New York Hosp, 57-67; asst vis physician, 2nd Med Div, Bellevue Hosp, 57-66, assoc vis physician, 66-; attend physician, Vet Admin Hosp, 58-; mem med adv bd, Kidney Found NY, 62-; actg chmn dept med, Cornell Univ, 65-67. *Mem:* Am Soc Clin Invest; Am Physiol Soc; Harvey Soc; Am Fedn Clin Res; Am Col Physicians. *Res:* Renal and electrolyte physiology. *Mailing Add:* New York Hosp Rm HT341 New York NY 10021

THOMPSON, DAVID FRED, b Columbus, Ohio, Mar 6, 41; m 66; c 4. EMISSIONS CONTROL, SPECIALTY MATERIALS. *Educ:* Ohio State Univ, BSc, 64, MSc, 64, PhD(ceramic eng), 68. *Prof Exp:* Proj engr, Edward Orton Jr Ceramic Found, 62-68; sect head, GTE Sylvania, 68-74; MGR TECH DEVELOP, ZIRCOA PROD, CORNING GLASS WORKS, 74-, SR DEVELOP ASSOC, CORNING, INC. *Mem:* Am Ceramic Soc; Am Soc Testing & Mat; Am Soc Metals; Soc Automotive Engrs. *Res:* Zirconia ceramics; solid electrolytes; processing technology; optimization of ceramic material properties; physical property measurement; test development; emissions control. *Mailing Add:* Corning Glass Works Sullivan Park Corning NY 14831

THOMPSON, DAVID J, b Danville, Ind, Apr 17, 34; m 68; c 2. GENETICS, PLANT BREEDING. *Educ:* Univ Idaho, BS, 54, MS, 56; Cornell Univ, PhD(plant breeding), 60. *Prof Exp:* Co-geneticist, 60-63, res dir genetics & plant breeding, 63-76, VPRES, RES DIV, FERRY-MORSE SEED CO, 76-, PRES, 86- *Mem:* Int Soc Hort Sci; Am Soc Hort Sci. *Res:* Genetics, cytology and physiology of male-sterility and self-incompatibility in plants. *Mailing Add:* 610 Harper PO Box 4938 Fletcher CA 73541

THOMPSON, DAVID JEROME, b Sand Creek, Wis, July 21, 37; m 62; c 2. MINERAL NUTRITION. *Educ:* Univ Wis-Madison, BS, 60, MS, 61, PhD(biochem), 63; Univ Chicago, MBA, 75. *Prof Exp:* Res assoc biochem, Univ Wis-Madison, 63-64; res biochemist, Int Minerals & Chem Corp, 64-69, mgr tech serv, 69-78, dir tech serv, 78-79, regional sales mgr, 79-81, vpres, sci & technol, 81-84, vpres & gen mgr, Sterwin Div, 84-86, vpres planning & develop, 87-89; DIR TECH PLANNING & EVAL, PITMAN-MOORE INC, 89- *Concurrent Pos:* Mem, Mineral Toxic Animals Subcomt, Nat Res Coun, Nat Acad Sci, 76-80; mem, Nutrit Sci External Adv Comt, Univ Ill, 85- *Mem:* NY Acad Sci; Am Chem Soc; AAAS; Coun Agr Sci & Technol. *Res:* Mineral nutrition of animals. *Mailing Add:* 421 E Hawley St Bldg F Mundelein IL 60060-2461

THOMPSON, DAVID JOHN, b Cincinnati, Ohio, Jan 11, 45; m 72; c 2. ASTROPHYSICS, GAMMA RAY ASTRONOMY. *Educ:* Johns Hopkins Univ, BA, 67; Univ Md, PhD(physics), 73. *Prof Exp:* Res assoc physics, Univ Md, 73; ASTROPHYSICIST, GODDARD SPACE FLIGHT CTR, NASA, 73- *Concurrent Pos:* Co-investr, Energetic Gamma Ray Exp Telescope, Gamma Ray Observ, 78- *Mem:* Am Astron Soc; Am Phys Soc. *Res:* Gamma ray astronomy and its relationship to cosmic ray physics and other aspects of astrophysics; high energy gamma ray detectors. *Mailing Add:* Code 662 Goddard Space Flight Ctr NASA Greenbelt MD 20771

THOMPSON, DAVID RUSSELL, b Cleveland, Ohio, Apr 4, 44; m 66; c 3. ENGINEERING, FOOD SCIENCE. *Educ:* Purdue Univ, West Lafayette, BS, 66, MS, 67; Mich State Univ, PhD(agr eng), 70. *Prof Exp:* From asst prof to prof food eng, Univ Minn, St Paul, 70-85; PROF & HEAD, AGR ENGR DEPT, OKLA STATE UNIV, STILLWATER, 85- *Concurrent Pos:* NSF grant, 72-73; sabbatical leave to work for Green Giant (Pillsbury) Co, LeSueur, Minn, 78-79; mem rev team, Dept Defense Food Res Develop, Test & Eng Prog, Nat Res Coun, 80. *Honors & Awards:* Young Researcher Award, Am Soc Agr Engrs, 83. *Mem:* Am Soc Agr Engrs; Inst Food Technol; Am Soc Eng Educ; Am Soc Heating, Refrig & Air Conditioning Engrs. *Res:* Food processing; modeling heat and mass transfer reaction kinetics of changes in nutrition; microbiological populations and organoleptic characteristics in food systems during heating and cooling processes; energy conservation in food systems. *Mailing Add:* Dept Agr Eng Okla State Univ Stillwater OK 74078-0497

THOMPSON, DAVID WALKER, b Philadelphia, Pa, Mar 21, 54. SPACE TECHNOLOGY. *Educ:* Mass Inst Technol, BS; Harvard Bus Sch, MBA; Calif Inst Technol, MS. *Prof Exp:* Proj mgr & engr, Marshall Space Flight Ctr, NASA; spec asst to pres, Missile Systs Group, Hughes Aircraft Co; CHMN BD, PRES & CHIEF EXEC OFFICER, ORBITAL SCI CORP, 82- *Concurrent Pos:* Dir, Am Astron Soc, Space Found, Nat Space Club, Found for Space Bus Res & Aurora Flight Sci Corp; mem, Com Prog Adv Comt, NASA; vchmn, Com Space Transp Adv Comt, Dept Transp; consult, Defense Sci Bd. *Honors & Awards:* Nat Medal of Technol, 91; Lawrence Sperry Award, Am Inst Aeronaut & Astronaut, 88; Lloyd V Brucker Award, Am Astronaut Soc, 89. *Mem:* Fel Am Astronaut Soc; assoc fel Am Inst Aeronaut & Astronaut. *Res:* Commercial space technology; develop, manufacture and market space transportation systems; commercial satellite applications. *Mailing Add:* Orbital Sci Corp 12500 Fair Lakes Circle Fairfax VA 22033

THOMPSON, DAVID WALLACE, b Chicago, Ill, Jan 27, 42; m 63; c 2. INORGANIC CHEMISTRY, ORGANOMETALLIC CHEMISTRY. *Educ:* Wheaton Col, BS, 63; Northwestern Univ, Evanston, PhD(chem), 68. *Prof Exp:* Asst prof, 67-70, ASSOC PROF CHEM, COL WILLIAM & MARY, 70- *Mem:* Am Chem Soc. *Res:* Coordination chemistry of group IV elements; use of transition elements to catalyze organic reactions. *Mailing Add:* Dept Chem Col William & Mary Williamsburg VA 23185-3647

THOMPSON, DONALD B, b Camden, NJ, Oct 27, 48; m 71; c 3. NUTRIENT BIOAVAILABILITY, INGREDIENT FUNCTIONALITY. *Educ:* Haverford Col, BA, 70; Univ Ill, MS, 80, PhD(food sci), 84. *Prof Exp:* Food scientist, R T French Co, 80-81; ASST PROF FOOD SCI, PA STATE UNIV, 84- *Mem:* Inst Food Technologists; Am Chem Soc; assoc Am Inst Nutrit. *Res:* Influence of food processing on bioavailability of minerals, especially iron and zinc; functionality of hydrocolloid materials in food systems. *Mailing Add:* 111 Borland Lab Pa State Univ University Park PA 16802

THOMPSON, DONALD LEO, b Keota, Okla, Dec 31, 43; m 65; c 3. PHYSICAL CHEMISTRY. *Educ:* Northeastern Okla State Univ, BS, 65; Univ Ark, Fayetteville, PhD(phys chem), 70. *Prof Exp:* Res assoc theoret chem, Univ Calif, Irvine, 70-71; mem staff theoret chem, Los Alamos Nat Lab, 71-83; PROF CHEM, OKLA STATE UNIV, 83- *Concurrent Pos:* Vis assoc prof physics, Univ Miss, 75-76; vis prof chem, Okla State Univ, 80-81. *Mem:* Am Chem Soc. *Res:* Theoretical molecular dynamics; reaction kinetics and intermolecular energy transfer. *Mailing Add:* Dept Chem Okla State Univ Stillwater OK 74078

THOMPSON, DONALD LEROY, b Highland Park, Mich, Nov 15, 32; m 62; c 1. HEALTH PHYSICS. *Educ:* City Col New York, BS, 61; Long Island Univ, MS, 66; St Johns Univ, PhD(phys chem), 72, cert health physics, 81. *Prof Exp:* Sr scientist med physics, Radiation Physics Lab, State Univ NY Downstate Med Ctr, 62-65, asst dir, 65-69, co-dir, 69-72; health physicist, 72-74, actg dep dir, Div Radioactive Mat Br, 75-78, DEP CHIEF NUCLEAR MED, BUR RADIOL HEALTH, 78- *Concurrent Pos:* Instr radiol, State Univ NY, 65-69, asst prof radiol sci, 69-72; instr, Found Advan Educ Sci, 73-81. *Mem:* Am Asn Physicists in Med; Health Physics Soc. *Res:* Radiation exposures related to medical and consumer products. *Mailing Add:* Ctr Devices & Radiol Health 5600 Fishers Lane Rockville MD 20857

THOMPSON, DONALD LORAINE, b SDak, Feb 10, 21; m 49; c 1. AGRONOMY. *Educ:* SDak State Col, BS, 47, MS, 49; Iowa State Col, PhD(corn breeding), 53. *Prof Exp:* Asst small grain breeding, SDak State Col, 47-49; corn breeding, Iowa State Col, 49-52; PROF CROP SCI & RES AGRONOMIST, USDA, NC STATE UNIV, 52- *Mem:* Am Soc Agron. *Res:* Practical and theoretical aspects of corn breeding relating to quantitative genetics; disease resistance; relationships among economic traits; maximum production; forage evaluation. *Mailing Add:* Dept Crop Sci NC State Univ 1236 Williams Raleigh NC 27650

THOMPSON, DONALD OSCAR, b Clear Lake, Iowa, Feb 27, 27; m 46; c 3. SOLID STATE PHYSICS. *Educ:* Univ Iowa, BA, 49, MS, 50, PhD(physics), 53. *Prof Exp:* Group leader elastic & anelastic effects & mem radiation effects group, Solid State Physics Div, Oak Ridge Nat Lab, 54-64; mgr tech staff & dir struct mat dept, Rockwell Int Sci Ctr, 64-79; PROF AEROSPACE ENG & ENG MECH, IOWA STATE UNIV, 79-, PRIN SCIENTIST & PROG DIR, AMES LAB, 79-, DIR, CTR NONDESTRUCTIVE EVAL, 79- *Mem:* Nat Acad Eng; Am Inst Mech & Mining Engrs; Inst Elec & Electronics Engrs; Sigma Xi; Am Soc Nondestructive Testing; fel Am Phys Soc. *Res:* Radiation damage in metals, particularly interaction of radiation-produced defects and dislocations; anharmonic and nonlinear effects in solids; materials research in support of nondestructive testing; nondestructive testing apparatus. *Mailing Add:* Ctr Nondestructive Eval Appl Sci Complex II Iowa State Univ 1915 Scholl Rd Ames IA 50011

THOMPSON, DONOVAN JEROME, b Stoughton, Wis, Jan 30, 19; m 42; c 2. BIOSTATISTICS. *Educ:* St Olaf Col, BA, 41; Univ Minn, MA, 47; Iowa State Col, PhD(statist), 52. *Prof Exp:* Mem, staff Dept Math, Univ Minn, 46-47 & Statist Lab, Iowa State Col, 47-53; prof biostatist, Grad Sch Pub Health, Univ Pittsburgh, 53-66; prof biostatist, Sch Pub Health & Community Med, Univ Wash, 66-83, chmn dept, 73-83; RETIRED. *Concurrent Pos:* Scientist, Fred Hutchinson Cancer Res Ctr, 85- *Mem:* AAAS; Biomet Soc; Am Pub Health Asn; Am Statist Asn; Inst Math Statist. *Res:* Statistical theory and methodology. *Mailing Add:* 4231 NE 73rd Seattle WA 98115

THOMPSON, DOUGLAS STUART, b Richmond, Calif, Dec 5, 39; m 70. PHYSICAL CHEMISTRY. *Educ:* Univ Calif, Berkeley, BS, 61; Mass Inst Technol, PhD(phys chem), 65. *Prof Exp:* NIH trainee, Univ Calif, San Diego, 66-67; res phys chemist, E I du Pont de Nemours & Co, Inc, 68-79; MEM STAFF, DEPT CHEM, HAMPDEN-SYDNEY COL, 79- *Mem:* Am Chem Soc; Sigma Xi; AAAS; Am Phys Soc. *Res:* Characterization of macromolecules; inelastic light scattering; electron spin resonance; nuclear magnetic resonance; phospholipid vesicles, micelles, spin labels and viscometry. *Mailing Add:* 2325 Thackery Scott Hill Oakland CA 94611-2439

THOMPSON, DUDLEY, b Florence, SC, Jan 9, 13; m 41. CHEMICAL ENGINEERING. *Educ:* Va Polytech Inst, BSChE, 35, MSChE, 41, PhD(chem eng), 50. *Prof Exp:* Personnel asst, NAm Rayon Corp, Tenn, 35-38; asst control & prod chemist, Am Bemberg Corp, 38-39; instr chem eng, Va Polytech Inst, 41-46, from asst prof to assoc prof, 46-55, dir, appl ultrasonic res, 50-55, adv, basic div eng & arch, 52-54, coordr, 54-55; prof chem eng & chmn, dept chem & chem eng, Mo Sch Mines, 55-64, dir ultrasonic res, 56-64, Atomic Energy comn training facil, 57-64, dean fac, 64-74, dir res admin, 64-65, actg chancellor, 73-74, vchancellor, 74-78, EMER PROF CHEM ENG & EMER VCHANCELLOR, UNIV MO, ROLLA, 78- *Mem:* Fel AAAS; Am Chem Soc; Am Soc Eng Educ; Nat Soc Prof Engrs; Am Inst Chem Engrs; Inst Elec & Electronics Engrs; Am Inst Mining, Metall & Petrol Engrs; Soc Am Mil Engrs; Am Inst Physics; Acoust Soc. *Res:* Applied ultransonics; chemical processing; industrial instrumentation. *Mailing Add:* 1200 Homelife Plaza Apt A-1 Rolla MO 65401-2595

THOMPSON, EARL RYAN, b Lenoir, NC, Jan 9, 39; m 60; c 3. METALLURGICAL ENGINEERING, MATERIALS SCIENCE. *Educ:* NC State Univ, BS, 60, MS, 62; Univ VA, DSc, 66. *Prof Exp:* Res scientist metal, Reynolds Metals Co, 61-62; sr res scientist high temp alloy res, 65-74, mgr mat sci, 74-85, ASST DIR RES MAT TECH, UNITED TECHNOL RES CTR, 85- *Concurrent Pos:* Newcomb fel, Univ Va, 64; mem, Solid State Sci Panel, Nat Res Coun, 76-79; mem, Nat Mat Adv Comt, 90-, Int Mat Rev Comt; mem adv bd mat proj, Energy Conversion & Utilization Technol Prog, Dept Energy, 84-87, adv comt, Metals & Ceramics Div, Oak Ridge Nat Lab, 87- *Honors & Awards:* Grossman's Auth Award, Am Soc Metals, 70. *Mem:* Fel Am Soc Metals; Am Inst Mining, Metall & Petrol Engrs; Am Ceramic Soc; Sigma Xi; Am Soc Testing & Mat; AAAS. *Res:* High temperature alloy research and development; composite materials; directional solidification; ceramics for gas turbine use; rapidly solidified alloys; high temperature coatings; plasma spray processing; machinery and tribology. *Mailing Add:* United Technol Res Ctr Silver Lane East Hartford CT 06108

THOMPSON, EDWARD IVINS BRADBRIDGE, b Burlington, Iowa, Dec 20, 33; m 57; c 2. CELL GENETICS, HORMONES & CANCER. *Educ:* Rice Inst, BA, 55; Harvard Med Sch, MD, 60. *Prof Exp:* Intern & resident med, Presby Hosp, Col Physicians & Surgeons, Columbia Univ, 60-62; res assoc neurochem, Lab Clin Sci, NIMH, 62-64, res scientist molecular biol, Lab Molecular Biol, Nat Inst Arthritis & Metab Dis, 64-69, sr res scientist molecular & cell biol, 69-73, head, Sect Biochem Gene Expression, Lab Biochem, Nat Cancer Inst, 73-84; PROF & CHMN, DEPT HUMAN BIOL CHEM & GENETICS, UNIV TEX MED BR, 84-, PROF MED, 84- *Concurrent Pos:* Corresp ed, J Steroid Biochem; sci adv biotechnol, Am Cancer Soc Rev Bd; assoc ed, Cancer Res; ed-in-chief, Molecular Endocrinol, 87-; coun res, Am Cancer Soc, 89- *Mem:* Endocrine Soc; Am Soc Cell Biol; Am Soc Biochem & Molecular Biol; Am Asn Cancer Res; Am Soc Microbiol; Tissue Cult Asn. *Res:* Endocrinology; regulation of gene expression in eukaryotic cells; mechanism of steroid hormone action; effects of steroids in malignant cells; cholesterol regulation; steroids; AIDS. *Mailing Add:* Dept Human Biochem & Genetics Univ Tex Med Br Galveston TX 77550

THOMPSON, EDWARD VALENTINE, b Sharon, Conn, Feb 6, 35; m 56; c 2. POLYMER MATERIAL SCIENCE, MEMBRANE TECHNOLOGY. *Educ:* Cornell Univ, AB, 56; Polytech Inst Brooklyn, PhD(phys chem), 62. *Prof Exp:* Chemist, Am Cyanamid Co, 56-57, res chemist, 61-66; PROF CHEM ENG, UNIV MAINE, ORONO, 66- *Mem:* Am Chem Soc; Am Inst Chem Engrs. *Res:* Polymer composite structures; membrane separation proceses; flow and compressibility characteristics of porous media. *Mailing Add:* Dept Chem Eng Univ Maine Jenness Hall Orono ME 04469

THOMPSON, EDWARD WILLIAM, b Twin Falls, Idaho, June 26, 51; m 75; c 3. ELECTRON MICROSCOPY, CARDIAC SURGERY. *Educ:* Macalester Col, BA, 73; Med Col Wis, PhD(anat), 82. *Prof Exp:* Res fel anat & cardiol, Sch Med, Temple Univ, 82-84, res instr anat, 83-84; RES ASSOC, HORMEL INST, UNIV MINN, 84-, MGR, ELECTRON MICRO LAB, 84- *Concurrent Pos:* Mem, Adv Comt & Selection Comt, Am Asn Anatomists, 84-88; instr anat, Mayo Clinic, 89- *Mem:* Am Asn Anatomists; Am Heart Asn; Am Soc Cell Biol; Electron Micros Soc Am; Soc Exp Biol Med. *Res:* Stereologic, ultrastructural & biochemical analysis of myocardium in diabetic cardiomyopathy, obesity and hypertrophy induced by pressure or volume overloading. *Mailing Add:* Electron Micro Lab Hormel Inst Univ Minn 801 16th Ave NE Austin MN 55912

THOMPSON, EMMANUEL BANDELE, b Zarla, Nigeria, Mar 15, 28. PHARMACOLOGY. *Educ:* Rockhurst Col, BS, 55; Univ Mo-Kansas City, BS, 59; Univ Nebr, Lincoln, MS, 63; Univ Wash, PhD(pharmacol), 66. *Prof Exp:* Hosp pharmacist, Univ Kans Med Ctr, 59-60; retail pharmacist, Cundiff Drug Store, 61; sr res pharmacologist, Baxter Labs Inc, Ill, 63-66; asst prof, 69-73, ASSOC PROF PHARMACOL, COL PHARM, UNIV ILL MED CTR, 73- *Concurrent Pos:* Univ Ill Grad Col grant, 69-70 & Exten, 70-71; USPHS grant, 72-74; prin res investr & consult, West Side Vet Admin Hosp, Chicago, 71- *Mem:* NY Acad Sci; Am Asn Cols Pharm; Am Pharmaceut Asn. *Res:* Cardiovascular pharmacology. *Mailing Add:* Col Pharm Univ Ill Med Ctr PO Box 6998 Chicago IL 60680

THOMPSON, EMMETT FRANK, b El Reno, Okla, Nov 6, 36; m 61; c 3. FOREST ECONOMICS. *Educ:* Okla State Univ, BS, 58; NC State Univ, MS, 60; Ore State Univ, PhD(forest econ), 66. *Prof Exp:* From asst prof to prof forestry, Va Polytech Inst, 62-73; prof forestry & head dept, Miss State Univ, 73-77; head dept, 77-84, PROF FORESTRY, AUBURN UNIV, 77-, DEAN, SCH FORESTRY, 85- *Mem:* Fel Soc Am Foresters; Forest Products Res Soc. *Res:* Economics of forest resource management. *Mailing Add:* Sch Forestry Auburn Univ Auburn AL 36849

THOMPSON, ERIC DOUGLAS, b Buffalo, NY, Mar 24, 34; m 60; c 3. SOLID STATE PHYSICS, SOLID STATE ELECTRONICS. *Educ:* Mass Inst Technol, SB & SM, 56, PhD(physics), 60. *Prof Exp:* NSF res fel, 62-63; from asst prof to assoc prof, Case Western Reserve Univ, 63-69, prof eng, 69-83; Chandler-Weaver prof & chmn, 83-86, PROF COMPUTER SCI & ELEC ENG, LEHIGH UNIV, 86- *Concurrent Pos:* Sr res assoc, Jet Propulsion Lab, 72-73; prog dir, NSF, 81-82. *Mem:* Fel Am Phys Soc; sr mem Inst Elec & Electronics Engrs. *Res:* Theory of magnetism; solid state microwave active devices; Josephson Junction devices; thermal atomic scattering; cryogenic electronics; optical mixing; molecular computing. *Mailing Add:* 75 Hilltop Rd Bethlehem PA 18015

THOMPSON, ERIK G(RINDE), b Dallas, Tex, May 3, 34; m 59; c 3. ENGINEERING MECHANICS. *Educ:* Southern Methodist Univ, BS, 57; Univ Tex, MS, 59, PhD(eng mech), 65. *Prof Exp:* From asst prof to assoc prof eng sci, Univ Idaho, 64-68; assoc prof, 68-76, PROF CIVIL ENG, COLO STATE UNIV, 76- *Mem:* Am Soc Eng Educ; Am Acad Mech; Sigma Xi. *Res:* Plasticity and creep in engineering materials; finite element method; metal forming analysis. *Mailing Add:* 1812 Yorktown Ft Collins CO 80526

THOMPSON, ERNEST AUBREY, JR, b Tyler, Tex, Nov 17, 45; m 67; c 2. BIOCHEMISTRY, MOLECULAR BIOLOGY. *Educ:* Southern Methodist Univ, BS, 68; Univ Tex, Dallas, PhD(biochem), 74. *Prof Exp:* Fel biochem, Med Ctr, Univ Calif, San Francisco, 74-77; from asst prof to assoc prof biol, Univ SC, 77-87; assoc prof, 87-89, PROF BIOL CHEM, UNIV TEX, 89- *Concurrent Pos:* fel, Am Cancer Soc 74-75, res prof; NIH fel, 76-77, grant, 78- *Res:* Hormonal control of cellular proliferation. *Mailing Add:* Dept Human Biol Univ Tex Med Sch Galveston TX 77550

THOMPSON, EVAN M, b Payson, Utah, Aug 7, 33; m 59; c 5. PHYSICAL ORGANIC CHEMISTRY. *Educ:* Brigham Young Univ, BA, 60, PhD(org chem), 65. *Prof Exp:* Charles F Kettering & Great Lakes Cols Asn teaching fel chem, Antioch Col, 64-65; from asst prof to assoc prof, 65-74, dean, Sch Natural Sci, 70-76, PROF CHEM, CALIF STATE UNIV, STANISLAUS, 74- *Mem:* Am Chem Soc. *Res:* Organic reaction mechanisms; kinetics; chemical education. *Mailing Add:* Dept Chem Calif State Univ Stanislaus Turlock CA 95380

THOMPSON, FAY MORGEN, b St Paul, Minn, Dec 13, 35; m 55; c 2. OCCUPATIONAL HEALTH, ENVIRONMENTAL CHEMISTRY. *Educ:* Univ Minn, BA, 63, PhD(org chem), 70; Am Bd Indust Hyg, cert, 77. *Prof Exp:* Instr chem, Macalester Col, 67-68; instr occup health, Sch Pub Health, 70-78, occup health chemist, Environ Health & Safety, 74-82, ASST PROF ENVIRON HEALTH, SCH PUB HEALTH, UNIV MINN, 78-, ASST DIR, ENVIRON HEALTH & SAFETY, 82- *Concurrent Pos:* Mem adv comt hazardous waste, Minn Waste Mgt Bd, 80-; mem, Comt Hazardous Substances in Lab, Nat Res Coun, 81-83; mem bd dirs, Am Lung Asn, 84- *Mem:* Am Indust Hyg Asn; Am Chem Soc; Am Conf Govt Indust Hygienists; Am Acad Indust Hyg; Sigma Xi; Air Pollution Control Asn. *Res:* Collection and analysis of selected air contaminants; waste minimization; hazardous waste disposal; laboratory safety; laboratory use of carcinogens. *Mailing Add:* Dept Environ Health & Safety Univ Minn Minneapolis MN 55455

THOMPSON, FRANCIS TRACY, b New York, NY, Nov 22, 30; m 55; c 3. ELECTRICAL ENGINEERING. *Educ:* Rensselaer Polytech Inst, BSEE, 52; Univ Pittsburgh, MS, 55, PhD(elec eng), 64. *Prof Exp:* Develop engr, Res Labs, 53-57, fel engr, New Prod Lab, 57-61, supvry engr, Res Labs, 61-64, mgr info & control circuitry, elec systs & power conditioning, 64-69, dir instrumentation & systs res, 69-72, dir electronics & electromagnetics res, 72-74, dir elec sci res, 74-77, div mgr Electronics Technol, 77-87, GEN MGR, ENG TECHNOL DIV, WESTINGHOUSE ELEC CORP, 87- *Concurrent Pos:* Mem bd dir, Siliconix, 80- *Honors & Awards:* B G Lamme Award. *Mem:* Fel Inst Elec & Electronics Engrs; Instrument Soc Am. *Res:* Digital computer development; control systems; instrumentation; television systems; solid-state circuitry; magnetic and mechanical systems; power electronic systems. *Mailing Add:* 3482 Treeline Dr Murrysville PA 15668

THOMPSON, FRED C, b Snow Shoe, Pa, Feb 26, 28; m 52; c 4. ELECTRONICS ENGINEERING. *Educ:* Pa State Univ, BS, 50, MS, 58. *Prof Exp:* Engr, Martin Co, Md, 52-54; engr, HRB-Singer, Inc, 54-58, div mgr receiving systs, 58-60, staff engr, 60-63, lab dir countermeasures equip, 63-66, staff asst to tech vpres, 66-69; vpres, 68-80, PRES, LOCUS, INC, 80- *Mem:* Sr mem Inst Elec & Electronics Engrs. *Res:* Very high frequency-ultra high frequency receiving systems; microwave devices. *Mailing Add:* Locus Inc PO Box 740 State College PA 16801

THOMPSON, FRED GILBERT, b Cleveland, Ohio, Nov 13, 34; m 57; c 1. MALACOLOGY. *Educ:* Univ Mich, BS, 58; Wayne State Univ, MA, 61; Univ Miami, PhD(zool), 64. *Prof Exp:* Res scientist, Univ Miami, 64-66; interim assoc cur, 66-71, assoc cur malacol, 71-81, CUR MALACOL & PROF ZOOL, FLA STATE MUS, UNIV FLA, 81- *Concurrent Pos:* NIH res grant systs Amnicolidae, 64-67. *Mem:* Am Malacol Union; Asn Syst Malacologists. *Res:* Systematics, ecology, land and freshwater mollusks. *Mailing Add:* Fla Mus Nat Hist Museum Rd Gainesville FL 32601

THOMPSON, FREDERICK NIMROD, JR, b Newport News, Va, Dec 9, 39; m 62; c 2. REPRODUCTIVE ENDOCRINOLOGY. *Educ:* Wake Forest Univ, BS, 61; Univ Ga, DVM, 65; Iowa State Univ, PhD(physiol), 73. *Prof Exp:* Instr physiol, Iowa State Univ, 67-73; ASST PROF PHYSIOL, UNIV GA, 73- *Mem:* Soc Study Reproduction; Sigma Xi; Am Vet Med Asn. *Res:* Hormonal control of parturition and cardiovascular effects of adrenal corticosteroids. *Mailing Add:* Dept Physiol & Pharmacol Col Vet Med Univ Ga Athens GA 30602

THOMPSON, GARY GENE, b Beach, NDak, Oct 18, 40; m 70, 82; c 1. GEOLOGY, PALYNOLOGY. *Educ:* Univ NDak, BS, 62; Mich State Univ, PhD(geol), 69. *Prof Exp:* Geologist, Shell Develop Co, 68-69 & Shell Oil Co, 69-70; from asst prof to assoc prof geol, Salem State Col, 71-81; assoc prof, 83-89, SEAGER PROF GEOL, ROCKY MOUNTAIN COL, 89- *Mem:* Soc Econ Paleontologists & Mineralogists; Am Asn Stratig Palynologists. *Res:* Cretaceous, tertiary and quaternary palynomorph biostratigraphy and paleoecology. *Mailing Add:* Dept Geol Rocky Mountain Col 1511 Poly Dr Billings MT 59102

THOMPSON, GARY HAUGHTON, b Long Beach, Calif, Mar 25, 35; m 64; c 3. PHYSICAL CHEMISTRY, INORGANIC CHEMISTRY. *Educ:* Univ Colo, Boulder, BS, 60; Univ Utah, PhD(chem), 69. *Prof Exp:* Engr, Hercules Powder Co, 60-63; res chemist, Savannah River Lab, E I du Pont de Nemours & Co, Inc, 69-76; group leader, Rocky Flats Plant, 76-81, PROCESS OPER MGR, ROCKWELL INT CORP, 81- *Mem:* Am Nuclear Soc; Am Chem Soc; Am Soc Metals. *Res:* Chromatography, including gas, liquid and ion exchange; radioiodine sorption, gas-solid and gas-liquid systems; radioactive waste management and solvent extraction processes. *Mailing Add:* 10729 Verese Lane Northglenn CO 80234

THOMPSON, GEOFFREY, b Stockton-on-Tees, Durham, Eng, Oct 18, 35; m 61; c 3. GEOCHEMISTRY, OCEANOGRAPHY. *Educ:* Univ Manchester, BSc, 61, PhD(geochem), 65. *Prof Exp:* Res chemist, Imp Chem Industs, UK, 58-59; geologist, Transvaal Gold Mines, SAfrica, 60; asst scientist, 65-70, assoc scientist, 70-78, SR SCIENTIST & CHMN, DEPT CHEM, WOODS HOLE OCEANOG INST, 78- *Concurrent Pos:* Res assoc dept mineral sci, Smithsonian Inst, 70-; assoc ed, Geochimica et Cosmochmica Acta, 73- & J Marine Res, 74- *Mem:* AAAS; Geochem Soc; Am Geophys Union; Soc Appl Spectros. *Res:* Origin, evolution and geochemistry of oceanic crust; geochemistry of ocean sediments and marine organisms. *Mailing Add:* Dept Chem Woods Hole Oceanog Inst Woods Hole MA 02543

THOMPSON, GEORGE ALBERT, b Swissvale, Pa, June 5, 19; m 44; c 3. GEOPHYSICS, GEOLOGY. *Educ:* Pa State Col, BS, 41; Mass Inst Technol, MS, 42; Stanford Univ, PhD(geol), 49. *Prof Exp:* Actg instr, 47-48, lectr, 48-49, from asst prof to assoc prof, 49-60, chmn, Dept Geol, 79-82, chmn, Dept Geophys, 67-86, Otto N Miller, prof, earth sci, 80-89, dean, Sch Earth Sci, 87- 89, PROF GEOPHYS, STANFORD UNIV, 60- *Concurrent Pos:* Geologist & geophysicist, US Geol Surv, 42-76; NSF fel, 56-57; Guggenheim fel, 63-64; G K Gilbert award seismic geol, 64; mem, Geodynamics Comt, Nat Res Coun, 75-78; consult Adv Comt Reactor Safeguards, Nuclear Reg Comt, 72-; coun mem, Geol Soc Am, 83-86; mem bd, Earth Sci, Nat Res Coun, 86-88; co-chmn, Sci planning & rev comt, Edge Deep Seismic Reflection Prog, 86-; sr external events rev group, Lawrence Livermore Nat Lab, 90-; exec comt, Inc Res Inst Seismol (IRIS), 90-; coun, Continental Sci Drilling, Nat Acad Sci, 90-; panel coupled processes, Yucca Mt, Nat Res Coun, 90- *Honors & Awards:* George P Woollard Award, Geol Soc Am, 83. *Mem:* Seismol Soc Am; Soc Explor Geophys; Soc Econ Geol; fel Geol Soc Am; fel Am Geophys Union; fel AAAS. *Res:* Structure and geophysics of Basin Range Province; crust-mantle structure from deep seismic reflection and refraction measurements; lunar traverse gravity experiment; geophysics of ultramafic rocks; geology of quicksilver deposits. *Mailing Add:* Dept Geophys Stanford Univ Stanford CA 94305-2215

THOMPSON, GEORGE REX, b Oakley, Idaho, July 24, 43; m 63; c 4. TOXICOLOGY, PHARMACOLOGY. *Educ:* Ore State Univ, BS, 65, PhD(toxicol & pharmacol), 69. *Prof Exp:* Res asst toxicol & pharmacol, Ore State Univ, 66-69; researcher toxicol, Mason Res Inst, Mass, 69-72; supvr toxicol, Biomed Res Lab, ICI Am Inc, 72-73; head sect gen toxicol, Abbott Labs, 73-77; mgr prod safety systs, 77-80, DIR CORP SAFETY ASSURANCE, INT FLAVORS & FRAGRANCES INC, 80- *Mem:* Soc Toxicol; Am Soc Pharmacol & Exp Therapeut; Environ Mutagen Soc; Inst Food Technol. *Res:* Toxicity of marihuana or tetrahydrocannabinol, cyclamate/cyclohexylamine, new drugs, anticancer compounds and pesticides; delineation of normal physiological parameters via the utilization of toxic materials; safety criteria for flavors and fragrance; computerized safety evaluations; employee health and safety; environmental protection. *Mailing Add:* Forum Sci Excellence 200 Wood Port Rd Sparta NJ 07871

THOMPSON, GEORGE RICHARD, b Ann Arbor, Mich, Apr 2, 30; m 57; c 3. INTERNAL MEDICINE, RHEUMATOLOGY. *Educ:* Univ Mich, Ann Arbor, BS, 50, MD, 54. *Prof Exp:* Intern, Ohio State Univ Hosp, 54-55; Dir Rheumatol Sect, Wayne County Gen Hosp, Eloise, 63-84; Vet Admin Hosp, Ann Arbor, 84-87; resident, Hosp, 55-58, from instr to assoc prof, 62-76, PROF INTERNAL MED, MED SCH, UNIV MICH, ANN ARBOR, 76-, UNIV MICH HEALTH SERV, 87- *Concurrent Pos:* USPHS fel rheumatol, Rackham Arthritis Res Unit, Univ Mich Hosp, Ann Arbor, 60-62, assoc physician, 63- *Mem:* Am Rheumatism Asn; fel Am Col Physicians; Am Fedn Clin Res; Cent Soc Clin Res. *Res:* Arthritis; mucopolysaccharide metabolism; gout and urate excretion; rubella-associated arthritis. *Mailing Add:* Dept Internal Med Univ Mich Ann Arbor MI 48109

THOMPSON, GERALD LEE, b Swea City, Iowa, Mar 16, 45; m 90; c 3. ORGANIC CHEMISTRY. *Educ:* Iowa State Univ, BS, 68; Ohio State Univ, PhD(chem), 72. *Prof Exp:* NIH fel, Harvard Univ, 72-74; sr org chemist, Eli Lilly & Co, 74-79, res scientist, 80, head chem res div, 80-84, head cancer res, 84-86, head process res, 86-89, DIR CHEM PROCESS RES & DEVELOP, ELI LILLY & CO, 89- *Mem:* Am Chem Soc. *Res:* Antitumor drug design; alkaloid synthesis; general medicinal chemistry. *Mailing Add:* Dept IC414 Bldg 110/1 Lilly Corp Ctr Indianapolis IN 46285

THOMPSON, GERALD LUTHER, b Rolfe, Iowa, Nov 25, 23; m 54; c 3. COMBINATORIAL OPTIMIZATION. *Educ:* Iowa State Col, BS, 44; Mass Inst Technol, SM, 48; Univ Mich, PhD(math), 53. *Prof Exp:* Instr math, Princeton Univ, 51-53; asst prof, Dartmouth Col, 53-58; prof, Ohio Wesleyan, 58-59; IBM PROF SYSTS & OPER RES, GRAD SCH INDUST ADMIN, CARNEGIE MELLON UNIV, 59- *Concurrent Pos:* Consult, Princeton Univ, Int Bus Mach Corp, Sandia Corp, Beth Steel Corp, Timken Co, Westinghouse Elec Co & McKinsey & Co, Gen Motors, PPG Co; Inst Mgt Sci rep, Math Div, Nat Res Coun, 71-73; mem, Sealift Readiness Comt, Nat Acad Sci, 74-75. *Mem:* Am Math Soc; Soc Indust & Appl Math; Inst Mgt Sci; Math Asn Am; Opers Res Soc Am; Math Prog Soc. *Res:* Applications of mathematics to management; mathematical economics; optimal control theory; graph theory and combinatorial problems; game theory; combinatorial optimization. *Mailing Add:* GSIA Carnegie Mellon Univ 5000 Forbes Ave Pittsburgh PA 15213-3890

THOMPSON, GORDON WILLIAM, b Vancouver, BC. EPIDEMIOLOGY. *Educ:* Univ Alta, DDS, 65; Univ Toronto, MScD, 67, PhD(epidemiol & biostatist), 71; FRCD(C), 75. *Prof Exp:* assoc dean & assoc prof dent, Univ Toronto, 69-77; dean fac dent, 77-89, CHMN, DEPT DENT HEALTH CARE, UNIV ALTA, 89- *Mem:* Biomet Soc; Int Asn Dent Res. *Res:* Biostatistical and computer applications in the fields of dental science and growth and development of humans with particular emphasis on the craniofacial complex and stomatognathic system. *Mailing Add:* Dept Dent Health Care Univ Alta Edmonton AB T6G 4G4 Can

THOMPSON, GRANT, b Ogden, Utah, Feb 26, 27; m 49; c 6. PROPELLANT CHEMISTRY. *Educ:* Univ Utah, BA, 50, PhD(chem), 53. *Prof Exp:* Res chemist, E I du Pont de Nemours & Co, 53-58; proj chemist, Thiokol Corp, 58-59, sr chemist, 59-60, supvr new propellants sect, 60-62, mgr propellant develop dept, Wasatch Div, 63-75, mgr, Res & Develop Labs, Wasatch Div, 75-85, DIR, RES & DEVELOP LABS, ADVAN TECHNOL, THIOKOL CORP, 85- *Mem:* Am Chem Soc. *Res:* Mechanism of propellant cure; curing agents and catalysts for hydrocarbon propellants; mechanism of hydrocarbon propellant aging; high energy oxidizers and propellants. *Mailing Add:* Res & Develop Labs Morton Thiokol Corp PO Box 707 MS 240 Brigham City UT 84302-0707

THOMPSON, GRANVILLE BERRY, b Sedalia, Mo, June 18, 29; m 58; c 3. AGRICULTURAL RESEARCH ADMINISTRATION, ANIMAL NUTRITION & MANAGEMENT. *Educ:* Univ Mo, BS, 51, MS, 55, PhD(animal nutrit), 58. *Prof Exp:* Asst exten agent, Mo Agr Exten Serv, 51-55; grad asst animal sci, Univ Mo, 53-55, from instr to prof, 55-76; RESIDENT DIR RES & PROF ANIMAL SCI, TEX A&M UNIV, 76- *Mem:* Sigma Xi; Am Soc Animal Sci; Am Forage & Grassland Coun; Am Registry Prof Animal Scientists; Coun Agr & Technol; Acad Vet Consults. *Res:* Beef cattle nutrition and management; beef production systems; research administration. *Mailing Add:* Syst Res & Exten Ctr Tex A&M Univ 6500 Amarillo Blvd W Amarillo TX 79106

THOMPSON, GUY A, JR, b Rosedale, Miss, May 31, 31; m 60; c 3. BIOCHEMISTRY. *Educ:* Miss State Univ, BS, 53; Calif Inst Technol, PhD(biochem), 59. *Prof Exp:* NSF res fel chem, Univ Manchester, 59-60; res assoc biochem, Univ Wash, 60-62, from instr to asst prof, 62-67; assoc prof, 67-74, PROF BOT, UNIV TEX, AUSTIN, 74- *Concurrent Pos:* NIH res career develop award, 63-67. *Mem:* Fel AAAS; Am Soc Biol Chem; Am Oil Chem Soc; Am Chem Soc; Sigma Xi. *Res:* Lipid metabolism; biochemistry of membranes. *Mailing Add:* Dept Bot Univ Tex Austin TX 78712

THOMPSON, HANNIS WOODSON, JR, b Salisbury, NC, Sept 3, 28; m 51; c 2. ELECTRICAL ENGINEERING, SOLID STATE PHYSICS. *Educ:* NC State Univ, BS, 53, MS, 59; Purdue Univ, PhD(elec eng), 63. *Prof Exp:* Engr missile systs, Western Elec Co, 53-57; PROF ELEC ENG, PURDUE UNIV, 63-, ASST HEAD EDUC, SCH ELEC ENG, 84- *Concurrent Pos:* Sr scientist, Navy Electronics Lab, 63, 64; consult, electronics & solid state, CTS Microelectronics, 62, 65- *Mem:* Inst Elec & Electronics Engrs; Am Phys Soc; Sigma Xi. *Res:* Solid state devices, oxide deposition, thin film and silicon technology. *Mailing Add:* 215 Timbercrest Rd West Lafayette IN 47906

THOMPSON, HARTWELL GREENE, JR, b Hartford, Conn, Aug 30, 24; m 55; c 4. NEUROLOGY. *Educ:* Yale Univ, BA, 46; Cornell Univ, MD, 50. *Prof Exp:* From asst to assoc neurol, Col Physicians & Surgeons, Columbia Univ, 57-59; from asst prof to assoc prof, Univ Wis, 59-64; prof & chmn dept, Sch Med, WVa Univ, 64-69; prof & assoc dean student affairs, Sch Med, Univ Pa, 69-73; prof neurol & dean, Charleston Div, WVA Univ Med Ctr, 73-76; PROG DIR, DEPT NEUROL, HARTFORD HOSP, 76- *Concurrent Pos:* Consult, Vet Admin Hosp, Clarksburg, WVa, 64- *Mem:* AMA; Am Acad Neurol; Sigma Xi. *Res:* Neurology training programs and undergraduate education in neurology; multiple sclerosis; motor neuron disease. *Mailing Add:* 12 Mountain Rd West Hartford CT 06107

THOMPSON, HARVEY E, b Valders, Wis, Oct 30, 20; m 53; c 2. AGRONOMY. *Educ:* Univ Wis, BS, 47, MS, 48, PhD(agron, econ entom), 51. *Prof Exp:* enten agronomist, Iowa State Univ, 50-; RETIRED. *Mem:* Am Soc Agron. *Res:* Forage and grain crop production. *Mailing Add:* 2104 Agron Bldg Iowa State Univ Ames IA 50011

THOMPSON, HAZEN SPENCER, b Frelighsburg, Que, Sept 10, 28; m 56; c 2. PLANT PATHOLOGY. *Educ:* McGill Univ, BSc, 50; Univ Toronto, MA, 52, PhD(plant path), 55. *Prof Exp:* Demonstr bot, Univ Toronto, 50-53; res off, Plant Res Inst, Res Br, Can Dept Agr, 51-65, fungicide liaison officer, Sci Info Sect, 65-73; SR EVAL OFFICER, PESTICIDES DIV, COM CHEM BR, ENVIRON CAN, 73- *Mem:* Can Phytopath Soc. *Res:* Chemical control of plant diseases; environmental fate and effects of pesticides. *Mailing Add:* Com Chem Br Conserv & Protection Environ Can Ottawa ON K1A 0H3 Can

THOMPSON, HENRY JOSEPH, b Mamaroneck, NY, Sept 5, 21; m 47; c 3. BIOLOGY. *Educ:* Whittier Col, AB, 47; Stanford Univ, MA, 48, PhD, 52. *Prof Exp:* Instr biol, Whittier Col, 48-49; actg instr, Stanford Univ, 51-52; instr bot, 52-54, from asst prof to assoc prof, 54-66, prof bot, Univ Calif, Los Angeles, 66-83; CONSULT, 83- *Mem:* Am Soc Plant Taxon; Soc Study Evolution. *Res:* Systematics and evolution. *Mailing Add:* AMC Cancer Res Ctr-Nutrit Ctr 1600 Pierce St Denver CO 80214

THOMPSON, HERBERT BRADFORD, b Detroit, Mich, Apr 22, 27; m 49. STRUCTURAL CHEMISTRY. *Educ:* Olivet Col, BS, 48; Oberlin Col, AM, 50; Mich State Col, PhD(chem), 53. *Prof Exp:* Res instr chem, Mich State Univ, 53-55; from asst prof to assoc prot, Gustavus Adolphus Col, 55-63; res assoc, Inst Atomic Res, Iowa State Univ, 63-65; res assoc, Univ Mich, 65-67; chmn dept chem, 68-69, 74-75, PROF CHEM, UNIV TOLEDO, 67- *Mem:* Am Chem Soc; Am Phys Soc. *Res:* Molecular structure and geometry; conformational analysis; data acquisition and computer applications in chemistry; electron diffraction; dipole moments. *Mailing Add:* 1604 Riverview Rd St Peter MN 56082

THOMPSON, HERBERT STANLEY, b China, June 12, 32; nat US; m 55; c 5. OPHTHALMOLOGY, NEUROLOGY. *Educ:* Univ Minn, BA, 53, MD, 61; Univ Iowa, MS, 66. *Prof Exp:* From instr to assoc prof, 67-76, PROF OPHTHAL, UNIV IOWA, 76- *Concurrent Pos:* Nat Inst Neurol Dis & Blindness spec fel clin neuro-ophthal, Univ Calif, San Francisco, 66-67; Nat Inst Neurol Dis & Blindness res career develop award, 68; coun mem, Int Neuro-Ophthal Soc, 76-; assoc ed, 81-84, book rev ed, Am J Ophthal, 84-; dir, Am Bd Ophthal, 89-; Ophthal consult, 26th ed Stedmans Med Dict. *Mem:* Asn Res Vision & Ophthal; Ophthal Soc UK; France Soc Ophthal; Am Acad Ophthal & Otolaryngol; Am Ophthal Soc; Int Neuro-Ophthal Soc. *Res:* Neuro-ophthalmology, especially of the autonomic nervous system. *Mailing Add:* Dept Ophthal Univ Iowa Hosps Iowa City IA 52242

THOMPSON, HOLLY ANN, US citizen. BIOTECHNOLOGY, CELLULAR & DEVELOPMENTAL BIOCHEMISTRY. *Educ:* Univ Del, BA, 76; Kans State Univ, PhD(cell & develop biol), 82. *Prof Exp:* Fel, NIH, 82-84; vis asst prof biochem, Univ Mont, 84-86; MGR RES & DEVELOP, CHROMATOCHEM, INC, 87- *Mem:* Am Soc Cell Biol. *Res:* Affinity chromatography, analytical immunoaffinity chromatography, novel quantitative solid phase assays, proteoglycan biosynthesis, extracellular matrix and morphogenesis, and mammalian development. *Mailing Add:* 2837 Ft Missoula Rd Missoula MT 59801

THOMPSON, HOWARD DOYLE, b Cedar City, Utah, Apr 17, 34; m 56; c 5. GAS DYNAMICS, LASER VELOCIMETRY. *Educ:* Univ Utah, BS, 57; Purdue Univ, MS, 62, PhD(mech eng), 65. *Prof Exp:* Asst eng sci, 61-62, asst mech eng, 62-65, from asst prof to assoc prof, 65-74, PROF MECH ENG, PURDUE UNIV, WEST LAFAYETTE, 74- *Concurrent Pos:* Consult, Dynetics, Inc, 66-72, Detroit Diesel Allison, 74-76, Univ Dayton, 78-80, McDonnell-Douglas, 80-85, Air Force Wright Aeronaut Lab, 80-, Arnold Eng Develop Ctr, 81-85 & Ballistic Missile Div, Redstone Arsenal, Ala; assoc res scientist, Pratt & Whitney Aircraft, 69-70; sr mech engr, Arnold Eng Develop Ctr, 80-81; vis scientist, WL/POP, Wright-Patterson AFB, 90-91. *Mem:* Assoc fel Am Inst Aeronaut & Astronaut; Am Soc Mech Engrs; Laser Inst Am. *Res:* Propulsion gas dynamics; nozzle design; three-dimensional supersonic flows; optimization of aerodynamic shapes; transonic and annular flows; laser doppler velocimetry; experimental and numerical fluid mechanics; fluid mechanics in turbomachinery. *Mailing Add:* Sch Mech Eng Purdue Univ West Lafayette IN 47907

THOMPSON, HOWARD K, JR, b Boston, Mass, May 19, 28; m 64; c 3. INTERNAL MEDICINE. *Educ:* Yale Univ, BA, 49; Columbia Univ, MD, 53. *Prof Exp:* Intern internal med, 1st Med Div, Bellevue Hosp, NY, 53-54, jr asst res, 55; clin fel, Mary I Bassett Hosp, Cooperstown, 54; cardiovasc res fel, Duke Hosp, Durham, NC, 58, sr asst res internal med, 58-59, cardiovasc res fel, 59-60, chief res, 60-61; biophys fel, Mass Inst Technol, 61-62; assoc med, Duke Univ, 62-65, asst prof biophys, 65-69, asst prof biomath, 66-69, assoc physiol, 63-69; prof med, Baylor Col Med, 71-78; prof med, Albany Med Col, 78-83; ASSOC, AM BD INTERNAL MED, 63- *Mem:* Fel Am Col Cardiol; NY Acad Sci. *Res:* Indicator dilution method of blood flow estimation; estimation of regional cerebral blood flow by radio-xenon inhalation; biostatistics; computers in medicine. *Mailing Add:* 416 Allison Dr Dallas TX 75208

THOMPSON, HUGH ALLISON, b Chattanooga, Tenn, Mar 24, 35; m 57; c 1. MECHANICAL ENGINEERING. *Educ:* Auburn Univ, BS, 56; Tulane Univ, MSc, 62, PhD(mech eng), 64. *Prof Exp:* Process engr, Mobil Oil Corp, 56-60; from instr to assoc prof mech eng, 63-71, PROF MECH ENG, TULANE UNIV, 71-, DEAN, 76- *Mem:* Am Soc Mech Engrs; Inst Elec & Electronics Engrs. *Res:* Dynamic response of bus conductor structures to short circuit loads, of pole-mounted electric transmission lines to hurricane winds and of transformer coils to through faults. *Mailing Add:* 1312 Eighth St New Orleans LA 70115

THOMPSON, HUGH ANSLEY, b Olympia, Wash, Feb 22, 36; m 66. MICROENGINEERING, APPLIED CAPILLARITY. *Educ:* Mass Inst Technol, BS & MS, 63. *Prof Exp:* Develop engr, 63-72, INVENTOR-SCIENTIST, PROCTER & GAMBLE CO, 73- *Concurrent Pos:* Prin, Thompson Eng, 82- *Mem:* Sigma Xi; Int Solar Energy Soc; Acad Appl Sci; Am Soc Mech Engrs. *Res:* Invention and development of microengineered capillary networks, textiles and related materials; absorbent products; solar and renewable energy development; 9 US patents. *Mailing Add:* 5777 Windermere Ln Fairfield OH 45014

THOMPSON, HUGH ERWIN, b Newport, RI, Aug 4, 17; m 46; c 4. ENTOMOLOGY. *Educ:* Univ RI, BS, 47; Cornell Univ, PhD(entom), 54. *Prof Exp:* Asst state entomologist, State Dept Agr & Conserv, RI, 47-48; entomologist, State Dept Agr, Pa, 53-56; from asst prof to prof entom, Kans State Univ, 56-86; RETIRED. *Mem:* Int Soc Arboricult; Arboricult Res & Educ Acad (pres, 78-79); Sigma Xi; Entom Soc Am; Soc Am Foresters. *Res:* Biology and control of insects attacking shade trees and ornamental plants; insect transmission of tree diseases. *Mailing Add:* Dept Entom Kans State Univ Manhattan KS 66506

THOMPSON, HUGH WALTER, b New York, NY, Dec 7, 36; m 64; c 1. ORGANIC CHEMISTRY. *Educ:* Cornell Univ, AB, 58; Mass Inst Technol, PhD(org chem), 63. *Prof Exp:* NIH res fel, Columbia Univ, 62-64; from asst prof to assoc prof chem, 64-72, PROF CHEM, RUTGERS UNIV, NEWARK, 72- *Mem:* Am Chem Soc. *Res:* Mechanisms and stereochemical courses of organic reactions; compounds of unusual symmetry and stereochemistry; development of new synthetic methods. *Mailing Add:* Dept Chem Rutgers Univ Newark NJ 07102

THOMPSON, IDA, paleobiology, marine biology, for more information see previous edition

THOMPSON, J G, MATHEMATICS. *Prof Exp:* PROF, MATH DEPT, UNIV CAMBRIDGE. *Mem:* Nat Acad Sci. *Mailing Add:* Math Dept Univ Cambridge Cambridge CB2 1SB England

THOMPSON, JAMES ARTHUR, b Sturgeon Bay, Wis, Aug 15, 31; m 55; c 3. ENVIRONMENTAL & INDUSTRIAL HYGIENE MANAGEMENT. *Educ:* St Olaf Col, BA, 55; Iowa State Univ, MS, 58. *Prof Exp:* Anal chemist, Ames Lab, AEC, 55-59; res chemist anal div, Alcoa Res Lab, 59-61, sr chemist, Warrick Opers, Ind, 61-70, chief chemist, Wenatchee Works, 70-75, NORTHWEST ENVIRON & INDUST HYG MGR, ALUMINUM CO AM, 75- *Mem:* AAAS; Am Chem Soc; Air Pollution Control Asn; Sigma Xi; Am Indust Hyg Asn. *Res:* Instrumental analysis; industrial hygiene; air and water pollution. *Mailing Add:* Aluminum Co Am PO Box 221 Wenatchee WA 98807-0221

THOMPSON, JAMES BURLEIGH, JR, b Calais, Maine, Nov 20, 21; m 57; c 1. PETROLOGY, GEOCHEMISTRY. *Educ:* Dartmouth Col, AB, 42; Mass Inst Technol, PhD, 50. *Hon Degrees:* DSc, Dartmouth Col, 75. *Prof Exp:* Instr geol, Dartmouth Col, 42; asst, Mass Inst Technol, 46-47, instr, 47-49; instr petrol, 49-50, asst prof petrog, 50-55, assoc prof mineral, 55-60, prof mineral, 60-77, STURGIS HOOPER PROF GEOL, HARVARD UNIV, 77- *Concurrent Pos:* Ford Found fel, 52-53; Guggenheim fel, 63. *Honors & Awards:* A L Day Medal, Geol Soc Am, 64; Roebling Medal, Mineral Soc Am, 78; V M Goldschmidt Medal, Geochem Soc, 85. *Mem:* Nat Acad Sci; AAAS; fel Geol Soc Am; fel Mineral Soc Am; fel Am Acad Arts & Sci. *Res:* Metamorphic petrology; geology of New England. *Mailing Add:* Dept Earth & Planetary Sci Harvard Univ Cambridge MA 02138

THOMPSON, JAMES CHARLES, b San Antonio, Tex, Aug 16, 28; m 67; c 5. SURGERY, PHYSIOLOGY. *Educ:* Agr & Mech Col Tex, BS, 48; Univ Tex, MD, 51, MA, 52; Am Bd Surg, dipl. *Prof Exp:* Intern, Univ Tex Med Br Galveston, 51-52; asst resident surg, Hosp, Univ Pa, 52-54 & 56-58, chief resident, 58-59; from asst surgeon to assoc surgeon, Pa Hosp, 59-63; head physician, Harbor Gen Hosp, 63-67, chief surg, 67-70; PROF SURG & CHMN DEPT, UNIV TEX MED BR GALVESTON, 70-, CHIEF SURG, HOSP, 70- *Concurrent Pos:* Fel, Harrison Dept Surg Res, Sch Med, Univ Pa, 52-54 & 56-57, Albert & Mary Lasker fel, 57-59; John A Hartford Found grants, 60-; NIH grants, 60-; asst instr surg, Sch Med, Univ Pa, 53-54 & 56-58, from instr to assoc instr, 58-61, asst prof, 61-63; from assoc prof to prof, Sch Med, Univ Calif, Los Angeles, 63-70. *Mem:* AAAS; Am Surg Asn; Am Physiol Soc; Am Gastroenterol Asn; Endocrine Soc; Soc Univ Surgeons. *Res:* Gastric physiology; general surgery; humoral control of gastric secretion; histamine metabolism; secretion in isolated tissue; radioimmunoassay and metabolism of gastrointestinal hormones, gastrin, cholecystokinin and secretin; molecular heterogeneity of gastrointestinal hormones; metabolic mechanisms (synthesis storage, release, transport, mechanisms of action and catabolisms) of gastrointestinal regulatory peptides and related substanstces and the effects of aging on these mechanisms; the role of gastro-intestros hormone in gastro-intestros cancer; gastrointestinal homormones receptors; intercellular mechanisms stimulated by gastrointestinal hormones. *Mailing Add:* Dept Surg Univ Tex Med Br Galveston TX 77550

THOMPSON, JAMES CHARLTON, b Leeds, UK, Jan 4, 41; div; c 2. INORGANIC CHEMISTRY. *Educ:* Cambridge Univ, BA, 62, PhD(chem), 65. *Prof Exp:* Fel, Rice Univ, 65-67; asst prof chem, 67-72, assoc chmn dept, 74-77, actg chmn, 79-80, ASSOC PROF CHEM, UNIV TORONTO, 72-, ASSOC CHMN DEPT, 82- *Mem:* Chem Inst Can. *Res:* Studies on the synthesis, structures and properties of silicon compounds, particularly those with fluorine or hydrogen bound to silicon. *Mailing Add:* Dept Chem Univ Toronto Toronto ON M5S 1A1 Can

THOMPSON, JAMES CHILTON, b Ft Worth, Tex, June 14, 30; m 55; c 3. PHYSICS. *Educ:* Tex Christian Univ, BA, 52; Rice Inst, MA, 54, PhD(physics), 56. *Prof Exp:* From asst prof to assoc prof, 56-67, PROF PHYSICS, UNIV TEX, AUSTIN, 67- *Mem:* Am Phys Soc. *Res:* Transport coefficients in solid and liquid metals; metal-ammonia solutions; metal-nonmetal transition; electrode-electrolyte interfaces. *Mailing Add:* Dept Physics Univ Tex Austin TX 78712

THOMPSON, JAMES EDWIN, b Maryville, Mo, Feb 2, 36; m 65; c 2. ORGANIC CHEMISTRY. *Educ:* Cent Methodist Col, AB, 56; Univ Mo, PhD(chem cyclopropanes), 61. *Prof Exp:* CHEMIST, PROCTER & GAMBLE CO, 61- *Mem:* Am Chem Soc. *Res:* Electronic effects in cyclopropanes; synthetic lipid and phospholipid, organo-phosphorus and organo-sulfur chemistry; radiochemical synthesis. *Mailing Add:* Procter & Gamble Co Miami Valley Labs Box 39175 Cincinnati OH 45247

THOMPSON, JAMES JARRARD, b Des Moines, Iowa; m 69; c 3. IMMUNOLOGY. *Educ:* Univ Iowa, BA, 65, MS, 68, PhD(microbiol), 70. *Prof Exp:* Instr microbiol & immunol, Dept Microbiol, Univ Iowa, 70-71; instr, Temple Univ, 71-72, asst prof, 72-74; actg head, 81-83, from asst prof to assoc prof, 74-90, PROF, DEPT MICROBIOL, LA STATE, UNIV MED CTR, NEW ORLEANS, 90- *Concurrent Pos:* Prin investr, Nat Sci Found, 75-77, Am Heart Asn, 77-78, Arthritis Found, 78-79, NIH, 80- *Mem:* Am Asn Immmunologists; Am Soc Microbiol; AAAS; Sigma Xi; Am Asn Univ Prof. *Res:* Immmunoassay of apolipoproteins; structure and function of apolipoproteins; humoral immune responses in periodontal diseases; mechanisms of complement activation; humoral mediators of adaptive host responses. *Mailing Add:* Dept Microbiol & Immunol La State Univ Med Ctr 1901 Perdido St New Orleans LA 70112-1393

THOMPSON, JAMES JOSEPH, b Waterbury, Conn, Oct 22, 40; m 63; c 3. HEALTH PHYSICS, INDUSTRIAL HYGIENE. *Educ:* Univ NMex, BA, 62; Purdue Univ, MS, 70, PhD(bionucleonics), 72; Am Bd Health Physics, cert, 76; Am Bd Indust Hyg, cert, 78; Bd Cert Safety Prof, cert, 78. *Prof Exp:* HEALTH PHYSICIST & INDUST HYGIENIST, LOVELACE INHALATION TOXICOL RES INST, 72- *Mem:* Health Physics Soc; Am Indust Hyg Asn; Am Nuclear Soc; Nat Fire Protection Asn; Int Radiation Protection Soc. *Res:* Personnel dosimetry, thermoluminescence, applied radiation protection and industrial hygiene; hazardous waste management. *Mailing Add:* Dept Microbiol La State Univ Med Ctr 1901 Perdido St New Orleans LA 70112

THOMPSON, JAMES LOWRY, b Syracuse, NY, Oct 5, 40; m 63; c 2. APPLIED MATHEMATICS, ENGINEERING SCIENCE. *Educ:* Brown Univ, AB, 62; Johns Hopkins Univ, PhD(mech), 68. *Prof Exp:* Asst prof math & eng sci, State Univ NY Buffalo, 68-74; mech engr, 74-82, CHIEF, SURVIVAL TECHNOL BR, US ARMY TANK AUTOMOTIVE COMMAND, 82- *Concurrent Pos:* NSF res grant, State Univ NY Buffalo, 71-73. *Mem:* AAAS; Am Math Soc; Soc Natural Philos; Soc Indust & Appl Math; Asn Comput Mach. *Res:* Analysis and optimization of complex systems; continuum mechanics. *Mailing Add:* 1448 Anita Ave Grosse Pointe Woods MI 48236

THOMPSON, JAMES MARION, b Findlay, Ohio, July 26, 26; m 53; c 3. PLANT BREEDING, PLANT GENETICS. *Educ:* Ohio State Univ, BSc, 50, MS, 54, PhD(agron), 63. *Prof Exp:* Fel corn breeding, Agron Dept, Ohio State Univ, 54-56; corn breeder, Steckley Hybrid Corn Co, 57-62; apple breeder, Blairsville, Ga, 63-70, APPLE & PLUM BREEDER, AGR RES SERV, USDA, BYRON, GA, 70- *Mem:* Am Genetic Asn; Am Pomol Soc; Am Soc Hort Sci; Apple Breeders Coop. *Res:* Apple breeding project designed to develop new varieties adapted in the Southern Coastal Plain and in the Southern Appalachian Mountains; plum breeding project; genetic research in pears, apples and plums. *Mailing Add:* Rte 2 Box 2993 Blairsville GA 30512

THOMPSON, JAMES NEAL, JR, b Lubbock, Tex, May 24, 46. GENETICS. *Educ:* Univ Okla, BS, 68, BA, 68; Univ Cambridge, PhD(genetics), 73. *Prof Exp:* Fel genetics, Univ Cambridge, 73-75; asst prof, 75-79, ASSOC PROF ZOOL, UNIV OKLA, 79-, CHMN DEPT ZOOL, 84- *Concurrent Pos:* Marshall scholar, Univ Cambridge, 70-73. *Mem:* Genetics Soc Am; Genetical Soc Gt Brit; Sigma Xi; Soc Study Evolution. *Res:* Development and genetics of quantitative characters; genetic determination of patterns; hybrid dysgenesis and mutator genes in natural populations. *Mailing Add:* Dept Zool Univ Okla 730 Van Vleet Oval Norman OK 73019

THOMPSON, JAMES R, b Charlotte, NC, Sept 14, 42. SOLID STATE PHYSICS, HIGH TEMPERATURE SUPERCONDUCTIVITY. *Educ:* Davidson Col, BS, 64; Duke Univ, PhD(physics), 69. *Prof Exp:* From asst prof to prof physics, Univ Tenn, Dept Physics, 71-88; RES & DEVELOP SCIENTIST, SOLID STATE DIV, OAKRIDGE NAT LAB, 78- *Mem:* Am Phys Soc; Sigma Xi; Mat Res Soc. *Res:* Experimental aspects of high temperature super conductors. *Mailing Add:* Dept Phys Univ Tenn Knoxville TN 37916

THOMPSON, JAMES ROBERT, b Memphis, Tenn, June 18, 38; m 67. MATHEMATICS, STATISTICS. *Educ:* Vanderbilt Univ, BE, 60; Princeton Univ, MA, 63, PhD(math), 65. *Prof Exp:* Asst prof, Vanderbilt Univ, 64-67; asst prof math, Ind Univ, Bloomington, 67-70; assoc prof, 70-77, PROF MATH SCI, RICE UNIV, 77-, PROF STATIST, 87- *Concurrent Pos:* Adj prof, Univ Tex, M D Anderson Cancer Ctr, 77-, Baylor Col Med, 88-, Univ Tex, Sch Pub Health, 90- *Mem:* Am Math Soc; fel Inst Math Statist; fel Am Statist Asn; fel Int Statist Inst. *Res:* Biomathematics; modelling. *Mailing Add:* Dept Statist Rice Univ Houston TX 77251-1892

THOMPSON, JEFFERY SCOTT, b Hartford, Conn, Mar 20, 52; m 74. INORGANIC CHEMISTRY. *Educ:* Trinity Col, BS, 74; Northwestern Univ, PhD(chem), 79. *Prof Exp:* Fel biochem, Med Sch, Harvard Univ, 78-80; CHEMIST, E I DU PONT DE NEMOURS & CO INC, 80- *Mem:* Am Chem Soc. *Res:* Investigation of the role of metal ions in biological processes, through the study of both native and model systems. *Mailing Add:* E I Du Pont de Nemours & Co Inc Exp Sta-E328-331B Wilmington DE 19880-0328

THOMPSON, JEFFREY MICHAEL, b Eau Claire, Wis, May 10, 50; m 84. BIOCHEMISTRY, ANIMAL PHYSIOLOGY. *Educ:* Mich State Univ, BS, 72; Fla State Univ, PhD(molecular biophys), 76. *Prof Exp:* Staff fel res, Nat Heart, Lung & Blood Inst, 77-78; staff fel res, 79-81, sr staff fel res, Nat Inst Aging, 81-82; asst prof dept anat sci & Col Med, Univ Ill, 82-88; ASST PROF DEPT BIOL, CALIF STATE UNIV, SAN BERNARDINO, 88- *Mem:* AAAS; Fedn Am Scientists; Int Soc Develop Neurosci; Soc Neurosci. *Res:* Mechanisms and specificity of synapse formation and their development patterns; synapse formation of isolated cells in culture detected by electrophysiological recording; neurochemical correlates of synapse behavior. *Mailing Add:* Dept Biol Calif State Univ San Bernardino CA 92407

THOMPSON, JERRY NELSON, b Cincinnati, Ohio, Apr 2, 39; m 65; c 2. GENETICS, BIOCHEMISTRY. *Educ:* Univ Cincinnati, BS, 64; Ind Univ, PhD(med genetics), 70. *Prof Exp:* Res asst teratology, Cincinnati Children's Hosp Res Found, 61-64; asst prof, 72-77, ASSOC PROF BIOCHEM & PEDIAT, MED CTR, UNIV ALA, BIRMINGHAM, 77- *Concurrent Pos:* USPHS fel, Univ Chicago, 70-72; Nat Found March Dimes Basil O'Connor starter res grant, Med Ctr, Univ Ala, Birmingham, 74-76. *Mem:* Am Soc Human Genetics; Am Soc Biol Chemists. *Res:* Biochemical and genetic studies of genetic lysosomal storage diseases. *Mailing Add:* Dept Med Genetics Univ Ala Sch Med Univ Sta Birmingham AL 35294

THOMPSON, JESSE CLAY, JR, b Hot Springs, Va, Sept 17, 26; m 50; c 3. SYSTEMATICS. *Educ:* Hampden-Sydney Col, BS, 49; Univ Va, PhD, 56. *Prof Exp:* From asst prof to assoc prof biol, Hollins Col, 55-63; prof & chmn dept, Hampden-Sydney Col, 63-67; prof, Queens Col, NC, 67-69; PROF & CHMN DEPT, ROANOKE COL, 69- *Concurrent Pos:* Mem, Va Inst Marine Sci, 61, Int Indian Ocean Exped, 63, Mt Lake Biol Sta, 65 & Eniwetok Marine Biol Lab, 66; res partic, Palmer Sta, Antarctica, 68-69. *Mem:* AAAS; Am Inst Biol Sci; Soc Protozool. *Res:* Morphology; systematics and geographical distribution of hymenostome ciliated protozoa. *Mailing Add:* Dept Biol Roanoke Col Salem VA 24153

THOMPSON, JESSE ELDON, b Laredo, Tex, Apr 7, 19; m 44; c 4. SURGERY, VASCULAR SURGERY. *Educ:* Univ Tex, BA, 39; Harvard Univ, MD, 43. *Prof Exp:* Instr surg, Boston Univ, 51-54; from asst prof to assoc prof, 54-68, CLIN PROF SURG, UNIV TEX HEALTH SCI CTR, DALLAS, 68-; CHIEF CONSULT PERIPHERAL VASCULAR SURG, BAYLOR UNIV MED CTR, DALLAS, 68-, CHIEF SURG, 82- *Concurrent Pos:* Rhodes scholar physiol & Fulbright fel, Oxford Univ, 49-50; chief surg & vascular surg, Baylor Univ Med Ctr, Dallas, Tex, 82-86. *Mem:* Int Soc Surg; Soc Vascular Surg; Am Surg Asn; Am Col Surgeons; AMA; Int Soc Cardiovasc Surg; Southern Surg Asn; Sigma Xi. *Res:* Vascular surgery; clinical investigation of hypertension, gastric physiology, peripheral vascular diseases and strokes; surgical management of vascular diseases. *Mailing Add:* 3600 Gaston Ave Dallas TX 75246

THOMPSON, JILL CHARLOTTE, b Detroit, Mich, Aug 18, 54. LASER SPECTROSCOPY, PARTICLE LEVITATION. *Educ:* NMex State Univ, BS, 80. *Prof Exp:* Biochemist, 80-81, PHYSICIST, ATMOSPHERIC SCI LAB, NMEX STATE UNIV, 81- *Concurrent Pos:* Geophysicist, GeoVann, 82-83; physicist, LaSen, 84-87 & PetroLaser, 87- *Mem:* Optical Soc Am; Am Meteorol Soc; Am Inst Biol Sci; Nat Orgn Female Exec. *Res:* Ultraviolet fluorescence spectroscopy; remote sensing, lidar and imaging; Raman spectroscopy; atmospheric physics and modeling; non-linear, 2-photon, research and particle characterization; ion spectroscopy. *Mailing Add:* PO Box 330 Organ NM 88052

THOMPSON, JOE DAVID, b Columbus, Ind, Oct 28, 47; m 67; c 2. SOLID STATE PHYSICS, LOW TEMPERATURE PHYSICS. *Educ:* Purdue Univ, BS, 69; Univ Cincinnati, MS, 71, PhD(physics), 75. *Prof Exp:* Fel, 75-77, staff mem condensed matter & thermal physics, 77-89, STAFF MEM, AFFIL CTR MAT SCI, LOS ALAMOS NAT LAB, 84-, DEP GROUP LEADER CONDENSED MATTER & THERMAL PHYSICS, 89- *Honors & Awards:* Japan Soc for the Promotion of Sci Award, 90. *Mem:* Am Phys Soc; AAAS; Sigma Xi. *Res:* Superconducting and magnetic materials; high pressure physics, heavy electron physics and correlated electron behavior. *Mailing Add:* 4873 Trinity Dr Los Alamos NM 87544

THOMPSON, JOE FLOYD, b Grenada, Miss, Apr 13, 39; m; c 2. COMPUTATIONAL FLUID DYNAMICS. *Educ:* Miss State Univ, BS, 61, MS, 63; Ga Inst Technol, PhD(aerospace eng), 71. *Prof Exp:* Aerospace engr, NASA Marshall Space Flight Ctr, 63-64; from asst prof to prof, 64-88, DISTINGUISHED PROF AEROSPACE ENG, MISS STATE UNIV, 88 - *Concurrent Pos:* consult, Numerical Grid Generation; assoc ed, Numerical Heat Transfer & sr assoc ed, Appl Math & Computation. *Honors & Awards:* Res Award, Am Soc Eng Educ, 75. *Mem:* Am Inst Aeronaut & Astronaut; Soc Indust & Appl Math; Inst Elec & Electronics Engrs. *Res:* Numerical grid generation; computational fluid dynamics. *Mailing Add:* NSF Eng Res Ctr Box 6176 Miss State Univ Mississippi State MS 39762

THOMPSON, JOHN ALEC, b Newton, Mass, Nov 27, 42. DRUG METABOLISM, TOXICOLOGY. *Educ:* Clark Univ, BA, 64; Univ Calif, Los Angeles, PhD(org chem), 69. *Prof Exp:* Fel chem, Univ Calif, Irvine, 69-70 & Syntex Corp, Mexico City, 70-71; res chemist, pharmacol, Vet Admin Hosp, Minneapolis, 71-73; res assoc pharmacol, Univ Colo Med Ctr, 73-76; asst prof, 77-80, ASSOC PROF MED CHEM, SCH PHARM, UNIV COLO, 80- *Concurrent Pos:* Res grants, NIH, 79-85, 86-89; Coun for Tobacco Res, 79-86; Chemex Pharmaceut Res Fel, 86-88. *Mem:* Am Chem Soc; Am Soc Mass Spectrometry; Am Soc Pharmacol & Exp Therapeut; Soc Toxicol. *Res:* Chemical and biochemical aspects of the metabolism of drugs and other xenobiotics; application of gas chromatographic/mass spectrometric techniques to studies in pharmacology and toxicology; studies of enzyme mechanisms. *Mailing Add:* Sch Pharm Univ Colo Campus Box 297 Boulder CO 80309-0297

THOMPSON, JOHN C, JR, b Thomas, WVa, Oct 4, 30; m 54; c 3. ENVIRONMENTAL HEALTH. *Educ:* Va Polytech Inst, BS, 51, MS, 58; Cornell Univ, PhD(agr econ), 62. *Prof Exp:* Res assoc phys biol, 61-65, asst prof environ radiation biol, 65-68, ASSOC PROF ENVIRON RADIATION BIOL, CORNELL UNIV, 68- *Concurrent Pos:* Instr, NY State Drinking Driver Prog. *Mem:* Health Physics Soc. *Res:* Radioactive contamination of the food chain, sampling techniques, controlled human studies, radionuclide deposition and cycling, world wide evaluation of fallout; biological costs of energy production; comparative environmental analyses and energy options; animal health and veterinary economics. *Mailing Add:* 792 Ridge Rd Lansing NY 14882

THOMPSON, JOHN CARL, b Toronto, Ont, Nov 28, 41; m 68; c 3. CIVIL ENGINEERING, APPLIED MECHANICS. *Educ:* Univ Toronto, BSc, 63, Univ Ill, MS, 65, PhD(civil eng), 69. *Prof Exp:* Res asst prof, 69-70, asst prof, 70-76, ASSOC PROF CIVIL ENG, UNIV WATERLOO, 76- *Mem:* Soc Exp Mech; Can Soc Civil Eng. *Res:* Optimization of experimental and numerical stress analysis techniques for stress concentration regions with or without cracks; forensic engineering. *Mailing Add:* Dept Civil Eng Univ Waterloo Waterloo ON N2L 3G1 Can

THOMPSON, JOHN DARRELL, b Mitchell, SDak, Sept 13, 33; m 57; c 4. PHYSICS, MOLECULAR BIOPHYSICS. *Educ:* Augustana Col, SDak, BA, 55; Iowa State Univ, MS, 62; Univ Wis, PhD(biophys), 67. *Prof Exp:* From instr to assoc prof, 57-78, PROF PHYSICS, AUGUSTANA COL, S DAK, 78- *Mem:* AAAS; Am Asn Physics Teachers. *Res:* Structure and function of Escherichia coli ribosomes; hormonal control of protein synthesis in the chick enbryo; microcomputers in the laboratory. *Mailing Add:* Dept Physics Augustana Col 29th St & S Summit Ave Sioux Falls SD 57197

THOMPSON, JOHN EVELEIGH, b Toronto, Ont, May 30, 41; m 65. BIOCHEMISTRY, PLANT PHYSIOLOGY. *Educ:* Univ Toronto, BSA, 63; Univ Alta, PhD(plant biochem), 66. *Prof Exp:* Fel med biochem, Univ Birmingham, 66-67; asst prof biol, 68-72, assoc prof, 72-77, PROF BIOL, UNIV WATERLOO, 77- *Mem:* Can Soc Plant Physiol; Am Soc Plant Physiol. *Res:* Membrane biochemistry and molecular biology; the molecular basis of membrane deterioration in aging tissues; plant membrane-hormone interactions; comparative aspects of senescence and stress including the role of hormones and the involvement of free radicals; molecular cloning of genes involved in senescence and stress. *Mailing Add:* Dept Biol Univ Waterloo Waterloo ON N2L 3G1 Can

THOMPSON, JOHN FANNING, b Ithaca, NY, May 24, 19; m 43; c 5. PLANT BIOCHEMISTRY. *Educ:* Oberlin Col, AB, 40; Cornell Univ, PhD(biochem), 44. *Prof Exp:* Instr biochem, Cornell Univ, 44-45; res assoc bot, Univ Chicago, 46-47; NIH fel, Univ Rochester, 47-49, res assoc, 49-50; from instr to asst prof bot, 50-55, ASSOC PROF BOT, CORNELL UNIV, 55-; PLANT PHYSIOLOGIST, PLANT, SOIL & NUTRIT LAB, USDA, 52- *Concurrent Pos:* NSF sr fel, 59-60. *Mem:* Am Soc Plant Physiol; Am Chem Soc; Am Soc Biol Chem; Sigma Xi; AAAS. *Res:* Nitrogen and sulfur metabolism and mineral nutrition of plants; chromatographic techniques; control mechanisms; seed storage proteins. *Mailing Add:* US Plant Soil & Nutrit Lab USDA Tower Rd Ithaca NY 14853

THOMPSON, JOHN FREDERICK, b Los Angeles, Calif, Mar 19, 47; m 80; c 1. FORENSIC PHARMACOLOGY, MEDICAL TOXICOLOGY. *Educ:* Calif State Univ, Los Angles, BSc, 69; Univ Southern Calif, PharmD, 73. *Prof Exp:* Resident clin pharmacol, Wadsworth Vet Admin Med Ctr, Los Angeles, 73-74; clin pharmacist, Los Angeles County-Univ Southern Calif Med Ctr, 74-75; asst clin prof, Med Sch, Loma Linda Univ, 80-83; ASST PROF CLIN PHARMACOL, UNIV SOUTHERN CALIF, 75- *Concurrent Pos:* Pres, Pharmanalysis Assoc, Inc, 78-; prin investr, Univ Southern Calif, 80-83; consult, Peer Standards Rev Orgn, Pasadena, 80-83. *Honors & Awards:* George F Archambalt, Am Soc Consult Pharm. *Mem:* Am Col Clin Pharmacol; Royal Soc Med. *Res:* Detection and analysis of adverse drug reactions and interactions; prescribing habits of physicians: analysis, and the appropriateness of drug therapy prescribing. *Mailing Add:* Sch Pharm Univ Southern Calif 1985 Zonal Ave Los Angeles CA 90033

THOMPSON, JOHN HAROLD, JR, b Amsden, Ohio, Jan 2, 21; m 43; c 3. PARASITOLOGY. *Educ:* Heidelberg Col, AB, 43; Ohio State Univ, MS, 48; Univ Minn, PhD, 52. *Prof Exp:* From instr path to asst prof clin path, 52-70, assoc prof clin path, 70-77, ASSOC PROF LAB MED, MAYO MED SCH, UNIV MINN, 77- *Concurrent Pos:* Consult, Mayo Clin, 52- *Mem:* AAAS; Am Soc Parasitol; Wildlife Dis Asn; Am Fedn Clin Res; Sigma Xi. *Res:* Blood coagulation. *Mailing Add:* Dept Lab Med Hilton 4 Mayo Found Mayo Grad Sch Med Second Ave S Rochester MN 55901

THOMPSON, JOHN LESLIE, b New Castle, Pa, July 11, 17; m 42; c 2. ENVIRONMENTAL SCIENCES. *Educ:* Slippery Rock State Teachers Col, BS, 40; Univ Wis, MS, 48, PhD(geog), 56. *Prof Exp:* Teacher gen sci, Sharpsville High Sch, Pa, 40-41; from asst prof to assoc prof geog, Miami Univ, 49-61, prof, 61-84; RETIRED. *Mem:* Asn Am Geog; Am Geog Soc; Nat Coun Geog Educ; Conserv Educ Asn. *Res:* Social aspects of environmental problems. *Mailing Add:* 6073 Conteras Oxford OH 45056

THOMPSON, JOHN MICHAEL, b Sacramento, Calif; m 72; c 3. GEOTHERMAL GEOCHEMISTRY, WATER CHEMISTRY. *Educ:* Univ Calif, Davis, BS, 68; Sacramento State Col, MS, 74. *Prof Exp:* Phys scientist, 71-74, chemist, 74-88, RES CHEMIST, US GEOL SURV, 88- *Concurrent Pos:* Agency Indust Sci & Technol fel, Japan, 89. *Mem:* AAAS; Am Geophys Union; Geothermal Resources Coun. *Res:* Chemical investigations of hot springs and geothermal well waters; active hydrothermal ore formations and effects of hydrothermal fluids in fresh water streams. *Mailing Add:* US Geol Surv MS 910 345 Middlefield Rd Menlo Park CA 94025

THOMPSON, JOHN N, b Pittsburgh, Pa, Nov 15, 51; m 73. EVOLUTIONARY ECOLOGY, COEVOLUTION. *Educ:* Wash & Jefferson Col, BA, 73; Univ Ill, Urbana, PhD(ecol), 77. *Prof Exp:* Vis asst prof entom, Univ Ill, 77-78; from asst prof to assoc prof, 78-87, PROF BOT & ZOOL, WASH STATE UNIV, 87- *Concurrent Pos:* Fulbright sr scholar, Australia, 91-92. *Mem:* Fel AAAS; Brit Ecol Soc; Ecol Soc Am; Soc Study Evolution; Am Soc Naturalists. *Res:* Coevolution of animals and plants; theory on the evolution of interspecific interactions. *Mailing Add:* Dept Zool & Bot Wash State Univ Pullman WA 99164

THOMPSON, JOHN R, b Beltrami, Minn, Oct 6, 18; m 47; c 4. AGRONOMY. *Educ:* Univ Minn, BS, 48, MS, 52; Iowa State Univ, PhD(crop prod), 64. *Prof Exp:* From instr to assoc prof agron, Univ Minn, 52-67; SUPT, HAWAII AGR EXP STA, UNIV HAWAII, 67-, AGRONOMIST, 74- *Mem:* Am Soc Agron; Corp Sci Soc Am. *Res:* Seed and crop production; plant breeding. *Mailing Add:* Hawaii Exp Station Univ Hawaii Hilo HI 96720

THOMPSON, JOHN ROBERT, b San Francisco, Calif, Dec 15, 51; m 74; c 2. PULSED POWER. *Educ:* Univ Calif, San Diego, BA, 74, PhD, 85. *Prof Exp:* SR STAFF SCIENTIST, MAXWELL LABS INC, 85- *Mem:* Am Inst Physics. *Res:* Research involving opening switch technology used in inductive energy storage based pulsed power systems. *Mailing Add:* 4476 Maryland St San Diego CA 92116

THOMPSON, JOHN S, b Lincoln, Nebr, Oct 29, 28; m 54, 72; c 5. INTERNAL MEDICINE, IMMUNOLOGY. *Educ:* Univ Calif, Berkeley, BA, 49; Univ Chicago, MD, 53. *Prof Exp:* Intern, Univ Chicago Hosps, 53-54; jr asst resident, Presby Hosp, New York, 54-55; sr asst resident med, Univ Chicago Hosps, 57-58, resident, 58-59; from instr to assoc prof, Sch Med, Univ Chicago, 59-69; vchmn vet affairs, Univ Iowa, 69-71, chmn dept, 71-77, prof med, 69-80; prof & chmn med, Univ Ky, 80-90; PROG MED & ASSOC CHIEF STAFF RES, VA MED CTR, LEXINGTON, KY, 90- *Concurrent Pos:* Nat Cancer Inst res fel, 58-60; Lederle med fac award, 66-68; chief med & chief sect allergy & clin immunol, Vet Admin Hosp, Iowa City. *Mem:* Fel Am Col Physicians; fel Am Acad Allergy; Transplantation Soc; Am Asn Immunologists; Sigma Xi. *Res:* HLA- and B-cell typing and genetics; natural monoclonal antibodies for immunotherapy and diagnosis. *Mailing Add:* Dept Med Univ Ky Med Ctr Lexington KY 40536

THOMPSON, JOHN STEWART, b Pittsburgh, Pa, Nov 19, 40. ELECTRICAL ENGINEERING. *Educ:* Lehigh Univ, BS, 62 & 63; Univ Rochester, MS, 65, PhD(elec eng), 67. *Prof Exp:* Mem tech staff signal processing res, 67-80, SUPVR SMALL BUS SYSTS DEVELOP, BELL LABS, 80- *Mem:* Inst Elec & Electronics Engrs. *Res:* Digital signal processing; digital source encoding; computer architecture and programming as applied to signal processing. *Mailing Add:* AT&T Info Systs Rm 30K23 11900 N Pecos St Denver CO 80432

THOMPSON, JON H, b NB, Can, Jan 30, 42. DIFFERENTIAL EQUATIONS & APPLICATIONS. *Educ:* Univ Toronto, PhD(math), 70. *Prof Exp:* From asst prof to assoc prof, 70-81, PROF MATH, UNIV NB, 81- *Mailing Add:* Dept Math Univ NB Fredericton NB E3B 5A3 Can

THOMPSON, JOSEPH GARTH, b Logan, Utah, Aug 15, 35; m 60; c 7. MECHANICAL ENGINEERING, AUTOMATIC CONTROLS. *Educ:* Brigham Young Univ, BES, 60; Purdue Univ, MSME, 62, PhD(mech eng), 67. *Prof Exp:* Design engr, Space Tech Labs, Thompson, Ramo, Wooldridge, Inc, 61-62; instr mech eng, Purdue Univ, 62-66; asst prof, Univ Tex, 66-71; dir, Ctr Res Comput Controlled Automation, 81-89, PROF MECH ENG, KANS STATE UNIV, 71- *Mem:* Simulation Coun; Am Soc Mech Engrs; Nat Soc Prof Engrs; Am Soc Eng Educ; Am Soc Heating, Refrig & Air Conditioning Engrs. *Res:* Modeling, design and compensation of nonlinear dynamic systems; design of regulator and feedback control systems; simulation and optimization of dynamic systems; application of microprocessors to automatic control; feature based design; integrated manufacturing systems. *Mailing Add:* Dept Mech Eng Kans State Univ Manhattan KS 66506

THOMPSON, JOSEPH KYLE, b Columbus, Ohio, Oct 2, 20; m 56; c 3. PHYSICAL INORGANIC CHEMISTRY. *Educ:* Sterling Col, BA, 42; Univ Kans, MA, 49, PhD, 50. *Hon Degrees:* DSc, Sterling Col, 67. *Prof Exp:* Chemist, US Naval Res Lab, 42-46 & 50-86; RETIRED. *Mem:* Am Chem Soc; Sigma Xi. *Res:* Kinetics and mechanisms of adsorption and filtration, particularly air cleaning devices; oxides of alkali and alkaline earth metals; nuclear magnetic resonance. *Mailing Add:* 6106 Baxter Dr Suitland MD 20746

THOMPSON, JOSEPH LIPPARD, b Newport News, Va, May 12, 32; m 55; c 5. RADIOCHEMISTRY, PHYSICAL CHEMISTRY. *Educ:* Va Polytech Inst, BS, 54; Pa State Univ, MS, 59, PhD(chem), 63. *Prof Exp:* Nat Res Coun res asst, Nat Bur Standards, Washington, DC, 63-64; from asst prof to assoc prof, 64-78, prof chem, Idaho State Univ, 78-82; LOS ALAMOS SCI LAB, 82- *Mem:* Am Chem Soc. *Res:* Radionuclide transport in the environment. *Mailing Add:* Los Alamos Sci Lab Inc 11 MS J514 Los Alamos NM 87545

THOMPSON, JULIA ANN, b Little Rock, Ark, Mar 13, 43. ELEMENTARY PARTICLE PHYSICS, HIGH ENERGY PHYSICS. *Educ:* Cornell Col, BA, 64; Yale Univ, MS, 66, PhD(physics), 69. *Prof Exp:* Res assoc physics, Brookhaven Nat Lab, 69-71; res assoc & assoc instr, Univ Utah, 71-72; from asst prof to assoc prof, 72-86, PROF PHYSICS, UNIV PITTSBURGH, 86- *Concurrent Pos:* Nat Acad Sci exchange, Novosibirsk, USSR, 89-90. *Mem:* Am Phys Soc. *Res:* Direct photon and lepton production; application of optoelectronic techniques to high energy physics pattern recognition problems, classification of hadronic jets; strange particle interactions and decays. *Mailing Add:* Dept Physics & Astron Allen Hall 107 Univ Pittsburgh 4200 Fifth Ave Pittsburgh PA 15260

THOMPSON, KENNETH DAVID, b Wimbeldon, NDak, Apr 13, 40; m 65; c 3. MICROBIOLOGY, IMMUNOLOGY. *Educ:* Univ NDak, BS, 63; MS, 67, PhD(microbiol), 70. *Prof Exp:* NIH fel, Temple Univ, 70-72, instr, 72-73, asst prof microbiol & immunol, Health Sci Ctr, 73-78; ASSOC PROF PATH & MICROBIOL, MED CTR, LOYOLA UNIV, 78- *Mem:* Am Soc Microbiol; NY Acad Sci. *Res:* Tumor immunology and clinical immunology. *Mailing Add:* Dept Path Loyola Univ Stritch Sch Med 2160 S First Maywood IL 60153

THOMPSON, KENNETH LANE, b New Orleans, La, Feb 4, 43; m 67; c 1. OPERATING SYSTEMS, NETWORKS. *Educ:* Univ Calif, Berkeley, BS, 65, MS, 66. *Prof Exp:* MEM STAFF, BELL LABS, 66- *Concurrent Pos:* Vis Mackay lectr comput sci, Univ Calif, Berkeley, 75-76. *Honors & Awards:* Piorie Award, Inst Elec & Electronics Engrs, 82 & Hamming Award. *Mem:* Nat Acad Sci; Nat Acad Eng; Asn Comput Mach. *Res:* Operating systems for telephone switching; computer chess. *Mailing Add:* Bell Labs Rm 2C519 Murray Hill NJ 07974

THOMPSON, KENNETH O(RVAL), b Fielding, Sask, Sept 12, 17; US citizen; m 78; c 2. MECHANICAL ENGINEERING, AERONAUTICAL ENGINEERING. *Educ:* Univ Minn, BAeroE & BBusAdmin, 53, MSAE, 58; Univ Ala, PhD(mech eng), 67. *Prof Exp:* Prin engr, Univ Minn, 53-62; res assoc, 62-67, ASSOC PROF ENG, RES INST, UNIV ALA, HUNTSVILLE, 67-, DIR, INST & RES SUPPORT SERV, 80-, ASSOC DEAN. *Mem:* Am Astronaut Soc; Am Inst Aeronaut & Astronaut; Am Soc Eng Sci. *Res:* Supersonic and hypersonic aerodynamics; gas dynamics; digital computers; design and operation of supersonic and hypersonic inlets, wind-tunnels and other research facilities. *Mailing Add:* Dean's Off Col Eng Univ Ala Huntsville AL 35899

THOMPSON, LANCELOT CHURCHILL ADALBERT, b Jamaica, West Indies, Mar 3, 25; US citizen; m 52; c 2. INORGANIC CHEMISTRY. *Educ:* Morgan State Col, BS, 52; Wayne State Univ, PhD(inorg chem), 56. *Prof Exp:* Instr chem, Wolmers Boys Sch, Jamaica, 55-56; Int Nickel Co fel, Pa State Univ, 57; from asst prof to assoc prof inorg chem, 58-66, asst dean col arts & sci, 64-66, PROF CHEM & DEAN STUDENT SERV, UNIV TOLEDO, 66-, VPRES STUDENT AFFAIRS, 68- *Concurrent Pos:* Consult, Owens-Ill Glass Co, Ohio, 62-64. *Mem:* AAAS; Am Chem Soc. *Res:* Determination of structure of coordination compounds; coordination polymers; solubility of hydrous oxides. *Mailing Add:* Dept Chem Univ Toledo Toledo OH 43606

THOMPSON, LARRY CLARK, b Hoquiam, Wash, June 13, 35; m 55; c 2. INORGANIC CHEMISTRY. *Educ:* Willamette Univ, BS, 57; Univ Ill, MS, 59, PhD(inorg chem), 60. *Prof Exp:* From asst prof to assoc prof, 60-68, head dept, 72-84, PROF CHEM, UNIV MINN, DULUTH, 68- *Concurrent Pos:* Vis prof, Univ Sao Paulo, 69, Fed Univ Ceara, 73 & 74, Fed Univ Pernambuco, 77. *Mem:* Am Chem Soc; Sigma Xi. *Res:* Coordination chemistry of the rare earth elements; high coordination numbers; ligands with unusual steric requirements. *Mailing Add:* Dept Chem Univ Minn Duluth MN 55812

THOMPSON, LARRY DEAN, b Warren, Ohio, Oct 16, 51; m 73; c 2. PHYSICAL METALLURGY, MATERIALS SCIENCE. *Educ:* Youngstown State Univ, BE, 73; Univ Calif, Berkeley, MS, 76, PhD(mat sci & eng), 78. *Prof Exp:* Res asst mat sci, Lawrence Berkeley Lab, 73-77; sr scientist mat sci, Gen Atomic Co, 77-81; PRES, PSI MET, 81- *Concurrent Pos:* Lectr, San Diego State Univ, 81- *Honors & Awards:* Achievement Award, Am Soc Metals, 77. *Mem:* Am Soc Metals; Am Inst Mining, Metall & Petrol Engrs. *Res:* Structural instability of high-temperature alloys and superalloys; high-temperature gaseous corrosion of metals; phase transformations in austenitic stainless steels; alloy design of stainless steels, fracture/mechanical properties of structural materials. *Mailing Add:* Dept Mech Eng San Diego State Univ 5300 Chapanile Dr San Diego CA 92182

THOMPSON, LARRY FLACK, b Union City, Tenn, Aug 31, 44; m 64; c 2. POLYMER CHEMISTRY. *Educ:* Tenn Technol Univ, BS, 66, MS, 68; Univ Mo-Rolla, PhD(chem), 71. *Prof Exp:* Mem tech staff chem & thin films, 70-78, HEAD ORG MAT & CHEM ENG, BELL LABS, 78- *Concurrent Pos:* Guest prof, Rutgers Univ, 72. *Mem:* Nat Acad Eng; Am Chem Soc; AAAS; Inst Elec & Electronics Engrs; Soc Photo-Optical Instrumentation Engrs; Am Inst Chem Engrs. *Res:* Electron beam polymer resist studies for microfabrication of integrated electronics; thin polymer films for use in microelectronic fabrication; materials and processes for optical fiber fabrication. *Mailing Add:* 6C302 AT&T Bell Labs 600 Mountain Ave Murray Hill NJ 07974

THOMPSON, LAWRENCE HADLEY, b Tyler, Tex, July 22, 41; div; c 2. DNA REPAIR, MUTAGENESIS. *Educ:* Univ Tex, Austin, BS, 63, MS, 67, PhD(biophys), 69. *Prof Exp:* Fel cell biol, Ont Cancer Inst, 69-71, staff physicist, 71-73; sr biomed scientist, Lawrence Livermore Nat Lab, 73-90, ADJ PROF, DEPT RADIATION ONCOL, SCH MED, UNIV CALIF, SAN FRANCISCO, 90- *Concurrent Pos:* Councilor, Environ Mutagen Soc, 85-88. *Mem:* Environ Mutagen Soc; AAAS. *Res:* Study mechanisms of somatic cell mutation and DNA repair through the cloning and analysis of mammalian DNA repair genes using cultured cells; develop DNA-repair deficient transgenic mice. *Mailing Add:* Biomed Sci Div Lawrence Livermore Nat Lab PO Box 5507 Livermore CA 94550

THOMPSON, LEE P(RICE), b Pastura, NMex, June 29, 13; m 36; c 4. MECHANICS. *Educ:* Ind Univ, BA, 36; Agr & Mech Col, Tex, MS, 38, PhD(eng), 49. *Prof Exp:* Asst, Agr & Mech Col, Tex, 36-38, from instr to prof mech eng, 38-55; PROF ENG, DEAN COL ENG & APPL SCI & DIR SCH ENG, ARIZ STATE UNIV, 55- *Concurrent Pos:* Mgr res & testing lab, AiResearch Corp, 44-46; partic, Am Soc Eng Educ-NSF vis engr prog. *Mem:* Am Soc Eng Educ; Am Inst Aeronaut & Astronaut. *Res:* Applied mechanics; aircraft cabin pressure systems; aircraft electronic equipment cooling research; vibrations; heat and mass transfer by electrical analogy; math studies for application of computers to solution of missile-satellite problems. *Mailing Add:* Dept Eng Ariz State Univ Tempe AZ 85287

THOMPSON, LEIF HARRY, b Chadron, Nebr, Dec 6, 43; c 1. REPRODUCTIVE PHYSIOLOGY. *Educ:* Univ Nebr, BS, 67; NC State Univ, MS, 70, PhD(animal sci), 72. *Prof Exp:* Asst prof reproductive physiol, Tex Tech Univ, 72-77; EXTEN SPECIALIST ANIMAL PHYSIOL, UNIV ILL, 78- *Mem:* Am Soc Animal Sci; Soc Study Fertil; Soc Study Reproduction. *Res:* Influence of environment, nutrition, development, hormonal therapy and selection on reproductive efficiency in swine, beef cattle and sheep and hormonal regulation of growth of feedlot animals. *Mailing Add:* Dept Animal Sci 319 Mumford House Univ Ill Urbana IL 61801

THOMPSON, LEITH STANLEY, b Margate, PEI, Aug 22, 34; m 58; c 4. ENTOMOLOGY, PLANT PATHOLOGY. *Educ:* McGill Univ, BSc, 56; Cornell Univ, PhD(entom), 61. *Prof Exp:* RES SCIENTIST ENTOM, AGR CAN, 56-, ASST DIR, SECT ENTOM & ADMIN, 77- *Mem:* Entom Soc Can; Entom Soc Am. *Res:* Forage insect studies; biological control of insects; potato insect studies. *Mailing Add:* Suffolk RR 3 Charlottetown PE C1A 7J7 Can

THOMPSON, LELAND E, engineering, for more information see previous edition

THOMPSON, LEWIS CHISHOLM, b Brechenridge, Tex, Jan 18, 26; m 55; c 2. NUCLEAR PHYSICS. *Educ:* Rice Univ, BA, 50, MA, 52, PhD(physics), 54. *Prof Exp:* Physicist, Naval Res Lab, 54-56; sr nuclear engr, Gen Dynamics/Convair, 56-59; asst prof physics, Univ Ga, 59-63; assoc prof, La Sierra Col, 65-70; prof, Loma Linda Univ, 70-77; PROF PHYSICS, OAKWOOD COL, 77- *Mem:* Am Asn Physics Teachers. *Res:* Energy levels of light nuclei; nuclear instruments; nuclear shielding; low energy particle accelerators. *Mailing Add:* 322 Farmstead Rd Huntsville AL 35806

THOMPSON, LILIAN UMALE, b Cavinti, Philippines; m 68; c 2. ANTINUTRIENTS, PROTEINS AND STARCH. *Educ:* Mapua Instit Technol, BSc, 60; Univ Philippines, MSc, 64; Univ Wis, PhD(food sci), 69. *Prof Exp:* Chemist, Inhelder Lab Inc, 61; res instr, chem, Univ Philippines, 64-65; res asst, Univ Wis, 65-69; from asst prof to assoc prof, 69-88, PROF NUTRIT SCI, UNIV TORONTO, 88- *Concurrent Pos:* Chmn, Nutrit Interest Group, Can Instit Food, 85-86; mem, expert comt plant prods, Can Comt food, Agr, Can, 81-84; chmn, Acad Panel, Ont grad scholarship, Ont Ministry Cols & Univ, 86-87; mem, grant selection Comm, Nat Sci & Eng Res Coun, 86-89. *Mem:* Am Instit Nutrit; Can Soc Nutrit Scis; Am Chem Soc; Am Assoc Cereal Chemists; Instit Food Technologists. *Res:* Bioavailability-digestibility of nutrients; diet and cancer; protein isolation and functionality. *Mailing Add:* Dept Nutrit Scis Fac Med Univ Toronto 150 College St Toronto ON M5S 1A8 Can

THOMPSON, LOREN EDWARD, JR, geophysics, for more information see previous edition

THOMPSON, LOUIS JEAN, b Big Spring, Tex, Apr 26, 25; m 46; c 4. CIVIL ENGINEERING, SOIL MECHANICS. *Educ:* Tex A&M Univ, BSCE, 49, MSCE, 51; Univ Va, DSc(civil eng), 66. *Prof Exp:* Engr, Lockwood & Andrews, Tex, 51-52; partner, Benson-Thompson-Nash, Engrs-Architects, 52-61; asst prof civil eng, Univ NMex, 61-64; assoc prof, 66-80, PROF CIVIL ENG, TEX A&M UNIV, 80- *Concurrent Pos:* Consult, Sandia Corp, NMex, 63- *Mem:* Am Soc Civil Engrs; Am Soc Eng Educ; Int Asn Bridge & Struct Eng; Sigma Xi. *Res:* High rate of deformation of earth materials; earth penetration; wave propagation in soils and rock; earth impact; cratering; drilling; tunnelling and design of earth structures. *Mailing Add:* 1216 Haines College Station TX 77840

THOMPSON, LYELL, b Rock Island, Ill, May 10, 24; m 46; c 5. SOILS. *Educ:* Okla State Univ, BS, 48; Ohio State Univ, PhD(soils), 52. *Prof Exp:* Asst prof, Ohio State Univ, 51-53; soil scientist, Noble Found, Okla, 53-58; assoc prof agron, 58-69, PROF AGRON, UNIV ARK, FAYETTEVILLE, 69- *Mem:* Am Soc Agron; Soil Sci Soc Am; Sigma Xi. *Res:* Effect of soil fertility, trace element availability and soil acidity upon crop production; increase of food production. *Mailing Add:* Dept Agron Univ Ark 115 Plant Sci Fayetteville AR 72703

THOMPSON, LYNNE CHARLES, b St Paul, Minn, Jan 30, 44; m; c 3. FOREST ENTOMOLOGY, BIOLOGICAL CONTROL. *Educ:* Kans State Univ, BS, 70; Univ Minn, MS, 73, PhD(entom), 76. *Prof Exp:* Res assoc forest entom, Univ Minn, 76-77; asst prof, Kansas State Univ, 77-80; ASSOC PROF, DEPT FOREST RESOURCES, UNIV ARK, 80- *Mem:* Entom Soc Am; Ecol Soc Am; Sigma Xi. *Res:* Conduct research on the biology and control of forest insects. *Mailing Add:* Dept Forest Resources Univ Ark Monticello AR 71655

THOMPSON, MAJOR CURT, b Cullman, Ala, May 25, 37; m 62; c 2. INORGANIC CHEMISTRY. *Educ:* Birmingham-Southern Col, BS, 59; Ohio State Univ, MS, 61, PhD(inorg chem), 63. *Prof Exp:* res assoc, Atomic Energy Div, Savannah River Lab, E I du Pont de Nemours & Co, Inc, 63-89; ADV SCIENTIST, WESTINGHOUSE, SAVANNAH RIVER CO, 89- *Mem:* Am Chem Soc; Sigma Xi. *Res:* Synthesis of binary compounds of actinide elements which are stable at high temperature; complexes of the actinides and lanthanides; solvent extraction. *Mailing Add:* 4950 Arroyo Chamisa NE Albuquerque NM 87111-3717

THOMPSON, MALCOLM J, b Baldwin, La, Feb 15, 27; m 53; c 3. ORGANIC CHEMISTRY. *Educ:* Xavier Univ, La, BS, 50, MS, 52. *Prof Exp:* Instr chem, Xavier Univ, La, 52-54; chemist, US Bur Mines, 54-55; org chemist, NIH, 55-60; res org chemist, Chem Warfare Labs, Army Chem Ctr, Md, 60-62; RES CHEMIST, INSECT & NEMATODE HORMONE LAB, USDA, 62- *Honors & Awards:* Hillebrand Award, 87. *Mem:* Fel AAAS; Am Chem Soc; NY Acad Sci. *Res:* Chemistry of steroids, sapogenins and natural products; synthesis and structural elucidations; insect hormones, isolation and structural elucidation of insect molting hormones; feeding stimulants; synthesis of compounds with insect hormonal activity and inhibitors of insect development and reproduction. *Mailing Add:* 3607 Cedardale Rd Baltimore MD 21215-7305

THOMPSON, MARGARET A WILSON, b Northwich, Eng, Jan 7, 20; Can citizen; wid; c 2. HUMAN & MEDICAL GENETICS. *Educ:* Univ Sask, BA, 43; Univ Toronto, PhD(human genetics), 48. *Prof Exp:* Lectr zool, Univ Toronto, 47-48 & Univ Western Ont, 48-50; lectr, Univ Alta, 50-59, asst prof human genetics, 59-62; vis investr, Jackson Lab, 62-63; res assoc pediat & lectr zool, 63-64, asst prof pediat & zool, 64, assoc prof zool, 65-70, assoc prof med cell biol, 69-72, from assoc prof to prof pediat, 66-85, prof med genetics, 73-85, EMER PROF, UNIV TORONTO, 85- *Concurrent Pos:* Muscular Dystrophy Asn Can res fel, 62-63; sr staff geneticist, Hosp for Sick Children, Toronto, 63-88, consult genetics, 88-; mem bd dirs, Am Soc Human Genetics, 75-78; mem bd trustees, Queen Elizabeth II Res Fund for Res Dis Children, 72-; hon res assoc, Dept Human Genetics & Biomet, Univ Col London, 77-78; Saul Lehman vis prof, Downstate Med Ctr, State Univ, NY, 81. *Mem:* Am Soc Human Genetics; Can Col Med Genetics (pres, 83-85); Genetics Soc Can (pres, 72-73); Order Can, 88. *Res:* Human and medical genetics; muscular dystrophy carrier detection; prenatal diagnosis; population agents. *Mailing Add:* Dept Genetics Hosp Sick Children Toronto ON M5G 1X8 Can

THOMPSON, MARGARET DOUGLAS, b Wilmington, Del, May 12, 47; m 83; c 2. GEOLOGY. *Educ:* Smith Col, BA, 69; Harvard Univ, MA, 74, PhD(geol sci), 76. *Prof Exp:* ASSOC PROF GEOL, WELLESLEY COL, 76- *Concurrent Pos:* Brachman-Hoffman fel, Wellesley Col, 81-83. *Mem:* Geol Soc Am; Sigma Xi. *Res:* Analysis of structure and tectonic evolution of the Boston Basin, Massachusetts. *Mailing Add:* Geol Dept Wellesley Col Wellesley MA 02181

THOMPSON, MARSHALL RAY, b Monterey, Ill, July 22, 38; m 60; c 2. CIVIL ENGINEERING. *Educ:* Univ Ill, Urbana, BS, 60, MS, 62, PhD(civil eng), 64. *Prof Exp:* Field engr, McCann & Co, Inc, 57-60; res asst, 60-63, from instr to assoc prof, 63-70, PROF CIVIL ENG, UNIV ILL, URBANA, 70- *Concurrent Pos:* Spec consult, Mil Asst Command, US Navy, Vietnam, 69-70; consult engr, Construct Eng Res Lab, US Army Corps Engrs, Caterpillar Tractor Co & var indust and govt agencies, 72-; Sect J comt rep, chmn lime stabilization comt & mem cement stabilization comt, Hwy Res Bd, Nat Acad Sci-Nat Res Coun. *Honors & Awards:* A W Johnson Mem Award, Hwy Res Bd, 70; Huber Res Prize, Am Soc Civil Engrs, 70. *Mem:* Am Soc Testing & Mat; Am Soc Eng Educ; Am Concrete Inst; Am Soc Civil Engrs; Sigma Xi. *Res:* Soil stabilization; highway materials; surficial soils; pavements. *Mailing Add:* Univ Ill 205 N Mathews Urbana IL 61801

THOMPSON, MARTIN LEROY, b Kindred, NDak, Jan 8, 35; m 63; c 4. INORGANIC CHEMISTRY. *Educ:* Concordia Col, Moorhead, Minn, BA, 56; Ind Univ, PhD(inorg chem), 64. *Prof Exp:* From instr to assoc prof, 62-78, PROF CHEM, LAKE FOREST COL, 79- *Mem:* Am Chem Soc; Sigma Xi. *Res:* Inorganic chemistry of silicon boron and phosphorus compounds; organometallic chemistry. *Mailing Add:* Dept Chem Johnson Sci Bldg Lake Forest Col Lake Forest IL 60045

THOMPSON, MARVIN P, b Troy, NY, June 22, 33; m 53; c 3. BIOCHEMISTRY. *Educ:* Kans State Univ, BS, 56, MS, 57; Mich State Univ, PhD(food sci), 60. *Prof Exp:* Biochemist, 60-71, chief, Milk Properties Lab, 71-74, res chemist, Dairy Lab, 74-80, res leader, Plant Sci Lab, 80-85, lead scientist, Plant & Soil Biophys, LEAD SCIENTIST MILK COMPONENTS, EASTERN REGIONAL RES CTR, 87- *Concurrent Pos:* Prof, Pa State Univ, 65- *Honors & Awards:* Borden Award, 70; Arthur S Flemming Award, 71; Super Serv Award, USDA, 71; Fed Lab Corsortium Award, 87. *Mem:* Am Chem Soc; Am Dairy Sci Asn; Am Soc Biol Chem; AAAS; Am Soc Plant Physiol; Inst Food Technol. *Res:* Isolation and properties of milk proteins; genetic polymorphism of milk proteins; structure of casein micelles; calcium binding proteins. *Mailing Add:* 548 S 49th St Philadelphia PA 19143

THOMPSON, MARVIN PETE, b Mackville, Ky, Sept 28, 41; m 62; c 4. WILDLIFE ECOLOGY, MAMMALOGY. *Educ:* Univ Ky, BS, 63; Kans State Univ, MS, 67; Southern Ill Univ, Carbondale, PhD(zool), 71. *Prof Exp:* From asst prof to assoc prof, 68-80, PROF BIOL, EASTERN KY UNIV, 80-, FAC RES GRANTS, 69- *Mem:* Wildlife Soc; Am Soc Mammal; Nat Wildlife Fedn. *Res:* Woodchuck ecology and physiology; ecology of pest mammals; wildlife restoration. *Mailing Add:* Dept Biol Moore 235 Eastern Ky Univ Richmond KY 40475

THOMPSON, MARY E, b Minneapolis, Minn, Dec 21, 28. PHYSICAL INORGANIC CHEMISTRY. *Educ:* Col St Catherine, BA, 53; Univ Minn, MS, 58; Univ Calif, Berkeley, PhD(chem), 64. *Prof Exp:* Instr sci & math, Derham Hall High Sch, 53-57 & 58-59; res asst chem, Lawrence Radiation Lab, Calif, Berkeley, 61-64; lab instr, Col St Catherine, 53-56, from asst prof to assoc prof, 64-78, chmn dept, 69-90, PROF CHEM, COL ST CATHERINE, 78- *Concurrent Pos:* Consult-Evaluator, N Cent Asn, 84-; Women Chemists Comt, Am Chem Soc, 86- *Mem:* AAAS; Am Chem Soc; The Chem Soc; Sigma Xi; Nat Sci Teachers Asn. *Res:* Hydrolytic polymerization in aqueous solutions; kinetics; magnetic susceptibility of solutions of transition metal polymers. *Mailing Add:* Dept Chem Col St Catherine 2004 Randolph Ave St Paul MN 55105

THOMPSON, MARY ELEANOR, b Cleveland, Ohio, Nov 5, 26. GEOCHEMISTRY. *Educ:* Boston Univ, BA, 48; Harvard Univ, MA, 63, PhD(geol), 64. *Prof Exp:* Mineralogist, US Geol Surv, 48-57; electrode chemist, EPSCO, Inc, Mass, 62-63; res assoc geochem, Dept Geol, Univ SC & electrode chem, Dept Geol & Sch Med, Stanford Univ, 64-67; res scientist & mgr chem limnol, Can Ctr Inland Waters, 67-; RETIRED. *Concurrent Pos:* Co-recipient, NSF grant, Univ SC, 65-66; res assoc, Dept Geol, McMaster Univ, 68-69. *Mem:* Int Asn Hydrol Sci; Geochem Soc; AAAS. *Res:* Low temperature aqueous geochemistry; specific-ion electrodes; chemical limnology; aquatic effects of acid precipitation. *Mailing Add:* 110 Regis Rd Falmouth MA 02536

THOMPSON, MARY ELINORE, b Winnipeg, Man, Sept 9, 44; m 68; c 3. STATISTICS. *Educ:* Univ Toronto, BSc, 65; UniY Ill, MS, 66, PhD(math), 69. *Prof Exp:* From lectr to assoc prof, 69-80, PROF STATIST, UNIV WATERLOO, 80- *Mem:* Am Math Soc; Can Math Soc; Inst Math Statist; Statist Soc Can; fel Am Statist Asn; Int Statist Inst. *Res:* Finite population sampling; probability. *Mailing Add:* Statist & Act Sci Univ Waterloo Waterloo ON N2L 3G1 Can

THOMPSON, MAX CLYDE, b Winfield, Kans, Jan 10, 36. ORNITHOLOGY, ORCHIDOLOGY. *Educ:* Southwestern Col, BA, 57; Univ Kans, MA, 64. *Prof Exp:* Zoologist, Univ Kans, 58 & USPHS, 58-60; ornithologist, Bernice P Bishop Mus, 62-63 & Univ Md, 63; cur res, Smithsonian Inst, 64-67; INSTR BIOL, SOUTHWESTERN COL, 67-; RES ASSOC, UNIV KANS, 70- *Mem:* Am Ornithologists Union; Wilson Ornith Soc; Cooper Ornith Soc; Brit Ornithologists Union; Am Hort Soc; Am Orchid Soc. *Res:* Migration and systematics of birds, particularly shorebirds, of Mexico, Argentina, Africa, Southeast Asia, Australia and the United States. *Mailing Add:* Dept Biol 100 Col Southwestern Col Winfield KS 67156-2499

THOMPSON, MAXINE MARIE, b Bloomington, Ill, Nov 3, 26; m 53; c 2. GENETICS, HORTICULTURE. *Educ:* Univ Calif, BS, 48, MS, 51, PhD(genetics), 60. *Prof Exp:* Jr specialist viticulture, Univ Calif, Davis, 62-63; asst prof biol, Wis State Univ, Oshkosh, 63-64; res assoc hort, 64-67, asst prof bot, 66-68, from asstprof to prof, 68-86, EMER PROF HORT, ORE STATE UNIV, 86- *Mem:* Am Soc Hort Sci. *Res:* Cytological and botanical studies related to horticultural problems, especially horticultural breeding; fruit breeding and genetics. *Mailing Add:* Dept Hort Ore State Univ Corvallis OR 97331

THOMPSON, MAYNARD, b Michigan City, Ind, Sept 8, 36; m 55; c 2. MATHEMATICAL MODELING. *Educ:* DePauw Univ, AB, 58; Univ Wis, MS, 59, PhD(math), 62. *Prof Exp:* Lectr, Ind Univ, 62-64, from asst prof to prof, 64-73, chmn dept, 74-77, assoc dean, Grad Sch, 81-84, assoc dean, 84-88, PROF MATH, IND UNIV, BLOOMINGTON, 73-, DEAN, BUDGETARY ADMIN, 88-, VCHANCELLOR, 90- *Concurrent Pos:* Res assoc, Univ Md, 70-71; sr res scientist, Gen Motors Res Labs, 78. *Mem:* Am Math Soc; Math Asn Am; Soc Indust & Appl Math. *Res:* Approximation theory; complex analysis; mathematical biology. *Mailing Add:* Dept Math Ind Univ Bloomington IN 47405

THOMPSON, MICHAEL BRUCE, b Kansas City, Mo, Aug 25, 39; m 67; c 2. EMBRYOLOGY. *Educ:* Baker Univ, BS, 63; Kans State Univ, MS, 67, PhD(biol), 69. *Prof Exp:* Head dept biol, 70-74, asst prof, 69-76, chmn div sci & math, 74-84, ASSOC PROF BIOL, MINOT STATE COL, 76- *Mem:* Soc Study Reproduction; Sigma Xi. *Res:* Developmental placentation. *Mailing Add:* Div Sci Cyril C Moore Hall Minot State Col Minot ND 58701

THOMPSON, MILTON AVERY, b Salem, Ore, July 5, 29; m 57; c 3. ENVIRONMENTAL MANAGEMENT. *Educ:* San Jose State Col, BA, 51; Ore State Univ, MS, 53, PhD(phys chem), 57. *Prof Exp:* From chemist to sr chemist, Dow Chem Co, 57-61, res supvr, 62-65, sr res mgr, 66-68, dir chem res & develop, 69-70, mgr environ sci, 70-74; mgr environ sci & waste control, Rockwell Int, 75-78; environ scientist, Stearns-Roger Inc, 78-79; environ scientist, Cyprus Mines Corp, 79-80; prin licensing engr & dir mining serv, Harding-Lawson Assoc, 81-82; MGR, BRECKENRIDGE SANIT DIST, 82- *Mem:* Am Chem Soc; Sigma Xi; Am Nuclear Soc. *Res:* Plutonium chemistry; plutonium processing, recovery and corrosion; nonaqueous plutonium chemistry. *Mailing Add:* PO Box 7215 Breckenridge CO 80424

THOMPSON, NANCY LYNN, b Charlotte, NC, Sept 28, 56. MEMBRANE BIOPHYSICS, FLUORESCENCE MICROSCOPY. *Educ:* Guilford Col, BS(physics) & BS(math), 77; Univ Mich, Ann Arbor, MS, 80, PhD(physics), 82. *Prof Exp:* Res fel chem, Stanford Univ, Damon Runyon-Walter Winchell Cancer Fund, 82-85; asst prof, 85-90, ASSOC PROF CHEM, UNIV NC, CHAPEL HILL, 90- *Honors & Awards:* Presidential Young Investr Award, NSF, 86; Margaret Oakley Dayhoff Award, Biophys Soc, 89. *Mem:* Biophys Soc; Am Phys Soc; Am Chem Soc; AAAS. *Res:* Cell membrane biophysics and cell-surface immunology; fluorescence microscopy and spectroscopy. *Mailing Add:* Dept Chem Venable Hall CB No 3290 Univ NC Chapel Hill NC 27599-3290

THOMPSON, NEAL PHILIP, b Brooklyn, NY, July 18, 36; m 58; c 5. PLANT PHYSIOLOGY, PLANT ANATOMY. *Educ:* Wheaton Col, Ill, BS, 57; Miami Univ, Ohio, MA, 62; Princeton Univ, PhD(biol), 65. *Prof Exp:* Asst prof plant physiol & asst plant physiologist, 65-72, assoc prof plant physiol & assoc plant physiologist, 72-77, prof plant physiol & plant physiologist, 77-80, asst dean res, 80-86, ASSOC DEAN RES, UNIV FLA, 87- *Mem:* Am Chem Soc. *Res:* Developmental structure of higher plants; translocation of materials, exogenously applied or endogenous, in higher plants, their effects on anatomical structure and their metabolism; pesticides in the environment, particularly as related to birds and fish. *Mailing Add:* Dept Food Sci & Nutrit Univ Fla Gainesville FL 32611

THOMPSON, NOEL PAGE, b San Francisco, Calif, Oct 22, 29; m 54; c 2. BIOMEDICAL ENGINEERING. *Educ:* Stanford Univ, BA, 51, MS, 61; Univ Calif, Los Angeles, MD, 55. *Prof Exp:* Intern med, Univ Hosps, Univ Wis, 55-56; chief bioeng & physiol div, Palo Alto Med Res Found, 58-73; CHIEF, MED INSTRUMENTATION LAB, PALO ALTO MED CLINIC, 64-; MEM STAFF, PHYSICIAN MED INST, 77- *Concurrent Pos:* Consult assoc prof, Stanford Univ, 61- & Univ Santa Clara, 62-68. *Mem:* AMA; Am Inst Ultrasonics in Med; Am Acad Family Physicians; sr mem Inst Elec & Electronics Engrs. *Res:* Theoretical and applied biomedical engineering; mathematics and electronic instruments as applied to research and in the practice of medicine. *Mailing Add:* Physician Med Inst Palo Alto Clin 300 Homer Ave Palo Alto CA 94301

THOMPSON, NORMAN STORM, b Ft William, Ont, Nov 10, 23; m 51; c 4. CHEMISTRY OF RADICALS, POLYSACCHARIDE CHEMISTRY. *Educ:* Univ Man, BSc, 50, MSc, 52; McGill Univ, PhD(wood chem), 54. *Hon Degrees:* MSc, Lawrence Univ, 87. *Prof Exp:* Res chemist, Rayonier, Inc, Wash, 53-60; res assoc, Inst Paper Chem, 60-69, prof chem & sr res assoc, 69-86; CONSULT, 85- *Mem:* AAAS; Am Chem Soc; Sigma Xi; Can Pulp & Paper Asn. *Res:* Location and composition of the constituents of wood and their behavior during pulping. *Mailing Add:* 6042 Rosewood Dr Appleton WI 54915

THOMPSON, OWEN EDWARD, b St Louis, Mo, Nov 20, 39; m 72; c 1. METEOROLOGY, ATMOSPHERIC PHYSICS. *Educ:* Univ Mo-Columbia, BS, 61, MS, 63, PhD(atmospheric sci), 66. *Prof Exp:* Instr physics & math, Stephens Col, 64-66; instr atmospheric sci, Univ Mo-Columbia, 66-68; from asst prof to assoc prof, 68-82, PROF METEROL, UNIV MD, COLLEGE PARK, 82- *Concurrent Pos:* Wallace Eckert vis scientist, IBM Thomas J Watson Res Ctr, Yorktown Heights, NY, 75-76; asst provost, Div Math & Phys Sci & Eng, Univ Md, 78-79; comnr educ & manpower, Am Meterol Soc, 80-86; dir educ affairs, Univ Corp Atmospheric Res, Boulder, Colo, 86-87. *Mem:* Am Meteorol Soc; Am Geophys Union. *Res:* Dynamical and physical meteorology; atmospheric waves and oscillations; micrometeorology and boundary layer studies; forest environment; satellite meteorology; meteorological instrumentation; science education. *Mailing Add:* Dept Meteorol Univ Md College Park MD 20742-2425

THOMPSON, PAUL DEVRIES, b Glen Cove, NY, Dec 6, 39; m; c 4. BIOMEDICAL ENGINEERING, AGRICULTURAL EQUIPMENT. *Educ:* Cornell Univ, BEE, 62; Univ Pa, PhD(biomed eng), 70. *Prof Exp:* Electronics engr, Mastitis Res, Agr Res Serv, USDA, 71-78; prod mgr, 78-80, vpres, 80-82, PRES, DAIRY EQUIP CO, 82- *Mem:* Am Inst Ultrasonics in Med; Am Soc Agr Engrs; Am Dairy Sci Asn; Inst Elec & Electronics Engrs; Am Solar Energy Soc; Human Factors Soc; Farm & Indust Equip Inst. *Res:* Biological flow measurements using electromagnetic and ultrasonic techniques; applications to blood flow in all species and to milk flow in cows. *Mailing Add:* Dairy Equip Co PO Box 8050 Madison WI 53708

THOMPSON, PAUL O, b Stoughton, Wis, Feb 12, 21; m 77; c 2. PSYCHOACOUSTICS, BIOACOUSTICS. *Educ:* St Olaf Col, BA, 43; Univ Southern Calif, MA, 50. *Prof Exp:* Res psychologist, US Navy Electronics Lab, 48-67; res psychologist, Naval Ocean Systs Ctr, 67-82; CONSULT, 82- *Mem:* Acoust Soc Am. *Res:* Speech intelligibility, intensity and pitch sensation and perception; thresholds; bioacoustics of marine mammals, particularly whales. *Mailing Add:* 3713 Kite St San Diego CA 92103

THOMPSON, PAUL WOODARD, b Manchester, NH, May 21, 09; m 36; c 2. CHEMISTRY. *Educ:* Univ Ill, BS, 30, MS, 32. *Prof Exp:* Chemist, State Water Surv, Ill, 30-32; res chemist, Sherwin Williams Paint & Varnish Co, Ill, 35-39, Acme White Lead & Color Works, Mich, 39-42 & Ethyl Corp, 42-71; RES ASSOC ECOL, CRANBROOK INST SCI, 56-, FEL, 66- *Concurrent Pos:* Mem ecol surv, Lake Mich Sand Dunes; lectr ecol communities, Leelanau Sch Mich, 83-84; tech comt, Mich Threatened & Endangered Plants. *Honors & Awards:* Oakleaf Award, The Nature Conserv, 75. *Mem:* AAAS; Am Chem Soc; Am Forest Asn; Natural Areas Asn; Wilderness Soc; Nature Conservancy. *Res:* Oxidation reactions of tetraethyl lead; stability of halogen compounds and fuels; ecology and flora of Michigan; petroleum chemistry; conservation of natural areas; ecological survey, Sleeping Bear Dunes National Lakeshore, Huron Mountains, Michigan and Michigan prairies & Fens, Keweenaw Peninsula, Michigan. *Mailing Add:* Cranbrook Inst Sci Bloomfield Hills MI 48013

THOMPSON, PETER ERVIN, b Urbana, Ill, Mar 20, 31; m 60; c 2. GENETICS. *Educ:* Purdue Univ, BS, 54, MS, 56; Univ Tex, PhD(genetics), 59. *Prof Exp:* NIH fel zool, Univ Calif, Berkeley, 59-60; res assoc biol, Oak Ridge Nat Lab, 60-61; from asst prof to assoc prof genetics, Iowa State Univ, 61-68; prof zool, Univ Ga, 68-72, head dept, 72-81. *Concurrent Pos:* Vis lectr, Univ Wis, 63, 64 & 66. *Mem:* Genetics Soc Am; Am Soc Nat. *Res:* Invertebrate and primate genetics; genetic control of protein synthesis; developmental regulation of gene activities; hemoglobin structures and evolution. *Mailing Add:* Dept Zool & Bio Sci Univ Ga Athens GA 30602

THOMPSON, PETER TRUEMAN, b Palmerton, Pa, Oct 15, 29; m 53; c 4. PHYSICAL CHEMISTRY. *Educ:* Johns Hopkins Univ, AB, 51; Univ Pittsburgh, PhD(phys chem), 57. *Prof Exp:* Res asst, Univ Pittsburgh, 51-56, res assoc & instr, 56-58; from instr to assoc prof, 58-73, chmn dept, 71-72, 77-78 & 81-86, PROF CHEM, SWARTHMORE COL, 73- *Concurrent Pos:* NSF sci fac fel, Cambridge Univ, 65-66; vis adj prof, Univ Del, 73-74 & 76-77; vis scientist, Nat Bur Standards, 85. *Mem:* Am Chem Soc; Sigma Xi. *Res:* Physical chemistry of solutions both aqueous and non-aqueous. *Mailing Add:* Dept Chem Swarthmore Col Swarthmore PA 19081

THOMPSON, PHEBE KIRSTEN, b Glace Bay, NS, Sept 5, 97; nat US; wid; c 4. ENDOCRINOLOGY, GERIATRICS. *Educ:* Dalhousie Univ, MD, CM, 23. *Prof Exp:* Asst biochem, Sch Pub Health, Harvard Univ, 24-26; res fel med, Thyroid Clin, Mass Gen Hosp, 26-29; asst endocrinol, metab dept, Rush Med Col, Univ Chicago & Centr Free Dispensary, Chicago, 30-46; med ed & writing, 46-53; ed, J Am Geriatric Soc, 54-82. *Concurrent Pos:* Managing ed, J Clin Endocrinol & Metab, Endocrine Soc, 54-61; consult ed, J Clin Endocrinol & Metab, 61-65 & Endocrinol, 61-65; freelance ed & writer, 61- *Honors & Awards:* Thewlis Award, Am Geriat Soc, 66; Cert of Appreciation, Am Thyroid Asn, 66. *Mem:* Fel Am Med Writers' Asn; Am Pub Health Asn; fel Am Geriat Soc; fel Geront Soc Am; Am Genetic Asn; Endocrine Soc; AAAS. *Res:* Medical writing and editing. *Mailing Add:* 40 Donald M Thompson 55 W Monroe No 3550 Chicago IL 60603

THOMPSON, PHILIP A, b Galesburg, Ill, Sept 10, 28; m 46; c 3. FLUID MECHANICS, THERMODYNAMICS. *Educ:* Rensselaer Polytech Inst, BS, 57, MS, 58; Mass Inst Technol, ScD(mech eng), 61. *Prof Exp:* From asst prof to assoc prof, 60-74, PROF MECH ENG, RENSSELAER POLYTECH INST, 74- *Concurrent Pos:* Ford Found resident, Large Steam Turbine-Generator Dept, Gen Elec Co, 64-65; vis scientist, Max-Planck Inst Stromungsforschung, 75-78 & 81 & 88; Alexander von Humboldt sci grant, 75; prog mgr conserv, US Dept Energy, 78-80. *Honors & Awards:* Alexander von Humboldt Award, WGer, 75. *Mem:* Nat Acad Eng; fel Am Soc Mech Engrs; Am Inst Aeronaut & Astronaut; Am Chem Soc; Sigma Xi; Am Phys Soc. *Res:* Fundamental gas dynamics; mechanics of dense fluids; fast phase changes in liquid-vapor states; thermodynamics of real gases; author of various publications; granted one patent. *Mailing Add:* Dept Mech Eng Aeronaut Eng & Mech Rensselaer Polytech Inst Troy NY 12180-3590

THOMPSON, PHILLIP EUGENE, b York, Pa, Nov 14, 46; m 73; c 2. SOLID STATE PHYSICS. *Educ:* Lebanon Valley Col, BS, 68; Univ Del, PhD(physics), 75. *Prof Exp:* Asst engr, York Div, Borg-Warner Corp, 69-70; asst prof physics, Lebanon Valley Col, 74-81; RES PHYSICIST, NAVAL RES LAB, 81-, HEAD, VACUUM DEPOSITION SECT, 88- *Concurrent Pos:* Consult physics educ, Annville-Cleona High Sch, 80-81. *Mem:* Am Phys Soc. *Res:* Million electron volts implantation of gallium arsenide; device structures using million electron volts implimentation; molecular beam epitanial growth of narrow band gap materials; indium antimonide, indium arsenide, indium arsonic antimonide. *Mailing Add:* Code 6823 Naval Res Lab 4555 Overlook Ave SW Washington DC 20375

THOMPSON, PHILLIP GERHARD, b Eagle Grove, Iowa, Jan 28, 30; m 55; c 2. INDOOR AIR QUALITY, HAZARDOUS WASTE MANAGEMENT. *Educ:* St Olaf Col, BA, 54; Cornell Univ, PhD(inorg chem), 59. *Prof Exp:* Fulbright scholar & Ramsay fel, Cambridge Univ, 59; sr chemist, Cent Res Labs, 3M Co, 59-64, sr res chemist, Contract Res Lab, Cent Res, 64-68, sr res scientist, Magnetic Prod Div, 68-70; tech dir & consult, Thompson Assocs, 70-77; patent adminr, Univ Minn, 77-83; PRES, THOMPSON ASSOCS, 84-; ASSOC PROF INDOOR AIR QUALITY, UNIV MINN, 87- *Concurrent Pos:* Mem, Metrop Coun Comprehensive Health Planning Bd, Metrop Coun, 68-70; 3M Indust lectr, 3M Co, 68-70; vis Ramsay res fel, Dept Physics & Chem, St Olaf Col, 70-71; mem & secy, St Paul Environ Qual Bd, 70-72; comprehensive environ health fel, Dept Environ Health, Sch Pub Health, Univ Minn, 72 & 73; comnr, St Paul Water Bd, 72-80, vpres, 74-80; consult, Environ Health & Safety, Univ Minn, 74, vis lectr, 74-77; dir, Minn Acad Sci, 76-80, consult, 77-; dir, Inventors & Technol Transfer Corp, 84-87; ed, Soc Univ Patent Admin, 85-86. *Mem:* Air Pollution Control Asn; AAAS; Am Chem Soc; Am Indust Hyg Asn; Am Water Works Asn; Am Soc Heating, Refrig & Air Conditioning Engrs. *Res:* environmental chemistry; air and respirable mass sampling; water quality; trace contaminants; transformation of environmental pollutants; new synthetic techniques; unusual oxygen fluorine compounds; propellants; high performance sealants; ferrites; Mossbauer spectroscopy; university-industry technology transfer; environmental law and regulations; laboratory fume hoods; cold climate housing moisture and indoor air quality problems; radon; statewide information center and data base for industrial air quality in residential housing; hazardous waste treatment methods and facilities; audio-visual training programs. *Mailing Add:* Thompson Assocs 229 Kennard St St Paul MN 55106-6817

THOMPSON, RALPH J, b Greenville, Tex, Apr 11, 30; m 59; c 2. PHYSICAL CHEMISTRY. *Educ:* ETex State Col, BS & MS, 54; Univ Tex, Austin, PhD(chem), 63. *Prof Exp:* Asst prof chem, Univ Tex, Arlington, 55-59; fel, Ind Univ, 63-65; assoc prof, 65-70, PROF CHEM, EASTERN KY UNIV, 70- *Mem:* Am Chem Soc. *Res:* Nuclear magnetic resonance of boron; nucleic acid research as related to brain function. *Mailing Add:* Dept Chem Eastern Ky Univ Richmond KY 40475

THOMPSON, RALPH J, JR, b Los Angeles, Calif, Jan 27, 28; m 48; c 3. SURGERY. *Educ:* La Sierra Col, BS, 50; Loma Linda Univ, MD, 51; Am Bd Surg, dipl, 61. *Prof Exp:* From instr to assoc prof, 64-77, PROF SURG, SCH MED, LOMA LINDA UNIV, 77- *Concurrent Pos:* Fel cancer surg, Mem Sloan-Kettering Cancer Ctr, 60-61. *Mem:* Am Soc Surg Oncol; James Ewing Soc; Am Soc Clin Oncol. *Res:* Cancer surgery. *Mailing Add:* Dept Surg OB-GYN Loma Linda Univ Loma Linda CA 92354

THOMPSON, RALPH LUTHER, b Niangua, Mo, Feb 6, 43; m 70. PLANT TAXONOMY, PLANT GEOGRAPHY. *Educ:* Southwest Mo State Univ, BS, 71, MA, 75; Northeast La Univ, MEd, 72; Southern Ill Univ, PhD(bot), 80. *Prof Exp:* Instr bot, Dept Life Sci, Southwest Mo State Univ, 74-75; spec asst, Dept Bot, Southern Ill Univ, 76-80; asst prof, 80-85, ASSOC PROF BOT & PLANT TAXON, DEPT BIOL, BEREA COL, 85- *Concurrent Pos:* Vis asst prof, Ohio State Univ, 80. *Mem:* Sigma Xi; AAAS. *Res:* Floristic and descriptive studies of the vascular flora of Kentucky; taxonomy and distributional history of nonindigenous plants of the United States; revisionary studies in the subfamily Mimosoideae of the fabaceae (Leguminosae); vegetation of coal surface mines. *Mailing Add:* CPO 2325 Berea Col Berea KY 40404

THOMPSON, RALPH NEWELL, b Boston, Mass, Mar 4, 18; m 42; c 3. CHEMICAL ENGINEERING. *Educ:* Mass Inst Technol, BS, 40. *Prof Exp:* Res engr paper mfg, Middlesex Prod, 40-42, tech dir, Falulah Paper Co, 45-48; staff engr paper chem, Calgon Corp, 48-54, res mgr water treat spec chem, 55-57, mgr res & develop, 58-63, dir res & eng, 63-67, vpres & gen mgr, Spec Chem Div, 67-70; vpres corp develop polymers, Pa Indust Chem Corp, 70-74; gen mgr, Chem Div, Thiokol Corp, 74-76, group vpres spec chem, 76-83; RETIRED. *Honors & Awards:* Goodreau Medal, Goodreau Mem Fund, 36. *Mem:* Tech Asn Pulp & Paper Indust; Soc Chem Indust; NY Acad Sci; fel Am Inst Chemists. *Res:* Colloid chemistry; polymer chemistry; industrial water treatment; chemical engineering. *Mailing Add:* 1006 Lehigh Dr Yardley PA 19067-2908

THOMPSON, RAMIE HERBERT, b St Johnsbury, Vt, Sept 26, 33; m; c 4. ELECTROMAGNETICS, ELECTROEXPLOSIVES. *Educ:* Univ Pa, BSEE, 61, MSEE, 65. *Prof Exp:* PRIN SCIENTIST ELECTROMAGNETICS, FRANKLIN RES CTR, 57- *Mem:* Inst Elec & Electronics Engrs; AAAS; Sigma Xi. *Res:* Interactions of electromagnetic energy and electroexplosives. *Mailing Add:* 1518 Noble Rd Rydal PA 19046

THOMPSON, RAYMOND G, b Birmingham, Ala, Dec 5, 52; m 72; c 3. JOINING OF MATERIALS. *Educ:* Univ Ala Birmingham, BSE, 74, MSE, 75; Vanderbilt Univ, PhD(mat sci), 79. *Prof Exp:* Asst prof mat eng, Clemson Univ, 78-81; asst prof to assoc prof mat eng, 81-89, PROF MAT SCI & ENG, UNIV ALA BIRMINGHAM, 89- *Concurrent Pos:* Res fac, Am Soc Eng Educ, 79 & 81; prin investr, Nat Aeronaut & Space Admin, 80-85 & NSF, 85-88 & 87-92. *Honors & Awards:* Adams Award, Am Welding Soc, 86. *Mem:* Am Welding Soc; Am Soc Metals; Metall Soc, Am Inst Mining Engrs. *Res:* Surface atomic structure and chemistry and their effects on material properties and processing. *Mailing Add:* Sch Eng Univ Ala Birmingham AL 35294

THOMPSON, RICHARD BAXTER, b Fresno, Calif, June 1, 26; m 50, 82; c 3. FISH BIOLOGY. *Educ:* San Jose State Col, BA, 50; Univ Wash, PhD(fisheries), 66. *Prof Exp:* Fishery res biologist, Fisheries Res Inst, Univ Wash, 50-54; fishery res biologist, Northwest Fisheries Ctr, 55-75, fishery res biologist, Northwest Regional Off, Nat Marine Fisheries Serv, Nat Oceanic & Atmospheric Admin, 75-88; RETIRED. *Mem:* Am Inst Fishery Res Biol. *Res:* Fish ethology; orientation and navigation of anadromous fishes; sensory perception of fishes; marine game fishery resources and utilization. *Mailing Add:* 9707 45th Ave NE Seattle WA 98115-6349

THOMPSON, RICHARD BRUCE, b Fargo, NDak, Oct 12, 39; m 61; c 2. MATHEMATICS. *Educ:* Univ Northern Iowa, BA, 61; Univ Wis, MS, 63, PhD(topology), 67. *Prof Exp:* Asst prof, 67-70, ASSOC PROF MATH, UNIV ARIZ, 70- *Concurrent Pos:* NSF res grants, 68-71. *Mem:* Am Math Soc; Math Asn Am. *Res:* Topological fixed point theory, particularly semicomplexes, quasi-complexes and local and global fixed point indices. *Mailing Add:* Dept Math Univ Ariz Tucson AZ 85721

THOMPSON, RICHARD CLAUDE, b Kansas City, Mo, Mar 12, 39. INORGANIC CHEMISTRY. *Educ:* Univ Chicago, BS, 61; Univ Md, PhD(chem), 65. *Prof Exp:* Resident res assoc, Chem Div, Argonne Nat Lab, 65-66; asst prof chem, Ill Inst Technol, 66-67; from asst prof to assoc prof, 67-77, PROF CHEM, UNIV MO-COLUMBIA, 77- *Concurrent Pos:* Consult, Argonne Nat Lab, 66-75. *Mem:* Am Chem Soc. *Res:* Kinetics and mechanisms of inorganic reactions. *Mailing Add:* Dept Chem Univ Mo Columbia MO 65211

THOMPSON, RICHARD E(UGENE), b Parsons, Kans, Oct 15, 29; m 51; c 2. PROCESS DESIGN. *Educ:* Okla State Univ, BS, 51, PhD(chem eng), 63; Colo Sch Mines, MS, 59. *Prof Exp:* Reactor engr, Atomic Energy Div, Phillips Petrol Co, 51-53; process engr, Tex Div, Dow Chem Co, 53-54; from asst prof to assoc prof chem eng, 62-75, chmn dept, 77-79, PROF CHEM ENG, UNIV TULSA, 75-, CHMN DEPT, 89- *Concurrent Pos:* Consult, Crest Eng, Inc, 64-82; sr consult, BWT Furlow-Philbeck Assocs, Inc, 75-82. *Mem:* Am Inst Chem Engrs; Soc Petrol Engrs; Sigma Xi. *Res:* Equilibrium-stage processes; oil and gas processing; computer simulation; computer-aided design. *Mailing Add:* Dept Chem Eng Univ Tulsa 600 S College Tulsa OK 74104

THOMPSON, RICHARD EDWARD, b Wichita, Kans, Oct 17, 46; m 71; c 2. BIOCHEMISTRY, BIOMETRICS- BIOSTATISTICS. *Educ:* Wichita State Univ, BS, 68, MS, 69; Okla State Univ, PhD(biochem), 74. *Prof Exp:* Fel biochem, Univ Cincinnati, 74-77; asst prof biochem, N Tex State Univ, 77-83; RES & DEVELOP MGR, ABBOTT LABS, 83- *Concurrent Pos:* Adj assoc prof biochem, Tex Col Osteop Med, 77-83; NIH fel, Univ Cincinnati, 75-77. *Mem:* Am Chem Soc; Sigma Xi; Am Asn Clin Chem. *Res:* Physical biochemistry; enzymology; regulation of cholesterol biosynthesis; analytical biochemistry; clinical chemistry; computational chemistry. *Mailing Add:* 1616 Pleasant Ct Libertyville IL 60048

THOMPSON, RICHARD FREDERICK, b Portland, Ore, Sept 6, 30; m 60; c 3. PHYSIOLOGICAL PSYCHOLOGY. *Educ:* Reed Col, BA, 52; Univ Wis, MS, 53, PhD(physiol psychol), 56. *Hon Degrees:* MS, Harvard Univ, 73. *Prof Exp:* NIH fel physiol, Univ Wis, 56-59; from asst prof to prof med psychol, Med Sch, Univ Ore, 59-67; prof, Univ Calif, Irvine, 67-73; prof psychol, Harvard Univ, 73-75; prof psychobiol, Univ Calif, Irvine, 75-80; prof psychol & Bing prof human biol, Stanford Univ, 80-87; KECK PROF PSYCHOL & BIOL SCI, UNIV SOUTHERN CALIF, 87-, DIR NEUROSCI PROG, 89- *Concurrent Pos:* Nat Inst Ment Health res career award, 62-73; mem adv panel psychobiol, NSF, 67-70; mem res scientist rev comt, Nat Inst Ment Health, 69-74; mem comt biol bases soc behav, Social Sci Res Coun, 72-80; mem US Nat Comt, Int Brain Res Orgn, 75-78. *Honors & Awards:* Commonwealth Award, 66; Distinguished Sci Contrib Award, Am Psychol Asn, 77. *Mem:* Nat Acad Sci; Soc Neurosci; Am Psychol Asn; fel AAAS; Am Physiol Soc; Am Acad Arts & Sci. *Res:* Neurophysiology; cerebral cortex and behavior; neural basis of learning. *Mailing Add:* Dept Psychol Neurosci Prog Univ Southern Calif Los Angeles CA 90089-2520

THOMPSON, RICHARD JOHN, b Chapman Ranch, Tex, Aug 9, 27; div; c 2. ANALYTICAL CHEMISTRY, AIR POLLUTION. *Educ:* Univ Tex, BS, 52, MA, 56, PhD(inorg chem), 59. *Prof Exp:* Asst prof chem, Lamar State Col, 57-58 & NTex State Univ, 59-68; from asst prof to assoc prof, Tex Tech Col, 62-68; chief metals & adv analytical unit, Air Qual & Emission Data Prog & supvry res chemist, Nat Air Pollution components and sampling of large quantities of sized respirable air-borne particulate matter. serv br, Div Air Qual & Emissions Data, Bur Criteria & Standards, 69-71; chief, Air Qual Analytical Lab Br, Div Atmospheric Surveillance, Environ Protection Agency, 71-73, chief, Qual Assurance & Environ Monitoring Lab, 73-75, chief, Analytical Chem Br, 75-78, actg dir, Environ Monitoring Div, 78-79, chief adv, Analytical Tech Br, Environ Monitoring Support Lab, Environ Res Ctr, 80; prof, Sch Pub Health, Univ Ala, Birmingham, 81-86; res sci & adj prof chem & civil eng, Arlington Univ, 86-88; training specialist, Tex A&M Univ, 88-89; RETIRED. *Concurrent Pos:* Res grants, Res Corp, 60-, Welch Found, 61-69 & NSF, 64-68, 74-75; adj prof, NC State Univ, 74-; consult, Lawrence Livermore Lab, 70-72, World Meteorol Orgn, 74-80, environ, prog assessment & develop, training & litigation support, indust hyg area. *Mem:* AAAS; Air Pollution Control Asn; Am Chem Soc; Sigma Xi; Soc Appl Spectros; fel Am Inst Chem; Am Indust Hyg Asn; Am Bd Indust Hyg. *Res:* Development of methods for collection and analysis of atmospheric pollutants, including analysis of trace elements, organics, non-metals inorganics, and precipitation components. *Mailing Add:* 12113 Shropshire Blvd Austin TX 78753

THOMPSON, RICHARD MICHAEL, b Thief River Falls, Minn, May 30, 45; m 67; c 2. BIOCHEMISTRY, ANALYTICAL CHEMISTRY. *Educ:* Univ Minn, BChem, 67; Univ Wis, PhD(biochem), 71. *Prof Exp:* Asst prof pediat, Sch Med, Ind Univ, Indianapolis, 73-78; PRIN RES SCIENTIST, COLUMBUS DIV, BATTELLE MEM INST, 78- *Concurrent Pos:* NIH trainee, Baylor Col Med, 71-72, Nat Heart Inst res associateship, 72-73. *Mem:* Am Soc Mass Spectrometry; Sigma Xi. *Res:* Biochemical genetics; gas phase analytical techniques; drug metabolism; identification of natural products. *Mailing Add:* Merrell Dow Res Inst PO Box 68470 Indianapolis IN 46268-0470

THOMPSON, RICHARD SCOTT, b Lubbock, Tex, May 24, 39; m 67; c 2. SUPERCONDUCTIVITY. *Educ:* Calif Inst Technol, BS, 61; Harvard Univ, AM, 62, PhD(physics), 65. *Prof Exp:* NSF fel, Ctr Nuclear Res, France, 66; mem, Inter-Acad Exchange Prog, Inst Theoret Physics, Moscow, 66-67 & 74-75; vis foreign scientist, Ctr Nuclear Res, France, 67-68; asst physicist, Brookhaven Nat Lab, 68-70; asst prof physics, 70-72, ASSOC PROF PHYSICS, UNIV SOUTHERN CALIF, 72- *Concurrent Pos:* Vis scientist, Univ Dortmund, Ger, 75-76. *Mem:* Am Phys Soc. *Res:* Superconductivity. *Mailing Add:* Dept Physics Univ Southern Calif Los Angeles CA 90089-0484

THOMPSON, ROBERT ALAN, b Catskill, NY, July 16, 37; m 71; c 3. MECHANICAL & PRODUCTION ENGINEERING. *Educ:* Bucknell Univ, BS, 60; Rensselaer Polytech Inst, MS, 62; Univ Rochester, PhD(mech & aerospace sci), 66. *Prof Exp:* Engr trainee truck develop, Ford Motor Co, Mich, 61-62; mech engr advan energy systs, Pratt & Whitney Aircraft Co, Conn, 62-63; mech engr appl mech unit, 66-70, MECH ENGR CORP RES & DEVELOP, PROCESS TECHNOL PROG, GEN ELEC CO, SCHENECTADY, 70- *Honors & Awards:* Blackall Machine Tool & Gage Award, Am Soc Mech Engrs. *Mem:* Fel Am Soc Mech Engrs; Sigma Xi. *Res:* Mechanical analysis of processes for process optimization and process automation; manufacturing process conception, process equipment design and development; technical areas: machine tool dynamics, flame hardening, residual stress, ceramic manufacturing process, shot peening and thermal barrier coating. *Mailing Add:* Box 44 Quaker St NY 12141

THOMPSON, ROBERT BRUCE, b Bryan, Tex, July 18, 41; m 67; c 2. NON-DESTRUCTIVE EVALUATION, ULTRASONICS. *Educ:* Rice Univ, BA, 64; Stanford Univ, MS, 65, PhD(appl physics), 71. *Prof Exp:* Mem tech staff, Rockwell Int Sci Ctr, 70-75, group leader, 75-80; dir metall & ceramics div, 86-89, SR SCIENTIST, AMES LAB, IOWA STATE UNIV, 80-, PROF MAT SCI, ENG & MECHANICS, 86-, ASSOC DIR, AMES LAB, 89- *Concurrent Pos:* Vchmn, Nondestructive Eval Working Group, 77-; consult, var indust & acad orgns, 80-; dir, Magnasonics Inc, Albuquerque, NMex, 84-;

dep dir, Nondestructive Eval Ctr, Iowa State Univ, 85-; ed, J Nondestructive Eval, 87-; vis fel, Wolfson col, Oxford Univ, 87. *Mem:* Inst Elec & Electronics Engrs; Am Soc Mech Engrs; Soc Eng Sci; Am Soc Mat. *Res:* Development of new ultrasonic techniques for characterizing the structure of materials and their flaws; specialties include the characterization of stresses and preferred grain orientation in metals; strengths of solid state metallic bonds; the prediction of the reliability of flaw detection. *Mailing Add:* 109 Off & Lab Iowa State Univ Ames IA 50011

THOMPSON, ROBERT CHARLES, b Winnipeg, Man, Apr 21, 31; m 60. MATHEMATICS. *Educ:* Univ BC, BA, 54, MA, 56; Calif Inst Technol, PhD(math), 60. *Prof Exp:* Defense sci officer, Defence Res Bd, Can, 56-57; from instr to asst prof math, Univ BC, 60-64; from asst prof to assoc prof, 64-69, PROF MATH, UNIV CALIF, SANTA BARBARA, 69- *Concurrent Pos:* Ed, J Linear & Multilinear Algebra. *Mem:* Am Math Soc; Math Asn Am; Soc Indust & Appl Math. *Res:* Algebra, especially linear algebra and number theory. *Mailing Add:* Dept Math Univ Calif Santa Barbara CA 93106

THOMPSON, ROBERT GARY, b Jewell, Iowa, Jan 20, 38; m 59; c 3. ENDOCRINOLOGY, LIPID METABOLISM. *Educ:* Univ Iowa, BA, 62, MD, 65. *Prof Exp:* Resident, Johns Hopkins Hosp, 65-68, fel, 68-71, asst prof pediat, 71-74; assoc prof, 74-80, PROF PEDIAT, UNIV IOWA, 80-, DIR PEDIAT ENDOCRINOL, 74- *Mem:* Endocrine Soc; Cent Soc Clin Res; Soc Pediat Res; Am Pediat Soc; Am Diabetes Asn. *Res:* Endocrinology and diabetes with emphasis of nutrition on insulin secretion and lipid homeostasis. *Mailing Add:* Dept Pediat Univ Iowa Hosps Iowa City IA 52242

THOMPSON, ROBERT GENE, b Hiddenite, NC, Dec 23, 31; m 53; c 2. PHYSICAL ORGANIC CHEMISTRY. *Educ:* Univ NC, BS, 52; Univ Tenn, MS, 54, PhD(chem), 56. *Prof Exp:* Res chemist, 56-60, sr res chemist, 60-61, supvr res, 61-63, tech, 63-64, sr supvr, 64-66, tech supt, 66-71, res mgr, 71-76, bus coordr, 76-78, mft mgr, 78-86, TECH MGR, E I DU PONT DE NEMOURS & CO, INC, 78- *Mem:* Am Chem Soc; Sigma Xi. *Res:* Vinyl and condensation polymers; synthetic textile fibers. *Mailing Add:* Textile Fibers Dept E I Du Pont de Nemours & Co Inc Wilmington DE 19898

THOMPSON, ROBERT HARRY, b Columbus, Ohio, May 2, 24; m 47. MATHEMATICS, COMPUTER SCIENCE. *Educ:* Sterling Col, BS, 45, DSc, 69; Univ Kans, MA, 51. *Prof Exp:* Prof math, Sterling Col, 47-67; assoc prof math, Washburn Univ, Topeka, 67-89; RETIRED. *Mem:* Math Asn Am. *Res:* General mathematics. *Mailing Add:* 3215 SE 156th Ave Vancouver WA 98684

THOMPSON, ROBERT JAMES, b Dayton, Ohio, Sept 21, 30; m 56; c 2. MATHEMATICS. *Educ:* Ohio State Univ, BSc, 52, MSc, 54, PhD(math), 58. *Prof Exp:* Asst math, Ohio State Univ, 52-53, from asst instr to instr, 53-58; mem staff, Sandia Nat Lab, 58-65, supvr appl math div II, 65-72, supvr numerical anal div, 72-75, supvr appl math div, 75-90, DISTINGUISHED MEM TECH STAFF, SANDIA NAT LAB, 90- *Mem:* Am Math Soc; Math Asn Am; Soc Indust & Appl Math. *Res:* Applied mathematics; numerical analysis. *Mailing Add:* 12500 Loyola Ave NE Albuquerque NM 87112

THOMPSON, ROBERT JOHN, JR, b San Francisco, Calif, Nov 10, 17; m 45; c 3. PHYSICAL CHEMISTRY, GUIDED MISSILE TECHNOLOGY. *Educ:* Univ Calif, Los Angeles, BS, 40; Univ Rochester, PhD(phys chem), 46. *Prof Exp:* Control chemist, Eastman Kodak Co, 37-41; res assoc, George Washington Univ, 43-46; sr res engr, M W Kellogg Co Div, Pullman, Inc, 46-53 & Bendix Aviation Corp, 53-54; vpres & dir res div, Rocketdyne Div, NAm Aviation, Inc, Calif, 54-71, vpres & gen mgr, Rocketdyne Solid Rocket Div, 71-72, sr staff scientist, Rocketdyne Div, NAm Rockwell Corp, 72-73; spec asst to dir, Appl Physics Lab, 74-80, supvr tech, info, 80-85, PRIN PROF STAFF, JOHNS HOPKINS UNIV, 85- *Concurrent Pos:* Mem subcomt rocket engines, NASA, 51-54 & subcomt aircraft fuels, 58, mem res adv comt energy processes, Md Gov Sci Adv Coun, 59- *Mem:* AAAS; Am Chem Soc; Am Inst Aeronaut & Astronaut; Am Inst Chem Engrs. *Res:* Guided missiles; rocket and jet propulsion; propellants and fuels; combustion; heat transfer and fluid flow; chemical processes, thermodynamics and kinetics; radiation and spectra; space science; energy processes, systems and applications. *Mailing Add:* 12912 Ruxton Rd Silver Spring MD 20904-5278

THOMPSON, ROBERT KRUGER, b Jeffersonville, Ohio, Jan 15, 22; m 43; c 3. ENTOMOLOGY. *Educ:* Ohio State Univ, BSc, 47, MSc, 48, PhD(entom), 50. *Prof Exp:* Field aide, Bur Entom & Plant Quarantine, US Dept Agr, 46-47; field res entomologist, Ortho Div, Chevron Chem Co, 50-51, field res supvr, 51-65, supvr, Biol Res Labs, 66-83; RETIRED. *Mem:* Entom Soc Am. *Res:* Chemical control of insects, weeds and plant diseases. *Mailing Add:* 1409 Dela Cruz Way Moraga CA 94556

THOMPSON, ROBERT LEE, b Canton, NY, Apr 25, 45; m 68; c 2. AGRICULTURAL POLICY, INTERNATIONAL TRADE. *Educ:* Cornell Univ, BS, 67; Purdue Univ, MS, 69, PhD(agr econ), 74. *Prof Exp:* From asst prof to assoc prof, 74-83, PROF AGR ECON, PURDUE UNIV, 83-, DEAN AGR, 87- *Concurrent Pos:* Vol agriculturist, Int Vol Serv, Laos, 68-70; vis prof agr econ, Fed Univ Vicosa, Brazil, 72-73 & Econ Res Serv, USDA, 79-80; res scholar, Int Inst Appl Systs Anal, Austria, 83; sr staff economist, President's Coun Econ Adv, 83-85; asst secy econ, USDA, 85-87; mem, Nat Comn Agr Trade & Export Policy, 85-86, Bd Agr, Nat Res Coun & Int Policy Coun Agr & Trade, 88-; chair adv coun, Nat Ctr Food & Agr Policy, 89- *Honors & Awards:* Qual of Commun Award, Am Agr Econ Asn, 79 & 91. *Mem:* Fel AAAS; Am Agr Econ Asn; Am Econ Asn; Int Asn Agr Economists (pres-elect); Sigma Xi. *Res:* Agricultural trade policy; US agricultural policy; world agricultural development. *Mailing Add:* 114 Agr Admin Bldg Purdue Univ West Lafayette IN 47907-1140

THOMPSON, ROBERT POOLE, b Winnipeg, Man, Feb 8, 23; m 53; c 5. DEVELOPMENTAL BIOLOGY. *Educ:* Univ Western Ont, BA, 49, MSc, 53; Univ Toronto, PhD(zool), 63. *Prof Exp:* Res officer entom, Can Dept Agr, 53-56; from instr to assoc prof biol, St Francis Xavier Univ, 56-67; assoc prof, 67-72, prof biol, State Univ Ny Col Brockport, 72-88; RETIRED. *Concurrent Pos:* Nat Res Coun Can grants, 63-68; State Univ NY Res Found fac res fel, 68-69. *Mem:* AAAS; Soc Develop Biol; Soc Study Reproduction; Am Soc Zoologists; Sigma Xi. *Res:* Homologous inhibition in the development of pattern in the embryo; time of eruption of the third molar tooth. *Mailing Add:* 40 College St Brockport NY 14420

THOMPSON, ROBERT QUINTON, b Morristown, NJ, Nov 3, 55; m 81; c 2. CLINICAL CHEMISTRY, CONTINUOUS FLOW ANALYSIS. *Educ:* Col Wooster, BA, 78; Mich State Univ, PhD(anal chem), 82. *Prof Exp:* Asst prof, 82-88, ASSOC PROF CHEM, OBERLIN COL, 88- *Concurrent Pos:* Prin investr, NSF grant, 85-87 & Petrol Res Fund grant, 88-91; appointee, Comt Prof Status, Am Chem Soc, 88-91; vis prof, Dept Chem, Univ Tenn, 89 & Dept Chem, Univ Cincinnati, 89-90. *Mem:* Am Chem Soc; Sigma Xi. *Res:* Use of immobilized biomolecules for the determination of clinically-important analytes; use of capillaries as sampling and reaction vessels. *Mailing Add:* Dept Chem Kettering Hall Oberlin Col Oberlin OH 44074

THOMPSON, ROBERT RICHARD, b Springfield, Mo, Mar 30, 31; m 55; c 2. ORGANIC GEOCHEMISTRY. *Educ:* Drury Col, BS, 53; Wash Univ, St Louis, 55-56, PhD(org chem), 57. *Prof Exp:* Sr res engr, Pan Am Petrol Corp, Standard Oil Co, Ind, 57- 61, tech group supvr, 61-65, staff res scientist, 65-71; res group supvr, 71-75, res sect dir, 75-86, RES CONSULT, AMOCO PROD CO, 86- *Mem:* Am Chem Soc; Geochem Soc. *Res:* Organic geochemistry; origin of oil; geochemical prospecting. *Mailing Add:* 5617 S Quebec Tulsa OK 74135-4231

THOMPSON, RODGER IRWIN, b Texarkana, Tex, Aug 9, 44; div; c 2. ASTROPHYSICS. *Educ:* Mass Inst Technol, SB, 66, PhD(physics), 70. *Prof Exp:* Asst prof optical sci, 70-71, asst prof astron, 71-74, assoc prof, 74-81, PROF ASTRON, STEWARD OBSERV, UNIV ARIZ, 81- *Mem:* Am Phys Soc; Am Astron Soc. *Res:* Theoretical astrophysics including molecular physics, stellar evolution and star formation; observational infrared and dispersive spectroscopy with Fourier transform spectrometers; principal investigator for an infrared instrument for the Hubble Space Telescope. *Mailing Add:* Steward Observ Univ Ariz Tucson AZ 85721-0655

THOMPSON, ROGER KEVIN RUSSELL, b Eng, Dec 12, 45; m 68; c 2. ANIMAL COGNITION, COMPARATIVE PSYCHOLOGY. *Educ:* Univ Auckland, BA, 70, MA, 71; Univ Hawaii, PhD(psychol), 76. *Prof Exp:* Asst prof psychol, 76-77, ASST PROF BIOL & PSYCHOL, FRANKLIN & MARSHALL COL, 77- *Mem:* AAAS; Animal Behav Soc; Am Primatological Soc; Am Asn Univ Prof; Sigma Xi. *Res:* Comparative analysis of animal memory and related cognitive processes; animal auditory and tonic immobility. *Mailing Add:* Dept Psychol Biol Franklin & Marshall Col PO Box 3003 Lancaster PA 17604

THOMPSON, RONALD G, b Texas City, Tex, Mar 15, 60. SOLUTION INTERACTIONS BETWEEN SURFACE ACTIVE POLYMERS & SUSPENDED INORGANIC PARTICLES, PARTICLE CHARACTERIZATION. *Educ:* Abilene Christian Univ, BS, 82; Colo State Univ, PhD(chem), 87. *Prof Exp:* ADVAN CHEMIST, PETROL TECHNOL CTR, MARATHON OIL CO, 87- *Mem:* Am Chem Soc; Sigma Xi. *Res:* Investigation of mechanisms of interaction between surface active polyelectrolytes and suspended inorganic particles in aqueous fluids; controlled release systems for delivery of surface active agents in field applications. *Mailing Add:* 612 Bowen St Longmont CO 80501

THOMPSON, RONALD HALSEY, b Brooklyn, NY, Apr 29, 26; m 51; c 2. PHYSIOLOGY. *Educ:* Adelphi Col, BA, 50; Columbia Univ, MA, 51; Univ Pa, PhD, 59. *Prof Exp:* Technician, Cornell Univ, 51-53 & Univ Pa, 53-55; physiologist, Nat Insts Health, 55-75, Sci Dir, 68-75; teacher, Northern Va Community Col, 74-80, asst prof mat sci, 80-85; ASSOC PROF MAT SCI, PASCO-HERNANDO COMMUNITY COL, 85- *Mem:* AAAS; Am Physiol Soc. *Res:* Temperature regulation; environmental physiology; instrumentation for physiology. *Mailing Add:* Pasco-Hernando Community Col 7025 Moon Lake Rd New Port Richey FL 34654

THOMPSON, RONALD HOBART, b Memphis, Tenn, Feb 21, 35; m 60; c 4. NUCLEAR CHEMISTRY. *Educ:* La Tech Univ, BS, 61, MS, 68; Univ Ark, PhD(chem), 72. *Prof Exp:* Chemist, Western Elec Corp, 68-70; from asst prof to assoc prof chem, 72-82, PROF CHEM ENG, LA TECH UNIV, 82-, DIR, NUCLEAR CTR & RADIATION SAFETY OFFICER, 86- *Mem:* Am Chem Soc; AAAS; Am Nuclear Soc; Am Health Physic Soc. *Res:* Cosmology; concentration of trace elements on the earth; migration rates of radionuclides in soils; medical applications of radioisotopes; radiation dosimetry. *Mailing Add:* Box 3015 TS Ruston LA 71272

THOMPSON, ROSEMARY ANN, b San Diego, Calif, May 15, 45; m 67; c 1. MARINE BIOLOGY, FRESHWATER BIOLOGY. *Educ:* Univ Mo-Columbia, BA, 67; Univ Calif, San Diego, PhD(marine biol), 72. *Prof Exp:* Res assoc marine biol, Univ Southern Calif, 72-73; sr biologist, Henningson, Durham & Richardson, 74-84; sr biologist, URS Consults, 84-89; PRES, SWIFT'S ENVIRON ANALYSIS, 85-; SR BIOLOGIST, SCI APPLICATIONS INT CORP, 89- *Concurrent Pos:* Consult environ scientist, EG&G Co, 74; consult, Hinningson, Durham & Richardson, 74 & 84. *Mem:* Sigma Xi; Am Fisheries Soc. *Res:* Marine and aquatic biology of fish and invertebrates as related to pollution and modification of the environment; preparation of environmental reports; field surveys; expert witness. *Mailing Add:* Swift's Environ Analysis 4634 Mint Lane Santa Barbara CA 93110

THOMPSON, ROY CHARLES, JR, b Kansas City, Mo, June 19, 20; m 76; c 4. RADIATION BIOLOGY, BIOCHEMISTRY. *Educ:* Univ Tex, BA, 40, MA, 42, PhD(biochem), 44. *Prof Exp:* Tutor, Univ Tex, 40-41, instr, 41-43, res assoc biochem, 43-44, asst prof chem, 47-50; res chemist, Manhattan Dist, US Army Engrs Plutonium Proj, Metall Lab, Univ Chicago, 44-46; res chemist, Radiation Lab, Univ Calif, 46-47; res chemist, Gen Elec Co, Washington, 50-65; staff scientist, Biol Dept, Pac Northwest Lab, Battelle

Mem Inst, 65-85; CONSULT, 85- *Concurrent Pos:* Mem comt int exposure, Int Comn Radiol Protection, 69-85; mem, Nat Coun Radiation Protection & Measurements, 76-88; assoc ed, Radiation Res, 80-83; hom mem, Nat Coun Radiation Protection & Measurements, 88- *Honors & Awards:* Distinguished Sci Achievement Award, Health Physics Soc, 85. *Mem:* AAAS; Radiation Res Soc; Health Physics Soc. *Res:* Radiochemical study of biochemical processes; evaluation of hazards from internally deposited radioisotopes; especially plutonium and other transuranium elements. *Mailing Add:* Biol Dept Pac Northwest Lab Battelle Mem Inst Richland WA 99352

THOMPSON, ROY LLOYD, b Minn, Apr 29, 27; m 54; c 3. AGRONOMY. *Educ:* Univ Minn, BS, 51, MS, 59; Pa State Univ, PhD(agron), 67. *Prof Exp:* Field supvr, Minn Crop Improv Asn, 49-51; agronomist, Univ Minn, Morris, 56-67 & Rockefeller Found, 67-72; exten agronomist, 72-78, ASST DIR, MINN AGR EXP STA, UNIV MINN, ST PAUL 78- *Mem:* Am Soc Agron; Crop Sci Soc Am; Int Soc Tropical Root Crops. *Res:* Applied crop physiology in the management of field crops for the development and improvement of crop production systems; bean improvement cooperative. *Mailing Add:* Agr Exp Sta Univ Minn St Paul MN 55108

THOMPSON, SAMUEL, III, b Dallas, Tex, Aug 12, 32. PETROLEUM GEOLOGY, STRATIGRAPHY. *Educ:* Southern Methodist Univ, BS, 53; Univ NMex, MS, 55. *Prof Exp:* Petrol geologist, Exxon Corp, 54-74; PETROL GEOLOGIST, NMEX BUR MINES & MINERAL RESOURCES, 74- *Mem:* Am Asn Petrol Geologists. *Res:* Regional evaluation of the potential for petroleum exploration in southwestern New Mexico; system for analysis of sedimentary units; physico-stratigraphy, chronostratigraphy and eustatic geochronology. *Mailing Add:* NMex Bur Mines & Mineral Resources Socorro NM 87801

THOMPSON, SAMUEL LEE, b Hopkinsville, Ky, Oct 24, 41; m 59; c 2. THEORETICAL PHYSICS. *Educ:* Murray State Univ, BS, 62; Univ Ky, PhD(physics), 66. *Prof Exp:* TECH STAFF MEM, SANDIA CORP, 66- *Mem:* Am Phys Soc. *Res:* Equation of state; hydrodynamics; radiation transport; molecular relaxation. *Mailing Add:* 1201 Arizona NE Albuquerque NM 87110

THOMPSON, SHELDON LEE, b Minneapolis, Minn, Oct 7, 38; m 62; c 3. CHEMICAL ENGINEERING. *Educ:* Univ Minn, Minneapolis, BS, 60, MS, 62. *Prof Exp:* Res engr, Sun Oil Co, 62-69, assoc engr, 69-70, chief, Eng Res Lab, 70-72, res prog mgr, Eng Res, 72-74, mgr, Venture Eng, Sun Ventures, 74-77, mgr chem, 77-80, dir, Appl Res & Develop Dept, 80-, VPRES TECH CHEM, LUBES, SUN REFINING & MKT CO. *Mem:* Am Inst Chem Engrs; Am Petrol Inst; Ind Res Inst. *Res:* Process engineering research and appropriate computer simulation leading to the design of new petroleum and chemical plants; analytical risk-related economic studies; petroleum product development. *Mailing Add:* Sun Refining & Mkt Co Ten Penn Ctr 1801 Market St Philadelphia PA 19103

THOMPSON, SHIRLEY JEAN, b Danville, Va, Dec 19, 37. NOISE-EFFECTS EPIDEMIOLOGY, REPRODUCTIVE EPIDEMIOLOGY. *Educ:* Med Col Va, BS, 60; Univ NC, Chapel Hill, MS, 65, PhD(epidemiol), 72. *Prof Exp:* Instr nursing, Roanoke Mem Hosp, 60-61; staff nurse & supvr, IVNA-CNS, Richmond, Va, 61-63; asst prof community health, Med Col Va, 65-67; res assoc epidemiol, Univ NC, Chapel Hill, 69-70; assoc prof edidemiol & nursing, Med Col Va, 72-76; ASSOC PROF EPIDEMIOL, SCH PUB HEALTH, UNIV SC, 77- *Concurrent Pos:* mem, Int Comn Biol Effects Noise. *Mem:* Am Col Epidemiol; Soc Epidemiol Res; Am Pub Health Asn; Geront Soc Am; Aerospace Med Asn; Biol Res Units Intensive Treatment Noise; Int Soc Environ Epidemiol; Soc Pediat Epidemiol Res. *Res:* Epidemiology of noise-related health effects and communicative disorders as they relate to aging; perinatal reproductive epidemiology. *Mailing Add:* 4514 Sylvan Dr Columbia SC 29206

THOMPSON, SHIRLEY WILLIAMS, b Laurens, SC, Oct 12, 41; m 69; c 3. MATHEMATICAL STATISTICS. *Educ:* Johnson C Smith Univ, BS, 63; Univ NC, MAEd, 71; Ga State Univ, PhD(career & math develop), 80. *Prof Exp:* Comput prog, Celanese Corp, Charlotte, NC, 67, Wyoming Hosp Med Serv, 68-69; instr math, C A Johnson High, Columbia, SC, 63-65; instr math, East Mechlenburg High, Charlotte, NC, 69-71, Cent Piedmont Community Col, 70, De Kalb Community Col, Ga, 75-80 & Ga State Univ, Atlanta, 76-77; ASST PROF MATH, MOREHOUSE COL, ATLANTA, 80- *Concurrent Pos:* Math consult, Proj Opportunity, Univ NC, 70, Richmond County Sch Syst, Augusta, Ga, 75-76. *Mem:* Math Asn Am; Nat Coun Teachers Math. *Res:* Testing the effects of career education awareness in mathematics on the career maturity, attitude toward mathematics and mathematics achievement of students in beginning algebra at a community college. *Mailing Add:* Dept Math Morehouse Col 830 Westview Dr SW Atlanta GA 30314

THOMPSON, STARLEY LEE, b Victoria, Tex, Jan 16, 54. CLIMATOLOGY, CLIMATE MODELING. *Educ:* Tex A&M Univ, BS, 76, MS, 77; Univ Wash, PhD(atmospheric sci), 83. *Prof Exp:* Fel, 82-83, SCIENTIST, NAT CTR ATMOSPHERIC RES, 83- *Mem:* Am Meterol Soc; Am Geophys Union. *Res:* Climate theory and numerical simulation of global climate; paleoclimatic theory; impact of man-made influences on future climates. *Mailing Add:* Nat Ctr Atmospheric Res PO Box 3000 Boulder CO 80307

THOMPSON, STEVEN RISLEY, b Hermiston, Ore, Dec 3, 38; m 68. ZOOLOGY, GENETICS. *Educ:* Portland State Col, BS, 61; Ore State Univ, MS, 64, PhD(zool), 66. *Prof Exp:* Res assoc genetics, Univ Notre Dame, 66-68; asst prof, 68-73, ASSOC PROF GENETICS, ITHACA COL, 73- *Mem:* Nat Asn Biol Teachers; Nat Sci Teachers Asn. *Res:* General and population genetics of Drosophila melanogaster; Canalizina selection. *Mailing Add:* Dept Biol Ithaca Col Ithaca NY 14850

THOMPSON, SUE ANN, b New Orleans, La, March 26, 38. REPRODUCTIVE ENDOCRINOLOGY, PROSTATIC CANCER. *Educ:* Univ Ala, BS, 59; La State Univ Med Ctr, MS, 71, PhD(physiol), 75. *Prof Exp:* Assoc biologist, Southern Res Inst, 59-65; res biologist, Med Sch, Tulane Univ, 65-82; assoc histol & anat, 75-78, ASST PROF ANAT, UNIV IOWA, 78-, RES BIOLOGIST, DEPT SURG, UNIV HOSP. *Mem:* Sigma Xi; Am Asn Anatomists. *Res:* Prolactin and testosterone interaction in the initiation and maintenance of prostatic cancer using in vivo and in vitro model systems. *Mailing Add:* Dept Surg 321 Med Lab Univ Hosp Univ Iowa Iowa City IA 52242

THOMPSON, THOMAS EATON, b San Mateo, Calif, Aug 10, 38; div; c 2. SOLID STATE PHYSICS. *Educ:* Univ Calif, Berkeley, AB, 60; Univ Pa, MS, 62, PhD(physics), 69. *Prof Exp:* Asst prof solid state electronics, Univ Pa, 70-76; SR RES PHYSICIST, SRI INT, 77- *Mem:* Am Phys Soc. *Res:* Electronic properties of metals and semiconductors; graphite intercalation compounds; photovoltaic materials; magnetic quantum effects in solids; ultrasonics; surface acoustic waves. *Mailing Add:* 446 Los Altos Ave Los Altos CA 94022-1603

THOMPSON, THOMAS EDWARD, b Cincinnati, Ohio, Mar 15, 26; m 53; c 4. BIOCHEMISTRY. *Educ:* Kalamazoo Col, BA, 49; Harvard Univ, PhD(biochem), 55. *Prof Exp:* From asst prof to assoc prof physiol chem, Sch Med, Johns Hopkins Univ, 58-66; chmn dept, 66-76, PROF BIOCHEM, SCH MED, UNIV VA, 66-, HARRY FLOOD BYRD, JR PROF BIOCHEM, 83- *Concurrent Pos:* NIH res fel biochem, Harvard Univ, 55-57; Swed-Am exchange fel, Am Cancer Soc, LKB-Produkter Fabrikasaktiebolog, Sweden, 57-58; hon res fel, Birmingham, Eng, 58; ed, Biophys J, 88. *Honors & Awards:* K C Cole Award, Biophys Soc, 80; Alexander von Humboldt Prize, 86. *Mem:* Am Chem Soc; Biophys Soc (pres, 76); Am Soc Biol Chemists; Sigma Xi. *Res:* Physical chemistry of proteins; lipid protein interactions; biological membrane structure. *Mailing Add:* Dept Biochem Univ Va Sch Med Charlottesville VA 22908

THOMPSON, THOMAS LEIGH, b Erie, Pa, Feb 18, 41; m 66; c 2. AGRICULTURAL ENGINEERING. *Educ:* Pa State Univ, BS, 62; Purdue Univ, MS, 64, PhD(agr eng), 67. *Prof Exp:* Grad asst, Dept Agr Eng, Purdue Univ, 62-63, res asst agr eng, 63-66; from asst prof to assoc prof, 66-73, PROF BIOL SYSTS ENG, UNIV NEBR, LINCOLN, 73- *Concurrent Pos:* Consult grain drying, comput simulation. *Mem:* Am Soc Agr Engrs. *Res:* Design support systems for agriculture; animal growth simulators. *Mailing Add:* L W Chase Hall Univ Nebr Lincoln NE 68583-0726

THOMPSON, THOMAS LEO, b Gering, Nebr, Dec 8, 22; m 46; c 3. BACTERIAL PHYSIOLOGY. *Educ:* Univ Nebr, AB, 48, MS, 50; Univ Tex, PhD, 53. *Prof Exp:* From asst prof to prof microbiol, Univ Nebr, Lincoln, 52-89, actg chmn dept, 70-76; RETIRED. *Mem:* Am Soc Microbiol. *Res:* Bacterial physiology, mainly resistance to antibiotics; bacterial genetics in relation to thermophily and the transformation phenomenon. *Mailing Add:* 370 Bruce Dr Lincoln NE 68510

THOMPSON, THOMAS LUMAN, b Boulder, Colo, Dec 25, 27; m 56; c 6. GEOLOGY. *Educ:* Univ Colo, BA, 50; Stanford Univ, PhD(geol), 62. *Prof Exp:* Geologist, Phillips Petrol Corp, 50-51; staff res scientist, Amoco Prod Co, 62-76; prof geol & geophys, Univ Okla, 76-81; PRES, GEO-DISCOVERY INC, 81- *Mem:* Geol Soc Am; Am Asn Petrol Geol; Am Geophys Union. *Res:* World tectonics; structure of continental margins; deep water petroleum and mineral exploration. *Mailing Add:* Geo-Discovery Inc 580 Euclid Ave Boulder CO 80302

THOMPSON, THOMAS LUTHER, b Houston, Tex, Feb 28, 38. GEOLOGY. *Educ:* Univ Kans, BS, 60, MS, 62; Univ Iowa, PhD(geol), 65. *Prof Exp:* GEOLOGIST, MO GEOL SURV & WATER RESOURCES, 65-, CHIEF AREAL GEOL & STRATIG, 71- *Mem:* Geol Soc Am; Paleont Soc; Soc Econ Paleont & Mineral; Int Paleont Union; Pander Soc. *Res:* Stratigraphy; biostratigraphy; micropaleontology; correlation of Paleozoic strata through the use of paleontology. *Mailing Add:* 580 Euclid Ave Boulder CO 80302

THOMPSON, THOMAS WILLIAM, b Canton, Ohio, May 25, 36; m 66; c 2. SPACE PHYSICS. *Educ:* Case Inst Technol, BS, 58; Yale Univ, ME, 59; Cornell Univ, PhD(elec eng), 66. *Prof Exp:* Engr, Sylvania Elec Prod, 59-61; res asst radar astron, Arecibo Observ, Cornell Univ, 63-64, 66-69; mem tech staff, Jet Propulsion Lab, Calif Inst Technol, 69-76; staff scientist, Planetary Sci Inst, Pasadena, Ca, 77-81; MEM TECH STAFF, JET PROPULSION LAB, CALIF INST TECHNOL, 81- *Mem:* Am Geophys Union; Am Astron Soc; Int Astron Union; Int Union Radio Sci; Inst Elec & Electronics Engrs. *Res:* Radar astronomy; mapping of lunar radar echoes; resolution of the delay-Doppler ambiguity. *Mailing Add:* 3043 Cloudcrest Rd La Crescenta CA 91214

THOMPSON, TIMOTHY J, b Alhambra, Calif, Dec 1, 49. RADIO ASTRONOMY. *Educ:* Calif State Univ, Los Angeles, BS, 78, MS, 85. *Prof Exp:* Mem tech staff, Ball Aerospace Corp, 82-84; MEM TECH STAFF, JET PROPULSION LAB, 84- *Mem:* AAAS; Am Phys Soc. *Res:* Primarily planetary radio astronomy; outer planet atmospheres and magnetic fields. *Mailing Add:* 1947 E Huntington Dr Apt C Duarte CA 91010

THOMPSON, TOMMY BURT, b Tucumcari, NMex, Apr 3, 38; m 58; c 4. ECONOMIC GEOLOGY, PETROLOGY. *Educ:* Univ NMex, BS, 61, MS, 63, PhD(geol), 66. *Prof Exp:* From asst prof to assoc prof geol, Okla State Univ, 66-73; assoc prof, 73-81, PROF GEOL, COLO STATE UNIV, 81- *Mem:* Geol Soc Am; Am Inst Mining, Metall & Petrol Engrs; Soc Econ Geologists. *Res:* Conceptual models for exploration of metallic resources; mineral resources of Colorado; igneous petrology; hydrothermal alteration of igneous rocks; exploration for gold in sedimentary and igneous rocks. *Mailing Add:* Dept Earth Resources Colo State Univ Ft Collins CO 80523

THOMPSON, TOMMY EARL, b Dublin, Tex, June 18, 44; m 72; c 3. PLANT GENETICS, PLANT BREEDING. *Educ:* Tex A&M Univ, BS, 66, MS, 70; Purdue Univ, West Lafayette, PhD(genetics & plant breeding), 73. *Prof Exp:* Field researcher agr econ, Tex A&M Univ, 65, res asst plant breeding, 66-67 & 69-70; res asst plant genetics, Purdue Univ, West Lafayette, 70-73, asst prof forage breeding, 73-74; res geneticist flax genetics, Agr Res Serv, NDak, 74-76, res geneticist sunflower genetics, Agr Res Ser, 76-79, RES GENETICIST PECAN GENETICS, AGR RES SERV, USDA, 79- *Concurrent Pos:* Sci adv, Flax Inst US, 74-76; nat tech adv sunflower prod, Sci & Educ Admin-Fed Res, USDA, 77-79; instr advan plant breeding, WTex State Univ, 78-79. *Mem:* Crop Sci Soc Am; Am Soc Agron; Am Genetic Asn. *Res:* Genetics; pecan research; yieldability of pecans; heritability of yield traits; heritability of dichogamy; insect resistance; disease resistance of pecans. *Mailing Add:* USDA W R Poage Pecan Field Sta 701 Woodson Rd Brownwood TX 76801

THOMPSON, TRUET B(RADFORD), b Noble, La, Mar 24, 17; m 41; c 3. ELECTRICAL ENGINEERING, ELECTRICAL SAFETY. *Educ:* La Polytech, BS, 47, BSEE, 48; Okla State Univ, MSEE, 50; Northwestern Univ, PhD(elec eng), 56. *Prof Exp:* From asst prof to assoc prof elec eng, Okla State Univ, 48-59; prof, 59-83, chmn dept, 60-67, EMER PROF ENG, ARIZ STATE UNIV, 83- *Concurrent Pos:* Forensic consult. *Mem:* Nat Soc Prof Engrs; Inst Elec & Electronics Engrs; Nat Fire Protection Asn; Am Nat Standards Inst; Am Soc Safety Engrs; Nat Acad Forensic Engrs. *Res:* Circuit theory; power systems; statistical communications theory. *Mailing Add:* 230 E Garfield St Tempe AZ 85281

THOMPSON, VINTON NEWBOLD, b Mt Holly, NJ, July 24, 47; m 75; c 2. EVOLUTIONARY GENETICS, ECOLOGICAL GENETICS. *Educ:* Harvard Univ, AB, 69; Univ Chicago, PhD(genetics), 74. *Prof Exp:* Indust hygienist, Ill Dept Labor, 75; indust hygienist, Occup Safety & Health Admin, US Dept Labor, 75-77; asst prof, 80-85, ASSOC PROF BIOL, ROOSEVELT UNIV, 85- *Mem:* Soc Study Evolution; Genetics Soc Am. *Res:* Drosophila population genetics; mathematical models of frequency-dependent selection; ecological genetics of spittlebugs. *Mailing Add:* Dept Biol Roosevelt Univ 430 S Michigan Ave Chicago IL 60605

THOMPSON, W(ILLIAM) E(DGAR), fluids engineering, turbomachinery, for more information see previous edition

THOMPSON, W P(AUL), b Elmira, NY, June 3, 34; m 77; c 4. CHEMICAL PHYSICS, LASERS. *Educ:* Yale Univ, BS, 55; Lehigh Univ, MS, 57, PhD(physics), 63. *Prof Exp:* Mem tech staff, Aerophys Dept, Aerospace Corp, 61-67, sect head, Exp Aerophys, Aerodyn & Propulsion Lab, 67-68, dir countermeasures, Reentry Systs Div, 68-74, assoc prin dir technol develop, Reentry Systs Div, 74-79, prin dir space technol planning, 79-81, dir, Aerophys Lab, 81-89, PRIN SCIENTIST, DEVELOP GROUP, AEROSPACE CORP, 89- *Concurrent Pos:* instr physics, Moravian Co, 57-58; astron, Los Angeles Trade-Tech Col, 67. *Mem:* Am Phys Soc; assoc fel Am Inst Aeronaut & Astronaut; Sigma Xi. *Res:* Shock tube gasdynamics; gas kinetics; microwave-plasma interactions; reentry physics; electronic and optical countermeasures; system engineering. *Mailing Add:* Aerospace Corp PO Box 92957 Mail Sta M1-008 Los Angeles CA 90009

THOMPSON, WARREN CHARLES, b Santa Monica, Calif, May 22, 22; m 48; c 4. OCEANOGRAPHY. *Educ:* Univ Calif, Los Angeles, BA, 43; Univ Calif, San Diego, MS, 48; Agr & Mech Col Tex, PhD, 53. *Prof Exp:* Asst, Scripps Inst, Univ Calif, San Diego, 46-47, 47-48; proj dir, Tex A&M Res Found, 50-52; assoc prof aerol & oceanog, 53-59, PROF OCEANOG, NAVAL POSTGRAD SCH, 59- *Concurrent Pos:* Assoc petrol engr, Humble Oil & Ref Co, La, 47; sci liaison officer, London Br, US Off Naval Res, 60-61; vpres, Oceanog Serv Inc, Santa Barbara, 65-66; consult, 50- *Mem:* AAAS; Geol Soc Am; Soc Econ Paleont & Mineral; Am Meteorol Soc; Am Asn Petrol Geol; Sigma Xi. *Res:* Shallow water processes; ocean waves. *Mailing Add:* 830 Dry Creek Rd Monterey CA 93940

THOMPSON, WARREN ELWIN, b Joliet, Ill, June 15, 30; m 62; c 2. SCIENCE ADMINISTRATION, PHYSICAL CHEMISTRY. *Educ:* Univ Wis, BS, 51; Harvard Univ, AM, 53, PhD(chem), 56. *Prof Exp:* Fulbright scholar, Kamerlingh Onnes Lab, Univ Leiden, 55-57; instr & asst prof chem, Univ Calif, Berkeley, 57-59; asst prof, Case Western Reserve Univ, 59-65; prog mgr, Div Int Progs, Nat Sci Found, 65-76, 77-83, policy analyst, Div Policy Res & Anal, 76-77, sr prog mgr, Div Int Progs, 83-91; GUEST RESEARCHER, NAT INST STANDARDS & TECHNOL, GAITHERSBURG, MD, 88- *Mem:* Am Chem Soc; AAAS. *Res:* Molecular spectroscopy; photochemistry; chemical studies related to astronomy. *Mailing Add:* 4509 Amherst Lane Bethesda MD 20814

THOMPSON, WARREN SLATER, b Utica, Miss, Aug 19, 29; m 53; c 4. WOOD SCIENCE & TECHNOLOGY. *Educ:* Auburn Univ, BS, 51, MS, 55; NC State Univ, PhD, 60. *Prof Exp:* Asst forester, Miss State Univ, 53-54; wood technologist, Masonite Corp, 57-59; asst & assoc prof forestry, La State Univ, 59-64; DIR, FOREST PROD LAB, MISS STATE UNIV, 64-, PROF WOOD SCI & TECHNOL & HEAD DEPT, 74- *Honors & Awards:* Fred Gottschalk Mem Award, Forest Prod Res Soc. *Mem:* Forest Prod Res Soc; Soc Wood Sci & Technol; Am Soc Testing & Mat; Tech Asn Pulp & Paper Indust; Am Wood-Preservers' Asn. *Res:* Wood pathology and preservation. *Mailing Add:* Dean Forest Res Miss State Univ PO Box 5328 Mississippi State MS 39762

THOMPSON, WAYNE JULIUS, b Chicago, Ill, Oct 18, 52; m; c 2. SYNTHESIS. *Educ:* Ill Inst Technol, BS, 74; Calif Inst Technol, PhD(chem), 78. *Prof Exp:* NIH FEL, MASS INST TECHNOL, 79-80. *Concurrent Pos:* Asst prof, Univ Calif, Los Angeles. *Mem:* Am Chem Soc. *Res:* Design and synthesis of pharmaceutical agents. *Mailing Add:* 2291 Locust Dr Lansdale PA 19446

THOMPSON, WESLEY JAY, b Alice, Tex, Dec 10, 47; m 86; c 1. NEUROMUSCULAR DEVELOPMENT. *Educ:* NTex State Univ, BS, 70, MA, 71; Univ Calif, Berkeley, PhD(molecular biol), 75. *Prof Exp:* Fel neurobiol, Inst Physiol, Univ Oslo, 75-77; fel, Sch Med, Wash Univ, 77-78; asst prof, 79-85, ASSOC PROF ZOOL, UNIV TEX, 85- *Concurrent Pos:* Muscular Dystrophy Asn fel, 75-76; NATO fel, 76-77; Searle scholar. *Honors & Awards:* Res Career Develop Award, NIH. *Mem:* AAAS; Soc Neurosci. *Res:* The formation and maintenance of synaptic connections at the neuromuscular junction. *Mailing Add:* Dept Zool Univ Tex Austin TX 78712

THOMPSON, WILEY ERNEST, b Murphysboro, Ill, June 30, 41; m 62; c 2. ELECTRICAL & SYSTEMS ENGINEERING. *Educ:* Southern Ill Univ, Carbondale, BS, 63; Mich State Univ, MS, 64, PhD(elec eng), 68. *Prof Exp:* Design engr, Olin Mathieson Chem Corp, 63; asst elec eng, Mich State Univ, 63-64, res asst systs, 64-68, asst prof elec eng & systs, 68; asst prof, 68-72, assoc prof, 72-80, PROF ELEC ENG, NMEX STATE UNIV, 80- *Concurrent Pos:* Consult, White Sands Missile Range, 69-; NSF initiation grant, NMex State Univ, 70-72. *Mem:* Inst Elec & Electronics Engrs; Am Soc Eng Educ; Sigma Xi. *Res:* Systems optimization, stability; mathematical modeling; computer-aided analysis and design; systems structure; guidance and control; stochastic systems; ecological systems. *Mailing Add:* Box 3-0 Dept Elec Eng NMex State Univ Las Cruces NM 88003

THOMPSON, WILLIAM, JR, b Hyannis, Mass, Dec 4, 36; m 59; c 4. ACOUSTICS, ELECTROACOUSTIC TRANSDUCERS. *Educ:* Mass Inst Technol, BS, 58; Northeastern Univ, MS, 63; Pa State Univ, PhD(eng acoust), 71. *Prof Exp:* Jr engr, Raytheon Co, 58-60; sr engr, Cambridge Acoust Assocs, Inc, 60-66; res asst transducer studies, 66-72, from asst prof to assoc prof acoust mech, Appl Res Lab, 72-80, assoc prof, 80-85, PROF ENG SCI, APPL RES LAB & DEPT ENG SCI & MECH, PA STATE UNIV, 85- *Mem:* Fel Acoust Soc Am; Inst Elec & Electronics Engrs; Soc Eng Sci. *Res:* Electroacoustic transducer design, construction and calibration; acoustic radiation and scattering; underwater acoustics. *Mailing Add:* 601 Glenn Rd State College PA 16803-3475

THOMPSON, WILLIAM A, b Moorestown, NJ, Oct 25, 36; m 61; c 1. SOLID STATE PHYSICS. *Educ:* Drexel Inst Tech, BS, 59; Univ Pittsburgh, PhD(physics), 64. *Prof Exp:* RES STAFF MEM PHYSICS, THOMAS J WATSON RES CTR, IBM CORP, 64- *Mem:* Am Phys Soc; Fedn Am Sci. *Res:* Electron tunneling properties of superconductors and magnetic semiconductors. *Mailing Add:* 1956 Glen Rock Rd Yorktown Heights NY 10598

THOMPSON, WILLIAM BALDWIN, b Meriden, Conn, Mar 28, 35; m 60; c 3. GEOPHYSICS. *Educ:* Mass Inst Technol, BS & MS, 58, PhD(geophys), 63. *Prof Exp:* Mem tech staff, Bellcomm, Inc, 63-72; head dept, Bell Labs, 72-75; dir, Am Bell Int, Inc, 75-77; MGR, AM TEL & TEL LONG LINES, 77- *Mem:* Am Geophys Union. *Res:* Electromagnetic cavity resonance phenomena in the earth's atmosphere; physical properties of lunar and planetary surfaces and atmospheres; scientific mission planning for Apollo lunar exploration and planetary missions; economic analyses of telephone network costs and investment; business economics. *Mailing Add:* Nine Oak Knoll Rd Mendham NJ 07945

THOMPSON, WILLIAM BELL, b Belfast, Northern Ireland, Feb 27, 22; m 52, 72; c 2. PLASMA PHYSICS. *Educ:* Univ BC, BA, 45, MA, 47; Univ Toronto, PhD(math), 50. *Hon Degrees:* MA, Oxford Univ, 62. *Prof Exp:* Sr fel, UK Atomic Energy Authority, Harwell, Eng, 50-53, sr scientist plasma theory, 53-60, dep chief scientist, Culham Lab, 60-62; prof, Oxford Univ, 63-65; PROF PHYSICS, UNIV CALIF, SAN DIEGO, 65- *Concurrent Pos:* Vis researcher, Plasma Physics Lab, Princeton Univ, 60; vis prof, Univ Calif, San Diego, 60-61 & Univ Colo, 72; ed, Advances in Plasma Physics; assoc ed, J Plasma Physics. *Honors & Awards:* Hulton Award, UK, 58. *Mem:* Fel Am Phys Soc; Can Asn Physicists; fel Royal Astron Soc; AAAS. *Res:* Theoretical physics; controlled thermonuclear fusion. *Mailing Add:* Dept Physics Univ Calif San Diego La Jolla CA 92093

THOMPSON, WILLIAM BENBOW, JR, b Detroit, Mich, July 26, 23; m 47, 58; c 3. OBSTETRICS & GYNECOLOGY. *Educ:* Univ Southern Calif, AB, 47, MD, 51; Am Bd Obstet & Gynec, dipl. *Prof Exp:* Intern, Harbor Gen Hosp, Los Angeles, 51-52; resident obstet & gynec, Galliinger Munic Hosp, Washington, DC, 52-53; from resident to sr resident, George Washington Univ Hosp, 53-55; asst, La State Univ, 55-56; clin instr, Univ Calif, Los Angeles, 56-62, asst clin prof, 62-64; assoc prof, Calif Col Med, 64-66; asst prof, Univ Calif, 66-73, assoc dean med student serv, 69-73, actg chmn, 73-77, ASSOC PROF OBSTET & GYNEC & DIR, GYNEC DIV, COL MED, UNIV CALIF, IRVINE, 77- *Concurrent Pos:* Dir obstet & gynec, Orange County Med Ctr, Orange, 67- *Mem:* Fel Am Col Obstetricians & Gynecologists; fel Am Col Surgeons. *Res:* Techniques in tubal sterilization. *Mailing Add:* UCI Fac Med Grp 101 The City Dr Orange CA 92668

THOMPSON, WILLIAM FRANCIS, III, b Seattle, Wash, Jan 21, 45; m 65; c 2. PLANT MOLECULAR BIOLOGY, PLANT DEVELOPMENT. *Educ:* Princeton Univ, AB, 66; Univ Wash, PhD(plant physiol), 70. *Prof Exp:* NSF fel biol, Harvard Univ, 70-72; asst prof bot, Univ Mass, 72-74; mem staff plant biol, Carnegie Inst Wash, 74-86; UNIV PROF BOT & GENETICS, NC STATE UNIV, 86- *Concurrent Pos:* Adj asst prof, Univ Mass, 74-79; vis scientist, Plant Breeding Inst, Cambridge, UK, 82-83; asst-assoc prof by courtesy, Stanford Univ, 74-86. *Mem:* Am Soc Plant Physiologists; Genetics Soc Am; Int Soc Plant Molecular Biol; AAAS. *Res:* Control of plant gene expression; plant molecular genetics. *Mailing Add:* Dept Bot NC State Univ Raleigh NC 27695

THOMPSON, WILLIAM HORN, b Somerville, NJ, Feb 9, 37; m 60; c 3. CHEMICAL ENGINEERING. *Educ:* Pa State Univ, BS, 61, MS, 62, PhD(chem eng), 66. *Prof Exp:* Res asst, Dept Chem Eng, Pa State Univ, 62-65, instr, 65-66; engr, Petrol Processing Dept, Shell Develop Co, Calif, 66-68; group leader, Technol Dept, Wood River Refinery, Shell Oil Co, 68-69, asst mgr, Lube Oil Dept, 69-70, mgr, Refinery Lab, 70-71 & Catalytic Cracking Dept, 71-72, supt opers light oil processing, 72-73, sr staff engr, 73-74, mgr, supply & qual lube, 74-75, mfg oper, 75-77, fuels logistics, 77-78, mgr, Gasoline Bus Ctr, 78-80; supt, 80-82, complex mgr, 82-85, MGR MFG MGT SYSTS, DEER PARK MFG COMPLEX, 85- *Mem:* Am Inst Chem Engrs. *Res:* Physical thermodynamic and transport properties of hydrocarbons and related substances. *Mailing Add:* Shell Oil Co PO Box 2463 Houston TX 77252

THOMPSON, WILLIAM LAY, b Austin, Tex, Feb 16, 30; m 58; c 3. VERTEBRATE ZOOLOGY. *Educ:* Univ Tex, BA, 51, MA, 52; Univ Calif, Berkeley, PhD(zool), 59. *Prof Exp:* From asst prof to assoc prof, 59-71, PROF BIOL, WAYNE STATE UNIV, 71- *Mem:* Fel AAAS; Animal Behav Soc; Am Ornith Union; Wilson Ornith Soc. *Res:* Animal behavior, especially communication and habitat selection in birds. *Mailing Add:* Dept Biol Wayne State Univ Detroit MI 48202

THOMPSON, WILLIAM OXLEY, II, b Richmond, Va, Apr 25, 41; m 63; c 2. MATHEMATICAL STATISTICS, TECHNICAL MANAGEMENT. *Educ:* Univ Va, BA, 63; Va Polytech Inst & State Univ, PhD(statist), 68. *Prof Exp:* Teacher high sch, Va, 63-64; asst prof & adj assoc prof statist, Univ Ky, 67-75; mgr statist serv group, Tech Serv Div, Agr Mkt Serv, 74-76, dir, Tech Serv Div, 76-79, STATIST & SYSTS COORDR RES, FOREST SERV, USDA, 79- *Concurrent Pos:* Consult, Clin Res Ctr, NIMH, 68-71, consult, Addiction Res Ctr, 70-72, math statistician, 72-74; adj prof statist, Va Polytech Inst & State Univ, 78- *Mem:* Am Statist Asn; Biomet Soc; Sigma Xi. *Res:* Experimental designs and analysis for estimating linear and nonlinear models; estimation of variance components; development of statistical methodology in fields of application; biostatistics. *Mailing Add:* 4260 Quail Springs Circle Augusta GA 30907

THOMPSON, WILMER LEIGH, b Shreveport, La, June 25, 38; m 57; c 1. CLINICAL PHARMACOLOGY, CRITICAL CARE MEDICINE. *Educ:* Col Charleston, BS, 58; Med Univ SC, MS, 60, PhD(pharmacol), 63; Johns Hopkins Univ, MD, 65; Am Bd Internal Med, dipl, 71. *Prof Exp:* Intern, Osler Med Serv, Johns Hopkins Hosp, 65-66, resident, 66-67 & 69-70; asst prof med & pharmacol, Sch Med, Johns Hopkins Univ, 70-74; assoc prof med & pharmacol, Sch Med, Case Western Reserve Univ, 74-80, prof, 80-82; dir, 82, exec dir, 82-87, vpres, 87-89, GROUP VPRES, LILLY RES LAB, ELI LILLY & CO, INDIANAPOLIS, 89- *Concurrent Pos:* Fels med, Sch Med, Johns Hopkins Univ, 65-67 & 69-70; staff assoc, Nat Cancer Inst, 67-69; asst physician & dir med intensive care unit, Johns Hopkins Hosp, 70-74; physician & dir clin pharmacol prog, Univ Hosps Cleveland, 74-; adj prof, Sch Libr Sci, Case Western Reserve Univ, 74-; adj assoc prof, Dept Pharmacol, Col Med, Ohio State Univ, 75-; Burroughs Wellcome scholar clin pharmacol, 75-80. *Mem:* Am Soc Pharmacol & Exp Therapeut; Soc Critical Care Med (pres, 81-82); Am Soc Clin Res; fel Am Col Physicians; Am Fedn Clin Res; Am Soc Clin Pharmacol & Therapeut. *Res:* Critical care medicine; medical decision analysis; clinical toxicology; regulatory affairs; artificial blood; large scale clinical trials. *Mailing Add:* Lilly Res Lab Lilly Corp Ctr Indianapolis IN 46285

THOMPSON, WYNELLE DOGGETT, b Birmingham, Ala, May 25, 14; m 38; c 4. BIOCHEMISTRY, ORGANIC CHEMISTRY. *Educ:* Birmingham-Southern Col, BS, 34, MS, 35; Univ Ala, MS, 56, PhD(biochem), 60. *Prof Exp:* Instr chem, Birmingham-Southern Col, 35-36; high sch instr gen sci, Ala, 36-37; jr chemist, Bur Home Econ, USDA, Washington, DC, 37-38; high sch instr gen sci, Ala, 40-41; instr chem, Birmingham-Southern Col, 41-44; instr, Exten Ctr, Univ Ala, 50-52 & 53-55; from asst prof to prof, 55-76, EMER PROF CHEM, BIRMINGHAM-SOUTHERN COL, 76- *Concurrent Pos:* Adj prof biochem, Med Col, Univ Ala, Birmingham, 76-78; adj prof chem, Jeff State Jr Col, Birmingham, 80-87. *Mem:* Am Chem Soc; Sigma Xi. *Res:* Protozoa growth and culture; enzymes, structure of and assays techniques; allosteric effects; fluorescence produced; circular dichroism. *Mailing Add:* 917 Valley Rd Pl Birmingham AL 35208-1020

THOMS, RICHARD EDWIN, b Olympia, Wash, June 5, 35; m 67. PALEONTOLOGY. *Educ:* Univ Wash, Seattle, BS, 57, MS, 59; Univ Calif, PhD(paleont), 65. *Prof Exp:* Teaching asst peleont, Univ Calif, 60-64; asst prof, 64-70, assoc prof, 70-80, PROF GEOL, PORTLAND STATE UNIV, 80- *Mem:* Paleont Soc. *Res:* West coast marine Tertiary biostratigraphy; ichnology. *Mailing Add:* Dept Geol Portland State Univ PO Box 751 Portland OR 97207

THOMSEN, HARRY LUDWIG, b Boise, Idaho, June 14, 11; m 35; c 3. GEOLOGY, OIL & GAS RESOURCE APPRAISAL. *Educ:* Oberlin Col, AB, 32, MA, 34. *Prof Exp:* Seismic party chief, Shell Oil Co, Calif, 35-38, seismologist, 38-41, div geophysicist, Calif & Rocky Mt area, 41-48, dist geologist, Colo, 48-51, area geologist, Okla, 51-53; spec assignment, Bataafse Petrol Maatschappij NV, Holland, 53-54; div explor mgr, Shell Oil Co, Mont, 54-60, mgr, explor econ, NY, 60-66, sr staff geologist, Shell Develop Co, Tex, 66-69, spec asst to vpres explor, Shell Oil Co, 69-70; geol consult, 70-74; geophysicist, US Geol Surv, 74-76, geol consult, 76-77; geol consult, Colo Sch Mines, 81-85; RETIRED. *Concurrent Pos:* Pres, Rocky Mt Sect, Am Asn Petrol Geol, 59-60, mem, Exec Adv Comt & vchmn, Indust & Acad Relations Comt, 64-65, mem, Energy Minerals Comt, 71-75, Petrol Resources Estimation Proj, 76-77; guest lectr, Univ Houston, 70 & Colo Sch Mines Grad Sch, 72. *Mem:* Am Asn Petrol Geol; sr fel Geol Soc Am; Soc Explor Geophys; Am Inst Prof Geologists. *Res:* Exploration for oil and gas; development and application of methods for evaluating petroleum exploration opportunities; estimation of undiscovered oil and gas resources. *Mailing Add:* 13717 E Marina Dr No B Aurora CO 80014-3775

THOMSEN, JOHN STEARNS, b Baltimore, Md, June 10, 21; m 52; c 4. ATOMIC PHYSICS. *Educ:* Johns Hopkins Univ, BE, 43, PhD(physics), 52. *Prof Exp:* Elec engr, Gen Elec Co, 43-45; asst prof physics, Univ Md, 50-51; res staff asst, Radiation Lab, Johns Hopkins Univ, 51-52, res scientist, 52-53; asst prof, Stevens Inst Technol, 53-54; res scientist, 54-55, asst prof mech eng, 55-61, res scientist, 62-70, FEL BY COURTESY PHYSICS, JOHNS HOPKINS UNIV, 70- *Concurrent Pos:* Mem comt fundamental constants, Nat Res Coun, 61-72, chmn, 69-71. *Mem:* Fel Am Phys Soc; Am Asn Physics Teachers. *Res:* Thermodynamics; irreversible processes; nonlinear electrical circuits; heat conduction with temperature dependent properties; statistical evaluation of atomic constants; x-ray wave lengths and precision experiments. *Mailing Add:* Dept Physics & Astron Johns Hopkins Univ Baltimore MD 21218

THOMSEN, LEON, b Tulsa, Okla, Oct 22, 42; m 65. ROCK PHYSICS, EXPLORATION GEOPHYSICS. *Educ:* Calif Inst Tech, BS, 64; Columbia Univ, PhD(geophys), 69. *Prof Exp:* Sr researcher, Ctr Nat Recherche Sci, 69-70; res fel, Calif Inst Tech, 70-72; from asst prof to assoc prof geophys, State Univ NY, 72-80; sr res scientist & staff res scientist, Amoco Corp, 80-86; RES ASSOC GEOPHYS, AMOCO CORP, 86- *Concurrent Pos:* Prof, Scuola Int Fisica Enrico Fermi, 70; consult, IBM, 70-72; vis res fel, Australian Nat Univ, 79; mem Mineral Phys Comt, Am Geophys Union, 83-; mem External Eval Comt, Div Earth Sci, Laurence-Berkeley Lab, 86-88; mem res comt, Soc Explor Geophys, 87- *Mem:* Am Geophys Union; Soc Explor Geophys; Am Physics Soc; AAAS. *Res:* Application of rock & mineral physics to exploration for hydrocarbons; anisotropic wave propagation in the sedimentary crust. *Mailing Add:* PO Box 27273 Tulsa OK 74149

THOMSEN, MICHELLE FLUCKEY, b Burlington, Colo, June 25, 50; m 73; c 2. SPACE PLASMA PHYSICS. *Educ:* Colo Col, BA, 71; Univ Iowa, MS, 74, PhD(physics), 77. *Prof Exp:* Res assoc, Univ Iowa, 77-80; res fel, Max Planck Inst Aeronomy, 80-81; STAFF SCIENTIST, LOS ALAMOS NAT LAB, 81- *Mem:* Am Geophys Union. *Res:* Physics of planetary magnetospheres and their interaction with the solar wind; Kinetic instabilities and wave-particle interactions in space plasmas; ion and electron heating and acceleration at collisionless shocks. *Mailing Add:* ESS-8 Mail Stop D438 Los Alamos Nat Lab Los Alamos NM 87545

THOMSEN, WARREN JESSEN, b Iowa, Mar 5, 22; m 43; c 3. MATHEMATICS. *Educ:* Iowa State Col, BA, 42; Univ Iowa, MS, 47, PhD(math), 52. *Prof Exp:* Instr math, Wis State Col, Whitewater, 52-53; res mathematician, Appl Physics Lab, Johns Hopkins Univ, 53-56; assoc prof math, Western Ill Univ, 56-57; prof & head dept, Mankato State Col, 57-65; head dept, 65-73, PROF MATH, MOORHEAD STATE COL, 65- *Mem:* Nat Coun Teachers Math; Math Asn Am. *Res:* Meteors; resistance of air to meteorites. *Mailing Add:* HCR 1 Box 273 Pequot Lakes MN 56472

THOMSON, ALAN, b Passaic, NJ, July 1, 28; m 53; c 6. GEOLOGY. *Educ:* WVa Univ, BS, 52, MS, 54; Rutgers Univ, PhD(geol), 57. *Prof Exp:* Asst geol, WVa Univ, 52-54; petrogr, Shell Oil Co, 57-65, sr geologist, 65, res geologist, 65-67, res assoc, 67-72, staff res geologist, 72-73, from staff geologist to sr staff geologist, 73-82, geol adv, 82-85; RETIRED. *Concurrent Pos:* Instr, Odessa Col, 58-62; geol consult & adj prof geol, Univ New Orleans, 85- *Mem:* Fel Geol Soc Am; Soc Econ Paleont & Mineral; Int Asn Sedimentol; Am Petrol Geologists; Soc Petrol Engrs. *Res:* Petrology of sedimentary rocks; determination of depositional environments; provenance of sandstones, sandstone diagenesis. *Mailing Add:* 36 Old Farmers Rd Long Valley NJ 07853

THOMSON, ALAN B R, b Toronto, Ont, May 29, 43; c 4. GASTROENTEROLOGY. *Educ:* Queen's Univ, BA, 65, MD, 67, MSc, 70, Phd, 71; FRCP(C), 74 & 75. *Prof Exp:* Asst prof med, Southwestern Med Sch, Dallas, 74-75; from asst prof to assoc prof, 75-82, PROF MED, UNIV ALTA, EDMONTON, 82- *Concurrent Pos:* Mem, med adv bd, Can Found Ileitis & Colitis, 77-82, med adv bd exec comt, 78-82; secy, med staff adv bd, Univ Alta Hosp, 81-82. *Mem:* Royal Soc Med; Can Asn Gastroenterol (vpres, 85-86); Am Gastroenterol Asn; NY Acad Sci; Am Geriat Soc; Can Physiol Soc; Can Soc Clin Pharmacol; Am Physiol Soc. *Res:* Mechanisms of intestinal adaptation in health and disease; alterations in intestinal transport, morphology and brush border membrane composition, following abdominal radiation, intestinal resection, chronic ethanol indigestion, aging, and in response to dietary manipulation. *Mailing Add:* Univ Alta 8-104 B Clin Sci Bldg Edmonton AB T6G 2E2 Can

THOMSON, ALAN JOHN, b Nov 4, 46; Brit & Can citizen; m 67; c 2. BIOLOGICAL SYSTEMS ANALYSIS, ECOLOGY. *Educ:* Glasgow Univ, BSc, 68; McMaster Univ, PhD(ecol), 72. *Prof Exp:* Res fel physiol ecol, Inst Animal Resource Ecol, Univ BC, 72-76; RES SCIENTIST BIOL SYST ANALYSIS, PAC FOREST RES CTR, CAN FORESTRY SERV, 76- *Concurrent Pos:* Killam res scholar, Univ BC, 74-76. *Mem:* Can Entom Soc. *Res:* Insect feeding behaviour; computer simulation of forest pest and disease dynamics and impact; effects of topography and weather on ecological processes; expert systems in resource management. *Mailing Add:* Pac Forest Res Ctr 506 W Burnside Rd Victoria BC V8Z 1M5 Can

THOMSON, ASHLEY EDWIN, b Regina, Sask, June 6, 21; m 47; c 7. MEDICINE, PHARMACOLOGY. *Educ:* Univ Sask, BA, 43; Univ Man, MD, 45, MSc, 48; FRCPS(C). *Prof Exp:* Prof med pharmacol & therapeut, 65-89, EMER PROF MED, UNIV MAN, 89- *Mem:* Can Soc Nephrology; Am Soc Nephrology; Int Soc Nephrology; Am Physiol Soc; Can Soc Clin Invest. *Res:* Hemodialysis and transplantation. *Mailing Add:* 172 Harvard Ave Winnipeg MB R3M 0K5 Can

THOMSON, DALE S, b Cleveland, Ohio, Feb 12, 34; m 56; c 3. DEVELOPMENTAL BIOLOGY. *Educ:* Cedarville Col, AB, 56; Ohio State Univ, MS, 62, PhD(zool), 65. *Prof Exp:* Teacher, Miami Christian Sch, 56-57; from asst prof to assoc prof biol, Cedarville Col, 57-67; assoc prof, 67-74, PROF BIOL, MALONE COL, 74-, CHMN, DIV SCI & MATH, 69- *Mem:* Am Sci Affil. *Res:* Developmental biology, especially cellular ultrastructure with respect to gland development in the chick. *Mailing Add:* Dept Biol Malone Col 515 25th St NW Canton OH 44709

THOMSON, DAVID JAMES, b Victoria, BC, June 25, 44; m 76. UNDERWATER ACOUSTICS. *Educ:* Univ Victoria, BS, 66, PhD(geophys), 73. *Prof Exp:* Fel physics, Univ Victoria, BC, 72-73 & Univ Alta, 73-74; DEFENCE SCIENTIST, DEFENCE RES ESTAB PAC, 74- *Mem:* Can Asn Physicists. *Res:* Geomagnetism; modelling of electric currents induced in non-uniform conductors; underwater acoustics; numerical modelling of underwater sound propagation in horizontally stratified and range dependent environments. *Mailing Add:* Defence Res Estab PAC FMO Victoria BC V0S 1B0 Can

THOMSON, DAVID M P, b Nov 4, 39; c 2. CANCER IMMUNOLOGY. *Educ:* Univ Western Ont, MD, 64; Univ London, Eng, PhD(immunol), 73. *Prof Exp:* Assoc physician, 72-82, SR PHYSICIAN, MONTREAL GEN HOSP, 83-, RES ASSOC, RES INST, 72- *Concurrent Pos:* From asst prof to assoc prof med, McGill Univ, 72-81, prof med, 81-, res assoc, McGill Cancer Ctr, 78- *Honors & Awards:* Med medal, Royal Col Physicians & Surgeons, 71; Presidential Award, New Eng Cancer Soc, 81. *Mem:* Can Soc Clin Invest; Am Asn Immunologists; Am Asn Clin Immunol & Allergies; Am Asn Cancer Res. *Res:* Human immune response to cancer; purification, definition and sequencing of putative tumor antigen; characterization of leukocytes responding to cancer antigens. *Mailing Add:* 1650 Cedar Ave Rm 7113 Montreal PQ H3G 1A4 Can

THOMSON, DENNIS WALTER, b New York, NY, Mar 14, 41; m 65; c 2. ENVIRONMENTAL EARTH & MARINE SCIENCES. *Educ:* Univ Wis-Madison, BS, 63, MS, 64, PhD(meteorol), 68. *Prof Exp:* Ger Acad Exchange Serv fel, Univ Hamburg, 68-69; from asst prof to assoc prof, 70-78, PROF METEOROL, COL EARTH & MINERAL SCI, PA STATE UNIV, UNIVERSITY PARK, 78- *Concurrent Pos:* Consult various industs & govt; res aviation panel, Nat Ctr Atmospheric Res, 71-77, Nat Sci Eval Comt, Univ Corp Atmospheric Res, 76, 81 & Nat Ocean Agency Hq, Environ Res Lab, 76, 85-; vis sci, Risoc Nat Lab, Denmark, 77-78; vist asst prof metorol, Univ Wis-Madison, 69-70; rev comt, Environ Res Div, Argonne Nat Lab, 85-87, 90-; G J Haltiner res chair prof, Naval Postgrad Sch, Monterey, Calif, 86; IPA sci officer, Off Chief Naval Res, Arlington, Va, 89-91. *Mem:* Fel Am Meteorol Soc; Sigma Xi. *Res:* Physical meteorology; indirect atmospheric sounding; meteorological measurements and instrumentation systems; marine meteorology. *Mailing Add:* 506 Walker Bldg University Park PA 16801

THOMSON, DONALD A, b Detroit, Mich, Apr 9, 32; m 57; c 4. MARINE ECOLOGY, ICHTHYOLOGY. *Educ:* Univ Mich, BS, 55, MS, 57; Univ Hawaii, PhD(zool), 63. *Prof Exp:* From asst prof to assoc prof zool, 63-77, PROF ECOL & EVOLUTIONARY BIOL, UNIV ARIZ, 77-, CUR FISHES, 66-, CHMN MARINE SCI PROG, ECOL & EVOLUTIONARY BIOL DEPT, 73- *Concurrent Pos:* Coordr, Ariz-Sonora Marine Sci Prog, 64-66; prin investr, Off Naval Res, 65-69; chief scientist, R/V Te Vega, Stanford Exped 16, 67; assoc investr, Off Saline Water, 68. *Mem:* Am Soc Naturalists; Am Soc Ichthyol & Herpet; Asn Syst Collections; Ecol Soc Am. *Res:* Community ecology, species diversity and stability of marine shore fishes in the Gulf of California. *Mailing Add:* Dept Ecol & Evolutionary Biol Univ Ariz Tucson AZ 85721

THOMSON, GEORGE WILLIS, b Seward, Ill, July 10, 21; m 45; c 3. FORESTRY. *Educ:* Iowa State Univ, BS, 43, MS, 47, PhD(silvicult, soils), 56. *Prof Exp:* Instr gen forestry, 47-52, from asst prof to assoc prof forest mgt, 52-60, actg head forestry dept, 67 & 75, chmn dept, 75-85, prof mensuration & photogram, 60-88, prof photogram & range mgt, 85-88, EMER PROF, DEPT FORESTRY, 88- *Mem:* Fel Soc Am Foresters; Soc Range Mgt; Am Soc Photogram. *Res:* Forest management employing aerial photogrammetry; forest regulation; range management; case studies in land management; remote sensing from LANDSAT-1 with emphasis on forest type delineation. *Mailing Add:* 3334 Morningside St Ames IA 50010

THOMSON, GERALD EDMUND, b New York, NY, June 6, 32; m 58; c 2. INTERNAL MEDICINE, NEPHROLOGY. *Educ:* Queens Col, BS, 55; Howard Univ, MD, 59. *Prof Exp:* Clin dir dialysis unit, Kings County Hosp, State Univ NY Downstate Med Ctr, 65-67; assoc dir med, Coney Island Hosp, Brooklyn, 67-70; chief div nephrology, 70-71, DIR MED, HARLEM HOSP CTR, 71-; PROF, COLUMBIA UNIV, 72- *Concurrent Pos:* NY Heart Asn fel renal dis, State Univ NY Downstate Med Ctr, 65-65; mem med adv bd, NY Kidney Found, 71-; mem, Health Res Coun City of New York, 72-75; mem, Health Res Coun State of subsurface injection; oilwell hypertension info comt & mem educ adv comt, NIH, 73; mem bd dirs, NY Heart Asn, 73-; chmn comt high blood pressure, NY Heart Asn, 75- *Mem:* Am Soc Nephrology. *Res:* Hypertension. *Mailing Add:* Dept Med Columbia Presby Med Ctr 622 W 168th St New York NY 10032-3784

THOMSON, GORDON MERLE, b Madison, Wis, May 3, 41; m 63; c 2. BIOSTATISTICS. *Educ:* Cornell Univ, BS, 63; Iowa State Univ, MS, 66, PhD(animal breeding statist), 68. *Prof Exp:* Assoc statist & comput sci, Iowa State Univ, 68-71; STATISTICIAN, RALSTON PURINA CO, 71-, MGR BIOL SERV, RES 900, 75- *Mem:* Am Statist Asn; Am Dairy Sci Asn; Am Soc Animal Sci; Sigma Xi. *Res:* Application of statistics to biological research data; conduct of biological experiments in the areas of toxicology and efficacy of feed additives. *Mailing Add:* 143 Gray Ave St Louis MO 63119

THOMSON, JAMES ALEX L, b Vancouver, BC, Aug 4, 28. ENGINEERING PHYSICS. *Educ:* Univ BC, BASc, 52; Calif Inst Technol, PhD(eng sci), 58. *Prof Exp:* Sr staff scientist, Space Sci Lab, Gen Dynamics-Convair, 58-67; mem tech staff, Sci & Technol Div, Inst Defense Anal, 67-68; prof eng sci, Res Inst Eng Sci & co-dir, Wayne State Univ, 68-72; vpres eng physics, Phys Dynamics, Inc, 69-; SCI APPL INT. *Mem:* Am Phys Soc. *Res:* Fluid dynamics; ionospheric physics; radiative transfer; oceanography. *Mailing Add:* 683 Longridge Rd Oakland CA 94610

THOMSON, JOHN ANSEL ARMSTRONG, b Detroit, Mich, Nov 23, 11; m 38; c 3. HORTICULTURAL IMPROVEMENTS WITH CARBON-HYDROGEN-OXYGEN COMPLEXES. *Educ:* Univ Southern Calif, AB, 57. *Hon Degrees:* BS, Calif Polytech State Univ, San Luis Opispo, 61. *Prof Exp:* BIOCHEMIST & OWNER, VITAPURE LABS, VITAMIN INST, 39- *Concurrent Pos:* Voc educ instr, US War Manpower Comn, 43-44. *Honors & Awards:* Sci & Indust Gold Medal, World Fair, 40. *Mem:* Int Soc Hort Sci; Am Inst Biol Sci; NY Acad Sci; Soc Nutrit Educ; AAAS; Am Hort Soc. *Res:* Biochemistry; vigorous longevity; nutrients' behavior-normalization; formulation, processing and utilization of products supplying life process substances, especially carbon-hydrogen- oxygen molecular groups for horticulture and environment improvements; human physical and mental health improvements through optimization of nutrients and of interference with their metabolism by adverse exposures; sole originator of over 100 products. *Mailing Add:* Vitamin Inst Box 230, 5411 Satsuma Ave North Hollywood CA 91603

THOMSON, JOHN FERGUSON, b Garrett, Ind, Apr 18, 20; m 43, 73; c 2. RADIATION BIOLOGY, PHARMACOLOGY. *Educ:* Univ Chicago, SB, 41, SM, 42, PhD(pharmacol), 47. *Prof Exp:* Res assoc anat, Univ Chicago, 43-45, from res assoc to asst prof pharmacol, 46-51, mem staff, Toxicity Lab, 43-51; assoc pharmacologist, Div Biol & Med Res, Argonne Nat Lab, 51-62, actg dir, 69-70 & 74-75, assoc dir, 70-74 & 75-79, sr biologist, Div Biol & Med Res, 62-82; RETIRED. *Concurrent Pos:* Adj prof, Northern Ill Univ, 71-83 & Univ Ill, Chicago Circle, 71-77. *Mem:* AAAS; Soc Exp Biol & Med; Sigma Xi; Am Inst Biol Sci; Radiation Res Soc. *Res:* Radiation biology; cytoenzymology; isotope toxicity. *Mailing Add:* Rte 1 Box 105-B Scales Mound IL 61075

THOMSON, JOHN OLIVER, b Cleveland, Ohio, Feb 2, 30; m 55; c 3. PHYSICS. *Educ:* Williams Col, AB, 51; Univ Ill, MS, 53, PhD, 56. *Prof Exp:* Fulbright grant, Univ Rome, 56-57 & Univ Padua, 57-58; from asst prof to assoc prof physics, 58-72, PROF PHYSICS, UNIV TENN, KNOXVILLE, 72- *Concurrent Pos:* Consult, Physics Div, Oak Ridge Nat Lab, 58-; assoc prof, Memphis State Univ, 66-68. *Mem:* Am Phys Soc. *Res:* Solid state and low temperature physics. *Mailing Add:* Dept Physics & Astron Univ Tenn Knoxville TN 37996-1200

THOMSON, KEITH STEWART, b Heanor, Eng, July 29, 38; m 63; c 2. ZOOLOGY, PALEONTOLOGY. *Educ:* Univ Birmingham, Eng, BSc, 60; Harvard Univ, Am, 61, PhD(biol), 63. *Prof Exp:* NATO sci fel & temporary lectr zool, Univ Col, Univ London, 63-65; asst prof & asst cur vert zool, 65-70, assoc cur vert zool, Peabody Mus Natural Hist & assoc prof biol, 70-76, dir, 77-79, prof biol & cur vert zool, Peabody Mus Natural Hist, Yale Univ, 76-87, dir, Sears Found Marine Res & Oceanog Hist, 77-88, dean, Grad Sch Arts & Sci, 79-86; PRES, ACAD NATURAL SCIS, PHILADELPHIA, 87- *Concurrent Pos:* Bd mem & trustee, Wetlands Inst, 75-, Woods Hole Oceanog Inst, 81-, Wistar Inst & Cent Philadelphia Develop Corp, 87- *Mem:* Am Soc Zool; Soc Nautical Res; fel Linnean Soc London; fel Zool Soc London; Soc Vert Paleont; Sigma Xi. *Res:* Vertebrate biology, especially of lower vertebrates; coelacanths, lungfishes, origin of tetrapods; origin of adaptations and of major groups; morphogenesis; history of science, Darwinism; popular science writing. *Mailing Add:* 41 Summit St Philadelphia PA 19118

THOMSON, KENNETH CLAIR, b Gunnison, Utah, Mar 5, 40; m 61; c 4. ECONOMIC GEOLOGY, PETROGRAPHY. *Educ:* Univ Utah, BS, 63, PhD, 70. *Prof Exp:* Illustrator-geologist, Utah Geol Surv, 59-68; from asst prof to assoc prof, 68-78, PROF GEOL, SOUTHWEST MO STATE UNIV, 78- *Mem:* Fel Nat Speleol Soc; Am Inst Prof Geologists; Nat Water Well Asn. *Res:* Karst hydrogeology; speleology; stratigraphy of Southwestern Missouri. *Mailing Add:* Dept Geosci Southwest Mo State Univ 901 S Nat Springfield MO 65804-0089

THOMSON, KERR CLIVE, b Toronto, Ont, Mar 2, 28; US citizen; m 55; c 2. GEOPHYSICS, SEISMOLOGY. *Educ:* Univ BC, BA, 52; Colo Sch Mines, DSc(geophys), 65. *Prof Exp:* Seismologist, Seismograph Serv Corp, Okla, 52-54; seismologist, Standard Oil Co Calif, 54-55, seismic party chief, 55-58; instr physics, Colo Sch Mines, 58-61; res physicist, Air Force Cambridge Res Labs, 61-65, br chief seismol, 65-75, dir terrestrial sci lab, 75-; PROF PHYSICS, BRYAN COL. *Concurrent Pos:* Lectr, Gordon Col, 66-69 & Boston Col, 72. *Mem:* Seismol Soc Am; Am Geophys Union; Soc Explor Geophys; Europ Asn Explor Geophys; Am Sci Affil. *Res:* Earthquake focal mechanism; theoretical seismology; model seismology; wave propagation in absorptive media; seismic radiation in tectonically stressed media; nuclear test detection; structural vibration; terrestrial gravity. *Mailing Add:* Dept Physics Bryan Col Dayton TN 37321-7000

THOMSON, MICHAEL GEORGE ROBERT, electron-optics, for more information see previous edition

THOMSON, QUENTIN ROBERT, b Lake Charles, La, Nov 14, 18; m 42; c 8. ENGINEERING. *Educ:* Ga Inst Technol, BS, 40; Univ Ariz, MS, 53; Calif Western Univ, PhD(eng econ), 75. *Prof Exp:* Exp test engr, Pratt & Whitney Aircraft Co, 40-41; engr & inspector, US Vet Admin, 51-53; from instr to assoc prof mech eng, Univ Ariz, 53-70, prof aerospace & mech eng, 70-81; CONSULT ENGR, 81- *Concurrent Pos:* Chief engr, Krueger Mfg Co, 63-67, consult, Krueger Div, Lear-Siegler, Inc, 67- & Shipley & Assoc. *Honors & Awards:* Ralph Teetor Award, Soc Automotive Engrs, 76; Am Soc Heating, Refrig & Air Conditioning Engrs Award, 76. *Mem:* Am Soc Mech Engrs; Am Soc Heating, Refrig & Air Conditioning Engrs; Am Mgt Asn; Soc Automotive Engrs. *Res:* Performance of gasoline as related to octane and distillation in automobiles. *Mailing Add:* 4730 Camino Luz Tucson AZ 85718

THOMSON, RICHARD EDWARD, b Comox, BC, Apr 14, 44. PHYSICAL OCEANOGRAPHY. *Educ:* Univ BC, BS, 67, PhD(physics & oceanog), 71. *Prof Exp:* RES SCIENTIST, INST OCEAN SCI, FISHERIES & OCEANS CAN, 71- *Concurrent Pos:* Vis scientist, Monash Univ, Australia, 74; vis researcher, Australian Inst Marine Sci, 82; vis lectr, Univ New South Wales,

Australia, 88. *Mem:* Can Meteorol & Oceanog Soc; Am Geophys Union; Am Soc Limnol & Oceanog. *Res:* Wave propagation in random media; energetics of planetary waves and internal gravity waves; vorticity mixing and redistribution in the ocean; baroclinic tides and inertial currents; generation and propagation of shelf waves; long-term sea level fluctuations; vortex streets in the atmosphere; rectilinear leads in arctic ice; mesoscale eddies; hydrothermal vents; tidal rectification. *Mailing Add:* Inst Ocean Sci 9860 W Saanich Rd Box 6000 Sidney BC V8L 4B2 Can

THOMSON, RICHARD N, chemical engineering; deceased, see previous edition for last biography

THOMSON, ROBB M(ILTON), b El Paso, Tex, Feb 4, 25; m 48; c 4. MATERIALS SCIENCE, SOLID STATE PHYSICS. *Educ:* Univ Chicago, MS, 50; Syracuse Univ, PhD(physics), 53. *Prof Exp:* Res assoc physics, Univ Ill, Urbana, 53-56, from asst prof to assoc prof metall, 56-60, prof physics & metall, 60-68; prof mat sci & chmn dept, State Univ NY Stony Brook, 68-71; SR RES SCIENTIST, NAT INST STANDARDS & TECHNOL, 71- *Concurrent Pos:* Dir mat sci, Adv Res Projs Agency, US Dept Defense, 65-68. *Mem:* Am Phys Soc; AAAS. *Res:* Theory of imperfections in solids; mechanical properties of solids. *Mailing Add:* Mat Sci & Eng Lab Nat Inst Standards & Technol Gaithersburg MD 20899

THOMSON, ROBERT FRANCIS, b Marine City, Mich, Feb 26, 14; m 37, 73; c 2. METALLURGICAL ENGINEERING. *Educ:* Univ Mich, BSE, 37, MSE, 40, PhD(metall eng), 41. *Prof Exp:* Metallurgist, Repub Steel, 37-38, Chrysler Corp, 42-45, Int Nickel, 45-50; metallurgist, Gen Motors Res, 50-62, tech dir, 62-69, exec dir, 69-77; RETIRED. *Concurrent Pos:* Trustee, Am Soc Metals, 62-64. *Honors & Awards:* John Penton Gold Medal, Am Foundrymen's Asn, 55; William Park Woodside Lectr, Am Soc Metals, 57, Campbell Mem Lectr, 63. *Mem:* Sigma Xi; Am Soc Metals; Soc Automotive Engrs; Am Inst Mining Metall & Petrol Engrs; Am Foundrymen's Soc. *Res:* Fundamental properties of metals and their application; foundry technology and its application. *Mailing Add:* 4645 Chamberlain Dr St Clair MI 48079

THOMSON, STANLEY, b Toronto, Ont, Dec 23, 23; m 46; c 5. GEOTECHNICS, ENGINEERING GEOLOGY. *Educ:* Univ Toronto, BASc, 50; Univ Alta, MSc, 55, PhD(soil mech, found), 63, BSc, 71. *Prof Exp:* From asst prof to prof, 61-84, EMER PROF CIVIL ENG, UNIV ALTA, 84- *Concurrent Pos:* Nat Res Coun Can oper res grants, 65-91. *Honors & Awards:* Eng Medal, Can Pac Rwy, 90. *Mem:* Eng Inst Can; Can Geotech Soc. *Res:* Foundation engineering; slope stability in highly over-consolidated soils; influence of geology in soil mechanics. *Mailing Add:* Dept Civil Eng Univ Alta Edmonton AB T6G 2G7 Can

THOMSON, TOM RADFORD, b Hachiman, Japan, Nov 7, 18; nat US; m 46; c 5. PHYSICAL ORGANIC CHEMISTRY. *Educ:* Univ Calif, BS, 39; Kans State Univ, MS, 40, PhD(chem), 45. *Prof Exp:* Asst chem, Kans State Univ, 39-40, instr, 42, chemist, 42-47; plant chemist, Cascade Frozen Foods, Wash, 41; from assoc prof to prof chem, Adams State Col, 47-61, head dept, 47-61; from assoc prof to prof, 61-81, EMER PROF CHEM, ARIZ STATE UNIV, 81- *Concurrent Pos:* Sigma Xi res grant, 54; res assoc, Univ Calif, 54; vis scholar, Utah, 59; Res Corp grant, 59-60; resident author, Addison-Wesley Publ Co, Calif, 67-68; chem consult, 81- *Mem:* Am Chem Soc. *Res:* Carbohydrates; starch chemistry; chemical education; relationship of structure to physical properties; theoretical chemistry; water-soluble polymers; food preservation. *Mailing Add:* 65 Ramon Sonoma CA 95476

THOMSON, WILLIAM ALEXANDER BROWN, biochemistry, analytical chemistry, for more information see previous edition

THOMSON, WILLIAM JOSEPH, b New York, NY, May 15, 39; m 64; c 4. CHEMICAL ENGINEERING. *Educ:* Pratt Inst, BChE, 60; Stanford Univ, MS, 62; Univ Idaho, PhD(chem eng), 69. *Prof Exp:* Asst scientist, Avco Res & Advan Develop Div, 61-62; assoc prof, 69-75, prof chem eng, Univ Idaho, 75-; DEPT CHEM ENG, WASH STATE UNIV. *Mem:* Am Inst Chem Engrs; Am Chem Soc. *Res:* Catalytic kinetics; chemical reactor development; fluidization. *Mailing Add:* Dept Chem Eng Wash State Univ Pullman WA 99163

THOMSON, WILLIAM TYRRELL, b Kyoto, Japan, Mar 24, 09; US citizen; m 41; c 3. ENGINEERING. *Educ:* Univ Calif, BS, 33, MS, 34, PhD(elec eng), 38. *Prof Exp:* Asst math & physics, Univ Calif, 34-35; instr mech, Kans State Col, 37-41; res engr, Boeing Airplane Co, Wash, 41; asst prof mech, Cornell Univ, 41-44; head of vibration & flutter, Ryan Aeronaut Co, Calif, 44-46; from assoc prof to prof mech, Univ Wis, 46-51; prof eng, Univ Calif, Los Angeles, 51-66; prof mech eng & chmn dept, 66-76, EMER PROF MECH ENG, UNIV CALIF, SANTA BARBARA, 76- *Concurrent Pos:* Consult, Space Technol Labs, 55-; Fulbright res prof, Kyoto Univ, 57-58; Guggenheim fel, Ger, 61-62. *Honors & Awards:* Den Hartog Award, Am Soc Mech Engrs, 89. *Mem:* Fel Am Soc Mech Engrs; assoc fel Am Inst Aeronaut & Astronaut. *Res:* Applied mathematics; vibrations; dynamics. *Mailing Add:* Dept Mech Eng Univ Calif Col Eng Santa Barbara CA 93106

THOMSON, WILLIAM WALTER, b Chico, Calif, Oct 11, 30; c 3. BOTANY, CYTOLOGY. *Educ:* Sacramento State Col, BA, 53, MA, 60; Univ Calif, Davis, PhD(bot), 63. *Prof Exp:* Asst res botanist, Air Pollution Res Ctr, 63-64, from asst prof to assoc prof, 64-74, PROF BIOL, UNIV CALIF, RIVERSIDE, 74- *Mem:* Am Soc Plant Physiol; Bot Soc Am. *Res:* Ultrastructure of chloroplasts and plant membranes as related to development, physiology and stress conditions. *Mailing Add:* Dept Botany & Plant Sci Univ Calif Riverside Riverside CA 92521

THON, J GEORGE, civil & structural engineering, hydroelectric projects; deceased, see previous edition for last biography

THONAR, EUGENE JEAN-MARIE, b Liege, Belg, Oct 4, 45; m 82; c 2. CARTILAGE BIOCHEMISTRY, CARBOHYDRATE BIOCHEMISTRY. *Educ:* Witwatersrand Univ, BSc, 69, BSc(hons), 70, PhD(biochem), 77. *Prof Exp:* Res asst orthop surg, Witwatersrand Univ, 73-76, res officer, 77-79; vis assoc biochem, Nat Inst Dent Res, 79-80; from asst prof to assoc prof, 80-88, PROF BIOCHEM & INTERNAL MED, RUSH PRESBY, ST LUKE'S MED CTR, 88-, GEORGE W STUPPY PROF ARTHRITIS, 90- *Concurrent Pos:* Prin investr, William Noble Lane, MRO, 80-88; specialist lectr, Cook County Grad Sch Med, 82-; consult, NIH pathobiochem study sect, 85 & 88, CIBA-Geigy Corp, 86-; mem med & sci adv comt, Arthritis Found, Ill chap. *Honors & Awards:* Carol Nachman Prize Rheumatology. *Mem:* Am Soc Biol Chemists; Orthop Res Soc; Sigma Xi; NIH Alumni Asn; AAAS; Am Rheumatism Asn. *Res:* Biochemistry of cartilage; proteoglycans of cartilage matrix in health and disease; age-related changes in structure of proteoglycans; blood assays for assessing cartilage disease. *Mailing Add:* 14503 Pheasant Lane Lockport IL 60441

THONNARD, NORBERT, b Berlin, Germany, Jan 22, 43; US citizen; m 64; c 4. ATOMIC PHYSICS, ASTROPHYSICS. *Educ:* Fla State Univ, BA, 64; Univ Ky, MS, 69, PhD(physics), 71. *Prof Exp:* Physicist, US Army Engr Res & Develop Labs, 64-66; consult astrophys, Battelle Pac Northwest Labs, Battelle Mem Inst, 72; fel, Carnegie Inst, 70-72, staff mem, Dept Terrestrial Magnetism, 72-84; dir technol, 83-85, vpres res & develop, 85-89, PRES, ATOM SCI, INC, 89- *Concurrent Pos:* Mem Users Comt, Nat Radio Astron Observ, 78-81; vis scientist, Kapteyn Lab, Univ Groningen, Neth, 82. *Mem:* Am Phys Soc; Am Astron Soc; AAAS; Int Astron Union; Am Chem Soc. *Res:* Development of laser-based ultra-sensitive element analysis techniques and instrumentation; analysis application developmenet of resonance ionization spectroscopy; application of one-atom detection; galaxy studies from Z1-cm and optical observations; development of instrumentation for radio and optical astronomy; observational cosmology. *Mailing Add:* Atom Sci 114 Ridgeway Ctr Oak Ridge TN 37830

THOR, EYVIND, b Oslo, Norway, Nov 24, 28; US citizen; m 56; c 3. FOREST GENETICS, SILVICULTURE. *Educ:* Univ Wash, Seattle, BS, 54, MS, 56; NC State Univ, PhD(forestry), 61. *Prof Exp:* Asst prof forestry res, 59-65, from assoc prof to prof, 65-82, EMER PROF FORESTRY, UNIV TENN, 82- *Concurrent Pos:* Elwood L Demmon res award, 61; US Forest Serv res grant, 66-68; grant, Inland Container Corp, 75-76. *Mem:* Soc Am Foresters. *Res:* Population genetics; breeding of trees for timber production (pines and hardwoods), Christmas trees (pine and spruce) and disease resistance (American chestnut). *Mailing Add:* Dept Forestry & Ecol Univ Tenn Knoxville TN 37916

THOR, KARL BRUCE, b Pittsburgh, Pa, Mar 30, 54; m 81; c 2. PHARMACOLOGY. *Educ:* Pa State Univ, BS, 76; Univ Pittsburgh, PhD(pharmacol), 85. *Prof Exp:* Pharmacol fel, Uniformed Serv Univ, 85-88; sr staff fel neurophysiol, Nat Inst Neurol Dis & Stroke, NIH, 88-90; SR PHARMACOLOGIST, LILLY RES LABS, 90- *Concurrent Pos:* Grant reviewer, Am Heart Asn, 87-90, Paralyzed Vet Am, 90-; adj asst prof pharmacol, Uniformed Serv Univ, 88-90, Ind Univ Med Sch, 91-, adj asst prof urol, 91-; ad hoc reviewer, J Urol, 90- *Mem:* Soc Neurosci; Soc Basic Urol Res. *Res:* Central control of lower urinary tract function; use of electrophysiology, neuroanatomy, receptor autoradiography and immunohistochemistry to determine neurotransmitter systems that regulate lower urinary tract function in adults, neonates and paraplegic patients. *Mailing Add:* Div CNS Res Lilly Res Labs Indianapolis IN 46285

THORBECKE, GEERTRUIDA JEANETTE, b Neth, Aug 2, 29; m 57; c 3. IMMUNOLOGY, EXPERIMENTAL PATHOLOGY. *Educ:* State Univ Groningen, MD, 50, PhD, 54. *Prof Exp:* Asst histol, State Univ Groningen, 48-54; asst path, State Univ Leiden, 56-57; from res assoc to assoc prof, 57-70, PROF PATH, SCH MED, NY UNIV, 70- *Concurrent Pos:* Foreign Opers Mission to Neth scholar, Lobund Inst, Ind, 54-56; USPHS res grants, 59-, USPHS res career develop award, 61-71; career scientist award, Health Res Coun, City of New York, 71-72; corresp mem, Royal Dutch Acad Sci, 80; coun, Am Asn Immunol, 84-91, pres, 89-90; coun, Fed Am Socs Exp Biol Bd, 89-91. *Mem:* Am Soc Exp Path; Soc Exp Biol & Med; fel NY Acad Sci; Reticuloendothelial Soc; Brit Soc Immunol; Am Asn Immunol; Dutch Soc Immunol; Harvey Soc. *Res:* Antibody formation; serum proteins; lymphoid tissues and lymphoid cell interaction; immunological tolerance; tumor immunity, effect of aging on immune response. *Mailing Add:* Dept Path NY Univ Sch Med New York NY 10016

THORBJORNSEN, ARTHUR ROBERT, b Winter, Wis, Sept 27, 36; m 62; c 2. ELECTRICAL ENGINEERING. *Educ:* Univ Wis-Madison, BSc, 62; Univ Ala, Huntsville, MSc, 68; Univ Fla, PhD(elec eng), 72. *Prof Exp:* Assoc engr elec eng, Boeing Co, Wash, 62-64, assoc res engr, Ala, 64-67; engr, Northrop Corp, 67-68; grad asst, Univ Fla, 68-72; from asst prof to assoc prof, 72-83, PROF ELEC ENG, UNIV TOLEDO, 83- *Concurrent Pos:* Prin investr res grant, Univ Toledo, 76 & 81; NSF grants, 74-76, 76, 76-79 & 86-87; US Air Force grant, 82-83. *Honors & Awards:* New Technol Award, Marshall Space Flight Ctr, NASA, 67; Centennial Award, Inst Elec & Electronics Engrs, 84. *Mem:* Inst Elec & Electronics Engrs; Sigma Xi. *Res:* Computer-aided design of electronic circuits; statistical simulation of semiconductor devices and integrated circuits. *Mailing Add:* Dept Elec Eng Univ Toledo Toledo OH 43606

THORESEN, ASA CLIFFORD, b Blenheim, NZ, Sept 9, 30; nat US; m 52; c 2. ORNITHOLOGY. *Educ:* Emmanuel Missionary Col, BA, 54; Walla Walla Col, MA, 58; Ore State Col, PhD, 60. *Prof Exp:* Vis prof biol, Biol Sta, Walla Walla Col, 60, 70; from asst prof to assoc prof, 60-67, chmn dept, 63-83, PROF BIOL, ANDREWS UNIV, 67- *Concurrent Pos:* Leader biol expeds Peru, 64-65, 68 & SPac & Australia, 72, 82, 85; NSF grant, NZ, 66-67. *Mem:* Electron Micros Soc Am; Cooper Ornith Soc; Am Ornith Union. *Res:* Life history and behavioral studies of oceanic birds, particularly of the family Alcidae; ultrastructural studies of avian tissues. *Mailing Add:* Dept Biol Andrews Univ Berrien Springs MI 49104-0410

THORGEIRSSON, SNORRI S, b Iceland, Dec 1, 41; nat US. CARCINOGENESIS RESEARCH. *Educ:* Univ Iceland, MD, 68; Univ London, PhD, 71. *Prof Exp:* Intern, Univ Hosp, Reykjavik, Iceland, 68-69; registrar & res fel, Dept Clin Pharmacol, Royal Postgrad Med Sch, London, Eng, 69-71; vis fel, Lab Chem Pharmacol, Nat Heart & Lung Inst, NIH, Bethesda, Md, 72-73, vis scientist, Sect Develop Pharmacol Neonatal & Pediat Med Br, Nat Inst Child Health & Human Develop, 74-75, chief, Sect Molecular Toxicol, 75-76, head, Biochem Pharmacol Sect, Lab Chem Pharmacol, DCT, Nat Cancer Inst, 76-81, CHIEF, LAB EXP CARCINOGENESIS DCE, NAT CANCER INST, NIH, 81- *Concurrent Pos:* Mem, Chem Selection Working Group, 78-82, Comt Occup Carcinogenesis, 79, Comt Amines, Nat Acad Sci; preceptor, Pharmacol Res Assoc Prog, Nat Inst Gen Med Sci, NIH, 77-; Nat Cancer Inst rep, Subcomt Testing & Test Method Validation Comt Coord Environ Health & Related Prog, 91- *Mem:* AAAS; Am Chem Soc; Am Asn Cancer Res; Am Soc Cell Biol; Am Soc Microbiol; Am Soc Pharmacol & Exp Therapeut; NY Acad Sci; Soc Toxicol. *Res:* Cellular and molecular biology of hepatocarcinogenesis; regulation and role of the multidrug resistance gene family; genetic control of growth and differentiation during normal and neoplastic development. *Mailing Add:* Nat Cancer Inst Div Cancer Etiol NIH Lab Exp Carcinogenesis Bldg 37 Rm 3C28 Bethesda MD 20892

THORHAUG, ANITRA L, b Chicago, Ill, June 1, 40. BIOPHYSICS, MARINE BOTANY. *Educ:* Univ Miami, BS, 63, MS, 65, PhD(marine sci), 69. *Prof Exp:* Environ Sci Serv Admin assoc biophys, Atlantic Labs, Miami, Fla, 68-69; res scientist, Algology Div Fish & Appl Estuarine Ecol, 69-71, res scientist microbiol, Sch Med, Univ Miami, 71-77; PROF BIOL, FLA INT UNIV, 77- *Concurrent Pos:* Fed Water Qual Admin grant, Univ Miami, 69-70, NSF inst grant, 70-71, Nat Oceanic & Atmospheric sea grant, 70-78 & NSF grant biophysics, 70-71, AEC grant, 70; Dept Energy grant, 71-; fel, Weizmann Inst Sci, 71-; vis scientist, Donner Lab, Univ Calif, Berkeley, 71; consult various orgns & govt, 69-; mem, XII Int Bot Cong, Leningrad, 75, nat adv panel, Energy Res & Develop Admin, 77, 1st Int Symp Trop Estuarine Pollution, Djakarta, 76, environ adv panel, Ocean Thermal Energy Conversion, Energy Res & Develop Admin, 77-78, conserv comt, Am Bot Soc, 77-78; prog mgr environ studies, Nat Solar Energy Consortium, Solar Energy Res Inst, 76-77; chmn res counr comt, Biscayne Bay, Univ Miami, 75-76, symp seagrasses, 2nd Int Ecol Cong, Jerusalem, 78, ed, Thalassia. *Honors & Awards:* Diamond Award, Am Bot Soc, 75. *Mem:* AAAS; Bot Soc Am; Phycol Soc Am; Ecol Soc Am; Brit Phycol Soc. *Res:* Membrane transport biophysics; membrane transport and physiology of giant algal cells; near-shore macro-plant ecology; physiology of tropical macro-flora-temperature, salinity, light, sediment. *Mailing Add:* PO Box 490559 Key Biscayne FL 33149

THORINGTON, RICHARD WAINWRIGHT, JR, b Philadelphia, Pa, Dec 24, 37; m 67; c 2. BIOLOGY. *Educ:* Princeton Univ, BA, 59; Harvard Univ, MA, 63, PhD(biol), 64. *Prof Exp:* Primatologist, New Eng Regional Primate Res Ctr, 64-69, assoc mammal, Mus Comp Zool, 64-69; CUR PRIMATES, BIOL PROG, 69-, CHMN, DEPT VERT ZOOL, SMITHSONIAN INST, 87- *Mem:* AAAS; Am Soc Mammal; Soc Study Evolution; Int Primatology Soc; Sigma Xi; Am Soc Primatology. *Res:* Form and function of mammals; thermoregulation and thermal effects on development; primate ecology and taxonomy. *Mailing Add:* 7714 Old Chester Rd Bethesda MD 20817

THORLAND, RODNEY HAROLD, b Lake Mills, Iowa, Feb 16, 41. SOLID STATE PHYSICS. *Educ:* Luther Col, BA, 64; Emory Univ, MSc, 69, PhD(physics), 71. *Prof Exp:* Vol, US Peace Corps, 64-67; asst prof physics, Kennesaw Jr Col, 71-72; res assoc physics, Emory Univ, 72-73; asst prof, 73-77, prof physics, Vol State Community Col, 77-81; SR, HONEY WELL CO, 82- *Mem:* AAAS; Am Phys Soc; Am Asn Physics Teachers; assoc Sigma Xi. *Res:* Nuclear magnetic resonance, electron paramagnetic resonance and far infrared spectroscopy. *Mailing Add:* 5808 Royal Oaks Dr Shoreview MN 55126

THORMAN, CHARLES HADLEY, b Albany, Calif, June 14, 36; m 57; c 2. GEOLOGY. *Educ:* Univ Redlands, BS, 58; Univ Wash, MS, 60, PhD(geol), 62. *Prof Exp:* Geologist, Humble Oil & Ref Co, 62-65 & Olympic Col, Wash, 65-68; vis asst prof geol, Univ Ore, 68-71; GEOLOGIST, US GEOL SURV, 71- *Mem:* Am Asn Petrol Geol; Geol Soc Am. *Res:* Tectonics of the eastern basin and range; structure of Liberia; tectonics of southwest Arizona and southeast New Mexico. *Mailing Add:* 1500 S Zang St Denver CO 80228

THORMAR, HALLDOR, b Iceland, Mar 9, 29; m 62; c 3. VIROLOGY. *Educ:* Copenhagen Univ, PhD(cell physiol), 56, DrPhil(virol), 66. *Prof Exp:* Res scientist, State Serum Inst, Copenhagen, Denmark, 60-62; from res scientist to assoc res scientist, Inst Exp Path, Keldur, Iceland, 62-67; chief res scientist virol, Inst Res Develop Disabilites, 67-88; PROF BIOL, UNIV ICELAND, 88- *Concurrent Pos:* Investr virol, Sci Res Inst, Caracas, Venezuela, 65-66; vis prof, Free Univ Brussels, Belgium, 75. *Mem:* Am Soc Microbiol. *Res:* Cell physiology; effect of temperature changes on cell growth and division; slow virus infections; isolation and study of visna virus and comparison of visna virus to other viruses; pathogenesis of subacute sclerosing panencephalitis in animal models; virus inhibitors. *Mailing Add:* Inst Biol Univ Iceland Grensasvegur 12 Reykjavik Iceland

THORN, CHARLES BEHAN, III, b Washington, Ind, Aug 14, 46; m 79; c 2. THEORETICAL PHYSICS. *Educ:* Mass Inst Technol, BS, 68; Univ Calif, Berkeley, MA, 69, PhD(particle theory), 71. *Prof Exp:* Res assoc, Europ Orgn Nuclear Res, NSF, 72; res assoc physics, Mass Inst Technol, 73, from asst prof, to assoc prof, 73-80; PROF PHYSICS, UNIV FLA, 80- *Concurrent Pos:* Alfred P Sloan fel, 74; vis, Dept Appl Math & Theoret Physics, Cambridge Univ, 76; invited prof, Ecole Normal Superieure, Paris, 85; mem, Inst Advan Study, Princeton, 86-87. *Mem:* Am Phys Soc. *Res:* Theory of elementary particles; theory of strongly interacting particles; theory of relativistic strings. *Mailing Add:* 215 Williamson Hall Univ Fla Gainesville FL 32611

THORN, DONALD CHILDRESS, b Ft Worth, Tex, June 2, 29; m 53; c 3. MICROWAVE DEVICES. *Educ:* Tex A&M Univ, BS, 51; Univ Tex, Austin, MS, 55, PhD(elec eng), 58. *Prof Exp:* Asst prof elec eng, Univ Tex, Austin, 56-58; from asst prof to prof, Univ NMex, 58-67; prof, Univ Akron, 67-87; CONSULT, 58- *Mem:* Sigma Xi; Nat Soc Prof Engrs; Nat Acad Forensic Engrs. *Res:* Microwave devices; microwave propagation in the atmosphere and other media. *Mailing Add:* PO Box 90045 San Antonio TX 78209-9045

THORN, GEORGE W, b Buffalo, NY, Jan 15, 06; m 31; c 1. ENDOCRINOLOGY. *Educ:* Univ Buffalo, MD, 29. *Hon Degrees:* Numerous from foreign & US univs, 42-87. *Prof Exp:* Asst med, Univ Buffalo, 32-34; assoc prof med, Johns Hopkins Med Sch, 38-42; prof, 42-72, EMER PROF, HOWARD MED SCH, 72- *Concurrent Pos:* Physician, Johns Hopkins Hosp, Baltimore, 40-42; physician-in-chief, Peter Bent Brigham Hosp, Boston, 42-72, emer physician-in-chief, 72-; emer chmn bd, Howard Hughes Med Inst; emer corp mem, Mass Inst Technol. *Mem:* Asn Am Physicians; Endocrine Soc; Clin & Climat Soc; Am Acad Arts & Sci. *Res:* Endocrinology, studies of the adrenal gland; kidney, development of dialysis and kidney transplant. *Mailing Add:* 16 Gurney St Cambridge MA 02138

THORN, RICHARD MARK, b New Castle, Pa, Mar 8, 47; m 69. IMMUNOBIOLOGY, VIRAL IMMUNOLOGY. *Educ:* Univ Calif, San Diego, BA, 69; Univ Pa, PhD(molecular biol), 74. *Prof Exp:* Fel immunol, Sch Med, Johns Hopkins Univ, 74-76; scientist II, Frederick Cancer Res Ctr, 76-79; asst pror, Vet Sch, Univ Pa, 79-; sr scientist, Cambridge Biosci Corp, 83-89; ASSOC DIR, DEPT IMMUNOL, BAXTER DIAG, INC, 89- *Concurrent Pos:* Prin investr, 79- *Mem:* Am Asn Immunologists. *Res:* Mechanism of target cell destruction by cytolytic T cells; mechanism of cytolytic T cell induction; role of immunity in the control of infectious disease and cancer; immunoassay research and development. *Mailing Add:* PO Box 520672 Miami FL 33152

THORN, ROBERT NICOL, b Coeur d'Alene, Idaho, Aug 31, 24; m 62; c 4. PHYSICS. *Educ:* Harvard Univ, AB, 48, AM, 49, PhD(physics), 53. *Prof Exp:* Mem staff, 53-71, leader theoret design div, 71-76, assoc dir, 76-79, actg dir, 79, DEP DIR, LOS ALAMOS NAT LAB, 79- *Concurrent Pos:* Mem sci adv group, Space Systs Div, US Air Force, 62-63, sci adv bd, 62-, nuclear panel, 64-; mem sci adv group, Defense Nuclear Agency & Defense Intel Agency. *Honors & Awards:* E O Lawrence Award, Atomic Energy Comn, 67. *Mem:* AAAS; Am Phys Soc; Sigma Xi. *Res:* Classical theoretical physics; quantum and nuclear physics; weapons systems analysis and design. *Mailing Add:* 155 Kimberly Los Alamos NM 87544

THORNBER, JAMES PHILIP, b Hebden Bridge, Eng, Dec 22, 34; m 60, 83; c 4. PLANT BIOCHEMISTRY. *Educ:* Cambridge Univ, BA, 58, MA, 61, PhD(biochem), 62. *Prof Exp:* Sci off plant biochem, Twyford Labs Ltd, Arthur Guinness, Son & Co, Ltd, 61-67; res assoc biol, Brookhaven Nat Lab, 67-69, asst scientist, 69-70; asst prof bot, 70-72, assoc prof biol, 72-75, PROF BIOL, UNIV CALIF, LOS ANGELES, 75-, chmn dept, 81-86. *Concurrent Pos:* NSF grant, Univ Calif, Los Angeles, 71-; Guggenheim mem fel, 76-77; USDA grant, 78- *Mem:* Am Soc Biol Chem; Am Soc Plant Physiol. *Res:* Photosynthesis; organization of chlorophyll in plants and bacteria; chlorophyll-protein complexes; photochemical reaction centers; membrane composition, structure and biogenesis. *Mailing Add:* Dept Biol Univ Calif Los Angeles CA 90024

THORNBERRY, HALBERT HOUSTON, b Corydon, Ky, Dec 28, 02; m 46; c 1. PLANT PATHOLOGY. *Educ:* Univ Ky, BS, 25, MS, 26; Univ Minn, PhD(plant path), 34. *Prof Exp:* Asst plant path, Univ Minn, 26-28 & Univ Ill, 28-31; fel, Rockefeller Inst Med Res, 31-35; jr plant pathologist, Citrus Exp Sta, Univ Calif, 35-36; asst pathologist, Univ Ky, 36-37; pathologist, Bur Plant Indust, US Dept Agr, 37-38; from asst prof to prof, 38-71, EMER PROF PLANT PATH, UNIV ILL, URBANA, 71- *Concurrent Pos:* Res award, Soc Am Florists, 68; consult plant health & mgt, 71- *Mem:* AAAS; Am Phytopath Soc; Am Soc Microbiol; Am Chem Soc. *Res:* Phytovirology; chemopathology; antibiotics; bacterial diseases. *Mailing Add:* 1602 S Hillcrest St PO Box 128 Urbana IL 61801

THORNBOROUGH, JOHN RANDLE, b Columbus, Ohio, Feb 2, 39; c 2. NEUROBIOLOGY, NEUROENDOCRINOLOGY. *Educ:* Ohio State Univ, BA, 60, MA, 61; NY Med Col, PhD(physiol), 72. *Prof Exp:* Instr biol, Denison Univ, 61-67; assoc med prof physiol, City Col NY-Sophie Davis Sch Biomed Educ, 80-86, dir acad affairs, 80-84; res med, 72-73, instr, 73-74, asst prof, 74-80, ADJ ASSOC PROF PHYSIOL, NY MED COL, 80-; ASSOC PROF PHYSIOL & ASSOC DEAN ACAD AFFAIRS, MT SINAI SCH MED, 87- *Concurrent Pos:* Adj prof physiol, Sarah Lawrence Col, 73-77; assoc prof med educ, Mt Sinai Sch Med, 87- *Mem:* Soc Neurosci. *Res:* Hypothalamic control of sodium and water metabolism. *Mailing Add:* Dept Physiol Mt Sinai Sch Med One Gustave Levy Pl New York NY 10029

THORNBURG, DAVID DEVOE, physical metallurgy, for more information see previous edition

THORNBURG, DONALD RICHARD, b Pittsburgh, Pa, Oct 16, 33; m 71; c 3. PHYSICAL METALLURGY, MAGNETISM. *Educ:* Rensselaer Polytech Inst, BMetE, 55; Carnegie Inst Technol, MS, 63; Carnegie-Mellon Univ, PhD(metall), 72. *Prof Exp:* Navigator, US Air Force, 55-57; sr engr magnetics, 58-73, FEL ENGR MAGNETICS, RES & DEVELOP CTR, WESTINGHOUSE ELEC CORP, 73- *Mem:* AAAS; Am Soc Metals; Am Inst Mining, Metall & Petrol Engrs; Sigma Xi. *Res:* Development of magnetic materials; calculation of texture development using crystal plasticity theory; methods for improving the properties of magnetic materials; corrosion and mechanical properties of zirconium-based materials. *Mailing Add:* 387 Barclay Ave Pittsburgh PA 15221

THORNBURG, JOHN ELMER, b Syracuse, Ind, Apr 15, 42; m 70; c 2. CLINICAL PHARMACOLOGY, TOXICOLOGY. *Educ:* Purdue Univ, BS, 65, MS, 68, PhD(pharmacol), 70; Mich State Univ, DO, 76. *Prof Exp:* Fel pharmacol, 70-72, ASST PROF PHARMACOL, DEPT PHARMACOL, TOXI-FAMILY MEDICINE COL & FAMILY MED, MICH STATE UNIV, 77- *Mem:* Soc Neurosci; AAAS; Am Osteop Asn. *Res:* Perinatal toxicity of organophosphates; sympathetic nervous system and platelet alpha adrenergic receptors in hypertensive subjects. *Mailing Add:* Dept Pharmacol & Toxicol Mich State Univ East Lansing MI 48824

THORNBURGH, DALE A, b Tiffin, Ohio, Dec 1, 31; m 61; c 3. FOREST ECOLOGY. *Educ:* Univ Wash, BS, 59, PhD(forestry), 69; Univ Calif, Berkeley, MS, 62. *Prof Exp:* Lectr silviculture, Univ Wash, 63-64; from asst prof to assoc prof forest ecol, 64-74, chmn dept, 77-80, PROF FOREST ECOL, HUMBOLDT STATE UNIV, 74- *Mem:* Ecol Soc Am; Soc Am Foresters. *Res:* Carrying capacity of subalpine meadows in wilderness areas; development of forest habitat types and successional models. *Mailing Add:* Dept Forestry Humboldt State Univ Arcata CA 95521

THORNBURGH, GEORGE E(ARL), b Blair, Nebr, Apr 16, 23; m 43; c 4. MECHANICAL ENGINEERING. *Educ:* Univ Nebr, BS, 44; Iowa State Col, MS, 50. *Prof Exp:* Instr mech eng, Iowa State Col, 48-50; proj engr oil combustion, Lennox Furnace Co, 50-52; from asst prof to prof mech eng, Ore State Univ, 52-66, dir facilities planning, 66-67, dir planning & instnl res, 67-74, prof mech eng, 71-84; RETIRED. *Concurrent Pos:* Mech engr, US Bur Mines, Ore; engr, Bechtel Power Corp, 74-75. *Mem:* Am Soc Heating, Refrig & Air Conditioning Engrs; Am Soc Mech Engrs; Am Soc Eng Educ. *Res:* Heating and air conditioning engineering; energy. *Mailing Add:* Dept Mech Eng Ore State Univ Corvallis OR 97331

THORNBURN, THOMAS H(AMPTON), b Urbana, Ill, June 29, 16; m 42; c 4. SOILS, ENGINEERING. *Educ:* Univ Ill, BS, 38; Mich State Col, PhD(soil sci), 41. *Prof Exp:* Res engr, State Hwy Dept, Mich, 41-42; res assoc civil eng, 45-48, from res asst prof to res prof, 48-57, prof civil eng, 57-75, EMER PROF, UNIV ILL, URBANA, 75- *Concurrent Pos:* Consult civil eng, 48- *Mem:* Am Soc Testing & Mat; Am Soc Civil Engrs; Am Soc Photogram; Geol Soc Am; Nat Soc Prof Engrs. *Res:* Construction of stabilized soil roads; highway and railroad subgrade stability; engineering properties of soils; preparation of soil maps; terrain analysis. *Mailing Add:* 1034 Tam O'Shanter Las Vegas NV 89109

THORNBURY, JOHN R, b Cleveland, Ohio, Mar 16, 29; m 55; c 2. RADIOLOGY. *Educ:* Miami Univ, AB, 50; Ohio State Univ, MD, 55. *Prof Exp:* From instr to asst prof radiol, Univ Colo, 62-63; asst prof, Univ Iowa, 63-66 & Univ Wash, 66-68; from assoc prof to prof radiol, Univ Mich, Ann Arbor, 68-79; prof & chief, diag radiol, Univ NMex, 79-84; actg chmn, 85-87, PROF RADIOL, UNIV ROCHESTER, 84- *Mem:* Asn Univ Radiologists (pres, 80-81); Radiol Soc NAm; Am Roentgen Ray Soc; Soc Med Decision Making; Soc Uroradiol (pres, 76-77); fel Am Col Radiol. *Res:* Urological radiology; application of decision theory and probability theory principles to improve radiologists' diagnostic performance; applications of computer technology to diagnostic radiology; visual perception and information processing in diagnostic radiology. *Mailing Add:* Dept Radiol Univ Hosp Univ Mich Med Ctr Ann Arbor MI 48109

THORNDIKE, ALAN MOULTON, b Montrose, NY, June 27, 18; m 42; c 5. HIGH ENERGY PHYSICS, COMPLEX SYSTEMS. *Educ:* Wesleyan Univ, BA, 39; Columbia Univ, AM, 40; Harvard Univ, PhD(chem physics), 47. *Prof Exp:* Asst physicist, Div War Res, Univ Calif, 41-42; res assoc, Div War Res, Columbia Univ, 42-43; field serv consult, Off Sci Res & Develop, 43-45; asst scientist, Brookhaven Nat Lab, 47-52, from assoc scientist to scientist, 52-58; vis prof, Johns Hopkins Univ, 58-59; scientist, Brookhaven Nat Lab, 59-65, sr physicist, 65-70, assoc chmn physics dept, 70-73, sr physicist, 73-81; RETIRED. *Concurrent Pos:* Consult, Off Technol Assessment, 74-78. *Mem:* Am Phys Soc; Int Fuzzy Systs Asn; World Future Soc; AAAS. *Res:* Primary cosmic radiation; meson physics; elementary particles; electronic data processing; high energy interactions; computer simulation. *Mailing Add:* Physics Dept Brookhaven Nat Lab Upton NY 11973

THORNDIKE, EDWARD HARMON, b Pasadena, Calif, Aug 2, 34; m 55; c 3. ELEMENTARY PARTICLE PHYSICS. *Educ:* Wesleyan Univ, AB, 56; Stanford Univ, MS, 57; Harvard Univ, PhD(physics), 60. *Prof Exp:* Res fel physics, Harvard Univ, 60-61; from asst prof to assoc prof, 61-72, PROF PHYSICS, UNIV ROCHESTER, 72- *Concurrent Pos:* NSF sr fel, Univ Geneva & Europ Orgn Nuclear Res, 70; fel, Guggenheim Found, 87-88. *Mem:* Am Phys Soc. *Res:* Nucleon-nucleon interactions; few nucleon problems; electron-positron colliding beam phenomena; energy and environment; high energy photoproduction processes; b-quark decay. *Mailing Add:* Dept Physics Univ Rochester Rochester NY 14627

THORNDIKE, EDWARD MOULTON, b New York, NY, Sept 25, 05; m 30; c 3. PHYSICS. *Educ:* Wesleyan Univ, BS, 26; Columbia Univ, MA, 27; Calif Inst Technol, PhD(physics), 30. *Prof Exp:* Fel physics, Calif Inst Technol, 30-31; instr, Polytech Inst Brooklyn, 31-38; from instr to asst prof, Queen's Col, NY, 38-43; assoc prof, Univ Southern Calif, 43-44; from asst prof to prof, 44-70, emer prof physics, Queen's Col, NY, 70; CONSULT, LAMONT-DOHERTY. *Concurrent Pos:* Physicist, Woods Hole Oceanog Inst, 42-43, consult, 44-45; consult, Columbia Univ,44, 45-48, physicist, r45, res assoc, 50- *Mem:* AAAS. *Res:* Optics; oceanogaphy. *Mailing Add:* College Hill Montrose NY 10548-0098

THORNE, BILLY JOE, b Chanute, Kans, Aug 19, 37; m 59; c 4. MINING ENGINEERING. *Educ:* Phillips Unvi, AB, 59; Kans State Univ, MA, 61; Univ NMex, PhD(math), 68. *Prof Exp:* Staff mem, Sandia Nat Labs, 61-66; instr math, Smith Col, Northampton, Mass, 67-68; staff mem, Sandia Nat Labs, 68-74; div mgr, Civil Eng Res Facil, Univ NMex, 74-77; vpres, Civil Systs Inc, Subsid Sci Appl Inc, 77-80, pres, 80-81; staff mem, 81-82 & 85-89, div mgr, 82-85, DISTINGUISHED MEM TECH STAFF, SANDIA NAT LABS, 89- *Concurrent Pos:* Asst vpres, Sci Appl Inc, 80-81. *Res:* The use of computational techniques for the solution of a wide range of physical and engineering problems, usually involving the numerical solution off non-linear systems of partial differential equations. *Mailing Add:* Sandia Nat Labs Div 6253 Albuquerque NM 87185-5800

THORNE, CHARLES JOSEPH, b Pleasant Grove, Utah, May 28, 15; m 42; c 2. APPLIED MATHEMATICS. *Educ:* Brigham Young Univ, AB, 36; Iowa State Col, MS, 38, PhD(math physics), 41. *Prof Exp:* Asst math, Iowa State Col, 38-41; instr, Univ Mich, 41-43; asst prof, La State Univ, 43-44; develop engr, Curtiss-Wright Corp, 44-45; from asst prof to prof math, Univ Utah, 45-55; res scientist, Res Dept, US Naval Ord Test Sta, 55-60, sr res scientist & head math div, 60-61; supvry mathematician, Pac Missile Range, Pac Missile Test Ctr, 61-65, sr opers res analyst, Naval Missile Ctr, 65-76, head assessment div, 75-79, head mgt systs div, 79-82; RETIRED. *Concurrent Pos:* Assoc prof, Univ Calif, Los Angeles, 48-49, lectr, 56-68; sr investr, US Navy Projs, 49-51; dir & prin investr, US Army Ord Projs, 51-55. *Mem:* Am Math Soc; Am Soc Mech Engrs; Math Asn Am; Soc Indust & Appl Math. *Res:* Differential equations; boundary value problems; elasticity; analysis; operations research; numerical analysis. *Mailing Add:* 1447 Sunrise Ct Camarillo CA 93010

THORNE, CHARLES M(ORRIS), b Seattle, Wash, June 27, 21; m 47; c 2. ENGINEERING. *Educ:* Univ Wash, BS, 50. *Prof Exp:* Design engr, Boeing Airplane Co, 50; proj engr, Naval Undersea Warfare Eng Sta, 51-52, sr proof engr, 52-56, electronics div head, Weapons Qual Eng Ctr, 56-60, dir eng, 60-63, tech dir, 64-81; RETIRED. *Concurrent Pos:* Trustee, Oceanog Inst Wash. *Mem:* Inst Elec & Electronics Engrs; Am Soc Qual Control. *Res:* Testing and evaluation of underwater missiles; proximity fuzes, high energy batteries, pyrotechnics; environmental testing; non-destructive testing; metrology and functional testing. *Mailing Add:* 1499 NW Bucklin Hill Rd Bremerton WA 98310

THORNE, CURTIS BLAINE, b Pine Grove, WVa, May 13, 21; m 59; c 1. MICROBIAL GENETICS. *Educ:* WVa Wesleyan Col, BS, 43; Univ Wis, MS, 44, PhD(biochem), 48. *Prof Exp:* Biochemist, US Army Biol Labs, 48-61; prof bact genetics, Ore State Univ, 61-63; biochemist, US Army Biol Labs, 63-66; PROF BACT GENETICS, UNIV MASS, AMHERST, 66- *Concurrent Pos:* Waksman hon lectr, 59. *Mem:* Am Soc Microbiol. *Res:* Bacterial genetics; transformation and transduction; genetics of Bacillus species. *Mailing Add:* Dept Microbiol Univ Mass Amherst MA 01003

THORNE, JAMES MEYERS, b Logan, Utah, June 3, 37; m 60; c 2. PHYSICAL CHEMISTRY. *Educ:* Utah State Univ, BS, 61; Univ Calif, Berkeley, PhD(chem), 66. *Prof Exp:* PROF CHEM, BRIGHAM YOUNG UNIV, 66- *Concurrent Pos:* Vis staff, Laser Div, Los Alamos Sci Lab, Univ Calif, 72- *Mem:* Am Chem Soc; Sigma Xi. *Res:* Applications of lasers to nuclear fusion, nonlinear optics; magneto and electrooptics. *Mailing Add:* 218 ESC Brigham Young Univ Provo UT 84601

THORNE, JOHN CARL, b Ft Dodge, Iowa, Feb 24, 43; m 70; c 2. PLANT BREEDING, GENETICS. *Educ:* Augustana Col, Ill, BA, 65; Iowa State Univ, MS, 67, PhD(plant breeding), 69. *Prof Exp:* Res assoc soybean breeding, Iowa State Univ, 67-69; plant breeder, 69-79, SOYBEAN RES DIR, NORTHRUP, KING & CO, 79- *Concurrent Pos:* Dir & past pres, Nat Coun Com Plant Breeders. *Mem:* Am Soc Agron; Am Soybean Asn. *Res:* Plant breeding and genetics related to soybean variety development. *Mailing Add:* Northrup King & Co PO Box 49 Washington IA 52353

THORNE, JOHN KANDELIN, nonferrous metallurgy, for more information see previous edition

THORNE, KIP STEPHEN, b Logan, Utah, June 1, 40; m 62, 84; c 2. ASTROPHYSICS, THEORETICAL PHYSICS. *Educ:* Calif Inst Technol, BS, 62; Princeton Univ, AM, 63, PhD(theoret physics), 65. *Hon Degrees:* DSc, Ill Col, 79; DHC, Moscow Univ, 81. *Prof Exp:* NSF postdoctoral fel physics, Princeton Univ, 65-66; res fel physics, 66-67, assoc prof, 67-70, PROF THEORET PHYSICS, CALIF INST TECHNOL, 70-, WILLIAM R KENAN JR PROF, 81- *Concurrent Pos:* Lectr, Enrico Fermi Int Sch Physics, Varenna, Italy, 65, 68; Fulbright lectr, Sch Theoret Physics, Les Houches, France, 66; Alfred P Sloan res fel, 66-70; Guggenheim fel, Inst Astrophys, Paris, France, 67-68; vis assoc prof, Univ Chicago, 68; vis prof, Moscow State Univ, 69, 75, 78, 81, 86, 88 & 90; gov comt, Div of High Energy Astrophys, Am Astron Soc, 70-72; adj prof, Univ Utah, 71-; mem Int Comt Gen Relativity and Gravitation, 71-80; vis sr res assoc, Cornell Univ, 77, Andrew D White Prof, 86-; mem comt, US-USSR Coop in Physics, 78-79; mem adv bd, Inst Theoret Physics, Univ Calif, Santa Barbara, 78-80, mem, Space Sci Bd, 80-84; mem, Inst Theoret Physics, Univ Calif, Santa Barbara, 81-82. *Honors & Awards:* Sci Writing Award, Am Inst Physics-US Steel Corp, 69; John Danz Lectr, Univ Wash, 81; Silliman Lectr, Yale Univ, 84; Karl Herzfeld Lectr, Cath Univ, 85; Hans Bethe Lectr, Cornell Univ, 86; Sigma Xi Centenial Lectr, 86; Bart Bok Mem Lectr, Astron Soc Pac, 87. *Mem:* Nat Acad Sci; fel Am Acad Arts & Sci; fel Am Phys Soc; Sigma Xi; fel AAAS; Am Astron Soc. *Res:* Theoretical physics; theoretical and relativistic astrophysics; gravitation physics. *Mailing Add:* 130-33 Calif Inst Technol Pasadena CA 91125

THORNE, MARLOWE DRIGGS, b Perry, Utah, Nov 4, 18; m 41; c 4. AGRONOMY. *Educ:* Utah State Agr Col, BS, 40; Iowa State Col, MS, 41; Cornell Univ, PhD, 48. *Prof Exp:* Soil physicist & head dept agron, Pineapple Res Inst, Univ Hawaii, 47-54; soil scientist & irrig work proj leader, Eastern Soil & Water Mgt Sect, Soil & Water Conserv Res Br, Agr Res Serv, US Dept Agr, 55-56; prof agron & head dept, Okla State Univ, 56-63; prof agron, Univ Ill, Urbana, 63-74, head dept, 63-70, emer prof agron, 74-84; RETIRED. *Concurrent Pos:* Water technol adv, GB Pant Univ, Pantnagar, India, 70-72; team leader, Inst Agr & Animal Sci, Rampur Nepal, 82-84. *Mem:* Soil Sci Soc Am; Am Soc Agron (pres, 77); Soil Conserv Soc Am. *Res:* Irrigation; mulching; tillage. *Mailing Add:* 1403 E Mumford Dr Urbana IL 61801

THORNE, MELVYN CHARLES, b San Francisco, Calif, Dec 27, 32; m 58; c 2. EPIDEMIOLOGY. *Educ:* Univ Calif, AB, 56; Harvard Univ, MD, 60; Johns Hopkins Univ, MPH, 68. *Prof Exp:* Epidemiologist, Field Epidemiol Res Sta, Nat Heart Inst, 61-63 & 65-66; Peace Corps physician, Morocco, 63-65; resident internal med, Mary Imogene Bassett Hosp, 66-67; rep, Pop Coun, Tunisia, 68-72; asst prof int health, 72-77, ASSOC PROF INT HEALTH, SCH HYG & PUB HEALTH, JOHNS HOPKINS UNIV, 77-, MEM STAFF HEALTH SERV ADMIN & POP DYNAMICS, 78- *Concurrent Pos:* Tech adv, Urban Life-Pop Educ Inst, 73-76; health syst adv, Overseas Develop Coun Seminars, Priv Vol Orgns, 74-76; consult, Health Educ, Porter Novelli & Assocs. *Mem:* Am Pub Health Asn; Pop Asn Am. *Res:* Simple, effective methods to introduce health services into populations currently deprived of them, and to introduce population education into school systems with focus on using information in decisions. *Mailing Add:* APHA Dept State Washington DC 20523

THORNE, RICHARD EUGENE, b Aberdeen, Wash, Apr 12, 43; m 64; c 2. FISHERIES, HYDROACOUSTICS. *Educ:* Univ Wash, BS, 65, MS, 68, PhD(fisheries), 70. *Prof Exp:* Sr res assoc, Fisheries Res Inst, Univ Wash, 70-75, res assoc prof, 76-80, res prof fisheries, 81-88; SR SCIENTIST, BIOSONICS, INC, 88- *Concurrent Pos:* Acoust expert, Food & Agr Orgn, UN, 71-78; prog coordr, Div Marine Resources, Univ Wash, 73-86, sr scientist, Appl Physics Lab, 78-86. *Honors & Awards:* Cert for Outstanding Res & Professionalism, Am Fisheries Soc, 86. *Mem:* Acoust Soc Am; Am Fisheries Soc; Marine Technol Soc; Am Soc Limnol & Oceanog. *Res:* Hydroacoustic techniques of fish detection and abundance estimation; ecology of fishes. *Mailing Add:* BioSonics Inc 3670 Stone Way N Seattle WA 98103

THORNE, RICHARD MANSERGH, b Birmingham, Eng, July 25, 42; m 63; c 2. SPACE PHYSICS, PLASMA PHYSICS. *Educ:* Univ Birmingham, BSc, 63; Mass Inst Technol, PhD(physics), 68. *Prof Exp:* from asst prof to assoc prof meteorol, 68-75, PROF ATMOSPHERIC PHYSICS, UNIV CALIF, LOS ANGELES, 75- *Concurrent Pos:* NSF grants, 71-; mem nat comt, Int Union Radio Sci, 71-; consult, Jet Propulsion Lab, Calif Inst Technol & Aerospace Corp; chmn, atmospheric sci, Univ Calif, Los Angeles 76-79. *Mem:* Am Geophys Union; Int Union Radio Sci. *Res:* Structure and stability of radiation belts; magnetosphere-ionosphere interactions; wave propagation in anisotropic media; space plasma physics. *Mailing Add:* Dept Atmospheric Sci Univ Calif Los Angeles CA 90024

THORNE, ROBERT FOLGER, b Spring Lake, NJ, July 13, 20; m 47; c 1. SYSTEMATICS, BIOGEOGRAPHY. *Educ:* Dartmouth Col, AB, 41; Cornell Univ, MS, 42, PhD(bot), 49. *Prof Exp:* Asst bot, Cornell Univ, 45-46, instr, 48-49; from asst prof to prof, Univ Iowa, 49-62; PROF BOT, CLAREMONT GRAD SCH, 62-; CUR & TAXONOMIST, RANCHO SANTA ANA BOT GARDEN, 62- *Concurrent Pos:* Fulbright res scholar, Univ Queensland, 59-60; NSF sr fel, 60; chmn adv coun, Flora NAm Proj. *Mem:* Bot Soc Am; Am Soc Plant Taxon (secy, 57-58, pres, 68); Int Soc Plant Morphol; fel Linnean Soc London; fel French Soc Biogeog; Int Soc Plant Taxonomists. *Res:* Phylogeny and geography of flowering plants; floristics; marine phanerogams and fresh-water aquatic plants. *Mailing Add:* Rancho Santa Ana Bot Garden 1500 N College Ave Claremont CA 91711

THORNER, JEREMY WILLIAM, b Quincy, Mass, Jan 18, 46; m. GENE REGULATION, CELL BIOLOGY. *Educ:* Harvard Col, BA, 67; Harvard Univ, PhD(biochem), 72. *Prof Exp:* Fel biochem, Sch Med, Stanford Univ, 72-74; asst prof, 74-80, from assoc prof to prof microbiol, 80-85, prof biochem, 85-89, PROF BIOCHEM & MOLECULAR BIOL, UNIV CALIF, BERKELEY, 89- *Concurrent Pos:* Prin investr res grant, Nat Inst Gen Med Sci, 75-; consult, Chron Corp, Berkeley, Calif, 82-, Syntex Corp, Palo Alto, Calif, 88-; merit award, NIH, 89- *Mem:* Am Soc Biol Chemists; Am Soc Microbiol; Am Chem Soc; AAAS; NY Acad Sci; Genetic Soc Am; Protein Soc. *Res:* Molecular and cellular basis of the interactions that control the conjugation response of the yeast Saccharomyces cerevisiae to provide information about the mechanisms of developmental gene regulation and morphogenic control in eukaryotic cells. *Mailing Add:* Div Biochem & Molecular Biol Univ Calif 2120 Oxford Berkeley CA 94720

THORNER, MICHAEL OLIVER, b Beaconsfield, Bucks, UK, Jan 14, 45; US citizen; m 66; c 2. ENDOCRINOLOGY, NEUROENDOCRINOLOGY. *Educ:* Univ London, BS & MS, 70; Royal Col Physicians, London, MRCP(med), 72. *Hon Degrees:* FRCP, Royal Col Physicians, London, 84. *Prof Exp:* Lectr chem path, St Bartholomew's Hosp, London, 74, lectr internal med, 75-77; assoc prof internal med, 77-82, PROF MED, MED CTR, UNIV VA, 82-, DIR, CLIN RES CTR, 84-, HEAD, DIV ENDOCRINOL & METAB, 86- *Concurrent Pos:* Mem, med adv bd, Nat Hormone & Pituitary Prog, 84-86; mem, Endocrinol & Metabolic Drugs Adv Comn, 84-88; mem, Biochem Endocrinol Study Sect, NIH, 85- *Honors & Awards:* Albion O Bernstein Award, NY Med Soc, 84. *Mem:* Fel Royal Col Physicians; Endocrine Soc; Am Physiol Soc; Am Soc Clin Invest; Soc Neurosci; Asn Am Physicians. *Res:* Neuroendocrinology, basic and clinical aspects of hypothalamic pituitary functions with particular emphasis on the regulation of prolactin secretion and regulation of growth hormone secretion by growth hormone releasing hormone. *Mailing Add:* Med Ctr Univ Va Box 511 Charlottesville VA 22908

THORNGATE, JOHN HILL, b Eau Claire, Wis, Dec 23, 35; m 56; c 3. PHYSICS. *Educ:* Ripon Col, BA, 57; Vanderbilt Univ, MS, 61, PhD, 76. *Prof Exp:* Inspector health physics, Oak Ridge Opers Off, US Atomic Energy Comn, 59; res group leader radiation dosimetry, Health Physics Div, Oak Ridge Nat Lab, 64-75, asst sect chief, 73-75, health physicist, 60-78; PHYSICIST, LAWRENCE LIVERMORE LAB, 78- *Mem:* Health Physics Soc; Am Phys Soc; Sigma Xi; Inst Elec & Electronics Engrs. *Res:* Radiation dosimetry and spectrometry. *Mailing Add:* Lawrence Livermore Lab L-383 PO Box 5505 Livermore CA 94550

THORNHILL, JAMES ARTHUR, b London, Ont, Feb 11, 51; m 76. ENDOCRINOLOGY, NEUROPHYSIOLOGY. *Educ:* Univ Western Ont, BSc, 74, MSc, 75, PhD(pharmacol), 78. *Prof Exp:* Fel, Univ Calgary, 78-80; asst prof, 80-83, ASSOC PROF ENDOCRINOL, UNIV SASK, 83- *Concurrent Pos:* Med Res Coun Fel, 78-80. *Mem:* Am Physiol Soc; Can Physiol Soc; Can Pharmacol Soc; Soc Neurosci; AAAS; Can Soc Neurosci. *Res:* Possible physiological role that endogenous opioid peptides (endorphins or enkephalins) have on feeding and temperature regulation; cardiovascular mechanisims of action of opiates and opioids. *Mailing Add:* Dept Physiol Col Med Univ Sask Saskatoon SK S7N 0W0 Can

THORNHILL, PHILIP G, b Maidstone, England, July 7, 18; Can citizen; m 45; c 3. METALLURGY. *Educ:* Univ Toronto, BASc, 50, MASc, 51. *Prof Exp:* Res engr, 51-53, res metallurgist, 54-59, supvr metall res, 60-68, mgr process metall, 68-69, dir metall res, 69-82, PROCESS CONSULT, FALCONBRIDGE NICKEL MINES LTD, 82- *Concurrent Pos:* Pres, Lakefield Res Can Ltd, 81-82. *Honors & Awards:* Technol Medal, Metall Soc, 74; Airey Award, Can Inst Mining & Metall, 76. *Mem:* Am Inst Mining, Metall & Petrol Engrs; Electrochem Soc; Can Inst Mining & Metall; The Chem Soc. *Res:* Hydrometallurgical processes for extraction and refining of nickel. *Mailing Add:* 330 Second St Newmarket ON L3Y 3W6 Can

THORNTHWAITE, JERRY T, b Huntsville, Ala, Aug 16, 48. ONCOLOGY. *Educ:* Fla State Univ, PhD(chem), 77. *Prof Exp:* SCI DIR, ONCOL RES LABS, CEDARS MED CTR, 83- *Mem:* Soc Anal Cytol; Histochem Soc; Am Asn Immunologists. *Res:* Solid tumors. *Mailing Add:* 7880 SW 139th Terr Miami FL 33158

THORNTON, ARNOLD WILLIAM, b Hull, Eng, Apr 20, 43; m 73; c 3. MANAGEMENT OF PRODUCT DEVELOPMENT & PRODUCT ASSURANCE IN THE BIOMEDICAL DEVICE INDUSTRY, TECHNOLOGY DEVELOPMENT IN THE BIOMEDICAL DEVICE INDUSTRY. *Educ:* Oxford Univ, Eng, BA, 64, MA, 68, DPhil, 68. *Prof Exp:* Res scientist, Dept Metall & Mat Sci, Univ Denver, Colo, 68-73; proj engr, Uniroyal Corp, Detroit, Mich, 73-75; dir res, Cardiac Pacemakers Inc, St Paul, Minn, 75-82; dir prod develop & prod assurance, Medtronic, Minneapolis, Minn, 83-90, dir leads technol, 90-91; MGR CATHETER DEVELOP, SCIMED LIFE SYSTS, MINNEAPOLIS, MINN, 91- *Concurrent Pos:* Lectr polymer & ceramic sci, Dept Metall & Mat Sci, Univ Denver, Colo, 70-73. *Res:* Application of polymer, ceramic and metallurgical science to the design and implementation of implantable devices for electrically stimulating the heart and devices for treating heart disease. *Mailing Add:* 1258 Willow Lane Roseville MN 55113

THORNTON, C G, b Detroit, Mich, Aug 3, 25; m 49; c 2. SEMICONDUCTORS, MICROELECTRONICS. *Educ:* Univ Mich, BS, 49, MS, 50, PhD(phys chem), 52. *Prof Exp:* Proj engr, Sylvania Elec Co, 51-52; from sect head to dir, Semiconductors Div, 52-60, dir, Res & Develop, Philco Corp, 60-72; DIR, ELEC TECHNOL & DEVICES LAB, DEPT ARMY, 72- *Concurrent Pos:* Chmn, numerous Gov Comts & study teams. *Honors & Awards:* Res & Develop Achievement Award, 76; Crozier Award; Centennial Medal, Inst Elec & Electronics Engrs. *Mem:* Fel Inst Elec & Electronics Engrs; Electrochem Soc; Armed Forces Commun & Electronics Asn; Electron Devices Soc. *Res:* Research in semiconductor devices, microelectronics, semiconductor processing; emphasis on high frequency devices and circuits and microfabrication techniques; very high speed integrated circuits and monolithic microwave/millimeter integrated circuits displays; frequency control and acoustic wave devices; power sources including pulse power. *Mailing Add:* Attn SLCET-D US Army Electronics Technol & Devices Lab Ft Monmouth NJ 07703-5000

THORNTON, CHARLES PERKINS, b Indianapolis, Ind, Jan 1, 27; m 54; c 2. PETROLOGY. *Educ:* Univ Va, AB, 49; Yale Univ, MS, 50, PhD(geol), 53. *Prof Exp:* Field geologist, State Geol Surv, Va, 50-52; from instr to asst prof petrog, Pa State Univ, 52-61; asst prof geol, Bucknell Univ, 61-63; assoc prof, 63-69, PROF GEOL, PA STATE UNIV, UNIVERSITY PARK, 69- *Mem:* Geol Soc Am. *Res:* Geology of central Shenandoah Valley, Virginia; petrography and petrology of volcanic rocks; volcanology. *Mailing Add:* Dept Geosci Pa State Univ University Park PA 16802

THORNTON, DONALD CARLTON, b Baltimore, Md, Apr 16, 47. ANALYTICAL CHEMISTRY. *Educ:* Univ Va, BS, 69; Pa State Univ, MS & PhD(chem), 76. *Prof Exp:* Res assoc chem, Univ Fla, 76-77; asst prof chem, 77-84, res scholar, 84-86, RES ASSOC PROF, DREXEL UNIV, 86- *Mem:* AAAS; Am Chem Soc; Sigma Xi; Am Geophys Union. *Res:* Measurements of sulfur compounds in the atmosphere; sulfur biogeochemical cycle; development of analytical instrumentation and techniques applicable to atmospheric chemistry. *Mailing Add:* Dept Chem Drexel Univ Philadelphia PA 19104

THORNTON, EDWARD RALPH, b Syracuse, NY, July 19, 35; m 69; c 1. ORGANIC CHEMISTRY, BIOLOGICAL CHEMISTRY. *Educ:* Syracuse Univ, BA, 57; Mass Inst Technol, PhD(org chem), 59. *Hon Degrees:* MA, Univ Pa, 71. *Prof Exp:* NIH fel, Mass Inst Technol, 59-60 & Harvard Univ, 60-61; from asst prof to assoc prof, 61-69, PROF CHEM, UNIV PA, 69- *Mem:* Fedn Am Sci; Am Chem Soc; Royal Soc Chem; Am Soc Biol Chemists. *Res:* Molecular interactions and selectivity: stereoselectivity, recognition, transition state structure. *Mailing Add:* Dept Chem Univ Pa Philadelphia PA 19104-6323

THORNTON, ELIZABETH K, b Brooklyn, NY, June 4, 40. PHYSICAL ORGANIC CHEMISTRY. *Educ:* Mt Holyoke Col, AB, 61; Univ Pa, PhD(org chem), 66. *Prof Exp:* Teaching asst chem, Univ Pa, 61-62, NIH fel org chem, 63-66; NATO fel, Swiss Fed Inst Technol, 66-68; asst prof, 68-75, ASSOC PROF CHEM, WIDENER UNIV, 75-, CURRIC COORDR, SCI DIV, 87- *Mem:* AAAS; Fedn Am Sci; Sigma Xi; Am Chem Soc. *Res:* Kinetic isotope effects and reaction mechanisms; organic and biochemistry. *Mailing Add:* Dept Chem Widener Univ Chester PA 19013

THORNTON, GEORGE DANIEL, b Elberton, Ga, Aug 10, 10; m 39. SOILS. *Educ:* Univ Ga, BS, 36, MS, 38; Iowa State Col, PhD(soil fertility), 47. *Prof Exp:* County agr agent, Ga Exten Serv, 36; asst soil surveyor, Ga State Col, 36, instr agron, 37-40; asst agronomist, Exp Sta, Univ Ga, 40-41; asst prof soils & asst soil microbiologist, Col Agr, Univ Fla, 41-45; asst, Iowa State Col, 45-47; assoc prof soils & assoc soil microbiology, 47-51, prof soils, 51-71, soil microbiologist, 51-56, asst dean col, 56-71, EMER PROF SOILS, COL AGR, UNIV FLA, 71- *Mem:* Fel Am Soc Agron; Soil Sci Soc Am. *Res:* Soil microbiology. *Mailing Add:* PO Box 833 Venice FL 34284

THORNTON, GEORGE FRED, b Newton, Mass, Mar 8, 33; m 63; c 2. INTERNAL MEDICINE, INFECTIOUS DISEASES. *Educ:* Harvard Univ, AB, 55; Boston Univ, MD, 59. *Prof Exp:* Instr clin med, Sch Med, Yale Univ, 64-65; instr med, Johns Hopkins Univ, 65-67; from asst prof to assoc prof clin med, 72-78, CLIN PROF MED, SCH MED, YALE UNIV, 78-; DIR MED SERV, WATERBURY HOSP, 72- *Concurrent Pos:* Fel allergy & infectious dis, Johns Hopkins Univ, 62-64; assoc clin prof med, Univ Conn, 75-81. *Mem:* AMA; Am Fedn Clin Res; fel Infectious Dis Soc Am; fel Am Col Physicians; Am Soc Microbiol. *Res:* Clinical epidemiology. *Mailing Add:* Med Serv Waterbury Hosp Waterbury CT 06720

THORNTON, HUBERT RICHARD, b Van Etten, NY, Nov 15, 32; m 59; c 2. CERAMIC ENGINEERING, METALLURGY. *Educ:* Alfred Univ, BS, 54, MS, 57; Univ Ill, PhD(ceramic eng), 63. *Prof Exp:* Res engr, Nat Bur Standards, 56-59; res assoc ceramic eng, Univ Ill, 59-63; proj standards engr, Gen Dynamics, Ft Worth, 63-67; assoc prof, 67-77, ASSOC PROF MECH ENG, TEX A&M UNIV, 77- *Mem:* Am Ceramic Soc; Am Soc Metals; Soc Aerospace Mat & Process Eng; Am Soc Mech Engrs; fel Am Inst Chem. *Res:* Explosive forming; fracture mechanics; failure modes in materials; fracture analysis; design optimization; teaching of materials and materials in design. *Mailing Add:* 2505 Willow Bend Dr Bryan TX 77802

THORNTON, JANICE ELAINE, b Vancouver, Wash, Apr 5, 52; m 85; c 1. NEUROENDOCRINOLOGY, BEHAVIORAL ENDOCRINOLOGY. *Educ:* Portland State Univ, BS, 76; Univ Wis-Madison, MS, 79, PhD(physiol, psychol), 83. *Prof Exp:* Postdoctoral fel, Rockefeller Univ, New York, NY & Rutgers Univ, Newark, NJ, 83-86; postdoctoral fel, Ore Health Sci Univ, Portland, Ore, 87-89; asst scientist, Ore Regional Primate Res Ctr, 89-90; ASST PROF NEUROSCI & BIOL, OBERLIN COL, OHIO, 90- *Mem:* Soc Neurosci; Soc Study Reproduction; AAAS; Sigma Xi. *Res:* Organizational and activational effects of gonadal hormones on brain and behavior, noradrenergic system and neuropeptides in the control of female reproductive behavior and ovulation. *Mailing Add:* Neurosci Prog Oberlin Col Oberlin OH 44074

THORNTON, JOHN IRVIN, b Sacramento, Calif, Jan 11, 41; m 75; c 4. FORENSIC CHEMISTRY. *Educ:* Univ Calif, Berkeley, BS, 62, MCriminalistics, 68, DCriminalistics, 74. *Prof Exp:* Criminologist, Contra Costa Count Sheriff's Dept, 63-72; from asst prof to assoc prof forensic sci, 74-82, vchmn, dept biomed & environ health sci, 81-84, PROF FORENSIC SCI, SCH PUB HEALTH, UNIV CALIF, BERKELEY, 82- *Concurrent Pos:* Mem proj adv comt, Nationwide Crime Lab Proficiency Testing Proj, Forensic Sci Found, 74-82. *Honors & Awards:* Paul Kirk Award, Am Acad Sci. *Mem:* Am Chem Soc; Am Acad Forensic Sci; Forensic Sci Soc Gt Brit; Sigma Xi. *Res:* Analysis, identification, and interpretation of physical evidence; author or coauthor of over 150 publications. *Mailing Add:* Pub Health Dept Univ Calif Berkeley CA 94720

THORNTON, JOHN WILLIAM, b Shawnee, Okla, Apr 21, 36; m 57; c 3. ZOOLOGY, CYTOLOGY. *Educ:* Okla State Univ, BS, 58; Univ Wash, PhD(zool), 64. *Prof Exp:* From asst prof to assoc prof, 60-74, PROF ZOOL, OKLA STATE UNIV, 74- *Concurrent Pos:* USPHS res grant, 65-68; staff biologist, Comn Undergrad Educ Biol Sci, 70-71; mem adv comt, Purdue Minicourse Proj. *Mem:* AAAS; Am Soc Zoologists; Am Inst Biol Sci. *Res:* Cellular ultrastructure; cell and tissue culture; undergraduate curricular improvement; investigative laboratories. *Mailing Add:* Dept Zool Okla State Univ Stillwater OK 74078

THORNTON, JOSEPH SCOTT, b Sewickley, Pa, Feb 6, 36; c 2. MATERIALS SCIENCE. *Educ:* Univ Tex, BS, 57; Carnegie Inst Technol, MS, 62; Univ Tex, Austin, PhD(mat sci), 69. *Prof Exp:* Design engr, Walworth Co, 57; res engr, Westinghouse Elec Co, 61-64; instr metall, Univ Tex, Austin, 64-66; group leader metals & composites, Tracor, Inc, 67-69; mgr mat, Horizons Inc, 67-73; dir appl sci, Tracor, Inc, 73-75; PRES, TEX RES INT, INC, 75- *Mem:* Am Soc Metals; Am Soc Testing & Mat; Am Soc Mech Engrs; Adhesion Soc. *Res:* Contract research administration; development and characterization of engineering materials; reliability of devices in adverse environments; failure analysis; accelerated life test development; recovery techniques for hazardous materials spills. *Mailing Add:* 9063 Bee Caves Rd Austin TX 78733-6201

THORNTON, KENT W, b Ames, Iowa, Apr 29, 44; m 66; c 1. AQUATIC ECOLOGY, SYSTEMS SCIENCE. *Educ:* Univ Iowa, BA, 67, MS, 69; Okla State Univ, PhD(ecol), 72. *Prof Exp:* Teaching asst zool, Univ Iowa, 68-69; lectr environ systs theory, Okla State Univ, 72; asst prof biol, Bowling Green State Univ, 73-74; SYSTS ECOLOGIST, WATERWAYS EXP STA, US ARMY ENGRS, 74- *Concurrent Pos:* NSF fel, Ctr Systs Sci, Okla State Univ, 72-73; mem methods ecosyst anal, Nat Comn Water Qual, 74; actg br chief, Waterways Exp Sta, US Army Engrs, 75. *Mem:* AAAS; Am Inst Biol Sci; Ecol Soc Am; NAm Benthological Soc; Int Soc Limnol. *Res:* Systems theoretical approach to the conceptualization, analysis and application of mathematical ecosystem models for watershed-reservoir planning and management; sampling theory approach to dynamic systems. *Mailing Add:* Five Clover Ct Little Rock AR 72209

THORNTON, LINDA WIERMAN, b Charleston, SC, Oct 16, 42; m 68; c 2. EARTH SCIENCES, COMPUTER SCIENCES. *Educ:* Univ Kansas, BA, 66. *Prof Exp:* From econ to sr econ, Hare & Hare, Inc, 64-74; sr economist, Res & Mgt Serv, 76-80, mgr res econ, 80-82, sect head, econ & financial analysis, 82-84, dir, Econ & Mgt Sci Dept, 84- 90, DIR, RES & MGT SERV, 90- *Mem:* Nat Asn Bus Econ. *Res:* Market analysis. *Mailing Add:* 425 Volker Blvd Kansas City MO 64110

THORNTON, MELVIN CHANDLER, b Sioux City, Iowa, July 2, 35; m 58; c 4. MATHEMATICS. *Educ:* Univ Nebr, BS, 57; Univ Ill, MS, 61, PhD(math), 65. *Prof Exp:* Asst prof math, Univ Wis, 65-69; asst prof, 69-73, ASSOC PROF MATH, UNIV NEBR, LINCOLN, 73- *Mem:* Am Math Soc; Math Asn Am. *Res:* General topology. *Mailing Add:* Dept Math Univ Nebr Lincoln NE 68588

THORNTON, MELVIN LEROY, b Billings, Mont, Nov 7, 28; m 52; c 2. BOTANY. *Educ:* Univ Denver, BA, 52; Tufts Univ, MA, 58; Univ Mont, PhD(bot), 69. *Prof Exp:* NSF fel, Birkbeck Col, Univ London, 69-70; from asst prof to assoc prof bot, Univ Mont, 71-80; RETIRED. *Mem:* Sigma Xi. *Res:* Dispersal of fungi; ecology of zoosporic fungi. *Mailing Add:* 810 First St Sultan WA 98294

THORNTON, PAUL A, b Campbell Co, Ky, June 29, 25; m 45; c 2. PHYSIOLOGY, NUTRITION. *Educ:* Univ Ky, BS, 49, MS, 53; Mich State Univ, PhD(nutrit), 56. *Prof Exp:* Asst & assoc prof nutrit, Colo State Univ, 56-62, assoc prof physiol, 63-64; vis prof, 62-63, assoc prof, 64-77, PROF PHYSIOL, UNIV KY, 77-; RES PHYSIOLOGIST, VET ADMIN HOSP, LEXINGTON, 64- *Mem:* Soc Exp Biol & Med; Am Inst Nutrit; Am Physiol Soc; Geront Soc. *Res:* Skeletal physiology and the influence of age on bone tissue change; endocrinological and other environmental factors which affect bone. *Mailing Add:* Dept Allied Health Univ Ky Col Med Lexington KY 40506

THORNTON, RICHARD D(OUGLAS), b New York, NY, Sept 24, 29; m 59; c 3. ELECTRICAL ENGINEERING, COMPUTER SCIENCE. *Educ:* Princeton Univ, SB, 51; Mass Inst Technol, MS, 54, ScD(elec eng), 57. *Prof Exp:* From asst prof to assoc prof, 57-68, PROF ELEC ENG & COMPUTER SCI, MASS INST TECHNOL, 68- *Concurrent Pos:* Chmn, Thornton Assocs, Inc. *Honors & Awards:* Baker Award, Inst Radio Eng, Inst Elec & Electronics Engrs, 59. *Mem:* Inst Elec & Electronics Engrs. *Res:* Maglev transportation systems; electronic circuits and computer aided engineering design; microcomputer controlled electromechanical systems. *Mailing Add:* Dept Elec Eng & Comp Sci Mass Inst Technol 77 Mass Ave Cambridge MA 02139

THORNTON, ROBERT MELVIN, b Auburn, Calif, Nov 14, 37; m 57, 71; c 2. BIOLOGY, PLANT PHYSIOLOGY. *Educ:* Calif Inst Technol, BS, 59; Harvard Univ, MA, 61, PhD(biol), 66. *Prof Exp:* Sr scientist, Biol/Eng, Appl Sci Corp, Calif, 61-63; instr, Ojai Valley Sch, 63-64; instr biol, Univ Calif, Santa Cruz, 66-67, asst prof, 67-68; from asst prof to assoc prof, 68-81, SR LECTR BOT, UNIV CALIF, DAVIS, 81- *Concurrent Pos:* NSF res grant, Univ Calif, Davis, 69-72. *Mem:* AAAS; Am Soc Plant Physiol; Bot Soc Am. *Res:* Regulatory mechanisms in morphogenesis of plants and fungi; educational research on critical thinking limitations with particular attention to memory management. *Mailing Add:* Dept Bot 218 Robbins Hall Univ Calif Davis CA 95616

THORNTON, ROGER LEA, b Wilmington, Del, Mar 9, 35; m 58; c 3. ORGANIC CHEMISTRY. *Educ:* Univ Del, BS, 57; Mass Inst Technol, PhD(org chem), 61. *Prof Exp:* Res chemist, Del, 61-65, Va, 65-67, sr res chemist, 67-86, LICENSING ASSOC, E I DU PONT DE NEMOURS & CO, INC, DEL, 86- *Res:* Vapor-phase catalytic reactions; thermally stable condensation polymers; emulsion polymerization; fluorinated compounds. *Mailing Add:* Chems E I Du Pont de Nemours & Co Inc Wilmington DE 19898

THORNTON, ROY FRED, b Upper Darby, Pa, Feb 27, 41; m 66; c 2. ELECTROCHEMICAL ENGINEERING, ELECTROCHEMISTRY. *Educ:* Johns Hopkins Univ, BS, 63, PhD(chem eng), 67. *Prof Exp:* Chem engr, Battery Bus Dept, 67-69, STAFF MEM ELECTROCHEM, GEN ELEC CO CORP RES & DEVELOP, 69- *Mem:* Am Inst Chem Engrs; Electrochem Soc. *Res:* Development of electrochemical energy storage systems and electrochemical processes. *Mailing Add:* Box Eight Corp Res & Develop Gen Elec Co Schenectady NY 12301

THORNTON, STAFFORD E, b Campbell Co, Va, July 29, 34; m 63; c 3. EDUCATION ADMINISTRATION. *Educ:* Univ Va, BCE, 59, MCE, 62. *Prof Exp:* Instr civil eng, Univ Va, 60-61, res engr, 62-63; from asst prof to prof civil eng, 63-86, head dept, 64-77, asst dean, 77-86, DIR TECH ASSISTANCE CTR, WVA INST TECHNOL, 87- *Concurrent Pos:* City Engr, Montgomery, WVa, 77-; dir, Am Soc Civil Engrs, 79-81; mem, WVa Regist Bd, 79-90. *Honors & Awards:* Roy D Koch Award, 86. *Mem:* Am Soc Civil Engrs (vpres, 82-84); Am Soc Eng Educ; Nat Soc Prof Engrs. *Res:* Structural design; concrete and steel; foundations; city planning; stress analysis of rotating disks and cylinders. *Mailing Add:* Tech Assistance Ctr WVa Inst Technol Montgomery WV 25136

THORNTON, STEPHEN THOMAS, b Kingsport, Tenn, Oct 2, 41; m 61, 79; c 5. EXPERIMENTAL NUCLEAR PHYSICS. *Educ:* Univ Tenn, Knoxville, BS, 63, MS, 64, PhD(physics), 67. *Prof Exp:* US Atomic Energy Comn fel, Univ Wis-Madison, 67-68; asst prof, 68-72, assoc prof, 72-82, PROF PHYSICS, UNIV VA, 82- *Concurrent Pos:* Consult, Physics Div, Oak Ridge Nat Lab, 72-77; Fulbright-Hays sr fel, Max Planck Inst, Heidelberg, 73-74 & 79-80. *Mem:* AAAS; Am Phys Soc; Sigma Xi; AAAS. *Res:* Electromagnetic nuclear physics; science education; textbook author. *Mailing Add:* Dept Physics Univ Va Charlottesville VA 22903

THORNTON, WILLIAM ALOYSIUS, b New York, NY, Sept 8, 38; m 63; c 2. STRUCTURAL MECHANICS. *Educ:* Manhattan Col, BCE, 60; Case Western Reserve Univ, MSEM, 64, PhD(eng mech), 67. *Prof Exp:* Asst prof civil eng, Clarkson Col Technol, 67-72, assoc prof civil & environ eng, 72-79; CHIEF ENGR, CIVES STEEL CO, ROSWELL, GA, 79-; PRES, CIVES ENG CORP, 87- *Concurrent Pos:* Lectr, AISC Regional Lect Ser, 84- *Mem:* Fel Am Soc Civil Engrs; Am Soc Testing & Mat; Nat Soc Prof Engrs; Sigma Xi; Am Welding Soc. *Res:* computer oriented structural analysis and design; behavior, analysis and design of structural connections. *Mailing Add:* 9475 Martin Rd Roswell GA 30076

THORNTON, WILLIAM ANDRUS, JR, b Buffalo, NY, June 16, 23; m 44; c 4. PHYSICS. *Educ:* Univ Buffalo, BA, 48; Yale Univ, MS, 49, PhD(physics), 51. *Prof Exp:* Res assoc labs, Gen Elec Co, 51-56; sr res engr, 56-59, fel res engr, 59-65, mgr phosphor res, 65-67, RES ENG CONSULT, WESTINGHOUSE ELEC CORP, BLOOMFIELD, 67- *Mem:* Am Phys Soc; Optical Soc Am; fel Illum Eng Soc. *Res:* Light and color. *Mailing Add:* 27 Harvard Rd Cranford NJ 07016

THORNTON, WILLIAM EDGAR, b Faison, NC, Apr 14, 29; c 2. MEDICINE, ASTRONAUTICS. *Educ:* Univ NC, BS, 52, MD, 63. *Prof Exp:* Chief engr, Electronics Div, Del Mar Eng Labs, Calif, 55-59; intern, Wilford Hall Hosp, Lackland AFB, US Air Force, Tex, 64, assigned to Aerospace Med Div, Brooks AFB, 65-67; scientist-astronaut, Johnson Space Ctr, NASA, 67-; CHMN, DEPT CRIMINAL JUSTICE, LOYOLA UNIV, NEW ORLEANS. *Concurrent Pos:* Instr, Dept Med, Univ Tex Med Br, Galveston. *Honors & Awards:* NASA Except Sci Achievement Award, 74. *Res:* Physics; biomedical instrumentation; cardiovascular; principle investigator on sky lab experiments including mass measurements in space, musculo-skeletal and cardiovascular investigations. *Mailing Add:* Dept Criminal Justice Loyola Univ 6363 St Charles Ave New Orleans LA 70118

THOROUGHGOOD, CAROLYN A, b Sept 1, 43; m 64; c 1. MARINE SCIENCES. *Educ:* Univ Del, BS, 65; Univ Md, MS, 66, PhD(nutrit), 68. *Prof Exp:* Asst prof food sci & nutrit, 68-72, assoc prof, 72-74, dir, Marine Adv Serv, 74-76, assoc dir, Del Sea Grant Col Prog, 76-78, assoc prof marine studies, food sci & nutrit, 78-84, assoc dean, 80-84, DEAN & PROF, COL MARINE STUDIES, UNIV DEL, 84- *Concurrent Pos:* Exec dir, Del Sea Grant Col Prog, Col Marine Studies, Univ Del, 78-84; chair, Nat Coun Sea Grand Dirs, 80-81, exec comt mem, 81-; dir, Del Sea Grant Col Prog, 84-; mem, Governor's Task Force Aquacult, 90- *Mem:* AAAS; Sigma Xi; Marine Technol Soc; Am Chem Soc. *Res:* Nutritional biochemistry of bivalve molluscs; aquaculture of bivalves; lipid biochemistry of marine organisms; marine education and the development of policy and materials required to enhance the general public's marine literacy. *Mailing Add:* PO Box 3693 Greenville DE 19807

THORP, BENJAMIN A, b Albany, NY, May 31, 38; m 87; c 1. FLUIDS, MATHEMATICS STATISTICS. *Educ:* Univ Md, BS, 64. *Prof Exp:* Mkt mgr, Huyck Formex Div, Huyck Corp, 71-74, vpres & gen mgr, 74-76, vpres & gen mgr, Huytech Systs Div, 76-80, vpres & dir res, 80-81; pres, Benjamin A Thorp Inc, 81-83 & Poyry-Bek Inc, Raleigh, NC, 83-85; vpres eng, B E & K Inc, Birmingham, Ala, 85-86; VPRES ENG, JAMES RIVER CORP, RICHMOND, VA, 86- *Mem:* Fel Tech Asn Pulp & Paper Indust; Paper Indust Mgt Asn; Exp Aircraft Asn. *Res:* Impact and relationship between human systems and hi-tech systems; development of high performance work systems. *Mailing Add:* James River Corp Tredegar St PO Box 2218 Richmond VA 23217

THORP, EDWARD O, b Chicago, Ill, Aug 14, 32; m 56; c 3. MATHEMATICS. *Educ:* Univ Calif, Los Angeles, BA, 53, MA, 55, PhD(math), 58. *Prof Exp:* Instr math, Univ Calif, Los Angeles, 58-59; C L E Moore instr, Mass Inst Technol, 59-61; from asst prof to assoc prof, NMex State Univ, 61-65; assoc prof, Univ Calif, Irvine, 65-67, prof math & finance, 67-82; PRES & CHMN, OAKLEY SUTTON MGT CORP, 72-; MANAGING GEN PARTNER, PRINCETON NEWPORT PARTNERS, 69- *Concurrent Pos:* Res grants, NSF, 62-64 & USAF Off Sci Res, 64-74. *Mem:* Fel Inst Math Stat. *Res:* Functional analysis; probability theory; game theory; statistics; mathematical finance. *Mailing Add:* Edward O Thorp & Assocs 620 Newport Center Dr Suite 1075 Newport Beach CA 92660

THORP, FRANK KEDZIE, b Denver, Colo, Apr 29, 36; m 65; c 1. BIOCHEMISTRY, PEDIATRICS. *Educ:* Mich State Univ, BA, 55; Univ Chicago, MD, 60, PhD(biochem), 62. *Prof Exp:* Intern pediat, Univ Chicago, 61-62; resident, Children's Hosp Med Ctr, Boston, 62-63; instr, 65-66, asst prof, 66-72, ASSOC PROF PEDIAT, UNIV CHICAGO, 72- *Concurrent Pos:* NIH res fel biochem, Children's Hosp Med Ctr, Boston, 63-65; Joseph P Kennedy, Jr scholar, 66-; Am Acad Pediat grant, 69-; dir ment develop clin, Joseph P Kennedy, Jr Ment Retardation Res Ctr, 71-74; dir clin serv, Wyler Children's Hosp, 77- *Mem:* Am Acad Pediat; Sigma Xi; NY Acad Sci. *Res:* Teaching and clinical work in pediatrics; metabolic and nutritional diseases of children; development of nutrition training and nutrition education programs. *Mailing Add:* Dept Pediat Box 308 Univ Chicago 5841 S Maryland Ave Chicago IL 60637

THORP, JAMES HARRISON, III, b Kansas City, Mo, July 23, 48; m 70; c 2. AQUATIC ECOLOGY, COMMUNITY ECOLOGY. *Educ:* Univ Kans, BA, 70; NC State Univ, MS, 73, PhD(zool), 75. *Prof Exp:* PROF BIOL & DIR WATER RESOURCES LAB, UNIV LOUISVILLE, 88- *Concurrent Pos:* Dir, Calder Conserv & Ecology Ctr, Fordham Univ, 85-88; educ prog dir, Savannah River Ecol Lab, 77-80; assoc ed, Freshwater Invertebrate Biol, 81-84; prin investr, numerous grants, ecology & education; proj dir, Indust Fac Res Partic, NSF grant, 79. *Mem:* Ecol Soc Am; NAm Benthological Soc Am; Soc Int Limnol; AAAS. *Res:* Experimental field studies of factors regulating structure in benthic macroinvertebrate communities within freshwater rivers and lakes; behavioral ecology and competitive interactions among crustaceans. *Mailing Add:* Biol Dept Univ Louisville Louisville KY 40292

THORP, JAMES SHELBY, b Kansas City, Mo, Feb 7, 37; m 59; c 2. ELECTRICAL ENGINEERING. *Educ:* Cornell Univ, BEE, 59, MS, 61, PhD(elec eng), 62. *Prof Exp:* From asst prof to assoc prof, 62-75, PROF ELEC ENG, CORNELL UNIV, 75- *Concurrent Pos:* Consult, ADCOM, Mass, 62, Gen Elec Co, 64 & Am Elec Power Serv Corp, 77- *Mem:* Inst Elec & Electronics Engrs; Sigma Xi. *Res:* Control systems and optimal control; power systems. *Mailing Add:* Phillips Hall Cornell Univ Ithaca NY 14850

THORP, JAMES WILSON, b Cohocton, NY, Mar 17, 42; m 65; c 3. DIVING MEDICINE, PEDIATRICS. *Educ:* Cornell Univ, BS, 63; Iowa State Univ, PhD(nutrit), 67; Georgetown Univ, MD, 75. *Prof Exp:* Res physiologist radiobiol, Armed Forces Radiobiol Res Inst, Bethesda, Md, 67-71; resident pediat, Naval Hosp, Bethesda, Md, 75-78; fel neonatology, Childrens Hosp, Washington, DC, 78-80; staff physician neonatology, Naval Hosp, Bethesda, Md, 80-87; RES PHYSICIAN DIVING MED, NAVAL MED RES INST, BETHESDA, MD, 87- *Concurrent Pos:* Assoc prof pediat, Uniformed Serv, Univ Health Sci, Bethesda, Md, 80-; staff neonatologist, Naval Hosp, Bethesda, Md, 87-; asst ed, Am J Clin Nutrit, 86- *Mem:* Am Soc Clin Nutrit; Am Acad Pediat; AMA. *Res:* Metabolism and nutrition requirements of humans working under hyperboric conditions and effects of extreme cold or heat. *Mailing Add:* Diving Med Naval Med Res Inst Bethesda MD 20889-5055

THORP, ROBBIN WALKER, b Benton Harbor, Mich, Aug 26, 33; m 54, 67; c 3. INSECT TAXONOMY, ECOLOGY. *Educ:* Univ Mich, BS, 55, MS, 57; Univ Calif, Berkeley, PhD(entom), 64. *Prof Exp:* Jr specialist, Univ Calif, Berkeley, 62-63, asst specialist, 63-64, asst res entomologist, 64; asst apiculturist, 64-72, assoc prof entom & assoc apiculturist, 72-78, PROF ENTOM & APICULTURIST, UNIV CALIF, DAVIS, 78- *Mem:* AAAS; Ecol Soc Am; Soc Syst Zool; Entom Soc Am; Soc Study Evolution; Sigma Xi. *Res:* Pollination ecology, especially bee and flower relationships; ecology and systematics of bees and ecology of their biotic enemies; coevolution and coadaptation of pollinating insects and entomophilous angiosperms. *Mailing Add:* Dept Entom Univ Calif Davis CA 95616

THORPE, BERT DUANE, b Spanish Fork, Utah, Sept 21, 29; m 55; c 6. IMMUNOLOGY, WILDLIFE DISEASES. *Educ:* Univ Utah, BS(chem) & BS(bact), 58, PhD(microbiol), 63. *Prof Exp:* Asst chem, Univ Utah, 55-57, asst bact, 57, res bacteriologist, Epizool Lab, 58-61, res instr ecol & epizool, 61-63, from asst res prof to assoc res prof, 63-68, dir epizool lab, 61-68, clin lectr microbiol, 67-68; prof microbiol, 68-77, prof zool, 77-88, EMER PROF, NORTHERN COLO, 88- *Concurrent Pos:* Lectr, Brigham Young Univ, 65-66; consult, Dept Defense, 67-; vpres nat comt, Int Northwestern Conf Dis Man. *Mem:* Am Chem Soc; Am Asn Immunol; Am Soc Microbiol; Am Soc Trop Med & Hyg; Soc Exp Biol & Med. *Res:* Host mechanisms of resistance to infectious diseases; new methods of detection and isolation of microorganisms; zoonoses; animal infections and human diseases; natural and acquired immunity. *Mailing Add:* Dept Biol Scis Univ Northern Colo Greeley CO 80639

THORPE, COLIN, b Grantham, Eng, May 1, 47; m 75; c 2. FLAVOPROTEINS, FLAVINS. *Educ:* Univ Cambridge, UK, BA, 69; Univ Kent, UK, PhD(chem), 72. *Prof Exp:* Fel biochem, Dept Biol Chem, Univ Mich, 72-78; from asst prof to assoc prof, 78-89, PROF BIOCHEM, DEPT CHEM & BIOCHEM, UNIV DEL, 89- *Mem:* Am Chem Soc; Biochem Soc; Am Soc Biochem & Molecular Biol. *Res:* Structure, function and mechanism of action of flavoproteins involved in fatty acid oxidation. *Mailing Add:* Dept Chem & Biochem Univ Del Newark DE 19716

THORPE, HOWARD A(LAN), b Joplin, Mo, Nov 21, 14; m 41. PHYSICS, ENGINEERING. *Educ:* Univ Ark, BA, 37; Tulane Univ, MS, 39. *Prof Exp:* Instr physics, Pa State Univ, 40-46, res assoc, 46-47; supvry physicist, US Navy Marine Eng Lab, 47-53, physicist, Bur Ships, 53-56; res engr, Lockheed Aircraft Corp, 56-57, supvry group engr, 57-61, sr res specialist, Lockheed-Calif Co, 61-67; sr res scientist, Stanford Res Inst, 68-69; sr res engr, Aerospace Group, Rohr Industs, Inc, 69-74, eng specialist, Rohr Marine, Inc, 74-81; CONSULT ACOUST, 81- *Mem:* Acoust Soc Am; Am Inst Aeronaut & Astronaut; Marine Technol Soc; Sigma Xi. *Res:* Acoustical physics; aircraft sonar system and component research; basic environmental effects on sonar performance; submarine, ship and aircraft noise measurement, prediction and control; propagation of sound in atmosphere. *Mailing Add:* 285 Moss No 44 Chula Vista CA 92011

THORPE, JAMES F(RANKLIN), b Sandusky, Ohio, Oct 2, 26; m 49; c 3. MECHANICAL ENGINEERING. *Educ:* Univ Cincinnati, ME, 52; Univ Ky, MS, 55; Univ Pittsburgh, PhD(mech eng), 60. *Prof Exp:* Proj engr, E W Buschman Co, Ohio, 52-53; instr mech eng, Univ Ky, 53-55; engr, Bettis Atomic Power Lab, Westinghouse Elec Corp, 55-57, sr instr reactor eng, 57-59, sr engr, 60-61; assoc prof mech eng, Univ Ky, 61-67; head dept, 70-79, PROF MECH ENG, UNIV CINCINNATI, 67- *Concurrent Pos:* Consult numerous co, govt & univs, 61-; Nat Endowment for Humanities ethics fel, 78. *Mem:* Am Soc Mech Engrs; Am Soc Eng Educ; Nat Soc Prof Engrs. *Res:* Fluid mechanics; non-traditional machining; vibrations; mechanical design; product liability and product safety. *Mailing Add:* 1478 Beech Grove Dr Cincinnati OH 45238

THORPE, JOHN ALDEN, b Lewiston, Maine, Feb 29, 36; m 59; c 2. GEOMETRY. *Educ:* Mass Inst Technol, SB, 58; Columbia Univ, AM, 59, PhD(math), 63. *Prof Exp:* Instr math, Columbia Univ, 63; C L E Moore instr, Mass Inst Technol, 63-65; asst prof, Haverford Col, 65-68; from assoc prof to prof, State Univ NY, Stony Brook, 68-87; PROF, VPROVOST UNDERGRAD EDUC & DEAN UNDERGRAD COL, STATE UNIV, NY, BUFFALO, 87- *Concurrent Pos:* Mem, Inst Adv Study, 67-68; prog dir, NSF, 84-86, dep div dir, 86-87; bd govs, Math Asn Am, 84-87, chair, sci policy comt, 87- *Mem:* Am Math Soc; Math Asn Am; AAAS. *Res:* Differential geometry; general relativity. *Mailing Add:* State Univ NY 544 Capen Hall Buffalo NY 14260

THORPE, MARTHA CAMPBELL, b Tullahoma, Tenn, Apr 28, 22; m 43; c 2. PHYSICAL ORGANIC CHEMISTRY. *Educ:* Vanderbilt Univ, BA, 44; Samford Univ, MA, 68. *Prof Exp:* Anal chemist, E I du Pont de Nemours & Co, Inc, 44-45; sr chemist molecular spectros, Southern Res Inst, 61-84; RETIRED. *Mem:* Am Chem Soc; Int Soc Magnetic Resonance. *Res:* H-1 and C-13 nuclear magnetic resonance spectroscopy of organic compounds. *Mailing Add:* 2161 Kent Way Birmingham AL 35226

THORPE, MICHAEL FIELDING, b Bromley, Eng, Mar 12, 44; m 68; c 2. THEORETICAL PHYSICS. *Educ:* Univ Manchester, BSc, 65; Oxford Univ, DPhil(physics), 68. *Prof Exp:* Res assoc physics, Brookhaven Nat Lab, 68-70; from asst prof to assoc prof physics, Yale Univ, 70-77; assoc prof, 76-80, PROF PHYSICS, MICH STATE, 80- *Concurrent Pos:* Yale jr fac fel & guest scientist, Max Planck Inst Solids, Stuttgart, 72-73; sr res fel, Oxford Univ, 78 & 82; vis prof, Brazil & Japan. *Mem:* Fel Am Phys Soc; fel Brit Inst Physics. *Res:* Theoretical condensed matter physics, including low temperature excitations, amorphous solids, glasses and networks. *Mailing Add:* Dept Physics & Astron Mich State Univ East Lansing MI 48824

THORPE, NEAL OWEN, b Wausau, Wis, Sept 8, 38; m 60; c 3. BIOCHEMISTRY. *Educ:* Augsburg Col, BA, 60; Univ Wis, Madison, PhD(physiol chem), 64. *Prof Exp:* USPHS fel, 65-66; Am Heart Asn adv res fel, 66-67; assoc prof, 67-80, PROF BIOL, AUGSBURG COL, 80- *Concurrent Pos:* Res Corp grant & Am Heart Asn grant, 69-71; regional dir Grants, Res Corp, 73-74. *Mailing Add:* Dept Biol Augsburg Col Minneapolis MN 55454

THORPE, RALPH IRVING, b Halls Harbour, NS, Feb 29, 36; m 68; c 2. ECONOMIC GEOLOGY. *Educ:* Acadia Univ, BSc, 58; Queen's Univ, Ont, MSc, 63; Univ Wis-Madison, PhD(econ geol), 67. *Prof Exp:* RES SCIENTIST MINERAL DEPOSITS, GEOL SURV CAN, 65- *Mem:* Can Geophys Union; Mineral Asn Can; Can Inst Mining & Metall; Geol Asn Can; Geochem Soc. *Res:* Genesis of metalliferous ore deposits; lead isotope interpretations; ore mineralogy. *Mailing Add:* Mineral Deposits Sect Geol Surv Can Ottawa ON K1A 0E8 Can

THORPE, RODNEY WARREN, b Boston, Mass, Sept 27, 35. APPLIED MATHEMATICS, OPERATIONS RESEARCH. *Educ:* Harvard Col, AB, 59, SM, 65, PhD(appl math), 70. *Prof Exp:* Programmer, Ling Proj, Dept Environ Syst, Harvard Univ, 59-65, comput programmer, 71-72; DIR ENVIRON SYST, APPL MATH, QEI, INC, 72- *Mem:* Sigma Xi. *Res:* Mathematical modeling of transportation problems; pollution problems; simulation operations research. *Mailing Add:* 58 Bradford St Needham MA 02192

THORPE, TREVOR ALLEYNE, b Barbados, WI, Oct 18, 36; m 63; c 2. PLANT PHYSIOLOGY. *Educ:* Allahabad Agr Inst, BScAgr, 61; Univ Calif, Riverside, MS, 64, PhD(plant sci & physiol), 68. *Prof Exp:* Nat Res Coun fel & res plant physiologist, Fruit & Vegetable Chem Lab, US Dept Agr, Calif, 68-69; from asst prof to assoc prof, 69-78, asst dean fac arts & sci, 74-76, PROF BOT, UNIV CALGARY, 78- *Concurrent Pos:* Convener, panel on plant biol & technol, Int Cell Res Organ, UNESCO, 79-85; adj prof, Forestry Sci, Univ Alta, 86-; affilated scientist, Nat Res Coun Can, Plant Biotechnol Inst, Sask, 87-, head biol sci, 88-89. *Mem:* Can Soc Plant Physiol; Am Soc Plant Physiol; Japanese Soc Plant Physiol; Scand Soc Plant Physiol; Tissue Cult Asn; Int Soc Plant Morphologists; Int Asn Plant Tissue Cult (chmn, 74-78). *Res:* Experimental plant morphogenesis; cytology, physiology and biochemistry of organ formation in tissue culture systems; plant propagation of conifers and selection of salt-tolerant variants by tissue culture methods. *Mailing Add:* Dept Biol Sci Univ Calgary Calgary AB T2N 1N4 Can

THORSELL, DAVID LINDEN, b July 6, 42; US citizen. PHYSICAL CHEMISTRY. *Educ:* Univ Minn, Duluth, BA, 60; Ohio State Univ, PhD(phys chem), 71. *Prof Exp:* Lectr & fel chem, Ohio State Univ, 72-74; asst prof chem, 74-79, actg dean, Sch Sci & Eng, 80-81, CHMN, DEPT CHEM, SEATTLE UNIV, 76-, ASSOC PROF CHEM, 79- *Mem:* AAAS; Am Chem Soc; Inter-Am Photochem Soc. *Res:* Electron paramagnetic reasonance of organic and inorganic single crystal systems; photochemistry; atmospheric chemistry. *Mailing Add:* Dept Chem Seattle Univ 900 12th & E Columbia Seattle WA 98122

THORSEN, ARTHUR C, b Portland, Ore, July 27, 34. EXPERIMENTAL SOLID STATE PHYSICS. *Educ:* Reed Col, BA, 56; Rice Inst, MA, 58, PhD(physics), 60; Calif Lutheran Col, MBA, 81. *Prof Exp:* Res physicist, Atomics Int Div, NAm Rockwell Corp, 60-63, mem tech staff, Sci Ctr, 63-67, Autonetics Div, 67-70, Res & Technol Div, Anaheim, Calif, 70-73, mem tech staff, 73-74, prog mgr independent res & develop, 74-78, dir, Corp Res Progs, 78-84, dir mat synthesis & processing, 84-87, sr res exec, Rockwell Int Sci Ctr, 87-90; RETIRED. *Mem:* Am Phys Soc. *Res:* Low temperature solid state physics; electronic structure of metals; transport properties of semiconductors; thin films; superconductivity. *Mailing Add:* 3263 W Sierra Dr Thousand Oaks CA 91362

THORSEN, RICHARD STANLEY, b New York, NY, Oct 6, 40; m 62; c 2. THERMAL SCIENCES, SOLAR ENERGY. *Educ:* City Univ New York, BS, 62; NY Univ, PhD(mech eng), 67. *Prof Exp:* PROF & HEAD, DEPT MECH & AEROSPACE ENG, POLYTECH INST NY, 74-, DIR, SOLAR ENERGY APPLNS CTR, 77- *Concurrent Pos:* Consult, Grumman Aerospace Corp, 68- *Mem:* Am Soc Mech Engrs; Am Inst Aeronaut & Astronaut; Int Solar Energy Soc; Am Soc Eng Educ. *Res:* Multi-phase heat transfer; solar energy. *Mailing Add:* Dept Res Grad Studies Polytechnic Inst NY 333 Jay St Brooklyn NY 11201

THORSETT, EUGENE DELOY, b Wadena, Minn, Nov 28, 48; m 70; c 3. ORGANIC CHEMISTRY. *Educ:* Univ Minn, BA, 70; Colo State Univ, PhD(org chem), 73. *Prof Exp:* Res assoc, Rice Univ, 73-75; sr res chemist, Merck, Sharp & Dohme Res Lab, Merck & Co, Inc, 75-88; SR SCIENTIST, GENENTECH, INC, 88- *Mem:* Am Chem Soc; AAAS. *Res:* Organic synthesis; peptidomimetic drug design; enzyme inhibitor design. *Mailing Add:* 571 Buena Vista Moss Beach CA 94038

THORSETT, GRANT OREL, b Shelton, Wash, Jan 25, 40; m 63; c 3. MOLECULAR BIOLOGY. *Educ:* Wash State Univ, BS, 62; Yale Univ, MS, 65, PhD(molecular biophys), 69. *Prof Exp:* From asst prof to assoc prof, 67-79, PROF BIOL, WILLAMETTE UNIV, 79-, CHMN DEPT, 84- *Concurrent Pos:* Instr, Proj Newgate, Ore State Penitentiary, 71-72; NSF res partic, Willamette Univ, 71-73. *Mem:* AAAS. *Res:* Bacterial transformation; biochemical systematics; bacterial biochemistry; use of computers in undergraduate curricula. *Mailing Add:* Dept Biol Willamette Univ Salem OR 97301

THORSON, JAMES A, b Chicago, Ill, Oct 8, 46; m 66; c 2. GERONTOLOGY, THANATOLOGY. *Educ:* Northern Ill Univ, Dekalb, BS, 67; Univ NC, Chapel Hill, MEd, 71; Univ Ga, Athens, EdD(adult educ), 75. *Prof Exp:* Asst, Exten Div, Univ NC, Chapel Hill, 70-71; from instr to asst prof adult educ, Univ Ga, Athens, 71-77, coordr continuing educ, 71-77; from asst prof to prof geront, Omaha, 77-89, assoc prof adult educ, Lincoln, 79-84, DIR GERONT, UNIV NEBR, OMAHA, 79-, ISAACSON PROF, 89-, PROF ADULT EDUC, LINCOLN, 84- *Concurrent Pos:* Prin investr, Career Training Prog, US Admin Aging, 78-81; adj assoc prof, dept theol studies, St Louis Univ, 79; vis assoc prof, dept educ, Concordia Univ, Montreal, 80; co-leader, US Sci Deleg, People's Repub China, 82-; trainer, Ghost Ranch Ctr, Abiquiu, NMex, 84, Nebr Scholars' Inst, Lincoln, 85-; ed consult, Prentice Hall, Univ Southern Calif Geront Ctr, 85; mem, Comn Prof Adult Educ; expert witness, numerous retirement home tax cases. *Honors & Awards:* John Tyler Mauldin Award Serv Aging. *Mem:* Fel Geront Soc Am. *Res:* Psychology of aging and spiritual well-being of the elderly; training and adult learning; attitudes toward the aged; death anxiety, personality, death education, nursing and social worth of the aged; aging in China; suicide; humor and death; lethal behaviors. *Mailing Add:* Chair Dept Geront Univ Nebr Omaha NE 68182

THORSON, JOHN WELLS, b Detroit, Mich, Feb 25, 33; m 64. NEUROPHYSIOLOGY, BIOPHYSICS. *Educ:* Rensselaer Polytech Inst, BS, 55, MS, 58; Univ Calif, Los Angeles, PhD(zool), 65. *Prof Exp:* Physicist, Gen Elec Co, 55-60; NIH trainee biophys, Univ Calif, Los Angeles, 60-65; NSF fel physiol, Max Planck Inst Biol, 65-66 & Oxford Univ, 66-67; asst prof neurosci, Univ Calif, San Diego, 67-68, res scientist, 68-69; vis lectr zool, Oxford Univ, 69-70; res fel, Max Planck Inst Physiol of Behav, 70-72; affil zool, Oxford Univ, 72-79; AFFIL, MAX PLANCK INST, 79- *Concurrent Pos:* Mass Inst Technol Neurosci Res Prog fel, Univ Colo, 66; prin investr, Air Force Off Sci Res grants, 67-69 & 69-70; consult, Max Planck Inst, 75-, Univ Calif, San Diego, 80 & J W Goethe Univ, Frankfurt, 80-85. *Mem:* Sigma Xi. *Res:* Experimental and theoretical analysis of the dynamics of biological systems; visual movement perception; macromolecular basis of muscle contraction; mathematics of distributed relaxation processes; computers in physiological analysis and control of experiments; methodology in behavioral experiments identifying mechanisms of sensory recognition. *Mailing Add:* The Old Marlborough Arms Combe Oxford OX7 2NQ England

THORSON, RALPH EDWARD, medical parasitology, veterinary parasitology; deceased, see previous edition for last biography

THORSON, THOMAS BERTEL, b Rowe, Ill, Jan 12, 17; m 41; c 2. ZOOLOGY. *Educ:* St Olaf Col, BA, 38; Univ Wash, Seattle, MS, 41, PhD(zool), 52. *Prof Exp:* Teacher high sch, Mont, 38-39 & Wash, 42-43; instr zool, Yakima Jr Col, Wash, 46-48, Univ Nebr, 48-50 & San Francisco State Univ, 52-54; asst prof, SDak State Univ, 54-56; from asst prof to prof, 56-82, chmn dept, 67-71, vdir, Sch Life Sci, 75-77, EMER PROF ZOOL, UNIV NEBR, LINCOLN, 82- *Concurrent Pos:* NSF-NIH & Off Naval Res grants, Field Expeds Cent Am, SAm & Nigeria, 60-; Nat Geog Soc grants, 74-82. *Mem:* Am Elasmobranch Soc; Am Soc Zoologists; Sigma Xi; Am Inst Biol Sci; fel Explorers Club; Am Soc Ichthyologists & Herpetologists; fel AAAS; Am Fisheries Soc. *Res:* Water economy of amphibians in relation to terrestrialism; ecological and phylogenetic significance of body water partitioning in vertebrates; osmoregulation freshwater adaptation; elasmobranch biology. *Mailing Add:* 21233 SW Martinazzi Tualatin OR 97062-9324

THORSON, WALTER ROLLIER, b Tulsa, Okla, Sept 3, 32. THEORETICAL CHEMISTRY. *Educ:* Calif Inst Technol, BS, 53, PhD(chem), 57. *Prof Exp:* NSF fel chem, Harvard Univ, 56-57; instr, Tufts Univ, 57-58; asst prof phys chem, Mass Inst Technol, 58-64, assoc prof chem, 64-68; PROF CHEM, UNIV ALTA, 68- *Mem:* Am Phys Soc; Can Asn Physicists. *Res:* Theory of atomic collisions; electronic structure of molecules and solids; quantum mechanics. *Mailing Add:* Dept Chem Univ Alta Edmonton AB T6G 2G2 Can

THORSTENSEN, THOMAS CLAYTON, b Milwaukee, Wis, Nov 29, 19; c 3. INDUSTRIAL EFFLUENT TREATMENT. *Educ:* Univ Minn, BS, 42; Lehigh Univ, MS, 47, PhD, 49. *Prof Exp:* Chemist, S B Foot Tanning Co, 42-44; asst, Lehigh Univ, 46-49, res assoc, 49-51; res chemist, J S Young Co, 51-55; proj dir, Res Found, Lowell Technol Inst, 55-59; CONSULT TO PRES, THORSTENSEN LAB, 59- *Concurrent Pos:* Vis prof, Lowell Technol Inst, 60-; consult aid to underdeveloped nations, UN & Dept State. *Mem:* Am Chem Soc; Am Leather Chem Asn (pres, 74). *Res:* Mineral tannages; chromium; iron; aluminum and zirconium; theory of mineral tannages; synthetic tanning agents; chemical waste streams. *Mailing Add:* 66 Little Ton Rd Box 426 Westford MA 01886

THORUP, JAMES TAT, b Salt Lake City, Utah, Dec 20, 30; m 58; c 5. AGRONOMY, SOIL FERTILITY. *Educ:* Brigham Young Univ, BA, 55; NC State Col, MS, 57; Univ Calif, Davis, PhD(soils, plant nutrit), 66. *Prof Exp:* Teacher high schs & city cols, Calif, 61-66; AGRONOMIST, CHEVRON CHEM CO, 66- *Mem:* Am Soc Agron; Soil Sci Soc Am. *Res:* Factors affecting plant growth in sodic soils; pH effect on plant growth and water uptake; plant nutrition. *Mailing Add:* 6001 Bollinger Canyon Rd San Ramon CA 94583

THORUP, OSCAR ANDREAS, JR, b Washington, DC, Mar 12, 22; m 44; c 1. MEDICINE. *Educ:* Univ Va, BA, 44, MD, 46; Am Bd Internal Med, dipl, 55. *Prof Exp:* From asst resident to resident internal med, Hosp, Univ Va, 50-52, asst to dean, Sch Med, 53-67, dir teacher's preventorium, 53-54, from instr to assoc prof internal med, 53-66; prof & head dept, Col Med, Univ Ariz, 66-74; prof internal med & assoc dean, Sch Med, Univ Va, 74-89. *Concurrent Pos:* Fel internal med, Hosp, Univ Va, 50; res fel, Univ NC, 52-53; AMA coun on continuing physician educ. *Mem:* AMA; Am Fedn Clin Res; fel Am Col Physicians; Am Clin & Climat Asn; Sigma Xi. *Res:* Hematology, particularly red blood cell enzymes and proteins. *Mailing Add:* 208 Colthurst Dr Charlottesville VA 22901

THORUP, RICHARD M, b Salt Lake City, Utah, Dec 20, 30; m 57, 80; c 6. AGRONOMY, SOIL FERTILITY. *Educ:* Brigham Young Univ, BA, 55; NC State Univ, MS, 57; Univ Calif, Davis, PhD(soil sci, plant nutrit), 62. *Prof Exp:* Agronomist, Chevron Chem Co, 60-61, field agronomist, 61-67, regional agronomist, 67-75, nat mgr agron, 75-83, chief agronomist, 83-86, consult, Fertilizer Div, 87-88; CONSULT, 88- *Concurrent Pos:* Consulting in Spain & Mexico; adj prof, Brigham Young Univ, 87- *Mem:* Am Soc Agron; Soil Sci Soc Am; Crop Sci Soc Am. *Res:* Chemistry of phosphates in the soil, including solubility and interrelationships with soil moisture; maximum fertility studies with field and tree crops; micronutrients; effect of fertilizers on environment; fertilizer response studies on numerous crops. *Mailing Add:* 1741 N 1500 E Provo UT 84604

THOULESS, DAVID JAMES, b Bearsden, Scotland, Sept 21, 34; m 58; c 3. STATISTICAL MECHANICS & CRITICAL PHENOMENA. *Educ:* Cambridge Univ, BA, 55, ScD, 86; Cornell Univ, PhD(theoret physics), 58. *Prof Exp:* Physicist, Lawrence Berkeley Lab, 58-59; res fel, Univ Birmingham, 59-61; lectr, Cambridge Univ, 61-65; prof math physics, Univ Birmingham, 65-78; prof appl sci, Yale Univ, 79-80; PROF PHYSICS, UNIV WASH, 80- *Concurrent Pos:* Prof, Queens Univ, Kingston, Ont, 78; royal soc res prof, Cambridge Univ, 83-85. *Honors & Awards:* Maxwell Medal, Inst Physics, 73; Holweck Medal, French Soc Physics, 81; Fritz London Award, Int Conf Low Temperature Physics, 84. *Mem:* Fel Am Acad Arts & Sci; fel Am Phys Soc. *Res:* Statistical mechanics and critical phenomena; theory of disordered electron and magnetic systems; quantum hall effect. *Mailing Add:* Dept Physics FM-15 Univ Wash Seattle WA 98195

THOURET, WOLFGANG E(MERY), b Berlin, Ger, Aug 27, 14; US citizen; m 62. PHYSICS, ELECTRICAL ENGINEERING. *Educ:* Tech Univ, Berlin, MS, 36; Univ Karlsruhe, D Ing, 52. *Prof Exp:* Physicist, Res Dept, OSRAM Corp, Ger, 36-40, sect mgr, 40-48; lab mgr, Quartz Lamp Soc, Germany, 49-52; develop engr, Lamp Div, Westinghouse Elec Corp, NJ, 52-57; assoc dir eng, 57-71, DIR ENG, DURO-TEST CORP, 71- *Mem:* Fel Am Illum Eng Soc; Am Phys Soc; NY Acad Sci; Inst Elec & Electronics Engrs; Soc Motion Picture & TV Engrs. *Res:* Gaseous discharges; spectroscopy; optical equipment, high intensity light and radiation sources; compact arc high pressure lamps; incandescent lamps; halogen quartz incandescent lamps; metal vapor additive lamps; alumina and sapphire metal vapor lamps. *Mailing Add:* Claridge House I Apt 402 Verona NJ 07044

THOURSON, THOMAS LAWRENCE, electrostatic imaging, for more information see previous edition

THRAILKILL, JOHN, b San Diego, Calif, Aug 31, 30; m 52. HYDROGEOLOGY, GEOCHEMISTRY. *Educ:* Univ Colo, AB, 53, MS, 55; Princeton Univ, PhD(geol), 65. *Prof Exp:* Geologist, Continental Oil Co, 55-61; from asst prof to assoc prof, 65-77, chmn dept, 74-77, PROF GEOL, UNIV KY, 69- *Mem:* AAAS; Nat Speleol Soc; Geol Soc Am; Geochem Soc; Am Geophys Union; Sigma Xi. *Res:* Hydrogeology and geochemistry of limestone terrains. *Mailing Add:* Dept Geol Sci Univ Ky Lexington KY 40506-0059

THRALL, ROBERT MCDOWELL, b Toledo, Ill, Sept 23, 14; m 36; c 3. MATHEMATICS. *Educ:* Ill Col, AB, 35; Univ Ill, AM, 35, PhD(math), 37. *Hon Degrees:* ScD, Ill Col, 60. *Prof Exp:* Instr math, Univ Mich, 37-40; mem, Inst Advan Study, 40-42; from asst prof to prof math, Univ Mich, Ann Arbor, 42-69; prof math sci & chmn dept, Rice Univ, 69-85; DECISION-MGT SCI, NSF, 85- *Concurrent Pos:* Res mathematician appl math group, Nat Defense Res Coun, Columbia Univ, 44-45; mem staff radiation lab & sect chief & ed-in-chief, Mars, Mass Inst Technol, 44-46; prof opers anal, Univ Mich, Ann Arbor, 56-69, head opers res dept, 57-60, res mathematician inst sci & technol, 60-69; consult, Rand Corp, Weapon Syst Eval Group, US Dept Defense & Math Steering Comt, Army Res Off, 58-; ed-in-chief, Mgt Sci, 61-69; adj prof, Dept Comput Sci & Inst Rehab & Res, Baylor Col Med, 71- & Univ Tex Sch Pub Health, Houston, 72-; pres, Robert M Thrall & Assoc, Inc; vis prof quant methods, Univ Houston & sr scientist NSF Industs Studies, 74-75. *Mem:* AAAS; Am Math Soc; Soc Indust & Appl Math; Opers Res Soc Am; Math Asn Am; Inst Mgt Sci (pres, 69-70); Sigma Xi. *Res:* Representations of groups; rings and lie rings; operations research linear and nonlinear programming and game theory; theory of application of mathematical models. *Mailing Add:* 12003 Pebble Hill Dr Houston TX 77024-6208

THRASHER, GEORGE W, b Bloomington, Ind, July 8, 31; m 53; c 1. ANIMAL NUTRITION. *Educ:* Purdue Univ, BS, 52, MS, 54, PhD, 58. *Prof Exp:* Asst animal nutrit, Chas Pfizer & Co, 54; voc agr teacher, Morgan County Schs, Ind, 54-56; asst animal nutrit, Purdue Univ, 56-58, exten swine specialist, 58-59; res animal nutritionist, Com Solvents Corp, 59-64; asst dir animal health prod res, 64-80, DIR ANIMAL HEALTH RES, PFIZER INC, 80- *Mem:* Am Soc Animal Sci; Sigma Xi. *Res:* Antibiotics; hormones; anthelmintics; antimicrobials; chemotherapeutics; minerals; vitamins. *Mailing Add:* Pfizer Inc 1107 S Missouri Dr Lees Summit MO 64063

THRASHER, L(AWRENCE) W(ILLIAM), b Gary, Ind, Dec 30, 22; m 52; c 2. MECHANICAL ENGINEERING. *Educ:* Purdue Univ, BSME, 43, MSME, 49, PhD(mech eng), 54. *Prof Exp:* Engr aircraft heating & vent systs, Martin Co, 46-47; oil field prod facilities, Arabian Am Oil Co, 49-50; res facil nuclear res, Calif Res & Develop Corp, 52, sect supvr prod tech oil field res, 54-65; sect supvr, Chevron Res Co, Chevron Oil Field Res Co, 65-68, mgr, Opers Technol Div, 68-74, vpres prod res, 74-83; RETIRED. *Concurrent Pos:* Chmn, Coord Subcomt for Study, Enhanced Oil Recovery, Nat Petrol Coun, 75-76. *Mem:* AAAS; assoc Am Soc Mech Engrs; Soc Petrol Engrs. *Res:* Oil field production; well drilling and stimulation; fluid mechanics; ocean wave forces; dynamics of anchored vessels; enhanced oil recovery. *Mailing Add:* 2424 E Clark Ave Fullerton CA 92631

THRASHER, TERRY NICHOLAS, PHYSIOLOGY, NEUROSCIENCE. *Educ:* Univ Fla, PhD(physiol), 76. *Prof Exp:* ASST PROF PHYSIOL, UNIV CALIF, SAN FRANCISCO, 76- *Mailing Add:* Dept Physiol Sch Med Univ Calif San Francisco CA 94143

THREADGILL, ERNEST DALE, b Tallassee, Ala, June 26, 42; m 67; c 2. BIOLOGICAL ENGINEERING. *Educ:* Auburn Univ, BS, 64, PhD(agr eng), 68. *Prof Exp:* From asst prof to assoc prof agr & biol eng, Miss State Univ, 68-75; assoc prof & head dept agr eng, 75-87, PROF & CHMN, DIV AGR ENG, COASTAL PLAIN EXP STA, UNIV GA, 87- *Mem:* Am Soc Agr Engrs. *Res:* Soil erosion and drainage; pesticide applications; plant microclimate; irrigation; soil tillage; chemigation. *Mailing Add:* Driftmier Eng Ctr Univ Ga Athens GA 30602

THREADGILL, W(ALTER) D(ENNIS), b Huron, Tenn, Mar 17, 22; m 59. CHEMICAL ENGINEERING. *Educ:* Vanderbilt Univ, BE, 50; Univ Mo, PhD(chem eng), 54. *Prof Exp:* Asst instr chem eng, Univ Mo, 50-52, asst & chem engr, Eng Exp Sta, 52-53; asst prof chem eng, 54-57, actg head dept, 56-57, head dept, 60-72, chmn dept, 75-80, PROF CHEM ENG, VANDERBILT UNIV, 57- *Mem:* Am Soc Eng Educ; Am Inst Chem Engrs. *Res:* Thermodynamics; chemical engineering unit operations. *Mailing Add:* Box 1821 Sta B Vanderbilt Univ Nashville TN 37235

THREEFOOT, SAM ABRAHAM, b Meridian, Miss, Apr 10, 21; m 54; c 3. CARDIOVASCULAR DISEASES. *Educ:* Tulane Univ, BS, 43, MD, 45; Am Bd Internal Med, dipl, 53. *Prof Exp:* Intern, Michael Reese Hosp, Chicago, 45-47; from instr to prof med, Tulane Univ, 48-70; prof med & asst dean, Med Col Ga, 70-76; chief staff, Forest Hills Div, Vet Admin Hosp, Augusta, 70-76; assoc chief staff res, 76-79, CHIEF OF STAFF, VET ADMIN HOSP, NEW ORLEANS, 79-; PROF MED, TULANE UNIV, 76- *Concurrent Pos:* Fel med, Sch Med, Tulane Univ, 47-49; from asst vis physician to sr vis physician, Charity Hosp La, New Orleans, 47-69, consult, 69-70 & 76-; consult, Lallie Kemp Charity Hosp, Independence, Mo, 51-53; clin asst, Touro Infirmary, 53-56, dir res & med studies, 53-63, jr staff mem, 56-60, sr assoc, 60-63, sr dept med, 63-70, dir res, Touro Res Inst, 53-70; mem exec comt, Coun on Circulation, Am Heart Asn, 68-75, from vchmn to chmn, 71-75, chmn credentials comt, 72-73; mem bd consult, Int Soc Lymphology, 70-76; mem bd dirs, Am Heart Asn, 66-70 & 72-75, mem exec comt, 69-70 & 73-75. *Honors & Awards:* Honors Achievement Award, Angiol Res Found, 68; Award of Merit, Am Heart Assoc, 76. *Mem:* Soc Nuclear Med; Am Fedn Clin Res; fel Am Col Physicians; Am Heart Asn; fel Am Col Cardiol. *Res:* Electrolyte turnover in congestive heart failure; anatomy and physiology of lymphatics as a transport system and their role in pathogenesis of disease. *Mailing Add:* 1601 Perdido St Vet Admin Med Ctr New Orleans LA 70146

THREET, RICHARD LOWELL, b Browns, Ill, Nov 17, 24; m 46; c 4. GEOLOGY. *Educ:* Univ Ill, BS & AB, 47, AM, 49; Univ Wah, Seattle, PhD(geol), 52. *Prof Exp:* From instr to asst prof geol, Univ Nebr, 51-57; asst prof, Univ Utah, 57-61; from asst prof to assoc prof, Calif State Univ, San Diego, 61-68,; prof, 68-80, EMER PROF GEOL, SAN DIEGO STATE UNIV, 80- *Concurrent Pos:* Vis prof, Ohio State Univ, 53, 63, 66-70 & 72-73, Col Southern Utah, 54-55 & Univ Ill, 57; chmn dept geol, San Diego State Univ, 72-73; vis prof, Western Washington State Univ, 80-82. *Mem:* Fel Geol Soc Am. *Res:* Colorado plateau geology; geomorphology; structural geology; photogeology. *Mailing Add:* 2801 17th St Anacortes WA 98221

THRELFALL, WILLIAM, b Preston, Eng, Oct 14, 39; m 65; c 2. PARASITOLOGY, ORNITHOLOGY. *Educ:* Univ Wales, BSc, 62, PhD(agr zool), 65. *Prof Exp:* From asst prof to assoc prof, 65-75, PROF BIOL, MEM UNIV NFLD, 75- *Mem:* Sci fel Zool Soc London; fel Linnean Soc London; fel Royal Soc Trop Med Hyg; Am Ornith Union; Brit Ornith Union. *Res:* Ecological and geographical aspects of parasitology; helminthology; breeding biology and migratory movements of marine birds. *Mailing Add:* Dept Biol Mem Univ Nfld St John's NF A1B 3X9 Can

THRELKELD, STEPHEN FRANCIS H, b Watford, Eng, Dec 27, 24; m 52; c 2. BIOLOGY, GENETICS. *Educ:* Univ Alta, BSc, 57, MSc, 58; St Catharine's Col, Cambridge, PhD(bot), 61. *Prof Exp:* From asst prof to assoc prof genetics, 61-71, chmn res unitbiochem, biophys & molecular biol, 64-68, assoc chmn dept biol, 66-68 chmn, deptbiol, 77-83, PROF GENETICS, MCMASTER UNIV, 71-,CHMN, DEPT BIOL, 86- *Mem:* Am Soc Naturalists; Genetics Soc Am; Genetics Soc Can; Chem Inst Can; Can Soc Cell Biol. *Res:* Neurospora; recombination; Drosophila behavioral genetics; insecticide resistance. *Mailing Add:* Dept Biol McMaster Univ Hamilton ON L8S 4K1 Can

THRIFT, FREDERICK AARON, b St George, Ga, Oct 6, 40; m 67; c 2. ANIMAL BREEDING. *Educ:* Univ Fla, BSA, 62; Univ Ga, MS, 65; Okla State Univ, PhD(animal breeding), 68. *Prof Exp:* Assoc prof, 67-78, PROF ANIMAL SCI, UNIV KY, 78- *Mem:* Biomet Soc; Am Soc Animal Sci. *Res:* Beef cattle and sheep breeding research. *Mailing Add:* Dept Animal Sci Univ Ky Lexington KY 40506

THRO, MARY PATRICIA, b St Charles, Mo, Mar 14, 38. PHYSICS, MATHEMATICS. *Educ:* Maryville Col, BA, 61; Fordham Univ, MS, 67; Wash Univ, PhD(educ), 76. *Prof Exp:* Teacher math, Villa Duchesne Elem & Sec Schs, St Louis, 61-64; instr physics, Maryville Col, 67-69; team mem admin, Relig of Sacred Heart, 69-73; asst prof physics, 73-77, ASSOC PROF PHYSICS & MATH, MARYVILLE COL, 77-, CHAIRPERSON DEPT, 74- *Concurrent Pos:* NSF grants, New Orleans, 69-71, Kansas City, 72-73 & Memphis, 78-79; eval chairperson, Acad Sacred Heart, 76; mem bd joint grad educ prog, Wash Univ-Maryville Col, 77-79; co-dir, Mo Jr Acad Sci, 77-; publ, J Educ Psychol, 78. *Mem:* AAAS; Sigma Xi; Am Asn Physics Teachers; Am Educ Res Asn. *Res:* Relationships between associative and content structure of physics concepts; chemical abundance of elements in the cosmos determined by analysis of ion damage recorded on meteorite crystals; spectral analysis of compounds. *Mailing Add:* Maryville Col 13550 Conway Rd St Louis MO 63141

THROCKMORTON, GAYLORD SCOTT, b Kansas City, Kans, Aug 7, 46; m 67; c 3. COMPARATIVE ANATOMY, BIOMECHANICS. *Educ:* Univ Kans, BA, 68; Univ Chicago, PhD(evolutionary biol), 74. *Prof Exp:* Asst prof human anat, Col Dent, Univ Ill Med Ctr, 73-75; asst prof, 75-81, ASSOC PROF HUMAN ANAT, UNIV TEX HEALTH SCI CTR, 81- *Concurrent Pos:* Prin investr, NSF, 78-81; co-investr, NIH, 80-83, Am Asn Oral & Maxillofacial Surgeons, 90-95; cur, Anat Teaching Mus, Univ Tex Southwestern Med Ctr, 79-; reviewer, Systs & Ecol Prog, NSF, 81- *Mem:* Am Asn Anatomists; Am Soc Zoologists; Am Soc Biomech; Int Asn Dent Res. *Res:* Form and function of the vertebrate feeding apparatus including biomechanical modeling of mandibular forces, mechanisms controlling mastication, and the effect of orthognatic surgery on function of the jaw muscles in humans. *Mailing Add:* Dept Cell Biol & Neurosci Univ Tex Southwestern Med Ctr Dallas TX 75235-9039

THROCKMORTON, JAMES RODNEY, b St John, Wash, Sept 4, 36; m 58; c 3. PESTICIDE CHEMISTRY. *Educ:* Univ Idaho, BS, 58, MS, 60; Univ Minn, PhD(org chem), 64. *Prof Exp:* Sr chemist, Imaging Res Lab, 64-66, Contract Res Lab, 66-71, RES SPECIALIST, 3M CO, 74- *Mem:* Am Chem Soc. *Res:* Organic fluorochemicals; imaging technology. *Mailing Add:* 7149 Aberdeen Curve St Paul MN 55125-1667

THROCKMORTON, LYNN HIRAM, b Loup City, Nebr, Dec 20, 27. ZOOLOGY. *Educ:* Univ Nebr, BS, 49, MS, 56; Univ Tex, PhD(zool), 59. *Prof Exp:* Instr zool, Univ Nebr, 56; spec instr, Univ Tex, 59-60; vis asst prof, Univ Calif, 60-61; res assoc, 61-62; from instr to assoc prof, 62-71, PROF BIOL, UNIV CHICAGO, 71- *Mem:* AAAS; Am Genetics Soc; Soc Syst Zool; Am Inst Biol Sci; Sigma Xi. *Res:* Taxonomy; phylogeny and biogeography of Drosophila and other Drosophilids; biochemical evolution and speciation of Drosophila; utilization and evaluation of computer methods in taxonomy. *Mailing Add:* Dept Biol Univ Chicago 1103 E 57th St Chicago IL 60637

THROCKMORTON, MORFORD C, b Waynesburg, Pa, July 28, 19. POLYMER CHEMISTRY. *Educ:* Grove City Col, BS, 40; Western Reserve Univ, MS, 41, PhD(phys chem), 44. *Prof Exp:* Res chemist, Texaco, Inc, 43-54, group leader, 54-59; sr res chemist, Firestone Tire & Rubber Co, 60-63; sr res chemist, Goodyear Tire & Rubber Co, 64-68, res scientist, 68-86; RETIRED. *Mem:* Am Chem Soc. *Res:* Catalysis; synthetic fuels; development of synthetic rubber; stereoregular polymerization; petrochemicals. *Mailing Add:* 967 Newport Rd Akron OH 44303-1319

THROCKMORTON, PETER E, b St Paul, Minn, Jan 20, 27; m 48; c 3. ORGANIC CHEMISTRY, CHEMICAL ENGINEERING. *Educ:* Univ Minn, BCHE, 48, MS, 55; Kans State Univ, PhD, 60. *Prof Exp:* Asst res engr, Tainton Co, 48-49; res engr, Glenn L Martin Aircraft Co, 49-52; chemist, Gen Mills Co, 52-56; asst, Kans State Univ, 56-59; assoc chemist, Midwest Res Inst, 59-65; sr res chemist, Archer, Daniels, Midland, 65-67, Ashland Chem Co, 67-86; CONSULT, THROCKMORTON CONSULT, 86- *Concurrent Pos:* China-US Sci Exchange, China Asn Sci & Technol, 84. *Mem:* Am Chem Soc; emer mem Sigma Xi; fel Am Inst Chemists. *Res:* Heterocyclic and organometallic compounds in organic chemistry; synthesis; process research; surfactants from cornstarch; sulfur derivatives; selective oxidation of hydrocarbons and catalysis; author of 44 technical publications. *Mailing Add:* 15943 Hawn Rd Plain City OH 43064

THRODAHL, MONTE C(ORDEN), b Minneapolis, Minn, Mar 25, 19; m 48; c 2. RESEARCH MANAGEMENT, ENVIRONMENTAL MANAGEMENT. *Educ:* Iowa State Univ, BS, 41. *Prof Exp:* Res chemist, Org Res Dept, Monsanto Co, 41-45, group leader, 45-50, asst res dir, 50-52, mgr rubber chem sect, Org Develop Dept, 52-54, asst dir develop, 54-56, dir develop, 56-60, dir res, Org Res Dept, 60-62, dir mkt, 62-64, asst gen mgr, Int Div, 64, gen mgr, 64-66, vpres technol & dir, 66-77, group vpres & sr vpres environ policy, 77-84; CONSULT, 84- *Mem:* Nat Acad Eng; fel AAAS; Com Develop Asn; fel Am Inst Chem Engrs; Soc Chem Indust. *Mailing Add:* 7811 Carondelet Ave Beck Bldg Rm 204 St Louis MO 63166

THRON, JONATHAN LOUIS, b Boulder, Colo, Nov 21, 54; m 87. NUCLEON DECAY EXPERIMENTS. *Educ:* Carleton Col, BA, 77; Yale Univ, MPhil, 79, PhD(physics), 83. *Prof Exp:* Postdoctoral, 83-85, asst physicist, 85-90, PHYSICIST, ARGONNE NAT LAB, 90- *Mem:* Am Phys Soc. *Res:* Construction and data analysis of the Soudan 2, high resolution, tracking calorimeter nucleon decay experiment. *Mailing Add:* Argonne Nat Lab 9700 S Cass Ave HEP Bldg 362 Argonne IL 60439-4815

THRON, WOLFGANG JOSEPH, b Ribnitz, Ger, Aug 17, 18; US citizen; m 53; c 5. MATHEMATICS. *Educ:* Princeton, AB, 39; Rice Inst, MA, 42, PhD(math), 43. *Prof Exp:* Instr math, Harvard Univ, 43-44; from instr to assoc prof, Wash Univ, St Louis, 46-54; from assoc prof to prof, 54-85, EMER PROF MATH, UNIV COLO, BOULDER, 85- *Concurrent Pos:* Vis prof, Free Univ Berlin, 51, Philippines, 66-67, Univ Erlangen, 70-71, Punjab Univ, India, 74-75 & Univ Trondheim, 78-79 & 82-83; res grant, Air Force Off Sci Res, Ger, 57-58; vis prof & Fulbright lectr, India, 62-63. *Mem:* Am Math Soc; Math Asn Am; Kongl Norske Videnskabers Selskab. *Res:* Complex variables, analysis of convergence and truncation errors of infinite processes in particular continued fractions; general topology, lattice of topologies, proximity, contiguity and nearness spaces, extensions of spaces. *Mailing Add:* Dept Math Univ Colo Campus Box 426 Boulder CO 80309-0426

THRONE, JAMES EDWARD, b Calembrone, Italy, Sept 10, 54; US citizen; m 85; c 1. BIOMATHEMATICS. *Educ:* Southeastern Mass Univ, BS, 76; Wash State Univ, MS, 78; Cornell Univ, PhD(entom), 83. *Prof Exp:* Res fel, NC State Univ, 83-85; RES ENTOMOLOGIST & ECOLOGIST, STORED PROD INSECTS RES & DEVELOP LAB, AGR RES SERV, USDA, 85- *Mem:* Entom Soc Am; Sigma Xi. *Res:* Bionomics and management of agricultural pests. *Mailing Add:* Stored Prod Insects Res & Develop Lab Agr Res Serv USDA PO Box 22909 Savannah GA 31403

THRONE, JAMES LOUIS, b Cleveland, Ohio, July 10, 37; m 59; c 2. CHEMICAL ENGINEERING. *Educ:* Case Inst Technol, BS, 59; Univ Del, MChE, 61, PhD(chem eng), 64. *Prof Exp:* Res engr, Eng Res Labs, E I du Pont de Nemours & Co, Inc, 63-64; assoc prof chem eng, Ohio Univ, 64-68; supvr plastics processing, Am Standard, NJ, 68-71; assoc prof energetics, Univ Wis-Milwaukee, 71-72; dir plastics res, Beloit Corp, Wis, 72-74; res assoc plastics processing, Amoco Chem Corp, 74-85; eng consult, 85-86, PRES, SHERWOOD TECHNOL, 86-; prof eng, Univ Akron, 86-89; fel, B F Goodrich, Avon Lake, Ohio, 89- *Concurrent Pos:* Vis prof chem eng, Univ Cincinnati, 65; adj prof plastics processing, Newark Col Eng, 68-71; prof food eng, Orgn Am States, Brazil, 72; consult, Sherwood Tech Serv, Beloit, Wis, 72-74; Am Soc Elec Eng fel, NASA, Hampton, Va, 88-89. *Mem:* Polymer Processing Soc; Soc Rheology; fel Brit Plastics & Rubber Inst; Soc Plastics Indust; fel Soc Plastics Engrs. *Res:* Plastics process engineering, with emphasis in thermoforming, rotational molding, foam processing and comparative process economics. *Mailing Add:* 158 Brookside Blvd Hinckley OH 44233-9676

THRONEBERRY, GLYN OGLE, b Rule, Tex, Nov 1, 27; m 48; c 1. PLANT PHYSIOLOGY. *Educ:* NMex State Univ, BS, 50; Iowa State Col, MS, 52, PhD(plant physiol), 53. *Prof Exp:* Plant physiologist, Kans State Col, USDA, 54-55; asst prof biol, 55-57, from asst prof to assoc prof, 57-67, PROF BOT & ENTOM, NMEX STATE UNIV, 67- *Mem:* AAAS; Am Soc Plant Physiol; Am Phytopath Soc; Sigma Xi. *Res:* Plant host-pathogen relationships; intermediary metabolism; fungus physiology; plant biochemistry. *Mailing Add:* 1778 Imperial Ridge Las Cruces NM 88001

THRONER, GUY CHARLES, b Minneapolis, Minn, Sept 14, 19; m 43; c 3. TERMINAL BALLISTICS, WEAPON SYSTEMS ENGINEERING. *Educ:* Oberlin Col, AB, 43. *Prof Exp:* Mgr res & develop, Ordnance Div, Aerojet Gen, 53-63, mgr res & develop, Tactical Weapon Systs Div, Aerojet Gen Corp, 63-64; vpres & div gen mgr res & develop, FMC Corp, 64-74; dir res & develop, Vacu-Blast & Tronic Corp, 76-78; vpres prod eng, Dahlman Inc, 78-79; sect mgr res & develop, Columbus Labs, 79-86, PRES, G C THRONER & ASSOC-TECH & MGT CONSULTS, BATTELLE MEM INST, 86- *Concurrent Pos:* Vpres & gen mgr res & develop, Steel Prod Div, FMC Corp, 64-74; exec vpres res & develop, Am Vidionetics Corp, 76-78; mem, Air Armament Bd, Bomb & Warhead Steering Comt & Underwater Weapons Steering Comt. *Honors & Awards:* IR-100 Award, 72; Bronze Medallion, Am Defense Preparedness Asn, 74, Simon Silver Medal, 85; Distinguished Inventor's Medallion, Battelle Mem Inst, 83. *Mem:* Sigma Xi; Sci Res Soc Am; Am Defense Preparedness Asn; Am Inst Aeronaut & Astronaut. *Res:* Metal forming; high speed instrumentation; medical devices; oil field equipment; agri-machinery; weapon and space technology; author or coauthor of over 80 publications. *Mailing Add:* 17992 Jayhawk Dr Penn Valley CA 95946-9206

THRONSON, HARLEY ANDREW, JR, b Madison, Wis, Oct 13, 48; m 78; c 2. STAR FORMATION, INTERSTELLAR MEDIUM. *Educ:* Univ Calif, Berkeley, BS, 71; Univ Chicago, MS, 73, PhD(astrophysics), 78. *Prof Exp:* Res assoc, astron, Steward Observ, Univ Ariz, 77-81; from asst prof to assoc prof, 81-91, PROF ASTRON, DEPT PHYSICS, UNIV WYO, 91-, PLANETARIUM DIR, 81- *Mem:* Int Astron Union; Am Astron Soc. *Res:* Formation of stars in the Milky Way and other galaxies; evolution of the interstellar medium in galaxies; history and sociology of astronomy. *Mailing Add:* 1902 Beaufort Laramie WY 82070

THROOP, LEWIS JOHN, b Detroit, Mich, June 18, 29; m 54; c 1. ANALYTICAL CHEMISTRY. *Educ:* Wayne State Univ, BS, 54, MS, 56, PhD(analytical chem), 57. *Prof Exp:* Chemist, Syntex, SA, Mex, 57-59; group leader analytical chem, Mead Johnson & Co, 59-64; dept head, 64-71, asst dir analytical chem, Inst Org Chem, 71-78, dir analytical res, 78-85, VPRES & DIR ANALYTICAL & ENVIRON RES, SYSTEX RES, SYNTEX CORP, 85- *Mem:* Am Chem Soc; Am Pharmaceut Asn. *Res:* Electrochemistry; organic structure characterization; laboratory automation. *Mailing Add:* Syntex Corp 3401 Hillview Ave Palo Alto CA 94304

THROW, FRANCIS EDWARD, b Ottumwa, Iowa, Oct 4, 12; wid; c 3. SCIENCE COMMUNICATIONS, GENERAL PHYSICS. *Educ:* Park Col, BA, 33; Univ Mich, MS, 36, PhD(physics), 40. *Prof Exp:* Instr physics, Milwaukee State Teachers Col, 39; instr physics & math, Polytech Inst PR, 40-41; instr physics, Altoona Undergrad Ctr, 41-42; ground sch instr math, physics & theory of flight, US Navy Pre-Flight Sch, Iowa Univ, 42-44; prof physics & head dept, Cornell Col, 44-52; chmn dept physics, Wabash Col, 52-56; asst dir physics div, Argonne Nat Lab, 56-73, asst dir radiol & environ res div, 73-77; RETIRED. *Concurrent Pos:* Consult, tech ed & writer, report on electronic control, Borg-Warner, 78, annual brochures, Argonne Univ Asn, 80, 81 & 82. *Mem:* Am Phys Soc; Am Asn Physics Teachers. *Mailing Add:* 3131 Simpson St Apt GW-206 Evanston IL 60201

THROWER, PETER ALBERT, b Norfolk, Eng, Jan 9, 38; m 85. PHYSICS, MATERIALS SCIENCE. *Educ:* Cambridge Univ, BA, 60, MA, 63, PhD(physics), 69. *Prof Exp:* Sci officer, Atomic Energy Res Estab, Eng, 60-65, sr sci officer, 65-69; PROF MAT SCI, PA STATE UNIV, UNIVERSITY PARK, 69- *Concurrent Pos:* Ed, Chem & Physics of Carbon, ed-in-chief, Carbon J. *Mem:* Am Soc Metals Int; Am Carbon Soc. *Res:* Structure and properties of carbon and graphite; irradiation damage to graphite; electron microscopy; mineral microstructures. *Mailing Add:* Dept Mat Sci & Eng Penn State Univ University Park PA 16802

THRUPP, LAURI DAVID, b Sask, Nov 30, 30; US citizen; m 52; c 4. INFECTIOUS DISEASES, MICROBIOLOGY. *Educ:* Stanford Univ, AB, 51; Univ Wash, MD, 55. *Prof Exp:* Asst chief & chief polio surveillance, Epidemiol Br, Nat Commun Dis Ctr, 56-58; resident physician & fel, Boston City Hosp, Thorndike Serv & Harvard Med Sch, 58-63; jr asst physician, Harvard Serv, 61-63; asst prof med & med microbiol, Sch Med, Univ Southern Calif, 63-66, asst prof med, 66-68; assoc prof med & head div infectious dis, 68-77, PROF MED, DIV INFECTIOUS DIS, UNIV CALIF, IRVINE, 77- *Concurrent Pos:* Life Ins Med Res Found res fel med & bact, Boston City Hosp & Harvard Med Sch, 60-63; asst chief commun dis, Los Angeles County Gen Hosp, 63-64, attend physician, 63-, med microbiologist, 64-65; consult, Los Angeles County Health Dept, 64- & Calif State Health Dept, 66-; chief infectious dis serv, Orange County Med Ctr, 68- *Mem:* Am Fedn Clin Res; Am Soc Microbiol; Am Pub Health Asn; Infectious Dis Soc Am; NY Acad Sci. *Res:* Clinical and experimental pyelonephritis, pathogenesis and immune response, role of bacterial L-forms; gram-negative hospital-acquired infections; meningitis, clinical and immunological; bacteriology. *Mailing Add:* Dept Infectious Dis Univ Calif Irvine CA 92717

THRUSTON, ALFRED DORRAH, JR, b Greenville, SC, Nov 3, 34; m 64; c 2. ANALYTICAL CHEMISTRY. *Educ:* Ga Inst Technol, BS, 57. *Prof Exp:* Chemist, US Food & Drug Admin, 59-66; RES CHEMIST, ATHENS ENVIRON RES LAB, ENVIRON PROTECTION AGENCY, 66- *Mem:* Am Chem Soc. *Res:* Liquid chromatographic, mass spectrometric and gas chromatographic analysis of water pollutants. *Mailing Add:* Environ Res Lab College Station Rd Athens GA 30613

THUAN, TRINH XUAN, b Hanoi, Vietnam, Aug 20, 48. ASTROPHYSICS, ASTRONOMY. *Educ:* Calif Inst Technol, BS, 70; Princeton Univ, PhD(astrophys), 74. *Prof Exp:* Res fel astrophys, Calif Inst Technol, 74-76; assoc prof, 76-90, PROF ASTRON, UNIV VA, 90- *Concurrent Pos:* Vis fel, Inst d'Astrophys, Paris, 78 & 87; vis prof, Ctr d'Etudes Nucleaires de Saclay, Paris, 81 & 87; Fulbright res scholar, 87. *Mem:* Am Astron Soc; Int Astron Union. *Res:* Study of the formation, clustering and evolution of galaxies; observational cosmology. *Mailing Add:* Dept Astron Univ Va PO Box 3818 Univ Sta Charlottesville VA 22903

THUBRIKAR, MANO J, b Nagpur, India, Aug 27, 47; m 73; c 2. HEART VALVE PROSTHETICS, ATHEROSCLEROSIS. *Educ:* Nagpur Univ, BE, 69; NY Univ, MS, 71, PhD(biomed eng), 75. *Prof Exp:* From asst prof to assoc prof, Univ Va, 75-91; HEINEMAN RES PROF, HEINEMAN MED RES CTR, 91- *Concurrent Pos:* Res career develop award, NIH, 80-85; dir surg res, Univ Va, 87-91, res prof, Div Atherosclerosis. *Mem:* Biomed Eng; Am Soc Artificial Internal Organs; Eng Med & Biol; Soc Cardiac Biol Implants. *Res:* Mechanism of atherogenesis; mechanics of arterial tissue; heart valve function; designing bioprosthetic and prosthetic heart valves. *Mailing Add:* 6401-7 Reafield Dr Charlotte NC 28226-3522

THUENTE, DAVID JOSEPH, b Decorah, Iowa, Mar 17, 45; m 67; c 2. OPERATIONS RESEARCH, COMPUTER SCIENCE. *Educ:* Loras Col, BS, 67; Univ Kans, MA, 69, PhD(math), 74. *Prof Exp:* Systs analyst transp, Jewel Co, Inc, 67; instr math, Univ Kans, 70-74; mem fac res staff nonlinear prog, Argonne Nat Lab, 76-77, 80, 83; asst prof math, 74-79, assoc prof math & computer sci, 80-89, PROF COMPUTER SCI, IND UNIV-PURDUE UNIV, FT WAYNE, 89- *Concurrent Pos:* Res grant, Purdue Univ, 75; fac res grant, Argonne Nat Lab, 76-77, 80 & 83, consult, 80-81; mem tech staff, Mitre Corp, 81-82; Comput Literacy grant, 84-; sr software engr, Magnavox Electronic Systs Co, 88- *Mem:* Soc Indust & Appl Math; Opers Res Soc Am; Math Prog Soc; Sigma Xi; Asn Comput Mach; Int Asn Math Modeling. *Res:* Simulation for distributed processing, Simulation languages and applications; nonlinear programming, optimization and mathematical modeling. *Mailing Add:* Dept Computer Sci 2101 Coliseum Blvd Ft Wayne IN 46805

THUERING, GEORGE LEWIS, b Milwaukee, Wis, Sept 2, 19; m 45, 75; c 1. INDUSTRIAL ENGINEERING. *Educ:* Univ Wis, BS, 41, ME, 54; Pa State Univ, MS, 49. *Prof Exp:* Mfg engr, Lockheed Aircraft Corp, Calif, 41-47, supvr plant layout & space control dept, Ga Div, 51-52; from instr to assoc prof, Pa State Univ, 47-56, dir, Dept Mgt Engr, 62-77, prof, 56-, emer prof indust eng; RETIRED. *Concurrent Pos:* Consult. *Mem:* Am Inst Indust Engrs; Am Soc Eng Educ; Am Soc Mech Engrs (vpres, 82-84); Sigma Xi. *Res:* Manufacturing science. *Mailing Add:* 436 Homan Ave State College PA 16801

THUESEN, GERALD JORGEN, b Oklahoma City, Okla, July 20, 38; m 60, 82; c 2. INDUSTRIAL ENGINEERING. *Educ:* Stanford Univ, BS, 60, MS, 61, PhD(indust eng), 68. *Prof Exp:* Engr, Pac Tel Co, 61-62; mgt engr comput systs, Atlantic Refining Co, 62-63; asst prof indust eng, Arlington State Col, 63-64 & Univ Tex, Arlington, 67-68; assoc prof, 68-76, PROF INDUST ENG, GA INST TECHNOL, 76- *Honors & Awards:* Eugene L Grant Award, Am Soc Eng Educ, 77, 89; Wellington Award, Inst Indust Engrs, 89, Outstanding Inst Indust Engrs Publ Award, 90. *Mem:* Fel Inst Indust Engrs; Opers Res Soc Am; Am Soc Eng Educ. *Res:* Decision analysis; engineering economy; capital budgeting; statistical decision theory. *Mailing Add:* Sch Indust & Systs Eng Ga Inst Technol Atlanta GA 30332

THUESON, DAVID OREL, b Twin Falls, Idaho, May 9, 47; m 69; c 5. IMMUNOPHARMACOLOGY, DRUG DESIGN. *Educ:* Brigham Young Univ, BS, 71; Univ Utah, Salt Lake City, PhD(pharmacol), 76. *Prof Exp:* Res scientist, Univ Tex Med Br, 76-77, asst prof immunol & pharmacol, 77-82; sr res assoc res, Parke-Davis Pharmaceut Res, 82-88; dept dir pharmacol, Immunetech Pharmaceut, 88-90; DEPT DIR IMMUNOPHARMACOL, TANABE RES LABS, USA, 90- *Concurrent Pos:* Postdoctoral fel, Univ Tex Med Br, 75-77, asst prof, Grad Fac, Biomed Sci, 77-82; young investr award, NIH, 78. *Mem:* Am Acad Allergy & Clin Immunol; Am Asn Immunologists; Am Thoracic Soc. *Res:* Drug discovery development of clinical agents for arthritis, autoimmune diseases, asthma and allergies; regulation of cell activation and mechanisms of inflammation; inhibition of mediator production/release from inflammatory effector cells. *Mailing Add:* Dept Immunopharmacol Tanabe Res Labs 11045 Roselle St San Diego CA 92121

THUM, ALAN BRADLEY, b Washington, DC, May 30, 43; m 66; c 2. MARINE ECOLOGY. *Educ:* Univ Redlands, BS, 65; Univ Pac, MS, 67; Ore State Univ, PhD(marine ecol), 71. *Prof Exp:* Lectr invert zool, Univ Cape Town, 71-75; sr scientist & environ consult, Lockheed Marine Biol Lab, Lockheed Aircraft Serv, 75-84, Kinnetic Labs, Inc, 84-90. *Concurrent Pos:* Consult, Bur Land Mgt, Univ Southern Calif, 76- *Mem:* Ecol Soc Am; Am Soc Limnol & Oceanog; Int Asn Meiobenthologists; Royal Soc SAfrica. *Res:* Ecology of interstitial meiofauna; ecology and systematics of turbellaria; reproductive ecology of marine benthic invertebrates; aquatic toxicology. *Mailing Add:* 1392 Peachwood Dr Encinitas CA 92024

THUMANN, ALBERT, b New York, NY, Mar 12, 42; m 66; c 1. ELECTRICAL & ENERGY ENGINEERING. *Educ:* City Col New York, BEE, 64; NY Univ, MSEE, 67, MSIE, 70. *Prof Exp:* Proj mgr eng construct, Bechtel, 64-77; EXEC DIR, ASN ENERGY ENGRS, 77- *Concurrent Pos:* Adj prof, Univ Louisville & lectr, Sch Continuing Educ, NY Univ, 76- *Mem:* Nat Soc Prof Engrs; Asn Energy Engrs (pres, 77); Am Soc Asn Execs; Am Soc Assoc Execs; Coun Eng; Sci Soc Execs. *Res:* Noise control; energy. *Mailing Add:* 931 Smoketree Dr Tucker GA 30084

THUMM, BYRON ASHLEY, b Malden, WVa, Jan 2, 23; m 56. ANALYTICAL CHEMISTRY. *Educ:* Morris Harvey Col, BS, 45; Duke Univ, PhD(chem), 51. *Prof Exp:* Res chemist, Am Viscose Div, FMC Corp, 51-63; emer prof chem, State Univ NY Col, Fredonia, 63-87; RETIRED. *Mem:* Am Chem Soc. *Res:* Solution kinetics and equilibrium; rayon spinning process; water analysis; formaldehyde complexes. *Mailing Add:* 24 Maple Ave Fredonia NY 14063

THUN, RUDOLF EDUARD, b Berlin, Germany, Jan 30, 21; nat US; m 44; c 2. ENGINEERING MANAGEMENT, MICROELECTRONICS. *Educ:* Univ Frankfurt, dipl, 54, PhD(physics), 55. *Prof Exp:* Physicist, German Gold & Silver Separation Plant, 51-55 & US Army Corps Engrs Res & Develop Labs, Ft Belvoir, Va, 55-59; var sci & managerial positions, Int Bus Mach Corp, 59-67; mgr microelectronics, Missile Systs Div, 67-83, DIR DEVICE TECHNOL, RAYTHEON CO, BEDFORD, 83- *Honors & Awards:* 1975 Contributions Award, Inst Elec & Electronics Engrs, 75. *Mem:* Fel Am Phys Soc; Optical Soc Am; fel Inst Elec & Electronics Engrs; Electron Microscope Soc; Int Soc Hybrid Microelectronics; NY Acad Sci. *Res:* Microelectronics; physical and optical properties of thin films; electron diffraction; metastable phases in binary metal systems; electronic packaging and dedicated computers; integrated circuit technology. *Mailing Add:* 228 Heald Rd Carlisle MA 01741

THUNING-ROBERSON, CLAIRE ANN, b Cincinnati, Ohio, Nov 17, 45; m 84. CANCER CHEMOTHERAPY, ONCOGENIC VIRUSES. *Educ:* St Mary-of-the-Woods Col, BA, 67; Nova Univ, MS, 77, PhD(biol), 82. *Prof Exp:* Sr res assoc, St Vincent Charity Hosp, 69-74; res assoc, 74-87, assoc dir, 88-90, DIR, GOODWIN INST CANCER RES, 90- *Concurrent Pos:* Dir grad studies, Goodwin Inst Cancer Res, 81- *Mem:* AAAS; Am Soc Clin Path; Am Asn Cancer Res. *Res:* Investigating the control of cancer using combined hyperthermia and chemotherapy; the use of oxygen immunosuppression in promoting xenogeneic tumor growth and blocking autoimmune disease; properties of recombinant herpes virus strains. *Mailing Add:* Goodwin Inst Cancer Res 1850 Northwest 69th Ave Plantation FL 33313

THURBER, CLIFFORD HAWES, b Doylestown, Pa, Aug 3, 54; m 76; c 2. SEISMOLOGY, VOLCANOLOGY. *Educ:* Cornell Univ, AB, 75; Mass Inst Technol, PhD(geophys), 81. *Prof Exp:* From asst prof to assoc prof geophys, State Univ NY, Stony Brook, 81-89; ASSOC PROF GEOPHYS, UNIV WIS-MADISON, 89- *Concurrent Pos:* Secy, Inc Res Insts Seismol, 86-88. *Mem:* Am Geophys Union; Seismol Asn Am; Sigma Xi. *Res:* Seismic imaging of earth structure, methods for earthquake location and seismic wave propagation; seismotectonics; volcanic earthquakes and magma transport; regional and global geophysical studies, especially gravity, stress and tectonics; seismic verification. *Mailing Add:* Dept Geol & Geophys Univ Wis Madison WI 53706-1692

THURBER, DAVID LAWRENCE, b Oneonta, NY, Dec 29, 34; m 64; c 1. GEOCHEMISTRY. *Educ:* Union Col, NY, BS, 56; Columbia Univ, MA, 58, PhD(geol), 63. *Prof Exp:* Res asst geochem, Lamont Geol Observ, Columbia Univ, 56-63, res scientist, 63-64, res assoc, 64-66; assoc prof, 66-70, PROF GEOL, QUEENS COL, NY, 70- *Concurrent Pos:* Am Geophys Union vis lectureship, 64; lectr, Queens Col, NY, 65-66; vis res assoc, Lamont Geol Observ, 66-70, vis sr res assoc, Lamont-Doherty Geol Observ, 70-74; vis prof, Fed Univ Bahia, 74-78. *Mem:* AAAS; Am Geophys Union; Geochem Soc. *Res:* General geochemistry; stable and radioisotope geochemistry; geochronology; hydrochemistry; soil chemistry. *Mailing Add:* Dept Geol Queens Col Flushing NY 11367

THURBER, JAMES KENT, b Utica, NY, Oct 29, 33. APPLIED MATHEMATICS. *Educ:* Brooklyn Col, BS, 55; NY Univ, PhD, 61. *Prof Exp:* Asst appl math, NY Univ, 57-61; asst prof math, Adelphi Univ, 61-64; assoc math, Brookhaven Nat Lab, 64-66, mathematician, 66-69; PROF MATH,

PURDUE UNIV, LAFAYETTE, 69- *Mem:* Am Math Soc; Am Nuclear Soc; Math Asn Am; Soc Indust & Appl Math; NY Acad Sci. *Res:* Neutron transport; kinetic theory of gases; asymptotic analysis; mathematical programming; applications of nonstandard analysis. *Mailing Add:* 1808 Charles St W Lafayette IN 47904

THURBER, ROBERT EUGENE, b Bayshore, NY, Oct 11, 32; m 53, 84; c 4. PHYSIOLOGY, BIOPHYSICS. *Educ:* Col of the Holy Cross, BS, 54; Adelphi Univ, MS, 61; Univ Kans, PhD(physiol), 65. *Prof Exp:* Res assoc radiation biol, Brookhaven Nat Lab, Assoc Univs Inc, 56-61; from instr to assoc prof physiol, Med Col Va, Va Commonwealth Univ, 64-69; assoc prof, Jefferson Med Col, Thomas Jefferson Univ, 69-70; PROF PHYSIOL & CHMN DEPT, SCH MED, ECAROLINA UNIV, 70- *Concurrent Pos:* Consult, US Vet Admin, Va State Bd Med Examr & NASA, Va, 66-69, US Naval Hosp, Portsmouth, 68-69 & Psychol Consult Inc, 68-70; mem bd dirs, NC Heart Asn, 72-, pres elect, 78-79, pres, 79-80; mem, Comt Regional & Nat Res, Am Heart Asn, 81-82. *Mem:* AAAS; Am Physiol Soc; NY Acad Sci; Sigma Xi; Asn Chmn Dept Physiol. *Res:* Nonequilibrium transfer and distribution of electrolytes; renal transport; radiation biology and carbohydrate metabolism. *Mailing Add:* Dept Physiol Sch Med ECarolina Univ Greenville NC 27858-4354

THURBER, WALTER ARTHUR, b East Worcester, NY, Nov 27, 08; m 34; c 2. ORNITHOLOGY. *Educ:* Union Col NY, BS, 33; NY State Col Teachers, MS, 38; Cornell Univ, PhD(nature study), 41. *Prof Exp:* Teacher NY schs, 26-29, 33-38; asst ed, Cornell Univ, 38-39; instr physics & phys sci, State Univ NY Teachers Col, Cortland, 40-43; asst prof physics & phys & earth sci, 43-48, prof sci, 48-58; vis prof sci educ, Syracuse Univ, 58-61, adj prof, 61-72; textbook writer, 61-80; RETIRED. *Concurrent Pos:* Consult, NY State Educ Dept, 43-55, NY pub schs, 47-52, Orgn Cent Am States, 66-, Ministry of Educ, 72 & Gen Direction Natural Resources, El Salvador, 74-; lab assoc, Lab Ornith, Cornell Univ, 70-; vis prof, Nat Univ El Salvador, 71-79. *Mem:* Fel AAAS; NY Acad Sci; Am Ornith Union; Wilson Ornith Soc; Cooper Ornith Soc; Sigma Xi. *Res:* Elementary and secondary science education; organization of syllabuses and textbooks; distribution and life histories Central American birds. *Mailing Add:* PO Box 16918 Temple Terrace FL 33687

THURBER, WILLIAM SAMUELS, b Ann Arbor, Mich, Mar 6, 22; m 43; c 4. ORGANIC CHEMISTRY. *Educ:* Mich State Univ, BS, 46, MS, 48. *Prof Exp:* Asst dir styrene polymerization lab, Dow Chem Co, 54-57, dir, Strosackers Res & Develop Group, 57-61, plant supt, 61-64, tech dir, Saginaw Bay Res Dept, 64-68, admin asst to div dir res, 68-70, sect mgr process develop & eng, 70-76, mgr, 76-81, mgr Equal Opportunity Employ Progs, Dow Chem USA, 81-83; RETIRED. *Mem:* Am Chem Soc. *Res:* Styrene polymers; antioxidants; polyglycols; surface active agents. *Mailing Add:* 11005 Elk Lake Rd Williamsburg MI 49690

THURBER, WILLIS ROBERT, b Butte, Nebr, July 10, 38; m 65; c 2. SEMICONDUCTORS, MATERIALS SCIENCE. *Educ:* Nebr Wesleyan Univ, AB, 60; Univ Md, MS, 63. *Prof Exp:* Res physicist, Solid State Physics Sect, 62-66, res physicist, Electron Devices Div, 66-80, res physicist, semiconductors, Semiconductor Mat & Processes Div, 80-85, SEMICONDUCTOR ELECTRONICS DIV, NAT BUR STANDARDS, 85- *Mem:* Am Soc Testing & Mat. *Res:* Electrical and optical properties of semiconductors, particularly silicon, including Hall effect, resistivity, mobility, lifetime, infrared transmission and absorption; deep level transient spectroscopy. *Mailing Add:* Nat Inst Standards & Technol Bldg 225 Rm A305 Gaithersburg MD 20899

THURBERG, FREDERICK PETER, b Weymouth, Mass, Aug 31, 42; m 64; c 2. PHYSIOLOGY, MARINE BIOLOGY. *Educ:* Univ Mass, BA, 64, MEd, 66; Univ NH, MS, 69, PhD(zool, physiol), 72. *Prof Exp:* Teacher pub schs, Mass, 64-67; PHYSIOLOGIST, NAT MARINE FISHERIES SERV, NAT OCEANIC & ATMOSPHERIC ADMIN, 71- *Mem:* Estuarine Res Fedn; Am Soc Zoologists; Nat Shellfisheries Asn. *Res:* Physiological ecology; effects of pollutants on marine organisms; invertebrate physiology; marine biotoxins; red tides. *Mailing Add:* Northeast Fisheries Ctr Nat Marine Fisheries Serv Milford CT 06460

THURESON-KLEIN, ASA KRISTINA, b Sweden, May 31, 34; m 61; c 2. BIOLOGY, PHARMACOLOGY. *Educ:* Univ Stockholm, MA, 58, PhD(biol), 68. *Prof Exp:* Instr bot, Univ Stockholm, 58-64; res assoc, 65-68, asst prof, 69-74, assoc prof, 74-78, PROF PHARMACOL, MED CTR, UNIV MISS, 78- *Concurrent Pos:* Pharmaceut Mfrs Asn Found fel pharmacol & morphol, Med Ctr, Univ Miss, 71-73; vis prof, Dept Physiol, Karolinska Inst, Sweden, 83 & 85. *Mem:* Am Soc Pharmacol & Exp Therapeut; Scand Soc Physiol; Soc Neurosci; Sigma Xi. *Res:* Effects of pharmacological agents on neurotransmitter release using combined biochemical and morphological methods; non-synaptic exocytosis from large neuropeptide containing vesicles; effects of leukotrienes B4 and C4 on the microcirculation. *Mailing Add:* Dept Pharmacol Univ Miss Med Ctr 2500 N State St Jackson MS 39216

THURLBECK, WILLIAM MICHAEL, b Johannesberg, Transvaal, SAfrica, Sept 7, 29; Can citizen; m 55; c 3. PULMONARY PATHOLOGY, PEDIATRIC PATHOLOGY. *Educ:* Univ Cape Town, BSc, 50, MB, ChB, 53. *Prof Exp:* From asst prof to prof path, McGill Univ, 61-73; sr investr path, Midhurst Med Res Inst, 73-75; prof path & head dept, Univ Man, 75-80; PROF PATH, UNIV BC, 80- *Concurrent Pos:* Adj prof, dept anat, Univ Calif, Davis, 84. *Honors & Awards:* Medallist, Am Col Chest Physicians, 76; Sommer Mem Lectr, Univ Ore, 81. *Mem:* Can Thoracic Soc (pres, 84); Am Asn Pathologists; Int Acad Path; Can Soc Clin Invest; Am Thoracic Soc; corresp mem Brit Thoracic Soc. *Res:* Human lung growth; manipulation and control of experimental lung growth; pathology and pathophysiology of chronic airflow obstruction. *Mailing Add:* Dept Path Fac Med Univ BC 2211 Wesbrook Mall Vancouver BC V6T 1W5 Can

THURLIMANN, BRUNO, b Gossau, Switz, Feb 6, 23; Swiss & US citizen; m 53; c 3. STRUCTURAL ENGINEERING, MASONRY. *Educ:* Swiss Fed Inst Technol, Zurich, Dipl, 46; Lehigh Univ, PhD(civil eng), 51. *Hon Degrees:* Dr, Univ Stuttgart, Fed Repub Germany, 83. *Prof Exp:* Res asst struct eng, Lehigh Univ, 49-50, res prof, 53-60; res assoc, Brown Univ, 51-52; res asst, 46-48, prof struct eng, 60-90, EMER PROF, SWISS FED INST TECHNOL, 90- *Honors & Awards:* Res Prize, Am Soc Civil Engrs, 60; Norman Medal, 63; Moisseff Award, 64; Howard Award, 86. *Mem:* Nat Acad Eng; fel Am Soc Civil Engrs; Swiss Engrs & Architects Soc; Ger Concrete Soc; Int Asn Bridge & Struct Eng (hon pres); hon mem Am Concrete Inst. *Res:* Concrete shell structures, plastic design of steel structures, stability, plate girders, composite construction; design of static and fatigue testing installation; plastic design of concrete structures; prestressed and partially prestressed structures; shear, torsion and combined actions; structural analysis; masonry construction. *Mailing Add:* Pfannenstiel-Str 56 Zurich CH-8132 Switzerland

THURMAIER, ROLAND JOSEPH, b Chicago, Ill, June 25, 28; m 55; c 4. ORGANIC CHEMISTRY, POLYMER CHEMISTRY. *Educ:* Bradley Univ, BS, 50; Univ Iowa, MS, 58, PhD(org chem), 60. *Prof Exp:* Plant chemist, Corn Prod Ref Corp, 51-55; res chemist, E I du Pont de Nemours & Co, 60-66; ASST PROF ORG CHEM, UNIV WIS-STEVENS POINT, 66- *Concurrent Pos:* Mem, Stevens Point Transit Comn, 70; mem, Study Comt Mass Transit, 72 & Environ Coun, Univ Wis- Stevens Point, 73-74. *Mem:* Am Chem Soc; Sigma Xi. *Res:* Polymers; stabilizers for polyurethanes. *Mailing Add:* Dept Chem Univ Wis-Stevens Pt Stevens Point WI 54481

THURMAN, GARY BOYD, b 1941; c 4. LYMPHOKINES, CELL MEDIATED IMMUNITY. *Educ:* Univ Utah, PhD(radiation biol), 71. *Prof Exp:* Assoc sci dir, Biotherapeut, 85-89; vis scholar, Dept Biochem, Vanderbilt Univ, 90. *Mem:* Am Asn Immunologists; Am Asn Cancer Res. *Mailing Add:* 1007 Mooreland Blvd Brentwood TN 37027-5224

THURMAN, HENRY L, JR, b Lawrenceville, Va, Jan 14, 27; m 52; c 1. ENGINEERING. *Educ:* Hampton Inst, BS, 47; Univ Ill, MS, 49. *Prof Exp:* Instr tech archit construct, 48-52, dir div indust technol, 53-56, dir technol & eng, 56-59, dean col eng, 59-72, PROF ARCHIT ENG, SOUTHERN UNIV, BATON ROUGE, 72- *Concurrent Pos:* Mem bd dirs, Comn Eng Educ, 67- *Mem:* Am Soc Eng Educ. *Res:* Structural and architectural design; shelter analysis. *Mailing Add:* Dept Archit A&M Col & Southern Univ Baton Rouge LA 70813

THURMAN, LLOY DUANE, b Oconto, Nebr, Sept 3, 33; m 57; c 4. PLANT ECOLOGY. *Educ:* Univ Nebr, BS, 59, MS, 61; Univ Calif, Berkeley, 66. *Prof Exp:* Asst prof biol, Southern Calif Col, 65-67; from asst prof to assoc prof biol, 67-72, chmn dept natural sci, 69-73, PROF BIOL, ORAL ROBERTS UNIV, 72- *Mem:* Sigma Xi; Soc Col Sci Teachers; Am Sci Affil; Nat Sci Teachers Asn. *Res:* Ecology of Orthocarpus; curricula and teaching methods in biology; computers in biology and nutrition; creation versus evolution controversy. *Mailing Add:* 2520 E 57 St Tulsa OK 74105

THURMAN, RICHARD GARY, b Wichita, Kans, Mar 1, 40; m 70. CHEMISTRY. *Educ:* NMex State Univ, BS, 62, MS, 65; Univ Ariz, PhD(chem), 71. *Prof Exp:* Asst prof chem, Univ Ariz, 70-71; from asst prof to assoc prof chem, Univ Nebr-Omaha, 71-77; MEM STAFF, UNION CAMP, 77- *Mem:* Am Chem Soc. *Res:* Liquid and gas chromatography-study of the separation processes; computer-controlled chemical instrumentation; use of computers in chemical education. *Mailing Add:* 2028 Old York Rd Burlington NJ 08016-9767

THURMAN, ROBERT ELLIS, II, b Springfield, Mo, Oct 15, 39; m 62; c 4. PHYSICS. *Educ:* Mo Sch Mines, BS, 62; Univ Wis, MS, 64; Univ Mo-Rolla, PhD(physics), 77. *Prof Exp:* Res asst physics, Univ Wis, 63-66; from instr to asst prof, 66-77, ASSOC PROF PHYSICS, SOUTHWEST MO STATE UNIV, 77- *Concurrent Pos:* Physicist, US Forest Prod Lab, 63-64; NSF sci fac fel, 70-71; res asst physics, Univ Mo-Rolla, 71-72; sabbatical leave, Cloud Physics Ctr, Univ Mo-Rolla, 81-82. *Mem:* AAAS; Am Asn Physics Teachers; Sigma Xi. *Res:* Atmospheric physics; electrical mobility of water molecule cluster ions; aerosol evolution. *Mailing Add:* 1851 S Maryland Springfield MO 65807

THURMAN, RONALD GLENN, b Carbondale, Ill, Nov 25, 41. BIOCHEMISTRY, PHARMACOLOGY. *Educ:* St Louis Col Pharm, BS, 63; Univ Ill, PhD(pharmacol), 68. *Prof Exp:* Asst prof biophys & phys biochem, Johnson Res Found, Univ Pa, 71-77; ASSOC PROF PHARMACOL, UNIV NC, CHAPEL HILL, 77- *Concurrent Pos:* Fel, Johnson Res Found, Univ Pa, 67-69; NATO fel, Inst Physiol Chem, Munich, Ger, 69-70; Alexander von Humboldt fel, 70-71; NIMH career develop award, 71-82. *Mem:* AAAS; Am Pharmaceut Soc. *Res:* Drug and alcohol metabolism. *Mailing Add:* Dept Pharmacol Univ NC Chapel Hill NC 27514

THURMAN, WAYNE LAVERNE, b Detroit, Mich, June 11, 23. SPEECH PATHOLOGY. *Educ:* Southeast Mo State Col, BA & BSE, 48; State Univ Iowa, MA, 49; Purdue Univ, PhD(speech path), 53. *Prof Exp:* Instr speech, Southeast Mo State Col, 49-51; prof speech path & chmn dept, Eastern Ill Univ, 53-84; RETIRED. *Mem:* Fel Am Speech-Lang-Hearing Asn. *Res:* Voice quality disorders severity scales; voice therapy procedures. *Mailing Add:* 238 El Caminito Carmel Valley CA 93924

THURMAN, WILLIAM GENTRY, b Jacksonville, Fla, July 1, 28; m 49; c 4. PEDIATRICS, ONCOLOGY. *Educ:* Univ NC, BS, 49; McGill Univ, MD & CM, 54. *Prof Exp:* Asst prof pediat, Tulane Univ, 60-61; assoc prof, Emory Univ, 61-62; prof, Cornell Univ, 62-64; prof & chmn dept, Sch Med, Univ Va, 64-73, dir, Ctr Delivery Health Care, 69-73; dean sch med, Tulane Univ, 73-75; prof pediat & provost, Univ Okla Health Sci Ctr, 75-79; PROF PEDIAT & PRES, OKLA MED RES FOUND, 79- *Concurrent Pos:* Fel hemat & oncol, 58-60; Markle scholar acad med, 59-64; consult, USAF, 59-, Comn on Cancer, 63- & Comn on Pediat Hemat, 65-; chmn dept pediat, Mem

Sloan-Kettering Cancer Ctr, 62-64; prog consult, Nat Found, 64-; prog consult pediat, VI, 66-; mem, Nat Rev Comt Regional Med Prog. *Mem:* Soc Pediat Res; Am Soc Hemat; Am Soc Human Genetics; Am Asn Cancer Res; Am Pediat Soc. *Res:* Immunologic abnormalities associated with malignancy in children; clinical management of children with malignancy; evaluation of various drugs; methods and models for delivery of health care. *Mailing Add:* Okla Med Res Found 825 NE 13th St Oklahoma City OK 73104

THURMAN-SWARTZWELDER, ERNESTINE H, medical entomology, research administration; deceased, see previous edition for last biography

THURMON, JOHN C, b Redford, Mo, Mar 4, 30; m 56; c 1. VETERINARY ANESTHESIOLOGY. *Educ:* Univ Mo, BS, 60, DVM, 62, MS, 67. *Prof Exp:* From instr to asst prof vet med, 62-70, assoc prof vet anesthesiol, 71-76, assoc prof physiol & pharmacol, 75-76, HEAD DIV, COL VET MED, UNIV ILL, URBANA, 71-, PROF VET ANESTHESIOL, VET ANAT, PHYSIOL & PHARMACOL, 76-, ASSOC PROF BIOENG, COL ENG, 72- *Concurrent Pos:* Nat Heart Inst fel, Baylor Col Med, 69; vis prof anesthesiol, Col Med, Univ Ill, Chicago, 70; consult, Bristol Labs, Syracuse, NY, 70-74, Affil Labs, Whitehall, Ill, 72-74, Bay Vet Corp, Shawnee Mission, Kans, 73-, Norden Labs, Lincoln, Nebr, 74- & Dept Med Sci, Southern Ill Univ, 75- *Mem:* Am Soc Vet Anesthesiol (pres elect, 73, pres, 74); Am Soc Anesthesiologists; Int Anesthesia Res Soc; Am Col Vet Anesthesiologists (pres, 75-77); Sigma Xi. *Res:* General anesthesia and its effects on homeostatic mechanisms of domestic and wild animals; development and design of equipment for use in veterinary anesthesia. *Mailing Add:* 2003 Cureton Dr Urbana IL 61801

THURMON, THEODORE FRANCIS, b Baton Rouge, La, Oct 20, 37; m 61; c 2. MEDICAL GENETICS. *Educ:* La State Univ, Baton Rouge, BS, 60; La State Univ, New Orleans, MD, 62. *Prof Exp:* Intern, Naval Hosp, Pensacola, Fla, 62-63, resident, Philadelphia, Pa, 63-65, pediatrician, cytogeneticist & cardiologist, St Albans, NY, 65-68; assoc prof, 69-77, PROF PEDIAT & DIR, PED GENETICS DIV, MED SCH, LA STATE UNIV, NEW ORLEANS, 77- *Concurrent Pos:* Fel med genetics, Johns Hopkins Hosp, Baltimore, 68-69; consult, La State Dept Hosps, 70- *Mem:* AAAS; Am Soc Human Genetics; Human Biol Coun; NY Acad Sci. *Res:* Genetic regulation. *Mailing Add:* 1542 Tulane Ave New Orleans LA 70112

THURMOND, JOHN TYDINGS, b Dallas, Tex, Oct 22, 41; m 69; c 3. VERTEBRATE PALEONTOLOGY, ENVIRONMENTAL GEOLOGY. *Educ:* St Louis Univ, BS, 63; Southern Methodist Univ, MS, 67, PhD(geol), 69. *Prof Exp:* Res asst to pres, Inst Study Earth & Man, Southern Methodist Univ, 70; asst prof geol, Birmingham-Southern Col, 70-77; ASSOC PROF EARTH SCI, UNIV ARK, LITTLE ROCK, 77-, CHAIRPERSON DEPT, 78- *Concurrent Pos:* Vis prof fac chem sci & pharm, San Carlos Univ, Guatemala, 71; geologist, Harbert Construct Co, Ala, 71-72; asst cordr coop univ upper-div prog, Univ Ala, Gadsden, 72. *Mem:* AAAS; Paleont Soc; Soc Vert Paleont. *Res:* Paleoecology, functional morphology, taxonomy of Mesozoic/Cenozoic fishes; Pleistocene paleoecology; Cretaceous marine reptiles; environmental geology; data retrieval in paleontology. *Mailing Add:* Dept Earth Sci 33rd & Univ Ave Univ Ark Little Rock AR 72204

THURMOND, WILLIAM, b Lodi, Calif, Apr 11, 26; m 49; c 2. DEVELOPMENTAL PHYSIOLOGY. *Educ:* Univ Calif, Berkeley, AB, 48, MA, 50, PhD(zool), 57. *Prof Exp:* Instr zool, San Mateo Community Col, Calif, 49-50; from instr to prof zool, 76, PROF BIOL SCI, CALIF POLYTECH STATE UNIV, SAN LUIS OBISPO, 76- *Concurrent Pos:* Vis prof, Univ Frankfort, 69-70. *Mem:* Fel AAAS; Am Soc Zool; Am Inst Biol Sci; Sigma Xi. *Res:* Development of the pituitary and the developmental interdependance with other endocrine glands and the hypothalamus. *Mailing Add:* Dept Biol Sci Calif Polytech State Univ San Luis Obispo CA 93407

THURNAUER, HANS, b Nurnberg, Ger, June 11, 08; nat US; m 35; c 3. CERAMICS. *Educ:* Tech Univ Berlin, Dipl, 31, DrEng, 58; Univ Ill, MS, 32. *Prof Exp:* Ceramic engr, Steatite-Magnesia Co, Ger, 32-33; ceramic lab asst, Steatite & Porcelain Prod, Ltd, Eng, 33-35; dir res & vpres, Am Lava Corp, Tenn, 35-55; head ceramic dept, Minn Mining & Mfg Co, 55-66; consult & tech adv, 67-73, INT EXEC SERV CORPS, COORS PORCELAIN CO, 74- *Concurrent Pos:* Dir, Israel Ceramic & Silicate Inst, Haifa, 64-66; consult mat comts, US Dept Defense & NASA. *Mem:* AAAS; fel Am Ceramic Soc; Electrochem Soc (pres, 57); Am Chem Soc. *Res:* Technical ceramics; solid state devices; refractory and abrasion resistant materials; nuclear ceramics. *Mailing Add:* 440 College Ave Boulder CO 80302

THURNAUER, MARION CHARLOTTE, b Chattanooga, Tenn; m 67. PHOTOSYNTHESIS, MAGNETIC RESONANCE. *Educ:* Univ Chicago, BA, 68, MS, 69, PhD(chem), 74. *Prof Exp:* Fel, 74-77, asst scientist, 77-81, SCIENTIST CHEM, ARGONNE NAT LAB, 81- *Mem:* Am Chem Soc; Biophys Soc; Sigma Xi; Asn Women Sci. *Res:* Electron paramagnetic resonance studies (continuous wave and pulsed) of radicals, radical pairs and triplet state; magnetic resonance studies of photosynthetic systems; time resolved magnetic resonance studies; pulsed electron paramagnetic resonance studies of photosynthetic systems. *Mailing Add:* 8511 Kearney Rd Downers Grove IL 60516

THURNER, JOSEPH JOHN, b Middletown, NY, Oct 26, 20; m 48; c 2. INORGANIC CHEMISTRY. *Educ:* Hartwick Col, BS, 49; Harvard Univ, MA, 51. *Prof Exp:* From instr to assoc prof, Colgate Univ, 51-69, chmn dept, 67-70, dir, Div Natural Sci & Math, 70-74, prof chem, 69-87; RETIRED. *Concurrent Pos:* Consult, Indium Corp Am, 52-62. *Mem:* AAAS; Am Chem Soc; Sigma Xi. *Res:* Organometallic compounds of germanium; organometallic and inorganic compounds; alloys of indium; synthesis and analysis of indium-bearing substances. *Mailing Add:* 50 Broad St Hamilton NY 13346

THUROW, GORDON RAY, b Aurora, Ill, Feb 13, 29; m 55; c 5. ZOOLOGY, ANATOMY. *Educ:* Univ Chicago, PhB, 48, BS, 50, MS, 51; Ind Univ, PhD, 55. *Prof Exp:* Asst zool & ornith, Ind Univ, 52 & 54, tech adv ed film, 55; assoc prof natural sci, Newberry Col, 57-59; fel anat, Med Col SC, 59-61; asst prof, Univ Kans, 61-66; ASSOC PROF BIOL, WESTERN ILL UNIV, 66- *Mem:* Am Soc Ichthyologists & Herpetologists; Soc Study Amphibians & Reptiles; Soc Study Evolution; Ecol Soc Am; Am Asn Anat; Sigma Xi. *Res:* Morphological, physiological and behavioral adaptations of vertebrates; transplantation; herpetology; ecology. *Mailing Add:* Dept Biol-Sci Western Ill Univ Macomb IL 61455

THURSTON, EARLE LAURENCE, b New York, NY, Jan 17, 43; m 62; c 3. BOTANY, CELL BIOLOGY. *Educ:* State Univ NY, Geneseo, BS, 64; Iowa State Univ, MS, 67, PhD(bot), 69. *Prof Exp:* NSF fel cell res inst, Univ Tex, Austin, 69-70; ASSOC PROF CELL BIOL & COORDR, ELECTRON MICROS CTR, TEX A&M UNIV, 70- *Mem:* Bot Soc Am; Electron Micros Soc; Am Soc Cell Biol. *Res:* Developmental morphology and ultrastructure; electron microprobe analysis. *Mailing Add:* RFD Three College Station TX 77843

THURSTON, GAYLEN AUBREY, b Garwin, Iowa, May 15, 29; m 51; c 5. ENGINEERING MECHANICS. *Educ:* Iowa State Col, BS, 50; Ohio State Univ, MS, 51; Cornell Univ, PhD(eng mech), 56. *Prof Exp:* Shell struct specialist, Jet Engine Dept, Gen Elec Co, 55-63; res scientist mech, Denver Div, Martin Marietta Corp, 63-69; prof, Univ Denver, 69-75; assoc prof mech, Univ Colo, Denver, 75-80. *Mem:* Sigma Xi; Am Soc Eng Educ. *Res:* Nonlinear mechanics; shell structures. *Mailing Add:* 506 Marlbank Dr Yorktown VA 23690

THURSTON, GEORGE BUTTE, b Austin, Tex, Oct 8, 24; m 47; c 2. BIOMEDICAL ENGINEERING, RHEOLOGY. *Educ:* Univ Tex, BS, 44, MA, 48, PhD(physics), 52. *Prof Exp:* Res scientist, Defense Res Lab, Tex, 49-52; asst prof physics, Univ Wyo, 52-53 & Univ Ark, 53-54; from assoc prof to prof, Okla State Univ, 54-68; PROF MECH ENG & BIOMED ENG, UNIV TEX, AUSTIN, 68- *Concurrent Pos:* Consult, US Naval Ord Test Sta, 54-55; res physicist, Univ Mich, 58-59; NSF fel, Macromolecules Res Ctr, Strasbourg, 63-64; vis prof, Helmholtz Inst, Rhineland-Westphalian Tech High Sch, Aachen, WGer, 75-76; guest prof, Nat Ctr Sci Res, CRM, Strasbourg, France, 83-84. *Honors & Awards:* Alexander von Humboldt US Sr Scientist Award, WGer, 75. *Mem:* Fel Am Phys Soc; fel Acoust Soc Am; Soc Rheology. *Res:* Acoustics; rheology; polymer science; macromolecules; optical and electrical properties of solutions. *Mailing Add:* Dept Mech Eng Univ Tex Austin TX 78712

THURSTON, HERBERT DAVID, b Sioux Falls, SDak, Mar 24, 27; m 51; c 3. PLANT PATHOLOGY. *Educ:* Univ Minn, MS, 53, PhD(plant path), 58. *Prof Exp:* Asst plant path, Univ Minn, 50-53, instr, 53-54; asst plant pathologist, Rockefeller Found, Colombia, 54-56; instr plant path, Univ Minn, 56-57; from assoc plant pathologist to plant pathologist, Rockefeller Found, Colombia, 58-67; PROF PLANT PATH, CORNELL UNIV, 67- *Mem:* Am Phytopath Soc; Potato Asn Am. *Res:* Diseases of potatoes; tropical plant pathology; nature of resistance to fungus diseases. *Mailing Add:* Dept Plant Path Plant Sci Bldg Cornell Univ Ithaca NY 14853

THURSTON, JAMES N(ORTON), b Murphysboro, Ill, May 6, 15; m 40; c 4. ELECTRICAL ENGINEERING. *Educ:* Ohio State Univ, BEE, 36; Mass Inst Technol, SM, 43, ScD(elec eng), 50. *Prof Exp:* Test engr, Gen Elec Co, 36-38; geophysicist, Mott-Smith Corp, 38-40; from asst to asst prof elec eng, Mass Inst Technol, 40-49; assoc prof, Univ Fla, 49-52 & Calif Inst Technol, 52-54; prof & head dept, Clemson Univ, 54-66, alumni prof elec eng, 66-80; RETIRED. *Concurrent Pos:* Consult, Ruge-Deforest, 45-49, Firestone Tire & Rubber Co, 53- & Nat Coun Eng Examr, 73-88. *Mem:* Am Soc Eng Educ; sr mem Inst Elec & Electronics Engrs. *Res:* Electronic measurement and control systems; communications systems. *Mailing Add:* 322 Woodlandway Clemson SC 29631

THURSTON, JOHN ROBERT, b Maumee, Ohio, May 6, 26; m 53; c 2. BACTERIOLOGY, IMMUNOLOGY. *Educ:* Ohio State Univ, BSc, 49, MSc, 51, PhD(bact), 55. *Prof Exp:* Res microbiologist, Ohio Tuberc Hosp, Columbus, 55-57; microbiologist, Trudeau Found, Inc, NY, 57-61; LEAD SCIENTIST, NAT ANIMAL DIS LAB, USDA, 61- *Mem:* Am Asn Vet Immunol. *Res:* Serologic investigations of mycobacteria, serologic diagnosis in tuberculosis; intradermal tuberculin testing of cattle; glycoprotein changes in tuberculous cattle; serology of nocardiosis and aspergillosis; effect of mycotoxins on the immune system; atrophic rhinitis. *Mailing Add:* 632 20th St Ames IA 50010

THURSTON, M(ARLIN) O(AKES), b Denver, Colo, Sept 20, 18; m 42; c 3. ELECTRICAL ENGINEERING. *Educ:* Univ Colo, BA, 40, MS, 46; Ohio State Univ, PhD(elec eng), 55. *Prof Exp:* Assoc prof elec eng & actg head dept, Air Force Inst Technol, 46-52; res assoc, Ohio State Univ, 52-55, from assoc prof to prof elec eng, 55-82, chmn dept, 65-77, dir, Electron Device Lab, 77-82; PRES, THURSTON-BELL ASSOCS, INC, 82- *Concurrent Pos:* Elec eng ed, Marcel Dekker, Inc, 75-; chmn bd, Nat Eng Consortium, Inc, 77-79; vpres, Neoprobe Corp, 83- *Mem:* Fel Inst Elec & Electronics Engrs; Am Phys Soc; Am Soc Eng Educ. *Res:* Solid state electron devices; biomedical engineering. *Mailing Add:* 3751 Kidka Ave Columbus OH 43220

THURSTON, ROBERT NORTON, b Kilbourne, Ohio, Dec 31, 24; m 49; c 4. WAVE PROPAGATION, ELECTROMECHANICAL INTERACTIONS. *Educ:* Ill Inst Technol, BS, 45; Ohio State Univ, MS, 48, PhD(physics), 52. *Prof Exp:* Teacher high sch, Ohio, 46-47; instr aeronaut eng, Ohio State Univ, 49-50; mem tech staff, Bell Tel Labs, Murray Hill, 52-83, BELL COMMUN RES, HOLMDEL, NJ, 84- *Mem:* Am Phys Soc; Acoust Soc Am; Inst Elec & Electronics Engrs; Soc Natural Philos; Soc Eng Sci; Optical Soc Am. *Res:* Mechanics; crystal physics; communications science; liquid crystals; guided waves and optoelectronics; physical acoustics; nonlinear propagation. *Mailing Add:* Bell Commun Res Rm 3X-223 Red Bank NJ 07701

THURSTON, RODNEY SUNDBYE, b Brooklyn, NY, Sept 17, 33; m 55; c 5. MECHANICAL ENGINEERING. *Educ:* Columbia Univ, AB, 55, BS, 56, MS, 58; Univ NMex, PhD(mech eng), 66. *Prof Exp:* Jr engr, Am Elec Power Co, NY, 56-57; teaching asst mech eng, Columbia Univ, 57-58 & Cornell Univ, 58-59; assoc group leader, 59-76, PROJ MGR, LOS ALAMOS SCI LAB, 76- *Mem:* Am Soc Mech Engrs; Sigma Xi. *Res:* Heat transfer; shock wave propagation; material damage induced by photon and neutron deposition. *Mailing Add:* Five Dakota Lane Los Alamos NM 87544

THURSTON, WILLIAM P, b Washington, DC, Oct 30, 46. GEOMETRY. *Educ:* New Col, Sarasota, BA, 67; Univ Calif, Berkeley, PhD, 72. *Hon Degrees:* Dr, Univ Paris-Sud, Orsay, 87. *Prof Exp:* Asst prof, Mass Inst Technol, 73; PROF, DEPT MATH, PRINCETON UNIV, 74- *Concurrent Pos:* Alfred P Sloan fel, 74; Ulam vis prof math, Univ Colo, 80-81; coun mem, Am Math Soc, 81-84. *Honors & Awards:* Oswald Veblen Prize in Geom, Am Math Soc, 76; Alan T Waterman Award, NSF, 79; Fields Medal, Int Math Union, 82. *Mem:* Fel Nat Acad Sci; fel Am Acad Sci; Am Math Soc (vpres, 87-88); Math Asn Am; Soc Indust & Appl Math; fel Am Acad Arts & Sci. *Mailing Add:* Dept Math Fine Hall Princeton Univ Washington Rd Princeton NJ 08544

THURSTON, WILLIAM R, b New York, NY, Nov 1, 15; m 37; c 1. GEOLOGY & SCIENCE ADMINISTRATION. *Educ:* Columbia Univ, AB, 38, AM, 43, PhD, 52. *Prof Exp:* Asst geologist, Cuban-Am Manganese Corp, Cuba, 38-39; asst geol, Brown Univ, 39-40; trustees asst, Columbia Univ, 40-42; geologist, US Geol Surv, 42-51, mining co, Mex, 51-52 & Nicaro Nickel Co, Cuba, 52-54; assoc prof, Sch Eng, La Polytech Inst, 54-55; exec secy div earth sci, Nat Acad Sci-Nat Res Coun, 55-59; staff geologist, US Geol Surv, 59-67; asst to sci adv, US Dept Interior, 67-70, spec asst to dir, US Geol Surv, 70-76; staff officer, Nat Acad Sci-Nat Res Coun, 76-77; consult geologist, 77-85; RETIRED. *Concurrent Pos:* Mem & nat del, Int Geol Cong, 64, 68 & 72. *Mem:* Fel AAAS; Geol Soc Am; Soc Econ Geol. *Res:* Economic geology; geology of pegmatites and fluorspar. *Mailing Add:* 200 Alhambra Sedona AZ 86336

THURSTONE, ROBERT LEON, b Chicago, Ill, July 29, 27; m 66. ELECTRICAL ENGINEERING. *Educ:* Ill Inst Technol, BS, 51; Univ Mo, MS 53; NC State Univ, PhD(elec eng), 65. *Prof Exp:* Instr elec eng, Duke Univ, 53-57 & NC State Univ, 57-61; from asst prof to assoc prof elec eng, Univ Ala, Huntsville, 65-88, chmn dept, 72-88; RETIRED. *Mem:* AAAS; Am Soc Eng Educ; Inst Elec & Electronics Engrs. *Res:* Energy transfer through biological tissue. *Mailing Add:* 1401 Toney Dr Huntsville AL 35802

THUT, PAUL DOUGLAS, b Exeter, NH, Feb 1, 43; m 66; c 3. PSYCHOPHARMACOLOGY. *Educ:* Hamilton Col, AB, 65; Univ RI, MS, 68; Dartmouth Col, PhD(pharmacol), 71. *Prof Exp:* Asst prof pharmacol, Col Med, Univ Ariz, 70-74; ASSOC PROF PHARM, SCH DENT, UNIV MD, 74- *Concurrent Pos:* Pharmaceut Mfrs Found grant, 72-73; consult, Vet Admin Hosp, Tucson, 72- *Mem:* Soc Neurosci; Am Soc Pharmacol & Exp Therapeut; Sigma Xi. *Res:* Psychomotor effects of l-dehydroxphenylalanine; competitive neuromuscular antagonists. *Mailing Add:* Dept Pharmacol Univ Md Sch Dent Baltimore MD 21201

THWAITES, THOMAS TURVILLE, b Madison, Wis, Aug 21, 31; m 53; c 2. NUCLEAR PHYSICS. *Educ:* Univ Wis, BS, 53; Univ Rochester, MA, 56, PhD(physics), 59. *Prof Exp:* Physicist, Stromberg Carlson Div, Gen Dynamics Corp, 55-56; from asst prof to assoc prof physics, Pa State Univ, University Park, 59-; RETIRED. *Mem:* Am Phys Soc. *Res:* Scattering of 200 million electron volts polarized protons; radioactive decay schemes; radiative capture of charged particles; scattering of charged particles. *Mailing Add:* Dept Physics 104 Davie Pa State Univ University Park PA 16802

THWAITES, WILLIAM MUELLER, b Madison, Wis, July 10, 33; m 55; c 3. GENETICS, SCIENCE EDUCATION. *Educ:* Univ Wis, BS, 55; Univ Mich, MS, 62, PhD(genetics), 65. *Prof Exp:* Asst prof, 65-70, ASSOC PROF BIOL, SAN DIEGO STATE UNIV, 70- *Concurrent Pos:* Prin investr, NSF res grant, 66-68 & 76-78; fel, Battelle Mem Inst, Richland, Washington, 72-73; vis prof, Instituto de Investigaciones Biomedicas, Mexico City, 73-74. *Mem:* Genetics Soc Am; AAAS. *Res:* General, microbial and biochemical genetics; evolution education. *Mailing Add:* Dept Biol San Diego State Univ San Diego CA 92182

THWEATT, JOHN G, b Norton, Va, Nov 21, 32; m 54; c 2. ORGANIC CHEMISTRY. *Educ:* Ga Inst Technol, BS, 54, PhD(org chem), 61. *Prof Exp:* Instr chem, Ga Inst Technol, 57-59; res chemist, Tenn Eastman Co, 60-63, sr res chemist, res labs, 63-75, sr chemist, develop dept, 75-76, tech staff, dyes dept, 76-83, tech staff, develop dept, 83-87, DEVELOP ASSOC, DEVELOP DEPT, ORG CHEM DIV, TENN EASTMAN CO, 87- *Mem:* Am Chem Soc. *Res:* Chemistry of enamines and other electron rich olefins; process development, photographically active compounds; process improvement; hydrogenation and other high pressure reactions. *Mailing Add:* Tenn Eastman Co Bldg 231 Kingsport TN 37662

THYAGARAJAN, B S, b Tiruvarur, India, July 14, 29; m 56; c 3. ORGANIC CHEMISTRY. *Educ:* Loyola Col, Madras, India, MA, 51; Presidency Col, Madras, MSc, 53, PhD(chem), 56. *Prof Exp:* Fel, Northwestern Univ, 56-58; fel, Univ Wis, 58-59; reader org chem, Madras Univ, 60-68; prof, Univ Idaho, 69-74; dir earth & phys sci div, 74-77, PROF CHEM, UNIV TEX, SAN ANTONIO, 74- *Honors & Awards:* Intrasci Res Award, 66. *Mem:* Fel Am Inst Chemists; NY Acad Sci; The Chem Soc; Sigma Xi; Soc Cosmetic Chemists. *Res:* Heterocyclic chemistry; aromaticity; molecular migrations; synthesis of organosulphur pesticides; novel reactions in sulphur chemistry; mass spectral rearrangements in organo-sulphur molecules. *Mailing Add:* Earth & Phys Sci Div Univ Tex San Antonio TX 78285

THYE, FORREST WALLACE, b Burlington, Iowa, Aug 24, 40; m 68; c 2. NUTRITION & EXERCISE. *Educ:* Cornell Univ, PhD(nutrit), 69. *Prof Exp:* ASSOC PROF NUTRIT & ENERGY METAB, VA POLYTECH INST & STATE UNIV, 81- *Mem:* Am Inst Nutrit. *Res:* Effect of diet on blood lipids and mineral balance. *Mailing Add:* Dept Human Nutrit & Foods Va Polytech Inst & State Univ Blacksburg VA 24061

THYER, NORMAN HAROLD, b Gloucester, Eng, Sept 3, 29; m 65; c 3. METEOROLOGY, SURVEYING ENGINEERING. *Educ:* Univ Birmingham, BSc, 57; Univ Wash, Seattle, PhD(meteorol), 62; BC Inst Technol, dipl, 83; Univ Calgary, MSc, 87. *Prof Exp:* Meteorol asst, Air Ministry, UK, 49-50 & Falkland Islands Dependencies Surv, 50-53; tech asst aeronaut eng, Gloster Aircraft Co, 54; micrometeorologist, Univ Wash, Seattle, 60, meteorol training expert, World Meteorol Orgn, 62-63; asst prof meteorol & physics, Univ BC, 63-66; res assoc meteorol, McGill Univ, 66-68; from asst prof to assoc prof, Notre Dame Univ, Nelson, 68-77. *Concurrent Pos:* Climatologist, 79-80; vis prof, Univ Vercruzana, Mex, 80-82; jr surv eng, Geodetic Surv Can, 88-89; forecaster, Oceanroutes (UK) Ltd, Aberdeen, Scotland, 89. *Honors & Awards:* Intera Kenting Award, Can Inst Surv & Mapping, 89. *Mem:* Royal Meteorol Soc; Am Meteorol Soc. *Res:* Studies of local winds near valleys and convectional storms; surveying engineering. *Mailing Add:* RR 2 Sproule Creek Rd Nelson BC V1L 5P5 Can

THYGESEN, KENNETH HELMER, b Cambridge, NY, June 30, 37; m 55; c 2. EXPERIMENTAL SOLID STATE PHYSICS & SURFACE PHYSICS. *Educ:* Washington & Lee Univ, 58; Clarkson Col Technol, MS, 60, PhD(physics), 67. *Prof Exp:* Instr physics, Clarkson Col Technol, 59-65; assoc prof, 67-77, chmn dept, 76-87, PROF PHYSICS, STATE UNIV NY COL POTSDAM, 77- *Concurrent Pos:* Adj prof physics, Univ Albany, NY. *Mem:* Am Phys Soc; Sigma Xi. *Res:* Atomic and molecular physics low energy ion implantation simulations; surface physics: x-ray standing wave studies of semiconductor-metal interfaces. *Mailing Add:* Dept Physics State Univ NY Col Potsdam NY 13676

THYGESON, JOHN R(OBERT), JR, b Boston, Mass, Sept 25, 24. CHEMICAL ENGINEERING. *Educ:* Drexel Inst, BS, 47; Univ Pa, MS, 55, PhD(chem eng), 61. *Prof Exp:* Res engr, Proctor & Schwartz, Inc, 47-58; asst prof, 61-63, ASSOC PROF CHEM ENG, DREXEL UNIV, 63- *Concurrent Pos:* Consult, Proctor & Schwartz, Inc, 63- *Mem:* Am Inst Chem Engrs; Am Chem Soc. *Res:* Transport phenomena; mechanisms of the reaction of hydrogen with carbon steels; simultaneous heat and mass transfer in packed beds of particulate solids. *Mailing Add:* 727 Seminole Ave Philadelphia PA 19111

THYR, BILLY DALE, b Kansas City, Kans, June 1, 32; m 58, 83; c 3. PLANT PATHOLOGY, MYCOLOGY. *Educ:* Univ Ottawa, Kans, BA, 59; Wash State Univ, PhD(plant path), 64. *Prof Exp:* Res plant pathologist, Agr Res Serv, USDA, 63-86, alfalfa res leader, 80-83; CONSULT, 83- *Mem:* Am Soc Agron; Am Phytopath Soc; Crop Sci Soc Am; Soc Nematologists. *Res:* Bacterial diseases of tomato and other vegetables; control of bacterial diseases through host resistance; alfalfa diseases; interaction of organisms infecting alfalfa; interaction of water relations and alfalfa disease. *Mailing Add:* 2230 King Edward Dr Reno NV 89503

THYSEN, BENJAMIN, b Bronx, NY, July 27, 32; m 75; c 2. LABORATORY MEDICINE, GYNECOLOGY. *Educ:* City Col New York, BS, 54; Univ Mo, MS, 63; St Louis Univ, PhD(biochem), 67. *Prof Exp:* Res asst obstet & gynec & biochem, Albert Einstein Col Med, 59-61; asst biochem, Sch Med, Univ Mo, 61-63; instr, St Louis Univ, 67-68; sr res scientist, Technicon Corp, 68-69, group leader, 69-70; asst prof, 71-86, DIR, ENDOCRINE LABS, 71-, ASSOC PROF, OBSTET & GYNEC & LAB MED, ALBERT EINSTEIN COL MED, 86- *Concurrent Pos:* Mem spec study sect Core Ctr Grants, Nat Inst Environ Health Sci, 86. *Mem:* AAAS; Endocrine Soc; Am Chem Soc; Am Asn Clin Chemists; Soc Study Reproduction; Sigma Xi. *Res:* Estrogen metabolism; testes; laboratory medicine; toxicology; methodology; reproductive endocrinology. *Mailing Add:* Dept Obstet Gynec & Lab Med Albert Einstein Col Med Bronx NY 10461

TIAB, DJEBBAR, b Ain Beida, Algeria, Sept 23, 50; m 75; c 1. PETROLEUM ENGINEERING. *Educ:* NMex Inst Mining & Technol, BSc, 74, MSc, 75; Univ Okla, PhD(petrol eng), 76. *Prof Exp:* Chief technician petrol geol, Alcore, Inc, Algeria, 69-71; res assoc petrol eng, NMex Inst Mining & Technol, 70-76, asst prof, 76-77; from asst prof to assoc prof, 77-88, PROF PETROL ENG, UNIV OKLA, 88- *Concurrent Pos:* NSF res scientist, 78-80; pres, United Petrol Tech, 79-83; sr res eng, Core Labs, 88-89; Hallibarton Lectureship Award, 87-92. *Honors & Awards:* Kerr-McGee Distinguished Lectr, 85. *Mem:* Assoc Soc Petrol Engrs. *Res:* Reservoir mechanics of enhanced oil recovery; well testing; natural gas technology; reservoir engineering; horizontal wells; rock properties; petrophysics. *Mailing Add:* Petrol & Geol Eng Sch Energy Ctr Rm T201 Univ Okla Norman OK 73019

TIAO, GEORGE CHING-HWUAN, b London, Eng, Nov 8, 33; Chinese citizen; m 58; c 4. MATHEMATICAL STATISTICS, ECONOMIC STATISTICS. *Educ:* Nat Taiwan Univ, BA, 55; NY Univ, MBA, 58; Univ Wis-Madison, PhD(econ), 62. *Prof Exp:* Asst prof statist & bus, Univ Wis-Madison, 62-65; vis assoc prof statist, Harvard Bus Sch, 65-66; assoc prof, 66-68, prof statist & bus, 68-81, BASCOM PROF STATIST & BUS, UNIV WIS-MADISON, 81- *Concurrent Pos:* Vis lectr, Post Col Prof Educ, Carnegie Mellon Univ, 66-73; statist consult to var indust co, 66-; vis prof, Univ Essex, 70-71; chmn dept statist, Univ Wis-Madison, 73-75; vis prof, Nat Taiwan Univ, 75-76; assoc ed, J Am Statist Asn, 76-81; vis Ford Found prof, Univ Chicago, Ill, 80-81. *Mem:* Fel Royal Statist Soc; fel Am Statist Asn; fel Inst Math Statist; AAAS; Int Statist Inst. *Res:* Bayesian methods in statistics; time series analysis; statistical analysis of environmental data; economic and business forecasting. *Mailing Add:* Dept Statist Univ Chicago 1118-32 E 58th St Chicago IL 60637

TIBBALS, HARRY FRED, III, b Jacksonville, Tex, Apr 7, 43; m 66; c 2. DATA SYSTEMS ENGINEERING, SIMULATION ANALYSIS. *Educ:* Baylor Univ, BS, 65; Univ Houston, PhD(chem), 70. *Prof Exp:* Appln consult phys sci & comput sci, Univ Glasgow, 72-74; acad staff, Comput Unit, Durham Univ, UK, 74-79; asst prof anal chem, NTex State Univ, 78-79; mem staff, Advan Anal Dept, Collins Commun Systs Div, Rockwell Int, 79-89; PRES, BIODIGITAL TECHNOL, INC, 89- *Concurrent Pos:* Sci Res Coun fel chem, Univ Leicester, 70-72; mem commun coord group, Scottish

Regional Comput Org, 72-74; systs programmer & adv nonnumerical appln, Comput Unit, Univ Durham, Eng, 74-; tutor & counr, Technol Fac, Open Univ, Eng, 76-78; analyst commun proj, NUmbrian Multi-Access Comput Org, UK, 75-78; fel & partic various advan study insts, NATO & Comn on Sci & Technol, 73-78; consult, Delta Mgt & Software Systs, 78-79; prin design engr, Mostek; transputer prod mgr, Inmos. *Mem:* Am Chem Soc; Asn Comput Mach; Inst Elec & Electronics Engrs; Brit Comput Soc; Royal Soc Chem. *Res:* Applications of computer systems to problems in chemical analysis; analysis of complex systems; communications, measurement and control systems; simulation; real-time interactive man-machine systems. *Mailing Add:* Biodigital Technol Inc 708 S College St Suite 204 McKinney TX 75069

TIBBETTS, CLARK, b Hartford, Conn, Feb 18, 47; m 67, 82; c 4. VIROLOGY, MOLECULAR GENETICS. *Educ:* Amherst Col, BA, 68; Calif Inst Technol, PhD(biophysics), 72. *Prof Exp:* Fel microbiol, Wallenberg Lab, Univ Uppsala, Sweden, 72-74; asst prof, Sch Med, Univ Conn, 74-80, assoc prof, 80-82; assoc prof, 82-87, PROF MICROBIOL, SCH MED, VANDERBILT UNIV, 88- *Mem:* Am Soc Biol Chemists; Am Soc Virologists; Am Soc Microbiol; AAAS. *Res:* Structure and function of viral DNA sequences which regulate gene expression; viral DNA encapsidation to transfer novel genes or arrangements of genes to animal cells in culture; large scale automated DNA sequencing technology. *Mailing Add:* Dept Microbiol Sch Med Vanderbilt Univ Nashville TN 37232

TIBBETTS, GARY GEORGE, b Omaha, Nebr, Oct 12, 39; m 64; c 3. PHYSICS. *Educ:* Calif Inst Technol, BS, 61; Univ Ill, Urbana, MS, 63, PhD(physics), 67. *Prof Exp:* Ger Res Asn grant, vis scientist, Munich Tech Univ, 67-69; SR RES PHYSICIST, GEN MOTORS RES LABS, 69-, SR STAFF RES SCIENTIST, 85- *Honors & Awards:* Graffin Lectr, Am Carbon Soc, 90. *Mem:* Am Phys Soc; Am Carbon Soc; Mat Res Soc. *Res:* Surface physics; plasma-surface interactions; adsorption of gases on surfaces; electronic and chemical properties of surfaces; carbon fibers; growth of whiskers; inorganic fibers. *Mailing Add:* Physics Dept Gen Motors Res Labs 30500 Mound Rd Warren MI 48090

TIBBETTS, MERRICK SAWYER, b Keene, NH, Dec 30, 25; m 50; c 4. ORGANIC CHEMISTRY. *Educ:* Univ NH, BS, 48, MS, 51; Stevens Inst Technol, PhD(org chem), 66. *Prof Exp:* Plant chemist, Rubber Div, Eberhard Faber Pencil Co, 51-53; sr chemist, Bendix Corp, 53-62; proj leader synthetic org chem, Int Flavors & Fragrances, 66-69; sr res chemist, Florasynth, Inc, 69-70; sr res scientist, Pepsico Inc, Long Island City, NY, 70-81; ADJ PROF CHEM, DEPT MATH & SCI, PRATT INST, 81-; ASST PROF CHEM, ST JOSEPH'S COL, BROOKLYN, NY, 86- *Concurrent Pos:* Vis prof chem, Monmouth Col, 85- *Mem:* Am Chem Soc; Sigma Xi. *Res:* Conformational analysis; organic synthesis; subjective-objective correlation; carbohydrate chemistry. *Mailing Add:* 16 N Cherry Lane Rumson NJ 07760

TIBBITS, DONALD FAY, b May 7, 43; US citizen. SPEECH PATHOLOGY. *Educ:* Univ Ark, BS, 64, MS, 68; Univ Mo-Columbia, PhD(speech path), 73. *Prof Exp:* Speech pathologist, Pub Schs, Ark, 63-70; teaching asst, Univ Mo-Columbia, 70-73; asst prof, Univ Ark, Fayetteville, 73-76; asst prof commun dis, Univ Tex, Dallas, 76-; PROF SPEECH PATH, CENT MO STATE UNIV, WARRENSBURG. *Concurrent Pos:* Mem, Coun Except Children. *Mem:* Am Speech & Hearing Asn; Asn Children Learning Disabilities. *Res:* Language acquisition and development in children and adolescents; langauge disabilities in children; gestural communications. *Mailing Add:* Dept Biol Univ Nev Reno NV 89557

TIBBITTS, FORREST DONALD, b Tacoma, Wash, Jan 23, 29; m 51; c 1. EMBRYOLOGY. *Educ:* East Wash Univ, BA, 51; Ore State Univ, MA, 55, PhD, 58. *Prof Exp:* Instr sci, Ore Col Educ, 57-59; asst prof zool, 59-69, PROF BIOL, UNIV NEV, RENO, 69-, PROF ANAT, MED SCH, 71- *Mem:* AAAS; Am Asn Anat. *Res:* Reproductive biology; placentation. *Mailing Add:* Dept Anat Sch Med Univ Nev Reno NV 89557

TIBBITTS, THEODORE WILLIAM, b Melrose, Wis, Apr 10, 29; m 55, 75, 86; c 2. ENVIRONMENTAL PHYSIOLOGY, LIFE SUPPORT IN SPACE. *Educ:* Univ Wis, BS, 50, MS, 52, PhD(hort, agron), 53. *Prof Exp:* From asst prof to assoc prof, 55-71, PROF HORT, UNIV WIS-MADISON, 71-, DIR BIOTRON, 87- *Concurrent Pos:* Res engr space biol, NAm Aviation, Inc, Calif, 65-66; bot adv, Manned Space Craft Ctr, Tex, 68-69; mem NASA rev panel, Skylab Biol Exps, 70-71 & Space Shuttle Biol Exp, 78; res leave, Lab Plant Physiol Ond, Netherlands, 74 & Climate Lab, Dept Sci & Indust Res, New Zealand, 81; vis scientist, Guelph, Can, 81; assoc ed, Am Soc Hort Sci, 87-90; co-prin investr, Commercialization Ctr Space Automation & Robotics, NASA, 87-, mem, CELSS Disciple Working Group, 89-; chmn, CIE TC6-22 Comt Light Terminology for Plants. *Honors & Awards:* Marion Meadows Award, Am Soc Hort Sci, 86. *Mem:* AAAS; fel Am Soc Hort Sci; Am Soc Plant Physiol; Int Soc Hort Sci; Am Inst Biol Sci; Am Soc Gravit Space Biol. *Res:* Environmental physiology of vegetable crops; bioregenerative life support systems for space; optimization of growth; air pollution and contaminants in enclosed environments; geophysical environment of plants; standardization in plant growth chambers; physiological breakdowns. *Mailing Add:* Biotron Univ Wis Madison WI 53706

TIBBLES, JOHN JAMES, b Toronto, Ont, Mar 16, 24; m 51; c 2. FISHERIES. *Educ:* Ont Agr Col, BSA, 51; Univ Wis, MS, PhD(fisheries), 56. *Prof Exp:* Assoc scientist, Fisheries Res Bd Can, Can Dept Fisheries & Oceans, 56-65, dir, Sea Lamprey Control Ctr, Pacific & Freshwater Fisheries, 66-86; RETIRED. *Mem:* Am Fisheries Soc. *Res:* Sea lamprey control. *Mailing Add:* 72 Bainbridge Sault Ste Marie ON P6C 2H1 Can

TIBBS, JOHN FRANCISCO, b Pacific Grove, Calif, Oct 12, 38; m 66. PROTOZOOLOGY. *Educ:* Fresno State Col, BA, 60; Univ Southern Calif, MS, 64, PhD(biol), 68. *Prof Exp:* From asst prof to assoc prof zool, 68-77, DIR BIOL STA, UNIV MONT, 70-, PROF ZOOL, 77-, CHMN, ZOOL DEPT. *Mem:* AAAS; Soc Protozool. *Res:* Ecology of Arctic and Antarctic protozoans and marine invertebrates; taxonomy of the Phaeodarina; ecology of the Sarcodina. *Mailing Add:* Dept Zool Univ Mont Missoula MT 59812

TIBBS, NICHOLAS H, b Windsor, Eng, May 31, 45; US citizen; m 68; c 3. ENVIRONMENTAL GEOLOGY. *Educ:* Univ Mo-Rolla, BS, 66, MS, 69, PhD(geol), 72. *Prof Exp:* Res fel environ geol, Univ Mo-Rolla, 72-73; asst prof geol, Paducah Community Col, Univ Ky, 73-75; geologist environ geol, Nuclear Raw Mat Br, Tenn Valley Authority, 75-78; from asst prof to assoc prof, 78-88, PROF GEOL, SOUTHEAST MO STATE UNIV, 88-, CHMN, EARTH SCI DEPT, 85- *Mem:* Sigma Xi; Geol Soc Am; Nat Asn Geol Teachers. *Res:* Environmental geology primarily in southeast Missouri. *Mailing Add:* Dept Earth Sci Southeast Mo State Univ Cape Girardeau MO 63701

TICE, DAVID ANTHONY, b Brooklyn, NY, Dec 31, 29; m 52, 85; c 4. THORACIC SURGERY, CARDIOVASCULAR SURGERY. *Educ:* Columbia Univ, BA, 51; NY Univ, MD, 55; Am Bd Surg, dipl, 61; Bd Thoracic Surg, dipl, 65. *Prof Exp:* Intern, Third Surg Div, Bellevue Hosp, 55-56, asst resident, 56-59; resident, Bellevue Hosp & asst surg, NY Univ, 59-60; res asst, Univ, 60-62, from instr to asst prof, 60-67, from asst attend to assoc attend, Univ Hosp, 60-68, assoc prof, Sch Med, 67-73, PROF SURG, SCH MED, NY UNIV, 73- *Concurrent Pos:* NY Heart Asn fel, Sch Med, NY Univ, 60-62; asst attend surg, Methodist Hosp Brooklyn, 60-63; from asst vis physician to assoc vis physician, Bellevue Hosp, 60-66, vis physician, 66-; consult cardiovasc surg, NY State Dept Health, 62 & Lutheran Med Ctr, 70; asst med examr, City of New York, 62-73; attend cardiovasc surg, New York Vet Admin Hosp, 63-66, attend thoracic surg, 64-66, chief surg serv, 67-74, chief thoracic & cardiovasc surg, 67-; attend surg, NY Univ Hosp, 68-; mem, Vet Admin Res & Educ Coun, Washington, DC, 72-73. *Mem:* Am Asn Thoracic Surg; fel Am Col Cardiol; fel Am Col Surgeons; Soc Univ Surgeons; Transplantation Soc; Soc Vascular Surg. *Res:* Thoracic and cardiovascular physiology and disease. *Mailing Add:* Dept Surg Va Med Ctr 408 First Ave New York NY 10010

TICE, LINWOOD FRANKLIN, b Salem, NJ, Feb 17, 09; m 29, 86; c 2. PHARMACEUTICAL CHEMISTRY. *Educ:* Philadelphia Col Pharm, BS, 33, MSc, 35; St Louis Col Pharm, DSc, 54. *Prof Exp:* Res fel, Wm R Warner & Co, 31-35; res fel, Edible Mfrs Res Soc, 35-38; asst prof, 38-40, dean, 59-75, dir sch pharm, 40-71, asst dean, 41-56, assoc dean, 56-59, prof pharm, 40-75, EMER DEAN, PHILADELPHIA COL PHARM, 75-, CONSULT, MEDICO-LEGAL, 75- *Concurrent Pos:* Res fel, Sharp & Dohme, 38-40; ed, Am J Pharm, 40-77; tech ed, El Farmaceutico, 41-59 & Pharm Int, 47-59; mem revision comt, US Pharmacopoeia, 40-60, bd trustees, 60-70, 70-75; dir, Am Found Pharmaceut Ed, 54-59; mem Am Coun Pharmaceut Ed, 60-66; mem, Philadelphia Med-Pharmaceut Sci, 63- *Honors & Awards:* Remington Honor Medal, Am Pharmaceut Asn, 71. *Mem:* AAAS (vpres, 71-72); Am Chem Soc; Am Pharmaceut Asn (pres elect, 65-66, pres, 66-67); Asn Cols Pharm (pres, 55-56); fel Am Inst Chem. *Res:* Pharmaceuticals; surfactants; proteins; emulsions. *Mailing Add:* Philadelphia Col Pharm & Sci 43rd St & Kingsessing Ave Philadelphia PA 19104

TICE, RAYMOND RICHARD, b Bridgeport, Conn, Jan 22, 47. GENETIC TOXICOLOGY, CYTOGENETICS. *Educ:* Univ Calif, San Diego, BA, 69; San Diego State Univ, MS, 72; Johns Hopkins Univ, PhD(human genetics), 76. *Prof Exp:* Res assoc, Brookhaven Nat Lab, 76-78, asst scientist genetic toxicol, 78-81, assoc scientist, 81-84, scientist, Dept Med, 85-90; DIR GENETICS TOXICOL & DATA MGT, INTEGRATED LAB SYSTS, 90- *Concurrent Pos:* Training grant, Dept Med, Brookhaven Nat Lab, 76-78. *Mem:* Genetic Toxicol Asn; Environ Mutagen Soc; Soc Risk Analysis. *Res:* Examination of mechanisms of cytogenetic manifestations of genotoxic damage in vivo, cytogenetic toxicological evaluation of environmental pollutants; detection of individuals at minimal risk to genotoxic agents. *Mailing Add:* Integrated Lab Systs PO Box 13501 Research Triangle Park NC 27709

TICE, RUSSELL L, b Parkersburg, WVa, Dec 5, 32; m 56; c 4. PHYSICAL CHEMISTRY. *Educ:* Marshall Col, BS, 60; Univ Calif, Los Angeles, PhD(chem), 65. *Prof Exp:* Res asst chem, Univ Calif, Los Angeles, 62-65; from asst prof to assoc prof, 65-75, PROF CHEM, CALIF POLYTECH STATE UNIV, 75- *Concurrent Pos:* NSF fel, Ind Univ, 66; vis prof, Purdue Univ, 76-77. *Mem:* Am Chem Soc. *Res:* Investigation of electron-atom and electron-molecule collision cross sections by cyclotron resonance; electron spin resonance of radicals. *Mailing Add:* Dept Chem Calif Polytech State Univ San Luis Obispo CA 93401

TICE, THOMAS E(ARL), b Florence, Ala, Jan 24, 24; m 49; c 2. ELECTRICAL ENGINEERING. *Educ:* Ohio State Univ, BEE, 47, MSc, 48, PhD(elec eng), 51. *Prof Exp:* Asst eng, Marshall Col, 43; asst math, Ohio State Univ, 47, asst elec eng, 47-48, res assoc & proj engr, Antenna Lab, 48-54, dir, 54-61, from asst prof to prof, 52-61; chief engr, Antenna & Microwave Group, Motorola, Inc, 61-67; PROF ELEC ENG & CHMN DEPT, ARIZ STATE UNIV, 67- *Mem:* Fel Inst Elec & Electronics Engrs; Sigma Xi. *Res:* Antennas; electronics; radar; microwave reflection and refraction. *Mailing Add:* 545 N Miller St Mesa AZ 85203

TICE, THOMAS ROBERT, b Rochester, NY. CONTROLLED RELEASE, MICROENCAPSULATION. *Educ:* Syracuse Univ, BS, 70, PhD(biophysics), 75. *Prof Exp:* Postdoctoral res, Univ Ala Birmingham, 75-79; res scientist microencapsulation res & develop, 79-82, sect head, 82-84, DIV HEAD CONTROLLED RELEASE RES & DEVELOP, SOUTHERN RES INST, 84- *Concurrent Pos:* Prin investr, NIH & Dept Defense, 82-; lectr, Ctr Prof Advan, 82- & Thomas Alva Edison Sci Found, 84-86; bd gov, Controlled Release Soc, 86-89; NIH, Study Sect, 86-; dir, CRS-APV Joint Workshop, 90. *Mem:* Am Asn Pharmaceut Scientists; Controlled Release Soc. *Res:* Microencapsulation and controlled-release technology as used for pharmaceutical, specialty chemicals, cosmetics, and consumer products; injectable LHRH microspheres to treat prostate and other cancers; author of various publications; granted several patents. *Mailing Add:* Southern Res Inst 2000 Ninth Ave S Birmingham AL 35205

TICHAUER, ERWIN RUDOLPH, b Berlin, Ger, Apr 27, 18; m 46. OCCUPATIONAL HEALTH & BIOMECHANICS. *Educ:* Technische Hochschule, dipl, 38; Albertus Univ, ScD(phys sci), 40. *Prof Exp:* Dep dir area team 1065, UNRRA, 46-47; engr in-chg trainiing & res, FAMIC Ltda, Chile, 47-50; works mgr design & develop, PMS Pty Ltd, Australia, 50-53; specialist lectr eng, Univ Queensland, 53-56; expert productivity, UN Tech Assistance Admin, 56-59; sr lectr indust eng, Univ New SWales, 60-64; prof, Tex Tech Univ, 64-67; res prof, 67-68, prof biomech & dir div, Ctr Safety & Inst Rehab Med, Med Ctr, 68-77, PROF BIOMECH & DIR PROGS ERGONOMICS & BIOMECH, DEPT OCCUP HEALTH & SAFETY, NY UNIV, 77- *Concurrent Pos:* Distinguished vis prof, San Marcos Univ, Lima, 58; hon consult, Royal S Sidney Hosp, 60-64; consult, UNICEF, 61, Major Corp, 64- & Waterbury Hosp, Conn, 69-; vis assoc prof, Tex Tech Univ, 63; mem, Australian Coun Rehab of Disabled, 64-; guest lectr, USPHS, 67; chmn subcomt biomech, Comt Z-94, Am Nat Stand Inst, 68-75; Am Soc Mech Engrs rep, US Nat Comt Eng, Med & Biol, Nat Acad Eng-Nat Res Coun, 69-72; mem bd trustees comt, NJ Inst Technol, 69-; chmn biomed eng, NY Acad Med, 71-72; mem comt prosthetics res in Vet Admin, Nat Res Coun, 75-77. *Honors & Awards:* Gilbreth Medal, Soc Advan Mgt; Golden Plate Award; Metrop Life Award, Nat Safety Coun. *Mem:* Fel NY Acad Sci; fel Royal Soc Health; fel Am Soc Mech Engrs; Am Inst Indust Engrs; Am Soc Eng Educ; Sigma Xi. *Res:* Occupational biomechanics, ergonomics; anatomy; medical education; medical thermography; preventive occupational medicine, safety and traumatology; functional anatomy and physiology applied to design of tasks, tools and equipment for both healthy and disabled workers; work stress on women. *Mailing Add:* 330 E 33 St Apt 12J New York NY 10016

TICHENOR, ROBERT LAUREN, b Ft Atkinson, Wis, Sept 1, 18; m 43; c 4. PHYSICAL CHEMISTRY. *Educ:* Mont State Univ, BS, 39; Harvard Univ, AM, 41, PhD(phys chem), 43. *Prof Exp:* Res chemist, Tenn Eastman Corp, 42-45; sect head, Thomas A Edison, Inc, 45-51; sr res chemist, E I du Pont de Nemours & Co, Inc, 51-85; RETIRED. *Mem:* AAAS; Electrochem Soc. *Res:* Durability of organic polymers to weathering; electrochemistry of alkaline iron and nickel oxide electrodes; development of acrylic textile fibers. *Mailing Add:* 437 Walnut Ave Waynesboro VA 22980

TICHO, HAROLD KLEIN, b Brno, Czech, Dec 21, 21; nat US. PARTICLE DETECTORS. *Educ:* Univ Chicago, PhD(physics), 49. *Prof Exp:* Asst, Univ Chicago, 42-48; from asst prof to prof physics, Univ Calif, Los Angeles, 48-83, chmn dept, 67-71, dean, Div Phys Sci, 74-83; vchancellor acad affairs, 83-89, PROF PHYSICS, UNIV CALIF, SAN DIEGO, 83- *Concurrent Pos:* Guggenheim fel, 66-67, 73-74. *Mem:* Fel Am Phys Soc; Sigma Xi. *Res:* High energy nuclear physics; elementary particles. *Mailing Add:* Dept Physics 0319 Univ Calif San Diego 9500 Gilman Dr La Jolla CA 92093

TICHY, ROBERT J, b Berwyn, Ill, June 6, 51. RELIABILITY BASED STRUCTURAL DESIGN, TOTAL QUALITY EDUCATION. *Educ:* Univ Ill, BS, 73; Colo State Univ, MS, 75; Wash State Univ, PhD(eng sci), 81. *Prof Exp:* Mgr eng res & develop, Western Wood Prod Asn, 79-86; prog mgr, Composites & Spec Prod Res & Develop, 86-89, SR ENGR, ENG SYSTS & PROD, WEYERHAEUSER CO, 89- *Mem:* Am Soc Civil Engrs; Am Soc Testing & Mat; Forest Prod Res Soc; Soc Wood Sci & Technol. *Res:* Engineering applications of materials; materials science with focus on wood and composites. *Mailing Add:* 1023 S 295th Pl Federal Way WA 98003-3716

TICKLE, ROBERT SIMPSON, b Norfolk, Va, July 31, 30; m 55, 86; c 1. NUCLEAR PHYSICS. *Educ:* US Mil Acad, BS, 52; Univ Va, MS, 58, PhD(nuclear physics), 60. *Prof Exp:* From instr to assoc prof, 60-68, PROF PHYSICS, UNIV MICH, ANN ARBOR, 68- *Mem:* Am Phys Soc; Am Asn Physics Teachers. *Res:* Measurement of photonuclear cross sections; accelerator design and development; study of nuclear structure with charged particle experiments; study of reaction mechanisms using heavy ions. *Mailing Add:* Dept Physics Univ Mich Ann Arbor MI 48109

TICKNOR, LELAND BRUCE, b Centralia, Wash, May 9, 22; m 52. PHYSICAL CHEMISTRY. *Educ:* Univ Wash, BS, 44; Mass Inst Technol, PhD(phys chem), 50. *Prof Exp:* Mem staff, dept metall, Mass Inst Technol, 50-52; instr chem, Swarthmore Col, 52-54; phys chemist & sect leader, Res & Develop Dept, Viscose Div, FMC Corp, 54-65; sr res chemist, Appl Res Lab, US Steel Corp, 65-68; sr res chemist, BASF Corp, 68-75, group leader, Acrylic Fibers Res Dept, Fibers Div, 75-85; instr chem, William & Mary Col, 85-88; RETIRED. *Mem:* Am Chem Soc. *Res:* Thermodynamics of solutions; retained energies of cold working in metals; cellulose chemistry; cellulose fibers; metal coatings; acrylic fibers. *Mailing Add:* 1202 Pinchurst Rd Staunton VA 24401

TICKNOR, ROBERT LEWIS, b Portland, Ore, Oct 26, 26; m 50; c 3. ORNAMENTAL HORTICULTURE. *Educ:* Ore State Univ, BS, 50; Mich State Univ, MS, 51, PhD(pomol), 53. *Prof Exp:* Asst hort, Mich State Univ, 50-53; from asst prof to assoc prof nursery culture, Univ Mass, 53-59; assoc prof, 59-69, PROF HORT, ORE STATE UNIV, 69- *Honors & Awards:* Gold Medal, Am Rhododendron Soc, 70; Jackson Dawson Gold Medal, Mass Hort Soc, 83; C J Alley Award, Int Plant Propagators Soc, 85. *Mem:* Am Rhododendron Soc (sec-treas, 64-71, pres, 71-73); Weed Sci Soc Am; Int Plant Propagators Soc; Am Hort Soc. *Res:* Chemical weed control; plant nutrition, propagation, breeding and materials. *Mailing Add:* North Willamette Exp Sta 15210 NE Miley Aurora OR 97002

TICKU, MAHARAJ K, b India, March 19, 48. PHARMACOLOGY. *Educ:* Birla Inst Technol & Sci, India, BS, 69; Univ Okla, MS, 72; State Univ NY, Buffalo, PhD(biochem), 76. *Prof Exp:* Teaching asst pharmacol, Univ Okla, 71-72; grad asst, State Univ NY, Buffalo, 72-75; fel, Univ Calif, Riverside, 76-78; ASST PROF PHARMACOL, UNIV HEALTH SCI CTR, SAN ANTONIO, 78-, ASST PROF PSYCHIAT, 81- *Concurrent Pos:* Res starter grant, Pharmaceut Mfg Asn, 79; travel award, NSF, 81. *Mem:* Am Soc Pharmacol & Exp Therapeut; Soc Neurosci; Europe Brain & Behav Soc; Brit Brain Res Asn; Sigma Xi. *Res:* Molecular pharmacology of synaptic transmission; molecular mechanisms of depressant and convulsant drugs; y-aminobutyic acid synaptic pharmacology; y-aminobutyic acid pharmacology; experimental hypertension; drug receptor studies. *Mailing Add:* Dept Pharmacol Univ Tex Health Sci Ctr 7703 Floyd Curl Dr San Antonio TX 78284

TIDBALL, CHARLES STANLEY, b Geneva, Switz, Apr 15, 28; US citizen; m 52. COMPUTER ASSISTED INSTRUCTION, INFORMATION RETRIEVAL. *Educ:* Wesleyan Univ, BA, 50; Univ Rochester, MS, 52; Univ Wis, PhD(physiol), 55; Univ Chicago, MD, 58. *Prof Exp:* Asst physiol, Univ Wis, 52-55; res asst surg, Univ Chicago, 55-56, asst physiol, 56-58, res asst, 57; intern, Madison Gen Hosp, Wis, 58-59; physician, Mendota State Hosp, 59; asst res prof, George Wash Univ, 59-63, from assoc prof to prof, 63-65, from actg chmn to chmn dept, 63-71, dir comput assisted educ, 73-78, res prof med, 72-80, Henry D Fry Prof physiol, 65-84, prof educ, Sch Educ, 81-84, PROF COMPUT MED, MED CTR, GEORGE WASHINGTON UNIV, 84-, PROF NEUROL SURG, 90- *Concurrent Pos:* USPHS fel, 60-61, USPHS res career develop award, 61-63; consult to var hosp, indust, col & govt orgn, 60- *Mem:* Asn Comput Mach; Am Physiol Soc; Asn Am Univ Prof. *Res:* Computer assisted education; medical computer utilization; medical education; computer literacy; information retrieval; small college database. *Mailing Add:* Dept Comput Med 2300 K Street NW Washington DC 20037

TIDBALL, M ELIZABETH PETERS, b Anderson, Ind, Oct 15, 29; m 52. MEDICAL PHYSIOLOGY, SCIENCE EDUCATION. *Educ:* Mt Holyoke Col, BA, 51; Univ Wis, MS, 55, PhD(physiol), 59; Wesley Theological Sem, MTS, 90. *Hon Degrees:* ScD, Wilson Col, 73; DSc, Trinity Col, 74, Cedar Crest Col, 77, Univ of the South, 78 & Goucher Col, 79; LHD, Mt Holyoke Col, 76, Skidmore Col, 84, Marymount Col, Tarrytown, 85, Converse Col, 85 & Mt Vernon Col, 86; LLD, St Joseph Col, 83; HHD, St Mary's Col, 77, Hood Col, 82; DLitt, Regis Col & Col St Catherine, 80; DSc, St Mary of the Woods Col, 86; Alverno Col, Litt D, 89. *Prof Exp:* Asst physiol, Univ Wis, 52-55, 58-59; asst histochem, Univ Chicago, 55-56, physiol, 56-58; USPHS fel, Nat Heart Inst, 59-61; staff pharmacologist, Hazelton Labs, Inc, 61-62; assoc physiol, George Washington Univ, 60-62, asst res prof pharmacol, 62-64, assoc res prof physiol, 64-70, res prof, 70-71, PROF PHYSIOL, GEORGE WASHINGTON UNIV, 71- *Concurrent Pos:* Shattuck Fel, 55-56; Mary E Woolley fel, Mt Holyoke Col, 58-59; postdoctoral fel, USPHS, 59-61; consult, Hazelton Labs, Inc, 62-63; assoc sci coordr, Food & Drug Admin, Sci Assocs Training Prog, 66-67; consult, Food & Drug Admin, 66-68; trustee, 68-73, vchmn, 72-73, Mt Holyoke Col & vchmn, 78-85, chmn trustee acad affairs comt, Sweet Briar Col, 82-85; trustee, 72-84 & 86-, exec comt mem, 78-84 & 88-, chmn trustee acad affairs comt, Hood Col, 74-82 & 88-; nat adv comt, NIH training progs & fels, Nat Acad Sci, 72-75; exec secy, 74-75, vchmn, 77-82, NAS-Nat Res Coun comn on human resources in comt educ & employment of women in sci & eng; consult, inst res, Wellesley Col, 74-75 & Woodrow Wilson Nat Fel Found, 74-; chmn task force on women, Am Physiol Soc, 73-80; consult & chmn rev panels, NSF, 74-; trustee, 79-85, chmn vd, 83-85, scholar in residence, Col Preachers, 89; mem gov bd, exec comt, Washington Cathedral Found, 83-85; Lucie Stern distinguished vis trustee prof natural sci, Mills Col, 80; nat panelist, Am Coun Educ, 83-; assoc res, Nat Resource Ctr, Girls Clubs Am, 83-; consult ed, J Higher Educ, 84; distinguished scholar in residence, Southern Methodist Univ, 85, Salem Col, 85; bd visitors, Salem Col, 86-; founder & dir, Summer Sem for Women, 87-; bd trustees, 88- & chair, acad affairs comt, 90- *Mem:* AAAS; Am Physiol Soc; Am Asn Higher Educ; Sigma Xi. *Res:* Neurotransmitters; environments for the education of women; institutional research; science literacy; human resources development. *Mailing Add:* 4100 Cathedral Ave NW Washington DC 20016

TIDMAN, DEREK ALBERT, b London, Eng, Oct 18, 30; US citizen; m 59; c 2. PLASMA MASS LAUNCH TECHNOLOGY. *Educ:* London Univ, BSc, 52, PhD(physics), 56. *Prof Exp:* Res fel cosmic radiation, Sydney Univ, Australia, 56-57; asst prof plasma physics, Fermi Inst, Univ Chicago, 57-60; res assoc prof, Inst Phys Sci & Technol, Univ Md, 61-64, res prof, 64-80; PRES, GT-DEVICES, SUBSID GEN DYNAMICS, 80- *Concurrent Pos:* Consult, Goddard Space Flight Ctr, NASA, 61-69 & Lawrence Livermore Lab, 77-80; vis scientist, Harvard Col Observ, 66; mem, Assoc Eval Comt, Nat Acad Sci, 66-69; assoc ed, Physics Fluids, 70-72 & J Math Physics, 72-74; prin investr, GT-Devices, Subsid Gen Dynamics, 89- *Mem:* Fel Am Phys Soc; Inst Elec & Electronics Engrs. *Res:* Plasma-driven mass launchers including both electrothermal and electromagnetic launchers; plasma physics generally including kinetic theory, turbulence, shock waves, instabilities, and applications to space physics and thermonuclear plasmas; co-author of three books and over 80 journal publications; granted ten US patents. *Mailing Add:* 6801 Benjamin St McLean VA 22101

TIDMORE, EUGENE F, b Ballinger, Tex, Oct 1, 40. ANALYSIS, LINEAR OPTIMIZATION. *Educ:* Okla State Univ, PhD(math), 68. *Prof Exp:* Asst prof math, Tex Tech, 67-71; assoc prof, 71-79, PROF MATH, BAYLOR UNIV, 80- *Mailing Add:* Baylor Univ Waco TX 76798

TIDWELL, EUGENE DELBERT, b Lehi, Utah, Sept 5, 26; m 48; c 8. INSTRUMENTATION, OPTICS. *Educ:* Brigham Young Univ, BS, 51. *Prof Exp:* Asst gas chemist copper smelt, Kennecott Copper Corp, 44-45 & 47-52; staff physicist spectros, Nat Bur Standards, 52-63; staff engr instrumentation, Aro, Inc, 63-85; GEN ENGR HIGH ENDOATMOSPHERIC INTERCEPTOR, US ARMY STRATEGIC DEFENSE COMMAND, 85- *Concurrent Pos:* Prog mgr flying optical platform, Infrared Instrumentation Syst. *Res:* Theoretical spectral analysis of light gases; environmental erosion research for aeroballistics and space engineering; missile boost and reentry phases infrared and visible signature data research. *Mailing Add:* 13002 Astalot Dr Huntsville AL 35803-1802

TIDWELL, THOMAS TINSLEY, b Atlanta, Ga, Feb 20, 39; m 71. ORGANIC REACTION MECHANISMS, SYNTHETIC METHODS. *Educ:* Ga Inst Technol, BS, 60; Harvard Univ, AM, 63, PhD(reaction mechanisms), 64. *Prof Exp:* NIH fel, Univ Calif, San Diego, 64-65; asst prof chem, Univ SC, 65-72; assoc prof chem, Scarborough Col, Univ Toronto, 72-77, assoc dean, 79-82, actg chmn phys sci, 88-89, PROF CHEM, SCARBOROUGH COL, UNIV TORONTO, 77- *Concurrent Pos:* vis prof, Stanford Univ, 79; Nato fel, Turkey, 81-82, France, 85-86; vis fel, Syntex Corp, 82-83; Nat Acad Sci fel, Bulgaria, 83, USSR, 89; Nat Sci & Eng Res Coun fel, Italy, 88; fel, Japan Soc Promising Sci, 89. *Mem:* Am Chem Soc; Chem Inst Can. *Res:* Steric strain; peroxides; enolates; carbonium ions; ketenes; fluorocarbons; oil from wood; organic conductors; radiochemistry. *Mailing Add:* Dept Chem Scarborough Campus Univ Toronto West Hill ON M1C 1A4

TIDWELL, WILLIAM LEE, b Greenville, SC, Jan 14, 26; m 46; c 1. MICROBIOLOGY. *Educ:* Univ SC, BS, 45; Univ Hawaii, MS, 48; Univ Calif, Los Angeles, PhD(microbiol), 51. *Prof Exp:* Asst bact, Univ Hawaii, 46-48 & Univ Calif, Los Angeles, 48-51; asst prof biol, Agr & Mech Col, Tex, 51-55, asst res bact, Eng Exp Sta, 52-55; asst prof biol, 55-58, assoc prof bact, 58-62, chmn microbiol area, 78-82, chmn dept biol sci, 83-85, prof, 62-88, EMER PROF MICROBIOL, SAN JOSE STATE UNIV, 88- *Concurrent Pos:* Consult, State Col Affairs, Calif State Employees Asn, 66-68; vpres, Calif Fac Asn, 81-85. *Mem:* Am Soc Microbiol; NY Acad Sci. *Res:* General bacteriology; water and sewage bacteriology. *Mailing Add:* 40649 Crystal Dr Three Rivers CA 93271

TIECKELMANN, HOWARD, b Chicago, Ill, Oct 29, 16; m 42; c 5. BIO-ORGANIC CHEMISTRY. *Educ:* Carthage Col, BA, 42; Univ Buffalo, PhD(org chem), 48. *Prof Exp:* Chemist, Armour Res Found, Ill, 46; from instr to prof, 46-61, chmn dept, 70-74, DISTINGUISHED TEACHING PROF CHEM, STATE UNIV NY BUFFALO, 75- *Concurrent Pos:* Lectr, Inst Teknologi Mara, Malaysia, 88-91. *Mem:* AAAS; Royal Soc Chem; Am Inst Chemists; Am Chem Soc. *Res:* Heterocyclic compounds; pyridines; pyrimidines, alkylations and rearrangements; natural products. *Mailing Add:* 59 Chaumont Dr Williamsville NY 14221

TIEDCKE, CARL HEINRICH WILHELM, b Lübeck, Ger, Mar 1, 03; nat US; m 48. CHEMISTRY, TOXICOLOGY. *Educ:* Hamburg Univ, PhD(chem), 28. *Prof Exp:* Consult, US Army, US Navy & USPHS; adv air pollution control, various industs & govts, SAm & Africa; RETIRED. *Concurrent Pos:* Concurrent. *Mem:* AAAS; Am Chem Soc; Am Microchem Soc; NY Acad Sci; Austrian Asn Microchem & Analytical Chem. *Res:* High vacuum distillation and countercurrent distribution; vegetable oils; coffee constituents; applied microchemistry; foods; pharmacology; nutrition. *Mailing Add:* Carrillo Hotel 420 Santa Barbara CA 93101

TIEDEMANN, ALBERT WILLIAM, JR, b Baltimore, Md, Nov 7, 24; m 53; c 4. QUALITY ASSURANCE. *Educ:* Loyola Col, Md, BS, 47; NY Univ, MS, 49; Georgetown Univ, PhD(chem), 58. *Prof Exp:* Instr chem, Mt St Agnes Col, 50-55; sr res chemist, Emerson Drug Co, 55-56, chief chemist, Emerson Drug Co Div, Warner-Lambert Pharmaceut Co, 56-60; supvr analytical group, Allegany Ballistic Lab, Hercules, Inc, 61-68, supt tech serv, Radford Army Ammunition Plant, Hercules Inc, 68-72; DIR, CONSOLIDATED LABS, COMMONWEALTH VA, 72- *Concurrent Pos:* Mem, drinking water lab working group, US Environ Protection Agency, 76-77 & 86-88; vchmn, Va Toxic Substance Adv Bd, 81-; chmn, Sci & Technol Comt, Asn Food & Drug Officials, 82-85, secy-treas, 85-87; lectr, lab qual assurance, Asn Off Analytical Chemists, 84-87, bd dirs, 86-90; exec bd, Cent Atlantic States Asn Food & Drug Officials, 77-84, vpres, 81-82, pres, 82-83. *Mem:* Am Soc Qual Control; fel Am Inst Chemists; Am Mgt Asn; Asn Food & Drug Officials (secy-treas, 85-87); Asn Official Analytical Chemists; Analytical Lab Mgr Asn; NY Acad Sci. *Res:* Development of analytical methods. *Mailing Add:* 10511 Cherokee Rd Richmond VA 23219

TIEDEMANN, HERMAN HENRY, b New York, NY, July 25, 17; m 44; c 3. ELECTROCHEMISTRY, PETROCHEMICALS. *Educ:* City Col New York, BChE, 39, MChE, 40. *Prof Exp:* From chem operator to group leader develop eng, Rohm & Haas Co, 40-48; sr chem engr, GAF Corp, 48-49, supv chem engr, 49-51, actg mgr chem eng, 51-52, supvr process engr, 52-57, supt ethylene chem prod, 58-60, prod mgr, heavy chems, 60-65, plant mgr, 66-70, div mgr prod control, 70-71; prod mgr, Linden Chlorine Prod, Inc, 72-74; vpres, PCF Machinery Corp, 74-75; CONSULT, 76- *Concurrent Pos:* eng instr, Univ Penn; chem & eng instr, Col Mainland, Tex City. *Mem:* Am Inst Chem Engrs. *Res:* Liquid-liquid extraction columns; use of wetting agents to improve column performance; dyes, detergent, beverages and culinary mixes; continuous anodes for chlorine cells; calcium hypochlorite. *Mailing Add:* 16 11th Ave N PO Box 28 Texas City TX 77592-0028

TIEDEMANN, WILLIAM HAROLD, b June 3, 43; c 3. ELECTROCHEMICAL ENGINEERING. *Educ:* Univ Calif, Los Angeles, BS, 66, MS, 68, PhD(electro chem eng), 71. *Prof Exp:* Sr electrochemist, Globe Union, Inc, 71-76, mgr electrochem res, 76-79; assoc dir res, 79-85, DIR CHEM & MAT RES, JOHNSON CONTROLS, INC, 85- *Mem:* Electrochem Soc. *Res:* Theoretical mathematical modeling; experimental investigations of a variety of electrochemical systems; system simulation and scale-up. *Mailing Add:* 5757 N Green Bay Ave Milwaukee WI 53201

TIEDERMAN, WILLIAM GREGG, JR, b Tulsa, Okla, Jan 29, 38; m 63; c 3. MECHANICAL ENGINEERING, FLUID MECHANICS. *Educ:* Stanford Univ, BS, 60, MS, 61, PhD(mech eng), 65. *Prof Exp:* Engr, Shell Develop Co, 65-68; from asst prof to prof mech eng, Okla State Univ, 68-78; PROF MECH ENG, PURDUE UNIV, WEST LAFAYETTE, 78-, ASST DEAN ENG, 89- *Mem:* Am Soc Mech Engrs; Am Phys Soc; Sigma Xi. *Res:* Turbulence; viscous fluid mechanics; laser velocimetry; solid-liquid and liquid-liquid separation; drag reduction. *Mailing Add:* 623 Ridgewood Dr West Lafayette IN 47906

TIEDJE, J THOMAS, b Sarnia, Ont, Feb 11, 51; m 77; c 3. SOLID STATE PHYSICS, ELECTRONICS ENGINEERING. *Educ:* Univ Toronto, BASc, 73; Univ BC, MSc, 75, PhD(solid state physics), 77. *Prof Exp:* From res physicist to group head, Corp Res Lab, Exxon Res & Eng Co, Annandale, NJ, 77-87; assoc prof, 87-89, PROF, DEPT PHYSICS & ELEC ENG, UNIV BC, 89- *Concurrent Pos:* Steacie fel, Natural Sci & Eng Res Coun, 90- *Honors & Awards:* Herzberg Prize, Can Asn Physicists, 89. *Mem:* Can Asn Physicists; fel Am Phys Soc; Inst Elec & Electronics Engrs. *Res:* Molecular beam epitaxy growth of III-V semiconductor materials and optoelectronic devices; scanning tunneling microscopy studies of surfaces in UHV and STM instrumentation; photoemission, x-ray absorption studies of semiconductor surfaces with synchrotron radiation. *Mailing Add:* Dept Physics Univ BC Vancouver BC V6T 1Z1 Can

TIEDJE, JAMES MICHAEL, b Newton, Iowa, Feb 9, 42; m 65; c 3. MICROBIAL ECOLOGY, SOIL MICROBIOLOGY. *Educ:* Iowa State Univ, BS, 64; Cornell Univ, MS, 66, PhD(soil microbiol), 68. *Prof Exp:* From asst prof to assoc prof, 68-78, PROF MICROBIAL ECOL, MICH STATE UNIV, 78-; DIR, SCI & TECHNOL CTR MICROBIAL ECOL, NSF, 88- *Concurrent Pos:* Eli Lilly career develop grant, 74; vis assoc prof, Univ Ga, 74-75; ed, Appl Microbiol, 74-, ed in chief, 80-; consult, NSF, 74-77; vis prof, Univ Calif, Berkeley, 81-82; Sigma Xi jr res award, 81; mem, Biotech Sci Adv Comt, GPA, 86-89, chair, Sci Adv Panel, GPA, 88-90; fel Int Inst Biotechnol. *Honors & Awards:* Res Award, Soil Sci Soc Am. *Mem:* Am Soc Microbiol; fel Am Soc Agron; Soil Sci Soc Am; fel AAAS; Sigma Xi; Ecol Soc Am. *Res:* Denitrification; microbial metabolism of organic pollutants; molecular microbiol ecology. *Mailing Add:* Dept Crop & Soil Sci Mich State Univ East Lansing MI 48824

TIEFEL, RALPH MAURICE, b Brazil, Ind, Sept 3, 28; m 66. BOTANY. *Educ:* Cent Mo State Col, BS, 53; Univ Mo, MA, 55, PhD(bot), 57. *Prof Exp:* Assoc prof, 57-59, PROF BIOL, CARTHAGE COL, 60- *Mem:* Soc Econ Bot; Torrey Bot Club. *Res:* Soil and plant relationships; meristems; history of science. *Mailing Add:* Dept Biol Lentz Hall Carthage Col 2001 Alford Dr Kenosha WI 53141

TIEFENTHAL, HARLAN E, b Kalamazoo, Mich, Aug 23, 22; m 42; c 4. ORGANIC CHEMISTRY. *Educ:* Kalamazoo Col, BA, 44; Univ Ill, MS, 48; Mich State Col, PhD(chem), 50; Univ Chicago, MBA, 65. *Prof Exp:* Fel chem, Mellon Inst, 50-54, group leader, Koppers Co, Inc, 54-58, mgr fine chem group, 58-62; asst res dir, Armour Indust Chem Co, 62-65, dir spec projs, 66-68; V PRES, MDM CHEM, INC, 76-; Pres, Western Springs Corp, 80-87. *Concurrent Pos:* Owner, Tiefenthal Assocs, 68-; exec dir, Assoc Cols Chicago Area, 69-75; lectr, George Williams Col, 72-75; pres, Deep Valley Chem Co, 73-89, UKiah Corp, 89- *Mem:* Am Chem Soc; Soc Tribologists & Lubricating Engrs. *Res:* Grignard reactions; dehydrogenation reactions; dehydrocyclizations; polymerization; alkylation; hydrogenation; organic synthesis; amination; aliphatic amine chemistry; corrosion inhibitors, metal working lubricants. *Mailing Add:* 4544 Grand Ave Western Springs IL 60558

TIEH, THOMAS TA-PIN, b Peking, China, May 2, 34; US citizen; m 62; c 4. MINERALOGY. *Educ:* Univ Ill, Urbana, BS, 58; Stanford Univ, MS, 60, PhD(geol), 65. *Prof Exp:* Geologist, Bear Creek Mining Co, 60; res assoc geol, Univ Hawaii, 62-63; asst cur mining & petrol, Stanford Univ, 65-66; asst prof, 66-71, assoc prof, 71-81, PROF GEOL, TEX A&M UNIV, 81- *Concurrent Pos:* Welch Found grant, Tex A&M Univ, 69-72. *Mem:* Brit Mineral Soc; Soc Econ Paleontologists & Mineralogists. *Res:* Petrology and geochemistry. *Mailing Add:* Dept Geol Tex A&M Univ College Station TX 77843

TIELEMAN, HENRY WILLIAM, b Rotterdam, Neth, May 26, 33; Can citizen; m 62; c 2. FLUID MECHANICS. *Educ:* Ont Agr Col, BSA, 61; Univ Toronto, BAS, 62; Univ Iowa, MS, 64; Colo State Univ, PhD(civil eng), 69. *Prof Exp:* Asst prof eng mech, 69-72, assoc prof, 72-81, PROF ENG SCI & MECH, VA POLYTECH INST & STATE UNIV, 81- *Mem:* Am Soc Mech Engrs; Sigma Xi. *Res:* Fluid mechanics; turbulent boundary layers; theory of turbulence; turbulence measurements. *Mailing Add:* PO Box 37 Riner VA 24149

TIEMAN, SUZANNAH BLISS, b Washington, DC, Oct 10, 43; m 69. DEVELOPMENTAL NEUROBIOLOGY, VISION. *Educ:* Cornell Univ, AB, 65; Stanford Univ, PhD(psychol), 74. *Prof Exp:* Nat Eye Inst fel, Dept Anat, Univ Calif Med Ctr, San Francisco, 74-77; res assoc, 77-90, SR RES ASSOC, NEUROBIOL RES CTR, STATE UNIV NY ALBANY, 90- *Concurrent Pos:* Adj asst prof biol, State Univ NY, Albany, 77-84, adj assoc prof, 84-87, res assoc prof, 87-90, res prof, 90-; prin investr, Nat Eye Inst res grant, 79-83; NSF res grants, 83-86, 88-91; assoc prof biomed sci, St Univ NY, Albany, 88. *Mem:* Am Asn Anatomists; Asn Res Vision & Ophthal; AAAS; Soc Neurosci; Asn Women in Sci. *Res:* Structure and function of the visual system; anatomical and behavioral effects of restricted early visual experience. *Mailing Add:* Neurobiol Res Ctr State Univ NY 1400 Washington Ave Albany NY 12222

TIEMANN, JEROME J, b Yonkers, NY, Feb 21, 32; m 57. APPLIED MATHEMATICS, SOLID STATE PHYSICS. *Educ:* Mass Inst Technol, ScB, 53; Stanford Univ, PhD(physics), 60. *Prof Exp:* Asst, Stanford Univ, 53-55 & 56-57; PHYSICIST, RES & DEVELOP CTR, GEN ELEC CO, 57- *Concurrent Pos:* Consult, Radiation Lab, Univ Calif, 55-57; Coolidge fel Gen Elec Corp Res & Develop Ctr. *Mem:* fel Inst Elec & Electronics Engrs; Nat Acad Eng. *Res:* Quantum mechanics; electronics; solid state electronic device phenomena; electronic circuit and system design; signal processing circuit and system design. *Mailing Add:* Gen Elec Co MS KWC 1307 PO Box 8 Schenectady NY 12301

TIEMSTRA, PETER J, b Chicago, Ill, Oct 7, 23; m 46; c 3. FOOD CHEMISTRY. *Educ:* Northwestern Univ, BS, 44; Univ Chicago, BS, 47, MS, 48. *Prof Exp:* Anal chemist, Swift & Co, 48-49, anal res chemist, 49-53, from asst head chemist to head chemist, 53-63, head res & develop, Stabilizer Div, 63-65; res chemist, Derby Foods, Inc, Ill, 65-66, tech dir, 66-69; dir res & qual assurance, Swift Grocery Prod Co, Chicago, 69-70, dir spec serv, 70-76; dir res & qual assurance, F&F Labs, Chicago, 77-78; CORP QUAL ASSURANCE LAB MGR, WM WRIGLEY JR CO, CHICAGO, 79- *Mem:* Am Chem Soc; Am Oil Chem Soc; Am Soc Qual Control; Inst Food Technologists; Am Asn Candy Technologists. *Res:* Quality control; product research and development. *Mailing Add:* 6543 Pontiac Dr LaGrange IL 60525-4351

TIEN, C(HANG) L(IN), b Wuhan, China, July 24, 35; nat US; m 59; c 3. HEAT TRANSFER. *Educ:* Nat Taiwan Univ, BS, 55; Univ Louisville, MME, 57; Princeton Univ, MA & PhD(mech eng), 59. *Prof Exp:* Actg asst prof mech eng, Univ Calif, Berkeley, 59-60, from asst prof to assoc prof, 60-68, chmn, Thermal Systs Div, 69-72, Dept Mech Eng, 74-81, vchancellor res, 83-85, PROF MECH ENG, UNIV CALIF, BERKELEY, 68-, CHANCELLOR, 90- *Concurrent Pos:* Consult, Lockheed Missile & Space Co, 63-80 & Gen Elec Co, 72-80; Guggenheim fel, 65-66; res assoc prof, Miller Inst Basic Res Sci, Univ Calif, 67-68; assoc ed, J Quant Spectros & Radiative Transfer, 71-; US scientist award, Alexander von Humboldt Found, 79; Japan Soc Promo Sci fel, 80; hon res prof, Inst Thermophysics, Chinese Acad Sci, 81; ed, Int J Heat & Mass Transfer, 81-, Int Commun in Heat & Mass Transfer, 81- & Exp Heat Transfer, 87-; vis Viola D Hank chair prof, Univ Notre Dame, 83; distinguished lectr, Am Soc Mech Engrs, 87-, vpres basic eng, 88-; mem, Int Affairs Adv Comt, Nat Acad Eng, 87-90 & Mech Eng Peer Comt, 87-90, chair, 88-90; exec vchancellor & distinguished prof, Univ Calif, Irvine, 88-90; consult, numerous indust, govt & educ orgn. *Honors & Awards:* Heat Transfer Mem Award, Am Soc Mech Engrs, 74, Gustus L Larson Mem Award, 75 & Max Jakob Mem Award, 81; Thermophysics Award, Am Inst Aeronaut & Astronaut, 77; Prince Distinguished Lectr, Ariz State Univ, 83; Hawkins Mem Lectr, Purdue Univ, 87. *Mem:* Fel Nat Acad Sci; Nat Acad Eng; fel Am Inst Aeronaut & Astronaut; fel Am Soc Mech Engrs; fel AAAS. *Res:* Heat transfer; radiative heat transfer; thermal insulation and enclosure convection; reactor safety heat transfer. *Mailing Add:* Off Chancellor Univ Calif 700 University Hall Berkeley CA 94720

TIEN, CHI, b Peking, China, Oct 8, 30; US citizen; m 60; c 2. CHEMICAL ENGINEERING. *Educ:* Nat Taiwan Univ, BSc, 52; Kans State Univ, MSc, 54; Northwestern Univ, Evanston, PhD(chem eng), 58. *Prof Exp:* Asst prof chem eng, Univ Tulsa, 57-59; from asst prof to assoc prof, Univ Windsor, 59-63; assoc prof, 63-66, PROF CHEM ENG, SYRACUSE UNIV, 66-, CHMN DEPT CHEM ENG & MAT SCI, 70- *Concurrent Pos:* Expert, US Army Cold Regions Res & Eng Lab, Hanover, NH, 59-64 & 69- *Mem:* AAAS; Am Inst Chem Engrs; Chem Inst Can. *Res:* Advanced treatment of waste water, especially by filtration and adsorption; heat transfer with phase change. *Mailing Add:* Dept Chem Eng Syracuse Univ 320 Hinds Hall Syracuse NY 13244

TIEN, H TI, b Peking, China, Feb 1, 28; US citizen; m 53; c 4. BIOPHYSICS. *Educ:* Univ Nebr, BS, 53; Temple Univ, MA, 60, PhD(chem), 63. *Prof Exp:* Proj engr, Allied Chem Corp, 56-57; med scientist, Eastern Pa Psychiat Inst, 57-63; assoc prof chem, Northeastern Univ, 63-66; PROF BIOPHYS & CHMN DEPT, MICH STATE UNIV, 66- *Concurrent Pos:* Grants, Res Corp, 64-65, NIH, 64-, Off Saline Water, US Dept Interior, 68-71 & Dept Energy, 80-82; hon prof, Academia Sinica. *Mem:* AAAS; Am Chem Soc; Biophys Soc. *Res:* Physical chemical investigations of membranes, particularly bilayer lipid membranes; photosynthesis and vision; solar energy conversion; specific electrodes; ion-exchange equilibria; bilayer lipid membranes as models of biological membranes; biomolecular electronics devices. *Mailing Add:* Dept Physiol Mich State Univ East Lansing MI 48824

T'IEN, JAMES SHAW-TZUU, b China, Mar 8, 42; m 67; c 2. COMBUSTION, PROPULSION. *Educ:* Nat Taiwan Univ, BS, 63; Purdue Univ, Lafayette, MS, 66; Princeton Univ, PhD(aeronaut & mech eng), 71. *Prof Exp:* Res assoc, 70-71, from asst prof to assoc prof, 71-82, PROF ENG, CASE WESTERN RESERVE UNIV, 82- *Mem:* Combustion Inst; Am Inst Aeronaut & Astronaut; Am Soc Mech Engrs. *Res:* Combustion and chemically-reacting flows; propulsion and fire research. *Mailing Add:* Dept Mech & Aerospace Eng Case Western Reserve Univ Cleveland OH 44106

TIEN, JOHN KAI, b Chungking, China, June 4, 40; US citizen; m 71; c 2. HIGH TEMPERATURE ALLOYS. *Educ:* Worcester Polytech Inst, BSME, 62; Yale Univ, MEng, 64; Carnegie-Mellon Univ, MS, 68, PhD(metall & mat sci), 69. *Prof Exp:* Res supvr metall, Chase Brass & Copper Co, 64-65; sr res assoc, Pratt & Whitney Aircraft, 68-71; from assoc prof to prof metall & mat sci, Columbia Univ, 71-84, Howe chair prof, 84-89, dir, Ctr Strategic Mat, 81-89; RASHID ENG REGENTS CHAIR & DIR, STRATEGIC MAT RES & DEVELOP LAB, UNIV TEX, 89- *Concurrent Pos:* Prin investr grants, NSF, NASA, Dept Energy, Air Force & Navy, 71-; consult, Industs & UN, 71-; mem, Nat Res coun, Solid State Sci Panel, Nat Acad Sci, 76- & Panels on Strategic Mat, 80- *Mem:* Am Inst Mining, Metall & Petrol Engrs; Am Soc Metals; Metall Soc; Am Soc Testing & Mat; Am Soc Mech Engrs; Nat Asn Corrosion Engrs. *Res:* Physical, mechanical and process metallurgy of high temperature superalloys and other materials for jet engine and turbine applications; hydrogen embrittlement and corrosion; supply and demand analysis of strategic materials. *Mailing Add:* Dept Mech Eng Univ Tex Austin TX 78712-1063

TIEN, P(ING) K(ING), b China, Aug 2, 19; m; c 2. MICROWAVE TECHNOLOGY, HIGH SPEED ELECTRONICS. *Educ:* Nat Cent Univ, China, BS, 42; Stanford Univ, MS, 48, PhD, 51. *Prof Exp:* Res assoc, Stanford Univ, 51-52; mem tech staff, Bell Tel Labs, Inc, 52-59, head, Dept Electron Physics Res, 59-84, head, Dept High Speed Electronics, 85-91, FEL, PHOTONICS RES LAB, BELL TEL LABS, INC, 91- *Honors & Awards:* Monis N Lieberman Award, Inst Elec & Electronics Engrs, 79. *Mem:* Nat Acad Sci; Nat Acad Eng; Am Inst Physics; fel Inst Elec & Electronics Engrs; fel Optical Soc Am. *Res:* Device physics; microwave electronics; electron dynamics; wave propagation; noise; ferrites; acoustics in solids; gas lasers; superconductivity; integrated optics. *Mailing Add:* Bell Tel Labs Inc Holmdel NJ 07733

TIEN, REX YUAN, b Hupei, China, Aug 4, 35; c 3. ORGANIC CHEMISTRY. *Educ:* Chung Hsing Univ, Taiwan, BS, 58; Univ RI, PhD(chem), 68. *Prof Exp:* NSF fel chem, State Univ NY Albany, 67-68; SR CHEMIST, AM HOECHST CORP, 68- *Mem:* Am Chem Soc. *Res:* Applied chemistry, dyes and pigments; photochemistry; organometallic chemistry, kinetics and instrumentation. *Mailing Add:* 129 Sturbridge Dr Warwick RI 02886-8630

TIEN, TSENG-YING, b Hopei, China, June 28, 24; US citizen; c 5. CERAMICS. *Educ:* Pa State Univ, MS, 60, PhD(ceramics), 65. *Prof Exp:* Sr scientist, Westinghouse Res Labs, 60-66; assoc prof mat, 66-73, PROF MAT, UNIV MICH, ANN ARBOR, 73- *Mem:* Am Ceramic Soc; fel Am Chem Soc. *Res:* Structure and physical properties of solid oxide materials. *Mailing Add:* Dept Mat Sci Univ Mich Main Campus Ann Arbor MI 48109

TIER, CHARLES, b Albany, NY, Sept 25, 47; m 74; c 2. ASYMPTOTIC METHODS, QUEUEING THEORY. *Educ:* Rensselaer Polytech Inst, BS, 69, MS, 71; NY Univ, PhD(appl math), 76. *Prof Exp:* Instr math, NY Univ, 72-76; PROF MATH, UNIV ILL, CHICAGO, 76- *Concurrent Pos:* Vis asst prof, Northwestern Univ, 81-82, res fel, 83-84. *Mem:* Soc Indust & Appl Math; AAAS; Asn Comput Mach. *Res:* Application of applied mathematics methods to analysis of stochastic models arising in queueing systems; biological models; chemical systems. *Mailing Add:* Dept Math M/C 249 Univ Ill Chicago IL 60680

TIERCE, JOHN FORREST, b Los Angeles, Calif, Aug 17, 42; m 67; c 2. ANIMAL GENETICS, BIOSTATISTICS. *Educ:* Okla State Univ, BS, 65, MS, 67; Iowa State Univ, PhD(animal genetics), 73. *Prof Exp:* Res assoc poultry sci, Iowa State Univ, 70-73; res geneticist poultry genetics, Perdue Farms, Inc, 73-75; DIR RES POULTRY GENETICS, INDIAN RIVER INT, 75- *Mem:* Poultry Sci Asn; World Poultry Sci Asn; Biometric Soc; Am Statist Asn. *Res:* Poultry genetics. *Mailing Add:* 3407 Kings Row Nacogdoches TX 75961

TIERNAN, ROBERT JOSEPH, b Boston, Mass, Dec 14, 35; m 59; c 4. PHYSICS, CERAMICS. *Educ:* Boston Col, AB, 57, MS, 59; Mass Inst Technol, PhD(ceramics), 69. *Prof Exp:* Res asst accelerators, Grad Sch, Boston Col, 57-59; solid state physicist, Naval Res Lab, 59-62; atomic physicist, Nat Bur Stand, 63; develop engr, Sylvania, 63-64; AEC res asst, Grad Sch, Mass Inst Technol, 64-69; sr res scientist, Raytheon Co, Waltham, 69-74; ENGR/SCIENTIST ADVAN RES & DEVELOP, GTE/SYLVANIA, 76- *Concurrent Pos:* Res assoc, Argonne Nat Lab, 75. *Mem:* Am Ceramic Soc; Int Soc Optical Engrs; Sigma Xi. *Res:* Solid state physics; electronic, optical and magnetic properties of ceramics; diffusion in ceramics; mechanical properties; sodium reaction in HID lamps; emissivity of alumina; thermal shock of alumina. *Mailing Add:* 224 North St Stoneham MA 02180

TIERNAN, THOMAS ORVILLE, b Chattanooga, Tenn, July 22, 36; m 61; c 1. CHEMICAL PHYSICS, ANALYTICAL CHEMISTRY. *Educ:* Univ Windsor, BSc, 58; Carnegie Inst Technol, MS, 60, PhD(chem), 66. *Prof Exp:* Ohio State Univ Res Found res chemist, Wright-Patterson AFB, 60-61, res chemist, Off Aeronaut Space Res, Aerospace Labs, 61-67, group leader high energy chem kinetics, 67-75; dir, Brehm Lab, 76-87, PROF CHEM, WRIGHT STATE UNIV, 75-, DIR, TOXIC CONTAMINANT RES PROG, 87- *Concurrent Pos:* Extenstive consulting on environ issues. *Mem:* AAAS; Am Chem Soc; Am Phys Soc; Am Soc Lubrication Engrs; fel Am Inst Chemists. *Res:* Mass spectrometry; gas phase kinetics; ion and electron impact-phenomena; plasma characterization and diagnostics; lasers; gaseous electronics; analytical methods development; environmental monitoring, materials characterization. *Mailing Add:* 6532 Senator Lane Dayton OH 45459

TIERNEY, DONALD FRANK, b Butte, Mont, May 24, 31; m 54; c 2. PHYSIOLOGY, MEDICINE. *Educ:* Univ Calif, Berkeley, BA, 53; Univ Calif, San Francisco, MD, 56. *Prof Exp:* Intern, Philadelphia Gen Hosp, 57; asst resident, Univ Pa, 58; asst resident, Univ Calif, San Francisco, 59, asst prof physiol, 65-68; assoc prof, 68-75, PROF MED, UNIV CALIF, LOS ANGELES, 75- *Concurrent Pos:* USPHS fels, Univ Calif, San Francisco, 59-65; mem pulmonary dis adv comt, Nat Heart & Lung Inst, 73-76; chief palmonary div, Univ Calif, Los Angeles Hosp; chmn Study Sect, Pediat Specialized Ctr Res, Nat Heart Lung & Blood Inst, NIH, 80-81; mem, Pulmonary Dis Adv Comt, NHLB, 74-78, appl physiol study sect, NIH, 87-91. *Mem:* Am Physiol Soc; Am Thoracic Soc (vpres, 76-77, pres, 78-79); Am Soc Clin Invest; Asn Am Physicians. *Res:* Pulmonary physiology, biochemistry and metabolism. *Mailing Add:* Dept Med Univ Calif Los Angeles CA 90024

TIERNEY, JOHN W(ILLIAM), b Oak Park, Ill, Dec 29, 23. CHEMICAL ENGINEERING, SEPARATION SYSTEMS. *Educ:* Purdue Univ, BS, 47; Univ Mich, MS, 48; Northwestern Univ, PhD(chem eng), 51. *Prof Exp:* Res engr, Pure Oil Co, 48-50, 51-54; asst prof chem eng, Purdue Univ, 54-56; dept mgr, Res Div, Univac Div, Sperry Rand Corp, 56-60; assoc prof chem eng, 60-62, PROF CHEM ENG, UNIV PITTSBURGH, 62- *Concurrent Pos:* Lectr, Univ Minn, 58-59; vis prof, Santa Maria Univ, Chile, 60-62; Fulbright lectr, Univ Barcelona, 68-69. *Res:* Application of computer techniques to chemical engineering; reaction engineering; distillation; filtration; direct and indirect coal liquefaction. *Mailing Add:* Dept Chem Eng Sch Eng Univ Pittsburgh Pittsburgh PA 15261

TIERNEY, WILLIAM JOHN, b New York, NY, Aug 17, 44; m 77; c 2. TOXICOLOGY, PHARMACOLOGY. *Educ:* Columbia Univ, BS, 68; St John's Univ, MS, 74, PhD(pharmacol), 77; Am Bd Toxicol, dipl, 81. *Prof Exp:* Toxicologist, Bio/Dynamics, Inc, 77-80, asst dir toxicol, 80-84, dir res, 84-89; PRES, TIERNEY & ASSOCS, 89- *Concurrent Pos:* Asst prof, St John's Univ, 80-82. *Mem:* Soc Toxicol; Europ Soc Toxicol; Am Col Toxicol; Int Soc Study Xenobiotics. *Res:* Preclinical safety evaluations of foods, drugs, chemicals, pesticides and cosmetics. *Mailing Add:* RD 3 Box 279 Pittstown NJ 08867-9416

TIERNO, PHILIP M, JR, b Brooklyn, NY, June 5, 43; m 67; c 2. CLINICAL MICROBIOLOGY. *Educ:* Brooklyn Col Pharm-Long Island Univ, BS, 65; New York Univ, MS, 74, PhD(microbiol), 77. *Prof Exp:* Clin microbiologist, Lutheran Med Ctr, 65-66; chief res microbiologist, US Vet Admin Hosp, 66-70; asst dir, Goldwater Mem Hosp, 70-75, dir, 75-81, DIR MICROBIOL, TISCH-UNIV HOSP, NEW YORK UNIV MED CTR, 81- *Concurrent Pos:* Asst prof microbiol, City Univ New York, 74-76 & Bloomfield Col Sch

Nursing, 75-82; consult microbiologist, Maimonides Med Ctr, 70-79; assoc prof microbiol, New York Univ Sch Med, 81-; co-founder & chmn bd, Found Sci Res Pub Interest, 85- *Mem:* Am Acad Microbiol; Am Soc Microbiol; AAAS; Am Pub Health Asn; Am Asn Univ Prof. *Res:* Clinical application of enzymic reactions to microorganisms especially as they relate to indentifications and typing systems; microbial ecology and epidemiology; role of staphylococci in human health and disease; Toxic Shock Syndrome; acquired immune deficiency syndrome (AIDS). *Mailing Add:* 30 Carter St Norwood NJ 07648

TIERS, GEORGE VAN DYKE, b Chicago, Ill, Mar 23, 27; m 50; c 1. ORGANIC MATERIALS SCIENCE, NON-LINEAR OPTICS. *Educ:* Univ Chicago, SB, 46, SM, 50, PhD(chem), 56. *Prof Exp:* Asst pharmacol, Univ Chicago, 45-46, chemist, Ord Res Proj, 48-49; chemist, 51-65, CORP SCIENTIST, CORP RES LABS, 3M CO, 65- *Honors & Awards:* Carbide Award, Am Chem Soc, 59. *Mem:* Am Chem Soc. *Res:* Nuclear magnetic resonance spectroscopy; fluorine, organic, dye, polymer and physical-organic chemistry; duplicating and imaging technology; non-linear-optical materials; materials science of organic compounds. *Mailing Add:* 3M Sci Res Lab PO Box 33221 St Paul MN 55133

TIERSTEN, HARRY FRANK, b Brooklyn, NY, Jan 4, 30; m 53; c 2. APPLIED MECHANICS. *Educ:* Columbia Univ, BS, 52, MS, 56, PhD(appl mech), 61. *Prof Exp:* Stress analyst, Grumman Aircraft Eng Corp, 52-53; struct designer, J G White Eng Corp, 53-56; instr civil eng, City Col New York, 56-60; res asst appl mech, Columbia Univ, 60-61; mem tech staff, Bell Tel Labs, 61-68; PROF MECH, RENSSELAER POLYTECH INST, 68- *Mem:* Am Phys Soc; Acoust Soc Am; Am Soc Mech Engrs; Soc Natural Philos; Inst Elec & Electronics Engr; Sigma Xi. *Res:* Elasticity; couple stress elasticity; electromagnetism; electrostriction; piezoelectricity; magnetism; magnetoelasticity; waves; vibrations. *Mailing Add:* 2288 Pinehaven Dr Schenectady NY 12309

TIERSTEN, MARTIN STUART, b Aug 7, 31; US citizen; m 53; c 2. PHYSICS. *Educ:* Queens Col, NY, BA, 53; Columbia Univ, AM, 58, PhD(theoret solid state physics), 62. *Prof Exp:* Tutor physics, 57-62, from instr to assoc prof, 62-83, PROF PHYSICS, CITY COL NEW YORK, 83- *Mem:* Am Phys Soc; Am Asn Physics Teachers. *Res:* Theoretical physics. *Mailing Add:* Dept Physics City Col New York Convent at 138th St New York NY 10031

TIESZEN, LARRY L, b Marion, SDak, Mar 2, 40; m 59; c 2. PLANT PHYSIOLOGY, PLANT ECOLOGY. *Educ:* Augustana Col, SDak, BA, 61; Univ Colo, PhD(bot), 65. *Prof Exp:* Kettering fel biol, Albion Col, 65-66; asst prof, Univ Minn, Duluth, 66; from asst prof to assoc prof, 66-75, PROF BIOL, AUGUSTANA COL, SDAK, 75-, CHMN DEPT, 77- *Concurrent Pos:* Res grants, Sigma Xi, 65-67, Arctic Inst NAm, 66-68 & NSF, 70-82; NSF vis prof, Univ Nairobi, Kenya, 75; prog assoc, NSF, 78-79; Fulbright fel, Univ Nairobi, 82; consult, Int Agencies. *Mem:* AAAS; Am Soc Plant Physiol; Ecol Soc Am; Arctic Inst NAm; Can Soc Plant Physiol. *Res:* Photosynthesis and pigments in arctic and alpine grasses; environmental influence of photosynthesis; growth under extreme conditions; physiological adaptation; United States tundra biome program; photosynthesis and water stress in finger millet and other tropical grasses; C3 and C4 photosynthesis; stable isotope ecology; tropical ecology; paleoecology; agroforestry. *Mailing Add:* Dept Biol Augustana Col 29th St & S Summit Sioux Falls SD 57197

TIETHOF, JACK ALAN, b Grand Rapids, Mich, July 23, 43; div; c 3. INORGANIC CHEMISTRY, SURFACE SCIENCE. *Educ:* Western Mich Univ, BS, 67, PhD(inorg chem), 71. *Prof Exp:* Res assoc inorg chem, Ohio State Univ, 71-73, lectr gen chem, 73-74; proj leader prod develop, 74-78, group leader new bus res, 78-81, GROUP LEADER CATALYST RES, ENGELHARD MINERALS & CHEM CORP, 81- *Concurrent Pos:* Nat Defense Educ Act fel, Western Mich Univ, 67-70; fel, Grad Sch, Ohio State Univ, 71-72. *Mem:* Am Chem Soc; Clay Mineral Soc. *Res:* Development of new catalysts for petroleum refining, particularly in fluidized bed cracking reforming catalysts; material research include zeolites and supported precious metals. *Mailing Add:* 24 Eton Ct Bedminster NJ 07921-1603

TIETJEN, JAMES JOSEPH, b New York, NY, Mar 29, 33; m 58; c 2. PHYSICAL CHEMISTRY. *Educ:* Iona Col, BS, 56; Pa State Univ, MS, 58, PhD(chem), 63. *Prof Exp:* Mem tech staff mat sci, 63-69, group head, 69-70, Dir Mat Res, RCA LABS, 70-77, STAFF VPRES, MAT & COMPONENTS RES, RCA CORP, 77- *Concurrent Pos:* Assoc ed, Mat Res Bull, 71; mem Solid State Sci Adv Panel, Nat Acad Sci, 73-; mem, NASA Space Systs & Technol Adv Comt, 78- *Honors & Awards:* David Sarnoff Outstanding Achievement Awards, 67 & 70. *Mem:* Am Chem Soc; Am Inst Mining Metall & Petrol Engrs; Electrochem Soc; Sigma Xi. *Res:* Displays, semiconductor materials and devices; insulators; metallic systems; optical phenomena; electron optics; negative electron affinity effects; luminescent materials; videodisc systems. *Mailing Add:* Three De Hart Dr Belle Mead NJ 08502

TIETJEN, JOHN H, b Jamaica, NY, June 19, 40; m 68; c 2. MARINE BIOLOGY, INVERTEBRATE ZOOLOGY. *Educ:* City Col New York, BS, 61; Univ RI, PhD(oceanog), 66. *Prof Exp:* From asst prof to assoc prof, 66-75, PROF BIOL, CITY COL NEW YORK, 75-; DIR, INST MARINE & ATMOSPHERIC SCI, CITY UNIV NY, 78- *Concurrent Pos:* Consult, Northeast Utilities, 68-, SW Res Inst, 78-79, other pvt & govt consult; res grants, NSF, 68-88, Nat Oceanic & Atmospheric Admin, 73-82, Off Naval Res, 80-82 & Brookhaven Nat Lab, 84-86; chmn biol, City Col NY, 81-87; vis scientist, Aust Inst Marine Sci, 87-88. *Honors & Awards:* Sigma Xi. *Mem:* AAAS; Am Soc Nematologists; Am Arbitrators Asn. *Res:* Estuarine ecology; physiological ecology of meiofauna; ecology; pollution ecology of benthos. *Mailing Add:* Dept Biol City Col NY New York NY 10031

TIETJEN, WILLIAM LEIGHTON, b Americus, Ga, Jan 3, 37; m 68; c 1. AQUATIC ECOLOGY. *Educ:* Univ Ga, BS, 58; Univ Tenn, PhD(radiation biol), 67. *Prof Exp:* Res asst marine biol, Marine Inst, Univ Ga, 60; from asst prof to assoc prof, 67-78, PROF & CHMN BIOL DEPT, GA SOUTHWESTERN COL, 78- *Mem:* AAAS; Am Inst Biol Sci; Am Soc Limnol & Oceanog; Ecol Soc Am; Entom Soc Am; NAm Beuthological Soc. *Res:* Arthropod metabolism and energy flow; environmental effects on physiology of arthropods; wetland ecology; nutrient flux in streams. *Mailing Add:* Dept Biol Ga Southwestern Col Americus GA 31709

TIETZ, NORBERT W, b Stettin, Ger, Nov 13, 26; US citizen; m 59; c 4. CLINICAL CHEMISTRY. *Educ:* Stuttgart Tech Univ, PhD(natural sci), 50. *Prof Exp:* Res fel biochem, Univ Munich, 51-54; res fel clin chem, Rockford Mem Hosp, 54-55 & Univ Chicago, 55-56; head div biochem, Reid Mem Hosp, Richmond, Ind, 56-59; assoc path, Chicago Med Sch-Univ Health Sci, 59-64, from asst prof to assoc prof clin path, 64-69, prof clin chem, 69-76; DIR CLIN CHEM, UNIV KY MED CTR & PROF PATH, COL MED, 76- *Concurrent Pos:* Dir clin chem, Mt Sinai Hosp Med Ctr, 59-76; consult, Dept Health, State of Ill, 67-76; consult, Vet Admin Hosp, Hines, Ill, 74-76 & Vet Admin Hosp, Lexington, Ky, 76- *Honors & Awards:* Clin Chemist Award, Am Asn Clin Chemists, 71, Award Outstanding Efforts Educ & Training, 76, Steuben Bowl Award, 78, Bernard F Gerulat Award, 88, Donald D Van Slyke Award, 89. *Mem:* Fel AAAS; Am Asn Clin Chem; Am Chem Soc; Am Soc Clin Path; fel Am Inst Chemists; Sigma Xi; Acad Clin Lab Physicians & Scientists. *Res:* Methodology related to clinical chemistry; enzyme chemistry; writing and editing in clinical chemistry. *Mailing Add:* Dept Path Univ Ky Med Ctr Lexington KY 40536

TIETZ, THOMAS E(DWIN), b Pittsburgh, Pa, July 22, 20; m 47; c 3. PHYSICAL METALLURGY. *Educ:* Univ Calif, BS, 44, MS, 51, PhD(phys metall), 54. *Prof Exp:* Design engr, Lane-Wells Co, 44-47; res engr, Inst Eng Res, Univ Calif, 47-50; sr metallurgist, Stanford Res Inst, 54-59; sr mem, Lockheed Palo Alto Res Lab, Lockheed Missiles & Space Co, 59-63, mgr, Metall & Composites Lab, 65-81; vis sr staff assoc, Nat Mat Adv Bd, Nat Acad Sci, 81-88; RETIRED. *Concurrent Pos:* Vis lectr, Stanford Univ, 57-58; mem, Comt Amorphous & Metastable Mat, Nat Mat Adv Bd, Nat Acad Sci, 78-79; guest scientist, Max-Planck-Inst Metall Res, Stuttgart, WGer, 79. *Mem:* Fel Am Soc Metals; Am Inst Mining, Metall & Petrol Engrs; Am Inst Aeronaut & Astronaut; Am Powder Metall Inst. *Res:* Deformation behavior of metals; strengthening mechanisms in metals; refractory metals; powder metallurgy. *Mailing Add:* 12640 Corte Madera Lane Los Altos Hills CA 94022

TIETZ, WILLIAM JOHN, JR, b Chicago, Ill, Mar 6, 27; div; c 3. NEUROPHYSIOLOGY. *Educ:* Swarthmore Col, BA, 50; Univ Wis, MS, 52; Colo State Univ, DVM, 57; Purdue Univ, PhD(physiol), 61. *Hon Degrees:* DSc, Purdue Univ, 82. *Prof Exp:* Instr physiol, Purdue Univ, 57-59, from instr to assoc prof physiol, 59-64; sect leader, Collab Radiol Health Lab, 64-67, assoc prof physiol & biophys, Univ, 64-67, chmn dept & prof physiol & biophys, 67-70, vpres student-univ rels, relations, 70-71, prof physiol & biophys, Colo State Univ, 67-77, dean, Col Vet Med & Biomed Sci, 71-77; prof vet specialties & pres, 77-90, EMER PRES, MONT STATE UNIV, 90- *Concurrent Pos:* Trustee, Yellowstone Asn, 81-, chmn, 87-; mem, Mont Sci & Technol Alliance, 85-88; mem, Gov Trade Comn, 85 & 90. *Mem:* AAAS; Am Physiol Soc; Am Vet Med Asn; Conf Res Workers Animal Dis. *Res:* Radiation biology; veterinary neurophysiology and neurosurgery; effects of ionizing radiation on early embryogenesis. *Mailing Add:* 2310 Spring Creek Dr Bozeman MT 59715

TIETZE, FRANK, b Manila, Philippines Aug 19, 24; nat US; m 54; c 4. BIOCHEMISTRY. *Educ:* Trinity Col, Conn, BS, 45; Northwestern Univ, MS, 47, PhD(biochem), 49. *Prof Exp:* USPHS fel, Duke Univ, 49-50; USPHS fel, Univ Wash, 50-51, instr biochem, 51-52; instr, Univ Pa, 52-56; RES BIOCHEMIST, NAT INST DIABETES & DIGESTIVE & KIDNEY DIS, 56- *Mem:* AAAS; Am Soc Biol Chem. *Res:* Inherited disorders of lysosomal transport. *Mailing Add:* Lab Molecular & Cellular Biol Nat Inst Diabetes & Digestive & Kidney Dis NIH Bethesda MD 20892

TIFFANY, CHARLES F, b Aitkin, Minn, Nov 23, 29; m 52; c 3. ENGINEERING. *Educ:* Univ Minn, BCE & BS, 52. *Prof Exp:* Spec tech adv, airframe & propulsion syst struct, aeronaut systs div, US Air Force, 72-79; vpres res & eng, Boeing Mil Airplane Co, 79-84, advan systs, 84-87, exec vpres prog & technol, 87-90; RETIRED. *Mem:* Nat Acad Eng; fel Am Inst Aeronaut & Astronaut; Am Defense Preparedness Asn; Unmanned Vehicle Soc. *Mailing Add:* Boeing Mil Airplane PO Box 7730 MS K15-08 Wichita KS 67227

TIFFANY, LOIS HATTERY, b Collins, Iowa, Mar 8, 24; m 45; c 3. PLANT PATHOLOGY, MYCOLOGY. *Educ:* Iowa State Univ, BS, 45, MS, 47, PhD, 50. *Prof Exp:* From instr to assoc prof, 50-65, PROF BOT & PLANT PATH, IOWA STATE UNIV, 65- *Honors & Awards:* William H Weston Award Mycol Soc Am, 80; William H Weston Award, 80. *Mem:* Am Phytopath Soc; Mycol Soc Am; Bot Soc Am. *Res:* Soil fungi and parasitic fungi of prairie plants; ascomycetes. *Mailing Add:* Dept Bot Iowa State Univ Ames IA 50011

TIFFANY, OTHO LYLE, b Flint, Mich, Nov 26, 19; m 42; c 5. PHYSICS. *Educ:* Univ Mich, BS, 43, MS, 46, PhD(physics), 50. *Prof Exp:* Mem staff, Radiation Lab, Mass Inst Technol, 43-45; res engr, Willow Run Res Ctr, Mich, 49-58; chief scientist, Aerospace Systs Div, Bendix Corp, 58-70, dir space & earth sci, 70-80; CONSULT, 80- *Mem:* Am Phys Soc; Am Geophys Union; Inst Elec & Electronics Engrs. *Res:* Space sciences, geophysics, oceanography, environmental research. *Mailing Add:* 1828 Vinewood Blvd Ann Arbor MI 48104

TIFFANY, WILLIAM JAMES, III, b Syracuse, NY, Sept 12, 44; m; c 2. MARINE BIOLOGY. *Educ:* Univ Miami, BS, 66; Fla State Univ, MS, 68, PhD(physiol), 72. *Prof Exp:* Instr biol, Fla State Univ, 70-71; asst prof, New Col, Univ SFla, 71-73, res fel marine biol, 73-74; res scientist, Marine Biomed

Inst, Med Br, Univ Tex, 74-75; environ scientist biol, Conserv Consults, 75-77; res scientist biomed, Mote Marine Lab, 77-80; environ spacialist, Health Dept, Manatee County, Fla, 80-84; DIR ENVIRON AFFAIRS, PORT MANATEE, 85- *Concurrent Pos:* Selby res fel, William G & Marie Selby Found, 73-74; mem marine adv bd, Sarasota County, Fla, 78-80 & Manatee County, 80-; res assoc, New Col, Univ SFla, 80-84; res fel, Health Res Found, Fla, 81-84. *Mem:* AAAS; Am Soc Zoologists; Fla Acad Sci. *Res:* Comparative animal physiology; renal physiology; invertebrate zoology; estuarine and intertidal ecology; benthic ecology. *Mailing Add:* Port Manatee Rte 1 Palmetto FL 34221

TIFFNEY, BRUCE HAYNES, b Sharon, Mass, July 3, 49. PLANT EVOLUTION, PALEOBIOLOGY. *Educ:* Boston Univ, BA, 71; Harvard Univ, PhD(biol), 77. *Prof Exp:* Assoc prof paleobiol, dept biol, Yale Univ, 77-86; ASSOC PROF GEOL, DEPT GEOL SCI, UNIV CALIF, 86- *Concurrent Pos:* Cur Herbarium, Paleobotanical Collections, Peabody Mus Natural Hist, 77-86. *Mem:* Bot Soc Am; fel Geol Soc Am; Int Asn Plant Taxonomists; Soc Study Evolution; Asn Trop Biol; Soc Syst Zool. *Res:* The fossil record and evolution of angiosperms with emphasis on the study of their fossilized fruiting remains; patterns and processes involved in the evolution of land plants as a group. *Mailing Add:* Dept Geol Sci Univ Calif Santa Barbara CA 93106

TIFFT, WILLIAM GRANT, b Derby, Conn, Apr 5, 32; m 65; c 6. GALAXIES, REDSHIFT. *Educ:* Harvard Univ, AB, 54; Calif Inst Technol, PhD, 58. *Prof Exp:* Hon res fel astron, Australian Nat Univ, 58-60; res assoc astron & physics, Vanderbilt Univ, 60-61; astronr, Lowell Observ, 61-64; assoc prof, 64-73, PROF ASTRON, UNIV ARIZ, 73- *Concurrent Pos:* NSF fel, 58-60, grant, 84-87; vis scientist, Inst Astron, Bologna, Italy, 78-79, Max-Planck Inst Radioastron, Bonn, Ger, 87, Nat Radio Astron Observ, Greenbank, WVa, 88. *Mem:* Am Astron Soc; Int Astron Union. *Res:* Optical stellar astronomy; galactic structure and extragalactic problems; interpretation of the redshift in galaxies; pairs and clusters of galaxies; precision of 21 cm redshifts; large scale structure, superclusters; large scale gravitation. *Mailing Add:* Dept Astron Univ Ariz Tucson AZ 85721

TIGCHELAAR, EDWARD CLARENCE, b Hamilton, Ont, Feb 10, 39; m 62; c 3. VEGETABLE CROPS, PLANT BREEDING. *Educ:* Univ Guelph, BScA, 62; Purdue Univ, MSc, 64, PhD(genetics, plant breeding), 66. *Prof Exp:* Asst prof hort, Univ Guelph, 66-67; from asst prof to assoc prof, 67-78, PROF HORT, PURDUE UNIV, WEST LAFAYETTE, 78- *Concurrent Pos:* Asst prof, Int Prog Agr, AID, Brazil, 67-69; consult. *Honors & Awards:* Asgrow Award; Marion Meapows Award. *Mem:* Fel Am Soc Hort Sci. *Res:* Tomato breeding; genetics of fruit ripening; physiological genetics and breeding of vegetables. *Mailing Add:* Dept Hort Purdue Univ West Lafayette IN 47907

TIGCHELAAR, PETER VERNON, b Chicago, Ill, May 15, 41; m 63; c 2. PHYSIOLOGY. *Educ:* Calvin Col, AB, 63; Univ Ill, Urbana, MS, 66, PhD(physiol), 69. *Prof Exp:* NIH fel endocrinol, Univ Ill, 70-71; asst prof physiol, Sch Med, Ind Univ-Purdue Univ, Indianapolis, 71-75; assoc prof, 75-79, PROF BIOL, CALVIN COL, 79- *Concurrent Pos:* NIH fel endocrinol, 69-70. *Mem:* AAAS; Endocrine Soc; Am Physiol Soc. *Res:* Mammalian reproductive physiology; synthesis, control and effects of mammalian gonadotrophic hormones. *Mailing Add:* Dept Biol Calvin Col Grand Rapids MI 49506

TIGERTT, WILLIAM DAVID, b Wilmer, Tex, May 22, 15; m 38; c 2. PATHOLOGY. *Educ:* Baylor Univ, MD, 37, AB, 38; Am Bd Path, dipl, 42. *Prof Exp:* Intern, Baylor Hosp, 37-38, instr path, Baylor Col Med, 38-40; pathologist, Brooke Gen Hosp, Med Corps, US Army, 40-43, commanding officer, 26th Army Med Lab, Southwest Pac, 44-46 & 406th Med Gen Lab, Tokyo, 46-49, asst commandant, Army Med Serv Grad Sch, 49-54, chief spec opers br, Walter Reed Army Inst Res, 54-56, commanding officer, Army Med Unit, Ft Detrick, Md, 56-61, med officer, Field Teams, Walter Reed Army Inst Res, 61-63, dir & commandant, 63-68, commanding gen, Madigan Army Hosp, 72; assoc prof med, Sch Med, Univ Md, Baltimore City, 56-68, prof exp med, 69-71, prof path, 71-81; mem sci adv bd, Armed Forces Inst Path, 81-83; ed, Am J Trop Med Hyg, 85-90; RETIRED. *Concurrent Pos:* Fel path, Baylor Col Med, 38-40; lab consult, Far East Command, US Army, 46-49, consult, Surgeon Gen, 60-68 & 73-82; consult, WHO, 61-62 & 69-71; dir comn malaria, Armed Forces Epidemiol Bd, 67-69; chmn, US Army Med Res & Develop Command Adv Panel, 73-82; mem bd dirs, Gorgas Mem Inst, 75-84. *Honors & Awards:* Gorgas Medal, 66. *Mem:* Am Asn Immunologists; fel Col Am Pathologists; fel Am Col Physicians; Am Soc Trop Med Hyg. *Res:* Infectious diseases; immunology. *Mailing Add:* 15 Charles Plaza 2203 Baltimore MD 21201

TIGGES, JOHANNES, b Rietberg, Ger, July 7, 31; m 59; c 2. NEUROANATOMY. *Educ:* Univ M nster, PhD(zool), 61. *Prof Exp:* Res assoc neuroanat, Max Planck Inst Brain Res, 61-62; res assoc vision in primates, Yerkes Primate Labs, Fla, 62-63; res assoc neuroanat & physiol, Max Planck Inst Brain Res, 63-65; neuroanatomist, Yerkes Regional Primate Res Ctr, 66-71, sr neuroanatomist, 71-79, from asst prof to assoc prof, 67-78, PROF ANAT, EMORY UNIV, 78-, PROF, DEPT OPHTHAL, 81-, RES PROF & CHIEF, DIV NEUROBIOLOGY, YERKES REGIONAL PRIMATE RES CTR, 79- *Mem:* Asn Am Anatomists; Int Primatological Soc; Soc Neurosci; Asn Res Vision & Opthal; Am Soc Primatologists; Int Brain Res Orgn; Cajal Club. *Res:* Light and electron microscopy of primate visual system; anatomical correlates of aging in primate brain. *Mailing Add:* Yerkes Regional Primate Res Ctr Emory Univ Atlanta GA 30322

TIGNER, JAMES ROBERT, b El Paso, Tex, June 17, 36; m 57; c 2. WILDLIFE BIOLOGY. *Educ:* Colo State Univ, BS, 58, MS, 60; Univ Colo, PhD(biol), 72. *Prof Exp:* Wildlife biologist, US Fish & Wildlife Serv, 60-61, proj leader, 62-70, sta leader animal damage control res, 73-81; ENVIRON IMPACT STUDY TEAM LEADER, BUR LAND MGT, DEPT INTERIOR, 81- *Mem:* Wildlife Soc; Sigma Xi. *Res:* Animal damage control, especially coyote. *Mailing Add:* 104 E Kendrick Rawlins WY 82301

TIGNER, MAURY, b Middletown, NY, Apr 22, 37; m 60; c 2. HIGH ENERGY PHYSICS. *Educ:* Rensselaer Polytech Inst, BS, 58; Cornell Univ, PhD(physics), 63. *Prof Exp:* Res assoc, Cornell Univ, 63-68, sr res assoc physics, 68-77, prof physics & mem staff, Lab Atomic & Solid State Physics, 77- *Mem:* Am Phys Soc; Am Vacuum Soc. *Res:* Design and development of new particle accelerators and improvement of existing designs. *Mailing Add:* 103 Campbell Ave Ithaca NY 14850

TIHANSKY, DIANE RICE, b Kansas City, Mo, Fev 20, 48; div; c 2. THEORY, SOFTWARE SYSTEMS. *Educ:* Marygrove Col, Detroit, Mich, BS, 69; Harvard Univ, MS, 72. *Prof Exp:* Programmer-analyst, City Alexandria, Va, 72 & US Dept Treas, 72-74; instr comput sci & math, Fla Int Univ, Miami, 77-78; instr, 83-84, CHMN, DEPT MATH & COMPUTER SCI, ST MARY'S COL, ORCHARD LAKE, MICH, 84- *Mem:* Math Asn Am; Am Math Soc. *Res:* Mathematics and computer anxiety; computer-assisted instruction for bilingual education and language learning. *Mailing Add:* Computer Sci Dept St Mary's Col Orchard Lake MI 48033

TIHEN, JOSEPH ANTON, b Harper, Kans, Nov 20, 18; m 40; c 7. ZOOLOGY. *Educ:* Univ Kans, AB, 40; Univ Rochester, PhD(zool), 45. *Prof Exp:* Asst instr zool, Univ Kans, 40-41; asst instr, Univ Rochester, 41-44, res assoc mouse genetics unit, Manhattan Proj, 44-46; asst prof zool, Tulane Univ, 46-47; asst prof, Univ Fla, 50-57, res assoc, AEC Proj, Sch Med, 57-58; asst prof zool, Univ Ill, 58-61; from asst prof to prof, 61-87, EMER PROF BIOL, UNIV NOTRE DAME, 87- *Mem:* Am Soc Zool; Soc Study Amphibians & Reptiles; Soc Vert Paleont; Am Soc Ichthyologists & Herpetologists; Sigma Xi. *Res:* Systematics, Cenozoic paleontology and phylogeny of reptiles and amphibians. *Mailing Add:* Dept Biol Univ Notre Dame Notre Dame IN 46556

TIHON, CLAUDE, b Shanghai, China, June 12, 44; US citizen; m 69; c 1. PATHOBIOLOGY, BIOCHEMISTRY. *Educ:* Univ Colo, BA, 65; Columbia Univ, PhD(path), 71. *Prof Exp:* Res assoc virol, Inst Molecular Virol, Sch Med, St Louis Univ, 72-74; sr scientist biochem, Frederick Cancer Res Ctr, 74-76; sr investr biochem, Nat Jewish Hosp & Res Ctr, 76-80; mem staff, Bristol Labs, Div Bristol Myers, 80-90; VPRES, DEPT RES & TECH ASSESSMENT, AM MED SYST, 90- *Concurrent Pos:* Asst prof path, Sch Med, Univ Colo, 77- *Mem:* Am Soc Cell Biol. *Res:* Eukaryotic gene regulatory mechanisms; molecular action of cyclic adenosine monophosphate in Chinese hamster ovary cells and Dictyostelium discoideum; RNA tumor virology. *Mailing Add:* Dept Res & Tech Assessment Am Med Syst 11001 Bren Rd East Minnetonka MN 55343

TIKOO, MOHAN L, b Sringagar, Kashmir, India, Feb 15, 43; m 69; c 1. EXTENSION THEORY, DIFFERENTIAL EQUATIONS. *Educ:* Univ Kashmir, BA, 60, MA, 63; Univ Kans, MA, 81, PhD(math), 84. *Prof Exp:* Asst prof math, Univ Kashmir, 66-78; asst instr, Univ Kans, 78-84; asst prof, 84-88, ASSOC PROF MATH, SOUTHWEST MO STATE UNIV, CAPE GIRADEAU, 88- *Concurrent Pos:* Reviewer, Math Rev; reviewer, Zentralblatt Math. *Honors & Awards:* Florence Black Award, Univ Kans, 82. *Mem:* Am Math Soc; Math Asn Am; Planetary Soc. *Res:* Topology; functional analysis; zeros of polynomials. *Mailing Add:* Dept Math Southeast Mo State Univ Cape Girardeau MO 63701

TIKSON, MICHAEL, b Campbell, Ohio, Nov 22, 24; m 57; c 4. MATHEMATICS. *Educ:* Youngstown Univ, BS, 48; Lehigh Univ, MA, 49; Mass Inst Technol, MS, 56. *Prof Exp:* Instr math, Lehigh Univ, 49-50; guest worker, Nat Bur Standards, DC, 51-52; sr mathematician, Wright Air Develop Div, US Air Force Res & Develop Command, 52-56, chief anal sect, Digital Comput Br, 56-58, br chief, 58-60; consult & head digital comput ctr, 60-66, assoc mgr, Systs & Electronics Dept, Columbus Labs, 66-70, mgr comput systs & applns, 70-74, MGR, COMPUT & INFO SYSTS DEPT, COLUMBUS LABS, BATTELLE MEM INST, 74-, COMPUT & INFO SYSTS. *Mem:* Simulation Coun; Asn Comput Mach. *Res:* Management of research and operations in computer and information systems. *Mailing Add:* Battelle Mem Inst 505 King Ave Columbus OH 43201

TIKU, MOTI, RHEUMATOLOGY & IMMUNOLOGY. *Prof Exp:* ASST PROF RHEUMATOLOGY, DIV RHEUMATOLOGY, UNIV ILL HOSP, CHICAGO. *Res:* Hematology. *Mailing Add:* Univ Med & Dent Robert Wood Johnson Pl CN-19 New Brunswick NJ 08903

TILBURY, ROY SIDNEY, b Ealing, Eng, Aug 7, 32; US citizen; m 61; c 2. RADIOCHEMISTRY, BIOPHYSICS. *Educ:* Univ London, BSc, 55; McGill Univ, PhD(radiochem), 63. *Prof Exp:* Asst exp officer chem, UK Atomic Energy Res Estab, 55-59; res scientist, Union Carbide Corp, 63-67; assoc biophys, Mem Sloan-Ketterring Cancer Ctr, 67-74, assoc mem, 74-81; PROF CHEM (NUCLEAR MED), UNIV TEX SYST CANCER CTR, HOUSTON, 81- *Concurrent Pos:* Asst prof biophys, Grad Sch Med Sci, Cornell Univ, 67-74, assoc prof, 74-81; adj prof med chem, Col Pharm, Univ Ky, 78-; assoc ed, J Nuclear Med, 85-89. *Mem:* Am Chem Soc; Soc Nuclear Med; Sigma Xi. *Res:* Radiochemicals and labeled compounds for use in medical research, especially cyclotron produced short-lived radionuclides. *Mailing Add:* Dept Nuclear Med Univ Tex Cancer Ctr 1515 Holcombe Houston TX 77030

TILDON, J TYSON, b Baltimore, Md, Aug 7, 31; m 55; c 2. BIOCHEMISTRY. *Educ:* Morgan State Col, BS, 54; Johns Hopkins Univ, PhD(biochem), 65. *Prof Exp:* Res asst chem, Sinai Hosp, Baltimore, Md, 54-59; asst prof, Goucher Col, 67-68; res asst prof biochem & pediat, 68-71, dir pediat res, 70-83, assoc prof pediat, 71-74, assoc prof biochem, 72-82, PROF PEDIAT, SCH MED, UNIV MD, BALTIMORE CITY, 74-, PROF BIOCHEM, 82- *Concurrent Pos:* Fulbright scholar, Univ Paris, 59-60; Helen Hay Whitney fel biochem, Brandeis Univ, 65-67; lectr, Antioch Col, Baltimore Campus, 72-; Josiah Macy, Jr Fac Scholar, State Univ Groningen, Neth, 75-76. *Mem:* AAAS; Am Soc Biol Chemists; Am Soc Neurochem; Am Chem Soc; Tissue Cult Asn. *Res:* Developmental biochemistry and metabolic control processes. *Mailing Add:* Dept Pediatrics Univ Md Med 655 W Baltimore St Rm 10-033 BRB Baltimore MD 21201

TILFORD, SHELBY G, b Grayson Co, Ky, Jan 11, 37; m 56; c 2. ATMOSPHERIC SCIENCES, EARTH & ENVIRONMENTAL SCIENCES. *Educ:* Western Ky Univ, BS, 58; Vanderbilt Univ, PhD(phys chem), 62. *Prof Exp:* Res assoc spectros, Naval Res Lab, 61-63, res chemist, 63-66, spectros consult physicist, 66-72, sect head extreme ultraviolet spectros, 72-76; space scientist solar physics, NASA Hq, 76, discipline chief upper atmosphere, 76-78, br chief atmospheric processes, 78-81, div dir environ opers, 81-84, DIV DIR EARTH SCI & APPLN, NASA HQ, 84- *Concurrent Pos:* Vis prof, Univ Md, 69-76. *Res:* High resolution vacuum ultra violet spectroscopy of atoms and molecules of atmospheric, laser, and astrophysical interest; mesospheric research; tropospheric and stratospheric air quality; upper atmospheric research; oceanic processes. *Mailing Add:* 8805 Church Field Lane Laurel MD 20708-2428

TILGHMAN, SHIRLEY MARIE, DEVELOPMENTAL BIOLOGY, EMBRYOLOGY. *Educ:* Temple Univ, PhD(biochem), 75. *Prof Exp:* PROF BIOL, PRINCETON UNIV, 86- *Mailing Add:* Dept Biol Princeton Univ Princeton NJ 08544

TILL, CHARLES EDGAR, b Can, June 14, 34; m; c 3. ENGINEERING PHYSICS. *Educ:* Univ Sask, BE, 56, MSc, 58; Univ London, PhD(reactor physics), 60. *Prof Exp:* Jr res officer physics, Nat Res Coun Can, 56-58; reactor physicist, Can Gen Elec, 61-63; asst physicist, Argonne Nat Lab, 63-65, assoc physicist, 65-66, sect head exp develop sect, 66-68, head critical exp anal sect, 68-72, mgr zero power reactor prog, 68-72, assoc dir, Appl Physics Div, 72-73, dir appl physics, 73-80, ASSOC LAB DIR, ENG RES, ARGONNE NAT LAB, 80- *Mem:* Nat Acad Res; Am Nuclear Soc. *Res:* Fast reactor research. *Mailing Add:* Bldg 208 9700 S Cass Ave Argonne IL 60439

TILL, JAMES EDGAR, b Lloydminster, Sask, Aug 25, 31; m 59; c 3. BIOPHYSICS, MEDICAL & HEALTH SCIENCES. *Educ:* Univ Sask, BA, 52, MA, 54; Yale Univ, PhD(biophys), 57. *Prof Exp:* Biophysicist, Ont Cancer Inst, 57-69, head, Biores Div, 69-82; assoc dean grad studies, 81-84, PROF MED BIOPHYS, UNIV TORONTO, 65-, HON PROF, 84-; SR SCIENTIST, ONT CANCER INST, 82- *Concurrent Pos:* Res fel microbiol, Connaught Med Res Labs, 56-57. *Honors & Awards:* Gairdner Found Award, 69. *Mem:* Soc Med Decision-Making; Judgment/Decision Making Soc; fel Royal Soc Can. *Res:* Various aspects of cancer research; research on judgment and decision problems in health care settings; public health; epidemiology. *Mailing Add:* Ont Cancer Inst Epidemiol & Statist Div 500 Sherbourne St Toronto ON M4X 1K9 Can

TILL, MICHAEL JOHN, b Independence, Iowa, July 30, 34; m 67; c 2. PEDIATRIC DENTISTRY. *Educ:* Univ Iowa, DDS, 61, MS, 63; Univ Pittsburgh, MEd & PhD(higher educ), 70. *Prof Exp:* Instr pedodontics, Univ Iowa, 61-63; pedodontist, Eastman Inst, Stockholm, Sweden, 63-64; asst prof, Royal Dent Col, Denmark, 64-66; asst prof, Univ Pittsburgh, 66-70; chmn dept, 70-90, PROF PEDODONTICS, UNIV MINN, MINNEAPOLIS, 70- *Concurrent Pos:* Pres, Minn Dent Asn, 88-89; dir, Proj Hope, Portugal, 89-90. *Mem:* Am Dent Asn; Int Asn Dent Res; Am Educ Res Asn; Am Soc Dent Children; Am Acad Pedodont. *Res:* Dental educational research; pedodontics. *Mailing Add:* Pediat Dent 6-150 Moos Univ Minn Minneapolis MN 55455

TILLAY, ELDRID WAYNE, b Yerington, Nev, Feb 26, 25; m 53; c 2. INORGANIC CHEMISTRY. *Educ:* Pac Union Col, BA, 50; Stanford Univ, MS, 52; La State Univ, PhD(inorg chem), 67. *Prof Exp:* Res asst chem, Stanford Univ, 52-57; instr, Sacramento City Col, 57-60; from asst prof to assoc prof, 60-72, PROF CHEM, PAC UNION COL, 72-, HEAD DEPT, 74- *Mem:* Am Chem Soc. *Res:* Organometallic chemistry and bio-inorganic chemistry. *Mailing Add:* 160 Cold Springs Rd Angwin CA 94508-9654

TILLER, CALVIN OMAH, b Richmond, Va, June 22, 25; m 52; c 2. PHYSICS. *Educ:* Col William & Mary, BS, 48; Syracuse Univ, MS, 50. *Prof Exp:* Qual engr, Eastman Kodak Co, 50-51; supvr, Optical Eng Dept, Otis Elevator Co, 51-55; physicist, Titmus Optical Co, 55-56; sr res physicist, Va Inst Sci Res, 56-68; res scientist, Res & Develop Ctr, Philip Morris, Inc, 68-88; RETIRED. *Honors & Awards:* J Sheldon Horsley Award, Va Acad Sci, 60; IR-100 Award, Indust Res, Inc, 72. *Mem:* NAm Thermal Analytical Soc; Am Phys Soc. *Res:* Solid state physics of thin metallic films; electron microscopy and diffraction; physical optics; thermal analysis of tobacco. *Mailing Add:* 1561 King William Woods Rd Midlothian VA 23113

TILLER, F(RANK) M(ONTEREY), b Louisville, Ky, Feb 26, 17; m 82; c 2. CHEMICAL ENGINEERING. *Educ:* Univ Louisville, BChE, 37; Univ Cincinnati, MS, 39, PhD(chem eng), 46. *Hon Degrees:* Dr, Univ Brazil, 62 & State Univ Rio de Janeiro, 67. *Prof Exp:* Technician, Charles R Long, Jr Co, 34-35; chemist, Durkee Famous Foods Div, Glidden Co, 36 & Colgate-Palmolive-Peet Co, 37; civil engr, US Corps Engrs, Ky, 39; chem engr, C M Hall Lamp Co, 40; instr chem eng, Univ Cincinnati, 40-42; from asst prof to assoc prof, Vanderbilt Univ, 42-51; dean eng, Lamar State Col, 51-55; prof chem & elec eng & dean eng, Univ Houston, 55-63, dir, int affairs, 63-67, Latin Am, 66-67, dir ctr study higher educ, Latin Am, 68-73, M D ANDERSON PROF CHEM ENG, UNIV HOUSTON, 63- *Concurrent Pos:* Indust consult, 43-; dir, Gupton-Jones Col Mortuary Sci, 45-51; vis prof, Inst Oleos, Rio de Janeiro, 52, Rice Univ, 72-73, Fed Univ, Rio de Janeiro & Sergipe, 79 & Univ PR, 85; lectr, Humble Oil Col, 58, Esso Res Labs, 62 & Nat Taiwan Univ, 82; Fulbright prof, Univ Guayaquil & Univ Cent Equador, 58; hon prof, Cent Univ Ecuador, Univ Guayaquil, 58, Pontifical Cath Univ, Rio de Janeiro, 63, Fed Univ Santa Mairi, 69, Fed Univ Espirito Santo, 72, Autonomous Univ Guadalajara, 75; AID Univ Contract dir, Univ Guayaquil, 60-64, consult, 64-67; mem, President's Sci Adv Comt, 61, Latin Am Sci Bd/Nat Acad Sci, 63-65 & Int Exchange Persons Comt, Conf Bd, Assoc Rec Coun, 67; consult, Int Coop Admin Mission, Ecuador, 61; titular prof, Cath Univ Rio de Janeiro, 62; grants, Orgn Am States, Univ Brazil, 62-63, NATO sr fac, Univ Col London, 69-70 & Japanese Soc Prom Sci, Nagoya Univ, 82-83; dir, univ contracts, Univ Brazil & Cath Univ Rio de Janeiro, 63-70 & Coun Rectors Brazilian Univs, 66-72; adv, Autonomous Univ Guadalajara, 64-67; consult, Union Tex Petrol contract, Univ Costa Rica, 64-68; joint prog study univ admin & finances, Off Cult Affairs, US Dept State-GULERPE, 66-67; pres, Int Consortium Filtration Res Groups, 71-76; hon pres, First World Cong Filtration, 73; vis prof, Nagoya Univ, 82 & 86, vis res, Commonwealth Sci & Indust Res Orgn, Canberra, Melbourne, 86; ed, Fluid/Particle Separation J, 88- *Honors & Awards:* Streng Award, Univ Louisville, 37; Colburn Award, Am Inst Chem Engrs, 50, Founder's Award, 87; Phillips Lectr, Okla State Univ, 77; Gold Medal, Filtration Soc, 78. *Mem:* Am Inst Chem Engrs; Fine Particle Soc; Am Filtration Soc; Am Soc Eng Educ; Int Consortium Filtration Res Groups; Sigma Xi; hon mem Filtration Soc Belg; Am Ceramic Soc; Water Pollution Control Fedn; Am Asn Drilling Engrs. *Res:* Flow through porous media; solid-liquid separation; thickening, filtration, centrifugation and expression; particle science. *Mailing Add:* Dept Chem Eng Univ Houston Houston TX 77204-4792

TILLER, RALPH EARL, b Birmingham, Ala, Nov 16, 25; m 49; c 3. PEDIATRICS. *Educ:* Birmingham-Southern Col, BS, 47; Tulane Univ, MD, 51; Am Bd Pediat, dipl, 56. *Prof Exp:* Intern, Fitzsimmons Army Hosp, Denver, 51-52; resident pediat, Tulane Univ, 53-54, chief resident, 54-55; pvt pract, Columbus, Ga, 55-67; assoc prof, 67-71, PROF PEDIAT, SCH MED, UNIV ALA, BIRMINGHAM, 71- *Concurrent Pos:* Consult, Martin Army Hosp, US Army Med Corps, Ft Benning, Ga, 59-67; dir, State Crippled Children's Seizure Clin, Columbus, 60-67; chief pediat serv, Med Ctr, Columbus, 62-67; consult, Muscogee Health Dept, 62-67; dir outpatient clin, Children's Hosp, Birmingham, 67-69, dir inpatient teaching serv, Cystic Fibrosis Care Teaching & Res Ctr & Pediat Chest Dis Clin, 69- *Mem:* Am Acad Pediat; Am Thoracic Soc. *Res:* Tuberculosis. *Mailing Add:* 1600 Seventh Ave S Birmingham AL 35233

TILLER, WILLIAM ARTHUR, b Toronto, Ont, Sept 18, 29; m 52; c 2. PHYSICS, PHYSICAL METALLURGY. *Educ:* Univ Toronto, BASc, 52, MASc, 53, PhD(phys metall), 55. *Prof Exp:* Res engr, Res Lab, Westinghouse Elec Corp, 55-57, adv physicist, 57-59, sect mgr crystallogenics, 59-64; exec head dept mat sci, 66-71, PROF MAT SCI, STANFORD UNIV, 64- *Concurrent Pos:* Guggenheim fel, Oxford Univ, 70-71; distinguished vis prof, Univ Del, 81-82. *Mem:* Fel AAAS. *Res:* Solidification and crystal growth; physics of metals; surfaces; properties of materials; solid state physics; biomaterials; psychoenergetics; semiconductor processing. *Mailing Add:* Dept Mat Sci Stanford Univ Stanford CA 94305-2205

TILLERY, BILL W, b Muskogee, Okla, Sept 15, 38; m 59, 81; c 3. SCIENCE EDUCATION. *Educ:* Northeastern Okla State Univ, BS, 60; Univ Northern Colo, MA, 65, EdD(sci educ), 67. *Prof Exp:* Teacher pub schs, Okla & Colo, 60-64; res assoc sci, Univ Northern Colo, 66-67; asst prof sci educ, Fla State Univ, 67-69; assoc prof & dir sci ctr, Univ Wyo, 69-73; assoc prof, 73-75, PROF PHYSICS, ARIZ STATE UNIV, 75- *Mem:* Nat Sci Teachers Asn; Asn Educ Teachers Sci; Nat Asn Res Sci Teaching. *Res:* Physical science textbook and curriculum writer; energy education; science for nonscience students. *Mailing Add:* Dept Physics & Astron Ariz State Univ Tempe AZ 85287-1504

TILLERY, MARVIN ISHMAEL, b Idabel, Okla, Oct 10, 36; m 61; c 3. AEROSOL PHYSICS, INHALATION TOXICOLOGY. *Educ:* Univ NMex, BS, 67; Univ Rochester, MS, 71. *Prof Exp:* Tech assoc aerosol technol, Univ Rochester, 68-70; staff scientist, 71-72, sect leader aerosol technol, 72-83, STAFF SCIENTIST, INDUST HYG GROUP, LOS ALAMOS NAT LAB, 83- *Mem:* Am Indust Hyg Asn; Am Asn Aerosol Res; Gesell Fur Aerosol Forschuns; Inst Environ Sci. *Res:* Aerosol coagulation; generation and characterization instrumentation; aerosol filtration; development of inhalation chambers for toxicology studies; characterization of inhalation hazards; industrial hygiene; respirators; protective equipment. *Mailing Add:* MS K499 Los Alamos Nat Lab PO Box 1663 Los Alamos NM 87545

TILLES, ABE, b New York, NY, Mar 9, 07; m 30; c 2. ELECTRICAL ENGINEERING, FORENSIC ENGINEERING. *Educ:* Univ Calif, BS, 28, MS, 32, PhD(elec eng), 34. *Prof Exp:* Elec tester, Los Angeles Bur Power & Light, 28-30; jr testing engr, State Hwy Testing & Res Labs, Calif, 30; assoc elec eng, Univ Calif, 30-32, from instr to asst prof, 32-45, lectr, Exten Div, 31 & 38, instr defense training, 41, sr elec engr, Manhattan Proj, Radiation Lab, 44-45; asst transmission engr, Pac Gas & Elec Co, 45-54; prof elec eng, Israel Inst Technol, 54-56; consult, R W Thomas & Assocs, Calif, 56-57; sr electronic engr, Lawrence Livermore Lab, Univ Calif, 58-73; CONSULT ENGR, 73- *Concurrent Pos:* Jr elec engr, Los Angeles Bur Power & Light, 36; elec designer, Pac Gas & Elec Co, 37-38; with L S Ready, 39; assoc elec engr, Mare Island Navy Yard, 41; chief elec design engr, Southwest Eng Co, 42; develop engr, Richmond Shipyard, Kaiser Co, 42-44; ed, San Francisco Engr, 47-49; chmn, San Francisco Eng Coun, 53; consult engr, 58- *Honors & Awards:* Recipient, Founder Socs Alfred Noble Prize, 36. *Mem:* Am Soc Eng Educ; fel Inst Elec & Electronics Engrs; Sigma Xi. *Res:* Sparkover; high voltage cable; steel shaft quality; specialized electromagnetic instruments; nuclear science; continuing engineering education; forensic engineering. *Mailing Add:* 2663 Pillsbury Ct Livermore CA 94550

TILLES, HARRY, b Buffalo, NY, Mar 22, 23; m 48; c 4. ORGANIC CHEMISTRY. *Educ:* Univ Buffalo, BA, 48; Univ Calif, PhD(org chem), 51. *Prof Exp:* Res chemist org synthesis, Nat Aniline Div, Allied Chem & Dye Corp, 51-53; res chemist, Stauffer Chem Co, 53-59, group leader indust chem group, 59-62, proj officer, Nat Cancer Inst contract, 62-64, sr res chemist, 64-68, res assoc, Western Res Ctr, 68-74, sr res assoc, De Guigne Tech Ctr, 74-86; RETIRED. *Mem:* Sigma Xi. *Res:* Synthesis of organic chemicals for agricultural screening; optimization of chemical synthesis processes. *Mailing Add:* 703 Balra Dr El Cerrito CA 94532

TILLEY, BARBARA CLAIRE, b San Rafael, Calif, Apr 26, 42. BIOMETRICS, BIOSTATISTICS. *Educ:* Calif State Univ, Northridge, BA, 72; Univ Wash, MS, 75; Univ Tex, PhD(biometry), 81. *Prof Exp:* Biostatistician, Child Develop & Ment Retardation Ctr, 73-74, Mayo Clinic,

74-77; fac assoc, Syst Cancer Ctr, Univ Tex, 78-80, asst prof biomath, Cancer Prevention & Biometry, 80-82; DIV HEAD, DIV BIOSTATIST, RES EPIDEMIOL & COMPUT, HENRY FORD HOSP, DETROIT, MICH, 83-; ADJ ASSOC PROF, UNIV MICH, ANN ARBOR, MICH, 88- *Concurrent Pos:* Mem, policy adv bd, Div Cancer Biol & Diagnostics, Nat Cancer Inst Data Monitoring Group, Minn Colon Cancer Screening Study, 86- *Mem:* Biometrics Soc; Am Statist Asn; Am Pub Health Asn; Am Women Sci. *Res:* Developing and improving a subset selection algorithm for categorical data analysis; carrying out biostatistical research relating to cancer prevention; clinical trials health services. *Mailing Add:* Dept Biostatists & Res Epidemiol Henry Ford Health Syst 23725 Northwestern Hwy Southfield MI 48075

TILLEY, BRIAN JOHN, b Croydon, Eng, Apr 28, 36; m 63; c 3. ELECTRICAL ENGINEERING. *Educ:* Univ Wales, BSc, 61, PhD(elec eng), 65. *Prof Exp:* Technician, Marconis Wireless Tel Co Ltd, Eng, 54-56, jr engr, 56-58; mem sci staff, RCA Res Labs, Montreal, Que, 65-68; mem sci staff, Semiconductor Div, TRW Inc, 68-70, mem sci staff, Systs Microwave Div, 70-71; SECT HEAD, HUGHES AIRCRAFT CO, CULVER CITY, 71- *Mem:* Inst Elec & Electronics Engrs. *Res:* Process and materials; large scale integrated circuits; high-speed, low-power and low-noise transistors; high reliability passive elements. *Mailing Add:* 29026 Indian Valley Rd Palos Verdes Peninsula CA 90274

TILLEY, DAVID RONALD, b Fuquay Springs, NC, Mar 10, 30; m 65; c 1. RADIATIVE CAPTURE, NUCLEAR DATA. *Educ:* Univ NC, BS, 52; Vanderbilt Univ, MS, 54; Johns Hopkins Univ, PhD(nuclear physics), 58. *Prof Exp:* Jr instr physics, Johns Hopkins Univ, 53-58; res assoc nuclear physics, Duke Univ, 58-61, asst prof, 61-66; assoc prof, 66-72, PROF PHYSICS, NC STATE UNIV, 72- *Concurrent Pos:* Staff physicist, Triangle Univs Nuclear Lab, 66- *Mem:* Am Phys Soc; AAAS; Sigma Xi. *Res:* Radiative capture; gamma ray spectroscopy; nuclear reactions. *Mailing Add:* Dept Physics NC State Univ Box 8202 Raleigh NC 27695

TILLEY, DONALD E, b Flushing, NY, July 6, 25; Can citizen; m 48; c 3. PHYSICS. *Educ:* McGill Univ, BSc, 48, PhD(physics), 51. *Hon Degrees:* DSc, Royal Mil Col Can, 88. *Prof Exp:* Res assoc physics, Radiation Lab, McGill Univ, 51-52; from asst prof to prof physics, Col Mil Royal, Quebec, 52-78, head dept, 61-71, dean sci & eng, 69-78; prin, Royal Mil Col Can, Kingston, 78-84; RETIRED. *Mem:* Am Phys Soc; Can Asn Physicists. *Res:* Physics of dielectrics; nuclear reactions; radioactive isotopes. *Mailing Add:* 44 Faircrest Blvd RR 1 Kingston ON K7L 4V1 Can

TILLEY, JEFFERSON WRIGHT, b Detroit, Mich, Dec 13, 46; m 70; c 2. ORGANIC CHEMISTRY, MEDICINAL CHEMISTRY. *Educ:* Harvey Mudd Col, BS, 68; Calif Inst Technol, PhD(chem), 72. *Prof Exp:* Sr chemist, 72-80, group chief, 80-85, RES LEADER, HOFFMAN-LA ROCHE, INC, 85- *Mem:* Am Chem Soc; NY Acad Sci. *Res:* Heterocyclic chemistry; peptide mimetics; design and synthesis of antiallergy agents such as PAF antagonists and leukotriene antagonists; design of novel drugs based on peptide hormones as lead compounds. *Mailing Add:* 19 Evergreen Dr North Caldwell NJ 07006

TILLEY, JOHN LEONARD, b New York, NY, June 4, 28; m 51. MATHEMATICS. *Educ:* Univ Pa, BS, 50; Univ Fla, MEd, 54, PhD(math), 61. *Prof Exp:* Teacher high sch, Fla, 54-56 & St Petersburg Jr Col, 56-58; instr math, Univ Fla, 60-61; from asst prof to assoc prof, Clemson Univ, 61-64; assoc prof, Miss State Univ, 64-69, actg head dept math, 71-72, prof math, 69-88, dir S D Lee hons prog, 69-81, EMER PROF MATH, MISS STATE UNIV, 88- *Concurrent Pos:* Secy-Treas & Newsletter Ed, La-Miss Sect, Math Asn Am, 78-89. *Honors & Awards:* Cert Meritorious Serv, Math Asn Am, 89. *Mem:* Nat Coun Teachers Math; Math Asn Am. *Res:* Classical methods of applied mathematics. *Mailing Add:* PO Box 391 Mississippi State MS 39762-0391

TILLEY, SHERMAINE ANN, b Shawnee, Okla, Feb 22, 52. MOLECULAR IMMUNOLOGY, IMMUNODEFICIENCY DISEASES. *Educ:* Okla City Univ, BA, 73; Johns Hopkins Univ, PhD(biochem), 80. *Prof Exp:* Fel cell biol, Albert Einstein Col Med, 80-85; ASST MEM, PUB HEALTH RES INST CITY NY, INC, 85-; RES ASST PROF PATH, NY UNIV SCH MED, 85- *Concurrent Pos:* Prin investr, Life & Health Ins Med Res Fund grant, Pub Health Res Inst, 86-, NIH pub health serv grant, 88-; lectr cell biol, NY Univ, 87- *Mem:* Am Soc Human Genetics; Am Asn Immunologists. *Res:* Isolation & characterization of human monoclonal antibodies against HIV-1; improvement of the human monoclonal antibody technology; human immunoglobulin gene rearrangement & expression in cells from normal &immunodeficient individuals; the effects of in vitro Epstein-Barr virus transformation on immunoglobulin gene expression. *Mailing Add:* Pub Health Res Inst 455 Fifth Ave Rm 1133 New York NY 10016

TILLEY, STEPHEN GEORGE, b Lima, Ohio, July 21, 43; div; c 2. EVOLUTIONARY BIOLOGY, HERPETOLOGY. *Educ:* Ohio State Univ, BS, 65; Univ Mich, Ann Arbor, MS, 67, PhD(zool), 70. *Prof Exp:* From asst prof to assoc prof, 70-82, chmn dept, 77-80, PROF BIOL SCI, SMITH COL, 83- *Mem:* Soc Study Evolution; Am Soc Ichthyol & Herpet; Soc Study Amphibians & Reptiles; AAAS; Ecol Soc Am. *Res:* Population biology and evolution of amphibians, especially desmognathine salamanders; genetic structures of populations. *Mailing Add:* Dept Biol Sci Smith Col Northampton MA 01063

TILLEY, T(ERRY) DON, b Norman, Okla, Nov 22, 54; m 85; c 2. COORDINATION POLYMERIZATIONS. *Educ:* Univ Tex, Austin, BS, 77; Univ Calif, Berkeley, PhD(chem), 82. *Prof Exp:* Postdoctoral inorg chem, Swiss Fed Inst Technol, 82-83; from asst prof to assoc prof, 83-90, PROF INORG CHEM, UNIV CALIF, SAN DIEGO, 90- *Concurrent Pos:* Alfred P Sloan Found award, 88; consult, Exxon Res & Develop Labs, 90- *Mem:* Am Chem Soc; Royal Soc Chem; Mat Res Soc; Sigma Xi; AAAS. *Res:* Synthetic, mechanistic and catalytic problems in inorganic and organometallic chemistry; coordination polymerizations and silicon-containing polymers; organometallic precursors to solid state materials. *Mailing Add:* Dept Chem 0506 Univ Calif La Jolla CA 92093-0506

TILLING, ROBERT INGERSOLL, b Shanghai, China, Nov 26, 35; US citizen; m 62; c 2. GEOLOGY, VOLCANOLOGY. *Educ:* Pomona Col, BA, 58; Yale Univ, MS, 60, PhD(geol), 63. *Prof Exp:* Geologist, US Geol Surv, Va, 62-72, Hawaiian Volcano Observ, 72-75, scientist-in-chg, Hawaiian Volcano Observ, 75-76, chief off geochem & geophys, Geol Div, 76-81, GEOLOGIST/VOLCANOLOGIST, US GEOL SURV, 82- *Mem:* Geol Soc Am; Mineral Soc Am; Geochem Soc; Int Asn Volcanology & Chem Earth's Interior. *Res:* Igenous petrology and volcanology. *Mailing Add:* US Geol Sur Mail Stop 910 345 Middlefied Rd Menlo Park CA 94025

TILLINGHAST, JOHN AVERY, b New York, NY, Apr 30, 27; m 48; c 3. ELECTRICAL ENGINEERING. *Educ:* Columbia Univ, BS, 48, MS, 49. *Prof Exp:* Mem staff, Am Elec Power Serv Corp, 49-67, exec vpres eng & construct, 67-72, sr exec vpres, 72-75, vchmn eng & construct, 75-79; sr vpres technol, Wheelabrator-Frye Inc, 79-81, chmn, Wheelabrator Utility Serv Inc, 81-86; PRES, TELTEC INC, PORTSMOUTH NH, 86- *Mem:* Nat Acad Eng; fel Am Soc Mech Engrs; Edison Elec Inst; Inst Elec & Electronics Engrs. *Res:* Generating unit control system. *Mailing Add:* 20 Ladd St Portsmouth NH 03801

TILLITSON, EDWARD WALTER, b Charlevoix Co, Mich, Jan 13, 03; m 30; c 3. DENTAL MATERIALS, CHEMICAL ENGINEERING. *Educ:* Univ Mich, BS, 29, MS, 32. *Prof Exp:* Res & develop chem engr, Whiting Swenson Co, 29-31; chief chemist, Iodent Chem Co, 32-35; sr res chemist, Parke Davis & Co, 35-42; head develop labs, Gelatin Prod Div, R P Scherer Corp, 42-46; assoc prof chem eng, Wayne State Univ, 46-63; res assoc, Sch Dent, Univ Mich, 63-73; CONSULT ENGR, 73- *Concurrent Pos:* Consult, 46- *Mem:* Am Chem Soc; Int Asn Dent Res. *Res:* Materials science; polymers; casualty investigation; custom research instrumentation; friction and wear; stress analysis. *Mailing Add:* 9945 Carter Rd Traverse City MI 49684

TILLMAN, ALLEN DOUGLAS, b Rayville, La, June 9, 16; m 45; c 5. ANIMAL HUSBANDRY. *Educ:* Southwestern La Univ, BS, 40; La State Univ, MS, 42; Pa State Univ, PhD, 52. *Prof Exp:* Asst, La State Univ, 40-42; instr animal nutrit, Pa State Univ, 46-48; asst prof, La State Univ, 48-52; from assoc prof to prof animal husb, Okla State Univ, 52-73; vis prof animal husb, Univ Gadjah Mada & field officer, Rockefeller Found, Indonesia, 73-81; CONSULT, 81- *Concurrent Pos:* Gen Educ Bd fel, 51-52; res partic, Oak Ridge Inst Nuclear Studies, 56-57; consult, AEC, 57-, gen chmn, Radioisotope Conf, 59; consult, USDA exchange team, USSR, 59; consult, Estab Labs for Food & Agr Orgn, Arg, 61 & nutrit study, Libya, 62; Fulbright lectr, Univ Col, Dublin, 62-63; Nat Feed Ingredients travel fel, Europe, 66; mem, Comt Animal Nutrit, Nat Res Coun-Nat Acad Sci, 67-73; Ford Found head animal prod div, EAfrica Agr & Res Orgn, Nairobi, Kenya, 69-71. *Honors & Awards:* Am Soc Animal Sci Award, 59; Tyler Award, 67; Int Animal Agr Award, 87. *Mem:* AAAS; Am Soc Animal Sci; Poultry Sci Asn; Am Inst Nutrit; Sigma Xi. *Res:* Metabolism of protein; energy and minerals; vitamin and mineral deficiencies; use of radioisotopes in mineral studies. *Mailing Add:* 523 Hamed Pl Stillwater OK 74075

TILLMAN, FRANK A, b Linn, Mo, July 22, 37; m 59; c 3. INDUSTRIAL ENGINEERING. *Educ:* Univ Mo, BS, 60, MS, 61; Univ Iowa, PhD(indust eng), 65. *Prof Exp:* Instr indust eng, Univ Mo, 60-61; oper res analyst, Standard Oil Ohio, 61-63; instr indust eng, Univ Iowa, 63-65; from asst prof to assoc prof, 65-69, head dept, 66-87, PROF INDUST ENG, KANS STATE UNIV, 69-; CHIEF EXEC OFFICER, HTX INT. *Concurrent Pos:* Res grants, Ford Found, 63-65, NSF, 65-67; NASA fel, 66-68; pres, Systs Res Corp, 68-; dir prog control, Price Comn, Exec Off of the President & consult, 72; consult, USDA; chief exec officer, HTX Int, 87, vpres, IBES, 86. *Mem:* Fel Am Inst Indust Engrs; Sigma Xi. *Res:* Operations research; engineering statistics and the applications of multiple criteria decision methods and optimization to a wide variety of applications. *Mailing Add:* HTX Int 115 N Fourth St Manhattan KS 66502

TILLMAN, J(AMES) D(AVID), JR, b Evansville, Ind, July 4, 21; m 56; c 3. ELECTRICAL ENGINEERING. *Educ:* Univ Tenn, BS, 47, MS, 50; Auburn Univ, PhD, 68. *Prof Exp:* From instr to prof elec eng, Univ Tenn, Knoxville, 47-, dir antenna projs, 56-; RETIRED. *Concurrent Pos:* Researcher, Eng Exp Sta, Univ Tenn, Knoxville, 50-51 & Navy & Air Force Projs, 51-64. *Mem:* Am Soc Eng Educ; Inst Elec & Electronics Engrs. *Res:* Antenna systems and arrays, especially circular symmetry; propagation studies; electronic scanning systems; scattering of pulses from long wires; transient response of antennas. *Mailing Add:* 8200 Fox Run Lane Knoxville TN 37919

TILLMAN, LARRY JAUBERT, b Bay St Louis, Miss, Aug 20, 48; m 70; c 1. ANATOMY. *Educ:* Univ Miss, BA, 70, MS, 72, PhD(histol, electron micros), 74. *Prof Exp:* Chief electron micros, Dept Path, Brooke Army Med Ctr, 74-77; ASST PROF ANAT, MED CTR, UNIV MISS, 78- *Concurrent Pos:* Clin appointee, Dept Anat, Micros Anat Sect, Univ Tex Health Sci Ctr & Med Sch, San Antonio, 75-77. *Mem:* Soc Armed Forces Med Lab Scientists; Electron Micros Soc Am; AAAS. *Res:* Correlation of the fine structure of the cells comprising the uriniferous tubules of Gallus domesticus to their function; use of transmission and scanning electron microscopy in the diagnosis of renal and tumor disease. *Mailing Add:* Dept Phys Ther Ga State Univ University Plaza Atlanta GA 30303

TILLMAN, MICHAEL FRANCIS, b Seattle, Wash, Feb 10, 43; m 65; c 3. CETACEAN BIOLOGY, POPULATION DYNAMICS. *Educ:* Univ Wash, BS, 65, MS, 68, PhD(fisheries), 72. *Prof Exp:* Fishery biologist res, Marine Fish & Shellfish Div, 72-74, leader, Cetaceans Task Unit, 74-76, leader, Cetaceans Res Unit, 76-78, dep dir, Marine Mammal Div, 78-79, dir, Nat Marine Mammal Lab, Nat Marine Fisheries Serv, 79-83, dir, Int Union for Conserv Nature & Natural Resources Conserv Monitoring Ctr, 83-87; chief conserv sci div, 87-90, DEPT ASST ADMIN, NAT MARINE FISHERIES SERV, 90- *Concurrent Pos:* Actg asst prof, Col Fisheries, Univ Wash, 72-73, affil asst prof, 73-78, affil assoc prof, 78-; sci adv to US Comnr, Int Whaling Comn, 74-, vchmn Int Whaling Comn Sci Comt, 79-82; US mem sci comt, Ad Hoc Comt Marine Mammals, Int N Pac Fisheries Comn, 82-85. *Mem:*

Am Fisheries Soc; Am Inst Fishery Res Biologists. *Res:* Biology, abundance, distribution, behavior, migrations and population dynamics of endangered cetacean species; assessments of exploited cetacean stocks. *Mailing Add:* Nat Marine Fisheries Serv 1335 East West Hwy Silver Spring MD 20910

TILLMAN, RICHARD MILTON, b Muskogee, Okla, Sept 7, 28; m 48; c 2. ORGANIC CHEMISTRY. *Educ:* Southern Methodist Univ, BS, 52, MS, 53. *Prof Exp:* From asst res chemist to sr res chemist, Continental Oil Co, 53-61, res group leader, 61-63, tech asst, 63-64, supvry res scientist, 64-67, mgr, Plant Foods Res Div, 67-72, assoc mgr, Petrol Prod Div, Res & Develop, 72-75, mgr, Res Serv Div, Res & Develop, 75-88; CONSULT, 88- *Mem:* Sr mem Am Chem Soc. *Res:* Hydrocarbon fuels; petroleum-based specialties; plant foods; phosphate rock; analytical chemistry; laboratory safety and fire protection; environmental chemistry. *Mailing Add:* 2400 Wildwood Ponca City OK 74604

TILLMAN, ROBERT ERWIN, b Hammondsport, NY, June 29, 37; m 60; c 2. WILDLIFE ECOLOGY. *Educ:* State Univ NY, Albany, AB, 59, MA, 61; Cornell Univ, PhD(environ educ), 72. *Prof Exp:* Teacher biol, Dundee Cent Sch, NY, 59-65; asst prof biol, Dutchess Community Col, 66-69, assoc prof nat res, 71-73; coordr wildlife res, Cary Arboretum, 71-78, CHMN ENVIRON ASSESSMENT, NY BOT GARDEN, 78- *Concurrent Pos:* Mem, NY State Forest Pract Bd, 74-, chmn, 80-; mem, Comnr's Adv Comn, NY State Environ Conserv, 76-78 & NY State Forest Resources Planning Comt. *Mem:* Ecol Soc Am; Inst Ecol; Wildlife Soc. *Res:* Powerline ecology, including vegetation management on powerline rights of way; hydroelectric generation sites, including transmission line rights of way. *Mailing Add:* 1971 Beach Rd Rock Stream NY 14878

TILLMAN, RODERICK W, b Macomb, Il, Feb 24, 34; m 57; c 4. SILICLASTIC SEDIMENTOLOGY, RESERVOIR DESCRIPTION. *Educ:* Univ Wisc, BS, 57, MS, 60; Univ Colo, PhD, 67. *Prof Exp:* Explor geologist, Conoco, 60-62; res geologist, Sinclair Oil Co, 67-69; res assoc, Cities Serv Oil & Gas Corp, 69-85; CONSULT SEDIMENTOLOGIST/STRATIGRAPHER, 85- *Concurrent Pos:* Res assoc, Univ Tulsa, 86-; comt mem, Distinguished lect select comt, Am Asn Petrol Geologist, 82-88; comt mem, Soc Petrol Engrs, 87. *Mem:* Am Asn Petrol Geologists; Soc Sedimentary Geol (pres, 90-91); fel Geol Soc Am; Soc Petrol Engrs; Am Inst Prof Geologists. *Res:* Sequence stratigraphy of silicastic deep water, shelf, shoreline and valley fill sandstones; petroleum reservoir geologic and production description and modeling. *Mailing Add:* 2121 E 51st St Suite 112 Tulsa OK 74105

TILLMAN, STEPHEN JOEL, b Springfield, Mass, Mar 31, 43; m 65; c 2. MATHEMATICS. *Educ:* Brown Univ, ScB, 65, PhD(math), 70; Lehigh Univ, MS, 78. *Prof Exp:* Instr math, Brown Univ, 69-70; asst prof, 70-75, ASSOC PROF MATH, WILKES COL, 75- *Mem:* Am Math Soc; Math Asn Am; Opers Res Soc Am. *Res:* Teaching and developing additional courses in operations research and related areas. *Mailing Add:* Dept Math & Computer Sci Wilkes Col 170 S Franklin St Wilkes-Barre PA 18703

TILLOTSON, JAMES E, b Cambridge, Mass, Feb 9, 29; m 56; c 2. FOOD SCIENCE. *Educ:* Harvard Univ, AB, 53; Boston Univ, MA, 56; Mass Inst Technol, PhD(food sci), 64; Univ Del, MBA, 69. *Prof Exp:* Teacher, Manter Hall Sch, 57-63; res asst nutrit & food sci, Mass Inst Technol, 61-63; training fel, Nat Inst Health, MIT, 63-64; res chemist indust & biochem dept, E I du Pont de Nemours & Co, 64-66, tech rep, Agr Tech Develop, 66-69; dir res & develop, Ocean Spray Cranberries Inc, 69-77, vpres tech res & develop, 77-89; DIR, FOOD POLICY INST, SCH NUTRIT, TUFTS UNIV, 89-, PROF, FOOD POLICY, 89- *Mem:* Fel Am Inst Chem; Am Chem Soc; prof mem Inst Food Technol; Soc Nutrit Educ; Nat Food Processors Asn; Grocery Mfg Am. *Res:* Commercial development of food products and agricultural chemicals; research management; technical forecasting; government regulation of agribusiness. *Mailing Add:* Food Policy Inst Sch Nutrit Tufts Univ 132 Curtis St Medford MA 02155

TILLOTSON, JAMES GLEN, b Brandon, Man, July 20, 23; m 48; c 3. PHYSICS. *Educ:* Univ Man, BSc, 45; Univ Western Ont, MSc, 47. *Prof Exp:* Asst prof physics, Univ NB, 47-53; dir appl physics sect, Can Armament Res & Develop Estab, 53-55; prof physics, Acadia Univ, 55-89; RETIRED. *Concurrent Pos:* Consult, NB Dept Health, 51-52; Defence Res Bd Can grant, 57-59. *Mem:* Inst Elec & Electronics Engrs. *Res:* Acoustic radiation. *Mailing Add:* RR 2 Comp C3 Site 8 Wolfville NS B0P 1X0 Can

TILLOTSON, JAMES RICHARD, b Berkeley, Calif, Oct 3, 33; m 82; c 2. INFECTIOUS DISEASES, INTERNAL MEDICINE. *Educ:* Lehigh Univ, BA, 55; Univ Calif, San Francisco, MD, 59. *Prof Exp:* Instr med, Med Sch, Harvard Univ, 67-68; asst prof, Sch Med, Univ Mich, 68-70; from assoc prof to prof, 70-86, head, Div Infectious Dis, 70-86, CLIN PROF, ALBANY MED COL & MED CTR HOSP, 86- *Concurrent Pos:* Clin res fel, Sch Med, Wayne State Univ, 64-66 & Med Sch, Harvard Univ-Boston City Hosp, 66-68; chief, Div Infectious Dis, Wayne County Gen Hosp, 68-70; consult, Albany Vet Admin Hosp & head, Div Infectious Dis, Albany Med Ctr Hosp, 70-86. *Mem:* Am Fedn Clin Res; Am Soc Microbiol; fel Infectious Dis Soc Am; fel Am Col Clin Pharmacol. *Res:* Pneumonia, especially gram-negative bacillary; prostatitis; antimicrobial activity of anti-tumor drugs; antimicrobial synergy; other areas of clinical microbiology and infectious diseases; medical education, especially instructional methods and decision making. *Mailing Add:* Albany Med Col New Scotland Ave Albany NY 12208

TILLSON, HENRY CHARLES, b Philadelphia, Pa, Sept 16, 23; m 50; c 2. RUBBER CHEMISTRY, POLYMER CHARACTERIZATION. *Educ:* Mass Inst Technol, SB, 44; Pa State Univ, MS, 48, PhD(org chem), 51. *Prof Exp:* Res chemist, Res Ctr, Hercules Inc, 50-52, tech rep, 52-54, res chemist, Res Ctr, 54-86; RETIRED. *Mem:* AAAS; Am Chem Soc. *Res:* Organic nitrogen compounds-nitramines; emulsion polymerization vinyl and condensation polymers, especially protective coatings; rubber compounding and crosslinking; polymer fractionation, mainly polyolifins. *Mailing Add:* Ten Dartmouth Rd Wilmington DE 19808

TILMAN, G DAVID, b Aurora, Ill, July 22, 49; m 71; c 4. RESOURCE COMPETITION THEORY, PLANT ECOLOGY. *Educ:* Univ Mich, BS, 71, PhD(zool), 76. *Prof Exp:* Lectr, 75-76, from asst prof to assoc prof, 76-84, PROF ECOL, UNIV MINN, 84- *Concurrent Pos:* Ed, Limnol & Oceanog, 78-81; mem, bd dirs, Orgn Trop Studies, 80-82 & Ecol Panel, NSF, 89-90; prin investr, Long Term Ecol Res, 81-; John Simon Guggenheim Found fel, 84-85; adv, Theoret Ecol, Lund Univ, Sweden, 85. *Honors & Awards:* W S Cooper Award, Ecol Soc Am, 89. *Mem:* Ecol Soc Am; Brit Ecol Soc; Am Soc Naturalists; Am Soc Limnol & Oceanog; fel AAAS. *Res:* Mechanistic models designed to predict the biodiversity, species composition and dynamics of ecological communities; role of environmental constraints; quantitative tradeoffs organisms face in dealing with several constraints. *Mailing Add:* Univ Minn 318 Church St SE Minneapolis MN 55455

TILNEY, LEWIS GAWTRY, ELECTROMICROSCOPY. *Educ:* Cornell Univ, PhD(biol), 64. *Prof Exp:* PROF BIOL, UNIV PA, 62- *Mailing Add:* Dept Biol Univ Pa Philadelphia PA 19104

TILNEY, NICHOLAS LECHMERE, KIDNEY TRANSPLANTATION. *Educ:* Cornell Univ, MD, 62. *Prof Exp:* PROF SURG, SCH MED, HARVARD UNIV & SURGEON, BRIGHAM & WOMEN'S HOSP, BOSTON, 73- *Res:* Peripheral vascular disease. *Mailing Add:* Dept Surg Harvard Med Sch 25 Shattuck St Boston MA 02115

TILSON, BRET RANSOM, b Yuba City, Calif, May 19, 37. MATHEMATICS, COMPUTER SCIENCE. *Educ:* Mass Inst Technol, BS, 60; Univ Calif, Berkeley, PhD(math), 69. *Prof Exp:* Asst prof math, Columbia Univ, 69-74; asst prof, 74-77, ASSOC PROF MATH, QUEEN'S COL, NY, 77- *Concurrent Pos:* Dir res & develop, Comarc Design Systs, San Francisco, 80-81. *Res:* Decomposition and complexity of finite semigroups; automata theory. *Mailing Add:* 1517 Holly St Berkeley CA 94703

TILSON, HUGH ARVAL, b Plainview, Texas, July 24, 46; m 81. BEHAVIORAL NEUROSCIENCE & TOXICOLOGY. *Educ:* Texas Tech Univ, BA, 68; Univ Minn, PhD, 72. *Prof Exp:* Res assoc, Michigan State Univ, 72-73; PHARMACOLOGIST, NAT INST ENVIRON HEALTH SCI, 76-88. *Concurrent Pos:* Consult, 83- *Mem:* Am Soc Pharm & Exp Therapeut; Sigma Xi; Int Neurotoxicol Asn; Soc Neurosci; Soc Toxicol. *Res:* Compensatory and modulatory processes associated with progressive degeneration. *Mailing Add:* Neurotoxicol Div US Environ Protection Agency MD-74B Res Triangle Park NC 27711

TILSWORTH, TIMOTHY, b Norfolk, Nebr, Apr 6, 39; m 84; c 2. GENERAL ENVIRONMENTAL SCIENCES, WATER RESOURCES. *Educ:* Univ Nebr-Lincoln, BS, 66, MS, 67; Univ Kans, PhD(environ health eng), 70. *Prof Exp:* Lab technician, Nitrogen Div, Allied Chem & Dye Corp, 60-61; civil engr technician, Scott Eng, 61-62; asst city engr, Norfolk, Nebr, 62-64; instr civil eng, Univ Nebr, 67; asst prof environ health eng, 70-74, head prog, 71-76, asst to pres, 76-77, assoc prof environ qual eng & civil eng, 74-84, PROF ENVIRON QUAL ENG & CIVIL ENG, UNIV ALASKA, FAIRBANKS, 84- *Concurrent Pos:* Spec consult, Philleo Eng & Archit Serv, 71- & Hill, Ingman & Chase & Co, 71-74; owner, Tilsworth & Assoc, 72-85, Alaska Arctic Environ Serv, 85- & DJT's Shelties Delight, 85-; proj mgr, State of Alaska Proposal for Superconducting Super Collider, 87; head, Dept Civil Eng, Univ Alaska, 89-, chmn, Grad Coun, 89- , Univ Alaska Chancellor Search Comt, 89- *Mem:* Am Soc Civil Engrs; Asn Prof Environ Eng; Water Pollution Control Fedn; Am Water Works Asn. *Res:* Environmental health engineering; pollution control; solid waste management; biological waste water treatment; physical and chemical treatment; water quality; air pollution control; environmental impact assessment. *Mailing Add:* Prog Environ Qual Eng Univ Alaska 306 Tanana Dr Fairbanks AK 99775

TILTON, BERNARD ELLSWORTH, pharmacology, for more information see previous edition

TILTON, GEORGE ROBERT, b Danville, Ill, June 3, 23; m 48; c 4. GEOCHEMISTRY. *Educ:* Univ Ill, BS, 47; Univ Chicago, PhD(chem), 51. *Hon Degrees:* DSc, Swiss Inst Technol, Zürich, 84. *Prof Exp:* Asst, Univ Chicago, 47-51; mem staff, Dept Terrestrial Magnetism, Carnegie Inst, 51-56, phys chemist, Geophys Lab, 56-65; PROF GEOCHEM, UNIV CALIF, SANTA BARBARA, 65- *Concurrent Pos:* Assoc ed, Geochimica Cosmochimica Acta, 74-; guest prof, Swiss Fed Inst Technol, Zurich, 71-72. *Honors & Awards:* Alexander von Humboldt Sr Scientist Award, 89. *Mem:* Nat Acad Sci; Geochem Soc (pres, 80-81); fel Am Geophys Union; fel Geol Soc Am; Meteoritical Soc; fel AAAS. *Res:* Geochemical studies applied to origin of volcanic and plutonic rocks, and the evolution of the Earth's crust and mantle; isotopic composition of lead in terrestrial and meteoritic materials; geologic age of minerals. *Mailing Add:* Dept Geol Sci Univ Calif Santa Barbara CA 93106

TILTON, VARIEN RUSSELL, b Willimantic, Conn, Aug 10, 43; m 76; c 1. PLANT BIOENGINEERING. *Educ:* Northern Ariz Univ, BS, 66, MS, 75; Iwo State Univ, PhD (bot), 78. *Prof Exp:* Fel exp plant embroyol, Rockefeller Found, Univ Minn, 78-79; res assoc develop bot, Univ Ariz, 79; vis asst prof bot & electron micros, Iowa State Univ, 79-80; vis asst prof biol, cytol & electron micros, Bowling Green State Univ, 80-81; RES SCIENTIST CELL BIOL, AGRIGENETICS CORP, 81-; PRIN SCIENTIST BIOTECHNOL, AGR RES CTR, DEL MONTE CORP. *Concurrent Pos:* Adj asst prof bot, Univ Wis, Madison, 81- & Iowa State Univ, 82- *Mem:* Am Inst Biol Sci; Am Soc Agron; Bot Soc Am; Electron Micros Soc Am; Sigma Xi; Int Asn Plant Molecular Biologist. *Mailing Add:* Agr Res Ctr Del Monte Corp PO Box 36 San Leandro CA 94577-0406

TIMASHEFF, SERGE NICHOLAS, b Paris, France, Apr 7, 26; nat US; m 53; c 1. PHYSICAL BIOCHEMISTRY. *Educ:* Fordham Univ, BS, 46, MS, 47, PhD(chem), 51. *Hon Degrees:* Dr, Univ Aix-Marseille, France, 90. *Prof Exp:* Instr chem, Fordham Univ, 47-49, lectr, 49-50; res fel, Calif Inst Technol, 51 & Yale Univ, 51-55; prin phys chemist, Eastern Regional Res Lab, USDA,

Pa, 55-66, head pioneering res lab, Mass, 66-73; PROF BIOCHEM, BRANDEIS UNIV, 66- *Concurrent Pos:* NSF sr res fel, Macromolecule Res Ctr, France, 59-60; head, phys-chem invest, Milk properties lab, Eastern Regional Res Lab, USDA, Philadelphia, 61-66, adj prof, Drexel Inst Technol, 63-64; vis prof, Univ Ariz, 66; mem, Fordham Univ Coun, 68-; mem, Biophys-Phys Biochem Study Sect, NIH, 68-72; Guggenheim fel, Inst Molecular Biol, Paris, 72-73; vis prof, Univ Paris, 72-73; co-ed, Biol Macromolecules; exec ed, Archives of Biochem & Biophys, 70-86, volume ed, Methods in Enzymology; vis prof, Duke Univ, 77, Univ Tech de Lille, 80, Pierre et Marie Curie Univ de Paris, 82-83, Univ de Paris, Orsay, 86-87; distinguished lectr, Univ Maine, 79; sr int fel, Fogarty Int Ctr, NIH, 86-87. *Honors & Awards:* Am Chem Soc Award, 63 & 66, Arthur H Flemming Award, 64; Frances Stone Burns Award, Am Can Soc, 74; Kelly Lectr, Purdue Univ, 85. *Mem:* Fel AAAS; Am Chem Soc; Am Soc Biol Chemists; Biophys Soc; Sigma Xi; Am Soc Biochem & Molecular Biol. *Res:* Structure and interactions of proteins and nucleic acids; physical methods of high polymer studies; solution thermodynamics of macromolecules; author of 226 research publications. *Mailing Add:* Dept Biochem Brandeis Univ Waltham MA 02154

TIMBERLAKE, JACK W, b Middletown, Ohio, May 26, 40; m 62; c 2. ORGANIC CHEMISTRY. *Educ:* Univ Ill, MS, 65, PhD, 67. *Prof Exp:* Fel org chem, Univ Calif, Irvine, 67-68; from asst prof to assoc prof, 68-78, PROF ORG CHEM, UNIV NEW ORLEANS, 78- *Concurrent Pos:* Petrol Res Found grant, Am Chem Soc, 75-77 & 76-79; Army Res Off grant, 76-78; consult med prog, Sch Med, Tulane Univ, 76-79; Diamond Shamrock crop grant, 77-79. *Mem:* Am Chem Soc; Sigma Xi. *Res:* Physical and synthetic organic chemistry; free radical reactions; small ring heterocycles; synthesis and screening of new anti-convulsants. *Mailing Add:* Dept Chem Univ New Orleans New Orleans LA 70122

TIMBERLAKE, JOSEPH WILLIAM, b Kansas City, Kans, Sept 5, 40; m 67. CLINICAL BIOCHEMISTRY. *Educ:* Univ Mo-Kansas City, BA, 63, MS, 69; Univ Kans Med Ctr, Kansas City, PhD(biochem), 74. *Prof Exp:* Technologist clin chem, Res Hosp & Med Ctr, 63-65; develop chemist, Univ Kans Med Ctr, Kansas City, 66-69; lab dir, Statlabs of Kans, Inc, 73-75; CLIN BIOCHEMIST, KANSAS CITY GEN HOSP-UNIV MO MED SCH, 75- *Mem:* Am Asn Clin Chemists. *Res:* Developmental clinical biochemistry. *Mailing Add:* Apt No 101 8325 E Harry Wichita KS 67207

TIMBERLAKE, WILLIAM EDWARD, b Washington, DC, May 2, 48; m 69; c 2. MICROBIOLOGY, MYCOLOGY. *Educ:* State Univ NY Col Forestry, BS, 70; State Univ NY Col Environ Sci & Forestry, MS, 72, PhD(biol), 74. *Prof Exp:* Assoc, Univ Geneva, 74; asst prof biol, Wayne State Univ, 74-79, assoc prof, 79-81; prof plant path, Univ Calif, Davis, 81-86; PROF GENETICS & PLANT PATH, UNIV GA, ATHENS, 86- *Mem:* AAAS; Genetics Soc Am; Mycol Soc Am; Sigma Xi; Am Soc Microbiol; Int Soc Plant Molecular Biol; Soc Develop Biol. *Res:* Genetic regulation of development in fungi. *Mailing Add:* Dept Genetics Univ Ga Athens GA 30602

TIMBERS, GORDON ERNEST, b Regina, Sask, Sept 14, 40; c 1. FOOD SCIENCE. *Educ:* Univ BC, BSA, 62, MSA, 64; Rutgers Univ, New Brunswick, PhD(food sci), 71. *Prof Exp:* Res scientist, 64-82, SR SCIENTIST, ENG & STATIST RES INST, CAN DEPT AGR, 83-, HEAD, FOOD ENG SECT, 80- *Concurrent Pos:* Adj prof, Univ Man, Univ Giuelph & McDonald Col, McGill Univ. *Honors & Awards:* Prix Innovation Technique, Indust Alimentaires et Agricole for Cryogran, 72; Royal Gordon Maybee Award, CIFST, 84; Inst Award, Can Inst Food Sci & Technol, 88. *Mem:* Can Inst Food Sci & Technol (pres, 82-83); Inst Food Technologists. *Res:* Food engineering; development of new processes and equipment for food processing; unit operations; thermal properties of food products. *Mailing Add:* Five Dallas Pl Ottawa ON K2G 3E2 Can

TIMBIE, PETER T, b Hartford, Conn, Aug 19, 57; m 89. COSMOLOGY, DETECTOR DEVELOPMENT. *Educ:* Harvard Univ, BA, 79; Princeton Univ, PhD(physics), 85. *Prof Exp:* Postdoctoral fel physics, Princeton Univ, 85-87 & Univ Calif, Berkeley, 87-90; ASST PROF PHYSICS, BROWN UNIV, 90- *Concurrent Pos:* NSF presidential young investr, 90. *Mem:* Am Phys Soc; Am Astron Soc. *Res:* Cosmic microwave background radiation and its anisotropy through the development of sensitive microwave and infrared detectors. *Mailing Add:* Dept Physics Brown Univ Box 1843 Providence RI 02912

TIMBLIN, LLOYD O, JR, b Denver, Colo, June 25, 27; m 50; c 1. WATER RESOURCES ENGINEERING, PHYSICAL SCIENCE. *Educ:* Univ Colo, BS, 50; Univ Denver, MS, 67. *Prof Exp:* Physicist, 50-58, head Spec Invests Lab Sect, 58-63, chief Chem Eng Br, 63-70, CHIEF APPL SCI BR, RES & LAB SERV DIV, US BUR RECLAMATION, 70- *Concurrent Pos:* Mem Colo adv coun, Sem Environ Arts & Sci; chmn US team, US/USSR Joint Study Plastic Films & Soil Stabilizers, 75-85; accredited corrosion specialist, Nat Asn Corrosion Engrs, 76-; mem, Nat Sanit Found Comt on Flexible Membrane Liners, 79-87; chmn, Comt Mat Embankment Dams, 81-86; chmn, Tech Prog Int Symp Geomembranes, Denver, Colo, 84; mem, Int Comn Large Dams Comt Environ, 86-; chmn, Comt Environ Effects, 88-; mem, US Comt Large Dams; US tech coordr, Sci & Tech Coop Study Water Resources, US Bur Reclamation & Ministry Agr, Israel. *Honors & Awards:* Meritorious Serv Award, US Dept Interior, 77. *Mem:* Am Phys Soc; Am Water Works Asn; Nat Asn Corrosion Engrs; Ecol Soc Am; Sigma Xi. *Res:* Water resources development and management; corrosion engineering; cathodic protection; radioisotopes applications; water quality and pollution control; reservoir and river ecology; materials analysis and development; water treatment and desalting; aquatic weed control; remote sensing; global climate change. *Mailing Add:* 355 Martin Dr Boulder CO 80303

TIMELL, TORE ERIK, b Stockholm, Sweden, Mar 31, 21; nat US; m 47; c 3. ORGANIC CHEMISTRY. *Educ:* Royal Inst Technol, Sweden, ChemE, 46, lic, 48, DrTech(cellulose chem), 50. *Prof Exp:* Chief asst, Royal Inst Technol, Sweden, 46-50; res assoc chem, McGill Univ, 50-51; res assoc, State Univ NY Col Forestry, Syracuse, 51-52; Hibbert Mem fel, McGill Univ, 52, res assoc, 53-62; PROF FOREST CHEM, STATE UNIV NY COL ENVIRON SCI & FORESTRY, SYRACUSE, 62- *Concurrent Pos:* Chemist, Pulp & Paper Res Inst Can, 53-59, res group leader, 60-62. *Honors & Awards:* Anselme Payen Award, Am Chem Soc, 71. *Mem:* Am Chem Soc; Chem Inst Can; Int Acad Wood Sci; Int Asn Wood Anat. *Res:* Chemistry of wood and bark; ultrastructure, cytology and physiology of wood and bark; reaction wood. *Mailing Add:* State Univ NY Col Environ Sci & Forestry Syracuse NY 13210

TIMIAN, ROLAND GUSTAV, b Langdon, NDak, Mar 5, 20; m 49; c 5. PLANT VIROLOGY. *Educ:* NDak Agr Col, BS, 49, MS, 50; Iowa State Col, PhD(plant path), 53. *Prof Exp:* Asst bot, NDak Agr Col, 47-49, fed agent pathologist, 49-50; path res asst, Iowa State Col, 50-53; res plant pathologist, Agr Res, NDak State Univ, USDA, 53-86; RETIRED. *Concurrent Pos:* Tech adv barley, NCent Region, Agr Res Serv, USDA, 73-80, mem nat barley improv comt, 78-83; plant pathol consult, NDak State Univ, USDA, 86- *Mem:* Am Phytopath Soc; Sigma Xi. *Res:* Cereal virus diseases; diseases in barley, especially virus diseases; serology; host-virus interaction. *Mailing Add:* Dept Plant Path NDak State Univ Fargo ND 58102

TIMIRAS, PAOLA SILVESTRI, b Rome, Italy, July 21, 23; nat US; m 46; c 2. DEVELOPMENTAL PHYSIOLOGY, NEUROENDOCRINOLOGY. *Educ:* Univ Rome, MD, 47; Univ Montreal, PhD(exp med, surg), 52. *Prof Exp:* Asst prof exp med & surg, Univ Montreal, 50-51, asst prof physiol, 51-53; asst prof pharmacol, Univ Utah, 54-55; asst physiologist, 55-58, from asst prof to assoc prof, 58-67, PROF PHYSIOL & CHMN DEPT, UNIV CALIF, BERKELEY, 67- *Mem:* AAAS; Endocrine Soc; Am Soc Pharmacol & Exp Therapeut; Am Physiol Soc; Geront Soc; Int Soc Develop Neurosci. *Res:* Endocrinology; environmental physiology; aging. *Mailing Add:* Dept Molecular & Cell Biol Univ Calif 221 Life Sci Add Berkeley CA 94720

TIMKO, JOSEPH MICHAEL, b Danville, Ill, May 20, 49. ORGANIC CHEMISTRY. *Educ:* Univ Ill, Urbana, BS, 71; Univ Calif, Los Angeles, PhD(chem), 75. *Prof Exp:* Fel org chem, Univ Wis-Madison, 75-77; RES CHEMIST ORG CHEM, UPJOHN CO, 77- *Concurrent Pos:* NIH fel, Univ Wis-Madison, 76-77. *Mem:* AAAS; Am Chem Soc. *Res:* Development of synthesis and transformations of medicinally important compounds. *Mailing Add:* 12368 Sprinkle Vicksburg MI 49097

TIMKOVICH, RUSSELL, b East Chicago, Ind. BIOCHEMISTRY. *Educ:* Mich State Univ, BS, 70; Calif Inst Technol, PhD(chem), 74. *Prof Exp:* Fel, Calif Inst Technol, 74-75; asst prof, 75-80, assoc prof, Ill Inst Technol, 80-; PROF, CHEM DEPT, UNIV ALA. *Mem:* Am Chem Soc; Am Soc Biol Chemists; Biophys Soc. *Res:* Biochemistry and biophysics of electron transport proteins and enzymes; bacterial electron transport systems. *Mailing Add:* Dept Chem Univ Ala Box 870336 Tuscaloosa AL 35487-0336

TIMM, DELMAR C, b Muscatine, Iowa, Aug 19, 40; m 62; c 3. CHEMICAL ENGINEERING. *Educ:* Iowa State Univ, BS, 62, MS, 65, PhD(chem eng), 67. *Prof Exp:* Chem engr, Esso Res & Eng Co, 62-63; from asst prof to assoc prof chem eng, 67-75, PROF CHEM ENG, UNIV NEBR, LINCOLN, 75- *Mem:* Am Chem Soc; Am Inst Chem Engrs. *Res:* Crystallization from solution; kinetics of polymerization; molecular characterization thermosets. *Mailing Add:* 2933 Jackson Dr Lincoln NE 68502

TIMM, GERALD WAYNE, b Brandon, Minn, Dec 9, 40; m 75; c 4. BIOMEDICAL ENGINEERING, NEURO-UROLOGIC DEVICES. *Educ:* Univ Minn, Minneapolis, BEE, 63, MS, 65, PhD(elec eng), 67. *Prof Exp:* Assoc prof neurol, Med Sch, Univ Minn, Minneapolis, 67-76, adj assoc prof mech eng, 78-83; exec vpres, Am Med Systs, Inc, 76-79; PRES, CHMN, CEO, DACOMED CORP, 80-; FOUNDER & CHMN, TIMM URO-CARE INST, 89- *Concurrent Pos:* Res fel elec eng, Univ Minn, Minneapolis, 67, NIH trainee, 70-72, prin investr, NIH grant, 74-76; mem grad fac, Univ Minn, 71-76; consult, Baylor Col Med, 72-76, Purdue Univ, 74- & Long Beach, Vet Admin Med Ctr, res assoc neurol, 83-85. *Honors & Awards:* Gold Medal, Urol Div AMA, 74. *Mem:* AAAS; Inst Elec & Electronics Engrs; sr mem Instrument Soc Am; NY Acad Sci. *Res:* Lower urinary tract and male erectile function; investigations of instrumentation systems and devices to diagnose and treat impaired genito-urinary and gastrointestinal function. *Mailing Add:* Dacomed Corp 1701 E 79th St Suite 17 Minneapolis MN 55425

TIMM, RAYMOND STANLEY, b Bay City, Mich, Nov 28, 18; m 44; c 8. ELECTRONICS ENGINEERING, OPERATIONS RESEARCH. *Educ:* Lawrence Inst Technol, BSc, 42. *Prof Exp:* Assoc sect head electronic systs, Naval Res Lab, 42-48; sr engr naval res, Off Naval Res, 48-51; asst to pres component res, Balco Corp, 51-55; sr analyst electronic systs, Westinghouse Elec Corp, 55-57; br head systs anal, Melpar Corp, 57-58; div mgr systs anal, Anal Serv Inc, 58-81, consult MIT Systs, 81-86; RETIRED. *Mem:* Inst Elec & Electronics Engrs; sr mem Opers Res Soc Am. *Res:* Analysis of electronic aides to navigation; research and development of antenna filter networks. *Mailing Add:* 9803 Singleton Dr Bethesda MD 20817

TIMMA, DONALD LEE, b Lebanon, Kans, Sept 1, 22; m 45; c 1. CHEMISTRY. *Educ:* Kans State Col, BS, 44; Ohio Univ, PhD(chem), 49. *Prof Exp:* Spectrographer, Tenn Eastman Co, 44-45; asst chem, Ohio Univ, 46-48; spectrographer, Monsanto Chem Co, 45-46, res physicist, Mound Lab, 49-50, res group leader, 50-51, res sect chief, 51-57; dir anal chem & methods develop, Mead Johnson & Co, 57-67, exec dir control lab, 67, vpres qual control, 67-76, mgr contract mfg, 76-87; RETIRED. *Mem:* Am Chem Soc. *Res:* Analytical instrumentation; emission and absorption spectroscopy. *Mailing Add:* 9402 Petersburg Rd Evansville IN 47711-1461

TIMME, ROBERT WILLIAM, b Victoria, Tex, July 22, 40; m 62; c 2. APPLIED PHYSICS. *Educ:* Tex A&M Univ, BS, 62; Rice Univ, MA, 69, PhD(physics), 70. *Prof Exp:* Res physicist, Ames Res Lab, NASA, 62-65, aerospace engr, Manned Spacecraft Ctr, 65-66; res assoc solid state physics, Rice Univ, 66-70; prin engr electromagnetics, Lockheed Electronics Co, Inc, 70-71; head mat res sect, 73-79, RES PHYSICIST, NAVAL RES LAB, 71-, HEAD, TRANSDUCER BR, 79- *Concurrent Pos:* Mem, Transducer Mat

Comt, Naval Sea Systs Command, 73-; mem oceanology adv group, Naval Res Lab, 75-78; prof oceanology, Fla Inst Technol, 75-78; program manager, Sonar Transducer Reliability Improvement Program, 78- *Honors & Awards:* NASA Achievement Awards, Manned Spacecraft Ctr, 66 & Hq, 73; Res Publ Award, Naval Res Lab, 76, 79 & 80, Performance Awards, 74, 78 & 80. *Mem:* Am Phys Soc; Acoust Soc Am. *Res:* Characterization of the effects of stress, temperature and time on the piezoelectric, magnetostrictive, elastic and acoustic properties of materials applicable to sonar systems; development of long life, reliable sonar transducers. *Mailing Add:* Naval Res Labs PO Box 568337 Orlando FL 32856-8337

TIMMER, KATHLEEN MAE, b Ellsworth, Mich, July 21, 42. SOFTWARE DEVELOPMENT. *Educ:* Calvin Col, BS, 64; Purdue Univ, MS, 66; Colo State Univ, PhD(math), 72. *Prof Exp:* Instr math, Calvin Col, 66-68; ASST PROF MATH, JACKSONVILLE UNIV, 72- *Mem:* Am Math Soc. *Mailing Add:* Dept Math Jacksonville Univ Jacksonville FL 32211

TIMMERHAUS, K(LAUS) D(IETER), b Minneapolis, Minn, Sept 10, 24; m 52; c 1. CHEMICAL ENGINEERING. *Educ:* Univ Ill, BS, 48, MS, 49, PhD(chem eng), 51. *Prof Exp:* Process design engr, Calif Res Corp, Standard Oil Co, Calif, 52-53; from asst prof to prof, Univ Colo, Boulder, 53-61, assoc dean eng & dir Eng Res Ctr, 63-86, PROF CHEM ENG, UNIV COLO, BOULDER, 61- *Concurrent Pos:* Lectr, Exten, Univ Calif, 53 & Univ Calif, Los Angeles, 61-62, 67, 70 & 78-85; ed, Advan in Cryogenic Eng, 54-81; consult numerous govt & indust orgn, 55-; mem & secy-treas, Cryogenic Eng Conf Bd, 56-66, 70-; co-ed, Int Cryogenic Monogr Ser; sect head, NSF, Washington, DC, 72-73; 3-M lectr award, Am Soc Eng Educ, 80; dir, Sigma Xi, 80-86 & Am Soc Eng Educ, 86-88; presidential teaching scholar, Univ Colo, Boulder, 89- *Honors & Awards:* S C Collins Award, 67; G Westinghouse Award, Am Soc Eng Educ, 68; Founders Award, Am Inst Chem Engrs, 78; Eminent Chem Eng Award, Am Inst Chem Engrs, 83, W K Lewis Award, 87; W T Pentzer Award, Int Inst Refrig, 89. *Mem:* Nat Acad Eng; fel Am Inst Chem Engrs (vpres & pres); Am Soc Eng Educ; Am Astronaut Soc; Austrian Acad Sci; fel AAAS; Am Acad Environ Engrs; Sigma Xi (pres elect, pres 86); Am Soc Heating Refrigerating & Air Conditioning Engrs; Int Inst Refrig (vpres, 79-87, pres, 87-). *Res:* Cryogenic processes; heat transfer; distillation; thermodynamic properties. *Mailing Add:* Eng Ctr Campus Box 424 Univ Colo Boulder CO 80309

TIMMERMANN, BARBARA NAWALANY, b Suffolk, Eng, May 30, 47; div; c 2. NATURAL PRODUCTS CHEMISTRY. *Educ:* Univ Nacional de Cordoba, Argentina, BA, 70; Univ Tex, Austin, MA, 77, PhD(bot), 80. *Prof Exp:* Res asst, Univ Nacional de Cordoba, 67-70; res asst, Univ Tex, Austin, 70-72, teaching asst biol, 70-72, instr, 80; res assoc, 81-85, asst res scientist, 85-87, asst prof, 87-90, ASSOC PROF, UNIV ARIZ, 90- *Mem:* Phytochem Soc NAm; Am Soc Pharmcognosy; Bot Soc Am; Europ Phytochem Soc; Int Soc Chem Ecol. *Res:* Chemical studies of North and South American desert plants, including the isolation and identification of natural products and evaluation of their potential uses as alternate sources of energy and chemical feedstocks; biocides and antitumor and antiviral agents. *Mailing Add:* Dept Pharmaceut Sci Univ Ariz Tucson AZ 85721

TIMMERMANN, DAN, JR, b New Braunfels, Tex, Oct 8, 33; m 55; c 2. BOTANY, CROP BREEDING. *Educ:* Tex A&M Univ, BS, 55, PhD(bot), 67; Ohio State Univ, MA, 62. *Prof Exp:* Teacher high sch, Tex, 58-61 & 62-63; asst prof, 67-70, ASSOC PROF BOT, ARK STATE UNIV, 70- *Concurrent Pos:* Consult plant breeding, Plant Res Div, Bryco, Inc, 74- *Mem:* Bot Soc Am. *Res:* Electron microscopy of plant cells; effects of gaseous air pollutants on the ultrastructure of specific cell organelles; cotton and soybean breeding. *Mailing Add:* Dept Biol Sci Ark State Univ Box 599 State University AR 72467

TIMMINS, ROBERT STONE, b Dallas, Tex, Aug 25, 33; m 55; c 3. CHEMICAL ENGINEERING. *Educ:* Univ Tex, BS, 55; Mass Inst Technol, SM, 57, ScD, 59. *Prof Exp:* Sr engr prod res, Sun Oil Co, 59-61, group leader reservoir anal, 61-62; sect chief energy transfer & nuclear effects, Avco Corp, 62-63, asst mgr mat dept, 63-64, mgr mat sci dept, 64-66; vpres res & develop, Abcor, Inc, 66-70, exec vpres, 70-71, pres, 72-77; sr vpres, Cobe Labs, Inc, 77-88; PRES & CHIEF EXEC OFFICER, ORGANON TEKNIKA, 88- *Mem:* Am Inst Chem Engrs; Am Chem Soc. *Res:* Separation and purification; materials; high temperature chemistry. *Mailing Add:* Organon Teknika 100 Akzo Ave Durham NC 27704

TIMMIS, GERALD C, b Apr 23, 30. CARDIOLOGY. *Educ:* Univ Detroit, BS, 51; Wayne State Univ, MD, 55; Am Bd Internal Med, cert, 62 & 65. *Prof Exp:* Intern, Harper Hosp, 55-56, resident internal med, 56-58; fel, Dept Med & Cardiol, Wayne State Univ, 58-60; chief cardiovasc serv, US Army Med Corps, Ft Belvoir, Va, 60-62; asst prof, 62-67, CLIN ASST PROF, DEPT PEDIAT, WAYNE STATE UNIV SCH MED, 67-; CLIN PROF HEALTH SCI & MED PHYSICS, OAKLAND UNIV, 85- *Concurrent Pos:* Co-dir, Cardiovasc Lab, Children's Hosp Mich, 61-67, attend staff, Cardiol Dept, 61-75; co-dir, Cardiovasc Lab, Harper Hosp, Detroit, Mich, 67-73 & Div Cardiovasc Dis, William Beaumont Hosp, Royal Oak, Mich, 67-87; assoc med dir res, William Beaumont Res Inst, Royal Oak, Mich, 77-83, dir, Cardiovasc Lab, Hosp, 80-, dir cardiovasc res, 87-, consult staff, Div Cardiol, Troy, Mich, 78-89; clin prof health sci & internal med, Oakland Univ, 83-84. *Mem:* Fel Am Col Cardiol; fel Am Col Angiol; sr fel Soc Cardiac Angiography; fel Am Col Physicians; fel Am Col Chest Physicians; Am Heart Asn; AMA; Am Fedn Clin Res; Am Col Nuclear Cardiol. *Res:* Author of numerous articles, books and chapters in the field of cardiovascular studies. *Mailing Add:* 26277 Dundalk Lane Farmington Hills MI 48018

TIMMONS, DARROL HOLT, b Little River, Kans, July 7, 40; m 62; c 2. NUCLEAR ENGINEERING. *Educ:* Kans State Univ, BS, 63, MS, 66, PhD(nuclear eng), 69. *Prof Exp:* Instr nuclear eng, Kans State Univ, 67-68; asst prof nuclear eng, Univ Mo-Columbia, 68-74, assoc prof, 74-79; sr engr, 79, mgr neutronics develop, 79-81, mgr Incore Monitoring, 81-83, staff engr, 83-85, MGR INCORE MONITORING SOFTWARE DEVELOP, EXXON NUCLEAR CO, 85- *Concurrent Pos:* Am Soc Eng Educ/Ford Found Eng Residency Prog partic, Commonwealth Edison Co, Ill, 71-72. *Mem:* Am Nuclear Soc. *Res:* Nuclear reactor physics, fuel management and incore monitoring. *Mailing Add:* 2521 Davison Ave Richland WA 99352

TIMMONS, RICHARD B, b Sherbrooke, Que, June 23, 38; m 63; c 3. PHYSICAL CHEMISTRY. *Educ:* St Francis Xavier Univ, BS, 58; Cath Univ Am, PhD(chem), 62. *Prof Exp:* Fel chem kinetics, Brookhaven Nat Lab, 62-64; asst prof chem, Boston Col, 64-65; from asst prof to prof chem, Cath Univ Am, 65-77; chmn dept, 77-90, PROF CHEM, UNIV TEX, ARLINGTON, 77- *Mem:* Am Chem Soc. *Res:* Chemical kinetics; kinetic isotope effects on reaction rates; heterogeneous catalysis; plasma chemistry. *Mailing Add:* Dept Chem Univ Tex Box 19088 Univ Tex Arlington Sta Arlington TX 76019

TIMMONS, THOMAS JOSEPH, b Webster City, Iowa. FISHERIES BIOLOGY, ICHTHYOLOGY. *Educ:* Iowa State Univ, BS, 72; Tenn Tech Univ, MS, 75; Auburn Univ, PhD(fisheries & aquacult), 79. *Prof Exp:* Asst prof fisheries, Auburn Univ, 79-82; ASSOC PROF BIOL, MURRAY STATE UNIV, 82- *Mem:* Am Fisheries Soc; Am Soc Ichthyologists & Herpetologists; Sigma Xi. *Res:* Life history and population dynamics of freshwater fishes in lakes and streams of the southeastern United States. *Mailing Add:* Hancock Biol Sta Murray State Univ Murray KY 42071

TIMMS, ROBERT J, b Chicago, Ill, Jan 2, 23; m 50; c 4. MICROWAVE ENGINEERING. *Educ:* Univ Md, BS, 49; Kans State Univ, MS, 51; Northwestern Univ, PhD(appl mech), 53. *Prof Exp:* Asst appl mech, Kans State Univ, 49-51; res asst, Northwestern Univ, 52-53; res sect head, Electronic Tube Div, 53-64, RES SECT HEAD, MICROWAVE ENG DEPT, UNISYS GYROSCOPE CO, 64- *Mem:* Inst Elec & Electronics Engrs. *Res:* Microwave components and related radar and communication systems; linear beam tubes; antennas; phased arrays; radar systems. *Mailing Add:* Res Sect Unisys Corp Marcus Ave & Lakeville Rd Great Neck NY 11020

TIMNICK, ANDREW, b Kremianka, Russia, Dec 29, 18; nat US; m 43; c 2. ANALYTICAL CHEMISTRY. *Educ:* Wartburg Col, BA, 40; Univ Iowa, MS, 42, PhD(phys chem), 47. *Prof Exp:* Asst, Univ Iowa, 40-42 & 46-47; control & develop chemist, Polymer Corp, Ont, 43-45; res chemist, Imp Oil, Ltd, 45-46; asst prof chem, WVa Univ, 47-49; from asst prof to assoc prof, 49-65, dir labs, 64-69, prof, 65-85, EMER PROF CHEM, MICH STATE UNIV, 86- *Concurrent Pos:* Sr chemist, Oak Ridge Nat Lab, 52, consult, 53-62; vis prof, Univ Newcastle, 68. *Mem:* Am Chem Soc. *Res:* Spectrophotometric, spectrofluorometric and electrometric methods of analysis. *Mailing Add:* 832 Knoll Rd East Lansing MI 48823

TIMON, WILLIAM EDWARD, JR, b Natchitoches, La, Jan 13, 24; m 46, 56; c 6. MATHEMATICAL STATISTICS. *Educ:* Northwestern State Col La, BS, 50; Tulane Univ, MS, 51; Okla State Univ, PhD, 62. *Prof Exp:* Chief comput, Western Geophys Co, 45-46; instr, Tulane Univ, 50-51; mathematician, Esso Stand Oil Co, 51-53; instr math, La State Univ, 53-54; asst prof, Northwestern State Col La, 54-56 & Southwestern La Inst, 56-57; from asst prof to prof, Northwestern State Col La, 57-65, head dept, 62-65; prof, Parsons Col, 65-74, chmn dept, 67-74; prof math, Glassboro State Col, 74-90; RETIRED. *Mem:* Math Asn Am. *Res:* Analysis of the slipped-block design. *Mailing Add:* 16 Lakeside Dr Natchitoches LA 71457

TIMONY, PETER EDWARD, b Orange, NJ, Dec 30, 43; m 66; c 4. ORGANIC CHEMISTRY. *Educ:* Fairleigh Dickinson Univ, BA, 67; Univ Notre Dame, PhD(org chem), 72. *Prof Exp:* Res chemist, Stauffer Chem Co, 71-75, sr res chemist, 75-76, supvr prod develop, 76-79, asst to dir, 79-81, mgr functional fluids, 81-82, mgr org res, 82-88; dir res, Alcolac, 88-90; DIR CHEM TECHNOL, LONZA, INC, 90- *Mem:* Am Chem Soc; Am Soc Lubrication Engrs. *Res:* Mechanistic organoboron chemistry; synthetic lubricants; organic synthesis; funtional monomers. *Mailing Add:* 446 Alder Trail Crownsville MD 21032-1624

TIMOSHENKO, GREGORY STEPHEN, b St Petersburg, Russia, Nov 1, 04; nat US; m 34; c 2. ELECTRICAL & ELECTRONIC MEASUREMENTS OF PHYSICAL PHENOMENA. *Educ:* Tech Univ, Dipl Ing, 29; Univ Mich, PhD(elec eng), 32. *Prof Exp:* Instr elec eng, Mass Inst Technol, 34-39; from asst prof to emer prof, Univ Conn, 39-71, head dept, 46-68; RETIRED. *Concurrent Pos:* Res & consult engr, Pratt & Whitney Aircraft, E Hartford, Conn, 42-46. *Mem:* Fel Am Phys Soc; Inst Elec & Electronics Engrs. *Res:* Conduction of electricity in gases, liquids and solids applied to aircraft ignition systems, arc reignition, ion sources, cathode sputtering and lasers; illuminating engineering applied to US air defense systems and to habitability of submarines. *Mailing Add:* 700 Ringling Blvd Apt 1112 Sarasota FL 34236

TIMOTHY, DAVID HARRY, b Pittsburgh, Pa, June 9, 28; m 53; c 3. PLANT GENETICS, PLANT BREEDING. *Educ:* Pa State Univ, BS, 52, MS, 55; Univ Minn, PhD(plant genetics), 56. *Prof Exp:* Asst geneticist, Rockefeller Found, 56-58, assoc geneticist, 58-61; assoc prof, 61-66, PROF CROP SCI, NC STATE UNIV, 66- *Concurrent Pos:* Consult, Latin Am Sci Bd, Nat Acad Sci, 64-65; mem, Adv Comt, Orgn Trop Studies, 68-70; mem, Exec Comt, Southern Pasture & Forage Crop Improvement Conf, 68-72, chmn, 71; mem, Nat Cert Grass Variety Rev Bd, 68-74 & Nat Found Seed Proj Planning Conf, 64-68; mem germplasm task force, Nat Plant Germ Plasm Syst, US Dept Agr, 81 & Germplasm Resources Info Prog Coord Comt, 81 & 82; assoc ed, Crop Sci, 82-84; consult, World Bank, 82; mem policy adv comt, USDA, Sci & Educ Res Grants Prog, 82-83; crop adv comt, Forage Grasses, 82-86; consult, Off Int Coop & Develop, US Agency Int Develop, 84-86; chief Scientist, Competive Res Grants Prog, USDA, Fiscal, 85, 86; mem bd dirs, exec comt & treas, Genetic Resources Commun Systs Inc, 85-; comt managing global genetic resource, agr imperatives, Nat Plant Germplasm Syst Working Group, Bd Agr, Nat Res Coun, 87-90; mem, Nat Plant Genetic Resources Bd, 88-90, vchmn, 91-92. *Mem:* Fel AAAS; Asn Trop Biol; fel Am Soc Agron; fel Crop Sci Soc Am; Am Inst Biol Sci. *Res:* Origin, race inter-relationships and evolution of maize; corn and forage grass breeding; germ plasm resources; evaluation and improvement methods in the Gramineae; cytotaxonomy; evolution in domesticated grasses and their wild relatives. *Mailing Add:* Dept Crop Sci Box 7620 NC State Univ Raleigh NC 27695-7620

TIMOTHY, JOHN GETHYN, b Ripley, Eng, Sept 23, 42. SPACE PHYSICS. *Educ:* Univ London, BS, 63, PhD(space physics), 67. *Prof Exp:* Res asst space physics, Mullard Space Sci Lab, Univ Col London, 67-71; physicist, Harvard Col Observ, 71-72, sr physicist, 73-78; res assoc, lab atmospheric & space physics, Univ Colo, 78-; AT CTR FOR SPACE SCI, STANFORD UNIV. *Mem:* Optical Soc Am; Am Geophys Union; Int Astron Union; Am Astron Soc. *Res:* Space astronomy; instrumentation for photometric measurements at extreme ultraviolet and soft x-ray wavelengths; photoelectric detector systems, imaging and nonimaging, for use at visible, ultraviolet and soft x-ray wavelengths. *Mailing Add:* Appl Physics GRI 315a Stanford Univ, ERL 315 Stanford CA 94305

TIMOURIAN, HECTOR, b Mex, Aug 24, 33; US citizen; m 58; c 2. DEVELOPMENTAL BIOLOGY. *Educ:* Univ Calif, Los Angeles, BA, 55, PhD(zool), 60. *Prof Exp:* Commonwealth Sci & Indust Res Orgn res fel immunol, Queensland Univ, 60-61; res zoologist, Univ Calif, Los Angeles, 61-62; NIH res fel biol, Calif Inst Technol, 62-64; asst prof biol, Calif State Univ, Northridge, 64-65; BIOLOGIST, LAWRENCE LIVERMORE LAB, UNIV CALIF, 65- *Mem:* AAAS; Sigma Xi. *Res:* Sperm morphology, activity and fertilization; environmental toxicology; genetic toxicology of effluents and products from energy technologies; science education. *Mailing Add:* 935 Lynn St Livermore CA 94550

TIMOURIAN, JAMES GREGORY, b New York, NY, May 5, 41. MATHEMATICS. *Educ:* Syracuse Univ, PhD(math), 67. *Prof Exp:* Asst prof math, Univ Tenn, 67-69; asst prof, 69-70, assoc prof, 70-77, PROF MATH, UNIV ALTA, 77- *Mem:* Sigma Xi. *Res:* Differential topology; singularities of maps on manifolds. *Mailing Add:* Dept Math Univ Alta Edmonton AB T6G 2G1 Can

TIMS, EUGENE F(RANCIS), b Madison, Wis, Oct 31, 21; m 49; c 4. ELECTRICAL ENGINEERING. *Educ:* La State Univ, BS, 43, MS, 49; Wash Univ, DSc(elec eng), 55. *Prof Exp:* From instr to asst prof elec eng, La State Univ, 46-50; lectr, Wash Univ, 51-55; engr sr staff, Appl Phys Lab, Johns Hopkins Univ, 55-58; sect chief engr, Martin Co, Fla, 58-59; staff engr, Fla Aero Div, Honeywell Inc, 59-60, from proj engr to prin staff engr, 60-64; prof elec eng, La State Univ, Baton Rouge, 64-83; CONSULT, ENGR, 83- *Concurrent Pos:* Pres, Nocon Corp, 70-89. *Mem:* AAAS; Inst Elec & Electronics Engrs; Inst Noise Control Eng; Acoust Soc Am; Nat Acad Forensic Engrs; Am Acad Forensic Sci; Int Asn Arson Investrs. *Res:* Noise control engineering; acoustics. *Mailing Add:* 4840 Newcomb Dr Baton Rouge LA 70808

TIMS, GEORGE B(ARTON), JR, b Purcell, Okla, June 23, 18; m 41; c 2. INDUSTRIAL ENGINEERING. *Educ:* Okla State Univ, BS, 47, MS, 49. *Prof Exp:* Asst instr mach shop, Okla State Univ, 39, instr nat defense training prog, 40, asst instr indust eng, 47, from instr to asst prof, 48-51; prof & head dept, Lamar Univ, 51-62; chief party, Western Mich Univ-AID Proj, Tech Col, Nigeria, 62-64; head dept, Lamar Univ, 64-66, assoc dean eng, 66-73, dir vet affairs, 73-74, dir coop educ, 74-79, prof indust eng, 64-, dir eng Coop Educ, 79-82; RETIRED. *Concurrent Pos:* Agr collabr, Indust Develop Serv, Okla State Univ, 51; Jefferson Amusement Co, 52 & Tex State Optical Co, 55-60. *Mem:* Am Soc Eng Educ; Am Inst Indust Engrs. *Res:* Management and organization. *Mailing Add:* Ridgeway 2520 IH 10E No 80 Beaumont TX 77703

TIMUSK, JOHN, b Narva, Estonia, Jan 2, 35; Can citizen; m 58; c 1. CIVIL ENGINEERING. *Educ:* Univ Toronto, BASc, 58, MASc, 61; Univ London, PhD(civil eng), 69. *Prof Exp:* Demonstr, 58-60, lectr, 60-63, from asst prof to assoc prof, 65-76, PROF CIVIL ENG, UNIV TORONTO, 76-, ASSOC, SYSTS BLDG CTR, 72- *Mem:* Am Concrete Inst. *Res:* Creep and shrinkage of portland cement concrete; development of materials for thermal insulation; plaster casts for orthopedic applications. *Mailing Add:* Dept Civil Eng Univ Toronto 35 St George St Toronto ON M5S 1A2 Can

TIMUSK, THOMAS, b Estonia, June 3, 33; Can citizen; m 57; c 2. PHYSICS. *Educ:* Univ Toronto, BA, 57; Cornell Univ, PhD(physics), 61. *Prof Exp:* Res assoc physics, Cornell Univ, 61-62; asst, Univ Frankfurt, 62-64; res asst prof, Univ Ill, Urbana, 64-65; from asst prof to assoc prof, 65-73, PROF PHYSICS, MCMASTER UNIV, 74- *Concurrent Pos:* Sloan fel, 66-68. *Mem:* Fel Am Phys Soc; Can Asn Physicists. *Res:* Solid state physics; localized vibrations; far infrared spectroscopy; excitons and electron-hole drops; superconductivity. *Mailing Add:* Dept Physics McMaster Univ 1280 Main St W Hamilton ON L8S 4L8 Can

TINANOFF, NORMAN, b Baltimore, Md, Mar 10, 45; m 67; c 1. PEDIATRIC DENTISTRY. *Educ:* Gettysburg Col, BA, 63; Univ Md, DDS, 71; Univ Iowa, MS, 73. *Prof Exp:* Instr pediat dent, Univ Iowa, 73-74; asst prof, Univ Md, 74-76; ASST PROF PEDIAT DENT, HEALTH CTR, UNIV CONN, 76- *Concurrent Pos:* Fel, Vet Admin Hosp, Iowa City, 73-74; spec proj officer, US Army Inst Dent Res, 74-76; contract prin investr, US Army, 77-80. *Mem:* Int Asn Dent Res; Am Dent Asn; Int Asn Dent Children; Am Acad Pedodont. *Res:* Fluorides; preventive dentistry; bacterial attachment; enamel structure. *Mailing Add:* Dept Pediat Dent Univ Conn Health Ctr 263 Farmington Ave Farmington CT 06032-1956

TINBERGEN, NIKOLAAS, zoology; deceased, see previous edition for last biography

TINCHER, WAYNE COLEMAN, b Frankfort, Ky, Jan 15, 35; m 57; c 3. PHYSICAL CHEMISTRY. *Educ:* David Lipscomb Col, BA, 56; Vanderbilt Univ, PhD(chem), 60. *Prof Exp:* Res chemist, Chemstrand Res Ctr, Monsanto Co, 60-65, group leader spectros, 65-71; assoc prof textile chem, 71-77, PROF TEXTILE ENG, A FRENCH TEXTILE SCH, GA INST TECHNOL, 77- *Mem:* AAAS; Am Chem Soc; Am Asn Textile Chemists & Colorists; Sigma Xi. *Res:* Nuclear magnetic resonance spectra and structure of polymers; mechanics of polymer degradation; fiber and fabric flammability; textile process water pollution control. *Mailing Add:* A French Textile Sch Ga Inst Technol Atlanta GA 30332

TINDALL, CHARLES GORDON, JR, b Trenton, NJ, Sept 3, 42; m 64; c 3. FORENSIC SCIENCE. *Educ:* Col Wooster, BA, 64; Ohio State Univ, MS, 67, PhD(org chem), 70. *Prof Exp:* Fel med chem, Nucleic Acid Res Inst, Int Chem & Nuclear Corp, 70-72; TECH SUPVR FORENSIC CHEM, NJ STATE POLICE SOUTH REGIONAL LAB, 72- *Concurrent Pos:* Adj prof, Stockton State Col, 74- & Ocean County Col, 75- *Mem:* Am Chem Soc; Royal Soc Chem; Forensic Sci Soc; Am Acad Forensic Sci. *Res:* Detection of accelerants in arson investigation; characterization of hair; toxicology and detection of drugs and poisons in blood and urine. *Mailing Add:* 283 Chicagami Trail Medford Lakes NJ 08055-2137

TINDALL, DONALD J, b Columbia, SC, May 16, 44; m 67; c 4. ANDROGEN ACTION, PROSTATE CANCER. *Educ:* Univ SC, BS, 66; Clemson Univ, MS, 70; Univ NC, PhD(biochem), 73. *Prof Exp:* Postdoctoral, Baylor Col Med, 74-76, from instr to assoc prof, 76-88; PROF, MAYO CLIN-FOUND, 89- *Concurrent Pos:* Consult, Mayo Clin-Found, 89-; assoc ed, Cancer Res, 91. *Mem:* Am Soc Biol Chemists; Endocrine Soc; Am Soc Androl; Am Asn Cancer Res; Am Soc Cell Biol. *Res:* Mechanism of androgen action in prostate cancer. *Mailing Add:* Dept Urol & Molecular Biol Mayo Clin 17 Guggenheim Rochester MN 55905

TINDALL, DONALD R, b Shelby County, Ky, June 29, 37; m 63; c 2. MARINE TOXINS, MARINE & FRESHWATER ECOLOGY. *Educ:* Georgetown Col, BS, 59, Univ Louisville, MS, 62 & PhD(bot), 66. *Prof Exp:* Instr bot, Univ Louisville, 60-61, researcher, 61-65; fel bot, Ind Univ, 65-66; from asst prof to assoc prof, 66-78, chmn, 79-86, PROF BOT, SOUTHERN ILL UNIV, 78-, ASSOC DEAN, COL SCI, 86- *Concurrent Pos:* Proj dir ciguatera res, dept bot, Southern Ill Univ, 78- *Mem:* Sigma Xi; Phycol Soc Am; Int Phycol Soc. *Res:* Identification, culture and ecology of ciguatoxigenic dinoflagellates and the extraction, purification and characterization of toxins responsible for ciguatera. *Mailing Add:* Dept Cell Biol Baylor Col Med One Baylor Plaza Houston TX 77030

TINDALL, GEORGE TAYLOR, b Magee, Miss, Mar 13, 28; m 47; c 4. NEUROSURGERY. *Educ:* Univ Miss, AB, 48; Johns Hopkins Univ, MD, 52. *Prof Exp:* Intern gen surg, Johns Hopkins Univ, 52-53; resident neurosurg, Med Ctr, Duke Univ, 55-61, from asst prof to assoc prof, 61-68; chief neurosurg serv, Vet Admin Hosp, 61-68; prof neurosurg & chief div, Univ Tex Med Br Galveston, 68-73; PROF NEUROSURG, SCH MED, EMORY UNIV, 73- *Mem:* Cong Neurol Surgeons; Soc Neurol Surgeons; Soc Univ Neurosurgeons (pres, 65). *Res:* Hypophysectomy and neuroendocrinology; measurement of the cerebral circulation; physiologic changes induced by increases in intracranial pressure, hemorrhage and effect of various pharmacologic agents; cranial aneurysms. *Mailing Add:* Div Surg Emory Univ 1365 Clifton Rd NE Atlanta GA 30322

TINDELL, RALPH S, b Tampa, Fla, Jan 16, 42; div; c 2. DISTRIBUTED SYSTEMS, GRAPH THEORY. *Educ:* Univ SFla, BA, 63; Fla State Univ, MS, 65, PhD(math), 67; Stevens Inst Technol, MEng, 84. *Prof Exp:* Res assoc math, Univ Ga, 66; asst prof, Univ Ga, 67-70; assoc prof, 70-78, prof math, 78-84, PROF COMPUTER SCI, STEVENS INST TECHNOL, 84- *Concurrent Pos:* Vis mem & grantee, Inst Advan Study, 66-67, res assoc, 69-70; vis prof, Univ der Saavlandes, 76 & Univ SFla, 82 & 84-85. *Mem:* Am Math Soc; Asn Mems Inst Advan Study; Asn Comput Mach. *Res:* Graph theory; theoretical computer science. *Mailing Add:* EECS Dept Stevens Inst Technol Castle Pt Sta Hoboken NJ 07030

TINDER, RICHARD F(RANCHERE), b Long Beach, Calif, Dec 17, 30; m 69; c 4. MATERIALS SCIENCE. *Educ:* Univ Calif, Berkeley, BS, 57, MS, 58, PhD(metall), 62. *Prof Exp:* From asst prof to assoc prof metall, 61-70, assoc prof elec eng, 70-73, PROF ELEC ENG, WASH STATE UNIV, 73- *Concurrent Pos:* Vis assoc prof mech eng, Univ Calif, Davis, 72-73. *Res:* Initiation of plastic flow in crystals; mechanism of surface ionization of heated filaments; shock studies of single crystals; peizothermoelectric effects in crystals; tensor properties of solids; direct energy conversion. *Mailing Add:* Dept Computer Sci Wash State Univ Pullman WA 99164-1210

TING, CHIH-YUAN CHARLES, b Tsingtao, China, Feb 1, 47; m 71; c 2. HIGH PERFORMANCE LIQUID & GAS CHROMATOGRAPHY. *Educ:* Fu-Jen Univ, Taiwan, BS, 70; Wilkes Col, MS, 73; Pa State Univ, PhD(anal chem), 78. *Prof Exp:* res specialist, Monsanto Co, 77-83; res scientist, G D Searle, 84-87; SR SCIENTIST, CALIF BIOTECHNOL, INC, 88- *Mem:* Am Chem Soc; Sigma Xi; Parental Drug Asn. *Res:* Electrochemistry of biological compounds; automated chromatographic instrumentation analysis of trace level of herbicide and waste effluents; product quality control of herbicides; peptide, protein and amino acid analysis; industrial hygiene; analytical specifications of biopharmaceuticals; analytical phospholipids; immunoassay of biopharmaceuticals in manufacturing process and serum. *Mailing Add:* Calif Biotechnol Inc 2450 Bayshore Pkwy Mountain View CA 94043

TING, CHOU-CHIK, b Fu-Yang, An-Hweir, China, Jan 7, 39; US citizen; m 68; c 2. IMMUNOLOGY, PATHOLOGY. *Educ:* Nat Taiwan Univ Med Sch, MD, 62. *Prof Exp:* Resident path, NY Med Col, Metrop Hosp, 63-64; resident path, Albert Einstein Col Med, Bronx Munic Hosp, 64-67, chief resident, 67-68; spec fel, 68-70, staff fel, 71-74, SR INVESTR IMMUNOL, NAT CANCER INST, NIH, 74- *Concurrent Pos:* China scholar, Nat Acad Sci, 81; adj prof, dept path, Uniformed Serv Univ Health Sci, 90- *Mem:* Am Asn Immunologists; Am Asn Cancer Res. *Res:* Tumor immunology to study specific humoral and cell mediator immunity to tumor associated antigens in animals; activation of killer cells by anti-T cell receptor antibody and cytokines and to determine the mechanism of activation and its implication in immunotherapy of cancer. *Mailing Add:* Nat Cancer Inst NIH Bldg 10 Rm 4B17 Bethesda MD 20892

TING, FRANCIS TA-CHUAN, b Tsingtao, China, Apr 26, 34; US citizen; m 66; c 2. GEOLOGY. *Educ:* Nat Taiwan Univ, BS, 57; Univ Minn, MS, 62; Pa State Univ, PhD, 67. *Prof Exp:* Fel, Pa State Univ, 67-68; asst prof geol, Macalester Col, 68-69; res assoc coal petrol, Pa State Univ, 69-70; from asst prof to assoc prof geol, Univ NDak, 70-74; assoc prof, 74-79, PROF GEOL, WVA UNIV, 79- *Concurrent Pos:* Fel, Univ Minn, 68-69; NATO sr fel, 73. *Mem:* Sigma Xi; Geol Soc Am; Bot Soc Am; Soc Econ Paleont & Mineral; Geochem Soc; Am Asn Petrol Geologists. *Res:* Coal petrology and chemistry; paleobotany. *Mailing Add:* Dept Geol & Geog WVa Univ Morgantown WV 26506

TING, IRWIN PETER, b San Francisco, Calif, Jan 13, 34; m 52; c 1. PLANT PHYSIOLOGY, METABOLISM. *Educ:* Univ Nev, BS, 60, MS, 61; Iowa State Univ, PhD(plant physiol, biochem), 64. *Prof Exp:* NSF fel, 64-65; plant physiologist, 65-66, from asst prof to assoc prof plant physiol & metab, 66-72, PROF BIOL, UNIV CALIF, RIVERSIDE, 72- *Concurrent Pos:* Vis fel, Australian Nat Univ, 72; exchange prof, Univ Paris, 74. *Mem:* AAAS; Bot Soc Am; Am Soc Plant Physiologists. *Res:* Gas transfer between plant surfaces and environment; carbon dioxide metabolism; plant isoenzymes. *Mailing Add:* Dept Bot & Plant Sci Univ Calif 900 Univ Ave Riverside CA 92521

TING, LU, b China, Apr 18, 25; nat US; div; c 3. APPLIED MATHEMATICS. *Educ:* Chiao Tung Univ, China, BS, 46; Mass Inst Technol, SM, 48; Harvard Univ, MS, 49; Univ, ScD, 51. *Prof Exp:* Res assoc aerodyn, NY Univ, 51-52; spec design engr, Foster Wheeler Corp, 52-55; res prof aerodyn, Polytech Inst Brooklyn, 55-64; prof aeronaut & astronaut, 64-68, PROF MATH, NY UNIV, 68- *Concurrent Pos:* Consult, Inst Comput Appln Sci & Eng NASA Langley Res Ctr, 77-79, 86- *Mem:* Soc Indust & Appl Math; Am Inst Aeronaut & Astronaut; Am Phys Soc; NY Acad Sci. *Res:* Shock deflections; boundary layer theory; supersonic wing-body interference; space mechanics; nonlinear wave propagations; perturbation methods; aeroacoustics. *Mailing Add:* Dept Math NY Univ Washington Sq New York NY 10012

TING, ROBERT YEN-YING, b Kwei-Yang, China, March 8, 42; US citizen; m 67; c 2. RHEOLOGY, APPLIED MECHANICS. *Educ:* Nat Taiwan Univ, BS, 64; Mass Inst Technol, MS, 67; Univ Calif, San Diego, PhD(eng sci), 71. *Prof Exp:* Res staff, Aerophys Lab, Mass Inst Technol, 67-68; mech engr surface chem, Chem Div, 71-76, sect head polymer mech, 77-80, HEAD MAT RES, UNDERWATER SOUND REFERENCE DIV, NAVAL RES LAB, 80- *Concurrent Pos:* Assoc prof lectr, George Washington Univ, 75; lectr appl rheology, Kent State Univ, 76-80; translation ed, Chinese Physics, Am Inst Physics, 81-; mem, Comt Composite Technol Transfer, Soc Plastics Indust, 77-81. *Mem:* Acoust Soc Am; Am Res Soc; Sigma Xi. *Res:* Research and management in underwater acoustical materials and transducers including piezoelectrics and ferroelectrics, elastomers, plastics, optical fibers, fluids, composites and sonar transducer development. *Mailing Add:* Naval Res Lab PO Box 568337 Orlando FL 32856-8337

TING, SAMUEL C C, b Ann Arbor, Mich, Jan 27, 36; m 60; c 2. PARTICLE PHYSICS. *Educ:* Univ Mich, BSE(physics) & BSE(math), 59, MS, 60, PhD(physics), 62. *Hon Degrees:* ScD, Univ Mich, 78. *Prof Exp:* Ford Found fel, Europ Coun Nuclear Res, Switz, 63-64; instr physics, Columbia Univ, 64-65, asst prof, 65-67; assoc prof, 67-69, PROF PHYSICS, MASS INST TECHNOL, 69-, THOMAS DUDLEY CABOT INST PROF, 77- *Concurrent Pos:* Ground leader, Deutches Electronen Synchrotronen, Hamburg, Ger, 66; assoc ed, Nuclear Physics B, 70; hon prof, Beijing Normal Col, China, 84, Jiatong Univ, 87. *Honors & Awards:* Nobel Prize in Physics, 76; Ernest Orlando Lawrence Award, US Dept Energy, 76; Eringen Medal, Soc Eng Sci, 77; DeGasperi Award Sci, Govt Italy, 88; Golden Leopard Award for Excellence, 88; Gold Medal Sci, 88. *Mem:* Nat Acad Sci; Europ Phys Soc; Ital Phys Soc; fel Am Phys Soc; fel AAAS; fel Am Acad Arts & Sci; foreign mem Pakistan Acad Sci; foreign mem Soviet Acad Sci. *Res:* Experimental particle physics; quantum electrodynamics; interactions of photons with matter. *Mailing Add:* Lab Nuclear Sci Mass Inst Technol Cambridge MA 02139

TING, SHIH-FAN, b Changteh, China, Sept 27, 17; US citizen; m 47; c 1. CHEMISTRY. *Educ:* Univ Chekiang, BS, 41; Univ Ala, MS, 57, PhD(chem), 60. *Prof Exp:* Asst prof chem, Fisk Univ, 60-65; fel, Duquesne Univ, 65-66; from assoc prof to prof, 69-83, EMER PROF CHEM, MILLERSVILLE STATE COL, 83- *Concurrent Pos:* NSF grant, 63-65. *Mem:* Am Chem Soc. *Res:* High-frequency titration; rhenium chemistry; coordination compounds; nuclear magnetic resonance studies of hydrogen bonding. *Mailing Add:* 777 E Valley Blvd No 102 Alhambra CA 91801-5267

TING, SIK VUNG, b Shanghai, China, Mar 3, 18; nat US; m 46; c 3. HORTICULTURE. *Educ:* Mich State Univ, BS, 41; Ohio State Univ, MS, 43, PhD(hort), 52. *Prof Exp:* Asst, Ohio State Univ, 41-43 & 49-52, Agr Exp Sta, 43-45; assoc prof hort, Nanking Univ, 47-49; asst horticulturist, Citrus Exp Sta, Fla State Citrus Comn, 52-60, assoc biochemist, 60-68, res biochemist, 68-83; prof biochem, Univ Fla, 68-83. *Concurrent Pos:* Consult, 83- *Mem:* Am Soc Hort Sci; Am Chem Soc; Inst Food Technologists. *Res:* Biochemistry of horticultural plants, especially chemical components of citrus fruit. *Mailing Add:* 1440 Seventh St SW Winterhaven FL 33880-4935

TING, THOMAS C(HI) T(SAI), b Taipei, Taiwan, Feb 9, 33; m 62; c 2. APPLIED MATHEMATICS, MECHANICS. *Educ:* Nat Taiwan Univ, BSc, 56; Brown Univ, PhD(appl math), 62. *Prof Exp:* Res asst appl math, Brown Univ, 59-62, res assoc eng, 62-63, asst prof, 63-65; from asst prof to assoc prof appl mech, 65-70, PROF APPL MECH, UNIV ILL, CHICAGO CIRCLE, 70- *Concurrent Pos:* Assoc mem, Ctr Advan Study, Univ Ill, 67-68; vis prof, Stanford Univ, 72-73; assoc ed, J Appl Mech, Am Soc Mech Engrs, 75-82. *Mem:* Math Asn Am; fel Am Soc Mech Engrs; Soc Indust & Appl Math; AAAS; Sigma Xi. *Res:* Continuum mechanics; viscoelasticity and viscoplasticity; wave propagations; numerical analysis; partial differential equations. *Mailing Add:* Univ Ill Chicago Box 4348 SEO-825 Chicago IL 60680

TING, TSUAN WU, b Anking, Anhwei, China, Oct 10, 22; nat US; m 57; c 2. MATHEMATICS. *Educ:* Nat Cent Univ, China, BS, 47; Univ RI, MS, 56; Ind Univ, MS, 59, PhD(math), 60. *Prof Exp:* Technician, China 60th Arsenal, 47-49, assoc engr, 49-53; res asst eng, Univ RI, 54-56; math, Ind Univ, 56-59; sr mathematician, Gen Motors Res Labs, 60-61; asst prof mech, Univ Tex, 61-63; vis mem, Courant Inst Math Sci, NY Univ, 63-64; assoc prof math, NC State Univ, 64-66; PROF MATH, UNIV ILL, URBANA, 66- *Mem:* Soc Indust & Appl Math; Tensor Soc; Am Math Soc; Soc Natural Philos. *Res:* Theory of partial differential equations; mathematical physics; principles of continuum mechanics and differential geometry. *Mailing Add:* 403 E Sherwin Dr Urbana IL 61801

TING, YU-CHEN, b Honan, China, Oct 3, 20; m 60; c 2. PLANT CYTOLOGY, PLANT GENETICS & TISSUE CULTURE. *Educ:* Nat Honan Univ, China, BS, 44; Cornell Univ, MSA, 52; La State Univ, PhD(hort genetics), 54. *Prof Exp:* Asst bot, Nat Honan Univ, 44-47; res fel, Harvard Univ, 54-62; from asst prof to assoc prof biol, 62-67, PROF BIOL, BOSTON COL, 67- *Concurrent Pos:* Hon prof, Genetics Inst, Beijing, China & Honan Univ, Honan, China; sr res fel, Nat Acad Sci, 79. *Mem:* AAAS; Genetics Soc Am; Am Genetic Asn; Bot Soc Am; Am Soc Hort Sci. *Res:* Cytology and genetics of Ipomoea batatas and related species; flower induction and site of synthesis of pigments in sweet potato plants; cytogenetics and tissue culture of maize and its relatives. *Mailing Add:* Dept Biol Boston Col Chestnut Hill MA 02167

TINGA, JACOB HINNES, b Wilmington, NC, Mar 9, 20; m 57; c 2. HORTICULTURE, PLANT PHYSIOLOGY. *Educ:* NC State Col, BS, 42; Cornell Univ, MS, 52, PhD, 56. *Prof Exp:* Assoc prof hort, Va Polytech Inst, 55-68; PROF HORT, UNIV GA, 68- *Mem:* Am Soc Hort Sci. *Res:* Effect of environment on growth of horticultural crops; ornamental and greenhouse crops; winter protection; container production; propagation of large cuttings of woody plants; nursery economics. *Mailing Add:* Dept Hort & Plant Sci Univ Ga Athens GA 30602

TING-BEALL, HIE PING, b Sibu, Malaysia, Dec 15, 40; US citizen; m 70; c 1. CELL PHYSIOLOGY, BIOPHYSICS. *Educ:* Greensboro Col, BS, 63; Tulane Univ, MS, 65, PhD(physiol), 67. *Prof Exp:* NIH fel biophys, Mich State Univ, 67-69, res assoc biochem, 70-72; NIH trainee phys biochem, Johnson Res Found, Univ Pa, 69-70; res assoc physiol & pharm, Duke Univ Med Ctr, 72-74, asst med res prof physiol, 75-83, asst med res prof anat, 75-88, asst med res prof, cell biol, 88-89, RES ASSOC, MECH ENG & MAT SCI, SCH ENG, DUKE UNIV, 89- *Concurrent Pos:* K E Osserman fel, Myasthenia Gravis Found, 73-74; NIH res grants, 82-88. *Mem:* Biophys Soc; Sigma Xi; Electron Micros Soc Am. *Res:* Structure and function of cell membranes; biophysics of bimolecular lipid membranes; ultrastructure of sodium, potassium-ATPase and calcium-ATPase; cell mechanics; cell culture of hybridomas; neutrophil activation. *Mailing Add:* Dept Mech Eng & Mat Sci Sch Eng Duke Univ Durham NC 27706

TINGELSTAD, JON BUNDE, b McVille, NDak, Jan 15, 35; m 60; c 3. PEDIATRICS, PEDIATRIC CARDIOLOGY. *Educ:* Univ NDak, BA, 57, BS, 58; Harvard Univ, MD, 60. *Prof Exp:* From asst prof to assoc prof pediat, Med Col Va, 67-76; PROF PEDIAT, SCH MED, ECAROLINA UNIV, 76-, CHMN DEPT, 77- *Concurrent Pos:* Intern pediat, Children's Hosp Med Ctr, Boston, 60-61, jr asst resident, 61-62; chief resident, Med Ctr, Univ Colo-Denver, 62-63; fel pediat cardiol, Children's Hosp, Buffalo, NY, 65-67. *Mem:* Am Acad Pediat; Am Col Cardiol; Am Inst Ultrasound Med; Ambulatory Pediat Asn. *Res:* Pediatric echocardiography and education. *Mailing Add:* Dept Pediat ECarolina Univ Sch Med Greenville NC 27834

TINGEY, DAVID THOMAS, b Salt Lake City, Utah, Jan 30, 41; m 68; c 5. PLANT PHYSIOLOGY, AIR POLLUTION. *Educ:* Univ Utah, BA, 66, MA, 68; NC State Univ, PhD(plant physiol), 72. *Prof Exp:* Plant physiologist, US Environ Protection Agency, NC State Univ, 68-73; PLANT PHYSIOLOGIST, ENVIRON PROTECTION AGENCY, CORVALLIS ENVIRON RES LAB, 73- *Concurrent Pos:* Asst prof bot, Ore State Univ, 74-80, assoc prof, 80-87, prof, 88- *Mem:* Am Soc Plant Physiol; Scand Soc Plant Physiol. *Res:* Studying the effects of atmospheric pollutants on plant physiology. *Mailing Add:* US Environ Protection Agency Corvallis Environ Res Lab Corvallis OR 97333

TINGEY, GARTH LEROY, b Woodruff, Utah, Apr 14, 32; m 53; c 8. NUCLEAR BYPRODUCTS, NUCLEAR GRAPHITE. *Educ:* Brigham Young Univ, BS, 54, MS, 59; Pa State Univ, PhD(phys chem), 63. *Prof Exp:* Sr scientist, Gen Elec Corp, 63-65; sr res scientist, 65-66, unit mgr mat, 66-70, RES ASSOC, PAC NORTHWEST LABS, BATTELLE MEM INST, 70- *Mem:* Am Chem Soc; Sigma Xi. *Res:* Radiation chemical studies of gaseous mixtures; inert gas sensitized radiolysis reactions; chemical kinetics; chemical studies of carbon and graphite; high temperature gas cooled nuclear reactor technology; food irradiation; nuclear waste management; nuclear graphite. *Mailing Add:* 1519 Alder Ave Richland WA 99352

TINGEY, WARD M, b Brigham City, Utah, Apr 9, 44; m 68; c 2. ENTOMOLOGY. *Educ:* Brigham Young Univ, BS, 66, MS, 68; Univ Ariz, PhD(entom), 72. *Prof Exp:* Asst res entomologist, Univ Calif, Davis, 72-74; asst prof entom, 74-80, assoc prof, 80-87, PROF ENTOM, CORNELL UNIV, 87- *Mem:* Entom Soc Am; AAAS; Potato Asn Am. *Res:* Genetic resistance of crop plants to insect pests; plant defense mechanisms against insects; pest management. *Mailing Add:* Dept Entom Comstock Hall Cornell Univ Ithaca NY 14853

TINGLE, FREDERIC CARLEY, b Meridian, Mass, Jan 28, 40. INSECT BEHAVIOR, INSECT-HOST PLANT INTERACTIONS. *Educ:* Miss State Univ, BS, 62, MS, 64. *Prof Exp:* RES ENTOMOLOGIST, BOLL WEEVIL RES LAB, AGR RES SERV, USDA, 64-72, 72- *Mem:* Int Soc Chem Ecol; Entom Soc Am; Am Registry Prof Entomologists. *Res:* Developing pest management strategies for Heliothis spp and other economic insect pests with emphasis on identification of the biological mechanisms that drive the mating process, host selection and oviposition behavior. *Mailing Add:* USDA-Agr Res Serv PO Box 14565 Gainesville FL 32604

TINGLE, MARJORIE ANNE, b Far Rockaway, NY, Oct 5, 38. MICROBIOLOGY. *Educ:* Brown Univ, AB, 60; Univ Wis-Madison, MS, 63, PhD(bacteriol), 66. *Prof Exp:* NIH trainee bacteriol, Univ Wis, 62-66; NIH fel, Dept Biol Sci, Purdue Univ, 67-68; Am Cancer Soc fel, Lab Enzymol, Nat Ctr Sci Res, France, 68-70; res assoc, Rosenstiel Ctr, Brandeis Univ, 71-75; health scientist adminr, Nat Ctr Res Resources, 75-86, CHIEF, BIOMED RES SUPPORT BR, NIH, 86- *Mem:* Am Soc Microbiol; AAAS; Genetics Soc Am; NY Acad Sci. *Res:* Microbial physiology; regulation of gene expression; health administration. *Mailing Add:* Nat Ctr Res Resources Westwood Bldg Rm 10A11 5333 Westbard Ave Bethesda MD 20892

TINGLE, WILLIAM HERBERT, b Parnassus, Pa, Aug 31, 17; m 45; c 2. INSTRUMENTATION, SPECTROSCOPY. *Educ:* Univ Pittsburgh, BS, 49. *Prof Exp:* Instr eng physics, Univ Pittsburgh, 49-50; spectroscopist anal chem, 51-62, sect head anal chem, 63-72, sci assoc equip develop, Alcoa Labs, 73-81; CONSULT, 82- *Concurrent Pos:* Chmn subcomt, Am Soc Testing & Mat, 62-73; adv, USA Adv Group, Int Stand Orgn Comt 79, 70-79; secretariat, Int Stand Orgn, 70-79. *Honors & Awards:* H V Churchill Award Meritorious Serv Spectro-Chem Anal, Am Soc Testing & Mat, 77. *Mem:* Sigma Xi; Optical Soc Am; Soc Appl Spectros; Am Chem Soc; Am Inst Physics. *Res:* Evaluating principles of measurement and control in chemical and metallurgical processes, atomic emission spectroscopy, plasma excitation, and optical instrumentation. *Mailing Add:* 3104 Leechburg Rd Lower Burrell PA 15068

TINGLEY, ARNOLD JACKSON, b Point de Bute, June 9, 20; m 46; c 2. MATHEMATICS. *Educ:* Mt Alison Univ, BA, 49; Univ Minn, PhD(math), 52. *Prof Exp:* Instr math, Univ Nebr, 52-53; head dept, Dalhousie Univ, 66-73, secy senate, 75-80, from asst prof to prof math, 53-85, registr, 73-85, bd gov, 80-86, EMER PROF, DALHOUSIE UNIV, 86- *Mem:* Can Math Soc. *Res:* Analysis. *Mailing Add:* Dept Math Statist & Computer Sci Dalhousie Univ Halifax NS B3H 3J5 Can

TINKER, DAVID OWEN, b Toronto, Ont, Jan 25, 40; m 62; c 4. BIOCHEMISTRY, PHYSICAL CHEMISTRY. *Educ:* Univ Toronto, BSc, 61; Univ Wash, PhD(biochem), 65. *Prof Exp:* Nat Res Coun Can fel, Univ London, 65-66; from asst prof to assoc prof, 66-82, PROF BIOCHEM, UNIV TORONTO, 82- *Concurrent Pos:* Assoc ed, Can J Biochem, 74-82. *Mem:* Am Chem Soc; Can Biochem Soc; Biophys Soc. *Res:* Structure, occurrence, metabolism and physical chemical properties of complex lipids from biological membranes; lipid-protein interactions. *Mailing Add:* Dept Biochem Fac Med Univ Toronto Toronto ON M5S 1A8 Can

TINKER, EDWARD BRIAN, chemical engineering, for more information see previous edition

TINKER, JOHN FRANK, b Wis, Mar 25, 22; m 48. CHEMISTRY, INFORMATION SCIENCE. *Educ:* Univ Va, BS, 43; Harvard Univ, PhD(chem), 51. *Prof Exp:* Instr chem, Harvard Univ, 50-52; INFO SCIENTIST, EASTMAN KODAK CO, 52- *Mem:* Am Chem Soc. *Res:* Chemical information. *Mailing Add:* 9843 47th Ave SW Seattle WA 98136

TINKER, SPENCER WILKIE, b Anamoose, NDak, Jan 29, 09; m 38; c 1. ICHTHYOLOGY, CONCHOLOGY & CETOLOGY. *Educ:* Univ Wash, BS, 31; Univ Hawaii, MS, 34. *Prof Exp:* Instr zool, Univ Hawaii, 33-34, ed educ, 35-55, dir, Waikiki Aquarium, 40-72, from asst prof to prof, Univ Hawaii, 55-72; RETIRED. *Res:* Indo-Pacific fish and molluscs; whales. *Mailing Add:* 1121 Hunakai St Honolulu HI 96816

TINKHAM, MICHAEL, b Green Lake Co, Wis, Feb 23, 28; m 61; c 2. SUPERCONDUCTIVITY. *Educ:* Ripon Col, AB, 51; Mass Inst Technol, MS, 51, PhD(physics), 54. *Hon Degrees:* ScD, Ripon Col, 76; MA, Harvard Univ, 66. *Prof Exp:* NSF fel, Clarendon Lab, Oxford Univ, 54-55; res physicist, Univ Calif, Berkeley, 55-57, lectr, 56-57, from asst prof to prof physics, 57-66; chmn dept physics, 75-78, prof physics, 66-80, GORDON MCKAY PROF APPL PHYSICS, HARVARD UNIV, 66-, RUMFORD PROF PHYSICS, 80- *Concurrent Pos:* Guggenheim fel, 63-64; NSF sr fel, Cavendish Lab, Cambridge Univ, 71-72; Humboldt sr scientist, Univ Karlsruhe, 78-79; vis Miller res prof, Univ Calif, Berkeley, 87; mem, Briefing Panel, High Temperature Superconductivity, Nat Acad Sci, 87. *Honors & Awards:* Buckley Prize, Am Phys Soc, 74. *Mem:* Nat Acad Sci; fel Am Acad Arts & Sci; fel Am Phys Soc; fel AAAS. *Res:* Superconductivity: energy gap, fluxoid quantization and macroscopic quantum interference, fluctuation effects, Josephson junctions, nonequilibrium effects including charge imbalance, and energy imbalance and macroscopic quantum tunneling, charging energy effects; microwave and far-infrared magnetic resonance. *Mailing Add:* Dept Physics Harvard Univ Cambridge MA 02138

TINKLEPAUGH, J(AMES) R(OOT), ceramic engineering; deceased, see previous edition for last biography

TINKLER, JACK D(ONALD), b Topeka, Kans, Apr 12, 36; m 60; c 2. CHEMICAL ENGINEERING. *Educ:* Univ Ill, BS, 58; Univ Del, MChE, 61, PhD(chem eng), 63. *Prof Exp:* Res engr, Res & Develop Dept, Sun Oil Co, 63-70, chief eng fundamentals, Corp Res Dept, 70-77; vpres eng, Occidental Res Corp, 77-80, Occidental Chem Co, 81-82, Occidental Petrol Corp, 83-85; mgr, Process Develop, Pennwalt Corp, 85-87, gen mgr, Cent Eng, 87-89; VPRES TECH, ATOCHEM N A, 90- *Mem:* Am Inst Chem Engrs; Am Chem Soc. *Res:* Kinetics; simulation; model building; process design and evaluation; technical economics. *Mailing Add:* 498 Meadow Lane King of Prussia PA 19406

TINLINE, ROBERT DAVIES, b Moose Jaw, Sask, Aug 4, 25; m 48; c 5. PLANT PATHOLOGY. *Educ:* Univ Sask, BA, 48; Univ Wis, MS, 52, PhD(plant path), 54. *Prof Exp:* Tech officer I, 48-49, tech officer II, 49-51, agr res officer, 51-67, RES SCIENTIST, AGR CAN, 67- *Mem:* Mycol Soc Am; Am Phytopath Soc; fel Can Phytopath Soc (pres, 77-78); Agr Inst Can. *Res:* Root and leaf diseases of cereals; variability and genetics of plant pathogenic fungi. *Mailing Add:* Agr Can Res Sta 107 Science Crescent Saskatoon SK S7N 0X2 Can

TINNEY, FRANCIS JOHN, b Brooklyn, NY, July 31, 38; m 74; c 4. MEDICINAL CHEMISTRY. *Educ:* St John's Univ, NY, BS, 59, MS, 61; Univ Md, PhD(aza steroids), 65; Univ Toledo, JD, 85. *Prof Exp:* Ortho Res Found fel chem, Univ Md, 65-66; assoc res chemist, 66-69, res chemist, 69-70, sr res chemist, Parke-Davis & Co, 70-78; res assoc, 78-87, sr patent atty, 87-89, ASST PATENT COUN, WARNER-LAMBERT/PARKE-DAVIS & CO, 87- *Mem:* Am Chem Soc; Am Pharmaceut Asn; NY Acad Sci. *Res:* Organic medicinal chemistry; steroids; heterocyclic steroids; heterocyclics; organic nitrogen containing compounds; peptides; antiallergy compounds. *Mailing Add:* Warner-Lambert/Parke-Davis & Co 2800 Plymouth Rd Ann Arbor MI 48105

TINNEY, WILLIAM FRANK, b Portland, Ore, May 5, 21; m 48; c 2. ELECTRIC POWER SYSTEMS. *Educ:* Stanford Univ, BS, 48, MS, 49. *Prof Exp:* Elec engr, Bonneville Power Admin, 50-63, head methods anal, 63-79; INDEPENDENT CONSULT, 79- *Mem:* Fel Inst Elec & Electronics Engrs. *Res:* Development of solution methods for large scale electric power network problems, energy management systems, optimal operation & control. *Mailing Add:* 9101 SW Eighth Ave Portland OR 97219

TINNIN, ROBERT OWEN, b Santa Barbara, Calif, Sept 6, 43; m 65; c 1. PLANT ECOLOGY. *Educ:* Univ Calif, Santa Barbara, BA, 65, PhD(ecol), 69. *Prof Exp:* From asst prof to assoc prof, 69-80, PROF BIOL, PORTLAND STATE UNIV, 80-4. *Mem:* Ecol Soc Am. *Res:* Host-parasite interactions and their effect on plant communities; natural systems under study are coniferous forest sites infected by arceuthobium species. *Mailing Add:* Dept Biol Portland Univ PO Box 751 Portland OR 97207

TINOCO, IGNACIO, JR, b El Paso, Tex, Nov 22, 30; m 51; c 1. PHYSICAL CHEMISTRY. *Educ:* Univ NMex, BS, 51, DSc, 72; Univ Wis, PhD(chem), 54. *Prof Exp:* Res fel chem, Yale Univ, 54-56; from instr to assoc prof, 56-66, chmn dept, 79-82, PROF CHEM, UNIV CALIF, BERKELEY, 66- *Concurrent Pos:* Guggenheim fel, 64. *Honors & Awards:* Calif Sect Award, Am Chem Soc, 65. *Mem:* Nat Acad Sci; Am Chem Soc; Am Phys Soc; Biophys Soc; Am Soc Biol Chemists. *Res:* Biophysical chemistry. *Mailing Add:* Dept Chem Univ Calif Berkeley CA 94720

TINOCO, JOAN W H, b Nome, Alaska, Oct 14, 32. LIPID METABOLISM. *Educ:* Univ Calif, Berkeley, BS, 58, PhD(nutrit), 62. *Prof Exp:* Technician II lipid res, dept home econ, Univ Calif, Berkeley, 58-59, res asst, 59-61, res fel, dept nutrit sci, 61-67, asst res biochemist, 67-73, res fel, dept chem, 73-75, assoc res biochemist, dept nutrit sci, 75-85. *Concurrent Pos:* Res fel, dept path, Cambridge Univ, Eng, 64; actg asst prof biochem, Ore State Univ, Corvallis, 70-71. *Mem:* Am Inst Nutrit; Am Oil Chemists' Soc; AAAS; Sigma Xi. *Res:* Functions of lipids; metabolism of cholesterol, fatty acids, and phospholipids; dietary requirements for essential fatty acids, especially n-3 fatty acids; behavior of cholesterol and phospholipids in monomolecular films (membrane model). *Mailing Add:* 1035 Spruce St Berkeley CA 94707

TINSLEY, BRIAN ALFRED, b Wellington, NZ, Apr 23, 37; m 61; c 4. SPACE PHYSICS. *Educ:* Univ Canterbury, BSc, 58, MSc, 61, PhD(physics), 63. *Prof Exp:* Res assoc atmospheric & space sci, Univ Tex, Dallas, 63-65, res scientist, 65-67, from asst prof to prof physics, 67-86; aeronomy prog dir, NSF, 86-88; PROF, UNIV TEX, DALLAS, 88- *Concurrent Pos:* Mem, Sci Comn Solar Terrestrial Physics, 76- & Aeronomy, 73-79; chmn, div II, Int Assoc Geomog, Assoc Educ Rev Geophys, 88-90. *Mem:* Am Geophys Union. *Res:* Optical observations of airglow and low latitude aurorae; design of optical instruments for measurement of airglow, and astronomical emissions; theoretical studies of atmospheres of earth and planets; effects of solar variability on weather and climate. *Mailing Add:* Physics Prog Univ Tex-Dallas PO Box 830688 Richardson TX 75083-0688

TINSLEY, IAN JAMES, b Sydney, Australia, Sept 23, 29; m 55; c 2. BIOCHEMISTRY. *Educ:* Univ Sydney, BSc, 50; Ore State Univ, MS, 55, PhD(food sci), 58. *Prof Exp:* Res officer, Commonwealth Sci & Indust Res Orgn, Australia, 50-53; from asst prof to assoc prof, 57-70, PROF BIOCHEM, ORE STATE UNIV, 70- *Honors & Awards:* Florasynth Award, Inst Food Technologists, 55. *Mem:* Am Inst Nutrit; Am Oil Chem Soc; Am Chem Soc. *Res:* Lipid metabolism; essential fatty acid nutrition; biochemical effects of pesticide ingestion; interactions of pesticides with lipids. *Mailing Add:* Dept Agr Chem Ore State Univ Corvallis OR 97331

TINSLEY, RICHARD STERLING, b Richmond, Va, Mar 7, 31; m 52; c 5. CHEMICAL ENGINEERING. *Educ:* Va Polytech Inst, BS, 52, MS, 53, PhD(chem eng), 55. *Prof Exp:* Res chem engr, Nitrogen Div, Allied Chem Corp, 55-58, sr chem engr, 58-61, supvry chem engr, 61-64, mgr chem eng, 64-67, start-up chief, Grismar Complex, 67-68, mgr prod, 68-69, PROJ MGR, FIBERS DIV, ALLIED CHEM CORP, 69- *Mem:* Am Chem Soc. *Res:* Chemical engineering design and development; fiber engineering. *Mailing Add:* 2009 Surreywood Ct Richmond VA 23235

TINSLEY, SAMUEL WEAVER, b Hopkinsville, Ky, July 15, 23; m 45; c 3. ORGANIC CHEMISTRY. *Educ:* Western Ky State Col, BS, 44; Northwestern Univ, PhD(org chem), 50. *Prof Exp:* Asst prof chem, Tex Tech Col, 49-50; res chemist, Union Carbide Corp, 50-60, asst dir org res, 60-64, assoc dir res & develop, 64-67, mgr new chem, 67-71, mgr corp res, 71-74, dir corp res, 74-77, dir corp technol, 77-85; RETIRED. *Mem:* Am Chem Soc; Indust Res Inst (past-pres); Com Develop Asn (past-pres). *Res:* Aromatic synthesis; peracids; epoxides. *Mailing Add:* 1739 Irish Blvd Sanford NC 27330

TINSMAN, JAMES HERBERT, JR, b Philadelphia, Pa, Apr 22, 30; m 56; c 4. PHYSICAL ANTHROPOLOGY. *Educ:* Univ Pa, AB, 56, AM, 60; Univ Colo, MA, 66, PhD(anthrop), 71. *Prof Exp:* From instr to assoc prof philos & econ, 59-71, chmn, dept Soc Sci, 74-77, chmn, dept Anthrop & Sociol, 85-86, PROF ANTHROP, KUTZTOWN STATE COL, 71-; PRES, ASN PA STATE COL & UNIV FACULTIES, 86- *Concurrent Pos:* Lectr anthrop, Univ Colo, 69-71; co-ed, Newsletter Pa Anthropologists; NSF fel, 64, 65-66

& 68-71. *Mem:* Fel Am Anthrop Asn; Soc Study Social Biol; Am Asn Phys Anthrop. *Res:* Contemporary human variation, anthropometric, anthroscopic and serological, as relates to human genetics, population structure and, ultimately, human evolution. *Mailing Add:* Dept Anthrop Kutztown Univ Kutztown PA 19530

TINT, HOWARD, b Philadelphia, Pa, Jan 22, 17; m 41; c 2. BIOCHEMISTRY, BIOLOGICALS. *Educ:* Univ Pa, AB, 37, PhD(mycol, plant path), 43. *Prof Exp:* Leader scouting crews, Bur Entom & Plant Quarantine, USDA, Washington, DC, 39-40, biol tech aide, Bur Plant Indust, 41, physiologist, Bur Agr & Indust Chem, 43-45; biochemist, Wyeth, Inc, 45-50, sr res biochemist, 50-56, supvr biologics lab, Wyeth Labs, 56-60, dir, Prod Develop Div, 60-63, DIR, BIOL & CHEM DEVELOP DIV, WYETH LABS, 63- *Concurrent Pos:* Microbiologist, Off Sci Res & Develop, Johnson Found, Univ Pa, 45. *Mem:* Am Chem Soc; Sigma Xi; NY Acad Sci. *Res:* Microbiology; physiology; virology; tissue culture; cancer immunology; chemical development and pilot-plant; fine-chemicals production. *Mailing Add:* 1219 W Wynnewood Rd Apt 615 Wynnewood PA 19096-2102

TINTI, DINO S, b San Bernardino, Calif, Feb 20, 41; m 62; c 3. PHYSICAL CHEMISTRY. *Educ:* Univ Calif, Riverside, BA, 62; Calif Inst Technol, PhD(chem), 68. *Prof Exp:* Res chemist, Univ Calif, Los Angeles, 67-70; from asst prof to assoc prof, 70-81, PROF CHEM, UNIV CALIF, DAVIS, 81- *Concurrent Pos:* Sloan fel, 74-78. *Mem:* Am Phys Soc; Am Chem Soc. *Res:* Electronic and magnetic resonance spectroscopy of excited electronic states. *Mailing Add:* Dept Chem Univ Calif Davis CA 95616

TINUS, RICHARD WILLARD, b Orange, NJ, Mar 26, 36; m 58; c 2. PLANT PHYSIOLOGY. *Educ:* Wesleyan Univ, BA, 58; Duke Univ, MF, 60; Univ Calif, Berkeley, PhD(plant physiol), 65. *Prof Exp:* Plant physiologist, Agr Res Serv, 65-68, PRIN PHYSIOLOGIST & PROJ LEADER, US FOREST SERV, USDA, 68- *Concurrent Pos:* Chmn working party, cold & drought hardiness, Int Union Forestry Res Orgn, ex-chmn working party, nursery opers. *Mem:* Am Soc Plant Physiologists; Int Plant Propagators Soc; Soc Am Foresters; Sigma Xi. *Res:* Development of greenhouse container systems for tree seedling production; vegetative propagation of pine; cold and drought resistance of trees. *Mailing Add:* Rocky Mt Forest & Range Exp Sta Forestry Sci Lab 700 S Knoles Dr Flagstaff AZ 86001

TIN-WA, MAUNG, b Rangoon, Burma, May 12, 40; US citizen; m 79. MEDICINAL CHEMISTRY, PHARMACOGNOSY. *Educ:* Univ Rangoon, Burma, BSc, 61; Ohio State Univ, Columbus, MSc, 65, PhD(pharmacog & natural prods), 69. *Prof Exp:* Teaching asst chem, Univ Rangoon, 61-62; teaching asst pharmacog, Univ Pittsburgh, 69; NSF fel plant sci, Univ Calif, Riverside, 70-71; NIH res assoc natural anticancer agents, Univ Ill Med Ctr, 71-72, asst prof pharmacog, 73-75; res dir, Res & Consult Assocs, 75-79; dir biosci, Sci Innovations, 80-87; PRES BIOSCI, ANPRA INC, 88- *Concurrent Pos:* Sr investr, Nat Cancer Inst, 72-74; dir, Nortech Labs Ltd, 82- *Mem:* Inst Food Technologists; Am Soc Pharmacog. *Res:* Development of synthetic and natural products with potential use for the treatment of cancer, immunodeficiency and cardiovascular diseases; ethnopharmacology and economic botany; food products; natural products chemistry. *Mailing Add:* Anpra Inc 30 Arroyo Way San Francisco CA 94127

TIO, CESARIO O, b Cebu, Philippines, Aug, 30, 32; m 62; c 2. ORGANIC SYNTHESIS, PHARMACOKINETICS. *Educ:* Univ Santo Tomas, Philippines, BS, 56; Georgetown Univ, Wash, MS, 61. *Prof Exp:* Res asst, Georgetown Univ Hosp, 78-86; res chemist II, 65-67, res chemist III, 67-78, UNIT SUPVR, WYETH-AYERST RES LABS, 65-, RES SCIENTIST, 86- *Res:* Drug disposition and metabolism; methods development for drug analysis; radiotracer synthesis; isolation and characterization of drug metabolites; synthesis of metabolites of drugs and pharmacokinetics. *Mailing Add:* Wyeth-Ayerst Res Labs Inc CN 8000 Princeton NJ 08543-8000

TIPEI, NICOLAE, b Calarasi, Romania, Apr 19, 13; US citizen; m 41; c 1. TRIBOLOGY, FLIGHT MECHANICS. *Educ:* Polytech Inst, Bucharest, MD, 36; Romanian Acad, DrEng, 68. *Prof Exp:* Asst prof airports & airplane oper, Polytech Inst, Bucharest, Romania, 36-48, assoc prof flight mech & rocket dynamics, 48-64, emer prof flight mech, rocket dynamics & aerodyn, 64-71; head solid & fluid mech, Gen Mech Div, Inst Appl Mech, Romanian Acad, 49-67, head tribology, Lubric Div, Inst Fluid Mech, 67-69; head, Tribology Div, Inst Sci Creation & Technol, Bucharest, 69-71; res fel engr res fluid mech, Gen Motors Corp Res Labs, 72-85; RETIRED. *Concurrent Pos:* Chief engr locomotive overhaul, Romanian Rwy, 37-38; chief engr airplane aerodyn & maintenance, Romanian Airlines, Romanian Air Transport, 38-39; pres, Nicolae Tipei Mach Trade Co, 39-43; chief engr, Romanian Mech & Chem Indust, 45-49; consult engr, Elec France & Total Co, 71-72. *Honors & Awards:* Mayo D Hersey Award, Am Soc Mech Engrs, 80. *Mem:* Am Inst Aeronaut & Astronaut; Am Soc Mech Engrs. *Res:* General mechanics; aerodynamics and flight mechanics; rocket dynamics and space flight; lubrication-tribology; fluid mechanics. *Mailing Add:* 105 Flora Dr Champaign IL 61821

TIPLER, FRANK JENNINGS, III, b Andalusia, Ala, Feb 1, 47; m 84; c 2. GLOBAL GENERAL RELATIVITY, COSMOLOGY. *Educ:* Mass Inst Technol, SB, 69; Univ Md, PhD(physics), 76. *Prof Exp:* Res mathematician, Univ Calif, Berkeley, 76-79; sr res fel, Oxford Univ, 79; res assoc, Univ Texas, Austin, 79-81; assoc prof, 81-87, PROF MATH & PHYSICS, TULANE UNIV, 87- *Concurrent Pos:* Vis fel, Astron Ctr, Univ Sussex, UK, 85, 87; Vis Prof, Univ Berne, Switz, 88, Univ Liege, Belg, 88. *Mem:* Am Phys Soc; Royal Astron Soc; Sigma Xi; Int Soc Gen Relativity & Gravitation. *Res:* Structure of singularities in classical general relativity; quantum cosmology wave function interpretation; extraterrestrial intelligent life; future evolution of universe; anthropic principle; wave function interpretation. *Mailing Add:* Dept Math & Physics Tulane Univ New Orleans LA 70118

TIPLER, PAUL A, b Antigo, Wis, Apr 12, 33; div; c 2. NUCLEAR PHYSICS. *Educ:* Purdue Univ, BS, 55; Univ Ill, PhD(physics), 62. *Prof Exp:* Asst prof physics, Wesleyan Univ, 61-62; from asst prof to assoc prof, Oakland Univ, 62-77, prof physics, 77-82; CONSULT, 82- *Mem:* Am Phys Soc; Am Asn Physics Teachers. *Res:* Low energy nuclear physics. *Mailing Add:* 1329 Arch St Berkeley CA 94708

TIPPENS, DORR E(UGENE) F(ELT), b Grand Rapids, Mich, Dec 28, 23; m 43; c 3. CHEMICAL ENGINEERING. *Educ:* Purdue Univ, BS, 48. *Prof Exp:* Engr in training, Oper Dept, Am Sugar Co, 49-50, asst supt refining dept, Baltimore Refinery, 50-53, refining supt, 53-56, asst to refining mgr, 56-58, process design engr, NY, 58-59, dept head new food prod develop, Res & Develop Div, 59-64, dir process develop, 64-68, sr chem engr, 68-76, proj mgr, Amstar Corp, 76-82; OWNER, DNE NO-LOAD MUTUAL FUND ADV SERV, 82- *Mem:* Am Chem Soc; Sugar Indust Tech; Am Inst Chem Engrs. *Res:* New process development in sugar refining and new food product development, directing and administrating. *Mailing Add:* Box Five Gibson Island MD 21056

TIPPER, DONALD JOHN, b Birmingham, Eng, July 21, 35; m 65; c 3. MICROBIOLOGY, MOLECULAR BIOLOGY. *Educ:* Univ Birmingham, BSc, 56, PhD(chem), 59. *Prof Exp:* Res assoc immunochem, Dept Surg Res, St Luke's Hosp, 59-60; chemist, Guinness's Brewery, Dublin, Ireland, 60-62; res asst pharmacol, Wash Univ, 62-64; res asst, Univ Wis-Madison, 64-65, from asst prof to assoc prof, 65-71; chmn dept, 71-80, PROF MICROBIOL & MOLECULAR GENETICS, MED SCH, UNIV MASS, 71- *Concurrent Pos:* USPHS career develop award, 68-71. *Mem:* Am Soc Microbiol; Brit Soc Gen Microbiol; Genetics Soc Am. *Res:* Structure and biosynthesis of bacterial cell walls; mode of action of penicillins; biosynthesis of bacillus spore-specific proteins; yeast dsRNA-coded killer system; G-protein coupled signal transduction in yeast. *Mailing Add:* Dept Molecular Genetics & Microbiol Univ Mass Med Sch Worcester MA 01655

TIPPER, RONALD CHARLES, b Sacramento, Calif, July 17, 42; m 64; c 2. BIOLOGICAL OCEANOGRAPHY. *Educ:* Ore State Univ, BS, 64, PhD(biol oceanog), 68. *Prof Exp:* Antisubmarine warfare officer, USS Coontz DLG-9, San Diego, Calif, 68-70; res plans officer oceanog res, US Naval Oceanog Off, 70-71; sci officer environ qual div, US Naval Oceanog Off, 71-72, aide & exec asst oceanog res & develop, 72-74, dir oceanic biol prog, Off Naval Res, 74-76 & Naval Ocean Res & Develop Activity, 76-77, cmndg officer, Oceanog Unit One, 77-78, dir, Oceanic Biol Prog, Off Naval Res, 78-81, dir Ocean Surv Prog, 81-83, res prog mgr, Off Naval Res, 83-86, off asst secy Navy Res & Systs, 86-87, cmndg officer Polar Oceanog Ctr, 87-89, dir, Naval Requirements D & D Hydrographic & Topographic Ctr, 89-91, DIR SATELLITE PROG, JOINT OCEANOG INST, INC, US NAVAL OCEANOG OFF, 91- *Mem:* Sigma Xi; Am Geophys Union; AAAS; Oceanog Soc. *Res:* Oceanic biology; marine biodeterioration; distribution and abundance of Micronekton and Zooplankton; benthic ecology; geophysics; cartography; navigation. *Mailing Add:* 9819 Summerday Dr Burke VA 22015

TIPPETT, JAMES T, b Oxford, NC, May 11, 31; m 65; c 2. COMPUTER SCIENCE, ELECTRONICS ENGINEERING. *Educ:* NC State Col, BSEE, 54, EE, 55. *Prof Exp:* ELECTRONIC ENGR, NAT SECURITY AGENCY, 55- *Mem:* AAAS; Inst Elec & Electronics Engrs; Asn Comput Mach. *Res:* Ultra high speed computer circuits; software techniques; computer languages; networks of computers; computer security; software engineering; telecommunications; general computer sciences. *Mailing Add:* Dept Defense Nat Security Agency 9800 Savage Rd Ft George G Meade MD 20755

TIPPING, RICHARD H, b Abington, Pa, Aug 31, 39. MOLECULAR PHYSICS. *Educ:* Pa State Univ, BSc, 63, MSc, 66, PhD(physics), 69. *Prof Exp:* From asst prof to assoc prof, Mem Univ Nfld, 69-77; assoc prof physics, Univ Nebr, Omaha, 77-80, prof, 80; PROF PHYSICS & ASTRON, UNIV ALA, 80- *Mem:* Am Phys Soc; Can Asn Physicists; Sigma Xi. *Res:* Molecular spectral line shapes; intensities and spectroscopic constants of diatomic molecules. *Mailing Add:* Box 870324 Tuscaloosa AL 35487-0324

TIPPLES, KEITH H, b Cambridge, Eng, Feb 4, 36; m 62; c 3. CEREAL CHEMISTRY. *Educ:* Univ Birmingham, BSc, 59, PhD(appl biochem), 62. *Prof Exp:* Nat Res Coun Can res fel, 63-64, res scientist, 64-79, DIR, GRAIN RES LAB, CAN GRAIN COMN, 79- *Honors & Awards:* C.W Brabender Award, 75; Gold Medal, Czechoslovak Acad Agr, 82. *Mem:* Am Asn Cereal Chem. *Res:* Basic and applied research on the quality of cereal grains and oilseeds aimed at understanding the chemistry and technology of end-use quality and development of improved methods for quality assessment. *Mailing Add:* Grain Res Lab Can Grain Comn 1404-303 Main St Winnipeg MB R3C 3G8 Can

TIPSWORD, RAY FENTON, b Beecher City, Ill, Sept 9, 31; m 52; c 2. PHYSICS. *Educ:* Eastern Ill Univ, BSEd, 53; Southern Ill Univ, MS, 57; Univ Ala, PhD(physics), 62. *Prof Exp:* Instr physics, Mo Sch Mines, 57-59; fel, Univ Ala, 62-63; asst prof, 63-68, assoc prof, 68-81, PROF PHYSICS, VA POLYTECH INST & STATE UNIV, 81- *Res:* Nuclear magnetic resonance and low temperature physics. *Mailing Add:* Dept Physics Va Polytech Inst Blacksburg VA 24061

TIPTON, C(LYDE) R(AYMOND), JR, b Cincinnati, Ohio, Nov 13, 21; m 42; c 2. SCIENCE COMMUNICATIONS. *Educ:* Univ Ky, BS, 46, MS, 47. *Prof Exp:* Asst, Univ Ky, 46-47; res engr, Battelle Mem Inst, 47-49; phys metallurgist, Los Alamos Sci Lab, 49-51; from asst div chief to sr tech adv, Physics Dept, Battelle Mem Inst, 51-62; dir res, Basic, Inc, 62-64; from staff mgr to asst dir, Pac Northwest Labs, 64-69, coordr corp commun, 69-73, vpres commun, 73-75, pres & trustee, Battelle Commons Co, 75-78, asst pres, 78-80, vpres & corp dir commun & pub affairs, Battelle Mem Inst, 80-88; CONSULT, 86- *Concurrent Pos:* Consult, Atomic Energy Comn, 58-61; secy, US Deleg 2nd UN Int Conf Peaceful Uses Atomic Energy, Geneva, Switz, 58; sr fel, Otterbein Col, 78; chmn, Prof Engrs Ind, 88-89. *Mem:* Am Soc Metals; Nat Soc Prof Engrs; Sigma Xi. *Res:* Reactor materials; technical and scientific information handling; industrial research. *Mailing Add:* 6475 Strathaven Ct W Worthington OH 43085

TIPTON, CARL LEE, b Collins, Iowa, July 26, 31; m 57; c 3. BIOCHEMISTRY. *Educ:* Univ Nebr, BS, 54, MS, 57; Univ Ill, PhD, 61. *Prof Exp:* Assoc & instr, 61-62, from asst to assoc prof, 62-78, PROF BIOCHEM, IOWA STATE UNIV, 78- *Concurrent Pos:* NIH sr fel, Univ Calif, Davis, 69-70. *Mem:* AAAS; Am Chem Soc; Am Soc Biochem & Molecular Biol. *Res:* Cholesterol metabolism. *Mailing Add:* Dept Biochem & Biophys Iowa State Univ Ames IA 50011

TIPTON, CHARLES M, b Evanston, Ill, Nov 29, 27; m 53; c 4. PHYSIOLOGY. *Educ:* Springfield Col, BS, 52; Univ Ill, MS, 53, PhD(physiol), 62. *Prof Exp:* High sch teacher, Ill, 53-55; teaching asst health educ, Univ Ill, 55-57, instr, 57-58, asst physiol, 58-61; asst prof, Springfield Col, 61-63; from asst prof to assoc prof physiol, Univ Iowa, 63-73, prof, 73-; PROF & HEAD, DEPT EXERCISE & SPORTS SCI, UNIV ARIZ. *Concurrent Pos:* Chmn study sect appl physiol & bioeng, NIH, 73-76; ed, Med Sci Sports Exercises, 79-; mem med adv comn, Am Inst Biol Sci, 80- *Mem:* Am Physiol Soc; fel Am Heart Asn; Am Col Sports Med (pres, 74-75). *Res:* Exercise physiology, including bradycardia of training, mechanisms of cardiac hypertrophy, ligamentous strength, diabetes, endocrines and training; exercise testing; hypertension; pharmacological differences and training. *Mailing Add:* Dept Phys Educ Univ Ariz INA E Gittings Bldg Tucson AZ 85721

TIPTON, KENNETH WARREN, b Belleville, Ill, Nov 14, 32; m 57; c 2. PLANT GENETICS, AGRONOMY. *Educ:* La State Univ, BS, 55, MS, 59; Miss State Univ, PhD(agron, genetics), 69. *Prof Exp:* Teaching asst, 58-59, from asst prof to prof agron, Agr Exp Sta, La State Univ, Baton Rouge, 59-75; prof & supt, Red River Valley Agr Exp Sta, Bossier City, 75-79; assoc dir, 79-89, VCHANCELLOR & DIR, LA AGR EXP STA, LA STATE UNIV, AGR CTR, 89- *Concurrent Pos:* Chmn, Comt Nine, 88, Exp Sta Comt Policy, 89-91. *Mem:* Am Soc Agron; Crop Sci Soc Am; Coun Agr Sci & Technol. *Res:* Varietal evaluation and adaptation of grain sorghum; varietal evaluation, breeding and genetics of small grains. *Mailing Add:* La Agr Exp Sta PO Box 25055 Baton Rouge LA 70894-5055

TIPTON, MERLIN J, b Watertown, SDak, Mar 23, 30; m 54; c 3. GEOLOGY. *Educ:* Univ SDak, BA, 55, MA, 58. *Prof Exp:* Geologist, 56-62, from asst state geologist to assoc state geologist, 62-82, STATE GEOLOGIST, SDAK GEOL SURV, 82-, LECTR EARTH SCI, UNIV SDAK, 76- *Mem:* Fel Geol Soc Am. *Res:* Pleistocene geology and hydrogeology. *Mailing Add:* Dept Earth Sci & Physics Univ SDak Vermillion SD 57069

TIPTON, VERNON JOHN, b Springville, Utah, July 12, 20; m 43; c 5. PARASITOLOGY, MEDICAL ENTOMOLOGY. *Educ:* Brigham Young Univ, BS, 48, MS, 49; Univ Calif, PhD, 59. *Prof Exp:* Entomologist, Walter Reed Army Inst Res, Med Serv Corps, US Army, 49-52, Calif, 52-54, commanding officer, 37th Prev Med Co, Korea, 54-55, chief entom sect, 5th Army Med Lab, St Louis, Mo, 56-59, chief environ health br, Off Surgeon, Ft Amador, CZ, 57-62, chief entom br dept prev med, Med Field Serv Sch, Ft Sam Houston, Tex, 62-66, chief dept entom, 406th Med Lab, 66-68; from assoc prof to prof zool, Brigham Young Univ, 68-86, dir, Ctr Health Environ Studies, 75-86; RETIRED. *Res:* Siphonaptera; mesostigmatid mites; systematics; bionomics. *Mailing Add:* 345 S Canyon Ave Springville UT 84663

TIRMAN, ALVIN, b Brooklyn, NY, Nov 27, 31. PHYSICAL CHEMISTRY, MATHEMATICS. *Educ:* Hofstra Col, AB, 53; Bowling Green State Univ, MA, 65; Carnegie-Mellon Univ, PhD(chem), 70. *Prof Exp:* From asst prof to assoc prof math, 69-88, PROF MATH, KINGSPORT UNIV CTR, ETENN STATE UNIV, 88- *Mem:* Math Asn Am; Sigma Xi. *Res:* History of mathematics; geometry; number theory. *Mailing Add:* Dept Math Kingsport Univ Ctr ETenn State Univ Kingsport TN 37662

TIRRELL, MATTHEW VINCENT, b Phillipsburg, NJ, Sept 5, 50. POLYMER SCIENCE, CHEMICAL ENGINEERING. *Educ:* Northwestern Univ, BSChE, 73; Univ Mass, PhD(polymer sci), 77. *Prof Exp:* from asst prof to assoc prof, 77-85, SHELL DISTINGUISHED PROF CHEM ENG, UNIV MINN-MINNEAPOLIS, 85- *Concurrent Pos:* Vis prof, Univ Guadalajara, Australian Nat Univ, Ecole Superieure de Physique & Chimie & Univ Nac'l del Sur. *Mem:* Am Chem Soc; Am Inst Chem Engrs; Soc Rheology; Am Phys Soc; AAAS. *Res:* Engineering applications of chemistry and physics of macromolecules; polymerization reactor design; flow-induced conformational and structural changes; rheology; diffusion and interfacial properties of polymers. *Mailing Add:* Dept Chem Eng & Mat Sci Univ Minn 421 Washington Ave SE Minneapolis MN 55455-0100

TISCHENDORF, JOHN ALLEN, b Lincoln City, Ind, July 22, 29; m 59; c 3. APPLIED STATISTICS, DATA ANALYSIS. *Educ:* Evansville Col, AB, 50; Purdue Univ, MS, 52, PhD(math statist), 55. *Prof Exp:* Mem tech staff, 57-59, supvr reliability & statist studies, Allentown, 59-64, supvr statist appln, 64-78, supvr eng data anal, Bell Tel Labs, Holmdel, 78-83; supvr eng data anal, AT&T-IS, 83-85; supvr, Prod Qual & Reliability, 85-89; MGR, NAT TECH SUPPORT CTR, 89- *Concurrent Pos:* Vis lectr, Rutgers Univ, 66 & 67 & Stanford Univ, 68-69. *Mem:* Am Statist Asn; Am Soc Qual Control. *Res:* Statistics applied to engineering and management problems; data analysis; statistical consulting; mathematical modeling; reliability; sampling; order statistics; market analysis; data base management; quality assurance. *Mailing Add:* 7798 Oakview Pl Castle Rock CO 80104

TISCHER, FREDERICK JOSEPH, b Plan, Austria, Mar 14, 13; US citizen; m 42. ELECTROPHYSICS, COMMUNICATIONS. *Educ:* Prague Tech Univ, MSc, 36, PhD(elec eng), 38. *Prof Exp:* Fel physics, Univ Berlin, 38; res assoc, Telefunken, Berlin, 38-42; owner, Tischer Phys Lab, Austria, 42-47; lectr microwaves, Royal Inst Technol, Sweden, 47-54; br chief guid lab, Ord Missile Lab, Ala, 54-56; assoc prof electromagnetics, Ohio State Univ, 56-62; asst dir res inst, Univ Ala, Huntsville, 62-64; prof electromagnetics, 64-78, EMER PROF, NC STATE UNIV, 78- *Concurrent Pos:* Vis lectr, Helsinki Inst Technol, 52; consult, Chance Vought Aircraft Co, Tex, 59-61 & Harry Diamond Labs, DC, 60-74; consult, NASA Hq, 62-63, expert consult, NASA-Goddard Space Flight Ctr, Md, 62; fel plasma physics, Princeton Univ, 62; consult, 76-; vis distinguished prof, Naval Postgrad Sch, 78-79; vis prof, Univ Bern, Switz, 80, Swiss Fed Inst Technol, Zurich, 81 & Sheffield, Eng, 83. *Mem:* Fel Inst Elec & Electronics Engrs; Optical Soc Am; Am Phys Soc; Sigma Xi. *Res:* Microwaves, waveguides, plasma physics, space communications; holography; millimeter waves; satellite communications. *Mailing Add:* 2313 Wheeler Rd Raleigh NC 27612

TISCHER, RAGNAR P(ASCAL), b Berlin, Ger, Apr 30, 22; m 61; c 2. PHYSICAL CHEMISTRY, METALLURGY. *Educ:* Univ Goettingen, dipl phys, 53; Stuttgart Tech Univ, Dr rer nat(phys chem), 57. *Prof Exp:* Res assoc electrochem, Dept Chem, Univ Ill, 57-59; asst metall, Gebr Boehler A G, Edelstahlwerk Dusseldorf, Ger, 59-60; res assoc electrochem, Inst Phys Chem, Bonn, 61-63; sr res chemist, Electrochem Dept, Res Labs, Gen Motors Corp, 63-66; prin res scientist assoc, Ford Motor Co, 66-68, staff scientist, res staff, 68-88; RETIRED. *Mem:* Electrochem Soc; Sigma Xi. *Res:* Electrode kinetics in aqueous solutions and in fused salts; fuel cells; corrosion of binary alloys; battery electrodes and metal deposition in organic solvents. *Mailing Add:* 1449 Suffield Ave Birmingham MI 48009-1041

TISCHER, THOMAS NORMAN, b Milwaukee, Wis, Apr 21, 34. PHOTOGRAPHIC CHEMISTRY. *Educ:* Marquette Univ, BS, 56, MS, 58; Univ Wis, PhD(anal chem), 61. *Prof Exp:* RES ASSOC, EASTMAN KODAK CO, 61- *Mem:* Am Chem Soc. *Res:* Photographic patent liaison; organic chemistry; materials science engineering; thermal-dye imaging and research. *Mailing Add:* 115 Heritage Circle Rochester NY 14615

TISCHFIELD, JAY ARNOLD, b New York, NY, June 15, 46; m 78; c 3. SOMATIC CELL GENETICS. *Educ:* Brooklyn Col, BS, 67; Yale Univ, MPhil, 69, PhD(biol), 73. *Prof Exp:* Asst prof biol, genetics & pediat, Case Western Reserve Univ, 72-78; prof anat, cell & molecular biol & pediat, Med Col Ga, 78-87; PROF MED GENETICS & DIR, DIV MOLECULAR GENETICS IND UNIV SCH MED, 87- *Concurrent Pos:* Fel, Univ Calif, San Francisco, 72-73; prin investr, US Pub Health grant, 73- & NSF grant, 82- *Mem:* Genetics Soc Am; Tissue Culture Asn; Am Soc Cell Biol; Am Soc Microbiol; NY Acad Sci; Am Soc Human Genetics. *Res:* Human genetics and mammalian somatic cell genetics, especially the effects of mutations on all levels of gene expression. *Mailing Add:* Dept Med Genetics Ind Univ Sch Med Indianapolis IN 46202-5251

TISCHIO, JOHN PATRICK, b Newark, NJ, Mar 17, 42; m 66, 88; c 2. BIO-PHARMACEUTICALS. *Educ:* Fairleigh Dickinson Univ, BS, 65; Univ Rochester, PhD(biochem), 71. *Prof Exp:* Asst prof biochem, Philadelphia Col Pharm & Sci, 70-77; scientist, drug metab, Ortho Pharm Corp, 77-78; SCIENTIST, DRUG METAB, IMMUNO BIOL RES INST, 88- *Concurrent Pos:* Lectr, Wagner Free Inst Sci, 71-77, Philadelphia Community Col, 76-77, Raritan Valley Community Col, 89- *Mem:* AAAS; Am Chem Soc; Sigma Xi; Am Pharmaceut Asn; Am Asn Pharmaceut Scientists. *Res:* Biotransformation and disposition of drugs; methods development for drugs in biological fluids; radio immune assay technology; bioavailability and pharmacokinetics in animals and humans. *Mailing Add:* Immunol/Biol Res Inst Rte 22 East Annandale NJ 08801

TISCHLER, HERBERT, b Detroit, Mich, Apr 28, 24; m 54; c 2. INVERTEBRATE PALEONTOLOGY. *Educ:* Wayne State Univ, BS, 50; Univ Calif, Berkeley, MA, 55; Univ Mich, PhD(geol), 61. *Prof Exp:* Instr geol, Wayne State Univ, 56-58; assoc prof earth sci, Northern Ill Univ, 58-65; chmn dept earth sci, 65-90, PROF GEOL, UNIV NH, 65- *Concurrent Pos:* NSF sci fac fel, Columbia Univ, 64-65. *Mem:* Fel Geol Soc Am; Paleont Soc; Am Asn Petrol Geol; Nat Asn Geol Teachers. *Res:* Paleoecology of marine invertebrates; ecology of benthic foraminifera; carbonate petrology; stratigraphy. *Mailing Add:* Dept Earth Sci James Hall Univ NH Durham NH 03824

TISCHLER, MARC ELIOT, b New York, NY, Nov 10, 49; m 79; c 2. INTERMEDIARY METABOLISM, HORMONAL REGULATION. *Educ:* Boston Univ, BA, 71; Univ SC, MS, 73; Univ Pa, PhD(biochem), 77. *Prof Exp:* Fel, Med Sch, Harvard Univ, 77-79; from asst prof to assoc prof, 79-90, PROF BIOCHEM, UNIV ARIZ, 90- *Concurrent Pos:* Estab investr, Am Heart Asn, 82-87. *Mem:* Am Physiol Soc; Am Soc Biol Chemists; Am Soc Gravitational & Space Biol (vpres, 89-90, pres, 90-91). *Res:* Regulation of muscle metabolism and responses to physiologic perturbations such as fasting, trauma and muscle disuse; regulation and role of insect protein turnover. *Mailing Add:* Dept Biochem Univ Ariz Tucson AZ 85724

TISCHLER, OSCAR, b New York, NY, Oct 7, 23; m 50; c 3. PROCESSING OF THIN FILMS, SUBSTRATES FOR THIN FILM DEPOSITION. *Educ:* Cooper Union, BChE, 51; Newark Col Eng, MS, 58. *Prof Exp:* Engr, Balco Corp, 51-56 & Daven Corp, 56-58; staff engr, Kearfott Div, KDI Electronics, 58-60, dir res, Pyrofilm Div, 60-; RETIRED. *Mem:* Am Chem Soc; Am Ceramics Soc; Mat Res Soc; Am Vacuum Soc. *Res:* Techniques used to deposit thin films of metals and alloys and non-metals or insulators; study of the properties and uses of these films. *Mailing Add:* 32 Holiday Dr West Caldwell NJ 07006

TISDALE, GLENN E(VAN), b Madison, Wis, July 4, 24; m 60; c 2. ELECTRICAL ENGINEERING. *Educ:* Yale Univ, BE, 44, MEng, 47, PhD(elec eng), 49. *Prof Exp:* Jr engr, Raytheon Mfg Co, 49; engr, Servo Corp Am, NY, 50 & Perkin-Elmer Corp, 51-52; sr res engr, Electro-Mech Res, Inc, 53-57, mgr systs eng dept, 57-63; pres, Sea Technol Corp, 63-64; gen mgr sea technol dept, 65, MGR INFO TECHNOL AEROSPACE DIV, WESTINGHOUSE ELEC CORP, 66- *Concurrent Pos:* Consult, Spencer-Kennedy Labs, 49-52. *Mem:* Inst Elec & Electronics Engrs; Sigma Xi. *Res:* Recognition logic; signal processing; data transmission systems; design of equipment for aerospace and underseas missions. *Mailing Add:* 13 Windward Dr Severna Park MD 21146

TISE, FRANK P, b Washington, DC, Dec 4, 51; m 81; c 1. ORGANIC PHOTOCHEMISTRY. *Educ:* Univ Md, BS, 73; Univ NC, PhD(org chem), 80. *Prof Exp:* RES SCIENTIST, RES CTR, HERCULES, INC, 81- *Mem:* Am Chem Soc. *Res:* Integrated circuit photoresists; phenolic-modified ink resins. *Mailing Add:* 5907 Stone Pine Rd Wilmington DE 19808

TISHBY, NAFTALI Z, b Jerusalem, Israel, Dec 28, 52; m 80; c 2. APPLICATION OF STATISTICAL MECHANICS TO LEARNING, DYNAMICAL SYSTEMS APPROACH TO PATTERN RECOGNITION. *Educ:* Hebrew Univ, Jerusalem, BS, 74; Tel-Aviv Univ, MS, 80; Hebrew Univ, PhD(phys). *Prof Exp:* Teaching asst physics, Tel Aviv Univ, 77-80; tutor & teaching asst physics, Hebrew Univ, Jerusalem, 81-85; vpres res, Sesame Systs Ltd, Israel, 85-86; fel physics, Mass Inst Technol, 86-87; MEM TECH STAFF PHYSICS & COMPUT, AT&T BELL LABS, 87- *Honors & Awards:* E Golomo Award, Israel Defense Ministry, 80. *Mem:* Am Phys Soc; Inst Elec & Electronics Engrs. *Res:* Machine learning based on statistical physics and information theory; dynamical systems theory and temporal pattern recognition; applying group theoretical methods to pattern recognition. *Mailing Add:* AT&T Bell Labs 2C-576 600 Mountain Ave Murray Hill NJ 07974-2070

TISHKOFF, GARSON HAROLD, b Rochester, NY, Aug 8, 23; m 59. MEDICINE, HEMATOLOGY. *Educ:* Univ Rochester, BS, 44, PhD(pharmacol), 51, MD, 53. *Prof Exp:* AEC assoc, Univ Rochester, 46-52; instr med, Med Sch, Tufts Univ, 56-57 & Harvard Med Sch, 57-60; from asst prof to assoc prof, Sch Med, Univ Calif, Los Angeles, 60-71; prof med, Mich State Univ & dir, Great Lakes Regional Blood Prog, Am Red Cross, 71-; RETIRED. *Concurrent Pos:* Nat Acad Sci-Nat Res Coun fel, New Eng Ctr Hosp, Tufts Univ, 55-56; assoc, Beth Israel Hosp, Boston, 57-60. *Mem:* AAAS; Am Soc Hemat; Int Soc Hemat; Int Soc Thrombosis & Haemostasis; Int Soc Blood Transfusion. *Res:* Internal medicine; hemolytic anemia; blood coagulation; blood transfusion. *Mailing Add:* 1850 Abbott Rd Apt B-10 E Lansing MI 48823

TISHLER, MAX, organic chemistry; deceased, see previous edition for last biography

TISHLER, PETER VERVEER, b Boston, Mass, July 18, 37; m 60; c 2. MEDICAL GENETICS. *Educ:* Harvard Univ, AB, 59; Yale Univ, MD, 63. *Prof Exp:* Staff assoc, Nat Inst Arthritis & Metab Dis, 66-68; house officer internal med II & IV, Harvard Med Serv, 63-66, res fel med, Thorndike Mem Lab & Channing Lab, 68-69, staff mem med & genetics, Channing Lab, Boston City Hosp & Harvard Med Sch, 69-77 & Peter Bent Hosp & Harvard Med Sch, 77-80; ASSOC CHIEF STAFF EDUC, VET ADMIN MED CTR, BROCKTON & W ROXBURY, MASS, 80- *Concurrent Pos:* Francis Weld Peabody fel med, Harvard Univ, 69-75; asst physician, Dept Pediat, Boston City Hosp, assoc vis physician, Dept Med, 71-77; assoc, Ctr Human Genetics, Harvard Med Sch, 71, asst prof med, 72; assoc med, Peter Bent Brigham Hosp, 76, assoc prof med, 83- *Mem:* AAAS; NY Acad Sci; Am Soc Human Genetics; Am Fedn Clin Res; Int Soc Twin Studies. *Res:* Genetics of chronic disease; biochemistry of diseases of porphyrin metabolism. *Mailing Add:* Vet Admin Med Ctr 940 Belmont St Brockton MA 02401

TISON, RICHARD PERRY, b Pontiac, Mich, June 23, 44; m 63; c 3. ELECTROCHEMISTRY. *Educ:* Gen Motors Inst, BME, 68; Rensselaer Polytech Inst, MS, 68. *Prof Exp:* Chem engr plant eng, Gen Motors Truck & Coach Div, 68-69, STAFF RES ENGR ELECTROCHEM, GEN MOTORS RES LABS, GEN MOTORS CORP, 69- *Mem:* Am Electroplaters' Soc. *Res:* Stress-strain analysis; adhesion failure of plated plastic automotive parts; methods for electrophoretic analysis of concentrated suspensions; improved methods for electrochemical recovery of metals from industrial process streams. *Mailing Add:* 1655 Nancy G Lane Orion MI 48359

TISONE, GARY C, b Boulder, Colo, Dec 24, 37; m 59; c 4. ATMOSPHERIC PHYSICS. *Educ:* Univ Colo, BS, 59, PhD(physics), 67. *Prof Exp:* Physicist, Vallecitos Atomic Lab, Gen Elec Co, 59-61; physicist, Nat Bur Standards, 61-62; res assoc atomic physics, Univ Colo, 66-67; STAFF MEM, SANDIA CORP, 67- *Mem:* Am Phys Soc. *Res:* Electron-negative ion collisions; photodetachment of negative ions; gaseous electronics; physics of the upper atmosphere; gas laser research; particle beam fusion. *Mailing Add:* Sandia Labs Org 1275 Albuquerque NM 87185

TISSERAT, BRENT HOWARD, b Oakland, Calif, Oct 2, 51; m 69; c 1. AGRICULTURAL SCIENCES. *Educ:* San Bernardino Jr Valley Col, AA, 71; Calif State Polytech Univ, BS, 73; Univ Calif, Berkeley, PhD(bot), 76. *Prof Exp:* Cabot fel Res, Cabot Found, Harvard Univ, 76-77; AGR RES GENETICIST, USDA, 77- *Mem:* Am Bot Soc; Crop Sci Soc Am; Palm Soc. *Res:* Palm and fruit tree crop morphogenesis; studying trees through tissue culture, protoplast, cryobiology and gene-enzyme techniques. *Mailing Add:* 2513 Euclid Crescent Upland CA 91756

TISSUE, ERIC BRUCE, b Hagerstown, Md, Feb 7, 55; m 80; c 2. TEST EVALUATION, VERIFICATION & VALIDATION. *Educ:* Towson State Univ, BS, 77; Mich State Univ, MS, 82. *Prof Exp:* SR PROG ANALYST MATH, SYSCON CORP, SUBSID HARNISCHFEGER INDUST, 83- *Concurrent Pos:* Sr lectr, Dept Math, Towson State Univ, 85- *Mem:* Soc Indust & Appl Math; Int Test Eval Asn. *Res:* Simulation and modeling in performing test and evaluation of complex systems, including computer software, and verification and validation activities over the software development life-cycle. *Mailing Add:* 1867 King Richard Rd Eldersburg MD 21784-6265

TISUE, GEORGE THOMAS, b Carroll, Iowa, Nov 25, 40; c 4. ENVIRONMENTAL CHEMISTRY, ANALYTICAL CHEMISTRY. *Educ:* Beloit Col, BS, 61; Yale Univ, PhD(org chem), 66. *Prof Exp:* Res chemist tech ctr, Celanese Chem Co, Tex, 66 & 68; NIH fel plant biochem, Univ Freiburg, 66-67; from asst prof to assoc prof chem, Beloit Col, 68-74; chemist, Argonne Nat Lab, 74-81; res assoc, Ctr Great Lakes Studies, Univ Wis-Milwaukee, 81; assoc prof, 82-91, PROF CHEM, CLEMSON UNIV, 91- *Concurrent Pos:* Vis prof, Dept Chem, Univ Wis-Milwaukee, 81, Shaw vis prof, 88. *Mem:* AAAS; Am Chem Soc; Int Asn Environ Anal Chem; Int Asn Great Lakes Res. *Res:* Biogeochemical cycling and effects of heavy metal pollutants in the Great Lakes and coastal waters; fluxes at the sediment-water interface. *Mailing Add:* 77 Broadway Newry SC 29665

TISZA, LASZLO, b Budapest, Hungary, July 7, 07; nat US; m 73. PHYSICS. *Educ:* Univ Budapest, PhD(physics), 32. *Prof Exp:* Res assoc, Phys Tech Inst, Kharkov, 35-37; res assoc, Col France, 37-40; from instr physics to prof physics, 41-73, EMER PROF PHYSICS, MASS INST TECHNOL, 73- *Concurrent Pos:* Guggenheim fel, 62-63; vis prof, Univ Paris, 62-63. *Mem:* Fel Am Phys Soc; fel Am Acad Arts & Sci; fel AAAS. *Res:* Theoretical physics; foundations of quantum mechanics; statistical thermodynamics. *Mailing Add:* Dept Physics Mass Inst Technol 6-108 Cambridge MA 02139

TITCHENER, EDWARD BRADFORD, b Cambridge, Mass, July 15, 27; m 52; c 2. BIOCHEMISTRY. *Educ:* Univ Mich, BS, 51; Ohio State Univ, MS, 54, PhD(physiol chem), 56. *Prof Exp:* Asst biochem, Ohio State Univ, 53-56; from asst prof to assoc prof, 58-74, PROF BIOCHEM, UNIV ILL COL MED, 74- *Concurrent Pos:* NIH trainee, Enzyme Inst, Univ Wis, 56-58. *Mem:* AAAS; Am Chem Soc; Am Soc Biol Chemists; NY Acad Sci; Sigma Xi. *Res:* Transfer RNA. *Mailing Add:* Dept Biochem Univ Ill 833 S Wood St Chicago IL 60612

TITCHENER, JAMES LAMPTON, b Binghamton, NY, Apr 9, 22; c 4. PSYCHIATRY, PSYCHOANALYSIS. *Educ:* Princeton Univ, AB, 46; Duke Univ, MD, 49; Chicago Psychoanal Inst, cert psychoanal, 64. *Prof Exp:* Intern psychiat, Walter Reed Gen Hosp, 49-50; resident, Cincinnati Gen Hosp, 50-54; from instr to assoc prof, 54-68, PROF PSYCHIAT, MED CTR, UNIV CINCINNATI, 68-; MEM FAC PSYCHOANAL, CINCINNATI PSYCHOANAL INST, 74- *Concurrent Pos:* Career investr, USPHS, 55-60 & NIMH, 57-62; training & supv analyst, Cincinnati Psychoanal Inst, 74-; attend physician, Cincinnati Gen Hosp. *Mem:* Fel AAAS; fel Am Psychiat Asn; Am Psychosom Soc; Am Psychoanal Asn; Sigma Xi. *Res:* Study of the effects of physical trauma on personality functioning; family and marital dynamics and marital therapy; psychological trauma resulting from disasters such as the Buffalo Creek disaster and Beverly Hills fire. *Mailing Add:* 4021 Rose Hill Ave Cincinnati OH 45229

TITELBAUM, SYDNEY, b Luck, Russia, Apr 24, 13; US citizen; m 39; c 1. ANIMAL PHYSIOLOGY, EVOLUTIONARY BIOLOGY. *Educ:* Univ Chicago, PhB, 33, PhD(physiol), 38, JD, 42. *Prof Exp:* Teaching asst physiol, Univ Chicago, 34-38; chief physiologist, Chicago Biol Res Lab, 39-59; lectr forensic med, Sch Med, Loyola Univ Chicago, 46-64; prof biol, City Cols Chicago, 60-77; PROF BIOL, FROMM INST, UNIV SAN FRANCISCO, 78- *Concurrent Pos:* Ed, Asn Off Racing Chem Jour, 56-58; consult-examr, NCent Asn Cols & Sec Schs, 66-78; dean, Bogan Col, 67-68. *Honors & Awards:* Award, Asn Off Racing Chem, 60. *Mem:* AAAS; Am Inst Biol Sci; Asn Off Racing Chem (pres, 57). *Res:* Physiology of sleep; history of science. *Mailing Add:* 3628 Fillmore St San Francisco CA 94123-1602

TITELER, MILT, b Lakewood, NJ, Nov 8, 49. PHARMACOLOGY, BIOCHEMISTRY. *Educ:* State Univ NY, Buffalo, BS, 72, MSc, 76; Univ Toronto, PhD(pharmacol), 78. *Prof Exp:* Res asst biochem, Univ Buffalo, 73-76; from res asst to res assoc pharmacol, 76-79, ASST PROF PHARMACOL, UNIV TORONTO, 79-; AT DEPT PHARMACOL, ALBANY MED COL. *Mem:* Soc Neurosci. *Res:* Neuronal receptor pharmacology; biological psychiatry; neurochemistry. *Mailing Add:* Dept Pharm & Toxicol Albany Med Col 47 New Scotland Ave Albany NY 12208

TITKEMEYER, CHARLES WILLIAM, b Rising Sun, Ind, Jan 14, 19; m 47; c 2. ANATOMY, BACTERIOLOGY. *Educ:* Ohio State Univ, DVM, 49; Mich State Univ, MS, 51, PhD, 56. *Prof Exp:* From instr to prof anat, Mich State Univ, 49-69; prof vet anat & head dept, Sch Vet Med, La State Univ, Baton Rouge, 69-85; RETIRED. *Concurrent Pos:* Consult vet, Univ Ky Contract Team, Bogor, Indonesia, 60-62; prof & head dept vet sci, Univ Nigeria, 66-68. *Mailing Add:* 1148 Aurora Pl Baton Rouge LA 70806

TITLEBAUM, EDWARD LAWRENCE, b Boston, Mass, Mar 23, 37; c 2. ELECTRICAL ENGINEERING. *Educ:* Northeastern Univ, BSEE, 59; Cornell Univ, MS, 63, PhD(elec eng), 65. *Prof Exp:* Engr, Avco Res & Adv Develop Labs, 59-61; from asst prof to assoc prof, 64-85, PROF ELEC ENG, UNIV ROCHESTER, 85- *Concurrent Pos:* Consult, Gen Dynamics Corp, 64-65; vis assoc prof, Johns Hopkins Univ, 70-71. *Mem:* AAAS; Inst Elec & Electronics Engrs. *Res:* Communication systems and signals; radar; sonar; echo-location systems in nature, acoustics and hearing; bat and dolphin sonar systems. *Mailing Add:* Dept Elec Eng Univ Rochester Wilson Blvd Rochester NY 14627

TITLEY, SPENCER ROWE, b Denver, Colo, Sept 27, 28; m 51; c 3. GEOLOGY, GEOCHEMISTRY. *Educ:* Colo Sch Mines, GeolE, 51; Univ Ariz, PhD(geol), 58. *Prof Exp:* Geologist, NJ Zinc Co, 51 & 53-55, explor geologist, 58-60; geologist, US Geol Surv, 63-73; from asst prof to assoc prof, 60-67, PROF GEOL, UNIV ARIZ, 67- *Concurrent Pos:* Consult occidental minerals, Placer prospecting, Cerro de Pasco, Australia & ConZincRioTinto Homestake Mining Co, 73-84; consult, NSF Adv Comt Appl Sci & Eng, 78-80. *Honors & Awards:* Thayer Lindsley Distinguished lectr, Soc Econ Geol, 85-86; Fulbright Sr Lectr, Fed Univ Para, Brazil, 86. *Mem:* Soc Econ Geol; fel Geol Soc Am; Soc Explor Geophys; Am Inst Mining, Metall & Petrol Engrs; Am Geophys Union; Australasian Inst Mining & Metall. *Res:* Mineral deposit geology, regional geology and tectonics; metallogenesis; geochemistry and petrology of hydrothermal ore systems. *Mailing Add:* Dept Geosci Univ Ariz Tucson AZ 85721

TITMAN, PAUL WILSON, b Lowell, NC, Aug 30, 20. BOTANY. *Educ:* Belmont Abbey Col, BS, 39; Univ NC, AB, 41, MA, 49; Harvard Univ, PhD(biol), 52. *Prof Exp:* Instr bot, Univ NC, 48-49; asst prof, Univ Louisville, 52-53; instr, Univ Conn, 53-55; from assoc prof to prof bot, Chicago State Univ, 55-88; RETIRED. *Concurrent Pos:* Res resident, Harvard Univ, 54; chmn coun faculties, Ill State Bd Cols & Univs; consult. *Mem:* Bot Soc Am; fel Royal Hort Soc. *Res:* Morphogenesis; systematic anatomy; microbiology; paleontology. *Mailing Add:* 1453 E 54th Chicago IL 60615

TITMAN, RODGER DONALDSON, b Montreal, Que, Can, Aug 26, 43; m 74; c 2. WATERFOWL ECOLOGY. *Educ:* McGill Univ, BSc, 65; Bishops Univ, MSc, 69; Univ NB, PhD(wildlife ecol), 73. *Prof Exp:* Asst dir, Delta Waterfowl Res Sta, 69-70; lectr wildlife biol, Univ NB, 72-73; asst prof, 73-79, ASSOC PROF WILDLIFE BIOL, MACDONALD COL, MCGILL UNIV, 79- *Concurrent Pos:* Exec, Wildlife Biologists Sect, Can Soc Zoologists, 78-83; chmn, dept renewable resources, McGill Univ, 83-88, assoc dean. *Mem:* Can Soc Zoologists; Animal Behav Soc; Wildlife Soc; Am Ornithologists Union; Cooper Ornith Soc; Wilson Ornith Soc; Sigma Xi. *Res:* Waterfowl ecology, behavior and energetics, particularly spacing systems in breeding ducks and waterfowl post-breeding and wintering ecology; avian groups; marsh ecology; vertebrate pest interaction with agriculture. *Mailing Add:* Dept Renewable Resources MacDonald Campus McGill Univ 21111 Lakeshore Rd Ste Anne de Bellevue PQ H9X 1C0 Can

TITONE, LUKE VICTOR, b Marsala, Italy, Oct 25, 11; nat US; m 39; c 4. PHYSICS, SCIENCE EDUCATION. *Educ:* NY Univ, MS, 40. *Prof Exp:* Instr physics, NY Univ, 40-51; Prof physics, Manhattan Col, 51-76; RETIRED. *Concurrent Pos:* Adj prof physics, Brooklyn Polytech Inst, 52-72. *Mem:* Am Phys Soc. *Res:* Atomic and nuclear physics. *Mailing Add:* 66 N Gate Circle Melville NY 11747

TITTEL, FRANK K(LAUS), b Berlin, WGer, Nov 14, 33; US citizen; m 65; c 2. PHYSICS, ELECTRICAL ENGINEERING. *Educ:* Oxford Univ, BA, 55, PhD(physics), 59. *Prof Exp:* Physicist, Gen Elec Co, 59-65; assoc prof physics, Am Univ Cairo, 65-67; assoc prof elec eng, 67-72, PROF ELEC ENG, RICE UNIV, 72- *Mem:* Fel Am Phys Soc; fel Optical Soc Am; fel Inst Elec & Electronics Engrs. *Res:* Precision measurements of atomic constants; quantum electronics; laser development; laser spectroscopy; nonlinear optics. *Mailing Add:* Dept Elec Eng & Comput Engr Rice Univ PO Box 1892 Houston TX 77251

TITTERTON, PAUL JAMES, b Copiague, NY, Feb 23, 40; m 63; c 3. ELECTROOPTICS. *Educ:* Boston Col, BS, 61; Brandeis Univ, MS, 63, PhD(physics), 67. *Prof Exp:* SR TECHNOLOSIST, GTE INC, 66- *Mem:* Optical Soc Am; Am Phys Soc. *Res:* Atmospheric effects on laser beams; precise optical ranging techniques; optimum optical communication methodology; sensitive optical receiver research. *Mailing Add:* 1412 Hamilton Ave Palo Alto CA 94301

TITTLE, CHARLES WILLIAM, b Bonham, Tex, Nov 11, 17; m 43; c 7. OPERATIONS RESEARCH. *Educ:* NTex State Univ, BS, 39, MS, 40; Mass Inst Technol, PhD(physics), 48. *Prof Exp:* Instr physics, NTex State Univ, 40-41; instr pre-radar, Southern Methodist Univ, 43; asst prof physics, NTex State Univ, 43-44; instr elec commun radar sch, Mass Inst Technol, 44-45; from assoc prof to prof physics, NTex State Univ, 48-51; head nuclear physics sect, Gulf Res & Develop Co, 51-55; dir western div, Tracerlab, Inc, 55-56, assoc tech dir, 56-57; prof nuclear eng, Southern Methodist Univ, 57-63, chmn dept mech eng, 61-65, chmn dept physics, 65-75, prof physics & mech eng, 63-88; VIS PROF, NUCLEAR ENG, NC STATE UNIV, 88- *Concurrent Pos:* Fel physics, Gulf Oil Corp, Mass Inst Technol, 46-48; consult, S W Marshall, Jr, 47, nuclear shielding proj, Mass Inst Technol, 48-51, Gulf Res & Develop Co, 50, Atlantic Refining Co, 57-58, Western Co, 58-77, Well Reconnaissance, Inc, 62-64, Ling-Temco-Vought, 63-64, Mobil Oil Corp, 63-74, 81-82 & 88, Nuclear-Chicago Corp, 64-66, Core Labs, 65-81 & Gearhart Industs, 81-; Amoco Prod Co, 87-88. *Mem:* Am Phys Soc; Am Asn Physics Teachers; Am Nuclear Soc; Soc Prof Well Log Analysts. *Res:* Neutron physics; nuclear well logging; applications of radioisotopes; boundary value problems; fundamental concepts of physics. *Mailing Add:* 6815 Lyre Lane Dallas TX 75214

TITTMAN, JAY, b Bayonne, NJ, Dec 28, 22; m 44; c 3. PETROPHYSICS, WELL LOGGING. *Educ:* Drew Univ, BA, 44; Columbia Univ, MA, 48, PhD(physics), 51. *Prof Exp:* Physicist, Schlumberger-Doll Res Ctr, 51-54, sect head nuclear physics, 54-66, sr staff scientist, 66-67, dept head physics res, 67-72, sci consult to dir res, 81-83, dept head eng physics, Schlumberger Well Serv, Houston, 72-78, dir develop eng, 78-81; CONSULT, JAY TITTMAN TECH CONSULT SERV, 83- *Mem:* Sigma Xi; fel Am Phys Soc; AAAS; Soc Petrol Engrs; Soc Prof Well Log Analysts. *Res:* Physical methods and instruments for subsurface geophysical exploration, borehole or well logging, in particular the application of neutron and gamma-ray physics and instruments to this field. *Mailing Add:* 11 Tanglewood Dr Danbury CT 06811

TITTMANN, BERNHARD R, b Moshi, Tanzania, Sept 15, 35; US citizen; m 66; c 5. SOLID STATE PHYSICS, ACOUSTICS. *Educ:* George Washington Univ, BS, 57; Univ Calif, Los Angeles, PhD(solid state physics), 65. *Prof Exp:* Mem tech staff res & develop aerospace & systs group, Hughes Aircraft Co, 57-61; res asst solid state physics, Univ Calif, Los Angeles, 61-65, asst prof & fel, 65-66; mem tech staff, Rockwell Int, 66-79, mgr earth & planetary sci group, 79-84, mgr Mat Characterization Dept, Sci Ctr, 84-90; BAYARD KUNCLE PROF ENG SCI & MECH, PA STATE UNIV, UNIVERSITY PARK, PA, 89- *Concurrent Pos:* Vis prof physics, Univ Paris VII, Paris, France, 77-78; res grant, Ecole Normale Superior, Paris, France, 82-83. *Mem:* Am Phys Soc; Am Geophys Union; fel Inst Elec & Electronics Engrs; Sigma Xi; Soc Explor Geophys. *Res:* Non-destructive evaluation and acoustic surface waves; acoustic properties of lunar and terrestrial rock; ultrasonic absorption of type I and type II superconductors; dislocation-electron interaction; superconductivity in high pressure polymorphs of semiconductors; ferromagnetic modes in epitaxial single crystal yttrium iron garnet; microwave conformal array antennas. *Mailing Add:* 2466 Sassafras Ct State College PA 16803-3366

TITUS, CHARLES JOSEPH, b Mt Clemens, Mich, June 23, 23; m 49; c 2. MATHEMATICS. *Educ:* Univ Detroit, BSc, 44; Brown Univ, ScM, 45; Syracuse Univ, PhD(math), 48. *Prof Exp:* Instr math, Syracuse Univ, 48; from instr to prof math, Univ Mich, Ann Arbor, 49-; RETIRED. *Concurrent Pos:* Vis prof, Univ Calif, 58-59; mem, Inst Defense Anal, 60-61. *Mem:* Am Math Soc. *Res:* Complex variables and generalizations; transformation semigroups; qualitative theory of differential equations; communications; geometric analysis; differential topology. *Mailing Add:* 1210 Arlington Ann Arbor MI 48104

TITUS, CHARLES O, b Augusta, Maine, Jan 26, 27; m 54; c 7. OPHTHALMOLOGY. *Educ:* Univ Ottawa, BA & BSc, 51, MD, 55. *Prof Exp:* Chief ophthal, Travis AFB, 59-63 & USAF Hosp Weisbaden, 63-66; pvt pract, Chevy Chase, Md, 66-74; CORP DIR MED AFFAIRS, BAUSCH & LOMB, 74. *Mem:* Am Acad Ophthal; Am Soc Cataract & Refractive Surg; Am Soc Contemp Ophthal; Am Acad Med Dirs; Asn Advan Med Instrumentation. *Res:* contact lens of ophthalmology. *Mailing Add:* 19 Foxbourne Rd Penfield NY 14526

TITUS, DONALD DEAN, b Worland, Wyo, Mar 22, 44; m 66; c 2. INORGANIC CHEMISTRY. *Educ:* Univ Wyo, BS, 66; Calif Inst Technol, PhD(chem), 71. *Prof Exp:* Asst prof, 71-77, ASSOC PROF CHEM, TEMPLE UNIV, 77- *Concurrent Pos:* Am Chem Soc Petrol Res Fund grant, 71-74. *Mem:* Am Chem Soc; Am Crystallog Asn. *Res:* Transition-metal complexes, non-rigidity in hydrides, the trans-influence and selenium complexes; x-ray crystal structures. *Mailing Add:* Dept Chem Temple Univ Broad & Montgomery Philadelphia PA 19122-2585

TITUS, DUDLEY SEYMOUR, b Ithaca, NY, Mar 18, 29; m 54. FOOD TECHNOLOGY. *Educ:* Cornell Univ, BS, 52; State Col Wash, MS, 54; Univ Ill, PhD(food tech), 57. *Prof Exp:* Asst food technol, State Col Wash, 52-54; food technologist, Merck & Co, Inc, 57-67; head food technol sect, Mallinckrodt Chem Works, 67-72, res & develop scientist, 72, mkt res specialist, Mallinckrodt, Inc, 72-77, dir planning, food, flavor & fragrance group, 77-86; CONSULT, 86- *Mem:* Am Asn Cereal Chem; Int Nutrit Anemia Consultative Group; Inst Food Technol. *Res:* Food microbiology and preservation; human nutrition; cereal products canned and frozen foods; dairy products; food additives; fragrances and olfaction. *Mailing Add:* 8343 Ardsley Dr St Louis MO 63121

TITUS, ELWOOD OWEN, b Rochester, NY, Sept 20, 19; m 51; c 2. ORGANIC CHEMISTRY. *Educ:* Williams Col, BA, 41; Columbia Univ, PhD(chem), 47. *Prof Exp:* Asst antimalarials, Div War Res, Columbia Univ, 43-46; res assoc antibiotics & metab prod, Squibb Inst Med Res, 46-50; chemist, Chem Pharmacol Lab, Nat Heart & Lung Inst, 50-77; dir div drug biol, Bur Drugs, Food & Drug Admin, 77-86; SR SCI CONSULT, LIFE SCI RES ORGN, FED NAM SOCS EXP BIOL, 90- *Concurrent Pos:* Mem, Nat Res Coun, 66-74; vis prof, Dept Pharmacol, Univ Bern, Switz, 86, Dept Clin Pharmacol, Karolinska Inst, Stockholm, Sweden, 87, Dept Physiol, Uniformed Serv Univ of the Health Sci, Bethesda, MD, 88- *Mem:* Am Chem Soc; Am Soc Biol Chem; Am Soc Pharmacol & Exp Therapeut; NY Acad Sci; Biophys Soc. *Res:* Organic synthesis; biochemistry; metabolism of biologically active compounds; application of counter-current distribution to metabolic studies on 4-amino-quinoline antimalarials; mechanism of action of steroids and catecholamines; lipid metabolism; biochemistry of cell membranes. *Mailing Add:* 9650 Rockville Pike Bethesda MD 20814

TITUS, HAROLD, b Detroit, Mich, Jan 10, 30; m 55; c 3. SYSTEMS DESIGN, CONTROL ENGINEERING. *Educ:* Univ Kans, BS, 52; Stanford Univ, MS, 57, PhD(eng mech), 62. *Prof Exp:* PROF ELEC ENG, US NAVAL POSTGRAD SCH, 62- *Concurrent Pos:* Consult, US Naval Ord Test Sta, 65-, Stanford Res Inst & US Naval Air Develop Ctr, 66- *Mem:* Inst Elec & Electronics Engrs; Sigma Xi. *Res:* Optimum filtering, identification and control applications to naval weapons problems. *Mailing Add:* Dept Electronic & Comput Eng Naval Postgrad Sch Code 0223 Monterey CA 93940

TITUS, JACK L, b South Bend, Ind, Dec 7, 26; m 49; c 5. CARDIOVASCULAR PATHOLOGY. *Educ:* Univ Notre Dame, BS, 48; Wash Univ, MD, 52; Univ Minn, PhD(path), 62. *Prof Exp:* Physician, Rensselaer, Ind, 53-57; assoc prof path, Mayo Grad Sch Med, Univ Minn, 61-72, prof, Mayo Clin, 71-72; prof path & chmn dept, Baylor Col Med, 72-87; DIR REGISTRY CARDIOVASC DIS, UNITED HOSP, ST PAUL, MINN, 87- *Concurrent Pos:* Fel path, Mayo Grad Sch Med, Univ Minn, 57-61; chief path serv, Methodist Hosp, 72-87; pathologist-in-chief, Harris County Hosp Dist, 72-87. *Honors & Awards:* Billings Gold Medal, AMA, 68, Hoektoen Gold Medal, 69. *Mem:* AMA; Am Asn Pathologists; Am Soc Clin Path; Int Acad Path; Col Am Pathologists; Sigma Xi. *Res:* Cardiac conduction system; cardiac anomalies; valvular heart disease; ischemic heart disease; atherosclerosis. *Mailing Add:* 4274 Pond View Dr White Bear Lake MN 55110

TITUS, JOHN ELLIOTT, b Iowa City, Iowa, July 24, 49; m 73; c 2. PLANT ECOLOGY. *Educ:* Oberlin Col, BA, 71; Univ Wis, MA, 73, PhD(bot), 77. *Prof Exp:* ASSOC PROF BIOL, STATE UNIV NY BINGHAMTON, 77- *Concurrent Pos:* Prin investr, NSF, 78-82, 85-89, Environ Protection Agency, 81-84. *Mem:* Am Soc Limnol & Oceanog; Am Inst Biol Sci; Sigma Xi; Ecol Soc Am; Torrey Bot Club. *Res:* Comparative physiological ecology of submersed macrophytes; importance of physico-chemical environment, especially acidic conditions, as determinants of plant distribution and abundance; community compositional change. *Mailing Add:* Dept Biol Sci State Univ NY Binghamton NY 13902

TITUS, JOHN S, b Mich, Apr 19, 23; m 46; c 4. POMOLOGY. *Educ:* Mich State Col, BS, 46, MS, 47; Cornell Univ, PhD(pomol), 51. *Prof Exp:* Instr hort, Mich State Col, 46-48; asst, Cornell Univ, 49-51; from instr to assoc prof pomol, 51-66, PROF POMOL, DEPT HORT, UNIV ILL, URBANA, 66- *Concurrent Pos:* Res assoc, Univ Calif, Davis, 62-63; Fulbright-Hays lectr,

Univ Col, Dublin, 71-72. *Honors & Awards:* Gourley Award, Am Soc Hort Sci, 74. *Mem:* Fel AAAS; Am Soc Hort Sci; Am Soc Plant Physiol; Soc Exp Biol & Med. *Res:* Mineral nutrition of deciduous fruit trees; soil fertility requirements of fruit trees; soil morphology in relation to fruit tree performance; translocation in woody plants; amino acid synthesis in higher plants. *Mailing Add:* 102 Hort Field Lab Univ Ill Urbana IL 61801

TITUS, RICHARD LEE, b Dayton, Ohio, Aug 6, 34. ORGANIC CHEMISTRY. *Educ:* DePauw Univ, BA, 56; Mich State Univ, PhD(chem), 64. *Prof Exp:* From instr to asst prof chem, Univ Toledo, 62-67; from asst prof to assoc prof, 67-73, PROF CHEM, UNIV NEV, LAS VEGAS, 73- *Mem:* AAAS; Am Chem Soc; Sigma Xi. *Res:* Synthesis of heterocyclic compounds. *Mailing Add:* Dept Chem Univ Nev Las Vegas NV 89154

TITUS, ROBERT CHARLES, b Paterson, NJ, Aug 9, 46. INVERTEBRATE PALEONTOLOGY. *Educ:* Rutgers Univ, BS, 68; Boston Univ, AM, 71, PhD(geol), 74. *Prof Exp:* Instr geol, Windham Col, 73-74; ASST PROF GEOL, HARTWICK COL, 74- *Mem:* Sigma Xi; Paleont Soc; Paleont Asn; Geol Soc Am; Soc Econ Paleontologists & Mineralogists. *Res:* Paleontology of Middle Ordovician fossil invertebrate benthic communities. *Mailing Add:* Dept Biol Hartwick Col Oneonta NY 13820

TITUS, WILLIAM JAMES, b Oakland, Calif, Dec 13, 41; m 67; c 1. LOW TEMPERATURE & COMPUTATIONAL PHYSICS, STATISTICAL MECHANICS. *Educ:* Univ Calif, Davis, BS, 63; Stanford Univ, MS, 65, PhD(physics), 68. *Prof Exp:* Res assoc physics, Univ Minn, 68-70; from asst prof to assoc prof, 70-85, PROF PHYSICS, CARLETON COL, 85- *Mem:* Am Phys Soc; Am Asn Physics Teachers; Sigma Xi. *Mailing Add:* Dept Physics & Astrom Carleton Col One N College St Northfield MN 55057-4025

TITZE, INGO ROLAND, b Hirschberg, Ger, July 8, 41; US citizen; m 69; c 4. VOICE ACOUSTICS, BIOMECHANICS OF THE LARYNX. *Educ:* Univ Utah, BSEE, 63, MS, 65; Brigham Young Univ, PhD(physics), 72. *Prof Exp:* Res engr, NAm Aviation, 65-66 & Boeing Co, 68-69; instr physics, Brigham Young Univ, 72-73; lectr, Calif State Polytech Univ, 73-74; asst prof, Univ Petrol & Minerals, Saudi Arabia, 74-76; asst prof speech commun, Gallaudet Col, 76-79; assoc prof, 79-84, PROF SPEECH SCI, UNIV IOWA, 84-; DIR RES, RECORDING & RES CTR, DENVER CTR PERFORMING ARTS, 83- *Concurrent Pos:* Mem, Tech Comt Musical Acoust, Acoust Soc Am, 73-82; consult speech sci, Bell Labs, 77-78; prin investr, NIH, 78-; dir publ, Int Asn Res in Singing, 82-; assoc ed, Nat Asn Teachers Singing J, 85-; panelist & site visitor, Nat Res Coun, Nat Acad Sci, 84-; pres, Voice Consults, 85-; mem, Collegium Medicorum Teatri, 88-; prin investr res award, Toward Standards for Voice Anal & Recording, NIH, 87- & Naturalness in Speech Synthesis, US West Advan Technol, 88; adj prof, Westminster Choir Col, Princeton, 89-; dir, Nat Ctr Voice & Speech, 90- *Honors & Awards:* Jacob Javits Neurosci Invest Award, NIH, 84; Gould Award, William & Harriot Gould Found, 84. *Mem:* Fel Acoust Soc Am; Nat Asn Teachers Singing; Am Speech, Hearing & Lang Asn; Am Asn Phonetic Sci; Int Asn Res in Singing; Int Platform Asn. *Res:* Acoustics and biomechanics of voice production understanding the physical and physiologic mechanisms by which voice is produced; voice characteristics of actors, singers, public speakers; effects of environment, fatigue, drugs, aging on human voice. *Mailing Add:* 330 SHC Univ Iowa Iowa City IA 52242

TIUS, MARCUS ANTONIUS, b Izmir, Turkey, Apr 18, 53; US citizen. SYNTHETIC ORGANIC CHEMISTRY. *Educ:* Dartmouth Col, BA, 75; Harvard Univ, MS, 77, PhD(chem), 80. *Prof Exp:* Asst prof, 80-84, ASSOC PROF CHEM, UNIV HAWAII, MANOA, 84- *Concurrent Pos:* Alfred P Sloan Fel, 87-89. *Mem:* Am Chem Soc. *Mailing Add:* Chem Dept Univ Hawaii 2545 Mall Honolulu HI 96822-1888

TIWARI, SURENDRA NATH, b Gorakhpur, India, Jan 1, 38; c 1. AEROSPACE ENGINEERING, ENVIRONMENTAL SCIENCES. *Educ:* Univ Allahabad, BS, 59; Univ Maine, MS, 62 & 64; State Univ NY, Stony Brook, PhD(eng sci), 69. *Prof Exp:* Syst analyst, Implements Factory, India, 55-57, res engr, 57-60; res asst agr eng, Univ Maine, 60-62, instr mech eng, 62-64; instr eng, State Univ NY, Stony Brook, 64-69, res assoc radiation, 69-70, asst prof eng, 70-71; assoc prof thermal eng, 71-77, prof mech eng & mech, 77-79, EMINENT PROF MECH ENG & MECH, OLD DOMINION UNIV, 79-, EMINENT SCHOLAR, 89- *Concurrent Pos:* Consult, Grumman Aerospace Corp, NY, 69-71; NASA grant, Old Dominion Univ, 71-; res consult, Langley Field, NASA, Va, 71-; dir Inst Computational & Appl Mech, Old Dominion Univ, 83- *Honors & Awards:* NASA Achievement Award. *Mem:* Fel Am Inst Aeronaut & Astronaut; fel Am Soc Mech Engrs; AAAS; Am Soc Eng Educ; Am Asn Univ Professors. *Res:* Radiation gas dynamics; boundary layer flows; multi-phase flows; combustion processes and flow of chemically reacting and radiating gases; high temperature gas kinetics; atmospheric radiation; computational fluid mechanics; planetary entry heating. *Mailing Add:* Dept Mech Eng Old Dominion Univ Hampton VA 23669

TIXIER, MAURICE PIERRE, b Clermont, France, Feb 1, 13; nat US; m 39; c 2. GEOPHYSICS. *Educ:* Ecole des Arts et Metiers d'Erquelinnes, Belg, Eng, 32; Ecole Superieure d'Electricite, Paris, Eng, 33. *Prof Exp:* Field engr, Societe de Prospection Electrique, Paris, 34-35; dist engr, Schlumberger Well Surv Corp, Tex, 35-39, area mgr, Colo, 41-49, chief petrol engr, Tex, 49-52, chief field develop engr, 52-57, mgr field develop, 57-66, dir field interpretation, 66-69, dir prod logging, 69-72, tech adv, 72-77; pres, Tixier Tech Corp, 78-83; CONSULT ENGR, 83- *Concurrent Pos:* Distinguished lectr, Soc Petrol Eng, 60. *Honors & Awards:* Gold Medal, Soc Prof Well Log Analysts, 70. *Mem:* Soc Explor Geophys; Am Asn Petrol Geol; Am Geophys Union; Am Inst Mining, Metall & Petrol Engrs; Soc Petrol Eng. *Res:* Electrical logging; electrochemistry; well bore geophysics. *Mailing Add:* 2319 Bolsover Rd Houston TX 77005

TIZARD, IAN RODNEY, b Belfast, Northern Ireland, Oct 27, 42; m 69; c 2. IMMUNOLOGY. *Educ:* Univ Edinburgh, BVMS, 65, BSc, 66; Cambridge Univ, PhD(immunol), 69. *Prof Exp:* Med Res Coun Can fel, Univ Guelph, 69-71; vet res officer, Animal Dis Res Asn, 71-72; from asst prof to prof vet immunol, Univ Guelph, 72-82; prof & head, Dept Vet Microbiol & Parasitol, 82-90, PROF, DEPT PATHOBIOL, TEX A&M UNIV, COLLEGE STATION, 90- *Mem:* Royal Col Vet Surg; Am Asn Immunologists. *Res:* Veterinary immunology. *Mailing Add:* Dept Vet Pathobiol Tex A&M Univ College Station TX 77843

TJEPKEMA, JOHN DIRK, b Madison, Wis, June 14, 43. NITROGEN FIXATION, NITROGEN CYCLE. *Educ:* Univ Mich, BA, 65, MA, 67, PhD(bot), 71. *Prof Exp:* Res assoc, Wash Univ, 72-73, Univ Wis, 73-74 & Ore State Univ, 74-75; res fel, EMBRAPA, Rio de Janeiro, 76; asst & assoc prof soil biol, Harvard Univ, 76-82; from asst prof to assoc prof, 82-89, PROF PLANT PHYSIOL, UNIV MAINE, 89- *Mem:* Am Soc Plant Physiologists; Bot Soc Am. *Res:* Physiology and ecology of nitrogen fixation by nodulated plants and their symbiotic bacteria; associative nitrogen fixation; nitrogen cycle of ecosystems, especially exchange of nitrogenous compounds with the atmosphere. *Mailing Add:* Dept Plant Biol & Path Univ Maine Orono ME 04469-0118

TJIAN, ROBERT TSE NAN, b Hong Kong, Sept 22, 49; Brit citizen; m 76. MOLECULAR BIOLOGY, BIOCHEMISTRY. *Educ:* Univ Calif, Berkeley, AB, 71; Harvard Univ, PhD(molecular biol), 76. *Prof Exp:* Staff investr molecular virol, Cold Spring Harbor Lab, 76-79; PROF BIOCHEM, UNIV CALIF, BERKELEY, 79- *Concurrent Pos:* Robertson fel, Cold Spring Harbor Lab, 78. *Mem:* Nat Acad Sci. *Res:* Oncogenic viruses and their interactions with the host cell; control of gene expression; simian virus 40, a small DNA containing oncogenic virus, tumor antigen, its structure and function. *Mailing Add:* Dept Biochem Univ Calif 401 Barker St Berkeley CA 94720

TJIO, JOE HIN, b Java, Indonesia, Feb 11, 19; US citizen; m 48; c 1. CYTOGENETICS. *Educ:* Univ Colo, PhD(biophys, cytogenetics), 60; Univ Zaragosa, Spain, Dr, 81. *Hon Degrees:* Dr, Univ Claude Bernard, France, 74, Sci-Univ, Zaragosa, Spain, 81. *Prof Exp:* Head cytogenetics, Estacion Exp de Aula Dei, 48-59; res biologist, 59-78, CHIEF CYTOGENETICS SECT LAB EXP PATH, NAT INST ARTHRITIS & METAB DIS, 74- *Concurrent Pos:* Vis scientist, Univ Lund, Sweden, 47-59, res assoc, Genetics Inst, 59-; res collabr, Univ Berkeley, 79-, Albany Med Col, 85-; fel award, Japan Soc Prom Sci, 84. *Honors & Awards:* Joseph P Kennedy Jr Found Award, 62. *Mem:* Genetics Soc Am; Am Soc Human Genetics; Am Genetic Asn; Am Soc Nat. *Res:* Plant, animal and human cytogenetics; mammalian cytogenetics and immunology. *Mailing Add:* NIH Bldg Ten Rm 4D-44 Bethesda MD 20014

TJIOE, DJOE TJHOO, b Medan, Indonesia, Oct 1, 37; US citizen; m 64; c 2. MAMMALIAN PHYSIOLOGY, PHARMACOLOGY. *Educ:* Sioux Falls Col, BSc, 65; Univ Wis, MSc, 67, PhD(physiol), 70. *Prof Exp:* Teaching asst physiol, Univ Wis, 65-70; ASSOC PROF PHYSIOL, CALIF STATE UNIV, LONG BEACH, 70- *Concurrent Pos:* Consult, Concept Media, 74-78. *Mem:* AAAS; Am Physiol Soc; Fedn Am Socs Exp Biol. *Res:* Cardiovascular, neuro and respiratory physiology. *Mailing Add:* Dept Anat & Physiol Calif State Univ 1250 Bellflower Blvd Long Beach CA 90840

TJIOE, SARAH ARCHAMBAULT, b Philadelphia, Pa, Oct 12, 44; m 67; c 1. PHARMACOLOGY. *Educ:* Univ Pa, BA, 66, PhD(pharmacol), 71. *Prof Exp:* Instr, 72-75, ASST PROF PHARMACOL, COL MED, OHIO STATE UNIV, 75- *Concurrent Pos:* NIH training grant pharmacol, Col Med, Ohio State Univ, 71-72, Pharmaceut Mfrs Asn Found fel pharmacol-morphol, 72-74. *Mem:* Am Soc Pharmacol & Exp Therapeut; Soc Neurosci. *Res:* Neuropharmacology; neurochemistry. *Mailing Add:* Dept Pharmacol Ohio State Univ Col Med 333 W Tenth Ave Columbus OH 43210

TJOSTEM, JOHN LEANDER, b Sisseton, SDak, June 6, 35; m 62; c 3. MICROBIOLOGY, PLANT PHYSIOLOGY. *Educ:* Concordia Col, Moorhead, Minn, BA, 59; NDak State Univ, MS, 62, PhD(bot), 68. *Prof Exp:* Instr biol, Luther Col, Iowa, 62-65; assoc prof, Concordia Col, Moorhead, Minn, 67-68; assoc prof, 68-80, PROF BIOL, LUTHER COL, IOWA, 80-, BIOL DEPT HEAD, 88- *Mem:* Am Soc Microbiol; Am Biol Teachers. *Res:* Metabolic pathways of certain carbohydrates in algae; groundwater pollution. *Mailing Add:* Dept Nursing/Baccalaureate Luther Col Decorah IA 52101

TKACHEFF, JOSEPH, JR, b Waterbury, Conn, Feb 13, 26; m 54; c 2. PHARMACY. *Educ:* RI Col Pharm, BSc, 46; Philadelphia Col Pharm, MSc, 47. *Prof Exp:* Pharmacist, E M Altman & Waterbury Drug Co, Conn, 45-46; control chemist, G F Harvey Co, 47-49, res & develop chemist, 49-50, plant chemist, 50-51, prod mgr, 51-58; SR RES PHARMACIST & GROUP LEADER PROD DEVELOP & RES, SOLID DOSAGE SECT, STERLING-WINTHROP RES INST, 58- *Mem:* Am Chem Soc; Am Pharmaceut Asn; Acad Pharmaceut Sci. *Res:* Pharmaceutical chemistry; physiological biochemistry. *Mailing Add:* 13 Russell Rd Greenfield Ctr NY 12833-1318

TKACHUK, RUSSELL, b Redwater, Alta, Dec 25, 30; m 57; c 3. CHEMISTRY. *Educ:* Univ BC, BA, 54, MSc, 56; Univ Sask, PhD(chem), 59. *Prof Exp:* RES SCIENTIST, GRAIN RES LAB, CAN GRAIN COMN, 59- *Concurrent Pos:* Adj prof, Univ Man; fel, St Vincent's Sch Med Res, Melbourne, Australia, 68-69. *Mem:* Am Asn Cereal Chem; Am Chem Soc; fel Chem Inst Can; Can Inst Chem; Sigma Xi. *Res:* Chemistry of amino acids and proteins in wheat; near infrared reflectance spectroscopy. *Mailing Add:* 1404-303 Main St Winnipeg MB R3C 3G8 Can

TKACZ, JAN S, b Kittery, Maine, Oct 24, 44; m 78. ANTIBIOTICS & NATURAL PRODUCTS. *Educ:* Univ NH, BA, 66; Rutgers Univ, PhD(microbiol), 71. *Prof Exp:* Fel biochem, Med Sch, Harvard Univ, 71-73; res assoc, M S Hershey Med Ctr, Pa State Univ, 73-74; asst res prof microbiol, Rutgers Univ, 75-78, asst prof microbiol, Waksman Inst Microbiol, 78-81; sr res investr, Dept Microbiol, The Squibb Inst Med Res, 81-87; SR RES FEL, MERCK SHARP & DOHME RES LAB, 87- *Concurrent Pos:* NSF res fel, Dept Biol Chem, Lab Carbohydrate Res, Med Sch, Harvard Univ, 71-72; vis prof microbiol, Rutgers Univ. *Mem:* Soc Complex Carbohydrates; Am Soc Microbiol; Am Chem Soc; AAAS. *Res:* Biosynthesis of cell wall and membrane constituents in eukaryotic microorganisms, antibiotics and natural products of fungi. *Mailing Add:* Fermentation Micro R80Y-330 Merck & Co Inc PO Box 2000 Rahway NJ 07065-0900

TOBA, H(ACHIRO) HAROLD, b Puunene, Hawaii, Aug 24, 32; m 58; c 3. APHID VECTORS, POTATO VIRUSES. *Educ:* Univ Hawaii, BS, 57, MS, 61; Purdue Univ, PhD(entom), 66. *Prof Exp:* RES ENTOMOLOGIST, AGR RES SERV, USDA, 65- *Mem:* AAAS; Entom Soc Am; Sigma Xi. *Res:* Elucidate vector-virus-plant host relationships of potato viruses and develop management strategies of aphid vectors for virus control. *Mailing Add:* Yakima Agr Res Lab USDA 3706 W Nob Hill Blvd Yakima WA 98902

TOBACH, ETHEL, b Miaskovka, Russia, Nov 7, 21; nat US; wid. COMPARATIVE PSYCHOLOGY. *Educ:* Hunter Col, BA, 49; NY Univ, MA, 52, PhD(comp psychol, physiol psychol), 57. *Hon Degrees:* DSc, Long Island Univ, 75. *Prof Exp:* Res assoc, Payne Whitney Psychiat Clin, New York, 49-53 & Pub Health Res Inst NY, Inc, 53-56; res fel comp & physiol psychol, Am Mus Natural Hist, 57-61; asst prof comp physiol & exp psychol, Sch Med, NY Univ, 61-65; assoc cur, Am Mus Natural Hist, 64-69, cur, Dept Animal Behav, 69-81, cur, Dept Mammal, 81-90, EMER CUR, AM MUS NATURAL HIST, 90- *Concurrent Pos:* Adj prof Biol & Psychol & Pres, City Univ New York, 69- *Honors & Awards:* Res Achievements Award, Asn Women Psychol. *Mem:* AAAS; fel Am Psychol Asn; fel Animal Behav Soc; Int Soc Comp Psychol. *Res:* Development and evolution of behavior; emotional behavior; social behavior; neurohormonal relationships; sensory processes; autonomic phenomena. *Mailing Add:* Am Mus Nat Hist Dept Mammal Cent Park West at 79th St New York NY 10024-5192

TOBACK, F(REDERICK) GARY, b Brooklyn, NY, Oct 23, 41; m 63; c 3. NEPHROLOGY, EPITHELIAL CELL GROWTH. *Educ:* Columbia Col, AB, 63; New York, Univ, MD, 67; Boston Univ, PHD(biochem), 74; Am Bd Int Med, dipl, 74; Am Bd Nephrol, dipl, 76. *Prof Exp:* Internship internal med, Cleveland Metrop Gen Hosp, 67-68, residency, 68-69; lieutenant med, US Navy Med Corps, 69-70; res assoc med & nephrol, Sch Med, Boston Univ, 70-73; clin fel nephrol, Harvard Med Sch, Beth Israel Hosp, 74; from asst prof to assoc prof, 80-85, PROF MED & CELL PHYSIOL, SCH MED, UNIV CHICAGO, 85- *Concurrent Pos:* Am Cancer Soc scholar, Salk Inst Biol Studies, 79-80; estab investr, Am Heart Asn, 80-85 & sci councils; co-chmn, NIDDK Workshop on Control of Renal Growth; NIDDK Spec Grants Rev Comt. *Mem:* Am Physiol Soc; Cent Soc Clin Res; Am & Int Soc Nephrol; Am Soc Clin Invest. *Res:* Growth of kidney cells in physiological and pathological states: regeneration after acute renal failure; potassium depletion nephropathy; kidney autocrine growth factors, signal transduction in growing kidney cells. *Mailing Add:* Dept Med Univ Chicago Pritzker Sch Med 5841 S Maryland Ave Box 453 Chicago IL 60637

TOBE, STEPHEN SOLOMON, b Niagra-on-the-Lake, Ont, Can, Oct 11, 44; m 69. INVERTEBRATE ENDOCRINOLOGY, REPRODUCTIVE PHYSIOLOGY. *Educ:* Queen's Univ, Kingston, Que, BSc, 67; York Univ, Downsview, Ont, MSc, 69; McGill Univ, Montreal, Ont, PhD(parasitol), 72. *Prof Exp:* Res fel invert endocrinol, Agr Res Coun, Univ Sussex, 72-74; from asst prof to assoc prof, 74-82, assoc dean sci, 88-92, PROF ZOOL & PHYSIOL, DEPT ZOOL, UNIV TORONTO, 82- *Concurrent Pos:* Ed Bull Can Soc Zoologists, 79-82; vis prof, dept entom sci, Univ Calif, Berkeley, 81; E W R Steacie Mem fel, Nat Sci & Eng Res Coun Can, 82-84. *Honors & Awards:* C Gordon Hewitt Award, Entom Soc Can, 82, Gold Medal, 90. *Mem:* Sigma Xi; fel Royal Entom Soc; Can Soc Zoologists; Am Soc Zoologists; Soc Exp Biol; AAAS; fel Royal Soc Can. *Res:* Invertebrate endocrinology; regulation of hormone biosynthesis, particularly juvenile hormone; regulation of hormone titres in arthropods; hormonal control of metamorphosis and reproduction in insects; mode of action of hormone agonists and antagonists. *Mailing Add:* Dept Zool Univ Toronto 25 Harbord St St George Campus Toronto ON M5S 1A1 Can

TOBERMAN, RALPH OWEN, b Harrisburg, Pa, Sept 23, 23; m 46; c 3. ANALYTICAL CHEMISTRY. *Educ:* Albright Col, BS, 49. *Prof Exp:* Control chemist, Pa Ohio Steel Corp, 49-51; res asst anal chem, Sharp & Dohme, Inc, 51-52, res assoc, 52-63, unit head anal res, 63-65, specifications assoc, 65-66, HEAD CONTAINER & PROD TESTING UNIT, MERCK SHARP & DOHME RES LABS, 66- *Mem:* Am Chem Soc; Drug Info Asn. *Res:* Analysis of pharmaceutical products; methods development; specifications for new products; package research and product testing. *Mailing Add:* 2939 Hickory Hill Dr Worcester Acres RD 1 Norristown PA 19403-4714

TOBES, MICHAEL CHARLES, b Detroit, Mich, Oct 7, 48; m 79; c 2. BIOCHEMICAL RADIOPHARMACOLOGY, ONCOLOGY. *Educ:* Mich State Univ, BS, 70; Univ Mich, MS, 72, PhD(biol chem), 76; Sch Med, Univ Miami, MD, 90. *Prof Exp:* Scholar, dept internal med, Univ Mich, Ann Arbor, 76-79, res investr, 79-81, dir, biochem res unit, 79-85, asst res scientist, 81-85; mem tech staff, AT&T Bell Labs, Middletown, NJ, 85-88; RESIDENT, INTERNAL MED, MILTON S HERSHEY MED CTR, HERSHEY, PA, 90- *Mem:* AMA. *Res:* Development of gamma and positron-labeled radiopharmaceuticals for diagnostic imaging; development of radiolabeled enzyme inhibitors for the diagnosis of cancer; APUDomas; amine metabolism; picture archiving and communication systems; biosensors. *Mailing Add:* 2028 Mill Plain Ct Harrisburg PA 17110

TOBEY, ARTHUR ROBERT, b Portland, Ore, Aug 4, 20; m 77; c 3. APPLIED PHYSICS. *Educ:* Yale Univ, BS, 42, MS, 46, PhD(physics), 48. *Prof Exp:* Mem staff radiation lab, Mass Inst Technol, 42-45; asst cosmic ray proj, Off Naval Res, Yale Univ, 47-48; asst prof physics, State Col Wash, 48-50; supvr physics, Armour Res Found, 50-52; sr scientist, 52-53; supvr TV res, Stanford Res Inst, 53-56, group head video systs lab, 56-59, STAFF SCIENTIST ENG DIV, SRI INT, 59- *Mem:* Am Phys Soc; Sigma Xi. *Res:* Radar and radar-type systems; neutron component of cosmic rays; electromechanical and electro-optical devices; electronic instrumentation; communication theory; man-computer systems; satellite systems modelling. *Mailing Add:* SRI Int Menlo Park CA 94025

TOBEY, FRANK LINDLEY, JR, b Coeur d'Alene, Idaho, Aug 28, 23; m 57; c 1. VISION, PHOTOMETRY. *Educ:* Univ Mich, BSCh, 47, MSCh, 48, MS, 50, PhD(physics), 62. *Prof Exp:* Res assoc physics, Univ Mich, 50-60; res physicist, Cornell Aeronaut Lab, 62-64; fel lab astrophys, Harvard Col Observ, 64-65; fel shocktube physics, McDonnell-Douglas Corp, 66-69; res instr spectros of vision, Med Sch, Wash Univ, 70-74; ASST PROF SPECTROS VISION, MED SCH, UNIV FLA, 74- *Concurrent Pos:* Asst prof physics & astron, Southern Ill Univ, Edwardsville, 70. *Mem:* AAAS; Am Phys Soc; Asn Res Vision & Ophthal; Sigma Xi. *Res:* Measurement of transition probabilities; atomic and molecular parameters; plasma diagnostics; optical properties of the retina and retinal receptors; spectroscopy; experimental determination of receptor waveguide properties; physics of vision. *Mailing Add:* 4114 NW 19th Pl Gainesville FL 32605

TOBEY, ROBERT ALLEN, b Owosso, Mich, May 26, 37; m 60; c 2. CELL BIOLOGY, CANCER. *Educ:* Mich State Univ, BS, 59; Univ Ill, PhD, 63. *Prof Exp:* STAFF MEM, GENETICS GROUP, LOS ALAMOS NAT LAB, 64- *Concurrent Pos:* Los Alamos Nat Lab Fel, 83. *Mem:* AAAS; Am Soc Cell Biol; Am Soc Biochem & Molecularþ Biol. *Res:* Factors controlling traverse of the life cycle; sequential biochemical markers in the mammalian cell cycle; control of mammalian cell proliferation; mechanisms of action and effects on mammalian cell growth and division of anticancer drugs; mammalian cell synchronization. *Mailing Add:* MS-M886 Los Alamos Nat Lab PO Box 1663 Los Alamos NM 87545

TOBEY, STEPHEN WINTER, b Chicago, Ill, Jan 9, 36; m 53; c 4. PHARMACEUTICAL CHEMISTRY. *Educ:* Ill Inst Technol, BS, 57; Univ Wis, MS, 59, PhD(inorg chem), 65. *Prof Exp:* Instr phys chem, WVa Wesleyan Col, 59-61; asst prof inorg chem, Purdue Univ, 64-65; res chemist, Eastern Res Lab, 65-68, sr res chemist, 68-70, res dir, 70-74, dir chem lab, 73-75, mgr chem process develop, 75-76, mgr pharmaceut, process develop, 76-84, sr assoc scientist, 78-84, RES SCIENTIST, MICH DIV, DOW USA, 84- *Concurrent Pos:* Vis prof, Harvard Univ, 65 & Saginaw Valley State Col, 83 & 84. *Mem:* Am Chem Soc. *Res:* Pharmaceutical product development; water soluble polymers. *Mailing Add:* Dow Chem USA Bldg 566 Midland MI 48667

TOBIA, ALFONSO JOSEPH, b Brooklyn, NY, June 19, 42; m 70; c 2. PHARMACOLOGY. *Educ:* St Louis Col Pharm, BS, 65; Purdue Univ, MS & PhD(pharmacol), 69. *Prof Exp:* Asst prof pharmacol, Univ Ga, 69-74; sr investr pharmacol, Smith Kline & French Labs, 74-77; group leader, Ortho Pharmaceut Corp, 77-81, sect head, 82-84, asst dir, cardiovascular immunopharmacol dermato-pharmacol, 85, dir pharmacol, Ortho Pharmaceut Corp, 85-89; SR DIR PHARMACOL, DRUG DISCOVERY RES, R W JOHNSON PRI, 90- *Concurrent Pos:* Ga Heart Asn res grant, Univ Ga, 71-72, Nat Heart & Lung Inst res grant hypertension, 71-74, mem coun high blood pressure coun circulation & coun basic sci, Am Heart Asn. *Mem:* Am Soc Pharmacol & Exp Therapeut; Sigma Xi, Am Heart Asn; NY Acad Sci; Soc Exp Biol Med; Am Chem Soc. *Res:* Drug discovery and development; cardiovascular, gastrointestinal, micro biology and biochemical pharmacology. *Mailing Add:* R W Johnson PRI Welsh & McKean Rds Spring House PA 19477-0776

TOBIAN, LOUIS, b Dallas, Tex, Jan 26, 20; m 51. INTERNAL MEDICINE. *Educ:* Univ Tex, BA, 40; Harvard Med Sch, MD, 44. *Prof Exp:* House officer med, Peter Bent Brigham Hosp, Boston, 44; asst resident, Univ Hosp, Univ Calif, 44-45 & Parkland Hosp, Tex, 45-46; asst prof med, Univ Tex Southwestern Med Sch Dallas, 54; assoc prof, 54-64, PROF MED, SCH MED, UNIV MINN, MINNEAPOLIS, 64- *Concurrent Pos:* Res fel med, Univ Tex Southwestern Med Sch Dallas, 46-51; res fel biochem, Harvard Med Sch, 51-54; estab investr, Am Heart Asn, 51-56, mem coun arteriosclerosis, 56-, chmn, Coun High Blood Pressure Res, 72; George Brown lectureship, 69; chmn task force hypertension, Nat Heart & Lung Inst, 72-73, mem adv comt hypertension res ctrs, 72-74; chmn comt hypertension & renal vascular dis, NIH Kidney Res Surv Group, 74-75. *Honors & Awards:* Ciba Award, 78; Karger Award, 79; Volhard Award, 88. *Mem:* Am Clin & Climat Asn; Asn Am Physicians; Am Soc Clin Invest; Am Physiol Soc; fel NY Acad Sci. *Res:* Hypertension; renal circulation; sodium excretion. *Mailing Add:* Dept Med Univ Minn Sch Med Box 285 UMHC Minneapolis MN 55455

TOBIAS, CHARLES W, b Budapest, Hungary, Nov 2, 20; nat US; m 50, 82; c 3. ELECTROCHEMISTRY, CHEMICAL ENGINEERING. *Educ:* Univ Tech Sci, Budapest, ChemEng, 42, PhD(phys chem), 46. *Prof Exp:* Res & develop engr, United Incandescent Lamp & Elec Co, Ltd, Hungary, 42-43, 45-47; instr chem eng, 47-48, lectr, 48-50, from asst prof to assoc prof, 50-60, chmn dept chem eng, 67-72, PROF CHEM ENG, UNIV CALIF, BERKELEY, 60-, FAC SR SCIENTIST, LAWRENCE BERKELEY LAB, 54- *Concurrent Pos:* Instr phys chem, Univ Tech Sci, Budapest, 45-46; res prof, Miller Inst, 58-59; consult. *Honors & Awards:* Acheson Medal & Prize, Electrochem Soc, 72, Henry B Linford Award, 82, Vittorio de Nora, Diamond Shamrock Award, 90; Alpha Chi Sigma Award, Am Inst Chem Engrs, 83. *Mem:* Nat Acad Eng; fel AAAS; Int Soc Electrochem (pres-elect, 75-76, pres, 77-78); hon mem Electrochem Soc (pres, 72); Am Chem Soc; Am Inst Chem Engrs. *Res:* Current distribution in electrolytic cells; mass transfer in electrode processes; electrolytic oxidation and reduction; electrodeposition; batteries; fuel cells; nonaqueous ionizing media. *Mailing Add:* Dept Chem Eng Univ Calif Berkeley CA 94720

TOBIAS, CORNELIUS ANTHONY, b Budapest, Hungary, May 28, 18; nat US; m 43; c 2. BIOPHYSICS. *Educ:* Univ Calif, Berkeley, MA, 40, PhD(nuclear physics), 42. *Prof Exp:* Physicist, 42-45, from instr to assoc prof biophys, 45-55, vchmn dept physics in-chg of med physics, 60-67, chmn, Div Med Physics, 67-71, chmn grad group biophys & med physics, 69-73, PROF MED PHYSICS, DONNER LAB, UNIV CALIF, BERKELEY, 55-, PROF ELEC ENG, 65-; PROF RADIOL, UNIV CALIF, SAN FRANCISCO, 77- *Concurrent Pos:* Fel med physics, Univ Calif, Berkeley, 45-47; Guggenheim fel, Karolinska Inst, Sweden, 56-57; vis prof, Harvard Univ, 60; mem subcomt, Nat Res Coun, mem comt radiol, Nat Acad Sci-Nat Res Coun; mem radiation study sect, NIH, 60-63; pres radiation biophys comn, Int Union Pure & Appl Physics, 69-72, coun mem, Int Union Pure & Appl Biophys, 69-75; Alexander von Humboldt US sr scientist award, 81; res assoc, BC Cancer Res Ctr, 85. *Honors & Awards:* Lawrence Mem Award, 63; Annual Award, Am Nuclear Soc Aerospace Div, 72. *Mem:* Am Asn Physicists in Med; NY Acad Sci; Am Phys Soc; Radiation Res Soc (pres, 62-63); Biophys Soc. *Res:* Biological effects of radiation; cancer research; space medicine. *Mailing Add:* 130 Summit Rd Walnut Creek CA 94598

TOBIAS, GEORGE S, b Portland, Ore, Apr 14, 16; m 53; c 5. CHEMICAL ENGINEERING. *Educ:* Ohio State Univ, BS, 38, MS, 39. *Prof Exp:* Chem engr, Eastman Kodak Co, 39-41; sr engr, Prod Res Div, Esso Res & Eng Co, 46-51, group head, 51-54, res assoc, 54-55, sect head, 55-58, asst dir prod res div, 58-61, mkt-res-tech serv coordr, Standard Oil Co, NJ, Esso Int Inc, 61-65; tech dir, Pittsburgh Activated Carbon Co, Calgon Corp, 65-70; PRES, ENVIROTROL, INC, 70- *Mem:* Am Chem Soc; Am Inst Chem Engrs; Am Inst Chemists; Soc Automotive Engrs. *Res:* Management of research, engineering and development; adsorption, activated carbon, liquid and gas phase; treatment of industrial and municipal waste and water; air pollution control and gas purification; motor and aviation fuels, lubricants and additives. *Mailing Add:* Backbone Rd RD 5 Sewickley PA 15143-9805

TOBIAS, JERRY VERNON, b St Louis, Mo, Oct 14, 29; c 2. PSYCHOACOUSTICS. *Educ:* Univ Mo, AB, 50; Univ Iowa, MA, 54; Western Reserve Univ, PhD(audition), 59. *Prof Exp:* Asst, Univ Iowa, 53-54; instr speech path & audiol, Ball State Teachers Col, Ind, 54-56; res assoc audition, Western Reserve Univ, 56-59; res scientist psychophys, Defense Res Lab & vis asst prof psychol, Univ Tex, 59-61; assoc prof psychol, Univ Okla, 63-71, prof, 71-80; chmn commun processes, Civil Aeromedical Inst, Fed Aviation Admin, 77-80; at Indust Audiol, 80-; AT AUDITORY & COMMUN SCI DEPT, NAVAL SUBMARINE MED RES LAB. *Concurrent Pos:* Mem comt hearing, bioacoust & biomech, Nat Acad Sci-Nat Res Coun; supvry psychologist, Civil Aeromed Inst, Fed Aviation Admin, 61-77. *Mem:* Fel Acoust Soc Am; Sigma Xi; fel Am Speech & Hearing Asn. *Res:* Physiological and psychological acoustics; audition; experimental phonetics; physiological psychology; psychophysics; sensation and perception; auditory time constants; binaural audition; noise hazards and control. *Mailing Add:* Six Huntington Way Ledyard CT 06339

TOBIAS, PHILIP E, b Harrison, NJ, July 15, 19; m 46; c 4. OPTICS. *Educ:* Cooper Union Inst Technol, BChE, 39, Polytechnic Inst, MChE, 41. *Hon Degrees:* ChE, Cooper Union Inst Tech, 46. *Prof Exp:* Civil serv examr, US Civil Serv Comn, 39-40; asst res eng, Nat Bur Standards, 40-42; head, instrument lab, Publicker Indust, 42-46; dir res eng, Edward Stern & Co, 45-56; PRES, TOBIAS ASSOC, INC, IVYLAND, PA, 56- & COMPUSTATICS, INC, PA, 73- *Concurrent Pos:* graphic arts consult, self employed, Philadelphia, 56-59. *Honors & Awards:* Tech Asn Graphic Arts Honors Award, 80. *Mem:* Tech Asn Graphic Arts (vpres, 58-59, pres 59-60); ACS; fel Am Inst Chemists; Sigma Xi; Graphic Arts Tech Found. *Res:* Developer of quality control approaches and devices used in the printing industry. *Mailing Add:* 50 Industrial Dr PO Box 2699 Ivyland PA 18974

TOBIAS, RUSSELL LAWRENCE, b Upper Darby, Pa, Dec 18, 48. MICROCOMPUTER DESIGN. *Educ:* Temple Univ, AB, 70; Univ Md, MS, 75, PhD(physics), 78. *Prof Exp:* Res asst physics, Dept Physics & Astron, Univ Md, 75-78; RES ANALYST, COMPUSTATICS, INC, 78- *Res:* Instrumentation design and development in graphic arts and radiology; management information system design and administration. *Mailing Add:* PO Box 2699 Warminster PA 18974-0347

TOBIASON, FREDERICK LEE, b Pe Ell, Wash, Sept 15, 36; m 61; c 3. PHYSICAL CHEMISTRY, POLYMER CHEMISTRY. *Educ:* Pac Lutheran Univ, BA, 58; Mich State Univ, PhD(phys chem), 63. *Prof Exp:* Res assoc nuclear magnetic resonance spectros, Emory Univ, 63-64; res chemist, Benger Lab, E I du Pont de Nemours & Co, Inc, 64-66; from asst prof to assoc prof, 66-73, PROF PHYS CHEM, PAC LUTHERAN UNIV, 73- *Concurrent Pos:* Consult, Reichhold Chem, Inc, 67-; chmn chem dept, Pac Lutheran Univ, 73-76 & regency prof, 75-76; consult, Bennett Lab, Inc, 80-82, US Commodities, 85; vis res prof, Univ Wash, 76, Univ Helsinki, 76, La State Univ, 82; vis lectr, Japan, 83, Yugoslavia, 87- *Mem:* Am Chem Soc. *Res:* Molecular structure; nuclear magnetic resonance spectroscopy; electric dipole moments; molecular characterization of polymers; wood chemistry; adhesives; polymer chain configuration calculations; phenolic resins; quantum mechanical calculations. *Mailing Add:* Dept Chem Pac Lutheran Univ Tacoma WA 98447

TOBIE, JOHN EDWIN, immunology, parasitology, for more information see previous edition

TOBIESSEN, PETER LAWS, b Philadelphia, Pa, Mar 30, 40; m 68. PLANT ECOLOGY. *Educ:* Wesleyan Univ, BA, 63; Pa State Univ, University Park, MS, 66; Duke Univ, PhD(bot), 71. *Prof Exp:* From asst prof to assoc prof, 77-83, PROF BIOL, UNION COL, NY, 83- *Mem:* Ecol Soc Am. *Res:* Physiological plant ecology. *Mailing Add:* 1377 Dean St Schenectady NY 12309

TOBIN, ALBERT GEORGE, b Boston, Mass, June 7, 38; m 63. PHYSICAL METALLURGY. *Educ:* Mass Inst Technol, BS, 60, MS, 63; Columbia Univ, PhD(metall), 68. *Prof Exp:* Staff metallurgist, Nuclear Metals Inc, 60-62; res asst metall, Mass Inst Technol, 62-63 & Columbia Univ, 63-68; res metallurgist, Union Carbide Corp, 68-69; STAFF SCIENTIST, GRUMMAN AEROSPACE CORP, 69- *Mem:* Am Soc Metals; Am Ceramic Soc. *Res:* Magnetic and dielectric materials; high temperature properties of materials; fracture and fatigue properties of titanium alloys; hydrogen embrittlement; application of materials to fusion reactors; metal matrix composites; radiation damage in ceramics; surface physics and chemistry; ceramic matrix composites; ceramic metal composites. *Mailing Add:* Grumman Aerospace Corp 26 1111 Stuart Ave Bethpage NY 11787

TOBIN, ALLAN JOSHUA, b Manchester, NH, Aug 22, 42; m 81; c 2. MOLECULAR NEUROBIOLOGY. *Educ:* Mass Inst Technol, BS, 63; Harvard Univ, PhD(biophys), 69. *Prof Exp:* USPHS fels, Weizman Inst Sci, 69-70 & Mass Inst Technol, 70-71; asst prof biol, Harvard Univ, 71-74; asst prof, 75-81, assoc prof, 81-84, PROF, UNIV CALIF, LOS ANGELES, 84-, CHAIR, INTERDEPARTMENTAL PROG NEUROSCI, 89. *Concurrent Pos:* Sci dir, Hereditary Dis Found, 79- *Mem:* AAAS; Soc Develop Biol; Am Soc Neurochem; Soc Neurosci. *Res:* Molecular neurobiology. *Mailing Add:* Dept Biol Univ Calif Los Angeles CA 90024-1606

TOBIN, ELAINE MUNSEY, b Louisville, Ky, Dec 23, 44; m 83; c 2. PLANT DEVELOPMENT, MOLECULAR BIOLOGY. *Educ:* Oberlin Col, BA, 66; Harvard Univ, PhD(biol), 72. *Prof Exp:* Fel biol, Brandeis Univ, 73-75; from asst prof to assoc prof, 75-85, PROF BIOL, UNIV CALIF, LOS ANGELES, 85- *Concurrent Pos:* Mem grant rev panel, USDA & NIH. *Mem:* Am Soc Plant Physiologists; AAAS; Int Soc Plant Molecular Biol. *Res:* Control of plant development by light. *Mailing Add:* Dept Biol Univ Calif Los Angeles CA 90024

TOBIN, GORDON ROSS, b Twin Falls, Idaho, Jan 6, 43; m 68; c 2. PLASTIC & RECONSTRUCTIVE SURGERY, MEDICAL RESEARCH. *Educ:* Whitman Col, AB, 65; Univ Calif, San Francisco, MD, 69. *Prof Exp:* From instr to assoc prof, 77-86, PROF SURG, UNIV LOUISVILLE, 86- *Concurrent Pos:* Prin investr, Paralyzed Vets Am, 75-76; mem var comts, Am Soc Plastic & Reconstruct Surgeons, 81-91; prin investr, Jewish Hosp Heart & Lung Inst, 91- *Mem:* Soc Univ Surgeons; Am Col Surgeons; Am Asn Plastic Surgeons; Plastic Surg Res Coun; Am Soc Plastic & Reconstructive Surgeons; Sigma Xi. *Res:* Tissue repair and wound healing; spinal cord wound healing and regeneration; muscle and skin anatomy and flap physiology; skin tissue culture and burn resurfacing; bone and soft tissue reconstruction. *Mailing Add:* 1505 Northwind Rd Louisville KY 40207-1636

TOBIN, JOHN ROBERT, JR, b Elgin, Ill, Dec 18, 16; m 42; c 1. INTERNAL MEDICINE, CARDIOLOGY. *Educ:* Univ Notre Dame, BS, 38; Univ Chicago, MD, 42; Univ Minn, MS, 50; Am Bd Internal Med, dipl, 52; Am Bd Cardiovasc Dis, dipl, 66. *Prof Exp:* Intern med & surg, Presby Hosp, Chicago, 42-43; resident path, Univ Chicago Clins, 46-47; asst staff, Mayo Clin, 47-51; staff, Rockwood Clin, Spokane, Wash, 51-55; dir adult cardiol, Cook County Hosp, Ill, 59-69; PROF MED, STRITCH SCH MED, LOYOLA UNIV CHICAGO, 62- *Concurrent Pos:* Fel med, Mayo Found, 47-50, NIH spec fel physiol, 63-64; assoc & attend staff med, Cook County Hosp, 55-69; physician-in-chief, Loyola Hosp, 69-82; chmn dept, Stritch Sch Med, Loyola Univ Chicago, 69-82, dean, 82-87. *Mem:* Fel Am Col Physicians; fel Am Col Cardiol. *Res:* Cardiovascular physiology and disease. *Mailing Add:* Loyola Med Ctr 2160 S First Ave Chicago IL 60153

TOBIN, MARTIN JOHN, b Kilkenny, Ireland, Apr 23, 51; m 77; c 2. REGULATION OF RESPIRATION, PULMONARY MECHANICS. *Educ:* Fac Med, Univ Col, Dublin, MB(med), BCh(surg) & BAO(obstet), 75; Am Bd Internal Med, dipl, 83, dipl pulmonary dis, 84, dipl critical care med, 88. *Prof Exp:* Registr, Med Professorial Unit, Univ Col-St Vincent's Hosp, Dublin, 78-79; Brit Thoracic Asn res fel, King's Col Hosp, London, 79-80, res & clin fel, div pulmonary dis, Mt Sinai Med Ctr, Miami Beach, 80-82 & Critical Care Training Prog, Presby-Univ Hosp, Univ Health Ctr, Pittsburgh, 82-83; from asst prof to assoc prof internal med, Div Pulmonary Med, Univ Tex Health Sci Ctr, Houston, 83-90; PROF MED, DIV PULMONARY & CRITICAL CARE MED, LOYOLA UNIV SCH MED, CHICAGO, 90- *Mem:* Am Physiol Soc; Am Thoracic Soc; Am Col Chest Physicians; Soc Critical Care Med; Brit Thoracic Soc; Royal Col Physicians Ireland. *Res:* Neuromuscular control of breathing; respiratory monitoring; cigarette smoking behavior; critical care medicine. *Mailing Add:* Div Pulmonary Med Hines Vet Admin Hosp 111 N Loyola Univ Hines IL 60141

TOBIN, MARVIN CHARLES, b St Louis, Mo, Jan 10, 23; wid; c 3. PHYSICAL CHEMISTRY, PHYSICS. *Educ:* Wash Univ, BA, 47; Ind Univ, MA, 49; Univ Conn, PhD(chem), 52. *Prof Exp:* Sr chemist, Arthur D Little, Inc, 52-53; group leader spectros, Olin Mathieson Chem Corp, 53-57; res chemist, Am Cyanamid Co, 57-60, group leader polymer physics, 60-62, solid state physics, 62-64; sr staff scientist, Perkin-Elmer Corp, 64-70; prof physics, Univ Bridgeport, 70-74; consult, 74-80; mat engr, Norden Systs, 80-89; RETIRED. *Res:* Polymer physics; light scattering; molecular and laser spectroscopy; materials science. *Mailing Add:* Five Clinton Ave Westport CT 06880

TOBIN, MICHAEL, b Russia, Jan 22, 13; nat US; m 44; c 2. INSTRUMENTATION, BIOENGINEERING. *Educ:* City Col New York, BSEE, 43; Polytech Inst Brooklyn, MSEE, 50. *Prof Exp:* Instr elec eng, Polytech Inst Brooklyn, 46-48; instr, instrumentation, Dept Neurol, Col Physicians & Surgeons, Columbia Univ, 48-50; clin asst prof biophys, State Univ NY Downstate Med Ctr, 50-65; res assoc, Dept Psychiat, Col Physicians & Surgeons, Columbia Univ, 65-77; RETIRED. *Concurrent Pos:* Res assoc, retired fac, Columbia Univ. *Mem:* AAAS; Inst Elec & Electronics Engrs. *Res:* Instrumentation; data communication, bioengineering, sensing, recording and analysis of physiologic data; logic design; interfacing to digital computers; stand-alone system design; microprocessor systems organization; history of technology and political economy. *Mailing Add:* 900 W 190th St Apt 5H New York NY 10040

TOBIN, RICHARD BRUCE, b Buffalo, NY, Mar 6, 25; m 47; c 5. PHYSIOLOGY. *Educ:* Union Univ, NY, BS, 45; Univ Rochester, MD, 49. *Prof Exp:* Instr physiol & med, Sch Med & Dent, Univ Rochester, 54-59, asst prof & asst physician, 59-63; assoc prof med, 66-68, assoc prof physiol, 67-68, PROF MED & BIOCHEM, UNIV NEBR MED CTR, OMAHA, 68-; STAFF PHYSICIAN, VET ADMIN HOSP, OMAHA, 68- *Concurrent Pos:* USPHS fel biochem, Univ Amsterdam, 63-65. *Mem:* Am Physiol Soc; Endocrine Soc; Am Inst Nutrit; fel Am Col Physicians; Am Diabetes Asn. *Res:* Cellular energetics; thyroid hormone regulation of intermediary metabolism. *Mailing Add:* Vet Admin Hosp 4101 Woolworth Ave Omaha NE 68105

TOBIN, ROGER LEE, b Grand Rapids, Mich, Oct 11, 40; m 85; c 2. ECONOMICS, TRANSPORTATION RESEARCH. *Educ:* Univ Mich, BSE, 64, MSE, 69, MA, 72, PhD(indust eng), 73. *Prof Exp:* Sr indust & process engr, Hoover Ball & Bearing, 64-68; sr res engr transp & traffic sci, Gen Motors Res Labs, Gen Motors Corp, 73-80; environ systs engr, environ res div, Argonne Nat Lab, 80-85; MEM TECH STAFF, DEPT ECON & STATIST, GTE LABS, 85- *Concurrent Pos:* Adj assoc prof, dept civil eng, Univ Pa, 83- *Mem:* Opers Res Soc Am; Inst Mgt Sci; Math Prog Soc; Am Econ Asn. *Res:* Economics, transportation and mathematical programming. *Mailing Add:* 29 Oxford St Arlington MA 02174

TOBIN, SARA L, b Springfield, Mass, Apr 27, 46; m 68. CONTRACTILE PROTEIN GENETICS, MOLECULAR GENETICS. *Educ:* Univ Wash, Seattle, BS, 68 & PhD(develop biol), 77. *Prof Exp:* Postdoc fel, dept genetics, Univ Calif Berkeley, 78-82, dept biochem & biophys, Univ Calif San Francisco, 78-81; asst res geneticist, Dept Genetics, Univ Calif Berkeley, 81-82; muscular dystrophy postdoc fel, 78-79; NIH postdoctoral fel, 79-81; sr postdoctoral fel, Calif Div, Am Cancer Soc, 81-82; asst prof, 83-89, ASSOC PROF, DEPT BIOCHEM & MOLECULAR BIOL, HEALTH SCI CTR, UNIV OKLA, 89- *Concurrent Pos:* Prin investr, NIH, 81, Am Heart Asn, 85. *Mem:* Am Soc Cell Biol; Sigma Xi; Am Soc Biochem & Molecular Biol; AAAS; Asn Women Sci. *Res:* Expression of individual actin genes during development of Drosophila melanogaster; structure, function and expression of the single Drosophila calmodulin gene; alteration by heat shock of Drosophila small nuclear ribonuclear protein particles. *Mailing Add:* Dept Biochem & Molecular Biol Univ Okla Health Sci Ctr PO Box 26901 Oklahoma City OK 73190

TOBIN, SIDNEY MORRIS, b Toronto, Ont, Jan 18, 23; m 49; c 4. OBSTETRICS & GYNECOLOGY. *Educ:* Univ Toronto, MD, 46; FRCS(C), 51. *Prof Exp:* Clin lectr, 61-71, ASST PROF OBSTET & GYNEC, UNIV TORONTO, 71- *Concurrent Pos:* Chmn, Res Adv Comt, Res Dept, Mt Sinai Hosp, Toronto, 70- *Mem:* Can Med Asn; Soc Obstetricians & Gynecologists Can; Royal Col Physicians & Surgeons Can; Can Oncol Soc; Can Fel Travel Soc. *Res:* Studies to elucidate the role of herpes simplex virus type II as a carcinogen or co-carcinogen in the etiology of squamous cell carcinoma of the cervix in humans. *Mailing Add:* 99 Avenue Rd Suite 206 Toronto ON M5R 2G5 Can

TOBIN, THOMAS, b Dublin, Ireland, Aug 7, 41; US citizen; div. TOXICOLOGY, PHARMACOLOGY. *Educ:* Univ Col Dublin, DVM, 64; Univ Guelph, MSc, 66; Univ Toronto, PhD(parmacol), 69; Am Bd Toxicol, cert, 80. *Prof Exp:* Res assoc pharmacol, Univ Toronto, 66-70; asst prof, Mich State Univ, 71-75; assoc prof vet sci, 75-78, prof vet sci & toxicol, 78-90, GRAD DIR, GRAD CTR TOXICOL, UNIV KY, 90- *Concurrent Pos:* NIH grant, 73-76; NSF grant, 74-76; Ky Equine Drug Res grant, 75-; consult equine medication & drug detection, 75- *Mem:* Am Col Vet Pharmacol & Therapeut; Am Soc Pharmacol & Exp Therapeut; Am Col Vet Toxicol; Equine Vet Asn Eng. *Res:* Equine medication; doping; drug analysis; drug disposition; pharmacokinetics; molecular pharmacology; toxicology. *Mailing Add:* 1242 Sommet Dr Lexington KY 40502

TOBIN, THOMAS VINCENT, b Plymouth, Pa, Apr 8, 26; m 47; c 1. BIOLOGY. *Educ:* King's Col, BS, 51; Boston Col, MS, 53. *Prof Exp:* Asst biol, Boston Col, 51-52; from instr to asst prof, King's Col, 52-62, chmn div natural sci, 71-76, chmn dept biol, 85-90, ASSOC PROF BIOL, KING'S COL, 62-, CHIEF HEALTH PROFESSIONS ADV, 89- *Res:* Chemoreception and proprioception in the crayfish; cold acclimitization; influence of drugs on animal behavior. *Mailing Add:* Dept Biol King's Col Wilkes-Barre PA 18711

TOBIS, JEROME SANFORD, b Syracuse, NY, July 23, 15; m 38; c 3. MEDICINE. *Educ:* City Col New York, BS, 36; Chicago Med Sch, MD, 43. *Prof Exp:* Consult phys med & rehab, Vet Admin Hosp, Bronx, NY, 48-53; prof phys med & rehab & dir, NY Med Col, 52-70; PROF PHYS MED & REHAB & CHMN DEPT, UNIV CALIF, IRVINE-CALIF COL MED, 70-, DIR DEPT, MED CTR, 77- *Concurrent Pos:* Baruch fel, Columbia Univ, 47; dir phys med & rehab, Metrop Hosp, 52-70 & Bird S Coler Hosp, 52-70; consult, City Hosp, 48 & Bur Handicapped Children, New York City Dept Health, 53-70; ed, Arch Phys Med & Rehab, 59-73; chief rehab med, Montefiore Hosp Med Ctr, 61-70; dir dept phys med & rehab, Orange County Med Ctr, 70-77. *Mem:* AAAS; AMA; Am Pub Health Asn; Am Cong Rehab Med; Am Acad Phys Med & Rehab. *Res:* Hemiplegia; cardiac rehabilitation; rehabilitation of handicapped children; geriatrics. *Mailing Add:* Dept Phys Comm Med Univ Calif Irvine CA 92717

TOBISCH, OTHMAR TARDIN, b Berkeley, Calif, June 18, 32; m 64; c 1. STRUCTURAL GEOLOGY. *Educ:* Univ of Calif, Berkeley, BA, 58, MA, 60; Univ London, PhD(struct geol), 63. *Prof Exp:* Fulbright fel, Innsbruck, Austria, 63-64; res geologist, US Geol Surv, 64-69; from asst prof to assoc prof earth sci, 69-78, PROF EARTH SCI, UNIV CALIF, SANTA CRUZ, 78- *Mem:* Geol Soc Am; fel Geol Soc London. *Res:* Polyphase deformation in orogenic belts; quantitative strain determination of deformed rocks; genesis of orogenic belts. *Mailing Add:* Dept Earth Sci Univ Calif Santa Cruz CA 95064

TOBKES, MARTIN, b New York, NY, Feb 8, 28; m 52; c 2. ORGANIC CHEMISTRY. *Educ:* City Col New York, BS, 48; Polytech Inst Brooklyn, MS, 54, PhD(org chem), 63. *Prof Exp:* Res & develop chemist, Nopco Chem Co, 49-54 & Charles Bruning Co, 54-56; res org chemist, Ethicon, Inc, 56-58; res fel org chem, Polytech Inst Brooklyn, 59-63; prin res chemist fermentation develop, Lederle Labs, Am Cyanamid Co, 63-90; RETIRED. *Mem:* Am Chem Soc. *Res:* Vitamin synthesis and stability; Perkin condensations; corrosion; light sensitive coatings; medicinal synthesis; polymers; antibiotics; steroids. *Mailing Add:* 98 Sutin Pl Chestnut Ridge NY 10977

TOBKES, NANCY J, b Somerville, NJ, Feb 12, 58; m 86. PROTEIN CHEMISTRY. *Educ:* Brandeis Univ, BA, 80; Columbia Univ, MA, 81, MPhil, 83, PhD(biochem), 86. *Prof Exp:* Postdoctoral fel cell biol, NY Univ Med Ctr, 86-89; SR SCIENTIST PROTEIN BIOCHEM, HOFFMANN-LA ROCHE, INC, 89- *Mem:* Protein Soc; Am Soc Cell Biol; Am Chem Soc; AAAS. *Res:* Structure and function relationships in membrane proteins; purifying and characterizing different recombinant forms of the human immunoglobulin E receptor and immunoglobulin E fragments to study their interactions. *Mailing Add:* Protein Biochem Dept Hoffmann-La Roche Inc 340 Kingsland St Nutley NJ 07110

TOBOCMAN, WILLIAM, b Detroit, Mich, Mar 14, 26; m 50; c 2. THEORETICAL NUCLEAR PHYSICS, INVERSE THEORY. *Educ:* Mass Inst Technol, SB, 50, PhD(physics), 53. *Prof Exp:* Res assoc, Cornell Univ, 53-54; mem, Inst Adv Study, 54-56; NSF fel, Univ Birmingham, 56-57; asst prof physics, Rice Univ, 57-60; assoc prof, 60-66, PROF PHYSICS, CASE WESTERN RESERVE UNIV, 66- *Concurrent Pos:* Sloan Found fel, 61-64; res fel, Weizmann Inst, 63-64. *Mem:* Am Phys Soc. *Res:* Theory of nuclear reactions; quantum mechanical many-body problem; classical wave scattering, inverse scattering; medical imaging. *Mailing Add:* Dept Physics Case Western Reserve Univ Cleveland OH 44106

TOBUREN, LARRY HOWARD, b Clay Center, Kans, July 9, 40; m 62; c 2. ATOMIC PHYSICS, MOLECULAR PHYSICS. *Educ:* Emporia State Univ, BA, 62; Vanderbilt Univ, PhD(physics), 68. *Prof Exp:* sr res scientist, 67-80, MGR RADIOL PHYSICS SECT, PAC NORTHWEST LABS, BATTELLE MEM INST, 80- *Concurrent Pos:* Adj assoc prof radiol sci, Univ Wash, 81- *Mem:* Fel Am Phys Soc; Radiation Res Soc; AAAS; Int Radiation Physics Soc. *Res:* Atomic and molecular collision processes; Auger electron studies; measurement of inner and outer shell ionization cross sections and continuum electron distributions resulting from charged particle impact. *Mailing Add:* Pac Northwest Labs Battelle Mem Inst PO Box 999 Richland WA 99352

TOBY, SIDNEY, b London, Eng, May 30, 30; m 53; c 2. PHYSICAL CHEMISTRY. *Educ:* Univ London, BSc, 52; McGill Univ, PhD, 55. *Prof Exp:* Fel photochem, Nat Res Coun, Can, 55-57; from instr to assoc prof, 57-69, PROF CHEM, RUTGERS UNIV, NEW BRUNSWICK, 69- *Mem:* AAAS; Am Chem Soc. *Res:* Kinetics of gaseous reactions; photochemistry; chemiluminescence. *Mailing Add:* Dept Chem Rutgers Univ New Brunswick NJ 08903

TOCCI, PAUL M, b Brooklyn, NY, Nov 11, 33; m 68; c 2. BIOCHEMISTRY, GENETICS. *Educ:* Johns Hopkins Univ, BA, 55; Univ Md, PhD(biochem), 64. *Prof Exp:* DIR, BIOCHEM GENETICS LAB, UNIV MIAMI, 64-, ASST PROF PEDIAT, SCH MED, 67- *Concurrent Pos:* pres, TLC Corp, 68-76. *Mem:* AAAS; Am Asn Clin Chemists; fel Am Inst Chemists; Am Soc Human Genetics; Am Fedn Clin Res; Int Fedn Clin Chem. *Res:* Amino acid metabolism; inborn errors of metabolism. *Mailing Add:* Dept Pediat R131 Univ Miami Sch Med 1600 NW Tenth Ave Miami FL 33101

TOCCO, DOMINICK JOSEPH, b New York, NY, Jan 25, 30; m 52; c 4. BIOCHEMISTRY, PHARMACOLOGY. *Educ:* St John's Univ, BS, 51, MS, 53; Georgetown Univ, PhD(chem), 60. *Prof Exp:* Biochemist, US Army Chem Ctr, Md, 53-55, Nat Heart Inst, 55-60, Merck Inst Therapeut Res, 60-66 & Shell Develop Co, 66-70; BIOCHEMIST, MERCK SHARP & DOHME, 70- *Mem:* Am Soc Pharmacol & Exp Therapeut. *Res:* Transport of drugs and natural substances across biological membranes; drug metabolism; pharmacodynamics; experimental enzyme kinetics. *Mailing Add:* Drug Metab Dept Merck Inst Therapeut Res West Point PA 19486

TOCHER, RICHARD DANA, b Oakland, Calif, Oct 8, 35; m 61; c 3. PLANT PHYSIOLOGY. *Educ:* Stanford Univ, AB, 57; Univ Wash, Seattle, MS, 63, PhD(bot), 65. *Prof Exp:* Nat Res Coun Can fel marine bot, Atlantic Regional Lab, 65-66; asst prof, 66-70, ASSOC PROF BIOL, PORTLAND STATE UNIV, 70- *Mem:* Am Soc Plant Physiol; Soc Econ Bot. *Res:* Plant physiology and biochemistry, especially marine algae and parasitic angiosperms. *Mailing Add:* Dept Biol Portland State Univ Box 751 Portland OR 97207

TOCK, RICHARD WILLIAM, b Centerville, Iowa, July 13, 40; m 65; c 3. CHEMICAL ENGINEERING, POLYMER SCIENCE. *Educ:* Univ Iowa, BS, 63, MS, 64, PhD(chem eng), 67. *Prof Exp:* From instr to asst prof chem eng, Univ Iowa, 65-67; res engr, Monsanto Co, 68-70; asst prof chem eng, Univ Iowa, 70-74; vis assoc prof, 74-75, assoc prof, 75-85, PROF, CHEM ENG, TEX TECH UNIV, 85- *Concurrent Pos:* Consult engr, 80- *Honors & Awards:* Ralph R Teetor Award, Soc Automotive Engrs, 80. *Mem:* Am Inst Chem Engrs; Soc Plastics Engrs; Am Soc Eng Educ; Soc Am Mil Engrs. *Res:* Membrane separation processes; polymer science, engineering materials and properties; biomass conversion; oxidations with ozone. *Mailing Add:* Dept Chem Eng Tex Tech Univ PO Box 4679 Lubbock TX 79409-3121

TOCUS, EDWARD C, b Youngstown, Ohio, Apr 22, 25; m 53; c 2. PHARMACOLOGY. *Educ:* Grinnell Col, AB, 50; Univ Chicago, MS, 56, PhD(pharmacol), 59. *Prof Exp:* Res asst radiol, Wash Univ, 51-53; res assoc med, Univ Chicago, 56-60; pharmacologist, Lederle Labs, Am Cyanamid Co, NY, 60-66; pharmacologist, Bur Med, Food & Drugs Admin, 66-70, supvry pharmacologist, Div Neuropharmacol Drugs, 70-72, chief of drug abuse staff, Bur Drugs, 72-; RETIRED. *Concurrent Pos:* Mem Am del, Int Atoms for Peace Conf, Geneva, 58; consult, WHO, Geneva, Switz, 75- *Mem:* Am Soc

Clin Pharmacol & Therapeut. *Res:* Development of new drugs and regulations controlling the national and international use of drugs; develop and implement programs to control illicit use of drugs. *Mailing Add:* 19142 Roman Way Gaithersburg MD 20879

TOCZEK, DONALD RICHARD, b LaPorte, Ind, Nov 21, 38; m 72; c 3. ENTOMOLOGY, BOTANY. *Educ:* Purdue Univ, BS, 61; NDak State Univ, MS, 63, PhD(entom), 67. *Prof Exp:* Res asst entom, NDak State Univ, 61-67; PROF BIOL, HILLSDALE COL, 67- *Mem:* Entom Soc Am; Ecol Soc Am; Sigma Xi. *Res:* Basic botany; invertebrate zoology. *Mailing Add:* 1771 N Lake Pleasant Rd Hillsdale MI 49242

TODARO, GEORGE JOSEPH, b New York, NY, July 1, 37; m 62; c 3. CANCER. *Educ:* Swarthmore Col, BA, 58; NY Univ, MD, 63. *Prof Exp:* Intern Path, NY Univ Sch Med, 62-63, fel, 64-65, asst prof path, 65-67; chief, Molecular Biol Sect, Viral Carcinogenesis Br, Nat Cancer Inst, NIH, Bethesda, 67-70, chief, Viral Leukemia & Lymphoma Br, 70-83, chief Lab Viral Carcinogenics, 76-83; sci dir, Oncogen, Seattle, 83-87; exec vpres & sci dir, Genetics Systs Corp, 87-88; pres & sci dir, Oncogen, 88-90; sr vpres, Pharmaceut Res Inst, Bristol-Myers Squibb, Princeton, NJ, 90; CHMN & PROF, DEPT PATHOBIOL, SCH PUB HEALTH & COMMUNITY MED, UNIV WASH, SEATTLE, 91- *Concurrent Pos:* Career develop award, USPHS, 67; affil prof path, Sch Med, Univ Wash, Seattle, 83-, adj prof path, 91-; spec lectr, Japanese Cancer Asn, 85; vis scientist, Fred Hutchinson Cancer Res Ctr, 90- *Honors & Awards:* Superior Serv Award, Dept Health Educ & Welfare, 71; Gustav Stern Award for Virol, 72; Parke-Davis Award, Am Soc Exp Path, 75; Walter Hubert Lectr, Brit Cancer Soc, 77; Solomon A Bernson Med Achievement Award, 81. *Mem:* Nat Acad Sci; Am Asn Cancer Res; Soc Exp Biol & Med; Am Soc Microbiol; Am Soc Biol Chemists; Am Soc Clin Invest; Am Physic Pluto Soc; Am Asn Pathologists; NY Acad Sci. *Res:* Virus, growth and genetic factors, oncogenes and growth inhibitors in cancer etiology; diagnostic tests for retroviral infections and for cancer susceptibility; viral genes and vaccines; population based and public health related effects of infectious disease and cancer in health care. *Mailing Add:* Fred Hutchins Cancer Res Ctr 1124 Columbia St Mail Stop EP-310-N Seattle WA 98104

TODARO, MICHAEL P, b NY, May 14, 42; m 74; c 1. POPULATION ECONOMICS, DEVELOPMENT ECONOMICS. *Educ:* Haverford Col, BA, 64; Yale Univ, MPhil, 66 & PhD(econ), 67. *Prof Exp:* Lectr econ, Mararere Univ, Uganda, 64-65; sr lectr econ, Univ Nairobi, Kenya, 68-70; assoc dir, Rockefeller Found, 70-74; vis prof econ, Univ Nairobi, Kenya, 74-76 & Univ Calif, Santa Barbara, 76; PROF ECON, NY UNIV, 78-; SR ASSOC, CTR POLICY STUDIES POP COUN, 78- *Concurrent Pos:* Mem, Coun Foreign Rel. *Mem:* Am Econ Asn; Int Union Sci Study Pop; Pop Asn Am. *Res:* Determinants and consequences of population movements in the international economy; economic problems of underdeveloped countries; interrelationships between population and development. *Mailing Add:* 150 E 61st St New York NY 10021

TODD, AARON RODWELL, b El Portal, Fla, Dec 25, 42; m 85; c 2. LINEAR TOPOLOGICAL SPACES, TOPOLOGY. *Educ:* Univ Mich, BS, 64; Univ Leeds, MSc, 68; Univ Fla, PhD(math), 72. *Prof Exp:* Asst prof math, Univ PR, 72-74; asst prof math, Brooklyn Col, 74-78; asst prof, 78-81, ASSOC PROF MATH, ST JOHN'S UNIV, NY, 78- *Concurrent Pos:* Vchmn, Sect Math, NY Acad Sci, 83-85. *Mem:* Am Math Soc; NY Acad Sci; Asn Comput Mach; Asn Women Math. *Res:* Functional analysis and especially properties related to the Baire category theorem; algorithms; numerical analysis. *Mailing Add:* Baruch Col City Univ New York 17 Lexington Ave New York NY 10010

TODD, DAVID BURTON, b Chester, Pa, Dec 21, 25; m 50; c 4. CHEMICAL ENGINEERING. *Educ:* Northwestern Univ, BS, 46, MS, 48; Princeton Univ, PhD(chem eng), 52. *Prof Exp:* Engr, Shell Develop Co, Calif, 52-62, supvr develop, 62-63; mgr eng, Podbielniak Div, Dresser Indust, Ill, 63-67; tech dir, Baker Perkins, Inc, 67-87; vpres tech, APV Chem Mach Inc, 87-88; PRES, TODD ENG, 89- *Concurrent Pos:* Fulbright fel, Delft Tech Univ, 50-51; mem adv comt eng & technol, Saginaw Valley State Col, 77-; mem indust adv comt, Mich Molecular Inst, 78-; lectr, Ctr Prof Advan & Plastics Inst Am, 80- & Polymer Processing Inst, 82- *Mem:* AAAS; fel Am Inst Chem Engrs; Am Chem Soc; Am Soc Safety Engrs; Soc Plastics Engrs. *Res:* Liquid-liquid extraction; fluidization; polymerization; catalysis; acid treating; equipment design; mixing. *Mailing Add:* 35-H Chicopee Dr Princeton NJ 08540

TODD, DAVID KEITH, b Lafayette, Ind, Dec 30, 23; m 48; c 2. CIVIL ENGINEERING, HYDROLOGY. *Educ:* Purdue Univ, BS, 48; NY Univ, MS, 49; Univ Calif, Berkeley, PhD(civil eng), 53. *Prof Exp:* Hydraul engr, US Bur Reclamation, 48-50; from instr to assoc prof civil eng, 50-62, PROF CIVIL ENG, UNIV CALIF, 62- *Concurrent Pos:* NSF fel, 57-58; NSF sr fel, 64-65; centennial prof, Am Univ Beirut, 67; Pres, David Keith Todd Consult Engrs, Inc, 78- *Honors & Awards:* Res Prize, Am Soc Civil Engrs, 60. *Mem:* Am Soc Civil Engrs; Am Geophys Union; Am Water Works Asn; Am Meteorol Soc; AAAS; Nat Water Well Asn. *Res:* Water resources planning, development and management; surface water and groundwater hydrology; precipitation, runoff and floods; saline water intrusion underground; groundwater pollution. *Mailing Add:* 2914 Domingo Ave Berkeley CA 94705

TODD, EDWARD PAYSON, b Newburyport, Mass, Jan 26, 20; m 50; c 3. ATMOSPHERIC PHYSICS, SCIENCE ADMINISTRATION. *Educ:* Mass Inst Technol, BS, 42; Univ Colo, PhD(physics), 54. *Prof Exp:* Res physicist, United Shoe Mach Corp, Mass, 46-49; supvr appl res, Pitney-Bowes, Inc, Conn, 54-57; mem res staff physics, Univ Colo, 57-59, tech dir upper air lab, 59-63; assoc prog dir atmospheric sci sect, NSF, 60-61, prog dir aeronomy, 63-65, actg sect head atmospheric sci sect, 63-64, spec asst to assoc dir res, 65-66, dep assoc dir res, 66-70, dep asst dir res, 70-75, dep asst dir astron, atmospheric, earth & ocean sci, 75-77, actg dir, Atmospheric Sci Div, 75-76, dir, Div Polar Progs, 77-84; RETIRED. *Concurrent Pos:* Mem fed comt meteorol serv & supporting res, 70-77, chmn interdept comt atmospheric sci, 73-78, chmn, Interagency Arctic Res Coord Comt, 77-78. *Honors & Awards:* Distinguished Serv Award, NSF, 71. *Mem:* Am Geophys Union; Am Meteorol Soc; Antarctican Soc (pres, 84-86); Oceanog Soc. *Mailing Add:* 312 Van Buren St Falls Church VA 22046-3655

TODD, ERIC E(DWARD), b Marlow, Eng, Apr 27, 06; nat US; m 36; c 3. CHEMICAL ENGINEERING. *Educ:* Univ BC, BASc, 29, MASc, 30; Stanford Univ, PhD(chem, physics), 34. *Prof Exp:* Chemist, B C Cement Co, Can, 28; chemist metall assays, Consol Mining & Smelting, 29; chemist petrol, Union Oil Co, Calif, 31-32; res chemist, Mich Alkali Co, Calif, 34-35; head chemist, Calif Testing Labs, 36-39; res chemist frozen foods, Calif Consumers Corp, 39-47; tech dir packaged foods, Lady's Choice Foods, Inc, 47-48; chem engr, Ventura Farms Frozen Foods, Inc, 48-59; res chemist, Milo Harding Co, 59-80; RETIRED. *Concurrent Pos:* Consult chem engr, M M H Corp, 48-, Oxnard Frozen Foods Coop & H L Hunt Foods, 59-80, Gen Elec Co & Case Swayne Co, 69-81. *Mem:* AAAS; Am Chem Soc; Nat Soc Prof Engrs; Inst Food Technol. *Res:* Frozen and specialty foods; chemistry; physics. *Mailing Add:* 28730 Grayfox St Malibu CA 90265-4251

TODD, EWEN CAMERON DAVID, b Glasgow, Scotland, Dec 25, 39; m 67; c 4. MICROBIOLOGY. *Educ:* Glasgow Univ, BSc, 63, PhD(bact taxon), 68. *Prof Exp:* Asst lectr bact, Glasgow Univ, 65-68; res scientist, 68-70, head, Methodology Sect, Food Microbiol, 71-73, HEAD, CONTAMINATED FOODS SECT, BUR MICROBIAL HAZARDS, HEALTH PROTECTION BR, DEPT NAT HEALTH & WELFARE, 74-, CHMN, FOOD-BORNE DIS REPORTING CTR, 77- *Mem:* Int Asn Milk, Food, & Environ, Sanitarians. *Res:* Development of methods for food microbiology; public health aspects of food; microbial taxonomy; food-borne disease statistics; costs of food-borne disease and food spoilage. *Mailing Add:* Res Div Bur Microbial Hazards Health & Welfare Can Health Protection Br Ottawa ON K1A 0L2 Can

TODD, FRANK ARNOLD, b Merrill, Iowa, Sept 11, 11; m 36; c 2. VETERINARY MEDICINE. *Educ:* Iowa State Col, DVM, 33; Yale Univ, MPH, 35; Am Bd Vet Pub Health, dipl. *Prof Exp:* Secy res & develop bd, Off Secy Defense, US Army, 49-51; consult vet serv, Fed Civil Defense Admin, 51-54; asst to adminstr, Agr Res Serv, USDA, 54-65; Wash rep, Am Vet Med Asn, 65-75; RETIRED. *Mem:* Am Vet Med Asn; Am Pub Health Asn; NY Acad Sci. *Res:* Epidemiology and epizootiology of animal diseases; exotic animal diseases; biological and chemical warfare defense; atomic energy. *Mailing Add:* 145 S Aberdeen St Arlington VA 22204

TODD, GLEN CORY, b Crawfordsville, Ind, May 10, 31; m 54; c 4. PATHOLOGY. *Educ:* Ind Cent Col, BA, 54; Univ Pa, VMD, 58; Cornell Univ, PhD(path), 65; Am Col Vet Path, dipl. *Prof Exp:* Vet, Agr Res Serv, USDA, 58-62; teaching assoc path, Cornell Univ, 62-65; pathologist, Agr Res Serv, USDA, 65-66 & Vet Res Div, Food & Drug Admin, 66-67; SR PATHOLOGIST, LILLY RES LABS, ELI LILLY & CO, 67- *Mem:* Am Vet Med Asn; NY Acad Sci; Int Acad Path. *Res:* Physiopathology of vitamin E and selenium; nutritional hepatic necrosis; aflatoxicosis in animals; induced myopathies; developmental and neoplastic diseases. *Mailing Add:* 1410 Bittersweet Dr Greenfield IN 46140

TODD, GLENN WILLIAM, b Kansas City, Mo, Sept 30, 27; m 51; c 4. PLANT PHYSIOLOGY. *Educ:* Univ Mo, AB, 49, MA, 50, PhD, 52. *Prof Exp:* Res fel plant physiol, Calif Inst Technol, 52-53; jr biochemist, Citrus Exp Sta, Riverside, Calif, 53-54, asst biochemist, 54-58; from asst prof to assoc prof, Okla State Univ, 58-66, dir sch biol sci, 72-81, head Bot Dept, 81-82, head Bot/Microbiol Dept, 82-90, PROF BOT, OKLA STATE UNIV, 67-, HEAD, BOT DEPT, 90- *Concurrent Pos:* Fulbright prof, Cairo Univ, 66-67 & Azerbaijan State Univ, USSR, 81; ed, Proc Okla Acad Sci, 80-89. *Mem:* Am Inst Biol Sci; Sigma Xi; Am Soc Plant Physiol; Scand Soc Plant Physiol. *Res:* Physiological responses of plants to environmental stress. *Mailing Add:* Bot Dept Okla State Univ Stillwater OK 74078

TODD, GORDON LIVINGSTON, b Princeton, WVa, Mar 17, 44; m 68; c 3. ANATOMY, ELECTRON MICROSCOPY. *Educ:* Kenyon Col, AB, 66; Med Col Ga, MS, 69, PhD(anat), 72. *Prof Exp:* ASST PROF ANAT, SCH MED, CREIGHTON UNIV, 72-; ASSOC PROF ANAT & PREV & STRESS MED, UNIV NEBR MED CTR, 75- *Concurrent Pos:* Nebr Heart Asn grant, Univ Nebr Med Ctr, 75-76, 77-78 & 78-79, Nat Heart, Lung & Blood Inst grant, 80-84, Marion Labs grant, 83-85. *Mem:* Int Soc Heart Res; Am Asn Anat; Sigma Xi; Am Heart Asn; NY Acad Sci; Electron Micros Soc Am. *Res:* Cardiac pathology; cardiac physiology; stress; catecholamines; electron microscopy; lymphatics; autonomic innervation. *Mailing Add:* Dept Prev Stress Med Univ Neb Med Ctr Omaha NE 68105

TODD, HAROLD DAVID, b Mt Vernon, Ill, Nov, 19, 44; m 69. CHEMISTRY. *Educ:* Univ Ill, Urbana, BS, 66; Johns Hopkins Univ, PhD(chem), 71. *Prof Exp:* ASST PROF CHEM & DIR COMPUT CTR, WESLEYAN UNIV, 71-, ADJ ASSOC PROF MATH & CHEM, COORDR ACAD COMPUT & DIR, COMPUT CTR, 80- *Mem:* Asn Comput Mach. *Res:* Theoretical chemistry with emphasis upon mathematical and computational problems. *Mailing Add:* Comput Ctr Wesleyan Univ Middletown CT 06457

TODD, HARRY FLYNN, JR, b Baton Rouge, La, Apr 9, 41. ANTHROPOLOGY, MEDICAL ANTHROPOLOGY. *Educ:* La State Univ, BS, 62; George Washington Univ, JD, 65; Univ Calif, Berkeley, MA, 70, PhD(anthrop), 72. *Prof Exp:* Fel, Ctr Study Law & Soc & actg asst prof anthrop, Univ Calif, Berkeley, 72-73; lectr & fel, Med Anthrop Prog, Dept Int Health, 74-76, adj asst prof, 76-78, asst prof, 78-81, ASSOC PROF ANTHROP IN RESIDENCE, UNIV CALIF, SAN FRANCISCO, 81- *Concurrent Pos:* Ed, Med Anthrop Newsletter. *Mem:* Fel Am Anthrop Asn; fel Geront Soc; Law & Soc Asn; Soc Med Anthrop; fel Soc Appl Anthrop. *Res:* Legal problems and behavior of urban residents, with a focus on both formal and informal mechanisms and agencies used for dispute settlement and conflict management. *Mailing Add:* 4431 19th San Francisco CA 94101

TODD, HOLLIS N, b Glens Falls, NY, Jan 28, 14; m 36; c 1. PHOTOGRAPHY, PHYSICS. *Educ:* Cornell Univ, BA, 34, MEd, 35. *Prof Exp:* Prof, 46-77, EMER PROF PHYSICS, ROCHESTER INST TECHNOL, 77- *Res:* Photographic physics, science and engineering. *Mailing Add:* 3213 Bay Berry Terr Sarasota FL 34237

TODD, JAMES WYATT, b Houston Co, Ala, Dec 16, 42; m 64; c 2. ECONOMIC ENTOMOLOGY, HOST PLANT RESISTANCE. *Educ:* Auburn Univ, BS, 66, MS, 68; Clemson Univ, PhD(entom), 73. *Prof Exp:* From field res asst to res asst entom, Auburn Univ, 64-68; from asst prof to prof res entom, Univ Ga, 68-90; CONSULT, 90- *Concurrent Pos:* Sr examr, Pest Mgt Rev Bd, Am Registry Prof Entomologists; peer panelist, Competitive Grants Prog, USDA; regional coordr, Pesticide Impact Assessment, 87-88; pres, Southeastern Br, Entom Soc Am, 90. *Mem:* Entom Soc Am. *Res:* Development of peanut insect pest management systems including host plant resistance; utilization of natural control agents; chemical and microbial pesticides; cultural control practices and economic injury thresholds. *Mailing Add:* Entom Univ Ga Coastal Plain Exp Sta Tifton GA 31793

TODD, JERRY WILLIAM, b La Crosse, Wis, Jan 7, 30; m 56; c 1. ANALYTICAL CHEMISTRY. *Educ:* Wis State Univ, Platteville, BS, 51; Univ Wis, PhD(anal chem), 60. *Prof Exp:* Teacher high sch, Wis, 51-52; chemist, Liberty Powder Defense Corp, Olin Mathieson Chem Corp, 52-56; sr chemist, Cent Res Lab, 60-69, sr res specialist, Agrichem Anal Lab, 69-83, SR RES SPECIALIST, MAGNETIC MEDIA ANALSIS LAB, 3M Co, 83- *Mem:* Am Chem Soc. *Res:* Gas and liquid chromatography, ur-vis-IR spectroscopy; titrimetry. *Mailing Add:* Minn Mining Mfg Co 3M Ctr 236-3D-02 St Paul MN 55144

TODD, JOHN, b Carnacally, Ireland, May 16, 11; nat US; m 38. NUMERICAL ANALYSIS. *Educ:* Queen's Univ, Belfast, BSc, 31. *Prof Exp:* Lectr, Queen's Univ, Belfast, 33-37 & King's Col, London, 37-49; expert appl math, Nat Bur Standards, 47-48, chief comput lab, 49-54, numerical anal, 54-57; PROF MATH, CALIF INST TECHNOL, 57- *Concurrent Pos:* Scientist, Brit Admiralty, 40-46; Fulbright prof, Univ Vienna, 65. *Mem:* Am Math Soc; Soc Indust & Appl Math; Math Asn Am. *Res:* Mathematical analysis; algebra. *Mailing Add:* Calif Inst Technol 253-37 Calif Inst Technol Pasadena CA 91125

TODD, KENNETH S, JR, b Three Forks, Mont, Aug, 25, 36. VETERINARY PARASITOLOGY. *Educ:* Mont State Univ, BS, 62, MS, 64; Utah State Univ, PhD(zool), 67. *Prof Exp:* Asst zool, Utah State Univ, 64-67; from asst prof to assoc prof, 67-76, PROF VET PARASITOL, UNIV ILL, URBANA, 76- *Concurrent Pos:* Chairperson, div vet parasitol, Univ Ill, Urbana, 83-90, prof vet progs in agr, 84-90, from asst head to actg head, 87-88, head, Dept Vet Pathbiol, 90-; affil prof scientist, natural hist surv, Urbana, Ill. *Honors & Awards:* Beecham Res Excellence Award, 85. *Mem:* Am Soc Parasitol; Am Vet Med Asn; Am Soc Trop Med & Hyg; Am Heartworm Soc; Am Asn Vet Parasitologists; Sigma Xi; Soc Protozoologists. *Res:* Parasites of wildlife and domestic animals; ecology of parasitism. *Mailing Add:* Dept Vet Pathbiol Univ Ill 2001 S Lincoln Ave Urbana IL 61801

TODD, LEE JOHN, b Denver, Colo, Nov 8, 36; m 79; c 3. BORON CHEMISTRY, SOLID STATE CHEMISTRY. *Educ:* Univ Notre Dame, BS, 58; Fla State Univ, MS, 60; Ind Univ, PhD(chem), 63. *Prof Exp:* Res assoc chem, Mass Inst Technol, 63-64; asst prof inorg chem, Univ Ill, Urbana, 64-68; assoc prof, 68-74, PROF CHEM, IND UNIV, BLOOMINGTON, 74- *Mem:* Am Chem Soc. *Res:* Synthetic chemistry of boron-containing materials; NMR; boron neutron capture therapy; propellants; biological activity of boron compounds; organometallic chemistry; coal chemistry; solid state chemistry. *Mailing Add:* Dept Chem Ind Univ A250C Chem Bldg Bloomington IN 47405

TODD, LEONARD, b Glasgow, Scotland, Feb 7, 40. THEORETICAL MECHANICS. *Educ:* Strathclyde Univ, BSc, 61; Cambridge Univ, PhD(appl math), 64. *Prof Exp:* C L E Moore instr appl math, Mass Inst Technol, 64-66; lectr, Strathclyde Univ, 66-69; PROF APPL MATH, LAURENTIAN UNIV, 69- *Concurrent Pos:* Nat Res Coun Can res grants, 69- *Res:* Theoretical fluid mechanics; asymptotic expansions with emphasis on applications. *Mailing Add:* Dept Math & Computer Sci Laurentian Univ Ramsey Lake Rd Sudbury ON P3E 2C6 Can

TODD, MARGARET EDNA, physiology, medical sciences; deceased, see previous edition for last biography

TODD, MARY ELIZABETH, b Kingston, Ont; m 72. ANATOMY, EMBRYOLOGY. *Educ:* Univ BC, BA, 57, MSc, 59; Univ Glasgow, PhD(zool), 62. *Prof Exp:* Lectr & head biol dept, United Col, Man, 62-63; from lectr to sr lectr anat, Univ Glasgow, 62-71; vis asst prof anat, Univ BC, 70-71; asst prof anat, Univ Western Ont, 72; sessional lectr oral biol & zool, 72-73, asst prof, 73-75, ASSOC PROF ANAT, UNIV BC, 75- *Mem:* Anat Soc Gt Brit; Can Asn Anatomists; Am Asn Anatomists; Sigma Xi. *Res:* Ultrastructure of vascular tissues; specifically developmental studies on components of the walls of vessels and on vasomotor innervation; morphometrics of vascular tissues from computer assisted reconstructions using serial sections; smooth muscle cell dimensional parameters in hypertension. *Mailing Add:* Dept Anat Univ BC 2075 Wesbrook Mall Vancouver BC V6T 1W5 Can

TODD, MICHAEL JEREMY, b Chelmsford, Eng, Aug 14, 47; m 71; c 1. NUMERICAL PROGRAMMING & ANALYSIS. *Educ:* Cambridge Univ, Eng, BA, 68; Yale Univ, PhD(admin sci), 72. *Prof Exp:* Asst prof opers res, Univ Ottawa, 72-73; from asst prof to prof opers res, 73-82, dir, Ctr Appl Math, LEON C WELCH PROF ENG, CORNELL UNIV, 88- *Concurrent Pos:* Prin investr, NSF, 74-76, 77-79, 79-82, 82-85 & 89; assoc ed, Math Opers Res, 78-, co-ed, Math Prog, 80-86, ed-in-chief, 86-89; sr vis, Dept Appl Math & Theoret Physics, Cambridge Univ, Eng, 80-81; Guggenheim fel, 80-81; Sloan fel, 81-85. *Honors & Awards:* Dantzig Prize, Math Prog Soc & Soc Indust Appl Math, 88. *Mem:* Opers Res Soc Am; Sigma Xi; Math Prog Soc; Soc Indust Appl Math. *Res:* Mathematical programming; combinatorial optimization; mathematical economics. *Mailing Add:* Sch Opers Res Cornell Univ Ithaca NY 14853-3801

TODD, NEIL BOWMAN, b Cambridge, Mass, Jan 3, 36. EVOLUTIONARY BIOLOGY, GENETICS. *Educ:* Univ Mass, BS, 59; Harvard Univ, PhD(biol), 63. *Prof Exp:* Geneticist, Animal Res Ctr, Med Sch, Harvard Univ, 63-68; DIR & GENETICIST, CARNIVORE GENETICS RES CTR, 68- *Concurrent Pos:* Geneticist, Bio-Res, Inst, Mass, 68-69; res dir, Faunalabs, Inc, 68-72; adj prof dept biol, Boston Univ, 71-86; Fulbright fel, 76-77. *Mem:* Am Genetic Asn. *Res:* Chromosomal mechanisms in the origin and evolution of mammals; population genetics and mutantallele frequencies in domestic cats. *Mailing Add:* 26 Walnut Pl Newtonville MA 02160

TODD, PAUL WILSON, b Bangor, Maine, June 15, 36; m 57; c 4. BIOPHYSICS, CHEMICAL ENGINEERING. *Educ:* Bowdoin Col, BA, 59; Mass Inst Technol, SB, 59; Univ Rochester, MS, 60; Univ Calif, Berkeley, PhD(biophys), 64. *Prof Exp:* Lectr med physics, Univ Calif, Berkeley, 64-66; from asst prof to prof biophys, Pa State Univ, 66-86, chmn, Grad Prog Genetics, 73-78,; dir, Bioprocessing & Pharmaceut Res Ctr, Philadelphia, 84-87; BIOPHYSICIST, NAT INST STANDARDS & TECHNOL, 88- *Concurrent Pos:* Eleanor Roosevelt Int Cancer res fel, 67-68; mem biomed steering comt, Los Alamos Meson Physics Facil, 70-77, chmn, 73-74 & vis staff mem, Los Alamos Sci Lab, 74, mem bd dir, Los Alamos Meson Physics Facil Users Group, 76-77; vis fel, Princeton Univ, 71-72; mem biol comt, Argonne Univ Asn, 73-75 & chmn, 75; consult, Oak Ridge Nat Lab, 75-; assoc ed, Radiation Res, 76-79, Cell Biophys, 78-, Cytometry, 81-88, Anal Quant Cytol, 81-86, J Biochem & Biophys Methods, 82-, Electrophoresis, 85-89 & Theoret & Appl Electrophoresis, 90-; vis prof med physics, Univ Calif, 79; mem, Discipline Working Group in Microgravity & Biotechnol, Univs Space Res Asn, 78-, chmn, 82-85; vis scientist, Univ Uppsala, 79, 84 & Oncol Sci Ctr, Moscow, 79; low level radiation expert, NIH, 80-83; mem, Health & Environ Res Adv Comt, Dept Energy, 83-87 & adv comt, Bevelac Biomed Prog, Lawrence Berkeley Lab, 83-87; mem, Space Appln Bd, Nat Res Coun, Nat Acad Sci, 85-88 & LifeSat Working Group, NASA, 90- *Honors & Awards:* R & D 100 Award, 90. *Mem:* Electrophoresis Soc; Soc Anal Cytol; Tissue Cult Asn; Radiation Res Soc; Am Soc Gravitational & Space Biol; Am Chem Soc. *Res:* Bioseparations; aqueous two-phase partitioning; radiation physics; cellular radiation biology; mammalian cell culture; cell electrophoresis; automated cytology; laser light scattering and interferometry; space biophysics and bioprocessing. *Mailing Add:* 2595 Vassar Dr Boulder CO 80303

TODD, PETER JUSTIN, b Lackawanna, NY, June 26, 49; m 72; c 1. ION OPTICS. *Educ:* Rensselaer Polytech Inst, BS, 71; Cornell Univ, MS, 77, PhD(chem), 80. *Prof Exp:* STAFF SCIENTIST, OAK RIDGE NAT LAB, 80- *Mem:* Am Chem Soc; Am Soc Mass Spectrometry; Sigma Xi. *Res:* Organic mass spectrometry, particularly in design and construction of spectrometers; collision and fragmentation of high energy polyatomic ions. *Mailing Add:* 147 Cumberland View Dr Oak Ridge TN 37830

TODD, PHILIP HAMISH, b Edinburgh, Scotland, Sept 28, 56. MATHEMATICS, COMPUTING GEOMETRY. *Educ:* Dundee Univ, PhD(math), 85. *Prof Exp:* Res asst math, Dundee Univ, 79-84; PRIN ENGR, TEKTRONIX INC, 86- *Res:* Developing techniques in computational geometry to mechanical engineering design. *Mailing Add:* 325 N First St Central Point OR 97502

TODD, ROBIN GRENVILLE, b Devon, Eng, July 24, 48; m 78. ENTOMOLOGY. *Educ:* Univ Lancaster, BA, 71; Univ Reading, PhD(entom), 79. *Prof Exp:* Res asst entom mosquito res & control, Grand Cayman Brit W Duties Unit, 79-80; ENTOMOLOGIST, INSECT CONTROL & RES, INC, 80- *Mem:* Am Mosquito Control Asn; Entom Soc Am; Inst Biol. *Res:* Minnows as potential control agents of mosquiotes in Caribbean; evaluation of pesticides against household and medically important insects and arachnids. *Mailing Add:* 10174 Green Clover Dr Ellicott City MD 21043

TODD, TERRY RAY, b De Kalb, Ill, Oct 9, 47; m 78; c 3. FUEL TECHNOLOGY & PETROLEUM ENGINEERING, ANALYTICAL CHEMISTRY. *Educ:* Northern Ill Univ, BS, 69; Pa State Univ, MS, 72, PhD(physics), 76. *Prof Exp:* Nat Bur Standards-Nat Res Coun fel physics, Gaithersburg, 76-78; staff scientist physics, Laser Anal Inc, 78-80; staff physicist, Exxon Res & Eng Co, 80-91; PRES, T R TODD ENTERPRISES, INC, 91- *Mem:* Optical Soc Am; Soc Appl Spectros; Int Soc Optical Eng; Int Standard Atmosphere. *Res:* High resolution molecular spectroscopy in the infrared region; optics; radiation thermometry; industrial applications of near infrared spectroscopy. *Mailing Add:* T R Todd Enterprises Inc 19 Third St Budd Lake NJ 07828-1614

TODD, WILBERT R, BIOCHEMISTRY. *Educ:* Univ Wis-Madison, PhD(biochem), 33. *Prof Exp:* Prof biochem, Sch Med, Ore Health Scis Univ, 62-72; RETIRED. *Mailing Add:* Dept Biochem Univ Ore Health Sci Ctr Sch Med 3181 SW Sam Jackson Park Rd Portland OR 97201

TODD, WILLIAM MCCLINTOCK, b Colon, Panama, July 17, 25; US citizen; m 47; c 2. MEDICAL MICROBIOLOGY, VIROLOGY. *Educ:* Univ Ga, BS, 50; Vanderbilt Univ, MS, 55, PhD(microbiol), 57. *Prof Exp:* Asst prof microbiol, Sch Med, Univ Miss, 57-63; ASSOC PROF MICROBIOL, MED UNITS, UNIV TENN, MEMPHIS, 63-, PROF, CTR FOR HEALTH SCI, 65- *Concurrent Pos:* USPHS fel biochem & res assoc, Vanderbilt Univ, 58-60. *Mem:* Am Soc Microbiol; Tissue Cult Asn. *Res:* Host-virus relationships. *Mailing Add:* 3821 Mariataria Memphis TN 38127

TODHUNTER, ELIZABETH NEIGE, b Christchurch, NZ, July 6, 01; nat US. NUTRITION. *Educ:* Univ NZ, BS, 26, MS, 28; Columbia Univ, PhD(chem), 33. *Prof Exp:* From asst prof to assoc prof nutrit, State Col Wash, 34-41; from assoc prof to prof & head dept, Univ Ala, 41-53, dean sch home econ, 53-66; NUTRIT CONSULT, 66-; VIS PROF NUTRIT, SCH MED, VANDERBILT UNIV, 67- *Concurrent Pos:* Consult to Surgeon Gen, USAF, 65-67; vis lectr & sci writing, 66- *Mem:* AAAS; Am Chem Soc; Am Dietetic Asn (pres, 57-58); fel Am Inst Nutrit. *Res:* Vitamin C in foods and body fluids; vitamin A in foods; human food consumption; dietary studies; nutritional status measurements; nutrition of elderly; history of nutrition. *Mailing Add:* 400 First American Ctr Nashville TN 37237

TODHUNTER, JOHN ANTHONY, b Cali, Colombia, Oct 9, 49; US citizen; m 72, 86; c 2. BIOCHEMISTRY, MOLECULAR BIOLOGY. *Educ:* Univ Calif, Los Angeles, BS, 71; Calif State Univ, Los Angeles, MS, 73; Univ Calif, Santa Barbara, PhD(chem), 76. *Prof Exp:* Instr, Dept Chem, Calif State Univ, Los Angeles, 72-73; teaching asst, Univ Calif, Santa Barbara, 74, res asst biochem, 74-76; fel, Roche Inst Molecular Biol, Hoffman-La Roche, Nutley, 76-78; asst prof biol & chmn prog biochem, Cath Univ Am, 78-81; asst adminr, pesticides & toxic substances, Environ Protection Agency, 81-83; Todhunter Assocs, 83-88; PRIN, SRS INT, 89- *Concurrent Pos:* Consult, Arral Industs, Encino, 72-; assoc, Andrulis Res Corp, Bethesda, 78-; mem, Hazardous Waste Siting Bd, State Md, 80-81. *Mem:* Am Chem Soc; AAAS; NY Acad Sci. *Res:* Mechanisms of drug and toxicant action; enzymology; dynamic behavior of biochemical systems; biochemistry of gene transcription and expression; science administration. *Mailing Add:* 1625 K St NW Suite 975 Washington DC 20006

TODOROVIC, PETAR N, probability, engineering, for more information see previous edition

TODREAS, NEIL EMMANUEL, b Peabody, Mass, Dec 17, 35. NUCLEAR ENGINEERING. *Educ:* Cornell Univ, BA, 53, MS, 58; Mass Inst Technol, PhD(nuclear eng), 66. *Prof Exp:* PROF NUCLEAR ENG, MASS INST TECHNOL, 70- *Mem:* Nat Acad Eng; fel Am Nuclear Soc; fel Am Soc Mech Engrs. *Mailing Add:* Dept Nuclear Eng Mass Inst Technol 77 Massachusetts Ave Cambridge MA 02139

TODSEN, THOMAS KAMP, b Pittsfield, Mass, Oct 21, 18; m 39; c 2. SCIENCE ADMINISTRATION, BOTANY. *Educ:* Univ Fla, BS, 39, MS, 42, PhD(org chem), 50. *Prof Exp:* Asst & instr, Univ Fla, 47-50; instr, NMex Col Agr & Mech Arts, 50-51; chief chemist, White Sands Missile Range, 51-53, chief warheads engr, 53-58, sci adv off, 58-59, land combat systs eval, 59-66, dir test opers, 66-69, dir SSMPO, 69-72, tech dir army missile test & eval, 72-78; res assoc, 77-79, ASST PROF, NMEX STATE UNIV, 79- *Mem:* AAAS; Am Chem Soc; Sigma Xi. *Res:* Naturally occurring plant constituents; plant taxonomy; plant distribution. *Mailing Add:* 2000 Rose Lane Las Cruces NM 88005

TODT, WILLIAM LYNN, b Columbus, Ohio, Dec 31, 54; m 78; c 4. CELL BIOLOGY. *Educ:* Berea Col, Ky, BA, 80; Univ Wis-Madison, PhD(anat), 87. *Prof Exp:* Researcher, Univ Calif, Irvine, 87-90; ASST PROF BIOL, CONCORDIA COL, MOOREHEAD, MINN, 90- *Mem:* AAAS; Soc Develop Biol; Am Soc Zoologists. *Res:* Cellular and tissue interactions that lead to the development of pattern formation and differentiation; vertebrate embryos. *Mailing Add:* Biol Dept Concordia Col Moorhead MN 56562

TOENISKOETTER, RICHARD HENRY, b St Louis, Mo, Mar 21, 31; m 53; c 6. ENVIRONMENTAL SCIENCES, INDUSTRIAL HYGIENE. *Educ:* Univ St Louis, BS, 52, MS, 56, PhD(chem), 58. *Prof Exp:* Res chemist, Union Carbide Corp, 57-67; sr res chemist, ADM Chem Co Div, Ashland Oil Co, 67-68, group leader, Ashland Chem Co, 68-70, mgr inorg chem res, 70-73, foundry res, 73-78, environ occup safety, 78-88, health & safety, 88-89, DIR HEALTH & SAFETY, ASHLAND CHEM CO, 89- *Honors & Awards:* Sci Merit Award, Am Foundrymen's Soc, 84. *Mem:* Am Chem Soc; Am Foundrymen's Soc; Sigma Xi. *Res:* Organic and inorganic polymer chemistry; coordination and fluorine compounds; boron hydrides; materials and foundry products research; environmental science; industrial hygiene; safety and product safety; toxicology. *Mailing Add:* 6771 Masefield St Worthington OH 43085

TOENNIES, JAN PETER, b Philadelphia, Pa, May 3, 30; m 66; c 2. MOLECULAR PHYSICS, SURFACE PHYSICS. *Educ:* Brown Univ, PhD(chem), 57. *Prof Exp:* Asst, 57-67, docent, 67-71, HON PROF, INST PHYSICS, UNIV BONN, 71-; DIR, MAX PLANCK INST FLUID DYNAMICS, 69- *Concurrent Pos:* Guest docent, Gothenburg Univ, 66-75; apl prof, Univ Goettingen, 72-; fel Am Phys Soc, 83. *Mem:* Am Phys Soc; Ger Phys Soc; Europ Phys Soc. *Res:* Molecular beam investigations of elastic, inelastic and reactive collisions; theory of inelastic scattering; chemical reactions in shock waves; molecular beam investigations of structure and dynamics of crystal surfaces. *Mailing Add:* Max Planck Inst Fluid Dynamics Bunsenstrasse 10 Goettingen Germany

TOENNIESSEN, GARY HERBERT, b Lockport, NY, July 9, 44; m 67. ENVIRONMENTAL SCIENCES. *Educ:* State Univ NY Buffalo, BA, 66; Univ NC, MS, 68, PhD(microbiol), 71. *Prof Exp:* Prog assoc, 71-72, asst dir natural & environ sci, 72-78, asst dir, 78-84, ASSOC DIR AGR SCI, ROCKEFELLER FOUND, 84- *Mem:* Am Soc Microbiol; Int Soc Plant Molecular Biol. *Res:* Plant molecular and cellular biology, plant breeding and environmental problems associated with agriculture. *Mailing Add:* 11 Fenimore Dr Harrison NY 10528

TOENSING, C(LARENCE) H(ERMAN), b St Paul, Minn, Aug 23, 15; m 45; c 1. METALLURGY, PHYSICAL CHEMISTRY. *Educ:* Macalester Col, AB, 37; Mont Sch Mines, MS, 39; Carnegie Inst Technol, ScD(phys chem), 47. *Prof Exp:* Instr chem, Mont Sch Mines, 37-39; asst, Carnegie Inst Technol, 39-44, instr, 44-49; sr technologist, US Steel Corp, 44-50; asst to mgr opers, Brush Beryllium Co, Ohio, 50, res engr, Brush Electronics Co, 50-51; res & develop engr, Lamp Metals & Components Dept, Gen Elec Co, 51-59; dir res, Firth Sterling, Inc, 59-64, dir tech serv, Carmet Co, 64-69; plant mgr, Valeron Corp, 69-81; RETIRED. *Mem:* Am Chem Soc; Am Soc Metals. *Res:* Gas-metal reactions; chemistry and metallurgy of tungsten and molybdenum; refractory carbides and materials. *Mailing Add:* 702 Via Zapata Riverside CA 92507

TOEPFER, ALAN JAMES, b Chicago, Ill, Oct 20, 41. PULSED POWER, FUSION. *Educ:* Marquette Univ, BA, 62; Univ Southern Calif, MS, 64, PhD(physics), 68. *Prof Exp:* Res asst, Royal Inst Technol, Stockholm, 64-65; tech staff, Aerospace Corp, 65-66; tech staff, Sandia Labs, 68-75, supvr, 75-79; dir, 79-81, vpres & dir, Res & Develop, Physics Int Co, 81-85; ASST VPRES, SCI APPL INT CORP, 85- *Mem:* Am Phys Soc; AAAS; Am Inst Aeronaut & Astronaut. *Res:* Inertial confinement fusion; pulsed power; intense relativistic electron beams; electromagnetic armor; radiation effects of nuclear weapons; neutral particle beams. *Mailing Add:* Sci Appl Intl Co 623 Cedar Hill Rd NE Albuquerque NM 87122

TOEPFER, RICHARD E, JR, b Chicago, Ill, Oct 9, 34; m 65; c 3. DATA PROCESSING, CONTROL SYSTEMS. *Educ:* Univ Ill, BSEE, 56, MSEE, 57, PhD(elec eng), 62. *Prof Exp:* Mem tech staff, Aerospace Corp, 61-63; res specialist, Autonetics Div, NAm Aviation, Inc, 63-64; adv engr, IBM Corp, Calif, 65-69; chief systs engr, Measurex Corp, 68-69; proj mgr, Data Systs Develop Div, Hewlett Packard, 70-71, sect mgr, 71-76, prod eng mgr, Data Systs Div, 76-80; opers mgr, Spectra Physics, 80-81; vpres mfg, Magnuson Comput, 81-82; mfg mgr, 82-84, dir pilot mfg, 84-88, TECHNOL DIR, DISTRIB CTR, CONVERGENT, INC, 87-; DIR MFG TECHNOL, UNISYS, COMPUTER SYSTS PROD GROUP, SAN JOSE, CALIF, 90- *Concurrent Pos:* Mem & subcomt chmn, Tech Adv Comt on Comput Systs, Bur East- West Trade, US Dept Com, 73-75; vpres develop & opers, Lasa Inc, San Jose, 88-90. *Mem:* Inst Elec & Electronics Engrs; Soc Indust & Appl Math; Sigma Xi. *Res:* Data processing systems; process control systems; control systems theory; video systems design. *Mailing Add:* 16300 Los Serenos Robles Los Gatos CA 95030-3026

TOETZ, DALE W, b Milwaukee, Wis, Sept 23, 37. FISH BIOLOGY, LIMNOLOGY. *Educ:* Univ Wis, BS, 59, MS, 61; Ind Univ, PhD(zool), 65. *Prof Exp:* Actg instr zool, Univ Wis, Milwaukee, 61-62; teaching asst, Ind Univ, 62-65; from asst prof to assoc prof, 65-79, PROF ZOOL, OKLA STATE UNIV, 80- *Concurrent Pos:* Assoc res biologist, Scripps Inst Oceanog, 74; tech consult, US Environ Protection Agency, 77; res affil, Inst Arctic & Alpine Res, 88-91. *Mem:* Am Fisheries Soc; Ecol Soc Am; Am Soc Limnol & Oceanog; Int Soc Theoret & Appl Limnol; Soc Wetland Sci. *Res:* Limnology of nitrogen; alpine limnology, wetlands; lake restoration. *Mailing Add:* Zool Dept Okla State Univ Stillwater OK 74078

TOEWS, ARREL DWAYNE, b Enid, Okla, July 22, 48; m 71; c 2. NEUROCHEMISTRY, BIOCHEMICAL NEUROPATHOLOGY. *Educ:* Tabor Col, BA, 70; Ohio State Univ, PhD(physiol chem), 74. *Prof Exp:* From res instr to res asst prof, 77-86, RES ASSOC PROF BIOCHEM, UNIV NC, CHAPEL HILL, 86-, RES SCIENTIST, BIOL SCI RES CTR, 79- *Mem:* Int Soc Neurochem; Am Soc Neurochem. *Res:* Neurochemistry; myelin metabolism; axonal transport; neurotoxicology; gene expression. *Mailing Add:* 321 BSRC CB 7250 Univ NC Chapel Hill NC 27599-7250

TOEWS, CORNELIUS J, b Altona, Man, Mar 22, 37; m 61; c 3. ENDOCRINOLOGY, BIOCHEMISTRY. *Educ:* Univ Man, BSc & MD, 63; Queen's Univ, Ont, PhD(biochem), 67; FRCPS(C), 69. *Prof Exp:* ASST PROF BIOCHEM & MED, MED SCH, MCMASTER UNIV, 71-, ASSOC PROF. *Concurrent Pos:* Med Res Coun Can Centennial res fel med, Joslin Res Lab, Harvard Univ, 68-71; jr assoc med, Peter Bent Brigham Hosp, Boston, 70-71; instr, Med Sch, Harvard Univ, 70-71. *Mem:* Can Soc Clin Invest; Can Biochem Soc; Am Diabetes Asn; Can Diabetes Asn. *Res:* Regulation of gluconeogenesis in the liver; regulation of glycolysis and intermediary metabolism in skeletal muscle. *Mailing Add:* Dept Med McMaster Univ Med Sch Hamilton ON L8N 3Z5 Can

TOEWS, DANIEL PETER, b Grande Prairie, Alta, Dec 18, 41; m 64; c 3. ANIMAL PHYSIOLOGY. *Educ:* Univ Alta, BSc, 63, MSc, 66; Univ BC, PhD(zool), 69. *Prof Exp:* Asst prof zool, Univ Alta, 69-71; assoc prof, 71-80, PROF BIOL, ACADIA UNIV, 80- *Concurrent Pos:* NSERC Grant Selection Comt. *Mem:* Can Soc Zool; Brit Soc Exp Biol. *Res:* Comparative respiration and circulation in fishes and amphibians. *Mailing Add:* Dept Biol Acadia Univ Wolfville NS B0P 1X0 Can

TOFE, ANDREW JOHN, b New York, NY, May 6, 40. NUCLEAR MEDICINE. *Educ:* Univ Dayton, BS, 63; Fla State Univ, PhD(nuclear chem), 69. *Prof Exp:* Gen mgr/sr vpres, Procter & Gamble Co, 70-72, scientist nuclear med, 72-80; MGR NUCLEAR MED, BENEDICT NUCLEAR PHARMACEUT, 81- *Mem:* Soc Nuclear Med; Am Chem Soc. *Res:* Manufacture and sales of radioisotopes for use in early detection of human disease or abnormalities. *Mailing Add:* 2195 Urban Dr Denver CO 80215

TOFFEL, GEORGE MATHIAS, b Greensburg, Pa, Jan 28, 11; m 38; c 4. CHEMISTRY. *Educ:* Vanderbilt Univ, BA, 35, MS, 36. *Prof Exp:* Instr chem, Vanderbilt Univ, 34-36; teacher high sch, Ga, 36-37; head sci dept, Marion Inst, 37-47; from asst prof to assoc prof, 47-77, EMER ASSOC PROF CHEM, UNIV ALA, TUSCALOOSA, 77- *Concurrent Pos:* Vis prof, Univ Hawaii, 64-65. *Mem:* Am Chem Soc; fel Am Inst Chem; AAAS. *Res:* Electro-organic chemistry; free radicals; resistor research; magnetic alloys of manganese; reaction mechanism studies with carbon-14 fatty acid esters; ketonization of fatty acids. *Mailing Add:* 303 Queen City Ave Tuscaloosa AL 35401

TOGASAKI, ROBERT K, b San Francisco, Calif, July 24, 32; m 59. PLANT PHYSIOLOGY, CELL BIOLOGY. *Educ:* Haverford Col, BA, 56; NIH fel & PhD(biochem), Cornell Univ, 64. *Prof Exp:* Res fel biol, Harvard Univ, 67, lectr, 67-68; from asst prof to assoc prof plant sci, 68-83, PROF BIOL, IND UNIV, BLOOMINGTON, 83- *Concurrent Pos:* NIH fel, 65-67. *Mem:* Am Soc Plant Physiol; Am Soc Cell Biol; Phycol Soc Am; Japan Soc Plant Physiol. *Res:* Photosynthetic carbon metabolism and its regulation; biochemical and genetic analysis of photosynthetic mechanisms and its regulation in Chlamydomonas reinhardi, a model eukaryotic photosynthetic organism. *Mailing Add:* Dept Biol Ind Univ Bloomington IN 47405

TOGLIA, JOSEPH U, b Pescopagano, Italy, Apr 24, 27; US citizen; c 3. NEUROLOGY. *Educ:* Liceo Scientifico, Avellino, Italy, BS, 45; Univ Rome, MD, 51. *Prof Exp:* Staff neurologist, Baylor Col Med, 60-63; prof neurol & otorhinol, Sch Med, Temple Univ, 66-74, prof neurol & chmn dept, 74-76; chief neurol, Philadelphia Gen Hosp, 66-70; RETIRED. *Concurrent Pos:* Attend physician, Temple Univ Hosp, 66-; consult, Vet Admin Hosp, 66-88 & NIH, 73-88. *Mem:* AMA; Am Acad Neurol; Am Acad Ophthal & Otolaryngol; Ital Med Asn; Ital Neurol Soc. *Res:* Electronystagmography; clinical vestibular physiology. *Mailing Add:* 1370 Duck Rd Kitty Hawk NC 27949

TOGURI, JAMES M, b Vancouver, BC, Sept 22, 30; m 57; c 5. CHEMISTRY. *Educ:* Univ Toronto, BASc, 55, MASc, 56, PhD(metall), 58. *Prof Exp:* Nat Res Coun Can fel metall, Imp Col Sci & Technol, Univ London, 58-59; fel inorg chem, Tech Univ Norway, 59-61; res assoc chem, Inst Metals, Univ Chicago, 61-62; group leader, Noranda Res Ctr, Que, 62-63, head dept, 63-66; assoc prof, 66-69, chmn dept, 76-81, PROF METALL & MAT SCI, UNIV TORONTO, 69-; CHAIR CHEM PROCESS METALL, INCO/NSERC, 88- *Concurrent Pos:* Royal Norweg Sci Coun fel, 60-61; grants, Nat Res Coun Can & Defence Res Bd Can, 66-; ed-in-chief, Can Metall Quart, 67-86. *Honors & Awards:* Alcan Award, 79; Extractive Metall Sci Award, Am inst Mining, Metall & Petrol Engrs, 81; Silver Medal, Can Inst Mining Metall, 86. *Mem:* Am Inst Mining, Metall & Petrol Engrs; Can Inst Mining & Metall; fel Am Soc Metals; Chem Inst Can. *Res:* High temperature chemistry; thermodynamic properties; kinetics high temperature; fused salt chemistry; nonferrous pyrometallurgy. *Mailing Add:* Dept Metall & Mat Sci Univ Toronto Toronto ON M5S 1A4 Can

TOHLINE, JOEL EDWARD, b Crowley, La, July 15, 53; m 74; c 1. MULTIDIMENSIONAL HYDRODYNAMICS. *Educ:* Centenary Col La, BS, 74; Univ Calif, Santa Cruz, PhD(astron), 78. *Prof Exp:* J W Gibbs instr astron, Yale Univ, 78-80; fel astrophysics, Los Alamos Nat Lab, 80-82; ASST PROF PHYSICS & ASTRON, LA STATE UNIV, 82- *Mem:* Am Astron Soc; Int Astron Union. *Res:* Computer modeling of multidimensional, hydrodynamic flows in astrophysical phenomena; star formation and gas dynamics in galaxies. *Mailing Add:* Dept Physics 255f Nicholson Hall La State Univ Baton Rouge LA 70803-4001

TOHVER, HANNO TIIT, b Tartu, Estonia, Dec 18, 35; Can citizen. PHYSICS. *Educ:* Queen's Univ, Ont, BS, 57, MS, 59; Purdue Univ, PhD, 68. *Prof Exp:* Asst, Purdue Univ, 60-67; asst prof, 68-71, ASSOC PROF PHYSICS, UNIV ALA, BIRMINGHAM, 71- *Mailing Add:* Dept Physics Univ Ala Univ Sta Birmingham AL 35294

TOIDA, SHUNICHI, b Shizuoka, Japan, Jan 8, 37; m 67; c 4. ELECTRICAL ENGINEERING. *Educ:* Univ Tokyo, BS, 59; Univ Ill, Urbana, MS, 66, PhD(elec eng), 69. *Prof Exp:* Comput engr, Mitsubishi Elec Co, 59-63; from asst prof to assoc prof, Systs Design, Univ Waterloo, 69-82; PROF COMPUTER SCI, OLD DOMINION UNIV, 82- *Concurrent Pos:* Nat Res Coun Can grant, 69-71 & 73-81; vis prof, Univ Dortmund, WGer, 75-76. *Mem:* Inst Elec & Electronic Engrs. *Res:* Linear graph theory and its applications; fault diagnosis of logic circuits; automatic reasoning. *Mailing Add:* Dept Computer Sci Old Dominion Univ Hampton Blvd Norfolk VA 23508

TOIVOLA, PERTTI TOIVO KALEVI, b Pori, Finland, Aug 15, 46; US citizen; m 73. MEDICAL PHYSIOLOGY, NEUROENDOCRINOLOGY. *Educ:* Univ Wash, BA, 68, PhD(physiol & biophys), 72. *Prof Exp:* Res asst physiol, Regional Primate Res Ctr, Univ Wash, 68-69, sr fel endocrinol, Dept Med, 73; ASST SCIENTIST, REGIONAL PRIMATE RES CTR, UNIV WIS, 73-, PROF PHYSIOL, SCH MED, 74- *Mem:* Endocrine Soc; Int Soc Neuroendocrinol; Am Physiol Soc. *Res:* Central nervous system regulation of the endocrine system. *Mailing Add:* Dept Lab Med Univ Wash Sch Med SB-10 Seattle WA 98195

TOJO, KAKUJI, b Osaka, Japan. COMPUTER SCIENCE, SOFTWARE SYSTEMS. *Educ:* Univ Osaka, BEng, 69, MEng, 71, DrEng, 75. *Prof Exp:* Asst prof chem eng, Univ Osaka, 71-77 & 79-82 & Kans State Univ, 77-78; asst prof, 83-84, ASSOC PROF PHARM, RUTGERS UNIV, 84- *Mem:* Am Asn Pharmaceut Sci; Am Inst Chem Engrs; Controlled Release Soc; NY Acad Sci; Pharm Soc Japan. *Res:* Controlled drug delivery; transdermal rate-controlled medications and enhanced bioavailability; ocular controlled drug administration; membrane permeability-molecular structure relationships. *Mailing Add:* Dept Pharmaceut Rutgers Univ Col Pharm New Brunswick NJ 08903

TOKAR, MICHAEL, b Elizabeth, NJ, Apr 27, 37; m 61; c 3. CERAMICS, METALLURGY. *Educ:* Univ Mich, BS, 61; Stevens Inst Technol, MS, 64; Rutgers Univ, PhD(ceramics), 68. *Prof Exp:* Metall trainee, Nat Castings Co, Ill, 61-62; res asst res & develop high temperature mat, Am Standard Corp Res Lab, 62-65; res asst ceramic sci, Rutgers Univ, 65-67; staff mem, Los Alamos Sci Lab, Univ Calif, 67-75; sr ceramist, 75-80, sr reactor fuels engr, 80-83, waste mgt mat engr, 83-85, SECT LEADER, US NUCLEAR REGULATORY COMN, 85- *Mem:* Am Ceramics Soc. *Res:* Analysis and evaluation of reactor fuel systems design; fabrication of ceramic nuclear fuels; measurement of physical and mechanical properties of ceramic materials; permanent magnet ferrites; high temperature oxidation-resistant coatings. *Mailing Add:* US Nuclear Regulatory Comn Washington DC 20555

TOKAY, ELBERT, b Brooklyn, NY, May 27, 16; m 40; c 2. BIOLOGY. *Educ:* City Col New York, AB, 36; Univ Chicago, PhD(physiol), 41. *Prof Exp:* Asst physiol, Univ Chicago, 39-41; instr, 41-43, from asst prof to assoc prof, 47-58, prof physiol, 58-77, PROF BIOL, VASSAR COL, 77- *Mem:* AAAS; Sigma Xi. *Res:* Drugs and other factors affecting central nervous potentials and metabolism. *Mailing Add:* Dept Biol Vassar Col Box 475 Poughkeepsie NY 12601

TOKAY, F HARRY, b St Paul, Minn, July 30, 36. AUDIOLOGY. *Educ:* St Cloud State Col, BS, 60; Mich State Univ, MA, 62, PhD(audiol), 66. *Prof Exp:* Asst prof audiol, Cent Mich Univ, 65-66 & Univ Mass, Amherst, 67-74; assoc prof audiol, 74-77, ASSOC PROF COMMUN DIS, UNIV NH, 77-, CHMN COMMUN DIS PROG, 74- *Mem:* Am Speech & Hearing Asn; Acoust Soc Am. *Res:* Audiology with children. *Mailing Add:* Dept Commun Dis Univ NH Durham NH 03824-3501

TOKER, CYRIL, surgical pathology, for more information see previous edition

TOKES, LASZLO GYULA, b Budapest, Hungary, July 7, 37; US citizen. ORGANIC CHEMISTRY. *Educ:* Univ Southern Calif, BA, 61; Stanford Univ, PhD(chem), 65. *Prof Exp:* Nat Ctr Sci Res, France fel, Univ Strasbourg, 65-66; fel, 66-67, dept head spectros, 67-79, ASST DIR ANALYSIS & ENVIRON RES, SYNTEX RES, SYNTEX CORP, 79- *Concurrent Pos:* Chmn award comt, Am Chem Soc, 73-74. *Mem:* Am Chem Soc; Am Soc Mass Spectrometry. *Res:* Mass spectrometric fragmentation mechanisms; isotope labeling studies; structure elucidations by using advanced spectroscopic methods; natural product chemistry with special interest in marine chemistry; environmental contaminants. *Mailing Add:* Analysis & Environ Res Syntex Res 3401 Hillview Ave Palo Alto CA 94304-1320

TOKES, ZOLTAN ANDRAS, b Budapest, Hungary, May 14, 40; m 72; c 2. BIOCHEMISTRY, DEVELOPMENTAL BIOLOGY. *Educ:* Univ Southern Calif, BSc, 64; Calif Inst Technol, PhD(biochem), 70. *Prof Exp:* Lectr biochem, Univ Malaya, 70-71; res immunol, Basel Inst Immunol, Hoffmann-LaRoche, Inc, 71-74; asst prof, 74-80, ASSOC PROF BIOCHEM, UNIV SOUTHERN CALIF & DIR CELL MEMBRANE & CELL CULTURE LABS, UNIV SOUTHERN CALIF CANCER CTR, 80- *Concurrent Pos:* Consult biotechnol. *Mem:* Am Soc Biol Chemists; Am Asn Pathologists; Am Asn Cancer Res; Int Soc Differentiation; Am Soc Cell Biologists. *Res:* Recognition of cell surface changes with differentiation; cell-cell interactions, tumor markers and drug delivery. *Mailing Add:* 2405 Kenilworth Ave Los Angeles CA 90039

TOKITA, NOBORU, b Sapporo, Japan, Feb 20, 23; m 53. PHYSICS. *Educ:* Hokkaido Univ, BS, 47, DrSci, 56. *Prof Exp:* Res mem, Kobayashi Inst Physics, Japan, 47-52; asst prof physics, Waseda Univ, Japan, 52-57; res assoc polymer sci, Duke Univ, 57-60; sr res physicist, 60-68, RES ASSOC RES CTR, UNIROYAL INC, 68- *Mem:* Am Chem Soc; Soc Rheology. *Res:* Polymer physics; vibration and sound; rheology of elastomer and plastics; tire technology. *Mailing Add:* Uniroyal Inc Res Ctr Middlebury CT 06749

TOKOLI, EMERY G, b Budapest, Hungary, June 6, 23, nat US; m 48; c 1. ORGANIC CHEMISTRY. *Educ:* E-tv-s Lorand Univ, Budapest, MS, 47. *Prof Exp:* Res chemist, Chinoin Co Ltd, Hungary, 45-47; dir res, Fine Orgs, Inc, NJ, 48-60; sr res chemist, Minn Mining & Mfg, 60-64; res scientist, Union-Camp Paper Corp, 64-66; SCIENTIST, RES & ENG DIV, XEROX CORP, 66- *Concurrent Pos:* Instr, Fairleigh Dickinson, 58-59. *Mem:* Am Chem Soc; fel Am Inst Chem. *Res:* Nucleophilic substitution reactions; Friedel Crafts alkylations and acylations; thermally stable polymers and intermediates; fluorocarbon and silicone chemistry; lignin chemistry; organic photoconductors; redox polymers. *Mailing Add:* 92 Torrington Dr Rochester NY 14618

TOKSOZ, MEHMET NAFI, b Antakya, Turkey, Apr 18, 34. GEOPHYSICS. *Educ:* Colo Sch Mines, GpE, 58; Calif Inst Technol, MS, 60, PhD(geophys, elec eng), 63. *Prof Exp:* Res fel geophys, Calif Inst Technol, 63-65; from asst prof to assoc prof, 65-71, PROF GEOPHYS, MASS INST TECHNOL, 71-, DIR, GEORGE WALLACE JR GEOPHYS OBSERV, DEPT EARTH & PLANETARY SCI, 75-, DIR, EARTH RESOURCES LAB, 82- *Mem:* AAAS; Am Geophys Union; Seismol Soc Am; Soc Explor Geophys; Soc Prof Well Log Analysts; Geol Soc Am; Am Asn Petrol Geologists; Royal Astron Soc. *Res:* Seismology; borehole geophysics; well logging; rock physics. *Mailing Add:* Earth Resources Lab Mass Inst Technol E34-440 Cambridge MA 02139

TOKUDA, SEI, b Ewa, Hawaii, Aug 17, 30; m 55; c 3. IMMUNOLOGY, MICROBIOLOGY. *Educ:* Univ Hawaii, BS, 53; Univ Wash, PhD(microbiol), 60. *Prof Exp:* Trainee microbiol, Univ Wash, 59-60, res instr immunol, 62-63; asst prof immunol & microbiol, Univ Vt, 63-66; from asst prof to assoc prof, 66-74, PROF IMMUNOL, SCH MED, UNIV NMEX, 74- *Concurrent Pos:* Res fel immunochem, Calif Inst Technol, 60-62; NIH fel, 61-62, NIH res grant, 64-; Am Cancer Soc res grant, 64; USPHS career develop award, 68-73; regional ed, Life Sci, 68-73; assoc ed, J Immunol, 79-81; mem Cancer Res Manpower rev comt, Nat Cancer Inst, 76-80. *Mem:* AAAS; Am Asn Immunologists; Am Soc Microbiol; Transplantation Soc; Sigma Xi. *Res:* Immune responses of mice against syngeneic tumors; transplantation immunity; hormonal regulation of the immune response; neuroimmuno endocrinology. *Mailing Add:* Dept Microbiol Univ NMex Sch Med Albuquerque NM 87131

TOKUHATA, GEORGE K, b Matsue, Japan, Aug 25, 24; US citizen; m 49. EPIDEMIOLOGY, PUBLIC HEALTH. *Educ:* Keio Univ, Japan, BA, 50; Miami Univ, MA, 53; Univ Iowa, PhD(behav sci), 56; Johns Hopkins Univ, DPH(epidemiol), 62. *Prof Exp:* Res assoc ment health, Mich State Dept Ment Health, 56-59; spec asst div chief, Div Chronic Dis, USPHS, 59-62, prin epidemiologist, 62-63; assoc prof prev med, Col Med, Univ Tenn & chief epidemiol, St Jude Children's Res Hosp, 63-67; RES DIR, DIV RES &

BIOSTATIST, PA STATE DEPT HEALTH, 67- *Concurrent Pos:* USPHS fel, Med Ctr, Johns Hopkins Univ, 59-60, NIH fel, 60-61; Mary Reynold Babcock Found res grant, 59-61; Food & Drug Admin res grant; US Consumer Prod Safety Comn grant, 73-74; Maternal & Child Health Serv grant, 74-; Nat Ctr Health Statist grant, 74-; prof epidemiol & biostatist, Grad Sch Pub Health, Univ Pittsburgh; assoc prof community med, Col Med, Temple Univ. *Mem:* Fel Am Pub Health Asn; fel Am Sociol Asn; fel Am Col Epidemiol. *Res:* Genetics aspects of cancer and other chronic diseases in adults and children; epidemiology of chronic diseases; environ health evaluation of public health programs research and development. *Mailing Add:* 410 Rupley Rd Camp Hill PA 17011

TOKUHIRO, TADASHI, physical chemistry, chemical physics, for more information see previous edition

TOKUNAGA, ALAN TAKASHI, b Puunene, Hawaii, Dec 17, 49; m 75. PLANETARY SCIENCE, INTERSTELLAR MEDIUM. *Educ:* Pomona Col, BS, 71; State Univ NY, Stony Brook, MS, 73, PhD(astron), 76. *Prof Exp:* Res assoc, Ames Res Ctr, NASA, 76-77; res assoc, Steward Observ, Univ Ariz, 77-79; asst astronr, 79-83, assoc astronr, 83-90, ASTRONR, UNIV HAWAII, 90- *Concurrent Pos:* Mem Kuiper Airborne Obs Time Allocation Comt, 88-90; mem planetary & optical panels, Astron & Astrophys Surv Comt, 89-90; mem div Planetary Sci Comt, Am Astron Soc, 90- *Mem:* Am Astron Soc; Astron Soc Pac. *Res:* Spectroscopy of comets, planetary satellites and interstellar medium; construction of ground-based infrared instrumentation. *Mailing Add:* Inst Astron 2680 Woodlawn Dr Honolulu HI 96822

TOKUYASU, KIYOTERU, b Nagasaki, Japan, Oct 16, 25; US citizen; m 57; c 2. CELL BIOLOGY, DEVELOPMENTAL BIOLOGY. *Educ:* Kyushu Univ, Japan, BS, 49, PhD(med sci), 57. *Prof Exp:* Assoc prof histol, Med Sch, Kurume Univ, Japan, 56-58; chief designer electron microscope, Hitachi Naka Works, Japan, 58-63; assoc res pathologist, 64-69, assoc res biologist, 69-73, res biologist, 73-77, PROF-IN-RESIDENCE CELL BIOL, UNIV CALIF, SAN DIEGO, 77- *Mem:* Am Soc Cell Biol; Electron Micros Soc Am; Histochem Soc Am; AAAS; NY Acad Sci. *Res:* Developmental, cell biological and molecular biological aspects of embryonic heart development; immunocytochemistry and autoradiography at the light and electron microscopic levels. *Mailing Add:* Dept Biol B-022 Univ Calif La Jolla CA 92093

TOLBERT, BERT MILLS, b Twin Falls, Idaho, Jan 15, 21; m 59; c 4. BIOCHEMISTRY, NUTRITION. *Educ:* Univ Calif, Berkeley, BS, 42, PhD(chem), 45. *Prof Exp:* Teaching asst, Univ Calif, Berkeley, 42-44, res chemist, Lawrence Radiation Lab, 44-57; assoc prof, 57-61, PROF CHEM, UNIV COLO, 61- *Concurrent Pos:* USPHS fel, 52-53; Int Atomic Energy Agency vis prof, Univ Buenos Aires, 62-63; biophysicist, US AEC, Washington, DC, 67-68; vis staff, Los Alamos Sci Lab, 70-80; consult, Surgeon Gen Off, US Army. *Mem:* Am Chem Soc; Am Soc Biochem & Molecular Biol; Am Inst Nutrit; Radiation Res Soc; Soc Exp Biol Med; AAAS. *Res:* Metabolism and function of ascorbic acid; radiation chemistry of proteins; catabolism of labeled compounds; application of C-14 and H-3 to biochemistry; instrumentation in radiochemistry; synthesis of labeled compounds; use of stable isotopes. *Mailing Add:* 444 Kalmia Ave Boulder CO 80304

TOLBERT, CHARLES RAY, b Van, WVa, Nov 14, 36; m 67; c 3. ASTRONOMY, EDUCATION. *Educ:* Univ Richmond, BS, 58; Vanderbilt Univ, MS, 60, PhD(physics, astron), 63. *Prof Exp:* Res assoc, Kapteyn Astron Lab, Netherlands, 63-67; res assoc ctr advan studies, 67-69, asst prof, 69-70, ASSOC PROF, UNIV VA, 70- *Concurrent Pos:* Educ officer, Am Astron Soc, 85-91. *Mem:* Am Astron Soc; Int Astron Union; Int Union Radio Sci; fel AAAS; Soc Sci Explor; Sigma Xi. *Res:* Photoelectric photometry; 21 centimeter radio-astronomical studies. *Mailing Add:* Box 3818 University Sta Univ Va Charlottesville VA 22903

TOLBERT, DANIEL LEE, b Fairview Heights, Ill, Oct 16, 46; m 72; c 2. NEUROSCIENCES. *Educ:* Quincy Col, BS, 68; St Louis Univ, MS, 72, PhD, 75. *Prof Exp:* Fel, Dept Neurosurg, Univ Minn, 75-78; from asst prof to assoc prof, 78-88, PROF, DEPT ANAT, ST LOUIS UNIV, 88-, ASST PROF, DEPT SURG, 78-, DIR, MURPHY NEUROANAT RES LAB, 78- *Mem:* Am Asn Anatomists; Soc Neurosci. *Res:* Anatomical and neurophysiological study of the development of the brain. *Mailing Add:* Dept Anat Sch Med St Louis Univ 1402 S Grand Blvd St Louis MO 63104

TOLBERT, GENE EDWARD, economic geology; deceased, see previous edition for last biography

TOLBERT, LAREN MALCOLM, b New Orleans, La, Sept 30, 49; m 68; c 4. ORGANIC SYNTHESIS, ORGANIC ELECTRONIC MATERIAL. *Educ:* Tulane Univ, BA, 70; Univ Wis, Madison, PhD(org chem), 75. *Prof Exp:* Res fel, Harvard Univ, 75-76; asst prof chem, Univ Ky, 76-81, assoc prof, 81-85; PROF CHEM, GA INST TECHNOL, 85. *Mem:* Am Chem Soc; AAAS. *Res:* Organic photochemistry, particularly of anions; photodehalogenation and radicalnucleophile interactions; organic polymer chemistry. *Mailing Add:* Sch Chem Ga Inst Technol Atlanta GA 30332

TOLBERT, MARGARET ELLEN MAYO, b Suffolk, Va, Nov 24, 43; c 1. BIOCHEMISTRY, ANALYTICAL CHEMISTRY. *Educ:* Tuskegee Inst, BS, 67; Wayne State Univ, MS, 68; Brown Univ, PhD(biochem), 74. *Prof Exp:* Instr, Opportunities Industrialization Ctr, 71-72; instr math, Tuskegee Inst, 69-70, res technician biochem, 69, asst prof chem, 73-76; assoc prof pharmaceut chem & assoc dean, Sch Pharm, Fla A&M Univ, 77-78; prof chem & dir, Assoc Provost Res & Develop, Carver Res Found, Tuskegee Univ, 79-87; ADM STAFF MEM, BUDGETS & CONTROL GROUP, BP AM INC, R&D CTR, 87- *Concurrent Pos:* NIH fel, 78-79; vis res scientist, Int Inst Cellular & Molecular Path, Cath Univ Louvain, Brussels, Belg & Brown Univ, RI; partic, Biomed Summer Inst, Lawrence Livermore Lab, 74. *Honors & Awards:* Award Sci & Med; Carver Res Asn Cert, Carver Res Found Fel. *Mem:* Sigma Xi; Am Chem Soc; Orgn Black Scientists; fel AAAS; Am Asn Univ Women; Int Develop Conf. *Res:* Metabolic studies involving isolated rat hepatic cells. *Mailing Add:* NSF 1800 G St NW Rm 1225 Washington DC 20550

TOLBERT, NATHAN EDWARD, b Twin Falls, Idaho, May 19, 19; m 52; c 3. BIOCHEMISTRY, PLANT SCIENCES. *Educ:* Univ Calif, BS, 41; Univ Wis, MS, 48, PhD(biochem), 50. *Prof Exp:* Asst chem dept viticult, Col Agr, Univ Calif, 41-43, biochemist, Radiation Lab, 50; res admin, US AEC, 50-52; sr biochemist, Oak Ridge Nat Lab, 52-58; PROF BIOCHEM, MICH STATE UNIV, 58- *Mem:* Nat Acad Sci; Am Chem Soc; Am Soc Biol Chem; Am Soc Plant Physiol. *Res:* Plant biochemistry and plant growth substances; glycolic acid metabolism, biosynthesis and function; photosynthesis and relation to plant growth; microbodies and peroxisomes. *Mailing Add:* Dept Biochem Mich State Univ East Lansing MI 48823

TOLBERT, ROBERT JOHN, b Pelican Rapids, Minn, Apr 16, 28; m 53; c 2. PLANT ANATOMY. *Educ:* Moorhead State Univ, BS & BA, 55; Rutgers Univ, PhD(bot), 59. *Prof Exp:* From asst prof to assoc prof biol, Univ WVa, 59-63; assoc prof, 63-65, PROF BIOL, MOORHEAD STATE UNIV, 65- *Mem:* AAAS; Bot Soc Am; Sigma Xi. *Res:* Anatomical investigation of vegetative shoot apices; comparative studies of shoot apices in the order Malvales. *Mailing Add:* Moorhead State Univ Moorhead MN 56560

TOLBERT, THOMAS WARREN, b Greenwood, SC, Dec 1, 45; m 78; c 1. PHYSICAL CHEMISTRY, MOLECULAR SPECTROSCOPY. *Educ:* Wofford Col, BS, 67; State Univ NY Binghamton, PhD(chem), 74. *Prof Exp:* Asst prof chem, Wofford Col, 72-74; proj scientist, 74-75, lab supvr, Parke-Davis Med Surg Div, 75-77, proj mgr res & develop, Parke Davis Deseret Div, 77-81; dir res & develop, Flexible Tubing Div, Automation Industs, 81-85; proj mgr res & develop, Springs Industs, 85-88, DIR, NEW VENTURES PRODS, RES & DEVELOP, 89-, DIR, AUTOMATION GROUP. *Mem:* Am Chem Soc. *Res:* Interaction characteristics of reacting molecules; industrial and manfacturing engineering; issued four patents. *Mailing Add:* 1823 Appletree Lane Ft Mill SC 29715

TOLBERT, VIRGINIA ROSE, b Scottsboro, Ala, July 16, 48; m; c 1. AQUATIC ECOLOGY, HYDROLOGY & WATER RESOURCES. *Educ:* ETenn State Univ, BS, 70; Univ Tenn, Knoxville, MS, 72, PhD(ecol), 78. *Prof Exp:* Fel, Dept Zool, Univ Tenn, Knoxville, 78-79; res assoc, 79-85, RES STAFF, ENVIRON SCI DIV, OAK RIDGE NAT LAB, TENN, 85- *Mem:* Ecol Soc Am; NAm Benthological Soc (treas, 88); Am Soc Surface Mining & Reclamation; Sigma Xi. *Res:* Effects of coal surface mining on aquatic communities; examination of the various aspects of pertarbation on aquatic systems; effects of energy related development on water and aquatic biota; assessment of the effects of chemical nerve agents on aquatic biota. *Mailing Add:* Rte 3 Box 77-1 Lenoir City TN 37771

TOLDERLUND, DOUGLAS STANLEY, b Newport, RI, Jan 14, 39; m 61; c 2. MARINE ECOLOGY, GLACIAL GEOLOGY. *Educ:* Brown Univ, BA, 60; Columbia Univ, PhD(marine geol), 69. *Prof Exp:* Sr ecologist, Raytheon Co, 69-70; assoc prof marine sci, USCG Acad, 70-77, chmn, Marine Sci Sect, 78-83, head dept sci, 83-91, PROF MARINE SCI, USCG ACAD, 77- *Mem:* Am Geophys Union; Am Fisheries Soc; Int Oceanog Found; Sigma Xi. *Res:* Estuarine ecology, especially water quality and finfish studies. *Mailing Add:* Dept Sci USCG Acad Mohegan Ave New London CT 06320

TOLE, JOHN ROY, b Washington, DC, Nov 6, 45; m 69; c 2. BIOENGINEERING. *Educ:* Drexel Inst Technol, BSEE, 68; Mass Inst Technol, SM, 70, ScD, 76. *Prof Exp:* Engr, Div Res Serv, NIH, 68; res asst, Mass Inst Technol, 68-70; lab supvr, Peter Bent Brigham Hosp, Boston, 70-72; NIH trainee, Mass Inst Technol, 72-76, res scientist, 76-81; ASSOC PROF BIOMED ENG, WORCESTER POLYTECH INST, 81- *Concurrent Pos:* Res affil, Mass Inst Technol, 81- *Mem:* Inst Elec & Electronics Engrs; Human Factors Soc. *Res:* Computer based medical instrumentation; neurological function testing; oculomotor system dynamics; man-machine interaction particularly in aviation and rehabilitation; stress physiology. *Mailing Add:* Stanford Telecommun 1761 Business Center Dr Reston VA 22090

TOLEDO, DOMINGO, b Hato Rey, PR, Apr 5, 45. DIFFERENTIAL GEOMETRY, ALGEBRAIC GEOMETRY. *Educ:* Stanford Univ, BS, 67; Cornell Univ, PhD(math), 72. *Prof Exp:* Mem, Inst Adv Studies, 72-74; J F Ritt asst prof math, Columbia Univ, 74-78; from asst prof to assoc prof, 78-86, PROF MATH, UNIV UTAH, 86- *Concurrent Pos:* Vis fel, Univ Warwick, 76; visitor, Inst Higher Sci Studies, 77 & 88-; Alfred P Sloan fel, 82-84; vis prof, Univ Bonn, 83-84. *Mem:* Am Math Soc. *Res:* Characteristic classes and coherent sheaves on complex manifolds; application of harmonic mappings to the geometry of Kiahler manifold and locally symmetric spaces. *Mailing Add:* Univ Utah Salt Lake City UT 84112

TOLEDO, ROMEO TRANCE, b Philippines, Apr 27, 41. CHEMICAL ENGINEERING, FOOD SCIENCE. *Educ:* St Augustine Univ, Philippines, BSChE, 60; Univ Ill, MS, 65, PhD(food sci), 67. *Prof Exp:* Chem engr, Calif Packing Corp, 60-62; instr chem eng, St Augustine Univ, Philippines, 62-63; res asst, Univ Ill, 63-65, teaching asst food sci, 65-67; res chem engr, Libby McNeill & Libby, 67-68; asst prof food sci, 68-76, assoc prof, 76-79, PROF FOOD SCI, UNIV GA, 79- *Concurrent Pos:* Consult, Mead Packaging Corp, 70-76. *Mem:* Assoc mem Am Inst Chem Engrs; Inst Food Technol; Am Soc Agr Engrs. *Res:* Engineering food processing and handling systems, heat transfer in food processing systems; rheological properties of food fluids and kinetics of food degradation and microbiological inactivation. *Mailing Add:* Dept Food Sci Univ Ga Athens GA 30601

TOLEDO-PEREYRA, LUIS HORACIO, b Nogales, Ariz, Oct 19, 43; m 74; c 2. BIOLOGICAL SCIENCES, MEDICINE. *Educ:* Colegio Regis, BS, 60; Univ Mex, MD, 67, MS, 70; Univ Minn, PhD(surg), 76, PhD(hist med), 84. *Prof Exp:* Dir surg res & co-dir transplantation, Henry Ford Hosp, 77-79; dir surg res, 79-84, CHIEF TRANSPLANTATION, MT CARMEL MERCY HOSP, 79-, DIR RES, 84- *Concurrent Pos:* Instr biochem, Univ Mex, 63, clin instr internal med, 68; adj prof, Sch Health Sci, Mercy Col Detroit, 83-; consult transplantation, Hutzel Hosp, 84- *Honors & Awards:* Resident Res Award, Asn Acad Surg, 74; Cecile Lehman Mayer Res Award, Am Col Chest Physicians, 75. *Mem:* NAm Soc Dialysis & Transplantation (pres, 82-84); Asn Acad Surg; Am Soc Transplant Surgeons; Am Diabetes Asn; Am Asn Immunologists. *Res:* Organ preservation and perfusion; modification of immune response by cyclosporine and lectins; islet cell transplantation; pancreas transplantation; new techniques to improve islet cell transplantation for diabetics; new ways of organ preservation for transplantation. *Mailing Add:* Mt Carmel Mercy Hosp 6071 W Outer Dr Detroit MI 48235

TOLER, J C, b Carthage, Ark, Jan 31, 36; c 2. ELECTRICAL ENGINEERING. *Educ:* Univ Ark, BSEE, 57; Ga Inst Technol, MSEE, 70. *Prof Exp:* DIR BIOENG CTR, 84-, CTR REHAB TECHNOL, 87-, PRIN RES ENGR, GA INST TECHNOL, 66- *Mem:* Fel Inst Elec & Electronics Engrs; Bioelectromagnetics Soc. *Res:* Interaction of electromagnetic waves with biological systems. *Mailing Add:* GIT/OIP 225 North Ave NW Centennial Bldg Rm 329 Atlanta GA 30332

TOLER, ROBERT WILLIAM, b Norphlet, Ark, Dec 15, 28; m 50; c 4. VIROLOGY, PLANT PATHOLOGY. *Educ:* Univ Ark, BS, 50, MS, 58; NC State Univ, PhD(plant virol), 62. *Prof Exp:* Res technician rice br exp sta, Univ Ark, 51-54, specialist & plant pathologist, Agr Mission, Panama, 55-57; res plant pathologist, coastal plain exp sta, Agr Res Serv, USDA, Univ Ga, 61-65; assoc prof, 69-74, PROF PLANT PATH, TEX A&M UNIV, 74-, CEREAL VIROLOGIST, 66- *Concurrent Pos:* Consult, Foy Pittman Rice Farms, Ark, 50-51, Campos Manola Arca, SA, Manziuillo, Cuba, 54 & Ford Found, Antonia Narro Col Agr, Coahuila, Mex, 66; dir plant protection lab, Remote Sensing Ctr, Tex A&M Univ, 71- *Mem:* Am Phytopath Soc; Sigma Xi. *Res:* Physiological effects and host response in plant pathology; cereal virology identification and transmission of viruses that cause cereal diseases. *Mailing Add:* Rte 5 Box 880 College Station TX 77845

TOLGYESI, EVA, b Budapest, Hungary. ORGANIC CHEMISTRY, POLYMER CHEMISTRY. *Educ:* Budapest Technol Univ, BSc, 53; Univ Leeds, PhD(textile chem), 59. *Prof Exp:* Asst prof chem, Budapest Technol Univ, 53-56; sr chemist, Harris Res Labs, 65-69, proj supvr polymer chem, 69-74, res supvr, 74-77, GROUP LEADER, GILLETTE RES INST, 77- *Mem:* Am Chem Soc; Soc Cosmetic Chemists; AAAS. *Res:* Keratin chemistry; chemical modification of wool; moth-proofing; cationic surfactants; fluoropolymers; silicones; hair cosmetics; hair removal; controlled release polymer systems. *Mailing Add:* 803 Tulane Pl Rockville MD 20850-1142

TOLIN, SUE ANN, b Montezuma, Ind, Nov 29, 38. PLANT VIROLOGY, PHYTOPATHOLOGY. *Educ:* Purdue Univ, BS, 60; Univ Nebr, MS, 62, PhD(bot), 65. *Prof Exp:* Res asst plant path, Univ Nebr, 60-65; res assoc bot & plant path, Purdue Univ, 65-66; from asst prof to assoc prof, 66-83, PROF PLANT PATH, VA POLYTECH INST & STATE UNIV, 83- *Concurrent Pos:* USDA rep, NIH Recombinant DNA Adv Comt, 79; spec consult, Coop State Res Serv, US Dept Agr, 79. *Mem:* Am Phytopath Soc; Am Soc Microbiol; Am Soc Virol; AAAS. *Res:* Identification, purification and molecular characterization of plant pathogenic viruses; electron microscopy; molecular genetic mechanisms of resistance of plants to viruses; biotechnology science policy. *Mailing Add:* Dept Plant Path Physiol & Weed Sci Va Polytech Inst & State Univ Blacksburg VA 24061-0330

TOLINE, FRANCIS RAYMOND, b Alliance, Nebr, Nov 3, 18; m 45; c 5. AEROSPACE ENGINEERING, NUCLEAR ENGINEERING. *Educ:* US Naval Postgrad Sch, BS, 52; Mass Inst Technol, SM, 53. *Prof Exp:* Br head air launched missile propulsion, Bur Naval Weapons, 59-60; from assoc prof to emer prof aerospace & nuclear eng, Tenn Technol, Univ 60-88; RETIRED. *Concurrent Pos:* NSF sci fac fel, 64-67. *Res:* Altitude simulation for environmental testing of nuclear rocket engines; ejector-diffuser systems used in altitude simulation. *Mailing Add:* 1856 E Spring St Cookeville TN 38501

TOLIVER, ADOLPHUS P, b St Louis, Mo, Apr 9, 31. DNA REPLICATION, REGULATION GENE EXPRESSION. *Educ:* Purdue Univ, PhD(molecular biol & biochem), 68. *Prof Exp:* Asst prof biochem, Univ Calif, Davis, 70-75; EXEC SECY, BIOCHEM STUDY SECT, DIV RES GRANTS, NIH, 75- *Mem:* Sigma Xi; Am Soc Biol Chemists. *Mailing Add:* Biochem Study Sect Div Res Grants NIH Westwood Bldg 318 Bethesda MD 20892

TOLIVER, MICHAEL EDWARD, b Albuquerque, NMex, Oct 1, 49; m 80. ENTOMOLOGY. *Educ:* Univ NMex, BS, 73; Univ Ill, MS, 77, PhD(entom), 79. *Prof Exp:* Entomologist, City Urbana, 76; res asst, Univ Ill, 77-79, asst prof entom, 80; field supvr, Macon Mosquito Abatement Dist, 81; ASSOC PROF BIOL, EUREKA COL, 81- *Concurrent Pos:* Mem, Lepidopter Res Found. *Mem:* Lepidopterists Soc; AAAS. *Res:* Evolution of mimetic color patterns; biogeography of southwestern lepidopter. *Mailing Add:* Div Sci & Math Eureka Col Eureka IL 61530

TOLK, NORMAN HENRY, b Idaho Falls, Idaho, Jan 9, 38; m 61; c 5. SPACE PHYSICS. *Educ:* Harvard Col, AB, 60; Columbia Univ, PhD(physics), 66. *Prof Exp:* Grad res asst, Radiation Lab, Dept Physics, Columbia Univ, 60-61, & 63-66, teaching asst, 61-63, res physicist, 66-67, lectr & staff mem, 67-68, adj asst prof physics, 68-69; mem tech staff, AT&T Bell Lab, 68-83, Bell Commun Res, 84; PROF PHYSICS, DEPT PHYSICS & ASTRON, VANDERBILT UNIV, 84-, DIR, VANDERBILT CTR ATOMIC & MOLECULAR PHYSIC SURFACE, 87- *Concurrent Pos:* Harvard Nat Fel & Gen Motors Fel, 56-60; Brattle Fel, Harvard Univ, 57; vis univ fel, Australian Nat Univ, Canberra, 75; vis scientist, Max Planck Inst Plasmaphysics, WGer, 78; NAm ed, Int J, Radiation & Solids; mem, div electron & atomic physics prog comt, Am Phys Soc, 76-77, int comt, Int Workshops on Inelastic Ion Surface Collisions, 82, Int Workshops Desorption Induced Electronics Transitions, adv bd, Int J Radiation Effects, exec comt, Vanderbilt Free Electron Laser Proj, exect comt, dept physics & astron, Vanderbilt Univ, 85-, comt atomic & molecular sci, Nat Res Coun, 84-; chmn, Int Workshop Desorption Induced Electronics Transitions, Williamsburg, 82, Int Conf Atomic Collision Solids, Washington, DC, 85, development comt & dir search comt, Free Electron Laser, patent comt, Vanderbilt Univ, 87-; consult, S-cubed Corp, Acurex Corp & Chem Div, Los Alamos Nat Lab, 86-, Lockheed Corp, 87; Eaton Corp, 87- Inst Defense Anal, 88-; US rep int comt, Int Conf Atomic Collisions Solids; thesis supvr, Mass Inst Technol, Univ Wis, Columbia & Vanderbilt Univ; invited speaker, Symp Surface Modification Directed Deposition Energy, Am Vacuum Soc, Physic Colloquium, Free Univ Berlin, Tech Univ Vienna, physics seminar, Univ Osnabruck, WGer, London Univ, Am Phys Soc Plasma Physics Meeting, Boston & Am Optical Soc Conf Luminescence, Madison, 84, Phys Chem Colloquium, Univ Houston, Rice Univ, Smithsonian Astrophys Seminar, Harvard Univ, Second Spacecraft Glow Workshop, Chem Seminar, Los Alamos Nat Lab, 86, Physics Colloquium, Univ Ark, 87, Inst Defense Anal, Washington, DC, Fourth Int Symp Resonance Ionization Spectors, Mass, 88 & other various seminars, symp & lect from different univ, inst & soc, 84-88. *Honors & Awards:* Alexander von Humboldt Sr Sci Award, 87. *Mem:* Fel Am Phys Soc; Am Vacuum Soc; Am Optical Soc. *Res:* Desorption induced by electronic transitions, processes associated with electron and photon bombardment on surfaces; interaction of neutral atoms, in particular, atomic oxygen with variety of surface; atomic and molecular collisions on surfaces; author of over 100 publications on optical physics and chemistry. *Mailing Add:* Dept Physics & Astron Univ Vanderbilt Box 1807 Sta B Nashville TN 37235

TOLL, JOHN SAMPSON, b Denver, Colo, Oct 25, 23; m 70; c 2. THEORETICAL PHYSICS, ELEMENTARY PARTICLE PHYSICS. *Educ:* Yale Univ, BS, 44; Princeton Univ, AM, 48, PhD(physics), 52. *Hon Degrees:* DSc, Univ Md, 73 & Univ Wroclaw, 75; LLD, Adelphi Univ, 78; Doctorate, Fudam Univ, China, 87; DHL, State Univ NY Stony Brook, 90. *Prof Exp:* Managing ed & actg chmn, Yale Sci Mag, 43-44; asst proctor fel, Princeton Univ, 46-49; theoret physicist, Los Alamos Sci Lab, 50-51; staff mem & assoc dir, Proj Matterhorn, Forrestal Res Ctr, Princeton Univ, 51-53; prof physics & chmn dept physics & astron, Univ Md, 53-65; prof physics & pres, State Univ NY Stony Brook, 65-78; pres & prof physics, 78-88, chancellor, 88-89, EMER CHANCELLOR & PROF PHYSICS, UNIV MD, 89-; PRES, UNIVS RES ASN, 89- *Concurrent Pos:* Guggenheim Mem Found fel, Inst Theoret Physics, Univ Copenhagen & Univ Lund, 58-59; US deleg & head sci secretariat, Int Conf High Energy Physics, 60; mem-at-large US nat comn, Int Union Pure & Appl Physics, 61-63; mem, Gov Adv Comt Atomic Energy, State of NY, 66-70; Nordita vis prof, Niels Bohr Inst Theoret Physics, Univ Copenhagen, 75-76; mem univ progs panel energy res bd, Dept Energy, 82-83, SBHE Adv Com, 83-89; bd dirs, Am Coun Educ, 86-89, bd trustees Aspen Inst Humanities, 87 & 89; chmn adv panel, tech risks & opportunities for US energy supply & demand, US Off Tech Assessment, 87, chmn adv panel, Int Collab in Defense Tech, US Off Tech Assessment, 89- *Mem:* AAAS; fel Am Phys Soc; Am Asn Physics Teachers; Nat Sci Teachers Asn; Fedn Am Scientists; Sigma Xi; NY Acad Sci; Philos Soc; Asn Higher Soc. *Res:* Elementary particle theory; scattering. *Mailing Add:* Univs Res Asn 1111 19th St NW Suite 400 Washington DC 20036

TOLLE, JON WRIGHT, b Mattoon, Ill, June 26, 39; m 64. MATHEMATICS, OPERATIONS RESEARCH. *Educ:* DePauw Univ, BA, 61; Univ Minn, PhD(math), 66. *Prof Exp:* Res assoc, Argonne Nat Lab, 63; instr math, Univ Minn, 66-67; asst prof, 67-73, assoc prof math & opers res, 73-78, chmn curric opers res, 74-79, PROF MATH & OPERS RES, UNIV NC, CHAPEL HILL, 78- *Concurrent Pos:* Vis prof, Grad Sch Bus, Univ Chicago, 75 & 81-82. *Mem:* Opers Res Soc Am; Soc Indust & Appl Math; Math Asn Am. *Res:* Mathematical programming; optimization theory; numerical analysis. *Mailing Add:* Dept Math Univ NC Chapel Hill NC 27599-3902

TOLLEFSON, CHARLES IVAR, b Moose Jaw, Sask, Oct 2, 18; m 40; c 2. BIOCHEMISTRY. *Educ:* Univ Sask, BSA, 40, MSc, 47; Univ Minn, PhD(agr biochem), 50. *Prof Exp:* Asst radioactive ruthenium, Univ Sask, 45-47; asst lactose, Univ Minn, 47-50; res biochemist, Stine Lab, E I du Pont de Nemours & Co, 50-57; res sect leader nutrit, 57-72, mgr res & develop, 72-83, CONSULT COMPUT PROGS, R T FRENCH CO, 84- *Mem:* AAAS; Am Chem Soc; Inst Food Technologists; NY Acad Sci. *Res:* Selenium in grains; adsorption and solvent distribution of ruthenium; nutritional effects of lactose; new growth factors; nutrition of cage birds; food dehydration. *Mailing Add:* 3655 E Elswood Dr Idaho Falls ID 83402-4736

TOLLEFSON, ERIC LARS, b Moose Jaw, Sask, Oct 15, 21; m 47; c 3. PHYSICAL CHEMISTRY, CHEMICAL ENGINEERING. *Educ:* Univ Sask, BA, 43, MA, 45; Univ Toronto, PhD(phys chem), 48. *Prof Exp:* Demonstr chem, Univ Sask, 41-43 & 45, asst, Directorate Chem Warfare, 43-44; jr res officer, Nat Res Coun Can, 45; lab asst, Univ Toronto, 45-47, lectr, 47-48; jr res officer, Nat Res Coun Can, 48-49, asst res officer, 50-51; chemist, Process Res Div, Stanolind Oil & Gas Co, Okla, 51-52, sr chemist, 53-56; head phys chem sect, Res Dept, Can Chem Co, Ltd, 56-66, supt chem develop dept, 65-66, tech mgr, Can Chem Co Div, Chem-Cell Ltd, Alta, 66-67; assoc prof chem eng, 67-70, actg head dept, 71, head dept, 72-81, PROF CHEM ENG, UNIV CALGARY, 70-, ALTA GIL SANDS TECHNOL & RES PROF, 82- *Mem:* Am Chem Soc; fel Chem Inst Can; Air Pollution Control Asn; Asn Prof Engrs. *Res:* Preparation of activated carbon from Alberta coals; recovery of hydrocarbons from aqueous wastes from oil sands; bitumen recovery operations; kinetics; atomic hydrogen with acetylene; oxidation of ethylene; Fischer-Tropsch synthesis; alcohol dehydrogenation; reduction of nitrogen oxides in stack gases; oxidation of low concentrations of hydrogen sulfide over activated carbon; biological oxidation, difficiency of waste gas flares. *Mailing Add:* Dept Chem & Petrol Eng Univ Calgary 2500 Univ Dr NW Calgary AB T2N 1N4 Can

TOLLEFSON, JEFFREY L, b Hampa, Idaho, July 30, 42; m 65; c 2. MATHEMATICS. *Educ:* Univ Idaho, BS, 65; Mich State Univ, MS, 66, PhD(math), 68. *Prof Exp:* NASA trainee, Mich State Univ, 65-68; asst prof math, Tulane Univ, 68-71 & Tex A&M Univ, 71-74; assoc prof, 74-77, PROF MATH, UNIV CONN, 77-, DEPT HEAD, 85- *Mem:* Am Math Soc. *Res:* Topology of manifolds. *Mailing Add:* Dept Math Univ Conn Storrs CT 06269

TOLLER, GARY NEIL, b 1950; m 87; c 1. GALACTIC RADIATION FIELD. *Educ:* Dickinson Col, BS, 72; State Univ NY, Stony Brook, PhD(astron), 81. *Prof Exp:* Astronr, Space Astron Lab, Univ Fla, 81-85; CONSULT, INT ASTRON UNION, 82-, MEM, 88-; PROJ MGR, GEN SCI CORP, 89- *Concurrent Pos:* res assoc, Appl Res Corp, 85-89. *Honors & Awards:* Group Achievement Award, NASA. *Mem:* Am Astron Soc; Int Astron Union. *Res:* Measurement of zodiacal light, integrated starlight, diffuse galactic light and extragalactic light at optical and infrared wavelengths for study of interplanetary dust and galactic structure; space-based experimentation. *Mailing Add:* 9364 Dewlit Way Columbia MD 21045

TOLLES, WALTER EDWIN, b Moline, Ill, Feb 1, 16; m 37; c 1. BIOPHYSICS, PHYSIOLOGY. *Educ:* Antioch Col, BS, 39; Univ Minn, MS, 41; State Univ NY Downstate Med Ctr, PhD(biophys & physiol), 69. *Prof Exp:* Asst, Kettering Found, Antioch Col, 37-39; physicist, Div War Res, Airborne Instruments Lab, Columbia Univ, 42-45; supvr, Airborne Instruments Lab, Inc, 45-54, head dept med & biol physics, 54-69; dir, Inst Oceanog & Marine Biol, 59-69; instr obstet & gynec, State Univ NY Downstate Med Ctr, 69- 70, assoc prof, 70-79; PRES, APPL SCI ASSOCS INC, 79- *Concurrent Pos:* Consult, NIH. *Mem:* AAAS; Biophys Soc; Am Soc Limnol & Oceanog; Am Phys Soc; Am Soc Cytol. *Res:* Magnetic techniques in undersea warfare; electronic countermeasures; high speed micro-scanning systems associated data handling system; physiological monitoring systems; clinical instrumentation; diagnostic computer methods. *Mailing Add:* Appl Sci Assocs Inc Lee Hwy Fairfield VA 24435

TOLLES, WILLIAM MARSHALL, b New Britain, Conn, June 30, 37; m 59; c 2. PHYSICAL CHEMISTRY. *Educ:* Univ Conn, BA, 58; Univ Calif, Berkeley, PhD(phys chem), 62. *Prof Exp:* Fel, Rice Univ, 61-62; from asst prof to prof sci & eng, Naval Postgrad Sch, 62-78, dean res & dean sci & eng, 78-84; supt chem, 84-88, ASSOC DIR RES, NAVAL RES LAB, WASHINGTON DC, 88- *Concurrent Pos:* Consult, Naval Weapons Ctr, China Lake, summers, 66-77. *Mem:* Optical Soc Am; Am Phys Soc; Am Chem Soc; Sigma Xi. *Res:* Microwave spectroscopy; rotational spectra of molecules; electron spin resonance; microwave properties of materials; non-linear molecular spectroscopy. *Mailing Add:* Assoc Dir Res Code 1003 Naval Res Lab Washington DC 20375

TOLLESTRUP, ALVIN V, b Los Angeles, Calif, Mar 22, 24; m 44; c 4. SUPERCONDUCTING MAGNET DESIGN. *Educ:* Univ Utah, BS, 44; Calif Inst Technol, PhD(physics), 50. *Prof Exp:* Res fel physics, Calif Inst Technol, 50-53, from asst prof to prof physics, 53-77; MEM STAFF, FERMI NAT ACCELERATOR LAB, 77- *Concurrent Pos:* NSF fel, Europe Nuclear Res, 57-58. *Honors & Awards:* Nat Medal Technol, 89; Wilson Prize, Am Phys Soc, 89. *Mem:* Fel Am Phys Soc; AAAS. *Res:* Design and operation of colliding beam detector; design and construction of superconducting magnets. *Mailing Add:* Fermi Nat Accelerator Lab PO Box 500 Batavia IL 60510

TOLLIN, GORDON, b New York, NY, Dec 26, 30; m 55; c 3. BIOPHYSICAL CHEMISTRY. *Educ:* Brooklyn Col, BS, 52; Iowa State Univ, PhD, 56. *Prof Exp:* Res assoc chem, Fla State Univ, 56; chemist, Lawrence Radiation Lab, Univ Calif, 56-59, NSF fel, 56-57; from asst prof to assoc prof, 59-67, PROF CHEM, UNIV ARIZ, 67-, PROF BIOCHEM, 78- *Concurrent Pos:* Sloan fel, 62-66. *Mem:* Am Soc Photobiol; Biophys Soc; Am Soc Biol Chemists. *Res:* Mechanism of redox protein action; photosynthesis; mechanisms of biological energy conversion; biological oxidation-reduction. *Mailing Add:* Dept Biochem Univ Ariz 1501 N Campbell Tucson AZ 85724

TOLLMAN, JAMES PERRY, b Chadron, Nebr, Nov 6, 04; m 29; c 3. PATHOLOGY. *Educ:* Univ Nebr, BSc, 27, MD, 29; Am Bd Path, dipl, 37. *Prof Exp:* Intern & resident, Peter Bent Brigham Hosp, Boston, Mass, 29-31; from asst prof clin path to prof path, 31-74, chmn dept path & bact, 48-54, dean, 52-64, EMER PROF PATH & EMER DEAN, COL MED, UNIV NEBR, 74- *Concurrent Pos:* Vis prof, Univ Chiengmai, 64-65. *Mem:* Am Soc Clin Path; Am Asn Pathologists & Bacteriologists; fel AMA; fel Col Am Path. *Res:* Effects of dusts on tissues; effect of toxic gases; tissue changes in endocrine disturbances. *Mailing Add:* 4441 E Sixth St Tucson AZ 85711

TOLMACH, L(EONARD) J(OSEPH), b New York, NY, Apr 18, 23; m 45; c 3. CELL BIOLOGY, RADIOBIOLOGY. *Educ:* Univ Mich, BS, 43; Univ Chicago, PhD(chem), 51. *Prof Exp:* Jr chemist, Manhattan Proj, 44-46; from instr to asst prof biophys, Sch Med, Univ Colo, 51-58; assoc prof, 58-64, PROF RADIATION BIOL, SCH MED, WASHINGTON UNIV, 64-, PROF ANAT, 69- *Concurrent Pos:* Mem comt molecular biol, Washington Univ, 61-72, chmn, 64-66; NSF sr fel, 63-64; mem biophys sci training comt, Nat Inst Gen Med Sci, 66-68. *Mem:* Radiation Res Soc; Biophys Soc; Am Asn Cancer Res; fel AAAS. *Res:* Effects of radiations and other toxic agents on cell proliferation. *Mailing Add:* Dept Anat & Radiol Wash Univ Sch Med 4511 Forest Park St Louis MO 63108

TOLMAN, CHADWICK ALMA, b Oct 11, 38; US citizen; m 62; c 3. INORGANIC CHEMISTRY & BIOCHEMISTRY, SCIENCE EDUCATION. *Educ:* Mass Inst Technol, BS, 60; Univ Calif, Berkeley, PhD(phys chem), 64. *Prof Exp:* Fel & res assoc chem, Mass Inst Technol, 64-65; CHEMIST, EXP STA, E I DU PONT DE NEMOURS & CO, INC, 65- *Concurrent Pos:* Vis prof, Ctr Nat Res Scientist, France, 81; exec on leave, precol educ, Sci Alliance, 70. *Mem:* Am Chem Soc; AAAS. *Res:* Mechanisms of homogeneous catalysis by transition metal complexes; hydrocarbon oxidations; kinetics and equilibria of organic and organometallic reactions; chemistry in zeolites, enzyme catalysis. *Mailing Add:* Exp Sta E I du Pont de Nemours & Co Inc Wilmington DE 19880-0328

TOLMAN, EDWARD LAURIE, b Chelsea, Mass, Oct 9, 42; m 67; c 1. PHARMACOLOGY. *Educ:* Univ Mass, BA, 64, MA, 65; State Univ NY Upstate Med Ctr, PhD(pharmacol), 70. *Prof Exp:* NIH fel physiol, Milton S Hershey Med Ctr, Pa State Univ, 69-71, res assoc, 71-72; sr res biologist, Lederle Labs, Am Cyanamid Co, 72-77, group leader, 77-80; sect head biochem res, Ortho Pharmacol Corp, 80-84, asst dir, biochem pharmacol/ microbiol, 84-87, dir, drug metab, 87-89; DIR, PRECLIN DEVELOP COORD, WORLDWIDE, R W JOHNSON PHARMACEUT RES INST, 89- *Mem:* Am Soc Pharmacol & Exp Therapeut; Am Diabetes Asn; AAAS; Am Soc Microbiol. *Res:* Disorders of carbohydrate and lipid metabolism; prostaglandins and inflammation. *Mailing Add:* R W Johnson Pharmaceut Res Ins Rte 202 Raritan NJ 08869

TOLMAN, RICHARD LEE, b Ames, Iowa, Nov 6, 41; m 64; c 3. HETEROCYCLIC CHEMISTRY, CARBOHYDRATE SYNTHESIS. *Educ:* Brigham Young Univ, BA, 65; Univ Utah, PhD(bioorg chem), 69. *Prof Exp:* Group leader, ICN Nucleic Acid Res Inst, 69-73; res fel synthetic org chem, 73-79, asst dir, 79-82, DIR, MERCK SHARP & DOHME RES LABS, 82- *Concurrent Pos:* Lectr, Merck Sharp & Dohme Res Labs, 80- *Mem:* Am Chem Soc; Am Peptide Soc; Int Soc Antiviral Res; Am Soc Microbiol; Int Soc Heterocyclic Chem. *Res:* Chemistry of antiviral agents, nucleosides, heterocydes, and peptides; synthetic conjugate vaccines; AIDS-derived peptides and bacterial polysaccharides; chemotherapy of metabolic diseases, diabetes and benign prostatic hypertrophy. *Mailing Add:* Merck Sharp & Dohme Res Labs Box 2000 R50G-204 Rahway NJ 07065-0900

TOLMAN, ROBERT ALEXANDER, b Springfield, Mass, Feb 28, 24; m 49; c 3. PHYSIOLOGY, ENDOCRINOLOGY. *Educ:* Univ Mass, BS, 49; Ind Univ, MA, 50, PhD(zool), 54. *Prof Exp:* Asst zool, Ind Univ, 50-53, asst chem embryol, 53-54; instr physiol & pharmacol, Col Osteop Med & Surg, 54-58; cardiovasc res trainee, Dept Physiol, Med Col Ga, 58-59; from instr to asst prof physiol & biophys, Sch Med, Univ Louisville, 59-67; grants assoc, NIH, 67-68, assoc myocardial infarction br, Nat Heart & Lung Inst, 68-69, ENDOCRINOL RES PROG DIR, DIABETES, ENDOCRINOL & METABOLIC DIS PROGS, NAT INST DIABETES & DIGESTIVE & KIDNEY DIS, 69- *Mem:* AAAS; Endocrinol Soc. *Res:* Thyroid, thyrotropic hormone interaction and thyroid-stimulating hormone assay; time of appearance of cardiac actin in chick embryo; ventricular pressure curves. *Mailing Add:* 14309 Briarwood Terr Rockville MD 20853

TOLNAI, SUSAN, b Budapest, Hungary, Nov 29, 28; Can citizen; m 50; c 2. CELL BIOLOGY, HISTOLOGY. *Educ:* Semmelweiss Med Univ, Budapest, MD, 53. *Prof Exp:* Bacteriologist, Lab Hyg, Dept Nat Health & Welfare, Can, 57-58, biologist, 58-62; lectr, 62-63, from asst prof to assoc prof, 63-71, PROF HISTOL & EMBRYOL, FAC MED, UNIV OTTAWA, 71- *Concurrent Pos:* Nat Acad Sci Hungary fel, 53-56. *Mem:* Can Soc Cell Biol; Can Asn Med Educ; Can Soc Immunol; Int Soc Heart Res. *Res:* Myocardial enzymes, proteinases; medical education. *Mailing Add:* Dept Anat Fac Med Univ Ottawa Ottawa ON K1H 8M5 Can

TOLSMA, JACOB, b Passaic, NJ, Mar 4, 23; m 48; c 5. CHEMICAL ENGINEERING, POLYMER CHEMISTRY. *Educ:* Newark Col Eng, BSChE, 56, MSChE, 65. *Prof Exp:* Chem engr, Res Ctr, Uniroyal, Inc, 56-65, res engr, 65-71; res engr, Weavenit Surg Corp, 71-80; MEADOX MEDICALS INC, 80- *Mem:* Am Chem Soc. *Res:* Chemical engineering development; polymers scale up; water pollution control. *Mailing Add:* 31 N Haledon Ave Haledon NJ 07508

TOLSON, ROBERT HEATH, b Portsmouth, Va, July 23, 35; m 78; c 2. ATMOSPHERIC SCIENCE, PLANETARY SCIENCES. *Educ:* Va Polytech Inst & State Univ, BS, 58, MS, 63. *Prof Exp:* Aerospace scientist lunar & planetary studies, Langley Res Ctr, NASA, 58-72, head, Planetary Physics Br, 72-75, head, Atmospheric Sci Br, 75-82, chief scientist, 82-84, head interdisplinary res off, 84-90, PROF JIHFS, LANGLEY RES CTR, NASA, GEORGE WASHINGTON UNIV, 91- *Honors & Awards:* H J E Reid, Langley Res Ctr, NASA, 78. *Mem:* Am Geophys Union; Am Inst Aeronaut & Astronaut. *Res:* Stratospheric minor constituent distributions and transport; regional tropospheric pollutant photochemistry, chemistry and transport. *Mailing Add:* Langley Res Ctr NASA George Washington Univ Mail Stop 269 Hampton VA 23665

TOLSTEAD, WILLIAM LAWRENCE, b Howard Co, Iowa, Nov 25, 09. BOTANY. *Educ:* Luther Col, Iowa, BS, 33; Iowa State Univ, MS, 36; Univ Nebr, PhD(plant ecol), 42. *Prof Exp:* Biologist, Conserv & Surv Div, Univ Nebr, 35-42; from assoc prof to prof, 57-76, EMER PROF BIOL, DAVIS & ELKINS COL, 76- *Concurrent Pos:* chmn dept, Davis & Elkins Col, 72-84. *Mem:* Am Rhododendron Soc; Am Hort Soc; Royal Hort Soc. *Res:* Plant breeding rhododendron. *Mailing Add:* Pheasant Mountain Rd Elkins WV 26241

TOLSTED, ELMER BEAUMONT, b Philadelphia, Pa, Apr 28, 20. MATHEMATICS. *Educ:* Univ Chicago, BS, 40, MS, 41; Brown Univ, PhD(math), 46. *Prof Exp:* Instr math, Brown Univ, 42-47; from asst prof to assoc prof, 47-61, prof, 61-80, Rueben C & Eleanor Winslow prof, 80-82, EMER PROF MATH, POMONA COL, 82- *Concurrent Pos:* Fulbright exchange prof, Eng, 49-50; instr, Claremont Inst Music, 51- *Honors & Awards:* Ford Award, Math Asn Am, 65. *Mem:* Am Math Soc; Math Asn Am. *Res:* Subharmonic functions. *Mailing Add:* Dept Math Pomona Col Claremont CA 91711

TOLSTOY, IVAN, b Baden-Baden, Ger, Mar 30, 23; nat US; m 47, 64; c 3. UNDERWATER SOUND, WAVE THEORY. *Educ:* Univ Sorbonne, Lic es sc, 45; Columbia Univ, MA, 47, PhD(geophys), 50. *Prof Exp:* Mem sci staff, Lamont Geol Observ, Columbia Univ, 48-51; sr res engr, Stanolind Oil & Gas Co, 51-53; res scientist, Hudson Lab, Columbia Univ, 53-60, sr res assoc, 62-67, assoc dir, 64-67, prof ocean eng, 67-68; prof geol & fluid dynamics, Geophys Fluid Dynamics Inst, Fla State Univ, 68-74; prof math, Univ Leeds, 73-76, distinguished vis prof acoust, Naval Postgrad Sch, 77-78; CONSULT,

79- *Concurrent Pos:* Chief scientist, MidAtlantic Ridge Exped, Columbia Univ, 50; consult, Gen Elec Co & Carter Oil Co, 60-62 & Schlumberger Tech Co, 62; sr vis fel, Dept Appl Math Studies, Leeds Univ, 71-72, vis prof, 73-76; hon mem, Sci Res Ctr, Far E Br, USSR Acad Nauk, 89. *Honors & Awards:* Pioneers of Underwater Acoust Medal, Acoust Soc Am, 90. *Mem:* AAAS; Am Phys Soc; fel Acoust Soc Am; Am Geophys Union; Sigma Xi; NY Acad Sci. *Res:* Theory of acoustic and elastic wave propagation; hydrodynamics; theoretical mechanics; seismology; submarine topography and geology; applied mathematics; author of two publications. *Mailing Add:* Knockvennie DG7-3PA Castle Douglas SW Scotland

TOM, BALDWIN HENG, b San Francisco, Calif, Sept 19, 40. IMMUNOLOGY. *Educ:* Univ Calif, Berkeley, BA, 63; Univ Ariz, MS, 67, PhD(microbiol), 70. *Prof Exp:* Fel, Stanford Univ Sch Med, 70-72, res assoc immunol, 72-73; instr, Sch Med, Northwestern Univ, 73-74, assoc, 74-75, asst prof surg & physiol, 75-77; asst prof, 77-84, ASSOC PROF BIOCHEM MOLECULAR BIOL & SURG, MED SCH, UNIV TEX, HOUSTON, 84-, ASSOC DIR, BIOPROCESSING RES CTR HOUSTON, HEALTH SCI CTR, 85- *Concurrent Pos:* Res career develop award, Nat Cancer Inst, 79-84; vis scientist, Prairie View A&M Univ, 81; Int Cancer Res exchange fel, Nottingham, Eng, 81. *Mem:* NY Acad Sci; Soc Exp Biol & Med; Tissue Cult Asn; Am Asn Immunol; Am Asn Cancer Res. *Res:* Dissection of the cellular and molecular bases for immune reactivities in tumor cell-lymphocyte interactions with liposomes and monoclonal antibodies; space bioprocessing. *Mailing Add:* Bioprocessing Res Ctr 1343 Moursund SH1 R127 Houston TX 77030

TOM, GLENN MCPHERSON, b Honolulu, Hawaii, Sept 1, 49; m 76; c 3. INORGANIC CHEMISTRY, ANALYTICAL CHEMISTRY. *Educ:* Univ Hawaii, BS, 71; Stanford Univ, PhD(inorg chem), 75. *Prof Exp:* Fel chem, Univ Chicago, 75-77; res chemist, Hercules Res Ctr, Hercules Inc, 77-86; VPRES RES, ADV TECHNOL MAT, 86- *Mem:* Am Chem Soc. *Res:* Electroanalytic chemistry; inorganic and organometallic chemistry; Zeigler-Natta polymerizations; gas purification. *Mailing Add:* ATM Inc 520-B Danbury Rd New Milford CT 06776-4342

TOMA, RAMSES BARSOUM, b Cairo, Egypt, Nov 9, 38; US citizen; m 69; c 2. PROTEIN METABOLISM, DAIRY SCIENCE & TECHNOLOGY. *Educ:* Ain Shams Univ, Cairo, BSc, 59, MSc, 65; La State Univ, PhD(food sci & nutrit), 71, Univ Minn, MPH, 80. *Prof Exp:* Chemist, Ministry Food Supplies, Cairo, 60-68; res assoc food sci, La State Univ, 69-71; asst prof, 72-74, assoc prof & chmn, 74-79, prof food sci & nutrit, Univ NDak, 79-; CALIF STATE UNIV, LONG BEACH. *Concurrent Pos:* Dir res & develop, Food Prod Div, Evangeline Foods, La, 72; prin investr, Red River Proj, Nat Potato Coun, 75-78; mem, NDak Trade Mission to Mid East, 75-78; vis assoc prof, Mansoura Univ, Egypt, 79-80; Egyptian-Am Scholars for Year 2000, 76- *Mem:* Am Chem Soc; Inst Food Technologists; Am Dietetic Asn; Am Pub Health Asn; Am Asn Cereal Chemists; fel Am Inst Chemists; NY Acad Sci. *Res:* Food analyses; protein composition; naturally occurring toxicants in foods; protein metabolism; dietary patterns of selected ethnic groups. *Mailing Add:* Home Econ Dept Calif State Univ-Long Beach 1250 Bellflower Blvd Long Beach CA 90840

TOMAJA, DAVID LOUIS, b Bridgeport, Conn, July 15, 46. MINI-COMPUTER SYSTEMS SUPPORT. *Educ:* Univ Conn, BA, 68; State Univ NY Albany, PhD(chem), 74. *Prof Exp:* Chemist, Gen Elec Res & Develop Ctr, 68-70; RES CHEMIST, PHILLIPS PETROL RES & DEVELOP CTR, 74- *Res:* Development and support of personal computer work stations; development and maintenance of software applications for technical needs. *Mailing Add:* Phillips Petrol Co Res & Develop Ctr 169-PL Bartlesville OK 74004

TOMALIA, DONALD ANDREW, b Owosso, Mich, Sept 5, 38; m 59, 86; c 5. PHYSICAL ORGANIC CHEMISTRY. *Educ:* Univ Mich, BA, 61; Bucknell Univ, MS, 62; Mich State Univ, PhD(phys org chem), 68. *Prof Exp:* Proj leader, Dow Chem Co, 66-68, group leader, 68-71, res mgr, 71-76, assoc scientist, 76-79, sr assoc scientist, 79-84, res scientist, Functional Polymers & Process Div, 84-90; RES PROF & SR RES SCIENTIST, MICH MOLECULAR INST, 90- *Concurrent Pos:* Distinguished lectr, Japan Soc Polymer Sci, Kyoto, 84. *Honors & Awards:* Global Proj Prize, Ministry Int Trade & Indust, Japan, 88. *Mem:* Am Chem Soc; Sigma Xi. *Res:* Functional monomers and polymers usually incorporating heterocyclic moieties; unusual cross linking devices; betaine surfactants and chelating agents; onium type chemistry; water soluble polymer systems; polyamines; cationic polymerization; starburst polymers; polymer topology; molecular morphogenesis. *Mailing Add:* Mich Molecular Inst 1910 St Andrews Rd Midland MI 48640

TOMAN, FRANK R, b Ellsworth, Kans, June 6, 39; m 62; c 2. PLANT BIOCHEMISTRY, PLANT METAL TOXICITY. *Educ:* Kans State Univ, BS, 61, MS, 63, PhD(biochem), 67. *Prof Exp:* From asst prof to assoc prof, 69-79, PROF BIOCHEM, WESTERN KY UNIV, 79- *Concurrent Pos:* Res assoc, Plant Res Lab, Mich State Univ-AEC, 73-74; vis prof, dept plant physiol, Univ Ky, 82; vis scientist, dept biochem, Kans State Univ, 84; Ogden Found fel instrnl improv, 84. *Mem:* Am Chem Soc; Sigma Xi. *Res:* Plant proteins and enzymes; metal toxicity in plants. *Mailing Add:* Dept Biol Western Ky Univ Bowling Green KY 42101

TOMAN, KAREL, b Pilsen, Czech, Mar 19, 24; m 48; c 1. CRYSTALLOGRAPHY. *Educ:* Prague Tech Univ, Ing Chem, 48, Dr Tech(chem), 51; Czech Acad Sci, DrSc(physics), 65. *Prof Exp:* Res officer, Inst Metals, Czech, 50-56; sr res officer, Inst Solid State Physics, 56-62; head lab crystallog, Inst Macromolecular Chem, 62-68; Sci Res Coun UK sr vis res fel, Univ Birmingham, 68-69; res fel, Inst Mat Sci, Univ Conn, 69-70; PROF CRYSTALLOG, WRIGHT STATE UNIV, 70- *Concurrent Pos:* Nicolet fel, McGill Univ, 66-67. *Mem:* Am Crystallog Asn; Mineral Soc Am. *Res:* Crystal structure and imperfections. *Mailing Add:* Dept Geol Wright State Univ Colonel Glenn Hwy Dayton OH 45435

TOMAN, KURT, b Vienna, Austria, Aug 11, 21; US citizen; m 58; c 3. IONOSPHERIC PHYSICS. *Educ:* Vienna Tech Univ, MS, 49; Univ Ill, Urbana, PhD(elec eng), 52. *Prof Exp:* Lab engr, Lecher Inst, Reichenau, Austria, 43-44 & Ctr Tube Res, Tanvald, Czech, 44-45; asst electronics, Univ Ill, Urbana, 49-52, res assoc, 52; re fel, Harvard Univ, 52-55; physicist, Air Force Cambridge Res Lab, 55-63, supvry res physicist & br chief ionospheric radio physics, 63-73, sr scientist, 73-76; PHYSICIST, IONOSPHERIC PROPAGATION, ROME AIR DEVELOP CTR, 76- *Concurrent Pos:* Mem nat comn G, Int Union Radio Sci, 61-; chmn, Inst Elec & Electronics Engrs Wave Propagation Standards Comt, 72-83; US mem working group, Int Electrotech Comn, 79- *Mem:* Inst Elec & Electronics Engrs; Am Geophys Union; Sigma Xi; Antenna & Propagation Soc. *Res:* Dynamics of the ionosphere; radio wave propagation; spectral analysis of internal ionospheric gravity waves; high-frequency ionospheric ducting; group and phase path studies; ionospheric Doppler analysis; method of determining ionospheric reflection height; theoretical and experimental studies of high frequency ducted propagation; propagation study for a tropospheric transhorizon radar RADC-TR-81-166; experimental radar system auroral clutter statistics RADC-TR-84-157; Christian Doppler and the Doppler effect. *Mailing Add:* 85 Tobey Rd Belmont MA 02178

TOMANA, MILAN, b June 7, 32; c 2. CLINICAL IMMUNOLOGY. *Educ:* Charles Univ Prague, PhD(biochem), 68. *Prof Exp:* ASSOC PROF BIOCHEM, DEPT CLIN IMMUNOL & RHEUMATOLOGY, UNIV ALA, 68- *Mem:* Am Asn Immunologists; Soc Complex Carbohydrates. *Res:* Immunoglobulin receptors on cells; glycosylation of immunoglobulins in various diseases; biological role of glycosyltransferases. *Mailing Add:* Div Clin Immunol & Rheumatology Univ Ala Birmingham AL 35294

TOMANEK, GERALD WAYNE, b Collyer, Kans, Sept 16, 21; m 45; c 3. BOTANY. *Educ:* Ft Hays State Univ, AB, 42, MS, 47; Univ Nebr, PhD, 51. *Prof Exp:* From asst prof to prof biol, Ft Hays State Univ, 47-87, chmn, Div Natural Sci & Math, 56-87, actg pres univ, 75-76, pres univ, 76-87; RETIRED. *Concurrent Pos:* Consult, Int Coop Admin Arg, 61. *Res:* Grassland ecology. *Mailing Add:* 1503 Oakmont Hays KS 67601

TOMAR, RUSSELL H, b Philadelphia, Pa, Oct 19, 37; m 65; c 2. IMMUNOLOGY, LABORATORY MEDICINE. *Educ:* George Washington Univ, BS, 59, MD, 63. *Prof Exp:* Resident, Barnes Hosp, Washington Univ, 63-65; surgeon, NIH, USPHS, 65-67; fel clin immunol, Univ Pa, 67-70, assoc, 70-71; ASST PROF MED, STATE UNIV NY UPSTATE MED CTR, 71-, DIR, IMMUNOPATH LAB, 74-, ASSOC PROF PATH, 76- *Concurrent Pos:* Mem immunopath test comt, Am Bd Path, 79- *Mem:* Am Asn Immunol; Am Acad Allergy; Am Soc Clin Path; Soc Exp Med & Biol; Reticuloendotheliol Soc. *Res:* Cellular immunity; mechanisms of cellular interactions, especially transfer factor; immunodeficiency diseases; tumor immunology. *Mailing Add:* Dept Lab Med Univ Wis Hosp & Clin B4-251 CSC 600 Highland Ave Madison WI 53792

TOMARELLI, RUDOLPH MICHAEL, b Pittsburgh, Pa, Jan 10, 17; m 52; c 4. BIOCHEMISTRY, NUTRITION. *Educ:* Univ Pittsburgh, BS, 38, MS, 42; Western Reserve Univ, PhD(biochem), 43. *Prof Exp:* Asst, Mellon Inst, 39-41; res chemist, Wyeth Inst Appl Biochem, 43-47; instr nutrit, Univ Pa, 47-50; res chemist, Wyeth Labs, Inc, 50-55, sr investr, Wyeth Inst Med Res, 55-70, mgr nutrit dept, Wyeth Labs, Inc, 70-79, dir nutrit sci, 79-87; RETIRED. *Concurrent Pos:* Asst prof, St Joseph's Col, Pa, 48-50. *Mem:* Am Chem Soc; Am Soc Biol Chemists; Am Inst Nutrit. *Res:* Infant nutrition. *Mailing Add:* Box 46 RD 3 Phoenixville PA 19460

TOMAS, FRANCISCO, b Montreal, Que, Dec 21, 30; m 81; c 2. SCIENTIFIC SYSTEMS ANALYST. *Educ:* Sir George Williams Univ, BSc, 56. *Prof Exp:* Cur physics, Lab, Sir George Williams Univ, 56-65, dir, 65-73; planning officer, 73-86, SCI SYST ANALYST, CONCORDIA UNIV, 86-, DIR, PHYSICS LAB, 90- *Concurrent Pos:* Consult, Int Youth Sci Week, Expo 67, 66-67. *Mem:* Can Asn Physicists; Can Info Processing Soc. *Res:* Laboratory instruction and administration; programming; computerization of physics laboratory student testing; physics laboratory student testing; analytical chemistry database; instrumentation calculation aids; budget control; typesetting and telidon, laboratory reservation. *Mailing Add:* Sci Syst Analyst Concordia Univ 1455 Demaisonnevue Blvd W Physics Dept Montreal PQ H3G 1M8 Can

TOMASCH, WALTER J, b Cleveland, Ohio, July 26, 30; m 55; c 2. EXPERIMENTAL SOLID STATE PHYSICS. *Educ:* Case Western Reserve Univ, BS, 52, PhD(physics), 58; Rensselaer Polytech Inst, MS, 55. *Prof Exp:* Sr physicist, Atomics Int Div, Rockwell Corp, 58-68; PROF PHYSICS, UNIV NOTRE DAME, 68- *Mem:* Fel Am Phys Soc. *Res:* High-temperature superconductivity; physics of metals and alloys. *Mailing Add:* Dept Physics Univ Notre Dame Notre Dame IN 46556

TOMASCHKE, HARRY E, b Kendall, NY, Apr 25, 29; m 53; c 4. OPTICS. *Educ:* Mich State Univ, BS, 56; Univ Ill, MS, 58, PhD(physics), 64. *Prof Exp:* Res assoc, Coord Sci Lab, Univ Ill, Urbana, 58-64; ASSOC PROF PHYSICS, GREENVILLE COL, 64- *Mem:* Am Asn Physics Teachers; Optical Soc Am; Sigma Xi. *Res:* Electrical breakdown in vacuum; absorption of gases under ultrahigh vacuum conditions; thin film optics. *Mailing Add:* 625 Eastern Ave Greenville IL 62246

TOMASELLI, VINCENT PAUL, b Weehawken, NJ, May 3, 41; m 66; c 3. PHYSICS. *Educ:* Fairleigh Dickinson Univ, BS, 62, MS, 64; NY Univ, PhD(physics), 71. *Prof Exp:* Res physicist, Uniroyal Res Ctr, 64-66; from instr to prof, 66-80, PROF PHYSICS, FAIRLEIGH DICKINSON UNIV, 74-, ASST DEAN RES & GRAD STUDIES, COL SCI & ENG, 84- *Concurrent Pos:* App Gov Sci Adv Comt, NJ State Panel Sci & Adv, 81-83; Fulbright Award, 82. *Mem:* Optical Soc Am. *Res:* Infrared and far-infrared spectroscopy; infrared optical properties of materials and instrumentation. *Mailing Add:* Dept Elec Eng Fairleigh Dickinson Univ Teaneck NJ 07666

TOMASETTA, LOUIS RALPH, b New York, NY, Nov 1, 48; m 73. ELECTRICAL ENGINEERING, SEMICONDUCTORS. *Educ:* Mass Inst Technol, BS & MS, 71, ScD(elec eng), 74. *Prof Exp:* Mem tech staff laser syst, Lincoln Lab, Mass Inst Technol, 74-77; group leader optical electronics, Rockwell Int, 77-84; PRES & CHIEF EXEC OFFICER, VITESSE-SEMICONDUCTOR CORP, 84- *Mem:* Inst Elec & Electronics Engrs; Sigma Xi. *Res:* Semiconductor devices; fiber optics; optical communication laser; laser systems; optics. *Mailing Add:* Vitesse-Semiconductor Corp 741 Calle Plano Camarillo CA 93012

TOMASHEFSKI, JOSEPH FRANCIS, b Plymouth, Pa, Dec 30, 22; m 49; c 3. PHYSIOLOGY. *Educ:* Hahnemann Med Col, MD, 47; Am Bd Prev Med, dipl & cert aerospace med. *Prof Exp:* Asst biol, Temple Univ, 41-43; intern, Wilkes-Barre Gen Hosp, Pa, 47-48, resident med, 48-49; resident pulmonary dis, Jefferson Med Col & Hosp, 49-51; asst prof med & physiol, Col Med, Ohio State Univ, 53-66, assoc prof prev med & physiol, 66-72; DIR, PULMONARY FUNCTION LAB & STAFF PHYSICIAN, DEPT PULMONARY DIS, CLEVELAND CLIN, 71-, HEAD DEPT, 73- *Concurrent Pos:* Dir res, Ohio Tuberc Hosp, 53-67; med res consult, Battelle Mem Inst, 59-67, med dir & med res adv, 67-17; dir pulmonary function labs, Univ Hosps, Ohio State Univ, 67; consult, USAF & Vet Admin; clin prof prev med, Col Med, Ohio State Univ, 72- *Mem:* Am Physiol Soc; Am Thoracic Soc; fel Am Col Chest Physicians; AMA; Aerospace Med Asn. *Res:* Pulmonary diseases; respiratory physiology; aviation medicine; pulmonary function testing; environmental medicine. *Mailing Add:* Dept Path 9500 Euclid Ave Cleveland OH 44106

TOMASHEFSKY, PHILIP, b Brooklyn, NY, May 4, 24; m 48; c 2. EXPERIMENTAL PATHOLOGY. *Educ:* City Col New York, BS, 46, MS, 51; NY Univ, MS, 63, PhD(biol), 69. *Prof Exp:* Chemist, Funk Found, 48-65 & US Vitamin Corp, 65; biochemist, Dept Urol, Col Physicians & Surgeons, Columbia Univ, 65-69, assoc, 69-71, asst prof path, 71-76, asst prof clin path, 76-89; RETIRED. *Mem:* Fel AAAS; NY Acad Sci; Sigma Xi; Am Inst Ultrasound Med. *Res:* Neoplastic and hyperplastic growth of the kidney and prostate and other urological tissues; chemotherapy and thermotherapy; metastasis. *Mailing Add:* 330 S Middletown Rd Pearl River NY 10965

TOMASI, GORDON ERNEST, b Denver, Colo, Dec 16, 30; m 54; c 3. BIOCHEMISTRY, ORGANIC CHEMISTRY. *Educ:* Colo State Col, BA, 57, MA, 58; Univ Louisville, PhD(biochem), 63. *Prof Exp:* From asst prof to assoc prof, 62-70, PROF CHEM, UNIV NORTHERN COLO, 70-, CHMN, DEPT CHEM, 81- *Mem:* Sigma Xi. *Res:* Bioenergetics; nutritional biochemistry; mechanisms of organic reaction and enzyme catalyzed reactions. *Mailing Add:* Dept Chem Univ Northern Colo Greeley CO 80639

TOMASI, THOMAS B, JR, b Barre, Vt, May 24, 27; m 48; c 3. IMMUNOLOGY, MOLECULAR BIOLOGY. *Educ:* Dartmouth Col, AB, 50; Univ Vt, MD, 54; Rockefeller Univ, PhD, 65. *Prof Exp:* Intern & resident internal med, Columbia Presby Hosp, 54-57, chief resident, 57-58; assoc prof med & chmn div exp med, Col Med, Univ Vt, Burlington, 60-65; Buswell prof, div immunol & rheumatic dis & res prof microbiol, State Univ NY, Buffalo, 65-73; prof med & chmn dept immunol, Mayo Med Sch, Rochester, Minn, 73-81, William H Donner prof immunol, 75-81; distinguished univ prof cell biol, chmn dept & dir cancer ctr & special diag lab, Univ NMex, 81-85; INST DIR, ROSWELL PARK CANCER INST, 85- *Concurrent Pos:* Sr investr, Arthritis & Rheumatism Found, 60-65; dir, NIH training grant arthritis & metab dis; chief med, DeGoesbriand Hosp, 61-65; mem gen med study sect, NIH; consult, Los Alamos Nat Lab, 82-; James Melville Cramer fel award, Dartmouth Col. *Honors & Awards:* Woodbury Prize, Univ Vt, 54. *Mem:* Arthritis & Rheumatism Asn; Asn Am Physicians; Am Asn Immunologists; Am Fedn Clin Res; Am Soc Clin Invest; Am Asn Cancer Res; Am Soc Cell Biol. *Res:* Internal medicine; immunological diseases; immunochemistry. *Mailing Add:* Roswell Park Cancer Inst Elm & Carlton Sts Buffalo NY 14263

TOMASI, THOMAS EDWARD, b San Diego, Calif, Aug 10, 55; m 78; c 2. PHYSIOLOGICAL ECOLOGY, ANIMAL ENERGETICS. *Educ:* Univ RI, BS, 76, MS, 78; Univ Utah, PhD(biol), 84. *Prof Exp:* ASST PROF ANIMAL PHYSIOL, DEPT BIOL, SOUTHWEST MO STATE UNIV, 86- *Concurrent Pos:* Vis lectr animal physiol, Univ Calif, Davis, 84-86; prin investr, NIH, 88-90 & 91-94; chair, Educ & Grad Stud Comt, Am Soc Mammalogists, 88-90. *Mem:* Am Soc Mammalogists; Am Soc Zoologists; AAAS; Sigma Xi; Int Soc Cryptozool. *Res:* Interspecific and intraspecific comparative energetics; comparative thyroid function of mammals and its relationship to metabolic rate. *Mailing Add:* Dept Biol Southwest Mo State Univ Springfield MO 65804-0095

TOMASOVIC, STEPHEN PETER, b Bend, Ore, Jan 5, 47; m 70. RADIATION BIOLOGY, CELL BIOLOGY. *Educ:* Ore State Univ, BS, 69, MS, 73; Colo State Univ, PhD(radiation biol), 77. *Prof Exp:* Fel, Dept Radiation Biol, Colo State Univ, 77; fel radiation biol, Dept Radiol, Med Ctr, Univ Utah, 78-79, asst prof, M D Anderson Cancer Ctr, 78-80; asst prof, 80-86, ASSOC PROF, DEPT TUMOR BIOL, M D ANDERSON CANCER CTR, 86-, CHIEF, SECT BIOL, 88- *Concurrent Pos:* Mem grad fac, Univ Tex Health Sci Ctr, Houston, 81-; mem, NAm Hyperthermia Group, 86-, Coun Biol & Chem, 90-92; dir, Interdisciplinary Studies in Cancer Biol, 87-90; assoc ed, Int J Radiation Oncol Biol Phys, 89-, Cancer Bull, 91- *Mem:* AAAS; Radiation Res Soc; Am Asn Cancer Res; Am Soc Cell Biol; Am Soc Therapeut Radiol & Oncol. *Res:* Molecular biology; experimental combined modality therapy; biology of the cell cycle; tumor biology; experimental tumor metastasis; hyperthermic biology; heat shock proteins. *Mailing Add:* Dept Tumor Biol Box 108 M D Anderson Ctr 1515 Holcombe Blvd Houston TX 77030

TOMASZ, ALEXANDER, b Budapest, Hungary, Dec 23, 30; US citizen; m 56; c 1. BIOCHEMISTRY, CELL BIOLOGY. *Educ:* Pazmany Peter Univ, Budapest, dipl, 53; Columbia Univ, PhD(biochem), 61. *Prof Exp:* Res assoc cytochem, Inst Genetics, Hungarian Nat Acad, 53-56; Am Cancer Soc fel & guest investr genetics, 61-63, from asst prof to assoc prof genetics & biochem, 63-77, PROF MICROBIOL & CHMN DEPT, ROCKEFELLER UNIV, 77- *Mem:* AAAS; Am Soc Microbiol; Am Soc Cell Biol; Harvey Soc. *Res:* Biosynthesis and functioning of cell surface structures; cell to cell interactions; control of cell division; molecular genetics. *Mailing Add:* Dept Microbiol Rockefeller Univ 1230 York Ave New York NY 10021

TOMASZ, MARIA, b Szeged, Hungary, Oct 18, 32; US citizen; m 88; c 2. MEDICINAL CHEMISTRY, MOLECULAR PHARMACOLOGY. *Educ:* Univ Eotvos Lorand, Budapest, dipl chem, 56; Columbia Univ, MA, 59, PhD(chem), 62. *Prof Exp:* Res assoc, Rockefeller Inst, 61-62, res assoc biochem, NY Univ, 62-64, instr, 64-66; from asst prof to assoc prof, 66-78, PROF CHEM, HUNTER COL, CITY UNIV NY, 79- *Mem:* NY Acad Sci; Fedn Am Soc Exp Biol; Am Chem Soc; Am Asn Cancer Res; Biophys Soc. *Res:* Chemistry of nucleic acids; chemical basis of action of antitumor agents and antibiotics. *Mailing Add:* Dept Chem Hunter Col New York NY 10021-5024

TOMBACK, DIANA FRANCINE, b Los Angeles, Calif, June 9, 49; m 86. EVOLUTIONARY ECOLOGY, PLANT-ANIMAL MUTUALISMS. *Educ:* Univ Calif, Los Angeles, BA, 70, MA, 72; Univ Calif, Santa Barbara, PhD(biol sci), 77. *Prof Exp:* Teaching asst biol sci, Univ Calif, Santa Barbara, 72-73, teaching assoc, 73-76; instr biol, Dept Zool, Brigham Young Univ, 77; vis asst prof zool, Pomona Col, 77-78; lectr biol, Univ Calif, Riverside, 78-79; postdoc fel zool, Colo State Univ, 79-81; asst prof biol, 81-86, actg dir, Environ Sci Grad Prog, 90- 91, ASSOC PROF, DEPT BIOL, UNIV COLO, DENVER, 86- *Mem:* Ecol Soc Am; Am Ornith Union; Soc Study Evolution; Am Soc Naturalists; AAAS. *Res:* Ecological relationship between Nucifraga columbiana and pines; avian foraging behavior. *Mailing Add:* Dept Biol Univ Colo PO Box 173364 Denver CO 80217-3364

TOMBALAKIAN, ARTIN S, b Jerusalem, Palestine, Nov 4, 29; Can citizen; m 59; c 3. CHEMICAL ENGINEERING, CHEMISTRY. *Educ:* Am Univ Beirut, BA, 52; Univ Toronto, MASc, 54, PhD(chem eng), 58. *Prof Exp:* Prof chem & chmn, dept chem & eng, 58-70, dir, Sch Eng, 68-79, PROF CHEM & CHEM ENG, LAURENTIAN UNIV, 66- *Mem:* Fel Chem Inst Can. *Res:* Ion-exchange; diffusion; mass transfer; electrochemistry; sorption of crude oil derivatives on Arctic terrain; treatment of industrial waste waters; recovery of metals by hydrometallurgical techniques. *Mailing Add:* Sch Eng Laurentian Univ Sudbury ON P3E 2C6 Can

TOMBAUGH, CLYDE W(ILLIAM), b Streator, Ill, Feb 4, 06; m 34; c 2. ASTRONOMY. *Educ:* Kans Univ, AB, 36, MA, 39; Ariz State Col, DSc, 60. *Prof Exp:* Asst observer, Lowell Observ, 29-38, asst astronr, 38-43; instr physics & navigation, Ariz State Col, 43-45; vis asst prof astron, Univ Calif, Los Angeles, 45-46; astronr, Ballistics Res Lab, White Sands Missile Range, 46-55, chief optical measurements sect, 47-51; astronr, phys sci lab, 55-59, res ctr, 59-70, from assoc prof to prof, 61-73, EMER PROF EARTH SCI & ASTRON, NMEX STATE UNIV, 73- *Concurrent Pos:* Consult, Off Ord Res Proj, Small Earth Satellites, 53-59; mem, planetary atmospheres study group, Space Sci Bd, Nat Acad Sci, 60-61; prin investr, NASA grant; mem comn planets & satellites, Int Astron Union; lectr, discovery of Pluto, 86-90. *Honors & Awards:* Jackson-Gwilt Medal, Royal Astron Soc, 31; Distinguished Serv Citation, Univ Kans, 66; Rittenhouse Award, 90. *Mem:* Am Astron Soc; Meteoritical Soc; fel Am Inst Aeronaut & Astronaut; Asn Lunar & Planetary Observers; Astron Soc Pac; Int Astron Union. *Res:* Trans-Neptunian planet search; discovered ninth planet, Pluto; planetary observations; optics applied to rocket ballistics; hypothetical geology of Mars; interpretation of Martian features. *Mailing Add:* PO Box 306 Mesilla Park NM 88047

TOMBAUGH, LARRY WILLIAM, b Erie, Pa, Jan 28, 39; m 60; c 2. FOREST ECONOMICS. *Educ:* Pa State Univ, BS, 60; Colo State Univ, MS, 63; Univ Mich, PhD(resource econ), 68. *Prof Exp:* Economist, NCent Forest Exp Sta, 66-69; prin economist, Southeastern Forest Exp Sta, 69-71; prog mgr, NSF, 71-75, dir div advan environ res & technol, 75-76, dep asst dir anal & planning, 76-78; CHMN, DEPT FORESTRY, MICH STATE UNIV, 78- *Concurrent Pos:* Lectr resource econ, Univ Mich, 66-69. *Mem:* Fel AAAS; Sigma Xi; fel Soc Am Foresters. *Res:* Economics of the forest products industry; relationship between research & development and technological change. *Mailing Add:* Col Forest Resources 2028 Biltmore Hall Box 80001 Raleigh NC 27695-8001

TOMBER, MARVIN L, b South Bend, Ind, Aug 4, 25; m 48; c 2. ALGEBRA. *Educ:* Univ Notre Dame, BS, 46; Univ Pa, PhD(math), 52. *Prof Exp:* Instr math, Amherst Col, 52-55; from asst prof to assoc prof, 55-65, PROF MATH, MICH STATE UNIV, 65- *Mem:* Am Math Soc; Math Asn Am. *Res:* Non-associative algebras. *Mailing Add:* Dept Math D306 Wells Hall Mich State Univ East Lansing MI 48824

TOMBES, AVERETT S, b Easton, Md, Sept 13, 32; m 57; c 4. INVERTEBRATE PHYSIOLOGY. *Educ:* Univ Richmond, BA, 54; Va Polytech Inst, MS, 56; Rutgers Univ, PhD, 61. *Prof Exp:* From asst prof to prof biol, Clemson Univ, 61-77, chmn zool, 68-71; prof biol, George Mason Univ, 77-81, dean, Grad Sch & Univ Res, 81-86; vpres res & grad studies, Wichita State Univ, 86-88; VPRES RES & ECON DEVELOP, NMEX STATE UNIV, 88- *Concurrent Pos:* Prin investr, NSF, NIH, USDA, Dept Defense, Environ Protection Agency & Water Resources Res Inst, 61-83; NSF consult, Indian Coun Educ, 68-69; lectr, Univ de Lille, France, 71-72; Am Coun Educ fel, 80-81; mem, Pub Affairs Comt, Am Soc Cell Biol, 83-86. *Mem:* Am Soc Cell Biol; Soc Res Adminr; Am Soc Zoologists; Sigma Xi; Am Asn Higher Educ; fel AAAS. *Res:* Invertebrate physiology principally hormonal influences on growth, reproduction and reduced metabolism. *Mailing Add:* NMex State Univ Box 30001 Dept 3RED Las Cruces NM 88003-0001

TOMBOULIAN, PAUL, b Rochester, NY, Oct 19, 34; m 57; c 3. TOXIC SUBSTANCE MANAGEMENT. *Educ:* Cornell Univ, AB, 53; Univ Ill, PhD(org chem), 56. *Prof Exp:* Res fel chem, Univ Minn, 56-59, instr, 57-58; from asst prof to assoc prof, 59-67, PROF CHEM, OAKLAND UNIV, 67-,

CHMN DEPT, 62-, COORDR ENVIRON STUDIES, 70-, DIR ENVIRON HEALTH, 75- *Mem:* Am Chem Soc. *Res:* Water resources; instrumental analysis; water quality studies. *Mailing Add:* Dept Chem Oakland Univ Rochester MI 48309

TOMBRELLO, THOMAS ANTHONY, JR, b Austin, Tex, Sept 20, 36; m 77; c 4. SURFACE SCIENCE, PLANETARY SCIENCE. *Educ:* Rice Univ, BA, 58, MA, 60, PhD(physics), 61. *Prof Exp:* Res fel physics, Calif Inst Technol, 61-63; from instr to asst prof, Yale Univ, 63-64; res fel, 64-65, from asst prof to assoc prof, 65-71, PROF PHYSICS, CALIF INST TECHNOL, 71- *Concurrent Pos:* NSF fel, 61-62; Alfred P Sloan Found fel, 71-73; consult, Schlumberger; assoc ed, Nuclear Physics, 72-, Nuclear Sci Appl, 79- & Radiation Effect, 85-88; distinguished vis prof, Univ Calif, Davis, 84; Alexander von Humboldt Award, 84-85; vpres & dir res, Schlumberger-Doll Res, 87-89. *Mem:* Fel Am Phys Soc; Mats Res Soc; AAAS. *Res:* Surface physics; applications of nuclear physics; space physics. *Mailing Add:* Calif Inst Technol 200-36 Pasadena CA 91125

TOMCUFCIK, ANDREW STEPHEN, b Czech, Oct 26, 21; nat US; m 54; c 5. ORGANIC CHEMISTRY. *Educ:* Fenn Col, BS, 43; Western Reserve Univ, MS, 48; Yale Univ, PhD(org chem), 51. *Prof Exp:* Res chemist uranium refining, Mallinckrodt Chem Works, 46; instr chem, Fenn Col, 46-47; res chemist uranium refining, Mallinckrodt Chem Works, 49; res chemist, Am Cyanamid Co, 50-53, res chemist, Lederle Labs, 53-55, dept head, 74-76, GROUP LEADER, LEDERLE LABS, AM CYANAMID CO, 55- *Mem:* AAAS; Am Chem Soc; NY Acad Sci; Royal Soc Chem; Sigma Xi. *Res:* Chemotherapy of cancer; parasitic infections; heterocyclic chemistry; cardiovascular-renal diseases. *Mailing Add:* 48 Dearborn Dr Old Tappan NJ 07675

TOMEI, L DAVID, b Williamsport, Pa, Apr 27, 45; m 77; c 5. CELLULAR PHYSIOLOGY, CELL CYCLE REGULATION. *Educ:* Canisius Col, BS, 68, MS, 70; State Univ NY, Buffalo, PhD(biochem & pharmacol), 74. *Prof Exp:* Chemist, Plum Island Animal Dis Ctr, USDA, 73-75; res scientist, Roswell Park Mem Inst, 75-81; RES SCIENTIST, 81-,OHIO STATE UNIV, DIR, BIOMED INSTRUMENTATION DEVELOP PROG, 84-, DIR, CHEMOSENSITIVITY TESTING LAB COMPREHENSIVE CANCER CTR, 84- *Concurrent Pos:* Pres, Optical Anal Inc, Scanning Laser Imaging Develop. *Mem:* Am Asn Cancer Res; Cell Kinetics Soc. *Res:* In vitro cell growth and function: cell cycle regulation during G1, and the perturbation of regulation by drugs believed to be tumor promoters; the role of plant derived tumor promoting agents in malignant transformation of human cells in vitro by viruses; the function of tumor promoters in modulation of cellular gene directed death related to trauma, irradiation and psychological stress in humans; design and construct new instrumentation based on computer assisted scanning laser imaging technology of which Dr Tomei is primary inventor. *Mailing Add:* Comprehensive Cancer Ctr Ohio State Univ Suite 302 410 W 12 Ave Columbus OH 43210

TOMER, KENNETH BEAMER, b New Kensington, Pa, Mar 13, 44; m 82; c 3. MASS SPECTROMETRY, ORGANIC CHEMISTRY. *Educ:* Ohio State Univ, BS, 66; Univ Colo, PhD(chem), 70. *Prof Exp:* Fel photochem, H C Orsted Inst, Univ Copenhagen, 70-71; fel mass spectrometry, Dept Chem, Stanford Univ, 71-73; asst prof chem, Brooklyn Col, 73-75; asst prof pediat, Med Sch, Univ Pa, 75-77; mem staff, Chem & Life Sci Div, Res Triangle Inst, 77-81; asst dir, Midwest Ctr Mass Spectrometry, Univ Nebr-Lincoln, 81-84, assoc dir & assoc res prof, 84-86; RES SCIENTIST, NAT INST ENVIRON HEALTH SCI, 86- *Mem:* Am Chem Soc; Am Soc Mass Spectrometry. *Res:* Investigations of fragmentation mechanisms of organic compounds in a mass spectrometer; clinical and biochemical applications of mass spectrometry; environmental analysis; application of tamdem mass spectrometry to structure determination; Czelms II Microdialysis-MS. *Mailing Add:* Nat Inst Environ Health Sci PO Box 12233 Res Triangle Park NC 27709

TOMES, DWIGHT TRAVIS, b Bowling Green, Ky, Sept 21, 46; m 66, 86; c 4. PLANT GENETICS, PLANT TISSUE CULTURE. *Educ:* Western Ky Univ, BS, 68; Univ Ky, PhD(crop sci), 75. *Prof Exp:* Asst prof crop sci, Univ Guelph, 75-80, assoc prof, 80-82; PIONEER HI-BRED INT, INC, JOHNSTON, IOWA, 82- *Mem:* Am Soc Agron; Sigma Xi; Tissue Culture Asn. *Res:* Cell Biology of Maize (FEA Mays L) including initiation of embryogenic cultures and plant regeneration; in vitro selection and selection of genetic transformants in maize and other crop species. *Mailing Add:* Pioneer Hi-Bred Int Inc PO Box 38 Johnston IA 50131-0038

TOMES, MARK LOUIS, b Ft Wayne, Ind, Nov 15, 17; m 44; c 2. GENETICS. *Educ:* Ind Univ, AB, 39; Agr & Mech Col, Tex, MS, 41; Purdue Univ, PhD(genetics), 52. *Prof Exp:* Chief plant res, Stokely-Van Camp, Inc, 46-48; asst geneticist, 48-53, ASSOC GENETICIST, EXP STA, PURDUE UNIV, WEST LAFAYETTE, 53-, PROF GENETICS, 59-, HEAD, DEPT BOT & PLANT PATH, 69-, ASSOC DIR, AGR EXP STA, 77- *Concurrent Pos:* Vis prof, Pa State Univ, 66-67. *Mem:* AAAS; Am Soc Hort Sci; Am Phytopath Soc; Genetics Soc Am; Am Genetic Asn; Sigma Xi. *Res:* Genetics and breeding of horticultural crops. *Mailing Add:* 1129 Glenway West Lafayette IN 47906

TOMETSKO, ANDREW M, b Mt Pleasant, Pa, Feb 13, 38; m 65; c 3. BIOCHEMISTRY, ORGANIC CHEMISTRY. *Educ:* St Vincent Col, BS, 60; Univ Pittsburgh, PhD(biochem), 64. *Prof Exp:* Res assoc biochem, Brookhaven Nat Lab, 64-65, asst scientist, 65-66; asst prof biochem, Sch Med & Dent, Univ Rochester, 66-76; PRES, LITRON LABS, LTD, 76- *Mem:* AAAS; NY Acad Sci; Sigma Xi; Am Chem Soc. *Res:* Chemical synthesis of polypeptides; protein structure and function; separation techniques in protein chemistry; photochemistry and photobiology. *Mailing Add:* Litron Labs Ltd 1351 Mt Hope Ave Suite 207 Rochester NY 14620

TOMEZSKO, EDWARD STEPHEN JOHN, b Philadelphia, Pa, Apr 9, 35; m 62; c 4. PHYSICAL CHEMISTRY. *Educ:* Villanova Univ, BS, 57; Pa State Univ, MS, 61, PhD(phys chem), 62. *Prof Exp:* Resident res assoc phys chem, Nat Bur Standards, 62-64; sr res chemist, Arco Chem Co Div, Atlantic Richfield Co, 64-71; asst prof chem, Pa State Univ, 71-78, assoc dir, 81-82, dir, 83-84, assoc dean acad affairs, Common Wealth Educ Syst, 84-85, asst vpres & dir, Div Technol, 85-86, ASSOC PROF CHEM, PA STATE UNIV, DEL COUNTY CAMPUS, 78-, CAMPUS EXEC OFFICER, 86- *Concurrent Pos:* Vis lectr, Johnson Res Found, Col Med Univ Pa, 79-80. *Mem:* Am Chem Soc; Catalysis Soc; Sigma Xi. *Res:* Thermodynamics; homogeneous and heterogeneous catalysis; inorganic and organic synthesis. *Mailing Add:* Four Prince Eugene Lane Media PA 19063-5211

TOMIC, ERNST ALOIS, b Vienna, Austria, Feb 1, 26; m 52; c 2. SYNTHETIC INORGANIC CHEMISTRY, CERAMICS BY SOL-GEL. *Educ:* Univ Vienna, PhD, 56. *Prof Exp:* Asst inorg, geochem & anal chem, Univ Vienna, 55-57; res chemist, Explosives Dept, 58-70, sr res chemist, Polymer Intermediates Dept, 70-74, staff res chemist, Polymer Intermediates Dept, 74-78, res assoc petrochem dept, Exp Sta, 78-85, SR RES ASSOC, DU PONT CHEM, E I DU PONT DE NEMOURS & CO, INC, 85- *Mem:* Sigma Xi; Am Chem Soc; Am Ceramic Soc; Mat Res Soc; Am Electrochem Soc. *Res:* Ion exchange; coordination chemistry; radiochemistry; inorganic synthesis; inorganic cements; molten salts; electrochemistry; hydrometallurgy; solution-derived ceramics. *Mailing Add:* 1430 Emory Rd Wilmington DE 19803

TOMICH, CHARLES EDWARD, b Gallup, NMex, Oct 23, 37; m 59; c 3. DENTISTRY, ORAL PATHOLOGY. *Educ:* Loyola Univ, La, DDS, 61; Ind Univ, Indianapolis, MSD, 68; Am Bd Oral Path, dipl. *Prof Exp:* Assoc prof, 69-78, PROF ORAL PATH, SCH DENT, IND UNIV, INDIANAPOLIS, 78-, DEPT CHMN, 85- *Concurrent Pos:* USPHS training grant, Sch Dent, Ind Univ, Indianapolis, 66-69; ed, Oral Surg, Oral Med & Oral Path, Oral Path Sect, 72-76. *Mem:* Fel Am Acad Oral Path; Sigma Xi. *Res:* In vivo hard tissue marking agents; salivary gland histochemistry; oral neoplasms. *Mailing Add:* Dept Oral Path Sch Dent DS 106 Ind Univ 1121 W Michigan St Indianapolis IN 46202

TOMICH, JOHN MATTHEW, b Baltimore, Md, May 29, 52; m 84; c 2. PROTEIN CHEMISTRY, ENZYMOLOGY. *Educ:* Univ Conn, BA, 74; Purdue Univ, MS, 75; Guelph-Waterloo Ctr Grad Work Chem, Waterloo, Ont, PhD(chem), 79. *Prof Exp:* FEL PROTEIN CHEM, CHEM DEPT, UNIV DEL, 79- *Concurrent Pos:* Res fel, Div Chem, Calif Inst Technol, 83-86. *Mem:* Sigma Xi. *Res:* Mechanisms by which enzymes exhibit specificity for both substrates and allosteriz effects, utilizing chemical and biophysical techniques; synthesis of ion channels. *Mailing Add:* 92 Grace Terr Pasadena CA 91105

TOMICH, PROSPER QUENTIN, b Orange Vale, Calif, Oct 11, 20; m 46; c 5. VERTEBRATE ZOOLOGY, ANIMAL ECOLOGY. *Educ:* Univ Calif, Berkeley, AB, 43, PhD(zool), 59. *Prof Exp:* Lab asst plague res, Hooper Found, Univ Calif, 43-44, res zoologist, Hastings Natural Hist Reservation, 47-52; assoc zool, Univ Calif, Davis, 56-59; Animal Ecologist, State Dept Health, Hawaii, 59-85. *Concurrent Pos:* Adv, Naval Med Res Unit, Egyptian Govt, 46-47; arbovirus res training grant, Pa State Univ, 66-67; mem, Island Ecosysts Proj, Int Biol Prog, Univ Hawaii, 69-75; chmn, Hawaii Natural Area Reserves Syst Comn, 76-84; pvt consult, 86- *Mem:* Fel AAAS; Ecol Soc Am; Am Soc Mammalogists; Wildlife Soc; Am Ornithologists Union; Marine Mammal Soc. *Res:* Rodents, fleas, and plague; field ecology of mule deer, ground squirrel, mongoose, and of Arctic birds and mammals; leptospirosis in populations of small mammals; history and adaptation of mammals in Hawaiian Islands; rehabilitation of depleted subtropical rain-forest ecosystems. *Mailing Add:* Biol Factors PO Box 675 Honokaa HI 96727

TOMIKEL, JOHN, b Cuddy, Pa, Apr 30, 28; m 49, 68; c 2. EARTH SCIENCE. *Educ:* Clarion State Col, BS, 51; Univ Pittsburgh, MLitt, 56, PhD(higher educ), 70; Syracuse Univ, MS, 62. *Prof Exp:* Sci teacher, Fairview High Sch, 51-63; asst prof earth sci, Edinboro State Col, 63-65; prof, 65-82, EMER PROF GEOG & EARTH SCI, CALIFORNIA STATE COL, PA, 65-82. *Mem:* Nat Asn Geol Teachers; Nat Wildlife Fedn. *Res:* Earth science education. *Mailing Add:* 19323 Elgin Rd Corry PA 16407

TOMITA, JOSEPH TSUNEKI, b Los Angeles, Calif, Mar 23, 38; m 63; c 2. CANCER DIAGNOSTICS, TUMOR MARKERS RESEARCH. *Educ:* Stanford Univ, BA, 59, PhD(physiol), 66. *Prof Exp:* Res asst biochem res, Dept Pediat, Stanford Med Ctr, 60-62, teaching asst immunophysiol, 62-66, actg instr, 66-67; Nat Multiple Sclerosis Soc fel, Dept Biol, Univ Calif, San Diego, 68-69, Nat Cystic Fibrosis Found fel, 69-70, asst res biologist neuroimmunol, 70-71; res immunologist tumor immunol, Dept Exp Biol, Abbott Labs, 71-76, head, Cancer Res Lab, Diag Div, 77-81, mgr cancer/cell biol, 81-84, DIR CANCER/IMMUNOL, DIAG DIV, ABBOTT LABS, 84- *Concurrent Pos:* Guest lectr, Dept Biol, Northwestern Univ, 72-73. *Mem:* Am Asn Immunol; Sigma Xi; AAAS. *Res:* Tumor antigen isolation and characterization, serum-based/immunocytochemical tumor marker assays; monoclonal Ab technology; oncogenes, growth factor receptors; autoimmune diseases. *Mailing Add:* Abbott Labs D-90C 1400 Sheridan Rd North Chicago IL 60064

TOMITA, TATSUO, b Tokyo, Japan, Apr 20, 39. ENDOCRINOLOGY. *Educ:* Tokyo Med & Dent Univ, Japan, BS, 61, MD, 65. *Prof Exp:* Intern rotating, US Naval Hosp, Yokosuka, Japan, 65-66; resident path, Wash Univ Sch Med, 70-74; fel, 74-75, from asst prof to assoc prof, 75-85, PROF PATH, UNIV KANS MED CTR, 85- *Mem:* Fel Am Diabetes Asn; fel AAAS; fel Sigma Xi; fel Am Asn Pathologists. *Res:* Experimental diabetes in regards to secretion of pancreatic hormones; radioimmunoassay of peptide hormones; immunohistochemistry of endocrine tumors; anatomic pathology. *Mailing Add:* Dept Path Univ Kans Med Ctr 39th at Rainbow Blvd Kansas City KS 66103

TOMIYASU, KIYO, b Las Vegas, Nev, Sept 25, 19; m 47. MICROWAVE ENGINEERING, ELECTRONICS ENGINEERING. *Educ:* Calif Inst Technol, BS, 40; Columbia Univ, MS, 41; Harvard Univ, MES, 47, PhD(eng sci, appl physics), 48. *Prof Exp:* Instr, Lyman Lab, Harvard Univ, 48-49; proj engr, Sperry Gyroscope Co, 49-52, head, Eng Sect, 52-55; consult engr, Microwave Lab, Calif, 55-60 & Res & Develop Ctr, NY, 60-69, CONSULT ENGR, SPACE DIV, GEN ELEC CO, 69- *Concurrent Pos:* Bd dirs, Inst Elec & Electronic Engrs, 85-86. *Honors & Awards:* C P Steinmetz Award, Gen Elec, 77; Microwave Career Award, Inst Elec & Electronics Engrs, 80, Centennial Medal, 84. *Mem:* Am Phys Soc; fel Inst Elec & Electronics Engrs. *Res:* Microwave radiometry; microwave scatterometry of sea surface; communications; atmospheric propagation; radar; tropospheric propagation; ionospheric propagation; synthetic aperture radar; satellites. *Mailing Add:* Valley Forge Space Ctr Gen Elec Co PO Box 8048 Philadelphia PA 19101

TOMIZUKA, CARL TATSUO, b Tokyo, Japan, May 24, 23; nat US; m 56; c 4. SOLID STATE PHYSICS. *Educ:* Univ Tokyo, BS, 45; Univ Ill, MS, 51, PhD(physics), 54. *Prof Exp:* Asst physics, Univ Ill, 51-54, res assoc, 54-55, res asst prof physics & elec eng, 55-56; asst prof physics, Inst Study Metals, Univ Chicago, 56-60; head dept, 70-77, assoc dean, 77-79, PROF PHYSICS, UNIV ARIZ, 60- *Mem:* Fel Am Phys Soc; Phys Soc Japan. *Res:* Solid state diffusion; anelasticity; high pressure; magnetism. *Mailing Add:* Dept Physics Univ Ariz Tucson AZ 85721

TOMKIEWICZ, MICHA, b Warsaw, Poland, May 25, 39; Israeli citizen; c 1. PHYSICAL CHEMISTRY. *Educ:* Hebrew Univ, MS, 63, PhD(chem), 69. *Prof Exp:* Instr phys chem, Hebrew Univ, 67-69; fel, Univ Guelph, 69-71; Nat Inst Gen Med Sci fel biophys, Univ Calif, Berkeley, 71-72, Nat Inst Gen Med Sci spec fel, 72-73; fel biophys, Thomas J Watson Res Ctr, IBM Corp, 73-76; res scientist, Union Carbide Corp, Tarrytown Tech Ctr, 76-80; MEM FAC, DEPT PHYSICS, BROOKLYN COL, 80- *Mem:* Am Phys Soc. *Res:* Using biophysical and electrochemical approaches to photolyse water with visible radiation for the purpose of converting solar energy to useful chemical energy. *Mailing Add:* Dept Physics Brooklyn Col Bedford Ave & Ave H Brooklyn NY 11210

TOMKINS, FRANK SARGENT, b Petoskey, Mich, June 24, 15; m 42, 63; c 1. ATOMIC SPECTROSCOPY. *Educ:* Kalamazoo Col, BS, 37; Mich State Col, PhD(phys chem), 42. *Prof Exp:* Physicist, Buick Motor Div, Gen Motors Corp, Ill, 41-43; physicist, Argonne Nat Lab, 43-46, sr scientist & group leader, 46-85; RETIRED. *Concurrent Pos:* Guggenheim fel, Nat Ctr Sci Res, France, 60-61; consult, Bendix Corp, 63-75; Argonne Nat Lab-Argonne Univs Asn distinguished appt, 75; fel Sci Res Coun (England), 75; assoc, Harvard Col Observ, 77- *Honors & Awards:* William F Meggars Award, Optical Soc Am, 77. *Mem:* AAAS; fel Optical Soc Am; assoc Am Phys Soc; assoc French Phys Soc. *Res:* Physical chemistry; optical spectroscopy. *Mailing Add:* 11714 83rd Ave Palos Park IL 60464

TOMKINS, MARION LOUISE, b Pembroke, NH, Mar 28, 26. X-RAY FLUORESCENT SPECTROSCOPY. *Educ:* Univ Tampa, BS, 52; Roosevelt Univ, MS. *Prof Exp:* Jr res chemist, Int Minerals & Chem Corp, 52-58; chief spectrographer, Martin Marietta Corp, 58-71; dir labs, H Kramer & Co, 71-89; RETIRED. *Mem:* Am Chem Soc; Soc Appl Spectros; Am Soc Testing & Mat. *Res:* Methods of analyses for high temperature alloys, geological materials and pollution control. *Mailing Add:* 9045 N Menard Morton Grove IL 60053

TOMKINS, ROBERT JAMES, b Ottawa, Ont, Aug 14, 45; m 67; c 2. STRONG LIMIT THEOREMS, MARTINGALE THEORY. *Educ:* Univ Saskatchewan, BA, 65 & 66; Purdue Univ, MS, 67, PhD(statist), 70. *Prof Exp:* From asst prof to assoc prof math, 69-76, PROF MATH, UNIV REGINA, 76-, DEPT HEAD MATH & STATIST, 85- *Concurrent Pos:* Mem scholar comt, 80-83, statist grant, Natural Sci & Eng Res Coun Can, 88-91. *Mem:* Statist Soc Can (secy, 82-85, prog secy, 86-); Inst Math Statist; Math Asn Am; Bernoulli Soc Probability & Statist. *Res:* Probability limit theory especially stability theorems, laws of iterated logarithm and laws of large numbers for independent random variables; sample maxima and martingales. *Mailing Add:* Dept Head Math & Statist Univ Regina Regina SK S4S 0A2 Can

TOMKOWIT, THADDEUS W(ALTER), b New York, NY, Sept 10, 18; m 43; c 3. CHEMICAL ENGINEERING. *Educ:* Columbia Univ, BS, 41, ChE, 42. *Prof Exp:* Semiworks supvr, E I du Pont de Nemours & Co, Inc, 42-43, develop engr, 43-53, gen engr in charge minor construct, Chambers Works, 54-65, gen supt process dept, 65-72, mgr logistics, Org Chem Dept, Freon Prod Div, 72-74, mgr logisitcs, Org Chem Dept, 74-78, mgr logistics & works supplies, Dyes & Pigments Dept, 78-82; RETIRED. *Concurrent Pos:* Mem exec bd, Salem County Voc Inst; mem deans adv comt, Cath Univ Am, 73-76, Lehigh Univ, 75-77 & Univ Tex, Austin, 76-78; mem NSF eval comt, Worcester Polytech Inst, 76-78; consult & pres, Eng Resources Inc. *Honors & Awards:* Achievement Award, Am Polish Descent Cult Soc, 68; Founders Award, Am Inst Chem Engrs. 75. *Mem:* Am Inst Chem Engrs (pres, 72); Sigma Xi. *Res:* Fluorine and fluorocarbons; detergents; Freon products; dyestuffs and intermediates; maintenance, design, chemical construction and physical distribution. *Mailing Add:* 511 Clearview Ave Woodside Hill Wilmington DE 19809

TOMLIN, ALAN DAVID, b Woking, Eng, Jan 27, 44; Can citizen; m 66; c 4. ENTOMOLOGY, ZOOLOGY. *Educ:* Univ Western Ont, BA, 65, MSc, 67; Rutgers Univ, PhD(entom), 72. *Prof Exp:* Res asst zool, Univ Western Ont, 65-67; res officer forestry, Dept Fisheries & Forestry, Can, 67-68; res asst entom, Rutgers Univ, 68-70; res officer forestry, Environ Can, 71-72; RES SCIENTIST AGR, AGR CAN, 72- *Concurrent Pos:* Fel, Rothamsted Exp Sta, Harpenden, Eng, 74-75; sci comt mem, Biol Surv Insects Can, 76-79; hon prof, Assoc Fac, Dept Environ Biol, Univ Guelph, 77-; hon lectr, Dept Zool, Univ Western Ont, 78-; hon prof, assoc fac, Dept Land Res Sci, Univ Guelph, 83- *Mem:* Entom Soc Am; Entom Soc Can; fel Royal Entom Soc. *Res:* Ecology of soil arthropods; effects of pesticides on soil fauna; morphology and taxonomy of soil arthropods; biology of earthworms. *Mailing Add:* London Res Ctr Agric Can 1400 Western Rd London ON N6G 2V4 Can

TOMLIN, DON C, b Meridian, Idaho, Aug 29, 32; m 58; c 1. ANIMAL NUTRITION. *Educ:* Calif State Polytech Col, BSc, 55; Univ Fla, MSc, 56, PhD(animal nutrit), 60. *Prof Exp:* Res asst, Univ Fla, 55-60; animal nutrit, Madera Milling Co, Calif, 60; fel forage eval, Ohio Agr Exp Sta, 61-62; res officer, Exp Farm, Can Dept Agr, BC, 62-65; res assoc range livestock nutrit, Utah Agr Exp Sta, 65-67; animal scientist, US Sheep Exp Sta, USDA, Idaho, 67-70; asst prof animal sci, Univ Alaska, Fairbanks, 70-75; consult, Agro-North Assoc, 75-88; NATURAL RESOURCES SPECIALIST, BUR INDIAN AFFAIRS, US DEPT INTERIOR, 88- *Mem:* Sigma Xi; Am Soc Animal Sci; Soc Range Mgt. *Res:* Ruminant nutrition; evaluation and utilization of native and domestic forages. *Mailing Add:* Box 6190-A HC-01 Palmer AK 99645

TOMLINSON, EVERETT PARSONS, science education; deceased, see previous edition for last biography

TOMLINSON, GEORGE HERBERT, b Fullerton, La, May 2, 12; m 37; c 3. CHEMISTRY. *Educ:* Bishop's Univ, Can, BA, 31; McGill Univ, PhD(chem), 35. *Hon Degrees:* DCL, Bishop's Univ, 86. *Prof Exp:* Res assoc cellulose & indust chem, McGill Univ, 35-36; chief chemist, Howard Smith Chem Ltd, 36-40; res dir, Howard Smith Paper Mills Ltd, 41-60; res dir, 61-70, vpres res & environ technol, 70-77, SR SCI ADV, DOMTAR LTD, 77- *Concurrent Pos:* Mem res adv bd, Int Joint Comn; mem, US/Can/Mex Tri-Acad Comt Acid Deposition. *Honors & Awards:* Medal, Tech Asn Pulp & Paper Indust, 69; Gold Medal, Can Pulp & Paper Asn, 91. *Mem:* AAAS; hon mem Can Pulp & Paper Asn; hon mem Tech Asn Pulp & Paper Indust; Am Chem Soc; fel Chem Inst Can; fel Royal Soc Can. *Mailing Add:* 920 Perrot Blvd N Ile Perrot PQ J7V 3K1 Can

TOMLINSON, GERALDINE ANN, b Vancouver, BC, Feb 5, 31; m 57. MICROBIOLOGY, BIOCHEMISTRY. *Educ:* Univ BC, BSA, 57, PhD(agr microbiol), 64; Univ Calif, Berkeley, MA, 60. *Prof Exp:* Res assoc comp biol, Kaiser Res Ctr Comp Biol, 60-61; develop pharmacol, Dept Pediat, State Univ NY Buffalo, 64-65; asst prof biol, Rosary Hill Col, 65-66; res assoc agr biochem, Stauffer Agr Ctr, Calif, 66-67; asst prof, 67-72, assoc prof, 72-82, actg chmn dept, 78-79, PROF BIOL, UNIV SANTA CLARA, 82- *Concurrent Pos:* Ames Res Ctr, NASA-Univ Santa Clara res grants, Ames Res Ctr, 68-; assoc dir summer prog planetary biol-microbiol ecol, NASA, 80 & 82. *Mem:* AAAS; Sigma Xi; Am Soc Microbiol. *Res:* Microbial physiology and biochemistry; extremely halophilic bacteria. *Mailing Add:* Dept Biol Santa Clara Univ Santa Clara CA 95053

TOMLINSON, HARLEY, b Tunbridge, Vt, July 20, 32; m 58; c 2. PLANT PATHOLOGY. *Educ:* Univ Vt, BS, 59, MS, 61, PhD(bot), 65. *Prof Exp:* Asst plant pathologist, Conn Agr Exp Sta, 65-73; mem staff, 73-84, MGR ENVIRON CONTROL, CHEM RES & DEVELOP, HUMPHREY CHEM CO, 84-; MGR ENVIRON CONTROL, CHEM RES & DEVELOP, HUMPHREY CHEM CO, 84- *Mem:* Am Chem Soc; Sigma Xi. *Res:* Chemical applications and plant sciences. *Mailing Add:* 191 Knob Hill Dr Hamden CT 06518

TOMLINSON, JACK TRISH, b Bakersfield, Calif, Aug 22, 29; m 63; c 3. INVERTEBRATE ZOOLOGY. *Educ:* Univ Calif, AB, 50, MA, 52, PhD(zool), 56. *Prof Exp:* Instr biol, Oakland City Col, 54-57; from instr to assoc prof, 57-68, chmn dept, 75-76, PROF BIOL, SAN FRANCISCO STATE UNIV, 68- *Mem:* AAAS; Animal Behav Soc; Soc Syst Zool; Crustacean Soc; Ecol Soc; Am Soc Zoologists. *Res:* Invertebrate physiology; animal behavior. *Mailing Add:* Dept Biol San Francisco State Univ San Francisco CA 94132

TOMLINSON, JAMES EVERETT, b Petersburg, Va, July 8, 42; m 65; c 3. RUMINANT NUTRITION, DAIRY MANAGEMENT. *Educ:* Va Polytech Inst & State Univ, BS, 64, MS, 68; Univ Ky, PhD(animal nutrit), 72. *Prof Exp:* Ruminant res assoc antibiotic res, Hoechst Pharmaceut Co, 72-73; dairy nutritionist, Ralston Purina Co, 73-75; from asst prof to assoc prof res & teaching, 75-86, PROF RES TEACHING & EXTEN, MISS STATE UNIV, 87- *Concurrent Pos:* Nutrit consult. *Mem:* Am Soc Animal Sci; Am Dairy Sci Asn; Sigma Xi; Am Registry Prof Animal Scientists; Am Forage & Grassland Coun. *Res:* Energy sources for lactating dairy cows; supplemental concentrate feeding systems; ensiled forage utilization by dairy cows; dairy heifer feeding and management; forage systems for lactating dairy cows and heifers; protein sources for lactating dairy cows. *Mailing Add:* Dept Dairy Sci Drawer DD Miss State Univ Mississippi State MS 39762

TOMLINSON, JOHN LASHIER, b Salem, Ore, Sept 15, 35; m 58; c 2. PHYSICAL METALLURGY. *Educ:* Loma Linda Univ, BA, 58; Univ Ore, MA, 61; Univ Wash, PhD(metall), 67. *Prof Exp:* Physicist, Naval Ord Lab, Calif, 60-63; res engr, Boeing Co, Wash, 63-64; res assoc metall eng, Univ Wash, 64-67; res physicist, Naval Weapons Ctr, Calif, 67-69; from asst prof to assoc prof, 69-76, PROF CHEM & MAT ENG, CALIF STATE POLYTECH UNIV, POMONA, 76- *Concurrent Pos:* Eve instr, Chaffey Col, 61-63; from asst prof to assoc prof, Sch Dent, Loma Linda Univ, 70-78, prof, 78-; consult, Naval Ocean Systs Ctr, San Diego, 71-78 & RSI Assocs, La Verne, Calif, 79- *Honors & Awards:* Charles Babbage Award, Inst Electronic & Radio Engrs, London, 76. *Mem:* Metall Soc; Am Phys Soc; Sigma Xi. *Res:* Failure analysis and product liability; physical metallurgy of electronic materials; electrical properties of liquid metals and semiconductors; properties and structure of thin films; dental materials. *Mailing Add:* Dept Chem & Mat Eng Calif State Polytech Univ Pomona CA 91768-4069

TOMLINSON, MICHAEL, b Leeds, Eng, Mar 30, 29; m 59; c 3. PHYSICAL CHEMISTRY. *Educ:* Univ Leeds, BSc, 49. *Prof Exp:* Asst exp officer radiation chem, Atomic Energy Res Estab, Harwell, Eng, 49-54; exp officer, Chalk River Nuclear Labs, Atomic Energy Can Ltd, 54-57; sr sci officer, Atomic Energy Res Estab, Harwell, Eng, 57-62; assoc res officer, Chalk River Nuclear Labs, Atomic Energy Can Ltd, 62-63, assoc res officer, Whiteshell Nuclear Res Estab, 63-71, head res chem br, 71-76, dir chem & mat sci div, 76-86, mgr, bus develop, 86-90, SAVANT GEN, WHITESHELL NUCLEAR RES ESTAB, ATOMIC ENERGY CAN LTD, 90- *Mem:* Fel Chem Inst Can. *Res:* Chemistry for nuclear power. *Mailing Add:* Whiteshell Nuclear Res Estab Pinawa MB R0E 1L0 Can

TOMLINSON, MICHAEL BANGS, mathematics, for more information see previous edition

TOMLINSON, PHILIP BARRY, b Leeds, Eng, Jan 17, 32; m 65; c 2. BOTANY. *Educ:* Univ Leeds, BSc, 53, PhD(bot), 55; Harvard Univ, AM, 71. *Prof Exp:* Fel bot, Univ Malaya, 55-56; lectr, Univ Col Ghana, 56-59 & Univ Leeds, 59-60; res scientist, Fairchild Trop Garden, Fla, 60-71; PROF BOT, HARVARD UNIV, 71-, E C JEFFREY PROF BIOL, 86- *Concurrent Pos:* Forest anatomist, Cabot Found, Harvard Univ, 65-71. *Res:* Morphology and anatomy of monocotyledons, especially palms; tropical botany. *Mailing Add:* PO Box 68 Petersham MA 01366

TOMLINSON, RAYMOND VALENTINE, b Smithers, BC, July 25, 27; m 57. BIOCHEMISTRY. *Educ:* Univ BC, BA, 54, MSc, 56; Univ Calif, Berkeley, PhD(biochem), 61. *Prof Exp:* Head technician, Children's Hosp, Vancouver, BC, 4953; res assoc, BC Neurol Inst, Vancouver, 56-57; fel, Univ BC, 61-64; asst res prof, State Univ NY Buffalo, 64-66; prin scientist, Syntex Corp Res Div, 66-86; RETIRED. *Res:* Intermediary metabolism of drugs. *Mailing Add:* 4087 Orme Palo Alto CA 94306

TOMLINSON, RICHARD HOWDEN, b Montreal, Que, Aug 2, 23; m 49. PHYSICAL CHEMISTRY, INORGANIC CHEMISTRY. *Educ:* Bishop's Univ, Can, BSc, 43; McGill Univ, PhD(chem), 48. *Prof Exp:* Nat Res Coun Can fel, Cambridge Univ, 49; asst prof phys & inorg chem, 52-58, chmn dept, 67-74, PROF PHYS CHEM, MCMASTER UNIV, 58- *Mem:* The Chem Soc; Sigma Xi. *Res:* Diffusion; mass spectrometry; polymerization; radiochemistry. *Mailing Add:* Dept Chem McMaster Univ 1280 Main St W Hamilton ON L8S 4L8 Can

TOMLINSON, WALTER JOHN, III, b Philadelphia, Pa, Apr 3, 38; m 61; c 1. INTEGRATED OPTICS, NONLINEAR OPTICS IN FIBERS. *Educ:* Mass Inst Technol, BS, 60, PhD(physics), 63. *Prof Exp:* Consult, Edgerton, Germeshausen & Grier, Inc, 60-63, sr scientist, 63; mem tech staff, Bell labs, 65-81, supvr, optical disk recording group, 81-83, DIST RES MGR, PHOTONIC COMPONENT RES, BELL COMMUN RES, 83- *Mem:* Am Phys Soc; fel Optical Soc Am. *Res:* Optical fiber components, integrated optics and, nonlinear optics; photochemistry; gaseous optical masers, atomic and molecular; magnetic field effects in optical masers; optical disk memories; isotope shifts in radioactive nuclei. *Mailing Add:* 22 Indian Creek Rd Holmdel NJ 07733

TOMLJANOVICH, NICHOLAS MATTHEW, b Susak, Yugoslavia, Mar 5, 39; US citizen; m 66; c 1. THEORETICAL PHYSICS, IONOSPHERIC PHYSICS. *Educ:* City Col New York, BS, 61; Mass Inst Technol, PhD(physics), 66. *Prof Exp:* Physicist, US Weather Bur, 62; teaching asst physics, Mass Inst Technol, 62-64; res asst elem particle physics, Lab Nuclear Sci, 64-66; MEM TECH STAFF, MITRE CORP, 66- *Concurrent Pos:* Woodrow Wilson fel, 62; physicist, Nat Bur Standards, 63. *Mem:* Am Phys Soc; Sigma Xi. *Res:* Radar detection theory and electromagnetic wave propagation; holography; modern optics; scattering theory; plasma physics. *Mailing Add:* 131 Nowell Farme Rd Carlisle MA 01741

TOMMERDAHL, JAMES B, b Adair, Iowa, Dec 2, 26; m 52; c 2. ELECTRICAL ENGINEERING. *Educ:* NC State Col, BS, 55, MS, 58. *Prof Exp:* Sr proj engr, Aeronaut Electronics, Inc, NC, 55-57; unit head & lead engr, Radiation Inc, Fla, 58-60; mem tech staff, 60-75, dir, Systs & Measurements Div, 75-83, VPRES ENVIRON SCI & ENG, RES TRIANGLE INST, 83- *Mem:* Inst Elec & Electronics Engrs. *Res:* Data acquisition systems and signal analysis; circuit and system performances modeling; design and development of very high frequency transceiver; high speed tracking systems and various instrumentation systems; design and development of airborne environmental monitoring systems. *Mailing Add:* Environ Sci & Eng Res Triangle Inst PO Box 12194 Research Triangle Park NC 27709

TOMOMATSU, HIDEO, b Tokyo, Japan, June 8, 29; m 68; c 1. ORGANIC CHEMISTRY, AGRICULTURAL CHEMISTRY. *Educ:* Waseda Univ, Japan, BEn, 53; Univ of the Pac, MSc, 60; Ohio State Univ, PhD(org chem), 65. *Prof Exp:* Res chemist, Hodogaya Chem Co Ltd, Japan, 53-59 & Texaco Chem Co, Inc, Tex, 65-71; supvr, 71-80, SR STAFF, RES LABS, QUAKER OATS CO, 80- *Concurrent Pos:* US ed, High Polymers, Japan, 79-81. *Mem:* Am Chem Soc; Inst Food Technologists; Soc New Food Mat (Japan). *Res:* Synthesis of new elastomers and new polymers; food science and technology. *Mailing Add:* Res Labs Quaker Oats Co 617 W Main St Barrington IL 60010

TOMONTO, JAMES R, b White Plains, NY, Apr 14, 32; m 56; c 5. NUCLEAR PHYSICS. *Educ:* Villanova Univ, BS, 54; Rensselaer Polytech Inst, MS, 59. *Prof Exp:* Engr, Airborne Instruments Lab, 57; anal physicist, Nuclear Power Eng Div, Alco Prod Inc, 58-59; exp physicist, Knolls Atomic Power Lab, 59-64; mgr nuclear eng dept, Gulf United Nuclear Fuels Corp, NY, 64-74; mgr, Nuclear Anal Dept, Fla Power & Light Co, 74-81, sr consult, 81-88; EXEC CONSULT, YFF, INC, 89- *Mem:* Fel Am Nuclear Soc. *Res:* Design and analysis of water moderated power and research reactors; development of analysis methods relating to use of uranium and plutonium as a fuel in thermal power and fast breeder reactors; economic analysis of energy systems; development of computer analysis systems; economic and technical analysis of high level nuclear waste treatment and disposal. *Mailing Add:* 14311 SW 74 Ct Miami FL 33158-1655

TOMOZAWA, YUKIO, b Iyo-City, Japan, Sept 3, 29; nat US; m 57; c 2. THEORETICAL HIGH ENERGY PHYSICS. *Educ:* Univ Tokyo, BSc, 52, DSc(physics), 61. *Prof Exp:* Asst physics, Univ Tokyo, 56-57, Tokyo Univ Educ, 57-59, Cambridge Univ, 59-60 & Univ Col, Univ London, 60-61; res assoc, Inst Physics, Univ Pisa, 61-64; mem, Inst Advan Study, 64-66; from asst prof to assoc prof, 66-72, PROF PHYSICS, UNIV MICH, ANN ARBOR, 72- *Mem:* Am Phys Soc. *Res:* Symmetries in elementary particle physics; theories of unified gauge particle; axiomatic field theory; quantum field theory. *Mailing Add:* Dept Physics Univ Mich Ann Arbor MI 48109

TOMPA, ALBERT S, b Trenton, NJ, Aug 26, 31; m 57; c 5. ANALYTICAL CHEMISTRY, PHYSICAL CHEMISTRY. *Educ:* St Joseph Univ, BS, 54; Fordham Univ, MS, 57, PhD(analytical chem), 60. *Prof Exp:* Lab instr analytical chem, Fordham Univ, 54-59; res chemist, 60-74, RES CHEMIST, NAVAL SURFACE WARFARE CTR, NAVAL ORD STA, 74- *Mem:* AAAS; Am Chem Soc. *Res:* Infrared and thermal analysis study of polymers and energetic materials; toxicity and demil of propellants and explosives. *Mailing Add:* Rte 2 Box 20 Laurel Acres Indian Head MD 20640

TOMPA, FRANK WILLIAM, b New York, NY, Nov 5, 48; m 72; c 3. DATA STRUCTURES, DATABASES. *Educ:* Brown Univ, ScB, 70, ScM, 70; Univ Toronto, PhD(computer sci), 74. *Prof Exp:* Lectr computer sci, Univ Toronto, 74; from asst prof to assoc prof, 74-89, PROF COMPUTER SCI, UNIV WATERLOO, 89- *Mem:* Asn Comput Mach. *Res:* Text-dominated databases; data structures design and specification; systems design for hypertexts; machine-readable dictionary systems. *Mailing Add:* Dept Computer Sci Univ Waterloo Waterloo ON N2L 3G1 Can

TOMPKIN, GERVAISE WILLIAM, b Vinton, Iowa, July 11, 24; m 48; c 4. PHYSICAL CHEMISTRY. *Educ:* Colo State Univ, BS, 47; Univ Colo, PhD(phys chem), 51. *Prof Exp:* Chemist, Mallinckrodt Chem Works, 51-53, develop chemist, 53-55, plant mgr nuclear fuels, Mallinckrodt Nuclear, 56-59, res dir indust div, Mallinckrodt Chem, 59-61; assoc prof phys chem, Colo State Univ, 61-70; dir, Western Opers, Basal Eng, 80-84; PRES, FARAD CORP, 63-; MGR, TECH TRANSFER, BUSINESS EXPANSION & RETENTION STAFF, GOV COUN SCI & HIGH TECH, AZ DEPT COM, 85- *Concurrent Pos:* Pres, Res Inst Colo, 78-84. *Mem:* Am Chem Soc. *Res:* Surface chemistry; development of industrial coatings and surface treatments. *Mailing Add:* 3800 N Central Suite 1500 Phoenix AZ 85012

TOMPKIN, ROBERT BRUCE, b Akron, Ohio, Apr 2, 37; m 61; c 3. FOOD MICROBIOLOGY. *Educ:* Ohio Univ, BSc, 59; Ohio State Univ, MSc, 61, PhD(microbiol), 63. *Prof Exp:* Res microbiologist, 64-65, head microbiol res div, 65-66, chief microbiologist, Swift & Co, 66-85, CHIEF MICROBIOLOGIST & DIR, ANALYTICAL SERV, SWIFT-ECKRICH, 85- *Concurrent Pos:* Chmn, Food Microbiol Sect, Am Soc Microbiol, 72, Food Microbiol Div, Inst Food Technologists, 77, Comt Microbiol Food, Nat Acad Sci-Nat Res Coun Comt, Nitrite & Alternative Curing Agents in Foods, 80-82; mem, Int Comn Microbiol Specif Foods, 81-, Nat Adv Comt Microbiol Criteria Foods, 88- *Mem:* Am Soc Microbiol; Am Meat Sci Asn; Inst Food Technologists; Int Asn Milk, Food & Environ Sanit; Am Acad Microbiol. *Res:* Prevention of food-borne diseases and food spoilage. *Mailing Add:* Swift-Eckrich Inc 3131 Woodcreek Dr Downers Grove IL 60515-5429

TOMPKINS, CURTIS JOHNSTON, b Roanoke, Va, July 14, 42; m 64; c 3. ENGINEERING MANAGEMENT, STATISTICAL ANALYSIS. *Educ:* Va Polytech Inst & State Univ, BS, 65, MS, 67; Ga Inst Technol, PhD(indust & systs eng), 71. *Prof Exp:* Indust eng, E I Du Pont de Nemours, Co Inc, 65-67 & South Fulton Hosp, Atlanta, 68; res eng, Health Systs Res Ctr, Ga Tech, 68-70, instr indust eng, Sch Indust & Syst Eng, 69-71; from asst prof to assoc prof quantitative methods, Grad Sch Bus Admin, Univ Va, 71-77; prof & chmn indust eng, dept indust eng, 77-80, DEAN, COL ENG, WVA UNIV, 80- *Concurrent Pos:* Vis lectr, Indust Col Armed Forces, 74-77; consult, Western Electric, 75-81; lectr, Nat Acad Voluntarism, 76-; mem, Comn Eng Educ, Nat Asn State Univ & Land Grant Col, 85-89. *Mem:* Inst Indust Engrs (pres, 88-89); Am Soc Eng Educ; Sigma Xi; Exec Comt Bd Gov; Am Asn Eng Soc. *Res:* Investigation of group theoretic structures in the fixed charge transportation problem; integer programming methods for the cutting-stock problem. *Mailing Add:* 1453 Anderson Ave Morgantown WV 26505

TOMPKINS, DANIEL REUBEN, b New York, NY, Oct 2, 31; m 64; c 2. HORTICULTURE, PLANT PHYSIOLOGY. *Educ:* Univ Md, BS, 59, MS, 62, PhD, 63. *Prof Exp:* Asst horticulturist, Western Wash Res & Exten Ctr, Wash State Univ, 62-68; from assoc prof to prof hort food sci, Univ Ark, Fayetteville, 69-75; PRIN HORTICULTURIST, COOP STATE RES SERV, USDA, 75- *Mem:* AAAS; fel Am Soc Hort Sci (vpres, 87-88); Sigma Xi; Plant Growth Regulator Soc Am. *Res:* Growth substances; physiology of horticultural plants and research administration. *Mailing Add:* Coop State Res Serv (CSRS) USDA Washington DC 20250

TOMPKINS, DONALD ROY, JR, b Calif, 32; m 61; c 5. PHYSICS. *Educ:* Univ NDak, BS, 55; Univ Colo, MS, 58; Univ Ariz, PhD(physics), 64. *Prof Exp:* Asst prof physics, La State Univ, 64-67 & Univ Ga, 67-70; assoc prof, Univ Wyo, 70-76; PRES, TERRENE CORP, 76- *Res:* Geophysics; mathematical physics. *Mailing Add:* Terrene Corp 604 Travis St Refugio TX 78377

TOMPKINS, GEORGE JONATHAN, b La Plata, Md, May 6, 44; c 2. ENTOMOLOGY, IMMUNOLOGY. *Educ:* Univ Md, BS, 67, MS, 68, PhD(entom microbiol), 79. *Prof Exp:* Med entomologist, 1st Army Med Lab, US Army Ft Meade, Md, 69-70, 71-73; res entomologist, Insect Path Lab, USDA, 73-89; ENTOMOLOGIST, ENVIRON GROUND WATER BR, US ENVIRON PROTECTION AGENCY, 89- *Concurrent Pos:* Prev med officer, 926 Med Detachment, Vietnam, 70-71. *Mem:* Am Inst Biol Sci; Soc Invertebrate Path; Entom Soc Am. *Res:* Evaluation of effects of passaging insect viruses in alternate hosts in terms of the enhancement of the virulence and also encapsulating these viruses with dyes to protect them from ultraviolet light; field evaluations to determine the effectiveness of these encapsulated viruses for control of insect pests of cole crops. *Mailing Add:* Environ Protection Agency 401 M St SW H-7507C Washington DC 20460

TOMPKINS, HOWARD E(DWARD), b Brooklyn, NY, Apr 19, 22; m 43; c 4. APPLIED COMPUTER SCIENCE, COMPUTER LANGUAGES. *Educ:* Swarthmore Col, BA, 42; Univ Pa, MS, 47, PhD(elec eng), 57. *Prof Exp:* Engr, Philco Corp, 42-47; instr & res asst, Moore Sch Elec Eng, Pa, 47-51, res assoc, 56-57, asst prof elec eng, 57-60; proj supvr, Burroughs Corp, 51-54, ed serv mgr, 55-56; prof elec eng, Univ NMex, 60-61; chief sect tech develop, Nat Inst Neurol Diseases & Blindness-NIMH, 61-63; prof elec eng & head dept, Univ Md, 63-67; dir info serv, Inst Elec & Electronics Engrs, NY, 67-71; chmn dept, Ind Univ Pa, 71-77, prof computer sci, 71-87; RETIRED. *Mem:* Asn Comput Mach; Inst Elec & Electronics Engrs. *Res:* Ease of learning and ease of use in computer languages; simple mapping. *Mailing Add:* 19 Congress Terr Milford MA 01757-4021

TOMPKINS, ROBERT CHARLES, b Bucyrus, Ohio, Aug 23, 24; m 57; c 3. ELECTROMAGNETIC THEORY. *Educ:* Ohio State Univ, BSc, 44. *Prof Exp:* Res engr chem, Battelle Mem Inst, 44-45; asst, Univ Chicago, 46-47; phys scientist, US Army Nuclear Defense Lab, 49071; res chemist, US Army Ballistic Res Lab, 71-86; RETIRED. *Concurrent Pos:* Mem adv comt civil defense, Nat Acad Sci-Nat Res Coun, 70-73; lectr, St Mary's Univ, Md, 75. *Mem:* Am Chem Soc; Sigma Xi; Am Phys Soc; NY Acad Sci. *Res:* Fallout from nuclear weapons; chemistry of propellants and explosives; high-pressure effects on electronic spectra; electromagnetic signatures. *Mailing Add:* 541 Valley View Rd Towson MD 21204

TOMPKINS, RONALD K, b Malta, Ohio, Oct 14, 34; m 56; c 3. SURGERY. *Educ:* Ohio Univ, BA, 56; Johns Hopkins Univ, MD, 60; Ohio State Univ, MS, 68. *Prof Exp:* NIH trainee surg & fel phys chem, Col Med, Ohio State Univ, 66-69, instr & fel phys chem, 66-69, instr surg, 68-69; from asst prof to assoc prof, 69-79, PROF SURG, SCH MED, UNIV CALIF, LOS ANGELES, 79-, ASSOC DEAN, 88- *Concurrent Pos:* NIH res grants, Inst Arthritis & Metab Dis, Dept Health Educ & Welfare, 68-71; res grants, John A Hartford Found, Inc, 70-78; consult, Sepulveda Vet Admin Hosp, 71-, Rand Corp Study Cholecystectomy, 76; mem, Prog Comt, Soc Univ Surgeons, 74-77, Asn Acad Surg, 75-77; hosp rep, Am Col Surgeons Southern Calif chap, 72-76, pres, 87-88; mem, Long Range Planning Comt, Soc Surg Alimentary Tract, 74-76 pres, 86-87; pres, Int Biliary Asn, 79-81. *Mem:* Soc Clin Surg; Am Surg Asn; Soc Univ Surgeons; Soc Surg Alimentary Tract (secy, 82-86, pres, 86-87); Am Gastroenterol Asn. *Res:* Biochemical and nutritional research related to diseases of the gastrointestinal tract, especially hepatobiliary and pancreatic diseases. *Mailing Add:* Dept Surg Sch Med Ctr Health Sci UCLA Rm 74-121 CHS Los Angeles CA 90024

TOMPKINS, STEPHEN STERN, b Portsmouth, Va, Nov 1, 38; m 61; c 2. MATERIALS INTERACTION WITH SERVICE ENVIRONMENT, ENGINEER MECHANICS SCIENCE. *Educ:* Va Polytech Inst & State Univ, BS, 62; Univ Va, MAE, 68; Old Dominion Univ, PhD(mech eng), 78. *Prof Exp:* Res engr heat & mass transfer, 62-72, res scientist, 72-78, SR RES SCIENTIST MAT, LANGLEY RES CTR, NASA, 78- *Mem:* Am Soc Mech Engrs; Soc Advan Mat & Process. *Res:* Analysis of the thermal and mechanical response of complex metallic and non-metallic composite materials over a wide range of environmental exposures and loading conditions; heat and mass transfer. *Mailing Add:* Mail Stop 188B Hampton VA 23665

TOMPKINS, VICTOR NORMAN, b Milbrook, NY, May 30, 13; m 38; c 4. PATHOLOGY. *Educ:* Cornell Univ, AB, 34; Union Univ, NY, MD, 38; Am Bd Path, dipl, 46. *Prof Exp:* Resident path, New Eng Deaconess Hosp, 39-40, Albany Hosp, 40-41 & Pondville Hosp, 41-42; sr pathologist, Div Labs & Res, State Dept Health, NY, 47-49, asst dir in charge diag labs, 49-56, assoc dir, 56-58, dir, 58-68; PROF PATH, ALBANY MED COL, 60- *Concurrent Pos:* Assoc prof path, Albany Med Col, 53-60. *Mem:* AAAS; Am Soc Human Genetics; Am Soc Clin Path; Am Soc Exp Path; AMA; Sigma Xi. *Res:* Immunology. *Mailing Add:* 312 N Florida Ave No G123 Tarpon Springs FL 34689

TOMPKINS, WILLIS JUDSON, b Presque Isle, Maine, July 20, 41; m 67; c 2. BIOMEDICAL ENGINEERING, ELECTRICAL ENGINEERING. *Educ:* Univ Maine, BS, 63, MS, 65; Univ Pa, PhD(biomed eng), 73. *Prof Exp:* Elec engr res & develop, Sanders Assoc Inc, 65-68; assoc biomed comput, Univ Pa, 73-74; PROF RES & TEACHING, UNIV WIS, 74- *Concurrent Pos:* Pres, Inst Elec E Electronics Engrs. *Mem:* Sr mem Inst Elec & Electronics Engrs; Biomed Eng Soc; Asn Adv Med Instrumentation; Am Soc Eng Educ. *Res:* Computers in medicine; microcomputer-based medical instruments; electrocardiography. *Mailing Add:* Dept Elec & Comput Eng Univ Wis 1425 Johnson Dr Madison WI 53706

TOMPSETT, MICHAEL F, b Eng, May 4, 39; m 67; c 4. ELECTRONICS. *Educ:* Cambridge Univ, BA, 62, MA & PhD(elec eng), 66. *Prof Exp:* Res asst mat sci, Cambridge Univ, 65-66; proj leader camera tubes, Eng Elec Valve Co, 66-69; SUPVR DATA CONVERSION DESIGN GROUP, AT&T BELL LABS, 69- *Mem:* Fel Inst Elec & Electronics Engrs. *Res:* Integrated circuits; charge coupled devices; imaging devices; integrated filters; analog circuits. *Mailing Add:* Data Conversion Design Group 67 Oakridge Ave Summit NJ 07901

TOMPSETT, RALPH RAYMOND, b Tidioute, Pa, Oct 8, 13; m 42; c 4. INTERNAL MEDICINE. *Educ:* Cornell Univ, AB, 34, MD, 39. *Prof Exp:* From instr to assoc prof med, Med Col, Cornell Univ, 46-57; PROF MED, UNIV TEX HEALTH SCI CTR, DALLAS, 57- *Concurrent Pos:* Attend physician, Med Ctr, Baylor Univ, 57- & Parkland Mem Hosp, 60- *Mem:* Am Soc Clin Invest; Asn Am Physicians; master Am Col Physicians; Am Fedn Clin Res. *Res:* Infectious diseases. *Mailing Add:* 3500 Gaston Ave Dallas TX 75246

TOMPSON, CLIFFORD WARE, b Mexico, Mo, Dec 12, 29; m 51; c 3. PHYSICS. *Educ:* Univ Mo, BS, 51, AM, 56, PhD(physics), 59. *Prof Exp:* Physicist, US Navy Electronics Lab, San Diego, 51-55; assoc prof, 59-72, PROF PHYSICS, UNIV MO-COLUMBIA, 72- *Mem:* Am Phys Soc; Am Asn Physics Teachers. *Res:* X-ray diffraction; neutron diffraction; structure of liquids; lattice vibrations. *Mailing Add:* Dept Physics Univ Mo Columbia MO 65201

TOMPSON, ROBERT NORMAN, b Adrian, Mich, Jan 7, 20; m 47; c 1. MATHEMATICAL ANALYSIS, APPLIED MATHEMATICS. *Educ:* Adrian Col, ScB, 41; Univ Nev, MS, 49; Brown Univ, PhD(math), 53. *Prof Exp:* Res inspector ord mat, US War Dept, 42-43; from asst to instr math, Univ Nev, 46-49; asst prof, Fla State Univ, 53-54; mem tech staff, Bell Tel Labs, Inc, 54-55; asst prof, Fla State Univ, 55-56; mathematician & programmer, Int Bus Mach Corp, 56; assoc prof, 56-64, PROF MATH, UNIV NEV, RENO, 64-, CHMN DEPT, 78- *Concurrent Pos:* Consult, NSF/AID Sci Asst to India Prog, 67-68. *Mem:* AAAS; Am Math Soc; Am Phys Soc. *Res:* Measure and integration theory; topology; systems theory; probability theory. *Mailing Add:* Dept Math Univ Nev Reno NV 89507

TOMSON, MASON BUTLER, b Syracuse, Kans, Nov 18, 46; m 68. PHYSICAL CHEMISTRY, ENVIRONMENTAL SCIENCES. *Educ:* Southwestern State Univ, BS, 67; Okla State Univ, PhD(chem), 72. *Prof Exp:* Teaching asst, Okla State Univ, 67-72; instr, Dept Chem, State Univ NY, Buffalo, 72-76, res asst prof, 76-77; asst prof, 77-81, ASSOC PROF ENVIRON SCI & ENG, RICE UNIV, 81- *Concurrent Pos:* Compiler, IUPAC Solubility Data Proj, 78-79. *Mem:* Am Chem Soc. *Res:* Kinetics and thermodynamics of precipitation and dissolution of sparingly soluble salts; solution equilibria of electrolytes; trace level organics, analysis and fate in water. *Mailing Add:* Dept Environ Sci & Eng Rice Univ PO Box 1892 Houston TX 77251

TOMUSIAK, EDWARD LAWRENCE, b Edmonton, Alta, Mar 3, 38; m 61; c 1. THEORETICAL NUCLEAR PHYSICS. *Educ:* Univ Alta, BSc, 60, MSc, 61; McGill Univ, PhD(theoret physics), 64. *Prof Exp:* NATO overseas fel, Oxford Univ, 64-66; from asst prof to assoc prof, 66-76, PROF PHYSICS, UNIV SASK, 76- *Mem:* Can Asn Physicists. *Res:* Nuclear structure calculations using realistic two-nucleon potentials; nuclear models and electromagnetic interactions with nuclei. *Mailing Add:* Dept Physics Univ Sask Saskatoon SK S7N 0W0 Can

TON, BUI AN, b Hanoi, Vietnam, Jan 23, 37; m 63; c 3. MATHEMATICAL ANALYSIS. *Educ:* Saigon Univ, BSc, 59; Mass Inst Technol, PhD(math), 64. *Prof Exp:* Res staff mathematician, Yale Univ, 63-64; vis asst prof, Math Res Ctr, Univ Wis, 64-65; Nat Res Coun Can fel, 65-66; asst prof math, Univ Montreal, 66-67; assoc prof, 67-71, PROF MATH, UNIV BC, 71- *Mem:* Can Math Cong. *Res:* Partial differential equations. *Mailing Add:* Dept Math Univ BC Vancouver BC V6T 1W5 Can

TONASCIA, JAMES A, b Los Banos, Calif, Mar 2, 44; m 65. BIOSTATISTICS. *Educ:* Univ San Francisco, BS, 65; Johns Hopkins Univ, PhD(biostatist), 70. *Prof Exp:* Asst prof, 70-77, ASSOC PROF BIOSTATIST, JOHNS HOPKINS UNIV, 77- *Mem:* Am Statist Asn; Biomet Soc; Inst Math Statist; Math Asn Am; Royal Statist Soc; Sigma Xi. *Res:* Biostatistical methods; epidemiology; statistical computing. *Mailing Add:* 4031 Deepwood Rd Baltimore MD 21213

TONDEUR, PHILIPPE, b Zurich, Switz, Dec 7, 32; m 65. MATHEMATICS. *Educ:* Univ Zurich, PhD(math), 61. *Prof Exp:* Res fel math, Univ Paris, 61-63; lectr, Univ Zurich, 63-64; res fel, Harvard Univ, 64-65; lectr, Univ Calif, Berkeley, 65-66; assoc prof, Wesleyan Univ, 66-68; PROF MATH, UNIV ILL, URBANA, 68- *Mem:* Am Math Soc; Math Soc France; Swiss Math Soc. *Res:* Geometry and topology. *Mailing Add:* Dept Math Univ Ill Urbana IL 61801

TONDRA, RICHARD JOHN, b Canton, Ohio, Jan 23, 43; m 66; c 2. MATHEMATICS. *Educ:* Univ Notre Dame, BS, 65; Mich State Univ, MS, 66, PhD(topology, manifold theory), 68. *Prof Exp:* Assoc prof, 68-80, PROF MATH, IOWA STATE UNIV, 80- *Mem:* Am Math Soc. *Res:* Topological and piecewise linear manifold theory. *Mailing Add:* 122 S Riverside Dr Ames IA 50010

TONE, JAMES N, b Grinnell, Iowa, Feb 9, 33; m 54; c 2. ANIMAL PHYSIOLOGY. *Educ:* Coe Col, BA, 54; Drake Univ, MA, 61; Iowa State Univ, PhD(animal physiol), 63. *Prof Exp:* PROF PHYSIOL, ILL STATE UNIV, 63- *Mem:* AAAS. *Res:* Physiological effects of gossypol on mammals. *Mailing Add:* Dept Biol Sci Ill State Univ Normal IL 61761

TONEGAWA, SUSUMU, Nagoya, Japan, Sept 5, 39; m; c 2. IMMUNOLOGY, IMMUNOBIOLOGY. *Educ:* Kyoto Univ, Japan, BS, 63; Univ Calif, San Diego, PhD(biol), 68. *Hon Degrees:* PhD physiol, Northwestern Univ, 87. *Prof Exp:* Res asst, Univ Calif, San Diego, 63-64, teaching asst, 64-68; staff mem immunol, Basel Inst Switzerland, 71-81; PROF BIOL, CTR CANCER RES & DEPT BIOL, MASS INST TECHNOL, 81- *Concurrent Pos:* Howard Hughes Med Inst investr, 88- *Honors & Awards:* Nobel Prize in Physiol & Med, 87; Albert & Mary Lasker Award, Albert & Mary Lasker Found, 87; Kihara Prize Japanese Soc Genetics, Kyoto, Japan, 88. *Mem:* Foreign assoc Nat Acad Sci; Scand Soc Immunol; Am Asn Immunol; hon mem Scand Soc Immunol; Am Acad Arts & Sci. *Res:* Molecular biology. *Mailing Add:* Ctr Cancer Res Bldg E17 Rm 353 Mass Inst Technol 77 Massachusetts Ave Cambridge MA 02139

TONELLI, ALAN EDWARD, b Chicago, Ill, Apr 14, 42; m 74; c 2. POLYMER PHYSICS. *Educ:* Univ Kans, BS, 64; Stanford Univ, PhD(polymer chem), 68. *Prof Exp:* MEM TECH STAFF POLYMER PHYSICS, BELL LABS, 68- *Mem:* Am Chem Soc; Am Phys Soc. *Res:* Study of the conformations and physical properties of synthetic and biological macromolecules. *Mailing Add:* 42 W Washington Rock Rd Green Brook NJ 08812-2598

TONELLI, GEORGE, b Tenafly, NJ, Feb 20, 21; m 55; c 4. VETERINARY MEDICINE. *Educ:* Parma Univ, DVM, 48. *Prof Exp:* Biologist, Peters Serum Co, Kans, 50; res vet, Animal Indust Sect, Am Cyanamid Co, 51-55, group leader & pharmacologist, Exp Therapeut Sect, 55-60, group leader endocrinol, Endocrine Res Dept, 60-68, group leader, Toxicol Dept, 68-73, mgr toxicol/pharmacol eval, Int Div, 73-77, mgr overseas toxicol, Med Res Div, Lederle Labs, 77-88; RETIRED. *Mem:* Am Soc Pharmacol & Exp Therapeut; Endocrine Soc. *Mailing Add:* 83 N Little Tor Rd New City NY 10956

TONELLI, JOHN P, JR, b Grove City, Pa, Nov 27, 46; c 1. REMOTE SENSING FOR FORESTRY, MATERIALS RESEARCH. *Educ:* Slippery Rock State Univ, BA, 68; State Univ NY, MA, 81. *Prof Exp:* Res chemist, Int Paper Corp Res, 68-69, sr res chemist, 69-79, res technologist, 79-83, res & develop assoc, 83-87, sr res & develop assoc, 88-91, RES SCIENTIST, INT PAPER CORP RES, 91- *Concurrent Pos:* Res & develop assoc, Com Uses Space, NASA, 85-89. *Mem:* Am Soc Photogram & Remote Sensing; Tech Asn Pulp & Paper Indust. *Res:* Use of remote sensing technologies to obtain information on land resources; impact of companies products on environment; determine wood property-pulp, paper and solid wood product property relationships; evaluate new technologies that can be applied to paper industry. *Mailing Add:* 160 Country Club Dr Florida NY 10921

TONER, JOHN JOSEPH, b Mineola, NY, Oct 12, 55; m 74. CONDENSED MATTER PHYSICS. *Educ:* Mass Inst Tech, BS, 77; Harvard Univ, MA, 79, PhD(physics), 81. *Prof Exp:* Postdoctoral fel, physics, Univ Chicago, 81-83; RES STAFF MEM, IBM THOMAS J WATSON RES CTR, 83- *Concurrent Pos:* Vis Scientist, IBM, TJ Watson Res Ctr, 82; vis scientist, Nordita, Copenhagen, 83; vis prof, Domiaine Univ, Bordeaux, 85. *Res:* Fluctuating hydrodynamics; hydrodynamics of novel condensed matter systems; macroscopic properties of quasicrystals and liquid crystals; phase transitions and critical phenomena; fractal structures in disk packings; glasses; fluctuating membranes; self-organized criticality; disordered super conductors. *Mailing Add:* IBM Res Div Thomas Watson Res Ctr Yorktown Heights NY 10598

TONER, RICHARD K(ENNETH), b Terre Haute, Ind, Jan 9, 13; m 37; c 1. CHEMICAL ENGINEERING. *Educ:* Rose-Hulman Inst Technol, BS, 34; Purdue Univ, MS, 36, PhD(chem eng), 39. *Hon Degrees:* DE, Rose-Hulman Inst Technol, 72. *Prof Exp:* Instr chem eng, Purdue Univ, 37-39 & Lehigh Univ, 39-40; instr, NY Univ, 40-42; from asst prof to prof, 42-81, assoc chmn dept, 66-81, EMER PROF CHEM ENG, PRINCETON UNIV, 81- *Concurrent Pos:* Consult chem engr, Textile Res Inst, 44-59, ed, Textile Res Jour & dir publ, 59-; vis prof, NY Univ, 42-46. *Mem:* Am Chem Soc; Am Soc Eng Educ; Am Inst Chem Engrs; Sigma Xi; hon fel Fiber Soc. *Res:* Thermodynamics. *Mailing Add:* Chem Eng Lab Princeton NJ 08540

TONEY, JOE DAVID, b Rosston, Ark, Aug 12, 42; m 64; c 2. INORGANIC CHEMISTRY. *Educ:* Univ Ark, Pine Bluff, BS, 64; Univ Ill, Urbana, PhD(chem), 69. *Prof Exp:* Teacher, 69-77, ASSOC PROF CHEM, CALIF STATE UNIV, FRESNO, 77- *Concurrent Pos:* Vis res assoc chem, Argonne Nat Lab, 72; lectr health manpower prog, Pac Col, 73. *Mem:* Am Chem Soc. *Res:* Syntheses and structural characterization of transition metal chelate compounds involving amino acid ligands; visible, infrared and Raman studies of bonding in chelate complexes. *Mailing Add:* Dept Chem Calif State Univ 6241 N Maple Ave Fresno CA 93740

TONEY, MARCELLUS E, JR, b Baltimore, Md, Dec 25, 20; m 45; c 1. CLINICAL MICROBIOLOGY. *Educ:* Va Union Univ, BS, 42; Meharry Med Col, MT, 46; Cath Univ Am, MS, 53, PhD(zool), 56. *Prof Exp:* Assoc prof zool, 57-71, PROF BIOL, VA UNION UNIV, 71- *Concurrent Pos:* Vis prof, Va State Col, 59-; chief premed adv, Va Union Univ, 63-, coordr biol, 74-; res grants, NSF & Nat Urban League; consult, Richmond Math & Sci Ctr, NSF & Friends Adoption Asn; US Dept Health, Educ & Welfare grant pesticide res, 72-77. *Mem:* AAAS; Am Inst Biol Sci; Am Asn Biol Teachers; NY Acad Sci; Sigma Xi. *Res:* Microbiology; endocrinology. *Mailing Add:* 2349 Brook Rd Richmond VA 23220

TONG, ALEX W, b Hong Kong, Apr 8, 52; US citizen; c 1. TUMOR IMMUNOLOGY, SOMATIC CRYLINDJATION & MONOCLONAL ANTIBODIES. *Educ:* Univ Ore, BA, 73; Ore Health Sci Univ, PhD(microbiol & immunol), 80. *Prof Exp:* Postdoctoral fel immunother, Surg Res Lab, Portland Vet Admin Med Ctr, 80-82; res assoc immunol, Dept Microbiol & Immunol, Sch Med, Ore Health Sci Univ, 81-82; res assoc cancer immunol, 82-86, ASSOC DIR CANCER IMMUNOL, IMMUNOL LAB, BAYLOR-SAMMONS CANCER CTR, 86-; ASST PROF BIOMED SCI, INST BIOMED STUDIES, BAYLOR UNIV, WACO, 88- *Concurrent Pos:* Adj fac, Immunol Grad Study Prog, Southwestern Med Ctr, Univ Tex, Dallas, 82-; prin investr, Monoclonal Antibody Res, Baylor Res Inst, 86- *Mem:* Am Asn Immunologists; Am Asn Cancer Res; Am Soc Hemat; Clin Immunol Soc; Int Soc Hemat; AAAS. *Res:* Development of murine monoclonal antibodies potentially useful for diagnosis and treatment of human cancers, study of tumor antigen expression in human lung, liver cancers and hematologic malignancies; characterization of pathogenetic mechanisms of human plasma cell dryserasia including oneogine expression, cytobine inodulation and antigenic stimulation. *Mailing Add:* Cancer Immunol Res Lab Baylor-Sammons Cancer Ctr Dallas TX 75246

TONG, BOK YIN, b Shanghai, China, Mar 5, 34; m 69; c 1. SOLID STATE PHYSICS, BIOPHYSICS. *Educ:* Univ Hong Kong, BSc, 57; Univ Calif, Berkeley, MA & MLS, 59; Univ Calif, San Diego, PhD(solid state physics), 67. *Prof Exp:* Asst librn, Oriental Collection, Univ Hong Kong, 59-61, asst lectr math, 61-65, lectr math, 65-67; from asst prof to assoc prof, 67-76, PROF PHYSICS, UNIV WESTERN ONT, 76- *Mem:* Am Phys Soc. *Res:* Theory of metals, surface physics and amorphous material; DNA molecules, muscle contraction and membrane activity; amorphous silicon; solar cells and devices; recrystallization of amorphous silicon. *Mailing Add:* Dept Physics Univ Western Ont London ON N6A 3K7 Can

TONG, JAMES YING-PEH, b Shanghai, China, Dec 8, 26; US citizen; m 51; c 3. PHYSICAL CHEMISTRY, FORENSIC CHEMISTRY. *Educ:* Univ Calif, Berkeley, BS, 50, MS, 51; Univ Wis, PhD(chem), 54. *Prof Exp:* Res chemist, Le Roy Res Lab, Durex Plastics & Chem, Inc, 53-54; res assoc phys inorg chem, Univ Ill, 54-57; from asst prof to assoc prof, 57-68, PROF CHEM, OHIO UNIV, 68-, DIR FORENSIC CHEM, 76-, DIR VIS SCHOLARS, 79- *Concurrent Pos:* Chmn, policy adv comt, Hocking River Basin, 80- *Mem:* Am Chem Soc; fel Am Inst Chemists. *Res:* Environmental water chemistry; forensic chemistry; chemistry and history of photography; equilibrium and kinetics studies. *Mailing Add:* 21 Rocking St Athens OH 45701

TONG, LONG SUN, b China, Aug 20, 15; US citizen; m 39. HEAT TRANSFER. *Educ:* Chinese Nat Inst Technol, BS, 40; Univ Fla, MS, 53; Stanford Univ, PhD(mech eng), 56. *Prof Exp:* Asst prof mech eng, Ord Eng Col, Taiwan, 47-52; sr engr, Atomic Power Dept, Westinghouse Elec Corp, 56-59, supvr thermal & hydraul design, Atomic Power Dept, 59-62, adv engr, 63-65, mgr thermal & hydraul design & develop, 65-66, mgr thermal & hydraul eng, 66-70, consult engr, PWR Syst Div, 70-72, sr consult, 72-73; asst dir, Div Reactor Safety Res, Nuclear Regulatory Comn, 73-81, chief scientist, 81-83; RETIRED. *Concurrent Pos:* Lectr, Univ Pittsburgh, 57-60, adj prof, 65-72; lectr, Carnegie Inst Technol, 61-67. *Honors & Awards:* Mem Award, Am Soc Mech Engrs, 73; Don Q Kern Award, Am Inst Chem Engrs, 81. *Mem:* Fel Am Soc Mech Engrs; fel Am Nuclear Soc. *Res:* Fluid flow; thermal and hydraulic design; analysis and development of pressurized water reactors; research in water reactor safety. *Mailing Add:* 9733 Lookout Pl Gaithersburg MD 20879

TONG, MARY POWDERLY, b New York, NY, May 24, 24; m 56; c 5. MATHEMATICS. *Educ:* St Joseph's Col, NY, BA, 50; Columbia Univ, MA, 51, PhD, 69. *Prof Exp:* Instr math, St Joseph's Col, NY, 51-54, City Col New York, 54 & Columbia Univ, 54-59; asst prof, Univ Conn, 60-66; from asst prof to assoc prof, Fairfield Univ, 66-70; prof math, William Paterson Col NJ, 70-84; RETIRED. *Concurrent Pos:* Delta Epsilon Sigma res fel, 68; NSF fac fel, 59-60. *Mem:* Am Math Soc; Math Asn Am; NY Acad Sci. *Res:* Topology; foundations of mathematics; applications of mathematics. *Mailing Add:* 725 Cooper Ave Oradell NJ 07649

TONG, PIN, b Kwang Tung, China, Dec 25, 37; m 63; c 4. STRUCTURAL MECHANICS. *Educ:* Nat Taiwan Univ, BS, 60; Calif Inst Technol, MS, 63, PhD(aeronaut, math), 66. *Prof Exp:* Sr res engr, Aeroelastic & Struct Res Lab, Mass Inst Technol, 66-67, from asst prof to assoc prof aeronaut, 67-74; adv, US Dept Transp, 74-78, Chief Struct & Dynamics Div, Transp Systs Ctr, 78-90; PROF & HEAD, MECH ENG DEPT, HONG KONG UNIV SCI & TECHNOL, 90- *Concurrent Pos:* Adj prof, Northeastern Univ, 71-73; vis prof appl mech, Univ Calif, San Diego, 73-74; hon prof, Acad Railway Sci, China. *Honors & Awards:* Von Karmen Mem Award Outstanding Contrib Aerospace Struct-Mat Technol, 74. *Res:* Continuum mechanics and its application to structural and vehicle dynamics; fracture mechanics; computational mechanics; finite element methods. *Mailing Add:* Transp Systs Ctr DTS-76 US Dept Transp Kendall Sq Cambridge MA 02142

TONG, STEPHEN S C, b Shanghai, China, May 3, 36; US citizen; m 67; c 2. ANALYTICAL CHEMISTRY, PHYSICAL CHEMISTRY. *Educ:* Univ Ottawa, BSc, 59; Mass Inst Technol, MS, 61; Cornell Univ, PhD(analytical chem), 66. *Prof Exp:* Res chemist, Rohm & Hass Co, 61-62; fel, Argonne Nat Lab, 66-68; SR RES CHEMIST, CORNING INC, 68-, RES ASSOC CHEMIST, RES & DEVELOP LAB, 80- *Mem:* Mat Res Soc; Am Chem Soc; Am Soc Mass Spectrometry; Sigma Xi; Soc App Spectros. *Res:* Trace analysis; spark source mass spectrometry; gas chromatography; neutron activation analysis; electron microprobe; secondary ion mass spectrometry; thermal analysis; gas chromatography-mass spectrometry. *Mailing Add:* Corning Inc SP-FR-41 Corning NY 14831

TONG, THEODORE G, b La Jolla, Calif, Oct 8, 42. PHARMACOLOGY, TOXICOLOGY. *Educ:* Univ Southern Calif, BS, 64; Ore State Univ, BS, 65; Univ Calif, PharmD, 69. *Prof Exp:* Instr nursing, Contra Costa Col, 72-74; clin instr pharm, Sch Pharm, Univ Calif, San Francisco, 69-73, asst clin prof, 73-78, assoc clin prof, 78-82; dir, San Francisco Bay Area Poison Control Ctr, 79-82; PROF PHARM, CLIN PHARMACOL & TOXICOL, COL PHARM, UNIV ARIZ, 82-, ASSOC DEAN ACAD AFFAIRS, COL PHARM, 87-; DIR, ARIZ POISON CONTROL SYST, 82- *Concurrent Pos:* Assoc, div clin Pharm & Toxicol, San Francisco Gen Med Ctr, 73-82; co-investr, Drug Dependence Clin Res Prog, Univ Calif, 79-82; ed, Poison Line, Am Asn Poison Control Ctrs, 82-85; sec, Am Asn Poison Control Ctrs, 86-; dip, Am Bd Appl Toxicol, 88-; mem, US Pharmacopeia Comn Rev, 90-95. *Mem:* Sigma Xi; fel Am Acad Clin Toxicol; Am Col Clin Pharm; Am Soc Clin Pharmacol & Therapeut; Am Asn Poison Control Ctrs; Am Pharm Asn. *Res:* Pharmacologic interactions of drugs and drugs of abuse, misuse and in overdose; toxicokinetics; poison control systems. *Mailing Add:* Col Pharm Univ Ariz Tucson AZ 85721

TONG, WINTON, b Los Angeles, Calif, May 3, 27; m 51; c 3. PHYSIOLOGY, BIOCHEMISTRY. *Educ:* Univ Calif, Berkeley, BS, 47, PhD(thyroid function), 53. *Prof Exp:* Res physiologist, Univ Calif, Berkeley, 47-62; from asst prof to assoc prof, 62-71, dir grad studies, 70-72, PROF PHYSIOL, SCH MED, UNIV PITTSBURGH, 71-, CHMN DEPT, 81- *Concurrent Pos:* USPHS res career develop award, 62-72; assoc ed, Am J Physiol, 77- *Honors & Awards:* Van Meter Prize, Am Thyroid Asn, 64. *Mem:* AAAS; Am Soc Biol Chemists; Am Physiol Soc; Endocrine Soc; Am Thyroid Asn. *Res:* Mechanisms in the biosynthesis of thyroid hormones and thyroglobulin; mechanism of action of thyrotropin; physiology of thyroid function; cultivation of thyroid cells in vitro. *Mailing Add:* Dept Physiol Scaite Hall 614 Univ Pittsburgh 4200 Fifth Ave Pittsburgh PA 15261

TONG, YULAN CHANG, b Nanking, China, Oct 21, 35. ORGANIC CHEMISTRY. *Educ:* Cheng Kung Univ, Taiwan, BS, 56; Univ Ill, MS, 58, PhD(org chem), 61. *Prof Exp:* Res assoc, Univ Mich, 61-62; org res chemist, Edgar C Britton Res Lab, Mich, 62-65, res chemist, Res Lab, Western Div, 66-70, sr res chemist, 70-72, res specialist, 72-78, sr research specialist, 78-80, RES ASSOC, RES LAB, WESTERN DIV, DOW CHEM CO, 80- *Mem:* Am Chem Soc; Int Soc Heterocyclic Chem; NY Acad Sci. *Res:* Heterocyclic chemistry. *Mailing Add:* 567 Monarch Ridge Dr Walnut Creek CA 94596-2797

TONG, YUNG LIANG, b Shantung, China, July 15, 35; m 65; c 3. STATISTICS, MATHEMATICS. *Educ:* Nat Taiwan Univ, BS, 58; Univ Minn, Minneapolis, MA, 63, PhD(statist), 67. *Prof Exp:* Asst prof statist, Univ Nebr, Lincoln, 67-69; vis asst prof, Univ Minn, 69-70; from asst prof to assoc prof statist, Univ Nebr-Lincoln, 70-76, prof, 76-; MATH DEPT, GA

TECH RES CORP. *Concurrent Pos:* Vis prof, Univ Calif, Santa Barbara, 78-79. *Mem:* Am Statist Asn; Inst Math Statist. *Res:* Mathematical and applied statistics; multivariate analysis; sequential analysis; probability inequalities. *Mailing Add:* Dept Math Ga Inst Technol Atlanta GA 30332

TONIK, ELLIS J, b Philadelphia, Pa, Jan 9, 21; m 48; c 3. MEDICAL MICROBIOLOGY. *Educ:* Roanoke Col, BS, 50. *Prof Exp:* Med bacteriologist diag bact, Dept Pub Welfare, Ill, Chicago State Hosp, 50-51, supvry bacteriologist, East Moline State Hosp, 51-52 & Kankakee State Hosp, 52; med bacteriologist, Process Res & Pilot Plants Div, Chem Corps Res & Develop Labs, US Dept Army, 52-60, supvry bacteriologist, Tech Eval Div, 60, actg chief animal path sect, 60-61, chief exp animal sect, Appl Aerobiol Div, 61-71, sr investr, Microbiol Res Div, 71-72; chief microbiol, Ft Howard Vet Admin Hosp, 72-79; RETIRED. *Mem:* Am Soc Microbiol; Sigma Xi. *Res:* Experimental respiratory diseases of laboratory animals; aerobiological research and technology; experimental and clinical pathology; immunology; chemotherapeutic agents; virulence of airborne particulates; laboratory diagnosis and assay methods; management of laboratory animals. *Mailing Add:* 526 Mary St Frederick MD 21701

TONKING, WILLIAM HARRY, b Newton, NJ, Apr 22, 27; m 64; c 2. MINING GEOLOGY. *Educ:* Princeton Univ, AB, 49, PhD(geol), 53. *Prof Exp:* Asst geol, Princeton Univ, 49-50 & 51-53; asst instr, Northwestern Univ, 50-51; geologist, Bear Creek Mining Co, 53-55 & Stand Oil Co, Tex, 55-62; dep mgr, Mohole Proj, 62-67, mgr spec projs, 67-75, SR MGR, MINING & GEOL & CHIEF GEOLOGIST, BROWN & ROOT INC, 75- *Honors & Awards:* Silver Medallist, Royal Soc Arts, 66. *Mem:* Geol Soc Am; Royal Soc Arts. *Res:* Petrology; volcanic rocks and ore deposits in the Southwest; petroleum geology; deep ocean engineering, geology and geophysics. *Mailing Add:* 12319 Rip Van Winkle Houston TX 77024

TONKS, DAVID BAYARD, b Edmonton, Alta, Aug 31, 19; m 46; c 2. BIOCHEMISTRY, CLINICAL CHEMISTRY. *Educ:* Univ BC, BA, 41; McGill Univ, PhD(org chem), 49. *Prof Exp:* Sr chemist, Clin Labs, Lab Hyg, Dept Nat Health & Welfare, Can, 48-57; asst biochemist, Biochem Dept & Res Inst, Hosp Sick Children, Toronto, 57-62; from asst prof to assoc prof, 62-75, PROF MED, MCGILL UNIV, 75-; DIR, DIV CLIN CHEM, DEPT MED, MONTREAL GEN HOSP, 62- *Concurrent Pos:* Tech dir, Seaforth Clin Labs, Montreal, 64-77; lab consult, Douglas Hosp, Verdun, 65-, Reddy Mem Hosp, Montreal, 66-, & Cybermedix Ltd, Toronto, 69-; mem bd dirs, Bio Res Labs, Point Claire, 65-72; Can nat rep, Int Comn Clin Chem, 66-70; secy sect clin chem, Int Union Pure & Appl Chem, 67-71, pres sect, 71-75, bur mem, 71-75, past-pres, 75-77, mem comn toxicol, 73-, mem Can nat comt, 74-; mem exec bd, Int Fedn Clin Chem, 67-75. *Honors & Awards:* Warner-Chilcott Award, Can Soc Lab Technologists, 67; Ames Award, Can Soc Clin Chem, 68; Ann Award, Que Corp Hosp Biochemists, 72. *Mem:* Am Asn Clin Chemists; Can Biochem Soc; fel Chem Inst Can; Can Soc Clin Chem (from secy to pres, 57-66, treas, 74-76); NY Acad Sci. *Res:* Development of synthetic antigens for serodiagnosis of syphilis; quality control and evaluation of laboratory precision in clinical chemistry laboratories; analytical methods in clinical chemistry. *Mailing Add:* 1295 Carson Ave Dorval PQ H9S 1M5 Can

TONKS, DAVIS LOEL, b Pocatello, Idaho, May 6, 47; m 71; c 5. PLASTIC FLOW IN METALS, FRACTURE. *Educ:* Brigham Young Univ, BSc, 72; Univ Utah, PhD(physics), 78. *Prof Exp:* researcher, dept physics, Ariz State Univ, 78-81; researcher, 81-82, code developer, 83-85, RESEARCHER PLASTICITY'S FRACTURE, LOS ALAMOS NAT LAB, 86- *Concurrent Pos:* Res assoc, Ariz State Univ, 78-80. *Mem:* Am Phys Soc. *Res:* Modeling and theory of high-strain-rate plasticity and fracture in metals. *Mailing Add:* T-1 MS B221 Los Alamos Nat Lab Los Alamos NM 87545

TONKS, ROBERT STANLEY, b Aberystwyth, Wales; m 53; c 4. PATHOLOGY, PHARMACY. *Educ:* Univ Wales, BS, 51, Med Sch, PhD(pharmacol), 54; Inst Biol, Eng, fel, 73. *Prof Exp:* Organon fel, Welsh Nat Sch Med, Cardiff, 53-54, Nat Health Serv fel, Nevill Hall Hosp, 54-56, sr fel, 56-58, univ lectr, Dept Mat Media & Pharmacol & Therapeut, 58-72, sr lectr, 72-73; prof & dir, Col Pharm, Dalhousie Univ, 73-77, dean, fac health professions, 77-89. *Concurrent Pos:* Chmn, Pharmaceut Sci Grant Comt, Med Res Coun Can, 76-77, Co-Chmn Northeast Can-Am Health Coun, 80-; mem, Personel Review Comt, Nat Health & Welfare Can, 79-81, 82-84, Med Res Coun, Joint Comt on Nursing Res, 88-; consult, NB Dept Health, 74-, Health & Welfare Can, 77-, Rector Riyadh Univ & Ministers Health & Educ, Saudi Arabia, 80, Pew Charitable Trusts,88. *Mem:* Brit Pharmacol Soc; Int Soc Thrombosis & Haemostasis; Can Soc Clin Invest. *Res:* Platelet micro-emboli in circulating human blood: their role in the production of primary pulmonary thrombosis and myocardiol infarction in man and during the immune response, explaining tissue and organ transplant refection; introduced aspirin for the treatment of recurrent myocardial infarction in the early sixties. *Mailing Add:* Dalhousie Univ Coburg Rd Halifax NS B3H 1R8 Can

TONKYN, RICHARD GEORGE, b Portland, Ore, Mar 26, 27; c 6. ORGANIC CHEMISTRY, POLYMER CHEMISTRY. *Educ:* Reed Col, BA, 48; Univ Ore, MA, 51; Univ Wash, PhD(org chem), 60. *Prof Exp:* Instr org chem, Univ Ore, 52; supvr res, Anal Labs, Titanium Metals Corp Am, Nev, 52-54; sr res anal chemist, Allegheny Ludlum Steel Corp, Pa, 54-55; res engr, Boeing Airplane Co, Wash, 55-59; NSF fel, Univ Col, Univ London, 60-61; chemist, Union Carbide Corp, NJ, 61-67; proj scientist, 67-69; sr res chemist, Betz Labs, Inc, 69-70, group leader, 70-72, mgr org res & process develop, 72-76; dir res & develop, Mogul Corp, 76-77, vpres res & develop, Mogul Div, Dexter Corp, 77-85; VPRES RES & DEVELOP, PETROLITE CORP, 85- *Mem:* Am Chem Soc; Sigma Xi; Cooling Tower Inst. *Res:* Monomer and polymer synthesis; oil and gas production chemicals for demulsification, corrosion, oilfieldwater treatment; polyelectrolyte synthesis; chemicals and processes for water treatment and water pollution control; corrosion and deposit control research; industrial cooling and boiler. *Mailing Add:* Petrolite Corp 369 Marshall Ave St Louis MO 63119

TONN, ROBERT JAMES, b Watertown, Wis, June 23, 27; m 61; c 2. RESEARCH ADMINISTRATION. *Educ:* Colo State Univ, BS, 49, MS, 50; Okla State Univ, PhD(entom), 59, MPH, 63. *Prof Exp:* Res assoc, Sch Med, La State Univ, 61-63; dir, Encephalitis Field Sta, Mass, 63-65; proj leader, WHO, 65-78, chief planning mgt, 83-87; regional adv entom, Pan Am Health Orgn, 78-83; ADJ PROF, UNIV TEX, EL PASO, 87- *Concurrent Pos:* Consult, Vector Biol & Control, USAID, 87- *Mem:* Am Soc Trop Med & Hyg; Soc Vector Ecol (pres, 84); Am Mosquito Control Asn; US-Mex Pub Health Border Asn. *Res:* Vector borne disease control. *Mailing Add:* 4277 Winchester Las Cruces NM 88001

TONNA, EDGAR ANTHONY, b Malta, May 10, 28; nat US; m 51; c 4. CELL PHYSIOLOGY, CELL CHEMISTRY. *Educ:* St John's Univ, NY, BS, 51; NY Univ, MS, 53, PhD(biol), 56. *Prof Exp:* Res collabr div exp path, Med Res Ctr, Brookhaven Nat Lab, 56-59, head histochem & cytochem res lab, 59-67; PROF HISTOL, GRAD SCH BASIC MED SCI, NY UNIV, 67-, DIR, LAB CELLULAR RES, 67-, DIR, INST DENT RES, COL DENT, 71-, CHMN, DEPT HISTOL & CELL BIOL, 81- *Concurrent Pos:* Res biochemist, Hosp Spec Surg, New York, 53-56, head histochem & cytochem res lab, 56-59; adj assoc prof, Grad Sch, Long Island Univ, 56-62; consult radiobiol, Inst Dent Res, NY Univ, 64-67; ed chief, Gerodontology, 81- *Mem:* Fel Geront Soc; fel NY Acad Sci; Histochem Soc; fel Royal Micros Soc; Sigma Xi. *Res:* Cellular contribution to skeletal and dental development, growth, repair and disease during aging; autoradiographic, cytochemical and cytological studies using optical analytical and electron microscopic techniques to determine biochemical and cell morphological changes in skeletal and dental cell parameters; cell gerontology. *Mailing Add:* Inst Dent Res NY Univ Col Dent 345 E 24th St New York NY 10010

TONNDORF, JUERGEN, physiology; deceased, see previous edition for last biography

TONNE, PHILIP CHARLES, b Chicago, Ill, Apr 2, 38; m 63; c 2. MATHEMATICS. *Educ:* Marquette Univ, BS, 60; Univ NC, MA, 63, PhD(math), 65. *Prof Exp:* Instr math, Univ NC, 65-66; asst prof, 66-71, ASSOC PROF MATH, EMORY UNIV, 71- *Mem:* Am Math Soc. *Res:* Classical analysis. *Mailing Add:* Dept Math Emory Univ 1364 Clifton Rd NE Atlanta GA 30322

TONNER, BRIAN P, b Jersey City, Sept 27, 53; m 84. SURFACE SCIENCE. *Educ:* Brown Univ, BSc, 76; Univ Pa, MSc, 78, PhD(physics), 82. *Prof Exp:* Fel physics, Univ Pa, 78-81; res assoc physics, Cornell Univ, 82-83; asst prof, 83-88, ASSOC PROF PHYSICS, UNIV WIS, MILWAUKEE, 88- *Mem:* Am Phys Soc; Am Vacuum Soc. *Res:* Applications of synchrotron radiation to condensed matter physics; photoemission, photo absorption and x-ray scattering. *Mailing Add:* Dept Physics Univ Wis PO Box 413 Milwaukee WI 53201

TONNIS, JOHN A, b Scottsburg, Ind, Apr 18, 39; m 72. ORGANIC CHEMISTRY. *Educ:* Hanover Col, BA, 61; Ind Univ, MS, 64, PhD(org chem), 68. *Prof Exp:* Res chemist, Reilly Tar & Chem Co, 64-65; fel org chem, Ind Univ, 68; asst prof, 68-70, assoc prof, 70-75, PROF CHEM, UNIV WIS-LA CROSSE, 75- *Mem:* AAAS; Am Chem Soc. *Res:* Preparation and use of sulfonamides as organic chelating reagents; new synthetic methods in organic chemistry. *Mailing Add:* Dept Chem Univ Wis 1725 St La Crosse WI 54601

TONRY, JOHN LANDIS, US citizen. EXTRAGALACTIC DYNAMICS, COSMOLOGY. *Educ:* Princeton Univ, AB, 75; Harvard Univ, MA, 76, PhD(physics), 80. *Prof Exp:* Mem, Inst Advan Study, 80-82; Bantrell fel, Calif Inst Technol, 82-85; asst prof, 85-90, ASSOC PROF, PHYSICS DEPT, MASS INST TECHNOL, 90- *Concurrent Pos:* Sloan Found fel, 86; NSF presidential young investr, 89. *Mem:* Am Astron Soc; Int Astron Union. *Res:* Observations of extragalactic objects intended to uncover the nature of dark matter; structure and evolution of the universe. *Mailing Add:* Dept Physics Rm 6-204 Mass Inst Technol Cambridge MA 02139

TON-THAT, TUONG, b Hue, Vietnam, Feb 1, 43; m 75; c 2. REPRESENTATIONS OF LIE GROUPS, MATHEMATICAL PHYSICS. *Educ:* Univ de Grenoble, France, Licence es Sciences, Maitrices es Sciences, 69; Univ Calif, MA, 71, PhD(math), 74. *Prof Exp:* Post doctoral res fel,math, Harvard Univ, 74-75; from asst prof to assoc prof, 75-83, PROF MATH, UNIV IOWA, 83- *Concurrent Pos:* Vis assoc prof, Univ Calif, Irvine, 81-82, Univ Southern Calif, 82-83. *Mem:* Am Math Soc; Math Asn Am. *Res:* Theory of representations of lie groups; harmonic analysis and the applications to quantum physics. *Mailing Add:* Dept Math Univ Iowa Iowa City IA 52242

TONZETICH, JOHN, b Nanaimo, BC, Oct 28, 41; m 78; c 2. GENETICS. *Educ:* Univ BC, BSc, 63; Duke Univ, MA, 67, PhD(zool), 72. *Prof Exp:* Asst prof, 70-77, ASSOC PROF BIOL, BUCKNELL UNIV, 77- *Concurrent Pos:* Vis prof, Univ Hawaii, 76-77, Univ BC, 85. *Mem:* Genetics Soc Am; Am Soc Zoologists. *Res:* Chromosomal inversions in Drosophila species; radiation induced aberrations; heterochromatin in Drosophila. *Mailing Add:* Dept Biol Bucknell Univ Lewisburg PA 17837

TOOHEY, RICHARD EDWARD, b Cincinnati, Ohio, Sept 2, 45; m 68. HEALTH PHYSICS. *Educ:* Xavier Univ, AB, 68; Univ Cincinnati, MS, 70, PhD(physics), 73. *Prof Exp:* Asst physicist, 73-79, biophysicist, 80-87, group leader, 83-87, health physics mgr, 87-90, DOSIMETRY & ANALYTICAL SERV MGR, ARGONNE NAT LAB, US DEPT ENERGY, 91- *Concurrent Pos:* Consult, Ill Emergency Serv & Disaster Agency, 77-; chmn, Plutonium Intercalibration Comt, US Dept Energy, 80-89. *Mem:* Health Physics Soc; Am Phys Soc; Sigma Xi. *Res:* Human radiobiology; whole-body counting; metabolism of radionuclides; radon measurements. *Mailing Add:* Environ Safety & Health Dept Argonne Nat Lab Argonne IL 60439-4831

TOOHIG, TIMOTHY E, b Lawrence, Mass, Feb 17, 28. EXPERIMENTAL HIGH ENERGY PHYSICS. *Educ:* Boston Col, BS, 51; Univ Rochester, MS, 53; Johns Hopkins Univ, PhD(physics), 62; Woodstock Col, STB, 64, STL, 65. *Prof Exp:* Asst dir admin res, Inst Natural Sci, Woodstock Col, 63-66; assoc physicist, Brookhaven Nat Lab, 67-70; assoc head neutrino lab sect, Nat Accelerator Lab, 70-74; asst to head res div, Fermi Nat Accelerator Lab, 74-76, head, Meson Dept, 77-78; res physicist, Joint Inst Nuclear Res, Dubna, USSR, 78-79; group leader construct & scheduling, Accelerator Div, 79-86, STAFF PHYSICIST, FERMI NAT ACCELERATOR LAB, 86-, DEP HEAD, CONVENTIONAL FAC DIV, URA SUPECONDUCTING, SUPERCOLLIDER, CENT DESIGN GROUP, 85- *Mem:* Am Phys Soc. *Res:* Elementary particle physics; investigation of the properties and interactions of elementary particles. *Mailing Add:* 2550 Beckleymeade Ave Dallas TX 75237-3946

TOOKE, WILLIAM RAYMOND, JR, b Atlanta, Ga, June 18, 25; m 48; c 3. CHEMICAL ENGINEERING. *Educ:* Ga Inst Technol, BChE, 49, MSChE, 55. *Prof Exp:* Res asst, Ga Technol Exp Sta, 49-55; process engr, Am Viscose Corp, 55-58; tech dir paint res, Oliver B Cannon & Sons, 58-60; sr res engr, Eng Exp Sta, Ga Inst Technol, 60-66; head indust prod br, 66-72, PRES, TOOKE ENG ASSOCS, 72- *Concurrent Pos:* Consult, Am Viscose Corp, 60-61, Southern Mills, Inc, 62-64 & Thomas Mfg Co, 64-; NIH grant, 63-67; mem comt A2GO2, Hwy Res Bd. *Mem:* Am Chem Soc; Am Inst Chem Engrs; Am Soc Testing & Mat; Nat Asn Corrosion Engrs; Sigma Xi. *Res:* Protective coatings and plastics technology; thermoplastic films, corrosion engineering; instrumentation and testing methods for coatings and plastics. *Mailing Add:* Tooke Eng Assocs PO Box 13804 Atlanta GA 30324

TOOKER, EDWIN WILSON, b Concord, Mass, May 9, 23; m 46; c 3. ECONOMIC GEOLOGY. *Educ:* Bates Col, BS, 47; Lehigh Univ, MS, 49; Univ Ill, PhD(geol), 52. *Prof Exp:* NSF fel, Univ Ill, 52-53; geologist, Base & Ferrous Metals Br & Pac Mineral Resources Br, 53-71, chief, Pac Mineral Resources Br, 71-72, chief, Off Mineral Resources, 72-76, RES GEOLOGIST MINERAL RESOURCES, WESTERN MINERAL RESOURCE BR, US GEOL SURV, 76- *Mem:* Soc Econ Geologists; Geol Soc Am; Sigma Xi. *Res:* Geology of base and precious metals ore deposits of Utah; metalogenesis and environment of ore deposition; wall rock alteration; base metal commodity resources; industrial rock & mineral resources. *Mailing Add:* 345 Middlefield Rd US Geol Surv MS 901 Menlo Park CA 94025

TOOKEY, HARVEY LLEWELLYN, b Hooper, Nebr, Dec 2, 22; m 50; c 4. BIOCHEMISTRY. *Educ:* Univ Nebr, AB, 44, MS, 50; Purdue Univ, PhD(biochem), 55. *Prof Exp:* Sr control chemist, Norden Lab, Nebr, 46-48; res chemist, USDA, 55-77, res leader natural toxicants res, Northern Regional Res Ctr, 77-83; RETIRED. *Mem:* Sigma Xi. *Res:* Enzymes of lipid metabolism; proteinases, enzymes acting on glucosinolates; chemistry of natural products; alkaloids; non-infectious diseases of cattle. *Mailing Add:* 5330 N Isabel Ave Peoria IL 61614

TOOLAN, HELENE WALLACE, PARVOVIRUS. *Educ:* Cornell Univ, PhD(path), 46. *Prof Exp:* Dir, Inst Med Res, Bennington, Vt, 62-85; RETIRED. *Mailing Add:* Three Monument Circle Bennington VT 05201

TOOLE, BRYAN PATRICK, b Clunes, Australia, Nov 6, 40; m 63; c 2. DEVELOPMENTAL BIOLOGY, CELL BIOLOGY. *Educ:* Univ Melbourne, BSc, 62; Monash Univ, Australia, MSc, 65, PhD(biochem), 68. *Prof Exp:* Instr med, Harvard Med Sch, 70-72, asst prof biochem, 72-75; asst biochemist med, Mass Gen Hosp, 72-78, assoc biochemist, 78-80; assoc prof anat, Med Sch, Harvard Univ, 75-80; PROF ANAT & CELLULAR BIOL, SCH MED, TUFTS UNIV, 80- *Concurrent Pos:* Estab investr, Am Heart Asn, 73-78, mem, basic sci coun, 76-; prin investr, Nat Inst Dent Res, 75- *Mem:* Complex Carbohydrate Soc; Soc Develop Biol; Am Soc Cell Biol. *Res:* Role of extracellular macromolecules in embryonic development and adult tissue remodelling and their influence on normal and aberrant cell behavior. *Mailing Add:* Dept Anat & Cell Biol Sch Med Tufts Univ 136 Harrison Ave Boston MA 02111

TOOLE, FLOYD EDWARD, b Moncton, NB, June 19, 38; m 61. PSYCHOACOUSTICS. *Educ:* Univ NB, BSc, 60; Univ London, PhD(elec eng) & DIC, 65. *Prof Exp:* Res officer, 65-80, SR RES OFFICER PHYSICS & ACOUST, NAT RES COUN CAN, 80- *Honors & Awards:* Publ Award, Audio Eng Soc, 88 & 90. *Mem:* Acoust Soc Am; fel Audio Soc Am; Can Acoust Asn; fel Audio Eng Soc. *Res:* Sound reproduction; loudspeaker systems; perception of sound and psychoacoustics; noise control; subjective measurements of listener preferences in loudspeakers and reproduced sound in rooms; relationship between these listener preferences and technical measurements on loudspeakers; establishing standard design and evaluation objectives for loudspeakers and rooms. *Mailing Add:* Inst Microstruct Sci Nat Res Coun Can Ottawa ON K1A 0R6 Can

TOOLE, JAMES FRANCIS, b Atlanta, Ga, Mar 22, 25; m 52; c 4. STROKE, NEUROLOGY. *Educ:* Princeton Univ, BA, 47; Cornell Univ, MD, 49; LaSalle Exten Univ, LLB, 63. *Prof Exp:* Intern, Univ Pa, 49-50, resident internal med, 53-55, resident neurol, 56-58, instr, 59-60, assoc, 60-61; flight surgeon, US Navy, 53-55; chmn, Dept Neurol, 62-83, WALTER C TEAGLE PROF NEUROL & STROKE CTR DIR, BOWMAN GRAY SCH MED, 83- *Concurrent Pos:* Fulbright fel, Nat Hosp, London, 55-56; mem res comt, Am Heart Asn, 65, chmn med ethics comt, 66-70, ed, Current Concepts Cerebrovasc Dis-Stroke, 68-72; chmn, Sixth & Seventh Princeton Conf Cerebrovasc Dis, 68 & 70; vis prof, Univ Calif, San Diego, 69-70; mem exam bd, Nat Bd Med Examr & Am Bd Psychol & Neurol; consult, WHO, Japan, 72, Moscow, 68, Abidjean, Ivory Coast, 77 & Switz, 74; FDA adv comt, 76-80; mem, Stroke Long Range Planning Comt, Nat Inst Neurol & Commun Disorders & Stroke, 81, adv comt, 80-83; adv bd, Am Soc NeuroImaging, 83, vpres, 90; counr, Am Neurol Asn; ed-in-chief, J Neurol Sci, 90- *Mem:* Am Neurol Asn (secy-treas, 78-82, pres-elect, 84-85, pres, 85-86); Am Clin & Climat Asn; World Fedn Neurol (secy-treas gen, 82-89); Am Osler Soc; Ger Soc Neurol. *Res:* Cerebral circulation and cerebrovascular diseases; physiology and pathology of the brain. *Mailing Add:* Dept Neurol Bowman Gray Sch Med 300 Hawthone Rd SW Winston-Salem NC 27103

TOOLES, CALVIN W(ARREN), b Burlington, Vt, June 3, 21; m 43; c 1. CIVIL ENGINEERING. *Educ:* Univ Vt, BS, 43; Iowa State Col, MS, 50. *Prof Exp:* Instr eng drawing, Univ Vt, 46-47; instr civil eng, Iowa State Col, 48-50; from asst prof to assoc prof, Va Polytech Inst, 51-60; assoc prof civil eng, Ga Inst Technol, 60-86; RETIRED. *Mem:* Am Soc Civil Engrs; Am Soc Eng Educ; Am Soc Photogram; Am Cong Surv & Mapping. *Res:* Geodetic and photogrammetric engineering. *Mailing Add:* 3632 Eaglerock Dr Doraville GA 30340

TOOLEY, F(AY) V(ANISLE), b Nokomis, Ill, May 4, 08; m 34; c 2. GLASS TECHNOLOGY, CERAMICS. *Educ:* Univ Ill, BS, 32, MS, 36, PhD(ceramic eng), 39. *Prof Exp:* Asst mineral wool & raw mat, State Geol Surv, Ill, 32-38; head dept glass res, Owens-Corning Fiberglas Corp, 39-46; prof glass technol, 46-72, EMER PROF GLASS TECHNOL, UNIV ILL, URBANA, 72- *Concurrent Pos:* Dir, Annual Conf Glass Probs, 46-75; mem, Int Comn Glass, 53-59, Inventor's Network, Columbus, Ohio; consult in glass control, prod & res & in gen res mgt. *Honors & Awards:* First Recipient, Phoenix Award, Glass Indust, 71. *Mem:* AAAS; fel Am Ceramic Soc; Am Soc Testing & Mat; Am Chem Soc; Soc Glass Technol Eng. *Res:* Glass research, production and control; general research management. *Mailing Add:* 2910 Silver St SW PO Box 301 Granville OH 43023

TOOLEY, RICHARD DOUGLAS, b Baltimore, Md, Apr 11, 32; m 84; c 3. GEOPHYSICS, SYSTEMS ENGINEERING. *Educ:* Mass Inst Technol, SB, 54, PhD(geophys), 58. *Prof Exp:* Res physicist, Calif Res Corp, 58-64; chief scientist mission & systs analysis unit, Northrop Systs Labs, Hawthorne, 64-69, Palos Verdes, 69-72, prin systs engr, Electro-Mech Div, 72-88, SR RES ENGR, ELECTRONIC SYST DIV, NORTHROP CORP, 88- *Mem:* Sigma Xi; Am Defense Preparedness Asn; Asn Old Crows. *Res:* Space science; solid body geophysics; rock physics; ultrasonic wave propagation; data processing; electrooptical systems; infrared systems; military systems analysis. *Mailing Add:* Electronic Syst Div Northrop Corp 2301 W 120th St CA Hawthorne CA 90251-5032

TOOLEY, WILLIAM HENRY, b Berkeley, Calif, Nov 18, 25. PEDIATRICS. *Educ:* Univ Calif, MD, 49. *Prof Exp:* From clin instr to asst clin prof, 56-61, from asst prof to assoc prof, 61-72, chief newborn serv, hosp, 62-71, PROF PEDIAT, MED CTR, UNIV CALIF, SAN FRANCISCO, 72-, CHIEF, DIV PEDIAT PULMONARY DIS, UNIV HOSPS, 71-, SR STAFF MEM, CARDIOVASC RES INST, 63- *Concurrent Pos:* Pvt pract, Calif, 56-58. *Mem:* Am Acad Pediat; Soc Pediat Res; Am Pediat Soc. *Res:* Cardiopulmonary disease; neonatal medicine. *Mailing Add:* Dept Pediat Univ Calif Med Ctr San Francisco CA 94122

TOOM, PAUL MARVIN, b Pella, Iowa, Apr 1, 42; m 65; c 2. BIOCHEMISTRY. *Educ:* Cent Col, Iowa, BA, 64; Colo State Univ, PhD(biochem), 69. *Prof Exp:* Res asst biochem, Colo State Univ, 67-68, asst prof, 69-70; from asst prof to assoc prof, 70-78, dir res & sponsored prog, 84-86, prof biochem, Univ Southern Miss, 78-86; assoc vpres, Grad Studies & Res, 86-89, EXEC DIR PLANNING & POLICY, SOUTHWEST MO STATE UNIV, 89- *Concurrent Pos:* Consult, Nat Marine Fisheries Serv, 76- *Mem:* AAAS; Am Chem Soc; Nat Coun Univ Res Adminr; Sigma Xi. *Res:* analytical biochemistry. *Mailing Add:* Southwest Mo State Univ Springfield MO 64804

TOOME, VOLDEMAR, b Estonia, Sept 10, 24; m 52. PHYSICAL CHEMISTRY. *Educ:* Univ Bonn, dipl, 48 & 52, Dr rer nat, 54. *Prof Exp:* Sci asst, Inst Phys Chem, Univ Bonn, 54; dept head control div, Merck & Co, Ger, 54-57; sr phys res chemist, 57-68, res fel, 67-70, group chief, 70-77, sr group chief, 78-85, RES LEADER, HOFFMANN-LA ROCHE, INC, 85- *Mem:* Am Chem Soc; NY Acad Sci. *Res:* Ultraviolet and infrared spectroscopy; optical rotatory dispersion and circular dichroism; instrumental analysis; polarography; electrochemistry; dissociation constants of organic acids and bases; physical organic chemistry; microanalysis; gas chromatography, fourier transform infrared spectroscopy; mass spectroscopy; nuclear magnetic resonance spectroscopy. *Mailing Add:* Dept Phys Chem Hoffmann-La Roche Inc Nutley NJ 07110

TOOMEY, DONALD FRANCIS, geology, stratigraphy-sedimentation, for more information see previous edition

TOOMEY, JAMES MICHAEL, b Boston, Mass, Mar 2, 30; m 54; c 6. OTORHINOLARYNGOLOGY. *Educ:* Col of Holy Cross, Mass, BS, 51; Harvard Sch Dent Med, DMD, 55; Boston Univ, MD, 58. *Prof Exp:* Intern surg, Wash Univ-Barnes Hosp, Med Ctr, 58-59, asst resident surg, 61-62, asst resident otolaryngol, 62-64, chief resident, 64-65, instr, 64-65; sr instr, Med Ctr, Univ Rochester, 65-66, asst prof, 66-68; assoc prof otolaryngol & head dept, Health Ctr, Univ Conn, 68-77; MEM FAC, SCH MED, WASHINGTON UNIV, 77- *Concurrent Pos:* NIH training grant, Sch Med, Wash Univ, 63-65; consult, Newington Vet Admin Hosp, Conn, 68-; lectr, Dept Speech, Univ Conn, 69-; consult, Hartford Hosp, 69-, St Francis Hosp, Hartford, 69- & Rocky Hill Vet Admin Hosp, Conn, 71- *Mem:* Am Laryngol, Rhinol & Otol Soc; Soc Univ Otolaryngol; Am Acad Facial Plastic & Reconstruct Surg; Am Soc Maxillofacial Surg. *Res:* Skin flap physiology; laryngeal physiology; experimental laryngeal surgery. *Mailing Add:* 150 S Huntington Ave Boston MA 02130

TOOMEY, JOSEPH EDWARD, b Somerville, NJ, Aug 8, 43; m 67; c 4. SYNTHETIC ORGANIC CHEMISTRY, STRUCTURAL CHEMISTRY. *Educ:* Rider Col, BS, 70; Purdue Univ, PhD(org chem), 76. *Prof Exp:* SECT HEAD, REILLY LAB, REILLY TAR & CHEM CO, 75- *Mem:* Am Chem Soc; AAAS. *Res:* Synthesis, isolation and structural determination of pyridine chemicals; electrochemistry of heterocyclic compounds; prediction of optical rotatory power, its relation to molecular structure and absolute configuration; unusual Diels-Adler condensation reactions; electroorganic synthesis; electroorganic synthesis. *Mailing Add:* 2035 Rosedale Dr Indianapolis IN 46227-4317

TOOMRE, ALAR, b Rakvere, Estonia, Feb 5, 37; US citizen; m 58; c 3. ASTRONOMY, APPLIED MATHEMATICS. *Educ:* Mass Inst Technol, BS(aeronaut eng) & BS(physics), 57; Univ Manchester, PhD(fluid mech), 60. *Prof Exp:* C L E Moore instr math, Mass Inst Technol, 60-62; fel astrophys, Inst Advan Study, 62-63; from asst prof to assoc prof, 63-70, PROF APPL MATH, MASS INST TECHNOL, 70- *Concurrent Pos:* Guggenheim fel astrophys, Calif Inst Technol, 69-70; Fairchild fel, 75; MacArthur fel, 84-89. *Mem:* Nat Acad Sci; AAAS; Am Acad Arts & Sci; Am Astron Soc; Int Astron Union. *Res:* Dynamical studies of galaxies; aerodynamics; rotating fluids. *Mailing Add:* Dept Math Rm 2-371 Mass Inst Technol Cambridge MA 02139

TOON, OWEN BRIAN, b Bethesda, Md, May 26, 47; m 68; c 1. ASTRONOMY, CLIMATOLOGY. *Educ:* Univ Calif, Berkeley, AB, 69; Cornell Univ, PhD(physics), 75. *Prof Exp:* Res assoc, Nat Res Coun, 75-77 & Cornell Univ, 77-78; RES SCIENTIST, AMES RES CTR, NASA, 78- *Honors & Awards:* Medal Except Sci Achievement, NASA, 83; Int Peace Garden Award, Univ NDak, 84; Leo Sziland Award, Am Physics Soc, 85. *Mem:* Fel Am Meterol Soc; Am Astron Soc; Am Geophys Soc. *Res:* Physics of terrestrial and planetary climates with emphasis on clouds, aerosols and radiative transfer; volcanos and climate, troposheric aerosols and climate change on Mars; clouds of Mars, Venus and Titan; nuclear winter; ozone hole physics and chemistry. *Mailing Add:* Earth Syst Sci Div M/S 245-3 Ames Res Ctr NASA Moffett Field CA 94035

TOONEY, NANCY MARION, b Ilion, NY, Feb 19, 39. BIOCHEMISTRY, BIOPHYSICS. *Educ:* State Univ NY Albany, BS, 60, MS, 61; Brandeis Univ, PhD(biochem), 66. *Prof Exp:* Teaching intern biochem, Dept Chem & Biol, Hope Col, 66-67; fel biophys, Childrens Cancer Res Found & Sch Med, Harvard Univ, 67-73; asst prof, 73-77, assoc dean arts & sci, 88-90, ASSOC PROF BIOCHEM, POLYTECH INST NY, 77-, ASST PROVOST ACAD AFFAIRS, 90- *Concurrent Pos:* NIH fel, 67-70, res career develop award, Nat Heart, Lung & Blood Inst, 75-80; environ consult. *Mem:* AAAS; Am Chem Soc; Biophys Soc; Sigma Xi; Asn Women Sci. *Res:* Protein chemistry, electron microscopy and optical methods of analysis; structure and function of fibronectin; the blood clotting protein, fibrinogen; biological macromolecules; environmental chemistry. *Mailing Add:* Dept Chem Polytech Univ 333 Jay St Brooklyn NY 11201

TOONG, TAU-YI, b Shanghai, China, Aug 15, 18; nat US; m 43; c 2. MECHANICAL ENGINEERING. *Educ:* Chiao-Tung Univ, BSME, 40; Mass Inst Technol, SM, 48, ScD(mech eng), 52. *Prof Exp:* Mgr, Nanyang Eng Corp, China, 42-45; plant supt, Shanghai Transit Co, 45-47; from instr to assoc prof, 51-63, PROF MECH ENG, MASS INST TECHNOL, 63- *Concurrent Pos:* Guggenheim fel, 59; consult, Joseph Kaye & Co, Stewart-Warner Corp, USAF, Churchill Lighting Corp, Thermo-Electron Eng, Corp, Foster-Miller Assocs, Dynatech Corp, Avco-Everett Res Lab, Kenics Corp, Steam Engine Systs Corp, Factory Mutual Eng, Firepro, Inc & Arthur D Little, Inc. *Honors & Awards:* Alfred P Sloan Award, 87. *Mem:* Am Soc Mech Engrs; Combustion Inst; Am Inst Aeronaut & Astronaut. *Res:* Combustion; propulsion; fluid mechanics; heat and mass transfer. *Mailing Add:* Dept Mech Eng Rm 31-165 Mass Inst Technol Cambridge MA 02139

TOOP, EDGAR WESLEY, b Chilliwack, BC, Feb 26, 32; m 59; c 4. ORNAMENTAL HORTICULTURE. *Educ:* Univ BC, BSA, 55; Ohio State Univ, MSc, 57, PhD(plant path), 60. *Prof Exp:* Res officer, Can Dept Agr, 55-56; instr bot, Ohio State Univ, 61-62; from asst prof to prof hort, Univ Alta, 62-87; RETIRED. *Mem:* Agr Inst Can; Can Soc Hort Sci (secy-treas, 78-81); Am Soc Hort Sci; Int Soc Hort Sci. *Res:* Greenhouse flower crops; herbaceous ornamentals; controlled environment studies. *Mailing Add:* 42 Lafonde Crescent St Albert AB T8N 2N6 Can

TOOR, ARTHUR, b Altadena, Calif, Aug 17, 38; c 4. RADIATION TRANSPORT, ASTROPHYSICS. *Educ:* Univ Calif, Berkeley, BA, 62. *Hon Degrees:* PhD, Univ Leister, Eng, 77. *Prof Exp:* Staff physicist, High Altitude Physics Group, Livermore, 63-76, leader, Laser Appln Prog, 76-78, Space Appln Proj, 78-81, Exp Non-Equilibrium Radiation Physics, 81-86, SCI ADV, NUCLEAR TESTING PROG, LAWRENCE LIVERMORE NAT LAB, 86-, SR SCIENTIST, NUCLEAR TESTING & EXP SCI PROG. *Concurrent Pos:* Consult, NASA, 78- *Honors & Awards:* Sci Excellence Award, Dept Energy, 85; Indust Res & Develop Award, Dept Energy, 87. *Mem:* Am Astron Soc; Am Phys Soc; Soc Optical Engrs; AAAS. *Res:* High-energy astrophysics; dense plasma physics; space physics; x-ray laser kinetics. *Mailing Add:* Nuclear Testing & Exp Sci Prog Lawrence Livermore Nat Lab PO Box 808 Livermore CA 94550

TOOR, H(ERBERT) L(AWRENCE), b Philadelphia, Pa, June 22, 27; m 50; c 3. CHEMICAL ENGINEERING. *Educ:* Drexel Inst Technol, BS, 48; Northwestern Univ, MS, 50, PhD(chem eng), 52. *Prof Exp:* Res chemist, Monsanto Chem Ltd, 52-53; from asst prof to assoc prof chem eng, Carnegie Mellon Univ, 53-61, head dept, 67-70, dean eng, 70-79, prof, 61-80, MOBAY PROF CHEM ENG, CARNEGIE MELLON UNIV, 80- *Concurrent Pos:* UNESCO prof, AC Col, Madras, India. *Honors & Awards:* Colburn Award, Am Inst Chem Engrs, 64. *Mem:* Nat Acad Eng; fel AAAS; fel Am Inst Chem Engrs; Am Chem Soc; Sigma Xi. *Res:* Transport phenomena; heat and mass transfer; chemical reactions with mixing. *Mailing Add:* Schenley Park Chem Eng Dept Carnegie Mellon Univ Pittsburgh PA 15213

TOOTHILL, RICHARD B, b Philadelphia, Pa, July 28, 36; m 59; c 2. ORGANIC CHEMISTRY. *Educ:* Lehigh Univ, BS, 58; Mass Inst Technol, MS, 60; Univ Del, PhD(org chem), 64. *Prof Exp:* Tech serv rep paper chem, Hercules Powder Co, 60-61, chemist, 61-62; res chemist, 64-67, group leader, Bound Brook Labs, 67-73, dyes res group leader, 73-78, elastomers res group leader, 78-80, TECH DIR COLOR TEXTILE, CHEM INTERMEDIATES DEPT & TEXTILE CHEM & PLASTICS ADDITIVES DEPT, AM CYANAMID CO, 81- *Mem:* Am Chem Soc. *Res:* Thiosemicarbazones; s-triazines; benzothiazoles; anthraquinone derivatives; polyurethanes; millable gum; synthetic rubber; wrinkle recovery agents; light absorbers; antioxidants; lead stabilizers; pigments. *Mailing Add:* 16 Sunrise Dr Warren NJ 07060-5030

TOP, FRANKLIN HENRY, JR, b Detroit, Mich, Mar 1, 36; m 61; c 3. MICROBIOLOGY, VIROLOGY. *Educ:* Yale Univ, BS, 57, MD, 61. *Prof Exp:* Residency & intern pediat, Univ Minn Hosp, 61-64, res fel, Univ Minn, 64-66; investr virol, Walter Reed Army Inst Res, 66-70; chief, Dept Virol, Seato Med Res Lab, 70-73; chief, Dept Virus Dis, Walter Reed Army Inst Res, 73-76, dir, Div Commun Dis & Immunol, 76-78, dep dir, 79-81; comndr, US Army Med Res Inst Chem Defense, 81-83, dir & commandant, 83-87; sr vpres med & regulatory affairs, Praxis Biologics, 87-88; EXEC VPRES & MED DIR, MED IMMONE INC, 88- *Concurrent Pos:* Mem, Microbiol & Infectious Dis Adv Comt, Nat Inst Allergy & Infectious Dis, 76-80; prof pediat, Sch Med, Uniformed Serv Univ, 78-87. *Mem:* Soc Pediat Res; Am Soc Trop Med & Hyg; Am Med Asn; Infectious Dis Soc Am; Am Asn Immunologists; fel Am Acad Pediat. *Res:* Epidemiology and prevention of respiratory infections; vaccine development. *Mailing Add:* 9922 Foxborough Circle Rockville MD 20850-4611

TOPAZIAN, RICHARD G, b Greenwich, Conn, Feb 2, 30; m 58; c 4. ORAL & MAXILLOFACIAL SURGERY. *Educ:* Houghton Col, BA, 51; McGill Univ, DDS, 55; Univ Pa, cert oral surg, 57; Am Bd Oral Surg, dipl, 64. *Prof Exp:* Lectr dent & oral surg, Christian Med Col, Vellore, India, 59-61, reader, 61-63; from asst prof to assoc prof oral surg, Col Dent, Univ Ky, 63-67; prof & chmn dept, Sch Dent & Sch Med, Med Col Ga, 67-75; PROF ORAL & MAXILLOFACIAL SURG & HEAD DEPT, SCH DENT MED & PROF SURG, SCH MED, UNIV CONN, FARMINGTON, 75- *Concurrent Pos:* Consult, USPHS Hosp, Lexington, Ky, 64-67, US Army, Ft Jackson, SC, 67-75, Vet Admin Hosps, Augusta, Ga, 67-75 & Newington, Conn, 75- & Coun Dent Educ, Am Dent Asn; mem adv comt, Am Bd Oral Surg, 67-75; sect ed, J Oral & Maxillofacial Surg, 82- *Mem:* AAAS; Am Dent Asn; Am Asn Oral & Maxillofacial Surgeons; Int Asn Dent Res; fel Am Col Dentists; fel Int Col Dentists. *Res:* Dental education; research in oral surgery; bone pathology and diseases of the temporomandibular joint. *Mailing Add:* 14 Rocklyn Dr West Simsbury CT 06092

TOPCIK, BARRY, b Passaic, NJ, Apr 7, 24; m 50. CHEMICAL ENGINEERING. *Educ:* Cooper Union, BChE, 52; Newark Col Eng, MS, 60. *Prof Exp:* Sr develop engr, Uniroyal, Inc, 44-56; chief chemist, Eberhard Faber, Inc, 56-62; proj leader butyl lab, Columbian Carbon Co, 62-64, asst mgr butyl lab, 64-67, mgr new appln lab, Cities Serv Co, Cranbury, 67-77; tech mat mgr, Wyrough & Loser, Trenton, 77-83; consult, Union Carbide Corp, Bound Brook, NJ, 84-87; CONSULT, 87- *Mem:* Am Chem Soc; Am Inst Chem Engrs; Am Mgt Asn. *Res:* Rubber technology, including formulation, engineering, and product development associated with laboratory research and application to production; plastics compounding, moisture cure, adhesives, TPE's, polymer alloying, wire and cable formulation. *Mailing Add:* 545 Spring Valley Dr Bridgewater NJ 08807

TOPEL, DAVID GLEN, b Lake Mills, Wis, Oct 24, 37; m. AGRICULTURAL RESEARCH. *Educ:* Univ Wis-Madison, BS, 60; Kans State Univ, MS, 62; Mich State Univ, PhD(food sci), 65. *Prof Exp:* Prof animal sci, Iowa State Univ, 73-79; prof & head animal sci, Auburn Univ, 79-88; DEAN & DIR COL AGR, IOWA STATE UNIV, 88- *Concurrent Pos:* Fulbright fel, Royal Vet & Agr Univ, Copenhagen, Denmark, 71-72; mem Gov's sci adv coun, State Iowa, 89-93. *Honors & Awards:* Res Award, Am Asn Animal Sci, 79. *Mem:* Am Meat Sci Asn; Am Soc Animal Sci. *Res:* Stress adaption with special emphasis on muscle physiology and metabolic functions with the pig. *Mailing Add:* 2630 Meadow Glen Rd Ames IA 50010

TOPHAM, RICHARD WALTON, b Montgomery, WVa, May 22, 43; m 67; c 1. BIOCHEMISTRY. *Educ:* Hampden-Sydney Col, BS, 65; Cornell Univ, PhD(biochem), 70. *Prof Exp:* NIH fel, Fla State Univ, 69-71; from asst prof to assoc prof, 71-82, PROF & CHMN, DEPT CHEM, UNIV RICHMOND, 82- *Concurrent Pos:* Res scientist, Res Corp grant, 75-77; NIH grant, 77- *Mem:* Am Chem Soc Res. *Res:* Role of copper-containing enzymes of blood serum in iron metabolism; mechanism and regulation of intestinal absorption; iron metabolism in marine organisms. *Mailing Add:* 11821 Young Manor Dr Rte 1 Midlothian VA 23113

TOPICH, JOSEPH, b Steubenville, Ohio, Apr 25, 48; m 71; c 2. INORGANIC CHEMISTRY. *Educ:* Columbia Univ, BA, 70; Case Western Reserve Univ, PhD(chem), 74. *Prof Exp:* Res assoc, Univ Chicago, 74-76; ASST PROF CHEM, VA COMMONWEALTH UNIV, 76- *Mem:* Am Chem Soc; Sigma Xi. *Res:* Synthesis and characterization of new molybdenum coordination complexes; chemical properties are correlated with ligand structure and molybdenum oxidation state. *Mailing Add:* Dept Chem Va Commonwealth Univ 923 W Franklin St Richmond VA 23284

TOPLISS, JOHN G, b Mansfield, Eng, June 3, 30; nat US; m 58; c 2. MEDICINAL CHEMISTRY. *Educ:* Univ Nottingham, BSc, 51, PhD(chem), 54. *Prof Exp:* Res fel chem, Royal Inst Technol, Stockholm, Sweden, 54-56 & Columbia Univ, 56-57; from res chemist to sr res chemist, Schering-Plough Corp, 57-66, sect leader, 66-68, from asst dir to assoc dir chem res, 68-73, dir chem res, 73-75, sr dir chem res, 75-79; dir, 79-83, VPRES CHEM, WARNER LAMBERT/PARKE DAVIS, 83-; ADJ PROF MED CHEM, UNIV MICH, 83- *Mem:* Am Chem Soc; Royal Soc Chem; NY Acad Sci; fel AAAS. *Res:* Design, synthesis, and structure-activity relationships of drugs. *Mailing Add:* 2800 Plymouth Rd Warner Lambert/Parke Davis Pharmaceut Ann Arbor MI 48105

TOPOFF, HOWARD RONALD, b New York, NY, May 7, 41. BIOLOGY, ANIMAL BEHAVIOR. *Educ:* City Col New York, BS, 64, PhD(biol), 68. *Prof Exp:* Lectr biol, City Col New York, 67-68; res fel animal behav, Am Mus Natural Hist, 68-70; asst prof, 70-76, assoc prof, 76-79, PROF PSYCHOL, HUNTER COL, 80-; RES ASSOC ANIMAL BEHAV, AM MUS NATURAL HIST, 70- *Mem:* AAAS; Animal Behav Soc; NY Acad Sci. *Res:* Behavioral development in social insects; insect communication, behavior and physiology. *Mailing Add:* PO Box 366 Portal AZ 85632

TOPOL, LEO ELI, b Boston, Mass, Apr 15, 26; m 48; c 2. PHYSICAL CHEMISTRY. *Educ:* Northeastern Univ, BS, 46; Univ Minn, Minneapolis, PhD(phys chem), 52. *Prof Exp:* Res chemist, Oak Ridge Nat Lab, 52-57; res specialist, Atomics Int Div, NAm Aviation, Inc, 57-63, mem tech staff, Sci Ctr, 63-69 & Atomics Int Div, 69-71, mem tech staff, Sci Ctr, NAm Rockwell Corp, 71-75, mem tech staff, Air Monitoring Ctr & mem tech staff & prog mgr, Environ Monitoring & Serv Ctr, Rockwell Int Corp, 75-84; mem tech staff & sr prog mgr combustion eng, Environ Monitoring & Serv, Inc, 84-89; CONSULT COMBUSTION ENG, ABB ENVIRON SERV, INC, 89- *Concurrent Pos:* Vchmn, qual assurance comt, Nat Atmospheric Deposition Prog, 80-81, chmn, 81-82; reviewer, Nat Acid Precipitation Assessment Prog, 86. *Mem:* Am Chem Soc; Sigma Xi. *Res:* Acid precipitation, air pollution, fugitive hydrocarbon emissions; quality assurance; electrolytes; new glass compositions, electrochemistry; thermodynamics and phase studies of fused salt systems and molten metal-metal salt solutions; high-conducting solid electrolytes; solid electrochemical gas pollutant sensors; materials damage from acid deposition; author EPA manuals on quality assurance and operations and maintenance for precipitation measurement systems; US patents; author of over 50 publications. *Mailing Add:* 23435 Strathern St West Hills CA 91304

TOPOLESKI, LEONARD DANIEL, b Wilkes-Barre, Pa, Apr 11, 35; m 58; c 3. PLANT BREEDING, VEGETABLE CROPS. *Educ:* Pa State Univ, BS, 57, MS, 59; Purdue Univ, PhD(genetics, plant breeding), 62. *Prof Exp:* Asst prof, 62-68, assoc prof, 68-78, PROF VEG CROPS, NY STATE COL AGR & LIFE SCI, CORNELL UNIV, 78- *Mem:* Am Soc Hort Sci. *Res:* Genetics; vegetative hybridization; physiology of interspecific incompatability; greenhouse vegetable production. *Mailing Add:* Dept Veg Crops Cornell Univ Ithaca NY 14853

TOPOREK, MILTON, b New York, NY, Apr 18, 20; m 42; c 3. BIOCHEMISTRY, LIVER METABOLISM. *Educ:* Brooklyn Col, BA, 40; George Washington Univ, MA, 48; Univ Rochester, PhD(biochem), 52. *Prof Exp:* Res assoc org chem, George Washington Univ, 48; res assoc biochem, Univ Rochester, 48-52; res chemist, Univ Mich, 52-57; from asst prof to assoc prof biochem, Jefferson Med Col, Thomas Jefferson Univ, 58-72, prof, 72-83; RETIRED. *Mem:* AAAS; Am Chem Soc; Am Inst Nutrit; Am Soc Biochem & Molecular Biol; NY Acad Sci; Sigma Xi. *Res:* Control of plasma protein synthesis by liver, relationship to disease states; vitamin B-12, intrinsic factor relationships. *Mailing Add:* 4667 Oak Forest Dr E Sarasota FL 34231-6416

TOPP, G CLARKE, b Canfield, Can, Nov 12, 37; m; c 3. SOIL SCIENCE, PHYSICS. *Educ:* Univ Toronto, BSA, 59; Univ Wis, MSc, 62, PhD(soils), 64. *Prof Exp:* Res asst soil physics, Univ Wis-Madison, 59-64; res assoc, Univ Ill, 64-65; RES SCIENTIST SOIL PHYSICS, CAN DEPT AGR, 65- *Concurrent Pos:* Adj prof, Dept Geog, Carleton Univ, 73-81; vis lectr, Dept Soil Sci, Univ Sask, 80. *Mem:* Fel Can Soc Soil Sci (secy, 69-72, pres, 77-78); Sigma Xi; fel Soil Sci Soc Am; Am Geophys Union. *Res:* Development of instrument to measure soil water content; soil water properties; microhydrology of soils; soil structure. *Mailing Add:* Land Resource Res Centre Agr Can Ottawa ON K1A 0C6 Can

TOPP, STEPHEN V, b Longview, Tex, Oct 19, 37; m 57; c 2. REACTOR PHYSICS, PROBABILISTIC RISK ASSESSMENT. *Educ:* Col William & Mary, BS, 59; Univ Va, MS, 60, PhD(physics), 62. *Prof Exp:* Res physicist, E I du Pont de Nemours & Co, Inc, 62-67, sr physicist, 67-69, asst chief supvr, Savannah River Lab, 69-75, res staff physicist, Advan Planning Sect, Savannah River Lab, 75- 85, PHYSICIST, REACTOR PROBABILISTIC RISK ASSESSMENT, WESTINGHOUSE SAVANNAH RIVER CO, 85- *Concurrent Pos:* Instr physics, Univ SC, 62-75. *Mem:* Am Phys Soc; Am Nuclear Soc; Opers Res Soc Am; Mat Res Soc. *Res:* Neutron polarization measurements, reactor criticality measurements and lattice calculations for heavy water reactors; application of computer techniques to weapons production system modeling; decision modeling for nuclear waste management; human reliability in PRA. *Mailing Add:* Savannah River Lab Westinghouse Savannah River Co Aiken SC 29808

TOPP, WILLIAM CARL, b Cleveland, Ohio, Feb 3, 48; m 84; c 2. CELL BIOLOGY. *Educ:* Oberlin Col, BA, 69; Princeton Univ, MA, 71, PhD(chem), 73. *Prof Exp:* Res assoc physics, Princeton Univ, 73, instr chem, 73-74, res assoc, 74; res assoc biol, Cold Spring Harbor Lab Quant Biol, 74-76, staff scientist, 76-78, sr staff scientist, 78-83; pres & chief operating officer, Otisville Bio Technol, 83-84; dir molecular biol, Agr Res Div, Am Cyanamid, 85-86; GEN MGR, BIOTECHNOL ASSOCS, 86- *Mem:* Sigma Xi; Am Phys Soc. *Res:* Virus/cell interactions; cell growth control. *Mailing Add:* RD 1 Box 271 Otisville NY 10963

TOPP, WILLIAM ROBERT, b Milwaukee, Wis, May 27, 39. MATHEMATICS. *Educ:* St Louis Univ, BA, 63, MA, 64; Univ Wash, MS, 67, PhD(math), 68. *Prof Exp:* Instr math, Univ Seattle, 67-68; asst prof, Marquette Univ, 69-70; from asst prof to assoc prof, 70-79, PROF MATH, UNIV PAC, 80- *Mem:* Math Asn Am; Asn Comput Mach; Opers Res Soc Am. *Res:* Rings; algebras. *Mailing Add:* Dept of Math Univ Pac 3601 Pacific Ave Stockton CA 95211

TOPPEL, BERT JACK, b Chicago, Ill, July 2, 26; m 50; c 2. REACTOR PHYSICS. *Educ:* Ill Inst Technol, BS, 48, MS, 50, PhD(physics), 52. *Prof Exp:* Instr physics, Ill Inst Technol, 49-51; assoc physicist, Brookhaven Nat Lab, 52-56; assoc physicist, 56-66, SR PHYSICIST, REACTOR ANAL DIV, ARGONNE NAT LAB, 66- *Mem:* Am Phys Soc; Am Nuclear Soc. *Res:* Nuclear reactions initiated by charged particles and neutrons; scintillation detector studies of gamma ray events; reactor critical facility experimentation; theoretical reactor physics calculations; reactor physics computer code development. *Mailing Add:* Reactor Analytical Div Argonne Nat Lab Argonne IL 60439

TOPPER, LEONARD, b New York, NY, Jan 11, 29. CHEMICAL ENGINEERING, SCIENCE POLICY. *Educ:* City Col New York, BChE, 48; NY Univ, MChE, 49; Cornell Univ, PhD(chem eng), 51. *Prof Exp:* Asst prof chem eng, Johns Hopkins Univ, 53-55; prog mgr, US Atomic Energy Comn, 57-73; sr policy analyst, Off Energy Res & Develop Policy, NSF, 73-75; dir, Div Technol Evaluation, Energy Res & Develop Admin, 75-76; sr policy analyst, Off Sci & Technol Policy, Exec Off President, 76-77; dir, Div Res Assessment, Off Energy Res, US Dept Energy, 77-79; CONSULT,79- *Mem:* Sigma Xi. *Res:* Energy technology; chemical engineering science; energy research policy. *Mailing Add:* 2126 Connecticut Ave NW Washington DC 20008

TOPPER, T(IMOTHY) H(AMILTON), b Kleinburg, Ont, May 20, 36; m 58; c 3. CIVIL ENGINEERING. *Educ:* Univ Toronto, BASc, 59; Cambridge Univ, PhD(fatigue), 62. *Prof Exp:* Lectr, 62-63, from asst prof to assoc prof, 63-69, assoc chmn dept, 66-72, chmn dept, 72-78, PROF CIVIL ENG, UNIV WATERLOO, 69- *Concurrent Pos:* Vis asst res prof, Univ Ill, 66, vis assoc prof, 68. *Mem:* Am Soc Metals; Am Soc Testing & Mat; Soc Automotive Engrs. *Res:* Mechanical behavior and fatigue of metals including applications to structures. *Mailing Add:* Dept Civil Eng Univ Waterloo Waterloo ON N2L 3G1 Can

TOPPER, YALE JEROME, b CHicago, Ill, Aug 11, 16; m 56; c 4. BIOCHEMISTRY. *Educ:* Northwestern Univ, BS, 42; Harvard Univ, MA, 43, PhD(chem), 47. *Prof Exp:* Assoc nutrit & physiol, Pub Health Res Inst, City of NY, Inc, 48-53; Am Heart Asn res fel, Biochem Res Lab, Mass Gen Hosp, 53-54; mem staff, 54-62, chief sect intermediary metab, 62-87, EMER SCIENTIST, NAT INST ARTHRITIS & METAB DIS, 87- *Mem:* Endocrine Soc; Am Soc Biol Chem. *Res:* Biochemistry of development and differentiation. *Mailing Add:* Develop Biol NIDDK-LBM-NIH Bldg 10 Rm 9B-18 Bethesda MD 20892

TOPPETO, ALPHONSE A, b Wheeling, WVa, Jan 7, 25; wid; c 3. ELECTRICAL ENGINEERING. *Educ:* Carnegie Inst Technol, BS, 48, MS, 49; Univ Mich, PhD(elec eng), 63. *Prof Exp:* from instr to asst prof elec eng, Univ Detroit, 50-60, assoc prof & vchmn dept, 60-63; res physicist, Aladdin Electronics Div, Aladdin Indust, Inc, 63-68, dir res, 66-68, dir res & eng, 68-76; res prof elec eng, Vanderbilt Univ, 76-77; mgr res & develop, Corcom Inc, 77-80 & appln eng, 80-89; RETIRED. *Concurrent Pos:* US deleg, Int Electrotech Comn, 74-89 & Int Comt Radio Interference, 81-89. *Mem:* Inst Elec & Electronics Engrs. *Res:* Characterization and application of ferrites; theoretical and practical design of filters and delay lines; computer aided design methods; development and design of RFI filters. *Mailing Add:* 32402 N Forest Dr Grayslake IL 60030-2523

TOPPING, JOSEPH JOHN, b Amsterdam, NY, Oct 9, 42; m 65; c 3. CHROMATOGRAPHY, SPECTROSCOPY. *Educ:* Le Moyne Col, NY, BS, 64; Univ NH, MS, 67, PhD(anal chem), 69. *Prof Exp:* Res fel chem, Ames Lab, Iowa State Univ, 69-70; from asst prof to assoc prof, 70-82, PROF CHEM, TOWNSON STATE UNIV, 82- *Concurrent Pos:* Adj prof chem, Univ Md, Baltimore County, 79-86; chmn-elect, Md sect, Am Chem Soc, 88, chmn, 89. *Mem:* AAAS; Am Inst Chemists; Sigma Xi; NY Acad Sci; Am Chem Soc; Am Asn Univ Prof. *Res:* Determination of trace metals and organics in environmental systems; the study of sample/substrate interactions in chromatography; chromatography. *Mailing Add:* Dept Chem Towson State Univ Towson MD 21204

TOPPING, NORMAN HAWKINS, b Flat River, Mo, Jan 12, 08; m 30; c 2. INFECTIOUS DISEASES. *Educ:* Univ Southern Calif, AB, 33, MD, 36. *Hon Degrees:* LLD, Occidental Col, Univ Calif, Los Angeles, 62, Loyola Univ Los Angeles, 68; DSc, Univ Southern Calif, 63; LHD, Hebrew Union Col, 66. *Prof Exp:* Intern, USPHS, Marine Hosps, San Francisco, Calif & Seattle, Wash, 36-37; mem staff med res viral & rickettsial dis, NIH, Md, 37-48, assoc dir, Insts, 48-52; vpres in-chg med affairs, Univ Pa, 52-58; pres, 58-70, chancellor, 70-80, EMER CHANCELLOR, UNIV SOUTHERN CALIF, 80- *Concurrent Pos:* Chmn res comt & mem comn virus res & epidemiol, Nat Found, 58-77. *Honors & Awards:* Ashford Award, 43; Medal, US Typhus Comn, 45. *Mem:* AAAS; Am Epidemiol Soc; Soc Exp Biol & Med; Asn Am Physicians. *Res:* Virus and rickettsial diseases. *Mailing Add:* Off Chancellor Univ Southern Calif Suite 1202 3810 Wilshire Blvd Los Angeles CA 90010

TOPPING, RICHARD FRANCIS, b Boston, Mass, Dec 19, 49; m; c 5. PRODUCT DEVELOPMENT, GAS APPLIANCE ENGINEERING. *Educ:* Tufts Univ, BS, 71; Mass Inst Technol, MS, 72. *Prof Exp:* Propulsion engr, USAF Aeropropulsion Lab, 72-75; res staff, Mass Inst Technol Energy Lab, 75-78; res engr, 78-86, unit mgr, 86-88, DIR TECHNOL & PROD DEVELOP, ARTHUR D LITTLE, INC, 88- *Concurrent Pos:* Chmn, Tech Comt 7-1, Am Soc Heating, Refrig & Air Conditioning Engrs, 90-92. *Mem:* Am Soc Mech Engrs; Am Soc Heating, Refrig & Air Conditioning Engrs. *Res:* Product design and commercialization of consumer products, major appliances and natural gas fired equipment; technology-based consulting to global manufacturers of appliances and housewares. *Mailing Add:* Arthur D Little Inc 20-529 Acorn Park Cambridge MA 02140

TOPUZ, ERTUGRUL S, b Sumnu, Bulgaria, Dec 11, 35; Turkish citizen; m 72; c 1. MINING ENGINEERING, MINERAL ECONOMICS. *Educ:* Istanbul Tech Univ, dipl eng, 59; Univ Calif, Berkeley, MEng, 72; Columbia Univ, DEngSc, 77. *Prof Exp:* Mining engr, Mineral Res & Explor Inst, 59-65, chief planning div, 65-67, asst gen dir, 67-69, mem sci bd, 69-70; asst prof, SDak Sch Mines & Technol, 76-77; ASSOC PROF MINING ENG, VA POLYTECH INST & STATE UNIV, 77- *Mem:* Am Inst Mining, Metall & Petrol Engrs; Am Inst Indust Engrs. *Res:* Mining evaluation and analysis; mineral economics; application of mathematical optimization techniques to problems of mining industry; mine ventilation. *Mailing Add:* Dept Mining & Minerals Eng Va Polytech Inst & State Univ Blacksburg VA 24060

TORACK, RICHARD M, b Passaic, NJ, July 23, 27; m 53; c 4. PATHOLOGY. *Educ:* Seton Hall Univ, BS, 48; Georgetown Univ, MD, 52. *Prof Exp:* Asst pathologist, Montefiore Hosp, 58-59, asst neuropathologist, 59-61; asst prof path, New York Hosp-Cornell Med Ctr, 62-65, assoc prof, 65-68, assoc attend pathologist, 62-68; assoc prof, 68-70, PROF PATH & ANAT, WASH UNIV, 70- *Concurrent Pos:* Nat Cancer Inst fel path, Montefiore Hosp, 58-59; NIH res fel, Yale Med Sch, 61-62; consult, Mem Hosp, 64-68; assoc attend, Barnes Hosp, 68- *Mem:* AAAS; Am Asn Neuropath; Am Asn Path & Bact; Histochem Soc; Am Neurol Asn. *Res:* Electron histochemistry of disease of the nervous system. *Mailing Add:* Dept Path Wash Univ 660 Euclid Ave St Louis MO 63110

TORALBALLA, GLORIA C, b Philippines, Jan 18, 15; nat US; m 46; c 1. CHEMISTRY, BIOCHEMISTRY. *Educ:* Univ Philippines, BS, 36, MS, 38; Univ Mich, PhD(chem), 42. *Prof Exp:* Fel, Univ Mich, 43-44; res asst org chem, Columbia Univ, 44-46; instr anal & org chem, Marquette Univ, 49-52; res assoc biochem, Columbia Univ, 54-57, asst prof chem, Barnard Col, 58-63; asst prof, Hunter Col, 64-70; assoc prof, 70-75, chmn dept, 75-78, prof, 75-80, EMER PROF CHEM, LEHMAN COL, 80- *Mem:* Am Chem Soc; Sigma Xi. *Res:* Structure and mechanism of action of porcine pancreatic amylase; analytical studies of metallo-biochemicals. *Mailing Add:* 143 Jeanne Ct Stamford CT 06903

TORBETT, EMERSON ARLIN, b Athens, Tenn, July 20, 39; m 61; c 3. OPERATIONS RESEARCH, MATHEMATICS. *Educ:* Ga Inst Technol, BS, 61; Univ Md, College Park, MA, 66; Stanford Univ, PhD(opers res), 72. *Prof Exp:* Instr, Ga Inst Technol, 60-61; mathematician, Nat Security Agency, 61-63; res engr, Adaptronics, Inc 63-64 & SRI Int, 64-73; prog mgr, Systs Control, Inc, 73-78; dept mgr, Western Develop Labs Div, Ford Aerospace & Commun Corp, 78-80; dir eng, Icot Corp, 80-83; PRES, CASCADE TECH, 83- *Concurrent Pos:* Lectr, Univ Calif, Berkeley, 69-70; lectr, Stanford Univ, 69-70, res asst, 70-72; adj prof, San Jose State Univ, 73- *Mem:* Opers Res Soc Am; Inst Elec & Electronics Engrs. *Res:* System effectiveness analysis; system simulation; digital filtering and prediction; queueing theory; decision analysis under uncertainty; optimal control of stochastic systems; mathematical system reliability; design/analysis of communication networks/systems. *Mailing Add:* 240 Mimosa Way Menlo Park CA 94028

TORBIT, CHARLES ALLEN, JR, b Fountain, Colo, Nov 8, 24; m 41; c 4. DEVELOPMENTAL BIOLOGY. *Educ:* Colo State Univ, BS, 62, PhD(cell biol), 72; Colo Col, MA, 66. *Prof Exp:* Fel physiol, Sch Med, Univ Kans, 72-74; embryologist, Codding Embryol Sci Inc, 74-75; cell biologist, Stanford Res Inst, 75-80; mem fac, Dept Obstet & Gynec, Vanderbilt Univ, 80-83; dir, IVF Lab, Presby Hosp, Dallas, 83-85; dir, IVF Lab, Univ Kans, Sch Med, Wichita, 85-88; DIR, IVF LAB, REPRODUCTION MED LAB, VA, 88- *Mem:* Sigma Xi; AAAS; Electron Micros Soc Am; Soc Study Reproduction; Int Embryo Transfer Soc. *Res:* Cell biology of ovulation, fertilization and early development of the mammalian egg and its hormonal control. *Mailing Add:* 506 N Main St Fountain CO 80817

TORCH, REUBEN, b Chicago, Ill, Dec 20, 26; m 49; c 3. PROTOZOOLOGY. *Educ:* Univ Ill, BS, 47, MS, 48, PhD(zool), 53. *Prof Exp:* Asst zool, Univ Ill, 47-53; from instr to assoc prof, Univ Vt, 53-65; from asst dean to actg dean, Col Arts & Sci, Oakland Univ, 66-73, prof biol, 65-80, dean, Col Arts & Sci, 73-80; vpres acad affairs & prof zool, Calif State Univ, Stanislaus, 80-88; RETIRED. *Mem:* Soc Protozoologists; Sigma Xi; Am Soc Cell Biol. *Res:* Taxonomy of marine psammophilic ciliates; nucleic acid synthesis and regeneration in ciliates. *Mailing Add:* 405 Robin Hood Dr Modesto CA 95350-1561

TORCHIA, DENNIS ANTHONY, b Reading, Pa, June 15, 39; m 67; c 3. BIOPHYSICS. *Educ:* Univ Calif, Riverside, BA, 61; Yale Univ, MS, 64, PhD(physics), 67. *Prof Exp:* NIH fel, Med Sch, Harvard Univ, 67-69; mem tech staff polymer chem, Bell Labs, 69-71; physicist, Polymers Div, Nat Bur Standards, 71-74; BIOPHYSICIST, NAT INST DENT RES, 74- *Mem:* Am Chem Soc; Am Phys Soc; Biophys Soc. *Res:* Solution and solid state magnetic resonance studies of the molecular conformaton and motion of proteins. *Mailing Add:* NIH Bldg 30 Rm 106 Bethesda MD 20014

TORCHIANA, MARY LOUISE, b Philadelphia, Pa, July 22, 29. PHYSIOLOGY, PHARMACOLOGY. *Educ:* Immaculata Col, Pa, BA, 51; Temple Univ, MS, 60; Boston Univ, PhD(physiol), 64. *Prof Exp:* Res assoc pharmacol, 52-58 & 63-65, sr res pharmacologist, 65, res fel pharmacol, 65-72, dir gastrointestinal res, 72-80, sr res fel pharmacol, Merck, Sharp & Dohme Res Labs, 81-84; RETIRED. *Mem:* Am Soc Pharmacol & Exp Therapeut. *Res:* Catecholamine distribution; cardiovascular physiology and pharmacology; pharmacology of the gastrointestinal tract. *Mailing Add:* 686 Hoover Rd Blue Bell PA 19422

TORCHINSKY, ALBERTO, b Buenos Aires, Arg, Mar 9, 44; m 69; c 2. MATHEMATICAL ANALYSIS. *Educ:* Univ Buenos Aires, Licenciado, 66; Univ Wis, Milwaukee, MS, 67; Univ Chicago, PhD(math), 71. *Prof Exp:* Asst prof, Cornell Univ, 71-75; asst prof, 75-77, assoc prof, 77-80, PROF MATH, IND UNIV, BLOOMINGTON, 80-, DEAN LATINO AFFAIRS, 81- *Concurrent Pos:* Mem, Inst Adv Studies, 77-78; fel, Ctr Math Analysis, Australia Nat Univ, 84. *Mem:* Am Math Soc. *Res:* Problems related to singular integrals; Hp spaces and applications to differential equations. *Mailing Add:* Dept Math Ind Univ Bloomington IN 47401

TORDA, CLARA, b Budapest, Hungary, Apr 1, 10; US citizen; m 42; c 1. BRAIN PHYSIOLOGY, PSYCHIATRY. *Educ:* Univ Budapest, PhD(philos), 33; Univ Milan, MD, 39. *Prof Exp:* Res assoc, Dept Biophysics, Univ Col, London, 38-39; res assoc biophysics, Univ Pa, 39-40; res assoc pharmacol, Med Col, Cornell Univ, NY, 40-42, res assoc neurol, 42-52; res psychiat, NY Stae Psychiat Inst, Columbia Univ, 52-55, sr res psychiat, 55-57; instr, Dept Psychiat, NY Med Col, NY, 60-65; STANFORD UNIV, CALIF, 65- *Concurrent Pos:* Assoc prof, Dept Psychiat, Downstate Med Ctr, State Univ NY, Brooklyn, 64-69 & Dept Neurol, Mt Sinai Sch Med, 69-71; with Bur Child Guidance, 64-; vis prof, Univ Argentina, 67, Theoret Study Prog, NASA, Ft Collins, Colo, 71, Rockefeller Univ, 72, Stanford Univ, 74, 75 & 80, Med Col, Univ Rochester, 76; US delegate, Int congress, 69, First World Congress Biol Psychiat, 74. *Mem:* Am Med Soc; Am Psychiat Soc; Acad Psychoanalysis; Soc Med Psychoanalysts; Soc Child Psychiat. *Res:* Artificial intelligence; computer science; neurophysiology; biophysics; biochemistry; author or coauthor of over 300 publications. *Mailing Add:* 111 Stedman Brookline MA 02146-3008

TORDOFF, HARRISON BRUCE, b Mechanicville, NY, Feb 8, 23; m 46; c 2. ZOOLOGY. *Educ:* Cornell Univ, BS, 46; Univ Mich, MA, 49, PhD(zool), 52. *Prof Exp:* Cur birds, Sci Mus, Inst Jamaica, BWI, 46-47; asst prof zool, Univ & asst cur birds, Mus, Univ Kans, 50-57, assoc prof zool & assoc cur birds, 57; from asst prof to prof zool, Univ & cur birds, Mus Zool, Univ Mich, Ann Arbor, 57-70; dir, Bell Mus Natural Hist, 70-82, PROF ECOL & BEHAV BIOL, UNIV MINN, MINNEAPOLIS, 70- *Concurrent Pos:* Ed, Wilson Bull, 52-54. *Mem:* Am Ornithologists' Union (pres, 78-80); Cooper Ornith Soc; Wilson Ornith Soc; fel Am Ornithologists Union. *Res:* Ornithology; systematics; paleontology; morphology; behavior; breeding biology. *Mailing Add:* Bell Mus Natural Hist Univ Minn Minneapolis MN 55455

TORDOFF, WALTER, III, b Newton, Mass, Jan 2, 43; m 65; c 3. POPULATION BIOLOGY. *Educ:* Univ Mass, BA, 65; Colo State Univ, MS, 67, PhD(zool), 71. *Prof Exp:* From asst prof to assoc prof, 70-81, PROF ZOOL SCI & CHMN, DEPT BIOL SCI, CALIF STATE UNIV, STANISLAUS, 81- *Concurrent Pos:* Res consult Bur Land Mgt, Calif Dept Fish & Game. *Mem:* Am Soc Ichthyologists & Herpetologists; Soc Study Evolution; Soc Study Amphibians & Reptiles; Sigma Xi. *Res:* Ecology and genetics of chapparal and montane populations of reptiles and amphibians, particularly Hydromantes brunus; Gambelia silus. *Mailing Add:* Dept Biol Sci Calif State Univ Stanislaus Turlock CA 95380

TORELL, DONALD THEODORE, b Mont, Oct 19, 26; m 50; c 2. ANIMAL SCIENCE. *Educ:* Mont State Col, BS, 49; Univ Calif, MS, 50. *Prof Exp:* Assoc animal husb, Univ Calif, 49-50, res asst beef cattle invest, 50-51; instr, Ariz State Col, 51; livestock specialist & lectr, Hopland Field Sta, Univ Calif, 51-81. *Concurrent Pos:* Fulbright res sr scholar, Uganda, 61-62; specialist, Univ Chile-Univ Calif Coop Prog, 69-70; livestock specialist & consult, 82- *Mem:* Am Soc Animal Sci. *Res:* Sheep nutrition, genetics, physiology and general sheep improvement. *Mailing Add:* 7950 Sanel Dr Ukiah CA 95482

TOREN, ERIC CLIFFORD, JR, b Chicago, Ill, Sept 16, 33; m 63; c 1. ANALYTICAL CHEMISTRY. *Educ:* Northwestern Univ, BS, 55; Univ Ill, MS, 60, PhD(chem), 61. *Prof Exp:* Instr chem, Duke Univ, 61-62, from asst prof to assoc prof, 62-70; assoc prof med & path, Med Ctr, Univ Wis-Madison, 70-75, prof, 75-77; PROF PATH, UNIV SALA, 77- *Concurrent Pos:* Consult, Lawrence Livermore Labs, 74- *Mem:* Am Chem Soc; Am Asn Clin Chemists; NY Acad Sci; Acad Clin Lab Physicians & Scientists; Sigma Xi. *Res:* Kinetic methods of analysis; analytical instrumentation and automation laboratory computing; high performance liquid chromatography of proteins. *Mailing Add:* Dept Path Univ SAla Mobile AL 36617-2293

TOREN, GEORGE ANTHONY, b Chicago, Ill, June 12, 24; m 49. ORGANIC CHEMISTRY. *Educ:* Hope Col, AB, 48; Purdue Univ, MS, 51, PhD(chem), 53. *Prof Exp:* Prod control specialtist, Minn Mining & Mfg Co, St Paul, 53-82; RETIRED. *Mem:* Am Chem Soc. *Res:* Boron and graphite advanced composites; pressure sensitive tapes; urethane foams and films. *Mailing Add:* 678 E Eldridge Ave Maplewood MN 55117-2210

TOREN, PAUL EDWARD, b Lincoln, Nebr, July 18, 23; wid; c 3. ANALYTICAL CHEMISTRY. *Educ:* Univ Nebr, AB, 47, MS, 48; Univ Minn, PhD(chem), 54. *Prof Exp:* Chemist, Phillips Petrol Co, 53-59; sr chemist, 59-67, res specialist, 67-80, SR RES SPECIALIST, CENT RES LABS, 3M CO, 80- *Mem:* Am Chem Soc; Electrochem Soc. *Res:* Electroanalytical chemistry; analytical instrumentation. *Mailing Add:* 805 Park Ave St Paul MN 55115

TORESON, WILFRED EARL, b Calif, Dec 25, 16; m 45; c 1. PATHOLOGY. *Educ:* McGill Univ, MD, 42, MSc, 48, PhD(path), 50; Am Bd Path, dipl & cert clin path, 53. *Prof Exp:* Lectr path, McGill Univ, 46-50, asst prof, 50; from instr to prof, Univ Calif, San Francisco & pathologist, Univ Hosp, 50-66; prof path, State Univ NY Downstate Med Ctr & dir labs, Univ Hosp, 66-70; prof path, Sch Med, Univ Calif, Davis, 70-83; RETIRED. *Concurrent Pos:* Dir labs, South Pac Hosp, Calif, 52-58, consult, 58-66; consult, Letterman Army Hosp, 58-66; attend, Ft Miley Vet Admin Hosp, 60-66. *Mem:* Am Asn Pathologists & Bacteriologists; AMA; Col Am Pathologists; Am Soc Clin Path; Int Acad Path. *Res:* Experimental diabetes; automation and computers in clinical pathology. *Mailing Add:* 2315 Stockton Blvd Sacramento CA 95817

TORGERSEN, PAUL E, b New York, NY, Oct 13, 31; m; c 3. MINING RESEARCH. *Educ:* Lehigh Univ, BS, 53; Ohio State Univ, MS, 56 & PhD(indust eng), 59. *Prof Exp:* Instr, Ohio State Univ, 56-59; from asst prof to assoc prof, Okla State Univ, 59-66; prof & head, Dept Indust Eng & Opers Res, Va Polytech Inst & State Univ, 67-70, dean, Col Eng, 70-90, John Grado prof indust eng & opers res, 87-89, interim pres, 88, JOHN W HANCOCK, JR, CHAIR ENG, VA POLYTECH INST & STATE UNIV, 89-, PRES, VA TECH CORP RES CTR, 90-; DIR, ROANOKE ELEC STEEL, 86- *Concurrent Pos:* Res assoc, Opers Res Group, Ohio State Univ, 56-59; consult, var co & univs, 61-; dir, Queuing Systs Simulation Res Proj, NSF, 64-65; mem, Coun Tech Div & Comts, Am Soc Eng Educ, 66-68, Dean's Inst Comt, 71-81, bd dirs, 79-81 & Lamme Medal Comt, 90-; mem, Task Force Ethical Pract, Am Inst Indust Engrs, 70-77, chmn, 72-77; inspector, Accreditation Bd Eng & Technol, 70-, served on Eng Accreditation Comn, 80-85; chmn, Eng Deans Coun, 79-81, Eng Comn, Nat Asn State Univ & Land-Grant Col, 79-83, comt construct, Eng Grad Record Exam, 84- 86;

vpres, Educ & Prof Develop, Int Indust Engrs; dir, Va Mining & Mineral Resources & Res Inst, 79-87. *Mem:* Nat Acad Eng; sr mem Am Inst Indust Engrs; Am Soc Eng Educ (vpres, 80-81); Soc Logistics Engrs; Int Asn Qual Circles; Am Arbitration Asn; Nat Soc Prof Engrs. *Res:* Author of five books on industrial engineering; published numerous technical papers. *Mailing Add:* Dept Indust & Systs Eng Va Polytech Inst & State Univ Blacksburgh VA 24061-0118

TORGERSON, DAVID FRANKLYN, b Winnipeg, Man, July 11, 42; m 66; c 3. WASTE MANAGEMENT. *Educ:* Univ Man, BSc, 65, MSc, 66; McMaster Univ, PhD(chem), 69. *Prof Exp:* Asst prof chem, Dept Chem, 69-70, res scientist, 70-74, sr scientist chem, Cyclotron Inst, Tex A&M Univ, iv, 74-76; res chemist, Atomic Energy Can Ltd, 76-78, sect head, 78-79, head chem, 79-84, dir, Appl Sci Div, 84-86, Reactor Safety Div, 86-89, VPRES ENVIRON SCI & WASTE MANAGEMENT, ATOMIC ENERGY CAN LTD, PINAWA, MAN, 89- *Mem:* Can Nuclear Soc; fel Chem Inst Can. *Res:* Waste management. *Mailing Add:* Atomic Energy Can Whiteshell Labs Pinawa MB R0E 1L0 Can

TORGERSON, RONALD THOMAS, b Minneapolis, Minn, Sept 20, 36; m 63; c 2. HIGH ENERGY PHYSICS, THEORETICAL PHYSICS. *Educ:* Col St Thomas, BS, 58; Univ Chicago, MS, 62, PhD(physics), 65. *Prof Exp:* Instr physics, Univ Notre Dame, 65-68; asst prof, Ohio State Univ, 68-73; res assoc physics, 73-77, PROGRAMMER ANALYST, UNIV ALTA, 77- *Mem:* Am Phys Soc. *Res:* Quantum field theory; high energy collisions; pi pi scattering; weak and electromagnetic interactions; hadron spectroscopy. *Mailing Add:* 3305-110 A St Edmonton AB T6G 3G3 Can

TORGESON, DEWAYNE CLINTON, b Ambrose, NDak, Oct 1, 25; m 59; c 3. PLANT PATHOLOGY. *Educ:* Iowa State Univ, BS, 49; Ore State Univ, PhD(plant path), 53. *Prof Exp:* Plant pathologist, Boyce Thompson Inst Plant Res, Inc, 52-85, prog dir Bioregulant Chem, 63-85, secy, 73-85; RETIRED. *Concurrent Pos:* Mem & chmn, Fed Insecticide, Fungicide & Rodenticide Act Sci Adv Panel, Environ Protection Agency, 76-81. *Mem:* AAAS; Am Inst Biol Sci; Am Phytopath Soc. *Res:* Fungicides; discovery and development of pesticides. *Mailing Add:* 106 Berkshire Rd Ithaca NY 13736

TORGOW, EUGENE N, b Bronx, NY, Nov 26, 25; m 50; c 3. MICROWAVES, ENGINEERING MANAGEMENT. *Educ:* Cooper Union, BSEE, 46; Polytech Inst Bklyn, MSEE, 80. *Prof Exp:* Mgr microwave lab, Allen B Dumont Labs, 51-53; res assoc, Microwave Res Inst, Polytech Inst Brooklyn, 53-59; mgr microwave prod, Dorne & Margolin Inc, 60-64; dir res, Rantec Div, Emerson Elec Co, 64-68; sr scientist, Missile Systs Group, Hughes Aircraft Co, 68-74, prog mgr, 74-81, assoc labs mgr, 81-85; RETIRED. *Concurrent Pos:* Teacher, Calif State Univ, Northridge, 85-; consult eng mgt & training, 85- *Mem:* Sigma Xi; fel Inst Elec & Electronics Engrs; fel Inst Advan Eng. *Res:* Microwave filters and equalizers; management of major aerospace electronic system development programs; published over a dozen papers. *Mailing Add:* 9531 Donna Ave Northridge CA 91324

TORIBARA, TAFT YUTAKA, b Seattle, Wash, Apr 10, 17; m 48; c 2. BIOPHYSICS, CHEMISTRY. *Educ:* Univ Wash, BS, 38, MS, 39; Univ Mich, PhD(chem), 42. *Prof Exp:* Res chemist, Dept Eng Res, Univ Mich, 42-48; scientist chem, Atomic Energy Proj, Univ Rochester, 48, from asst prof to assoc prof, 50-63, prof radiobiol & biophys, 63-87, prof, 87-89, EMER PROF TOXICOL BIOPHYS, MED SCH, UNIV ROCHESTER, 89- *Concurrent Pos:* Nat Inst Gen Med Sci spec res fel, Univ Tokyo, 60-61. *Mem:* AAAS; Am Chem Soc. *Res:* Binding of ions and small molecules to serum proteins; analytical chemistry of trace materials in biological systems; trace element analysis along a single strand of hair by X-ray fluorescence; measurement of environmental pollutants. *Mailing Add:* Dept Biophys Univ Rochester Med Ctr Rochester NY 14642

TORIDIS, THEODORE GEORGE, b Istanbul, Turkey, Sept 7, 32; US citizen; m 61; c 2. STRUCTURAL DYNAMICS, APPLIED MECHANICS. *Educ:* Robert Col, Istanbul, BS, 54; Mich State Univ, MS, 61, PhD(civil eng), 64. *Prof Exp:* Design engr, EMC-RAR Contractors, 54-56; asst div engr, Raymond Concrete Pile Co, 56-57; res asst, Mich State Univ, 59-64; assoc prof eng mech, 64-77, PROF ENG & APPL SCI, GEORGE WASHINGTON UNIV, 77- *Concurrent Pos:* NSF res grants, 65-67, prin investr, 70-; co-investr, David Taylor Model Basin Res contract & prin investr, Naval Ship Res & Develop Ctr contract, 66-69; sr res scientist, Nat Biomed Res Found, 66-70. *Mem:* AAAS; Am Soc Civil Engrs. *Res:* Elastoinelastic response of beams to moving loads; improved vibration analysis of beams and plates; biomechanics, stress analysis of a bone; seismic analysis of structures; nonlinear deformations of framed structures. *Mailing Add:* Sch Eng George Washington Univ 2121 Eye St NW Washington DC 20006

TORIO, JOYCE CLARKE, b Biddeford, Maine, Oct 1, 34; m 55. SCIENCE ADMINISTRATION, INFORMATION SCIENCE. *Educ:* Rutgers Univ, BS, 56, MS, 61, PhD(hort, soils), 65. *Prof Exp:* Ed biochem, hort & soils, Chem Abstr Serv, Am Chem Soc, 65-69; staff officer, Bd Agr & Renewable Resources, Nat Acad Sci, 69-74; HEAD INFO SERV, INT RICE RES INST, 74- *Concurrent Pos:* Consult, World Bank. *Mem:* Am Soc Hort Sci; Am Chem Soc; Am Inst Biol Sci; Sigma Xi. *Res:* Pomology, mineral nutrition and plants; soil fertility and analysis; plant physiology and pathology; rice culture and associated multiple cropping systems research-information management. *Mailing Add:* 10723 West Dr Apt 202 Fairfax VA 22030

TORKELSON, ARNOLD, b Thompson, NDak, Oct 28, 22; m 44; c 4. ORGANOMETALLIC CHEMISTRY. *Educ:* Univ NDak, BSc, 46; Purdue Univ, MS, 48, PhD, 50. *Prof Exp:* Asst, Purdue Univ, 46-48; prod develop chemist, Silicone Prod Dept, Gen Elec Co, 50-58, mgr anal chem, 58-65, fluid prod develop, 65-72, specialities develop, 72-76, & fluids resins & specialities prod develop, 76-88; RETIRED. *Mem:* AAAS; Am Chem Soc; fel Am Inst Chem. *Res:* Synthesis of organosilicon compounds; rate studies on the cleavage of silicon-carbon bond; product development and research on silicone fluids, resins and specialty products. *Mailing Add:* River Rd Box 60 M RD 2 West Lebanon NH 03784

TORLEY, ROBERT EDWARD, b Monmouth, Ill, Jan 28, 18; m 41; c 2. CHEMISTRY. *Educ:* Monmouth Col, BS, 39; Univ Iowa, MS, 41, PhD(chem), 42. *Prof Exp:* Chemist, Am Cyanamid Co, Conn, 43-51, chemist, Chem Processing Plant, Idaho, 51-53, gen supt, 53, asst plant mgr, Bridgeville Plant, Pa, 53-56, asst plant mgr, Res Div, NY, 56-58, contract mgr govt solid rocket propellant contract, Stamford Labs, 58-62, dir, Contract Res Dept, 62-63, dir, Physics Dept, 63-71, dir, Sci Serv Dept, 71-73; vpres & dir technol, T&E Ctr, Evans Prod Co, 73-76; RETIRED. *Mem:* Fel AAAS; Am Chem Soc; Am Phys Soc. *Mailing Add:* 1820 NW Woodland Dr Corvallis OR 97330

TORMANEN, CALVIN DOUGLAS, b Litchfield, Minn, Nov 24, 46; m 71; c 2. ENZYMOLOGY. *Educ:* Univ Minn, BA, 68, PhD(biochem), 74. *Prof Exp:* Res fel, Univ NMex, 74-75; asst prof chem, Ambassador Col, 75-78; res asst prof, Univ NMex, 78-81; asst prof, 81-86, ASSOC PROF BIOCHEM, CENT MICH UNIV, 86-; ASSOC PROF BIOCHEM, CENT MICH UNIV, 86- *Mem:* Am Chem Soc; Sigma Xi; Am Inst Chemists; AAAS. *Res:* Regulation of creatine biosynthesis by transamidinase in rat kidney and in chicken liver and kidney. *Mailing Add:* Dept Chem Cent Mich Univ Mt Pleasant MI 48859

TORMEY, DOUGLASS COLE, b Madison, Wis, Sept 2, 38. ONCOLOGY, INTERNAL MEDICINE. *Educ:* Univ Wis, BS, 60, MD, 64, PhD(oncol), 69. *Prof Exp:* Intern med, Med Ctr, Univ Calif, San Francisco, 64-65, resident, 65-66; fel oncol, Univ Wis, Madison, 66-69 & Roswell Park Mem Inst, 69-70; staff oncologist, Walter Reed Gen Hosp, 70-72; head , Med Breast Cancer Serv, Nat Cancer Inst, NIH, 72-76; assoc prof, 76-82, PROF HUMAN ONCOLOGY & MED, MED SCH, UNIV WIS, MADISON, 82- *Concurrent Pos:* Sr investr, Cancer & Leukemia Group B, 70-76, Nat Cancer Inst, NIH, 72-76 & Eastern Coop Oncol Group, 72-; consult, Dept Biol Sci, George Washington Univ, 74-76, William S Middleton Mem Vet Hosp, 79-, US-Japan Coop Breast Cancer Prog, 79-81; assoc dir, Clin Trials, Wis Clin Cancer Ctr, 85-; chief exec officer, Eastern Coop Oncol Group, 85-90, chmn, 90- *Honors & Awards:* Borden Award, Med Sch, Univ Wis, 64. *Mem:* Am Asn Cancer Res; Am Soc Clin Oncol; Cell Kinetics Soc; Am Soc Hematol; Int Asn Breast Cancer Res. *Res:* Treatment and biology of breast cancer, and the clinical utility of the role of biomarkers in the disease. *Mailing Add:* K4-632 CSC 600 Highland Ave Madison WI 53792

TORMEY, JOHN MCDIVIT, b Baltimore, Md, Oct 7, 34; div; c 2. PHYSIOLOGY, CELL BIOLOGY. *Educ:* Loyola Col, Md, BS, 56; Johns Hopkins Univ, MD, 61. *Prof Exp:* Fel ophthal, Johns Hopkins Univ, 61-62, instr, 62-63; res fel biol, Harvard Univ, 63-64; asst prof anat & ophthal, Johns Hopkins Univ, 64-66; staff assoc phys biol, Nat Inst Arthritis & Metab Dis, 66-68; asst prof, 68-70, assoc prof, 70-78, PROF PHYSIOL, UNIV CALIF, LOS ANGELES, 78- *Concurrent Pos:* Nat Inst Neurol Dis & Blindness fel, 61-63, spec fel, 63-66, res grants, 65-66 & 68- *Mem:* Microbeam Analytical Soc; Am Physiol Soc; Am Soc Cell Biol; Am Asn Anat. *Res:* Relationship between structure and function of body tissues, especially epithelia and muscle; development of methods for localizing transport functions; electron microprobe analysis. *Mailing Add:* Dept Physiol 53-170 CHS Univ Calif 405 Hilgard Ave Los Angeles CA 90024

TORNABENE, THOMAS GUY, b Cecil, Pa, May 6, 37; m 62, 77; c 6. MICROBIOLOGY. *Educ:* St Edward's Univ, BS, 59; Univ Houston, MS, 62, PhD(biol chem), 67. *Prof Exp:* Instr biol, Univ Houston, 62-65; fel biochem, Nat Res Coun, Ottawa, Can, 67-68; from asst prof to assoc prof microbiol, Colo State Univ, 68-78, prof, 78-80, group mgr, solar Energy Res Inst, 80; dir, Dept Appl Biol, 81-90, ASSOC DEAN, COL SCI, GA INST TECHNOL, 90- *Mem:* Am Soc Microbiol; Am Oil Chem Soc. *Res:* Biogenesis and distribution of microbial hydrocarbons; microbial lipids and carbohydrates; metabolic pathways and mechanisms of synthesis of biochemical compounds; cell immobilization; fermentation processes. *Mailing Add:* Col Sci Ga Inst Technol 225 N Ave NW Atlanta GA 30332

TORNG, HWA-CHUNG, b Yangchow, China, Aug 12, 32; US citizen; m 60; c 2. COMPUTER & ELECTRICAL ENGINEERING. *Educ:* Nat Taiwan Univ, BS, 55; Cornell Univ, MS, 58, PhD(elec eng), 60. *Prof Exp:* From asst prof to assoc prof, 60-71, PROF ELEC ENG, CORNELL UNIV, 71- *Concurrent Pos:* Mem tech staff, Switching Div, Bell Tel Labs, 66-67 & 80-81. *Mem:* Inst Elec & Electronics Engrs; Asn Comput Mach. *Res:* Very-large-scale intergration systems; computer structures and design; microprocessor systems; digital systems; telecommunications. *Mailing Add:* Sch Elec Eng Phillips Hall Cornell Univ Ithaca NY 14853-5401

TORNHEIM, PATRICIA ANNE, b Chicago, Ill, June 12, 39. ANATOMY. *Educ:* Rosary Col, BA, 61; Univ Ill, MS, 64; Univ Kans, PhD(anat), 73. *Prof Exp:* Instr anat, Univ Kans, 68-69; instr, Kansas City Col Osteop Med, 69-73; asst prof, 73-80, ASSOC PROF ANAT, UNIV CINCINNATI, 80- *Mem:* Fel AAAS; Am Asn Anatomists; Cajal Club. *Res:* Traumatic cerebral edema; metabolic cerebral edema; cerebrospinal fluid pathways. *Mailing Add:* Col Biol & Anat 521 Gob3 Msb Univ Cincinnati Col Med Cincinnati OH 45267

TORNQVIST, ERIK GUSTAV MARKUS, b Lund, Sweden, Jan 13, 24; m 69; c 1. POLYMER CHEMISTRY, BIOCHEMISTRY. *Educ:* Royal Inst Technol, Sweden, MSc, 48; Univ Wis, MS, 53, PhD(biochem), 55. *Prof Exp:* First res asst, Div Food Chem, Royal Inst Technol, Sweden, 49-51; res asst, Dept Biochem, Univ Wis, 51-55; res chemist, Chem Res Div, Esso Res & Eng Co, 55-58, res assoc, 58-66, SR RES ASSOC, ENJAY POLYMER LABS, LINDEN, EXXON RES & ENG CO, 66- *Mem:* AAAS; Am Chem Soc; NY Acad Sci; Swedish Asn Eng & Archit; Sigma Xi. *Res:* Organometallic chemistry and catalysis; polymer chemistry, especially synthesis and mechanisms of polymerization; biotechnical production of protein, fat, vitamins and antibiotics. *Mailing Add:* 38 Mareu Dr Scotch Plains NJ 07076

TORO, RICHARD FRANK, b South Amboy, NJ, Nov 11, 39; m 63; c 3. ENVIRONMENTAL ENGINEERING, CHEMICAL ENGINEERING. *Educ:* Lafayette Col, BS, 61; Univ Del, MChE, 64. *Prof Exp:* Process engr chem plants, Pullman-Kellogg, 62-65; process engr mgr eng & vpres environ eng & testing & chem res & develop, Princeton Chem Res, 65-73; VPRES & CONSULT ENVIRON ENG & TESTING, RECON SYST INC, 73- *Concurrent Pos:* Chmn, Intersoc Comt Methods Air Sampling & Analyisi, 76- *Mem:* Air Pollution Control Asn; Am Inst Chem Engrs; Am Chem Soc; Sigma Xi. *Res:* Chemical processing; environmental science. *Mailing Add:* 589 Foothill Rd Bridgewater NJ 08807

TORO-GOYCO, EFRAIN, physical chemistry, biochemistry, for more information see previous edition

TOROK, ANDREW, JR, b Hopewell, Va, Oct 30, 25; m 51; c 2. CHEMISTRY. *Educ:* Pa State Univ, BS, 49; Stevens Inst Technol, MS, 56. *Prof Exp:* Anal chemist, William P Warner, Inc, 49-51; res chemist, Venus Pen & Pencil Co, 51-53, chief chemist, 53-57, tech dir, 57-61; prod develop mgr, Ga Kaolin Co, 61-74; DIR RES & DEVELOP, FABER-CASTELL CORP, 74- *Mem:* Am Chem Soc; Am Ceramic Soc; fel Am Inst Chemists; NY Acad Sci; Fine Particle Soc (treas, 70-73). *Res:* Clays and clay products, especially application in new fields. *Mailing Add:* 44 Long Ridge Rd RD 3 Dover NJ 07801

TOROK, NICHOLAS, b Budapest, Hungary, June 13, 09; US citizen; m 39. OTOLARYNGOLOGY. *Educ:* E-tv-s Lorand Univ, Budapest, 34. *Prof Exp:* From instr to asst prof otolaryngol, E-tv-s Lorand Univ, Budapest, 40-47; from instr to assoc prof, 50-68, PROF OTOLARYNGOL, UNIV ILL COL MED, 68- *Concurrent Pos:* Consult, Chicago Read Hosp, Ill State Psychiat Inst, Ill Hosp Sch & Michael Reese Hosp; pres, Chicago Laryngol & Otolarynol Soc, 74. *Honors & Awards:* Award, Am Acad Ophthal & Otolaryngol, 69; NASA Skylab Achievement Award. *Mem:* Am Acad Ophthal & Otolaryngol; Am Laryngol, Rhinol & Otol Soc; Am Neurotol Soc (pres, 73-74, 76); Am Acad Cerebral Palsy; affil Royal Soc Med. *Res:* Otology; neuro-otology; vestibular studies. *Mailing Add:* 42 Portwine Rd Lake Hinsdale Village Clarendon Hills IL 60514

TOROK, THEODORE ELWYN, b Pine Grove, Pa, Aug 16, 31; m 61; c 3. METALLURGICAL ENGINEERING. *Educ:* Univ Idaho, BS, 54; Lehigh Univ, MS, 62, PhD(metall eng), 65. *Prof Exp:* Metallurgist, Bendix Aviation Corp, 57-58; metallurgist, Convair/Astronaut, Gen Dynamics Corp, 58-59, group leader chem & metall, 59; metallurgist, Eng Res Ctr, Western Elec Co, NJ, 65-68; RES ENGR, HOMER RES LABS, BETHLEHEM STEEL CORP, 68- *Mem:* Am Soc Metals; Am Welding Soc; Sigma Xi. *Res:* Dilatometry; weldability; delayed cracking; hot dipped coatings; phase transformations; wire patenting and drawing. *Mailing Add:* PO Box 326 Springtown PA 18081

TOROP, WILLIAM, b New York, NY, Jan 12, 38; m 60; c 2. INORGANIC CHEMISTRY, SCIENCE EDUCATION. *Educ:* Univ Pa, AB, 59, MS, 61, EdD(sci educ), 68. *Prof Exp:* Prof employee chem, Upper Darby Sr High Sch, Pa, 60-68; asst prof chem & sci educ, St Joseph's Col, Pa, 68-71; PROF CHEM, WEST CHESTER STATE UNIV, 71- *Concurrent Pos:* Elem sci consult, Interboro Sch Dist, Pa, 69-70; elem sci consult, Marple Newtown Sch Dist, 70-72; dir, Del Valley Inst Sci Educ, 71-; Commonwealth distinguished teaching chair. *Mem:* Am Chem Soc. *Res:* Use of written laboratory reports in high school chemistry; trivalent basic polyphosphates; evaluation of elementary science programs; computer managed and computer assisted instruction. *Mailing Add:* Dept Chem West Chester State Univ West Chester PA 19383

TOROSIAN, GEORGE, b Racine, Wis, Jan 1, 36; m 64; c 2. PHARMACY, PHARMACOLOGY. *Educ:* Univ Wis-Madison, BS, 62, MS, 64, PhD(pharm), 66. *Prof Exp:* Sr pharm chemist, Menley & James Labs Div, Smith Kline & French Labs, Inc, 66-69; assoc prof pharm, Col Pharm, Univ Fla, 69-81; MGR MFG TECHNOL, DUPONT-MERCK, 81- *Mem:* Am Pharmaceut Asn; Am Asn Cols Pharm; Acad Pharmaceut Sci; Am Asn Pharmaceut Sci. *Res:* Product development and design; biopharmaceutics; solution kinetics. *Mailing Add:* 801 Gen Cornwallis Dr West Chester PA 19382

TORP, BRUCE ALAN, b Duluth, Minn, Sept 5, 37; m 60; c 3. INORGANIC CHEMISTRY, COMPUTER SCIENCE. *Educ:* Univ Minn, BA, 59; Iowa State Univ, MS, 62, PhD(inorg chem), 64. *Prof Exp:* Sr chemist, 64-68, supvr, Inorg Chem Res Group, 68-71, lab mgr, Physics & Mat Res Lab, 71-74, dir, Mat & Electronics Res Lab, Cent Res Labs, 75-77, dir, Data Rec Prod Div Labs, 3M Co, 77-86, TECH DIR, TEL COMM PROD 3M CO, 86- *Mem:* Am Chem Soc. *Res:* Coordination, transition metal and solid state chemistry; semiconductor research; magnetic materials research; magnetic media development. *Mailing Add:* 2924 Gabriel View Dr Georgetown TX 78628

TORQUATO, SALVATORE, b Falerna, Italy, Feb 10, 54; US citizen; m 75; c 1. STATISTICAL PHYSICS. *Educ:* Syracuse Univ, BS, 75; State Univ NY, Stony Brook, MS, 77, PhD(mech eng), 80. *Prof Exp:* Asst prof mech eng, Gen Motors Inst, 81-82; from asst prof to assoc prof, 82-91, PROF MECH & AEROSPACE ENG, NC STATE UNIV, 91- *Concurrent Pos:* Prin investr, Nat Sci Found, 82-88 & US Dept Energy, 86-; vis prof, Courantinstitute Math Sci, NY Univ, 90-91. *Mem:* Am Sec Mech Engrs; Am Inst Chem Engrs; Soc Eng Sci; Am Inst Physics; Soc Appl & Indust Math. *Res:* The relationship of mechanical, electrical, thermal and chemical properties of composite and other heterogenous materials to their microstructures; author of 80 publications. *Mailing Add:* Dept Mech & Aero Eng Box 7910 NC State Univ Raleigh NC 27695-7910

TORRANCE, DANIEL J, b Peking, China, Nov 14, 21; US citizen; m 51; c 2. RADIOLOGY. *Educ:* Univ Wash, BSc, 44; Johns Hopkins Univ, MD, 49. *Prof Exp:* Intern med, Johns Hopkins Hosp, 49-50, fel path, 50-51, asst resident radiol, 51-53, from instr to assoc prof, Sch Med, Johns Hopkins Univ, 53-63; head div, Scripps Clin & Res Found, La Jolla, Calif, 63-66; assoc prof, Sch Med, Wash Univ, 66-68; prof, Univ Calif, Los Angeles, 68-72; chief radiologist, Bay Harbor Hosp, Calif, 72-76; radiologist & chief, Chest & Gen Radiol Sect, Harbor Gen Hosp, 76-85; ADJ PROF RADIOL, UNIV CALIF, LOS ANGELES, 76-; RADIOLOGIST, BAY SHORE MED CLIN, 85- *Concurrent Pos:* Consult, USPHS Hosp, Baltimore, Md, 54-; radiologist, Johns Hopkins Hosp, 55-63; assoc radiologist, Mallinckrodt Inst Radiol, Barnes Hosp, St Louis, 66-; chief dept radiol, Harbor Gen Hosp, Torrance, Calif; clin prof radiol, Univ Calif, Los Angeles, 72-; consult, Vet Admin Hosp, Long Beach, Calif, 80- *Mem:* Am Col Radiol. *Res:* Chest radiograph in connection with the pulmonary circulation; problems in the radiography of pulmonary atelectasis; radiographic manifestations of pulmonary edema. *Mailing Add:* 502 Torrance Blvd-Rad Redondo Beach CA 90277

TORRANCE, JERRY BADGLEY, JR, b San Diego, Calif, July 20, 41; div; c 2. SYNTHETIC INORGANIC & ORGANOMETALLIC CHEMISTRY. *Educ:* Stanford Univ, BS, 63; Univ Calif, Berkeley, MA, 66; Harvard Univ, PhD(appl physics), 69. *Prof Exp:* Res physicist, Thomas J Watson Res Ctr, 69-76, mgr phys properties organic solids group, 74-76, MGR, ALMADEN RES CTR, IBM CORP, 76-, MGR MAGNETICS, 81- *Mem:* Am Chem Soc; fel Am Phys Soc; Mat Res Soc. *Res:* High temperature superconductivity; synthesis of new materials; organic furomagnets; organic solids. *Mailing Add:* IBM Almaden Res Ctr K32/803 650 Harry Rd San Jose CA 95120

TORRANCE, KENNETH E(RIC), b Minneapolis, Minn, Aug 23, 40; m 62; c 3. PHOTOREALISTIC IMAGES, BOILING HEAT TRANSFER. *Educ:* Univ Minn, Minneapolis, BS, 61, MSME, 64, PhD(mech eng), 66. *Prof Exp:* Factory Mutual Eng Co res assoc, Nat Bur Standards, 66-68; from asst prof to assoc prof, 69-81, assoc dean, Col Eng, 83-86, PROF MECH & AEROSPACE ENG, CORNELL UNIV, 81- *Concurrent Pos:* Sr fel, Nat Ctr Atmospheric Res, 74-75. *Mem:* AAAS; Am Inst Aeronaut & Astronaut; fel Am Soc Mech Engrs; Am Phys Soc. *Res:* Heat transfer; fluid mechanics; numerical computations; computer graphics. *Mailing Add:* 37 Deerhaven Dr Ithaca NY 14850

TORRE, DOUGLAS PAUL, b New Orleans, La, Feb 6, 19; m 54, 77; c 2. CRYOSURGERY, DERMATOLOGIC SURGERY. *Educ:* Tulane Univ, BS, 40, MD, 43. *Prof Exp:* PVT PRACT DERMATOLOGIST, 50- *Concurrent Pos:* Clin prof med, Med Col, Cornell Univ. *Honors & Awards:* Gold Award, Am Acad Dermat, 77. *Mem:* Am Soc Dermat Surg (pres, 80-81); Am Col Cryosurg (pres, 83-84); Am Acad Dermat; AMA. *Res:* Development of instrumentation for and clinical application of cryosurgery. *Mailing Add:* 320 E 65th St New York NY 10021

TORRE, FRANK JOHN, b Newark, NJ, Oct 6, 44; m 68; c 1. PHYSICAL CHEMISTRY. *Educ:* Monmouth Col NJ, BS, 67; Rutgers Univ, PhD(phys chem), 71. *Prof Exp:* Res chem, Bell Tel Labs, 67-68; fel, Univ Rochester, 71-73; asst prof, 73-80, ASSOC PROF CHEM, SPRINGFIELD COL, 80- *Mem:* Am Chem Soc. *Mailing Add:* Dept Chem Springfield Col 263 Alden St Springfield MA 01109-3788

TORRE-BUENO, JOSE ROLLIN, b Tucson, Ariz, Nov 20, 48; m 69; c 1. PHYSIOLOGY. *Educ:* State Univ NY Stony Brook, BS, 70; Rockefeller Univ, PhD(physiol), 75. *Prof Exp:* Res assoc physiol, 75-78, med res asst prof, Dept Physiol, Med Ctr, 78-84, PRES, AM INNOVISION, DUKE UNIV, 84- *Mem:* Sigma Xi. *Res:* Respiratory physiology and energetics particularly during hypoxia. *Mailing Add:* Amer Innovision 9581 Ridgehaven Ct San Diego CA 92123

TORREGROSSA, ROBERT EMILE, b Bogalusa, La, Oct 24, 51; m 75; c 2. MICROBIAL PHYSIOLOGY, FERMENTATION TECHNOLOGY. *Educ:* La Tech Univ, BS, 73; Univ Ga, PhD(microbiol), 77. *Prof Exp:* Res microbiologist fermentations, CPC Int Inc, 78-80; mem staff, Chem Div, Tethyl Corp, 80-82, Enzyme Technol Corp, 82-84, Phillips Petrol Co, 84-89; MGR PROCESS ENG, TRIAD TECHNOL INC, 89- *Concurrent Pos:* NSF fel, Mass Inst Technol, 77-78. *Mem:* Am Soc Microbiol; Am Chem Soc; Parenteral Drug Asn; Int Soc Pharmaceut Eng. *Res:* Biochemistry; enzyme processes; immobilized cells and enzymes; biotransformations; hydrocarbon oxidation; rDNA fermentation scale-up; biopharmaceuticals. *Mailing Add:* Triad Tech Inc 101 Centerpoint Blvd New Castle DE 19720-4180

TORRENCE, PAUL FREDERICK, b New Brighton, Pa, April 22, 43; m 67; c 2. INTERFERON, NUCLEIC ACIDS. *Educ:* Geneva Col, Pa, BS, 65; State Univ NY, PhD(chem), 69. *Prof Exp:* Staff fel, Nat Inst Arthritis & Metab Dis, 69-71, sr staff fel, Nat Inst Arthritis, Metab & Digestive Dis, 71-74, RES CHEMIST, NAT INST ARTHRITIS, DIABETES & DIGESTIVE & KIDNEY DIS, NIH, 74- *Concurrent Pos:* Ad Hoc consult, Spec Proj Adv Comt, Nat Cancer Inst, 80 & Nat Inst Allergy Infectious Dis, 79- *Mem:* Am Chem Soc; Soc Exp Biol & Med; Am Soc Microbiol; AAAS. *Res:* Mechanisms of induction and action of interons; the role played by double-standard RNA and 2,5-oligoadenlates in these mechanisms and how this information may be used to design antiviral or antitumor agents. *Mailing Add:* Dept Chem NIH Bldg 8 Rm B2A02 Washington DC 20205

TORRENCE, ROBERT JAMES, b Pittsburgh, Pa, June 7, 37; m 59. THEORETICAL PHYSICS. *Educ:* Carnegie-Mellon Univ, BS, 59; Univ Pittsburgh, PhD(physics), 65. *Prof Exp:* Res assoc physics, Syracuse Univ, 65-67; adj prof, Ctr Advan Studies, Nat Polytech Inst, Mex, 67-68; asst prof, 68-70, chmn div appl math, 75-77, ASSOC PROF MATH, UNIV CALGARY, 70- *Mem:* Am Phys Soc. *Res:* General relativity with emphasis on gravitational radiation. *Mailing Add:* Dept Math & Statist Univ Calgary Calgary AB T2N 1N4 Can

TORRES, ANDREW M, b Albuquerque, NMex, Jan 20, 31; m 55; c 4. BOTANY. *Educ:* Univ Albuquerque, BS, 52; Univ NMex, MS, 58; Ind Univ, PhD(bot), 61. *Prof Exp:* Instr biol, Wis State Univ, Oshkosh, 60-61; asst prof bot & genetics, Univ Wis, Milwaukee, 61-64; assoc prof, 64-70, assoc dean grad sch, 69-72, chmn dept, 79-81, PROF BOT & GENETICS, UNIV

KANS, 70- *Concurrent Pos:* Ford Found sr adv, Univ Oriente, Venezuela, 66-67; chief party, Aid to higher educ, Dominican Republic, 68; Calif Avocado Adv Bd, 78; vis prof, Univ Calif, Riverside, 78, Univ Nat Del Sur, 81; consult, Agr Res Orgn, Bet Dagan, Israel, 81 & 85; sum vis scientist, Div Hort, Commonwealth Sci Indust Res Orgn, Adelaide, Australia, 84 & 85; sum vis prof, Univ Colo, 86-87. *Mem:* AAAS; Bot Soc Am; Soc Study Evolution; Genetics Soc Am; Sigma Xi. *Res:* Cytogenetics; alcohol dehydrogenase isozymes of sunflowers; genetics; paleozoic algae. *Mailing Add:* Dept Bot Univ Kans Lawrence KS 66044

TORRES, ANTHONY R, b Trinidad, Colo, Apr 1, 43; m 75; c 7. BIOCHEMISTRY, MOLECULAR BIOLOGY. *Educ:* Univ Utah, BS, 70, MD, 74. *Prof Exp:* Res assoc, Nat Cancer Inst, 75-78, expert consult, 78-81; resident, Yale Univ, 81-83, asst prof lab med, 83-86; MEM STAFF, HYCLONE LAB INC, 86- *Concurrent Pos:* Adj fac mem, Dept Med, Univ Utah, 90- *Mem:* NY Acad Sci. *Res:* Separation of biologically active proteins from complex mixtures using capacity and high resolution displacement chromatography. *Mailing Add:* 4289 Noal Dr Salt Lake City UT 84124

TORRES, FERNANDO, b Paris, France, Nov 29, 24; m 55; c 1. NEUROLOGY, NEUROPHYSIOLOGY. *Educ:* Ger Col, Colombia, BA, 41; Nat Univ Colombia, MD, 48; Am Bd EEG, dipl, 51; Am Bd Psychiat & Neurol, dipl, 61. *Prof Exp:* Asst neurosurg, Inst Cancer, Buenos Aires, Arg, 49-50; resident neurol, Montefiore Hosp, New York, 53-55; from instr to assoc prof, 56-64, PROF NEUROL, UNIV MINN, MINNEAPOLIS, 64- *Concurrent Pos:* Fel, Johns Hopkins Hosp, 50-52, NIH res fel, 52-53; NIH spec fel, LaSalpetriere Hosp, Paris, 63-64; asst, Columbia Univ, 54-55; consult prof, Univ PR, 61-64. *Mem:* AAAS; fel Am Acad Neurol; Am Neurol Asn; Soc Neurosci; Am Electroencephalog Soc. *Res:* Electroencephalography; clinical neurophysiology; epilepsy; cerebrovascular physiology; developmental cerebral physiology. *Mailing Add:* Dept Neurol Univ Minn Hosp & Clinic Box 28 Minneapolis MN 55455

TORRES, JOSEPH CHARLES, CARDIAC ELECTROPHYSIOLOGY, CARDIAC MECHANICS. *Educ:* Boston Univ, PhD(physiol), 61. *Prof Exp:* PROF CARDIOVASC PHYSIOL & VCHMN, DEPT PHYSIOL & BIOPHYS, HAHNEMANN UNIV, 69- *Mailing Add:* Hahnemann Univ MS409 235 N 15th St Philadelphia PA 19102

TORRES, LOURDES MARIA, b Havana, Cuba, Sept 26, 57; US citizen; m 85; c 2. BIOINORGANIC CHEMISTRY. *Educ:* Univ PR, BS, 78, PhD(phys-inorg chem), 82. *Prof Exp:* Postdoctoral fel phys chem, Univ Utah, 82-84; sr res chemist, Union Carbide Corp, 84-88; instr chem, Ga State Univ, 88-89; NSF vis prof chem, Emory Univ, 89-91; RES SCIENTIST, SCI & TECHNOL CTR, CLARK ATLANTA UNIV, 91- *Mem:* Am Chem Soc; Am Catalysis Soc. *Res:* Synthesis and characterization of inert transition metal-nucleotide and polynucleotide complexes; metal-catalyzed phosphate hydrolysis: mechanistic and structural studies using NMR spectroscopy; interaction of environmentally relevant metals with DNA. *Mailing Add:* Chem Dept Clark Atlanta Univ James P Brawley Dr at Fair St Atlanta GA 30314

TORRES-ANJEL, MANUEL JOSE, b Bogota, Colombia, Apr 15, 42; US citizen; m 64; c 4. CLINICAL EPIDEMIOLOGY & CLINICAL TRIALS, SCIENTIFIC MASS MEDIA OF COMMUNICATIONS-PRINTED & BROADCASTING. *Educ:* Nat Univ Colombia, DVM, 64; Tulane Univ & Mich State Univ, MSc, 68; Univ Calif, Davis, MPVM, 72, PhD(comp path), 74. *Prof Exp:* Instr to prof microbiol, Nat Univ Colombia, 63-78; int scientist pub health, World (Pan Am) Health Orgn, 79-83; assoc scientist virol, Wistar Inst, Univ Pa, 83-84; assoc prof epidemiol, Univ Mo Columbia, 84-88; health sci admin clin epidemiol, Nat Inst Allergy Infectious Dis, 88-90, Molecular Biol, Nat Cancer Inst, 90, HEALTH SCI ADMIN CLIN TRIALS, CANCER CLIN INVEST RES COMT, NAT CANCER INST, NIH, 90- *Concurrent Pos:* Lectr & vis prof epidemiol, Univ Calif Davis, 71-78, postgrad res, 71-74; consult, Nestle Res Co, Vevey, Switz, 77-78, Food & Agr Orgn, United Nations, & WHO, 87-90; res award, Col Vet Med Asn Meeting, 77, Col Vet Med Asn, 78, Angel Escobar Found, 79; adj prof int health, Dept Prev Med, Cornell Univ, 79-83; adj prof epidemiol, Col Vet Med, Univ Pa, 83-84. *Mem:* Am Vet Med Asn; Am Asn Path; Am Col Epidemiol. *Res:* Virus induced cachexy/wasting syndrome; AIDS; hypothalamic-hypophyseal-thymic axis; diarrheal disease. *Mailing Add:* PO Box 1284 Rockville MD 20849-1284

TORRES-BLASINI, GLADYS, b Ponce, PR; m; c 2. MICROBIOLOGY. *Educ:* Univ PR, BS, 48; Univ Mich, Ann Arbor, MS, 52, PhD(bact), 53; Duke Univ, cert mycol, 54. *Prof Exp:* Teaching asst bact, Univ Mich, 52; from assoc prof to asst prof bact, Univ PR, 53-62, from assoc prof to prof mycol, 62-77, prof & head, Dept Microbiol & Med Zool, 77-84, DIR, DEPT MICROBIOL & MED ZOOL, SCH MED, UNIV PR, 84- *Concurrent Pos:* USPHS fel mycol, 54; Hoffmann-La Roche grant fungistatic drugs, 54, Trichophyton species, 56; Vet Admin Hosp grant, 56 & 57; Univ PR Sch Med & NIH grant, 58; NIH grant, 58, 60, 64 & 68-70; lectr, Hahnemann Med Sch, 62; Univ PR Med Sch Gen Res Funds grant, 64; mem, Study Sect Res, Vet Admin Hosp, San Juan, PR, 70; mem, Acad Senate, Univ PR Sch Med, 77-84; mem, Int Activities Comt, Am Soc Microbiol, 78-81. *Mem:* Soc Am Bacteriologists; Sigma Xi; Tissue Cult Asn. *Res:* Bacteriology; comparison of phagocytosis of various candida species; medical education; medical mycology. *Mailing Add:* Dept Microbiol Univ PR Med Sci Box 5067 San Juan PR 00936

TORRES-MEDINA, ALFONSO, b Colombia, Aug 28, 45; m 71; c 2. VETERINARY MEDICINE, VIROLOGY. *Educ:* Nat Univ Colombia, DVM, 68; Univ Nebr, Lincoln, MS, 71; Univ Nebr, Omaha, PhD(med microbiol), 73. *Prof Exp:* Asst instr vet path, Nat Univ Colombia, 69, fel, 69-71; from instr to asst prof vet sci, Univ Nebr, Lincoln, 73-75; new prod mgr diag, Miles Labs Inc, Cali, Colombia, 76-78; from asst prof to assoc prof vet sci, Univ Nebr, 78-87; prof mgr, Biol Res & Develop, Norden Labs Inc, 87-90; MGR, VIROL DEVELOP, SMITHKLINE BEECHAM ANIMAL HEALTH, 90- *Mem:* Sigma Xi; Am Soc Microbiologist; Am Vet Med Asn; Am Asn Indust Vet; Am Soc Virol. *Res:* Neonatal infectious diseases of bovine and swine, especially in regard to viral diarrheas of calves and piglets; use of gnotobiotic calves and piglets for the study of experimental infectious diseases; vaccine development. *Mailing Add:* 601 W Cornhusker Hwy Smithkline Beecham Animal Health Lincoln NE 68521-3596

TORRES-PINEDO, RAMON, b Burgos, Spain, Apr 3, 29; US citizen; m 57; c 4. PEDIATRICS, GASTROENTEROLOGY. *Educ:* Univ Granada, BS, 48; Univ Madrid, MD, 56. *Prof Exp:* Intern, San Juan City Hosp, PR, 58-59; resident pediat, San Juan City Hosp & Univ Hosp, 59-61; assoc, Univ Hosp, Univ PR, 63-65, from asst prof to prof, 65-75, asst dir pediat res, Clin Res Ctr, Sch Med, 63-75, prof physiol & head dept, 66-75; PROF PEDIAT & CHIEF PEDIAT GASTROENTEROL, 75- DIR, CLIN RES CTR, OKLA CHILDREN'S MEM HOSP, OKLAHOMA CITY, 78- *Concurrent Pos:* Fels pediat res, Michael Reese Hosp & Med Ctr, Univ Ill, 61-63; consult physician, San Juan City Hosp, 64-75. *Mem:* Am Fedn Clin Res; Am Inst Nutrit; Am Pediat Soc; Soc Pediat Res; Sigma Xi. *Res:* Pediatric research; electrolyte transport; intermediary metabolism; nutrition. *Mailing Add:* Children's Mem Hosp PO Box 26307 Oklahoma City OK 73126

TORREY, HENRY CUTLER, b Yonkers, NY, Apr 4, 11; m 37; c 2. NUCLEAR MAGNETIC RESONANCE, RELAXATION PHENOMENA. *Educ:* Univ Vt, BSc, 32; Columbia Univ, MA, 33 & PhD(physics), 37. *Hon Degrees:* DSc, Univ Vt, 65. *Prof Exp:* From instr to asst prof, Penn State Univ, 37-42; from assoc prof to prof physics, Rutgers Univ, 46-76, dean, grad sch, 65-74, dir res arts & sci, 65-74; RETIRED. *Concurrent Pos:* Consult, Calif Res Corp, 52-65; prin investr, NSF & Off Naval Res, 48-65. *Mem:* Sigma Xi; AAAS; Am Phys Soc; Am Asn Physic Teachers. *Res:* Radio frequency spectra; spin diffusion and relaxation. *Mailing Add:* The Colony House 1050 George St Apt 12D New Brunswick NJ 08901

TORREY, JOHN GORDON, b Philadelphia, Pa, Feb 22, 21; m 49; c 5. PLANT PHYSIOLOGY. *Educ:* Williams Col, BA, 42; Harvard Univ, MA, 47, PhD, 50. *Prof Exp:* Harvard Univ traveling fel, Cambridge Univ, 48-49; from instr to assoc prof bot, Univ Calif, 49-60; dir, Cabot Found, Harvard Univ, 66-75, prof bot, 60-84, dir, Harvard Forest, 84-90, BULLARD PROF FORESTRY, HARVARD UNIV, 84- *Concurrent Pos:* Guggenheim fel, 65-66; hon sr res fel, Univ Glasgow, 73; res collabr, Div Plant Indust, Commonwealth Sci & Indust Res Orgn, Australia, 80; Fulbright sr scholar, Univ Aberdeen, Scotland, 84. *Mem:* Nat Acad Sci; Am Acad Arts & Sci; Bot Soc Am; Am Soc Plant Physiol; Soc Develop Biol (pres, 63). *Res:* Physiology of root growth; physiology and biochemistry of tissue differentiation; root nodules in legumes and non-legumes; biology of Frankia. *Mailing Add:* Harvard Forest Harvard Univ Petersham MA 01366

TORREY, RUBYE PRIGMORE, b Sweetwater, Tenn, Feb 18, 26; m 57; c 2. RADIATION CHEMISTRY, ANALYTICAL CHEMISTRY. *Educ:* Tenn State Univ, BS, 46, MS, 48; Syracuse Univ, PhD(chem), 68. *Prof Exp:* Res assoc & instr chem, 48-57, from asst prof to assoc prof, 57-72, PROF CHEM, TENN STATE UNIV, 72- *Concurrent Pos:* Asst lectr, Syracuse Univ, 63-68; US AEC res grant & res collabr, Brookhaven Nat Lab, 70- *Mem:* AAAS; Am Chem Soc. *Res:* Electro-analytical chemistry; gas phase reaction mechanisms using alpha radiolysis and high-pressure impact mass spectrometry; effects of various factors on polarographic diffusion coefficients using chronopotentiometric technique. *Mailing Add:* Box 150234 Nashville TN 37215-0234

TORRIANI GORINI, ANNAMARIA, b Milan, Italy, Dec 19, 18; nat US; m 60; c 1. BACTERIAL PHYSIOLOGY. *Educ:* Univ Milan, PhD(natural sci), 42. *Prof Exp:* Res asst physiol & bact, Pasteur Inst, Paris, 48-55; Fulbright fel microbiol, NY Univ, 55-58; res assoc, Biol labs, Harvard Univ, 58-59; from res assoc to assoc prof, 59-75, PROF MICROBIOL, MASS INST TECHNOL, 75- *Concurrent Pos:* NIH res career award, 63-73. *Mem:* Am Soc Biol Chemists; Am Soc Microbiol; AAAS; Genetics Soc Am. *Res:* Control of protein synthesis; bacterial genetics; bacterial spores germination. *Mailing Add:* Dept Biol Mass Inst Technol Cambridge MA 02139

TORTONESE, MARCO, b Rome, Italy, Dec 14, 62. INTEGRATED CIRCUIT MANUFACTURING, MICROFABRICATION. *Educ:* Univ Genova, Italy, Laurea, 86; Stanford Univ, MSEE, 90. *Prof Exp:* RES ASST APPL PHYSICS, GINZTON LAB, STANFORD UNIV, 89- *Res:* Microfabrication of probes for scanning probe microscopy; atomic force microscopy; scanning ion conductance microscopy; near field scanning optical microscopy. *Mailing Add:* E Ginzton Lab Stanford Univ Stanford CA 94305

TORTORELLO, ANTHONY JOSEPH, b Chicago, Ill, Sept 26, 45; m 71; c 3. SYNTHETIC ORGANIC CHEMISTRY. *Educ:* St Joseph's Col, BS, 67; Loyola Univ, Chicago, MS, 70, PhD(chem), 75. *Prof Exp:* Res scientist chem, Am Can Co, Barrington, 74-77; sr res chemist, 77-87, MGR, POLYMER DEVELOP DEPT, DESOTO, INC, DES PLAINES, ILL, 87- *Concurrent Pos:* Adj fac, Elmhurst Col, Loyola Univ; lectr, Gordon Res Conf. *Mem:* Am Chem Soc; Fed Soc Coatings Technol. *Res:* Polymer synthesis; organic coatings; polymer structure property relationships; radiation cure coatings. *Mailing Add:* 449 East Ct Elmhurst IL 60126

TORÚN, BENJAMÍN, b Guatemala City, Guatemala, Dec 8, 39; m 67; c 3. CLINICAL NUTRITION, HUMAN METABOLISM. *Educ:* San Carlos Univ, Guatemala, MD, 65; Harvard Univ, MA, 69; Mass Inst Technol, PhD(nutrit biochem & metab), 72. *Prof Exp:* Fel physiol, Kellogg Found, Harvard Univ, 67-69; res assoc nutrit, Mass Inst Technol, 69-72; med officer clin & exp nutrit, Inst Nutrit Cent Am & Panama, 65-67, med officer human nutrit, 72-80, div chief nutrit & health, 80-85, PROF BASIC & HUMAN NUTRIT, INST NUTRIT CENT AM & PANAMA, SAN CARLOS UNIV, 73-, PROG HEAD METAB & CLIN NUTRIT, 85- *Concurrent Pos:* Dir, Clin Res Ctr, Inst Nutrit Cent Am & Panama, 72-83; prin investr, various res projs, 73-; lectr numerous hosps, univs & insts in more than 20 countries, 73-;

ed, J Guatemalan Med Col, 75-76 & 86-87; vis prof, Dept Nutrit & Food Sci, Mass Inst Technol, 77-79; consult-adv, WHO, UN Food & Agr Orgn, UN Univ, Int Dietary Energy Consultancy Group, Pan Am Health Orgn, 81-; vis scientist, Dept Appl Biol Sci, Mass Inst Technol, 84-85. *Honors & Awards:* Nestlé Award, Guatemalan Pediat Asn, 89. *Mem:* Am Soc Clin Res; Am Inst Nutrit; Am Col Nutrit. *Res:* Protein-energy malnutrition; energy, protein and amino acid requirements; protein-energy interactions; nutrition and work capacity; physical activity and growth; nutrient absorption; nutritional status evaluation; diarrheal diseases. *Mailing Add:* Inst Nutrit Cent Am & Panama Apartado Postal 1188 Guatemala

TORVIK, PETER J, b Fergus Falls, Minn, Dec 6, 38; m 58; c 2. AERONAUTICAL & ASTRONAUTICAL ENGINEERING. *Educ:* Univ Minn, BS, 60, MS, 62, PhD(eng mech), 65; Wright State Univ, BA, 80. *Prof Exp:* Asst mech & mat, Univ Minn, 60-62, res fel & instr, 62-64; from asst prof to assoc prof, 64-73, head, Dept Aeronaut & Astronaut, 80-90, PROF AEROSPACE ENG & ENG MECH, AIR FORCE INST TECHNOL, 73- *Concurrent Pos:* Vis prof, Ohio State Univ, 79. *Honors & Awards:* Fel, Am Soc Mech Engrs. *Mem:* Am Acad Mech; Am Soc Mech Engrs; Acoust Soc Am; fel Am Inst Aeronaut & Astronaut; Am Soc Eng Educr; Sigma Xi; Soc Engr Sci; Soc Rheology. *Res:* Elasticity and wave propagation; material behavior; effects of high power lasers. *Mailing Add:* 2000 Harvard Blvd Dayton OH 45406

TORY, ELMER MELVIN, b Vermilion, Alta, Dec 10, 28; m 56; c 2. APPLIED MATHEMATICS, CHEMICAL ENGINEERING. *Educ:* Univ Alta, BSc, 52; Purdue Univ, PhD(chem eng), 61. *Prof Exp:* Res chemist, Aluminium Labs Ltd, 54-58; asst prof chem eng, McMaster Univ, 60-63; assoc chem engr, Brookhaven Nat Lab, 63-65; assoc prof, 65-73, PROF MATH, MT ALLISON UNIV, 73- *Mem:* Can Appl Math Soc; Can Soc Chem Eng; AAAS; Can Math Soc. *Res:* Stochastic modelling of the slow sedimentation of small particles in a viscous fluid; stability of sedimenting clusters of identical spheres; computer simulation of random packing of spheres. *Mailing Add:* Dept Math & Computer Sci Mt Allison Univ Sackville NB E0A 3C0 Can

TOSATO, GIOVANNA, IMMUNITY OF EPSTEIN-BARR VIRUS IN MEN. *Educ:* State Univ Rome, Italy, MD, 73. *Prof Exp:* SR STAFF FEL, FOOD & DRUG ADMIN-NIH, 84- *Res:* Suppression in the defense against Epstein-Barr virus; B-cell activation by the Epstein-Barr virus. *Mailing Add:* Dept Biochem & BiophysCcs Bur Biologics FDA 8800 Rockville Pike Bldg 29 Rm 520 Bethesda MD 20892

TOSCANO, WILLIAM AGOSTINO, JR, b Santa Monica, Calif, June 26, 45; m 69; c 1. BIOCHEMISTRY, CELL BIOLOGY. *Educ:* Indiana Univ, Pa, BSc, 68, MSc, 72; Univ Ill, Urbana, PhD(biochem), 78. *Prof Exp:* Chemist, Gulf Res & Develop Co, Pittsburgh, Pa, 68-71; res assoc biochem, Univ Ill, Urbana, 78; res fel pharmacol, Univ Wash, Seattle, 78-80; from asst prof to assoc prof toxicol, Harvard Univ, 80-90; ASSOC PROF TOXICOL, UNIV MINN, 90- *Honors & Awards:* Pincus Award, Fedn Am Soc Exp Biol, 77. *Mem:* Am Chem Soc; Am Soc Biol Chemists; Am Soc Microbiol; Am Soc Cell Biol; Biophys Soc; Sigma Xi. *Res:* Regulation of cyclic nucleotide metabolism in mammalian systems and in sperm; regulation of differentiation and proliferation of human cells. *Mailing Add:* Dept Environ Occup Health Univ Minn Box 197 Mayo Mem Bldg Minneapolis MN 55455

TOSCANO, WILLIAM MICHAEL, b Santa Barbara, Calif, June 22, 45; m 82; c 3. THERMODYNAMICS, HEAT TRANSFER. *Educ:* Univ Calif, Berkeley, BS, 67; Mass Inst Technol, SM, 69, ME,71, PhD(eng), 73; Boston Univ, MBA, 80. *Prof Exp:* Res engr thermodyn, Western Elec Eng Res Ctr, 69; res assoc cryog, Mass Inst Technol, 73-74; res & develop mgr cryogenic eng, Helix Technol Corp, 74-77; div mgr mech eng & consult, Foster-Miller, Inc, 77-82; div mgr thermal systs, ORFMA, 82-84; mkt dir software systs, Intermetrics, Inc, 84-87; PRES, SYSTS DESIGNERS SOFTWARE, INC, 87- *Concurrent Pos:* Consult, 73-; co-founder & treas, Aspen Systs, Inc, 84-; chmn, Cryog Comt, Am Soc Mech Engrs, 79-82 & paper review chmn, Proces Industs Div, 77-79. *Mem:* Sigma Xi; Am Soc Mech Engrs; AAAS. *Res:* Cryogenic engineering research; heat pumps and heat engines; compressors and expanders; energy conservation; energy conversion and generation; appliances; burner technology; thermal systems; thermodynamics and heat transfer; refrigeration systems. *Mailing Add:* 82 Old Garrison Rd Sudbury MA 01776

TOSCH, WILLIAM CONRAD, b Lee's Summit, Mo, Jan 19, 34; m 57; c 3. OIL PRODUCTION. *Educ:* Univ SDak, AB, 57; Purdue Univ, MS, 60, PhD(phys chem), 62. *Prof Exp:* Teaching asst chem, Purdue Univ, 57-61, res asst, 58-62; res scientist, 62-68, RES DEPT MGR, MARATHON OIL CO-DENVER RES CTR, 68- *Concurrent Pos:* Asst prof, Arapahoe Community Col, 66-67; lectr, Casper Col. *Mem:* Soc Petrol Engrs. *Res:* Recovery of petroleum products from water-depleted reservoirs, including interfacial phenomena, rock and fluid interactions, rheology, surfactancy,polymer chemistry, carbon dioxide flooding and thermal oil recovery. *Mailing Add:* Assoc Prof/Petrol Pa State Univ 202 Munerat Sci Bldg University Park PA 16802

TOSH, FRED EUGENE, b Bemis, Tenn, Feb 13, 30; m 55; c 3. MEDICINE, EPIDEMIOLOGY. *Educ:* Univ Tenn, MD, 54; Univ Calif, MPH, 63. *Prof Exp:* Intern, Baptist Hosp, Memphis, Tenn, 54-55; med epidemiologist, Commun Dis Ctr, USPHS, 55-57; pvt pract, Tenn, 57-58; med epidemiologist, Kansas City Field Sta, Ctr Dis Control, USPHS, 58-64, chief pulmonary mycoses unit, 64-66, dep dir ecol invests prog, 67-73, dir div qual & stand, 73-78, dir off regional health planning, USPHS Regional Off, Colo, 78-80; DIR, WICHITA-SEDGWICK COUNTY DEPT COMMUNITY HEALTH, 80- *Concurrent Pos:* Resident, Mo State Sanitorium, 60-61; instr med, Univ Kans, 64-70, asst clin prof, 70-73. *Mem:* AMA; Am Epidemiol Soc. *Res:* Public health. *Mailing Add:* Wichita-Sedgwick County Community Health 1900 E Ninth St Wichita KS 67214

TOSHIO, MAKIMOTO, b Hiroshima, Japan, Aug 30, 16; m 41; c 3. MICROWAVE & MILLIMETERWAVE CIRCUITS & ANTENNAS. *Educ:* Tokyo Inst Technol, Bachelor, 41, Dr(eng), 56. *Prof Exp:* From asst prof to prof electronics, Inst Sci & Res, Osaka Univ, 48-80, dean, Fac Eng Sci, 74-78; prof electronics, Fac Eng, 80-88, vpres, 85-88, PRES, SETSUNAN UNIV, 85-; EMER PROF, OSAKA UNIV, 80- *Concurrent Pos:* Chmn, Prof Group Microwaves, Inst Electronics & Commun Engrs, 73-75; co-leader, Spec Proj Res Optical Guided-Wave Electronics, Japanese govt, 77-80. *Mem:* Fel Inst Elec & Electronics Engrs; Inst Electronics Info & Commun Engrs; Inst Elec Engrs; Soc Radiation Sci; Inst Laser Engrs. *Res:* Microwave and millimeterwave circuits and antennas; optical electronics and fundamental research on optical guided-wave circuits. *Mailing Add:* Setsunan Univ Ikeda-Nakamachi 17-8, Neyagawa Osaka 572 Japan

TOSI, JOSEPH ANDREW, JR, b Worcester, Mass, July 1, 21; m 48; c 3. ECOLOGY. *Educ:* Mass State Col, BS, 43; Yale Univ, MF, 48; Clark Univ, PhD(geog), 59. *Prof Exp:* Forester & ecologist, Northern Zone, Inter-Am Inst Agr Sci, 51-52, Andean Zone, 52-63; resident staff geogr, Cent Am Field Prog, Assoc Cols Midwest, 64-67; land-use ecologist & adminr, Trop Sci Ctr, 67-90; RETIRED. *Concurrent Pos:* Consult, Forest Surv, Venezuela, 55; consult ecol surv, Colombia, 59-60; consult to comn on environ landscape planning, Int Union for Conserv Nature & Natural Resources, 73- *Mem:* Soc Am Foresters; Sigma Xi. *Res:* Tropical ecology; bioclimatology; land utilization, especially tropical rural areas; biogeography; tropical forest management; life zone ecological theory; economic botany. *Mailing Add:* Trop Sci Ctr Apt 83870 San Jose Costa Rica

TOSI, OSCAR I, b Trento, Italy, June 17, 29; US citizen. AUDIOLOGY, ACOUSTICS. *Educ:* Univ Buenos Aires, ScD; Ohio State Univ, PhD, 65. *Prof Exp:* Assoc prof physics, Univ Buenos Aires, 51-62; res assoc voice commun, Ohio State Univ, 63-65; from asst prof to assoc prof, 65-70, PROF AUDIOL & SPEECH SCI & PHYSICS, MICH STATE UNIV, 70-, DIR, INST VOICE IDENTIFICATION, 74- *Concurrent Pos:* Dept Justice-Mich State Police grant, 68-71; expert witness on voice identification, Fed & State Courts, US, Can & Europe, 68-; vpres, Int Asn Voice Identification, Inc, 72; elected staff mem voice commun tech comt, Acoust Soc Am, 74-77. *Mem:* Acoust Soc Am; Am Asn Physics Teachers; Am Speech & Hearing Asn; Int Asn Logopedics & Phoniatrics; Int Col Exp Phonology. *Res:* Voice spectrography and identification; low levels of human acoustical energy; voice identification; articulatory pauses. *Mailing Add:* Dept Audiol & Speech Sci Mich State Univ East Lansing MI 48824

TOSKEY, BURNETT ROLAND, b Seattle, Wash, May 27, 29. MATHEMATICS. *Educ:* Univ Wash, BS, 52, MA, 58, PhD(algebra), 59. *Prof Exp:* From instr to assoc prof, 58-69, PROF MATH, SEATTLE UNIV, 69- *Mem:* Math Asn Am. *Res:* Abelian groups; ring theory; homological algebra; additive groups of rings. *Mailing Add:* Dept of Math Seattle Univ 900 12th & E Columbia Seattle WA 98122

TOSNEY, KATHRYN W, b Sept 3, 46. AXONAL GUIDANCE, MORPHOGENESIS. *Educ:* Univ Ore, BS, 75; Stanford Univ, PhD(biol), 80. *Prof Exp:* Postdoctoral fel, Yale Univ, 80-82, Univ Conn, 83-84; asst prof develop, 84-89, ASSOC PROF DEVELOP & PROF WRITING, UNIV MICH, 89-, ASSOC CHAIR, DEPT BIOL, 91- *Concurrent Pos:* Assoc ed, J Morphol, 85-90; mem, Neurol B2 Study Sect, NIH, 88-; assoc dir, Develop Neurosci Prog, Univ Mich, 91- *Mem:* Soc Neurosci; AAAS; Soc Develop Biol. *Res:* Ellucidate the mechanisms that control the guidance of axons; specific migration of neural crest cells; morphogenesis of muscle during development. *Mailing Add:* Nat Sci Bldg Univ Mich Ann Arbor MI 48109-1048

TOSTESON, DANIEL CHARLES, b Milwaukee, Wis, Feb 5, 25; m 49, 69; c 6. PHYSIOLOGY, BIOPHYSICS. *Educ:* Harvard Univ, MD, 49. *Hon Degrees:* DSc, Univ Copenhagen, 79, Univ Liege, 83 & Med Col Wis, 84. *Prof Exp:* Intern med, Presby Hosp, 49-51; res fel, Dept Med, Brookhaven Nat Lab, 51-53, Lab Kidney & Electrolyte Metab, Nat Heart Inst, 53-55 & 57-58, Biol Isotope Res Lab, Univ Copenhagen, 55-56 & Physiol Lab Cambridge, Eng, 56- 57; assoc prof physiol, Sch Med, Wash Univ, 58-61; prof physiol & pharmacol & chmn dept, Sch Med, Duke Univ, 61-75, James B Duke distinguished prof, 71-75; dean, Div Biol Sci & Pritzker Sch Med, vpres, Med Ctr & Lowell T Coggeshall prof med sci, Univ Chicago, 75-77; DEAN FAC MED & CAROLINE SHIELDS WALKER PROF PHYSIOL, PRES, MED CTR, HARVARD UNIV, 77- *Concurrent Pos:* Mem, Molecular Biol Panel, NSF, Sci Rev Comt, NIH, 64-67, Off Technol Assessment, 76, Ethics Adv Bd, HEW, 77-80, Nat Adv Coun, Nat Inst Gen Med Sci, 82-86 & Panel Gen Prof Educ Physician, Asn Am Med Cols, 82-84; chmn, Coun Acad Socs, Asn Am Med Cols, 73-74, Flexner Award Comt, 78 & Task Force Physician Supply, 87-90. *Mem:* Inst Med-Nat Acad Sci; Nat Inst Med; Am Physiol Soc; Soc Gen Physiologists (pres, 68-69); Biophys Soc; fel AAAS; Am Acad Arts & Sci. *Res:* Membrane physiology. *Mailing Add:* Med Sch Harvard Univ 25 Shattuck St Boston MA 02115

TOSTEVIN, JAMES EARLE, b Mandan, NDak, June 28, 38; m 65; c 2. PAPER CHEMISTRY. *Educ:* Carleton Col, BA, 60; Inst Paper Chem, MS, 62, PhD(paper chem), 66. *Prof Exp:* Group leader analysis, Columbia Cellulose Co, Ltd, 66-69; RES CHEMIST & GROUP LEADER ANALYSIS, ITT RAYONIER INC, 69- *Mem:* Am Soc Testing & Mat; Tech Asn Pulp & Paper Indust. *Res:* Research into application and properties of natural cellulose fibers. *Mailing Add:* 1043 Connection St Shelton WA 98584

TOTH, BELA, b Pecs, Hungary, Oct 26, 31; US citizen; m 63; c 4. PATHOLOGY, ONCOLOGY. *Educ:* Univ Vet Sci, Budapest, DVM, 56. *Prof Exp:* From res asst to res assoc oncol, Chicago Med Sch, 59-63, asst prof, 63-66; fel exp biol, Weizmann Inst, 66-67; assoc prof path, 68-72, PROF PATH, EPPLEY INST RES CANCER, COL MED, UNIV NEBR, OMAHA, 72- *Concurrent Pos:* USPHS trainee path, 61-63, res career develop award, 69; Eleanor Roosevelt int cancer res fel, 67-68. *Mem:* AAAS; Am Asn Cancer Res; Am Soc Exp Path; NY Acad Sci; Am Asn Path & Bact. *Res:* Experimental oncology; chemical carcinogenesis; leukemogenesis. *Mailing Add:* Eppley Inst Res Cancer Univ Nebr Med Ctr 600 S 42nd St Omaha NE 68198

TOTH, EUGENE J, b Csikvand, Hungary, Jan 27, 32; US citizen; m 55; c 2. NEUROCHEMISTRY. *Educ:* State Col Peis Hungary, BS, 54; NY Med Sch, MS, 72, PhD(biochem), 78. *Prof Exp:* Chemist, NY State Res Inst, 61-66, res scientist, 66-77; SR RES SCIENTIST, CTR NEUROCHEM, ROCKLAND RES INST, 77- *Mem:* Am Neurochem Soc; Int Neurochem Soc. *Res:* Transport and metabolism of amino acids in the central nervous system and their functional roles in health and in various neurological and mental disorders; synthesis and tumors of brain proteins; influences on protein metabolisms. *Mailing Add:* 12 Crossbar Rd Hastings on Hudson NY 10706

TOTH, JOZSEF, b Bekes, Hungary, June 22, 33; Can citizen; c 2. HYDROGEOLOGY. *Educ:* Univ Utrecht, BSc, 58, MSc, 60, PhD(hydrogeol), 65. *Prof Exp:* From jr res officer to sr res officer hydrogeol, 60-68, HEAD GROUND WATER, DIV HYDROGEOL, RES COUN ALTA, 68- *Concurrent Pos:* Mem subcomt hydrol, assoc comt geod & geophys, Nat Res Coun Can, 63-68; lectr, Univ Alta, 66-71; vis prof, Univ Calgary, 78-79. *Honors & Awards:* O E Meinzer Award, Geol Soc Am, 65. *Res:* Hydrogeology; theoretical and practical investigations of the interaction between groundwater and the geologic environment and hydrogeological applications in water resources, soil mechanics, agriculture and mineral exploration. *Mailing Add:* Dept Geol Univ Alta Edmonton AB T6G 2M7 Can

TOTH, KENNETH STEPHEN, b Shanghai, China, Mar 17, 34; m 56; c 2. NUCLEAR PHYSICS. *Educ:* San Diego State Univ, AB, 54; Univ Calif, PhD(chem), 58. *Prof Exp:* Asst chem, Univ California, 54-55, asst, Lawrence Radiation Lab, 55-58; Fulbright fel, Inst Theoret Physics, Denmark, 58-59; NUCLEAR CHEMIST, OAK RIDGE NAT LAB, 59- *Concurrent Pos:* Guggenheim fel, Niels Bohr Inst, Copenhagen, Denmark, 65-66; exchange physicist joint inst for nuclear res, Dubna, USSR, Nat Acad Sci, 75. *Mem:* Am Chem Soc; Am Phys Soc. *Res:* Nuclear properties of radioactive isotopes in rare earth region; low energy; heavy-ion nuclear reactions; nuclei far from stability. *Mailing Add:* Physics Div Oak Ridge Nat Lab MS-6371 Oak Ridge TN 37831-6371

TOTH, LOUIS MCKENNA, b Lexington, Ky, Aug 27, 41; m 62; c 3. PHYSICAL CHEMISTRY. *Educ:* La State Univ, Baton Rouge, BS, 63; Univ Calif, Berkeley, PhD(chem), 67. *Prof Exp:* CHEMIST, OAK RIDGE NAT LAB, 67- *Mem:* Am Chem Soc. *Res:* High temperature molten salt chemistry; infrared, Raman and Ligand-Field spectroscopy of molten salt systems; gas phase kinetics; aqueous actinide photochemistry. *Mailing Add:* 1040 W Outer Dr Oak Ridge TN 37830-4898

TOTH, PAUL EUGENE, b Welland, Ont, July 11, 20; US citizen; wid; c 3. INDUSTRIAL HYGIENE. *Educ:* Lawrence Inst Technol, BSChE, 49. *Prof Exp:* Supvr chem, Qual Control Lab, Briggs Mfg Co, 42-49, indust hygienist, 49-54; indust hygienist, Chrysler Corp, 54-60; occup health engr indust hyg, Dept Health Mich, 60-61; Mgr Indust Hyg & Toxicol, Ford Motor Co, 61-; AT PAUL E TOTH & ASSOC. *Concurrent Pos:* Chmn, Am Bd Indust Hyg, 75-76. *Mem:* Am Indust Hyg Asn (treas, 67-70, pres, 78-79); Acoust Soc Am; Am Soc Safety Engrs; Am Indust Hyg Found (pres, 81-82); Am Acad Indust Hyg. *Mailing Add:* Paul E Toth & Assoc 9611 Manor Ave Allen Park MI 48101

TOTH, ROBERT ALLEN, b Richmond, Ind, Aug 10, 39; div; c 3. PHYSICS. *Educ:* Earlham Col, AB, 62; Fla State Univ, MS, 66, PhD(physics), 69. *Prof Exp:* Physicist, Infrared Spectros, Nat Bur Standards, 62-66; instr Earlham Col, 66-67; assoc, Fla State Univ, 69-70; RES SCIENTIST, INFRARED SPECTROS & REMOTE SENSING, JET PROPULSION LAB, 70- *Mem:* Fel Optical Soc Am. *Res:* Infrared spectroscopy; high resolution, its application to laboratory and theoretical data and to remote sensing of the atmosphere. *Mailing Add:* MS-183-301 Jet Propulsion Lab Pasadena CA 91109

TOTH, ROBERT S, b Detroit, Mich, Sept 4, 31; m 53; c 3. SOLID STATE PHYSICS. *Educ:* Wayne State Univ, AB, 54, MS, 55, PhD(physics), 60. *Prof Exp:* Res assoc physics, Wayne State Univ, 55-60; sr scientist, sci lab, Ford Motor Co, 60-69; vpres, Sensors, Inc, 69-76; PRES, DEXTER RES CTR, INC, 78- *Mem:* Am Phys Soc. *Res:* Metal oxide semi-conductors; crystal structure theory of alloy phases; magnetic structure of metals and alloys; thin film physics; epitaxy; thermoelectricity; infrared physics. *Mailing Add:* 8495 Mast Rd Dexter MI 48130

TOTH, STEPHEN JOHN, b Elizabeth, NJ, Feb 19, 12; m 46; c 1. SOIL CHEMISTRY. *Educ:* Rutgers Univ, BS, 33, MS, 35, PhD(soil chem), 37. *Prof Exp:* Specialist forest soils, 37-39, instr agr chem, 39-42, asst soil chemist, NJ Agr Exp Sta, 39-81, from asst prof to prof soils, 46-81, assoc res specialist, 47-81, EMER PROF SOIL COLLOIDS, RUTGERS UNIV, NEW BRUNSWICK, 81- *Mem:* Fel AAAS; Am Chem Soc; Soil Sci Soc Am; fel Am Inst Chem; fel Am Geog Soc; Sigma Xi. *Res:* Soil chemistry; colloids; nutrition; radioisotopes; fertilizers; water quality; bottom sediments; wildlife crops; composts. *Mailing Add:* 187 Echo Ave Edison NJ 08837-2632

TOTH, WILLIAM JAMES, b Carteret, NJ, Jan 20, 36; m 67; c 3. POLYMER CHEMISTRY. *Educ:* Rutgers Univ, New Brunswick, BA, 68; Princeton Univ, MS, 71, PhD(chem), 72. *Prof Exp:* Sr res chemist, Mobil Chem Co, Mobil Oil Corp, 63-76; SR DEVELOP ASSOC, ICI AM INC, HOPEWELL, VA, 76- *Mem:* Am Chem Soc; Soc Rheology. *Res:* Chemical mechanical, dielectric, rheological and physical properties of new polymers; physical chemistry of liquid crystals; characterization of monomeric and polymeric liquid crystals. *Mailing Add:* 735 Yager Rd Clinton OH 44216

TOTO, PATRICK D, b Niles, Ohio, Jan 6, 21; m 45; c 3. ORAL PATHOLOGY. *Educ:* Kent State Univ, BS, 48; Ohio State Univ, DDS, 48, MS, 50; Am Bd Oral Path, dipl, 48. *Prof Exp:* Asst prof, 50-53 & 55-57, clin dir, 55-57, assoc prof, dir res & coordr grad studies, 57-76, PROF ORAL PATH & CHMN DEPT, SCH DENT, LOYOLA UNIV CHICAGO, 71- *Concurrent Pos:* Consult, Vet Admin Hosps, Hines, Ill, 53 & Chicago, 61- *Mem:* Int Asn Dent Res; Am Soc Clin Path; Am Acad Oral Path; NY Acad Sci; Am Dent Asn. *Res:* Lectin binding to premalignant and malignant oral neoplasms; induction of oral cancer; histiocyte modulation; immunopathology oral mucosa; pathogenesis of periodontitis. *Mailing Add:* 2160 S First Ave Loyola Univ Maywood IL 60153

TOTON, EDWARD THOMAS, b Philadelphia, Pa, Dec 6, 42; m 70. ASTROPHYSICS. *Educ:* St Joseph's Col, Pa, BS, 64; Univ Md, College Park, PhD(physics), 69. *Prof Exp:* Air Force Off Sci Res fel, Inst Advan Study, 69-70; NSF fel, Inst Theoret Physics, Univ Vienna, 70-71; res assoc & assoc instr astrophys, Univ Utah, 71-72; res assoc physics, Univ Pa, 72-74; RES PHYSICIST, NAVAL SURFACE WEAPONS CTR, 74- *Concurrent Pos:* Vis asst prof physics, St Joseph's Col, Pa, 74-75; consult, Naval Res Lab, Washington, DC, 75- *Mem:* Am Phys Soc; AAAS. *Res:* Astrophysical studies related to structure of neutron stars, nature of radiation from galaxies, nature of universe at moment of creation; research in combustion physics, including flame propagation, ignition, quenching and noise generation; detonation physics. *Mailing Add:* 10296 Wayover Way Columbia MD 21046

TOTTA, PAUL ANTHONY, b Middletown, NY, May 17, 30; m 54; c 2. MATERIALS SCIENCE, METALLURGY. *Educ:* Rensselaer Polytech Inst, BMetE, 52. *Prof Exp:* Metallurgist, Gen Elec Co, 54-58; staff metallurgist, Handy & Harman, 58-59; sr engr, IBM Corp, 59-72, mgr metal & insulator technol, 72-81, sr mem tech staff, 81-87. *Concurrent Pos:* App Fel, IBM, 87. *Honors & Awards:* Tech Achievement Award, Int Soc Hybrid Microelectronics, 84; App Fel, IBM, 87. *Mem:* Am Vacuum Soc; Am Soc Metals; Sigma Xi; Int Soc Hybrid Microelectronics. *Res:* Thin film metallurgy deposited by vacuum evaporation or sputtering for use as conductors in monolithic integrated semiconductor devices; metal and insulator technology for integrated circuit semiconductors and electronic packaging. *Mailing Add:* 29 Sandi Dr Poughkeepsie NY 12603

TOTTEN, JAMES EDWARD, b Saskatoon, Sask, Aug 9, 47; m 68; c 1. GEOMETRY. *Educ:* Univ Sask, BA, 67; Univ Waterloo, MATH 69, PhD(geom), 74. *Prof Exp:* Nat Res Coun Can fel geom, Univ Math Inst, Tubingen, WGer, 74-76; asst prof math, St Mary's Univ, NS, 76-78; vis asst prof math, Univ Sask, 78-79; INSTR MATH & COMPUT, CARIBOO COL, KAMLOOPS, BC, 79- *Mem:* Math Asn Am; Can Math Soc. *Res:* Linear spaces, a set of elements called points and distinguished subsets of points called lines, such that two points determine a unique line and all lines have at least two points. *Mailing Add:* Dept Computer Sci & Math Cariboo Col Kamloops BC V2C 5N3 Can

TOTTEN, STANLEY MARTIN, b Lodi, Ohio, July 15, 36; m 58; c 5. GEOLOGY. *Educ:* Col Wooster, BA, 58; Univ Ill, MS, 60, PhD(geol), 62. *Prof Exp:* From asst prof to assoc prof, 62-71, PROF GEOL, HANOVER COL, 71- *Concurrent Pos:* NSF fel, Univ Birmingham, 68-69. *Mem:* Fel Geol Soc Am; Soc Econ Paleont & Mineral; Nat Asn Geol Teachers; Am Quaternary Asn. *Res:* Glacial geology; Pleistocene and Paleozoic stratigraphy; sedimentary petrology; history of geology. *Mailing Add:* Dept Geol Hanover Col Hanover IN 47243

TOTTER, JOHN RANDOLPH, b Saragosa, Tex, Jan 7, 14; m 38; c 3. BIOCHEMISTRY. *Educ:* Univ Wyo, AB, 34, AM, 35; Univ Iowa, PhD(biochem), 38. *Prof Exp:* Instr chem, Univ Wyo, 35-36; asst biochem, Univ Iowa, 36-38; instr, Univ WVa, 38-39; instr, Sch Med, Univ Ark, 39-42, from asst prof to assoc prof, 42-52; biochemist, Oak Ridge Nat Lab, 52-56; biochemist, USAEC, 56-58; biochemist, Univ of the Repub, Uruguay, 58-60; prof chem & chmn div biol sci, Univ Ga, 60-62; assoc dir res, Div Biol & Med, USAEC, 63-67, dir, 67-72; assoc dir biomed & environ sci, Oak Ridge Nat Lab, 72-74, biochemist, 74-78; SCIENTIST, OAK RIDGE ASSOC UNIV, 78- *Concurrent Pos:* Nutrit biochemist, Univ Alaska, 47; prof biochem, Univ Tenn, 75- *Mem:* Am Chem Soc; Soc Exp Biol & Med; Am Soc Biol Chemists; Am Soc Nat; Am Soc Photobiol. *Res:* Radiation effects; amino acid and formate metabolism; synthesis and metabolism of pterins; luminescence; cancer epidemiology. *Mailing Add:* 109 Wedgewood Dr Oak Ridge TN 37830

TOTUSEK, ROBERT, b Garber, Okla, Nov 3, 26; m 47; c 3. ANIMAL NUTRITION. *Educ:* Okla Agr & Mech Col, BS, 49; Purdue Univ, MS, 50, PhD(animal nutrit), 52. *Prof Exp:* Asst, Purdue Univ, 49-50, instr, 50-52; from asst prof to assoc prof, 52-60, PROF ANIMAL HUSB, OKLA STATE UNIV, 60- & HEAD, DEPT ANIMAL SCI, 77- *Mem:* Am Soc Animal Sci. *Res:* Range cow nutrition and management. *Mailing Add:* 2201 Countryside St Stillwater OK 74074

TOU, JAMES CHIEH, b Su-yang, China, Apr 25, 36; US citizen; m 64; c 3. ANALYTICAL CHEMISTRY, PHYSICAL CHEMISTRY. *Educ:* Taiwan Norm Univ, BSc, 61; Univ Utah, PhD(chem), 66. *Prof Exp:* Teaching asst chem, Taiwan Norm Univ, 60-61; res asst, Univ Utah, 62-65; from res chemist to sr res chemist, Chem Physics Res Lab, 65-71, sr analytical specialist chem, 75-78, assoc scientist, 78-81, sr assoc scientist, 81-84, RES SCIENTIST, ANALYTICAL LAB, DOW CHEM CO, 84- *Honors & Awards:* V A Stenger Analytical Sci Award, Dow Chem Co, 75. *Mem:* NAm Thermal Analytical Soc; Am Chem Soc; Am Soc Testing & Mat; Am Soc Mass Spectrometry. *Res:* Organic mass spectrometry; chemical ionization; electron impact and field ionization; gas-chromatography-mass spectrometry; chemical property and analysis of bis-chloromethyl ether and chloromethyl methyl ether; thermal analysis; thermokinetics and high temperature chemistry; mass spectrometry; membrane permeation. *Mailing Add:* 1910 Wilmington Dr Midland MI 48640-3831

TOU, JEN-SIE HSU, b Shantung, China, Sept 17, 36; US citizen; m 65. LIPID BIOCHEMISTRY, METABOLISM OF PHOSPHOLIPIDS. *Educ:* Nat Taiwan Univ, BS, 59; Baylor Univ, MS, 64; Tulane Univ, PhD(biochem), 68. *Prof Exp:* Teaching asst biochem, Med Col, Baylor Univ, 62-64; res asst, Med Sch, Harvard Univ, 64-65; trainee, 65-68, res fel, 68-71, from instr to res asst prof, 71-88, ASSOC PROF BIOCHEM, MED SCH, TULANE UNIV,

88- *Concurrent Pos:* Spec res fel award, NIH, 72-74, res carrer develop award, 75-80; Albert L Hyman Res grant, Am Heart Asn, La, 85. *Mem:* Am Soc Biochem & Molecular Biol. *Res:* Regulation of the metabolism of phospholipids in human leukocytes; endocrinology. *Mailing Add:* Dept Biochem Tulane Univ Med Sch 1430 Tulane Ave New Orleans LA 70112

TOU, JULIUS T(SU) L(IEH), b Shanghai, China, Aug 15, 26; m 56; c 4. ELECTRICAL ENGINEERING, COMPUTER & INFORMATION SCIENCE. *Educ:* Chiao Tung Univ, BS, 47; Harvard Univ, MS, 50; Yale Univ, DEng, 52. *Prof Exp:* Proj engr, Philco Corp, 52-55; asst prof elec eng, Univ Pa, 55-57; assoc prof, Purdue Univ, 57-61, vis prof, 61-62; prof & dir comput sci lab, Northwestern Univ, 61-64; dir info sci res, Battelle Mem Inst, 64-67; GRAD RES PROF ELEC ENG, UNIV FLA, 67-, DIR, CTR RES INFO, 77- *Concurrent Pos:* Consult, Philco Corp, 55-57, Barber Colman Co, 56, Int Bus Mach Corp, 60, Gen Elec Co, 61, McDonnell Douglas, 69 & Martin Marietta, 81; adj prof, Ohio State Univ, 64-67; adv, Nat Inst Health, 83-87. *Honors & Awards:* Achievement Award, Comput & Automated Systs Asn. *Mem:* Am Soc Eng Educ; fel Inst Elec & Electronics Engrs; Am Inst Mgt; Int Soc Cybernet Med; Comput & Automated Systs Asn; Academia Sinica. *Res:* Control and information systems; computer vision; digital control; artificial intelligence; pattern recognition; computer-based automation; knowledge engineering. *Mailing Add:* Ctr Info Res 339 Larsen Hall Univ Fla Gainesville FL 32611

TOUBA, ALI R, b Tabriz, Iran, Apr 25, 25; m 57; c 4. FOOD TECHNOLOGY. *Educ:* Rutgers Univ, BSc, 51, MSc, 52; Univ Ill, PhD(food technol), 56. *Prof Exp:* Asst food microbiol, Univ Ill, 53-56; assoc technologist food res, Res Ctr, Gen Foods Corp, 56-63; proj mgr food res, Tronchemics Res, Inc, 63-65; res assoc explor food res, 65-76. head explor food res, 76-81, DEPT HEAD, BETTY CROCKERS DIV, GEN MILLS, INC, 81- *Concurrent Pos:* Tech consult, Teheran, Iran, 60-63. *Mem:* Am Chem Soc; Inst Food Technologists; Am Asn Cereal Chemists; Am Soc Microbiol; Sigma Xi. *Res:* Food texture; fabricated foods; gums; space foods; freeze drying beverages; flavors; cereals and snacks; desserts; dehydrated products; fruit products; technical management. *Mailing Add:* 4609 Island View Dr Mound MN 55364

TOUBASSI, ELIAS HANNA, b Jaffa, Israel, May 28, 43; US citizen; m 67; c 2. MATHEMATICS. *Educ:* Bethel Col (Kans), AB, 66; Lehigh Univ, MS, 69, PhD(math), 70. *Prof Exp:* Sr tech aide prog design, Bell Tel Labs, 66-67; res assoc, 70-71, from asst prof to assoc prof, 70-84, assoc head, Dept Math, 77-90, PROF MATH, UNIV ARIZ, 85- *Mem:* Math Asn Am; Nat Coun Teachers Math. *Res:* Algebra, specifically infinite abelian groups; math education. *Mailing Add:* Dept Math Univ Ariz Tucson AZ 85721

TOUCHBERRY, ROBERT WALTON, b Manning, SC, Oct 27, 21; m 48; c 4. ANIMAL BREEDING. *Educ:* Clemson Col, BS, 45; Iowa State Col, MS, 47, PhD(animal breeding, genetics), 48. *Prof Exp:* Asst dairy sci, Univ Ill, Urbana, 48-49, asst prof dairy cattle genetics, 49-55, assoc prof genetics in dairy sci, 55-59, prof, 59-70; prof animal sci & head dept, Univ Minn, St Paul, 70-82; PROF & CHAIR, DEPT ANIMAL SCI, UNIV CALIF, DAVIS, 82- *Concurrent Pos:* Fulbright res fel, Denmark, 56-57; geneticist, Div Biol & Med, US AEC, 67-68. *Honors & Awards:* Animal Breeding & Genetics Award, Am Soc Animal Sci, 71. *Mem:* Fel AAAS; fel Am Soc Animal Sci; Genetics Soc Am; Am Diary Sci Asn; Am Genetic Asn. *Res:* Population genetics; quantitative genetics; effects of crossbreeding on the growth and milk production of dairy cattle; effects of x-irradiation on quantitative traits of mice and fruit flies; statistical studies of animal records; quantitative genetics of levels of hormones in the blood of dairy cattle. *Mailing Add:* Dept Animal Sci Univ Calif Davis Davis CA 95616

TOUCHSTONE, JOSEPH CARY, b Soochow, China, Nov 27, 21; US citizen; m 55; c 3. CLINICAL CHEMISTRY, BIOCHEMISTRY. *Educ:* Stephen F Austin State Univ BS, 43; Purdue Univ, MS, 46; St Louis Univ, PhD(biochem), 53. *Prof Exp:* Asst, Purdue Univ, 43-45; res assoc, Univ Tex, Southwestern Med Sch, 46-49; res assoc med, 52-56, assoc, Pepper Lab Clin Chem, Univ Hosp, 52-56, asst res prof obstet & gynec & res assoc, Harrison Dept Surg Res, Univ, 56-63, res assoc prof, 63-67, assoc prof res surg, 63-68, RES PROF OBSTET & GYNEC, SCH MED, UNIV PA, 67-, DIR, STEROID LAB & PROF RES SURG, 68- *Concurrent Pos:* NIH res career award, 61-71; pres & co-founder, Chromatog Forum, 66-67, pres, 71-72, exec comt, 66- *Honors & Awards:* Chromatography Forum Del Valley Award, 82. *Mem:* Am Chem Soc; Endocrine Soc; Am Soc Biol Chemists; Am Asn Clin Chemists; Am Acad Forensic Sci. *Res:* Steroid chemistry; organic synthesis; isolation and metabolism of steroid hormones; chromatography; adrenal physiology; chromatography of lipids; forensic and environmental methodology; phospolipids; thin layer gas and liquid chromatography; published 300 articles in various journals and 14 books. *Mailing Add:* Univ Pa Hosp 3400 Spruce St Philadelphia PA 19104

TOUGER, JEROLD STEVEN, b Brooklyn, NY, Aug 6, 45; m 69; c 2. PHYSICS. *Educ:* Cornell Univ, BA, 66; City Univ New York, PhD(physics), 74. *Prof Exp:* From asst prof to assoc prof, 74-85, PROF PHYSICS, CURRY COL, 85- *Concurrent Pos:* Proj dir, NSF grant, 80-82; adj prof physics, Univ Mass, Amherst, 87. *Mem:* Am Asn Physics Teachers (pres New Eng Sect, 86-87). *Res:* Thermoelectric power; transport properties in magnetic alloys; cognitive science and structural linguistics applied to scientific and mathematical discourses and to research in physics and mathematics education; curriculum development in physics, calculus and integrated science. *Mailing Add:* Div Sci & Math Curry Col Milton MA 02186

TOUGH, JAMES THOMAS, b Chicago, Ill, May 4, 38; m 60; c 2. LOW TEMPERATURE PHYSICS. *Educ:* Univ Ill, BS, 60; Univ Wash, PhD(liquid helium), 64. *Prof Exp:* Res assoc, 64-65, from asst prof to assoc prof, 65-76, PROF PHYSICS, OHIO STATE UNIV, 68-, VCHMN GRAD STUDIES, 85- *Concurrent Pos:* Hon vis prof, Univ St Andrews, Scotland, 73-74; prin investr, NSF, 75- *Mem:* Am Phys Soc, Div Fluid Dynamics & Condensed Matter; fel Am Phys Soc. *Res:* Hydrodynamics and turbulence in liquid helium II. *Mailing Add:* Dept Physics Ohio State Univ 174 W 18th Ave Columbus OH 43210

TOUHILL, CHARLES JOSEPH, b Newark, NJ, Aug 27, 38; m 60; c 4. ENVIRONMENTAL ENGINEERING. *Educ:* Rensselaer Polytech Inst, BCE, 60, PhD(environ eng), 64; Mass Inst Technol, SM, 61; Univ Wash, dipl, 70; Am Acad Environ Engrs, dipl, 70. *Prof Exp:* With, Gen Elec Co, 64-65 & Battelle Pac Northwest Labs, 65-71; officer in consult firms, 71-77; pres, Baker/TSA, Inc, 77-90; GROUP SR VPRES, ICF KAISER ENGRS, INC, 90- *Concurrent Pos:* US deleg, Int Asn Water Pollution Res, 70-71 & 79-83; mem bd trustees, Am Acad Environ Engrs, 71-78 & 83-86; mem ed adv bd, Environ Sci & Technol, 75-77 & Environ Progress, 78- *Mem:* Am Inst Chem Engrs; Am Chem Soc; Am Water Works Asn; Water Pollution Control Fedn. *Res:* Management of water, wastewater, solids and toxic and hazardous materials. *Mailing Add:* 2206 Almanack Ct Pittsburgh PA 15237

TOULMIN, PRIESTLEY, b Birmingham, Ala, June 5, 30; m 52; c 2. GEOLOGY. *Educ:* Harvard Univ, AB, 51, PhD(geol), 59; Univ Colo, MS, 53. *Prof Exp:* Geologist, US Geol Surv, 53-56, chief br exp geochem & mineral, 66-72, geologist, 58-86; RETIRED. *Concurrent Pos:* Lectr vis geol scientist prog, Am Geol Inst, 64; adj assoc prof, Columbia Univ, 66; scientist, Proj Viking, NASA, 68-81, team leader inorg chem invest, 72-81; ed J Translations, Geochem Soc, 65-68; assoc ed Am Mineralogist, J Mineral Soc Am, 74-76; res assoc geochem, Calif Inst Technol, 76-77; dir, Petrogenesis & Mineral Resources Prog, NSF, 85. *Mem:* fel Mineral Soc Am; Geochem Soc; fel Geol Soc Am; Geochem Soc; Mineral Soc Great Britain; Am Geophys Union; Mineral Asn Can; Soc Econ Geologists; Sigma Xi. *Res:* Igneous and sulfide petrology; phase equilibria and thermochemistry of ore minerals; mineralogy and geochemistry of Mars. *Mailing Add:* PO Box 183 Alexandria VA 22313-0183

TOULOUSE, JEAN, b Paris, France, Mar 30, 48; US citizen; m 73; c 4. RELAXATION PHENOMENA & STRUCTURAL PHASE TRANSITIONS, FERROELECTRICITY. *Educ:* Univ Paris France, MS, 71; Columbia Univ NY, MS, 77, PhD(solid state sci), 82. *Prof Exp:* Staff econ statist, Nat Inst Statist & Econ Studies, 71-73 & Usinor Steel Corp, NY, 73-75; res asst, Columbia Univ NY, 75-82, postdoctoral res assoc, 82-84; asst prof, 84-89, ASSOC PROF PHYSICS, LEHIGH UNIV, 89- *Concurrent Pos:* Prin investr, Dept Energy grant, 86- & Off Naval Res grant, 90-; guest ed, Ferroelectrics, 91; prin organizer, Conf Fundamental Exp in Ferroelectrics, 91. *Mem:* Am Phys Soc; Mat Res Soc. *Res:* Dielectric, ultrasonic, raman and neutron studies of structural phase transitions in disordered crystals and amorphous systems. *Mailing Add:* Physics Dept Lehigh Univ Bethlehem PA 18015

TOUPIN, RICHARD A, b Miami, Fla, Aug 20, 26; m 50; c 3. MATHEMATICAL PHYSICS, CONTINUUM MECHANICS. *Educ:* Univ SC, BS, 46; Univ Hawaii, MS, 49; Syracuse Univ, PhD(physics), 61. *Prof Exp:* Instr physics, Univ Hawaii, 49-50; res asst theoret mech, US Naval Res Lab, 50-62; res asst appl math, res ctr, 62-74, dir math sci, 74-81, MEM STAFF SCI/ENG COMPUT, IBM SCI CTR, HEIDELBERG, 82- *Mem:* Am Math Soc; Soc Natural Philos (secy, 65-67). *Res:* Elasticity and electromagnetic theories; relativity mechanics; dielectrics; differential geometry. *Mailing Add:* 1276 Copper Peak Lane San Jose CA 95120

TOUR, JAMES MITCHELL, b New York, NY, Aug 18, 59; m 82; c 3. CONDUCTING POLYMERS, HETEROGENEOUS CATALYSIS. *Educ:* Syracuse Univ, BS, 81; Purdue Univ, PhD(chem), 86. *Prof Exp:* Postdoctoral org chem, Univ Wis-Madison, 86-87 & Stanford Univ, Calif, 87-88; ASST PROF ORG/POLYMER, UNIV SC, COLUMBIA, 88- *Concurrent Pos:* Postdoctoral fel, NIH, 87; Vis lectr, IBM Almaden Res Ctr, San Jose, 88; young investr award polymer chem, Off Naval Res, 89-91; consult, Ethyl Corp, Baton Rouge, La, 90-; presidential young investr award polymer chem, NSF, 91-96. *Mem:* Am Chem Soc; Mat Res Soc. *Res:* Synthesis of conducting polymers; polymers for nonlinear optical applications; heterogeneous catalysis; metal deposition in sol-gel materials. *Mailing Add:* Dept Chem Univ SC Columbia SC 29208

TOURGEE, RONALD ALAN, b Wakefield, RI, May 2, 38; c 3. MATHEMATICAL STATISTICS, COMPUTER SCIENCE. *Educ:* Univ RI, BS, 60, MS, 62; Univ SFla, PhD(math), 75. *Prof Exp:* Teacher math, Keene State Col, 64-66 & Mt Holyoke Col, 66-68; TEACHER MATH & COMPUT SCI, KEENE STATE COL, 68- *Mem:* Am Math Soc; Am Statist Asn. *Res:* Stochastic systems; mathematical statistics; applied probability. *Mailing Add:* Dept Math Keene State Col 229 Main St Keene NH 03431

TOURIAN, ARA YERVANT, b Jerusalem, May 19, 33; US citizen; m 59; c 3. BIOCHEMICAL GENETICS. *Educ:* Am Univ Beirut, BS, 55; Iowa State Univ, MD, 58. *Prof Exp:* Intern med, Washington Hosp Ctr, DC, 58-59; resident neurol, NY Univ Med Ctr, 62-63, chief resident, 64-65; instr & fel biophys & neurol, Med Ctr, Univ Colo, 65-69; ASSOC PROF MED, MED CTR, DUKE UNIV, 69- *Concurrent Pos:* NIH res career develop award, Med Ctr, Duke Univ; vis scientist cell biol, Dept Zool, Cambridge Univ, 75-76. *Mem:* AAAS; Am Soc Neurochem; Am Acad Neurol; Cambridge Philos Soc; NY Acad Sci; Int Soc Neurochem. *Res:* Biochemical genetics of Huntington's chorea, tissue culture/protein glycosylation and the control of hexosawine metabolism; metabolic and genetic control mechanisms of phenylalanine hydroxlase; the treatment of human pain in nervous system damaged patients. *Mailing Add:* Duke Univ Med Ctr Durham NC 27710

TOURIN, RICHARD HAROLD, b New York, NY, Dec 4, 22; m 48; c 2. ENERGY CONVERSION, ENERGY CONSERVATION. *Educ:* City Col New York, BS, 47; NY Univ, MS, 48. *Prof Exp:* Res physicist, 48-51, chief physicist, 51-59, mgr res lab, 59-63, div mgr, Control Instrument Div, Warner & Swasey Co, 63-71; dir mkt, Klinger Sci Apparatus Corp, 71-73; dir new prog develop, NY State Energy Res & Develop Authority, 73-78; mgr indust mktg, Stone & Webster Eng Corp, 78-81; dir develop, Syska & Hennessy, Inc, 81-83; proj mgr, New York City Energy Off, 84-87; CONSULT, 87- *Concurrent Pos:* Adj instr, Cooper Union, 55-60; US mem joint comt, Int Flame Res Found, 66-68; vchmn, eng sect, NY Acad Sci, 84-86 & 89-91; chmn, eng sect, NY Acad Sci, 87-88; mem, energy comt, Inst Elec &

Electronics Engrs, 85-89. *Mem:* Fel Optical Soc Am; Combustion Inst; NY Acad Sci. *Res:* Energy conversion, spectroscopic gas temperature measurement; optical physics; remote sensing of environment; combined heat and power generation; rapid-scan spectroscopy; infrared spectra of hot gases; fuel utilization and the environment. *Mailing Add:* 195-10A 67th Ave Flushing NY 11365

TOURLENTES, THOMAS THEODORE, b Chicago, Ill, Dec 7, 22; m 56; c 3. PSYCHIATRY, HOSPITAL ADMINISTRATION. *Educ:* Univ Chicago, BS, 45, MD, 47, Am Bd Psychiat & Neurol, dipl, 53. *Prof Exp:* Resident psychiat, Northwestern Univ-Vet Admin Prog, 48-51; captain, US Army, Ft Carson, Colo, 52-54; supt psychiat, Galesburg State Res Hosp, 54-71; exec dir psychiat, Franciscan Ment Health Ctr, 71-85; chief psychiat serv, Vet Admin Outpatient Clin, Peoria, Ill, 85-88; CONSULT, 89- *Concurrent Pos:* Clin prof, dept psychiat, Col Med, Univ Ill, 55-91; examnr, Am Bd Psychiat & Neurol, 60-85; regional dir psychiat, Ill Dept Ment Health, 62-71; trustee, Ill Hosp Asn, 65-67. *Mem:* Fel AAAS; fel NY Acad Sci; fel Am Psychiat Asn; fel Am Asn Psychiat Adminrs (pres, 79-80); fel Am Col Psychiatrists; fel Am Col Ment Health Adminrs; fel Central Neuropsychiat Asn (pres, 86). *Res:* Brain-behavior interface and psychopharmacology; design and implementation of comprehensive systems of mental health care. *Mailing Add:* PO Box 251 RR 2 Valley View Rd Galesburg IL 61401-9544

TOURNEY, GARFIELD, b Quincy, Ill, Feb 6, 27; m 50; c 3. PSYCHIATRY. *Educ:* Univ Ill, BS, 46, MD, 48; State Univ Iowa, MS, 52. *Prof Exp:* Asst prof psychiat, Sch Med, Univ Miami, 54-55; from asst prof to prof, Sch Med, Wayne State Univ, 55-67; prof, Univ Iowa, 67-71; co-chmn dept, Sch Med, Wayne State Univ, 71-73, prof psychiat, 71-78, chmn dept, 73-78; PROF PSYCHIAT, SCH MED, UNIV MISS, 78-, VCHMN DEPT, 81- *Concurrent Pos:* Assoc examr, Am Bd Psychiat & Neurol, 67- *Mem:* Fel Am Psychiat Asn; fel Am Col Psychiat. *Res:* Biochemical and clinical studies of schizophrenia and depressive illnesses; history of psychiatry. *Mailing Add:* Dept Psychiat & Human Behav Univ Miss Med Ctr 2500 N State St Jackson MS 39216

TOURTELLOTTE, CHARLES DEE, b Kalamazoo, Mich, Aug 28, 31; m 55; c 4. BIOCHEMISTRY. *Educ:* Johns Hopkins Univ, AB, 53; Temple Univ, MS & MD, 57; Am Bd Internal Med, dipl. *Prof Exp:* Intern med, Univ Mich, 57-58, resident & jr clin instr, 58-60; instr med & biochem, 63-65, from asst prof to assoc prof, 65-72, res asst prof biochem, 65-71, actg chief sect rheumatol, 66-67, PROF MED, SCH MED, TEMPLE UNIV, 72-, CHIEF SECT RHEUMATOL, SCH MED & UNIV HOSP, 67- *Concurrent Pos:* USPHS trainee rheumatol, Temple Univ, 60-61; Helen Hay Whitney Found fel biochem, Rockefeller Univ, 61-63; Arthritis Found fel, 63-66; mem, Gov Bd, Arthritis Found; consult, St Christopher's Hosp Children, Philadelphia. *Mem:* Fel Am Col Physicians; Am Fedn Clin Res; fel Am Col Rheumatology. *Res:* Biochemistry and physiology of connective tissue; endochondral ossification; amino acid metabolism; histidine; heritable disorders of bone and connective tissues; rheumatic diseases; medical education. *Mailing Add:* 3400 N Broad St Temple Univ Sch Med Philadelphia PA 19140-5192

TOURTELLOTTE, MARK ETON, b Worcester, Mass, Oct 25, 28; m 53; c 3. BIOCHEMISTRY. *Educ:* Dartmouth Col, BA, 50; Univ Conn, MS, 53, PhD(microbiol), 60. *Prof Exp:* From asst instr to instr bact, Univ Conn, 53-60; res assoc biophys, Yale Univ, 60-62; assoc prof, 63-67, PROF ANIMAL PATH, UNIV CONN, 67- *Mem:* AAAS; Am Soc Microbiol; Am Asn Avian Path; fel Am Inst Chem; NY Acad Sci. *Res:* Immunology; diagnostic bacteriology; lipids; chemistry and biosynthesis in mycoplasma; structure and function of biomembranes; mycoplasma toxins; mechanisms of pathogenesis. *Mailing Add:* Dept Pathbiol Univ Conn 61 N Eaglevill Storrs CT 06268

TOURTELLOTTE, WALLACE WILLIAM, b Great Falls, Mont, Sept 13, 24; m 53; c 4. NEUROLOGY. *Educ:* Univ Chicago, PhB & BS, 45, PhD(biochem neuropharmacol), 48, MD, 51; Am Bd Psychiat & Neurol, dipl, 60. *Prof Exp:* Res assoc & instr pharmacol, Univ Chicago, 48-51; intern med, Sch Med, Univ Rochester, 51-52; resident neurol, Med Sch, Univ Mich, 54-57; from asst prof to prof, 57-71; PROF NEUROL & VCHMN DEPT, UNIV CALIF, LOS ANGELES, 71-; CHIEF NEUROL SERV & DIR, NEUROL TRAINING PROG, VET ADMIN WADSWORTH MED CTR, LOS ANGELES, 71- *Concurrent Pos:* Consult, Vet Admin Hosp, Ann Arbor, Mich, 58-71; chief neurol serv, Wayne County Gen Hosp, Detroit, 59-71; mem, Multiple Sclerosis Res Comt, Int Comn Correlation Neurol & Neurochem, World Fedn Neurol, 59-; vis assoc prof, Washington Univ, 63-64; asst examr, Am Bd Psychiat & Neurol, 64-; mem, Med Adv Bd, Nat Multiple Sclerosis Soc, 68-; exchange biomed investr, Vet Admin-Fr NIH & Med Res, Paris, 72; mem, Cerebrospinal Fluid & Immunol Comns, World Fedn Neurol; dir, Nat Neurol Res Bank, 71- *Honors & Awards:* Mitchell Award, Am Acad Neurol, 59- *Mem:* AAAS; Am Neurol Asn; Am Soc Pharmacol & Exp Therapeut; Asn Res Nerv & Ment Dis; Am Acad Neurol. *Mailing Add:* Neurol Serv Bldg 212 Rm 31 Vet Admin Wadsworth Hosp Ctr Los Angeles CA 90073

TOURTELOT, HARRY ALLISON, b Lincoln, Nebr, June 15, 18; m 40, 65, 77; c 6. GEOLOGY. *Educ:* Univ Nebr, AB, 40. *Prof Exp:* Proj technician, State Geol Surv, Ala, 40-42; GEOLOGIST, US GEOL SURV, 42- *Mem:* Fel Geol Soc Am; Geochem Soc; Soc Econ Paleont & Mineral; Clay Minerals Soc; Am Asn Petrol Geol. *Res:* Stratigraphy of continental tertiary rocks; geologic structure of Central Wyoming; geochemistry of sedimentary rocks; petrology of shale; environmental geochemistry; geochemistry and health. *Mailing Add:* Geol Survey Mail Stop 911 Box 25046 Fed Ctr Denver CO 80225

TOURYAN, KENELL JAMES, b Beirut, Lebanon, Dec 2, 36; US citizen; m 63; c 3. SOLAR ENERGY TECHNOLOGIES. *Educ:* Univ Southern Calif, BS, 58, MS, 59; Princeton Univ, MA, 60, PhD(aerospace), 62. *Prof Exp:* Supvr reentry studies, Sandia Labs, 65-68, mgr, Aerothermodynamics Res, 68-75, mgr, Fluid & Plasmadynamics, 75-77, mgr, Fluid & Thermal Sci, 77-78; assoc dir res, Solar Energy Res Inst, 78-80, dep dir, 80-81; SR VPRES, RES & TECHNOL, FLOW INDUST, 81-, TETRA CORP, 87- *Concurrent Pos:* Adj prof, Dept Nuclear Eng, Univ NMex, 66-72; assoc ed, Am Inst Aeronaut & Astronaut J, 75-78, J Energy, 78-; high level expert, Panel UN & World Bank Renewable Energy Utilizaiton, 82-; Fulbright scholar USSR, 86. *Honors & Awards:* Energy Systs Award, Am Inst Aeronaut & Astronaut, 82. *Mem:* Am Inst Aeronaut & Astronaut; Am Phys Soc; AAAS; fel Am Sci Affil; Sigma Xi. *Res:* Fluid dynamics; plasmadynamics; solar and renewable energy; pulsed power technology. *Mailing Add:* Tetra Corp 3701 Hawkins NE Albuquerque NM 87109

TOUSEY, RICHARD, b Somerville, Mass, May 18, 08; m 32; c 1. PHYSICS. *Educ:* Tufts Univ, AB, 28; Harvard Univ, AM, 29, PhD(physics), 33. *Hon Degrees:* ScD, Tufts Univ, 62. *Prof Exp:* Instr physics, Harvard Univ, 33-36, tutor, 34-36, Cutting fel, 35-36; res instr, Tufts Univ, 36-41; head, Instrument Sect, 42-45, head, Micron Waves Br, 45-48, head, Rocket Spectros Br, 58-78, PHYSICIST, US NAVAL RES LAB, 41- *Concurrent Pos:* Darwin lectr, Royal Astron Soc, 63; Russell lectr, Am Astron Soc, 66; mem comt vision, Armed Forces-Nat Res Coun. *Honors & Awards:* Hulburt Award, 48; Medal, Photog Soc Am, 59; Ives Medal, Optical Soc Am, 60; Prix Ancel, Photog Soc France, 62; Draper Medal, Nat Acad Sci, 63; Eddington Medal, Royal Astron Soc, 64; Except Sci Achievement Medal, NASA, 74; George Ellery Hale, Am Astron Soc, 90. *Mem:* Nat Acad Sci; fel Am Phys Soc; fel Optical Soc Am; Am Astron Soc (vpres, 64-66); fel Am Acad Arts & Sci. *Res:* Optical properties of the atmospheres; spectroscopy from rockets; physiological optics; photographic photometry; vacuum ultraviolet. *Mailing Add:* Code 4107 US Naval Res Lab Washington DC 20375-5000

TOUSIGNANT, MICHEL, b Montreal, Que, May 4, 45; m 77; c 2. MEDICAL ANTHROPOLOGY. *Educ:* Univ Montreal, MA, 68; Univ Chicago, PhD(human develop), 74. *Prof Exp:* Res assoc epidemiol, Univ Sherbrooke, 72-74; PROF PSYCHOL, UNIV QUE, MONTREAL, 75- *Concurrent Pos:* Chmn, Lab Social & Human Ecol, 85-88. *Mem:* Am Psychol Asn; Am Anthrop Asn; Soc Med Anthrop; Can Psychol Asn. *Res:* Epidemiology of suicide, mental problems and alcoholism. *Mailing Add:* Lab Social & Human Ecol Univ Que CP 8888 Montreal PQ H3C 3P8 Can

TOUSIGNAUT, DWIGHT R, b Ironwood, Mich, Dec 4, 33; m 64; c 3. PHARMACY. *Educ:* Univ Mich, BS, 59; Univ Calif, Pharm D, 61. *Prof Exp:* Residency hosp pharm, San Francisco Med Ctr, Univ Calif, 59-61; pharmacist, Queen Elizabeth Hosp & Royal Perth Hosp, Australia, 62-63; pharmacist, Stanford Med Ctr, 63-64; Fulbright prof hosp pharm, Cairo, 64-66; dir dept prof pract, 66-72, assoc dir bur communs & publs, 72-81, VPRES, AM SOC HOSP PHARMACISTS, 82-, DIR, DIV DATA BASE SERV, 82- *Concurrent Pos:* Past mem nomenclature adv comt, Nat Libr Med; ed, Int Pharmaceut Abstracts, 66-91; mem bd dir, Nat Fedn Abstracting & Info Serv, 87. *Honors & Awards:* Bristol Award, 59. *Mem:* Am Pharmaceut Asn; Am Soc Hosp Pharmacists; Drug Info Asn (vpres, 73-74, pres, 77-80); Fedn Int Pharm. *Res:* Drug information processing and searching; pharmacy education and practice standards; drug absorption from implanted or injected routes of administration; griseofulvin solubility studies; plastic drug sorption studies. *Mailing Add:* Am Soc of Hosp Pharmacists 4630 Montgomery Ave Bethesda MD 20814

TOUSSIENG, POVL WINNING, b Nysted, Denmark, Sept 5, 18; US citizen. PSYCHIATRY. *Educ:* Copenhagen Univ, MD, 45. *Prof Exp:* Resident gen psychiat, Menninger Sch Psychiat, 50-53, John Harper Seeley fel child psychiat, Children's Div, Menninger Clin, 53-55, staff psychiatrist, 55-65; assoc prof child psychiat & pediat, 65-69, PROF CHILD PSYCHIAT, HEALTH SCI CTR, COL MED, UNIV OKLA, 69- *Concurrent Pos:* Consult, Kans Indust Sch Boys, 53-61; mem fac, Menninger Sch Psychiat, 53-65; consult, Kans Neurol Inst, 64-65, Minn Dept Ment Health, Minneapolis, 65-66 & Spec Subcomt Indian Educ, US Senate Comt Labor & Pub Welfare, 69; mem, Nat Drafting Comt Juv Studies Proj, 73-; mem bd, Psychiat Outpatients Ctr Am, 73- *Mem:* Fel Am Psychiat Asn; fel Am Orthopsychiat Asn; Soc Res Child Develop. *Res:* Childhood autism; coping devices of normal and disturbed children; various modalities of psychotherapy; adolescent experience in changing times; delinquency; adoption; delivery systems of help. *Mailing Add:* Dept Psych & Behav Sci Univ Okla Health Sci Ctr Box 26901 Oklahoma City OK 73190

TOUSTER, OSCAR, b New York, NY, July 3, 21; m 44; c 1. MOLECULAR BIOLOGY, BIOCHEMISTRY. *Educ:* City Col New York, BS, 41; Oberlin Col, MA, 42; Univ Ill, Urbana, PhD(biochem), 47. *Prof Exp:* Chemist, Atlas Powder Co, 42-43; res biochemist, Abbott Labs, 44-45; from instr to assoc prof biochem, 47-58, PROF BIOCHEM, VANDERBILT UNIV, 58-, PROF MOLECULAR BIOL, 73-, CHMN DEPT MOLECULAR BIOL, 63- *Concurrent Pos:* Guggenheim fel, Oxford Univ, 57-58; H Hughes investr, Vanderbilt Univ & Oxford Univ, 57-60; consult, NIH, 61-70; mem, Subcomt Metab Intermediates, Nat Res Coun, 66-76; mem, Bd Dirs, Oak Ridge Assoc Univs, 73-, vpres, 74-76, pres, 76-; mem, Sci Adv Bd, Eunice Kennedy Shriver Ctr Ment Retardation, Waltham, Mass, 74-; mem, Sci Adv Bd, St Jude Children's Res Hosp, 84-87. *Honors & Awards:* Theobald Smith Award Med Sci, AAAS, 56. *Mem:* Fel AAAS; Am Soc Biol Chemists; Am Chem Soc; Sigma Xi. *Res:* Lysosome biochemistry; golgi enzymes; carbohydrate metabolism; glycoproteins; glycosidases. *Mailing Add:* Dept Molecular Biol Vanderbilt Univ Nashville TN 37235

TOUTANT, JEAN-PIERRE, b Paris, France, Nov 13, 44; m 71. NEUROSCIENCES, BIOCHEMISTRY. *Educ:* Univ Nantes, France, PhD(biol), 79; Univ Paris VII, France, PhD(biochem), 85. *Prof Exp:* Assoc prof neurobiol, Secondary Teachers Training Col, Paris, 81-85; postdoctoral neurobiol, Case Western Reserve Univ, Cleveland, Ohio, 87-89; res assoc neurobiol, 85-87, DIR RES NEUROBIOL, NAT INST AGRON RES, MONTPELLIER, FRANCE, 89- *Concurrent Pos:* Prof biochem, Univ Montpellier II, 90- *Mem:* Am Soc Biochem & Molecular Biol; Am Soc Chem Biol; Int Soc Develop Neurosci. *Res:* Structure, expression and regulation of cholinesterases in higher vertebrates and in the nematode Caenorhabditis elegans with the methods of biochemistry and molecular biology. *Mailing Add:* Animal Physiol INRA Pl Viala Montpellier 34060 France

TOVE, SAMUEL B, b Baltimore, Md, July 29, 21; m 45; c 3. BIOCHEMISTRY, NUTRITION. *Educ:* Cornell Univ, BS, 43; Univ Wis, MS, 48, PhD(biochem), 50. *Prof Exp:* Asst, Univ Wis, 46-50; from asst res prof to assoc res prof animal sci, 50-60, PROF BIOCHEM, NC STATE UNIV, 60-, head dept, 75-88. *Concurrent Pos:* William Neal Reynolds prof biochem, NC State Univ, 75. *Mem:* AAAS; Am Chem Soc; Soc Exp Biol & Med; Am Inst Nutrit; Am Soc Biol Chemists. *Res:* Lipid and intermediary metabolism. *Mailing Add:* Dept Biochem NC State Univ Box 7622 Raleigh NC 27695

TOVE, SHIRLEY RUTH, b New York, NY, Jan 31, 25; m 45; c 3. BACTERIOLOGY, BIOCHEMISTRY. *Educ:* Cornell Univ, BS, 45; Univ Wis, MS, 48, PhD(bact, biochem), 50. *Prof Exp:* Instr chem, NC State Univ, 50-51, bact, 51-52; vis teacher biol, NC Col Durham, 64-65; assoc prof, 65-72, chmn dept, 65-75, consult planning, 65, prof biol, Div Natural Sci & Math, Shaw Univ, 72-77; assoc chief, 76-82, CHIEF, BIOL SCI PROG, CHEM & BIOL SCI DIV, US ARMY RES OFF, 82- *Mem:* Am Soc Microbiol; Sigma Xi. *Res:* Biochemistry of nitrogen fixation. *Mailing Add:* US Army Res Off PO Box 12211 Research Triangle Park NC 27709

TOVELL, WALTER MASSEY, b Toronto, Ont, June 25, 16; m 72. GEOLOGY. *Educ:* Univ Toronto, BA, 40, PhD, 54; Calif Inst Technol, MS, 42. *Prof Exp:* Geologist, Calif Standard Co, 42-46; lectr geol, Univ Toronto, 49-50, assoc prof, Univ & Col Educ, 59-64; from asst prof to assoc prof geol, Univ Toronto, 62-81; RETIRED. *Concurrent Pos:* Mus asst, Royal Ont Mus, 46-48, cur geol dept, 48-72, assoc dir, 71-73, dir pro tem, 72-73, dir, 73-77; mem & vchmn info & educ comt, Met Toronto & Region Conserv Authority, 68-74, chmn, 75- *Mem:* Fel Geol Asn Can (secy-treas, 60-62); Mus Dirs Asn Can. *Res:* Stratigraphy and Pleistocene geology; research on geology history of Great Lakes with special emphasis on Georgian Bay. *Mailing Add:* Box 14 Violet Hill RR 4 Shelburne ON L0N 1S8 Can

TOVERUD, SVEIN UTHEIM, b Oslo, Norway, Dec 14, 29; m 54; c 3. PHARMACOLOGY, ENDOCRINOLOGY. *Educ:* Harvard Univ, DMD, 54; Norweg State Dent Sch, DDS, 56; Univ Oslo, PhD, 64. *Prof Exp:* Instr physiol, Univ Oslo, 62-63, res assoc, 63-64, from asst prof to assoc prof, 65-70; assoc prof pharmacol, Sch Med, & assoc prof oral biol, Sch Dent, 69-76, PROF PHARMACOL, SCH MED, UNIV NC, CHAPEL HILL, 76-, PROF ORAL BIOL, SCH DENT, 76- *Concurrent Pos:* Res fel physiol & biochem, Univ Oslo, 56-62; USPHS int fel, Sch Dent Med, Harvard Univ, 64-66. *Mem:* Fel AAAS; US Endocrine Soc; Am Soc Pharmacol & Exp Therapeut; Am Soc Bone Mineral Res. *Res:* Hormonal regulation of calcium and bone metabolism, especially during lactation and the neonatal period; developmental aspects of vitamin D toxicity; purification, characterization and function of acid phosphatases. *Mailing Add:* Dent Res Ctr CB No 7455 Univ NC Chapel Hill NC 27599

TOW, JAMES, b Canton, China, June 25, 36; US citizen; m 68; c 1. ELECTRICAL ENGINEERING. *Educ:* Univ Calif, Berkeley, BS, 60, MS, 62, PhD(elec eng), 66. *Prof Exp:* MEM TECH STAFF, AT&T BELL LABS, 66- *Mem:* Inst Elec & Electronics Engrs. *Res:* Computer aided network and circuit analysis and design; active filter realization; digital signal processing; microprocessor applications. *Mailing Add:* AT&T Bell Labs Crawford Corners Rd H02D-223 Holmdel NJ 07733

TOWBIN, EUGENE JONAS, b New York, NY, Sept 18, 18; m 49; c 4. INTERNAL MEDICINE, PHYSIOLOGY. *Educ:* NY Univ, BA, 41; Univ Colo, MS, 42; Univ Rochester, MD & PhD(physiol), 49. *Prof Exp:* Asst psychologist, Univ Rochester, 42-44; asst physiologist, Duke Univ, 44-47, intern med, 49-50, resident, 50-52; clin asst prof med, 55-56, from asst prof to assoc prof, 56-69, PROF MED & PHYSIOL, SCH MED, UNIV ARK, LITTLE ROCK, 69-, ASSOC DEAN, 68-; CHIEF OF STAFF, VET ADMIN HOSP, 68- *Concurrent Pos:* Fel cardiol, Duke Univ, 52; ward physician, Vet Admin Hosp, 55-58, exec secy & mem res comt, 56-58, asst dir prof serv for res & educ, 58-61, assoc chief of staff for res & educ, 61-72. *Mem:* Am Fedn Clin Res; Geront Soc Am; Am Col Physicians; Am Physiol Soc; Soc Exp Biol & Med; Am Asn Med Syst & Informatics. *Res:* Water and electrolyte metabolism; physiological regulation of thirst and hunger. *Mailing Add:* Vet Admin Med Ctr 4300 W Seventh St Little Rock AR 72205

TOWE, ARNOLD LESTER, b Patterson, Calif, July 25, 27; wid. PHYSIOLOGY, BIOPHYSICS. *Educ:* Pac Lutheran Col, BA, 48; Univ Wash, PhD(psychol, physiol), 53. *Prof Exp:* Res assoc, 53-54, from instr to asst prof anat & physiol, 54-58, from asst prof to assoc prof physiol & biophys, 58-65, PROF PHYSIOL & BIOPHYS, SCH MED, UNIV WASH, 65- *Concurrent Pos:* Mem, NIH Study Sect, 66-70, 78-82, 90-94. *Res:* Neurophysiology, particularly analysis of sensory and motor systems, including gross potentials and single unit activity; cortical physiology, brain evolution. *Mailing Add:* Dept Physiol & Biophys SJ-40 Univ Wash Seattle WA 98195

TOWE, GEORGE COFFIN, b Passaic, NJ, Nov 28, 21; m 47; c 1. PHYSICS, SCIENCE EDUCATION. *Educ:* Hamilton Col (NY), BS, 43; Univ Mich, MS, 47, PhD(chem), 54. *Prof Exp:* Physicist, US Naval Ord Lab, 43-45; res assoc, Eng Res Inst, Univ Mich, 46-53; res engr, Sci Lab, Ford Motor Co, 53-55; from asst prof to assoc prof physics, Mont State Col, 55-61; prof physics, head dept & chmn div natural sci, Findlay Col, 61-62; assoc prof physics, 62-65, chmn dept, 65-72, chmn, Div Spec Progs, 74-77, prof physics, 65-84, EMER PROF PHYSICS, ALFRED UNIV, 84- *Concurrent Pos:* Lectr, Univ Wyo, 59; vis scientist, Atomic Energy Res Estab, Eng, 67-68; consult, Oak Ridge Inst Nuclear Studies, 66-72 & consult educ, Australian Univs, 77-78. *Mem:* Am Asn Physics Teachers; Astron Soc Pacific. *Res:* Radioactivity; radiation; solid state diffusion; nuclear activation analysis. *Mailing Add:* 4089 Normandy Way Eugene OR 97405-4919

TOWE, KENNETH MCCARN, b Jacksonville, Fla, Jan 31, 35; m 75; c 2. GEOLOGY, PALEOBIOLOGY. *Educ:* Duke Univ, AB, 56; Brown Univ, MSc, 59; Univ Ill, PhD(geol), 61. *Prof Exp:* Res assoc electron micros, Univ Ill, 61-62; Ford Found res fel geol, Calif Inst Technol, 62-64; RES GEOLOGIST, DEPT PALEOBIOL, SMITHSONIAN INST, 64- *Concurrent Pos:* Vis prof Geol-Paleont Inst, Univ Tubingen, 73; assoc ed Am Mineralogist, 76-81; judge, AAAS-Westinghouse Sci Jour Awards, 77-86; ed, J Foraminiferal Res, 81-85, Clays & Clay Minerals, 90- *Mem:* AAAS; fel Geol Soc Am; Clay Minerals Soc (treas, 81-); fel Mineral Soc Am; Int Asn Study Clays; fel Cushman Found Foraminiferal Res (pres, 89-90); Geol Soc Wash (pres, 84-85). *Res:* Biomineralogy; clay mineralogy; application of electron microscopy to geology and paleontology; Precambrian paleobiology. *Mailing Add:* Smithsonian Inst Natl Mus Washington DC 20560

TOWELL, DAVID GARRETT, b Fillmore, NY, May 30, 37; m 60; c 2. GEOCHEMISTRY. *Educ:* Pa State Univ, BS, 59; Mass Inst Technol, PhD(geochem), 63. *Prof Exp:* Res fel geochem, Calif Inst Technol, 63-64; asst prof, 64-68, ASSOC PROF GEOCHEM, IND UNIV BLOOMINGTON, 68-, ASSOC DIR, GEOL FIELD STA, 82- *Concurrent Pos:* Assoc chair, dept geol, 87-90. *Mem:* Geochem Soc; Geol Soc Am; Sigma Xi. *Res:* General inorganic, rare-earth, trace element and isotope geochemistry; igneous and metamorphic geology of the Northern Rocky Mountains. *Mailing Add:* Dept Geol Sci Ind Univ Bloomington IN 47405

TOWELL, WILLIAM EARNEST, b St James, Mo, June 11, 16; m 40; c 2. FOREST POLICY, CONSERVATION. *Educ:* Univ Michigan, BSF, 38, MF, 38. *Hon Degrees:* DSc, Univ Mo, 82. *Prof Exp:* Dir, Conservation, Mo Dept Conservation, 57-67; exec vpres, conservation, Am Forestry Assoc, 67-79; RETIRED. *Concurrent Pos:* Pres, Internat Assoc Fish & Wildlife Agencies, 65-66; ad, prof, NC State Univ, 79-88; dir, Nat Wildlife Fedn, 79-87; dir, Forest Hist Soc, 80-88; pres, Soc Am Foresters, 85-86. *Honors & Awards:* Schlich Medal, Soc Am Foresters, 75; Wander Medal, Am Forestry Assoc, 80. *Mem:* Soc Am Foresters (pres, 85). *Mailing Add:* Four Village Green Southern Pines NC 28387-6603

TOWER, DONALD BAYLEY, b Orange, NJ, Dec 11, 19; m 47; c 1. RESEARCH ADMINISTRATION. *Educ:* Harvard Univ, AB, 41, MD, 44; McGill Univ, MSc, 48, PhD(exp neurol), 51. *Hon Degrees:* DSc, McGill Univ, 84. *Prof Exp:* Intern surg, Univ Minn Hosps, 44-45; asst resident neurosurg, Montreal Neurol Inst, McGill Univ, 48-49, assoc neurochemist, 51-53, lectr exp neurol, Fac Med, Univ, 51-52, asst prof, 52-53; chief, Sect Clin Neurochem, 53-60, chief, Lab Neurochem, 61-73, dir, 73-81, emer dir, nat inst neurol & commun dis & stroke, 81-; asst surgeon gen, USPHS, 75-81; RETIRED. *Concurrent Pos:* Res fel neurochem, Montreal Neurol Inst, McGill Univ, 47-51; Markle scholar med sci, 51-53; clin clerk, Nat Hosp, London, Eng, 51; assoc prof, Sch Med & consult, Georgetown Univ, 53-81; mem, Neurol Study Sect, Div Res Grants, NIH, 54-61; mem, US Bd Civil Serv Exam, 61-67; chmn, Neurochem Deleg to USSR, US-USSR Exchange Prog Health & Med Sci, 69; mem, Basic Res Task Force, Adv Comt Epilepsies, USPHS, 69-73; chief ed, J Neurochem, 69-73; mem, Neurochem Panel, Int Brain Res Orgn, mem, Cent Coun, 74-; temp adv neurosci, World Health Orgn, 76-84. *Honors & Awards:* Hist Prize, Justus-Liebig-Univ, Giessen, WGer, 84. *Mem:* Am Acad Neurol; Am Neurol Asn; Am Soc Biol Chemists; Int Soc Neurochem; Am Soc Neurochem (treas, 70-75); hon mem Am Neurol Asn; hon men Peruvian Soc Psychiat, Neurol & Neurosurg; AAAS; Can Neurol Soc; Can Physiol Soc. *Res:* Neurochemistry of epilepsy; cerebral amino acids and electrolytes; history of neurochemistry. *Mailing Add:* 7105 Brennon Lane Chevy Chase MD 20815

TOWERS, BARRY, b Toledo, Ohio, July 20, 38; m 63. FOREST PATHOLOGY, MYCOLOGY. *Educ:* Thiel Col, BA, 61; Duke Univ, MF, 61, DF(forest path), 65. *Prof Exp:* Res asst forest path, Duke Univ & Southern Forest Dis & Insect Res Coun, 62; fel phytotoxic air pollutants, Sch Pub Health, Univ NC, Chapel Hill, 65-68; forest pathologist, 68-87, CHIEF, DIV FOREST PEST MGT, PA DEPT ENVIRON RESOURCES, 87- *Concurrent Pos:* Adj asst prof plant path, Dept Plant Path, Pa State Univ, Univ Park; mem, Beech Bark Dis Working Group, Int Union Forestry Res Orgn. *Mem:* AAAS; Am Phytopath Soc; Soc Am Foresters; Int Soc Arboriculture. *Res:* Diseases of forest trees and coniferous plantations, particularly root rot diseases; phytotoxicity of air pollutants. *Mailing Add:* Div of Forest Pest Mgt 34 Airport Dr HIA Middletown PA 17057

TOWERS, BERNARD, b Preston, Eng, Aug 20, 22; c 4. PSYCHIATRY, ANATOMY. *Educ:* Univ Liverpool, MB, ChB, 47; Cambridge Univ, MA, 54; Royal Col Physicians, Licentiate, 47. *Prof Exp:* House surgeon, Royal Infirmary, Liverpool, 47; asst lectr anat, Bristol Univ, 49-50; lectr anat & histol, Univ Wales, 50-54; lectr anat, Cambridge Univ, 54-70, dir med studies, 64-70; prof pediat, 71-84, co-dir, prog med, Law & Human Values, 77-84, PROF ANAT, SCH MED, UNIV CALIF, LOS ANGELES, 71-, PROF PSYCHIAT, 83- *Concurrent Pos:* Fel med, Jesus Col, Cambridge Univ, 57-70; ed, Brit Abstr Med Sci, 54-56; chmn, Teilhard Ctr Future Man, London, 66-69; consult, Inst Human Values in Med, 71-; dir med soc Forum, 74- *Mem:* Anat Soc Gt Brit & Ireland; Soc Health & Human Values (pres, 77-78); Am Asn Anatomists; fel Royal Soc Med; Brit Soc Hist Med; Am Asn Study Metall; Inst Asn Study Pain. *Res:* Fetal and neonatal lung; development of the heart and congenital anomalies; early detection of myocardial ischemia; primate evolution, especially human; medical humanities; medical history; medical ethics; somato psycho therapy. *Mailing Add:* Dept Anat UCLA Ctr Health Sci Los Angeles CA 90024

TOWERS, GEORGE HUGH NEIL, b Bombay, India, Sept 28, 23; div; c 6. PHYTOCHEMISTRY, PHOTOBIOLOGY. *Educ:* McGill Univ, BSc, 50, MSc, 51; Cornell Univ, PhD, 54. *Prof Exp:* From asst prof to assoc prof bot, McGill Univ, 53-62; sr res officer, Nat Res Coun Can, 62-64; head dept, 64-70, PROF BOT, UNIV BC, 70- *Honors & Awards:* Lalor Found Award, 55. *Mem:* Fel Royal Soc Can; Can Soc Plant Physiol (pres, 65-66); Phytochem Soc NAm. *Res:* Ethnopharmacology; phytochemistry; photobiology. *Mailing Add:* Dept Bot Univ BC 2075 Westbrook Mall Vancouver BC V6T 1W5 Can

TOWILL, LESLIE RUTH, b Milwaukee, Wis, Nov 19, 44. PLANT PHYSIOLOGY, PHOTOBIOLOGY. *Educ:* Univ Wis-Milwaukee, BS, 66, MS, 68; Univ Mich, Ann Arbor, PhD(bot), 73. *Prof Exp:* Sr fel biochem, Univ Wash, 73-75; ASST PROF BOT, ARIZ STATE UNIV, 75- *Concurrent Pos:* Prin investr, NSF grant, 77-79. *Mem:* Am Soc Plant Physiologists; Am Soc Photobiol; Bot Soc Am. *Res:* Mechanism of light action on initial biochemical events in plant development. *Mailing Add:* Dept Bot & Microbiol Ariz State Univ Tempe AZ 85287

TOWLE, ALBERT, b Stockton, Calif, May 10, 25; m 46; c 3. INVERTEBRATE PHYSIOLOGY, SCIENCE EDUCATION. *Educ:* Col of Pac, AB, 46; San Jose State Col, MA, 53; Stanford Univ, PhD(biol), 62. *Prof Exp:* Chmn, Dept Math & Sci, James Lick High Sch, Calif, 50, chmn, Dept Sci, 51, teacher biol, physiol & chem, 53-57 & 58-9, chmn, Dept Sci, 61-64 & 65-66; asst biol, Stanford Univ, 57-58; lectr biol, 64-65, chmn, Dept Marine Biol, 72-75, PROF BIOL, CALIF STATE UNIV, SAN FRANCISCO, 66- *Concurrent Pos:* Lectr, NSF partic, San Jose State Col, 60 & Purdue Univ, 65, partic, NSF Social Psychol Workshop & Conf for Dirs, DC, 68, dir, NSF-Nat Asn Sec Sch Prin Inst Sec Sch Adminr, 68; Nat Sch Teachers Asn Res Antarctica, 70; Smithsonian sponsored attendance and presentation of paper on Sipunculida, Yugoslavia, 70; res in the Galapagos, 74-75. *Mem:* AAAS; Am Inst Biol Sci; Nat Asn Biol Teachers; Nat Sci Teachers Asn; Sigma Xi. *Res:* Teaching of high school biology; behavior and ecology of terrestrial isopods; reproductive physiology of marine invertebrates; distribution of Sipunculida in the Galapagos Archepelago. *Mailing Add:* 11304 Timber Ct Auburn CA 95603

TOWLE, DAVID WALTER, b Concord, NH, May 26, 41; m 74; c 3. MARINE PHYSIOLOGY, MEMBRANE BIOLOGY. *Educ:* Univ NH, BS, 65, MS, 67; Dartmouth Col, PhD(biol sci), 70. *Prof Exp:* From asst prof to assoc prof, 70-82, prof biol, Univ Richmond, 82-88; PROF & CHAIRPERSON BIOL, LAKE FOREST COL, 88- *Concurrent Pos:* Prin investr, Mt Desert Island Biol Lab, 83-88 & 90; Res Corp, Jeffress Trust, NSF grants; prog off, Am Soc Zoologists, 87-89. *Mem:* Am Physiol Soc; Sigma Xi; Am Soc Zoologists; fel AAAS. *Res:* Biochemistry and physiology of osmoregulation in marine and estuarine organisms; ion transport by membrane vesicles; molecular biology of transport proteins. *Mailing Add:* Dept Biol Lake Forest Col Lake Forest IL 60045

TOWLE, HOWARD COLGATE, b Philadelphia, Pa, June 15, 47; m 76; c 3. MOLECULAR ENDOCRINOLOGY. *Educ:* Mich State Univ, BA, 69, PhD(biochem), 74. *Prof Exp:* Fel cell biol, Baylor Col Med, 74-76; asst prof med, 77-79, asst prof biochem, 79-83, ASSOC PROF BIOCHEM, UNIV MINN, 84- *Mem:* Endocrine Soc; Am Soc Biol Chemists; Sigma Xi. *Res:* Intracellular mechanism of action of thyroid hormones; regulation of messenger RNA production by hormonal and dietary factors. *Mailing Add:* Dept Biochem Univ Minn 4-225 Millard Hall Minneapolis MN 55455

TOWLE, JACK LEWIS, organic chemistry, for more information see previous edition

TOWLE, LOUIS WALLACE, b Frog Mountain, Ala, Nov 21, 08; m 31; c 2. ANALYTICAL CHEMISTRY. *Educ:* Univ Ariz, BS, 30, MS, 32. *Prof Exp:* Instr chem, Ariz State Col, 33; teacher high sch, Ariz, 33-36; res chemist, 37-44, tech serv supvr, 44-51, tech dir, 51-54, gen supt, 54-65, vpres, 69-71, gen mgr, 65-79, pres, 71-80, CHMN BD, APACHE POWDER CO, 80- *Mem:* Am Chem Soc; fel Am Inst Chem; Am Inst Mining Metall & Petrol Engrs. *Res:* Preparation and uses of acetylene di-carboxylic acid; analysis of albumen and globulin blood proteins; nitroglycerin blasting explosives; manufacturing heavy chemicals; nitric and sulphuric acids; ammonium nitrate; anhydrous ammonia; ammonium nitrate blasting agents. *Mailing Add:* PO Box 1300 Benson AZ 85602-1300

TOWLER, MARTIN LEE, b Hockley, Tex, Sept 18, 10; m 40; c 5. NEUROLOGY, PSYCHIATRY. *Educ:* Univ Tex, MD, 35; Am Bd Psychiat & Neurol, dipl, 42. *Prof Exp:* Intern, Med Br, Univ Tex, 36, resident neurol & psychiat, 39, instr, 39-41; PROF NEUROL & PSYCHIAT, UNIV TEX MED BR GALVESTON, 46- *Concurrent Pos:* Rockefeller Found fel, Sch Med, Univ Colo, 41-42; pvt pract; consult, Surg Gen, US Army, 49; consult, Lackland AFB Hosp, 54- *Mem:* Am EEG Soc; fel Am Psychiat Asn; AMA; Asn Am Med Cols; fel Am Acad Neurol. *Res:* Effect of drugs on the electroencephalograph pattern; clinical value and limitation of antidepressant drugs. *Mailing Add:* 5115 Ave T Galveston TX 77550

TOWNE, DUDLEY HERBERT, b Schenectady, NY, Nov 7, 24. THEORETICAL PHYSICS. *Educ:* Yale Univ, BS, 47; Harvard Univ, MA, 49, PhD(physics), 54. *Prof Exp:* From instr to assoc prof, 52-63, PROF PHYSICS, AMHERST COL, 63- *Concurrent Pos:* Staff mem, Rockefeller Found, 63-64. *Mem:* Am Asn Physics Teachers; Am Phys Soc. *Res:* Scattering of electromagnetic radiation; broadening of spectral lines; wave propagation in inhomogeneous media. *Mailing Add:* Dept Physics Amherst Col Amherst MA 01002

TOWNE, JACK C, b New York, NY, Apr 23, 27; m 50; c 2. BIOCHEMISTRY, RADIOCHEMISTRY. *Educ:* Univ Calif, Los Angeles, BS, 50; Univ Wis, MS, 52, PhD(biochem), 55. *Prof Exp:* USPHS fel, 54-56; dir biochem lab, Inst Psychosom & Psychiat Res & Training, 56-58; prin scientist biochem, Vet Admin Hosp, Tucson, Ariz, 58-70; PROF CHEM, UNIV DALLAS, 70-, CHMN DEPT, 72- *Concurrent Pos:* Asst prof, Med Sch, Northwestern Univ, 58-65; holder & co-investr, NIH grants, 59-65; lectr, Univ of the Andes, Venezuela, 64; res assoc, Col Med, Univ Ariz, 65-70. *Mem:* AAAS; Am Chem Soc; NY Acad Sci; Sigma Xi. *Res:* Enzymology; intermediary and amine metabolism; radiometric syntheses and analyses. *Mailing Add:* Dept Chem Univ Dallas 1845 E Northgate Irving TX 75015

TOWNER, HARRY H, b Albany, NY, Sept 4, 49; m 72; c 1. PHYSICS, NUCLEAR ENGINEERING. *Educ:* State Univ NY Albany, BS, 72; Univ Ill, Urbana, MS, 75. *Prof Exp:* MEM PROF TECH STAFF CONTROLLED NUCLEAR FUSION, PLASMA PHYSICS LAB, PRINCETON UNIV, 75- *Mem:* Am Phys Soc; Am Nuclear Soc. *Res:* Controlled nuclear fusion and in particular the plasma physics aspects. *Mailing Add:* Princeton Plasma Physics Lab Princeton NJ 08543

TOWNER, HOWARD FROST, b Los Angeles, Calif, Aug 10, 43; m 65; c 2. BIOLOGY, ECOLOGY. *Educ:* Univ Calif, Riverside, AB, 65; Stanford Univ, PhD(biol), 70. *Prof Exp:* NIH fel, Univ Calif, Los Angeles, 70-71, asst res neurologist, Ctr Health Sci, 71-72; PROF BIOL & DEPT CHAIR, LOYOLA MARYMOUNT UNIV, 87- *Concurrent Pos:* Consult, Wadsworth Hosp, US Vet Admin, 72- *Mem:* AAAS; Ecol Soc Am; Sigma Xi. *Res:* Ecology of desert organisms; cytogenetics and plant evolution. *Mailing Add:* Biol Dept Loyola Marymount Univ Loyola Blvd W80th Los Angeles CA 90045

TOWNER, IAN STUART, b Hastings, Eng, May 24, 40; m 66; c 3. THEORETICAL NUCLEAR PHYSICS. *Educ:* Univ London, BSc, 62, PhD(nuclear physics), 66. *Prof Exp:* Res assoc, Nuclear Physics Lab, Oxford Univ, 65-70; RES OFFICER NUCLEAR PHYSICS, CHALK RIVER NUCLEAR LABS, ATOMIC ENERGY CAN LTD, 70- *Mem:* Can Asn Physicists. *Res:* Nuclear structure; models. *Mailing Add:* Physics Div Chalk River Nuclear Labs Chalk River ON K0J 1J0 Can

TOWNER, O W, b Peterson, Iowa, Mar 29, 03. RADIO COMMUNICATIONS. *Educ:* Univ Kans, BS, 27, EE, 33. *Prof Exp:* Chief, Broadcasting Sta Opers, WREN, Lawrence, Kans, 26-27; mem staff, Broadcasting Sta Transmitter Develop Dept, Bell Tel Labs, NY, 27-38; dir eng, Expansion Opers Personnel, AM/FM & FAX, WHAS Inc, Louisville, Ky, 38-42, dir broadcasting, 45-68; asst dir surv, War Res Proj, Airborne Instr Lab, Columbia Univ, NY, 42-45; engr consult, Cover Thack Proj AM/FM & TV, covering shock wave, Anchorage, Ky, 68-80; RETIRED. *Mem:* Fel Inst Elec & Electronics Engrs. *Res:* Construction of broadcasting facsimile and insulation of technical equipment AM/FM and television included; field intensity measurements and performance. *Mailing Add:* 1404 Chesterfield Ct Eustis FL 32726

TOWNER, R(AYMOND) J(AY), b Norwich, NY, Dec 23, 25; m 57; c 3. METALLURGICAL ENGINEERING. *Educ:* Rensselaer Polytech Inst, BMetE, 50, MMetE, 51; Univ Pittsburgh, PhD(metall eng), 58. *Prof Exp:* Student aide trainee metall eng, Phys Metall Div, US Naval Res Lab, 49; res asst, Rensselaer Polytech Inst, 50-51; res engr, Alcoa Res Labs, Aluminum Co Am, 51-63; fel scientist, Aerospace Elec Div, Westinghouse Elec Corp, Ohio, 63-69, FEL ENGR, BETTIS ATOMIC POWER LAB, WESTINGHOUSE ELEC CORP, 69- *Mem:* Am Soc Metals; Am Inst Mining Metall & Petrol Engrs; Brit Inst Metals; Sigma Xi. *Res:* Physical metallurgy; development of aluminum powder metallurgy products; high strength aluminum alloys; high temperature ceramic to metal seals; dispersion strengthened magnetic materials; development and application of materials for nuclear power plants. *Mailing Add:* Westinghouse Elec Corp PO Box 79 West Mifflin PA 15122-0079

TOWNER, RICHARD HENRY, b Gunnison, Colo, Oct 7, 48; m 83. ANIMAL GENETICS. *Educ:* Colo State Univ, BS, 70; Univ Wis-Madison, MS, 73, PhD(genetics, meat & animal sci), 75. *Prof Exp:* Geneticist, H&N Inc, 75-82; dir res, Tatum Farms, 82-86; DIR GENETICS RES, H&N INT, 86- *Concurrent Pos:* Affil asst prof, Col Ocean & Fisheries Sci, Univ Wash, 78- *Mem:* Am Soc Animal Sci; Am Poultry Sci Asn; Worlds Poultry Sci Asn. *Res:* Develop and improve the H & N white and brown egg layers utilizing the existing genetic differences between birds. *Mailing Add:* 24300 NE 193rd Pl Woodinville WA 98072

TOWNES, ALEXANDER SLOAN, b Birmingham, Ala, June 19, 29; m 51; c 5. INTERNAL MEDICINE, RHEUMATOLOGY. *Educ:* Vanderbilt Univ, BA, 50, MD, 53. *Prof Exp:* From instr to assoc prof, Johns Hopkins Univ, 61-72; chief sect rheumatol, 72-75, PROF MED, COL MED, UNIV TENN, MEMPHIS, 72-; CHIEF OF STAFF, VET ADMIN HOSP NASHVILLE, TENN, 87- *Concurrent Pos:* Fel med, Sch Med, Johns Hopkins Univ, 59-61; asst physician in chief, Baltimore City Hosps, 63-70; chief med serv, Memphis Vet Admin Hosp, 75-85, chief of staff, 85-87. *Mem:* AAAS; fel Am Rheumatism Asn; Am Fedn Clin Res; fel Am Col Physicians; Am Asn Immunologists. *Res:* Clinical medicine and rheumatology; role immune reactions in pathogenesis of rheumatic diseases; correlation of clinical findings with immunologic changes and effects of therapy; auto immune arthritis in experimental animals. *Mailing Add:* Vet Admin Hosp 1310 24th Ave S Nashville TN 37212-2637

TOWNES, CHARLES HARD, b Greenville, SC, July 28, 15; m 41; c 4. PHYSICS. *Educ:* Furman Univ, BA & BS, 35; Duke Univ, MA, 37; Calif Inst Technol, PhD(physics), 39. *Hon Degrees:* Twenty-four from US & foreign univs & cols, 60-78. *Prof Exp:* Asst physics, Calif Inst Technol, 37-39; mem tech staff, Bell Tel Labs, 39-47; from assoc prof to prof physics, Columbia Univ, 48-61, chmn dept, 52-55, exec dir radiation lab, 50-52; prof physics & provost, Mass Inst Technol, 61-66, inst prof physics, 66-67; univ prof, 67-86, EMER UNIV PROF PHYSICS, UNIV CALIF, BERKELEY, 86- *Concurrent Pos:* Adams fel, 50; Guggenheim fel, 55-56; Nat lectr, Sigma Xi, 50-51, Fulbright lectr, Univ Paris, 55-56 & Univ Tokyo, 56, Fulbright fel lectr, Col France & Ecole Normale Superieure, 87; lectr, Enrico Fermi Int Sch Physics, 55 & 60, dir, 63; mem, Coun am Phys soc, 59-62, NY City, 65-71; mem, President's Sci Adv Comt, Panel on strategic Weapons, 60-63, chmn, Strategic Weapons Panel, Dept Defense, 61-62; vpres & dir res, Inst Defense Anal, 59-61; trustee, Salk Inst Biol Studies, 63-68, Carnegie Inst, 65-, Rand Corp, 65-70 & Calif Inst Technol, 79-; mem bd dirs, Perkin-Elmer Corp, 66-85, Gen Motors Corp, 73-78, astron Soc Pac, 87-90; mem corp, Woods Hole Oceanog Inst, 69-, trustee, 71-74; mem sci adv bd, USAF, 58-61; chmn sci & tech adv comt manned space flight, NASA, 64-70, mem space prog adv coun, 71-77; mem, President's Sci Adv Comt, 66-70, vchmn, 67-69; chmn,

President-Elect's Task Force on Space, 68, mem, President's Task Force Nat Sci Policy, 69; mem coun, Nat Acad Sci, 69-72 & 78-81, chmn space sci bd, 70-73, comt int security & arms control, 80-89; chmn sci adv comt, Gen Motors Corp, 71-73; mem, President's Comt Sci & Technol, 76-77; mem Bd Trustees, Gen Motors Cancer Res Found, 78-86; chmn, Comt Scholarly Exchanges, Peoples' Repub China, 80-83; chmn, Comt MX Basing, 81; mem, Pontifical Acad Sci, Rome, 81-; mem, defense sci bd, Dept Defense, 82-86; trustee, Calif Acad Sci, San Francisco, 87-; fac res lectr, Univ Calif, Berkeley, 85-86. *Honors & Awards:* Nobel Prize in Physics, 64; Res Corp Award, 58; Morris Liebmann Mem Prize, Inst Radio Engrs, 58, David Sarnoff Award, Inst Elec & Electronics Engrs, 61 & Medal of Honor, Inst Elec & Electronics Engrs, 67; Comstock Prize, Nat Acad Sci, 59, John C Carty Medal, 62; Stuat Ballantine Medal, Franklin Inst, 59 & 62; Rumford Premium Award, Am Acad Arts & Sci, 61; Beckman Award, Instrument Soc Am, 61; Thomas Young Medal & Prize, Brit Inst Physics & Phys Soc, 63; Joseph Priestley Award, Dickinson Col, 66; C E K Mees Medal, Optical Soc Am, 68; Michelson-Morley Award, 70; Wilhem-Exner Award, 70; Earle K Plyler Prize, Am Phys Soc, Niels Bohr Int Gold Medal, Richtmeyer lectr, 59, Scott lectr, Cambridge Univ, 63, Centennial lectr, Univ Toronto, 67, Jansky lectr, Nat Radio Astron Observ, 71, Halley lectr, Oxford Univ, 76, Michelson Mem lectr, US Naval Acad, 82. *Mem:* Nat Acad Sci; fel Am Phys Soc (pres, 67); hon mem Optical Soc Am; foreign mem Royal Soc London; Am Astron Soc; Am Acad Arts & Sci; Am Philos Soc; foreign fel Nat Acad Sci India. *Res:* Molecular and nuclear structure; masers; lasers; radio and infrared astronomy; microwave spectroscopy; optics; quantum electronics. *Mailing Add:* Dept Physics Univ Calif Berkeley CA 94720

TOWNES, GEORGE ANDERSON, b Augusta, Ga, Oct 30, 43; m 68; c 2. NUCLEAR FUEL REPROCESSING, NUCLEAR WASTE. *Educ:* Ga Inst Technol, BME, 65, MSME, 67. *Prof Exp:* Engr, USPHS, 68-70; sr engr & proj mgr, Allied-Gen Nuclear Serv, 71-83; VPRES & OWNER, BE, INC, 83- *Concurrent Pos:* Chmn, Nuclear Lifting Devices, Am Nat Standards Inst, 75- *Res:* Investigations in the reprocessing and waste aspects of the nuclear fuel cycle including remote processing and handling of spent fuel; consolidation of spent fuel to enhance existing storage and transportation capacities. *Mailing Add:* PO Box 381 Barnwell SC 29812

TOWNES, HARRY W(ARREN), b Machias, Maine, Oct 12, 37. MECHANICAL ENGINEERING. *Educ:* Brown Univ, BS, 59; Calif Inst Technol, MS, 60, PhD(mech eng), 65. *Prof Exp:* Asst prof mech eng, 65-71, assoc prof aerospace & mech eng, 71-77, prof aerospace, 77-80, PROF MECH ENG, MONT STATE UNIV, 77- *Mem:* Am Soc Mech Engrs. *Res:* Heat transfer; fluid mechanics; robotics. *Mailing Add:* Dept Mech Eng Mont State Univ Bozeman MT 59717

TOWNES, HENRY KEITH, JR, entomology; deceased, see previous edition for last biography

TOWNES, MARY MCLEAN, b Southern Pines, NC, July 12, 28; m 54; c 2. CELL PHYSIOLOGY. *Educ:* NC Univ Durham, BS, 49, MSPH, 50; Univ Mich, MS, 53, PhD(cell physiol), 62. *Prof Exp:* From instr to assoc prof, 50-68, dean, grad sch arts & sci, 79-86, PROF BIOL, NC CENT UNIV, 68-, DEAN, SCH GRAD STUDIES, 86- *Concurrent Pos:* Consult biol improv prog, NSF & NC Acad Sci Prog High Sch Teachers Biol, 65-66; consult minority access to res careers, Nat Inst Gen Med Sci, NIH, 75-79. *Mem:* AAAS; Soc Gen Physiologists; NY Acad Sci; Sigma Xi; Am Soc Zoologists. *Res:* pH relations of contractility of glycerinated stalks of Vorticella convallaria; contractile properties of glycerinated stalks of Vorticella. *Mailing Add:* Grad Off NC Cent Univ Durham NC 27707

TOWNES, PHILIP LEONARD, b Salem, Mass, Feb 18, 27; m 56. GENETICS. *Educ:* Harvard Univ, AB, 48; Univ Rochester, PhD(zool), 53, MD, 59; Am Bd Pediat, dipl, 80; Am Bd Genetics, dipl, 82. *Prof Exp:* Asst biol, Sch Med & Dent, Univ Rochester, 48-51, from instr to prof anat, 52-79, from asst prof to prof pediat, 65-79, prof genetics, 66-79; PROF PEDIAT, SCH MED, UNIV MASS, 79- *Mem:* Soc Pediat Res; Am Pediat Soc; Am Soc Human Genetics. *Res:* Experimental embryology; cell movements; biochemical aspects of development; enzymes; proteins; metabolic inhibitors; physiology of development; human genetics and embryology. *Mailing Add:* Dept Pediat Med Ctr Univ Mass Lake Ave N Worcester MA 01605

TOWNLEY, CHARLES WILLIAM, b East Liverpool, Ohio, Oct 27, 34; m 57; c 2. PHYSICAL CHEMISTRY, RESEARCH ADMINISTRATION. *Educ:* Ohio State Univ, BSc, 56, PhD(nuclear chem), 59. *Prof Exp:* Sr chemist, Battelle Mem Inst, 59-62, fel, 62-65, chief chem physics res, 65-67, chief struct physics res, 67-70, mgr mat sci, 70-73, mgr info & commun systs, 73-74, mgr, William F Clapp Labs, 74-77, sr prog mgr, Toxic Substance Res, 77-80, prog mgr, toxic & hazardous mat res, Columbus Labs, 80-91; RETIRED. *Mem:* Fel Am Inst Chem. *Res:* Environmental monitoring; research management. *Mailing Add:* 5441 Rockport St Columbus OH 43235

TOWNLEY, JOHN LEWIS, III, petroleum, for more information see previous edition

TOWNLEY, JUDY ANN, b San Antonio, Tex, Sept 19, 46. INFORMATION SCIENCE. *Educ:* Univ Tex, Austin, BA, 68; Harvard Univ, SM, 69, PhD(appl math), 73. *Prof Exp:* Dir Info Sci Prog, 75-79, RES FEL APPL MATH, HARVARD UNIV, 73-; SR SCIENTIST, SOFTWARE OPTIONS INC, 81- *Mem:* Asn Comput Mach; Inst Elec & Electronics Engrs. *Res:* Programming languages; environments (software systems). *Mailing Add:* c/o Software Options 22 Hilliard St Cambridge MA 02138

TOWNLEY, ROBERT WILLIAM, b Lampasas, Tex, Apr 28, 07; m 29; c 2. CHEMISTRY. *Educ:* Austin Col, BA, 29; Univ Tex, MA, 35, PhD(phys chem), 38. *Prof Exp:* Anal chemist, First Tex Chem Mfg Co, 31-33; from asst to instr chem, Univ Tex, 35-37; bacteriologist, State Dept Health, Tex, 37-38, from chemist to chief chemist, 38-41; res chemist, Humble Oil & Ref Co, 41-42; indust hyg engr, USPHS, Md, 42-44; res chemist, Ciba Pharmaceut Prod, Inc, NJ, 44-50; assoc prof chem, Drew Univ, 50-54; head res dept, Personal Prod Corp, 54-57; dir Townley Res & Consult, 57-73; assoc prof Fairleigh Dickinson Univ, 58-59; RETIRED. *Concurrent Pos:* Consult, 73-85. *Mem:* Am Chem Soc. *Res:* Foods; drugs; water; corrosion; air and water pollution; industrial hygiene; microbiology. *Mailing Add:* Clinton Gardens Apts No 3 School St Clinton NJ 08809

TOWNLEY-SMITH, THOMAS FREDERICK, b Scott, Sask, Aug 27, 42; m 63; c 3. PLANT BREEDING. *Educ:* Univ Sask, BSA, 64, MSc, 65; Univ Guelph, PhD(plant breeding), 69. *Prof Exp:* Res asst plant breeding, Univ Guelph, 65-68; res scientist wheat breeding, Res Sta, Can Dept Agr, 68-85; dir, Res Sta, Regina, 85-88; RES SCIENTIST, WINNIPEG RES STA, 88- *Concurrent Pos:* Sr wheat breeder, Plant Breeding Sta, Can Int Develop Agency, Njoro, Kenya, EAfrica, 72-74. *Mem:* Genetics Soc Can; Can Soc Agron. *Res:* Breeding durum wheat; genetics and cytogenetics of wheat; breeding triticale; physiology of drought resistance. *Mailing Add:* 195 Dafoe Rd Winnipeg MB R3T 2M9 Can

TOWNS, CLARENCE, JR, b Little Rock, Ark, July 22, 16; m 44; c 3. PATHOLOGY, HISTOLOGY. *Educ:* Cent YMCA Col, BS, 42; Univ Ill, DDS, 45, MS, 74; Am Bd Endodontic, dipl, 57. *Prof Exp:* Instr histol, 68-69, asst instr basic sci, 70-72, ASST PROF HISTOL, COL DENT, UNIV ILL, 74- *Mem:* Sigma Xi; Am Acad Forensic Dent; Am Soc Oral Med; Am Asn Endodontics; Am Dent Asn. *Res:* Exfoliative cytology of the oral mucosa in the male Negro nonsmoker and smoker; ultra structures study of oral mucosa; comparison of normal and hyperkerotic human oral mucosa. *Mailing Add:* Col Dent 808 S Wood St Chicago IL 60612

TOWNS, DONALD LIONEL, b Sioux City, Iowa, Mar 8, 35; m 60; c 2. PHYSICAL ORGANIC CHEMISTRY, CHEMICAL ENGINEERING. *Educ:* Ga Inst Technol, BCheE, 57; Univ Wis, PhD(chem), 63. *Prof Exp:* Process res chemist, Agr Chem FMC Corp, 62-72, sr process res chemist, Niagara Chem Div, 72-73, process eng group leader, Indust Chem Div, 73-74, tech mgr, 74-75, Furadan prod mgr, 75-76, gen operating supt, 76-77, mgr pesticide formulation & delivery res, 77-80; proj mgr, Herzog-Hart Corp, 80-84; proj mgr, Carlson Assoc, 84-88; proj mgr, Facil Group, 88-90; PROJ MGR, HERZOG-HART CORP, 90- *Mem:* Am Chem Soc; Royal Chem Soc; Am Inst Chem Engrs. *Res:* Technical improvement, environmental protection and production of the insecticide Furadan. *Mailing Add:* Two Happy Hollow Rd Wayland MA 01778

TOWNS, ROBERT LEE ROY, b Bartlesville, Okla, Oct 27, 40; m 60; c 2. CHEMISTRY. *Educ:* Univ New Orleans, BS, 65; Univ Tex, Austin, PhD(phys chem), 69. *Prof Exp:* Vis asst prof chem, Univ New Orleans, 68-70; asst prof, Tex A&M Univ, 70-73; assoc prof, 73-77, PROF CHEM, CLEVELAND STATE UNIV, 78-, CHMN DEPT CHEM, 87- *Concurrent Pos:* NATO Advan Study Inst, Univ York, Eng, 71. *Mem:* Am Chem Soc; Am Crystallog Asn; Am Inst Physics. *Res:* X-ray fluorescence; trace and ultratrace metal analysis in human tissues and body fluids; instrument design, development and automation; x-ray crystallography; molecular structure determination. *Mailing Add:* Dept Chem Cleveland State Univ Cleveland OH 44115

TOWNSEND, ALDEN MILLER, b Tulsa, Okla, Mar 4, 42; m 68; c 2. PLANT GENETICS, TREE PHYSIOLOGY. *Educ:* Pa State Univ, University Park, BS, 64; Yale Univ, MF, 66; Mich State Univ, PhD(plant genetics & physiol), Mich State Univ, 69. *Prof Exp:* RES LEADER, US NAT ARBORETUM, 84- *Concurrent Pos:* Res Geneticist, Nursery Crops Res Lab, USDA, 70-84; exec dir, Metrop Tree Improvement Alliance, 82-85. *Mem:* Soc Am Foresters; Am Phytopath Soc; Int Soc Arboricult. *Res:* Genetic improvement of urban trees with emphasis on disease and insect resistance of maples; physiological genetics of Acer rubrum and elm hybrids. *Mailing Add:* US Nat Arboretum 3501 New York Ave NE Washington DC 20002

TOWNSEND, CHARLEY E, b Decatur Co, Kans, July 2, 29; m 59; c 1. PLANT BREEDING, GENETICS. *Educ:* Kans State Univ, BS, 50, MS, 51; Univ Wis, PhD(agron, plant path), 56. *Prof Exp:* RES GENETICIST, AGR RES SERV, USDA & COLO STATE UNIV, 56- *Mem:* Am Soc Agron; Crop Sci Soc Am. *Res:* Breeding legumes for western ranges and pastures. *Mailing Add:* 638 Gregory Rd Ft Collins CO 80524

TOWNSEND, CRAIG ARTHUR, b Chicago, Ill, Aug 19, 47; m 71; c 2. BIOORGANIC CHEMISTRY, ORGANIC CHEMISTRY. *Educ:* Williams Col, BA, 69; Yale Univ, PhD(org chem), 74. *Prof Exp:* Int exchange fel bio-org chem, Swiss Fed Inst Technol, 74-76; from asst prof to assoc prof, 76-85, PROF CHEM, JOHNS HOPKINS UNIV, 85- *Concurrent Pos:* Res fel, Alfred P Sloan Found, 82-86. *Honors & Awards:* Stuart Pharmaceut Award in Chem, 86. *Mem:* Am Chem Soc; Royal Soc Chem; AAAS. *Res:* Biosynthesis of natural products; stereochemical and mechanistic studies of enzyme action; application of spectroscopic techniques to the solution of biological problems. *Mailing Add:* Dept Chem Johns Hopkins Univ Baltimore MD 21218-2680

TOWNSEND, DAVID WARREN, b Westbrook, Maine, Jan 14, 52; m 75; c 2. PLANKTON ECOLOGY, MARINE FISHES. *Educ:* Univ Maine, BA, 74, PhD(oceanog), 81; Long Island Univ, MS, 77. *Prof Exp:* RES SCIENTIST, BIGELOW LAB OCEAN SCI, 81- *Mem:* Am Soc Limnol & Oceanog; Estuarine Res Fedn; Sigma Xi; AAAS; Oceanog Soc. *Res:* Biological oceanography of shelf seas; physical-biological coupling; fisheries oceanography including plankton trophodynamics. *Mailing Add:* Box 255 West Boothbay Harbor ME 04575

TOWNSEND, DOUGLAS WAYNE, b Covington, Ky, Aug 7, 48; m 72. MATHEMATICS, STATISTICS. *Educ:* Ohio State Univ, BS, 70; Univ Ill, MSc, 75, PhD(math), 76. *Prof Exp:* ASST PROF MATH, IND-PURDUE UNIV, 76- *Mem:* Am Math Soc; Math Asn Am; Sigma Xi. *Res:* Complex analysis, primarily Nevanlinna theory for complex valued functions of a single complex variable. *Mailing Add:* Mathematics Dept Indiana Univ-Purdue Univ 2101 Coliseum Blvd E Ft Wayne IN 46805

TOWNSEND, EDWIN C, b Vienna, WVa, July 7, 36; m 58; c 3. BIOMETRY. *Educ:* Univ WVa, BS, 58, MS, 64; Cornell Univ, PhD(biomet), 68. *Prof Exp:* Staff asst comput, 61-62, res assoc, 62-63, assoc prof statist, 68-74, ASSOC DIR, WVA AGR & FORESTRY EXP STA, WVA UNIV, 75- *Mem:* Biomet Soc; Am Statist Asn. *Mailing Add:* Dept Statist Knapp Hall WVa Univ Morgantown WV 26506

TOWNSEND, FRANK MARION, b Stamford, Tex, Oct 29, 14; m 51; c 2. PATHOLOGY. *Educ:* Tulane Univ, MD, 38. *Prof Exp:* USAF, 40-65, pathologist, Sch Med, Washington Univ, 45-47, instr clin path, Col Med, Univ Nebr, 47-48, assoc pathologist, Scott & White Clin, Temple Tex, 49, regional consult path, Vet Admin Hosp, Tex & La, 50, chief lab serv, Lackland AFB, 50-54, dep dir, Armed Forces Inst Path, 55-59, dir, 59-63, vcommander, Aerospace Med Div, Air Force Systs Command, Brooks AFB, Tex, 63-65; pathologist, Tex State Dept Health, 65-69; clin prof path, 69-72, prof & chmn dept, 72-87, EMER PROF & CHMN, UNIV TEX HEALTH SCI CTR, SAN ANTONIO, 87- *Concurrent Pos:* Assoc prof, Med Br, Univ Tex, 49-58, lectr, 58-63, assoc prof, Postgrad Sch, 53-54; consult, Surgeon Gen, 54-63, Bd Gov, Armed Forces Inst Path, 84-, NASA, 67-74, Armed Forces Epidemiol Bd, 83-; mem, Joint Comt Aviation Path, 56-63, chmn, 60-62; mem, Armed Forces Comt Bioastronaut, Nat Res Coun, 59-60, Nat Adv Cancer Coun, 59-63, Exp Adv Panel Cancer, WHO, 58-81 & Proj Mercury Recovery Team, NASA, 60-63; regional comnr lab accreditation, SCent Region, Col Am Pathologists, 70-84, adv, Coun Educ, 85-; chair path, Univ Tex, San Antonio, 89. *Honors & Awards:* Moseley Award, Aerospace Med Asn, 62; Founders Medal, Asn Mil Surgeons US, 62; Ward Burdick Award, Am Soc Clin Pathologists, 83. *Mem:* Am Asn Pathol; fel Am Soc Clin Path; fel AMA; fel Am Col Physicians; fel Am Col Path; Int Acad Aviation & Space Med. *Res:* Aerospace and respiratory disease pathology. *Mailing Add:* Dept Path Univ Tex Health Sci Ctr 7703 Floyd Curl Dr San Antonio TX 78632

TOWNSEND, GORDON FREDERICK, apiculture; deceased, see previous edition for last biography

TOWNSEND, HERBERT EARL, JR, b Bristol, Pa, July 1, 38; m 63; c 3. CORROSION, METALLURGICAL ENGINEERING. *Educ:* Drexel Univ, BS, 63; Univ Pa, PhD(metall eng), 67. *Prof Exp:* Res engr, 67-72, res supvr, 72-84, res mgr, 84-85, SR RES FEL, HOMER RES LABS, BETHLEHEM STEEL CORP, 85- *Honors & Awards:* D J Blickwede Award Res Excellence, 83. *Mem:* Am Soc Metals; Nat Asn Corrosion Engrs; Soc Automotive Engrs. *Res:* Corrosion-resistant low-alloy steels; metallic and organic coatings for protection of steel; corrosion testing. *Mailing Add:* Homer Res Labs Bethlehem Steel Corp Bethlehem PA 18016

TOWNSEND, HOWARD GARFIELD, JR, b Rochester, NY, Sept 10, 38; m 64; c 2. ENTOMOLOGY. *Educ:* Cornell Univ, BS, 60; Va Polytech Inst, MS, 63; Pa State Univ, PhD(entom), 70. *Prof Exp:* Res asst entom, Va Polytech Inst, 60-62; experimentalist II, NY Agr Exp Sta, Geneva, 63-65; instr, Pa State Univ, 65-69; RES ENTOMOLOGIST & RES PROF ENTOM, STATE FRUIT EXP STA, SOUTHWEST MO STATE UNIV, 70- *Mem:* Entom Soc Am; Am Registry Prof Entomologists. *Res:* Insect and mite pests of pome and stone fruits, grapes and small fruits; pest management. *Mailing Add:* State Fruit Exp Sta Southwest Mo State Univ Mountain Grove MO 65711-9201

TOWNSEND, J(OEL) IVES, b Greenwood, SC, Aug 20, 20. GENETICS. *Educ:* Univ SC, BS, 41; Columbia Univ, PhD(zool), 52. *Prof Exp:* Lectr, Columbia Univ, 50-51; asst prof zool, Univ Tenn, 52-60; asst prof genetics, 60-62, ASSOC PROF HUMAN GENETICS, MED COL VA, VA COMMONWEALTH UNIV, 62- *Concurrent Pos:* Prof, Univ Rio Grande do Sul, Brazil, 54. *Mem:* Fel AAAS; Soc Study Evolution; Genetics Soc Am; Soc Study Social Biol; Am Genetic Asn; Sigma Xi. *Res:* Population genetics; genetics of human isolates; genetics and cytology of marginal populations of Drosophila. *Mailing Add:* Dept Human Genetics Med Col Va Va Commonwealth Univ Box 33 MCV Sta Richmond VA 23219

TOWNSEND, JAMES SKEOCH, b Belwood, Ont, Apr 12, 34; m 60; c 4. AGRICULTURAL ENGINEERING. *Educ:* Ont Agr Col, BSA, 56; Univ Toronto, BASc, 57; Cornell Univ, MS, 65, PhD(agr eng), 69. *Hon Degrees:* Dr, Khon Kaen Univ, Thailand, 88. *Prof Exp:* Teacher, St Catharines Collegiate Inst, 60-62; instr agr eng, Cornell Univ, 64-66; PROF AGR ENG, UNIV MAN, 68- *Concurrent Pos:* Nat Res Coun & Can Dept Agr grants, Univ Man, 69-72; prof, Khon Kaen univ, Thailand, 72-74; team leader, Int Rice Res Inst, Rangoon, Burma, 79-82; consult, Can Int Develop Agency, Ottawa, 82-, Int Ctr Maize & Wheat Res, Mex, 85-86, CUSO, Ottawa, 87, Int Rice Res Inst, 89-90. *Mem:* Am Soc Agr Engrs; Can Soc Agr Eng; Am Soc Eng Educ; Can Soc Mech Eng; Sigma Xi. *Res:* Mechanics of machine milking; agricultural mechanization; heat transfer problems in greenhouse operations; crop and animal modeling. *Mailing Add:* Dept Agr Eng Univ Man Winnipeg MB R3T 2N2 Can

TOWNSEND, JAMES WILLIS, b Evansville, Ind, Sept 9, 36; m 58; c 1. ULTRASTRUCTURAL PATHOLOGY, TOXICOLOGY. *Educ:* Ball State Univ, Ind, BS, 62; Iowa State Univ, PhD(zool), 70. *Prof Exp:* Instr gen biol, anat, cell biol & physiol, Ind State Univ, Ind, 67-70, asst prof genetics, cell physiol & electron micros, 70-71; consult electron micros, Dept Path & Toxicol, Mead Johnson Res Ctr, Ind, 71-72; mgr, Dept Anat & Physiol, Lab Neurosci & Ultrastructure Res, Col Vet Med, Kans State Univ, 74-76; ASST PROF PATH, ELECTRON MICROS & CELL PATH, UNIV ARK MED SCI, 76- *Concurrent Pos:* Lectr, Am Soc Clin Path, Ill, 80-81, Nat Soc Histotechnol, 84; dir, Electron Micros Div, Path Serv Proj, Nat Ctr Toxicol Res, Food & Drug Admin, Ark, 76-81. *Mem:* Electron Micros Soc Am; Sigma Xi. *Res:* Ultrastructural pathology of humans and laboratory animals; ultrastructural effects of long-term, low dose exposure to toxins and carcinogens; applications of energy-dispersive X-ray microanalysis and stereology to problems in toxicology and pathology. *Mailing Add:* Path Dept Slot 517 Univ Ark Med Sci 4301 W Markham St Little Rock AR 72205-7101

TOWNSEND, JANE KALTENBACH, b Chicago, Ill, Dec 21, 22; m 66. HISTOLOGY, ENDOCRINOLOGY. *Educ:* Beloit Col, BS, 44; Univ Wis, MA, 46; Univ Iowa, PhD(zool), 50. *Prof Exp:* Teaching asst zool, Univ Wis, 44-47; teaching asst & instr, Univ Iowa, 47-50; res asst & proj assoc path, Univ Wis, 50-53; Am Cancer Soc res fel, Wenner-Grens Inst, Univ Stockholm, 53-56; asst prof zool, Northwestern Univ, 56-58; from asst prof to assoc prof, 58-70, chmn dept, 80-86, PROF BIOL, MT HOLYOKE COL, 70- *Concurrent Pos:* Mem, Corp Marine Biol Lab. *Mem:* Fel AAAS; Am Asn Anatomists; Am Inst Biol Sci; Am Soc Zoologists; Sigma Xi. *Res:* Thyroxine and other hormonal controls of amphibian metamorphosis; immunohistochemistry. *Mailing Add:* Dept Biol Sci Mt Holyoke Col South Hadley MA 01075

TOWNSEND, JOHN FORD, b Kansas City, Mo, Jan 14, 36; m 59; c 3. PATHOLOGY. *Educ:* Univ Mo-Columbia, AB, 58, MD, 61; Am Bd Path, dipl, 67. *Prof Exp:* Intern, Med Br, Univ Tex, 61-62; resident, 62-66, from asst prof to assoc prof, 68-75, vchmn dept, 75-77, interim chmn dept, 77-78, PROF PATH, SCH MED, UNIV MO-COLUMBIA, 75-, CHMN DEPT, 78- *Concurrent Pos:* Chief path serv, Vet Admin Hosp, Columbia, Mo, 72-75. *Res:* Electrical energy transport through tissue; uterine peroxidase; diabetes using animal model Mystromys albicandatus. *Mailing Add:* 39 King Arthur Way Mansfield MA 02048

TOWNSEND, JOHN MARSHALL, b Amarillo, Tex, Sept 1, 41; m 84; c 3. MEDICAL ANTHROPOLOGY. *Educ:* Univ Calif, Berkeley, BA, 63; Univ Calif, Santa Barbara, MA, 67, PhD(anthrop), 72. *Prof Exp:* Asst prof anthrop, Univ Mont, 72-73; asst prof, 73-76, ASSOC PROF ANTHROP, SYRACUSE UNIV, 76- *Mem:* Am Anthrop Asn; fel Soc Appl Anthrop; Soc Med Anthrop. *Res:* Cross-cultural mental health; labeling theory; health care delivery and human fertility; human sexuality and mate selection. *Mailing Add:* Dept Anthrop 308 Browne Hall Syracuse Univ Syracuse NY 13244-1200

TOWNSEND, JOHN ROBERT, b Brooten, Minn, Oct 26, 25; m 48; c 2. PHYSICS. *Educ:* Cornell Univ, BS, 45, PhD(physics), 51. *Prof Exp:* Physicist, Hanford Atomic Prod Opers, Gen Elec Co, Wash, 51-54; from instr to prof physics, Univ Pittsburgh, 54-; RETIRED. *Mem:* Am Phys Soc; Am Asn Physics Teachers. *Res:* Radiation effects in solids; defects in metals. *Mailing Add:* Dept Physics & Astron 100 Allen Hall Univ Pittsburgh 3941 O'Hara Pittsburgh PA 15260

TOWNSEND, JOHN WILLIAM, JR, b Washington, DC, Mar 19, 24; m 48; c 4. PHYSICS. *Educ:* Williams Col, BA, 47, MA, 49. *Hon Degrees:* DSc, Williams Col, 61. *Prof Exp:* Asst physics, Williams Col, 47-49; physicist, US Naval Res Lab, DC, 49-50, unit head, 50-52, sect head, 52-53, asst br head, 53-55, head rocket sonde br, 55-58; chief space sci div, NASA, 58-59, asst dir, Goddard Space Flight Ctr, 59-65, dep dir, 65-68; dep adminstr, Environ Sci Serv Admin, Nat Oceanic & Atmospheric Admin, 68-70, assoc adminr, 70-77; pres, Fairchild Space Co, Fairchild Industs, 77-85, corp vpres, 79-85, dir, Spacecom, Inc, 84-85, chmn bd, Am Satellite Co, 85, sr vpres & group exec, Aerospace, 86, exec vpres, 87; chief, Space Sci Div, NASA, 58-59, asst div, Space Sci & Satellite Appln, Goddard Space Flight Ctr, 59-65, dep dir, 65-68, dir, 87-90; RETIRED. *Concurrent Pos:* Mem comt aeronomy, Int Union Geod & Geophys; mem tech panel on rocketry, US Nat Comt, Int Geophys Year; exec secy, US Rocket & Satellite Res Panel, 58-; mem, Int Acad Astronaut & Int Astronaut Fedn; trustee, Int Acad Aeronaut; mem, Nat Res Coun, Prog Rev Comt, NASA, 81-87; dir & trustee, Telos Corp, Santa Monica, Calif, 90-; mem adv bd, Loral Corp, 91. *Honors & Awards:* Outstanding Leadership Medal, NASA, 62; Arthur S Fleming Award, 63. *Mem:* Nat Acad Eng; Am Phys Soc; fel Am Meteorol Soc; Sigma Xi; Am Geophys Union; fel AAAS; fel Aerospace Industs Assoc Am. *Res:* Space science and space applications; aeronomy; upper atmosphere physics; composition of the upper atmosphere; mass spectrometry; scientific, meteorological and communications satellites; design and development of sounding rockets; author of various publications. *Mailing Add:* 15810 Comus Rd Clarksburg MD 20871

TOWNSEND, JONATHAN, b Colo, July 17, 22; m 55; c 1. ELECTRONICS ENGINEERING. *Educ:* Univ Denver, BS, 43; Washington Univ, MA, 48, PhD(physics), 51. *Prof Exp:* Engr, Gen Elec Co, 43-44; physicist, Carbide & Carbon Chem Co, 45-46; from asst prof to assoc prof, 51-87, EMER PROF PHYSICS, WASH UNIV, 87- *Mem:* AAAS; Am Phys Soc; Am Asn Physics Teachers. *Res:* Nuclear and paramagnetic resonance; electronics. *Mailing Add:* Dept Physics Wash Univ One Brookings Dr St Louis MO 63130-4899

TOWNSEND, LAWRENCE WILLARD, b Jacksonville, Fla, May 13, 47; m 69; c 3. THEORETICAL NUCLEAR PHYSICS, SPACE RADIATION TRANSPORT. *Educ:* US Naval Acad, BS, 69; US Naval Postgrad Sch, MS, 70; Univ Idaho, PhD(physics), 80. *Prof Exp:* Res asst prof, Old Dom Univ, 80; res scientist, 81-88, SR RES SCIENTIST, NASA LANGLEY RES CTR, 88- *Concurrent Pos:* Adj fac prof physics, Old Dominion Univ, 81-; adj fac math, Christopher Newport Col, 88-; mem, sci subcomn F 2, Comt Space Res & Sci Comt, 75, Nat Coun Radiation Protection & Measurements, 90-; prin investr, NASA Space Radiation Protection Prog, Langley Res Ctr. *Mem:* Am Phys Soc; Radiation Res Soc; Comt Space Res. *Res:* Nuclear and particle physics (heavy ion collisions, nuclear theory, pion production mechanisms and interactions, relativistic nucleon-nucleus interactions); space radiation interactions. *Mailing Add:* Langley Res Ctr NASA Mail Stop 493 Hampton VA 23665-5225

TOWNSEND, LEROY B, b Lubbock, Tex, Dec 20, 33; m 53; c 2. MEDICINAL CHEMISTRY. *Educ:* NMex Highlands Univ, BA, 55, MS, 57; Ariz State Univ, PhD(chem), 65. *Prof Exp:* Res assoc chem, Ariz State Univ, 60-65; res assoc, Univ Utah, 65-67, asst res prof, 67-69; asst prof, 69-71, asst res prof chem, 67-79, from assoc prof med chem to prof, 71-79; prof chem & med chem, 79-86, ALBERT B PRESCOTT PROF MED CHEM, UNIV MICH, ANN ARBOR, 79- *Concurrent Pos:* Consult, Heterocyclic Chem Corp, 69-74, Warner Lambert-Parke Davis, 80- & Baxter Travenol, 81-83.

Mem: Am Chem Soc; Royal Soc Chem; Int Soc Heterocyclic Chem (treas, 73-77, pres-elect, 78-79, pres, 80-81). *Res:* Nitrogen heterocycles, for example pyrrole, pyrimidine, imidazo(4,5-c)pyridine, pyrazole, pyrazolo(3,4-d)pyrimidine, purine, pyrrolo(2,3-d)pyrimidine, pyrazolo(4,3-d)pyrimidine imidazole and the nucleosides arabinofuranosides, ribopyranosides, 2'-deoxyribofuranosides and ribofuranosides of these systems with biological and chemotherapeutic interest as well as structure elucidation and chemical synthesis of certain antibiotics. *Mailing Add:* Col Pharm Univ Mich Ann Arbor MI 48109

TOWNSEND, MARJORIE RHODES, b Washington, DC, Mar 12, 30; m 48; c 4. AEROSPACE ENGINEERING, ELECTRONICS ENGINEERING. *Educ:* George Washington Univ, BEE, 51. *Prof Exp:* Electronics engr basic & appl sonar res, Naval Res Lab, 51-59; sect head design & develop elec instruments, Goddard Space Flight Ctr, NASA, 59-65, tech asst to chief syst div, 65-66, proj mgr small astron satellites, 66-75, proj mgr appl explorer mission, 75-76, mgr preliminary syst design group advan syst design, 76-80; consult, 80-90; DIR, SPACE SYSTS ENG, BDM INT, 90- *Honors & Awards:* Except Serv Medal, NASA, 71, Outstanding Leadership Medal, 80; Knight, Italian Repub Order, 72; Fed Women's Award, 73. *Mem:* Fel Inst Elec & Electronics Engrs; assoc fel Am Inst Aeronaut & Astronaut; AAAS; Am Geophys Union; Soc Women Engrs; corresp mem Int Acad Astronaut. *Res:* Advanced space and ground systems design for a large variety of missions in space and terrestrial applications and in space sciences; new applications for use of the space shuttle. *Mailing Add:* 3529 Tilden St NW Washington DC 20008-3194

TOWNSEND, MILES AVERILL, b Buffalo, NY, Apr 16, 35; m 57; c 5. MECHANICAL ENGINEERING, BIOMECHANICS. *Educ:* Univ Mich, BS, 58; Univ Ill, Urbana, advan cert, 63, MS, 67; Univ Wis-Madison, PhD(mech eng), 71. *Prof Exp:* Res engr, Sundstrand Corp, Ill, 59-63; design engr, Twin Disc, Inc, 63-64; sr engr, Westinghouse Elec Corp, Calif, 64-66; proj engr, Twin Disc, Inc, Ill, 66-68; lectr mech eng, Univ Wis-Madison, 68-69; assoc prof, Univ Toronto, 71-74; prof mech eng, Vanderbilt Univ, 74-81; WILSON PROF & CHMN MECH & AERO ENG, UNIV VA, 82- *Concurrent Pos:* US rep educ, Int Fedn Theory of Mach & Mech, 77-83; mem, comt on productivity, coun engrs, Am Soc Mech Eng. *Honors & Awards:* Am Soc Eng Educ Outstanding Award, 78. *Mem:* Am Soc Mech Engrs; AAAS; Sigma Xi. *Res:* Optimal design, optimal and adaptive control; biomechanics; modeling, dynamics. *Mailing Add:* Dept Mech & Aero Eng Univ Va Charlottesville VA 22901

TOWNSEND, PALMER W, b New York, NY, Aug 1, 26; m 49; c 5. CHEMICAL ENGINEERING. *Educ:* Dartmouth Col, AB, 47; Columbia Univ, BS, 47, MS, 48, PhD(chem eng), 56. *Prof Exp:* Instr chem eng, Columbia Univ, 48-53; sr engr, Pilot Plant Div, Cent Res Labs, Air Reduction Co, Inc, 53-56, sect head, Chem Eng Div, 57-61, asst dir, 61-64, mgr exp eng, Cent Eng Dept, 64-66, asst to group vpres, 66-67, dir commercial develop, Airco Chem & Plastics Div, 67-70; dir commercial develop, Plastics Div, Allied Chem Corp, 70-72; CONSULT, 72- *Mem:* Am Chem Soc; Am Inst Chem Engrs; Soc Plastics Engrs; Soc Plastics Indust; Commercial Develop Asn; Sigma Xi; Asn Consult Chemists & Chem Engrs (pres, 90-91); fel Am Inst Chemists. *Res:* Plastics research and development in products and processes; chemicals and monomers; applications and market research; materials selection; energy development; fluorochemical specialty chemicals. *Mailing Add:* Nine Bristol Ct Berkeley Heights NJ 07922-1306

TOWNSEND, R REID, b Atlanta, Ga, Dec 8, 51. PLASMA GLYCOPROTEINS. *Educ:* Tulane Univ, MD, 76, PhD(biochem), 82. *Prof Exp:* Teaching fel hemat, 78-82, ASSOC RES SCIENTIST, JOHNS HOPKINS UNIV, 82- *Mem:* Am Soc Cell Biol; Soc Complex Carbohydrates; Math Asn Am; Am Chem Soc. *Mailing Add:* Pharm Chem Dept Calif Univ San Francisco CA 94143-0446

TOWNSEND, RALPH N, b Normal, Ill, May 20, 31; m 58; c 2. MATHEMATICS. *Educ:* Ill Wesleyan Univ, BS, 53; Univ Ill, MS, 55, PhD(math), 58. *Prof Exp:* Asst math, Univ Ill, 54-58; asst prof math, San Jose State Col, 58-60; from asst prof to assoc prof, 60-71, asst dean, Col Arts & Sci, 69-75, PROF MATH, BOWLING GREEN STATE UNIV, 71-, ASSOC DEAN, COL ARTS & SCI, 75- *Mem:* Am Math Soc; Math Asn Am; Sigma Xi. *Res:* Analysis, including Schwartz distributions; complex analysis. *Mailing Add:* Dept Math Bowling Green State Univ Bowling Green OH 43403

TOWNSEND, SAMUEL FRANKLIN, b Montague, Mich, Mar 22, 35; m 58; c 2. BIOLOGY, ANATOMY. *Educ:* Kalamazoo Col, AB, 57; Univ Mich, MS, 59, PhD(anat), 61. *Prof Exp:* From asst prof to assoc prof biol, Kalamazoo Col, 61-69, chmn dept, 66-69; assoc prof anat, Col Med, Univ Cincinnati, 69-81; PROF BIOL & CHMN, DIV NATURAL SCI & MATH, HILLSDALE COL, MICH, 81- *Concurrent Pos:* Partic, NSF Res Participation Prog Col Teachers, 64-66. *Mem:* Am Asn Anatomists; AAAS. *Res:* Cellular differentiation in the adult rat; healing and control of experimental ulcers; diabetes. *Mailing Add:* Dept Biol Hillsdale MI 49242

TOWNSHEND, JOHN LINDEN, b Hamilton, Ont, Feb 6, 26; m 53; c 2. PLANT PATHOLOGY. *Educ:* Univ Western Ont, BSc, 51, MSc, 52; Imp Col, Univ London, dipl, 63. *Prof Exp:* Tech officer, Forest Path Unit, Forest Prod Lab, Dept Northern Affairs & Nat Resources, Can, 52; RES SCIENTIST, RES BR, CAN DEPT AGR, 52- *Honors & Awards:* Hoechst Award, Can Soc Hort Sci, 85. *Mem:* Soc Nematol; Can Phytopath Soc; Europ Soc Nematol. *Res:* Ecology of plant parasitic nematodes; forage nematodes; host-parasite relationships-ultrastructures. *Mailing Add:* Res Sta Agr Can Vineland Station ON L0R 2E0 Can

TOWNSLEY, JOHN D, b London, Eng, July 19, 33. DENTAL RESEARCH. *Educ:* Univ Leeds, Eng, BS, 58, PhD(biochem), 61. *Prof Exp:* CHIEF, CRANIOFACIAL ANOMALIES, PAIN CONTROL & BEHAV RES BR, NAT INST DENT RES, NIH, 85- *Mem:* Am Asn Dent Res; Biochem Soc Eng. *Mailing Add:* Pain Control & Behav Res Br Nat Inst Dent Res Craniofacial Anomalies NIH Westwood Bldg Rm 506 5333 Westbard Ave Bethesda MD 20892

TOWNSLEY, PHILIP MCNAIR, b Vancouver, BC, Nov 21, 25; m 52; c 3. INDUSTRIAL MICROBIOLOGY. *Educ:* Univ BC, BSA, 49; Univ Calif, Berkeley, MS, 50, PhD(comp biochem), 56. *Prof Exp:* Biochemist, Dept Agr, Govt Can, 56-61, group leader, Process & Prod Group, Fishery Res Bd, 61-63; group leader biochem, BC Res Coun, 63-67; PROF INDUST MICROBIOL, UNIV BC, 67- *Concurrent Pos:* Mem bd dirs, John Dunn Agencies Ltd & Pac Micro-Bio Cult Ltd. *Mem:* Inst Food Technol; Can Inst Food Sci & Technol; Int Asn Plant Tissue Cult; Can Soc Microbiol. *Res:* Practical application of basic research. *Mailing Add:* Dept Food Sci Univ BC 2075 W Brook Pl Vancouver BC V6T 1W5 Can

TOWNSLEY, SIDNEY JOSEPH, b Colorado Springs, Colo, Aug 6, 24; m 50; c 5. RADIOBIOLOGY. *Educ:* Univ Calif, AB, 47; Univ Hawaii, MS, 50; Yale Univ, PhD(zool), 54. *Prof Exp:* Asst prof marine zool & asst, Marine Lab, Univ Hawaii, 54-60, assoc prof marine biol, 60-66, prof, 66-70, prof marine zool, 70-; RETIRED. *Mem:* Am Soc Zoologists; Am Soc Limnol & Oceanog; Ecol Soc Am. *Res:* Ecology of radioisotopes in marine organisms; systematics; stomatopod Crustacea and cephalopod mollusks; histochemistry and physiology of heavy metals. *Mailing Add:* Dept Zool Univ Hawaii 2538 The Mall Honolulu HI 96822

TOWSE, DONALD FREDERICK, b Somerville, Mass, Dec 5, 24; m 45; c 6. GEOLOGY. *Educ:* Mass Inst Technol, BS, 48, PhD(geol), 51. *Prof Exp:* Geologist, Amerada Petrol Corp, 48-49, 50-51; asst prof geol, Univ NDak, 51-54; geologist, State Geol Surv, NDak, 51-54; consult geologist, 54-56; vis assoc prof, Univ Calif, Los Angeles, 56-57; proj geologist, Kaiser Aluminum & Chem Corp, 57-60, sr proj geologist, Kaiser Cement & Gypsum Corp, 60-71, sr geologist, Kaiser Explor & Mining Co, 71-73; geologist, Lawrence Livermore Nat Lab, Univ Calif, 74-80, sect leader, 80-82, sr scientist, 83-90; CONSULT, 90- *Concurrent Pos:* Managing dir, Delta Res Inst, 79- *Honors & Awards:* Pres Award, Am Asn Petrol Geologists, 52. *Mem:* Am Asn Petrol Geologists (pres, Energy Min Div, 89-90); Am Inst Mining, Metall & Petrol Engrs. *Res:* Stratigraphy and petroleum resources; carbonate oil reservoirs; cement raw materials; laterites; uranium ores; geothermal resources; geology and economics of geothermal energy deposits; geologic disposal of nuclear and hazardous waste; environmental and groundwater geology. *Mailing Add:* 2420 Ruby Ave San Jose CA 95148-1749

TOY, ARTHUR DOCK FON, b Canton, China, Sept 13, 15; nat US; m 42; c 3. INDUSTRIAL CHEMISTRY. *Educ:* Univ Ill, BS, 39, MS, 40, PhD(chem), 42. *Prof Exp:* Res chemist, Victor Chem Works Div, Stauffer Chem Co, 42-53, dir org res, 53-59, assoc dir res, 59-63, dir res, 63-65; vis scientist, Cambridge Univ, 65-66; sr scientist, Stauffer Chem Co, Westport, Conn, 66-68, sr scientist & actg mgr, Chem Res Dept, 68-70, sr scientist & actg mgr, Specialties Dept, 70-72, chief scientist, 72-74, dir, Eastern Res Ctr, 75-78, dir res, 79-80; CONSULT, STAMFORD, CONN, 80- *Mem:* Am Chem Soc; Royal Soc Chem; fel AAAS; NY Acad Sci; Sigma Xi. *Res:* Organic phosphorus compounds for plastic applications and insecticides; allyl aryl-phosphonate flame resistant plastic; economic process for synthesis of phosphorus insecticides and insecticide intermediates; aquo ammono phosphoric acids; organic reaction mechanisms; new reactions leading to the formation of carbon to phosphorus bonds. *Mailing Add:* 14 Katydid Lane Stamford CT 06903

TOY, ARTHUR JOHN, JR, b Pasadena, Calif. RADIATION BIOPHYSICS, HEALTH PHYSICS. *Educ:* Calif State Univ, Hayward, BS, 69; Univ Kans, MS, 71, PhD(radiation biophys), 73. *Prof Exp:* RES & DEVELOP SCIENTIST ENVIRON SCI & HEALTH PHYSICIST, LAWRENCE LIVERMORE LAB, 69- *Mem:* Sigma Xi; Health Physics Soc. *Res:* Radiation accidents. *Mailing Add:* 6800 Telsa Rd Livermore CA 94550

TOY, MADELINE SHEN, b Shanghai, China, Nov 6, 28; US citizen; m 51; c 1. FLUORINE CHEMISTRY, POLYMER CHEMISTRY. *Educ:* Col St Teresa, Minn, BS, 49; Univ Wis, MS, 51; Ohio State Univ, MS, 57; Univ Pa, PhD(org chem), 59. *Prof Exp:* Mgr org lab, Freelander Res & Develop Div, Dayco Corp, Calif, 59-60; asst prof, Calif State Univ, Northridge, 60-61; staff mem, Int Tel & Tel Corp, Fed Labs, 61-63; res scientist & sect chief, Astropower Lab, McDonnell Douglas Corp, 64-69, head polymer sci, Douglas Adv Res Labs, 69-70; sr polymer chemist, Stanford Res Inst, Menlo Park, 71-75; sr scientist, 75-88, HEAD CHEM LAB, SCI APPLN INT CORP, 88-; SR SCIENTIST, ANALYTICAL LAB INC, CAMPBELL, CALIF. *Concurrent Pos:* Mem exec comt, Div Fluorine Chem, Am Chem Soc; reviewer of polymer progs, NSF, Adv Chem Series, J Org Chem & J Phys Chem; prin investr, NASA, Jet Propulsion Lab, 69-70, Lewis, 72, Wright Patterson AFB, 72-73, Air Force Off Sci Res, 73-79, NASA, Joint Strategic Comt, 75-77, Aerospace Corp, 77-78, Edwards AFB, 79, Navy 79 & 82, Elec Power Res Inst, 81-86 & 87-90, Naval Weapons Ctr, 83-86, NASA Marshall Flight Ctr, 91. *Mem:* Am Chem Soc; fel Royal Soc Chem. *Res:* Optically active polymers; fire retardant polyurethanes; high temperature plastics; thermoplastic films; high energy perfluorinated salts; surface polymerizations on metal substrates; fluoroelastomers; multifunctional fluoropolymers; perfluoropolymer-forming reactions; flame resistant surface treatments; organic chemistry; fuel cell electrolytes; photooxidative degradation studies of aromatic polyamide; photosonolysis for synthetic and degradation studies; aqueous chemistry. *Mailing Add:* 4190 Manuela Ave Palo Alto CA 94306

TOY, TERRENCE J, b Sidney, Ohio, Aug 10, 46; m 68. DISTURBED LAND RECLAMATION, EROSION. *Educ:* State Univ NY Buffalo, BA, 69, MA, 70; Univ Denver, PhD(geog), 73. *Prof Exp:* From asst prof to assoc prof, 75-86, PROF GEOG, UNIV DENVER, 86-, DEPT CHAIR, 87- *Concurrent Pos:* Prin investr, US Geol Survey, 75-79 & Northern Energy Resource Co, 80-85. *Mem:* Soil Conservation Soc Am; Am Soc Surface Mining & Reclamation; Asn Am Geographers. *Res:* Geomorphology of disturbed lands; author of several publications and one book. *Mailing Add:* 2657 S Linden Ct Denver CO 80222

TOY, WILLIAM W, b Chicago, Ill, Mar 11, 50. STOCHASTIC PROCESSES. *Educ:* Mass Inst Technol, SB, 73, SMEE, 73, PhD(physics), 78. *Prof Exp:* Mem tech staff, Bell Labs, Inc, 78-84; VPRES, GOLDMAN SACHS & CO, 84- *Mem:* Am Phys Soc; Inst Elec & Electronics Engrs; Sigma Xi. *Res:* Modeling of stochastic processes; financial applications; mathematical modeling. *Mailing Add:* 211 Chelsea Circle Lake Villa IL 60046

TOY, WING N, digital communications; deceased, see previous edition for last biography

TOZER, E T, b Kirkland, Ont, Mar 2, 24. GEOLOGY. *Educ:* Univ Toronto, BA, 47, MA, 49, PhD, 55. *Prof Exp:* RES SCIENTIST, GEOL SURV CAN, 52- *Mem:* Geol Soc Am; Geol Asn Can; Paleont Soc Can. *Mailing Add:* Cordilleran Div 100 W Pender St Vancouver BC V6B 1R8 Can

TOZER, THOMAS NELSON, b San Diego, Calif, July 4, 36; m; c 4. PHARMACEUTICAL CHEMISTRY, PHARMACY. *Educ:* Univ Calif, San Francisco, BS & PharmD, 59, PhD(pharmaceut chem), 63. *Prof Exp:* Lectr chem & pharmaceut chem, Univ Calif, San Francisco, 63; NIMH fel, 63-65; from asst prof to assoc prof, 65-81, PROF PHARM & PHARMACEUT CHEM, UNIV CALIF, SAN FRANCISCO, 81- *Concurrent Pos:* Pharmacist, 59-; fel, Lab Chem Pharmacol, Nat Heart Inst, 63-65. *Mem:* AAAS; Am Pharmaceut Asn; Am Asn Pharmaceut Sci; Am Chem Soc; NY Acad Sci; Sigma Xi. *Res:* Toxicokinetics; colon specific drug delivery; pharmacokinetics and metabolic fate of contrast agents for magnetic resonance imaging; saturable first-pass metabolism; effect of disease states on drug pharmacokinetics. *Mailing Add:* 361 Avalon Dr S San Francisco CA 94080

TOZIER, JOHN E, b DuBois, Pa, Nov 3, 42; m 65; c 1. EXPERIMENTAL DATA ANALYSIS WITH MACINTOSH-HP COMPUTERS, EXPERIMENTAL PREDICTIONS BASED ON MATH ANALYSIS. *Educ:* Bucknell Univ, BS, 64. *Prof Exp:* Engr, Clevite Corp, 65-69, Bailey Controls, 69-71; sr engr, GTE Sylvania, 75-81; SR SCIENTIST, BACHARACH INC, 81- *Res:* Industrial research; catalytic combustion sensors; solid state hydrogen-sulphide sensors. *Mailing Add:* 286 Courtney Pl Wexford PA 15090

TOZZI, SALVATORE, ALLERGIC DISORDERS. *Educ:* NY Inst, DC, 54. *Prof Exp:* AT SCHERING CORP. *Mailing Add:* 60 Orange St Schering Corp Bloomfield NJ 07003

TRABANT, EDWARD ARTHUR, b Los Angeles, Calif, Feb 28, 20; m 43; c 3. APPLIED MATHEMATICS. *Educ:* Occidental Col, AB, 41; Calif Inst Technol, PhD(appl math), 47. *Prof Exp:* From instr to prof math & eng sci, Purdue Univ, 47-60; dean sch eng, State Univ NY Buffalo, 60-66; vpres acad affairs, Ga Inst Technol, 66-68; PRES, UNIV DEL & PROF ENG SCI, COL ENG, 68- *Concurrent Pos:* Consult, Allison Div, Gen Motors, Ind, 50-55, Argonne Nat Lab, Ill, 55-61; Carborundum Corp, NY, 64-68 & Army Sci Adv Panel, 66-71. *Mem:* Am Soc Eng Educ; Am Soc Mech Eng; Am Math Soc; Am Nuclear Soc; Sigma Xi. *Mailing Add:* 102 Bellant Circle Wilmington DE 19807

TRABER, DANIEL LEE, b Victoria, Tex, Apr 28, 38; m 59. PHYSIOLOGY, PHARMACOLOGY. *Educ:* St Mary's Univ, Tex, BA, 59; Univ Tex, MA, 62, PhD(physiol), 65. *Prof Exp:* Asst physiol, Med Br, Univ Tex, 60-65; fel pharmacol, Col Med, Ohio State Univ, 65-66; asst prof physiol & res asst prof anesthesiol, 66-70, dir, Interdisciplinary Labs, 70-72, assoc prof, 70-74, PROF ANESTHESIOL & PHYSIOL, UNIV TEX MED BR GALVESTON, 74-, DIR, INTEGRATED FUNCTIONAL LAB, 72- *Concurrent Pos:* Chief, Div Anesthesia Res, Shriners Burn Inst, Galveston, 71-78. *Mem:* AAAS; Am Physiol Soc; Am Burn Asn; Soc Exp Biol & Med; Am Soc Pharmacol & Exp Therapeut. *Res:* Shock, endotoxemia and sepsis. *Mailing Add:* Dept Anesthesiol Suite 2A-Rte E91 Univ Tex Med Br Galveston TX 77551

TRABER, MARET G, b Stockton, Calif, Nov 4, 50; m 71; c 1. LIPOPROTEIN METABOLISM, VITAMIN E. *Educ:* Univ Calif, Berkeley, BS, 72, PhD(nutrit), 76. *Prof Exp:* Instr, dept nutrit, Rutgers Univ, 76-77; asst res scientist, NY Univ Sch Med, 77-80, assoc res scientist, 80-82, res scientist, 82-86, res asst prof, 86-89, RES ASSOC PROF, DEPT MED, NY UNIV SCH MED, 89- *Mem:* Am Heart Asn; Am Inst Nutrit; Am Soc Clin Nutrit; Am Soc Biol Chemists. *Res:* Lipids and lipoprotein metabolism; vitamin E deficiency in humans. *Mailing Add:* Dept Med NY Univ Med Ctr 550 First Ave New York NY 10016

TRACE, ROBERT DENNY, b Zanesville, Ohio, Oct 27, 17; m 50; c 2. GEOLOGY. *Educ:* Southern Methodist Univ, BS, 40; Univ Calif, Los Angeles, MA, 47. *Prof Exp:* Geologist, US Geol Surv, 42-77; sr geologist, Ky Geol Surv, 77-81; RETIRED. *Concurrent Pos:* Instr, Hopkinsville Community Col, Ky, 66-70. *Mem:* Fel Geol Soc Am; Soc Econ Geol. *Res:* Stratigraphic and structural geology of fluorspar-zinc-lead deposits. *Mailing Add:* RR One Box 143 A Lot A-9 Edinburg TX 78539-9705

TRACEY, DANIEL EDWARD, b Boston, Mass, Mar 31, 47; div; c 2. IMMUNOLOGY, INFLAMMATION RESEARCH. *Educ:* Johns Hopkins Univ, PhD(biochem), 73. *Prof Exp:* Sr res scientist, 83-87, ASSOC DIR, DEPT HYPERSENSITIVITY DIS RES, UPJOHN CO, 87- *Mem:* Am Asn Immunol; NY Acad Sci. *Res:* Arthritis. *Mailing Add:* Dept Hypersensitivity Dis Res Upjohn Co 301 Henrietta St Kalamazoo MI 49007

TRACEY, JOSHUA IRVING, JR, b New Haven, Conn, May 2, 15; m 46; c 2. GEOLOGY. *Educ:* Yale Univ, BA, 37, MS, 43, PhD(geol), 50. *Prof Exp:* Geologist, 42-86, EMER SCIENTIST, US GEOL SURV, 86-; RES ASSOC, DEPT PALEOBIOL, NAT MUS NATURAL HIST, 86- *Concurrent Pos:* Chmn, Geol Names Comt, US Geol Surv, 81-86. *Mem:* Fel AAAS; fel Geol Soc Am; Am Asn Petrol Geol; Am Geophys Union; Sigma Xi. *Res:* Bauxite; geology and ecology of coral reefs; geology and resources of the Pacific Islands; tertiary stratigraphy of the fossil basin in southwestern Wyoming. *Mailing Add:* Nat Mus Natural Hist E308 Smithsonian Inst Washington DC 20560

TRACEY, MARTIN LOUIS, JR, b Boston, Mass, Mar 3, 43; m 66; c 2. GENETICS. *Educ:* Providence Col, AB, 65; Brown Univ, PhD(biol), 71. *Prof Exp:* Fel genetics, Univ Calif, Davis, 71-73; dir genetics, Bodega Marine Lab, Calif, 73-74; asst prof biol, Brock Univ, 74-77; from asst prof to assoc prof, 77-85, chmn dept, 79-85, PROF BIOL, FLA INT UNIV, 85- *Mem:* AAAS; Genetics Soc Am; Am Soc Naturalists; Can Genetics Soc; Soc Study Evolution. *Res:* Speciation and genetic differentiation; recombination; polymorphism; DNA fingerprinting. *Mailing Add:* Dept Biol Sci Fla Int Univ Miami FL 33199

TRACHEWSKY, DANIEL, b Montreal, Que, Nov 4, 40; US citizen; m 67; c 2. MOLECULAR BIOLOGY, ENDOCRINOLOGY. *Educ:* McGill Univ, BSc, 61, MSc, 63, PhD(biochem), 66. *Prof Exp:* Res assoc biochem of reproduction, Biomed Div Pop Coun, Rockefeller Univ, 65-67; dir molecular biol, Montreal Clin Res Inst, 68-75; from asst prof to assoc prof med, McGill Univ, 69-75; from asst prof to assoc prof med, Univ Montreal, 69-75; assoc prof med, biochem & molecular biol, 75-80, PROF MED, BIOCHEM & MOLECULAR BIOL, & PHYSIOL & BIOPHYSICS, HEALTH SCI CTR, UNIV OKLA, 80- *Concurrent Pos:* Prin investr on grants, Can Med Res Coun, Que Med Res Coun, Que Heart Found & Montreal Clin Res Inst, 68-75, NIH, 76-; off referee, review grant applns, Can Med Res Coun, 69-75; reviewer, Res & Develop Comt, Vet Admin, 78-; mem res comt, Am Heart Asn, 80-83, 87-; vis prof, Univ Calif, San Francisco, 83-84; sr fel, NIH, 84; mem bd dirs, Biomed Res Labs Inc, Okla City, 84- *Mem:* NY Acad Sci; Endocrine Soc; Am Fedn Clin Res; Am Soc Biochem & Molecular Biol; Soc Exp Biol & Med; Sigma Xi. *Res:* Mechanism of action of steroid and peptide hormones at the molecular level employing recombinant DNA and other molecular biological strategies; pathophysiology and pharmacological management of hypertension. *Mailing Add:* Health Sci Ctr Dept Med Univ Okla PO Box 26307 Oklahoma City OK 73126

TRACHMAN, EDWARD G, b New York, NY, Apr 10, 46; m 68; c 2. TRIBOLOGY, COMPUTER SIMULATION. *Educ:* Cooper Union, BE, 66; Calif Inst Technol, MS, 67; Northwestern Univ, Evanston, PhD(mech eng), 71; Mich State Univ, MBA, 83. *Prof Exp:* Assoc sr res engr, Res Labs, Gen Motors Corp, 71-75; mem tech staff, RCA Labs, 75-77; head eng anal dept, Vadetec Corp, 77-81; chief engr Appl Res, Rockwell Int Corp, 81-85, dir, Res & Develop, 85-89, dir, Prod Eng, Automotive Opers, 89-90; VPRES PORD ENG, HARMAN-MOTIVE, INC, 90- *Concurrent Pos:* Fac asst, Lawrence Inst Technol, 71-75. *Mem:* Am Soc Mech Engrs; Sigma Xi; Soc Automotive Engrs. *Res:* Computer simulation of mechanical systems; tribology; traction drives; mechanical design; solid and fluid mechanics. *Mailing Add:* 31326 Downing Place Birmingham MI 48009

TRACHTENBERG, EDWARD NORMAN, b New York, NY, Dec 8, 27; m 54; c 3. ORGANIC CHEMISTRY. *Educ:* NY Univ, AB, 49; Harvard Univ, AM, 51, PhD(org chem), 53. *Prof Exp:* Instr chem, Columbia Univ, 53-58; from asst prof to assoc prof, 58-70, PROF CHEM, CLARK UNIV, 70- *Concurrent Pos:* NSF fel, Univ London, 67-68. *Mem:* Am Chem Soc; Royal Soc Chem. *Res:* Mechanism of organic reactions; organic synthesis; selenium dioxide oxidation of organic compounds; 1,3-dipolar cycloadditions. *Mailing Add:* Dept Chem Clark Univ Worcester MA 01610

TRACHTENBERG, ISAAC, b New Orleans, La, Aug 20, 29; m 86; c 2. ELECTROCHEMISTRY. *Educ:* Rice Inst, BA, 50; La State Univ, MS, 52, PhD(chem), 57. *Prof Exp:* From assoc chemist to chemist, Am Oil Co, Tex, 57-60, sr chemist & group leader, 60; mem tech staff, Tex Inst, 60-63, br head basic electrochem, 63-66, chem kinetics, 66-68, systs anal process control, 69-70, br head environ monitoring, 71-72, br head sensors res & develop, 72-74, mgr process control, Semiconductor Group, 74-75, mgr qual & reliability assurance, Semiconductor Group, 74-80, solar energy proj, 80-84; from adj prof to prof, 84-90, PAUL D & BETTY ROBERTSON MEEK CENTENNIAL PROF CHEM ENG, CHEM ENG DEPT, UNIV TEX, AUSTIN, 90- *Concurrent Pos:* VChmn & chmn, Gordon Res Con Electrochem, 68-69; ed, J Battery Div, Electrochem Soc, 69-72. *Mem:* Am Chem Soc; Electrochem Soc. *Res:* Semiconductors; semiconductor device quality and reliability; energy conversion; electrochemistry; semiconductor device processing, plasma etching, aerosol etching; microelectronics packaging, scanning, tunneling and electrochemical microscopy, electrochemistry and energy conversion. *Mailing Add:* 4711 Spicewood Springs Rd No 278 Austin TX 78759-8402

TRACHTENBERG, MICHAEL CARL, neurophysiology, cell biology, for more information see previous edition

TRACHTMAN, MENDEL, b May 6, 29; US citizen; m 50; c 3. RADIATION CHEMISTRY, PHOTOCHEMISTRY. *Educ:* Temple Univ, AB, 51; Drexel Inst Technol, MS, 56; Univ Pa, PhD(chem), 61. *Prof Exp:* Res chemist, Frankford Arsenal, Philadelphia, Pa, 51-67; PROF CHEM, PHILADELPHIA COL TEXTILES & SCI, 67- *Concurrent Pos:* Res Corp res grant, Philadelphia Col Textiles & Sci, 72-74. *Mem:* Am Chem Soc. *Res:* Determination of quenching cross sections of aromatic molecules; fading properties of various dye molecules. *Mailing Add:* 1705 Tustin St Philadelphia PA 19152

TRACTON, MARTIN STEVEN, b Brockton, Mass, Feb 9, 45; m 66; c 1. METEOROLOGY. *Educ:* Univ Mass, Amherst, BS, 66; Mass Inst Technol, MS, 69, PhD(meteorol), 72. *Prof Exp:* Asst prof meteorol, Naval Postgrad Sch, 72-75; RES METEOROLOGIST, NAT METEOROL CTR, NAT OCEANIC & ATMOSPHERIC ADMIN, 75- *Mem:* Am Meteorol Soc; Sigma Xi. *Res:* Synoptic-dynamic aspects of the role of cumulus convection in the development of extratropical cyclones; test and evaluation of numerical prediction models. *Mailing Add:* 13011 Rhame Dr Ft Washington MD 20744

TRACY, C RICHARD, b Glendale, Calif, May 24, 43; div; c 1. ECOLOGY. *Educ:* Calif State Univ, Northridge, BA, 66, MS, 68; Univ Wis, PhD(zool), 72. *Prof Exp:* Res assoc environ studies, Univ Wis, 72-73, lectr, 73-74, asst scientist, 73-74; from asst prof to assoc prof, 74-83, PROF ZOOL, COLO

STATE UNIV, 83- *Concurrent Pos:* Prof biol, Univ Mich Biol Sta, 74-79; nat adv comt person, Univ Wis, Biotron, 74-; ed, Ecol & Ecological Monogr, 78-; Guggenheim fel, 80-81; vis distinguished prof, Pepperdine Univ, 86. *Mem:* AAAS; Am Soc Naturalists; Ecol Soc Am; Sigma Xi; Am Soc Ichthyologists & Herpetologists; Herpetologist League; Am Soc Zoologists. *Res:* Biophysical ecological method to study evolutionary ecological questions of adaptations to physical environments; dispersal; dispersion; habitat selection and space utilization in animals. *Mailing Add:* Dept Zool Colo State Univ Ft Collins CO 80523

TRACY, CRAIG ARNOLD, b London, Eng, Sept 9, 45; US citizen; m 70; c 2. STATISTICAL MECHANICS. *Educ:* Univ Mo, BS, 67; State Univ NY, Stony Brook, PhD(physics), 73. *Prof Exp:* Woodrow Wilson fel, 67; res assoc physics, Univ Rochester, 73-75 & Inst Theoret Physics, State Univ NY, Stony Brook, 75-78; from asst prof to assoc prof math, Dartmouth Col, 78-84; PROF MATH, UNIV CALIF, DAVIS, 84- *Mem:* Am Math Soc; Soc Indust & Appl Math. *Res:* Exactly solvable lattice models in statistical mechanics; theory of phase transitions; theory of completely integrable systems. *Mailing Add:* Dept Math Univ Calif Davis CA 95616

TRACY, DAVID J, b Covington, Ky, Jan 22, 37. ORGANIC CHEMISTRY. *Educ:* Villa Madonna Col, AB, 59; Univ Ill, MS, 61, PhD(chem), 64. *Prof Exp:* Res specialist, GAF Corp, 64-69, tech assoc, 69-84, group leader, 84-87, sect mgr surfactants, 87-90, DIR SYNTHESIS RES, GAF CORP, 90- *Mem:* Am Chem Soc. *Res:* Acetylenics; photographic couplers; surfactants. *Mailing Add:* 37-03 Ravens Crest Rd Plainsboro NJ 08536

TRACY, DERRICK SHANNON, b Mirzapur, India, July 1, 33; Can citizen; m 69. MATHEMATICAL STATISTICS, MULTIVARIATE STATISTICAL ANALYSIS. *Educ:* Univ Lucknow, BSc, 51, MSc, 53; Univ Mich, MS, 60, ScD(math), 63. *Prof Exp:* Lectr math, Ewing Col, Allahabad, 53-55; sr lectr math statist, Govt Col Bhopal, 56-59; res asst math, Univ Mich, 61-63; asst prof statist, Univ Conn, 63-65; assoc prof, 65-70, PROF MATH & STATIST, UNIV WINDSOR, 70- *Concurrent Pos:* Nat Res Coun Can res grants, 66-; Defence Res Bd Can res grants, 68-74; consult, Bell Tel Co Can & Ont Inst Studies Educ, 67-; consult, Walter Reed Army Res Inst, 68; vis prof, Univ Calif, Riverside, 72-73, Univ Waterloo, 74; assoc ed Can J Statist, 73-77; Can Coun travel grant to Poland, 75 & India, 77; vis prof, Univ Fed do Rio de Janeiro, 76 & 78; Ministry External Affairs travel grant, Brazil & India, 76; Nat Res Coun exchange scientist, Brazil, 78; vis prof, Univ Amsterdam, 79, Indian Statist Inst, 80; consult, UN Food & Agr Orgn, Indian Agr Statist Inst, 87; vis prof, Tex A&M Univ, 89. *Mem:* Am Statist Asn; Sigma Xi; Statist Soc Can; Math Asn Am; Int Statist Inst; Statist Soc Australia. *Res:* Products of generalized k-statistics; finite moment formulae; symmetric functions; finite sampling; matrix derivatives in multivariate analysis, multiresponse permutation procedures. *Mailing Add:* Dept Math & Statist Univ Windsor Windsor ON N9B 3P4 Can

TRACY, JAMES FRUEH, b Isle of Pines, July 29, 16; US citizen; m 57; c 2. NUCLEAR PHYSICS. *Educ:* Univ Ill, BS, 40; Univ Calif, MS, 49, PhD(physics), 53. *Prof Exp:* Elec engr, Gen Elec Co, NY, 40-46; opers analyst, Broadview Res & Develop, Calif, 53; physicist, Lawrence Livermore Lab, Univ Calif, 53-83; RETIRED. *Mem:* Am Phys Soc; AAAS. *Res:* Physics design of nuclear weapons; application of nuclear explosives to industry and science. *Mailing Add:* 1262 Madison Ave Livermore CA 94550

TRACY, JOSEPH CHARLES, b Wilkes Barre, Pa, Jan 15, 43; m 66; c 3. SOLID STATE PHYSICS, SURFACE PHYSICS. *Educ:* Rensselaer Polytech Inst, BEE, 64; Cornell Univ, PhD(appl physics), 68. *Prof Exp:* Fel surface physics, NAm Rockwell Sci Ctr, Calif, 68-69, mem tech staff, 69-70; mem tech staff, Bell Tel Labs, Murray Hill, 70-73; group leader, Gen Motors Res Lab, 73-75, asst dept head, Dept Physics, 75-82; chief engr, Intergrated Circuits, 85-87, DEPT HEAD, ELECTRONICS DEPT & CHIEF ENGR, AUDIO SYSTS, DELCO ELECTRONICS, 87- *Concurrent Pos:* Chmn, Nat Acad Sci Eval Panel Nat Bur Standards Ctr Thermodyn & Molecular Sci, 79-80, Eval Panel Ctr Chem Physics, Nat Bur Standards, 80-82; panel mem, Nat Acad Sci Eval Panel, Nat Bur Standards Nat Measurement Lab, 79-82. *Mem:* AAAS; Soc Automotive Engrs; fel Am Phys Soc; Am Vacuum Soc; Sigma Xi; Inst Elec & Electronics Engrs. *Res:* Surface physics; electron spectroscopy; semiconductor physics; management of research. *Mailing Add:* R225 Delca Electronics Corp Kokomo ID 46904-0055

TRACY, JOSEPH WALTER, b Seattle, Wash, June 22, 24; m 50; c 5. INORGANIC CHEMISTRY, MEAT SCIENCE. *Educ:* Univ Wash, BS, 51, MS, 54, PhD(chem), 60. *Prof Exp:* Assoc res engr, Boeing Airplane Co, 54-58; prof chem, Northwest Nazarene Col, 60-70; chemist-in-chg, Meat Inspection Lab, Idaho State Dept Agr, 70-81; RETIRED. *Mem:* Am Chem Soc. *Res:* Meat analysis; x-ray crystallography; crystal structures of simple inorganic compounds. *Mailing Add:* 823 Ninth Ave S Nampa ID 83651

TRACY, RICHARD E, b Klamath Falls, Ore, Apr 30, 34; m 62; c 2. PATHOLOGY. *Educ:* Univ Chicago, BA, 55, MD & PhD(path), 61. *Prof Exp:* Intern, Presby Hosp, Denver, Colo, 61-62; res assoc path, Univ Chicago, 62-64, instr, 64-65; asst prof, Med Sch, Univ Ore, 65-67; asst prof, 67-73, assoc prof, 73-81, PROF PATH, SCH MED, LA STATE UNIV, NEW ORLEANS, 81- *Concurrent Pos:* USPHS trainee, 62-65. *Honors & Awards:* Bausch & Lomb Medal, 61; Joseph A Capps Prize, 65. *Mem:* AAAS; Am Heart Asn; Am Soc Exp Path. *Res:* Arteriosclerotic and hypertensive cardiovascular and renal disease; biostatistics. *Mailing Add:* Dept Path La State Univ Sch Med New Orleans LA 70112-1393

TRACY, WILLIAM E, b Memphis, Tenn, Aug 16, 34; m 56; c 3. PEDODONTICS. *Educ:* Univ Tenn, DDS, 61; Am Bd Pedodont, dipl, 71. *Prof Exp:* Res asst dent mat, Dent Sch, 60-61, instr pediat, Med Sch & instr dent, Dent Sch, 61-62, asst prof dent, 62-68, ASSOC PROF DENT, DENT SCH, UNIV ORE, 68-, INSTR PEDIAT & DENT MED, MED SCH, 67-, MEM STAFF, CHILD STUDY CLIN, 71- *Concurrent Pos:* Nat Inst Dent Res fel growth & develop, Dent Sch, Univ Ore, 61-62, spec fel, 65-; Nat Inst Child Health & Human Develop career develop award, 66-67, res grant, 66-71. *Mem:* AAAS; Am Acad Pedodont; Am Soc Dent Children; Am Dent Asn; Int Asn Dent Res. *Res:* Growth and development of children; dentofacial growth. *Mailing Add:* 12909 SW Timera Tigard OR 97224

TRACY, WILLIAM FRANCIS, b Cambridge, Mass, Oct 11, 54; m 77; c 1. PLANT BREEDING. *Educ:* Univ Mass, Amherst, BS, 76, MS, 79; Cornell Univ, PhD(plant breeding), 82. *Prof Exp:* Res scientist, Int Plant Res Inst, 82-83; plant breeder, Cargill Inc, 83-84; ASST PROF AGRON, DEPT AGRON, UNIV WIS-MADISON, 84- *Mem:* Am Soc Agron; Crop Sci Soc; AAAS; Am Genetics Asn; Nat Sweet Corn Breeders Asn. *Res:* Sweet corn breeding and genetics; use of non-sweet corn germplasm in the improvement of quality, pest resistance and agronomic character. *Mailing Add:* Univ Wis 1575 Linden Dr Madison WI 53706

TRACZ, WILL, b Salamanca, NY, Sept 8, 50. COMPUTER SCIENCE. *Educ:* State Univ NY, BA, 72; Pa State Univ, MS, 74; Syracuse Univ, MS, 79. *Prof Exp:* ADVAN PROGRAMMER, FED SECTOR DIV, IBM, 81- *Concurrent Pos:* Chmn, Int Tech Liaison Microprocessors, IBM Corp, 83-85 & Microprog Comt, Asn Comput Mach, 87-; ed, Inst Elec & Electronics Engrs Computer Mag, 91- *Mem:* Asn Comput Mach; Inst Elec & Electronics Engrs. *Mailing Add:* Int Bus Mach Corp MD 0210 Owego NY 13827

TRAEXLER, JOHN F, b Brooklyn, NY, Feb 21, 30; m 55; c 4. MECHANICAL ENGINEERING, ELECTRICAL ENGINEERING. *Educ:* Pratt Inst Technol, BME, 51; Univ Pa, MSME, 54, PhD(eng mech), 65. *Prof Exp:* Design engr, Westinghouse Elec Corp, 51-57, sr engr, 57-62, fel engr, 62-67, mgr mech develop, 67-70, mgr technol develop, 70-77, adv engr, 77-78, mgr turbine develop, 78-80, mgr steam turbine generator eng, 80-88; RETIRED. *Concurrent Pos:* Mem, Pressure Vessel Res Comt, 59-79. *Mem:* Am Soc Mech Engrs. *Res:* Application of plasticity and creep to turbomachinery mechanical design; dynamics and vibrations problems in turbomachinery. *Mailing Add:* 319 Santiago Dr Winter Park FL 32789

TRAFTON, LAURENCE MUNRO, b Boston, Mass, July 31, 38. PLANETARY ATMOSPHERES. *Educ:* Calif Inst Technol, BS, 60, MS, 61, PhD(astron & physics), 65. *Prof Exp:* Proj officer, Weapons Lab, US Air Force, 65-68, proj scientist, 68-69; spec res assoc, 69-72, RES SCIENTIST, UNIV TEX, AUSTIN, 72- *Concurrent Pos:* Team mem, Instrument Develop Team High Resolution Spectrog Space Telescope, 77; assoc ed, Icarus, 79- *Mem:* Int Astron Union; Am Astron Soc; fel AAAS. *Res:* State, structure, composition, dynamics and energy balance in the atmospheres of the major planets and satellites; spectroscopic observations from infrared to ultraviolet; theoretica investigations; investigations from spacecraft, including space telescope. *Mailing Add:* Astron Dept Univ Tex Austin TX 78712

TRAGER, WILLIAM, b Newark, NJ, Mar 20, 10; m 35; c 3. PARASITOLOGY. *Educ:* Rutgers Univ, BS, 30; Harvard Univ, AM, 31; PhD(biol), 33. *Hon Degrees:* ScD, Rutgers Univ, 65, Rockefeller Univ, 87. *Prof Exp:* Nat Res Coun fel med, 33-34; fel, Rockefeller Inst, 34-35, asst, 35-40, assoc, 40-50, assoc mem, 50-59, from assoc prof to prof, 59-80, EMER PROF PARASITOL, ROCKEFELLER UNIV, 80- *Concurrent Pos:* Ed J, Soc Protozool, 53-65; mem study sect parasitol & trop med, Nat Inst Allergy & Infectious Dis, 54-58 & 66-70, microbiol & infectious dis adv comt, 78-80; mem training grant comt, 61-64; mem malaria comn, Armed Forces Epidemiol Bd, 65-70; guest investr, W African Inst Trypanosomiasis Res, 58-59; vis prof, Fla State Univ, 62, Med Sch, Puerto Rico, 63 & Med Sch, Nat Univ Mex, 65; Guggenheim found fel, 73; mem malaria immunol steering comt, WHO; Avivah Zuckerman fel, Kuvin Ctr Infectious & Tropical Dis, Hebrew Univ, Jerusalem, 82; hon pres, Asia & Pac Conf Malaria, 85. *Honors & Awards:* S T Darling Medal & Prize, World Health Orgn, 80; Leuckart Medal, Ger Soc Parasitol, 82; Manson Medal, Royal Soc Trop Med & Hyg, 86; Rameshwardas Birla International Award in Tropical Medicine, 82. *Mem:* Nat Acad Sci; Am Soc Parasitol (pres, 74); Soc Protozool (pres, 60-61); Am Soc Trop Med & Hyg (vpres, 64-65, pres, 78-79); NY Acad Sci. *Res:* Insect physiology; physiology of parasitisms; cultivation of intracellular parasites; malaria. *Mailing Add:* Rockefeller Univ York Ave & 66th St New York NY 10021

TRAGER, WILLIAM FRANK, b Winnipeg, Man, Oct 17, 37; m 60; c 3. PHARMACEUTICAL CHEMISTRY, ORGANIC CHEMISTRY. *Educ:* Univ San Francisco, BSc, 60; Univ Wash, PhD(pharmaceut chem), 65. *Prof Exp:* NIH fel pharm, Chelsea Col Sci & Technol, London, 65-67; fel, Univ Wash, 67; asst prof chem & pharmaceut chem, Sch Pharm, Univ San Francisco, 67-72; assoc prof, 72-77, PROF PHARMACEUT CHEM, SCH PHARM, UNIV WASH, 77- *Mem:* Am Chem Soc. *Res:* Drug metabolism studies and mass spectroscopy. *Mailing Add:* Dept Med Chem & Pharmaceut BG20 Univ Wash Seattle WA 98195-0001

TRAHAN, DONALD HERBERT, b North Adams, Mass, Mar 14, 30; m 61, 76; c 1. MATHEMATICS. *Educ:* Univ Vt, BS, 52; Univ Nebr, MA, 54; Univ Pittsburgh, PhD, 61. *Prof Exp:* Instr math, Univ Mass, 56-59; asst prof, Univ Pittsburgh, 61-65; asst prof & chmn dept, Chatham Col, 65-66; ASSOC PROF MATH, NAVAL POSTGRAD SCH, 66- *Concurrent Pos:* Hays-Fulbright lectr, Nat Univ Ireland, 63-64. *Mem:* Am Math Soc; Math Asn Am. *Res:* Complex variables; univalent function theory; real analysis. *Mailing Add:* Dept Math Naval Postgrad Sch Code 0223 Monterey CA 93940

TRAHAN, RUSSELL EDWARD, JR, b New Orleans, La, Jan 30, 49; m 70; c 2. CONTROL SYSTEMS, OPTIMIZATION. *Educ:* Univ New Orleans, BS, 70, MS, 73; Univ Calif, Berkeley, PhD(elec eng), 77. *Prof Exp:* Design engr, Tano Corp, 71-73; PROF ELEC ENG, UNIV NEW ORLEANS, 77- *Concurrent Pos:* Vis prof, Univ Calif, Berkeley, 80; consult, Litton Data Systs, 80-86, Naval Oceanog & Atmospheric Lab, 86- *Mem:* Inst Elec & Electronics Engrs; Inst Elec & Electronics Engrs Control Systs Soc. *Res:* Optimization applied to speech processing; optimal identification of control systems; optimization applied to ocean data acquisition systems. *Mailing Add:* 20501 Alba E New Orleans LA 70129

TRAHANOVSKY, WALTER SAMUEL, b Conemaugh, Pa, June 15, 38; m 67; c 3. ORGANIC CHEMISTRY. *Educ:* Franklin & Marshall Col, BS, 60; Mass Inst Technol, PhD(chem), 63. *Prof Exp:* NSF fel chem, Harvard Univ, 63-64; from instr to assoc prof, 64-74, PROF CHEM, IOWA STATE UNIV, 74- *Concurrent Pos:* A P Sloan fel, 70-72. *Mem:* AAAS; Am Chem Soc; Royal Soc Chem. *Res:* Physical-organic chemistry, including the study of oxidations and reductions of organic compounds; oxidative cleavages; cyclization reactions; free radicals; carbocations; arene tricarbonylchromium complexes; flash vacuum pyrolysis; methylenecyclobutenones; tropolone derivatives. *Mailing Add:* Dept Chem A202 Gilman Iowa State Univ Ames IA 50011

TRAIL, CARROLL C, b Forney, Tex, Dec 25, 27; m 51; c 3. NUCLEAR PHYSICS. *Educ:* Agr & Mech Col Tex, BS, 49, MS, 51, PhD, 56. *Prof Exp:* Asst physics, Argonne Nat Lab, 56-60, assoc physicist, 60-64; assoc prof, 64-68, chmn dept, 69-76, PROF PHYSICS, BROOKLYN COL, 68- *Concurrent Pos:* Vis prof, Linear Accelerator Lab, Orsay, France, 70-71; vis scientist, Mass Inst Technol, 77-78; fel, Sci Fac Imp Prog, NSF, 77-78. *Mem:* Fel Am Phys Soc. *Res:* Low energy; nuclear experimentation. *Mailing Add:* Dept Physics Brooklyn Col Bedford Ave & Ave H Brooklyn NY 11210

TRAIN, CARL T, b Lindsborg, Kans, Jan 19, 39; m 60; c 3. PARASITOLOGY, INVERTEBRATE ZOOLOGY. *Educ:* Bethany Col, Kans, BS, 61; Kans State Univ, MS, 63, PhD(parasitol), 67. *Prof Exp:* Asst zool, Kans State Univ, 61-64, instr, 66-67; from asst prof to assoc prof, 67-79, PROF BIOL, SOUTHEAST MO STATE UNIV, 79-, CHMN DEPT, 84- *Mem:* Am Soc Parasitol; Am Micros Soc; Sigma Xi. *Res:* General parasitology, biology and reproduction of nematodes; life cycles of parasites; parasites of aquatic birds. *Mailing Add:* Dept Biol Southeast Mo State Univ Cape Girardeau MO 63701

TRAINA, PAUL J(OSEPH), b New York, NY, Mar 8, 34; m 55; c 5. CIVIL ENGINEERING. *Educ:* Manhattan Col, BCE, 55; Univ Mich, MS, 60. *Prof Exp:* Chief water resources, Southeast Region, USPHS, 60-64; asst dir comprehensive planning, Fed Water Pollution Control Admin, 64-67, dir tech progs, Southeast Region, 67-71, dir off water progs, 71-73, dir enforcement, 73-79, dir, water div, Region IV, Environ Protection Agency, 79-85; ENVIRON CONSULT, 85- *Mem:* Am Soc Civil Engrs; Water Pollution Control Fedn; NY Acad Sci; Am Acad Environ Engrs. *Res:* Water supply and water quality control; regional water programs; hazardous waste management. *Mailing Add:* 2366 Wood Creek Tucker GA 30084

TRAINA, VINCENT MICHAEL, b Oceanside, NY, May 8, 43; c 2. TOXICOLOGY. *Educ:* Rutgers Univ, BA, 65, MS, 70, PhD(physiol), 73; Am Bd Toxicol, dipl, 81. *Prof Exp:* Res investr toxicol, Squibb Inst Med Res, 66-74; mgr toxicol, 74-79, assoc dir,toxicol & path, 79-82, dir toxicol, 82-84, EXEC DIR TOXICOL & PATH, CIBA-GEIGY CORP, 84- *Concurrent Pos:* Industrial relations comt, Ciba-Geigy. *Honors & Awards:* Prinomide Phase III Award. *Mem:* Sigma Xi; Environ Mutagen Soc; Am Inst Biol Sci; Am Col Toxicol; Soc Toxicol; Soc Comp Ophthal. *Res:* Body fluid volumes and concentrations and electrolyte changes during periods of prolonged starvation; all phases of toxicology. *Mailing Add:* Ciba-Geigy Corp Toxicol-Pathol Div Morris Ave Summit NJ 07901

TRAINER, DANIEL OLNEY, b Chicago, Ill, July 13, 26; m 55; c 2. WILDLIFE DISEASES. *Educ:* Ripon Col, BS, 50; Univ Wis-Madison, MS, 55, PhD(bact), 61. *Prof Exp:* Res virologist, Fromm Labs, 55-56; pathologist, Wis Conserv Dept, 56-62; from asst prof to prof vet sci, Univ Wis-Madison, 62-71; dean, Col Natural Resources, 71-80, VCHANCELLOR, UNIV WIS-STEVENS POINT, 80- *Mem:* AAAS; Wildlife Dis Asn (vpres, 66-68, pres, 68-70); Wildlife Soc; Am Inst Biol Sci; Soc Am Foresters. *Res:* Ecology of disease, especially diseases of wild or natural populations. *Mailing Add:* Col Natural Resources Univ Wis Stevens Point WI 54481

TRAINER, DAVID GIBSON, b Allentown, Pa, Mar 11, 45; m 74; c 2. PHYSIOLOGY. *Educ:* Washington & Jefferson Col, AB, 67; Univ Maine, MSc, 69; Univ NH, PhD(zool), 75. *Prof Exp:* From asst prof to assoc prof, 75-82, PROF BIOL, E STROUDSBURG UNIV, 78-, DEPT CHAIRPERSON, 90- *Mem:* AAAS; Am Soc Zoologists; Sigma Xi. *Res:* Physiology of invertebrates. *Mailing Add:* Dept Biol E Stroudsburg State Col East Stroudsburg PA 18301

TRAINER, FRANK W, b Manchester, Eng, Dec 2, 21; m; c 2. GEOLOGY. *Educ:* Univ Va, BA, 43; Harvard Univ, PhD(geol), 54. *Prof Exp:* Geologist, US Geol Soc, 48-81; RETIRED. *Mem:* Asn Groundwater Scientists & Engrs; fel Geol Soc Am. *Mailing Add:* PO Box 1735 Corrales NM 87048

TRAINER, JOHN EZRA, JR, b Allentown, Pa, Aug 31, 43; m 67; c 3. HELMINTHOLOGY, BIOLOGICAL CHEMISTRY. *Educ:* Muhlenburg Col, BS, 65; Wake Forest Univ, MA, 67; Univ Okla, PhD(zool), 71. *Prof Exp:* Teaching asst, Wake Forest Univ, 65-67; teaching asst, Univ Okla, 67-69; asst prof, 71-78, ASSOC PROF BIOL, JACKSONVILLE UNIV, 78-, VPRES & DEAN FAC, 81- *Concurrent Pos:* Teaching asst biol sta, Mich State Univ, 66 & Univ Okla, 67; consult, Environ Ctr. *Mem:* Am Soc Parasitol; Am Soc Zool; Am Inst Biol Sci; AAAS. *Res:* Ecology, biochemistry and ultrastructure of the Pentastomida and other parasitic helminths. *Mailing Add:* President Lenoir Phyne Col Hickory NC 28603

TRAINER, JOHN EZRA, SR, b Allentown, Pa, Feb 8, 14; m 39; c 2. ORNITHOLOGY. *Educ:* Muhlenberg Col, BS, 35; MS, Cornell Univ, 38, PhD(ornith), 46. *Prof Exp:* Teacher biol, Tenn State Teachers Col, 38-39; from instr to sr prof, 39-80, EMER PROF BIOL, MUHLENBERG COL, 80- *Mem:* Wilson Ornith Soc; Am Ornith Union. *Res:* Hearing ability of birds; auditory acuity of certain birds; respiration of birds; vertebrate morphology; led three trips to East Africa. *Mailing Add:* HCR Box 165 Lakeville PA 18438

TRAINER, MICHAEL NORMAN, b Lehighton, Pa, Jan 26, 52; m 72. LIGHT SCATTERING, FIBER OPTICS. *Educ:* Lehigh Univ, BS, 73, MS, 91; Univ Rochester, MS, 91. *Prof Exp:* Physicist, NJ Zinc Co, 73-75; PRIN SCIENTIST, LEEDS & NORTHRUP CO, 76- *Mem:* Optical Soc Am; Soc Photo-Optical Instrumentation Engrs. *Res:* Fiber optic sensors; optical data storage; optical instrumentation for size analysis of small particles; interferometry; signal processing; optical and electrical engineering. *Mailing Add:* 186 Fretz Rd Telford PA 18969

TRAINOR, FRANCIS RICE, b Pawtucket, RI, Feb 11, 29; m 56. PHYCOLOGY. *Educ:* Providence Col, BS, 50; Vanderbilt Univ, MA, 53, PhD(biol), 57. *Prof Exp:* From instr to assoc prof, 57-67, PROF BOT, UNIV CONN, 67- *Concurrent Pos:* Consult, Elec Boat Div, Gen Dynamics Corp, Conn, 58-61; Fulbright res scholar, Stockholm, 70; Fulbright lectr, Greece & Yugoslavia, 71. *Honors & Awards:* Distinguished Fac Award, Univ Conn, 62; Darbaker Award, Bot Soc Am, 65. *Mem:* AAAS; Phycol Soc Am (vpres, 68, pres, 69); Brit Phycol Soc; Int Phycol Soc; Am Inst Biol Sci; Sigma Xi. *Res:* Sexual reproduction in unicellular algae; algal nutrition and morphogenesis; eutrophication. *Mailing Add:* Biol Sci U-42 Univ Conn Storrs CT 06268

TRAINOR, GEORGE L, b Staten Island, NY, Aug 19, 52; m 75; c 1. BIO-ORGANIC CHEMISTRY. *Educ:* Stevens Inst Technol, BS, 74; Harvard Univ, MA & PhD(chem), 79. *Prof Exp:* Res asst org chem, Stevens Inst Technol, 71-74 & Harvard Univ, 75-79; res assoc, Columbia Univ, 79-81; chemist org chem, 81-85, proj leader, 85-87, RES LEADER, CENT RES & DEVELOP, E I DU PONT DE NEMOURS & CO, INC, 87- *Concurrent Pos:* Teaching asst org chem, Stevens Inst Technol, 73-74, Harvard Univ, 74-77. *Mem:* Sigma Xi; Am Chem Soc. *Res:* Organic chemistry of biologically relevant processes. *Mailing Add:* Nine Carillon Ct Wilmington DE 19803

TRAINOR, JAMES H, b Lancaster, NH, Aug 22, 35. NUCLEAR RADIATION DAMAGE, COMPUTER SYSTEMS. *Educ:* Univ NH, BC, 58, MS, 59, PhD(physics), 64. *Prof Exp:* Sr physicist & div chief cosmic ray physics, 64-83, dep dir sci, 83-86, DIR SPACE & EARTH SCI, GODDARD SPACE FLIGHT CTR, NASA, 86- *Mem:* Am Geophys Union; fel Inst Elec & Electronics Engrs. *Mailing Add:* 2803 Lakehurst Ave District Heights MD 20747

TRAINOR, LYNNE E H, b Chamberlain, Sask, Dec 4, 21; div; c 3. THEORETICAL PHYSICS. *Educ:* Univ Sask, BA, 46, MA, 47; Univ Minn, PhD(physics), 51. *Prof Exp:* Fel, Nat Res Coun Can, 51-52; asst prof physics, Queen's Univ, Ont, 52-55; vis prof, Univ BC, 55-56; from asst prof to prof, Univ Alta, 56-63; PROF PHYSICS, UNIV TORONTO, 63- *Concurrent Pos:* Chmn, North York Bd Educ, 70-72. *Mem:* Am Phys Soc; Can Asn Physicists (secy, 66-68, hon secy-treas, 68-70); Biophys Soc; Can Soc Clin Ecol. *Res:* Statistical mechanics of Bose-Einstein systems; properties of thin helium films; field approach to structuralism in theoretical biology; theoretical biology and biophysics; pattern formation and morphology of developmental systems; structure and function of DNA; transmembrane transport; neurophysics; brain theory. *Mailing Add:* Dept Physics Univ Toronto Toronto ON M5S 1A7 Can

TRAINOR, ROBERT JAMES, b Bell, Calif, June 15, 44. PHYSICS. *Educ:* Calif State Polytech Univ, BS, 66; Univ Calif, Riverside, MS, 69, PhD(physics), 74. *Prof Exp:* Res assoc physics, Argonne Nat Lab, 74-76; physicist, Lawrence Livermore Nat Lab, 76-82; GROUP LEADER, HIGH ENERGY-DENSITY PHYSICS GROUP, LOS ALAMOS NAT LAB, 83- *Mem:* Am Phys Soc. *Res:* Experimental condensed matter physics; properties of matter at extreme high pressures and temperatures; laser-matter interactions; properties of hot, dense plasmas. *Mailing Add:* MS E526 Los Alamos Nat Lab Los Alamos NM 87544

TRAISMAN, HOWARD SEVIN, b Chicago, Ill, Mar 18, 23; m 56; c 3. PEDIATRICS. *Educ:* Northwestern Univ, BS, 43, BM, 46, MD, 47. *Prof Exp:* Intern, Cook County Hosp, Chicago, 46-47; resident, Children's Mem Hosp, 49-51; from instr to assoc, 52-57, from asst prof to assoc prof, 62-73, PROF PEDIAT, MED SCH, NORTHWESTERN UNIV, CHICAGO, 73- *Concurrent Pos:* Attend pediatrician, Children's Mem Hosp, 52- & Northwestern Mem Hosp, 67- *Mem:* Am Acad Pediat; Am Pediat Soc; Endocrine Soc; Lawson Wilkins Pediat Endocrine Soc; Am Diabetes Asn. *Res:* Juvenile diabetes mellitus. *Mailing Add:* 1325 W Howard St Evanston IL 60202

TRAITOR, CHARLES EUGENE, b West Frankfort, Ill, Jan 28, 34; m 57; c 2. PHARMACOLOGY, TOXICOLOGY. *Educ:* St Louis Col Pharm, BS, 60; Purdue Univ, MS, 63, PhD(pharmacol), 65. *Prof Exp:* RES TOXICOLOGIST, MED RES DIV, AM CYANAMID CO, 65- *Mem:* Am Col Toxicol; NY Acad Sci; Sigma Xi. *Res:* Toxicology of drugs on various organ systems. *Mailing Add:* Toxicol Evaluation Dept Am Cyanamid Co Lederle Labs Med Res Div Pearl River NY 10965

TRAJMAR, SANDOR, b Bogacs, Hungary, Sept 7, 31; US citizen; m 57; c 1. ATOMIC & MOLECULAR PHYSICS. *Educ:* Debrecen Univ, dipl, 55; Univ Calif, Berkeley, PhD(phys chem), 61. *Prof Exp:* Chemist, N Hungarian Chem Works, 55-57; chemist, Stauffer Chem Co, 57-58; mem technol staff, Calif Inst Tech, 61-80, head, Electron Col Physics Group, 70-89, mgr planetary atmospheric sect, 83-85, SR RES SCIENTIST, JET PROPULSION LAB, CALIF INST TECHNOL, 80- *Concurrent Pos:* Teaching asst, Dept Chem, Univ Calif, 58-59, res asst, Lawrance Radiation Lab, 59-61; res fel, Div Chem & Chem Eng, Calif Inst Technol, 64-66, sr res fel, 69-72; adj prof physics, Univ Southern Calif, 82-84; adj prof physics, Univ Calif, Riverside, 86- *Honors & Awards:* NASA Medal Except Sci Achievement, 73. *Mem:* Am Chem Soc; fel Am Phys Soc. *Res:* High temperature chemistry; molecular spectroscopy; low-energy electron scattering; atomic physics. *Mailing Add:* Jet Propulsion Lab 4800 Oak Grove Dr Pasadena CA 91109

TRAMA, FRANCESCO BIAGIO, b Philadelphia, Pa, Dec 13, 27; m 54. LIMNOLOGY. *Educ:* Temple Univ, AB, 48, MA, 50; Univ Mich, PhD(zool), 57. *Prof Exp:* Asst limnol, Acad Natural Sci Philadelphia, 50-53; asst prof zool, Chicago Teachers Col, 57-60; asst prof, 60-63, ASSOC PROF ZOOL, RUTGERS COL, RUTGERS UNIV, NEW BRUNSWICK, 63-, ASSOC DEAN, 73- *Concurrent Pos:* Res assoc, Great Lakes Res Inst, Univ Mich, 57-61. *Mem:* Fel AAAS; Am Soc Limnol & Oceanog; Ecol Soc Am; Sigma Xi. *Res:* Trophic dynamics and energy transfer in aquatic ecosystem; primary productivity in fresh water. *Mailing Add:* 4490 Plumosa St Spring Hill FL 34607

TRAMBARULO, RALPH, b East Longmeadow, Mass, Jan 24, 25; m 55; c 4. PHYSICS, PHYSICAL CHEMISTRY. *Educ:* Yale Univ, BS, 44, PhD(phys chem), 49. *Prof Exp:* Res assoc, Duke Univ, 49-52; asst prof physics, Pa State Col, 52-53 & Univ Del, 53-56; MEM TECH STAFF, BELL TEL LABS, 56- *Mem:* Inst Elec & Electronics Engrs; Sigma Xi. *Res:* Microwave physics and spectroscopy; microwave integrated circuits. *Mailing Add:* 20 Irving Pl Red Bank NJ 07701-1710

TRAMELL, PAUL RICHARD, b El Centro, Calif, Mar 10, 43; m 67; c 1. PHARMACOLOGY, BIOCHEMISTRY. *Educ:* Fresno State Col, BA, 65, MA, 67; Rice Univ, PhD(biochem), 70. *Prof Exp:* Dir instrumentation, Cent Calif Med Labs, 65-67; biochemist, Abbott Labs, 71-72; SR PHARMACOLOGIST, ALZA CORP, 72- *Concurrent Pos:* Nat Inst Dent Res fel, Rice Univ, 70-; NIH fel pharmacol, Med Sch, Stanford Univ, 70-71. *Mem:* AAAS. *Res:* Enzymology; drug metabolism; pharmacokinetics. *Mailing Add:* 2040 Delmar Ave San Marino CA 91108

TRAMMEL, KENNETH, b Skipperville, Ala, Oct 30, 37; m 58; c 4. ENTOMOLOGY. *Educ:* Univ Fla, BS, 60, PhD(entom), 65. *Prof Exp:* Res assoc citrus pest control, Citrus Exp Sta, Univ Fla, 64-65, asst entomologist, 65-67; entomologist, Ciba Agrochem Co, Fla, 67-69; assoc prof entom, NY State Agr Exp Sta, Cornell Univ, 69-75; OWNER & PRES, AGR CHEM DEVELOP SERV, INC, 75- *Mem:* Entom Soc Am; Weed Sci Soc Am. *Res:* Apple and pear pest management; field application of sex pheromones for monitoring and control of pests; pesticides on fruit, vegetables and field crops. *Mailing Add:* One Lester Rd Phelps NY 14532

TRAMMELL, GEORGE THOMAS, b Marshall, Tex, Feb 5, 23; m 45; c 4. THEORETICAL PHYSICS. *Educ:* Rice Inst, BA, 44; Cornell Univ, PhD, 50. *Prof Exp:* Physicist, Oak Ridge Nat Lab, 50-61; PROF PHYSICS, RICE UNIV, 61- *Mem:* Fel Am Phys Soc. *Res:* Solid state theory. *Mailing Add:* Dept Physics Box 1892 Rice Univ Box 1892 Houston TX 77251

TRAMMELL, GROVER J(ACKSON), JR, b Attalla, Ala, July 17, 19; m 45; c 5. MECHANICAL ENGINEERING. *Educ:* Tulane Univ, BS, 49, MS, 50. *Prof Exp:* Instr math, La State Univ, 50-52, asst prof eng mech, 52-57; prof mech eng, La Tech Univ, 57-84, dir continuing educ, 67-84; RETIRED. *Mem:* Am Soc Eng Educ; Am Soc Mech Engrs; Nat Soc Prof Engrs. *Res:* Vibrations; solid mechanics; thermodynamics. *Mailing Add:* 911 James St Ruston LA 71270

TRAMMELL, REX COSTO, b Sevier County, Tenn, Nov 9, 38. NUCLEAR SCIENCE, SEMICONDUCTOR MATERIALS. *Educ:* Univ Tenn, BS, 65, MS, 69. *Prof Exp:* Staff physicist radiation detectors, 65-71, staff physicist semiconductor mat, 71-78, SR SCIENTIST, EG&G ORTEC INC, 78- *Honors & Awards:* EG&G Germeshausen Award, 78. *Mem:* Sr mem Inst Elec & Electronics Engrs. *Res:* Germanium crystal growth technology as applied to germanium gamma-ray spectrometers. *Mailing Add:* EG&G Ortec Inc 100 Midland Rd EG&G Ortec Inc 100 Midland Lane Oak Ridge TN 37830

TRAMONDOZZI, JOHN EDMUND, b Malden, Mass, Aug 28, 42. ORGANIC CHEMISTRY. *Educ:* Boston Col, BS, 64, PhD(chem), 72. *Prof Exp:* From asst prof to assoc prof, 69-83, PROF CHEM, CURRY COL, 83-, CHMN, 87- *Mem:* Am Chem Soc; Sigma Xi. *Res:* Reactions and syntheses of organic sulfur compounds, especially sufinates and sulfones; organic reactions in fused salt media; chemical conservation of artistic objects. *Mailing Add:* Dept Chem Sci Div Curry Col Milton MA 02186

TRAMPUS, ANTHONY, b Cleveland, Ohio, July 22, 27. MATHEMATICS. *Educ:* Case Inst Technol, BS, 51, PhD(math), 57; George Washington Univ, MS, 53. *Prof Exp:* Mathematician, Nat Bur Standards, DC, 51-53, Firestone Tire & Rubber Co, 53-56 & Gen Elec Co, 57-63; staff mathematician, Interstate Electronics Corp, Calif, 63- 70; res mathematician, Univ Dayton Res Inst, 70-72; mathematician consult, Nat Space Technol Labs, Sperry Corp, 72-77; mathematical statistician, 77-80, OPER RES ANALYST, FED ENERGY REGULATORY COMN, DEPT ENERGY, 80- *Mem:* Math Asn Am. *Res:* Function theory in linear algebra; mathematical analysis in science and engineering. *Mailing Add:* 2001 Columbia Pike Apt 214 Arlington VA 22204

TRAN, LONG TRIEU, b Saigon, Vietnam, Oct 10, 56; US citizen; m 88. AUTOMATION, COMPUTER-INTEGRATED MANUFACTURING. *Educ:* Univ Kansas, BS, 76; Mass Inst Technol, MS, 80. *Prof Exp:* Prod engr, Gen Elec, Cleveland, 80-81, adv mfg engr, Louisville, 81-82, qual systs engr, 82-84, qual control engr, 84-86, sr qual info equip engr, 86-89, SR QUAL INDUST ENGR, GEN ELEC, LOUISVILLE, 90- *Concurrent Pos:* Exec adv, Jr Achievement Inc, Louisville, 83-84; pvt consult, 88- *Mem:* NY Acad Sci; Sigma Xi; Am Soc Qual Control; Am Soc Mech Engrs; sr mem Soc Mfg Engrs; AAAS. *Res:* Grinding processes and surface analysis; computer-integrated testing in manufacturing process control; real-time data collection and feedback systems in monitoring plant production and quality; manufacturing planning and management. *Mailing Add:* Gen Elec Appliances 3410 Fountain Dr Louisville KY 40218

TRAN, NANG TRI, b Binh Dinh, Viet Nam, Jan 2, 48; m; c 4. THIN FILM TECHNOLOGY, SOLID STATE DEVICES. *Educ:* Kyushu Inst Technol, BEE, 73, MEE, 75; Univ Osaka, PhD(elec eng), 78. *Prof Exp:* Res scientist, Sharp Electronic Inc, 79-80; sr res scientist, Arco Solar Indust, 80-84; SR RES SPECIALIST & GROUP LEADER, 3M, 85- *Concurrent Pos:* Instr, Japaneses Cult Indust. *Mem:* Japan Soc Appl Physics; Am Vacuum Soc; Inst Elec & Electronics Engrs. *Res:* Zinc sulfide thin film; electroluminescent displays, amorphous silicon semiconductors and devices based on these materials. *Mailing Add:* 3M Co 201-IE-16 St Paul MN 55144

TRANEL, DANIEL T, b Colfax, Wash, Oct 20, 57; m 81; c 2. NEUROPSYCHOLOGY, BEHAVIORAL NEUROLOGY. *Educ:* Univ Notre Dame, BA, 79; Univ Iowa, MA, 81, PhD(clin psychol), 82. *Prof Exp:* Fel neurophsychol, 82-83, assoc, 84-85, asst res scientist behav neurol, 85-86, ASST PROF NEUROL, BEHAV NEUROL, UNIV IOWA HOSP & CLINS, 86- *Mem:* AAAS; Int Neuropsychol Soc; Am Psychol Asn; Sigma Xi; Behav Neurol Soc; Soc Neurosci. *Res:* Brain behavior relationships; agnosia, aphasia, dementia, amnesia and neuroanatomy; autonomic measures to investigate non-conscious brain processes; prosopagnosia and auditory agnosia. *Mailing Add:* Dept Neurol 2155 RCP Univ Iowa Hosp & Clins Iowa City IA 52242

TRANIELLO, JAMES FRANCIS ANTHONY, b Somerville, Mass, Aug 24, 52; m 78; c 2. BEHAVIORAL ECOLOGY, SOCIOBIOLOGY. *Educ:* Boston Univ, AB, 74; Univ Mass, MS, 76; Harvard Univ, PhD(biol), 80. *Prof Exp:* Lectr biol & ethol, 80-81, RES ASSOC ENTOM, HARVARD UNIV, 81-; ASSOC PROF BIOL, BOSTON UNIV, 81- *Concurrent Pos:* Consult & dir, Monadnock Res Inst; NSF Grant recipient, 82-85 & 86-90, Whitehall Found grant recipient, 86-91. *Mem:* Int Union Study Social Insects; AAAS; Animal Behav Soc; Entom Soc Am; Behav Ecol Soc. *Res:* Behavioral ecology and sociobiology of insects, including communication, foraging behavior, defensive behavior and caste evolution. *Mailing Add:* Dept Biol Boston Univ Five Cummington St Boston MA 02215

TRANK, JOHN W, b Minneapolis, Minn, July 24, 28; m 52; c 3. PHYSIOLOGY, ELECTRICAL ENGINEERING. *Educ:* Univ Minn, BEE, 51, MS, 56, PhD(physiol), 61. *Prof Exp:* From instr biophys to instr physiol, Univ Minn, 54-61; lectr, McGill Univ, 61-63, asst prof, 63-64; from asst prof to assoc prof, Med Ctr, 64-82, PROF PHYSIOL, UNIV KANS MED SCH, 82- *Concurrent Pos:* Dep dir univ surg clin, Montreal Gen Hosp, 61-64. *Mem:* Am Physiol Soc; Inst Elec & Electronics Engrs; Biophys Soc. *Res:* Engineering analysis of cardiovascular control and instrumentation for biological research; bioengineering; mechanics of muscle contraction. *Mailing Add:* Dept Physiol Univ Kans Med Ctr 39th & Rainbow Kansas City KS 66103

TRANNER, FRANK, b Chelsea, Mass, May 22, 22; m 49; c 4. PHARMACY, CHEMISTRY. *Educ:* Mass Col Pharm, BS, 43. *Prof Exp:* Develop & control chemist, Hat Corp Am, 44-46; develop chemist, Remington Rand Inc, 47-54 & Rilling-Dermetics Inc, 54-56; asst chief chemist, Germaine Monteil Cosmetiques, 57-59; develop chemist & sect head, Chesebrough-Pond's Inc, 60-69, res mgr, 69-72, dir tech serv, 72-76, dir appl technol, 76-84; CONSULT, COSMETICS & TOILETRIES, 84- *Mem:* Fel Am Inst Chemists; Am Chem Soc; Am Pharmaceut Asn; Soc Cosmetic Chem; NY Acad Sci. *Res:* Cosmetic, food, toiletry and pharmaceutical products and processes. *Mailing Add:* 23 Beech Tree Circle Trumbull CT 06611

TRANQUADA, ROBERT ERNEST, b Los Angeles, Calif, Aug 27, 30; m 51; c 3. INTERNAL MEDICINE. *Educ:* Pomona Col, BA, 51; Stanford Univ, MD, 55; Am Bd Internal Med, cert, 63. *Hon Degrees:* DSc, Worcester Polytech Inst, 85. *Prof Exp:* Intern in med, Univ Calif Los Angeles, 55-57, asst resident, 56-57; intermediate resident med, Los Angeles VA Hosp, 57-58; from instr to assoc prof med, Univ Southern Calif Sch Med, Dept Med, 59-68, chmn, Dept Community Med & Pub Health, 66-69, prof community med & pub health, 68-75, assoc dean, 69-75; assoc dean regional & postgrad, Med Educ, Univ Calif Los Angeles, Sch Med, 76-79, prof med, 76-79, prof pub health, 76-79, vchmn, Dept Med, 77-79; chancellor & dean, Univ Mass Med Ctr, 79-86, prof med, 79-86; DEAN & PROF MED, UNIV SOUTHERN CALIF SCH MED, 86- *Concurrent Pos:* Fel, Univ Calif Los Angeles Med Ctr, USPHS, 58-59, Univ Southern Calif, 59-60; attend physician med, Huntington Mem Hosp, Pasadena, Calif, 65-77; consult med, White Mem Hosp, Los Angeles, Calif, 65-77; assoc proj dir, SCentral Multipurpose Health Serv Ctr, 66-68; from attend physician to hon sr attend physician, Los Angeles County-Univ Southern Calif Med Ctr, 69-76; mem bd dirs, Charles R Drew Univ Med & Sci, 66-75 & 86-; mem, Prev Med & Dent Rev Comt, Bureau of Health Manpower Educ, Div Allied Health Manpower, Prog Asst Br, NIH, Bethesda, Md, 70-74; mem bd dir, Nat Med Fel, Inc, 71- chmn, 80-86; fac med teacher's training course, Paradeniya, Sri Lanka, 70; consult, Fourth World Conf Med Educ, World Med Asn, Inc, 72; nat adv, Pub Health Training Coun, 75-77, Comt Vet Med, Univ Minn, 75; consult, Robert Wood Johnson Found, 75-78; attend internist, Harbor Gen Hosp, Torrance, Calif, 76-79, UCLA Hosp & Clin, 76-79; treas bd gov, Health Systs Agency, Los Angeles County, Inc, 77-79; hon staff, Huntington Mem Hosp, Pasadena, Calif, 77-79; attend physician, Olive View Med Ctr, 77-79; active staff, Univ Mass Med Ctr, 79-86; mem rev panel, Div Med, Family Med, Bureau of Health Prof, 80-83; chmn admin bd, Counc Deans, Asn Am Med Col, 90-91; rev med educ, Inst Med, Nat Acad Sci, 82-83; mem, Mass Statewide Health Coord Coun, 82-83; corporator, Greater Worcester Community Found, Inc, 82-86. *Mem:* Inst Med Nat Acad Sci; NY Acad Sci; Am Diabetes Asn; Am Pub Health Asn; NY Acad Sci; fel AAAS; fel Am Antiquarian Soc; Asn Teachers Prev Med; Sigma Xi. *Res:* Fibrosis in scleroderma; phenethylbiquamide on human hepatic metabolism as measured; effect of insulin on lactic acid production by rat adipose tissue; author of forty publications; spontaneous hypoglycemia; medical care research. *Mailing Add:* Univ Southern Calif Sch Med 1975 Zonal Ave KAM 500 Los Angeles CA 90033

TRANSUE, LAURENCE FREDERICK, b Summerfield, Kans, Apr 2, 14; m 58. PHYSICAL CHEMISTRY. *Educ:* Tarkio Col, AB, 36; Univ Nebr, AM, 39, PhD(phys chem), 41. *Prof Exp:* Asst, Univ Nebr, 37-41; res chemist, E I du Pont de Nemours & Co, Inc, 41-50, supvr, 50-52, supt & res mgr, Photoprod Dept, 52-79; RETIRED. *Mem:* Am Chem Soc; Soc Photog Sci & Eng. *Res:* Surface films; chemistry and physics of photography. *Mailing Add:* 110 Wendover Rd Rochester NY 14610

TRANSUE, WILLIAM REAGLE, b Pen Argyl, Pa, Nov 30, 14; m 36; c 3. MATHEMATICAL ANALYSIS. *Educ:* Lafayette Col, BS, 35; Lehigh Univ, MA, 39, PhD(math), 41. *Hon Degrees:* ScD, Kenyon Col, 82. *Prof Exp:* Asst, Inst Advan Study, 42-43; assoc physicist, Ord Dept, US Dept Army, 43-44, physicist, 44-45; assoc prof math, Kenyon Col, 45-48; asst, Inst Advan Study, 48-49; prof math, Kenyon Col, 49-66; prof math, State Univ NY Binghamton, 66-81; RETIRED. *Concurrent Pos:* Fulbright scholar, Italy, 51-52; NSF fac fel, Paris, 60-61. *Mem:* Am Math Soc; Math Asn Am. *Res:* Functional analysis; theory of measure and integration. *Mailing Add:* RD 2 Box 2370 Mt Bethel PA 18343

TRAPANI, IGNATIUS LOUIS, b San Francisco, Calif, Nov 19, 25; m 52; c 2. PHYSIOLOGY, IMMUNOLOGY. *Educ:* Univ San Francisco, BS, 48, MS, 50; Stanford Univ, PhD(physiol), 56. *Prof Exp:* Asst physiol, Stanford Univ, 51-54, jr res assoc, 54-56; USPHS fel, Calif Inst Technol, 56-58, res fel immunochem, 56-60; asst chief dept exp immunol, Nat Jewish Hosp, 60-69, actg chief, 63-69; asst prof microbiol, Univ Colo Med Ctr, Denver, 65-69; prof chem, 69-85, chmn, dept sci & math, 75-82, instrnl supvr arts & sci, 82-85, EMER PROF CHEM, COLO MOUNTAIN COL, 85- *Mem:* AAAS; Am Physiol Soc; Am Asn Immunol; Soc Exp Biol & Med; NY Acad Sci. *Res:* Physiological and physico-chemical properties of plasma substitutes; physiology and immunology of animals at high altitude and low temperatures; antigen-antibody complexes, immunophysiological parameters of antibody formation; appropriate technology; science and man; energy and its societal implications. *Mailing Add:* 0060 Ptarmigan Lane Glenwood Springs CO 81601

TRAPANI, ROBERT-JOHN, immunology, for more information see previous edition

TRAPIDO, HAROLD, medical entomology; deceased, see previous edition for last biography

TRAPP, ALLAN LAVERNE, b Stockbridge, Mich, July 20, 32; m 55; c 4. VETERINARY PATHOLOGY. *Educ:* Mich State Univ, BS, 54, DVM, 56; Iowa State Univ, PhD(vet path), 60. *Prof Exp:* Vet livestock investr, Animal Dis Eradication Br, USDA, 56-57; res assoc animal dis res, Iowa State Univ, 57-60; asst prof, Ohio Agr Exp Sta, Ohio State Univ, 60-65, assoc prof, Ohio Agr Res & Develop Ctr, 65-66; assoc prof animal dis res, 66-70, PROF ANIMAL DIS DIAG WORK & TEACHING, MICH STATE UNIV, 70- *Concurrent Pos:* Mem, Med Adv Coun, Detroit Zoo, 69- *Mem:* Wildlife Dis Asn; Am Vet Med Asn; Am Asn Vet Lab Diag. *Res:* Respiratory diseases of cattle; gastrointestinal diseases of cattle and swine; naturally occuring diseases in fishes; diseases of wild and pet birds. *Mailing Add:* Animal Health Diag Lab PO Box 30076 Lansing MI 48909

TRAPP, CHARLES ANTHONY, b Chicago, Ill, July 9, 36; m 58; c 2. PHYSICAL CHEMISTRY. *Educ:* Loyola Univ, Ill, BS, 58; Univ Chicago, MS, 60, PhD(chem), 63. *Prof Exp:* NSF fel physics, Oxford Univ, 62-63; asst prof chem, Ill Inst Technol, 63-69; assoc prof, 69-74, PROF CHEM, UNIV LOUISVILLE, 74- *Concurrent Pos:* Petrol Res Fund starter grant, 63-64, type A res grant, 65-68; NSF res grant, 64-67; consult, Argonne Nat Lab, 69-72; Res Corp grant, 70-72. *Mem:* Am Phys Soc. *Res:* Magnetic properties of matter; electron spin resonance in transition metal compounds, organic free radicals and biologically important compounds; electrical conductivity studies of nonmetals. *Mailing Add:* Dept Chem Univ Louisville Box 35260 Louisville KY 40292

TRAPP, GENE ROBERT, b Hammond, Wis, June 16, 38; m 60. MAMMALOGY. *Educ:* Wash State Univ, BS, 60; Univ Alaska, MS, 62; Univ Wis, PhD(zool), 72. *Prof Exp:* Res asst, Coop Wildlife Res Unit, Univ Alaska, 60-62; teaching asst, dept biol, Univ NMex, 62-63; wildlife biologist, US Soil Conserv, Honesdale, Pa, 63-64 & Br River Basin Studies, US Fish & Wildlife Serv, Tex, 64-65; teaching asst, dept zool, Univ Wis-Madison, 65-70; collabr carnivore res, US Nat Park Serv, Zion Nat Park, 67-69; from asst prof to assoc prof, 70-77, PROF BIOL, CALIF STATE UNIV, SACRAMENTO, 77- *Mem:* Am Soc Mammalogists; Animal Behav Soc; Wildlife Soc; Sigma Xi. *Res:* The behavioral ecology of mammals, especially carnivores. *Mailing Add:* Dept Biol Sci Calif State Univ Sacramento CA 95819

TRAPP, GEORGE E, JR, b Pittsburgh, Pa, June 30, 44; m 68; c 2. APPLIED MATHEMATICS. *Educ:* Carnegie-Mellon Univ, BS, 66, MS, 67, PhD(math), 70. *Prof Exp:* Assoc prof, 70-76, PROF COMPUT SCI, WVA UNIV, 76-; CONSULT, WESTINGHOUSE ELEC CORP, 70- *Concurrent Pos:* Consult, Brookhaven Nat Lab, 83. *Mem:* Soc Indust & Appl Math; Sigma Xi; Am Math Soc; Math Asn Am. *Res:* Algebraic analysis of electrical networks and numerical analysis, and product data modeling. *Mailing Add:* Dept Statist & Comput Sci WVa Univ Morgantown WV 26506

TRAPP, ROBERT F(RANK), b Taylorville, Ill, Nov 4, 32; m 53; c 3. ENGINEERING PHYSICS, NUCLEAR ENGINEERING. *Educ:* Univ Ill, BS, 54. *Prof Exp:* Asst, Cyclotron Lab, Univ Ill, 54; assoc engr, Douglas Aircraft Co, Inc, 56-57, adv design coordr, 57-60, supvr adv propulsion, 60-61, chief proj engr, 61-62; br chief life sci, NASA, 62-68; chief res div, Weapons Eval & Control Bur, Arms Control & Disarmament Agency, Va, 68-72; mfrs rep, TDM, Inc, Fairfax, 72-76; dir tech res, Panasonic Co, 76-77; pres, Microelectronics Technol Corp, 77-82; vpres & gen mgr, Western Plant, Matsushita Elec Components Co, 82-89; CONSULT, 89- *Mem:* Int Soc Hybrid Microelectronics. *Res:* High temperature and thermal properties of materials for reactor cores and missile nosecones; engineering design and supervision of all aspects of nuclear rockets and aerospace life sciences; electronic components and subsystems applicable to consumer and industrial products; areas include hybrid circuits, monolytic IC's, sensors, displays, controllers. *Mailing Add:* 46990 Ocotillo Ct Fremont CA 94539

TRAPPE, JAMES MARTIN, b Spokane, Wash, Aug 16, 31; m 63; c 4. MYCOLOGY, FOREST ECOLOGY. *Educ:* Univ Wash, BS, 53, PhD(forest bot), 62; State Univ NY, MS, 55. *Prof Exp:* Forester, Colville Nat Forest, 53-56, res forester, Pac Northwest Forest & Range Exp Sta, US Forest Serv, 56-65, proj leader & prin mycologist, Pac Northwest Forest & Range Exp Sta, 65-86; PROF BOT & FOREST SCI, ORE STATE UNIV, 76- *Concurrent Pos:* Assoc prof bot, Ore State Univ, 65-76; NSF grants, 66-71, 74, 76, 78-91; Am Philos Soc grants for mycol res, Univ Torino, Italy, 67-68, Nat Polytech Inst, Mex, 72; Japan Soc Prom Sci res fel, 75; mem, Joint Comn on Rural Reconstruct, Repub China, 77, Kuwait Inst Sci Res, 79; Commonwealth Sci & Ind Res Orgn, Perth, Australia, 82, 87-91; Indo-US Sci & Technol Initiative, New Delhi & Bangalore, 84-91; Sweden spec prof, Uppsala, 90-91; Weyerhaueser Found fel, 55-56. *Honors & Awards:* Milestone Res Award, Univ Hannover, Ger, 85. *Mem:* Fel AAAS; Brit Mycol Soc; Mex Soc Mycol; Mycol Soc Am (pres, 86-87); Sigma Xi; NAm Truffling Soc (pres, 70). *Res:* Taxonomy of fungi, especially hypogeous species, Mycorrhizae. *Mailing Add:* Dept Forest Sci Ore State Univ Corvallis OR 97331-5705

TRAQUAIR, JAMES ALVIN, b London, Ont, Aug 1, 47. MYCOLOGY, PLANT PATHOLOGY. *Educ:* Univ Western Ont, BSc, 70; Univ Alta, PhD(mycol), 74. *Prof Exp:* Nat Res Coun Can fel plant path, Univ Western Ont, London, 74-77; res assoc mycol, Erindale Col, Univ Toronto, 77-78; plant pathologist forage path, Lethbridge, Alta, 78-81, TREE FRUIT PATHOLOGIST, RES STA, AGR CAN, HARROW, ONT, 81- *Mem:* Can Phytopath Soc; Can Bot Asn; Brit Mycol Soc; Mycol Soc Am. *Res:* Biosystematics of polyporaceous basidiomycetes; microbial ecology; hyperparasitism of fungi; ultrastructure of fungi; host-parasite relations and epidemiology of snowmold diseases of winter cereals, forage legumes and grasses; biocontrol of cytospora canker, root rots and plant disorders; tree fruit mycorrhizae. *Mailing Add:* Res Sta Agr Can Harrow ON N0R 1G0 Can

TRASHER, DONALD WATSON, b Malverne, NY, Nov 1, 37; m 59; c 4. MATHEMATICS. *Educ:* Houghton Col, BA, 59; Univ Buffalo, MA, 62. *Prof Exp:* Instr math, Univ Buffalo, 61-64; vis instr, Univ Ind, 69-70; asst prof, 64-68 & 70-78, CHMN DEPT MATH, STATE UNIV, NY, GENESEO, 78- *Concurrent Pos:* Dir, Math Enrichment Prog, Cheektowaga Central Sch, NY, 62-64; consult, GEM Prog for mathematically gifted, York Central Sch, NY, 79-80. *Mem:* Math Asn Am. *Res:* Geometry and the history of mathematics. *Mailing Add:* Dept Math State Univ NY Col Geneseo NY 14454

TRASK, CHARLES BRIAN, b Bar Harbor, Maine, June 13, 44; m 69; c 2. SEDIMENTOLOGY, SEDIMENTARY PETROLOGY. *Educ:* Amherst Col, AB, 66; Univ Tex, Austin, MA, 72; Syracuse Univ, PhD(geol), 76. *Prof Exp:* Res geologist, Gulf Sci & Technol Co, Gulf Oil Corp, 76-77; assoc geologist coal, 81-87, ASSOC GEOLOGIST ENVIRON STUDIES & ASSESSMENT, ILL STATE GEOL SURV, 87- *Mem:* Geol Soc Am; Int Asn Sedimentologists; Soc Econ Paleontologists & Mineralogists; Sigma Xi; Soc Sedimentary Geol. *Res:* Sedimentology, engineering geology environmental geology & geology for planning. *Mailing Add:* 615 E Peabody Dr Champaign IL 61820

TRASKOS, RICHARD THOMAS, b New Britain, Conn, Jan 3, 40; m 64; c 2. POLYMER SCIENCE, CHEMICAL ENGINEERING. *Educ:* Univ Notre Dame, BS, 61; Mass Inst Technol, DSc(chem eng), 66. *Prof Exp:* RES ENGR, ROGERS CORP, 69- *Mem:* Am Chem Soc; Soc Photog Sci & Eng. *Res:* Development of polymer-based printing plates; study of lithography. *Mailing Add:* 43 Herrick Rd Brooklyn CT 06234-1413

TRASLER, DAPHNE GAY, b Iquique, Chile, July 2, 26; Can citizen; m 51; c 2. GENETICS. *Educ:* McGill Univ, BSc, 48, MSc, 54, PhD, 58. *Prof Exp:* Demonstr genetics, McGill Univ, 47; chief asst plant genetics, Inst Cotton Genetics, Peru, 49-51; demonstr genetics, 52-53, res assoc develop genetics & teratol, 58-70, ASSOC PROF BIOL, MCGILL UNIV, 70-, ASSOC PROF, CTR HUMAN GENETICS, 79- *Concurrent Pos:* Grants, NSF, 59-62, Asn Aid Crippled Children, 65-66, NIH, 66-69 & Nat Res Coun Can, 70-76 & Med Res Coun, 76; mem study sect, Div Res Grants, NIH, 80-82 & 84-88; vis fel, physiol dept, Fitzwilliam Col, Cambridge Univ, Eng, 81-82 & Human Anat Dept, Oxford Univ, Eng, 90. *Mem:* Genetics Soc Can; Teratology Soc (pres, 72-73). *Res:* Gene-teratogen interaction and embryonic mechanisms in mouse neural tube defects; mouse teratology. *Mailing Add:* Dept Biol McGill Univ 1205 Ave Docteur Penfield Montreal PQ H3A 1B1 Can

TRASS, O(LEV), b Estonia, Oct 9, 31; m 61; c 2. CHEMICAL ENGINEERING. *Educ:* Princeton Univ, BSE, 55; Mass Inst Technol, ScD(chem eng), 58. *Prof Exp:* From asst prof to assoc prof, 58-68, PROF CHEM ENG & APPL CHEM, UNIV TORONTO, 68- *Concurrent Pos:* Vis prof, Swiss Fed Inst Technol, 68-69; dir & consult, Chem Eng Res Consults, Ltd Can; mem, Grants Comt on Chem & Metall Eng, Nat Res Coun Can, 72-73, co-chmn, 73-74, chmn, 74-75; assoc chmn, Div Eng Sci, Univ Toronto, 74-77; vpres, Gen Comminution, Inc, 75-86, dir & chmn, 86-; vis prof, Ecole Nat Superieur Indust Chimiques, Nancy, France, 78; speaker, fac coun, Fac Appl Sci & Eng, Univ Toronto, 78-82. *Mem:* Am Chem Soc; Am Inst Chem Engrs; fel Chem Inst Can; Can Soc Chem Eng; Asn Advan Baltic Studies; Fine Particle Soc. *Res:* Fluid flow and mass transfer; solid-fluid interface phenomena, particularly rough surface phenomena; high temperature chemical reactions in shock tubes; comminution and particle dynamics; coal slurry fuels technology; coal beneficiation technology. *Mailing Add:* Dept Chem Eng & Appl Chem Univ Toronto Toronto ON M5S 1A4 Can

TRAUB, ALAN CUTLER, b Hartford, Conn, Jan 20, 23; m 51; c 3. ELECTRO-OPTICS. *Educ:* Trinity Col, Conn, BS, 47; Univ Cincinnati, MS, 49, PhD(physics), 52. *Prof Exp:* Res physicist, Am Optical Co, 52-56; res physicist, Fenwal, Inc, 56-61, chief res engr, 61-63; mem tech staff, Mitre

Corp, Mass, 63-70; chief scientist, Foto-Mem, Inc, 70-71; consult, 71-73; prod develop engr, Identicon Corp, 73-74; advan develop mgr, Vanzetti Infrared & Comput Systs, Inc, 74-80, advan develop mgr, Vanzetti Systs, Inc, 80-89; RETIRED. *Concurrent Pos:* Res fel, Tufts Univ, 72-73. *Honors & Awards:* Soc Tech Writers & Publ Award of Excellence, 70. *Res:* Spectrophotometry; colorimetry; thin optical films; optical and thermal sensors; visual perception; three-dimensional displays; fiber optics; aerospace electro-optical instrumentation; atmospheric optical propagation; optical communications systems; laser applications; optical memories. *Mailing Add:* 56 Donna Rd Framingham MA 01701

TRAUB, JOSEPH FREDERICK, b WGer, June 24, 32; nat US; m 69; c 2. COMPUTER SCIENCE. *Educ:* City Col New York, BS, 54; Columbia Univ, PhD(appl math), 59. *Prof Exp:* Mem tech staff, Bell Tel Labs, Inc, NJ, 59-70; prof computer sci & math & head, Dept Computer Sci, Carnegie-Mellon Univ, 71-79; Edwin Howard Armstrong prof computer sci & chmn, Dept Computer Sci & prof math, Columbia Univ, 79-89. *Concurrent Pos:* Chmn, Award Comt, Asn Comput Mach, 72-76; mem adv comt fed judicial ctr; mem sci coun, Inst de Res d'Info et d'Automatique, Paris, 76-80, cent steering comt, Comput Sci & Eng Res Study, NSF & liaison to panel on theoret comput sci & panel on numerical anal, 74-80; mem adv comt, Carnegie-Mellon Inst Res, 78-79, Inst Defense Anal, 76-79 & mem adv comt, math & comput sci, NSF, 78-; mem, Conf Bd Math Sci, Math Sci Res Inst, Berkeley, Calif, 85-86; dir, NY State Ctr Comput & Info Systs, 84-; ed, J Complexity, Acad Press & Ann Rev Comput Sci, Ann Rev, Inc, 85-; chmn, Computer Sci Technol Bd, Nat Acad Sci, 86-90, Computer Sci Telecommun Bd, 90. *Honors & Awards:* Emanuel R Piore Award, Inst Elec & Electronics Engrs, 91. *Mem:* Nat Acad Sci; Nat Acad Eng; Soc Indust & Appl Math; Asn Comput Mach; Am Math Soc; Inst Elec & Electronics Engrs. *Res:* Computational complexity; parallel computation; algorithmic analysis; numerical mathematics; large scientific problems. *Mailing Add:* 96 Battle Rd Princeton NJ 08540

TRAUB, RICHARD KIMBERLEY, b Bessemer, Mich, Mar 13, 34; m 56; c 2. MEDICAL RESEARCH. *Educ:* Johns Hopkins Univ, BES, 56; Towson State Col, MA, 71; Univ Del, MS, 75. *Prof Exp:* Res engr, Plastics Dept, Du Pont Co, 59-62; chief of group, Defense Develop & Eng Lab, 62-72, Human Factors Eng, 72-74, prog dir prophylaxis & ther, Biomed Lab, 74-80, PRIN INVESTR, NEUROTOX & EXP THERMAL BR, US MED RES INST DEFENSE, 80- *Concurrent Pos:* Mem, Pyrotech Comt, Am Ordnance Asn, 62-72; clin psychologist, Harford County, State Md, 71-; intel specialist, Surgeon Gen, US Army & mem, Acad Coun, Edgewood Arsenal, 73- *Mem:* Am Psychol Asn; Soc for Neurosci. *Res:* Therapy and prophylaxis against chemical warfare agents including basic mechanisms. *Mailing Add:* 1900 Park Beach Dr Aberdeen MD 21001

TRAUB, ROBERT, b New York, NY, Oct 26, 16; m 39; c 2. MEDICAL ENTOMOLOGY. *Educ:* City Col New York, BS, 38; Cornell Univ, MS, 39; Univ Ill, PhD(med entom), 47. *Prof Exp:* Asst entom, Univ Ill, 39-41; chief dept parasitol, Med Ctr, US Army, 47-55, commanding officer, Med Res Unit, Malaya, 55-59, chief, Prev Med & Entom Res Br, Med Res & Develop Command, 59-62; prof, 62-90, EMER PROF MICROBIOL, SCH MED, UNIV MD, 90- *Concurrent Pos:* Parasitologist, 4th Hoogstraal Exped, Mex, 41; field dir, Army Med Res Units, Malaya, NBorneo & Labrador, 47-55; Comn Hemorrhagic Fever, Korea, 52 & 53; Univ Md Sch Med Res Units, Pakistan, Ethiopia, Burma, Australia, Thailand & New Guinea; hon assoc, Field Mus, Ill, Smithsonian Inst, DC, Carnegie Mus Natural Hist, Pittsburgh & Bishop Mus, Honululu; consult ectoparasite-borne dis, WHO, US Army & USN; mem, Comn Immunization, 48-53, Comn Riskettsial Dis, 64-73 & Armed Forces Epidemiol Bd. *Mem:* AAAS; Entom Soc Am; fel Am Soc Parasitol; fel Am Soc Trop Med & Hyg; Soc Syst Zool; Sigma Xi. *Res:* Ecology and control of vectors and reservoirs of disease; systematics of Siphonaptera and trombiculid mites. *Mailing Add:* 5702 Bradley Blvd Bethesda MD 20814

TRAUB, WESLEY ARTHUR, b Milwaukee, Wis, Sept 25, 40; m 63; c 1. STRATOSPHERIC COMPOSITION. *Educ:* Univ Wis-Milwaukee, BS, 62; Univ Wis(Madison), MS, 64, PhD(physics), 68. *Prof Exp:* PHYSICIST, CTR ASTROPHYS, SMITHSONIAN & HARVARD COL OBSERVS, 68-; LECTR ASTRON, HARVARD UNIV, 76- *Concurrent Pos:* Res assoc eng & appl physics, Harvard Univ, 68-74. *Mem:* AAAS; Am Astron Soc; Int Astron Union; Optical Soc Am; Sigma Xi; Am Geophys Union. *Res:* Far-infrared spectroscopy of the terrestrial atmosphere to determine molecular abudances; ground-based Michelson spatial interferometry; far-infrared laboratory spectroscopy of molecules of stratospheric interest; coherent arrays of optical telescopes for astronomy from space, design studies; Fourier-transform infrared spectrometers; astrophysical high angular resolution imaging. *Mailing Add:* Smithsonian & Harvard Col Observs 60 Garden St Cambridge MA 02138

TRAUGER, DAVID LEE, b Ft Dodge, Iowa, June 16, 42; div; c 2. ZOOLOGY, ECOLOGY. *Educ:* Iowa State Univ, BS, 64, MS, 67, PhD(animal ecol), 71. *Prof Exp:* Instr zool & entom, Iowa State Univ, 67-70, asst prof zool & entom & exec secy environ coun, 70-72; wildlife res biologist, Northern Prairie Wildlife Res Ctr, Jamestown, NDak, US Dept Interior, Washington, DC, 72-75, asst dir, 75-79, CHIEF, DIV WILD LIFE ECOL RES SERV, PATUXENT WILDLIFE RES CTR, FISH & WILDLIFE SERV, US DEPT INTERIOR, 79-, DIR. *Concurrent Pos:* Wildlife technician, Northern Prairie Wildlife Res Ctr, 66-70. *Mem:* Wildlife Soc; Am Ornith Union; Cooper Ornith Union; Wilson Ornith Soc. *Res:* Breeding biology, population dynamics, habitat requirements of waterfowl, particularly diving ducks in prairie parklands and subarctic taiga. *Mailing Add:* Fish & Wildlife Serv Patuxent Wildlife Res Ctr Rte 197 Laurel MD 20708

TRAUGER, DONALD BYRON, b Exeter, Nebr, June 29, 20; m 45; c 2. PHYSICS. *Educ:* Nebr Wesleyan Univ, AB, 42. *Hon Degrees:* DSc, Wesleyan Univ, Nebr, 74 & Wesleyan Col, Tenn, 77. *Prof Exp:* Physicist, Manhattan Proj, Columbia Univ, 42-44; engr, Union Carbide Corp, 44-54; head irradiation eng dept, 54-64, dir gas-cooled reactor prog, 64-70, assoc dir Nuclear & Eng Technol, 70-84, SR TECH ASST TO DIR, OAK RIDGE NAT LAB, 84- *Mem:* AAAS; fel Am Nuclear Soc; Am Phys Soc; Sigma Xi. *Res:* Reactor technology; gas cooled reactor fuels; nuclear irradiation tests of fuels and materials; liquid metals; behavior of gases; isotope separation. *Mailing Add:* 510 Delaware Ave Oak Ridge TN 37830

TRAUGER, FREDERICK DALE, b Lindsay, Calif, Apr 21, 16; m 51. GEOLOGY. *Educ:* Fresno State Col, BA, 39. *Prof Exp:* Geologist, US Bur Reclamation, 46; hydrologist, US Geol Surv, 47-73; geohydrologist, Geohydrology Consults Inc, 73-85; RETIRED. *Mem:* Fel Geol Soc Am; Am Geol Union; AAAS; Am Water Well Asn. *Res:* Quantitative and qualitative evaluation of ground-water resources for public supplies for municipal, county and state governments; groundwater recharge; movement of ground-water contaminants. *Mailing Add:* 1206 Field Dr NE Albuquerque NM 87112

TRAUGH, JOLINDA ANN, b Detroit, Mich; c 1. BIOCHEMISTRY, MOLECULAR BIOLOGY. *Educ:* Univ Calif, Davis, BS, 60; Univ Calif, Los Angeles, PhD(microbiol), 70. *Prof Exp:* Res asst, Gerber Baby Foods, 60-62 & Univ Calif, Berkeley, 62-64; USPHS res fel molecular biol, Univ Calif, Davis, 71-73; from asst prof to assoc prof, 73-82, chmn dept, 81-86, PROF BIOCHEM, UNIV CALIF, RIVERSIDE, 82- *Concurrent Pos:* Resident scholar, Study & Conf Ctr, Rockefeller Found, Belliago, Italy, 79; biochem study sect, NIH, 76-80; bd sci coun, Nat Inst Environ Health Sci; sci adv comt biochem & endocrinol, Am Cancer Soc, 91- *Mem:* AAAS; Fedn Am Soc Exp Biol; Biophys Soc. *Res:* Regulation of protein synthesis; protein kinases. *Mailing Add:* Dept Biochem Univ Calif Riverside Riverside CA 92521

TRAUGOTT, STEPHEN C(HARLES), b Frankfurt, Ger, Dec 10, 27; nat US; m 55; c 2. FLUID DYNAMICS, HEAT TRANSFER. *Educ:* Johns Hopkins Univ, BES, 49, MSE, 51, DEng, 56. *Prof Exp:* Sr scientist, Res Dept, Martin Co, 57-66, chief aerophys res staff, 60-62, prin res scientist & mgr mech, Martin Marietta Labs, 66-85; PROG DIR, NSF, 85- *Concurrent Pos:* Adj prof, Drexel Inst Technol, 59; vis lectr, Johns Hopkins Univ, 62-66; adj prof, Univ Md, 66-67; vis assoc prof, Cornell Univ, 67-68. *Mem:* Am Inst Aeronaut & Astronaut; Am Phys Soc; Soc Natural Philos; Sigma Xi. *Res:* High speed flow; radiation gas dynamics; dynamics of atmospheres; aluminum smelting; hydrodynamics. *Mailing Add:* 11219 Old Calliage Rd Glen Arm MD 21057

TRAUMANN, KLAUS FRIEDRICH, b Schweinfurt, Ger, Mar 23, 24; nat US; m 58. ORGANIC CHEMISTRY. *Educ:* Univ Heidelberg, PhD(org chem), 54. *Prof Exp:* Res chemist, Carothers Lab, E I Du Pont de Nemours & Co, Inc, 54-66, sr res chemist, Textile Res Lab, 67-73, develop assoc, Textile Res Lab, 73-86; RETIRED. *Mem:* Am Chem Soc. *Res:* Synthetic organic fibers. *Mailing Add:* Sharpley 502 Whitby Dr Wilmington DE 19803

TRAURIG, HAROLD H, b Chicago, Ill, July 28, 36; m 59; c 3. NEUROBIOLOGY, ENDOCRINOLOGY. *Educ:* Mankato State Col, BS, 58; Univ Minn, PhD(anat), 63. *Prof Exp:* From instr to assoc prof anat, 63-75, PROF ANAT & NEUROL, MED CTR, UNIV KY, 75- *Concurrent Pos:* NIH res grants, 64-67 & 68-72; res fel neurochem, Ohio State Univ, 70-71; vis res fel, Univ Graz, Austria, 83 & Univ Ulm, Ger, 91. *Mem:* Am Asn Anatomists; Soc Exp Biol & Med; Soc Study Reproduction; Neurosci Soc. *Res:* Cytology; neurobiology; reproductive biology; cell proliferation. *Mailing Add:* Dept Anat Univ Ky Med Ctr Lexington KY 40536

TRAURING, MITCHELL, b Brooklyn, NY, Mar 8, 22; m 43; c 2. OPERATIONS RESEARCH. *Educ:* Brooklyn Col, BA, 41; Johns Hopkins Univ, MA, 47; Univ Calif, Los Angeles, MS, 76. *Prof Exp:* Physicist, Nat Adv Comt Aeronaut, 41-46; optical engr, Bur Ships, USN, 49; ballistician & sect head, Ballistics Res Labs, Ord Corps, US Dept Army, 49-53; sect head, Guided Missiles Div, Repub Aviation Corp, 53-57; asst sect head, Ground Systs Group, Hughes Aircraft Co, 57-59, sr staff physicist, Res Labs, 59-63, sr staff engr, Aerospace Corp, Calif, 63-68; sr mem tech staff, Data Systs Div, Litton Industs, Inc, 68-72; consult opers anal, 72-78; sr scientist, Hughes Aircraft Co, 78-87; CONSULT, OPERS ANALYSIS & FINANCE, 87- *Mem:* Sigma Xi; Opers Res Soc Am. *Res:* Weapon, electronic and space systems; automatic recognition; business and international economics. *Mailing Add:* 1645 Comstock Ave Los Angeles CA 90024

TRAUT, ROBERT RUSH, b Utica, NY, Oct 21, 34; m 62; c 1. BIOCHEMISTRY, MOLECULAR BIOLOGY. *Educ:* Haverford Col, AB, 56; Rockefeller Univ, PhD(biochem), 62. *Prof Exp:* Res asst, Inst Molecular Biol, Univ Geneva, 64-68, Am Heart Asn estab investr, 68-70; assoc prof, 70-76, PROF BIOL CHEM, SCH MED, UNIV CALIF, DAVIS, 76- *Concurrent Pos:* Jane Coffin Childs Mem Fund fel molecular biol, Med Res Coun Lab Molecular Biol, Cambridge Univ, 62-64; Am Heart Asn estab investr, Univ Calif, Davis, 70-73. *Mem:* AAAS; Am Soc Microbiol; NY Acad Sci; Am Soc Biol Chem; Sigma Xi. *Res:* Mechanism and regulation of protein synthesis; structure and function of ribosomes. *Mailing Add:* Dept Biol Chem Sch Med Univ Calif Davis CA 95616

TRAUT, THOMAS WOLFGANG, b Mar 17, 43. ENZYME REGULATION, PROTEIN STRUCTURE. *Educ:* Univ Southern Calif, PhD(molecular biol), 74. *Prof Exp:* ASSOC PROF BIOCHEM, SCH MED, UNIV NC, 78- *Mem:* Am Soc Biochem & Molecular Biol; Protein Soc; AAAS. *Res:* Working with allosteric enzymes in nucleotide metabolism; the importance of enzyme polymer dissociation to change in specific activity; the relation of subunit size to enzyme function; we are exploring the hypothesis that enzymes are composed of modules that specifically bind some ligand. *Mailing Add:* Dept Biochem 430 Fac Lab Off Bldg Sch Med Univ NC Chapel Hill NC 27599-7260

TRAUTMAN, JACK CARL, b Cushing, Okla, Dec 7, 29; m 57; c 1. DAIRY SCIENCE, BIOCHEMISTRY. *Educ:* Univ Idaho, BS, 51; Univ Calif, MS, 53; Univ Wis, PhD(dairy tech), 58. *Prof Exp:* Asst dairy indust, Univ Calif, 51-53; asst dairy & food indust, Univ Wis, 56-58, from instr to asst prof dairy indust, 58-59; asst prof dairy technol, Ohio State Univ, 59-60; supvr indust prod & processes, 60-72, mgr biol res, Oscar Mayer & Co, 72-83; PRES, INT

BIOCHEM TECHNOL, INC, 83- *Concurrent Pos:* Mem sci adv comt, Fats & Protein Res Found; Mfg Enzymolgists. *Mem:* Am Chem Soc; Am Meat Sci Asn; Inst Food Technol. *Res:* Process development of animal biologicals; biologicals needed in human disease states; engineering scale-ups; enzymes, anti-coagulants and blood fractions. *Mailing Add:* 3017 Dianne Dr Middleton WI 53562

TRAUTMAN, MILTON BERNHARD, b Columbus, Ohio, Sept 7, 99; m 40; c 1. ZOOLOGY. *Hon Degrees:* DSc, Col Wooster, 51 & Ohio State Univ, 78. *Prof Exp:* Asst, Bur Sci Res, Ohio Div Conserv, 30-34; asst cur, Mus Zool, Univ Mich, 34-39; res biologist, Stone Lab, Ohio State Univ, 39-40, res assoc, Stone Inst Hydrobiol, 40-55, lectr zool & cur vert collections, Dept Zool & Entom, 55-69, prof fac biol, Col Biol Sci, 69-72, emer prof zool & emer cur birds, 72-; RETIRED. *Concurrent Pos:* Asst dir inst fisheries res, State Conserv Dept, Mich, 34-35, res assoc, 35-36; mem, Univ Mich Zool Exped, Yucatan, 36; consult, US Bur Comm Fish, 59 & 61; mem, Ohio State Univ Exped, Inst Polar Studies, Alaska, 65; res assoc biol, John Carroll Univ, 72-74. *Honors & Awards:* Wildlife Soc Award, 58. *Mem:* AAAS; Wilson Ornith Soc (treas, 43-45); Am Soc Ichthyol & Herpet (vpres, 46-49); Am Ornith Union. *Res:* Factors affecting animal distribution and abundance; animal behavior; changes in animal and plant distribution and abundance in Ohio since 1750; endangered vertebrates; Great Lakes wetlands; author of over 140 publications. *Mailing Add:* Mus Zool Ohio State Univ 1813 N High St Columbus OH 43210

TRAUTMAN, RODES, biophysics, microcomputer education; deceased, see previous edition for last biography

TRAVELLI, ARMANDO, b Rome, Italy, Feb 6, 34; m 70; c 3. NUCLEAR ENGINEERING & SCIENCE, TECHNICAL MANAGEMENT. *Educ:* Univ Rome, Dr Ing, 58; Rensselaer Polytech Inst, PhD(nuclear eng, sci), 63; Univ Chicago, MBA, 84. *Prof Exp:* Res asst nuclear eng & sci, Rensselaer Polytech Inst, 61-63; asst prof nuclear eng, Mass Inst Technol, 63-65; asst nuclear engr, Argonne Nat Lab, 65-68, nuclear engr, 68-69, head, Fast Flux Test Facil Sect, 69-74, head, Physics Reactor Safety Sect, 74-76, assoc dir, Safety Test Facil Proj, 76-78, SR NUCLEAR ENGR, ENG, PHYSICS DIV, ARGONNE NAT LAB, 87-, MGR ARMS CONTROL PROG, 89- *Concurrent Pos:* Ford Found res fel eng, 63-65; mgr, Rertr Prog, Eng Physics Div, Argonne Nat Lab, 78- *Honors & Awards:* IR-100 Award, 85; Lab Consortium Award for Excellence, 86. *Mem:* Am Nuclear Soc; AAAS; Sigma Xi. *Res:* Arms control verification technology; neutron transport and high-order perturbation theories; analysis of neutron waves and pulses; analyses and design of research reactors, safety test facilities, critical experiments and fast breeder reactors; patent on research reactor fuel. *Mailing Add:* Eng Physics Div Argonne Nat Lab 9700 S Cass Ave Argonne IL 60439

TRAVER, ALFRED ELLIS, b New York, NY, Dec 17, 39; m 66; c 2. MECHANICAL ENGINEERING, SYSTEMS ENGINEERING. *Educ:* Mass Inst Technol, BS, 61; Iowa State, MS, Univ Tex, Austin, PhD(mech eng), 68. *Hon Degrees:* JD, Univ Tex, 80. *Prof Exp:* Researcher, Mass Inst Technol-Harvard Joint Ctr Urban Studies, 60-61; aerosysts engr, Gen Dynamics Corp, 63-64; engr-scientist, Tracor, Inc, 66-70; prof mech eng & actg chmn dept systs eng, Tenn Technol Univ, 70-78; DEPT MECH ENG, UNIV TEX, AUSTIN, 78- *Concurrent Pos:* consult engr; consult atty. *Mem:* Am Soc Mech Engrs; Opers Res Soc Am; Inst Elec & Electronics Engrs; Am Soc Eng Educ; Nat Soc Prof Engrs; Soc Mfg Engrs. *Res:* Simulation and modeling of dynamic systems; control systems; design studies; interface of technology and the law. *Mailing Add:* Dept Mech Eng Univ Tex Austin TX 78712

TRAVER, JANET HOPE, biochemistry; deceased, see previous edition for last biography

TRAVERS, WILLIAM BRAILSFORD, b Long Beach, Calif, June 13, 34; m 58; c 3. GEOLOGY. *Educ:* Stanford Univ, BS, 56, MS, 59; Princeton Univ, PhD(geol), 72. *Prof Exp:* Geologist, Standard Oil Co, Calif, 59-61; geologist, Santa Fe Drilling Co, 61-63; chief geologist, Santa Fe Int, Inc, 63-67; asst instr geol, Princeton Univ, 67-71; asst prof, 72-78, ASSOC PROF GEOL SCI, CORNELL UNIV, 78- *Concurrent Pos:* Consult petrol geologist; vpres, Anacapa Oil Co, 67-; vis prof, Stanford Univ, 79; res vis, Oxford Univ, 80. *Mem:* AAAS; fel Geol Soc Am; Am Asn Petrol Geol; Am Geophys Union; fel Geol Asn Can; Sigma Xi. *Res:* Problems of mountain building; deformation of continental margins; structural geology and sedimentology; continental rifting; tectonism in Italy, western United States and western Canada. *Mailing Add:* 671 Sand Point Rd Carpinteria CA 93013

TRAVERSE, ALFRED, b Port Hill, Prince Edward Island, Sept 7, 25; nat US; m 51; c 4. PALYNOLOGY, PALEOBOTANY. *Educ:* Harvard Univ, SB, 46, AM, 48, PhD(paleobot), 51; Episcopal Theol Sem Southwest, MDiv, 65. *Prof Exp:* Coal technologist, Lignite Res Lab, US Bur Mines, NDak, 51-55, head fuels micros lab, Colo, 55; geologist, Shell Develop Co, 55-62; palynological consult, Tex, 62-65; asst prof geol, Univ Tex, 65-66; assoc prof, 66-70, PROF PALYNOLOGY, PA STATE UNIV, UNIV PARK, 70- *Concurrent Pos:* Mem, Int Comn Palynology, 73-76, pres, 77-80, archivist, 84-; vis prof, Swiss Fed Tech Inst, Zurich, 80-81; adj prof geobiol, Juniata Col, Huntingdon, Pa, 77-85; secy, Comm Fossil Plants, Int Asn Plant Taxon, 69- *Mem:* Am Asn Stratig Palynologists (secy-treas, 67-70, pres, 70-71); AAAS; Geol Soc Am; Bot Soc Am; Int Asn Plant Taxon; Soc Econ Paleontologists & Mineralogists. *Res:* Palynology of Cenozoic and older rocks; theory of palynology; plant evolution. *Mailing Add:* Dept Geosci 435 Pa State Univ Deike 435 University Park PA 16802

TRAVIS, DAVID M, b Nashville, Tenn, June 6, 26; m 53; c 3. INTERNAL MEDICINE, PHARMACOLOGY. *Educ:* Vanderbilt Univ, BA, 47, MD, 51; Am Bd Internal Med, dipl, 60, recert, 77. *Prof Exp:* Intern & resident med, Boston City Hosp, Harvard Univ, 51-54; from asst prof to prof pharmacol & med, Col Med, Univ Fla, 58-80; PROF MED & PHARMACOL, UNIV NDAK, 80- *Concurrent Pos:* Teaching fel, Harvard Med Sch, 52-54; Nat Heart Inst res fel, Peter Bent Brigham Hosp, Boston, 56-58; Nat Heart & Lung Inst sr res fel, Harvard Univ, 71-72; mem corp, Marine Biol Lab, Woods Hole, 62- *Honors & Awards:* Borden Res Award, 51. *Mem:* Am Fedn Clin Res; Am Col Physicians; Am Physiol Soc; Soc Gen Physiol; Am Soc Pharmacol & Exp Therapeut. *Res:* Respiratory physiology and pharmacology; biological role of respiratory gases in health and disease. *Mailing Add:* VA Med Ctr Fargo ND 58102

TRAVIS, IRVEN, b McConnelsville, Ohio, Mar 30, 04; m 35; c 3. ELECTRICAL ENGINEERING, MATHEMATICS. *Educ:* Drexel Inst Technol, BS, 26; Univ Pa, MS, 28. *Hon Degrees:* DEng, Drexel Inst Technol, 62; DSc, 38. *Prof Exp:* Prof elec eng, Univ Pa, 28-49, supvr res, Moore Sch Elec Eng, 46-48; dir res, Burroughs Corp, 49-52, vpres, 52-69, mem bd dirs, 50-71; RETIRED. *Concurrent Pos:* Consult, Gen Elec Co, 38-40, Reeves Instrument Corp, 46-48 & Burroughs Corp, 48-49; chmn bd eng educ & vpres, IPAC, Pa State Univ; trustee, Detroit Inst Technol; mgt consult, 69- *Mem:* Fel AAAS; Am Soc Eng Educ; Am Soc Naval Engrs; Am Ord Asn; fel Inst Elec & Electronics Engrs. *Res:* Applied mathematics; weapons systems, especially antiaircraft fire control and missiles; information processing systems; communications. *Mailing Add:* 121 S Valley Rd Paoli PA 19301

TRAVIS, J(OHN) C(HARLES), b Cheboygan, Mich, Aug 6, 27; m 51; c 2. ELECTRICAL ENGINEERING. *Educ:* Purdue Univ, BSEE, 47, MSEE, 50, PhD(elec eng), 55. *Prof Exp:* From instr to asst prof, Purdue Univ, 47-56; mem tech staff, Hughes Aircraft Co, 56-60 & Aerospace Corp, 60-66; sr staff engr, TRW Systs, Calif, 66-69; assoc group dir, Off of Systs Requirements, 69-70, MEM TECH STAFF, OFF FOR DEVELOP, AEROSPACE CORP, 70- *Concurrent Pos:* Lectr, Univ Calif, Los Angeles, 57 & 60- *Mem:* Inst Elec & Electronics Engrs. *Res:* Automatic control; circuit theory; systems engineering. *Mailing Add:* Aerospace Corp 2350 El Segundo Blvd El Segundo CA 90245

TRAVIS, JAMES, b Winnipeg, Can, Nov 11, 35; m 60, 89; c 4. BIOCHEMISTRY. *Educ:* Univ Man, BSc, 58, MSc, 60; Univ Minn, PhD(biochem), 64. *Prof Exp:* Fel biochem, Johns Hopkins Univ, 64-66; asst prof, Univ Md, 66-67; from asst prof to prof, 67-87, RES PROF BIOCHEM, UNIV GA, 87- *Honors & Awards:* Merit Award, NIH. *Mem:* Am Soc Biol Chem. *Res:* Protein structure and function. *Mailing Add:* Dept Biochem Univ Ga Athens GA 30602

TRAVIS, JAMES ROLAND, b Iowa City, Iowa, Dec 20, 25; m 50. PHYSICS, EXPLOSIVES. *Educ:* Tufts Univ, BS, 49; Johns Hopkins Univ, PhD(physics), 56. *Prof Exp:* Res assoc spectros, Johns Hopkins Univ, 56-57; MEM STAFF, LOS ALAMOS NAT LAB, 57- *Mem:* Am Phys Soc; Sigma Xi; Combustion Inst. *Res:* Physics of detonation and shock waves; atomic and molecular spectroscopy. *Mailing Add:* 9420 Avenida De La Luna NE Albuquerque NM 87111

TRAVIS, JOHN RICHARD, b Billings, Mont, Sept 3, 42; div; c 2. COMPUTATIONAL FLUID DYNAMICS. *Educ:* Univ Wyo, BS, 65; Purdue Univ, MS, 69, PhD(nuclear eng), 71. *Prof Exp:* Asst scientist reactor anal & safety, Argonne Nat Lab, 71-73; staff mem numerical fluid dynamics, Los Alamos Nat Lab, 73-90; SR SCIENTIST, SCI APPLICATIONS INT CORP, 90- *Mem:* Am Soc Mech Engr; Am Nuclear Soc; Sigma Xi. *Res:* Develop computational fluid dynamics and transport phenomena models for analyzing safety issues involving nuclear reactors. *Mailing Add:* Sci Applications Int Corp 2109 Air Park Rd SE Albuquerque NM 87106

TRAVIS, JOSEPH, b Philadelphia, Pa, Dec 22, 53; m 83; c 2. ECOLOGICAL GENETICS, POPULATION BIOLOGY. *Educ:* Univ Pa, BA, 75; Duke Univ, PhD(zool), 80. *Prof Exp:* From asst prof to assoc prof, 80-90, PROF BIOL, FLA STATE UNIV, 90- *Concurrent Pos:* Fel, Univ Va, 80-81, vis fac, 82-90; vis fac, Orgn Trop Studies, 86, Kellogg Biol Sta, 87; adv panel mem, Prog Pop Biol, NSF, 86-89; coun mem, Ecol Soc Am, 85-91 & Soc Study Evolution, 90-92. *Mem:* Soc Study Evol; Europ Soc Evol Biol; Ecol Soc Am; Am Soc Zoologists; AAAS; Am Soc Naturalists; Society for Study of Evol. *Res:* Factors that influence the dynamics of numbers of individuals and genetic variation in natural populations; ecological determinants of selection and genetic determinants of response to selection. *Mailing Add:* Dept Biol Sci Fla State Univ Tallahassee FL 32306-2043

TRAVIS, LARRY DEAN, b Burlington, Iowa, July 29, 43. SPACE PHYSICS. *Educ:* Univ Iowa, BA, 65, MS, 67; Pa State Univ, University Park, PhD(astron), 71. *Prof Exp:* Asst prof physics, Pa State Univ, Worthington Scranton Campus, 71-73; MEM STAFF, INST SPACE STUDIES, 73- *Honors & Awards:* Except Sci Achievement Medal, NASA, 80. *Mem:* AAAS; Am Astron Soc; Am Geophys Union. *Res:* Planetary atmospheres. *Mailing Add:* Inst for Space Studies 2880 Broadway New York NY 10025

TRAVIS, LUTHER BRISENDINE, b Atlanta, Ga, May 25, 31; m 80; c 6. PEDIATRIC NEPHROLOGY, PEDIATRIC DIABETES. *Educ:* NGa Col, BS, 51; Med Col Ga, MD, 55. *Prof Exp:* Intern, Med Col Va, 55-56; resident pediat, Wyeth Labs, Col Med, Baylor Univ, 58-60; from asst prof to assoc prof pediat, 62-73, co-dir pediat nephrology, 64-71, PROF PEDIAT, UNIV TEX MED BR, GALVESTON, 73-, DIR PEDIAT NEPHROLOGY & DIABETES, 71- *Concurrent Pos:* Nat Inst Arthritis & Metab Dis fel pediat nephrology, Univ Tex Med Br, Galveston, 60-62; vis prof, William Beaumont Army Hosp, El Paso, Tex, 66-85; consult, NIH, 78-85, Food & Drug Asn, 80-82, Tex Diabetes Coun, 82-86. *Honors & Awards:* Diabetes in Youth Award, Am Diabetes Asn, 82. *Mem:* Am Soc Nephrology; Am Soc Pediat Nephrology; Am Acad Pediat; Am Fedn Clin Res; Soc Pediat Res; Am Diabetes Asn. *Res:* Medical diseases of the kidney in children, particularly glomerulonephritis, nephrosis and pyelonephritis; juvenile Diabetes Mellitus; author or coauthor of over 150 publications. *Mailing Add:* Dept Pediat Univ Tex Med Br Galveston TX 77550

TRAVIS, RANDALL HOWARD, b Curdsville, Ky, July 11, 24; div; c 2. PHYSIOLOGY. *Educ:* Univ Chicago, BS, 47; Case Western Reserve Univ, MD, 52. *Prof Exp:* Intern, Univ Hosps, Cleveland, 52-53, jr asst resident, 53-54, asst resident, 54-55; sr instr physiol & med, 59-63, asst prof physiol, 63-68, ASSOC PROF PHYSIOL, CASE WESTERN RESERVE UNIV, 68-, ASST PROF MED, 63-; DIR ENDOCRINOL, CLEVELAND METROP GEN HOSP, 74- *Concurrent Pos:* Nat Heart Inst res fel, 55-57; Am Heart Asn res fel, 57-59; estab investr, Am Heart Asn, 59-64; asst physician, Univ Hosps, Cleveland, 59-, assoc dir employees clin & consult endocrinol, 67-; attend physician, Wade Park Vet Admin Hosp, 59-; asst phys, Cuyahoga County Hosp, 73-75, assoc, 75- *Mem:* Endocrine Soc. *Res:* Experimental endocrinology, adrenal and renal hormones relating to cardiovascular system and to nervous system. *Mailing Add:* 355 E 266th St Cleveland OH 44132

TRAVIS, ROBERT LEROY, b Oakland, Calif, Oct 7, 40; m 63; c 2. PLANT PHYSIOLOGY, AGRONOMY. *Educ:* Univ Calif, Davis, BS, 64, MS, 66, PhD(plant physiol), 69. *Prof Exp:* Fel plant physiol, Univ Ga, 69-73, asst prof bot, 73-74; res agronomist herbicides, US Borax Res Corp, 74-76; asst prof, 76-79, ASSOC PROF AGRON, UNIV CALIF, DAVIS, 79- *Mem:* Am Soc Plant Physiologists. *Res:* Development of plasma membrane in higher plants; protein synthesis; photosynthetic efficiency. *Mailing Add:* Dept Agron Univ Calif Davis CA 95616

TRAVIS, ROBERT VICTOR, b Ames, Iowa, Aug 6, 33; m 55; c 5. ENTOMOLOGY, ENVIRONMENTAL SCIENCES. *Educ:* Cornell Univ, BS, 55; Univ Md, MS, 57, PhD(entom), 61. *Prof Exp:* Horticulturist, Agr Res Serv, USDA, 55-60; teacher sci & chmn dept, Gwynn Park High Sch, Md, 60-63; assoc prof biol, Mansfield State Col, 63-66; PROF BIOL, WESTMINSTER COL, PA, 66- *Concurrent Pos:* Owner, Garden Pest Control Co, 58-61. *Mem:* Nat Asn Biol Teachers; Entom Soc Am; Sigma Xi. *Res:* Computer modeling; use of insects and microorganisms in teaching; biological clocks; insect diapause. *Mailing Add:* Biol Dept Westminster Co New Wilmington PA 16172

TRAVIS, RUSSELL BURTON, b San Francisco, Calif, June 18, 18; m 40, 60; c 6. STRUCTURAL GEOLOGY. *Educ:* Colo Sch Mines, GeolE, 43; Univ Calif, PhD(geol), 51. *Prof Exp:* Geologist, Standard Oil Co Calif, 46; teaching asst, Univ Calif, 47-50; asst prof geol, Univ Idaho, 51; sr geologist, Int Petrol Co, Ltd, 51-53; asst prof geol, Colo Sch Mines, 53-56; sr geologist, Int Petrol Co, Ltd, Peru, 56-62, 67-68, Fla, 62-63 & Colombia, 63-67 & 69-73; consult petrol geol, 77-80; tech adv geol/comput appl, Petroleos Del Peru, 73-77 & 80-86; consult, Systs Design, 86-89; RETIRED. *Mem:* Am Inst Prof Geologists; fel Geol Soc Am; Am Asn Petrol Geologists. *Res:* Computer science in exloration/production in petroleum and minerals. *Mailing Add:* 5043 Shoshone Dr Pensacola FL 32507

TRAVNICEK, EDWARD ADOLPH, b Morse Bluff, Nebr, Mar 19, 36; m 66; c 2. ORGANIC CHEMISTRY, CHEMICAL ENGINEERING. *Educ:* Univ Nebr, Lincoln, BS, 58, MS, 60; Kans State Univ, PhD(chem eng), 68. *Prof Exp:* Testing engr chem eng, Atomic Energy Div, Phillips Petrol Co, 59-61; mgr chem lab, TRW Capacitor, Nebr, 61-63, consult, 63-68; sr res chem engr, Monsanto Co, Mo, 68-70 & Mass, 70-72; sr res engr, Am Optical Corp, Southbridge, Ma, 72-86; VISION-EASE CORP, FT LAUDERDALE, FLA, 86- *Concurrent Pos:* Lab asst, Kans State Univ, 66. *Mem:* Am Inst Chem Engrs; Am Chem Soc; Sigma Xi. *Res:* Polymer processing and formulation; dye and monomer synthesis, purification and analysis; gas chrom analytical techniques; mechanical, electrical and thermoelectric devices; liquid diffusion. *Mailing Add:* 9920 NW 39th Ct Coral Springs FL 33065-1527

TRAWICK, WILLIAM GEORGE, b Sandersville, Ga, Aug 16, 24; m 48; c 2. CLINICAL CHEMISTRY, PHYSICAL CHEMISTRY. *Educ:* Ga Inst Technol, BS, 48; PhD(phys chem), 55; Am Bd Clin Chem, dipl, 76. *Prof Exp:* Chemist, Union Carbide Nuclear Co, Tenn, 54-58; assoc prof phys chem, La Polytech Inst, 58-61; prof chem, Ga State Univ, 61-85, chmn dept, 62-74; RETIRED. *Concurrent Pos:* Bd dirs, Nat Registry Clin Chem, 80-89; consult, 85- *Mem:* Am Chem Soc; Sigma Xi; Am Asn Clin Chemists. *Res:* Microcomputer applications. *Mailing Add:* 2479 Burnt Leaf Lane Decatur GA 30033-2806

TRAWINSKI, BENON JOHN, b Poland, Oct 20, 24; c 1. MATHEMATICAL STATISTICS. *Educ:* McMaster Univ, BSc, 58; Va Polytech Inst & State Univ, PhD(math statist), 61. *Prof Exp:* Asst prof statist, Va Polytech Inst & State Univ, 60-61; from asst prof to assoc prof biostatist, Tulane Univ, 61-65; assoc prof statist, Univ Ky, 65-66; ASSOC PROF BIOSTATIST, MED CTR, UNIV ALA, BIRMINGHAM, 66-, DIR GRAD PROG, 69- *Concurrent Pos:* NIH fel, 62; Nat Res Coun Can grant, 63; vis lectr, NSF, 71-; univ statist & math sci curric consult. *Mem:* AAAS; Inst Math Statist; Am Statist Asn; Am Math Soc; NY Acad Sci. *Res:* Theoretical and applied research in statistics, especially order and nonparametric statistics and decision theory. *Mailing Add:* Dept Biostatist Univ Ala Birmingham AL 35233

TRAWINSKI, IRENE PATRICIA MONAHAN, b Bayonne, NJ, Mar 17, 29; m 63; c 1. MATHEMATICAL STATISTICS. *Educ:* Rutgers Univ, BSc, 50; Univ Ill, MS, 51; Va Polytech Inst, PhD(math statist), 61. *Prof Exp:* From instr to prof math, Keuka Col, 51-63, head dept, 57-63; assoc prof, La State Univ, New Orleans, 64-66; ASSOC PROF BIOSTATIST, UNIV ALA, BIRMINGHAM, 66- *Mem:* Math Asn Am; Inst Math Statist; Biomet Soc; Sigma Xi. *Res:* Multivariate analysis. *Mailing Add:* Dept Biostatist Univ Ala Birmingham AL 35233

TRAXLER, JAMES THEODORE, b Le Center, Minn, Oct 17, 29; m 56; c 5. SYNTHETIC ORGANIC CHEMISTRY, PESTICIDE CHEMISTRY. *Educ:* St John's Univ, Minn, BA, 51; Univ Notre Dame, PhD(org chem, biochem), 56. *Prof Exp:* Res chemist, Cent Res Labs, Armour & Co, Ill, 55-60 & Am Cyanamid Co, Conn, 60-62; head org lab, Durkee Foods Div, Glidden Co, Ill, 62-66; sr res chemist, Peter Hand Found, Ill, 66-69; org res specialist, Growth Sci Ctr, Int Minerals & Chem Corp, Libertyville, 69-74; SR SCIENTIST, SANDOZ CROP PROTECTION CORP, 74- *Concurrent Pos:* Scientific translator. *Mem:* AAAS; Am Chem Soc; Sigma Xi. *Res:* Amino acids and alcohols; pesticides; heterocycles; natural products; polycyclic aromatics; antimalarials; 14 research publications, 11 US & foreign patents. *Mailing Add:* 917 Forest Ave Evanston IL 60202

TRAXLER, RICHARD WARWICK, b New Orleans, La, July 25, 28; m 52; c 3. BACTERIAL PHYSIOLOGY. *Educ:* Univ Tex, BA, 51, MA, 55, PhD(bact), 58. *Prof Exp:* Asst serologist, Port Arthur Health Dept, 49-52; asst, Univ Tex, 54-58; asst prof bact, Univ Southwestern La, 58-62, from assoc prof to prof microbiol, 62-71; prof plant path, entom & microbiol & chmn dept, 71-83, prof microbiol, 83-85, PROF FOOD SCI & MICROBIOL, UNIV RI, 85- *Mem:* Am Soc Microbiol; Soc Indust Microbiol. *Res:* Microbial physiology, especially degradation; aliphatic hydrocarbons and related molecules; aerobic and anaerobic fermentation, bacterial resistance to lead and other metals. *Mailing Add:* Dept Food Sci & Nutrit Univ RI Kingston RI 02881

TRAYLOR, MELVIN ALVAH, JR, b Chicago, Ill, Dec 16, 15; m 41, 70; c 2. ORNITHOLOGY. *Educ:* Harvard Univ, AB, 37. *Prof Exp:* Assoc, Div Birds, 40-48, res assoc, 48-55, assoc cur, 55-62, cur birds, 72-77, chmn dept zool, 77-80, EMER CUR, FIELD MUS NATURAL HIST, 81- *Concurrent Pos:* Mem expeds, Yucatan, Mex, 39-40, Galapagos Island, 41, US, 41, Mex, 48 & Africa, 61-62; pelagic fishing surv, Oper Crossroads, Bikini Atoll, 46. *Mem:* Wilson Ornith Soc; fel Am Ornith Union; hon mem Soc Orinthol France; Brit Ornith Union. *Res:* Taxonomy of Neotropical and African birds; biogeography of South America. *Mailing Add:* Field Mus Natural Hist Chicago IL 60605

TRAYLOR, PATRICIA SHIZUKO, b San Francisco, Calif, Jan 21, 30; m 59; c 2. CHEMISTRY, BIOCHEMISTRY. *Educ:* Univ Calif, Berkeley, AB, 51; Univ Wis, MS, 53; Harvard Univ, PhD(chem), 63. *Prof Exp:* Res biochemist, Univ Calif, Berkeley, 53-55; chemist, Dow Chem Co, 55-59; NIH res fel, 63-66; from asst prof to prof chem, 66-87, CHMN DEPT, UNIV SAN DIEGO, 87- *Mem:* AAAS; Am Chem Soc; NY Acad Sci. *Res:* Mechanisms of reactions. *Mailing Add:* Dept Chem Univ San Diego Alcala Park San Diego CA 92110

TRAYLOR, TEDDY G, b Sulphur, Okla, May 21, 25; m 59; c 6. ORGANIC CHEMISTRY. *Educ:* Univ Calif, Los Angeles, AB, 49, PhD(chem), 52. *Prof Exp:* Sr res chemist, Dow Chem Co, Calif, 52-59; fel, Harvard Univ, 59-61, instr chem, 61; from asst prof to assoc prof, 61-68, PROF CHEM, UNIV CALIF, SAN DIEGO, 68- *Concurrent Pos:* Consult, Rohm and Haas Res Labs, Pa, 63-74; Guggenheim fel, 76. *Mem:* Am Chem Soc. *Res:* Organometallic chemistry; autoxidation; oxygen transport; bioorganic chemistry. *Mailing Add:* Dept Chem D-006 Univ Calif San Diego La Jolla CA 92093

TRAYNHAM, JAMES GIBSON, b Broxton, Ga, Aug 5, 25; m 48, 80; c 2. ORGANIC CHEMISTRY, HISTORY OF CHEMISTRY. *Educ:* Univ NC, BS, 46; Northwestern Univ, PhD(chem), 50. *Prof Exp:* Instr chem, Northwestern, 49-50; asst prof, Denison Univ, 50-53; from asst prof to prof, La State Univ, Baton Rouge, 53-63, chmn dept, 68-73, vchancellor advan studies & res & dean, Grad Sch, 73-81, prof chem, 63-88, EMER PROF, CHEM, LA STATE UNIV, BATON ROUGE, 88- *Concurrent Pos:* Res assoc, Ohio State Univ, 51-53; Am Chem Soc Petrol Res Fund int award, Swiss Fed Inst Technol, 59-60; NATO sr fel sci, Univ Saarland, 72; chmn, div hist chem, Am Chem Soc, 88. *Mem:* Am Chem Soc; Sigma Xi. *Res:* Mechanisms of reactions; halogenations; ipso aromatic substitutions; history of organic chemistry. *Mailing Add:* Dept Chem La State Univ Baton Rouge LA 70803-1804

TRAYNOR, LEE, b Flint, Mich, July 9, 38; m 60; c 2. ORGANIC CHEMISTRY, POLYMER CHEMISTRY. *Educ:* Mich State Univ, BS, 60; Univ Mich, PhD(org chem), 64. *Prof Exp:* Res chemist, 64-67, sr res chemist, 67-72, RES ASSOC, B F GOODRICH CO, 72- *Mem:* Am Chem Soc. *Res:* Organic reaction mechanisms; heterogeneous catalysis; new methods in vinyl polymerization. *Mailing Add:* 2824 Yellow Creek Rd Akron OH 44313-2208

TRAYSTMAN, RICHARD J, b Brooklyn, NY, April 5, 42; m 71. CARDIOVASCULAR & PULMONARY PHYSIOLOGY. *Educ:* Long Island Univ, BS, 63, MA, 66; Johns Hopkins Univ, PhD(cardiopulmonary physiol), 71. *Prof Exp:* Teaching & res fel physiol, Brooklyn Col Med, 63-66; instr, Bowman Gray Sch Med, 71-72; from asst prof to assoc prof environ physiol, 73-85, assoc prof anesthesiol, 80-85, PROF ENVIRON PHYSIOL, SCH HYG, JOHNS HOPKINS UNIV, 85-, PROF ANESTHESIOL & CRITICAL CARE MED, SCH MED, 85-, DIR RES LABS, ANESTHESIOL & CRITICAL CARE MED, 80- *Concurrent Pos:* Fel, Bowman Gray Sch Med, 71-72. *Mem:* Am Physiol Soc; Microcirculatory Soc; Am Soc Anesthesiol; Am Thoracic Soc. *Res:* Control of cerebral circulation in fetal, neonatal and adult animals; physiological interrelationships between the circulatory system. *Mailing Add:* 1510 Berwick Rd Ruxton MD 21204

TREADO, PAUL A, nuclear physics; deceased, see previous edition for last biography

TREADWAY, WILLIAM JACK, JR, b Johnson City, Tenn, Feb 22, 49; m 71. CHEMISTRY. *Educ:* Univ Ill, Urbana-Champaign, BS, 72; Loyola Univ Chicago, PhD(biochem), 76. *Prof Exp:* Teaching asst biochem, Loyola Univ Chicago Med Ctr, 71-73, res asst, 73-75; res assoc immunochem, Jefferson Med Col, 75-77; res assoc immunol, Sch Med, Temple Univ, 77-78; res instr, Bowman Gray Sch Med, 78-80, res asst prof med, 80-81; INSTR CHEM & BIOCHEM, PARKLAND COL, 81- *Concurrent Pos:* Chmn, Staff Develop Comt, Parkland Col, 85. *Honors & Awards:* Bausch & Lomb Sci Award, 67. *Mem:* Am Chem Soc; AAAS. *Res:* Immunology; immunochemistry; biochemistry. *Mailing Add:* 2303 Southwood Dr Champaign IL 61821-5418

TREADWELL, CARLETON RAYMOND, biochemistry; deceased, see previous edition for last biography

TREADWELL, ELLIOTT ALLEN, b Rockford, Ill, May 14, 47; m 80. COMPUTER SCIENCE, PHYSICAL CHEMISTRY. *Educ:* Cent State Univ Ohio, BS, 69; State Univ NY Stony Brook, MA, 71; Cornell Univ, MS, 73, PhD(exp high energy physics), 78. *Prof Exp:* Res asst high energy physics, Wilson Synchrotron, Cornell Univ, 73-78; res assoc, 78-81, ASSOC SCIENTIST ACCELERATOR & HIGH ENERGY PHYSICS, FERMI NAT ACCELERATOR LAB, 81-; PROF PHYSICS, CHICAGO STATE UNIV, 90- *Concurrent Pos:* Ford Found Panel Postdoctoral Fel Minorities, NSF grantee. *Mem:* Am Phys Soc; Nat Soc Black Physicists. *Res:* Neutrino-nucleon scattering at 400 billion electron volts, utilizing a 15 foot neon-hydrogen bubble chamber. *Mailing Add:* 1306 Naperville Rd Wheaton IL 60187

TREADWELL, GEORGE EDWARD, JR, b Selma, Ala, Dec 22, 41. BOTANY, BIOCHEMISTRY. *Educ:* King Col, BA, 64; Iowa State Univ, MS, 67, PhD(biochem, plant physiol), 70. *Prof Exp:* From asst prof to assoc prof, 70-85, PROF BIOL, EMORY & HENRY COL, 85- *Mem:* Am Chem Soc; Bot Soc Am. *Res:* Plant physiology; vitamin B-2; chromatography. *Mailing Add:* PO Drawer DDD Emory VA 24327

TREADWELL, KENNETH MYRON, b Cleveland, Ohio, May 5, 23; m 51; c 1. MECHANICAL ENGINEERING, NUCLEAR ENGINEERING. *Educ:* US Naval Acad, BS, 48; US Naval Postgrad Sch, BS, 54; Mass Inst Technol, SM, 55. *Prof Exp:* Engr thermal design, Westinghouse Elec Corp, 55-59, supvr, 59-64, mgr thermal & fuel design, 64-68, mgr reactor anal, 68-70, mgr reactor eng, 70-72, mgr naval fuel element develop, 72-77, mgr fuel element develop & statist, Bettis Atomic Power Lab, 77-82; PRES, TREADWELL CONSULT, 82- *Concurrent Pos:* sr consult, Westinghouse Elec, 82-87, O'Donnell & Assocs, 87-90. *Mem:* Am Soc Mech Engrs; Am Soc Naval Engrs; Sigma Xi; Am Nuclear Soc. *Res:* Analysis, engineering design and operation of naval and power nuclear reactors; engineering and materials science. *Mailing Add:* 4983 Parkvue Dr Pittsburgh PA 15236

TREAGAN, LUCY, b Novosibirsk, Russia, July 20, 24; US citizen; m 42; c 2. MICROBIOLOGY. *Educ:* Univ Calif, Berkeley, AB, 45, PhD(bact), 60. *Prof Exp:* Lectr, Col Holy Names, Calif, 61-66; from asst prof to prof biol, Univ San Francisco, 66-87; RETIRED. *Concurrent Pos:* Lectr, Univ San Francisco, 62-66. *Res:* Viral inhibitors; immunological study of interferons; effect of metals on the immune response. *Mailing Add:* Dept Biol Univ San Francisco Ignation Heights San Francisco CA 94117

TREANOR, CHARLES EDWARD, b Buffalo, NY, Oct 22, 24; m 50; c 5. PHYSICS, AERODYNAMICS. *Educ:* Univ Minn, BA, 47; Univ Buffalo, PhD(physics), 55. *Prof Exp:* Instr physics, Univ Buffalo, 53; physicist, Cornell Aeronaut Lab, Inc, 54-68, head aerodyn res dept, 68-78, vpres phys sci group, 78-82, VPRES & CHIEF SCIENTIST, CALSPAN CORP, 82- *Mem:* Fel Am Phys Soc; fel Am Inst Aeronaut & Astronaut; Combustion Inst. *Res:* High temperature gases; spectroscopy; hypersonic flows; molecular interactions. *Mailing Add:* 140 Segsbury Dr Williamsville NY 14221

TREAT, CHARLES HERBERT, b Cambridge, Mass, Dec 9, 31; m 56; c 3. NUMERICAL ANALYSIS, HEAT TRANSFER. *Educ:* Purdue Univ, BSCE, 53, MSE, 59; Univ NMex, PhD(mech eng), 68. *Prof Exp:* Instr eng graphics, Purdue Univ, 56-59; staff mem heat transfer, Sandia Corp, 59-61; instr mech eng, Univ NMex, 61-68; asst prof eng sci, 68-72, asst prof comput & info sci, 72-75, assoc prof comput & info sci, 75-78, assoc prof, 78-81, PROF & CHMN ENG SCI DEPT, TRINITY UNIV, TEX, 81- *Concurrent Pos:* Res assoc, United Nuclear Corp, 63 & Los Alamos Sci Lab, 64; consult, Sch Aerospace Med, Brooks AFB, Tex, 68, Q-Dot Corp, NMex, 69, Southwest Res Inst, 81-82 & small comput systs, 81-; dir instrumentation, Solar Heating & Cool Proj, Trinity Univ, 75-78. *Mem:* Am Soc Mech Eng. *Res:* Solar Energy. *Mailing Add:* Dept Eng Sci Trinity Univ 715 Stadium Dr Box 48 San Antonio TX 78284

TREAT, DONALD FACKLER, b Hartford, Conn, Feb 14, 25; m 49; c 4. MEDICAL EDUCATION. *Educ:* Univ Mich, Ann Arbor, BA, 46, MD, 49; Am Bd Family Pract, dipl. *Prof Exp:* Intern gen pract, Univ Hosp, Ann Arbor, Mich, 49-50, resident, 50-52; pvt pract, Springfield, Vt, 54-69; actg dir, Family Med Prog & dir Grad Educ Family Med, Univ Rochester, 78-81, assoc prof family med, Sch Med & Dent, 69-89, chmn dept, 81-89; RETIRED. *Concurrent Pos:* Vis prof, Univ Vt, 78, Univ Colo, 80 & Case Western Univ, 83. *Mem:* Soc Teachers Family Med; Am Acad Family Practioners. *Res:* Family medicine; medical audit and peer review; measurement of attitudinal change in residents; defining patterns of medical care. *Mailing Add:* Main St Holcomb NY 14469

TREAT, JAY EMERY, JR, b Trinidad, Colo, Nov 16, 20; m 45; c 4. PHYSICS. *Educ:* Univ Ariz, BS, 42; Cornell Univ, PhD, 54. *Prof Exp:* Mem staff magnetrons, Radiation Lab, Mass Inst Technol, 42-45; asst gen physics, Cornell Univ, 45-48, cosmic rays and nuclear physics, 48-51; asst prof nuclear physics, 51-58, ASSOC PROF PHYSICS, UNIV ARIZ, 58- *Mem:* Am Phys Soc; Am Asn Physics Teachers; Sigma Xi. *Res:* Cosmic rays; electromagnetic theory; elementary particles. *Mailing Add:* 1910 E Hawthorne Tucson AZ 85719

TREAT-CLEMONS, LYNDA GEORGE, b Wooster, Ohio, May 23, 46; m 79; c 1. GENETICS, BIOLOGY. *Educ:* Ohio State Univ, BS, 73, MS, 74, PhD(genetics), 78. *Prof Exp:* Assoc genetics, Ohio State Univ, 73-78; sr chemist, Biotrack, Inc, Mountain View, Calif, 84-86; postdoctoral fel, Stanford Univ, 86-89; CHROMATOGRAPHY COMMUN & APPLN CHEMIST, INTERACTION CHEMICALS INC, MOUNTAIN VIEW, CALIF, 89- *Mem:* Genetics Soc Am; AAAS; Soc Develop Biol; Sigma Xi. *Res:* Biochemistry. *Mailing Add:* 18781 Newsom Ave Cupertino CA 95014

TREBLE, DONALD HAROLD, b Liverpool, Eng, Apr 14, 34; m 59; c 2. BIOCHEMISTRY. *Educ:* Bristol Univ, BSc, 55; Univ Liverpool, PhD(biochem), 59. *Prof Exp:* From asst prof to assoc prof, 63-75, PROF BIOCHEM, ALBANY MED COL, 75- *Concurrent Pos:* Res fel biochem, Inst Animal Physiol, Babraham, Eng, 58-59, Cambridge Univ, 59-61 & Harvard Med Sch, 61-63. *Mem:* Am Soc Biol Chemists. *Res:* Lipid metabolism. *Mailing Add:* Dept Biochem Albany Med Col 43 New Scotland Ave Albany NY 12208

TREDICCE, JORGE RAUL, b Buenos Aires, Arg, Jan 11, 53; Ital citizen; m 76; c 2. OPTICS. *Educ:* Univ Buenos Aires, Arg, Lic en Fisica, 76; Univ Firenze, Italy, Laurea Fisica, 80. *Prof Exp:* Staff mem, Ctr Invest Appl Lasers, 76-77; staff mem, Nat Inst Ottica, 77-80, sr researcher, 80-85; from asst prof to assoc prof, 85-91, PROF, DREXEL UNIV, 91- *Concurrent Pos:* Vis prof, Drexel Univ, 85, Univ Nice, 91. *Mem:* Optical Soc Am; Am Phys Soc. *Res:* Laser physics; non linear dynamics; spatio-temporal structures in lasers. *Mailing Add:* Dept Physics Drexel Univ Philadelphia PA 19104

TREE, DAVID R, b Wanship, Utah, July 18, 36; m 58; c 5. ENERGY USAGE. *Educ:* Brigham Young Univ, BES, 62, MS, 63; Purdue Univ, PhD(mech eng), 66. *Prof Exp:* Assoc prof, 66-74, PROF MECH ENG, PURDUE UNIV, 74- *Concurrent Pos:* US Dept Health, Educ & Welfare fel, Univ Southampton, 70-71. *Mem:* Am Soc Heating, Refrig & Air-Conditioning; Am Soc Eng Educ. *Res:* Two-phase flow; air-conditioners and heat pumps; energy usage in residential buildings; thermal systems modeling; numerical methods applied to thermal systems. *Mailing Add:* Sch Mech Eng Purdue Univ West Lafayette IN 47907

TREECE, JACK MILAN, b Findlay, Ohio, Dec 19, 32; m 54; c 3. CLINICAL BIOCHEMISTRY. *Educ:* Ohio State Univ, BS, 54, MSc, 55, PhD(biochem genetics), 60. *Prof Exp:* Res asst biol res, Ohio Agr Exp Sta, 54-60; tech mgr genetics, Cent Ohio Breeding Asn, 60-62; Nat Acad Sci-Nat Res Coun res fel biochem, Animal Protein Pioneering Lab, Eastern Regional Res Utilization Lab, Agr Res Serv, USDA, 62-63; asst prof, Sch Vet Med, Univ Pa, 63-66; asst prof biochem, Univ Del, 66-70; dir, Blood Plasma & Components, Inc, 70-77; sr res biochemist, E I du Pont de Nemours & Co, Inc, 77-81; assoc prof clin biochem, Univ Fla, 81-85; dir mfg eng clin diag, Technicon Inst Corp, 85-87; sr mgr process develop & prod design, Coulter Diagnostics Div, Coulter Electronics Corp, 87-89; CONSULT, 89- *Concurrent Pos:* Consult, govt & indust. *Mem:* Sigma Xi; Am Chem Soc; Am Asn Clin Chemists; NY Acad Sci. *Res:* Clinical biochemistry and related product development. *Mailing Add:* 3842 Jeb Dr Winchester PA 22601

TREECE, ROBERT EUGENE, b Bluffton, Ohio, Oct 1, 27; m 55; c 2. ECONOMIC ENTOMOLOGY. *Educ:* Ohio State Univ, BS, 51, MS, 53; Cornell Univ, PhD(econ entom), 57. *Prof Exp:* Asst exten specialist entom, Rutgers Univ, 56-58; asst prof entom & asst entomologist, Agr Exp Sta, Ohio State Univ, 58-64, from assoc prof to prof, Ohio Agr Res & Develop Ctr, 64-73, ASSOC CHMN ENTOM DEPT, OHIO AGR RES & DEVELOP CTR, OHIO STATE UNIV, 73- *Mem:* Entom Soc Am; AAAS; Coun Agr Sci & Technol. *Res:* Bionomics and control of insect pests of livestock and forage crops. *Mailing Add:* Dept Entom Ohio Agr Res & Develop Ctr Wooster OH 44691

TREFETHEN, JOSEPH MUZZY, b Kent's Hill, Maine, May 27, 06; m 31; c 3. GEOLOGY. *Educ:* Colby Col, AB, 31; Univ Ill, MS, 32; Univ Wis, PhD(geol), 35. *Prof Exp:* Instr geol, Univ Mo, 35-38; from asst prof to prof, 38-71, EMER PROF, UNIV MAINE, 71-; CONSULT GEOLOGIST, 71- *Concurrent Pos:* Field geologist, State Geol Surv, Maine, 29-32, dir, 42-55; dir, State Geol Surv, Wis, 35; vis lectr, Stephens Col, 35-36. *Mem:* Fel Geol Soc Am; Sigma Xi. *Res:* Structural and applied geology; economic geology of the non-metallics; engineering geology. *Mailing Add:* PO Box 99 Friendship ME 04547

TREFETHEN, LLOYD MACGREGOR, b Boston, Mass, Mar 5, 19; m 44; c 2. FLUID MECHANICS. *Educ:* Webb Inst Naval Archit, BS, 40; Mass Inst Technol, MS, 42; Cambridge Univ, PhD, 50. *Prof Exp:* Appln engr, Gen Elec Co, 40-44; sci consult, US Off Naval Res, Eng, 47-50; physicist, tech aide to chief scientist & exec secy, US Naval Res Adv Comt, Washington, DC, 50-51; exec secy, Nat Sci Bd & tech aide to dir, NSF, 51-54; asst prof mech eng, Harvard Univ, 54-58; chmn dept, 58-69, 81, PROF MECH ENG, TUFTS UNIV, 58- *Concurrent Pos:* NSF fel, Cambridge Univ, 56; vis prof, Univ Sydney, 65, 72 & 79, Stanford Univ, 79; vis fel, Seattle Res Ctr, Battelle Mem Inst, 71; hon res assoc, Harvard Univ, 78; Russell Springer prof mech eng, Univ Calif, Berkeley, 86. *Honors & Awards:* Golden Eagle Award, Comn Int Nontheatrical Events, 67; Prix de Physique, Inst Sci Film Festival, 68. *Mem:* Am Soc Mech Engrs. *Res:* Fluid mechanics. *Mailing Add:* Dept Mech Eng Tufts Univ Medford MA 02155

TREFETHEN, LLOYD NICHOLAS, b Boston, Mass, Aug 30, 55. NUMERICAL ANALYSIS. *Educ:* Harvard Col, BA, 77; Stanford Univ, MS, 80, PhD(comput sci), 82. *Prof Exp:* Adj asst prof comput sci, Courant Inst, New York Univ, 82-84; ASST PROF MATH, MASS INST TECHNOL, 84- *Honors & Awards:* Fox Prize, Inst Math & Appln, UK, 85. *Mem:* Soc Indust & Appl Math; Am Math Soc; Inst Elec & Electronics Engrs. *Res:* Numerical analysis; solution of differential equations; numerical conformal mapping; approximation theory. *Mailing Add:* Dept Math Rm 2-383 Mass Inst Technol Cambridge MA 02139

TREFFERS, HENRY PETER, b New York, NY, Aug 21, 12; wid; c 2. MICROBIOLOGY. *Educ:* Columbia Univ, AB, 33, PhD(chem), 37. *Hon Degrees:* Yale Univ, MA, 50. *Prof Exp:* Asst chem, Univ Exten, Columbia Univ, 33-36, instr biochem, 36-42; asst prof comp path & biochem, Harvard Med Sch, 42-44; asst prof immunol, Sch Med, Yale Univ, 44-46, assoc prof immunochem, 46-49, prof microbiol & Davenport Col fel, 49-69, chmn dept microbiol, 50-61, prof path, 69-83; RETIRED. *Concurrent Pos:* Fulbright res scholar, Univ Otago, NZ, 54; USPHS spec fel, Wright-Fleming Inst, St Mary's Hosp, London, Eng, 61-62; consult, USPHS, 48-50, 58-61 & 66-69; asst, Presby Hosp, 36-42; assoc mem, Comn Immunization, US Army Epidemiol Bd, 47-55; vchmn, Sect Microbiol, Chem & Biol Coord Ctr, Nat Res Coun, 48-53; mem, Microbiol Panel, US Off Naval Res, 51-54; ed, J Am Asn Immunol, 52-57; vis instr, Univ Calif, San Diego, 69-70. *Mem:* Am Asn Immunologists. *Res:* Chemistry of non-aqueous solutions; quantitative chemistry of immune reactions; antibiotic resistance; bacterial genetics. *Mailing Add:* 550 Amity Rd Bethany CT 06525

TREFFERS, RICHARD ROWE, b Bethany, Conn, Nov 7, 47; m 72; c 2. ASTRONOMY. *Educ:* Yale Univ, BA, 69; Univ Calif, Berkeley, PhD(astron), 73. *Prof Exp:* Asst res astronomer, Univ Ariz, 74-76; asst res astronomer, Ames Res Ctr, NASA, 76; ASST RES ASTRONOMER, UNIV CALIF, 76- *Mem:* Am Astron Soc. *Res:* Infrared astronomy and spectroscopy as applied to planets and interstellar matter. *Mailing Add:* Dept Astron Univ Calif 601 Campbell Hall Berkeley CA 94720

TREFFERT, DAROLD ALLEN, b Fond du Lac, Wis, Mar 12, 33; m 55; c 4. MEDICINE, PSYCHIATRY. *Educ:* Univ Wis, BA, 55, MD, 58. *Hon Degrees:* DH, Marion Col, 87. *Prof Exp:* Intern, Sacred Heart Gen Hosp, Eugene, Ore, 58-59; resident psychiat, Univ Wis Hosp, Madison, 59-62; dir, Child-Adolescent Unit, Winnebago Ment Health Inst, 62-64, dir, 64-79; EXEC DIR, FOND DU LAC COUNTY HEALTH CARE CTR, 79- *Concurrent Pos:* Assoc clin prof, Dept Psychiat, Univ Wis Med Sch, Madison, 65-78; chmn, Controlled Substances Bd Wis, 70-; mem, Adv Coun Health Probs Educ, Dept Pub Instr Wis, 73- & Tech Adv Comt, Joint Comn Accreditation Hosps. *Mem:* Am Asn Psychiat Adminr (pres, 83-85); Am Psychiat Asn; Am Col Psychiat; AMA. *Res:* Infantile autism; alcoholism and drug abuse; rights of the mentally ill; the Savant Syndrome (idiot savant), islands of genius in severely handicapped persons; implications for understanding normal brain function, especially memory. *Mailing Add:* Brookside Med Ctr 481 E Division St Fond du Lac WI 54935

TREFIL, JAMES S, b Chicago, Ill, Sept 10, 38; m 72; c 5. PARTICLE PHYSICS. *Educ:* Univ Ill, BS, 60; Oxford Univ, BA & MA, 62; Stanford Univ, MS & PhD(physics), 66. *Prof Exp:* Res assoc physics, Stanford Linear Accelerator Ctr, 66; Air Force Off Sci Res fel, Europ Ctr Nuclear Res, 66-67; res assoc, Mass Inst Technol, 67-68; asst prof, Univ Ill, Urbana, 68-70; assoc prof & fel, Ctr Advan Studies, 70-75, PROF PHYSICS, UNIV VA, 75-; CLARENCE J ROBINSON PROF PHYSICS, GEORGE MASON UNIV. *Concurrent Pos:* John Simon Guggeheim fel. *Mem:* Fel Am Phys Soc. *Res:* Theoretical studies of mass extinctions. *Mailing Add:* Dept Physics George Mason Univ Fairfax VA 22030

TREFNY, JOHN ULRIC, b Greenwich, Conn, Jan 28, 42; div; c 1. PHYSICS. *Educ:* Fordham Univ, BS, 63, Rutgers Univ, New Brunswick, PhD(physics), 68. *Prof Exp:* Res assoc physics, Cornell Univ, 67-69; asst prof physics, Wesleyan Univ, 69-77; from asst prof to assoc prof, 77-84, PROF PHYSICS, COLO SCH MINES, 84-, HEAD DEPT, 90- *Concurrent Pos:* Consult, Inst for Future, 70, Solar Energy Res Inst, 78-81, 86- *Mem:* Am Phys Soc; Sigma Xi; Am Asn Physics Teachers. *Res:* Experimental studies of crystalline solids and amorphous semiconductors; superconductivity general topics in low temperature physics; thermoelectric materials. *Mailing Add:* Dept Physics Colo Sch Mines Golden CO 80401

TREFONAS, LOUIS MARCO, b Chicago, Ill, June 21, 31; m 57; c 6. STRUCTURAL CHEMISTRY, PHYSICAL BIOCHEMISTRY. *Educ:* Univ Chicago, BA, 51, MS, 54; Univ Minn, PhD, 59. *Prof Exp:* From asst prof to prof chem, Univ New Orleans, 59-81, chmn dept, 64-80; assoc vpres, res, 81-86, ASSOC VPRES ACAD AFFAIRS & GRAD DEAN, UNIV CENT FLA, 86- *Concurrent Pos:* NIH spec fel, 72-73; hon res assoc, Harvard Univ, 72-73. *Mem:* Am Chem Soc. *Res:* Molecular structure studies by x-ray diffraction; small ring nitrogen compounds; structures of biologically interesting compounds by x-ray diffraction techniques. *Mailing Add:* Off Grad Studies Univ Cent Fla Orlando FL 32816

TREGILLUS, LEONARD WARREN, physical chemistry; deceased, see previous edition for last biography

TREGLIA, THOMAS A(NTHONY), engineering; deceased, see previous edition for last biography

TREHAN, RAJENDER, b Bihar, India. SOLID STATE SURFACE SCIENCE. *Educ:* Panjab Univ, BSc, 73, MSc, 74; Univ Utah, PhD(chem), 85. *Prof Exp:* Sci officer radiochemistry, Bhabha Atomic Res Ctr, 75-78; postdoctoral fel phys chem, Pa State Univ, 85-87; lectr, Rutgers Univ, 87-88; RES ASSOC CHEM, CTR SUPERCONDUCTIVITY, UNIV HOUSTON, 88- *Mem:* Am Phys Soc; Am Chem Soc. *Res:* Electron and ion beam spectroscopy; high temperature superconductivity. *Mailing Add:* Dept Chem Univ Houston Houston TX 77004

TREHU, ANNE MARTINE, b Princeton, NJ, Jan 7, 55. SEISMOLOGY, MARINE GEOPHYSICS. *Educ:* Princeton Univ, BA, 75; Mass Inst Technol, PhD(marine seismol), 82. *Prof Exp:* ASSOC, US GEOL SURV, 82-; AT COL OCEANOG, ORE STATE UNIV. *Mem:* Am Geophys Union. *Res:* Interpretation of data recorded by ocean bottom seismometers to study offshore crystal structure and earthquake activity. *Mailing Add:* Col Oceanog Ore State Univ Corvallis OR 97331

TREHUB, ARNOLD, b Malden, Mass, Oct 19, 23; m 50; c 3. PSYCHOPHYSIOLOGY. *Educ:* Northeastern Univ, AB, 49; Boston Univ, MA, 51, PhD(clin psychol), 54. *Prof Exp:* Clin psychologist, Northampton Vet Admin Med Ctr, 54-59, dir psychol res lab, 59-82; MEM GRAD FAC, UNIV MASS, AMHERST, 70-, ADJ PROF, 71- *Concurrent Pos:* Vis lectr, Univ Mass, 59-66; lectr, Clark Univ, 60-62. *Mem:* AAAS; NY Acad Sci; Soc Neurosci; Cognitive Sci Soc. *Res:* Electrophysiology of brain; biomathematics; artificial intelligence; neurophysiology. *Mailing Add:* 145 Farview Way Amherst MA 01002

TREI, JOHN EARL, b Freeport, Ill, Sept 19, 39; m 66; c 2. RUMINANT NUTRITION, PHYSIOLOGY. *Educ:* Univ Ill, BS, 61; Univ Ariz, MS, 63, PhD(agr biochem, nutrit), 66. *Prof Exp:* Res & teaching assoc animal sci, Univ Ariz, 63-66; sr scientist animal nutrit, Smith Kline & French Labs, 66-71, sr investr, Smith Kline Corp, 71-73, asst mgr develop animal health prod, 73-74; assoc prof, 74-80, PROF ANIMAL SCI, CALIF STATE POLYTECH UNIV, 80-, SUPVR, BEEF UNIT & FEEDMILL, 75- *Concurrent Pos:* Actg chair, Animal Sci Dept; chmn, steering comt, Calif Animal Nutrit Conf, 86-87. *Mem:* Am Soc Animal Soc; Am Inst Biol Sci; NY Acad Sci; Am Dairy Sci Asn; Sigma Xi; Am Registry Prof Animal Scientists. *Res:* Energy metabolism; feed additive evaluations. *Mailing Add:* 2117 Villa Maria Claremont CA 91711

TREICHEL, PAUL MORGAN, JR, b Madison, Wis, Dec 4, 36; m 61; c 2. INORGANIC CHEMISTRY. *Educ:* Univ Wis, BS, 58; Harvard Univ, AM, 60, PhD(chem), 62. *Prof Exp:* Teaching asst chem, Harvard Univ, 61-62; NSF fel, Queen Mary Col, London, 62-63; from asst prof to assoc prof, 63-72, PROF CHEM, UNIV WIS-MADISON, 72-, CHMN, CHEM, 86- *Mem:* Am Chem Soc; Royal Soc Chem. *Res:* Organometallic chemistry, including metal carbonyls, cyclopentadienyls, alkyls and related compounds; organophosphorus and organoboron chemistry. *Mailing Add:* Dept Chem Univ Wis Madison WI 53706

TREICHEL, ROBIN STONG, b Billings, Mont, Feb 13, 50; m 85. TUMOR IMMUNOLOGY, IMMUNOTHERAPY. *Educ:* Macalester Col, BA, 72; Univ Wis-Madison, MS, 78, PhD(genetics), 81. *Prof Exp:* Postdoctoral tumor immunol, Univ Minn Med Sch, 81-84; postdoctoral molecular immunol, Southwest Found Biomed Res, San Antonio, Tex, 85-86; res scientist oncol, Childrens Hosp Orange County, Calif, 86-87; ASST PROF BIOL, OBERLIN COL, OHIO, 87- *Concurrent Pos:* Instr, Trinity Univ, San Antonio, Tex, 85-86; adj asst prof, Univ Tex Health Sci Ctr, San Antonio, Tex, 85-86. *Mem:* Sigma Xi; Transplantation Soc; AAAS. *Res:* Immunotherapy for the treatment of human leukemia; investigation of the susceptibility of multidrug-resistant leukemic cells to lysis mediated by natural cytotoxic cells and by antibody and complement; development of immunotoxins for purging autologous bone marrow grafts for subsequent transplantation into patients following high dose chemoradiotherapy. *Mailing Add:* Dept Biol Oberlin Col Oberlin OH 44074

TREICHLER, RAY, b Rock Island, Ill, Sept 10, 07; m 42. AGRICULTURAL CHEMISTRY, BIOLOGICAL CHEMISTRY. *Educ:* Pa State Col, BS & MS, 29; Univ Ill, PhD, 39. *Prof Exp:* Chemist, Exp Sta, Agr & Mech Col, Tex, 29-37, 39-40; spec asst animal nutrit, Univ Ill, 37-38; vis scientist, Univ Ill, 40-41; chemist, US Fish & Wildlife Serv, 41-42, technologist, 42-43, chemist, 43-44; res & develop div & head biol activities sect, Off Qm Gen, US Dept Army, 45-52, head chem & biol br, 53; chief agts br, Res & Eng Command, US Army Chem Corps, 53-55, asst to dir med res, 55-58; scientist, USAF, 58-68; TECH SERV MGR, H D HUDSON MFG CO, 68- *Mem:* NY Acad Sci; Am Soc Trop Med & Hyg; Entom Soc Am; Am Mosquito Control Asn; Am Soc Agr Engrs; Am Chem Soc; Soc Vector Ecol; Am Soc Testing & Mat; Sigma Xi. *Res:* Biology of natural products; environmental pollution; research and development of pesticides; application equipment for pesticides; pesticide dissemination systems. *Mailing Add:* Essex Apt 402 4740 Conn Ave NW Washington DC 20008

TREICK, RONALD WALTER, b Scotland, SDak, June 8, 34; m 55; c 2. MICROBIAL PHYSIOLOGY. *Educ:* Univ SDak, BA, 56, MA, 57; Ind Univ, PhD(bact), 65. *Prof Exp:* Med technician, Univ SDak, 57-58; res asst infectious dis, Upjohn Co, 58-62; from asst prof to assoc prof, 65-77, PROF MICROBIOL, MIAMI UNIV, 77- *Concurrent Pos:* Res grants, Miami Univ, 65-66, 68-69, 70-71 & 72- *Mem:* AAAS; Am Soc Microbiol; Sigma Xi. *Res:* Antimicrobial agents; effect of metabolic inhibitors on bacterial macromolecular synthesis; inhibition of bacterial luminescence; bacteria lipid metabolism. *Mailing Add:* Dept Microbiol Miami Univ Oxford OH 45056

TREIMAN, SAM BARD, b Chicago, Ill, May 27, 25; m 52; c 3. THEORETICAL PHYSICS, ELEMENTARY PARTICLES. *Educ:* Univ Chicago, PhD(physics), 52. *Prof Exp:* Res assoc physics, Univ Chicago, 52; from instr to assoc prof, 52-63, PROF PHYSICS, PRINCETON UNIV, 63- *Concurrent Pos:* Chmn, Physics Dept, Princeton Univ, 81-87, Univ Res Bd, 88- *Honors & Awards:* Oersted Medal, Am Asn Physics Teachers. *Mem:* Nat Acad Sci; Am Phys Soc; Am Acad Arts & Sci. *Res:* Cosmic ray physics; fundamental particles; field theory; elementary particle theory, especially weak interactions, unified field theories and connections with cosmology. *Mailing Add:* Dept Physics Princeton Univ Princeton NJ 08540

TREISTMAN, STEVEN NEAL, b New York, NY, May 3, 45; m 67; c 2. NEUROBIOLOGY. *Educ:* State Univ NY, Binghamton, BA, 67; Univ NC, Chapel Hill, PhD(neurobiol), 72. *Prof Exp:* Teacher biol, Heuvelton Cent Sch, 67-69; neurobiol trainee, Sch Med, Univ NC, 69-72; fel, Sch Med, NY Univ, 72-75; res assoc, Friedrich Miescher Inst, Basel, 75-76; asst prof biol, Bryn Mawr Col, 76-80; staff scientist exp biol, 80-84, SR SCIENTIST, WORCESTER FOUND EXP BIOL, 84- *Concurrent Pos:* Scottish Rite Found Schizophrenia res grant, 73-75; NSF grant, 77-80; NIH res grant, 79-82, NSF res grant, 81-; corp mem, Marine Biol Lab, Woods Hole, Mass, 80-; adj assoc prof pharmacol, Univ Mass Med Ctr, 82- *Mem:* Soc Neurosci; AAAS; Biophys Soc. *Res:* Neurobiology of simple systems; biochemical correlates of long-term changes in electrical membrane characteristics. *Mailing Add:* Worcester Found Exp Biol 222 Maple Ave Shrewsbury MA 01545

TREITEL, SVEN, b Freiburg, Ger, Mar 5, 29; US citizen; m 58; c 4. GEOPHYSICS. *Educ:* Mass Inst Technol, BS, 53, MS, 55, PhD(geophys), 58. *Prof Exp:* Geophysicist, Standard Oil Co Calif, 58-60; res assoc commun theory, Pan Am Petrol Corp, 60-65, group supvr, 66-71, mgr res sect, 71-74, sr res assoc, 74-77, RES CONSULT, AMOCO PROD CO, 77- *Honors & Awards:* Fessenden Medal, Soc Explor Geophys, 69. *Mem:* Fel Inst Elec & Electronics Engrs; hon mem Soc Explor Geophys; Sigma Xi; Europ Asn Explor Geophys; Seismol Soc Am. *Res:* Application of statistical communication theory to seismic analysis. *Mailing Add:* Amoco Prod Co Res Ctr PO Box 3385 Tulsa OK 74102

TREITERER, JOSEPH, b Grafrath, Ger, Dec 7, 18; m 53; c 2. TRANSPORTATION ENGINEERING. *Educ:* Munich Tech Univ, Dipl Ing, 49. *Hon Degrees:* DSc, Munich Tech Univ, 58. *Prof Exp:* Sci asst civil engr & lectr munic eng & city traffic, Munich Tech Univ, 48-55; chief res officer, Nat Inst Rd Res Coun Sci & Indust Res, SAfrica, 56-63; from asst prof to assoc prof civil eng, 63-72, PROF CIVIL ENG, OHIO STATE UNIV, 72- *Concurrent Pos:* Consult, Water Supply, Sewage & Traffic, Munich, 50-55; mem, Ger Res Coun Hwy Pract, 55; lectr, Pretoria Univ, 60-63; mem tech comt, Inter-Provincial Adv Bd Rd Traffic Legis, Repub SAfrica, 60-63; chmn steering comt transp planning, Natural Resources Develop Coun, 61-63; mem, Hwy Res Bd, Nat Acad Sci-Nat Res Coun, 64. *Res:* Automatic control and guidance of motor vehicles; theory of traffic flow; aerial photogrammetry techniques for traffic surveys; transportation systems. *Mailing Add:* 171 Medick Way Worthington OH 43085

TRELA, EDWARD, physical metallurgy; deceased, see previous edition for last biography

TRELA, JOHN MICHAEL, b Greenfield, Mass, Nov 15, 42; m 68; c 3. BIOCHEMISTRY, MICROBIOLOGY. *Educ:* Tufts Univ, BS, 65; Univ Mass, Amherst, MS, 67; State Univ NY, Stony Brook, PhD(biochem), 71. *Prof Exp:* Instr biol sci, State Univ NY, Stony Brook, 70-72; asst prof, 72-78, ASSOC PROF BIOL SCI, UNIV CINCINNATI, 78- *Mem:* Am Soc Microbiol; Am Asn Univ Prof. *Res:* Microbial physiology, with special emphasis on micromolecular synthesis in thermophilic bacteria. *Mailing Add:* Dept Biol Sci Univ Cincinnati Main Campus Cincinnati OH 45221

TRELA, WALTER JOSEPH, b Pawtucket, RI, May 31, 36; m 62; c 2. LOW TEMPERATURE PHYSICS. *Educ:* Brown Univ, BS, 58; Stanford Univ, PhD(physics), 67. *Prof Exp:* Asst prof physics, Haverford Col, 67-72; staff mem, Los Alamos Nat Lab, 73-80, asst div leader, 80-83, prog mgr, 83-88, STAFF MEM, LOS ALAMOS NAT LAB, 88- *Mem:* Am Phys Soc; AAAS. *Res:* Low temperature physics with primary emphasis on liquid helium and superconductivity; quantum fluids and solids; atomic, molecular and solid state physics; x-ray optics; synchrotron radiation. *Mailing Add:* 137 San Juan Los Alamos NM 87544

TRELAWNY, GILBERT STERLING, b Cincinnati, Ohio, Nov 12, 29; m 50; c 3. MICROBIAL PHYSIOLOGY. *Educ:* Delaware Valley Col, BS, 57; Lehigh Univ, MS, 60, PhD(biol), 66. *Prof Exp:* From instr to asst prof microbiol, Delaware Valley Col, 57-66; PROF BIOL, JAMES MADISON UNIV, VA, 66-, HEAD DEPT, 71- *Concurrent Pos:* NSF res grant, 67-69. *Res:* Microbial metabolism and nutrition. *Mailing Add:* Dept Biol James Madison Univ Harrisonburg VA 22807

TRELEASE, RICHARD DAVIS, b Chicago, Ill, Sept 23, 17; m 42; c 6. FOOD CHEMISTRY. *Educ:* Univ Ill, BS, 40. *Prof Exp:* Chemist, Swift & Co, 40-42, res chemist poultry, 43-44; food technician, Qm Food & Container Inst, 44-45; asst to vpres res, Swift & Co, 48-50, head frozen food res, 50-63, mgr processed meats res, 63-71, mgr meat res, 71-72, gen mgr, processed meats res, 72-73, mgr contract res, 73-80, mgr processed meats res, 80-83; CONSULT, 83- *Mem:* Am Chem Soc; Inst Food Technol. *Res:* Food processing and preservation; poultry products; frozen foods; cured meats. *Mailing Add:* Trelease Consults Inc 504 Chicago Ave Downers Grove IL 60515-3811

TRELEASE, RICHARD NORMAN, b Las Vegas, Nev, Nov 6, 41; m 65; c 2. PLANT CELL BIOLOGY, PLANT PHYSIOLOGY. *Educ:* Univ Nev, Reno, BS, 63, MS, 65; Univ Tex, Austin, PhD(cell biol), 69. *Prof Exp:* NIH fel, Univ Wis-Madison, 69-71; from asst prof to assoc prof, 71-81, PROF BIOL, ARIZ STATE UNIV, 81- *Concurrent Pos:* Adv panel mem, NSF Cell Biol Prog, 78-80, Cellular Biochem Prog, 88-90. *Mem:* AAAS; Am Soc Cell Biol; Am Soc Plant Physiol; Sigma Xi. *Res:* Oilseed metabolism, especially cottonseeds, aspects of maturation and germination; protein and membrane intracellular trafficking; enzymology; cell fractionation. *Mailing Add:* Dept Bot Ariz State Univ Tempe AZ 85287-1601

TRELFORD, JOHN D, b Toronto, Ont, Feb 7, 31; US citizen; c 3. OBSTETRICS & GYNECOLOGY, ONCOLOGY. *Educ:* Univ Toronto, MD, 56; FRCS(C), 64; FRCOG. *Prof Exp:* Asst prof obstet & gynec, Med Sch, Ohio State Univ, 65-70; assoc prof, 70-75, PROF OBSTET & GYNEC, SCH MED, UNIV CALIF, DAVIS, 75- *Concurrent Pos:* Grants, Univ Calif, Davis, 71-72; dir, Oncol Serv, Sacramento Med Ctr, 72-; consult, Vet Admin Hosp, Martinez, Calif, 72-; dir, Am Cancer Soc, Yolo County. *Mem:* Fel Am Col Surgeons; Soc Obstet & Gynec Can; fel Am Col Obstet & Gynec. *Res:* Antigenicity of the trophoblastic cell and its relationship to cancer. *Mailing Add:* 3420 Lakeview Dr El Macero CA 95618

TRELKA, DENNIS GEORGE, b Lorain, Ohio, June 11, 40; m 66; c 2. EXERCISE PHYSIOLOGY. *Educ:* Kent State Univ, BA, 67, MA, 68; Cornell Univ, PhD(animal physiol), 72. *Prof Exp:* Res asst invert zool, Kent State Univ, 67-68; res asst animal physiol, Cornell Univ, 68-72; from asst prof to assoc prof, 77-78, PROF BIOL, WASH & JEFFERSON COL, 88- *Mem:* Zool Soc Am; AAAS; Sigma Xi; Nat Strength & Conditioning Asn. *Res:* Research in areas of strength training; histological and pharmacological studies on the central nervous system of common garden slugs. *Mailing Add:* Dept Biol Wash & Jefferson Col Washington PA 15301-4801

TRELOAR, ALAN EDWARD, b Melbourne, Australia, Sept 27, 02; nat US; m 29, 49; c 3. MEDICAL ANTHROPOLOGY, BIOMETRICS. *Educ:* Univ Sydney, BSc, 26; Univ Minn, MS, 29, PhD(agr biochem), 30. *Prof Exp:* Demonstr geol, Univ Sydney, 24-25; from instr to assoc prof biomet, Univ Minn, 29-47, prof biostatist, Sch Pub Health, 47-56; asst dir res, Am Hosp Asn, 56-59; chief statist & anal br, Div Res Grants, NIH, 59-61, spec asst to dir for biomet, Nat Inst Neurol Dis & Blindness, 61-66, chief reproduction anthropometry sect, Nat Inst Child Health & Human Develop, 66-74; dir mensuration & reproduction hist res prog, Ruth E Boynton Health Serv, Univ Minn, Minneapolis, 74-77; from assoc dir to dir to consult, menstrual & reproductive health res prog, Dept Obstet & Gynec, Univ NC, 77-81; vpres, Ctr Advan Reproductive Health, Chapel Hill, NC, 82-85; RETIRED. *Concurrent Pos:* Consult, USPHS, 53-58; dir, Hosp Res & Educ Trust, 57-59; auxiliary res prof, Col Nursing, Univ Utah, 81- *Mem:* Human Biol Coun; fel AAAS; fel Am Statist Asn; Pop Asn Am; Am Fertil Soc. *Res:* Biometry of the menstrual cycle and gestation period; relationship of menstrual history to illness; effects of oral contraceptives. *Mailing Add:* 4064 Army St San Francisco CA 94131-1919

TRELSTAD, ROBERT LAURENCE, b Redding, Calif, June 16, 40; m 61; c 4. EMBRYOLOGY, PATHOLOGY. *Educ:* Columbia Univ, BA, 61; Harvard Univ, MD, 66. *Prof Exp:* Intern path, Mass Gen Hosp, 66-67; NIH res assoc embryol, 67-69; ASST PATHOLOGISTS, MASS GEN HOSP, SHRINERS BURN INST, 72-, CHIEF PATH, 75-; CHMN & PROF, PATH DEPT, ROBERT WOOD JOHNSON MED SCH. *Concurrent Pos:* Helen Hay Whitney Found res fel path, Mass Gen Hosp, 69-72; Am Cancer Soc fac res award, 72-77; asst prof path, Harvard Med Sch, 72-77; assoc prof path, Harvard Med Sch, 77-; mem cell biol study sect, NIH, 77-; assoc ed, Develop Biol, 78- *Mem:* Soc Develop Biol; Am Soc Cell Biol; Am Soc Zoologists; Int Soc Develop Biol. *Res:* Biological function of connective tissues in normal growth and development and in disease. *Mailing Add:* Pathol Dept Univ Med & Dent NJ Robert Wood Johnson Med Sch Piscataway NJ 08854

TREMAINE, JACK H, b Galt, Ont, June 15, 28; m 56; c 3. PLANT VIROLOGY. *Educ:* McMaster Univ, BSc, 51, MSc, 53; Univ Pittsburgh, PhD(virol), 57. *Prof Exp:* Plant pathologist, Can Dept Agr, 52-64, plant virologist, 64-91; RETIRED. *Mem:* Am Phytopath Soc; Can Phytopath Soc. *Res:* Structure of plant viruses; in vitro reassembly stabilizing interactions using monoclonal antibodies as structural probes of plant viruses; studies of plant viruses in pseudorecombinant studies. *Mailing Add:* 3936 W 22nd Vancouver BC V6S 1K1 Can

TREMAINE, PETER RICHARD, b Toronto, Ont, Dec 20, 47; m 72; c 2. PHYSICAL CHEMISTRY. *Educ:* Univ Waterloo, Ont, BSc, 69; Univ Alta, PhD(phys chem), 74. *Prof Exp:* Nat Res Coun Can fel surface chem, Pulp & Paper Res Inst Can, Montreal, 74-75; asst res officer phys chem, Whiteshell Nuclear Res Estab, Atomic Energy Can, 75-80; assoc res officer, Alta Res Coun, 80-85, sr res officer phys chem & mgr fund res, Oil Sands Res Dept, 85-90; PROF & HEAD, DEPT CHEM, MEM UNIV NF, 90- *Concurrent Pos:* Vis assoc prof, Geol Dept, Univ Alt, 84; adj prof geol, Univ Alta, 87-90, chem eng, 89-90. *Mem:* Chem Inst Can; Am Chem Soc. *Res:* Thermodynamics of high temperature, high pressure aqueous systems; hydrothermal geochemistry. *Mailing Add:* Dept Chem Mem Univ Mfld St John's NF A1B 3X7 Can

TREMAINE, SCOTT DUNCAN, b Toronto, Ont, May 25, 50. ASTROPHYSICS. *Educ:* McMaster Univ, BSc, 71; Princeton Univ, MA, 73, PhD(physics), 75. *Prof Exp:* Res fel, Calif Inst Technol, 75-77 , Cambridge Univ, 77-78 & Inst Advan Studt, 78-81; assoc prof, Mass Inst Technol, 81-85; DIR, CAN INST THEORET ASTROPHYSICS, 85- *Concurrent Pos:* E W R Steacie Mem fel. *Honors & Awards:* Helen B Warner Prize; Steacie Prize; Rutherford Medal in Physics; Carlyle S Beals Award. *Mem:* Am Astron Soc; Int Astron Union; Can Astron Soc. *Res:* Theoretical studies in dynamics; galactic structure; cosmology; planetary rings. *Mailing Add:* CITA McLennan Lab Univ Toronto Toronto ON M5S 1A1 Can

TREMBA, EDWARD LOUIS, cratering, geotechnical engineering, for more information see previous edition

TREMBLAY, ANDRÉ -MARIE, b Montreal, Que, Jan 2, 53; m 75; c 2. SUPERCONDUCTIVITY. *Educ:* Univ Montreal, BSc, 74; Mass Inst Technol, PhD(physics), 78. *Prof Exp:* Postdoc fel res, Cornell Univ, 78-80; PROF RES TRAINING, UNIV SHERBROOKE, 80- *Concurrent Pos:* Assoc prof, Univ Provence, France, 82; vis scientist, Cornell Univ, 86-87; Steacie fel, Natural Sci & Eng Res Coun, 87. *Honors & Awards:* Herzberg Medal, Can Asn Physicists, 86. *Mem:* Am Phys Soc; Can Asn Physicists. *Res:* Scaling and universality in disordered systems; fractal and multifractal properties of percolating systems; analogies with critical phenomena; 1/f noise; properties of stongly correlated electronic systems; superconductivity; quantum Monte Carlo simulations. *Mailing Add:* Physics Dept Univ Sherbrooke Sherbrooke PQ J1K 2R1 Can

TREMBLAY, GEORGE CHARLES, b Pittsfield, Mass, Oct 13, 38; c 4. BIOCHEMISTRY. *Educ:* Mass Col Pharm, BS, 60; St Louis Univ, PhD(biochem), 65. *Prof Exp:* Am Cancer Soc fel biol chem, Harvard Univ, 65-66; from asst prof to assoc prof, 66-75, PROF BIOCHEM, UNIV RI, 75-, CHMN DEPT BIOCHEM & BIOPHYS, 84- *Concurrent Pos:* Res grants, Nat Inst Child Health & Human Develop, Nat Inst Arthritis & Metab Dis, Nat Cancer Inst, NSF, & Nat Inst Arthritis Metab & Diag Dis. *Mem:* AAAS; Am Soc Biol Chemists; Soc Study Inborn Errors Metab. *Res:* Regulatory mechanisms in metabolism. *Mailing Add:* Dept Biochem & Biophysics Univ RI 117 Morrill Hall Kingston RI 02881

TREMBLAY, GILLES, b Montreal, Que, Apr 18, 28; m 75; c 3. PATHOLOGY. *Educ:* Univ Montreal, BA, 48, MD, 53. *Prof Exp:* Resident path, Hotel-Dieu Hosp, 54-55; resident path, New Eng Deaconess Hosp, Boston, Mass, 55-57; pathologist, Hotel-Dieu Hosp, Montreal, 59-61; pathologist, Notre-Dame Hosp, 61-64; prof path & chmn dept, Fac Med, Univ Montreal, 64-70; PROF PATH, McGILL UNIV, 70-, SR PATHOLOGIST, ROYAL VICTORIA HOSP, 77- *Concurrent Pos:* Nat Res Coun Can med res fel, Hotel-Dieu Hosp, Montreal, 53-54; Can Cancer Soc Allan Blair Mem res fel histochem, Postgrad Med Sch, Univ London, 57-59. *Mem:* Am Asn Cancer Res; Am Asn Pathologists; Can Asn Path; Int Acad Path. *Res:* Experimental studies on tumor cell-host cell interaction; ultrastructural and cytochemical studies on breast carcinoma. *Mailing Add:* Dept Path McGill Univ 3775 University St Montreal PQ H3A 2B4 Can

TREMELLING, MICHAEL, JR, physical organic chemistry; deceased, see previous edition for last biography

TREMMEL, CARL GEORGE, b Lakewood, Ohio, June 25, 33; m 54; c 3. PHOTOGRAPHIC ANALYTICAL CHEMISTRY METHODS DEVELOPMENT & EDITING, BREATH & BLOOD ALCOHOL TESTING. *Educ:* Kent State Univ, BS, 55; Iowa State Univ, MS, 58. *Prof Exp:* Res chemist, 58-75, photog anal chemist, 75-86, TECH CONSULT, EASTMAN KODAK CO, 86-, FORENSIC CHEMIST, 89- *Concurrent Pos:* Tech consult, PRC, 86- *Mem:* Am Chem Soc. *Res:* Photographic chemistry; physical chemistry of color photography; electrochemistry; photographic analytical chemistry; forensic chemistry; breath and blood alcohol analysis. *Mailing Add:* 130 Northside Dr Burlington VT 05401-1272

TREMOR, JOHN W, b East Aurora, NY, Jan 24, 32; m 59; c 2. COMPARATIVE PHYSIOLOGY, ENVIRONMENTAL BIOLOGY. *Educ:* Univ Buffalo, BA, 53, MA, 57; Univ Ariz, PhD(zool), 62. *Prof Exp:* Res asst biochem, Vet Admin Hosp, Buffalo, NY, 58; teaching asst radiation biol, NSF Inst Biol Sta, Mont State Univ, 59; instr biol & bot, Phoenix Col, 62; asst prof physiol & embryol, Humboldt State Col, 62-63; group leader gen biol, Biosatellite Proj, 63-72, dep chief earth sci applns off, Ames Res Ctr, NASA, 72-74; proj scientist, biomed exp sci satellite & joint USSR/US biol satellite prog, 74-76; MGR SPACE SHUTTLE PLANT EXP & DEP PROJ SCIENTIST, AMES LIFE SCI PAYLOAD SPACELAB III & PRIN INVESTR, SPACELAB IV, AMES RES CTR, NASA, 77- *Mem:* AAAS. *Res:* Development and implementation of biological experiments for space flight; developmental biology of amphibians. *Mailing Add:* 13525 Surrey Lane Saratoga CA 95070

TRENBERTH, KEVIN EDWARD, b Christchurch, NZ, Nov 8, 44; m 70; c 2. METEOROLOGY. *Educ:* Univ Canterbury, BSc(hons), 66; Mass Inst Technol, ScD, 72. *Prof Exp:* Meteorologist res, NZ Meteorol Serv, 66-77, supt dynamic meteorol, 77; from assoc prof to prof meteorol, Univ Ill, 77-84; scientist, 84-86, leader toga activities, 86-87 & empirical studies group, 87, SR SCIENTIST, NAT CTR ATMOSPHERIC RES, 86-, HEAD CLIMATE ANAL SECT, 87- *Concurrent Pos:* Mem, NZ Working Group Global Data Processing Syst, Comn Basic Syst, World Meteorol Orgn, 75-77; NSF res grant, 78-88 & Nat Oceanic & Atmospheric Admin res grant, 86-; consult, Nat Oceanic & Atmospheric Admin, 80-, NASA, 81-; ed, Monthly Weather Rev, Am Meteorol Soc, 81-87, assoc ed, 87-88 & J Climate, 87-; NZ res fel, 68-72; NASA res grant, 89-; mem, Comt Earth Scis, Space Sci Bd, 82-85, Climate Res Comt, Bd Atmospheric Scis & Climate, 87-90, Nat Oceanic & Atmospheric Admin Panel Climate & Global Change, 87-; US Trop Oceans Global Atmosphere adv panel, 84-87; Polar Res Bd, 86-90; mem, Int Sci Steering Group, Trop Oceans Global Atmospheric Programme, 89- *Honors & Awards:* Am Meteorol Soc Editors Award, 89. *Mem:* Fel Am Meteorol Soc; AAAS; Royal Soc NZ; Meteorol Soc NZ. *Res:* Dynamics of climate and climate change; meteorology of the southern hemisphere; numerical weather prediction; dynamics of the stratosphere; published over 100 scientific articles or papers; global warming; El Niño-Southern Oscillation; climate system modeling; 1988 North American drought. *Mailing Add:* Nat Ctr Atmospheric Res PO Box 3000 Boulder CO 80307-3000

TRENCH, ROBERT KENT, b Belize City, Brit Honduras, Aug 3, 40; nat US; m 68. MARINE BIOLOGY, BIOCHEMISTRY. *Educ:* Univ West Indies, BSc, 65; Univ Calif, Los Angeles, MA, 67, PhD(invert zool), 69. *Prof Exp:* UK Sci Res Coun fel, Oxford Univ, 69-71; instr, Yale Univ, 71-72, asst prof, 72-76; assoc prof biol, 76-80, PROF BIOL & GEOL, UNIV CALIF, SANTA BARBARA, 80- *Mem:* AAAS; Am Soc Limnol & Oceanog; Am Soc Cell Biol; Soc Exp Biol UK; Sigma Xi. *Res:* Coral reef biology and ecology; biochemical integration of plasmids in autotroph-heterotroph endosymbioses; intercellular recognition phenomena. *Mailing Add:* 253 San Napoli Goleta CA 93117

TRENCH, WILLIAM FREDERICK, b Trenton, NJ, July 31, 31; m 54, 81; c 4. DIFFERENTIAL EQUATIONS, NUMERICAL LINEAR ALGEBRA. *Educ:* Lehigh Univ, BA, 53; Univ Pa, MA, 55, PhD(math), 58. *Prof Exp:* Instr res, Moore Sch Elec Eng, Univ Pa, 53-56; appl mathematician, Missile & Space Vehicle Div, Gen Elec, 56-57; eng specialist, Philco Corp, Philadelphia, 57-59; class AA engr, Radio Corp Am, Moorestown, NJ, 59-64; from assoc prof to prof math, Drexel Univ, 64-86; ANDREW G COWLES DISTINGUISHED PROF MATH, TRINITY UNIV, SAN ANTONIO, TEX, 86- *Mem:* Am Math Soc; Soc Indust & Appl Math; Math Asn Am; Int Linear Algebra Soc. *Res:* Asymptotic theory of solutions of differential equations and in numerical linear algebra; published papers in special functions, smoothing and prediction and elementary number theory. *Mailing Add:* Math Dept Trinity Univ San Antonio TX 78212

TRENHOLM, ANDREW RUTLEDGE, b Charleston, SC, May 1, 42; m 66; c 2. CHEMICAL CHEMISTRY. *Educ:* Clemson Univ, BS, 64; Ga Inst Technol, MS, 68. *Prof Exp:* Sanitary engr, Health Dept, Dade Couty, Fla, 64-67; chem engr, Union Carbide Corp, 68-70; environ engr, US Environ Protection Agency, 70-79; SECT HEAD, MIDWEST RES INST, 79- *Res:* Multimedia environmental research in air pollution, water pollution, solid waste and hazardous waste. *Mailing Add:* Midwest Res Inst 401 Harrison Oaks Blvd Cary NC 27513-2413

TRENHOLM, HAROLD LOCKSLEY, b Amherst, NS, July 24, 41; m 68; c 2. AGRICULTURE. *Educ:* McGill Univ, BSc, 63; NY State Vet Col, Cornell Univ, PhD(biochem), 68. *Prof Exp:* Asst physiol, State Univ NY Vet Col, Cornell, 63-67; res scientist, Toxicol Dept, Health Protection Br, Can Dept Health & Welfare, 67-73, dir res bur, Nonmed Use Drugs Directorate, 73-77; prog chmn, Animal Feed Safety & Nutrit, 77-88; SR SCIENTIST ADV, FEED & FOOD SAFETY & QUAL, ANIMAL RES CTR, CAN, 88- *Concurrent Pos:* Computer processing. *Mem:* Am Chem Soc; Can Soc Res Toxicol; Can Soc Animal Sci; Sigma Xi. *Res:* Mycotoxins; toxicology; food safety. *Mailing Add:* Animal Res Ctr Agr Can Ottawa ON K1A 0C6 Can

TRENHOLME, JOHN BURGESS, b Portland, Ore, Feb 4, 39; m 69; c 1. LASER ENGINEERING, NUMERICAL ANALYSIS. *Educ:* Calif Inst Technol, BS, 61, MS, 62, PhD(mat sci), 69. *Prof Exp:* Scientist lasers & light sources, Naval Res Lab, Washington, DC, 69-72; physicist laser fusion, 72-78, SR SCIENTIST SOLID STATE LASER PROG, LAWRENCE LIVERMORE LAB, 78- *Mem:* AAAS. *Res:* Laser fusion; laser design and engineering; numerical simulation of laser pumping and propagation; nonlinear optics; optical design. *Mailing Add:* 1212 Upper Happy Valley Rd Lafayette CA 94549-2723

TRENKLE, ALLEN H, b Alliance, Nebr, July 23, 34; m 56; c 3. NUTRITION, ENDOCRINOLOGY. *Educ:* Univ Nebr, BSc, 56; Iowa State Univ, MSc, 58, PhD(nutrit), 60. *Prof Exp:* NIH res fel, Univ Calif, Berkeley, 61-62; from asst prof to prof 62-83, DISTINGUISHED PROF ANIMAL SCI, IOWA STATE UNIV, 83- *Mem:* Am Soc Animal Sci; Am Chem Soc; Soc Exp Biol & Med; Endocrine Soc; Am Inst Nutrit. *Res:* Physiology of growth hormone and insulin secretion; endocrinology studies with ruminants; influence of hormones on growth and development of mammals; molecular biology of growth; protein metabolism in ruminants. *Mailing Add:* 301 Kildee Hall Iowa State Univ Ames IA 50011

TRENT, DENNIS W, b Bend, Ore, Oct 17, 35; m 55; c 7. IMMUNOLOGY, BIOCHEMISTRY. *Educ:* Brigham Young Univ, BS, 59, MS, 61; Univ Okla, PhD(med sci), 64. *Prof Exp:* From asst prof to assoc prof bact, Brigham Young Univ, 67-69; from asst prof to assoc prof microbiol, Univ Tex Med Sch, San Antonio, 69-74; chief molecular virol, Branett-Borne Dis Div, Ctr Dis Control, USPHS, 74-89; CHIEF MOLECULAR BIOL, DIV VECTOR-BORNE INFECTIOUS DIS, CTR DIS CONTROL, USPHS, 89- *Concurrent Pos:* NIH res grants, 66-72 & US Army contracts, 80-90. *Mem:* Am Soc Microbiol; Am Soc Virol; Am Soc Trop Med Hyg. *Res:* Biochemistry of togavirus and flavivirus replication; virus immunology; viral nucleic acids; vaccines. *Mailing Add:* Div Vector-Borne Infectious Dis Ctr for Dis Control PO Box 2087 Ft Collins CO 80522

TRENT, DONALD STEPHEN, b Cloverdale, Ore, Mar 29, 35; m 58; c 3. MECHANICAL ENGINEERING, APPLIED MATHEMATICS. *Educ:* Ore State Univ, BS, 62, MS, 64, PhD(mech eng), 72. *Prof Exp:* Develop engr, Battelle Mem Inst, 64-67; lectr, Ore State Univ, 67-68, res asst, 68-70; sr res engr, Pac Northwest Lab, Battelle Mem Inst, 70-73, res assoc, 73-75, res & develop mgr, 76-87, CHIEF SCIENTIST, PAC NORTHWEST LABS, BATTELLE MEM INST, 87- *Concurrent Pos:* Mech engr, Environ Protection Agency, Ore, 68-70; instr, Columbia Basin Col, 70-72; standard chmn, Am Nat Standard Inst, 75-; consult, Am Soc Reactor Safety, 82-85; courtesy prof, Ore State Univ, 87-; res affil, Mass Inst Technol, 89- *Mem:* Am Soc Mech Engrs; Sigma Xi; NY Acad Sci. *Res:* Computational heat transfer and fluid flow in natural convecting systems; numerical modeling of free turbulence, thermal plumes and air-sea interactions; reactor heat transfer; thermal hydraulics of advanced energy systems. *Mailing Add:* 721 Lynnwood Loop Richland WA 99352

TRENT, JOHN ELLSWORTH, b Wabash, Ind, Sept 22, 42; m 68; c 2. ANALYTICAL CHEMISTRY. *Educ:* Manchester Col, BA, 64; Ohio State Univ, PhD(org chem), 70. *Prof Exp:* RES CHEMIST, AMOCO CORP, IND, 71- *Mem:* Am Chem Soc. *Res:* Analytical instrumentation. *Mailing Add:* PO Box 3011 Naperville IL 60566-7011

TRENT, WALTER RUSSELL, chemistry, for more information see previous edition

TRENTELMAN, GEORGE FREDERICK, b Amsterdam, NY, Apr 27, 44. PHYSICS. *Educ:* Clarkson Col Technol, BS, 66; Mich State Univ, MS, 68, PhD(physics), 70. *Prof Exp:* Res assoc nuclear physics, Mich State Univ, 70-71; asst prof, 71-76, assoc prof, 76-82, PROF PHYSICS, NORTHERN MICH UNIV, 82- *Concurrent Pos:* Vis physicist, Inst Nuclear Sci, Univ Tokyo, 72-73; res fel, Jet Propulsion Lab, 81, 82. *Mem:* Soc Photo Optical Instrument Eng. *Res:* Nuclear physics; optics. *Mailing Add:* Dept Physics Northern Mich Univ Marquette MI 49855

TRENTHAM, DAVID R, b Solihull, Eng, Sept 22, 38. ENZYMOLOGY, MUSCLE PHYSIOLOGY. *Educ:* Cambridge Univ, BA, 61, PhD(chem), 64. *Prof Exp:* Lectr biochem, Biochem Dept, Univ Bristol, 66-77; chmn biochem & biophysics & Edwin M Chance prof biophysics, Univ Pa, 77-84; HEAD, DIV PHYS BIOCHEM & PHYSIOL & NEURAL MECHANISMS GROUP, NAT INST MED RES, LONDON, 84- *Honors & Awards:* Colworth Medal, Biochem Soc UK, 75; Wilhelm Feldberg Prize, 91. *Mem:* Fel Royal Soc; Biophys Soc; Biophys Soc UK; Am Soc Biochem & Molecular Biol; Physiol Soc UK; Biochem Soc UK; Am Chem Soc. *Res:* Mechanism of muscular contraction and its regulation studied through a variety of transient kinetic techniques including laser pulse photolysis of caged compounds to introduce biochemical metabolites rapidly into organized cellular preparations. *Mailing Add:* Div Phys Biochem Nat Inst Med Res The Ridgeway Mill Hill London NW7 1AA England

TRENTHAM, JIMMY N, b Dresden, Tenn, Jan 7, 36; m 65; c 2. MICROBIOLOGY. *Educ:* Univ Tenn, Martin, BS, 58; Vanderbilt Univ, PhD(microbiol), 65. *Prof Exp:* From asst prof to assoc prof microbiol, Univ Tenn, Martin, 65-73, chmn dept biol, 69-73, provost, 73-79, vchancellor, 76-79, PROF, UNIV TENN, MARTIN, 73-, ALUMNI DISTINGUISHED PROF, 90- *Concurrent Pos:* NSF res fel, 67-69. *Mem:* AAAS; Am Soc Microbiologists; Am Inst Biol Sci; Sigma Xi. *Res:* Microbial ecology, including qualitative and quantitative fluctuations of bacterial populations in freshwater and effects of temperature and nutritional factors on the structure of bacterial communities in freshwater. *Mailing Add:* 429 Raven Martin TN 38237

TRENTIN, JOHN JOSEPH, b Newark, NJ, Dec 15, 18; m 46; c 2. EXPERIMENTAL BIOLOGY & HEMATOLOGY. *Educ:* Pa State Univ, BS, 40; Univ Mo, AM, 41, PhD(endocrinol), 47. *Prof Exp:* Res asst, Univ Mo, 47-48; Childs fel anat, Sch Med, Yale Univ, 48-51, from instr to asst prof, 51-54; from assoc prof to prof, 54-60, actg chmn dept anat, 58-60, PROF EXP BIOL & HEAD DIV, BAYLOR COL MED, 60- *Concurrent Pos:* Assoc prof anat, Univ Tex Dent Br, 54-60; mem, Adv Comt Pathogenesis Cancer, Am Cancer Soc, 58-60, Comt Tissue Transplantation, Nat Res Coun, 60-70 & Comt Med Res & Educ, Vet Admin Hosp, Houston, 60-65; consult, Univ Tex M D Anderson Hosp & Tumor Inst, 59-62; mem, Bd Sci Counrs, Nat Cancer Inst, 63-65, chmn, 65-67; mem, Spec Animal Leukemia Ecol Studies Comt, NIH, 64-67; mem bd dirs, Am Asn Cancer Res, 70-73 & Int Soc Exp Hemat, 74-75, pres, 78-79. *Mem:* Am Asn Immunol; Soc Exp Biol & Med; Am Soc Microbiol; Transplant Soc; Am Asn Cancer Res; Int Soc Exp Hemat (pres, 78-79). *Res:* Cancer research; cancer viruses; immunology; tissue transplantation; experimental hematology; cancer immunity; bone marrow transplantation and control of stem cell differentiation; graft-versus-host disease; immunological tolerance; atherogenesis; gene therapy of genetic diseases. *Mailing Add:* Div Exp Biol Baylor Col Med One Baylor Plaza Houston TX 77030

TRENTO, ALFREDO, b Cittadella, Italy, July 3, 50. SURGERY. *Educ:* Univ Padua, Italy, MD, 75. *Prof Exp:* Asst prof surg, Univ Pittsburgh Sch Med, 85-88, dir, Ecmo Prog, Children's Hosp, 87-88; DIR, HEART TRANSPLANT PROG, DEPT THORACIC & CARDIOVASC SURG, CEDARS-SINAI MED CTR, 88- *Concurrent Pos:* Lectr, Inst Giannina Gaslini, Italy, 87. *Res:* Cardiac and lung transplantation. *Mailing Add:* Dept Thoracic & Cardiovasc Surg Suite 6215 Cedars-Sinai Med Ctr 8700 Beverly Blvd Los Angeles CA 90048

TREPANIER, PIERRE, b Montreal, Que, Oct 29, 53. VIROLOGY. *Educ:* Univ Ottawa, BSc, 75; Univ Montreal, BSc, 77; Armand Fruppier Inst, MSc, 79, PhD(virol), 82. *Prof Exp:* Res fel, Nat Sci & Eng Res Coun, Univ Melbourne, Australia, 82-84; assoc prof virol, Ctr Virol, 84-86, RES & DEVELOP MGM, DIV VIRAL VACCINES, ARMAND-FRAPPIER INST, 87- *Mem:* Can Soc Microbiol; Am Soc Virol. *Res:* High density cell culture for the production of biologicals virus production, monoclonals; characterization of proteins antigenic determinants. *Mailing Add:* 525 Blvd Des Prairies Laval PQ H7M 4O2 Can

TREPKA, ROBERT DALE, b Crete, Nebr, Dec 16, 38; m 61; c 3. PHYSICAL ORGANIC CHEMISTRY. *Educ:* Grinnell Col, BA, 61; Univ Calif, Los Angeles, PhD(org chem), 65. *Prof Exp:* NATO fel org chem, Munich, 65-66; SR RES SPECIALIST, DIV PRINTING & PUBL SYSTS, 3M CO, 66- *Mem:* Am Chem Soc; Soc Photog Sci & Eng. *Res:* Physical organic chemistry; organic stereochemistry; photographic chemistry; synthetic organic chemistry. *Mailing Add:* 6381 Birchwood Rd Woodbury MN 55125

TREPKA, WILLIAM JAMES, b Crete, Nebr, Apr 17, 33; m 55; c 3. POLYMER CHEMISTRY, RUBBER CHEMISTRY. *Educ:* Doane Col, BA, 55; Iowa State Univ, PhD(chem), 60. *Prof Exp:* RES CHEMIST, PHILLIPS PETROL CO, 60- *Mem:* Am Chem Soc. *Res:* Polymer and synthesis. *Mailing Add:* Dept Res & Develop Phillips Petrol Co Bartlesville OK 74004

TREPTOW, RICHARD S, b Chicago, Ill, Feb 8, 41; m 68; c 2. SCIENCE EDUCATION. *Educ:* Blackburn Col, BA, 62; Univ Ill, Urbana, MS, 64, PhD(chem), 66. *Prof Exp:* Staff chemist, Procter & Gamble Co, 66-72; from asst prof to assoc prof, 72-75, PROF CHEM, CHICAGO STATE UNIV, 78- *Mem:* Am Chem Soc. *Res:* Transition metal compounds; forensic chemistry; physical biochemistry. *Mailing Add:* Dept Chem & Physics Chicago State Univ Chicago IL 60628

TRESER, GERHARD, b Offenbach, West Ger, June 14, 29. NEPHROLOGY. *Educ:* Johann Wolfgang Goethe Univ, Frankfurt, Germany, MD, 56. *Prof Exp:* DIR MED, LINCOLN HOSP, 79-; PROF MED, NY MED COL, 80-, VCHMN, 83- *Mailing Add:* Dept Med Nephrology Lincoln Hosp 234 E 149th St New York NY 10451

TRESHOW, MICHAEL, b Copenhagen, Denmark, July 14, 26; nat US; m 51, 82; c 2. PLANT PATHOLOGY, AIR POLLUTION BIOLOGY. *Educ:* Univ Calif, Los Angeles, BS, 50; Univ Calif, Davis, PhD(plant path), 54. *Prof Exp:* Sr lab technician, Univ Calif, 52-53; plant pathologist, Columbia-Geneva Steel Div, US Steel Corp, 53-61; from asst prof to assoc prof bot, 61-67, assoc prof biol, 67-70, PROF BIOL, UNIV UTAH, 70- *Mem:* Air Pollution Control Asn; Am Phytopath Soc. *Res:* Environmental pathology, particularly diseases caused by air pollutants; environmental stress; impact of air pollution on forest species; study of growth trends based on dendrochronological methods. *Mailing Add:* Dept Biol Univ Utah Salt Lake City UT 84112

TRESSLER, RICHARD ERNEST, b Bellefonte, Pa, June 14, 42; m 65; c 3. CERAMICS, MATERIALS SCIENCE. *Educ:* Pa State Univ, BS, 63, PhD(ceramic sci), 67; Mass Inst Technol, SM, 64. *Prof Exp:* Sr scientist mat res, Tem Pres Res, Inc, Pa, 67; nuclear res officer, US Air Force, McClellan AFB, 67-68, mat scientist, Mat Lab, Wright-Patterson AFB, 68-70, tech area mgr, 70-71; NSF fel, Univ Essex, 71-72; from asst prof to assoc prof, Pa State Univ, University Park, 72-81, prof & head ceramic sci & eng, 81-86, dir, Ctr Adv Mat, 86-91, HEAD DEPT MAT SCI & ENG, PA STATE UNIV, 91- *Concurrent Pos:* Adj asst prof, Univ Cincinnati, 69-71; adv, US Arm Off Res, Mat Div, 79-80. *Honors & Awards:* Sci Achievement Medal, Systs Command, USAF, 70; Pace-Schwartz Walter Award, Am Ceramic Soc; IR-100 Award. *Mem:* Fel Am Ceramic Soc; Metall Soc; Nat Inst Ceramic Engrs; Electrochem Soc. *Res:* Fabrication and mechanical behavior of structural ceramic and composite materials; fracture and strengthening mechanisms; IC processing and properties. *Mailing Add:* Dept Mat Sci & Eng Pa State Univ University Park PA 16802

TRETIAK, OLEH JOHN, b Podkamen, Ukraine, Jan 18, 39; US citizen; m 65; c 1. ELECTRICAL ENGINEERING, BIOMEDICAL ENGINEERING. *Educ:* Cooper Union, BS, 58; Mass Inst Technol, SM, 60, ScD, 63. *Prof Exp:* Asst prof elec eng, Mass Inst Technol, 63-66, res assoc biomed, 66-73; proj engr storage tubes, Image Instruments, 66; assoc prof elec eng, 73-84, PROF ELEC ENG, DREXEL UNIV, 84-, MEM BIOMED ENG & SCI INST, 73-, DIR, IMAGING & COMP VISION CTR, 81- *Concurrent Pos:* Consult, Raytheon Corp, 64-65, Cognos Corp, 70-71, E G & G Inc, 71-72, Kulicke & Soffa, 74-80 & RCA, 81-82, FAA, 86, SAIC, 87; Soviet Union exchange scholar, Nat Acad Sci, 72; adj prof, radiol dept, Sch Med, Univ Pa, 81- *Mem:* Inst Elec & Electronics Engrs; Asn Comput Mach; Soc Mfg Engrs; AAAS. *Res:* Medical imaging systems; ultrasonography, image processing & machine vision; pictorial pattern recognition; mathematical modeling of images; systems for image analysis of autoradiograms; image pattern recognition; quantitative ultrasonic imaging. *Mailing Add:* Drexel Univ 32nd & Chestnut Sts Philadelphia PA 19104

TRETTER, STEVEN ALAN, b Greenbelt, Md, May 28, 40; m 68; c 1. ELECTRICAL ENGINEERING. *Educ:* Univ Md, Col Park, BSEE, 62; Princeton Univ, MA, 64, PhD(elec eng), 66. *Prof Exp:* Mem tech staff elec eng, Hughes Aircraft Co, 65-66; ASSOC PROF ELEC ENG, UNIV MD, COL PARK, 66- *Concurrent Pos:* Suppl engr, Ctr Explor Studies, IBM Corp, 68; consult, Vitro Corp Am, 69-70, Case Commun Inc, 70-, US Naval Res Lab, 74-86, Penril Data Comm, 86- *Mem:* Inst Elec & Electronics Engrs. *Res:* Statistical communication theory; error correcting codes; digital filtering and communications. *Mailing Add:* 601 Hawkesbury Terr Silver Spring MD 20904

TREU, JESSE ISAIAH, b New York, NY, Apr 10, 47; m 70; c 1. BIOPHYSICS, OPTICS. *Educ:* Rensselaer Polytech Inst, BS, 68; Princeton Univ, MA, 71, PhD(physics), 73. *Prof Exp:* Physicist optics & immunol, 73-75, liaison scientist components & mat group, Gen Elec Res & Develop Ctr, 75-77; prog mgr adv develop, automated equip for hemat & histol, Technicon Corp, 77-, AT CW GROUP INC. *Mem:* Am Phys Soc; Am Asn Physicists Med; Biophys Soc. *Mailing Add:* Domain Assocs One Palmer Sq Princeton NJ 08542

TREU, SIEGFRIED, b Hohenwalde, Prussia, Ger, Jan 25, 40; US citizen; m 62; c 4. HUMAN-COMPUTER INTERACTION, INTERFACE STRUCTURES. *Educ:* Univ Wis-Madison, BS, 62, MS, 63, Univ Pittsburgh, Pa, PhD(info processing), 70. *Prof Exp:* Info systs specialist, Goodyear Aerospace Corp, 65-67; postdoctoral res assoc, Nat Res Coun, Nat Bur Standards, Washington, DC, 70-71; asst dir, Computer Ctr, Univ Pittsburgh, 67-70, from asst prof to assoc prof computer sci, 71-89, chmn, Dept Computer Sci, 84-86, PROF COMPUTER SCI, UNIV PITTSBURGH, 89-, CO-DIR, CTR PARALLEL, DISTRIBUTED & INTEL SYSTS, 90- *Concurrent Pos:* Computer scientist & consult, Inst Computer Sci & Technol, Nat Bur Standards, Washington, DC, 71-81; chair, Spec Interest Group Off Info Systems, Asn Comput Mach, 83-87, area dir, Spec Interest Group Bd, 86- *Mem:* Asn Comput Mach. *Res:* Methodologies for structure-based design, measurement, and evaluation of adaptive, network-oriented human-computer interface systems. *Mailing Add:* Dept Comp Sci Univ Pittsburgh 324 Alumni Hall Pittsburgh PA 15260

TREUMANN, WILLIAM BORGEN, b Grafton, NDak, Feb 26, 16; m 45, 48; c 3. PHYSICAL CHEMISTRY. *Educ:* Univ NDak, BS, 42; Univ Ill, MS, 44, PhD(phys chem), 47. *Prof Exp:* Asst chem, Univ Ill, 42-46, asst math, 46; from asst prof to prof phys chem, NDak State Univ, 46-55; from assoc prof to prof, 60-81, assoc dean acad affairs, 68-70, dean, 70-81, EMER PROF PHYS CHEM & DEAN FAC MATH & SCI, MOORHEAD STATE UNIV, 81- *Mem:* AAAS; Am Chem Soc; Am Inst Chemists. *Res:* Complex ion formation. *Mailing Add:* 433 Eighth Ave S Fargo ND 58103-2827

TREVATHAN, LARRY EUGENE, b Phoenix, Ariz, Apr 13, 47; m 70; c 2. PLANT PATHOLOGY, PLANT PHYSIOLOGY. *Educ:* Univ Tenn, BS, 69; Va Polytech Inst & State Univ, PhD(plant path), 78. *Prof Exp:* ASST PROF & ASST PLANT PATHOLOGIST, MISS STATE UNIV, 78- *Mem:* Sigma Xi. *Res:* Physiology of pathogenesis of diseases of forage crops; genetic basis and inheritance of resistance in forage grasses. *Mailing Add:* PO Box 2645 Mississippi State MS 39762

TREVELYAN, BENJAMIN JOHN, b Beamsville, Ont, Nov 8, 22; m 53; c 4. PULP & PAPER TECHNOLOGY. *Educ:* McMaster Univ, BA, 44; McGill Univ, PhD(chem), 51. *Prof Exp:* Chemist, Aluminum Co Can, Ltd, Que, 44-47; asst res dir, Fraser Co, Ltd, NB, 51-53, res dir, 53-60; res scientist, WVa Pulp & Paper Co, 60-62; dir res, Celfibe Div, Johnson & Johnson, 62-65; mgr pulp res & eng, Kimberly-Clark Corp, 65-69, dir pulp & wood prep, Res & Eng, 69-73; indust liaison officer, Pulp & Paper Res Inst Can, 73-77; consult, Sync-Rust Ltd, Montreal, 77-78; assoc dir develop, ITT Rayonier, 78-80 & Rust Int, 80-85; RETIRED. *Mem:* Tech Asn Pulp & Paper Indust; Am Chem Soc; Can Pulp & Paper Asn. *Res:* Pulp and paper technology; all phases of the production of pulp from northern wood species; pulping of southern pines; pulping of vegetable fibers such as bagasse, flax and hemp; the application of fibers to the production of paper, dry-formed fiber products and cellulose products; application of research to the development and use of new processes on an industrial scale; financial analyses of industrial processes; mill evaluations and the evaluation of new equipment developments. *Mailing Add:* 200 Rockgreen Pl Santa Rosa CA 95409

TREVES, DAVID, b Milan, Italy, June 28, 30; Israeli citizen; m 54; c 3. MAGNETIC MATERIALS & MAGNETIZATION PROCESSES. *Educ:* Technion, Haifa, Israel, BSc, 53, MSc, 56, DSc(elec eng), 58. *Prof Exp:* Mem staff, Bell Tel Labs, Murray Hill, NJ, 60-62, Res Lab, Ampex Corp, Redwood City, Calif, 65-67, Xerox Palo Alto Res Ctr, 81-82, Optimem, Mountain View, 86-87; DIR, KOMAG INC, MILIPITAS, CALIF, 87- *Concurrent Pos:* Mem staff, Appl Physics, Weizmann Inst Sci, Rehovot, Israel, 57-86, chmn, Dept Electronics, 77-86. *Mem:* Am Optical Soc; Am Phys Soc; Soc Photo-Optical Instrumentation Engrs; fel Inst Elec & ElectronicsR Engrs. *Res:* Basic studies of magnetic materials and magnetization processes; high resolution magnetooptics for research and memory applications; extensive work on lasers, holography, coherent optics and microwave printed-array antennae. *Mailing Add:* Komag 591 Yosemite Dr Milpitas CA 95035

TREVES, JEAN FRANCOIS, b Brussels, Belg, Apr 23, 30; m 62; c 2. PURE MATHEMATICS. *Educ:* Univ Sorbonne, Lic, 53, Dr(math), 58. *Prof Exp:* Asst prof math, Univ Calif, Berkeley, 58-60; assoc prof, Yeshiva Univ, 61-64; prof, Purdue Univ, 64-71; PROF MATH, RUTGERS UNIV, NEW BRUNSWICK, 71- *Concurrent Pos:* Sloan fel, 60-64; vis prof, Univ Sorbonne, 65-67, Univ Paris, 74-75; mem, Mission Orgn Am States In Brazil, 61; Guggenheim fel, 77-78. *Honors & Awards:* Chauvenet Prize, Am Math Soc, 71. *Mem:* Am Math Soc; Math Soc France. *Res:* Partial differential equations; functional analysis. *Mailing Add:* Dept Math Rutgers Univ New Brunswick NJ 08903

TREVES, S T, b Ramallo, Arg, Aug 3, 40; US citizen; m; c 2. NUCLEAR MEDICINE. *Educ:* Nat Col Ill, Buenos Aires, BA, 59; Univ Buenos Aires, MD, 66. *Prof Exp:* Asst physician, Ctr Nuclear Med, Hosp Clin, Univ Buenos Aires, 66; res fel, Inst Med & Exp Surg, Univ Montreal, 67; resident nuclear med, Royal Victoria Hosp, McGill Univ, 67-68; clin fel & res assoc nuclear med & clin fel med, Yale-New Haven Hosp, Sch Med, Yale Univ, 68-70; INSTR & PROF RADIOL, HARVARD MED SCH, 70-; CHIEF, DIV NUCLEAR MED, CHILDREN'S HOSP. *Mem:* Soc Nuclear Med; Sigma Xi; Am Heart Asn; Soc Pediat Radiol. *Res:* Development of ultrashort lived radionuclide generators for angiocardiography in children; development of computer software for nuclear medicine functional studies. *Mailing Add:* Div Nuclear Med Children's Hosp 300 Longwood Ave Boston MA 02115

TREVES, SAMUEL BLAIN, b Detroit, Mich, Sept 11, 25; m 60; c 2. VOLCANOLOGY, PETROLOGY. *Educ:* Mich Col Mining & Technol, BS, 51; Univ Idaho, MS, 53; Ohio State Univ, PhD(geol), 58. *Prof Exp:* Geologist, Ford Motor Co, 51, State Bur Mines & Geol, Idaho, 52 & Otago Catchment Bd, NZ, 53-54; from instr to assoc prof geol, 59-55; chmn dept, 64-70 & 75-89, assoc dean, Col Arts & Sci, 89, PROF GEOL, UNIV NEBR, LINCOLN, 66- *Concurrent Pos:* Fulbright scholar, Univ Otago, NZ, 53-54; chief scientist, Antartica Expeds, 60-75 & Greenland Exped, 62-64. *Mem:* AAAS; fel Geol Soc Am; Am Mineral Soc. *Res:* Precambrian geology; antarctic geology. *Mailing Add:* 1710 B St Lincoln NE 68502

TREVILLYAN, ALVIN EARL, b Moline, Ill, Apr 12, 36; m 58; c 4. ORGANIC CHEMISTRY. *Educ:* Augustana Col, BA, 57; Purdue Univ, MS, 59, PhD(organometallics), 62. *Prof Exp:* Res chemist, Sinclair Res Inc, 62-69; res chemist, Amoco Chem, 69-76, Supvr, 76-79, dir, 79-82, mgr, 82-85, GEN MGR, AMOCO CHEM, 85- *Mem:* Am Chem Soc. *Res:* Organometallics; petrochemicals; Terephthalic acid; Olefins; environmental research; Xylene; engineering design construction; computer operations and support. *Mailing Add:* 2552 River Woods Dr Naperville IL 60565-6321

TREVINO, DANIEL LOUIS, b Edinburg, Tex, May 15, 43; m 83; c 2. NEUROPHYSIOLOGY. *Educ:* Univ Tex, Austin, BA, 65, PhD(physiol), 70. *Prof Exp:* Instr physiol, Sch Med Univ NC, Chapel Hill, 72-73, asst prof, 73-79; assoc dir student affairs & res asst prof anat, Sch Med, Univ NMex, Albuquerque, 79-81; res assoc prof anat & asst dean student affairs, Sch Med, Univ Tex Med Br, Galveston, 81-87; ASSOC PROF APPLIED PHYSIOL, DIR OFF MINORITY PROGS, 87- *Concurrent Pos:* USPHS fel, Univ Tex Southwest Med Sch Dallas, 70 & Marine Biomed Inst, Galveston, 70-72; fel neurobiol, Sch Med, Univ NC, Chapel Hill, 72- *Mem:* AAAS; Sigma Xi; Soc Neurosci. *Res:* Sensory neurophysiology; pain. *Mailing Add:* Pa State Univ 111 Old Main University Park PA 16802-0001

TREVINO, GILBERTO STEPHENSON, infectious diseases, research administration, for more information see previous edition

TREVINO, SAMUEL FRANCISCO, b San Antonio, Tex, Apr 2, 36; m 58; c 5. SPECTROSCOPY & SPECTROMETRY. *Educ:* St Mary's Univ, Tex, BS, 58; Univ Notre Dame, PhD(physics), 63. *Prof Exp:* RES PHYSICIST, ARDEC, PICATINNY ARSENAL, NJ, 64- *Concurrent Pos:* Guest physicist, Reactor Radiation Div, Nat Bur Stand, 71- *Honors & Awards:* Paul A Siple Award, Dept Army, 70. *Mem:* AAAS; Am Physics Soc. *Res:* Spectroscopy using inelastic scattering of low energy neutrons, particularly vibrational properties of molecular crystals; computer molecular dynamics simulation of condensed matter chemistry in both equilibrium and non-equilibrium environments via shock and detonation waves. *Mailing Add:* Reactor Radiation Div Nat Inst Standards & Technol Gaithersburg MD 20899

TREVITHICK, JOHN RICHARD, b St Thomas, Ont, Nov 30, 38; c 1. BIOCHEMISTRY. *Educ:* Queen's Univ, Ont, BSc, 61; Univ Wis-Madison, PhD(physiol chem), 65. *Prof Exp:* Nat Res Coun fel, Univ BC, 65-67; asst prof, 67-72, assoc prof, 73-80, PROF BIOCHEM, UNIV WESTERN ONT, 80- *Concurrent Pos:* Ed, J Can Fedn Biol Socs, 71-72 & 77-80. *Mem:* Can Biochem Soc; Can Fedn Biol Socs; Am Chem Soc. *Res:* Biochemistry of development and differentiation, mechanism of cortical cataract formation, especially diabetic and microwave cataracts; extracellular enzymes of microorganisms; histones and nuclear proteins; the lens. *Mailing Add:* Dept Biochem Univ Western Ont Med Sch London ON N6A 5C1 Can

TREVOR, ANTHONY JOHN, b London, Eng, Dec 27, 34; m 63; c 4. BIOCHEMICAL PHARMACOLOGY, NEUROCHEMISTRY. *Educ:* Univ Southampton, BSc, 60; Univ London, PhD(biochem), 63. *Prof Exp:* Lectr, 64-65, from asst prof to assoc prof, 65-77, PROF PHARMACOL, SCH MED, UNIV CALIF, SAN FRANCISCO, 77-, ACTG CHMN, 78- *Concurrent Pos:* NSF res grants, 72-74 & 75-; USPHS fel neuropharmacol, Sch Med, Univ Calif, San Francisco, 63-64, NIH res grants, 66-75. *Mem:* AAAS; Am Soc Pharmacol & Exp Therapeut; Am Soc Neurochem; Int Soc Neurochem; Soc Toxicol. *Res:* Brain enzymes purification, acetylcholinesterase, mechanisms of action and biodisposition of anesthetics. *Mailing Add:* Dept Pharmacol & Exp Therapeut Univ Calif Sch Med San Francisco CA 94143

TREVORROW, LAVERNE EVERETT, b Moline, Ill, Nov 1, 28; m 50; c 2. INORGANIC CHEMISTRY, PHYSICAL CHEMISTRY. *Educ:* Augustana Col, Ill, AB, 50; Okla State Univ, MS, 52; Univ Wis, PhD(chem), 55. *Prof Exp:* Asst, Okla State Univ, 50-52; asst, Univ Wis, 52-55; asst chemist, 55-59, ASSOC CHEMIST, ARGONNE NAT LAB, 59- *Mem:* Am Chem Soc. *Res:* Chemistry of fluorine and metal fluorides; uranium, neptunium, plutonium and fission elements; molten salt batteries; analysis of radioactive waste disposal systems and nuclear fuel cycle systems; disposal of low-level radioactive waste and nuclear byproduct wastes; environmental impact assessment. *Mailing Add:* Argonne Nat Lab CMT Bldg 9700 S Cass Ave Argonne IL 60439

TREVORS, JACK THOMAS, b Berwick, NS, Feb 24, 53; m 79; c 1. PLASMID BIOLOGY. *Educ:* Acadia Univ, BSc, 78, MSc, 79; Univ Waterloo, PhD(microbiol), 82. *Prof Exp:* Asst prof, 82-86, ASSOC PROF MICROBIOL, UNIV GUELPH, 86- *Concurrent Pos:* Consult, Health & Welfare Can, 87 & Agriculture Can, 88. *Mem:* Am Soc Microbiol; Soc Environ Toxicol & Chem; Can Soc Microbiol. *Res:* Physiology and genetics of metal-resistance mechanisms in bacteria; gene transfer between bacteria in soil and aquatic environments; methods for isolating DNA; development of electroporation protocols for transforming gram-negative and gram-positive bacteria; environmental microbiology; biotechnology. *Mailing Add:* Dept Environ Biol Univ Guelph Guelph ON N1G 2W1 Can

TREVOY, DONALD JAMES, b Saskatoon, Sask, Jan 27, 22; m 46; c 4. SOLID STATE CHEMISTRY, ENERGY CONVERSION. *Educ:* Univ Sask, BE, 44, MSc, 46; Univ Ill, PhD(chem eng), 49. *Prof Exp:* Chem engr, Nat Res Coun Can, 44-46; SR LAB HEAD, EASTMAN KODAK CO, 49- *Mem:* Am Chem Soc; AAAS; NY Acad Sci; Electrochem Soc; Sigma Xi. *Res:* Liquid-liquid extraction; thermal diffusion; high vacuum evaporation of liquids; physical chemistry of lithography; antistatic agents; photochemistry; organic semiconductors; conducting coatings; batteries; solar energy conversion. *Mailing Add:* 13 County Side Rd Fairport NY 14450

TREWELLA, JEFFREY CHARLES, NUCLEAR MAGNETIC RESONANCE SPECTROSCOPY, MASS SPECTROMETRY. *Educ:* Lock Haven State Col, Pa, BS, 75; Pa State Univ, PhD(chem), 79. *Prof Exp:* Res chemist, 79-81, SR RES CHEMIST, MOBIL RES DEVELOP CORP, 81- *Mem:* Am Chem Soc; Soc Appl Spectrosc. *Res:* Hydrogen bonding in liquids by nuclear magnetic resonance techniques; developed liquid chromatographic and high resolution mass spectrometric methods for the characterization of shale oils; developed inductively coupled plasma methods for multielemental analysis of both agreous and organic samples. *Mailing Add:* Mobil Res & Develop Corp PO Box 250 Edison NJ 08818-0250

TREWILER, CARL EDWARD, b Sheridan, NY, Sept 17, 34; m 54; c 4. POLYMER CHEMISTRY. *Educ:* Alfred Univ, BA, 56; Univ Akron, PhD(polymer sci), 66. *Prof Exp:* Resin chemist, Durez Plastics Div, Hooker Chem Co, 56-58; develop engr, Laminated Prod Bus Dept, Gen Elec Co, Ohio, 58-64; staff chemist polymerization, Akron Univ, 64-66; polymer chemist, 66-78, MGR PAPER & COMPOSITE PROD DEVELOP, LAMINATED PROD BUS DEPT, GEN ELEC CO, 78- *Mem:* Am Chem Soc. *Res:* Thermosetting resins and polymers in laminate applications. *Mailing Add:* 1979 Walnut Coshocton OH 43812

TREWYN, RONALD WILLIAM, b Edgerton, Wis, Aug 24, 43; m 70; c 1. CARCINOGENESIS, CELL BIOLOGY. *Educ:* Wis State Univ, BS, 70; Ore State Univ, PhD(microbiol), 74. *Prof Exp:* Res assoc biochem, Univ Colo Health Sci Ctr, 74-77, instr, 77-78; asst prof, 78-83, ASSOC PROF PHYSIOL CHEM, OHIO STATE UNIV, 83- *Concurrent Pos:* Res scientist, Comp Cancer Ctr, Ohio State Univ, 78-, dir, Tumor Procurement Serv, 81- *Mem:* Am Asn Cancer Res; Am Soc Biol Chem. *Res:* Cellular changes in the transfer RNA metabolism associated with carcinogenesis in an effort to elucidate the role of these changes in the neoplastic process and differentiation. *Mailing Add:* Physiol Chem 309 Cancer Ctr Ohio State Univ 410 W 12th Ave Columbus OH 43210

TREXLER, DAVID WILLIAM, geology, paleontology, for more information see previous edition

TREXLER, DENNIS THOMAS, b Compton, Calif, Aug 6, 40; m 66. REMOTE SENSING, GEOTHERMAL. *Educ:* Univ Southern Calif, BS, 65, MS, 68. *Prof Exp:* Eng geologist, State Calif Dept Water Resources, 66-67, Geolabs, Inc, 67-68; staff geologist, Space Div, Aerojet-Gen Corp, 68-70; mgr, Microwave Sensor Systs Div, Spectran Inc, 70-71; res assoc, Mackay Sch Mines, Univ Nev, 71-74 & Nev Bur Mines & Geol, 74-81; DIR & PRIN INVESTR, DIV EARTH SCI, ENVIRON RES CTR, UNIV NEV, LAS VEGAS, 81- *Concurrent Pos:* Consult, Am Mus Natural Hist, 74- *Mem:* Am Asn Petrol Geologists; Geothermal Resources Coun; Sigma Xi. *Res:* Assessment and application of geothermal resources; interpretation and application of remote sensing techniques; earthquake hazards assessment; geoarcheology to determine holocene climatic conditions. *Mailing Add:* 115 Blair Pl Reno NV 89509

TREXLER, FREDERICK DAVID, b Rahway, NJ, Feb 24, 42; m 64; c 2. GEOLOGY, MICROCOMPUTER INTERFACING. *Educ:* Houghton Col, BS, 64; Pa State Univ, University Park, PhD(solid state sci), 71. *Prof Exp:* From asst prof to assoc prof physics, 69-78, dept head, 78-82, PROF PHYSICS, HOUGHTON COL, 78-, DEPT HEAD, 88- *Concurrent Pos:* Radio eng, 75- *Mem:* Am Asn Physics Teachers; Sigma Xi. *Res:* Micro-electronics. *Mailing Add:* RD 1 Box H-6 Houghton NY 14744

TREXLER, JOHN PETER, b Allentown, Pa, Nov 8, 26; m 50; c 2. GEOLOGY, STRATIGRAPHY. *Educ:* Lehigh Univ, BA, 50, MS, 53; Univ Mich, PhD(geol), 64. *Prof Exp:* Asst geologist, Lehigh Portland Cement Co, 50-52; geologist, Fuels Br, US Geol Surv, 53-59; assoc prof geol, 62-69, chmn dept, 62-69, chmn sci div, 67-70, PROF GEOL, JUNIATA COL, 69-, CHMN DEPT, 74- *Concurrent Pos:* NSF fel, Princeton Univ, 70-71; Vis prof,

Univ Edinburgh, 84. *Mem:* Fel Geol Soc Am; Am Asn Petrol Geologists. *Res:* Field geology; stratigraphy and structural geology of sedimentary rocks of central and eastern Pennsylvania, particularly in the anthracite region. *Mailing Add:* Box 294 Rd Two Huntingdon PA 16652

TREYBIG, LEON BRUCE, b Yoakum, Tex, Aug 29, 31; m 57; c 3. MATHEMATICS. *Educ:* Univ Tex, BA, 53, PhD(math), 58. *Prof Exp:* Spec instr math, Univ Tex, 54-58; instr, Tulane Univ, 58-59, res assoc, 59-60, from asst prof to prof, 60-70; PROF MATH, TEX A&M UNIV, 70- *Mem:* Am Math Soc. *Res:* Integer programming; separability in metric spaces; continuous images of ordered compacta; knot theory. *Mailing Add:* Dept Math Tex A&M Univ College Station TX 77843

TREZEK, GEORGE J, b Chicago, Ill, July 10, 37; m 62; c 3. MECHANICAL ENGINEERING. *Educ:* Gen Motors Inst, BME, 61; Univ Ill, MS, 62, PhD(mech eng), 65. *Prof Exp:* Asst prof mech eng, Northwestern Univ, 65-66; from asst prof to assoc prof, 66-74, PROF MECH ENG, UNIV CALIF, BERKELEY, 74- *Concurrent Pos:* Consult engr. *Mem:* Am Soc Mech Engrs; Am Soc Testing & Mat. *Res:* Solid and hazardous waste management; size reduction, material and energy recovery from wastes; development of toxic waste treatment technology; power plant waste-heat dispersal and utilization problems; spray cooling system design. *Mailing Add:* 6185 Etcheverry Hall TSD Berkeley CA 94720

TRIA, JOHN JOSEPH, JR, b Shelby, NC, May 26, 46; m 76. PHYSICAL CHEMISTRY, THERMOCHEMISTRY. *Educ:* Duke Univ, BS, 68; Fla State Univ, PhD(phys chem), 77. *Prof Exp:* Res fel phys chem, dept chem, Fla State Univ, 77-78 & Ohio State Univ, 78-79; sr res chemist, 79-82, res specialist, 83-87, SR RES SPECIALIST, MONSANTO CO, 87- *Mem:* Am Chem Soc; Sigma Xi; Am Soc Testing & Mat; NAm Thermal Anal Sci; AAAS. *Res:* Thermochemistry; phsico-chemical property measurements; chemical instrumentation. *Mailing Add:* Monsanto Co 800 N Lindbergh S4B St Louis MO 63122

TRIANDAFILIDIS, GEORGE EMMANUEL, b Istanbul, Turkey, Nov 6, 22; US citizen; m 56; c 2. CIVIL ENGINEERING, SOIL MECHANICS. *Educ:* Robert Col, Istanbul, BS, 45; Univ Ill, Urbana, MS, 47, PhD, 60. *Prof Exp:* From instr to asst prof civil eng, Univ Ill, Urbana, 57-61; asst prof, Rice Univ, 61-64; assoc prof, 64-68, PROF CIVIL ENG, UNIV NMEX, 68- *Concurrent Pos:* Consult, 57- & Eric H Wang Civil Eng Res Facility, 61-64; mem, Hwy Res Bd, Nat Acad Sci-Nat Res Coun. *Mem:* Am Soc Civil Engrs; Am Soc Testing & Mat; Int Soc Soil Mech & Found Engrs; Sigma Xi. *Res:* Soil mechanics and foundations, especially dynamic properties of earth materials and soil-structure interaction phenomena; soil and rock mechanics; foundations and pavements. *Mailing Add:* 4611 Glenwood Hills Dr NE Albuquerque NM 87111

TRIANO, JOHN JOSEPH, b Bath, NY, Oct 20, 49; m 83; c 1. NEUROPHYSIOLOGY, APPLIED MECHANICS. *Educ:* Logan Col, BS & DC, 83; Webster Col, MA, 84; Univ Mich, PhD(appl mech), 89. *Prof Exp:* Inst physiol, Logan Col Chiropractic, 74-77, chair basic sci, 76-77, asst prof diag, 76-77; res assoc neurophysiol, Univ Colo, 77-78; pvt pract, Baseline Chiropractic Clin, 78-83; assoc prof, 83-87, PROF ERGONOMICS, NAT COL CHIROPRACTIC, 88- *Concurrent Pos:* Consult, Occup Safety & Health Admin, US Dept Labor & Spec Study Sect, Orthop Small Bus Innovation Res, NIH, 77-78. *Mem:* Am Soc Biomech; Int Chiropractic Acad Study Backpain (pres, 87-88); fel Am Back Soc; Am Col Chiropractic Clin Consult; Nat Asn Disability Eval Physicians; NAm Spine Soc. *Res:* Evaluation of spinal biomechanics; undertaking study of muscle behavior and tissue loads; primary methods of investigation are measures of muscle actions, load amplitudes and biomechanical computer model. *Mailing Add:* 604 Cottage Rd Batavia IL 60510

TRIANTAPHYLLOPOULOS, DEMETRIOS, b Athens, Greece, July 8, 20; m 54. PHYSIOLOGY, INTERNAL MEDICINE. *Educ:* Athens Sch Med, MD, 46. *Prof Exp:* From asst prof to assoc prof physiol, Univ Alta, 60-65; assoc prof, Wayne State Univ, 65-69; SR RES SCIENTIST & HEAD COAGULATION DEPT, AM NAT RED CROSS BLOOD RES LAB, 69- *Concurrent Pos:* Res assoc, Med Res Coun Can, 60-66. *Mem:* Am Physiol Soc; Can Physiol Soc; Am Soc Hemat; Int Soc Hemat; Am Heart Asn. *Res:* Blood coagulation; immunohematology. *Mailing Add:* 11625 Danville Dr Rockville MD 20852

TRIANTAPHYLLOPOULOS, EUGENIE, b Astros, Greece, Nov 27, 21; US citizen; m 54. CLINICAL TRIALS EVALUATION. *Educ:* Nat Univ Athens, MD, 47; Univ Alta, PhD(biochem), 57. *Prof Exp:* Nat Cancer Inst Can res fel, Univ Alta, 57-60, res assoc blood coagulation, 60-62; Can Heart Found res fel, Univ Alta & Wayne State Univ, 62-67; res scientist, Mt Carmel Mercy Hosp, Detroit, Mich, 68-69; NIH grant & chief coagulation res, Washington Hosp Ctr, 69-73; MED OFFICER, CTR DRUG EVALUATION & RES, FOOD & DRUG ADMIN, 73- *Concurrent Pos:* Asst prof, Univ Alta, 65-66. *Mem:* AAAS; Am Physiol Soc; Can Biochem Soc; Am Soc Clin Pharmacol & Therapeut; NY Acad Sci. *Res:* Blood coagulation; fibrinolysis; hemolysis. *Mailing Add:* Parklawn 10-92 5600 Fishers Lane Rockville MD 20857

TRIANTAPHYLLOU, ANASTASIOS CHRISTOS, b Amaliapolis-Volou, Greece, Nov 30, 26; m 60; c 1. CYTOGENETICS, NEMATODES. *Educ:* Athens Superior Sch Agr, Greece, BS, 49, MS, 50; NC State Univ, PhD(plant path, bot), 59. *Prof Exp:* Nematologist, Benaki Phytopath Inst, Greece, 50-60; asst geneticist, 60-62, from asst prof to assoc prof, 62-68, PROF GENETICS, NC STATE UNIV, 68- *Concurrent Pos:* Ed, J Nematology, 83-87, Revue de nematologie, 84-90. *Mem:* AAAS; Genetics Soc Am; fel Soc Nematologists; Am Inst Biol Sci. *Res:* Cytogenetics, evolution, mode of reproduction and sexuality of nematodes; genetics and cytology of plant parasitic nematodes. *Mailing Add:* Dept Genetics NC State Univ Raleigh NC 27695

TRIANTAPHYLLOU, HEDWIG HIRSCHMANN, b Fuerth, Bavaria, Ger, Jan 16, 27; nat US; m 60; c 1. PLANT NEMATOLOGY. *Educ:* Univ Erlangen, PhD(zool), 51. *Prof Exp:* Tech asst, 54-55, res instr, 55-57, from asst prof to assoc prof, 57-67, PROF PLANT PATH, NC STATE UNIV, 67- *Concurrent Pos:* Ed, J , Soc Nematol, 69-73; ed, Nematologica Soc Europ Nematologists, 75-90. *Honors & Awards:* Res Award, Sigma Xi, 62. *Mem:* Soc Europ Nematologists; fel Soc Nematologists; Sigma Xi. *Res:* Taxonomy and biology of free-living and plant-parasitic nematodes; ultrastructure of plant parasitic nematodes. *Mailing Add:* Dept Plant Path Box 7616 NC State Univ Raleigh NC 27695

TRIBBEY, BERT ALLEN, b Moorpark, Calif, Aug 8, 38; m 61; c 2. ZOOLOGY, ECOLOGY. *Educ:* Univ Calif, Santa Barbara, AB, 61; Univ Tex, PhD(zool), 65. *Prof Exp:* From asst prof to assoc prof, 65-73, chmn dept, 72-78, PROF BIOL, CALIF STATE UNIV, FRESNO, 73- *Concurrent Pos:* Vis prof, Mich State Univ, 78-79 & Univ Wis, 78. *Mem:* Am Fisheries Soc; Ecol Soc Am. *Res:* Structure and succession of aquatic communities; ecology of temporary ponds; physiological ecology of freshwater invertebrates. *Mailing Add:* Dept Biol Calif State Univ Fresno CA 93740

TRIBBLE, LELAND FLOYD, b Oxnard, Calif, July 12, 23; wid; c 4. ANIMAL SCIENCE. *Educ:* Univ Mo, BS, 49, MS, 50, PhD(agr), 56. *Prof Exp:* Instr animal husb, Univ Mo, 49-56, assoc prof, 56-67; PROF ANIMAL SCI, TEX TECH UNIV, 67- *Concurrent Pos:* Vis prof, Kans State Univ, 65-66; nonruminant nutritionist, USDA, Washington, DC, 74-75. *Mem:* Am Soc Animal Sci. *Res:* Animal nutrition; swine production and management. *Mailing Add:* Tex Tech Univ 6613 Norfolk Lubbock TX 79413

TRIBBLE, ROBERT EDMOND, b Mexico, Mo, Jan 7, 47; m 86; c 1. NUCLEAR PHYSICS. *Educ:* Univ Mo, Columbia, BS, 69; Princeton Univ, PhD(nuclear physics), 73. *Prof Exp:* Instr physics, Princeton Univ, 73-75; from asst prof to assoc prof, 75-82, head dept, 79-87, PROF PHYSICS, TEX A&M UNIV, 82- *Concurrent Pos:* A P Sloan fel, 76-80; vis scientist, Max Planck Inst, Heidelburg, 77-78; assoc fac fel, Western Univ, 87-88. *Mem:* fel Am Phys Soc. *Res:* Experimental research to determine reaction rates, masses and radioactive decays pertinent to nuclear astrophysics; weak interaction experiments that will test the standard model of electro-weak interactions. *Mailing Add:* Physics Dept Tex A&M Univ College Station TX 77843

TRIBUS, MYRON, b San Francisco, Calif, Oct 30, 21; m 45; c 2. ENGINEERING, QUALITY MANAGEMENT. *Educ:* Univ Calif, Berkeley, BS, 42, Los Angeles, PhD(eng), 49. *Prof Exp:* From lectr to assoc prof eng, Univ Calif, Los Angeles, 46-58, prof, 58-61; dean, Thayer Sch Eng, Dartmouth Col, 61-69; asst secy commerce for sci & technol, 69-71; sr vpres res & eng, Info Technol Group, Xerox Corp, 71-75; dir, Ctr Advan Eng Study, Mass Inst Technol, 75-86; DIR, EXERGY INC, HAYWARD, CALIF, 86- *Concurrent Pos:* Consult, Gen Elec Co, 50- & NATO, 53; dir icing res, Univ Mich, 51-54; Presidential appointee, Nat Adv Comn Oceans & Atmosphere, 72- *Honors & Awards:* Wright Bros Medal, Soc Automotive Engrs, 45; Bane Award, Inst Aerospace Sci, Am Inst Aeronaut & Astronaut, 46; Alfred Noble Prize, 52. *Mem:* Nat Acad Eng; Am Soc Mech Engrs; Am Inst Aeronaut & Astronaut; Soc Automotive Engrs. *Res:* Heat transfer; thermodynamics; decision theory. *Mailing Add:* Exergy Inc 22320 Foothill Blvd Hayward CA 94538

TRICE, VIRGIL GARNETT, JR, b Indianapolis, Ind, Feb 3, 26; m 58; c 3. NUCLEAR WASTE MANAGEMENT. *Educ:* Purdue Univ, BS & MS, 45; Ill Inst Technol, MS, 70. *Prof Exp:* Chem engr, Argonne Nat Lab, 49-71; nuclear waste mgt engr, Energy Res & Develop Admin, 71-77; sr prog analyst, 77-81, PROG MGR, US DEPT ENERGY, 81- *Mem:* Sigma Xi; Am Nuclear Soc. *Res:* Development of pyrochemical processes for nuclear fuel processing. *Mailing Add:* Dept Energy EM 34 Washington DC 20545

TRICE, WILLIAM HENRY, b Geneva, NY, Apr 4, 33; m 55; c 2. RESEARCH ADMINISTRATION. *Educ:* State Univ NY Col Forestry, BS, 55; Inst Paper Chem, MS, 60, PhD(phys chem), 63. *Prof Exp:* Res scientist, 63-66, group leader res & develop, 66-68, sect leader, 68-72, tech dir bleached div, 72-74, vpres res & develop, 74-79, sr vpres, 79-85, EXEC VPRES RES & DEVELOP, UNION CAMP CORP, 85-, CORP TECH DIR, 74- *Concurrent Pos:* Pres, Paper Technol Found, Western Mich Univ, 79-81; chmn, Res Adv Comt, Inst Paper Chem, 82-83; bd dir, Syracuse Pulp & Paper Found, 84- *Mem:* Fel Tech Asn Pulp & Paper Indust; AAAS. *Res:* Fatty acid and terpene chemistry; pulp and paper chemistry; engineering in fields related to pulp and paper. *Mailing Add:* Union Camp Corp 1600 Valley Rd Wayne NJ 07470

TRICHE, TIMOTHY J, b San Angolo, Tex, June 4, 44; c 3. CANCER DIAGNOSIS & TREATMENT, CHILDHOOD CANCER. *Educ:* Cornell Univ, AB, 66; Tulane Univ, PhD(med), 71. *Prof Exp:* Sect chief cell biol & histol, Lab Path, Nat Cancer Inst, NIH, 75-88; CHMN & PATHOLOGIST-IN-CHIEF, DEPT PATH & MED LAB, CHILDRENS HOSP, LOS ANGELES, 88- *Concurrent Pos:* Vchmn & prof, Dept Path, Univ Southern Calif, 88-; chmn, Path Discipline Comt, Childrens Cancer Study Group, 90- *Mem:* Am Soc Cell Biol; Int Acad Path; Soc Pediat Pathologists; Am Asn Pathologists; Am Soc Clin Pathologists; Am Asn Cancer Res. *Res:* Biologic behavior and genetic basis of childhood cancer using tumor cell lines and tumor tissue from childhood cancer patients. *Mailing Add:* Childrens Hosp Los Angeles 4650 Sunset Blvd Los Angeles CA 90027

TRICK, CHARLES GORDON, b Medicine Hat, Alta, Jan 29, 54; m 79; c 1. MICROBIAL BIOCHEMISTRY, BIOGEOCHEMISTRY. *Educ:* Univ Man, BSc, 75; Acadia Univ, MSc, 77; Univ BC, PhD(oceanog), 82. *Prof Exp:* Teaching fel biochem, Univ Calif, Berkeley, 82-84; asst prof environ studies, Inst Environ Studies, Univ Toronto, 84-86; PROF PLANT SCI, UNIV WESTERN ONT, 86- *Mem:* Am Soc Limnol & Oceanog; Am Soc Microbiol; Phys Soc Am; Int Asn Great Lakes Res. *Res:* Mechanisms of microbial adaptation; use of molecular genetic techniques in environmental studies; plant and microbial biotechnology. *Mailing Add:* Dept Plant Sci Univ Western Ont London ON N6A 5B7 Can

TRICK, GORDON STAPLES, b Winnipeg, Man, May 6, 27; m 52; c 4. TECHNOLOGY COMMERCIALIZATION. *Educ:* McGill Univ, BS, 49, PhD(chem), 52; Univ Western Ont, MS, 50. *Prof Exp:* Ramsay Mem fel, London, 53; sci serv officer, Defence Res Bd Can, 53-56; sr res chemist & res scientist, Goodyear Tire & Rubber Co, 56-71; exec dir, Man Res Coun & Technol Div, Govt Man, 71-85; SCI & TECHNOL CONSULT, 85- *Concurrent Pos:* Adj prof, Univ Manitoba, 72-80; mem adv bd sci & tech info, Nat Res Coun Can, 72-78, Standards Coun Can, 78-85 & Strategic Res, 84-86. *Mem:* Sigma Xi; fel Chem Inst Can; Can Res Mgt Asn. *Res:* Gas and liquid phase kinetics; phase transitions in polymers; stress-strain properties; technology transfer; innovation development; science policy; economic analysis; manufacturing productivity. *Mailing Add:* Box 101 St Germain MB R0G 2A0 Can

TRICK, TIMOTHY NOEL, b Dayton, Ohio, July 14, 39; m 58; c 6. ELECTRICAL ENGINEERING. *Educ:* Univ Dayton, BEE, 61; Purdue Univ, MSEE, 62, PhD(elec eng), 66. *Prof Exp:* From asst prof to assoc prof elec eng, 65-75, dir, Coord Sci Lab, 83-85, PROF ELEC ENG, UNIV ILL, URBANA, 75-, DEPT HEAD ELEC & COMP ENG, 85- *Concurrent Pos:* Summer fac fel, Am Soc Eng Educ-NASA, 70 & 71; consult, Rome Air Develop Ctr, 73-75, IBM, 82-83 & Cadence, 89-; vis assoc prof, Univ Calif, Berkeley, 73-74; vis lectr, Nat Inst Astrophysics, Optics & Electronics, Mex, 75; vpres publ, Inst Elec & Electronics Engrs, 88-90, mem bd dirs, 86-90. *Honors & Awards:* Guillemin-Cauer Award, Inst Elec & Electronics Engrs, 76; Centennial Medal, 84, Meritorious Serv Award, 88. *Mem:* Fel Inst Elec & Electronics Engrs; Sigma Xi. *Res:* Numerical methods and computer algorithms for the analysis and design of electrical circuits. *Mailing Add:* Everitt Lab Univ Ill 1406 W Green St Urbana IL 61801-2991

TRICKEY, SAMUEL BALDWIN, b Detroit, Mich, Nov 28, 40; div; c 2. THEORETICAL PHYSICS, SOLID STATE PHYSICS. *Educ:* Rice Univ, BA, 62; Tex A&M Univ, MS, 66, PhD(physics), 68. *Prof Exp:* Physicist, Mason & Hanger, Silas Mason Corp, Tex, 62-64; prof & chmn dept physics, Tex Tech Univ, 77-79; vis asst prof, 68-70, from asst prof to assoc prof, 70-77, dir comput & commun res, Col Lib Arts & Sci, 86-90, PROF PHYSICS & CHEM, UNIV FLA, 79- *Concurrent Pos:* Vis staff mem, Los Alamos Sci Lab, 71-91; consult, Quantum Physics Group, Redstone Arsenal, Ala, 72-76; consult phys sci, IBM Res Labs, San Jose, Calif, 75-76, Theoretical Div, Los Alamos Nat Lab, 84-; dir comput quantum theory proj, Univ Fla, 82-, Inst Cond Matter Studies, Mich Tech Univ, 85-; vis scientist, Max Planck Inst, Munich, 85- *Mem:* Fel Am Phys Soc; Am Asn Physics Teachers. *Res:* Theory of structure and energetics of crystals and thin films, computational physics, density functional theory; quantum chemistry; high performance computing & networking. *Mailing Add:* Dept Physics Univ Fla Gainesville FL 32611

TRICOLES, GUS P, b San Francisco, Calif, Oct 18, 31; m 53; c 2. OPTICS, ELECTRICAL ENGINEERING. *Educ:* Univ Calif, BA, 55, MS, 62, PhD, 71; San Diego State Col, MS, 58. *Prof Exp:* Asst res engr, Gen Dynamics/Convair, 55, res engr, 55-59, sr res engr, 59; physicist, Smyth Res Assocs, 59-61 & Univ Calif, 62; SR ENG STAFF SPEC, ELECTRONICS DIV, GEN DYNAMICS CORP, 62- *Mem:* Fel Inst Elec & Electronics Engrs; fel Optical Soc Am; Int Sci Radio Union; NY Acad Sci. *Res:* Microwave optics; diffraction; physical optics; holograms; antennas; waves. *Mailing Add:* PO Box 85227 Mail Zone 7-61 Gen Dynamics Electronics Div San Diego CA 92186-5227

TRICOMI, VINCENT, b New York, NY, Sept 16, 21; m 49; c 6. OBSTETRICS & GYNECOLOGY. *Educ:* Syracuse Univ, AB, 42; State Univ NY Downstate Med Ctr, MD, 50. *Prof Exp:* From instr to assoc prof, 55-69, clin prof, 69-74, asst dean, 70-74, ASSOC DEAN, STATE UNIV NY-HSCB, 70-, PROF OBSTET & GYNEC, 74-; VPRES MED AFFAIRS, BROOKLYN HOSP CTR, 74- *Concurrent Pos:* Brooks scholar, 54-55; consult, Lutheran Med Ctr, 69- & Kings County Med Ctr, 71-; chmn, Dept Obstet/Gynec, Brooklyn Hosp Ctr, 65- *Mem:* Soc Gynec Invest; Sigma Xi. *Res:* Human cytogenetics. *Mailing Add:* Brooklyn Hosp Ctr 121 DeKalb Ave Brooklyn NY 11201

TRIEBWASSER, JOHN, b Emery, SDak, Feb 16, 36; m 57; c 2. INTERNAL MEDICINE, CARDIOLOGY. *Educ:* Univ Mo, MD, 61; Am Bd Internal Med, dipl, 68. *Prof Exp:* Residency internal med, Wilferd Hall, USAF Med Ctr, Lackland AFB, Tex, 65; chief internal med sect, Wright Patterson USAF Med Ctr, 65-67; chief metab sect, Clin Sci Div, Sch Aerospace Med, Brooks AFB, Tex, 67-71, chmn dept med cardiovasc res, 73-78; fel cardiol, Univ Tex Southwestern Med Ctr, Dallas, 71-73; chmn dept med, Wright Patterson USAF Med Ctr, 78-80; dir Med Surveillance, Health & Environ Sci, Dow Chem USA, 80-83, med dir, Mich Div, 83-85; physician-in-chg, Cent Med Serv, 85-87, DIR, OCCUP HEALTH & SAFETY, FORD MOTOR CO, 87- *Concurrent Pos:* Clin asst prof med, Univ Tex Med Sch, San Antonio, 69-71; clin assoc prof, Wright State Univ, Dayton, Ohio, 78- *Honors & Awards:* Paul Dudley White Award, Asn Mil Surgeons US, 78. *Mem:* Fel Am Col Physicians; fel Am Col Preventive Med; fel Aerospace Med Asn; Soc Physicians (pres, 77); AMA; fel Am Col Occup Med. *Res:* Exercise stress testing and the use of radionuclide imaging techniques for the detection of subclinical coronary artery disease; automated electrocardiogram analysis using mathematical approaches to signal analysis. *Mailing Add:* 1376 W Ann Arbor Tr Plymouth MI 48170-1504

TRIEBWASSER, SOL, b New York, NY, Aug 16, 21; m 41; c 2. PHYSICS. *Educ:* Brooklyn Col, AB, 41; Columbia Univ, MA, 48, PhD, 52. *Prof Exp:* Instr physics, Brooklyn Col, 47-50; asst, Radiation Lab, Columbia, 51; mem tech staff, IBM Corp, 52-69, asst dir appl res, 69-81, prog mgr, Res Ctr, 81-89, DIR TECH J & PROF RELATIONS, IBM CORP, 89- *Mem:* Fel Am Phys Soc; Sigma Xi; fel Inst Elec & Electronics Engrs; fel AAAS. *Res:* Atomic structure; microwave spectroscopy; solid state physics; ferroelectricity; photoconductivity; semiconductors; microelectronics. *Mailing Add:* IBM Corp 500 Columbus Ave Thornwood NY 10594

TRIEFF, NORMAN MARTIN, b Brooklyn, NY, May 11, 29; div; c 4. ENVIRONMENTAL CHEMISTRY, TOXICOLOGY. *Educ:* Polytech Inst Brooklyn, BS, 50; Univ Iowa, MS, 55; NY Univ, PhD(chem), 63. *Prof Exp:* Res asst biochem, Atran Labs, Mt Sinai Hosp, New York, 54-55; instr chem, Cooper Union, 55-58; res assoc biochem, Isaac Albert Res Ctr, Jewish Chronic Dis Hosp, Brooklyn, NY, 61-62; supvr blood res, Blood Derivatives Sect, Mich Dept Pub Health, 63-65; asst prof chem, Drexel Inst, 65-68; asst prof toxicol & phys chem, 68-70, from asst prof to assoc prof, 70-72, PROF PREV MED & COMMUNITY HEALTH, UNIV TEX MED BR GALVESTON, 77- *Concurrent Pos:* Nat Ctr for Air Pollution Control grant, 66-68; consult, US Army Corps Engr, 67-68 & US Coast Guard, 70-71; Robert A Welch Found res grant, 70-; consult, La Conroe, Tex Pac Co, & Heat Systs Ultrasonics, Inc, Plainview, NY, 75- *Mem:* Fel Am Inst Chemists; Am Chem Soc; Air Pollution Control Asn; Am Indust Hyg Asn; Am Acad Indust Hyg; NY Acad Sci. *Res:* Environmental chemistry; development of analytic methods of drugs, toxins and environmental pollutants; odor analysis and olfaction; structure-activity relations. *Mailing Add:* Dept Prev Med 207 Ewing Bldg J09 Univ Tex Med Sch 301 Univ Blvd Galveston TX 77550

TRIEMER, RICHARD ERNEST, PHYLOGENY, PHYCOLOGY. *Educ:* Univ NC, Chapel Hill, PhD(bot), 75. *Prof Exp:* PROF BIOL SCI, RUTGERS UNIV, 75- *Mem:* Phycological Soc Am; Soc Protozoologists; Int Soc Evolutionary Protistology. *Res:* Ultrastructure and molecular biology of protists; systematics. *Mailing Add:* Dept Biol Sci Rutgers Univ Piscataway NJ 08855-1059

TRIER, JERRY STEVEN, b Frankfurt, Ger, Apr 12, 33; US citizen; m 57; c 3. INTERNAL MEDICINE, GASTROENTEROLOGY. *Educ:* Univ Wash, MD, 57. *Hon Degrees:* AM, Harvard Univ, 73. *Prof Exp:* Intern med, Univ Rochester, 57-58, asst resident, 58-59; clin assoc, Nat Cancer Inst, 59-61; trainee gastroenterol, Univ Wash, 61-63; asst prof med, Univ Wis, 63-67; assoc prof, Univ NMex, 67-69; assoc prof med & anat, Sch Med, Boston Univ, 69-73; assoc prof, 73-76, PROF MED, HARVARD MED SCH, 76- *Concurrent Pos:* USPHS grant, 64-67; USPHS grant, Univ NMex, 67-69; USPHS grants, Sch Med, Boston Univ, 69-; assoc physician, Univ Hosp, Boston & Boston City Hosp, 69-73; consult, Vet Admin Hosp, Boston, 69-, Chelsea Naval Hosp, 71-74 & US Vet Admin Cent Off, Washington, DC, 71-; sr physician & dir div gastroenterol, Brigham & Women's Hosp, Boston, 73-; NIH gen med & study sect, 74-78. *Mem:* Asn Am Physicians; Am Soc Clin Invest; Am Gastroenterol Asn; Am Fedn Clin Res; Am Soc Cell Biol. *Res:* Functional morphology of the gastrointestinal tract of humans in health and disease; developmental morphology of the intestine; cell renewal in the alimentary tract. *Mailing Add:* Div Gastroenterol Brigham & Women's Hosp 75 Francis St Boston MA 02115

TRIFAN, DANIEL SIEGFRIED, b Cleveland, Ohio, Dec 23, 18; m 48; c 3. PHYSICAL CHEMISTRY, ORGANIC CHEMISTRY. *Educ:* Baldwin-Wallace Col, BS, 40; Western Reserve Univ, MS, 41; Harvard Univ, MA, 46, PhD(chem), 48. *Prof Exp:* Res fel phys & org chem, Univ Calif, Los Angeles, 48-49, instr org chem, 49-50; asst prof, Bowling Green State Univ, 50-51; asst prof phys, org & polymer chem & head chem sect, Plastics Lab, Princeton Univ, 51-64; prof chem, Fairleigh Dickenson Univ, 64-89; RETIRED. *Mem:* Am Chem Soc. *Res:* Organic reaction mechanisms; polymer chemistry. *Mailing Add:* 466A Liberty St Apt 303 Little Ferry NJ 07666

TRIFAN, DEONISIE, applied mathematics; deceased, see previous edition for last biography

TRIFARÓ, JOSÉ MARÍA, b Mercedes, Arg, Nov 29, 36; m 64; c 2. PHARMACOLOGY. *Educ:* Liceo Militar Gen San Martín, BA, 54; Univ Buenos Aires, MD, 61. *Prof Exp:* Demonstr anat, Univ Buenos Aires, 57-58, instr pharmacol, 61-62, lectr physiol, 62-64; lectr, 67-68, from asst prof to assoc prof, 72-78, prof pharmacol, McGill Univ, 78-86; CHMN PHARMACOL, UNIV OTTAWA, 86- *Concurrent Pos:* A Thyssen Found res fel, Arg, 62-63; res fel, Nat Res Coun, Arg, 63-64; Rockefeller Found res fel, US, 64-66; NIH res fel, 66-67; Med Res Coun Can scholar, 68. *Mem:* Am Soc Pharmacol & Exp Therapeut; Pharmacol Soc Can; Int Soc Neurochem. *Res:* Cellular and molecular mechanism of hormone and neurotransmitter release; role of contractile proteins in secretory cell functions. *Mailing Add:* Dept Pharmacol Univ Ottawa 451 Smyth Rd Ottawa ON K1H 8M5 Can

TRIFFET, TERRY, b Enid, Okla, June 10, 22; m 46; c 3. MATHEMATICAL MODELING & MECHANICS. *Educ:* Univ Okla, BA, 45; Univ Colo, BS, 48, MS, 50; Stanford Univ, PhD(struct mech), 57. *Prof Exp:* Instr eng, Univ Colo, 47-50; gen engr rocket & guided missile res, US Naval Ord Test Sta, 50-55; gen engr radiol res & head radiol effects br, US Naval Radiol Defense Lab, 55-59; assoc prof appl mech, Mich State Univ, 59-63, prof mech & mat sci, 63-76; assoc dean res, 76-87, actg dean, 87, PROF MATS SCI, COL ENG, UNIV ARIZ, 88-; DIR, VA/NASA SPACE ENG RES CTR, 88- *Concurrent Pos:* Mem apex comt, US Naval Res Labs, 59-65; consult, US Dept Defense, 59-65, Battelle Mem Inst, 65-68, Lear-Siegler, Inc, 65-, US Dept Energy, 78-, NASA, 82-, NSF, 83-; Australian Res Grants Comt res fel math physics, Univ Adelaide, 66-67 & 72-73, distinguished vis scientist, 85-86. *Honors & Awards:* Maxwell Prize, Fifth Int Math Model Conf. *Mem:* Am Phys Soc; Am Math Soc; Soc Eng Sci; Soc Indust & Appl Math; Inst Elec & Electronics Engrs; Am Inst Aeronaut & Astronaut; Sigma Xi. *Res:* Mathematical modeling of neural systems. *Mailing Add:* 6935 Stardust Circle Tucson AZ 85718

TRIFUNAC, ALEXANDER DIMITRIJE, b Yugoslavia, July 29, 44; US citizen; m 67. PHYSICAL CHEMISTRY, MAGNETIC RESONANCE. *Educ:* Columbia Univ, BA, 66; Univ Chicago, PhD(chem), 71. *Prof Exp:* Res asst chem, Univ Chicago, 66-71; res assoc chem, Univ Notre Dame, 71-72 & Univ Chicago, 72; asst scientist, 74-77, scientist, 77-86, SR SCIENTIST & GROUP LEADER RADIATION & PHOTOCHEM, CHEM DIV, ARGONNE NAT LAB, 82- *Concurrent Pos:* Presidential intern, Argonne Nat Lab, 72-73. *Mem:* Am Chem Soc; Sigma Xi; Radiation Res Soc. *Res:*

Chemistry and physics of transient reactive intermediates in radiation and photochemistry; development of novel time resolved magnetic resonance methods for study of transient radicals and radical ions in liquids. *Mailing Add:* Chem Div Argonne Nat Lab 9700 S Cass Argonne IL 60439

TRIFUNAC, NATALIA PISKER, b Budapest, Hungary, June 26, 42; US citizen; m 66; c 1. BIOCHEMISTRY. *Educ:* Univ Belgrade, BS, 65; Calif Inst Technol, PhD(chem), 69. *Prof Exp:* Res assoc dept chem, Calif Inst Technol, 69-70; res assoc, Columbia Univ, 70-71, res assoc dept biochem, Col Physicians & Surgeons, 71-72; instr biochem, 72-75, asst prof, Dept Obstet & Gynec, Sect Reprod Biol, 75-78, ASST PROF RES BIOCHEM DEPT PATH, SCH MED, UNIV SOUTHERN CALIF, 78- *Concurrent Pos:* Consult, Jet Propulsion Lab, Calif Inst Technol, 78- *Res:* Cell surface changes during lymphocyte transformation; reproductive biology; human spermatozoa metabolism and survival. *Mailing Add:* 1458 Old House Rd Pasadena CA 91107

TRIGG, GEORGE LOCKWOOD, b Washington, DC, Sept 30, 25; wid; c 2. THEORETICAL PHYSICS. *Educ:* Washington Univ, AB, 47, AM, 50, PhD(physics), 51. *Prof Exp:* Asst physics, Washington Univ, 47-50; asst prof, Knox Col, 51-54, actg chmn dept, 51-52; from asst prof to assoc prof, Ore State Univ, 54-62; from asst ed to ed, 62-85, pub ed, Phys Rev Lett, 85-88, FREE-LANCE ED, AM PHYS SOC, 88- *Concurrent Pos:* NSF fel, 57-58; asst ed, Phys Rev & Phys Rev Letters, 58; consult, Funk & Wagnall's Dictionary, 66-71 & Am Heritage Dictionary, 67-70; mem comt symbols, units & terminology, Nat Res Coun, 71-; consult, Wiley Int Dictionary Med & Biol, 81- *Mem:* Fel AAAS; fel Am Phys Soc; Fedn Am Sci; Soc Scholarly Publ. *Res:* Elementary particle theory; fundamentals of quantum theory; history of physics. *Mailing Add:* Phys Rev Letters One Research Rd Box 1000 Ridge NY 11961

TRIGG, WILLIAM WALKER, b Little Rock, Ark, Dec 4, 31; m 57; c 2. INORGANIC CHEMISTRY. *Educ:* Univ Ark, BSChE, 56, MS, 60; La State Univ, PhD(inorg chem), 66. *Prof Exp:* Control chemist, Niagara Chem Div, Food Mach & Chem Corp, 56-57; from instr to assoc prof, 59-75, PROF CHEM, ARK POLYTECH COL, 75-, HEAD DEPT, 66- *Mem:* Am Chem Soc; Am Inst Chem Engrs. *Res:* Precipitation from homogeneous solution and studies of ion solvent effects in solvents of low dielectric constant. *Mailing Add:* Rte 3 Box 4 Russellville AR 72801-9305

TRIGGER, KENNETH JAMES, b Carsonville, Mich, Sept 6, 10; m 39; c 3. MECHANICAL ENGINEERING. *Educ:* Mich State Univ, BS, 33, MS, 35, ME, 43. *Prof Exp:* Asst, Mich State Col, 33-34, instr mech eng, 35-36; instr mech eng, Swarthmore Col, 37-38 & Lehigh Univ, 38-39; assoc, 39-40, from asst prof to prof, 40-77, EMER PROF MECH ENG, UNIV ILL, URBANA, 77- *Concurrent Pos:* Consult, Nuclear Div, Union Carbide Corp, Continental Can Co, Aeroprojects Inc & Atlantic Richfield Co. *Honors & Awards:* Blackall Award, Am Soc Mech Engrs, 57; Medal, Soc Mfg Eng, 59. *Mem:* Fel Am Soc Mech Engrs; fel Am Soc Metals; Am Soc Eng Educ; fel Soc Mfg Engrs. *Res:* Metal cutting and machinability; physical metallurgy; cutting temperatures and temperature distribution in cutting of metals; mechanism of tool wear. *Mailing Add:* 705 W Columbia Ave Champaign IL 61820

TRIGGER, KENNETH ROY, b Chicago, Ill, Mar 7, 24; m 46; c 4. COMPUTER SIMULATIONS, SHOCK HYDRODYNAMICS. *Educ:* Stanford Univ, BS, 49, MS, 50, PhD(physics), 56. *Prof Exp:* Staff scientist, Los Alamos Sci Lab, 55-56, Hughes Res & Develop Labs, 56-58, Lawrence Livermore Lab, 58-62, Stanford Linear Accelerator Ctr, 62-64, Lawrence Livermore Lab, 64-65 & Appl Theory, Inc, 65-67; STAFF SCIENTIST, LAWRENCE LIVERMORE NAT LAB, 68- *Mem:* Am Phys Soc. *Res:* Chemical and nuclear explosive devices. *Mailing Add:* 1440 Via Loma Walnut Creek CA 94598

TRIGGIANI, ROBERTO, b Bolzano, Italy, Mar 28, 42. APPLIED MATHEMATICS, CONTROL THEORY. *Educ:* Univ Padova, Dr, 67; Univ Minn, PhD(math control theory), 73. *Prof Exp:* From asst prof to assoc prof math, Iowa State Univ Ames, 75-80; from assoc prof to prof Math Univ Fla, Gainesville, 80-87; PROF MATH, UNIV VA, CHARLOTTESVILLE, 87- *Concurrent Pos:* Vis prof, Univ Calif, Los Angeles, 79-80; prin co-investr, Nat Sci Found, 81, Air Force Off Sci Res, 76-; assoc ed, Appl Math & Optimization, 85. *Res:* Mathematical control theory with emphasis on boundary control theory for partial differential equations. *Mailing Add:* Dept Appl Math Univ Va Thornton Hall Charlottesville VA 22903

TRIGGLE, DAVID J, b London, Eng, May 4, 35; m 59; c 2. PHARMACOLOGY, MEDICINAL CHEMISTRY. *Educ:* Univ Southampton, BSc, 56; Univ Hull, PhD(org chem), 59. *Prof Exp:* Fel org chem, Univ Ottawa, 59-61; res fel, Bedford Col, London, 61-62; asst prof biochem pharmacol, 62-65, assoc prof biochem pharmacol & theoret biol, 65-69, chmn dept biochem pharmacol, 71-85, PROF BIOCHEM PHARMACOL & THEORET BIOL, SCH PHARM, STATE UNIV NY, BUFFALO, 69-, DEAN , 85- *Concurrent Pos:* NIH res grants, 64- *Mem:* Am Chem Soc; Royal Soc Chem; Am Soc Pharmacol & Exp Therapeut; Brit Pharmacol Soc; Can Pharmacol Soc; fel AAAS. *Res:* Molecular pharmacology of adrenergic and cholinergic systems; organic reaction mechanisms; synthesis of organic heterocyclic systems; ion translocation and cell membranes; molecular basis of neurotransmitter action. *Mailing Add:* Sch Pharm State Univ 126 Cooke Buffalo NY 14260

TRIGIANO, ROBERT NICHOLAS, b Johnstown, Pa, Nov 30, 53; m 83. MICROPROPAGATION, FUNGAL PHYSIOLOGY. *Educ:* Juniata Col, BS, 75; Pa State Univ, MS, 77; NC State Univ, PhD(bot & plant path), 83. *Prof Exp:* Assoc agronomist, Green Giant Co, 77-79; mushroom grower, Rol-Land Farms Ltd, 79-80; res assoc tissue cult, 84-86, asst prof, 87-91, ASSOC PROF TISSUE CULT, UNIV TENN, 91- *Concurrent Pos:* Assoc ed, Am Soc Hort Sci, 90-91. *Honors & Awards:* Distinguished Young Scientist Award, Inst Agr, Univ Tenn, 91. *Mem:* Mycol Soc Am; Am Soc Hort Sci; Int Soc Plant Morphologists; Int Soc Hort Sci; Sigma Xi. *Res:* Cell biology; development plant biology; somatic embryogenesis; plant pathology-diseases of woody ornamentals; transformation. *Mailing Add:* Dept Ornamental Hort & Landscape Design Univ Tenn Knoxville TN 37901-1071

TRIGLIA, EMIL J, b Lucca, Italy, Aug 26, 21; US citizen; m 48; c 5. ANALYTICAL CHEMISTRY. *Educ:* City Col New York, BS, 43. *Prof Exp:* Chemist, Ledoux & Co, Inc, 46-51; group leader anal chem, Chem Construct Corp & Am Cyanamid Co, 51-56; sr res chemist, Minerals & Chem, Philipp Corp, 56-61, res group supvr, 61-68; group leader anal & phys testing, Engelhard Minerals & Chem Corp, Menlo Park, 68-74, mgr anal & phys measurements, 74-81; res assoc, Ledoux & Co, 82-83; SR RES ASSOC, ESSEX CHEM CORP, 83- *Mem:* Am Chem Soc; Tech Asn Pulp & Paper Indust; Am Soc Testing & Mat; Acids Indust Chem. *Res:* Analyses of minerals, ores and rocks by chemical and instrumental methods; development of methods for the analysis of elements and metals in industrial chemicals by instrumental methods primarily spectrophotometric and atomic absorption by flame and flameless techniques. *Mailing Add:* 12 Sharon Ct Metuchen NJ 08840-1730

TRILLING, CHARLES A(LEXANDER), b Wiesbaden, Ger, Apr 13, 23; nat US; m 49; c 3. CHEMICAL ENGINEERING, NUCLEAR ENGINEERING. *Educ:* Calif Inst Technol, BS, 44; Mass Inst Technol, ScD(chem eng), 49. *Prof Exp:* Res assoc, Chem Warfare Serv Develop Lab, Mass Inst Technol, 44-45, res assoc chem eng, 47-49; chem engr, Dennison Mfg Co, 49-50; sr res engr, Atomic Energy Res Dept, N Am Aviation, Inc, 50-52; process engr, United Engrs & Constructors, Inc, 52-54; sr res engr, Atomics Int Div, NAm Aviation, Inc, Rockwell Int, Canoga Park, 54-55, proj engr, 55-59, group leader org reactors develop, 56-59, chief proj engr, 59-61, assoc dir, Org Reactors Dept, 61-62, proj mgr org cooled reactors, 62-65, mgr heavy water org cooled reactors prog off, 65-67, mem tech staff, Fast Breeder Prog Off, Atomics Int Div, NAm Rockwell Corp, 67-69, pollution control technol, 69-72, proj mgr coal gasification, Rockwell Int, 73-78, prog mgr, Advan Technol Process Design, Environ & Energy Systs Div, 78-83; CONSULT, ENERGY-RELATED FIELDS, 83- *Mem:* AAAS; Am Chem Soc; Am Nuclear Soc; Am Inst Chem Engrs. *Res:* Nuclear power reactor development and engineering; organic cooled reactor concept for economic generation of electric power; air pollution control; coal gasification and liquefaction. *Mailing Add:* 5254 Melvin Ave Tarzana CA 91356-2940

TRILLING, GEORGE HENRY, b Bialystok, Poland, Sept 18, 30; nat US; m 55; c 3. ELEMENTARY PARTICLE PHYSICS. *Educ:* Calif Inst Technol, BS, 51, PhD(physics), 55. *Prof Exp:* Res fel physics, Calif Inst Technol, 55-56; Fulbright res fel, Polytech Sch, Paris, 56-57; from asst prof to assoc prof, Univ Mich, 57-60; assoc prof, 60-64, chmn dept, 68-72, PROF PHYSICS, UNIV CALIF, BERKELEY, 64- *Concurrent Pos:* NSF sr fel, Europ Orgn Nuclear Res, 66-67, Guggenheim fel, 73-74. *Mem:* Nat Acad Sci; fel Am Phys Soc. *Res:* Properties of elementary particles produced by high energy accelerators and colliders. *Mailing Add:* Dept Physics Univ Calif Berkeley CA 94720

TRILLING, LEON, b Poland, July 15, 24; nat US; m 46; c 2. AERONAUTICS & ASTRONAUTICS, SCIENCE TECHNOLOGY & SOCIETY. *Educ:* Calif Inst Technol, BS, 44, MS, 46, AeroEng, 47, PhD, 48. *Prof Exp:* Physicist, US Naval Ord Test Sta, 44-46; asst, Calif Inst Technol, 46-48, res fel, 48-49, instr, 49-50; Fulbright scholar, Univ Paris, 50-51; res assoc, 51-54, from asst prof to assoc prof, 54-62, PROF AERONAUT ENG, MASS INST TECHNOL, 62-, MEM FAC, COL SCI, TECHNOL & SOC, 78- *Concurrent Pos:* Guggenheim fel & vis prof, Sorbonne Univ, 63-64; vis prof, Aeronaut Dept, Delft Tech Univ, Neth, 74-75, Carleton Col, 87; consult. *Mem:* Fel AAAS. *Res:* Aerodynamics; gas dynamics; kinetic theory of gases; gas surface interactions. *Mailing Add:* 180 Beacon St Boston MA 02116

TRIM, CYNTHIA MARY, b England, Apr 12, 47; m 78. ANIMAL PHYSIOLOGY, PHARMACOLOGY. *Educ:* Univ Liverpool, BVSc, 70. *Prof Exp:* Res asst vet anesthesiol, Univ Cambridge, Eng, 70-72; pvt practitioner vet med, Nixon & Partners, Suffolk, Eng, 72-74; asst prof anesthesiol, Univ Guelph, Can, 74-76 & Univ Ill, Urbana, 76-77; assoc prof, Univ Mo, Columbia, 77-80; assoc prof, 81-85, PROF ANESTHESIOL, UNIV GA, ATHENS, 85- *Mem:* Royal Col Vet Surgeons; Am Vet Med Asn; Asn Vet Anaesthetists Gt Brit & Ireland; Am Col Vet Anesthesiologists (pres, 85); Brit Vet Asn; Int Anesthesia Res Soc. *Res:* Cardiopulmonary physiology and pharmacology relating to anesthesiology of all species; physiology and treatment of equine endotoxemia. *Mailing Add:* Dept Large Animal Med Col Vet Med Univ Ga Athens GA 30602

TRIMBERGER, GEORGE WILLIAM, b Neilsville, Wis, Dec 8, 09; m 38; c 3. DAIRY SCIENCE. *Educ:* Univ Wis, BS, 33; Univ Nebr, MS, 42, PhD(zool), 48. *Prof Exp:* Herd supt, Univ Nebr, 34-40, instr dairy prod, 40-44; from asst prof to prof, 44-75, EMER PROF DAIRY HUSB, CORNELL UNIV, 75- *Concurrent Pos:* Vis prof & proj leader, Univ Philippines, 55-57 & 66-67 & Ahmadu Bello Univ, Nigeria, 75-77. *Mem:* AAAS; Am Sci Animal Sci; Am Diary Sci Asn; Am Genetic Asn. *Res:* Artificial insemination; reproduction and nutrition in dairy cattle. *Mailing Add:* Morrison Hall Cornell Univ Ithaca NY 14850

TRIMBLE, DONALD E, b Westhope, NDak, Mar 6,16. FIELD MAPPING GEOLOGY. *Educ:* Wash State Univ, BS, 38; Univ Minn, BA. *Prof Exp:* US Army, 41-45; field mapping geologists, US Geol Surv, 46-80; RETIRED. *Mem:* Sr fel Geol Soc Am; Col Sci Soc. *Res:* Strategy of structure. *Mailing Add:* 2640 Lamar Denver CO 80214

TRIMBLE, JOHN LEONARD, b Detroit, Mich, Feb 27, 44. HUMAN FACTORS ENGINEERING, SENSORY PSYCHOPHYSICS. *Educ:* Univ Ill, Chicago, BS, 68; Med Ctr, Univ Ill, PhD(physiol, bioeng), 72. *Prof Exp:* Res assoc & asst prof, Eye Res Labs, Univ Chicago, 72-79; eng dir, 79-83, DIR, REHAB RES & DEVELOP CTR, HINES VET ADMIN HOSP, 83-; PRES, IMT, INC, 84- *Concurrent Pos:* Vis assoc biol, Calif Inst Technol, 77-79; consult, Nat Ctr Health Care Technol, NIH, 80-, Off Productivity, Technol & Innovation & Ctr Utilization fed Technol, US Dept Com, 84-; clin assoc prof, Dept Orthop Surg, Loyola Med Sch, 83-; adj assoc prof, Pritzker Inst Med Eng, 84- *Mem:* Inst Elec & Electronics Engrs; Sigma Xi; Soc Neurosci; AAAS; Asn Res Vision & Ophthal. *Res:* Neural modelling; biological signal processing; intelligent systems. *Mailing Add:* Rehab Res & Develop Ctr Va Hines Hosp PO Box 20 Hines IL 60141

TRIMBLE, MARY ELLEN, b Englewood, NJ, Nov 1, 36. PHYSIOLOGY, BIOLOGY. *Educ:* Wellesley Col, AB, 58; Syracuse Univ, MA, 59; Case Western Reserve Univ, PhD(biol), 69. *Prof Exp:* Res assoc develop biol, Brown Univ, 68-70; res physiologist, Vet Admin Hosp, Syracuse, 71-86; asst prof, 71-84, ASSOC PROF, HEALTH SCI CTR, SYRACUSE, STATE UNIV NY, 84- *Mem:* AAAS; Am Soc Nephrology; Am Physiol Soc; Int Soc Nephrology; Am Soc Renal Biochem & Metab. *Res:* Renal physiology, particularly metabolic aspects of ion transport, lipid metabolism and fatty acid binding proteins. *Mailing Add:* Dept Physiol State Univ NY Health Sci Ctr-Syracuse 750 E Adams St Syracuse NY 13210

TRIMBLE, ROBERT BOGUE, b Baltimore, Md, July 2, 43; wid; c 1. CELL BIOLOGY, GLYCOPROTEIN BIOSYNTHESIS. *Educ:* Rensselaer Polytech Inst, BS, 65, MS, 67, PhD(microbiol), 69. *Prof Exp:* Health Res Inc fel, 69-70, res scientist, 70-72, sr res scientist, 73-80, RES SCIENTIST V, WADSWORTH CTR LABS & RES, NY STATE DEPT HEALTH, 81-; PROF BIOMED SCI, SCH PUB HEALTH, STATE UNIV NY, ALBANY, 88- *Concurrent Pos:* Mem, US Pub Health Serv-NIH Sci Rev Group, 85-89, Am Cancer Soc Personnel Rev Comt, 91- *Mem:* AAAS; Am Soc Microbiol; Am Soc Biochem & Molecular Biol; Sigma Xi; Soc Complex Carbohydrates. *Res:* Cell biology and biochemistry of glycoprotein biosynthesis and secretion in yeast and mammalian cells; N-linked glycan structure and function; oligosaccharide processing pathways; glycohydralases and glycosyl transferases; oligosaccharide and protein structure. *Mailing Add:* NY State Dept Health Wadsworth Ctr Labs & Res Box 509 Albany NY 12201-0509

TRIMBLE, RUSSELL FAY, b Montclair, NJ, Feb 23, 27; m 50; c 3. COORDINATION COMPOUNDS, CHEMICAL LITERATURE. *Educ:* Mass Inst Technol, BS, 48, PhD(chem), 51. *Prof Exp:* Instr chem, Univ Rochester, 51-54; from asst prof to assoc prof, 54-70, PROF CHEM, SOUTHERN ILL UNIV, CARBONDALE, 70- *Concurrent Pos:* Abstractor, Chem Abstr, 54-, Metals Abstr, 84-, tech translator, 59-; vis lectr, Univ Ill, 63-64. *Mem:* AAAS; Am Chem Soc; Am Inst Chemists; Am Translators Asn; Sigma Xi. *Res:* Coordination compounds; inorganic synthesis. *Mailing Add:* Dept Chem & Biochem Southern Ill Univ Carbondale IL 62901

TRIMBLE, VIRGINIA LOUISE, b Los Angeles, Calif, Nov 15, 43; m 72. ASTRONOMY, ASTROPHYSICS. *Educ:* Univ Calif, Los Angeles, BA, 64; Calif Inst Technol, MS, 65, PhD(astron), 68; Cambridge Univ, MA, 69. *Prof Exp:* Res fel astrophys, Inst Theoret Astron, Cambridge Univ, 68; asst prof astron, Smith Col & Four Cols Observ, 68-69; NATO sr fel, Inst Theoret Astron, Cambridge Univ, 69-70, vis fel, 70-71; from asst prof to assoc prof physics, 71-80, PROF PHYSICS, UNIV CALIF, IRVINE, 80- *Concurrent Pos:* From vis asst prof to vis assoc prof astron, Univ Md, College Park, 72-80, vis prof, 80-; Sloan fel, 72-74; nat lectr, Sigma Xi, 74-77; ed, Comments Astrophys 87-; assoc ed, Astrophys J, 89- *Honors & Awards:* Sci Reviewing Award, Nat Acad Sci, 86. *Mem:* Am Astron Soc; Royal Astron Soc; Europ Phys Soc; Int Astron Union; Int Soc Gen Relativity & Gravitation. *Res:* Late phases of stellar evolution; supernovae; white dwarfs; neutron stars; galactic evolution; binary stars. *Mailing Add:* Dept Physics Univ Calif Irvine CA 92717

TRIMITSIS, GEORGE B, b Assiut, Egypt, Nov 28, 39; m 64; c 1. ORGANIC CHEMISTRY. *Educ:* Am Univ Cairo, BSc, 64; Va Polytech Inst & State Univ, PhD(org chem), 68. *Prof Exp:* Res assoc org res, Ohio State Univ, 68-69; asst prof org chem, Western Mich Univ, 69-76; MEM FAC, DIV NAT SCI, UNIV PITTSBURGH, 76- *Mem:* Am Chem Soc. *Res:* Formation, study and synthetic applications of carbanions. *Mailing Add:* Div Nat Sci Univ Pittsburgh Johnstown PA 15904-2990

TRIMMER, ROBERT WHITFIELD, b Binghamton, NY, Dec 13, 37; m 65; c 2. INDUSTRIAL ORGANIC CHEMISTRY. *Educ:* Hope Col, Mich, AB, 60; Rensselaer Polytech Inst, PhD(org chem), 73. *Prof Exp:* Asst res polymer chem, Schenectady Chem Co, 60; asst res med chemist, Sterling Winthrop Res Inst, Sterling Drug, Inc, 64-69; supvr lab prof develop & org chemist, Sumner Div, 73-77 & Corp Div, 77-80, SUPVR LAB PROD DEVELOP & ORG CHEMIST, AMES DIV DIAG GROUP, MILES LABS, INC, 80- *Mem:* Am Chem Soc; AAAS; Org Reactions Catalysis Soc. *Res:* Hydrogenation technology; specialty products; pharmaceuticals; aryl and alkyl amines as intermediates and polymer catalysts; quaternary ammonium compounds, citric acid derivatives; heterocycles; organic photochemistry; development of new synthetic methods; peptide chemistry. *Mailing Add:* 20410 Bargene Way Germantown MD 20874-1160

TRIMMER, WILLIAM S(TUART), b Long Beach, Calif, Apr 12, 43; m 76; c 2. MICROMECHANICS. *Educ:* Occidental Col, BA, 62; Wesleyan Univ, PhD(physics), 72. *Prof Exp:* Asst prof, Montclair State Col, NJ, 71-72; asst prof & chmn, Physics Dept, Col Wooster, Ohio, 72-77; staff physicist, Singer Corp Res, 77-79; sr scientist, Johnson & Johnson, 79-82; res micromech & robotic syst develop, AT&T Bell Labs, 82-90; OWNER, BELL MEAD RES, 90- *Concurrent Pos:* Adj prof, NJ Inst Technol, 90-; vis fel, Princeton Univ, 90- *Mem:* Am Phys Soc; Inst Elec & Electronics Engrs; Am Soc Mech Engrs. *Res:* Help start the micro electr mechanical systems (or micromechanics, microdynamics) field; co-invented harmonic motor; co-developed sacrificial method for making micro gears, etc; nine US patents. *Mailing Add:* 58 Riverview Terr Belle Mead NJ 08502

TRINDLE, CARL OTIS, b Des Moines, Iowa, Aug 26, 41; m 62; c 1. THEORETICAL CHEMISTRY. *Educ:* Grinnell Col, BA, 63; Tufts Univ, PhD(phys chem), 67. *Prof Exp:* NSF fel theoret chem, Yale Univ, 67-68; res assoc, Argonne Nat Lab, 68-69; asst prof, Univ Va, 69-73, ASSOC PROF THEORET CHEM, UNIV VA, 73-; dir, Curr One Prog, 81-86, DIR STUDIES, MONROE HILL COL, 86- *Concurrent Pos:* Consult, Argonne Nat Lab, 69-; vis asst prof, Mideast Tech Univ, 71; Sloan Found fel, 71-73; vis assoc prof, Mideast Tech Univ, 73; vis scientist, Israel Inst Technol, 76; Nat Acad Sci exchange fel, Yugoslavia, 79, Univ Ga, 81, Amherst Col, 86, Am Chem Soc Tour Lectr, 82, 87, 88, 89, 90. *Mem:* AAAS; Am Chem Soc; Sigma Xi. *Res:* Impact of orbital topology on organic stereo-chemistry; localized description of charge distributions; group theory of easily rearranged systems; computer algebra systems; artificial intelligence applications in chemistry. *Mailing Add:* Dept Chem Univ Va Charlottesville VA 22901

TRINKAUS, JOHN PHILIP, b Rockville Ctr, NY, May 23, 18; m 63; c 3. CELL BIOLOGY, DEVELOPMENTAL BIOLOGY. *Educ:* Wesleyan Univ, BA, 40; Columbia Univ, MA, 41; Johns Hopkins Univ, PhD(embryol), 48. *Prof Exp:* From instr to prof, 48-68, EMER PROF BIOL, YALE UNIV, 88- *Concurrent Pos:* Mem staff embryol, Marine Biol Lab, Woods Hole, 53-57, 70 & 78; master, Branford Col, Yale Univ, 66-73, dir grad studies biol, 65-66; Guggenheim fel, Lab Exp Embryol, Col France, 59-60; chmn, Gordon Conf Cell Contact & Movement, 79. *Honors & Awards:* Merit Award, NIH, 87. *Mem:* Soc Develop Biol; Am Soc Zoologists; Am Soc Cell Biol; Int Inst Embryol. *Res:* Cytodifferentiation; mechanism of morphogenetic cell movements; teleost development; contact behavior and locomotion of normal and transformed cells. *Mailing Add:* 1010 KBT Yale Univ New Haven CT 06520

TRINKAUS-RANDALL, VICKERY E, b Albuquerque, NMex, Jan 11, 53. OPTHALMOLOGY. *Educ:* Kenyon Col, AB, 74; Univ Wis-Madison, PhD(zool), 81. *Prof Exp:* Fel, Marine Biol Lab, 81; RES FEL, EYE RES INST RETINA FOUND, 81- *Mem:* Sigma Xi; Am Res Vision & Ophthal. *Res:* Trabecular meshwork of the eye for the presence of non-muscle contractile filaments to examine their role in the aqueous outflow; the role of divalent cations on hemidesmosome formation in the cornea. *Mailing Add:* Eye Res Inst Retina Found 20 Staniford St Boston MA 02114

TRINKLEIN, DAVID HERBERT, b Jefferson City, Mo, July 3, 47. FLORICULTURE, PLANT GENETICS. *Educ:* Lincoln Univ, BS, 69; Univ Mo, Ms, 71, PhD(hort), 74. *Prof Exp:* Res asst hort, Univ Mo, 70-75; res scientist plant sci, Farmland Industs, 75-76; exten horticulturist, Lincoln Univ, 76-77; ASST PROF HORT, UNIV MO, 77- *Mem:* Sigma Xi. *Res:* Physiology and genetics of flower crops with allied interest in energy conservation in the greenhouse. *Mailing Add:* Hort 1-40 Agr Bldg Univ Mo Columbia MO 65211

TRINKO, JOSEPH RICHARD, JR, b Washington DC, Dec 20, 39; m 65; c 2. NUCLEAR ENGINEERING, ELECTRICAL ENGINEERING. *Educ:* Univ Tenn, BS, 63, MS, 65, PhD(nuclear eng), 67. *Prof Exp:* Asst nuclear engr breeder, Ebasco Serv, 70-73; SR NUCLEAR ENGR REACTOR ANALYSIS, MIDDLE SOUTH SERV, 73- *Mem:* Elec Power Res Inst. *Res:* Advanced methods of reactor core analysis; stress corrosion mechanisms in nuclear systems. *Mailing Add:* 303 Brett Dr Gretna LA 70056

TRINLER, WILLIAM A, b Louisville, Ky, Dec 24, 29; m 62; c 1. ORGANIC CHEMISTRY. *Educ:* Univ Louisville, BS, 55, PhD(org chem), 59. *Prof Exp:* Chemist, E I du Pont de Nemours & Co, 59-60; from asst prof to assoc prof, 60-74, PROF ORG CHEM, IND STATE UNIV, TERRE HAUTE, 74- *Mem:* Am Chem Soc; Am Inst Chemists; Sigma Xi. *Res:* Synthesis and polymerization of vinyl monomers; liquid chromatography. *Mailing Add:* Dept Chem Ind State Univ Terre Haute IN 47809

TRIOLO, ANTHONY J, b Philadelphia, Pa, Aug 8, 32; m 59; c 3. PHARMACOLOGY. *Educ:* Philadelphia Col Pharm, BS, 59; Jefferson Med Col, MS, 62, PhD(pharmacol), 64. *Prof Exp:* From instr to asst prof, 67-72, assoc prof, 72-81, PROF PHARMACOL, JEFFERSON MED COL, 81- *Res:* Neuropharmacological effects of tremorine on motor reflex activity; toxicological interactions between organochlorine or organophosphate insecticides and Benzo-(a)-pyrene carcinogenesis. *Mailing Add:* 1025 Walnut St Philadelphia PA 19107

TRIONE, EDWARD JOHN, b Ill, Mar 10, 26; m 49; c 3. BIOCHEMISTRY. *Educ:* Chico State Col, BA, 50; Ore State Col, PhD(bot), 57. *Prof Exp:* Chem technician, Ore State Col, 53-54, asst, 54-57; res plant pathologist, Univ Calif, 57-59; PLANT BIOCHEMIST, ORE STATE UNIV, 59-, PROF BOT & PLANT PATH, 76- *Mem:* Am Soc Plant Physiologists; Bot Soc Am; Am Phytopath Soc; Can Soc Plant Physiologists; Japanese Soc Plant Physiologists; Scandinavian Soc Plant Physiol. *Res:* Biochemistry and physiology of reproduction in higher plants and fungi; biochemistry of host-pathogen interactions. *Mailing Add:* Dept Crop Sci Ore State Univ Corvallis OR 97331

TRIPARD, GERALD EDWARD, b Saskatoon, Sask, Apr 18, 40; m 63; c 2. NUCLEAR PHYSICS. *Educ:* Univ BC, BSc, 62, MSc, 64, PhD(physics), 67. *Prof Exp:* Nat Res Coun Can fel, Swiss Fed Inst Technol, 67-69; asst prof physics, 69-76, ASSOC PROF PHYSICS, WASH STATE UNIV, 76- *Res:* Neutron scattering; final state interactions; stopped pions. *Mailing Add:* Dept Physics Wash State Univ Pullman WA 99164-2814

TRIPATHI, BRENDA JENNIFER, b Rochford, Eng, July 5, 46; m 69; c 2. VISUAL SCIENCE, CELL BIOLOGY. *Educ:* Univ London, BSc, 67, PhD(med), 71. *Prof Exp:* Res asst path, Univ Col Hosp, London, 67-69; res asst, Inst Ophthal, Univ London, 69-72, lectr, 72-77; asst prof lectr, 79-84, RES ASSOC & ASSOC PROF OPHTHAL, UNIV CHICAGO, 77-, ASSOC PROF LECTR, BIOL SCI COL DIV, 84- *Concurrent Pos:* Ocular microbiologist, Eye Path Lab, Univ Chicago, 77-; sr lectr, Organismal Biol & Anat, Univ Chicago, 90- *Honors & Awards:* Cert Honor, Murgoci Award, Bedford Col, London, 65; Int Prize, Alcon Res Inst, 87; Spec Invited Lectr, Col Univ Chicago, 85; Serv Award, Am Acad Ophthal, 87, Honor Award, 88. *Mem:* Royal Soc Med, London; Asn Res Vision & Ophthal; Sigma Xi; AAAS; Am Acad Ophthal; Int Soc Eye Res; NY Acad Sci; Tissue Culture Asn; Soc Exp Biol & Med; Int Soc Ocular Toxicol; Am Soc Cell Biol. *Res:* Visual science; anatomy; physiology; experimental pathology; cell biology; electron microscopy; tissue culture; biochemistry; immunology; microbiology; cerebrospinal fluid physiology and pathology; molecular biology; over 270 articles to scientific publications. *Mailing Add:* 939 E 57th St Chicago IL 60637

TRIPATHI, GOVAKH NATH RAM, b Gorakhpur, India, Jan 1, 44; m 62; c 3. CHEMICAL PHYSICS, SOLID STATE PHYSICS. *Educ:* Univ Gorakhpur, India, BSc, 60, MSc, 62, PhD(physics), 68. *Prof Exp:* Fel physics, Univ Gorakhpur, India, 62-65, sr lectr, 65-76; vis sr lectr & sr acad staff fel,

Dept Physics, Univ Manchester & Manchester Inst Sci & Technol, 76-77; SR SCIENTIST RADIATION RES, RADIATION LAB, UNIV NOTRE DAME, 78- *Res:* Time resolved resonance raman studies of structure and reactions of transient free radicals and excited states; physical and chemical properties of excited radical species; early chemical steps in radical reactions. *Mailing Add:* Radiation Lab Univ Notre Dame Notre Dame IN 46556

TRIPATHI, RAMESH CHANDRA, b Jamira, India, July 1, 36; US citizen; m 69; c 2. CLINICAL OPHTHALMOLOGY, GLAUCOMA. *Educ:* Lucknow Christian Col, India, ISc, 54; Univ Agra, MBBS(MD), 59; Univ Lucknow, MS, 63; DORCS&P, 65; Univ London, PhD(med), 70; Royal Col Pathologists, FRCPath, 74; Col Ophthalmologists, London, FCOphth, 88. *Hon Degrees:* FACS, Am Col Surgeons; FICS, Int Col Surgeons; FNASc, Nat Acad Sci, India. *Prof Exp:* Resident ophthal, Med Col Kanpur, Lucknow Univ, 59-64; res fel, Univ Ghent, 64-65; registrar, S W Middlesex Hosp, London, 65-67; lectr, Inst Ophthal, London Univ, 67-70, sr lectr, 70-77; PROF OPHTHAL & VISUAL SCI, UNIV CHICAGO, 77-, PROF, THE COL, 79- *Concurrent Pos:* Attend opthalmologist & asst surgeon, Div Railway Hosp, Govt India, 63-64; hon registrar, Charing Cross Hosp, 65-67; Hayward res fel, Inst Ophthal, Univ London, 67-68; sr registrar, Moorfields Eye Hosp, London, 67-68, chief clin asst, 68-72, consult ophthal & pathologist, 72-77; attend physician & electron microscopist, Univ Chicago Med Ctr, 77-; ocular pathologist & dir, Eye Path Lab & NIH grant, 77-, attend ophthalmologist, Oak Forest Hosp, 86-; sect ed & exec ed, Exp Eye Res; life fel, Nat Acad Sci, India, 87; vis prof to many nat & int acad insts. *Honors & Awards:* Ophthalmologic Prize, Royal Soc Med, 71; Royal Eye Hosp Prize, Ophthal Soc, UK, 76; Honor Award, Am Acad Ophthal, 84; Honor Award, Asn Indians Am, 84; Litchfield lectr, Univ Oxford, 86; Honor Award, Nat Fedn Asian Indians, 86; Int Prize, Alcon Res Inst, 87. *Mem:* Asn Res Vision & Ophthal; Am Acad Ophthal; Royal Soc Med; Royal Col Pathologists; Sigma Xi; Physiol Soc London; AMA; Contact Lens Asn Ophthalmologists; Fedn Am Socs Exp Biol; Am Asn Pathologists. *Res:* Ophthalmology; visual science; anatomy; physiology; surgical pathology; experimental pathology; clinical research; cell biology; electron microscopy; tissue culture; biochemistry; immunology; microbiology; micrography; ophthalmic microsurgery; laser surgery; physiology and pathology of cerebrospinal fluid; author of more than 350 publications and 30 monographs and chapters in scientific journals and books. *Mailing Add:* Visual Sci Ctr Univ Chicago 939 E 57th St Chicago IL 60637

TRIPATHI, SATISH CHANDRA, b Gwalior, Madhya Pradesh, India, May 15, 56; m. CELL CYCLE CONTROL, ONCOGENE EXPRESSION. *Educ:* Jiwaji Univ, Gwalior, BS, 75, MS, 77; Bhopal Univ, MPhil, 78; Univ Glasgow, PhD(cell biol), 84. *Prof Exp:* Lectr bot, Govt Postgrad Col, Khargone, Madhya Pradesh, 78-80; asst prof genetics, Jawaharlal Nehru Krishi Univ, Jabalpur, 80-85; res scientist cell biol, Med Col Wis, Milwaukee, 85-87; vis scientist cell biol, Emory Univ, Atlanta, 87-89; RES FEL CELL BIOL, MASS INST TECHNOL, 89-; STAFF BIOLOGIST CELL BIOL, RES INST, ILL INST TECHNOL, CHICAGO, 90- *Concurrent Pos:* Consult, Biotechnol Consult Int, 89-; policy consult biotechnol, Mass Ctr Excellence Biotechnol, Boston, 89- *Honors & Awards:* Maitland Award, Univ Glasgow, 83. *Mem:* Am Soc Cell Biol; Cell Kinetics Soc; AAAS; Brit Soc Cell Biol. *Res:* Tension control of cell cycle and proto oncogene activation and expression; protein chemistry; recombinant DNA technology; mutational spectrometry; visual cortex; PCR technology; bioreactor technology for efficient production of macromolecules. *Mailing Add:* Life Sci Dept Res Inst Ill Inst Technol Ten W 35th St Chicago IL 60616

TRIPATHI, SATISH K, b Patna, Faizabad, India, Jan 20, 51; m 70; c 2. PERFORMANCE EVALUATION, NETWORKING. *Educ:* Banaras Hindu Univ, BSc, 68; Univ Alta, MS, 74; Univ Toronto, MS, 76, PhD(computer sci), 79. *Prof Exp:* From asst prof to assoc prof, 78-88, PROF COMPUTER SCI, UNIV MD, COLLEGE PARK, 88-, CHMN, DEPT COMPUTER SCI, 89- *Concurrent Pos:* Vis prof, Univ Paris, Sud, 84-85, Univ Erlangen, Nunenburg, 85. *Mem:* Sr mem Inst Elec & Electronics Engrs; Asn Computer Mach. *Res:* Performance evaluation of distributed, real-time systems; resource a-location in multi-processor systems; high speed networking. *Mailing Add:* Dept Computer Sci Univ Md College Park MD 20742

TRIPATHI, UMA PRASAD, b Lumbini, Nepal, Apr 29, 45; c 2. INORGANIC CHEMISTRY, COSMETIC CHEMISTRY. *Educ:* Univ Gorakhpur, India, BSc, 63; Univ Allahabad, India, MSc, 65; Mont State Univ, PhD(inorg chem), 72. *Prof Exp:* Anal chemist, Nepal Bur Mines, Kathmandu, 65; asst prof chem, Trichandra Col, Kathmandu, 65-67; sr res chemist, Cosmetics, Chesebrough-Pond's Inc, 72-75; sr chemist, Gillette Co, 75-76; group leader, 76-78, sect mgr, 78-81, ASSOC DIR, AM CYANAMID CO, 82- *Mem:* Soc Cosmetic Chemists; Am Chem Soc. *Res:* Elucidation of molecular and crystal structures of organometallics; inorganic synthesis; inorganic reaction mechanisms; emulsion technology; surface active agents; hair science; insecticides; household products. *Mailing Add:* Playtex Inc 215 College Rd PO Box 728 Paramus NJ 07653

TRIPATHI, VIJAI KUMAR, b Kanpur, India, Dec 23, 42; US citizen; m 68; c 2. ELECTRICAL ENGINEERING. *Educ:* Agra Univ, BSc, 58; Univ Allahabad, MScTech, 61; Univ Mich, Ann Arbor, MS, 64, PhD(elec eng), 68. *Prof Exp:* Sr res asst elec eng, Indian Inst Technol, Bombay, 61-63; res asst, Electron Physics Lab, Univ Mich, Ann Arbor, 63-65, res assoc, 65-67; asst prof elec eng, Univ Okla, 68-74; from asst prof to assoc prof, 74-85, PROF ELEC ENG, ORE STATE UNIV, 85- *Concurrent Pos:* Guest prof, Chalmers Tech Univ, Gothenburg, Sweden, 81-82 & Duisburg Univ, WGermany, 82. *Mem:* Inst Elec & Electronics Engrs. *Res:* Microwave circuits and devices; solid state devices; electromagnetic fields. *Mailing Add:* Dept Elec & Comput Eng Ore State Univ Corvallis OR 97331

TRIPATHY, DEOKI NANDAN, b Dwarahat, India, July 1, 33; m 69; c 2. VETERINARY MICROBIOLOGY, MOLECULAR BIOLOGY. *Educ:* Utter Pradesh Agr Univ, India, BVSc & AH, 64; Univ Ill, Urbana, MS, 67, PhD(vet microbiol), 70; Am Col Vet Microbiologists, dipl, 72. *Prof Exp:* Fel, 64-65, res asst, 65-70, assoc, 70-73, asst prof, 73-77, assoc prof vet microbiol, PROF, UNIV ILL, URBANA, 83- *Mem:* Am Vet Med Asn; Am Asn Avian Pathologists; Am Soc Microbiol; Am Col Vet Microbiologists; US Animal Health Asn; Conf Res Workers Animal Dis. *Res:* Diseases of food producing animals; poxviruses of veterinary importance; leptospirosis in domestic animals; infection laryngotracheitis virus, recombinant vaccines; development of fowlpox virus as eucaryotic cloning and expression vector for genes from poultry pathogens. *Mailing Add:* Dept Vet Pathobiol Univ Ill Urbana IL 61801

TRIPATHY, SUKANT K, b Chakradharpur, India, Aug 4, 52; m 81; c 2. STRUCTURE PROPERTY, POLYMER THEORY. *Educ:* Indian Inst Technol, India, BSc, 72, MSc, 74; Case Western Reserve Univ, PhD(macromolecular sci), 81. *Prof Exp:* Mem tech staff, 81-83, res mgr, GTE Labs Inc, 83-86; PROF CHEM, UNIV LOWELL, 86- *Concurrent Pos:* Consult & lectr, Case Western Reserve Univ, 77-81. *Mem:* Am Phys Soc; Am Chem Soc; AAAS; Sigma Xi. *Res:* Characterization and understanding of polymer microstructure using physical methods and theoretical modelling, establishing structure property relationships; new organic materials and novel polymeric applications. *Mailing Add:* Dept Chem Univ Lowell Lowell MA 01854

TRIPLEHORN, CHARLES A, b Bluffton, Ohio, Oct 27, 27; m 49; c 2. ENTOMOLOGY. *Educ:* Ohio State Univ, BS, 49, MS, 52; Cornell Univ, PhD(entom), 57. *Prof Exp:* Asst prof entom, Univ Del, 52-54, Ohio Agr Exp Sta, 57-62 & Ohio State Univ, 62-64; entomologist, US AID, Brazil, 64-66; assoc prof, 66-67, PROF ENTOM, OHIO STATE UNIV, 67- *Mem:* Entom Soc Am (pres, 85); Coleopterists Soc (pres, 76); Royal Entom Soc London. *Res:* Taxonomy of Coleoptera; animal ecology; herpetology. *Mailing Add:* Dept Entom Ohio State Univ 1735 Neil Ave Columbus OH 43210

TRIPLEHORN, DON MURRAY, b Bluffton, Ohio, July 24, 34; m 57; c 4. GEOLOGY. *Educ:* Ohio Wesleyan Univ, BA, 56; Ind Univ, MA, 57; Univ Ill, PhD(geol), 61. *Prof Exp:* Instr geol, Col Wooster, 60-61; res geologist, Tulsa Res Ctr, Sinclair Oil & Gas Co, 61-69; from assoc prof to prof geol, Univ Alaska, 69-76; US GEOL SURV, DENVER, COLO, 76- *Mem:* Geol Soc Am; Int Asn Sedimentologists; Am Asn Petrol Geologists; Soc Econ Paleontologists & Mineralogists; Am Asn Geol Teachers. *Res:* Glauconite; coal geology; clay mineralogy; shale petrology; diagenesis; Cretaceous-Tertiary boundary. *Mailing Add:* Dept Geol & Geophys Univ Alaska Fairbanks AK 99775

TRIPLETT, EDWARD LEE, b Denver, Colo, July 14, 30; m 51; c 3. ZOOLOGY. *Educ:* Stanford Univ, BS, 51, PhD, 56. *Prof Exp:* Actg instr biol, Stanford Univ, 54-55; from instr to assoc prof, 55-73, PROF BIOL, UNIV CALIF, SANTA BARBARA, 73- *Mem:* AAAS; Soc Develop Biol. *Res:* Development of the nervous system; cell differentiation of neural plate cells; control of protein synthesis in developing systems. *Mailing Add:* Dept Biol Sci Univ Calif Santa Barbara CA 93106

TRIPLETT, GLOVER BROWN, JR, b Miss, June 2, 30; m 51; c 1. AGRONOMY. *Educ:* Miss State Univ, BS, 51, MS, 55; Mich State Univ, PhD(farm crops), 59. *Prof Exp:* From asst prof to assoc prof, 59-67, prof agron, Ohio Agr Res & Develop Ctr, 67-83; PROF AGRON, MISS STATE UNIV, 83- *Concurrent Pos:* Agronomist, 59- *Mem:* Am Soc Agron; Weed Sci Soc Am; Sigma Xi. *Res:* Crop production and management; no-tillage cropping systems. *Mailing Add:* PO Box 2463 Starkville MS 39759

TRIPLETT, KELLY B, b Cincinnati, Ohio; m; c 2. CERAMICS FROM CHEMICALS, ZIEGLER-NATTA PROCESSES. *Educ:* Northwestern Univ, BA, 68; Univ Mich, PhD(chem), 74. *Prof Exp:* Res assoc, Mich State Univ, 74-76; res chemist, Stauffer Chem Co, 76-78, supvr, 78-82, tech mgr, 82-84, bus mgr, 84-87; prog mgr & bus mgr, 87-90, RES CTR MGR, AKZO CHEM INC, 90- *Mem:* Am Ceramic Soc; Indust Res Inst. *Res:* Investigation of chemical routes to advanced ceramics; new polymerization products and process; organometallics and transition metal chemistries; research and development administration; development of new research and development methodologies. *Mailing Add:* Akzo Chem Inc One Livingstone Ave Dobbs Ferry NY 10522

TRIPODI, DANIEL, b Cliffside Park, NJ, May 13, 39; m 63. IMMUNOCHEMISTRY, MICROBIOLOGY. *Educ:* Univ Del, BS, 61, MS, 63; Temple Univ, PhD(microbiol, immunol), 66. *Prof Exp:* Asst microbiol, Univ Del, 63; sr scientist, Ortho Res Found, 66-69, dir div immunol & diag res, 69-74; res asst prof, Sch Med, Temple Univ, 69-74, res assoc prof microbiol microbiol, 73-77; gen mgr, biomed div, New Eng Nuclear Corp, 74-77; dir, tech planning group, Becton, Dickinson & Co, 77-80; dir tech serv, Ortho Diagnostics, 80-83; corp dir biotechnol, Johnson & Johnson, 83-87; VPRES RES, THERAKOS, INC, 87- *Mem:* Am Soc Microbiol; Am Asn Immunol; Sigma Xi. *Res:* Physical and chemical aspects of structures which display serological reactivity. *Mailing Add:* Tricon Consult 35A Burlinghoff Lane Lebanon NJ 08833

TRIPP, JOHN RATHBONE, b Barberton, Ohio, Oct 18, 39; m 71. INVERTEBRATE ZOOLOGY, EMBRYOLOGY. *Educ:* Ore State Univ, BSc, 62; Ohio State Univ, MSc, 65, PhD(zool), 70. *Prof Exp:* Res specialist, Introductory Biol Prog, Ohio State Univ 69-70; ASST PROF DEPT BIOL, FLA SOUTHERN COL, 71- *Concurrent Pos:* Prin investr, Cottrell Found Col Sci grant, 75. *Mem:* Am Soc Zoologists; Am Inst Biol Sci; AAAS; Am Arachnological Soc. *Res:* Spider embryology and marine invertebrate zoology. *Mailing Add:* Dept Biol Fla Southern Col 111 Lake Hollingsworth Dr Lakeland FL 33801

TRIPP, JOHN STEPHEN, b Salina, Kans, Aug 3, 38. ELECTRONICS ENGINEERING, COMPUTER SCIENCE. *Educ:* Kans State Univ, BS, 61, MS, 62; Univ Mich, Ann Arbor, MS, 67, PhD(comput sci), 71. *Prof Exp:* AEROSPACE TECHNOLOGIST, LANGLEY RES CTR-NASA, 62- *Concurrent Pos:* Lectr, George Washington Univ, 81- *Mem:* Inst Elec & Electronics Engrs. *Res:* Analysis and simulation of physical systems; synthesis of control techniques for automation of aerospace test facilities; application of digital computers to automatic control. *Mailing Add:* Langley Res Ctr 238 Data Systs NASA Hampton VA 23665

TRIPP, LEONARD L, b Los Angeles, Calif, Oct 21, 41; m 63; c 6. SOFTWARE DESIGN, SOFTWARE ENGINEERING STANDARDS. *Educ:* Brigham Young Univ, BS, 65, MS, 67. *Prof Exp:* Appl programmer, Boeing Co, 67-71,; appl programmer, Boeing Comput Serv, 71-73, software technol specialist, 73-79, sr comput scientist, 79-89, SR COMPUT SCIENTIST, BOEING COM AIRPLANE, 89- *Concurrent Pos:* Chmn, working group for IEEE Standard 1002, 83-87; chmn, Inst Elec & Electronic Engrs Comput Sci Coord Comt, 90-91. *Mem:* Asn Comput Mach; Math Asn Am; Inst Elec & Electronics Engrs; Am Soc Testing & Mat. *Res:* Development and implementation of techniques, methodologies, standards and tools for systems engineering software requirements definition, software design, software verification, validation and testing and software project management. *Mailing Add:* 28632 Eighth Pl S Federal Way WA 98003

TRIPP, MARENES ROBERT, b Poughkeepsie, NY, Aug 20, 31; m 55; c 4. INVERTEBRATE PATHOLOGY. *Educ:* Colgate Univ, AB, 53; Univ Rochester, MS, 56; Rutgers Univ, PhD(zool), 58. *Prof Exp:* Res fel trop pub health, Harvard Univ, 58-60; from asst prof to assoc prof, 60-71, PROF BIOL SCI, UNIV DEL, 71-, from interim dir to dir, Sch Life & Health Sci, 81-87. *Mem:* AAAS; Am Soc Parasitol; Soc Invert Path; Am Soc Zoologists; Int Soc Develop Comp Immunol. *Res:* Invertebrate defense mechanisms. *Mailing Add:* Sch Life & Health Sci Univ Del Newark DE 19716

TRIPP, R(USSELL) MAURICE, b Holton, Kans, July 12, 16; m 37; c 7. RADIOLOGY, GEOPHYSICS. *Educ:* Colo Sch Mines, GeolE, 39, MGeophysEng, 43; Mass Inst Technol, ScD(geol), 48. *Prof Exp:* Geophysicist, Geotech Corp, Tex, 36-41, asst to pres-in-chg res, 43-46; instr geol & geophys, Colo Sch Mines, 41-43; sr scientist, Bur Ships, US Dept Navy, 46; consult geologist & geophysicist, 47-49; vpres & dir res, Res Inc, 49-53; pres, Explor, Inc, 52-58; PRES, TRIPP RES CORP, 55-, SKIA CORP, 72-, AKTINA CORP, 84- *Concurrent Pos:* Managing partner, Tripp Lead & Zinc Co, 52-, pres, Tripp Prod, Inc, 62-; dir, Sonic Res Corp, 56-; consult, Bostwick Propecting Co, 56- & Archilithic Corp, 58-; managing partner, Saratoga Develop Co. *Mem:* Soc Info Display; Soc Photog Scientists & Engrs; Soc Photo Optical Instrumentation Engrs; AAAS; Sigma Xi. *Res:* Geophysical and geochemical exploration for minerals; relation between clay minerals, organic matter and radioelements in sediments; genesis of uranium ore bodies; mineral benefication; free viewing depth-perception radiography; nuclear medicine instrumentation; animal genetics; ophthalmic surgical instrumentation. *Mailing Add:* SKIA Corp 5181 Lafayette St Santa Clara CA 95054

TRIPP, ROBERT D, b Oakland, Calif, Jan 9, 27; m 64; c 3. PHYSICS. *Educ:* Mass Inst Technol, BS, 49; Univ Calif, PhD, 55. *Prof Exp:* From asst prof to assoc prof, 60-66, PHYSICIST, LAWRENCE BERKELEY LAB, UNIV CALIF, BERKELEY, 55-, PROF PHYSICS, 66- *Concurrent Pos:* Physicist, AEC, France, 59-60; NSF sr fel, Europ Orgn Nuclear Res, Switz, 64-65, vis scientist, 71-72. *Mem:* Am Phys Soc. *Res:* Elementary particle physics. *Mailing Add:* Dept Physics Univ Calif 2120 Oxford St Berkeley CA 94720

TRIPPE, ANTHONY PHILIP, b Buffalo, NY, Aug 30, 43; m 65; c 3. NUCLEAR ENGINEERING, TECHNICAL MANAGEMENT. *Educ:* Rochester Inst Technol, BS, 66; Fairleigh Dickinson Univ, MS, 72; US Int Univ, DBA, 82. *Prof Exp:* Lab tech anal chem, Eastman Kodak Co, 63-66; chemist polymers, Smith Kline & French Labs, 66-67; reliability engr quality assurance, US Army Picatinny Arsenal, 67-72; prin engr, Calspan Corp, 72-76; mgr prog develop, IRT Corp, 77-84, mkt mgr, 84-88; VPRES SURVIVABILITY PRODS & EM PRODUCTS MGR, MAXWELL LABORATORIES, INC, 88- *Concurrent Pos:* Fire protection engr, Allstate Insurance Co, 74-75. *Mem:* Am Defense Preparedness Asn; Asn Old Crows; Am Mgt Asn; Armed Forces Commun & Electronics Asn; Inst Elec & Electronics Engrs. *Res:* Nondestructive testing using nuclear techniques; study of systems using computerized analytical models; application of pattern recognition to automated control; nuclear weapons effects on military systems. *Mailing Add:* 5590 Antigua Blvd San Diego CA 92124

TRIPPE, THOMAS GORDON, b Los Angeles, Calif, Nov 17, 39; m; c 3. EXPERIMENTAL HIGH ENERGY PHYSICS. *Educ:* Univ Calif, Los Angeles, PhD(physics), 68. *Prof Exp:* Physicist, Univ Calif, Los Angeles, 68-69; physicist, Europ Orgn Nuclear Res, 69-70 & 73; PHYSICIST, LAWRENCE BERKELEY LAB, 71- *Concurrent Pos:* NSF fel, 69-70. *Mem:* Am Phys Soc. *Res:* Experimental weak interactions; particle properties. *Mailing Add:* Lawrence Berkeley Lab Particle Data Group Berkeley CA 94720

TRIPPODO, NICK CHARLES, b Galveston, Tex, Sept 27, 45; m 67; c 2. CARDIOVASCULAR PHYSIOLOGY, HYPERTENSION. *Educ:* Stephen F Austin State Univ, BS, 68, MS, 69; Univ Tex, Galveston, PhD(physiol), 74. *Prof Exp:* Fel cardiovascular physiol, Univ Miss Med Ctr, 74-76; sci staff hypertension, Div Res, Alton Ochsner Med Found, 76-79, res coordr, 79-88, head vascular physiol, 88; head sect peptide pharmacol, Squibb Inst Med Res, 88-91, PRIN SCIENTIST, PHARMACOL DEPT, BRISTOL-MYERS SQUIBB PHARMACEUT RES INST, 91- *Concurrent Pos:* Assoc prof physiol, La State Univ Sch Med, 76-88; prin investr grant awards, Public Health Serv, 79-88; consult, Abbott Lab, 87-88; adj assoc prof physiol, Tulane Univ Sch Med, 88. *Mem:* Am Physiol Soc; Int Soc Hypertension; Am Fedn Clin Res; Int Soc Health Res. *Res:* Fluid and electrolyte homeostasis; arterial blood pressure regulation; blood volume; vascular capacitance; cardiovascular actions of the hormone; atrial natriuretic factor. *Mailing Add:* Five Water Lilly Way Newtown PA 18940

TRISCARI, JOSEPH, b Italy, Apr 17, 45; US citizen; m 66; c 3. OBESITY, LIPID METABOLISM. *Educ:* Cornell Univ, BS, 71; Fairleigh Dickinson Univ, MS, 75; Columbia Univ, PhD(nutrit), 80. *Prof Exp:* Lab asst, Dept Biochem, Cornell Univ, 68-71; res investr, neurobiol & obesity, anti-infarctives, Hoffmann La Roche, 71-89; ASST DIR, BRISTOL-MYERS SQUIBB, 89- *Concurrent Pos:* Guest lectr, Columbia Univ, Iowa State Univ & New York Univ, 82, Univ Ga, 86. *Honors & Awards:* Gold Bond Award, Am Chem Soc. *Mem:* Am Inst Nutrit; AAAS; NY Acad Sci; Soc Study Ingestive Behav. *Res:* Regulation of lipid and carbohydrate metabolism; endocrinology and pharmacology of insulin and CCK; appetite regulation and energy balance; pharmacologic treatment of obesity and hyperlipidemia; models of obesity; atherosclerosis; clinical pharmacokinetics. *Mailing Add:* Dept Cardiovasc Clin Res Metab Bristol-Myers Squibb PO Box 4000 Princeton NJ 08543

TRISCHAN, GLENN M, Milwaukee, Wis. ATOMIC SPECTROSCOPY, MOLECULAR SPECTROSCOPY. *Educ:* Marquette Univ, BS, 72; Univ Iowa, MS, 74 & PhD(anal chem), 77. *Prof Exp:* Res asst, dept chem, Univ Iowa, 73-76; assoc anal chemist, Midwest Rest Inst, 76-80; sr anal chemist, Johnson Controls Inc, 80-86, MGR MAT TEST LAB, 86- *Mem:* Am Chem Soc; Sigma Xi; Am Indust Hygiene Asn; Soc Appl Spectroscopy; Am Lab Mgrs Asn. *Res:* Track element analysis in metal materials; analysis of aerosols and vapors in compressed gas systems; surface, reaction layer characterization; environmental analysis of process effluents from waste to energy; surface reactions in low temperate flames. *Mailing Add:* Johnson Controls Inc MS G-Three PO Box 591 Milwaukee WI 53201

TRISCHKA, JOHN WILSON, b Bisbee, Ariz, Dec 30, 16; m 46; c 2. PHYSICS. *Educ:* Univ Ariz, BS, 37; Cornell Univ, PhD(physics), 43. *Prof Exp:* Test engr, Gen Elec Co, 37-38; asst physics, Cornell Univ, 39-42, instr, 42-45; res physicist, Los Alamos Sci Lab, Univ Calif, 45; assoc physics, Columbia Univ, 46-48; from asst prof to prof, 48-84, EMER PROF PHYSICS, SYRACUSE UNIV, 85- *Mem:* AAAS; Am Phys Soc. *Res:* Radio frequency spectroscopy of molecules; molecular beams; surface physics; atmospheric physics. *Mailing Add:* Dept Physics Syracuse Univ Syracuse NY 13210

TRISCHLER, FLOYD D, b Pittsburgh, Pa, Aug 31, 29; m 51; c 6. ORGANIC CHEMISTRY, POLYMER CHEMISTRY. *Educ:* Univ Pittsburgh, BS, 51. *Prof Exp:* Res chemist, Pa Indust Chem Corp, 53-55, asst lab mgr, 55-56; develop & tech chemist, Tar Prod Div, Koppers Co, 56-58, asst group leader, Res Dept, 58-63; sr res chemist, Narmco Res & Develop, Whittaker Corp, Calif, 63-65, prog mgr, 65-69; mgr mkt & admin asst to pres, Mat Systs Corp, Calif, 69-72; exec vpres, Taylor Bldg Corp, Ind, 72-73; pres, Guadalupe Builders, Ind, 73-78; PRES, IROQUOIS REALTY, INC, 78- *Mem:* Am Chem Soc. *Res:* Polymer applications and research; pulp and paper; organic coatings; fluorine compounds; elastomers; adhesives; high performance polymers. *Mailing Add:* 8249 Filly Lane Plainfield IN 46168

TRISLER, JOHN CHARLES, b Eva, La, Dec 24, 33; m 53; c 2. ORGANIC CHEMISTRY. *Educ:* La Polytech Inst, BS, 56; Tex Tech Univ, PhD(org chem), 59. *Prof Exp:* From asst prof to assoc prof, 59-66, PROF CHEM, LA TECH UNIV, 66-, HEAD DEPT, 78- *Mem:* Am Chem Soc. *Res:* Organic reaction mechanisms. *Mailing Add:* PO Box 3044 TS Ruston LA 71270

TRISTAN, THEODORE A, b Mexico City, Mex, Oct 5, 24; US citizen; m 48; c 4. RADIOLOGY. *Educ:* Univ Nebr, BS, 47, MD & MSc, 50; Univ Pa, MS, 58; Am Bd Radiol, dipl, 57. *Prof Exp:* Intern, Hosp, Univ Pa, 50-51; instr radiol, Med Ctr, Univ Rochester, 56-59, sr instr, 59; from asst prof to assoc prof, Univ Pa, 59-65, lectr, 65-70; prof radiol & anat, Hershey Med Ctr, Pa State Univ, 70-86, prof radiol & clin prof anat, 86-89; RETIRED. *Concurrent Pos:* Fel radiol, Univ Pa Hosp, 53-56; Am Cancer Soc fel, 53-54; NIH grants, Univ Rochester, 58-59 & Univ Pa, 59-63; Nat Res Coun Picker Found grant, Univ Pa, 58-65; mem & task group chmn, Int Comn Radiation Units & Measurements, 64-67; assoc prof radiol, Hahnemann Med Col, 70-73. *Honors & Awards:* Gold Medal, Radiol Soc NAm. *Mem:* Radiol Soc NAm (secy, 75-79, pres elect, 81 & pres, 82); Am Roentgen Ray Soc; AMA; fel Am Col Radiol; Sigma Xi; Asn Univ Radiol. *Res:* Image intensification, image quality and information development, storage and retrieval; analysis of the function of motion in cinefluorography with regard to its value in development of diagnostic criteria in clinical radiology. *Mailing Add:* 353 N 28th St Camp Hill PA 17011-2837

TRISTRAM-NAGLE, STEPHANIE ANN, b New York, NY, Nov 21, 48; m 80; c 2. MEMBRANE THERMODYNAMICS, MEMBRANE STRUCTURE. *Educ:* Douglass Col, BA, 70; Univ Calif, Berkeley, PhD(comp biochem), 81. *Prof Exp:* Postdoctoral res membrane thermodyn, 82-86, RES BIOLOGIST STRUCT MEMBRANES, CARNEGIE MELLON UNIV, PITTSBURGH, 86- *Concurrent Pos:* Reviewer, Birchimica Biophysica Acta (Elsevier), NHolland Biomed Press, 85- & Biochem, Univ Wash, Seattle, 87-; postgrad biol & chem, Univ Mass, Boston. *Mem:* Biophys Soc; Sigma Xi; Asn Women Sci. *Res:* Structure and thermodynamic properties primarily of biomimetic membranes composed of purified lipids; x-ray diffraction of both oriented and dispersed lipids; differential dilatometry; differential scanning calorimetry; neutral buoyancy centrifugation. *Mailing Add:* Dept Biol Sci 4400 Fifth Ave Pittsburgh PA 15213-3890

TRITCHLER, DAVID LYNN, b Portage, Wis, Dec 17, 44; m 75. BIOMETRICS, BIOSTATISTICS. *Educ:* Univ Nebr, BA, 70, MS, 75; Harvard Sch Pub Health, ScD, 80. *Prof Exp:* Asst prof, Cornell Univ, 81; ASST PROF BIOSTATIST, HARVARD SCH PUB HEALTH, 81- *Concurrent Pos:* Assoc, Mem Sloan Kettering Cancer Ctr, 80-81; asst prof, Sidney Farber Cancer Inst, 81- *Mem:* Am Statist Asn; Biometric Soc; Soc Indust & Appl Math. *Res:* Computational algorithms for nonparametric statistics; survival analysis; discrete data analysis. *Mailing Add:* 71 Rochester Rd Carver MA 02330

TRITES, RONALD WILMOT, b Moncton, NB, July 17, 29; m 56; c 5. PHYSICAL OCEANOGRAPHY. *Educ:* Univ NB, BSc, 50; Univ BC, MA, 52, PhD(physics), 55. *Prof Exp:* Phys oceanogr, Fisheries Res Bd Can, 50-55 & 56-66 & Defence Res Bd Can, 55-56; head appl oceanog, Bedford Inst Oceanog, 66-70; adv Atlantic, Fisheries Res Bd Can, 70-71; head coastal oceaonog div, 71-75, sr oceanogr, Marine Ecol Lab, 75-87, RES SCIENTIST, COASTAL OCEANOG DIV, BEDFORD INST OCEANOG, 87- *Concurrent Pos:* Hon lectr, Dalhousie Univ, 60-77. *Res:* Coastal and estuarine circulation, mixing and dispersion processes; role of physical processes in production of fish stocks, including mariculture. *Mailing Add:* Two Eljay Dr Dartmouth NS B2W 2C1 Can

TRITSCHLER, LOUIS GEORGE, b St Louis, Mo, Jan 24, 27; m 47; c 2. VETERINARY MEDICINE, VETERINARY SURGERY. *Educ:* Univ Mo, BSAgr, 49, DVM, 60, MS, 62. *Prof Exp:* From instr to asst prof, 60-72, assoc prof, 72-77, PROF VET MED & SURG, UNIV MO-COLUMBIA, 77-, DIR, EQUINE CTR, 79- *Mem:* Am Vet Med Asn; Am Asn Equine Practrs; Am Asn Vet Clinicians. *Res:* Use of estrone in the treatment of anestrus in cattle; evaluation of bulls for breeding soundness; wound treatment and fracture repair in equine. *Mailing Add:* Vet Hosp Surger Connaway Hall Univ Mo Columbia MO 65211

TRITTON, THOMAS RICHARD, b Lakewood, Ohio, Dec 20, 47; m 77; c 2. BIOPHYSICAL CHEMISTRY, MOLECULAR PHARMACOLOGY. *Educ:* Ohio Wesleyan Univ, BA, 69; Boston Univ, PhD(chem), 73. *Prof Exp:* Fel biophys chem, Sch Med, Yale Univ, 73-75, asst prof, 75-80, assoc prof pharmacol, 80-84; PROF PHARMACOL & DEP DIR, VT REGIONAL CANCER CTR, UNIV VT, 85- *Concurrent Pos:* NIH fel, Yale Univ, 74-75, res career develop award, 80-85, exp ther study sect, 88- *Mem:* Am Chem Soc; Sigma Xi; Am Soc Biol Chemists; Am Asn Cancer Res. *Res:* Membrane mediated cytotoxicity; drug resistance; membrane dynamics and the cell surface as a target for antineoplastic agents; biophysical approaches to pharmacology. *Mailing Add:* Dept Pharmacol Univ Vt Med Col Given Bldg Burlington VT 05405

TRITZ, GERALD JOSEPH, b Sioux City, Iowa, Apr 12, 37; m 66; c 2. MICROBIOLOGY, GENETICS. *Educ:* Utah State Univ, BS, 62; Colo State Univ, MS, 65; Univ Tex Med Sch Houston, PhD(biomed res), 70. *Prof Exp:* Res microbiologist, USPHS, 65-67; NIH fel microbiol, Univ Tex M D Anderson Hosp & Tumor Inst, Houston, 70; asst prof microbiol, Univ Ga, 70-76; assoc prof & chmn dept, 76-81, PROF & CHMN MICROBIOL & IMMUNOL, KIRKSVILLE COL OSTEOP MED, 81- *Concurrent Pos:* NSF grant, 71-78; Am Osteop Asn res grant, 79-85; citizen ambassador & vis scientist, Peoples Repub China, 83; Fermature Found grant, 85-89, Commerce Found grant, 87; vis scientist, minority student sci careers support prog, Am Soc Microbiol, 87-; ed, cellular & molecular biol. *Honors & Awards:* M G Michael Award. *Mem:* Am Soc Microbiol; Am Acad Microbiol. *Res:* Pyridine nucleotide metabolism and its control. *Mailing Add:* Dept Microbiol & Immunol Kirksville Col Osteop Med Kirksville MO 63501

TRIVEDI, KISHOR SHRIDHARBHAI, b Bhavnagar, India, Aug 20, 46; m 73; c 2. COMPUTER SCIENCES. *Educ:* Indian Inst Technol, Bombay, BTech, 68; Univ Ill, Urbana-Champaign, MS, 72, PhD(comput sci), 74. *Prof Exp:* Assoc customer engr, Int Bus Mach World Trade Corp, Bombay, 68-70; res asst comput sci, Univ Ill, Urbana-Champaign, 70-74, res assoc, 74-75; from asst prof to assoc prof, 75-83, PROF COMPUT SCI & ELEC ENG, DUKE UNIV, 83- *Concurrent Pos:* Consult, various Res Inst & Indust Labs, 79-; Asn Comput Mach, nat lectr, 81-82; ed, Inst Elec & Electronics Engrs Transactions Comput, 83-87. *Mem:* Comput Soc India; Am Asn Comput Mach; Inst Elec & Electronics Engrs Comput Soc. *Res:* Algorithms for construction and performance evaluation of computer operating systems; reliability and performance analysis of fault-tolerant multiple processor systems. *Mailing Add:* Dept Comput Sci Duke Univ Durham NC 27706

TRIVEDI, MOHAN MANUBHAI, b Wardha, India, Oct 4, 53; m 82. SENSOR DRIVEN ROBOTICS. *Educ:* Birla Inst Tech & Sci, Pilani, India, BE, 74; Utah State Univ, ME, 76, PhD(elec eng), 79. *Prof Exp:* Teaching asst, Utah State Univ, 75-76, res asst, Space Dynamics Lab, 76-79; assoc prof comput eng, La State Univ, 79-86; assoc prof, 86-90, PROF ELEC & COMPUT ENG, UNIV TENN, 90- *Concurrent Pos:* Prin investr, var govt & indust res orgn, 79- consult, 83-; chmn, Pattern Recognition, Image Processing & Comput Vision Comt, Inst Elec & Electronics Engrs, 86- & Robotics Comt, 87 & assoc ed, Transactions of Systs, Man & Cybernetics, Machine Vision & Appln J, Int J Approx Reasoning, 88-; lectr, Int Soc Optical Engrs, 87- *Honors & Awards:* Pioneer Award, Inst Elec & Electronics Engrs, 89, Meritorious Serv Award, 89. *Mem:* Comput Soc; sr mem Inst Elec & Electronics Engrs; fel Int Soc Optical Engrs; Sigma Xi. *Res:* Design and engineering of computer vision systems including image processing, pattern recognition, knowledge representation, control, parallel architecture issues; development of sensor-driven intelligent robots including, sensor design, information processing, sensor fusion, image tactile, thermal infrared information acquisition and processing path planning. *Mailing Add:* Elec & Comput Eng Dept Univ Tenn Ferris Hall Knoxville TN 37996-2100

TRIVEDI, NAYAN B, b Kapadwanj, India, Feb 13, 47; US citizen; m 70; c 2. TECHNOLOGY MANAGEMENT, NEW FOOD PRODUCTS. *Educ:* Gujarat Univ, India, Sc, 67; Univ Baroda, India, MSc, 69; Univ Southwest La, PhD(microbiol), 73. *Prof Exp:* Dir labs, Arroyo Pharmaceut Corp, Chase Chem Co, 73-76; sr scientist, Schering Corp, 76; mgr, genetics-biochem, Standard Brands & Nabisco Brands, Inc, 76-81; DIR, RES & DEVELOP, UNIVERSAL FOODS CORP, 81- *Concurrent Pos:* Asst prof biochem, Univ Conn, Danbury, 75. *Mem:* Am Soc Microbiol; Am Chem Soc; Soc Indust Microbiol; Am Asn Cereal Chemists; Am Soc Bakery Engrs; Inst Food Technologists. *Res:* Fermentation research including isolation, selection, mutation and genetic manipulation of microorganisms; recovery processes for fermentation products; yeast, mold actinomycetes and bacterial fermentations covering from cell mass and alcohol production to emulsifiers, antibiotics, vitamins and special enzymes using r-DNA technology. *Mailing Add:* 33 Parkview Rd Cranbury NJ 08512

TRIVEDI, ROHIT K, b Bhavnagar, India, Mar 8, 39; m; c 3. METALLURGY, MATERIALS SCIENCE. *Educ:* Indian Inst Technol, Kharagpur, BS, 60; Carnegie Inst Technol, MS, 64, PhD(metall, mat sci), 66. *Prof Exp:* Res scientist, Sci Lab, Ford Motor Co, 63-64; res assoc metall, Inst Atomic Res, 65-66, asst prof, 66-70; assoc prof metall, 70-75, PROF MAT SCI & ENG, IOWA STATE UNIV, 75-; SR METALLURGIST, AMES LAB, US DEPT ENERGY, 75- *Concurrent Pos:* Vis assoc prof, Stanford Univ, 69-70; guest prof, Tech Univ Braunschweig, WGer, 77-78; invited prof, Fed Polytech Sch Lausanne, Switz, 84-85; invited prof, Université Aix-Marseille III, Marseille, France, 88, 89. *Honors & Awards:* Sr Alexander von Humboldt Award, 77; Sr Fulbright Res Award, 77. *Mem:* Fel Am Soc Metals; Am Inst Mining, Metall & Petrol Engrs. *Res:* Solidification; stability of interphase boundaries. *Mailing Add:* Dept Mat Sci & Eng Iowa State Univ Ames IA 50011

TRIVELPIECE, ALVIN WILLIAM, b Stockton, Calif, Mar 15, 31; m 53; c 3. MICROWAVE DEVICES, PARTICLE ACCELERATOR. *Educ:* Calif Polytech State Univ, BS, 53; Calif Inst Technol, MS, 55, PhD(elec eng & physics), 58. *Prof Exp:* Fulbright scholar, Tech Hogesch, 58-59; from asst prof to assoc prof elec eng, Univ Calif, 59-66; prof physics, Univ Md, 66-76; vpres eng & res, Maxwell Labs, Inc, 76-78; corp vpres, Sci Appln, Inc, 78-81; dir, Off Energy Res, US Dept Energy, 81-87; exec dir, AAAS, 87-88; VPRES, MARTIN MARIETTA ENERGY SYSTS, INC, 89-; DIR, OAK RIDGE NAT LAB, 89- *Concurrent Pos:* Guggenheim fel, Guggenheim Found, 66; asst dir res, Div Controlled Thermonuclear Res, US AEC, 73-75; chmn, Div Plasma Physics, Am Phys Soc, 75; mem bd dirs, Bausch & Lomb, 90; chmn, Math Sci Educ Bd, 90-92 & Coord Coun Educ, Nat Res Coun, 91-92. *Honors & Awards:* Secretarial Gold Medal, US Dept Energy, 87. *Mem:* Fel Am Phys Soc; Am Nuclear Soc; Inst Elec & Electronics Engrs; Nuclear & Plasma Sci Soc; fel AAAS; Am Asn Physics Teachers; Am Asn Univ Professors. *Res:* Plasma physics; Controlled thermonuclear research; particle accelerators; granted several patents; author of various publications. *Mailing Add:* Eight Rivers Run Way Oak Ridge TN 37830

TRIVETT, TERRENCE LYNN, b Madison, Tenn, Oct 3, 40; m 65; c 2. BACTERIOLOGY. *Educ:* Southern Missionary Col, BS, 64; Univ Ore, PhD(microbiol), 69. *Prof Exp:* From asst prof to assoc prof, 69-77, PROF BIOL, PAC UNION COL, 77- *Mem:* Am Soc Microbiol; AAAS. *Res:* Carbohydrate metabolism of Listeria monocytogenes; pathogenesis of Neisseria meningitidis. *Mailing Add:* Dept Biol Pac Union Col Angwin CA 94508-9707

TRIVISONNO, CHARLES F(RANCIS), b Cleveland, Ohio, Dec 30, 24; m 58; c 3. ANALYTICAL CHEMISTRY, ORGANIC CHEMISTRY. *Educ:* Case Inst Technol, BS, 45, MS, 49. *Prof Exp:* Squad trainee, Goodyear Tire & Rubber Co, 48, res chemist, 48-53, group leader uranium chem, Goodyear Atomic Corp, 53-65, supvr chem anal, 65-70, supvr, Chem Dept, 70-73, supvr, Chem Anal Dept, 73-82, supt, Anal Serv Subdiv, 82-85; RETIRED. *Mem:* Am Chem Soc; Sigma Xi. *Res:* Preparation characterization and study of physical properties of condensation polymers; chemical development and chemical analyses, uranium, trace constituents, environmental, and industrial hygiene, related to operation of uranium isotope enrichment plant. *Mailing Add:* 2226 Micklethwaite Portsmouth OH 45662

TRIVISONNO, JOSEPH, JR, b Cleveland, Ohio, Feb 28, 33; m 57; c 4. SOLID STATE PHYSICS. *Educ:* John Carroll Univ, BS, 55, MS, 56; Case Western Reserve Univ, PhD(physics), 61. *Prof Exp:* Instr physics & math, John Carroll Univ, 55-58 & physics, Case Western Reserve Univ, 58-61; from asst prof to assoc prof, 61-69, chmn dept, 79-89, PROF PHYSICS, JOHN CARROLL UNIV, 69- *Concurrent Pos:* Vis prof, Univ Ariz, 74; Prog dir, Solid State Physics, NSF, 86-87. *Mem:* Inst Elec & Electron Engrs Sonics & Ultrasonics; Am Phys Soc; Mat Res Soc; Coun Undergrad Res. *Res:* Elastic constants of metals; magnetoacoustic studies; ultrasonics; low temperature physics; superconductivity. *Mailing Add:* Dept Physics John Carroll Univ Cleveland OH 44118

TRIX, PHELPS, organic chemistry, waste disposal, for more information see previous edition

TRIZNA, DENNIS BENEDICT, b Joilet, Ill, Oct 25, 41; m 63; c 3. RADAR, PHYSICS. *Educ:* Ill Benedictine Col, BS, 63; Iowa State Univ, MS, 66, PhD(physics), 70. *Prof Exp:* RES PHYSICIST RADAR, NAVAL RES LAB, 70- *Honors & Awards:* Res Publ Award, Naval Res Lab, 72, 77 & 83. *Mem:* Am Phys Soc; Inst Elec & Electronics Engrs; Am Geophys Union; Int Union Radio Scientists. *Res:* Remote sensing of geophysical phenomena with radar; remote sensing of ocean waves using high frequency radar; air-sea interaction and the marine boundary layer; general wave propagation. *Mailing Add:* 6130 Virgo Ct Burke VA 22015

TRKULA, DAVID, b Patton Twp, Pa, Aug 19, 27; m 54; c 3. VIROLOGY. *Educ:* Univ Pittsburgh, BS, 49, MS, 55, PhD(biophys), 59. *Prof Exp:* Instr physics, Johnstown Col, Univ Pittsburgh, 57-59; asst biophysicist, M D Anderson Hosp & Tumor Inst, 59-61; physicist, US Army Biol Labs, Ft Detrick, 61-68; asst prof biophys, Baylor Col Med, 68-85; INSTR BIOLOGY, HOUSTON COMMUNITY COL SYST, 85- *Mem:* Am Soc Microbiol; Sigma Xi. *Res:* Biophysics of viruses including mammalian tumor viruses and herpes viruses; biophysics of bacterial toxin proteins. *Mailing Add:* 10603 Del Monte Dr Houston TX 77042

TROBAUGH, FRANK EDWIN, JR, medicine; deceased, see previous edition for last biography

TROEH, FREDERICK ROY, b Grangeville, Idaho, Jan 23, 30; m 51, 87; c 3. SOIL SCIENCE. *Educ:* Univ Idaho, BSAgr, 51, MSAgr, 52; Cornell Univ, PhD(soil sci), 63. *Prof Exp:* Soil scientist, Soil Conserv Serv, 52-59; from asst prof to assoc prof, 63-76, PROF AGRON, IOWA STATE UNIV, 76- *Mem:* Am Soc Agron; Soil Sci Soc Am; Soil Conserv Soc Am. *Res:* Soil formation and classification; measuring the rate of soil creep; soil permeability relationships with microbial activity. *Mailing Add:* Dept Agron Iowa State Univ Ames IA 50011

TROELSTRA, ARNE, b Zelhem, Neth, Mar 30, 35; m 59; c 1. PHYSICS, BIOENGINEERING. *Educ:* State Univ Utrecht, BS, 55, MS, 58, PhD(med physics), 64. *Prof Exp:* Res assoc vision res, Inst Perception, Nat Defense Res Orgn, Neth, 60-65; assoc prof bioeng, Univ Ill, Chicago, 65-69; from assoc prof to prof elec eng, Rice Univ, 69-80; CONSULT, 80- *Concurrent Pos:* Assoc biomed eng, Presby-St Luke's Hosp, Chicago, 66-69; consult, Biosysts Div, Whittaker Corp, 66-69; lectr ophthalmol, Univ Tex Med Sch, Houston, 74-80. *Mem:* Inst Elec & Electronics Engrs. *Res:* Vision research; biological control systems; systems analysis of biological systems; electroretinography; biomedical instrumentation. *Mailing Add:* 501 NE 15th St Oklahoma City OK 73104

TROEN, PHILIP, b Portland, Maine, Nov 24, 25; wid; c 3. ENDOCRINOLOGY, ANDROLOGY. *Educ:* Harvard Univ, AB, 44, MD, 48; Am Bd Internal Med, dipl. *Prof Exp:* Intern, Boston City Hosp, 48-49, asst resident med, 49-50; asst resident, Beth Israel Hosp, 50 & 52-53, resident, 53-54; asst, Harvard Univ, 53-54, instr, 56-59, assoc, 59-60, asst prof, 60-64; physician-in-chief, 64-90, EMER PHYSICIAN-IN-CHIEF, MONTEFIORE UNIV HOSP, 90-; PROF MED, SCH MED, UNIV PITTSBURGH, 64- *Concurrent Pos:* Chief med serv, US Army Hosp, Japan, 50-52; Teaching fel, Harvard Univ, 52-53, res fel, 55-56; Ziskind teaching fel, Beth Israel Hosp, 55-60; fel endocrinol & metab, Mayo Clin, 54-55, Kendall-Hench res fel, 55; Guggenheim fel, Stockholm, Sweden, 60-61; from asst to assoc, Beth Israel Hosp, 55-64, asst vis physician, 59-64; chmn, Contract Rev Comt, Contraceptive Develop Br, Nat Inst Child Health & Human Develop; assoc chmn dept, Univ Pittsburgh, 69-79, vchmn dept, 79-90; mem, Nat Med Serv, Res Merit Rev Bd Endocrinol, Vet Admin, 79-82; mem, 84-87, chmn, Endocrinol & Metab Drugs Adv Comt, Food & Drugs Admin, 87-88; mem publ comt, Endocrine Soc, 84-87, chmn, 87-90; mem, prog organizing comts, Third & Fourth Int Cong Andrology, 85 & 89 & Seventh Int Cong Hormonal Steroids, 86; chmn publ comt, Am Soc Andrology, 90- *Honors & Awards:* Distinguished Andrologist, Am Soc Andrology, 91. *Mem:* Am Soc Clin Invest; Asn Am Physicians; Endocrine Soc; Am Soc Biochem & Molecular Biol; Am Soc Andrology (vpres & pres, 79-81); Int Soc Andrology (secy, 80-89, pres, 89-); Cent Soc Clin Res; Am Fedn Clin Res. *Res:* Endocrinology; internal medicine; andrology. *Mailing Add:* Montefiore Univ Hosp 3459 Fifth Ave Pittsburgh PA 15213

TROESCH, BEAT ANDREAS, b Bern, Switz, Mar 2, 20; nat US; m 48; c 4. APPLIED MATHEMATICS. *Educ:* Swiss Fed Inst Technol, Zurich, dipl, 47, PhD(math), 52. *Prof Exp:* Asst mech & physics, Swiss Fed Inst Technol, Zurich, 47-52; res assoc appl math, Inst Math, NY Univ, 52-56; head appl math sect, Comput Ctr, Ramo-Wooldridge Corp, 56-58 & Space Tech Labs, 58-61; mgr comput sci dept, Aerospace Corp, 61-66; prof, 66-90, EMER PROF AEROSPACE ENG & MATH, UNIV SOUTHERN CALIF, 90- *Concurrent Pos:* Consult, Aerospace Corp, 66- *Mem:* Am Math Soc. *Res:* Applied mathematics and numerical analysis in hydrodynamics and gas dynamics; elliptic and hyperbolic partial differential equations. *Mailing Add:* 523 N Elm Dr Beverly Hills CA 90210-3418

TROFIMENKOFF, FREDERICK N(ICHOLAS), b Veregin, Sask, Aug 10, 34; m 57; c 3. ELECTRICAL ENGINEERING, PHYSICS. *Educ:* Univ Sask, BSc, 57, MSc, 59; Univ London, PhD(elec eng) & DIC, 62. *Prof Exp:* Jr res off, Div Bldg Res, Nat Res Coun Can, 57-59; asst prof elec eng, Univ Sask, 62-66; assoc prof, 66-68, head dept, 68-78, PROF ELEC ENG, UNIV CALGARY, 68- *Concurrent Pos:* Consult to indust. *Honors & Awards:* Publ Award, Brit Inst Elec Engrs, 66. *Mem:* Inst Elec & Electronics Engrs; Can Asn Physicists; Am Soc Eng Educ; Eng Inst Can; Asn Prof Eng. *Res:* Electronics and instrumentation related to oil and gas industry. *Mailing Add:* Univ Calgary 2500 University Dr NW Calgary AB T2N 1N4 Can

TROFIMOV, BORIS ALEXANDROVICH, b Tchita, USSR, Oct 2, 38; m 61; c 1. ORGANIC & PHYSICAL CHEMISTRY. *Educ:* Irkutsk State Univ, MS, 61, PhD(org chem), 65; Leningrad State Univ, DSc, 70. *Prof Exp:* Mem res staff, 61-66, sr scientist, 66-70, LAB HEAD, IRKUTSK INST ORG CHEM, 70-, PROF, 71-, VDIR, 90- *Mem:* Corresp mem USSR Acad Sci. *Res:* Chemistry, physical chemistry and structure of vinylic and acetylenic heteroatomic compounds; pyrroles; organic reactions under superbasic conditions; new synthetic methods; synthesis of monomers and useful compounds. *Mailing Add:* Irkutsk Inst Org Chem USSR Acad Sci One Favorsky St Irkutsk SU 664033 USSR

TROGDON, WILLIAM OREN, b Anadarko, Okla, Nov 1, 20; m 42; c 2. SOILS. *Educ:* Okla State Univ, BS, 42; Ohio State Univ, PhD(soil fertility), 49. *Prof Exp:* Asst agronomist, Agr Exp Sta, Univ Tex, 48; soil scientist, Res Div, Soil Conserv Serv, USDA, 49; chmn dept agr & dir soils lab, Midwestern Univ, 49-53; agronomist, Olin Mathieson Chem Corp, 53-58; prof agron & head dept, Tex A&M Univ, 58-63; exec vpres, Best Fertilizers Co, Tex, 63-65; dir agron & mkt develop, Occidental Agr Chem Corp, 65-66; pres, 66-82, EMER PRES & PROF AGR, TARLETON STATE UNIV, 82- *Mem:* Am Soc Agron; Crop Sci Soc Am; Soil Sci Soc Am; Sigma Xi. *Res:* Soil fertility and management, especially fertilizer usage; fertilizer technology; salinity control and water quality; polyphosphate fertilizers; academic administration. *Mailing Add:* 105 Sandra Palmer Stephenville TX 76401

TROGLER, WILLIAM C, CHEMISTRY. *Prof Exp:* FAC MEM, UNIV CALIF, SAN DIEGO. *Mailing Add:* 92093-0332 Univ Calif San Diego La Jolla CA 92093

TROIANO, A(LEXANDER) R(OBERT), b Boston, Mass, Sept 5, 08; m 38; c 2. PHYSICAL METALLURGY. *Educ:* Harvard Univ, AB, 31, ScD(metall), 39; Mass Inst Technol, MS, 37. *Prof Exp:* Instr physics, Middlesex Col, 31-35; asst instr phys metall, Harvard Univ, 37-39; from assoc prof to prof, Univ Notre Dame, 39-49; assoc prof, 49-53, prof & head dept, 53-67, repub steel distinguished prof phys metall, 67-78, SR RES SCIENTIST, CASE WESTERN RESERVE UNIV, 78- *Concurrent Pos:* Keynote lectr, Brit Iron & Steel Res Inst, Harrogate, Eng, 62 & Int Conf Heat Treatment, Bremen, Ger, 66; distinguished vis prof, NY Univ, 65-66; keynote lectr, Int Conf Hydrogen in Metals, Paris, France, 72. *Honors & Awards:* Hunt Award, Am Inst Mining, Metall & Petrol Engrs, 40; Howe Medal, Am Soc Metals, 57, Sauveur Achievement Medal, 68; Le Chatlier Gold Medal, 80; Nat Acad Engrs, 86. *Mem:* Am Soc Metals; Am Soc Testing & Mat; Am Foundrymen's Soc; Am Inst Mining, Metall & Petrol Engrs. *Res:* Heat treatment of steels; phase transformations in solid state; gases in metals; stress corrosion; materials for geothermal energy. *Mailing Add:* Dept Metall Case Western Reserve Univ Cleveland OH 44106

TROITSKY, MICHAEL S(ERGE), b St Petersburgh, Russia, Sept 20, 17; Can citizen; m 52; c 1. CIVIL ENGINEERING. *Educ:* Univ Belgrade, Dipl eng, 40, DSc(struct eng), 43. *Prof Exp:* Asst prof civil eng, Univ Belgrade, 43-47; proj br engr, Ministry Transp, Belg, 48-50; bridge engr & group leader, Found Co, Can, 51-62; vis asst prof & Ford Found grant, Univ Calif, Los Angeles, 62-63; asst prof civil eng & appl mech, McGill Univ, 64-68; assoc prof, 68-77, PROF ENG, SIR GEORGE WILLIAMS CAMPUS, CONCORDIA UNIV, 77- *Concurrent Pos:* Prof, Univ Montreal, 60-62. *Mem:* Am Soc Civil Engrs; Int Asn Bridge & Struct Engrs; Eng Inst Can; Am Soc Eng Educ; Int Asn Shell & Spat Struct. *Res:* Structures; bridges; stiffness of plates and shells; applied elasticity; structural dynamics; tubular steel structures; prestressed steel structures. *Mailing Add:* Sir George Williams Campus 1455 de Maisonneuve Montreal PQ H3G 1M8 Can

TROJAN, PAUL K, b Chicago, Ill, Sept 8, 31; m 53; c 3. MATERIALS ENGINEERING. *Educ:* Univ Mich, BS, 55, MS, 56, PhD(metall eng), 61. *Prof Exp:* Engr-trainee, Engine & Foundry Div, Ford Motor Co, 52-53; instr metall eng, Univ Mich, 58-61, from asst prof to assoc prof, 61-70, chmn, Div Eng, 62-64, actg dean, Sch Eng, 80-82, PROF MAT & METALL ENG, UNIV MICH-DEARBORN, 70- *Concurrent Pos:* Mich Mem Phoenix res grant, 64-67; sr eng mfg & develop, Ford Motor Co, 64; prog develop mgr, Rocketdyne Div, Rockwell Int Corp, 83-86; environ consult, 78-, hazardous mats consult, Appl Geosci, Inc, 88- *Honors & Awards:* Howe Medal, Am Soc Metals, 63; Thomas Pangborn Gold Medal, Am Foundrymen's Soc, 78. *Mem:* Am Foundrymen's Soc; Am Soc Metals; Am Soc Eng Educ. *Res:* Cast metals; liquid metal processing solidification; relationship of processing and service performance; structure of engineering materials. *Mailing Add:* Mat & Metall Eng Univ Mich 4901 Evergreen Dearborn MI 48128-1491

TROLINGER, JAMES DAVIS, b Shelbyville, Tenn, Mar 2, 40. LASER APPLICATIONS, HOLOGRAPHY. *Educ:* Univ Tenn, BS, 63, PhD(physics), 67; La State Univ, 64. *Prof Exp:* Scientist, Serdrup Technol, 58-63; assoc prof optics, Univ Tenn, 67-75; vpres, Spectron Inc, 75-88; PARTNER, METROLASER, 88- *Concurrent Pos:* Scientist, Sci Appl Inc, 73-75; consult, NATO, 85-88; prin invest, Int Microgravity Space Lab, 87- *Res:* Coherent optics and holography; applications in flow diagnostics and non destructive evaluation. *Mailing Add:* 3917 Wimbledon Way Costa Mesa CA 92626

TROLL, JOSEPH, b Paterson, NJ, May 5, 20; m 43; c 2. SOIL SCIENCE, AGROSTOLOGY. *Educ:* Univ RI, BS, 54, MS, 57; Univ Mass, PhD(nematol), 65. *Prof Exp:* Asst, Univ RI, 54-57; asst prof agron, 57-65, assoc prof plant & soil sci, 65-71, PROF PLANT & SOIL SCI, UNIV MASS, AMHERST, 71- *Res:* Turf management; plant pathology and nematology. *Mailing Add:* 34 Comins Rd RFD No 2 Hadley MA 01035

TROLL, RALPH, b Reinheim, Ger, Oct 8, 32; US citizen; m 58; c 3. DEVELOPMENTAL BIOLOGY, VERTEBRATE ZOOLOGY. *Educ:* Univ Ill, BS, 57, MS, 58; Univ Minn, PhD(parasitol), 65. *Prof Exp:* From inst to assoc prof, 59-72, chmn dept, 68-77, PROF BIOL, AUGUSTANA COL, ILL, 72-, chmn, Div Natural Sci, 83-86. *Mem:* Am Inst Biol Sci; Am Fern Soc; Soc Conserv Biol; Nat Asn Biol Teachers. *Res:* Biology of ferns; Johann Wolfgang Von Goethe's contribution to science; plant-animal interactions. *Mailing Add:* Dept Biol Augustana Col Rock Island IL 61201

TROLL, WALTER, b Vienna, Austria, Oct 25, 22; nat US; m 44; c 2. BIOCHEMISTRY, ORGANIC CHEMISTRY. *Educ:* Univ Ill, BS, 44; Pa State Univ, MS, 46; NY Univ, PhD(biochem), 51. *Prof Exp:* Instr biochem, Univ Cincinnati, 51-52, asst prof, 52-54; assoc, Cancer Res Inst, New Eng Deaconess Hosp, 54-56; from asst prof to prof indust med, 56-76, PROF ENVIRON MED, SCH MED, NY UNIV, 76- *Concurrent Pos:* Invited partic, Int Symp, Princess Takamatsu Cancer Res Fund, Japan, 73, 75 & 83, Coloquium der Gesellschaft für Biologische Chemie, Mosbach, Ger, 79; invited chmn & partic, Third Int Conf Environ Mutagens, Japan, 81. *Mem:* Am Soc Biol Chem; Am Chem Soc; NY Acad Sci; AAAS; Harvey Soc; Am Asn Cancer Res. *Res:* Assay of amino acids; synthetic substrates for enzymes involved in blood clotting; metabolism of aromatic amines and its relation to carcinogenesis; role of free oxygen radicals in tumor promotion; chemopreventive agents, protease inhibitors and vitamin A suppress oncogene expression. *Mailing Add:* Dept Environ Med NY Univ Sch Med New York NY 10016-6402

TROLLER, JOHN ARTHUR, b Hartford, Wis, Apr 17, 33; m 56; c 4. MICROBIOLOGY. *Educ:* Univ Wis, BS, 55, MS, 56, PhD(bact), 62. *Prof Exp:* GROUP LEADER MICROBIOL, WINTON HILL TECH CTR, PROCTER & GAMBLE CO, 62-, SR RES SCIENTIST, 78- *Mem:* AAAS; Am Soc Microbiol; Inst Food Technologists; Soc Indust Microbiol; Brit Soc Appl Bact. *Res:* Food technology and microbiology; water relations of microorganisms; mechanism of action of food preservatives; staphylococcal food poisoning and other food-borne diseases; food hygiene and sanitation. *Mailing Add:* 314 Ritchie Ave Cincinnati OH 45215

TROMANS, D(ESMOND), b Birmingham, Eng, Mar 7, 38; Can citizen; m 61; c 2. PHYSICAL METALLURGY, CORROSION. *Educ:* Leeds Univ, BSc, 60, PhD(metall), 63. *Prof Exp:* Res assoc, 63-66, from asst prof to assoc prof, 66-81, PROF METALL ENG, UNIV BC, 81- *Concurrent Pos:* Sr indust fel, Nat Res Coun, MacMillan Bloedel Res Ltd, Vancouver, 76; bd dirs, Can Comt Res Strength & Fracture Mat, 81-; fac assoc, Pulp & Paper Res Inst Can, 86- *Honors & Awards:* Campbell Award, Nat Asn Corrosion Engrs, 65. *Mem:*

Am Soc Metals; Nat Asn Corrosion Engrs; Brit Inst Metals; Electrochem Soc. *Res:* Stress corrosion; fracture of metals; electron diffraction and microscopy; corrosion; surface chemistry and electrochemistry; fatigue of metals. *Mailing Add:* Metals Mat & Eng Univ BC Stores Rd Vancouver BC V6T 1W5 Can

TROMBA, ANTHONY JOSEPH, b Brooklyn, NY, Aug 10, 43; m 86. PURE MATHEMATICS. *Educ:* Cornell Univ, BA, 65; Princeton Univ, MA, 67, PhD(math), 68. *Prof Exp:* Asst prof math, Stanford Univ, 68-69; vis prof, Univ Pisa, Italy, 70; assoc prof, 70-77, PROF MATH, UNIV CALIF, SANTA CRUZ, 77- *Concurrent Pos:* Woodrow Wilson & NSF fels; vis prof, Univ Calif, Stony Brook, 74 & Bonn Unvi; mem, Inst Advan Study, 75. *Mem:* Am Math Soc. *Res:* Topological methods in non-linear analysis; minimal surfaces Teichmuller theory. *Mailing Add:* Dept Math Univ Calif Santa Cruz CA 95064

TROMBETTA, LOUIS DAVID, b New York, NY, Sept 8, 46; m 73. CELL BIOLOGY, PATHOLOGY. *Educ:* Fordham Univ, BS, 68, MS, 69, PhD(biol), 74. *Prof Exp:* Res assoc path, Issac Albert Res Inst, 73-80; ASSOC PROF, ST JOHN 'S UNIV, 80- *Concurrent Pos:* Adj res assoc path, Kingsbrook Jewish Med Ctr. *Mem:* NY Acad Sci; Electron Micros Soc Am; Sigma Xi. *Res:* Insect development and endocrinology by electron microscopy and histochemistry, cell pathology, neuropathology and histogenesis. *Mailing Add:* Dept Pharmacol St Johns Univ Grand Central & Utopia Pkwy Jamaica NY 11439

TROMBKA, JACOB ISRAEL, b Detroit, Mich, Jan 7, 30; m 52; c 3. RADIATION PHYSICS, SPACE PHYSICS. *Educ:* Wayne State Univ, BS, 52, MS, 54; Univ Mich, PhD(nuclear sci), 62. *Prof Exp:* Res physicist, Oak Ridge Inst Nuclear Studies, 54-56; res assoc nuclear eng & fel gamma ray spectros, Univ Mich, 56-62; sr scientist, Jet Propulsion Lab, Calif Inst Technol, 62-64; prog scientist, Hq, 64-65, STAFF SCIENTIST, GODDARD SPACE FLIGHT CTR, NASA, 65- *Concurrent Pos:* Mem panel in-flight exp, NASA, 63-66, mem working group, Manned Space Flight Exp Bd, 64-66, secy & mem, Geochem Working Group Planetology Subcomt, 65, mem, Terrestrial Bodies Sci Working Group, 77-; adj prof, Law Sch, Georgetown Univ, 67-; co-investr, Apollo 15 & 16 x-ray, gamma ray & alpha particle spectrometer exp, 68-; mem Apollo sci working panel, 71-; prin investr, Apollo 17 & Apollo-Soyuz Crystal Activation Exp, 71 & 75-; vis prof, dept chem, physics & geol, Univ Md, 76-; mem, Coop Planetory Explor Prog, Nat Acad Sci, 83; mem, Instr Design Sci Team, Planetary Explor Prog, NASA, 83, team mem, Mars Observer Gamma Ray Spectrometer, 87 & proj scientist, Gamma Ray Spectrometer Mars Observer, 87. *Honors & Awards:* John Lindsay Mem Award, Goddard Space Flight Ctr, NASA, 72, Except Sci Achievement Medal, 73. *Mem:* Am Phys Soc; Am Nuclear Soc; Sigma Xi; NY Acad Sci. *Res:* Gamma ray spectroscopy; techniques in activation analysis, dosimetry and tracer techniques; planetary physics; gamma ray astrophysics; gamma-ray, x-ray and neutron-gamma ray in situ and remote sensing methods; gamma and x-ray imaging. *Mailing Add:* 11703 Farmland Dr Rockville MD 20852

TROMBLE, JOHN M, b Lincoln, Kans, Jan 26, 32; m 52; c 2. HYDROLOGY, SOIL SCIENCE. *Educ:* Utah State Univ, BS, 61; Univ Ariz, MS, 64, PhD(watershed hydrol), 73. *Prof Exp:* Res assoc watershed hydrol, Univ Ariz, 64-67; asst dir, Southwest Watershed Res Ctr, 67-74, RES HYDROLOGIST, JORNADA EXP RANGE, AGR RES SERV, USDA, 74- *Mem:* Crop Sci Soc Am; Soc Range Mgt; Am Soc Agron; Soil Sci Soc Am; Int Soil Sci Soc. *Res:* Watershed hydrological studies; runoff, erosion, infiltration, simultaneous transfer of heat and water in soils; evaluation of consumptive use of water by native vegetation. *Mailing Add:* Box 173 Fair Acres NM 88033

TROMBULAK, STEPHEN CHRISTOPHER, US citizen; m 83; c 1. RODENT ECOLOGY, ANNUAL CYCLES OF RODENTS. *Educ:* Univ Calif, Los Angeles, BA, 77; Univ Wash, PhD(zool), 83. *Prof Exp:* Instr biol, zool & environ studies, Univ Wash, 83-85; ASST PROF BIOL, MIDDLEBURY COL, 85- *Mem:* Am Soc Mammalogists; AAAS; Ecol Soc Am; Am Soc Zoologists; fel Sigma Xi. *Res:* Annual cycles of small mammals; environmental effects on reproductive effort and hibernation. *Mailing Add:* Dept Biol Middlebury Col Middlebury VT 05753

TROPF, CHERYL GRIFFITHS, b Newark, NJ, Oct 15, 46; m 68; c 1. APPLIED MATHEMATICS. *Educ:* Col William & Mary, BS, 68; Univ Va, MAM, 72, PhD(appl math), 73, Georgetown Univ, MS, 83. *Prof Exp:* Sr mathematician, Appl Physics Lab, Johns Hopkins Univ, 73-80; cong sci fel, US Senate Subcomt Sci, Technol & Space, 80-81; proj mgr operating reactors, US Nuclear Regulatory Comn, 81-82; asst prof, Univ Baltimore, 83-85; SELF EMPLOYED CPA, 84- *Concurrent Pos:* Instr, Johns Hopkins Evening Col, 74- *Mem:* Sigma Xi; Soc Indust & Appl Math. *Res:* Application of analytical mathematical techniques to the modelling of physical systems; scientific/technological management and policy. *Mailing Add:* 13060 St Patricks Ct Highland MD 20777

TROPF, WILLIAM JACOB, b Chicago, Ill, Jan 14, 47; m 68; c 1. OPTICS, MILITARY SYSTEMS. *Educ:* Col William & Mary, BS, 68; Univ Va, PhD(physics), 73. *Prof Exp:* Proj dir, B-K Dynamics, Inc, Rockville, 73-76; PRIN PHYSICIST, SUPVR, ELECTRO-OPTICS SYSTS GROUP, APPL PHYSICS LAB, JOHNS HOPKINS UNIV, LAUREL, 77- *Mem:* Am Phys Soc; Optical Soc Am; Sigma Xi. *Res:* Analysis and modelling of missile guidance and control systems; optical properties of materials. *Mailing Add:* 13060 St Patricks Ct Highland MD 20777

TROPP, BURTON E, b New York, NY, Aug 8, 40; m 65; c 3. BIOCHEMISTRY. *Educ:* Brooklyn Col, BS, 61; Harvard Univ, PhD(biochem), 66. *Prof Exp:* NIH fel bacteriol, Harvard Med Sch, 65-67; asst prof biochem, Richmond Col, NY, 67-70; from asst prof to assoc prof, 70-76, PROF BIOCHEM, QUEEN'S COL, CITY UNIV NEW YORK, 76- *Concurrent Pos:* Vis scientist, Weizmann Inst, 78-79. *Mem:* AAAS; Am Chem Soc; Am Soc Microbiol; Am Soc Biochem & Molecular Biol. *Res:* Regulation of lipid metabolism. *Mailing Add:* Dept Chem & Biochem Queen's Col Flushing NY 11367

TROPP, HENRY S, b Chicago, Ill, July 15, 27; m 54; c 3. MATHEMATICS. *Educ:* Purdue Univ, BS, 49; Ind Univ, MS, 53. *Prof Exp:* Instr math, Mont Sch Mines, 55-57; from asst prof to assoc prof, Humboldt State Col, 57-72; prin investr, Comput Hist Proj, Smithsonian Inst, 71-74; PROF MATH, HUMBOLDT STATE UNIV, 74- *Concurrent Pos:* Vis lectr, Asn Comput Mach, 73-; mem, Prog Comt, Int Res Conf Hist Comput. *Mem:* Math Asn Am; Hist Sci Soc; Can Soc Hist & Philos Math; Asn Comput Mach. *Res:* History of mathematics; history of computers. *Mailing Add:* Dept Math Humboldt State Univ Arcata CA 95521

TROREY, A(LAN) W(ILSON), b London, Eng, May 8, 26; m 50; c 2. ENGINEERING, PHYSICS. *Educ:* Univ BC, BASc, 49; Stanford Univ, MS, 51, PhD(electronic eng), 54. *Prof Exp:* Sr res assoc, Chevron Oil Field Res Co, Standard Oil Co Calif, 54-86; RETIRED. *Concurrent Pos:* Founding ed, Trans Geosci Electronics, Inst Elec & Electronics Engrs, 65-67; designed & installed seismic data process syst, Standard Oil Co, Calif, 67. *Mem:* Soc Explor Geophys. *Res:* Exploration seismology; elastic wave propagation; classical physics; mathematics of linear systems; digital computer systems. *Mailing Add:* 25295 Carmel Knoll Dr Carmel CA 93923-8827

TROSCINSKI, EDWIN S, b Chicago, Ill, May 8, 28; m 52; c 7. CHEMICAL ENGINEERING. *Educ:* Univ Ill, BS, 51. *Prof Exp:* Chem engr, Gen Labs, US Rubber Co, 51-53; process engr, Corn Prods Refining Co, 53-56; res engr, 56-57, proj engr, 57-59, proj supvr, Standard Oil Co, Inc, 59-62; sr res chemist, 62-67, group leader, 67-69, tech dir, 69-72, mgr mkt & res, 72-74, mgr corp mkt servs, 74-76, mgr admin & planning servs, 76- 81, proj mgr Synfuels, 81-82, mgr mkt, 82-84, MGR ADMIN, NALCO CHEM CO, 84- *Mem:* Am Inst Chem Engrs; Nat Asn Corrosion Engrs; Soc Petrol Engrs. *Res:* Corrosion phenomena with respect to metal and alloys and of special environments; water technology, particularly stabilization and corrosion phenomena; oil field emulsion breaking. *Mailing Add:* Nalco Chem Co One Nalco Ctr Naperville IL 60566-1024

TROSKO, JAMES EDWARD, b Muskegon, Mich, Apr 2, 38; m 60; c 1. GENETICS, ONCOLOGY. *Educ:* Cent Mich Univ, BA, 60; Mich State Univ, MS, 62, PhD(radiation genetics), 63. *Prof Exp:* Fel, Oak Ridge Nat Lab, 63-64 & Am Cancer Soc, 64-65; res scientist radiation biophys, Biol Div, Oak Ridge Nat Lab, 65-66; asst prof sci & philos, Dept Natural Sci, 66-70, assoc prof carcinogenesis & med ethics, 70-77, PROF CARCINOGENESIS & MED ETHICS, DEPT HUMAN DEVELOP, MICH STATE UNIV, 77- *Concurrent Pos:* Career develop award, Nat Cancer Inst, 72; consult, Oak Ridge Nat Lab, 70-72; vis prof oncol, McArdle Lab Cancer Res, Univ Wis-Madison, 72-73; consult, Wis Res & Develop Ctr Cognitive Learning, 73-74; mem biol comt, Argonne Nat Lab, 76-79; Pancreas Cancer Working Group, Nat Cancer Inst; chief res, Radiation Effects Res Found, Hiroshima, Japan, 90-92. *Honors & Awards:* Searle Award, UK Environ Mutagen Soc Award, 79; Sigma Xi Sr Scientist Award, Mich State Univ, 85. *Mem:* AAAS; Genetics Soc Am; Am Asn Cancer Res; Tissue Cult Asn; Soc Toxicol. *Res:* Molecular basis for genetic and environmental influences on carcinogenesis and aging; integration of science and human values; chemical modulation of intercellular communication & its consequences in teratogenesis/carcinoogenesis/neurotocology & reproductive dysfunction. *Mailing Add:* Dept Human Develop Col Human Med Mich State Univ East Lansing MI 48824

TROSMAN, HARRY, b Toronto, Ont, Dec 9, 24; nat US; m 52; c 3. PSYCHIATRY, PSYCHOANALYSIS. *Educ:* Univ Toronto, MD, 48. *Prof Exp:* Intern, Grace Hosp, Detroit, Mich, 48-49; resident, Psychopath Hosp, Iowa City, Iowa, 49-51; resident, Cincinnati Gen Hosp, Ohio, 51-52; from asst prof to assoc prof, 54-74, PROF PSYCHIAT, PRITZKER SCH MED, UNIV CHICAGO, 75- *Concurrent Pos:* Psychoanal training, Chicago Inst Psychoanal, 54-62, fac mem, 65-, training & supv analyst, 74- *Honors & Awards:* Franz Alexander Prize, Chicago Inst Psychoanal, 65. *Mem:* Am Psychiat Asn; Int Psychoanal Asn; fel Am Col Psychoanalysts; Am Psychoanal Asn. *Res:* Psychoanalysis and the arts; applied psychoanalysis; creativity. *Mailing Add:* Dept Psychiat Univ Chicago Sch Med Chicago IL 60637

TROSPER, JAMES HAMILTON, b Indianapolis, Ind, May 26, 44; m 68. MEDICAL ENTOMOLOGY. *Educ:* Ind Univ, AB, 67; Univ Ga, MS, 71, PhD(entom), 74. *Prof Exp:* Entomologist, Dis Vector Ecol Control Ctr, US Navy, 74-76; entomologist, US Naval Med Res Unit 2, 76-79; mem staff, Defensive Post Mgt Info Anal Ctr, 81-; AT MALARIA BR, INFECTIOUS DIS PROG CTR, NAVAL MED RES INST. *Mem:* Entom Soc Am; Sigma Xi. *Res:* Factors which control susceptibility of mosquitoes to malaria parasites; isolation of arboviruses and determination of their primary vectors; control of medically important insects. *Mailing Add:* 3209 Verona Dr Silver Springs MD 20906

TROSS, RALPH G, b Bad Kreuznach, Ger, Jan 17, 23; US citizen; m 47; c 2. MATHEMATICAL PHYSICS. *Educ:* Sophia Univ, Japan, BS, 52; Mo Sch Mines, BS, 59; Univ Mo-Rolla, MS, 66, PhD(physics), 68. *Prof Exp:* NASA fel physics, Univ Mo-Rolla, 64-67, instr math, 67-68, res assoc physics, 68; asst prof, Univ Ottawa, 68-69, actg chmn dept math, 70-71, chmn comput comt, 71-75, dir continuing educ, 78-82, ASSOC PROF MATH, UNIV OTTAWA, 69- *Concurrent Pos:* Nat Res Coun grant, 68-; mem adv comt, Algonquin Col, Ottawa, 71-75 & Univ Ottawa, St Lawrence, 78-81; educ develop grant, Comt Ont Univs, 74; sci secy pure & appl math grant selection comt, Nat Sci & Engr Res Coun, 90- *Mem:* Am Phys Soc; Can Asn Physicists; Am Math Soc; Soc Indust & Appl Math; Sigma Xi; Can Math Soc; NY Acad Sci. *Res:* Statistical mechanics; Ising model; cooperative phenomena. *Mailing Add:* Dept Math Univ Ottawa Ottawa ON K1N 6N5 Can

TROST, BARRY M, b Philadelphia, Pa, June 13, 41; m; c 2. ORGANIC CHEMISTRY. *Educ:* Univ Pa, BA, 62; Mass Inst Technol, PhD(org chem), 65. *Prof Exp:* From asst prof to prof, Univ Wis-Madison, 65-76, Helfaer prof, 76-82, chmn, Dept Chem, 80-82, Vilas Res Prof, 82-87; PROF CHEM, STANFORD UNIV, 87-, TAMAKI PROF, HUMANITIES & SCI, 90- *Concurrent Pos:* Assoc ed, J Am Chem Soc & adv, NSF Chem Sect, 73-; Sloan fel; Dreyfuss Found teacher-scholar grant; consult, E I du Pont de Nemours & Co; Am-Swiss Found fel; consult, Merck Co & E I du Pont de Nemours & Co; mem comt chem sci, Nat Acad Sci, 80-; centenary lectr, Chem Soc London, 82; chmn, med chem study sect, NIH, 82-86. *Honors & Awards:* Award, Am Chem Soc, 77, 81, 89 & 90; Baekland Medal, 81; Merit Award, NIH, 88; Janssen Prize, 90. *Mem:* Nat Acad Sci; Royal Soc Chem; Am Chem Soc; fel AAAS; Am Acad Arts & Sci. *Res:* Development of new synthetic methods; synthesis of natural products and theoretically important systems; investigations of model biogenetic systems. *Mailing Add:* Dept Chem Stanford Univ Stanford CA 94305

TROST, CHARLES HENRY, b Erie, Pa, Apr 4, 34; m 60; c 1. VERTEBRATE ZOOLOGY, PHYSIOLOGICAL ECOLOGY. *Educ:* Pa State Univ, BS, 60; Univ Fla, MS, 64; Univ Calif, Los Angeles, PhD(zool), 68. *Prof Exp:* Grad fac grant, 69-70, asst prof biol, 68-81, ASSOC PROF BIOL, IDAHO STATE UNIV, 81- *Mem:* Am Ornithologists Union; Cooper Ornith Soc; Wilson Ornith Soc; Am Soc Zoologists; Am Soc Mammalogists. *Res:* Water balance and energetics of birds and mammals; relation of behavior to the adaptations of animals to their environment. *Mailing Add:* Dept Biol Idaho State Univ Pocatello ID 83209-0009

TROST, HENRY BIGGS, b Lancaster, Pa, Aug 18, 20; m 43; c 3. ORGANIC CHEMISTRY. *Educ:* Franklin & Marshall Col, BS, 42. *Prof Exp:* Anal chemist, Org Anal Group, Hercules Inc, 42-44. shift supvr, Acid Lab, Badger Ord Works, 44-45, Org Anal Lab, 45-46, Size & Solvents Anal Lab, Naval Stores Res Div, 55-61, mem staff, Indust Labs Div, 61-82; RETIRED. *Mem:* Am Chem Soc. *Res:* Product application, development, formulation and sales service type work on water soluble polymers and surface active agents. *Mailing Add:* 2603 Orchard Ave Wilmington DE 19805-2274

TROSTEL, LOUIS J(ACOB), JR, b Baltimore, Md, May 21, 27; m 53; c 3. CERAMICS ENGINEERING. *Educ:* Ohio State Univ, BCerE & MS, 51, PhD(ceramic eng), 55. *Prof Exp:* Res assoc ceramic eng, Res Found, Ohio State Univ, 50-51 & 52-55; ceramic res engr, Advan Ceramics, Norton Co, Worcester, 55-65, sr res engr, 65-69, res assoc, 68-78, res mgr, res & develop dept, 78-90; CONSULT, 90- *Mem:* Fel Am Ceramic Soc; Nat Inst Ceramic Engrs; fel Am Soc Testing & Mat; Sigma Xi. *Res:* Ceramic applications and development of cermets; refractory coatings; special refractories. *Mailing Add:* PO Box 199 Princeton MA 01541

TROTT, GENE F, b Louisville, Ky, May 27, 29; m 55; c 4. RUBBER CHEMISTRY. *Educ:* Univ Louisville, BA, 54, MS, 68, PhD, 71. *Prof Exp:* Chemist, Pillsbury Co, 54-56; res chemist & group leader, Am Synthetic Rubber Corp, 56-66; chemist, Gen Elec Co, Louisville, Ky, 66-73; mgr res & develop, Burton Rubber Processing Co, 73-77; TECH DIR, AM SYNTHETIC RUBBER CORP, LOUISVILLE, 77- *Mem:* Am Chem Soc; Soc Automotive Engrs. *Res:* Polymer chemistry; polymer characterization; biopolymeric interactions; polymerization kinetics. *Mailing Add:* 2301 St Andrews Rd Jeffersonville IN 47130-6763

TROTT, WINFIELD JAMES, underwater acoustics; deceased, see previous edition for last biography

TROTTA, PAUL P, b Brooklyn, NY, Sept 10, 42. BIOCHEMISTRY. *Educ:* Columbia Col, BA, 64, Downstate Med Ctr, State Univ NY, PhD(biochem), 68. *Prof Exp:* Res fel, Sloan-Kettering Inst, 68-70, assoc cancer res, 74- 81; from instr to asst prof, 70-82, ASSOC PROF BIOCHEM, MED COL, CORNELL UNIV, 82- *Concurrent Pos:* Fac res award, Am Cancer Soc, 79; sect chief, protein chem, Schering Corp, 84, res fel, 87, dir, biotechnol-biochem. *Mem:* Am Soc Biol Chemists; Am Chem Soc; NY Acad Sci; Am Asn Cancer Res; AAAS; Interferon Soc; Protein Soc. *Res:* Structure-function relations and regulatory properties of proteins and polypeptides, especially cytokines; physico-chemical characterization of macromolecules; pre-clinical biology of cytokines. *Mailing Add:* Dept Biotechnol-Biochem Schering Corp 60 Orange St Bloomsfield NJ 07003

TROTTER, GORDON TRUMBULL, b Washington, DC, Aug 27, 34; m 65; c 2. COMPUTER SCIENCES. *Educ:* Univ Md, BS, 56; Johns Hopkins Univ, MS, 72. *Prof Exp:* Mathematician, Nat Bur Standards, 56-58; from assoc mathematician to mathematician appl physics lab, Johns Hopkins Univ, 58-66; res mathematician & supvr comput opers, IIT Res Inst, 66-67; COMPUTER SCIENTIST & SUPVR APPLN DEVELOP, APPL PHYSICS LAB, JOHNS HOPKINS UNIV, 67- *Concurrent Pos:* Asst, Grad Sch, Univ Md, 56-58. *Mem:* Asn Comput Mach. *Res:* Information storage and retrieval systems; text processing; software management; programming theory. *Mailing Add:* App Physics Lab Johns Hopkins Univ John Hopkins Rd Laurel MD 20723

TROTTER, HALE FREEMAN, b Kingston, Ont, May 30, 31. MATHEMATICS. *Educ:* Queen's Univ, Ont, BA, 52, MA, 53; Princeton Univ, PhD(math), 56. *Prof Exp:* Fine instr math, Princeton Univ, 56-58; asst prof, Queen's Univ, Ont, 58-60; vis assoc prof, 60-62, assoc prof, 63-69, PROF MATH, PRINCETON UNIV, 69-, ASSOC DIR COMPUT CTR, 62- *Mem:* Am Math Soc; Can Math Cong; Math Asn Am; Asn Comput Mach. *Res:* Knot theory; computing. *Mailing Add:* Dept Math Princeton Univ Princeton NJ 08540

TROTTER, JAMES, b Dumfries, Scotland, July 15, 33; m 57; c 2. PHYSICAL CHEMISTRY. *Educ:* Univ Glasgow, BSc, 54, PhD(chem), 57. *Hon Degrees:* DSc Univ Glasgow, 63. *Prof Exp:* Asst lectr chem, Univ Glasgow, 54-57, Nat Res Coun Can fel physics, 57-59, Imp Chem Indust fel chem, 59-60; from asst prof to assoc prof, 60-65, PROF CHEM, UNIV BC, 65- *Mem:* AAAS; Am Crystallog Asn; Chem Inst Can; Royal Soc Chem; Royal Inst Chemists; Royal Soc Can. *Res:* Chemistry; crystallography. *Mailing Add:* Dept Chem Univ BC 2075 Wesbrook Mall Vancouver BC V6T 1W5 Can

TROTTER, JOHN ALLEN, b Robstown, Tex, May 26, 45; m 78; c 1. ULTRASTRUCTURE, CYTOCHEMISTRY. *Educ:* Johns Hopkins Univ, BA, 69; Univ Wash, PhD(biol structure), 76. *Prof Exp:* Guest worker, NIH, 77-78; asst prof, 78-89, PROF ANAT, MED SCH, UNIV NMEX, 89- *Concurrent Pos:* Assoc ed, Anatomical Record. *Mem:* Am Soc Cell Biol; Am Asn Anatomists; Biophys Soc; Orthop Res Soc; AAAS; Sigma Xi. *Res:* Functional morphology of muscle; structure, function, development and pathology of the muscle-tenden junction; molecular organization of collagenous tissues; morphometry. *Mailing Add:* Dept Anat Univ NMex Sch Med Albuquerque NM 87131

TROTTER, JOHN WAYNE, b Clifton, Tex, Jan 8, 48. ORGANIC CHEMISTRY. *Educ:* Tex Lutheran Col, BS, 70; Tex A&M Univ, PhD(chem), 75. *Prof Exp:* Sr chemist, 76-80, PROD SUPVR, MOBAY CHEM CORP, 80- *Mem:* Am Chem Soc; AAAS. *Mailing Add:* 2001 Holcombe Blvd No 3103 Houston TX 77030-4219

TROTTER, MILDRED, anatomy, for more information see previous edition

TROTTER, NANCY LOUISA, b Monaca, Pa, July 26, 34. CYTOLOGY, ELECTRON MICROSCOPY. *Educ:* Oberlin Col, AB, 56; Brown Univ, ScM, 58, PhD(cytol), 60. *Prof Exp:* From instr to asst prof histol, Col Physicians & Surgeons, Columbia Univ, 61-68; assoc prof, 68-90, EMER ASSOC PROF ANAT, JEFFERSON COL MED, THOMAS JEFFERSON UNIV, 90- *Concurrent Pos:* USPHS trainee, 60, fel, 61, res grant, 62-68. *Mem:* Am Asn Anatomists. *Res:* Hepatomas; liver cytology; partial hepatectomy. *Mailing Add:* 1617 Huron Ave New Castle PA 16101

TROTTER, PATRICK CASEY, b Longview, Wash, Jan 26, 35; div; c 2. PULP & PAPER CHEMISTRY. *Educ:* Ore State Univ, BS, 57; Inst Paper Chem, MS, 59, PhD(chem), 61. *Prof Exp:* Sect leader, Paperboard & Coatings Group, Pulp & Paperboard Res Dept, 61-68, DEPT MGR, FIBER PROD RES & DEVELOP DIV, WEYERHAEUSER CO, 68- *Mem:* Tech Asn Pulp & Paper Indust; Am Chem Soc. *Res:* Long range and basic research on pulping, bleaching, papermaking and properties of paper and paperboard. *Mailing Add:* 4926 26th Ave S Seattle WA 98108-2018

TROTTER, PHILIP JAMES, b Jackson, Mich, Jan 31, 41. PHYSICAL CHEMISTRY. *Educ:* Ill Inst Technol, BS, 64; Univ Colo, PhD(phys chem), 67. *Prof Exp:* Fel molecular complexes, New Eng Inst & Univ Conn, 67-69; res chemist Raman spectra & surface studies, Shell Res, Holland, 69-72; sr res chemist Raman & infrared spectra, Eastman Kodak Co, 73-; AT COHERENT INC. *Mem:* Am Chem Soc; Soc Appl Spectros. *Res:* Reaction systems and dye structures; laser-Raman and infrared spectroscopic applications; surface reactions; instrument development. *Mailing Add:* 332 Canyon Dr Portola Valley CA 94025-7811

TROTTER, ROBERT RUSSELL, b Morgantown, WVa, Apr 23, 15; wid. OPHTHALMOLOGY. *Educ:* WVa Univ, AB & BS, 40; Temple Univ, MD, 42. *Prof Exp:* Asst ophthal res, Howe Lab, Harvard Med Sch, 47, instr, 52-55, instr ophthal, 56-60; clin assoc prof surg, Med Ctr, WVa Univ, 61-63, chmn div ophthal, 61-80, from assoc prof to prof surg, 63-88, emer prof ophthal, 80-88; RETIRED. *Concurrent Pos:* Fel, Harvard Med Sch, 48-49; resident ophthal, Mass Eye & Ear Infirmary, 49-51, dir glaucoma consult serv & asst to chief ophthal, 55-60. *Mem:* AMA; Am Acad Ophthal & Otolaryngol; Am Col Surgeons; Asn Res Vision & Ophthal. *Res:* Glaucoma; testing vision of pre-school children. *Mailing Add:* 281 Dormont St Morgantown WV 26505-6503

TROTZ, SAMUEL ISAAC, b Chattanooga, Tenn, Nov 6, 27; m 55; c 2. CHEMISTRY, ENVIRONMENTAL SCIENCE. *Educ:* Univ Chattanooga, BS, 48; Univ Tenn, MS, 51; St Louis Univ, PhD, 56. *Prof Exp:* Asst chem, Univ Tenn, 48-50 & Univ St Louis, 52-55; res chemist, Olin Mathieson Chem Corp, 55, sr res chemist & group leader, 55-59, proj supvr org div, 59-66, sect mgr, 66-70, tech mgr chem group, 70-74, mgr res & develop, 74-82, consult scientist, Olin Corp, 83-88; CONSULT, 88- *Mem:* AAAS; Am Chem Soc; Sigma Xi. *Res:* Synthesis; product development; custom chemicals process chemistry; oxyhalogens; water chemistry; antimicrobial agents; organic chemistry; organometallics; boranes; light metal hydrides; heterocyclics; polymers; microbiology. *Mailing Add:* 17 Gunning Point Falmouth MA 02540-1864

TROUBETZKOY, EUGENE SERGE, b Clamart, Seine, France, Apr 7, 31; US citizen; m 58; c 3. THEORETICAL PHYSICS. *Educ:* Univ Paris, B es Sc, 49, lic es Sc, 53; Columbia Univ, PhD(physics), 58. *Prof Exp:* Res asst physics, Columbia Univ, 53-58; sr scientist, United Nuclear Corp, NY, 58-64, adv scientist, 64-69; sr res assoc, Div Nuclear Sci & Eng, Columbia Univ, 69-70; consult, Magi Corp, 70-87; DIR RES & DEVELOP, CGI, 88- *Concurrent Pos:* Consult, 87- *Mem:* Am Phys Soc; Asn Comput Mach. *Res:* Software systems; computer graphics; neutrons; radiation transport theory and calculations applied to shielding and reactor calculations. *Mailing Add:* 42 Summit Dr Hastings-on-Hudson NY 10706-1215

TROUP, STANLEY BURTON, b Minneapolis, Minn, Feb 9, 25; m 49; c 2. INTERNAL MEDICINE, HEMATOLOGY. *Educ:* Univ Minn, Minneapolis, BS, 48, BM, 49, MD, 50; Mass Inst Technol, MS, 72; Am Bd Internal Med, dipl, 57. *Prof Exp:* Intern med, Strong Mem Hosp, Univ Rochester, 49-50, intern path, 50-51, asst resident med, 51-52; resident, Beth Israel Hosp, Harvard Univ, 52-53; from instr to prof med, Strong Mem Hosp, Univ Rochester, 58-74; prof med, dir med ctr & vpres, 74-82, PROF MED, HEALTH CARE & HUMAN VALUES, UNIV CINCINNATI, 82- *Concurrent Pos:* Fel path, Strong Mem Hosp, Univ Rochester, 50-51, fel hemat, 55-58; NIH spec fel, Kocher Inst, Univ Bern, 61-62; Alfred P Sloan fel mgt, Mass Inst Technol, 71-72; chief med, Rochester Gen Hosp, 65-74; consult, Genesee, St Mary's & Highland Hosps, Rochester & Vet Admin Hosp, Bath, NY, 65-; consult spec ctrs res, NIH, 71-72; vis prof health mgt, Mass Inst Technol, 80. *Mem:* Fel Am Col Physicians; Am Fedn Clin Res; Am Soc Hemat. *Res:* Bleeding disorders; hemolytic anemia; medical education and management. *Mailing Add:* 234 Goodman St Cincinnati OH 45267

TROUSDALE, WILLIAM LATIMER, b Littleton, NH, Nov 10, 28; m 55; c 4. PHYSICS. *Educ:* Trinity Col, Conn, BS, 50; Rutgers Univ, PhD, 56. *Prof Exp:* Asst prof physics, Trinity Col, Conn, 55-61; res assoc, Univ Pa, 61-62; assoc prof physics, Wesleyan Univ, 66-89; RETIRED. *Concurrent Pos:* Consult, United Aircraft Corp, 56-61; vis scientist, Brookhaven Nat Lab, 66-67. *Mem:* Am Phys Soc. *Res:* Mossbauer effect; magnetism; low temperature physics; physical electronics; holography. *Mailing Add:* 534 Pine St Middletown CT 06457

TROUSE, ALBERT CHARLES, b Hanford, Calif, May 19, 21; m 47; c 4. AGRONOMY. *Educ:* Univ Calif, BS, 43, MS, 48; Univ Hawaii, PhD(soil physics), 64. *Prof Exp:* Soil scientist, USDA, Nev, Calif & Hawaii, 46-51; assoc agronomist, Exp Sta, Hawaiian Sugar Planters Asn, 51-57, sr agronomist, 58-63; soil scientist, 64-83, COLLABR, NAT TILLAGE MACH LAB, AGR RES SERV, SOUTHERN REGION, USDA, 83- *Concurrent Pos:* Pres, Tilth Int, 83- *Mem:* Int Soil Tillage Res Orgn. *Res:* Soil physical properties; requirements for plant root bed and seed bed with respect to various crops and climatic situations, soil strength and aeration; interactions of roots of various species to each other. *Mailing Add:* PO Box 449 Auburn AL 36830

TROUT, DAVID LINN, b Ann Arbor, Mich, Aug 2, 27; m 59; c 3. GASTROINTESTINAL PHYSIOLOGY, CARBOHYDRATE NUTRITION. *Educ:* Swarthmore Col, AB, 51; Duke Univ, MA, 54, PhD(physiol, pharmacol), 58. *Prof Exp:* Lab technician, Baxter Labs, 51-52; asst physiol & pharmacol, Duke Univ, 54-58; sr biochemist, Cent Ref Lab, Vet Admin Hosp, Durham, NC, 58-61; res physiologist, US Air Force Sch Aerospace Med, Brooks AFB, Tex, 61-66; res physiologist, Human Nutrit Res Div, Agr Res Serv, 66-71, RES PHYSIOLOGIST, BELTSVILLE HUMAN NUTRIT RES CTR, AGR RES SERV, US DEPT AGR, BELTSVILLE, MD, 71- *Concurrent Pos:* Adj prof, Dept of Food, Nutrit & Instnl Admin, Univ Md, College Park, MD, 76. *Mem:* AAAS; Am Physiol Soc; Am Inst Nutrit; Soc Exp Biol & Med; Int Soc Chronobiology. *Res:* Carry out and publish researches exploring how carbohydrates in foods and the pattern of eating may influence intermediary metabolism and endocrine status, principally through effects on gastrointestinal physiology. *Mailing Add:* Carbohydrate Nutrit Lab Beltsville Human Nutrit Res Ctr Agr Res Serv USDA Beltsville MD 20705

TROUT, DENNIS ALAN, b Washington, DC, July 26, 47; m 75. AIR POLLUTION METEOROLOGY, ENVIRONMENTAL ENGINEERING. *Educ:* Pa State Univ, BS, 68, MS, 69, PhD(meteorol, air pollution), 73. *Prof Exp:* Res asst, Dept Meteorol, Pa State Univ, 68-70, res asst, Ctr Air Environ Studies, 70-71; environ/syst analyst, Environ Tech Appl Ctr, US Air Force, 71-73; scientist, Battelle Mem Inst, Columbus, Ohio, 73-76; sr scientist & mgr, Environ Studies Div, Environ Res & Technol, Inc, 76-78; regional meteorologist & nat expert air pollution dispersion modeling, Region V, 78-84, RES PROG MGR, ACID DEPOSITION & GLOBAL CHANGE RES PROGS, OFF RES & DEVELOP, US ENVIRON PROTECTION AGENCY, WASHINGTON, DC, 84- *Concurrent Pos:* Lectr air pollution & meteorol, Ohio State Univ, 75. *Mem:* Am Meteorol Soc; Am Geophys Union; AAAS; Sigma Xi. *Res:* Ambient air quality measurements and analysis; computer modeling of atmospheric dispersion of air pollutants; development of emission control strategies; assessment of trace contaminant emissions and resulting ambient concentrations; assessment of emissions, transport, chemical transformation, and deposition of acidic substances and radiatively important trace gasses and their precursors; research program management: development, evaluation and application of regional and global atmospheric models for use in assessing ecological and other environmental effects. *Mailing Add:* 6121 N 18th St Arlington VA 22205-2117

TROUT, JEROME JOSEPH, MEMBRANE TRANSPORT, LYSOSOMAL SYSTEM. *Educ:* Iowa State Univ, PhD(molecular, cellular & develop biol), 77. *Prof Exp:* RES ASST PROF, SCH MED, NORTHWESTERN UNIV, 82- *Mailing Add:* Res 151 MSB Vet Admin Lakeside Med Ctr 400 E Ontario St Chicago IL 60611

TROUT, THOMAS JAMES, b Bluffton, Ohio, Mar 30, 49; m 76; c 2. IRRIGATION WATER MANAGEMENT, SOIL PHYSICS. *Educ:* Case Western Reserve Univ, BS, 72; Colo State Univ, MS, 75, PhD(agr), 79. *Prof Exp:* Res assoc, Colo State Univ, 77-79, res asst prof, 79-82; AGR ENGR, USDA-AGR RES SERV, KIMBERLY, IDAHO, 82- *Concurrent Pos:* Consult, Univ Wis, 79 & World Bank & US AID, 80. *Mem:* Am Soc Agr Engrs; Am Soc Civil Engrs; Sigma Xi. *Res:* Methods of managing surface irrigation water at the farm level including both soil management and methods of water application. *Mailing Add:* Snake River Conserv Res Ctr USDA-Agr Res Serv Rte 1 Box 186 Kimberly ID 83341

TROUT, WILLIAM EDGAR, III, b Staunton, Va, Apr 21, 37. GENETICS. *Educ:* Univ Richmond, BS, 59; Ind Univ, AM, 64, PhD(genetics), 65. *Prof Exp:* USPHS res fel radiation genetics, Biol Div, Oak Ridge Nat Lab, 65-66; res scientist, biol dept, City of Hope Med Ctr, 66-82; PRES, AM CANAL SOC, 85- *Concurrent Pos:* Canal archaeol consult. *Mem:* Behav Genetics Asn; Sigma Xi. *Res:* Behavior genetics of Drosophila melanogaster. *Mailing Add:* 35 Towana Rd Richmond VA 23226

TROUTMAN, JAMES SCOTT, b Hannibal, Mo, Mar 7, 30; m 52; c 2. ENGINEERING, OPERATIONS RESEARCH. *Educ:* US Naval Acad, BS, 52. *Prof Exp:* Sr engr, Corvey Div, Melpar, Inc, 56-58; weapon syst analyst, 58-60, corp secy, 60-76, vpres, 76-83, sr vpres, Anser, Arlington, Va, 84-88; CONSULT, 88- *Mem:* Fel Am Astronaut Soc; Nat Mil Intel Asn; Sigma Xi. *Res:* Electronics and communications systems; seismology; electronic parts for extreme environments; weapon systems analysis and operations research; intelligence systems; research and development management. *Mailing Add:* 5624 Newington Ct Bethesda MD 20816-3315

TROUTMAN, RICHARD CHARLES, b Columbus, Ohio, May 16, 22; c 3. OPHTHALMOLOGY. *Educ:* Ohio State Univ, BA, 42, MD, 45; Am Bd Ophthal, dipl, 51. *Prof Exp:* Intern ophthal, New York Hosp, 45-46; resident, Cornell Med Ctr, 48-50, from instr to asst prof, Med Col, Cornell Univ, 52-55; prof & chmn div, 55-83, EMER PROF OPHTHAL, STATE UNIV NY DOWNSTATE MED CTR, 83- *Concurrent Pos:* Instr, Manhattan Eye, Ear & Throat Hosp, 51-55, mem courtesy staff, 55-, surgeon dir, 61-; consult hosps, 54-; vis surgeon, Kings County Hosp, 55-; mem courtesy staff, Cornell Med Ctr, New York Hosp, 55-, attend surgeon, 71-; mem courtesy staff, New York Eye & Ear Infirmary, 55-; mem ophthal postgrad training comt, Nat Inst Neurol Dis & Blindness, 59-63; consult neurol & blindness div, Bur State Serv, Dept Health, Educ & Welfare, 63-67; mem, Bd Dirs, Baraquer Inst, 63-73. *Mem:* Am Asn Res Vision & Ophthal; fel Am Col Surgeons; fel Am Acad Ophthal & Otolaryngol; fel NY Acad Med; Am Ophthal Soc; hon mem Asn Advan Ophthal Oper Techniques. *Res:* Orbital surgery and surgery of the anterior segment of the eye; microsurgery and refractive surgery. *Mailing Add:* 755 Park Ave New York NY 10021

TROUTMAN, RONALD R, US citizen; m 66; c 2. ELECTRICAL ENGINEERING. *Educ:* Mass Inst Technol, BS, 62; NY Univ, MS, 63, PhD(elec eng), 66. *Prof Exp:* Proj officer, Off Naval Res, 66-68; staff engr, Gen Technol Div, IBM, 69-73, adv engr, 73-81, sr engr, 81-87, SR ENGR, T J WATSON RES CTR, IBM, 87- *Concurrent Pos:* Fel, Ctr Advan Eng Studies, Mass Inst Technol, 83-84. *Mem:* Fel Inst Elec & Electronics Engrs; Soc Info Display; NY Acad Sci. *Res:* Research contract administration; random access memory design; semiconductor technology development; analysis and characterization of short channel effects; subthreshold behavior; hot electron phenomena in field-effect transistors; author of over 55 publications; granted 14 patents. *Mailing Add:* 38 Deer Hill Dr Ridgefield CT 06877

TROUTNER, DAVID ELLIOTT, b Eolia, Mo, Oct 11, 29; m 55; c 3. NUCLEAR CHEMISTRY. *Educ:* Washington Univ, AB, 52, PhD(chem), 59; Univ Mo-Rolla, MS, 56. *Prof Exp:* From asst prof to assoc prof, Univ Mo-Rolla, 59-61; assoc prof, 61-69, chmn dept, 71-78, PROF CHEM, UNIV MO-COLUMBIA, 69- *Concurrent Pos:* Vis scientist, Oak Ridge Nat Lab, 67-68. *Mem:* Am Chem Soc; Soc Nuclear Med. *Res:* Radioisotopes in medicine; radiochemical studies of nuclear fission. *Mailing Add:* Dept Chem Univ Mo Columbia MO 65201

TROUTT, LOUISE LEOTTA, b New York, NY, Oct 22, 58; m 83. MICROTUBULE-BASED CELL MOTILITY, ELECTRON MICROSCOPY. *Educ:* Adelphi Univ, BS, 80; Syracuse Univ, MA, 82; Univ Calif, Berkeley, PhD(physiol), 88. *Prof Exp:* Teaching assoc physiol, human anat & histol, Dept Physiol-Anat, Univ Calif, Berkeley, 84-88; res scientist I, Dnax Res Inst, Palo Alto, Calif, 88-89; POSTDOCTORAL FEL, SCH BOT, UNIV MELBOURNE, AUSTRALIA, 89- *Mem:* Am Soc Cell Biol. *Res:* Microtubule polarity orientation and its relation to function, organization of a microtubule array; role of microtubules in cell elongation, in pigment granule transport and in chromosome congression during prometaphase; reactivation of microtubule-based functions in lysed cell systems. *Mailing Add:* Sch Bot Univ Melbourne Parkville Victoria 3052 Australia

TROW, JAMES, b Chicago, Ill, Apr 21, 22; m 47; c 1. GEOLOGY. *Educ:* Univ Chicago, SB, 43, SM, 45, PhD(geol), 48. *Prof Exp:* From asst prof to assoc prof geol, 47-59, PROF GEOL, MICH STATE UNIV, 59- *Concurrent Pos:* Consult, US Geol Surv, 48-52, Cleveland Cliffs Iron Co, 58-60, Anaconda, 61-68, Elcor Corp, 69-70, Chevron Resources, 71-76, Mich Geol Surv, 77-79, Falconbridge Ltd, 81, & Exmin Corp, 86-88. *Mem:* Soc Mining Engrs; Can Inst Mining & Metall; Geol Soc Am; Am Geophys Union; Prospectors & Developers Asn Can. *Res:* Inductive electrostatic gradiometry; chemical thermodynamics, structural geology and Pleistocene sedimentology applied to mining exploration for uranuim, gold, diamonds, copper and iron. *Mailing Add:* Dept Geol Sci Mich State Univ East Lansing MI 48824

TROWBRIDGE, DALE BRIAN, b Glendale, Calif, May 17, 40; m 66; c 2. ORGANIC CHEMISTRY. *Educ:* Whittier Col, AB, 61; Univ Calif, Berkeley, MS, 64, PhD(org chem), 70. *Prof Exp:* Chemist, Aerojet Gen Corp, 61-62; teacher chem high sch, Calif, 64-66; from asst prof to assoc prof, 69-77, PROF CHEM, CALIF STATE COL, SONOMA, 77- *Concurrent Pos:* Vis prof, Univ Calif-Berkeley, 71, 72, 74 & 88; vis res assoc, Cambridge Univ, 78. *Mem:* Sigma Xi; Am Chem Soc. *Res:* Preparation and study of organo-phosphorus compounds of biological interest. *Mailing Add:* Dept Chem Sonoma State Univ 1801 E Cotati Ave Rohnert Park CA 94928

TROWBRIDGE, FREDERICK LINDSLEY, b Newark, NJ, June 8, 42; m 70; c 2. NUTRITION, EPIDEMIOLOGY. *Educ:* Princeton Univ, BA, 64; Harvard Univ, MD, 68; London Sch Hyg & Trop Med, MSc, 74. *Prof Exp:* Med epidemiologist nutrit epidemiol, Ctr Dis Control, USPHS, 71-77; ASST PROF INT HEALTH EPIDEMIOL, SCH HYG & PUB HEALTH, JOHNS HOPKINS UNIV, 77-; DIR, NUTR DIV, CTR HEALTH PROM, CTR DIS CONTROL. *Mem:* Int Health Soc; Am Pub Health Asn. *Res:* Field assessment of nutrition status; epidemiologic studies in nutrition; methods of nutriontal surveillance. *Mailing Add:* Ctr Health Prom Ctr Dis Control 1600 Clifton Rd Atlanta GA 30333

TROWBRIDGE, GEORGE CECIL, b Delta, Colo, May 6, 38; m 58; c 2. COMPUTER SCIENCE, MATHEMATICS. *Educ:* Western State Col, Colo, BA, 60; Univ Ill, MS, 64. *Prof Exp:* Teacher math, Lamar High Sch, 60-61; teacher, Delta County Jct Sch Dist number 50, 61-63 & 64-65; teacher, Western State Col, Colo, 65-67; res mathematician electro-magnetic pulse, 67-68, sr res mathematician & proj leader, 68-71, dep prog mgr, Dikewood Industs Inc, 71-79, DIR COMPUT SERV, HANCOCK/DIKEWOOD SERV INC, 79- *Res:* Application of computers to health care information; computer and mathematical applications in the physical sciences. *Mailing Add:* 7412 Gila Rd Albuquerque NM 87109

TROWBRIDGE, IAN STUART, b Apr 13, 47; m 70; c 2. MONOCLONAL ANTIBODIES, CELL SURFACE RECEPTORS. *Educ:* Oxford Univ, Eng, BA, 69, PhD(immunol), 72. *Prof Exp:* DIR CANCER BIOL, SALK INST, 72- *Mem:* Asn Am Immunologists; Asn Am Cell Biologists. *Res:* Analysis of the structure-function relationships of cell surface receptors involved in cell growth or in the regulation of the immune response. *Mailing Add:* Salk Inst 110010 N Torrey Pines Rd La Jolla CA 92038

TROWBRIDGE, LEE DOUGLAS, b Akron, Ohio, Oct 25, 49; m 70; c 3. PHYSICAL CHEMISTRY. *Educ:* Mich State Univ, BS, 70; Harvard Univ, MA, 71, PhD(chem physics), 78. *Prof Exp:* Develop assoc, Oak Ridge Gaseous Diffusion Plant, Union Carbide, 78-81, develop staff member, Nuclear Div, 81-85; staff mem, Oak Ridge Nat Lab, 85-87; SECT HEAD, ENRICHMENT TECH ORGN, 87- *Concurrent Pos:* Mem staff, fossil energy progs, Dept Energy, 85-87. *Mem:* Am Phys Soc; Am Vacuum Soc; Am Chem Soc. *Res:* Gas phase and surface reactions of uranium hexaflouride; materials and chemistry studies for the uranium enrichment program. *Mailing Add:* PO Box 2003 MS 7266 Bldg K 1004L MS 7266 Oak Ridge TN 37831-7266

TROWBRIDGE, LESLIE WALTER, b Curtiss, Wis, May 21, 20; m 46; c 4. SCIENCE EDUCATION. *Educ:* Wis State Univ, Stevens Point, BS, 40; Univ Chicago, MS, 48; Univ Wis, MS, 53; Univ Mich, PhD(sci educ), 61. *Prof Exp:* Teacher jr high sch, Wis, 41 & 46, instr high sch, 46-54; univ scholar, Univ Mich, 54-62; from asst prof to prof, 62-70, chmn dept sci educ, 66-72 & 81-83, PROF SCI EDUC, UNIV NORTHERN COLO, 70- *Concurrent Pos:* Fel, NY Univ, 69-70. *Mem:* Nat Sci Teachers Asn (pres, 73-74); Nat Asn Res Sci Teaching. *Mailing Add:* Dept Earth Sci Univ Northern Colo Greeley CO 80639

TROWBRIDGE, RICHARD STUART, b Cambridge, Mass, Apr 3, 42; m 65; c 1. MAMMALIAN CELL CULTURE, MAMMALIAN VIROLOGY. *Educ:* Univ Mass, Amherst, BS, 64, MS, 66, PhD(microbiol & mammalian virol), 71. *Prof Exp:* Sanitarian, Natick Bd Health, Natick, Mass, 64-65; lab instr public health bact, Univ Mass, 64-67; bacteriologist, Amherst Bd Health, 64-67; sanitarian, Mass Dept Pub Health, 65-66, bacteriologist, 66-68; res scientist, NY State Inst Basic Res Ment Retardation, 70-72, sr res scientist, 72-77, grants adminr, 77-80, res scientist III, 80-81; RES SCIENTIST IV, NY STATE INST BASIC RES DEVELOP DISABILITIES, 81- *Concurrent Pos:* Bd mem, Res Found Ment Hyg, Inc, 77-80; reviewer, Teratol J Abnorm Develop, 78-80. *Mem:* Am Soc Microbiol; Tissue Cult Asn; NY Acad Sci; Am Soc Virol. *Res:* Elucidation of the virus-host cell interactions that lead to pathogenesis in slow-virus infections of the central nervous system. *Mailing Add:* Inst Basic Res Develop Disabilities 1050 Forest Hill Rd Staten Island NY 10314

TROWER, W(ILLIAM) PETER, b Rapid City, SDak, May 25, 35; m 57, 63; c 3. EXPERIMENTAL PHYSICS. *Educ:* Univ Calif, Berkeley, AB, 57; Univ Ill, Urbana, MS, 63, PhD(physics), 66. *Prof Exp:* Physicist, Lawrence Radiation Lab, Univ Calif, Berkeley, 60-62; res asst, Digital Comput Lab & Dept Physics, Univ Ill, Urbana, 62-66; PROF PHYSICS, VA POLYTECH INST & STATE UNIV, 70- *Concurrent Pos:* Chmn, Gordon Res Conf Multiparticle Prod Processes, 73; dir, Coun Munic Performance, NY, 73-75; co-chmn, Physics in Collision Res Conf, 80, Magnetic Monopole Res Conf, 82. *Honors & Awards:* Bronze Medal, Int Film & TV Festival NY, 73. *Mem:* Fel Am Phys Soc; fel AAAS. *Res:* Experimental nuclear and particle physics; scientific computer applications. *Mailing Add:* Dept Physics Va Polytech Inst Blacksburg VA 24061

TROXEL, BENNIE WYATT, b Osawatomie, Kans, Aug 9, 20; m 46; c 2. GEOLOGY. *Educ:* Univ Calif, Los Angeles, BA, 51, MA, 58. *Prof Exp:* Geologist, Calif Div Mines & Geol, 52-71; sci ed, Geol Soc Am, 71-75; mem fac, Calif State Univ, Sacramento, 75-77; mem fac, Univ Calif, Davis, 77-88; RETIRED. *Concurrent Pos:* Mem staff, Calif Div Mines, 75-77; consult geologist, 77- *Mem:* AAAS; Geol Soc Am. *Res:* Geology of Death Valley region; geologic factors that influence slope stability in urban areas of California; mineral resources; Precambrian stratigraphy and faults in Death Valley, California. *Mailing Add:* 2961 Redwood Rd Napa CA 94558

TROXEL, DONALD EUGENE, b Trenton, NJ, Mar 11, 34; m 63; c 3. ELECTRICAL ENGINEERING. *Educ:* Rutgers Univ, BS, 56; Mass Inst Technol, SM, 60, PhD(elec eng), 62. *Prof Exp:* From instr to assoc prof, 61-85, PROF ELEC ENG, MASS INST TECHNOL, 85- *Concurrent Pos:* Ford Found fel, 62-64. *Mem:* Inst Elec & Electronics Engrs. *Res:* Digital systems applications; communications; electronics; computers; computer aided fabrication of integrated circuits. *Mailing Add:* Dept Elec Eng Mass Inst Technol Cambridge MA 02139

TROXELL, HARRY EMERSON, JR, wood science & technology; deceased, see previous edition for last biography

TROXELL, TERRY CHARLES, b Allentown, Pa, Jan 1, 44; m 64; c 2. PHYSICAL CHEMISTRY. *Educ:* Muhlenberg Col, BS, 65; Cornell Univ, PhD(biophys chem), 71. *Prof Exp:* Res assoc phys chem, Univ Ore, 71-74; sr phys chemist, Eli Lilly & Co, 74-77; SCI ADMINR, FOOD & DRUG ADMIN, 77- *Mem:* Am Chem Soc; AAAS. *Res:* Scientific policy and administration; food safety, especially regulation of food additives and food packaging. *Mailing Add:* 1522 Powells Tavern Pl Herndon VA 22070-2831

TROXELL, WADE OAKES, b Ft Collins, Colo, Nov 15, 56; m 81; c 1. BEHAVIOR-BASED ROBOTICS, ENGINEERING DESIGN SUPPORT SYSTEMS. *Educ:* Colo State Univ, BS, 80, MS, 82, PhD(mech eng), 87. *Prof Exp:* Proj engr, Eastman Kodak Co, Colo Div, 82-85; res asst & teaching asst, 80-82, instr, 85-88, ASST PROF MECH ENG, COLO STATE UNIV, 88- *Concurrent Pos:* Colo Inst Artificial Intel, 85-91; dir, Mfg & Robotic Systs Lab, 86-; NATO postdoctoral fel artificial intel, Univ Edinburgh, 87-88; dir, Mfg Excellence Ctr, Colo State Univ, 88-; examr, Coun Nat Acad Awards, UK, 90- *Mem:* Am Soc Mech Engrs; Nat Soc Prof Engrs; Inst Elec & Electronics Engrs; Am Asn Artificial Intel; Soc Mfg Engrs. *Res:* Engineering of robust task-achieving robots; formulation of robot control methodologies; artificial intelligence-based engineering design support systems; robot programming and high level robot programming languages. *Mailing Add:* Mech Eng Colo State Univ Ft Collins CO 80523

TROXLER, RAYMOND GEORGE, b New Orleans, La, Sept 21, 39; m 63; c 2. PATHOLOGY, PUBLIC HEALTH. *Educ:* Univ Southwestern La, BS, 64; La State Univ, MD, 64; Univ Tex Health Sci Ctr, MPh, 87. *Prof Exp:* chief clin path, Clin Sci Div, USAF Sch Aerospace Med, 71-83; coordr patient care, Med Ctr Hosp, San Antonio, Texas, 83-85; CONSULT CARDIOVASC DIS, SOUTHWEST RES INST, 85- *Concurrent Pos:* Clin assoc prof, Univ Tex Health Sci Ctr, San Antonio, 72-82; consult clin path, USAF Surg Gen. *Honors & Awards:* Edward Rhodes Stitt Award, Best Res Lab Med, 78. *Res:* Early detection of latent coronary artery disease by laboratory screening of blood and serum. *Mailing Add:* 10318 Willowick San Antonio TX 78217

TROXLER, ROBERT FULTON, b Santa Monica, Calif, July 11, 38; m 64. BIOCHEMISTRY. *Educ:* Grinnell Col, BS, 60; Pa State Univ, MS, 62; Univ Chicago, PhD(bot), 65. *Prof Exp:* Res assoc bot, Univ Chicago, 65-66; res assoc med, 66-68, from asst prof to assoc prof, 68-80, PROF BIOCHEM, SCH MED, BOSTON UNIV, 80- *Mem:* Am Soc Biol Chemists; AAAS; Am Soc Plant Physiologists; Int Asn Dent Res. *Res:* Porphyrin and bile pigment chemistry and biochemistry; structure and function of proteins; molecular biology. *Mailing Add:* Dept Biochem 80 E Concord St Boston MA 02118

TROY, DANIEL JOSEPH, b St Louis Co, Mo, Feb 2, 32; m 55; c 6. MATHEMATICS. *Educ:* St Louis Univ, BS, 53, MS, 58, PhD(math), 61. *Prof Exp:* Instr math, St Louis Univ, 58-61; asst prof, Ohio State Univ, 61-67; ASSOC PROF MATH, PURDUE UNIV, 67- *Mem:* Am Math Soc; Math Asn Am. *Res:* Complex variable. *Mailing Add:* 17042 Evans Dr South Holland IL 60473

TROY, FREDERIC ARTHUR, b Evanston, Ill, Feb 16, 37; m 59; c 2. BIOCHEMISTRY, ONCOLOGY. *Educ:* Washington Univ, BS, 61; Purdue Univ, West Lafayette, PhD(biochem), 66. *Prof Exp:* Am Cancer Soc res fel physiol chem, Sch Med, Johns Hopkins Univ, 66-68; from asst prof to assoc prof, 68-79, PROF BIOL CHEM, SCH MED, UNIV CALIF, DAVIS, 80-, CHMN, 91- *Concurrent Pos:* USPHS res grant, Sch Med, Univ Calif, Davis, Nat Cancer Inst res grant, 71-83; Nat Cancer Inst career res develop award, 75-80; co-dir tumor biol training grant, Nat Cancer Inst, 72-; Am Cancer Soc-Eleanor Roosevelt-Int Cancer fel, Int Union Against Cancer, Stockholm, 76-77; consult, NIH, NSF & Basic Sci Rev bd, Vet Admin; ed, J Biol Chem, 88-; pres-elect & vpres, Bd Dirs, Soc Complex Carbohydrates, 91- *Mem:* Am Soc Microbiol; Am Inst Chemists; Am Soc Biol Chemists; Am Chem Soc; Brit Biochem Soc; Am Soc Cancer Res; Am Soc Enologists. *Res:* Membrane chemistry, biosynthesis of glycoconjugates; chemistry and biosynthesis of bacterial capsular polysaccharides; nuclear magnetic resonance studies of the conformation and dynamics of glycosyl carrier polyisoprenoids; membranes biogenesis of multi-enzyme complexes; role of polysialic acid in neural development and tumor metastasis. *Mailing Add:* Dept Biol Chem Univ Calif Sch Med Davis CA 95616

TROY, WILLIAM CHRISTOPHER, b Rochester, NY, July 7, 47. APPLIED MATHEMATICS. *Educ:* St John Fisher Col, BS, 69; State Univ NY Buffalo, MA, 70, PhD(math), 74. *Prof Exp:* From asst prof to assoc prof, 74-86, PROF MATH, UNIV PITTSBURGH, 87- *Mem:* Am Math Soc. *Res:* Application of the theory of differential equations to mathematical problems arising in biology, chemistry and engineering (fluid mechanics). *Mailing Add:* Dept Math Univ Pittsburgh 4200 Fifth Ave Pittsburgh PA 15260

TROYER, ALVAH FORREST, b LaFontaine, Ind, May 30, 29; m 50; c 4. PLANT BREEDING, CORN BREEDING. *Educ:* Purdue Univ, BS, 54; Univ Ill, MS, 56; Univ Minn, PhD(plant breeding & genetics), 64. *Prof Exp:* Res assoc, Univ Ill, Urbana, 55-56; res fel, Univ Minn, St Paul, 56-58; sta mgr, Pioneer Corn Co, Mankato, 58-65; res coordr, Northern Res, Pioneer Hi-Bred Int, Inc, 65-77, Eastern Res, 71-77; dir, Res & Develop, Pfizer Genetics, Inc, St Louis, 77-81, vpres, 81-82; VPRES, RES & DEVELOP, DEKALB PFIZER GENETICS, DEKALB, ILL, 82- *Mem:* Sigma Xi; fel AAAS; fel Am Soc Agron; NY Acad Sci; fel Crop Sci Soc Am; Am Genetics Soc; Genetic Soc Am. *Res:* Increasing the effectiveness and efficiency of corn breeding in all possible ways so that superior products for agriculture will be developed. *Mailing Add:* 611 Joanne Lane Dekalb IL 60115

TROYER, JAMES RICHARD, b Goshen, Ind, Feb 26, 29; m 51; c 3. HISTORY OF BOTANY. *Educ:* DePauw Univ, BA, 50; Ohio State Univ, MS, 51; Columbia Univ, PhD(bot), 54. *Prof Exp:* Vis asst prof biol, Univ Ala, 54-55; instr plant physiol, Sch Forestry, Yale Univ, 55-57; from asst prof bot to assoc prof, 57-69, PROF BOT, NC STATE UNIV, 69- *Concurrent Pos:* Fel, Biomath Training Prog, NC State Univ, 64-66. *Mem:* AAAS; Am Soc Plant Physiol; Bot Soc Am; Sigma XI; Soc Hist Natural Hist; Hist Sci Soc. *Res:* Mathematical plant physiology; flavonoid substances of plants; history of botany. *Mailing Add:* Dept Bot Box 7612 NC State Univ Raleigh NC 27695-7612

TROYER, JOHN ROBERT, b Princeton, Ill, Feb 5, 28; m 56; c 4. ANATOMY. *Educ:* Syracuse Univ, AB, 49; Cornell Univ, PhD(histol, embryol), 55. *Prof Exp:* Asst histol & embryol, Cornell Univ, 49-54; from instr to assoc prof anat, 54-69, actg chmn dept, 71-72, vchmn dept in chg teaching, 73-79, PROF ANAT, SCH MED, TEMPLE UNIV, 69-, CHMN, ANAT DEPT, 79- *Mem:* AAAS; Am Asn Anat; Am Soc Mammal; NY Acad Sci; Sigma Xi. *Res:* Liver glycogen, porphyrin synthesis and neurosecretion in the hibernating bat; gross anatomy, normal development and abnormal development of the human heart; medical education. *Mailing Add:* Dept Anat Temple Univ Sch Med Broad & Ont St Philadelphia PA 19140

TROYER, ROBERT JAMES, b Sturgis, Mich, Sept 21, 28; m 54; c 4. MATHEMATICS. *Educ:* Ball State Univ, BS, 50; Ind Univ, MAT, 56, PhD(math), 60. *Prof Exp:* Instr math, Ind Univ, 60-62, asst prof, 62-65; vis fel, Dartmouth Col, 65-66; from asst prof to assoc prof, Univ NC, Chapel Hill, 66-68; vis assoc prof, 68-69, assoc prof, 69-70, PROF MATH, LAKE FOREST COL, 70- *Concurrent Pos:* Vis scholar, Northwestern Univ, 74-75. *Mem:* Math Asn Am. *Res:* Geometry; history of mathematics. *Mailing Add:* Dept Math Lake Forest Col Lake Forest IL 60045

TROYER, STEPHANIE FANTL, b Jan 30, 44; US citizen; m 65; c 1. DIFFERENTIAL TOPOLOGY. *Educ:* Swarthmore Col, BA, 65; Northeastern Univ, MS, 67, PhD(math), 73. *Prof Exp:* Instr, 70-74, ASST PROF MATH, UNIV HARTFORD, 74- *Concurrent Pos:* Vis assoc prof math, Brown Univ, 82; consult, Computopia Corp, Set Logic Assocs & AAAS consult prog for minority insts; NSF working sci grants. *Mem:* Am Math Soc; Math Asn Am; Asn Women Math. *Res:* Extension problems in low codimension, particularly for curves and surfaces. *Mailing Add:* Dept Math Univ Hartford 200 Bloomfield Ave West Hartford CT 06117

TROZZOLO, ANTHONY MARION, b Chicago, Ill, Jan 11, 30; m 55; c 6. ORGANIC CHEMISTRY, PHOTOCHEMISTRY. *Educ:* Ill Inst Technol, SB, 50; Univ Chicago, SM, 57, PhD(chem), 60. *Prof Exp:* Asst chemist, Chicago Midway Labs, 52-53; assoc chemist, Armour Res Found, 53-56; mem tech staff, Bell Tel Labs, 59-75; HUISKING PROF CHEM, UNIV NOTRE DAME, IND, 75- *Concurrent Pos:* Adj prof, Columbia Univ, 71; assoc ed, J Am Chem Soc, 75-77; ed, Chem Rev, 77-84; Chevron lectr, Univ Nev, Reno, 83; mem bd trustees, Gordon Res Conf; vis prof, Univ Colo, 81; fac sci lectr, Leuven, Belg, 83; vis lectr, Acad Sinica, 84 & 85; vis prof, Max Planck Inst Radiation Chem, Mülheim, Ruhr, Ger, 90. *Honors & Awards:* Am Inst Chemists Award, 50; Halpern Photochemistry Award, NY Acad Sci, 80; Phillips lectr, Univ Okla, 71; Reilly lectr, Univ Notre Dame, 72; Brown lectr, Rutgers Univ, 75; Faraday lectr, Northern Ill Univ, 76; Butler lectr, SDak State Univ, 78; Hesburgh lectr, Univ Nev, Reno, 86. *Mem:* Fel AAAS; Am Chem Soc; fel Am Inst Chemists; fel NY Acad Sci; Sigma Xi. *Res:* Free radicals; carbenes; charge transfer complexes; electron spin resonance; organic solid state; singlet molecular oxygen; polymer stabilization; chemically-induced dynamic nuclear polarization; laser spectroscopy. *Mailing Add:* Dept Chem Univ Notre Dame Notre Dame IN 46556

TRPIS, MILAN, b Mojsova Lucka, Czechoslovakia, Dec 20, 30; US citizen; m 56; c 3. MEDICAL ENTOMOLOGY. *Educ:* Comenius Univ, Bratislava, Prom Biol, 56; Charles Univ, Prague, Dr rer nat(zool, med entom), 60. *Prof Exp:* Res asst entom, Faunistic Lab, Slovak Acad Sci, 53-56, sci asst, Dept Biol, 56-60, scientist, 60-62, independent scientist & head, Dept Ecol Physiol Insects, Inst Landscape Biol, 62-65; res assoc med entom, Univ Ill, Urbana, 66-67; res assoc, Can Dept Agr, Alta, 67-68; independent scientist & head, Dept Ecol Physiol Insects, Slovak Acad Sci, 68-69; entomologist-ecologist, EAfrica Aedes Res Unit, WHO, UN, Tanzania, 69-71; from asst fac fel to assoc fac fel, Vector Biol Labs, Dept Biol, Univ Notre Dame, Ind, 71-74; assoc prof, 74-78, PROF MED ENTOM, LABS MED ENTOM, DEPT PATHOBIOL, JOHNS HOPKINS UNIV, 78- *Concurrent Pos:* Dir, Biol Res Inst Am, 72-77; proj dir epidemiol river blindness (onchocerciasis), Liberia & Sierra Leone, 81-84. *Honors & Awards:* First Prize Award, Slovak Acad Sci, Bratislava, 61. *Mem:* AAAS; Am Soc Trop Med & Hyg; Am Soc Parasitologists; Entom Soc Am; Am Mosquito Control Asn. *Res:* Parasitic insects, particularly their population dynamics, ecological genetics of populations, behavior and behavioral genetics; embryonic development of insects; biological and genetic control of vectors; ecology of vector-borne diseases. *Mailing Add:* 1504 Ivy Hill Rd Cockeysville MD 21030

TRUAX, DONALD R, b Minneapolis, Minn, Aug 29, 27; m 50; c 4. MATHEMATICAL STATISTICS. *Educ:* Univ Wash, BS, 51, MS, 53; Stanford Univ, PhD(statist), 55. *Prof Exp:* Res fel math, Calif Inst Technol, 55-56; asst prof, Univ Kans, 56-59; from asst prof to assoc prof, 59-69, PROF MATH, UNIV ORE, 69- *Concurrent Pos:* Managing ed, Inst Math Statist, 75-81. *Mem:* Am Statist Asn; fel Inst Math Statist. *Res:* Testing statistical hypotheses; multiple decision problems. *Mailing Add:* Dept Math Univ Ore Eugene OR 97403

TRUAX, ROBERT LLOYD, b Gillett, Ark, May 8, 28; m 52; c 2. MATHEMATICS. *Educ:* Ark State Teachers Col, BSE, 50; Univ Miss, MA, 58; Okla State Univ, EdD(math educ), 64. *Prof Exp:* Coordr math, Pub Schs, Ark, 50-62; assoc prof math, Southern State Col, 63-65; assoc prof, Northeast La State Col, 65-67; ASSOC PROF MATH, UNIV MISS, 67- *Mem:* Math Asn Am. *Res:* Multivariable function approximations; statistical analysis of research related to paper industry. *Mailing Add:* Dept Math/Sec Ed Univ Miss University MS 38677

TRUBATCH, JANETT, b New York, NY, Oct 13, 45; c 4. TECHNICAL MANAGEMENT, NEUROSCIENCE. *Educ:* Polytech Inst Brooklyn, BSc, 62; Brandeis Univ, MA, 64, PhD(physics), 68; Univ Miami, MBA, 91. *Prof Exp:* Asst prof physics, Calif State Univ, Los Angeles, 67-68; res fel biol, Calif Inst Technol, 68-74; asst prof physiol, NY Med Col, 74-77; prog dir neurobiol, NSF, 77-79; health sci admin, Neurol Dis Prog, NIH, 79-85; ASSOC VPRES RES, UNIV CHICAGO, 85- *Mem:* Soc Neurosci; Am Physiol Soc; AAAS; Asn Univ Technol Mgrs. *Res:* Mechanisms of synaptic transmission; neural basis of memory and learning; synapse formation; mathematical modeling of biological systems. *Mailing Add:* 14385 SW 87th Ave Miami FL 33176

TRUBATCH, SHELDON L, b Brooklyn, NY, Mar 12, 42; m 62; c 4. THEORETICAL PHYSICS, BIOPHYSICS. *Educ:* Polytech Inst Brooklyn, BS, 62; Brandeis Univ, MA, 64, PhD(physics), 68;. *Hon Degrees:* JD, Columbia Univ, 77. *Prof Exp:* Assoc prof physics, Calif State Univ, Long Beach, 67-77; mem staff, Off Gen Coun, US Nuclear Regulatory Comn, 77-80, spec asst technol affairs, 80-85; atty off staff coun, Commonwealth Edison, 85-89; ATTY, SIDLEY & AUSTIN, 89- *Mem:* Am Nuclear Soc. *Res:* Non-relativistic field theory; sensory physiology. *Mailing Add:* 1722 I St Washington DC 20006

TRUBEK, MAX, b New York, NY, Nov 28, 98; m 37; c 1. MEDICINE. *Educ:* Johns Hopkins Univ, AB, 22; Univ Md, MD, 26. *Prof Exp:* House physician, Bellevue Hosp, 27-29; asst pathologist, Newark City Hosp, 29-31; assoc prof med, NY Univ-Bellevue Med Ctr, 44-56, PROF CLIN MED, MED SCH, NY UNIV, 56- *Concurrent Pos:* Vis physician, Bellevue Hosp, 46-; attend physician, Univ Hosp, 52- *Res:* Clinical medicine. *Mailing Add:* Dept Clin Med Sch Med NY Univ 550 First Avenue New York NY 10016

TRUBERT, MARC, b Soissons, France, Feb 7, 27; US citizen; m 89; c 2. APPLIED MECHANICS, ELECTRONICS. *Educ:* Univ Paris, BS, 50; Spec Sch Pub Works, Paris, BS, 51; Univ Fla, MS, 60, PhD(eng mech), 62. *Prof Exp:* Res engr, Nat Off Aeronaut Study & Res, Paris, 53-57; develop engr, Compagnie Generale de TSF, Paris, 57-58; asst in res eng mech, Univ Fla, 58-62, asst prof, 62-63; asst prof, Univ Calif, Berkeley, 63-65; res engr, 65-66, mem tech staff, 66-70, group supvr, 70-80, SR MEM TECH STAFF, JET PROPULSION LAB, 80- *Concurrent Pos:* Instr, Exten Sch, Univ Calif, Los Angeles, 65-70. *Honors & Awards:* Except Eng Achievement Medal, NASA. *Mem:* Acoust Soc Am; Am Inst Aeronaut & Astronaut; Sigma Xi. *Res:* Structural dynamics; vibration, random vibration, frequency domain method and analog methods; attitude control of spacecraft; furlable antennas; load analysis and structural testing. *Mailing Add:* PO Box 2064 Covina CA 91722

TRUBEY, DAVID KEITH, b Coldwater, Mich, Apr 23, 28; m 50; c 2. RADIATION PHYSICS, RADIATION PROTECTION. *Educ:* Mich State Univ, BS, 53; Univ Tenn, MS, 88. *Prof Exp:* Physicist, 53-54 & 55-56, mgr radiation shielding info ctr, 66-70, MEM STAFF, OAK RIDGE NAT LAB, 72- *Concurrent Pos:* Lectr, Oak Ridge Sch Reactor Technol, 60-62; chmn, Standards Subcomt ANS-6, Am Nuclear Soc, 72- & Radiation Protection & Shielding Div, 74; mem, Comt N13 Radiation Protection, Am Nat Standards Inst, 82-85 & N17 Res Reactors, Reactor Physics & Shielding, 85- *Honors & Awards:* Tech Achievement Award, Radiation Protection & Shielding Div, Am Nuclear Soc, 83. *Mem:* fel Am Nuclear Soc; Health Physics Soc. *Res:* Radiation shielding, transport and dosimetry. *Mailing Add:* Oak Ridge Nat Lab PO Box 2008 Oak Ridge TN 37831-6362

TRUCE, WILLIAM EVERETT, b Chicago, Ill, Sept 30, 17; m 40; c 2. CHEMISTRY. *Educ:* Univ Ill, BS, 39; Northwestern Univ, PhD(chem), 43. *Prof Exp:* Instr chem, Wabash Col, 43-44; res chemist, Swift & Co, Ill, 44-46; from asst prof to assoc prof, 46-56, PROF CHEM, PURDUE UNIV, WLAFAYETTE, 56- *Concurrent Pos:* Guggenheim fel, Oxford Univ, 57. *Mem:* Am Chem Soc; Royal Soc Chem. *Res:* Organic sulfur chemistry; acetylenes; vinylic halides; organic theory and its relationship to synthetic organic chemistry. *Mailing Add:* Dept Chem Purdue Univ West Lafayette IN 47907-9980

TRUCHARD, JAMES JOSEPH, b Sealy, Tex, June 25, 43; m 66; c 4. ACOUSTICS, ELECTRONICS ENGINEERING. *Educ:* Univ Tex, BS, 64, MA, 67, PhD(elec eng), 74. *Prof Exp:* Lab res asst, Appl Res Labs, Univ Tex, Austin, 63-65, res scientist acoust electronics, 65-80; WITH NAT INSTRUMENTS, 80- *Mem:* Acoust Soc Am. *Res:* Transducer measurement systems; digital signal processing of acoustic signals; nonlinear acoustics; parametric receiving arrays for acoustic signals. *Mailing Add:* 4406 Aqua Verde Austin TX 78746

TRUCKER, DONALD EDWARD, photographic science, organic chemistry; deceased, see previous edition for last biography

TRUDEL, MICHEL D, b Montreal, Que, Feb 5, 44; Can citizen; m 67, 83; c 2. VIROLOGY. *Educ:* Univ Montreal, BA, 65, BSc, 68, MSc, 70; Univ Sherbrooke, PhD(cell biol), 73. *Prof Exp:* Fel cell membranes, Nat Cancer Inst, 73-74; asst prof, 74-75, PROF VIROL INST ARMAND-FRAPPIER, 75-, DIR, CTR VIROL. *Concurrent Pos:* Grants, Formation Researcher, Ministry of Educ, Que, 76-78 & 88-91, Health & Welfare, Can, 76-78, Nat Res Coun Can, 78-81 & 84-93 & Can Med Res Coun, 79-91, conseil de recherches en peche et agroalimentaire, Quebec, 85-89. *Mem:* Am Soc Virologists; Int Soc Antiviral Res; Can Asn Clin Microbiol Infectious Dis; Can Soc Microbiologists. *Res:* Viral subunit vaccines and the implication of the physical form of the viral proteins that induce the immune response; neutralization epitopes rubella, respiratory syncytial virus, bovine herpes virus 1; synthetic peptides; recombinant vaccines. *Mailing Add:* Inst Armand-Frappier 531 Boul des Prairies Ville De Laual PQ H7V 1B7 Can

TRUDEN, JUDITH LUCILLE, b Duluth, Minn, Sept 29, 31. VIROLOGY. *Educ:* Wayne State Univ, BA, 53, MS, 55; Univ Miami, PhD(microbiol), 67. *Prof Exp:* Technician virol, Henry Ford Hosp, Detroit, Mich, 55-59; USPHS res fel, Pub Health Res Inst, New York, NY, 68-71; res instr, Med Col Wis, 71-74, res assoc, 74; scholar molecular biol, Univ Mich, Ann Arbor, 75-78; RES ASSOC, WAYNE STATE UNIV, DETROIT, 78- *Mem:* Sigma Xi; Am Soc Microbiol. *Res:* Immunological and biochemical aspects of liver fibrosis in schistosomiasis. *Mailing Add:* 12700 Greiner Detroit MI 48205

TRUE, NANCY S, b Waterbury, Conn, Sept 30, 51. PHYSICAL CHEMISTRY. *Educ:* Univ Conn, PhD(phys chem), 77. *Prof Exp:* Assoc chem, Univ Conn, 77-78, Univ Col London, 78-79 & Cornell Univ, 79-80; from asst prof to assoc prof, 80-89, PROF CHEM, UNIV CALIF, DAVIS, 89- *Honors & Awards:* Presidental Young Invest, 85; Alfred P Sloan Fel, 86-88. *Mem:* Am Phys Soc; Am Chem Soc; Sigma Xi. *Mailing Add:* Dept Chem Univ Calif Davis CA 95616

TRUE, WILLIAM WADSWORTH, b Rockland, Maine, Dec 27, 25; m 54; c 4. PHYSICS, NUCLEAR STRUCTURE. *Educ:* Univ Maine, BS, 50; Univ RI, MS, 52; Ind Univ, PhD(physics), 57. *Prof Exp:* Instr physics, Princeton Univ, 57-60; from asst prof to assoc prof, 60-69, PROF PHYSICS, UNIV CALIF, DAVIS, 69- *Mem:* Am Phys Soc. *Res:* Theoretical nuclear physics. *Mailing Add:* Dept Physics Univ Calif Davis CA 95616

TRUEBLOOD, KENNETH NYITRAY, b Dobbs Ferry, NY, Apr 24, 20; m 70. CHEMISTRY. *Educ:* Harvard Univ, AB, 41; Calif Inst Technol, PhD(chem), 47. *Prof Exp:* Asst chem, Calif Inst Technol, 43-46, res fel, 47-49; from instr to assoc prof, 49-60, dean, Col Lett & Sci, 71-74, chmn dept chem, 65-70 & 90-91, PROF CHEM, UNIV CALIF, LOS ANGELES, 60- *Concurrent Pos:* Fulbright award, 56-57; mem, US Nat Comt Crystallog, 60-65; vis prof, Ibadan, 64-65; vis scientist, Inst Elemento-Org Compounds, Moscow, 65; Guggenheim fel, 76. *Mem:* Am Chem Soc; Am Crystallog Asn (pres, 61). *Res:* X-ray studies of molecular and crystal structure. *Mailing Add:* Dept Chem & Biochem Univ Calif Los Angeles CA 90024-1569

TRUELOVE, BRYAN, b Bradford, Eng. WEED SCIENCE, HERBICIDE RESISTANCE. *Educ:* Sheffield Univ, BSc, 55, PhD(plant physiol), 61. *Prof Exp:* Asst lectr bot, Manchester Univ, 60-62, lectr, 62-67; assoc prof bot, 68-75, PROF BOT & MICROBIOL, AUBURN UNIV, 75- *Concurrent Pos:* Vis asst prof, Univ Ill, 65. *Mem:* Soc Exp Biol & Med; Am Soc Plant Physiologists; Weed Sci Soc Am. *Res:* The mechanism of herbicide resistance in plants; mode of action of herbicides and effects on plant physiological processes and metabolism. *Mailing Add:* Dept Bot Plant Path & Microbiol Auburn Univ Auburn AL 36849

TRUEMAN, RICHARD E(LIAS), operations research, for more information see previous edition

TRUEMAN, THOMAS LAURENCE, b Media, Pa, Sept 24, 35; m 61; c 2. MANAGEMENT OF HIGH ENERGY, NUCLEAR PHYSICS RESEARCH. *Educ:* Dartmouth Col, AB, 57; Univ Chicago, MS, 58, PhD(physics), 62. *Prof Exp:* Res assoc physics, Brookhaven Nat Lab, 62-64, from asst physicist to physicist, 65-74, dep chmn physics dept, 80-86, assoc dir, 88-91, SR PHYSICIST & GROUP LEADER, BROOKHAVEN NAT LAB, 74- *Concurrent Pos:* Guggenheim fel, Oxford Univ, 72-73; vis prof, Univ D'Aix-Marseille, 78-79, 86-87. *Mem:* Fel Am Phys Soc. *Res:* High energy theory; finite temperature field theory. *Mailing Add:* Brookhaven Nat Lab Upton NY 11973

TRUESDALE, LARRY KENNETH, b San Mateo, Calif, Jan 3, 47. CHEMICAL SYNTHESIS. *Educ:* Univ Calif, San Diego, AB, 69; Univ Calif, Los Angeles, PhD(chem), 74. *Prof Exp:* Res fel chem, Mass Inst Technol, 74-75; res chemist, Allied Chem Corp, 75-78; SR SCIENTIST CHEM, HOFFMANN-LAROCHE INC, 79- *Concurrent Pos:* Adj prof organometall chem, Rutgers Univ, Newark, 81. *Mem:* Am Chem Soc. *Res:* Chemical synthesis; synthetic methods; oxidations; and organometallics. *Mailing Add:* 148 Crossbow Lane North Andover MA 01845

TRUESDELL, ALFRED HEMINGWAY, b Washington, DC, Sept 10, 33; m 64. GEOLOGY, CHEMISTRY. *Educ:* Oberlin Col, AB, 57; Harvard Univ, AM, 61, PhD(geol), 62. *Prof Exp:* Geochemist, US Geol Surv, 55-91; CONSULT, 91- *Concurrent Pos:* Res assoc, Stanford Univ, 64- *Mem:* AAAS; Am Mineral Soc; Geochem Soc. *Res:* Application of physical chemistry to the study of geologic processes; electrochemistry of membranes; ion exchange equilibria and energetics; solution geochemistry; chemistry and physics of geothermal systems. *Mailing Add:* 700 Hermosa Way Menlo Park CA 94025

TRUESDELL, CLIFFORD AMBROSE, III, b Los Angeles, Calif, Feb 18, 19; m 39, 51; c 1. MECHANICS. *Educ:* Calif Inst Technol, BS, 41, MS, 42, Princeton Univ, PhD(math), 43. *Hon Degrees:* Dr Eng, Milan Polytech Inst, 64; DSc, Tulane Univ, 76; Fil Dr, Uppsala Univ, 79; Dr Phil, Basel Univ, 79. *Prof Exp:* Asst math & hist, Calif Inst Technol, 41-42; asst mech, Brown Univ, 42; instr math, Princeton Univ, 42-43 & Univ Mich, 43-44; mem staff, Radiation Lab, Mass Inst Technol, 44-46; chief theoret mech subdiv, Naval Ord Lab, 46-48, head theoret mech sect, Naval Res Lab, 48-51; prof math, Ind Univ, 50-61; prof, 61-89, EMER PROF RATIONAL MECH, JOHNS HOPKINS UNIV, 89- *Concurrent Pos:* From lectr to assoc prof, Univ Md, 46-50; consult, Naval Res Lab, 51-55, Nat Bur Standards, 59-62, Sandia Corp, 66, Ga Inst Technol, 73-74 & US Nuclear Regulatory Comn, 75-83; ed, J Rational Mech & Anal, 52-56 & Arch Hist Exact Sci, 60-; Guggenheim fel, 57; ed, Arch Rational Mech & Anal, 57-66, 85-90, co-ed, 67-85; co-ed, Results Appl Math, 57-62; NSF sr res fel, Univ Bologna & Univ Basel, 60-61; ed, Springer Tracts Natural Philos, 62-66 & 79-, co-ed, 67-78; Walker-Ames prof, Univ Wash, 64; distinguished vis prof, Syracuse Univ, 65; 75th Anniversary lectr, Drexel Inst Technol, 66-67; chmn, Soc Natural Philos, 67-68 & 83-84; Lincean prof, Ital Acad Lincei, Rome, 70, 73 & 74; lectr, Fed Univ Rio de Janeiro, 72; vis res scholar, Japan Soc Prom Sci, Kyoto, 80. *Honors & Awards:* Bingham Medal, Soc Rheol, 63; Panetti Int Medal & Prize, Acad Sci Turin, 67; Birkhoff Prize, Am Math Soc & Soc Indust & Appl Math, 78; Ritt lectr, Dept Math, Columbia Univ, 82; Page-Barbour lectr, Univ Va, 85; Humbolt Prize, US Sr Scientist Award, 85. *Mem:* Soc Natural Philos(secy, 63-65, 70-71 & 80-81); Int Acad Hist Sci; hon mem Polish Soc Theoret & Appl Mech; Int Acad Philos Sci; foreign mem Lincean Acad Sci; corresp mem Acad Brasileira Ciencias. *Res:* Rational mechanics. *Mailing Add:* Johns Hopkins Univ Baltimore MD 21218

TRUESDELL, SUSAN JANE, b Oak Park, Ill, Mar 22, 45; m 78. MOLECULAR BIOLOGY. *Educ:* Mich State Univ, BS, 67; Univ Calif, Los Angeles, PhD(molecular biol), 71. *Prof Exp:* Am Cancer Soc fel, Univ Mich, 71-73; MICROBIOLOGIST, PFIZER, INC, 73- *Concurrent Pos:* Instr introductory microbiol, Conn Col, 77 & 78. *Mem:* Am Soc Microbiol; Soc Indust Microbiol. *Res:* Genetics and physiology of penicillin production; viruses that infect penicillium chrysogenum; microbiol enzyme production; biotransformations; antibiotic and vitamin fermentations. *Mailing Add:* Fermentation Process Res Pfizer Inc Groton CT 06340

TRUEX, RAYMOND CARL, b Norfolk, Nebr, Dec 11, 11; m 38; c 2. ANATOMY. *Educ:* Nebr Wesleyan Univ, AB, 34; St Louis Univ, MS, 36; Univ Minn, PhD(anat), 39. *Prof Exp:* From instr to assoc prof anat, Col Physicians & Surgeons, Columbia Univ, 38-48; prof & head div, Hahnemann Med Col, 48-61; PROF ANAT, SCH MED, TEMPLE UNIV, 61-, AT DEPT NEUROSURG. *Concurrent Pos:* USPHS award prof, 61-; consult, NIH, 62-66 & Nat Bd Med Examrs, 67-71 & 75-79. *Honors & Awards:* AMA Awards, 52 & 58. *Mem:* AAAS; Geront Soc; Am Asn Anatomists (pres, 71); Harvey Soc; Am Vet Med Asn. *Res:* Histological changes with age and pathology of the human nervous system; histology and physiology of the conduction system and circulation of the heart. *Mailing Add:* Dept Neurosurg Health Sci Campus Temple Univ Broad & Ont Philadelphia PA 19140

TRUEX, TIMOTHY JAY, b Goshen, Ind, June 11, 45. INORGANIC CHEMISTRY. *Educ:* Hanover Col, BS, 67; Mass Inst Technol, PhD(inorg chem), 72. *Prof Exp:* Res scientist inorg chem, Ford Motor Co, 72-80. *Mem:* Am Chem Soc. *Res:* Atmospheric environmental chemistry; catalysis chemistry; chemistry of surface coatings. *Mailing Add:* Johnson Matthey Tech Centre Bounts Ct Sowning Commons Reading RG4 9NH England

TRUFANT, SAMUEL ADAMS, b New Orleans, La, May 24, 19; m 45; c 4. MEDICINE, NEUROLOGY. *Educ:* Tulane Univ, BS, 40, MD, 43; Am Bd Psychiat & Neurol, dipl, 51. *Prof Exp:* Asst gross anat, Tulane Univ, 40-41; Rockefeller fel neurol, Washington Univ, 47-49, USPHS res fel, 49-50; from asst prof to prof, 50-78, from asst dean to assoc dean, Col Med, 51-62, EMER PROF NEUROL, COL MED, UNIV CINCINNATI, 78-; PROF NEUROL, TULANE UNIV, 79-; STAFF NEUROLOGIST, VET ADMIN MED CTR, LA, 79- *Concurrent Pos:* Consult, Wright-Patterson AFB, Ohio, 52-; dir child neurol, Children's Hosp, 59-71; dir, Am Bd Psychiat & Neurol, 66-73, vpres, 71, pres, 72-; proj officer, India Neurol & Sensory Dis Serv Prog, USPHS, 66-; mem, Residency Rev Comt Psychiat & Neurol, 67-72, chmn, 70-72; ed, Trans, Am Neurol Asn, 68-73. *Mem:* Am Neurol Asn (vpres, 65, secy-treas, 68-73, pres, 75); Asn Res Nerv & Ment Dis; Asn Am Med Cols (asst secy, 59-64); Am Acad Neurol. *Res:* Clinical neurology; electroencephalography. *Mailing Add:* Dept Psychiat & Neurol Tulane Univ Sch Med 1415 Tulane Ave New Orleans LA 70112

TRUHLAR, DONALD GENE, b Chicago, Ill, Feb 27, 44; m 65; c 2. CHEMICAL DYNAMICS, COMPUTATIONAL SCIENCE. *Educ:* St Mary's Col, Minn, BA, 65; Calif Inst Technol, PhD(chem), 70. *Prof Exp:* Student aide chem, Argonne Nat Lab, 65; from asst prof to assoc prof chem, 69-76, PROF CHEM & CHEM PHYSICS, UNIV MINN, MINNEAPOLIS, 76-, DIR, MINN SUPERCOMPUTER INST, 88- *Concurrent Pos:* Sr vis fel, Battelle Mem Inst, Columbus, 73; & Sloan res fel, 73-77; vis fel, Joint Inst Lab Astrophysics, Boulder, Colo, 75-76. *Mem:* Fel Am Phys Soc; Am Chem Soc. *Res:* Theory and computations for collision processes involving atoms, molecules and electrons; potential energy surfaces for reactive and inelastic collisions; theory of molecular spectroscopy. *Mailing Add:* Dept Chem Univ Minn 139 Smith Hall Minneapolis MN 55455

TRUITT, EDWARD BYRD, JR, b Norfolk, Va, Aug 23, 22; m 49; c 2. PHARMACOLOGY. *Educ:* Med Col Va, BS, 43; Univ Md, PhD(pharmacol), 50. *Prof Exp:* Asst prof pharmacol, Bowman Gray Sch Med, Wake Forest Col, 50-55; from assoc prof to prof, Sch Med, Univ Md, Baltimore City, 55-67; sr res fel, Columbus Labs, Battelle Mem Inst & prof pharmacol, Col Med, Ohio State Univ, 67-72; res prof pharmacol, Sch Med, George Washington Univ, 73-76; prog chief & prof pharmacol, 76-84, RES PROF PHARMACOL, COL MED, NORTHEASTERN OHIO UNIV, 84- *Concurrent Pos:* Robins Co fel, Bowman Gray Sch Med, Wake Forest Col, 50-55. *Mem:* Res Soc Alcoholism; Am Chem Soc; Am Soc Pharmacol & Exp Therapeut; Soc Exp Biol & Med; Sigma Xi; Int Soc Biol Res Alcoholism. *Res:* Neuropharmacology; psychopharmacology; drug metabolism; alcoholism; marijuana and drug abuse research; study of alcohol-acetaldehyde metabolism as a possible genetic and/or chemical risk marker for alcoholism. *Mailing Add:* Dept Pharmacol Northeastern Ohio Univ Col Med 4209 S Rte 44 Rootstown OH 44272-0095

TRUITT, MARCUS M(CCAFFERTY), b Enid, Okla, Oct 10, 21; m 42; c 2. CIVIL ENGINEERING. *Educ:* Okla State Univ, BS, 47; Harvard Univ, MS, 48; Stanford Univ, Engr, 51; Johns Hopkins Univ, PhD, 68. *Prof Exp:* Soil mechanic, United Fruit Co, Cent Am, 48-49; asst prof civil eng, Univ Ala, 49-50; from asst prof to prof eng, Tex A&I Univ, 51-77, chmn dept civil & mech eng, 68-77, prof civil & mech eng, 77-84; RETIRED. *Concurrent Pos:* Consult, 53-; city engr, Kingsville, Tex, 53-; res assoc, Johns Hopkins Univ, 66- *Mem:* Am Soc Civil Engrs; Am Soc Eng Educ. *Res:* Municipal engineering; environmental engineering; application of computer simulation techniques to hurricane track predictions. *Mailing Add:* 504 Ailsie Kingsville TX 78363

TRUITT, ROBERT LINDELL, b Carbondale, Ill, July 26, 46; m 67; c 5. TUMOR IMMUNOLOGY, TRANSPLANTATION BIOLOGY. *Educ:* Southern Ill Univ, Carbondale, BA, 68, PhD(microbiol), 73. *Prof Exp:* Fel germfree syst, Lobund Lab, Univ Notre Dame, 72-74; res assoc, Winter Res Lab, Mt Sinai Med Ctr, 74-77, sr scientist, 77-83, assoc dir, 83-84; res assoc prof, 84-88, RES PROF, DEPT PEDIAT, MED COL WIS, MILWAUKEE, 88- *Concurrent Pos:* Fel, United Cancer Coun, 72-73 & Damon Runyon Mem Fund Cancer Res, 73-75; NIH/Nat Cancer Inst res grants, 75, 77, 79, 83 & 86; spec fel, Leukemia Soc Am, 76-78, scholar, 78-83; assoc scientist, Univ Wis Med Sch, Milwaukee, 83-84; NIH/Nat Inst Allergy & Infectious Dis res grant, 84. *Mem:* Am Asn Immunologists; Fedn Am Soc Exp Biol; AAAS; Am Soc Microbiol; Int Soc Exp Hemat; Asn Gnotobiotics; Int Asn Gnotobiol; Soc Biol Ther. *Res:* Germfree animal systems; tumor immunology; bone marrow transplantation; transplantation biology. *Mailing Add:* Med Col Wis Dept Pediat Milwaukee WI 53226

TRUJILLO, EDUARDO E, b Horconcitos, Panama, Apr 22, 30. PLANT PATHOLOGY. *Educ:* Univ Ark, BSA, 56, MS, 57; Univ Calif, PhD(plant path), 62. *Prof Exp:* Res asst plant path, Univ Calif, 57-62; asst plant pathologist, 62-65, asst prof plant path & asst specialist, 65-67, assoc prof, 67-74, prof plant path, 74-80, PROF BOT SCI, UNIV HAWAII, 80-, ASSOC PLANT PATHOLOGIST, 67- *Concurrent Pos:* Consult, Pac Southwest Forest & Range Exp Sta, US Forest Serv, 63- *Mem:* AAAS; Am Phys Soc; Am Soc Hort Sci. *Res:* Aspects of research dealing with soil borne pathogens, mainly ecology and epidemiology of Pythium and Phytophthoras in tropical environments; biology of Fusarium species. *Mailing Add:* Dept Plant Path St John 310 Univ Hawaii at Manoa 2500 Campus Rd Honolulu HI 96822

TRUJILLO, EDWARD MICHAEL, b Los Angeles, Calif, Apr 14, 47. CHEMICAL ENGINEERING. *Educ:* Univ Ariz, BS, 69; Calif Inst Technol, MS, 70; Univ Utah, PhD(chem eng), 75. *Prof Exp:* Prod engr, Kimberly-Clark Corp, 70-71; instr chem eng, Univ Utah, 71-75; res eng prod, Marathon Oil Co, 75-84; ASSOC PROF CHEM ENG, UNIV UTAH, 84- *Mem:* Am Inst Chem Engrs; Soc Petrol Engrs; Am Inst Mining, Metall & Petrol Engrs; Am Chem Soc. *Res:* Biochemical engineering; polymer rheology; flow through porous media; ultrafiltration of protein solutions; immobilized enzyme reactor systems; rheology of nanoparticle suspensions. *Mailing Add:* Dept Chem Eng Univ Utah 3297 Merrill Eng Bldg Salt Lake City UT 84112

TRUJILLO, PATRICIO EDUARDO, b Santa Fe, NMex, Jan 21, 37; m 62; c 4. ANALYTICAL CHEMISTRY, GEOCHEMISTRY. *Educ:* Univ NMex, BS, 60. *Prof Exp:* Chemist, NMex Bur Revenue, 59-60; lieutenant missile tech, USAF, 61-64; chief chemist, Eberline Instrument Corp, 64-67; STAFF MEM CHEM, LOS ALAMOS NAT LAB, 67- *Res:* Analytical chemistry related to geothermal energy. *Mailing Add:* 1069 Calle Largo Santa Fe NM 87501

TRUJILLO, RALPH EUSEBIO, b Embudo, NMex, Sept 22, 40; m 70. BIOCHEMISTRY. *Educ:* Univ NMex, BS, 62; Ind Univ, PhD(biochem), 67. *Prof Exp:* Mem Peace Corps, Ecuador, 62-64; USPHS fel biochem, Univ Tex M D Anderson Hosp & Tumor Inst, 67-69; MEM TECH STAFF, SANDIA LABS, 69- *Res:* Response of macromolecular systems to thermal, chemical and radiation environments. *Mailing Add:* 5725 El Rito Ave NW Albuquerque NM 87105

TRUJILLO, STEPHEN MICHAEL, b Culver City, Calif, Mar 5, 32; m 59. ATOMIC PHYSICS, ELECTRICAL ENGINEERING. *Educ:* Univ Kans, BSc, 58; Univ London, DPHil(physics), 75. *Prof Exp:* Staff scientist physics, Convair Div, Gulf Oil Corp, 69-72; prin physicist, IRT Corp, 72-80; sr scientist, Inesco Inc, 80-83; Staff scientist, S-Cubed div, Maxwell Labs, Inc, 83-89; PRIN PHYSICIST, IRT CORP, 90- *Concurrent Pos:* Vis res fel physics, Univ London, 70-75. *Mem:* Am Phys Soc; Inst Physics, Eng; Inst Elec & Electronics Engrs. *Res:* Molecular physics; scientific instrumentation; electro optics; x-ray technology. *Mailing Add:* 5931 Bellevue Ave La Jolla CA 92037

TRULSON, MICHAEL E, MICROANATOMY, DRUG ABUSE. *Educ:* Univ Iowa, PhD(biopsychol), 74. *Prof Exp:* PROF ANAT, TEX A&M UNIV, 84- *Mailing Add:* 701 Auburn Dr Richardson TX 75081

TRUMAN, JAMES WILLIAM, b Akron, Ohio, Feb 5, 45; m 70. INVERTEBRATE NEURODEVELOPMENT. *Educ:* Univ Notre Dame, BS, 67; Harvard Univ, MA, 69, PhD(biol), 70. *Prof Exp:* Harvard Soc Fels jr fel, Harvard Univ, 70-73; from asst prof to assoc prof, 73-78, PROF ZOOL, UNIV WASH, 78- *Concurrent Pos:* John Simon Guggenheim Fel, 86; NIH res career develop award, 76-81; McKnight Neurosci develop award, 82-84. *Honors & Awards:* Newcomb Cleveland Prize, AAAS, 70. *Mem:* Entom Soc Am; Soc Neurosci; Am Soc Zoologists. *Res:* Physiological aspects of circadian rhythms; interaction of hormones with the nervous system; nervous system development. *Mailing Add:* Dept Zool NJ-15 Univ Wash Seattle WA 98195

TRUMBO, BRUCE EDWARD, b Springfield, Ill, Dec 12, 37. STATISTICS. *Educ:* Knox Col, Ill, AB, 59; Univ Chicago, SM, 61, PhD(statist), 65. *Prof Exp:* Asst prof math, San Jose State Col, 63-64; from asst prof to assoc prof, 65-72, chmn dept, 70-75, PROF STATIST, CALIF STATE UNIV, HAYWARD, 72- *Concurrent Pos:* Consult, 64-70; vis assoc prof, Stanford Univ, 67-69, 71 & 81; coun fel, Acad Admin Internship Prog, Am Coun Educ, 68-69; prog dir statist res, NSF, 74-75, 78-79 & 85-86. *Mem:* Am Statist Asn; fel Inst Math Statist. *Res:* Application of statistical methods to social, behavioral and biological sciences; probability; statistical graphics. *Mailing Add:* Dept Statist Calif State Univ Hayward CA 94542

TRUMBORE, CONRAD NOBLE, b Denver, Colo, Feb 17, 31; m 55; c 2. PHYSICAL CHEMISTRY. *Educ:* Dickinson Col, BS, 52; Pa State Univ, PhD(chem), 55. *Prof Exp:* Fulbright grant, Inst Nuclear Res, Netherlands, 55-56; asst scientist, Argonne Nat Lab, 56-57; instr chem, Univ Rochester, 57-60; asst prof, 60-66, assoc prof chem, 66-, DIR GRAD STUDIES & ASST CHAIR, UNIV DEL, 89- *Concurrent Pos:* USPHS spec fel, Inst Cancer Res, Sutton, Eng, 67-68. *Mem:* AAAS; Radiation Res Soc; Am Chem Soc. *Res:* Primary chemical processes in radiation chemistry of aqueous solutions; correlations between photochemistry and radiation chemistry; biological radiation chemistry; pulse radiolysis and flash photolysis; behavior of diffusing molecules in shear fields; characterization of mixtures by flow properties in capillaries. *Mailing Add:* 113 Dallas Ave Newark DE 19711

TRUMBORE, FORREST ALLEN, b Denver, Colo, Dec 28, 27; m 51; c 2. PHYSICAL CHEMISTRY. *Educ:* Dickinson Col, BS, 46; Univ Pittsburgh, PhD(chem), 50. *Prof Exp:* Aeronaut res scientist thermodyn alloys, Lewis Flight Propulsion Lab, Nat Adv Comt Aeronaut, 50-52; mem tech staff, Bell Labs, AT&T Inc, 52-89; CONSULT, 89- *Mem:* AAAS; Electrochem Soc. *Res:* Solubilities and electrical properties of impurities in semiconductors; crystal growth; photoluminescence and electroluminescence in semiconductors; battery materials. *Mailing Add:* 30 Glen Oaks Ave Summit NJ 07901

TRUMBULL, ELMER ROY, JR, b Lawrence, Mass, Apr 5, 24; m 54; c 3. ORGANIC CHEMISTRY. *Educ:* Dartmouth Col, AB, 44; Univ Ill, PhD(org chem), 47. *Prof Exp:* Asst chem, Univ Ill, 44-46; res assoc, Mass Inst Technol, 47-48, Du Pont fel, 51-52; instr, Tufts Col, 48-51; asst prof chem, Brown Univ, 52-58; from asst prof to assoc prof, Colgate Univ 58-63, dir div natural sci & math, 64-70, prof chem, 63-87, chmn dept, 70-87; RETIRED. *Concurrent Pos:* NSF fac fel, Univ Ariz, 66-67; res assoc, Eidgenoessische Technische Hochschule, Zurich, 71-72; vis prof, Ore State Univ, 78-79, Univ Ga, 86-87. *Mem:* Am Chem Soc; Royal Soc Chem. *Res:* Elimination reactions; natural products. *Mailing Add:* Rt One Box 200 B Arnoldsville GA 30619

TRUMMEL, J(OHN) MERLE, b Maroa, Ill, Dec 28, 16; m 42; c 4. MECHANICAL ENGINEERING. *Educ:* Univ Ill, BS, 39; Iowa State Univ, MS, 40; Univ Iowa, PhD(mech eng), 60. *Prof Exp:* Instr mech eng, 41-44, asst prof, 44-58, instr, 58-59, assoc prof, 60-61, actg chmn dept, 65-66, PROF MECH ENG, UNIV IOWA, 61- *Concurrent Pos:* Consult, Oak Ridge Nat Lab, 54-58, Hawkeye Prod Corp, Iowa, 60-61 & Pioneer-Cent Div Bendix Corp, 61-63. *Mem:* Am Soc Mech Engrs; Am Soc Eng Educ. *Res:* Heat transport by unsteady flows; design and analysis of mechanical systems; mechanical engineering measurement and instrumentation. *Mailing Add:* Dept Mech Eng Univ Iowa Iowa City IA 52242

TRUMMER, MAX JOSEPH, b Bogota, Colombia, Aug 12, 24; US citizen; m 45; c 1. THORACIC SURGERY. *Educ:* Univ Ill, MD, 48; Univ Pa, MS, 65; Am Bd Surg, dipl, 58; Bd Thoracic Surg, dipl, 61. *Prof Exp:* Resident thoracic surg, US Naval Hosp, St Albans, NY, Med Corps, USN, 58-60, chief thoracic surgeon, US Naval Hosp, San Diego, 67-70; chief thoracic and cardiac surg, Los Angeles County-Olive View Med Ctr, 70-71; dir surg teaching prog, Mercy Hosp & Med Ctr, San Diego, 71-84; CLIN PROF THORACIC SURG, UNIV SOUTHERN CALIF, 84- *Concurrent Pos:* Assoc clin prof thoracic surg, Univ Southern Calif, 69-84 & Univ Calif, San Diego, 70- *Mem:* Fel Am Col Surgeons; fel Am Col Chest Physicians; Soc Thoracic Surg; Am Asn Thoracic Surg; fel Am Col Cardiol. *Res:* Lung transplantation; open-heart surgery; cardiopulmonary physiology. *Mailing Add:* 711 Cornish Dr San Diego CA 92107-4225

TRUMMER, STEVEN, analytical chemistry, for more information see previous edition

TRUMP, BENJAMIN FRANKLIN, b Kansas City, Mo, July 23, 32; m 61; c 2. PATHOLOGY, CELL BIOLOGY. *Educ:* Univ Mo-Kansas City, BA, 53; Univ Kans, MD, 57. *Prof Exp:* Intern path, Med Ctr, Univ Kans, 57-58, resident, 58-59; resident anat, Sch Med, Univ Wash, 59-60, resident-trainee, 60-61, investr exp path, Armed Forces Inst Path, 61-63; asst prof path, Sch Med, Univ Wash, 63-65; from assoc prof to prof, Med Ctr, Duke Univ, 65-70; PROF & CHMN DEPT PATH, SCH MED, UNIV MD, BALTIMORE, 70- *Concurrent Pos:* Fel, Med Ctr, Univ Kans, 58-59; US Food & Drug Admin fel, Univ Assoc for Res & Educ in Path; NIH fel & training grant, Univ Md; mem bd dirs, Univ Assoc for Res & Educ in Path, 70-; mem, Md Post Mortem Exam Comn, 70-; consult, US Food & Drug Admin, 71- & Vet Admin Hosp, Baltimore, Md, 72-; docent, Dept Cell Biol, Univ Jyvaskyla, Finland, 73-; Am Cancer Soc prof oncol, & dir, Md Cancer Prog, 77- *Mem:* AAAS; Am Asn Path & Bact; Am Soc Exp Path; Am Soc Cell Biol; Am Soc Microbiol. *Res:* Cellular and subcellular pathology; membrane structure and functions; lysosome structure and function; chemical carcinogenesis; kidney pathophysiology; fish physiology and pathology; environmental pathology. *Mailing Add:* Dept Path Univ Md Sch Med Ten S Pine St Baltimore MD 21201

TRUMPLER, PAUL R(OBERT), turbomachinery dynamics, film bearings; deceased, see previous edition for last biography

TRUMPOWER, BERNARD LEE, b Chambersburg, Pa, July 20, 43; m 64. BIOCHEMISTRY, YEAST MOLECULAR GENETICS. *Educ:* Univ Pittsburgh, BS, 65; St Louis Univ, PhD(biochem), 69. *Prof Exp:* From asst prof to assoc prof biochem, 72-82, PROF BIOCHEM, DARTMOUTH MED SCH, 82- *Concurrent Pos:* NIH fel, Cornell Univ, 69-71; vis scientist, Whitehead Inst; chmn, Phys Biochem Study Sect, NIH, 81-84; estab investr, Am Heart Asn. *Honors & Awards:* Humboldt Prize, Fed Rep Ger, 85. *Mem:* Fed Am Socs Exp Biol; NY Acad Sci; Am Soc Cell Biol; Genetic Soc Am. *Res:* Genetic aspects of organelle assembly with particular emphasis on genetic aspects of mammalian cell respiration and energy transduction. *Mailing Add:* Dept Biochem Dartmouth Med Sch Hanover NH 03756

TRÜMPY, D(ANIEL) RUDOLF, b Glarus, Switz, Aug 16, 21; m 48; c 2. HISTORY OF SCIENCES. *Educ:* Swiss Fed Inst Technol, Dip (ing geol), 45, Dr, 47. *Hon Degrees:* DSc, Univ Pierre & Marie Curie, Paris, 77, Univ Lausanne, 87. *Prof Exp:* Asst dept geol, Univ Lausanne, 47-53; prof, Swiss Fed Inst Technol, 53-86, prof, Univ Zurich, 56-86, EMER PROF GEOL, UNIV ZURICH & SWISS FED INST TECHNOL, 86- *Concurrent Pos:* Treas, Int Union Geol Scis, 64-68, pres, 76-80; vis prof, Yale Univ, 81. *Honors & Awards:* Von Buch Medal, Deutsche Geol Gesellschaft; Penrose Medal, Geol Soc Am, 85; Suess Medal, Oesterreichische Geol Gesellschaft, 87. *Mem:* Foreign assoc Nat Acad sci; hon mem geol Soc Am; Am Assoc Petroleum Geologist; foreign assoc Am Acad Arts & Sci; foreign assoc Acad Scis, Paris; German Acad Leopolidina. *Res:* Interrelation between sedimentary and structural evolution of Alps; other areas: Morocco, Spain, France, Greece; Traissic and Jurassic strutigraphy (Alps, Greenland); history of Geology. *Mailing Add:* Allmen 19 Kusnacht-8700 Switzerland

TRUNK, GERARD VERNON, b Baltimore, Md, May 9, 42. ELECTRICAL ENGINEERING. *Educ:* Johns Hopkins Univ, BES, 63, PhD(elec eng), 67. *Prof Exp:* Elec engr radar, Naval Res Lab, 67-74; vis fel elec eng, Johns Hopkins Univ, 74-75; head radar anal staff, 75-79, HEAD, RADAR ANALYTICAL BR, NAVAL RES LAB, 79- *Concurrent Pos:* Sabatical lectr, Johns Hopkins Univ, 74- *Mem:* Sr mem Inst Elec & Electronics Engrs; Sci Res Soc Am. *Res:* Radar systems; detection and estimation theory; simulation; and pattern recognition. *Mailing Add:* Head Radar Analytical Br Naval Res Lab Code 5310 Washington DC 20375

TRUNZO, FLOYD F(RANK), b Apollo, Pa, Dec 3, 17; m 46; c 3. CHEMICAL ENGINEERING. *Educ:* Univ Ala, BS, 41, MS, 47. *Prof Exp:* Observer, Process Control, Carnegie Ill Steel Corp, 41; jr engr pilot resin develop, Westinghouse Res Labs, 47-48, assoc engr wire insulation develop, 48-53, res engr, 53-62, sr res engr, 62-76, sr res scientist, 76-77, fel engr, 77-81; RETIRED. *Mem:* Am Chem Soc. *Res:* Wire insulation for motors, transformers and magnet wires; composite insulating materials for electrical power and generating equipment. *Mailing Add:* 106 Jamison Lane Monroeville PA 15146-2316

TRUONG, XUAN THOAI, b French Cochin-China, Nov 17, 30; US citizen. PHYSICAL MEDICINE, PHYSIOLOGY. *Educ:* West Liberty State Col, BS, 52; Columbia Univ, MD, 56; Univ Louisville, PhD(physiol), 64; Am Bd Phys Med & Rehab, dipl, 65; Am Bd Electrodiagnostic Med, 89. *Prof Exp:* Intern surg, Ind Univ Med Ctr, 56-57; resident phys med & rehab, Univ Louisville Hosps, 58-61, instr, Sch Med, Univ, 62-64; clin & res consult, Inst Phys Med & Rehab, 64-68; asst prof, Baylor Col Med, 68-71; DIR RES & EDUC, INST PHYS MED & REHAB, 72- *Concurrent Pos:* Clin investr, Vet Admin Hosp, Houston, 69-71; asst prof, Peoria Sch Med, Univ Ill, 72-; dir, Muscular Dystrophy Asn Clin, 78- *Mem:* Am Physiol Soc; Am Acad Phys Med & Rehab; Inst Elec & Electronics Engrs; Am Asn Electrodiag Med; Am Med Asn; Am Soc Clin Evoked Potentials. *Res:* Mechanical properties of muscle tissue; electrophysiology of neuro-muscular system; rehabilitation engineering. *Mailing Add:* Inst Phys Med & Rehab 6501 N Sheridan Rd Peoria IL 61614-2932

TRUPIN, JOEL SUNRISE, b Brooklyn, NY, Mar 15, 34; m 57; c 1. BIOCHEMISTRY, NUTRITION. *Educ:* Cornell Univ, BS, 54, MNS, 56; Univ Ill, PhD(biochem), 63. *Prof Exp:* Asst prof microbiol, Sch Med, St Louis Univ, 66-71; ASSOC PROF BIOCHEM, MEHARRY MED COL, 71- *Concurrent Pos:* Am Cancer Soc res fel biochem genetics, Nat Heart Inst, 63-66. *Mem:* AAAS; Am Chem Soc; Am Soc Microbiol; Am Soc Cell Biol. *Res:* Biochemistry and regulation of amino acid biosynthesis; transfer RNA; protein and amino acid nutrition; biochemistry and molecular biology of growth and metabolism of human dermal fibroblasts and of fibroblast functions relevant to normal and abnormal wound healing. *Mailing Add:* Div Biomed Sci Meharry Med Col Nashville TN 37208

TRURAN, JAMES WELLINGTON, JR, b Brewster, NY, July 12, 40; m 65; c 3. ASTROPHYSICS. *Educ:* Cornell Univ, BA, 61; Yale Univ, MS, 63, PhD(physics), 66. *Prof Exp:* Resident res assoc, Goddard Inst Space Studies, NASA, NY, 65-67; res fel physics, Calif Inst Technol, 68-69; from assoc prof to prof physics, Belfer Grad Sch Sci, 70-73; prof astron, Univ Ill, Urbana-Champaign, 73-91; PROF ASTRON ASTROPHYS, UNIV CHICAGO, CHICAGO, 91- *Concurrent Pos:* Mem bd contribr, Comments Astrophys & Space Physics, 73-74; ed, Contributions Nuclear & High-energy Astrophys, Physics Letters B, 74-80; trustee, Aspen Ctr Physics, 79-85; assoc, Ctr Advan Sci, Univ Ill; sr vis fel, Inst Astron, Univ Cambridge, England & Guggenheim Mem Found fel, 79-80; Alexander von Humboldt-Stiftung, Sr Scientist, Max-Planck-Inst Astrophysics, Munich, WGer, 86-87; vpres, Aspen Ctr Physics, 85-88. *Mem:* Am Phys Soc; Am Astron Soc; Int Astron Union. *Res:* Nucleosynthesis; nuclear reactions in stars; mechanisms of nova and supernova explosions; stellar evolution; galactic evolution; origin of cosmic rays; white dwarfs; binary evolution. *Mailing Add:* Dept Astron Univ Ill 1002 W Green Urbana IL 61801

TRUS, BENES L, b Tyler, Tex, May 9, 46; m 72; c 2. IMAGE PROCESSING. *Educ:* Tulane Univ, BS, 68; Calif Inst Technol, PhD(phys chem), 72. *Prof Exp:* Jane Coffin Childs fel, Calif Inst Technol, 72-75; staff fel, Lab Biochem, Nat Inst Dent Res, 75-77, sr staff fel, 77-80, RES CHEMIST, COMPUT SYST LAB, DIV COMPUT RES & TECHNOL, NIH, 80- *Honors & Awards:* Dirs Award, NIH, 87. *Mem:* Sigma Xi; Electron Micros Soc Am; NY Acad Sci. *Res:* Fibrous protein and virus; computer analysis of chemical or biochemical problems, and electron microscopy; image processing; molecular graphics and computer modeling. *Mailing Add:* Rm 2053 Bldg 12A NIH Bethesda MD 20892

TRUSAL, LYNN R, b Williamsport, Pa, Dec 4, 45. CELLULAR PHYSIOLOGY, ELECTRON MICROSCOPY. *Educ:* Pa State Univ, PhD(physiol), 75. *Prof Exp:* RES SCIENTIST, DIV PATH, US ARMY MED RES INST, 81- *Mem:* Am Soc Cell Biol; Electron Micros Soc Am; Int Soc Toxinology. *Mailing Add:* c/o Armed Forces Med Intel Ctr Ft Detrick Frederick MD 21702

TRUSCOTT, FREDERICK HERBERT, b Meredith, NY, Mar 16, 26; m 54; c 2. PLANT PHYSIOLOGY. *Educ:* State Univ NY Albany, AB, 50; Rutgers Univ, PhD(bot), 55. *Prof Exp:* Asst bot, Rutgers Univ, 52-55; res fel, Jackson Mem Lab, Bar Harbor, Maine, 55-56; instr bot, Univ RI, 56-58; from asst prof to assoc prof, 58-65, chmn dept, 72-75, PROF BIOL, STATE UNIV NY ALBANY, 65- *Mem:* Bot Soc Am; Am Soc Plant Physiologists; Am Inst Biol Sci. *Res:* Morphogenesis; photophysiology. *Mailing Add:* Dept Biol Sci State Univ NY 1400 Washington Ave Albany NY 12222

TRUSCOTT, ROBERT BRUCE, b Winnipeg, Man, July 9, 28; m 53; c 5. VETERINARY MICROBIOLOGY. *Educ:* Univ Toronto, BSA, 50, MSA, 53, DVM, 62; Univ Waterloo, PhD(microbiol physiol), 66. *Prof Exp:* Supvr, Animal House, Univ Western Ont, 50; fermentation supvr, Merck & Co Ltd, Que, 50-51; res asst microbiol, Ont Agr Col, Guelph, 51-53; lectr poultry path, Ont Vet Col, Univ Guelph, 53-76, assoc prof vet microbiol, 69-76; res scientist, Health Animals Br, Can Dept Agr, Animal Path Lab, Sackville, NB, 77-85; res scientist, Animal Path Lab, Guelph, Ont, 85-89; RETIRED. *Mem:* Microbiol; Am Soc Microbiol; Am Asn Avian Pathologists; Can Vet Med Asn. *Res:* Avian salmonella; food microbiology. *Mailing Add:* 106 Dovercliffe Rd Guelph ON N1G 3A6 Can

TRUSELL, FRED CHARLES, analytical chemistry, for more information see previous edition

TRUSK, AMBROSE, analytical chemistry, for more information see previous edition

TRUSSELL, HENRY JOEL, b Atlanta, Ga, Feb 3, 45; m 68; c 2. COMPUTER SCIENCE, ELECTRICAL ENGINEERING. *Educ:* Ga Inst Technol, BS, 67; Fla State Univ, MS, 68; Univ NMex, PhD(elec eng & comput sci), 76. *Prof Exp:* Mem staff, Los Alamos Sci Lab, 69-80; assoc prof, 80-86, PROF ELEC ENG, NC STATE UNIV, 86- *Concurrent Pos:* Adj prof, Univ NMex, 77; vis prof, Heriot-Watt Univ, Edinburgh, Scotland, 78-79. *Honors & Awards:* Sr Paper Award, Inst Elec & Electonics Engrs, 86. *Mem:* Inst Elec & Electronics Engrs. *Res:* Image processing; restoration, enhancement and pattern recognition. *Mailing Add:* ECE Dept NC State Univ Raleigh NC 27695-7911

TRUSSELL, PAUL CHANDOS, b Vancouver, BC, July 4, 16; m 43; c 2. BACTERIOLOGY. *Educ:* Univ BC, BSA, 38; Univ Wis, MS, 42, PhD(agr bact), 43. *Prof Exp:* Chief res microbiologist, Ayerst, McKenna & Harrison, Ltd, Que, 44-47; head div appl biol, 47-61, dir, BC Res Coun, 61-80; CONSULT, 80- *Mem:* Sigma Xi. *Res:* Agricultural bacteriology; industrial fermentations; antibiotics; marine borer control; industrial coatings; food spoilage; water pollution; bacteriological leaching of ores; forestry; research administration. *Mailing Add:* 15270 17th Ave 202 White Rock BC V4A 1T9 Can

TRUST, RONALD I, b Philadelphia, Pa, May 23, 47; m 69; c 2. CLINICAL STUDY MONITORING, QUALITY ASSURANCE. *Educ:* Drexel Univ, BS, 69; Calif Inst Technol, PhD(org chem), 74; Fairleigh Dickenson, MBA, 86. *Prof Exp:* Res & develop chemist, Ciba Geigy Corp, 73-74; res chemist, Am Cyanamid Co, 74-77, sr res chemist, 77-80, mgr overseas clin mat, 80-84, clin trials mat coordr, 84-86, clin trials coordr, 86-88, mgr, Clin Res Assocs, 88-90, ASST DIR, GLOBAL COORD ONCOL, AM CYANAMID CO, 90- *Mem:* Am Chem Soc; Assocs Clin Pharmacol; Drug Info Asn; AAAS; Int Photodynamic Asn. *Res:* Taking chemical compounds from laboratory curiosity to medical reality; involved in laboratory testing, manufacturing and human testing of chemical compounds; photodynamic therapy; global coordination; oncology. *Mailing Add:* Am Cyanamid Co Pearl River NY 10965

TRUST, TREVOR JOHN, b Melbourne, Australia, June 24, 42. MICROBIOLOGY. *Educ:* Univ Melbourne, BSc, 64, MSc, 66, PhD(microbiol), 69. *Prof Exp:* Lectr microbiol, Royal Melbourne Inst Technol, 69; from asst prof to assoc prof, 69-80, PROF MICROBIOL, UNIV VICTORIA, BC, 81- *Concurrent Pos:* Vis res fel, Southampton Fac Med, 77-78. *Mem:* Can Soc Microbiol; Am Soc Microbiol; Brit Soc Gen Microbiol. *Res:* Fish diseases; molecular basis for bacterial virulence and immunogenicity; campylobacter, salmonella and enteric pathogens. *Mailing Add:* Dept Biochem & Microbiol Univ Victoria Box 1700 Victoria BC V8W 2Y2 Can

TRUSZKOWSKA, KRYSTYNA, b Stare Rochowice, Poland. SURFACE SCIENCE PHYSICS. *Educ:* Univ Wroclaw, Poland, MS, 70; Tech Univ Wroclaw, PhD(solid state physics), 77. *Prof Exp:* Asst prof physics, Inst Physics, Tech Univ Wroclaw, Poland, 77-80 & Inst Physics, Univ Nat Autonoma, Mex, 81-83; res assoc, Dept Physics & Lab Surface Studies, Univ Wis-Milwaukee, 83-86; SR ANALYSIS ENGR, G E MED SYST CORP, MILWAUKEE, 86- *Concurrent Pos:* Lectr, dept physics, Univ Wis-Milwaukee, 84-85. *Mem:* Am Phys Soc; Am Vacuum Soc. *Res:* New materials for optical interference coatings; structural characterization of thin metal films by electron microscopy; absorption and reactions of gases at solid surfaces; optics of thin solid films; mechanical engineering; physics engineering. *Mailing Add:* Med Sys Group GE Corp 4511 N Newhall Milwaukee WI 53201

TRUXAL, FRED STONE, b Great Bend, Kans, Feb 20, 22; m 43; c 2. SCIENCE ADMINISTRATION. *Educ:* Univ Kans, AB, 47, MA, 49, PhD(entom), 52. *Prof Exp:* Agent, Bur Entom & Plant Quarantine, USDA, 42; asst instr biol & entom, Univ Kans, 47-52; cur entom, 52-61, chief cur, 61-82, EMER CHIEF CUR, LIFE SCI DIV, LOS ANGELES COUNTY MUS NATURAL HIST, 82- *Concurrent Pos:* Asst to state entomologist, Kans, 48-51; asst prof, Ottawa Univ, Kans, 51-52; adj prof, Univ Southern Calif, 57-82; biol consult, Pac Horizons; mem expeds, Mex, Cent Am, Brazil, Peru, Australia & Africa. *Mem:* Fel AAAS; Entom Soc Am; Soc Syst Zool; Sigma Xi. *Res:* Biology, ecology and taxonomy of aquatic Hemiptera. *Mailing Add:* Life Sci Div 900 Expos Blvd Los Angeles Co Mus Natural Hist Los Angeles CA 90007

TRUXAL, JOHN G(ROFF), b Lancaster, Pa, Feb 19, 24; m 49; c 2. ELECTRICAL ENGINEERING, COMMUNICATIONS. *Educ:* Dartmouth Col, AB, 44; Mass Inst Technol, BS, 47, DSc, 50. *Hon Degrees:* DE, Purdue Univ, 64 & Ind Inst Technol, 71. *Prof Exp:* Teaching & res asst, Mass Inst Technol, 46-50; from asst prof to assoc prof, Purdue Univ, 50-54; assoc prof, Polytech Inst Brooklyn, 54-57, prof & head dept, 57-61, vpres ed develop, 61-64, dean eng, 64-66, provost, 66-71; dean eng & appl sci, State Univ NY, Stony Brook, 72-76, prof, 76-77, distinguished teaching prof, 77-91, DISTINGUISHED TEACHING EMER PROF, STATE UNIV NY, STONY BROOK, 91- *Concurrent Pos:* Prin investr res grants, NSF, Alfred P Sloan Found & other govt & pvt agencies, 62-; co-dir, Eng Concepts Curric Proj, 65-74; mem, President's Sci Adv Comt, 70-72; mem, Vis Comt, Nat Bur Standards, 72-76, var comts, Nat Acad Eng & Nat Res Coun & World Bk Phys Sci Comt, 83-90; dir, Nat Coord Ctr Curric Develop, State Univ NY, Stony Brook, 76-85 & Stony Brook Ctr New Liberal Arts Prog & prin investr, State Univ NY Res Found, 91-; chmn, Comt Pub Understanding Sci, AAAS, 79-82; mem, Coun Understanding Technol Human Affairs, chmn, 82-84. *Honors & Awards:* Lanchester Award, Opers Res Soc Am; Westinghouse Award, Am Soc Eng Educ. *Mem:* Nat Acad Eng; fel AAAS; Am Soc Eng Educ; fel Inst Elec & Electronics Engrs; fel Instrument Soc Am (pres, 65-66); Sigma Xi. *Res:* Network theory; feedback control systems; author of various publications. *Mailing Add:* Col Eng State Univ NY Stony Brook NY 11794-2250

TRUXILLO, STANTON GEORGE, b New Orleans, La, June 23, 41; m 65; c 2. GEOPHYSICS. *Educ:* Loyola Univ, La, BS, 63; La State Univ, Baton Rouge, PhD(physics), 69. *Prof Exp:* Fel, Coastal Studies Inst, La State Univ, Baton Rouge, 68-70; asst prof physics, Univ Tampa, 70-73, assoc prof, 73-79, prof, 79-81; PETROL GEOPHYSICIST, AMOCO PROD CO, 81- *Concurrent Pos:* Am Coun Educ Fel, 77-78. *Mem:* Am Phys Soc; Soc Explor Geophysicists; Southeastern Geophys Soc; Sigma Xi. *Res:* Time-dependent rotating fluid dynamics; fluid dynamics of circulatory system; acoustics. *Mailing Add:* 18219 Farnsfield Dr Houston TX 77084-2364

TRYFIATES, GEORGE P, b Mesolongi, Greece, Feb 26, 35; US citizen; m 59; c 4. BIOCHEMISTRY, CANCER. *Educ:* Univ Toledo, BS, 58; Bowling Green State Univ, MA, 59; Rutgers Univ, PhD(biochem), 63. *Prof Exp:* Teaching & res asst biol, Bowling Green State Univ, 58-59; res asst biochem, Rutgers Univ, 59-62; res assoc, Grad Sch Med, Univ Pa & Sch Med, Temple Univ, 62-64; res biochemist, P Lorillard Co, NC, 64-66; instr pharmacol, Sch Med, Duke Univ, 66-67, assoc, 67; from asst prof to assoc prof, 67-83, PROF BIOCHEM, SCH MED, WVA UNIV, 83- *Concurrent Pos:* USPHS trainee, Grad Sch Med, Univ Pa & Sch Med, Temple Univ, 62-64, USPHS grants, Nat Cancer Inst; ed, J Nutr Growth & Cancer. *Mem:* AAAS; Am Chem Soc; Soc Exp Biol & Med; Am Inst Nutrit; Am Soc Biochem & Molecular Biol; Am Asn Cancer Res. *Res:* Enzyme regulation in vivo and in vitro; novel tumor products and synthesis; nutritional oncology; tumor markers; cancer detection. *Mailing Add:* Dept Biochem WVa Univ Sch Med Morgantown WV 26506

TRYON, EDWARD POLK, b Terre Haute, Ind, Sept 4, 40. HIGH ENERGY PHYSICS, COSMOLOGY. *Educ:* Cornell Univ, AB, 62; Univ Calif, Berkeley, PhD(physics), 67. *Prof Exp:* Res assoc physics, Columbia Univ, 67-68, asst prof, 68-71; from asst prof to assoc prof, 71-79, PROF PHYSICS, HUNTER COL, CITY UNIV NEW YORK, 79- *Concurrent Pos:* Vis mem, Inst Advan Study, Princeton, 77-78; Sigma Xi Nat Lectr, 82-84. *Mem:* Am Phys Soc; NY Acad Sci; Sigma Xi. *Res:* Gravitational interactions; high energy theory; pion-pion interaction; origin of universe. *Mailing Add:* Dept Physics Hunter Col Hunter Col City Univ NY 695 Park Ave New York NY 10021

TRYON, JOHN G(RIGGS), b Washington, DC, Dec 18, 20; m 48; c 2. ENGINEERING PHYSICS. *Educ:* Univ Minn, BS, 41; Cornell Univ, PhD(eng physics), 52. *Prof Exp:* Mem tech staff, Bell Tel Labs, Inc, 51-58; prof elec eng & head dept, Univ Alaska, 58-69; prof, Tuskegee Inst, 69-75; chmn dept, 80-84, prof, 75-86, EMER PROF ELEC ENG, UNIV NEV, LAS VEGAS, 86- *Mem:* Am Phys Soc; Inst Elec & Electronics Engrs; Am Solar Energy Soc; Sigma Xi; Am Soc Eng Educ. *Mailing Add:* 631 Ave I Boulder City NV 89005

TRYON, ROLLA MILTON, JR, b Chicago, Ill, Aug 26, 16; m 45. BOTANY. *Educ:* Univ Chicago, BS, 37; Univ Wis, PhM, 38; Harvard Univ, MS, 40, PhD(bot), 41. *Prof Exp:* Lab technician, Chem Warfare Serv, Mass Inst Technol, 42; instr bot, Dartmouth Col, 42, lab technician, 43-44; instr bot, Univ Wis, 44-45; asst prof plant taxon, Univ Minn, 45-48; assoc prof, Wash Univ, St Louis, 48-57; cur, herbarium & cur ferns, Gray Herbarium Harvard Univ, 58-89, prof biol, 72-89; CUR, HERBARIUM UNIV SFLA, 89- *Concurrent Pos:* Cur, Herbarium, Minneapolis, Minn, 46-48; asst cur herbarium, Mo Bot Garden, 48-57. *Mem:* Am Soc Plant Taxon; Am Fern Soc; Bot Soc Am. *Res:* Taxonomy of pteridophytes; Doryopteris; Pteridium; ferns and fern allies of Wisconsin, Minnesota and Peru. *Mailing Add:* Herbarium Univ SFla Tampa FL 33620-5150

TRYON, WARREN W, b 1944; m 70; c 1. PSYCHOLOGY. *Educ:* Ohio Northern Univ, BA, 66; Kent State Univ, MA, 69, PhD, 70; Am Bd Prof Psychol, dipl. *Prof Exp:* From asst prof to assoc prof, 70-83, PROF PSYCHOL, FORDHAM UNIV, 83- *Mem:* Am Psychol Asn; Asn Advan Behav Ther. *Res:* Measurement of human activity under ambulatory conditions; therapy based on operant and respondent conditioning principles; tremor and other dyskinesias. *Mailing Add:* Dept Psychology Fordham Univ Bronx NY 10458-5198

TRYPHONAS, HELEN, b Greece, June 21, 39; m 62; c 2. IMMUNOTOXICOLOGY, FOOD ALLERGY. *Educ:* Univ Sask, BSc, 68, MSc, 72. *Prof Exp:* Dept asst microbiol, West Col Vet Med, Univ Sask, 72-74; supvr microbiol, 74-77, biologist II immunol, 77-82, SCIENTIST II IMMUNOTOXICOL, HEALTH & WELFARE CAN, 82- *Res:* Immunotoxicity studies of food additives and environmental contaminants in man and experimental animals; the casual relationship of food allergies to behavioral abnormalities in children. *Mailing Add:* Toxicol Res Div Bur Food Chem Health Protection Br Health & Welfare Tunney's Pasture Ottawa ON K1A 0L2 Can

TRYTTEN, GEORGE NORMAN, b Pittsburgh, Pa, Apr 21, 28; m 52; c 5. MATHEMATICS. *Educ:* Luther Col, Iowa, AB, 51; Univ Wis, MS, 53; Univ Md, PhD(math), 62. *Prof Exp:* Asst math, Univ Wis, 51-53; asst, Inst Fluid Dynamics & Appl Math, Univ Md, 53-57; mathematician, US Naval Ord Lab, 57-62; res assoc math, Inst Fluid Dynamics & Appl Math, Univ Md, College Park, 62-63, from res asst prof to res assoc prof, 63-69, assoc dean sponsored res & fels, Grad Sch, 67-69; vpres, Math Sci Group, Inc, 69-70; free-lance filmstrip producer, 70-71; prof math & chmn dept, Hood Col, 71-72; assoc prof, 72-75, PROF MATH, LUTHER COL, IOWA, 75- *Mem:* AAAS; Am Math Soc; Math Asn Am. *Res:* Partial differential equations; fluid dynamics; numerical solution of partial differential equations; celestial mechanics. *Mailing Add:* Dept Math Luther Col 700 College Dr Decorah IA 52101

TRYTTEN, ROLAND AAKER, b Tower City, NDak, Oct 15, 13; m 42; c 6. CHEMISTRY. *Educ:* St Olaf Col, AB, 35; Univ Wis, PhD(chem), 41. *Prof Exp:* Control chemist, Kimberly-Clark Corp, 41-42; instr chem, Ripon Col, 42-45; instr, Cent State Teachers Col, 45-51, chmn dept, 49-51; chmn dept, Univ Wis-Stevens Point, 51-72, prof chem, 51-83; RETIRED. *Mem:* Am Chem Soc. *Res:* Foams; foaming tendency of aqueous aliphatic alcohol solutions; sulfur dioxide determination in polluted air. *Mailing Add:* 2809 Algoma St Stevens Point WI 54481

TRZASKOMA, PATRICIA POVILITIS, b New York, NY, 40; m 61; c 2. ELECTROCHEMISTRY, PHYSICAL CHEMISTRY. *Educ:* Barnard Col, BA, 61; Am Univ, MS, 72, PhD(chem), 76. *Prof Exp:* Anal chemist, Gen Dynamics, 61-62; res chemist fuel cell res, Apollo Prog, Pratt & Whitney Aircraft Corp, 62-65; asst prof chem, George Mason Univ, 77-85; RES CHEMIST, NAVAL RES LAB, WASHINGTON, DC, 79- *Mem:* Electrochem Soc; Am Chem Soc. *Res:* Electrochemistry with emphasis on kinetics of electrode surface reactions; passivation and inhibition; fast reaction kinetics; corrosion behavior of aluminum; ion implanted materials; metal matrix composites. *Mailing Add:* Code 6322 Naval Res Lab Washington DC 20375

TRZCIENSKI, WALTER EDWARD, JR, b Montague City, Mass, Sept 19, 42; m 65; c 1. GEOLOGY, MATH. *Educ:* Bowdoin Col, AB, 65; McGill Univ, PhD(geol), 71. *Prof Exp:* NASA fel, Princeton Univ, 71; asst prof geol, Brooklyn Col, 71-72; from asst prof to assoc prof, Ecole Polytech, Montreal, 72-81; PROF GEOL, UNIV MONTREAL, 81- *Mem:* Am Math Soc; Geol Soc Am; Mineral Soc Am; Mineral Asn Can; Mineral Soc; Am Geophys Union; Soc Indust & Appl Math. *Res:* Metamorphic and igneous petrology and mineralogy-geochemistry, especially in the northern Appalachians. *Mailing Add:* 2001 Breezeway Ave Dorval PQ H9S 1C7 Can

TSAGARIS, THEOFILOS JOHN, b Fernandina, Fla, June 27, 29; m 54; c 3. INTERNAL MEDICINE, CARDIOLOGY. *Educ:* Univ Fla, BS, 50; Emory Univ, MD, 54. *Prof Exp:* Chief cardiol, Vet Admin Hosp, Wood, Wis, 62-65; from asst prof to assoc prof, 65-75, PROF INTERNAL MED, UNIV UTAH, 75-, RES ASST PROF PHYSIOL, 76-; CHIEF CARDIOL, VET ADMIN HOSP, SALT LAKE CITY, 65- *Concurrent Pos:* Fel cardiol, Emory Univ, 59-60 & Univ Utah, 60-62; asst prof, Marquette Univ, 62-65. *Mem:* Am Fedn Clin Res. *Res:* Hemodynamics; coronary blood flow. *Mailing Add:* Dept Cardiol/Med Univ Utah BA Med Ctr Heart Sta Salt Lake City UT 84148

TSAHALIS, DEMOSTHENES THEODOROS, fluid mechanics, fluid-structure interaction, for more information see previous edition

TSAI, ALAN CHUNG-HONG, b Chang-hua Hsien, Taiwan, June 18, 43; m 69; c 2. NUTRITION. *Educ:* Taiwan Prov Chung Hsing Univ, BS, 66; Wash State Univ, MS, 69, PhD(nutrit), 72. *Prof Exp:* Res assoc nutrit, Mich State Univ, 72-73; asst prof, 73-76, ASSOC PROF NUTRIT, UNIV MICH, ANN ARBOR, 76- *Mem:* Am Inst Nutrit; NY Acad Sci. *Res:* Cholesterol feeding associated metabolic alterations including enzyme activities; microsomal activities; tissue lipid peroxidation and insulin metabolism; nutrition of dietary fiber, its effect on cholesterol metabolism and gastrointestinal functions; interaction of Vitamin E and the function of thyroid hormones; metabolic effects of exercise training and detraining. *Mailing Add:* Human Nutrit Prog Univ Mich Ann Arbor MI 48109

TSAI, BILIN PAULA, b Seattle, Wash, May 23, 49. CHEMICAL PHYSICS. *Educ:* Univ Chicago, BS, 71; Univ NC, Chapel Hill, PhD(chem physics), 75. *Prof Exp:* Res assoc chem physics, Univ Nebr, Lincoln, 75-76; asst prof, 76-82, ASSOC PROF CHEM, UNIV MINN, DULUTH, 82-, ASSOC DEAN, COL SCI & ENG, 83- *Mem:* Sigma Xi; Am Chem Soc; Am Soc Mass Spectrometry; Am Phys Soc. *Res:* Physical chemistry; gas phase ion fluorescence spectroscopy; photoionization; chemical dynamics. *Mailing Add:* Dept Chem Univ Minn Duluth MN 55812

TSAI, BOH CHANG, b Taiwan, China, Jan 2, 44; US citizen; m 69; c 2. POLYMER SCIENCE, CHEMICAL ENGINEERING. *Educ:* Nat Taiwan Univ, BS, 67; Univ Akron, MS, 70, PhD(polymer sci), 73. *Prof Exp:* Staff mat engr polymer processing, Acushnet Co, 73-78; SR SUPVR MATS SCI, AM CAN CO, 82- *Concurrent Pos:* Fel, Inst Polymer Sci, Univ Akron, 73-78; vis lectr polymer sci, Southeastern Mass Univ, 75-76. *Mem:* Am Chem Soc; Rheology Soc. *Res:* Interaction of structure-property-processing of polymeric materials. *Mailing Add:* 5502 Silent Brook Lane Rolling Meadows IL 60008

TSAI, CHESTER E, b Amoy, China, Mar 7, 35; m 62; c 2. ALGEBRA. *Educ:* Nat Taiwan Univ, BA, 57; Marquette Univ, MS, 61; Ill Inst Technol, PhD(math), 64. *Prof Exp:* Instr math, Ill Inst Technol, 62-64; from asst prof to assoc prof, 64-78, PROF MATH, MICH STATE UNIV, 78- *Concurrent Pos:* Consult, World Bank Proj. *Mem:* Am Math Soc; Math Asn Am. *Res:* Non-associative algebra. *Mailing Add:* Dept Math D216 Wells Hall Mich State Univ East Lansing MI 48824

TSAI, CHIA-YIN, b Taichung, Taiwan, Dec 15, 37; m 67; c 2. GENETICS, BIOCHEMISTRY. *Educ:* Nat Taiwan Univ, BS, 60; Purdue Univ, Lafayette, PhD(genetics), 67. *Prof Exp:* Res asst genetics, 63-67, res assoc, 67-69, asst prof, 69-74, assoc prof, 74-80, PROF GENETICS, PURDUE UNIV, WEST LAFAYETTE, 80- *Concurrent Pos:* Vis prof, Inst Bot, Acad Sinica, Taiwan, 76; NIH panel mem, 81-85. *Mem:* Am Soc Plant Physiologists; Crop Sci Soc Am. *Res:* Carbohydrate metabolism and storage protein synthesis in maize; nutritional quality of protein and grain yield potential of maize; ammonium assimilation in maize roots; maize hybrid response to nitrogen fertility. *Mailing Add:* Dept Bot & Plant Path Purdue Univ West Lafayette IN 47907

TSAI, CHING-LONG, b Kaohsiung, Taiwan, May 25, 45; m 73; c 3. SOLID STATE PHYSICS. *Educ:* La State Univ, PhD(physics), 76. *Prof Exp:* Res & teaching asst physics, La State Univ, 71-76; res assoc, Ill Inst Technol, 76-78; physicist, Nat Standards Co, 78-80; sr scientist, Northeastern Univ, 80-84; RES SPECIALIST, 3M, 84- *Concurrent Pos:* Vis scientist, Mass Inst Technol, 82-84. *Mem:* Am Phys Soc; Am Inst Metall Engrs. *Res:* Low temperature solid state physics; material preparations and characterizations; rapid quenching technology. *Mailing Add:* 8010 Somerset Road Woodbury MN 55125

TSAI, CHISHIUN S, b Chia-yi, Taiwan, Dec 19, 33; m 65. BIOCHEMISTRY, ENZYMOLOGY. *Educ:* Nat Taiwan Univ, BS, 56; Purdue Univ, MS, 61, PhD(biochem), 63. *Prof Exp:* Fel chem, Cornell Univ, 63; fel biosci, Nat Res Coun Can, 63-64, asst res officer, 64-65; from asst prof to assoc prof, 65-76, PROF CHEM & BIOCHEM, CARLETON UNIV, 76- *Concurrent Pos:* Res assoc, Univ Tex, 72; vis sci, Foxchase Cancer Res Inst, 73. *Mem:* Am Chem Soc; Can Biochem Soc; Am Soc Biol Chemists. *Res:* Function and reactivity of enzymes in relation with the structures of enzymes; substrates and inhibitors. *Mailing Add:* Dept Chem & Biochem Carleton Univ Colonel By Dr Ottawa ON K1S 5B6 Can

TSAI, CHUANG CHUANG, b Taipei, Taiwan; US citizen; c 2. SOLID STATE PHYSICS, SEMICONDUCTOR TECHNOLOGY. *Educ:* Nat Taiwan Univ, BS, 72; Univ Chicago, MS, 73, PhD(physics), 79. *Prof Exp:* MEM RES STAFF, XEROX CORP, PALO ALTO RES CTR, 78- *Mem:* Am Phys Soc; Mat Res Soc; Inst Elec & Electronics Engrs. *Res:* Material science and thin film technology. *Mailing Add:* Xerox Palo Alto Res Ctr 3333 Coyote Hill Rd Palo Alto CA 94304

TSAI, CHUN-CHE, b Chiayi, Taiwan, China, Sept 17, 37; US citizen; m 63; c 3. PHYSICAL CHEMISTRY, BIOCHEMISTRY. *Educ:* Cheng-Kung Univ, Taiwan, BS, 60; Ind Univ, Bloomington, PhD(chem), 68. *Prof Exp:* Fel crystallog, Univ Pa, 68-69; fel bio-crystallog, Cornell Univ, 69-71; res chemist phys chem, Univ Colo, Boulder, 71-72; sr res assoc biophys chem & molecular biol, Univ Rochester, 72-76; asst prof chem, 76-80, ASSOC PROF CHEM, KENT STATE UNIV, 80- *Concurrent Pos:* Res fac mem, Dept Microbiol & Immunol, Col Med, Northeastern Ohio Univ, 78- *Mem:* Am Chem Soc; Am Biophys Soc; Am Crystallog Asn; AAAS; Sigma Xi; Int Soc Antiviral Res. *Res:* Biological crystallography; drug-nucleic acid interactions; structure and function of nucleic acids; molecular associations and interactions in biological systems; quantitative structure-activity relationships. *Mailing Add:* Dept Chem Kent State Univ Kent OH 44242

TSAI, CHUNG-CHIEH, b Taiwan, Oct 1, 48; c 3. RESIN TRANSFER MOLDING. *Educ:* Nat Cheng Kung Univ, Taiwan, BS, 71; Univ Lowell, MS, 77, PhD(chem), 80. *Prof Exp:* Lab instr chem, Nat Cheng Kung Univ, 73-74; res asst polymer, Univ Lowell, 75-79; res chemist polymer, Stauffer Chem Co, 79-84; PROD DEVELOP SPECIALIST, ASHLAND OIL CO, 84- *Mem:* Am Chem Soc; Soc Plastic Engrs. *Res:* Polyelectrolytes; unsaturated polyesters; epoxy adhesives; polyurethane; thermosets; synthesis of polymers for electronic application; reactive polymers. *Mailing Add:* PO Box 45 Dublin OH 43017

TSAI, FRANK Y, b China; US citizen; m 64; c 2. TECHNICAL MANAGEMENT, SCIENCE ADMINISTRATION. *Educ:* Univ Minn, MS, 60, PhD(eng). *Prof Exp:* Res fel, St Anthony Falls Hydraul Lab, 60-68; asst prof, Iowa State Univ, 68-72; sect chief, Ebasco Serv Inc, 72-75; sr engr, HUD, 75-79, SR TECH ADVISOR, FED EMERGENCY MGT AGENCY, FED INS ADMIN, 79- *Mem:* Sigma Xi. *Res:* Fluid mechanics; hydromechanics; ocean and coastal engineering. *Mailing Add:* 522 S Larrimore St Arlington VA 22204

TSAI, JAMES HSI-CHO, b Fuzhou, Fujian, China, June 10, 34; US citizen; m 60; c 2. MYCOPLASMOLOGY, PLANT VIROLOGY. *Educ:* Nat Chung Hsing Univ, Taiwan, BS; Mich State Univ, MS, PhD(entom). *Prof Exp:* Scientist entom, Int Inst Trop Agr, Nigeria, 69-70; res assoc entom, Mich State Univ, 70-72; from asst prof to assoc prof, 73-84, PROF ENTOM, UNIV FLA, 84- *Concurrent Pos:* Vis prof, Shaanxi Acad Agr Sci, China, 81 & Chinese Acad, Beijing, 84. *Mem:* Entom Soc Am; Am Phytopath Soc; Int Orgn Citrus Virologists; Int Orgn Mycoplasmology. *Res:* Plant prokaryotic diseases; vector-pathogen-host relationships; ecology and control of vector borne diseases. *Mailing Add:* Ft Lauderdale Res & Educ Ctr 3205 College Ave Ft Lauderdale FL 33314

TSAI, JIR SHIONG, INTERNAL MEDICINE, ENDOCRINOLOGY. *Educ:* Nat Taiwan Univ, MD, 65. *Prof Exp:* ASSOC PROF CLIN MED, DEPT MED, NY UNIV MED CTR, 75- *Mailing Add:* Dept Med NY Med Ctr 550 First Ave New York NY 10016

TSAI, KUEI-WU, b Taiwan, Jan 22, 41; m 69; c 2. SOILS, FOUNDATION ENGINEERING. *Educ:* Nat Taiwan Univ, BSCE, 62; Princeton Univ, MSCE & MA, 65, PhD(soils eng), 67. *Prof Exp:* From asst prof to assoc prof, 67-76, PROF CIVIL ENG, SAN JOSE STATE UNIV, 76-, CHMN DEPT, 81- *Concurrent Pos:* Proj engr, Dames & Moore, 69-74, sr engr, 74-75, consult, 75- *Mem:* Fel Am Soc Civil Engrs. *Res:* Shear strength of clays; slope stability; settlement; land reclamation and development; penetrometer method; difficult foundation problems; design of dams; soils investigation and foundation recommendation; site improvement. *Mailing Add:* Dept Civil Eng San Jose State Univ Washington Sq San Jose CA 95192

TSAI, LIN, b Hong Kong, May 30, 22; US citizen. ORGANIC CHEMISTRY. *Educ:* Chinese Nat Southwest Assoc Univ, BSc, 46; Univ Ore, MA, 49; Fla State Univ, PhD(org chem), 54. *Prof Exp:* Res assoc chem, Ohio State Univ, 54-57; res scientist, Worcester Found Exp Biol, 57-59; vis scientist, 59-62, ORG CHEMIST, NAT HEART, LUNG & BLOOD INST, 62- *Mem:* Am Chem Soc; Royal Soc Chem; Am Soc for Biochem & Molecular Biol. *Res:* Syntheses, reactions and microbial degradations of heterocyclic compounds; stereochemistry of enzymic reactions. *Mailing Add:* Nat Heart Lung & Blood Inst Rm 110 Bethesda MD 20892

TSAI, LUNG-WEN, b Taipei, Taiwan, Feb 20, 45; US citizen; m 71; c 2. ROBOT MANIPULATORS & MACHINE DESIGN. *Educ:* Nat Taiwan Univ, BS, 67; State Univ NY, MS, 70; Stanford Univ, PhD(mech eng), 73. *Prof Exp:* Develop engr, Hewlett-Packard Corp, 73-78; sr staff res engr, Power Syst Res Dept, Gen Motors Corp, 78-86; assoc prof, 86-90, PROF MECH ENG, UNIV MD, 90- *Honors & Awards:* Melville Medal, Am Soc Mech Engrs, 85, Mechanisms Comt Award, 84; Campbell Award, Gen Motors Res Labs, 86; A T Colwell Merit Award, Soc Automotive Engrs, 88; Award of Merit, Proctor & Gamble, 89. *Mem:* Mem Am Soc Mech Engrs. *Mailing Add:* Mech Engr Dept Univ Md College Park MD 20742

TSAI, MING-DAW, b Taiwan, Repub China, Sept 1, 50; US citizen; m 76; c 3. BIOCHEMISTRY, ORGANIC CHEMISTRY. *Educ:* Nat Taiwan Univ, BS, 72, Purdue Univ, PhD(biochem), 78. *Prof Exp:* Postdoctoral assoc, Purdue Univ, 78-79, vis asst prof, 79-80; asst prof chem, Rutgers Univ, 80-81; from asst prof to assoc prof, 86-90, PROF CHEM, OHIO STATE UNIV, 90- *Concurrent Pos:* Alfred P Sloan fel, 83-85; Camille & Henry Dreyfus Teacher Scholar, Dreyfus Found, 85-90; mem, Phys Biochem Study Sect, NIH, 88-92. *Mem:* Am Chem Soc; Am Soc Biochem & Molecular Biol; AAAS; Sigma Xi. *Res:* Chemical basis of enzyme catalysis; structure-function relationship of proteins; mechanism of adenylate kinase and phospholipase A2; synthesis of phosphatidylinositides. *Mailing Add:* Dept Chem 100 New Chem Bldg 120 W 18th Ave Columbus OH 43210-1173

TSAI, MING-JER, b Taichung, Taiwan, Nov 3, 43; m 71; c 2. BIOCHEMISTRY, MOLECULAR BIOLOGY. *Educ:* Nat Taiwan Univ, BS, 66; Univ Calif, Davis, PhD(biochem), 71. *Prof Exp:* Damon Runyon fel, Univ Tex M D Anderson Hosp & Tumor Inst Houston, 71-73; from instr to assoc prof, 73-87, PROF CELL BIOL, BAYLOR COL MED, 88- *Mem:* Am Soc Cell Biol; Sigma Xi; Am Soc Microbiol; Endocrinol Soc. *Res:* Hormonal regulation of gene expression; chromatin structure; precursors of mRNA and their processing; initiation of RNA synthesis by DNA-dependent RNA polymerases; development and differentiation. *Mailing Add:* Dept Cell Biol Baylor Col Med One Baylor Plaza Houston TX 77030

TSAI, MING-JONG, b Taiwan, Aug 25, 50; US citizen; c 1. OPTOELECTRONICS, SEMICONDUCTORS. *Educ:* Nat Tsing Hwa Univ, Taiwan, BS, 72; Stanford Univ, MS, 77, PhD(mat sci & eng), 79. *Prof Exp:* Engr res & develop, Am Microsyst Inc, 79-80; engr & supvr res & develop, Hewlett-Packard, 80-84; dir eng, Compound Semiconductor, Inc, 84-86, vpres eng, 86-88; VPRES, RES & DEVELOP DIV, LITE-ON INC, 88- *Mem:* Inst Elec & Electronics Engrs. *Res:* Compound semiconductor materials and devices for the optoelectronics applications. *Mailing Add:* Lite-On Inc 720 S Hillview Dr Milpitas CA 95035

TSAI, STEPHEN W, b Beijing, China, July 6, 29; US citizen; m 54; c 2. MECHANICS. *Educ:* Yale Univ, BE, 52, DEng(mech), 61. *Prof Exp:* Proj engr, Foster Wheeler Corp, 52-58; dept mgr mat res, Aeronutronic Div, Philco Corp, 61-66; prof eng, Washington Univ, 66-68; chief scientist, 68-76, SCIENTIST, AIR FORCE MAT LAB, WRIGHT-PATTERSON AFB, 76- *Concurrent Pos:* Lectr, Univ Calif, Los Angeles, 65-66; ed-in-chief, J Composite Mat, 66-; ed, Int J Fibre Sci & Technol, 68-; affil prof, Washington Univ, 68-; Battelle vis prof, Ohio State Univ, 69. *Mem:* Am Inst Aeronaut & Astronaut; Sigma Xi; Am Phys Soc; Soc Rheol. *Res:* Mechanics of composite materials for structural applications. *Mailing Add:* Dept Aeronaut & Astronaut Durand Bldg Rm 250 Stanford CA 94305-4035

TSAI, TSUI HSIEN, b Taichung, Formosa, Nov 19, 35; m 63; c 2. PHARMACOLOGY. *Educ:* Nat Taiwan Univ, BS, 58; WVa Univ, MS, 63, PhD(pharmacol), 65. *Prof Exp:* Sect head autonomic pharmacol, Merrell-Nat Labs, Ohio, 65-71; sr res pharmacologist, Wellcome Res Labs, 71-77; asst dir res support & biomet, Merrell Nat Labs, 77-80; assoc dir clin invest, Merrel Dow Pharmaceut, Inc, 80-87, dir, med & lab liasion, 87-89, dir, Hirakata Ctr, Merrell Dow Res Inst, 89-91, DIR CLIN RES, MARION MERRELL DOW, TOKYO, 91- *Concurrent Pos:* Fel, Harvard Med Sch, 64-65. *Mem:* Am Soc Pharmacol & Exp Therapeut; Am Soc Clin Pharmacol & Therapeut; Am Col Allergy & Immunol. *Res:* Autonomic and cardiovascular pharmacology. *Mailing Add:* Merrell Dow Pharmaceut Inc Global Med 2110 E Galraith Rd Cincinnati OH 45215

TSAI, Y(U)-M(IN), b Taiwan, Formosa, Mar 31, 37; m 63; c 1. ENGINEERING MECHANICS. *Educ:* Taipei Inst Technol, Taiwan, dipl civil eng, 57; Univ Tenn, ScM, 62; Brown Univ, ScM, 64, PhD(eng), 67. *Prof Exp:* Res assoc eng, Brown Univ, 66-67; from asst prof to assoc prof eng mech, 67-77, PROF ENG MECH, IOWA STATE UNIV, 77- *Concurrent Pos:* Nat Sci Found res initiation grants, 69-70. *Mem:* Sigma Xi. *Res:* Elasticity; stress waves; fracture mechanics. *Mailing Add:* Dept Eng Mech Iowa State Univ Ames IA 50011

TSAI, YUNG SU, b Yuli, Taiwan, Feb 1, 30, US citizen; m 61; c 2. THEORETICAL PHYSICS, ELEMENTARY PARTICLE PHYSICS. *Educ:* Nat Taiwan Univ, BS, 54; Univ Minn, MS, 56, PhD(physics), 58. *Prof Exp:* Res assoc theoret physics, Stanford Univ, 59-61, asst prof, 61-63, SR STAFF MEM THEORET PHYSICS, STANFORD LINEAR ACCELERATOR CTR, STANFORD UNIV, 63- *Mem:* Am Phys Soc; Sigma Xi. *Res:* Passage of particles through matter; production of particles; energy loss and straggling due to ionization; bremsstrahlung and pair productions; radiative corrections to scatterings; properties of tay leptons; physics of electron-positron collision. *Mailing Add:* Stanford Univ PO Box 4349 Stanford Univ Stanford CA 94309

TSAKONAS, STAVROS, b Resht, Iran, July 11, 20; nat US; m 54; c 1. ENGINEERING MECHANICS. *Educ:* Nat Univ Greece, BS, 44; Columbia Univ, MS, 52, PhD(eng, eng mech), 56. *Prof Exp:* Design engr, Defense Dept, Greece, 45-50; asst hydraul, Sch Eng, Columbia Univ, 52-56; lectr math & staff scientist, Davidson Lab, 56-60, chief fluid dynamics sect, 60-72, assoc prof appl math, 60-77, RES ASSOC PROF OCEAN ENG, STEVENS INST TECHNOL, 77- *Mem:* Am Soc Civil Engrs; Sigma Xi. *Res:* Applied mechanics; fluid dynamics; applied mathematics. *Mailing Add:* Davidson Lab Stevens Inst Technol Castle Point Hoboken NJ 07030

TSALIOVICH, ANATOLY, b USSR, 1936; US citizen. ELECTROMAGNETIC COMPATIBILITY, ELECTRONIC CABLES. *Educ:* Odessa Elec Telecommun Inst, USSR, BSEE & MSEE, 60; Leningrad Elec Telecommun Inst, USSR, PhD(transmission lines & electromagnetic shielding), 66. *Prof Exp:* Sr scientist & group mgr, Cent Res Telecommun Inst, USSR, 66-79; prod develop engr, Belden Corp, 80-84; fel engr, Thomas & Betts Corp, 84-85; DISTINGUISHED MEM TECH STAFF, AT&T BELL LABS, 85- *Concurrent Pos:* Assoc prof, Leningrad Elec Telecommun Inst, USSR, 63-79; tech translator, Scripta Technika, US & Kabelnaya Technika, USSR, 64-89; lectr & consult, Ctr Prof Advan, US-Neth, 85- *Honors & Awards:* Richard R Stoddard Award, Inst Elec & Electronics Engrs, Electromagnetic Compatibility Soc, 89. *Mem:* Inst Elec & Electronics Engrs. *Res:* Electromagnetic compatibility; electromagnetic shielding; electronic and telephone cables; telecommunications lines and carrier systems; operations research application to telecommunications; radio frequency absorbers. *Mailing Add:* 13 Branton Dr East Brunswick NJ 08816

TSAN, MIN-FU, b Taiwan, Jan 27, 42; m 75; c 2. HEMATOLOGY, NUCLEAR MEDICINE. *Educ:* Nat Taiwan Univ, MB, 67; Harvard Univ, PhD(physiol), 71; Am Bd Internal Med, dipl, 75, dipl hemat, 78; Am Bd Nuclear Med, dipl, 76. *Prof Exp:* Intern med, Nat Taiwan Univ Hosp, 66-67; med officer, Chinese Navy, 67-68; med intern, Boston Vet Admin Hosp, 71-72; med resident, 72-73, fel hematol, Johns Hopkins Hosp, 73-75; asst prof med, radiol & radiol sci, Med Sch & asst prof environ health, Sch Pub Health & Hyg, Johns Hopkins Univ, 75-79, assoc prof med, radiol & radiol sci, Med Sch & assoc prof environ health, Sch Pub Health & Hyg, 79-82; ASSOC CHIEF STAFF RES & DEVELOP, ALBANY VA MED CTR, 82-, PROF MED & PHYSIOL, ALBANY MED COL, 82- *Honors & Awards:* Sci Award, Chinese Am Med Soc, 90. *Mem:* AAAS; Am Fedn Clin Res; Am Soc Hematol; Am Soc Nuclear Med. *Res:* Oxygen radical-induced tissue injury. *Mailing Add:* Res Serv Dept Vet Affairs Med Ctr 113 Holland Ave Albany NY 12208

TSANDOULAS, GERASIMOS NICHOLAS, b Preveza, Greece, Aug 14, 39; m 64, 74. ELECTRICAL ENGINEERING. *Educ:* Harvard Univ, BA, 61, BS, 63; Univ Pa, PhD(elec eng), 67. *Prof Exp:* Engr, Kel Corp, Mass, 63-64; staff mem antenna res, Mass Inst Technol, 67-74, staff scientist, EM Tech Group, 74-77, syst engr, Airborne Radar Group, Lincoln Lab, 77-79; asst leader & leader, ARPA Lincoln C-Band Observable Radar, US Army, Krems & Kwajalein, Marshall Islands, 79-81; GROUP LEADER, SPACE RADAR TECHNOL GROUP, LINCOLN LAB, MASS INST TECHNOL, 82- *Mem:* AAAS; fel Inst Elec & Electronics Engrs; Sigma Xi. *Res:* Electromagnetic scattering and diffraction; antennas and arrays; wave propagation; radar systems; airborne moving target indicator radar; space based radar technology. *Mailing Add:* Lincoln Lab Mass Inst Technol Lexington MA 02173

TSANG, CHARLES PAK WAI, Can citizen; c 3. ENDOCRINOLOGY. *Educ:* McGill Univ, BSc, 61, MSc, 65, PhD(steroid biochem), 68. *Prof Exp:* Staff scientist, Worcester Found Exp Biol, 68-70; fel steroid & cyclic necleotide protein-binding assays, Queen Mary Hosp, Montreal, 70-71; RES SCIENTIST REPRODUCTIVE PHYSIOL, POULTRY ENDOCRINOL ANIMAL RES CTR, 71- *Concurrent Pos:* Res grant, Japan Soc Prom Sci, Nagoya Univ. *Mem:* Poultry Science Asn; Can Soc Animal Sci; World's Poultry Sci; fel Japan Soc Prom Sci. *Res:* Hormonal control of egg production and shell quality in the hen; estrogens, vitamin D and calcium metabolism; gene expression for calcium binding protein synthesis. *Mailing Add:* Animal Res Ctr Ottawa ON K1A 0C6 Can

TSANG, DEAN ZENSH, b Detroit, Mich, Aug 13, 52. OPTOELECTRONICS. *Educ:* Mass Inst Technol, SB, 74, ScD, 81; Univ Ill Urbana-Champaign, MS, 76. *Prof Exp:* STAFF MEM, LINCOLN LAB, MASS INST TECHNOL, 81- *Concurrent Pos:* Chmn, Cent New Eng, Inst Elec & Electronics Engrs Lasers & Electro-Optics Soc, 87-88, mem chapters comt, 88-, eng comt, 89-, chmn mem comt, 91-; bd electors, Sigma Xi. *Mem:* Sr mem Inst Elec & Electronics Engrs; Optical Soc Am. *Res:* Optical interconnections; semiconductor diode lasers, especially high speed dynamics; q-switched diode lasers; optoelectronic devices. *Mailing Add:* Mass Inst Technol Lincoln Lab Rm C-213 244 Wood St Lexington MA 02173

TSANG, GEE, b Macao, Mar 29, 38; Can citizen; m 70; c 2. FLUID MECHANICS, ICE ENGINEERING. *Educ:* Univ New South Wales, BE, 63, MEngSc, 65; Univ Waterloo, PhD(fluid mech), 68. *Prof Exp:* Asst engr, New South Wales Water Conserv & Irrig Comn, 63-64; tech officer, Water Res Lab, Univ New South Wales, 64-65; instr fluid mech, Univ Guelph, 68-69; res assoc air pollution, Mass Inst Technol, 69-70; asst prof fluid mech & dynamics, Univ Guelph, 70-72; res scientist, Can Ctr Inland Waters, 72-90; RES SCIENTIST, NAT HYDROL RES INST, 90- *Concurrent Pos:* Vis scholar, Univ Cambridge, 86-87. *Mem:* Int Asn Hydraul Res; NY Acad Sci. *Res:* Air and water pollution; atmospheric diffusion; hydraulics of cold regions; ice mechanics; micrometeorology; oil spill containment and recovery technology; ice and hydraulics; cold weather hydraulic instrument development; lake rehabilitation. *Mailing Add:* Nat Hydrol Res Inst 11 Innovation Blvd Saskatoon SK S7N 3H5 Can

TSANG, JAMES CHEN-HSIANG, b New York, NY, June 1, 46. SOLID STATE PHYSICS. *Educ:* Mass Inst Technol, BS & MS, 68, PhD(elec eng), 73. *Prof Exp:* RES STAFF MEM SEMICONDUCTOR PHYSICS, IBM RES CTR, IBM CORP, 73- *Mem:* Fel Am Phys Soc; fel Soc Photo-Optical Instrumentation Engrs. *Res:* Raman spectroscopy of solids and surfaces; optical spectroscopy of solids; time resolved spectroscopy. *Mailing Add:* IBM TJ Watson Res Ctr PO Box 218 Yorktown Heights NY 10598

TSANG, JOSEPH CHIAO-LIANG, b Hong Kong, Oct 11, 36; US citizen; m 69. MICROBIAL BIOCHEMISTRY. *Educ:* Grantham Teachers Col, Hong Kong, dipl, 58; Univ Okla, BS, 62, MS, 65, PhD(biochem), 68. *Prof Exp:* Res asst, Okla Med Res Found, 64-68; from asst prof to assoc prof, 68-78, PROF CHEM & BIOCHEM, ILL STATE UNIV, 78-; HEAD DEPT APPL BIOL & CHEM TECHNOL, HONG KONG POLYTECH, KOWLOON, 87- *Concurrent Pos:* Adj prof, Peoria Sch Med, Univ Ill, 73-; prof biochem, Med Sch, Jinan Univ, China, 80-; vis prof biochem, Univ Hong Kong, 83-84 & 85-86. *Mem:* Am Chem Soc; Roayl Soc Chem; Am Soc Microbiol; fel Am Inst Chemists. *Res:* Biochemical and pharmacological studies of Serratia marcescens as a bacterium causing nosocomial diseases: role of the pigment, prodigiosin, in the stability of the cell envelope and transfer of R-plasmids; interactions of antibiotics and surfactants with the outer membrane components such as lipopolysaccharides and phospholipids. *Mailing Add:* Dept Appl Biochem Tech Hong Kong Poly Hung Hom Kowloon Hong Kong

TSANG, KANG TOO, b Hong Kong. PLASMA PHYSICS. *Educ:* Chinese Univ Hong Kong, BSc, 70; State Univ NY Stony Brook, MA, 71; Princeton Univ, PhD(plasma physics), 74. *Prof Exp:* Res staff plasma physics, Oak Ridge Nat Lab, 74-81; SR RES PHYSICIST, SCI APPLICATIONS, INC, 81- *Concurrent Pos:* Adj prof, Nuclear Eng Dept, NC State Univ, Raleigh, 78. *Mem:* Am Phys Soc. *Res:* Theoretical investigation of equilibrium, stability and transports in thermonuclear plasma. *Mailing Add:* 8820 Charles Hawkins Way Annadale VA 22003

TSANG, LEUNG, b Hong Kong, China, July 5, 50; US citizen; m 78; c 1. ELECTROMAGNETICS, OPTICAL PHYSICS. *Educ:* Mass Inst Technol, BS, 71, MS, 73, PhD(elec eng), 76. *Prof Exp:* Res engr acoust, Schlumberger Doll Res Ctr, 76-78; res assoc elec eng, Mass Inst Technol, 78-80; prof elec eng, Tex A&M Univ, 80-83; PROF ELEC ENG, UNIV WASH, 83- *Concurrent Pos:* Assoc ed, Trans Geosci & Remote Sensing, Inst Elec & Electronics Engrs, 87- & Radio Sci, Am Geophys Union, 88- *Mem:* Fel Inst Elec & Electronics Engrs; Am Geophys Union; Optical Soc Am; Soc Photo-optical Instrumentation Engrs. *Res:* Developing theoretical models of wave propagation in random media and application to remote sensing; study of nonlinear optical properties of compound semiconductors; study of electromagnetic and optical propagation and scattering. *Mailing Add:* Elec Eng Dept Univ Wash FT-10 Seattle WA 98195

TSANG, REGINALD C, b Hong Kong, Sept 20, 40; US citizen; m 66; c 2. NEONATOLOGY, NUTRITION. *Educ:* Univ Hong Kong, MBBS, 66. *Prof Exp:* Intern med & surg, Queen Mary Hosp, Hong Kong Univ, 64-65, resident pediat, 65-66; resident psychiat, Hong Kong Psychiat Hosp, 65; intern pediat & med, Michael Reese Hosp, Chicago, 66-67, resident pediat, 67-68, fel neonatology, 68-69; fel, Cincinnati Gen Hosp & Childrens Hosp, 69-71; from asst prof to assoc prof, 71-79, dir, Fels Div Pediat Res, 74-76, PROF PEDIAT & OBSTET & GYNEC, UNIV CINCINNATI, 79-, DAVID G & PRISCILLA R GAMBLE PROF NEONATOLOGY, 87- *Concurrent Pos:* Attend pediatrician, Childrens Hosp, Cincinnati & NIH grant neonatal mineral metab, Nat Inst Child Health & Human Develop, 71-; attend pediat, Cincinnati Gen Hosp, 71-; dir, perinatal neonatology training, NIH, 78-, Diabetes prog PPG, 78-, Div Neonatology, 83-, Perinatal Res Inst, 86-, chief, Ctr Growth Retardation PERC, 85- *Mem:* Am Fedn Clin Res; Soc for Pediat Res; Am Soc Clin Nutrit; Am Pediat Soc; Perinatal Res Soc; Am Soc for Bone Mineral Res; NY Acad Sci; Nat Perinatal Asn; Endocrine Soc; fel Am Col Nutrit (secy-treas, 79-85, vpres, 85-86, pres, 89-91). *Res:* Pathophysiology of disturbances in calcium-phosphate-magnesium homeostasis in the neonate; examination of parathyroid hormone, vitamin D, glucagon and calcitonin; diabetic pregnancy; pediatric hyperlipoproteinemia; identification and prevention of premature atherosclerosis. *Mailing Add:* Dept Pediat Univ Cincinnati Bethesda Ave Cincinnati OH 45267-0541

TSANG, SIEN MOO, photochemistry; deceased, see previous edition for last biography

TSANG, TUNG, b Shanghai, China, Aug 17, 32; US citizen; m 57; c 1. PHYSICAL CHEMISTRY, SOLID STATE PHYSICS. *Educ:* Ta-Tung Univ, China, BS, 49; Univ Minn, MS, 52; Univ Chicago, PhD(chem), 60. *Prof Exp:* Chemist, Minneapolis-Honeywell Regulator Co, 52-56; asst chemist, Argonne Nat Lab, 60-64, assoc chemist, 64-67; phys chemist, Nat Bur Standards, 67-69; assoc prof, 69-75, PROF PHYSICS, HOWARD UNIV, 75- *Concurrent Pos:* Nat Acad Sci-Nat Res Coun sr res fel, NASA Goddard Space Flight Ctr, 75-76; consult, Naval Res Lab, 83- *Mem:* Am Phys Soc. *Res:* Magnetic resonance and susceptibility; statistical physics; photoelectron spectroscopy. *Mailing Add:* Dept Physics Howard Univ Washington DC 20059

TSANG, VICTOR CHIU WANA, b Hong Kong, June 20, 47; US citizen; m 72; c 1. IMMUNOLOGY. *Educ:* ETex Baptist Col, BS, 67; Stephen F Austin State Univ, MS, 69; Notre Dame Univ, MS, 72; Univ Ga, Athens, PhD(parasitic biochem immunol), 76. *Prof Exp:* Postdoctoral parasitic biochem immunol, Univ Ga, Athens, 76-77; res chemist, 77-88, CHIEF IMMUNOL & MOLECULAR BIOL PARASITIC BIOCHEM, CTR DIS CONTROL, CTR INFECTIOUS DIS, DIV PARASITIC DIS, 88- *Concurrent Pos:* Adj prof, Dept Biol, Natural Sci Col, Univ PR, 86-89; collab scientist, Div Pathobiol, Yerkes Regional Primate Res Ctr, Emory Univ, 86-; assoc, Dept Int Health, Sch Hyg & Pub Health, Johns Hopkins Univ, 89-; adj prof, Dept Chem, Ga State Univ, 89- *Honors & Awards:* Distinguished Serv Award, US Dept Health & Human Serv, 89. *Mem:* Am Asn Immunol; Am Soc Parasitol; Am Soc Trop Med & Hyg. *Res:* Quantitative immunological assay systems; diagnostic assay invention for infectious diseases; human immunovirus; helminthis diseases; immunochemistry; schistosomiasis; cysticercosis; biologically active co-polymers. *Mailing Add:* Ctr Dis Control CID DPD POB Mail Stop F13 Atlanta GA 30333

TSANG, WAI LIN, b Toishan, Kwangtung, June 3, 51; US citizen; m 85; c 2. SATELLITE NAVIGATION, HIGH PERFORMANCE COMPUTING. *Educ:* Northeastern Univ, BS, 74, MS, 75, PhD(elec eng), 81. *Prof Exp:* Asst prof, Northeastern Univ, 80-81; systs engr navig, Intermetrics Inc, 81-82; res engr commun, GTE Sylvania CSD, 82-86; mem tech staff C3I systs, Mitre Corp, 86-88; SR ENGR NAVIG, SCI APPLN INT CORP, 88- *Concurrent Pos:* Consult, Intermetrics Inc, 82-83; mem, Radio Tech Comn Aeronaut, 86-, Airlines Electronic Eng Comt, 89-, Air Traffic Tech Comt, 91- *Mem:* sr mem Am Inst Aeronaut & Astronaut; sr mem Inst Elec & Electronics Engrs; Inst Navig. *Res:* Satellite navigation and communications; technology assessment studies on supercomputer; biotechnology; high definition television; lithography; development of signal processing algorithms with applications to synthetic aperture radar, estimation/filtering, and electromagnetic interference. *Mailing Add:* 1710 Goodridge Dr McLean VA 22102

TSAO, CHEN-HSIANG, b Shanghai, China, Jan 21, 29; US citizen; m 57; c 2. PHYSICS. *Educ:* Univ Wash, Seattle, BS, 53, MS, 56, PhD(physics), 61. *Prof Exp:* Res assoc particle physics, Univ Wash, Seattle, 56-60, res instr, 60-61; res assoc high energy physics, Enrico Fermi Inst Nuclear Studies, Univ Chicago, 61-65; Nat Acad Sci-Nat Res Coun resident res assoc fel, 65-66, RES PHYSICIST, E O HULBURT CTR SPACE RES, US NAVAL RES LAB, 66- *Concurrent Pos:* Chap pres, Naval Res Lab, 86-87. *Mem:* Am Phys Soc; Sigma Xi. *Res:* Particle physics; high energy interactions; astrophysics; radiation effects. *Mailing Add:* Space Sci Div Code 4154 US Naval Res Lab Washington DC 20375-5000

TSAO, CHIA KUEI, b China, Jan 14, 22; m 52; c 4. MATHEMATICAL STATISTICS. *Educ:* Univ Ore, MA, 50, PhD(math statist), 52. *Prof Exp:* Asst math, Univ Ore, 48-52; from instr to assoc prof, 52-63, PROF MATH, WAYNE STATE UNIV, 63- *Mem:* Am Math Soc; Math Asn Am; Inst Math Statist. *Res:* Nonparametric statistics. *Mailing Add:* 2254 Belmont Rd Ann Arbor MI 48104

TSAO, CHING H, b China, Nov 16, 20; US citizen; m 52; c 2. MECHANICAL ENGINEERING. *Educ:* Chiao Tung Univ, BS, 41; Mich State Univ, MS, 48; Ill Inst Technol, PhD(eng mech), 52. *Prof Exp:* Asst prof civil eng, Univ Southern Calif, 53-55; res engr, Hughes Aircraft Co, 55-61; head stress anal sect, Aerospace Corp, 61-65; assoc prof mech eng, 65-69, PROF MECH ENG, CALIF STATE UNIV, LONG BEACH, 69- *Mem:* Am Soc Eng Educ; Soc Exp Stress Anal; Am Inst Aeronaut & Astronaut; Sigma Xi. *Res:* Theoretical and experimental stress. *Mailing Add:* Dept Mech Eng Calif State Univ Long Beach CA 90840

TS'AO, CHUNG-HSIN, b Nanking, China, 33; m 62; c 3. PHYSIOLOGY, EXPERIMENTAL PATHOLOGY. *Educ:* Tunghai Univ, Taiwan, BS, 60; Ind Univ, Bloomington, MA, 61; Yale Univ, PhD(physiol), 66. *Prof Exp:* Res assoc hemat, Montefiore Hosp & Med Ctr, 66-67; res assoc physiol, Sch Med, Univ Chicago, 67-72, asst prof path, 68-72; asst prof, 73-75, assoc prof, 75-80, PROF PATH, MED SCH, NORTHWESTERN UNIV, CHICAGO, 80- *Concurrent Pos:* Dir path, Coagulation Lab, Chicago Northwestern Mem Hosp, 73- *Mem:* AAAS; Am Soc Hemat; NY Acad Sci; Int Soc Thrombosis & Haemostasis; Am Soc Exp Path. *Res:* Experimental thrombosis; vascular morphology and function. *Mailing Add:* Dept Path Northwestern Univ Sch Med Chicago IL 60611

TSAO, CONSTANCE S, b Hong Kong, China, July 27, 34; US citizen; m 75. BIOCHEMISTRY. *Educ:* Tex Woman's Univ, BS, 62; Cornell Univ, PhD(chem), 67. *Prof Exp:* Res fel, Harvard Med Sch, 69-70; res sci, Tyco Lab, Inc, 70-73, Arthur D Little, Inc, 73-75; sr res sci, 76-79, DIR, DEPT PHYSIOCHEM, LINUS PAULING INST SCI & MED, 79- *Concurrent Pos:* Postdoctoral fel, Brandeis Univ, 67-69. *Mem:* NY Acad Sci; Am Chem Soc. *Res:* Metabolism of vitamin C; detoxification action of ascorbic acid; medical application and nutrition of vitamin C and vitamin B12; cancer research. *Mailing Add:* Linus Pauling Inst 440 Page Mill Rd Palo Alto CA 94306-2025

TSAO, FRANCIS HSIANG-CHIAN, b China, July 22, 36. BIOCHEMISTRY, ORGANIC CHEMISTRY. *Educ:* Taiwan Chung Hsing Univ, BS, 61; Dalhousie Univ, MS, 66; Iowa State Univ, PhD(biochem), 72. *Prof Exp:* Anal chemist qual control, Biotech Indust, London, Ont, 67-68; fel lipolytic enzymes, Univ Chicago, 72-74; fel pulmonary dis, 74-76, asst scientist, 76-83, ASSOC SCIENTIST RES LUNG SUFACTANT, UNIV WIS-MADISON, 83- *Concurrent Pos:* Pediat award, Spec Ctr Res, NIH, 81-86. *Mem:* Sigma Xi; Am Chem Soc; Am Soc Biol Chemists. *Res:* Metabolism of lung surfactant phospholipids. *Mailing Add:* Neonatal Res Labs Madison Gen Hosp 202 S Park St Madison WI 53715

TSAO, GEORGE T, b Nanking, China, Dec 4, 31; m 60; c 3. CHEMICAL ENGINEERING, MICROBIOLOGY. *Educ:* Nat Taiwan Univ, BSc, 53; Univ Fla, MSc, 56; Univ Mich, PhD(chem eng), 60. *Prof Exp:* Asst prof physics, Olivet Col, 59-60; chem engr, Merck & Co, Inc, 60-61; res chemist, Tenn Valley Authority, 61-62; sect leader hydrolysis & fermentation, Res Dept, Union Starch & Refining Co, Inc Div, Miles Labs, Inc, 62-65, asst res dir, 65-66; from assoc prof to prof chem eng, Iowa State Univ, 66-77; PROF CHEM ENG, PURDUE UNIV, 77- *Mem:* Am Chem Soc; Am Inst Chem Engrs; Am Soc Eng Educ. *Res:* Biological technology; fermentation; agricultural and natural products utilization; waste disposal; organic synthesis; industrial carbohydrates; process development; enzyme engineering. *Mailing Add:* Sch Eng Chem Purdue Univ West Lafayette IN 47907-9980

TSAO, JEFFREY YEENIEN, b Santa Monica, Calif, May 27, 55; m 79; c 1. CRYSTAL GROWTH, THIN FILM MATERIALS. *Educ:* Stanford Univ, BS, 77, MS, 77; Harvard Univ, MS, 81, PhD(appl physics), 81. *Prof Exp:* Mem tech staff, Lincoln Lab, Mass Inst Technol, 81-84; MEM TECH STAFF, SANDIA NAT LABS 84- *Mem:* Am Phys Soc; Mat Res Soc; Am Vacuum Soc. *Res:* Fundamental aspects of crystal growth; economics of scientific research. *Mailing Add:* Orgn 1141 Sandia Nat Labs PO Box 5800 Albuquerque NM 87185-5800

TSAO, KEH CHENG, b Kiangsu, China, Apr 20, 23; m 57; c 2. MECHANICAL ENGINEERING. *Educ:* Nat Chung Cheng Univ, BS, 46; Ill Inst Technol, MS, 56; Univ Wis, PhD(mech eng), 61. *Prof Exp:* Assoc prof mech eng, SDak Sch Mines & Technol, 61-67; assoc prof, 67-74, PROF MECH ENG, UNIV WIS-MILWAUKEE, 74- *Concurrent Pos:* Consult, BPS Inc, Beloit, Wis & Wis Elec Power Co, Milwaukee; consult prof, Xian Jiaotong Univ, Peoples Rep China; adv prof, Harbin Inst Technol, Harbin, Peoples Rep China. *Mem:* AAAS; Am Soc Eng Educ; Am Soc Mech Engrs; Soc Automotive Engrs. *Res:* High temperative measurement in combustion and arc heated gases; diesel combustion and ignition delays; energy recovery; air quality and modeling; coal combustion and hot gas cleaning. *Mailing Add:* 13640 N Lake Shore Dr Mequon WI 53092

TSAO, MAKEPEACE UHO, b Shanghai, China, Aug 28, 18; nat US; m 47; c 4. CHEMISTRY. *Educ:* Univ Tatung, BS, 37; Univ Mich, MS, 41, PhD(pharmaceut chem), 44. *Prof Exp:* Wm S Merrill Co fel, Univ Mich, 44-45, sr biochemist, 46-48, head biochemist, 48-52, from asst prof to assoc prof biochem, 52-67; prof, 67-82, EMER PROF SURG, UNIV CALIF, DAVIS, 82- *Mem:* AAAS; Am Chem Soc; Am Soc Biol Chemists; Biomet Soc; NY Acad Sci. *Res:* Synthetic medicinals; physiological chemistry of premature and newborn infants; biochemical analytical methods; multiple molecular forms of enzymes; carbohydrate metabolism; experimental diabetes; Neurospora crassa. *Mailing Add:* 533 Antioch Dr Davis CA 95616

TSAO, NAI-KUAN, b Shanghai, China, June 25, 39; US citizen; m 65; c 1. NUMERICAL ANALYSIS, ERROR ANALYSIS. *Educ:* Nat Taiwan Univ, BSEE, 61; Nat Chiao Tung Univ, MSEE, 64; Univ Hawaii, MS, 66, PhD(elec eng), 70. *Prof Exp:* Vis asst prof elec eng, Univ Hawaii, 71; res assoc, Aerospace Res Labs, 71-73, res analyst, 73-74; asst prof, 74-80, ASSOC PROF COMPUT SCI, WAYNE STATE UNIV, 80- *Concurrent Pos:* Prin investr, Air Force Off Sci Res, 76-81. *Mem:* Asn Comput Mach; Math Asn Am; Soc Indust & Appl Math; Inst Elec & Electronics Engrs Comput Soc. *Res:* Numerical algorithms in linear algebra and fast Fourier transform using sequential or parallel machines. *Mailing Add:* Dept Comput Sci Wayne State Univ 460 State Hall Detroit MI 48202

TSAO, PETER HSING-TSUEN, b Shanghai, China, Mar 22, 29; nat US; m 56; c 1. MYCOLOGY. *Educ:* Univ Wis, BA, 52, PhD(plant path), 56. *Prof Exp:* Jr plant pathologist, 56-58, asst plant pathologist, 58-64, assoc prof plant path, 64-70, PROF PLANT PATH, UNIV CALIF, RIVERSIDE, 70- *Concurrent Pos:* Guggenheim fel, 66-67; consult, UN Food & Agr Orgn, Thailand, 73-74 & 81, Malaysia, 82 & Indonesia, 83; res consult, Nat Sci Coun, Taiwan, 74-; lectr, Du Pont, 80; vis prof, Peking Agr Univ, 80, 84 & 86. *Mem:* Am Phytopath Soc; Mycol Soc Am; Brit Mycol Soc; Am Inst Biol Sci; Int Soc Citriculture; fel Am Phytopath Soc. *Res:* Phytophthora and other soil fungi; root diseases of citrus, avocado, black pepper, vanilla, and other subtropical and tropical crops. *Mailing Add:* Dept Plant Path Univ Calif Riverside CA 92521-0122

TSAO, SAI HOI, b Hong Kong, Oct 19, 36; Can citizen; m 61; c 3. ELECTRICAL METROLOGY, ELECTRONIC INSTRUMENT DESIGN. *Educ:* Univ BC, BASc, 59, MASc, 61; Univ Birmingham, UK, PhD, 65. *Prof Exp:* Assoc res off, 61-79, SR RES OFFICER, NAT RES COUN CAN, 79- *Concurrent Pos:* Vis lectr, China & Taiwan. *Mem:* Fel Inst Elec & Electronics Engrs; Prof Inst Can; Instrument & Measurement Soc. *Res:* High precision electrical standards, measurements and instrumentation. *Mailing Add:* Inst Nat Measurement Standards Nat Res Coun Ottawa ON K1A 0R6 Can

TSAO, UTAH, b Shanghai, China, June 6, 13; nat US; m 40; c 2. CHEMICAL ENGINEERING. *Educ:* Tatung Univ, China, BSc, 33; Univ Mich, MSc, 37, DSc(chem eng), 40. *Prof Exp:* Plant engr, Audubon Sugar Factory, La, 39-40; res engr, Eng Res Dept, Univ Mich, 40-41; petrol engr, Nat Resources Comn, China, 41-42; process engr, 42-53, staff process engr, 53-64, mgr chem plant design, 64-78, CONSULT, LUMMUS CO, 78- *Mem:* Fel Am Inst Chem Engrs; Chinese Inst Engrs (vpres, Am Sect, 48, pres, 67). *Res:* Design of ethylene oxide and glycol, vinyl acetate, polyvinyl alcohol, vinyl chloride, chloroform, carbon tetrachloride, urea, acetylene from hydrocarbons, polyvinyl pyrolidone, phenol and acetone plants, ethylene cracking heaters, caprolactum, chlorine-caustic, formaldehyde, styrene, aluminum chloride, aromatic nitriles; process improvements. *Mailing Add:* 1887 Kennedy Blvd Jersey City NJ 07305

TSAO-WU, NELSON TSIN, b Tientsin, China, Sept 9, 34; US citizen; m 59; c 3. ELECTRONICS ENGINEERING. *Educ:* Loughborough Col Technol, DLC, 57; Univ London, BSc, 57; Northeastern Univ, MSEE, 65, PhD(elec eng), 68. *Prof Exp:* Engr, Rediffusion Ltd, Hong Kong, 57-59; audio engr, Far East Broadcasting Co, Hong Kong, 59-60; asst educ officer, Hong Kong Govt, 61-63; biomed engr, Harvard Med Sch & Peter Bent Brigham Hosp, 65-67; MEM TECH STAFF, BELL TEL LABS, 68- *Mem:* Inst Elec & Electronics Engrs. *Res:* Switching networks; coding theory; mathematical modeling; signal theory; software engineering and development. *Mailing Add:* Tong Bldg Bell Labs Singapore Si 1 Basking Ridge NJ 07920

TSAROS, C(ONSTANTINE) L(OUIS), b East Chicago, Ind, Sept 6, 21; m 50; c 4. CHEMICAL ENGINEERING, ENGINEERING ECONOMICS. *Educ:* Purdue Univ, BS, 43; Univ Mich, MS, 48. *Prof Exp:* Chem engr, Armour & Co, Ill, 43-46; assoc chem engr, Inst Gas Technol, Ill Inst Technol, 48-51; chem engr, Standard Oil Co, Ind, 51-57; chem engr, Inst Gas Technol, Ill Inst Technol, 57-64, supvr process econ, 64-70, mgr process econ, 70-83; RETIRED. *Mem:* Am Inst Chem Engrs; Am Asn Cost Engrs; Sigma Xi. *Res:* Chemical engineering design and process development; process economics in petroleum processing, hydrocarbon conversion, energy studies and synthetic fuels. *Mailing Add:* 5328 Lawn Ave West Springs IL 60558

TSAUR, BOR-YEU, b Taiwan, June 8, 55. ELECTRONICS ENGINEERING. *Educ:* Nat Taiwan Univ, BS, 77; Calif Inst Technol, MS, 78, PhD(elec eng), 80. *Prof Exp:* Vis scientist, IBM Res Ctr, 79; RES STAFF MEM, LINCOLN LAB, MASS INST TECHNOL, 80- *Mem:* Am Phys Soc; Inst Elec & Electronics Engrs; Mat Res Soc; Metall Soc-Am Inst Metal Engrs. *Res:* Development of novel crystal growth technique for preparing large-area single-crystal semiconductor sheets on insulating substrates for integrated electronic devices; novel electronic devices or circuits; processing technologies; device physics based on the semiconductor on insulator structures. *Mailing Add:* Lincoln Lab Mass Inst Technol Lexington MA 02173

TSCHANG, PIN-SENG, b Penang, Malaysia, May 14, 34; m 62; c 2. ELECTRICAL ENGINEERING. *Educ:* Ore State Univ, BS, 58, MS, 59; Newark Col Eng, DEngSci(elec eng), 67. *Prof Exp:* Instr elec eng, Newark Col Eng, 59-62; res engr, Electronics Res Labs, Columbia Univ, 62-65, sr res engr, 65-67; res assoc, Res Labs, Eastman Kodak Co, 67-90; PRES, ACTIVE TEXT INC, 90- *Mem:* Inst Elec & Electronics Engrs; Brit Inst Elec Engrs. *Res:* Electronics and scientific instrumentation; television; digital signal processing; electronic memories. *Mailing Add:* Active Text Inc PO Box 17400 Rochester NY 14617

TSCHANTZ, BRUCE A, b Akron, Ohio, Sept 15, 38; m 62; c 2. CIVIL ENGINEERING, WATER RESOURCES. *Educ:* Ohio Northern Univ, BSCE, 60; NMex State Univ, MSCE, 62, ScD(civil eng), 65. *Prof Exp:* Civil engr, Facilities Div, White Sands Missile Range, 65; from asst prof to assoc prof civil eng, 69-76, PROF CIVIL ENG, UNIV TENN, KNOXVILLE, 76-

Concurrent Pos: Consult, Exec Off Pres, Off Sci & Technol Policy, Washington, DC, 77-79; chief, Fed Dam Safety, Fed Emergency Mgt Agency, Washington, DC, 80. *Honors & Awards:* Dow Chem Co Award, Am Soc Eng Educ, 69. *Mem:* Am Soc Eng Educ; Am Soc Civil Engrs. *Res:* Remote sensing of the environment, particularly water resources; analysis of the safety of dams; unsteady open channel flow; hydrologic impact of coal strip mining. *Mailing Add:* 1508 Meeting House Rd Knoxville TN 37931

TSCHANZ, CHARLES MCFARLAND, b Mackay, Idaho, July 9, 26; m 58; c 3. GEOLOGY. *Educ:* Univ Idaho, BS, 49; Stanford Univ, MS, 51. *Prof Exp:* Geologist, Colo, US Geol Surv, 49-50, Pioche, Nev, 51-53, chief uranium-copper proj, NMex, 53-55, geochem researcher, 55-56, chief mapping proj, Lincoln County, Nev, 56-60, advisor, Bolivian Mineral Resources, US Opers Mission, USAID, 60-65, geol consult, Nat Mineral Inventory, Colombia, 65-69, proj chief, Boulder Mountains Mapping Proj, Idaho, 69-70, proj chief mineral eval, Sawtooth Nat Recreation Area, Idaho, 71-74, proj chief, Boulder Mountains, Idaho, 74-82; RETIRED. *Mem:* Geol Soc Am; Geochem Soc; Soc Econ Geol. *Res:* Regional mapping and economic evaluation as an integrated project; geochemistry, especially distribution of minor elements in igneous rocks; geology of eastern Nevada, Colorado Plateau, Bolivian Altiplano and Sierra Nevada of Santa Marta, Colombia. *Mailing Add:* 876 S Moore Denver CO 80226

TSCHANZ, CHRISTIAN, pharmacology, clinical chemistry, for more information see previous edition

TSCHINKEL, WALTER RHEINHARDT, b Lobositz, Czech, Sept 15, 40; US citizen; m 68; c 1. BIOLOGY, INSECT BEHAVIOR. *Educ:* Wesleyan Univ, BA, 62; Univ Calif, Berkeley, MA, 65, PhD(comp biochem), 68. *Prof Exp:* Fel biol, Dept Neurobiol & Behav, Cornell Univ, 68-70; lectr entom, Rhodes Univ, 70; from asst prof to assoc prof 70-80, assoc chmn, 85-87, PROF BIOL SCI, FLA STATE UNIV, 80- *Concurrent Pos:* NSF grants, 71-; consult, Environ Protection Agency, 73-75; vis assoc prof, Entom Dept, Univ Calif, Berkeley, 77; mem Panel Regulatory Biol, NSF, 77- & consult, Integrated Basic Res, 78-80; guest prof, Univ Pretoria, SAfrica, 82; Fla Pesticide Rev Coun, 84- *Mem:* AAAS; Entom Soc Am; Int Union Study Soc Insects. *Res:* Insect behavior and chemical communication; biology of ants; biology of tenebrionid beetles. *Mailing Add:* Dept Biol Sci Fla State Univ Tallahassee FL 32306

TSCHIRGI, ROBERT DONALD, b Sheridan, Wyo, Oct 9, 24; m 83. PHYSIOLOGY. *Educ:* Univ Chicago, BS, 45, MS, 47, PhD, 49, MD, 50. *Prof Exp:* Asst physiol, Univ Chicago, 45-48, from instr to asst prof, 48-53; from assoc prof to prof, Sch Med, Univ Calif, Los Angeles, 53-66; vchancellor acad planning, 66-67, vchancellor acad affairs, 67-68, PROF NEUROSCI, UNIV CALIF, SAN DIEGO, 66- *Concurrent Pos:* Dir med educ study, Univ Hawaii, 63-64; univ dean planning, Univ Calif, 64-66; consult, NSF, NIH, Greek Govt, World Health Orgn, Brit Med Asn & var univs & med schs. *Mem:* Int Brain Res Orgn; fel AAAS; Am Physiol Soc; Biophys Soc. *Res:* Intracranial fluids and barriers; direct current potentials in central nervous system; neurophysiology of perception. *Mailing Add:* Dept Neurosci Sch Med M-008 Univ Calif San Diego La Jolla CA 92093

TSCHIRLEY, FRED HAROLD, b Ethan, SDak, Dec 19, 25; m 48; c 5. ECOLOGY. *Educ:* Univ Colo, BA, 51, MA, 54; Univ Ariz, PhD, 63. *Prof Exp:* Res asst, Univ Ariz, 52-53, instr, 53-54; range scientist, Crops Res Div, Agr Res Serv, USDA, 54-68, asst br chief, Crops Protection Res Br, 68-71, asst coordr environ qual activ sci & educ, 71-73, coordr environ qual activ, Off Secy, 73-74; chmn dept, bot & plant path, Mich State Univ, 74-80, prof, 74-84; exec dir, Mich Agr/Bus Coun, 84-88; RETIRED. *Mem:* AAAS; Weed Sci Soc Am; Soc Range Mgt; Ecol Soc Am. *Res:* Woody plant control; pesticides. *Mailing Add:* 3401 Placitadel Emblema Green Valley AZ 85614

TSCHOEGL, NICHOLAS WILLIAM, b Zidlochovice, Czech, June 4, 18; m 46; c 2. PHYSICAL CHEMISTRY. *Educ:* New South Wales, BSc, 54, PhD(chem), 58. *Prof Exp:* Sr res officer, Bread Res Inst, Australia, 58-61; proj assoc dept chem, Univ Wis, 61-63; sr phys chemist, Stanford Res Inst, 63-65; assoc prof mat sci, 65-67, prof, 67-85, EMER PROF CHEM ENG, CALIF INST TECHNOL, 85- *Concurrent Pos:* Consult, Phillips Petrol Co, 67-83; Alexander von Humboldt Found award, 70. *Mem:* Am Phys Soc; Am Chem Soc; Soc Rheol; Brit Soc Rheol. *Res:* Polymer rheology; physical chemistry of macromolecules; mechanical properties of polymeric materials. *Mailing Add:* 228 Spalding Calif Inst Technol Pasadena CA 91125

TSCHUDY, DONALD P, b Palmerton, Pa, Nov 8, 26; m 51; c 2. INTERNAL MEDICINE, BIOCHEMISTRY. *Educ:* Princeton Univ, AB, 46; Columbia Univ, MD, 50; Am Bd Internal Med, dipl, 61. *Prof Exp:* Intern med, Presby Hosp, New York, 51-52, asst resident, 52-53; asst resident, Francis Delafield Hosp, 53-54; clin assoc, Clin Ctr, NIH, 54-55, sr investr, Metab Serv, Clin Ctr, Nat Cancer Inst, 55-85; RETIRED. *Mem:* Am Soc Clin Invest; Am Fedn Clin Res; Am Soc Biol Chemists; AMA. *Res:* Clinical and biochemical research on porphyrin metabolism and the porphyrias; research on tumor-host relationships. *Mailing Add:* 3905 Woodlawn Rd Chevy Chase MD 20815

TSCHUIKOW-ROUX, EUGENE, b Kharkov, USSR, Jan 16, 36; US citizen; m 59; c 1. CHEMICAL KINETICS, THERMODYNAMICS. *Educ:* Univ Calif, Berkeley, BS, 57, PhD(chem), 61; Univ Wash, Seattle, MS, 58. *Prof Exp:* Sr scientist, Jet Propulsion Lab, Calif Inst Technol, 60-65; Nat Acad Sci-Nat Res Coun res assoc chem, Nat Bur Standards, 65-66; assoc prof, 61-71, head chmn dept, 73-77, PROF CHEM, UNIV CALGARY, 71- *Concurrent Pos:* Vis scholar, Univ Calif, Santa Barbara, 72; consult, Jet Propulsion Lab, Calif Inst Technol, 73, 87; fel, Alexander von Humboldt Found, IRG, 80. *Mem:* Am Chem Soc; Am Phys Soc; Can Inst Chem. *Res:* Gas phase reaction kinetics; high temperature shock tube studies; kinetic isotope effects; photochemistry; reaction dynamics; unimolecular reactions; gas phase ion-molecule reactions; thermodynamics. *Mailing Add:* Dept Chem Univ Calgary 2500 University Dr NW Calgary AB T2N 1N4 Can

TSCHUNKO, HUBERT F A, b Weidenau, Austria, Sept 9, 12; US citizen; m 46; c 2. PHYSICS, OPTICS. *Educ:* Darmstadt Tech Univ, Diplom-Ing, 35. *Prof Exp:* Develop engr aeronaut indust, Europe, 36-45; res assoc astron, Astron Observ, Heidelberg, 45-50; engr pvt indust, WGer, 51-57; physicist, USAF, Wright-Patterson AFB, 57-65; opticist Electronics Res Ctr, NASA, 65-70, Goddard Space Flight Ctr, 70-87; RETIRED. *Honors & Awards:* Apollo Achievement Award, NASA, 69. *Mem:* Optical Soc Am; Ger Soc Aeronaut & Astronaut. *Res:* Wave optics; performances of large and space optical systems, as space telescopes, space cameras, space energy collectors and space energy transmitters. *Mailing Add:* 12501 N Point Lane Laurel MD 20708

TSE, FRANCIS LAI-SING, b Hong Kong, Jan 20, 52; US citizen; m 79; c 1. PHARMACOKINETICS, DRUG METABOLISM. *Educ:* Univ Wis-Madison, BS, 74, MS, 75, PhD(pharmaceut), 78. *Prof Exp:* Asst prof pharmacokinetics, Rutgers Univ, 78-80; sr scientist & unit head, 81-84, group leader, 85-90, ASST DIR, DRUG METAB, SANDOZ RES INST, 91- *Concurrent Pos:* Prin investr, Nat Inst Drug Abuse, 80-81; vis asst prof pharmaceut, Rutgers Univ, 81-, mem grad fac, 80-81. *Mem:* Am Pharmaceut Asn; NY Acad Sci; Sigma Xi; Int Soc Study Xenobiotics; Am Asn Pharmaceut Scientists; Am Soc Clinical Pharmacol & Therapeut; fel Acad Pharmaceut Res & Sci, 90; fel Am Col Clin Pharmacol, 88. *Res:* Absorption, distribution, metabolism and excretion of therapeutic agents in laboratory animals and humans; influence of various environmental and physiological factors on drug pharmacokinetics. *Mailing Add:* Drug Metab Dept Sandoz Res Inst East Hanover NJ 07936

TSE, FRANCIS S, b Canton, China, Dec 15, 19; US citizen; m 52; c 2. MECHANICAL ENGINEERING. *Educ:* Univ Hong Kong, BSc, 41; Purdue Univ, Lafayette, MSME, 42; Univ Pa, MBA, 49; Ohio State Univ, PhD(mech eng), 57. *Prof Exp:* Jr engr, Baldwin Locomotive Works, Pa, 42-45; engr, Fairbanks Morse, Wis, 46-47; instr & res assoc mech eng, Ohio State Univ, 47-57; from asst prof to assoc prof, Mich State Univ, 57-62; NSF fel, Purdue Univ, Lafayette, 62-63; PROF MECH ENG, UNIV CINCINNATI, 63- *Concurrent Pos:* NSF grants undergrad educ, Univ Cincinnati, 64-66. *Mem:* Am Soc Eng Educ; Am Soc Mech Engrs; Instrument Soc Am; Sigma Xi. *Res:* Vibrations; control theory; measurement and instrumentation; machine dynamics. *Mailing Add:* 10167 Lochcrest Dr Cincinnati OH 45231

TSE, HARLEY Y, b China, July 17, 47; US citizen; m 79; c 3. NEUROIMMUNOLOGY, IMMUNOGENETICS. *Educ:* Calif Inst Technol, BS, 72; Univ Calif, San Diego, PhD(immunol), 77; Rutgers Univ, MBA, 86. *Prof Exp:* Fel, NIH, 77-80; sr immunologist, Merck & Co, 80-86; ASSOC PROF, SCH MED, WAYNE STATE UNIV, 86- *Concurrent Pos:* Adj assoc prof, Columbia Univ Col Surgeons & Physicians, 81-83; prin investr res grants, NIH, 88-, Nat Multiple Sclerosis Soc, 88-; coun mem, Immunol Autumn Conf, 90- *Mem:* Am Asn Immunologists; Int Soc Neuroimmunol; Soc Chinese Bioscientists Am. *Res:* T cell functions in autoimmune diseases; molecular biology and genetics of T cell receptor genes; in vivo T cell trafficking; T cell activation signals. *Mailing Add:* Dept Immunol & Microbiol Wayne State Univ Sch Med Detroit MI 48201

TSE, ROSE (LOU), b Shanghai, China, July 27, 27; US citizen; m 53. ORGANIC CHEMISTRY, MEDICINE. *Educ:* St John's Univ, China, BS, 49; Mt Holyoke Col, MA, 50; Yale Univ, PhD(org chem), 53; Med Col Pa, MD, 60; Am Bd Internal Med, dipl & cert rheumatology. *Prof Exp:* Instr, Ohio State Univ, 53-55; res assoc, Univ Pa, 55-56; intern, Philadelphia Gen Hosp, 60-61, resident internal med, 61-64; assoc med, 68-71, asst prof clin med, 71-75, ASSOC PROF MED, SCH MED, UNIV PA, 75- *Concurrent Pos:* Attend physician, Philadelphia Gen Hosp, 64-68, sr attend physician, 68-77, assoc chief spec ward cardiol, 68-71, chief rheumatology sect, 71-77; clin instr internal med, Med Col Pa, 64-68; assoc chief med, Philadelphia Gen Hosp, 75-77; chief rheumatology, West Park Hosp, 77-86; dir reheumatology, Jefferson Park Hosp, 87- *Mem:* Fel Am Col Physicians; fel Am Inst Chemists; fel Am Col Angiol; Am Heart Asn; Am Rheumatism Asn; fel Am Col Reumatology. *Res:* Reaction mechanisms; organic synthesis; electrocardiology; inflammatory mediators; non-steroidal inflammatory agents; cardiology; rheumatology; catecholamines; cyclic adenosine monophosphate; prostaglandin; crystal-induced synovitis. *Mailing Add:* 191 Presidential Condominium 191 Presidential Blvd Bala-Cynwyd PA 19004

TSE, WARREN W, b Hong Kong, Mar 13, 39; m 68; c 2. HUMAN PHYSIOLOGY. *Educ:* Univ Cincinnati, BS, 65; Univ Wis, MS, 67, PhD(physiol), 70. *Prof Exp:* Lectr physiol, Univ Wis, 69-70, fel physiol, 70-72; res assoc cardiol, Albany Med Col, 72-75; ASST PROF PHYSIOL, DEPT PHYSIOL & BIOPHYS, CHICAGO MED SCH, UNIV HEALTH SCI, 75- *Mem:* Am Physiol Soc. *Res:* Electrophysiological properties of single fibers of atrioventricular node and purkinje fibers of dog hearts. *Mailing Add:* Dept Physiol & Biophys Univ Chicago Med Sch 3333 Green Bay Rd North Chicago IL 60064

TSEN, CHO CHING, biochemistry, nutrition; deceased, see previous edition for last biography

TSENG, CHARLES C, b Fuchow, Fukien, China, Dec 20, 32; m 65; c 2. MOLECULAR BIOLOGY, CELL BIOLOGY. *Educ:* Taiwan Norm Univ, BS, 55; Taiwan Univ, MS, 57; Univ Calif, Los Angeles, PhD(plant sci), 65. *Prof Exp:* From asst prof to assoc prof bot, Windham Col, 65-75; assoc prof, 75-80, PROF BIOL, PURDUE UNIV, CALUMET, 80- *Concurrent Pos:* Dir med genetics prog, Northwest Ctr Med Educ, Indiana Univ, Gary, Ind. *Mem:* AAAS; Bot Soc Am; Am Inst Biol Sci; Am Soc Plant Taxon. *Res:* Acid phosphatase isoenzymes of vertebrates and plants; plant molecular systematics; human karyology. *Mailing Add:* Dept Biol Purdue Univ Calumet 2233 171st St Hammond IN 46323

TSENG, CHIEN KUEI, b Tao Yuan, Taiwan, Feb 21, 34; m 66; c 2. ORGANIC CHEMISTRY. *Educ:* Cheng Kung Univ, Taiwan, BS, 57; WVa Univ, MS, 64; Ill Inst Technol, PhD(chem), 68. *Prof Exp:* USPHS fels, Ill Inst Technol, 67-68; from res chemist to sr res chemist, Stauffer Chem Co, 68- 71, supvr anal chem, 71-81; MGR SPECTROS & PHYS CHEM, ICI AMERICAS INC, 81- *Mem:* Am Chem Soc. *Res:* Nuclear magnetic resonance; structure determination; stereochemistry; phosphorus chemistry; infrared and mass spectroscopy; environmental fate studies. *Mailing Add:* ICI Americas Inc 1200 S 47th St Richmond CA 94804

TSENG, FUNG-I, b Pingtung, Taiwan, Jan 12, 36; m 65; c 3. ELECTRICAL ENGINEERING, ELECTROMAGNETICS. *Educ:* Nat Taiwan Univ, BS, 58; Chiao Tung Univ, MS, 60; Syracuse Univ, PhD(elec eng), 66. *Prof Exp:* Res engr antennas, Syracuse Univ, 66-69; from asst prof to assoc prof, 69-82, PROF ELEC ENG, ROCHESTER INST TECHNOL, 83- *Mem:* Inst Elec & Electronics Engrs; Optical Soc Am. *Res:* Optimization of antenna arrays in noisy environments subject to random fluctuations. *Mailing Add:* Dept Elec Eng Rochester Inst Technol Lamb Memorial Dr Rochester NY 14623

TSENG, HSIANG LEN, pathology; deceased, see previous edition for last biography

TSENG, LEON L F, b Tainan City, Taiwan, Nov 20, 37; US citizen; m 65; c 4. NEUROPEPTIDES, CATECHOLAMINES. *Educ:* Nat Taiwan Univ, Taipei, BS, 61, MS, 64; Univ Kans, PhD(pharmacol), 70. *Prof Exp:* Instr pharmacol, Nat Taiwan Univ, 66-67; asst res pharmacologist, dept pharmacol, Univ Calif, San Francisco, 72-75, adj asst prof, 75-78; from asst prof to assoc prof, 78-89, PROF PHARMACOL, MED COL WIS, 89-; CONSULT, VET ADMIN MED CTR, WOOD, WIS, 78- *Mem:* Am Soc Pharmacol & Exp Therapeut; Soc Neurosci. *Res:* Neuronal mechanism of opiates and opioid peptides on the production of analgesic actions. *Mailing Add:* Dept Pharmacol & Toxicol Med Col Wis Milwaukee WI 53226

TSENG, LINDA, b China, Sept 29, 36; US citizen; m 63; c 1. PHYSICAL CHEMISTRY, BIOCHEMISTRY. *Educ:* Cheng Kung Univ, Taiwan, BS, 55; Univ NDak, PhD(chem), 68. *Prof Exp:* Assoc prof biochem, 76-80, ASSOC PROF OBSTET & GYNEC, HEALTH SCI CTR, STATE UNIV NY STONY BROOK, 80- *Honors & Awards:* Irma Hirsh Res Award, Irma Hirsh Inc, 76. *Mem:* AAAS; Endocrine Soc. *Res:* Steroid biochemistry; hormone action. *Mailing Add:* Dept Obstet & Gynec State Univ NY Health Sci Ctr Stony Brook NY 11794

TSENG, MICHAEL TSUNG, b Chungking, China, Jan 25, 44; US citizen; m 70; c 1. EXPERIMENTAL PATHOLOGY, ONCOLOGY. *Educ:* Iowa State Univ, BS, 67; State Univ NY, Buffalo, PhD(exp path), 73. *Hon Degrees:* Dr, Henan Med Univ, China. *Prof Exp:* Res assoc biochem, Ore Regional Primate Res Ctr, 73-74; asst prof anat, Upstate Med Ctr, State Univ NY, 74-78; ASSOC PROF ANAT & ONCOL ASSOC CANCER CTR, HEALTH SCI CTR, UNIV LOUISVILLE, 78-, DIR, TUMOR EVAL LABS, 82-, PROF ANAT, 87- *Concurrent Pos:* Instr anat, Sch Med, Univ Ore, 73-74; prin investr, Univ Award, State Univ NY, 73-76; prin investr, Am Cancer Soc, 78-82; actg dir, Anesthesia Critical Care Res Unit, Univ Louisville, 83-84; prin investr, NIOSH, 88-90; hon prof, Henan Med Univ, China. *Mem:* Am Asn Anatomists; Am Soc Cell Biol; Am Asn Cancer Res; NY Acad Sci; Endocrine Soc; Int Soc Prev Oncol. *Res:* Photodynamic therapy; vitreoretinal research; environmental toxicology; neurobiology. *Mailing Add:* Dept Anat/Neurobiol Univ Louisville Louisville KY 40292

TSENG, SAMUEL CHIN-CHONG, b Tainan, Taiwan, Mar 6, 33; m 57. SOLID STATE ELECTRONICS. *Educ:* Ching-Kung Univ, Taiwan, BS, 56; Chiao Tung Univ, MS, 60; Yale Univ, ME, 61; Univ Calif, Berkeley, PhD(solid state electronics), 66. *Prof Exp:* Res engr, Electron Tube Div, Litton Industs, Inc, Calif, 66-68; staff engr, 68-69, MEM RES STAFF, T J WATSON RES CTR, INT BUS MACH CORP, 69- *Mem:* Am Phys Soc; sr mem Inst Elec & Electronics Engrs. *Res:* Excitation, propagation and amplification of surface elastic waves in piezoelectric crystals and ceramics; application of surface waves to delay lines, matched filters, binary sequence recognitions and signal processing in general. *Mailing Add:* Int Bus Mach Corp 650 Harry Rd San Jose CA 95120-6099

TSENG, SHIN-SHYONG, b Tainan, Taiwan, Nov 24, 38; US citizen; m 67; c 2. ORGANIC CHEMISTRY, COMPUTER APPLICATION IN CHEMISTRY. *Educ:* Nat Taiwan Univ, BS, 61; Kent State Univ, MA, 64; Univ Chicago, PhD(org chem), 69. *Prof Exp:* Fel, Univ Chicago, 69-70; fel, Syva Res Inst, 70-72; res assoc, Ames Res Ctr, NASA, 72-73; assoc res scientist, Inst Environ Med, Med Ctr, NY Univ, 73-77; SR RES CHEMIST, AM CYANAMID CO, 77- *Mem:* Am Chem Soc. *Res:* Chemiluminescence; photochemistry; chemical carcinogenesis; chemistry of electron-rich olefins; organic synthesis; charge-transfer complexes; heterocyclic chemistry; pharmaceuticals (CNS-geriatric agents); computer application in chemical database management. *Mailing Add:* 232 Windmill Ct Bridgewater NJ 08807

TSERNOGLOU, DEMETRIUS, b Mytilene, Greece, Feb 10, 35. BIOPHYSICS, CRYSTALLOGRAPHY. *Educ:* Univ London, BSc, 60; Dalhousie Univ, MSc, 62; Yale Univ, MS, 64, PhD(molecular biophys), 66. *Prof Exp:* Fel molecular biophys, Yale Univ, 67-69; res asst prof biochem, Sch Med, Wash Univ, 69-71; from asst prof to prof, Dept Biochem, Sch Med, Wayne State Univ, 71-82; SR SCIENTIST, EUROP MOLECULAR BIOL LAB, 82- *Concurrent Pos:* Lectr biochem, Yale Univ, 68-69; vis prof, Univ Athens, 75-76 & Biozentrum Basel, 80; vis fel, All Souls Col, Oxford, 80-81; Josiah Macy Found Fel, 80-81. *Res:* Crystallographic study of structure and function of proteins; pore-forming proteins; neurotoxins. *Mailing Add:* Europ Molecular Biol Lab Postfach 102209 Heidelberg D-6900 Germany

TSIATIS, ANASTASIOS A, b New York, NY, July 12, 48; m 70; c 1. BIOMETRICS, BIOSTATISTICS. *Educ:* Mass Inst Technol, BS, 70; Univ Calif, Berkeley, PhD(statist), 74. *Prof Exp:* Asst prof statist, Univ Wis-Madison, 74-79; assoc mem biostatist, St Jude Children's Res Hosp, 79-81; ASSOC PROF BIOSTATIST, SCH PUB HEALTH, HARVARD UNIV & SIDNEY FARBER CANCER INST, 81- *Mem:* Am Statist Asn; Biometrics Soc; Inst Math Statist. *Res:* Application of survival analysis in clinical trials with specific emphasis on sequential rules for stopping a trial early if large treatment differences occur. *Mailing Add:* Dept Biostatist Sch Pub Health Harvard Univ 677 Huntington Ave Boston MA 02115

TSIBRIS, JOHN-CONSTANTINE MICHAEL, b Jannina, Greece, Dec 22, 36; m 69; c 2. BIOCHEMISTRY, ENDOCRINOLOGY. *Educ:* Nat Univ Athens, BSc, 59; Cornell Univ PhD(biochem), 65. *Prof Exp:* Vis scientist, Univ Fla, 69-71, asst prof biochem, 71-77, assoc prof, Dept Obstet & Gynec, 77-79; ASSOC PROF, DEPT OBSTET/GYNEC & PHYSIOL/BIOPHYSICS, MED CTR, UNIV ILL, CHICAGO, 79- *Concurrent Pos:* NIH trainee biophys chem, 67-68, grant, 69-78, career develop award, 69-77; Am Diabetes Asn, 80-81. *Mem:* AAAS; NY Acad Sci; Am Soc Biol Chem; Am Chem Soc; Soc Gynecologic Invest; Sigma Xi. *Res:* Chemical carcinogenesis; steroid and peptide hormone receptors; diabetes; fertility control. *Mailing Add:* Obstet-Gynec Dept Univ SFla Four Columbia Dr Harbor Side Med Tampa FL 33606-3500

TSICHRITZIS, DENNIS, computer science, for more information see previous edition

TSIEN, HSIENCHYANG, b Nanking, China, July 26, 39; m 66; c 2. MICROBIOLOGY. *Educ:* Nat Taiwan Univ, BS, 61; Cath Univ Louvain, Belg, DrS(microbiol & biochem), 67. *Prof Exp:* Asst, Cath Univ Louvain, 67-68 & 70-72; res instr microbiol, Temple Univ, 72-75; res fel, 68-70, ASST PROF MICROBIOL, UNIV MINN, 75- *Mem:* Am Soc Microbiol; AAAS. *Res:* Soil microbiology; microorganism-plant symbiotic nitrogen fixation; microbial ecology; microbial physiology; structure and function of bacterial cell membranes and cell walls. *Mailing Add:* Dept Microbiol Box 196 Univ Minn 420 Delaware St SE Minneapolis MN 55455

TSIEN, RICHARD WINYU, b Tating, China, Mar 3, 45; US citizen; m 71; c 3. ION CHANNELS. *Educ:* Mass Inst Technol, SB, 65, SM, 66; Oxford Univ, DPhil, 70. *Prof Exp:* Lect fel, Balliol Col, Oxford, 69-70; from asst prof to prof, dept physiol, Yale Univ Sch Med, 70-88; GEORGE D SMITH PROF & CHMN DEPT MOLECULAR & CELLULAR PHYSIOL, STANFORD UNIV, 88- *Concurrent Pos:* Weir jr res fel, Univ Col, Oxford, 66-70; estab investr, Am Heart Asn, 74-79, Javits investr, Nat Inst Neurol & Communicative Disorders & Stroke, 86-93; SKF lectr, Univ Col, London & Bass lectr, Vanderbilt Univ, distinguished lectr, Stanford Univ, Sterling lectr, Columbia Univ; mem, US Nat Comt Int Union Pure & Appl Biophys. *Honors & Awards:* Kenneth S Cole Award, 85; Otsuka Award, Int Soc Heart Res, 85. *Mem:* Biophys Soc; Soc Gen Physiologists (pres, 87-88). *Res:* Mechanisms of calcium delivery to neurons and muscle cells and their modulation by neurotransmitters, hormones and drugs; the function of heart, smooth muscle, neurons and their electrical activity and such functions as contraction and neurotransmitter release; author of numerous articles and papers. *Mailing Add:* Dept Molecular & Cellular Physiol Beckman Ctr Stanford Univ Med Ctr Stanford CA 94305-5426

TSIGDINOS, GEORGE ANDREW, inorganic chemistry; deceased, see previous edition for last biography

TSIN, ANDREW TSANG CHEUNG, b Hong Kong, July 19, 50; Can citizen; m 79; c 1. VISUAL BIOCHEMISTRY, NEUROBIOLOGY. *Educ:* Dalhousie Univ, BSc, 73; Univ Alta, MSC, 76, PhD(zool), 79. *Prof Exp:* From asst prof to assoc prof, Div Life Sci, Univ Tex, San Antonio, 81-90, from adj asst prof to adj assoc prof, Dept Ophthal, Univ Tex Health Sci Ctr, San Antonio, 84-90, PROF, DIV LIFE SCI, UNIV TEX, SAN ANTONIO, 90-, ADJ PROF, DEPT OPHTHAL, UNIV TEX HEALTH SCI CTR, 90- *Concurrent Pos:* Ed consult vision res, Exp Eye Res, J Exp Biol, Arch Biochem & Biophys, Physiol & Behav & Brain Res Bull; reviewer and panelist, NIH & NSF; prin investr, NIH grant, NSF res grant; consult, Technol Inc, San Antonio, 85-86, Vision Res & Develop, Lubbock, 86-87 & Alcon Lab, Ft Worth, 89- *Mem:* Am Soc Zoologists; NY Acad Sci; AAAS; Asn Researchers Vision & Opthal; Soc Neurosci; Biophys Soc; Soc Neurosci. *Res:* Metabolism of vitamin A; rhodopsin biosynthesis; comparative physiology and biochemistry; disease mechanism of the eye. *Mailing Add:* Div Life Sci Univ Tex San Antonio TX 78249-0662

TSIPIS, KOSTA M, b Athens, Greece, Feb, 12, 34; US citizen; m 70; c 3. NUCLEAR PHYSICS. *Educ:* Rutgers Univ, BSc, 58, MSc, 60; Columbia Univ, PhD(nuclear physics), 66. *Prof Exp:* Res assoc particle physics, 66-68, asst prof, 68-71, res fel biophys, 71-73, res fel, Ctr Int Studies,73-78, res assoc Physics, 78-79, res sci Physics 80-81, prin res sci, physics, 81-83, DIR, PROG SCI & TECHNOL INT SECURITY, MASS INST TECHNOL, 78- & PRIN RES SCI, SCI,TECHNOL & SOC, 83- *Concurrent Pos:* Sr consult, Stockholm Int Peace Res Inst, 73-77; adv, govt Greece, 77-; sci adv, World Coun Churches, 83- *Honors & Awards:* Szilard Award, Am Phys Soc, 84. *Mem:* Fel Am Phys Soc; fel Asn Advan Sci; fel NY Acad Sci. *Res:* Particle physics; technical and scientific aspects of national defense policy. *Mailing Add:* Prog Sci & Tech Int Security Mass Inst Technol 20A-011 Cambridge MA 02139

TSIVIDIS, YANNIS P, b Piraeus, Greece, 46. ELECTRICAL ENGINEERING, ELECTRONICS. *Educ:* Univ Minn, Minneapolis, BEE, 72; Univ Calif, Berkeley, MS, 73, PhD(eng), 76. *Prof Exp:* Engr electronics, Motorola Semiconductor, Phoenix, Ariz, 74; lectr elec eng, Univ Calif, Berkeley, 76; from asst prof to assoc prof, 76-84, PROF ELEC ENG, COLUMBIA UNIV, 84- *Concurrent Pos:* Mem tech staff, Bell Labs, Am Telephone & Telegraph Co, 77, resident vis, 77-; vis assoc prof, Mass Inst Technol, 80. *Honors & Awards:* Baker Prize, Inst Elec & Electronics Engrs. *Mem:* Fel sr mem Inst Elec & Electronics Engrs; Sigma Xi. *Res:* Design, analysis and simulation of integrated circuits; electronics; semiconductor device modelling; signal processing; circuit theory. *Mailing Add:* Dept Elec Eng Columbia Univ New York NY 10027

TSO, MARK ON-MAN, b Hong Kong, China, Oct 19, 36; US citizen; m 64; c 2. OPHTHALMOLOGY, PATHOLOGY. *Educ:* Univ Hong Kong, MB, BS, 61. *Prof Exp:* Res assoc, Armed Forces Inst Path, 71-76; PROF OPHTHAL, DIR GEORGIANA THEOBALD OPHTHALMIC PATH LAB & DIR MACULAR CLIN, EYE & EAR INFIRMARY, UNIV ILL, 76- *Concurrent Pos:* Fel ophthal path, Armed Forces Inst Path, 67-68, res fel, 68-69; assoc res prof ophthal, Med Ctr, George Washington Univ, 73-76; William Friedkin scholar, 76; mem, Visual Disorder Study Sect, NIH, 78-82 & bd scientific counrs, Nat Eye Inst; vpres, Asn Res Vision & Opthalmol, 84; pres, Am Asn Opthalmic Pathologists, 84-85. *Honors & Awards:* Friedenwald Award, Asn Res Vision & Ophthal, 89. *Mem:* Fel Am Acad Ophthal & Otolayrngol; Asn Res Vision & Ophthal. *Res:* Experimental pathology; clinical ophthalmology, especially diseases of retina and macula; electron microscopy; tissue culture; ocular oncology. *Mailing Add:* Eye & Ear Infirmary 1855 W Taylor St Chicago IL 60612

TS'O, PAUL ON PONG, b Hong Kong, July 17, 29; m 55; c 3. BIOPHYSICAL CHEMISTRY. *Educ:* Lingnan Univ, China, BS, 49; Mich State Univ, MS, 51; Calif Inst Technol, PhD, 55. *Prof Exp:* Res fel biol, Calif Inst Technol, 55-61, sr res fel, 61-62; assoc prof biophys chem, 62-67, PROF BIOPHYS CHEM, JOHNS HOPKINS UNIV, 67-, DIR DIV BIOPHYSICS SCH HYG & PUB HEALTH, 73- *Concurrent Pos:* Consult, Nat Cancer Inst; assoc ed, Molecular Pharmacol, 64-, Biochem, 66-74, Biophys J, 69-72, Biochem Biophys Acta, 71-81, Cancer Review, 73-, Cancer Res, 75-, J Environ Health Sci, 76-; mem, Biophys Study Sect, NIH, 76-80. *Mem:* Am Chem Soc; Am Soc Biol Chemists; Biophys Soc; Am Asn Cancer Res; Am Soc Microbiol; Am Soc Cell Biol. *Res:* Biophysics, chemistry and biology of nucleic acids; chemical carcinogenesis and mutagenesis; magnetic resonance; cell biology of differentiation and aging; interferon induction and function; DNA rearrangement and control of gene expression. *Mailing Add:* Sch Hyg & Pub Health Johns Hopkins Univ 615 N Wolfe St Baltimore MD 21205-2122

TSO, TIEN CHIOH, b Hupeh, China, July 25, 17; nat US; m 49; c 2. PHYTOCHEMISTRY. *Educ:* Nanking Univ, China, BS, 41, MS, 44; Pa State Univ, PhD(agr biochem), 50. *Prof Exp:* Supt exp farm, Ministry Social Affairs, China, 44-46; secy, Tobacco Improv Bur, 46-47; chemist res lab, Gen Cigar Co, 50-51; chemist div tobacco & spec crops, USDA, 52; asst prof & res assoc agron, Univ Md, 53-59; res plant physiologist, Tobacco & Sugar Crops Res Br, USDA, 59-62, sr plant physiologist, 62-64, prin plant physiologist, 64-66, leader tobacco qual invests, Tobacco & Sugar Crops Res Br, 66-72, chief tobacco lab, 72-83, sr exec serv, 79-83, COLLABR TOBACCO & HEALTH, BELTSVILLE AGR RES CTR, AGR RES SERV, USDA, 84- *Concurrent Pos:* Mem, Tobacco Chem Res Conf, Tobacco Workers Conf, World Conf Tobacco & Health, Tobacco Working Group, Lung Cancer Task Force & Int Tobacco Working Group; exec dir, Int Develop & Educ in Agr & Life Sci, 83-; sr consult, China Nat Tobacco Corp, 84-, US & foreign tobacco indust, 84- *Honors & Awards:* Coresta Prize, 78; US Presidential Rank Award, 84. *Mem:* Fel AAAS; fel Am Agron Soc; Am Chem Soc; fel Am Inst Chem; Am Soc Plant Physiol. *Res:* Plant physiology; tobacco alkaloids; biochemistry; culture; radio elements; health related components; tobacco production research relating to smoking and health. *Mailing Add:* Agr Res Ctr W Rm 009 Bldg 005 USDA Beltsville MD 20705

TS'O, TIMOTHY ON-TO, b Hong Kong, Nov 9, 34; US citizen; m 63; c 2. NEUROPHARMACOLOGY, BEHAVIORAL TOXICOLOGY. *Educ:* Univ Hong Kong, MB & BS, 59; Stanford Univ, PhD(neuropharmacol & psychopharmacol), 68; Am Bd Psychiat & Neurol, dipl, 79. *Prof Exp:* Demonstr, Dept Path, Fac Med, Univ Hong Kong, 60-62; sr res specialist pharmacol, Dow Chem Co, Mich, 68-74; res prof human biol, Saginaw Valley State Col, Mich, 74-75; lab chief neuropharmacol, Long Island Res Inst, 75-79; psychiatrist, Vet Admin Med Ctr, Northport, NY, 77-79; assoc prof psychiat & behav sci, Univ Health Sci, Chicago Med Sch, 79-83; PSYCHIATRIST, VET ADMIN MED CTR, NORTH CHICAGO, 79- *Concurrent Pos:* House physician & surgeon, Govt Surg Unit & Univ Med Unit, Queen Mary Hosp, Hong Kong, 59-60; resident, Dept Psychiat, State Univ NY Stony Brook, 75-7, res asst prof, 77-79; psychiatrist, pvt pract, 81- *Mem:* NY Acad Sci; AAAS; Biophys Soc; Am Psychiat Asn; AMA; Am Pharm Asn; Am Med Soc Alcohol & Drug Dependancies. *Res:* Effects of central nervous system drugs and toxicants on performance, memory and learning; computer system applications in behavioral pharmacology and toxicology, psychiatry and in electroencephalogram analysis; biomathematics; broncho-genic carcinoma; congenital tumors. *Mailing Add:* 1110 W Golf Rd Libertyville IL 60048

TSOKOS, CHRIS PETER, b Greece, Mar 25, 37; US citizen; c 3. APPLIED MATHEMATICS. *Educ:* Univ RI, BS & MS, 61; Univ Conn, PhD(math statist & probability), 67. *Prof Exp:* Consult opers res anal bur naval weapons, US Naval Air Sta, 64; proj engr elec boat div, Gen Dynamics Corp, 61-63; asst prof math, Univ RI, 63-69; assoc prof statist, Va Polytech Inst, 69-71; prof math & statist, 71-76, DIR GRAD PROG STATIST & STOCHASTIC SYSTS, UNIV SFLA, 71- *Concurrent Pos:* RI Res Coun res grants 65-66 & 67-68; NSF lectr, Univ RI, 65-68; consult, US Army Electronics Command Ctr, Ft Monmouth, NJ; dir contracts, AFSOR, 74-79, NASA, 75-79 & Bur Land Mgt, 74-77; vpres, Robert M Thrall & Assocs, Houston. *Mem:* AAAS; Am Math Soc; fel Am Statist Asn; Opers Res Soc Am. *Res:* Statistical theory and applications; stochastic integral equations; stochastic systems theory; biomathematics; stochastic modeling; time series; stochastic differential games; Bayesian reliability theory and simulation. *Mailing Add:* Dept Math Univ SFla 4202 Fowler Ave Tampa FL 33620

TSOLAS, ORESTES, b Istanbul, Turkey, Dec 5, 33; US citizen; m 86. IMMUNOLOGY. *Educ:* Robert Col, Istanbul, BSc, 54; Cambridge Univ, BA, 57, MA, 61; Albert Einstein Col Med, PhD(molecular biol), 67. *Prof Exp:* Asst prof molecular biol, Albert Einstein Col Med, 70-72; asst mem, Roche Inst Molecular Biol, 72-78; PROF & HEAD LAB BIOCHEM, SCH MED, UNIV IOANNINA, 78-, DIR, UNIV CLIN CHEM LAB, 81- *Concurrent Pos:* Prof, Univ Sao Paulo, 69-; adj prof, Rutgers Univ, 74-78; vis lectr, Rotterdam Med Fac, Neth, 72; vis asst prof microbiol & immunol, Albert Einstein Col Med, 72-79; dean Med Sch, Univ Ioannina, 80-82; sci coun, Hellenic Pasteur Inst, 83-85; Greek Nat Adv Coun Res, 85-; exec coun, Hellenic Biochem & Biophys Soc, 84-85. *Mem:* Am Soc Biochem & Molecular Biol; Am Chem Soc; Biochem Soc Gt Brit; Am Asn Clin Chem; Nat Acad Clin Chem; Hellenic Biochem & Biophys Soc (vpres, 80-81). *Res:* Enzymes and immunopeptides. *Mailing Add:* Lab Biol Chem Univ Ioannina Med Sch Ioannina GR-451 10 Greece

TSONG, IGNATIUS SIU TUNG, b Hong Kong, Jan 4, 43; Australian citizen; m 70; c 1. PHYSICS, MATERIALS SCIENCE. *Educ:* Univ Leeds, BSc, 66, MSc, 67; Univ London, PhD(physics), 70. *Prof Exp:* Fel physics, Univ Essex, 70-73; sr tutor, Monash Univ, 73-76; res assoc, Pa State Univ, 76-78, asst prof mat res, 78-79, res assoc, 79-81; AT PHYSICS DEPT, ARIZ STATE UNIV, 81- *Mem:* Am Phys Soc; Mineral Soc Am; Am Ceramic Soc; Am Vacuum Soc; AAAS; Electrochem Soc. *Res:* Sputter-induced optical emission; surface characterization using ion beam techniques; analysis of hydrogen in solids; physics of particle-solid interactions. *Mailing Add:* Physics Dept Ariz State Univ Tempe AZ 85287

TSONG, TIAN YOW, b Taiwan, Sept 6, 34, US citizen; m 71; c 2. BIOPHYSICAL CHEMISTRY, BIOCHEMISTRY. *Educ:* Chung Hsing Univ, Taiwan, BS, 64; Yale Univ, MS, 67, MPh, 68, PhD(phys biochem), 69. *Prof Exp:* From asst prof to assoc prof, physiol chem, Sch Med Johns Hopkins Univ, 72-87; PROF BIOCHEM, UNIV MINN, 88- *Concurrent Pos:* Fel, Stanford Univ, 70-72; NSF res grant, Sch Med, Johns Hopkins Univ, 73-; NIH res grant, 75-; mem, NIH Study Sect, 80-84; Off Naval Res Contract, 87- *Mem:* Am Chem Soc; Biophys Soc; Am Soc Biol Chemists; AAAS. *Res:* Physical chemistry of proteins and membrane lipids, and its correlation to biological functions. *Mailing Add:* Dept Biochem Univ Minn 1479 Gortner Ave St Paul MN 55108

TSONG, TIEN TZOU, b Taiwan, China, Sept 6, 34; m 64; c 3. SOLID STATE PHYSICS. *Educ:* Taiwan Norm Univ, BSc, 59; Pa State Univ, MS, 64, PhD(physics), 66. *Prof Exp:* Res assoc physics, 67-69, from asst prof to assoc prof, 69-74, PROF PHYSICS, PA STATE UNIV, 75- *Concurrent Pos:* Fel, Japan Soc Prom Sci. *Mem:* Fel Am Phys Soc; Am Vacuum Soc; fel Japan Soc Promotion Sci; Mat Res Soc. *Res:* Surface physics; field effect on metal surface; field ionization; field desorption and field ion microscopy; atomic processes on solid surfaces. *Mailing Add:* Dept Physics Pa State Univ 104 Davey Lab University Park PA 16802

TSONG, YUN YEN, b Taiwan, China, Jan 15, 37; m 67; c 2. BIOCHEMISTRY, ORGANIC CHEMISTRY. *Educ:* Nat Taiwan Univ, BS, 60; Univ Wis-Madison, PhD(biochem), 68. *Prof Exp:* Sr med chemist, Smith Kline & French Labs, 68-70; res assoc biochem, 70-72, SCIENTIST BIOCHEM, POP COUN, ROCKEFELLER UNIV, SCIENTIST, 74- *Mem:* AAAS; Am Chem Soc; Endocrine Soc; Am Fertil Soc; NY Acad Sci. *Res:* Mechanism of action of steroid and peptide hormones; metabolism and microbial transformation of steroids; contraceptive developmemt. *Mailing Add:* 33 Evergreen Dr N Caldwell NJ 07006

TSONOPOULOS, CONSTANTINE, b Megalopolis, Greece, Sept 5, 41; US citizen; m 69; c 2. CHEMICAL ENGINEERING, THERMODYNAMICS. *Educ:* Ga Inst Technol, BChemEng, 64, MS, 65; Univ Calif, Berkeley, PhD(chem eng), 70. *Prof Exp:* Engr appl thermodynamics, 70-71, res engr, 71-74, sr res engr, 74-78, sr staff engr, 78-80, eng assoc, 80-82, SR ENG ASSOC, EXXON RES & ENG CO, SUBSID EXXON CORP, 82- *Concurrent Pos:* Adj prof, NJ Inst Technol, 76 & 78; mem, DOE peer rev panels, 84, 87, NRC panel chem eng, 88-90, NSF adv cmt chem & process eng, 83-84. *Mem:* Am Inst Chem Engrs; Am Chem Soc; Am Petrol Inst; Sigma Xi; AAAS. *Res:* Thermodynamics of fluid-phase equilibria; properties of polar systems and electrolyte solutions (hydrocarbon/water/weak electrolytes); synthetic liquids (from coal or shale). *Mailing Add:* Exxon Res & Eng Co PO Box 101 Florham Park NJ 07932

TSOU, CHEN-LU, b Wuxi, China, May 17, 23; m 49; c 1. ENZYME MECHANISM, PROTEIN FOLDING. *Educ:* Nat SW Univ, China, BSc, 45; Univ Cambridge, Eng, PhD(biochem), 51. *Prof Exp:* Res assoc, Inst Biochem, Academia Sinica, 51-56, res prof, 56-91, vdir, 77-83, DIR, NAT LAB BIOMACROMOLECULES, INST BIOPHYS, ACADEMIA SINICA, 88- *Concurrent Pos:* Ed, Fed Am Socs Exp Biol J, 88; vis prof, Harvard Univ Med Sch, 81-82; Fogarty scholar, NIH, 86-90. *Mem:* Hon mem Am Soc Biochem & Molecular Biol; NY Acad Sci. *Res:* Comparison of activity and conformational changes during enzyme folding and unfolding; mechanism and regulation of enzyme action; kinetics of irreversible inactivation of enzymes; effect of chemical modification on the activities of proteins. *Mailing Add:* Inst Biophys Academia Sinica Beijing 100080 China

TSOU, F(U) K(ANG), b Kiangsu, China, May 25, 22; m 50. HEAT TRANSFER, FLUID MECHANICS. *Educ:* Nat Cent Univ, China, BS, 43; Univ Toronto, MS, 58; Univ Minn, PhD(mech eng), 65. *Prof Exp:* Engr, Taiwan Mach Mfg Co, 51-57; res asst & fel mech eng, Univ Minn, 58-62; from asst prof to assoc prof, 65-75, PROF MECH ENG, DREXEL UNIV, 75- *Mem:* Fel Am Soc Mech Engrs; Am Acad Mech. *Res:* Convective heat transfer. *Mailing Add:* 32 Eastwood Dr West Berlin NJ 08091

TSOULFANIDIS, NICHOLAS, b Ioannina, Greece, May 6, 38; m 64; c 2. NUCLEAR ENGINEERING, PHYSICS. *Educ:* Nat Univ Athens, BS, 60; Univ Ill, Urbana, MS, 65, PhD(nuclear eng), 68. *Prof Exp:* From instr to prof nuclear eng, 68-81, chmn dept, 81-85, interim vchancellor acad affairs, 85-86, PROF & ASST DEAN, UNIV MO-ROLLA, 87- *Concurrent Pos:* Sabbatical leave, Nuclear Res Ctr, Cadarache, France, 86-87; consult. *Mem:* Nat Soc Prof Engrs; Am Nuclear Soc; Health Physics Soc. *Res:* Neutron and gamma transport; health physics; nuclear fuel cycle; author of two books. *Mailing Add:* Dept Nuclear Eng Univ Mo Rolla MO 65401

TSU, T(SUNG) C(HI), b Haining, China, Aug 27, 15; nat US; m 48; c 1. PROJECT PLANNING & EVALUATION. *Educ:* Chiao Tung Univ, BSc, 37; Univ Toronto, MASc, 41; Mass Inst Technol, ScD(aeronaut eng), 44. *Prof Exp:* Asst engr, Bur Aeronaut Res, China, 39-40; engr, Div Indust Coop, Mass Inst Technol, 44-45; consult engr, Gen Mach Corp, Ohio, 45-47; res assoc, Pa State Col, 47-49, assoc prof, 50-52; sr engr, Aviation Gas Turbine Div, Div Westinghouse Elec Corp, 52-55, res engr, Res Labs, 55-57, adv engr, 57-82; RETIRED. *Concurrent Pos:* Adj prof, Drexel Inst Technol, 53-54. *Mem:* AAAS; Am Soc Mech Engrs. *Res:* Space vehicles propulsion; magnetohydrodynamics; fluid mechanics; thermodynamics; energy conversion; design optimization. *Mailing Add:* 3540 Ridgewood Dr Pittsburgh PA 15235-5230

TSUANG, MING TSO, b Tainan, Taiwan, Nov 16, 31; US citizen; m 58; c 3. PSYCHIATRY, PSYCHIATRIC EPIDEMIOLOGY. *Educ:* Nat Taiwan Univ, MD, 57; Univ London, PhD(psychiat), 65, DSc, 81. *Hon Degrees:* MA, Brown Univ, 83; AM, Harvard Univ, 87. *Prof Exp:* Lectr psychiat & sr psychiatrist, Dept Neurol & Psychiat, Nat Taiwan Univ Hosp, 61-63; vis res worker psychiat, Med Res Coun Psychiat Genetics Res Unit, Maudsley Hosp & Inst Psychiat, Univ London, 63-65; lectr psychiat & sr psychiatrist, Dept Neurol & Psychiat, Nat Taiwan Univ Hosp, 65-68, assoc prof psychiat & sr psychiatrist, 68-71; vis assoc prof psychiat & staff psychiatrist, Barnes & Renard Hosp, Sch Med, Wash Univ, 71-72; assoc prof psychiat & staff psychiatrist, Iowa Psychiat Hosp, Col Med, Univ Iowa, 72-75, prof psychiat, 75-82; prof psychiat & vchmn dept psychiat, Brown Univ, 82-85; DIR PSYCHIAT EPIDEMIOL, HARVARD UNIV, 85-, PROF PSYCHIAT, MED SCH, 85-; CHIEF PSYCHIAT SERV, BROCKTON-WEST ROXBURY VA CTR, 85- *Concurrent Pos:* Res fel, Nat Coun Sci Develop, Repub China, 60-70; fel, Sino-Brit Fel Trust, UK, 63-65; collab investr, Int Pilot Study Schizophrenia, WHO, Geneva, Switz, 66-71; consult psychiatrist, Vet Admin Hosp, Iowa City, Iowa, 72-82; chief staff psychiatrist E Ward, Psychiat Hosp, Univ Iowa, 72-82; vis prof, Dept Psychiat, Univ Oxford, Eng, 79-80; Josiah Macy fac scholar award, 79-80; chief psychiat, Psychiat Serv, Brockton-West Roxbury Vet Admin Med Ctr, 85-; NIMH Merit Award on Psychopath & Heterogeneity of Schizophrenia, 88-93; mem, Med Res Serv Planning Coun, Vet Health Serv & Res Admin, Vet Admin Cent Off, 90-, Extramural Sci Adv Bd, NIMH. *Honors & Awards:* Clin Res Award, Am Acad Clin Psychiatrists, 83; Rema Lapouse Award, Am Pub Health Asn, 84; Stanley Dean Award for Res in Schizophrenia, Am Col Psychiatrists, 89. *Mem:* Psychiat Res Soc; AAAS; Am Psychopath Asn; Behav Genetics Asn; Am Psychiat Asn; Sigma Xi. *Res:* Long-term follow-up and family studies of schizophrenia, mania, depression and atypical psychoses; diagnostic classification of mental disorder; psychiatric genetics; clinical psychopharmacological research; heterogeniety of schizophrenia; genetic linkage studies of schizophrenia. *Mailing Add:* Psychiat Serv 116A Brockton/W Roxbury VA Med Ctr 940 Belmont St Brockton MA 02401

TSUBOI, KENNETH KAZ, b Seno, Japan, Feb 7, 22; nat US; m 47; c 2. BIOCHEMISTRY. *Educ:* St Thomas Col, BS, 44; Univ Minn, MS, 46, PhD(biochem), 48. *Prof Exp:* Asst physiol chem, Univ Minn, 44-47; res assoc path, Washington Univ, 48; res assoc oncol, Univ Kans Med Ctr, Kansas City, 48-51; res assoc biochem, Columbia Univ, 51-55; asst prof, Med Col, Cornell Univ, 55-60; assoc prof pediat, 60-66, sr res assoc, 66-73, adj prof, 73-82, PROF PEDIAT, SCH MED, STANFORD UNIV, 82- *Concurrent Pos:* Estab investr, Am Heart Asn, 59-64; vis prof, Univ Tokyo, 67. *Mem:* Am Soc Biol Chemists; Biophys Soc; Am Asn Cancer Res. *Res:* Cellular and muscle biochemistry; enzymology. *Mailing Add:* Dept Pediat Med Ctr S 232 Stanford Univ Stanford CA 94305

TSUCHIYA, HENRY MITSUMASA, bacteriology, for more information see previous edition

TSUCHIYA, MIZUKI, b Matsuyama, Japan, May 2, 29; m 56; c 2. PHYSICAL OCEANOGRAPHY. *Educ:* Univ Tokyo, BS, 53, DSc, 62. *Prof Exp:* Res asst geophys, Univ Tokyo, 55-60; instr oceanog, Meteorol Col, Japan Meteorol Agency, 60-64; res assoc, Johns Hopkins Univ, 64-67; lectr, Univ Tokyo, 67-69; from asst res oceanogr to assoc res oceanogr, 69-79, RES OCEANOGR, SCRIPPS INST OCEANOG, 79- *Honors & Awards:* Okada Takematsu Prize, Oceanog Soc Japan, 67. *Mem:* Am Geophys Union; Oceanog Soc. *Res:* Circulation and distributions of water characteristics in the ocean. *Mailing Add:* Scripps Inst Oceanog A-030 La Jolla CA 92093

TSUCHIYA, TAKUMI, b Oita-Ken, Japan, Mar 10, 23; US citizen; m 53; c 2. PLANT CYTOLOGY, PLANT GENETICS. *Educ:* Gifu Univ, BAgr, 43; Kyoto Univ, BAgr, 47, DAgr(genetics), 60. *Prof Exp:* Asst prof biol, Beppu Univ, 50-57; cytogeneticist, Kihara Inst Biol Res, 57-63; Nat Res Coun Can fel plant cytogenetics, Univ Man, 63-64; cytogeneticist, Children's Hosp, Winnipeg, 64-65; res assoc plant cytogenetics, Univ Man, 65-68; assoc prof plant cytogenetics, 68-73, PROF GENETICS, COLO STATE UNIV, 73- *Concurrent Pos:* Rockefeller Found travel grant insts & univs, US & Can, 61; ed, Barley Genetics Newletters, 69-88; coordr genetic & linkage studies barley, Int Barley Genetics Symp, 70-86; chmn, Int Comt Nomenclature & Symbolization Barley Genes, 70-86; chmn barley genetics comt, Am Barley Res Workers' Conf, 71-; standing collabr, Cytologia, Int Soc Cytol, Tokyo, 83-87. *Honors & Awards:* Crop Sci Res Award, Crop Sci Soc Am, 86. *Mem:* Fel Am Soc Agron; fel Crop Sci Soc Am; Am Genetics Asn; Genetics Soc Can; fel Japan Soc Prom Sci; hon foreign mem Genetics Soc Japan; Soc Econ Bot. *Res:* Genetics of barley, sugar beet, triticale, rye; cytogenetic and evolutionary studies of species of Gramineae; cytotaxonomy of tree species in Taxodiaceae; cytology of ornamental crops. *Mailing Add:* Dept Agron Colo State Univ Ft Collins CO 80523

TSUDA, ROY TOSHIO, b Honolulu, Hawaii, Dec 25, 39; m 59; c 3. PHYCOLOGY. *Educ:* Univ Hawaii, BA, 63, MS, 66; Univ Wis-Milwaukee, PhD(bot), 70. *Prof Exp:* Instr biol, Univ Guam, 67-68, asst prof dept biol & marine lab, 68-70, assoc prof marine biol, 70-74, dir, marine lab, 74-76, dean, Grad Sch & Res, 78-84, prof marine biol, 74-89, acad vpres, 84-89, EMER PROF MARINE BIOL, UNIV GUAM, 89-; CHIEF ENVIRON SERV, DUENAS & SWAVELY, INC, 90- *Concurrent Pos:* gen ed, Micronesica, Univ Guam, 72-76, chmn, Coral Reef Comt, Pac Sci Asn, 75-81; chmn coral reef comt, Int Asn Biol Oceanog, 79-81; mem comt ecology, Int Union Conserv Nature & Natural Resources, 81-84, Survival Serv comt, 77-83. *Mem:* Am Soc Limnol & Oceanog; Asn Trop Biol; Phycol Soc Am; Int Soc Reef Studies; Int Phycol Soc. *Res:* Taxonomy and ecology of tropical marine algae; primary productivity. *Mailing Add:* Univ Guam Sta PO Box 5316 Mangilao GU 96923

TSUEI, YEONG GING, b China, Feb 25, 32; m 62; c 4. APPLIED MECHANICS. *Educ:* Cheng Kung Univ, Taiwan, BSCE, 56; Colo State Univ, MCE, 60, PhD(fluid mech), 63. *Prof Exp:* Asst, Cheng Kung Univ, Taiwan, 56-58; from instr to assoc prof, 61-76, PROF MECH, UNIV CINCINNATI, 76- *Mem:* Am Inst Aeronaut & Astronaut; Am Soc Civil Engrs; Am Soc Eng Educ. *Res:* Modal analysis and engineering mechanics. *Mailing Add:* Dept Mech & Indust Eng Univ Cincinnati Cincinnati OH 45221-0072

TSUI, BENJAMIN MING WAH, b Hong Kong, June 1, 48; m 75; c 2. MEDICAL PHYSICS & IMAGING. *Educ:* Chung Chi Col, BSc, 70; Dartmouth Col, AM, 72; Univ Chicago, PhD(med physics), 77. *Prof Exp:* Res assoc med physics, Univ Chicago, 77-79, asst prof, dept radiol & Franklin McLean Res Inst, 79-82; ASSOC PROF RADIOL & BIOMED ENG, UNIV NC, CHAPEL HILL, 82- *Honors & Awards:* Sci Res Award, Eastman Kodak Co, 77. *Mem:* Am Asn Physicists Med; Soc Nuclear Med; Soc Magnetic Resonance in Med; Inst Elec & Electronics Engrs; AAAS; Sigma Xi. *Res:* Theory and instrumentation in radiation detection; image formation and recording in nuclear medicine; emission computed tomography imaging; magnetic resonance imaging. *Mailing Add:* Dept Radiol & Biomed Eng CB 7575 Univ NC 152 MacNider Hall Chapel Hill NC 27599-7575

TSUI, DANIEL CHEE, b Henan, China, Feb 28, 39; US citizen; m 64; c 2. SOLID STATE PHYSICS. *Educ:* Augustana Col, BA, 61; Univ Chicago, MS & PhD(physics), 67. *Prof Exp:* Res assoc, Univ Chicago, 67-68; mem tech staff, Bell Labs, Murray Hill, NJ, 68-82; ARTHUR LEGRAND DOTY PROF ELEC ENG, PRINCETON UNIV, 82- *Honors & Awards:* Oliver E Buckley Condensed Matter Physics Prize, Am Phys Soc, 84. *Mem:* Nat Acad Sci. *Res:* Electronic properties of metals, surface properties of semiconductors; low temperature physics. *Mailing Add:* Elec Eng Dept Princeton Univ Princeton NJ 08544

TSUI, JAMES BAO-YEN, b Shantung, China; US citizen. ELECTRICAL ENGINEERING. *Educ:* Nat Taiwan Univ, BS, 57; Marquette Univ, MS, 61; Univ Ill, PhD(elec eng), 65. *Prof Exp:* From asst prof to assoc prof elec eng, Univ Dayton, 65-73; ELECTRONICS ENGR, AVIONICS LAB, WRIGHT-PATTERSON AFB, 73- *Concurrent Pos:* Scientist, Labtron Corp Am, 68-69. *Mem:* Am Soc Eng Educ; Inst Elec & Electronics Engrs. *Res:* Rare earth cobalt permanent magnets and their applications; microwave receivers. *Mailing Add:* Air Force Avionics Lab AFWAL-AAWP Wright-Patterson AFB OH 45433

TSUI, Y(AW) T(ZONG), physics, engineering science, for more information see previous edition

TSUJI, FREDERICK ICHIRO, b Honolulu, Hawaii, Aug 23, 23. BIOCHEMISTRY. *Educ:* Cornell Univ, AB, 46, MS, 48, PhD(biochem), 50. *Prof Exp:* Asst biochem & nutrit, Cornell Univ, 48-49; res biochemist, Children's Fund Mich, 49-50; asst prof biochem & pharmacol, Duquesne Univ, 50-52; res asst biol, Princeton Univ, 52-55; tech dir res lab, Vet Admin Hosp, Pittsburgh, Pa, 55-72; biochemist, Brentwood Vet Admin Hosp, Los Angeles, 72-76; prog dir biochem, NSF, 76-78; assoc res biochemist, 79-82, RES BIOCHEMIST, SCRIPPS INST OCEANOG, UNIV CALIF, SAN DIEGO, 82- *Concurrent Pos:* Res assoc, Mercy Hosp, 51-52; Anathan fel inst res, Montefiore Hosp, 52; investr, Marine Biol Lab, Woods Hole, 53-54; lectr, Univ Pittsburgh, 56-65, from adj assoc prof to adj prof, 65-72; sr scientist, Te Vega Exped Pac Ocean, Hopkins Marine Sta, Stanford Univ, 66; mem, Alpha Helix Biol Exped to New Guinea, Scripps Inst Oceanog, Univ Calif, San Diego, 69; Hancock fel, Univ Southern Calif, 72-76; vis prof dept med chem, Fac Med, Kyoto, Univ, Japan, 74; vis res prof, Univ Southern Calif, 78-82, res prof biol, 82-; biochemist, Brentwood Vet Admin Hosp, Los Angeles, 78-87; head, Dept Enzymes & Metab, Osaka Biosci Inst, Japan, 87- *Mem:* Am Soc Biol Chem; Am Chem Soc; Biophys Soc; Am Asn Immunol; Soc Gen Physiol. *Res:* Bioluminescence; enzyme reactions. *Mailing Add:* Marine Biol Res Div A-002 Scripps Inst Oceanog Univ Calif San Diego La Jolla CA 92093

TSUJI, GORDON YUKIO, b Honolulu, Hawaii, July 31, 42; m 67; c 3. SOIL PHYSICS. *Educ:* Univ Hawaii, BS, 65, MS, 67; Purdue Univ, Lafayette, PhD(soil physics), 71. *Prof Exp:* Asst soil scientist dept agron & soil sci, 71-74, proj mgr, Benchmark soils proj, 74-83, PROJ MGR, INT BENCHMARK SITES NETWORK FOR AEROTECH TRANSFER & US AID SUBGRANT ON TROPSOILS & SOIL MGT, UNIV HAWAII/US AID, 83- *Concurrent Pos:* AID fel, Univ Hawaii. 71- *Mem:* Am Soc Agron; Sigma Xi; Int Soc Soil Sci; Soil Sci Soc Am; NZ Soc Soil Sci. *Res:* Water movement in soils; infiltration of water into soils; water distribution under drip irrigation; tropical meteorology; United States soil taxonomy and agricultural development; agrotechnology transference; systems analysis; crop modeling; decision support systems. *Mailing Add:* Dept Agron & Soil Sci Univ Hawaii 1910 Eastwest Rd Honolulu HI 96822

TSUJI, KIYOSHI, b Kyoto, Japan, May 31, 31; m 58; c 3. MICROBIOLOGY, ANALYTICAL CHEMISTRY. *Educ:* Kyoto Univ, BS, 54; Univ Mass, MS, 56, PhD(food technol), 60. *Prof Exp:* Fel food sci, Rutgers Univ, 59-60; staff microbiol, Nat Canners Asn, Calif, 60-63; res assoc food sci, Mass Inst Technol, 63-64; res assoc analytical res & develop, 64-71, sr res scientist, 71-74, SR SCIENTIST, CONTROL ANALYTICAL RES & DEVELOP, UPJOHN CO, 74- *Concurrent Pos:* Ed, J Lab Robotics & Automation & J Radiation Sterlization. *Honors & Awards:* William E Upjohn Prize, Upjohn Co, 71 & 88; ed, GLC & HPLC Determination of Therapeut

Agents, 78; Pioneer Lab Robotics, 85; Nicholas Copernicus Award, Qual Control Acad, 86. *Mem:* Am Chem Soc; Am Soc Microbiol; Inst Food Technologists. *Res:* Analytical microbiology; microbioassay automation; analysis of antibiotics by gas-liquid chromatography and high-performance liquid chromatography; sterility test; bacterial endotoxin analysis of vitamins; endotoxin detection by Limulus amebocyte lysate; environmental microbiology; application of expert system; laboratory automation; vision systems; oligonucleotide probe; biotechnology; analysis of recombinant proteins by capillary electrophoresis. *Mailing Add:* Upjohn Co 7831-41-1 Kalamazoo MI 49001

TSUK, ANDREW GEORGE, b Budapest, Hungary, July 11, 32; US citizen; wid; c 2. PHYSICAL CHEMISTRY, POLYMER CHEMISTRY. *Educ:* Budapest Polytech Inst, dipl chem eng, 54; Polytech Inst Brooklyn, PhD(chem), 64. *Prof Exp:* Engr, Indust Fermentations, Hungary, 54-56; res chemist, Schwarz Biores, Inc, 58-62, tech asst to pres, 64-65, dir radiochem div, 65-66; sr res chemist, W R Grace & Co, Md, 66-72; group leader pharmaceut develop, Ayerst Labs Inc, 72-74, res assoc, 74-76; SR SCIENTIST, BIOTEC INC, WOBURN, MASS, 86- *Concurrent Pos:* Lectr, Polmer Sci, Northeastern Univ, Boston, Mass, 86, 90-91. *Mem:* Am Chem Soc. *Res:* Polymers in pharmaceutical dosage forms; physical chemistry of polymers; polyelectrolytes; biomedical materials; biological macromolecules; ion-exchange and radioactive tracers. *Mailing Add:* 145 Robbins Rd Arlington MA 02174

TSUKADA, MATSUO, b Nagano, Japan, Jan 4, 30; m 56; c 2. ECOLOGY, PALEOECOLOGY. *Educ:* Shinshu Univ, Japan, BS, 53; Osaka City Univ, MA, 58, PhD(biol), 61. *Prof Exp:* Japan Acad Sci fel, Osaka City Univ, 61; Seessel fel, Yale Univ, 61-62, res assoc palynology, 62-66, lectr & res assoc biol, 66-68; assoc prof, 69-71, PROF BOT, UNIV WASH, 71-, DIR LAB PALEOECOL, 69-, ADJ PROF GEOL & QUATERNARY STUDIES, 76- *Concurrent Pos:* Sigma Xi res grant, Yale Univ, 63-64, Am Philos Soc res grant, 64-65; prin investr NSF res grants, Univ Wash, 70- *Mem:* Ecol Soc Am; Am Soc Limnol & Oceanog; Am Quaternary Asn; Am Asn Stratig Palynologists; Bot Soc Japan; Sigma Xi. *Res:* Present and past environmental changes on a global scale, mainly by means of modern and fossil plants, including pollen and also animals, chemicals and heavy metals, such as lead and cadmium. *Mailing Add:* 13809 SE 20th Bellevue WA 98005

TSUNG, YEAN-KAI, b Taiwan, July 30, 43; US citizen; m 72; c 2. SOMATIC CELL GENETICS, TUMOR BIOLOGY. *Educ:* Nat Taiwan Univ, BS, 67; Univ Ill, MS, 71, PhD(plant path), 74. *Prof Exp:* Asst prof genetics, Univ Mich, 75-76; trainee immunol, Div Immunol, Duke Univ Med Ctr, 76-77; asst geneticist, E K Shriver Ctr Ment Retardation, 77-81; ASST PROF HUMAN GENETICS, SCH MED, BOSTON UNIV, 82- *Concurrent Pos:* Consult, Brain Res, Inc, 80- *Mem:* AAAS. *Res:* Immunological identification of human cell membrane components by the development of cross membrane transport, defective mutants and monoclonal antibodies distinguishing the mutants from wild type population. *Mailing Add:* Ctr Human Genetics Boston Univ Sch Med 80 E Concord St Boston MA 02118

TSURUTANI, BRUCE TADASHI, b Los Angeles, Calif, Jan 29, 41. SPACE PLASMA PHYSICS. *Educ:* Univ Calif, Berkeley, BA, 63, PhD(physics), 72. *Prof Exp:* RES SCIENTIST & MEM TECH STAFF PHYSICS, JET PROPULSION LAB, CALIF INST TECHNOL, 72- *Mem:* AAAS; Am Geophys Union; Int Union Radio Sci; Sigma Xi. *Res:* Plasma instabilities; wave-particle interactions; interplanetary and planetary magnetic fields; particle acceleration processes; magnetospheric and heliospheric physics; x-ray sources. *Mailing Add:* Jet Propulsion Lab 4800 Oakgrove Dr Pasadena CA 91103

TSUTAKAWA, ROBERT K, b Seattle, Wash, Mar 28, 30; m 61; c 3. STATISTICS. *Educ:* Univ Chicago, BS, 56, MS, 57, PhD(statist), 63. *Prof Exp:* Res specialist, Boeing Co, 58-60 & 63-65; res assoc statist, Univ Chicago, 65-68; assoc prof, 68-78, PROF STATIST, UNIV MO-COLUMBIA, 78- *Mem:* Am Statist Asn; Inst Math Statist. *Res:* Statistical inference. *Mailing Add:* Dept Statist 222 Math Sci Bldg Univ Mo Columbia MO 65211

TSUTSUI, ETHEL ASHWORTH, b Geneva, NY, May 31, 27; m 56; c 1. BIOCHEMISTRY. *Educ:* Keuka Col, BA, 48; Univ Rochester, PhD(biochem), 54. *Prof Exp:* Res assoc med sch med & dent, Univ Rochester, 53-55; Nat Cancer Inst fel, Sloan-Kettering Inst Cancer Res, 55-56; lectr, Tokyo Med & Dent Univ, Japan, 56-57; asst prof biol & res biochemist, C F Kettering Found, Antioch Col, 57-60; res assoc Inst Cancer Res, Col Physicians & Surgeons, Columbia Univ, 60-63, res assoc dept biochem, 63-65; asst prof biol sci, Hunter Col, 65-69; assoc prof biol, 69-71, assoc prof, 71-80, PROF BIOCHEM & BIOPHYS, TEX A&M UNIV, 80- *Mem:* AAAS; Am Asn Cancer Res; Am Chem Soc; fel NY Acad Sci; Harvey Soc. *Res:* Enzymatic methylation of nucleic acids; biochemistry of cancer cells; tRNA metabolism during insect development. *Mailing Add:* 413 E Brookside Bryan TX 77801

TSUZUKI, JUNJI, virology, tumorigenesis, for more information see previous edition

TTERLIKKIS, LAMBROS, b Beirut, Lebanon, Oct 17, 34; US citizen; m 60; c 2. PHYSICS. *Educ:* Walla Walla Col, BSc, 59; Univ Denver, MSc, 62; Univ Calif, Riverside, PhD(physics), 68. *Prof Exp:* Res asst solid state physics, Denver Res Inst, Univ Denver, 59-62; assoc physicist, IBM Corp, 62-63; NIH res assoc biophys, Inst Molecular Biophys, Fla State Univ, 68-70; asst prof, 70-74, ASSOC PROF PHYS PHARMACEUT, FLA A&M UNIV, 74- *Mem:* AAAS; Soc Nuclear Med; Am Pharmaceut Asn. *Res:* Pharmacokinetics of drug metabolism; physicochemical properties of drugs; solid state physics; optical properties of biopolymers. *Mailing Add:* Sch Pharm Fla A&M Univ PO Box 367 Tallahassee FL 32304

TU, ANTHONY T, b Taipei, Formosa, Aug 12, 30; US citizen; m 57; c 5. BIOCHEMISTRY. *Educ:* Nat Taiwan Univ, BS, 53; Univ Notre Dame, MS, 56; Stanford Univ, PhD(biochem), 60. *Prof Exp:* Res assoc biochem, Yale Univ, 61-62; asst prof, Utah State Univ, 62-67; assoc prof, 67-70, PROF BIOCHEM, COLO STATE UNIV, 70- *Concurrent Pos:* NIH career develop award, 69-73. *Honors & Awards:* Merit Award, NIH, 87. *Mem:* Am Chem Soc; Am Soc Biol Chem & Molecular Biol. *Res:* Snake venom toxins and enzymes; metal-nucleotide interaction; raman spectroscopy; structure-function relationship of snake neurotoxins, hemorrhagic and myonecrotic toxins; application of raman spectroscopy to biological compounds. *Mailing Add:* Dept Biochem Colo State Univ Ft Collins CO 80523

TU, CHARLES WUCHING, b 1951; m 76; c 2. MOLECULAR BEAM EPITAXY, HIGH SPEED ELECTRONIC DEVICES. *Educ:* McGill Univ, BSc, 71; Yale Univ, MPhil, 72, PhD(appl phys), 78. *Prof Exp:* Lectr physics, Yale Univ, 78-80; mem tech staff, AT&T Labs, 80-87, dist MTS, 87-88; ASSOC PROF, UNIV CALIF, SAN DIEGO, 88- *Mem:* Am Vacuum Soc; Mats Res Soc; Inst Elec & Electron Eng. *Res:* Molecular beam epitaxy of compound semiconductor heterostructures; high speed and high frequency electronic devices based on semiconductor heterostructures; property of quantum wells & superlattices. *Mailing Add:* ECE Dept Mail Code R-007 Univ Calif San Diego LaJolla CA 92093

TU, CHEN CHUAN, b Husin, Oct 5, 18; nat US; m 47; c 2. CHEMISTRY. *Educ:* Chinese Nat Col Pharm, dipl, 42; Purdue Univ, MS, 49, PhD, 51. *Prof Exp:* Asst, Chinese Nat Chekiang Univ, 42-47; asst, Purdue Univ, 47-51, res fel, 51-52; res asst, Inst Paper Chem, 52-56; res assoc, 56-57, SR SCIENTIST, EXP STA, HAWAIIAN SUGAR PLANTERS' ASN, 57- *Mem:* Am Chem Soc; Int Soc Sugarcane Technol. *Res:* Biochemistry; natural products; sugar technology; food science. *Mailing Add:* 1644 Ulueo St Kailua HI 96734-4459

TU, CHEN-PEI DAVID, b Taipei, China, Nov 23, 48; US citizen. GENE EXPRESSION, TRANSPOSABLE ELEMENTS. *Educ:* Nat Taiwan Univ, Taipei, BS, 70; Cornell Univ, PhD(biochem & molecular biol), 76. *Prof Exp:* Fel biochem genetics, Med Sch, Stanford Univ, 76-80; from asst prof to assoc prof, 80-90, PROF BIOCHEM & MOLECULAR BIOL, PA STATE UNIV, 90- *Concurrent Pos:* Fel, Am Cancer Soc, 76-78; res career develop award, USPHS, 85-90; mem, Environ Health Sci Rev Comt, Nat Inst Environ Health Sci, 89-93. *Mem:* Am Soc Biochem & Molecular Biologists; Am Soc Microbiologists. *Res:* Gene Regulation, structure and function of glutathione S- transferal; gene expression. *Mailing Add:* Dept Molecular & Cell Biol 108 Althouse Lab Pa State Univ University Park PA 16802

TU, CHIN MING, b Hsinchu, Taiwan, Dec 14, 32; Can citizen; m 62; c 3. MICROBIOLOGY, BIOCHEMISTRY. *Educ:* Chung Hsing Univ, Taiwan, BSc, 56; Univ Sask, MSc, 63; Ore State Univ, PhD(microbiol), 66. *Prof Exp:* Asst org chem & soil fertil dept agr chem, Chung Hsing Univ, Taiwan, 57-60; asst soil sci, Univ Sask, 60-62; res fel microbiol, Ore State Univ, 63-66; RES SCIENTIST, RES CTR, AGR CAN, LONDON, ONT, 66- *Mem:* Am Soc Microbiol; Can Soc Microbiol; Soc Invert Path; Sigma Xi. *Res:* Interaction between pesticides and soil microorganisms; soil science; insect pathology; pesticide pollution; pesticide degradation; nitrogen fixation; rhizobia-leguminous plants symbiosis-pesticide interaction; microbial control of insect pests; soil fertility. *Mailing Add:* Res Ctr Agr Can 1400 Western Rd London ON N6G 2V4 Can

TU, JUI-CHANG, b Tainan, Taiwan, Aug 14, 36; Can citizen; m 64; c 2. PHYTOPATHOLOGY. *Educ:* Nat Taiwan Univ, BSc, 59, MSc, 61; Wash State Univ, PhD(plant path), 66. *Prof Exp:* Res assoc, Iowa State Univ, 67-69; res scientist, Univ Alta, 69-70, from asst prof to assoc prof & asst dir biol & electron micros, 70-78; RES SCIENTIST, HARROW RES STA, AGR CAN, 78- *Concurrent Pos:* Adj prof, Dept Biol, Univ Windsor. *Honors & Awards:* Bailey Award, Can Phytopathol Soc, Outstanding Res Award. *Mem:* Fel Am Phytopath Soc; Can Phytopath Soc; Sigma Xi; Can Seed Growers Asn. *Res:* Diseases of legume crops and their control. *Mailing Add:* Harrow Res Sta Harrow ON N0R 1G0 Can

TU, KING-NING, b Canton, China, Dec 30, 37; m 65; c 2. MATERIALS SCIENCE. *Educ:* Nat Taiwan Univ, BS, 60; Brown Univ, MS, 64; Harvard Univ, PhD(appl physics), 68. *Prof Exp:* Res asst mat sci, Brown Univ, 63-64; res asst appl physics, Harvard Univ, 66-68, res fel, 68; sr mgr thin film sci, 78-84, mat sci, 84-86, RES STAFF MEM PHYS SCI, IBM WATSON RES CTR, 68- *Concurrent Pos:* Sci Res Coun sr vis fel, Cavendish Lab, Cambridge Univ, Eng, 75-76. *Honors & Awards:* Appl to Practice Award, Metall Soc, 88. *Mem:* Fel Am Phys Soc; Mat Res Soc (pres, 81); fel Metall Soc. *Res:* Phase transformations in alloys; kinetics in thin solid films; electrical properties of metal-silicon interfaces and compounds; device metallurgy; diffusion in superconducting oxides. *Mailing Add:* 44 Whitlaw Close Chappaqua NY 10598

TU, SHIAO-CHUN, b Henan, China, Dec 29, 43; US citizen; m 70; c 1. BIOCHEMISTRY, BIOPHYSICS. *Educ:* Nat Taiwan Univ, BS, 66; Cornell Univ, MNS, 69, PhD(biochem), 73. *Prof Exp:* Res assoc biochem, Grad Sch Nutrit, Cornell Univ, 73; res fel biol, Biol Labs, Harvard Univ, 73-77; from asst prof to assoc prof, 77-85, PROF BIOCHEM & BIOPHYSICS, DEPT BIOCHEM & BIOPHYS SCI, UNIV HOUSTON, 85-, CHMN DEPT, 89- *Concurrent Pos:* Tutor biol, Harvard Univ, 74-75; NIH fel, 75-77, & mem study sect, phys biochem, 84-88; assoc ed, Photochem & Photobiol, 85-; res career develop award, 81-86. *Honors & Awards:* Sigma Xi Award, Univ Houston, 82. *Mem:* Sigma Xi; AAAS; Am Soc Photobiol; Am Chem Soc; Am Soc Biochem & Molecular Biol. *Res:* Mechanisms of biological oxidation; structure-function relationships of flavin-and pyridine nucleotide-dependent enzymes; enzyme biotechnology; bioluminescence. *Mailing Add:* Dept Biochem & Biophys Sci Univ Houston Houston TX 77004

TU, SHU-I, b Chungking, China, Jan 3, 43; m 69; c 1. BIOPHYSICAL CHEMISTRY. *Educ:* Nat Taiwan Univ, BS, 65; Yale Univ, MPhil, 68, PhD(chem), 69. *Prof Exp:* Res assoc biochem, Yale Univ, 69-72; res asst prof, State Univ NY Buffalo, 72-74; asst prof chem, State Univ NY Stony Brook, 74-81; res chemist, 81-88, SUPVRY RES CHEMIST, EASTERN REGIONAL RES CTR, AGR RES SERV, USDA, PHILADELPHIA, 88- *Mem:* Am Chem Soc; Biophys Soc; Sigma Xi. *Res:* Bioenergetics of ion transport in plant root system,; H; interactions between soil and roots. *Mailing Add:* USDA Ars E Reg Res Ctr 600 E Mermaid Lane Philadelphia PA 19118

TU, YIH-O, b Jiangxi, China, Jan 8, 20; m 60; c 1. APPLIED MATHEMATICS. *Educ:* Col Ord Eng, Chungking, China, BS, 46; Carnegie Inst Technol, MS, 54; Rensselaer Polytech Inst, PhD(math), 59. *Prof Exp:* Designer mech eng, Rockwell Mfg Co, Pa, 53-55; STAFF MATHEMATICIAN, IBM CORP, 59- *Mem:* Am Math Soc; Am Phys Soc; Am Soc Mech Eng; Soc Indust & Appl Math. *Res:* Fluid mechanics; elasticity; vibration and elastic stability; continuum mechanics. *Mailing Add:* 6716 Bret Harte Dr San Jose CA 95120-6099

TUAN, DEBBIE FU-TAI, b Kiangsu, China, Feb 2, 30; m 87. PHYSICAL CHEMISTRY, CHEMICAL PHYSICS. *Educ:* Taiwan Univ, BS, 54, MS, 58; Yale Univ, MS, 60, PhD(chem), 61. *Prof Exp:* Teaching asst chem, Taiwan Univ, 54-55; NSF res fel, Yale Univ, 61-64; NASA res grant & proj assoc, theoret chem inst, Univ Wis, 64-65; from asst prof to assoc prof, 65-73, summer res fels, 66, 68 & 71, PROF CHEM, KENT STATE UNIV, 73- *Concurrent Pos:* Vis scientist, Belfer Grad Sch Sci, Yeshiva Univ, 66 & Stanford Res Inst Int, 81; vis prof, Academia Sinica of China, Nat Taiwan Univ & Nat Tsing-Hwa Univ, 67; res fel, Harvard Univ, 70; res assoc, Cornell Univ, 83. *Mem:* Am Phys Soc; Am Chem Soc; Sigma Xi. *Res:* Many electron theory of atoms and molecules; perturbation theory; other applications of quantum mechanics to chemical problems. *Mailing Add:* Dept Chem Kent State Univ Kent OH 44242-0001

TUAN, HANG-SHENG, b Hankow, Hupei, China, Oct 23, 35; m 65. ELECTRICAL ENGINEERING. *Educ:* Nat Taiwan Univ, BS, 58; Univ Wash, MS, 61; Harvard Univ, PhD(appl physics), 65. *Prof Exp:* Res fel electronics, Harvard Univ, 65; asst prof elec sci, 65-69, assoc prof, 69-80, PROF ELEC ENG, STATE UNIV NY, STONY BROOK, 80- *Mem:* Inst Elec & Electronics Engrs. *Res:* Electromagnetic theory; antenna and wave propagation; plasma physics, microwave acoustics. *Mailing Add:* Dept Elec Sci State Univ NY Stony Brook NY 11794

TUAN, ROCKY SUNG-CHI, b Hong Kong, Mar 5, 51. CELL BIOLOGY, DEVELOPMENTAL BIOLOGY. *Educ:* Berea Col, BA, 72; Rockefeller Univ, PhD(life scis), 77; Univ PA, MA, 87. *Prof Exp:* Res fel, med & orthop surg, Harvard Med Sch, Mass Gen Hosp & Children's Hosp, 77-80; from asst prof to assoc prof biol, Univ Pa, 80-86; PROF ORTHOPED SURG, BIOCHEM & MOLECULAR BIOL, DIR, ORTHO RES LAB, THOMAS JEFFERSON UNIV, 88- *Mem:* Soc Develop Biol; Am Soc Cell Biol. *Res:* Biochemistry and molecular biology of embryonic calcium metabolism; biology of cell formation; cell differentiation and extracellular matrix; cardiovascular functions during development. *Mailing Add:* Dept Orthopaedic Surg Thomas Jefferson Univ Philadelphia PA 19107

TUAN, SAN FU, b Tientsin, China, May 14, 32; m 63; c 4. THEORETICAL PHYSICS, APPLIED MATHEMATICS. *Educ:* Oxford Univ, BA, 54, MA, 58; Univ Calif, PhD(appl math), 58. *Prof Exp:* Res assoc, Univ Chicago, 58-60; asst prof, Brown Univ, 60-62; assoc prof, Purdue Univ, 62-65; vis prof, Univ Hawaii, 65-66; mem inst adv study, Princeton Univ, 66-72; PROF THEORET PHYSICS, UNIV HAWAII, 66- *Concurrent Pos:* Mackinnon scholar, Magdalen Col, Oxford Univ, 51-54; consult, Argonne Nat Lab, 63-70; John S Guggenheim fel, 65-66; dir & co-ed proc, Second, Third, Fifth, Sixth & Seventh Hawaii Topical Conf Particle Physics, 67, 69, 73, 75 & 77; vis lectr, Bariloche Atomic Ctr, Argentina & Univ Buenos Aires, 69-70; vis lectr, US-China Sci Coop Prog, 70-71; vis prof, Peking Univ & Inst Theoret Physics, 79-80. *Mem:* Am Math Soc; fel Am Phys Soc. *Res:* Mathematical physics; theory of elementary particles; superconductivity; political science. *Mailing Add:* Dept Physics & Astron Wat 232 Univ Hawaii 2505 Correa Rd Honolulu HI 96822

TUAN, TAI-FU, b Tientsin, China, Sept 7, 29; US citizen; m 68. THEORETICAL PHYSICS, ATMOSPHERIC DYNAMICS. *Educ:* Cambridge Univ, BA, 51; La State Univ, MS, 53; Univ Pittsburgh, PhD(physics), 59. *Prof Exp:* Instr physics, Northwestern Univ, 59-60; univ res fel, Univ Birmingham, 61-64, Dept Sci & Indust Res res fel, 64-65; from asst prof to assoc prof, 65-71, PROF PHYSICS, UNIV CINCINNATI, 71- *Concurrent Pos:* USAF res grant, 69-, NSF grant, 85- *Mem:* Am Phys Soc; Am Geophys Union; NY Acad Sci. *Res:* Scattering theory; atmospheric physics; research in airglow, gravity waves and magnetohydrodynamic models for magnetosphere; application of scattering theory for atmospheric dynamics; investigation of instability in gravity waves gravity-wave ducting through atmospheric structure; inhomogeneous dissipation; Brunt-Doppler mechanism; non-linear response of airglow to linearized waves. *Mailing Add:* Dept Physics Univ Cincinnati Cincinnati OH 45221

TUBA, I STEPHEN, b Hungary, Jan 22, 32; US citizen; m 55; c 2. ELEVATED TEMPERATURE DESIGN, FAILURE ANALYSIS. *Educ:* Tech Univ Budapest, BSME, 56; Carnegie-Mellon Univ, MSME, 60; Univ Pittsburgh, PhD(mech eng), 64. *Prof Exp:* Res engr & mgr, Anal Mech Res & Develop, Westinghouse, 57-70; PRES, BASIC TECHNOL, INC, 70-; EXEC DIR, INT TECHNOL INST, 76- *Concurrent Pos:* Sr lectr math & mech eng, Carnegie-Mellon Univ, 64-70; adj prof mech eng, Univ Pittsburgh, 64-70; lectr, univs, industs & tech socs, 70- *Mem:* Am Soc Mech Engrs; Sigma Xi; Int Technol Inst. *Res:* Solid mechanics: elasticity, plasticity, creep, fatique; engineering analysis: finite element methods, non-linear, combustion; failure analysis: machinery, equipment, systems; technology transfer: international design, development. *Mailing Add:* Int Technol Inst 7125 Saltsburg Rd Pittsburgh PA 15235

TUBB, RICHARD ARNOLD, b Weatherford, Okla, Dec 18, 31; m 57; c 2. LIMNOLOGY, FISHERIES. *Educ:* Okla State Univ, BS, 58, MS, 60, PhD(zool), 63. *Prof Exp:* Res asst aquatic biol lab, Okla State Univ, 60-62; asst prof biol, Univ NDak, 63-66; asst leader fisheries, SDak Coop Fishery Unit, 66-67; leader, Ohio Coop Fishery Unit, 67-75; PROF & HEAD DEPT FISHERIES & WILDLIFE MGT, ORE STATE UNIV, , 75- *Concurrent Pos:* Exec dir, Consortium Int Fisheries & Aquacult Develop, 85- *Mem:* Am Soc Limnol & Oceanog; Am Fisheries Soc; Am Inst Fishery Res Biol. *Res:* Herbivorous insect population in oil refinery effluent holding pond series; investigations of whirling disease of trout; freshwater bivalves as stream monitors for pesticides; environmental impact of nuclear power plants on fresh water fish; colonization of artificial reefs by fish. *Mailing Add:* Dept Fisheries & Wildlife Mgt Ore State Univ Corvallis OR 97331

TUBBS, ELDRED FRANK, b Buffalo, NY, Mar 31, 24; m 49; c 3. ATOMIC SPECTROSCOPY. *Educ:* Carnegie Inst Technol, BS, 49; Johns Hopkins Univ, PhD(physics), 56. *Prof Exp:* Sr physicist res ctr, Am Optical Co, 55-58; res physicist microwave physics lab, Sylvania Elec Prod Inc, 58-60; res physicist, WCoast Br, Gen Tel & Electronics Labs, Inc, 60-63; from asst prof to assoc prof physics, Harvey Mudd Col, 63-72, prof, 72-79; MEM TECH STAFF, JET PROPULSION LAB, 79- *Concurrent Pos:* Consult, Mech Universe. *Honors & Awards:* Prize, Am Asn Physics Teachers, 67. *Mem:* Optical Soc Am; Soc Photo-Optical Instrumentation Engrs; Sigma Xi. *Res:* Optical instruments and metrology; interferometry; absolute f-values; teaching apparatus and techniques. *Mailing Add:* 730 W 11th St Claremont CA 91711-3748

TUBBS, RAYMOND R, b Ithaca, NY, Aug 23, 46; m 69; c 2. HEMATOPATHOLOGY, IMMUNOPATHOLOGY. *Educ:* Bob Jones Univ, BS, 68; Kirksville Col Osteop Med, DO, 73; Am Bd Path, dipl, 79 & 83. *Prof Exp:* Intern & resident med, 73-75, resident lab med, 75-79, STAFF PATHOLOGIST, CLEVELAND CLIN FOUND, 79- *Concurrent Pos:* Clin assoc immunopath, Cleveland Clin Found, 79, fel, 79-80. *Mem:* Col Am Pathologists; Am Immunologists; Am Asn Pathologists; Am Asn Cancer Res; Int Acad Path; Am Soc Hemat. *Res:* Hematopathology; nephropathology; immunotyping support for biologic response modifiers research program. *Mailing Add:* Dept Path LM 2 Cleveland Clin Found 9500 Euclid Ave Cleveland OH 44106

TUBBS, ROBERT KENNETH, b Gary, Ind, Nov 25, 36; m 56; c 3. COLLOID CHEMISTRY, SURFACE CHEMISTRY. *Educ:* Ohio State Univ, BS, 58, PhD(colloid chem), 62. *Prof Exp:* Res chemist electrochem dept, E I du Pont de Nemours & Co, Inc, Del, 62-67, staff scientist, 67-68, res supvr, 68-70, gen tech supt indust chem dept, NY, 70-73, sr res supvr, Plastics Dept, 73-74, prod mgr, 74-75, develop mgr plastics dept, 75-76, com develop mgr pharmaceut, 76-77, prod mgr, 78-85; RETIRED. *Concurrent Pos:* Consult chemist, 85-; vis prof, Bloomsburg Univ, 87, asst prof chem, 89- *Mem:* Am Chem Soc. *Res:* Structure and interactions of macromolecules; kinetics of polymerization; molecular biology; emulsion polymerization; coatings; adhesives. *Mailing Add:* RD 2 Box 140 A Gnoga Lakes Benton PA 17814

TUBIS, ARNOLD, b Pottstown, Pa, Mar 28, 32; m 59; c 2. MUSICAL & PHYSIOLOGICAL ACOUSTICS. *Educ:* Mass Inst Technol, BS, 54, PhD(theoret physics), 59. *Prof Exp:* Res asst, Mass Inst Technol, 54-57; asst prof physics, Worcester Polytech Inst, 58-60; res assoc, 60-62, from asst prof to assoc prof, 62-69, asst head dept, 66-73, actg head dept, 88, PROF PHYSICS, PURDUE UNIV, 69-, HEAD DEPT, 89- *Concurrent Pos:* Asst physicist, Brookhaven Nat Lab, 59; res assoc, Argonne Nat Lab, 61; vis physicist, Lawrence Radiation Lab, 63, Stanford Linear Accelerator Ctr, 71, Los Alamos Sci Lab, 72, Naval Weapons Ctr, China Lake, 85-88. *Mem:* AAAS; Sigma Xi; Acoust Soc Am; fel Am Phys Soc; Am Asn Physics Teachers; Assoc Res Otolaryngol. *Res:* Physical, physiological and musical acoustics; theory of atomic structure; theory of interactions of nuclei and elementary particles. *Mailing Add:* Dept Physics Purdue Univ West Lafayette IN 47907

TUBIS, MANUEL, b Philadelphia, Pa, July 14, 09; m 36; c 1. NUCLEAR MEDICINE. *Educ:* Philadelphia Col Pharm, BSc, 31; Univ Pa, MSc, 32; Univ Tokyo, PhD(pharmaceut sci), 66. *Prof Exp:* Chemist, US Food & Drug Admin, 35-44; pharmaceut res chemist, Wyeth, Inc, 44-46; res chemist, Dartell Labs, 46-47; tech dir, Am Biochem Corp, 47-48; biochemist, 48-70, chief biochemist, Nuclear Med Serv, 70-79, CONSULT RES CHEMIST, VET ADMIN HOSP, WADSWORTH MED CTR, 79- *Concurrent Pos:* Asst prof sch med, Univ Calif, Los Angeles, 53-68; consult nuclear med & radiopharm, Int Atomic Energy Agency, Vienna; adj prof biomed chem & co-dir radiopharm prog, Sch Pharm, Univ Southern Calif, 74-, consult radiopharmaceut & nuclear med, 74- *Honors & Awards:* Super Performance Award, US Vet Admin, 75; Distinguished Scientist Award, Soc Nuclear Med, 72. *Mem:* Fel & hon mem Am Inst Chem; Am Chem Soc; Soc Nuclear Med. *Res:* Application of radioisotopes to biochemistry, medicine and pharmacy; research and development of new labeled radiopharmaceuticals; supervision of preparation, production and quality control. *Mailing Add:* 5371-3E Punta Alta Laguna Hills CA 92653

TUCCI, EDMOND RAYMOND, b Pawtucket, RI, May 31, 33; m 63. CATALYSIS CONSULTING. *Educ:* RI Sch Design, BS, 55; St Louis Univ, MS, 57; Duquesne Univ, PhD(phys chem), 61. *Prof Exp:* Res chemist, Olin Mathieson Chem Corp, 57-58; chemist, Gulf Res & Develop Co, 61-62, res chemist, 62-66, sr res chemist, 66-73; supvr catalysis, Engelhard Minerals & Chem Co, 73-75; mgr com develop, Johnson Matthey, Inc, 76-80; TECH CONSULT, CATALYSTS & CATALYTIC SYSTS, 80- *Mem:* Com Develop Asn. *Res:* Homogeneous and heterogeneous catalysis; organometallics; electrochemical syntheses; petrochemical process development; methanation of coal gas; fuel cell development; catalytic combustion. *Mailing Add:* Catalysts & Catalytic Systs Sandy Ridge Rd Box 85 RD Three Stockton NJ 08559

TUCCI, JAMES VINCENT, b Hollis, NY, Feb 13, 39; m 62; c 1. CHEMICAL PHYSICS. *Educ:* Hofstra Univ, BA, 62; Univ Mass, MS, 66, PhD(chem), 67. *Prof Exp:* Instr physics, 66-67, from asst prof to assoc prof, 67-73, actg chmn dept, 67-71, chmn dept, 72-90, PROF PHYSICS, UNIV BRIDGEPORT, 73-, CHMN, DIV SCI & MATH, 90- *Concurrent Pos:* NSF grant, 67-70; Conn Res Comn res grant, 68-70. *Mem:* Optical Soc Am; Am Phys Soc. *Res:* Laser Raman spectroscopy; vibrational spectroscopy. *Mailing Add:* Dept Physics Univ Bridgeport Bridgeport CT 06601

TUCCIARONE, JOHN PETER, b New York, NY, Apr 9, 40; m 63; c 3. MATHEMATICS. *Educ:* Fordham Univ, BS, 61; St John's Univ, NY, MA, 63, JD, 66; NY Univ, PhD(math), 69. *Prof Exp:* Asst prof math, St John's Univ, NY, 62-74; assoc prof, 74-80, PROF MATH, MERCY COL, 80-; DIR, PROB ANALYSIS CORP, 71- *Concurrent Pos:* Attorney pvt practice, NY, 70- *Mem:* Math Asn Am. *Res:* Numerical analysis; computer science; application of computers to instruction. *Mailing Add:* 390 Bedford Rd Pleasantville NY 10570

TUCHINSKY, PHILIP MARTIN, b Philadelphia, Pa, June 17, 45. COMPUTER SCIENCE, GRAPHICAL USER INTERFACES. *Educ:* Queens Col, BA, 66; Courant Inst, NY Univ, MS, 68, PhD(math), 71. *Prof Exp:* Instr math, NY Univ, 69-70; adj instr, Cooper Union Advan Sci & Art, 70-71; asst prof, Kalamazoo Col, 71-72; asst prof math, Ohio Wesleyan Univ, 72-78; res engr, Comput Sci Dept, Ford Res & Eng Ctr, 78-82, sr res engr, res staff, 82-85, PRIN RES ENGR ASSOC, RES STAFF, FORD MOTOR CO, 85- *Concurrent Pos:* Woodrow Wilson fel grad study, 66-67; NDEA Title IV fel, NY Univ, 67-69, res fel, 69-70; Great Lakes Col Asn teaching fel, Lilly Found Grant, 75; vis lectr, Math Asn Am, 76-; mem, Ford Aerospace Intelligent Systs Consortium, 86-90, Tech Training Adv Bd, Addison-Wesley Publ Co, 87-; co-chair, Expert Systs Conf & Expository Planning Comt, 88-89, exec chair, 89-90. *Res:* Vehicle design and engineering software systems; advanced user interfaces. *Mailing Add:* 48371 Bayshore Dr Belleville MI 48111

TUCHMAN, AVRAHAM, b Brooklyn, NY, July 1, 35; m 57; c 4. PHYSICS, THERMAL PHYSICS. *Educ:* Yeshiva Univ, BA, 56; Mass Inst Technol, PhD(physics), 63. *Prof Exp:* Staff scientist Res & Develop Div, Avco Corp, 63-64, group leader propulsion, 64-66, sect chief Plasma Physics, Space Systs Div, 66-70, sr consult scientist, Systs Div, 70-73, prin staff scientist physics, 73-85, chief scientist, Sensors, Res & Guidance Lab, 85-88, CHIEF SCIENTIST, SENSORS & GUIDANCE LAB, TEXTRON DEFENSE SYSTS, AVCO CORP, 88- *Concurrent Pos:* Vis prof, Weizmann Inst Sci, Israel, 74, 78 & 82; mem, Sensor Systs Tech Comt, Am Inst Aeronaut & Astronaut, 90- *Mem:* Asn Orthodox Jewish Scientists; Asn Old Crows; Sigma Xi; Am Defense Preparedness Asn; Am Inst Aeronaut & Astronaut. *Res:* Plasma, quantum and elementary particle physics; optical and infrared sources; infrared optical design; missile countermeasures; optical, infrared and laser countermeasures; infrared target signatures; knowledge-based classification; millimeter wave sensors; satellite detection; high energy lasers; artificial intelligence; expert systems. *Mailing Add:* Textron Defense Systs 201 Lowell St Wilmington MA 01887

TUCHOLKE, BRIAN EDWARD, b Hot Springs, SDak, Mar 19, 46; m 68; c 2. MARINE GEOLOGY, GEOPHYSICS. *Educ:* SDak Sch Mines & Technol, BS, 68; Mass Inst Technol & Woods Hole Oceanog Inst, PhD(oceanog), 73. *Prof Exp:* fel, Lamont-Doherty Geol Observ, Columbia Univ, 73-74, from res assoc to sr res assoc, 74-79; assoc scientist, 79-87, SR SCIENTIST, MARINE GEOL, WOODS HOLE OCEANOG INST, 87-; ADJ SR RES SCIENTIST, LAMONT-DOHERTY GEOL OBSERV, COLUMBIA UNIV, 82- *Concurrent Pos:* Corp mem, Woods Hole Oceanog Inst, 76-79; prin investr grants & contracts, NSF, Off Naval Res & US Dept Energy, 76-; Western NAtlantic proj co-leader, Decade NAm Geol, Geol Soc Am, 80-86; mem, Joint Oceanog Insts Deep Earth Sampling Passive Margin Panel, 81-83 & Atlantic Regional Panel, 84-87; mem, Interim US Sci Adv Comt, Ocean Drilling Prog, 83; assoc ed, Geol Soc Am Bull, 85-88; mem, Am Asn Petrol Geologists Comt on Marine Geol, 87-90; mem, Geophys Union Comt on Paleoceanog, 87-88, Joint Oceanog Insts Deep Earth Sampling Planning Comt, 87-; Seward Johnson chair oceanog, Woods Hole Oceanog Inst, 86-88. *Honors & Awards:* Silver Trout Award, Trout Unlimited, 88. *Mem:* Am Asn Petrol Geologists; Am Geophys Union; fel Geol Soc Am; fel Geol Asn Can; AAAS. *Res:* Seismic stratigraphy; oceanic rock stratigraphy; paleo-oceanography; benthic boundary layer processes; physical properties of sediments; sedimentology; tectonic framework of Atlantic Ocean crust; structure/evolution of Atlantic passive margins. *Mailing Add:* Woods Hole Oceanog Inst Woods Hole MA 02543

TUCK, DENNIS GEORGE, b UK, Apr 8, 29; m 56; c 3. INORGANIC CHEMISTRY, ORGANOMETALLIC CHEMISTRY. *Educ:* Univ Durham, BSc, 49, PhD(chem), 56. *Hon Degrees:* DSc, Univ Durham, 71. *Prof Exp:* Brit Coun fel, Inst du Radium, Paris, France, 52-53; sci officer chem, Windscale Works, UK Atomic Energy Auth, Eng, 53-56; lectr inorg chem, Univ Nottingham, 59-65; from assoc prof to prof chem, Simon Fraser Univ, 66-72; prof chem, 72-87, UNIV PROF, UNIV WINDSOR, 87- *Concurrent Pos:* Res fel chem, Univ Manchester, 56-59; res fel lab nuclear sci, Cornell Univ, 57-58; vis expert, Concepcion Univ, Chile, 64; mem chem grant selection comt, Nat Res Coun Can, 72-74; dir, Can Patents & Develop Ltd, 81-86; mem bd, Can Soc Chem, 85-89. *Honors & Awards:* Main Group Element Award, Royal Soc Chem, 86; Montreal Medal, Chem Inst Can, 87; Alcan Award, Can Soc Chem, 88. *Mem:* Royal Soc Chem; fel Chem Inst Can. *Res:* Coordination chemistry of non-transition metals, especially indium; complexes in solution; use of electrochemical methods in inorganic and organometallic chemistry. *Mailing Add:* Dept Chem Univ Windsor Windsor ON N9B 3P4 Can

TUCK, LEO DALLAS, b San Francisco, Calif, Oct 12, 16; m 53; c 2. PHYSICAL CHEMISTRY. *Educ:* Univ Calif, AB, 39, PhD(chem), 48. *Prof Exp:* Res assoc radiation lab, Univ Calif, 42; res assoc chem dept, Univ Chicago, 42-43; res assoc radio res lab, Harvard Univ, 43-45; lectr & res asst chem, 48-50, instr, 50-51, from asst prof to assoc prof, 51-63, vchancellor acad affairs, 71-73, PROF CHEM & PHARMACEUT CHEM, SCH PHARM, UNIV CALIF, SAN FRANCISCO, 63-, ASSOC DEAN, 80- *Mem:* Fel AAAS; Am Chem Soc; Am Phys Soc; Am Pharmaceut Asn. *Res:* Thermodynamics and electrochemistry; thermodynamics of nonisothermal systems; chemistry of boron and uranium compounds; electrolytes in aqueous and nonaqueous solutions; microwave electronics; chemistry of free radicals; magnetic resonance spectroscopy. *Mailing Add:* Sch Pharm Univ Calif Med Ctr San Francisco CA 94143

TUCKER, ALAN, b Worthing, Eng, Sept 17, 47; US citizen; m 68; c 2. CARDIOPULMONARY PHYSIOLOGY. *Educ:* Univ Calif, Santa Barbara, BA, 68, PhD(biol), 72. *Prof Exp:* Res assoc physiol, Cardiovasc Pulmonary Res Lab, Med Ctr, Univ Colo, 73-74, fel, 74-76; asst prof physiol, Sch Med, Wright State Univ, 76-79; assoc prof, 79-86, PROF PHYSIOL, COLO STATE UNIV, 86- *Concurrent Pos:* Mem, Circulation Coun & Cardiopulmonary Coun, Am Heart Asn. *Honors & Awards:* Res Serv Award, Nat Heart & Lung Inst, 74. *Mem:* Am Physiol Soc; Soc Exp Biol & Med; Am Heart Asn. *Res:* Control of the pulmonary circulation; hypoxia and high altitude physiology; pharmacology of pulmonary and systemic vasculature; respiratory physiology; environmental and exercise physiology. *Mailing Add:* Dept Physiol Colo State Univ Ft Collins CO 80523

TUCKER, ALAN CURTISS, b Princeton, NJ, July 6, 43; m 68; c 2. MATHEMATICS. *Educ:* Harvard Univ, BA, 65; Stanford Univ, MS, 67, PhD(math), 69. *Prof Exp:* Vis asst prof math, Math Res Ctr, Univ Wis-Madison, 69-70; from asst prof to prof, 70-89, chmn, 78-88, DISTINGUISHED TEACHING PROF APPL MATH & STATIST, STATE UNIV NY, STONY BROOK, 89- *Concurrent Pos:* Res consult, Rand Corp, 64-71; vis assoc prof comput sci, Univ Calif, San Diego, 76-77; chmn panel gen math sci prog, 77-81, chmn publ comt, 82-87, chmn educ coun, 90-; mem, US Comn on Math Educ, 80-82; vis prof oper res, Stanford Univ, 83-84. *Mem:* Am Math Soc; Soc Indust & Appl Math; AAAS; Oper Res Soc Am; Math Asn Am (first vpres, 88-90); Sigma Xi. *Res:* Extremal characterization problems in graph theory; zero-one matrices; combinatorial algorithms. *Mailing Add:* 36 Woodfield Rd Stony Brook NY 11790

TUCKER, ALBERT WILLIAM, b Oshawa, Ont, Nov 28, 05; m 38, 64; c 3. LINEAR & NON-LINEAR PROGRAMMING, GAME THEORY. *Educ:* Univ Toronto, BA, 28, MA, 29; Princeton Univ, PhD(math), 32. *Hon Degrees:* DSc, Dartmouth Col, 61. *Prof Exp:* Nat Res Coun fel math, Cambridge Univ, Harvard Univ, Univ Chicago, 32-33; from instr to prof, 33-54, chmn dept, 53-63, Albert Baldwin Dod prof, 54-74, EMER ALBERT BALDWIN DOD PROF MATH, PRINCETON UNIV, 74- *Concurrent Pos:* Ed, Princeton Math Ser, 38-69 & Ann Math Studies, 40-49; assoc dir fire control proj, Nat Defense Res Comt, Off Sci Res Develop, Princeton Univ, 41-45; prin investr, Logistics Res Proj, Off Naval Res, Princeton Univ, 48-72; consult, Rand Corp, 49063, IBM Res Ctr, 57-64; vis prof, Stanford Univ, 49-50, Dartmouth Col, 63, Ariz State Uhiv, 71 & 77, Cornell Univ, 74, Univ Western Australia, 75, 78 & 81; lectr, Haverford Col, 53-54 & 58-59, Orgn Europ Econ Coop, 58-59; chmn comn math, Col Entrance Exam Bd, 55-59; mem, Phys Sci Prog Comt, Alfred P Sloan Found, 55-59, President's Comt, Nat Medal Sci, 62-66; Fulbright lectr, Australian Univs, 56; chmn, Math Prog Soc, 77-80. *Honors & Awards:* John von Neumann Theory Prize, Opers Res Soc Am & Inst Mgt, 80. *Mem:* AAAS (vpres, 57, 66); Am Math Soc; Math Asn Am (pres, 61-62); Math Prog Soc; Can Math Soc; Australian Math Soc; Am Acad Arts & Sci 87. *Res:* Differential geometry; combinatorial topology; theory of games; linear and non-linear programming. *Mailing Add:* 37 Lake Lane Princeton NJ 08540-7222

TUCKER, ALLEN B, b Worcester, Mass, Feb 19, 42; m 65; c 2. SOFTWARE ENGINEERING, PROGRAMMING LANGUAGES. *Educ:* Wesleyan Univ, BA, 63; Northwestern Univ, MS, 68, PhD(comput sci), 70. *Prof Exp:* Systs analyst, Norton Co, 63-67; asst prof comput sci, Univ Mo-Rolla, 70-71; from asst prof to assoc prof comput sci, Georgetown Univ, 71-83, chmn dept & dir acad comput, 76-83; chmn dept, 83-86, Macarthur Prof Comput Sci, Colgate Univ, 83-88, assoc dean fac, 86-88; PROF COMPUT SCI, BOWDOIN COL, 88- *Concurrent Pos:* Prin investr, NSF grants, 71, 84 & 86 & var found grants, 80-; consult, WHO, 73-80, var cols & univ, 82- & Smithsonian Inst, 84; assoc ed, J Comput Lang, 78-; vis assoc prof comput sci, overseas prog Heidelberg, Ger, Boston Univ, 82 & 84. *Mem:* Asn Comput Mach; Sigma Xi. *Res:* Formal theory of languages; automata theory; computer applications; natural language analysis. *Mailing Add:* Comput Sci Dept Bowdoin Col Brunswick ME 04011

TUCKER, ALLEN BRINK, b Highland, Ind, Oct 12, 36; m 63; c 2. NUCLEAR PHYSICS. *Educ:* Mass Inst Technol, BS, 58; Stanford Univ, PhD(physics), 65. *Prof Exp:* Asst prof physics, San Jose State Col, 63-65 & Iowa State Univ, 65-70; assoc prof, 70-77, PROF PHYSICS, SAN JOSE STATE UNIV, 77- *Mem:* Am Phys Soc; Am Asn Physics Teachers. *Res:* Neutron scattering; nuclear spectroscopy; delayed neutrons. *Mailing Add:* Dept Physics San Jose State Univ Washington Sq San Jose CA 95192

TUCKER, ANNE NICHOLS, b Ashland, Ky, Oct 16, 42; m 65; c 1. IMMUNOTOXICOLOGY. *Educ:* Univ Ky, BS, 64, PhD(toxicol), 76. *Prof Exp:* Res assoc biochem, Univ Ky, 65-74; fel toxicol, 76-78, asst prof, Med Col Va, 78-86; IPA Scientist, Nat Inst Environ Health Sci, 83-87. *Concurrent Pos:* Nat Cancer Inst-Nat Res Serv Award fel, 76; spec consult, Food & Drug Admin, 78-; consult, Indust Health Found, 78-79. *Mem:* Reticuloendothelial Soc. *Res:* Immunotoxicology; the role of arachidonic acid metabolism in cellular proliferation. *Mailing Add:* 10703 Old Squaws Lane Chesterfield VA 23832

TUCKER, BILLY BOB, b Cheyenne, Okla, Jan 13, 28; m 49; c 3. AGRONOMY. *Educ:* Okla State Univ, BS, 52, MS, 53; Univ Ill, PhD, 55. *Prof Exp:* Soil scientist, Agr Res Serv, USDA, 55-56; from asst prof to prof, Okla State Univ, 56-79, regents prof agron, 79-; RETIRED. *Mem:* Am Soc Agron; Soil Sci Soc Am; Soil Conserv Soc Am. *Res:* Soil management, especially improvement and maintenance of soil productivity; soil chemistry, plant nutrition and fertilizer technology. *Mailing Add:* 1212 N Jardot Stillwater OK 74074

TUCKER, CHARLES EUGENE, b Montgomery, Ala, July 2, 33; m 58; c 2. BIOLOGY. *Educ:* Huntingdon Col, BA, 59; Univ Ala, MS, 65, PhD(biol), 67. *Prof Exp:* Assoc prof biol, 67-80, PROF BIOL, LIVINGSTON UNIV, 80-, CHMN, DIV NATURAL SCI & MATH, 70-, ASSOC DEAN, GEN STUDIES HEALTH-RELATED PROG, 75- *Mem:* Am Soc Ichthyologists & Herpetologists; Sigma Xi. *Res:* Vertebrate field zoology; ichthyology; survey of fishes. *Mailing Add:* PO Box 665 Livingston AL 35470

TUCKER, CHARLES L, b Durham, NC, July 29, 53; m 75; c 2. POLYMER PROCESSING, COMPOSITE MATERIALS. *Educ:* Mass Inst Technol, SB, 75, SM, 77, PhD(mech eng), 78. *Prof Exp:* From asst prof to assoc prof 78-89, PROF MECH ENG, UNIV ILL, 89- *Concurrent Pos:* Consult, Dow Chem Co, 83-; Postdoctoral Award in Mfg Eng, TRW Found, 84; vis eng dept, Cambridge Univ, 84-85; vis fel commoner, Churchill Col, Cambridge, 84-85; mem adv comt, Mech & Struct Systs Div, NSF, 88-90. *Mem:* Am Soc Mech Engr; Soc Rheology; Polymer Processing Soc. *Res:* Processing of polymers and polymer-matrix composites; modeling and numerical simulation of flow, heat transfer, reaction, structure development; flow-induced fiber orientation; compression and injection molding; mechanical properties of composites. *Mailing Add:* Dept Mech & Ind Eng Univ Ill Urbana IL 61801

TUCKER, CHARLES LEROY, JR, b Winston-Salem, NC, May 19, 21; m 49; c 2. PLANT CHEMISTRY. *Educ:* The Citadel, BS, 43; NC State Col, MS, 52. *Prof Exp:* Jr analytical chemist div tests & mat, NC State Hwy Comn, 47-48, analytical chemist dept conserv & develop water resources div, 48-50; chemist analytical res, Liggett & Myers Tobacco Co, 52-58; res chemist, Lorillard Res Ctr, 58-67, mgr leaf & flavor res, 67-68, mgr prod develop, 68-80, dir prod develop, 80-87; RETIRED. *Mem:* Am Chem Soc. *Res:* Chemical and physical constitution of tobacco, especially as related to final tobacco product characteristics; analytical methods peculiar to tobacco and tobacco smoke. *Mailing Add:* 903 Caswell Dr Greensboro NC 27408

TUCKER, CHARLES THOMAS, b Laredo, Tex, Aug 6, 36; m 64. MATHEMATICS. *Educ:* Tex A&M Univ, BS & BA, 58; Univ Tex, MA, 62, PhD(math), 66. *Prof Exp:* Chem engr, Tracor, Inc, Tex, 60-62, engr & scientist, 63-66; asst prof math, 66-73, ASSOC PROF MATH, UNIV HOUSTON, 73- *Mem:* Am Math Soc; Math Asn Am. *Res:* Sonar signal processing; pure mathematics. *Mailing Add:* Dept Math Univ Houston 4800 Calhoun Rd Houston TX 77204

TUCKER, DAVID PATRICK HISLOP, b Trinidad, West Indies, Oct 26, 34; m 66; c 1. HORTICULTURE. *Educ:* Univ Birmingham, BSc, 58; Univ Calif, PhD(plant sci), 66. *Prof Exp:* Agronomist, Dept Agr, Brit Honduras, 60-63; agronomist, 66-74, assoc prof & assoc horticulturist, 74-80, PROF FRUIT CROPS & HORTICULTURIST, CITRUS RES & EDUC CTR, UNIV FLA, 80- *Mem:* Am Soc Hort Sci. *Res:* All aspects of citrus production. *Mailing Add:* Citrus Res Educ Ctr Univ Fla Lake Alfred FL 33850

TUCKER, DON HARRELL, b Brown County, Tex, Jan 21, 30; m 51; c 5. MATHEMATICS. *Educ:* WTex State Univ, BA, 51; Univ Tex, MA, 55, PhD(math), 58. *Prof Exp:* Res scientist mil physics res lab, Balcones Res Ctr, Univ Tex, 52-53, instr math, Univ, 53-58; from asst prof to assoc prof, 58-67, PROF MATH, UNIV UTAH, 67- *Concurrent Pos:* Vis prof, Cath Univ Am, 68-69; guest prof, Univ Marburg, 66 & 69, vis lectr, Math Asn Am, 71- *Mem:* Am Math Soc; Math Asn Am. *Res:* Functional analysis; abstract summability theory; differential equations. *Mailing Add:* 66 N Wolcutt St Salt Lake City UT 84103

TUCKER, EDMUND BELFORD, b NS, Can, May 6, 22; m 46; c 3. PHYSICS. *Educ:* Mt Allison Univ, BSc, 43; Oxford Univ, BA, 48; Yale Univ, MS, 49, PhD(physics), 51. *Prof Exp:* Res assoc physics, Univ Minn, 50-52, asst prof, 53, res assoc, 53-55; physicist res & develop ctr, 55-66, mgr personnel & admin info sci lab, 66-69, consult educ rels, NY, 69-71, consult, Conn, 71-75, MGR SCI & TECHNOL SUPPORT PROG, GEN ELEC CO, CONN, 75- *Mem:* AAAS; Am Soc Eng Educ; Sigma Xi; Am Phys Soc. *Res:* Linear accelerators; magnetic resonance; microwave ultrasonics; crystal fields; energy related problems. *Mailing Add:* 237 Strawberry Hill Stamford CT 06902

TUCKER, GARY EDWARD, b Michigan Valley, Kans, Aug 17, 41; m 60; c 2. BOTANY. *Educ:* Kans State Teachers Col, BA, 64; Univ NC, Chapel Hill, MA, 67; Univ Ark, PhD(bot), 76. *Prof Exp:* Asst prof, 66-77, PROF BIOL & CUR HERBARIUM, ARK TECH UNIV, 77- *Concurrent Pos:* Consult, var state & fed agencies. *Mem:* Int Asn Plant Taxonomists; Sigma Xi. *Res:* Endangered and threatened plant species of Southeastern states; endangered and threatened plant species of Arkansas; woody flora of Arkansas. *Mailing Add:* 422 S Denver Russellville AR 72801

TUCKER, GARY JAY, b Cleveland, Ohio, May 6, 34; m 56; c 2. PSYCHIATRY. *Educ:* Oberlin Col, AB, 56; Western Reserve Univ, MD, 60. *Prof Exp:* Asst med dir, Acute Psychiat Inpatient Div, Yale-New Haven Hosp, Yale Univ, 67-68, from asst prof to assoc prof psychiat, Sch Med, 67-71, med dir, Psychiat Inpatient Div, Med Ctr, 68-71, attend psychiatrist, 69-71, asst chief psychiat, 70-71; from assoc prof to prof psychiat, Sch Med, Dartmouth Univ, 71-85, dir residency training, 71-78, chmn dept, 78-85; PROF PSYCHIAT & CHMN DEPT, SCH MED, UNIV WASH, SEATTLE, 85- *Concurrent Pos:* Fel psychiat, Sch Med, Yale Univ, 61-64; consult, Norwich State Hosp, Conn, 67-68, Univ Conn, 68-70, Off Aviation Med, Fed Aviation Admin, Dept Transp, 68-70, Vet Admin Hosp, West Haven, Conn, 70-71 & White River Junction, Vt, 71-85. *Mem:* Fel Am Psychiat Asn. *Res:* Behavioral implications of neurologic functions; psychopathology and hospital psychiatry. *Mailing Add:* Dept Psychiat Sch Med Univ Wash Seattle WA 98195

TUCKER, HARVEY MICHAEL, b New Brunswick, NJ, Nov 27, 38; m 81; c 3. OTOLARYNGOLOGY, SURGERY. *Educ:* Bucknell Univ, BS, 60; Jefferson Med Col, MD, 64. *Prof Exp:* Resident, Jefferson Med Col, 69; asst otolaryngol, Barnes Hosp, Wash Univ, 69-70; assoc prof otolaryngol, State Univ NY Upstate Med Ctr, 70-75; CHMN, DEPT OTOLARYNGOL & COMMUN DIS, CLEVELAND CLIN FOUND, 75- *Concurrent Pos:* Fel head & neck surg, Barnes Hosp, Wash Univ, 69-70. *Honors & Awards:* Benjamin Shuster Award, Am Acad Plastic & Reconstruct Surg, 70. *Mem:* Fel Am Col Surgeons; fel Am Acad Facial Plastic & Reconstruct Surg; Am Soc Surg; fel Am Acad Head & Neck Surg; fel Am Acad Ophthal & Otolaryngol. *Res:* Laryngeal reinnervation and transplantation; cancer surgery of the head and neck. *Mailing Add:* 9500 Euclid Ave Cleveland OH 44106

TUCKER, HERBERT ALLEN, b Milford, Mass, Oct 25, 36; m 59; c 3. ANIMAL PHYSIOLOGY, ENDOCRINOLOGY. *Educ:* Univ Mass, BS, 58; Rutgers Univ, MS, 60, PhD(animal physiol), 63. *Prof Exp:* From asst prof to assoc prof, 62-75, PROF MAMMARY PHYSIOL, MICH STATE UNIV, 75- *Concurrent Pos:* NIH spec fel, Univ Ill, 69. *Honors & Awards:* Borden Award, Am Dairy Sci Asn, 79; Upjohn Physiol Award, Am Dairy Sci Asn & Cyanamid Animal Physiol & Endocrinol Award, Am Soc Animal Sci, 83; L E Casida Award, Am Soc Animal Sci, 87. *Mem:* AAAS; Am Soc Animal Sci; Am Dairy Sci Asn; Soc Exp Biol & Med; Am Physiol Soc; Endocrine Soc. *Res:* Endocrinology of mammary development and lactation; environmental control hormones, growth, lactation; radioimmunoassay of hormones; endocrinology of reproduction; hormone binding to mammary cells. *Mailing Add:* Dept Animal Sci Mich State Univ East Lansing MI 48824

TUCKER, HOWARD GREGORY, b Lawrence, Kans, Oct 3, 22; m 46; c 4. MATHEMATICS. *Educ:* Univ Calif, AB, 48, MA, 49, PhD(math), 55. *Prof Exp:* Instr math, Rutgers Univ, 52-53; asst prof, Univ Ore, 55-56; from asst prof to prof, Univ Calif, Riverside, 56-68; PROF MATH, UNIV CALIF, IRVINE, 68- *Mem:* Am Math Soc; Math Asn Am; Inst Math Statist; Am Statist Asn; Biometric Soc. *Res:* Probability theory; mathematical statistics. *Mailing Add:* Dept Math Univ Calif Irvine CA 92717

TUCKER, IRWIN WILLIAM, b New York, NY, Oct 30, 14; m 73; c 1. ORGANIC CHEMISTRY, ENVIRONMENTAL ENGINEERING. *Educ:* George Washington Univ, BS, 39; Univ Md, PhD(org chem), 48. *Prof Exp:* Chemist, USDA, Washington, DC, 36-45; asst, Univ Md, 45-48; res chemist, Ligget & Meyers Tobacco Co, NC, 48-51; indust specialist, Indust Eval Bd, US Dept Com, 51-53; dir res, Brown & Williamson Tobacco Corp, 53-59, mem, Bd Dirs, 56-59; prof eng res, 66-81, prof environ eng, 77-81, EMER PROF ENG RES & ENVIRON ENG, UNIV LOUISVILLE, 81- *Concurrent Pos:* Dir, Inst Indust Res, 66-72; pres, Coun Environ Balance, 73- *Mem:* Am Chem Soc; Air Pollution Control Asn; Inst Food Technologists. *Res:* Synthetic and determination of structure; fermentation chemistry; enzyme chemistry; chemistry of natural products; air pollution and solid wastes; energy resources, foods, agriculture, agricultural chemicals and product safety. *Mailing Add:* 1810 Crossgate Lane Louisville KY 40222

TUCKER, JOHN MAURICE, b Yamhill Co, Ore, Jan 7, 16; m 42; c 3. BOTANY. *Educ:* Univ Calif, Berkeley, AB, 40, PhD(bot), 50. *Prof Exp:* Botanist, Univ Calif Exped, El Salvador, Cent Am, 41-42; asst bot, Univ Calif, Berkeley, 46-47; assoc, Univ Calif, Davis, 47-49, instr & jr botanist, 51-63, assoc, Exp Sta, 47-49, prof bot & botanist, 63-86, dir, Arboretum, 72-84; RETIRED. *Concurrent Pos:* Guggenheim fel, 55-56. *Mem:* Int Asn Plant Taxon; Bot Soc Am; Soc Study Evolution; Am Soc Plant Taxon. *Res:* Systematics and evolution of oaks of North America; classification of Fagaceae of the New World. *Mailing Add:* 1101 Ovejas Ave Davis CA 95616

TUCKER, JOHN RICHARD, b Baltimore, Md, Aug 22, 48; m 88. ANALYSIS & FUNCTIONAL ANALYSIS, APPLIED STATISTICS. *Educ:* Wash Col, Chestertown, Md, BA, 70; George Washington Univ, Wash, DC, MPh, 76, PhD(math), 80. *Prof Exp:* Math analyst & computer specialist, Chi Assocs, Inc, Arlington, Va, 76-80; asst prof appl math, Dept Math, Va Commonwealth Univ, Richmond, 80-83 & Mary Washington Col, Fredericksburg, 83-88; STAFF OFFICER, BD MATH SCI, NAT ACAD SCI, WASHINGTON, DC, 89- *Concurrent Pos:* Nat Res Coun prog officer, Nat Security Agency's Math Sci Prog, Nat Acad Sci, Washington, DC, 89-91, Nat Res Coun staff officer, Comt Appl & Theoret Statist, 90- *Mem:* Sigma Xi; Am Math Soc; Math Asn Am; Soc Indust & Appl Math; Am Statist Asn. *Res:* Nonlinear dynamical systems; fractals-chaos-turbulence; quasicrystals; self-organized criticality; solitons; applied mathematics; mathematics applied to environmental and biological sciences; history of mathematics. *Mailing Add:* Bd Math Sci Rm NAS 312 Nat Acad Sci 2101 Constitution Ave NW Washington DC 20418

TUCKER, KENNETH WILBURN, b Santa Barbara, Calif, Aug 8, 24; m 53. APICULTURE. *Educ:* Univ Calif, BS, 50, PhD(entom), 57. *Prof Exp:* Res fel, Univ Minn, 54-60; instr biol, Lake Forest Col, 60-63; asst res apiculturist, Univ Calif, Davis, 63-66; res entomologist, USDA, 66-85; RETIRED. *Mem:* Entom Soc Am; Int Bee Res Asn; Genetics Soc Am; Am Genetic Asn. *Res:* Genetics of honey bees. *Mailing Add:* 1022 Baird Dr Baton Rouge LA 70808

TUCKER, PAUL ARTHUR, b Albemarle, NC, May 14, 41; m 65. TEXTILES, MICROSCOPY. *Educ:* NC State Univ, BS, 63, MS, 66, PhD(fiber & polymer sci), 73. *Prof Exp:* Instr textiles, NC State Univ, 64-71, asst prof, 73; NATO vis fel, Dept Textile Indust, Univ Leeds, 74; asst prof textiles, 75-80, ASSOC PROF TEXTILE MAT & MGT, SCH TEXTILES, NC STATE UNIV, 80- *Mem:* Royal Micros Soc. *Res:* Seeking the basic materials science underlying fibrous materials and relating applied technology to this science; polymer fine structure; microscopy; yarn processing; particulate analyses. *Mailing Add:* Dept Textile Mat Sch Textiles NC State Univ Raleigh NC 27695

TUCKER, RAY EDWIN, b Somerset, Ky, Dec 31, 29; m 53; c 3. ANIMAL SCIENCE. *Educ:* Univ Ky, BS, 51, MS, 66, PhD(animal nutrit), 68. *Prof Exp:* Asst prof animal sci, Va Polytech Inst & State Univ, 68-69; asst prof, 69-80, ASSOC PROF ANIMAL SCI, UNIV KY, 80- *Mem:* Am Soc Animal Sci. *Res:* Ruminant nutrition; starch utilization; urea utilization; magnesium deficiency; vitamin A antagonists; poultry litter as a feedstuff for ruminants. *Mailing Add:* Dept Animal Sci Univ Ky Lexington KY 40506

TUCKER, RICHARD FRANK, b New York, NY, Dec 25, 26. RESEARCH ADMINISTRATION. *Educ:* Cornell Univ, BChE, 50. *Prof Exp:* Esso Standard Oil Co, 50-55; Caltex Oil Co, 55-61; mem staff, Mobil Oil Corp, 61-, pres, 86-; RETIRED. *Mem:* Nat Acad Eng. *Mailing Add:* c/o Corp Secy Mobil Corp 3225 Gallows Rd Fairfax VA 22037

TUCKER, RICHARD LEE, b Wichita Falls, Tex, July 19, 35; m 56; c 2. CIVIL ENGINEERING, CONSTRUCTION. *Educ:* Univ Tex, BS, 58, MS, 60, PhD(civil eng), 63. *Prof Exp:* Proj engr, Eng-Sci Consult, 58-60; instr civil eng, Univ Tex, 60-62; from asst prof to prof, Univ Tex, Arlington, 62-74, assoc dean eng, 67-74; vpres, Luther Hill & Assoc, Inc, Dallas, 74-76; C T WELLS PROF PROJ MGT, 76-, DIR, CONSTRUCT INDUST INST, UNIV TEX, AUSTIN, 83- *Honors & Awards:* R L Peurifoy Res Award, Am Soc Civil Engrs, 86; Thomas F Rowland Prize, 87. *Mem:* Am Soc Eng Educ; Am Soc Civil Engrs; Nat Soc Prof Engrs; Am Soc Testing & Mat; Soc Exp Stress Anal. *Res:* Construction engineering; project management. *Mailing Add:* 3208 Red River No 300 Univ Tex Austin TX 78705

TUCKER, ROBERT C, JR, b Kansas City, Mo. RESEARCH & DEVELOPMENT MANAGEMENT, FAILURE ANALYSIS. *Educ:* NDak State Univ, BS, 57; Iowa State Univ, MS, 64, PhD(metall), 67. *Prof Exp:* Chemist, Ames Lab, AEC, Iowa State Univ, 57-58 & 61-67; sr res metallurgist, 67-72, mgr mat develop, 72-81, ASSOC DIR TECHNOL, UNION CARBIDE CORP, 81- *Concurrent Pos:* Mem numerous adv comts, Am Soc Metals Int, Am Vacuum Soc, Nat Asn Corrosion Engrs & Am Soc Testing & Mat, 86-; vis prof, Univ Ill, Urbana, 89- *Mem:* Fel Am Soc Metals Int; Am Vacuum Soc; Nat Asn Corrosion Engrs; Am Soc Testing & Mat; Sigma Xi; Am Inst Mining Metall & Petrol Engrs. *Res:* Metallic, ceramic and cermet coatings wear and corrosion resistance at low and high temperature; thermal barrier materials using thermal spray; chemical vapor deposition and physical vapor deposition. *Mailing Add:* 61 Ridgeway Dr Brownsburg IN 46112

TUCKER, ROBERT GENE, b Springfield, Mo, Sep 14, 18; m 46; c 3. MINERAL METABOLISM, NEW DRUGS & MEDICAL DEVICES. *Educ:* Southwestern Mo State Univ, AB, 40; Univ Minn, MS, 48, PhD(biochem), 50. *Prof Exp:* Chemist, MFA Milling Co, Springfield, Mo, 40-44; chemist, US Army, 44-45; biochemist, Vet Admin Hosp, Nashville Tenn, 51-55; sr scientist, Smith Kline Corp, Philadelphia, 55-60; mgr sci services, Baxter Labs, Deerfield, Ill, 60-76; RETIRED. *Mem:* Am Chem Soc; Am Inst Nutrit; Sigma Xi. *Res:* Mineral metabolism; new drugs and medical devices. *Mailing Add:* Rte 2 Box 112 Noel MO 64854

TUCKER, ROBERT WILSON, CELL BIOLOGY, MEDICAL ONCOLOGY. *Educ:* Harvard Univ, MD, 70. *Prof Exp:* ASST PROF ONCOL & CELL BIOL, SCH MED, JOHNS HOPKINS UNIV, 79- *Mailing Add:* Dept Oncol Johns Hopkins Oncol Ctr 600 N Wolfe St Baltimore MD 21205

TUCKER, ROY WILBUR, b Exeter, Calif, Jan 25, 27; m 54; c 2. MATHEMATICS. *Educ:* Stanford Univ, BS, 51, MA, 53, MS, 54. *Prof Exp:* Instr math, Colo Col, 54-55; instr, Modesto Jr Col, 55-59; from asst prof to assoc prof, 59-71, coordr comput ctr, 64-65 & consult, 65-66, PROF MATH, HUMBOLDT STATE UNIV, 71- *Mem:* Math Asn Am; Soc Indust & Appl Math; Asn Comput Mach. *Res:* Numerical analysis; linear algebra. *Mailing Add:* Dept Math Humboldt State Univ Arcata CA 95521

TUCKER, RUTH EMMA, b Warrensburg, Ill, Feb 17, 01. NUTRITION. *Educ:* Univ Ill, AB, 23, MS, 25; Univ Chicago, PhD(nutrit, food chem), 48. *Prof Exp:* Asst, Univ Ill, 23-25; instr food & nutrit, Kans State Col, 25-37; prof home econ, Univ Alaska, 37-42; prof & res prof food & nutrit, 44-72, EMER PROF FOOD & NUTRIT, UNIV RI, 72 - *Mem:* AAAS; fel Am Pub Health Asn; Am Dietetic Asn; Am Home Econ Asn. *Res:* Food chemistry. *Mailing Add:* 160 Linden Dr Kingston RI 02881-1920

TUCKER, SHIRLEY COTTER, b St Paul, Minn, Apr 4, 27; m 53. BOTANY. *Educ:* Univ Minn, Minneapolis, BA, 49, MS, 51; Univ Calif, Davis, PhD, 56. *Prof Exp:* Instr bot, Univ Minn, 60; res fel biol, Northwestern Univ, 61-63; res fel bot, Univ Calif, 63-66; from asst prof to prof bot, 68-82, BOYD PROF, DEPT BOT, LA STATE UNIV, BATON ROUGE, 82- *Honors & Awards:* Merit Award, Bot Soc Am. *Mem:* Bot Soc Am (pres-elect, 86-87, pres, 87-88); Am Bryol & Lichenological Soc; Brit Lichen Soc; Sigma Xi; fel Linnean Soc London. *Res:* Developmental anatomy of flower and vegetative shoots; determinate growth; plant anatomy; morphology; lichenology. *Mailing Add:* Dept Bot La State Univ Baton Rouge LA 70803

TUCKER, THOMAS CURTIS, b Hanson, Ky, Nov 1, 26; m 47; c 3. SOILS, PLANT NUTRITION. *Educ:* Univ Ky, BS, 49; Kans State Univ, MS, 51; Univ Ill, PhD, 55. *Prof Exp:* Asst soils, Kans State Univ, 49-51, instr, 51; asst soil fertility, Univ Ill, 51-55; asst prof soils, Miss State Univ, 55-56; from assoc prof to prof agr chem & soils, 56-74, PROF SOILS, WATER & ENG & SOIL SCIENTIST, AGR EXP STA, UNIV ARIZ, 74- *Concurrent Pos:* Vis prof, NC State Univ, 66-67; vis scientist, Univ Ariz, Tuscon, 77-78. *Mem:* Soil Sci Soc Am; Am Soc Agron; Am Chem Soc. *Res:* Agronomy; soil fertility and chemistry; analytical chemistry; soil-plant relationships; soil nitrogen transformations, fixation, fertilizer use efficiency using 15-nitrogen labelled materials. *Mailing Add:* Dept Soils & Water Engr Univ Ariz Tucson AZ 85721

TUCKER, THOMAS WILLIAM, b Princeton, NJ, July 15, 45; m 68; c 2. TOPOLOGY, COMBINATORICS. *Educ:* Harvard Univ, AB, 67; Dartmouth Univ, PhD(math), 71. *Prof Exp:* Instr math, Princeton Univ, 71-73; from asst prof to assoc prof, 73-82, chmn, 82-86, PROF MATH, COLGATE UNIV, 83- *Concurrent Pos:* Consult, Col Bds & Educ Testing Serv, 73-, Inst Defense Anal, 74, 75, 78, 79, 84 & 85; prin investr, various NSF res grants; vis assoc prof, Darmouth Col, 78-79; chmn, AP Calculus Comt Col Bd, 83-87; Nat Adv Comt, NSF, 86, Col Bd, 86-, coun chief, State Sch Officers, 87-88, Westat Assessment, NSF/ILI Prog, 88-90, Nat Assessment Educ Prog, 88-, Nat Fac, 89-, co-prin invester, Harvard Calculus Consortium, 89-, ed, NSF/MAA Surv Calculus Projs, 90; Math Comt, Nat Assessment Educ Progress, 87-; chmn, CUPM Comt Math Asn Am, 88- *Mem:* Math Asn Am (first vpres, 90-); Am Math Soc. *Res:* Three-dimensional topology and combinatorics, especially topological graph theory; author of topological graph theory. *Mailing Add:* Dept Math Colgate Univ Hamilton NY 13346

TUCKER, VANCE ALAN, b Niagara Falls, NY, Apr 4, 36; m. COMPARATIVE PHYSIOLOGY. *Educ:* Univ Calif, Los Angeles, BA, 58, PhD(zool), 63; Univ Wis, MS, 60. *Prof Exp:* NSF fel zool, Univ Mich, 63-64; from asst prof to assoc prof zool, 64-73, PROF ZOOL, DUKE UNIV, 73- *Concurrent Pos:* NSF res grants & Duke Univ Coun Res grants, 65-91. *Mem:* AAAS. *Res:* Vertebrate locomotion, respiration, circulation, energy metabolism; avian aerodynamics. *Mailing Add:* Dept Zool Duke Univ Durham NC 27706

TUCKER, W(ILLIAM) HENRY, b Seaford, Del, July 7, 20; m 85; c 2. CHEMICAL & ENVIRONMENTAL ENGINEERING, ENERGY & MASS TRANSFER. *Educ:* Univ Va, BS, 42; Mass Inst Technol, MS, 46, ScD(chem eng), 47. *Prof Exp:* Res assoc, Manhattan Proj, 44-45; res engr & supvr eng res, Servel, Inc, 47-53; assoc prof chem eng, Purdue Univ, Lafayette, 53-69; prof chem eng & head dept, 69-84, dir, Energy Anal & Diag Ctr, US Dept Energy, Tri-State Univ, 84, EMER PROF CHEM, TRI-STATE UNIV, 84- *Concurrent Pos:* Adv, Cheng Kung Univ, Taiwan, 58-59; vis teacher, Swiss Fed Inst Technol, 59; consult, Whirlpool Res, 62-69; Am Inst Chem Eng traveling fel study co-op educ, Gt Brit, 69; mid-career fel, Lilly Endowment, Inc, 78-79; chmn, Theol Dialogue Sci & Technol, United Ministries Educ, 79-81, Chem Plant Safety Workshops, Taiwan, 85. *Mem:* Sigma Xi; Am Inst Chem Engrs. *Res:* Chemical heat pump; process design and economics; energy conservation; absorption refrigeration. *Mailing Add:* 4175 Stratus Ct S Salem OR 97302

TUCKER, WALLACE HAMPTON, b McAlester, Okla, Nov 4, 39; m 57; c 2. HIGH ENERGY ASTROPHYSICS, X-RAY ASTRONOMY. *Educ:* Univ Okla, BS, 61, MS, 62; Univ Calif, San Diego, Phd(physics), 66. *Prof Exp:* Res assoc, Cornell Univ, 66-67; asst prof space sci, Rice Univ, 67-69; sr staff scientist, Dept Am Sci & Eng, Harvard-Smithsonian Observ, 69-72, consult, 72-76; ASTROPHYSICIST, SMITHSONIAN ASTROPHYS OBSERV, 76- *Concurrent Pos:* Vis lectr physics, Univ Calif, Irvine, 80-86, vis prof, 86-89. *Mem:* Am Astron Soc; AAAS; Int Astron Union. *Res:* High energy astrophysics, particularly x-ray astronomy; writings in astronomy and cosmology for general audiences. *Mailing Add:* PO Box 266 Bonsall CA 92003

TUCKER, WALTER EUGENE, JR, b Atlanta, Ga, Aug 7, 31; m 56; c 3. VETERINARY PATHOLOGY. *Educ:* Univ Ga, DVM, 56; Am Col Vet Pathologists, dipl, 62. *Prof Exp:* From resident to staff mem, Vet Path Div, Armed Forces Inst Path, 58-62; sr pathologist, Dow Chem Co, 62-68; mgr path sect, Wyeth Labs Inc, 68-74; DIR, DIV TOXICOL & PATH, BURROUGHS WELLCOME CO, 74- *Mem:* Int Acad Pathologists; Soc Toxicol Pathologists; Am Col Vet Pathologists. *Res:* Research and development of pharmaceutical and agricultural chemicals with emphasis on characterization and safety evaluation of toxicopathologic responses in laboratory and domestic animals following administration of candidate human and veterinary biological and chemotherapeutic agents. *Mailing Add:* Burroughs Wellcome Co 3030 Cornwallis Rd Research Triangle Park NC 27709

TUCKER, WILLIAM PRESTON, b Louisville, Ky, Sept 23, 32; m 59; c 3. ORGANIC CHEMISTRY. *Educ:* Wake Forest Col, BS, 57; Univ NC, MA, 60, PhD(chem), 62. *Prof Exp:* NIH fel, Univ Ill, 62-63; from asst prof to assoc prof chem, 63-72, PROF CHEM, NC STATE UNIV, 72- *Mem:* Am Chem Soc. *Res:* Chemistry of organic compounds of divalent sulfur; natural products. *Mailing Add:* Dept Chem NC State Univ Raleigh NC 27695-8204

TUCKER, WILLIE GEORGE, b Tampa, Fla, Nov 26, 34. ORGANIC CHEMISTRY. *Educ:* Tuskegee Inst, BS, 56, MS, 58; Univ Okla, PhD(org chem), 62. *Prof Exp:* Head dept, 69-89, PROF CHEM, SAVANNAH STATE COL, 62- *Concurrent Pos:* Dir coop educ, Savannah State Col, 83-89. *Mem:* AAAS; Am Chem Soc; Sigma Xi. *Res:* Chlorination with cupric chloride; halogenation of pyridine; iodination of lactate dehydrogenase isoenzymes. *Mailing Add:* 1523 Cathy St Savannah GA 31401

TUCKER, WOODSON COLEMAN, JR, b Halsey, Ky, Sept 17, 08; m 32, 73; c 2. ACADEMIC ADMINISTRATION, PHYSICAL CHEMISTRY. *Educ:* Univ Fla, BS, 29, MS, 30, PhD(chem), 53. *Prof Exp:* Chemist, Superior Earth Co, 31-36; asst gen mgr in chg prod, United Prod Co, 36-38; chemist & tech adv to supt, Edgar Plastic Kaolin Co, 40-41; interim instr chem, Univ Fla, 46-51; instr, 51-52, from asst prof to prof, 52-68, asst vchancellor acad affairs, 69-76, EMER PROF CHEM, UNIV NEW ORLEANS, 76- *Mem:* Am Chem Soc; Sigma Xi. *Res:* Physical and thermodynamic properties of terpenes; chemical education. *Mailing Add:* 315 E Hathaway Dr San Antonio TX 78209-6414

TUCKERMAN, MURRAY MOSES, b Boston, Mass, July 19, 28; m 48; c 4. PHARMACEUTICAL QUALITY ASSURANCE, DRUG REGULATORY AFFAIRS. *Educ:* Yale Univ, BS, 48; Temple Univ, BS, 53; Rensselaer Polytech Inst, PhD(chem), 58. *Prof Exp:* Asst anal chem, Sterling-Winthrop Res Inst, 53-55, res assoc, 55-58; assoc prof chem, Temple Univ, 58-62, dir radiol health specialist training prog, 63-69, head dept, 61-72, prof chem, Sch Pharm, 62-90, dir progs for Pharmaceut Indus, 87-90, EMER PROF, TEMPLE UNIV, 90- *Concurrent Pos:* Resident res assoc, Argonne Nat Lab, 59; mem bd revision, US Pharmacopeia, 60-90; consult clin ctr, NIH, 59-65; sci adv, Food & Drug Admin, 67-71; consult, Drug Regulatory Affairs, 71-; temp hon mem secretariat Europ pharmacopeia, Coun Europe, 74. *Mem:* Fel Am Inst Chem; fel Am Acad Pharmaceut Sci; Int Pharmaceut Fedn; Am Soc Qual Control. *Res:* Drug standards; pharmaceutical quality assurance. *Mailing Add:* Turtleneck Enterprises Monomonac Rd W Winchendon Springs MA 01477-0079

TUCKETT, ROBERT P, b Salt Lake City, Utah, March 11, 43; m 69; c 2. NEUROPHYSIOLOGY, SOMATOSENSORY PHYSIOLOGY. *Educ:* Univ Utah, BS, 65, PhD(biophysics & bioeng), 72. *Prof Exp:* Biophysicist, Artificial Heart Test Ctr, 71-72; fel neurophysiol, 72-77, res assoc, 77-79, res instr, 79-81, ASST RES PROF, UNIV UTAH, 81- *Concurrent Pos:* Prin investr, NIH grant, 79- *Mem:* Am Physiol Soc; Soc Neurosci. *Res:* Study of cutaneous receptor behavior and transfer of information in somatosensory pathways; mechanism by which the sensation of itch is transmitted to central nervous system. *Mailing Add:* Univ Utah Med Sch 410 Chipeta Way Res Park Salt Lake City UT 84108

TUCKEY, STEWART LAWRENCE, b Browns Valley, Minn, Aug 24, 05; m 36; c 1. DAIRY TECHNOLOGY. *Educ:* Univ Ill, Urbana, BS, 28, MS, 30, PhD(dairy tech), 37. *Prof Exp:* Asst dairy mfg, 28-30, instr, 30-32, assoc, 32-37, from asst prof to assoc prof, 37-57, prof dairy technol, 57-72, EMER PROF DAIRY TECHNOL, UNIV ILL, URBANA, 72- *Concurrent Pos:* Sabbatical, Neth Inst Dairy Res, Ede, 67. *Honors & Awards:* Borden Award, 39; Charles E Pfizer Award, Am Dairy Sci Asn, 69. *Mem:* Am Chem Soc; Am Dairy Sci Asn; Sigma Xi. *Res:* Biochemical and microbiological changes in cheese; microbial clotting enzyme for cheese. *Mailing Add:* 919 W Charles St Champaign IL 61821

TUCKSON, COLEMAN REED, JR, dentistry, for more information see previous edition

TUDBURY, CHESTER A, b Warwick, RI, Jan 3, 13. MECHANICAL ENGINEERING. *Educ:* Mass Inst Technol, BSEE & MSEE, 34. *Prof Exp:* Asst prof elec eng, Cleveland State Univ, 36-50; supvr develop lab, Ohio Crankshaft, Cleveland, 40-46, mgr eng, 48-57; supvr design, Budd Corp, Detroit, 46-48; self employed, consult, 79-83; RETIRED. *Concurrent Pos:* Tech adv to the pres, Thermatool, Stanford, Conn, 58-79. *Mem:* Fel Inst Elec & Electronics Engrs. *Res:* Author of a book. *Mailing Add:* Heartlands Apt 110 3004 N Ridge Rd Ellicott City MD 21043

TUDDENHAM, W(ILLIAM) MARVIN, b Salt Lake City, Utah, July 8, 24; m 45; c 4. PHYSICAL CHEMISTRY, FUEL TECHNOLOGY. *Educ:* Univ Utah, BA, 47, MS, 48, PhD(fuel technol), 54. *Prof Exp:* Res anal chemist, Eastman Kodak Co, 48-50; res lab technician, 53-55, sr scientist, 55-59, head chem, Phys Methods & Spec Studies Sect, 59-72, mgr anal tech dept, Metal Mining Div Res, 72-78, dir anal serv, Utah Copper Div, 78-80, mgr prod qual projs, Kennecott Minerals Co, 80-83; INDEPENDENT CONSULT, 83- *Mem:* Am Chem Soc; Am Inst Mining, Metall & Petrol Engrs; Sigma Xi. *Res:* Application of infrared and x-ray in process control; application of solar energy for high temperature research; role of catalysis in the oxidation of carbon; quality control in copper production. *Mailing Add:* 1828 Lincoln St Salt Lake City UT 84105-3308

TUDDENHAM, WILLIAM J, radiology, roentgenology, for more information see previous edition

TUDOR, DAVID CYRUS, b Wildwood, NJ, May 10, 18; m 41; c 1. POULTRY PATHOLOGY, PIGEON DISEASES. *Educ:* Rutgers Univ, BS, 40; Univ Pa, VMD, 51. *Prof Exp:* Instr high sch, Voc Agr, NJ, 40-44; res asst poultry path, Rutgers Univ, 51-59, assoc res specialist, 59-66, res prof poultry path, 66-78; RETIRED. *Concurrent Pos:* Instr, poultry path, animal path & gen biol, Cook Col, 51-78; assoc ed, Poultry Sci, Poultry Sci Asn, 73-75; small animal pract, 78-91. *Honors & Awards:* Helyar House Award, 85. *Mem:* Am Vet Med Asn; Am Asn Avian Path; World Poultry Sci. *Res:* Poultry science; infectious nephrosis; Salmonella and chronic respiratory disease; mycoplasma; pox; pet bird diseases; pigeon diseases. *Mailing Add:* 29 Station Rd Cranbury NJ 08512

TUDOR, JAMES R, b Ft Smith, Ark, Mar 26, 22; m 72; c 1. ELECTRICAL ENGINEERING. *Educ:* Univ Mo, BS, 48, MS, 50; Ill Inst Technol, PhD(elec eng), 60. *Prof Exp:* Sr asst engr, Union Elec Co, 50-52; from asst prof to assoc prof elec eng, 52-65, PROF ELEC ENG, UNIV MO-COLUMBIA, 65-, MO ELEC UTILITIES PROF POWER SYSTS ENG, 69- *Concurrent Pos:* Proj dir, Signal Corps, US Army, 56-57 & 58. *Mem:* Inst Elec & Electronics Engrs. *Res:* Transmission lines; energy conversion; electric circuits; power systems; computer control of electric power systems. *Mailing Add:* Dept Elec Eng 202 Ee Bldg Univ Mo Columbia MO 65211

TUEL, WILLIAM GOLE, JR, b Indianapolis, Ind, Apr 16, 41. COMPUTER GRAPHICS, DATA BASES. *Educ:* Rensselaer Polytech Inst, BEE, 62, MEE, 64, PhD(elec eng), 65. *Prof Exp:* Mem res staff, 65-76, mgr power syst studies group, 70-72, mgr exp comput studies, 73-76, MEM SCI STAFF, IBM SCI CTR, 76-, MGR EXPLOR GRAPHIC, 84- *Mem:* Inst Elec & Electronics Engrs; Asn Comput Mach. *Res:* Interactive computer graphics; geographic data bases; computer programming environments. *Mailing Add:* 6042 Monteverde Dr San Jose CA 95120

TUELLER, PAUL T, b Paris, Idaho, July 30, 34; m 63; c 4. PLANT ECOLOGY, RANGE MANAGEMENT. *Educ:* Idaho State Univ, BS, 57; Univ Nev, MS, 59; Ore State Univ, PhD, 62. *Prof Exp:* From asst prof to assoc prof range sci, 62-73, head, Div Renewable Natural Resources, 80, PROF RANGE SCI, UNIV NEV, RENO, 73- *Mem:* Soc Range Mgt; Soc Am Foresters; Am Soc Photogram. *Res:* Range ecology, especially vegetation-soil relationships; management of big game populations; remote sensing of renewable natural resources. *Mailing Add:* Dept Range Wildlife Forestry Univ Nev Reno NV 89557

TUEMMLER, WILLIAM BRUCE, inorganic & polymer chemistry; deceased, see previous edition for last biography

TUERPE, DIETER ROLF, b Chemnitz, Ger, Dec 29, 40; US citizen; m 79; c 1. CONVENTIONAL ORDNANCE, NUCLEAR PHYSICS. *Educ:* Polytech Inst Brooklyn, BS, 62; Univ Calif, Berkeley, MA, 66; Univ Calif, Davis, PhD(appl sci), 73. *Prof Exp:* Physicist, Lawrence Livermore Lab, 67-80; AT PHYSICS INT CO, 80- *Mem:* Am Phys Soc. *Res:* Nuclear Hartree-Fock calculation; atmospheric boundary layer models; equations of state; social and economic systems modelling; general ordnance research and development; shaped charge design. *Mailing Add:* Physics Int Co 2700 Merced St San Leandro CA 94577

TUESDAY, CHARLES SHEFFIELD, b Trenton, NJ, Sept 7, 27; m 52; c 3. RESEARCH ADMINISTRATION. *Educ:* Hamilton Col, NY, AB, 51; Princeton Univ, MA & PhD(phys chem), 55. *Prof Exp:* Res chemist panelyte div, St Regis Paper Corp, 50-51; res asst, Princeton Univ, 51-55; sr res chemist fuels & lubricants dept, 55-64, supvry res chemist, 64-67, from asst head to head, 67-72, head environ sci dept & actg head phys chem dept, 72-74, tech dir, 74-88, EXEC DIR, RES LABS, GEN MOTORS CORP, 88- *Concurrent Pos:* Consult vapor-phase org air pollutants panel, Nat Res Coun, 72-76; mem, Air Pollution Res Adv Comt, Coord Res Coun, 70-82, chmn, 78-80; mem comt mat substitution methodology, Nat Res Coun, 79-81; mem, Comt Corp Assocs, Am Chem Soc, 85-; mem bd dir, Coord Res Coun, 87-; chair, Indust Sci Sect, AAAS, 90-91. *Mem:* AAAS; Am Chem Soc; Soc Automotive Eng; Sigma Xi. *Res:* Physical chemistry; molecular energy exchange; environmental science; polymers; metallurgy; fuels & lubricants; biomedical science; analytical chemistry. *Mailing Add:* 69900 Henry Ross Dr Romeo MI 48065

TUFARIELLO, JOSEPH JAMES, b Brooklyn, NY, Oct 3, 35; m 60; c 4. ORGANIC CHEMISTRY. *Educ:* Queens Col, NY, BS, 57; Univ Wis-Madison, PhD(chem), 61. *Prof Exp:* Assoc org chem, Purdue Univ, Lafayette, 62; NIH fel, Cornell Univ, 62-63; assoc prof, 63-80, PROF ORG CHEM, STATE UNIV NY, BUFFALO, 80-, CHMN DEPT CHEM, 84- *Mem:* AAAS; Am Chem Soc; The Chem Soc. *Res:* Organic synthesis; synthesis and reactivity of strained or otherwise unique carbocyclic systems; synthesis of natural products; organometallic chemistry; chemistry of 1,3-dipolar compounds. *Mailing Add:* Dept Chem State Univ NY Buffalo NY 14214

TUFF, DONALD WRAY, b San Francisco, Calif, May 4, 35; m 55; c 3. PARASITOLOGY, TAXONOMY. *Educ:* San Jose State Col, BA, 57; Wash State Univ, MS, 59; Tex A&M Univ, PhD(entomol), 63. *Prof Exp:* From asst prof to assoc prof biol, 63-73, PROF BIOL, SOUTHWEST TEX STATE UNIV, 73- *Mem:* Entom Soc Am; Soc Syst Zool; Wildlife Dis Asn. *Res:* Taxonomy of avian Mallophaga; parasites of wildlife. *Mailing Add:* Dept Biol Southwest Tex State Univ San Marcos TX 78666

TUFFEY, THOMAS J, US citizen. ENVIRONMENTAL SCIENCES. *Educ:* Rutgers Univ BS, 68, MS, 72, PhD(environ sci), 73. *Prof Exp:* VPRES RESOURCES ENG, ROY F WESTON, INC, 76- *Concurrent Pos:* Mem fac, Col Eng, Rutgers Univ, 76, vis asst prof, 77. *Res:* Surface water monitoring systems and water quality modeling simulations; nitrogen cycle-nitrication-dentrification; siting, restoration and environmental assessments of lakes and reservoirs; watershed management; land management of waste sludges. *Mailing Add:* Roy F Weston Inc Weston Way West Chester PA 19380

TUFFLY, BARTHOLOMEW LOUIS, b Houston, Tex, Apr 9, 28; m 58; c 3. PHYSICAL CHEMISTRY, ANALYTICAL CHEMISTRY. *Educ:* Univ Tex, BA, 48, MA, 50, PhD(chem), 52. *Prof Exp:* Chemist, Carbide & Carbon Chem Co, 52-60; chemist, Rocketdyne Div, NAm Rockwell Corp, 60-73, Rockwell Int Corp, 73-84; PVT CONSULT, ENVIRON SERV, 84- *Mem:* Water Pollution Control Fedn. *Res:* Mass spectrometry; pollution; water management; environmental control systems; rocket propellants; Technical analysis for the handling, disposing, and treatment of hazardous substances; compliance with regulations pertaining to hazardous substances. *Mailing Add:* 4709 Dunman Ave Woodland Hills CA 91364

TUFTE, MARILYN JEAN, b Iron Mountain, Mich, Nov 20, 39; m 72. IMMUNOLOGY & MOLECULAR BIOLOGY. *Educ:* Northern Mich Univ, AB, 61; Univ Wis, Madison, MS, 65, PhD(bact), 68. *Prof Exp:* Trainee & fel, Univ Wis, Madison, 68; asst prof biol, 68-71, assoc prof biol, 71-76, PROF BIOL, UNIV WIS-PLATTEVILLE, 76- *Mem:* Am Soc Microbiol; Sigma Xi. *Res:* Electron microscopic analysis of guinea pig peritoneal phagocytes infected with strains of Brucella abortus of different degrees of virulence; effects of vitamin A calcium and zinc gluconate on colon carcinoma in mice. *Mailing Add:* Dept Biol Univ Wis Platteville WI 53818

TUFTE, OBERT NORMAN, b Northfield, Minn, May 30, 32; m 56; c 4. SOLID STATE PHYSICS, ELECTRON DEVICE PHYSICS. *Educ:* St Olaf Col, BA, 54; Northwestern Univ, PhD(physics), 60. *Prof Exp:* Asst physics, Northwestern Univ, 54-59; res scientist, 60-70, dept mgr, Technol Ctr, 70-84, res fel, 84-87, CHIEF SCIENTIST, HONEYWELL INC, 87- *Concurrent Pos:* Honeywell res fel, 84-87. *Mem:* Am Phys Soc; Inst Elec & Electronics Engrs. *Res:* Semiconductors; electrical and optical properties of solids; solid state devices; integrated circuit technology. *Mailing Add:* Honeywell 10701 Lyndale Ave Bloomington MN 55420

TUFTS, DONALD WINSTON, b Yonkers, NY, Mar 5, 33; m 56; c 3. ELECTRICAL ENGINEERING. *Educ:* Williams Col, BA, 55; Mass Inst Technol, BS & MS, 58, ScD(elec eng), 60. *Prof Exp:* Asst prof appl math, Harvard Univ, 62-67; PROF ELEC ENG & COMPUT SCI, UNIV RI, 67- *Mem:* AAAS; fel Inst Elec & Electronics Engrs. *Res:* Information theory; communication theory; computer science; signal processing; underwater sound; data transmission. *Mailing Add:* Dept Elec Eng Univ RI Kelley Hall Kingston RI 02881

TUGWELL, PETER, b Mar 30, 44; Can citizen; m 71; c 2. CLINICAL EPIDEMIOLOGY, INTERNAL MEDICINE. *Educ:* Univ London, MBBS, 69, MD, 76; FRCP(C), 76; McMaster Univ, MSc, 77. *Prof Exp:* House officer med, Royal Free & WMiddlesex Hosps, London, 69-70; sr house officer, Whittington Hosp, London, 70-71; res fel & registr, Ahmadu Bello Univ, Nigeria, 71-74; chief resident internal med, 75-76, CLIN EPIDEMIOLOGIST & ATTENDING PHYSICIAN, MCMASTER UNIV,

77-, CHMN, DEPT EPIDEMIOL & BIOSTATISTICS, 79-, PROF, 84-, PROF MED, 84- *Mem:* Royal Col Physicians & Surgeons Can; Soc Epidemiologic Res; Am Rheumatology Asn; Am Asn Clin Res. *Res:* Rheumatology; effectiveness studies; educational evaluation; evaluation of quality of care. *Mailing Add:* McMaster Univ 1200 Main St W Hamilton ON L8N 3Z5 Can

TUITE, JOHN F, phytopathology; deceased, see previous edition for last biography

TUITE, ROBERT JOSEPH, b Rochester, NY, Aug 28, 34; m 58; c 3. RESEARCH MANAGEMENT, PHOTOGRAPHIC CHEMISTRY. *Educ:* St John Fisher Col, BS, 56; Univ Ill, PhD(org chem), 60. *Prof Exp:* Asst chem, Univ Ill, 56-57, asst pub health serv, 59; from res chemist to sr research chemist, Eastman Kodak Co, 59-67, res assoc res labs, 67-70, head color photo chem lab, 70-73, from head to sr head, Color Reversal Systs Lab, 73-74, asst dir, 74-76, dir Color Photog Div, 76-78, dir Color Instant Photog Div, 78-81, asst to the dir, Res Labs, 81-82, dir New Opportunity Develop, corp staff, 82-89; LOANED EXEC, HIGH TECHNOL, ROCHESTER, 89- *Concurrent Pos:* Mem, AMA coun res & develop mgt. *Mem:* Am Chem Soc; Soc Imaging Sci & Technol. *Res:* Applied photochemistry; color photographic imaging chemistry; color photographic systems design; parametrization of color photographic system response and correlation with molecular structure and other systems variables; new business development. *Mailing Add:* High Technol Rochester 55 St Paul St Rochester NY 14604

TUITES, DONALD EDGAR, b Saginaw, Mich, Dec 27, 25; m 50; c 4. POLYMER CHEMISTRY. *Educ:* Univ Rochester, BS, 49; Clarkson Tech Univ, MS, 52; Cornell Univ, PhD(org chem), 56. *Prof Exp:* Chemist, NY, 55-63, chemist electrochem dept, Del, 63-71, chemist plastics prod & resins dept, E I Du Pont de Nemours & Co, Inc, Del, 72-80; RETIRED. *Mailing Add:* 2515 Kittiwake Dr Wilmington DE 19805

TUITES, RICHARD CLARENCE, b Rochester, NY, Oct 31, 33; m 54; c 4. ORGANIC CHEMISTRY, POLYMER CHEMISTRY. *Educ:* Univ Rochester, BS, 55; Univ Ill, PhD(org chem), 59. *Prof Exp:* Res chemist, E I du Pont de Nemours & Co, Inc, 58-62; from chemist to sr chemist, 62-72, RES ASSOC, EASTMAN KODAK CO, 72- *Mem:* Am Chem Soc; Soc Photog Scientists & Engrs. *Res:* Photographic chemistry. *Mailing Add:* Rushford Lake Caneadea NY 14717

TUKEY, HAROLD BRADFORD, JR, b Geneva, NY, May 29, 34; m 55; c 3. HORTICULTURE. *Educ:* Mich State Univ, BS, 55, MS, 56, PhD(hort), 58. *Hon Degrees:* Dr, Portuguese Asn Hort, 85. *Prof Exp:* Res asst hort, South Haven Exp Sta, Mich State Univ, 55, AEC, 55-58; NSF fel hort, Calif Inst Technol, 58-59; from asst prof to prof hort, Cornell Univ, 59-80; PROF URBAN HORT, UNIV WASH, 80- *Concurrent Pos:* Consult, PR Nuclear Ctr, Rio Piedras, 56-66, Int Bonsai Mag, 80-, Elec Power Res Inst; NSF grant, 62 & 75, Bot Soc Am, 64; dir, eastern region Int Plant Propagators Soc, 69-71, Am Soc Hort Sci, 70-71, Am Hort Soc, 72-81, Bot Soc Am, Northwest Hort Soc & Arboretum Found, 80-, Ctr Urban Hort & Arboreta, Univ Wash, 80-; US deleg, Coun Int Soc Hort Sci, 71-90, chmn, Comm Amateur Hort, 74-83, chmn, Comm Urban Hort, 90-; regional vpres, Int Plant Propagators Soc, 72; vis prof, Univ Calif, Davis, 73; mem var comts, Nat Acad Sci; mem adv comt, Seattle Univ Wash Arboretum & Bot Garden, 80-, vchmn, 82, chmn, 86-87; bd dirs, Arbor Fund Bloedel Res, 80-, pres, 83-84; mem, Nat Adv Comt, USDA, 90- *Honors & Awards:* Citation Merit, Am Hort Soc, 81. *Mem:* Int Soc Hort Sci (vpres, 78-82, pres, 82-86); Int Plant Propagators Soc (pres, 73, int pres, 76); Am Hort Soc (vpres, 78-80); fel Am Soc Hort Sci; Bot Soc Am Northwest Hort Soc; Sigma Xi. *Res:* Physiology of horticultural plants; urban horticulture; uptake and loss of substance through plant foliage. *Mailing Add:* Ctr Urban Hort GF-15 Univ Wash Seattle WA 98195

TUKEY, JOHN WILDER, b New Bedford, Mass, June 16, 15; m 50. STATISTICS, STATISTICAL ANALYSIS. *Educ:* Brown Univ, ScB, 36, ScM, 37; Prineton Univ, MA, 38, PhD(math), 39. *Hon Degrees:* ScD, Case Inst Technol, 62, Brown Univ, 65, Yale Univ, 68, Univ Chicago, 69 & Temple Univ, 78. *Prof Exp:* Instr math, Princeton Univ, 39-41, res assoc, Fire Control Res Off, 41-45; mem tech staff, Bell Labs, 45-58, asst dir res commun prin, 58-61, assoc exec dir res commun, Princeton Div, 61-85; prof statist, 65-85, Donner prof sci, 76-85, EMER PROF & SR RES STATISTICIAN & EMER DONNER PROF SCI, PRINCETON UNIV, 85- *Concurrent Pos:* From asst prof to prof math, Princeton Univ, 41-65, chmn dept statist, 65-70; Guggenheim fel, 49-50; fel, Ctr Advan Study Behav Sci, 57-65; visitor, Commonwealth Sci & Indust Res Orgn, Canberra, Australia, 71-79, & Stanford Linear Accelerator Ctr, Calif, 72, 79; mem, US deleg, Tech Working Group 2 Conf Discontinuance Nuclear Weapon Tests, Geneva, Switz, 59 & UN Conf Human Eviron, Stockholm, Sweden, 72; mem, President's Sci Adv Comt, Off Sci & Technol, 60-63, chmn, Panel Environ Pollution, 64-65 & Panel Chem & Health, 71-72; mem, President's Air Qual Adv Bd, 68-71 & President's Comn Fed Statist, 70-71; mem, Sci Info Coun, NSF, 62-64; chmn anal adv comt, Nat Assessment Educ Progress, 63-73 & chmn sci panel anal adv comt, 73-82; mem coun, Nat Acad Sci, 69-71 & 75-78, chmn class III, 69-72 & chmn climatic impact comt, 75-79; mem, Nat Adv Comt Oceans & Atomsphere, 75-77; mem, Nat Ctr Atmospheric Res, Boulder, Colo, 78 & Dept Sci & Indust Res, Wellington, NZ, 79. *Honors & Awards:* S S Wilks Medal, Am Statist Asn, 65; Nat Medal Sci, 73; Shewhart Medal, Am Soc Qual Control, 77, Deming Medal, 83; Medal of Honor, Inst Elec & Electronics Engrs, 82; James Madison Medal, Princeton Univ, 84; Monei Ferst Award, Sigma Xi, 89. *Mem:* Nat Acad Sci; Am Philos Soc; Am Acad Arts & Sci; Int Statist Inst; hon mem Royal Statist Soc. *Res:* Theoretical, applied and mathematical statistics; point set topology; fire control equipment; military analysis. *Mailing Add:* 408 Fine Hall Princeton Univ Washington Rd Princeton NJ 08544-1000

TUKEY, LOREN DAVENPORT, b Geneva, NY, Dec 4, 21; m 52; c 2. POMOLOGY & PLANT PHYSIOLOGY. *Educ:* Mich State Univ, BS, 43, MS, 47; Ohio State Univ, PhD(hort), 52. *Prof Exp:* Asst hort, Ohio State Univ, 47-50; asst prof & assoc prof, 50-66, PROF POMOL, PA STATE UNIV, 66- *Concurrent Pos:* Mem, Coop Fruit Res Prog, Inst Nat Tech Agr, Arg, 65-70; assoc ed, J Hort Sci, Eng, 78-; ed, Pa State Hort Reviews, 62-; bus mgr, Am Pomol Soc, 68-90; res consult, Inst InterAm Coop Agr, Arg, 88. *Honors & Awards:* Paul Howe Shepard Award, Am Pomol Soc, 64; Serv & Leadership Award, Am Dwarf Fruit Tree Asn, 88; Milo Gibson Award, NAm Fruit Explorers, 89. *Mem:* Fel AAAS; fel Am Soc Hort Sci; Plant Growth Regulator Soc Am; Am Soc Plant Physiol; Int Soc Hort Sci; Am Pomol Soc (secy, 68-84, treas, 68-90); Brit Soc Plant Growth Regulation; corresp mem Acad Agr France. *Res:* Pomology; growth and development-plant growth regulators; plant-environmental relationships; developer of Penn State low trellis hedgerow system; intensive orchard systems; land productivity and rootstocks. *Mailing Add:* 103 Tyson Bldg Dept Hort Pa State Univ University Park PA 16802

TULAGIN, VSEVOLOD, b Leningrad, Russia, June 16, 14; US citizen; m 38. CHEMISTRY. *Educ:* Calif Inst Technol, BS, 37; Univ Calif, Los Angeles, MA, 41, PhD(chem), 43. *Prof Exp:* Chemist, Wesco Water Paints, Inc, Calif, 37-39; Nat Defense Res Comt asst, Univ Calif, Los Angeles, 42-43; res chemist, Gen Aniline & Film Corp, Pa, 43-47, group leader, 47-52, res specialist & group leader, Ansco Div, 52-57; res supvr, Minn Mining & Mfg Co, 57-60, mgr res photo prod, 60-66; res mgr, Xerox Corp, 66-69, chief scientist, 69-77; RETIRED. *Honors & Awards:* Kosar Mem Award, Soc Photog Sci & Eng. *Mem:* Am Chem Soc; Soc Photog Sci & Eng. *Res:* Paint technology; chemical synthesis; color photography; polychrome photoelectrophoresis. *Mailing Add:* 106 E Olive Ct Pine Knoll Shores NC 28512

TULCZYJEW, WLODZIMIERZ MAREK, mathematical physics, for more information see previous edition

TULECKE, WALT, b Detroit, Mich, Feb 10, 24; c 4. BOTANY. *Educ:* Univ Mich, BA, 46, MS, 50, PhD(bot), 53. *Prof Exp:* Asst prof bot, Ariz State Col, 53-55; res assoc, Brooklyn Bot Garden, 55-57; res botanist, Chas Pfizer & Co, 57-59; assoc plant physiologist, Boyce Thompson Inst, 59-67; PROF BIOL, ANTIOCH COL, 67- *Concurrent Pos:* Res plant physiologist, Univ Calif, Davis, 83-85. *Mem:* AAAS. *Res:* Plant tissue culture; Ginkgo; walnut. *Mailing Add:* Dept Biol Antioch Col Yellow Springs OH 45387

TULEEN, DAVID L, b Oak Park, Ill, Sept 19, 36; m 60; c 4. ORGANIC CHEMISTRY. *Educ:* Wittenberg Univ, BS, 58; Univ Ill, PhD(org chem), 62. *Prof Exp:* Fel, Pa State Univ, 62-63; asst prof, 63-68, ASSOC PROF ORG CHEM, VANDERBILT UNIV, 68- *Mem:* Am Chem Soc; Royal Soc Chem. *Res:* Sulfur chemistry. *Mailing Add:* Dept Chem Vanderbilt Univ Nashville TN 37240

TULENKO, JAMES STANLEY, b Holyoke, Mass, June 1, 36; m 65; c 3. COMPUTER AIDED ENGINEERING. *Educ:* Harvard Univ, BA, 58, MA, 60; Mass Inst Technol, MS, 63; George Washington Univ, MBA, 80. *Prof Exp:* Mgr nuclear develop, United Nuclear Corp, NY, 63-70; mgr physics, Nuclear Mat & Equip Corp, 70-71; mgr physics, Nuclear Power Generation Div, 71-74, mgr nuclear fuel eng, Babcock & Wilcox, 74-81; gen mgr, B & W Comput Serv, 81-83; mgr, Corp Eng Automation Serv, B & W/McDermott, 83-86; CHMN & PROF, NUCLEAR ENG SCI DEPT, UNIV FLA, 86- *Concurrent Pos:* Adj prof systs anal, George Washington Univ; chmn, spec comt waste mgt, Am Nuclear Soc; chmn-elect, Nuclear Dept Heads Orgn, 91; chmn, Nuclear Eng Div, Am Soc Eng Educ, 91. *Honors & Awards:* Am Nuclear Soc Silver Award, 81. *Mem:* Fel Am Nuclear Soc; Am Soc Mfrg Engrs; Soc Mfrg Engrs; Inst Elec & Electronics Engrs Comput Soc; Am Soc Eng Educ. *Res:* Mechanical and material design nuclear fuel; radioactive waste management; reactor physics; fuel management of nuclear reactors; fuel cycle economics of nuclear power plants; robotics for nuclear maintenance; engineering information flow; storage and retrieval information. *Mailing Add:* Nuclear Engr Sci Dept Univ Fla 202 Nuclear Sci Ctr Gainesville FL 32611

TULENKO, THOMAS NORMAN, b Pittsburgh, Pa, Dec 2, 42; m 73; c 1. VASCULAR DISEASE, HYPERTENISON RESEARCH. *Educ:* Grove City Col, BS, 66; Duquesne Univ, MS, 68; Botson Univ, PhD, 72. *Prof Exp:* Lectr biol, Gwynedd Mercy Col, 74-76; ASST PROF OBSTET & GYNEC, MED COL PA, 81-, ASSOC PROF PHYSIOL, 82- *Concurrent Pos:* Dir, City-Wide Bd Rev, 81-; consult, McGraw-Hill Pub, 81-; appointee, Nat Bd Med Examrs, 84- *Mem:* Am Physiol Soc; AAAS; Am Heart Asn. *Res:* Regulation of arterial activity in human blood vessels and the nature of their involvement in various disease states; atherosclerosis, vasospasm, heart attack, certain forms of hypertension and intrauterine growth retardation. *Mailing Add:* Dept Physiol & Biochem Med Col Pa 3300 Henry Ave Philadelphia PA 19129

TULER, FLOYD ROBERT, b Chicago, Ill, May 24, 39; m 61; c 2. MATERIALS SCIENCE, MECHANICAL METALLURGY. *Educ:* Univ Ill, Urbana, BS, 60, MS, 62; Cornell Univ, PhD(mat sci & eng), 67. *Prof Exp:* Tech staff mem, Sandia Labs, 66-69; vpres tech, Effects Technol Inc, 69-74; assoc prof mat sci, Hebrew Univ, Jerusalem, Israel, 74-80; PROF MAT ENG, WORCESTER POLYTECHNIC INST, MA, 81- *Concurrent Pos:* Panel mem nat mat adv bd ad hoc comt, Nat Acad Sci, 68-69; vis sci-assoc prof, Ctr for Policy Alternative & Dept of Mat Sciences & Eng, MIT, 78-81; mgr process develop, ManLabs Div, Alcan Aluminum, 90-91. *Mem:* AAAS; Sigma Xi; Am Soc Metals Inst. *Res:* Mechanical properties of metals and reinforced composite materials, materials processing; fracture; fatigue initiation and propagation; impact and impulsive loading; mechanical testing and nondestructive testing techniques; failure prediction and analysis; risk analysis and management. *Mailing Add:* 241 Perkins St Unit No F402 Boston MA 02130

TULEYA, ROBERT E, b York, Pa, Feb 22, 47; m 72. GEOPHYSICAL FLUID DYNAMICS. *Educ:* Pa State Univ, BS, 69, MS, 71. *Prof Exp:* Res asst, Dept Meteorol, Pa State Univ, University Park, 69-71; meteorologist, Geophys Fluid Dynamics Lab, 71-72, RES METEOROLOGIST, GEOPHYS FLUID DYNAMICS LAB, NAT OCEANIC ATMOSPHERIC ADMIN, PRINCETON UNIV, 72- *Honors & Awards:* Banner Miller Award, Am Meteorol Soc, 84. *Mem:* Am Meteorol Soc. *Mailing Add:* 107 Ironmaster Rd Cherry Hill NJ 08034

TULI, JAGDISH KUMAR, b India, Aug 7, 41; US citizen; m 75; c 2. NUCLEAR SPECTROSCOPY. *Educ:* Delhi Univ, MS, 65; Ind Univ, MS, 69, PhD(physics), 71. *Prof Exp:* Res assoc, Nat Acad Sci, Ind Univ, 71-73; physicist, Dept Atomic Energy, India, 73-75, Lawrence Berkeley Lab, 76-77; PHYSICIST, BROOKHAVEN NAT LAB, 77- *Concurrent Pos:* Ed, Nuclear Data Sheets, 81- *Mem:* Am Phys Soc. *Mailing Add:* Brookhaven Nat Lab 197D Upton NY 11973

TULIN, LEONARD GEORGE, b Mozyr, Russia, May 20, 20; US citizen; m 48; c 1. STRUCTURAL MECHANICS. *Educ:* Univ Colo, BSCE, 50, MSCE, 52; Iowa State Univ, PhD(theoret & appl mech), 65. *Prof Exp:* From instr to assoc prof, 50-61, chmn dept, 72-77, prof, 61-90, EMER PROF CIVIL ENG, UNIV COLO, BOULDER, 90- *Honors & Awards:* Wason Award, Am Concrete Inst, 64. *Mem:* Am Soc Civil Engrs; Am Concrete Inst; Am Soc Testing & Mat; Soc Exp Stress Anal; Masonry Soc. *Res:* Mechanics and materials; construction practices in reinforced masonry; design of timber structures; glass fiber reinforced concrete. *Mailing Add:* Dept Civil Environ & Archit Eng Campus Box 428 Univ Colo Boulder CO 80309-0428

TULIN, MARSHALL P(ETER), b Hartford, Conn, Mar 14, 26; m 55; c 2. WAVES, CAVITATIONS. *Educ:* Mass Inst Technol, BS, 46, MS, 49. *Prof Exp:* Aeronaut res scientist, Nat Adv Comt Aeronaut, 46-50; physicist & head turbulence & frictional sect, David W Taylor Model Basin, US Dept Navy, 50-54, aeronaut res engr, Mech Br, Off Naval Res, 54-57, sci liaison officer, London Br Off, 57-59; vpres & dir, Hydronautics, Inc, 59-71, chief exec officer & bd chmn, 71-82; PROF MECH & ENVIRON ENG & DIR, OCEAN ENG LAB, UNIV CALIF, SANTA BARBARA, 82- *Concurrent Pos:* Bd chmn, Hydronautics-Israel, Ltd. *Mem:* Nat Acad Eng; NY Acad Sci; Am Geophys Union. *Res:* Supercavitating, turbulent, stratified and polymer flows; hydrofoil and propeller theory; wakes. *Mailing Add:* Ocean Eng Lab Univ Calif Santa Barbara CA 93106

TULINSKY, ALEXANDER, b Philadelphia, Pa, Sept 25, 28; m 55; c 4. BLOOD PROTEINS. *Educ:* Temple Univ, AB, 52; Princeton Univ, PhD(chem), 56. *Prof Exp:* Res assoc, Protein Struct Proj, Polytech Inst Brooklyn, 55-59; asst prof chem, Yale Univ, 59-65; assoc prof chem, 65-67, PROF CHEM, MICH STATE UNIV, 68- *Concurrent Pos:* vis prof chem, Univ SC, Columbia, 82; chmn, Spec Interest Group Biol Macromolecules, Am Crystallog Asn, 77-78. *Honors & Awards:* Alberta Heritage Found vis scientist award, Univ Calgary, 81. *Mem:* Am Crystallog Asn; Am Chem Soc. *Res:* X-ray crystallographic structure determination of biological molecules; structure and function of enzymes; blood clotting proteins: thrombin, gla and tringle domains, protein inhibitor-enzyme complexes, other blood protein structural domains. *Mailing Add:* Dept Chem Mich State Univ East Lansing MI 48823

TULIP, THOMAS HUNT, b Anchorage, Alaska, Nov 16, 52; m 78. ORGANOTRANSITION METAL CHEMISTRY. *Educ:* Univ Vt, BS, 74; Northwestern Univ, MS, 75, PhD(chem), 78. *Prof Exp:* Res chemist, Cent Res & Develop, 78-85, CHEM SUPVR IMMUNOPHARMACEUT, BIOMED PROD, E I DU PONT DE NEMOURS & CO, INC, 85- *Concurrent Pos:* Vis scientist, Osaka Univ, 74. *Mem:* Am Chem Soc; Soc Nuclear Med; AAAS. *Res:* Organometallic and bioorganic chemistry; metals in medicine; protein modification; attachment of binding agents to antibodies; radiochemistry; immunochemistry. *Mailing Add:* Du Pont Med Prdts BMP 25/1174 PO Box 80024 Wilmington DE 19880-0024

TULK, ALEXANDER STUART, b Hamilton, Ont, Feb 25, 18; m 46; c 4. APPLIED CHEMISTRY. *Educ:* McMaster Univ, BSc, 44, MSc, 45; Pa State Univ, PhD(inorg chem), 51. *Prof Exp:* Lectr, McMaster Univ, 45-47; asst, Pa State Univ, 47-50; sr engr, GTE Prods Corp, 50-52, engr in-chg, 52-55, adv develop engr, 55-56, eng specialist, 56-57, eng mgr, 57-67, eng mgr inorg mat, 67-71, sr eng specialist, Electronic Mat, 71-73, sr eng specialist, Spec Proj, 73-83; RETIRED. *Honors & Awards:* Sullivan Award, Am Chem Soc, 74. *Mem:* Am Chem Soc; Electrochem Soc; Chem Inst Can. *Res:* Chemical warfare; photosensitive materials; liquid bright gold; fluorocarbon chemistry; electroplating of precious metals; germanium; silicon and gallium arsenide preparation; semiconductor measurements; chemistry of tungsten, molybdenum, rare earths, tantalum and niobium. *Mailing Add:* 510 Poplar St Towanda PA 18848-1120

TULL, JACK PHILLIP, b Jackson, Mich, Dec 2, 30; m 52; c 3. MATHEMATICS. *Educ:* Univ Ill, PhD(math), 57. *Prof Exp:* From instr to asst prof, 56-61, ASSOC PROF MATH, OHIO STATE UNIV, 61- *Concurrent Pos:* Vis sr lectr, Univ Adelaide, 63-64; prof & head dept, Univ Zambia, 70-71, dean humanities & social sci, 71. *Mem:* Am Math Soc; Math Asn Am; London Math Soc. *Res:* Analytic theory of numbers. *Mailing Add:* 6323 21st Ave NE Seattle WA 98115-6915

TULL, JAMES FRANKLIN, b New York, NY, May 26, 47; m 67; c 2. STRUCTURAL GEOLOGY, GEOTECTONICS. *Educ:* Univ NC, BS, 69; Rice Univ, PhD(geol), 73. *Prof Exp:* Asst prof geol, Univ Ala, 73-78, assoc prof, 78-81; assoc prof geol, 81-84, PROF GEOL & DEPT CHMN, FLA STATE UNIV, 84- *Concurrent Pos:* Consult, Geol Surv Ala, 75-76, E I du Pont de Nemours & Co, Inc, 78-, Amoco Petrol Co, 81-84, Champlin Petrol Co, 81 & J M Huber Co, 84, Hecla Mining Co, 87- *Mem:* Fel Geol Soc Am; Geol Soc Norway. *Res:* Structural evolution of mountain systems, particularly the development of metamorphic and igneous terraines; studying metamorphism associated with orogenesis and relationships between structural and metamorphic events. *Mailing Add:* Dept Geol Fla State Univ Tallahassee FL 32306

TULL, ROBERT GORDON, b Jackson, Mich, May 1, 29; m 52; c 4. ASTRONOMY, INSTRUMENTATION. *Educ:* Univ Ill, BS, 52, MS, 57; Univ Mich, PhD(astron), 63. *Prof Exp:* Res assoc, 61, from instr to asst prof, 62-70, res scientist astron, 70-83, SR RES SCIENTIST, MCDONALD OBSERV, UNIV TEX, AUSTIN, 83- *Concurrent Pos:* NSF grants, 63-66, 74-78, 78-82 & 88-91; mem high resolution spectrograph instrument definition team, NASA Large Space Telescope, 73-76; consult, Electronic Vision Co Div of Sci Appln Inc, 74-78; consult Europ Southern Observ, 78-; consult, Asiago Observ, Italy & Wise Observ, Israel, 75-; vis scientist, Chinese Acad Sci, 81; mem, Steering Comt, Nat New Technol Telescope, 81-83, High Resolution Spectrog Working Group, 85; proj scientist, 300-Inch Telescope, Univ Tex, 83-85, high resolution spectrograph, 2.7-m telescope, McDonald Observ, 88- *Mem:* Int Soc Optical Eng; Am Astron Soc; Int Astron Union; Optical Soc Am. *Res:* Photoelectric spectrophotometry of astronomical sources; astronomical instrumentation; development and application of multichannel image detectors for astronomical spectrophotometry; design of large telescopes. *Mailing Add:* Dept Astron Univ Tex Austin TX 78712

TULL, WILLIAM J, b Ontario, Can. ELECTRICAL ENGINEERING. *Educ:* Univ Mich, BSEE, 42. *Prof Exp:* Vpres & dir res & eng, Gen Porcelain Corp, NY, 64-69; pres, Tull Aviation, Armonk, NY, 69-78; vpres, Northrop Corp, Kansas City, MO, 78-82; RETIRED. *Honors & Awards:* Thurlow Award, Inst Navig, 59. *Mem:* Inst Navig; Fel Inst Elec & Electronics Engrs. *Res:* Inventor of the Doppler microwave navigation equipment. *Mailing Add:* 12 Oyster Rake Lane Hilton Head Plantation Hilton Head Island SC 29926

TULLER, ANNITA, b New York, NY, Dec 30, 10; wid; c 2. MATHEMATICS. *Educ:* Hunter Col, BA, 29; Bryn Mawr Col, MA, 30, PhD(math), 37. *Prof Exp:* Substitute math, Hunter Col, 30-31; teacher high sch, NY, 31-35; from tutor to assoc prof math, Hunter Col, 37-68; prof, 68-71, EMER PROF MATH, LEHMAN COL, 71- *Mem:* AAAS; Am Math Soc; Hist Sci Soc; Math Asn Am. *Res:* Differential geometry; ergodic theory. *Mailing Add:* 139-62 Pershing Crescent Jamaica NY 11435

TULLER, HARRY LOUIS, b Apr 2, 45; c 2. ELECTRONIC CERAMICS, SOLID STATE IONICS. *Educ:* Columbia Univ, New York, NY, BS, 66, MS, 67, EngScD, 73. *Prof Exp:* Res assoc physics, Technion-Israel Inst Technol, 74-75; PROF, MAT SCI & ENG, MASS INST TECHNOL, 75- & DIR, CRYSTAL PHYSICS & OPTICAL ELECTRONICS LAB. *Concurrent Pos:* Mem, Nat Mat Adv Bd Comt Fuel Cell Mat Tech Vehicular Propulsion, 82-83 & Mass Inst Technol Comn Indust Productivity, 87-88; vis fel, Imperial Col, London, 83; vis scientist, Raychem Corp, Menlo Park, Calif, 83; vis prof, Univ Pierre Marie Curie, Paris, 90. *Mem:* Fel Am Ceramic Soc; Am Phys Soc; Electrochem Soc; Inst Elec & Electronics Engrs; Mat Res Soc. *Res:* Charge and mass transport in ceramics, glasses and across interfaces; defect theory; optical properties; materials development for sensors and for electrochemical and photoelectrochemical energy conversion. *Mailing Add:* Dept Mat Sci & Eng Mass Inst Technol Rm 13-3126 Cambridge MA 02139

TULLIER, PETER MARSHALL, JR, operations research, systems analysis, for more information see previous edition

TULLIO, VICTOR, b Philadelphia, Pa, May 29, 27; m 51; c 2. CHEMISTRY. *Educ:* Univ Pa, BS, 48; Univ Ill, PhD(org chem), 51. *Prof Exp:* Asst, Univ Ill, 48-49; res chemist, 51-61, supvr new dye eval, 64-69, tech asst textile dyes, 69-71, supvr new dye eval, 71-72, SR RES CHEMIST, E I DU PONT DE NEMOURS & CO, INC, 72- *Mem:* Am Asn Textile Chemists & Colorists; Am Chem Soc. *Res:* Organic chemistry; dyes; dyeing of synthetic fibers. *Mailing Add:* 1304 Chadwick Rd Wilmington DE 19803-4116

TULLIS, J PAUL, b Ogden, Utah, July 24, 38; m 58; c 4. CIVIL ENGINEERING. *Educ:* Utah State Univ, BS, 61, PhD(civil eng), 66. *Prof Exp:* Gen contractor commercial construct, Paul & Milo Tullis Gen Contractors, 61-63; asst prof civil eng, Colo State Univ, 66-70, assoc prof, 70-80; PROF CIVIL ENG, UTAH WATER RES LAB, UTAH STATE UNIV, 80- *Res:* Cavitation research, viscous drag reduction and hydraulic modeling. *Mailing Add:* Dept Civil & Environ Eng Utah State Univ Logan UT 84322-4110

TULLIS, JAMES EARL, b Cincinnati, Ohio. GENETICS. *Educ:* Miami Univ, BS, 51; Ohio State Univ, MS, 54, PhD(genetics), 61. *Prof Exp:* Instr zool, Ohio Univ, 56-59; asst prof, Wash State Univ, 61-65; from asst prof to assoc prof, 65-85, PROF, IDAHO STATE UNIV, 85- *Mem:* AAAS; Genetics Soc Am. *Mailing Add:* Dept Biol Sci Idaho State Univ Pocatello ID 83209-0009

TULLIS, JAMES LYMAN, b Newark, Ohio, June 22, 14; m 37; c 4. BIOCHEMISTRY. *Educ:* Duke Univ, MD, 40; Am Bd Internal Med, dipl, 48. *Prof Exp:* Intern med, Roosevelt Hosp, New York, 40-41, sr intern & resident physician, 41-42; from assoc dir to dir blood characterization & preservation lab, 51-55, RES ASSOC BIOCHEM, HARVARD MED SCH, 54-, SR INVESTR, PROTEIN FOUND, 56-, DIR CYTOL LABS, 60-, PROF MED, 75- *Concurrent Pos:* Donner Found res fel, Harvard Med Sch, 45-48; asst, Peter Bent Brigham Hosp, Boston, 46-50, assoc, 55-58, sr assoc, 58-; attend physician, West Roxbury Vet Admin Hosp, 48-; attend physician & hematologist, New Eng Deaconess Hosp, 49-, chief hemat & chemother clin, 57-, chmn gen med div, 60-, chmn dept med, 64-; consult, Panel Mil & Field Med, Div Med Sci, Res & Develop Bd, US Secy Defense, 52-54; consult, Cambridge City Hosp, 54-60; vpres & treas, Int Cong Hemat, 55-56; assoc clin prof med, Harvard Med Sch, 70-75. *Honors & Awards:* Glycerol Producers Award, 57; Hektoen Medal, AMA, 58; Katsunuma Award, Int Soc Hemat, 59. *Mem:* Fel Am Soc Hemat (pres, 58-59); fel AMA; fel Am Col Physicians; fel NY Acad Sci; Int Soc Hemat (vpres, 56-58, secy gen, western hemisphere, 58-). *Res:* Chemical interactions between blood cells and plasma proteins. *Mailing Add:* 110 Francis St Boston MA 02215

TULLIS, JULIA ANN, b Swedesboro, NJ, Feb 21, 43; div. GEOLOGY. *Educ:* Carleton Col, AB, 65; Univ Calif, Los Angeles, PhD(geol), 71. *Prof Exp:* Res asst geol, Inst Geophys, Univ Calif, Los Angeles, 69-70; from asst res prof to assoc res prof, 71-79, assoc prof, 79- 87, PROF GEOL SCI, BROWN UNIV, 88- *Mem:* Am Geophys Union; Mineral Soc Am; Sigma Xi. *Res:* Experimental rock deformation; deformation mechanisms of crustal rocks and minerals. *Mailing Add:* Dept Geol Sci Brown Univ Providence RI 02912

TULLIS, RICHARD EUGENE, b Long Beach, Calif, Apr 26, 36; m 62; c 3. COMPARATIVE PHYSIOLOGY, COMPARATIVE ENDOCRINOLOGY. *Educ:* Univ Wash, BS, 63; Univ Hawaii, MS, 68, PhD(zool), 72. *Prof Exp:* Res asst, Univ Hawaii, 70-71; asst prof, 72-76, assoc prof, 76-81, PROF PHYSIOL, CALIF STATE UNIV, HAYWARD, 81- *Concurrent Pos:* Partic guest, Biomed Div, Lawrence Livermore Lab, 73-; NSF sci equip grant, 74. *Mem:* Sigma Xi; Am Soc Zoologists; AAAS. *Res:* Neuroendocrine control of hydromineral regulation in crustaceans including isolation of neuroendocrine substances, enzyme regulation mechanisms and target organ identification; basic physiological invertebrate functions affected by environmental and man-made substances. *Mailing Add:* Dept Biol Sci Calif State Univ Hayward CA 94542

TULLIS, TERRY EDSON, b Rapid City, SDak, July 21, 42; m 65. STRUCTURAL GEOLOGY, GEOPHYSICS. *Educ:* Carleton Col, AB, 64; Univ Calif, Los Angeles, MS, 67, PhD(struct geol), 71. *Prof Exp:* Actg instr geol, Univ Calif, Los Angeles, 69-70; asst prof, 70-76, ASSOC PROF GEOL, BROWN UNIV, 76- *Concurrent Pos:* Sloan res fel, Brown Univ, 73-75; vis fel res, Sch Earth Sci, Australian Nat Univ, 76; geologist, US Geol Surv, 77. *Mem:* Am Geophys Union; Geol Soc Am; AAAS. *Res:* Experimental rock deformation; tectonophysics; plate tectonics; origin of slaty cleavage and schistosity; thermodynamic systems under nonhydrostatic stress; rheology of rocks at high temperature and pressure; study of in situ stress; rock friction. *Mailing Add:* Dept Geol Sci Brown Univ Brown Sta Providence RI 02912

TULLOCH, ALEXANDER PATRICK, organic chemistry; deceased, see previous edition for last biography

TULLOCH, GEORGE SHERLOCK, b Bridgewater, Mass, Aug 3, 06; m 31, 77; c 2. PARASITOLOGY. *Educ:* Mass Col, BS, 28; Harvard Univ, MS, 29, PhD(entom), 31. *Prof Exp:* Asst biol, Harvard Univ, 28-31; from instr to prof, 32-65, emer prof biol, Brooklyn Col, 65-87; res dir, George S Tulloch & Assocs, 70-87; RETIRED. *Concurrent Pos:* Asst, Radcliffe Col, 29-30; assoc entomologist, PR Insect Pest Surv, 35-36; chief entomologist, State Mosquito Surv, Mass, 39; Rockefeller Found fel, Brazil, 40-41; consult, Arctic Aeromed Lab, USAF, 54-55, res scientist, 65-67; vis prof, Univ Queensland, 61; res scientist, USAF Sch Aerospace Med, 67-69; vis investr, Southwest Found Res & Educ, Tex, 70-80; consult, Dept Path & Labs, Nassau County Med Ctr, NY, 71- *Honors & Awards:* Entom Soc Am Award, 86. *Mem:* AAAS; Am Soc Parasitol; Entom Soc Am; Am Soc Trop Med & Hyg; Am Heartworm Soc; Am Soc Vet Parasitologists. *Res:* Arthropod carried and parasitic diseases of man and animals; life cycle and therapy of canine dirofilariasis; collection of parasitic antigens for skin tests. *Mailing Add:* 6146 Sunset Haven San Antonio TX 78249

TULLOCK, ROBERT JOHNS, b Atascadero, Calif, Oct 3, 40; m 62; c 3. SOIL CHEMISTRY. *Educ:* Calif State Polytech Col, San Luis Obispo, BS, 67; Purdue Univ, West Lafayette, MS, 70, PhD(soil chem), 72. *Prof Exp:* Asst prof soil sci, Univ Calif, Riverside, 72-73; asst prof soil sci, Ore State Univ, 74-76; ASST PROF SOIL SCI, CALIF STATE POLYTECH UNIV, POMONA, 76- *Mem:* Am Soc Agron; Soil Sci Soc Am; Clay Minerals Soc. *Res:* Physicochemical properties of colloidal surfaces. *Mailing Add:* Dept Soil Sci Calif State Polytech Univ 3801 W Temple Ave Pomona CA 91768

TULLY, EDWARD JOSEPH, JR, b Brooklyn, NY, Jan 22, 30. MATHEMATICS. *Educ:* Fordham Univ, AB, 51, MS, 52; Tulane Univ, PhD(math), 60. *Prof Exp:* Instr math, St John's Univ, Minn, 56-57; asst, Tulane Univ, 57-60; NSF fel, Calif Inst Technol, 60-61; lectr, Univ Calif, Los Angeles, 61-63; lectr, 63-64, asst prof, 64-68, ASSOC PROF MATH, UNIV CALIF, DAVIS, 68- *Mem:* Am Math Soc; Math Asn Am. *Res:* Algebraic theory of semigroups; ordered algebraic systems; application of algebra to linguistics. *Mailing Add:* Dept Math Univ Calif Davis CA 95616

TULLY, FRANK PAUL, b Hartford, Conn, Apr 27, 46; m 81; c 2. PHYSICAL CHEMISTRY. *Educ:* Clark Univ, BA, 68; Univ Chicago, MS, 69, PhD(chem), 73. *Prof Exp:* Res asst chem, Clark Univ, 65-68; res asst, Univ Chicago, 68-73; fel, Univ Toronto, 73-74; NSF fel chem, Mich State Univ, East Lansing, 74-76; mem fac eng exp sta, Appl Sci Div, Ga Inst Technol, 76-80; MEM TECH STAFF, SANDIA NAT LABS, LIVERMORE, CALIF, 80- *Concurrent Pos:* NSF energy related fel, 75. *Mem:* Am Chem Soc. *Res:* Use of the crossed molecular beam method in studies of elastic, inelastic and reactive scattering; photoionization; gas-phase reaction kinetics; laser photochemistry. *Mailing Add:* Sandia Nat Labs MS 8353 Livermore CA 94551

TULLY, JOHN CHARLES, b New York, NY, May 17, 42; m 71; c 3. CHEMICAL PHYSICS, SOLID STATE PHYSICS. *Educ:* Yale Univ, BS, 64; Univ Chicago, PhD(chem), 68. *Prof Exp:* NSF fel chem, Univ Colo, 68-69 & Yale Univ, 69-70; mem tech staff, 70-85, HEAD, PHYS CHEM RES DEPT, BELL LABS, 85- *Concurrent Pos:* Vis prof chem, Princeton Univ, 81-82; mem, Nat Sci Found Adv Comt for Chem, 87- & Petrol Res Fund Adv Bd, 88-; distinguished vis lectr, Univ Tex, Austin, 87. *Mem:* Am Chem Soc; fel Am Phys Soc. *Res:* Theory of chemical rate processes, molecular collisions and gas-surface interactions. *Mailing Add:* 11 Dell Lane Berkeley Heights NJ 07922-1492

TULLY, JOSEPH GEORGE, b Sterling, Colo, July 14, 25; m 57; c 1. MEDICAL MICROBIOLOGY. *Educ:* Portland Univ, BS, 49; Brigham Young Univ, MS, 51; Cincinnati Univ, PhD(microbiol), 55. *Hon Degrees:* Dr, Univ Bordeaux, France, 80. *Prof Exp:* Asst prof microbiol, Col Med, Cincinnati Univ, 55-57; microbiologist, Walter Reed Army Inst Res, 57-61, chief, Dept Microbiol, 61-62; res microbiologist, 62-68, HEAD MYCOPLASMA SECT, NAT INST ALLERGY & INFECTIOUS DIS, 68- *Concurrent Pos:* China med bd fel, Cent Am, 56; attend microbiologist, Cincinnati Gen Hosp, Ohio, 56-57; mem bd, Food & Agr Orgn/WHO Prog on Comp Mycoplasmology, 69-78, chmn bd, 72-77; mem, Int Subcomt Taxon of Mollicutes, 70-; chmn, Int Org Mycoplasmology, 75-78; mem adv bd, Bergey's Manual Syst Bacteriol, 80-81; assoc ed, Int J Syst Bact, 83-90. *Honors & Awards:* Klieneberger-Nobel Award, Int Orgn Mycoplasmology; J Roger Porter Award, Am Soc Microbiol, 82. *Mem:* Fel AAAS; Am Asn Immunol; fel Am Acad Microbiol; Am Soc Microbiol; Soc Exp Biol & Med; Int Orgn Mycoplasmology. *Res:* Bacillary dysentery; immunology and pathogenesis of enteric diseases; typhoid infection in primates; basic biology of the mycoplasmas; murine mycoplasmas; spiroplasmas; urogenital mycoplasmas. *Mailing Add:* Mycoplasma Sec Nat Inst Allergy & Infectious Dis Bldg 550 Frederick Cancer Res Ctr Frederick MD 21702-1201

TULLY, PHILIP C(OCHRAN), b Grand Island, Nebr, Jan 11, 23; m 45; c 4. CHEMICAL & NUCLEAR ENGINEERING. *Educ:* Iowa State Univ, BS, 47; Univ Pittsburgh, MS, 55; Okla State Univ, PhD(chem eng), 65. *Prof Exp:* Process engr, Plastics Div, Koppers Co, Inc, 47-54, mgr training, 54-56, asst prod supt polyethylene, 56-59, chief plant engr, 59-61; asst chem eng, Okla State Univ, 61-64; proj leader, Phase Equilibrium, Helium Res Ctr, 64-71, chem engr, Tech Servs Unit, Helium Opers, 71-76, helium technologist, Helium Opers, 76-83, CHIEF, PLANNING & ANALYSTS, US BUR MINES, 83- *Mem:* Am Chem Soc; Am Inst Chem Engrs; Sigma Xi. *Res:* Cryogenic, high pressure phase equilibria research and physical properties determinations on helium bearing systems. *Mailing Add:* Rte 4 Box 10 Amarillo TX 79119-9762

TULLY, RICHARD BRENT, b Toronto, Ont, Can, Mar 9, 43; m 72; c 3. EXTRAGALACTIC ASTRONOMY. *Educ:* Univ BC, BSc, 64; Univ Md, PhD(astron), 72. *Prof Exp:* Fel astron, Univ Toronto, 73 & Marseille Observ, 73-75; asst astronr, 75-77, assoc astronr, 77-82, ASTRONR, INST ASTRON, UNIV HAWAII, 82- *Concurrent Pos:* Mem, Sci Adv Comt, Can-France-Hawaii Telescope, 77-81; vis sr scientist, Cerro Tololo InterAm Observ, Chile, 82-83 & Meudon Observ, France, 83; mem, Sci Steering Comt, Keck Telescope, 85-87; vis sr scientist, Instituto di Radioastronomia CNR, Bologna, Italy. *Mem:* Am Astron Soc; Int Astron Union. *Res:* Extragalactic astronomy; formation, evolution and dynamics of ordinary galaxies; environment of galaxies; formation of large scale structure; the distance scale and age of universe; the distribution of matter in the universe. *Mailing Add:* Inst Astron 2680 Woodlawn Dr Honolulu HI 96822

TULSKY, EMANUEL GOODEL, b Philadelphia, Pa, Dec 6, 23; m 50; c 2. RADIOLOGY. *Educ:* Jefferson Med Col, MD, 48. *Prof Exp:* Attend radiologist, Delafield Hosp, New York, 52-53; assoc radiologist, Sch Med & Univ Hosp, Temple Univ, 53-55; assoc radiol, Div Grad Med & asst prof radiol, Univ Pa, 55-65; RADIOLOGIST & DIR DIV RADIATION THER & NUCLEAR MED, ABINGTON MEM HOSP, PA, 65- *Concurrent Pos:* Assoc radiologist & dir, Tumor Clin, Hosp Univ Pa, 55-67; asst prof radiol, Med Col Pa, 62- *Mem:* Radiol Soc NAm; AMA; Am Col Radiol. *Res:* Intracavitary radiation dosimetry; isotopic studies of gastrointestinal absorption; cancer therapy; synergistic action of radiation and cytotoxics. *Mailing Add:* AMA-ACR-RSNA 24 W Butler Ave Chalfont PA 18914

TULUNAY-KEESEY, ULKER, b Ankara, Turkey, Oct 18, 32. OPHTHALMOLOGY, PSYCHOPHYSICS. *Educ:* Mt Holyoke Univ, BA, 55; Brown Univ, MA, 57, PhD(physiol), 59. *Prof Exp:* Instr, 62-65, asst prof, 65-70, assoc prof, 70-75, PROF, DEPT OPHTHAL, UNIV WIS-MADISON, 75- *Concurrent Pos:* Prin investr, NIH & Nat Eye Inst, 62- *Mem:* Fel Optical Soc Am; Asn Res Vision & Ophthal. *Res:* Influence of eye measurements in vision, specifically contrast sensitivity, acuity and motion dedection, light adaption and contrast sensitivity in vision disorders. *Mailing Add:* Dept Ophthal Rm 573 Waisman Ctr Univ Wis Madison WI 53705

TUMA, DEAN J, b Howells, Nebr, Oct 20, 41; m 64; c 3. BIOLOGICAL CHEMISTRY. *Educ:* Creighton Univ, BS, 64, MS, 68; Univ Nebr, PhD(biochem), 73. *Prof Exp:* Instr, 73-75, ASST PROF INTERNAL MED & BIOCHEM, COL MED, UNIV NEBR, 75-; RES CHEMIST BIOCHEM, VET ADMIN HOSP, OMAHA, 64- *Mem:* Am Asn Study Liver Dis; Am Fedn Clin Res. *Res:* Investigation of the role of ethanol, drugs and nutrition in liver metabolism and liver disease. *Mailing Add:* Vet Admin Hosp 4104 Woolworth Ave Omaha NE 68105

TUMA, GERALD, b Oklahoma City, Okla, July 19, 14; m 38; c 2. ELECTRICAL ENGINEERING. *Educ:* Univ Okla, BS, 39, MEE, 41. *Prof Exp:* From instr to prof elec eng, 40-66, chmn dept, 58-62, dir sch, 66-77, DAVID ROSS BOYD PROF ELEC ENG, UNIV OKLA, 66- *Mem:* Am Soc Eng Educ; Inst Elec & Electronics Engrs; Sigma Xi. *Res:* Communications; feedback control systems; analog simulation; digital computers. *Mailing Add:* Univ Okla 202 W Boyd St Rm 219 Norman OK 73019

TUMA, JAN J, civil engineering; deceased, see previous edition for last biography

TUMAN, VLADIMIR SHLIMON, b Kermanshah, Iran, May 21, 23; US citizen; m 51; c 3. PHYSICS, GEOPHYSICS. *Educ:* Univ Birmingham, BSc, 48; Univ London, DIC, 49; Stanford Univ, PhD(geophys), 64. *Prof Exp:* Geophysicist, Anglo Iranian Oil Co, SW Iran, 50-52; actg chief petrol, Nat Iranian Oil Co, 52-55, engr trainee, Europe & USA, 55-56; sr petrol physicist, Nat Iranian Consortium, 56-57; res physicist, Atlantic Refining Oil Co, Dallas, Tex, 57-59; assoc prof petrol eng, Univ Ill, Urbana, 59-62; res assoc geophys, Stanford Univ, 62-65, res physicist, Res Inst, 65-66; assoc prof physics, 66-67, chmn dept phys sci, 66-71, PROF PHYSICS, CALIF STATE COL, STANISLAUS, 67- *Concurrent Pos:* Calif Res Corp grant, 60-61; consult, Esso Res Lab, 61, Schlumberger Well Logging Co, 61 & Sinclair Oil Co, 61; Am Petrol Inst grants, 61-63; consult, Comput Symp, Stanford Univ, 63-64, res assoc, Physics Dept, 65-; lectr, Varian Assoc, Palo Alto, Calif, 66. *Mem:* fel Royal Astron Soc; Am Phys Soc; Am Asn Physics Teachers; Am Geophys Union; Soc Explor Geophys. *Res:* Development of cryogenic gravity meter to study earth eigen vibrations and detection gravitational radiation; evolution of Babylonian and Assyrian astronomy. *Mailing Add:* 1401 E Toolumne Turlock CA 95380

TUMBLESON, M(YRON) E(UGENE), b Mountain Lake, Minn, Mar 13, 37; m 83; c 3. BIOCHEMISTRY, NUTRITION. *Educ:* Univ Minn, BS, 58, MS, 61, PhD(nutrit), 64. *Prof Exp:* From res assoc to asst prof animal sci, Univ Minn, 64-66; asst prof vet physiol & pharmacol, Univ Mo-Columbia & res assoc med biochem, Sinclair Comp Med Res, 66-69; assoc prof vet anat & physiol, Univ Mo, Columbia & Sinclair Comp Med res, 66-69, res assoc, 64-80, res prof, 80-86; PROF, COL VET MED, UNIV ILL, 86- *Mem:* Am Inst Nutrit; Soc Exp Biol & Med; Am Soc Neurochem; Am Soc Biol Chemists; Sigma Xi. *Res:* Protein-calorie malnutrition; alcoholism and aging, using miniature swine as biomedical research subjects. *Mailing Add:* Col Vet Med Univ Ill 2001 S Lincoln Urbana IL 61801

TUMELTY, PAUL FRANCIS, b Boston, Mass, May 9, 41; m 85; c 1. MAGNETIC & OPTICAL PROPERTIES OF MATERIALS, TRANSPORT PROPERTIES. *Educ:* Boston Col, BS, 62; Univ Iowa, MS, 64, PhD(physics), 70. *Prof Exp:* Prin res engr, Honeywell, Inc, 70-73; physicist, Allied Corp, 73-80, sr res physicist, 80-82, res assoc, 82-88; CONSULT, 88- *Mem:* Am Phys Soc; Inst Elec & Electronics Engrs; Sigma Xi. *Res:* Superconductivity; magnetic and optical properties of epitaxially-grown oxide films; liquid-phase expitaxy of oxide films; transport properties of semiconductors; low-temperature physics. *Mailing Add:* One East Dr Convent Station NJ 07961

TUMEN, HENRY JOSEPH, b Philadelphia, Pa, Apr 7, 02; m 26; c 1. MEDICINE. *Educ:* Univ Pa, AB, 22, MD, 25; Am Bd Internal Med, dipl, 52. *Prof Exp:* Prof clin gastroenterol, 54-60, prof med & chmn dept, 60-71, EMER PROF MED, GRAD SCH MED, UNIV PA, 71- *Concurrent Pos:* Consult, Walter Reed Army Med Ctr, DC, 58- & Albert Einstein Med Ctr, 60-; chmn, Am Bd Internal Med, 58-60. *Mem:* Am Soc Gastrointestinal Endoscopy; Am Gastroenterol Asn; Sigma Xi. *Res:* Gastroenterology. *Mailing Add:* 1830 Rittenhouse Square Philadelphia PA 19103

TUMMALA, RAO RAMAMOHANA, b Nandamuru, India, Feb 15, 42; US citizen; m 66; c 3. ELECTRON PACKAGING, MULTICHIP MODULE. *Educ:* Loyola Col, BSc, 61; Indian Inst Sci, BE, 63; Queen's Univ, Can, MS, 65; Univ Ill, PhD(ceramics), 68. *Prof Exp:* Process engr ceramics, Norton Co, 64-66; res asst, Univ Ill, 66-68; staff engr metall, IBM Corp, 68-70, adv engr ceramics, 70- 76, sr engr glass-ceramics, 76-84, IBM fel computer packaging, 84-86, dir packaging, 86-88, IBM FEL PACKAGING, IBM CORP, 88- *Concurrent Pos:* Mem adv bd, Univ Ill, 80-84; chmn adv bd, Univ Calif, Berkeley, 86-88, Mass Inst Technol, 90-91; vis prof, Polytech Inst NY, 87-88. *Honors & Awards:* Gilpin Mem Lectr, Clarkson Univ, 91. *Mem:* Nat Acad Eng; fel Am Ceramic Soc; Inst Elec & Electronics Engrs; Int Electron Packaging Soc; Int Soc Hybrid Microelectronics; Mat Res Soc. *Res:* Ceramics; glass; glass-ceramics; powder metallurgy; thin films; polymers; composites; interfaces; mechanical behavior of materials; ferroelectrics. *Mailing Add:* IBM Corp East Fishkill NY 12533

TUMOSA, NINA JEAN, b Dover-Foxcroft, Maine, Oct 12, 51. ELECTROPHYSIOLOGY, IMMUNOCYTOCHEMISTRY. *Educ:* Rensselaer Polytech Inst, BS, 73, MS, 74; State Univ NY, Albany, PhD(neurosci), 82. *Prof Exp:* Fel anat, Med Sch, Univ Calgary, 82-84; fel neurosci, Univ Wis, 85-87, AAAS, 88-89; ASST PROF, UNIV MO, 89- *Concurrent Pos:* Res scientist, 87-88. *Mem:* Asn Res Vision & Opthal; AAAS; Am Women in Sci; Nat Asn Female Exec; Women in Neurosci; Soc Neurosci. *Res:* Integrated neuroscientific approach to the development of visual processing in vertebrates, with an emphasis on the effects of altered visual experience on the maintenance of binocularity. *Mailing Add:* Sch Optom Univ Mo St Louis MO 63121

TUMULTY, PHILIP A, medicine; deceased, see previous edition for last biography

TUNA, NAIP, b Constanta, Romania, Aug 18, 21; m 49; c 2. INTERNAL MEDICINE, CARDIOVASCULAR DISEASES. *Educ:* Istanbul Univ, MD, 47; Univ Minn, PhD(med), 58. *Prof Exp:* Asst med, Therapeut Clin, Istanbul Univ, 49-52; resident med, St Joseph's Hosp, Lexington, Ky, 52-53; from instr to asst prof, 57-64, assoc prof, 64-80, PROF MED, MED SCH, UNIV MINN, MINNEAPOLIS, 80- *Concurrent Pos:* Am Heart Asn res fel, Univ Minn, Minneapolis, 58-59, advan res fel, 59-61. *Mem:* Fel Am Col Physicians; fel Am Col Cardiol; fel Am Heart Asn; Am Fedn Clin Res. *Res:* Electro and vector cardiography; cardiology. *Mailing Add:* Dept Med Univ Minn Hosp Box 481 Minneapolis MN 55455

TUNC, DEGER CETIN, b Izmir, Turkey, Apr 2, 36; m 63; c 2. PHYSICAL CHEMISTRY, BIOPOLYMERS. *Educ:* Columbia Univ, BS, 63; Fairleigh Dickinson Univ, MA, 66; Rutgers Univ, PhD(phys chem), 72. *Prof Exp:* Res chemist cellulose, Eastern Res, ITT Rayonier, 63-66; res scientist surg dressings, Johnson & Johnson, 66-74, sr res scientist bioeng & biochem, Cent Res, 74-81, res assoc Orthop Res, 81-85, prin scientist, 85-91, GROUP LEADER POLYMER RES, JOHNSON & JOHNSON ORTHOP, 91- *Concurrent Pos:* UN guest lectr, Dept Bioeng, Ege Univ, Izmir, Turkey, 79, Dept Chem, Tubitak, Izmit, Turkey, 81, Yale Univ Med Sch, 84, Cornell Univ, 88 & Clemson Univ, 90. *Mem:* Am Chem Soc; Soc Biomat; Soc Turkish Architects Engrs & Scientists Am (pres, 72-74); Orthop Res Soc. *Res:* Polyelectrolytes; body absorbable polymers; health care products in general; wound healing; orthopedics; gastrointestinal problems and drugs; controlled release membranes; biomedical devices; absorbable internal bone fixation devices; blood compatible polymers; 13 US patents and 69 international patents; 20 published articles. *Mailing Add:* Six Springfield Rd E Brunswick NJ 08816

TUNE, BRUCE MALCOLM, b New York, NY, Aug 26, 39; m 69; c 2. PEDIATRIC NEPHROLOGY, CLINICAL RENAL PHYSIOLOGY. *Educ:* Stanford Univ, AB, 63, MD, 65. *Prof Exp:* Intern med & pediat, Univ Rochester, 65-66; resident pediat, Stanford Univ, 66-67; res assoc renal physiol, Nat Heart Inst, NIH, 67-69, clin assoc renal dis, 68-69; chief resident pediat, Stanford Univ, 69-70, fel pediat renal dis, 70-71, from asst prof to assoc prof, 71-83, PROF PEDIAT, SCH MED, STANFORD UNIV, 83- *Concurrent Pos:* Dir pediat nephrol, Stanford Univ Hosp & Lucile Salter Packard Children's Hosp at Stanford, 71- *Mem:* Am Soc Nephrol; Int Soc Nephrol; Am Pediat Soc; Soc Pediat Res; Am Soc Pharmacol & Exp Therapeut; Am Soc Renal Biochem & Metab. *Res:* Investigation of molecular mechanisms of drug-induced injury to the kidney focused on the attack by Beta-Lactam antibiotics on renal tubular cell mitochondrial substrate transporters; identifying effective modes of therapy of the nephrotic syndrome caused by focal sclerosing glomerulonephritis. *Mailing Add:* Dept Pediat Stanford Univ Stanford CA 94305

TUNG, CHE-SE, b Nanking, China, Nov 19, 48; m 75; c 2. AUTONOMIC NERVOUS SYSTEM, NEURO-PSYCHOPHARMACOLOGY. *Educ:* Nat Defense Med Ctr, MD, 75; Vanderbilt Univ, PhD(pharmacol), 83. *Prof Exp:* Asst instr physiol, Dept Biophys, 75-78, asst instr pharmacol, Dept Pharmacol, 78-79, from assoc prof to prof pharmacol, Dept Pharmacol, 84-90, PROF PHYSIOL, DEPT PHYSIOL & BIOPHYS, NAT DEFENSE MED CTR, 90-; JOINTED PROF PHYSIOL, DEPT BIOMED ENG, CHUNG-YUAN CHRISTIAN UNIV, REPUB CHINA, 91- *Concurrent Pos:* Res assoc, Dept Med & Pharmacol, Vanderbilt Univ, 83-84; bd trustee, Chinese Soc Pharmacol, 84-; basic res awards med sci, Dept Health, Repub China, 84-; consult, Bur Drug, Dept Health, Repub China, 85-88; student adv, Dr Med Sci, Grad Fac Coun, Nat Defense Med Ctr, 85-88, teaching comt mem, Educ Fac Coun, 85-, chief secy, S C Wang Found Neurosci Res, 90-; vis scientist, Dept Pharmacol, Karolinska Institutet, Sweden, 88-89; mem, Coun Basic Sci, Am Heart Asn, 83- *Mem:* Am Fedn Clin Res; Am Soc Pharmacol & Exp Therapeut; NY Acad Sci. *Res:* Brain exerts control over the motivated behaviors and cardiovascular functions through the central monoaminergic neuronal pathways; physiological functions and pharmacological control of schedule-induced drinking, one of animal displacement behaviors, using in vivo as well as in vitro techniques. *Mailing Add:* Dept Physiol Nat Defense Med Ctr PO Box 90048-503 Taipei 107 Taiwan

TUNG, CHI CHAO, b Shanghai, China, Mar 24, 32; US citizen; m 64; c 2. STRUCTURAL MECHANICS. *Educ:* Tung-Chi Univ, China, BS, 53; Univ Calif, Berkeley, MS, 61, PhD(struct eng & struct mech), 64. *Prof Exp:* Asst prof struct eng, Univ Ill, Urbana, 64-69; assoc prof, 69-76, PROF STRUCT ENG, NC STATE UNIV, 76- *Mem:* Am Soc Civil Engrs; Am Soc Eng Educ. *Res:* Application of probability and statistics to civil engineering problems; ocean engineering. *Mailing Add:* Dept Civil Eng NC State Univ PO Box 7908 Raleigh NC 27695-7908

TUNG, FRED FU, b Manchouli, Inner Mongolia, July 23, 34; m 71; c 2. ENZYMOLOGY, PROTEIN CHEMISTRY. *Educ:* Taiwan Prov Col Agr, BS, 56; Univ Vt, MS, 63; Univ Mich, MS, 66; Univ Mo, PhD(biochem), 70. *Prof Exp:* Res asst agr chem, Taiwan Prov Col Agr, 58-59; fel biochem, State Univ NY Albany, 70-73; biochemist, 74-78, CLIN HEALTH SCIENTIST, MICH DEPT PUB HEALTH, 78- *Mem:* Am Chem Soc; AAAS; NY Acad Sci. *Res:* Use of plasmin to modify immune serum globulin for intravenous administration; isolation and purification of anticancer drugs; research and development of bacterial vaccines (pertussis). *Mailing Add:* Mich Dept Pub Health PO Box 30035 Lansing MI 48909

TUNG, JOHN SHIH-HSIUNG, b Keelung, Taiwan, July 19, 28; m 54; c 4. MATHEMATICS. *Educ:* Taiwan Norm Univ, BA, 50; Pa State Univ, MA, 60, PhD(math), 62. *Prof Exp:* Asst civil eng, Taihoku Imp Univ, Taiwan, 44-45; asst instr math, Taipei Inst Technol, 50-53; asst math & indust educ, Taiwan Norm Univ, 53-56, instr, 56-58; asst math, Pa State Univ, 58-60 & 61-62; asst prof, 62-66, assoc prof, 66-76, PROF MATH, MIAMI UNIV, OHIO, 76- *Mem:* Math Asn Am; Am Math Soc. *Res:* Theory of functions of a complex variable; infinite and orthogonal series. *Mailing Add:* Dept Math Miami Univ Oxford OH 45056

TUNG, KA-KIT, b Canton, China, Dec 6, 48; US citizen; m 76; c 3. DYNAMIC METEOROLOGY, FLUID MECHANICS. *Educ:* Calif Inst Technol, BSc & Msc, 72; Harvard Univ, PhD(appl math), 77. *Prof Exp:* Postdoctoral, Harvard Univ, 77-79; asst prof, Mass Inst Technol, 79-84, assoc prof appl math, 84-86; prof math & computer sci, Clarkson Univ, 86-88; PROF APPL MATH, UNIV WASH, 89- *Concurrent Pos:* Res fel, Harvard Univ, 77-79; consult, Dynamics Technol Inc, 80-81; John Simon Guggenheim Found fel, 85-86. *Mem:* Am Meteorol Soc; Am Geophys Union; Meteorol Soc Japan; Royal Meteorol Soc. *Res:* Large scale wave motions in the earth's atmosphere; internal waves in the ocean; modeling of tracer transport in the stratosphere. *Mailing Add:* Dept Appl Math FS-20 Univ Wash 77 Mass Ave Seattle WA 98195

TUNG, LU HO, b Tientsin, China, Dec 7, 23; US citizen; m 67; c 3. POLYMER CHEMISTRY. *Educ:* Tsing Hua Univ, China, BS, 48; Univ Ill, MS, 50, PhD(chem eng), 51. *Prof Exp:* Phys chemist, 53-59, assoc scientist, 59-70, RES SCIENTIST, DOW CHEM CO, 70- *Mem:* AAAS; Sigma Xi; Am Chem Soc. *Res:* Polymer physical chemistry. *Mailing Add:* Cent Res Walnut Creek CA 94598

TUNG, MARVIN ARTHUR, b Sask, Can, Nov 9, 37; c 3. FOOD SCIENCE. *Educ:* Univ BC, BSA, 60, teaching cert, 61, MSA, 67, PhD(food sci), 70. *Prof Exp:* Teacher sch bd, 61-70; from asst prof to prof food sci, Univ BC, 70-87; HEAD, DEPT FOOD SCI TECHNOL, UNIV NS, 87- *Honors & Awards:*

William J Eva Award, Can Inst Food Sci Technol, 85. *Mem:* Can Inst Food Sci & Technol; Inst Food Technologists; Brit Inst Food Sci & Technol; Can Soc Agr Eng; Micros Soc Can; Inst Thermal Processing Specialists; Soc Rheology. *Res:* Food rheology; microstructure of food systems; food processing and packaging. *Mailing Add:* Dept Food Sci Technol Univ NS PO Box 1000 Halifax NS B3J 2X4 Can

TUNG, MING SUNG, b Taiwan, Feb 25, 42; US citizen; m 70; c 2. DENTAL CHEMISTRY. *Educ:* Cheng-Kung Univ, Taiwan, BS, 64; Brown Univ, PhD(chem), 73. *Prof Exp:* Fel, Univ Md, 72-74, vis prof chem, 74; PROJ LEADER, DENT CHEM, AM DENT ASN HEALTH FOUND, PAFFENBARGER RES CTR, 74- *Honors & Awards:* E H Hatton Award, Int Asn Dent Res, 76. *Mem:* Am Chem Soc; Am Asn Dent Res; Int Asn Dent Res; Sigma Xi. *Res:* Calcium phosphate and fluoride chemistry as applied to dental and bone sciences; physical chemistry of biological systems; study of physical properties of biopolymers such as DNA, polypeptides, proteins and enzymes. *Mailing Add:* 15233 Falconbridge Terr Gaithersburg MD 20878

TUNG, PIERRE S, cell biology, reproductive & developmental biology, for more information see previous edition

TUNG, WU-KI, b Kunming, China, Oct 16, 39; m 63; c 2. THEORETICAL PHYSICS, ELEMENTARY PARTICLE PHYSICS. *Educ:* Univ Taiwan, BS, 60; Yale Univ, PhD(physics), 66. *Prof Exp:* Res assoc theoret physics, Inst Theoret Physics, State Univ NY Stony Brook, 66-68; mem staff, Inst Adv Study, Princeton, 68-70; asst prof physics & mem staff, Dept Physics, Enrico Fermi Inst, Univ Chicago, 70-75; assoc prof, 75-79, chmn Dept, 81-84, PROF PHYSICS, ILL INST TECHNOL, 79- *Mem:* Fel, Am Phys Soc. *Res:* High energy theoretical physics. *Mailing Add:* Dept Physics Ill Inst Technol Chicago IL 60616

TUNG, YEOU-KOUNG, b Taiwan, Repub China, Mar 4, 54; US citizen; m 77; c 3. RELIABILITY ANALYSIS, WATER RESOURCE SYSTEMS ANALYSIS. *Educ:* Tamkang Univ, BS, 76; Univ Tex, Austin, MS, 78, PhD(water resources), 80. *Prof Exp:* Res assoc water resources, Univ Tex, Austin, 80; asst prof civil eng, Univ Nev, Reno, 81-84; asst prof, 85-87, ASSOC PROF STATIST, UNIV WYO, 87- *Concurrent Pos:* Control mem, Task Comt Risk & Reliability Anal Water Distrib Systs, Am Soc Civil Engrs, 84-87 & Comt Probabilistic Approaches Hydraul, 84-90, chmn, 89; chmn, Subcomt Uncertainty & Reliability Anal Design Hydraul Struct, 89-91; sabbatical asst prof civil eng, Nat Chiao-Tung Univ, 91- *Honors & Awards:* Collingwood Prize, Am Soc Civil Engrs, 87. *Mem:* Am Geophys Union; Am Soc Civil Engrs; Am Statist Asn; Int Asn Hydraul Res. *Res:* Water resource systems analysis; probabilistic analysis of hydrologic and hydraulic systems; decision making under uncertainty. *Mailing Add:* Wyo Water Res Ctr Univ Wyo Laramie WY 82071

TUNHEIM, JERALD ARDEN, b Claremont, SDak, Sept 3, 40; m 63; c 3. SOLID STATE PHYSICS. *Educ:* SDak State Univ, BS, 62, MS, 64; Okla State Univ, PhD(physics), 68. *Prof Exp:* From asst prof to prof physics, SDak State Univ, 68-85, head dept, 80-85; dean sch math sci & technol, Eastern Wash Univ, 85-87; PRES, DAKOTA STATE UNIV, 87- *Mem:* Am Phys Soc; Am Asn Physics Teachers; Nat Soc Prof Engrs. *Res:* Alpha particle model of sulphur nucleus; electron spin resonance measurements of transition metal ions in stannic oxide; surface effects on conductivity of stannic oxide; application and development of models for remote sensing application; electrolytic capacitors; science education. *Mailing Add:* Pres Dakota State Univ Madison SD 57042-1799

TUNIK, BERNARD D, b New York, NY, Dec 22, 21; m 49; c 3. PHYSIOLOGY. *Educ:* Univ Wis, BA, 42; Columbia Univ, MA, 51, PhD(zool), 59. *Prof Exp:* Asst cytol, Sloan-Kettering Inst, 50-52; asst zool, Columbia Univ, 52-55; instr anat, Sch Med, Univ Pa, 58-60; assoc prof biol sci, State Univ NY Stony Brook, 60-90, dep chmn dept, 62-63 & 64-65; RETIRED. *Concurrent Pos:* Nat Inst Arthritis & Metab Dis spec fel, Dept Polymer Sci, Weizmann Inst, 66-67; vis scholar, Dept Zool, Univ Calif, Berkeley, 75. *Res:* Cellular physiology; mechanochemical aspects of muscle contraction; triggers of muscle hypertrophy. *Mailing Add:* Dept Neurobiol & Behav State Univ NY Stony Brook NY 11794

TUNIS, C(YRIL) J(AMES), b Montreal, Que, July 31, 32; m 51; c 3. ELECTRICAL ENGINEERING. *Educ:* McGill Univ, BEng, 54, MSc, 56; Univ Manchester, PhD(elec eng), 58. *Prof Exp:* Staff engr, 58-65, SR ENGR, COMPUT DESIGN, IBM CORP, ENDICOTT, 65- *Concurrent Pos:* Lectr, Harpur Col, State Univ NY, 59-60 & Lehigh Univ, 61-62; vis prof, Stanford Univ, 66-67. *Honors & Awards:* Babbage Award, Brit Inst Elec Engrs. *Mem:* Fel Inst Elec & Electronics Engrs. *Res:* Design of advanced digital computer systems. *Mailing Add:* Dir Corp Tech Inst IBM Corp Tec Inst 500 Columbus Ave Thornwood NJ 10594

TUNIS, MARVIN, b New York, NY, Apr 18, 25; m 52; c 4. BIOCHEMISTRY. *Educ:* Hunter Col, AB, 50; Univ Ill, MS, 51, PhD(biochem), 54. *Prof Exp:* Res assoc, Univ Ill, 54-55; USPHS fel, Col Physicians & Surgeons, Columbia Univ, 55-56; sr cancer res scientist, Roswell Park Mem Inst, 57-68; assoc prof, 68-77, PROF CHEM, STATE UNIV NY COL BUFFALO, 77- *Mem:* Am Chem Soc; Am Soc Biol Chemists; Am Asn Cancer Res. *Res:* Biochemistry and metabolism of nucleic acids, proteins, glycoproteins and mucopolysaccharides; enzymology. *Mailing Add:* Dept Chem State Univ NY Col 1300 Elmwood Ave Buffalo NY 14222

TUNKEL, STEVEN JOSEPH, b New York, NY, Jan 15, 29; m 52; c 3. CHEMICAL ENGINEERING, PHYSICAL CHEMISTRY. *Educ:* Polytech Inst Brooklyn, BS, 51; Newark Col Eng, MS, 56. *Prof Exp:* Res engr pilot plants, Allied Chem Corp, 51-56; mgr res eng aerospace, Thiokol Chem Corp, 56-68; mgr mkt res jet engines, Austenal Div, Howmet Corp, 68-70; mgr process res new plant start-up, Celanese Chem Corp, 70-72; CHIEF CHEM ENGR FIRE & EXPLOSION, HAZARDS RES CORP, 72- *Concurrent Pos:* Course lectr, Am Inst Chem Engrs & Nat Safety Coun. *Mem:* Am Chem Soc; Am Inst Chem Engrs; Combustion Inst; Nat Soc Prof Engrs. *Res:* Fire and explosion hazard evaluation of chemicals and chemical processes; vapor cloud explosions; metallic hydrode safety. *Mailing Add:* 37 Woodcrest Rd Whippany NJ 07981

TUNNELL, WILLIAM C(LOTWORTHY), b Knoxville, Tenn, May 19, 15; m 42; c 2. MECHANICAL ENGINEERING. *Educ:* Univ Tenn, BS, 40. *Prof Exp:* Engr, Blue Ridge Glass Corp, 40-43, Tenn Eastman Co, 43-47, Oak Ridge Nat Lab, 47-68 & Nuclear Div, Union Carbide Corp, 68-79; RETIRED. *Concurrent Pos:* Consult engr, 79- *Mem:* Nat Soc Prof Engrs. *Res:* Electromagnetic separation of isotopes; high temperature components of atomic power reactors; critical assemblies; nuclear reactions; liquid metals; pulse reactors; environmental statements for nuclear power plants. *Mailing Add:* 104 Ditman Lane Oak Ridge TN 37830

TUNNICLIFF, DAVID GEORGE, b Ord, Nebr, Sept 18, 31; m 59; c 2. BITUMINOUS ENGINEERING. *Educ:* Univ Nebr, BSCE, 54; Cornell Univ, MS, 58; Univ Mich, PhD(civil eng), 72. *Prof Exp:* Sr engr, Nebr Dept Roads, 58-60; prof civil eng, Wayne State Univ, 60-67; chief engr, Warren Bros Co, 67-79; CONSULT ENGR, 79- *Concurrent Pos:* Dir, Asn Asphalt Paving Technologists, 76-78. *Honors & Awards:* Distinguished Serv Award, Transp Res Bd. *Mem:* Hon mem Am Soc Testing & Mat; Asn Asphalt Paving Technologists; Transp Res Bd; Am Soc Civil Engrs; Nat Soc Prof Engrs. *Res:* Use of antistripping methods and introduction of lime into asphalt concrete mixtures. *Mailing Add:* 9624 Larimore Ave Omaha NE 68134

TUNNICLIFF, GODFREY, b Malvern, Eng, Jan 6, 41; m 71; c 2. NEUROCHEMISTRY. *Educ:* Univ Col Wales, BSc, 64; Univ Southampton, MSc, 67, PhD(biochem), 69. *Prof Exp:* Res biochemist, Liebig's Extract Meat Co Ltd, London, 64-66; fel, City of Hope Nat Med Ctr, Duarte, Calif, 69-71; fel, Univ Sask, 71-72, asst prof biochem, 72-74; dir, Lab Neurochem, Clin Res Inst Montreal, Que, 74-77; asst prof, 78-81, ASSOC PROF BIOCHEM, SCH MED, IND UNIV, 81- *Mem:* Can Biochem Soc; Int Soc Neurochem. *Res:* Role of gamma-aminobutyric acid in the functioning of the central nervous system. *Mailing Add:* Ind Univ Sch Med 8600 Univ Blvd Evansville IN 47712

TUNNICLIFFE, PHILIP ROBERT, b Derby, Eng, May 3, 22; Can citizen; m 46; c 2. PHYSICS. *Educ:* Univ London, BSc, 42. *Prof Exp:* Res staff, Telecommun Res Estab, Malvern, Eng, 42-46, UK Atomic Energy Auth, Ont, Can, 46-49 & Atomic Energy Res Estab, Harwell, Eng, 49-51; res staff, Atomic Energy Can Ltd, 51-61, br head electronics, 61-63, br head appl physics, 63-67, br head accelerator physics, 67-78; RETIRED. *Concurrent Pos:* Consult, Los Alamos Sci Lab, 79-80. *Mem:* Am Phys Soc. *Res:* Microwave tubes; neutron, low energy nuclear, reactor and accelerator physics; reactor control and instrumentation. *Mailing Add:* Six Beach Ave Deep River ON K0J 1P0 Can

TUNNICLIFFE, VERENA JULIA, b Deep River, Ont, June 6, 53; m 87; c 1. EVOLUTIONARY BIOLOGY, BIOGEOGRAPHY. *Educ:* McMasters Univ, BSc, 75; Yale Univ, MPhil, 78, PhD(biol), 80. *Prof Exp:* Asst prof, 82-87, ASSOC PROF BIOL, UNIV VICTORIA, 88- *Concurrent Pos:* Vis researcher, Ifremer, France, 88-89. *Mem:* Sigma Xi; Am Geophys Union; Can Meteorol & Oceanog Soc; Paleont Soc Am. *Res:* Ecology and evolution of marine communities with present emphasis on hydrothermal vent fauna and the development of biota on offshore seamounts. *Mailing Add:* Univ Victoria PO Box 1700 Victoria BC V8W 2Y2 Can

TUNSTALL, LUCILLE HAWKINS, b Thurber, Tex, Jan 17, 22; m 44; c 2. MICROBIOLOGY, IMMUNOLOGY. *Educ:* Univ Colo, BS, 43; Wayne State Univ, MS, 59, PhD(biol, microbiol), 63. *Prof Exp:* Med technologist, Med Sch, Univ Colo, 43-45, Presby Hosp Colo, 45-47, Evangel Deaconess Hosp, 50-52, Sinai Hosp Detroit, 52-55 & Brent Gen Hosp, 55-58; res & tech asst biol, Wayne State Univ, 58-62; asst prof, Delta Col, 62-65; assoc prof, Saginaw Valley Col, 65-67; prof & chmn dept, Bishop Col, 67-71; assoc dir, United Bd Col Develop, 71-72; prof biol & dir, Allied Health Prog, Clark Col, 72-75; PROF BIOL, ATLANTA UNIV, 72-; CHMN, ALLIED HEALTH PROFESSIONS DEPT, CALIF STATE UNIV, 75-, CLIN PROF, ALLIED HEALTH SCH, 78- *Concurrent Pos:* Consult, United Bd Col Develop, 72-; consult, Nat Urban League, Moton Consortium Admis & Financial Aid & Univ Assocs, 72-74; spec consult, Nat Inst Gen Med Sci, 75- *Mem:* AAAS; Am Soc Clin Path; Am Soc Microbiol; Am Soc Cell Biol; NY Acad Sci. *Res:* L-variation; frequency of occurrence and pathogenicity of organisms; immunological studies of blood and pleural fluid in patients with coronary thrombosis; cell wall-deficient bacteria in microbial ecology. *Mailing Add:* Dept Biol Clark Col Atlanta GA 30314

TUNTURI, ARCHIE ROBERT, neurophysics, communication; deceased, see previous edition for last biography

TUOMI, DONALD, b Willoughby, Ohio, Sept 12, 20; m 45; c 2. THERMOELECTRIC ENERGY CONVERSION. *Educ:* Ohio State Univ, BS, 43, PhD(phys chem), 52. *Prof Exp:* Res scientist, Columbia Univ, SAM Lab, 43-45 & Carbide & Carbon Chem Corp, 45-46; fel res assoc photoemissive surfaces, Res Found, Ohio State Univ, 50-53; mem staff semiconductor devices, Lincoln Lab, Mass Inst Technol, 53-54; res chemist, Baird Assocs, Inc, 54-55; res scientist, Thomas A Edison Res Lab, McGraw Edison, 55-61; staff scientist, RC Ingersoll Res Ctr, Borg Warner, 61-63, mgr solid state physics, 63-78, sr scientist, 78-83; PRES, DONALD TUOMI, PHD & ASSOC LTD, 83- *Honors & Awards:* Battery Div Award, Electrochem Soc, 68. *Mem:* Fel AAAS; Electrochem Soc; Am Phys Soc; Am Chem Soc; Am Crystal Soc; Sigma Xi; fel Am Inst Chem; Am Crystal Growth Soc. *Res:* Correlation of materials processing and composition variables to system performance and properties through solid state structural chemistry in polymer deformation-failure, thermoelectric energy conversion, and plated plastics. *Mailing Add:* 626 S Kaspar Ave Arlington Heights IL 60005

TUOMINEN, FRANCIS WILLIAM, b Floodwood, Minn, Mar 1, 43; m 64; c 4. BIOCHEMISTRY. *Educ:* Univ Minn, Duluth, BS, 65; Univ Minn, Minneapolis, MS, 68, PhD(biochem), 70. *Prof Exp:* NSF fel carcinogenesis, Oak Ridge Nat Lab, 70-71; sect leader, Gen Mills Chem Inc, 71-76, mgr res & develop, 76-78; dir res & develop chem, Henkel Corp, 78-85; VPRES CORP TECHNOL, ECOLAB, 85- *Concurrent Pos:* Consult, Sch Pub Health, Univ Minn, 65-66 & Oak Ridge Nat Lab, 71-72. *Mem:* Am Chem Soc; Am Soc Microbiol; Indust Res Inst; AAAS; Chem Specialities Mfrs Asn. *Res:* New products, processes and applications for specialty chemicals. *Mailing Add:* Ecolab Inc Ecolab Ctr St Paul MN 55102

TUOMY, JUSTIN M(ATTHEW), b Bemidji, Minn, Mar 7, 14; m 48; c 1. FOOD TECHNOLOGY, CHEMICAL ENGINEERING. *Educ:* Univ Minn, BChE, 38. *Prof Exp:* Chem engr, Northern Regional Res Lab, USDA, 40-48; food technologist, Oscar Mayer & Co, 48-49, head qual control, 49-55, plant mgr, 55-56; tech sales mgr equip, L C Spiehs Co, 56-57; plant mgr, Horton Fruit Co, 57-58; food technologist, Qm Food & Container Inst, 58-59, supvr food technol, 59-63; supvr food technol, US Army Naticks Lab, 63-77, chief, Animal Prod Group, US Army Res & Develop Command, 77-80, chief, Food Technol Divs, US Army Res & Develop Labs, 80-82. *Mem:* Am Chem Soc; Am Soc Qual Control; Inst Food Technol; Sigma Xi. *Res:* Dairy, poultry, fish, meat and combination products for military and stress subsistence; development of food and food systems for use in space; program management for central food preparation systems. *Mailing Add:* 83 Davidson Rd Framingham MA 01701

TUOVINEN, OLLI HEIKKI, b Helsinki, Finland, April 8, 44; m 72; c 3. APPLIED & INDUSTRIAL MICROBIOLOGY. *Educ:* Univ Helsinki, Finland, MSc, 69, LicSc, 70; London, Eng, PhD(microbiol), 73. *Prof Exp:* Fel agr biochem, Waite Agr Res Inst, SAustralia, 73-76; assoc prof microbiol, Univ Helsinki, 76; res assoc biotechnol, Univ Minn, 78; from asst prof to assoc prof, 80-85, PROF MICROBIOL, OHIO STATE UNIV, 85- *Concurrent Pos:* Chief investr, Int Atomic Energy Agency, Austria, 73-78; sr res fel, Acad Sci & Univ Helsinki, 76; prin investr, Ministry Trade & Indust, Finland, 78-; vis assoc prof microbiol, Ohio State Univ, 78-80. *Mem:* Soc Int Limnol; Soc Gen Microbiol; Am Soc Microbiol; Soc Indust Microbiol. *Res:* Biodegradation of pesticides; biohydrometallurgy; microbiological corrosion; drinking water microbiology; oligotrophic bacteria. *Mailing Add:* Dept Microbiol Ohio State Univ 484 W 12th Ave Columbus OH 43210-1292

TUPIN, JOE PAUL, b Comanche, Tex, Feb 17, 34; m 55; c 3. PSYCHIATRY. *Educ:* Univ Tex, Austin, BS, 55; Univ Tex Med Br, Galveston, MD, 59. *Prof Exp:* Resident psychiatrist, Univ Tex Med Br, Galveston, 60-62; resident, NIMH, 63-64; NIMH career teaching award, Group Advan Psychiat, 64-66; assoc prof psychiat & assoc dean, Univ Tex Med Br, Galveston, 68-69; assoc prof, 69-71, vchmn dept, 70-76, chmn dept, 76-84, PROF PSYCHIAT, SCH MED, UNIV CALIF, DAVIS, 71-, MED DIR, 84- *Concurrent Pos:* Fel, Group Advan Psychiat, 60-62; dir, Psychiat Consult Serv, Sacramento Med Ctr, 69-; consult, Calif Med Facil, Vacaville, 69-, Twin & Sibling Study, NIMH, 69- & Dept Corrections, Calif, 71; chmn, Clin Psychopharmacol Rev Comt, NIMH, 75-77; mem, Comt Psychiat & Criminal Law, Am Bar Asn, 75-81 & Task Force Recertification, Am Psychiat Asn, 77-; consult & grant receiver, Orphan Drugs-Violence, FDA, 82-85. *Mem:* Fel Am Psychiat Asn; Soc Biol Psychiat; Am Col Psychiat; Soc Health & Human Values; Am Psychosomatic Soc. *Res:* Teaching of medical education; psychopharmacology; identification and treatment of violent behavior. *Mailing Add:* Davis Med Ctr Univ Calif 2315 Stockton Blvd Sacramento CA 95817

TUPPER, CHARLES JOHN, b Miami, Ariz, Mar 7, 20; m 42; c 2. INTERNAL MEDICINE. *Educ:* San Diego State Col, BS, 43; Univ Nebr, MD, 48. *Hon Degrees:* DSc, Univ Nebr, 86. *Prof Exp:* Asst prof internal med, Med Sch, Univ Mich, 56-59, secy med sch, 57, assoc prof & asst dean, 59-66; prof med & dean, Sch Med, 66-80, prof internal med, Community Health & Family Practice, Sch Med, 80-90, ACTING CHAIR, DEPT COMMUNITY HEALTH, UNIV CALIF, DAVIS, 90- *Concurrent Pos:* Consult, St Joseph Mercy Hosp, 56-; pres, Calif Med Asn, 79-80; trustee, AMA, 85- *Honors & Awards:* Billings Bronze Medal, AMA, 55. *Mem:* Am Soc Internal Med; AMA (pres, 90-91); Am Col Health Asn; Asn Am Med Cols; fel Am Col Physicians. *Res:* Medical education; application of principles of preventive medicine to care of the individual patient through periodic health examination; evaluation of diagnostic procedures for effectiveness and reliability; geriatrics. *Mailing Add:* Dept Community Health Univ Calif Sch Med Davis CA 95616

TUPPER, W R CARL, b New Glasgow, NS, Feb 15, 15; m 43; c 3. OBSTETRICS & GYNECOLOGY. *Educ:* Dalhousie Univ, BSc, 39, MD, CM, 43; FRCOG; FRCS(C). *Prof Exp:* Head dept, Dalhousie Univ, 59-77, prof obstet & gynec, 59-81, mem fac, 81-88. EMER PROF OBSTET & GYNEC, DALHOUSIE UNIV, 88- *Mem:* Am Col Surgeons; Am Col Obstet & Gynec; Can Soc Obstet & Gynec; Int Col Surgeons. *Mailing Add:* Dept Obstet & Gynec Grace Maternity Hosp Halifax NS B3H 1W3 Can

TURBAK, ALBIN FRANK, b New Bedford, Mass, Sept 23, 29; m 52; c 2. TEXTILES. *Educ:* Southeastern Mass Technol Inst, BS, 51; Inst Textile Tech, MS, 53; Ga Inst Technol, PhD, 57. *Prof Exp:* Res chemist, Esso Res Co, 57-63; corp res dir, Teepak Inc, 63-72; mgr basic res, ITT Rayonier Co, 72-82; prof & dir, Sch Textiles, Ga Inst Technol, 82-88; DIR, S TECH APPL RES CTR, 88- *Concurrent Pos:* Int consult mgt & res. *Mem:* AAAS; Am Chem Soc; Am Asn Textile Chem & Colorists; NY Acad Sci; fel Royal Soc Dyers & Colorists; Sigma Xi; fel Am Inst Chemists; Tech Asn Pulp & Paper Indust. *Res:* Cellulose, natural and synthetic polymer research; polymer modification; new process and methods research; phosphorus chemistry; dyeing and finishing of textiles; food products research; paper products and wood research. *Mailing Add:* 7140 Brandon Mill Rd Sandy Springs GA 30328

TURBYFILL, CHARLES LEWIS, b Newland, NC, Feb 27, 33; m 55; c 2. RESEARCH ADMINISTRATION. *Educ:* Univ Ore, BA, 55, MS, 57; Univ Ga, PhD(zool), 64. *Prof Exp:* Prin investr, Worcester Found Exp Biol, 64-66, Armed Forces Radiobiol Res Inst, 66-72; health sci adminr, Nat Heart, Lung & Blood Inst, NIH, 72-76, head instnl training, Nat Cancer Inst, 75-76, CHIEF, CTR & SPECIAL PROJ SECT, REV BD, NAT HEART, LUNG & BLOOD INST, NIH, 76- *Res:* Primate cardiovascular physiology, atherosclerosis. *Mailing Add:* 5333 Westbard Ave Rev Br Nat Heart Lung Blood Inst Westwood Bldg Rm 553 Bethesda MD 20892

TURCHAN, OTTO CHARLES, b Ostrava, Czech, Dec 30, 25; nat US; m 52; c 2. PHYSICS, ENGINEERING. *Educ:* Tech Univ Brunn, Ger, Dipl Ing, 45; Charles Univ, Prague, RNDr(physics), 47; Detroit Univ, BS, 50, MS, 53. *Prof Exp:* Res engr, Junkers Airplane Works, Ger, 43; dir res & develop, Turchan Follower Mach Co, Mich, 46-55; mem tech staff & group head inertial systs develop, Res & Develop Labs, Hughes Aircraft Co, Calif, 55-61; mem tech staff & sect head spec projs, Systs Res Labs, Space Technol Labs, Inc, 61-62; mem tech staff, Spec Studies Directorate, Satellite Systs Div, Aerospace Corp, 62-65; sr staff engr & tech consult spacecraft eng, Space Systs Div, Lockheed Missiles & Space Co, 65-66; sr tech specialist, Apollo Syst Develop Dept, Space Systs Div, NAm Aviation, Inc, 66-67; prog develop engr advan systs, Strategic Missile Systs Autonetics Div, NAm Rockwell Corp, 67-71; prin engr, Bedford Labs, Raytheon Co, 71-72; sr proj engr, Bechtel Power Corp, 72-85; PRIN SCIENTIST, NUCLEAR ENERGY SYSTS, 85- *Mem:* AAAS; Am Phys Soc; Am Nuclear Soc; Am Geophys Union; Am Inst Aeronaut & Astronaut; NY Acad Sci. *Res:* Nuclear and plasma physics; astrophysics; celestial dynamics; space physics; space vehicle systems; aeronautical-astronautical navigation and guidance; astrionics systems; automatic control systems in nuclear power and propulsion; advanced nuclear and thermonuclear power systems; plasma systems; nuclear systems design; nuclear fusion power development; combustion technology and high energy fuels development. *Mailing Add:* 458 El Camino Dr Beverly Hills CA 90212

TURCHI, JOSEPH J, b Philadelphia, Pa, Feb 16, 33; m 59; c 4. INTERNAL MEDICINE, ONCOLOGY. *Educ:* Univ Pa, BA, 54; Jefferson Med Col, MD, 58. *Prof Exp:* Head clin hemat & cancer chemother, US Naval Hosp, Bethesda, Md, 62-64; sr investr med, Hahnemann Med Col, 64-69; clin asst prof med, 69-80, CLIN ASSOC PROF MED, THOMAS JEFFERSON UNIV, 80- *Concurrent Pos:* Nat Cancer Inst grant, Misericordia Hosp, 66-; assoc dept path, Hemat Sect & attend physician dept med, Misericordia Hosp, 64-, Nat Cancer Inst prin investr hemat res, 66- *Mem:* Am Col Physicians; Am Soc Clin Oncol. *Mailing Add:* Township & Belfield Ave Havertown PA 19083

TURCHI, PETER JOHN, b New York, NY, Dec 30, 46; m 67; c 2. INDUCTIVE ENERGY SYSTEMS, PLASMA DYNAMICS. *Educ:* Princeton Univ, BSE, 67, MA, 69, PhD(aero & mech sci), 70. *Prof Exp:* Res asst, Guggenheim Propulsion Labs, Princeton Univ, 63-70; plasma physicist, Air Force Weapons Lab, 70-72; res physicist, Naval Res Lab, 72-77, chief, Plasma Technol Br, 77-80; staff scientist, 80-81, DIR, WASHINGTON RES LAB, RES & DEVELOP ASSOCS, INC, 81-; LECTR DEPT CONTINUING ENG EDUC, GEORGE WASH UNIV, 86-; ADJ PROF, DEPT AERONAUT & ASTRONAUT ENG, OHIO STATE UNIV, 88- *Concurrent Pos:* NSF Grad fel, 67-70; lectr, 4th Sch Plasma Physics, Novosibirsk, USSR, 74, Christophilos Mem Sch Plasma Physics, Greece, 77 & Air Force Pulsed Power Lectr Series, 80; chmn, 2nd Int Conf Megagauss Magnetic Field Generation, 79, Conf Prime-Power High Energy Space Systs, 82; tech chmn, Inst Elec & Electronics Engrs, Pulsed Power Conf, 85 & gen chmn, Pulsed Power Conf, 87, int chmn, 18th & 19th Am Inst Aeronauts & Astronauts Elec Propulsion Conf, 85,87; ed, Megagauss Physics & Technol, Plenum Press, 80; chmn adv coun, Mech & Aerospace Eng Dept, Princeton Univ, 88-; mem, Am Inst Aeronauts & Astronauts Tech Comt on Plasmadynamics & Lasers, 84-87, Tech Comt on Elecpropulsion, 87-, Inst Elec & Electronics Engrs Plasmasci & Applns Exec Comt, 85- *Mem:* Am Phys Soc; Sigma Xi; Am Inst Aeronaut & Astronaut; sr mem, Inst Elec & Electronics Engrs; Am Soc Engr Educ. *Res:* Electromagnetic energy to create high energy density systems for rocket propulsion; controlled nuclear fusion; nuclear weapons simulation. *Mailing Add:* Dept Aero/Astro Eng Ohio State Univ 2036 Neil Ave Mall Columbus OH 43210-1276

TURCHINETZ, WILLIAM ERNEST, b Winnipeg, Man, Nov 18, 28; m 54; c 2. PHYSICS. *Educ:* Univ Man, BSc, 52, MSc, 53, PhD(physics), 55. *Prof Exp:* Asst prof physics, Univ Man, 55-56; res fel, Australian Nat Univ, 56-59; Sloan Found fel, 59-69, mem res staff, 60-65, lectr, 65-68, head opers, Bates Linear Accelerator, 73-80, SR RES SCIENTIST, MASS INST TECHNOL, 68-, ASSOC DIR, BATES LINEAR ACCELERATOR, 80- *Concurrent Pos:* Chmn, Gordon Conf Photonuclear Reactions, 69-71. *Mem:* Am Phys Soc; AAAS. *Res:* Nuclear physics; particle accelerators; science education. *Mailing Add:* Bates Linear Accelerator Mass Inst Technol PO Box 846 Middleton MA 01949-2846

TURCK, MARVIN, b Chicago, Ill, June 13, 34; m 56; c 4. INTERNAL MEDICINE, INFECTIOUS DISEASE. *Educ:* Univ Ill, BS, 57, MD, 59; Am Bd Internal Med, dipl. *Prof Exp:* Intern med, Res & Educ Hosp, Ill, 59-60; fel, Univ Wash, 60-62; resident & asst, Res & Educ Hosp, Ill, 62-63, chief resident, Cook County Hosp Serv, 63-64; head, Div Infectious Dis & prog dir, Res Infectious Dis Lab, King County Hosp, 64-68; chief med, USPHS Hosp, 68-72; PROF MED, UNIV WASH, 72-, PHYSICIAN-IN-CHIEF DEPT MED, HARBORVIEW MED CTR, 72- *Concurrent Pos:* Instr, Univ Ill, 63-64; from asst prof to assoc prof, Univ Wash, 64-72; attend physician, King County Hosp, Wash, 64-; attend physician & consult, Univ Wash Hosp, 66; attend physician, USPHS Hosp, 66. *Mem:* Am Fedn Clin Res; Infectious Dis Soc Am; fel Am Col Physicians. *Res:* Laboratory and clinical aspects of pyelonephritis; investigation of new antibiotics. *Mailing Add:* 325 Ninth Ave Seattle WA 98104

TURCO, CHARLES PAUL, b Brooklyn, NY, Sept 23, 34; m 55; c 4. PARASITOLOGY, NEMATOLOGY. *Educ:* St John's Univ, NY, BS, 56, MS, 58, MS, 60; Tex A&M Univ, PhD(biol), 69. *Prof Exp:* Teacher, High Sch, NY, 56-64; dir univ develop, 71-74, assoc prof, 65-81, dir develop, 74-81, PROF BIOL, LAMAR UNIV, 81-, DIR RES & PROGS, 81- *Honors & Awards:* Sigma Xi Res Award, 68. *Mem:* Am Soc Parasitol; Am Inst Biol Sci; Am Soc Nematol; Sigma Xi. *Res:* Nematodes of rice and associated insect pests; nematode parasites associated with man's domestic animals. *Mailing Add:* 113 Briggs Beaumont TX 77707

TURCO, JENIFER, b Morgantown, WVa, July 24, 50. RICKETTSIOLOGY & RICKETTSIAL DISEASES. *Educ:* Marywood Col, Scranton, Pa, BS, 72; WVa Univ, MS, 75, PhD(med microbiol), 78. *Prof Exp:* fel, 78-82, from instr to asst prof, 83-90, ASSOC PROF DEPT MICROBIOL & IMMUNOL, COL MED, UNIV SALA, 90- *Concurrent Pos:* Nat Needs fel, NSF, 79-80; Nat Res Serv Award, Nat Inst Allergy & Infectious Dis, NIH, 80-82; co-prin investr, USPHS grant, 83-; counr-at-large, Am Soc Rickettsiology & Rickettsial Dis. *Mem:* Am Soc Microbiol; Am Soc Rickettsiology & Rickettsial Dis; Am Asn Immunologists; Int Soc Interferon Res. *Res:* Interaction of obligate intracellular bacteria, particularly Rickettsia Prowazekii, with host cells; mechanisms of action of interferon-gamma and other lymphokines in host defense against rickettsiae; role of antibody in host defense against rickettsiae. *Mailing Add:* Dept Microbiol & Immunol Lab Molecular Biol Col Med Univ SAla Mobile AL 36688

TURCO, RICHARD PETER, b New York, NY, Mar 9, 43; div; c 1. AERONOMY, CLIMATOLOGY. *Educ:* Rutgers Univ, New Brunswick, BS, 65; Univ Ill, Urbana, MS, 67, PhD(elec eng), 71. *Prof Exp:* NSF res grant, Space Sci Div, Ames Res Ctr, NASA, Moffett Field, Calif, 71; atmospheric scientist, R&D Assocs, 71-88; PROF, UNIV CALIF, LOS ANGELES, 88- *Concurrent Pos:* Assoc ed, J Geophys Res, 81-; mem, comt Causes & Effects of Ozone Change, Nat Acad Sci, 83-84, comt Atmospheric Effects Nuclear Explosions, 84-85, comt Environ Effects of Nuclear War, Int Coun Sci Unions, 83-88; vis prof, Univ Calif, Los Angeles, 84-; mem comt, Ozone Trends, NASA, 86-88; pres-elect, Atmospheric Sci Div, Am Geophys Union, 90-; H Julian Allen Award, NASA, 83-88. *Honors & Awards:* H Julian Allen Award, NASA, 83; Leo Szilard Award, Am Phys Soc, 85; MacArthur Found Award, 86. *Mem:* Am Geophys Union; Sigma Xi. *Res:* Ozone photochemistry and the ozone hole; nuclear winter; climate change; air pollution; volcanic climate; global change; planetary atmospheres and clouds. *Mailing Add:* Dept Atmospheric Sci Univ Calif Los Angeles CA 90024-1565

TURCO, SALVATORE J, b Philadelphia, Pa, Mar 4, 32; m 57; c 2. PHARMACY. *Educ:* Philadelphia Col Pharm & Sci, BSc, 59, MSc, 66, PharmD, 67. *Prof Exp:* Instr sterile prod, Philadelphia Col Pharm & Sci, 66-67; from instr to asst prof, 67-73, assoc prof, 73-78, PROF PHARM, SCH PHARM, TEMPLE UNIV, 78- *Concurrent Pos:* Indust res grants, Temple Univ, 69, Roche award, 72, univ grant, 72-73. *Mem:* Am Pharmaceut Asn; Am Soc Hosp Pharmacists. *Res:* Parenteral products; particulate matter in parenterals; hospital pharmacy. *Mailing Add:* Dept Biochem Univ Ky Col Med 800 Rose St Lexington KY 40536

TURCOTTE, DONALD LAWSON, b Bellingham, Wash, Apr 22, 32; m 57; c 2. GEOPHYSICS, FLUIDS. *Educ:* Calif Inst Technol, BS, 54, DPh(aeronaut eng), 58; Cornell Univ, MAeroE, 55. *Prof Exp:* Asst prof aeronaut eng, US Naval Postgrad Sch, 58-59; from asst prof to prof aeronaut eng, 59-72, prof geol sci, 72-84, chmn geol sci, 80-84, MAXWELL UPSON PROF ENG, 84- *Concurrent Pos:* NSF fel, Oxford Univ, 65-66, Guggenheim fel, 72-73. *Honors & Awards:* Day Medal, Geol Soc Am, 81; Regents Medal, NY State, 84. *Mem:* Nat Acad Sci; Geol Soc Am; Am Geophys Union; Seismol Soc Am; Am Phys Soc. *Res:* Mantle convection; geophysical heat transfer; behavior of faults; distribution of stress; evolution of sedimentary basins. *Mailing Add:* Snee Hall Cornell Univ Ithaca NY 14853

TURCOTTE, EDGAR LEWIS, b Duluth, Minn, June 7, 29. PLANT GENETICS. *Educ:* Univ Minn, BA, 51, MS, 57, PhD(genetics), 58. *Prof Exp:* PLANT GENETICIST, USDA, 58- *Honors & Awards:* Cotton Genetics Res Award, USDA, 76. *Mem:* AAAS; Am Genetic Asn; Am Soc Agron; Genetics Soc Can. *Res:* Breeding, genetics and speciation of Gossypium barbadense. *Mailing Add:* Maricopa Agr Ctr 37860 W Smith Enke Rd Maricopa AZ 85239

TURCOTTE, JEREMIAH G, b Detroit, Mich, Jan 20, 33; m 58; c 4. SURGERY. *Educ:* Univ Mich, BS, 55, MD, 57; Am Bd Surg, dipl, 64. *Prof Exp:* Resident, 58-63, from instr to assoc prof, 63-71, PROF SURG, MED SCH, UNIV MICH, ANN ARBOR, 71-, CHMN DEPT, 74- *Concurrent Pos:* Co-investr, USPHS Res Grant, 64-; consult, Ann Arbor Vet Admin Hosp & Wayne County Gen Hosp. *Mem:* Fel Am Col Surg; Transplantation Soc; Soc Univ Surgeons; Soc Surg Alimentary Tract; Asn Acad Surg; Am Soc Transplant Surgeons; Cent Surg Asn (pres, 91). *Res:* Portal hypertension; organ transplantation. *Mailing Add:* Dept Surg Univ Mich Ann Arbor MI 48109-0331

TURCOTTE, JOSEPH GEORGE, b Boston, Mass, Dec 25, 36; m 62; c 5. ORGANIC CHEMISTRY, MEDICINAL CHEMISTRY. *Educ:* Mass Col Pharm, BS, 58, MS, 60; Univ Minn, PhD(med chem), 67. *Prof Exp:* Sr biochemist, Spec Lab Cancer Res & Radioisotope Serv, Vet Admin Hosp, Minneapolis, 65-67; from asst prof to assoc prof med chem, 67-77, PROF MED CHEM, COL PHARM, UNIV RI, 77- *Concurrent Pos:* Res comt grants, Univ RI, 67-69; res grants, Nat Cancer Inst, 67-70, RI Water Resources Ctr, 68-69, RI Heart Asn, 70-72 & Nat Heart & Lung Inst, 71- *Mem:* AAAS; NY Acad Sci; Am Chem Soc; Am Pharmaceut Asn. *Res:* Synthesis of potential medicinal agents, including phospholipids, anticancer agents, antihypertensives, molluscicides, parasympathomimetic and parasympatholytic agents. *Mailing Add:* Dept Med Chem Univ RI Kingston RI 02881

TURCOTTE, WILLIAM ARTHUR, b Stambaugh, Mich, May 6, 45; m 66; c 2. IRON ORE FLOTATION & PROCESSING, MINERALS PROCESSING. *Educ:* Mich Technol Univ, BS, 68. *Prof Exp:* Metall engr, Res Lab, 68-71, metall engr, Res Pilot Plant, 71-74, plant metallurgist, Tilden Mine, 74-79, asst chief metallurgist, 79-83, chief metallurgist, Res & Develop, 83-87, DIR, RES & DEVELOP, CLEVELAND-CLIFFS IRON CO, 87- *Mem:* Soc Mining Engrs. *Res:* Research and development in iron ore and other mineral beneficiation, pelletizing and environmental related areas; iron ore plant operation. *Mailing Add:* Cleveland-Cliffs Iron Co Division St Ishpeming MI 49849

TUREK, ANDREW, b Lemberg, Poland, Nov 11, 35; Can citizen; m 58; c 2. GEOCHEMISTRY. *Educ:* Univ Edinburgh, BSc, 57; Univ Alta, MSc, 62; Australian Nat Univ, PhD(geophys), 66. *Prof Exp:* Chemist, Scottish Agr Industs, 57-58; mine geologist, Lake Cinch Mines Ltd, Can, 58-60 & Sherritt-Gordon Mines, 62-63; res coordr geol, Man Dept Mines & Natural Resources, 66-69; assoc prof, Northern Ill Univ, 69-71; assoc prof geol & chem, 71-77, PROF GEOL, UNIV WINDSOR, 77- *Concurrent Pos:* Vis scientist, US Geol Survey, Denver, 78-79; adj prof geophys, Univ Western Ont, 86- *Mem:* Geol Asn Can; Spectros Soc Can; Geochem Soc. *Res:* Economic geology; geochronology and isotope geology; analytical geochemistry, geostatistics. *Mailing Add:* 6115 Diusputed Rd Windsor ON N9H 1X6 Can

TUREK, FRED WILLIAM, b Detroit, Mich, July 31, 47; m; c 2. REPRODUCTIVE ENDOCRINOLOGY, CIRCADIAN RHYTHMS. *Educ:* Mich State Univ, BS, 69; Stanford Univ, PhD(biol sci), 73. *Prof Exp:* Fel reproductive biol, Univ Tex, Austin, 73-75; asst prof reproductive endocrinol, 75-80, assoc prof, 80-83, PROF NEUROBIOL & PHYSIOL, NORTHWESTERN UNIV, 83-, CHMN, 87- *Concurrent Pos:* NIH fel, 73; res career develop award, 78-83; vis asst prof, Dept Anat, Univ Calif, Los Angeles, 79; vis scientist, Dept Zool, Univ Bristol, Eng, 81; vis prof, Free Univ Brussels, Belg. *Honors & Awards:* Fogarty Sr Int Fel Award, 86; Curt Richter Psychoneuroendocrinology Award, 87. *Mem:* Soc Neurosci; Soc Study of Reproduction; Am Physiol Soc; Endocrine Soc; AAAS. *Res:* Role of the photoperiod in regulating the hypothalamo-pituitary-gonadal axis in mammals; neural basis for the generation of circadian rhythms; importance of biological rhythm for human health and performance. *Mailing Add:* Dept Neurobiol & Physiol Northwestern Univ 633 Clark St Evanston IL 60201

TUREK, WILLIAM NORBERT, b St Paul, Minn, June 30, 31; m 66. THREE-D COMPUTER GRAPHICS. *Educ:* Univ St Thomas, BS, 53; Univ Md, PhD (org chem), 58. *Prof Exp:* From asst prof to assoc prof, 63-75, chmn dept, 77-84, PROF CHEM, ST BONAVENTURE UNIV, 75- *Mem:* Am Chem Soc. *Res:* Tutorials using computer graphics; substituted furans. *Mailing Add:* Dept Chem St Bonaventure Univ St Bonaventure NY 14778

TUREKIAN, KARL KAREKIN, b New York, NY, Oct 25, 27; m 62; c 2. GEOCHEMISTRY. *Educ:* Wheaton Col, Ill, AB, 49; Columbia Univ, MA, 51, PhD, 55. *Hon Degrees:* MAH, Yale Univ, 65; DSc, SUNY Stony Brook, 89. *Prof Exp:* Lectr geol, Columbia Univ, 53-54, res assoc geochem, Lamont Geol Observ, 54-56; from asst prof to prof geol, 56-72, Henry Barnard Davis prof geol & geophys, 72-85, chmn Dept Geol, 82-88, BENJAMIN SILLIMAN PROF GEOL & GEOPHYS, YALE UNIV, 85-, DIR, CTR STUDY GLOBAL CHANGE, 89- *Concurrent Pos:* Guggenheim fel, 62-63; consult, President's Comn Marine Sci Eng & Resources, 67-68 & Oceanog Panel, NSF, 68-71; ed, J Geophys Res, 69-75 & Earth & Planetary Sci Letters, 75-89, Global Biogeochem Cycles, 90-; mem, US Nat Comn Geochem, 70-73, Climate Res Bd, 77-80 & Ocean Sci Bd, 79-82; mem, Comn Phys Sci Mat Res, Nat Acad Sci, Nat Res Coun, 86-90, Comn Geosci Environ Res, 90-92; group experts Sci Aspects Marine Pollution, UN; Sherman Fairchild Distinguished Scholar, Caltech, 88. *Honors & Awards:* VM Goldschmidt Medal, Geochem Soc, 89. *Mem:* Nat Acad Sci; fel Geol Soc Am; Geochem Soc (pres, 75-76); fel Am Geophys Union; fel Meteoritical Soc; fel AAAS. *Res:* Marine geochemistry; geochemistry of radionuclides and trace elements; planetary evolution; atmospheric chemistry; geochemical archeology. *Mailing Add:* Dept Geol & Geophys Yale Univ Box 6666 New Haven CT 06511

TUREL, FRANZISKA LILI MARGARETE, b Berlin, Ger, Jan 10, 24; nat Can. PLANT PHYSIOLOGY. *Educ:* Swiss Fed Inst Tech, dipl, 47, Dr sc nat, 52. *Prof Exp:* Asst to prof, Inst Appl Bot, Swiss Fed Inst Tech, 47-50; asst to dir, Swiss Fed Exp Sta, Waedenswil, 50-53; fel, Prairie Regional Lab, Nat Res Coun Can, 53-55, asst res officer, 55-58; res assoc bact, 58-59, lectr, 65-69, asst prof, 69-75, ASSOC PROF PLANT PHYSIOL, UNIV SASK, 75-, RES ASSOC PLANT PHYSIOL, 60- *Mem:* AAAS. *Res:* Plant pathology; physiology of host-parasite relationships. *Mailing Add:* Dept Biol Univ Sask Saskatoon SK S7N 0W0 Can

TURER, JACK, b New York, NY, Mar 18, 12; m 38; c 2. ORGANIC CHEMISTRY, PHYSICAL CHEMISTRY. *Educ:* City Col New York, BS, 34; Fairleigh Dickinson Univ, MAS, 69. *Prof Exp:* Res chemist, US Pub Rds Admin, 36-39; res chemist, USDA, 39-41, Eastern Regional Res Labs, 41-45; chief chemist, Va-Carolina Chem Corp, 45-52; tech dir, Textile Chem, Witco Chem Corp, 52-66, tech mgr automotive lubricants & petrol prod, 66-72, corp dir labeling govt regulations & chem adv, 72-77; chem indust specialist toxic substances control act & fed regulations expert, Off Toxic Substances, 77-82, CONSULT, PESTICIDES ENVIRON PROTECTION AAGENCY, 82- *Concurrent Pos:* Abstr, Chem Abstracts, 53-; consult; co rep for Witco Corp at Chem Mfrs Asn, Chem Specialties Mfrs Asn & Petrol Packaging Asn. *Mem:* Am Chem Soc; Am Soc Test & Mat; Am Asn Textile Chem & Colorists; Am Inst Chem; Am Soc Lubrication Eng. *Res:* Soils; chemurgy; electrochemistry; oils and fats; proteins for synthetic textile fibers; textile chemicals and finishes; automotive lubricants and petrochemicals; pollution control; labeling; government regulations. *Mailing Add:* Apt 434N 1600 S Eads St Arlington VA 22202

TURESKY, SAMUEL SAUL, b Portland, Maine, Feb 22, 16; m 52; c 5. DENTISTRY. *Educ:* Harvard Univ, AB, 37; Tufts Col, DMD, 41. *Prof Exp:* Res assoc oral path & periodont, 47-55, from asst prof to assoc prof, 55-71, PROF PERIDONT, SCH DENT MED, TUFTS UNIV, 71- *Concurrent Pos:* Dent consult, Gillette Corp. *Mem:* Int Asn Dent Res. *Res:* Histochemistry of gingiva; calculus and plaque formation and prevention. *Mailing Add:* 1758 Beacon St Brookline MA 02146

TURGEON, JEAN, b Montreal, Que, May 8, 36. MATHEMATICS. *Educ:* Univ Toronto, MA, 65, PhD, 68. *Prof Exp:* Asst prof math, Univ Montreal, 69-73; assoc prof math, Concordia Univ, 73-; AT DEPT MATH & STATIST, UNIV MONTREAL. *Mem:* AAAS; Am Math Soc; Math Asn Am; Can Math Cong. *Res:* Geometry. *Mailing Add:* Dept Math & Statist Univ Montreal C P 6128 Succursale A Montreal PQ H3C 3J7 Can

TURGEON, JUDITH LEE, b Topeka, Kans, Mar 19, 42. CELLULAR ENDOCRINOLOGY, NEUROENDOCRINOLOGY. *Educ:* Washburn Univ, BA, 65; Univ Kans, PhD(anat), 69. *Prof Exp:* Res assoc physiol, Sch Med, Univ Md, 69-71, asst prof, 71-75; from asst prof to assoc prof, 75-86, PROF HUMAN PHYSIOL, SCH MED, UNIV CALIF, DAVIS, 86- *Concurrent Pos:* Mem ed bd, Am J Physiol, Endocrinol & Metab, 82-, Endocrinol, 83-87, Biol of Reproduction, 83-86; vis scientist, Inst Animal Physiol, Animal Res Ctr, Cambridge, Eng, 83; mem study sect, NIH, 85-89. *Mem:* Soc Study Reprod; Endocrine Soc; Am Physiol Soc. *Res:* Hypothalmic control of gonadotrophin secretion by the anterior pituitary; signal transfer mechanisms in hormone action; exocytosis. *Mailing Add:* Dept Human Physiol Sch Med Univ Calif Davis CA 95616

TURI, PAUL GEORGE, b Battonya, Hungary, Apr 16, 17; US citizen; m 41; c 1. INDUSTRIAL PHARMACY, ANALYTICAL CHEMISTRY. *Educ:* Pazmany Peter Univ, Budapest, MS, 40, PhD(pharm), 46. *Prof Exp:* Mgr, Szanto Pharm Labs, Budapest, 46-48; res coord, Pharmaceut Indust Ctr, 48-49; dep dir, Pharm Res Inst, 50-53, head pharm res & develop, 55-56; dep mgr qual control dept, Chinoin Chem Works, 53-55; anal chemist, Chase Chem Co, NJ, 57-59; sr scientist, 59-60, group leader anal res, 60-63, head anal labs, 63-70, mgr pharm res, 70-74, ASSOC SECT HEAD, SANDOZ PHARMACEUT, SANDOZ INC, 75- *Concurrent Pos:* Hon asst prof, Pazmany Peter Univ, Budapest, 46-48, hon adj prof, 48-56; lectr, Budapest Tech Univ, 52-55. *Mem:* Am Pharmaceut Asn; Acad Pharmaceut Sci; Int Pharmaceut Fedn; Am Chem Soc. *Res:* Pharmaceutical analysis; pharmacy research and development. *Mailing Add:* Five Oxford Dr Livingston NJ 07039-1406

TURI, RAYMOND A, b Cleveland, Ohio, Sept 1, 57. SEMI CONDUCTOR PHYSICS, NON VOLATILE DEVICES. *Educ:* Case Western Reserve Univ, BSc, 79. *Prof Exp:* Engr, non volatile technol develop, 79-81, proj leader, 81-86; SECT HEAD, EETROM & ANALOG PROCESS INTERGRATION GROUP, NCR CORP, 86- *Mem:* Inst Elec & Electronic Engrs; Electrochem Soc; Am Phys Soc. *Mailing Add:* Microelectronics Div NCR Corp 1635 Aeroplaza Dr Colorado Springs CO 80916

TURIN, GEORGE L, b New York, NY, Jan 27, 30. ELECTRICAL ENGINEERING. *Educ:* Mass Inst Technol, BS & MS, 52, ScD, 56. *Prof Exp:* Prof elec eng, Univ Calif, Berkeley, 60-80, chmn, dept elec eng & comput sci, 80-83, dean, Sch Eng, Los Angeles, 83-86, PROF ELEC ENG & COMPUT SCI, UNIV CALIF, BERKELEY, 86-; VPRES, TEKUEKRON CORP, MENLO PARK, CALIF, 88- *Concurrent Pos:* Guggenheim fel. *Mem:* Nat Acad Eng; fel Inst Elec & Electronics Engrs. *Mailing Add:* Dept Elec Eng & Computer Sci Univ Calif Berkeley CA 94720

TURINO, GERARD MICHAEL, b New York, NY, May 16, 24; m 51; c 3. MEDICINE. *Educ:* Princeton Univ, AB, 45; Columbia Univ, MD, 48. *Prof Exp:* Mem staff, Div Med Sci, Nat Res Coun, 51-53; resident med, Bellevue Hosp, New York, 53-54; assoc, 56-60, from asst prof to prof, 60-83, JOHN H KEATING SR PROF MED, COL PHYSICIANS & SURGEONS, COLUMBIA UNIV, 83; DIR, DEPT MED, ST LUKES-ROOSEVELT HOSP, NEW YORK, NY, 83- *Concurrent Pos:* Nat Found Infantile Paralysis fel, Col Physicians & Surgeons, Columbia Univ, 54-56, NY Heart Asn sr fel, 56-60; asst physician, Presby Hosp, New York, 56-61, from asst attend physician to assoc attend physician, 61-72, dir, Cardiovasc Lab, 66-, attend physician, 72-; consult, Vet Admin Hosp, East Orange, NJ; mem, Career Invest Health Res Coun, New York, 61; vpres coun, Am Heart Asn, 78-81; chmn, Pulmonary Dis Adv Comt, Nat Heart, Lung & Blood Inst, NIH. *Mem:* AAAS; Harvey Soc; Asn Am Physicians; Am Soc Clin Invest; Am Physiol Soc; Am Thoracic Soc (pres, 86-87). *Res:* Internal medicine; cardio-pulmonary physiology. *Mailing Add:* Dept Med Columbia Univ St Luke's Roosevelt Hosp Ctr Amsterdam Ave at 114th St New York NY 10025

TURINSKY, JIRI, b Prague, Czech, Apr 9, 35; m 64; c 2. PHYSIOLOGY, BIOCHEMISTRY. *Educ:* Charles Univ, Prague, MD, 59, PhD(physiol), 62. *Prof Exp:* Instr & res assoc physiol, Med Sch, Charles Univ, Prague, 59-66, asst prof, 68-69; from asst prof to assoc prof, 70-79, PROF PHYSIOL, ALBANY MED COL, 79- *Concurrent Pos:* Res fel surg, Med Sch, Univ Pa, 66-68. *Mem:* Am Physiol Soc; Am Diabetes Asn; Am Burn Asn. *Res:* endocrine control of metabolism; control of metabolism after trauma. *Mailing Add:* Dept Physiol Albany Med Col 43 New Scotland Ave Albany NY 12208

TURINSKY, PAUL JOSEF, b Hoboken, NJ, Oct 20, 44; m 66; c 2. COMPUTATIONAL ENGINEERING. *Educ:* Univ RI, BS, 66; Univ Mich, MSE, 67, PhD(nuclear eng), 70; Univ Pittsburgh, MBA, 79. *Prof Exp:* Asst prof nuclear eng, Rensselaer Polytech Inst, 70-73; mgr core develop, Water Reactor Div, Westinghouse Elec Corp, 73-80; dept head, 80-88, PROF NUCLEAR ENG, NC STATE UNIV, 88- *Concurrent Pos:* Tech expert, Int Atomic Energy Agency, 78-; consult, Elec Power Res Inst, 80-, Duke Power Co, 85-, Sci Appln Int, Corp, 90-; prin investr, NSF & Dept Energy, 80- *Honors & Awards:* Mark Mills Award, Am Nuclear Soc, 70; Glenn Murphy Award, Am Soc Eng Educ, 90. *Mem:* Fel Am Nuclear Soc; Am Soc Eng Educ; Soc Indust & Appl Math; Comput Soc. *Res:* Developing and applying numerical solution algorithms for reactor physics problems which take maximum advantage of computers with parallel architectures; applications include nuclear fuel management optimization, core physics benchmarks, and on-line digital control of nuclear power plants. *Mailing Add:* NC State Univ PO Box 7909 Raleigh NC 27695-7909

TURITTO, VINCENT THOMAS, b New York, NY, June 4, 44; m 70; c 4. BIOMATERIALS, HEMOSTASIS. *Educ:* Manhattan Col, BChemE, 65; Columbia Univ, DEngSc, 72. *Prof Exp:* Prof bioeng, Univ Rio de Janeiro, 72-73; res asst, F Hoffmann La Roche & Co, Ltd, 73-74; res assoc, St Lukes-Roosevelt Hosp & Columbia Univ, 74-84; ASSOC PROF MED, MT SINAI MED SCH, 84- *Concurrent Pos:* Prin investr, Am Heart Asn, 77-80 & NIH, 82-; established investr, Am Heart Asn, 77-82; chmn, Int Comt Thrombosis & Hemostasis, Subcomt Rheology, 82-85; chmn, sect biomed eng, NY Acad Med, 88-89. *Mem:* NY Acad Sci; AAAS; Sigma Xi; Soc Rheology; Int Soc Thrombosis & Hemostasis; NY Acad Med; Am Heart Asn. *Res:* Application of engineering principles for understanding blood and surface interactions as they pertain in hemostasis and thrombosis. *Mailing Add:* Dept Biomed Eng Memphis State Univ Memphis TN 38152

TURK, AMOS, b New York, NY, Feb 28, 18; m 41; c 3. ATMOSPHERIC CHEMISTRY & PHYSICS. *Educ:* City Col New York, BS, 37; Ohio State Univ, MA, 38, PhD(chem), 40. *Prof Exp:* Res assoc, Explosives Res Lab, Pa, 42-44 & Allegany Ballistics Lab, Md, 44-46; instr org chem, City Col New York, 46-49; dir res & develop, Connor Eng Corp, 49-54; from asst prof to assoc prof, 56-66, prof, 67-86, EMER PROF CHEM, CITY COL NEW YORK, 86-; CONSULT CHEMIST, 54- *Mem:* Am Chem Soc; Am Soc Test & Mat; Am Soc Heating Refrigerating & Air-conditioning Engrs; Air Pollution Control Asn; NY Acad Sci. *Res:* Organic synthesis; activated carbon; air analysis and purification; odors. *Mailing Add:* Seven Tarrywile Lake Dr Danbury CT 06810

TURK, DENNIS CHARLES, b New York, NY, Mar 15, 46; m 69; c 2. BEHAVIORAL SCIENCE. *Educ:* Univ Fla, BA, 67; Univ Waterloo, Can, MA, 75, PhD(clin psych), 78. *Prof Exp:* Asst prof psychol, Yale Univ, 77-82, assoc prof psychol, 82-85; PROF PSYCHIAT, UNIV PITTSBURGH, 85-, PROF ANESTHESIOL, 88-; MEM, PITTSBURGH CANCER INST, 86- *Concurrent Pos:* Attend, W Haven Vet Admin Med Ctr, 79-85, Newington Vet Admin Med Ctr, 82-85; consult, Boehringer-Mannheim Diagnostics Inc, 83-88, Nat Ctr Health Stats, 87-, SSA, comn Eval of Pain, 87; vis prof, Univ Tubingen, WGer, 88. *Mem:* Am Psychol Asn; Am Pain Soc; Int Asn Study Pain; Soc Behav Med; Acad Behav Med Res. *Res:* Multiaxial assessment and treatment of chronic pain with emphasis on the physiological mechanisms of cognitive variables; the integration of medical, psychosocial, and behavioral data, and the prescription of treatment based on the empirically-derived taxonomy of chronic pain patients. *Mailing Add:* Pain Eval & Treat Inst Univ Pittsburgh Sch Med Baum Blvd & Craig St Pittsburgh PA 15213

TURK, DONALD EARLE, b Dryden, NY, Sept 4, 31. NUTRITION, BIOCHEMISTRY. *Educ:* Cornell Univ, BS, 53, MNS, 57; Univ Wis, PhD(biochem, nutrit), 60. *Prof Exp:* From asst prof to assoc prof poultry sci, 60-74, from assoc prof to prof food sci, 74-88, EMER PROF FOOD SCI, CLEMSON UNIV, 88- *Concurrent Pos:* From vchmn to chmn, SC Nutrit Comt, 75-77 & 83-85. *Mem:* AAAS; Poultry Sci Asn; Am Chem Soc; Am Inst Nutrit; Sigma Xi. *Res:* Protein and energy relationships in the fowl; trace mineral metabolism; digestive tract disease and nutrient absorption; chronic pyridoxine intoxication in rats, microflora of the digestive tract. *Mailing Add:* Food Sci Dept Clemson Univ Clemson SC 29634-0371

TURK, FATEH (FRANK) M, b June 11, 24; Can citizen; m 60; c 2. AGRICULTURE, BIOLOGY. *Educ:* Univ Bombay, BSc, 47; Univ Sind Pakistan, MSc, 52; Univ Minn, PhD(plant path), 56. *Prof Exp:* Res scientist & lectr, Govt Res Sta & Sci Col, 47-52; res asst, Univ Minn, 53-55; tech adv, A M Lotia Chem Co, Pakistan, 56-57; microbiologist, Gallowhur Chem Can, Montreal, 60-78; SR EVAL OFFICER PESTICIDES, AGR CAN, 61- *Concurrent Pos:* Adv, Western Comt Plant Dis Control, 77-81. *Mem:* Can Phytopath Soc; Can Pest Mgt Soc. *Res:* Fungicides and nematicides. *Mailing Add:* 3039 Highway 16 Nepean ON K1V 8R7 Can

TURK, GREGORY CHESTER, b Elizabeth, NJ, Nov 16, 51; m 74; c 3. LASER SPECTROSCOPY, ATOMIC SPECTROMETRY. *Educ:* Rutgers Col, BA, 73; Univ Md, PhD(anal chem), 78. *Prof Exp:* RES CHEMIST, NAT INST STANDARDS & TECHNOL, 76- *Mem:* Am Chem Soc; Soc Applied Spectros. *Res:* Application of lasers for spectroscopic chemical analysis, in particular the development of laser-enhanced ionization spectrometry in flames. *Mailing Add:* Bldg 222 Rm-A223 Chem Sci & Technol Nat Inst Standards & Technol Gaithersburg MD 20899

TURK, KENNETH LEROY, b Mt Vernon, Mo, July 14, 08; m 34. DAIRY HUSBANDRY. *Educ:* Univ Mo, BS, 30; Cornell Univ, MS, 31, PhD(animal husb), 34. *Prof Exp:* Asst animal husb, Cornell Univ, 31-34, from exten instr to exten asst prof, 34-38; prof dairy husb, Univ Md, 38-44, head dept, 40-44; prof animal husb, 44-74, dir int agr develop, 63-74, head dept, 45-63, EMER PROF ANIMAL SCI, CORNELL UNIV, 74- *Concurrent Pos:* Vis prof, Univ Philippines, 54-55; consult, Rockefeller Found, 58-62; mem, Expert Panel Dairy Educ, Food & Agr Orgn, 65-71, Expert Panel Animal Husb Educ, 69-71; tech adv comt, Inst Nutrit Cent Am & Panama, 67-69; mem, Working Group Agr Res, US-USSR Joint Comn Sci & Tech Coop, 72. *Mem:* AAAS; Am Soc Animal Sci; Am Dairy Sci Asn (pres, 59-60); Asn US Univ Dir Int Agr Progs (pres, 71-72). *Res:* Nutrition of dairy calves and cows; nutritive value of hay and pasture crops; dairy cattle breeding; international agricultural development; animal science. *Mailing Add:* 259 Morrison Hall Cornell Univ Ithaca NY 14850

TURK, LELAND JAN, b Tulare, Calif, July 18, 38; m 56, 84; c 3. GEOLOGY, HYDROLOGY. *Educ:* Fresno State Col, BA, 61; Stanford Univ, MS, 63 & 67, PhD(geol), 69. *Prof Exp:* Jr geologist, Mobil Oil Libya Ltd, Tripoli, Libya, 63-65, geologist, 65-66; from asst prof to prof geol, Univ Tex, Austin, 68-81, adj prof, 81-88; RES GEOLOGIST, FAILURE ANALYSIS, 88- *Concurrent Pos:* Prof engr, Tex; ed-in-chief, Environ Geol, 74-78; partner, Turk-Kehle & Assocs, Consult Geologists, 74-86; pres, Oiltex Int Ltd, 81-85, OSIRIS Petrol, Inc, 85-87; consult, Hall Southwest Water Consults, Inc, 86-89. *Mem:* Geol Soc Am; Nat Water Well Asn; Am Inst Prof Geologists. *Res:* Hydrogeology; environmental geology; petroleum geology. *Mailing Add:* Failure Analysis 149 Commonwealth Dr Menlo Park CA 94025

TURKANIS, STUART ALLEN, b Everett, Mass, Dec 15, 36; m 64; c 2. ELECTROPHYSIOLOGY. *Educ:* Mass Col Pharm, BS, 58, MS, 60; Univ Utah, PhD(pharmacol), 67. *Prof Exp:* From asst prof to assoc prof, 67-85, PROF PHARMACOL, SCH MED, UNIV UTAH, 85- *Concurrent Pos:* USPHS fel, Univ Col, Univ London, 67-69. *Mem:* Am Soc Pharmacol & Exp Therapeut; Soc Neurosci. *Res:* Pharmacology and physiology of synaptic transmission; mechanisms of action of antiepileptic drugs; pharmacology of marijuana and other drugs of abuse. *Mailing Add:* Dept Pharmacol Univ Utah Sch Med 2C219 Med Ctr Salt Lake City UT 84132

TURKDOGAN, ETHEM TUGRUL, b Istanbul, Sept 12, 23; m 50; c 2. METALLURGY. *Educ:* Univ Sheffield, BMet, 47, MMet, 49, PhD, 51, DMet, 85. *Prof Exp:* Res metallurgist, Res & Develop, Brit Oxygen Co, Eng, 50-51; head phys chem sect, Brit Iron & Steel Res Asn, 51-59; staff scientist, US Steel Tech Ctr, 59-64, mgr chem metall, 64-72, sr res consult, 72-86; CONSULT, PYROMETALL & THERMOCHEM, 86- *Honors & Awards:* Brunton Medal, Univ Sheffield, 51; Andrew Carnegie Silver Medal, Brit Iron & Steel Inst, 53; Robert W Hunt Medal, Am Inst Mining, Metall & Petrol Engrs, 67; Mathewson Gold Medal, Metall Soc, 75; Chipman Award, Iron & Steel Soc, 78; Kroll Medal, Metals Soc, 78. *Mem:* Fel Am Inst Mining, Metall & Petrol Engrs; fel Inst Metals; fel Inst Mining & Metall. *Res:* Chemical metallurgy; physical chemistry of high temperature reactions of interest to metallurgical processes and related subjects. *Mailing Add:* 5820 Northumberland Pittsburgh PA 15217

TURKEL, RICKEY M(ARTIN), b New York, NY, Apr 12, 43; wid; c 2. ORGANIC CHEMISTRY, ABSTRACTING. *Educ:* Hofstra Univ, BA, 63; Mass Inst Technol, PhD(org chem), 68; Ohio State Univ, MA, 76. *Prof Exp:* Fel org chem, Hebrew Univ, Jerusalem, 68-69 & Tulane Univ, 69-70; assoc abstractor, 70-71, assoc ed, 71-76, sr assoc ed, 77-82, sr ed, 82-87, SR ED, DOCUMENT ANALYSIS, CHEM ABSTR SERV, 87- *Concurrent Pos:* Consult, translator Russian, Serbocroatian & Hebrew tech literature. *Honors & Awards:* Award, Am Chem Soc, 63. *Mem:* Am Chem Soc; Sigma Xi. *Res:* Synthetic organic chemistry; organometallic chemistry; chemistry literature. *Mailing Add:* 150 S Cassingham Rd Columbus OH 43209-1845

TURKELTAUB, PAUL CHARLES, b Brooklyn, NY, Jan 10, 44; m 67; c 2. ALLERGY, CLINICAL IMMUNOLOGY. *Educ:* Brooklyn Col, BS, 65; Univ Pittsburgh, MD, 69. *Prof Exp:* RES INVESTR ALLERGENIC PROD BR, BUR BIOLOGICS, 77-; CHIEF, LAB ALLERGENIC PROD, FOOD & DRUG ADMIN. *Concurrent Pos:* Consult, Nat Ctr Health Statist, 78-; instr, Johns Hopkins Univ Sch Med, 78- *Honors & Awards:* Commendation Medal, USPHS. *Mem:* Fel Am Acad Allergy; fel Am Col Chest Physicians. *Res:* Investigating the potency, safety and efficacy of allergenic products used for the diagnosis and treatment of allergic diseases. *Mailing Add:* Bldg 29 Rm 212 8800 Rockville Pike Bethesda MD 20892

TURKEVICH, ANTHONY, b New York, NY, July 23, 16; m 48; c 2. NUCLEAR CHEMISTRY, SPACE CHEMISTRY. *Educ:* Dartmouth Col, BA, 37; Princeton Univ, PhD(phys chem), 40. *Hon Degrees:* DSc, Dartmouth Col, 71. *Prof Exp:* Res assoc molecular spectra, Dept Physics, Univ Chicago, 40-41, res chemist, Metall Lab, 43-45; res chemist, SAM Labs, Columbia Univ, 42-43; res physicist, Los Alamos Sci Lab, 45-46; from asst prof to assoc prof, 46-50, James Franck prof, 65-71, James Franck distinguished serv prof, 71-86, PROF CHEM, 50-, EMER PROF CHEM, ENRICO FERMI INST & CHEM DEPT, UNIV CHICAGO, 86- *Concurrent Pos:* NSF fel, Europ Orgn Nuclear Res, Switz, 61-62; J W Kennedy Mem lectr, Univ Wash, St Louis, 64; NSF fel, Orsay, France, 70; consult, Labs, US AEC, DOE. *Honors & Awards:* E O Lawrence Award, US AEC, 62; Atoms for Peace Award, 69; Nuclear Applns Award, Am Chem Soc, 72; Pregel Award, NY Acad Sci. *Mem:* Nat Acad Sci; AAAS; Am Chem Soc; Am Acad Arts & Sci; Royal Soc Arts; NY Acad Sci. *Res:* Reactions of energetic particles with complex nuclei; radioactivity in meteorites; chemical composition of the moon and meteorites. *Mailing Add:* Enrico Fermi Inst Univ Chicago 5640 S Ellis Ave Chicago IL 60637

TURKEVICH, JOHN, b Minneapolis, Minn, Jan 20, 07; m 34; c 2. PHYSICAL CHEMISTRY. *Educ:* Dartmouth Col, BS, 28, MA, 30; Princeton Univ, AM, 32, PhD(chem), 34. *Hon Degrees:* DD, McMurry Col; DSc, Dartmouth Col. *Prof Exp:* Instr, Dartmouth Col, 28-31; from instr to prof, 36-55, Eugene Higgins prof, 55-75, EMER PROF CHEM, PRINCETON UNIV, 75-; MEM STAFF, FRICK CHEM LABS, 77-, EMER PROF, 75- *Concurrent Pos:* Consult, M W Kellogg Co Div, Pullman, Inc, NY, 36-, Radio Corp Am Labs, 43-, Brookhaven Nat Lab, 47, US Energy Res & Develop Admin, 50- & US Dept State; chmn, US Deleg Educrs, USSR, 58; rep, US Sci Am Nat Exhib, Moscow, 59; actg sci attache, US Embassy, Moscow, 60, sci attache, 61; lectr, US Army War Col & US Air War Col; Phi Beta Kappa vis scholar, 61. *Honors & Awards:* Mfg Chemists Asn Award, 56. *Mem:* Am Chem Soc; Am Phys Soc; Sigma Xi. *Res:* Catalysis; molecular structure; synthesis and characterization of monodisperse noble metals, sulica, alumina and zeolites for heterogeneous and homogeneous catalysis; synthesis of drugs for cancer and study of biochemistry of drug action and cancer. *Mailing Add:* Frick Chem Labs Princeton NJ 08544

TURKINGTON, ROBERT (ROY) ALBERT, b Portadown, Northern Ireland, Apr 8, 51; Brit & Can citizen; m 75; c 2. PLANT ECOLOGY. *Educ:* New Univ Ulster, BSc, 72; Univ Col North Wales, dipl ecol, 74, PhD(plant ecol), 75. *Prof Exp:* Fel plant ecol, Univ Western Ont, 76-77; from asst prof to assoc prof, 77-90, PROF BOT, UNIV BC, 90- *Mem:* Brit Ecol Soc; Can Bot Asn; Ecol Soc Am. *Res:* Plant population biology, with special reference to neighbor relationships among grasses and legumes; structure and dynamics of herbaceous vegetation in the Boreal Forest. *Mailing Add:* Dept Bot Univ BC Vancouver BC V6T 2B1 Can

TURKINGTON, ROGER W, b Manchester, Conn, Jan 13, 36; m 60; c 1. MEDICINE, BIOCHEMISTRY. *Educ:* Wesleyan Univ, AB, 58; Harvard Med Sch, MD, 63; Am Bd Internal Med, dipl. *Prof Exp:* Intern med, Duke Univ Hosp, 63-64, resident, 64-65; res assoc, NIH, 65-67; res assoc, Sch Med, Duke Univ, 67-68, asst prof med, 68-69, asst prof med & biochem, 69-71; assoc prof med, Univ Wis-Madison, 71-73; CHIEF ENDOCRINOL, ST LUKES HOSP, 73- *Concurrent Pos:* Chief endocrinol, Vet Admin Hosp, Durham, NC, 67-71. *Mem:* Am Soc Clin Invest; Am Soc Cancer Res; Endocrine Soc; Am Soc Biol Chem; Am Fedn Clin Res. *Res:* Biochemistry of development and cancer; mechanisms of hormone action. *Mailing Add:* 3237 16th St Milwaukee WI 53215

TURKKI, PIRKKO REETTA, b Laitila, Finland, Aug 27, 34; m 57; c 2. NUTRITION, FOOD SCIENCE. *Educ:* Helsinki Home Econ Teacher's Col, dipl, 57; Univ Mass, Amherst, MS, 62; Univ Tenn, Knoxville, PhD(nutrit, food sci), 65. *Prof Exp:* Res assoc biochem, Univ Tenn, 65-66; NIH res fel, Albany Med Col, 66-68; from asst prof to assoc prof, 68-78, asst dean, Col Human Develop, 75-79, PROF NUTRIT, SYRACUSE UNIV, 78- *Mem:* Am Dietetic Asn; Am Inst Nutrit. *Res:* Phospholipid metabolism in choline deficiency; role of diet in hyperlipemia; riboflavin status in protein/energy deprivation; vitamin utilization during negative energy balance. *Mailing Add:* 034 Slocum Hall Col Human Develop Syracuse Univ Syracuse NY 13244-1250

TURKOT, FRANK, b Woodlynne, NJ, Sept 29, 29; m 59; c 4. PHYSICS. *Educ:* Univ Pa, BA, 51; Cornell Univ, PhD(physics), 59. *Prof Exp:* Res assoc particle physics, Lab Nuclear Studies, Cornell Univ, 59-60; from asst physicist to sr physicist, Brookhaven Nat Lab, NY, 60-74; PHYSICIST, FERMILAB, 74- *Mem:* Fel Am Phys Soc. *Res:* Experiments in elementary particle physics to study photoproduction processes and high energy collisions of strongly-interacting particles; particle accelerator research. *Mailing Add:* Fermilab PO Box 500 Batavia IL 60510

TURKSTRA, CARL J, b Hamilton, Ont, Oct 29, 36; m 64; c 2. CIVIL ENGINEERING, STRUCTURAL ENGINEERING. *Educ:* Queen's Univ, Ont, BSc, 58; Univ Ill, MS, 60; Univ Waterloo, PhD(struct safety), 63, McGill Univ, dipl, 79; Univ Montreal, DEA, 80. *Prof Exp:* Engr, F R Harris, Consult Engrs, 62-63; lectr civil eng, Univ Col, Univ London, 63-65; from asst prof to prof civil eng, McGill Univ, 65-82; head dept, 82-85, PROF, POLYTECH UNIV, BROOKLYN, 82- *Concurrent Pos:* Res grants, Nat Res Coun Can, 65-82, NSF, 83-87, Nat Ctr Earthquake Eng Res, 87-, City New York, 87-89, Nat Tran Res Ctr, City Univ NY, 88-; mem, Int Joint Comt on Struct Safety, 76-; chmn, Comt Safety Bldg, Am Soc Civil Engs, 87-90 & Nat Bldg Code Can, 77-; consult struct engr, 68- *Honors & Awards:* State of the Art Award, Am Soc Civil Engrs, 72 & 88. *Mem:* Fel Am Soc Civil Engrs; Am Acad Mech; fel Can Soc Civil Eng; Int Asn Bridge & Struct Eng. *Res:* Choice of structural safety levels based on probabilistic and decision theory concepts; structural masonry; optimum structural design; earthquake risk analysis; bridge management systems. *Mailing Add:* Dept Civil Eng Polytech Univ 333 Jay St Brooklyn NY 11201

TURLAPATY, PRASAD, b Vijayawada, India, June 1, 42; m 72; c 1. HYPERTENSION, CLINICAL RESEARCH. *Educ:* Andhra Univ, India, BSc, 60, BPharm, 64, MPharm 65; Univ Hawaii, PhD(pharmacol), 71. *Prof Exp:* Fel pharmacol, Univ Tex Health Sci Ctr, San Antonio, 72-74; scientist pharmacol, Postgrad Med Inst, Pondicherry, India, 75-77; from instr to asst prof physiol, Downstate Med Ctr, State Univ NY, Brooklyn, 77-80; clin invest assoc, Ives Med Lab, New York, 80-81, sr clin invest assoc med, 82-; asst dir clin res, Med Dept, Am Critical Care; ASSOC MED DIR, DEPT MED, DUPONT MERCK PHARMACEUTICALS. *Mem:* Am Physiol Soc; NY Acad Sci; AAAS; Fel Am Col Clin Pharmacol; Am Soc Clin Pharmacol Ther; Am Soc Pharmacol & Ther. *Res:* Conduct of Phase I to Phase IV clinical trials towards new drugs development, design and development of clinical protocols in hypertension, stable & unstable angina, acute myocardial infraction, arrhythmias, preparation of medical summaries for USDA review towards new drug application. *Mailing Add:* DuPont Merck Pharmaceut Barley Mill Plaza Bldg 26/1370 Wilmington DE 19880-0026

TURLEY, JUNE WILLIAMS, b Boston, Mass, Apr 12, 29; m 50; c 3. X-RAY CRYSTALLOGRAPHY, TSCA REGULATORY COMPLIANCE. *Educ:* Wilkes Col, BS, 50; Pa State Univ, MS, 51, PhD(biochem), 57. *Prof Exp:* Res asst, molecular struct via x-ray, Pa State Univ, 53-56, res assoc, 56-57; res specialist, molecular struct & x-ray crystallog, 57-71, res mgr, Anal Labs, 71-74, sr econ planner, Capital Projs, Bus Develop & Futures Studies, 74-84, SR RES ASSOC REGULATORY COMPLIANCE MGT, DOW CHEM CO, 84- *Mem:* Am Chem Soc; World Future Soc; fel AAAS; Am Women Sci; Nat Resources Defense Coun; fel Sigma Xi. *Res:* Health and environmental regulatory issues, TSCA regulatory compliance; business analysis; computer applications; long range economic planning; single-crystal x-ray structure analysis; chemical identification using x-ray methods. *Mailing Add:* Regulatory Compliance Health & Environ Sci The Dow Chem Co Bldg 1803 Midland MI 48674

TURLEY, KEVIN, b New York, NY, May 21, 46. CARDIAC SURGERY, PHYSIOLOGY. *Educ:* Fordham Univ, BA, 68; Med Col Wis, MD, 72. *Prof Exp:* Intern surgery, Ohio State Univ, 72-73; gen surgery resident, Univ SFla, 73-75, chief resident, 75-76; chief resident, 76-78, INSTR CARDIAC SURGERY, UNIV CALIF, SAN FRANCISCO, 78- *Concurrent Pos:* Fel, Cardiovascular Res Inst, 78- *Res:* Cardiovascular research; right ventricular function pulmonary hypertension both chronic and acute; aortic insuffiency and deep hypothermia and total circulatory arrest. *Mailing Add:* Dept Surg Univ Calif M-896 San Francisco CA 94143-0118

TURLEY, RICHARD EYRING, b El Paso, Tex, Dec 29, 30; m 54; c 7. OPERATIONS RESEARCH, NUCLEAR & SYSTEMS ENGINEERING. *Educ:* Univ Utah, BS, 55, MS, 58; Iowa State Univ, PhD(nuclear eng), 66. *Prof Exp:* Engr, Convair Div, Gen Dynamics Corp, Tex, 55-56 & El Paso Natural Gas Co, 56-57; asst prof mech & nuclear eng, Univ Utah, 57-62; asst prof eng sci & nuclear eng, Iowa State Univ, 63-67; sect mgr systs anal, Battelle-Northwest, 67-71, res assoc systs eng, 71-72; assoc prof, 72-81, PROF MECH & INDUST ENG, UNIV UTAH, 81- *Concurrent Pos:* Mem adv bd, Wash/Alaska Regional Med Prog, 70-72; alt state rep, Western Interstate Nuclear Bd, 72-78; exec dir, Utah Nuclear Energy Comn, 72-73; sci adv, State of Utah, 73-77; consult to indust; pres, Mex Hermosillo Mission, 83-85. *Mem:* Inst Indust Engrs. *Res:* Discrete-event simulation; operations research; technology assessment; energy conservation; engineering economics; energy; management and systems engineering; simulation and modeling; waste management. *Mailing Add:* Univ Utah 2000 MEB Salt Lake City UT 84112

TURLEY, SHELDON GAMAGE, b Pa, June 13, 22; m 50; c 3. PHYSICS. *Educ:* Pa State Univ, BS, 50, MS, 51, PhD(physics), 57. *Prof Exp:* Res assoc physics of aerosols, Pa State Univ, 52-53; res physicist, 57-62, SR RES PHYSICIST, DOW CHEM CO, 62- *Mem:* Sigma Xi; Am Phys Soc. *Res:* High polymer physics, especially with dynamic mechanical and electrical properties of polymers and their relationship to molecular structure. *Mailing Add:* 1208 Wakefield Midland MI 48640

TURMAN, ELBERT JEROME, b Granite, Okla, Jan 26, 24; m 49; c 2. ANIMAL SCIENCE. *Educ:* Okla State Univ, BS, 49; Purdue Univ, MS, 50, PhD(physiol), 53. *Prof Exp:* From instr to asst prof animal husb, Purdue Univ, 50-55; from asst prof to prof animal sci, Okla State Univ, 55-87; RETIRED. *Mem:* Soc Study Reproduction; Am Soc Animal Sci. *Res:* Physiology of reproduction of beef cattle, sheep and swine. *Mailing Add:* 3401 W 24th St Stillwater OK 74074

TURNBLOM, ERNEST WAYNE, b Boston, Mass, Nov 1, 46; m 73; c 2. ORGANIC CHEMISTRY. *Educ:* Worcester Polytech Inst, BS, 68; Columbia Univ, PhD(chem), 72. *Prof Exp:* Instr org chem, Princeton Univ, 72-74; res chemist, Eastman Kodak Co, 74-77, res lab head, 77-83, mkt intelligence, 83-84, dir, planning & regulatory affairs, Bio-Prod Div, 85-88, dir, fine chem strategy, Lab & Res Prods Div, 88-89, MGR, TECHNOL DEVELOP, GRAPHICS IMAGING SYSTS DIV, EASTMAN KODAK CO, 89- *Concurrent Pos:* Adv coun, Mat Sci Prog, Roch Inst Tech. *Mem:* Am Chem Soc; Com Develop Asn. *Res:* Organophosphorus chemistry; pentaalkylphosphoranes and related compounds; chemistry of other main group elements; electrophotographic processes and materials; novel imaging systems. *Mailing Add:* 26 Stonefield Pl Honeoye Falls NY 14472-1158

TURNBULL, BRUCE FELTON, b Cleveland, Ohio, Mar 2, 28; m 51; c 3. PHYSICAL CHEMISTRY. *Educ:* Case Inst Technol, BS, 50; Faith Theol Sem, BD, 54; Western Reserve Univ, MS, 55, PhD(phys chem), 63. *Prof Exp:* Chemist, Gen Motors Corp, Ohio, 50-51; asst prof chem, Cedarville Col, 55-63; from asst prof to assoc prof, 63-70, PROF CHEM, CLEVELAND STATE UNIV, 70-, OMBUDSMAN, 71- *Mem:* AAAS; Am Chem Soc; Am Asn Physics Teachers. *Res:* Thermodynamics; molecular structure studies with infrared spectroscopy. *Mailing Add:* 2503 Dover Center Rd Westlake OH 44145

TURNBULL, BRUCE WILLIAM, b Purley, Eng, Oct 8, 46; m 72; c 2. BIOMETRICS, BIOSTATISTICS. *Educ:* Cambridge Univ, BA, 67; Cornell Univ, MS, 70, PhD(statist), 71. *Prof Exp:* Asst prof statist, Stanford Univ, 71-72; lectr math, Oxford Univ, 72-76; PROF OPER RES STATIST, CORNELL UNIV, 76- *Honors & Awards:* Snedecor Award, Am Statist Asn, 79,. *Mem:* Biomet Soc; fel Am Statist Asn, 85; Oper Res Soc Am; Royal Statist Soc; Statist Soc Can. *Res:* Statistical design and analysis of long-term animal studies; general biomedical statistics; survival analysis; reliability and life testing; quality control. *Mailing Add:* Dept Opers Res Cornell Univ Ithaca NY 14853-3801

TURNBULL, CRAIG DAVID, b Reading, Pa, Aug 28, 40; m 61; c 4. BIOSTATISTICS, PUBLIC HEALTH. *Educ:* Albright Col, BA, 62; Univ NC, Chapel Hill, MPH, 65, PhD(biostatist), 71. *Prof Exp:* Statistician pub health, Pa State Health Dept, 62-64; biostatistician & clin instr, Health Res Found, State Univ NY, Buffalo, 65-68; from instr to asst prof, 71-77, ASSOC PROF BIOSTATIST, UNIV NC, 78- *Concurrent Pos:* Statist consult, Dept Pub Health, NC, 70-76, Am Col Obstet & Gynec, 71-74 & Sch Nursing, Univ NC, Chapel Hill, 72-75; statist consult & adv bd mem, Asn Schs Pub Health, 74-79; prin investr, Doctoral & Postdoctoral Res Training, 76-; prof dir, BSPH Biostatist, 76-; co-dir, Health Admin Postdoctoral Training, 78-80. *Mem:* Am Statist Asn; Am Pub Health Asn; Sigma Xi; Biomet Soc. *Res:* Public health statistics; perinatal mortality; mental health problems; complications of pregnancy; congenital malformations. *Mailing Add:* Dept Biostat CB No 7400 Univ NC Chapel Hill NC 27599

TURNBULL, DAVID, b Kewanee, Ill, Feb 18, 15; m 46; c 3. PHYSICAL CHEMISTRY, MATERIALS SCIENCES. *Educ:* Monmouth Col, BS, 36; Univ Ill, PhD(phys chem), 39. *Hon Degrees:* ScD, Monmouth Col, 58, Case Western Reserve Univ, 90. *Prof Exp:* Instr phys chem, Case Inst Technol, 39-43, asst prof, 43-46, res proj leader, 45-46; res assoc, Res Lab, Gen Elec Co, 46-51, mgr, Chem Metall Sect, 51-58, phys chemist, 58-62; Gordon McKay prof appl physics, EMER PROF PHYSICS, HARVARD UNIV, 85- *Honors & Awards:* Acta Metallurgica Gold Medal, 79; Von Hipple Prize, Mat Res Soc, 79; Prize for New Mat, Am Physic Soc, 83; Hume-Rothery Award, 86; Japan Prize in Mat Sci & Technol, 86; Franklin Medal, 90; Bruce Chalmers Award, 90. *Mem:* Nat Acad Sci; AAAS; fel Am Acad Arts & Sci; Am Chem Soc; fel Am Phys Soc. *Res:* Thermionic emission; thermodynamic properties of gases at high pressures; corrosion in non-aqueous media; diffusion in metals; kinetics of nucleation in solid state transformation; solidification; theory of liquids; glass; crystal growth. *Mailing Add:* Div Appl Sci Pierce Hall Harvard Univ Cambridge MA 02138

TURNBULL, G(ORDON) KEITH, b Cleveland, Ohio, Nov 10, 35; m 57; c 5. METALLURGY. *Educ:* Case Inst Technol, BS, 57, MS, 59, PhD, 62. *Prof Exp:* Res engr, Aluminum Co Am, 62-65, sr res engr, 65-69, group leader, 69-71, sect head, Pa, 71-77, div mgr, Alcoa Res Labs, 77-79, asst dir, 79-80, mgr, Corp Planning, 80-82, dir, technol planning, 82-86, VPRES, TECHNOL PLANNING, ALUMINUM CO AM, PA, 86- *Mem:* Am Soc Metals; Am Foundrymen's Soc; Am Inst Mining, Metall & Petrol Engrs; Sigma Xi; Am Soc Eng Educ; Soc Mfg Engrs. *Res:* Solidification of metals; grain refinement of solidifying metals; aluminum alloy development; control of residual stresses in metals; aluminum and titanium forging research; ingot casting, melting, energy and fabricating; technology planning. *Mailing Add:* 550 Fairview Rd Fox Chapel PA 15238

TURNBULL, KENNETH, b Edinburgh, Scotland, July 12, 51; m 85; c 2. MESOIONIC CHEMISTRY, ORGANOSULFUR CHEMISTRY. *Educ:* Heriot Watt Univ, Scotland, BSc, 73, PhD(org chem), 76. *Prof Exp:* Res assoc, Erindale Col, Univ Toronto, 76-78; asst prof chem, Grinnell Col, 78-80; asst prof, 80-86, ASSOC PROF ORG CHEM, WRIGHT STATE UNIV, 86- *Mem:* Am Chem Soc. *Res:* Preparation of novel, fused ring mesoionic compounds; photochromic sydnones; sulfur analogues of N-nitro samines; unusual compounds containing the S equal S linkage in stable configuration; polymer supported reagents. *Mailing Add:* Chem Dept Wright State Univ Dayton OH 45435

TURNBULL, ROBERT JAMES, b Washington, DC, July 26, 41; m 65; c 2. ELECTRICAL ENGINEERING. *Educ:* Mass Inst Technol, BS & MS, 64, PhD(elec eng), 67. *Prof Exp:* From asst prof to assoc prof, 67-77, PROF ELEC ENG, UNIV ILL, URBANA, 77- *Mem:* Inst Elec & Electronics Engrs; Am Phys Soc. *Res:* Electromechanics; electrohydrodynamics; heat transfer; fluid mechanics; controlled thermonuclear fusion; ion sources. *Mailing Add:* Dept Elec & Comp Eng Univ Ill 1406 W Green St Urbana IL 61801

TURNEAURE, JOHN PAUL, b Yakima, Wash, Jan 16, 39; m 68; c 2. LOW TEMPERATURE PHYSICS. *Educ:* Univ Wash, BS, 61; Stanford Univ, PhD(physics), 67. *Prof Exp:* Res assoc, Stanford Univ, 66-69, res physicist, 69-75, actg asst prof, 70-73, sr res assoc, 75-88, PROF PHYSICS RES, STANFORD UNIV, 88- *Concurrent Pos:* Vis prof physics, Univ Wuppertal, Ger, 79. *Mem:* Am Phys Soc; Sigma Xi. *Res:* Study of radio frequency properties of superconductors, time variations of the fundamental physical constants and general relativistic effects; development of ultra-stable superconducting cavity oscillators; development of gyroscopes for measurement of general relativistic effects in earth orbit. *Mailing Add:* Hansen Exp Physics Lab Stanford Univ Stanford CA 94305-4085

TURNER, ALBERT JOSEPH, JR, b Arcadia, Fla, June 21, 38. SOFTWARE ENGINEERING. *Educ:* Ga Inst Technol, BS, 61, MS, 66; Univ Md, MS, 72, PhD(computer sci), 76. *Prof Exp:* Asst prof math, WGa Col, 64-70; asst prof math sci, 75-78, from asst prof to assoc prof, 78-84, PROF COMPUTER SCI, CLEMSON UNIV, 84-, HEAD DEPT, 79- *Concurrent Pos:* Chair, Educ Bd, Asn Comput Mach, 88-; chmn, Computer Sci Accreditation Comn, 88-90. *Mem:* Asn Comput Mach; Inst Elec & Electronics Engrs Computer Soc; Am Asn Univ Professors. *Res:* Computer science education; software engineering. *Mailing Add:* Dept Computer Sci Clemson Univ Clemson SC 29634-1906

TURNER, ALMON GEORGE, JR, b Detroit, Mich, June 9, 32; m 64; c 3. ATMOSPHERIC CHEMISTRY, CHEMICAL DYNAMICS. *Educ:* Univ Mich, BS, 55; Purdue Univ, MS, 56, PhD(inorg chem), 58. *Prof Exp:* Assoc prof inorg chem, NDak Agr Col, 58-59; instr & fel chem, Carnegie Inst Technol, 59-61; asst prof inorg chem, Polytech Inst, NY, 61-66; assoc prof, 66-74, PROF CHEM, UNIV DETROIT, 74- *Concurrent Pos:* Distinguished vis prof, USAF Acad, 81-82. *Mem:* Am Chem Soc; Am Phys Soc; Am Geophys Union. *Res:* Chemical bonding; electronic structure of molecules; nitrogen-sulfur chemistry; atmospheric chemistry. *Mailing Add:* Dept Chem Univ Detroit 4001 W McNichols Rd Detroit MI 48221

TURNER, ALVIS GREELY, b Manheim, Pa, Feb 26, 29; m 56; c 2. ENVIRONMENTAL HEALTH, ENVIRONMENTAL TOXICOLOGY. *Educ:* Univ NC, BA, 52, MSPH, 58, PhD(environ sci), 70; Am Intersoc Acad Cert Sanit, dipl, 70. *Prof Exp:* Sr sanitarian, Caswell County Health Dept, 54-58; supvr prev med, Arabian Am Oil Co, 58-66; from asst prof to assoc prof environ sci, 66-84, PROF ENVIRON TOXICOL, UNIV NC, CHAPEL HILL, 84- *Mem:* Nat Environ Health Asn; Am Pub Health Asn; Sigma Xi. *Res:* Hazardous waste management; environmental risk assessment. *Mailing Add:* 802 Emory Dr Chapel Hill NC 27514

TURNER, ANDREW, b Glasgow, Scotland, Dec 24, 22; nat US; m 50; c 2. CHEMICAL ENGINEERING. *Educ:* Univ Mich, BS, 49, MS, 50, PhD(chem eng), 58. *Prof Exp:* Chem engr, Union Carbide Corp, 53-71; sect chief, 72-78, asst off chief, 78-83, DIV CHIEF, OHIO ENVIRON PROTECTION AGENCY, 83- *Mem:* Am Chem Soc; Am Inst Chem Engrs; Water Pollution Control Conf. *Res:* Synthetic resins and plastics. *Mailing Add:* 417 Montreal Pl Westerville OH 43081

TURNER, ANDREW B, b Lock Haven, Pa, Dec 23, 40. ORGANIC CHEMISTRY. *Educ:* Franklin & Marshall Col, AB, 62; Bucknell Univ, MS, 65; Univ Va, PhD(chem), 68. *Prof Exp:* Interim asst prof chem, Univ Fla, 68-69; asst prof chem, Lycoming Col, 69-74; asst prof chem, St John Fisher Col, 74-80; from asst prof to assoc prof chem, 80-91, PROF CHEM & CHMN DEPT, ST VINCENT COL, 91- *Mem:* Am Chem Soc; Roy Soc Chem. *Res:* Aziridines; synthetic tropane alkaloid analogs; nuclear magnetic resonance spectroscopy. *Mailing Add:* Chem Dept St Vincent Col Latrobe PA 15650-2690

TURNER, ANNE HALLIGAN, b Columbus, Ohio, Feb 3, 41; m 66; c 2. MAGNETIC RESONANCE. *Educ:* Middlebury Col, AB, 63; Univ Rochester, PhD(chem), 69. *Prof Exp:* Lectr chem, Prince George's Commun Col, 69-79; res assoc, 79-85, asst prof, dept chem, 85-87, NMR LAB MGR, HOWARD UNIV, 87- *Concurrent Pos:* Instr chem, Grad Sch, US Dept Agr, 70-75. *Mem:* Am Chem Soc. *Res:* Application of nuclear magnetic resonance to problems of structure, kinetics and/or conformation in chemistry or biochemistry. *Mailing Add:* Dept Chem Howard Univ Washington DC 20059

TURNER, ARTHUR FRANCIS, b Detroit, Mich, Aug 8, 06; wid; c 2. OPTICS, PHYSICS. *Educ:* Mass Inst Technol, BS, 29; Univ Berlin, PhD(physics), 35. *Prof Exp:* From asst to instr physics, Mass Inst Technol, 35-39; from physicist to head optical physics dept, Bausch & Lomb, Inc, 39-71; prof, 71-78, PROF EMER OPTICAL SCI, UNIV ARIZ, 71- *Concurrent Pos:* Mem US Nat Comt, Int Comn Optics, 60-65. *Honors & Awards:* Frederic Ives Medalist, Optical Soc Am, 71. *Mem:* Fel Am Phys Soc; fel Optical Soc Am (pres, 68); Brit Inst Physics & Phys Soc. *Res:* Optical physics; evaporated thin films; infrared. *Mailing Add:* Ecanto Norte 3790 E 3rd St Tucson AZ 85716

TURNER, BARBARA BUSH, b Los Angeles, Calif; m 72; c 2. NEUROENDOCRINOLOGY, ENDOCRINOLOGY. *Educ:* Immaculate Heart Col, BA, 67, MA, 70; Univ Calif, Los Angeles, PhD(neurosci), 74. *Prof Exp:* Fel B S McEwen Lab, Rockefeller Univ, 74-76; res assoc, Ment Health Res Inst, Univ Mich, 76-78; vis asst prof biol, Va Polytech Inst & State Univ, 79-; ASSOC PROF, DEPT PHYSIOL, QUILLEN-DISHNER COL MED, ETENN STATE UNIV. *Concurrent Pos:* Res fel, NIH & Nat Inst Neurol & Commun Dis & Stroke, 74-76; NIMH fel res training biol sci, 76-77; vis asst prof psychol, Va Polytech Inst & State Univ, 78-81. *Mem:* Soc Neurosci; Endocrine Soc; Int Soc Psychoneuroendocrinol; NY Acad Sci; Int Soc Develop Neurosci. *Res:* Brain-hormone interactions; glucocorticoid binding in the brain; corticoid receptor regulation; regulation of neuronal activity by glucocorticoids. *Mailing Add:* Dept Physiol James H Quillen Col Med ETenn State Univ Box 19780A Johnson City TN 37614

TURNER, BARBARA HOLMAN, b Evergreen, Ala, Aug 31, 26; m 50; c 2. ECOLOGY. *Educ:* Miss State Col for Women, BS, 47; Univ Kans, MT, 48; Vanderbilt Univ, MA, 66, PhD(ecol), 72. *Prof Exp:* Asst prof biol, George Peabody Col, 66-72; from asst prof to prof micro-med technol, Miss State Univ, 73-90, asst dept head, 82-90; RETIRED. *Mem:* Am Inst Biol Sci; Am Soc Microbiology; Sigma Xi; Am Soc Med Technologists. *Res:* Interactions between microorganisms and higher plants. *Mailing Add:* 505 Briarwick Dr Starkville MS 39759

TURNER, BARRY EARL, b Victoria, BC, Sept 8, 36; m 62. RADIO ASTRONOMY. *Educ:* Univ BC, BSc, 59, MSc, 62; Univ Calif, Berkeley, PhD(astron), 67. *Prof Exp:* Res off elec eng, Nat Res Coun Can, 62-64; from res assoc radio astron to assoc scientist, 67-74, SCIENTIST, NAT RADIO ASTRON OBSERV, 74- *Concurrent Pos:* mem site rev team, Space Telescope, Assoc Univ, Inc, 80 & NSF, 81; titulaire & chmn vis comt, Observ Paris, 81. *Mem:* Int Astron Union; Am Astron Soc; Union Radio Sci Int. *Res:* Theoretical and observational studies of interstellar molecules, interstellar chemistry, physics of the interstellar medium. *Mailing Add:* Nat Radio Astron Observ Edgemont Rd Charlottesville VA 22901

TURNER, BILLIE LEE, b Yoakum, Tex, Feb 22, 25; div; c 4. SYSTEMATIC BOTANY. *Educ:* Sul Ross State Col, BS, 49; Southern Methodist Univ, MS, 50; Wash State Univ, PhD(bot), 53. *Prof Exp:* From instr to assoc prof, 53-58, chmn dept, 67-74, PROF BOT & DIR PLANT RESOURCES CTR, UNIV TEX, AUSTIN, 59- *Concurrent Pos:* Vis prof & NSF sr fel, Univ Liverpool, 65-66; assoc investr ecol study African veg, 56-57. *Honors & Awards:* NY Bot Garden Award, 65; Merit Award, Bot Soc Am, 88. *Mem:* AAAS; Bot Soc Am (secy, 58-59 & 60-64, vpres, 65); Am Soc Plant Taxon; Soc Study Evolution; Int Asn Plant Taxon; Sigma Xi. *Res:* Plant geography; chromosomal studies of higher plants; flora of Texas and Mexico; biochemical systematics. *Mailing Add:* Dept Bot Univ Tex Austin TX 78712

TURNER, BRUCE JAY, b Brooklyn, NY, Sept 19, 45; m 72; c 2. EVOLUTIONARY BIOLOGY, ICHTHYOLOGY. *Educ:* City Univ New York, Brooklyn, BS, 66; Univ Calif, Los Angeles, MA, 67, PhD(biol), 71. *Prof Exp:* Fel biochem genetics, Ment Health Res Unit, Neuropsychiat Inst, Univ Calif, 72-74; res assoc, Rockefeller Univ, 74-76; vis res scientist evolutionary biol, Mus Zool, Univ Mich, 76-78; asst prof, 78-83, ASSOC PROF BIOL, VA POLYTECH INST & STATE UNIV, 83- *Mem:* Soc Study Evolution; Am Soc Naturalists; Am Soc Ichthyologists & Herpetologists; Soc Syst Zool; Am Fisheries Soc; Genetics Soc Am; Am Genetic Assoc. *Res:* Evolutionary and ecological genetics of fish populations (polymorphism and divergence interspecific hybridization, thelytoky); emphasis on biochemical genetic, molecular and chromosomal studies; general ichthyology and systematics. *Mailing Add:* Dept Biol Va Polytech Inst & State Univ Blacksburg VA 24061-0406

TURNER, CARLTON EDGAR, b Choctaw Co, Ala, Sept 13, 40; c 2. CHEMISTRY, PHARMACOGNOSY. *Educ:* Univ Southern Miss, BS, 66, MS, 69, PhD(org chem), 70. *Hon Degrees:* DSc, Albany Col Pharm, Union Col, 85. *Prof Exp:* Fel pharmacog & dir res, Inst Pharmaceut Sci, Sch Pharm, Univ Miss, 70-71, proj supvr & coord marihuana proj, dept pharmacog, 71-72, dir Cannabis Proj, 72, assoc dir res, 72-80. *Concurrent Pos:* Dir & asst to Pres Reagan, Drug Abuse Policy Off, The White House; pres & chief exec officer, Princeton Diag Lab Am, Inc. *Honors & Awards:* Armstrong Lectr, Aerospace Med Soc, 82; Nat Award, Nat Parent Resource Inst Drug Educ, 82; Pres Award, New Eng Narcotic Enforcement Officers Asn, 82; Pres Award, Int Asn Lions Club, 83. *Mem:* AAAS; Am Soc Pharmacog; Soc Econ Bot; Am Chem Soc; Aerospace Med Asn. *Res:* Organo-silicone compounds hypotensive agents; chemistry of cannabinoids; gas chromatographic and thin layer chromatographic techniques for separation of natural products; phyto chemistry; drugs; biological fluids; pharmacokinetics. *Mailing Add:* 8406 Riverside Dr Alexandria VA 22308-1545

TURNER, CHARLIE DANIEL, JR, b Birmingham, Ala, Feb 24, 46; m 67; c 2. AEROELASTICITY, STRUCTURAL DYNAMICS. *Educ:* Univ Ala, BS, 71; Va Polytech Inst & State Univ, MS, 76, PhD(aerospace eng), 80. *Prof Exp:* Aerospace engr structural dynamics, Air Force Armament Technol Lab, 71-78; group leader dynamics, Cessna Aircraft, 78-79; dynamics engr structural dynamics, Beech Aircraft, 79-81; asst prof aeroelasticity, NC State Univ, 81-83; sr res engr, NTI, 83-87; pres & consult, Aeroelastic Anal Inc, 87-89; SR STAFF ENGR, NICHOLS RES CORP, 89- *Concurrent Pos:* Adj prof, Wichita State Univ, 80-81; consult, Accident Reconstruct Anal Corp, 81 & Lewis, Wilson, Lewis & Jones, 81-82. *Mem:* Am Inst Aeronaut & Astronaut. *Res:* Analytical and experimental subcritical flutter analysis; feedback system approach for subcritical flight flutter testing; wing and control surface; tab aeroelastic analysis; wing and store aeroelastic analysis. *Mailing Add:* Rte 2 Box 217 Ardmore AL 35739

TURNER, CHRISTY GENTRY, II, b Columbia, Mo, Nov 28, 33; m 57; c 3. DENTAL ANTHROPOLOGY. *Educ:* Univ Ariz, BA, 57, MA, 58; Univ Wis, Madison, PhD(phys anthrop), 67. *Prof Exp:* Actg asst prof phys anthrop, Univ Calif, Berkeley, 63-66; from asst prof to assoc prof, 66-75, asst dean, Grad Col, 72-76, PROF PHYS ANTHROP, ARIZ STATE UNIV, 75- *Concurrent Pos:* Am Dent Asn & NIH dent epidemiol & biomet trainee; collabr phys anthrop, US Nat Park Serv, 69-; fel, Ctr Advan Study Behav Sci, Stanford Univ, 70-71; Nat Geog Soc grants, 72, 73 & 79-85, 88-90; Irex to USSR, 80-81, 84; NSF grants, 83-84 & 85; Nat Acad Scis to USSR, 87; Wenner-Gren Found grant, 91. *Mem:* Am Asn Phys Anthropologists; Sigma Xi; Soc Am Archaeol; AAAS; Am Quarternary Asn; Indo-Pac Prehist Asn. *Res:* Co-evolution of human biology and culture; dental morphology, genetics and related behavior; biology and culture of New World peoples, especially southwestern United States Indians and Alaskan Aleuts; origins of peoples of Pacific and New World; dental anthropology; origin of modern humans. *Mailing Add:* Dept Anthrop Ariz State Univ Tempe AZ 85287-2402

TURNER, DANIEL SHELTON, b Montgomery, Ala, Dec 9, 45; m 78; c 4. TRAFFIC ENGINEERING, ROADWAY TORT LIABILITY DEFENSE. *Educ:* Univ Ala, BS, 68, MS, 70; Tex A&M Univ, PhD(civil eng), 80. *Prof Exp:* Civil eng officer, USAF, 69-73; asst prof civil eng technol, Ga Southern Col, 73-76; assoc prof civil eng technol & actg dir, 76-84, PROF CIVIL ENG & DEPT HEAD, UNIV ALA, 84- *Concurrent Pos:* Mem staff, A C Parker & Son & Consult Engrs, 64-69; owner, Turner-Meadows Land Surv Co, 75-77; consult, 76-; asst res engr, Tex Transp Inst, Tex A&M, 79; subcomt chair, Transp Res Bd, 87-90; vchair, educators coun, Inst Transp Engrs, 90-91, chair, 91-92. *Honors & Awards:* Hensley Award, 91. *Mem:* Transp Res Bd; fel Inst Transp Engrs; Am Soc Civil Engrs; Nat Safety Coun; Sigma Xi. *Res:* Highway research: 46 projects and over 160 publications; traffic engineering, traffic safety, and governmental defense of roadway tort liability. *Mailing Add:* Dept Civil Eng Univ Ala PO Box 870205 Tuscaloosa AL 35487-0205

TURNER, DANIEL STOUGHTON, b Madison, Wis, Feb 8, 17; m 44; c 4. PETROLEUM, GEOLOGY. *Educ:* Univ Wis, PhB, 40, PhM, 42, PhD(geol), 48. *Prof Exp:* Field geologist, Buchans Mining Co, Nfld, 47; geologist, US AEC, Colo, 48; asst prof geol, Univ Wyo, 49-50; geologist, Carter Oil Co, Okla, 51 & Petrol Res Co, Colo, 52-53; div geologist, Wm R Whittaker Co, Ltd, 53; geol consult, 54-63; consult, Earth Sci Curric Proj, Boulder, 63-65; prof, 65-85, EMER PROF GEOL, EASTERN MICH UNIV, 85- *Mem:* Am Inst Petrol Geologists; Am Asn Petrol Geol. *Res:* Arctic geology and glaciation; permafrost; thermal activity of Yellowstone National Park; hydrodynamics of oil; Rocky Mountain petroleum exploration; earth science education. *Mailing Add:* 7175 S Poplar Way Englewood CO 80112

TURNER, DANNY WILLIAM, b Shelby, NC, Oct 8, 47; m 70; c 1. CLUSTER ANALYSIS, MULTIVARIATE GRAPHICS. *Educ:* Clemson Univ, BS, 69, PhD(math), 73. *Prof Exp:* Asst prof math, Baylor Univ, 73-76; software analyst, Tex Instruments, 76; asst prof, 76-79, ASSOC PROF MATH, BAYLOR UNIV, 79- *Mem:* Classification Soc; Math Asn Am; Am Statist Asn. *Res:* Cluster analysis and in particular multivariate graphics. *Mailing Add:* Dept Math Baylor Univ PO Box 7328 Waco TX 76798

TURNER, DAVID GERALD, b Toronto, Ont, Dec 13, 45; m 70; c 1. OBSERVATIONAL ASTRONOMY, ASTROPHYSICS. *Educ:* Univ Waterloo, BSc, 68; Univ Western Ont, MSc, 70, PhD(astron), 74. *Prof Exp:* Postdoctorate fel astron, David-Dunlap Observ, 74-76; asst prof, Laurentian Univ, 76-78; asst prof astron, Univ Toronto, 78-80; from asst prof to assoc prof, dept physics & astron, Laurentian Univ, 80-84; assoc prof, 84-91, PROF ASTRON, ST MARY'S UNIV, HALIFAX, NS, 91- *Concurrent Pos:* Connaught fel, Univ Toronto, 75; Nat Sci & Eng Res Coun, Univ res fel, Laurentian Univ, 80-84, St Mary's, 84-90. *Mem:* Can Astron Soc; Am Astron Soc; Royal Astron Soc Can; Int Astron Union; Int Planetarium Soc. *Res:* Interstellar extinction; star clusters and associations; variable stars and galactic structure; stellar spectroscopy. *Mailing Add:* Dept Astron St Mary's Univ Halifax NS B3H 3C3 Can

TURNER, DAVID L(EE), b Afton, Wyo, Nov 20, 49; m 69; c 4. STATISTICAL ANALYSIS. *Educ:* Colo State Univ, BS, 71, MS, 73, PhD(statist), 75. *Prof Exp:* From res asst to instr statist, Colo State Univ, 71-75; from asst prof to assoc prof statist, 75-81, Utah State Univ, 75-89; MATH STATISTICIAN, USDA FOREST SERV, 89- *Concurrent Pos:* Vis Ariz State Univ,85. *Mem:* Sigma Xi; Am Statist Asn; Biomet Soc. *Res:* Analysis of unbalanced data, tolerance intervals and bands for regression; statistical computing, monitoring. *Mailing Add:* Intermountain Res Sta 324 25th St Ogden UT 84401

TURNER, DEREK T, physical chemistry, polymer chemistry; deceased, see previous edition for last biography

TURNER, DONALD LLOYD, b Richmond, Calif, Dec 21, 37; c 3. GEOLOGY, GEOCHRONOLOGY. *Educ:* Univ Calif, Berkeley, AB, 60, PhD(geol), 68. *Prof Exp:* Nat Res Coun res assoc, Isotope Geol Br, US Geol Surv, Colo, 68-69, Calif, 69-70; assoc prof, 70-80, PROF GEOL, GEOPHYS INST & DEPT GEOL, UNIV ALASKA, FAIRBANKS, 80- *Res:* Geochronology, K-Ar and fission track dating applied to problems of regional tectonics; radiometric calibration of paleontological time scales and geothermal systems; geological and geophysical exploration for geothermal resources. *Mailing Add:* Geophys Inst Univ Alaska Fairbanks AK 99775

TURNER, DONALD W, b St Louis, Mo, Jan 24, 32; m 57; c 6. INFLAMMATION, IMMUNE RESPONSES. *Educ:* Washington Univ, St Louis, BS, 57, DDS, 60; Md Univ, College Park, PhD(microbiol), 74. *Prof Exp:* Dentist, 60-64; dentist, Naval Dent Res Inst, 64-70, dental researcher immunol, 74-81, commanding officer, 81-82; asst prof, 82-87, PROF DEPT PERIODONT, NORTHWESTERN UNIV, 87- *Concurrent Pos:* Asst prof microbiol, Univ Md, 75-; mem, Oral Med & Biol Study Sect, Nat Inst Dent Res, 76-81; lectr, Sch Dent, Northwestern Univ, 81-; vis prof, Sch Dent, Univ Ill & Chicago Med Sch, Univ Health Sci, 81-, Okla Univ, Oklahoma City. *Mem:* Am Dent Asn; Am Asn Dent Res; Int Asn Dent Res; Am Soc Microbiol; Int Col Dentists. *Res:* The immunologic basis for dental diseases, especially in periodontal disease; transplantation immunology; inflammatory process in periodontal lesions. *Mailing Add:* Dept Periodont Sch Dent Northwestern Univ 240 E Huron Chicago IL 60611

TURNER, DOUGLAS HUGH, b Staten Island, NY, July 24, 46; m; c 1. BIOPHYSICAL CHEMISTRY. *Educ:* Harvard Col, AB, 67; Columbia Univ, PhD(phys chem), 72. *Prof Exp:* Fel biophys chem, Univ Calif, Berkeley, 73-74; asst prof, 74-81, assoc prof, 81-86, PROF CHEM, UNIV ROCHESTER, 86- *Concurrent Pos:* Tech collabr, Brookhaven Nat Lab, 70-; Alfred P Sloan fel, 79-83; vis prof, Univ Colo, Boulder, 84-85. *Mem:* Am Chem Soc; AAAS. *Res:* Structure and function of nucleic acids; laser temperature jump kinetics. *Mailing Add:* Dept Chem Univ Rochester Rochester NY 14627

TURNER, EDWARD C, b Princeton, NJ, Mar 16, 43. ALGEBRA & TOPOLOGY. *Educ:* Univ Rochester, BA, 65; Univ Calif, Los Angeles, PhD(math), 68. *Prof Exp:* Instr math, Mass Inst Technol, 69-71; from asst prof math to assoc prof math, 71-85, PROF MATH, STATE UNIV NY, ALBANY. *Mem:* Am Math Soc; Math Asn Am. *Res:* Combinatorial group theory. *Mailing Add:* 25 Providence St Albany NY 12203

TURNER, EDWARD FELIX, JR, b Newport News, Va, Apr 21, 20; m 45; c 3. PHYSICS. *Educ:* Washington & Lee Univ, BS & BA, 50; Mass Inst Technol, MS, 52; Univ Va, PhD(physics), 54. *Prof Exp:* Asst prof physics, George Washington Univ, 54-57; assoc prof, 57-59, PROF PHYSICS, WASHINGTON & LEE UNIV, 59-, CHMN DEPT, 61- *Concurrent Pos:* Consult & physicist, Diamond Ord Fuze Lab, 57-58; consult, US Off Educ, 65- *Mem:* AAAS; Am Phys Soc; Am Asn Physics Teachers. *Res:* Solid state physics; electrical properties of solids; electronics; astronomy. *Mailing Add:* 869 Parkridge Dr Wallingford PA 19063

TURNER, EDWARD HARRISON, b Cleveland, Ohio, Dec 21, 20; m 45; c 4. MAGNETISM. *Educ:* Harvard Univ, SB, 42, AM, 47, PhD, 50. *Prof Exp:* Mem staff, Radiation Lab, Mass Inst Technol, 42-45; MEM TECH STAFF UNDERSEA OPTICAL CABLE, BELL TEL LABS, INC, 49- *Mem:* Am Phys Soc; sr mem Inst Elec & Electronics Engrs. *Res:* Optical wave guide propagation; electro-magnetic theory in gyrotropic media; spin waves in ferrimagnetic materials. *Mailing Add:* One N Cherry Lane Rumson NJ 07760

TURNER, EDWARD V, b Belmont, Ohio, May 19, 13; m 39; c 1. PEDIATRICS. *Educ:* Ohio Univ, AB, 34; Harvard Med Sch, MD, 38. *Prof Exp:* From asst prof to assoc prof, 48-68, PROF PEDIAT, COL MED, OHIO STATE UNIV, 68- *Mem:* Fel Am Acad Pediat. *Res:* Clinical pediatrics; medical education. *Mailing Add:* 3341 E Livington Ave Columbus OH 43227

TURNER, EDWIN LEWIS, b Knoxville, Tenn, May 3, 49; m 71; c 2. EXTRAGALACTIC ASTRONOMY, COSMOLOGY. *Educ:* Mass Inst Technol, SB, 71; Calif Inst Technol, PhD(astron), 75. *Prof Exp:* Res fel physics & astron, Inst Advan Study, 75-77; asst prof astron, Harvard Col Observ, Harvard Univ, 77-78; asst prof, 78-81, assoc prof astrophys sci, Princeton Univ Observ, 81-85, PROF, 85- *Concurrent Pos:* Alfred P Sloan res fel, 80; mem bd dir, Asn Univ Res Astron, 80- *Mem:* Am Astron Soc; Int Astron Union. *Res:* Dynamics of galaxies and clusters of galaxies; quasars and active galactic nuclei; the cosmic x-ray background; image processing; gravitational lenses. *Mailing Add:* Peyton Hall Princeton Univ Observ Ivy Lane Princeton NJ 08544

TURNER, ELLA VICTORIA, b Columbia, Mo, Jan 23, 46; m 66; c 2. IMMUNOLOGY. *Educ:* Univ Ark, BA, 67; Univ Louisville, PhD(microbiol), 73. *Prof Exp:* Res asst, Univ Louisville, 73-74, res assoc, Dept Microbiol & Immunol, 74-78, adj asst prof, 78-80; DIR HISTOCOMPATIBILITY LAB, ST JUDE CHILDREN'S RES HOSP, MEMPHIS, TN, 83- *Concurrent Pos:* Consult, Tissue Typing Lab, Jewish Hosp, Louisville, 77-80. *Mem:* Am Asn Histocompatability & Immunogenetics; Am Soc Microbiol. *Res:* Immunologic parameters of tumor growth and rejection; effects of mediators of inflammation and of specific immunological responses on tumor growth; regulation of immune responses; biology of engraftment. *Mailing Add:* 6380 Candlewood Memphis TN 38119

TURNER, ERNEST CRAIG, JR, b West Jefferson, NC, June 15, 27; m 53; c 3. ENTOMOLOGY. *Educ:* Clemson Col, BS, 48; Cornell Univ, PhD(entom), 53. *Prof Exp:* Asst econ entom, Cornell Univ, 48-53; assoc prof entom, 53-65, PROF ENTOM, VA POLYTECH INST & STATE UNIV, 65-, ASSOC ENTOMOLOGIST, AGR EXP STA, 53- *Mem:* Am Mosquito Control Asn; Entom Soc Am; Sigma Xi. *Res:* Medical and veterinary entomology. *Mailing Add:* Dept Entom Va Polytech Inst Blacksburg VA 24061-0319

TURNER, EUGENE BONNER, b Wolf Point, Mont, Oct 6, 22; m 46; c 3. PHYSICS. *Educ:* Mont State Col, BS, 44; Univ Mich, MS, 50, PhD(physics), 56. *Prof Exp:* Mem tech staff, Ramo-Wooldridge Corp, Calif, 56-58 & Space Tech Labs, Inc, 58-60; mem tech staff, 60-68, head lasers & optics dept, Electronics Res Lab, 68-69, sr staff engr, Develop Planning Div, 69-73, staff engr, 73-75, STAFF SCIENTIST, LABS DIV, AEROSPACE CORP, 75- *Mem:* Am Phys Soc; Optical Soc Am; Soc Photo-Optical Instrument Eng (pres, 70); Sigma Xi; Am Inst Aeronaut & Astronaut. *Res:* Plasma physics; spectroscopy; optical and photographic instrumentation; lasers; optical systems; strategic space systems. *Mailing Add:* 23216 Juniper Torrance CA 90505

TURNER, FRED, JR, agricultural chemistry, soils, for more information see previous edition

TURNER, FRED ALLEN, b Chicago, Ill, Mar 16, 33; m 64; c 2. ORGANIC CHEMISTRY. *Educ:* Univ Ill, BS, 55, MS, 58, PhD(pharmaceut chem), 63. *Prof Exp:* Instr, Univ Ill, Chicago, 58-59 & 61-62; from instr to assoc prof, 62-74, PROF CHEM, ROOSEVELT UNIV, 74- *Concurrent Pos:* NSF fel, 59-61. *Mem:* Am Chem Soc; Sigma Xi. *Res:* Synthesis of medicinal compounds; study of organic halogenating agents; synthesis of heterocyclic systems. *Mailing Add:* Dept Chem Roosevelt Univ 430 S Michigan Ave Chicago IL 60605-1394

TURNER, FREDERICK BROWN, b Carlinville, Ill, Feb 4, 27; div. VERTEBRATE ZOOLOGY. *Educ:* Univ Calif, AB, 49, MA, 50, PhD(zool), 57. *Prof Exp:* Asst zool, Univ Calif, 50-52 & 55-56; instr, Ill Col, 52-53; seasonal park naturalist, Death Valley Nat Monument, Calif, 53-55; instr biol, Wayne State Univ, 56-60, univ res coun res fel, 60; vis asst prof zool, Univ Calif, Los Angeles, 60-61, mem staff, Lab Biomed & Environ Sci, 61-87; RETIRED. *Concurrent Pos:* NSF partic, Inst Desert Biol, Ariz State Univ, 59. *Res:* Population ecology of reptiles; ecosystem analysis; environmental effects of energy development in arid environments. *Mailing Add:* 900 Veteran Ave Los Angeles CA 90024

TURNER, GEORGE CLEVELAND, b Spokane, Wash, Jan 26, 25; m 52; c 1. SCIENCE EDUCATION, RESEARCH ADMINISTRATION. *Educ:* Stanford Univ, BA, 47; Utah State Univ, MS, 50; East Wash State Col, MEd, 52; Ariz State Univ, EdD(sci ed), 64. *Prof Exp:* Teacher high schs, Wash, 52-53 & Calif, 53-60; assoc prof, 60-67, chmn dept sci educ, 64-77, assoc vpres univ res, 77-82, PROF BIOL & SCI EDUC, CALIF STATE UNIV, FULLERTON, 67- *Concurrent Pos:* Lectr, Claremont Grad Sch, 57-60; dir, NSF grants for adv topics inst for high sch biol teachers, 65-67, intern-master's degree prog for sci teachers, 67-, Human Ecol Inst, 69- & Urban Sci Intern Teaching Proj, 72-; dir, Off Educ grant for biol sci curriculum study test eval team, 66-67; pres, Calif Intersci Coun, 73-75; dir, Energy and the Environ Inst, 75-, Calif Statewide Energy Consortium, 73- *Mem:* Nat Sci Teachers Asn; Nat Asn Biol Teachers. *Res:* Ecology of rodents, Wasatch Mountains, Utah; scientific enquiry. *Mailing Add:* Energy Consortium Calif State Univ Fullerton CA 92634

TURNER, GODFREY ALAN, chemical engineering, for more information see previous edition

TURNER, HOWARD S(INCLAIR), b Jenkintown, Pa, Nov 27, 11; m 36; c 3. CHEMICAL ENGINEERING. *Educ:* Swarthmore Col, AB, 33; Mass Inst Technol, PhD(org chem, chem eng), 36. *Prof Exp:* Res chemist & supvr, E I du Pont de Nemours & Co, Del, 36-47; dir res & develop div, Pittsburgh Consol Coal Co, 47-54; vpres res & develop, Jones & Laughlin Steel Corp, 54-65; pres & dir, Turner Construct Co, 65-70, chmn, 71-78, chmn exec comt, 78-82, dir, 52-88; RETIRED. *Concurrent Pos:* Dir, Ingersoll-Rand Co & Dime Savings Bank NY. *Mem:* Nat Acad Eng; Am Chem Soc. *Res:* Processes of iron ore beneficiation, reduction and steelmaking. *Mailing Add:* Dunwoody Village-Ch 125 3500 Westchester Pike Newtown Square PA 19073

TURNER, HUGH MICHAEL, b Marianna, Fla, Sept 20, 42; m 72. AQUATIC INVERTEBRATE ZOOLOGY. *Educ:* Ga Southern Univ, BS, 65, MS, 72; La State Univ, PhD(zool), 77. *Prof Exp:* PROF ZOOL, PARASITOL HISTOL, DEPT BIOL, & ENVIRON SCI, MCNEESE STATE UNIV, 77- *Mem:* Am Soc Parasitologists; Sigma Xi; Am Micros Soc; Soc Study Evol. *Res:* Host-parasite interactions including histopathology and immunity in the hosts of cestodes and digenetic trematodes; ecology, systematics and life histories of digenetic trematode parasites; use of animal parasites as indicators of water quality in both freshwater and marine habitats. *Mailing Add:* Dept Biol & Environ Soc McNeese State Univ Lake Charles LA 70609

TURNER, J HOWARD, b San Diego, Calif, Oct 16, 27; m 49; c 4. HUMAN GENETICS. *Educ:* Utah State Univ, BS, 57, MS, 59; Univ Pittsburgh, ScD(human genetics), 62. *Prof Exp:* Instr zool & res asst genetics, Utah State Univ, 57-59; res assoc cytogenetics, 61-62, asst prof human genetics, 62-66, assoc prof, 66-71, PROF BIOSTATIST, UNIV PITTSBURGH, 71- *Concurrent Pos:* Res assoc, Magee Womens Hosp, 64- *Mem:* AAAS; Soc Human Genetics; Am Genetic Asn; Genetics Soc Am; Am Pub Health Asn. *Res:* Developmental cytogenetics; somatic cell genetics; statistical genetics. *Mailing Add:* Rm 316 Grad Sch Pub Health Univ Pittsburgh Pittsburgh PA 15260

TURNER, JACK ALLEN, b Milner, Colo, Feb 2, 42; m 66; c 2. ECOLOGY. *Educ:* Colo State Univ, BS, 68; SDak State Univ, MS, 71; Univ Okla, PhD(ecol), 74. *Prof Exp:* Asst bact, SDak State Univ, 69-72; asst prof, 74-77, ASSOC PROF BIOL, UNIV SC, SPARTANBURG, 77- *Mem:* Am Soc Microbiol; Ecol Soc Am; Sigma Xi. *Res:* The antimicrobial activity of various types of textile finishes. *Mailing Add:* Dept Biol Univ SC Spartanburg SC 29303

TURNER, JAMES A, b Anna, Ill, Jan 9, 48. ORGANIC CHEMISTRY, PESTICIDE CHEMISTRY. *Educ:* Murray State Univ, BS, 71; Fla State Univ, PhD(org chem), 76. *Prof Exp:* Sr res chemist energy res, Hydrocarbons & Energy Lab, 76-78, SR RES CHEMIST AGR CHEM SYNTHESIS, AGR PROD RES, DOW CHEM CO, 78- *Mem:* Am Chem Soc. *Res:* Synthesis of new agricultural chemicals. *Mailing Add:* DOW Chem USA Wal Creek Res Ctr 2800 Mitchell Dr Walnut Creek CA 94598-1604

TURNER, JAMES DAVID, b Bristol, Va, Aug 23, 52. NUCLEAR PHYSICS. *Educ:* Wake Forest Univ, BS(physics) & BA(phil), 73; Duke Univ, PhD(physics), 78. *Prof Exp:* Asst prof physics, Eastern Ky Univ, 78-79; asst prof, 79-84, ASSOC PROF PHYSICS, FURMAN UNIV, 84- *Mem:* Am Phys Soc. *Res:* Studies of giant dipole resonances via polarized and unpolarized proton capture reactions. *Mailing Add:* Dept Physics Furman Univ Greenville SC 29613

TURNER, JAMES EDWARD, b Norfolk, Va, Feb 12, 30; m 55; c 3. RADIATION PHYSICS, HEALTH PHYSICS. *Educ:* Emory Univ, BA, 51; Harvard Univ, MS, 53; Vanderbilt Univ, PhD(physics), 56. *Prof Exp:* Instr physics, Yale Univ, 56-58; physicist, US Atomic Energy Comn, 58-62; PHYSICIST, OAK RIDGE NAT LAB, 62-, CORP FEL, 88- *Concurrent Pos:* Prof physics, Univ Tenn, 81-; mem, Nat Coun Radiation Protection & Measurements, 77-83; ed, Health Physics, 74-79; assoc ed, Radiation Res, 80-83; mem bd dirs, Health Physics Soc, 80-83. *Mem:* Fel Health Physics Soc; fel Am Phys Soc; fel AAAS; Radiation Res Soc. *Res:* Interaction of radiation with matter; early physical & chemical events; atomic & molecular interactions; interaction of metal ions with nucleic acids & proteins; chemical dosimetry. *Mailing Add:* Oak Ridge Nat Lab Bldg 4500-S MS-6123 PO Box 2008 Oak Ridge TN 37831-6123

TURNER, JAMES ELDRIDGE, b Richmond, Va, Oct 1, 42; m 67; c 3. NEUROBIOLOGY, ELECTRON MICROSCOPY. *Educ:* Va Mil Inst, BA, 65; Univ Richmond, MS, 67; Univ Tenn, PhD(zool), 70. *Prof Exp:* Teaching asst zool, Univ Tenn, 67-69; NIH res trainee neurobiol & electron micros, Dept Anat, Sch Med, Case Western Reserve Univ, 71,res fel, 72-74; asst prof biol, Va Mil Inst, 71-72; from asst prof to assoc prof, 74-83, PROF ANAT, BOWMAN GRAY SCH MED, 83-; VIS DISTINGUISHED PROF BIOL, VA MIL INST, 90- *Concurrent Pos:* Vis prof, dept neurochem, Max Planck Inst, Munich, WGermany, 80-81; Basil Oconnor Starter res fel, March Dimes, 75-78; Res Career Develop award, NIH, 78-83; assoc vis res opthal, Int Cong Eye Res. *Honors & Awards:* Victory in Sight Award, Retinitis Pigmentosa Int, 88. *Mem:* AAAS; Soc Neurosci; Am Soc Zool; Am Asn Anatomists; Sigma Xi. *Res:* Nerve injury, repair and regeneration; neurotrophic phenomenia; neuronal transplantation. *Mailing Add:* Dept Anat Bowman Gray Sch Med Winston-Salem NC 27103

TURNER, JAMES HENRY, b Stuart, Va, June 13, 22. PARASITOLOGY. *Educ:* Univ Md, BS, 47, MS, 52, PhD(zool), 57. *Prof Exp:* Entomologist, Div Insects, Dept Zool, US Nat Mus, 48; from jr parasitologist to sr res parasitologist, Animal Dis Parasite Res Div, Agr Res Serv, USDA, 48-62; prin res parasitologist, Beltsville Parasitol Lab & McMaster Health Lab, Commonwealth Sci & Indust Res Orgn, Australia, 62-64; HEALTH SCIENTIST ADMINSTR, IMMUNOBIOL STUDY SECT, DIV RES GRANTS, NIH, 64- *Concurrent Pos:* Tutorial lectr, Univ Md, 58-61; Fulbright res fel, Australia, 62-63. *Honors & Awards:* Ransom Mem Award, 60. *Mem:* Am Soc Parasitol; Am Soc Trop Med & Hyg; Transplantation Soc; Am Phys Soc. *Res:* Pathogenesis and immunological aspects of parasitic infections. *Mailing Add:* 19 Potter St Brunswick ME 04011

TURNER, JAMES HOWARD, b Colorado Springs, Colo, July 18, 12. ANALYTICAL CHEMISTRY. *Educ:* Colo Col, AB, 33; Univ Iowa, MS, 36. *Prof Exp:* Asst chem, Colo Col, 31-33 & Calif Inst Technol, 33-34; anal chemist, SW Shattuck Chem Corp, Colo, 41-44; asst chem, Univ Iowa, 45; chemist, Colo Fuel & Iron Corp, 47-48; res chemist, Holly Sugar Corp, 48-54; anal chemist, Holloman Air Force Base, NMex, 55-59 & US Bur Mines, 59-61; anal chemist, US Geol Surv, 61-78; RETIRED. *Mem:* Emer mem Am Chem Soc; assoc Cooper Ornith Soc; Am Ornith Union. *Res:* Spectrophotometric methods of analysis; sugar analysis; sugar beet by-products; rarer metal analysis; infrared analysis. *Mailing Add:* 807 N Wahsatch Ave Colorado Springs CO 80903

TURNER, JAMES MARSHALL, b Washington, DC, Aug 20, 44; m 67, 81; c 5. PLASMA PHYSICS. *Educ:* Johns Hopkins Univ, BA, 66; Mass Inst Technol, PhD(physics), 71. *Prof Exp:* Lectr physics, Lesley Col, 70-71; res staff mem, Mass Inst Technol, 66-71; asst prof, Southern Univ, 71-73; assoc prof physics, Morehouse Col, 73-77; mem staff, 77-88, DIR WEAPONS PROG SAFETY, US DEPT ENERGY, 88- *Mem:* Am Phys Soc; Am Geophys Union; Sigma Xi. *Res:* Instabilities in weakly ionized plasmas; MHD structures and plasma properties in the solar wind; structure of hemoglobin S. *Mailing Add:* 12517 Hialeah Way Gaithersburg MD 20878

TURNER, JAN ROSS, b Okla, Sept 25, 37; m 62; c 2. MICROBIOLOGY. *Educ:* Ore State Univ, BS, 61, MS, 63, PhD(microbial physiol), 65. *Prof Exp:* Res asst microbial physiol, Ore State Univ, 61-65; AEC fel microbiol, Biol Div, Pac Northwest Labs, Battelle Mem Inst, 65-67, sr res scientist, 67-73; SR MICROBIOLOGIST, LILLY RES LABS, 73 - *Mem:* AAAS; Am Chem Soc; Am Soc Microbiol; Sigma Xi. *Res:* Microbial physiology; sterol biosynthesis; biochemical genetics; aromatic amino acid metabolism; medical mycology; clinical microbiology; antibiotic mechanisms; antibiotic biosynthesis. *Mailing Add:* Natural Prod Res Lilly Res Labs 307 E McCarty St Indianapolis IN 46285-1535

TURNER, JANICE BUTLER, b Lincolnton, Ga, Dec 1, 36; m 58; c 1. MOLECULAR SPECTROSCOPY. *Educ:* Ga State Col for Women, AB, 58; Emory Univ, MS, 59; Univ SC, PhD(chem), 70. *Prof Exp:* From instr to assoc prof, 59-77, PROF CHEM, AUGUSTA COL, 77-, CHMN, 76- *Mem:* Sigma Xi; Coblentz Soc; Am Chem Soc. *Res:* Preparation and structure determination of organogermanes-microwave studies. *Mailing Add:* Dept Chem & Physics Augusta Col 2500 Walton Way Augusta GA 30910

TURNER, JOHN CHARLES, b Houston, Tex, Sept 28, 49; m 71; c 2. STATISTICS. *Educ:* Rice Univ, BA, 72; Princeton Univ, MA, 74, PhD(statist), 76. *Prof Exp:* Asst prof math, Univ SFla, 76-78; ASSOC PROF MATH, US NAVAL ACAD, 78- *Mem:* Am Statist Asn. *Res:* Applied statistics. *Mailing Add:* 418 Blue Bonnett Tyler TX 75701

TURNER, JOHN DEAN, b Pasadena, Calif, Oct 2, 30. MEDICINE. *Educ:* Univ Calif, Berkeley, BA, 52; McGill Univ, MD, CM, 56. *Prof Exp:* Intern med, Mass Gen Hosp, 56-57, asst resident, 57-58; attend physician, Nat Heart Inst, 61-65; asst prof med, Baylor Col Med, 65-71; assoc prof med, Sch Med, Univ Calif, San Diego, 71-; MEM STAFF, COTTON, INC, NC. *Concurrent Pos:* Paul Dudley White fel cardiol, Mass Gen Hosp, 59-60, teaching fels, 59-61; res fel med, Peter Bent Brigham Hosp, Harvard Univ, 58-59; staff assoc, President's Comn Heart Dis, Cancer & Stroke, 64; fel coun epidemiol, Am Heart Asn, 65-; mem staff, Vet Admin Hosp, La Jolla, 77- *Mem:* Am Heart Asn; fel Am Col Physicians; Am Oil Chem Soc; Am Pub Health Asn. *Res:* Cardiovascular hemodynamics and epidemiology; angiocardiography in the diagnosis of valvular and congenital heart disease; external scintillation counting in detection of intracardiac shunts; lipid composition in human erythrocytes and plasma; lipid and lipoprotein metabolism in human plasma. *Mailing Add:* 209 Westborne St La Jolla CA 92037

TURNER, JOHN K, b Fairfield, Ill, May 7, 23; m 46; c 5. ANIMAL PHYSIOLOGY. *Educ:* Millikin Univ, AB, 47; Univ Ill, MA, 49; Univ Wis, PhD(physiol), 60. *Prof Exp:* Physiologist, Army Med Res Lab, Ft Knox, Ky, 59-61 & Vet Admin Hosp, Long Beach, Calif, 61-64; assoc prof biol, sci, Western Ill Univ, 65-; RETIRED. *Concurrent Pos:* Lectr, Sch Med, Univ Calif, Los Angeles, 62-65. *Res:* Reflex regulation of respiration and circulation. *Mailing Add:* 6409 Alisha Circle Las Vegas NV 89130

TURNER, JOHN LINDSEY, b Birmingham, Ala, Sept 9, 49; m 70; c 2. ENGINEERING MECHANICS. *Educ:* Auburn Univ, BSME, 71, MS, 72; Univ Ill, PhD(mech), 75. *Prof Exp:* Res scientist, Firestone Res Labs, 75-77; asst prof mech eng, 77-81, ASSOC PROF AGR ENG, AUBURN UNIV, 81- *Mem:* Soc Exp Stress Anal; Am Soc Agr Eng; Sigma Xi. *Res:* Experimental and numerical methods of stress analysis and structural mechanics; combined applications of experimental and computer based techniques for improved methods of analysis and design. *Mailing Add:* Dept Mech Eng Univ SC Columbia SC 29028

TURNER, JUDITH ANN, b Houston, Tex, Nov 19, 52; m 82; c 2. CLINICAL PSYCHOLOGY, CHRONIC PAIN. *Educ:* Vanderbilt Univ, BA, 70; Univ Calif Los Angeles, MA, 71 & PhD(clin psychol), 79. *Prof Exp:* Asst prof, 80-85, ASSOC PROF, UNIV WASH SCH MED, 85- *Mem:* Int Asn Study Pain; Am Pain Soc; Soc Behav Med; Am Psychol Asn; Asn Advan Behav Ther. *Res:* Psychological aspects of chronic pain problems. *Mailing Add:* Dept Psychiat & Behav Sci Univ Wash RP 10 Seattle WA 98195

TURNER, KENNETH CLYDE, b Mt Vernon, Wash, Apr 6, 34; m 55; c 3. RADIO ASTRONOMY, RADIO INTERFEROMETRY. *Educ:* Portland Univ, BS, 57; Princeton Univ, PhD(physics), 62. *Prof Exp:* Instr physics, Princeton Univ, 62; Carnegie fel, 62-64; res staff assoc, 64-66, Carnegie Inst Washington Dept Terrestrial Magnetism, res staff mem, 66-78; sr res assoc, 78-87, PRES INNOVATIVE SYSTS, NAT ASTRON & IONOSPHERE CTR, ARECIBO, PR, 86- *Concurrent Pos:* Dir, Arg Inst Radio Astron, 70-73; vis prof, Nat Univ La Plata, 72-73; vis prof physics & computer sci, Inter Am Univ, San German, PR, 87- *Mem:* AAAS; Am Phys Soc; Am Astron Soc; Int Union Radio Sci; Int Astron Union; Am Asn Physics Teachers. *Res:* Radio instrumentation; hydrogen line and galactic continuum radio astronomy; observational cosmology; experimental foundations of relativity; astrophysicial jets. *Mailing Add:* NSF 1800 G St NW Washington DC 20550

TURNER, LEAF, b Brooklyn, NY, Mar 23, 43; m 66; c 4. THEORETICAL PHYSICS. *Educ:* Cornell Univ, AB, 63; Univ Wis, Madison, MS, 64, PhD(theoret physics), 69. *Prof Exp:* Weizmann Inst Sci fel, Rehovot, Israel, 69; fel physics, Univ Toronto, 69-71; asst scientist, Space Sci & Eng Ctr, Univ Wis, Madison, 71-72, proj assoc, 72-74; STAFF MEM, LOS ALAMOS NAT LAB, UNIV CALIF, 74- *Concurrent Pos:* Instr physics, Scarborough Col, Univ Toronto, 70-71; NSF fel, Inst Theoret Physics, Brandeis Univ, 70; vis prof, dept physics, Univ Wis, Madison, 84. *Mem:* Am Phys Soc; Sigma Xi. *Res:* Current algebra; phenomenological lagrangians; interactions of mesons and baryons; quantum field theory; symmetries in high energy physics; high energy phenomenology; scattering theory; plasma kinetic theory; magnetohydrodynamics; optics; plasma astrophysics; nonneutral plasma physics. *Mailing Add:* Los Alamos Nat Lab PO Box 1663 Mail Stop F647 T-DOT Los Alamos NM 87545

TURNER, LINCOLN HULLEY, b Chicago, Ill, June 30, 28. MATHEMATICS. *Educ:* Univ Chicago, MS, 48; Purdue Univ, PhD(math), 57. *Prof Exp:* Mathematician, Space Tech Labs, Inc, 58-60; prof math, Univ Minn, 60-63; PROF MATH, UNIV TENN, KNOXVILLE, 63- *Mem:* Am Math Soc; Math Asn Am; Soc Indust & Appl Math. *Res:* Real variables; calculus of variations; applied mathematics. *Mailing Add:* 5709 Lyons View Pk 2308 Knoxville TN 37919

TURNER, MALCOLM ELIJAH, JR, b Atlanta, Ga, May 27, 29; m 48, 68; c 8. MATHEMATICAL BIOLOGY, STATISTICS. *Educ:* Duke Univ, BA, 52; NC State Univ, MES, 55, PhD(statist), 59. *Prof Exp:* Asst biostatist, Univ NC, 53-54; sr res assoc biomet, Med Sch, Univ Cincinnati, 55, asst prof, 56-58; assoc prof & Williams res fel, Va Commonwealth Univ, 58-63, chmn div biomet, Dept Biophys & Biomet, 59-63; prof statist & biomet & chmn dept, Emory Univ, 63-69, prof math, 66-69; prof biostatist & biomath & assoc prof physiol & biophys, 70, chmn dept, 75-82, prof biomath, 72-82, PROF BIOSTATIST & BIOMATH, UNIV ALA, BIRMINGHAM, 82 - *Concurrent Pos:* Asst statistician, NC State Univ, 57-58; managing ed, Biometrics, 62-69; vis prof & chmn, Dept Biomet, Med Ctr, Univ Kans, 68-69; consult, Southern Res Inst, 74 - *Honors & Awards:* Smith Kline & French lectr, Med Ctr, Univ Kans, 65. *Mem:* Fel AAAS; Soc Indust & Appl Math; hon fel Am Statist Asn; Biometric Soc; Sigma Xi. *Res:* Application of mathematics and statistics to biological research. *Mailing Add:* Dept Biostatist & Biomath Univ Ala-Birmingham Birmingham AL 35294

TURNER, MANSON DON, b Pleasanton, Tex, Nov 15, 28; m 53; c 3. PHYSIOLOGY. *Educ:* Baylor Univ, BS, 50, MS, 51; Univ Tenn, PhD(physiol), 55. *Prof Exp:* Lab instr, Baylor Univ, 50-51; res asst, Univ Tenn, 51-54, instr clin physiol, 54-55; res asst prof surg, Sch Med, Univ Miss, 55-57, asst prof biochem, 58-61, assoc prof res surg & asst prof physiol & biophys, 61-65; supvry res physiologist, US Air Force Sch Aerospace Med, 65; assoc prof surg & physiol, 65-69, ASSOC PROF PHYSIOL & BIOPHYS & RES PROF SURG, SCH MED, UNIV MISS, 69- *Concurrent Pos:* Attend in physiol, Vet Admin Hosp, Jackson, 58-; consult, Oak Ridge Inst Nuclear Studies, 62-65. *Mem:* Transplantation Soc; Cryobiol Soc; Am Heart Asn; Am Physiol Soc. *Res:* Organ preservation; cardiovascular physiology. *Mailing Add:* Med Admissions Off Univ Miss Sch Med Jackson MS 39216

TURNER, MATTHEW X, b Rockaway, NY, July 27, 28; m 58; c 5. CYTOLOGY, IMMUNOLOGY. *Educ:* Fordham Univ, BS, 56, MS, 62, PhD(cytol), 68. *Prof Exp:* Jr bacteriologist, NY State Conserv Dept, 56-58; jr isotope officer, State Univ NY Downstate Med Ctr, 58-60, radiation officer, 60-62, res asst anat & cytol, 62-66, NIH fel cytol & immunol, 67-68; microbiologist, US Naval Appl Sci Lab, NY, 66-67; asst prof, 68-72, ASSOC PROF BIOL, JERSEY CITY STATE COL, 72-, CHMN BIOL DEPT, 75- *Mem:* Reticuloendothelial Soc; Am Soc Microbiol. *Res:* Effect of antigen upon the cellular and humoral responses in normal and immunized animals; determination of x-ray and drugs upon antigenic stimulation in normal and immunized mice. *Mailing Add:* Dept Biol Jersey City State Col 2039 Kennedy Blvd Jersey City NJ 07305

TURNER, MICHAEL D, b Weston, Eng, Oct 26, 27; m 57; c 4. MEDICINE, BIOCHEMISTRY. *Educ:* Bristol Univ, MB, ChB, 50, MD, 59; Univ Rochester, PhD(biochem), 64. *Prof Exp:* Tutor med, Postgrad Med Sch, Univ London, 54-57; lectr, Royal Free Hosp Med Sch, 60-63; consult med, 63-64, from assoc prof to prof, 64-72, Segal-Watson prof gastroenterol, Sch Med & Dent, Univ Rochester, 72-; AT DEPT MED, BROWN UNIV. *Concurrent Pos:* Lederle med fac fel, 63-66. *Mem:* Am Fedn Clin Res; Am Gasteroenterol Asn; Brit Soc Gastroenterol. *Res:* Liver diseases and portal hypertension; chemistry and immunology of gastric macromolecules. *Mailing Add:* Dept Med VA Hosp Davis Park Providence RI 02908

TURNER, MICHAEL STANLEY, b Los Angeles, Calif, July 29, 49; m 88; c 1. COSMOLOGY, PARTICLE PHYSICS. *Educ:* Calif Inst Technol, BS, 71; Stanford Univ, MS, 73, PhD(physics), 78. *Prof Exp:* Instr physics, Stanford Univ, 78; Enrico Fermi fel, Enrico Fermi Inst, 78-80, from asst prof to assoc prof, 80-85, PROF, UNIV CHICAGO, 85-; DEPT HEAD ASTROPHYS, FERMILAB, 89- *Concurrent Pos:* Sloan fel, 83. *Honors & Awards:* Quantrell Prize, 83; Helen B Warner Prize, 84. *Mem:* Sigma Xi; fel Am Phys Soc; Am Astron Soc; Int Astron Union. *Res:* Cosmology; study of earliest history of the universe; interplay of cosmology and particle physics. *Mailing Add:* Fermilab MS 209 Box 500 Batavia IL 60510

TURNER, MORTIMER DARLING, b Greeley, Colo, Oct 24, 20; m 45, 65; c 4. ECONOMIC GEOLOGY, ENGINEERING GEOLOGY. *Educ:* Univ Calif, Berkeley, BS, 43, MS, 54; Univ Kans, PhD, 72. *Prof Exp:* Mech & elec engr, Aberdeen Proving Grounds, US Dept Army, 46; asst geol sci, Univ Calif, Berkeley, 48; from jr mining geologist to asst mining geologist, Calif State Div Mines, 48-54; state geologist, Econ Develop Admin, PR, 54-58; phys sci administr & asst to dir, Antarctic Res Prog, NSF, 59-61; res assoc geol, Univ Kans, 62-65; prog dir, Antarctic Earth Sci, Off Antarctic Progs, NSF, 65-70, prog mgr, Polar Earth Sci, Div Polar Progs, 70-85; assoc prof lectr geol, George Washington Univ, 72-85; RETIRED. *Concurrent Pos:* Consult, 55-59; mem, Orgn Comt, 1st Conf Clays & Clay Technol, 52 & Caribbean Geol Conf, 59; mem, US Planning Comt, 2nd Int Conf Permafrost, Nat Acad Sci; mem, 19th, 20th (vpres), 21st, 22nd, 24th (US Govt deleg), and 25th Int Geol Congs; adj prof quaternary studies, Ctr Study Early Man, Inst Quaternary Studies, Univ Maine, Orono, 85- *Mem:* Geol Soc Am; Am Geog Soc; Antarctican Soc; Geol Asn PR; Am Polar Soc. *Res:* Tectonics, economic geology and engineering geology of California, Puerto Rico, Caribbean area and Antarctica; geology of early man in North America and China; engineering geology of coastal zones. *Mailing Add:* Instaar Campus Box 450 Univ Colo Boulder CO 80309-0450

TURNER, NOEL HINTON, b Redlands, Calif, Dec 24, 40; m 66; c 2. PHYSICAL CHEMISTRY. *Educ:* Univ Calif, Berkeley, BS, 62; Univ Rochester, PhD(phys chem), 68. *Prof Exp:* RES CHEMIST, US NAVAL RES LAB, 68- *Mem:* AAAS; Am Chem Soc; Sigma Xi; Am Vacuum Soc. *Res:* Surface chemistry; analytical chemistry; electron spectroscopy for chemical analysis and auger electron spectroscopy; gas-solid and gas-liquid adsorption. *Mailing Add:* Chem Div Code 6170 Naval Res Lab Washington DC 20375

TURNER, PAUL JESSE, spectroscopy, computer software, for more information see previous edition

TURNER, R JAMES, Chatham, Ont, Aug 26, 45. MEMBRANE TRANSPORT, MEMBRANE BIOLOGY. *Educ:* Univ Toronto, PhD(theoret physics), 71. *Prof Exp:* Vis assoc epithelial physiol, NIH, 79-83; assoc prof med, Univ Toronto, 83-85; SECT CHIEF, EXOCRINE PHYSIOL, NIH, 85- *Mem:* Am Physiol Soc; Am Soc Nephrology; Biophys Soc. *Res:* Study of characteristics and regulation of transport related and intercellular events associated with fluid secretion in exocrine glands, especially salivary glands. *Mailing Add:* NIH Bldg Ten Rm 1A06 Bethesda MD 20892

TURNER, RALPH B, b Lynchburg, Va, Mar 1, 31; m 53; c 2. BIOCHEMISTRY. *Educ:* Va Polytech Inst, BS, 52; Univ Tex, PhD(chem), 63. *Prof Exp:* Res biochemist, Entom Res Div, USDA, 63-68; RES BIOCHEMIST, DEPT BOT & ENTOM, NMEX STATE UNIV, 68- *Mem:* Fel AAAS; fel Am Inst Chem; Am Chem Soc; Sigma Xi; NY Acad Sci; Am Soc Biol Chem. *Res:* Insect nutrition and development; insect embryogenesis; molecular basis of metamorphosis. *Mailing Add:* 300 Capri ARC Las Cruces NM 88005

TURNER, RALPH WALDO, b Blakely, Ga, Nov 9, 38. PHYSICAL CHEMISTRY, MATHEMATICS. *Educ:* J C Smith Univ, BS, 59; Univ Pittsburgh, PhD(phys chem), 65. *Prof Exp:* Res asst chem, Univ Pittsburgh, 60-64; res engr, Gen Tel & Electronics, 65-66, advan res engr, 66-67; prof sci educ, 67-68, prof chem & physics, 68-69, dir div basic studies, 69-77, PROF CHEM, FLA A&M UNIV, 69-, CHMN CHEM DEPT, 86- *Concurrent Pos:* US Off Educ grant, Fla A&M Univ, 62-; dir 13 col prog, Fla A&M Univ, 68-69; consult, Inst Servs Educ, 68-72, Am Chem Soc, 80-; proposal reviewer, NIH & NSF, 87-; dir, Marc Prog, 86-, Fla Comprehensive State Ctr for Minorities, 89- *Mem:* AAAS; Am Chem Soc; Am Crystallog Asn; Sigma Xi. *Res:* Structure of metal ion aromatic complexes using x-ray crystallography; 113Cd NMR and structural studies of ceruloplasmin. *Mailing Add:* Dept Chem Fla A&M Univ Tallahassee FL 32307

TURNER, RAYMOND MARRINER, b Salt Lake City, Utah, Feb 25, 27; m 49; c 3. ECOLOGY. *Educ:* Univ Utah, BS, 48; State Col Wash, PhD(plant ecol), 54. *Prof Exp:* Instr range mgt, Univ Ariz, 54-56, asst prof, 56, instr bot, 56-57, asst prof, 57-62; RES BOTANIST, US GEOL SURV, 62- *Concurrent Pos:* NSF res grant, 57-60. *Mem:* AAAS; Ecol Soc Am; Bot Soc Mex. *Res:* Ecology of arid and semi-arid regions. *Mailing Add:* US Geol Surv 301 W Congress Tucson AZ 85701-1393

TURNER, REX HOWELL, b Birmingham, Ala, Aug 22, 41; m 60; c 2. CELLULOSE CHEMISTRY. *Educ:* Univ SAla, BS, 68; Univ Ga, PhD(chem), 73. *Prof Exp:* Res asst org chem, Univ Ga, 68-73; develop chemist, Millmaster Onyx Corp, 73-74; SR RES ASSOC INT PAPER CO, 74- *Mem:* Am Chem Soc; Am Asn Textile Chemists & Colorists. *Res:* Isolation and enrichinhomogeneities; macrocyclic synthesis via thermochemical and photochemical decomposition of ketone peroxides; ozonolysis mechanism non-chlorine bleaching of wood pulp; acetylation of wood pulp. *Mailing Add:* 6400 Poplar Ave Memphis TN 38119-4835

TURNER, ROBERT, b Boston, Mass, Jan 23, 25; m 56; c 3. ELECTRICAL ENGINEERING. *Educ:* Mass Inst Technol, SB, 45; Harvard Univ, SM, 48. *Prof Exp:* Proj engr, Sperry Gyroscope Co, NY, 48-53; analyst, Opers Eval Group, Mass Inst Technol, Washington, DC, 53-55; sr engr, Johns Hopkins Univ, 55-80, prin staff engr, Appl Physics Lab, 80-89; RETIRED. *Mem:* Am Phys Soc. *Res:* Servomechanism design; systems analysis; plasma physics; gas lasers; light scattering. *Mailing Add:* Appl Physics Lab Johns Hopkins Rd Laurel MD 20707

TURNER, ROBERT ALEXANDER, b Englewood, NJ, Oct 12, 37; m 60; c 3. RHEUMATOLOGY. *Educ:* Univ NC, BA, 59; Med Col Ala, Birmingham, MD, 66. *Prof Exp:* From asst prof to prof med, Bowman Gray Sch Med, Winston-Salem, NC, 71-89, chief, Rheumatology Sect, 79-89; DIR, ARTHRITIS CTR, HUMANA HOSP, 89- *Mem:* Fel Am Col Phys; Am Asn Immunologists; Am Pain Soc; Am Fedn Clin Res; Am Rheumatism Asn; Am Soc Clin Pharmacol & Therapeut. *Mailing Add:* Arthritis Ctr Humana Hosp Palm Beaches 201-203 Gould Prof Bldg 2151 45th St West Palm Beach FL 33407

TURNER, ROBERT ATWOOD, IMMUNOHISTOCHEMISTRY. *Educ:* Sam Houston Univ, MS, 52. *Prof Exp:* Dir, Surg Path Electron Microscope Lab, Scott & White Hosp, Temple, 70-; RETIRED. *Mailing Add:* 3914 Berie St Scott & White Temple TX 76501

TURNER, ROBERT DAVISON, b Goose Creek, Tex, Oct 16, 29; m 50; c 4. COMMAND & CONTROL SYSTEMS, SENSOR SYSTEMS. *Educ:* St Lawrence Univ, SB, 49; Harvard Univ, AM, 51, PhD(appl math), 54. *Prof Exp:* Commun theory analyst & consult anal & synthesis, Advan Electronics Ctr, Gen Elec Co, 54-63; staff mem, Inst Defense Anal, 63-68, asst to pres, 68-69, staff mem, Sci & Technol Div, 69-74; staff specialist, Net Tech Assessment, Off Dir Defense Res & Eng, Off Secy Defense, US Dept Defense, 74-76, spec asst tech plans & res, Off Asst Secy Defense, Commun, Command, Control & Intel, 77-81, dir strategic & theater nuclear forces command, control & commun, Off Secy Defense, 82-84, STAFF MEM, SYS EVAL DIV, INST DEFENSE ANALYSIS, 86- *Concurrent Pos:* Vis asst prof, Cornell Univ, 55; mem sci adv bd, USAF, 67-71, consult, 71-72; consult, Defense Sci Bd, Off Secy Defense, 71-78; mem, Naval Res Adv Comt, Lab Adv Bd Undersea Warfare, 71-74; mem, Chief Naval Opers Command Control Commun Adv Comt, 73-76. *Res:* Synthesis and evaluation of communication networks and sensor information handling systems; military technology assessment; analysis and planning of military command systems. *Mailing Add:* 9200 Quintana Dr Bethesda MD 20817

TURNER, ROBERT E L, b Montclair, NJ, Nov 15, 36; m 60; c 3. MATHEMATICS. *Educ:* Cornell Univ, BEngPhys, 59; NY Univ, PhD(math), 63. *Prof Exp:* From asst prof to assoc prof, 63-71, PROF MATH, UNIV WIS-MADISON, 71- *Concurrent Pos:* Woodrow Wilson hon fel. *Mem:* Am Math Soc. *Res:* Functional analysis; differential equations; hydrodynamics. *Mailing Add:* Dept Math Van Vleck Hall Univ Wis 480 Lincoln Dr Madison WI 53706

TURNER, ROBERT HAROLD, b San Francisco, Calif, Sept 20, 41; m 69; c 3. SOLAR ENERGY, FORENSIC ENGINEERING. *Educ:* Univ Calif, Berkeley, BS, 64, MS, 65; Univ Calif, Los Angeles, PhD(heat transfer), 71. *Prof Exp:* Engr, Rohr Corp, 64, AiResearch, 65-67 & Nus Corp, 72-74; consult, Spectrolab Co, 67-71; head, Thermal Sect, Southern Res Inst, 71-72; engr & supvr, Jet Propulsion Lab, 75-83; ASSOC PROF THERMAL SCI, UNIV NEV, RENO, 83- *Concurrent Pos:* Adj prof, Calif State Univ, 76-83; consult, 79- *Mem:* Am Soc Mech Engrs; Am Soc Heating Refrig & Air Conditioning Engrs; Sigma Xi. *Res:* Applied thermal sciences; ice formation conditions; mathematical modeling; solar energy; thermal systems; forensic engineering. *Mailing Add:* Dept Mech Eng Mail Stop 312 Univ Nev Reno NV 89557

TURNER, ROBERT JAMES, b Loda, Ill, Nov 15, 21; m 47; c 2. ORGANIC CHEMISTRY. *Educ:* Univ Ill, BS, 47; Univ Wis, MS, 49, PhD(org chem), 50. *Prof Exp:* Asst, Alumni Res Found, Univ Wis, 47-49; res chemist, Mallinckrodt Chem Works, 50-55; group leader, 55; group leader, 56-58, res supvr, 58-64, asst dir res, 64-68, dir res, 68-78, TECH DIR, RES, MORTON CHEM CO, MORTON-NORWICH PROD, INC, 78- *Mem:* Am Chem Soc. *Res:* Synthetic organic chemistry; hydrogenation; hydroformylation; polymers; surface coatings; adhesives; emulsions and dispersions. *Mailing Add:* Rte Four Box 2410 Lake Geneva WI 53147

TURNER, ROBERT LAWRENCE, b Chicago, Ill, Nov 4, 45; m 66; c 2. ORGANIC POLYMER CHEMISTRY. *Educ:* Albion Col, BA, 67; Mich State Univ, PhD(org chem), 72. *Prof Exp:* Teaching asst org chem, Mich State Univ, 67-69, res asst, 71-72; res chemist, 72-74, staff chemist, 74-76, RES SUPVR, E I DU PONT DE NEMOURS & CO INC, 76- *Res:* Development of new polymer-catalyst systems for use in low energy and nonpolluting protective organic coatings for industrial use. *Mailing Add:* DuPont Co PO Box 3886 Philadelphia PA 19146

TURNER, ROBERT STUART, b Red Oak, Iowa, June 25, 12; m 42; c 2. NEUROANATOMY. *Educ:* Dartmouth Col, AB, 33; Yale Univ, PhD(anat), 38. *Prof Exp:* Instr biol, Dartmouth Col, 33-35; asst resident in neuroanat, Med Sch, Yale Univ, 37-38; from instr to prof anat, 38-77, EMER PROF ANAT, STANFORD UNIV, 77 - *Concurrent Pos:* NIH spec fel, 48-49; consult, Agnews State Hosp & Vet Admin Hosp, 59-69. *Mem:* AAAS; Am Asn Anat; corresp mem Mex Soc Anat; Pan Am Soc Anat. *Res:* Functional anatomy of nervous system; comparative neurology and neurophysiology. *Mailing Add:* 345 Iris Way Palo Alto CA 94303

TURNER, RUTH DIXON, b Melrose, Mass, Dec 7, 14. MALACOLOGY. *Educ:* Bridgewater Teachers Col, BS, 36; Cornell Univ, MA, 43; Radcliffe Col, PhD, 54. *Hon Degrees:* DSC, New Eng Col, 79. *Prof Exp:* Teacher high sch, Vt, 36-37 & jr high sch, Mass, 37-40; asst ed, Boston Soc Natural Hist, 40-42, asst cur birds, 41-42; instr ornith, Vassar Col, 43-44; biologist, William F Clapp Labs, Mass, 44-45; res mollusks, Mus Comp Zool, 45-55, res assoc malacol & Agassiz fel oceanog & zool, 55-75, PROF BIOL, HARVARD UNIV & CUR MALACOL, MUS COMP ZOOL, 75- *Concurrent Pos:* Res assoc, Inst Marine Biol, PR, 56-; consult, William F Clapp Labs, 57- *Mem:* AAAS; Soc Syst Zool; Am Malacol Union; Sigma Xi. *Res:* Marine boring and fouling mollusks; taxonomy and biology of mollusks, particularly Western Atlantic marine and North American freshwater. *Mailing Add:* Mus Comp Zool Harvard Univ Cambridge MA 02138

TURNER, S RICHARD, b Nashville, Tenn, Oct 17, 42; m 69; c 2. SYNTHETIC POLYMER CHEMISTRY. *Educ:* Tenn Technol Univ, BS, 64, MS, 66; Univ Fla, PhD(org polymer chem), 71. *Prof Exp:* Fel polymer chem, Tech Univ, Parmstadt, WGer, 71-72; scientist, Webster Res, Xerox Corp, 72-79 & Corp Res, Exxon Corp, 80-82; RES ASSOC, CORP RES, EASTMAN KODAK CO, 82- *Concurrent Pos:* Vis adj prof polymer chem, Chem Dept, Univ Rochester, 82-; prog chmn, Div Polymer Mat, Am Chem Soc, 85-88, vchmn, 90, chmn elect, 91. *Mem:* Am Chem Soc. *Res:* Synthetic polymer chemistry directed toward new polymer forming reactors; new routes to controlling polymer architecture; novel photopolymers and ionomers; interacting polymers. *Mailing Add:* Corp Res Labs Eastman Kodak Co Rochester NY 14650-2110

TURNER, TERRY EARLE, physics, for more information see previous edition

TURNER, TERRY TOMO, b Moultrie, Ga, 45; m 65; c 2. TESTIS FUNCTION, MALE INFERTILITY. *Educ:* Univ Ga, BSA, 67, MS, 72, PhD(reprod biol), 74. *Prof Exp:* Rockefeller postdoctoral fel, Univ Tex, Med Sch, San Antonio, 75; asst prof, 76-83, ASSOC PROF UROL, UNIV VA, SCH MED, 83- *Concurrent Pos:* Prin investr NIH grants, 82-, & Studies Sect, 83 & 88. *Mem:* Soc Study Reprod; Am Androl Soc; Am Soc for Cell Biol; Am Physiol Soc; NY Acad Sci; Am Fertil Soc. *Res:* Investigations of the epithelial functions of the male reproductive tract, how those functions are regulated, how they are altered in pathophysiological conditions and the consequences of those functions on the development of fertile spermatozoa. *Mailing Add:* Dept Urol Univ Va Sch Med Charlottesville VA 22908

TURNER, THOMAS BOURNE, b Prince Frederick, Md, Jan 28, 02; m 27; c 2. MICROBIOLOGY. *Educ:* St John's Col, Md, BS, 21; Univ Md, MD, 25. *Hon Degrees:* ScD, Univ Md, 66. *Prof Exp:* Intern, Hosp for Women of Md, 25-26; resident, Mercy Hosp, 26-27; Loeb fel, Sch Med, Johns Hopkins Univ, 27-28, instr med, 28-31, assoc, 31-32; mem staff, Int Health Div, Rockefeller Found, 32-39, clin dir, Jamaica Yaws Comn, 32-34, labs, 34-36; lectr med & pub health admin, Sch Hyg & Pub Health, 36-39, prof microbiol, Sch Med, 39-68, dean, 57-68, EMER DEAN, SCH MED, JOHNS HOPKINS UNIV, 68- *Concurrent Pos:* Consult, Surgeon Gen, US Army; vchmn comt virus res & epidemiol, Nat Found, 49-67; coord, Regional Med Prog for Md, 67-68; physician, out-patient dept, Johns Hopkins Hosp; mem bd visitors, St John's Col, Md; pres, Alcoholic Beverage Med Res Found, 82- *Mem:* Am Soc Clin Invest; Asn Am Physicians; Asn Am Med Cols (past pres); Am Med Soc Alcoholism. *Res:* Spirochetal diseases; poliomyelitis; tetanus; internal medicine. *Mailing Add:* Animal Sci 2029 Fyffect No 110d Ohio St Univ Columbus OH 43210

TURNER, THOMAS JENKINS, b Albany, Ga, Sept 11, 26; m 48; c 4. SOLID STATE PHYSICS. *Educ:* Univ NC, BS, 47; Clemson Col, MS, 49; Univ Va, PhD(physics), 51. *Prof Exp:* Instr physics, Clemson Col, 47-49; asst prof, Univ NH, 52; from asst prof to prof, Wake Forest Univ, 53-78, chmn dept, 56-74, vpres & dean, 78-81, PROVOST, STETSON UNIV, 81- *Mem:* Fel Am Phys Soc; Am Asn Physics Teachers. *Res:* Defects in crystalline materials; color centers; internal friction. *Mailing Add:* 1629 N Ridge Dr Arkadelphia AR 71923

TURNER, VERAS D, mathematics, for more information see previous edition

TURNER, VERNON LEE, JR, chemistry, for more information see previous edition

TURNER, WALTER W(EEKS), b Augusta, Maine, Aug 3, 22; m 49; c 4. ELECTRICAL ENGINEERING. *Educ:* Mass Inst Technol, BS & MS, 47. *Prof Exp:* From instr to assoc prof, 47-65, PROF ELEC ENG, UNIV MAINE, ORONO, 65- *Mem:* Am Soc Eng Educ; sr mem Inst Elec & Electronics Engrs; Nat Soc Prof Engrs. *Res:* Instrumentation, communications and control. *Mailing Add:* Dept Elec Eng Univ Maine 109 Barrons Orono ME 04469

TURNER, WAYNE CONNELLY, b West Point, Va, May 7, 42; m 69; c 2. ENGINEERING ECONOMY. *Educ:* Va Polytech Inst & State Univ, BS, 64, MS, 69, PhD(indust eng), 71. *Prof Exp:* Mgr mfg engr, Modine Mfg Co, 64-67; grad asst indust eng, Va Polytech Inst & State Univ, 67-69, instr, 69-71, asst prof, 71-74; from assoc prof to prof, Okla State Univ, 74-85; PROF & HEAD & INDUST MGT ENG, MONT STATE UNIV, 85- *Concurrent Pos:* Consult, wide variety of govt & pvt firms, 70-; prin investr, US Dept Energy, 74-85; ed, Prentice Hall Ser on Energy, Prentice Hall Int, 80-85. *Mem:* Inst Indust Engrs; Am Soc Eng Educr; Asn Energy Engrs; Am Inst Plant Engrs. *Res:* Energy, water and hazardous material management in industrial and other institutions. *Mailing Add:* Dept Indust Eng Okla State Univ Stillwater OK 74078

TURNER, WILLIAM DANNY, b Moberly, Mo, Feb 14, 39; m 75; c 4. ENERGY CONSERVATION, MATERIALS SCIENCE. *Educ:* Univ Tex, Austin, BS, 61, MS, 62; Univ Okla, Norman, PhD(mech eng), 69. *Prof Exp:* Reliability engr, Ling-Temco-Vought Astronaut Div, 62-64; design engr, Gen Dynamics, Ft Worth, 64-66; from asst prof to assoc prof mat sci & heat transfer, Tex A&I Univ, 69-76; assoc prof, Univ Ark, Fayetteville, 76-81; from assoc prof to prof thermodyn & energy systs, 81-89, interim dept head mech eng, 85-87, ASSOC DEAN, TEX A&M UNIV, COLLEGE STA, 89- *Concurrent Pos:* Consult, Argonne Nat Labs, 81-83 & Ministry Petrol Indust, People's Repub China, 86-87; dir, Dept Mech Eng, Energy Systs Lab, Tex A&M Univ, 84- *Mem:* Fel Am Soc Mech Engrs; Am Soc Eng Educ (vpres, 88-90); Asn Energy Engrs; Am Solar Energy Soc. *Res:* Energy conservation; cogeneration; residential radiant barrier testing; metallurgy. *Mailing Add:* 9102 Waterford Dr College Station TX 77845

TURNER, WILLIAM JOSEPH, b Canandaigua, NY, Mar 7, 27; m 51; c 4. SOLID STATE PHYSICS. *Educ:* Villanova Univ, BS, 49; Cath Univ, PhD(physics), 55. *Prof Exp:* Physicist, Naval Res Lab, Washington, DC, 51-52 & Nat Bur Standards, 52-56; assoc physicist, Phys Res Dept, IBM Corp, 56-57, staff physicist, Res Lab, 57-58, proj physicist, 58-59, develop physicist & res staff mem semiconductor physics, 59-61, res staff mem optical properties semiconductors, 61-64, mgr res staff opers, Thomas J Watson Res Ctr, 64-86, dir tech journ & prof rels, sci & tech staff, IBM Corp, 86-89; CONSULT, 89- *Mem:* Fel Am Phys Soc. *Res:* Use of optical absorption, reflection and luminescence measurements to study the intrinsic, lattice and extrinsic properties of semiconductors. *Mailing Add:* 507 Croton Lake Rd RD 4 Mt Kisco NY 10549

TURNER, WILLIAM JUNIOR, b Bell, Calif, June 27, 40; m 62; c 2. ZOOLOGY, BOTANY-PHYTOPATHOLOGY. *Educ:* Univ Calif, Berkeley, AB, 63, MS, 66, PhD(entom), 71. *Prof Exp:* Lab technician entom, Univ Calif, Berkeley, 63-64, from asst to assoc entomologist, 64-67, NIH trainee, 67-70; asst prof & asst entomologist, 70-76, ASSOC PROF ENTOM & ZOOL & ASSOC ENTOMOLOGIST, WASH STATE UNIV, 76- *Concurrent Pos:* Dir, James Entomol Collection, Wash State Univ, 70-85. *Mem:* Entom Soc Am; Sigma Xi; Soc Syst Zool. *Res:* Insect biosystematics; syst zoogeography, phylogeny and biology of Diptera; swarming behavior in insects; medical entomology, especially biting flies as vectors of pathogens. *Mailing Add:* Dept Entom Wash State Univ Pullman WA 99164-6382

TURNER, WILLIAM RICHARD, b Drexel Hill, Pa, June 26, 36; m 60; c 3. ANALYTICAL CHEMISTRY. *Educ:* Philadelphia Col Pharm, BS, 58; Univ Conn, MS, 61, PhD(anal chem), 63. *Prof Exp:* Chemist, Borden Chem Co, 58-59; fel, Univ Mich, 64; sr res chemist, ICI Am, 65-91; RETIRED. *Mem:* Am Chem Soc; Sigma Xi. *Res:* Liquid chromatography; electroanalytical chemistry, especially organic polarography, votammetry at solid electrodes, coulometric titrimetry and amperometric titrimetry. *Mailing Add:* RR 1 Box 484 M Avondale PA 19311

TURNER, WILLIE, b Suffolk, Va, Feb 1, 35; m 64; c 2. VIROLOGY, IMMUNOLOGY. *Educ:* Md State Col, BS, 57; Ohio State Univ, MS, 59, PhD(microbiol), 61. *Prof Exp:* NIH fel, Naval Med Res Inst, 61-62, Nat Inst Allergy & Infectious Dis grant, 62-64; asst prof microbiol, Meharry Med Col, 62-66; head microbiol sect, Viral Biol Br, Nat Cancer Inst, 70-71; PROF MICROBIOL & CHMN DEPT, COL MED, HOWARD UNIV, 71- *Concurrent Pos:* NIH staff fel oncol virol, Nat Cancer Inst, 66-69, sr fel, 69-70. *Mem:* AAAS; Am Soc Microbiol; Am Asn Immunologists; Am Asn Cancer Res; Soc Exp Biol & Med. *Res:* Oncogenic virology, especially interaction of oncogenic and nononcogenic viruses in vitro and in vivo; immunology of murine oncogenic virus as well as the immunology involved with tumors induced by these agents in vivo. *Mailing Add:* Dept Microbiol Howard Univ Col Med 520 West St NW Washington DC 20059

TURNEY, TULLY HUBERT, b Lakewood, Ohio, Sept 7, 36; m 62; c 1. ZOOLOGY. *Educ:* Oberlin Col, AB, 58; Univ NC, PhD(zool), 63. *Prof Exp:* Fel, Oak Ridge Nat Labs, 63-64; instr zool, Univ NC, 64-65; assoc prof biol, 65-74, chmn dept, 67-76, PROF BIOL, HAMPDEN-SYDNEY COL, 74- *Mem:* AAAS; Am Inst Biol Sci. *Res:* Cell control mechanisms; molecular biochemistry; physiology; bioethics; computerized instruction. *Mailing Add:* Dept Biol Hampden-Sydney Col Hampden-Sydney VA 23943-0001

TURNIPSEED, GLYN D, b Hazlehurst, Miss, Dec 19, 42; m 67; c 1. PLANT PHYSIOLOGY. *Educ:* Delta State Univ, BS, 66; Miss State Univ, PhD(bot), 73. *Prof Exp:* Sci teacher biol chem, Jackson Pub Schs at Wingfield High Sch, 66-70; asst bot, Miss State Univ, 70-73; asst prof, 73-80, ASSOC PROF BIOL, ARK POLYTECH COL, 80- *Concurrent Pos:* Ark Nat Heritage Comn. *Mem:* Soc Study of Amphibians & Reptiles; Am Mus Nat Hist; Audubon Soc. *Res:* Nitrogen source preference of Oophila ambystomatis; feeding efficiency of larval anurans; population density of larval anurans; Arkansas amphibian and reptillian distributions. *Mailing Add:* Dept Biol Ark Tech Univ Russellville AR 72801

TURNIPSEED, MARVIN ROY, b Carrollton, Miss, Nov 11, 34; m 57; c 2. REPRODUCTIVE PHYSIOLOGY, BIOCHEMISTRY. *Educ:* Miss State Univ, BS, 56, MEd, 63; Univ Ga, PhD(zool), 69. *Prof Exp:* NIH fel steroid biochem, biochem dept, Med Sch, Univ Minn, Minneapolis, 69-71, res assoc steroid biochem, pediat dept, 71-77; from asst prof to assoc prof, 78-84, PROF BIOL, QUINNIPIAC COL, HAMDEN, 84- *Concurrent Pos:* Res affil, dept obstet & gynec, Yale Univ, 80-85. *Mem:* Endocrine Soc; AAAS; Sigma Xi. *Res:* Placental aromatase action upon testosterone stearate and testosterone linolenate; anatomy and physiology and radiation biology; receptors involved in reproduction of female mammals; ovarian production of steroid hormones in response to trophic hormones; plasma and urinary estrogens and adrenal metabolism in newborn babies; endocrinology. *Mailing Add:* Box 92 Quinnipiac Col Hamden CT 06518

TURNLUND, JUDITH RAE, b St Paul, Minn, Sept 28, 36; m 57; c 3. NUTRITION, CHEMISTRY. *Educ:* Gustavus Adolphus Col, BS, 58; Univ Calif, Berkeley, PhD(nutrit), 78. *Prof Exp:* Fel human nutrit, Univ Calif, Berkeley, 78-80; RES NUTRIT SCIENTIST, USDA, 80- *Concurrent Pos:* Lectr, Univ Calif, Berkeley, 81-82, 84-, adj assoc prof, 89-; vis prof, Am Univ Beirut, Lebanon, 79 & 80. *Honors & Awards:* Cert of Merit, USDA, 84. *Mem:* Am Inst Nutrit; Am Soc Clin Nutrit; Am Dietetics Asn; Am Chem Soc. *Res:* Enriched stable isotopes of minerals to study mineral metabolism and bioavailability in humans. *Mailing Add:* Western Human Nutrit Res Ctr PO Box 29997 Presidio of San Francisco CA 94129

TURNOCK, A C, b Winnipeg, Can, Sept 11, 30. GEOLOGIST. *Educ:* Univ Manitoba, BS, 53, MS, 56; Johns Hopkins Univ, PhD(geol), 60. *Prof Exp:* PROF GEOL, UNIV MANITOBA, 65- *Mem:* Geol Soc Am; Sigma Xi; Minerol Soc Am; Minerol Soc Can; Geol Asn Can; Minerol Soc Gt Brit. *Mailing Add:* Dept Earth Sci Univ Manitoba Ctr for Precambrian Studies Winnipeg MB R3T 2N2 Can

TURNOCK, WILLIAM JAMES, b Winnipeg, Man, May 17, 29; m 58; c 3. POPULATION ECOLOGY, BIOLOGICAL CONTROL. *Educ:* Univ Man, BSA, 49; Univ Minn, MS, 51, PhD(entom), 59. *Prof Exp:* Res scientist, Div Forest Biol, Can Dept Agr, 49-61 & Forest Entom & Path Br, Can Dept Forestry, Man, 61-70; sci adv, Can Ministry State for Sci & Technol, 70-72; SECT HEAD INTEGRATED PEST CONTROL, RES BR, CAN DEPT AGR, 72- *Concurrent Pos:* Hon prof, Grad Sch, Univ Man, 65-70; vis scientist, Dept Zool, Agric Univ, Wageningen, Neth, 66-67, Dept Pure & Appl Biol, Imp Col, Silwood Park, Eng, 83-84; Can del, Int Coord Coun Man & Biosphere, UNESCO, 71. *Mem:* Sigma Xi; fel Entom Soc Can (pres, 79-80). *Res:* Integrated and biological control of agriculture pests. *Mailing Add:* Agr Res Sta 195 Dafoe Rd Winnipeg MB R3T 2M9 Can

TURNQUIST, CARL RICHARD, b Midland, Mich, May 12, 44; m 69; c 3. POLYMER CHEMISTRY. *Educ:* Westminister Col, BA, 66; Univ Wis-Madison, PhD(chem), 72. *Prof Exp:* Res chemist textile fibers, E I du Pont de Nemours, Co, Inc, 72-75; sr chemist, C R Bard Inc, 75-78, res prog supvr, 78, chem & polymer supvr, Cardiol & Radiol Prod, 78-82, staff engr, USCI Cardiol & Radiol Div, 82, eng mgr, USCI Div, 82-84, Bard Crit Care Div, 84-86, TECHNOL DEVELOP MGR, BARD VASCULAR SYSTS DIV, C R BARD INC, 86- *Concurrent Pos:* Secy, div-in-formation, Med Plastics Div, Soc Plastics Engrs, 77-80, secy, 80-83, mem div bd dirs, 83-86. *Mem:* Am Chem Soc; Soc Plastics Engrs; AAAS; Sigma Xi. *Res:* Development of medical devices (implantable artificial arteries and heart catheters); analysis; material and process development; biotesting; physician interactions. *Mailing Add:* 106 Kenney Lane Concord MA 01742-2702

TURNQUIST, MARK ALAN, b July 26, 49; US citizen; m 71; c 2. LOGISTICS, TRANSPORTATION ENGINEERING. *Educ:* Mich State Univ, BS, 71; Mass Inst Technol, SM, 72, PhD(transp systs analysis), 75. *Prof Exp:* Asst prof transp eng, Dept Civil Eng & transp Ctr, Northwestern Univ, 75-79; assoc prof transp eng, Sch Civil/Environ Eng, 79-86, assoc dean comput, Col Eng, 84-86, PROF TRANSP ENG, SCH CIVIL/ENVIRON ENG, CORNELL UNIV, 86- *Concurrent Pos:* Consult, John Hamburg & Assocs, Inc, Chicago, Ill, 77-78, City of Chicago, 79-81, Cent Ohio Transit Authority, Columbus, Ohio, 82-85, Gen Motors, 84-, CSX Inc, 88- *Mem:* Opers Res Soc Am; Transp Res Bd. *Res:* Logistics and production systems design, operations and control; vehicle routing and scheduling; risk analysis. *Mailing Add:* Sch Civil/Environ Eng Cornell Univ Ithaca NY 14853

TURNQUIST, PAUL KENNETH, b Lindsborg, Kans, Jan 3, 35; m 62; c 3. AGRICULTURAL ENGINEERING. *Educ:* Kans State Univ, BS, 57; Okla State Univ, MS, 61, PhD(agr eng), 65. *Prof Exp:* Res engr, Caterpillar Tractor Co, 57; from instr to asst prof agr eng, Okla State Univ, 58-62; assoc prof, SDak State Univ, 64-70; engr in residence, Caterpillar Tractor Co, 70-71; prof agr eng, SDak State Univ, 71-76; PROF AGR ENG & HEAD DEPT, AUBURN UNIV, 77- *Concurrent Pos:* Vis prof, Purdue Univ, 85. *Mem:* Fel Am Soc Agr Engrs; Am Soc Eng Educ; Sigma Xi; Nat Soc Prof Engrs. *Res:* Agricultural power and machinery for tillage and harvesting; operator safety; environmental control. *Mailing Add:* Dept Agr Eng Auburn Univ Auburn AL 36849

TURNQUIST, RALPH OTTO, b Lindsborg, Kans, Aug 10, 28; m 65; c 2. MECHANICAL ENGINEERING, INSTRUMENTATION. *Educ:* Kans State Univ, BS, 52, MS, 61; Case Inst Technol, PhD(fluid controls), 65. *Prof Exp:* Engr, Aircraft Gas Turbine Div, Westinghouse Elec Co, Mo, 54-59; instr mech eng, Kans State Univ, 59-62; res asst fluid control systs, Case Inst Technol, 62-65; from asst prof to assoc prof, 65-77, PROF MECH ENG, KANS STATE UNIV, 77- *Mem:* Am Soc Mech Engrs; Am Soc Eng Educ; Instrument Soc Am; Fluid Power Soc. *Res:* Turbojet engine fuel distribution and atomization; automatic control theory; fluid control systems; fluidics; hydrostatic and power shift transmissions. *Mailing Add:* Dept Mech Eng Kans State Univ Manhattan KS 66506

TURNQUIST, RICHARD LEE, b Rugby, NDak, Aug 12, 44; m 66. PHYSIOLOGY, BIOCHEMISTRY. *Educ:* Concordia Col, BA, 66; Utah State Univ, PhD(physiol), 71. *Prof Exp:* Fel biochem, Utah State Univ, 71-74; ASST PROF BIOL, AUGUSTANA COL, 74- *Mem:* AAAS; Sigma Xi. *Res:* Ultrastructural changes and responses during detoxication of xenobiotics in mammals and insects. *Mailing Add:* Dept Biol Augustana Col 639 38th St Rock Island IL 61201

TURNQUIST, TRUMAN DALE, b Kipling, Sask, Apr 8, 40; m 64; c 3. ANALYTICAL CHEMISTRY. *Educ:* Bethel Col, Minn, BA, 61; Univ Minn, PhD(anal chem), 65. *Prof Exp:* Assoc prof, 65-76, PROF CHEM, MT UNION COL, 76- *Mem:* Am Chem Soc. *Res:* Metal complex formation; solvent extraction of metal complexes; spectrophotometry. *Mailing Add:* Dept Chem Mt Union Col Alliance OH 44601

TURNROSE, BARRY EDMUND, b New Britain, Conn, June 3, 47; m 70; c 2. SPACE ASTRONOMY, DIGITAL IMAGE PROCESSING. *Educ:* Wesleyan Univ, BA, 69; Calif Inst Technol, PhD(astron), 76. *Prof Exp:* Resident res assoc astron, NASA-Johnson Space Ctr, Houston, 75-77; resident astronomer image processing, 77-80, sect mgr, 80-82, asst dept mgr, 82-83, Dept Mgr, Int Ultraviolet explorer, Comput Sci, Goddard Space Flight Ctr, NASA, 83-86; OPER MGR, ASTRON PROG, COMPUTER SCI CORP, 86- *Mem:* Am Astron Soc; Sigma Xi. *Res:* Extragalactic astronomy; spectrophotometry of galaxies; stellar content of galaxies; surface photometry of extragalactic objects; astronomical image processing; ultraviolet astronomy. *Mailing Add:* 12008 Maddox Lane Bowie MD 20715

TUROCZI, LESTER J, b Jersey City, NJ, Nov 13, 42. GENETIC TOXICOLOGY, DEVELOPMENTAL BIOLOGY. *Educ:* Rutgers Univ, BA, 65, MS, 67, PhD(zool), 72. *Prof Exp:* From asst prof to assoc prof, 72-83, PROF BIOL, WILKES COL, WILKES-BARRE, PA, 84-, CHMN DEPT, 77- *Concurrent Pos:* Vis prof, Pa State Univ, Lehman, 75-76. *Mem:* AAAS; Am Soc Zoologists; NY Acad Sci; Genetic Toxicol Asn; Biol Photogr Asn. *Res:* Utilization of the Ames Salmonella-microsomal mutagenicity assay and the Mouse Micronucleus Test as genetic toxicology methods for the study of various nutrients, food additives and natural products. *Mailing Add:* Dept Biol Wilkes Univ Wilkes-Barre PA 18766

TUROFF, MURRAY, b San Francisco, Calif, Feb 13, 36; m 61; c 2. COMPUTER SCIENCE, OPERATIONS RESEARCH. *Educ:* Univ Calif, Berkeley, BA, 58; Brandeis Univ, PhD(physics), 65. *Prof Exp:* Syst engr, IBM Corp, 61-64; mem prof staff syst anal, Inst Defense Anal, 64-68; opers res & info systs, Off Emergency Preparedness, 68-73; PROF COMPUT & INFO SCI, NJ INST TECHNOL, 73- *Concurrent Pos:* Lectr, Am Univ, 70-73. *Mem:* AAAS; Inst Mgt Sci; Opers Res Soc Am; Asn Comput Mach; Am Soc Cybernet. *Res:* Delphi design; computerized conferencing systems; information systems design; technology assessment and forecasting; gaming, simulation and modeling; policy analyses. *Mailing Add:* Dept Computer & Info Sci NJ State Technol 323 High St Newark NJ 07102

TURPEN, JAMES BAXTER, b Sheridan, Wyo, Sept 27, 45; m; c 3. DEVELOPMENTAL BIOLOGY, IMMUNOLOGY. *Educ:* Univ Denver, BS, 67, MS, 69; Tulane Univ, PhD(biol), 73. *Prof Exp:* USPHS fel immunol, Med Ctr, Univ Rochester, 74-76; asst prof, 86-88, ASSOC PROF, DEPT ANAT, UNIV NEBR COL MED, 88-, ASST PROF BIOL, PA STATE UNIV, 76- *Concurrent Pos:* Prin investr, USPHS-NIH grant, 77-85; USPHS res career develop award, 80-85; mem, Basel Inst Immunol, 85. *Honors & Awards:* Masua hon lectr, 87. *Mem:* Am Soc Zoologists; Am Asn Anatomists; AAAS; Int Soc Study Differentiation; Am Asn Immunol; Int Soc Develop & Comp Immunol; Reticuloendothelial Soc. *Res:* Development of hematopoietic cells; comparative immunology. *Mailing Add:* Dept Anat Univ Nebr Col Med 600 S 42nd St Omaha NE 68198-6395

TURPIN, FRANK THOMAS, b Troy, Kans, June 4, 43; m 70; c 2. ECONOMIC ENTOMOLOGY. *Educ:* Washburn Univ, BS, 65; Iowa State Univ, PhD(entom), 71. *Prof Exp:* From asst prof to assoc prof entom, 71-82, PROF ENTOM PURDUE UNIV, 82- *Concurrent Pos:* Consult, US Environ Protection Agency, 73-, Pesticide Industs, 75- & USAID, Tanzania, 81. *Mem:* Entom Soc Am. *Res:* Biology; ecology; population dynamics and control of insects attacking corn. *Mailing Add:* 1158 Entom Hall Purdue Univ West Lafayette IN 47907-1158

TURRELL, BRIAN GEORGE, b Shoreham-by-Sea, Eng, May 6, 38; m 62, 70; c 3. PHYSICS. *Educ:* Oxford Univ, BA, 59, MA & PhD(nuclear orientation), 63. *Prof Exp:* Asst lectr physics, Univ Sussex, 63-64; from asst prof to assoc prof, 64-70, PROF PHYSICS, 76-, HEAD DEPT, UNIV BC, 87- *Concurrent Pos:* Sr fel, Sci Res Coun, Oxford, 71-72; vis prof, Univ New S Wales, Duntroon, Australia, 78-79; vis scientist, Walther Meissner Inst & Tech Univ Munich, 85-86. *Mem:* Can Asn Physicists. *Res:* Nuclear orientation; nuclear magnetic resonance; use of these techniques to study hyperfine interactions in magnetic materials; cryogenic detectors. *Mailing Add:* Dept Physics Univ BC Vancouver BC V6T 1W5 Can

TURRELL, EUGENE SNOW, b Hyattsville, Md, Feb 27, 19; m 88; c 1. PSYCHIATRY. *Educ:* Ind Univ, BS, 39, MD, 47; Am Bd Psychiat & Neurol, dipl, 53. *Prof Exp:* Asst physiol, Ind Univ, 39-42; res assoc, Fatigue Lab, Harvard Univ, 42-43; res assoc physiol, Ind Univ, 43-44, asst biochem, Sch Med, 45-47; med house officer, Peter Bent Brigham Hosp, Boston, 47-48; resident psychiat, Kankakee State Hosp, Ill, 48-49; clin asst, Sch Med, Univ Calif, 49-52; asst prof, Sch Med, Ind Univ, 52-53; assoc prof, Sch Med, Univ Colo, 53-58, asst dean sch, 57; prof & chmn dept, Sch Med, Marquette Univ, 58-63, clin prof, 63-69; sr psychiatrist, 69-72, dir, Ctr Spec Probs, Community Ment Health Serv, City & County of San Francisco, 72-75; STAFF PSYCHIATRIST, MIDTOWN COMMUNITY MENT HEALTH CTR, WISHARD MEM HOSP, 75- MED DIR, 86-, MED DIR FORENSIC SERV, 87- *Concurrent Pos:* Resident, Langley Porter Clin, 49-50; chief psychiat consult serv, Robert W Long Hosp, Indianapolis, Ind, 52-53; med dir, Colo Psychopath Hosp, 53-54; assoc attend, Denver Gen Hosp, 53-57, dir psychiat serv, 57-58; attend, Vet Admin Hosp, Denver, 53-58; mem staff psychosom div, Colo Gen Hosp, 54-57; dir psychiat serv, Milwaukee Sanitarium Found, 58-65; consult, Hosp Ment Dis, Milwaukee, 58-69, Vet Admin Hosp, Wood, Wis, 58-69, Columbia Hosp, 59-69 & Milwaukee Children's Hosp, 60-69; assoc prof, dept psychiat, Ind Univ Sch Med, 75- *Mem:* AAAS; fel Am Psychiat Asn; AMA; Sigma Xi. *Res:* Psychosomatic medicine; psychotherapy. *Mailing Add:* 13322 Mango Dr Del Mar CA 92014

TURRELL, GEORGE CHARLES, b Portland, Ore, June 19, 31; m 54; c 4. PHYSICAL CHEMISTRY. *Educ:* Lewis & Clark Col, BA, 50; Ore State Univ, MS, 52, PhD(phys chem), 54. *Prof Exp:* Asst, Ore State Univ, 50-52; mem tech staff, Electron Device Dept, Bell Tel Labs, Inc, 54-56; res assoc, Metcalf Res Lab, Brown Univ, 56-57, instr chem, 57-58; Guggenheim fel, Bellevue Labs, Nat Ctr Sci Res, France, 58-59; asst prof chem, Howard Univ, 59-62, assoc prof, 62-67; exchange prof & Fulbright fel, Infrared Spectros Lab,

Univ Bordeaux, 66-67, vis prof, 67-70; prof, Nat Univ Zaire, Kisangani, 70-71 & Kinshasa, 71-72; actg ed, Can Jour Spectroscopy, 72; vis prof chem, Univ Montreal, 72-74 & McGill Univ, 73-75; res prof chem, Univ Laval, Que, 75-79; engr, Bomem, Inc, 79-80; vis prof, Molecular Physics Lab, Univ Limoges, 80-81; PROF, INFRARED & RAMAN SPECTROS LAB, UNIV SCI TECH LILLE, FRANCE, 81- *Concurrent Pos:* Vis scientist, IBM Almaden Res Lab, 89. *Mem:* Am Phys Soc; Coblentz Soc; Can Spectroscopy Soc; Sigma Xi; French Chem Soc. *Res:* Molecular spectroscopy; studies of molecular interactions in solids and liquids using infrared and Raman spectroscopy. *Mailing Add:* Univ Sci Tech Lille LASIR Bat C5 Villeneuve d'Ascq 59655 France

TURRO, NICHOLAS JOHN, b Middletown, Conn, May 18, 38; m 60; c 2. ORGANIC CHEMISTRY. *Educ:* Wesleyan Univ, BA, 60; Calif Inst Technol, PhD(chem), 63. *Hon Degrees:* DSc, Wesleyan Univ, 84. *Prof Exp:* Instr, 64-65, from asst prof to assoc prof, 65-69, PROF ORG CHEM, COLUMBIA UNIV, 69- *Concurrent Pos:* NSF fel, Harvard Univ, 63-64; consult, E I du Pont de Nemours & Co, 64-; vis prof, Pa State Univ, 66; Sloan Found fel, 66-68; W P Schweitzer prof chem, Columbia Univ, 81, chmn, chem dept, 81-84; Sherman Fairchild fel, Calif Inst Technol, 84-85; Guggenheim fel, Oxford Univ, 85. *Honors & Awards:* Pure Chem Award, Am Chem Soc, 74; Halpern Award, NY Acad Sci, 77; Ernest Orlando Lawrence Award, Dept Energy, 83; Harrison Howe Award, Am Chem Soc, 86, Cope Scholar Award, 87, Norris Award, 88; Photochem Award, Inter-Am Photochem Soc, 91. *Mem:* Nat Acad Sci; Am Chem Soc; Royal Soc Chem; fel NY Acad Sci; Am Acad Arts & Sci. *Res:* Photochemistry; electronic energy transfer in fluid solution. cycloaddition reactions; thermal rearrangements; dioxetane chemistry; chemiluminescence; micellar chemistry; application of laser techniques to organic photochemistry; emulsion polymerization; magnetic field effects; polymer photochemistry and photo-physics; magnetic isotope and magnetic field effects on organic reactions; organic reactions on porous silica and zeolites. *Mailing Add:* 125 Downey Dr Tenafly NJ 07670

TURSE, RICHARD S, b Jersey City, NJ, Mar 24, 35; m 64; c 3. ANALYTICAL CHEMISTRY, SPECTROCHEMISTRY. *Educ:* Rutgers Univ, BS, 56, MS, 58, PhD(anal chem), 60. *Prof Exp:* Instr anal chem, Rutgers Univ, 56-58; sr res chemist, 60-87, RES ASSOC, COLGATE-PALMOLIVE CO, 87- *Mem:* Soc Appl Spectros; Am Pharm Asn. *Res:* Atomic absorption methods for determination of trace metals; X-ray diffraction and emission techniques for sample identification, secondary ion mass spectroscopy, electron spectroscopy chemical analysis and auger surface analysis; scanning electron microscopy. *Mailing Add:* Colgate-Palmolive Co 909 River Rd Piscataway NJ 08855-1343

TUSING, THOMAS WILLIAM, b New Market, Va, Feb 2, 20; m 49; c 5. PHARMACOLOGY. *Educ:* George Washington Univ, BS, 42; Med Col Va, MD, 50. *Prof Exp:* Med dir, Hazleton Labs, Inc, 51-58, dir res & vpres, 58-69; med dir pharmaceut res & develop div, Mallindkrodt Inc, 69-85; RETIRED. *Mem:* AMA; Indust Med Asn; Soc Toxicol. *Mailing Add:* 1876 Lake Francis Apopka FL 32712

TUSTANOFF, EUGENE RENO, b Windsor, Ont, Jan 30, 29; m 54; c 4. BIOCHEMISTRY. *Educ:* Assumption Col, BA, 52; Detroit Univ, MS, 54; Western Ont Univ, PhD(biochem), 59. *Prof Exp:* Res assoc biochem, Western Ont Univ, 55-59; fel, Western Reserve Univ, 59-61; Life Ins Med Res fel, Oxford Univ, 61-62; asst scientist, Hosp for Sick Children, Univ Toronto, 62-64; assoc biochem, Univ Toronto, 63-65, asst prof pharmacol, 64-65; assoc prof biochem, McMaster Univ, 65-67; assoc prof path chem, 67-72, assoc prof clin biochem, 72-75, PROF CLIN BIOCHEM & DEPT ONCOL, UNIV WESTERN ONT, 75- *Mem:* Can Biochem Soc; Brit Biochem Soc; Am Soc Biol Chem; Am Asn Clin Chem; Can Soc Clin Chem; Can Acad Clin Biochem. *Res:* Biogenesis and control of mitochondria; biogenesis of membranes; biochemistry of tumour model systems; steroid receptors in breast cancer. *Mailing Add:* Dept Biochem Victoria Hosp Div Clin Biochem Univ Western Ont London ON N6A 4G5 Can

TUSTING, ROBERT FREDERICK, b Alameda, Calif, May 11, 33. OCEAN ENGINEERING, INSTRUMENTATION. *Educ:* Univ Calif, Berkeley, BS, 61, MS, 65. *Prof Exp:* Engr, Lawrence Radiation Lab, Berkeley, 61-64, design engr, 65-68; chief engr & design eng mgr, Delco Electronics, Gen Motors, Santa Barbara, 68-72; res prof marine sci, Rosenstiel Sch, Univ Miami, 72-81; SR ENGR, HARBOR BR OCEANOG INST, 82- *Concurrent Pos:* Adj prof, Indian River Community Col, 86-89. *Mem:* Marine Technol Soc; Inst Elec & Electronics Engrs. *Res:* Ocean instrumentation; scientific sampling equipment design; optical and physical measurements; acoustic instrumentation; laser-based quantitative analysis; awarded six US patents; author of 23 technical papers. *Mailing Add:* Harbor Br Oceanog Inst Inc 5600 Old Dixie Hwy Ft Pierce FL 34946

TUSTISON, RANDAL WAYNE, b Ft Wayne, Ind, Dec 4, 47; m 75; c 2. THIN FILM TECHNOLOGY, OPTICAL MATERIALS. *Educ:* Purdue Univ, BS, 70; Univ Ill, MS, 72, PhD(metall eng) 76,. *Prof Exp:* Res assoc physics, Mass Inst Technol, 76-78; thin film scientist mat sci, Res Div, 3M Co, 81-82; sr scientist, 78-81, PRIN SCIENTIST MAT SCI, RES DIV, RAYTHEON CO, 82- *Concurrent Pos:* Lectr, Northeastern Univ, 83-86; Nat Prog Comt, Am Vacuum Soc, 86-91. *Mem:* Am Vacuum Soc; Am Phys Soc. *Res:* Thin film deposition processes; infrared optical materials and coatings; magnetic and ferroelectric materials and coatings; mechanical property characterization of thin films. *Mailing Add:* Raytheon Co 131 Spring St Lexington MA 02173

TUSZYNSKI, ALFONS ALFRED, b Poland, Apr 30, 21; US citizen; m 45; c 2. ELECTRONICS, MICROELECTRONICS. *Educ:* Univ London, BSc, 52; Newark Col Eng, MSc, 62, DEngSc(integrated circuits), 69. *Prof Exp:* Mem staff various co, Gt Brit & Can, 52-59; creative specialist, Am Optical Co, 60-63; proj engr, Singer Gen Precision, 63-66; mgr applns, Sprague Elec Co, 66-67; consult, Sylvania & Singer Gen Precision, 67-69; mgr res & develop, Solitron Devices Inc, 69-70; prof elec eng, Univ Minn, Minneapolis, 70-87; RETIRED. *Concurrent Pos:* Consult, Control Data Corp, 75-85, UNIVAC, 80-81, 3M, 82-83. *Mem:* Sr mem Inst Elec & Electronics Engrs. *Res:* Large scale integration and micropower techniques pertaining to linear and digital systems; microcircuits. *Mailing Add:* Dept Elec Eng San Diego CA 92182

TUSZYNSKI, GEORGE P, PROTEIN CHEMISTRY, CELL BIOLOGY. *Educ:* Univ Pa, PhD(biochem), 73. *Prof Exp:* ASSOC INVESTR PALETELET HEMAT, LANKENAU MED RES CTR, 84- *Res:* Paletes in coagulation in tumor cell metastasis. *Mailing Add:* Dept Med Med Col Penn 3300 Henry Ave Philadelphia PA 19129

TUSZYNSKI, JACK A, b Poznan, Poland, July 24, 56; Can citizen; m 80; c 1. PHASE TRANSITIONS, BIOPHYSICS. *Educ:* Univ Poznan, Poland, MSc, 80; Univ Calgary, PhD(solid state hysics), 83. *Prof Exp:* Fel chem, Univ Calgary, 83; asst prof physics, Mem Univ Nfld, 83-87; ASST PROF PHYSICS, UNIV ALTA, 88- *Concurrent Pos:* Hon asst prof physics, Mem Univ Nfld, 88. *Mem:* Can Asn Physicists. *Res:* Applications of nonlinear differential equations to the linetics of phase transitions and other nonlinear phenomena; crystal lattice dynamics magnetic phase transitions. *Mailing Add:* Dept Physics Univ Alta Edmonton AB T6G 2M7 Can

TUTEUR, FRANZ BENJAMIN, b Frankfurt, Ger, Mar 6, 23; nat US; m 52; c 2. ELECTRICAL ENGINEERING. *Educ:* Univ Colo, BS, 44; Yale Univ, MEng, 49, PhD(elec eng), 54. *Prof Exp:* From instr to assoc prof elec eng, 50-66, prof eng & appl sci, 66-81, PROF ELEC ENG, YALE UNIV, 81- *Concurrent Pos:* Guest lectr & consult, Israel Inst Technol, 57; consult, Tech Res Group, NY, 57-58, Sikorsky Aircraft Co, Conn, 58, Melpar, Inc, Mass, 59-61, Rand Corp, Calif, 62-77, Analytical Sci Corp, Mass, 70- & Perkin-Elmer Corp, 72-77; vis prof, Univ Calif, Berkeley, 65-66; vis res marine physicist, Scripps Inst Oceanog, San Diego, Calif, 78; vis prof, Tel Aviv Univ, 82, Ruhr Universität, Bochum WGer, 86, Ecole Nationale Superieure de Telecommunication, Paris, France, 87. *Mem:* Inst Elec & Electronics Engrs. *Res:* Communications theory; sonar; control systems; adaptive systems; medical sinalprocessing. *Mailing Add:* Elec Eng Dept Yale Univ 519 Becton New Haven CT 06520-2157

TUTHILL, ARTHUR F(REDERICK), b Cutchogue, NY, Dec 18, 16; m 41; c 5. MECHANICAL ENGINEERING, THERMODYNAMICS. *Educ:* Carnegie Inst Technol, BS, 38; Univ Wis, MS, 39. *Prof Exp:* Instr mech eng, Cooper Union, 39-42; from instr to prof mech eng, 46-81, EMER PROF, UNIV VT, 81- *Mem:* Am Soc Mech Engrs; Am Soc Eng Educ. *Res:* Air distribution, environmental engineering. *Mailing Add:* 947 Williston Rd Williston VT 05495

TUTHILL, HARLAN LLOYD, b Fillmore, NY, Nov 24, 17; m 41, 71; c 2. TECHNICAL MANAGEMENT, RESOURCE MANAGEMENT. *Educ:* Houghton Col, BS, 39; Cornell Univ, PhD(phys chem), 43. *Prof Exp:* Res chemist, Rohm & Haas Co, 43-46; head phys chem sect, Smith Kline & French Labs, 46-48, tech dir, 48-54, sci dir int div, 54-57, asst dir res & develop labs, 57-62, vpres, SmithKline Instruments, Inc, 62-65; dir prod develop & assoc dir, Squibb Inst Med Res, Squibb Corp, 65-70; dir health & med res, Int Paper Co, 70-77, assoc dir Sci & Technol Lab, 77-84; RETIRED. *Concurrent Pos:* Consult univ/indust technol transfer, 85- *Mem:* Am Chem Soc; Sigma Xi. *Res:* Health and environmental sciences; analytical and materials sciences. *Mailing Add:* 1183 Candlewood Dr Pen Argyl PA 18072-9689

TUTHILL, SAMUEL JAMES, b San Diego, Calif, Sept 6, 25; m 52; c 3. GEOLOGY, HYDROGEOLOGY. *Educ:* Drew Univ, AB, 51; Syracuse Univ, MS, 60; Univ NDak, MA, 63, PhD(geol), 69. *Prof Exp:* Geologist, NDak Geol Surv, 63-64; asst prof geol, Muskingum Col, 64-68; asst state geologist, Iowa Geol Surv, 68-69, dir & state geologist, 69-75; vpres energy resources/utilization, Environ & Res, Iowa Elec Light & Power Co, 77-78, sr vpres energy prod, 78-88; pres, 88-91, SR SCI ASSOC, TUTHILL INC, 91- *Concurrent Pos:* NSF grants, Muskingum Col expeds Alaska, 65-66, 67-69; adj prof, Univ Iowa, 69-; adminr, Iowa Oil & Gas Admin, 69-; mem, Iowa Natural Resources Coun, 70-; mem, Iowa Land Rehab Coun, 70-; secy, Iowa State Map Adv Coun, 72; sci adv to US Secy Interior, 75-; spec asst energy policy to chmn Presidents Energy Resources Coun, 75-76; dir, Off Indust Energy Conserv, US Dept Com, 75-76, Iowa Acad Sci, 87-90; adv, US Fed Energy Admin, 75-76, asst adminr conserv & environ, 76-77; comnr, Upper Miss River Basin Comn, Iowa, 80-82; chmn, Enerex Exec Comt; trustee, Herbert Hoover Presidential Libr Asn, Inc. *Mem:* Am Water Well Asn; fel Geol Soc Am; Asn Am State Geol; fel Explorers Club; Sigma Xi. *Res:* Research management; water; minerals; environmental protection; waste management; remote sensing; resources managemental development. *Mailing Add:* PO Box 983 Grand Marais MN 55604

TUTHILL, SAMUEL MILLER, b Rocky Point, NY, Jan 7, 19; m 41; c 3. ANALYTICAL CHEMISTRY. *Educ:* Wesleyan Univ, BA, 39, MA, 41; Ohio State Univ, PhD(chem), 48. *Prof Exp:* Lab asst chem, Wesleyan Univ, 39-41; anal chemist, Mallinckrodt Chem Works, 41-45; lab asst chem, Ohio State Univ, 45-46, from asst instr to instr, 47-48; anal lab supvr, Mallinckrodt Inc, 48-56, dir qual control, 56-70, corp dir qual control, 70-72, corp dir qual assurance, 72-76, dir corp anal serv, 76-81, TECH COMT & QUAL STANDARD CONSULT, MALLINCKRODT INC, 81- *Concurrent Pos:* Former mem, comt revision, US Pharmacopeia; consult, comt reagent specif, Am Chem Soc; mem, Food Chem Codex Comt; mem, E-15 comt, Am Soc Testing & Mat. *Mem:* Am Chem Soc; Am Soc Qual Control; Parenteral Drug Asn. *Res:* Methods of analysis of pharmaceutical and reagent chemicals; separation by electrodeposition; instrumental methods of analysis; separation of rhodium from iridium by electrolysis with control of the cathode potential; determination of rare earths in steels; analysis of opium and narcotics; good manufacturing practice in manufacture of drugs. *Mailing Add:* Mallinckrodt Inc Mallinckrodt & Second St PO Box 5439 St Louis MO 63147

TUTIHASI, SIMPEI, b Tokyo, Japan, Mar 2, 22; US citizen; m 47; c 2. SOLID STATE PHYSICS. *Educ:* Kyoto Univ, BSc, 46, DSc(physics), 56. *Prof Exp:* Res assoc solid state physics, Inst Optics, Univ Rochester, 56-59; sr engr, Sylvania Elec Prods, Inc, 59-64; scientist, res labs, Xerox Corp, 64-87; RETIRED. *Mem:* Am Phys Soc; Optical Soc Am. *Res:* Solid state spectroscopy; spectroscopy of ordered crystals, photoconductivity, luminescence. *Mailing Add:* 274 Sudden Valley Bellingham WA 98226

TUTTE, WILLIAM THOMAS, b Newmarket, Eng, May 14, 17; m 49. COMBINATORICS. *Educ:* Cambridge Univ, PhD(math), 48. *Hon Degrees:* Doctor Math, Univ Waterloo, 87. *Prof Exp:* Lectr, Univ Toronto, 48-52, from asst prof to assoc prof, 52-62; prof math, 62-85, PROF EMER, DEPT COMBINATORICS & OPTIMIZATION, UNIV WATERLOO, 85- *Honors & Awards:* Henry Marshall Tory Medal, Royal Soc Can, 75; Killam Prize, Can Coun, 82. *Mem:* Am Math Soc; Math Asn Am; fel Royal Soc Can; Can Math Soc; London Math Soc; fel Royal Soc London. *Res:* Graph theory; matroid theory. *Mailing Add:* Dept Combinatorics & Optimization Univ Waterloo Waterloo ON N2L 3G1 Can

TUTTLE, DAVID F(EARS), b Briarcliff Manor, NY, July 5, 14; m 44; c 2. ELECTRICAL ENGINEERING. *Educ:* Amherst Col, AB, 34; Mass Inst Technol, SB & SM, 38, ScD(elec eng), 48. *Prof Exp:* Mem tech staff, Bell Tel Labs, NY, 38-42; prof, 48-79, EMER PROF ELEC ENG, STANFORD UNIV, 79- *Concurrent Pos:* Fulbright lectr, France, 54-55 & Spain, 61-62; assoc prof, Univ Aix-Marseille, 68-69; vis prof math, Ga Inst Tech, 81-82, Amherst Col, 83-84. *Mem:* Fel AAAS; fel Inst Elec & Electronics Engrs. *Res:* Network theory. *Mailing Add:* 713 Alvarado Row Stanford CA 94305-1010

TUTTLE, DONALD MONROE, b Bay City, Mich, Feb 1, 17; m 47; c 3. ENTOMOLOGY. *Educ:* Mich State Univ, BS, 40, MS, 47; Univ Ill, PhD(entom), 52. *Prof Exp:* Lab asst entom, Mich State Col, 40; instr, Univ Maine, 47-49; asst, Univ Ill, 49-52; prof & res entomologist, 52-83, EMER PROF, UNIV ARIZ, 83- *Concurrent Pos:* Consult & pest control adv. *Mem:* Entom Soc Am; Acarological Soc Am; Sigma Xi; hon mem Future Farmers Am; Native Plant Soc; Geneal Soc. *Res:* Citrus, alfalfa, melon and turf insects; systematics and biology of Tetranychoidea. *Mailing Add:* Univ Ariz Farm 6425 W Eighth St Yuma AZ 85364-9737

TUTTLE, ELBERT P, JR, b Ithaca, NY, Sept 1, 21; m 52; c 5. PHYSIOLOGY, INTERNAL MEDICINE. *Educ:* Princeton Univ, AB, 42; Harvard Univ, MD, 51. *Prof Exp:* Asst med, Harvard Med Sch & Mass Gen Hosp, 54-56; from asst prof to assoc prof, 57-66, PROF MED, SCH MED, EMORY UNIV, 66- *Concurrent Pos:* Nat Heart Inst res fel, 53-56; Am Heart Asn res fel, 57; chair cardiovasc res, Ga Heart Asn, 58-72. *Mem:* Am Fedn Clin Res. *Res:* Inorganic metabolism; renal and circulatory physiology; hypertension; nephrology. *Mailing Add:* Dept Med Emory Univ 1364 Clifton Rd NE Atlanta GA 30322

TUTTLE, ELIZABETH R, b Boston, Mass, Dec 5, 38. MECHANISMS. *Educ:* Univ NH, BS, 60; Univ Colo, MS, 61, MS, 87, PhD(physics), 64. *Prof Exp:* From asst prof to prof physics, 64-86, PROF ENG, UNIV DENVER, 86- *Mem:* Am Phys Soc; Am Asn Physics Teachers; Nat Soc Prof Engrs; Am Soc Mech Engrs; Am Soc Eng Educ; Soc Indust & Appl Math. *Res:* Type synthesis of planar mechanisms; laminar flow in curved pipes. *Mailing Add:* Dept Eng Univ Denver Denver CO 80208

TUTTLE, KENNETH LEWIS, b Toledo, Ore, Apr 4, 44; m 67; c 3. COMBUSTION & GASIFICATION OF WOOD, AIR POLLUTION FROM HEAT ENGINES. *Educ:* US Naval Acad, BS, 67; Ore State Univ, MS, 74, PhD(mech eng), 78. *Prof Exp:* Grad asst, Ore State Univ, 72-77; res engr, Weyerhaeuser Co, 77-81; prin, Solid Fuel Energy Assocs, 81-83; PROF, NAVAL ACAD, 83- *Concurrent Pos:* Prin, Tuttle Energy Conversion & Exchange, 72-77 & Solid Fuel Res, 84-; dir, Marine Propulsion Labs, Naval Acad, 83-90; chmn, Ocean & Marine Eng Div, Am Soc Eng Educ, 89-91. *Mem:* Am Soc Eng Educ; Combustion Inst; Am Gear Mfr Asn. *Res:* Bio-mass energy; awarded one patent for a method of firing to reduce air pollution; internal combustion engines including gasohol, air pollution, and fuels; closed loop engines for deep submergence vehicles; reliability centered maintenance; computers in education. *Mailing Add:* 1098 Broadview Dr Annapolis MD 21401

TUTTLE, MERLIN DEVERE, b Honolulu, Hawaii, Aug 26, 41. POPULATION ECOLOGY & BEHAVIOR, MAMMALOGY. *Educ:* Andrews Univ, BA, 65; Univ Kans, MA, 69, PhD(pop ecol), 74. *Prof Exp:* Co-dir, Smithsonian Venezuelan Res Proj, Smithsonian Inst, 65-67; res assoc pop ecol, Univ Minn, 72; CUR MAMMALS, MILWAUKEE PUB MUS, 75- *Concurrent Pos:* Consult endangered bats, Tenn Valley Authority, 76-; mem, Recovery Team Endangered Indiana & Gray Bats, US Fish & Wildlife Serv, 79-; pres, Bat Conserv Int, 82- *Mem:* Am Soc Mammalogists; Am Soc Naturalists; Ecol Soc Am; Soc Study Evolution; Nat Speleol Soc. *Res:* Predator/prey interaction, communication; foraging behavior in refuging species and the energetics of thermoregulation, hibernation and migration. *Mailing Add:* Brackenridge Field Lab Univ Tex Austin TX 73712

TUTTLE, RICHARD SUNESON, b Pottsville, Pa, Aug 18, 30; m 60; c 1. PHYSIOLOGY, PHARMACOLOGY. *Educ:* State Univ NY, PhD(pharm), 60. *Prof Exp:* Res fel, 60-64, RES ASSOC NEUROPHYSIOL, MASONIC MED RES FOUND, 64- *Concurrent Pos:* USPHS fel, 60-61; NIH grants, 63-; Fogarty fel, Sweden, 78; vis chair pharmacol, Div Astra, Hasssle Res Inst. *Mem:* Am Physiol Soc; Am Soc Pharmacol & Exp Therapeut. *Res:* Pharmacology of cardiac glycosides and electrolytes; neurophysiology of vasomotor regulation; centrally evoked histamine release; role of histamine in control of cardiovascular tone; cardiovascular effects of imidazoles and mesenteric Pacinian baroreceptors. *Mailing Add:* Masonic Med Res Lab 2150 Bleeker St Utica NY 15302

TUTTLE, ROBERT LEWIS, b Boston, Mass, July 26, 22; m 42; c 2. MICROBIOLOGY. *Educ:* Univ NH, BS, 43; Univ Rochester, MD, 47. *Prof Exp:* Asst trop med, Bowman Gray Sch Med, 48-50, from instr to assoc prof microbiol, 50-70, chmn dept, 55-62, assoc dean, 62-69, acad dean, 69-70; prof microbiol, Univ Tex Med Sch Houston, 70-80, assoc dean acad affairs, 70-75, dean, 75-80. *Mailing Add:* 174 Wednesday Hill Rd Durham NH 03824

TUTTLE, RONALD RALPH, b Colorado Springs, Colo, July 10, 36; m 63; c 1. PHARMACOLOGY, PHYSIOLOGY. *Educ:* Colo Col, BA, 60; Univ Man, MS, 64, PhD(pharmacol), 66. *Prof Exp:* Fel pharmacol, Emory Univ, 66-67; sr pharmacologist, Lilly Res Labs, 67-71, res scientist, 71-74, res assoc pharmacol. 74-78, res adv, 78-; VPRES & DIR, NEW DRUG DEVELOP, KEY PHARMACEUT INC. *Concurrent Pos:* Mem coun circulation fel, Am Heart Asn. *Mem:* Am Soc Pharmacol & Exp Therapeut. *Res:* Cardiovascular pharmacology. *Mailing Add:* Dey Pharmaceuticals PO Box 694307 Miami FL 33169

TUTTLE, RUSSELL HOWARD, b Marion, Ohio, Aug 18, 39; m 68; c 2. PRIMATOLOGY, PALEOANTHROPOLOGY. *Educ:* Ohio State Univ, BSc, 61, MA, 62; Univ Calif, Berkeley, PhD(anthrop), 65. *Prof Exp:* PROF ANTHROP, UNIV CHICAGO, 64-, ASSOC SCIENTIST PRIMATOLOGY, YERKES PRIMATE RES CTR, 70- *Concurrent Pos:* Wenner-Gren res grants, Univ Chicago, 65, 66 & 69, NSF res grants, 66-77 & 85-89, USPHS res career develop award, 68-73; vis res prof, Japan Soc Promotion Sci, 74 & 80; Guggenheim fel, 85-86; managing ed, Int J Primatology, 89- *Honors & Awards:* Medal of Foundation, Singer-Polignac. *Mem:* AAAS; Am Anthrop Asn; Am Asn Phys Anthropologists; Int Primatological Soc; Sigma Xi. *Res:* Behavior and comparative functional morphology of anthropoid primates and available fossils in order to elucidate the evolution of human bipedalism, tool use and other subsistence behaviors. *Mailing Add:* Dept Anthrop Univ Chicago 1126 E 59th St Chicago IL 60637

TUTTLE, SHERWOOD DODGE, b Medford, Mass, June 8, 18; m 41; c 3. GEOMORPHOLOGY. *Educ:* Univ NH, BS, 39; Wash State Univ, MS, 41; Harvard Univ, MA & PhD(geol), 53. *Prof Exp:* Instr geol, Wash State Univ, 41 & 46-48; from asst prof to prof geol, Univ Iowa, 52-88, chmn dept geol, 63-68, assoc dean col lib arts, 70-84, EMER PROF GEOL, UNIV IOWA, 88- *Concurrent Pos:* Res assoc, Woods Hole Oceanog Inst, 59-66; Fulbright lectr, Chinese Univ Hong Kong, 68-69. *Mem:* Col Fel Geol Soc Am; Nat Asn Geol Teachers; Am Inst Prof Geologists. *Res:* Fluvial geomorphology; geology of national parks and national military parks. *Mailing Add:* 21 George St Iowa City IA 52246-1901

TUTTLE, THOMAS R, JR, b Somerville, Mass, Mar 28, 28; m 54; c 3. PHYSICAL CHEMISTRY. *Educ:* Northeastern Univ, BS, 53, MS, 55; Wash Univ, St Louis, PhD, 57. *Prof Exp:* Actg asst prof chem, Stanford Univ, 57-60; from asst prof to assoc prof, 60-83, PROF CHEM, BRANDEIS UNIV, 83- *Mem:* Am Chem Soc; Am Phys Soc; AAAS; Am Inst Chem; NY Acad Sci. *Res:* Determination of structures of chemical species in liquid solutions; thermodynamics of electrolytic solutions; properties of solvated electrons and metal solutions in polar solvents. *Mailing Add:* Dept Chem Brandeis Univ Waltham MA 02254

TUTTLE, WARREN WILSON, b Fulton, Mo, Aug 2, 30; m 52; c 4. NEUROPHARMACOLOGY. *Educ:* Univ Mo, BA, 52; Univ Kans City, BS, 58, MS, 60; Univ Calif, San Francisco, PhD(pharmacol), 66. *Prof Exp:* Asst prof pharmacol, Univ Mo-Kans City, 65-68; asst prof med pharmacol & therapeut, Sch Med, Univ Calif, Irvine, 68-69; asst prof pharmacol, Univ Mo-Kans City, 69-72; Actg chmn dept, 72-73, assoc prof & chmn dept, 72-83, PROF & CHMN DEPT PHARMACOL, COL OSTEOP MED, UNIV HEALTH SCI, 83- *Mem:* Sigma Xi. *Res:* Drug metabolism; electroencephalographic investigation into the sites of action of various drugs in the central nervous system. *Mailing Add:* Dept Pharmacol Col Osteop Med Univ Health Sci Kansas City MO 64124

TUTUPALLI, LOHIT VENKATESWARA, b Guntur, Andhra Pradesh, India, Aug 10, 45; m 74. PHARMACOGNOSY, PHYTOCHEMISTRY. *Educ:* Andhra Univ, BS, 63; Bombay Univ, BS, 66, MS, 68; Univ of the Pac, PhD(pharmacog), 74. *Prof Exp:* Res pharmacist product develop, M/S Pfizer (India), Ltd, Bombay, 68-69; from teaching asst to instr pharmacog, Univ of the Pac, 69-74; RES SCIENTIST, CALIF CEDAR PROD RES LAB, 74- *Mem:* Am Pharmaceut Asn; Am Chem Soc; NY Acad Sci; Am Soc Pharmacog; Forest Prod Res Soc. *Res:* Investigating the economic uses of forest products. *Mailing Add:* 1131 Stanton Way Stockton CA 95207-2536

TUTWILER, GENE FLOYD, b Peoria, Ill, Sept 19, 45; m 68; c 2. BIOCHEMISTRY, ENDOCRINOLOGY. *Educ:* Western Ill Univ, BS, 67; Univ Mich, Ann Arbor, PhD(biochem), 70. *Prof Exp:* Sr scientist, McNeil Pharmaceut, Inc, 70-74, group leader, 74-77, res fel, 77-80, sect head endocrinol & metab, Dept Biol Res, 80-84; dir biochem, Ayerst Res Labs, Inc, 84-86; dir, Biol Res, 86-87, DIR, PROJ MGT, MCNEIL PHARMACEUT, 87- *Concurrent Pos:* Teaching fel biochem, Univ Mich, Ann Arbor, 67-70; vis asst prof, Temple Univ, 73-74 & Bucks County Community Col, 74- & Univ Pa, 77-78; adj asst prof, Temple Univ, 79-83 & Univ Pa, 80-83; adj assoc prof, Temple Univ, 83- & Univ Pa, 83- *Mem:* Am Endocrine Soc; Am Diabetes Asn; Soc Exp Biol & Med; Am Soc Biochem Molecular Genetics; Am Soc Pharmacol & Exp Therapeut. *Res:* Diabetes; obesity; protein purification; isolation pituitary proteins; free fatty acid metabolism; carbohydrate metabolism; atherosclerosis. *Mailing Add:* McNeil Pharmaceut, New Bus Develop Welsh & McKean Rds Springhouse PA 19477-0776

TUUL, JOHANNES, b Tarvastu, Estonia, May 23, 22; US citizen; div; c 3. PHYSICS. *Educ:* Stockholm Univ, BS, 55, MA, 56; Brown Univ, ScM, 57, PhD(physics), 60. *Prof Exp:* Instr elec eng, Stockholm Tech Inst, 47-49; res engr, Elec Prospecting Co, Sweden, 49-53; elec engr, L M Ericsson Tel Co, 54-55; res asst physics, Brown Univ, 55-57, 58-60, res assoc, 60; res physicist, Stamford Res Labs, Am Cyanamid Co, Conn, 60-62; sr res physicist, Bell & Howell Res Ctr, Calif, 62-65; from asst prof to assoc prof, 65-74, chmn dept physics & earth sci, 71-75, PROF PHYSICS, CALIF STATE POLYTECH UNIV, 74- *Concurrent Pos:* Consult, Bell & Howell Res Ctr, 65 & Teledyne Inc, 68; vis assoc prof, Pahlavi Univ, Iran, 68-70; resident dir, Calif State Univ & Col Int Prog, Sweden & Denmark, 77-78. *Mem:* AAAS; Am Phys Soc; Am Chem Soc; Am Vacuum Soc; Am Asn Physics Teachers; NY Acad Sci. *Res:* Adsorption of gases on solids; low-energy electron diffraction studies of effects of adsorption and ion bombardment on initially clean surfaces; ultra-high vacuum technology; physics education; energy conservation and new energy technologies. *Mailing Add:* Dept Physics Calif State Polytech Univ Pomona CA 91768

TUVE, RICHARD LARSEN, b Canton, SDak, Feb 1, 12; m 36; c 2. PHYSICAL CHEMISTRY, INORGANIC CHEMISTRY. *Educ:* Am Univ, BA, 35. *Hon Degrees:* DSc, Carleton Col, 61. *Prof Exp:* Asst chemist, Res Assocs, Inc, 35-38; asst chemist, US Naval Res Lab, 38-40, assoc phys chemist, 40-41, head, Eng Res Br, 41-70; consult, Appl Physics Lab, Johns Hopkins Univ, Silver Spring, 70-80; RETIRED. *Concurrent Pos:* Mem comt fire res conf, Nat Acad Sci-Nat Res Coun, 55-68. *Mem:* Fel AAAS; Am Chem Soc; Am Inst Chem Eng; fel Am Inst Chem; Inst Chem Eng. *Res:* Chemistry of fire extinguishment; foam extinguishment methods and materials; flame propagation; surface chemistry; special explosives; fire fighting equipment design; combustion inhibition. *Mailing Add:* 9211 Crosby Rd Silver Spring MD 20910

TUVESON, ROBERT WILLIAMS, b Chicago, Ill, Aug 30, 31. GENETICS, BOTANY. *Educ:* Univ Ill, BS, 54, MS, 56; Univ Chicago, PhD(bot), 59. *Prof Exp:* Asst prof biol, Wayne State Univ, 59-61; asst prof bot, Univ Chicago, 61-68; assoc prof, 68-77, PROF BOT, UNIV ILL, URBANA-CHAMPAIGN, 77- *Concurrent Pos:* NIH spec fel, Dartmouth Col, 65-66. *Mem:* AAAS; Mycol Soc Am; Bot Soc Am; Genetics Soc Am. *Res:* Fungal genetics; viruses of fungi; radiation sensitivity. *Mailing Add:* Dept Bot Univ Ill 505 S Goodwin Ave Urbana IL 61801

TUZAR, JAROSLAV, b Czech, Mar 25, 15; nat US; m 48; c 1. MATHEMATICS. *Educ:* Charles Univ, Prague, MA, 39 & 45, ScD(math), 48. *Prof Exp:* Asst prof, State Tech Col, Prague, 45-48; Rockefeller Found fel, Univ Chicago, 48-50; dir control & res lab, Salerno-Megowen Biscuit Co, 50-70; from assoc prof to prof, 70-84, EMER PROF MATH, NORTHWESTERN UNIV, 84- *Concurrent Pos:* Lectr, Northwestern Univ, 60-70. *Mem:* Math Asn Am; Am Statist Asn. *Res:* Mathematical probability and statistics; pedagogy of mathematics. *Mailing Add:* 8929 Elmore St Niles IL 60648

TUZSON, JOHN J(ANOS), b Budapest, Hungary, Apr 29, 29; US citizen; m 60; c 3. FLUID MECHANICS, ROTATING MACHINERY. *Educ:* Conserv Nat Arts et Metiers, France, MMech Eng, 55; Mass Inst Technol, ScD(mech eng), 59. *Prof Exp:* Res asst mech eng, Mass Inst Technol, 56-58; res engr fluid dynamics, Whirlpool Corp, Mich, 58-61; assoc prof, Mich State Univ, 61; res engr fluid mech, IIT Res Inst, 62-63; asst dir, Res Ctr, Borg-Warner Corp, Des Plaines, 63-79; mgr fluid technol, Allis-Chalmers, Milwaukee, 79-83; PROG MGR, GAS RES INST, CHICAGO, 83- *Concurrent Pos:* Teaching asst, Ill Inst Technol, 62-72; ed, Proc Nat Conf Fluid Power, 62-72. *Mem:* Am Soc Mech Engrs. *Res:* Swirling flow; two phase flow; turbomachinery; hydraulic, pneumatic and fluidic controls; process equipment; slurry; industrial power generation; gas turbines; rotary engines. *Mailing Add:* 1220 Maple Ave Evanston IL 60202

TUZZOLINO, ANTHONY J, b Chicago, Ill, July 1, 31; m 54; c 2. PHYSICS, SOLID STATE PHYSICS. *Educ:* Univ Chicago, MS, 55, PhD(physics), 58. *Prof Exp:* PHYSICIST, UNIV CHICAGO, 58- *Res:* Semiconductor nuclear particle and photon detectors. *Mailing Add:* 6615 N Knox Lincolnwood IL 60646

TWARDOCK, ARTHUR ROBERT, b Normal, Ill, July 20, 31; m 54; c 4. VETERINARY PHYSIOLOGY, NUCLEAR MEDICINE. *Educ:* Univ Ill, BS, 54, DVM, 56; Cornell Univ, PhD(animal physiol), 61. *Prof Exp:* Vet practioner, Hillcrest Animal Hosp, 56-57; res asst radiobiol, Cornell Univ, 57-60, res assoc phys biol, 60-62; from asst prof to assoc prof, Univ Ill, Urbana, 62-70, actg head, Dept Vet Physiol & Pharm, 74-77, assoc dean acad affairs, 73-86, PROF, VET PHYS, UNIV ILL, URBANA, 70-, ACTG DEAN, COL VET MED, 89- *Mem:* AAAS; Am Physiol Soc; Am Soc Vet Physiol & Pharmacol; Am Vet Med Asn; Conf Res Workers Animal Diseases; Sigma Xi; Soc Nuclear Med. *Res:* Mineral metabolism; placental transfer of mineral elements; applications of radioisotope techniques in veterinary nuclear medicine. *Mailing Add:* Col Vet Med Univ Ill 220 Large Animal Clin 1102 Hazelwood Urbana IL 61801

TWARDOWSKI, ZBYLUT JOZEF, b Stanislawice, Poland, June 2, 34; US citizen; m 58; c 2. NEPHROLOGY, ARTIFICIAL INTERNAL ORGANS. *Educ:* Copernicus Med Acad, Krakow, Poland, MD, 59, PhD(med), 64; Silesian Med Acad, Katowice, Poland, 75. *Prof Exp:* Dir internal med, Hospital for Miners, Bytom, Poland, 63-76; dir dept nephrology, Med Acad, Lublin, Poland, 76-82; PROF, DEPT MED, UNIV MO, 85- *Concurrent Pos:* Mem Transplantation & Dialysis Comt, Ministry of Health & Social Servs, 66-82; consult, Baxter-Travenol, 82-, prin investr, peritoneal dialysis, 83-; dir, Outpatient Peritoneal Dialysis Prog, Univ Mo, 85-; mem prog comt, Am Soc Artificial Internal Organs, 87- *Mem:* Europ Dialysis & Transplant Asn; Int Soc Nephrology; Am Col Physicians; Am Soc Artificial Internal Organs; Biomed Eng Soc; Int Peritoneal Dialysis Soc, mem-founder, 84- *Res:* Kinetics of peritoneal dialysis and modifications of peritoneal dialysis to achieve adequate treatment of renal failure; introduced two major modifications; access (catheter) for peritoneal dialysis and hemodialysis; developed new peritoneal dialysis catheters which when used are associated with markedly decreased complications. *Mailing Add:* 304 Devine Ct Columbia MO 65203

TWAROG, BETTY MACK, b New York, NY, Aug 28, 27; m 47; c 1. PHYSIOLOGY, NEUROSCIENCES. *Educ:* Swarthmore Col, AB, 48; Tufts Col, MS, 49; Radcliffe Col, PhD(biol), 52. *Prof Exp:* Asst, Harvard Univ, 52 & Res Div, Cleveland Clin, 52-53; res assoc & instr, Tufts Col, 53-55; res fel, Harvard Univ, 55-58; instr, 58-60; USPHS trainee, Oxford Univ, 60-61; asst prof physiol & biophys, Sch Med, NY Univ, 61-65; res fel, Harvard Univ, 65-66; prof biol, Tufts Univ, 66-75; prof physiol & anat sci, State Univ NY, Stony Brook, 75-81; prof & chmn biol, Bryn Mawr Col, 82-85; res scientist, Bockus Res Inst, 84-91; RES SCIENTIST BIGELOW LAB OCEAN SCI, 90- *Concurrent Pos:* John Simon Guggenheim Mem fel, Melbourne Univ, Australia, 72-73; prog dir, NSF, 77-78 & 90; adj prof physiol, Univ Pa, 84-91. *Mem:* Soc Gen Physiologists (pres, 78-79); Biophys Soc; Am Physiol Soc; fel AAAS; Am Soc Zoologists. *Res:* Physiology and pharmacology of smooth muscle; neurophysiology; neuropharmacology; control of contraction and relaxation in molluce catch muscle; developmental and environmental regulation of vascular smooth muscle phenotype; marine neurotoxins associated with algal blooms; effects on vestor species. *Mailing Add:* Bigelow Lab Ocean Sci West Boothbay Harbor ME 04575

TWAROG, BRUCE ANTHONY, b Chester, Pa, Aug 10, 52; m 77; c 1. PHOTOMETRY. *Educ:* Case Western Reserve Univ, BS, 74; Yale Univ, MS, 78, PhD(astron), 80. *Prof Exp:* Asst prof astron, Univ Tex, Austin, 80-82; asst prof, 82-87, ASSOC PROF ASTRON, UNIV KANS, 87- *Concurrent Pos:* Mem time allocation comt, Cerro Tololo InterAm Observ, 82-86; mem sci adv comt, NASA-Univ Ariz Astrometric Space Telescope, 85-86. *Honors & Awards:* Trumpler Prize, Astron Soc Pac, 82. *Mem:* Am Astron Soc; Int Astron Union; Sigma Xi. *Res:* Galactic evolution and stellar populations; stellar photometry. *Mailing Add:* Dept Physics & Astron Univ Kans Lawrence KS 66045

TWAROG, ROBERT, b Lowell, Mass, Mar 17, 35; m 58; c 2. MICROBIAL BIOCHEMISTRY. *Educ:* Univ Conn, BS, 56, MS, 58; Univ Ill, PhD(microbiol), 62. *Prof Exp:* Lab officer, USAF Epidemiol Lab, San Antonio, Tex, 62-65; asst prof, 65-74, ASSOC PROF BACT, SCH MED, UNIV NC, CHAPEL HILL, 74- *Mem:* AAAS; Am Soc Microbiol. *Res:* Control mechanisms of microbial processes; molecular biology; microbiology. *Mailing Add:* Dept Microbiol & Immunol Med Sch Univ NC Chapel Hill NC 27514

TWAY, PATRICIA C, b Worcester, Mass, Sept 29, 45; m 78; c 2. CHROMATOGRAPHY, MASS SPECTROMETRY. *Educ:* Mt Holyoke Col, BA, 67; Rutgers Univ, MS, 69; Seton Hall Univ, PhD(anal chem), 80. *Prof Exp:* Staff chemist anal res, 69-70, res chemist, 70-73, sr res chemist, 73-78, res fel, 78-83, assoc dir, 84-86, dir, 86-88, sr dir, 88-90, EXEC DIR, MERCK SHARP & DOHME RES LAB, MERCK & CO, 90- *Mem:* Am Chem Soc; Am Soc Mass Spectrometry. *Res:* Analytical methodology to support pharmaceutical development process research; test methods and specifications for raw materials, process intermediates and bulk drug. *Mailing Add:* Merck & Co Inc PO Box 2000 Rahway NJ 07065

TWEDT, ROBERT MADSEN, b Rochester, Minn, July 4, 24; m 72. MICROBIOLOGY. *Educ:* Univ Minn, BS, 45; Univ Colo, MS, 49, PhD(microbiol), 52. *Prof Exp:* Res fel microbiol, Western Reserve Univ, 52-54; asst scientist, Univ Minn, 54-56, res fel physiol chem, 56-59; from asst prof to assoc prof biol, Univ Detroit, 59-67; res microbiologist, Nat Ctr Urban & Indust Health, 67-69, from microbiologist to asst chief, Food Microbiol Br, 69-77, actg chief, 78-79, CHIEF, BACTERIAL PHYSIOL BR, FOOD & DRUG ADMIN, 80- *Concurrent Pos:* Am Cancer Soc fel, 52-53; microbiologist, Fed Water Pollution Control Admin, 66; adj prof, Univ Detroit, 68-70. *Mem:* Am Soc Microbiol; fel Am Acad Microbiol; AAAS; Int Asn Milk, Food & Environ Sanitarians; Sigma Xi. *Res:* Research and field investigations to identify, evaluate and resolve microbiological problems of public health significance associated with foods. *Mailing Add:* Food & Drug Admin 200 C St SW Washington DC 20204

TWEED, DAVID GEORGE, b Troy, NY. SYSTEMS DESIGN. *Educ:* Mass Inst Technol, BS, 66. *Prof Exp:* Engr, Draper Labs, 66-67; sr engr, Electronic Image Systs, 67-72; prod mgr, Brattle Instruments, 72-77; VPRES ENG, GEN SCANNING, 77- *Res:* Research and development of advance laser based electro-photographic techniques. *Mailing Add:* 180 South St Boston MA 02111

TWEED, JOHN, b Greenock, Scotland, Mar 29, 42; m 66; c 2. APPLIED MATHEMATICS. *Educ:* Univ Strathclyde, MSc, 65; Univ Glasgow, PhD(appl math), 68, DSc, 81. *Prof Exp:* From asst lectr to lecr math, Univ Glasgow, 65-69, vis asst prof, NO State Univ, 69-70; Univ Glasgow, 70-73; vis prof, NC State Univ, 73-74; assoc prof math, 74-77, PROF MATH & COMPUT SCI, OLD DOMINION UNIV, 77- *Mem:* Soc Indust & Appl Math; fel Brit Inst Math & Appln; Am Acad Mech. *Res:* Applications of transform techniques and integral equations to the solution of mixed boundary value problems in fracture mechanics. *Mailing Add:* Dept Math Old Dominion Univ 5215 Hampton Blvd Norfolk VA 23508

TWEEDDALE, MARTIN GEORGE, b Bristol, Eng, Aug 22, 40. CLINICAL PHARMACOLOGY. *Educ:* King's Col, Univ London, BSc, 62, PhD(pharmacol), 65; Westminister Med Sch, Univ London, MB, BS, 67; FRCPS(C), 72. *Prof Exp:* Asst prof, 73-78, ASSOC PROF MED CLIN PHARMACOL, FAC MED, MEM UNIV NFLD, 78-; ASSOC PHYSICIAN INTERNAL MED, GEN HOSP, ST JOHN'S, NFLD, 73-, CHMN INTENSIVE CARE UNIT, 80- *Concurrent Pos:* Develop grant, Can Found Advan Clin Pharmacol, 73-77. *Mem:* Can Soc Clin Invest; Can Crit Care Soc; Can Pharmacol Soc; Am Soc Clin Pharmacol & Therapeut; Soc Critical Care Med. *Res:* Plasma inhibitors of pulmonary surfactant and their role in human pulmonary disorders; isotopic investigation of physiological fluid volumes in the critically ill; the effects of and indications for high frequency jet ventilation. *Mailing Add:* Vancouver Gen Hosp - ICC 855 W 12th Ave Vancouver BC V5Z 1M9 Can

TWEEDELL, KENYON STANLEY, b Sterling, Ill, Mar 28, 24; m 56; c 5. DEVELOPMENTAL BIOLOGY, ONCOLOGY. *Educ:* Univ Ill, BS, 47, MS, 49, PhD(zool,physiol), 53. *Prof Exp:* Res assoc biol, Control Systs Lab, Univ Ill, 51-54; from instr to asst prof zool, Univ Maine, 54-58; from asst prof to assoc prof, 58-68, PROF BIOL, UNIV NOTRE DAME, 68- *Concurrent Pos:* Corp mem, Marine Biol Lab, Woods Hole. *Mem:* AAAS; Int Soc Develop Biol; Am Soc Zool; Sigma Xi; Soc Develop Biol. *Res:* Experimental pathology; transmissable tumors of Amphibia; developmental biology, especially oogenesis, ovulation, regeneration in invertebrates, cytodifferentiation in normal and malignant cells. *Mailing Add:* Dept Biol Sci Univ Notre Dame Notre Dame IN 46556-0369

TWEEDIE, ADELBERT THOMAS, b Saginaw, Mich, Jan 5, 31; m 53; c 3. POLYMER CHEMISTRY. *Educ:* Univ Mich, BSCh, 53; Univ Ill, PhD(org chem), 56. *Prof Exp:* Res chemist, Dow Chem Co, Mich, 56-58 & Aerojet-Gen Corp, Calif, 58-62; tech supvr rocket propellants, Union Carbide Corp, WVa, 62-64; mgr mat eng, Space Div, Gen Elec Co, King of Prussia, 64-89, chief scientist, 88-89; RETIRED. *Concurrent Pos:* Mem Mat Adv Comt, NASA. *Mem:* Am Inst Aeronaut & Astronaut; Am Chem Soc; AAAS. *Res:* Behavior of materials in the space environment; vibration damping; application of materials to spacecraft; polymers; solar energy; fracture mechanics of viscoelastic materials. *Mailing Add:* 218 Chester Rd Devon PA 19333

TWEEDIE, VIRGIL LEE, b Norborne, Mo, Feb 18, 18; m 43; c 3. ORGANIC CHEMISTRY. *Educ:* Univ Mo, AB, 41, MA, 43; Univ Tex, PhD(chem), 51. *Prof Exp:* Res chemist, Commercial Solvents Corp, 42-46; asst prof, Baylor Univ, 46-48, assoc prof, 50-53, prof chem, 53-88; RETIRED. *Mem:* Sigma Xi; Am Chem Soc; Nat Asn Adv Health Professions. *Res:* Allylic compounds; organometallics; complex metal hydrides and alkides; hydrogenolysis. *Mailing Add:* 7720 Tallahassee Rd Waco TX 76712

TWEEDLE, CHARLES DAVID, b Astoria, Ore, Jan 22, 44; m 82. NEUROBIOLOGY. *Educ:* Univ Ore, BA, 66, MA, 67; Mich State Univ, PhD(zool), 70. *Prof Exp:* NIH fel, Yale Univ, 70-72; NIH res grant & res assoc, 72-73; asst prof, Dept Biomech & Zool, 73-78, assoc prof anat & zool, 78-84, PROF ANAT, MICH STATE UNIV, 84- *Mem:* AAAS; Soc Neurosci; Am Asn Anatomists. *Res:* Developmental neurobiology; trophic effects of nerves; neuromuscular development; nerve plasticity; nerve/glia interactions. *Mailing Add:* Dept Anat Mich State Univ East Lansing MI 48824

TWEEDY, BILLY GENE, b Cobden, Ill, Dec 31, 34; m 57; c 3. PLANT PATHOLOGY. *Educ:* Univ Southern Ill, BS, 56; Univ Ill, MS, 59, PhD(plant path), 61. *Prof Exp:* Asst plant pathologist, Boyce Thompson Inst Plant Res, 61-65; asst prof plant path, Univ Mo-Columbia, 65-69, from assoc prof to prof, 69-73; mgr residue invest, 73-78, DIR, BIOCHEM DEPT, CIBA-GEIGY CORP, 78- *Concurrent Pos:* USDA, 71-72. *Mem:* AAAS; Am Phytopath Soc; Am Chem Soc; Weed Sci Soc Am; Sigma Xi. *Res:* Degradation of pesticides; integrated pest control; fungus physiology; fruit pathology. *Mailing Add:* CIBA Geigy PO Box 18300 Greensboro NC 27419-8300

TWEEDY, JAMES ARTHUR, b Cobden, Ill, Nov 29, 39; m 64; c 2. HORTICULTURE. *Educ:* Southern Ill Univ, BS, 62; Mich State Univ, MS, 64, PhD(hort), 66. *Prof Exp:* From asst prof to assoc prof, 66-74, asst dean, Sch of Agr, 74-75, asst vpres acad affairs & res, 76-78, dean, Col Agr, 86-88, PROF PLANT INDUST, PLANT & SOIL SCI DEPT, SOUTHERN ILL UNIV, 74- *Mem:* Weed Sci Soc Am. *Res:* Influence of herbicides on plant physiological processes; evaluation of herbicides for weed control in agronomic and horticultural crops. *Mailing Add:* Col Agr Southern Ill Univ Carbondale IL 62901-4416

TWEET, ARTHUR GLENN, b Aberdeen, SDak, Sept 20, 27; m 50; c 3. SOLID STATE PHYSICS. *Educ:* Harvard Univ, AB, 48; Univ Wis, MS, 49, PhD(physics), 53. *Prof Exp:* Physicist, Semiconductor Sect, Res Labs, Gen Elec Corp, 53-59, liaison scientist, 59-61, physicist, Biol Studies Sect, 61-64; mgr phys imaging br, Advan Imaging Technol Lab, 64-68, mgr, Imaging Res Lab, 68-72, mgr technol planning, Info Technol Group, 72-75, MGR, TECH STRATEGIC PLANNING, XEROX CORP, 75- *Concurrent Pos:* Adj prof, Rensselaer Polytech Inst, 62. *Mem:* Fel Am Phys Soc. *Res:* Energy transfer in excited molecules; unconventional photographic systems; dye sensitized reactions; electrical and surface properties of polymers; surface chemistry of chromophores; optical properties of large molecules; imperfections in semiconductors and insulators; nonaqueous electrochemistry; decision analysis; technological forecasting; operations analysis and modelling. *Mailing Add:* Xerox Corp Xerox Sq 280 Weymouth Dr Rochester NY 14625

TWELVES, ROBERT RALPH, b Chicago, Ill, Nov 4, 27; m 53; c 3. ORGANIC CHEMISTRY. *Educ:* Univ Utah, BS, 50, MA, 52; Univ Minn, PhD(org chem), 57. *Prof Exp:* Asst, Univ Utah, 50-51; process develop chemist, Merck & Co, 52-53; asst, Univ Minn, 53-55; res chemist, E I du Pont de Nemours & Co, Inc, 57-88; RETIRED. *Mem:* Am Chem Soc; Sigma Xi. *Res:* Fluorochemicals; dyes and intermediates; synthetic organic chemistry; chemicals for elastomers; analytical chemistry of polymers. *Mailing Add:* 236 Duncan Ave McDaniel Cr Wilmington DE 19803

TWENHOFEL, WILLIAM STEPHENS, b Madison, Wis, Aug 30, 18; m 51; c 5. GEOLOGY. *Educ:* Univ Wis, BA, 40, PhD(geol), 52. *Prof Exp:* Geologist, US Geol Surv, Washington, DC, 42-45; physicist, US Naval Res Lab, 45-46; geologist, US Geol Surv, 47-64, chief, Br Spec Projs, 64-81. *Concurrent Pos:* Consult, 81. *Honors & Awards:* Meritorious Serv Award, US Dept Interior, 73 & Distiguished Serv Award, 79. *Mem:* AAAS; Geol Soc Am; Asn Eng Geol. *Res:* Geology of Alaska; growth of artificial crystals; geologic disposal of radioactive waste; structural and uranium geology; engineering geology of underground nuclear explosions. *Mailing Add:* 820 Estes St Lakewood CO 80215

TWENTE, JOHN W, b Lawrence, Kans, Dec 18, 26; m 53; c 1. ZOOLOGY. *Educ:* Univ Kans, AB, 50; Univ Mich, MS, 52, PhD(zool), 54. *Prof Exp:* Interim instr biol, Univ Fla, 54-55; instr zool, Col Pharm, Univ Ill, 55-56; instr, Univ Utah, 57-58, asst prof, 58-62, res biologist, 62-66; ASSOC PROF BIOL & INVESTR, DALTON RES CTR, UNIV MO-COLUMBIA, 66- *Mem:* Am Soc Zool; Ecol Soc Am; Am Soc Mammal; fel AAAS. *Res:* Physiological ecology and behavior; hibernation physiology. *Mailing Add:* Bio Sci 105 Tucker Hall Univ of Mo Columbia MO 65211

TWERSKY, VICTOR, b Poland, Aug 10, 23; nat US; m 50; c 3. MATHEMATICAL PHYSICS, MULTIPLE SCATTERING THEORY. *Educ:* City Col New York, BS, 47; Columbia Univ, AM, 48; NY Univ, PhD(physics), 50. *Prof Exp:* Assoc, Guid Device Proj, Biol Dept, City Col New York, 46-49; asst physics, NY Univ, 49, res assoc electromagnetic theory, Inst Math Sci, 50-53; assoc, Nuclear Develop Assocs, 51-53; specialist theoret physics, Electronic Defense Labs, Sylvania Electronic Systs-West, Sylvania Elec Prod, Inc Div, Gen Tel & Electronics Corp, 53-58, sr specialist & lab consult, 58-60, sr scientist, 60-66, head res, Electronics Defense Labs, 58-66 & Sylvania Electronic Systs-West, 64-66; PROF MATH, UNIV ILL, CHICAGO CIRCLE, 66 - *Concurrent Pos:* Mem tech res comt, Am Found Blind, 47-49; lectr math, Stanford Univ, 56-58; assoc ed, J Optical Soc, 61-68, subj ed, Scattering & Radiation, J Acoust Soc, 63-66, Electromagnetic Theory, I R E Trans Antennas & Propagation, 65-66, J Geophysical Res, 65-68, J Math Physics, 77-79, J Appl Math, 72-83; vis prof or vis scholar, Technion-Israel Inst Technol, 62-63, Courant Inst Math Sci, 63, Stanford Univ, 67-88, Hebrew Univ Jerusalem, 72 & Weizmann Inst Sci, Rehovoth & Ben-Gurion Univ Negev, Beer Sheva, 79; feature ed, Microwaves & Optics, J Appl Optics, 65; consult, Sylvania Elec Prod Inc Div, Gen Tel & Electronics Corp, 66; mem ctr advan study, Univ Ill, 69-70; Guggenehim fel, 72-73 & 79-80; mem at large, Conf Bd Math Sci, 75-77; mem US Comn B, Int Union Radio Sci. *Mem:* Fel AAAS; fel Am Phys Soc; fel Acoust Soc Am; fel Optical Soc Am; fel Inst Elec & Electronics Engrs; Am Math Soc; Soc Ind & Appl Math. *Res:* Multiple scattering of electromagnetic and acoustic waves; rough surfaces; gratings; scattering and propagation in random distributions; radiative diagnostics of biological cells; relativistic scattering; applied mathematics; obstacle perception by the blind. *Mailing Add:* Dept Math Univ Ill Box 4348 Chicago IL 60680

TWETO, OGDEN, b Abercrombie, NDak, June 10, 12; m 40; c 2. ECONOMIC GEOLOGY. *Educ:* Univ Mont, AB, 34, MA, 37; Univ Mich, PhD(geol), 47. *Prof Exp:* Instr, Univ NC, 39-40; geologist, 40-61, chief, South Rockies Br, 61-65, asst chief geologist & chief off econ geol, 65-68, RES GEOLOGIST, US GEOL SURV, 68- *Mem:* AAAS; Geol Soc Am; Soc Econ Geol; Mineral Soc Am; Am Inst Mining, Metall & Petrol Eng. *Res:* Geology and mineral deposits of Southern Rocky Mountains. *Mailing Add:* 1995 Taft Dr Denver CO 80215

TWIDWELL, LARRY G, b Jackson, Mo, July 5, 39; m 61; c 3. EXTRACTIVE & CHEMICAL METALLURGY. *Educ:* Mo Sch Mines, BS, 61, MS, 62; Colo Sch Mines, DSc, 66. *Prof Exp:* Develop engr, Rocky Flats Div, Dow Chem Co, 62-63, PROF METALL ENGR & PRES, MONLANES ENVIRON, 83-; assoc metallurgist, Monsanto Res Corp, 65-67; staff mem nuclear reactors, Reactor Develop Div, Sandia Corp, 67-69; assoc prof, Mont Col Mineral Sci & Technol, 69-76, prof metall eng & head dept, 76-82. *Concurrent Pos:* Proj officer, Environ Protection Agency, 78-80. *Mem:* Am Inst Mining, Metall & Petrol Engrs; Am Soc Eng Educ; Am Inst Mining & Metall Engrs. *Res:* Containment of radioisotopes at elevated temperatures; reactor development; thermodynamics of metallic solutions; nonferrous and ferrous process metallurgy; environmental and thermodynamics of metallic solutions; self-paced instruction; treatment of wastes, hydromettalurgy. *Mailing Add:* 54 Apple Orchard Rd Butte MT 59701

TWIEG, DONALD BAKER, b Port Arthur, Tex, Dec 8, 44; m 69; c 2. BIOMEDICAL NUCLEAR MAGNETIC RESONANCE IMAGING. *Educ:* Rice Univ, BA, 68, MS, 71; Southern Methodist Univ, PhD(biomed eng), 77. *Prof Exp:* Engr, Boeing Co, 72-73; asst prof, Dept Radiol, Univ Tex Health Sci Ctr, Dallas, 77-87; AT PHILIPS MED CTR , SAN FRANCISCO VET ADMIN MED CTR, 88- *Concurrent Pos:* Fac mem, Biomed Eng Grad Prog, Univ Tex Health Sci Ctr, Dallas, 77-, Radiol Sci Grad Prog, 82- *Mem:* Inst Elec & Electronics Engrs; AAAS; Soc Magnetic Resonance Imaging. *Res:* Theoretical investigations of nuclear magnetic resonance imaging processes; modeling of nuclear magnetic resonance imaging performance in biomedical applications; applications of optimal estimation and sampling theory in nuclear magnetic resonance imaging. *Mailing Add:* San Francisco Vet Admin Med Ctr 4150 Clement St San Francisco CA 94121

TWIEST, GILBERT LEE, b Grand Rapids, Mich, Apr 23, 37; m 58; c 3. ORNITHOLOGY, SCIENCE EDUCATION. *Educ:* Mich State Univ, BS, 61, MS, 63; Univ Toledo, PhD(sci educ), 68. *Prof Exp:* Instr biol, Kellogg Community assoc prof, 68-76, PROF SCI EDUC, CLARION UNIV, 76- *Concurrent Pos:* Ed, Newsletter, Coun Elementary Sci Int; NDEA fel, 65-68. *Honors & Awards:* Outstanding Leadership, Coun Elem Sci Int. *Mem:* Nat Asn Res Sci Teaching; Nat Sci Teachers Asn; Coun Elem Sci Int (pres, 83-86); Wilson Ornith Soc. *Res:* Methods of teaching students the processes needed to solve problems. *Mailing Add:* Dept Biol Clarion Univ Clarion PA 16214

TWIGG, BERNARD ALVIN, b Cumberland, Md, Oct 15, 28; m 51; c 4. HORTICULTURE, FOOD SCIENCE. *Educ:* Univ Md, BS, 52, MS, 54, PhD, 59. *Prof Exp:* Asst hort, 52-54, from instr to assoc prof, 54-69, PROF HORT, UNIV MD, COL PARK, 69-, CHMN DEPT, 75- *Mem:* Am Soc Hort Sci; fel Inst Food Technologists. *Res:* Objective evaluation of food products; statistical quality control; food chemistry and physics; horticultural food processing. *Mailing Add:* 3537 Duke College Park MD 20740

TWIGG, HOMER LEE, b Westminster, Md, Apr 10, 26; m 55; c 6. MEDICINE, RADIOLOGY. *Educ:* Univ Md, MD, 51; Am Bd Radiol, dipl, 56. *Prof Exp:* Intern med, USPHS Hosp, Boston, 51-52, intern surg serv outpatient clin, 52, resident radiol, New Orleans, 53-55, Baltimore, 55-56, chief radiol, Detroit, 56-57; actg chmn & dir dept, 67-69, from asst prof to assoc prof, 57-70, chmn dept, 69-79, PROF RADIOL, GEORGETOWN UNIV HOSP, 70- *Concurrent Pos:* Spec assignment, US Dept Interior, 52; consult, Vet Admin Hosp, DC, 60- *Mem:* AMA; fel Am Col Radiol; Am Roentgen Ray Soc; Radiol Soc NAm; Soc Thoracic Radiol. *Mailing Add:* Dept Radiol Georgetown Univ Hosp Washington DC 20007

TWILLEY, IAN CHARLES, b London, Eng, Apr 4, 27; m 54; c 4. CHEMISTRY. *Educ:* Univ London, BSc, 53, FRSC, 62. *Prof Exp:* Sr chemist, Nelsons Silk Ltd, Eng, 53-56 & Micanite & Insulators, Ltd, 56-57; sr develop chemist, Textile Fibers Div, Du Pont of Can, 57-59; sect leader moulding polymers, Nat Aniline Div, Fibers & Plastics Co, Allied Corp, Petersburg, 59-60, group leader polymer res, 60-61, res supvr, 61-65, process develop supvr, 65-66, mgr systs eng, 66-68, mgr eng res, Fibers Div, 68-69, asst chief engr, 69, tech dir polyester, 69-72, tech dir home furnishings, 72-76, mgr govt & indust affairs & liaison, 76-77, mgr advan technol, Fibers Div, Allied Chem Corp, 77-81, dir advan technol, 81-87; DIR, TEXTILE PERFORMANCE & COMFORT RES LABS, UNIV MD, 88- *Mem:* Am Chem Soc; Am Inst Chem Eng; Can Soc Chem Eng; fel Chem Inst Can; Royal Soc Chem; fel Plastics & Rubber Inst. *Res:* Polymeric and textile processes and products; production and design problems; polymer and textile chemistry; polyamide, polyolesine, and polyester technology. *Mailing Add:* 12625 Merry Dr Chester VA 23831

TWINING, LINDA CAROL, b Paterson, NJ, July 8, 52; m 80; c 2. IMMUNOPARASITOLOGY. *Educ:* William Paterson Col NJ, BA, 72; Rutgers Univ, MS, 75; Univ Ill Urbana-Champaign, PhD(zool), 82. *Prof Exp:* Teaching asst biol zool, Rutgers Univ, 72-74, immunol & parasitol, Univ Ill, 74-81; ASSOC PROF IMMUNOL & PARASITOL, NORTHEAST MO STATE UNIV, 81 - *Mem:* Am Soc Microbiol; AAAS. *Res:* Immunoparasitology, particularly the immune response to Plasmodium and other protozoans, both the in vivo and in vitro response are of interest, especially the roles of the various immune cells; phenomenon of immunosuppression in protozoal and helminth infections. *Mailing Add:* Dept Sci Northeast Mo State Univ Kirksville MO 63501

TWINING, SALLY SHINEW, b Bowling Green, Ohio, July 28, 47; m 71; c 1. PROTEASES, PROTEASE INHIBITORS. *Educ:* Bowling Green State Univ, BS, 69, MA, 71; Ohio State Univ, PhD(physiol chem), 76. *Prof Exp:* Instr chem, Bowling Green State Univ, 71-73; Mayo fel, immunol, Mayo Clin, 76-78; res assoc, Med Col Wis, 79-80, instr, 80-83, asst prof, 83- 89, ASSOC PROF BIOCHEM, MED COL WIS, 89- *Mem:* Sigma Xi; Am Chem Soc; AAAS; Am Soc Biol Chem; Asn Res Vision & Ophthalmol. *Res:* Role of proteases and protease inhibitors in corneal degradation; role of the immune system in Pseudomonas keratitis. *Mailing Add:* Dept Biochem Med Col Wis 8701 Watertown Plank Rd Milwaukee WI 53226

TWISS, PAGE CHARLES, b Columbus, Ohio, Jan 2, 29; m 54; c 3. GEOLOGY. *Educ:* Kans State Univ, BS, 50, MS, 55; Univ Tex, PhD(geol), 59. *Prof Exp:* From Instr to assoc prof, 53-69, head dept, 68-77, PROF GEOL, KANS STATE UNIV, 69- *Concurrent Pos:* Co-investr, NSF grants, 60-62, 66, 67-68; res scientist, Univ Tex, 66-67; fel, Pan Am Petrol Found, 57-58, Shell Found, 58-59; geologist, Agr Res Serv, USDA, 66-68. *Mem:* AAAS; fel Geol Soc Am; Am Asn Petrol Geol; Soc Econ Paleont & Mineral; Clay Minerals Soc; Soil Sci Soc Am; Am Soc Agron; Int Soc Soil Sci; Int Asn Sedimentologist; Int pour l'Etude des Argiles. *Res:* Sedimentary and igneous petrology; clay mineralogy; stratigraphy, tectonics, petrology and geochemistry of Mesozoic and Cenozoic rocks of Trans-Pecos Texas and Chihuahua, Mexico; grass phytoliths; petrology of recent dust deposits. *Mailing Add:* 2327 Bailey Dr Manhattan KS 66502

TWISS, ROBERT JOHN, b Baltimore, Md, May 12, 42; m 68; c 2. ROCK DEFORMATION MECHANISMS & STRUCTURES. *Educ:* Yale Univ, BS, 64; Princeton Univ, MA, 68, PhD(geol), 70. *Prof Exp:* NATO fel geol, Australian Nat Univ, 70-71; asst prof, 71-78, ASSOC PROF GEOL, UNIV CALIF, DAVIS, 78- *Mem:* Am Geophys Union; Geol Soc Am; Sigma Xi. *Res:* Continuum mechanics theory applied to understanding the behavior of geologic materials; deformation mechanisms in silicates; structural analysis of tectonites. *Mailing Add:* Dept Geol Univ Calif Davis CA 95616-8605

TWITCHELL, PAUL F, b Somerville, Mass, Mar 7, 32; m 56; c 4. METEOROLOGY. *Educ:* Boston Col, BS, 53, MS, 62; Pa State Univ, BS, 54; Univ Wis-Madison, PhD, 76. *Prof Exp:* Weather officer, US Air Force, 53-57; res engr, Res Dept, Melpar, Inc, Mass, 57-60; sr res engr, Appl Sci Div, 60-62; phys sci coordr, Boston Br, Off Naval Res, 62-72, phys sci adminr, 72-81; vis prof, US Naval Acad, 81-82; dir, Environ Progs, Hq Naval Air Systs Command, 82-86; mgr, Off Naval Res, Washington, 86-88; CONSULT, ENVIRON SCIS, 88- *Mem:* Am Meteorol Soc; Am Geophys Union. *Res:* Physical processes in the terrestrial atmosphere. *Mailing Add:* 36 Laurel Ave Wellesley Hills MA 02181

TWITCHELL, THOMAS EVANS, b Springfield, Ohio, Sept 4, 23; m 56; c 4. NEUROLOGY. *Educ:* Univ Mich, MD, 46. *Prof Exp:* Res fel physiol, Med Sch, Yale Univ, 47; intern neurol, Boston City Hosp, 47-48, res fel, Harvard Med Sch & Boston City Hosp, 48-49; asst resident med, New Eng Ctr Hosp, 54, chief resident neurol, 54-55; from instr to assoc prof, 55-83, actg chmn dept, 83-85, PROF NEUROL, SCH MED, TUFTS UNIV, 83- *Concurrent Pos:* USPHS res fel, Yale Univ, 49-51; res assoc, Mass Inst Technol, 63-; neurologist, New Eng Med Ctr Hosps, 63- *Mem:* AAAS; Asn Res Nerv & Ment Dis; Am Neurol Asn; AMA; Am Fedn Clin Res; Sigma Xi; Am Acad Neurol. *Res:* Neurophysiology of primate motor function; physiologic nature of development of behavior in infants; sensory mechanisms in movement; clinical neurology; applied neurophysiology; neuropsychology. *Mailing Add:* 54 Longfellow Rd Wellesley Hills MA 02181

TWOHY, DONALD WILFRED, b Clackamas, Ore, Sept 9, 24; m 55; c 1. PARASITOLOGY. *Educ:* Ore State Col, BS, 48, MS, 51; Johns Hopkins Univ, ScD, 55. *Prof Exp:* Aquatic biologist, Ore State Game Comn, 49-51; res asst, Sch Hyg & Pub Health, Johns Hopkins Univ, 55-56; asst prof zool, Okla State Univ, 56-60; instr, 60-62, asst prof microbiol, 62-77, ASSOC PROF MICROBIOL & PUB HEALTH, MICH STATE UNIV, 77- *Mem:* Am Soc Parasitologists; Am Soc Trop Med & Hyg; Soc Protozool; fel AAAS. *Res:* Parasitic protozoa; cellular immunity. *Mailing Add:* Dept Microbiol 57A Giltner Hall Mich State Univ East Lansing MI 48824

TWOMBLY, JOHN C, b Denver, Colo, Nov 26, 21. ELECTRICAL ENGINEERING. *Educ:* Univ Colo, BS, 44, PhD(elec eng), 59; Stanford Univ, MS, 50. *Prof Exp:* Engr, Manhattan Dist, Los Alamos Sci Labs, 45-46; instr elec eng, 46-49, res assoc electronics, 51-58, assoc prof elec eng, 59-62, PROF ELEC ENG, UNIV COLO, BOULDER, 62- *Concurrent Pos:* Fac study & res fel from Univ Colo, Swiss Fed Inst Technol, 65-66. *Mem:* Inst Elec & Electronics Engrs. *Res:* Electron devices; space-charge dynamics; network theory. *Mailing Add:* Dept Elec Eng Box 425 Univ Colo Campus Boulder CO 80309-0425

TWOMEY, JEREMIAH JOHN, b Co Cork, Ireland, July 30, 34; m 77; c 3. CELLULAR IMMUNOLOGY. *Educ:* Nat Univ Ireland, MB, BCh, BAO, 58. *Prof Exp:* From asst prof to prof med, Baylor Col Med, 67-; PVT PRACT. *Mem:* Fel Am Col Physicians; Am Fedn Clin Res; fel Royal Irish Acad Med; Am Soc Hemat; Am Asn Immunologists. *Res:* Immune regulation-physiology and pathophysiology; lymphomas-immune responses; thymic hormone; immunobiology of aging. *Mailing Add:* 4242 Southwest Freeway Suite 207 Houston TX 77027

TWOMEY, SEAN ANDREW, b Cork, Ireland; US citizen. ATMOSPHERIC CHEMISTRY & PHYSICS. *Educ:* Nat Univ Ireland, BSc, 48, MSc, 49, PhD(exp physics), 55. *Prof Exp:* Sr res officer, Radiophys Div, Commonwealth Sci & Indust Res Orgn, Australia, 50-59; atmospheric physicist, Nat Oceanic & Atmosphere Admin, Washington, DC, 59-63; consult physicist, Naval Res Lab, Washington, DC, 63-68; chief res scientist, Div Cloud Physics, 68-76; PROF ATMOSPHERIC PHYSICS, INST ATMOSPHERIC PHYSICS, UNIV ARIZ, 76- *Concurrent Pos:* Vis prof, dept atmospheric sci, Univ Ariz, 73-74; vis res fel, div radiophys,. *Honors & Awards:* C G Rossby Medal, Am Meteorol Soc, 80. *Res:* Extension of mathematical inversion techniques to enable solution for large number of unknowns; influence of pollution on clouds as it relates to climate change. *Mailing Add:* Inst Atmospheric Physics Univ Ariz Tucson AZ 85721

TYAGI, SURESH C, b Bankhandra, India, Jan 1, 55; Can citizen; m 80. REGULATION OF HIV & NEUTROPHIL PROTEASES, MECHANISM OF TRANSCRIPTION BY RNA POLYMERASE. *Educ:* Aligash Univ, PhD(chem), 80. *Prof Exp:* Sr demonstr chem, Univ Cork, Ireland, 80-82; postdoctoral fel biochem, Univ BC, Can, 83-85; res assoc pharmacol, 85-87, res assoc path, 87-89, INSTR BIOCHEM, STATE UNIV NY, 89- *Honors & Awards:* Travel Award, Am Soc Biochem & Molecular Biol, 91. *Mem:* Am Soc Biochem & Molecular Biol; Can Biochem Soc. *Res:* Viral and human inflammatory cell proteases: biochemical and biophysical studies; prokaryote and eukaryotic RNA polymerases: mechanism of transcription. *Mailing Add:* Dept Biochem State Univ NY Life Sci Bldg Stony Brook NY 11794-5215

TYAN, MARVIN L, b Los Angeles, Calif, Nov 29, 26; m 50; c 2. INTERNAL MEDICINE, EXPERIMENTAL BIOLOGY. *Educ:* Univ Calif, Berkeley, BA, 49; Univ San Francisco, MD, 52. *Prof Exp:* Intern med, Boston City Hosp, Mass, 52-53; resident, Boston Vet Admin Hosp, 53-54; sr asst resident, San Francisco County Hosp, Calif, 54-55; pvt pract, 56-61; sr investr exp path, US Naval Radiol Defense Lab, 61-68 & Stanford Res Inst, Calif, 68-71; prof bact, Immunol & Oral Biol, Dent Res Ctr, Univ NC, Chapel Hill, 71-77; PROF MED, UNIV CALIF, LOS ANGELES, 77- *Concurrent Pos:* Fel hemat, Stanford Lane Hosp, San Francisco, 55-56; fel, Tumor Biol Inst, Karolinska Inst, Sweden, 63-64. *Mem:* Am Asn Immunol. *Res:* Ontogeny of immune system of the mouse; processes involved in transplantation immunity. *Mailing Add:* Dept Med Vet Admin Wadsworth Hosp Ctr Wilshire & Sawtelle Blvds Los Angeles CA 90073

TYBERG, JOHN VICTOR, b Grantsburg, Wis, May 4, 38; m 60, 79; c 1. CARDIOVASCULAR PHYSIOLOGY. *Educ:* Bethel Col, Minn, BA, 60; Univ Minn, PhD(physiol), 67, MD, 72. *Prof Exp:* Res assoc med, Harvard Med Sch, 69; res physiologist, Riverside Res Inst, 69-71; res scientist cardiol, Cedars-Sinai Med Ctr, 71-73; asst prof med & physiol & mem,, Cardiovasc Res Inst, Med Ctr Univ Calif, San Francisco, 74-81; PROF, DEPT MED & MED PHYSIOL, UNIV CALGARY HEALTH SCI CTR, 81- *Concurrent Pos:* Lectr, Med Ctr, Univ Calif, San Francisco, 69-71; mem, Basic Sci Coun, Am Heart Asn, 74- *Mem:* Am Heart Asn; Am Physiol Soc; fel Am Col Cardiol. *Res:* Mechanics of ischemic myocardium; diastolic dynamics; pericardiol physiology; venous hemodynamics. *Mailing Add:* Dept Med & Med Physiol Univ Calgary Health Sci Ctr 3330 Hosp Dr NW Calgary AB T2N 4N1 Can

TYBOR, PHILIP THOMAS, b Fredericksburg, Tex, Oct 3, 48; m 69; c 4. FOOD SCIENCE. *Educ:* Tex A&M Univ, BS, 70, PhD(food sci), 73. *Prof Exp:* Dir protein res, 73-80, food prod develop dir, 80-86, HEAD, EXTEN FOOD SCI DEPT, CENT SOYA CO, INC, 86- *Concurrent Pos:* Mem, Ga Agribus Coun. *Mem:* Inst Food Technologists; Am Asn Cereal Chemists; Am Chem Soc; Am Dairy Sci Asn. *Res:* Administration of research programs pertaining to the development of food products for food service, protein applications development and the exploration of new food technologies; technical service and instruction on food quality, safety, and sanitation. *Mailing Add:* Cooperative Extension Serv Univ Ga Athens GA 30602

TYCE, FRANCIS ANTHONY, b South Wales, Eng, Oct 31, 17; US citizen; m 52; c 1. PSYCHIATRY. *Educ:* Univ Durham, BS & MD, 52; Univ Minn, MS, 64; Am Bd Psychiat & Neurol, dipl, 64. *Prof Exp:* House surgeon, Teaching Hosp, Durham, Eng, 52-53; rotating intern, St Vincents Hosp, Erie, Pa, 53-54;

gen pract, Seaham Harbor, 54-56; actg supt, Rochester State Hosp, 60-61; asst prof psychiat, Mayo Grad Sch Med, Univ Minn, 69-73, assoc prof, 73-77, PROF PSYCHIAT, MAYO MED SCH, ROCHESTER, MINN, 77-; SUPT, ROCHESTER STATE HOSP, 71-; pvt practice. *Concurrent Pos:* Fel psychiat, Mayo Clin, 56-60; lectr, Mayo Found, 65-; consult, WHO, 67-; vpres, Zumbro Valley Med Soc, Rochester, Minn, 71-72, pres, 72-73; task force comt, Psychiat Rehab in Correctional Systs, 73-; Field Rep Accreditation Coun for Psychiat Facil-Jt Comn on Accreditation of Hosps, 73-; mem, Juvenile Delinquency-Nat Adv Comt on Criminal Justice Stand & Goals, 75- *Mem:* Fel Psychiat Soc; Asn Med Supt Ment Hosp (pres elect, 67-68 & pres, 68-); chmn, Am Psychiat Asn. *Res:* Neurophysiology, especially electrical stimulation of the brain in rats; psychiatric program design in mental hospitals. *Mailing Add:* Am Asn Psychiat 929 SW 11th St Rochester MN 55901

TYCE, GERTRUDE MARY, b Wark, Eng, Mar 26, 27; m 52; c 1. BIOLOGY. *Educ:* Univ Durham, BSc, 48, PhD(plant physiol & biochem), 52. *Prof Exp:* Instr chem, Villa Maria Col, Pa, 52-53; instr biol & chem, Nottingham & Dist Tech Col Eng, 54-56; res asst biochem, Mayo Clin, 58-63, res assoc, 63-71, assoc consult biochem, 71-76, assoc prof biochem, 76-81, PROF PHYSIOL, MAYO MED SCH, MAYO CLIN & FOUND, 81- *Concurrent Pos:* Asst ed, News in Physiol Sci, 88-; consult, reviewal of grants, NSF; asst ed, Soc Exp Biol & Med, 90- *Mem:* AAAS; Am Chem Soc; Am Soc Exp Path; Int Soc Neurochem; Soc Neurosci; Am Soc Neurochem; Sigma Xi; Soc Exp Biol & Med; NY Acad Sci. *Res:* Metabolism of glucose, amino acids and biogenic amines in brain and liver. *Mailing Add:* Dept Physiol Mayo Med Sch Rochester MN 55901

TYCE, ROBERT CHARLES, b San Diego, Calif, July 9, 47. MARINE PHYSICS, OCEAN ENGINEERING. *Educ:* Univ Calif, San Diego, BA, 69, PhD(appl ocean sci), 77. *Prof Exp:* Asst programmer ocean instrumentation, Scripps Inst Oceanog, 69-70, res asst marine physics, Marine Phys Lab, 70-76; sr acoustics engr ocean vehicle instrumentation, Hydro Prod, 77-78; asst res engr marine physics, Marine Phys Lab, Scripps Inst Oceanog, 78-; RES ENGR, GRAD SCH OCEANOG, UNIV RI. *Concurrent Pos:* Physicist, Scripps Inst Oceanog, 76-77, res fel, 77-78; consult, Hydro Prod, 78- *Mem:* Acoust Soc Am; assoc mem Soc Explor Geophysicists; Marine Technol Soc; Am Geophys Union; Inst Elec & Electronics Engrs. *Res:* Underwater acoustics and geophysics; deep ocean engineering and instrumentation; vehicle technology; computer science; digital signal processing. *Mailing Add:* Grad Sch Oceanog-Ocean Eng Univ RI Narragansett RI 02882

TYCHSEN, PAUL C, b Chicago, Ill, Nov 1, 16; m 43; c 2. GEOLOGY. *Educ:* Carleton Col, BA, 41; Univ Nebr, MSc, 49, PhD(geol), 54. *Prof Exp:* Geolgist, US Geol Surv, 46-49; Proj rep Australia, Earth Sci Curric Proj, Am Geol Inst, 72-73; chmn dept geol, 52-82, EMER PROF, UNIV WIS-SUPERIOR, 82- *Mem:* AAAS; fel Geol Soc Am; Nat Asn Geol Teachers. *Res:* Stratigraphy; field-mapping; geomorphology. *Mailing Add:* 6603 Tower Ave Superior WI 54880

TYCKO, DANIEL H, b Los Angeles, Calif, Nov 14, 27; m 52; c 3. CYTOMETRY, LABORATORY INSTRUMENTATION. *Educ:* Univ Calif, Los Angeles, BA, 50; Columbia Univ, PhD(physics), 57. *Prof Exp:* Sr res assoc, Nevis Labs, Columbia Univ, 57-66; assoc prof physics, Rutgers Univ, 66-67; assoc prof comput sci, State Univ NY, Stony Brook, 67-70, prof, 70-79, prof, State Univ NY, New Paltz, 79-80; consult, Technicon Instruments Corp, 70-80, prin scientist, 80-82; RETIRED. *Concurrent Pos:* Res collab, Saclay Nuclear Res Ctr, France, 61-62; consult, appl physics & comput sci, 82- *Mem:* Am Phys Soc; Asn Comput Mach; Inst Elec Electronic Engrs; AAAS; Soc Anal Cytol. *Res:* Image processing by computers; pattern recognition; applications to cytology and hematology; flow cytometry instrumentation; application of light scattering to the measurement of the properties of cells. *Mailing Add:* PO Box 1033 Stony Brook NY 11790

TYCZKOWSKI, EDWARD ALBERT, b Providence, RI, May 15, 24; m 50. ORGANIC CHEMISTRY. *Educ:* Brown Univ, ScB, 49; Duke Univ, PhD(chem), 53. *Prof Exp:* Res assoc, US Army Off Ord Res, Duke Univ, 52-53; res chemist, Gen Chem Div, Allied Chem & Dye Corp, 53-56 & Pennsalt Chem Corp, 56-62; process chemist, Air Prod & Chem Corp, 62-63; vpres, Hynes Chem Res Corp, 63-66; sr res chemist, Fibers Div, Beaunit Corp, 66-67, res assoc, 67-72; pres, Armageddon Chem Co, 72-86; vpres, 86-89, PRES, FLURA CORP, 89- *Mem:* AAAS; Am Chem Soc; Sigma Xi. *Res:* Organic fluorine chemistry; reactions of elementary fluorine with organic compounds; flame reactions; explosions; reactor design; organic synthesis; polymer chemistry; fiber structure. *Mailing Add:* 3216 Landmark Dr Morristown TN 37814

TYE, ARTHUR, pharmacology; deceased, see previous edition for last biography

TYE, BIK-KWOON, b Hong Kong, Jan 7, 47; m 71; c 2. MOLECULAR BIOLOGY, MOLECULAR GENETICS. *Educ:* Wellesley Col, BA, 69; Univ Calif, San Francisco, MSc, 71; Mass Inst Technol, PhD(microbiol), 74. *Prof Exp:* Helen Hay Whitney Found fel biochem, Sch Med, Stanford Univ, 74-77; asst prof, 77-84, assoc, prof biochem, 84-90, PROF, CORNELL UNIV, 91- *Mem:* Genetics Soc Am. *Res:* DNA replication; transcription regulation; chromosome structure and function. *Mailing Add:* Dept Biochem Molecular & Cell Biol Biotechnol Bldg Cornell Univ Ithaca NY 14853

TYE, SZE-HOI HENRY, b Shanghai, China, Jan 15, 47; m 71; c 2. THEORETICAL & ELEMENTARY PARTICLE PHYSICS. *Educ:* Calif Inst Technol, BS, 70; Mass Inst Technol, PhD(physics), 74. *Prof Exp:* Res assoc physics, Stanford Linear Accelerator Ctr, Stanford Univ, 74-77, Fermi Nat Accelerator Lab, 77-78; sr res assoc, 78-87, PROF PHYSICS, NEWMAN LAB NUCLEAR STUDIES, CORNELL UNIV, 87- *Mem:* Am Phys Soc. *Mailing Add:* Newman Lab Nuclear Studies Cornell Univ Ithaca NY 14853

TYERYAR, FRANKLIN JOSEPH, b Frederick, Md, Apr 29, 35; m 59; c 3. MEDICAL MICROBIOLOGY. *Educ:* Univ Md, BS, 60, MS, 62, PhD(microbiol), 68. *Prof Exp:* Microbiologist, US Bur Mines, US Dept Interior, 62-63, US Dept Army, Ft Detrick, 63-71 & Dept Microbiol, Naval Med Res Inst, 71-73; microbiologist, Nat Inst Allergy & Infectious Dis, 73-90, chief, Develop & Applns Br, 84-90; RETIRED. *Honors & Awards:* Leroy Fothergill Sci Award, Sci Res Soc Am, 70; Dirs Award, NIH, 79. *Res:* Development and testing of bacterial and viral vaccines for clinical use; clinical evaluation of viral vaccines for efficacy; persistent viral infections; prevention and control of infectious diseases. *Mailing Add:* 7104 Autumn Leaf Lane Frederick MD 21702

TYHACH, RICHARD JOSEPH, b New York, NY, Aug 14, 49; m 73; c 2. BIOCHEMISTRY. *Educ:* Queens Col, BA, 70, MA, 73; City Univ New York, PhD(biochem), 76. *Prof Exp:* Lectr chem, Queens Col, City Univ New York, 71-76; Damon Runyon-Walter Winchell Cancer Fund fel biol chem, Harvard Med Sch, 76-78; res scientist biochem, Miles Labs Inc, 78-81, sr res scientist, 81-85, res & develop supvr, 85-87, res & develop mgr, Diagnostics Div, 87-89, DIR, URINE CHEM PROD RES & DEVELOP, MILES LABS INC, 89- *Mem:* Am Chem Soc; Am Asn Clin Chem; Sigma Xi. *Res:* Dry reagent chemistry; clinical biochemistry and immunoassay technology. *Mailing Add:* Diagnostics Div Miles Inc PO Box 70 Elkhart IN 46515

TYKOCINSKI, MARK L, b Lakewood, NJ, Nov 26, 52; m 78; c 4. DEVELOPMENTAL BIOLOGY, HEMATOLOGY & ONCOLOGY. *Educ:* Yale Univ, BA, 74; NY Univ, MD, 78. *Prof Exp:* Resident internal med, Columbia-Presby Med Ctr, 78-79; resident anat path, NY Univ Med Ctr, 79-81; med staff fel immunogenetics, Nat Inst Allergy & Infectious Dis, NIH, 81-83; ASSOC PROF PATH, SCH MED, CASE WESTERN RESERVE UNIV, 83-; STAFF PHYSICIAN, UNIV HOSPS CLEVELAND, 83- *Mem:* Am Asn Pathologists; Am Asn Cancer Res. *Res:* Study of human hematopoietic differentiation and leukemic cell induction using recombinant DNA technology; T cell biology; gene transfer technology. *Mailing Add:* Inst Path Case Western Reserve Univ 2085 Adelbert Rd Cleveland OH 44106

TYKODI, RALPH JOHN, b Cleveland, Ohio, Apr 18, 25; m 55; c 3. PHYSICAL CHEMISTRY. *Educ:* Northwestern Univ, BS, 49; Pa State Univ, PhD, 54. *Prof Exp:* Instr, Ill Inst Technol, 55-57, from asst prof to assoc prof, 57-65; assoc prof chem, 65-68, assoc dean, Col Arts & Sci, 69-72, PROF, SOUTHEASTERN MASS UNIV, 68- *Mem:* Am Chem Soc; Am Phys Soc. *Res:* Equliibrium and non-equilibrium thermodynamics. *Mailing Add:* Dept Chem Southeastern Mass Univ North Dartmouth MA 02747

TYLER, ALBERT VINCENT, b Philadelphia, Pa, June 25, 38; m 60; c 3. FISHERIES. *Educ:* Univ Pa, BA, 60; Univ Toronto, MA, 64, PhD(synecol), 68. *Prof Exp:* Scientist, Fisheries Res Bd Can, 64-74,; from assoc prof to prof fisheries, Ore State Univ, 74-82; RES SCIENTIST, GROUNDFISH RES SECT, PAC BIOL STA, BRIT COL, 82- *Mem:* Am Fisheries Soc; Can Soc Zool; Sigma Xi. *Res:* Physiological and ecological energetics; competitive and predatory relationships among fishes; population dynamics. *Mailing Add:* Groundfish Res Sect Pac Biol Sta Nanaimo BC V9R 5K6 Can

TYLER, AUSTIN LAMONT, b Provo, Utah, July 21, 36; m 60; c 5. CHEMICAL ENGINEERING. *Educ:* Univ Utah, BS, 61, PhD(chem eng), 65. *Prof Exp:* Supvr & mem staff, Semiconductor Processing, Bell Tel Labs, 65-70; PROF & CHMN CHEM ENG, UNIV UTAH, 70- *Mem:* Am Inst Chem Engrs. *Res:* Particle behavior in acceleration gas streams; fabrication processes for silicon semiconductors; kinetics of oil shale retorting processes; fluid bed reactors. *Mailing Add:* Dept Chem Eng Univ Utah Salt Lake City UT 84112

TYLER, BONNIE MORELAND, b New York, NY, Jan 5, 41; m 63; c 3. MICROBIAL PHYSIOLOGY, MOLECULAR BIOLOGY. *Educ:* Wheaton Col, BA, 62; Mass Inst Technol, PhD(biol), 68. *Prof Exp:* Res assoc bact, Univ Calif, Davis, 68-70; RES ASSOC, DEPT BIOCHEM, CORNELL UNIV. *Mem:* Sigma Xi; Am Soc Microbiol; Am Soc Biol Chemists. *Res:* Regulation of transcription and translation in bacteria; studies with whole cells and with purified transcription systems on factors regulating transcription. *Mailing Add:* Dept Human Develop Univ Md College Park MD 20742-1131

TYLER, CHRISTOPHER WILLIAM, b Leicester, UK, Dec 16, 43; m 85; c 1. OPTOMETRY, PSYCHOLOGY. *Educ:* Univ Leicester, UK, BA, 66; Univ Aston, UK, MSc, 67; Univ Keele, UK, PhD(communication), 70. *Prof Exp:* Asst prof, Northeastern Univ, Boston, 72-73; res fel, dept psychol, Univ Bristol, UK, 73-74 & dept sensory & perception progs, Bell Lab, 74-75, scientist, Smith-Kettlewell Inst Visual Sci, 75-80, sr scientist, William A Kettlewell Chair Res, 84-85; ASSOC DIR, SMITH-KETTLEWELL EYE RES INST, 90- *Concurrent Pos:* Vis prof, Jules Stein Inst, Univ Calif, Los Angeles, 85-88, sch optom, Univ Calif, Berkeley, 86-; assoc ed, Perception & Psychophysics. *Mem:* Asn Res Vision Ophthal; Optical Soc Am; Psychonomy Soc; NY Acad Sci; AAAS; Fed Am Scientists. *Res:* Retinal diseases; how light is turned into electrical energy in the receptors of the retina; effects of poor eye coordination; stereoscopic depth perception. *Mailing Add:* Smith-Kettlewell Eye Res Inst 2232 Webster St San Francisco CA 94115

TYLER, DAVID E, b Carlisle, Iowa, July 12, 28; m 52; c 2. VETERINARY PATHOLOGY. *Educ:* Iowa State Univ, BS, 53, DVM, 57, PhD(vet path), 63; Purdue Univ, MS, 60. *Prof Exp:* Instr, Purdue Univ, 57-60; asst prof, Iowa State Univ, 60-64, assoc prof, 64-66; prof & head dept, 66-79, PROF VET PATH, COL VET MED, UNIV GA, 79- *Concurrent Pos:* Mem, Conf Res Workers Animal Dis. *Honors & Awards:* Stange Award, 87. *Mem:* Am Vet Med Asn; Am Col Vet Path; Am Asn Vet Med Educr. *Res:* Epidemiology, pathology and immunology of the bovine mucosal disease-virus diarrhea complex; pathogenesis of equine colic; pathogenesis of porcine and equine salmonellosis. *Mailing Add:* Dept Path Col Vet Med Univ Ga Athens GA 30602

TYLER, DAVID RALPH, b Willimantic, Conn, Apr 26, 53; m 73; c 2. PHOTOCHEMISTRY, REACTION MECHANISMS. *Educ:* Purdue Univ, BS, 75; Calif Inst Technol, PhD(chem), 79. *Prof Exp:* Asst prof chem, Columbia Univ, 79-85; assoc prof, 85-90, PROF CHEM, UNIV ORE, 90- *Concurrent Pos:* Sloan Found fel, 86. *Mem:* Am Chem Soc. *Res:* Mechanisms of inorganic and organometallic photochemical reactions; electron-transfer, hypervalent reaction intermediates and metal oxide complexes. *Mailing Add:* Dept Chem Univ Ore Eugene OR 97403

TYLER, FRANK HILL, b Villisca, Iowa, Jan 5, 16; m 41; c 3. MEDICINE. *Educ:* Willamette Univ, BA, 38; Johns Hopkins Univ, MD, 42. *Prof Exp:* Intern med, Johns Hopkins Hosp, 42-43; from asst resident to resident, Peter Bent Brigham Hosp, Boston, 43-47; research instr, 47-54, from asst prof to assoc prof, 50-59, PROF MED, MED SCH, UNIV UTAH, 59- *Mem:* Am Soc Clin Invest; Endocrine Soc; Am Fedn Clin Res; Asn Am Physicians; master Am Col Physicians. *Res:* Disease of the muscle; human inheritance; metabolism of steroids and metabolic disorders. *Mailing Add:* 50 N Medical Dr Salt Lake City UT 84132

TYLER, GEORGE LEONARD, b Bartow, Fla, Oct 18, 40; m 77; c 2. PLANETARY EXPLORATION, RADAR ASTRONOMY. *Educ:* Ga Inst Technol, BS, 63; Stanford Univ, MS, 64, PhD(elec eng), 67. *Prof Exp:* Prof elec eng, Ctr Radar Astron, Stanford Univ, 67-79, res engr, 69-71, sr res assoc, 71-74; team leader, Voyager Radio Sci Team, 79-90, TEAM LEADER, MARS OBSERVER RADIO SCI TEAM, NASA, 86- *Concurrent Pos:* Consult, NASA & other res orgns; mem comt planetary explor; prin investr on numerous res projs; fel, NSF, 64-66. *Mem:* Fel Inst Elec & Electronics Engrs; Am Astron Soc; Am Geophys Union; Int Astron Union; Int Union Radio Sci. *Res:* Radio propagation experiments in space including theory and experiment; radio occultation measurements of planetary rings and atmospheres; radar astronomy including observation and interpretation of radiowave scatter from planetary surfaces using both spacecraft and ground based techniques; terrestrial applications of remote sensing. *Mailing Add:* Elec Eng Res Dept Stanford Univ Stanford CA 94305-4055

TYLER, H RICHARD, b Kings County, NY, Oct 16, 27; m 51; c 4. HISTORY NEUROLOGY. *Educ:* Syracuse Univ, BA, 47; Wash Univ, BS, 51, MD, 51. *Hon Degrees:* MA, Harvard Univ, 90. *Prof Exp:* Teaching fel neurol, 53-54, asst instr neurol, 56-59, Inst neurol, 59-61, assoc neurol, 61-64, asst prof neurol, 64-68, assoc prof, 69-73, PROF NEUROL, HARVARD MED SCH, 74-; PROF HEALTH SCI, MASS INST TECHNOL, 87- *Concurrent Pos:* Head, sect neurol, Brigham & Womens Hosp, 56-88, sr physician neurol, 88-; sr neurologist, Beth Israel Hosp, 86- *Mem:* Am Acad Neurol; Am Neurol Assoc; Am Assos Res Nervous & Mental Dis. *Res:* Neurological complications of medical diseases, especially renal disease; work on higher cortical visual disorder, Amyotrophic Laterel Sclerosis. *Mailing Add:* Brigham & Women Hosp 75 Francis St Boston MA 02115

TYLER, JACK D, b Snyder, Okla, July 18, 40; m 69. ORNITHOLOGY, ECOLOGY. *Educ:* Southwestern State Col, BS, 62; Okla State Univ, MS, 65; Univ Okla, PhD(zool), 68. *Prof Exp:* Teaching asst gen zool, Univ Okla, 64-66; from instr to assoc prof biol, 67-78, PROF BIOL, CAMERON UNIV, 78- *Concurrent Pos:* Ed, Bull Okla Orinth Soc, 72- *Mem:* Am Soc Mammal; Wilson Ornith Soc; Am Orinth Union. *Res:* Ecological relationships between certain birds in southwest Oklahoma and among vertebrates in prairie dog towns; mammals of Oklahoma. *Mailing Add:* Dept Biol Cameron Univ 2800 Gore Blvd Lawton OK 73505

TYLER, JAMES CHASE, b Shanghai, China, Mar 31, 35; US citizen; m 58; c 2. ICHTHYOLOGY. *Educ:* George Washington Univ, BS, 57; Stanford Univ, PhD(biol), 62. *Prof Exp:* Actg instr gen biol, Stanford Univ, 61; asst curichthyol, Acad Natural Sci Philadelphia, 62-66, assoc curr, 67-72; asst dir, Lerner Marine Lab, Am Mus Natural Hist, 72-73, dir, 73-75; prog mgr, Endangered Species, Nat Marine Fisheries Serv, 76-79; prog dir biol res resources, Div Environ Biol, NSF, Washington, DC, 80-84; assoc dir, Nat Mus Natural Hist, 85-87, DEP DIR, SMITHSONIAN INST, 87- *Concurrent Pos:* NSF grants, 63-72. *Mem:* Am Soc Ichthyol & Herpet. *Res:* Ichthyology, especially the anatomy and phylogeny of tetradontiform fishes and their classification; behavior and ecology of coral reef fishes. *Mailing Add:* Dep Dir Nat Mus Natural Hist Smithsonian Inst Washington DC 20560

TYLER, JEAN MARY, b Sheffield, Eng, Apr 7, 28. PITUITARY, PANCREAS. *Educ:* Univ London, BSc, 49, PhD(org chem), 55. *Prof Exp:* Asst lectr org chem, Univ London, 51-55; Imp Chem Industs fel, Univ Edinburgh, 55-59; res fel, Univ WI, 59-60; res assoc immunochem, Inst Microbiol, Rutgers Univ, 60-64; NATO res fel org chem, Univ Newcastle, 64-65; sr hosp biochemist, Dept Med, Royal Free Hosp, London, Eng, 65-66; res fel chem, Imp Col Sci & Technol, Univ London, 66-68; from asst res prof to assoc res prof, 68-82, ASSOC PROF MED, MED COL GA, 82- *Concurrent Pos:* Prin investr, NIH awards, 68-71, 72-75 &, 77-80. *Mem:* Royal Soc Chem; Endocrine Soc; Am Diabetes Asn; NY Acad Sci. *Res:* Radioimmunoassay of peptide hormones; regulation of pancreatic and pituary hormone secretion; carbohydrate chemistry; immunochemistry of microbiol. *Mailing Add:* Div Metabolic & Endocrine Dis Dept Med Med Col Ga Augusta GA 30912

TYLER, JOHN HOWARD, b Madison, Wis, Aug 29, 35; m 75. GEOLOGY. *Educ:* Univ Wis, BS, 58; Va Polytech Inst, MS, 60; Univ Mich, PhD(geol), 63. *Prof Exp:* Res asst geol, Va Polytech Inst, 58-60; res asst geol & paleont, Univ Mich, 60-63; tech asst, US Geol Surv, Calif, 63-64; fel, Univ of Wales, Swansea, 64-65; air photo interpreter, Itek Corp, Calif, 65-66 & Mark Systs Inc, 66; PROF GEOL, SAN FRANCISCO STATE UNIV, 66- *Mem:* Geol Soc Am; Soc Econ Paleont & Mineral. *Res:* Structural geology; remote sensing; stratigraphy. *Mailing Add:* Dept Geosci San Francisco State Univ San Francisco CA 94132

TYLER, KENNETH LAURENCE, b Boston, Mass, May 6, 53; m 79; c 2. NEUROVIROLOGY, CNS INFECTIONS. *Educ:* Harvard Univ, AB, 74; Johns Hopkins Univ Sch Med, MD, 78. *Prof Exp:* Intern med, Peter Bent Brigham Hosp, 78-79, resident, 79-80; resident neurol, Mass Gen Hosp, 80-82, chief resident, 82-83; instr microbiol & molecular genetics, 84-86, ASST PROF NEUROL & NEUROSCI, HARVARD MED SCH, 86-; ASST NEUROLOGIST, MASS GEN HOSP, 86- *Concurrent Pos:* Alfred P Sloan res fel, 86-89, Milton fel, Harvard Univ, 86-88; co-prin investr, Nat Inst Neurol & Communicative Dis & Stroke Prog Proj Grant, 87-92; vis prof, Univ Ala Sch Med, 88, Med Univ SC, 90; assoc ed, Jour Neurol Sci, 90- *Honors & Awards:* S Weir Mitchell Award, Am Acad Neurol, 83; Physician-Scientist Award, Nat Inst Allergy & Infectious Dis, 84. *Mem:* Am Soc Neurol Invest (secy-treas, 83-84 & pres, 84-85); fel Am Acad Neurol (pres hist sect, 89-91); Am Soc Virol; Am Fedn Clin Res; fel Am Col Physicians. *Res:* Study of how viruses produce diseases involving the central nervous system; identifying the role(s) played by specific viral genes at distinct stages in the pathogenesis of CNS infection; spread of virus to the CNS, tropism of virus for specific cell populations; role of the immune system on pathogenesis of viral CNS infection. *Mailing Add:* Dept Microbiol & Molecular Genetics D-1 Bldg Harvard Med Sch 25 Shattuck St Boston MA 02115

TYLER, LESLIE J, b Salamanca, NY, Nov 2, 19; m 47; c 7. ORGANIC CHEMISTRY. *Educ:* Univ Scranton, BS, 42; Pa State Univ, MS, 47, PhD(org chem), 48. *Prof Exp:* Res chemist silicon chem, Dow Corning Corp, 48-51, lab supvr resin res, 51-62, dir develop, 62-68, bus mgr fluids, 68-73, dir res, 73-75, vpres res & develop, 75-80; RETIRED. *Mem:* Am Chem Soc; Res Soc Am; Sigma Xi. *Res:* Silicon research; silica research; organosilicon research. *Mailing Add:* PO Box 177 Grawn MI 49637

TYLER, MARY STOTT, b Princeton, NJ, Apr 1, 49. DEVELOPMENTAL BIOLOGY. *Educ:* Swarthmore Col, BA, 71; Univ NC, Chapel Hill, MS, 73, PhD(zool), 75. *Prof Exp:* NSF-NATO fel develop biol, Dalhousie Univ, 75-76; asst prof, 76-80, ASSOC PROF ZOOL, UNIV MAINE, ORONO, 80- *Concurrent Pos:* Fac res grant, Univ Maine, 77-78; NIH res grant, 78-81. *Mem:* Soc Develop Biol; AAAS; Int Asn Dent Res; Sigma Xi. *Res:* Interacting systems in the developing vertebrate embryo; light-microscopic and ultrastructural aspects of tissue interactions; developmental capabilities of epithelial and mesenchymal tissues in experimental in vitro systems. *Mailing Add:* Dept Zool Murray Hall Univ Maine Orono ME 04469

TYLER, R(ONALD) A(NTHONY), b Burnham, Eng, June 4, 20; nat Can; m 42; c 1. THERMODYNAMICS, AERODYNAMICS. *Educ:* Univ London, BSc, 40. *Prof Exp:* Asst, Univ London, 40-42; engr, Bristol Aeroplane Co, Eng, 42-46; mathematician, Valve Res Labs, Standard Tel & Cables, Ltd, 46-47; asst res officer, Nat Res Coun, Can, 47-51, assoc res officer, 51-55, sr res officer, 55-61, head, Gas Dynamics Lab, 77-85, PRIN RES OFFICER, DIV MECH ENG, NAT RES COUN CAN, 61- *Mem:* Assoc fel Am Inst Aeronaut & Astronaut; assoc fel Can Aeronaut & Space Inst. *Res:* Turbomachinery, particularly aircraft turbines; locomotive gas turbine power plants; vertical take off and landing lift-propulsion systems. *Mailing Add:* 728 Lonsdale Rd Ottawa ON K1K 0K2 Can

TYLER, SETH, b Chicago, Ill, Feb 26, 49; m 70; c 2. INVERTEBRATE ZOOLOGY, ELECTRON MICROSCOPY. *Educ:* Swarthmore Col, BA, 70; Univ NC, Chapel Hill, PhD(zool), 75. *Prof Exp:* Killam fel anat, Dalhousie Univ, 75-76; from asst prof to assoc prof, 76-90, PROF ZOOL, UNIV MAINE, 90- *Concurrent Pos:* Prin investr, NSF grants, Univ Maine, 77-90; mem bd reviewers, Trans Am Micros Soc, 79-; assoc prof zool & dir, Electron Micros Ctr, Wash State Univ, 80; vis prof, Bermuda Biol Sta, 85. *Mem:* Am Soc Zoologists; Am Microscopical Soc; Sigma Xi; Electron Micros Soc Am; Int Asn Meiobenthologists; AAAS. *Res:* Comparative ultrastructure of lower metazoans; phylogeny of invertebrates; meiobenthology. *Mailing Add:* Dept Zool Univ Maine Orono ME 04469

TYLER, TIPTON RANSOM, b Milwaukee, Wis, Jan 3, 41; m 62; c 3. TOXICOLOGY, RISK ASSESSMENT. *Educ:* Colo State Univ, BS, 63, PhD(nutrit), 68; NC State Univ, MS, 65; Am Bd Toxicol, dipl, 80. *Prof Exp:* Sr res chemist, Merck Sharp & Dohme Res Labs, 68-74; asst prof animal sci, Univ Ill, Urbana-Champaign, 74-75; sr scientist chem hyg fel, Carnegie-Mellon Univ, 76-81; asst coord dir appl toxicol, 81-89, ASSOC DIR APPL TOXICOL, UNION CARBIDE CORP, 90- *Concurrent Pos:* Adj assoc prof toxicol, Univ Pittsburgh, 79-82 & Col Grad Studies, WVa Univ, 82-85. *Mem:* AAAS; Am Chem Soc; Soc Toxicol; Am Soc Pharmacol & Exp Therapeut; Am Col Toxicol. *Res:* Toxicology; drug metabolism; residues in tissues; disposition and clearance from animals; toxicology; risk assessment. *Mailing Add:* 39 Old Ridgebury Rd P2592 Danbury CT 06817-0001

TYLER, VARRO EUGENE, b Auburn, Nebr, Dec 19, 26; m 47; c 2. PHARMACOGNOSY. *Educ:* Univ Nebr, BS, 49; Univ Conn, MS, 51, PhD(pharmacog), 53. *Hon Degrees:* DSc, Univ Nebr, 87. *Prof Exp:* Assoc prof pharmacog, Univ Nebr, 53-57; assoc prof pharmacog, Univ Wash, 57-61, prof, 61-66, chmn dept pharmacog & dir drug plant gardens, 57-66; dean, Sch Pharm & Pharmacol Sci, 66-86, Sch Pharm, Nursing & Health Sci, 79-86, exec vpres, Acad Affairs, 86-91, PROF PHARMACOG, SCH PHARM & PHARMACAL SCI, PURDUE UNIV, 91- *Concurrent Pos:* Vis prof, Univ Gottingen, Ger, 84. *Honors & Awards:* Found Res Award, Am Pharmaceut Asn, 66. *Mem:* Am Asn Cols Pharm (pres, 70-71); Am Soc Pharmacog (pres, 59-61); Am Pharmaceut Asn; Am Coun Pharmaceut Educ (pres, 74-78); fel Acad Pharmaceut Sci; fel Am Asn Pharmaceutical Scientists. *Res:* Alkaloid biosynthesis; drug plant cultivation; phytochemical analysis; medicinal and toxic constituents of higher fungi; herbal medicine. *Mailing Add:* Sch Pharm & Pharmacal Sci Purdue Univ West Lafayette IN 47907

TYLER, WALTER STEELE, b Caspian, Mich, Nov 2, 25; m 49; c 2. ANATOMY. *Educ:* Mich State Col, DVM, 51; Univ Calif, Davis, PhD(comp path), 56. *Prof Exp:* Lectr vet sci, 52-57, from asst prof to assoc prof vet med, 57-67, actg chmn dept, 65-67, chmn dept, 67-70, jr vet, Exp Sta, 52-56, asst vet, 56-62, prof vet med, 67-76, PROF ANAT, UNIV CALIF, DAVIS, 76-,

ANATOMIST, 62- *Concurrent Pos:* Fel, Postgrad Med Sch, Univ London, 63-64; dir, Calif Primate Res Ctr, 72- *Mem:* Am Asn Anatomists; Am Asn Vet Anat (secy-treas, 66-67); Am Physiol Soc; Am Soc Zool; Am Vet Med Asn; Sigma Xi. *Res:* Relationship of structure to function in health and disease; pulmonary anatomy; emphysema; air pollution; histochemistry; scanning and transmission; electron microscopy; comparative anatomy; primate morphology. *Mailing Add:* Dept Anat Univ Calif Davis CA 95616

TYLER, WILLARD PHILIP, b Newton, Mass, Nov 8, 09; m 35, 55; c 3. RUBBER & PLASTICS, CHEMISTRY & TECHNOLOGY. *Educ:* Ore State Col, BS, 31, MS, 33; Univ Ill, PhD(anal chem), 38. *Prof Exp:* Instr, Clark Jr Col, 34-36; res analyst, BF Goodrich Co, 38-45, sect leader, Res Ctr, 45-74; RETIRED. *Mem:* Am Chem Soc. *Res:* Classical and instrumental chemical analysis; absorption spectroscopy; x-ray diffraction; gas chromatography; electroanalytical methods; high polymers. *Mailing Add:* 8471 Whitewood Rd Brecksville OH 44141

TYLUTKI, EDMUND EUGENE, b Chicago, Ill, Nov 6, 26; m 56; c 6. MYCOLOGY. *Educ:* Univ Ill, BS, 51, MS, 52; Mich State Univ, PhD(mycol), 55. *Prof Exp:* Asst bot, Univ Ill, 51-52; bot & plant path, Mich State Univ, 52-55; actg asst prof plant path & actg asst plant pathologist, Wash State Univ, 56; asst prof, 56-65, ASSOC PROF BOT, UNIV IDAHO, 65- *Concurrent Pos:* Ed-in-chief, J Idaho Acad Sci, 65-, pres, 70-71; actg chmn, Dept Biol Sci, Univ Idaho, 75-76. *Mem:* Mycol Soc Am; Classifaction Soc; Int Asn Plant Taxonomists; Sigma Xi. *Res:* Computer applications to fungal taxonomy; taxonomy of fleshy fungi of the Pacific northwest. *Mailing Add:* Dept Bot Univ Idaho Moscow ID 83844

TYMCHATYN, EDWARD DMYTRO, b Leoville, Sask, Nov 11, 42. MATHEMATICS, TOPOLOGY. *Educ:* Univ Sask, BA, 63, Hons, 64; Univ Ore, MA, 65, PhD(math), 68. *Prof Exp:* Asst prof, 68-71, assoc prof, 71-76, PROF MATH, UNIV SASK, 76- *Concurrent Pos:* Natural Sci & Eng Res Coun res grant, Univ Sask, 69-; vis assoc prof, Univ Ore, 75; vis prof, Ctr de Invest del Inst Politec Nac, Mexico City, 79 & Univ Ala, Birmingham, 81-82; mem fel comt, Natural Sci & Eng Res Coun, 79-82. *Mem:* Am Math Soc; Can Math Soc. *Res:* General topology and point set topology; continua; low dimensional spaces; partially ordered spaces and topological semigroups. *Mailing Add:* Dept Math Univ Sask Saskatoon SK S7N 0W0 Can

TYNDALL, JESSE PARKER, b Jones Co, NC, Jan 9, 25; m 65; c 1. BIOLOGY, SCIENCE EDUCATION. *Educ:* Atlantic Christian Col, AB, 45; Univ NC, Chapel Hill, MA, 49; Univ Fla, EdD, 56. *Prof Exp:* Teacher, Jones County Bd Educ, NC, 45-47; from instr to assoc prof biol & sci educ, 49-52, prof sci educ, 52-73, PROF BIOL, ATLANTIC CHRISTIAN COL, 52-, CHMN DEPT SCI, 54- *Concurrent Pos:* Mem bd dirs, Joint Comn Nursing Educ, NC Bd Educ & Bd Gov, 69-72, chmn, 72-73. *Mem:* Fel AAAS. *Res:* Genetics; nursing education. *Mailing Add:* Dept Sci Atlantic Christian Col Wilson NC 27893

TYNDALL, JOHN RAYMOND, b Greensboro, NC, Feb 25, 42; m 84; c 2. PHYSICAL CHEMISTRY, STATISTICS. *Educ:* Univ NC, Chapel Hill, BS, 63; Ill Inst Technol, Chicago, PhD(chem), 68; Univ Calif, San Diego, MA, 75. *Prof Exp:* Asst prof chem, Univ Pac, 70-74, sr chemist, Brin-Mont Chemicals, 75-78, Ethox Chemicals, 78-79; sr chemist, 80-82, SR SCIENTIST, GLYCO CHEMIST, 82-; CHEMTRONICS INC, 82- *Mem:* Am Chem Soc; Am Defense Preparedness Asn. *Res:* Specialty organics; electrically conducting polymers; ethoxylates and propoxylates; high temperature stable explosives; ultra-violet absorbers; flame retardants. *Mailing Add:* Pisgah Lab Box 567 Pisgah Forest NC 28768-0567

TYNER, DAVID ANSON, b Berrien Co, Mich, Feb 19, 22; m 49; c 4. ORGANIC CHEMISTRY. *Educ:* Univ Mich, BS, 44, MS, 49, PhD(org chem), 52. *Prof Exp:* Res chemist, G D Searle & Co, 52-86; RETIRED. *Concurrent Pos:* Civilian res chemist, Manhattan Proj, 44-45. *Mem:* Am Chem Soc. *Res:* Total and partial synthesis of steroids; peptides. *Mailing Add:* 909 Glendale Rd Glenview IL 60025

TYNER, GEORGE S, b Omaha, Nebr, Oct 9, 16; c 2. OPHTHALMOLOGY, LOW VISION. *Educ:* Univ Nebr, BS, 40, MD, 42; Univ Pa, MS, 52; Am Bd Ophthal, dipl, 50. *Prof Exp:* Intern, Philadelphia Gen Hosp, 42-43, resident ophthal, 47-48; resident, Hosp Univ Pa, 48-51, asst instr ophthal, Sch Med, Univ Pa, 48-52; from asst instr to asst clin prof, Sch Med, Univ Colo, 52-61, assoc dean & asst to vpres med affairs, 63-71, chief glaucoma clin & assoc prof ophthal, 64-71; dean, 74-81, PROF OPHTHAL, SCH MED, TEX TECH UNIV, 71-, EMER DEAN, 81- *Concurrent Pos:* Res fel, Univ Pa, 48-51; asst abstr ed, Am J Ophthal, 52-57; pvt pract, Colo, 52-61; mem, Colo State Bd Basic Sci Exam, 62-67; mem consult staff, Children's Hosp & Denver Gen Hosp; mem courtesy staff, St Luke's Hosp; mem hon staff, St Mary's Hosp, 73; active staff, Lubbock Gen Hosp. *Mem:* Fel Am Col Surg; Am Acad Ophthal & Otolaryngol; AMA. *Res:* Low vision rehabilitation. *Mailing Add:* 4006 Flint St Lubbock TX 79413

TYNER, MACK, b Laurel Hill, Fla, Feb 1, 16; m 46. CHEMICAL ENGINEERING. *Educ:* Univ Fla, BSChE, 38; Univ Cincinnati, MS, 40, PhD(phys chem), 41. *Prof Exp:* Tech control engr, Kimberly Clark Corp, NY, 41-42; res engr, Univ Wis, 42-43; asst chem engr, Armour Res Found, Ill, 42-45; prof chem eng, Univ Fla, 45-82; RETIRED. *Mem:* Instrument Soc Am; Am Inst Chem Engrs. *Res:* Process dynamics and control; systems engineering; instrumentation; chemical reaction engineering. *Mailing Add:* 1421 SW 13th Ave Gainesville FL 32608-1117

TYNES, ARTHUR RICHARD, b Great Falls, Mont, Oct 1, 26; m 46; c 5. OPTICS. *Educ:* Mont State Univ, BS, 50; Ore State Univ, MS, 53, PhD(physics), 63. *Prof Exp:* Physicist, US Bur Mines, 51-54; instr physics, Ore State Univ, 54-61; MEM TECH STAFF, AT&T BELL LABS, 61- *Concurrent Pos:* Consult, 88- *Mem:* Optical Soc Am. *Res:* Spectroscopy; plasma diagnositics; physical optics; applications of lasers to optical measurements; light scattering and light transmission; fiber optics. *Mailing Add:* 2734 Fern Dr Great Falls MT 59404

TYOR, MALCOLM PAUL, b New York, NY, Apr 20, 23; m 47; c 4. MEDICINE. *Educ:* Univ Wis, AB, 44; Duke Univ, MD, 46; Am Bd Internal Med, dipl, 57. *Prof Exp:* Intern, Madison Gen Hosp, Univ Wis-Madison, 46-47; resident med, Bowman Gray Sch Med, Wake Forest Col, 49-51, fel gastroenterol, 51-52; clinician, Med Div, Oak Ridge Inst Nuclear Studies, 52-54; physician, pvt pract, 54-55; assoc, 55-57, from asst prof to assoc prof, 57-62, PROF MED, MED SCH, DUKE UNIV, 62-, CHIEF DIV GASTROENTEROL, MED CTR, 65- *Concurrent Pos:* Asst chief med serv & chief radioisotope serv & gastroenterol, Vet Admin Hosp, Durham, 55- *Mem:* AAAS; AMA; Am Gastroenterol Asn; Am Fedn Clin Res; Am Soc Clin Invest. *Res:* Gastroenterology. *Mailing Add:* Med/Box 3902 Med Ctr Duke Univ Durham NC 27710

TYREE, BERNADETTE, PROTEIN CHEMISTRY, CELL CULTURE. *Educ:* Ill Inst Technol, PhD(biochem), 78. *Prof Exp:* Staff scientist, Howard Univ Cancer Ctr, 83-; MEM STAFF NAVAL RES INST. *Mailing Add:* Naval Med Res Inst MS 18 Bethesda MD 20814

TYREE, MELVIN THOMAS, b Santa Ana, Calif, Nov 15, 46; Can citizen; m 70; c 1. WHOLE PLANT PHYSIOLOGY, STRESS PHYSIOLOGY. *Educ:* Pomona Col, BA, 68; Cambridge Univ, PhD(biophys), 72. *Prof Exp:* Lectr, 71-72, from asst prof to prof plant physiol, Univ Toronto, 72-85; RES PROF TREE PHYSIOL, UNIV VT, 85- *Concurrent Pos:* Proj leader, N Eastern Forest Exp Sta, US Forest Serv, 88- *Honors & Awards:* CD Nelson Award, Can Soc Plant Physiologists, 79; Killam Award, 79. *Mem:* Am Soc Plant Physiologists; Can Soc Plant Physiologists; Scand Soc Plant Physiologists; Soc Exp Biol. *Res:* Water stress physiol; multiple stress physiol; translocation in phloem & symplasm; mathematical models of transport in plants; membrane transport; cavitation and embolism in xylem. *Mailing Add:* Dept Bot Univ Vt-Agr Col 120 MLS Bldg Burlington VT 05405

TYREE, SHEPPARD YOUNG, JR, inorganic chemistry; deceased, see previous edition for last biography

TYRER, HARRY WAKELEY, b Palmira, Colombia, Sept 20, 42; US citizen; m 68; c 3. ELECTRICAL ENGINEERING, BIOLOGY. *Educ:* Univ Miami, BSEE, 65; Duke Univ, MS, 69, PhD(elec eng), 72. *Prof Exp:* Asst prof elec eng, NC A&T Univ, 71-72; res & teaching assoc, Duke Univ, 67-72; instrumentation engr, Becton Dickinson Res Ctr, 73-76, prin investr automated cytol, 76-79; dir, biomed eng & biophys, Cancer Res Ctr, 79-87; PROF, DEPT ELEC ENG, UNIV MO, COLUMBIA, 85- *Concurrent Pos:* Adj lectr, Sch Med, Johns Hopkins Univ, 78- *Mem:* Inst Elec & Electronics Engrs; Asn Comput Mach; Soc Anal Cytol. *Res:* Real-time operating systems, software engineering, computer design; high resolution image analysis and processing; acoustic imaging; computer networks; computer aided electronic design; flow systems, applications and development. *Mailing Add:* Dept Elec & Computer Eng Univ Mo Columbia MO 64110

TYREY, LEE, b Chicago, Ill, Oct 26, 37; m 61; c 3. NEUROENDOCRINOLOGY, REPRODUCTION. *Educ:* Univ Ill, Urbana, BSc, 63, MSc, 64, PhD(physiol), 69. *Prof Exp:* Assoc obstet, gynec & anat, Duke Univ Sch Med, 70-72, from asst prof to assoc prof obstet & gynec, 72-83, from asst prof to assoc prof anat, 76-88, PROF OBSTET & GYNEC, DUKE UNIV SCH MED, 83-, ASSOC PROF NEUROBIOL, 88- *Concurrent Pos:* NIH res fel, Med Ctr, Duke Univ, 69-70, Duke Endowment res grant, 71-73, Pop Coun res grant, 74-76; dir, Gynec Endocrinol Lab, Duke Univ Med Ctr, 70-, mem Comprehensive Cancer Ctr, 78-; NC United Community Serv grant, 75-76; USPHS res grant, Nat Inst Drug Abuse, 78-89, mem, 83-87, chmn, Biomed Res Rev Comt, NIDA, 85-87; consult, NC Alcoholism Res Authority, 85- *Mem:* Endocrine Soc; Soc Study Reproduction; AAAS. *Res:* Neural control of gonadotropin secretion; effects of drugs of abuse on reproductive function; radioimmunoassay of protein hormones. *Mailing Add:* Dept Obstet-Gynec Duke Univ Med Ctr Box 3244 Durham NC 27710

TYRL, RONALD JAY, b Lawton, Okla, June 16, 43; m 65; c 3. PLANT TAXONOMY. *Educ:* Park Col, BA, 64; Ore State Univ, MS, 67, PhD(syst bot), 69. *Prof Exp:* Herbarium asst taxon, Ore State Univ, 65-69; asst prof biol, Park Col, 70-72; asst prof, 72-77, ASSOC PROF BOT & CUR HERBARIUM, OKLA STATE UNIV, 77- *Mem:* Am Soc Plant Taxon. *Res:* Plant biosystematics; evolutionary mechanisms; cytogenetic patterns. *Mailing Add:* Dept Bot Okla State Univ Stillwater OK 74078

TYROLER, HERMAN A, b New York, NY, Sept 5, 24. MEDICINE. *Educ:* Ohio Univ, AB, 43; NY Univ, MD, 47. *Prof Exp:* Resident physician internal med, Cornell Univ, 48-49; fel, NY Med Col & Metrop Hosp, 49-51; capt, Med Corps, USAF, Patrick AFB, 51-53; med consult & dir, Occup Health Serv, Asheville, NC, 53-58; res dir, Health Res Found, 58-60; from assoc prof to prof, 60-79, ALUMNI DISTINGUISHED PROF EPIDEMIOL, UNIV NC, CHAPEL HILL, 80- *Concurrent Pos:* Consult occup & cancer epidemiol, Champion Paper Co, 55-68, NY Times, 55-72, NASA, 65-70, Upjohn Co, 72-73 & Reynolds Aluminum, 77-82; mem, Res Training Grants Comt, Commun Health Serv, USPHS, 65-67 & 67-69, Health Serv Res Study Sect, HEW, 70-73, Lipid Res Clin Prog, Nat Heart, Lung & Blood Inst, 72-, US Polish Steering Comt Collab Studies, 79-, US Fed Repub Ger Steering Comt Collab Training, 81; chmn, Family Studies Comt, Nat Heart, Lung & Blood Inst, 72-81, Coun Epidemiol, Am Heart Asn, 83-85. *Honors & Awards:* Wade Hampton Frost Lectr, Am Pub Health Asn, 85, John Snow Award, 88. *Mem:* Inst Med-Nat Acad Sci; Sigma Xi; Am Pub Health Asn; Soc Epidemiol Res; Int Epidemiol Asn; Am Epidemiol Soc. *Res:* Cardiovascular epidemiology; atherosclerosis risk in communities. *Mailing Add:* Dept Epidemiol Sch Pub Health Univ NC Chapel Hill NC 27514

TYRRELL, ELIZABETH ANN, b Pittsfield, Mass, Oct 16, 31. MICROBIOLOGY. *Educ:* Simmons Col, BS, 53; Univ Mich, MS, 56, PhD(bact), 62. *Prof Exp:* Res asst virol, Parke, Davis & Co, Mich, 53-55; from instr to asst prof, 60-71, assoc prof microbiol, 71-79, PROF BIOL SCI, SMITH COL, 79- *Mem:* AAAS; Am Soc Microbiol. *Res:* Concentrated culture of microorganisms; autolysis in bacteria. *Mailing Add:* Burton Hall Smith Col Northampton MA 01063

TYRRELL, HENRY FLANSBURG, b Gloversville, NY, Aug 4, 37; m 69. NUTRITION, BIOMETRY. *Educ:* Iowa State Univ, BS, 59; Cornell Univ, MS, 64, PhD(nutrit), 66. *Prof Exp:* Asst prof animal sci, Cornell Univ, 66-69; res dairy husbandman, Energy Metab Lab, Animal Husb Res Div, Agr Res Serv, USDA, 69-72, res animal scientist, Ruminant Nutrit Lab, Nutrit Inst, 72-; RES SCI, USDA CSRS, AEROSPACE CTR. *Mem:* Am Dairy Sci Asn; Am Soc Animal Sci. *Res:* Utilization of energy by domestic animals; nitrogen utilization by ruminant animals; energy requirements for growthl lactation in cattle. *Mailing Add:* USDA CSRS Aerospace Ctr Suite 330-A 901 D St Washington DC 20250-2200

TYRRELL, JAMES, b Kilsyth, Scotland, Apr 19, 38; m 80; c 1. THEORETICAL CHEMISTRY. *Educ:* Univ Glasgow, BS, 60, PhD(chem), 63. *Prof Exp:* Teaching fel chem, McMaster Univ, 63-65; fel, Div Pure Physics, Nat Res Coun, 65-67; from asst prof to assoc prof, 67-80, actg chmn chem, 80-81, chmn, 82-90, PROF CHEM, SOUTHERN ILL UNIV, 80- *Mem:* Am Chem Soc; Sigma Xi; fel Am Inst Chem. *Res:* Theoretical calculations on atoms and molecules with particular application to transition metal systems and to the study of internal rotation. *Mailing Add:* Dept Chem & Biochem Southern Ill Univ Carbondale IL 62901

TYRRELL, WILLIS W, JR, b Mobile, Ala, Feb 12, 30; m 52; c 4. GEOLOGY. *Educ:* Fla State Univ, BS, 52; Yale Univ, MS, 54, PhD(geol), 57. *Prof Exp:* Asst, Fla State Univ, 52-53; field asst, Texaco, 54; geologist, Pan Am Petrol Corp, 55-64, res group supvr, 64-68; sr staff geologist, Amoco Prod Co, New Orleans, 68-77, sr geol assoc, Chicago, 77-81, sr consult geologist, Houston, Tex, 81-89; CONSULT GEOLOGIST, HOUSTON, TEX, 89- *Mem:* Geol Soc Am; Am Asn Petrol Geol. *Res:* Stratigraphy; petroleum exploration; seismic stratigraphy; carbonate petrology. *Mailing Add:* 807 Harvest Moon Houston TX 77077

TYSON, GRETA E, b Medford, Mass, Nov 2, 33; m 86. ZOOLOGY. *Educ:* State Teachers Col Bridgewater, BS, 55; Univ NH, MS, 57; Univ Calif, Berkeley, PhD(zool), 67. *Prof Exp:* NIH fel biol struct, Univ Wash, 67-69, NIH fel path, 69-70, instr, 70-72; asst prof, Univ Md, Baltimore, 72-76; assoc prof, 76-80, PROF, MISS STATE UNIV, 80- *Concurrent Pos:* Head, Electron Microscope Ctr, 76- *Mem:* Am Soc Cell Biol; Am Soc Zool; Am Micros Soc; Electron Micros Soc Am; Crustacean Soc. *Res:* Comparative renal morphology; structure and function of microtubules and microfilaments; ultrastructure of crustacean organs. *Mailing Add:* Dept Entom Box EM Miss State Univ PO Box 5328 Mississippi State MS 39762

TYSON, J ANTHONY, b Pasadena, Calif, Apr 5, 40; m 81; c 1. ASTROPHYSICS, IMAGING. *Educ:* Stanford Univ, BS, 62; Univ Wis, MS, 64, PhD(physics), 67. *Prof Exp:* Nat Res Coun-Air Force Off Sci Res fel, Univ Chicago, 67-68; vis lectr physics, Sussex Univ & Hebrew Univ, Jerusalem, 68-69; MEM TECH STAFF, AT&T BELL LABS, 69- *Concurrent Pos:* vis prof, Univ Calif, Berkeley, 75-76; chair, Adv Develop Prog, Nat Optical Astron Observ, 85-88, Dirs Adv Comt, 86-87; adj prof astrophys, Princeton Univ, 87-; adj staff mem, Lowell Observ, 87-; vchair, Astrophys Div, Int Union Pure & Appl Physics, 88- *Honors & Awards:* IR 100 Award, 85. *Mem:* Am Astron Soc; Am Phys Soc; AAAS; Int Astron Union. *Res:* Optical astronomy; astrophysics; experimental gravitation and relativity; gravitational radiation; radio astronomy; CCD imaging systems and automated imaging processing software. *Mailing Add:* 1D335 AT&T Bell Labs Murray Hill NJ 07974

TYSON, JOHN EDWARD ALFRED, b Hamilton, Ont, May 27, 35; c 3. REPRODUCTIVE ENDOCRINOLOGY. *Educ:* Univ Western Ont, MD, 56 & 60; Am Bd Obstet & Gynec, dipl, 72 & 79. *Prof Exp:* Fel gynec & obstet, Sch Med, Johns Hopkins Univ, 66-68, instr, 68-69, asst prof, 69-71, assoc prof, 71-78; PROF & CHMN, UNIV MAN, 78-, OBSTET & GYNEC CHIEF, 78-, PROF PHYSIOL, 79- *Concurrent Pos:* Mem med adv bd, Planned Parenthood Md, 70-78; chmn, Comt Int Reference Prep Placental Lactogen 69; ed-in-chief, Current Topics Obstet & Gynec, 72-78; mem, Adv Bd Educ TV, Md State Bd Educ, 67-69, Am Asn Planned Parenthood Physicians, 69-71, Nat Primate Cent Comt, Ottawa, 78-, Fertil & Maternal Clin Trials, Med Res Coun, Can, 80-; consult, Planned Parenthood Md, 67-78, reprod, Nat Zoo, Smithsonian Inst, 74-78, res, Agency Int Develop, US Dept State, 74-78, Prog Pub Health Educ, Can Broadcasting Corp, 76-79, field consult, Inst Nutrit Cent Am & Panama, 77-78, Nat Heart, Lung & Blood Inst, NIH, 78- *Mem:* Fel Am Col Obstetricians & Gynecologists; Am Diabetes Asn; Soc Gynec Invest; Endocrin Soc; Perinatal Res Soc. *Res:* Reproductive endocrinology, principally in those areas having to do with gestational diabetes, infertility, endocrinology of breastfeeding and neuroendocrinology of reproduction and prolactin physiology. *Mailing Add:* 2338 Hurontario St Mississauga ON L5B 1N1 Can

TYSON, JOHN JEANES, b Abington, Pa, Dec 12, 47; m 69; c 4. THEORETICAL BIOLOGY, CELL BIOLOGY. *Educ:* Wheaton Col, BS, 69; Univ Chicago, PhD(chem physics), 73. *Prof Exp:* NATO fel theoret biol, Max Planck Inst Biophys Chem, 73-74; asst prof math, State Univ NY, Buffalo, 74-75; Nat Cancer Inst fel cell biol, Inst Biochem & Cancer Res, Univ Innsbruck, 76-77; PROF BIOL, VA POLYTECH INST & STATE UNIV, 78- *Concurrent Pos:* Co-investr, NSF grant, 75- 77, prin investr, 80-; prin investr, NIH grant, 80-; guest prof, Univ Utah, 85; sr vis res fel, Oxford Univ, 84, 86. *Mem:* AAAS; Soc Math Biol. *Res:* Control of cell cycle events; chemical oscillations and traveling waves. *Mailing Add:* Dept Biol Va Polytech Inst & State Univ Blacksburg VA 24061

TYSON, RALPH ROBERT, b Philadelphia, Pa, Dec 14, 20; m 45; c 3. SURGERY. *Educ:* Dartmouth Col, AB, 41; Univ Pa, MD, 44; Am Bd Surg, dipl, 52; Pan-Am Med Asn, dipl. *Prof Exp:* From instr to assoc prof surg, Temple Univ, 52-62, chief sect vascular surg, 62-73, prof surg, Sch Med, 62-86, chmn dept & div, 73-83, EMER PROF SURG, SCH MED, TEMPLE UNIV, 86- *Concurrent Pos:* Mem Nat Bd Med Exam. *Mem:* AMA; fel Am Col Surg; Am Surg Soc; Soc Vascular Surg; Int Cardiovasc Soc. *Mailing Add:* RD 1 Box 179B Rome PA 18837

TYSON, WILLIAM RUSSELL, b Bourlamaque, Que, Sept 5, 39; m 73; c 3. METALLURGY & PHYSICAL METALLURGICAL ENGINEERING. *Educ:* Univ Toronto, BASc, 61; Cambridge Univ, PhD(metall), 65. *Prof Exp:* Fel metall, Univ Toronto, 66-67; fac mem physics, Trent Univ, 67-73; res scientist mat sci, 73-80, head, Eng Metal Phys Sect, 80-89, PROG COORDR, PMRL CANMET, 89- *Mem:* Am Soc Testing & Mat; fel Am Soc Metals; Can Inst Mining & Metall. *Res:* Hydrogen in metals; fracture mechanics. *Mailing Add:* 95 Kenora St Ottawa ON K1Y 3K9 Can

TYSVER, JOSEPH BRYCE, b Hazen, NDak, Mar 2, 18; m 45; c 1. APPLIED STATISTICS, OPERATIONS RESEARCH. *Educ:* Wash State Univ, BA, 42, MA, 48; Univ Mich, PhD(statist), 57. *Prof Exp:* Assoc res engr, Eng Res Inst, Univ Mich, 51-57; res specialist, Boeing Airplane Co, 57-63; statistician, Stanford Res Inst, 63-67; ASSOC PROF STATIST, NAVAL POSTGRAD SCH, 67-, ASSOC PROF OPER RES, 73- *Concurrent Pos:* Statistician, Litton Sci Support Lab, 68. *Mem:* Inst Math Statist; Am Math Soc; Math Asn Am; Inst Indust & Appl Math; Sigma Xi. *Res:* Statistical applications for engineering and military systems; sensitivity testing. *Mailing Add:* Box 2465 Carmel CA 93921

TYZBIR, ROBERT S, m; c 2. NUTRITIONAL BIOCHEMISTRY. *Educ:* Univ RI, PhD(biochem), 71. *Prof Exp:* ASSOC PROF NUTRIT BIOCHEM & CHMN, DEPT NUTRIT SCI, COL AGR & LIFE SCI, UNIV VT, 73- *Mem:* Sigma Xi; Am Inst Nutrit; Brit Nutrit Soc. *Res:* Energy metabolism; brown adipose tissue metabolism. *Mailing Add:* Dept Nutrit Sci Terrill Bldg Rm 315 Univ Vt Burlington VT 05405-0148

TYZNIK, WILLIAM JOHN, b Milwaukee, Wis, Apr 26, 27; m 50; c 5. ANIMAL NUTRITION. *Educ:* Univ Wis, BS, 48, MS, 49, PhD(nutrit), 51. *Prof Exp:* Asst nutrit, Univ Wis, 48-51; from asst prof to assoc prof animal nutrit, 51-59, PROF ANIMAL SCI & VET PREV MED, OHIO STATE UNIV, 59- *Concurrent Pos:* Pres, TIZCO Inc. *Mem:* Am Soc Animal Sci; Am Dairy Sci Asn; fel Soc Animal Sci. *Res:* Ruminant and monogastric nutrition; digestive physiology and mineral nutrition of equines and nutrition of zoological animals. *Mailing Add:* Dept Animal Sci Ohio State Univ Columbus OH 43210

TZAFESTAS, SPYROS G, b Corfu, Greece, Dec 3, 39; m 65; c 2. ROBOTICS & CONTROL, EXPERT SYSTEMS IN FAULT DIAGNOSTICS. *Educ:* Athens Univ, BSc, 63, dipl electronics, 65; Univ London, MSc & DIC(elec eng), 67; Southampton Univ, PhD(control), 69. *Hon Degrees:* DSc, Southampton Univ, 87, Int Univ Found, 89. *Prof Exp:* Dir computer control, Nat Res Coun, Demokritos, Athens, 69-73; prof systs control, Patras Univ, Greece, 73-84; dir, Control Systs Lab, Nat Tech Univ, 74-84, Systs & Control Div, 82-84 & Computer Sci Div, 86-88, PROF ROBOTICS & CONTROL & DIR, INTEL ROBOTICS & CONTROL UNIT, NAT TECH UNIV, ATHENS, 85- *Concurrent Pos:* Res fel, Nuclear Safety & Control, Nat Res Coun Demokritos, Athens, 75-84; vis prof fel, Southampton Univ, 75, Imp Col, Univ London, 81 & Denmark Techniske Hó4jskole, 83; vis prof adaptive control, Galabria Univ, Italy, 85, robotics, 87 & Tech Univ Delft, Neth, 91; chmn, Tech Comt Control Systs & Robotics, IFAC, Austria, 85- , co-chmn, 25th CDC, Inst Elec & Electronics Engrs, 86; mem, Systs Eng Comt, Hellenic Artificial Intel Soc; external prof, Grande Ecole d'Ingenieurs Lille, France, 87- *Mem:* Fel Inst Elec & Electronics Engrs; Am Soc Mech Engrs; Int Asn Math & Computers Simulation. *Res:* System estimation, simulation, optimization and control; distributed and multidimensional systems; queing systems and sequential machines; adaptive, robust and intelligent control; robotics and CIM; automated input and knowledge based system; author of 20 books and 150 publications. *Mailing Add:* Intel Robotics & Control Unit Nat Tech Univ 15773 Zografou Athens Greece

TZAGOURNIS, MANUEL, b Youngstown, Ohio, Oct 20, 34; m 58; c 5. MEDICINE, ENDOCRINOLOGY. *Educ:* Ohio State Univ, BS, 56, MD, 60, MMS, 65. *Prof Exp:* Intern med, Philadelphia Gen Hosp, Pa, 60-61; resident internal med, Univ Hosp, 61-62 & 64-65, chief resident internal med, 66-67, asst prof med, Col Med, 67-70, assoc prof med, 70-74, assoc dean, 75-76, PROF MED, COL MED, OHIO STATE UNIV, 74-, ASST DEAN RES & CONTINUING MED EDUC, 76- *Concurrent Pos:* USPHS fel endocrinol & metab, Ohio State Univ Hosp, 65-66. *Mem:* Am Fedn Clin Res; Am Diabetes Asn; AMA. *Res:* Diabetes, glucose metabolism and insulin secretion, especially as they relate to lipid disorders and coronary atherosclerosis. *Mailing Add:* Dean Med Admin Med Col Ohio State Univ 370 W Ninth Ave Columbus OH 43210

TZANAKOU, M EVANGELIA, b Athens, Greece. NEUROPHYSIOLOGY, NEURAL NETWORKS. *Educ:* Univ Athens, BS, 68; Syracuse Univ, MS, 74, PhD(physics), 77. *Prof Exp:* Fel biophysics, Physics Dept, Syracuse Univ, 77-80; from asst prof to assoc prof, 81-90, PROF BIOMED ENG & DEPT CHAIR, DEPT ELEC ENG, RUTGERS UNIV, 90- *Concurrent Pos:* Consult, Eye Defect & Vision Res Found, 78-80, biophysics, Syracuse Univ, 80-81; assoc ed, Inst Elec & Electronics Trans Neural Networking. *Honors & Awards:* Outstanding Adv Award, Inst Elec & Electronics Engrs, 85. *Mem:* Soc Neurosci; Asn Res Vision & Opthalmol; Biophys Soc; Sigma Xi; Inst Elec & Electronics Engrs. *Res:* Information processing in the visual system is examined by computer controlled techniques; recordings are done both in animals and in humans with a response feedback method where the information flow is reversed and a feature extractor becomes a feature generator; pattern recognition; digital signal processing of biological signals; computer application in biomedical engineering; neural networks; data compression, image reconstruction. *Mailing Add:* Dept Biomed Eng Rutgers Univ PO Box 909 Piscataway NJ 08855-0909

TZENG, CHU, b Tainan, Taiwan, Sept 19, 40; US citizen; m 69; c 3. FOOD SCIENCE & TECHNOLOGY. *Educ:* Nat Taiwan Univ, BS, 63, MS, 66. *Hon Degrees:* Scd, Mass Inst Technol, 72. *Prof Exp:* Res assoc biochem eng, Mass Inst Technol, 73-74; dir res & develop food indust, Milbrew Inc, 74-80; dir, fermentation technol, Abcor Inc, 80-83; chmn, Hansen Lab, Inc, 83-87; DAIRYLEA COOP, INC, 88- *Honors & Awards:* Res Awards, Environ

Protection Agency, NSF & NY State Res & Develop. *Mem:* Inst Food Technologists; Am Chem Soc; Am Soc Microbiol; Am Inst Chem Engrs; Soc Indust Microbiol. *Res:* Fermentation engineering in scale up; product recovery processing; utilization and development of food ingredients; product development and quality assurance, regulatory compliance. *Mailing Add:* 4827 Hyde Rd Manlius NY 13104

TZENG, KENNETH KAI-MING, b Kaifeng, China, Aug 6, 37; m 61; c 2. ELECTRICAL ENGINEERING, COMPUTER SCIENCES. *Educ:* Nat Taiwan Univ, BS, 59; Univ Ill, MS, 62, PhD(elec eng), 69. *Prof Exp:* Jr engr, IBM Corp, 62-63; elec res engr, Nat Cash Register Co, 63-65; res asst, Univ Ill, 65-69; from asst prof to assoc prof elec eng, 69-77, PROF ELEC ENG & COMPUTER SCI, LEHIGH UNIV, 77- *Concurrent Pos:* NSF res initiation grant, 70-71, res grants, 73-79 & 88-; mem tech staff, Bell Labs, 69, instr, In-Hour Continuing Educ Prog, 72-80, consult, 81-82; vis asst prof, Univ Ill, 72; fac fel, NASA, 76. *Mem:* Inst Elec & Electronics Engrs. *Res:* Error control in digital computing and communication systems; computer networks; fault-tolerant computing. *Mailing Add:* Dept Comput Sci & Elec Eng Lehigh Univ Packard Lab Bldg 19 Bethlehem PA 18015-3084

TZENG, WEN-SHIAN VINCENT, b Taipei, Taiwan, May 7, 43; m 72; c 2. METALLURGICAL ENGINEERING. *Educ:* Cheng Kung Univ, Taiwan, BSE, 66; Univ Conn, PhD(metall), 75. *Prof Exp:* Sr res scientist, Firestone Tire & Rubber Co, 75-80; DIR MAT SCI, CHOMERICS, W R GRACE, 80- *Mem:* Am Soc Metals; Am Vacuum Soc; Electron Micros Am; Am Powder Metall Inst; Microbeam Anal Soc. *Res:* Metal powder surface chemistry and physics for electrical conductive behavior and their industrial application; electron microscopy; surface analysis; elastomer compounding and testing. *Mailing Add:* 54 Eastway Reading MA 01867

TZIANABOS, THEODORE, b Manchester, NH, Feb 12, 33; m; c 2. VIROLOGY. *Educ:* Univ NH, BA, 55, MS, 59; Univ Mass, PhD(microbiol), 65; Am Bd Med Microbiol, dipl. *Prof Exp:* Res instr microbiol, Dept Poultry Dis, Univ NH, 55-57; microbiologist, Diagnostic Virol, State Mass, 59-60; res instr microbiol, Dept Vet Sci, Univ Mass, 60-65; resident, Ctr Dis Control, 65-67; res microbiologist, Med Sci Div, Ft Detrick, 67-70; microbiologist, Beckman Instruments, 70-71; RES MICROBIOLOGIST VIROL, CTR DIS CONTROL, 71- *Concurrent Pos:* Mem bd trustees, Am Type Cult Collection, Rockville, Md, 78-82. *Mem:* Am Soc Microbiologists; Am Soc Trop Med & Hyg; Res Soc Am; Am Soc Rickettsiology. *Res:* Development and research on rickettsial products involving serologic tests, including fluorescent microscopy; purification and protein composition of rickettsiae. *Mailing Add:* 6959 Lockridge Dr Doraville GA 30360

TZODIKOV, NATHAN ROBERT, b Brooklyn, NY, Feb 28, 52. BIORGANIC CHEMISTRY, RADIOCHEMISTRY. *Educ:* State Univ NY, Stony Brook, BS, 73; Mass Inst Technol, PhD(org chem), 77. *Prof Exp:* Sr chemist lubricants, Texaco Inc, 77-78; chemist radiochem res, 78-80, GROUP LEADER RADIOCHEM RES, NEW ENG NUCLEAR, 80- *Concurrent Pos:* Mkt mgr, Sadtler Div Bio Rad. *Mem:* AAAS; Am Chem Soc. *Res:* Organic synthesis; radiochemical research; radiochemical synthesis. *Mailing Add:* Sadtler Div Bio Rad 3316 Spring Garden St Philadelphia PA 19104-2552

TZOGANAKIS, COSTAS, b Chania, Crete, Greece, May 14, 60; m 88. POLYMER PROCESSING & RHEOLOGY, COMPUTER-AIDED DESIGN IN POLYMER ENGINEERING. *Educ:* Univ Thessaloniki, Greece, dipl eng, 83; McMaster Univ, Can, PhD(chem eng), 89. *Prof Exp:* Res asst chem eng, McMaster Univ, 83-88; res scientist, Du Pont Can, Inc, 88-90; ASST PROF CHEM ENG, UNIV WATERLOO, 90- *Mem:* Sigma Xi; Can Soc Chem Eng; Am Inst Chem Eng; Soc Plastics Engrs; Polymer Processing Soc; Soc Rheology. *Res:* Reactive processing of polymers; melt phase polymer reaction kinetics; mathematical modelling and computer simulations of polymer processing; rheology of polymer melts; expert systems in polymer processing. *Mailing Add:* Dept Chem Eng Univ Waterloo Waterloo ON N2L 3G1 Can

U

UBAN, STEPHEN A, b Waterloo, Iowa, May 10, 50. FILTRATION THEORY & PRACTICE, WATER TREATMENT. *Educ:* Iowa State Univ, BS, 73. *Prof Exp:* Res & develop engr, Smith & Loveless Div Ecodyne, 74-77, sr res & develop engr, 77-78; spec projs engr, Microfloc Div Neptune, 78-80, proj eng mgr, 80-83, dir res & develop, 83-84; spec projs mgr, Johnson Filtration Systs, WTI Corp, 84-87, dir res & develop, 87-90; DIR & MGR RES & DEVELOP, JOHNSON FILTRATION & WHEELABRATOR CLEAN WATER SYSTS, 90- *Mem:* Am Water Works Asn; Prod Develop Mgt Asn; Am Filtration Soc; Am Mgt Asn. *Res:* Dept filtration; dissolved air flotation; underdrains and distributors; adsorption clarification; crossflow microfiltration; oil well screen design. *Mailing Add:* 1104 S Fourth St Stillwater MN 55082

UBELAKER, DOUGLAS HENRY, b Horton, Kans, Aug 23, 46; m 75; c 2. PHYSICAL ANTHROPOLOGY. *Educ:* Univ Kans, Lawrence, BA, 68, PhD(anthrop), 73. *Prof Exp:* CUR PHYS ANTHROP & CHMN, DEPT ANTHROP, SMITHSONIAN INST, 73- *Mem:* AAAS; Am Asn Phys Anthrop; Am Anthrop Asn; Am Acad Forensic Sci; Soc Am Archaeol. *Res:* Physical anthropology of North America, Latin America; skeletal biology; prehistoric demography; forensic anthropology. *Mailing Add:* Nat Mus Natural Hist Smithsonian Inst Washington DC 20560

UBELAKER, JOHN E, b Everest, Kans, Mar 21, 40. PARASITOLOGY. *Educ:* Univ Kans, BA, 62, MA, 65, PhD(zool), 67. *Prof Exp:* Fel parasitol, Emory Univ, 67-68; asst prof biol, 68-71, assoc prof, 71-74, PROF BIOL, SOUTHERN METHODIST UNIV, 74- *Mem:* Am Soc Parasitol; Am Soc Zool; Am Micros Soc; Wildlife Dis Asn; Sigma Xi. *Res:* Helminthology; transmission, scanning electron microscopy; helminth reproduction; pathophysiology of lungworm infections. *Mailing Add:* 2733 Westminster Dallas TX 75205

UBERALL, HERBERT MICHAEL, b Neunkirchen, Austria, Oct 14, 31; US citizen; m 81; c 2. ACOUSTICS, RADAR. *Educ:* Univ Vienna, PhD(theoret physics), 53; Cornell Univ, PhD(theoret physics), 56. *Hon Degrees:* DSc, Univ LeHavre, France, 87. *Prof Exp:* Res fel physics, Univ Liverpool, 56-57; Ford Found fel, Europ Org Nuclear Res, Geneva, Switz, 57-58; res physicist, Carnegie Inst Technol, 58-60; asst prof, Univ Mich, 60-64; assoc prof, 64-65, PROF PHYSICS, CATH UNIV AM, 65- *Concurrent Pos:* Sr res physicist, Conductron Corp, Mich, 61-64; consult, Naval Res Lab, Washington, DC, 66- & Naval Surface Weapons Ctr, White Oak, Md, 76-81; vis prof, Univ Paris, 84-85. *Mem:* Fel Am Phys Soc; fel Acoust Soc Am; Am Asn Univ Prof; fel Inst Elec & Electronics Engrs; Int Union Radio Sci. *Res:* Scattering and radiation theory; electromagnetic and acoustic waves; underwater acoustics; theoretical nuclear physics; wave mechanics; author, co-author and co-editor of 4 books and 300 articles in professional journals. *Mailing Add:* Dept Physics Cath Univ Am Washington DC 20064

UBEROI, M(AHINDER) S(INGH), b Delhi, India, Mar 13, 24; nat US. AERONAUTICAL ENGINEERING, ASTRONAUTICS. *Educ:* Punjab Univ, India, BSc, 44; Calif Inst Technol, MS, 46; Johns Hopkins Univ, DEng, 52. *Prof Exp:* Res asst aeronaut eng, Johns Hopkins Univ, 52-53; res assoc aerospace eng, Univ Mich, 53-56, from assoc prof to prof, 56-63; chmn dept aerospace eng sci, 63-77, PROF AEROSPACE ENG SCI & FEL, JOINT INST LAB ASTROPHYS, UNIV COLO, BOULDER, 63- *Concurrent Pos:* Guggenheim fel, 58-59. *Mem:* Am Inst Aeronaut & Astronaut; Am Phys Soc; Am Soc Eng Educ. *Res:* Turbulent flows; statistical analysis of random functions; magnetohydrodynamics; aerothermodynamics. *Mailing Add:* Dept Eng Univ Colo Campus Box 422 Boulder CO 80309

UCCI, POMPELIO ANGELO, b Warwick, RI, Jan 15, 22; m 49; c 4. PHYSICAL CHEMISTRY, MATHEMATICS. *Educ:* Univ RI, BS, 43. *Prof Exp:* Res chemist, Celanese Corp Am, 43-44 & 46-52; sr res chemist, 52-54, group leader synthetic fibers, 54-58, sect head, 58-69, site mgr, New Enterprise Div, 69-71, sr res specialist, 71-75, ENG FEL, TEXTILES DIV, MONSANTO CO, 75- *Mem:* AAAS; Am Chem Soc. *Res:* Solution and melt properties of natural and synthetic fiber forming polymers; fundamental mechanical and engineering properties; statistics and quality control; paper making from synthetic fibers; testing equipment and procedures. *Mailing Add:* 4070 Aiken Rd Pensacola FL 32503

UCHIDA, IRENE AYAKO, b Vancouver, BC, Apr 8, 17. CYTOGENETICS. *Educ:* Univ Toronto, PhD(human genetics), 51. *Prof Exp:* Res assoc, Hosp Sick Children, Toronto, Ont, 51-59; proj assoc, Univ Wis, 59-60; lectr pediat, Univ Man, 60-62, from asst prof to assoc prof, 63-69, asst prof anat, 67-69; prof, 69-85, EMER PROF PEDIAT & PATH, 85-, DIR, REGIONAL CYTOGENETICS LAB, MED CTR, MCMASTER UNIV, 69- *Concurrent Pos:* Ramsay Wright scholar, Univ Toronto, 47; fel, Rockefeller Found, 59; vis prof, Univ Ala Med Sch, 68, Med Res Coun Can, 73; vis scientist, Med Res Coun Can, 69, mem, Sci coun Can & grant comt, 70-73 & vis prof, 73; Med Res Coun vis prof, Univ Western Ont, 73; consult, Int Prog Radiation Genetics, Nuclear Energy Agency, Orgn Econ Coop & Develop, Paris, 73 & Am Bd Med Genetics, 80-82; mem adv comt genetic serv, Ont Ministry Health, 79-85 & mem task force high technol diag procedures & equip, Ont Coun Health, 80-81; mem, Ment Retardation Comt, Nat Inst Child Health & Human Develop, 80-84; chmn, Genetic Cell Culture Comt, Ont Med Asn Lab Proficiency Testing Prog, 81-; Basic Sci Adv Comt, NY State Inst, 81-; Queen Elizabeth II speaker, Winnipeg, Man, 71. *Mem:* AAAS; Am Soc Human Genetics (pres, 68); Can Col Med Geneticists; Genetics Soc Can; Genetics Soc Am. *Res:* Human genetics; cytogenetics. *Mailing Add:* Dept Pediat McMaster Univ Hamilton ON L8N 3Z5 Can

UCHIDA, RICHARD NOBORU, b Honolulu, Hawaii, Sept 4, 29; m 55; c 3. MARINE SCIENCES. *Educ:* Univ Wash, Seattle, BS, 51. *Prof Exp:* Fishery res biologist, US Dept Com, Nat Oceanic & Atmospheric Admin, Nat Marine Fisheries Serv, Southwest Fisheries Ctr, Honolulu Lab, 54-76, supvry biologist, 76-84; RETIRED. *Concurrent Pos:* Counr, Hawaiian Acad Sci, 75-76; consult, UN Food & Agr orgn, 77. *Mem:* Fel Am Inst Fishery Res Biol. *Res:* Distribution, life history and relative abundance of demersal and pelgic fishes, mollusks, and crustaceans in waters surrounding central and western Pacific islands and overlying seamounts. *Mailing Add:* 1586 Hoolehua St Pearl City HI 96782

UCHUPI, ELAZAR, b New York, NY, Oct 31, 28. GEOLOGY. *Educ:* City Col New York, BS, 52; Univ Southern Calif, MS, 54, PhD(geol), 62. *Prof Exp:* Res asst geol, Univ Southern Calif, 55-62; res asst 62-64, assoc scientist, 64-79, SR SCIENTIST, WOODS HOLE OCEANOG INST, 79- *Honors & Awards:* Shepard Medal. *Mem:* Geol Soc Am; Am Asn Petrol Geol; Soc Econ Paleont & Mineral; Am Geophys Union; Sociedad Geologica de España; Am Archeol Inst. *Res:* Sedimentation; submarine geomorphology; tectonics; geologic development of oceanic basins. *Mailing Add:* Dept Geol & Geophys Woods Hole Oceanog Inst Woods Hole MA 02543

UCKO, DAVID A, b New York, NY, July 9, 48; m 72; c 1. SCIENCE MUSEUM ADMINISTRATION. *Educ:* Columbia Col, BA, 69; Mass Inst Technol, PhD(chem), 72. *Prof Exp:* Asst prof chem, Hostos Community Col, 72-76; from asst prof to assoc prof, Antioch Col, 76-79; res coordr, Mus Sci & Indust, Chicago, 79-80, sci dir, 81-87, prog div chmn, 85-86, vpres prog, 86-87; DEP DIR, CALIF MUS SCI & INDUST, 87-; VPRES PROG, CALIF MUS FOUND, 87- *Concurrent Pos:* NIH fel, Columbia Univ, 72; fac res fel, Res Found, State Univ NY, 75; adj staff scientist, C F Kettering Res Lab, 78;

res assoc & assoc prof, Dept Educ, Univ Chicago, 82-87. *Mem:* Am Chem Soc; Sigma Xi; Royal Soc Chem; Am Asn Mus; AAAS. *Res:* Enhancing public science literacy through informal education. *Mailing Add:* Kansas City Mus 3218 Gladstone Blvd Kansas City MO 64123-1199

UDALL, JOHN ALFRED, b Holbrook, Ariz, Sept 11, 29; m 50; c 3. INTERNAL MEDICINE, CARDIOLOGY. *Educ:* Brigham Young Univ, BS, 51; Temple Univ, MD, 58. *Prof Exp:* Resident internal med & cardiol, Med Ctr, Univ Calif, San Francisco, 59-62; dir med educ, Maricopa County Gen Hosp, Phoenix, Ariz, 62-66; asst prof, 66-69, ASSOC PROF MED & CARDIOL, COL MED, UNIV CALIF, IRVINE, 69-; CARDIOLOGIST, MULLIKIN MED CTR. *Concurrent Pos:* Consult cardiovasc dis, Long Beach Vet Admin Hosp, Calif, 66- & Fairview State Hosp Retarded Children, Costa Mesa, 69-; dir med serv, Orange County Med Ctr, Calif, 71- *Mem:* AAAS; fel Am Col Cardiol; fel Am Col Physicians; Am Fedn Clin Res; Asn Hosp Med Educ. *Res:* Pervenous pacemaker electrode endocardial implantation; clinical research in all aspects of oral anticoagulant therapy, especially the problem of the lack of stability of long-term therapy. *Mailing Add:* 17821 S Pioneer Blvd Artesia CA 90701

UDALL, JOHN NICHOLAS, JR, b Washington, DC, Dec 30, 40; m 67; c 4. MEDICINE, NUTRITIONAL BIOCHEMISTRY. *Educ:* Brigham Young Univ, BSc, 65; Temple Univ, MD, 69; Am Bd Pediat, dipl; Mass Inst Technol, PhD(nutrit biochem), 80. *Prof Exp:* Intern med & pediat, Los Angeles County/Univ Southern Calif Med Ctr, 69-70; general med officer, Indian Health Serv, USPHS, 70-72; resident pediat, Los Angeles County/Univ Southern Calif Med Ctr, 72-74; fel pediat nutrit & gastroenterol, Baylor Col Med, 74-76; fel clin nutrit, Mass Inst Technol, Mass Gen Hosp & Children's Hosp Med Ctr, 77-80; asst dir, Clin Res Ctr, Mass Inst Technol, 80-85; ASST PROF PEDIAT, HARVARD MED SCH, 81- *Concurrent Pos:* Nutrit consult eval acceptability & tolerance to purified single cell protein for adult human feeding, Dir, Nevin S Scrinshaw, 77-78; co-pirin investr, NIH grant, 78-80. *Mem:* Fel Am Acad Pediat; fel Am Bd Nutrit; NAm Soc Pediat Gastroenterol. *Res:* Effect of early nutrition on gastrointestinal development; obesity in infancy and childhood; pathophysiology of cholera. *Mailing Add:* Dept Pediat Harvard Med Sch 25 Shattuck St Boston MA 02115

UDANI, KANAKKUMAR HARILAL, b Rajkot, India, Dec 4, 36; US citizen; m 66; c 2. FOOD SCIENCE, CHEMICAL ENGINEERING. *Educ:* Gujarat Univ, India, BSc, 57; Univ Bombay, BSc, 59; Univ Ill, Urbana, MS, 61, PhD(food sci), 65. *Prof Exp:* Res fel chem eng, Univ Bombay, 59; supvr qual control, Accent Int, Div Int Minerals & Chem Corp, 61-62; sr food technologist, H J Heinz Co, 65-68; sr scientist oil prod, Res & Develop Div, Kraftco Corp, 68-79, res coordr, 79-87, sr res scientist, Kraft Inc, Glenview, 87-90; MGR PROG, DEPT ADVAN SOLUTIONS, BAXTER HEALTH CARE CORP, ROUNDLAKE, ILL, 90- *Mem:* Inst Food Technologists. *Res:* Dehydration of foods; oils, fats and starch technology; aseptic systems; emulsifiers and stabilizers; processed food rheology; flavor science; product, process and market development. *Mailing Add:* Baxter Health Care Corp Rte 120 & Wilson Rd Roundlake IL 60073

UDANI, LALIT KUMAR HARILAL, b Rajkot, India, Aug 19, 27; US citizen; m 70; c 1. CHEMICAL ENGINEERING. *Educ:* Univ Bombay, BSc, 49; Univ Nagpur, BSc, 52; Univ Mich, Ann Arbor, MSE, 56, ScD(chem eng), 62. *Prof Exp:* Develop engr, Kordite Co, NY, 61-63; specialist process develop, Chem Div, Gen Elec Co, Mass, 63-67; res engr, Org Chem Div, FMC Corp, Md, 67-69; supvry process engr, Catalytic Inc, Philadelphia, 69-80, sr process specialist, 80-87; CHEM ENGR CONSULT, 88- *Mem:* Am Chem Soc; Sigma Xi. *Res:* Organic chemicals and polymer process research; water pollution abatement for industrial systems; processes for clean fuels and electrode coke from coal, flue gas desulfurization; corporate planning and business development for synthetic fuels projects; coal-water slurry; composites. *Mailing Add:* 40 Masters Circle Marlton NJ 08053

UDD, JOHN EAMAN, b Rochester, NY, June 18, 37; Can citizen; m 70; c 2. ROCK MECHANICS, MINING ENGINEERING. *Educ:* McGill Univ, BEng, 59, MEng, 60, PhD(mining eng), 70. *Prof Exp:* Lectr, 61-64, asst prof, 64-70, ASSOC PROF MINING ENG, MCGILL UNIV, 70-, DIR, MINING PROG, 78- *Mem:* Eng Inst Can; Can Geotech Soc; Can Inst Mining & Metall. *Res:* Applications of methods of stress analysis to mining engineering; studies of stresses in the earth's crust; engineering properties of geological materials; ground control and mine stability. *Mailing Add:* Canmet Mining Res Lab 555 Booth St Ottawa ON K1A 0G6 Can

UDELSON, DANIEL G(ERALD), b New York, NY, Mar 7, 29; m 67; c 2. AEROSPACE & MECHANICAL ENGINEERING. *Educ:* George Washington Univ, AB, 53; Harvard Univ, AM, 54, PhD, 61. *Prof Exp:* Asst prof aero & mech eng, Col Indust Tech, 60-64, assoc prof, Col Eng, 64-70, PROF, COL ENG, BOSTON UNIV, 70-, CHMN AEROSPACE MECH ENG, 81- *Concurrent Pos:* Consult, Avco Missile Systs Div, Ctr Nuclear Studies, French Atomic Energy Comn, Urology Dept, Boston Univ Med Sch & Cambridge Air Force Res Labs. *Mem:* Assoc fel Am Inst Aeronaut & Astronaut; Am Soc Eng Educ; Am Soc Mech Engrs. *Res:* Fluid and applied mechanics. *Mailing Add:* 237 Marlborough St Boston MA 02116

UDEM, STEPHEN ALEXANDER, b New York, NY, Apr 4, 44. VIROLOGY, INFECTIOUS DISEASES. *Educ:* City Col New York, BS, 64; Albert Einstein Col Med, PhD(genetics), 71, MD, 72. *Prof Exp:* Intern internal med, Bronx Munic Hosp Complex, 72-73, resident, 73-74; NIH fel infectious dis, Montefiore/Albert Einstein/Jacobi Hosps & Albert Einstein Col Med, 74-76; asst prof, 76-83, ASSOC PROF MED & DEPT MICROBIOL & IMMUNOL, ALBERT EINSTEIN COL MED, 83- *Concurrent Pos:* Comt infectious dis, NY Acad Med. *Mem:* Am Soc Microbiol; Infectious Dis Soc. *Res:* Investigation of persistent viral infections and their relationship to the production of chronic disease, particularly chronic neurological and rheumatic diseases. *Mailing Add:* 1300 Morris Park Ave New York NY 10461

UDEN, PETER CHRISTOPHER, b Southampton, Eng, May 19, 39; m 67; c 3. CHROMATOGRAPHY. *Educ:* Bristol Univ, BSc, 61, PhD(chem), 64. *Prof Exp:* Instr chem, Univ Ill, Urbana, 65-66; ICI fel, Univ Birmingham, 66-67, lectr, 67-70; asst prof, 70-72, assoc prof, 72-78, PROF ANALYTICAL CHEM, UNIV MASS, AMHERST, 78- *Concurrent Pos:* Mallinckrodt Chem Corp res assoc, Univ Ill, Urbana, 64-66. *Mem:* Am Chem Soc; Royal Soc Chem; UK Chromatographic Soc; Sigma Xi; Int Union Pure & Appl Chem. *Res:* Analytical and inorganic chemistry; separation and thermal methods; gas and liquid chromatography; mass spectrometry; metal complexes. *Mailing Add:* GRC Tower A Dept Chem Univ Mass Amherst MA 01003-0035

UDENFRIEND, SIDNEY, b New York, NY, Apr 5, 18; m 43; c 2. BIOCHEMISTRY. *Educ:* City Col New York, BS, 39; NY Univ, MS, 42, PhD(biochem), 48. *Hon Degrees:* DSc, New York Med Col, 74, Col Med & Dent NJ, 79 & Mt Sinai Sch Med City Univ NY, 81. *Prof Exp:* Lab asst bact, City Dept Health, New York, 40-42; asst chemist, Res Div, NY Univ, 42-45, asst biochem, Col Med, 45-48; instr, Sch Med, Wash Univ, 48-50; chief sect cellular pharmacol, Chem Pharmacol Lab, Nat Heart Inst, 50-56, chief lab clin-biochem, 56-58; dir, Roche Inst Molecular Biol, 68-83, HEAD, LAB MOLECULAR NEURO-BIOL & DISTINGUISHED MEM, ROCHE INST MOLECULAR BIOL, 83- *Concurrent Pos:* US Nat Comt Int Union Biochem; mem bd trustees, Wistar Inst, 68-71; adj prof dept biochem, City Univ New York, 68-; adj prof dept human genetics & develop, Columbia Univ Col Physicians & Surgeons, 69-; mem panel narcotics, Off Sci & Technol, 72-73; mem sci adv bd, Scripps Clin & Res Found, 74-78 & Inst Cellular & Molecular Path, 74-; mem adv comt to dir, NIH, 76-78; mem adv bd, Weizmann Inst Sci, 78- *Honors & Awards:* Flemming Award, 58; Van Slyke Award, 67; Gairdner Award, 67; Hillebrand Award, Am Chem Soc, 62; Ames Award, Am Asn Clin Chem, 69; Torald Sollman Award, Am Soc Pharmacol & Exp Therapeut, 75; City of Hope Res Award, 75; Heinrich Waelsch Lectr neurosci, 77; Rudolf Virchow Gold Medal Award, 79; Chauncey Leake lectr, Univ Calif, San Francisco, 80; Wis Biol Div Hilldale Lectureship Award, Univ of Wis, 85. *Mem:* Nat Acad Sci; AAAS; Am Chem Soc; Am Soc Biol Chemists; Am Soc Pharmacol & Exp Therapeut (secy, 62-64); Soc Exp Biol & Med; Asn Clin Chemist; NY Acad Sci; Harvey Soc; Int Union Biochem; Int Soc Neurochemistry; Am Col Neuropsychopharmacol; Am Acad Arts & Sci; hon mem Japanese Pharmacol Soc, 77; hon mem Czech Pharmacol Soc, 77. *Res:* Peptide and protein biochemistry; neurochemistry. *Mailing Add:* Roche Inst Molecular Biol Nutley NJ 07110

UDIN, SUSAN BOYMEL, b Philadelphia, Pa, Aug 11, 47; m 67; c 2. DEVELOPMENT NEUROBIOLOGY. *Educ:* Mass Inst Technol, BS, 69, PhD(life sci), 75. *Prof Exp:* Sr staff mem, Nat Inst Med Res, Mill Hill, 78-79; asst prof, 79-85, ASSOC PROF, NEUROBIOL, STATE UNIV NY BUFFALO, 85- *Concurrent Pos:* Fel, Psychol Dept, Mass Inst Technol, 75-77, Nat Inst Med Res, Mill Hill, 77-78; prin investr, NY State Health Res Coun grant, 80-81; prin investr, Nat Eye Inst res grant, 80-91; prin investr, March of Dimes, basic res grant, 83-85 & 89-91. *Mem:* AAAS; Soc Neurosci; Sigma Xi; Asn Women Sci. *Res:* Effects of early visual experience on formation of connections in the brain and the role of glutamate receptors in control of plasticity. *Mailing Add:* Dept Physiol 327 Cary Hall State Univ NY Buffalo NY 14214

UDIPI, KISHORE, b Udipi, SIndia, May 19, 40; m 73; c 1. POLYMER CHEMISTRY. *Educ:* Univ Bombay, BSc Hons, 59, MSc, 63; Univ Akron, PhD(polymer chem), 72. *Prof Exp:* Works mgr paints & polymers, Bombay Paints, India, 63-68; fel, Princeton Univ, 72-73; res chemist polymer chem, Phillips Petrol Co, 73-80; res specialist, 80-89, SR RES SPECIALIST & SCI FEL, MONSANTO PLASTICS & RESINS, 89- *Mem:* Am Chem Soc. *Res:* Polymer synthesis; study of polymer microstructure and chemical modifications of polymers. *Mailing Add:* Monsanto Plastics & Resins 730 Worchester St Indian Orchard MA 01151

UDLER, DMITRY, b Kharkov, USSR, Dec 15, 54. SOLID STATE PHYSICS, APPLIED MATHEMATICS. *Educ:* Inst Solid State Physics, Acad Sci Chernogslovka, USSR, Cand Sci Physics & Math, 88. *Prof Exp:* Res scientist, Inst Solid State Physics, Chernogslovka, USSR, 89; RES ASSOC, DEPT MAT SCI & ENG, NORTHWESTERN UNIV, 90- *Mem:* Mat Res Soc. *Res:* Computer simulations at internal interfaces in solids and related phenomena. *Mailing Add:* Dept Mat Sci & Eng Northwestern Univ 2145 Sheridan Rd Evanston IL 60208-3108

UDO, TATSUO, b Tokyo, Japan, Sept 15, 25; m 57; c 2. ELECTRIC POWER TRANSMISSION LINE INSULATION DESIGN, HIGH VOLTAGE TESTING FACILITIES. *Educ:* Tokyo Univ, Bachelor, 49, Dr(eng), 60. *Prof Exp:* Engr, Japan Generation & Transmission Co Inc, 49-52 & Bonneville Power Admin, 66-68; researcher high voltage, Cent Res Inst Elec Power Indust, 52-64, sect chief, Syst Insulation, 64-66, div mgr, 68-75, vpres, Elec Eng Lab, 75-83, vpres, Adiko Res Lab, 83-87, sr adv, 87-89; PRES, DCC LTD, 89- *Concurrent Pos:* Vis lectr, Fac Eng, Nagoya Univ, 74-79 & Tokyo Univ, 85-86. *Mem:* Inst Elec & Electronics Engrs. *Res:* Insulation and electrical breakdown of extra high voltage and ultra high voltage power transmission systems. *Mailing Add:* Denryoku Comput Ctr Ltd 13-27 2-Chome Honkomagome Bunkyo-Ku Tokyo 113 Japan

UDOLF, ROY, b New York, NY, Aug 7, 26; m 50; c 4. PSYCHOLOGY, ENGINEERING. *Educ:* NY Univ, BEE, 50, Brooklyn Law Sch JD, 54; Hofstra Univ, MA, 63; Adelphi Univ, PhD(psychol), 71; Am Bd Forensic Psycol, dipl. *Prof Exp:* Test engr, Am Bosch Arma Corp, 56-63; asst dept head eng, Gyrodyne Co Am, 63-67; assoc prof, 67-80, PROF PSYCHOL, HOFSTRA UNIV, 80- *Concurrent Pos:* Human factors consult, Litcom Div, Litton Industs, 71-73. *Mem:* Am Psychol Asn; Am Psychol-Law Soc. *Res:* Human engineering; hypnosis; forensic psychology. *Mailing Add:* New Col Hofstra Univ Hempstead NY 11550

UDOVIC, DANIEL, b Cleveland, Ohio, July 9, 47; m 68; c 2. COMPUTERS IN EDUCATION, INTELLIGENT TUTORING SYSTEMS. *Educ:* Univ Tex, Austin, BA, 70; Cornell Univ, PhD(entom), 74: Univ Ore, MS. *Prof Exp:* Asst prof, 73-81, ASSOC PROF BIOL, UNIV ORE, 81-, DEPT HEAD, 89- *Mem:* Asn Comput Mach. *Res:* Computers in biology education, with emphasis on development of intelligent tutoring systems for use in university level science courses. *Mailing Add:* Dept Biol Univ Ore Eugene OR 97403

UDRY, JOE RICHARD, b Covington, Ky, Oct 12, 28; m 50; c 2. DEMOGRAPHY. *Educ:* Northwestern Univ, BS, 50; Calif State Univ, Long Beach, MA, 56; Univ Southern Calif, PhD(sociol), 60. *Prof Exp:* Instr sociol, Chaffey Col, 60-62; asst prof, Calif State Polytech Col, 62-65; assoc prof, 65-69, PROF MATERNAL & CHILD HEALTH & SOCIOL, UNIV NC, CHAPEL HILL, 69-, DIR, CAROLINA POP CTR, 77- *Mem:* Am Pub Health Asn; Am Sociol Asn; Pop Asn Am; Nat Coun Family Rels. *Res:* Demography; family and sexual behavior; interaction of social and biological processes. *Mailing Add:* Carolina Pop Ctr Univ NC Chapel Hill NC 27514

UDVARDY, MIKLOS DEZSO FERENC, b Debrecen, Hungary, Mar 23, 19; nat US; m 51; c 3. ZOOLOGY. *Educ:* Debrecen Univ, PhD, 42. *Prof Exp:* Asst biologist, Hungarian Inst Ornith, 42-45; res assoc, Biol Res Inst, Hungarian Acad Sci, 45-48; res fel zool, Univ Helsinki, 48-49 & Univ Uppsala, 49-50; asst cur, Swedish Mus Natural Hist, 51; vis lectr ecol, Univ Toronto, 51-52; lectr zool, Univ BC, 52-53, from asst prof to assoc prof, 53-66; prof, 66-90, EMER PROF BIOL SCI, CALIF STATE UNIV, 90- *Concurrent Pos:* Asst scientist, Fisheries Res Bd Can, 52-55; vis prof, Univ Hawaii, 58-59; vis spec lectr, Univ Calif, Los Angeles, 63-64; vis prof, Univ Bonn, 70-71; Fulbright lectr, Honduras, 71-72; mem, Int Protecting Bd, Biol Sta Wilhelmiberg, Austria & Point Reyes Bird Observ; consult, Int Union Conserv Nature, 80- *Mem:* Fel AAAS; Cooper Ornith Soc; Nat Audubon Soc; Wilson Ornith Soc; Sigma Xi; corresp mem, Finnish Ornithol Soc; cooresp mem, Argentinian Ornithol Soc. *Res:* Biogeography, especially distributional and ornithology. animal ecology and behavior; ornithology. *Mailing Add:* Dept Biol Sci Calif State Univ Sacramento CA 95819

UDVARHELYI, GEORGE BELA, b Budapest, Hungary, May 14, 20; US citizen; m 56; c 3. NEUROSURGERY. *Educ:* St Stephen's Col, Hungary, BA, 38; Pazmany Peter Univ, MD, 44; Univ Buenos Aires, MD, 52. *Prof Exp:* Resident neurol, Univ Budapest Hosp, 44-46; fel, Univ Vienna, 46-47; fel neuropath & psychiat, Univ Berne, 47-48; resident neurosurg, Hosp Espanol, Cordoba, 48-50; resident & surgeon, Univ Buenos Aires, 50-52; registr, Univ Edinburgh, 53-55; from instr neurosurg to asst prof neurosurg & radiol, 56-63, ASSOC PROF RADIOL, SCH MED, JOHNS HOPKINS UNIV, 63-, PROF NEUROSURG, 69- *Concurrent Pos:* WGer fel neurosurg, Univ Cologne, 52-53; Brit Coun scholar, Univ Edinburgh, 53-55; NIH fel neurosurg, Sch Med, Johns Hopkins Univ, 55-56; NIH res fel, 57-58; consult, Baltimore City Hosp, 58-, Danville Hosp, 58-71 & Harrisburg State Hosp, 59-71; dir, Off Cult Affairs, Johns Hopkins Med Inst, 76-; chief, Vet Admin Hosp, Baltimore, 79-81. *Honors & Awards:* Lincoln Award, Am Asn Neurol Surg, Humanitarian Award. *Mem:* AMA; Am Asn Neuropath; Am Asn Neurol Surg; Cong Neurol Surg; Soc Fr Speaking Neurosurgeons. *Res:* Clinical neurosurgery; neuroradiology; cerebral circulation; pediatric neurosurgery; pituitary surgery. *Mailing Add:* Johns Hopkins Med Inst 550 N Broadway Suite 407 Baltimore MD 21205

UEBBING, JOHN JULIAN, b Chicago, Ill, July 7, 37; m 66; c 2. ELECTRICAL ENGINEERING, ELECTRONICS. *Educ:* Univ Notre Dame, BS, 60; Mass Inst Technol, MS, 62; Stanford Univ, PhD(elec eng), 67. *Prof Exp:* Staff engr, Gen Motors Defense Res Labs, 62-63; staff engr surface sci & electron spectros, Varian Assocs, 66-72; staff engr electronics, Electromagnetic Systs Lab, 72-73; develop engr, 73-75, sect mgr optoelectronics, 75-80, proj mgr laser addressed crystal displays, 80-83, proj mgr obj oriented software for test measurements, 83-87, PROJ MGR LED PRINTHEAD, OPTOELECTRONICS DIV, HEWLETT-PACKARD, 88- *Concurrent Pos:* Consult, Stanford Univ, 74-77 & Radiologic Sci Inc, 75-77; mem, Comput Soc, Inst Elec & Electronics Engrs. *Mem:* Inst Elec & Electronics Engrs; Soc Info Display. *Res:* Light emitting diode products; plastic optics; electron spectroscopy; surface science. *Mailing Add:* Dev Engr Hewlett Packard 370 W Trimble Rd San Jose CA 95131

UEBEL, JACOB JOHN, b Chicago, Ill, Dec 25, 37; m 58; c 3. ORGANIC CHEMISTRY. *Educ:* Carthage Col, BA, 59; Univ Ill, MA, 62, PhD(chem), 64. *Prof Exp:* Res assoc, Univ Mich, 64; from asst prof to assoc prof, Univ NH, 64-73, prof org chem, 73-80; res scientist, 80-83, LAB HEAD, EASTMAN KODAK CO, ROCHESTER, NY, 83- *Concurrent Pos:* Vis prof, Univ Calif-Riverside, 71-72; vis scientist, Eastman Kodak Co, Rochester, NY & NSF sci fac fel, 78-79. *Mem:* Am Chem Soc; Sigma Xi. *Res:* Organic magnetic resonance and conformational analysis. *Mailing Add:* Five Landmark Lane Pittsford NY 14534

UEBELE, CURTIS EUGENE, b Kenosha Co, Wis, Dec 3, 35; m 58; c 6. POLYMER CHEMISTRY. *Educ:* Carroll Col, Wis, BS, 58; Univ Kans, PhD(chem), 65. *Prof Exp:* Proj leader, 65-75, res assoc, 75, SUPVR POLYMER RES, B P AM, CLEVELAND, 78- *Mem:* Am Chem Soc. *Res:* Formulation and evaluation of polyvinyl and polyolefin resins. *Mailing Add:* 4440 Warrensville Ctr Rd Cleveland OH 44128

UEBERSAX, MARK ALAN, b Baltimore, Md, Feb 13, 48; m 74. AGRICULTURE, FOOD SCIENCE. *Educ:* Delaware Valley Col, BS, 70; Mich State Univ, MS, 72, PhD(food sci), 77. *Prof Exp:* Food scientist prod develop, R T French Co, 73; MEM FAC, DEPT FOOD SCI & HUMAN NUTRIT, MICH STATE UNIV, 77- *Mem:* Inst Food Technologists; Am Soc Hort Sci; Am Asn Cereal Chemists; Sigma Xi. *Res:* Chemical and physical evaluation of processed fruits and vegetables; evaluation of processing techniques on nutrient retention, yield and overall quality of processed foods. *Mailing Add:* Dept Food Sci & Human Nutrit Mich State Univ East Lansing MI 48824

UECKER, FRANCIS AUGUST, b Ft Wayne, Ind, Dec 18, 30; m 61; c 4. MYCOLOGY. *Educ:* Quincy Col, BS, 56; Univ Ill, Urbana, MS, 59, PhD(bot), 62. *Prof Exp:* Asst prof bot, Univ Ill, Urbana, 62-63, biologist, 63; asst prof, Winona State Col, 63-65; RES MYCOLOGIST, BELTSVILLE AGR RES CTR, PLANT SCI INST, USDA, 65- *Mem:* AAAS; Sigma Xi; Mycol Soc Am; Am Inst Biol Scientists. *Res:* Development, cytology and taxonomy of fungi, especially pyrenomycetes and Fungi Imperfecti. *Mailing Add:* USDA Agr Res Ctr BARC-W Beltsville MD 20705

UEDA, CLARENCE TAD, b Kansas City, Mo, July 6, 42; m 71; c 2. PHARMACOKINETICS, BIOPHARMACEUTICS. *Educ:* Contra Costa Col, AA, 63; Univ Calif, San Francisco, PharmD, 67, PhD(bipharmaceut), 74. *Prof Exp:* From asst prof to assoc prof, 74-85, dir clin pharmacokinetics, Cardiovasc Ctr, 76-80, chmn dept, 76-87, PROF PHARMACEUT SCI, 85-, DEAN, UNIV NEBR, 87- *Concurrent Pos:* Consult, Sandoz Ltd, 82-85, Comt Revisions, USP, 90-95. *Mem:* Am Pharmaceut Asn; Acad Pharmaceut Sci; Am Soc Clin Pharmacol & Therapeut; Am Col Clin Pharmacol; Am Asn Pharmaceut Sci; Am Asn Cols Pharm. *Res:* Intestinal lymphatic drug absorption; pharmacokinetics of cardiovascular drugs; placental drug transfer pharmacokinetics; in vitro and in vivo forecasting of drug pharmacokinetics. *Mailing Add:* Dept Pharmaceut Sci Univ Nebr Med Ctr 600 S 42nd St Omaha NE 68198-6000

UEHARA, HIROSHI, b Kobe City, Japan, Mar 7, 23; m 47; c 2. MATHEMATICS. *Educ:* Univ Tokyo, MS, 49; Osaka Univ, DSc, 54. *Prof Exp:* Instr math, Nagoya Univ, 49-51; asst prof, Math Inst, Kyushu Univ, 53-56; prof, Univ of the Andes & Nat Univ Colombia, 56-58; asst prof, Univ Southern Calif, 58-60; assoc prof, Univ Iowa, 60-64; PROF MATH, OKLA STATE UNIV, 64- *Concurrent Pos:* Lectr, Nat Univ Mex, 58. *Mem:* Am Math Soc; Math Soc France; Math Soc Japan. *Res:* Algebraic topology. *Mailing Add:* Dept Math Okla State Univ Stillwater OK 74075

UELAND, KENT, b Chicago, Ill, May 27, 31; m 54; c 4. OBSTETRICS & GYNECOLOGY. *Educ:* Carleton Col, BA, 53; Univ Ill, BS & MD, 57; Am Bd Obstet & Gynec, dipl, 67; cert maternal-fetal med, 77. *Prof Exp:* Intern, Med Sch, Univ Ore, 57-58; asst resident obstet & gynec, King County & Univ Hosps, 60-61; from asst instr to prof obstet & gynec, Univ Wash, 61-77, dir obstet, Univ Hosp, 68-77; PROF OBSTET & GYNEC & CHIEF MATERNAL FETAL MED, STANFORD UNIV, 77- *Concurrent Pos:* Res fel cardiovasc res, Univ Wash, 63; res fel med, Univ Ore, 63-64; resident, King County & Univ Hosps, 61-62, chief resident, 63; consult, Santa Clara Valley Med Ctr, San Jose, Calif, Naval Regional Med Ctr, Oakland, Calif, Childbirth Educ Asn & Nat Found March of Dimes. *Mem:* Perinatal Res Soc; Soc Obstet, Anesthesia & Perinatal; Am Col Obstet & Gynec; Soc Gynec Invest; Soc Perinatal Obstet. *Res:* Pregnancy and cardiovascular dynamics; toxemia of pregnancy; control of labor; pregnancy and heart disease. *Mailing Add:* 2750 Kalapu Dr Lahaina HI 96761

UEMURA, YASUTOMO J, b Tokyo, Japan, Nov 22, 53; m 83. MAGNETISM, RANDOM SYSTEMS. *Educ:* Univ Tokyo, BSc, 77, MSc, 79, DSc(physics), 82. *Prof Exp:* Fel physics, Univ Tokyo, 82-83; Japan Soc Prom Sci spec res fel, 83-85, ASSOC PHYSICIST, BROOKHAVEN NAT LAB, 85- *Res:* Condensed matter physics; muon spin relaxation methods; neutron scattering. *Mailing Add:* Brookhaven Nat Lab Bldg 510B Upton NY 11973

UENG, CHARLES E(N) S(HIUH), b Kiangtu, Kiangsu, China, Sept 8, 30; m 62; c 2. SOLID MECHANICS. *Educ:* Cheng Kung Univ, BS, 53; Kans State Univ, MS, 60, PhD(appl mech), 63. *Prof Exp:* Struct engr, Taiwan Power Co, China, 53-58; res assoc appl mech, Kans State Univ, 62-63, asst prof, 63-64; from asst prof to assoc prof, 64-67, PROF ENG MECH, GA INST TECHNOL, 77- *Concurrent Pos:* Consult, Reliance Elec Co, Westinghouse Elec Corp, Combustion Eng, Inc, Electro-Mech Co, Victoreen Instruments & Wolfe & Mann Mfg Co; res grants, NASA & NSF. *Mem:* Am Soc Civil Engrs; Am Acad Mech; Soc Eng Sci; Am Soc Eng Educ. *Res:* Composite structures; vibration; elastic stability; variational principles; earthquake engineering. *Mailing Add:* Sch Civil Eng Ga Inst Technol Atlanta GA 30332-0355

UENO, HIROSHI, b Osaka, Japan, Dec 9, 50; m 78; c 1. PROTEIN CHEMISTRY, ENZYMOLOGY. *Educ:* Kyoto Univ, BE, 74; Brandeis Univ, MA, 76; Iowa State Univ, PhD(biochem), 82. *Prof Exp:* Res assoc, 82-83, Rockefeller Found fel, 84-85, ASST PROF BIOCHEM, ROCKEFELLER UNIV, 86- *Concurrent Pos:* Summer investr, Woods Hole Marine Biol Lab, 84-; vis scientist, Pop Coun, 84-; Ad hoc comt, Nat Heart, Lung & Blood Inst, NIH, 87-; vis prof, Kumamoto Univ, Japan, 90. *Mem:* Am Chem Soc; Am Soc Biol Chemists; Harvey Soc; NY Acad Sci. *Res:* Chemistry of gossypol, transaminases, hemoglobins. *Mailing Add:* Biochem Lab Rockefeller Univ 1230 York Ave New York NY 10021-6399

UETZ, GEORGE WILLIAM, b Philadelphia, Pa, Dec 8, 46. ECOLOGY, BEHAVIOR. *Educ:* Albion Col, BA, 68; Univ Del, MS, 70; Univ Ill, PhD(ecol), 76. *Prof Exp:* Res asst entom, Univ Del, 68-70; teacher biol, Sanford Sch, 70-72; res asst zool, Univ Ill, 72-75, teaching asst ecol, 75-76; from asst prof to assoc prof, 76-87, PROF BIOL SCI, UNIV CINCINNATI, 87- *Concurrent Pos:* Assoc cur, Cincinnati Mus Natural Hist, 76-84; Sigma Xi grant, 76-77; Nat Geog Soc grant, 78-79, 85-86 & 91-92; Elec Power Res Inst grant, 78-80; NSF grant, 87-89, 89-91, 90-92. *Mem:* Am Arachnological Soc; Brit Arachnological Soc; Ecol Soc Am; Entom Soc Am; Animal Behavior Soc. *Res:* Ecology and behavior of spiders; social behavior of colonial spiders; behavioral reproductive isolation in wolf spiders; spider community structure. *Mailing Add:* Dept Biol Sci Univ Cincinnati Cincinnati OH 45221-0006

UFFEN, ROBERT JAMES, b Toronto, Ont, Sept 21, 23; m 49; c 2. GEOPHYSICS, SANTITARY & ENVIRONMENTAL ENGINEERING. *Educ:* Univ Toronto, BASc, 49, MA, 50; Univ Western Ont, PhD(physics), 52. *Hon Degrees:* DSc, Univ Western Ont, 70, Queen's Univ, Ont, 67, Royal Mil Col Can, 78 & McMaster Univ, 83. *Prof Exp:* Lectr physics & geol, Univ

Western Ont, 51-53, from asst prof to assoc prof geophys, 53-58, prof & head dept, 58-61, actg head dept physics, 60-61, asst prin, Univ Col Arts & Sci, 60-61, prin, 61-65, dean col sci, 65-66; mem, Defence Res Bd Can, 63-66, vchmn, 66-67, chmn, 67-69; chief sci adv to cabinet, Can Govt, 69-71; dean fac appl sci, 71-80, prof geophys, 71-89, EMER PROF GEOPHYS, QUEEN'S UNIV, ONT, 89- *Concurrent Pos:* Consult, Kennco Explor Ltd, 52-59; res fel, Inst Geophys Univ Calif, Los Angeles, 53; Can deleg, Int Union Geod & Geophys, Rome, 54, Toronto, 57, Helsinki, 60, Tokyo, 62 & Int Union Geol Sci, New Delhi, 64; consult, Utah Construct Co, 55-58; mem Nat Adv Comt Res Geol Sci, 58-61; mem, Nat Res Coun Can, 63-66; mem, Nat Feasibility Comt Proj Oilsand, 59; chmn, Can Sci Comt, Int Upper Mantle Proj, 60-65; ed, Earth & Planetary Sci Lett, 65-69 & Tectonophysics, 67-70; mem, Coun Regents for Cols Appl Arts & Technol, Prov Ont, 66-69 & 72-75; mem Sci Coun Can, 67-71; chmn Can Eng Manpower Coun, 72-75; mem Ctr Resource Studies, 73-85; counr, Assoc Prof Engrs Ont, 75-79; mem Fisheries Res Bd Can, 75-78; vchmn bd, Ont Hydro, 74-79; comnr, Ont Royal Comn Asbestos, 80-84 & Ont Comn Truck Safety, 81-83; chmn, Ont Explor Tech Develop Fund, 81-83; consult, Ministry State, Sci & Technol, 83-84, nuclear energy, Elec & Electronics Comn on Energy, 87-88; coordr, Summer Prog Sci Teachers, Queens Univ, 86-89. *Honors & Awards:* Waddell Lectr, Royal Mil Col Can, 82; Centenial Medal Can, 67; Sigma Xi Medal, 80; Officer, Order of Can, 83; Pub Serv Medal, Asn Prof Engrs, Ont, 85. *Mem:* Fel AAAS; fel Royal Soc Can; Am Geophys Union; fel Geol Soc Am; Am Inst Mining, Metall & Petrol Eng; Can Geophys Union. *Res:* Geothermometry; internal constitution of the earth; paleomagnetism; science policy; radioactive waste management; occupational safety. *Mailing Add:* Dept Geol Queen's Univ Kingston ON K7L 3N6 Can

UFFEN, ROBERT L, b Oxnard, Calif, Dec 2, 37; m 78; c 2. MICROBIOLOGY. *Educ:* Stanford Univ, BA, 62; Univ Mass, Amherst, MA, 64, PhD(microbiol), 68. *Prof Exp:* NIH fel, Univ Ill, Urbana, 68-70; from asst prof to assoc prof, 70-84, PROF MICROBIOL, MICH STATE UNIV, 84- *Concurrent Pos:* Guggenheim fel. *Mem:* AAAS; Am Soc Microbiol; Am Chem Soc. *Res:* General microbiology and microbial physiology; regulation during cell differentiation; hydrogen gas and carbon monoxide metabolism; ecology. *Mailing Add:* Dept Microbiol Mich State Univ East Lansing MI 48824

UGARTE, EDUARDO, b Santa Ana, El Salvador, Oct 22, 35; m 63; c 4. BIOCHEMISTRY, MICROBIOLOGY. *Educ:* Nat Univ Mex, lic biochem, 62; Univ El Salvador, Dr(biochem), 65; Univ Rio de Janeiro, dipl oral microbiol, 66; Inter-Am Inst Agr Sci, El Salvador, dipl agr sci educ, 69. *Prof Exp:* Assoc prof biochem, Sch Dent, Univ El Salvador, 65-67, secy dept basic sci, 66, secy curriculum comn, 66-67, prof chem, Sch Agron Sci, 67-68, chief prof biochem, 68-70; researcher biochem, United Med Labs, 70-74; dir, Analytico Lab, 75; vpres, Page Biochem Labs, Inc, 75-76; tech dir, Nat Health Labs, Ft Lauderdale, 76-77; mem staff res cancer tissue, Oncolnovairul, Life Sci Res Inc, St Petersburg Fla, 78-; AT INTERSCI, INC. *Concurrent Pos:* Fel bact & virol, Life Labs, Ecuador, 63-64; fel oral microbiol, Pan-Am Health Orgn, Brazil, 66; lab clin supvr, Tampa Gen Hosp, Fla, 79-80; lab dir, St Petersburg Gen Hosp, Fla, 80-81; pres, Hemato Control Prod, Intersci Inc, Portland, Ore, 81-82. *Mem:* Am Chem Soc; Microbiol Soc El Salvador. *Res:* Relation of the mechanism of different hormones between pituitary and thyroid glands; development of the radioimmunoassays for these hormones; radiolabeling materials for different fractions. *Mailing Add:* 938 NE 173rd Ave Portland OR 97230

UGENT, DONALD, b Chicago, Ill, Dec 20, 33; m 62; c 3. PLANT TAXONOMY, ECONOMIC BOTANY. *Educ:* Univ Wis, Madison, BS, 56, MS, 61, PhD(bot), 66. *Prof Exp:* From asst prof to assoc prof, 68-82, PROF BOT, SOUTHERN ILL UNIV, CARBONDALE, 82-, CUR, HERBARIUM, 68- *Concurrent Pos:* NSF Proj assoc, 66-67; mem trop studies comt, Assoc Univs Int Educ, 70-76; reviewer grants, NSF, 70- *Mem:* Soc Econ Bot; Sigma Xi. *Res:* Biosystematics of the wild and cultivated species of Solanum, section tuberarium, potatoes; ethnobotany; phytogeography and archaeo-botany. *Mailing Add:* Dept Bot Southern Ill Univ Carbondale IL 62901

UGINCIUS, PETER, b Geniai, Lithuania, Feb 23, 36; US citizen; m 62; c 4. GEODESY. *Educ:* Kalamazoo Col, BA, 58; Ind Univ, MS, 61; Cath Univ Am, PhD(physics), 68. *Prof Exp:* Res physicist, 61-70, supvry geodist, 71-80, RES ASSOC, NAVAL SURFACE WARFARE CTR, 81- *Concurrent Pos:* Adj prof, Va Polytech Inst & State Univ, 72- *Mem:* Am Geophys Union. *Res:* Physical geodesy; earth gravitational field. *Mailing Add:* Naval Surface Warfare Ctr Code K-405 Dahlgren VA 22448

UGLEM, GARY LEE, b Grand Forks, NDak, Sept 19, 41; m 64; c 2. PARASITOLOGY, ZOOLOGY. *Educ:* Univ NDak, BS, 66, MS, 68; Univ Idaho, PhD(zool), 72. *Prof Exp:* Instr parasitol, Univ Idaho, 71; NIH fel, Rice Univ, 72-74; asst prof, 74-78, ASSOC PROF BIOL SCI, UNIV KY, 78- *Mem:* Am Micros Soc; Am Soc Parasitologists; Am Inst Biol Sci. *Res:* Physiology of host-parasite relations; membrane transport in parasitic helminths; Acanthocephalan life histories. *Mailing Add:* Dept Biol Sci Univ Ky Lexington KY 40506

UGOLINI, FIORENZO CESARE, b Florence, Italy, Jan 16, 29; m 63; c 2. SOILS. *Educ:* Rutgers Univ, BS, 57, PhD(soils), 60. *Prof Exp:* Arctic Inst NAm fel, Rutgers Univ, 60-61, asst prof soils, 61-64; asst prof & res assoc, Ohio State Univ, 64-66; from assoc prof to prof soils, Univ Wash, 66-90; PROF SOIL GENESIS, CLASSIFICATION & CARTOGRAPHY, UNIV FLORENCE, ITALY, 90- *Concurrent Pos:* NATO prof, Univ Milan. *Mem:* AAAS; Am Polar Soc; Am Soc Agron; Int Soc Soil Sci; fel Arctic Inst NAm; Int Union Quaternary Res. *Res:* Soil formation and weathering in the cold regions, including Arctic, Antarctica and Alpine environments; soil development and the impact of time on glacial deposits, chronologically different; forest soils; paleosoils. *Mailing Add:* Dept Soil Genesis, Classification & Cartography Univ Florence Piazzale delle Caseine 15 Firenze 50144 Italy

UGURBIL, KAMIL, b Tire, Turkey, July 11, 49. NUCLEAR MAGNETIC RESONANCE SPECTROSCOPY. *Educ:* Columbia Univ, NY, AB, 71, PhD(chem), 77. *Prof Exp:* Fel res biophysics, Bell Labs, 76-79; asst prof biochem, Columbia Univ, 79-85; mem staff, Gray Freshwater Bio Inst, 86-90; DIR, MAGNETIC RESONANCE RES CTR, UNIV MINN, 90- *Concurrent Pos:* Consult, Bell Labs, 80- *Mem:* Am Chem Soc; Biophys Soc; AAAS. *Res:* Applications of nuclear magnetic resonance spectroscopy to biological problems, primarily to studies of intact cells and organs. *Mailing Add:* Magnetic Resonance Res Ctr Univ Minn 385 E River Rd Minneapolis MN 55455

UHART, MICHAEL SCOTT, b Sacramento, Calif, Nov 25, 48; m 76; c 2. AGRICULTURAL METEOROLOGY, CLIMATOLOGY. *Educ:* Univ Calif, San Diego, BA, 70; Fla State Univ, MS, 76; Univ Okla, PhD(meteorol), 85. *Prof Exp:* Meteorologist, 76-79, forecaster, 79-82, METEOROLOGIST, PROG MGT, NAT WEATHER SERV, 83- *Concurrent Pos:* Adj prof, Montgomery Col, Md, 89-; mem, Agr Meteorol Working Group. *Mem:* Agr Res Inst; World Meteorol Orgn. *Res:* Agricultural and fire weather, climatology and air pollution; economic impacts of climate change, drought and weather modification. *Mailing Add:* 14620 Gallant Fox Lane Darnestown MD 20878

UHDE, THOMAS WHITLEY, b Louisville, Ky; m 77; c 2. PSYCHOPHARMACOLOGY, BEHAVIORAL PHARMACOLOGY. *Educ:* Duke Univ, BS, 71; Univ Louisville, MD, 75. *Prof Exp:* Resident, dept psychiat, Yale Univ, 75-79, chief resident, Clin Res Unit, 78; med staff fel, 79-81, chief, unit anxiety & affective dis, 82-90, CHIEF SECT/ANXIETY & AFFECTIVE DIS, BIOL PSYCHIAT BR, NIMH, 90-; ATTEND STAFF, NIH, 81-; ASSOC CLIN PROF PSYCHIAT, SCH MED, UNIFORMED SERV UNIV HEALTH SCI, 85- *Concurrent Pos:* Asst clin prof psychiat, Sch Med, Uniformed Serv Univ Health Sci, Bethesda, Md, 82-85; consult, develop rev comt, Dept Health & Human Serv, NIMH, 83-; vis prof, dept psychiat, Sch Med, Univ Hawaii, 85. *Honors & Awards:* Nat Res Serv Award, NIMH, 79; A E Bennett Neuropsychiat Res Found Award, Soc Biol Psychiat, 84. *Mem:* Am Psychiat Asn; Soc Biol Psychiat; AAAS; Pavlovian Soc; Am Col Neuropsychopharmacol; Sleep Res Soc. *Res:* Phenomenology, longitudinal course, neurobiology, pharmacological treatment of mood and anxiety disorders; psychobiological relationship of panic disorder to major affective disorders using neurophysiological, pharmacologic, biochemical and receptor binding techniques; chemical models of human anxiety and the influence of diet and nutrition in animals. *Mailing Add:* NIH Bldg Ten Rm 3S 239 9000 Rockville Pike Bethesda MD 20814

UHER, RICHARD ANTHONY, b McKeesport, Pa, June 8, 39; m 61; c 3. TRANSPORTATION RESEARCH, ENERGY RESEARCH. *Educ:* Carnegie-Mellon Univ, BS, 61, MS, 63, PhD(physics), 66. *Prof Exp:* Engr res, Westinghouse Elec Corp, 68-70, proj mgr, 70-75; sr engr res, 75-80, DIR RES, RAIL SYSTS CTR, CARNEGIE-MELLON UNIV, 80- *Concurrent Pos:* Sr lectr phyics & elec eng, Carnegie-Mellon Univ, 75-80; consult, 80- *Mem:* Sr mem Inst Elec & Electronics Engrs; Am Inst Physics; Am Railway Eng Asn; Transp Res Bd. *Res:* Modeling energy use in transportation systems; modeling railroad operations and safety in the area of tranportation. *Mailing Add:* Mellon Inst 4400 Fifth Ave Pittsburgh PA 15213

UHERKA, DAVID JEROME, b Wagner, SDak, June 2, 38; m 65; c 2. NUMERICAL ANALYSIS. *Educ:* SDak Sch Mines & Technol, BS, 60; Univ Utah, MA, 63, PhD(math), 64. *Prof Exp:* Mathematician, US Naval Radiol Defense Lab, Calif, 62; mathematician & programmer, US Army Natick Labs, Mass, 64-66; assoc prof, 68-76, PROF MATH, UNIV NDAK, 76- *Concurrent Pos:* Consult, US Army Natick Labs, 66-68; sabbatical leave, scientist in residence, Agronne Nat Lab, 81-82; acad consult, Comput Ctr, Univ NDak, 82-84. *Mem:* Math Asn Am; Am Asn Univ Profs; Am Soc Eng Educ; Soc Indust & Appl Math. *Res:* Functional and numerical analysis; computer applications; computer aided instruction. *Mailing Add:* Math Dept Univ NDak Grand Forks ND 58202-8162

UHERKA, KENNETH LEROY, b Wagner, SDak, May 30, 37; m 67; c 4. ADVANCED ENERGY SYSTEMS, THERMAL SCIENCES. *Educ:* SDak Sch Mines & Technol, BS, 59; Univ Ariz, MS, 61; Purdue Univ, Lafayette, PhD(aeronaut & eng sci), 67. *Prof Exp:* Instr aerospace eng, Purdue Univ, Lafayette, 61-67; assoc scientist, Res Inst, Ill Inst Technol, 67-68; asst prof, Energy Eng Dept, Univ Ill, Chicago Circle, 68-73; PROJ MGR, ARGONNE NAT LAB, 73- *Mem:* Am Soc Mech Engrs; Sigma Xi. *Res:* Advanced nuclear reactor power systems; stirling engines and other energy conversion systems; energy conservation technologies; Maglev trains and other super conductor applications; fluid mechanics; heat transfer; atmospheric and plasma physics; environmental pollution control. *Mailing Add:* Argonne Nat Lab MCT Div Argonne IL 60439

UHL, ARTHUR E(DWARD), b Chicago, Ill, Nov 11, 29; div; c 3. ENERGY SYSTEMS ENGINEERING, RESOURCE DEVELOPMENT PROGRAM PLANNING. *Educ:* Southwestern La Inst, BS, 50; Univ Tulsa, BS, 52, MPE, 53. *Prof Exp:* Asst prof petrol eng, Univ Southwestern La, 58-60; res assoc prod-pipeline-reservoir eng, Inst Gas Technol, 60-65, asst prof, Ill Inst Technol, 60-65, asst chmn Dept Gas Technol, 63-65; supvry engr, Bechtel Inc, 65-66, sr supvry engr, 66-68, asst chief engr, 68-69, chief engr, 69-71, prin engr, 71-72, asst mgr, Asia-Pac Opers, 72-74, mgr projs, 75-77, mgr resource develop progs, 78-79, eng mgr, Planning & Develop, 79-80, mgr spec proj opers, 81-82; gen mgr, Conserv Comt Calif, 83; PRIN PARTNER & PRES, PROJ MGT ASSOCS, 84- *Concurrent Pos:* Consult, McGraw-Hill Publ Co, Inc, 58-60, Layne & Bowler, 59, Tenn Gas Transmission Co, 59, Oil Ctr Res, 60, Natural Gas Pipeline Co Am, Columbia Gas Syst, 63-65, S & Q Corp, 85, AMX Mgt Group, 86-88, Contractors Licensing Cols Calif, 88, Christenson Eng Corp, 88-89 & EPA SuperLund Oper, 89-; mem Gas Piping Standards Comn, Am Nat Standards Inst-Am Soc Mech Engrs, 70-74; mem liquified natural gas task Force, Nat Energy Surv, Fed Power Comn, 71-72, mem tech adv subcomt, Comt Conserv Energy, 72-73; distinguished lectr, Soc Petrol Engrs, Am Inst Mining, Metall & Petrol

Engrs. *Res:* Flow behavior in conduits; liquified natural gas systems; fuel energy production and delivery systems; energy resources development; regional and national energy planning. *Mailing Add:* 130 Harvard Ave Mill Valley CA 94941

UHL, CHARLES HARRISON, b Schenectady, NY, May 28, 18; m 45; c 4. BOTANY. *Educ:* Emory Univ, BA, 39, MS, 41; Cornell Univ, PhD(bot), 47. *Prof Exp:* Asst, 41-42, 45-46, from instr to prof, 46-85, EMER PROF BOT, CORNELL UNIV, 85- *Mem:* AAAS; Bot Soc Am; fel Cactus & Succulent Soc Am; Int Asn Plant Taxon. *Res:* Chromosomes and evolution, especially of Crassulaceae. *Mailing Add:* Dept Plant Biol Cornell Univ Ithaca NY 14583-5908

UHL, JOHN JERRY, JR, b Pittsburgh, Pa, June 27, 40. MATHEMATICS. *Educ:* Col William & Mary, BS, 62; Carnegie Inst Technol, MS, 64, PhD(math), 66. *Prof Exp:* Asst prof, 68-72, assoc prof, 72-75, PROF MATH, UNIV ILL, URBANA-CHAMPAIGN, 75- *Mem:* Am Math Soc; Math Asn Am. *Res:* Functional analysis and integration theory; calculus reform. *Mailing Add:* Dept Math 273 Altgeld Hall Univ Ill Urbana IL 61801

UHL, V(INCENT) W(ILLIAM), b Philadelphia, Pa, May 16, 17. CHEMICAL ENGINEERING. *Educ:* Drexel Inst Technol, BS, 40; Lehigh Univ, MS, 49, PhD(chem eng), 52. *Prof Exp:* Develop engr, Sun Oil Co, 40-44; asst mgr heat transfer div, Downingtown Iron Works, 44-46; instr chem eng, Lehigh Univ, 47-51; mgr process equip div, Bethlehem Foundry & Mach Co, 51-54; assoc prof chem eng, Villanova Univ, 54-57; from assoc prof to prof, Drexel Inst Technol, 57-63; Union-Camp prof & chmn dept, 63-74, PROF CHEM ENG, UNIV VA, 74- *Concurrent Pos:* Consult, Bethlehem Corp, 54-70, R M Armstrong Co, 58-61, Atlantic Res Corp, 66-68, Joy Mfg Co, 70-73 & Philadelphia Gear Corp, 73-; NSF fac fel, Mass Inst Technol, 62-63; sr chem eng adv, US Environ Protection Agency, NC, 76-78. *Mem:* Am Chem Soc; Am Soc Eng Educ; fel Am Inst Chem Engrs. *Res:* Heat transfer, mixing, especially viscous fluids; technical economics; cost-benefit analysis; uncertainty analysis. *Mailing Add:* 14 Cheshire Ct Charlottesville VA 22901-3730

UHLENBECK, GEORGE EUGENE, theoretical physics; deceased, see previous edition for last biography

UHLENBECK, KAREN K, b Cleveland, Ohio, Aug 24, 42; div. MATHEMATICS. *Educ:* Univ Mich, BA, 64; Brandeis Univ, PhD(math), 68. *Prof Exp:* Instr math, Mass Inst Technol, 68-69; lectr, Univ Calif, Berkeley, 69-71; asst prof math, Univ Ill, Urbana, 71-76; from assoc prof to prof, Univ Ill, Chicago Circle, 76-83; prof, Univ Chicago, 83-88; SID RICHARDSON CENTENNIAL CHAIR, UNIV TEX, 88- *Concurrent Pos:* Sloan fel, 74-76; MacArthur fel, 83-88. *Mem:* Nat Acad Sci; Asn Women Math; Am Math Soc; Am Acad Arts Sci. *Res:* Calculus of variations; global analysis; gauge theories. *Mailing Add:* Dept Math Univ Tex Austin TX 78712

UHLENBECK, OLKE CORNELIS, b Ann Arbor, Mich, Apr 20, 42. BIOPHYSICAL CHEMISTRY. *Educ:* Univ Mich, Ann Arbor, BS, 64; Harvard Univ, PhD(biophys), 69. *Prof Exp:* Miller fel, Univ Calif, Berkeley, 69-71; asst prof, 71-75, assoc prof, 76-79, prof biochem & chem, Univ Ill, Urbana, 79-86. *Concurrent Pos:* NIH res grant, Univ Ill, Urbana, 71- *Res:* Nucleic acid interactions; structure and function of RNA. *Mailing Add:* Dept Biochem Box 215 Univ Colo Boulder CO 80309-0215

UHLENBROCK, DIETRICH A, b Schweinfurt, Ger, Oct 13, 37. APPLIED MATHEMATICS, MATHEMATICAL PHYSICS. *Educ:* Univ Cologne, Vordiplom, 59; NY Univ, MS, 62, PhD(physics), 63. *Prof Exp:* Adj asst prof theoret physics, Univ & res assoc, Courant Inst, NY Univ, 63-64; vis mem, Inst Adv Study, NJ, 64-66; from asst prof to assoc prof, 66-76, PROF MATH & PHYSICS, UNIV WIS-MADISON, 76- *Concurrent Pos:* Prof math, Free Univ Berlin, 73-78; vis prof math, Nat Autonomous Univ Mex, 81. *Res:* Classical and quantum statistical physics; quantum field theory; mathematical physics. *Mailing Add:* Dept Math & Physics Univ Wis 715 Eb Van Vleck Madison WI 53706

UHLENHOPP, ELLIOTT LEE, b Hampton, Iowa, Dec 8, 42; m 67; c 2. BIOCHEMISTRY. *Educ:* Carleton Col, BA, 65; Columbia Univ, PhD(biochem), 71. *Prof Exp:* Res chemist, Univ Calif, San Diego, 73-74; asst prof chem, Whitman Col, 75-78; asst prof, 78-80, ASSOC PROF CHEM, GRINNELL COL, 80-, CHMN DEPT, 85- *Concurrent Pos:* Fel, Damon Runyon Mem Fund, Cancer Res Inc, 71-73; IPA fel, NIH, 82-84. *Mem:* Sigma Xi; Biophys Soc; Am Chem Soc; AAAS. *Res:* Viscoelastic characterization of high molecular weight native and denatured DNA from bacterial and eukaryotic cells; chromosome structure; DNA damage and repair. *Mailing Add:* Dept Chem & Biochem Campus Box 25 Univ Colo Boulder CO 80309-0215

UHLENHUTH, EBERHARD HENRY, b Baltimore, Md, Sept 15, 27; m 52; c 3. PSYCHIATRY, PSYCHOPHARMACOLOGY. *Educ:* Yale Univ, BS, 47; Johns Hopkins Univ, MD, 51. *Prof Exp:* USPHS fel psychiat, Johns Hopkins Univ, 52-56, from instr to assoc prof, 56-68; chief, Adult Psychiat Clin, Univ Chicago, 68-76, from assoc prof to prof psychiat, 68-85, prof clin pharmacol, 75-85; PROF PSYCHIAT, UNIV MEX, 85- *Concurrent Pos:* Consult, Patuxent Inst, 56-57; asst psychiatrist chg, Outpatient Dept, Johns Hopkins Hosp, 56-61, consult, Div Plastic Surg, 59-60, psychiatrist chg, Outpatient Dept, 61-62; USPHS career teacher trainee, Johns Hopkins Univ, 57-59, USPHS career res develop awards, 62-68; consult, Ill State Psychiat Inst, 68-75; Clin Psychopharmacol Res Rev Comt, Psychopharmacol Br, NIMH, 68-72, consult, Ment Health Task Force, Mid-Southside Planning Orgn, 70-75, Woodlawn Ment Health Ctr, 71-73, & US Food & Drug Admin, 71-74; mem psychopharmacol adv comt, 74-78. *Honors & Awards:* Assoc Clin Psychiatrists' Award, 58; USPHS Res Sci Award, 76-81. *Mem:* Fel Am Col Neuropsychopharmacol; fel Am Psychiat Asn; Col Int Neuro-Psychopharmacol; AAAS; Psychiat Res Soc. *Res:* Clinical psychopharmacology; anxiety and depression; life stress; evaluation of treatments; consumption of psychotherapeutic drugs. *Mailing Add:* 1420 Ridge Crest Loop Albuquerque NM 87108

UHLER, MICHAEL DAVID, b San Bernardino, Calif, Oct 3, 56; m 77; c 2. SIGNAL TRANSDUCTION, REGULATION GENE EXPRESSION. *Educ:* Seattle Univ, BS(chem), 77, BS(clin chem), 77; Univ Ore, PhD(biochem), 82. *Prof Exp:* Asst prof biochem, Ore Health Sci Univ, 86-88; ASST PROF BIOCHEM, UNIV MICH, 88- *Mem:* AAAS; Am Soc Microbiol; Am Soc Biochem & Molecular Biol; Soc Neurosci. *Res:* Genes which code for signal transduction molecules of the nervous system and are required for intercellular coordination in proper nervous system function. *Mailing Add:* Ment Health Res Inst 205 Zina Pitcher Pl Ann Arbor MI 48109-0720

UHLHORN, KENNETH W, b Mankato, Minn, Sept 24, 33; m 54; c 2. SCIENCE EDUCATION. *Educ:* Mankato State Col, BS, 54; Univ Minn, MA, 60; Univ Iowa, PhD(biol, sci ed), 63. *Prof Exp:* Teacher pub sch, Minn, 56-60; instr biol & physics, Univ Iowa, 60-63, asst prof biol, 63; assoc prof, 63-68, PROF SCI EDUC, IND STATE UNIV, TERRE HAUTE, 68-, DIR SCI TEACHING CTR, 66- *Mem:* Nat Asn Res Sci Teaching; Nat Sci Teachers Asn. *Res:* Preparation, use and application of science experience inventories and their role in teaching science; science for elementary education majors; children's concept of science and scientists based on their drawings of scientists; the physics and perception of photography to undergraduate and graduate students. *Mailing Add:* Sci Teaching Ctr Ind State Univ Terre Haute IN 47809

UHLIG, HERBERT H(ENRY), b Haledon, NJ, Mar 3, 07; m 41; c 3. MATERIALS SCIENCE, ELECTROCHEMISTRY. *Educ:* Brown Univ, ScB, 29; Mass Inst Technol, PhD(phys chem), 32. *Prof Exp:* Phys chemist, Rockefeller Inst, 32-33; res chemist, Lever Bros Co, 34-36, asst chief chemist, 36-37; res assoc, Corrosion Lab, Mass Inst Technol, 37-40 & Gen Elec Co, 40-46; from assoc prof to prof, 46-72, In Charge Corrosion Lab, 46-75, EMER PROF METALL, MASS INST TECHNOL, 72- *Concurrent Pos:* Guggenheim fel, Max Planck Inst Phys Chem, 61; chmn, Int Corrosion Coun, 75-78. *Honors & Awards:* Whitney Award, Nat Asn Corrosion Engrs, 51; Palladium Medal, Electrochem Soc, 61, Acheson Medal, 88; U R Evans Award, Inst Corrosion Sci & Technol, Gt Brit, 80. *Mem:* Hon mem Electrochem Soc (pres, 55-56); Am Chem Soc; Am Soc Metals; Nat Asn Corrosion Engrs; Am Acad Arts & Sci; Meteoritical Soc. *Res:* Corrosion and passivity; stainless steels; meteorites; history and philosophy of science. *Mailing Add:* Dept Mat Sci & Eng Mass Inst Technol Cambridge MA 02139

UHLIR, ARTHUR, JR, b Chicago, Ill, Feb 2, 26; m 54; c 3. ELECTRICAL ENGINEERING, PHYSICS. *Educ:* Ill Inst Technol, BS, 45, MS, 48; Univ Chicago, PhD(physics), 52. *Prof Exp:* Asst engr mech, Armour Res Found, 45-48; mem tech staff transistors, Bell Labs, 51-58; vpres & dir semiconductors, Microwave Assocs, Inc, 58-69; dir res, Comput Metrics Inc, 69-73; chmn elec eng, 70-75, dean eng, 75-80, PROF ELEC ENG, COL ENG, TUFTS UNIV, 70- *Concurrent Pos:* Dir, Harvard Apparatus Corp, 75-79. *Mem:* Fel Inst Elec & Electronics Engrs; Am Phys Soc; Sigma Xi; AAAS. *Res:* Microwave devices and measurements. *Mailing Add:* Col Eng Tufts Univ Medford MA 02155

UHLMANN, DONALD ROBERT, b Chicago, Ill, Sept 22, 36; m 58; c 6. GLASS TECHNOLOGY, POLYMER SCIENCE. *Educ:* Yale Univ, BS, 58; Harvard Univ, PhD(appl physics), 63. *Prof Exp:* Res fel appl physics, Harvard Univ, 63-65; from asst prof to assoc prof ceramics, Mass Inst Technol, 65-75, prof ceramics & polymers, 75-83, Cabot prof mat, 83-86; PROF & DEPT HEAD, UNIV ARIZ, 86- *Concurrent Pos:* Convenor sci coun mat processing, Univ Space Res Asn, 77-78, mem sci coun, 78-80; Guggenheim fel, 81-82. *Honors & Awards:* F H Norton Award, 79; George W Morey Award, Am Ceramic Soc, 81; Sosman Lectr, 82. *Mem:* Am Phys Soc; Am Inst Mining, Metall & Petrol Engrs; fel Am Ceramic Soc; fel Brit Soc Glass Technol; Ceramic Educ Coun; Soc Photogeog & Inst Engrs. *Res:* Structure and properties of glasses; crystallization phenomena; structure and properties of polymers; sol-gel produced materials. *Mailing Add:* Skyline Country Club Estates 5309 E Mission Hill Dr Tucson AZ 85718

UHR, JONATHAN WILLIAM, b New York, NY, Sept 8, 27; m 54; c 2. MEDICINE, IMMUNOLOGY. *Educ:* Cornell Univ, AB, 48; NY Univ, MD, 52; Am Bd Internal Med, dipl, 60. *Prof Exp:* Dazian fel, Dept Microbiol, NY Univ Med Ctr, 55-56, instr microbiol, Sch Med, NY Univ, 57-58, from asst prof to prof med, 58-72, dir, Irvington House Inst Rheumatic Fever & Allied Dis, 62-72; PROF & CHMN, DEPT MICROBIOL, PROF INTERNAL MED & MARY NELL & RALPH ROGERS PROF IMMUNOL, UNIV TEX SOUTHWESTERN MED SCH, 72-, RAYMOND & ELLEN WILLIE DISTINGUISHED CHAIR CANCER RES, 90- *Concurrent Pos:* Intern, Mt Sinai Hosp, 52-53, resident path, 53-54, asst resident med, 54-55, chief resident med, 56-57; assoc mem, Comn Immunization, Armed Forces Epidemiol Bd, 59-65, mem, 65-73, dep dir, 69-73; assoc vis physician, Bellevue Hosp, 59-72; Commonwealth fel, Walter & Eliza Hall Inst Med Res, 61-62; USPHS career develop award, 62; assoc attending physician, Univ Hosp, 63-72; consult internal med, Manhattan Vet Hosp, 64-74; assoc ed, J Immunol, 65-73, Advan Immunol, 84-; mem, Panel Regulatory Biol, NSF, 66-68; mem, Allergy & Immunol Study Sect, USPHS, 69-73; vis prof, Dept Microbiol, Yale Univ, 70-71; adv ed, Immunogenetics, 74-80; counr, Am Asn Immunologists, 78-85; mem sci rev bd, Howard Hughes Med Inst, 80-89, Scripps Clin, 83-87; US rep, Int Union Immunol Socs, 80-86; mem, US-Japan Panel Coop Sci Prog Immunol, 81-84, head, 84. *Honors & Awards:* Newcomb Cleveland Prize, AAAS, 63; Squibb Award, Infectious Dis Soc Am, 71. *Mem:* Nat Acad Sci; Am Asn Immunologists (pres, 83-84); Am Soc Clin Invest; Am Asn Pathologists; Transplantation Soc; Asn Am Physicians; fel AAAS. *Res:* Immunology. *Mailing Add:* Dept Microbiol Southwestern Med Sch Univ Tex Dallas TX 75235-9048

UHR, LEONARD MERRICK, b Philadelphia, Pa, June 26, 27; m 49; c 2. COMPUTER SCIENCE, PSYCHOLOGY. *Educ:* Princeton Univ, BA, 49; Johns Hopkins Univ, MA, 51; Univ Mich, PhD(psychol), 57. *Prof Exp:* Assoc prof psychol, Univ Mich & res psychologist & coordr psychol sci, Ment Health Res Inst, 57-65; PROF COMPUTER SCI, UNIV WIS-MADISON,

65- *Mem:* Am Psychol Asn; Asn Comput Mach; Inst Elec & Electronics Engrs. *Res:* Dynamic computer models of perceptual and cognitive processes; perception; learning; computers and education; intelligent systems. *Mailing Add:* 211 Lathrop St Madison WI 53705

UHRAN, JOHN JOSEPH, JR, ELECTRICAL ENGINEERING, COMPUTER SCIENCES. *Educ:* Manhattan Col, BEE, 57; Purdue Univ, MSEE, 63, PhD(elec eng), 67. *Prof Exp:* Sr engr, Hazeltine Res Corp, 57-61; instr, Purdue Univ, 63-66; asst prof, 66-69, assoc prof, 69-79, PROF ELEC ENG, UNIV NOTRE DAME, 79- *Concurrent Pos:* Mem staff, Lincoln Lab, Mass Inst Technol, 78-79; consult indust; assoc ed, Transactions on Commun. *Mem:* Inst Elec & Electronics Engrs; Assoc Comput Mach; Soc Comput Simulation; Sigma Xi. *Res:* Hardware/software development; microcomputer applications; modelling and simulation techniques; communication theory; digital signal processing. *Mailing Add:* Dept Elec & Computer Eng Univ Notre Dame Notre Dame IN 46556

UHRICH, DAVID LEE, b Buffalo, NY, Jan 5, 39; m 61; c 3. PHYSICS. *Educ:* Canisius Col, BS, 60; Univ Pittsburgh, PhD(physics), 65. *Prof Exp:* Fels, Univ Pittsburgh, 66 & Iowa State Univ, 66-67; asst prof, 67-71, assoc prof, 71-77, PROF PHYSICS, KENT STATE UNIV, 77-, MEM LIQUID & CRYSTAL INST, 77- *Mem:* Am Phys Soc. *Res:* Mössbauer effect studies of liquid crystals and solids. *Mailing Add:* 5754 Caranor Dr Kent OH 44240

UHRIG, JEROME LEE, b Pittsburgh, Pa. PERFORMANCE MODELING & ANALYSIS, SIGNAL PROCESSING. *Educ:* Ohio Univ, BS, 63; Carnegie Inst Technol, MS, 64, PhD(systs & commun sci), 66. *Prof Exp:* MEM TECH STAFF, AT&T BELL LABS, 66- *Mem:* Inst Elec & Electronics Engrs; Asn Comput Mach. *Res:* Systems engineering and partitioning methodologies; operational requirements analysis for real-time computer processing applications; performance modeling analysis; real-time computer architectures; optimization theory and application; signal processing applications. *Mailing Add:* AT&T Bell Labs Whippany Rd Whippany NJ 07981

UHRIG, ROBERT EUGENE, b Raymond, Ill, Aug 6, 28; m 54; c 7. MECHANICAL ENGINEERING, COMPUTER SCIENCE. *Educ:* Univ Ill, BS, 50; Iowa State Univ, MS, 50, PhD(eng mech), 54. *Prof Exp:* Instr eng mech, West Pt Mil Acad, NY, 54-56; assoc prof nuclear eng, Iowa State Univ, 56-60; dept chmn nuclear eng, 60-68, dean, Col Eng, Univ Fla, 68-73; vpres, Fla Power & Light Co, 73-86; DISTINGUISHED PROF ENG, UNIV TENN, 86-; DISTINGUISHED SCIENTIST, OAK RIDGE NAT LAB, 86- *Concurrent Pos:* Instr eng mech, Iowa State Univ, 48-51, res engr, Ames Lab, 51-54; asst dir res, US Dept Defense, 67-68; chmn, NSF Eng Adv Comt, 72-73, Nuclear Div, Am Soc Eng Educ, 73-88; mem, Nat Acad Sci/NRC Shuttle Criticality Rev & Hazards Anal Audit, 86-88; bd dir, Am Nuclear Soc, 65-68, Eng Coun Prof Develop, 68-72; mem, Nuclear Safety Res Comt, Nuclear Regulatory Comn, 89- *Honors & Awards:* Richards Mem Award, Am Soc Mech Engrs, 69; Am Soc Eng Educ Res Award, 62. *Mem:* Fel Am Nuclear Soc; fel Am Soc Mech Engrs; fel AAAS; Am Soc Eng Educ; Int Neural Network Soc; Am Asn Artificial Intel; Inst Elec & Electronics Engrs. *Res:* Application of artificial intelligence (expert systems & neural networks) to enhance the operation & safety of nuclear power plants; application of random noise techniques to nuclear reactor systems; neutron wave propagation in nuclear systems; approximately 150 publications in scientific or technical journals and author of one book. *Mailing Add:* 113 Connors Dr Oak Ridge TN 37830-7662

UICKER, JOHN JOSEPH, JR, b Derry, NH, July 11, 38; div; c 6. MECHANICAL ENGINEERING. *Educ:* Univ Detroit, BME, 61; Northwestern Univ, Evanston, MS, 63, PhD(mech eng), 65. *Prof Exp:* From asst prof to assoc prof, Univ Wis-Madison, 67-75, dir, Comput-Aided Eng Ctr, 81-88, chair, Gen Eng Dept, 88-90, PROF MECH ENG, UNIV WIS-MADISON, 75- *Concurrent Pos:* Mem, US Coun Int Fedn for Theory of Mach & Mechanisms, 69-, ed-in-chief, Mechanism & Mach Theory, 73-78; prin res engr assoc & Am Soc Eng Educ resident fel, Advan Anal Technol Dept, Ford Motor Co, Mich, 72-73; Sr Fulbright lectr, Cranfield, England, 78-79. *Honors & Awards:* Ralph R Teeter Award, Soc Automotive Engrs, 68. *Mem:* Am Soc Mech Engrs; Am Soc Eng Educ; Asn Comput Mach. *Res:* Kinematic and dynamic analysis of mechanical systems; spatial linkage analysis; computer-aided design; computational geometry. *Mailing Add:* Dept Mech Eng 1513 Univ Ave Madison WI 53706

UITTO, JOUNI JORMA, b Helsinki, Finland, Sept 15, 43; m 65, 82; c 3. DERMATOLOGY, BIOCHEMISTRY. *Educ:* Univ Helsinki, MB, 65, MD & PhD(med biochem), 70. *Prof Exp:* Intern med, surg & med biochem, Univ Helsinki Cent Hosp, 69; instr med biochem, Univ Helsinki, 70-71; clin asst dermat, Univ Cent Hosp, Univ Copenhagen, 71; from instr to asst prof biochem, Rutgers Med Sch, Col Med & Dent NJ, 73-75; instr med & resident fel dermat, Sch Med, Washington Univ, 75-78, asst prof med & biochem, 78-80; from assoc prof to prof med & assoc chief & dir res, Div Dermat, Harbor Med Ctr, Univ Calif, Los Angeles, 80-86; PROF DERMAT, BIOCHEM & MOLECULAR BIOL & CHMN DEPT DERMAT, JEFFERSON MED COL, THOMAS JEFFERSON UNIV, PHILADELPHIA, PA, 86- *Concurrent Pos:* Fel, Gen Clin Res Ctr, Philadelphia Gen Hosp & Dept Dermat, Univ Pa, 71-72. *Honors & Awards:* William Montagna Lectr, Soc Invest Dermat, 87; Hermann Pinkens Mem Lectr, Am Soc Dermat Physicians, 88. *Mem:* Soc Invest Dermat; Am Soc Clin Invest; Am Soc Biol Chem; Am Acad Dermat; Am Chem Soc; Am Dermat Asn; Asn Am Physicians. *Res:* Biochemistry and molecular biology of connective tissues; collagen metabolism; investigative dermatology. *Mailing Add:* Dermat Jefferson Med Col 1020 Locust St Philadelphia PA 19107

UKELES, RAVENNA, b New York, NY, Aug 1, 29. MICROBIOLOGY. *Educ:* Hunter Col, BS, 49; NY Univ, MSc, 56, PhD(biol), 60. *Prof Exp:* RES MICROBIOLOGIST, LAB EXP BIOL, NAT MARINE FISHERIES SERV, 59- *Mem:* Soc Protozool; Phycol Soc Am; Am Soc Microbiol; NY Acad Sci; Sigma Xi. *Res:* Nutrition of protozoa and algae; growth of organisms in mass culture; algae as potential food sources for invertebrates. *Mailing Add:* Lab Exp Biol Nat Marine Fisheries Serv 212 Rogers Ave Milford CT 06460

UKLEJA, PAUL LEONARD MATTHEW, b Chicago, Ill, Nov 22, 46; m 67; c 1. PHYSICS, LIQUID CRYSTAL PHYSICS. *Educ:* New Col, Fla, BA, 67; Univ Chicago, MS, 69; Kent State Univ, PhD(physics), 76. *Prof Exp:* Peace Corps vol sci & math, Govt Malta, 70-73; res assoc & fel, Kent State Univ, 76-78; asst prof, 78-85, ASSOC PROF PHYSICS, SOUTHEASTERN MASS UNIV, 85- *Concurrent Pos:* Assoc mem, Liquid Crystal Inst, 88. *Mem:* Sigma Xi; Am Phys Soc; Am Asn Physics Teachers. *Res:* Liquid crystal physics, orientational ordering, and self diffusion in partially ordered systems. *Mailing Add:* Dept Physics SE Mass Univ Old Westport Rd North Dartmouth MA 02747

UKRAINETZ, PAUL RUVIM, b Erwood, Sask, Dec 28, 35; wid; c 3. FLUID POWER. *Educ:* Univ Sask, BS, 57; Univ BC, MASc, 60; Purdue Univ, PhD(mech eng), 62. *Prof Exp:* Eng trainee, Bristol Aeroplane Co Ltd, 57-59; from asst prof to assoc prof, 62-71, dept head, 74-82, PROF MECH ENG, UNIV SASK, 71- *Concurrent Pos:* Vis prof, Monash Univ, Melbourne, Australia, 82-83 & Univ Bath, Eng, 89. *Honors & Awards:* Ralph R Teetor Award, Soc Automotive Engrs, 74. *Mem:* Soc Automotive Engrs; Am Soc Eng Educ; Can Soc Mech Engrs; Eng Inst Can. *Res:* Investigations into problems associated with fluid power control systems; study of the dynamics of cables with particular reference to power transmission line vibration; experimental stress analysis. *Mailing Add:* Dept Mech Eng Univ Sask Saskatoon SK S7N 0W0 Can

ULABY, FAWWAZ TAYSSIR, b Damascus, Syria, Feb 4, 43; US citizen; m 68; c 3. ELECTRICAL ENGINEERING. *Educ:* Am Univ Beirut, BS, 64; Univ Tex, Austin, MSEE, 66, PhD(elec eng), 68. *Prof Exp:* From asst prof to prof elec eng, 68-80, dir remote sensing lab, Ctr Res, Univ Kans, 76-84, J L Constant Distinguished Prof, 80-84; PROF ELEC ENG & COMPUTER SCI, UNIV MICH, 84-; DIR, CTR SPACE TERAHERTZ TECHNOL, NASA, 88- *Concurrent Pos:* Res grants, US Army, Univ Kans, 68-, NSF, 72- & 76-, Sandia Labs & Eglin AFB, 78- & NASA, 70-; assoc dir remote sensing lab, Ctr Res, Inc, Kans, 69-71; consult, govt & univ. *Honors & Awards:* Distinguished Achievement, Inst Elec & Electronics Engrs, 83, Centennial Medal, 84; Kuwait Prize Appl Sci, 87. *Mem:* Fel Inst Elec & Electronics Engrs; Int Soc Photogram; Int Union Radio Sci. *Res:* Millimeter wave propagation; remote sensing; microwave radiometry; radar systems; optical data processing. *Mailing Add:* Dept Elec & Computer Eng 3228 EECS Ann Arbor MI 48109

ULAGARAJ, MUNIVANDY SEYDUNGANALLUR, b Tuticorin, India, Mar 4, 44; Can citizen; m 70; c 2. DIGITAL SIGNAL PROCESSING, SPEECH PROCESSING. *Educ:* Madras Univ, India, BSc, 64; Indian Agr Res Inst, New Delhi, MSc, 70; Univ Fla, Gainesville, PhD, 74. *Prof Exp:* Res asst, Tamil Nadu Govt, Coimbatore, India, 64-68; res & teaching asst, Univ Fla, Gainesville, 71-74, assoc, 74-75; res assoc, McGill Univ, MacDonald Campus, Montreal, 75-80; lectr, Comput Sci Dept, Concordia Univ, Montreal, 80-81; prog/systs anal, 81-89, RESEARCHER, RES & DEVELOP, CORP PLANNING & MGT DEPT, AGT LTD, EDMONTON, 89- *Concurrent Pos:* Res assoc, Univ Fla, Gainesville, 74; res assoc, Purdue Univ, WLafayette, Ind, 75; lectr, Comput Sci Dept, Univ Alta, Edmonton, 84- *Mem:* AAAS; Acoust Soc Am; Inst Elec & Electronics Engrs Signal Processing Soc; Inst Elec & Electronics Engrs Commun Soc. *Res:* Speech recognition, speech synthesis, speech processing, spoken language systems; acoustics, software engineering, subscriber loop in telecommunication; electronic data processing teaching; personal computers. *Mailing Add:* AGT Ltd Res & Develop Suite 500 Capitol Sq Bldg 10065 Jasper Ave Edmonton AB T5J 3B1 Can

ULANOWICZ, ROBERT EDWARD, b Baltimore, Md, Sept 17, 43; m 67; c 3. THEORETICAL ECOLOGY, NETWORK THEORY. *Educ:* Johns Hopkins Univ, BES, 64, PhD(chem eng), 68. *Prof Exp:* Res asst phys chem, Univ Gottingen, 64; res asst chem eng, Johns Hopkins Univ, 64-68; asst prof, Cath Univ Am, 68-70; res asst prof, Natural Resources Inst, 70-75, assoc prof, 75-80, PROF, CTR ENVIRON & ESTUARINE STUDIES, UNIV MD, 80- *Concurrent Pos:* Mem Sci Comt Oceanic Res, Int Coun Sci Unions, Working Group No 59, Biol Models in Oceanog & Working Group No 73, Ecosystems Theory in Relation to Biol & Oceanog; Marine Sci Panel, working group monitoring Chesapeake Bay, Nat Res Coun; orgn comn, Venice Summer Sch. *Mem:* Int Soc Ecol Modelling; Estuarine Res Fedn. *Res:* Mass and energy transfer in ecosystems; ecosystem network analysis; thermodynamics of ecosystems; information theory in ecology. *Mailing Add:* Ctr Environ & Estuarine Studies Univ Md Solomons MD 20688-0038

ULBERG, LESTER CURTISS, b Wis, Dec 2, 17; m 45; c 1. REPRODUCTIVE PHYSIOLOGY. *Educ:* Univ Wis, BS, 48, MS, 49, PhD(reprod physiol), 52. *Prof Exp:* Asst reprod physiol, Univ Wis, 47-50, instr, 50-52; instr animal husb, Miss State Col, 52-55, assoc prof, 55-57; assoc prof, 57-60, PROF ANIMAL HUSB, NC STATE UNIV, 60- *Concurrent Pos:* Agent, USDA, 50-52. *Mem:* AAAS; Am Soc Animal Sci; Am Dairy Sci Asn; Brit Soc Study Fertil. *Res:* Early embryonic development; hormone control of ovarian activity. *Mailing Add:* Dept Animal Sci NC State Univ Raleigh NC 27650

ULBRECHT, JAROMIR JOSEF, b Ostrava, Czech, Dec 16, 28; US citizen; m 52; c 2. ENGINEERING RHEOLOGY. *Educ:* Czech Inst Technol, Prague, Ing, 52; Inst Chem Technol, Prague, PhD(chem eng), 58. *Prof Exp:* Assoc dir chem eng, Rubber Res Inst, 57-62; head eng rheology lab, Czech Acad Sci, 62-68; prof chem eng, Univ Salford, 68-78; prof & chmn dept chem eng, State Univ NY, Buffalo, 78-84; chief chem eng, 84-88, DEP DIR, OFF TECHNOL EVAL & ASSESSMENT, NAT INST STANDARDS & TECHNOL, 88- *Concurrent Pos:* Res prof, Inst Chem Technol, Prague, 66-68; vis prof Univ Technol, Aachen, Germany, 67-68; Alexander von

Humboldt fel, 67; ed-in-chief, Chem Eng Commun, 77-84 & Chem Eng-Concepts & Rev, 84-; adj prof, Univ Md, 85- *Mem:* Am Inst Chem Eng; Brit Inst Chem Eng; Soc Rheology; Brit Soc Rheology; Sigma Xi; Am Chem Soc. *Res:* Momentum and mass transfer in rheologically complex and multiphase systems; mixing and gas-liquid and liquid-liquid contacting; stirred chemical; biochemical reactors. *Mailing Add:* Nat Inst Standards & Technol Bldg 411 Rm A115 Gaithersburg MD 20899

ULBRICH, CARLTON WILBUR, b Meriden, Conn, Oct 1, 32; m 62; c 3. ATMOSPHERIC PHYSICS. *Educ:* Univ Conn, BSME, 60, MS, 62, PhD(physics), 65. *Prof Exp:* Res asst physics, Univ Conn, 61-65; asst prof, Wittenberg Univ, 65-66; asst prof, 66-74, assoc prof, 74-79, PROF PHYSICS, CLEMSON UNIV, 79- *Mem:* Am Meteorol Soc; Am Geophys Union. *Res:* Radar meteorology; atmospheric physics. *Mailing Add:* Dept Physics Clemson Univ Clemson SC 29631

ULDRICK, JOHN PAUL, b Donalds, SC, Apr 11, 29; m 54; c 1. MECHANICAL ENGINEERING, THEORETICAL MECHANICS. *Educ:* Clemson Univ, BCE, 50, BME, 56, MS, 58; Univ Fla, PhD(eng mech), 63. *Prof Exp:* Bridge inspector, SC Hwy Dept, 50-51; res engr, J E Sirrine Co, Engrs, 53-54; instr eng, Clemson Univ, 55-58, asst prof eng mech, 58-60; instr, Univ Fla, 60-61; assoc prof, Clemson Univ, 63-66; from assoc to prof eng, US Naval Acad, 66-70, prof mech eng & chmn dept, 70-73, prof & dir eng mech, mat & design, 73-79, prof, 80-90; CONSULT, 90- *Concurrent Pos:* Fac assoc, Lockheed Aircraft Corp, Ga, 57; NSF initiation eng res grant, 64-66; NASA res grant, 68-69; consult, D W Taylor Naval Ship Res & Develop Ctr, Annapolis, 76-90. *Mem:* Sigma Xi. *Res:* Hydrodynamics of fish locomotion; optical data processing and information storage; flow noise, structural vibrations and time series analysis. *Mailing Add:* PO Box 592 Belton SC 29627

ULERY, DANA LYNN, b East St Louis, Mo, Jan 2, 38; m 59, 80; c 2. QUALITY MANAGEMENT, STATISTICAL COMPUTING. *Educ:* Grinnell Col, BA, 59; Univ Del, MS, 72, PhD(comput sci), 75. *Prof Exp:* Res engr, Jet Propulsion Lab, NASA, 60-63; programmer, Getty Oil Co, 63-64; lectr comput sci, Dept Comput Sci, Univ Del, 70-75, postdoctoral res assoc, 75-76; sr software engr, 77-82, consult supvr, 82-87, consult, 87-90, SR CONSULT, QUAL MGT & TECHNOL CTR, E I DU PONT DE NEMOURS & CO, 90-; ADJ PROF, DEPT MATH SCI, UNIV DEL, 89- *Concurrent Pos:* Vis lectr, Inst Statist & Comput Sci, Cairo Univ, 76; consult, Am Univ, Cairo, 76; coun rep, Am Nat Standard Inst Accredited Comt, Electronic Data Interchange, 86-; rep, Adv Comt NAm Rapporteur, UN EDIFACT Bd, 88-; mem, Accredited Comt Qual Assurance, Am Nat Standards Inst, 88-; rep, Tech Comt Qual Mgt & Qual Assurance, Int Orgn Standardization, 88-; rep, Automotive Indust Action Group, 88- *Mem:* Sigma Xi; Inst Elec & Electronics Engrs Computer Soc; Asn Comput Mach; Am Soc Qual Control; Am Statist Asn; AAAS. *Res:* Industrial quality management systems; quality management computer systems in industry; standards for electronic data interchange of quality information; statistical computing; symbolic and numeric algorithms for partial differential equations. *Mailing Add:* 18 Squirrel Lane Newark DE 19711

ULEVITCH, RICHARD JOEL, b Cleveland, Ohio, Apr 4, 44; m 74; c 1. BIOCHEMISTRY, IMMUNOPATHOLOGY. *Educ:* Washington & Jefferson Col, BA, 66; Univ Pa, PhD(biochem), 71. *Prof Exp:* Fel, Univ Minn, 71-72, 72-75, MEM STAFF, SCRIPPS CLIN & RES FOUND, 75- *Concurrent Pos:* NIH fel, 78-81, 75-77. *Mem:* Am Asn Immunologists; Am Asn Path. *Res:* Biochemical mechanisms of inflammatory disease processes. *Mailing Add:* Dept Immunol Scripps Clin 10666 N Torrey Pines Rd La Jolla CA 92037

ULFELDER, HOWARD, obstetrics & gynecology, surgery; deceased, see previous edition for last biography

ULICH, BOBBY LEE, b Bryan, Tex, Aug 13, 47; m 65; c 1. LARGE TELESCOPE TECHNOLOGY, RADIO ASTRONOMY. *Educ:* Tex A&M Univ, BS, 69; Calif Inst Technol, MS, 70; Univ Tex, Austin, PhD(elec eng), 73. *Prof Exp:* Head telescope oper, Tucson Div, Nat Radio Astron Observ, Assoc Univs Inc, 73-79; ASST DIR, MULTIPLE MIRROR TELESCOPE OBSERV, UNIV ARIZ, 79- *Concurrent Pos:* Mem, US Nat Comt Int Union Radio Sci, Comn J. *Mem:* Inst Elec & Electronics Engrs; Am Astron Soc; Int Astron Union. *Res:* Millimeter wavelength instrumentation and calibration techniques; solar system astronomy; interstellar molecules. *Mailing Add:* 5055 E Broadway Suite C104 5055 E Broadway No C214 Tucson AZ 85711-3694

ULICH, WILLIE LEE, b Somerville, Tex, Nov 10, 20; m 39; c 2. AGRICULTURAL ENGINEERING. *Educ:* Tex A&M Univ, BS, 43, MS, 47; Harvard Univ, PhD(pub admin), 51. *Prof Exp:* Res asst, Tex A&M Univ, 46-47, exten engr, 48-61; farm labor supvr, Fed Exten Serv, 47-48; prof agr eng & chmn dept, Tex Tech Univ, 61-84; RETIRED. *Concurrent Pos:* Mem, Tex Air Control Bd, 69, 71- *Honors & Awards:* James F Lincoln Found Award, 49-50. *Mem:* Am Soc Agr Engrs; Am Soc Eng Educ. *Res:* Covance systems and particulate control in cotton gins; efficiency studies of irrigation well pumping plants; confined animal odor and waste disposal systems; soil surface modification for soil and water conservation; farm machinery design; brush harvesting. *Mailing Add:* Rte 4 Box 119 Caldwell TX 77836

ULINSKI, PHILIP STEVEN, b Detroit, Mich, Feb 17, 43. NEUROANATOMY. *Educ:* Mich State Univ, BS, 64, MS, 67, PhD(zool), 69. *Prof Exp:* Asst prof biol, Oberlin Col, 69-70; asst prof anat, Sch Dent, Loyola Univ Chicago, 70-74; from asst prof to assoc prof, 75-86, PROF CHMN ANAT, UNIV CHICAGO, 86- *Concurrent Pos:* NIH fel, Univ Chicago, 75-81. *Mem:* Am Soc Zool; Soc Neurosci; Am Asn Anat. *Res:* Comparative anatomy of reptilian nervous systems. *Mailing Add:* Dept Organismal Biol & Anat Univ Chicago 1025 E 57th St Chicago IL 60637

ULLIMAN, JOSEPH JAMES, b Springfield, Ohio, July 7, 35; m 61; c 3. REMOTE SENSING, PHOTO INTERPRETATION. *Educ:* Univ Dayton, BA, 58; Univ Minn, MF, 68, PhD(forestry remote sensing),71. *Prof Exp:* Instr aerial photog interpretation, Col Forestry, Univ Minn, 68-71, asst prof, 71-74; PROF REMOTE SENSING, COL FORESTRY, WILDLIFE & RANGE SCI, UNIV IDAHO, 74- *Concurrent Pos:* Consult, USAID, 78-84; chmn, Int Soc Photogram & Remote Sensing, 80-84; dir, Forestry & Wildlife Remote Sensing Ctr, 81-; co-dir, remote sensing res unit, Univ Idaho, 81- *Mem:* Int Soc Photogram; Am Soc Photogram; Soc Am Foresters. *Res:* Large format camera system acquisition; Manual and digital analysis of aerial and satellite data for mapping, inventory, and geographic information systems. *Mailing Add:* Forest Resources Forest Wildlife & Range Sci Univ Idaho Moscow ID 83843

ULLMAN, ARTHUR WILLIAM JAMES, b New York, NY, Dec 30, 36. MATHEMATICS. *Educ:* Univ Miami, BS, 57; Rice Univ, MS, 64, PhD(math), 66. *Prof Exp:* Instr math, Rice Univ, 61-62; asst prof, Tex Christian Univ, 64-66; asst prof philos, Tex A&M Univ, 66-67; prof & chmn dept, State Univ NY Col New Paltz, 68-71, resident dir, State Univ NY prog, Int Documentation Ctr, Cuernavaca, Mex, 71-72, PROF MATH, STATE UNIV NY COL NEW PALTZ, 72- *Concurrent Pos:* Founder & pres, MH Hedging Analysts, Inc, 74; contract to produce educ modules on health, Empire State Col, State Univ NY, 75. *Mem:* Am Math Soc; Asn Symbolic Logic. *Res:* Logic applying algebraic and analytic techniques to the study of logical theories; mathematical analysis of various commodity markets; potential conflict of interest between the medical profession and the public interest. *Mailing Add:* Dept Math & Comput Sci State Univ NY Col New Paltz New Paltz NY 12561

ULLMAN, EDWIN FISHER, b Chicago, Ill, July 19, 30; m 54; c 2. CLINICAL CHEMISTRY, IMMUNOCHEMISTRY. *Educ:* Reed Col, AB, 52; Harvard Univ, AM, 54, PhD(org chem), 56. *Prof Exp:* Res chemist, Lederle Labs, Am Cyanamid Co, 55-60, group leader, Cent Res Div, 60-66; sci dir, Synvar Assocs, 66-70, VPRES & DIR RES, SYVA CO, 70- *Concurrent Pos:* Adv bd, J Org Chem, 70-75, J Immunoassay, 80- & J Clin Lab Anal, 85-87. *Honors & Awards:* Mallinckrodt Award, Clin Ligand Assay Soc, 81; Van Slyke Award, NY Sect, Am Asn Clin Chem, 84. *Mem:* Fel AAAS; Am Chem Soc; Royal Soc Chem; Am Asn Clin Chemists; Clin Ligand Assay Soc; Am Soc Biochem & Molecular Biol. *Res:* Enzyme chemistry; immunochemical assay methods; effects of ligand-receptor binding on chemical reactivity; methods for detection of specific nucleic acid sequences. *Mailing Add:* Syva Co PO Box 10058 900 Arastradero Rd Palo Alto CA 94303

ULLMAN, FRANK GORDON, b New York, NY, Dec 14, 26; m 51; c 3. SOLID STATE PHYSICS. *Educ:* NY Univ, BA, 49; Polytech Inst Brooklyn, PhD(physics), 58. *Prof Exp:* Jr engr, Sylvania Elec Prod, Inc, 51-54; asst physics, Polytech Inst Brooklyn, 54-57, res assoc, 57-58; sr physicist, Nat Cash Register Co, 58-66; PROF ELEC ENG, UNIV NEBR, LINCOLN, 66-, & ASSOC CHEM ELEC ENG, 87- *Mem:* Am Phys Soc; sr mem Inst Elec & Electronics Engrs; Sigma Xi. *Res:* Light scattering; ferroelectricity; photoconduction; luminescence. *Mailing Add:* 209 N Walter Scott Engr Ctr Univ Nebr Lincoln NE 68588-0511

ULLMAN, JACK DONALD, b Chicago, Ill, Sept 5, 29; m 72; c 2. SOLAR NEUTRINOS, MONOPOLE SEARCHES. *Educ:* Univ Ill, BS, 51, MS, 56, PhD(physics), 60. *Prof Exp:* Physicist, Bur Ships, Dept Navy, 51-53 & US Bur Standards, 60-61; res assoc, Dept Nuclear Physics, Univ Strasbourg, 61-62; res assoc physics, Columbia Univ, 62-70; assoc prof, 70-81, PROF PHYSICS, LEHMAN COL, 81- *Mem:* Am Phys Soc. *Res:* Solar neutrino detection. *Mailing Add:* Dept Physics & Astron Lehman Col Bedford Park Blvd Bronx NY 10468

ULLMAN, JEFFREY D(AVID), b New York, NY, Nov 22, 42. COMPUTER SCIENCE. *Educ:* Columbia Univ, BS, 63; Princeton Univ, PhD(elec eng), 66. *Prof Exp:* Mem tech staff, Bell Tel Labs, NJ, 66-69; assoc prof elec eng, Princeton Univ, 69-74, prof elec eng & comput sci, 74-79; PROF COMPUTER SCI, STANFORD UNIV, 79-, CHMN DEPT, 90- *Concurrent Pos:* Vis lectr, Columbia Univ, NY, 66-68, Princeton Univ, 68, vis prof, Univ Calif, Berkeley, 73; Ed, J Comput & Systs Sci, 74-, Theoret Comput Sci, 74-, J Computing, Soc Indust Appl Math, 75, J Asn Comput Mach, 77-84, J Computer Lang, 74-81, J Parallel & Distributing Comput, 84, J Logic Programming, 86-; mem adv panel computer sci, NSF, 74-77, chmn, coord exp res prog rev panel, 83, mem adv panel parallel software res, 85, adv panel info, robotics & intel systs, 86-89; mem vis comt, Dept Computer Sci, State Univ NY, Albany, 78, Dept Elec Eng & Computer Sci, Univ Calif, San Diego, 84, Dept Computer Info Systs, Univ Calif, Santa Cruz, 85, Dept Computer Sci, Univ Southern Calif, 89; Einstein fel, Israeli Acad Sci, 84; mem tech adv bd, Atherton Technol, 87-, Nucleus Int Corp, 89-; Guggenheim fel, 88-89. *Mem:* Nat Acad Eng; Asn Comput Mach; Asn Logic Programming. *Res:* Compilers; data bases; theory of algorithms. *Mailing Add:* Dept Comput Sci Bldg 460 Rm 332 Stanford Univ Stanford CA 94305

ULLMAN, JOSEPH LEONARD, b Buffalo, NY, Jan 30, 23. MATHEMATICS. *Educ:* Univ Buffalo, BA, 42; Stanford Univ, PhD(math), 49. *Prof Exp:* From instr to assoc prof, 49-66, PROF MATH, UNIV MICH, ANN ARBOR, 66- *Mem:* Am Math Soc; Math Asn Am. *Res:* Complex variable theory; potential theory; approximation theory in the complex plane. *Mailing Add:* Dept Math Univ Mich Ann Arbor MI 48104

ULLMAN, NELLY SZABO, b Vienna, Austria, Aug 11, 25; US citizen; m 47; c 4. APPLIED MATHEMATICS, BIOSTATISTICS. *Educ:* Hunter Col, BA, 45; Columbia Univ, MA, 48; Univ Mich, Ann Arbor, PhD(biostatist), 69. *Prof Exp:* Res assoc, Radiation Lab, Mass Inst Technol, 45; res assoc, Microwave Res Inst, Polytech Inst Brooklyn, 45-46, instr math, 45-63; instr, 63-64, asst prof, 64-66 & 68-71, assoc prof, 71-78, PROF MATH, EASTERN MICH UNIV, 78- *Mem:* Am Asn Univ Prof (treas, 71-); Am Math Asn; Am Statist Soc; Biomet Soc. *Res:* Integral equations; mathematical models in biological data; mathematical statistics. *Mailing Add:* 2360 St Francis Dr Ann Arbor MI 48104-4807

ULLMAN, ROBERT, b New York, NY, Nov 21, 20; m 47; c 4. POLYMER CHEMISTRY, PHYSICAL CHEMISTRY, POLYMER PHYSICS. *Educ:* City Col New York, BS, 41; Polytech Inst Brooklyn, MS, 46, PhD(chem), 50. *Prof Exp:* Res chemist, Ridbo Lab, 41-42, Columbia Univ, 42-45 & Carbide & Chem Co, 45-46; asst & instr chem, Polytech Inst Brooklyn, 46-47, from instr to assoc prof math, 47-59, assoc prof chem, 59-63; staff scientist, Ford Motor Co, 63-85; RETIRED. *Concurrent Pos:* Fulbright fel, Univ Groningen, 52; assoc prof, Univ Strasbourg, 61-62, Guggenheim fel, Ctr Res on Macromolecules, Strasbourg, France, 61-62; Fulbright fel, 77-78; adj res scientist nuclear eng, Univ Mich, 74-, adj prof chem, 75 & 80-83; vis prof, Polytech Inst Milan, 84, Univ Mainz, 88. *Mem:* AAAS; Am Chem Soc; fel Am Phys Soc; Sigma Xi. *Res:* Macromolecules viscosity; chain statistics; surface chemistry; light scattering; magnetic resonance and relaxation; molecular motion in liquids and solutions; neutron scattering from macromolecules; rubber elasticity; diffusion polymer blends. *Mailing Add:* 2360 St Francis Dr Ann Arbor MI 48104

ULLMANN, JOHN E, b Vienna, Austria, Dec 25, 23; US citizen; m 53; c 2. MILITARY-INDUSTRIAL RELATIONSHIPS, TECHNICAL INNOVATION. *Educ:* Univ London, BSc, 48; Columbia Univ, MS, 51, PhD(indust eng), 59. *Prof Exp:* Asst proj engr, Bechtel Corp, 50-54; proj engr, Bulova Res & Develop Labs, 54-58; asst prof indust eng, Stevens Inst, 58-61; PROF MGT, HOFSTRA UNIV, 61- *Concurrent Pos:* Consult engr, 55- *Res:* Author of 30 books and monographs and over 100 publications on industrial development and innovation, statistics and quantitative methods; political, economic and technical aspects of military production; environmental issues; industrial history. *Mailing Add:* 2518 Norwood Ave North Bellmore NY 11710-1705

ULLOM, STEPHEN VIRGIL, b Washington, DC, Nov 9, 38; m 66; c 2. MATHEMATICS. *Educ:* Am Univ, BA, 62; Harvard Univ, MA, 64; Univ Md, PhD(math), 68. *Prof Exp:* From asst prof to assoc prof, 70-78, PROF MATH, UNIV ILL, URBANA, 78- *Concurrent Pos:* NSF fel, Math Inst, Karlsruhe, WGer & King's Col, Univ London, 68-69; mem, Inst Advan Study, 69-70. *Mem:* Am Math Soc. *Res:* Algebraic number theory; galois groups; galois module structure. *Mailing Add:* Dept Math Univ Ill 337 Illini Hall Urbana IL 61801

ULLREY, DUANE EARL, b Niles, Mich, May 27, 28; m 61, 76; c 3. FISH & WILDLIFE SCIENCES. *Educ:* Mich State Univ, BS, 50, MS, 51; Univ Ill, PhD(animal nutrit), 54. *Prof Exp:* Asst animal sci, Univ Ill, 51-54; instr physiol & pharmacol, Okla State Univ, 54-55; from asst prof to assoc prof, 56-68, PROF ANIMAL SCI, MICH STATE UNIV, 68-, PROF FISH & WILDLIFE, 73- *Concurrent Pos:* Moorman fel nutrit res, 70; mem comt animal nutrit, Nat Res Coun; adv comt life sci, Coun Int Exchange Scholars Nat Acad Sci-Nat Res Coun, 74-77; dir Mich State Univ Inst Nutrit, 74-76; res assoc, San Diego Zoo, 78-; consult, Int Study Vitamin E, WHO. *Honors & Awards:* Am Feed Mfrs Asn Nutrit Res Award, 67; G Bohstedt Mineral Res Award, 69; Sr Res Award, Sigma Xi, 80. *Mem:* Am Am Soc Animal Sci; Am Inst Nutrit; Am Asn Zoo Veterinarians; Sigma Xi; Equine Nutrit Physiol Soc. *Res:* Nutrient requirements of swine; normal development of the swine fetus; hematology of domestic animals; nutrition of wild animals; mineral and vitamin metabolism. *Mailing Add:* Dept Animal Sci Mich State Univ East Lansing MI 48823

ULLRICH, DAVID FREDERICK, b Waterbury, Conn, Sept 10, 37. MATHEMATICS. *Educ:* Rensselaer Polytech Inst, BS, 59; Case Western Reserve Univ, MS, 62; Carnegie-Mellon Univ, PhD(differential equations), 67. *Prof Exp:* ASST PROF MATH, NC STATE UNIV, 66- *Mem:* Am Math Soc; Math Asn Am. *Res:* Non-linear ordinary differential equations. *Mailing Add:* Dept Math Box 8205 NC State Univ Raleigh NC 27695-8205

ULLRICH, FELIX THOMAS, b Elizabeth, NJ, June 1, 39; m 68; c 1. PHYSICS, ELECTRICAL ENGINEERING. *Educ:* Rutgers Col, BA, 61; Univ Pittsburgh, PhD(physics), 70. *Prof Exp:* Mem tech staff physics, Riverside Res Inst, 70 & GTE Labs, 70-72; advan res & develop engr, GTE Sylvania, 72-75; prof tech staff mem elec eng, Plasma Physics Lab, Princeton Univ, 75-85; DOWTY RFL INDUSTS, 85- *Mem:* Am Phys Soc. *Mailing Add:* 43 Holiday Dr Hopatcong NJ 07843

ULLRICH, ROBERT CARL, b Dumont, NJ, Aug 4, 40; m 64; c 2. FUNGAL GENETICS. *Educ:* Univ Minn, BSc, 68; Harvard Univ, AM, 69, PhD(biol), 73. *Prof Exp:* From asst prof to assoc prof, 74-86, PROF BOT, UNIV VT, 86- *Mem:* Genetics Soc Am; Mycological Soc Am. *Res:* Fungal genetics of Basidiomycetes; molecular genetics of fungi including transformation and ribosomal RNA genes; molecular, classical and population genetic studies of sexuality and mating types in Basidiomycetes. *Mailing Add:* Dept Bot Marsh Life Sci Univ Vt Burlington VT 05405

ULLRICH, ROBERT LEO, b Ottumwa, Iowa, Sept 8, 47. PATHOLOGY, RADIATION CARCINOGENESIS. *Educ:* Creighton Univ, BS, 69, MS, 71; Univ Rochester, PhD(radiation biol), 75. *Prof Exp:* Res assoc, 74-76, HEAD RADIATION CARCINOGENESIS UNIT, BIOL DIV, OAK RIDGE NAT LAB, 76- *Concurrent Pos:* Mem, biol basis radiation protection criteria, Nat Coun Radiat Protection & Measurements Sci Comt, 77-; consult, Comt Fed Res Biol & Health Effects Ionizing Radiation, Nat Acad Sci, 80-81. *Mem:* Am Asn Cancer Res; Radiation Res Soc; AAAS; NAm Late Effects Group. *Res:* Mechanisms of radiation carcinogenesis and cocarcinogenesis. *Mailing Add:* Dept Radiation Ther 310 Gail Borden F-56 Galveston TX 77550

ULLRICK, WILLIAM CHARLES, b Evanston, Ill, June 6, 24; m 48; c 3. BIOPHYSICS. *Educ:* Northwestern Univ, BS, 49; Univ Ill, MS, 51, PhD(physiol), 55. *Prof Exp:* Lab asst zool, comp anat & embryol, Northwestern Univ, 48-49; asst physiol, Col Med, Univ Ill, 50-54; from instr to assoc prof, 54-63, prof, 63-87, EMER PROF PHYSIOL, SCH MED, BOSTON UNIV, 87- *Concurrent Pos:* USPHS career res develop awards, 59-; mem bd dirs, Harvard Apparatus Co, Mass, 58-61. *Mem:* AAAS; Biophys Soc; Am Physiol Soc; NY Acad Sci. *Res:* Muscle physiology and biophysics. *Mailing Add:* Dept Physiol Boston Univ Sch Med 80 E Concord St Boston MA 02118

ULLYOT, GLENN EDGAR, b Clark Co, SDak, Mar 11, 10; m 35. ORGANIC CHEMISTRY, MEDICINAL CHEMISTRY. *Educ:* Univ Minn, BChem, 33; Univ Ill, MS, 35, PhD(org chem), 38. *Prof Exp:* Res chemist, Smith Kline & French Labs, 37-45, head org chem sect, 45-50, dir chem labs, 50-57, assoc dir res, 57-67, dir sci liaison res & develop div, 67-75; CONSULT, 75- *Concurrent Pos:* Mem ad hoc comt anticonvulsants, Nat Inst Neurol & Commun Dis & Stroke, 69-72, epilepsy adv comt, 72-76 & 81-84; consult, Epilepsy Br, Neurol Dis Progs, 75, Franklin Inst Res Lab, Sci Liaison, 76-80 & Biosearch Inc, 81-82. *Mem:* AAAS; Am Chem Soc; fel Am Inst Chem; Am Soc Pharmacol & Exp Therapeut. *Res:* Research and development administration; synthetic medicinal agents; central nervous system active agents; diuretics; structure-biological activity relation. *Mailing Add:* Box 686 Kimberton PA 19442-0686

ULM, EDGAR H, b McKeesport, Pa, July 23, 42; m 65; c 1. BIOCHEMISTRY. *Educ:* Ind Univ Pa, BA, 65; Ohio Univ, MS, 67; Purdue Univ, PhD(biochem), 72. *Prof Exp:* Res assoc biochem, St Louis Univ Med Sch, 71-73; sr res biochemist, 73-76, RES FEL, MERCK INST, 76- *Concurrent Pos:* NIH res fel, 72-73. *Mem:* Am Chem Soc; AAAS. *Res:* Biochemistry renin angiotensin; biochemistry of hypertension; rational drug design based on specific alterations of enzymatic activities; metabolism and disposition of enalaprilmaleate. *Mailing Add:* CG Corp CPD/Pharm 444 Saw Mill River Rd Ardsley NY 10502

ULM, LESTER, JR, b Palm Harbor, Fla, Aug 5, 22; m 45; c 2. ELECTRICAL ENGINEERING. *Educ:* Ga Inst Technol, BEE, 49. *Prof Exp:* Test engr, Gen Elec Co, 49-50; student engr, Tampa Elec Co, 50, sr inspector, 50-51, from jr engr to sr engr, 51-57, from assoc engr to engr, 57-58, gen engr, 58-62, dir eng, 62-74, dir methods & procedures, 74-80, vpres serv, 80-; RETIRED. *Mem:* Am Soc Eng Educ; Inst Elec & Electronics Engrs. *Res:* Electrical utility engineering. *Mailing Add:* 3006 Schiller St Tampa FL 33629

ULMER, GENE CARLETON, b Cincinnati, Ohio, Jan 28, 37; m 60; c 4. THERMODYNAMICS, ORE DEPOSITS. *Educ:* Univ Cincinnati, BS, 58; Pa State Univ, PhD(geochem), 64. *Prof Exp:* Asst geochem, Pa State Univ, 59-62, staff res asst, Col Mineral Industs, 62-64; engr, Homer Res Labs, Bethlehem Steel Corp, Pa, 64-69; assoc prof, 69-74, chmn dept, 74-77, PROF GEOL, TEMPLE UNIV, 74- *Concurrent Pos:* Geol fieldwork, Italy, Ger, SAfrica, Iceland & Mont. *Mem:* AAAS; Mineral Soc Am; Am Ceramic Soc; Am Geophys Union; Sigma Xi. *Res:* High temperature phase equilibria; oxide systems; oxidation reduction reactions, equilibria and kinetics; experimental petrology; materials research, especially spinels, diamonds and silicates, ultramafics in Africa and basalts in Idaho-Oregon; high pressure-high temperature research; nuclear waste management; platinum petrogenesis. *Mailing Add:* Rm 307 Buery Hall Temple Univ Dept Geol Philadelphia PA 19122

ULMER, MELVILLE PAUL, b Washington, DC, Mar 12, 43; m 68; c 1. X-RAY ASTRONOMY, GAMMA RAY ASTRONOMY. *Educ:* Johns Hopkins Univ, BA, 65; Univ Wis, PhD(physics), 70. *Prof Exp:* Res assoc physics, Univ Calif, San Diego, 70-74; astrophysicist, Smithsonian Astrophys Observ, 74-76; asst prof, 76-82, assoc prof, Physics & Astron Prog, 82-87, PROF PHYSICS & ASTRON, NORTHWESTERN UNIV, 87- *Mem:* Am Astron Soc; Int Astron Union; fel Am Phys Soc; Royal Astron Soc. *Res:* Gamma ray astrophysics, x-ray astronomy instrumentation measurement of spectra and positions of galactic and extragalactic x-ray sources; application of x-ray and gamma ray astronomy to studies in cosmology, galactic structure, pulsars and the interstellar medium. *Mailing Add:* Dept Physics & Astron Northwestern Univ Evanston IL 60208

ULMER, MILLARD B, b Demopolis, Ala, June 9, 46; div; c 2. MATHEMATICS. *Educ:* Univ Ala, PhD(math), 72. *Prof Exp:* From asst prof to assoc prof, 72-83, PROF MATH, UNIV SC, 83- *Mem:* Math Asn Am; Nat Coun Teachers Math. *Res:* Queneing models; several papers. *Mailing Add:* Univ SC-Spartanburg 800 University Way Spartanburg SC 29303-9395

ULMER, RAYMOND ARTHUR, b Chicago, Ill, Nov 15, 23; div; c 1. PATIENT COMPLIANCE RESEARCH EDUCATION, PATIENT COMPLIANCE CONTINUING. *Educ:* Univ Chicago, MS, 49; La State Univ, PhD(psychol), 65. *Prof Exp:* Asst prof speech path, Univ Calif, Los Angeles, 64-65; asst prof psychol, Calif State Univ, 65-68; asst proj dir psychother, Univ Southern Calif, 68-69; asst proj dir res, Camarillo State Ment Health Ctr, 69-72; assoc prof patient compliance, Drew Med Sch, 73-81; proj dir cancer patient compliance, Univ Calif-Los Angeles Jonsson Comprehensive Cancer Ctr, 82-85; DIR PATIENT COMPLIANCE PROBS, NONCOMPLIANCE INST LOS ANGELES, 76- *Concurrent Pos:* Ed, J Compliance Health Care, 85-; consult, Boehringer Ingelheim Pharmaceut, 86-, Univ Calif-Los Angeles Hispanics, Cocaine & Health, Nat Inst Drug Abuse grant, 88-, Am Lung Asn, Los Angeles, 81-86, Univ Calif Los Angeles, dept pulmonarry med, 85-86. *Mem:* Am Psychol Asn; Am Pub Health Asn; Am Med Writers Asn; Asn Behav Anal; Geront Soc Am; Soc Behav Med. *Res:* Patient compliance issues; compliance problems of tuberculars, cancer patients and cocaine abusers; author of 27 publications. *Mailing Add:* Noncompliance Inst Los Angeles 6411 W Fifth St PO Box 48555 Los Angeles CA 90048

ULMER, RICHARD CLYDE, b Lancaster, Ohio, July 4, 09; m 36; c 1. PHYSICAL CHEMISTRY. *Educ:* Ohio State Univ, AB, 30, PhD(chem), 36. *Prof Exp:* Chief chemist, Columbus and Southern Ohio Elec Co, 30-33; asst head chem div, Res Dept, Detroit Edison Co, 36-45; tech dir, E F Drew & Co, Inc, 45-53; mgr res, Combustion Eng, Inc, Windsor, 53-66, exec engr, 66-74; RETIRED. *Mem:* Am Chem Soc; Am Soc Mech Eng. *Res:* Water treatment and corrosion in the power boiler and electric utility fields. *Mailing Add:* 3508 Rue de Fleur Columbus OH 43221-1548

ULOTH, ROBERT HENRY, b Valley City, NDak, Mar 17, 27; m 50; c 2. ORGANIC CHEMISTRY. *Educ:* Valley City State Col, BS, 49; Univ NDak, MSc, 54. *Prof Exp:* Assoc chemist, 54-58, chemist, 59-60, sr scientist, 60-68, res assoc, 68-69, patent coordr, 69-70, PATENT AGENT, MEAD JOHNSON & CO, 70- *Mem:* Am Chem Soc; Sigma Xi. *Res:* Synthetic pharmaceutical drugs. *Mailing Add:* 411 Westmore Dr Evansville IN 47712

ULPIAN, CARLA, b Bacau, Romania, May 10, 47; Can citizen; m 69; c 2. NEUROLEPTIC DRUGS. *Educ:* Babes-Bolyai-Cluj, Romania, MS, 69, McGill Univ, Montreal, MS, 77. *Prof Exp:* Chemist, Res Inst, Bucharest, Romania, 69-73; res asst pharmacol, McGill Univ, 73-77; RES ASST PHARMACOL, UNIV TORONTO, 77- *Res:* Action of neuroleptic drugs in preventing delusions and hallucinations in paranoid schizophrenia; blockade of dopamine receptors by neuroleptics; measuring density of dopamine receptors in post-mortem brain tissue from schizophrenics. *Mailing Add:* Dept Pharmacol Med Sci Bldg Univ Toronto Toronto ON M5S 1A8 Can

ULREY, STEPHEN SCOTT, b Wilmington, Del, July 29, 46; m 69; c 2. INDUSTRIAL ORGANIC CHEMISTRY. *Educ:* WVa Univ, BS, 68; Ohio State Univ, PhD(org chem), 73. *Prof Exp:* DEVELOP CHEMIST, AM CYANAMID CO, 74-; AT ABBOTT LABS. *Mem:* Am Chem Soc; Sigma Xi. *Res:* Catalysis in organic chemistry, especially enzyme model compounds; industrial process development. *Mailing Add:* Dept 453 Abbott Labs 1400 Sheridan Rd North Chicago IL 60064

ULRICH, AARON JACK, b Benton, Ill, Feb 27, 21; m 54; c 2. NUCLEAR REACTOR PHYSICS. *Educ:* Univ Chicago, BS, 43, MS, 50. *Prof Exp:* Instr physics, Univ Chicago, 43-44; jr physicist, Oak Ridge Nat Lab, 44-46; asst physicist, 50-51, assoc physicist, 51-72, physicist, 72-85, SCIENTIST APPOINTEE, ARGONNE NAT LAB, 86- *Mem:* Am Phys Soc; Am Nuclear Soc; Sigma Xi. *Res:* Nuclear fission power reactors, safety test reactor design, planning and analysis of critical experiments and reactor safety tests; energy conversion; plasma physics and thermonuclear power reactors. *Mailing Add:* 1677 Linstead Dr Lexington KY 40504

ULRICH, BENJAMIN H(ARRISON), JR, b Olean, NY, Nov 5, 22; m 45; c 3. AERONAUTICAL ENGINEERING. *Educ:* Pa State Univ, BS, 44, MS, 49. *Prof Exp:* Aeronaut res scientist, Nat Adv Comt Aeronaut, 44 & 46-47; asst prof mech, Pa State Univ, 49-50; educ specialist, US Marine Corps Inst, 50-51; asst prof aeronaut eng, Univ WVa, 51-55, assoc prof aerospace eng, 56-66; prof aerospace eng & eng sci & chmn Dept, Parks Col Aeronaut Technol, St Louis Univ, 66-87; RETIRED. *Concurrent Pos:* Stress analyst, NAm Aviation, Inc, 53; flight test engr, Boeing Aircraft Co, 55; ed, Aero Div J, Am Soc Eng Educ, 59-62. *Mem:* Am Soc Eng Educ; Soc Exp Stress Anal; Am Inst Aeronaut & Astronaut. *Res:* Theories of failure and structural loading. *Mailing Add:* 4846 Chapel Hill Rd St Louis MO 63128

ULRICH, DALE V, b Wenatchee, Wash, Mar 1, 32; m 53; c 3. PHYSICS. *Educ:* La Verne Col, BA, 54; Univ Ore, MS, 56; Univ Va, PhD(physics), 64. *Prof Exp:* Instr, 58-61, from asst prof to assoc prof, 64-67, PROF PHYSICS & DEAN COL, BRIDGEWATER COL, 85- *Concurrent Pos:* NSF res grant, 64-68. *Mem:* Am Asn Physics Teachers; Am Phys Soc. *Res:* Partial specific volume studies on biological macromolecules; density studies in the critical region of single component systems. *Mailing Add:* Bridgewater Col Bridgewater VA 22812

ULRICH, FRANK, b Frankfurt-am-Main, Ger, Aug 30, 26; nat US; m 57; c 3. IMMUNOLOGY. *Educ:* Univ Calif, BA, 48, PhD, 52. *Prof Exp:* Jr res endocrinologist, Univ Calif, 52-53; estab investr, Am Heart Asn, 57-62; asst prof, Yale Univ, 62-67; sr res assoc, Grad Sch Nutrit, Cornell Univ, 67-69; ASSOC PROF PHYSIOL, DEPT SURG, SCH MED, TUFTS UNIV, 69- *Concurrent Pos:* Brown Mem fel physiol, Sch Med, Yale Univ, 53-54; Nat Cancer Inst fel, 54-56; Arthritis & Rheumatism Found fel, 56-57. *Mem:* Am Soc Cell Biol; Am Physiol Soc; Brit Biochem Soc; Soc Exp Biol Med. *Res:* Ion transport in mitochondria, ions and cell respiration; enzyme kinetics; macrophage physiology. *Mailing Add:* Surg Res Unit Vet Admin Med Ctr 150 S Huntington Ave Boston MA 02130

ULRICH, GAEL DENNIS, b Devils Slide, Utah, Oct 29, 35; m 58; c 5. CHEMICAL ENGINEERING. *Educ:* Univ Utah, BS, 59, MS, 62; Mass Inst Technol, DSc(chem eng), 64. *Prof Exp:* Sr researcher chem eng, Atomics Int Div, NAm Aviation, 64-65; res engr, Billerica Res Ctr, Cabot Corp, 65-70; from asst prof to assoc prof, 70-81, PROF CHEM ENG, UNIV NH, 82- *Mem:* Combustion Inst. *Res:* Particle formation in flames; furnace combustion and incineration; chemical engineering process design. *Mailing Add:* Dept Chem Eng Kingsbury Hall Univ NH Durham NH 03824

ULRICH, HENRI, b Rheinsberg, Germany, May 4, 25; nat US; m 54; c 4. ORGANIC POLYMER CHEMISTRY. *Educ:* Univ Berlin, dipl, 52, Dr rer nat, 54. *Prof Exp:* Instr org chem, Univ Berlin, 53-54; res assoc, Res Found, Ohio State Univ, 55-59; group leader org res, Carwin Co, 59-62, head org res, Donald S Gilmore Res Lab, 62-65, mgr chem res & develop, 65-76, dir, 76-81, vpres, Donald S Gilmore Res Labs, Upjohn Co, 82-85; dir, NHaven Labs, Dow Chem, USDA, 85-88; CONSULT, 88- *Mem:* AAAS; Am Chem Soc; Soc Ger Chem. *Res:* Isocyanates; polyurethanes; agricultural chemicals; light sensitive chemicals. *Mailing Add:* 28 Coginchaug Ct Guilford CT 06437

ULRICH, JOHN AUGUST, b St Paul, Minn, May 15, 15; m 40, 86; c 6. MICROBIOLOGY, BACTERIOLOGY. *Educ:* St Thomas Col, BS, 38; Univ Minn, PhD(bact), 47; Am Bd Microbiol, dipl, 63. *Prof Exp:* Teacher, High Sch, Minn, 38-41; asst bact, 41-45 & Hormel Inst, 45-46; first asst, Mayo Clin, 49-50, consult, 50-65, assoc prof bact, Mayo Grad Sch Med, Univ Minn, 65-69, assoc prof microbiol, Univ, 66-69; prof microbiol, Med Sch, Univ NMex, 69-84; RETIRED. *Concurrent Pos:* Res fel, Hormel Inst, Univ Minn, 46-49; consult, Econ Labs, Mo, 45, Hormel Packing Plant, Minn, 47-, NIH, 56-, Nat Commun Dis Ctr, 63-, NASA, 65-, Vet Admin, 69-, Sandia Labs, 70- & Midwest Res Inst, 71-, Int Chem Inc, 79, 3M, 80 - *Honors & Awards:* Silver Beaver; Bishops Medal. *Mem:* AAAS; Am Soc Microbiol; Mycol Soc Am; Am Chem Soc; Am Acad Microbiol. *Res:* Skin bacteriology; hospital epidemiology; surgery air recirculation; infected wounds; chemotherapy; food preservation; low temperature; bacterial metabolism; medical mycology and bacteriology. *Mailing Add:* 3807 Columbia Dr Longmont CO 80501

ULRICH, MERWYN GENE, b Norfolk, Nebr, July 14, 36; m 58. ZOOLOGY. *Educ:* Westmar Col, BA, 58; Univ SDak, MA, 62; Univ Southern Ill, PhD(zool), 66. *Prof Exp:* Asst prof, Westmar Col, 66-68, assoc prof, 68-74. *Concurrent Pos:* Vis prof biol, Silliman Univ, Phillippines, 72-73; aquacult specialist, World Bank Bangkok, Thailand, 74-75; adj prof biol, Westmar Col, 74- *Mem:* Am Fisheries Soc; Sigma Xi. *Res:* Fisheries management and culture. *Mailing Add:* 1600 Hiawatha Trail Sioux City IA 51104

ULRICH, ROGER STEFFEN, b Birmingham, Mich, Feb 20, 46; m 70; c 2. RESEARCH DEVELOPMENT IN ENVIRONMENTAL DESIGN, PLANNING & CONSTRUCTION. *Educ:* Univ Mich, BA, 68, MA, 71, PhD(geog), 73. *Prof Exp:* From asst prof to assoc prof geog, Univ Del, 74-88; PROF LANDSCAPE & URBAN PLANNING, TEX A&M UNIV, 88-, ASSOC DEAN RES, COL ARCHIT, 88- *Concurrent Pos:* Vis prof, Sch Archit, Lund Inst Technol, Sweden, 77-78; vis researcher, Dept Clin Psychol, Uppsala Univ, Sweden, 84-85. *Honors & Awards:* Nat Award for Exemplary Team Leadership in Higher Educ, Am Asn Univ Adminr, 90. *Mem:* Asn Am Geographers. *Res:* Effects of viewing natural and built environments on human physiological systems, behavior and health; stressful and stress reducing influences of designed and natural environments; relationships between healthcare facility design and patient stress, compliance and health. *Mailing Add:* Col Archit Res Off Tex A&M Univ College Station TX 77843-3137

ULRICH, VALENTIN, b Palmerton, Pa, Aug 5, 26. GENETICS. *Educ:* Rutgers Univ, AB, 53, PhD, 61. *Prof Exp:* From asst prof to assoc prof, 57-68, chmn develop biol fac, 74-77, PROF GENETICS, WVA UNIV, 68-, GENETICIST, 77- *Concurrent Pos:* Consult, Nat Tech Adv Comt Pesticides in Water Environ, 70 & Environ Protection Agency, 70-75. *Mem:* AAAS; Genetic Soc Am. *Res:* Biochemistry and genetics of heterosis. *Mailing Add:* 1096 Agric Sci Bldg WVa Univ Morgantown WV 26506

ULRICH, WERNER, b Munich, Ger, Mar 12, 31; US citizen; m 59; c 2. PATENT LAW, ELECTRICAL ENGINEERING. *Educ:* Columbia Univ, BS, 52, MS, 53, EngScD(elec eng), 57; Univ Chicago, Grad Sch Bus, MBA, 75; Loyola Univ Sch Law, Chicago, JD, 85. *Prof Exp:* Mem tech staff, 53-58, supvr, Switching Syst Develop, 58-64, dept head, 64-68, dir advan switching technol, 68-77, head, Systs Reliability Design Dept, 77-81, mem legal & patent staff, 81-85, PATENT ATTORNEY, AT&T BELL LABS, 85- *Concurrent Pos:* Vis lectr, Univ Calif, Berkeley, 66-67. *Mem:* Fel Inst Elec & Electronics Engrs. *Res:* System design and development of program controlled electronic telephone switching systems. *Mailing Add:* AT&T Bell Labs Naperville IL 60566

ULRICH, WILLIAM FREDERICK, b Pinckneyville, Ill, Nov 4, 26; m 51; c 3. INORGANIC CHEMISTRY, ANALYTICAL CHEMISTRY. *Educ:* Southern Ill Univ, BS, 49; Univ Ill, PhD, 52. *Prof Exp:* Supvr spectrochem group, Shell Develop Co, 52-56; supvr appln eng, Beckman Instruments, Inc, 55-66, mgr appl res, Sci Instruments Dir, 66-74, mgr clin mkt develop, 74-86; DIR PROD DEVELOP, SMITHKLINE DIAGS, INC, 86- *Mem:* Am Chem Soc; Soc Appl Spectros (treas, 67-69); Spectros Soc Can; Am Asn Clin Chemists. *Res:* Analytical instrumentation, ultraviolet and infrared spectrophotometry; atomic absorption; gas chromatography; electrochemistry; radioimmunoassay; enzyme immunoassay. *Mailing Add:* Smithkline Diag Inc 225 Bay Pointe San Jose CA 95143

ULRICHSON, DEAN LEROY, b Alma, Nebr, Mar 6, 37; m 61; c 3. CHEMICAL ENGINEERING. *Educ:* Univ Nebr, Lincoln, BSc, 62; Univ Ill, Urbana, MSc, 63; Iowa State Univ, PhD(chem eng), 70. *Prof Exp:* Res engr, E I du Pont de Nemours & Co, Inc, 63-68; res asst chem eng, 69-70, from asst prof to assoc prof, 70-81, PROF CHEM ENG, IOWA STATE UNIV, 81- *Honors & Awards:* Fulbright lectr, 87-88. *Mem:* Am Inst Chem Engrs; Am Soc Eng Educ; Nat Soc Prof Engrs. *Res:* Modelling and simulation of chemical processes. *Mailing Add:* Dept Chem Eng Iowa State Univ 231 Sweeney Hall Ames IA 50011

ULRYCH, TADEUSZ JAN, b Warsaw, Poland, Aug 9, 35; Can citizen; m 58; c 2. GEOPHYSICS. *Educ:* Univ London, BSc, 57; Univ BC, MSc, 61, PhD(geophys), 63. *Prof Exp:* Asst prof geophys, Univ Western Ont, 61-64; Nat Res Coun fel, Oxford Univ, 64-65 & Bernard Price Inst Geophys, 65; asst prof, Univ BC, 65-67; vis prof, Petrobras, Salvador, Brazil, 67-68; assoc prof, 68-74, PROF GEOPHYS, UNIV BC, 74- *Concurrent Pos:* Consult var cos; lectr var insts & univs. *Mem:* Soc Explor Geophys. *Res:* Applications of communication theory to geophysics and astronomy; inverse theory. *Mailing Add:* Dept Geophys & Astron Univ BC 2075 Westbrook Mall Vancouver BC V6T 1W5 Can

ULSAMER, ANDREW GEORGE, JR, b Yonkers, NY, Nov 13, 41; m 65; c 3. BIOCHEMISTRY, TOXICOLOGY. *Educ:* Siena Col, BS, 63; Albany Med Col, PhD(biochem), 67. *Prof Exp:* USPHS fel, Nat Heart Inst, Md, 67-68, staff fel, 68-70; res biochemist, Div Toxicol, US Food & Drug Admin, 70-73; res biochemsist, 73-75, chief Biochem Br, 75-80, dir Div Health Effects, 80-87, ASSOC EXEC DIR, HEALTH SCI, CONSUMER PROD SAFETY COMN, 87- *Concurrent Pos:* Proj adv group, Food & Drug Admin, Dept Housing & Urban Develop & Nat Inst Drug Abuse. *Mem:* Soc Toxicol; Asn Govt Toxicologist; Sigma Xi. *Res:* Evaluation of the hazards posed by toxic substances found in consumer products, with particular emphasis on indoor air pollutants. *Mailing Add:* 17 Duke St S Rockville MD 20850

ULSTROM, ROBERT, b Minneapolis, Minn, Feb 23, 23; m 46; c 3. MEDICINE. *Educ:* Univ Minn, BS, 44, MD, 46. *Prof Exp:* Intern & resident, Strong Mem Hosp, Rochester, NY, 46-48; from instr to asst prof pediat, Univ Minn, 50-53; asst prof, Univ Calif, Los Angeles, 53-56; from assoc prof to prof, Sch Med, Univ Minn, 56-64; prof & chmn dept, Sch Med, Univ Calif, Los Angeles, 64-67; assoc dean med sch, 67-70, PROF PEDIAT, MED SCH, UNIV MINN, MINNEAPOLIS, 67- *Concurrent Pos:* Markle scholar, 54-59; consult, Hennepin County Gen Hosp, 56-; mem study sect, NIH, 64-68;

examr, Am Bd Pediat, 71-, mem bd, 80-86. *Mem:* AAAS; Soc Pediat Res; Endocrine Soc; Lawson Wilkins Pediat Endocrine Soc; Am Pediat Soc. *Res:* Metabolism of children, particularly the endocrine aspects of the neo-natal period; developmental endocrinology. *Mailing Add:* 4616 Sunset Ridge Minneapolis MN 55416

ULTEE, CASPER JAN, b Noordwyk, Neth, Apr 5, 28; nat US; m 50; c 4. MOLECULAR SPECTROSCOPY, PHYSICAL CHEMISTRY. *Educ:* Hope Col, BA, 50; Purdue Univ, PhD(chem), 54. *Prof Exp:* Res chemist, Linde Co Div, Union Carbide Corp, 54-60; chemist, Res & Adv Develop Div, Avco Corp, 60-61; sr res scientist, United Aircraft Corp, 61-67, prin scientist, Res Lab, 67-76, mgr chem physics, 76-81, mgr chem physics & combustion sci, United Technol Res Ctr, 81-90; RETIRED. *Concurrent Pos:* Lectr chem, St Joseph Col, W Hartford, Conn. *Mem:* Am Phys Soc; Am Chem Soc. *Res:* Infrared, Raman and electron spin resonance spectroscopy; molecular structure; chemical kinetics; spectroscopy of high temperature arcs; gas phase and chemical lasers. *Mailing Add:* 55 Harvest Lane Glastonbury CT 06033

ULTMAN, JAMES STUART, b Chicago, Ill, Oct 24, 43; m 67; c 3. CHEMICAL ENGINEERING, BIOENGINEERING. *Educ:* Ill Inst Technol, BS, 65; Univ Del, MChE, 67, PhD(chem eng), 69. *Prof Exp:* From asst prof to assoc prof, 70-91, PROF CHEM ENG, PA STATE UNIV, UNIVERSITY PARK, 91- *Concurrent Pos:* Instr, Univ Del, 69, NIH fel, 69-70; Fulbright-Hays lectr, Technion-Israel Inst Technol, 77-78; vis res prof, Duke Univ Med Ctr, 89-90. *Mem:* Am Inst Chem Engrs; Biomed Eng Soc. *Res:* Mass transfer; heat transfer; air pollution. *Mailing Add:* 106 Fenske Lab Pa State Univ University Park PA 16802

ULTMANN, JOHN ERNEST, b Vienna, Austria, Jan 6, 25; US citizen; m 52; c 3. HEMATOLOGY, ONCOLOGY. *Educ:* Columbia Univ, MD, 52; Am Bd Internal Med, dipl, 60. *Hon Degrees:* MD, Heidelberg Univ, 87. *Prof Exp:* Intern, NY Hosp-Cornell Med Ctr, 52-53, asst resident med, 53-54, asst med & resident hemat, 54-55; instr med, Col Physicians & Surgeons, Columbia Univ, 56-61, assoc, 61-62, asst prof, 62-68; assoc prof, 68-70, dean res & develop, 78-88, PROF MED, SCH MED, UNIV CHICAGO, 70-, DIR CANCER RES CTR, 73-; DEAN, RES DEVELOP, UNIV CHICAGO, 78- *Concurrent Pos:* Nat Cancer Inst trainee, NY Hosp-Cornell Med Ctr, 53-55; Am Cancer Inst Soc fel hemat, Col Physicians & Surgeons, Columbia Univ, 55-56; from asst vis physician to vis physician, Francis Delafield Hosp, 56-68; career scientist, Health Res Coun City New York, 59-68; asst physician, Presby Hosp, 59-65, asst attend physician, 65-68; clin asst vis physician, 1st Med Div, Bellevue Hosp, 61-62, asst vis physician, 63-68; consult, Harlem Hosp, 66-68; hon prof, Cancer Inst Chinese Acad Med Sci, Peoples Repub China, 88; mem, numerous comts & bds, 57- *Honors & Awards:* Seventeenth Kretschmer mem lectr, 71. *Mem:* Fel Am Col Physicians; Am Asn Cancer Res; Am Soc Hemat; Am Soc Clin Oncol; Int Soc Hemat; NY Acad Sci; Am Fedn Clin Res; Am Asn Univ Prof; Am Soc Hemat; AAAS; Harvey Soc; Am Soc Clin Oncol; Cent Soc Clin Res. *Res:* Chemotherapy cancer, lymphoma and leukemia; pathophysiology of anemia of cancer; immune defects of patients with lymphoma. *Mailing Add:* Dept Med & Cancer Res Ctr Univ Chicago Box 444 Chicago IL 60637-1470

ULUG, ESIN M, US citizen. COMMUNICATIONS & INFORMATION SYSTEMS. *Educ:* Univ Durham, BSc, 53, MSc, 64; Carleton Univ, PhD(elec eng), 73. *Prof Exp:* Engr switching equip, Bell Can, 54-56; mgr eng, Can Gen Elec, 56-70; mgr tech develop, subsyst technol, Microsyst Int, 70-72; staff engr spec proj, Bell Can, 72-74; prof syst eng & comput sci, Carleton Univ, 74-78; mgr commun syst, Tex Instruments, 78-85; MGR COMMUN SYST, GEN ELEC CORP, 85-, ELEC ENGR, RES & DEVELOP. *Concurrent Pos:* Consult, Bell Can, Telesat Can & Govt Can, 74-78, Tex Instruments, 77-78. *Mem:* Inst Elec & Electronics Engrs; Inst Elec Eng UK. *Res:* Relational databases; local area networks; communication protocols; packet switched networks; artificial intelligence; satellite communications; queueing theory; algebraic topological methods; information theory. *Mailing Add:* 1537 E Hillsboro Blvd No 342 Deerfield Beach FL 33441

ULVEDAL, FRODE, b Oslo, Norway, Nov 20, 32; US citizen; m 57; c 1. PHYSIOLOGY, ENDOCRINOLOGY. *Educ:* St Svithun's Col, Norway, BS, 51; Drew Univ, BA, 55; Emory Univ, PhD(physiol), 59. *Prof Exp:* Chief physiol support div, Laughlin AFB, Tex, 59-60, aviation physiologist, USAF Scj Aerospace Med, 60-62, chief adv res unit, SMBE, 62-65, chief chem sect, 65-66, chief sealed environ sect & task scientist, 66-68, chief sealed environ br, 68-72; chief pulmonary dis br, Nat Heart & Lung Inst, 72-74; actg dir, 74-77, 82-84, supvry toxicologist & sr health scientist, Health Effects Div, Environ Protection Agency, 77-89. *Concurrent Pos:* Fel, Emory Univ, 66. *Mem:* Am Physiol Soc; Endocrine Soc; Am Col Toxicol. *Res:* Effects of altered atmospheric conditions like altitude and gaseous composition on man and other animals during prolonged exposures in space cabin environments; oxygen toxicity at decreased pressure, especially in regard to endocrinology, hematology and biochemistry; toxicology of environmental pollutants. *Mailing Add:* 1505 Dulcimer Ct Vienna VA 22182

UMAN, MARTIN A(LLAN), b Tampa, Fla, July 3, 36; m 62; c 3. ELECTRICAL ENGINEERING. *Educ:* Princeton Univ, BSE, 57, MA, 59, PhD(elec eng), 61. *Prof Exp:* Assoc prof elec eng, Univ Ariz, 61-65; fel scientist, Westinghouse Res Labs, Pa, 65-71; PROF ELEC ENG, UNIV FLA, 71- *Concurrent Pos:* Pres, Lightning Location & Protection, Inc, 75-85, vpres & chief consult scientist, 85- *Mem:* Fel Inst Elec & Electronics Engrs; fel Am Geophys Union; fel Am Meteorol Soc; Int Union Geophys & Geod. *Res:* Lightning physics; electromagnetic field theory. *Mailing Add:* Dept Elec Eng Univ Fla Gainesville FL 32611

UMAN, MYRON F, b Tampa, Fla, Oct 13, 39; m 63; c 1. ELECTRICAL ENGINEERING, PLASMA PHYSICS. *Educ:* Princeton Univ, BSEE, 61, MA, 66, PhD(elec eng), 68; Univ Ill, Urbana, MS, 62. *Prof Exp:* Res asst, Plasma Physics Lab, Princeton Univ, 64-68; asst prof elec eng, Univ Calif, Davis, 68-74; prog mgr, NSF, 74-75; staff dir, Environ Studies Bd, 81-86, SR STAFF OFFICER, NAT ACAD SCI, 75-, DIR REDESIGN SPACE SHUTTLE BOOSTER ROCKET, 86-, ASSOC EXEC OFFICER, NAT RES COUN, 90- *Concurrent Pos:* Chmn fac, Col Eng, Univ Calif, Davis, 70-71; Am Soc Eng Educ-Ford Found resident fel, E Fishkill Facility, IBM Systs Prods Div, NY, 72-73; Sloan resident fel, Nat Acad Sci, 73-74; spec asst to dir, US Geol Surv, 78; assoc dir, comt on sci, eng & pub policy, Nat Acad Sci, 87-88, dir, comn phys sci, math & resources, 88-90; lectr, George Mason Univ, 89- *Mem:* AAAS; Am Phys Soc; NY Acad Sci; Sigma Xi. *Res:* Applications of science and technology to public policy decision making; environmental, energy and mineral resource policy; acid deposition; engineering and management of aerospace systems. *Mailing Add:* 2101 Constitution Ave NW Washington DC 20418

UMANS, ROBERT SCOTT, b New York, NY, Dec 17, 41. BIOPHYSICAL CHEMISTRY, CELLULAR PHYSIOLOGY. *Educ:* Columbia Univ, AB, 62; Yale Univ, MS, 63, PhD(chem), 66. *Prof Exp:* Res assoc biophys chem, Johns Hopkins Univ, 66-68, NIH fel, 67-68; res assoc, Inst Biophys & Biochem, Paris, 68-69; Mass Div, Am Cancer Soc res grant, Boston Univ, 71-72, asst prof chem, 69-75; asst prof chem, Boston Col, 76-77; asst prof chem, Wellesley Col, 77-80; res fel cancer biol, 80-83, res assoc, Sch Pub Health, Harvard Univ, 83-86; asst prof chem, Wellesley Col, 86-90; CONSULT, 90- *Mem:* Am Chem Soc. *Res:* Biochemical studies of the role in cancer causation and treatment of steroid hormones and chemicals. *Mailing Add:* 23 Chauncy St Cambridge MA 02138

UMANZIO, CARL BEEMAN, b Thompsonville, Conn, Apr 9, 07; m 29; c 1. MICROBIOLOGY. *Educ:* Univ Boston, AM, 35; Wash Univ, St Louis, PhD, 50. *Prof Exp:* Asst med mycol, Sch Med, Univ Boston, 34-35; educ adv, Civilian Conserv Corps, 35-37; teacher, high sch, Mass, 37-39; mem staff pharm, Franklin Tech Inst, 39-47; prof bact & parasitol & chmn dept microbiol, 47-74, EMER PROF MICROBIOL, KIRKSVILLE COL OSTEOP MED, 74- *Concurrent Pos:* Mem staff pharm, Cambridge Jr Col, 42-47; consult, Microbiol & Allergic Dis, 47- *Mem:* Sigma Xi; Am Soc Microbiol; Mycol Soc Am; Am Soc Trop Med & Hyg; Nat Asn Biol Teachers. *Res:* Various publications, principally on parasitology, medical mycology, allergy and antibiotic hypersensitivity; parasitology; hypersensitivity; antibiotics. *Mailing Add:* D 222 6647 E Colegio Rd Goleta CA 93117

UMBARGER, H EDWIN, b 1921. BIOCHEMISTRY. *Prof Exp:* PROF BIOCHEM, PURDUE UNIV, RIGHT DISTINGUISHED PROF BIOL SCI. *Mem:* Nat Acad Sci; Am Chem Soc. *Mailing Add:* Biol Sci Dept Purdue Univ West Lafayette IN 47907

UMBERGER, ERNEST JOY, b Burke, SDak, Aug 5, 09; m 32; c 3. PHARMACOLOGY, ENDOCRINOLOGY. *Educ:* George Washington Univ, BS, 37, MA, 41; Georgetown Univ, PhD(biochem), 48. *Prof Exp:* Asst sci aide fermentation sect, Bur Agr Chem & Eng, USDA, 37-39, jr chemist, Div Allergen Invests, 39-42, asst chemist, Naval Stores Res Div, 42-43; asst chemist, Div Pharmacol, US Food & Drug Admin, 43-44, pharmacologist, 44-47, chief endocrine sect, Drug Pharmacol Br, 57-67, chief drug anal br, Div Pharmaceut Sci, Bur Sci Consumer Protection & Environ Health Serv, 67-70, dir div drug biol, Off Pharmaceut Res & Testing, Bur Drugs, 70-71; RETIRED. *Concurrent Pos:* Consult endocrinol & pharmacol, 71- *Mem:* Emer mem Am Chem Soc; Soc Exp Biol & Med; emer mem Endocrine Soc. *Res:* Fermentation; chemistry of allergenic proteins; rosin esters; absorption of calomel from ointments; bioassay of estrogens; androgens, gonadotropins and adrenocorticotropic hormone; metabolism of steroid hormones; central nervous systems endocrine relationships. *Mailing Add:* 4811 Flanders Ave Kensington MD 20895

UMBREIT, GERALD ROSS, b Minneapolis, Minn, June 17, 30; m 53; c 3. ANALYTICAL CHEMISTRY. *Educ:* Augustana Col, SDak, BA, 54; Iowa State Univ, PhD(anal chem), 57. *Prof Exp:* Res asst, Ames Lab, AEC, Iowa, 54-57; res assoc, Upjohn Co, Mich, 58-63; res scientist, Lockheed Missiles & Space Co, Calif, 63-64; appln lab mgr, F & M Sci Div, Hewlett-Packard Co, Pa, 64-66; PRES, GREENWOOD LABS, 66- *Honors & Awards:* Delaware Valley Chromatography Forum Award, 85. *Mem:* Am Chem Soc. *Res:* Chromatography; ion exchange. *Mailing Add:* Greenwood Labs 903 E Baltimore Pike Kennett Square PA 19348

UMBREIT, WAYNE WILLIAM, b Marksan, Wis, May 1, 13; m 37; c 3. BACTERIOLOGY, BIOCHEMISTRY. *Educ:* Univ Wis, BA, 34, MS, 36, PhD(bact, biochem), 39. *Prof Exp:* Asst bact & biochem, Univ Wis, 34-37, instr bact & chem, 38-41, asst prof bact, 41-44; instr soil microbiol, Rutgers Univ, 37-38; assoc prof bact, Cornell Univ, 44-46, prof, 46-47; head enzyme chem dept, Merck Inst Therapeut Res, 47-56, assoc dir, 56-58; prof bact & head dept, Rutgers Univ, 58-75; S Br Watershed Asn, 75-90; RETIRED. *Honors & Awards:* Lilly Award, 47; Waksman Award, 57; Carski Award, 68. *Mem:* AAAS; Am Soc Microbiol; Am Chem Soc; Am Soc Biol Chem; fel Am Acad Microbiol (vpres, 60). *Res:* Mode of action of antibiotics; nature of autotrophic bacteria; transformations of morphine by microorganisms. *Mailing Add:* 18 Dogwood Dr Flemington NJ 08822

UMEDA, PATRICK KAICHI, MUSCLE DEVELOPMENT CARDIAC HYPERTROPHY. *Educ:* Univ Chicago, PhD(biochem), 80. *Prof Exp:* ASST PROF MED, UNIV CHICAGO, 82- *Mailing Add:* Dept Med/Cardiol Univ Ala Zeigler 302 Birmingham AL 35294

UMEN, MICHAEL JAY, b Jamaica, NY, Feb 10, 48; m 69; c 3. GENERAL MEDICAL SCIENCES, IMMUNOLOGY. *Educ:* Queens Col, BA, 69; Mass Inst Technol, PhD(org chem), 73. *Prof Exp:* Res scientist, sr scientist org med chem, Johnson & Johnson, 74-77, mgr res info serv dept, McNeil Pharmaceut, 77-80, dir, 80-81; INDEPENDENT CONSULT, MICHAEL UMEN & CO INC, 81- *Mem:* Am Chem Soc; Proprietary Asn; Am Med Writers Asn; Drug Info Asn. *Res:* Pharmaceutical and medical device development; preparation of regulatory documents, literature reviews, publication manuscripts and technology assessments; evaluation of new product and licensing opportunities; coordination of preclinical and clinical scientists; management consulting; regulatory affairs. *Mailing Add:* 352 N Easton Rd Glenside PA 19038

UMEZAWA, HIROOMI, b Saitama-Ken, Japan, Sept 20, 24; m 58; c 2. THEORETICAL PHYSICS, SOLID STATE PHYSICS. *Educ:* Nagoya Univ, BS, 46, DSc(physics), 51. *Prof Exp:* Assoc prof physics, Nagoya Univ, 52-55; from assoc prof to prof, Univ Tokyo, 55-64; prof, Univ Naples, 64-66; prof, Univ Wis-Milwaukee, 66-67, distinguished prof, 67-75; KILLAM MEM PROF SCI, PROF PHYSICS, UNIV ALTA, 75- *Concurrent Pos:* Imp Chem Indust fel, Univ Manchester, 53-55; vis prof, Univ Wash, 56, Univ Md, 57, Univ Iowa, 57 & Univ Aix Marseille, 59-60; leading mem Naples group of struct of matter, Ctr Nat Res, Italy, 64-66. *Mem:* Fel & life mem Am Phys Soc; life mem Phys Soc Japan; emer mem NY Acad Sci. *Res:* Theoretical research on quantum field theory, high energy particle physics and solid state physics. *Mailing Add:* Dept Physics Univ Alta Edmonton AB T6G 2E2 Can

UMHOLTZ, CLYDE ALLAN, b Du Quoin, Ill, Dec 20, 47. ARTIFICIAL INTELLIGENCE, INDUSTRIAL ENGINEERING. *Educ:* Univ Ill, BS, 69, MBA, 71; Univ Melbourne, DSc, 75. *Prof Exp:* Eng planning analyst, Chem Mfg Group, W R Grace & Co & Subsidiaries; mgr, Res Serv, Ctr Indust Consult, Memphis State Univ, 78-84; DEP ADMIN, DATA PROCESSING SYSTS, SHELBY CO GOVT, 84- *Concurrent Pos:* Consult, var comput energy, chem & nuclear industs, 78-; vis lectr, Eng Sch, var univs, 79-; adj prof, Memphis State Univ, 80-; partner, Custom Data Systs Inc, Memphis, Tenn, 87- & Western Technologies Inc, 88-; adj prof, Univ Tenn, Memphis, 85-; bd dirs, Am Tech Inst, Memphis, 89- *Mem:* Nat Acad Sci; Am Chem Soc; Data Processing Mgt Asn; Am Inst Mfg Engrs; Am Inst Chem Engrs; NY Acad Sci. *Res:* Angle trisector; energy considerations of Haber cycle; comprehensive engineering and economic studies of the sulfur, sulfuric acid and phosphate manufacturing industries; cost and materials science studies for the nuclear industry; structured methodology in the development of artificial intelligence systems; distillation with vapor recompression; prototyping in development of computerized financial systems; context analysis in system design; computer aided software systems for engineering analysis and design methodologies. *Mailing Add:* 3580 Hanna Dr Memphis TN 38128

UMMINGER, BRUCE LYNN, b Dayton, Ohio, Apr 10, 41; m 66; c 2. CELLULAR BIOSCIENCES, COMPARATIVE PHYSIOLOGY. *Educ:* Yale Univ, BS, 63, MS, 66, MPhil, 68, PhD(biol), 69. *Prof Exp:* From asst prof to prof biol sci, Univ Cincinnati, 69-81, actg head dept, 73-75, admin intern, Off Develop, 74-75, fel, Grad Sch, 77-81, dir grad affairs biol sci, 78-79; dir, Regulatory Biol Prog, NSF, 79-84, dep dir, Div Cellular Biosci, 84-89, DIR, 89-; SR ADV HEALTH POLICY, OFF INT HEALTH POLICY, DEPT STATE, 88- *Concurrent Pos:* Trainee, NASA, 64-67 & NSF, 67-69; NSF fel, 64 & NSF grant, Univ Cincinnati, 71-79; assoc ed, J Exp Zool, 77-79; mem, NSF rev panels, US-India Exchange Scholars Prog, 79-81, US-India Coop Res Prog, 81-82; chmn, Cong Sci Fel Prog Comt, Am Soc Zoologists, 86-89, mem, 91; exec secy, Nat Sci Bd Comt Ctr & Individual Investr, 86-88; chairperson, Sect G-Biol Sci, AAAS, 88-89, steering group, Sect Comt G, 87-90; mem, Group, Nat Experts on Safety in Biotechnol, Organ Econ Coop & Develop, 88-89, panel Study Biol Diversity, Bd Sci & Technol in Int Develop, Nat Res Coun, 89, planning comt NIH Develop Conf Modeling Biomed Res, 88-89, working group, Int Biotechnol, 88-; chmn, Cell & Develop Biol Discipline Working Group, Space Biol Prog, NASA, 90-, Gravitational Biol Panel, NASA Specialized Ctrs Res & Training, 90- *Honors & Awards:* Sigma Xi Distinguished Res Award, 73. *Mem:* Fel AAAS; Am Physiol Soc; Am Soc Zoologists (secy, 79-81); fel NY Acad Sci; Am Inst Biol Sci; Sigma Xi. *Res:* Research administration of programs in cellular biosciences; comparative physiology, biochemistry and endocrinology of fish; low temperature biology. *Mailing Add:* Div Cellular Biosci NSF Washington DC 20550

UN, CHONG KWAN, b Seoul, Korea, Aug 25, 40; US citizen; m 66; c 3. DIGITAL COMMUNICATIONS, DIGITAL SIGNAL PROCESSING. *Educ:* Univ Del, BSEE, 64, MSEE, 66, PhD(elec eng), 69. *Prof Exp:* Asst prof elec eng, Univ Maine, 69-73; mem tech staff, SRI Int, 73-77; dean eng, 81-82, PROF ELEC ENG, KOREA ADVAN INST SCI & TECHNOL, 77-; DIR, CTR SPEECH INFO RES, 90- *Concurrent Pos:* Consult, Gold Star Elec Co, 77-87 & Digicom Inst Telematics, 87-; adv, Ministry Commun, Korea, 82-90 & Korea Inst Defense Anal, 85- *Mem:* Fel Inst Elec & Electronics Engrs. *Res:* Speech coding and processing; speech recognition and synthesis; adaptive filtering and signal processing; packet switching; voice-data integration. *Mailing Add:* Dept Elec Eng Korea Adv Inst Sci & Technol PO Box 150 Chongyangni Seoul Repub Korea

UN, HOWARD HO-WEI, b Hong Kong, June 8, 38; m 67; c 2. ORGANIC POLYMER CHEMISTRY. *Educ:* Beloit Col, BS, 60; Univ Mich, MSCh, 63, PhD(org chem), 65. *Prof Exp:* Res chemist, Exp Sta Lab, E I du Pont de Nemours & Co, Inc, 65-69, tech rep, Chestnut Run Lab, 69-71, tech rep, Fluorocarbons Mkt, 71-74, mkt rep, fluorocarbons sales, 74-76, sr mkt rep, 76-77, supvr & prod coord, Bus Servs Div, Plastic Prods & Resins Dept, 77-79, export mkt mgr, Int Mkt, Polymer Prod Dept, 79-82; develop mgr, Fluoropolymers, 82-84, develop mgr, 84-86, BUS DEVELOP MGR, ASIA-PAC WIRE & CABLE, 86- *Mem:* Am Chem Soc. *Res:* Fluorocarbon chemistry and polymers; thermally stable polymers; wire and cable materials. *Mailing Add:* 3314 S Rockfield Dr Wilmington DE 19810

UNAKAR, NALIN J, b Karachi, Pakistan, Mar 26, 35; m 62; c 2. CELL BIOLOGY. *Educ:* Gujarat Univ, India, BSc, 55; Univ Bombay, MSc, 61; Brown Univ, PhD(biol), 65. *Prof Exp:* Res asst biol, Indian Cancer Res Ctr, 55-61; res assoc path, Univ Toronto, 65-66; from asst prof to assoc prof, Oakland Univ, 69-74, prof & chmn dept biol sci, 74-87, PROF, DEPT BIOL SCI, 74-, ADJ PROF BIOMED SCI, OAKLAND UNIV, 83- *Concurrent Pos:* Nat Cancer Inst Can fel, 65-66; NIH res grant, 71-; mem visual sci A study sect, NIH, 82-86; adj prof biol sci, Wayne State Univ, 83- *Mem:* AAAS; Am Soc Cell Biol; Asn Res Vision & Opthal. *Res:* Cell ultrastructure and function; control of cell division; human and experimental cataracts; wound healing. *Mailing Add:* Dept Biol Sci Oakland Univ Rochester MI 48309

UNAL, AYNUR, b Ankara, Turkey, Jun 24, 46; m 74; c 2. NONLINEAR ACOUSTICS, NONLINEAR DYNAMICAL SYSTEMS. *Educ:* Middle East Tech Univ, BS, 68; Stanford Univ, MS, 70 & PhD(eng mechs), 73. *Prof Exp:* Asst prof mech eng, Univ Moncton, Que, 73-75; assoc prof mech eng, Alexandria Univ, 75-80; sr res assoc acoust, Stanford Univ, 80-83; prof acoust & dynamics, dept mech eng, Santa Clara Univ, 83-87; RES SCIENTIST ACOUSTICS, AERO FLIGHT DYNAMICS DYRECTORATE, NASA, 87- *Concurrent Pos:* Adj prof eng, Stanford Univ, 80-; NRC fel, Nat Res Coun, 80; mem Nat Curric Comt, 87-; prin invest, NASA grants nonlinear dynamical systs, 80-87,; vis prof, Univ Lyon, France,88. *Mem:* Am Inst Astron & Aeronaut; Am Soc Mech Eng; Soc Nat Philos; Soc Ind & Appl Math; Math Asn Am; Am Soc Mech Eng. *Res:* Nonlinear acoustics; nonlinear dynamical systems; fluid-structure interactions; biomedical dynamics; differential geometric interpretation of Navier-Stokes. *Mailing Add:* Dept Mech Eng Santa Clara Univ Santa Clara CA 95053

UNANGST, PAUL CHARLES, b Fountain Hill, Pa, Apr 19, 44; m 69; c 2. MEDICINAL CHEMISTRY. *Educ:* Lehigh Univ, BS, 65; Carnegie-Mellon Univ, MS, 68, PhD(org chem), 70. *Prof Exp:* Res chemist, Ozone Systs Div, Welsbach Corp, 70-72; assoc chem, Lehigh Univ, 72-73; SR RES ASSOC, WARNER-LAMBERT CO, 73- *Mem:* Am Chem Soc. *Res:* Medicinal chemistry; heterocycles; antiallergy and antiinflammatory agents. *Mailing Add:* Warner-Lambert/Parke-Davis 2800 Plymouth Rd Ann Arbor MI 48105

UNANUE, EMIL R, b Havana, Cuba, Sept 13, 34; m 65; c 3. IMMUNOLOGY. *Educ:* Inst Sec Educ, BSc, 52; Univ Havana Sch Med, MD, 60; Harvard Univ, MA, 74. *Prof Exp:* Intern path, Presby Univ Hosp, Pittsburgh, Pa, 61-62; res fel exp path, Scripps Clin & Res Found, 62-66, assoc, 68-70; res fel, Immunol Div, Nat Inst Med Res, London, Eng, 66-68; from asst prof to assoc prof path, Harvard Med Sch, 70-74, Mallinckrodt prof immunopath, 74-84; MALLINCKRODT PROF & CHMN, DEPT PATH, SCH MED, WASHINGTON UNIV, ST LOUIS, MO, 85- *Concurrent Pos:* Prof path, Harvard Med Sch; mem, Path A Study Sect, NIH, 73-77, Allergy & Immunol Res Comt, 80-85; consult path, Brigham & Women's Hosp, Boston, Mass, 77-84; Guggenheim fel, Biol Labs, Harvard Univ, 80-81; vis prof, Sch Med, Kuwait Univ, 82 & Royal Postgrad Med Sch, London Eng, 83; mem, Nat Sci Adv Coun, Nat Jewish Ctr Immunol & Respiratory Med, Denver, Colo, 85-86; pathologist-in-chief, Barnes & Allied Hosps & St Louis Children's Hosp, 85-; counr, Am Asn Pathologist, Inc, 85-88; sci adv bd, Harold C Simmons Arthritis Res Ctr, 87-; bd dirs, Barnes Hosp, 89- *Honors & Awards:* T Duckett Jones Award, Helen Hay Whitney Found, 68; Parke-Davis Award, Am Soc Exp Path, 73; Ecker Lectr, Western Reserve Univ, 79; D Allan Harmon Lectr, Okla Med Res Found, 85; Henry Kunkel Lectr, Johns Hopkins Univ, 85; Gerald Rodnan Lectr, Am Rheumatism Asn, 86; Albert H Coons Lectr, Harvard Med Sch, 88. *Mem:* Nat Acad Sci; Am Asn Path; Am Asn Immunologists; Brit Soc Immunol; Reticuloendothelial Soc; Am Soc Cell Biol. *Res:* The cellular basis of the immune response; regulatory mechanisms in immunity. *Mailing Add:* Dept Path 660 S Euclid PO Box 8188 St Louis MO 63110

UNBEHAUN, LARAINE MARIE, b Kearney, Nebr, May 4, 40; m 65. PLANT PATHOLOGY. *Educ:* Kearney State Col, BAEd, 61; Univ Northern Colo, MA, 64; Va Polytech Inst & State Univ, PhD(plant path), 69. *Prof Exp:* Teaching assoc biol, Univ Colo, Boulder, 64-65; asst prof, 69-71, assoc prof, 71-80, PROF BIOL, UNIV WIS-LA CROSSE, 80- *Concurrent Pos:* Dir, Gen Honors Prog, Univ Wis, La Crosse, 86- *Mem:* Am Phytopath Soc; Sigma Xi. *Res:* Pectic enzyme production by Thielaviopsis basicola grown on synthetic and natural media; enzyme purification; characterization of pectic enzymes produced in black root rot diseased tobacco. *Mailing Add:* Dept Biol Cowley Hall Univ Wis La Crosse WI 54601

UNDERDAHL, NORMAN RUSSELL, b Minn, June 5, 18; m 48; c 1. BACTERIOLOGY, VIROLOGY. *Educ:* St Olaf Col, BA, 41; Univ Minn, MS, 48. *Prof Exp:* Asst scientist, Hormel Inst, Univ Minn, 46-55; from asst prof to prof vet sci, Univ Nebr, Lincoln, 55-85; RETIRED. *Concurrent Pos:* Mem, Conf Res Workers Animal Dis. *Mem:* Assoc Am Vet Med Asn; Asn Gnotobiotics; Am Soc Microbiol. *Res:* Elimination of swine diseases by repopulation with disease-free pigs; isolation of causative agents of swine diseases using antibody-devoid, disease-free pig, obtained by surgery, as the host animal. *Mailing Add:* 935 N 67th St Lincoln NE 68505

UNDERDOWN, BRIAN JAMES, b Montreal, Que, Mar 23, 41; m 65; c 5. IMMUNOLOGY. *Educ:* McGill Univ, BSc, 64, PhD(immunol), 68. *Prof Exp:* Med Res Coun Can fel, Sch Med, Washington Univ, 68-70; prof med & grad secy, Inst Immunol, Univ Toronto, 70-; ASSOC DEAN RES, FAC HEALTH SCI, MCMASTER UNIV. *Mem:* Can Soc Immunol. *Res:* Studies of the IgA immune response; studies of the structure of antibody molecules. *Mailing Add:* Dept Path McMaster Univ Hamilton ON L8N 3Z5 Can

UNDERHILL, ANNE BARBARA, b Vancouver, BC, June 12, 20. ASTROPHYSICS, HOT MASSIVE STARS. *Educ:* Univ BC, BA, 42, MA, 44; Univ Chicago, PhD(astrophys), 48. *Hon Degrees:* DSc, York Univ, 69. *Prof Exp:* Nat Res Coun Can fel, Copenhagen Observ, 48-49; astrophysicist, Dom Astrophys Observ, 49-62; prof astrophys, State Univ Utrecht, 62-70; chief, Lab Optical Astron, Goddard Space Flight Ctr, NASA, 70-77, sr scientist, Astron & Solar Physics Lab, 77-85; HON PROF, DEPT GEOPHYS & ASTRON, UNIV BC, 85- *Concurrent Pos:* Vis lectr, Harvard Univ, 55-56; vis prof, Univ Colo, 67; Inst Astrophysics, Paris, 78-79. *Mem:* Am Astron Soc; Astron Soc Pac; Royal Astron Soc Can; Royal Astron Soc; Int Astron Union; fel Royal Soc Can, Acad III; Can Astron Soc. *Res:* Atmospheres of hot stars; Wolf-Rayet stars; model atmospheres; ultraviolet spectra of stars. *Mailing Add:* 4696 W Tenth Ave No 301 Vancouver BC V6R 2J5 Can

UNDERHILL, EDWARD WESLEY, b Regina, Sask, Jan 28, 31; m 54; c 3. PLANT BIOCHEMISTRY. *Educ:* Univ Sask, BScP, 54, MSc, 56; Univ RI, PhD(pharmacog), 60. *Prof Exp:* Lectr pharm, Univ Sask, 54-55, asst prof, 60-61; from asst res off to assoc res off, 61-74, SR RES OFF, NAT RES COUN CAN, 74- *Mem:* Am Entomol Soc; Can Entomol Soc. *Res:* Isolation and characterization of insect sex pheromones and attractants, paricularly of Lepidoptera; application of sex pheromones for monitoring and controling insect populations; studies on plant-insect interaction, particularly plant derived semiochemical affecting insect behavior. *Mailing Add:* 770 Cormorant St No 705 Victoria BC V8W 3J3 Can

UNDERHILL, GLENN, b Trenton, Nebr, Oct 30, 25; m 58; c 5. THEORETICAL PHYSICS, ASTRONOMY. *Educ:* Nebr State Col, BS, 55; Univ Nebr, MA, 57, PhD(physics), 63. *Prof Exp:* Instr physics, Univ Nebr, 59-61; assoc prof, 63-68, PROF PHYSICS, KEARNEY STATE COL, 68-, HEAD DEPT PHYSICS & PHYS SCI, 71- *Mem:* Am Phys Soc; Am Asn Physics Teachers; Am Sci Affiliation; Sigma Xi. *Res:* Structure of beryllium-9 nucleus; interaction of radiation with matter. *Mailing Add:* PO Box 70 Riverdale NE 68870

UNDERHILL, JAMES CAMPBELL, b Duluth, Minn, June 8, 23; m 43; c 3. ZOOLOGY. *Educ:* Univ Minn, BA, 49, MA, 52, PhD(zool), 55. *Prof Exp:* From asst prof to assoc prof zool, Univ SDak, 55-59; from asst prof to assoc prof, 59-69, coordr gen zool prog, 70-77, PROF ZOOL & BEHAV BIOL, UNIV MINN, MINNEAPOLIS, 69- *Mem:* Am Soc Ichthyol & Herpet; Ecol Soc Am; Soc Study Evolution; Am Fisheries Soc; Am Soc Limnol & Oceanog; Sigma Xi. *Res:* Ecology of minnows and darters; variation in fishes; aquatic ecology. *Mailing Add:* Bell Mus Natural Hist Univ Minn Minneapolis MN 55455

UNDERKOFLER, WILLIAM LELAND, b Ames, Iowa, Nov 10, 36; m 61; c 3. ANALYTICAL CHEMISTRY. *Educ:* Iowa State Univ, BS, 58; Univ Wis, PhD(anal chem), 64. *Prof Exp:* Staff chemist, 63-71, ADV CHEMIST, IBM CORP, 71- *Mem:* Am Chem Soc. *Res:* Electrochemistry; electrochemical analysis; electroplating; general chemical analysis. *Mailing Add:* 340 Raylene Dr Vestal NY 13850

UNDERWOOD, ARTHUR LOUIS, JR, b Rochester, NY, May 18, 24; m 48; c 3. BIOCHEMISTRY. *Educ:* Univ Rochester, BS, 44, PhD(biochem), 51. *Prof Exp:* Res assoc, Atomic Energy Proj, Univ Rochester, 46-51; res assoc anal chem, Mass Inst Technol, 51-52; from asst prof to assoc prof, 52-62, PROF CHEM, EMORY UNIV, 62- *Concurrent Pos:* Res fel biochem, Univ Rochester, 48-51; res assoc, Cornell Univ, 59-60; vis prof chem, Mont State Univ, 79-85. *Mem:* Fel AAAS; Am Chem Soc; Sigma Xi. *Res:* Ionic micelles with organic counterions; effects of counterion structure upon charge, critical micelle concentration and aggregation number; role of counterion in micellar solubilization; aqueous micellar systems as solvents in analytical chemistry. *Mailing Add:* Dept Chem Emory Univ Atlanta GA 30322

UNDERWOOD, BARBARA ANN, b Santa Ana, Calif, Aug 24, 34. NUTRITION, BIOCHEMISTRY. *Educ:* Univ Calif, Santa Barbara, BA, 56; Cornell Univ, MS, 58; Columbia Univ, PhD(nutrit biochem), 62. *Prof Exp:* Res asst nutrit, Cornell Univ, 58-59; res asst nutrit biochem, Columbia Univ & St Luke's Hosp, 59-61; res assoc, Inst Int Med, Univ Md, 62-64, asst prof, 64-66; res assoc, Columbia Univ, 66-68, asst prof nutrit sci, 68-72; assoc prof nutrit, Pa State Univ, 72-77, dir div biol health, 74-76; assoc prof nutrit, Mass Inst Technol, 78-82; spec asst, Nutrit Res & Int Progs, 82-89, ASST DIR INT PROGS, NAT EYE INST, NIH, 89- *Concurrent Pos:* Nutrit Found future leaders grant, 67-69; adv comt mem, Am Found Overseas Blind, 73-; mem malnutrit panel, US-Japan Med Res Comt, 74-85; mem Nat Acad Sci-Int Nutrit Progs Comt, 74-79; Int Vitamin A consult group, 75-; US deleg, World Health Assembly, 75-; mem adv comt, Off Sci & Technol Policy, 78-79; assoc ed, Am J Clin Nutrit, 84-; mem, bd trustees, Helen Keller Int, 85-90 & bd dirs, Int Eye Found. *Mem:* Am Inst Nutrit; Am Pub Health Asn; NY Acad Sci; Am Soc Clin Nutrit. *Res:* Malnutrition children; lipid metabolism; absorption and metabolism fat soluble vitamins in cystic fibrosis; vitamin A; breast feeding and child development; nutrition and nation development in developing countries. *Mailing Add:* Nat Eye Inst Off Dir Bldg 31-6A-17 NIH 9000 Rockville Pike Bethesda MD 20892

UNDERWOOD, DONALD LEE, b Grand Rapids, Mich, Apr 20, 28; m 56; c 2. COSMETIC CHEMISTRY. *Educ:* Wheaton Col, BS, 50; Princeton Univ, MA, 53. *Prof Exp:* Res chemist, Personal Care Div, 54-55, from res supvr to sr res supvr, 55-73, prin engr, 73-80, asst dir res, Advan Technol Lab, 80-85, CONSULT & PAT SEARCH, GILLETE CO, 85- *Mem:* Sigma Xi. *Res:* Sorption and diffusion of salt, acid and water in human hair; hair cosmetics; physics; appliance engineering and development. *Mailing Add:* 125 Marilyn St Holliston MA 01746-2035

UNDERWOOD, DOUGLAS HAINES, b Ravenna, Ohio, Nov 29, 34; m 58; c 2. MATHEMATICS. *Educ:* Case Inst Technol, BS, 56; Univ Calif, Berkeley, MA, 58; Univ Wis-Madison, PhD, 68. *Prof Exp:* Assoc prof, 58-74, PROF MATH, WHITMAN COL, 74- *Mem:* Am Math Soc; Math Asn Am. *Res:* Commutative rings. *Mailing Add:* Dept Math Whitman Col Walla Walla WA 99362

UNDERWOOD, ERVIN E(DGAR), b Gary, Ind, Jan 30, 18; c 3. PHYSICAL METALLURGY. *Educ:* Purdue Univ, BS, 49; Mass Inst Technol, SM, 51, ScD(metall), 54. *Prof Exp:* Metall observer, Open Hearth Dept, US Steel Corp, Inc, 40-41; res asst metall res, Mass Inst Technol, 49-54; metall consult, Battelle Mem Inst, 54-62; staff scientist, Lockhead Missiles & Space Co, 62-65, res dir mat, Lockheed-Ga Co, 65-71; Alcoa prof mech eng, 71-74, prof metall, 74-88, EMER PROF, FRACTURE & FATIGUE RES LAB, SCH MAT ENG, GA INST TECHNOL, 88- *Concurrent Pos:* Vis scientist, Max Planck Inst Metal Res, Stuttgart, 74-75; consult legal, sci & indust mat; prog dir metall, Div Mat Res, NSF, 87-88. *Honors & Awards:* Sorby Award, Int Metallographic Soc; Viella Award, Am Soc Testing & Mat. *Mem:* Am Inst Mining, Metall & Petrol Engrs; Am Soc Metals; Int Soc Stereol (pres, 71-75). *Res:* High temperature deformation of metals and alloys; fracture and fatigue; superplasticity; stereology, quantitative fracto; quantitative characterization of microstructures for applications based on mechanical behavior, mechanisms, patents, failure analysis and product liability; stereological study of fracture and fatigue leading to new area of quantitative fractography. *Mailing Add:* Rm 122 Sch Mat Eng Ga Inst Technol Atlanta GA 30332-0245

UNDERWOOD, HERBERT ARTHUR, JR, b Austin, Tex, Sept 4, 45; m 81. BIOLOGICAL RHYTHMS. *Educ:* Univ Tex, Austin, BA, 67, MA, 68, PhD(zool), 72. *Prof Exp:* Fel zool, Max-Planck Inst, 72-73; fel, Univ Tex, Austin, 73-75; from asst prof to assoc prof, 80-85, PROF ZOOL, NC STATE UNIV, 85- *Honors & Awards:* Res Career Develop Award, NIH, 79-84. *Mem:* Am Soc Zoologists; AAAS; Sigma Xi; Soc Res Biol Rhythms. *Res:* Role of the eyes and extraretinal photoreceptors in the control of the biological clock of lizards and birds; vertebrate photoperiodism; behavioral thermoregulation in poikilotherms; involvement of the pineal system in vertebrate circadian rhythms. *Mailing Add:* Dept Zool NC State Univ Raleigh NC 27695-7617

UNDERWOOD, JAMES HENRY, b Minster, Eng, Apr 18, 38; m 81. SOLAR PHYSICS, OPTICS. *Educ:* Univ Leicester, BSc, 59, PhD(physics), 63. *Prof Exp:* Res assoc space res, Nat Acad Sci, NSF, 63-66; aerospace scientist, Goddard Space Flight Ctr, NASA, 66-72; staff scientist space res, Aerospace Corp, 72-77; mem staff, Inst Plasma Res, Stanford Univ, 77-80; res scientist, Jet Propulsion Lab, 80-83; LAWRENCE LIVERMORE NAT LAB UNIV CALIF. *Mem:* Int Astron Union; Am Astron Soc; Optical Soc Am. *Res:* Application of x-ray techniques, in particular x-ray optics and crystal spectroscopy, to the study of hot plasmas, in particular the solar corona and other celestial x-ray sources. *Mailing Add:* Lawrence Livermore Nat Lab Univ Calif One Cyclotron Rd Berkeley CA 94720

UNDERWOOD, JAMES ROSS, JR, b Austin, Tex, May 15, 27; m 61; c 3. PLANETARY GEOLOGY, GEOLOGICAL EDUCATION. *Educ:* Univ Tex, Austin, BS, 48, BS, 49, MA, 56, PhD(geol), 62. *Prof Exp:* Petrol eng trainee, Sohio Petrol Co, 49-50, jr petrol engr, 50-51, petrol engr, 53-54; instr geol, Univ Tex, 56-57, Univ Tex-Agency Int Develop Prog, asst prof, Univ Baghdad, 62-65; temp asst prof, Univ Fla, 65-67; from assoc prof to prof geol, WTex State Univ, 67-77; head dept, 77-85, PROF GEOL, KANS STATE UNIV, 77- *Concurrent Pos:* On leave as Exxon-sponsored prof, Univ Libya, 69-71; discipline scientist, NASA Planetary Geol & Geophysics Prog, Washington, DC, 87-89. *Mem:* Geol Soc Am; Am Asn Petrol Geol; Soc Econ Paleontologists & Mineralogists; Am Geophys Union; Am Inst Mining, Metall & Petrol Engrs; AAAS. *Res:* Structural geology; geomorphology; planetary geology, especially Mercury, Mars and moons of Jupiter; terrestrial impact structures; geology of Trans-Pecos Texas, northern Chihuahua and the Middle East; petroleum engineering, especially drilling and production. *Mailing Add:* Dept Geol Kans State Univ Thompson Hall Manhatten KS 66506-3201

UNDERWOOD, LAWRENCE STATTON, b Kansas City, Kans, July 29, 36; m 64; c 2. PHYSIOLOGICAL ECOLOGY. *Educ:* Univ Kans, BA, 59; Syracuse Univ, MS, 68; Pa State Univ, University Park, PhD(zool), 71. *Prof Exp:* Teacher pub schs, Kans, Mo & Alaska, 57-67; asst biol, Pa State Univ, University Park, 68-71; asst prof zool, Univ Conn, West Hartford Br, 71-73; asst dir sci, Naval Arctic Res Lab, 73-76; res analyst, 76-80, SUPVR INFO SCI SERV, ARCTIC ENVIRON INFO & DATA CTR, 80- *Concurrent Pos:* Int Biol Prog, NSF & Arctic Inst NAm grants, Naval Arctic Res Lab, 71-72. *Mem:* AAAS; Am Inst Biol Sci; Ecol Soc Am; fel Arctic Inst NAm; Int Soc Biometeorol. *Res:* Physiological and behavioral adjustments in cold acclimatized mammals to varying environmental conditions; environmental and information transfer; establishing a system of ecological reserves in Alaska. *Mailing Add:* 12000 Waterside View Dr No 31 Reston VA 22094-1703

UNDERWOOD, LLOYD B, b Jackson, Mich, May 9, 19. HYDROLOGY, RECLAMATION. *Educ:* Mich State Univ, BS, 42. *Prof Exp:* Chief eng geologist, 42-80; cosult, BC Hydrol Corp & US Bur Reclamation, 80-; RETIRED. *Honors & Awards:* Burwell Award, Geol Soc Am, 69; Clare Holdred Award, Asn Econ Geologists, 68. *Mem:* Geol Soc Am; Am Asn Petrol Geologists; Sigma Xi. *Mailing Add:* 15682 Leavenworth St Omaha NE 68118

UNDERWOOD, LOUIS EDWIN, b Danville, Ky, Feb 20, 37; m 60; c 3. PEDIATRICS, ENDOCRINOLOGY. *Educ:* Univ Ky, AB, 58; Vanderbilt Univ, MD, 61. *Prof Exp:* From intern to asst resident pediat, Vanderbilt Univ, 61-63; asst resident, Univ NC, 63-64; instr, Vanderbilt Univ, 64-65; attend pediat, US Naval Hosp, Chelsea, Mass, 65-67; from instr to assoc prof, 69-80, PROF PEDIAT, SCH MED, UNIV NC, CHAPEL HILL, 80- *Concurrent Pos:* USPHS fel endocrinol, 67-70. *Mem:* AAAS; Am Fedn Clin Res; Endocrine Soc; Soc Pediat Res; Lawson Wilkins Pediat Endocrine Soc; Am Pediat Soc. *Res:* Pediatric endocrine diseases; growth problems and hormonal control of growth. *Mailing Add:* Dept Pediat CB 2770 Univ NC Sch Med Chapel Hill NC 27599

UNDERWOOD, REX J, b Eugene, Ore, Nov 13, 26; m 50; c 2. MEDICINE. *Educ:* Stanford Univ, AB, 50; Univ Ore, MS & MD, 55. *Prof Exp:* From instr to asst prof anesthesiol, Med Sch, Univ Ore, 58-67; asst dir anesthesiol, 67-68, DIR ANESTHESIOL, BESS KAISER HOSP, 68- *Mem:* AMA; Am Soc Anesthesiol; Sigma Xi. *Res:* Anesthesiology, particularly related to cardiovascular physiology. *Mailing Add:* 3272 NE Davis St Portland OR 97232

UNDERWOOD, ROBERT GORDON, b Nashville, Tenn, Feb 3, 45; m 68; c 2. APPLIED MATHEMATICS, OPERATIONS RESEARCH. *Educ:* Univ NC, Chapel Hill, BS, 67; Univ Va, PhD(appl math), 74. *Prof Exp:* Health Serv Officer, USPHS, 68-70; asst prof, Univ SC, 74-78; ASST PROF MATH, COLO SCH MINES, 78- *Mem:* Soc Indust & Appl Math; Am Math Asn; Sigma Xi. *Res:* Applied mathematics; optimization, game theory and control theory. *Mailing Add:* Dept Math Colo Sch Mines Golden CO 80401

UNDEUTSCH, WILLIAM CHARLES, b Hamilton, Ohio, Oct 6, 25; m 54; c 4. ORGANIC CHEMISTRY. *Educ:* Univ Cincinnati, BS, 48, MS, 50; Univ Del, PhD(chem), 53. *Prof Exp:* Lab technician, Children's Hosp Res Found, 48, 50; res chemist, Photo Prods Dept, E I du Pont de Nemours & Co, Inc, 53-58; SR ED CHEM ABSTR SERV, AM CHEM SOC, 58- *Res:* Chemical literature. *Mailing Add:* 102 N Harris Ave Columbus OH 43204

UNG, MAN T, b South Vietnam, Apr 1, 38; US citizen; m 60; c 2. COMPUTER SCIENCE, MATHEMATICS. *Educ:* Univ Wis, BS, 59; Ill Inst Technol, MS, 62; Univ Southern Calif, PhD(elec eng), 70. *Prof Exp:* Asst res engr, IIT Res Inst, 61-63; mgr educ & training, Electronic Assocs, Inc, 63-70; mgr comput & math serv, Dillingham Environ Co, 70-71; asst prof elec eng, 72-76, adj asst prof, 76-79, ADJ ASSOC PROF ELEC ENG, UNIV SOUTHERN CALIF, 79- *Concurrent Pos:* Ed, Analog/Hybrid Comput Educ Soc, 65-68. *Mem:* Simulation Coun; Marine Technol Soc; Sr mem Soc Comput Simulation. *Res:* Modeling of ecological systems; hybrid computer applications in engineering. *Mailing Add:* Dept Elec Eng SAL Rm 316 Univ Southern Calif University Park Los Angeles CA 90087-0781

UNGAR, EDWARD WILLIAM, b New York, NY, Feb 6, 36; m 78; c 3. TECHNICAL MANAGEMENT, RESEARCH ADMINISTRATION. *Educ:* City Col New York, BME, 57; Ohio State Univ, MSc, 59, PhD(mech eng), 66. *Prof Exp:* Res engr, 57-63, prog dir, Fluid & Thermal Mech Div, 63-66, assoc div chief, 66, chief fluid & gas dynamics div, 66-70, mgr eng physics sect, 70-71, mgr eng physics dept, 71-74, mgr energy & environ processes res dept, 74-76, assoc dir, 76-78, dir, Columbus Div, 78-85, vpres, Battelle Mem Inst, 78-86; PRES, TARATEC CORP, 86- *Mem:* Am Soc Mech Engrs; AAAS; Sigma Xi. *Res:* Physical, life and social sciences. *Mailing Add:* 5981 Whitman Rd Columbus OH 43213

UNGAR, ERIC E(DWARD), b Vienna, Austria, Nov 12, 26; nat US; m 51; c 4. STRUCTURAL DYNAMICS, VIBRATION AND NOISE CONTROL. *Educ:* Wash Univ, BSME, 51; Univ NMex, MS, 54; NY Univ, DEngSc(mech eng), 57. *Prof Exp:* Aero-ord engr, Sandia Corp, 51-53; from instr to asst prof mech eng, NY Univ, 53-58, res scientist, 58; sr eng scientist & mgr appl physics dept, 58-68, assoc div dir, 68-76, prin engr, 76-84, CHIEF CONSULT ENGR, BOLT BERANEK & NEWMAN, 85- *Honors & Awards:* Centennial Medallion, Am Soc Mech Engrs, 81; Trent-Crede Medal of Acoust, Soc Am, 83. *Mem:* Fel Am Soc Mech Engrs; assoc fel Am Inst Aeronaut & Astronaut; fel Acoust Soc Am; Inst Noise Control Eng. *Res:* Structural and machinery dynamics; vibrations and noise; stress analysis; machine design. *Mailing Add:* BBN Systs & Technol Ten Moulton St Cambridge MA 02238

UNGAR, FRANK, b Cleveland, Ohio, Apr 30, 22; m 48; c 4. BIOCHEMISTRY, ENDOCRINOLOGY. *Educ:* Ohio State Univ, BA, 43; Western Reserve Univ, MSc, 48; Tufts Univ, PhD(biochem, physiol), 52. *Prof Exp:* Res staff mem, Cleveland Clin, 47-48; res staff mem, Worcester Found Exp Biol, 51-58; from assoc prof to prof, 58-90, EMER PROF BIOCHEM, MED SCH, UNIV MINN, MINNEAPOLIS, 90- *Concurrent Pos:* Fulbright sr scholar, Univ Col, Cork, 74-75; asst vis prof chem, Clark Univ, 56-58; consult cancer chemother group, NIH, 65-72; Fogarty Int Sr fel, Weizmann Inst, Rehovot, Israel, 82-83. *Mem:* AAAS; Am Chem Soc; Endocrine Soc; Am Soc Biol Chem. *Res:* Regulation of hormone action; metabolism of steroid hormones. *Mailing Add:* Dept Biochem 4-217 OWRE Univ Minn Med Sch Minneapolis MN 55455

UNGAR, GERALD S, b Wilkes-Barre, Pa, Jan 27, 41; m 59; c 3. MATHEMATICS. *Educ:* Franklin & Marshall Col, BA, 61; Rutgers Univ, MS, 63, PhD(topol), 66. *Prof Exp:* Asst prof math, La State Univ, Baton Rouge, 66-68 & Case Western Reserve Univ, 68-70; PROF MATH, UNIV CINCINNATI, 70- *Mem:* Am Math Soc; Math Asn Am. *Res:* Fiber maps; local homogeneity. *Mailing Add:* Math Dept Univ Cincinnati 810 E Old Chem Cincinnati OH 45221

UNGAR, IRWIN A, b New York, NY, Jan 21, 34; m 59; c 3. PLANT ECOLOGY. *Educ:* City Col New York, BS, 55; Univ Kans, MA, 57, PhD(bot), 61. *Prof Exp:* Instr bot, Univ RI, 61-62; asst prof, Quincy Col, 62-66; from asst prof to assoc prof, 66-73, PROF BOT, OHIO UNIV, 74-, chmn dept, 83-88. *Concurrent Pos:* Sigma Xi res awards, 59 & 66; NSF grants, 63-65, 67-69, 74-75, 76-78 & 80-83, panelist, 66; res grant, Ohio Univ, 66-68, 74-75, 78-79, 83-85, & 88-89, John C Baker res grant, 72-73; res grant, Ohio Biol Surv, 70-71 & 74-75; res assoc, Ctr Nat Res Sci grant, France, 72-73; Res Inst fel, Ohio Univ, 74. *Mem:* AAAS; Ecol Soc Am; Bot Soc Am; Sigma Xi. *Res:* Vegetation-soil relations on acid and saline soils; ecology of halophytes; studies in salt tolerance and demography of species under field conditions. *Mailing Add:* Dept Bot Ohio Univ Athens OH 45701

UNGER, HANS-GEORG, b Braunschweig, Ger, Sept 14, 26; m 55; c 2. OPTICAL COMMUNICATION, MICROWAVES. *Educ:* Tech Univ Braunschweig, Dipl Ing, 51, Dr-Ing, 54. *Hon Degrees:* Dr-Ing Eh, Tech Univ Munich, 85. *Prof Exp:* Develop engr, Siemens Ag, Munich, 51-54, dept head, 54-55; mem staff, US Spec Proj 63, 55-56; mem tech staff, Bell Labs, Holmdel, 56-59, dept head, Res Commun Technol, 59-60 & Guided Wave Res, 61; PROF ELEC ENG & DIR, INST HIGH FREQUENCY TECHNOL, TECH UNIV BRAUNSCHWEIG, 60- *Concurrent Pos:* Dean, Elec Eng, Tech Univ Braunschweig, 61-64; vis prof, Univ Wis, 64-65, Univ Ghent, 68, Univ Rennes & Tech Univ Delft, 73 & Univ Tokyo, 75; int chmn, Comn B, Int Union Radio Sci, 81-84. *Honors & Awards:* Heinrich Hertz Medal, Inst Elec & Electronics Engrs, 88. *Mem:* Fel Inst Elec & Electronics Engrs. *Res:* Optical fiber guides and planar optical wave guides; waveguides components and circuits for optical communication, integrated optics and integrated optoelectronics; microwave electronics; microwave antennas. *Mailing Add:* Inst High Frequency Technol Tech Univ Braunschweig Braunschweig 3300 Germany

UNGER, ISRAEL, b Tarnow, Poland, Mar 30, 38; Can citizen; m 64. CHEMISTRY. *Educ:* Sir George Williams Univ, BSc, 58; Univ NB, MSc, 60, PhD(chem), 63. *Prof Exp:* Fel, Univ Tex, 63-65; from asst prof to assoc prof, 65-74, PROF CHEM, UNIV NB, FREDERICTON, 74-, DEAN SCI, 86- *Concurrent Pos:* Adv bd, Can Inst Sci & Technol Info. *Mem:* Chem Inst Can; Inst Res Pub Policy. *Res:* Kinetics and photochemistry of small organic molecules; photochemistry of pesticides. *Mailing Add:* Dept Chem Univ NB Fredericton NB E3B 6E2 Can

UNGER, JAMES WILLIAM, b Marshfield, Wis, Apr 1, 21; m 47; c 1. PLANT MORPHOLOGY. *Educ:* Wis State Col, Stevens Point, BS, 42; Univ Wis, MS, 47, PhD(bot), 53. *Prof Exp:* Asst prof bot, Hope Col, 47-51; prof bot & chmn dept biol, 53-67, PROF BIOL, UNIV WIS-OSHKOSH, 67- *Mem:* Am Inst Biol Sci; Bot Soc Am; Sigma Xi. *Res:* Anatomical considerations of gymnosperm tissue cultures; anatomical studies of stem apices and stem to root vascular transitions; membrane permeability studies; tissue culture of orange. *Mailing Add:* 1212 E New York Ave Oshkosh WI 54901

UNGER, JOHN DUEY, b Harrisburg, Pa, Mar 2, 43; m 66; c 3. GEOPHYSICS, SEISMOLOGY. *Educ:* Mass Inst Technol, BS & MS, 67; Dartmouth Col, PhD(geol), 69. *Prof Exp:* Geophysicist, Hawaiian Volcano Observ, US Geol Surv, Hawaii Nat Park, 69-74, geophysicist, Nat Ctr for Earthquake Res, 74-76, GEOPHYSICIST, BR ENG GEOL & TECTONICS, US GEOL SURV NAT CTR, RESTON VA, 76- *Mem:* AAAS; Geol Soc Am; Am Geophys Union; Seismol Soc Am. *Res:* Reflection seismology; deep crustal studies and tectonics; general microearthquake seismology; volcano geophysics. *Mailing Add:* US Geol Surv Nat Ctr MS-922 12201 Sunrise Valley Dr Reston VA 22092

UNGER, LLOYD GEORGE, b Stickney, SDak, Feb 24, 18; m 47; c 4. PHYSICAL CHEMISTRY. *Educ:* Yankton Col, BA, 39; Pa State Univ, MS, 41, PhD(phys chem), 45. *Prof Exp:* Lab asst, Pa State Univ, 39-44; res chemist, CPC Int Inc, Ill, 44-68; assoc prof, 68-80 PROF PHYS SCI & CHEM, WRIGHT COL, 80- *Mem:* Fel AAAS; Am Chem Soc; Sigma Xi. *Res:* Cereal proteins; textile chemicals. *Mailing Add:* 99 Lawton Rd Riverside IL 60546

UNGER, PAUL WALTER, b Winchester, Tex, Sept 10, 31; m 60; c 6. SOIL SCIENCE, AGRONOMY. *Educ:* Tex A&M Univ, BS, 61; Colo State Univ, MS, 63, PhD(soil sci), 66. *Prof Exp:* SOIL SCIENTIST, AGR RES SERV, USDA, 65- *Concurrent Pos:* Assoc ed, Soil Sci Soc Am J, 77-82; assoc ed, Iowa State J Res, 80-86; consult, Food Agr Orgn UN, 86; co-ed, proceedings, Int Conf Dry Land Farming, 88; div chmn, Am Soc Soil Sci, 86. *Mem:* fel Am Soc Agron; fel Soc Soil Sci Am; Soil & Water Conserv Soc; Coun Agr Sci & Technol; Int Soil Sci Soc; Int Soil Tillage Res Orgn. *Res:* Soil management and moisture conservation, especially tillage and crop residue management as they relate to soil structure, water movement and water storage in the soil. *Mailing Add:* Conserv & Prod Res Lab USDA Agr Res Serv PO Drawer 10 Bushland TX 79012

UNGER, ROGER HAROLD, b New York, NY, Mar 7, 24; m 46; c 3. INTERNAL MEDICINE. *Educ:* Yale Univ, BS, 44; Columbia Univ, MD, 47; Am Bd Internal Med, dipl, 56. *Hon Degrees:* Dr, Univ Geneva, 76, Univ Liege, 78. *Prof Exp:* From intern to resident med, Bellevue Hosp, New York, 47-51; dir, Dallas Diabetes Unit, Tex, 51-52; clin instr, 52-59, from asst prof to assoc prof, 59-70, PROF MED, UNIV TEX HEALTH SCI CTR DALLAS, 71-; DIR, CTR DIABETES RES, UNIV TEX SOUTHWESTERN MED SCH, 85- *Concurrent Pos:* Clin instr, Postgrad Med Sch, NY Univ, 51-56; chief gastroenterol sect, Vet Admin Hosp, Dallas, 58-64, chief metab sect, 64-74, dir res, 65-75; mem, Res & Educ Comt, Vet Admin Cent Off, 71-; vis prof, Univ Geneva, 72-73; vpres, Solomon A Berson Fund Med Res, Inc, 73 & pres, 74-76; sr med investr, Vet Admin Med Ctr, Dallas; consult, Nat Comn Diabetes, 75-77; mem, Subcomt to Nat Comt Diabetes, Vet Admin, 75-77 & Comt Sci Progs, Am Diabetes Asn, 76-77; prin investr, NIH contract, 76-82; assoc ed, Diabetes, 79-83. *Honors & Awards:* Lilly Award, Am Diabetes Asn, 64; Tinsley Harrison Award, 67; Middleton Award, Vet Admin, 69; Solomon A Bernson Mem Lectr, NIH, 75, Int Diabetes Fedn & Am Diabetes Asn, 82; Banting Medal & Mem Lectr, Am Diabetes Asn, 75; David Rumbough Award, Juvenile Diabetes Found, 75; Francis D W Lukens Hon Lectr, Univ Pa, 75; Sandoz Award & Lectr, Can Soc Endocrinol & Metab, 76; Woodyatt Mem Lectr, Chicago, Ill, 77; Joslin Mem Lectr, Harvard Univ, 79; Claude Bernard Mem Medal & Lectr, Europ Asn Study Diabetes, 80; Arthur R Colwell Mem Lectr, Northwestern Univ, 81; Fred Conrad Koch Award, Endocrine Soc, 83; Mosenthal Lectr, NY Diabetes Asn, 88. *Mem:* Nat Acad Sci; Am Fedn Clin Res; emer mem Am Soc Clin Invest; Endocrine Soc; Asn Am Physicians; Am Diabetes Asn; Int Diabetes Found; AMA; Cent Soc Clin Res; fel Am Col Physicians. *Res:* Diabetes. *Mailing Add:* Dept Internal Med Univ Tex Southwestern Med Sch Dallas TX 75235

UNGER, S(TEPHEN) H(ERBERT), b New York, NY, July 7, 31; m; c 2. ELECTRICAL ENGINEERING, COMPUTER SCIENCE. *Educ:* Polytech Inst Brooklyn, BEE, 52; Mass Inst Technol, SM, 53, ScD, 57. *Prof Exp:* Asst elec eng, Res Lab Electronics, Mass Inst Technol, 54-57; mem tech staff, Bell Tel Labs, Inc, 57-61; assoc prof elec eng, 61-68, prof elec eng, 68-79, PROF COMPUT SCI, COLUMBIA UNIV, 79- *Concurrent Pos:* Adj asst prof, Columbia Univ, 60-61; consult, Res Labs, RCA Corp, 64-69; Guggenheim fel, 67; consult, Western Elec Eng Res Ctr, 71 & IBM, 82-85; vis prof comput sci, Danish Tech Univ, 74-75. *Mem:* Fel AAAS; Am Asn Univ Professors; fel Inst Elec & Electronics Engrs; Asn Comput Mach. *Res:* Switching circuit theory; digital computer systems; programming theory; pattern recognition; technological aids to the democratic process; programming languages; computer conferencing; engineering ethics. *Mailing Add:* Dept Computer Sci Columbia Univ Broadway & N 116th St New York NY 10027

UNGER, VERNON EDWIN, JR, b Easton, Md, Dec 14, 35; m 58; c 2. INDUSTRIAL ENGINEERING, OPERATIONS RESEARCH. *Educ:* Johns Hopkins Univ, BES, 57, MS, 64, PhD(opers res), 68. *Prof Exp:* Staff asst, AAI, Inc, 57-65; prof indust eng, Ga Inst Technol, 68-79, assoc dir res, 74-79; PROF INDUST ENG & HEAD DEPT, AUBURN UNIV, 79- *Mem:* Opers Res Soc Am; Am Inst Indust Eng. *Res:* Operations research and engineering economics. *Mailing Add:* Dept Indust Eng Auburn Univ Auburn AL 36849

UNGLAUBE, JAMES M, b Milwaukee, Wis, Apr 13, 42; m 64; c 1. ORGANIC CHEMISTRY. *Educ:* Carthage Col, BA, 63; Univ Iowa, MS, 66, PhD(org chem), 68. *Prof Exp:* Teaching asst org chem, Univ Iowa, 63-64; asst prof chem, 67-70, assoc prof chem & acad dean, Lenoir Rhyne Col, 70-77; asst dir, Dept Higher Educ, Lutheran Church Am, 77-82, dir, 82-87; DIR, COL & UNIV, DIV EDUC, EVANGEL LUTHERAN CHURCH, 87- *Mem:* Am Asn Higher Educ; Am Chem Soc; Sigma Xi. *Res:* Chemical education; organic chemistry syntheses, including heterocyclic nitrogen compounds. *Mailing Add:* Evangelical Luthern Church 8765 W Higgens Rd Chicago IL 60631-4177

UNGVICHIAN, VICHATE, b Bangkok, Thailand, May 22, 49. ANTENNA DESIGNS, COMPUTER MODELLING. *Educ:* Khon-Kaen Univ, Thailand, BSEE, 67; Ohio State Univ, MS, 74; Ohio Univ, PhD(elec eng), 81. *Prof Exp:* Fel electromagnetics, elec eng dept, Ohio Univ, 81-; RES ENGR, DEPT ELEC ENG, FLA ATLANTIC UNIV. *Mem:* Inst Elec & Electronics Engrs; Sigma Xi. *Res:* Electromagnetic fields as applied to instrument landing systems; computer models to calculate the ground scattering by using the uniform theory of diffraction, including all rays up to the second and most of the third order types of rays. *Mailing Add:* Dept Elec Eng Fla Atlantic Univ Boca Raton FL 33431

UNIK, JOHN PETER, b Chicago, Ill, May 18, 34; m 57; c 2. NUCLEAR CHEMISTRY. *Educ:* Ill Inst Technol, BS, 56; Univ Calif, Berkeley, PhD(chem), 60. *Prof Exp:* From asst chemist to assoc chemist, 60-74, sect head, 72-74, assoc dir chem div, 74-82, dir Sci Support, 83-84, SR CHEMIST, 74-, ASSOC DIR SUPPORT SER, ARGONNE NAT LAB, 84- *Concurrent Pos:* Consult, Oak Ridge Nat Lab, 74-81, US Nuclear Data Comt, 78-82 & Lawrence Berkeley Lab, 79-82. *Mem:* Am Phys Soc; Am Chem Soc. *Res:* Nuclear fission and heavy ion reactions. *Mailing Add:* Argonne Nat Lab B201/241 SSD 9700 S Cass Ave Argonne IL 60439

UNKLESBAY, ATHEL GLYDE, b Byesville, Ohio, Feb 11, 14; m 40; c 4. GEOLOGIC EDUCATION, EDUCATION ADMINISTRATION. *Educ:* Marietta Col, AB, 38; State Univ Iowa, MA, 40 & PhD(paleont), 42. *Hon Degrees:* SDc, Marietta Col, 77. *Prof Exp:* Geologist, US Geol Surv, 42-45, Iowa Geol Surv, 45-46; asst prof geol, Colgate Univ, 46-47; from asst prof to prof geol, Univ Mo, 47-67, vpres admin, 67-79; exec dir, Am Geol Inst, 79-85; RETIRED. *Mem:* Geol Soc Am; Am Asn Petrol Geologists; Paleontol Soc; Nat Asn Geol Teachers. *Res:* General geology for the layman and of Missouri. *Mailing Add:* 37 G Broadway Village Columbia MO 65201-8662

UNKLESBAY, NAN F, b North Vancouver, BC, May 28, 44; m 74. FOOD SCIENCE, NUTRITION. *Educ:* Univ BC, BHE, 66; Univ Wis, Madison, MS, 71, PhD(food sci), 73. *Prof Exp:* Dietary consult, Dept Health, Govt Nfld & Labrador, 67-70; asst prof food systs mgt, 73-76, from asst prof to assoc prof, 76-85, PROF FOOD SCI & NUTRIT, UNIV MO, COLUMBIA, 85- *Concurrent Pos:* Consult, NSF, 75-76. *Mem:* Inst Food Technologists; Am Dietetic Asn; Can Dietetic Asn. *Res:* Major amounts of energy utilization within the food industry; optimization of resource utilization in food services while maintaining microbial safety, quality and nutritional value. *Mailing Add:* Dept Food Sci-Nutrit Univ Mo 122 Eckles Hall Columbia MO 65211

UNLAND, MARK LEROY, b Jacksonville, Ill, Mar 17, 40; m 62; c 2. PHYSICAL CHEMISTRY. *Educ:* MacMurray Col, AB, 62; Univ Ill, Urbana, MS, 64, PhD(chem), 66. *Prof Exp:* RES CHEMIST, MONSANTO CO, 66- *Mem:* Am Chem Soc; Sigma Xi; Sci Res Soc NAm. *Res:* Molecular structure; catalyst development, fundamental studies of heterogeneous catalysts and catalysis mechanisms. *Mailing Add:* 12903 Mayerling Dr Creve Coeur MO 63146

UNLU, M SELIM, b Sinop, Turkey, June 12, 64. HIGH SPEED ELECTRONIC DEVICES, OPTOELECTRONIC DEVICES. *Educ:* Middle East Tech Univ-Ankara, BS, 86; Univ Ill-Urbana-Champaign, MS, 88. *Prof Exp:* Res engr, Aselsan, Mil Electronics Inc, Ankara, Turkey, 84-86; RES ASST, COORD SCI LAB, UNIV ILL, 86- *Mem:* Inst Elec & Electronics Engrs; Lasers & Electrooptics Soc. *Res:* Design and characterization of high speed electronic devices; optoelectronic devices for optical communication systems; analysis and modeling of heterojunction semiconductor devices; basic semiconductor phenomena. *Mailing Add:* Coord Sci Lab 1101 W Springfield Urbana IL 61801

UNNAM, JALAIAH, b Tangutur, India, Dec 1, 47; m 73; c 2. MATERIALS SCIENCE. *Educ:* Indian Inst Technol, BTech, 70; Va Polytech Inst & State Univ, MS, 72, PhD(mat sci), 75. *Prof Exp:* Instr metall, Va Polytech Inst & State Univ, 74 & 75, res assoc mat sci, 75-77; res fel, George Washington Univ, 77-78; sr scientist, Vira Inc, 82-83; PRES, ANALYTICAL SERV & MAT, INC, 83- *Concurrent Pos:* Vis prof mat sci, Va Polytech Inst & State Univ, 78-82. *Mem:* Am Soc Metals; AIME; Sigma Xi. *Res:* Solid state diffusion; x-ray diffraction; composite materials; numerical analysis and computer programming; electron microprobe; scanning electron microscope; quantitative metallography; mechanical testing; electroplating; oxidation. *Mailing Add:* Analytical Serv & Mat 107 Research Dr Hampton VA 23666

UNO, HIDEO, b Tokyo, Japan, Nov 28, 29; m 56; c 2. COMPARATIVE PATHOLOGY, PHARMACO-MORPHOLOGY. *Educ:* Yokohama Med Col, MD, 55, PhD(path), 60. *Prof Exp:* Intern gen med, Tokyo Munic Hiroo Hosp, 55-56; resident path, Yokohama Med Col, 56-60; asst prof path, Sch Med, Yokohama City Univ, 60-64; instr path, Jefferson Med Col, Philadelphia, Pa, 64-66; vis scientist cutaneous biol, Ore Primate Res Ctr, Beaverton, 66-68; assoc prof path, Sch Med, Yokohama City Univ, 68-70; SR SCIENTIST, WIS REGIONAL PRIMATE RES CTR, UNIV WIS-MADISON, 70- *Concurrent Pos:* Scientist path, Ore Primate Res Ctr, 70-79; adj assoc prof, dept path & lab med, Sch Med, Univ Wis- Madison, 72-, adj prof, 89- *Mem:* Fedn Am Soc Exp Biol; Int Acad Path; Soc Invest Dermat; Japanese Soc Pathologists. *Res:* Comparative pathology of aging, spontaneous cancer, diabetes, cardiovascular disorders and aging brain disorders of rhesus monkeys; pharmaco-pathological study of dexamethasone to monkey fetal brain; effect of minoxidil and antiandrogen on macaque and human baldness; stress and brain damage. *Mailing Add:* Regional Primate Res Ctr Univ Wis Sch Med 1223 Capitol Ct Madison WI 53706

UNOWSKY, JOEL, b St Paul, Minn, Dec 11, 38; m 75; c 4. INDUSTRIAL MICROBIOLOGY, MICROBIAL GENETICS. *Educ:* Univ Minn, Minneapolis, BA, 61; Northwestern Univ, Evanston, PhD(bact), 66. *Prof Exp:* Res fel, E I du Pont de Nemours & Co, Inc, 65-66; sr microbiologist, Hoffmann-La Roche Inc, 66-79, asst res group, 79, res group chief, 79-84, RES LEADER, HOFFMANN-LA ROCHE INC, 87- *Concurrent Pos:* Adj assoc prof, Seton Hall Univ, 76. *Mem:* AAAS; Am Soc Microbiol; Sigma Xi; NY Acad Sci. *Res:* Resistance transfer factor; antibiotic strain development; antibiotic screening; mutation and genetics; chemotherapy; immunology; cytokines; immunmodulators; absorption of pharmaceuticals. *Mailing Add:* Hoffmann-La Roche Inc 340 Kingsland Rd Bldg 58 Nutley NJ 07110

UNRATH, CLAUDE RICHARD, b Benton Harbor, Mich, Nov 29, 41; m 76; c 2. HORTICULTURE. *Educ:* Mich State Univ, BS, 63, MS, 66, PhD(hort), 68. *Prof Exp:* From asst prof to assoc prof, tree fruit physiol, 68-84, PROF HORT/RES POMOLOGIST, NC STATE UNIV, 84- *Honors & Awards:* Raw Prod Res Award, Nat Canners Asn, 70. *Mem:* Am Soc Hort Sci; Am Pomol Soc. *Res:* Applied tree fruit physiology, apple research, growth regulator physiology, environmental control and modification and cultural improvement and efficiency. *Mailing Add:* Dept Hort Sci Mtn Crops Res Sta NC State Univ 2016 Fanning Bridge Rd Fletcher NC 28732

UNRAU, ABRAHAM MARTIN, biochemistry, plant physiology; deceased, see previous edition for last biography

UNRAU, DAVID GEORGE, b Leamington, Ont, July 21, 38; m 62; c 4. BIOCHEMISTRY. *Educ:* Univ Toronto, BSA, 62; Purdue Univ, MS, 65, PhD(biochem), 67. *Prof Exp:* Res scientist, Union Camp Corp, NJ, 67-71; RES CHEMIST, ITT RAYONIER INC, WHIPPANY, 71-, RES GROUP LEADER, 74- *Mem:* Am Chem Soc. *Res:* Carbohydrate modification for industrial uses; flame retardants for cellulosics; viscose; new rayon fiber development. *Mailing Add:* 20 Glenside Dr Budd Lake NJ 07828

UNRUG, RAPHAEL, b Cracow, Poland; US citizen. BASIN ANALYSIS, REGIONAL TECTONICS. *Educ:* Tech Univ Mining & Metall, Cracow, MSc, 57; Jagiellonian Univ, Cracow, PhD(geol), 62, DSc (geodynamics), 68. *Prof Exp:* From asst prof to assoc prof geol, Dept Geol, Jagiellonian Univ, 62-74, from assoc prof to prof geol & chair dept, Inst Geol Sci, 74-79; prof, Dept Geol, Univ Zambia, 80-83; chair dept, 84-89, PROF GEOL, DEPT GEOL SCI, WRIGHT STATE UNIV, 84- *Concurrent Pos:* Sr geologist, Cekop Consults Accra, Ghana, 62-63; vis res scientist, Univ Reading, UK & Univ Paris, 66; consult geologist, Polservice Consult Engrs, Tripoli, Libya, 74-76, Consol Oil Industs, Warsaw, 76-79 & Geoexplorers Int Inc, Denver, 82-84; vis prof, Univ Granada, Spain & Univ Bologna, Italy, 76; proj leader, Int Geol Correlation Prog Proj 288. *Mem:* Geol Soc Am; Am Asn Petrol Geologists. *Res:* Sedimentology, stratigraphy and basin analysis; regional tectonics; economic geology base and precious metals; oil geology. *Mailing Add:* Dept Geol Sci Wright State Univ Dayton OH 45435

UNRUH, HENRY, JR, b Greensburg, Kans, Dec 31, 26; wid; c 2. PHYSICS. *Educ:* Wichita State Univ, AB, 50; Kans State Univ, MS, 52; Case Inst Technol, PhD(physics), 60. *Prof Exp:* Instr physics, Fenn Col, 54-57; asst prof, Colo State Univ, 59-61; assoc prof, 61-72, PROF PHYSICS, WICHITA STATE UNIV, 72- *Mem:* Am Phys Soc; Sigma Xi. *Res:* Magnetic properties of solids; simple liquids; many body problems. *Mailing Add:* Dept Physics Wichita State Univ Wichita KS 67208

UNRUH, JERRY DEAN, b Colorado Springs, Colo, Nov 4, 44; div; c 2. INDUSTRIAL ORGANIC CHEMISTRY. *Educ:* Colo State Univ, BS, 66; Ore State Univ, PhD(org chem), 70. *Prof Exp:* Res chemist, Corpus Christi Tech Ctr, Celanese Chem Co, 70-73, sr res chemist, 73-78, staff chemist, 78-81, res assoc, 81-82, sect leader, 82-86, RES ASSOC, HOECHST-CELANESE CHEM GROUP, 86- *Mem:* Sigma Xi; AAAS; Am Chem Soc. *Res:* Organic free radicals; linear free energy relationships; molecular orbital theory; organometallic chemistry; homogeneous catalysis. *Mailing Add:* 622 Bradshaw Corpus Christi TX 78412-3002

UNRUH, WILLIAM GEORGE, b Winnipeg, Man, Aug 28, 45; m 74; c 1. COSMOLOGY, QUANTUM NOISE. *Educ:* Univ Man, BSc, 67; Princeton Univ, MA, 69, PhD(physics), 71. *Prof Exp:* Nat Res Coun Can fel physics, Birkbeck Col, Univ London, 71-72; Miller fel physics, Miller Inst Basic Res & Univ Calif, Berkeley, 73-74; asst prof appl math, McMaster Univ, 74-76; from asst prof to assoc prof, 76-82, PROF PHYSICS, UNIV BC, 82- *Concurrent Pos:* Fel, Rutherford Mem, Royal Soc Can, 71, Nat Res Coun Teaching, 71-72, Miller Res, Univ Calif, Berkeley, 73-74, Alfred P Sloan Res, Univ BC, 78-80, Steacie, Nat Sci & Eng Res Coun, 84-86; chmn, Theoret Physics Div, Can Asn Physicists, 85-86. *Honors & Awards:* Rutherford Medal, Royal Soc Can, 82 & Rutherford Lectr, Royal Soc London, 85; Hertzberg Medal, Can Asn Physics, 83; Steacie Medal, 84; Gold Medal, BC Sci Coun, 90. *Mem:* Fel Royal Soc Can; Can Asn Physicists. *Res:* Relation between quantum mechanics and gravitation; quantum gravity wave detectors; the early universe. *Mailing Add:* Dept Physics Univ BC 2075 Westbrook Mall Vancouver BC V6T 1Z2 Can

UNSWORTH, BRIAN RUSSELL, b London, Eng, July 30, 37; m 66; c 2. BIOCHEMISTRY. *Educ:* Univ London, BSc, 61, PhD(biochem), 65. *Prof Exp:* Proj assoc biochem, Univ Wis-Madison, 65-67; USPHS fel biol, Univ Calif, San Diego, 67-69; asst prof, 69-76, ASSOC PROF BIOL, MARQUETTE UNIV, 76- *Mem:* Am Col Sports Med. *Res:* Biochemical alterations in muscle with use and disuse; myogenesis in the rat; structural analysis of myosin light and heavy chains in various animal model systems; changes in sarcoplasmic reticulum structure and function during development. *Mailing Add:* Dept Biol Marquette Univ 1515 W Wisconsin Ave Milwaukee WI 53233

UNT, HILLAR, b Tallinn, Estonia, Mar 17, 35; US citizen; m 61. MECHANICAL ENGINEERING. *Educ:* Univ Southern Calif, BS, 58, MS, 60, PhD, 69. *Prof Exp:* Lectr mech eng, Univ Southern Calif, 58-60; asst prof, 60-65, assoc prof & chmn dept, 65-67, PROF MECH ENG, CALIF STATE UNIV, LONG BEACH, 70-, CHMN DEPT, 74- *Mem:* Am Soc Mech Engrs; Am Soc Metals; Am Inst Aeronaut & Astronaut; Acoust Soc Am; Soc Exp Stress Anal. *Res:* Mechanical vibration; bio-engineering; engineering education. *Mailing Add:* Dept Mech Eng Calif State Univ 1250 Bellflower Blvd Long Beach CA 90840

UNTCH, KARL GEORGE, b Cleveland, Ohio, Apr 24, 31; m 53; c 2. ORGANIC CHEMISTRY. *Educ:* Oberlin Col, BA, 53; Univ NDak, MS, 55; Columbia Univ, MA, 57, PhD(org chem), 59. *Prof Exp:* Du Pont teaching fel, Columbia Univ, 57-58; fel org chem, Univ Wis, 58-60; Fundamental Res staff fel, Mellon Inst, 60-66; assoc prof chem, Belfer Grad Sch Sci, Yeshiva Univ, 66-68; dept head, Syntex Res, 68-80, prin scientist, 80-81; instr, Evergreen Valley Col, 82-84, CONSULT CHEM, 81-; Dir, Ctr Chem Res, Dept Chem, Columbia Univ, 88-90; CONSULT, 90- *Honors & Awards:* Alfred P Sloan fel, 67-69. *Mem:* Am Chem Soc. *Res:* Aromaticity; chemistry of unsaturated medium sized ring compounds; synthesis and structure of natural products; synthesis of heterocyclic systems; total synthesis of prostaglandins; synthesis of anti-infammatory and cardiovascular agents; enzyme immunodiagnostics. *Mailing Add:* 7203 Via Carrizo San Jose CA 95135

UNTERBERGER, ROBERT RUPPE, b New York, NY, Apr 27, 21; m 44; c 3. PHYSICS. *Educ:* State Univ NY Col Forestry, Syracuse, BS, 43; Syracuse Univ, BS, 43; Duke Univ, PhD(physics), 50. *Prof Exp:* Lab instr physics, Syracuse Univ, 43; electronics engr, Watson Labs, NJ, 46; res physicist, Chevron Res Co, Standard Oil Co, Calif, 50-52, sr res physicist, 52-54, tech asst mgr, 54-56, res assoc, 56-59, supvr res physicist, 59-65, sr res assoc, 65-68; PROF GEOPHYS, TEX A&M UNIV, 68- *Concurrent Pos:* Electronic engr, White Sands Proving Ground, 46; consult to many salt and potash companies in US, Can, Europe & SAm. *Mem:* Am Phys Soc; Inst Elec & Electronics Engrs; Soc Explor Geophysists; Asn Explor Geophysists. *Res:* Geophysics; acoustics; electronic instrumentation; microwave spectroscopy; structure of molecules; secondary frequency standards for K-band and higher frequencies; electron spin resonance; optical pumping; high sensitivity magnetometry; electromagnetic wave and acoustic wave propagation in rocks; ground probing radar studies. *Mailing Add:* Dept Geophys Tex A&M Univ College Station TX 77843

UNTERHARNSCHEIDT, FRIEDRICH J, b Essen, Ger, July 17, 26; m; c 1. NEUROLOGY, PSYCHIATRY. *Educ:* Univ Munster, MD, 53, Venia Legendi, 61; Ger Bd Neurol & Psychiat, cert, 57. *Prof Exp:* Asst neurol & psychiat, Univ Hosp Neurol & Psychiat, Bonn, 52-57; assoc, Neuropath Inst, Univ Bonn, 57-61; res assoc, Ger Res Inst Psychiat, Max Planck Inst, Munich, 61-66; res prof neuropath & chief div neuropath & exp neurol, Univ Tex Med Br Galveston, 66-72; mem staff, Naval Aerospace Med Res Lab, 72-77; mem staff, 77-80, DIR, REHAB CLIN, BISCHOFSWIESEN, 80-, CONTRACTOR, NEUROSCI, 80- *Concurrent Pos:* Pres, Neurosci, Inc, New Orleans. *Mem:* Ger Neurol Soc; Ger Asn Neuropath; Japanese Soc Neurol & Psychiat; Ger Soc Scientists & Physicians; Am Asn Neuropath; Cerv Spine Res Soc; Scand Soc Neurosurg. *Res:* Mechanics and pathomorphology of central nervous system traumas; virus-induced tumors; neurovirology, especially safety tests of polio and measle vaccines; general and special neuropathology; malformations of the nervous system; diseases of the spinal cord. *Mailing Add:* 3512 Camp New Orleans LA 70115

UNTERMAN, RONALD DAVID, recombinant dna, environmental microbiology, for more information see previous edition

UNTERSTEINER, NORBERT, b Merano, Italy, Feb 24, 26. GLACIOLOGY, RESEARCH ADMINISTRATION. *Educ:* Innsbruck Univ, PhD(geophys), 50. *Prof Exp:* Asst prof meteorol, Univ Vienna, 51-56, res meteorologist, Cent Estab Meteorol & Geodyn, Vienna, Austria, 57-62; res assoc prof, 63-67, PROF ATMOSPHERIC SCI & GEOPHYS, UNIV WASH, 67- *Concurrent Pos:* Docent, Univ Vienna, 61; consult, Rand Corp, Calif, 65-72; mem, Int Comn Polar Meteorol, World Meteorol Orgn, 66-74; mem comt polar res, Nat Acad Sci, 70-77; vpres, Int Comn Snow & Ice, Asn Sci Hydrol, Int Union Geod & Geophys, 61-76; proj dir, Div Marine Resources, 70-78; sci adv polar affairs, Off Naval Res, Washington, DC, 78-80; dir res & develop, Off Ocean Prog, Nat Oceanic & Atmospheric Admin, Washington, DC, 80-81; chmn dept atmospheric sci, Univ Wash, 88- *Honors & Awards:* Austrian Hon Cross Arts & Sci, 60. *Mem:* Am Geophys Union; Ger Polar Soc; Norweg Polar Soc; Int Glaciol Soc. *Res:* Heat and mass budget of glaciers; physical properties of sea ice; sea-air interactions in polar regions; polar climatology. *Mailing Add:* Dept Atmospheric Sci Univ Wash Seattle WA 98195

UNTI, THEODORE WAYNE JOSEPH, b Kenosha, Wis, Mar 11, 31; m 70; c 3. OPTICS, PLASMA PHYSICS. *Educ:* Marquette Univ, BS, 54; Univ Pittsburgh, MS, 60, PhD(gen relativity physics), 64. *Prof Exp:* Sr physicist, Am Optical Co, Pa, 64-66; sr physicist, Jet Propulsion Labs, 66-81; PHYSICIST, HUGHES AIRCRAFT CO, 83- *Concurrent Pos:* Physicist & consult optics, Fairchild Space-Sci Div, Calif, 66; vis prof physics, Nat Univ, Costa Rica, 77. *Res:* Diffraction theory; scattering; statistical and spectral analysis; mathematics peripheral to lens design; computer science; solar wind; shape of magnetosphere. *Mailing Add:* PO Box 8624 La Crescenta CA 91224

UNTRAUER, RAYMOND E(RNEST), b Nevada, Iowa, Feb 9, 26; m 49; c 7. STRUCTURAL ENGINEERING. *Educ:* Iowa State Col, BS, 48; Univ Colo, MS, 51; Univ Ill, PhD(civil eng), 61. *Prof Exp:* Surveyor eng dept, Mo Pac RR, Co, 48-49; instr civil eng, Univ Colo, 49-51; struct designer, C F Braun & Co, Calif, 51-52; from asst prof to assoc prof civil eng, Univ Ark, 52-57; res assoc, Univ Ill, 57-61; assoc prof, Iowa State Univ, 61-65, prof civil eng & dir struct res lab, Eng Res Inst, 65-72; head dept civil eng, Pa State Univ, 72-79, prof, 72-88; RETIRED. *Mem:* Am Soc Civil Engrs; Am Soc Eng Educ; Am Concrete Inst. *Res:* Reinforced and prestressed concrete; structural dynamics; analysis by numerical methods. *Mailing Add:* 510 Sierra Lane State College PA 16801

UNWIN, STEPHEN CHARLES, b Bromley, Eng, Sept 8, 53. RADIO ASTRONOMY. *Educ:* Univ Cambridge, BA, 76, PhD(radio astron), 79. *Hon Degrees:* MA, Univ Cambridge, 80. *Prof Exp:* Researcher, Dept Physics, Univ Cambridge, 76-79; res fel radio astron, Owens Valley Radio Observ, 79-82, MEM PROF STAFF, CALIF INST TECHNOL, 82- *Mem:* Fel Royal Astron Soc; Am Astron Soc; Int Astron Union; Int Radio Sci Union. *Res:* Radio astronomy using interferometric techniques; mapping of compact radio sources using very long baseline interferometric methods; kinematics and spectra of variable sources. *Mailing Add:* Owens Valley Radio Observ Calif Inst Technol Mail Code 105-24 Pasadena CA 91125

UNZ, HILLEL, b Darmstadt, Ger, Aug 15, 29; nat US; m 75; c 3. WAVE PROPAGATION, QUANTUM WAVE MECHANICS. *Educ:* Israel Inst Technol, BS, 53, dipl, 61; Univ Calif, Berkeley, MS, 54, PhD(elec eng), 57. *Prof Exp:* Res asst, Microwave Res Lab, Univ Calif, Berkeley, 53-57; from asst prof to assoc prof elec eng, 57-62, PROF ELEC ENG, UNIV KANS, 62- *Concurrent Pos:* Consult, 58-; NSF fel, Cavendish Lab, Univ Cambridge, 63-64; vis prof, Israel Inst Technol, 64. *Mem:* AAAS. *Res:* Electromagnetic theory; antenna arrays; propagation in the ionosphere; plasma dynamics; moving plasmas; theory of plates; quantum wave mechanics. *Mailing Add:* Dept Elec & Comput Eng Univ Kans Lawrence KS 66045

UNZ, RICHARD F(REDERICK), b Syracuse, NY, Sept 15, 35; m 65; c 1. ENVIRONMENTAL MICROBIOLOGY. *Educ:* Syracuse Univ, BS, 57, MS, 60; Rutgers Univ, PhD(environ sci), 65. *Prof Exp:* Res chemist, Metrop Sanit Dist Greater Chicago, 65-66; from asst prof to assoc prof sanit microbiol, 66-81, PROF ENVIRON MICROBIOL, PA STATE UNIV, 81- *Concurrent Pos:* Ed, J Appl & Environ Microbiol, 91- *Mem:* Am Soc Microbiol; Water Pollution Control Fedn; Am Acad Microbiol; Int Asn Water Pollution Res & Control. *Res:* Microbiological flocculation and zoogloeal bacteria; microbiology of acid mine drainage; ecology and physiology of filamentous microorganisms in waste waters; disinfection of water and wastewater; wetland microbiology and geochemistry. *Mailing Add:* Dept Civil Eng Pa State Univ 116 Sackett Bldg University Park PA 16802

UNZICKER, JOHN DUANE, b Harvey, Ill, May 8, 38; m 71; c 1. ENTOMOLOGY. *Educ:* Univ Ill, BS, 62, MS, 63, PhD(entom), 66. *Prof Exp:* Res assoc Trichoptera, Ill Nat Hist Surv, 66-67, asst taxonomist, 67-76, assoc taxonomist, 76-85; asst prof zool, 72-75, ASST PROF AGR ENTOM, UNIV ILL, URBANA, 78-; AFFIL RES SCIENTIST, ILL NATURAL HIST SURV, 86- *Concurrent Pos:* NSF grant reviewer, 68-; consult, St John's Hosp Regional Poison Resource Ctr, Springfield, 78- *Mem:* Sigma Xi; Entom Soc Am. *Res:* Biosystematics of aquatic insects. *Mailing Add:* Ctr Biodiversity 93 Nat Resources Bldg 607 E Peabody St Champaign IL 61820

UOTILA, URHO A(NTTI KALEVI), b Poytya, Finland, Feb 22, 23; nat US; m 49; c 6. GEOPHYSICS. *Educ:* Finland Inst Tech, BS, 46, MS, 49; Ohio State Univ, PhD(geod), 59. *Prof Exp:* Surveyor & geodesist, Finnish Govt, 44-46 & 46-51; res asst, Ohio State Univ, 52-53, res assoc, 53-58, lectr geod, 55-57, from asst prof to prof, 59-88, chmn dept geod sci, 64-84, res supvr, 59-88, EMER PROF GEOD & CHMN, OHIO STATE UNIV, 89- *Concurrent Pos:* Geodesist, Swedish Govt, 46; mem, Solar Eclipse Exped Greenland, 54; mem, pres spec study group 5.30, 67-71 & sect V, Int Asn Geod, 71-75; mem geod adv panel, Nat Acad Sci to US Coast & Geod Surv, 64-66; mem geod & cartog working group, Space Sci Steering Comt, NASA, 65-67, geod & cartog adv subcomt, 67-72; vis prof, Tech Univ Berlin, 67 & Univ Fed Parana, Brasil, 73-75, 77 & 81; mem bd dirs, Int Gravity Bur, France, 75-83; mem bd trustees, Univ Space Res Asn, 73-75; mem ad hoc comt NAm datum, div earth sci, Nat Acad Sci-Nat Acad Eng, 68-70; mem comt geod, Nat Acad Sci, 75-78. *Honors & Awards:* Kaarina & W A Heiskanen Award, 62; Apollo Achievement Award, NASA, 69; Award for Except Serv, Am Cong Surv & Mapping, 80; Distinguished Serv Award, Survr Inst Sri Lanka. *Mem:* Fel Am Geophys Union (vpres geod sect, 64-68, pres, 68-70); fel Am Cong Surv & Mapping (vpres, 77-78, pres-elect, 78-79, pres, 79-80); Am Soc Photogram; Can Inst Surv; foreign mem Finnish Nat Acad Sci; Am Asn Geod Surv (pres, 84-86). *Res:* Geometric and physical geodesy and statistical analysis of data. *Mailing Add:* Dept Geod Sci & Surv Ohio State Univ 1958 Neil Ave Columbus OH 43210-1247

UPADHYAY, JAGDISH M, b Jambusar, Gujerat, India, July 2, 31; m 63; c 1. MICROBIOLOGY, BIOCHEMISTRY. *Educ:* Gujerat Univ, India, BPharm, 51; Univ Mich, MS, 57; Wash State Univ, PhD(bact), 63. *Prof Exp:* Chemist, Sarabhai Chem, India, 51-55; grant, Univ Tex, 63-65; asst prof microbiol, 65-67, ASSOC PROF MICROBIOL, LOYOLA UNIV, LA, 68- *Concurrent Pos:* NIH grants, Schlieder Found, 70-72. *Mem:* Am Soc Microbiol; Brit Soc Gen Microbiol; Sigma Xi. *Res:* Growth and metabolism of psychrophilic microorganisms and soil amebas; lytic enzymes; cell-wall composition; thermophilic microorganisms; carotenoid pigments. *Mailing Add:* Dept Microbiol Loyola Univ 6363 St Charles Ave New Orleans LA 70118

UPADHYAYA, BELLE RAGHAVENDRA, b Mangalore, India; US citizen; m 79; c 1. NEURAL NETWORKS APPLICATIONS, SIGNAL PROCESSING TECHNOLOGY. *Educ:* Regional Eng Co, Suratkal, India, BE, 65; Univ Toronto, MASc, 68; Univ Calif, San Diego, PhD(syst sci), 75. *Prof Exp:* From asst prof to assoc prof, 78-89, PROF NUCLEAR ENG, UNIV TENN, KNOXVILLE, 89- *Concurrent Pos:* Adj res assoc, Oak Ridge Nat Lab. *Mem:* Inst Elec & Electronics Engrs; Am Nuclear Soc; Instrument Soc Am; Am Soc Eng Educ; Sigma Xi; Am Phys Soc. *Res:* Digital signal processing and system dynamic analysis with application to power plant monitoring and diagnotics. *Mailing Add:* 1223 Hamstead Ct Knoxville TN 37922

UPATNIEKS, JURIS, b Riga, Latvia, May 7, 36; US citizen; m 68; c 2. COHERENT OPTICS. *Educ:* Univ Akron, BS, 60; Univ Mich, MS, 65. *Prof Exp:* Instr microwaves, Ord Sch, US Army, 61-62; from res asst to res engr holography, Willow Run Labs, Univ Mich, 60-72; RES ENGR COHERENT OPTICS, ENVIRON RES INST MICH, 73- *Concurrent Pos:* Consult, var pvt & govt orgn, 66-; adj assoc prof, dept elec eng & comput sci, Univ Mich, 74- *Honors & Awards:* Robert Gordown Mem Award, Soc Photographic Instr Eng, 65; R W Wood Prize, Optical Soc Am, 75; Holley Medal Am Soc Mech Engrs, 76. *Mem:* Optical Soc Am; Inst Elec & Electronics Engrs; Soc Photo-Optical Instrumentation Engrs. *Res:* Holography; optical data processing; holographic optical elements. *Mailing Add:* Environ Res Inst Mich PO Box 8618 Ann Arbor MI 48107

UPCHURCH, JONATHAN EVERETT, b Chicago, Ill, Jan 2, 51; m 71; c 1. TRANSPORTATION ENGINEERING. *Educ:* Univ Ill, BS, 71, MS, 75; Univ Md, PhD(civil eng), 82. *Prof Exp:* Transp engr, Harland Bartholomew & Assocs, 72-76; dir tech affairs, Inst Transp Engrs, Washington, DC, 76-80; res engr, Off Res, Fed Hwy Admin, Washington, DC, 81-82; asst prof, 82-91, ASSOC PROF CIVIL ENG, COL ENG & APPL SCI, ARIZ STATE UNIV, 86- *Concurrent Pos:* Consult, Am Asn State Hwy & Transp Officials, 80-88; exec secy, Nat Comt Uniform Traffic Control Devices, 79-88. *Mem:* Transp Res Bd; Inst Transp Engrs (int pres, 91); Am Soc Civil Engrs. *Res:* Traffic engineering; traffic operations; development of standards for traffic control devices which result in the most economical operation for the motoring public. *Mailing Add:* Dept Civil Eng Ariz State Univ Tempe AZ 85287-5306

UPCHURCH, ROBERT PHILLIP, b Raleigh, NC, Feb 9, 28; m 48; c 3. PLANT PHYSIOLOGY, WEED SCIENCE. *Educ:* BS & MS, NC State Univ, 49; Univ Calif, PhD(plant physiol), 53. *Prof Exp:* Instr crop sci, NC State Univ, 49-51, from asst prof to prof, 53-65; sr res group leader, Monsanto Co, 65-70, res mgr, 70-73, mgr res, 73-75; head, Dept Plant Sci, 75-81, dir agr develop & assoc dir, Agr Exp Sta, 81-83, assoc dean, 83-88, dir agr alumni affairs, 88-90, DIR AGR DEVELOP, UNIV ARIZ, 90- *Concurrent Pos:* Consult, Shell Develop Co, 62-65 & Eli Lilly & Co, 79-88; mem, Weeds Subcomt, Nat Acad Sci, 64-68; ed, Southern Weed Conf, 66-69; adv, Am Coun & Health, 83-; dir, Boyce Thompson Southwest Arboretum, 85- *Honors & Awards:* Sigma Xi Res Award, NC State Univ, 63. *Mem:* Am Soc Plant Physiol; Am Soc Agron; Crop Sci Soc Am; fel Weed Sci Soc Am (pres, 72-); Plant Growth Regulator Soc (chmn, 72). *Res:* Response of plants to phytoactive chemicals and the influence of soil and climate factors on the expression of such responses. *Mailing Add:* Col Agr Univ Ariz Tucson AZ 85721

UPCHURCH, SAM BAYLISS, b Murfreesboro, Tenn, June 30, 41; m 64; c 2. GROUND-WATER CHEMISTRY, SEDIMENTOLOGY. *Educ:* Vanderbilt Univ, AB, 63; Northwestern Univ, Evanston, MS, 66, PhD(geol), 70. *Prof Exp:* Resident in res marine geol, Northwestern Univ, Evanston, 67-68; res phys scientist chem limnol, Lake Surv Ctr, Nat Oceanic & Atmospheric Admin, 68-71; asst prof geol, Mich State Univ, 71-74; assoc prof, 74-80, PROF GEOL & CHMN DEPT, UNIV SFLA, 81- *Concurrent Pos:* Mem limnol work group, Great Lakes Basin Comn, 68-75, mem bd tech adv, 69-75; consult hydrol & geoarcheol, 73-; mem, Fla Bd Prof Geologists, 88-91. *Mem:* Soc Econ Paleont & Mineral; fel Geol Soc Am; Am Water Resources Asn; Am Asn Petrol Geologists; Southeastern Geol Soc (vpres, 79-80, pres, 80-81); Nat Water Well Asn. *Res:* Trace element-sediment interaction; ground-water chemistry; carbonate sediment genesis and diagenesis; chert petrology; mathematical geology; geohydrology; land-use planning. *Mailing Add:* Dept Geol Univ SFla Tampa FL 33620

UPDEGRAFF, DAVID MAULE, b Woodstock, NY, Dec 19, 17; m 43; c 3. MICROBIOLOGY. *Educ:* Univ Calif, Los Angeles, AB, 41; Univ Calif, PhD(microbiol), 47. *Prof Exp:* Res assoc & actg dir, Am Petrol Inst Res Proj, Scripps Inst, Univ Calif, 46; sr res chemist, Field Res Labs, Magnolia Petrol Co, 47-50, sr res technologist, 50-55; res microbiologist, Cent Res Dept, Minn Mining & Mfg Co, Minn, 55-68; head microbiol sect, Chem Div, Denver Res Inst, Univ Denver, 68-72; head basic res dept, Cawthron Inst, 72-75; vpres, Resource Industs Int, Ltd, 75-77; prof chem & geochem, 77-87, RES PROF, COLO SCH MINES, 88- *Res:* Biochemistry of carotenoid pigments; bacterial physiology; marine and petroleum microbiology; applied microbiology; fermentations; ecology of water pollution and waste treatment; bioremediation. *Mailing Add:* Dept Chem & Geochem Colo Sch Mines Golden CO 80401

UPDEGROVE, LOUIS B, b Kingsville, Tex, Sept 10, 28; m 52; c 2. CHEMICAL ENGINEERING. *Educ:* Mass Inst Technol, SB, 53. *Prof Exp:* Chem engr, Chemstrand Corp, Fla, 53-56; res engr, Sci Labs, Ford Motor Co, 56-57; lab develop supvr, Kordite Corp, NY, 57-62; tech dir, Standard Packaging Corp, 62-65; process develop mgr, Alcolac Chem Corp, 65; vpres & mgr res & develop, Vogt Mfg Corp, NY, 65-69; vpres eng, J H Day Co, Cincinnati, 69-77; PRES, TYLER SCOTT INC, TERRACE PARK, 77- *Mem:* Am Chem Soc; Soc Plastics Engrs. *Res:* Applied plastics research and development; packaging; coatings; emulsion polymers; polymer processing; solids drying; pigment dispersion; research management. *Mailing Add:* 313 Rugby Ave Terrace Park OH 45174

UPDIKE, OTIS L(EE), JR, b Roanoke, Va, Feb 12, 20; m 45; c 4. BIOMEDICAL & CHEMICAL ENGINEERING. *Educ:* Univ Va, BChE, 41; Univ Ill, PhD(chem eng), 44. *Prof Exp:* Asst, Eng Exp Sta, Univ Ill, 41-44; chem engr res dept, Westvaco Chlorine Prod Corp, 44-46; assoc prof chem eng, 46-60, res engr, Eng Exp Sta, 51-59, PROF CHEM ENG & MEM PARTIC FAC, RES LABS, ENG SCI, UNIV VA, 60-, PROF CHEM & BIOMED ENG, 67- *Concurrent Pos:* Consult, Philip Morris & Co, 52-59; tech consult, US Naval Air Missile Test Ctr, 54-55; vpres, Jefferson Res Labs, 55-58; NSF sci fac fel & vis assoc, Calif Inst Technol, 59-60; chmn, Joint Automatic Control Conf, 69. *Mem:* AAAS; Am Inst Chem Engrs; Biomed Eng Soc; Inst Elec & Electronics Engrs; Instrument Soc Am. *Res:* Instrumentation and automatic control; computer applications in process and biomedical engineering; dynamics of process, physiological and instrumentation systems; data acquisition and interpretation. *Mailing Add:* Box 377 Med Ctr Univ Va Charlottesville VA 22908

UPESLACIS, JANIS, b Bad-Rothenfelde, Ger, Jan 12, 46; US citizen; m 68; c 2. PHARMACEUTICAL CHEMISTRY. *Educ:* Univ Nebr-Lincoln, BS, 67; Harvard Univ, MA, 69, PhD(org chem), 75. *Prof Exp:* Asst nuclear physics, Walter Reed Army Med Ctr, US Army, 69-71; res chemist, Lederle Labs, Am Cyanamid Co, 75-80, group leader, 81-86, proj mgr, biotechnol chem, 86-89, DEPT HEAD, ONCOL & IMMUNOL RES SECT, LEDERLE LABS, AM CYNAMID CO, 89- *Mem:* Am Chem Soc. *Res:* Synthetic applications of carbohydrates; antiatherogenic agents; inhibitors of complement-mediated diseases; antineoplastic agents; synthetic vaccines; monoclonal antibodies as carriers of drugs. *Mailing Add:* Lederle Labs Pearl River NY 10965

UPGREN, ARTHUR REINHOLD, JR, b Minneapolis, Minn, Feb 21, 33; m 67; c 1. ASTRONOMY. *Educ:* Univ Minn, BA, 55; Univ Mich, MS, 58; Case Inst Technol, PhD(astron), 61. *Prof Exp:* Res assoc astron, Swarthmore Col, 61-63; astronr, US Naval Observ, 63-66; asst prof astron, 66-73, chmn dept, 68-86, adj prof, 66-73, dir, Van Vleck Observ, 73-, assoc prof astron, 73-81, JOHN MONROE VAN VLECK PROF ASTRON, WESLEYAN UNIV, 81- *Concurrent Pos:* Vis lectr, Univ Md, 64-66, George Washington Univ, 65-66 & Yale Univ, 67-68; vpres & exec officer, Fund Astrophys Res, Inc, 72-; pres, Comn 24, Int Astron Union, 85-88; adj prof, Univ Fla, 84-; NSF grant, 67-, Am Philos Soc grant, 62-; chmn, Sci Orgn Comt Symp, "Calibration Stellar Ages," 88. *Mem:* Am Astron Soc; Royal Astron Soc; Int Astron Union. *Res:* Galactic structure; photographic astrometry. *Mailing Add:* Dept Astron Van Vleck Observ Wesleyan Univ Middletown CT 06457

UPHAM, ROY HERBERT, b Boston, Mass, Feb 9, 20; m 45; c 8. ORGANIC CHEMISTRY. *Educ:* Boston Col, BS, 41, MS, 48. *Prof Exp:* Chemist, Howe & French, Inc, 41-42, Cities Serv Oil Co, 42, Rock Island Arsenal, 44-45 & E I du Pont de Nemours & Co, Inc, 47-49; prof chem, St Anselm Col, 49-86, chmn dept, 82-86; RETIRED. *Mem:* Am Chem Soc. *Res:* Reaction mechanisms; teaching methods; computer applications. *Mailing Add:* 735 Montgomery St Manchester NH 03102-3027

UPHAM, ROY WALTER, b Ogden, Kans, Apr 11, 20; m 69. VETERINARY MEDICINE, FOOD TECHNOLOGY. *Educ:* Kans State Univ, DVM, 43; Mass Inst Technol, MS, 60; Am Bd Vet Pub Health, dipl. *Prof Exp:* Instr food technol, US Army Med Serv Sch, 54-56, proj off radiation of foods prog, US Army, 56-58, mil adv, Food Prog for Vietnam, 62-63, chief lab br food testing, Defense Personnel Supply Command, 63-65, chief standardization br, Mil Specifications, Natick Army Lab, 65-66; chief, Div Food, Drugs & Dairies, Regulatory Agency, Ill Dept Pub Health, Springfield, 66-83. *Mem:* Inst Food Technologists; Asn Food & Drug Officials (pres, 81-82). *Res:* Application of controlled food processing to safeguard and protect public health. *Mailing Add:* RR 2 Rochester IL 62563

UPHOFF, DELTA EMMA, b Brooklyn, NY, Jan 23, 22. GENETICS. *Educ:* Russell Sage Col, AB, 44; Univ Rochester, MS, 47. *Hon Degrees:* DSc, Russell Sage Col, 82. *Prof Exp:* Asst radiation genetics, Manhattan Proj, Rochester, 46-47; zoologist, Radiation Lab, Univ Calif, 47-48; geneticist, Mound Lab, Monsanto Chem Co, Ohio, 48-49; RES BIOLOGIST, NAT CANCER INST, 49- *Mem:* Fel AAAS; Genetics Soc Am; Radiation Res Soc; Am Genetic Asn; Am Asn Cancer Res; Int Soc Exp Hemat; Transplantation Soc. *Res:* Immunogenetics of bone marrow transplantations; radiation biology. *Mailing Add:* NIH 10-6B18 Nat Cancer Inst 9000 Wisconsin Ave Bethesda MD 20892

UPHOLT, WILLIAM BOYCE, b Orlando, Fla, Sept 14, 43; m 80; c 2. MOLECULAR BIOLOGY. *Educ:* Pomona Col, BA, 65; Calif Inst Technol, PhD(chem), 71. *Prof Exp:* Res fel molecular biol, Damon Runyon Mem Fund Cancer Res, Biochem Lab, Univ Amsterdam, 71-73; res fel, Dept Embryol, Carnegie Inst Wash, Baltimore, 73-75; res assoc molecular biol, Dept Pediat & Biochem, Univ Chicago, 75-85; assoc prof, 85-90, PROF, DEPT BIOSTRUCT & FUNCTION, HEALTH CTR, UNIV CONN, 90- *Res:* Physical chemistry of nucleic acids, organization of genetic material, developmental biology, control of gene expression during chick limb cartilage differentiation; type II collagen gene structure and regulation. *Mailing Add:* Dept Biostruct & Function Univ Conn Health Ctr Farmington CT 06030

UPMEIER, HARALD, b Mainz, Ger, Dec 29, 50. FUNCTIONAL ANALYSIS, COMPLEX ANALYSIS. *Educ:* Univ Tubingen, PhD(math), 75. *Prof Exp:* Asst prof math, Univ Pa, 82-84; assoc prof math, 84-86, PROF MATH, UNIV KANS, 86- *Mailing Add:* Dept Math Univ Kans Lawrence KS 66045-2142

UPPULURI, V R RAO, b Machilipattanam, India, Feb 22, 31; m 60; c 1. MATHEMATICAL STATISTICS. *Educ:* Andhra Univ, India, MA, 54; Ind Univ, Bloomington, PhD(math), 63. *Prof Exp:* Res asst math & statist, Tata Inst Fundamental Res, 54-57; res asst math, Ind Univ, Bloomington, 57-61; res assoc statist, Mich State Univ, 61-62, biophys & statist, 62-63; sr math statistician, Oak Ridge Nat Lab, 63-74; mem staff, Nuclear Div, Union Carbide Corp, 74-77, SR RES STAFF MEM, UNION CARBIDE NUCLEAR DIV, 77- *Concurrent Pos:* Lectr, Oak Ridge Traveling Lect Prog, Oak Ridge Assoc Univs & Oak Ridge Nat Lab, 64-; sr engr & scientist, Douglas Aircraft Co, 66; vis lectr prog, Comt Statist Southern Regional Educ Bd, 67-; adj prof, Univ Tenn, 67-; consult, Syst Develop Corp, 67- & med div, Oak Ridge Assoc Univ, 67-; vis prof, Univ Sao Paulo, 70 & Univ Minn, 71. *Mem:* Fel AAAS; fel Am Statist Asn; Am Math Soc; Sigma Xi. *Res:* Probability; statistics; stochastic approach for a better understanding of the structure of physical and natural phenomena; limit theorems in random difference equations and applications of probability. *Mailing Add:* 130 Indian Lane Oak Ridge TN 37830

U'PRICHARD, DAVID C, b Rothesay, Scotland, May 27, 48; m 85; c 2. NEUROPHARMACOLOGY, NEUROCHEMISTRY. *Educ:* Univ Glasgow, BSc, 70; Univ Kans, MS, 74, PhD(pharmacol), 75. *Prof Exp:* Res scientist biochem, UK Ministry of Technol Lab, Torry, Scotland, 68; res scientist pharmacol, Ethicon, Gmbh, Hamburg, Ger, 70; res assoc, Johns

Hopkins Univ, 75-78; asst prof pharmacol, Northwestern Univ, 78-80, assoc prof pharmacol, neurobiol & physiol, 81-83; SR VPRES & SCI DIR, NOVA PHARMACEUT CORP, 83- *Concurrent Pos:* NIH fel, USPHS, 75-77. *Mem:* AAAS; Brit Asn Advan Sci; Soc Neurosci; Europ Neurosci Asn; Brit Pharmacol Soc; Am Soc Pharmacol Exp Therapeut; Am Col Neuropsychopharmacol. *Res:* Sympathetic nervous system and central catecholaminergic neurotransmission mechanisms; identification, characterization and isolation of catecholaminergic receptors; cellular mechanism of action of psychoactive drugs and antihypertensive agents; endogenous opioid peptides and receptors in cultured cells. *Mailing Add:* ICI Pharmaceut Group ICI Americas Wilmington DE 19897

UPSON, DAN W, b Hutchinson, Kans, July 30, 29; m 59; c 3. PHARMACOLOGY, PHYSIOLOGY. *Educ:* Kans State Univ, DVM, 52, MS, 62, PhD(physiol), 69. *Prof Exp:* Vet, Pvt Pract, 52-59; instr pharmacol & physiol, 59-69, assoc prof pharmacol, 69-73, asst dean, 72-73, PROF PHARMACOL, COL VET MED, KANS STATE UNIV, 73-, DIR TEACHING RESOURCES, 75- *Concurrent Pos:* Consult, Tevcon Ind Inc, 70- *Mem:* Am Vet Med Asn; Am Acad Vet Pharmacol & Therapeut; Am Acad Vet Consult. *Res:* Veterinary pharmacology and clinical pharmacology. *Mailing Add:* Dept Animal Sci Call Hall Kans State Univ Manhattan KS 66506

UPTHEGROVE, W(ILLIAM) R(EID), b Ann Arbor, Mich, Nov 10, 28; m 53; c 5. METALLURGICAL ENGINEERING, ENGINEERING EDUCATION. *Educ:* Univ Mich, BSE, 50, MSE, 54, PhD(metall eng), 57. *Prof Exp:* From asst prof to assoc prof metall eng, Univ Okla, 56-62, chmn sch, 56-62; sect leader powder metall res & develop, Res Labs, Int Nickel Co, Inc, NJ, 62-64; prof mech eng & chmn dept, Univ Tex, Austin, 64-70; prof metall & mech eng & dean, Col Eng, 70-81, REGENTS PROF ENG, UNIV OKLA, 81- *Concurrent Pos:* Indust consult, 64-; consult educ planning, US Overseas Inst, 70- & failure anal & prod liability. *Mem:* Fel Am Soc Metals; Am Soc Mech Engrs; Nat Soc Prof Engrs; Am Soc Eng Educ; fel AAAS; Soc Antomotive Engrs. *Res:* Powder metallurgy; diffusion; design and materials properties. *Mailing Add:* Aerospace-Mech-Nuclear-Fh 212 Univ Okla 660 Parington Oval Norman OK 73019

UPTON, ARTHUR CANFIELD, b Ann Arbor, Mich, Feb 27, 23; m 46; c 3. EXPERIMENTAL PATHOLOGY. *Educ:* Univ Mich, BA, 44, MD, 46. *Prof Exp:* Intern, Univ Hosp, Univ Mich, 47, resident path, Med Sch, 48-50, instr, 50-51; pathologist, Biol Div, Oak Ridge Nat Lab, 51-54, chief, Path-Physiol Sect, 54-69; chmn dept path, Health Sci Ctr, 69-70, dean, Sch Basic Health Sci, 70-75, prof path, State Univ NY, Stony Brook, 69-77, attend pathologist, Med Dept, Brookhaven Nat Lab, 69-77, dir, Nat Cancer Inst, NIH, 77-79; dir, Inst Environ Med, Med Sch, NY Univ, 80-91; RETIRED. *Concurrent Pos:* Mem comt biol effectiveness of radiation & long term effects of radiation, Nat Acad Sci, Nat Res Coun, 57-; Ciba Found Lectr, 59, Failla Lectr, Radiation Res Soc, 77, IBM- Princess Takamatsu Cancer Res Found, 81, Failla Mem Lectr, Health Physics Soc, 83, Fourth Ann Martin Schneider Mem Lectr, Univ Tex Med Br, 84; rep, USA Nat Comt Int Union against Cancer, 72-; mem, Int Comn Radiol Protection, 74-; mem comt, Int Comn Radiol Protection, 64-; mem, Nat Coun Radiation Protection & Measurements, 65-, adv comt, Ctr Human Radiobiol, Argonne Nat Lab, 72-76, sci adv group, US-Japan Coop Cancer Res Prog, 74-77, sci adv bd, Nat Ctr for Toxological Res, 74-77, sci coun, Int Agency Res Cancer, WHO, 77-77; chmn, Health Res Coun, NY State, 81-; Sigma Xi nat lectr, 90-91. *Honors & Awards:* Lawrence Award, 65; Comfort Crookshank Award for Cancer Res, 78; Claude M Fuess Award, 80; Sarah L Poilley Award, 83; CHUMS Physician of the Yr Award, 85; Basic Cell Res in Cytol Lectureship Awards, 85; Fred W Stewart Award, 86; Ramazzini Award, 86. *Mem:* Inst Med Nat Acad Sci; AAAS; Sigma Xi; Radiation Res Soc (pres, 65-66); Am Soc Exp Path (pres, 67-68); Am Asn Cancer Res (pres, 63-64); hon mem Peruvian Oncol Soc; hon mem Japan Cancer Asn; fel NY Acad Sci. *Res:* Pathology of radiation injury and endocrine glands; cancer; carcinogenesis; experimental leukemia; aging. *Mailing Add:* Inst Environ Med NY Univ Med Ctr 550 First Ave New York NY 10016

UPTON, G VIRGINIA, b New Haven, Conn, Oct 17, 29; wid; c 3. PHYSIOLOGY, BIOCHEMISTRY. *Educ:* Albertus Magnus Col, BA, 51; Yale Univ, MS, 61, PhD(physiol), 64. *Prof Exp:* NIH fel peptide chem, Yale Univ, 63-66, chief endocrine & polypeptide lab, Vet Admin Hosp, Yale Univ, 66-74, sr res assoc med, Sch Med, 71-74; assoc dir clin res, 74-78, dir Wyeth Int Ltd, 84-86, ASSOC MED, WYETH INT LTD, 78-, DIR CLIN DEVELOP, WYETH AYERST RES, 88- *Concurrent Pos:* Asst prof comp endocrinol, Eve Div, South Conn State Col, 66-67; assoc prof med, Med Col, PA, 82- *Mem:* Am Asn Cancer Res; Endocrine Soc; NY Acad Sci; Int Soc Neuroendocrinol; Am Physiol Soc; Am Fertil Soc. *Res:* Neuroendocrinology; hypothalamic-pituitary-adrenal relationships; isolation of pituitary peptides and tumor peptides with hormonal activity; relationship between endocrine disorders and hypothalamic dysfunction. *Mailing Add:* 8401 Roosevelt Blvd No C22 Philadelphia PA 19152

UPTON, RONALD P, b Boston, Mass, May 6, 41; m 65; c 4. ANALYTICAL CHEMISTRY. *Educ:* New Bedford Inst Tech, BS, 63; Univ Del, PhD(anal chem), 68. *Prof Exp:* Staff chemist, Res Lab, 67-70, mgr qual control, 70-84, dir, Qual Assurance Miles Labs, 84-85; mgr anal chem, Pennwalt Corp, 88-90; DIR, QUAL ASSURANCE/QUAL CONTROL, NASKA PHARMACOL, 90- *Mem:* Am Chem Soc; Sigma Xi. *Res:* Quality control; instrumental analysis; infrared, ultraviolet, atomic absorption and nuclear magnetic spectroscopies; gas and high pressure liquid chromatography, in vitro diagnostics and pharmaceutical analysis. *Mailing Add:* 1587 Nottingham Dr Newton NC 28658

UPTON, THOMAS HALLWORTH, b Dallas, Tex, Apr 14, 52; m 78; c 2. CATALYSIS, SURFACE SCIENCE. *Educ:* Stanford Univ, BS, 74; Calif Inst Technol, PhD(theoret), 80. *Prof Exp:* Res chemist, 80-90, SECT HEAD, EXXON RES & ENG CORP, 90- *Mem:* Am Chem Soc. *Res:* Theoretical investigations of mechanisms in homogeneous, heterogeneous catalysis and surface science. *Mailing Add:* Exxon Res & Eng Prod Res Div PO Box 51 Linden NJ 07036

URALIL, FRANCIS STEPHEN, b Kerala, India, June 3, 50; m 80; c 1. POLYMER PHYSICS. *Educ:* Univ Kerala, BS, 69; Marquette Univ, MS, 72; Univ Del, PhD(physics), 76. *Prof Exp:* Res assoc polymer physics, Case Inst Technol, 76-78 & physics, Schlumberger, 78-79; SCIENTIST & GROUP LEADER POLYMER PHYSICS, BATTELLE INST, 79- *Mem:* Am Phys Soc; Am Chem Soc. *Res:* Structure-property relationships; dynamic mechanical behavior of materials; fracture and fatigue studies; dielectrics and composites. *Mailing Add:* 4503 Mobile Dr Columbus OH 43220

URANO, MUNEYASU, b Osaka, Japan, Apr 21, 36; US citizen; m 63; c 2. HYPERTHERMIA, RADIATION DOSE FRACTIONATION. *Educ:* Kyoto Prefectural Univ, MD, 61, PhD(radiation biol), 68. *Prof Exp:* Res fel radiation biol, M D Anderson Hosp, 66-68; asst prof radiol, Kyoto Prefectural Univ Med, 68-70; sr researcher radiol biol, Nat Inst Radiol Sci, 70-77; from asst prof to assoc prof radiation biol, Mass Gen Hosp, 77-89; PROF RADIATION BIOL, COL MED, UNIV KY, 89- *Mem:* Radiation Res Soc; Am Soc Therapeut Radiol & Oncol; NAm Hyperthermia Soc; AAAS; Am Asn Cancer Res. *Res:* Effect of hyperthermia given alone or in combination with radiation and/or chemotherapeutic agents in animal tumor and normal tissues and in cultured cell. *Mailing Add:* 2205 Broadhead Pl Lexington KY 40515

URBACH, FREDERICK, b Vienna, Austria, Sept 6, 22; nat US; wid; c 3. DERMATOLOGY. *Educ:* Univ Pa, BS, 43; Jefferson Med Col, MD, 46; Am Bd Dermat, dipl, 53. *Hon Degrees:* Dr Med(hon), Univ Güttingen, WGer, 87. *Prof Exp:* Asst instr dermat, Med Sch, Univ Pa, 49-50, instr, 50-52, assoc, 52-54; chief dermat serv, Roswell Park Mem Inst, 54-58; assoc prof, Temple Univ, 58-60, prof res, 60-67, prof dermat & chmn dept, Sch Med, 67-89; med dir, Skin & Cancer Hosp, 67-89, EMER PROF DERMAT, PHILADELPHIA, 89- *Concurrent Pos:* Fel, Hosp Univ Pa, 49-52; Damon Runyon res fel clin cancer, Univ Pa, 51-53; asst vis physician, Philadelphia Gen Hosp, 51-54; mem, Int Cong Dermat, London, 52; from asst med dir to assoc med dir, Skin & Cancer Hosp, Temple Univ, 58-67; mem, US Nat Comt Photobiol, Nat Res Coun, 73-80; dir, Ctr Photobiol, 77-89. *Honors & Awards:* Hellerstrom Medal, Swed Derm Soc, 78; Ritter Medal, Polish Derm Soc, 79, Germ Derm Soc, 80, Austrian Dermat Soc, 84. *Mem:* Fel AAAS; Soc Invest Dermat; Soc Exp Biol & Med; Am Asn Cancer Res; Am Soc Photobiol (pres, 77); Asn Int Photobiol (pres, 80-84). *Res:* Blood supply of cancer; biologic effects of ultraviolet radiation; photobiology; epidemiology of cancer. *Mailing Add:* Temple Med Pract 220 Commerce Dr Ft Washington PA 19054

URBACH, FREDERICK LEWIS, b New Castle, Pa, Nov 21, 38; m 60; c 2. INORGANIC CHEMISTRY. *Educ:* Pa State Univ, University Park, BS, 60; Mich State Univ, PhD(chem), 64. *Prof Exp:* Res assoc & fel, Ohio State Univ, 64-66; asst prof, 66-74, assoc prof, 74-80, PROF CHEM, CASE WESTERN RESERVE UNIV, 80- *Mem:* Am Chem Soc. *Res:* Chemistry of metal chelates containing multidentate ligands; stereochemistry and optical activity; role of transition metal ions in biological processes. *Mailing Add:* Dept Chem Case Western Reserve Univ Cleveland OH 44106

URBACH, HERMAN B, b New York, NY, Jan 19, 23; m 56; c 2. THERMODYNAMICS OF HEAT ENGINES, ELECTROCHEMISTRY. *Educ:* Univ Ind, AB, 48; Columbia Univ, MA, 50; Case Western Reserve Univ, PhD, 54; George Washington Univ, MS, 76. *Prof Exp:* Group leader, Olin Mathieson Chem Corp, 53-59; res scientist, Res Lab, United Aircraft Corp, 59-65, consult electrochem, Pratt & Whitney Aircraft Div, 60-63; SCI STAFF ASST POWER & PROPULSION, POWER SYSTS DIV, US NAVAL RES & DEVELOP CTR, ANNAPOLIS, 65- *Concurrent Pos:* Guest prof mech eng, US Naval Acad, 81-82. *Mem:* Am Inst Aeronaut & Astronaut; Am Soc Mech Engrs; NY Acad Sci; Sigma Xi; Am Chem Soc; Electrochem Soc. *Res:* Kinetics of the oxygen electrode; fuel cells; theory of porous electrodes; ozone and plasma kinetics; atomic reactions; boranes; magnetohydrodynamics; biphase turbines; gas and steam turbines. *Mailing Add:* Power Systs Div Code 272T David Taylor Res Ctr Annapolis MD 21402

URBACH, JOHN C, b Vienna, Austria, Feb 18, 34; US citizen; m 56; c 3. IMAGE SCIENCE, ELECTRONIC IMAGING. *Educ:* Univ Rochester, BS, 55, PhD(optics), 62; Mass Inst Technol, MS, 57. *Prof Exp:* Assoc physicist, Int Bus Mach Corp, 57-58; NATO fel sci, Royal Inst Tech, Sweden, 61-62; scientist, Xerox Corp, NY, 63-66, sr scientist, 66-67, mgr optical & imaging anal br, 67-68, mgr optical sci br, 68-70, mgr optical sci area, 70-75, mgr Optical Sci Lab, Palo Alto Res Ctr, 75-85, mgr Color Systs Technol Res Group, 85-87; VPRES, DEVELOP, STRATA SYSTS INC, 87- *Concurrent Pos:* Chmn tech group info processing & holography, Optical Soc Am, 76-77; mem, US Nat Comt, Int Comn Optics, 75-77. *Mem:* Fel Optical Soc Am; Soc Photog Sci & Eng; Soc Photo-Optical Instrument Eng; Sigma Xi. *Res:* Optical techniques for information storage and retrieval; effects of recording materials upon holographic imaging; unconventional photography, especially electrophotography; evaluation of optical and photographic image quality; laser scanning. *Mailing Add:* 142 Crescent Ave Portola Valley CA 94028

URBACH, KARL FREDERIC, b Vienna, Austria, Nov 9, 17; nat US; m 52; c 2. HOSPITAL ADMINISTRATION. *Educ:* Reed Col, BA, 42; Northwestern Univ, PhD(chem), 46, MD, 51. *Prof Exp:* Asst, Eve Sch, Northwestern Univ, 42-46, asst chem, Dent Sch, 43-45, asst pharmacol, Med Sch, 45-47, lectr chem, Univ, 46-47, instr chem & pharmacol, Univ & Med Sch, 47-50; resident anesthesiol, USPHS Hosp, Staten Island, NY, 52-54; actg chief, USPHS Hosp, San Francisco, 54-55; chief anesthesiol, USPHS Hosp, Staten Island, 55-69; chief med educ & res, USPHS Hosp, San Francisco, 69-70, dir, 70-79; RETIRED. *Concurrent Pos:* NIH res fel, USPHS, 51-52. *Mem:* AAAS; Soc Exp Biol & Med; AMA; Am Soc Anesthesiol; Asn Mil Surg US. *Res:* Synthesis of vasopressors and testing, local anesthetics and testing; histamine methods for identification; actions; metabolism; pharmacology; evaluation of coronary dilators; anesthesia; hypnotics. *Mailing Add:* Two Atalaya Terr San Francisco CA 94117

URBAIN, WALTER MATHIAS, b Chicago, Ill, Apr 8, 10; m 39; c 2. FOOD SCIENCE, FOOD IRRADIATION. *Educ:* Univ Chicago, SB, 31, PhD(chem), 34. *Prof Exp:* Phys chemist, Swift & Co, 33-50, assoc dir res, 50-59, dir eng res & develop dept, 59-65; prof, 65-75, EMER PROF FOOD SCI, MICH STATE UNIV, 75- *Concurrent Pos:* Mem comn radiation preservation of foods, Nat Res Coun, 56-62, chmn, 59-62, mem adv bd mil personnel supplies, 58-75; mem adv comt isotopes & radiation develop, AEC, 64-66; sci ed, Food Technol & J Food Sci, Inst Food Technologists, 66-70; consult, US Food & Drug Admin, 74-80; vis prof food science & technol, Univ Calif, Davis, 82, 85; chmn adv comt radiation pasteurization foods, Am Inst Biol Sci, 71-; lectr, Miss State Univ, 88; lectr, consult & writer, Int Atomic Energy Agency, 67-90. *Honors & Awards:* Indust Achievement Award, Inst Food Technologists, 63; Food Eng Award, Dairy & Food Industs Supply Asn & Am Soc Agr Engrs. *Mem:* Am Chem Soc; fel Inst Food Technologists; Optical Soc Am. *Res:* Activity coefficients; detergent action of soaps; meat pigments; color standards; egg processing; meat packaging and processing; spectrochemical analysis of foods; x-ray diffraction; instrumentation; microwave heating; treatment of foods with ionizing radiation; spun protein foods. *Mailing Add:* 10645 Welk Dr Sun City AZ 85373

URBAN, EMIL KARL, b Milwaukee, Wis, May 27, 34; m 63; c 1. ORNITHOLOGY, VERTEBRATE ZOOLOGY. *Educ:* Univ Wis, BS, 56, PhD(zool), 64; Univ Kans, MA, 58. *Prof Exp:* From asst prof to assoc prof, Haile Sellassie I Univ, 64-75; assoc prof zool, Univ Ark, 75-76; PROF & CHMN BIOL, AUGUSTA COL, 76- *Concurrent Pos:* Vis prof, Univ Miami, 71. *Honors & Awards:* Louis K Bell Alumni Res Award, Augusta Col, 83. *Mem:* AAAS; Am Ornith Union; Brit Ornith Union; Cooper Ornith Soc; Wilson Ornith Soc. *Res:* Birds of Africa; biology of African pelicans, cormorants, ibises and cranes; monitoring of ciconiid colonies in Southeast United States. *Mailing Add:* Dept Biol Augusta Col Augusta GA 30910

URBAN, EUGENE WILLARD, b Omaha, Nebr, Apr 20, 35; m 60; c 8. PHYSICS. *Educ:* Harvard Univ, BS, 57; Univ Ala, MS, 63, PhD(physics), 70. *Prof Exp:* Physicist space res, US Army Ballistic Missile Agency, 59-60; physicist, 60-70, SUPVRY PHYSICIST SPACE RES, GEORGE C MARSHALL SPACE FLIGHT CTR, NASA, 70- *Mem:* Am Phys Soc; Sigma Xi. *Res:* Low temperature physics; superconducting instruments for space experiments; properties and applications of superfluid liquid helium in space; superfluid helium systems for space experiment cooling. *Mailing Add:* Rte 8 Box 497 Fayetteville TN 37334

URBAN, JAMES EDWARD, b Dime Box, Tex, Jan 5, 42; m 63; c 2. MICROBIAL PHYSIOLOGY. *Educ:* Univ Tex, Austin, BA, 65, PhD(microbiol), 68. *Prof Exp:* NSF fel, 68-70, asst prof, 70-77, ASSOC PROF BIOL, KANS STATE UNIV, 77- *Mem:* Am Soc Microbiol. *Res:* Regulation of cell division; medium and growth rate influences on the bacterial cell cycle; bacteroid morphogenesis in the Rhizobium legume symbiosis. *Mailing Add:* Div Biol Ackert Hall Kans State Univ Manhattan KS 66506-4901

URBAN, JOSEPH, b Brooklyn, NY, Mar 6, 21; m 47; c 4. CHEMISTRY, METALLURGY. *Educ:* Cooper Union, BChE, 43; Polytech Inst Brooklyn, MChE, 47, DrChE, 50. *Prof Exp:* Res chemist, Manhattan Proj, Columbia Univ, 43-44; res engr, Kellex Corp, 44-45; petrol res, M W Kellogg Co Div, Pullman, Inc, 45-47; instr chem, Adelphi Col, 47-50; dean, 69-83, prof, 48-86, EMER PROF SCI, WEBB INST NAVAL ARCHIT, 86- *Concurrent Pos:* Consult, Gen Elec Co, 50-52, A Pollak, 51-53 & J L Finck Labs, 53-56. *Res:* Catalysis; petroleum. *Mailing Add:* 69-64 64th St Glendale NY 11385-5248

URBAN, PETER ANTHONY, group technology, computer integrated manufacturing, for more information see previous edition

URBAN, RICHARD WILLIAM, b Newark, NY, July 30, 45; m 67; c 2. MICROBIOLOGY, MOLECULAR BIOLOGY. *Educ:* Ariz State Univ, BS, 70; Univ Hartford, MA, 73; Univ Colo, PhD(biol), 78. *Prof Exp:* Vis asst prof biol, microbiol & immunol, Metrop State Col, 78-79; res fel immunol, Dept Surg, Univ Colo Med Ctr, 78-79; RES ASST PROF TUMOR IMMUNOL, UTAH STATE UNIV, 79- *Concurrent Pos:* Researcher, Oncol Unit, Univ Colo Med Ctr & Dept of EPO Biol, Boulder Campus, 75-; res assoc, Dept Surg, Univ Calif Med Ctr, 80-; prin investr tumor immunother, New Agent Develop, 80-; independent investr tumor immunol, Dept Surg/EPO biol, Univ Colo, 80- *Mem:* Am Soc Microbiol; Sigma Xi; AAAS; NY Acad Sci. *Res:* Development and clinical evaluation of new tumor immunotherapeutic agents. *Mailing Add:* Cell Technol Inc 1668 Valtec Lane Boulder CO 80301

URBAN, THEODORE JOSEPH, physiology; deceased, see previous edition for last biography

URBAN, THOMAS CHARLES, b Vineland, NJ, Oct 11, 44; m 77; c 5. EXPERIMENTAL GEOPHYSICS, COMPUTER APPLICATIONS. *Educ:* Marist Col, BA, 67; Univ Rochester, PhD(geophys), 71. *Prof Exp:* Postdoctoral fel, Univ Rochester, 70-73; GEOPHYSICIST, US GEOL SURV, 73- *Mem:* Am Geophys Union; Soc Indust & Appl Math; NY Acad Sci; Sigma Xi. *Res:* Analysis of hydrothermal environments by means of borehole geophysical methods and computer modeling; study of local and regional crustal structure using gravity, magnetic, thermal and seismic methods. *Mailing Add:* 11384 Paul's Dr Conifer CO 80433-8208

URBAN, WILLARD EDWARD, JR, b Chicago, Ill, Sept 16, 36; m 57; c 3. BIOMETRICS, ANIMAL BREEDING. *Educ:* Va Polytech Inst & State Univ, BS, 58; Iowa State Univ, MS, 60, PhD(animal breeding), 63. *Prof Exp:* Animal husbandman, Animal Husb Res Div, Agr Res Serv, USDA, 58-63; from asst prof to assoc prof, Univ NH, 63-85, statistician, 63-72, from asst dir to assoc dir, 72-86, coordr info systs, Agr Exp Sta, 86-88, PROF BIOMET, UNIV NH, 85- *Concurrent Pos:* Adv, Int Crops Res Inst Semi-Arid Tropics, Hyderabad, India, 86-87 & 89. *Mem:* Biomet Soc; Am Soc Animal Sci. *Res:* Role of heredity and environment in economic traits of livestock; statistical methods for analyzing non-orthogonal data. *Mailing Add:* Biomet Pette Hall Univ NH Durham NH 03824

URBANEK, VINCENT EDWARD, b Chicago, Ill, Jan 2, 27; m 49; c 2. PROSTHODONTICS, DENTISTRY. *Educ:* Northwestern Univ, Evanston, BS, 51, MA, 52; Univ Ill, Chicago, DDS, 57. *Prof Exp:* Instr removable prosthodontics, Col Dent, Univ Ill Med Ctr, 57-63, asst prof, 63-67, assoc prof oral diag & oral med, 67-70; prof removable prosthodontics, Sch Dent, Med Col Ga, 70-89; RETIRED. *Concurrent Pos:* Mem attend staff, Eugene Talmadge Mem Hosp, 70- *Mem:* Fel Am Col Dent; Am Dent Asn; Am Prosthodont Soc; Am Acad Oral Med. *Res:* Removable prosthodontics; oral diagnosis; mandibular dysfunction as related to the temporomandibular joints, neuromuscular components and dental occlusion; effects of corticosteroids on vesiculobullous lesions of the oral mucosa. *Mailing Add:* 3231 Ramsgate Rd Augusta GA 30909

URBANIK, ARTHUR RONALD, b Union City, NJ, Apr 17, 39; m 66; c 2. QUALITY CONTROL. *Educ:* St Vincent Col, BS, 61; WVa Univ, MS, 63, PhD(org chem), 67. *Prof Exp:* Sr res chemist, Bjorksten Res Labs, 67; sr res chemist, Dan River, Inc, 67-90; SR RES CHEMIST, HICKSON DANCHEM CORP, 90- *Concurrent Pos:* Lectr chem, Stratford Col, 72-73. *Mem:* Am Chem Soc; assoc Sigma Xi; Am Asn Textile Chem & Colorists. *Res:* Quality control; organic chemical process development. *Mailing Add:* 709 Brightwell Dr Danville VA 24540

URBAS, BRANKO, b Zagreb, Yugoslavia, July 24, 29; m 56; c 1. ORGANIC CHEMISTRY, BIOCHEMISTRY. *Educ:* Univ Zagreb, Diplom Chem, 53, DSc(org chem), 60. *Prof Exp:* Group leader, Synthetic Org Chem, Pliva Chem & Pharmaceut Works, Zagreb, Yugoslavia, 52-60; sect leader polymer chem, Org Chem Indust, 60-61; Nat Res Coun Can fel carbohydrates, Ottawa Univ, Ont, 61-63; asst prof & NIH grant dept biochem, Purdue Univ, Lafayette, 63-65; res scientist, Res Br, Can Dept Agr, 65-68; sr res chemist, Moffett Tech Ctr, CPC Int, Inc, ARGO, 68-87; SR RES CHEMIST, AKZO CHEMICALS INC, 87- *Mem:* Am Chem Soc; Croatian Chem Soc; Sigma Xi. *Res:* Synthetic organic and carbohydrate chemistry. *Mailing Add:* 813 Belair Dr Darien IL 60559-4007

URBATSCH, LOWELL EDWARD, b Osage, Iowa, July 5, 42; m 65; c 2. SYSTEMATIC BOTANY. *Educ:* Univ Northern Iowa, BA, 64; Univ Ga, PhD(bot), 70. *Prof Exp:* Asst prof bot, Chadron State Col, 70-71; asst prof plant syst & biol, Univ Tex, Austin, 71-75; asst prof, 75-79, ASSOC PROF PLANT SYSTS, LA STATE UNIV, BATON ROUGE, 79- *Mem:* Bot Soc Am; Am Soc Plant Taxon. *Res:* Cytological and biochemical systematics of genera in the Compositae. *Mailing Add:* Dept Bot La State Univ Baton Rouge LA 70803-1705

URBSCHEIT, NANCY LEE, b Viroqua, Wis, Sept 7, 46. RESPIRATORY PHYSIOLOGY, PHYSICAL MEDICINE. *Educ:* State Univ NY Buffalo, BS, 68, MA, 70, PhD(physiol), 73. *Prof Exp:* From instr to asst prof physiol, State Univ NY, Buffalo, 73-76, asst to vpres health sci, 73-76; asst prof phys ther educ, Univ Iowa, Iowa City, 76-79; staff phys therapist, St Lawrence Hosp, Lansing, Mich, 79; chief phys therapist, Palo Alto County Hosp, 80-81; ASSOC PROF PHYS THER, ECAROLINA UNIV, 81- *Res:* Motor unit discharge patterns in elderly man reciprocal inhibition in hemiparetic man; mapping the activity of intercostal muscles in anesthetized cats during mechanical loading, in particular, positive and negative pressure breathing, threshold loading and elastic loading and chemical loading of respiration. *Mailing Add:* Dept Health Occup Univ Louisville Louisville KY 40292

URCH, UMBERT ANTHONY, b San Francisco, Calif, Aug 15, 46. FERTILIZATION. *Educ:* Univ Calif, Davis, PhD(biochem), 76. *Prof Exp:* ASST RES BIOCHEM, UNIV CALIF, DAVIS, 83- *Mailing Add:* 1309 Westwood Way Woodland CA 95695

URDAL, DAVID L, RECEPTOR BIOCHEMISTRY, LYMPHOKINE PURIFICATION. *Educ:* Univ Wash, PhD(biochem-encol), 80. *Prof Exp:* HEAD DEPT MEMBRANE BIOCHEM, IMMUNEX CORP, 83- *Mailing Add:* Immunex Corp 51 University St Seattle WA 98101

URDANG, ARNOLD, b Brooklyn, NY, Feb 10, 28; m 52; c 2. PHARMACEUTICAL CHEMISTRY, PHARMACY. *Educ:* Long Island Univ, BS, 49; Columbia Univ, MS, 52; Univ Conn, PhD(pharmaceut chem), 55. *Prof Exp:* Asst prof phys pharm, Long Island Univ, 55-58; res dir pharm, E Fougera & Co, NY, 58-64; asst dir clin res, Winthrop Prod Inc, Div, Sterling Drug Inc, 64-67, asst dir new prod, 67-69, dir tech coord, Winthrop Labs, 69-77, mgr sci affairs, Pharmaceut Group, Sterling Drug, Inc, 77-84; VPRES, SCI AFFAIRS, SANOFI INC, NY, 84- *Concurrent Pos:* Res Corp grant, 56-58. *Mem:* AAAS; Am Pharmaceut Asn; Am Chem Soc. *Res:* Development of new pharmaceuticals both from the aspect of new product development as well as the investigation of potential new chemical compounds. *Mailing Add:* 165 Murray Dr Oceanside Rockville Center NY 11570

URDY, CHARLES EUGENE, b Georgetown, Tex, Dec 27, 33; m 62; c 1. X-RAY CRYSTALLOGRAPHY, INORGANIC CHEMISTRY. *Educ:* Huston-Tillotson Col, BS, 54; Univ Tex, Austin, PhD(chem), 62. *Prof Exp:* Prof chem, Huston-Tillotson Col, 61-62; assoc prof, NC Col, Durham, 62-63; prof, Prairie View Agr & Mech Col, 63-72; PROF CHEM, HUSTON-TILLOTSON COL, 72- *Concurrent Pos:* Robert A Welch Found fel, Univ Tex, Austin, 62. *Mem:* Fel Am Inst Chem; Am Crystallog Asn; Am Chem Soc; Sigma Xi. *Res:* Determination of the crystal structures of coordination compounds of the transition metals by x-ray diffraction methods. *Mailing Add:* 7311 Hartnell Dr Austin TX 78723

URELES, ALVIN L, b Rochester, NY, Aug 8, 21; m 53; c 3. MEDICINE. *Educ:* Univ Rochester, MD, 45; Am Bd Internal Med, dipl; Am Bd Nuclear Med, dipl. *Prof Exp:* Intern, Beth Israel Hosp, Boston, 45-46, from resident to chief resident med, 48-51; from clin instr radiol & med to clin asst prof med, 51-64, assoc prof med, 64-67, clin assoc radiol, 61-67, PROF MED, SCH MED & DENT, UNIV ROCHESTER, 69-; ASSOC HEAD ENDOCRINOL & METAB DIV & CHIEF MED, GENESEE HOSP, 67- *Concurrent Pos:* Asst physician, Strong Mem Hosp, Rochester, NY, 51-58, sr assoc physician, 64- *Mem:* Fel Am Col Physicians; Int Soc Internal Med; Am Soc Internal Med; Endocrine Soc; AMA; Am Thyroid Asn. *Res:* Thyroid disease. *Mailing Add:* 56 Oxford St Rochester NY 14607

URENOVITCH, JOSEPH VICTOR, b Freeland, Pa, Nov 21, 37; m 59; c 4. EXPLOSIVES CHEMISTRY & ENGINEERING. *Educ:* Univ Pa, BA, 59, PhD(chem), 63. *Prof Exp:* Asst chem, Univ Wis, 62-63; res chemist, res & develop, Olin Mathieson Chem Corp, 64-65; section mgr corp res & develop, Air Prod & Chem Inc, 65-70, gen mgr specialty chem, 70-80; VPRES RES & DEVELOP, ATLAS POWDER CO, 80- *Mem:* Am Chem Soc; Soc Explosives Engrs. *Res:* Development and commercialization of new commercial explosives products, new pharmaceutical intermediates, new pesticides and products for the aerospace industry. *Mailing Add:* Box 136 RD 2 Tamaqua PA 18252-0136

URESK, DANIEL WILLIAM, b Price, Utah, July 18, 43; m 71; c 2. RANGE SCIENCE, WILDLIFE MANAGEMENT. *Educ:* Univ Utah, BS, 65, MS, 67; Colo State Univ, PhD(range sci), 72. *Prof Exp:* Res scientist, Battelle Pac Northwest Labs, 73-77; pres intern, 72-73, SUPVRY RES BIOLOGIST RANGE-WILDLIFE, ROCKY MOUNTAIN FOREST & RANGE EXP STA, FOREST SERV, USDA, 77- *Concurrent Pos:* Adj prof, SDak Sch Mines, SDak State Univ, fac affil & Colo State Univ, 78; referee, J Range Mgt, J Wildlife Mgt, Great Basin Naturalist, Prarie Naturalist, Northwest Sci, 78. *Mem:* Sigma Xi; Soc Range Mgt. *Res:* Basic range ecology; plant production; livestock grazing; dietary analysis; plant-animal wildlife relationships; nutrition studies. *Mailing Add:* 4406 Ridgewood Rapid City SD 57701

URETSKY, JACK LEON, b Great Falls, Mont, Mar 17, 24; div; c 2. APPLIED MATH, MUON ASTRONOMY. *Educ:* Mass Inst Technol, SB, 45, SM, 52 & PhD(theoret physics), 56, JD, Univ Chicago, 75. *Prof Exp:* Physicist, Univ Calif Radiation Lab, Berkeley, 56-58; fel, Imp Col, London, 59; asst prof physics, Purdue Univ, 60-61; assoc physicist, 61-72, GUEST PHYSICIST, ARGONNE NAT LAB, 73-; INSTR PHYSICS, COL DU PAGE, 87- *Concurrent Pos:* Prof physics, Northern Ill Univ, 69-70; Nat Acad Sci exchange fel to Romania, 70; vis physicist, DESY, Hamburg, Ger, 70; practiced law, 75-85; assoc prof math, Elmhurst Col, Ill, 85-86. *Mem:* Fel Am Phys Soc; AAAS; NY Acad Sci; Am Bar Asn. *Res:* Elementary particle theory; high energy physics. *Mailing Add:* 206 N Grant Hinsdale IL 60521

URETSKY, MYRON, b New York, NY, May 28, 40; m 81; c 4. COMPUTER SCIENCE, DATA PROCESSING. *Educ:* City Col NY, BBA, 61; Ohio State Univ, MBA, 62, PhD(acct), 62. *Prof Exp:* Asst prof acct, Univ Ill, 64-67; assoc prof bus, Columbia Univ, 67-70; PROF INFO SYST, NY UNIV, 70- *Concurrent Pos:* Fulbright scholar, 79; consult, 80- *Mem:* Inst Mgt Sci; Am Inst Cert Pub Acct; Asn Comput Mach. *Res:* Impact of computers on society; management fraud; simulation and gaming; East-West trade; technology assessment. *Mailing Add:* Dept Bus Info Syst NY Univ Washington Sq New York NY 10003

URETZ, ROBERT BENJAMIN, b Chicago, Ill, June 27, 24; m 55; c 2. BIOPHYSICS. *Educ:* Univ Chicago, BS, 47, PhD(biophys), 54. *Prof Exp:* Asst cosmic rays, 48-50, from instr to assoc prof biophys, 54-64, chmn dept biophys, 66-69; asst cosmic rays, Univ Chicago, 48-50, from instr to assoc prof biophys, 54-64, prof, 64-73, chmn, Dept Biophys, 66-69, assoc dean, 69-70, dep dean basic sci, 70-76, assoc vpres, Med Ctr & dep dean acad affairs, 76, actg vpres, Med Ctr & actg dean, Div Biol Sci & Pritzker Sch Med, 76-77, vpres, Med Ctr & dean, Div Biol Sci & Pritzker Sch Med, 77-82, RALPH W GERARD PROF BIOPHYS & THEORET BIOL, DIV BIOL SCI & PRITZKER SCH MED, UNIV CHICAGO, 73- *Mem:* Radiation Res Soc; Biophys Soc; Am Soc Cell Biol; Am Asn Med Cols; AMA. *Res:* Mechanism of biological effects of various radiations; optical analysis of biological structure. *Mailing Add:* Univ Chicago CLSC 920 E 58th St Chicago IL 60637

URIBE, ERNEST GILBERT, b Sanger, Calif, Nov 25, 35; m 57; c 3. PLANT BIOCHEMISTRY. *Educ:* Fresno State Col, AB, 57; Univ Calif, Davis, MS, 62, PhD(plant physiol), 65. *Prof Exp:* Res asst bot & biochem, Univ Calif, Davis, 58-65; res assoc biochem, Johns Hopkins Univ, 65-66 & Cornell Univ, 66-67; asst prof biol, Yale Univ, 67-74; assoc prof, 74-84, PROF BOT, WASH STATE UNIV, 85- *Concurrent Pos:* Res fels, NSF, 65-66 & NIH, 66-67. *Mem:* AAAS; Am Soc Plant Physiologists; Biophys Soc; Am Soc Biol Chemists; Sigma Xi. *Res:* Membrane transport in higher plants; energy conversion in photosynthesis mechanism of photosynthetic phosphorylation. *Mailing Add:* Dept Bot Wash State Univ Pullman WA 99164

URICCHIO, WILLIAM ANDREW, b Hartford, Conn, Apr 21, 24; m 50; c 5. BIOLOGY, MICROBIOLOGY. *Educ:* Cath Univ, BA, 49, MS, 51, PhD(zool), 53. *Prof Exp:* Asst prof, 53, PROF BIOL & CHMN DEPT, CARLOW COL, 53- *Concurrent Pos:* Guest prof, Carnegie-Mellon Univ, 63-68, NSF vis prof, 67-68; mem bd dirs, Duquesne Univ; pres, Int Fedn Family Life Prom. *Honors & Awards:* Bishop-Wright Award, 62; Knight of St Gregory the Great. *Mem:* Fel AAAS; Sigma Xi; Nat Asn Sci Teachers. *Res:* Sexuality; science education; natural family planning. *Mailing Add:* 1402 Murray Ave Pittsburgh PA 15217

URICK, ROBERT JOSEPH, b Brooklyn, NY, Apr 1, 15; m 55; c 3. UNDERWATER ACOUSTICS. *Educ:* Brooklyn Col, BS, 35; Calif Inst Technol, MS, 39. *Prof Exp:* Asst seismologist, Shell Oil Co, Tex, 36-38; chief computer, Tex Co, 39-42; physicist, Radio & Sound Lab, US Dept Navy, Calif, 42-45, Naval Res Lab, Wash, DC, 45-55; physicist, Ord Res Lab, Pa State Univ, 55-57, Mine Defense Lab, Fla, 57-60, US Naval Ord Res Lab, 60-74 & Naval Surface Weapons Ctr, 74-75; ACOUST CONSULT, 75- *Concurrent Pos:* Adj prof, Cath Univ Am, 77-88. *Honors & Awards:* Pioneers Medal, Acoust Soc Am. *Mem:* Fel Acoust Soc Am. *Res:* Underwater sound; sonar; sound propagation. *Mailing Add:* 11701 Berwick Rd Silver Spring MD 20904

URITAM, REIN AARNE, b Tartu, Estonia, Apr 11, 39; US citizen; m 70. ELEMENTARY PARTICLE PHYSICS. *Educ:* Concordia Col, Moorhead, Minn, BA, 61; Oxford Univ, BA, 63; Princeton Univ, MA, 65, PhD(physics), 68. *Prof Exp:* Res assoc physics, Princeton Univ, 67-68; asst prof, 68-74, ASSOC PROF PHYSICS, BOSTON COL, 74-, CHMN DEPT, 82- *Mem:* Am Phys Soc; Philos Sci Asn; Hist Sci Soc. *Res:* Theory of elementary particles; weak interactions; current algebra; high-energy hadron collisions; history and philosophy of science. *Mailing Add:* Dept Physics Boston Col 1410 Commonwealth Ave Chestnut Hill MA 02167

URIU, KIYOTO, b Berryessa, Calif, May 25, 17; m 49; c 4. POMOLOGY, PLANT PHYSIOLOGY. *Educ:* Univ Calif, BS, 48, MS, 50, PhD(plant physiol), 53. *Prof Exp:* Prin lab technol, 53-55, jr pomologist, 55-56, asst pomologist, 56-63, assoc specialist, 63-64, assoc pomologist, 64-70, POMOLOGIST, UNIV CALIF, DAVIS, 70-, LECTR, 62- *Mem:* Am Soc Hort Sci; Am Soc Plant Physiol; Sigma Xi. *Res:* Mineral nutrition, especially microelements of deciduous fruit trees; water relations of deciduous fruit trees. *Mailing Add:* Dept Pomol Univ Calif Davis CA 95616

URKOWITZ, HARRY, b Philadelphia, Pa, Oct 1, 21; m 46; c 2. ELECTRICAL ENGINEERING. *Educ:* Drexel Inst Technol, BS, 48; Univ Pa, MS, 54, PhD, 72. *Prof Exp:* From jr engr to sr engr, Res Div, Philco Corp, 48-53, proj engr, 53-56, sect engr, 56-58, sr res specialist, 58-64; sr eng specialist, Gen Atronics Corp, 64-70; STAFF SCIENTIST, GOVT ELECTRONIC SYSTS DIV, GEN ELEC CO, 70- *Concurrent Pos:* Adj prof, Drexel Univ, 52- *Mem:* Fel Inst Elec & Electronics Engrs. *Res:* Signal detection theory; signal processing; radar. *Mailing Add:* 9242 Darlington Rd Philadelphia PA 19115

URNESS, PHILIP JOEL, b Wenatchee, Wash, Jan 18, 36; m 56; c 2. WILDLIFE. *Educ:* Wash State Univ, BS, 58, MS, 60; Ore State Univ, PhD(range sci), 66. *Prof Exp:* Res biologist wildlife sci, Utah Wildlife Resources, 60-62; res biologist, Ore Game Comn, 62-65; res scientist, Rocky Mountain Forest & Range Exp Sta, US Forest Serv, 65-73; assoc prof, 73-88, PROF RANGE SCI, UTAH STATE UNIV, 88- *Concurrent Pos:* Vis assoc prof, Univ Calif, Davis, 80; consult, Chihuahua, Mex, 87; Fulbright res scholar, Nat Forest Corp, Chile, 88; proj leader res, Utah Div Wildlife Resources, 73-88. *Mem:* Soc Range Mgt; Wildlife Soc. *Res:* Wildlife interactions on wildlands of western North America including foraging behavior, nutritional ecology and competitive vs complemental relationships. *Mailing Add:* Dept Range Sci Utah State Univ Logan UT 84322-5230

URONE, PAUL, b Pueblo, Colo, Nov 29, 15; m 43; c 2. CHEMISTRY, AIR POLLUTION CONTROL. *Educ:* Western State Col Colo, AB, 38; Ohio State Univ, MS, 47, PhD, 54. *Prof Exp:* Teacher, High Sch, Colo, 38-42; chemist, Colo Fuel & Iron Corp, 42-45; chief chemist, Div Indust Hyg, State Dept Health, Ohio, 47-55; from asst prof to prof chem, Univ Colo, Boulder, 55-70; prof atmospheric chem, 71-81, EMER PROF, CHEM & ENVIRON ENG SCI, UNIV FLA, 81- *Concurrent Pos:* Consult, Martin Co, 60-61; Univ Colo fac fel, Univ Calif, Los Angeles, 61-62; sci adv, Food & Drug Admin, 66-71; Dept HEW air pollution fel, Univ Fla, 67; mem, Air Pollution Nat Manpower Adv Comt; sulfur oxides subcomt, Intersoc Comt Methods Sampling & Anal; Inter Govt Personnel Act Fel, Denver Fed Ctr, 77-78 & Air Pollution Control, Repub Korea, 81. *Mem:* Am Chem Soc; Am Indust Hyg Asn; Am Conf Govt Indust Hygienists; Air Pollution Control Asn. *Res:* Polarography of hydrocarbon combustion products; chemical analysis of air contaminants in industrial hygiene and air pollution; theoretical and applied gas chromatography; thermal and photochemical reactions of sulfur dioxide in air; air pollution control; air flow measurements. *Mailing Add:* 3726 SW Sixth Pl Gainesville FL 32607

URONE, PAUL PETER, b Pueblo, Colo, Feb 11, 44; m 65; c 2. NUCLEAR PHYSICS, MEDICAL PHYSICS. *Educ:* Univ Colo, BA, 65, PhD(physics), 70. *Prof Exp:* Teaching asst, Dept Physics, Univ Wash, 65-66; res asst nuclear physics, Univ Colo, 66-70; staff physicist, Kernfysisch Versneller Inst, Univ Groningen, Neth, 70-71; nuclear info res assoc, State Univ NY Stony Brook, 71-73; from asst prof to assoc prof, 73-82, PROF PHYSICS, CALIF STATE UNIV & COL, 82- *Concurrent Pos:* Consult, Calif State Univ & Cols, 74-78, Crocker Nuclear Lab, Univ Calif, Davis, 74-80 & Univ Calif, Davis, Med Ctr, 85- *Mem:* Am Phys Soc; Sigma Xi; Am Asn Physics Teachers. *Res:* Optical model of nucleus, nuclear data compilations, basic and applied neutron physics, medical physics and radiology. *Mailing Add:* Dept Physics Calif State Univ 6000 J St Sacramento CA 95819

URQUHART, ANDREW WILLARD, b Burlington, Vt, Aug 24, 39; m 64; c 2. COMPOSITES, MATERIALS SCIENCE. *Educ:* Dartmouth Col, BA, 61, MS, 64, PhD(metall), 71. *Prof Exp:* Engr, Div Naval Reactors, USAEC, 62-67, CREARE, Inc, 67-68; metallurgist, 71-75, br mgr inorg mat, Gen Elec Corp Res & Develop, 75-84; vpres res develop & engr, 84-89, SR VPRES TECHNOL, LANXIDE CORP, 89- *Mem:* Am Soc Metals Int; Am Ceramic Soc; Sigma Xi. *Res:* Management of research, development, and engineering for ceramic matrix and metal matrix composites. *Mailing Add:* 48 Bridleshire Rd Newark DE 19711

URQUHART, JOHN, III, b Pittsburgh, Pa, Apr 24, 34; m 57; c 3. PHYSIOLOGY, ENDOCRINOLOGY. *Educ:* Rice Univ, BA, 55; Harvard Univ, MD, 59. *Prof Exp:* Intern surg, Mass Gen Hosp, 59-60, asst resident, 60-61; investr exp cardiovasc dis, NIH, 61-63; from asst prof to prof physiol, Sch Med, Univ Pittsburgh, 63-70; prof biomed eng, Univ Southern Calif, 70-71; prin scientist & dir biol res, Alza Corp, 71-74, pres, 74-78, chief scientist, 78-82, sr vpres, 78-86, prin scientist, 82-86; pres & co-found, 86-88, CHMN & CHIEF SCIENTIST, APREX CORP, 88- *Concurrent Pos:* Josiah Macy, Jr fel obstet, Harvard Med Sch, 59-61; USPHS res career develop award, Nat Heart Inst, 63-70; NIH grants, 63-71 & 66-71; consult, Physiol Training Comt, NIH, 70-73; trustee, GMI Eng & Mgt Inst, Flint, Mich; adj prof pharm, Univ Calif Med Ctr; vis prof pharmaco-epidemiol, Rijksuniversiteit Limburg, Maastricht, Neth; dir adv comt, NIH, 85-88. *Honors & Awards:* Upjohn Award, Endocrine Soc, 62; Bowditch lectr, Am Physiol Soc, 69; Plenary lectr, UPHAR, 90. *Mem:* Am Physiol Soc; Endocrine Soc; Biomed Eng Soc (pres, 76-77). *Res:* Dynamics of drug and hormone action; patent compliance with prescribed drug requirements. *Mailing Add:* APREX Corp 47777 Warm Springs Blvd Fremont CA 94539

URQUHART, N SCOTT, b Columbia, SC, Mar 15, 40; m 59; c 7. STATISTICS. *Educ:* Colo State Univ, BS, 61, MS, 63, PhD(statist), 65. *Prof Exp:* From asst prof to assoc prof biol statist, Cornell Univ, 65-70; assoc prof, 70-75, prof exp statist, NMex State Univ, 75-; MEM FAC, ORE STATE UNIV. *Mem:* Am Statist Asn; Biomet Soc. *Res:* Development and dissemination of statistical methods used in biological research; teaching and development of teaching techniques for statistical methods. *Mailing Add:* 2360 NW Rolling Green Dr No 84 Corvallis OR 97330

URQUIDI-MACDONALD, MIRNA, 1946; m 85; c 1. THERMAL CONTAMINATION SHORE WATER COOLING POWER PLANTS. *Educ:* Inst Tech de Monterrey, BA, 69; Univ Paris-Sud, Orsay, MA, 70, PhD(plasma physics), 72. *Prof Exp:* Postdoctoral, x-spectroscopy, Lab de Chimie-Physique, Paris, 72-75; scientist, math model, Instituto Imgenieria, UNAM, Mex, 75-78; scientist math model, CFE-PNLV, Mex, 78-79; scientist, solar energy math model, SAHOP, Mexicali, Mex, 79-80; scientist, IEE Mexicali, Mex, 80-83; scientist, corrosion sci math model, Fontana Corrosion Ctr, Ohio, 83-84; scientist math model expert systs SRI chem, 84-85, SCIENTIST, CORROSION SCI, SRI INFO TECHNOL CTR, MENLO PARK, CA, 85. *Concurrent Pos:* Artificial Intel. *Mem:* Electrochem Soc; Math Soc; Mat Res Soc. *Res:* Expert systems and mathematical modeling of physic, chemical and engineering problem simulations; oxide passive film; breakdown of film; pitting corrosion; stress corrosion cracks, concrete corrosion, asphalt oxidation; corrosion, oxidation modeling & expert systems. *Mailing Add:* SRI Int Info Technol Ctr Menlo Park CA 94025

URQUILLA, PEDRO RAMON, b San Miguel, El Salvador, July 28, 39; m 64; c 2. CLINICAL PHARMACOLOGY. *Educ:* Cath Inst of the East, BA, 57; Univ El Salvador, MD, 65. *Prof Exp:* Instr pharmacol, Univ El Salvador, 64-65, Pan Am Health Orgn fel, 66-68, NIH fel, 68-69, assoc prof, 69-72, asst dean, Sch Med, 71-72; assoc prof pharmacol, Univ Madrid, 72-73; asst prof pharmacol, WVa Univ, 73-75, assoc prof, 75-79; assoc dir med res, Miles Lab, 79-81; ASSOC DIR CLIN RES, PFIZER INC, 81- *Mem:* Am Soc Pharmacol & Exp Therapeut; Am Soc Clin Pharmacol Therapuet; AMA. *Res:* Analysis of the pharmacological receptors of cerebral arteries; pharmacological studies on cerebral vasospasm. *Mailing Add:* Clin Res Bristol-Myers Five Research Pkwy PO Box 5100 Wallingford CT 06492

URRY, DAN WESLEY, b Salt Lake City, Utah, Sept 14, 35; c 4. MOLECULAR BIOPHYSICS, BIOCHEMISTRY. *Educ:* Univ Utah, BA, 60, PhD(phys chem), 64. *Prof Exp:* Fel, Univ Utah, 64; Harvard Corp Fel, 64-65; vis investr, Chem Biodynamics Lab, Univ Calif, 65-66; prof lectr, Dept Biochem, Univ Chicago, 67-70; dir div molecular biophys, Lab Molecular Biol, 70-72, PROF BIOCHEM, UNIV ALA, BIRMINGHAM, 70-, DIR LAB MOLECULAR BIOPHYS, 72- *Concurrent Pos:* Assoc mem, Inst Biomed Res, AMA, 65-69, mem, 69-70; vis prof, Univ di Padova, Centro di Studi sui Biopolimeri, 77; vchmn, Southern Region Res Review & Cert Subcomt, Am Heart Asn, 79-80, chmn, 81-82; Alexander von Humboldt Found award, Ger, 79-80; mem biophysics & biophys chem B study sect, Div Res Grants, NIH, 80-84. *Mem:* AAAS; Am Soc Biol Chem; Am Chem Soc; Biophys Soc; Am Inst Biol Chem. *Res:* Methods of absorption, optical rotation and nuclear magnetic resonance spectroscopies to study polypeptide conformation and its relation to biological function; emphasis on membrane structure, mechanism of ion transport, membrane active polypeptides, elastin, and atherosclerosis. *Mailing Add:* Dept Biochem Univ Ala Sch Med University Station AL 35294

URRY, GRANT WAYNE, b Salt Lake City, Utah, Mar 12, 26; m 46; c 4. INORGANIC CHEMISTRY, CHEMICAL BONDING. *Educ:* Univ Chicago, SB, 47, PhD(chem), 53. *Prof Exp:* Asst bot, Univ Chicago, 46-47, res assoc, 47-48, asst chem, 49-52, res assoc, 53-55; asst prof, Wash Univ, 55-58; from assoc prof to prof, Purdue Univ, 58-68; prof, 68-70, chmn dept, 68-73, ROBINSON PROF CHEM, TUFTS UNIV, 70- *Concurrent Pos:* Sloan fel, 56-58; consult, E I du Pont de Nemours & Co, Inc. *Mem:* AAAS; Am Chem Soc; Fedn Am Scientists. *Res:* Chemistry of convalently bonded inorganic compounds and electron spin resonance; equilibrium chemistry of carbon. *Mailing Add:* Dept Chem Tufts Univ Medford MA 02155

URRY, LISA ANDREA, b Chicago, Ill, July 27, 53; m 91. EXTRACELLULAR MATRIX & CELL-CELL ADHESION, INDUCTION OF CELL TYPES. *Educ:* Tufts Univ, BS, 75; Mass Inst Technol, PhD(biol), 90. *Prof Exp:* Res assoc, New Eng Aquarium, 77-82; postdoctoral fel, Brigham & Women's Hosp, Harvard Med Sch, 90-91; RES ASSOC, DEPT BIOL, TUFTS UNIV, 91- *Res:* Analysis at the molecular and cellular levels of developmental events such as morphogenesis and differentiation in the sea urchin embryo, as well as later phenomena associated with metamorphosis. *Mailing Add:* 29 Elm St Winchester MA 01890

URRY, RONALD LEE, b Ogden, Utah, June 5, 45; m 71; c 1. REPRODUCTIVE PHYSIOLOGY, UROLOGY. *Educ:* Weber State Col, BS, 70; Utah State Univ, MS, 72, PhD(physiol), 73. *Prof Exp:* Teaching & res asst physiol, Dept Biol, Utah State Univ, 70-72, NDEA fel, 72-73; dir urol res lab & asst prof urol surg, Sch Med & Dent, Univ Rochester, 73-76; ASSOC PROF ZOOL, BRIGHAM YOUNG UNIV & RES ASSOC PROF SURG, DIV UROL, UNIV UTAH MED CTR, 76-, PROF SURG, DIV UROL & DEPT OBSTET & GYNEC, SCH MED, SALT LAKE CITY, 84- *Mem:* Soc Study Reproduction; AAAS; Am Fertil Soc; Endocrine Soc; Am Andrology Soc; Am Urol Asn. *Res:* Relationship of stress and biogenic amines to male reproduction; testicular tissue culture, testicular perfusion, male infertility studies, vasectomy and vasovasostomy, and testicular physiology and endocrinology; in vitro fertilization, gamese freezing, fertilization. *Mailing Add:* Dept Surg Urol Univ Utah Sch Med 50 N Medical Dr Salt Lake City UT 84132

URRY, WILBERT HERBERT, b Salt Lake City, Utah, Nov 5, 14; m 36; c 4. CHEMISTRY. *Educ:* Univ Chicago, BS, 38, PhD(chem), 46. *Prof Exp:* Lectr sci, Mus Sci & Indust, Chicago, 37-40, cur & consult chem, 40-44; from instr to prof, 44-80, emer prof chem, Univ Chicago, 80-; RETIRED. *Concurrent Pos:* Daines Mem lectr, Univ Kans, 54; Carbide & Carbon Prof fel, 56; vis prof, Univ Calif, 58-59; consult, Monsanto Co, Com Solvents Corp, Wyandotte Chem Corp & US Naval Weapons Ctr, Calif, 51- *Mem:* AAAS; Am Chem Soc. *Res:* Reaction of free radicals in solution; rearrangements of free radicals and anions; organic photochemistry; homogeneous catalysis via complex ions; chemistry of hydrazines; structure and synthesis of natural products. *Mailing Add:* 23 E Redondo Tempe AZ 85282

URSELL, JOHN HENRY, b Leeds, Eng, June 9, 38. MATHEMATICS. *Educ:* Oxford Univ, BA, 59, MA & DPhil(math), 63. *Prof Exp:* Asst prof math, Pa State Univ, 62-63; assoc prof, State Univ NY Col Fredonia, 63-64; ASST PROF MATH, QUEEN'S UNIV, ONT, 64- *Mem:* Am Math Soc; Can Math Cong; Math Asn Am; fel Royal Asiatic Soc Gt Brit & Ireland; Am Pharmacog Soc. *Res:* Topological semigroups; algebra; graph theory; comparative religions; mathematical sociology; statistics. *Mailing Add:* Dept Math Queen's Univ Kingston ON K7L 3N6 Can

URSENBACH, WAYNE OCTAVE, b Lethbridge, Alta, Dec 4, 23; m 44; c 7. CHEMISTRY. *Educ:* Brigham Young Univ, BSc, 47, MSc, 48. *Prof Exp:* Asst chemist, Dept Agr Res, Am Smelting & Refining Co, 42-43 & 46-51; lab supvr health physics, Dow Chem Co, Colo, 51-52; asst, Explosives Res Group, Univ Utah, 52-55, res assoc, 55-59, asst res prof, Inst Metals & Explosives, 59-61; mgr prod & res develop, Inter-Mountain Res & Eng Co, 61-65, asst res dir, 65-66; prod mgr, Ireco Chem, 66-68, dir res, 68-69, mgr planning, 69-70; res assoc, Utah Eng Exp Sta, Univ Utah, 71-74, asst gen mgr, Res Inst, 74-75, vpres & dir, Appl Technol Div, 75-84; CONSULT, 84- *Concurrent Pos:* Consult air pollution effects, fires, explosions & explosives, 71- *Mem:* Am Chem Soc; AAAS; Air Pollution Control Asn; NY Acad Sci; Nat Fire Protection Asn. *Res:* Air pollution; agricultural chemistry; health physics; theory of detonation; explosion and long range blast effects; terminal ballistics; seismic effects of explosions; causes of fires and accidental explosions. *Mailing Add:* 4635 S 1175 E Salt Lake City UT 84117

URSIC, STANLEY JOHN, b Milwaukee, Wis, Apr 2, 24; m 50; c 3. WATERSHED MANAGEMENT, FOREST HYDROLOGY. *Educ:* Univ Minn, BS, 49; Yale Univ, MF, 50. *Prof Exp:* Res forester, Univ Ill, 50-51; proj leader, Southern Forest Exp Sta, USDA Forest Serv, 51-90; RETIRED. *Concurrent Pos:* Mem, Int Union Forestry Res Orgn; assoc ed, Southern J Appl Forestry. *Mem:* Soc Am Foresters. *Res:* Effects of forestry practices, including rehabilitation of eroding lands, on water quality, yields and distribution; flow processes on forested lands; effects of atmospheric deposition on forests and water quality. *Mailing Add:* 1031 Zilla Avent Dr PO Box 947 Oxford MS 38655

URSILLO, RICHARD CARMEN, b Lawrence, Mass, Oct 26, 26; m 66; c 2. PHARMACOLOGY. *Educ:* Tufts Univ, BS, 49, PhD(pharmacol), 54; Univ Calif, MS, 52. *Prof Exp:* From instr to asst prof pharmacol, Univ Calif, Los Angeles, 54-62; sect head pharmacol, Lakeside Labs, 62-66, dir pharmacol dept, 66-75; head dept pharmacol, Merrell Res Ctr, 75-84; dir pharmacol sci, Merrell Dow Res Inst, 84-87, sr dir res sci, 87-89, dir res admin, 88-89, VPRES RES ADMIN, MERRELL DOW RES INST, 89- *Concurrent Pos:* USPHS spec fel, Inst Sanita, Italy, 60-61. *Mem:* Am Soc Pharmacol & Exp Therapeut. *Res:* Pharmacology of autonomic and central nervous systems. *Mailing Add:* Merrell Dow Res Inst 2110 E Galbraith Rd Cincinnati OH 45215

URSINO, DONALD JOSEPH, b Toronto, Ont, Nov 11, 35; m 60; c 4. PLANT PHYSIOLOGY. *Educ:* Pomona Col, BA, 56; Queen's Univ, Ont, MSc, 64, PhD(biol), 67. *Prof Exp:* Teacher sci, High Sch, 57-63; Nat Res Coun Can fel, Milan, 67-69; asst prof biol, 69-72, chmn dept biol sci, 74-77, ASSOC PROF BIOL, BROCK UNIV, 72- *Mem:* Am Col Sports Med. *Mailing Add:* Dept Biol Sci Brock Univ St Catherines ON L2S 3A1 Can

URSINO, JOSEPH ANTHONY, b Brooklyn, NY, Feb 28, 39. ORGANIC CHEMISTRY. *Educ:* St John's Univ, NY, BS, 60, MS, 62, PhD(chem), 67. *Prof Exp:* From asst prof to assoc prof, 66-71, PROF CHEM, STATE UNIV NY AGR & TECH COL FARMINGDALE, 71- *Mem:* Am Chem Soc; NY Acad Sci; Am Soc Eng Educ. *Res:* Synthesis and properties of heterocyclic organotin compounds. *Mailing Add:* 2299 Narwood Ct Merrick NY 11566-3928

URSO, PAUL, b Sicily, Italy, Aug 3, 25; US citizen; m 52; c 2. IMMUNOLOGY, ZOOLOGY. *Educ:* St Francis Col, NY, BS, 50; Marquette Univ, MS, 52; Univ Tenn, PhD(zool), 61. *Prof Exp:* Asst zool, Marquette Univ, 50-52; instr biol, Cardinal Stritch Col, 52-53; biologist, Nat Cancer Inst, 53-55; jr biologist, Oak Ridge Nat Lab, 55-57, assoc biologist, 58-59; from asst prof to assoc prof biol, Seton Hall Univ, 61-71; sr scientist, Med Div, Oak Ridge Assoc Univs, 71-81; asst prof, 81-88, ASSOC PROF, DEPT MICROBIOL & IMMUNOL, MOREHOUSE SCH MED, 88- *Concurrent Pos:* Consult, Biol Div, Oak Ridge Nat Lab, 61-63, Med Div, Oak Ridge Assoc Univ, 63-71 & Water & Health Sanitation Proj, USAID, 89-; res partic, Oak Ridge Inst Nuclear Studies, 63-; prin investr, NIH, 64-67, 83-87, 88-90 & 90-, USAID, 85-86, 87-88 & 89-91 & US Environ Protection Agency, 89-92; co-prin investr, NIH, 64-66; co-investr, Dept Energy, Environ Protection Agency, 78-81; lectr, Univ Tenn, 78-81; vis res scientist Am Soc Microbiol, Univ PR Med Ctr & Lib Arts Col, San Juan, 88 & Savannah State Col, 89. *Mem:* Transplantation Soc; Radiation Res Soc; Am Asn Immunologists; Reticuloendothelial Soc; Soc Exp Hemat; Int Soc Comp & Develop Immunol; Am Asn Cancer Res; Clin Immunol Soc. *Res:* Antibody formation; immune cell interactions; immunologic recovery of chimeras; transplantation in chimeras; immunocarcinogenesis. *Mailing Add:* Dept Microbiol & Immunol Morehouse Sch Med Atlanta GA 30310

URSPRUNG, JOSEPH JOHN, organic chemistry, medicinal chemistry, for more information see previous edition

URTASUN, RAUL C, Can citizen. ONCOLOGY, EXPERIMENTAL RADIOBIOLOGY. *Educ:* Univ Buenos Aires, MD, 60; FRCP(C), 67; Am Bd Radiol, dipl radiother, 67. *Prof Exp:* Res fel oncol, Harvard Med Sch, 63-64; instr radiation oncol, Johns Hopkins Univ, 66-68; asst prof, McGill Univ, 68-70; PROF RADIATION ONCOL, UNIV ALTA, 70- *Mem:* Am Soc Therapeut Radiologists; Am Soc Clin Oncologists; Am Soc Cancer Res; Radiator Res Soc; Royal Col Physicians & Surgeons Can. *Res:* Clinical radiobiology; radiosensitizers; combined modalities in the treatment of cancer; high linear energy transfer particle radiation. *Mailing Add:* 26 Wellington Crescent Edmonton AB T5N 3V2 Can

URTIEW, PAUL ANDREW, b Nish, Yugoslavia, Feb 23, 31; US citizen; m 61; c 2. THERMODYNAMICS, PHYSICS. *Educ:* Univ Calif, Berkeley, BS, 55, MS, 59, PhD(mech eng), 64. *Prof Exp:* Res assoc, Detonation Lab, 59-64, asst res engr, Propulsion Dynamics Lab, 64-67, engr, Physics Dept, 67-73, ENGR, CHEM & MAT SCI DEPT, LAWRENCE LIVERMORE LAB, UNIV CALIF, 73- *Concurrent Pos:* Consult, Hiller Aircraft Corp, Calif, 63-64 & MB Assocs, 66, Chevron, 90- *Mem:* Combustion Inst; Am Inst Aeronaut & Astronaut. *Res:* High pressure physics of shocked solid materials; nonsteady wave dynamics; wave interaction processes in reactive and nonreactive media; graphical and experimental techniques applicable in research; diagnostic of high speed processes. *Mailing Add:* Lawrence Livermore Nat Lab L282 PO Box 808 Livermore CA 94550

URY, HANS KONRAD, b Berlin, Ger, Nov 4, 24; US citizen; m 55. BIOSTATISTICS, MATHEMATICAL STATISTICS. *Educ:* Univ Calif, Berkeley, AB, 45 & 55, MA, 64, PhD(statist), 71. *Prof Exp:* Res asst statist, Inst Eng Res, Univ Calif, Richmond, 61-62; res assoc, Stanford Univ, 62-63; biostatistician, Calif State Dept Pub Health, 63-66; statist consult, Comput Ctr, San Francisco Med Ctr, Univ Calif, 67; spec consult, Calif State Dept Pub Health, 67-68; res specialist biostatist, 68-71; biostatistician & consult statist, Permanente Med Group, Oakland, 71-75, sr statistician, Med Methods Res Dept, 75-84; PVT CONSULT, 85- *Concurrent Pos:* Instr, Exten Div, Univ Calif, Berkeley, 67-80; statist consult, Environ Resources, Inc, Calif, 68-70; chmn, San Francisco Bay Area Biostatist Colloquium, 74-75; vis prof, Div Statist, Univ Calif, Davis, 83-84. *Mem:* Am Statist Asn; Biomet Soc; fel Royal Statist Soc. *Res:* Nonparametric statistics; statistical techniques for evaluating environmental health studies; chronic disease epidemiology; application of computers to biostatistics; multiple comparison methods; statistical efficiency comparisons; sample size determination for comparing rates or proportions. *Mailing Add:* 2050 Drake Dr Oakland CA 94611

USBORNE, WILLIAM RONALD, b Rochester, NY, Nov 22, 37; m 62; c 2. MEAT & FOOD SCIENCE. *Educ:* Cornell Univ, BS, 59; Univ Ill, Urbana, MS, 61; Univ Ky, PhD(meat & animal sci), 67. *Prof Exp:* Res asst meat sci, Univ Ill, 59-61; res assoc, Cornell Univ, 61-62; res asst, Univ Ky, 63-64, instr, 64-65, res asst, 65-66; fel meat chem, Tex A&M Univ, 66-67; asst prof meat sci, Univ Minn, St Paul, 68-69; assoc prof meat sci, 69-79, PROF & CHMN FOOD SCI, UNIV GUELPH, 79- *Concurrent Pos:* Welch Found fel, 67-68; fel, Tex A&M Univ, 68. *Mem:* Am Soc Animal Sci; Am Meat Sci Asn; Am Inst Food Technologists; Can Inst Food Sci & Technol; Agr Inst Can. *Res:* Meat chemistry and technology; meat animal evaluation techniques; meat processing and quality. *Mailing Add:* Dept Food Sci Univ Guelph Guelph ON N1G 2W1 Can

USCAVAGE, JOSEPH PETER, bacteriology, mycology, for more information see previous edition

USCHOLD, RICHARD L, b Buffalo, NY, Sept 10, 28; m 51; c 9. MATHEMATICS. *Educ:* Canisius Col, BS, 53; Univ Notre Dame, MS, 55; State Univ NY Buffalo, PhD(math), 63. *Prof Exp:* Instr math, Nazareth Col, NY, 55-56; from instr to asst prof, 56-64, chmn dept, 66-72, ASSOC PROF MATH, CANISIUS COL, 64- *Mem:* Math Asn Am. *Mailing Add:* Dept Math Canisius Col Buffalo NY 14208

USDIN, VERA RUDIN, b Vienna, Austria, May 31, 25; nat US; m 49; c 4. BIOCHEMISTRY. *Educ:* Sterling Col, BS, 45; Duke Univ, MA, 47; Ohio State Univ, PhD(biochem), 51. *Prof Exp:* Res assoc physiol chem, Grad Sch Med, Univ Pa, 51-56; chemist, Res Labs, Rohm and Haas Co, 56-59; assoc res prof physiol chem, NMex Highlands Univ, 59-62; head physiol chem br, Melpar, Inc, Va, 62-67; proj supvr, Gillette Res Inst, Inc, 67-69, res supvr, 69-73, group leader, 73-79, prin scientist, 79-83, consult, 83-85; pres, Biotran Corp, 85-88; RETIRED. *Mem:* Am Chem Soc; Am Soc Cell Biol; Soc Invest Dermat. *Res:* Biochemistry of skin; enzyme inhibition; salivary proteins; dental plaque. *Mailing Add:* Biotran Corp Six Stevens Ct Rockville MD 20850-1919

USENIK, EDWARD A, b Eveleth, Minn, Jan 16, 27; m 55; c 2. VETERINARY MEDICINE. *Educ:* Univ Minn, BS, 50, DVM, 52, PhD(vet med, path), 57. *Prof Exp:* Instr vet surg & radiol, Col Vet Med, Univ Minn, 52-57, asst prof, 57-59; med assoc exp path, Med Dept, Brookhaven Nat Lab, 59-60; assoc prof vet surg & radiol, 60-64, PROF VET SURG & RADIOL, COL VET MED, UNIV MINN, ST PAUL, 64- *Concurrent Pos:* Collabr med dept, Brookhaven Nat Lab, 60-; Rockefeller consult, Col Vet Med, Lima, 65-66; mem adv coun, Inst Lab Animal Resources, Nat Res Coun-Nat Acad Sci; on leave to fac vet sci, Nat Univ, Neirobi, Kenya, 72-74. *Res:* Gastrointestinal diseases; anesthesia in veterinary medicine. *Mailing Add:* Fac Vet Med Dept Clin Studies Univ Zimbabwe PO Box MP 197 Mt Pleasant Harare Zimbabwe

USHER, DAVID ANTHONY, b Harrow, Eng, Nov 1, 36; m 74. BIO-ORGANIC CHEMISTRY. *Educ:* Victoria Univ, NZ, BSc, 58, MSc, 60; Univ Cambridge, PhD(chem), 63. *Prof Exp:* Res fel chem, Harvard Univ, 63-65; asst prof, 65-70, ASSOC PROF CHEM, CORNELL UNIV, 70- *Concurrent Pos:* NIH career develop award, 68-73; vis prof, Oxford Univ, 71-72. *Honors & Awards:* NZ Inst Chem Prize, 58. *Mem:* Am Chem Soc; AAAS. *Res:* Chemical evolution; chemical reactions of nucleic acids; enzyme action. *Mailing Add:* Dept Chem Cornell Univ Ithaca NY 14853-0001

USHER, PETER DENIS, b Bloemfontein, SAfrica, Oct 27, 35; US citizen; m 61; c 1. ASTRONOMY. *Educ:* Univ of the Orange Free State, BS, 56, MS, 59; Harvard Univ, PhD(astron), 66. *Prof Exp:* Fel, Harvard Col Observ, 66-67; sr scientist, Am Sci & Eng, Inc, Mass, 67-68; from asst prof to assoc prof, 68-85, PROF ASTRON, PA STATE UNIV, UNIVERSITY PARK, 86- *Concurrent Pos:* Sabbatical leave, Hale Observ, 75-76 & Royal Observ, Edinburgh, 86; ed, Astron Quarterly, 80-88. *Mem:* Int Astron Union; Am Astron Soc; Royal Astron Soc. *Res:* Perturbation theory; stellar structure; faint blue objects; quasars. *Mailing Add:* Dept Astron Pa State Univ University Park PA 16802

USHER, W(ILLIA)M MACK, b Devol, Okla, Nov 10, 27; m 52. MATHEMATICAL STATISTICS, EDUCATIONAL ADMINISTRATION. *Educ:* Okla State Univ, BS, 52, MS, 58. *Prof Exp:* Asst registr, Okla State Univ, 54-58; statistician, Tex Instruments, Inc, 58-59, opers res analyst, 59-61, mgr, 61-63, corp systs develop mgr, 63-67; dir instnl res, 67-69, DIR COMPUT & INFO SYST, OKLA STATE UNIV, 69- *Mem:* Am Statist Asn; Asn Comput Mach; Asn Instnl Res. *Res:* Development of management information systems. *Mailing Add:* Dir Inst Res Okla State Univ Main Campus Stillwater OK 74078

USHERWOOD, NOBLE RANSOM, b Atlanta, Ill, Jan 13, 38; m 63; c 3. SOIL FERTILITY, PLANT NUTRITION. *Educ:* Southern Ill Univ, BS, 59, MS, 60; Univ Md, PhD(soils, plant physiol), 66. *Prof Exp:* Res asst soil fertil & test correlation, Univ Md, 60-66; asst prof soil fertil & plant nutrit, Univ Del, 66-67; midwest agronomist, Ill, 67-69, dir, Potash Res Asn Northern Latin Am, Guatemala, 69-71, dir Fla & Latin Am Potash Inst, 71-77, VPRES, POTASH & PHOSPHATE INST, 77- *Concurrent Pos:* Assoc ed, J Agron Educ; chmn bd, Agron Sci Found. *Mem:* Fel Am Soc Agron; fel Soil Sci Soc Am; Brazilian Soc Sci; fel Crop Sci Soc Am. *Res:* Nitrogen, phosphorus, potassium, magnesium and manganese soil fertility and plant nutrition; subsurface irrigation feasability studies. *Mailing Add:* Potash & Phosphate Inst 2801 Buford Hwy NE Suite 401 Atlanta GA 30329

USHIODA, SUKEKATSU, b Tokyo, Japan, Sept 18, 41; m 85; c 3. ELECTRONICS ENGINEERING. *Educ:* Dartmouth Col, AB, 64; Univ Pa, MS, 65, PhD(physics), 69. *Prof Exp:* From asst prof to prof physics, Univ Calif, Irvine, 69-85; PROF, TOHOKU UNIV, 85- *Concurrent Pos:* Adv bd, J Mat Sci & Eng. *Mem:* Optical Soc Am; Am Phys Soc; Phys Soc Japan. *Res:* Solid state physics; Raman spectroscopy; surface physics; electron spectroscopy. *Mailing Add:* Res Inst Elec Comm Tohoku Univ Sendai 980 Japan

USINGER, WILLIAM R, b Chicago Ill, Mar 20, 51. IMMUNOLOGICAL ASPECTS OF BACTERIAL PRODUCTS. *Educ:* Univ Wis-Madison, PhD(immunol), 80. *Prof Exp:* NIH FEL & ASSOC RES SCIENTIST, DEPT IMMUNOL, UNIV CALIF, BERKELEY, 80- *Mem:* AAAS; Am Asn Immunologists; Reticuloendothelial Soc. *Mailing Add:* Dept Immunol Immusine Labs Inc 25 North Lane Orinda CA 94563

USISKIN, ZALMAN P, b Chicago, Ill, Jan 1, 43; m 79; c 2. MATHEMATICS EDUCATION, CURRICULUM. *Educ:* Univ Ill, BS(educ) & BS(math), 63; Harvard Univ, MAT, 64; Univ Mich, PhD(educ), 69. *Prof Exp:* From asst prof to assoc prof, 69-82, PROF EDUC, UNIV CHICAGO, 82- *Concurrent Pos:* Dir Chicago Univ Sch Math Proj, 87-; mem Nat Acad Math Sci Educ Bd, 88- *Mem:* Math Asn Am. *Res:* All aspects of math education with emphasis on matters related to curriculum & instruction; policy making; selection & organization of content. *Mailing Add:* 5835 S Kimbark Chicago IL 60637

USLENGHI, PIERGIORGIO L, b Turin, Italy, Aug 31, 37; m 78; c 3. THEORETICAL PHYSICS, ELECTRICAL ENGINEERING. *Educ:* Turin Polytech Inst, Laurea, 60; Univ Mich, MS, 64, PhD(physics), 67. *Prof Exp:* Asst prof elec eng, Turin Polytech Inst, 61; assoc res engr, Conductron Corp, 62-63; res physicist, Univ Mich, Ann Arbor, 63-70; assoc prof info eng, 70-74, assoc dean eng, 82-87 PROF ELEC ENG, UNIV ILL, CHICAGO, 74- *Concurrent Pos:* NASA, NSF & Dept Defense grants. *Mem:* Inst Elec & Electronics Engrs; Int Union Radio Sci; Sigma Xi. *Res:* Antennas; radars; quantum electronics; nonlinear phenomena. *Mailing Add:* Dept Elec Eng & Comput Sci Univ Ill PO Box 4348 Chicago IL 60680

USMANI, RIAZ AHMAD, b Farrukhabad, India, Nov 1, 34; m 54; c 3. NUMERICAL ANALYSIS. *Educ:* Aligarh Muslim Univ, India, BSc, 56, MSc, 57; Univ BC, PhD(numerical anal), 67. *Prof Exp:* Lectr math, Col Eng & Technol, Aligarh Muslim Univ, Indian, 57-61; teaching asst, Univ BC, 61-65; sessional lectr, Univ Calgary, 65-66; from asst prof to assoc prof comput sci, 66-76, assoc prof, 76-79, PROF APPL MATH, UNIV MAN, 79- *Mem:* Am Math Soc; Can Appl Math Soc. *Res:* Numerical integration of differential equations; initial and boundary value problems in ordinary differential equations. *Mailing Add:* Dept Appl Math Univ Man Winnipeg MB R3T 2N2 Can

USSELMAN, MELVYN CHARLES, b Ottawa, Ont, Jan 5, 46; m 79; c 4. BIOGRAPHY, CREATIVITY & ENTREPRENEURSHIP. *Educ:* Univ Western Ont, BSc, 68, PhD(chem), 72, MA, 75. *Prof Exp:* Asst prof chem, 75-81, asst prof, Dept Hist Med & Sci, 76-81, ASSOC PROF CHEM, UNIV WESTERN ONT, 81- *Mem:* Hist Sci Soc; Am Chem Soc; Brit Soc Hist Alchemy & Chem; Can Soc Hist & Philos Sci; Can Sci & Technol Hist Asn. *Res:* History of chemistry, 18th-20th centuries; the work of W H Wollaston; platinum metallurgy, origin of chemical laws. *Mailing Add:* Dept Chem Univ Western Ont London ON N6A 5B7 Can

USSELMAN, THOMAS MICHAEL, b Bismarck, ND, Aug 9, 47; m 73; c 1. GEOPHYSICS, GEOCHEMISTRY. *Educ:* Franklin & Marshall Col, BA, 69; Lehigh Univ, MS, 71, PhD(geol), 73. *Prof Exp:* Nat Res Coun res assoc geochem, Johnson Space Ctr, NASA, 73-75, fel, Lunar Sci Inst, 75-76; vis prof geol, State Univ NY, Buffalo, 76-78; SR STAFF SCIENTIST, NAT ACAD SCI, 78- *Concurrent Pos:* Co-editor, 7th Lunar Sci Conf Proc, 76. *Mem:* Am Geophys Union; fel Geol Soc Am. *Res:* Experimental geochemistry and geophysics including study of planetary interiors, crystallization of igneous melts, and effect of volatiles. *Mailing Add:* Bd Earth Sci & Resources Nat Acad Sci 2101 Constitution Ave NW Washington DC 20418

UTAGIKAR, AJIT PURUSHOTTAM, b Pune, Maharashtra, Sept 28, 67. ENGINEERING PHYSICS. *Educ:* Banaras Hindu Univ, BTech, 89. *Prof Exp:* Res asst plasma processing, Univ Kans, 89-91, teaching asst Fortran 77, 91; MAT ENGR, INTEL CORP, 91- *Mem:* Am Inst Chem Engrs. *Res:* Study of r-f plasmas; development of a method to determine the charged particle concentrations in low frequency glow discharges. *Mailing Add:* Intel Corp CH4-52 5000 W Chandler Blvd Chandler AZ 85224

UTECH, FREDERICK HERBERT, b Merrill, Wis, Apr 19, 43; m; c 1. SYSTEMATIC BOTANY, BOTANY. *Educ:* Univ Wis-Madison, BS, 66, MS, 68; Wash Univ, PhD(biol), 73. *Prof Exp:* Lectr bot, Univ Wis, Marshfield, 70; vis scientist, US-Jap Coop Sci Prog, 74-75; fel bot, Jap Soc Prom Sci, 75-76; CUR BOT, CARNEGIE MUS NATURAL HIST, 76- *Concurrent Pos:* NSF fel, Wash Univ, 71-72 & 72-73; fel, Univ Wis-Madison & res assoc, Wash Univ, 74-76; adj res scientist, Hunt Inst Bot Doc, 77-; M Graham Netting Res Fund grant, Carnegie Mus, 77-87. *Mem:* Bot Soc Am; Int Asn Plant Taxon; AAAS; Am Soc Plant Taxonomists. *Res:* Biosystematic investigations of the living elements of the Arcto-Tertiary geoflora in the Northern hemisphere; floral vascular anatomy of the Liliaceae; cytotaxonomy and systematics of the Liliaceae. *Mailing Add:* Sect Bot Carnegie Mus Natural Hist 4400 Forbes Ave Pittsburgh PA 15213

UTERMOHLEN, VIRGINIA, b New York, NY, June 17, 43; div; c 2. IMMUNOLOGY. *Educ:* Wash Univ, BS, 64; Columbia Univ, MD, 68; Am Bd Pediat, dipl. *Prof Exp:* Intern, resident & chief resident pediat, St Luke's Hosp, NY, 68-71; fel immunol, Rockefeller Univ, 71-74; asst prof biochem, Cornell Univ, 74-77, asst prof nutrit sci, 77-81; asst prof, 77-80, ASSOC PROF, NY STATE COL VET MED, 80-; ASSOC PROF NUTRIT SCI, CORNELL UNIV, 81- *Concurrent Pos:* NIH fel, 71-72; NY Heart Asn fel, 72-74; guest investr immunol, Rockefeller Univ, 74-76; asst dean, NY State Col Human Ecol, 86-87. *Mem:* Harvey Soc; Am Asn Immunologists; Am Med Women's Asn. *Res:* Nutrition and cell-mediated immunity. *Mailing Add:* N204B Van Rensselaer Hall Cornell Univ Ithaca NY 14853

UTGAARD, JOHN EDWARD, b Anamoose, NDak, Jan 22, 36; m 61; c 4. GEOLOGY, PALEOZOOLOGY. *Educ:* Univ NDak, BS, 58; Ind Univ, AM, 61, PhD(geol), 63. *Prof Exp:* Res assoc paleont, US Nat Mus, Smithsonian Inst, 63-65; from asst prof to assoc prof geol, 65-73, PROF GEOL, SOUTHERN ILL UNIV, CARBONDALE, 73- *Concurrent Pos:* Smithsonian fel evolutionary & syst biol, Smithsonian Inst, 72. *Mem:* Geol Soc Am; Am Asn Petrol Geologists; Paleont Soc; Brit Palaeont Asn; Soc Econ Paleont & Mineral; Sigma Xi. *Res:* Fossil bryozoans; carboniferous paleoecology and depositional environments; paleobiology of Paleozoic bryozoans; paleoecology of Late Paleozoic fossil communities. *Mailing Add:* Dept Geol Southern Ill Univ Carbondale IL 62901

UTGARD, RUSSELL OLIVER, b Star Prairie, Wis, July 30, 33; m 56; c 3. GEOLOGY, SCIENCE EDUCATION. *Educ:* Wis State Col, River Falls, BS, 57; Univ Wis, MS, 58; Ind Univ, Bloomington, MAT, 66, EdD(sci), 69. *Prof Exp:* Instr geol, Joliet Jr Col, Ill, 58-67; asst prof, 69-72, ASSOC PROF GEOL, OHIO STATE UNIV, 72- *Concurrent Pos:* Teaching asst, Ind Univ, Bloomington, 65-66. *Mem:* Nat Asn Geol Teachers; Nat Sci Teachers Asn; Geol Soc Am. *Res:* Environmental geology. *Mailing Add:* Dept Geol Sci Ohio State Univ Columbus OH 43210

UTGOFF, VADYM V, b Sevastopol, Russia, Aug 3, 17; m 46; c 3. AEROSPACE ENGINEERING. *Educ:* US Naval Acad, BS, 39; US Naval Postgrad Sch, BS, 48; Mass Inst Technol, MS, 49. *Prof Exp:* Assoc prof, 64-83, PROF EMER AEROSPACE ENG, US NAVAL ACAD, 84- *Mem:* Am Inst Aeronaut & Astronaut; Am Helicopter Soc. *Res:* Flight dynamics; rotary wing aerodynamics. *Mailing Add:* Two Ridge Rd Wardour Annapolis MD 21401-1201

UTHE, JOHN FREDERICK, b Saskatoon, Sask, Feb 27, 38; m 63; c 3. FISHERIES. *Educ:* Univ Sask, BA, 59, Hons, 60, MA, 61; Univ Western Ont, PhD(biochem), 68. *Prof Exp:* Res scientist, Fresh Water Inst, Fisheries Res Bd, 63-72, res mgr, Technol Br, 72-79, HEAD, INORGANIC CONTAMINANTS & STABLE ISOTOPES SECT, HALIFAX, FISHERIES & OCEANS CAN, 87- *Res:* Analytical chemistry applied to toxic residues and the biochemical effects of such residues on fish. *Mailing Add:* 32 Simcoe Pl Halifax NS B3M 1H3 Can

UTHE, P(AUL) M(ICHAEL), JR, nuclear engineering, materials science engineering; deceased, see previous edition for last biography

UTKE, ALLEN R, b Moline, Ill, Feb 5, 36; m 57; c 3. INORGANIC CHEMISTRY. *Educ:* Augustana Col, Ill, BS, 58; Univ Iowa, MS, 61, PhD(inorg chem), 63. *Prof Exp:* Sr res chemist, Chem Div, Pittsburgh Plate Glass Co, Tex, 62-64; assoc prof, 64-78, PROF CHEM, UNIV WIS-OSHKOSH, 78- *Mem:* Am Chem Soc. *Res:* Chemistry of the alkali and alkaline earth metals and their reactions in liquid ammonia. *Mailing Add:* Dept Chem Univ Wis 800 Olgoma Blvd Oshkosh WI 54901

UTKHEDE, RAJESHWAR SHAMRAO, b Kalmeshwar, India, Apr 18, 39; c 2. PLANT BREEDING, MICROBIOLOGY. *Educ:* Nagpur Univ, BSc, 61; Indian Agr Res Inst, PhD(genetics), 68. *Prof Exp:* Res assoc genetics, Rockefeller Found, 67-70; millet breeder, Haryana Agr Univ, 71-72; corn breeder, Ministry Agr, Tanzania, 72-75; fel plant path, Simon Fraser Univ, 75-77, res assoc, 77-80; PLANT PATHOLOGIST, AGR CAN, 80- *Concurrent Pos:* Res assoc, Can Ministry Manpower & Immigration, 75-76; consult, Food & Agr Orgn, UN, 84. *Mem:* Genetics Soc Can; Indian Soc Genetics; Can Soc Phytopath; Am Phytopath Soc. *Res:* Breeding for disease resistance; control of soilborne diseases by integrated use of resistance, microbial antagonist and chemical treatment. *Mailing Add:* Res Sta Agr Can Summerland BC V0H 1Z0 Can

UTKU, BISULAY BEREKET, b Bandirma, Turkey, Apr 28, 40; m 64; c 2. ARCHITECTURAL ENGINEERING, ARCHITECTURAL DESIGN. *Educ:* Instanbul Tech Univ, dipl arch ing, 62, PhD(arch eng), 78, docent arch, 82; NC State Univ, MArch, 73. *Prof Exp:* Teaching asst struct anal, Istanbul Tech Univ, 62-65, teaching asst arch struct, 74-78, asst prof, 78-82, assoc prof, 82-86; res assoc energy conserv, 79-80, ADJ ASSOC PROF ARCH ENG, DUKE UNIV, 87- *Mem:* Nat Asn Arch Engrs. *Res:* Earthquake resistance of masonry structures; passive climate control; energy conservation in buildings. *Mailing Add:* Dept Civil & Environ Eng Duke Univ Durham NC 27706

UTKU, SENOL, b Suruc, Turkey, Nov 23, 31; m 64; c 2. ENGINEERING, SIMULATIONS. *Educ:* Istanbul Tech Univ, MS, 54; Mass Inst Technol, MS, 59, ScD(struct eng), 60. *Prof Exp:* Res engr, Math & Appln Dept, Int Bus Mach Corp, 59-60; asst prof struct, Mass Inst Technol, 60-62; assoc prof, Mid East Tech Univ, Ankara, 62-63; exec chief, Comput Ctr, Istanbul Tech Univ, 63-65; sr res engr, Jet Propulsion Lab, Calif Inst Technol, 65-68, mem tech staff, 68-70; from assoc prof to prof civil eng, 70-78, PROF CIVIL ENG & COMPUT SCI, DUKE UNIV, 78- *Concurrent Pos:* Consult, Math & Appln Dept, Int Bus Mach Corp & Lincoln Lab, Mass Inst Technol, 60-61, Mitre Corp, 61-62, Westinghouse Res & Develop Ctr, 70, Langley Res Ctr, NASA, 71 & Jet Propulsion Lab, Calif Inst Technol, 71-; lectr, Istanbul Tech Univ, 62-63, Univ Southern Calif, 66-70, Univ Wash, 68 & Duke Univ, 70-; mem, NATO Study Group Comput Sci, Brussels, 68-70; partic, Tokten Prog, UN, 79, 80 & 81. *Mem:* Am Soc Civil Engrs; Am Acad Mech; Soc Indust & Appl Math. *Res:* Engineering and structural mechanics; applied mathematics; numerical analysis; concurrent processing; optimal control; adaptive structures. *Mailing Add:* 1843 Woodburn Rd Durham NC 27705

UTLAUT, WILLIAM FREDERICK, b Sterling, Colo, July 26, 22; m 46; c 3. TELECOMMUNICATIONS. *Educ:* Univ Colo, BSEE, 44, MSEE, 50, PhD(elec eng), 66. *Prof Exp:* Engr large motor design, Gen Elec Co, 46-48; instr elec eng, Univ Colo, 48-54; electronic engr radio propagation, Dept Commun, Nat Bur Standards, 54-64; dir Ionospheric Telecommun Lab, Environ Sci & Services Admin, Dept Com, 64-67; DIR TELECOMMUN RES & ENG, INST TELECOMMUN SCI, NAT TELECOMMUN & INFO ADMIN, DEPT COM, 67- *Concurrent Pos:* Chmn, spectrum utilization & monitoring, US Study Group I, Int Radio Consult Comt, 72-; chmn, integrated serv digital networks tech subcomt of TI, 84-88; chair, Serv architectures & signaling tech subcont of TI, 88-, Joint Working Party on ISDN, 87-89, US Study Group B, Consultative Comt Int Tel & Telephony, 89- *Mem:* Int Union Radio Sci; fel Inst Elec & Electronics Engrs. *Res:* Radiowave propagation and ionospheric modification by high-powered ground-based radio frequency transmitters; radio spectrum utilization studies. *Mailing Add:* Inst Telecommun Sci Nat Telecommun & Info Admin Boulder CO 80303-3328

UTLEY, JOHN FOSTER, III, b Detroit, Mich, June 23, 44; m 71. PLANT SYSTEMATICS, BIOLOGY. *Educ:* Univ SFla, BA, 68; Duke Univ, PhD(bot), 77. *Prof Exp:* Cur bot, Div Natural Hist, Mus Nac de Costa Rica, 73-76; fel, US Nat Herbarium, Smithsonian Inst, 76-77; ASST PROF BIOL, UNIV NEW ORLEANS, 78- *Mem:* Sigma Xi; Soc Study Evolution; Int Asn Plant Taxonomists. *Res:* Systematics; evolution and ecology of epiphytic angiosperms. *Mailing Add:* Dept Biol Sci Univ New Orleans New Orleans LA 70148

UTLEY, PHILIP RAY, b Ill, Dec 18, 41; m 63; c 3. ANIMAL SCIENCE. *Educ:* Southern Ill Univ, Carbondale, BS, 64; Univ Mo-Columbia, MS, 67; Univ Ky, PhD(animal sci), 69. *Prof Exp:* Asst instr animal sci, Southern Ill Univ, Carbondale, 63-65; assoc prof, 70-78, PROF ANIMAL SCI, UNIV GA, COASTAL PLAINS EXP STA, TIFTON, 78- *Mem:* Am Soc Animal Sci. *Res:* Beef cattle nutrition and management. *Mailing Add:* Dept Animal Sci Tifton GA 31793

UTRACKI, LECHOSLAW ADAM, b Poland, Aug 1, 31; Can citizen; m 56; c 2. POLYMER ENGINEERING, RHEOLOGY. *Educ:* Polytech Lodz, Poland, BS, 53, MEng, 56. *Hon Degrees:* DSc, Polytechnic Lodz, Poland, 60; Habitation, 63. *Prof Exp:* Adj phys chem polymers, Polish Acad SCi, 56-65; vis scientist, Univ Southern Calif, 65-67; vis prof macromol sci, Case Western Reserve Univ, 67-68; researcher polymer eng, Gulf Oil Can Ltd, 68-71; vis scientist polymer rheology, McGill Univ, 71-73; group leader, CIL Inc, 73-80; SR RES OFFICER RHEOLOGY, NAT RES COUN CAN, 80- *Concurrent Pos:* Postdoctorate fel, Univ Southern Calif, 60-62; dir, Plastic Eng, 82-; chmn, Versailles Proj Advan Mat & Sci, 85- *Mem:* Can Rheology Soc (pres, 84-86); Polymer Processing Soc (pres, 87-89); Am Chem Soc; Soc Plastic Eng; Chem Inst Can. *Res:* Preparation properties and performance of polymer alloys, blends and composites. *Mailing Add:* Indust Mat Inst Nat Res Coun Can 75 de Mortagne Boucherville PQ J4B 6Y4 Can

UTTER, FRED MADISON, b Seattle, Wash, Nov 25, 31; m 58; c 3. BIOCHEMICAL GENETICS. *Educ:* Univ Puget Sound, BSc, 54; Univ Wash, MSc, 64; Univ Calif, Davis, PhD(genetics), 69. *Prof Exp:* Serologist, Biol Lab, US Bur Com Fisheries, 59-60, chemist, 60-80, geneticist, 80-83, supvry geneticist, 83-88; RETIRED. *Concurrent Pos:* Affiliate asst prof, Univ Wash, 71, affil assoc prof, 76, affil prof, 82; consult, appl pop genetics, 88-; vis prof, Autonomous Univ Barcelona, 88- & Univ Oviedo, 90- *Mem:* Fel Am Fisheries Soc; Am Inst Fishery Res Biol. *Res:* Use of biochemical methods for the detection of genetic variations in fish for use in studies of fish populations; induced gynogenesis and polyploidy in salmon. *Mailing Add:* 19424 Tenth NE Seattle WA 98155

UTTERBACK, DONALD D, b 1904. OIL & GAS SULPHUR. *Educ:* Univ Ill, BS, MS, PhD(geol). *Prof Exp:* Sr geologist, Tex Co; dist geologist, Houston Oil Co; vpres, Exp & Develop, Freeport Oil Co; consult, Exp Oil, Gas & Sulphur, Guatemala, Iran, Brazil, 60-87; RETIRED. *Mem:* Petrol Eng Soc/ Am Inst Mech Eng; fel Geol Soc Am; Soc Eng Geologists; Am Inst Petrol Geologists; Am Asn Petrol Geologists. *Mailing Add:* 1703 Pere Marquette Bldg New Orleans LA 70112

UTTERBACK, NYLE GENE, b Oskaloosa, Iowa, Jan 19, 31; m 58; c 3. EXPERIMENTAL PHYSICS. *Educ:* Iowa State Univ, BS, 53, PhD(physics), 57. *Prof Exp:* Res asst, Ames Lab, AEC, 51-57; Fulbright scholar, Ger, 57-58; sr physicist, Ord Res Lab, Univ Va, 58-59; asst prof & res physicist, Denver Res Inst, 59-63; staff scientist, Gen Motors Corp, Calif, 63-72; staff scientist, Mission Res Corp, Santa Barbara, 72-75; consult, TRW, Redondo Beach, Calif, 76-81, CONSULT, SANSUM MED RES FOUND, SANTA BARBARA, 81- *Mem:* Am Phys Soc. *Res:* Laser technology; atomic and molecular reaction kinetics; diabetic microangiopathy. *Mailing Add:* 718 Willowglen Rd Santa Barbara CA 93105

UTZ, JOHN PHILIP, b Rochester, Minn, June 9, 22; m 47; c 5. MEDICINE. *Educ:* Northwestern Univ, BS, 43, MD, 47; Georgetown Univ, MS, 49. *Prof Exp:* Researcher, Lab for Infectious Dis, Nat Inst Allergy & Infectious Dis, 47-49, chief infectious dis serv, 52-65; prof med & chmn div immunol & infectious dis, Med Col Va, 65-73; dean fac, 73-78, PROF MED, GEORGETOWN UNIV, 73- *Concurrent Pos:* Fel, Mayo Found, 49-52; intern, Evans Mem Hosp, Boston, 46-47; consult, E I du Pont de Nemours & Co; pres, Nat Found Infectious Dis, 72-75. *Mem:* Am Fedn Clin Res; Am Col Physicians; Am Col Chest Physicians; Soc Exp Biol & Med; Am Thoracic Soc; Am Soc Clin Invest; Asn Am Phys. *Res:* Clinical investigations in infectious diseases. *Mailing Add:* Georgetown Univ Hosp Washington DC 20007

UTZ, WINFIELD ROY, JR, b Boonville, Mo, Nov 17, 19; m 41; c 3. MATHEMATICAL ANALYSIS. *Educ:* Cent Col, Mo, AB, 41; Univ Mo, MA, 42; Univ Va, PhD(math), 48. *Prof Exp:* Asst instr math, Univ Mo, 42; instr, Univ Notre Dame, 42-43; instr, Univ Va, 44-48; instr, Univ Mich, 48-49; from asst prof to assoc prof, 49-69, PROF MATH, UNIV MO, COLUMBIA, 69-, CHMN DEPT, 70- *Concurrent Pos:* Mem, Inst Advan Study, Princeton Univ, 55-56; vis scholar, Univ Calif, Berkeley, 62-63; vis prof, Brown Univ, 69. *Mem:* Am Math Soc; Math Asn Am; London Math Soc. *Res:* Surface dynamics; topological dynamics; differential equations. *Mailing Add:* Dept Math Univ Mo Columbia MO 65201

UWAYDAH, IBRAHIM MUSA, b Qualqiliya, Jordan, Sept 18, 43; US citizen; m 68; c 4. ANTI-INFLAMMATORY, ANTIALLERGY. *Educ:* Am Univ Beirut, BSc, 67; Univ Kans, Lawrence, PhD(med chem), 74. *Prof Exp:* Fel res assoc pharmacol, Med Col Va, 74-77; RES ASSOC, A H ROBINS CO, 77- *Concurrent Pos:* Adj assoc prof pharmacol, Med Col Va, 78-; adj asst prof medicinal chem, Med Col Va, 86- *Mem:* Am Chem Soc; NY Acad Sci; Sigma Xi; Int Soc Heterocyclic Chem. *Res:* Design and synthesis of potentially active bioactive agents; inflammation area; H2-antagonists (gastrointestinal); central nervous system area (analgesics); B-blockers. *Mailing Add:* A H Robins Co 1211 Sherwood Ave Richmond VA 23220

UY, WILLIAM CHENG, b Manila, Philippines, Feb 11, 40; US citizen; m 69; c 4. ENGINEERING, CHEMICAL ENGINEERING. *Educ:* De La Salle Col, Philippines, BSChE, 65; Northwestern Univ, MS, 67, PhD(chem eng), 70. *Prof Exp:* Res engr, 70-74, SR RES ENGR, E I DU PONT DE NEMOURS & CO, INC, 74-, RES ASSOC, 88- *Mem:* Soc Rheology; Am Chem Soc. *Res:* Polymer characterization; fiber fatigue resistance; fiber finishes; melt and wet fiber spinning; rheology; polymerization. *Mailing Add:* Pioneering Res Lab Du Pont Exp Sta Wilmington DE 19880-0302

UYEDA, CARL KAORU, b San Bernardino, Calif, July 11, 22; m 76; c 2. ANATOMY, CYTOPATHOLOGY. *Educ:* Syracuse Univ, BA, 47, MS, 49; Univ Md, PhD(anat), 66. *Prof Exp:* Div head cancer cytol dept, Md State Dept Health, 50-58; instr cytopath, Sch Med, Univ Md, 58-64, instr anat, 62-67; asst prof path & anat, Univ Ark, Little Rock, 67-73, assoc dir sch cytotechnol & dir cytopath Lab, Med Sci, 67-77, assoc prof path, 73-77; cytopathologist, Path Lab, Los Gatos, Calif, 77-84; cytopathologist, Lab Serv, San Jose, Calif, 85-89; RETIRED. *Concurrent Pos:* Res assoc, Sch Med, Johns Hopkins Univ, 56-65, sr cytologist, 65-67; sr cytologist, Ark State Dept Health, 65-67; contractor, Nat Ctr Toxicol Res, Food & Drug Admin, Ark, 71-77; consult, Vet Admin Hosp, Little Rock, 72-77. *Mem:* Am Soc Cytol; Int Acad Cytol; Pan-Am Cancer Cytol Soc; Am Asn Anat; NY Acad Sci; Sigma Xi. *Res:* Cytogenetics; abnormal cytogenetic changes; clinical carcinoma and congenital anomalies; spontaneous leukemic C3H and C57 mice; cytopathology, refinement of interpretation in structural change of cancer; neoplasm of mice bladder; circadian rhythmicity in bronchogenic carcinoma and mice tissue. *Mailing Add:* 9808 Catskill Rd Little Rock AR 72207-5525

UYEDA, CHARLES TSUNEO, b Penryn, Calif, Feb 20, 29; m 56; c 2. MEDICAL MICROBIOLOGY. *Educ:* San Jose State Col, BA, 51; Miami Univ, MA, 52; Stanford Univ, PhD(med microbiol), 56. *Prof Exp:* Officer-in-charge diag microbiol, 406th Med Gen Lab, US Army, Japan, 56-57; bacteriologist microbiol, 6th US Army Area Lab, Ft Baker, 57-58, lab serv, Vet Admin Hosp, Oakland, 58-63; MICROBIOLOGIST, LAB SERV, VET ADMIN MED CTR, PALO ALTO, 63- *Concurrent Pos:* Clin lab officer, US Army Med Serv Corps, 58-; res assoc, Sch Med, Stanford Univ, 63-78, clin asst prof path, 78-; instr, San Francisco State Univ, 79- *Mem:* Am Soc Microbiol; Sigma Xi. *Res:* Rapid and automated methods in the diagnosis and treatment of infectious diseases as well as in the immunology of multiple sclerosis; serology and immunology. *Mailing Add:* 875 Norfolk Pine Ave Sunnyvale CA 94087

UYEDA, KOSAKU, b Kokawa Naga-gun, Japan, Mar 15, 32; US citizen; m 57; c 2. BIOCHEMISTRY. *Educ:* Ore State Univ, BS, 55, MS, 57; Univ Calif, Berkeley, PhD(biochem), 62. *Prof Exp:* From asst prof to assoc prof, 67-81, PROF BIOCHEM, UNIV TEX HEALTH SCI CTR, DALLAS, 82-; RES CHEMIST, VET ADMIN HOSP, 67-, CHIEF CELLULAR REGULATION, 71- *Concurrent Pos:* Fel, Univ Calif, Berkeley, 62; NIH fel, Pub Health Res Inst NY, 62-64; scholar, Univ Calif, Berkeley, 64-67. *Honors & Awards:* William S Middleton Award, 84. *Mem:* Am Chem Soc; Am Soc Biol Chem. *Res:* Elucidation of the mechanism of action of enzymes and allosteric enzymes and their roles in regulation of carbohydrate metabolism. *Mailing Add:* Dept Biochem Univ Tex Southwestern Med Sch 4500 S Lancaster Rd Dallas TX 75216-7167

UYEHARA, OTTO A(RTHUR), b Hanford, Calif, Sept 9, 16; m 45; c 3. MECHANICAL ENGINEERING. *Educ:* Univ Wis, BS, 42, MS, 43, PhD(chem eng), 45. *Prof Exp:* Alumni Res Found fel, 45-46, res assoc, 46-47, from asst prof to prof, 47-82, EMER PROF MECH ENG, UNIV WIS-MADISON, 82-; CONSULT, 82- *Concurrent Pos:* Invited lectr, India, 68, Japan, Internal Combustion Engines, 76 & Norway Technol Inst, 77; instr combustion, Nat Cheng Kung Univ, Taiwan, 85. *Honors & Awards:* Benjamin Smith Reynolds Award, 67; Horning Award, Soc Automotive Engrs, 67 & 69, Colwell Awards; Dugold Clerk Award, Brit Inst Mech Engrs, 71. *Mem:* Fel Soc Automotive Engrs; Am Soc Mech Engrs; hon mem Japan Soc Mech Eng. *Res:* Emission and combustion in internal combustion engines; instantaneous flame temperature indicators; influence of operating variables; fuel droplet vaporization in transient state; compression temperature measurement in internal combustion engines. *Mailing Add:* 544 S Bond St Anaheim CA 92805

UYEKI, EDWIN M, b Seattle, Wash, Mar 12, 28; m 51; c 3. PHARMACOLOGY, RADIOBIOLOGY. *Educ:* Kenyon Col, AB, 49; Univ Chicago, PhD(pharmacol), 53. *Prof Exp:* Instr pharmacol, Univ Chicago, 53-54; instr, Sch Med, Western Reserve Univ, 54-60, sect assoc radiation biol, 54-60; sr scientist, Hanford Labs Gen Elec Co, 60-65; assoc prof, 65-70, PROF PHARMACOL, MED CTR, UNIV KANS, 70- *Mem:* AAAS; Am Soc Pharmacol & Exp Therapeut; Am Soc Cell Biol; Radiation Res Soc. *Res:* Immunopharmacology; immunosuppressants on antibody formation; bone marrow transplantation in radiation chimeras; radiation effects; short term tissue culture. *Mailing Add:* NIX 3008 Univ Kans G51 Health 39th St & Rainbow Blvd Kansas City KS 66103

UYEMOTO, JERRY KAZUMITSU, b Fresno, Calif, May 27, 39; m 65; c 1. PLANT PATHOLOGY, PLANT VIROLOGY. *Educ:* Univ Calif, Davis, BS, 62, MS, 64, PhD(plant path), 68. *Prof Exp:* Lab technician, Univ Calif, Davis, 63-67; from asst prof to assoc prof virol, Cornell Univ, 68-77; from assoc prof to prof, Kans State Univ, 77-83; PROF, USDA AGR RES SERV, UNIV CALIF, DAVIS, 86- *Mem:* Asn Appl Biologists; Am Phytopath Soc. *Res:* Epidemiology and control of plant virus diseases. *Mailing Add:* Dept Plant Path USDA Agr Res Serv Univ Calif Davis CA 95616

UYENO, EDWARD TEISO, b Vancouver, BC, Mar, 31, 21; m 69. PHARMACOLOGY, BEHAVIORAL PSYCHOPHARMACOLOGY. *Educ:* Univ Toronto, BA, 47, MA, 52, PhD(psychol), 58. *Prof Exp:* Res assoc psychol, Stanford Univ, 58-61, RES PSYCHOLOGIST & PHARMACOLOGIST, STANFORD RES INST, 61- *Concurrent Pos:* NIH grants, 63-66, 68-70, 71-73, 75-78 & 81-83. *Mem:* Am Soc Pharmacol & Exp Therapeut; Am Psychol Asn; Psychonomic Soc; Can Psychol Asn. *Res:* Behavioral psychopharmacology; effects of drugs on learning, retention, and reproduction of animals; interaction effects of drugs; self-administration of alcohol and narcotics by rats; analgesics and narcotic antagonists; behavioral toxicology; bioassay of peptides, anxiolytics, anti-convulsants, stimulants, depressants, and hallucinogens; author of over 90 scientific articles and of several chapters in books. *Mailing Add:* Life Sci Div Stanford Res Inst Menlo Park CA 94025

UYS, JOHANNES MARTHINUS, b Heidelberg, Rep SAfrica, Oct 17, 25; m 52; c 3. THERMODYNAMICS IRON & STEEL MAKING, METAL WORKING. *Educ:* Univ Pretoria, BSc, 46, MSc, 50; Mass Inst Technol, ScD(metall eng), 59. *Prof Exp:* Supvr, Ludlum Steel Corp, 59-61; asst dir res, Youngstown Sheet and Tube Co, 61-65, asst vpres opers, 65-70, dir res & develop, 70-71, dir tech servs, 71, vpres tech serv, Youngstown Sheet & Tube Co, 72-78; dir qual control, Jones & Laughlin Steel Corp, 79-83; dir, tech servs, McLouth Steel Prod Corp, 83-85; vpres technol, Sharon Steel Corp, 87-89; DIR TECHNOL, WHEELING-PITTSBURGH STEEL CORP, 89- *Mem:* Fel Am Soc Mat; Asn Iron & Steel Eng; Am Soc Testing & Mat; Am Iron & Steel Inst. *Res:* Improvements of the processes used to produce iron and steel. *Mailing Add:* 2009 Guadalupe Ave Youngstown OH 44504

UZER, AHMET TURGAY, b Samsun, Turkey, Feb 1, 52. QUANTUM DYNAMICS OF ATOMS & MOLECULES, COMPUTATIONAL PHYSICS. *Educ:* Mid East Tech Univ, Ankara, Turkey, BSc, 74; Harvard Univ, AM, 77, PhD(chem physics), 79. *Prof Exp:* Res asst, Dept Theoret Chem, Oxford Univ, UK, 79-81; res fel, Noyes Lab, Calif Tech, 82-83; res assoc, Joint Inst Lab Astrophys, Boulder, Colo, 83-85; asst prof, 85-90, ASSOC PROF, SCH PHYSICS, GA INST TECHNOL, 90- *Mem:* Am Phys Soc; Am Chem Soc; Sigma Xi. *Res:* Dynamics of intramolecular energy transfer; quantization of nonlinear systems; quantum mechanics of chaotic systems; computational physics; semiclassical theories; matter-laser interactions. *Mailing Add:* Sch Physics Ga Inst Technol Atlanta GA 30332-0430

UZES, CHARLES ALPHONSE, b Downey, Calif, Dec 14, 39; m 67; c 1. THEORETICAL PHYSICS. *Educ:* Calif State Univ, Long Beach, BS, 62; Univ Calif, Riverside, MA, 64, PhD(physics), 67. *Prof Exp:* Asst prof, 67-73, ASSOC PROF PHYSICS, UNIV GA, 73- *Mem:* Am Phys Soc. *Res:* The use of non-perturbative calculational methods in the nonlinear classical and quantum mechanical problems of field and many body theory, and in solid state physics. *Mailing Add:* Dept Physics & Astron Univ Ga Athens GA 30602

UZGIRIS, EGIDIJUS E, b Lithuania, Jan 11, 41; m 67; c 2. STRUCTURAL BIOLOGY, BIOTRANSFORMATIONS. *Educ:* Univ Ill, BS, 62; Harvard Univ, MS, 64, PhD(physics), 68. *Prof Exp:* Res assoc, Harvard Univ, 68-69; res assoc, Joint Inst Lab Astrophys, Univ Colo, 69-70; PHYSICIST, GEN ELEC RES & DEVELOP CTR, 70- *Concurrent Pos:* Vis scientist, INSERN, Nancy, France, 78 & Sch Med, Stanford Univ, 81-82. *Mem:* Am Phys Soc; Biophys Soc; NY Acad Sci; AAAS. *Res:* Structural biology, protein crystallization on membranes; light scattering and hydrodynamic measurements; cellular immunology and cell surface change measurements; frequency and wavelength standards; non linear laser spectroscopy; biophysics. *Mailing Add:* Gen Elec Res & Develop Ctr PO Box Eight Schenectady NY 12301

UZIEL, MAYO, b Seattle, Wash, May 3, 30; m 67; c 2. BIOLOGICAL & CLINICAL CHEMISTRY, CELL BIOLOGY. *Educ:* Univ Wash, BSc, 52, PhD(biochem), 55. *Prof Exp:* Nat Found Infantile Paralysis fel, Rockefeller Inst, 55-57; asst prof biochem, Sch Med, Tufts Univ, 57-62; biochemist, Mass Eye & Ear Infirmary, 62-64; biochemist, Biol Div, 64-81, BIOCHEMIST, HEALTH & SAFETY RES DIV, OAK RIDGE NAT LAB, 81- *Concurrent*

Pos: Mem subcomt specification & criteria nucleotides & related compounds, NSF-Nat Res Coun, 68-75; prof, Univ Tenn, 71- *Mem:* AAAS; Am Soc Biochem & Molecular Biol; Am Chem Soc. *Res:* Structure and function of biological macromolecules; risk assessment; bioindicators of injury and disease; human genonie. *Mailing Add:* 102 Newton Lane Oak Ridge TN 37830

UZODINMA, JOHN E, b Onitsha, Nigeria, July 26, 29; m 57; c 4. PREVENTIVE MEDICINE, MICROBIOLOGY. *Educ:* Grinnell Col, BA, 54; Univ Iowa, MS, 56, PhD(prev med, microbiol), 65. *Prof Exp:* Bacteriologist, State Hyg Labs, Iowa, 57-58; microbiologist, Broadlawns County Hosp, Des Moines, 58-61; PROF BIOL, JACKSON STATE UNIV, 64-, CHMN DEPT, 67- *Mem:* AAAS; Am Soc Clin Path; Am Inst Biol Sci; Am Soc Microbiol; Royal Soc Trop Med & Hyg; Sigma Xi. *Res:* Host-parasite relationships; effect of insulin, serotonin and thyroxine on the penetration of tissue culture cells by Toxoplasma gondii; effect of certain drugs on Trypanosoma equiperdum infections in mice. *Mailing Add:* Biol Dept Jackson State Univ PO Box 18723 Jackson MS 39209

UZZELL, THOMAS, b Charleston, SC, Apr 6, 32; m 75; c 2. SYSTEMATIC BIOLOGY, VERTEBRATE BIOLOGY. *Educ:* Univ Mich, BA, 53, MS, 58, PhD(zool), 62. *Prof Exp:* From instr to asst prof biol, Univ Chicago, 62-67; asst prof & asst cur herpet, Peabody Mus, Yale Univ, 67-72, fel, Berkeley Col, 70-72; assoc cur, Acad Natural Sci 72-77, cur herpet, 77-85; DIR, MUS NATURAL HIST, 85-; ASSOC PROF ECOL, ETHOLOGY & EVOLUTION, UNIV ILL, 85- *Concurrent Pos:* Adj assoc prof, Univ Pa, 74-85. *Mem:* AAAS; Am Soc Ichthyol & Herpet; Soc Study Evolution; Soc Syst Zool. *Res:* Origin and evolution of hybrid species of vertebrates; determination of the generic and specific limits of South American lizards of the family Teiidae. *Mailing Add:* 438 NHB 1301 W Green St Urbana IL 61801

V

VAALER, JEFFREY DAVID, b Grand Forks, NDak, May 2, 48; m 70; c 1. ANALYTIC NUMBER THEORY, DIOPHANTINE APPROXIMATION. *Educ:* Lawrence Univ, Appleton, Wis, BS, 70; Univ Ill, Urbana, MS, 71, PhD(math), 74. *Prof Exp:* Res instr math, Calif Inst Technol, 74-76; asst prof math, Univ Tex, Austin, 76-82; mem, Inst Advan Study, Princeton, 82-83; assoc prof, 83-87, PROF MATH, UNIV TEX, AUSTIN, 87- *Concurrent Pos:* Prin investr, NSF, 76- *Mem:* Am Math Soc. *Res:* Diophantine approximation; diophantine equations; fourier analysis; analytic number theory; applications of analysis to number theory. *Mailing Add:* Dept Math Univ Tex Austin TX 78712

VACCA, LINDA LEE, b Paterson, NJ, Mar 10, 47. NEUROSCIENCE, HISTOCHEMISTRY. *Educ:* Col William & Mary, BS, 68; Tulane Univ, MS, 71, PhD(biol), 73. *Prof Exp:* Res asst ecol, Marine Biol Lab, 69, teaching asst, comp physiol, 69-70, biol, Tulane Univ, 70-71; res assoc histochem, La State Univ Med Sch, 71-73; vis fel psychiat res, Dept of Path, Col Physicians & Surgeons NY State Psychiat Inst, 73-74; instr neuroanat, Dept of Physiol, NY Univ Med Ctr, 74-76; asst prof, dept physiol, Med Col Ga, 76-80, asst prof neurosci res, dept path & anat, 76-; AT DEPT ANAT, UNIV KANS HEALTH SCI CTR. *Concurrent Pos:* Mem Biol Stain Comn, 72-; NIMH fel, NY State Psychiat Inst, 73-74; dir histopath, Dept of Path, Med Col Ga, 76-, biomed res grant, Med Col Ga, 78-79 & 79-80; mem, Comt Combat Huntingtons Dis, 80-81 & 81-82; NSF travel fel, VI Int Cong Histochem & Cytochem, England, 80; invited speaker, Winter Conf Brain Res, 82. *Mem:* Sigma Xi; Histochem Soc; Soc Neurosci; Am Asn Anat. *Res:* Histochemical identification of neurotransmitters; ultrastructural evaluation of neural circuits in central nervous system; immunocytochemical tracing of pain pathways in spinal cord which contain substance P, methionine enkephalin, and other peptides; peptides in basal ganglia and Huntington's disease. *Mailing Add:* Dept Anat Univ Kans Health Sci Ctr 39th & Rainbow Blvd Kansas City KS 66103

VACHON, RAYMOND NORMAND, b Lawrence, Mass, Jan 14, 40; m 70; c 5. ORGANIC & POLYMER CHEMISTRY, FIBER LUBRICANTS. *Educ:* Lowell Technol Inst, BS, 63; Princeton Univ, PhD(chem), 67. *Prof Exp:* Res grant, Inst Sci & Technol, Univ Manchester, 67-69; sr res chemist, Burlington Industs, Inc, 69-72; res chemist, 72-74, sr res chemist, Tenn Eastman Co, 74-86, SR TECH REP, MAT SAFETY PROG REGULATORY AFFAIRS, TENN EASTMAN CO 86- *Mem:* Am Chem Soc. *Mailing Add:* 440 Forest Hills Rd Kingsport TN 37663-2220

VACHON, REGINALD IRENEE, b Norfolk, Va, Jan 29, 37; m 60; c 2. MECHANICAL ENGINEERING. *Educ:* Auburn Univ, BME, 58, MSNS, 60; Okla State Univ, PhD(mech eng), 63; Jones Law Sch, LLB, 69. *Prof Exp:* Instr physics, Auburn Univ, 58-59, assoc prof mech eng, 63-67, alumni prof, 67-78, prof, 78-80, assoc researcher heat transfer, Res Found, 60-61; chief oper off, Thacker Orgn Inc, 81-91; pres, Vachon Nix & Assocs, 80-82, PRES VNA SYSTS, INC, 82- *Concurrent Pos:* Mem, Southern Interstate Nuclear Bd; chmn bd, Optimal Systs Int Inc, 69- *Mem:* Am Soc Mech Engrs; Am Inst Aeronaut & Astronaut; Am Soc Eng Educ; Nat Soc Prof Engrs; Am Bar Asn. *Res:* Thermoscience; conduction in solids; gas dynamics; boiling and convection heat transfer; power and energy systems; systems approach management; systems design of machine vision and artificial intelligence. *Mailing Add:* VNA Systs Inc PO Box 467069 Atlanta GA 30346

VACIK, JAMES P, b North Judson, Ind, Nov 30, 31; m 67; c 6. PHARMACEUTICAL CHEMISTRY, BIONUCLEONICS. *Educ:* Purdue Univ, BS, 55, MS, 57, PhD(bionucleonics), 59. *Prof Exp:* Asst prof & res fel bionucleonics, Purdue Univ, 59-60; assoc prof pharmaceut chem & chmn dept, NDak State Univ, 60-63, prof pharmaceut chem & bionucleonics & chmn dept, 63-76; ASSOC PROF PHARMACOL & DIR ENVIRON SAFETY, UNIV S ALA, 76- *Mem:* AAAS; Health Physics Soc; Am Pharmaceut Asn; Am Conf Gov & Ind Hyg. *Res:* Bionucleonics including metabolism, uptake and distribution of radioisotope tracers and large animal biosynthesis; synthesis of benzodioxans; antiviral agents. *Mailing Add:* 1220 Vendome Dr W Mobile AL 36688

VACIRCA, SALVATORE JOHN, b Bronx, NY, July 20, 22; c 2. RADIATION PHYSICS, HEALTH PHYSICS. *Educ:* City Col New York, BS, 48; NY Univ, MS, 54, PhD(radiation physics), 70; Am Bd Health Physics, dipl, 60; Am Bd Med Physics, dipl, 89. *Prof Exp:* Res asst nuclear electronics instrumentation, Sloan Kettering Inst Cancer Res, 48-54, asst physicist, Mem-Sloan Kettering Ctr Cancer & Allied Dis, 54-56, asst attend physicist, 56-61; co-dir nuclear med lab, State Univ NY Downstate Med Ctr, 65-70, chmn radiation technol prog, 69-71, from asst prof to assoc prof radiol, 73-78, dir radiation physics lab & radiation safety off, 61-78; DIR MED PHYSICS DIV, DEPT RADIOL, NORTH SHORE UNIV HOSP, 78-; PROF CLIN RADIOL, CORNELL MED SCH, 78- *Concurrent Pos:* Instr & res assoc, Sloan Kettering Div, Med Col, Cornell Univ, 55-61; dir radiol physics & radiation safety off, Kings County Hosp Ctr, 61-78, sr med physicist, Prof Staff, 72; consult physicist, Col Health Related Professions, State Univ NY, 71-; consult physicist, Coney Island Hosp, 72; adj assoc prof, York Col, NY, 72 & City Col New York, 74. *Mem:* Am Asn Physicists Med; Health Physics Soc. *Res:* Radiation dosimetry as applied to therapy; film and thermoluminescent synergistic dosimetry system used as a method of mapping dose distribution; health physics problems as applied to hospital environment. *Mailing Add:* North Shore Univ Hosp 300 Community Dr Manhasset NY 11030

VACQUIER, VICTOR, b Leningrad, Russia, Oct 13, 07; nat US; m 66; c 1. GEOPHYSICS. *Educ:* Univ Wis, BS, 27, MA, 28. *Prof Exp:* Asst instr physics, Univ Wis, 27-30; geophysicist, Gulf Res & Develop Co, Pa, 30-42; mem staff airborne instruments lab, Columbia Univ, 42-44; marine instruments engr, Sperry Gyroscope Co, 44-53; prof geophys & prin geophysicist, NMex Inst Mining & Technol, 53-57; prof geophys, Scripps Inst Oceanog, 57-74, EMER PROF GEOPHYS, UNIV CALIF, SAN DIEGO, 74- *Honors & Awards:* Witherill Medal, Franklin Inst; Fessendem Award, Soc Explor Geophys; J A Fleming Medal, Am Geophys Union. *Mem:* Geol Soc Am; Am Geophys Union; Franklin Inst. *Res:* Geomagnetism; airborne magnetometry; terrestrial heat flow. *Mailing Add:* Scripps Inst Oceanog Univ Calif San Diego La Jolla CA 92093-0205

VACQUIER, VICTOR DIMITRI, b Pittsburgh, Pa, July 20, 40; m 73; c 2. DEVELOPMENTAL & CELL BIOLOGY. *Educ:* San Diego State Univ, BA, 63; Univ Calif, Berkeley, PhD(zool), 68. *Prof Exp:* Researcher, Intern Lab, Genetics & Biophys, Naples, Italy, 68-69; mem staff, Hopkins Marine Sta, Stanford Univ, 70-71 & Scripps Inst Oceanog, Univ Calif, San Diego, 71-73; from asst prof to assoc prof zool, Univ Calif, Davis, 73-78; assoc prof, 78-80, PROF MARINE BIOL, SCRIPPS INST OCEANOG, UNIV CALIF, SAN DIEGO, 80- *Mem:* Fel AAAS; Am Soc Cell Biol; Soc Develop Biol; Int Soc Develop Biol. *Res:* Biochemistry of fertilization. *Mailing Add:* Marine Biol Res Div Univ Calif La Jolla CA 92093-0202

VADAS, PETER, b Can, Aug 5, 53. CLINICAL IMMUNOLOGY & ALLERGY, REGULATION OF INFLAMMATION. *Educ:* Univ Toronto, BSc, Hons, 76, PhD(exp path), 80, MD, 83; Am Bd Internal Med, dipl, 89; FRCP(C), 90. *Prof Exp:* Med Res Coun postdoctoral fel, Dept Med, Div Immunol, Wellesley Hosp, Toronto, 84-86; res assoc, 82-84, lectr, 87-89, ASST PROF, DIV IMMUNOL, DEPT MED, FAC MED, UNIV TORONTO, 90-, ASST PROF, DEPT IMMUNOL, FAC MED, 91- *Concurrent Pos:* Ont-Que exchange fel, 80-81; med scientist award, Can Heart Found, 80-83; Samuel Castrilli award, 81-83; Walter F Watkins scholar, 82; F M Hill res award, 87; lectr, Dept Path, Univ Toronto, 90-; consult, Div Immunol & Gen Internal Med, Wellesley Hosp, Toronto, 90-; Med Res Coun scholar, 91- *Mem:* NY Acad Sci; Am Rheumatism Asn; Can Soc Clin Invest; Royal Col Physicians & Surgeons; Inflammation Res Asn; Can Soc Immunol; Shock Soc. *Res:* Role of phospholipase A2 in the pathogenesis of local and systemic inflammation; awarded two patents. *Mailing Add:* Wellesley Hosp Rm 238 Jones Bldg 160 Wellesley St E Toronto ON M4Y 1J3 Can

VADAS, ROBERT LOUIS, b New Brunswick, NJ, Aug 5, 36; m 61; c 3. MARINE ECOLOGY, PHYCOLOGY. *Educ:* Utah State Univ, BS, 62; Univ Wash, PhD(bot), 68. *Prof Exp:* Asst prof bot, 67-72, asst prof zool, 68-72, assoc prof, 72-77, PROF BOT, OCEANOG & ZOOL, UNIV MAINE, ORONO, 77-, CHMN DEPT BOT & PLANT PATH, 83- *Concurrent Pos:* Maine Yankee Nuclear Atomic Power Co study grant, 69-74; Off Water Resources grants, 70-72 & 72-75. *Mem:* Ecol Soc Am; Am Soc Naturalists; Phycol Soc Am; Brit Phycol Soc; Int Phycol Soc. *Res:* Ecology of kelp communities; marine plant-herbivore interactions and biogeography; algal distributions; population genetics of marine organisms; ecological studies of Ascophyllum Nodosum. *Mailing Add:* Dept Bot & Oceanog Univ Maine Orono ME 04473

VADHWA, OM PARKASH, b Mandi Maklot Ganj, India, May 10, 41; m 67; c 1. AGRONOMY. *Educ:* Rajasthan Univ, India, BS, 61; Punjab Agr Univ, MS, 63; Utah State Univ, PhD(agron), 71. *Prof Exp:* Lectr agron, Punjab Agr Univ, Hissar Campus, 63-65; fel agron & plant sci, Utah State Univ, 70-71; assoc prof natural resources, Ala A&M Univ, 71-72; ASST PROF AGRON, ALCORN STATE UNIV, 72- *Mem:* Am Soc Agron; Am Asn Univ Prof. *Res:* Forage crops; crop production; soil fertility and plant nutrition; vegetable crops. *Mailing Add:* Dept Agr Alcorn State Univ Lorman MS 39096

VADLAMUDI, SRI KRISHNA, b Moparru, Tenali, AP, India, Aug 15, 27; US citizen; m 54; c 4. MICROBIOLOGY, IMMUNOLOGY. *Educ:* Madras Univ, DVM, 52; Univ Wis-Madison, MS, 59, PhD(microbiol), 63. *Prof Exp:* Head cancer chemother res, Microbiol Asn, Inc, 66-74; sci expert cancer, Smithsonian Sci Info Exch, 75; EXEC SECY, PANELS IMMUNOL, CTR DEVICES & RADIOL HEALTH, FOOD & DRUG ADMIN, HHS, 75-,

IMMUNOL BR CHIEF, 77- , SUPVR MICROBIOLOGIST, 91- *Concurrent Pos:* Vet surgeon, Animal Husbandry Dept, Govt AP, India, 52-55; res vet, Govt NVD Lab, Guntur, India, actg supt, Govt Livestock Res Sta, 57; proj asst, Dept Vet Sci, Univ Wis-Madison, 57-62; sr microbiologist, Dept Infectious Dis, Abbott Lab, 62-65; chief, Viral Chemother Div, Microbiol Asn, Inc, 65-66. *Mem:* Am Soc Microbiol; Am Asn Cancer Res; Am Asn Path; Am Vet Med Asn; Sigma Xi. *Res:* Cancer chemotherapy and immunotherapy; tumor biology and metastasis; immunotoxicology; epizootiology and epidemiology of viral diseases, particularly arbor viruses; infection and immunity; pharmacology. *Mailing Add:* Ctr Radiol Health & Med Devices Food & Drug Admin 1390 Piccard Dr Rockville MD 20850

VAFAKOS, WILLIAM P(AUL), b Brooklyn, NY, Oct 12, 27; m 62; c 3. STRESS ANALYSIS, VIBRATIONS. *Educ:* Polytech Inst Brooklyn, BME, 51, MME, 55, PhD(appl mech), 60; Brooklyn Law Sch, JD, 76. *Prof Exp:* Engr, Westinghouse Elec Corp, 51-53; sr engr, Ford Instrument Co, 53-57; res assoc, 57-60, from asst prof to assoc prof, 60-68, PROF MECH ENG, POLYTECH UNIV, 68- *Mem:* Am Soc Mech Engrs. *Res:* Thin-walled structures. *Mailing Add:* Polytechnic Univ 333 Jay St Brooklyn NY 11201

VAFOPOULO, XANTHE, b Thessaloniki, Greece, Aug 22, 49; US citizen; m 72. DEVELOPMENTAL BIOLOGY, INSECT ENDOCRINOLOGY. *Educ:* Aristotelian Univ, BA, 72; Bridgewater State Col, MA, 78; Univ Conn, PhD(develop biol), 80. *Prof Exp:* Teaching asst, Univ Conn, 76-80, asst prof residence biol & res assoc, biol dept, 81-86; RES ASSOC, DEPT BIOL, YORK UNIV, 86- *Concurrent Pos:* Lectr-instr, W Alton Jones Cell Sci Ctr & Tissue Cult Asn, Inc, 80-81. *Honors & Awards:* Young Investr Award, Soc Develop Biol, 80. *Mem:* Can Soc Zoologists; Soc Develop Biol; Am Soc Zoologists; Can Fedn Biol Socs. *Res:* Circadian control of synthesis and release of developmentally significant hormones in insects; regulation of development by rhythmic release of hormones. *Mailing Add:* Dept Biol York Univ 4700 Keele St North York ON M3J 1P3 Can

VAGELATOS, NICHOLAS, b Kefallinia, Greece, Mar 8, 45; m 79; c 3. PENETRATING RADIATION INSPECTION TECHNOLOGY. *Educ:* Univ Mich, BSE, 67, MSE, 69, PhD(nuclear eng), 73. *Prof Exp:* Res assoc, Neutron spectros, Nat Bur Stand, US Dept of Com, 73-75; sr scientist, 76-78, prin scientist, 78-83, mgr appl develop, 83-85, mgr tech support, 85-86, tech mkt mgr, automation systs group, 86-87, MGR, PROG DEVELOP, NUCLEAR SYSTS DIV, IRT CORP, 87- *Concurrent Pos:* Nat Res Coun fel, Nat Bur Standards, US Dept Com, 73-75. *Mem:* Am Soc Nondestructive Testing; Am Soc Testing & Mat. *Res:* Interaction of radiation with matter; radiation detection and measurement; nuclear technology applications in natural resources evaluation; penetrating radiation nondestructive inspection technology for quality and process control. *Mailing Add:* 13474 Black Hills Rd San Diego CA 92129

VAGELOS, P ROY, b Westfield, NJ, Oct 8, 29; m 55; c 4. LIPID CHEMISTRY, ENZYME CHEMISTRY. *Educ:* Univ Pa, AB, 50; Columbia Univ, MD, 54. *Hon Degrees:* DSc, Wash Univ, 80, Brown Univ, 82, Univ Med & Dent NJ, 84, NY Univ, 89 & Columbia Univ, 90; LLD, Princeton Univ, 90. *Prof Exp:* Intern med, Mass Gen Hosp, Boston, 54-55, asst resident, 55-56; sr asst surgeon, Lab Cellular Physiol, Nat Heart Inst, 56-59, surgeon, 59-61, actg chief sect enzymes, 59-60, sr surgeon, Lab Biochem, 61-62, sr surgeon & res chemist, 63-64, head sect comp biochem, 64-66; sr surgeon, Pasteur Inst, Paris, 62-63; chmn dept biol chem, Sch Med, Wash Univ, 66-75, dir div biol & biomed sci, 73-75; sr vpres res, Merck Sharp & Dohme Res Labs Div, 75-76, pres, 76-84, sr vpres, Merck & Co, Inc, 82-84, exec vpres, 84-85, PRES & CHIEF EXEC OFFICER, MERCH & CO, INC, 85-, CHMN BD DIRS, 86- *Concurrent Pos:* NIH & NSF grants; Sloan vis prof chem, Harvard Col, 73, mem, vis comt biochem & molec biol dept, 81-; mem bd trustees, Rockefeller Univ, 76-, Danforth Found, 78-, Univ Pa, 88- & Partnership NJ, 89-; dir, TRW, Inc, 87-, Prudential Ins Co Am, 89- & NJ Ctr Performing Arts, 89-; mem, Conf Bd, Bus Coun, Policy Comt of Bus Roundtable & Bd Managing dirs, Metropolitan Opera Asn, Inc; vis comts & adv bds, Mass Inst Technol, Col Physicians & Surgeons, Columbia Univ, Cleveland Clin Found, NJ Ctr Advan Biotechnol & Med & Beckman Ctr Hist Chem. *Honors & Awards:* Enzyme Chem Award, Am Chem Soc, 67. *Mem:* Nat Acad Sci; Nat Inst Med; AAAS; Am Soc Biol Chem; Am Acad Arts & Sci. *Res:* Mechanism of lipid biosynthesis; involvement of acyl carrier protein in fatty acid biosynthesis. *Mailing Add:* Merck & Co Inc PO Box 2000 Rahway NJ 07065

VAGNINI, LIVIO L, b North Bergen, NJ, Apr 26, 17; m 49; c 3. CHEMISTRY. *Educ:* Fordham Col, BS, 38. *Prof Exp:* Chemist, H A Wilson Co Div, Englehard Industs, Inc, NJ, 40-42; chief forensic chemist, US Army Criminal Invest Lab, France, 44-46, chief chemist, US Army Graves Regist Lab, Belg, 46-48, chief forensic chemist, Ger, 48-60; microanalyst, US Food & Drug Admin, Washington, DC, 60-63; sr chemist, Cent Intel Agency, 63-73; tech staff mem, Mitre Corp, Va, 73-75; consult criminalist, 75; tech staff, Planning Res Corp, Va, 75-77; prog dir, L Miranda & Assoc, 78-81; RETIRED. *Concurrent Pos:* Consult criminalist, 82- *Mem:* Fel Am Inst Chem; Am Chem Soc; Asn Off Anal Chem; fel Am Acad Forensic Sci; Int Soc Forensic Toxicol. *Res:* Forensic chemistry; microchemistry; serology of dried blood factors; analysis of narcotics; optical crystallography of drugs; microanalysis of foods and drugs; x-ray spectrometry. *Mailing Add:* 26069 Mesa Dr Carmel CA 93923

VAGNUCCI, ANTHONY HILLARY, b Terni, Italy, July 9, 28; US citizen; m 62; c 3. MEDICINE, PHYSIOLOGY. *Educ:* Univ Genoa, MD, 54. *Prof Exp:* Intern med, Wesson Mem Hosp, Springfield, Mass, 57-58; resident, NY Univ-Bellevue Med Ctr, 58-60; Am Heart Asn res fel renal physiol, Med Sch, NY Univ, 60-62; advan res fel, Hypertension Unit, Dept Med, Peter Bent Brigham Hosp, Boston, 62-63 & advan res fel endocrinol, 63-64; jr assoc med & assoc dir endocrinol metab unit, Peter Bent Brigham Hosp, 64-65; from asst prof to assoc prof, 65-79, PROF MED, SCH MED, UNIV PITTSBURGH, 79-; HEAD ADRENAL UNIT, MONTEFIORE HOSP, 65-, HEAD ENDOCRINE UNIT, 77- *Concurrent Pos:* Res assoc, Harvard Med Sch, 64-65; adj prof, elec eng, Univ Pittsburgh, 89. *Mem:* Am Fedn Clin Res; Endocrine Soc; NY Acad Sci; sr fel Inst Elec & Electronics Engrs. *Res:* Circadian physiopath pituitary-adrenal/receptors/pattern recognition. *Mailing Add:* Montefore Hosp 3459 Fifth Ave Pittsburgh PA 15213

VAHALA, GEORGE MARTIN, b Tabor, Czech, Mar 26, 46; Australian citizen; m 70. MAGNETOHYDRODYNAMICS, PLASMA PHYSICS. *Educ:* Univ Western Australia, BSc Hons, 67; Univ Iowa, MS, 69, PhD(physics), 72. *Prof Exp:* Res assoc plasma physics, Univ Tenn, Knoxville, 72; res scientist magnetohydrodynamics, Courant Inst Math Sci, NY Univ, 72-74; asst prof, 74-80, PROF PHYSICS, COL WILLIAM & MARY, 80- *Res:* Magnetohydrodynamics and guiding-center stability of containment devices; spectral theory and its interpretation in plasma physics as well as in magnetohydrodynamics; transport effects in plasmas; nonlinear dynamics. *Mailing Add:* Dept Physics Col William & Mary Williamsburg VA 23185

VAHAVIOLOS, SOTIRIOS J, b Apr 16, 46; c 3. PHYSICAL ACOUSTICS. *Educ:* Fairleigh Dickinson Univ, BS, 70; Columbia Univ, MS, 72, MPh, 75 PhD(elec eng), 76. *Prof Exp:* Mem res staff integrated circuit bonding & automated handling oper, Western Elec Co, 70-71, proj leader, 72-74, res leader, 74-78; PRES, CHIEF EXEC OFFICER & DIR, PHYS ACOUST CORP, 78- *Concurrent Pos:* Adv comt, Indust Elec Cent Instrumental Soc. *Honors & Awards:* Centennial Award, Inst Elec & Electronics Engrs, 84. *Mem:* Fel Inst Elec & Electronics Engrs; Am Soc Testing & Mat; Sigma Xi; NY Acad Sci; Am Soc Nondestructive Testing; sr mem Indust Electronics Soc; sr mem Instrument Soc Am; Soc Exp Stress Anal; fel Acoust Emission Working Group. *Res:* Analog and digital design skills and thorough knowledge of microprocessor and microcomputer technology; author of over 25 publications. *Mailing Add:* Seven Ridgeview Rd Princeton NJ 08540-7601

VAHEY, DAVID WILLIAM, b Youngstown, Ohio, Nov 21, 44; m 77; c 2. OPTICAL PHYSICS, ULTRASONICS. *Educ:* Mass Inst Technol, BS, 66; Calif Inst Technol, MS, 67, PhD(elec eng), 73. *Prof Exp:* Fel physics, Battelle Mem Inst, 73-74, res scientist, 74-75, prin res scientist, 75-81; staff physicist, 81-82, PRIN PHYSICIST, COMBUSTION ENG, 83- *Mem:* Sigma Xi. *Res:* Optical inspection and measurement techniques; on-line optical and ultrasonic measurements of paper properties. *Mailing Add:* Int Paper Long Meadow Rd Tuxedo Park NY 10987

VAHEY, MARYANNE T, CELL BIOLOGY, ELECTRON MICROSCOPY. *Educ:* Univ Mass, PhD(molecular biol), 81. *Prof Exp:* Sr fel, Nat Heart, Lung & Blood Inst, NIH, 82- *Mailing Add:* 19315 Dimona Dr Brookville MD 20833

VAHLDIEK, FRED W(ILLIAM), b Eilsleben, Ger, Feb 5, 33; US citizen; m 59; c 1. CHEMICAL ENGINEERING, CHEMISTRY. *Educ:* Univ Halle, BSc, 53, MSc, 54. *Prof Exp:* Analytical chemist, Iron & Steel Co, Ger, 54-55 & E F Drew Co, NJ, 55-56; RES MAT ENGR & GROUP LEADER, AIR FORCE MAT LAB, WRIGHT-PATTERSON AFB, 59- *Mem:* AAAS; Am Chem Soc; Sigma Xi; fel Am Inst Chemists. *Res:* High pressure-high temperature research on refractory, monmetallic materials; electron microscopy studies on metallic and nonmetallic high temperature materials; oxidation on metallic and refractory solids; high temperature x-ray of solids. *Mailing Add:* 5851 Barrett Dr Dayton OH 45431

VAHOUNY, GEORGE V, cardiovascular diseases; deceased, see previous edition for last biography

VAICAITIS, RIMAS, b Sakei, Lithuania, Apr 30, 41; US citizen; m 65; c 2. AERONAUTICAL ENGINEERING. *Educ:* Univ Ill, Urbana, BS, 67, MS, 68, PhD(aero eng), 70. *Prof Exp:* Res asst eng, Univ Ill, 67-70; from asst prof to prof, 70-80, DIR INST FLIGHT STRUCT, COLUMBIA UNIV, 77- *Concurrent Pos:* Res engr NASA, Langley Res Ctr, 76-77; consult, USAF, 75-, US Army, 76-, Rockwell Int, 78-, Modern Anal, Inc, 74- *Mem:* Am Inst Aeronaut & Astronaut; Am Soc Civil Eng. *Res:* Fluid-solid interactions; random vibrations; structural acoustics. *Mailing Add:* Dept Civil Eng Columbia Univ New York NY 10027

VAIDHYANATHAN, V S, b Madras, India, Dec 15, 33; m 65; c 3. BIOPHYSICS. *Educ:* Annamalai Univ, Madras, BSC, 53, MA, 54; Ill Inst Technol, PhD(chem), 61. *Prof Exp:* Res assoc chem, Univ Kans, 60-62; chief math & statist sect, Southern Res Support Ctr, 62-63, chief theoret sci sect, 63-66; assoc prof theoret biol, State Univ NY, Buffalo, 66-70, assoc prof biophys, 67-72, assoc prof pharmaceut & biophys, 72-80, PROF BIOPHYS SCI, STATE UNIV NY, 80- *Concurrent Pos:* Consult, Vet Admin Hosp, New Orleans; vis prof theoret biol, State Univ NY, Buffalo, 65; Europ Molecular Biol Orgn fel, 69. *Mem:* AAAS; Am Chem Soc; Biophys Soc. *Res:* Statistical mechanics; active transport; nerve potentials; biophysics of membranes; regulation X control. *Mailing Add:* Dept Biophys & Sci R 114B Cory Hall State Univ NY Health Sci Ctr 3435 Main St Buffalo NY 14214

VAIDYA, AKHIL BABUBHAI, b Gondal, India, Oct 24, 47; m 73; c 2. MOLECULAR PARASITOLOGY, CELL BIOLOGY. *Educ:* Univ Bombay, BSc, 67, PhD(appl biol), 72. *Prof Exp:* Res asst ultrastruct, Cancer Res Inst, Bombay, 70-72; res assoc molecular biol, Inst Med Res, 72-75, assoc, 75-77; from asst prof to assoc prof, 77-89, PROF MICROBIOL, HAHNEMANN UNIV, 89- *Concurrent Pos:* Mem, Spec Study Sect, NIH; expert reviewer, NSF. *Mem:* Am Soc Microbiol; AAAS. *Res:* Molecular biology of malarial parasites; organization and expression of organelle genomes; molecular evolution; Control of eukaryotic gene expression; Ribsomal functioning viruses. *Mailing Add:* Dept Microbiol Immunol Hahnemann Univ Broad & Vine Philadelphia PA 19102-1192

VAIL, CHARLES BROOKS, b Bessemer, Ala, Apr 29, 23; m 44; c 2. PHYSICAL CHEMISTRY. *Educ:* Birmingham-Southern Col, BS, 45; Emory Univ, MS, 47, PhD(chem), 51. *Prof Exp:* Instr chem, Armstrong Col, 48-49; chemist, Southern Res Inst, 51-53; prof phys sci, Coker Col, 53-56; assoc prof chem, Agnes Scott Col, 56-57; prof chem & acad dean, Hampden-Sydney Col, 57-65; assoc exec secy, Comn on Cols, 65-68; dean, Sch Arts & Sci, Ga State Univ, 68-73; PRES, WINTHROP COL, 73- *Mem:* Am Chem Soc. *Res:* Thermal diffusion in liquids; heats of vaporization; photochemistry. *Mailing Add:* PO Box 26 Newland NC 28657-0026

VAIL, CHARLES R(OWE), b Glens Falls, NY, Oct 16, 15; m 39; c 3. ELECTRICAL ENGINEERING, ACADEMIC ADMINISTRATION. *Educ:* Duke Univ, BSEE, 37; Univ Mich, MS, 46, PhD(elec eng), 56. *Prof Exp:* Engr, Gen Elec Co, 37-39; from instr to prof elec eng, Duke Univ, 39-67, exec officer, dept elec eng, 53-56, chmn dept, 56-64, assoc dean grad study & res, Sch Eng, 64-67, actg dean, Sch Eng, 65; prof elec eng & electronic sci, Southern Methodist Univ, 67-73, assoc dean eng, 67-70, vpres, Univ, 70-73; assoc dean, Col Eng, Ga Inst Technol, 73-79, prof elec eng, 73-83, dir, Dept Continuing Educ, 79-83, EMER ASSOC DEAN ENG, COL ENG, GA INST TECHNOL, 83- *Concurrent Pos:* Consult, Chem War Res Proj, Duke Univ, 45, Solid State Div, US Naval Res Lab, 52-58, Gen Elec Co, 56-58 & NC Res Triangle Inst, 60-63; mem, Gov's Sci Adv Comt, NC, 61-64; mem tech utilization adv bd, NC Bd Sci & Technol, 65-67; agent & mem corp sect, Asn Media-Based Continuing Educ for Engrs, Inc, 76-83; mem, Fac Coun, Univ Ctr, Ga, 73-83. *Mem:* AAAS; fel Inst Elec & Electronics Engrs; Nat Soc Prof Engrs; Sigma Xi. *Res:* High voltage phenomena; dielectric materials; superconducting circuitry; thin-film properties. *Mailing Add:* 2669 Peppermint Dr Tucker GA 30084

VAIL, EDWIN GEORGE, b Toledo, Ohio, July 25, 21; m 46; c 5. MEDICAL PHYSIOLOGY, BIOENGINEERING. *Educ:* Univ Toledo, BSc, 47; Ohio State Univ, MSc, 48, PhD(aviation physiol), 53. *Prof Exp:* Proj engr, Aerospace Med Lab, Wright Air Develop Ctr, Ohio, 51-53, chief respiration sect, 53-54, proj scientist, 54-60, chief personnel protection equip & crew escape group X-20 syst prog officer, 60-62, asst chief, Bioastronaut Div, 63-64; chief human eng & space suit res & develop, Hamilton Standard, United Aircraft Corp, Conn, 64-69; mem staff physiol, Naval Coastal Systs Lab, Panama City, 70-78; pres, Vail Appl Res Co, Inc, 73-83; RETIRED. *Mem:* Aerospace Med Asn; Undersea Med Soc; Sigma Xi. *Res:* Aerospace and oceanographic physiology, including respiratory, cardiovascular, environmental stress tolerance; space-pressure suits, diving equipment and life support system research and development of medical devices. *Mailing Add:* 4502 Vista Lane Lynn Haven FL 32444

VAIL, JOHN MONCRIEFF, b Winnipeg, Man, Oct 17, 31; m 54; c 1. SOLID STATE PHYSICS. *Educ:* Univ Man, BSc, 55, MSc, 56; Brandeis Univ, PhD(physics), 60. *Prof Exp:* IBM res asst, Brandeis Univ, 57-59; Nat Res Coun Can fel, McGill Univ, 60-61; Leverhulme fel, Univ Liverpool, 61-62; from asst prof math physics to assoc prof physics, 62-72, PROF PHYSICS, UNIV MAN, 72- *Concurrent Pos:* Vis lectr, St Andrews Univ, 68-69; vis res assoc, AERE Harwell, 75-76 & 82-83; hon res fel chem, Univ Col London, UK, 82-83. *Mem:* Can Asn Physicists; Am Phys Soc; Brit Inst Physics; Mat Res Soc. *Res:* Solid state theory, applied to properties of localized defects in crystalline materials. *Mailing Add:* Dept Physics Univ Man Winnipeg MB R3T 2N2 Can

VAIL, PATRICK VIRGIL, b Pasadena, Calif, Nov 16, 37; m 84; c 3. INSECT PATHOLOGY, MICROBIAL CONTROL. *Educ:* Calif State Univ, Fresno, BA, 60, MS, 62; Univ Calif, Riverside, PhD(entom), 67. *Prof Exp:* Res entomologist, Agr Res Serv, USDA, 62-69, res leader, 70-75; sect head, UN-Int Atomic Energy Agency, 75-78; res entomologist, 78-82, LAB DIR, AGR RES SERV, USDA, 82- *Concurrent Pos:* Adj prof entom, Univ Ariz, 70-75. *Mem:* Entom Soc Am; Soc Invert Path; AAAS. *Res:* Insect pathology, virology and microbial control; quarantine treatments; insect behavior, biology and ecology; microbial control of insects infesting dried fruits and nuts. *Mailing Add:* Hort Crops Res Lab Agr Res Serv USDA 2021 S Peach Ave Fresno CA 93727

VAIL, PETER R, b New York, NY, Jan 13, 30. SEISMIC STRATIGRAPHIC INTERPRETATION. *Educ:* Dartmouth Col, AB, 52; Northwestern Univ, MS & PhD, 56. *Prof Exp:* From res geologist to sr res scientist, Exxon Prod Res Co, 56-86; W MAURICE EWING PROF OCEANOG, RICE UNIV, 86- *Concurrent Pos:* Mem US Geodynamics Comt, Nat Acad Sci, 87-, US Dept Energy, 87-, Ocean Sci Bd, Nat Acad Sci, 79-82; Gallagher vis scientist, Univ Galgary, 80; vis scientist, Woods Hole Oceanog Inst, Mass, 76. *Honors & Awards:* Virgil Kauffman Gold Medal Award, 76; William Smith Lectr, London Geol Soc, 78, William Smith Medal, 86; Burwell Lectr, UK Geophys Soc, 83. *Mem:* Sigma Xi; fel Geol Soc Am; Am Asn Petrol Geologists; hon mem Soc Explor Geophysicists; Soc Econ Paleontologists & Mineralogists; fel AAAS; Europ Asn Explor Geophysicists. *Res:* Stratigraphic mapping; well log correlation; computer applications to geology; the stratigraphic and structural interpretation of seismic data the sequence stratigraphy of outcrops and well logs. *Mailing Add:* 3745 Del Monte Houston TX 77019

VAIL, SIDNEY LEE, b New Orleans, La, Aug 10, 28; m 53; c 4. ORGANIC CHEMISTRY TEXTILES, SEED GERMINATION. *Educ:* Tulane Univ, BS, 49, PhD(org chem), 65; La State Univ, MS, 51. *Prof Exp:* Org chemist, Dow Chem Co, 51-53; sr chemist, Am Cyanamid Co, 55-59; proj leader, USDA, 59-72, res leader, Cotton Textile Chem Lab, 72-83, chief, 76-83, res leader, Crop Protection Chem, Southern Regional Res Ctr, 83-87, res chemist, Textile Finishing Chem, 87- 88; RETIRED. *Concurrent Pos:* Exchange scientist, Shirley Inst, Eng, 65-66; adj prof textile chem, Sch Textiles, NC State Univ, Raleigh, 79-83; consult, 88- *Mem:* Am Chem Soc; Am Asn Textile Chemists & Colorists; Sigma Xi. *Res:* Petrochemicals, synthesis and process chemistry; textile chemistry, organic synthesis and mechanisms; nuclear magnetic resonance; seed germination and parasitic weed chemistry; chemical modification of cotton. *Mailing Add:* 10137 Hyde Pl River Ridge LA 70123-1523

VAILLANCOURT, REMI ETIENNE, b Maniwaki, Que, June 16, 34. MATHEMATICS. *Educ:* Univ Ottawa, BA, 57, BSc, 61, BTh, 63, MSc, 64, MTh, 65; NY Univ, PhD(math), 69. *Prof Exp:* Instr, NY Univ, 68-69; Off Naval Res res assoc, Univ Chicago, 69-70; chmn dept, 72-76, ASSOC PROF MATH, UNIV OTTAWA, 70- *Mem:* Am Math Soc; Math Asn Am; Can Math Soc; French-Can Asn Advan Sci. *Res:* Partial differential equations; pseudo-differential operators; finite difference and finite element methods. *Mailing Add:* Dept Math Univ Ottawa Ottawa ON K1N 6N5 Can

VAILLANT, GEORGE EMAN, b New York, NY, June 16, 34; m 71; c 5. PSYCHIATRY. *Educ:* Harvard Univ, AB, 55, MD, 59. *Prof Exp:* Resident, Mass Ment Health Ctr, 60-63; staff psychiatrist, USPHS, Lexington, Ky, 63-65; from asst prof to assoc prof psychiat, Sch Med, Tufts Univ, 66-71; assoc prof, 71-76, PROF PSYCHIAT, HARVARD MED SCH, 76- *Concurrent Pos:* Dir study adult develop, Harvard Univ, 72-; dir training, Dept Psychiat, Mass Ment Health Ctr, 81-; consult, Div Manpower & Training, NIMH, 76-79; fel, Ctr Advan Study Behav Sci, 78-79; NIH res sci award, 81- *Mem:* Fel Am Psychiat Asn; Soc Life Hist Res in Psychopath; Int Soc Study Behav Develop. *Res:* Long term follow up in adult development and in psychopathology. *Mailing Add:* Dartmouth Med Sch Hanover NH 03751

VAILLANT, HENRY WINCHESTER, b New York, NY, Dec 17, 36; m 58; c 3. POPULATION BIOLOGY. *Educ:* Harvard Univ, AB, 58, MD, 62, SMHyg, 69. *Prof Exp:* Intern med, Boston City Hosp, 62-63, resident, 63-64; res assoc, Nat Inst Child Health & Human Develop, 64-66; resident, Boston City Hosp, 66-67; res fel obstet & gynec, Harvard Med Sch, 67-68, ASST PROF POP STUDIES, SCH PUB HEALTH, HARVARD UNIV, 68- *Concurrent Pos:* Consult, Cancer Control Prog, USPHS, 68; pres-elect, Emerson Hosp Med Staff, pres, 87- *Mem:* Am Pub Health Asn. *Res:* Clinical human reproductive physiology. *Mailing Add:* 321 Main Acton MA 01720

VAIRAVAN, KASIVISVANATHAN, b Madras, India, July 9, 39; m 67; c 2. ELECTRICAL ENGINEERING, COMPUTER SCIENCE. *Educ:* Univ Madras, BE, 62; George Washington Univ, MS, 65; Univ Notre Dame, PhD(elec eng), 68. *Prof Exp:* Jr elec engr, Madras State Elec Bd, 62-63; asst prof elec eng, 68-71, assoc prof elec eng & comput sci, 71-77, PROF ELEC ENG & COMPUT SCI, UNIV WIS-MILWAUKEE, 77- *Concurrent Pos:* Mem tech staff, Bell Tel Labs, 70; vis researcher, Hiroshima Univ, 80; vis lectr, var res ctrs in Japan. *Mem:* Inst Elec & Electronics Engrs. *Res:* Parallel computation; distributed processing; software science; computer organization. *Mailing Add:* Dept Elec Eng & Comput Sci Univ Wis PO Box 413 Milwaukee WI 53201

VAISEY-GENSER, FLORENCE MARION, b Winnipeg, Man, Apr 3, 29; m 53, 76; c 2. FOOD SCIENCE & TECHNOLOGY. *Educ:* Univ Man, BSc, 49; McGill Univ, MSc, 51. *Prof Exp:* Head food acceptance, Defense Res Med Labs, 51-53; metab dietician, Victoria Gen Hosp, 53-54; lectr foods & nutrit, Ore State Col, Corvallis, 56-60; asst prof, Univ Guelph, 62-65; from asst prof to assoc prof, Univ Man, 65-73, head dept, 78-80, assoc dean, Fac Grad Studies, 81-83, PROF FOOD & NUTRIT, UNIV MAN, 73-, ASSOC VPRES RES, 83- *Concurrent Pos:* Adv Bd, Can Food Prod Develop Ctr, 75-83; res grant, Agr Can, 75-77 & 80-, Alberta Agr, 79-82, Man Res Coun, 71-77, Canola Coun Can, 70-85 & Fisheries Res Bd, 66-69; guest prof, Swiss Fed Inst Technol, 71-72; chmn, Man Res Coun, 83-87; bd mem, assoc adv sci in Can, 84-86, Fisheries & Oceans Res Adv Coun, 86-89, Natural Sci & Engr Res Coun Scholar Comt, 88-, Nat Adv Bd Sci & Technol, 88-; mem, Can Res Mgr Asn, 88-90; mem, WPG 2000 Leaders Comn, 90- *Honors & Awards:* W J Eva Award. *Mem:* Can Inst Food Sci & Technol (vpres, 78-79, pres, 80-81); Can Dietetic Asn; Can Home Econ Asn; Am Asn Cereal Chem; Inst Food Technol; Soc Res Admin. *Res:* Sensory evaluation of foods and their components; rapeseed oil; plant proteins; amino acids. *Mailing Add:* 208 Admin Bldg Univ Man Winnipeg MB R3T 2N2 Can

VAISHNAV, RAMESH, soft tissue biomechanics, artificial intelligence; deceased, see previous edition for last biography

VAISHNAVA, PREM P, b Jodhpur, India, Oct 10, 42; US citizen; m 72; c 3. SUPERCONDUCTIVITY, MOSSBAUER SPECTROSCOPY. *Educ:* Jodhpur Univ, India, BSc, 63, MSc, 65, PhD(physics), 76. *Prof Exp:* Asst prof physics, Jodhpur Univ, 65-78; res assoc chem, Heriot-Watt Univ, Edinburgh, Scotland, 78-80; asst prof physics, WVa Univ, Morgantown, 80-83; res assoc mat sci, Argonne Nat Lab, 83-86; assoc prof physics, Northern Ill Univ, 83-86; assoc prof, 86-91, PROF PHYSICS, GMI ENG & MGT INST, FLINT, MICH, 91- *Concurrent Pos:* Guest scientist, Brookhaven Nat Lab, 86-; acad affil fel, Mich State Univ, 88-91. *Mem:* Am Phys Soc; Mat Res Soc. *Res:* Investigate electronic, magnetic and lattice vibrational behavior of high tech super conductors; Mössbauer spectroscopy; scanning transmission electron microscope and SOVID magnetometer for these investigations. *Mailing Add:* Dept Sci & Math GMI Eng & Mgt Inst Flint MI 48504-4898

VAISNYS, JUOZAS RIMVYDAS, b Kaunas, Lithuania, Mar 12, 37; US citizen. PHYSICAL CHEMISTRY. *Educ:* Yale Univ, BS, 56; Univ Calif, Berkeley, PhD(chem), 60. *Prof Exp:* ASSOC PROF APPL SCI, YALE UNIV, 67- *Mem:* Am Phys Soc; Sigma Xi; Asn Advan Baltic Studies; Am Chem Soc. *Res:* Evolution; ecology; biological dynamics. *Mailing Add:* Kline Geol Lab Box 2161 Yale Sta New Haven CT 06520

VAITKEVICIUS, VAINUTIS K, b Kaunas, Lithuania, Jan 12, 27; US citizen; m 51; c 6. ONCOLOGY. *Educ:* Univ Frankfurt, MD, 51. *Prof Exp:* Intern med, Grace Hosp, Detroit, 51-52, resident, 55-56; resident internal med, Detroit Gen Hosp, 56-58; assoc physician, Henry Ford Hosp, Detroit, 59-62; clin dir oncol, Detroit Inst Cancer Res, 62-66; from asst prof to prof med & dir oncol, 66-73, prof oncol & chmn dept, 73-82, PROF & CHMN, DEPT INTERNAL MED, SCH MED, WAYNE STATE UNIV, 82- *Concurrent Pos:* Fel cancer res, Detroit Inst Cancer Res, Mich, 58-59; consult, Beaumont Hosp, Harper Grace Hosps, Mt Carmel Mercy Hosp, Sinai Hosp & Vet Admin Hosp; clin dir, Comprehensive Cancer Ctr Detroit, 78-82. *Mem:* AAAS; Am Col Physicians; Am Asn Cancer Res; Am Asn Cancer Educ; Am Soc Hemat. *Res:* Mechanism of metastases; pharmacology of cytostatic drugs. *Mailing Add:* Dept Internal Med Harper Hosp 3990 John R St Detroit MI 48201

VAITUKAITIS, JUDITH L, b Hartford, Conn, Aug 29, 40. MEDICINE. *Educ:* Tufts Univ, BS, 62; Boston Univ, MD, 66. *Prof Exp:* Attend physician, Nat Inst Child Health & Human Develop, Clin Ctr, NIH, Bethesda, Md, 70-75; head, Sect Endocrin & Metab, Boston City Hosp, Mass, 74-86; assoc prof med, Sch Med, Boston Univ, 74-77, co-prog dir, Gen Clin Res Ctr, 75-77, prog dir, 77-86, prof med, 77-86, prof physiol, 80-86; dir Gen Clin Res tr Prog, 86-91, actg dep dir, 90-91, DEP DIR, EXTRAMURAL RES RESOURCES, NIH, 91- *Concurrent Pos:* Sr investr & med officer, Reprod Res Br, Nat Inst Child Health & Human Develop, NIH, 73-74, guest worker, 74-75, assoc dir clin res, Nat Ctr Res Resources, 86-91, actg dir, Biol Models & Mat Res Prog, Nat Ctr Res Resources, 90; assoc prof physiol, Sch Med, Boston Univ, 75-80, assoc prof obstet & gynec, 77-80. *Honors & Awards:* Mallinckrodt Award Investigative Res, Clin Radioassay Soc, 80. *Mem:* Endocrine Soc; Am Soc Clin Investigators; Asn Am Physicians; Soc Study Reproduction; Am Fedn Clin Res. *Res:* Reproductive endocrinology; subcellular action of glycoprotein hormones; neuroendocrinology of reproduction and structure-function studies of gonadotropins; laboratory and clinical biomedical research, including complex biotechnologies, models of human disease science education and career development. *Mailing Add:* Nat Ctr Res Resources Bldg 12A Rm 4011 NIH Bethesda MD 20892

VAJK, J(OSEPH) PETER, b Budapest, Hungary, Aug 3, 42; US citizen; m 70; c 4. PHYSICS, SPACE INDUSTRIALIZATION. *Educ:* Cornell Univ, AB, 63; Princeton Univ, MA, 65, PhD(physics), 68. *Prof Exp:* Sr physicist, Lawrence Livermore Lab, Univ Calif, 68-76; consult, 76-79, SR SCIENTIST, SCI APPLNS INC, CALIF, 77- *Mem:* Am Phys Soc; Sigma Xi; Am Inst Aeronaut & Astronaut. *Res:* Relativistic astrophysics and cosmology; general relativity theory; evolution of relativistic cosmological models; theory of electromagnetic pulses from nuclear explosions; world dynamics; socioeconomic implications of space industrialization and colonization; space technology assessment; atmospheric photochemistry and dynamics modeling; alternative futures research. *Mailing Add:* Suite 148 PO Box 8027 Walnut Creek CA 94596-8027

VAKILI, NADER GHOLI, b Bushir, Iran, Jan 14, 27; nat US; m 53; c 6. PLANT PATHOLOGY, PLANT GENETICS. *Educ:* Northwestern Univ, BS, 52; Univ Chicago, MS, 53; Purdue Univ, PhD, 58. *Prof Exp:* Pathologist, United Fruit Co, 58-65; asst plant pathologist, Everglades Exp Sta, Univ Fla, 65-67; area agron adv, US Agency Int Develop, US Dept Agr, 67-69; mem staff, 69-79, RES PLANT PATHOLOGIST, IOWA STATE UNIV, USDA, 79- *Mem:* Am Phytopath Soc; Am Soc Agron; Am Inst Biol Sci; Sigma Xi. *Res:* Genetics of pathogenicity; taxonomy and genetics of disease resistance in musa; vegetable diseases. *Mailing Add:* 421 Bessey Hall Iowa State Univ Ames IA 50011

VAKILZADEH, JAVAD, b Esfahan, Iran, June 26, 27. PUBLIC HEALTH, EPIDEMIOLOGY. *Educ:* Sharaf Col, Iran, BS, 48; Univ Teheran, DVM, 52; Univ Pittsburgh, CPH, 58; Univ NC, MPH, 59. *Prof Exp:* Epidemiologist, Int Coop Admin, Iran, 53-57; res scientist, NC Sanitorium Syst, 59-63, from asst to actg dir res respiratory dis, 63-68; mem fac, Med Ctr, Duke Univ, 68-72; epidemiologist, Int Fertil Res Prog, Pop Ctr, Univ NC, Chapel Hill, 72-80; CONSULT EPIDEMIOLOGIST, HAITIAN-AM TUBERC INST, 63- *Mem:* Fel Am Pub Health Asn; Am Vet Med Asn; Am Vet Epidemiol Soc. *Res:* Research and teaching of epidemiology; environmental health; communicable disease control. *Mailing Add:* 400 S Elliott Rd Chapel Hill NC 27514-5823

VALA, MARTIN THORVALD, JR, b Brooklyn, NY, Mar 28, 38; m 66; c 3. PHYSICAL CHEMISTRY, SPECTROCHEMISTRY. *Educ:* St Olaf Col, BA, 60; Univ Chicago, SM, 62, PhD(chem), 64. *Prof Exp:* NSF fel chem, Copenhagen Univ, 65-66; US-Japan Coop Sci Prog fel, Univ Nagoya, 66-67; from asst prof to assoc prof, 68-78, PROF CHEM, UNIV FLA, 78- *Concurrent Pos:* Merck Found fel, 70-71; vis prof, Advan Sch Physics & Chem, 73-74; NATO fel, 73-74; Fulbright sr fel, Franco-Am Scholar Exchange Comn, 73-74; vis scientist, US-India Exchange Scientists Prog, India, 77; vis scientist, Univ Poznan, Poland, 81. *Mem:* Am Phys Soc; Am Chem Soc; Am Inst Chem; InterAm Photochem Soc. *Res:* Optical and magnetic properties of organic molecules and transition metal atoms, dimers and complexes; cluster spectroscopy. *Mailing Add:* Dept Chem Univ Fla Gainesville FL 32611-2046

VALACH, MIROSLAV, b Hnusta, Czech, Sept 12, 26; US citizen; m 52; c 1. INFORMATION & COMPUTER SCIENCE. *Educ:* Prague Tech Univ, ME, 51; Czech Acad Sci, PhD(math & physics), 58. *Prof Exp:* Mgr peripheral equip, Res Inst Mach Mach, Czech, 51-64; consult engr, Gen Elec Co, Ariz, 65-69; prof into & computer sci, Ga Inst Technol, 69-74; res dir, Karsten Mfg Corp, 74-80; FRIDAY COMPUT INC, 80- *Mem:* Asn Comput Mach; Inst Elec & Electronics Engrs. *Res:* Switching theory; cybernetics; artifical intelligence; computer hardware; linguistics. *Mailing Add:* San Jose State Univ One Washington Square San Jose CA 95192

VALANIS, BARBARA MAYLEAS, b Harrisburg, Pa, Oct 4, 42; m 78; c 1. EPIDEMIOLOGY. *Educ:* Cornell Univ, BS, 65; Columbia Univ, MEd, 71, DrPH(epidemiol), 75. *Prof Exp:* Dist nurse, Vis Nurse Serv NY, 65-69; instr pub health nursing, Columbia Univ Sch Nursing, 69-71; asst prof, Fairleigh Dickinson Univ, 72-73; res worker, Columbia Univ, 73-75, staff assoc, 75-77, res assoc epidemiol, 77-78; assoc prof, 78-80, PROF, COL NURSING & HEALTH, UNIV CINCINNATI, 80-, ASST PROF EPIDEMIOL, COL MED, 78- *Concurrent Pos:* Columbia Univ Cancer Res Ctr fel, 75-78; vpres & prog eval consult, Eval Assoc, Inc, 76-78. *Mem:* Am Pub Health Asn; Soc Epidemiol Res; Sigma Xi; Nat League Nursing. *Res:* Social epidemiology; cancer epidemiology; reproductive epidemiology; health services evaluation; stress; occupational exposures. *Mailing Add:* 8605 NW Lake Crest Ct Vancouver WA 98665

VALASEK, JOSEPH, b Cleveland, Ohio, Apr 27, 97; m 24; c 2. PHYSICS, ELECTROMAGNETISM. *Educ:* Case Inst Technol, BS, 17; Univ Minn, MA, 20, PhD(physics), 21. *Hon Degrees:* DSc, Univ Minn, 83. *Prof Exp:* Asst physicist, Heat Div, Nat Bur Standards, 17-19; teaching asst, 19-20, from instr to prof, 20-65, EMER PROF PHYSICS, UNIV MINN, 65- *Concurrent Pos:* Nat Res Coun fel, 21. *Mem:* Sigma Xi; fel Am Phys Soc; Optical Soc Am; AAAS; Am Asn Physics Teachers; Am Asn Univ Professors. *Res:* Discovered ferroelectricity, obtaining the first dielectric hysteresis loops; discovered the Curie temperatures at which the ferroelectric property of Rochelle salt disappears. *Mailing Add:* 300 Seymour Pl SE Minneapolis MN 55414

VALASSI, KYRIAKE V, b Salonika, Greece, May 15, 17; US citizen. HYPERALIMENTATION NUTRITION ASSESSMENT. *Educ:* Syracuse Univ, BS, 50; Cornell Univ, MS, 51; Ore State Univ, PhD(nutrit & biochem), 56. *Prof Exp:* PROF FOODS & NUTRIT, CATH UNIV AM, 56- *Concurrent Pos:* Consult, Off Int Res, 61-67, Pan Am Health Orgn, 72-74, Food & Drug Admin, 78, Nat Coun Dis Control, 67-74, Nutrit Cancer NIH, 79; lectr, Corner Univ, 68, Ore State Univ, 69. *Mem:* Am Dietetic Asn; Soc Nutrit Educ; Am Inst Nutrit; Am Soc Parental & Enteral Nutrit. *Res:* Assessment of nutritional status of population groups; dietary survey methodology and training in foreign countries; nutritional management of cancer patients in a variety of therapeutic regiments; enteral hyperalimentation to patients undergoing treatment for burns. *Mailing Add:* Cath Univ Am Washington DC 20064

VALBERG, LESLIE S, b Churchbridge, Sask, June 3, 30; m 54; c 3. MEDICINE. *Educ:* Queen's Univ, Ont, MD, 54, MSc, 58; FRCPS(C), 60. *Prof Exp:* Lectr med, Queen's Univ, Ont, 60-61; res assoc, Med Res Coun, Can, 61-65; from asst prof to prof med, Queen's Univ, Ont, 61-75; prof med & chmn dept, 75-85, DEAN, FAC MED, UNIV WESTERN ONT, 85- *Concurrent Pos:* Consult, Univ Hosp, London, 75- *Mem:* Am Gastroenterol Asn; Am Fedn Clin Res; fel Am Col Physicians; fel Royal Col Physicians & Surgeons Can; Can Soc Clin Invest. *Res:* Iron metabolism; absorption of metals; hemochromatosis. *Mailing Add:* Dean Fac Med Univ Western Ont London ON N6A 5C1 Can

VALDES, JAMES JOHN, b San Antonio, Tex, Apr 25, 51. NEUROTOXICOLOGY. *Educ:* Loyola Univ, Chicago, BS, 73; Trinity Univ, MS, 76; Tex Christian Univ, PhD(neurosci), 79. *Prof Exp:* Res fel, Johns Hopkins Univ, 79-82, lectr, 80-86; phys scientist, toxicol div, 82-85, pharmacologist, Biotechnol Div, 85-90, SCI ADV BIOTECHNOL, US ARMY CHEM RES & DEVELOP CTR, 90- *Concurrent Pos:* Instr, Hood Col, 82-83; ed, Neurobehav Toxicol & Teratol, 82-; chmn, Ann Conf Receptor Res, Johns Hopkins Appl Physics Lab, 85, 86 & 87; res assoc, Johns Hopkins Univ, 83-87, adj assoc prof, 87-90; adj prof, UTSA, 90- *Mem:* Soc Neurosci; Soc Toxicol; Sigma Xi; Brit Brain Res Asn; Int Brain Res Asn; Europ Brain & Behav Soc. *Res:* Interaction of neurotoxins with receptor and ion channel proteins; synaptic mechanisms of neurotoxicity; biotechnology. *Mailing Add:* Sci Adv Biotechnol Attn: SMCCR-TDB US Army Chem Res & Develop Ctr Aberdeen Proving Ground MD 21010-5423

VALDES-DAPENA, MARIE A, b Pottsville, Pa, July 14, 21; div; c 11. PEDIATRICS. *Educ:* Immaculate Col, Pa, BS, 41; Temple Univ, MD, 44. *Prof Exp:* St Christopher's Hosp Children grant, 59-76; instr, Grad Sch Med, Univ Pa, 48-55, vis lectr, 60-; consult pediat path, Div Med Exam, Dept Pub Health, Philadelphia, 67-70; mem, Perinatal Biol & Infant Mortality Res & Training Comt Nat Inst Child Health & Human Develop, 71-75; consult, Lankenau Hosp, Philadelphia, 71-76; consult & lectr, US Naval Hosp, Philadelphia, 72; PROF PATH & PEDIAT, SCH MED, UNIV MIAMI, 76- *Mem:* Soc Pediat Path; Int Acad Path. *Res:* Causes of neonatal mortality; sudden infant death syndrome; gynecologic pathology in infancy and childhood; iatrogenic diseases in the perinatal period. *Mailing Add:* Dept Path (D-33) Sch Med Univ Miami PO Box 016960 Miami FL 33101

VALDIVIA, ENRIQUE, PATHOLOGY & ELECTROMICROSCOPY. *Educ:* Univ Chile, MD, 49. *Prof Exp:* PROF PATH & PREV MED, UNIV WIS-MADISON, 58- *Res:* Enzymology. *Mailing Add:* Dept Path & Prev Med Univ Wis 502 N Walnut Madison WI 53705

VALDIVIESO, DARIO, b Fusagasuga, Colombia, Dec 12, 36; US citizen. MEDICAL MICROBIOLOGY, PARASITOLOGY. *Educ:* Univ Andes, Colombia, BS, 60, MS, 62; Univ PR Sch Med, PhD(med zool), 67. *Prof Exp:* Coordr biochem, Seneca Col, Toronto, 68-70; resident med microbiol, US Nat Ctr Dis Control, Atlanta, 70-72; instr microbiol, Univ Tex Sch Med, San Antonio, 72-73; RES ASSOC MAMMAL, ROYAL ONT MUS, 73-; INT TECH REP, GIBCO DIV, DEXTER CORP, OHIO, 81- *Concurrent Pos:* Specialist microbiologist pub health & med lab microbiol, Am Acad Microbiol, 74-; vis prof microbiol, Pontificia Univ Javeriana, Colombia, & Fulbright-Hays Latin Am teaching fel, 75-77. *Mem:* Am Soc Microbiologists; Am Soc Trop Med & Hyg; Sigma Xi. *Res:* Comparative biochemistry of proteins; microbiology of human pathogens; immunochemistry of human parasitic and mycotic diseases. *Mailing Add:* 2903 Cedarview Dr Austin TX 78704

VALDSAAR, HERBERT, b Tallinn, Estonia, Dec 6, 25; nat US; m 59; c 2. HIGH TEMPERATURE CHEMISTRY, METAL CHLORIDES. *Educ:* Aachen Tech Univ, Dipl, 50; Univ Maine, MS, 52; Univ Fla, PhD(chem), 56. *Prof Exp:* Res chemist, 56-71, sr res chemist, 71-90, RES ASSOC, E I DU PONT DE NEMOURS & CO, INC, 91- *Res:* High temperature inorganic reactions; preparation of high purity silicon; metal chlorides; environmental control of heavy metals; pigment coatings. *Mailing Add:* Jackson Lab Chambers Works E I du Pont de Nemours & Co Inc Deepwater NJ 08023

VALEGA, THOMAS MICHAEL, b Linden, NJ, May 23, 37; m 58; c 4. ORGANIC CHEMISTRY, DENTISTRY. *Educ:* Rutgers Univ, BS, 59, PhD(org chem), 63. *Prof Exp:* Chemist, Pesticide Chem Res Br, Entom Res Div, Agr Res Serv, USDA, 63-67; grants assoc, NIH, 67-68, health scientist adminr, Nat Inst Environ Health Sci, 68-69, coordr contracts artificial kidney-chronic uremia prog, Nat Inst Arthritis & Metab Dis, 69-72; prog analyst, Off Categorical Progs, Prog Planning & Eval, Environ Protection Agency, DC, 72; health scientist adminr, periodontal dis prog, 72-74, chief restorative mat prog, extramural progs, 74-83, SPEC ASST MANPOWER DEVELOP & TRAINING, NAT INST DENT RES, NIH, 84- *Honors & Awards:* Spec Award, Soc Biomat, 84. *Mem:* Int Asn Dent Res; AAAS; Am Chem Soc; Nat Audubon Soc; Soc Biomaterials. *Res:* Peroxide and carbamate chemistry; insecticide and insecticide synergist chemistry; insect attractant and insect pheromone chemistry; medicinal chemistry; dental materials bioengineering; biomaterials; health sciences administration; oral biology; dental chemistry. *Mailing Add:* NIH-Nat Inst Dent Res Westwood Bldg Rm 510 Bethesda MD 20892

VALENCIA, MAURO EDUARDO, b Nogales Sonora, Mex, Jan 15, 49; m 73; c 2. NUTRITIONAL EVALUATION OF PROTEIN QUALITY, CHEMICAL & BIOLOGICAL METHODS. *Educ:* Univ Sonora, Mex, BS, 73; Univ Ariz, MS, 75, PhD(nutrit & food sci), 78. *Prof Exp:* Res asst, Univ Ariz, 75-78; res scientist nutrit & foods, 78-82, dept head, 82-86, PROF NUTRIT & TEACHING, CTR INVEST ALIMENT & DESARROLLO, 84-, DIV DIR NUTRIT & FOODS, 86-; PROF NUTRIT, UNIV SONORA, 78- *Concurrent Pos:* Postdoctoral fel, Rowett Res Inst, Aberdeen, UK, 87-88, vis scientist, 89- *Honors & Awards:* Agr Toward 2000 Medal, Foreign Agr Orgn, 85; Nat Award Food Sci & Technol, Coca-Cola Export, 86 & Conasupo, 87. *Mem:* Nutrit Soc; Am Soc Clin Nutrit; Am Soc Cereal Chemists; Latin Am Soc Nutrit. *Res:* Evaluation of nutritional status in the community by clinical, anthropometric and dietary methodologies; the effect of environmental factors on energy expenditure and utilization and body composition in adults and children, especially parasitic infections; the use of indirect calorimetry and stable isotopes for the latter. *Mailing Add:* Apt Postal 1735 Hermosillo Sonora 83000 Mexico

VALENCICH, TRINA J, b Long Beach, Calif, Feb 3, 43; c 1. PHYSICAL CHEMISTRY. *Educ:* Univ Calif, Irvine, BA, 68, PhD(chem), 74. *Prof Exp:* Adj asst prof chem, Univ Calif, Los Angeles, 73-76; asst prof, Tex A&M Univ, 76; ASST PROF CHEM, CALIF STATE UNIV, LOS ANGELES, 77- *Mem:* Am Chem Soc; Am Physics Soc. *Res:* Classical trajectory simulation of microscopic physical and chemical processes. *Mailing Add:* Dept Chem Univ Calif 5151 State University Dr Los Angeles CA 90032-4202

VALENSTEIN, ELLIOT SPIRO, b New York, NY, Dec 9, 23; m 47; c 2. NEUROSCIENCES, PSYCHOLOGY. *Educ:* City Col New York, BS, 49; Univ Kans, MA, 53, PhD(psychol), 54. *Prof Exp:* Asst anat & psychol, Univ Kans, 51, asst, Endocrinol Lab, 53-54, USPHS res fel anat, 54-55; chief lab neuropsychol, Walter Reed Army Inst Res, Walter Reed Army Med Ctr, 59-61; from assoc to prof psychol & sr res assoc, Fels Res Inst, Antioch Col, 61-70; PROF PSYCHOL, NEUROSCI LAB, UNIV MICH, ANN ARBOR, 70- *Concurrent Pos:* Mem, Exp Psychol Study Sect, Nat Sci Adv Bd, 64-66; vis prof, Univ Calif, Berkeley, 69-70; mem, Neurobiol Rev Panel, NSF, 71-72; mem, Exp Psychol Study Sect, NIH, 75-79; mem, Maternal & Child Health Res Comn, Nat Inst Child Health Develop, 80-; Kenneth Craik res award, Cambridge Univ, 80-81; pres, Div Comp & Physiol Psychol, Am Psychol Asn, 75-76. *Mem:* Fel AAAS; fel Am Psychol Asn; Int Brain Res Orgn; NY Acad Sci; Soc Exp Psychol. *Res:* Hormones and behavior; development of behavioral capacities; physiological and comparative psychology; nervous system and motivation. *Mailing Add:* Dept Psychol Univ Mich Ann Arbor MI 48109-1687

VALENTA, ZDENEK, b Havlickuv Brod, Czech, June 14, 27; Can citizen; m 57; c 3. ORGANIC CHEMISTRY. *Educ:* Swiss Fed Inst Technol, Dipl Ing Chem, 50; Univ NB, MSc, 52, PhD(chem), 53. *Prof Exp:* Spec lectr chem, Univ NB, 53-54, lectr, 54-56; Univ NB fel & res assoc, Harvard Univ, 56-57; from asst prof to assoc prof, 57-63, chmn dept, 63-72, prof chem, 63-90, RES PROF, UNIV NB, 90- *Honors & Awards:* Merck, Sharp & Dohme Lect Award, Chem Inst Can, 67. *Mem:* Chem Inst Can; fel Royal Soc Can; Am Chem Soc. *Res:* Total synthesis of organic molecules of biological and pharmaceutical interest; study of organic reactions and stereochemistry. *Mailing Add:* Dept Chem Univ NB Fredericton NB E3B 6E2 Can

VALENTEKOVICH, MARIJA NIKOLETIC, b Dubrovnik, Yugoslavia, Feb 5, 32; m 62; c 2. CHEMISTRY. *Educ:* Univ Zagreb, MSChE, 57, PhD(chem), 63. *Prof Exp:* Res assoc, Rudjer Boskovic Inst, Zagreb, Yugoslavia, 57-65; fel & res assoc, Radiocarbon Lab, Univ Ill, Urbana, 65-67 & Univ Southern Calif, 67-68; sr chemist, Cyclo Chem Co, 68-69, dir qual control, 69-73; head dept radioisotopes, Curtis Nuclear Co, 73-74; dir qual control, Nichols Inst, 74-76, dir radiochem, 76-79; prin develop chemist, Beckman Instrument Inc, 79-83; vpres immunochem, Innotron Diag, 83-86; OPER MGR, MEREL INC, 87- *Mem:* Am Chem Soc; Am Asn Clin Chem; Clin Ligand Assay Soc. *Res:* Clinical diagnostics, particularly immunoassays; radiolabelling of peptides and hormones; development and evaluation of new immunoassay techniques. *Mailing Add:* 33 Silver Spring Dr Rolling Hills Estates CA 90274-2312

VALENTICH, JOHN DAVID, CELL PHYSIOLOGY. *Educ:* Med Col Pa, PhD(pathobiol), 77. *Prof Exp:* ASST PROF CELL BIOL & MED PHYSIOL, UNIV TEX HEALTH SCI CTR, 84- *Res:* Epithelial transport; cell differentiation. *Mailing Add:* Dept Physiol & Cell Biol Univ Tex Health Sci Ctr PO Box 20708 Houston TX 77225

VALENTINE, BARRY DEAN, b New York, NY, June 6, 24; m 53; c 2. SYSTEMATICS, WEEVIL BIOLOGY. *Educ:* Univ Ala, BS, 51, MS, 54; Cornell Univ, PhD(entom), 60. *Prof Exp:* Asst prof biol, Miss Southern Col, 55-57, actg head dept, 57; from asst prof to assoc prof zool & entom, 60-74, PROF ZOOL, OHIO STATE UNIV, 74- *Concurrent Pos:* Consult, Standard Fruit Co, 56, Lerner Marine Lab, Am Mus Natural Hist, 65, Dames & Moore Inc, 72 & US Army CEngr, 72 & Rockefeller Brothers Fund, 78; entomologist zool exped, Haiti & Jamaica, 56, Cent Am, 56, Mexico, 59, Bahama Islands, 65, 72, 82 & 83, Kenya & Tanzania, 71, 74, 75 & 86, Costa Rico, 87, Galapagos Islands, 87 & 88; Entom Soc Am travel grant, London, Eng, 64; Ohio State Univ develop fund travel grant, London, Copenhagen, Stockholm & Paris, 70; vis prof, Univ Okla Biol Sta, 65, 67-69; vis cur of Coleoptera, Am Mus Natural Hist, 77; Nat Geog Soc res grant, Kenya & Seychelles Islands, 86; res assoc, Fla Dept Agr, 86- *Mem:* Entom Soc Am; Coleopterists Soc; Am Soc Ichthyologists & Herpetologists; Soc Study Amphibians & Reptiles; Herpetologists League. *Res:* Theory and practice of systematics and zoogeography, especially the weevil family Anthribidae of the world and salamanders of Eastern United States; comparative grooming behavior of arthropods. *Mailing Add:* Dept Zool Ohio State Univ Columbus OH 43210

VALENTINE, DONALD H, JR, b Orange, NJ, Nov 7, 40; m 66. ORGANOMETALLIC CHEMISTRY, PHOTOCHEMISTRY. *Educ:* Wesleyan Univ, BA, 62; Calif Inst Technol, PhD(photochem), 66. *Prof Exp:* NSF fel, Stanford Univ, 65-66; asst prof chem, Princeton Univ, 66-71; sr chemist, Hoffmann-La Roche Inc, 71-74, res fel, 74, group chief, 75-80; sr res assoc, Catalytica Asn 80, tech dir, 81-83; proj mgr, 84, tech dir, 85-86, DIR, AM CYANAMID, 87- *Concurrent Pos:* Lectr, Bell Tel Labs, 70-71; lectr, Exten Div, Rutgers Univ, 72 & 76; ed, Molecular Photochem, 72-77; adj prof, Rutgers Univ, 79; fel, Hydrocarbon Res Inst, Univ Southern Calif, 80. *Mem:* Sigma Xi. *Res:* Redox reactions; spectroscopy; homogeneous catalysis; asymmetric synthesis; electronic chemicals; bioseparations. *Mailing Add:* 1937 W Main St Am Cyanamid Stamford CT 06904

VALENTINE, FRED TOWNSEND, b Detroit, Mich, Sept 1, 34; m 64; c 2. IMMUNOLOGY, INFECTIOUS DISEASES. *Educ:* Harvard Univ, AB, 56, MD, 60. *Prof Exp:* Asst prof, 69-75, ASSOC PROF MED, SCH MED, NY UNIV, 75- *Concurrent Pos:* Attend med, Manhattan Vet Admin Hosp, 70-; assoc attend med, Univ Hosp & assoc attend physician, Bellevue Hosp, NY, 76- *Mem:* Am Asn Immunologists; Infectious Dis Soc Am; Transplant Soc; Harvey Soc. *Res:* Cellular immunology, immunological defenses against infectious agents and against neoplasia. *Mailing Add:* Dept Med Sch Med NY Univ 550 First Ave New York NY 10016

VALENTINE, FREDRICK ARTHUR, b Detroit Lakes, Minn, June 26, 26; m 66; c 3. FOREST GENETICS. *Educ:* St Cloud State Teachers Col, BS, 49; Univ Wis, MS, 53, PhD(genetics), 57. *Prof Exp:* Instr genetics, Univ Wis, 54-56; from asst prof to assoc prof forest bot, 56-69, PROF ENVIRON & FOREST BIOL, STATE UNIV NY COL ENVIRON SCI & FORESTRY, 69- *Mem:* Genetics Soc Am; Am Genetic Asn. *Res:* Genetic control of growth and wood properties in Populus tremuloides, the genetics of Hypoxylon mammatum susceptibility to canker in Populus spp; genetics of resistance to verticillium wilt in urban maple trees. *Mailing Add:* Dept Environ & Forest Biol State Univ NY Col Environ Sci & Forestry Syracuse NY 13210

VALENTINE, JAMES K, b Culver City, Los Angeles, Calif, Oct 18, 42; m 85; c 6. ELECTRONICS DESIGN, SYSTEMS INTERGRATION. *Educ:* USAF Inst, BS, 65. *Hon Degrees:* MSIE, Loyola, Paris, France, 72. *Prof Exp:* Res assoc, Temple Univ Los Alamos Sci Labs, 74-80; process engr, Solid State Sci, 80-81; reliability mgr, Tracor Aerospace/Failure Anal, 81-82; component engr, Motorola Commun, 82-83; engr, Plain Elec, 83; test engr, Kirk Mayer Eng Consult Sandia Labs, 83-86 & Allied Signal Sandia Labs, 86-88; TEST ENGR, ALLIANCE ANALYSIS, 88- *Concurrent Pos:* Eng technician, Tex Nuclear, 66 & Tracor Aerospace Avionics, 66-67; engr, RCA Govt Serv, Goddard Space Flight Ctr, 67-68 & Gen Dynamics, 68-72; elec engr, Los Alamos Sci Labs, 72-74. *Res:* Gain stabilization of photo multiplier tubes using calibation techniques and blue LED diodes for timing and synthesizing beam experiments. *Mailing Add:* PO Box 199 McIntosh NM 87032-0199

VALENTINE, JAMES WILLIAM, b Los Angeles, Calif, Nov 10, 26; m 57, 87; c 5. GEOLOGY. *Educ:* Phillips Univ, BA, 51; Univ Calif, Los Angeles, MA, 54, PhD(geol), 58. *Prof Exp:* Asst geol, Univ Calif, Los Angeles, 52-55, asst geophys, 57-58; from asst prof to assoc prof geol, Univ Mo, 58-64; assoc prof, Univ Calif, Davis, 64-68, prof geol, 68-77; prof geol sci, Univ Calif, Santa Barbara, 77-90; PROF INT BIOL, UNIV CALIF, BERKELEY, 90- *Concurrent Pos:* Fulbright res scholar, Australia, 62-63; Guggenheim fel. *Mem:* Nat Acad Sci; AAAS; Geol Soc Am; Ecol Soc Am; Paleont Soc (pres, 73-74); Am Acad Arts & Sci. *Res:* Evolutionary paleoecology. *Mailing Add:* Dept Int Biol Univ Calif Berkeley CA 94720

VALENTINE, JIMMIE LLOYD, b Shreveport, La, Oct 18, 40; m 63, 84; c 3. PHARMACOLOGY. *Educ:* Centenary Col La, BS, 62 & 64; Univ Miss, MS, 66, PhD(med chem), 68. *Prof Exp:* Sr scientist med chem, Mallinckrodt Pharmaceut, 68-73; asst prof, Univ Mo, Kansas City, 73-78; assoc prof pharmacol, Oral Roberts Univ, 78-84, prof & chmn, 84-90; PROF PEDIAT, UNIV ARK COL MED, 90- *Concurrent Pos:* Consult, var pvt & pub orgn, 73-; adj prof, Univ Mo, 78-81. *Mem:* Am Chem Soc; Am Pharmaceut Asn; Am Soc Mass Spectrometry; AAAS; Am Soc Pharmacol & Exp Therapeut. *Res:* Analysis of physiological specimens; human breath analysis; drug effects on endocrine function; laser analysis of drugs and physiological fluids; gas liquid chromatography and high pressure liquid chromatography-mass spectrometry of physiological fluids; alcoholism. *Mailing Add:* Univ Ark Med Sci Ark Children's Hosp 800 Marshall St Little Rock AR 72202-3591

VALENTINE, JOAN SELVERSTONE, bioinorganic chemistry, inorganic chemistry; deceased, see previous edition for last biography

VALENTINE, JOSEPH EARL, b Kansas City, Kans, Apr 6, 33; m 55; c 2. MATHEMATICS. *Educ:* Southwest Mo State Col, BSEd, 58; Univ Ill, Urbana, MS, 60; Univ Mo, Columbia, PhD(distance geom), 67. *Prof Exp:* Teacher high sch, Mo, 58-59; teacher & prin high sch, Mo, 60-61; from instr to asst prof math, Southwest Mo State Col, 61-68; asst prof, Utah State Univ, 68-70, assoc prof math, 70-77; RETIRED. *Concurrent Pos:* Fulbright fel, Univ Jordan, 71-72. *Mem:* Am Math Soc; Math Asn Am. *Res:* Distance geometry; non-euclidean geometry. *Mailing Add:* Rte 2 Verona MO 65769

VALENTINE, MARTIN DOUGLAS, b Greenwich, Conn, Apr 13, 35; m 57; c 4. ALLERGY, CLINICAL IMMUNOLOGY. *Educ:* Union Col, BS, 56; Tufts Univ, MD, 60; Am Bd Internal Med, cert, 72; Am Bd Allergy & Immunol, cert, 74. *Prof Exp:* Instr, Sch Med, Harvard Univ, 68-70; from asst prof to assoc prof, 70-85, PROF, SCH MED, JOHNS HOPKINS UNIV, 85- *Concurrent Pos:* Staff physician allergy, Lahey Clin Found, Boston, 68-70; physician allergy, Lahey Clin Found, Boston, 68-70; physician allergy, Johns Hopkins Hosp & Good Samaritan Hosp, Baltimore, 70-; chmn comt insects, Am Acad Allergy, 75-77, chmn res coun, 77-80, vpres, 89-90; chmn study group hymenotera venoms, Nat Inst Allergy & Infectious Dis, 77- *Honors & Awards:* Philip S Norman Lectr, Am Acad Allergy Immunol, 90. *Mem:* Am Acad Allergy Immunol; Am Asn Immunologists; Am Thoracic Soc. *Res:* Mechanisms of immediate hypersensitivity reactions; allergy to hymenoptera venoms; allergic reaction to foods. *Mailing Add:* Johns Hopkins Asthma/Allergy Ctr 301 Bayview Blvd Baltimore MD 21224

VALENTINE, RAYMOND CARLYLE, b Piatt Co, Ill, Sept 20, 36; m 58; c 1. BIOCHEMISTRY. *Educ:* Univ Ill, Urbana, BS, 58, MS, 60, PhD(microbiol), 62. *Prof Exp:* Asst microbiol, Univ Ill, Urbana, 58-62; fel, Rockefeller Inst, 62-64; asst prof biochem, Univ Calif, Berkeley, 64-70; asst prof in residence microbial biochem, Univ Calif, San Diego, 72-74; mem staff, 74-77, ASSOC PROF PLANT GROWTH LAB, UNIV CALIF, DAVIS, 77- *Mem:* Am Soc Microbiol; fel Am Soc Biol Chemists. *Res:* Nitrogen fixation; ferredoxin; microbial biochemistry and genetics. *Mailing Add:* Dept Agron & Range Sci Univ Calif Davis CA 95616

VALENTINE, WILLIAM NEWTON, b Kansas City, Mo, Sept 29, 17; m 40; c 3. MEDICINE, HEMATOLOGY. *Educ:* Tulane Univ, MD, 42; Am Bd Internal Med, dipl, 49. *Prof Exp:* Intern med, Strong Mem Hosp, Rochester, NY, 42-43, asst resident, 43, chief resident, 43-44; instr, Sch Med, Univ Rochester, 47-48, head sect hemat, AEC Proj, 47-53; asst clin prof, Sch Med, Univ Calif, Los Angles, 49-50, from asst prof to prof, 50-88, chmn dept, 63-71, EMER PROF MED, SCH MED, UNIV CALIF, LOS ANGELES, 88- *Concurrent Pos:* Assoc, St John's Hosp, Santa Monica, Calif, 47, hon consult, 52; sr attend, Harbor Hosp, Torrance, 50; consult, AEC Proj, 53; consult, Hemat Study Sect, NIH, 55-58, mem coun, Inst Arthritis & Metab Dis, 66-70; mem, Am Bd Internal Med, 64-67; mem, Gov Adv Coun Southern Calif, Am Col Physicians, 64-74 & 81-85; mem adv coun, Am Soc Hemat, 69-74, counr, 76-77; outstanding fac res lectr, Dept Med, Univ Calif, Los Angeles, 74-75 & 53rd ann fac res lectr, 78. *Honors & Awards:* Mayo Soley Award for Excellence in Res, Western Soc Clin Res, 78; Henry Stratton Medalist, Am Soc Hemat, 78; John Phillips Mem Award for Distinguished Achievements in Internal Med, Am Col Physicians, 79. *Mem:* Nat Acad Sci; AAAS; Am Soc Clin Invest (vpres, 62); Asn Am Physicians; master Am Col Physicians; fel Am Acad Arts & Sci; AMA; fel Am Soc Hemat; fel Int Soc Hemat (vpres, 76-80); Western Soc Clin Res. *Res:* Hematology. *Mailing Add:* Div Hemat & Oncol 37-068-Ctr Health Sci Bldg Univ Calif Sch Med 10833 LeConte Ave Los Angeles CA 90024-1678

VALENTINI, JAMES JOSEPH, b Martins Ferry, Ohio, Mar 20, 50; m 81; c 2. CHEMICAL DYNAMICS, SPECTROSCOPY. *Educ:* Univ Pittsburgh, BS, 72; Univ Chicago, MS, 73; Univ Calif, Berkeley, PhD(chem), 76. *Prof Exp:* Chaim Weizmann fel, Harvard Univ, 77-78; J Robert Oppenheimer fel, Los Alamos Nat Lab, 78-80, staff mem, 80-84; prof, Univ Calif, Irvine, 84-90; PROF, COLUMBIA UNIV, 90- *Mem:* Am Phys Soc; Sigma Xi; Am Chem Soc; AAAS. *Res:* Chemical dynamics; experimental studies employing molecular beam and laser spectroscopic methods. *Mailing Add:* Chem Dept Columbia Univ New York NY 10027

VALENTY, STEVEN JEFFREY, b Minneapolis, Minn, May 28, 44; m 69; c 2. POLYMER SURFACE CHEMISTRY, PHOTOCHEMISTRY. *Educ:* Lewis Col, BA, 66; Pa State Univ, PhD(org chem), 71. *Prof Exp:* Fel, Royal Inst London, 71-72; STAFF SCIENTIST, CORP RES & DEVELOP CTR, GEN ELEC CO, NY, 72- *Mem:* Am Chem Soc; AAAS. *Res:* Design and develop new materials and processes using the theory and tools of polymer surface science. *Mailing Add:* 7855 S River Pkwy Suite 100 Temple AZ 85284-1825

VALENTY, VIVIAN BRIONES, b Tarlac, Philippines, Dec 15, 44; US citizen; m 69; c 2. ORGANIC CHEMISTRY. *Educ:* Mapua Inst Technol, BS, 64; Pa State Univ, PhD(chem), 71. *Prof Exp:* Res asst cereal chem, Int Rice Res Inst, 64-66; asst prof chem, Skidmore Col, 75-77; res assoc, Div Labs & Res, NY State Dept Health, 7;-81; res chemist, A E Staley Mfg Co, 81-83; prog leader, Gen Elec Co, 83-88, microsci, 88-90; VPRES, ANALYZE INC, 90- *Mem:* Am Chem Soc; Sigma Xi; AAAS. *Res:* Organic chemistry applied to biomolecules; chemicals from carbohydrates; polyimide siloxanes. *Mailing Add:* 7918 S Kenwood Lane Temple AZ 85284

VALENZENO, DENNIS PAUL, b Cleveland, Ohio, June 15, 49; m 72, 86; c 4. MEMBRANE BIOPHYSICS, PHOTOSENSITIZATION. *Educ:* Case Western Reserve Univ, BS, 71, MS, 75, PhD(physiol), 76. *Prof Exp:* NIH fel, Emory Univ, 76-80; asst prof, 80-86, ASSOC PROF PHYSIOL, UNIV KANS MED CTR, 86- *Concurrent Pos:* Porter vis lectr, Spelman Col, 78-79; instr, Emory Univ, 79-80; prin investr, Am Heart Asn, 80, 83, Am Lung Asn, 84; vis prof, Univ L'Aquila, Italy, 85; ed, Am Soc Photobiol Newslett, 87-; mem, Am Soc Photobiol Educ Comt, 86-, Publ Comt, 87-, Pub Affairs Comt, 87- *Mem:* Am Soc Photobiol; Biophys Soc. *Res:* Membrane biophysics in nerve axons, cardiac cells, and red blood cells, particularly sensitization of membrane functions to visible light by photosensitizing dyes. *Mailing Add:* Dept Physiol Univ Kans Med Ctr 39th Rainbow Blvd Kansas City KS 66103

VALENZUELA, GASPAR RODOLFO, b Coelemu, Chile, Jan 6, 33; US citizen; m 58; c 2. AIR-SEA INTERACTION, NONLINEAR DYNAMICS. *Educ:* Univ Fla, BSEE, 54, MSE, 55; Johns Hopkins Univ, DrEng, 65. *Prof Exp:* Assoc eng, Westinghouse Elec Corp, Md, 55-57; assoc sr res staff, Appl Physics Lab, Johns Hopkins Univ, 57-59, res staff asst, Carlyle Barton, 59-64, sr res staff, Appl Physics Lab, 64-68; RES ELECTRONIC ENGR, NAVAL RES LAB, WASHINGTON, DC, 68- *Concurrent Pos:* Assoc ed, J Geophys Res Oceans, 82-85; rep, Sci Comt Oceanog Res, Int Union Radio Sci, 84-90; mem ad-hoc working group global change, Int Union Radio Sci, 87-90. *Mem:* Sr mem Inst Elec & Electronics Engrs; Am Geophys Union; Am Meteorol Soc; Int Union Radio Sci; Oceanog Soc. *Res:* Electromagnetic theory; rough surface scattering theory; interaction of electromagnetic waves with ocean; oceanography; nonlinear interactions in geophysics; hydrodynamics; wave dynamics; radio-oceanography; remote sensing. *Mailing Add:* Ctr Advan Space Sens US Naval Res Lab Code 4234 4555 Overlook Ave SW Washington DC 20375-5000

VALENZUELA, REINALDO A, b San Antonio, Chile, Aug 30, 52; m 75; c 2. WIRELESS COMMUNICATIONS, SIGNAL PROCESSING. *Educ:* Univ Chile, BSc, 75; Imp Col London, PhD(elec eng) & DIC, 81. *Prof Exp:* Sr engr data commun, Databit Ltd, 83-84; res asst signal processing, Imp Col, Univ London, 80-82; mgr voice res, CODEX Corp, 88-90; mem tech staff signal processing, Commun Methods Res Dept, 84-88, MEM TECH STAFF, COMMUN RES DEPT, AT&T BELL LABS, 90- *Concurrent Pos:* Ed, Inst Elec & Electronics Engrs Trans on Commun, 90- *Mem:* Sr mem Inst Elec & Electronics Engrs. *Res:* Propagation modeling and system design for wireless communication systems and personal communication networks. *Mailing Add:* 17 Partridge Run Holmdel NJ 07733-0400

VALEO, ERNEST JOHN, b New London, Conn, Aug 6, 45; m 70; c 3. PLASMA PHYSICS. *Educ:* Rensselaer Polytech Inst, BS, 67; Princeton Univ, MA, 69, PhD(astrophys sci), 71. *Prof Exp:* Res assoc plasma physics lab, Princeton Univ, 72-73; physicist, Lawrence Livermore Lab 73-76; RES PHYSICIST, PLASMA PHYSICS LAB, PRINCETON UNIV, 77- *Concurrent Pos:* Consult, Lab Laser Energetics, Univ Rochester, 77- & Lawrence Livermore Lab, Univ Calif, 77- *Mem:* Am Phys Soc. *Res:* Theoretical plasma physics, especially as related to controlled thermonuclear fusion research. *Mailing Add:* 23 Monterey Princeton Jct NJ 08550

VALERIOTE, FREDERICK AUGUSTUS, b Montreal, Que, May 19, 41; m 66; c 3. BIOPHYSICS. *Educ:* Univ Toronto, BSc, 62, MA, 64, PhD(med biophys), 66. *Prof Exp:* Can Cancer Soc fel, Ont Cancer Inst, 66-67; Med Res Coun Can fel, NIH, 67-68, USPHS vis fel, 68-69; assoc prof, Edward Mallinckrodt Inst Radiol, Med Sch, Wash Univ, 69-76, prof radiol, 69-, assoc dir, Dept Radiation Oncol, 76-; STAFF MEM, MICH CANCER FOUND. *Concurrent Pos:* Mem ed bd, Cell & Tissue Kinetics. *Mem:* Am Asn Cancer Res; Cell Kinetics Soc (pres); Am Asn Cancer Educ. *Res:* Cancer research; experimental cancer chemotherapy; cell population kinetics. *Mailing Add:* Med Oncol 3800 Woodward Wayne State Univ Sch Med 540 E Canfield Detroit MI 48201

VALI, GABOR, b Budapest, Hungary, Oct 22, 36; US citizen; m 56; c 3. ATMOSPHERIC PHYSICS. *Educ:* Sir George Williams Univ, BSc, 61; McGill Univ, MSc, 64, PhD(physics), 68. *Prof Exp:* Lectr agr physics, Macdonald Col, McGill Univ, 65-68, asst prof, 68-69; from asst prof to assoc prof, 69-76, PROF ATMOSPHERIC SCI, UNIV WYO, 76- *Mem:* Am Meteorol Soc; Am Asn Aerosol Res; Sigma Xi; Royal Meteorol Soc; Int Asn Aerobiol. *Res:* Ice nucleation; development of ice elements in clouds; physics of precipitation; weather modification; atmospheric aerosols; human impact; biogenic ice nuclei. *Mailing Add:* Dept Atmospheric Sci Univ Wyo PO Box 3038 Univ Laramie WY 82070

VALIAVEEDAN, GEORGE DEVASIA, b Erattupetta, India, Mar 29, 32; m 65; c 2. ORGANIC CHEMISTRY. *Educ:* Univ Madras, BSc, 52; Georgetown Univ, MS, 60, PhD(org chem), 62. *Prof Exp:* Lab instr chem, St Joseph's Col, India, 52-53; high sch teacher, Ceylon, 54-55; instr, St Sylvester's Jr Col, 56-57; res asst steroid & carbohydrate chem, Georgetown Univ, 60-62; res fel, Sch Chem, Univ Minn, Minneapolis, 62-65; res chemist, 65-74, SR RES CHEMIST, PHOTO PROD DEPT, E I DU PONT DE NEMOURS & CO, INC, PARLIN, 74- *Mem:* Am Chem Soc. *Res:* Steroids; carbohydrates; microbial metabolites; aromatic hydrocarbons; polymer chemistry; dyes and pigments; photochemistry; photographic processes. *Mailing Add:* Nine Coventry Dr Freehold NJ 07728

VALINSKY, JAY E, EMBRYONIC CELL MIGRATION. *Educ:* Brandeis Univ, PhD(biochem), 74. *Prof Exp:* ASST DIR RES & DEVELOP, NY BLOOD CTR, 83- *Res:* Cell-surface phenotypes. *Mailing Add:* Dept Flow Cytometry NY Blood Ctr 310 E 67th St New York NY 10021

VALK, HENRY SNOWDEN, b Washington, DC, Jan 26, 29; m 68; c 4. THEORETICAL NUCLEAR PHYSICS. *Educ:* George Washington Univ, BS, 53, MS, 54; Wash Univ, PhD(physics), 57. *Prof Exp:* Asst, Wash Univ, 54-56; asst prof physics, Univ Ore, 57-59; asst prog dir physics, NSF, 59-60; from asst prof to prof physics, Univ Nebr, Lincoln, 60-70, chmn dept, 66-70; dean, Col Sci & Liberal Studies, 70-82, PROF PHYSICS, GA INST TECHNOL, 70- *Concurrent Pos:* Prog dir theoret physics, NSF, 65-66; vis prof, Univ Frankfurt, 70, Rensselaer Polytech Inst, 82, Cath Univ Am, 82-83. *Honors & Awards:* Order of Brit Empire, 85. *Mem:* Fel Am Phys Soc; Am Math Soc; Am Asn Physics Teachers; Math Asn Am. *Res:* Theoretical atomic and nuclear physics. *Mailing Add:* 3032 St Helena Dr Tucker GA 30084

VALK, WILLIAM LOWELL, b Muskegon, Mich, Aug 23, 09; m 37; c 2. SURGERY. *Educ:* Univ Mich, AB, 34, MD, 37. *Prof Exp:* Instr surg, Med Sch, Univ Mich, 40-43; assoc prof, 46-47, PROF SURG, UNIV KANS MED CTR, 47- *Mem:* Soc Univ Surg; Clin Soc Genito-Urinary Surg; Am Surg Asn; Am Urol Asn; AMA. *Res:* Urological surgery; physiology of kidney. *Mailing Add:* Med Ctr Univ Kans 39th & Rainbow Blvd Kansas City KS 66208

VALLABHAN, C V GIRIJA, b Trichur, India, May 24, 35; m 61; c 3. CIVIL ENGINEERING, ENGINEERING MECHANICS. *Educ:* Univ Kerala, BSc, 57; Univ Mo-Rolla, MS, 60; Univ Tex, Austin, PhD(civil eng), 67. *Prof Exp:* Jr engr, Kerala Pub Works Dept, India, 57; lectr civil eng, Eng Col, Trichur, India, 58-59, asst prof, 60-64; asst prof, 66-76, assoc prof, 76-80, PROF CIVIL ENG, TEX TECH UNIV, 80- *Mem:* Am Soc Civil Engrs. *Res:* Finite element technique for solving elasticity and plasticity problems in structural and soil mechanics; deterministic and probabilistic analysis of soil-structure interaction problems. *Mailing Add:* Dept Civil Eng Tex Tech Univ Lubbock TX 79409

VALLANCE, MICHAEL ALAN, advanced composites, thermoplastic elastomers, for more information see previous edition

VALLBONA, CARLOS, b Barcelona, Spain, July 29, 27; m 56; c 4. PEDIATRICS, COMMUNITY MEDICINE. *Educ:* Univ Barcelona, BA & BS, 44, MD, 50. *Prof Exp:* Physician, Sch Child Health, Spain, 51-52; intern & resident, Sch Med, Univ Louisville, 53-55; from instr to assoc prof pediat & physiol, 56-67, from instr to assoc prof rehab, 57-67, PROF REHAB, BAYLOR COL MED, 67-, PROF & CHMN DEPT COMMUNITY MED, 69-; CHIEF COMMUN MED SERV, HARRIS COUNTY HOSP DIST, 69- *Concurrent Pos:* Fel, Children's Int Ctr, Univ Paris, 52-53; consult, Nat Heart, Lung & Blood Inst, Nat Ctr Health Servs Res, Nat Ctr Health Care Technol,

Nat Ctr Health Statist & Off Health Resources Opportunity. *Mem:* AAAS; Soc Pediat Res; Am Col Chest Physicians; Sigma Xi; AMA. *Res:* Pediatric rehabilitation; cardiorespiratory physiology in disabled persons and the newborn; application of electronic data processing techniques in health care; rehabilitation community medicine; community medicine; prevention of hypertension. *Mailing Add:* Dept Community Med Baylor Col Med Houston TX 77030

VALLEAU, JOHN PHILIP, b Toronto, Ont, Jan 17, 32; m 57, 86; c 2. STATISTICAL MECHANICS, CHEMICAL PHYSICS. *Educ:* Univ Toronto, BA, 54, MA, 55; Cambridge Univ, PhD(theoret chem), 58. *Prof Exp:* Nat Res Coun Can fel, 58-60; from asst prof to assoc prof, 61-74, PROF CHEM, UNIV TORONTO, 74- *Concurrent Pos:* Res visitor, Fac Sci, Orsay, France, 68-69, Norweg Inst Technol, Trondheim, 78-79, Dipartimento Fisica, Univ Roma, 85-86. *Res:* Theory of liquids and phase changes and of solutions; Monte Carlo and molecular dynamic computations; theory of surface phenomena. *Mailing Add:* Lash Miller Lab Univ Toronto Toronto ON M5S 1A1 Can

VALLEE, BERT L, b Hemer, WGer, June 1, 19; nat US; m 47. BIOCHEMISTRY, BIOPHYSICS. *Educ:* Univ Bern, BS, 38; NY Univ, MD, 43. *Hon Degrees:* MA, Harvard Univ, 60; Dr Med hon causa, Karolinska Inst, Sweden, 87. *Prof Exp:* Res fel med, Harvard Med Sch, 45-49, from res assoc to assoc, 49-55, from asst prof to prof, 55-65, Paul C Cabot prof biol chem, 65-80, Paul C Cabot prof biochem sci, 80-89, head, Ctr Biochem & Biophys Sci & Med, 80-89, distinguished sr prof, 89-90, EDGAR M BRONFMAN DISTINGUISHED SR PROF, HARVARD MED SCH, 90- *Concurrent Pos:* Nat Res Coun Can sr fel, Mass Inst Technol, 48-51; Hughes fel, Harvard Med Sch, 51-64; mem staff, Div Indust Coop, Mass Inst Technol, 45-48, res assoc biol, 48-; Merck Sharpe & Dohme prof, Univ Wash, 62; mem adv bd, La Trinidad Health Care Facil, Caracas, Venezuela, 70- & Metrop Univ, Caracas, 70-; mem bd gov, Tel Aviv Univ, 72-; Tracy & Ruth Storer vis prof life sci, Univ Davis, Calif, 79, vis prof, Oberlin Col, Ohio, 79; chmn, Sect Biochem, Nat Acad Sci, 81; chmn, US Nat Comt Int Union Biochem, 76; Arthur K Watkins vis prof chem & life sci, Wichita State Univ, 78; vis prof, Univ Zurich & Fed Inst Technol, Zurich, 78; head, Ctr Biochem & Biophys Sci & Med, Brigham & Women's Hosp, 80-89, biochemist-in-chief, Div Clin Chem, 80-89; chmn, Sect Biochem, Nat Acad Sci, 81-84; vis prof med chem, Mass Inst Technol, 86; Wellcome vis prof biol chem, Emory Univ Sch Med, 87; vis lectr, Biochem Fundamental to Med, Weizmann Inst Sci, Rehovot, Israel, 87; hon prof, Tsinghua Univ, Beijing, China, 87. *Honors & Awards:* Warner-Chilcott Award, Am Asn Clin Chem, 69; Arthur Kelley lectr, Purdue Univ, 63; DuPont lectr, Univ SC, 71; Venable lectr, Univ NC, 72; Bauchman lectr, Calif Inst Technol, 76; Linderstrom-Lang Medal & Award, 80; Willard Gibbs Medal & Award, Am Chem Soc, 81; William C Rose Award Biochem, 82; Messenger Lectr, Cornell Univ, 88. *Mem:* Nat Acad Sci; Biochem Soc; Am Soc Clin Invest; Am Chem Soc; Optical Soc Am. *Res:* Composition, conformation, structure, function and mechanism of action of metalloenzymes; local conformation of enzymes; enzyme kinetics; physical chemistry; emission; atomic absorption; absorption spectroscopy; circular dichroism; magnetic circular dichroism; physics of spectrographic sources. *Mailing Add:* Ctr Biochem & Biophys Sci & Med Med Sch Harvard Univ Seeley G Mudd Bldg 1st Floor Boston MA 02115

VALLÉE, JACQUES P, b Verdun, Que, Sept 5, 45. ASTROPHYSICS, RADIO ASTRONOMY. *Educ:* Univ Montreal, BA, 65, BSc, 68, MSc, 69; Univ Toronto, PhD(astron), 73. *Prof Exp:* Res fel astrophy, Sterrewacht te Leiden, Holland, 73-75; res asst, Herzberg Inst, Ottawa, 75-76; res assoc astrophys, Queen's Univ, Kingston, 76-80; RES OFFICER, HERZBERG INST, OTTAWA, 80- *Concurrent Pos:* Mem, comt scientist observ astron, Mont Megantic, 79-80 & 84-86; vis lectr radio astron, Univ Montreal, 85; sabbat res, observatory, Univ Grenoble, France, 88; secondment astron, Royal Observ, Edinburgh, 89-91. *Mem:* Can Astron Soc; Am Astron Soc; Royal Astron Soc Can; Int Astron Union. *Res:* Molecular clouds, interstellar magnetic fields, ionized emission regions in our galaxy; nearby galaxies, clusters of galaxies, cosmology; computer modeling of astronomical processes; search for extraterrestial intelligence; submillimeter radio astronomy. *Mailing Add:* Inst Herzberg 100 Sussex Dr Ottawa ON K1A 0R6 Can

VALLEE, RICHARD BERT, b New York, NY. CELL BIOLOGY, MOLECULAR BIOLOGY. *Educ:* Swarthmore Col, BA, 67; Yale Univ MPh & PhD(biol), 74. *Prof Exp:* Fel molecular biol, Lab Molecular Biol, Univ Wis-Madison, 74-78; STAFF SR & PRIN SCI, WORCESTER FOUND EXP BIOL, 78-, CO-DIR CANCER CTR, 87- *Concurrent Pos:* Instr, Marine Biol Lab, Woods Hole, Ma, 84-88; ed, Methods Enzym; assoc ed, Cell Motility & Cytoskeleton. *Mem:* Biophys Soc; AAAS; Am Soc Cell Biol; Am Soc Biochem & Molecular Biol. *Res:* Proteins associated with cytoplasmic microtubles; enzymology; motility; post translational modification; molecular cloning; ultrastructure; immunological characterization. *Mailing Add:* Worcester Found Exp Biol Shrewsbury MA 01545

VALLEE, RICHARD EARL, b Cincinnati, Ohio, June 21, 28; m 51. PHYSICAL CHEMISTRY, INORGANIC CHEMISTRY. *Educ:* Univ Cincinnati, BS, 51, MS, 52, PhD(chem), 62. *Prof Exp:* Chemist, Procter & Gamble Co, 47-48; asst, Univ Cincinnati, 51; chemist, Monsanto Chem Co, 52-59, group leader, 59-60, ASF fel, 61-62, group leader, Monsanto Res Corp, 62-63, sect mgr, 63-67, mgr nuclear technol, 67-69, mgr non-weapons progs, 69-72, MGR TECHNOL APPLN & DEVELOP, MONSANTO RES CORP, 72- *Mem:* AAAS; Am Chem Soc. *Res:* Preparation, evaluation and handling of radioactive compounds; high temperature compounds; vacuum technology; isotope separation. *Mailing Add:* 619 S Bourbon St Blanchester OH 45107

VALLENTINE, JOHN FRANKLIN, b Ashland, Kans, Aug 1, 31; m 50; c 3. RANCH MANAGEMENT, GRAZING MANAGEMENT. *Educ:* Kans State Univ, BS, 52; Utah State Univ, MS, 53; Tex A&M Univ, PhD(range mgt, animal nutrit), 59. *Prof Exp:* Res aide, Rocky Mountain Forest & Range Exp Sta, US Forest Serv, 52; range conservationist, US Bur Land Mgt, 55-56; res asst range mgt, Exp Sta, Tex A&M Univ, 56-58; exten range specialist, Utah State Univ, 58-62; assoc prof range exten & res, Univ Nebr, 62-68; PROF RANGE SCI, BRIGHAM YOUNG UNIV, 68- *Mem:* Soc Range Mgt; Am Soc Animal Sci. *Res:* Range science and agricultural bibliography; range seeding; range improvements. *Mailing Add:* 425 WIDB Brigham Young Univ Provo UT 84602

VALLERA, DANIEL A, b E Liverpool, Ohio, Oct 31, 51; m; c 2. IMMUNOTOXINS, BONE MARROW TRANSPLANTATION. *Educ:* Ohio State Univ, Columbus, BS, 73, MS, 75, PhD(microbiol), 78. *Prof Exp:* Fel, 78-79, res assoc, 79-80, from asst prof to assoc prof radiol & lab med, 81-87, assoc prof sec cancer immunol, 87-89, DIR, SEC CANCER IMMUNOL, DEPT THERAPEUT RADIOL, UNIV MINN, 87-, PROF SEC CANCER IMMUNOL, 89- *Concurrent Pos:* Leukemia Soc Am Scholar, 83; mem adv comt, Am Cancer Soc, 90- *Honors & Awards:* Am Cancer Soc Jr Fac Award, 83; Hubert H Humphrey Cancer Res Award, 81. *Mem:* Am Asn Immunologists; Transplantation Soc; AAAS; Am Soc Hemat; Am Asn Cancer Res. *Res:* Linkage of potent toxins and radionuclides to antibodies raised against human tumor cells to derive potent anti-tumor agents effective against a wide variety of human cancer cells; utilization of these agents in autologous and allogenic bone marrow transplantation for the treatment of leukemia; cellular and molecular biology of bone marrow transplantation. *Mailing Add:* Box 367 Mayo Mem Bldg Univ Minn 420 Delaware St Minneapolis MN 55455

VALLERGA, BERNARD A, ENGINEERING ADMINISTRATION. *Educ:* Univ Calif, Berkeley, BS, 43, MS, 48. *Prof Exp:* Mat testing engr, Hershey Inspect Bur, Oakland, Calif, 46-48; asst prof civil eng, Univ Calif, Berkeley, 48-53; managing engr, Pac Coast Div, Asphalt Inst, San Francisco, Calif, 53-60; vpres prod develop & mkt, GBO Div, Witco Chem Co, Los Angeles, Calif, 60-64; pres & chief exec officer, Mat Res & Develop, Inc, Oakland, Calif, 64-72; vpres & managing prin, Woodward-Clyde Consults, San-Francisco-Oakland, Calif, 68-76; PRES, B A VALLERGA, INC, CONSULT CIVIL ENG, OAKLAND, CALIF, 77- *Concurrent Pos:* Chmn, Triaxial Inst Struct Design Pavements, 50-52; mem, bd dirs, Asn Asphalt Paving Technologists, 60-62 & 80-82 & bd dirs, Woodward-Clyde Consults; mem, Bd Dirs, & vpres, Asphalt Inst, 62-64; vpres, Design Div, Am Road Builders Asn, 68-70; chmn bd dirs, Woodward-Envicon, 69-72 & Subcomt Asphalt Durability, Transp Res Bd, 80-; mem, Airfield Pavement Comt, Am Soc Civil Engrs, 72-79, Air Transp Publ Comt, 72-79; gen consult, Off Energy Related Inventions, Bur Standards, Dept Com, 80- *Honors & Awards:* Provost Hubbard Award, Am Soc Testing & Mat, 89; Recognition Award, Asn Asphalt Paving Technologists, 88. *Mem:* Nat Acad Eng; fel Am Soc Civil Engrs; Asn Asphalt Paving Technologists; Transp Res Bd; Am Soc Testing & Mat; Int Soc Asphalt Pavements; Sigma Xi. *Res:* Pavement design; construction; rehabilitation and maintenance of pavements; wind and water erosion and soil stabilization; hydraulic revetments; author of various publications. *Mailing Add:* B A Vallerga Inc 1330 Broadway Suite 1044 Oakland CA 94612

VALLE-RIESTRA, J(OSEPH) FRANK, b Oakland, Calif, Nov 12, 24; m 48; c 2. CHEMICAL ENGINEERING. *Educ:* Univ Calif, BAS, 45; Calif Inst Technol, BS, 48, MS, 49. *Prof Exp:* Air pollution chemist, Los Angeles County Air Pollution Control Dist, Calif, 46-47; air pollution chemist, Truesdail Labs, 48; res asst chem eng, Calif Inst Technol, 48-49; res & develop engr, Dow Chem Co, 49-62, sr res engr, 62-75, res specialist, Western Div, 70-75, sr res specialist, 75-78, assoc scientist, 78-81, sr assoc scientist, Western Div, 81-83, res scientist, 83-86; RETIRED. *Concurrent Pos:* Lectr chem eng, Univ Calif, Berkeley, 74- *Honors & Awards:* Chem Eng Pract Award, Am Inst Chem Engrs, 84. *Mem:* Fel Am Inst Chem; fel Am Inst Chem Engrs; Am Soc Eng Educ. *Res:* Transport phenomena in electrochemical and high temperature gaseous systems; high temperature kinetics; physical chemistry of graphite; solvent extraction; secondary fiber technology; food process engineering; project evaluation methodology. *Mailing Add:* 140 Cora St Walnut Creek CA 94596

VALLESE, FRANK M, b New York, NY, July 12, 50; m 76; c 2. AGRICULTURAL & FOOD CHEMISTRY. *Educ:* Wagner Col, Staten Island, NY, BS, 72; Rutgers Univ, NJ, MS, 76, PhD(food sci), 78. *Prof Exp:* Assoc chemist analytical chem, Lipton, Inc, 78-80, proj coordr flavor res, 80-81; sr res chemist anal chem, M&M/Mars, 81-82, mgr, 82-85, dir res lab serv chem microbiol, sensory eval, 85-89, DIR FUNDAMENTAL RES CHEM, MICROBIOL, BIOTECHNOL, PROD EVAL SERV, BASIC RES, INFOSERV, M&M/MARS, 89- *Mem:* Int Food Technologists; Am Oil Chemists Soc; Am Chem Soc; Asn Cereal Chemists. *Res:* Biotechnology; flavor research; analytical chemistry; corporate microbiology; microbiology research; sensory evaluation; marketing research; environmental research and progress. *Mailing Add:* Res Lab Serv M&M/Mars Res & Develop High St Hackettstown NJ 07840

VALLESE, LUCIO M(ARIO), b Naples, Italy, Sept 27, 15; nat US; m 54; c 2. ELECTRICAL ENGINEERING, PHYSICS. *Educ:* Univ Naples, Dott Elec Eng, 37; Carnegie Inst Technol, DSc(elec eng), 48. *Prof Exp:* Res assoc, Navy Inst Elec Commun, Leghorn, Italy, 37-39; from instr to asst prof elec eng, Univ Rome, 39-47; from asst prof to assoc prof physics, Duquesne Univ, 48-51; from asst prof to assoc prof elec eng, Polytech Inst Brooklyn, 51-59; sr scientist, Labs, Int Tel & Tel Corp, 59-68; pres, Electrophys Corp, Nutley, 69-80. *Concurrent Pos:* Adj prof, Polytech Inst Brooklyn, 59-65. *Mem:* Am Phys Soc; Optical Soc Am; Inst Elec & Electronics Engrs. *Res:* Electromagnetic theory; circuit theory; solid state devices; quantum electronics. *Mailing Add:* 340 Ridgewood Ave Glen Ridge NJ 07028

VALLETTA, ROBERT M, b Waterbury, Conn, Nov 10, 31; m 55; c 4. PHYSICAL CHEMISTRY. *Educ:* Univ Conn, BA, 53, MS, 56; Iowa State Univ, PhD(phys chem), 59. *Prof Exp:* Res physicist, Cent Res Lab, Am Mach & Foundry Co, 59-63; staff chemist, Components Div, Vt, 63-68, develop engr, 68-69, sr engr, 69-75, SR ENGR, GEN PROD DIV, IBM CORP, CALIF, 75- *Mem:* Am Chem Soc. *Res:* Solid state chemistry and physics. *Mailing Add:* 1224 Serene Valley Ct San Jose CA 95120

VALLEY, LEONARD MAURICE, b Little Falls, Minn, July 3, 33; m 58; c 3. HOLOGRAPHY, TEACHING. *Educ:* St John's Univ, Minn, BA(physics) & BA(math), 55; Iowa State Univ, PhD(physics), 60. *Prof Exp:* From asst prof to assoc prof physics, 60-72, PROF PHYSICS, ST JOHN'S UNIV, MINN, 72-, CHMN DEPT, 70- *Concurrent Pos:* Vis assoc prof, Univ Denver, 67-68, 83; prof, Ind Univ, Malaysia Coop prog, 87-89. *Mem:* Am Phys Soc; Am Asn Physics Teachers. *Res:* Vibrational, rotational and translational relaxation in gas molecules undergoing collisions; relaxation time; education. *Mailing Add:* Dept Physics St John's Univ Collegeville MN 56321

VALLEY, SHARON LOUISE, b Bay City, Mich, Oct 18, 41; m 75. PHARMACOLOGY, INFORMATION SCIENCE. *Educ:* Univ Mich, BS, 63, PhD(pharmacol), 67. *Prof Exp:* Intern pharm, Schulz Pharm, 60-62 & Health Serv Pharm, Univ Mich, 62-63; PHARMACOLOGIST SCI INFO, NAT LIBR MED, NIH, 67- *Concurrent Pos:* Mem comt user educ, Nat Fedn Abstracting & Indexing, 78-79; bd gov, Col Pharm Alumni Soc, Univ Mich, 78- *Honors & Awards:* Plaque and Gavel Award, Am Pharmaceut Asn, 62. *Mem:* Drug Info Asn; Am Pharmaceut Asn. *Res:* Cardiovascular pharmacology; toxicology. *Mailing Add:* Specialized Info Serv Nat Lib Med 8600 Rockville Pike 38A B1N28M Bethesda MD 20894

VALLIER, TRACY L, b Oakland, Iowa, Sept 19, 36; m 57; c 4. MARINE GEOLOGY. *Educ:* Iowa State Univ, BS, 62; Ore State Univ, PhD(geol), 67. *Prof Exp:* Assoc prof geol, Ind State Univ, 66-72; geologist, Deep Sea Drilling Proj, Scripps Inst Oceanog, 72-75; MARINE GEOLOGIST, US GEOL SURV, 75- *Mem:* Geol Soc Am; Am Soc Petroleum Geol; Am Geophys Union. *Res:* Geology of the Aleutian Island Arc; geology of the Tonga Island Arc; igneous petrology; geology of Hells Canyon, Oregon and Idaho. *Mailing Add:* US Geol Surv 345 Middlefield Rd Menlo Park CA 94025

VALLOTTON, WILLIAM WISE, medicine; deceased, see previous edition for last biography

VALLOWE, HENRY HOWARD, b Pittsburgh, Pa, Nov 18, 24; m 48. ENDOCRINOLOGY. *Educ:* Pa State Teachers Col, BS, 49; Univ Chicago, MS, 50, PhD(zool), 54. *Prof Exp:* Instr biol, Wright Jr Col, Ill, 52-56; assoc prof zool, Ohio Univ, 56-67; PROF BIOL, IND UNIV, PA, 67- *Mem:* AAAS; Am Soc Zoologists. *Res:* Endocrines of poikilotherms; sexual physiology; phylogeny of endocrines; circadian rhythms. *Mailing Add:* 14617 Turtle Creek Cir-704 Lutz FL 33549-6505

VALLS, ORIOL TOMAS, b Barcelona, Spain, Oct 15, 47; US citizen; m 74; c 2. CONDENSED MATTER PHYSICS. *Educ:* Univ Barcelona, BSc, 69; Brown Univ, MSc, 72, PhD(physics), 75. *Prof Exp:* Res assoc physics, James Franck Inst, Univ Chicago, 75-77; Miller fel, Univ Calif, Berkeley, 77-78; from asst prof to assoc prof, 78-88, PROF PHYSICS, UNIV MINN, 88- *Concurrent Pos:* Exchange prof, Univ Paris, 83; vis scientist, Argonne Nat Lab, 84-85; prof invité, Univ Paris, 88. *Mem:* Am Phys Soc; Sigma Xi. *Res:* Properties of quantum fluids; systems far from equilibrium; superconductivity. *Mailing Add:* Tate Lab Physics 116 Church St SE Minneapolis MN 55455

VALOCCHI, ALBERT JOSEPH, b Coatesville, Pa, Aug 20, 53. ENVIRONMENTAL ENGINEERING. *Educ:* Cornell Univ, BS, 75; Stanford Univ, MS, 76, PhD(civil eng), 81. *Prof Exp:* Asst prof, 81-86, ASSOC PROF CIVIL ENG, UNIV ILL, URBANA-CHAMPAIGN, 86- *Mem:* Am Soc Civil Engrs; Am Geophys Union; Asn Groundwater Scientist & Engrs; Sigma Xi. *Res:* Study of the transport and fate of contaminants in groundwater and soils. *Mailing Add:* Dept Civil Eng Univ Ill 205 N Mathews Urbana IL 61801

VALSAMAKIS, EMMANUEL, b Istanbul, Turkey, May 11, 33; US citizen; m 60; c 2. SOLID STATE ELECTRONICS. *Educ:* Robert Col, Istanbul, BSEE, 55; Rensselaer Polytech Inst, MEE, 58, PhD(plasma physics), 63. *Prof Exp:* Instr elec eng, Rensselaer Polytech Inst, 56-61, res asst, 61-62; res scientist, Grumman Aircraft Eng Corp, 62-67; adv physicist, 67-76, mem res staff, 76-80, adv engr, 80-89, SR ENGR, IBM CORP, 90- *Concurrent Pos:* Lectr, State Univ NY, New Paltz, 85- *Mem:* Inst Elec & Electronics Engrs; Am Phys Soc; NY Acad Sci. *Res:* Cryogenic tunneling device design and circuit analysis; experimental investigation of plasmas from pulsed plasma sources; bipolar, field effect transistor modeling, device design and circuit analysis for memory and logic applications. *Mailing Add:* 2685 Hilltop Dr Yorktown Heights NY 10598

VALTIN, HEINZ, b Hamburg, Ger, Sept 23, 26; nat US; m 53; c 2. RENAL PHYSIOLOGY, NEPHROLOGY. *Educ:* Swarthmore Col, AB, 49; Cornell Univ, MD, 53. *Prof Exp:* From instr to prof, 57-73, ANDREW C VAIL PROF PHYSIOL, DARTMOUTH MED SCH, 73-, CHMN DEPT, 77- *Concurrent Pos:* Consult, Hitchcock Clin, 61- *Mem:* Am Physiol Soc; Am Fedn Clin Res; Am Soc Clin Invest; Am Soc Nephrology; Int Soc Nephrology; Int Union Physiol Sci (treas, 84-). *Res:* Kidney, electrolyte and water metabolism; neuroendocrinology. *Mailing Add:* Dept Physiol Dartmouth Med Sch Hanover NH 03756

VALVANI, SHRI CHAND, b Mar 20, 40; US citizen; c 3. PHARMACEUTICAL CHEMISTRY, PHARMACEUTICS. *Educ:* Univ Saugar, BPharm, 65; Univ Mich, Ann Arbor, MS, 69, PhD(pharmaceut chem), 71. *Prof Exp:* From res scientist to sr res scientist, Upjohn Co, 70-83, res head, 83-85, assoc dir, 85-88, dir, drug delivery res & develop, 88-90, DIR, CONTROL DEVELOP, UPJOHN CO, 90- *Honors & Awards:* Prof Schroff Gold Medal, Indian Pharmaceut Asn, 65. *Mem:* Acad Pharmaceut Sci; Am Chem Soc; Am Pharmaceut Asn; Am Asn Pharmaceut Scientists. *Res:* Thermodynamics of solution process and its effect on drug design, physico-chemical and biochemical parameters influencing performance of various drug dosage forms and drug delivery systems, computer applications in stability testing in drug delivery research and development. *Mailing Add:* Control Develop Upjohn Co Kalamazoo MI 49001

VALVASSORI, GALDINO E, b Milan, Italy, July 16, 26; US citizen; m 55; c 4. MEDICINE, RADIOLOGY. *Educ:* Univ Milan, MD, 50; Am Bd Radiol, dipl, 59. *Prof Exp:* Resident radiol, Univ Milan, 51-53; resident, Mem Hosp, NY, 54-56; asst prof, Univ Chicago, 60-65; assoc prof, 65-67, PROF RADIOL, UNIV ILL MED CTR, 67- *Concurrent Pos:* Dir dept radiol, Ill Eye & Ear Infirmary, Chicago, 65-; consult, Grant Hosp, Chicago, 66, 71; pres, Int Collegium Radiol in Otolaryngol, 79-82; consult, MacNeal Hosp, Berwyn, Ill. *Mem:* AMA; Am Col Radiol; Am Roentgen Ray Soc; Radiol Soc NAm; Am Acad Ophthal & Otolaryngol. *Res:* Radiology of the head and neck; development and refinement of new radiographic techniques for study of temporal bone in pathological conditions of the ear. *Mailing Add:* 55 E Washington Chicago IL 60602

VALYI, EMERY I, b Murska Sobota, Yugoslavia, July 14, 11; nat US; m 39; c 2. APPLIED MECHANICS, MATERIAL SCIENCE. *Educ:* Fed Inst Technol, Zurich, ME, 33, DSc(phys metall, appl mech), 37. *Prof Exp:* Res engr, Swiss Fed Inst Testing Mat, 34-37; metall engr, Injecta Ltd, 37-40; mgr, die casting mach div, Hydraul Press Mfg Co, 40-42; vpres, Sam Tour & Co, 43-45; pres, ARD Corp, 45-62; consult, Ford Motor Co, Int Harvester Co, Continental Can Co, Owens-Ill, Inc, Olin Corp & Molins Mach Co, 62-71; consult, Nat Can Corp, 71-81; PRES, TPT MACH CORP, 75- *Mem:* Am Soc Metals; Am Inst Mining, Metall & Petrol Engrs; Soc Plastics Engrs; fel Am Inst Chemists. *Res:* Materials technology; rheology; metal casting; plastic molding; container technology. *Mailing Add:* 19 Moseman Ave Katonah NY 10536

VAMOS, TIBOR, b Budapest, Hungary, June 1, 26; m 50, 73; c 1. PATTERN RECOGNITION, EPISTEMOLOGY. *Educ:* Budapest Tech Univ, MA, 50; Hungarian Acad Sci, PhD(power control), 58, Dr Sc(power systs), 64. *Hon Degrees:* Dr, Tallinn Tech Univ, 86. *Prof Exp:* Dir construct, Power Plant Construct Co, 50-54; head dept, Power Syst Control, Power Res Inst, 57-62; dir res, 62-85, CHMN BD, COMPUTER & AUTOMATION INST, HUNGARIAN ACAD SCI, 86- *Concurrent Pos:* Prof, Budapest Tech Univ, 65-, mem bd, 83-; adv, Int Fedn Automatic Control, 87. *Mem:* Int Fedn Automatic Control (pres, 80-84); fel Inst Elec & Electronics Engrs. *Res:* Process control; robot vision, pattern recognition; artificial intelligence, especially expert systems combining logic and pattern-like features of cognitive psychology related to those epistemic problems of computer science. *Mailing Add:* Computer & Automation Inst Hungarian Acad Sci Victor Hugo U 18-22 Budapest 1132 Hungary

VAMPOLA, ALFRED LUDVIK, b Dwight, Nebr, July 10, 34; m 56; c 8. SPACE PHYSICS. *Educ:* Creighton Univ, BS, 56; St Louis Univ, MS, 58, PhD(physics), 61. *Prof Exp:* Sr physicist, Convair Div, Gen Dynamics Corp, 61-62; staff scientist space physics, Aerospace Corp, 62-78, sr scientist, 78-90; CONSULT, SPACE ENVIRON, 90- *Concurrent Pos:* Assoc ed, J Spacecraft & Rockets, 84-87, 90-; vis fel, Univ Otago, Dunedin, NZ, 86. *Mem:* Am Geophys Union; assoc fel Am Inst Aeronaut & Astronaut. *Res:* Magnetospheric physics; solar particles; spacecraft environment interactions. *Mailing Add:* Vampola PO Box 10225 Torrance CA 90505

VANABLE, JOSEPH WILLIAM, JR, b Providence, RI, May 29, 36; m 62; c 2. DEVELOPMENTAL BIOLOGY. *Educ:* Brown Univ, AB, 58; Rockefeller Inst, PhD(biol), 62. *Prof Exp:* From asst prof to assoc prof, 61-82, PROF BIOL, PURDUE UNIV, WEST LAFAYETTE, 82- *Concurrent Pos:* NIH spec fels, Univ Ore, 69 & Yale Univ, 69-70. *Honors & Awards:* Yasuda Award, Bioelec Repair & Growth Soc, 85. *Mem:* Soc Develop Biol; Am Soc Zool. *Res:* Regeneration and wound healing; the role of endogenous electrical fields; visual mutants. *Mailing Add:* Dept Biol Sci Purdue Univ West Lafayette IN 47907

VAN ALFEN, NEAL K, b Ogdan, Utah, July 17, 43; div; c 4. HOST-PATHOGEN RELATIONS. *Educ:* Brigham Young Univ, BS, 68; MS, 69; Univ Calif, Davis, PhD(plant path), 72. *Prof Exp:* Asst plant path, Conn Agr Exp Sta, New Haven, 72-75; asst prof biol, Utah State Univ, 75-78, assoc prof, 78-82, PROF BIOL, UTAH STATE UNIV, 82- *Concurrent Pos:* Consult, Kennecot Corp, 76-, US Dept Agr, 80-82 & 86; sr ed, Phytopath, 85- *Mem:* Am Phytopath Soc; Am Soc Microbiol; Am Soc Plant Physiologists; AAAS. *Res:* Mechanisms of virulence expression by plant pathogens; methods of reducing pathogen virulence for biological control of plant disease. *Mailing Add:* Dept Biol Utah State Univ Logan UT 84322-5305

VAN ALLEN, JAMES ALFRED, b Mt Pleasant, Iowa, Sept 7, 14; m 45; c 5. ASTRONOMY. *Educ:* Iowa Wesleyan Col, BS, 35; Univ Iowa, MS, 36, PhD(physics), 39. *Hon Degrees:* ScD, Iowa Wesleyan Col, 51, Grinnell Col, 57, Coe Col, 58, Cornell Col, 59, Univ Dubuque, 60, Univ Mich, 61, Northwestern Univ, 61, Ill Col, 63, Butler Col, 66, Boston Col, 66, Southampton Col, 67, Augustana Col, 69, St Ambrose Col, 82, Univ Bridgeport, 87. *Prof Exp:* Carnegie res fel nuclear physics, Dept Terrestrial Magnetism, Carnegie Inst, 39-41, physicist, 41-42; physicist, appl physics lab, Johns Hopkins Univ, 42 & 46-50; prof physics, Univ Iowa, 51-72, head, Dept Physics & Astron, 51-85, Carver prof, 72-85, REGENT DISTINGUISHED PROF PHYSICS, UNIV IOWA, 85- *Concurrent Pos:* Mem, Int Sci Radio Union; mem, Rocket & Satellite Res Panel, 46-, chmn, 47-48, mem exec comt, 58-; mem Space Sci Bd, Nat Acad Sci, 58-70, 80-83, chmn Ad Hoc Panel Small Planetary Probes, 66; leader sci exped, Cent Pac, 49, Gulf Alaska, 50, Arctic, 52, Int Geophys Year, Arctic, Atlantic, Cent Pac, SPac & Antarctic, 57; Guggenheim Mem Found fel, Brookhaven Nat Lab, 51; res assoc, Princeton Univ, 53-54; mem tech panel earth satellite prog, Int Geophys Year, 55-58, chmn working group internal instrumentation, 56-58, tech panel rocketry, 55-58, tech panel cosmic rays, 56-58, tech panel aurora & airglow, 57-58; adv comt nuclear physics, Off Naval Res, 57-59; adv comt physics, NSF, 57-60; mem, Space Sci Bd, Nat Acad Sci, 58-70 & 80-83, chmn Ad Hoc Panel Small Planetary Probes, 66; mem panel sci & technol, Comt Sci & Astronaut, US House Rep, 59-72; consult, President's Sci Adv Comt, 57-60; consult particles & fields subcomt, Nat Aeronaut & Space Admin, 61-, mem ad hoc sci adv comt, 66; chmn, Iowa's Int Coop Year Comt Sci & Advan Technol, 65; lectr, NATO Conf, Bergen, Norway, 65; mem, Planetary

Missions Bd, 67-71; foreign mem, Royal Swedish Acad Sci, 81. *Honors & Awards:* Hickman Medal, Rocket Soc, 49, First Annual Res Award, 61, Hill Award, Inst Aerospace Sci, 60; Space Flight Award, Am Astronaut Soc, 58; Space Flight Award, Int Acad Astronaut, 61; Elliot Cresson Medal, Franklin Inst, 61; Golden Omega Award, Elec Insulation Conf, 63; John A Fleming Awards, Am Geophys Union, 63, William Bowie Medal, 77; Gold Medal, Royal Astron Soc, London, 78; Space Sci Award, Am Inst Aeronaut & Astronaut, 82; Cospar Space Sci Award, 84; Nat Medal Sci, 87; Abelson Prize, AAAS, 86; Proctor Prize, Sigma Xi, 87; Crafoord Prize, Royal Swed Acad Sci, 89; Nansen Medal, Norweg Acad Sci & Lett, 90. *Mem:* Nat Acad Sci; fel Am Phys Soc; fel Am Geophys Union (pres, 82-84); fel Inst Elec & Electronics Engrs; fel AAAS; Royal Astron Soc. *Res:* Planetary magnetosphere; cosmic rays; use of rockets in physical research; satellites and space probes in planetary and solar physics. *Mailing Add:* Dept Physics & Astron Univ Iowa Iowa City IA 52242-1410

VAN ALLEN, MAURICE WRIGHT, neurology; deceased, see previous edition for last biography

VAN ALLER, ROBERT THOMAS, b Mobile, Ala, June 18, 33; m 59; c 2. ORGANIC CHEMISTRY, BIOCHEMISTRY. *Educ:* Univ Ala, BS, 60, MS, 62, PhD(sulfonyl halides), 65. *Prof Exp:* Res assoc chem, Univ Miss, 65-67; aerospace technologist, Marshall Space Flight Ctr, NASA, 67-68; chmn dept chem, 68-70, dean, Col Sci, 70-71, DEAN, GRAD SCH, UNIV SOUTHERN MISS, 71- *Mem:* Am Chem Soc. *Res:* Reactions of aliphatic sulfonyl halides; biosynthesis of phytosterols and monocyclic monoterpenes. *Mailing Add:* SS Box 5024 Southern Sta Univ Southern Miss Southern Sta Box 500 Hattiesburg MS 39401

VAN ALSTINE, JAMES BRUCE, b Whitefish Bay, Wis, July 30, 49; m 70; c 2. PALEOECOLOGY. *Educ:* Winona State Univ, BA, 70; Univ NDak, MS, 74, PhD(geol), 80. *Prof Exp:* From instr to asst prof, 74-84, ASSOC PROF GEOL, UNIV MINN, MORRIS, 85- *Mem:* Soc Econ Paleontologists & Mineralogists; Sigma Xi; Geol Soc Am. *Res:* Paleoecology of nonmarine and brackish water faunas of Cretaceous and Paleocene ages. *Mailing Add:* Dept Math & Sci Univ Minn Morris MN 56267

VAN ALSTYNE, JOHN PRUYN, b Albany, NY, Sept 12, 21; m 44; c 3. MATHEMATICS. *Educ:* Hamilton Col, BS, 44; Columbia Univ, MA, 52. *Prof Exp:* Vis instr math, Hamilton Col, 43-44, from instr to assoc prof, 48-61; assoc prof, 61-66, actg head dept, 68-71, PROF MATH, WORCESTER POLYTECH INST, 66-, DEAN ACAD ADVISING, 71- *Mem:* Am Math Soc; Math Asn Am; Sigma Xi. *Res:* Functional equations; linear algebra. *Mailing Add:* Four Lakeshore Lane Asheville NC 78804-2359

VAN ALTEN, LLOYD, b East Grand Rapids, Mich, Jan 2, 24; m 51; c 2. INORGANIC CHEMISTRY. *Educ:* Calvin Col, AB, 45; Purdue Univ, West Lafayette, MS, 48; Univ Wash, PhD(chem), 54. *Prof Exp:* Teacher, Lynden Christian High Sch, 48-50; Olin-Mathieson fel, Univ Wash-Boron Chem, 54-55; from asst prof to assoc prof, 55-68, PROF CHEM, SAN JOSE STATE UNIV, 68- *Mem:* Am Chem Soc; Sigma Xi. *Res:* Boranes; carboranes; absolute intensities in infrared spectroscopy; environmental mercury. *Mailing Add:* Dept Chem San Jose State Univ Washington Sq San Jose CA 95152

VAN ALTEN, PIERSON JAY, b Grand Rapids, Mich, Feb 21, 28; m 53; c 2. EMBRYOLOGY, IMMUNOLOGY. *Educ:* Calvin Col, AB, 50; Mich State Univ, MS, 55, PhD(zool, physiol), 58. *Prof Exp:* From asst prof to assoc prof, 60-73, PROF ANAT, UNIV ILL MED CTR, 73- *Concurrent Pos:* NIH fel exp embryol & immunobiol, Univ Calif, Los Angeles, 58-60; Am Cancer Soc scholar, Univ Bern, 73-74; vis assoc prof pediat, Univ Minn, 66-67; guest prof immunobiol, Univ Bern, 73-74; mem, Res Comt, Ill Div Am Cancer Soc, 75-91; bd trustees, Calvin Col & Sem, 90- *Mem:* AAAS; Leukocyte Biol (treas, 78-80); Am Asn Immunol; Am Asn Anat; Soc Develop Biol; Transplantation Soc; Sigma Xi. *Res:* Immunological ability of gut-associated lymphoid tissue; experimental embryology and immunobiology; phagocytosis-promoting activity of plasma and cell surface fibronectins; graft-versus-host disease and immunological competence of chicken embryo; development of brain antigens in hamsters; culture of lymphocytes and human myeloma cells. *Mailing Add:* Dept Anat MC 512 Univ Ill Col Med PO Box 6998 Chicago IL 60680

VAN ALTENA, WILLIAM F, b Hayward, Calif, Aug 15, 39; c 2. ASTRONOMY. *Educ:* Univ Calif, Berkeley, BA, 62, PhD(astron), 66. *Prof Exp:* From asst prof to assoc prof astron, Yerkes Observ, Univ Chicago, 66-74; dir observ, 72-74, chmn dept, 75-81, PROF ASTRON, YALE UNIV, 74- *Concurrent Pos:* Mem comt photographic plates & films, Am Nat Standards Inst. *Mem:* Am Astron Soc; Int Astron Union; Sigma Xi; Int Astron Union (vpres, 85-88, pres, 88-91). *Res:* Trigonometric parallaxes and proper motions. *Mailing Add:* Yale Univ Observ PO Box 6666 New Haven CT 06551

VANAMAN, SHERMAN BENTON, b Lexington, Ky, July 25, 28; m 55; c 2. MATHEMATICS. *Educ:* Univ Louisville, BA, 49; Univ Ky, MS, 51; Univ Md, PhD(math educ), 67. *Prof Exp:* Instr math, Univ Ky, 51-55; assoc prof & actg head dept, 56-66, PROF MATH & CHMN DEPT, CARSON-NEWMAN COL, 66- *Mem:* Math Asn Am; Nat Coun Teachers Math. *Res:* Learning theory, especially mathematics-education. *Mailing Add:* Dept Math Box 1994 Carson-Newman Col Jefferson City TN 37760

VANAMAN, THOMAS CLARK, b Louisville, Ky, Aug 12, 41; m 62, 83; c 4. BIOCHEMISTRY, MICROBIOLOGY. *Educ:* Univ Ky, BS, 64; Duke Univ, PhD(biochem), 68. *Prof Exp:* From asst prof to prof microbiol & immunol, Med Ctr, Duke Univ, 70-83, dir cancer ctr basic res, 81-83; PROF & CHMN, DEPT BIOCHEM, UNIV KY MED CTR, 83- *Concurrent Pos:* Am Can Soc fel, Med Ctr, Stanford Univ, 69-70, NSF, 71-73, 87-90 & NIH res grant, 71-90; Josiah Macy, Jr Found fac scholar, 77-78. *Mem:* NY Acad Sci; Am Soc Biol Chemists; Am Soc Microbiologists; Am Chem Soc; Am Soc Neurobiol; Protein Soc. *Res:* Study of the structure, function and evolution of proteins; evolution, structure and function of calcium dependent regulatory proteins; mechanisms of stimulus-response coupling. *Mailing Add:* Dept Biochem Univ Ky Med Ctr 800 Rose St Lexington KY 40536

VANAMBURG, GERALD LEROY, b Hunter, Kans, Dec 17, 41; m 63; c 2. PLANT ECOLOGY, BOTANY. *Educ:* Ft Hays Kans State Col, BS, 64, MS, 65; Tex A&M Univ, PhD(plant ecol), 69. *Prof Exp:* From asst prof to assoc prof, 69-86, PROF BIOL, CONCORDIA COL, MOORHEAD, MINN, 76- *Concurrent Pos:* Expert, Int Atomic Energy Agency, 70-71. *Mem:* Ecol Soc Am; Am Inst Biol Sci; Soc Range Mgt. *Res:* Soil-vegetation relationships; biogeochemical cycling; prairie wetland ecosystems. *Mailing Add:* Dept Biol Concordia Col Moorhead MN 56560

VANAMEE, PARKER, b Portland, Maine, Aug 9, 19; m 53; c 6. PHYSIOLOGY. *Educ:* Yale Univ, BS, 42; Cornell Univ, MD, 45. *Prof Exp:* Intern, RI Hosp, Providence, 45-46; res fel, Sloan-Kettering Inst Cancer Res, 51-54, res assoc, 55-58, assoc, 58-60, assoc mem, 60-84; asst prof, 56-61, ASSOC PROF MED, MED COL, CORNELL UNIV, 61- *Concurrent Pos:* From asst resident to chief resident med, Mem Hosp Cancer & Allied Dis, 51-54, clin asst, 54-57, from asst attend physician to assoc attend physician, 57-69, attend physician, 69-, chief clin physiol & renal serv, Dept Med, 70-; asst vis physician, James Ewing Hosp, NY, 57-61, assoc vis physician, 61-; consult, Urol Dept, USPHS Hosp, Staten Island, NY; adj assoc prof clin pharm, Brooklyn Col Pharm, 77-80. *Mem:* AAAS; fel Am Col Physicians; Am Fedn Clin Res; AMA; Am Soc Clin Nutrit. *Res:* Medicine; clinical physiology. *Mailing Add:* Mem Sloan-Kettering Cancer Ctr 1275 York Ave New York NY 10021

VAN ANDEL, TJEERD HENDRIK, b Rotterdam, Neth, Feb 15, 23; US citizen; m 88; c 6. MARINE GEOLOGY & GEOLOGICAL ARCHAEOLOGY. *Educ:* State Univ Groningen, BSc, 46, MSc, 48, PhD(geol), 50. *Prof Exp:* Asst prof geol, State Agr Univ, Wageningen, 48-50; sedimentologist, Royal Dutch Shell Res Lab, 50-53; sr sedimentologist, Cia Shell de Venezuela, 53-56; assoc res geologist, Scripps Inst Oceanog, Univ Calif, 57-64, res geologist, 64-68, lectr geol, 57-68; prof geol, Sch Oceanog, Ore State Univ, 68-76; Wayne Loel prof earth science, dept geol, Stanford Univ, 76-88; HON PROF EARTH SCI, DEPT EARTH SCI, CAMBRIDGE UNIV, ENG, 88- *Concurrent Pos:* Vis prof, Univ Calif, Berkeley, 63; sr fel, Woods Hole Oceanog Inst, 63; sci adv, Deep Sea Drilling Proj, 64-68; res assoc geol, Scripps Inst Oceanog, Univ Calif, 68-72; mem geodynamics comt, Nat Acad Sci, 70-75; managing consult, Int Ocean Explor, NSF, 71-72; vis prof geophys, Stanford Univ, 74-75; group chmn, Sci Comt Ocean Res, UNESCO, 75-78; co-ed, UNESCO World Ocean Atlas Comn. *Honors & Awards:* F P Shepard Medal, 78; N B Watkins Award, 80; Waterschoot Van der Gracht Medal, 84. *Mem:* AAAS; Soc Field Archaeologists; Soc Econ Paleont & Mineral; Geol Soc Am; Am Geophys Union; fel Royal Neth Acad Sci. *Res:* Recent sediments of continents and oceans; origin and nature of the continental shelf; geology and geophysics of mid-ocean ridges; paleoceanography; deep-diving research submersibles; geo-archeology; neotectonics; paleoclimates. *Mailing Add:* Dept Earth Sci Cambridge Univ Downing St Cambridge CB2 3EQ England

VANANTWERP, CRAIG LEWIS, b Binghampton, NY, Feb 24, 50; m 72; c 1. NUCLEAR MAGNETIC RESONANCE SPECTROSCOPY. *Educ:* Juniata Col, BS, 72; Stanford Univ, PhD(org chem), 77. *Prof Exp:* Asst prof chem, Rochester Inst Technol, 77-80; software develop engr, Nicolet Magnetics Corp, Div Nicolet Instruments, 80-89; MGR, SOFTWARE ADVAN PROD DEVELOP, VARIAN ASSOC, 89- *Mem:* Am Chem Soc. *Res:* Develop and implement state of the art research capabilities into nuclear magnetic resonance spectrometer products. *Mailing Add:* 48031 Chann Ct Fremont CA 94539

VAN ANTWERP, WALTER ROBERT, b Franklin, Ind, Aug 16, 25; m 46; c 2. SOLID STATE PHYSICS, HEALTH PHYSICS. *Educ:* Ind Univ, AB, 49; Univ Md, MS, 58. *Prof Exp:* Res physicist, Chem Res & Develop Lab, US Dept Army, 50-58 & Nuclear Defense Lab, Edgewood Arsenal, 58-60, chief solid state physics br, 60-64, chief nuclear physics div, 64-69; chief, Exp Physics Br, Ballistics Res Labs, Aberdeen Proving Ground, 69-81; consult, 81-86; HEALTH PHYSICIST, MD DEPT ENVIRON, 86- *Concurrent Pos:* Instr & researcher, Univ Del, 81-86, health physics, 86- *Mem:* AAAS; Am Phys Soc; Health Physics Soc. *Res:* Solid state and nuclear physics; radiation damage and detection; gamma-ray spectroscopy; thin films. *Mailing Add:* 110 Woodland Dr Bel Air MD 21014

VAN ARMAN, CLARENCE GORDON, b Detroit, Mich, Dec 29, 17; m 43, 69; c 3. PHARMACOLOGY. *Educ:* Univ Chicago, SB, 39; Northwestern Univ, MS, 48, PhD(pharmacol), 49. *Prof Exp:* Chemist, Price Extract Co, 39-40; chemist, G D Searle & Co, Ill, 40-41, pharmacologist, 50-60; res assoc pharmacol, Med Sch Northwestern Univ, 49-50; chief pharmacologist, Chem Therapeut Res Labs, Miles Labs, Inc, 60-61; dir pharmacol res dept, Chas Pfizer & Co, 63-65; sr res fel, 65-75; sr investr, Merck Inst Therapeut Res, 75-77; mgr pharmacol eval sect, Wyeth Labs, Inc, 61-63, dir biol res, 77-83; RETIRED. *Concurrent Pos:* Lectr, Med Sch, Northwestern Univ, 54-61. *Mem:* AAAS; Am Soc Pharmacol & Exp Therapeut; Soc Exp Biol & Med; Am Soc Clin Pharmacol & Therapeut; Am Rheumatism Asn. *Res:* Inflammation; polypeptides; analgesics; diuretics; cardiac drugs; glomerulonephritis. *Mailing Add:* 6020 Cannon Hill Rd Ft Washington PA 19034

VAN ARSDEL, JOHN HEDDE, b Chicago, Ill, Nov 22, 21; m 63; c 5. ENGINEERING PSYCHOLOGY. *Educ:* Purdue Univ, BS, 50; Commonwealth Univ, MA, 55, PhD(psychol), 56; Denver Univ, MA, 58. *Prof Exp:* Human factors engr, US Army Electronic Proving Ground, Ariz, 58-63; sr develop engr advan syst, Goodyear Aerospace Corp, Litchfield Park, Ariz, 63-66; sr human factors engr Minute Man III, Bell Aerosyst Corp, Niagara Falls, NY, 66-68; supvr comput lab, Col Bus Admin, Denver Univ, 68-73; human factors scientist syst eng, Syst Develop Corp, Colo Springs, Colo, 73-75; teacher & supvr acad & sci, Northrop, 76-80; writer med training manuals, Acad Health Sci, US Army, Ft Sam Houston, Tex, 81, OPERS RES ANALYST & ELECTRONIC ENGR, US ARMY, FT HUACHUCA, ARIZ, 81- *Concurrent Pos:* Consult, opers res anal & eng psychol. *Res:* Hypno-therapy in clinical treatment of demerol addiction; effectiveness and survival in hostile environments; launching system accessibility requirements; maintenance program human factors analysis. *Mailing Add:* 2364 Sonoita Dr Sierra Vista AZ 85635

VAN ARSDEL, PAUL PARR, JR, b Indianapolis, Ind, Nov 4, 26; m 50; c 2. MEDICINE. *Educ:* Yale Univ, BS, 48; Columbia Univ, MD, 51. *Prof Exp:* Intern med, Presby Hosp, New York, 51-52, asst resident, 52-53; asst, Sch Med, Univ Wash, 53-55; asst, Presby Hosp, New York, 55-56; from instr to assoc prof, 56-69, PROF MED, SCH MED, UNIV WASH, 69-, HEAD SECT ALLERGY, 56- *Concurrent Pos:* Res fel med, Sch Med, Univ Wash, 53-55, Mass Mem Hosp & Sch Med, Boston Univ, 55 & Presby Hosp, NY, 55-56; mem, Growth & Develop Training Comt, Nat Inst Child Health & Human Develop, 70-74; consult, Vet Admin Hosp, USPHS Hosp, King's County Hosp, Children's Hosp & Univ Hosp; chief of staff, Univ Hosp, 83-85. *Mem:* AAAS; fel Am Col Physicians; Am Asn Immunol; Asn Am Med Cols; Am Acad Allergy (secy, 64-68, pres, 71-72). *Res:* Hypersensitivity; human immunology; histamine release and metabolism; drug sensitivity; autoantibodies. *Mailing Add:* Dept Med Mail Stop 13 Univ Hosp Seattle WA 98195

VAN ARSDEL, WILLIAM CAMPBELL, III, b Indianapolis, Ind, June 27, 20; div. TOXICOLOGY, ANIMAL EKG. *Educ:* Ore State Col, BS, 49, MS, 51, PhD(physiol, zool), 59; Univ Ore, MS, 54. *Prof Exp:* Mem staff, Prod Control Lab, US Rubber Co, Ind, 41-45; lab asst pharmacol, Ore Health Sci Univ Sch Med, 51-54; teaching fel zool & res animal husb, Ore State Univ, 54-59, jr animal psychologist, 59-60, asst in animal physiol, 60-63; PHARMACOLOGIST, CTR DRUG EVAL & RES, FOOD & DRUG ADMIN, 63- *Concurrent Pos:* Mem, Comt Adv Scientific Educ, Neurotoxicity Study Recommendations Comt & Fed Drug Admin Cardiovasc Task Force Comt; assoc, Staff Col, Ctr Drug Eval & Res, Food & Drug Admin, 90. *Mem:* AAAS; NY Acad Sci; Sigma Xi; Soc Exp Biol & Med; Undersea & Hyperbaric Med Soc; Am Chem Soc. *Res:* Toxicology; teratology; electrocardiology; physiology; marine biology; anthropometry. *Mailing Add:* Div Cardio-Renal Drug Prod HFD-110 Ctr Drug Eval & Res Food & Drug Admin 5600 Fishers Lane Rockville MD 20857

VAN ARTSDALEN, ERVIN ROBERT, b Doylestown, Pa, Nov 13, 13; m 45. PHYSICAL CHEMISTRY, INORGANIC CHEMISTRY. *Educ:* Lafayette Col, BS, 35; Harvard Univ, AB, 39, PhD(phys chem), 41. *Prof Exp:* Asst chem, Harvard Univ, 36-40; from instr to asst prof, Lafayette Col, 41-45; res scientist, Los Alamos Sci Lab, 45-46; asst prof chem, Cornell Univ, 46-51; prin chemist, Oak Ridge Nat Lab, 51-56; asst dir res, Parma Res Ctr, Union Carbide Corp, 56-63; John W Mallet Prof chem & chmn dept, Univ Va, 63-68; head dept, 68-72, prof, 68-84, EMER PROF CHEM, UNIV ALA, 84- *Concurrent Pos:* Res assoc, Nat Defense Res Comt, Sch Med, Johns Hopkins Univ, 43-44; res assoc, Carnegie Inst Technol, 45; lectr, Western Reserve Univ, 59; mem coun, Oak Ridge Assoc Univs, 63-69, mem bd dir, 69-75, staff mem, Inst Energy Anal, 75-76, Radiation Adv Bd Health, State Dept Health, Ala, 80. *Mem:* AAAS; Am Chem Soc; Am Phys Soc; fel Am Inst Chemists. *Res:* Energy analysis; adsorption indicators; photochemistry; bond strengths; reaction kinetics; thermodynamics and structure of inorganic systems; high temperature chemistry; fused salts; atomic energy; Mössbauer spectroscopy; radiochemistry; nuclear chemistry. *Mailing Add:* Univ Ala Box 870336 Tuscaloosa AL 35487-0336

VAN ASDALL, WILLARD, b Knox, Ind, Apr 29, 34. PLANT ECOLOGY. *Educ:* Valparaiso Univ, AB, 56; Purdue Univ, SM, 58; Univ Chicago, PhD(bot), 61. *Prof Exp:* Instr biol, Knox Col, Ill, 61; asst prof, Duquesne Univ, 62-63; from asst prof to assoc prof bot, Univ Ariz, 62-76, assoc prof gen biol, 76-91; RETIRED. *Mem:* Ecol Soc Am; Bot Soc Am. *Res:* Physiological ecology of desert plant species, especially winter-spring desert ephemerals. *Mailing Add:* 4479 N Summer Set Loop Univ Ariz Tucson AZ 85715

VANASSE, GEORGE ALFRED, b Woonsocket, RI, Oct 8, 24; m 61; c 3. OPTICS, SPECTROSCOPY. *Educ:* Univ RI, BS, 50; Boston Col, MS, 52; Johns Hopkins Univ, PhD(physics), 58. *Prof Exp:* Physicist spectros, Air Force Cambridge Res Lab, 52; teaching asst physics, Johns Hopkins Univ, 52-55, res asst, 55-58, res staff asst, 58-59; vis lectr, Goucher Col, 59; RES PHYSICIST, AIR FORCE GEOPHYS LAB, BEDFORD, 59- *Concurrent Pos:* Vis lectr, Lowell Technol Inst, 64-65. *Mem:* Am Phys Soc; fel Optical Soc Am. *Res:* Spectrometric techniques. *Mailing Add:* 71 Old Stage Rd Chelmsford MA 01824

VAN ASSENDELFT, ONNO WILLEM, b Brummen, Neth, Aug 23, 32; US citizen; m 60; c 5. CLINICAL PATHOLOGY, CHEMICAL PHYSIOLOGY. *Educ:* Univ Groningen, Neth, MD, 59, PhD(med & physiol), 70. *Prof Exp:* Med officer, Royal Dutch Army Med Corps, 59-61; sr res asst & asst prof physiol, Lab Chem Physiol, Univ Groningen, Neth, 61-76, secy & actg dean, Med Sch, 73-75; chief gen hemat, Ctr Dis Control, Bur Labs, 76-80 & Ctr Infectious Dis, 80-88, CHIEF, CLIN MED BR, DIV HOST FACTORS, DIV IMMUNOL, ONCOL, HEMAT DIS, CTRS DIS CONTROL, CTR INFECTIOUS DIS, 88- *Concurrent Pos:* Consult, Food & Drug Admin, 78- & Col Am Pathologists Hemat Resource Comt, 84-90; mem bd & secy, Int Coun Standardization Hemat, 79-; mem bd & pres-elect, Nat Comt Clin Lab Standards, 83- *Honors & Awards:* Russel J Eilers Award, Nat Comt Clin Lab Standards, 87. *Mem:* AAAS; Am Soc Hemat; NY Acad Sci. *Res:* Reflection and transmission oximetry and spectrophotometry of hemoglobin derivatives, including standardization of hemoglobin determination; standards development of general hematology and clinical chemistry methods and quality control. *Mailing Add:* Ctr Dis Control Bldg I-1403 MS D02 1600 Clifton Rd Atlanta GA 30333

VAN ATTA, CHARLES W, b New London, Conn, Feb 24, 34; m 58; c 1. FLUID MECHANICS. *Educ:* Univ Mich, BS, 58, MS, 59; Calif Inst Technol, PhD(aeronaut), 65. *Prof Exp:* Scientist, Jet Propulsion Lab, Calif Inst Technol, 64-65; from asst prof to assoc prof, 65-75, PROF ENG SCI & OCEANOG, UNIV CALIF, SAN DIEGO, 75- *Concurrent Pos:* USSR Exchange fel, Nat Acad Sci, 72-73; Guggenheim fel, 72-73. *Honors & Awards:* Fel Am Phys Soc. *Res:* Transition and turbulence in fluid flow; geophysical fluid mechanics. *Mailing Add:* Ames 0411 Univ Calif San Diego CA 92093

VANATTA, JOHN CROTHERS, III, b Lafayette, Ind, Apr 22, 19; m 44; c 2. PHYSIOLOGY. *Educ:* Ind Univ, AB, 41, MD, 44; Am Bd Internal Med, dipl, 53. *Prof Exp:* Intern, Wayne County Gen Hosp, 44-45, asst resident med, 46-47, fel physiol, Southeastern Med Col, 47-49; instr physiol & pharmacol, 49-50, from asst prof to assoc prof physiol, 50-57, PROF PHYSIOL, UNIV TEX HEALTH SCI CTR DALLAS, 57- *Concurrent Pos:* Fel exp med, Univ Tex Southwestern Med Sch Dallas, 48-49; consult, Div Nuclear Educ & Training, USAEC, 64-67; adj prof physiol, Inst Technol, Southern Methodist Univ, 69-81. *Mem:* Am Physiol Soc; Soc Exp Biol & Med; AMA. *Res:* Sodium metabolism; transport functions of urinary bladder of toad; epithelial transport, toad urinary bladder and frog skin. *Mailing Add:* Dept Physiol Southwest Med Sch Dallas TX 75235-9040

VAN ATTA, JOHN R, biophysics, for more information see previous edition

VAN ATTA, LESTER CLARE, b Portland, Ore, Apr 18, 05; m 29; c 3. PHYSICS, ELECTRICAL ENGINEERING. *Educ:* Reed Col, BA, 27; Wash Univ, MS, 29, PhD(physics, math), 31. *Prof Exp:* Res asst physics, Princeton Univ, 31-32; res assoc & asst prof, Mass Inst Technol, 32-40, sr staff mem radar, Radiation Lab, 40-45; br leader, Naval Res Lab, Washington, DC, 45-50; dir microwave lab & dir res labs physics & electronics, Hughes Aircraft Co, Culver City, Calif, 50-62; chief scientist, Lockheed Missiles & Space, Sunnyvale, Calif, 62-64; asst dir, Res Ctr NASA, Cambridge, MA, 64-70; assoc dean & prof elec eng, Univ Mass, 70-73, adj prof, 73-76; RETIRED. *Concurrent Pos:* spec asst arms control, Defense Develop Res & Eng, 59-60. *Mem:* Fel Am Phys Soc; fel Inst Elec & Electronics Engrs; assoc fel Inst Aeronaut & Astronaut. *Res:* High voltage generation; high energy bombardment; radar; microwave antennas and systems. *Mailing Add:* PO Box 5 Laytonville CA 95454

VAN ATTA, ROBERT ERNEST, analytical chemistry, for more information see previous edition

VAN AUKEN, OSCAR WILLIAM, b Morristown, NJ, Dec 7, 39; m 61; c 3. PHYSIOLOGICAL ECOLOGY, PLANT ECOLOGY. *Educ:* High Point Col, BS, 62; Univ Utah, MS, 65, PhD(biol), 69. *Prof Exp:* Asst prof biol, Southwest Tex State Univ, 69-71; sr res scientist, Southwest Res Inst, 71-73, mgr environ biol, 75-76; assoc found scientist, Southwest Found Res & Educ, 73-75; ASSOC PROF, DIV LIFE SCI, UNIV TEX, 76- *Concurrent Pos:* Assoc ed, Plant Ecol, Southwestern Asn Naturalists, 84-86. *Mem:* Ecol Soc Am; AAAS; Bot Soc Am. *Res:* Plant, animal and environmental interaction; community ecology. *Mailing Add:* Div Life Sci Univ Tex 6700 N FM 1604 W San Antonio TX 78285-0662

VAN AUSDAL, RAY GARRISON, b Cincinnati, Ohio, Sept 16, 43. MEDICAL & HEALTH PHYSICS, MUSICAL ACOUSTICS. *Educ:* Miami Univ, AB, 64, MA, 66; Univ Mich, Ann Arbor, PhD(physics), 72. *Prof Exp:* Asst prof physics, Northern Mich Univ, 72-73 & Kalamazoo Col, 73-74; ASSOC PROF PHYSICS, UNIV PITTSBURGH, JOHNSTOWN, 74- *Res:* Development of effective teaching methods and materials in physics. *Mailing Add:* Dept Physics Univ Pittsburgh Johnstown PA 15914

VAN BAAK, DAVID ALAN, b Tokyo, Japan, July 13, 52; m 80; c 1. MICROWAVE SPECTROSCOPY, ATOMIC BEAM SPECTROSCOPY. *Educ:* Calvin Col, BS, 73; Harvard Univ, MA, 75, PhD(physics), 79. *Prof Exp:* Nat Res Coun & Nat Bur Standards fel physics, Joint Inst Lab Astrophys, 79-80; from asst prof to assoc prof, 80-87, PROF PHYSICS, CALVIN COL, 87- *Concurrent Pos:* Vis assoc prof phys, Notre Dame Univ, 86-87. *Mem:* Am Phys Soc; Am Sci Affil. *Res:* Fine and hyperfine structure in simple atoms; quantum-electrodynamical effects; gavitation. *Mailing Add:* Dept Physics Calvin Col Grand Rapids MI 49506

VAN BAVEL, CORNELIUS H M, agronomy, biology, for more information see previous edition

VAN BEAUMONT, KAREL WILLIAM, b Amsterdam, Netherlands, Sept 26, 30; US citizen; m 59; c 2. PHYSIOLOGY. *Educ:* Acad Phys Educ, The Hague, BS, 55; Cath Univ Louvain, MS, 57; Univ Ill, MS, 62; Ind Univ, PhD(physiol), 65. *Prof Exp:* Instr physiol, Univ Ind, 64-66; res physiologist, Miami Valley Labs, Procter & Gamble Co, Ohio, 66-68; asst prof, 68-73, ASSOC PROF PHYSIOL, SCH MED, ST LOUIS UNIV, 73- *Concurrent Pos:* Olympic Coach, 80. *Res:* Temperature regulation; neural control systems; high altitude physiology; physiology of exercise; acceleration stress; biometeorology; body fluids and electrolytes; hematology. *Mailing Add:* Dept Physiol St Louis Univ Sch Med St Louis MO 63104

VAN BELLE, GERALD, b Enschede, Netherlands, July 23, 36; Can citizen; m 63; c 5. STATISTICS. *Educ:* Univ Toronto, BA, 62, MA, 64, PhD(math), 67. *Prof Exp:* Statistician, Connaught Med Res Labs, Univ Toronto, 57-62; from asst prof to assoc prof statist, Fla State Univ, 67-74, dir statist consult ctr, 71-74; vis assoc prof biostatist, 74-75, assoc prof, 75-76, PROF BIOSTATIST, UNIV WASH, 76- *Mem:* Fel Am Statist Asn; Biomet Soc; Sigma Xi; Int Statist Inst; fel AAAS. *Res:* Application of statistics to biological and health-related problems. *Mailing Add:* Dept Biostatist SC 32 Univ Wash Seattle WA 98195

VAN BERGEN, FREDERICK HALL, b Minneapolis, Minn, Sept 21, 14; c 4. ANESTHESIOLOGY. *Educ:* Univ Minn, MB, 41, MD, 42, MS, 52. *Prof Exp:* From instr to assoc prof, 48-57, assoc dir, 53-54, actg dir, 54-55, head dept, 55-78, prof, 57-78, EMER PROF ANESTHESIOL, MED SCH, UNIV MINN, MINNEAPOLIS, 78- *Mem:* Am Soc Anesthesiol; Int Anesthesia Res Soc; AMA; Acad Anesthesiol. *Res:* Development and testing of respirators and respiratory assistors; evaluation of effects of respiratory patterns upon cardiovascular function; evaluation of pulmonary compliance under conditions of anesthesia; gas mass spectrometer. *Mailing Add:* 2005 Argonne Dr Minneapolis MN 55421

VAN BIBBER, KARL ALBERT, b New London, Conn, Dec 5, 50. HIGH ENERGY ELECTRON SCATTERING. *Educ:* Mass Inst Technol, BS & MS, 72, PhD(physics), 76. *Prof Exp:* Instr physics, Mass Inst Technol, 76-77; asst prof, Dept Physics, Stanford Univ, 80-85; res assoc, Lawrence Berkeley Lab, 77-79; SR PHYSICIST, E-DIV PHYSICS, LAWRENCE LIVERMORE NAT LAB, 85- *Concurrent Pos:* Lectr, Dept Nuclear Eng, Univ Calif, Berkeley, 77-78. *Mem:* Am Phys Soc. *Res:* High energy electron scattering from the nucleon and atomic nuclei; searches for the axion; dark matter of the universe. *Mailing Add:* L-288 Lawrence Livermore Nat Lab PO Box 808 Livermore CA 94550

VANBLARICOM, GLENN R, b Shelton, Wash, Apr 16, 49. MARINE ECOLOGY, BIOLOGY OF MARINE MAMMALS. *Educ:* Univ Wash, Seattle, BS(zool) & BS(oceanog), 72; Univ Calif, San Diego, PhD(oceanog), 78. *Prof Exp:* WILDLIFE RES BIOLOGIST, FISH & WILDLIFE SERV, US DEPT INTERIOR, 77- *Concurrent Pos:* Res assoc, Inst Marine Sci, Univ Calif, Santa Cruz, 84- *Mem:* Am Soc Naturalists; Ecol Soc Am; Soc Marine Mammal. *Res:* Studies of relationships of sea otters to nearshore marine benthic communities in California and Alaska; studies of impacts of offshore oil development and transport on sea otters in California. *Mailing Add:* US Fish & Wildlife Serv Univ Calif Appl Sci Bldg Santa Cruz CA 95064

VANBLARIGAN, PETER, b Jersey City, NJ, Apr 17, 52; m 82; c 3. KINEMATICS, ELECTROMECHANICS. *Educ:* Va Polytech Inst & State Univ, BS, 74; Univ Calif, Berkeley, MS, 75, DEng(mech eng), 79. *Prof Exp:* Scientist, Lawrence Livermore Nat Lab, 76-81; SCIENTIST, SANDIA NAT LAB, 81- *Res:* Development of novel simplifying solutions to complex engineering problems. *Mailing Add:* PO Box 2564 Truckee CA 96160

VAN BREEMAN, CORNELIS, b Singapore, Jan 9, 36. VASCULAR & CELLULAR PHYSIOLOGY. *Educ:* Univ Alta, PhD(pharm), 65. *Prof Exp:* PROF, SCH MED, UNIV MIAMI, 77- *Mem:* Am Soc Pharmacol & Exp Therapeut. *Mailing Add:* Dept Pharm Sch Med Univ Miami PO Box 016189 Miami FL 33101

VANBRUGGEN, ARIENA H C, b Delfzyl, Neth, Dec 7, 49. EPIDEMIOLOGY, SOIL MICROBIOLOGY. *Educ:* Agr Univ, Wageningen, Neth, BSc, 72, MSc, 76; Cornell Univ, PhD(plant path), 85. *Prof Exp:* Assoc expert plant path, Food & Agr Orgn, UN, Ethiopia, 76-80; grad res asst plant path, Cornell Univ, Ithaca, 80-84; postdoctoral assoc environ biol, Boyce Thompson Inst Plant Res, Ithaca, 84-86; ASST PROF PLANT PATH, UNIV CALIF, DAVIS, 86- *Concurrent Pos:* Prin investr, Low Input Sustainable Agr, USDA, 88-; assoc ed, Plant Dis, Am Phytopath Soc, 89- *Mem:* Am Phytopath Soc; Am Soc Microbiol; AAAS. *Res:* Etiology, epidemiology and control of vegetable diseases; integrated approach to disease management, using sustainable agricultural practices and systems analysis. *Mailing Add:* Dept Plant Path Univ Calif Davis CA 95616

VAN BRUGGEN, JOHN TIMOTHY, biochemistry; deceased, see previous edition for last biography

VAN BRUGGEN, THEODORE, b Hawarden, Iowa, Jan 26, 26; m 48; c 2. BOTANY. *Educ:* Buena Vista Col, BS, 48; Univ SDak, MA, 50; Univ Iowa, PhD(bot), 58. *Prof Exp:* Instr biol, Northwestern Col, 50-55; from asst prof to prof bot & chmn dept, 58-59, ASSOC DEAN COL ARTS & SCI, UNIV SDAK, VERMILLION, 69- *Concurrent Pos:* Asst prog dir, NSF, 65-66. *Mem:* AAAS; Bot Soc Am; Mycol Soc Am; Am Soc Plant Taxon. *Res:* Plant systematics; flora of South Dakota, especially identification of vascular plants; flora of the Great Plains. *Mailing Add:* Col Arts & Sci Univ SDak 414 E Carls St Vermillion SD 57069

VAN BRUNT, RICHARD JOSEPH, b Jersey City, NJ, May 11, 39; m 72; c 1. ATOMIC PHYSICS, MOLECULAR PHYSICS. *Educ:* Univ Fla, BS, 61, MS, 64; Univ Colo, Boulder, PhD(physics), 69. *Prof Exp:* Res asst atomic & molecular physics, Univ Fla, 64 & Joint Inst Lab Astrophys, Univ Colo, Boulder, 64-69; res assoc, 69-71, asst prof atomic & molecular physics, Univ Va, 71-76; physicist & res assoc basic atomic physics, Joint Inst Lab Astrophys, Univ Colo, Boulder, 75-78; PHYSICIST GAS DISCHARGES, NAT INST STANDARDS & TECHNOL, WASHINGTON, DC, 78- *Concurrent Pos:* Univ grant, Univ Va, 72-73; NASA fac fel, NASA/Goddard Space Flight Ctr, Greenbelt, Md, 74; vis mem, Joint Inst Lab Astrophys, Univ Colo, Boulder, 75-77. *Honors & Awards:* Bronze Medal, US Dept Com, 84; R&D-100 Award, 90. *Mem:* AAAS; Am Phys Soc; Inst Elec & Electronics Engrs; Am Asn Physics Teachers. *Res:* Experimental and theoretical studies of ionization and excitation by electron impact and photon absorption; measurement of electron-atom scattering and ion-molecule reaction rates; fundamental studies of high-voltage and rf gas discharge phenomena, corona and plasma chemistry. *Mailing Add:* Nat Inst Standards & Technol Bldg 220 Rm B344 Gaithersburg MD 20899

VAN BUIJTENEN, JOHANNES PETRUS, b Neth, May 8, 28; nat US; m 63; c 3. FOREST GENETICS. *Educ:* State Agr Univ, Wageningen, BS, 52; Univ Calif, Berkeley, MS, 55; Tex A&M Univ, PhD(genetics), 56. *Prof Exp:* Forest geneticist, Inst Paper Chem, 56-60, Tex Forest Serv, 60-66 & Northeastern Forest Exp Sta, NH, 66-68; assoc prof, 68-71, prin geneticist, Tex Forest Serv, 68-85, PROF FOREST GENETICS, TEX A&M UNIV, 71-, HEAD, REFORESTATION DEPT, TEX FOREST SERV, 85- *Concurrent Pos:* NSF travel grant, 63; consult, Tex Forest Serv, 66-68. *Mem:* AAAS; Tech Asn Pulp & Paper Indust; Soc Am Foresters; fel Int Acad Wood Sci. *Res:* Genetic improvement of forest trees for growth rate, drought resistance, wood quality, disease resistance; physiology of forest trees as related to forest tree improvement. *Mailing Add:* Forest Genetics Lab Tex Forest Serv College Station TX 77843-2131

VAN BUREN, ARNIE LEE, b Reynoldsburg, Ohio, Nov 28, 39; m 90; c 3. ACOUSTICS. *Educ:* Birmingham-Southern Col, BS, 61; Univ Tenn, PhD(physics), 67. *Prof Exp:* Res assoc acoustics, Univ Tenn, 67-68; res physicist, Naval Res Lab, Washington, DC, 68-76; res physicist, 76-79, head methods sect, 79-82, HEAD MEASUREMENTS BR, NAVAL RES LAB, ORLANDO, FLA, 82- *Concurrent Pos:* Chmn working group, Acoust Transducer Calibration, Am Nat Standards Inst, 83-88. *Mem:* Fel Acoust Soc Am. *Res:* Underwater acoustic measurements; acoustic radiation; nonlinear acoustics. *Mailing Add:* Code 5980 PO Box 8337 Orlando FL 32856

VAN BUREN, JEROME PAUL, b Brooklyn, NY, Oct 17, 26; m 53; c 3. BIOCHEMISTRY. *Educ:* Cornell Univ, BS, 50, MNS, 51, PhD, 54. *Prof Exp:* Proj leader cereal chem, Gen Mills, Inc, 54-57; from asst prof to assoc prof biochem, 57-69, PROF BIOCHEM, CORNELL UNIV, 69- *Concurrent Pos:* Consult food & agr, USPHS, 62-65; vis prof, Swiss Fed Inst Technol, 64-65; vis prof, Agr Univ, Holland, 71-72, Imp Col, London, 80, Inst Food Res, Norwich, 90. *Mem:* Am Chem Soc; Inst Food Technologists. *Res:* Protein interactions in food; anthocyanins and polyphenols; wine chemistry; effects of salts on vegetable texture; pectic substances; food color and pigments. *Mailing Add:* Dept Food Sci & Technol NY State Agr Exp Sta Geneva NY 14456

VAN BURKALOW, ANASTASIA, b Buchanan, NY, Mar 16, 11. MEDICAL GEOGRAPHY. *Educ:* Hunter Col, BA, 31; Columbia Univ, MA, 33, PhD(geomorphol), 44. *Prof Exp:* Res asst geomorphol, Columbia Univ, 34-37, Kemp fel geol, 37-38; res & ed asst geog, Am Geog Soc, 45-48; from instr to prof, 38-45 & 48-75, EMER PROF GEOL & GEOG, HUNTER COL, NY, 75- *Concurrent Pos:* Consult geologist, E I du Pont de Nemours & Co, Wilmington, Del, 45-59. *Mem:* Fel Geol Soc Am; fel AAAS; fel NY Acad Sci; Am Geophys Union; Asn Am Geographers; Sigma Xi. *Res:* Angle of repose of loose material; water resources; medical geography. *Mailing Add:* 160 E 95th St New York NY 10128-2511

VAN CALSTEREN, MARIE-ROSE, b Brussels, Belg, Mar 26, 58; Can citizen. NUCLEAR MAGNETIC RESONANCE, STRUCTURE-ACTIVITY RELATIONSHIPS. *Educ:* Univ Sherbrooke, BSc, 80, MSc, 83; Univ Ottawa, PhD(chem), 91. *Prof Exp:* Res asst, 87-88, chemist & nuclear magnetic resonance spectroscopist, 88-91, RES SCIENTIST, AGR CAN, 91- *Concurrent Pos:* Demonstr, Univ Sherbrooke, 80-82 & Univ Ottawa, 82-87. *Mem:* Chem Inst Can; Can Soc Chem; Spectros Soc Can. *Res:* Application of nuclear magnetic resonance and other spectroscopic techniques to structure determination of natural products, to structure-activity relationships and to biomolecular interactions. *Mailing Add:* Food Res & Develop Ctr Agr Can 3600 Casavant Blvd W St-Hyacinthe PQ J2S 8E3 Can

VAN CAMP, W(ILLIAM) M(ORRIS), heat transfer, thermodynamics; deceased, see previous edition for last biography

VAN CAMPEN, DARRELL R, b Two Buttes, Colo, July 15, 35; m 58; c 2. NUTRITION, BIOCHEMISTRY. *Educ:* Colo State Univ, BS, 57; NC State Univ, MS, 60, PhD(nutrit), 62. *Prof Exp:* NIH fel biochem, Cornell Univ, 62-63; res chemist, US Plant, Soil & Nutrit Lab, USDA, 63-80; asst prof animal nutrit, Cornell Univ, 72-80; LAB DIR, US PLANT, SOIL & NUTRIT LAB, USDA, 80-; ASSOC PROF ANIMAL NUTRIT, CORNELL UNIV, 80- *Mem:* AAAS; Am Inst Nutrit; Soc Exp Biol & Med; NY Acad Sci. *Res:* Mineral metabolism; absorption and utilization of trace minerals. *Mailing Add:* Plant Soil & Nutrit Lab USDA Tower Rd Ithaca NY 14853

VANCE, BENJAMIN DWAIN, b Cave City, Ark, May 7, 32; m 52; c 4. PLANT PHYSIOLOGY. *Educ:* Tex Tech Col, BS, 58; Univ Mo, AM, 59, PhD(bot), 62. *Prof Exp:* Asst prof biol, Tex Tech Col, 62-63; asst prof, 63-70, ASSOC PROF BIOL, NTEX STATE UNIV, 70- *Concurrent Pos:* Res grants, NTex Fac Res-Tex Col & Coord Bd, 66-67, Nat Commun Dis Ctr, 66-68, Su Corps, Inc, 78-80 & Robert Welch Found. *Mem:* Aquatic Plant Mgt Soc. *Res:* Physiology of blue-green algae; phytohormones; aquatic angiosperm physiology. *Mailing Add:* NTex Sta PO Box 5218 Denton TX 76203

VANCE, DENNIS E, b St Anthony, Idaho, July 14, 42; US & Can citizen; m 67; c 2. PHOSPHOLIPID METABOLISM, LIPOPROTEIN METABOLISM. *Educ:* Dickinson Col, BS, 64; Univ Pittsburgh, PhD(biochem), 68. *Prof Exp:* From asst prof to prof biochem, Univ BC, 73-86, assoc dean med, 78-81, head dept biochem, 82-86; PROF BIOCHEM & DIR, LIPID & LIPOPROTEIN RES GROUP, UNIV ALTA, 86- *Concurrent Pos:* Adv bds, Biochem J, 84- & J Lipid Res, 89-; ed, Phosphatidylcholine Metab, 89; co-ed, Biochem Lipids, Lipoproteins & Membranes, 92 & Phospholipid Biosynthesis in Methods in Enzym, 92. *Honors & Awards:* Bristol Lipoprotein Res Award, Can Lipoprotein Conf, 85; Boehringer-Mannheim Can Prize, Can Biochem Soc, 89. *Mem:* Can Biochem Soc (vpres, 91-92, pres 92-93); Am Soc Biochem & Molecular Biol; Brit Biochem Soc; AAAS. *Res:* Regulation of phosphatidycholine biosynthesis and catabolism in animal cells and the role of phosphaitdylcholine in the assembly and secretion of lipoproteins from hepatocytes. *Mailing Add:* Lipid Res Group Univ Alta Edmonton AB T6G 2S2

VANCE, DENNIS WILLIAM, b Quincy, Ill, Nov 20, 38; m 61; c 1. ELECTROOPTICS. *Educ:* St Lawrence Univ, BS, 60; Univ Fla, MS, 62, PhD(physics), 65. *Prof Exp:* SCIENTIST, XEROX RES LABS, 65-, MGR DISPLAY TECHNOL AREA, 75- *Mem:* Am Phys Soc. *Res:* Surface physics; display systems engineering and technology. *Mailing Add:* 3970 Los Alasdas Rd Paso Robles CA 93446

VANCE, EDWARD F(LAVUS), b Mansfield, Tex, Sept 1, 29; m 56; c 4. ELECTRICAL ENGINEERING. *Educ:* Univ Calif, Los Angeles, BS, 54; Univ Denver, MSEE, 58. *Prof Exp:* Design engr, NAm Aviation, Inc, 54-56; from instr to asst prof elec eng, Univ Denver, 56-59; res engr, Stanford Res Inst, 59-67, sr res engr, 67-76, prog mgr, 76-80, staff scientist, 80-83, SR STAFF SCIENTIST, SRI INT, 83- *Concurrent Pos:* Several tech adv groups, DOD; Electromagnetic Pulse fel, 86; fel, Inst Elec & Electronics Engrs, 90. *Mem:* Inst Elec & Electronics Engrs. *Res:* Electromagnetic coupling; cable shields and transmission lines; high voltage phenomena; electrical discharges; aircraft and rocket electrification. *Mailing Add:* SRI Int 333 Ravenswood Ave Menlo Park CA 94025-3493

VANCE, ELBRIDGE PUTNAM, b Cincinnati, Ohio, Feb 7, 15; m 75; c 4. MATHEMATICS. *Educ:* Col Wooster, AB, 36; Univ Mich, MA, 37, PhD(math), 39. *Prof Exp:* Dir statist lab, Univ Mich, 38; from instr to asst prof math, Univ Nev, 39-43; lectr, 43-46, from asst prof to prof, 46-83, chmn Dept 48-77, actg dean fac, 65-66 & 70-71, EMER PROF MATH, OBERLIN COL, 83- *Concurrent Pos:* NSF fel, Stanford Univ, 60-61; Columbia Univ & US AID consult, Ranchi Univ, India, 65; math assoc, Univ Auckland, 67; instr, Glenville High Sch & Phillips Acad, 73; ed, Am Math Monthly, 49-57 & 64-67. *Mem:* AAAS; assoc Am Math Soc; assoc Math Asn Am; Nat Coun Tchrs Math; Sigma Xi. *Res:* Continuous transformations; foundations of mathematics; topology. *Mailing Add:* Dept Math Oberlin Col Oberlin OH 44074

VANCE, HUGH GORDON, b Forest, Ont, Sept 18, 24; m 53; c 1. ANALYTICAL CHEMISTRY, BIOCHEMISTRY. *Educ:* Univ Western Ont, BSc, 50, PhD(path chem), 56. *Prof Exp:* Nat Res Coun Can fel, Dept Anat, McGill Univ, 55-57; res assoc biochem, Sinai Hosp of Baltimore, Ind, Md, 57-61, Nat Cancer Inst fel dept med, 60-61; from asst prof to assoc prof chem, Morgan State Col, 65-86; RETIRED. *Mem:* Am Chem Soc. *Res:* Investigations of the nature and content of mucopolysaccharides in skin; structural determination of the carbohydrate moiety of glycoporteins. *Mailing Add:* 2908 Mayfield Ave Baltimore MD 21207

VANCE, IRVIN ELMER, b Mexico, Mo, Apr 8, 28; m 58; c 3. MATHEMATICS EDUCATION. *Educ:* Wayne State Univ, BS, 57; Washington Univ, MA, 59; Univ Mich, Ann Arbor, DEduc(math), 67. *Prof Exp:* From asst prof to assoc prof math, Mich State Univ, 66-71; from assoc prof to prof math, NMex State Univ, 71-89; PROF MATH, MICH STATE UNIV, 89- *Concurrent Pos:* Consult, Morel Lab Math Proj, 67; asst dir, Grand Rapids Math Lab Proj, 68-69; dir, Inner City Math Proj, Mich State Univ, 69-72, Sch-Community Outreach Proj One, 73-75 & Mich Minority Math Proj, 89-; dir, Elem Teachers Math Proj, NMex State Univ, 77-80. *Mem:* AAAS; Nat Coun Teachers Math; Am Math Soc; Math Asn Am; Nat Asn Mathematicians. *Res:* Finite projective planes; inductive learning and teaching of mathematics; laboratory techniques at school level and individualized instruction at the college level. *Mailing Add:* Dept Math Mich State Univ East Lansing MI 48824

VANCE, JOHN MILTON, b Houston, Tex, Oct 5, 37; m 58, 83; c 4. MECHANICAL ENGINEERING. *Educ:* Univ Tex, Austin, BS, 60, MS, 63, PhD(mech eng), 67. *Prof Exp:* Mech engr, Houston Div, Armco Steel Corp, 60-62; res engr, Res & Tech Dept, Texaco Inc, 63-64; group leader missile decoy anal, Tracor, Inc, 67-69; asst prof mech eng, Univ Fla, 69-77; PROF, TEX A&M UNIV, 78- *Concurrent Pos:* Consult, Pratt & Whitney, 70-; sci adv, Air Mobility Res & Develop, US Army, 71-72; US Army res grant, Univ Fla, 72-73, Southwest Res Inst, 75-76; Dresser Industs assoc prof, Tex A&M Univ, 79-80, Shell, 81-82. *Mem:* Am Soc Mech Engrs; Am Soc Eng Educ. *Res:* Dynamics of rotating machinery; dynamic stability of mechanical systems; bearings; mechanical design synthesis; turobmachinery; helicopters. *Mailing Add:* Dept Mech Eng Tex A&M Univ College Station TX 77843

VANCE, JOSEPH ALAN, b Aberdeen, Wash, Mar 15, 30; m 49; c 3. GEOLOGY. *Educ:* Univ Wash, BSc, 51, PhD, 57. *Prof Exp:* Asst prof geol, 57-68, ASSOC PROF GEOL, UNIV WASH, 68- *Mem:* Geol Soc Am; Mineral Soc Am. *Res:* Igneous and metamorphic petrology; structure and stratigraphy; geology of the Pacific Northwest. *Mailing Add:* Dept Geol Univ Wash Seattle WA 98195

VANCE, JOSEPH FRANCIS, b Kansas City, Mo, July 24, 37. MATHEMATICS, STATISTICS. *Educ:* Southwest Tex State Col, BS, 59; Univ Tex, Austin, MA, 62, PhD(math), 67. *Prof Exp:* Teacher high sch, Tex, 60-61; spec instr math, Univ Tex, Austin, 66-67, asst prof, 67-68; from asst prof to assoc prof, 68-74, PROF MATH, ST MARY'S UNIV, TEX, 74- *Mem:* Am Math Soc. *Res:* Analysis; specialty, integration theory. *Mailing Add:* Dept Math St Mary's Univ Camino Santa Maria San Antonio TX 78284

VANCE, MILES ELLIOTT, b Findlay, Ohio, Jan 2, 32; m 55; c 3. PHYSICAL OPTICS, GEOMETRICAL OPTICS. *Educ:* Bowling Green State Univ, BA, 53; Ohio State Univ, PhD(physics), 62. *Prof Exp:* Instr physics, Ohio State Univ, 61-62; res physicist, Res & Develop Lab, Corning Glass Works, NY, 62-68, sr res physicist, Electronics Res Lab, Raleigh NC, 68-73, sr res physicist, Biomed Tech Ctr, 73-76. SR RES PHYSICIST, RES & DEVELOP LAB, CORNING GLASS WORKS, 76- *Mem:* Optical Soc Am; Inst Elec & Electronics Engrs; Sigma Xi. *Res:* Applied optics; interferometry; clinical instruments; microscopy; spectroscopy; optical communications. *Mailing Add:* 71 E Third St Corning NY 14830

VANCE, OLLIE LAWRENCE, b Birmingham, Ala, Feb 5, 37; m 56; c 2. ENGINEERING. *Educ:* Auburn Univ, BSME, 59, MSME, 61; Univ Tex, Austin, PhD(mech eng), 67. *Prof Exp:* Instr mech eng, Auburn Univ, 59-61; eng design analyst, Pratt & Whitney Inc, 61-62; asst prof mech eng, Auburn Univ, 62-64; ASSOC PROF ENG, UNIV ALA, BIRMINGHAM, 67-; CONSULT ENG, 70- *Mem:* Am Soc Mech Engr. *Res:* Vibrations; linear elasticity. *Mailing Add:* 410 Summerchase Dr Birmingham AL 35244

VANCE, PAUL A(NDREW), JR, b Ft Wayne, Ind, Feb 17, 30; m 53; c 3. ELECTRICAL ENGINEERING. *Educ:* Univ Ill, BS, 51, MS, 52, PhD(elec eng), 54. *Prof Exp:* Res engr, Eng Res Lab, Exp Sta, E I du Pont de Nemours & Co, Inc, 54-58, res proj engr, 58-61, sr res engr, Mech Res, Eng Res & Develop Labs, 61-65, sr res engr, Orchem Dept, Jackson Lab, 65-77, sr engr, Photo Prod Dept, Instrument Prod Div, 77-79, sr engr, Photo Prod Dept, Clin Syst Div, 79-85; RETIRED. *Mem:* Sigma Xi. *Res:* Optics; electronics; electromechanics. *Mailing Add:* 201 Country Club Dr Newark DE 19711

VANCE, ROBERT FLOYD, b Columbus, Ohio, May 12, 26; m 53; c 3. INORGANIC CHEMISTRY. *Educ:* Otterbein Col, BS, 49; Univ Ill, MS, 50, PhD(chem), 52. *Prof Exp:* Res chemist, Battelle Mem Inst, 52-58; develop supvr, Girdler Catalysts Div, Chemetron Corp, 58-60; sr chemist, Gen Elec Co, 60-80; MGR ENVIRON LAB, STATE OF KY, 81- *Mem:* Am Chem Soc. *Res:* Gas chromatography; mass spectrometry; thermal analysis. *Mailing Add:* 7502 Tudor Ct Louisville KY 40222-4142

VANCE, VELMA JOYCE, b Wilder, Idaho, May 13, 29. VERTEBRATE ZOOLOGY. *Educ:* Col Idaho, BS, 51; Univ Ariz, MS, 53; Univ Calif, Los Angeles, PhD(zool), 59. *Prof Exp:* Asst, Crookham Co, 52-54 & Univ Calif, Los Angeles, 54-58; instr biol, Occidental Col, 59; from asst prof to assoc prof zool, 59-71, PROF ZOOL, CALIF STATE UNIV, LOS ANGELES, 71- *Concurrent Pos:* Animal behavior; vertebrate biology. *Mailing Add:* Dept Biol Calif State Univ 5151 St Univ Dr Los Angeles CA 90032

VANCE, WILLIAM HARRISON, b Phoenix, Ariz, Nov 6, 34; m 73; c 2. FLUID DYNAMICS, HEAT TRANSFER. *Educ:* Univ NMex, BS, 56; Univ Wash, PhD(chem eng), 62. *Prof Exp:* NSF fel, Swiss Fed Inst Technol, 62-63; sr engr, 64-76, FEL ENGR, BETTIS ATOMIC POWER LAB, WESTINGHOUSE ELEC CORP, 77- *Mem:* Am Inst Chem Engrs. *Res:* Experimentation and analysis in fluid dynamics and heat transfer. *Mailing Add:* 205 Conover Rd Pittsburgh PA 15208

VAN CITTERS, ROBERT L, b Alton, Iowa, Jan 20, 26; m 49; c 4. CARDIOVASCULAR PHYSIOLOGY. *Educ:* Univ Kans, AB, 49, MD, 53. *Hon Degrees:* DSc, Northwestern Col, Iowa, 78. *Prof Exp:* Intern, Med Ctr, Univ Kans, 53-54, resident internal med, 55-58; res assoc, Scripps Clin Res Found, Univ Calif, 61-62; from asst prof to assoc prof physiol & biophys, 63-70, Robert L King chmn cardiovasc res, 63-68, mem staff, Regional Primate Res Ctr, 64-68, assoc dean, Sch Med, 68-70, dean, 70-81, PROF PHYSIOL, BIOPHYS & MED, SCH MED, UNIV WASH, 70- *Concurrent Pos:* Nat Heart Inst res fel, Univ Kans, 55-56 & trainee, 56-57, spec res fel, 58-59; res fel, Sch Med, Univ Wash, 59-62; NIH career res award, 62; res grants, Nat Heart Inst, 62-67, Am Heart Asn, 62-67 & US Air Force, 67; exchange scientist, Joint US-USSR Sci Exchange, 62; mem, Bd Trustees, Wash State Heart Asn; mem Admin Bd, Coun Deans, 72-, Exec Coun, Asn Am Med Col, 72-78; mem spec med adv group, Vet Admin, 74-78, chmn, 77-78; mem, gen res support prog adv comt, NIH, 75-78 & nat adv res resources coun, 78-84; mem Liaison Comt Med Educ, 81-84; chmn, mech circulatory assistance working group, Nat Heart, Lung & Blood Inst, 83- & mem, clin applns & prev adv comt, 85- *Honors & Awards:* Cummings Medal, 70. *Mem:* Inst Med-Nat Acad Sci; Am Physiol Soc; Am Heart Asn; Am Fedn Clin Res; NY Acad Sci; fel AAAS; distinguished serv mem Asn Am Med Col. *Res:* Cardiovascular physiology, left ventricular function and control, regional flow distribution, exercise and diving; development of instrumentation and techniques for studying cardiovascular dynamics in healthy subjects during spontaneous activity; prevention of cardiovascular disease. *Mailing Add:* Univ Wash Sch Med RG-22 Seattle WA 98195

VANCKO, ROBERT MICHAEL, b Johnson City, NY, Sept 15, 42; m 66; c 3. HISTORY OF MATHEMATICS. *Educ:* Pa State Univ, BA, 64, MA, 65, PhD(math), 69. *Prof Exp:* Lectr math, Univ Man, 67-69; asst prof, 69-89, ASSOC PROF MATH, OHIO UNIV, 89- *Mem:* Am Math Soc; Math Asn Am. *Res:* Universal algebra; general algebraic systems; history of mathematics. *Mailing Add:* Dept Math Ohio Univ Athens OH 45701

VANCLEAVE, ALLAN BISHOP, b Medicine Hat, Alta, Aug 19, 10; m 34; c 4. PHYSICAL CHEMISTRY. *Educ:* Univ Sask, BSc, 31, MSc, 33; McGill Univ, PhD, 35; Cambridge Univ, PhD, 37. *Hon Degrees:* LLD, Univ Regina, 80. *Prof Exp:* 1851 Exhib scholar, Cambridge Univ, 35-37; from asst prof to prof chem, Univ Sask, 37-62; chmn div natural sci, Univ Regina 62-69, dir sch grad studies, 65-69, dean grad studies & res, 69-76; RETIRED. *Mem:* Fel Chem Inst Can; Royal Soc Can. *Res:* Active hydrogen; viscosity of gases; catalysis; accomodation coefficient method of measuring gas adsorption; reactions of cyanogen halides; radiation chemistry; properties of Saskatchewan volcanic ashes; beneficiation of low grade uranium ores; flotation characteristics of minerals; x-ray fluorescence analysis. *Mailing Add:* 301-222 Saskatchewan Crescent E Saskatoon SK S7N 0K6 Can

VAN CLEAVE, HORACE WILLIAM, b Cherryvale, Kans, July 9, 31; m 56; c 2. ENTOMOLOGY. *Educ:* Tex A&M Univ, BS, 52, MS, 58; Okla State Univ, PhD(entom), 69. *Prof Exp:* Teacher high sch, Tex, 54-56; surv entomologist, Okla State Univ, 58-61, instr, 62-64; from asst prof to prof entom, 64-89, PROF & ASSOC HEAD ACAD PROGS, TEX A&M UNIV, 89- *Mem:* Entom Soc Am; Am Registry Prof Entomologists (pres, 88). *Res:* Insect pests of pecans; taxonomy of aphids; economic entomology. *Mailing Add:* Dept Entom Tex A&M Univ College Station TX 77843

VAN CLEVE, JOHN WOODBRIDGE, b Kansas City, Mo, Nov 22, 14; m 47; c 3. CARBOHYDRATE CHEMISTRY. *Educ:* Antioch Col, BS, 37; Univ Minn, PhD(biochem), 51. *Prof Exp:* Jr res chemist, Aluminum Co Am, 43-45, assoc res chemist, 45-48; RES CHEMIST, CEREAL CROPS LAB, NORTHERN REGIONAL RES LAB, USDA, 51- *Mem:* AAAS; Am Chem Soc; Sigma Xi; NY Acad Sci. *Res:* Carbohydrate chemistry; coorelation of anomeric configuration of glycosides; methylation analysis of dextrans; synthesis of derivitives of erythrose and glucose. *Mailing Add:* 903 W Meadows Peoria IL 61604

VAN COTT, HAROLD PORTER, b Schenectady, NY, Nov 16, 25; m 53; c 3. ENGINEERING SCIENCE, ERGONOMICS. *Educ:* Univ Rochester, BA, 48; Univ NC, MA, 52, PhD(psychol), 53. *Prof Exp:* Prog dir, Am Inst Res, 55-58, dir, Inst Human Performance, 64-68, dir human ecol, 74-75; develop engr, IBM Corp, 58-64; dir, Off Commun, Am Psychol Asn, 68-74; chief human factors, Nat Bur Standards, 75-78, chief consumer sci, 78-81; chief scientist, Biotechnol, Inc, 81-87; PRIN STAFF OFFICER, NAT RES COUN/NAT ACAD SCI, 81- *Concurrent Pos:* Consult to various pvt & govt orgn, 64- *Mem:* Fel Am Psychol Asn; fel Human Factors Soc; fel AAAS; Sigma Xi. *Res:* Ergonomics as it is applied to the design and evaluation of products and process control systems. *Mailing Add:* 8300 Still Spring Ct Bethesda MD 20817

VANDAM, LEROY DAVID, b New York, NY, Jan 19, 14; m 39; c 2. ANESTHESIOLOGY. *Educ:* Brown Univ, PhB, 34; NY Univ, MD, 38. *Hon Degrees:* MA, Harvard Univ, 68. *Prof Exp:* Fel surg, Sch Med, Johns Hopkins Univ, 45-47; asst prof anesthesia, Sch Med, Univ Pa, 52-54; from assoc clin prof to clin prof, 54-67, prof, 67-79, EMER PROF ANESTHESIA, HARVARD MED SCH, 79- *Concurrent Pos:* Consult, Valley Forge Army Hosp & Philadelphia Naval Hosp, 52-54, Children's Boston Lying In, Chelsea Naval, West Roxbury Vet Admin, Rutland Vet Admin, Winchester, Burbank & Nantucket Cottage Hosps, 54-; chmn adv panel anesthesiol, US Pharmacopoeia, 54-60; ed-in-chief, J Anesthesiol, 64; chmn comn anesthesia, Nat Acad Sci-Nat Res Coun, 65; pres, Boston Med Libr, 79-85. *Honors & Awards:* Distinguished Serv Award, Am Soc Anesthesiol. *Mem:* Am Soc Anesthesiol; AMA; Sigma Xi. *Res:* Pharmacology, physiology and biochemistry of surgery and anesthesia. *Mailing Add:* Ten Longwood Dr 268 Westwood MA 02090

VANDE BERG, JERRY STANLEY, b Sheldon, Iowa, June 1, 40; m 79; c 3. ELECTRON MICROSCOPY, TISSUE CULTURE. *Educ:* Univ Nebr, BSc, 64, MSc, 65; Va Polytech Inst & State Univ, PhD(physiol), 69. *Prof Exp:* Instr entom, Va Polytech Inst & State Univ, 67-69; asst prof neurophysiol, Wayne State Univ, 69-74; res assoc, Univ Wis-Madison, 74-79; asst prof electron micros, Old Dominion Univ, 75-79; dir core res, 79-82, DIR CLIN & RES, ELECTRON MICROS LAB, VET ADMIN MED CTR, 82- *Concurrent Pos:* Prin investr, 75-; Lectr, Univ Calif, San Diego Med Sch, 79-80, Eastern Va Med Sch, 78-79. *Mem:* Electron Micros Soc Am. *Res:* Cellular mechanisms of contraction in wound healing. *Mailing Add:* Vet Admin Med Ctr 151 3350 La Jolla Village Dr San Diego CA 92161

VANDEBERG, JOHN LEE, b Appleton, Wis, June 14, 47; m 75; c 2. GENETICS. *Educ:* Univ Wis-Madison, BS, 69; La Trobe Univ, Melbourne, BS, 70; Macquarie Univ, PhD(genetics), 75. *Prof Exp:* Tutor genetics, Macquarie Univ, Sydney, 73-75; res assoc genetics, Univ Wis-Madison, 75-79; asst scientist genetics, Wis Regional Res Ctr, Univ Wis- Madison, 79-80; from asst prof to assoc prof, Dept Cell & Struct Biol & Dept Path, Univ Tex Health Sci Ctr, San Antonio, 80-86; assoc scientist genetics, 80-85, CHMN GENETICS, SOUTHWEST FOUND BIOMED RES, 82-, SCIENTIST GENETICS, 85-; PROF CELL & STRUCT BIOL & PATH, UNIV TEX HEALTH SCI CTR, SAN ANTONIO, 86- *Concurrent Pos:* Fulbright fel, 69-70; fel, Pop Coun, 76-77; NIH trainee, 77-78 Biol & Dept Path, Univ Tex Health Sci Ctr, San Antonio, 80-86. *Mem:* AAAS; Genetics Soc Am; Sigma Xi; Am Soc Human Genetics; Am Heart Asn; Res Soc Alcoholism; Int Soc Biomed Res Alcoholism; Int Soc Animal Genetics; Am Soc Biochem & Molecular Biol; Am Soc Primatologists; Int Primatological Soc. *Res:* Genetic aspects of heart disease and alcoholism; sex chromosome evolution and dosage compensation; marsupial and primate models for biomedical research. *Mailing Add:* Dept Genetics Southwest Found Biomed Res P O Box 28147 San Antonio TX 78228-0147

VANDEBERG, JOHN THOMAS, b Great Falls, Mont, Aug 27, 39; c 2. SYNTHETIC INORGANIC & ORGANOMETALLIC CHEMISTRY. *Educ:* Carroll Col, BA, 62; Loyola Univ, MS, 66, PhD(chem), 69; Northwestern Univ, MM, 88. *Prof Exp:* Sect leader, DeSoto Inc, 69-73, mgr, Res Serv, 73-78, mgr, Polymer Develop, 78-84, dir, 84-89, dir, Technol, 89-90; VPRES TECHNOL, DSM DESOTECH INC, 90- *Concurrent Pos:* Mem bd dirs, Fedn Socs Coatings Technol. *Honors & Awards:* Mat Mkt Asn Award, Fedn Socs Coatings Technol, 80; Res & Develop IR 100 Award, 86. *Mem:* Sigma Xi; Indust Res Inst; Am Chem Soc; Fedn Socs Coatings Technol. *Res:* Polymer chemistry; coatings for optical fibers; spectroscopy; analytical chemistry; tetraanglborates and additives for concrete; author of 18 publications and two books; holder of two patents. *Mailing Add:* DSM Desotech Inc 1700 S Mt Prospect Rd Des Plaines IL 60017

VANDE BERG, WARREN JAMES, b Orange City, Iowa, Sept 28, 43; m 62; c 3. PHYCOLOGY. *Educ:* Iowa State Univ, BS, 66; Ind Univ Univ, Bloomington, PhD(phycol), 70. *Prof Exp:* From asst prof to assoc prof, 70-81, PROF BIOL, NORTHERN MICH UNIV, 81- *Mem:* Phycol Soc Am. *Res:* Control of cellular development in Volvox. *Mailing Add:* Dept Biol Northern Mich Univ Marquette MI 49855

VAN DE CASTLE, JOHN F, b New York, NY, Sept 30, 33; m 57; c 4. ORGANIC, POLYMER & PETROLEUM CHEMISTRY. *Educ:* St John's Col, BS, 55; Univ Md, PhD(org chem), 60. *Prof Exp:* Chemist, Hoffmann-La Roche, Inc, 55, Nat Bur Standards, 56 & Am Cyanamid Co, 57; group leader elastomers, Esso Res & Eng Co, NJ, 59-65, investment planning & mkt adv, Esso Chem Co, Inc, NY, 65-71; asst vpres, Englehard Minerals & Chem Co, 71-72, mgr com develop, 72-77; mgr technol acquision, 77-81, VPRES ARCO TECHNOL, ARCO CHEM CO, 81- *Res:* Synthesis and characterization of ethylene propylene copolymers and terpolymers, polybutadienes; chemical modification of polymers; Ziegler and organometallic catalytic studies; technology licensing for chemical, petroleum and petrochemical industries; chemical petroleum and petrochemical processing; hydrogenation processes; hydrocarbon isomerization processes. *Mailing Add:* 20 Fox Chase Dr Plainfield NJ 07060

VANDEGAER, JAN EDMOND, b Tienen, Belg, July 28, 27; nat US; m 51; c 4. BIOMEDICAL CHEMISTRY, ENVIRON COAGULANTS. *Educ:* Cath Univ Louvain, BS, 48, MS, 50, PhD(phys chem), 52. *Prof Exp:* Asst res proteins, Cath Univ Louvain, 52-54; fel, Nat Res Coun Can, 54-55; res chemist, Dow Chem Co, 55-59; sr chemist & proj leader, J T Baker Chem Co, 59-60, sr chemist & group leader, 60-62; mgr polymer res, Wallace & Tiernan, Inc, 62-65, dir cent res, 65-70; dir lab res, Chem Group, Dart Industs Inc, 70-74; dir res, 74-77, vpres, 75-77, INDEPENDENT CONSULT, CHEM PROD DEVELOP, LEGAL EXPERT, 77- *Concurrent Pos:* Scholar, Biochem Inst, Finland, 53 & Cambridge, 54. *Mem:* Asn Res Dirs; Soc Plastics Engrs; Am Chem Soc. *Res:* Physical chemistry of proteins; polymer chemistry; correlation of physical properties of high polymers and molecular structure; biomedical applications; plastics engineering; microencapsulation; polyolefins, surfactants, cosmetics and emulsion polymerization; research administration; technical planning. *Mailing Add:* 427 Auds Lane Pasadena MD 21122

VAN DE GRAAFF, KENT MARSHALL, b Ogden, Utah, May 21, 42; m 62; c 4. GROSS ANATOMY, MAMMALOGY. *Educ:* Weber State Col, BS, 65; Univ Utah, MS, 69; Northern Ariz Univ, PhD(zool), 73. *Prof Exp:* Asst prof vet sci, Univ Minn, St Paul, 73-75; ASST PROF HUMAN ANAT, BRIGHAM YOUNG UNIV, 75- *Honors & Awards:* A Brazier Howell Honoriarum, Am Soc Mammalogists, 72. *Mem:* Am Soc Mammalogists; Am Soc Zoologists; Am Soc Vet Anatomists; Sigma Xi. *Res:* Functional morphological aspects of mammalian posture and locomotion. *Mailing Add:* Dept Zool 575 Widtsoe Bldg Brigham Young Univ Provo UT 84602

VANDEGRIFT, ALFRED EUGENE, b Chanute, Kans, Nov 10, 37; m 59; c 4. RESEARCH ADMINISTRATION. *Educ:* Univ Kans, BS, 59; Univ Calif, Berkeley, PhD(chem eng), 63. *Prof Exp:* Assoc chem engr, Midwest Res Inst, 63-65, sr chem engr, 65-69, head, environ sci sect, 69-73, asst dir, phys sci div, 73-74, dir, N Star Div, Minneapolis, 74-77, dir, econ & mgt sci div, MRI Ventures, 77-80, vpres, social & eng systs, 80-84, pres, 84-87; pres, Ruf corp, 87-88; CONSULT & DIR, TECHNITRAN INT, 88- *Concurrent Pos:* Lectr gen eng, Univ Mo, Kansas City, 67-72; adj prof indust eng, Univ Mo, Columbia, 78-; adj prof eng mgt, Kansas Univ, 81- *Mem:* AAAS; Sigma Xi. *Res:* Management sciences, pollution control and environmental assessment; mathematical modeling of physical and social phenomena; surface chemistry-oscillating jet studies; assist small technology-based companies solve technical and management problems. *Mailing Add:* 637 E 115th Terr Kansas City MO 64131

VANDEGRIFT, VAUGHN, b Jersey City, NJ, Dec 7, 46; m 69; c 3. BIOCHEMISTRY. *Educ:* Montclair State Col, BA, 68, MA, 70; Ohio Univ, PhD(biochem), 74. *Prof Exp:* Teacher chem, River Dell Regional High Sch, Oradell, NJ, 68-70; asst prof chem & biol, Ill State Univ, 74-76; from asst prof to assoc prof chem, Murray State Univ, 76-84, prof & chmn, Dept Chem, 82-88; PROF & DEAN, SCH MATH & SCI, MONTCLAIR SATE COL, 88- *Concurrent Pos:* Vis assoc prof biochem, Southern Ill Univ, Carbondale, 80-81; vis scientist, Sch Med, Ohio Univ, Athens, 82. *Mem:* Sigma Xi; Am Chem Soc; Am Asn Univ Adminrs; Am Asn Higher Educ. *Res:* Protein-nucleic ascid interactions; structure and function of genes. *Mailing Add:* Sch Math & Natural Sci Montclair State Col Upper Montclair NJ 07043

VANDEHEY, ROBERT C, b Hollandtown, Wis, July 19, 24. ENTOMOLOGY. *Educ:* Univ Notre Dame, MS, 57, PhD(biol), 61. *Prof Exp:* Instr high sch, Pa, 49-55; from instr to asst prof, 55-69, ASSOC PROF BIOL, ST NORBERT COL, 69- *Concurrent Pos:* NIH res fel, Inst Genetics, Johannes Gutenberg Univ, Mainz, 63-64; res assoc, Mt St Mary's Col, Calif, 71-72. *Mem:* AAAS; Entom Soc Am; Am Mosquito Control Asn; Sigma Xi. *Res:* Genetic variability and control of the mosquitoes, especially Aedes aegypti and Culex pipiens. *Mailing Add:* 4841 S Woodlawn Ave Chicago IL 60615

VAN DE KAMP, PETER CORNELIS, b Plainfield, NJ, Aug 25, 40; m 64; c 2. GEOLOGY, GEOCHEMISTRY. *Educ:* Lehigh Univ, BA, 62; McMaster Univ, MSc, 64; Univ Bristol, PhD(geochem), 67. *Prof Exp:* Geologist, Shell Develop Co, Tex & Calif, 67-71 & Shell Oil Co, Colo, 71-73; res prof, Univ Man, 73-74; vpres, Geo-Logic, Inc, 74-77; independent petrol explor consult, 77-79; partner, Georesources assoc, 79-85; INDEPENDENT CONSULT, 85- *Mem:* Soc Econ Paleont & Mineral; Geol Soc Am; Am Asn Petrol Geol. *Res:* Stratigraphy; sedimentary, metamorphic and igneous petrology and geochemistry. *Mailing Add:* 1750 Cabernet Lane St Helena CA 94574

VAN DE KAR, LOUIS DAVID, b Amsterdam, Neth, Apr 15, 47; m 78. NEUROPHARMACOLOGY, NEUROENDOCRINOLOGY. *Educ:* Univ Amsterdam, BS, 71, MS, 74; Univ Iowa PhD(pharmacol), 78. *Prof Exp:* Res assoc pharmacol, Neth Cent Inst Brain Res, 74-75; fel physiol, Univ Calif, San Francisco, 79-81; asst prof pharmacol, 81-87, ASSOC PROF PHARMACOL, STRITCH SCH MED, LOYOLA UNIV, 87- *Concurrent Pos:* Fel, Fulbright-Hays Found, 75. *Mem:* Int Soc Neuroendocrinol; Am Soc Pharmacol Exp Ther; Am Physiol Soc; Soc Neurosci; Endocrine Soc. *Res:* Role of serotonergic and catecholamineroic neurons in the regulation of renin carticosterone, oxytocin, vasopressin and prolactin secretion; neuroanatomy and physiology of brain serotonin; role of brain serotonin in cardiovascular homeostasis; stress and neuroendocrine function. *Mailing Add:* Dept Pharmacol Stritch Sch Med Loyola Univ 2160 S First Ave Maywood IL 60153

VANDE KIEFT, LAURENCE JOHN, b Grand Rapids, Mich, May 14, 32; m 62; c 2. SOLID STATE PHYSICS, EXPLOSIVES. *Educ:* Calvin Col, BA, 53; Univ Conn, MS, 55, PhD(physics), 68. *Prof Exp:* Sr physicist, Bendix Res Labs, 58-62; teaching & res asst, Univ Conn, 62-63 & 64-68; res physicist, Signature & Propagation Lab, 68-72, chief, Optical & Microwave Systs Br, Concepts Anal Lab, 72-75, dep prog mgr, 75-79, EXPLOSIVES FORMULATION TEAM LEADER, TERMINAL BALLISTICS DIV, BALLISTIC RES LAB, US ARMY ABERDEEN PROVING GROUND, MD, 79- *Mem:* Am Phys Soc. *Res:* Electron paramagnetic resonance investigations of radiation effects in single crystals; laser interactions with the atmosphere; laser semiactive terminal homing; computer simulation; development of safer energetic materials; patent on PEG/PMVT energetic binder for explosives. *Mailing Add:* US Army Ballistic Res Labs Attn SLCBR-TB-E Aberdeen Proving Ground MD 21005-5066

VAN DEMARK, DUANE R, b Elida, Ohio, Feb 26, 36; div; c 2. SPEECH PATHOLOGY. *Educ:* Hiram Col, BA, 58; Univ Iowa, MA & PhD(speech path, audiol), 62. *Prof Exp:* Instr speech path, Ind Univ, 62-65; asst prof, 65-68, assoc prof speech path & otolaryngol, 68-75, PROF SPEECH PATH & OTOLARYNGOL & MAXILLOFACIAL SURG, UNIV IOWA, 75- *Concurrent Pos:* Ind Univ Found res grant, 65; Am-Scand Found George Marshall fel, 70; Nat Inst Dent Res spec fel, 70; partic, Int Cong Cleft Palate, 67-69 & 73, mem prog comt, 68, secy & asst to secy gen, 69; sect ed, Cleft Palate J. *Mem:* Am Cleft Palate Educ Found; Am Speech & Hearing Asn; Am Cleft Palate Asn (vpres, 80, pres, 82). *Res:* Cleft palate research. *Mailing Add:* Dept Otolaryngol Univ Iowa Iowa City IA 52242

VANDEMARK, NOLAND LEROY, b Columbus Grove, Ohio, July 6, 19; m 40; c 3. PHYSIOLOGY. *Educ:* Ohio State Univ, BS, 41, MS, 42; Cornell Univ, PhD, 48. *Prof Exp:* Asst animal husb, Ohio State Univ, 41-42; vitamin chemist, State Dept Agr, Ohio, 42; asst animal husb, Cornell Univ, 42-44, 48; livestock specialist, US Dept Army, Austria, 46-47; from asst prof to prof physiol, Univ Ill, Urbana, 48-64; prof dairy sci & chmn dept, Col Agr & Home Econ, Ohio State Univ & Ohio Agr Res & Develop Ctr, 64-73, mem fac, Coop Exten Serv, Univ, 64-73; dir res, NY State Col Agr & Life Sci, 74-81; distinguished bicentennial prof, Univ Ga, 85; dir, Agr Exp Sta, 74-81, prof, 74-83, EMER PROF ANIMAL SCI, CORNELL UNIV, 83- *Honors & Awards:* Borden Award, 59. *Mem:* AAAS; Am Soc Animal Sci; Am Dairy Sci Asn; Am Physiol Soc; Brit Soc Study Fertil. *Res:* Physiology and biochemistry of reproductive processes in cattle, especially semen production; sperm metabolism; female reproductive processes; artificial insemination and fertility-sterility problems; research management and creativity; author of one book. *Mailing Add:* 8801 Leesville Rd Raleigh NC 27613-1012

VAN DEN AKKER, JOHANNES ARCHIBALD, b Los Angeles, Calif, Dec 5, 04; m 30, 58, 90; c 1. OPTICS, THERMODYNAMICS. *Educ:* Calif Inst Technol, BS, 26, PhD(physics), 31. *Prof Exp:* Instr physics, Wash Univ, 30-35; res assoc & chmn dept, Inst Paper Chem, Lawrence Univ, 35-36, sr res assoc physics & chmn dept physics & math, 56-70, res counsr, 65-70, EMER PROF PHYSICS, INST PAPER CHEM, LAWRENCE UNIV, 70- *Concurrent Pos:* Sr Fulbright lectr, Univ Manchester Inst Sci & Technol, 61-62; consult res & develop, Am Can Co, Neenah, 71-82, James River Corp, 82-85; gen consult, 85- *Honors & Awards:* Res & Develop Award, Tech Asn Pulp & Paper Indust, 67, Gold Medal, 68. *Mem:* Fel AAAS; fel Am Phys Soc; fel Optical Soc Am; Am Asn Physics Teachers; fel Tech Asn Pulp & Paper Indust. *Res:* Spatial distribution of x-ray photoelectrons; optical properties of paper; spectrophotometry and color measurement; instrumentation of all properties of paper; paper and fiber physics. *Mailing Add:* 1101 E Glendale Ave Appleton WI 54911

VAN DEN AVYLE, JAMES ALBERT, b South Bend, Ind, Sept 13, 46; m 68. METALLURGY. *Educ:* Purdue Univ, BS, 68; Mass Inst Technol, SM, 69, PhD(metall), 75. *Prof Exp:* STAFF MEM, SANDIA LABS, 68- *Mem:* Am Soc Metals, Am Soc Testing & Mat. *Res:* Fracture of metals, mechanical properties of metals. *Mailing Add:* Org 1832 Sandia Nat Labs Albuquerque NM 87185

VANDENBERG, JOANNA MARIA, b Heemstede, Netherlands, Jan 24, 38; div; c 2. HIGH RESOLUTION X-RAY DIFFRACTION,. *Educ:* Leiden State Univ, BS, 59, MS, 62, PhD(structural chem), 64. *Prof Exp:* Teaching asst, Lab Crystallog, Univ Amsterdam, 62-64; res chemist, Royal Dutch & Shell Lab, 64-68; fel, 68-69, MEM TECH STAFF, BELL LABS, MURRAY HILL, 72- *Concurrent Pos:* Consult, Bell Labs, 72. *Mem:* Am Phys Soc. *Res:* High-resolution x-ray analysis of semiconductor superlattices and epitacial films; superconductivity of thin film materials; molecular beam epitacy. *Mailing Add:* 49 Oak Ridge Ave Summit NJ 07901

VAN DEN BERG, L, b Hattem, Neth, Mar 14, 29; Can citizen; m 53; c 4. BIOCHEMICAL ENGINEERING, FOOD SCIENCE. *Educ:* State Agr Univ, Wageningen, MSc, 53; Univ Man, MSc, 55. *Prof Exp:* Sr res officer food technol, 56-80, prin res officer, Biol Prod Fuels, Inst Bio Sci, 80-84, asst dir & head carbohydrate lab, 84-90, DIR, NAT RES COUN CAN, 90- *Mem:* Am Soc Microbiol; AAAS; Inst Food Technologists; Can Inst Food Sci & Technol. *Res:* Application of refrigeration to food preservation, including freezing and frozen storage of meat and vegetables and storage of fresh vegetables; anaerobic digestion of food plant waste; methanogenesis from biomass; fermenter design. *Mailing Add:* Inst Biol Sci Nat Res Coun Can Ottawa ON K1A 0R6 Can

VANDENBERG, STEVEN GERRITJAN, b Den Helder, Netherlands, July 7, 15; nat US; m 48. BEHAVIORAL GENETICS, PSYCHOLOGY. *Educ:* Univ Groningen, DrsJur, 46; Univ Mich, PhD, 55. *Prof Exp:* Psychologist, Ionia State Hosp for Criminally Insane, 47-48; asst psychol, Univ Mich, 48-50, psychologist, Inst Human Biol, 51-57, assoc dir schizophrenia study, Ment Insane, 47-48; psychologist, Sch Med, Univ Louisville, 60-67; PROF PSYCHOL, UNIV COLO, BOULDER, 67- *Concurrent Pos:* Psychologist, State Child Guid Clin, Mich, 49-50; lab schs, Eastern Mich Univ & pub schs, Mich, 51; exec ed, Behav Genetics, 70-77; Nat Inst Ment Health res career develop award, 62-67; vis prof genetics, Univ Hawaii, 74-75 & Univ Amsterdam, 80-81. *Mem:* Am Soc Human Genetics; Soc Res Child Develop; Soc Personality Assessment; Am Psychol Asn; Brit Psychol Soc. *Res:* Objective measures of personality; behavior genetics; factor analysis and test theory; computer applications in the behavior sciences. *Mailing Add:* Inst Behav Genetics Univ Colo Boulder CO 80309

VANDENBERGH, DAVID JOHN, b Athens, Ohio, May 7, 59; m 85; c 2. MOLECULAR BIOLOGY, NEUROSCIENCES. *Educ:* Univ NC, BS, 81; Pa State Univ, PhD(biochem), 87. *Prof Exp:* Grad res asst biochem, Pa State Univ, 81-87; postdoctoral res fel, biol, Calif Inst Technol, 87-90; SR RES FEL, LAB MOLECULAR NEUROBIOL, ADDICTION RES CTR, NAT INST DRUG ABUSE, 90- *Mem:* AAAS. *Res:* Control of gene expression in neurons, particularly genes in drug abuse and the dopaminergic neurons. *Mailing Add:* Nat Inst Drug Abuse Addiction Res Ctr PO Box 5180 Baltimore MD 21224

VANDENBERGH, JOHN GARRY, b Paterson, NJ, May 5, 35; m 58; c 2. ANIMAL BEHAVIOR, ENDOCRINOLOGY. *Educ:* Montclair State Col, AB, 57; Ohio Univ, MA, 59; Pa State Univ, PhD(zool), 62. *Prof Exp:* Res biologist, Nat Inst Neurol Dis & Blindness, 62-65; res scientist, NC Dept Ment Health, 65-76; head dept zool, 76-90, PROF ZOOL, NC STATE UNIV, 76- *Concurrent Pos:* NIMH grant, 67-82 & 85-, NSF grant, 82-85; mem, nat primate adv comt, NIH, rev comt basic behav processes, NIMH; mem psychobiol panel, NSF; mem, comt conserv non-human primates, Nat Res Coun-Nat Acad Sci, comt rewrite guide care & use lab animals; assoc ed, Am J Primatology; mem adv comt, Calif Primate Res Ctr, Caribbean Primate Ctr & Wis Regional Primate Res Ctr; mem, NASA Space Biol Rev Comt; dir, Triangle Consortium Reproductive Biol. *Mem:* Am Soc Zoologists; Animal Behav Soc (pres, 82-83); Am Soc Mammal; Soc Study Reproduction; Int Primatology Soc; Am Soc Primatology; AAAS. *Res:* Environmental control of reproduction; endocrine basis of behavior; pheromones and reproduction; rodent and primate social behavior. *Mailing Add:* Dept Zool NC State Univ Box 7617 Raleigh NC 27695-7617

VAN DEN BERGH, SIDNEY, b Wassenaar, Holland, May 20, 29; wid; c 3. ASTRONOMY. *Educ:* Princeton Univ, AB, 50; Ohio State Univ, MSc, 52; Univ Gottingen, Dr rer nat(astron), 56. *Prof Exp:* Asst prof astron, Ohio State Univ, 56-58; from lectr to prof astron, Univ Toronto, 58-77; dir, 77-86, ASTRONR, DOMINION ASTROPHYS OBSERV, 86- *Concurrent Pos:* Pres & chmn bd, Can, France & Hawaii Telescope Corp, 82; pres, Can Astron Soc, 90-92. *Honors & Awards:* Beals Award, Can Astron Soc, 88 & Isaac Killam Prize, 90; Russell Lectr, Am Astron Soc, 90. *Mem:* Am Astron Soc; fel Royal Soc Can; Royal Astron Soc; Int Astron Union (vpres, 76-82); Royal Soc London. *Res:* Extragalactic nebulae; star clusters; variable stars; supernovae. *Mailing Add:* 418 Lands Ends Rd Sydney BC V8M 4R4 Can

VAN DEN BOLD, WILLEM AALDERT, b Amsterdam, Neth, Mar 30, 21; m 50; c 5. GEOLOGY, PALEONTOLOGY. *Educ:* State Univ Utrecht, PhD, 46. *Prof Exp:* Micropaleontologist, Royal Dutch Shell Group, 46-58; assoc prof geol, 58-59, PROF GEOL, LA STATE UNIV, BATON ROUGE, 59- *Mem:* Paleont Res Inst; Soc Econ Paleontologists & Mineralogists. *Res:* Post-Paleozoic ostracoda; planktonic Foraminifera; Cenozoic correlation, Caribbean. *Mailing Add:* 4646 Bennett Dr La State Univ Baton Rouge LA 70808

VANDEN BORN, WILLIAM HENRY, b Rhenen, Netherlands, Nov 17, 32; Can citizen; m 58; c 5. WEED SCIENCE, PLANT PHYSIOLOGY. *Educ:* Univ Alta, BSc, 56, MSc, 58; Univ Toronto, PhD(plant physiol), 61. *Prof Exp:* Fel, Dept Plant Sci, 60-61, from asst prof to assoc prof, 61-72, chmn dept, 70-75, 82-87, PROF WEED SCI & CROP ECOL, UNIV ALTA, 72- *Concurrent Pos:* Assoc ed, Weed Sci, 81-86 & Can J Plant Sci, 81-84. *Mem:* Fel Weed Sci Soc Am; Can Soc Plant Physiol; Can Soc Agron; Am Sci Affil; Am Soc Agron; Am Inst Biol Sci. *Res:* Herbicide physiology; physiology of herbicide action: absorption, translocation, mechanism of action, metabolism of systemic herbicides; weed biology and control, both annuals and perennials, chiefly in grain crops. *Mailing Add:* Dept Plant Sci Univ Alta Edmonton AB T6G 2P5 Can

VANDENBOSCH, ROBERT, b Lexington, Ky, Dec 12, 32; m 56; c 2. NUCLEAR CHEMISTRY. *Educ:* Calvin Col, AB, 54; Univ Calif, PhD(chem), 57. *Prof Exp:* From asst chemist to assoc chemist, Argonne Nat Lab, 57-63; PROF CHEM, UNIV WASH, 63- *Mem:* Fel Am Phys Soc; Am Chem Soc. *Res:* Heavy ion nuclear reactions and nuclear fission. *Mailing Add:* Dept Chem Univ Wash Seattle WA 98195

VANDEN BOUT, PAUL ADRIAN, b Grand Rapids, Mich, June 16, 39; m 61; c 2. INTERSTELLAR MEDIUM, RADIO ASTRONOMY. *Educ:* Calvin Col, AB, 61; Univ Calif, Berkeley, PhD(physics), 66. *Prof Exp:* Teaching asst physics, Univ Calif, Berkeley, 61-66; teaching fel, Columbia Univ, 67-68, asst prof, 68-70; from asst prof to prof astron, Univ Tex, Austin, 70-84; DIR, NAT RADIO ASTRON OBSERV, 85- *Mem:* Fel Am Phys Soc; Am Astron Soc; Int Astron Union; AAAS; Int Radio Sci Union. *Res:* Spectroscopic study of diffuse matter in space, particularly using radio techniques, interstellar matter, star formation isotopic abundances, radio instrumentation and interferometry. *Mailing Add:* Nat Radio Astron Observ Edgemont Rd Charlottesville VA 22903-2475

VANDEN EYNDEN, CHARLES LAWRENCE, b Cincinnati, Ohio, June 25, 36; m 67; c 2. NUMBER THEORY. *Educ:* Univ Cincinnati, BS, 58; Univ Ore, MA, 60, PhD(math), 62. *Prof Exp:* NSF fel, Univ Mich, 62-63; asst prof math, Univ Ariz, 63-65 & Miami Univ, 65-67; asst prof math, Miami Univ,65-67; vis asst prof, Pa State Univ, 67-68; asst prof, Ohio Univ, 68-69; assoc prof, 69-74, PROF MATH, ILL STATE UNIV, 74- *Concurrent Pos:* Vis prof, Portland State Univ, 77-78. *Mem:* Am Math Soc; Math Asn Am. *Res:* Diophantine approximation; elementary number theory; combinatorial theory; sequences of integers. *Mailing Add:* Dept Math Ill State Univ Normal IL 61761

VANDENHAZEL, BESSEL J, b Neth, 27; Can citizen; m; c 3. UNDERWATER ARCHEOLOGY, ARTIFACT PRESERVATION. *Educ:* State Col Agr, Neth, L Ing, 53; Univ Western Ont, BA, 60; Northern Ill Univ, MSc, 69. *Prof Exp:* Instr physics, biol & environ studies, Ont Sec Schs, 55-75; prof sci & environ studies, Nipissing Univ Col, North Bay, 75-; RETIRED. *Concurrent Pos:* Assoc ed, Crucible, Sci Teachers Asn Ont, 70-; prin investr, Lake Nipissing Underwater Archeol Proj, 80- *Honors & Awards:* Rolex Montres Award, Geneva, Switz, 87. *Res:* Role of science in society; underwater archeology; preservation of waterlogged wood and corroded iron recovered from fresh water. *Mailing Add:* 20 Young St St Thomas ON N5R 4W5 Can

VANDEN HEUVEL, WILLIAM JOHN ADRIAN, III, b Brooklyn, NY, Mar 7, 35; m 60; c 3. DRUG METABOLISM, ENVIRONMENTAL IMPACT STUDIES. *Educ:* Princeton Univ, AB, 56, AM, 58, PhD(org chem), 60. *Hon Degrees:* DSc, Bucknell Univ, 83. *Prof Exp:* Instr chem, Lipid Res Ctr, Col Med, Baylor Univ, 62, asst prof, Dept Biochem, 62-64; sr res biochemist, 64-67, res fel, Dept Biochem, 67-72, sr res fel, Dept Drug Metab, 72-76, sect dir, Animal Drug Metab & Radiochem, 76-79, SR INVESTR, MERCK SHARP & DOHME RES LABS, 79- *Concurrent Pos:* Sr asst scientist, Nat Heart Inst, 60-62; vis scientist, Bucknell Univ, 79-88. *Mem:* AAAS; Am Chem Soc; Am Soc Mass Spectrometry; Am Soc Pharm Exp Therapeut; Soc Environ Toxicol & Chem; Am Col Clin Pharmacol; Int Soc Study Xenobiotics. *Res:* Identification and quantification of drugs, metabolites and natural products; environmental impact studies; drug residue studies; use of radioactive and stable isotopes in metabolism, residue and environmental studies. *Mailing Add:* Merck Sharp & Dohme Res Labs Rahway NJ 07065

VAN DEN NOORT, STANLEY, b Lynn, Mass, Sept 8, 30; m 54; c 5. NEUROLOGY. *Educ:* Dartmouth Col, AB, 51; Harvard Univ, MD, 54. *Prof Exp:* Asst prof neurol, Sch Med, Case Western Reserve Univ, 65-71; prof med (neurol) & dean, Col Med, Univ Calif, Irvine, 73-85. *Mailing Add:* Med Dean Off Univ Calif Irvine Irvine CA 92717

VANDE NOORD, EDWIN LEE, b Pella, Iowa, Sept 10, 38; m 62; c 2. SPACE PHYSICS. *Educ:* Grinnell Col, BA, 60; Univ NMex, MS, 63, PhD(physics), 68. *Prof Exp:* Res asst physics, Grinnel Col, 60-61; asst, Univ NMex, 61-68, res assoc, 68-69; res scientist, Douglas Advan Res Labs, Calif, 69-70; staff scientist, 70-75, mgr advan progs, 75-78, asst dir, 78-80, dir, 80-84, vpres, Space Systs, 84-88, PRES ELECTRO-OPTICS & CRYOGENICS DIV, BALL AEROSPACE SYSTS GROUP, 88- *Mem:* Am Inst Aeronaut & Astronaut; Sigma Xi. *Res:* Photometry of zodiacal light; interplanetary dust; infrared Fourier transform spectroscopy; space instrumentation; remote sensing; earth radiation budget instrumentation. *Mailing Add:* Ball Aerospace Systs Group PO Box 1062 Boulder CO 80306

VAN DEN SYPE, JAAK STEFAAN, b Dendermonde, Belg, July 15, 35. METALLURGICAL ENGINEERING. *Educ:* Univ Louvain, MetEng, 59; Univ Pa, MSc, 61, PhD(metall eng), 65. *Prof Exp:* Instr metall, Sch Metall Eng, Univ Pa, 66-67, asst prof, 67-70; SR RES METALLURGIST, TARRYTOWN TECH CTR, LINDE DIV, UNION CARBIDE CORP, 70- *Mem:* Am Phys Soc; Am Inst Mech Engrs; Am Soc Metals. *Res:* Surface physics; phase transformations; mechanical properties; materials science. *Mailing Add:* Union Carbide Corp Tarrytown Tech Ctr Tarrytown NY 10591

VANDEPOPULIERE, JOSEPH MARCEL, b Parkville, Mo, June 21, 29; m 53; c 4. NUTRITION, BIOCHEMISTRY. *Educ:* Cent Mo State Univ, AB, 51; Univ Mo, Columbia, MS, 54; Univ Fla, PhD(animal husb), 60. *Prof Exp:* Res asst agr chem, Univ Mo, Columbia, 51-54; asst mgr, Biol Lab, Ralston Purina, 54-57; res asst animal husb, Univ Fla, 57-60; asst mgr broiler res, Ralston Purina, 60-62, mgr broiler res, 62-69, mgr field res & tech serv US, Can & Mex, 69-72; ASSOC PROF DEPT POULTRY HUSB, UNIV MO, COLUMBIA, 72- *Concurrent Pos:* Travel grant, World Poultry Sci Asn, 78. *Mem:* Poultry Sci Asn (secy-treas, 77-); World Poultry Sci Asn; Can Feed Mfg Nutrit Coun; Sigma Xi. *Res:* Efficient conversion of agriculture and industrial residuals to human food through the use of a monogastric-polygastric biological team; insect and ectoparasite control in the avian species. *Mailing Add:* S-138 Animal Sci Ctr Univ Mo Columbia MO 65211

VANDER, ARTHUR J, b Detroit, Mich, Dec 28, 33; m 55; c 3. PHYSIOLOGY. *Educ:* Univ Mich, BA, 55, MD, 59. *Prof Exp:* Intern med, New York Hosp-Cornell Med Ctr, 59-60; from instr to assoc prof, 60-69, PROF PHYSIOL, UNIV MICH, ANN ARBOR, 69- *Mem:* AAAS; Am Physiol Soc; Am Soc Nephrol; Soc Exp Biol & Med. *Res:* Renal physiology. *Mailing Add:* Dept Physiol M7744 Med Sci 2 Univ Mich 1301 Catherine Rd Ann Arbor MI 48109-0622

VANDER BEEK, LEO CORNELIS, b The Hague, Netherlands, Aug 28, 18; nat US; m 44; c 4. PLANT PHYSIOLOGY. *Educ:* Western Mich Univ, AB, 52; Univ Mich, MS, 53, PhD(bot), 56. *Prof Exp:* Asst, Univ Mich, 52-53, 54-55; from asst prof to assoc prof biol, Western Mich Univ, 56-62, prof, 62-; RETIRED. *Mem:* AAAS; Am Soc Plant Physiol; Am Inst Biol Sci. *Res:* Plant growth regulators; effect of pesticides on plants. *Mailing Add:* Dept Biol Sci Western Mich Univ Kalamazoo MI 49008

VANDERBERG, JEROME PHILIP, b New York, NY, Feb 5, 35; m 67; c 2. PARASITOLOGY, CELL PHYSIOLOGY. *Educ:* City Col New York, BS, 55; Pa State Univ, MS, 57; Cornell Univ, PhD(med entom), 61. *Prof Exp:* Fel biol, Johns Hopkins Univ, 62-63; from asst prof to assoc prof, 63-74, PROF PARASITOL SCH MED, NY UNIV, 74- *Mem:* Am Soc Trop Med & Hyg; Am Soc Parasitol; Soc Protozool. *Res:* Cellular physiology of insects and of host-parasite complex; malariology; invasion of host cells by malaria parasite. *Mailing Add:* Dept Med & Molecular Parasitol Med Sch NY Univ 550 First Ave New York NY 10016

VAN DER BIJL, WILLIAM, b Alphen aan den Rijn, Neth, Aug 15, 20; nat US; m 46; c 2. METEOROLOGY. *Educ:* Vrije Univ, Neth, BSc, 41, MSc, 43; Univ Utrecht, PhD(meteorol), 52. *Prof Exp:* Res assoc climat, Royal Netherlands Meteorol Inst, 46-56; assoc prof physics & meteorol, Kans State Univ, 56-61; ASSOC PROF METEOROL, NAVAL POSTGRAD SCH, 61- *Concurrent Pos:* Fel statist & meteorol, Univ Chicago, 54-55. *Mem:* Am Meteorol Soc; Am Geophys Union; Sigma Xi. *Res:* Physics of the atmosphere; statistical treatment of data; statistical analysis of geophysical data. *Mailing Add:* Dept Meteorol Naval Postgrad Sch Monterey CA 93943-5000

VANDERBILT, DAVID HAMILTON, b Huntington, NY, Aug 20, 54; m 81; c 2. THEORY OF SURFACES AND INTERFACES OF SEMICONDUCTORS. *Educ:* Swarthmore Col, Ba, 76; Mass Inst Technol, PhD(physics), 81. *Prof Exp:* Fel, Univ Calif, Berkeley, 81-84; at Lyman Lab Physics, Harvard Univ, 84-90; AT DEPT PHYSICS & ASTRON, RUTGERS UNIV, 91- *Mem:* Am Phys Soc; Sigma Xi. *Res:* Theoretical solid state physics; electronic structure of amorphous or other non-periodic systems, particularly semiconductors; chalcogenide glasses. *Mailing Add:* Dept Physics & Astron Rutgers Univ PO Box 849 Piscataway NJ 08855-0849

VANDERBILT, JEFFREY JAMES, b Sheboygan, Wis, July 18, 51. ORGANIC CHEMISTRY. *Educ:* Calvin Col, BS, 73; Univ Mich, MS, 76, PhD(org chem), 78. *Prof Exp:* Res chemist, 78-80, SR CHEMIST ORG CHEM, TENN EASTMAN CO, 80- *Mem:* Am Chem Soc. *Res:* Process research and development. *Mailing Add:* 2403 Wood Hollow Ct Long View TX 75604-2150

VANDERBILT, VERN C, JR, b Indianapolis, Ind, Mar 29, 20; m 42; c 3. ENGINEERING. *Educ:* Purdue Univ, BS, 42, MS, 47, PhD(electronic instrumentation, servo control), 54. *Prof Exp:* Asst prof aeronaut, Purdue Univ, 46-52; private pract, 52-54; res engr, Gen Elec Co, 54-55; chief res engr in charge, Gen Res & Road Testing Depts, Dynamometer Lab & Electronics Div, Perfect Circle Corp, 56-62, mgr electronics div, 62-64; pres, Dynamic Precision Controls Corp, 64-74; PRES, VANDERBILT ASSOCS, 74- *Concurrent Pos:* Adj prof, Purdue Prog, Ind Univ E, 75- *Mem:* Inst Elec & Electronics Engrs; Soc Automotive Engrs; Am Soc Metals; Nat Soc Prof Engrs. *Res:* Electronic instrumentation and servo control as applied to internal combustion engines and automotive equipment. *Mailing Add:* Vanderbilt Assocs PO Box 31 Hagerstown IN 47346

VANDERBORGH, NICHOLAS ERNEST, b Bay Shore, NY, June 24, 38; m; c 3. ELECTROCHEMISTRY. *Educ:* Hope Col, AB, 60; Southern Ill Univ, MS, 62, PhD(chem), 64. *Prof Exp:* Asst prof chem, Univ Minn, 64-66; from asst prof to assoc prof, Univ NMex, 66-75; staff mem, 75-77, ALT GROUP LEADER, LOS ALAMOS SCI LAB, 77- *Concurrent Pos:* Staff mem, Sandia Labs, 66-69; guest staff mem, Univ New Castle Upon Tyne, 73; res fels, Mead John & Assoc Western Univs. *Mem:* Mat Res Soc; Electrochem Soc; Am Chem Soc; Am Inst Chem Engrs; Soc Petrol Eng. *Res:* Development of electrochemical power systems for transportation; ionic transport and electrocatalysis; hydrogen production from hydrocarbon fuels. *Mailing Add:* MEE-13 MS J576 Los Alamos Nat Lab Los Alamos NM 87545

VANDERBURG, CHARLES R, b Langley, Va, Oct 14, 56; m 80; c 3. DEVELOPMENTAL BIOLOGY, CELLULAR BIOLOGY. *Educ:* Univ Pittsburgh, BSc, 78; Seton Hall Univ, MSc, 84; Univ Med & Dent NJ, PhD(cell biol), 89. *Prof Exp:* Res technician, Western Psychial Inst & Clin, Pittsburgh, 78-79; res & develop proj leader, Biotech Capital Corp, NY, 79-82; prod mgr, Clin Sci Inc, Whippany, NJ, 82-84; RES ASSOC, HARVARD MED SCH, BOSTON, 89- *Concurrent Pos:* Biomed consult, 83-86; instr anat, NJ Med Sch, Newark, 87-89; legis intern, NJ State Assembly Dist 24, 88-89. *Honors & Awards:* Jan Langman Award, 87. *Mem:* Am Asn Anatomists; Am Soc Cell Biologists; Electron Micros Soc Am; Sigma Xi; Microbeam Anal Soc. *Res:* Molecular biology of cells during development including cell motility and transformation, growth factors, electron microscopy; translational control, RNA processing and regulation; clinical diagnostic assay development. *Mailing Add:* Dept Anat & Cellular Biol Harvard Med Sch 220 Longwood Ave Boston MA 02115

VAN DER BURG, SJIRK, b Makkum, Netherlands, Mar 23, 26; nat US; c 3. RUBBER CHEMISTRY, TIRE TECHNOLOGY. *Educ:* Univ Groningen, Drs, 55. *Prof Exp:* Asst, Univ Groningen, 52-55; res chemist, Rubber Found, Delft Univ Technol, 55-56 & Res Ctr, US Rubber Co, NJ, 56-61; mgr mat res, US Rubber Tire Co, 61-66, develop mgr, Uniroyal Europ Tire Develop Ctr, Ger, 66-67, dir, 67-80; dir, Tyre Tech Div, Dunlop Ltd, 80-86; TIRE CONSULT, 86- *Mem:* Am Chem Soc; Royal Neth Chem Soc; Plastics & Rubber Inst. *Res:* Rubber chemistry and technology; plastics. *Mailing Add:* 39 Meadowview Rd PO Box 578 West Chatham MA 02669

VANDERBURG, VANCE DILKS, b Grand Rapids, Mich, July 22, 37; m 66; c 2. NUCLEAR ENGINEERING. *Educ:* Syracuse Univ, BS, 60; Purdue Univ, MS, 63, PhD(high energy physics), 65. *Prof Exp:* Physicist, US AEC, 65-67; asst physicist, Brookhaven Nat Lab, 67-69, assoc physicist, 69-73, physicist, 73; nuclear engr, Am Elec Power Serv Corp, 74-75; plant nuclear engr, Donald C Cook Nuclear Plant, 75-80, tech dept prod supvr, 80-84; sr scientist, 84-89, PRIN SCIENTIST, AM ELEC POWER SERV CORP, 89- *Mem:* Am Phys Soc; Am Nuclear Soc; Sigma Xi. *Res:* Experimental particle physics. *Mailing Add:* 30 Spring Creek Dr Westerville OH 43081

VANDER BURGH, LEONARD F, physical organic chemistry, for more information see previous edition

VANDERGRAAF, TJALLE T, b 's Gravenmoer, Neth, Sept 3, 36; Can citizen; m 66; c 4. ANALYTICAL CHEMISTRY, RADIOCHEMISTRY. *Educ:* Calvin Col, BS, 63; Pa State Univ, PhD(anal chem), 69. *Prof Exp:* RES SCIENTIST GEOCHEM RES BR, WHITESHELL NUCLEAR RES ESTAB, ENERGY CAN LTD, 69-, HEAD, GEOCHEMISTRY SECT, 84- *Concurrent Pos:* Consult, OECD/NEA, 86- *Mem:* Fel Chem Inst Can; Can Nuclear Soc; Sigma Xi; Int Asn Geochem & Cosmochem. *Res:* Radionuclide interaction with geological materials; redox reactions of multivalent radionuclides in geological environments; migration of radionuclides through crystalline rock formations; low temperature rock/water reactions; autoradiography of sorbed radionuclide distributions. *Mailing Add:* Geochem Res Br AECL Res Whiteshell Lab Pinawa MB R0E 1L0 Can

VANDERGRAFT, JAMES SAUL, b Gooding, Idaho, Apr 29, 37. NUMERICAL ANALYSIS, NUMERICAL SOFTWARE. *Educ:* Stanford Univ, BS, 59, MS, 63; Univ Md, PhD(math), 66. *Prof Exp:* Programmer, Lawrence Radiation Lab, 59-60; number analyst,Bellcomm Inc, Wash DC, 63-64; asst prof comput sci, Univ Md, 66-73, assoc prof, 73-79; asst dir, Automated Sci Group Inc, Silver Spring, Md, 79-84; staff analyst, Bus & Technol Systs, Inc, Greenbelt, Md, 84-88; Computational Eng Inc, Laurel, Md, 88-90; COLEMAN RES CORP, LAUREL, MD, 90- *Concurrent Pos:* Tech consult, Apollo Proj, Bellcomm Inc, 69-70; guest prof math, Swiss Fed Inst Zurich, 74. *Mem:* Soc Indust & Appl Math; Math Asn Am; Asn Comput Mach; Inst Elec & Electronics Engrs. *Res:* Numerical solution of linear and nonlinear systems of equations; numerical algorithms and numerical software; modeling and simulation. *Mailing Add:* 772 11th St SE Washington DC 20003

VANDER HART, DAVID LLOYD, b Rehoboth, NMex, May 20, 41; m 64; c 2. PHYSICAL CHEMISTRY, POLYMER CHEMISTRY. *Educ:* Calvin Col, AB, 63; Univ Ill, Urbana, PhD(phys chem), 68. *Prof Exp:* Fel microwave spectros, Univ Ill, 68-69; RES CHEMIST, NAT INST STANDARDS & TECHNOL, 69- *Mem:* Am Phys Soc. *Res:* Nuclear magnetic resonance, particularly carbon-13, application to the characterization of polymer solids. *Mailing Add:* Div 440 Nat Inst Standards & Technol Gaithersburg MD 20899

VANDERHEIDEN, GREGG, b Norway, Mich, Oct 27, 49; m 85; c 1. REHABILITATION ENGINEERING TECHNOLOGY. *Educ:* Univ Wis, Madison, BS, 72, MS, 74, PhD(technol commun rehab), 84. *Prof Exp:* ASSOC PROF TECHNOL & DISABILITY AGING, HUMAN FACTORS DIV, INDUST ENG, UNIV WIS, 86- *Concurrent Pos:* Dir, Trace Res & Develop Ctr, Univ Wis, 71-, mem staff, Commun Aids & Systs Clin, 80-; mem coun for Exceptional Children. *Honors & Awards:* Isabelle & Leonard Goldenson Award Outstanding Res Med & Technol, 78; Distinguished Serv Award, Rehab Eng Soc NAm, 78, 85 & 89; Clin Achievement Award, Am Social Health Asn, 85; 3rd Ann Award, Nat Coun Commun Dis, 85. *Mem:* Rehab Eng Soc NAm (secy, 88, pres-elect, 91); Int Soc Alternative & Augmentative Commun; Am Speech-Language-Hearing Asn; Asn Comput Mach; Inst Elec & Electronics Engrs; Inst Indust Engrs. *Res:* Use of technology in a rehabilitation; access to standard computers and electronics devices by persons with disabilities. *Mailing Add:* S-151 Waisman Ctr 1500 Highland Ave Madison WI 53705-2280

VAN DER HEIJDE, PAUL KAREL MARIA, b N Holland, Neth, May 29, 47; m 76; c 2. GEOHYDROLOGY, WATER RESOURCES MODELLING. *Educ:* Tech Univ Delft, Neth, MSc, 77. *Prof Exp:* Geohydrologist, Orgn Appl Sci Res TNO, Inst Appl Geosci, Delft, Neth, 77-85; DIR, INT GROUND WATER MODELING CTR, HOLCOMB RES INST, BUTLER UNIV, IND, 81-, DIR, WATER SCI PROG, 85-, ACTG DEAN, 89- *Concurrent Pos:* Mem sci adv bd, Nat Ctr Ground Water Res, 86-; mem Groundwater Modeling Assessment Comt, Nat Res Coun; prin investr, various US Environ Protection Agency grants, 91- *Mem:* Am Geophys Union; Royal Inst Engrs Neth; Am Water Resources Asn; Nat Asn Groundwater Scientists & Engrs; Am Soc Civil Engrs. *Res:* Application of groundwater hydrology; advancing the use of quality assured modelling methodologies in the management of groundwater resources; development of the technology transfer methods in groundwater science. *Mailing Add:* 7815 Pineview Ct Indianapolis IN 46250

VAN DER HELM, DICK, b Velsen, Netherlands, Mar 16, 33; m 60; c 6. PHYSICAL CHEMISTRY. *Educ:* Univ Amsterdam, Drs, 56, DSc(x-ray diffraction), 60. *Prof Exp:* Res assoc x-ray diffraction, Ind Univ, 57-59 & Inst Cancer Res, Philadelphia, 59-62; from asst prof to prof, 62-77, GEORGE LYNN CROSS RES PROF PHYS CHEM, UNIV OKLA, 77- *Concurrent Pos:* NIH Develop Award, 69-74. *Mem:* Am Chem Soc; Am Crystallog Asn; AAAS. *Res:* Molecular structure determination by means of x-ray diffraction of natural products, siderophores and peptides. *Mailing Add:* Dept Chem Univ Okla 620 Parrington Oval Norman OK 73019

VANDERHOEF, LARRY NEIL, b Frazee, Minn, Mar 20, 41; m 63; c 2. HORMONE PHYSIOLOGY, DEVELOPMENT. *Educ:* Univ Wis-Milwaukee, BS, 64, MS, 65; Purdue Univ, Lafayette, PhD(plant physiol), 69. *Prof Exp:* Nat Res Coun fel, Univ Wis-Madison, 69-70; assoc prof, Univ Ill, Urbana, 70-76, head biol progs, 74-76, prof plant develop, 76-80, head bot dept, 77-80; provost, Div Agr & Life Sci, Univ Md, College Park, 80-84; EXEC VCHANCELLOR, UNIV CALIF, DAVIS, 84- *Concurrent Pos:* Consult, Fed granting agencies; bd trustees, Am Soc Plant Physiol; Eisenhower fel, 87. *Mem:* AAAS; Am Soc Plant Physiol. *Res:* Plant hormones and nucleic acid metabolism; nitrogen fixation. *Mailing Add:* 573 Mrak Hall Chancellors Off Univ Calif Davis CA 95616

VANDERHOEK, JACK YEHUDI, b Hilversum, Neth, Jan 1, 41; US citizen; m 66; c 3. BIOCHEMISTRY, ORGANIC CHEMISTRY. *Educ:* City Col New York, BS, 60; Mass Inst Technol, PhD(org chem), 66. *Prof Exp:* Sr res chemist, Gen Mills, Inc, 66-68, group leader vitamin E and sterol synthesis, 68-69; NIH spec fel, Princeton Univ, 69-70 & Univ Fla, 71; res assoc biochem, Univ Mich, Ann Arbor, 71-72, instr, 72-74; lectr, Hadassah Univ Hosp, Hebrew Univ Med Sch, Jerusalem, Israel, 74-76; res assoc med & pharmacol, Univ Conn Health Ctr, Farmington, 77-78; asst prof, 78-82, ASSOC PROF BIOCHEM, GEORGE WASHINGTON SCH MED, WASHINGTON, D C, 83- *Mem:* Am Chem Soc; NY Acad Sci; AAAS; Fedn Am Socs Exp Biol. *Res:* Lipid metabolism including leukotrienes, prostaglandins and thromboxanes; lipid mediators in allergic and inflammatory disease. *Mailing Add:* Dept Biochem George Washington Univ Washington DC 20037

VAN DER HOEVEN, THEO A, b Indonesia, Feb 13, 33; US citizen; m 60; c 3. PHARMACOLOGY. *Educ:* Brooklyn Col, BS, 65; Columbia Univ, PhD(biochem), 71. *Prof Exp:* Instr biochem, Univ Mich, 71-74; asst prof med chem, Univ Md, Baltimore, 74-77; assoc prof pharmacol & exp therapeut, 77-80, assoc prof biochem, 80-83, ASSOC PROF PHARMACOL & TOXICOL, ALBANY MED COL, 83- *Mem:* AAAS; Am Soc Pharmacol Exp Therapeut. *Res:* Hormonal regulation of drug metabolism; membrane bound enzymes. *Mailing Add:* Dept Pharmacol & Toxicol Albany Med Col Albany NY 12208

VANDERHOFF, JOHN W, b Niagara Falls, NY, Aug 2, 25; m 50; c 2. POLYMER CHEMISTRY, COLLOID CHEMISTRY. *Educ:* Niagara Univ, BS, 47; Univ Buffalo, PhD(phys chem), 51. *Prof Exp:* Chemist, Phys Res Lab, Dow Chem Co, 50-56, proj leader, 56-58, assoc scientist, 58-70, Plastics Dept Res Lab, 63-70; assoc prof, 70-74, DIR, NAT PRINTING INK RES INST, ASSOC DIR COATINGS, CTR SURFACE & COATINGS RES, LEHIGH UNIV, 70-, PROF CHEM, 74-, CO-DIR EMULSION POLYMERS INST, 75- *Concurrent Pos:* Participant, Dow Career Scientist Assignment Prog, van't Hoff Lab, Utrecht, 65-66. *Mem:* AAAS; fel Am Inst Chem; Am Chem Soc; Sigma Xi. *Res:* Polymerization kinetics; solution properties of polymers; mechanism of emulsion polymerization; latex properties; foamed plastics; mechanism of latex film formation; colloidal properties of latexes; monodisperse latexes; printing inks; deinking of wastepaper. *Mailing Add:* Emulsion Polymers Inst Lehigh Univ 11 Research Dr Mountaintop Campus Bldg A Bethlehem PA 18015

VANDERHOLM, DALE HENRY, b Villisca, Iowa, Mar 28, 40; m 67; c 2. AGRICULTURAL WATER QUALITY. *Educ:* Iowa State Univ, BS, 62, MS, 69; Colo State Univ, PhD(agr eng), 72. *Prof Exp:* Watershed planning engr, Iowa Soil Conserv Serv, USDA, 62-63; instr, Iowa State Univ, 67-68, asst prof agr eng water supply & waste treat, 72-73; from asst prof to prof agr eng water, supply & waste treat, Univ Ill, Urbana-Champaign, 78-83, asst dir, Ill Agr Exp Sta, 81-83; ASSOC DEAN, AGR RES DIV, UNIV NEBR, LINCOLN, 83- *Concurrent Pos:* Res award, Great Plains Livestock Comt, 77; vis res fel, NZ Agr Inst, Lincoln Col, Cantebury, 79-80. *Mem:* Am Soc Agr Engrs; Soil Conserv Soc Am; AAAS; Coun Agr Sci & Technol. *Res:* Treatment and handling of livestock wastes, particularly land application systems and feedlot runoff control systems; treatment of domestic sewage by recirculating sand filters and aerobic package systems. *Mailing Add:* Agr Res Div Univ Nebr 207 Agr Hall Lincoln NE 68583-0704

VAN DER HULST, JAN MATHIJS, b 's-Gravenhage, Neth, Jan 26, 48; m 70; c 3. ASTROPHYSICS. *Educ:* Univ Groningen, Drs, 73, PhD(astron), 77. *Prof Exp:* Res assoc astron, Nat Radio Astron Observ, 77-78; asst prof astron, Univ Minn, 79-82; res scientist, astron, Neth Found Radio Astron, 82-88; ASSOC PROF ASTRON, UNIV GRONINGEN, 88- *Mem:* Am Astron Soc; Nederlandse Astron Club. *Res:* Structure and evolution of galaxies, gas content of galaxies; dynamics of galaxies; nuclear activity in galaxies. *Mailing Add:* Kapteyn Astron Inst Postbus 800 NL-9700 AV Groningen Netherlands

VAN DE RIJN, IVO, b Sept 30, 46; m; c 1. MICROBIOLOGY & IMMUNOLOGY. *Educ:* Univ Fla, PhD(microbiol), 72. *Prof Exp:* ASSOC PROF MICROBIOL & IMMUNOL & ASSOC, DEPT MED, DIV INFECTIOUS DIS, BOWMAN GRAY SCH MED, 82- *Res:* Bacterial pathogenesis; bacterial endocarditis; hyaluronic acids synthesis. *Mailing Add:* Dept Microbiol & Immunol Wake Forest Univ Med 300 S Hawthorne Rd Winston-Salem NC 27103

VAN DERIPE, DONALD R, b Lafayette, Ind, Feb 13, 34; m 64; c 1. PHARMACOLOGY. *Educ:* Purdue Univ, BS, 56, MS, 58; Northwestern Univ, PhD(pharmacol), 63. *Prof Exp:* Fel cardiovasc pharmacol, Emory Univ, 63-65; res pharmacologist, Mallinckrodt Chem Works, 65-75, DIR TECH EVAL, MALLINCKRODT, INC, 75- *Res:* Radiopaque diagnostic agents; cardiovascular and radionuclide pharmacology. *Mailing Add:* Mallinkrodt Med Inc 675 McDonnell Blvd PO Box 5840 St Louis MO 63134

VANDER JAGT, DAVID LEE, b Grand Rapids, Mich, Jan 13, 42; m 67. BIOCHEMISTRY. *Educ:* Calvin Col, AB, 63; Purdue Univ, Lafayette, PhD(chem), 67. *Prof Exp:* NIH fel biochem, Northwestern Univ, 67-69; asst prof, 69-74, ASSOC PROF BIOCHEM, UNIV NMEX, 74- *Concurrent Pos:* Res career develop award, Nat Cancer Inst, 74- *Mem:* AAAS; Am Chem Soc; Am Soc Biol Chemists. *Res:* Enzyme, coenzyme reaction mechanisms of glutathione requiring enzymes, especially glyoxalase; metabolic role of methylglyoxal; biomedical applications of 13-C; metabolism of chemical carcinogens. *Mailing Add:* Dept Biochem Univ NMex Sch Med Albuquerque NM 87131

VANDERJAGT, DONALD W, b Muskegon, Mich, Feb 25, 38; m 58; c 4. COMBINATORICS. *Educ:* Hope Col, AB, 59; Fla State Univ, MS, 61; Western Mich Univ, PhD(math), 73. *Prof Exp:* Instr math, Cent Univ Iowa, 62-64; from asst prof to assoc prof, 64-75, PROF MATH, GRAND VALLEY STATE COLS, 75- *Mem:* Am Math Soc; Math Asn Am; Asn Comput Mach; Nat Coun Teachers Math. *Res:* Graph theory, local properties, degree sets, Hamiltonian properties, generalized Ramsey theory. *Mailing Add:* Dept Math & Comput Sci Grand Valley State Cols Col Landing Allendale MI 49401

VANDER KLOET, SAM PETER, b Heidenschap, Frisia, Feb 18, 42; Can citizen; m 70; c 2. PLANT TAXONOMY, PLANT ECOLOGY. *Educ:* Queen's Univ, BA, 68, PhD(biol), 72. *Prof Exp:* PROF BIOL & CUR E C SMITH HERBARIUM, ACADIA UNIV, 72- *Mem:* Bot Soc Am; Can Bot Asn; Int Asn Plant Taxon. *Res:* Biosystematics of Vaccinium. *Mailing Add:* Dept Biol Acadia Univ Wolfville NS B0P 1X0 Can

VAN DER KLOOT, ALBERT PETER, b Chicago, Ill, Jan 22, 21; m 48; c 2. FOOD CHEMISTRY. *Educ:* Mass Inst Technol, SB, 42. *Prof Exp:* Chemist flower preserv, Flower Foods, Inc, 46-47; chemist biscuits & crackers, Independent Biscuit Mfg Tech Inst, 47-52; chief chemist brewing & foods, 52-57, PRES, WAHL-HENIUS INST, INC, 57- *Mem:* Am Soc Brewing Chem; Am Chem Soc; Inst Food Technol; AAAS. *Res:* Commercial application of plant tissue culture; soda cracker fermentation; freeze drying of microorganisms; vapor pressure-moisture relationships; statistical analysis of brewing process; gas chromatography; alcoholic beverages; trace components of food. *Mailing Add:* Wahl-Henius Inst Inc 4206 N Broadway Chicago IL 60613

VAN DER KLOOT, WILLIAM GEORGE, b Chicago, Ill, Feb 18, 27; m 84; c 2. PHYSIOLOGY. *Educ:* Harvard Univ, SB, 48, PhD(biol), 52. *Prof Exp:* Nat Res Coun fel, Cambridge Univ, 52-53; instr biol, Harvard Univ, 53-56; from asst prof to assoc prof zool, Cornell Univ, 56-58; prof pharmacol & chmn dept, 58-61, prof physiol & chmn dept physiol & biophys, Sch Med, NY Univ, 67-71; chmn dept, 71-86, PROF PHYSIOL & BIOPHYS, STATE UNIV NY, STONY BROOK, 71- *Concurrent Pos:* Consult, NSF, 59-65 & NIH, 68-; ed, Biosci, 81-84. *Mem:* AAAS; Sigma Xi; Am Physiol Soc; Physiol Soc UK; Soc Neurosci. *Res:* Neurophysiology and pharmacology. *Mailing Add:* Dept Physiol & Biophys Health Sci Ctr State Univ NY Stony Brook NY 11794

VANDERKOOI, JANE M, b Rochester, NY, Feb 28, 44. BIOCHEMISTRY, BIOPHYSICS. *Educ:* Cent Univ Iowa, BA, 67; St Louis Univ, PhD(biochem), 71. *Prof Exp:* Fel biophys, 71-73, res assoc, Johnson Found, 73-75, from asst prof to assoc prof, 75-87, PROF BIOCHEM, DEPT BIOCHEM & BIOPHYS, UNIV PA, 88- *Concurrent Pos:* NIH fel, 75. *Mem:* Biophys Soc; Am Soc Biol Chem; Am Chem Soc. *Res:* Membrane structure and function; excited state reactions. *Mailing Add:* 36 & Hamilton Walk Univ Pa Philadelphia PA 19104

VANDER KOOI, LAMBERT RAY, b Lynden, Wash, Jan 8, 35; m 65; c 1. ELECTRICAL ENGINEERING. *Educ:* Univ Mich, BSE, 58, MSE, 61, PhD(elec eng), 68; Calvin Col, BS, 59. *Prof Exp:* Engr, Systs Div, Bendix Corp, 58-61; assoc res engr, Radar & Optics Lab, Inst Sci & Technol, Univ Mich, 61-68; engr, Res & Develop Div, Kelsh Instrument Co, 68; radar systs engr, Polhemus Assocs, Inc, Mich, 69-70; ASSOC PROF ELEC ENG & TECHNOL, WESTERN MICH UNIV, 70- *Mem:* Inst Elec & Electronics Engrs. *Res:* Stochastic control systems; synthetic aperture radar systems; digital circuits and systems; active networks. *Mailing Add:* Dept Elec Eng Western Mich Univ Kalamazoo MI 49008

VANDERKOOI, WILLIAM NICHOLAS, b Paterson, NJ, Dec 19, 29; m 54; c 3. INDUSTRIAL CHEMISTRY. *Educ:* Calvin Col, AB, 51; Purdue Univ, MS, 53, PhD(phys chem), 55. *Prof Exp:* Lab asst chem & physics, Calvin Col, 48-51, asst, Off Naval Res, 50-51; chemist, Purdue Univ, 51-52, USAF contract, 52, instruments, 52-53; res chemist, C C Kennedy Res Lab, 55-64, group leader, Polymer & Chem Res Lab, 64-69, assoc scientist, Hydrocarbon & Monomers Res Lab, 69-81, assoc scientist, Functional Polymers & Process Res Lab, 81-88, ASSOC SCIENTIST, APPL ORG LAB, DOW CHEM CO, 88- *Mem:* Sigma Xi; Am Chem Soc. *Res:* Polymerization kinetics; radiation grafting; polymer synthesis and properties; metal chelation and purification; hydrocarbon analyses; high temperature reactions; hydrocarbon pyrolysis; acrylamide monomer process and acrylamide polymerization; processes; catalysis; data processing; plant simulation; miniplants; computer programming; artificial intelligence expert system development. *Mailing Add:* 677 Bldg Dow Chem Co Midland MI 48667

VANDERKOOY, JOHN, b Neth, Jan 1, 41; Can citizen; m 65; c 1. SOLID STATE PHYSICS, ELECTRO-ACOUSTICS. *Educ:* McMaster Univ, BEng, 63, PhD(physics), 67. *Prof Exp:* Nat Res Coun Can fel physics, Cambridge Univ, 67-69; res assoc, 69-70, asst prof, 70-79, ASSOC PROF PHYSICS, UNIV WATERLOO, 79- *Concurrent Pos:* Nat Res Coun Can grants, 69-81. *Mem:* Can Asn Physicists; Audio Eng Soc. *Res:* Low temperature solid state physics of metals; audio, transducer design and measurement. *Mailing Add:* Dept Physics Univ Waterloo Waterloo ON N2L 3G1 Can

VANDERLAAN, MARTIN, b San Francisco, Calif, Nov 6, 48. CYTOLOGY, ONCOLOGY. *Educ:* Univ Calif, Santa Barbara, BA, 70; NY Univ, MS, 72, PhD(environ health), 75. *Prof Exp:* Assoc res scientist oncol, Inst Environ Med, NY Univ, 72-76; BIOMED SCIENTIST CELL BIOL, LAWRENCE LIVERMORE LAB, 76- *Mem:* Am Asn Cancer Res; AAAS. *Res:* Cancer cytology; enzyme and immuno-cytochemistry; carcinogenesis; hybridomas. *Mailing Add:* Biomed Div Lawrence Livermore Lab L452 PO Box 5507 Livermore CA 94550

VANDERLAAN, WILLARD PARKER, b Muskegon, Mich, June 5, 17; m 44; c 3. ENDOCRINOLOGY. *Educ:* Harvard Univ, MD, 42. *Prof Exp:* Intern med, Boston City Hosp, 42-43, asst resident path, 43; instr pharmacother, Harvard Med Sch, 44-45; fel endocrinol, J H Pratt Diag Hosp, Boston, 45-47; from asst prof to assoc prof med, Med Sch, Tufts Univ, 47-56; HEAD LUTCHER BROWN CTR FOR DIABETES & ENDOCRINOL, SCRIPPS CLIN & RES FOUND, 56-; PROF MED, UNIV CALIF, SAN DIEGO, 68- *Concurrent Pos:* Fel med, Thorndike Mem Lab, Boston City Hosp, 44; consult, Boston Vet Admin Hosp, 54-56; mem endocrinol study sect, NIH, 71-75, chmn, 74-75. *Mem:* Am Thyroid Asn; Endocrine Soc; Am Soc Clin Invest. *Res:* Growth hormone; prolactin; thyroid physiology. *Mailing Add:* 9894 Genesee Ave La Jolla CA 92037

VANDERLASKE, DENNIS P, b Mineola, NY, Jan 15, 48; m 78; c 1. ELECTROOPTICS. *Educ:* State Univ NY, Stony Brook, BE, 69; George Washington Univ, MS, 75; Cent Mich Univ, MA, 81. *Prof Exp:* Jr engr, 69-73, proj engr, 73-79, prog mgr, 79-82, chief, Target Signature Team, Visionics Div, 82-85, CHIEF, STANDARD ADVAN INFRARED SENSOR TEAM, NIGHT VISION & ELECTRO-OPTICS LABS, US ARMY ELECTRONICS RES & DEVELOP COMMAND, 85- *Mem:* Inst Elec & Electronics Engrs. *Res:* Infrared imaging, primarily for military applications and relation with other electro-optical technologies and field evaluation of this technology; solid state electronics as applied to future generation thermal imaging system technology. *Mailing Add:* Ctr Night Vision & Electrooptics AMSEL-RD-NV-D Ft Belvoir VA 22060-5677

VANDERLIND, MERWYN RAY, b Grand Rapids, Mich. PHYSICS, POLYMER CHEMISTRY. *Educ:* Hope Col, BA, 58; Ohio Univ, MS, 60, PhD(physics), 64. *Prof Exp:* Nuclear eng, Atomics Int, 60-61; res polymer physics, Rohm & Haas, 64-66; res physicist, Battelle Mem Inst, 66-73, mgr, Phys Sci Sect, 73-79, mgr, Venture Develop, 80-81, assoc dir, Corp Tech Develop, 82-83, dir, Indust Technol Ctr, Geneva, Switz, 83-84, vpres, Electronic & Defense Systs, 85-88, GROUP VPRES & GEN MGR, DEFENSE SYSTS & TECHNOL, BATTELLE MEM INST, 89- *Concurrent Pos:* Consult, Nat Acad Sci & Nat Acad Eng, 68-72; mem alumni grad coun res, Ohio Univ, 76- *Mem:* Am Phys Soc; Sigma Xi. *Res:* Radiation transport; nuclear weapons effects; polymer physics; laser applications; reactor engineering. *Mailing Add:* Battelle Mem Inst 505 King Ave Columbus OH 43201

VANDERLINDE, RAYMOND E, b Newark, NY, Feb 28, 24; m 48; c 3. CLINICAL CHEMISTRY. *Educ:* Syracuse Univ, AB, 44, MS(educ), 45, MS(chem), 47, PhD(biochem), 50; Am Bd Clin Chem, dipl, 60. *Prof Exp:* Teacher high sch, NY, 45-46; from asst prof to assoc prof biochem, Sch Med, Univ Md, 50-57; asst prof, Col Med, State Univ NY Upstate Med Ctr, 57-62; assoc lab dir & clin chemist, Mem Hosp Cumberland, Md, 62-65; dir labs clin chem, Div Labs & Res, NY State Dept Health, 65-77; PROF & DIR, DIV CLIN CHEM, DEPT PATH & MED & DIV CLIN BIOCHEM, DEPT BIOL CHEM, HAHNEMANN MED COL, 77- *Concurrent Pos:* Lab admin dir & clin biochemist, Syracuse Mem Hosp, 57-62; consult, Madison County Lab, NY, 59-62, Rome City Lab, 61-62 & Meyersdale Community Hosp Lab, Pa, 64-65; clin asst prof, Med Ctr, Univ WVa, 64-65; adj assoc prof, Albany Med Col, 70-; mem diag prods adv comt, Food & Drug Admin, 72-75; pres, Nat Comn Accreditation Clin Chem, 85-88; mem, NIH Lipid Standardization Panel, 86-88; chmn, Coun Nat Ref Systs Clin Lab, 88-89. *Honors & Awards:* Fisher Award, 85. *Mem:* Am Chem Soc; fel Am Asn Clin Chemists; Asn Clin Sci; assoc Am Soc Clin Path; Acad Clin Lab Physicians & Scientists. *Res:* Clinical chemistry; clinical enzymology and diabetes. *Mailing Add:* 573 Weadley Rd Wayne PA 19087

VAN DER LINDE, REINHOUD H, b Amsterdam, Holland, July 14, 29; US citizen; m 58; c 5. MATHEMATICS. *Educ:* NY Univ, BA, 53, MS, 56; Rensselaer Polytech Inst, PhD(math), 68. *Prof Exp:* Res asst, NY Univ, 54-56; PROF MATH, BENNINGTON COL, 56- *Mem:* Math Asn Am. *Res:* Random Eigenvalue problems; differential equations; functional analysis. *Mailing Add:* Dept Math Bennington Col Bennington VT 05201

VANDERLINDEN, CARL R, b Pella, Iowa, Sept 26, 23; m 45; c 2. MECHANICAL ENGINEERING. *Educ:* Univ Wash, BS, 44; Iowa State Univ, PhD(chem eng), 50. *Prof Exp:* Instr chem eng, Iowa State Univ, 46-50; dir res & develop shelter prod, 69-73, dir res & develop govt contracts, 73-75, vpres dir, 75-79, vpres res & develop contracts & appl tech, Johns Mainville Sales Corp,79-81, Manville Serv Corp , 81-86; PRES, VANDERLINDEN & ASSOC CONSULT, 87- *Concurrent Pos:* Chmn tech comt, Perlite Inst, 64-70, dir, 70-72, vpres, 72-74 & pres 74-76; mem, Bldg Futures Coun, Nat Inst Bldg Sci; mem bd dirs, Am Inst Chem Engrs, 79-81 & Bldg Thermal Envelope Coord Coun, 85-88; chmn, Advan Indust Mat Guid Eval Bd, US Dept Energy, 88-89. *Honors & Awards:* Lewis Lloyd Award, Perlite Inst, 83. *Mem:* Am Inst Chem Engrs; Am Chem Soc; Nat Inst Bldg Sci; Soc Am Mil Engrs. *Res:* Minerals including diatomite and perlite; synthetic silicates; filtration; water treatment; building materials and construction systems. *Mailing Add:* VanderLinden & Assoc Consult Five Brassie Way Littleton CO 80123

VANDERLIP, RICHARD L, b Woodston, Kans, May 6, 38; m 60; c 3. AGRONOMY. *Educ:* Kans State Univ, BS, 60; Iowa State Univ, MS, 62, PhD(agron), 65. *Prof Exp:* From asst prof to assoc prof, 64-76, PROF AGRON, KANS STATE UNIV, 76- *Mem:* Am Soc Agron; Soil Sci Soc Am; Crop Sci Soc Am; Sigma Xi. *Res:* Ecology of crop plants, especially climatic interrelationships. *Mailing Add:* Dept Agron Throckmorton Hall Kans State Univ Manhattan KS 66506

VANDER LUGT, ANTHONY, b Dorr, Mich, Mar 7, 37; m 59; c 2. ELECTROOPTICS & ELECTRICAL ENGINEERING. *Educ:* Calvin Col, BS, 59; Univ Mich, BSEE, 59, MSEE, 62; Univ Reading, PhD, 69, DSc, 89. *Prof Exp:* Res asst, Radar & Optics Lab, Willow Run Labs, Inst Sci & Technol, Univ Mich, 59-63, res assoc, 63-64, assoc res engr, 64-65, res engr & asst head optics group, 65-69; mgr res & develop, Electro-Optics Ctr, Radiation Inc, 69-73; dir, Electro-Optics Dept, 73-79, sr scientist, Advan Technol Dept, Govt Commun Systs Div, Harris Corp, 79-86; PROF, ELEC & COMPUT ENG, NC STATE UNIV, RALEIGH, NC, 88- *Concurrent Pos:* Consult, Gen Elec Co, 64-66, Tex Instruments, Inc, 65-66, Bendix Res Lab, Harris Corp, 86- & Syracuse Res Corp, 87-; Am ed, Optica Acta, 69-75; assoc ed, Wave Electronics, 81-84; topical ed, Appl Optics, 90- *Mem:* Fel Optical Soc Am; Inst Elec & Electronics Engrs; fel Photo Instrumentation Engrs. *Res:* Optical data-processing; complex spatial filtering and optical matched filtering; modulation transfer functions of recording media; holography; acousto-optic devices; optical storage and retrieval; optical signal processing. *Mailing Add:* 108 O'Kelly Lane Cary NC 27511-5530

VANDER LUGT, KAREL L, b Pella, Iowa, Apr 25, 40; m 64; c 1. SOLID STATE PHYSICS. *Educ:* Hope Col, BA, 62; Wayne State Univ, PhD(exp solid state physics), 67. *Prof Exp:* Nat Res Coun assoc physics, Naval Res Lab, Washington, DC, 67-68; ASST PROF PHYSICS, AUGUSTANA COL, SDAK, 69- *Mem:* AAAS; Am Asn Physics Teachers; Hist Sci Soc. *Res:* Radiation damage in crystals; experimental solid state physics. *Mailing Add:* 1201 W 38 St Sioux Falls SD 57105

VAN DER MAATEN, MARTIN JUNIOR, b Alton, Iowa, Aug 6, 32; m 56; c 2. VETERINARY VIROLOGY. *Educ:* Iowa State Univ, DVM, 56, PhD(vet bact), 64. *Prof Exp:* Vet practice, 58-60; res asst vet virol, Iowa State Univ, 60-61, Nat Inst Allergy & Infectious Dis fel, 61-64, asst prof, 64-67; VET LAB OFFICER, NAT ANIMAL DIS CTR, US DEPT AGR, 67- *Mem:* Am Vet Med Asn; Am Soc Microbiol; Am Soc Virol. *Res:* Virological and serological studies of bovine lymphosarcoma and bovine leukemia virus; bovine viral reproductive disease; vesicular stomatitis virus; bovine lentivirus. *Mailing Add:* Nat Animal Dis Ctr US Dept Agr PO Box 70 Ames IA 50010

VANDERMEER, CANUTE, b Xiamen, Fujian, China, Apr 2, 30; US citizen; m 55; c 2. IRRIGATION WATER MANAGEMENT. *Educ:* Hope Col, BA, 50; Univ Mich, MA, 56 & PhD(geog), 62. *Prof Exp:* Instr geog, Univ Mich, 60-61; instr, 61-62, Univ Wis, Milwaukee, 61-62, from asst prof to assoc prof, 62-73, actg chmn dept, 70-71; chmn dept, 73-85, PROF GEOG, UNIV VT, 73- *Concurrent Pos:* Vis prof geog, Univ Hawaii, 72, Calif State Univ, Hayward, 74, Univ Tsukuba, Japan, 86-87. *Honors & Awards:* Fulbright-Hays Lectr, Dept Agr Econ, Univ Philippines, 73. *Mem:* Asn Am Geographers; Asn Asian Studies; Am Asn Univ Prof. *Res:* Irrigation water distribution procedures and problems where hundreds or thousands of farmers receive water from a single irrigation system; farmers perceptions of irrigation problems. *Mailing Add:* Dept Geog Univ Vt Burlington VT 05405-0114

VANDERMEER, JOHN H, b Chicago, Ill, July 21, 40; m 69; c 1. POPULATION BIOLOGY. *Educ:* Univ Ill, Urbana, BS, 61; Univ Kans, MA, 64; Univ Mich, Ann Arbor, PhD(ecol), 69. *Prof Exp:* Sloan Found fel, Univ Chicago, 69-70; asst prof ecol, State Univ NY, Stony Brook, 70-71; asst prof, 71-74, ASSOC PROF ZOOL, BIOL DEPT, UNIV MICH, ANN ARBOR, 74- *Mem:* Am Soc Ichthyologists & Herpetologists; Am Soc Naturalists; Ecol Soc Am. *Res:* Role of population processes as determiners of the structure of biological communities. *Mailing Add:* Dept Biol Univ Mich Ann Arbor MI 48109-1048

VAN DER MEER, JOHN PETER, b Netherlands, June 25, 43; Can citizen; m 66; c 2. PHYCOLOGY, MOLECULAR GENETICS. *Educ:* Univ Western Ont, BSc, 66; Cornell Univ, PhD(genetics), 71. *Prof Exp:* Fel biochem, Charles H Best Inst, Univ Toronto, 71-74; RES OFFICER GENETICS, NAT RES COUN CAN, 74-, ASST DIR, INST MARINE BIOSCI, 89- *Mem:* Genetics Soc Can; Phycol Soc Am; Int Phycol Soc (treas, 87-90); Can Soc Plant Molecular Biol. *Res:* Genetics of marine red algae. *Mailing Add:* Nat Res Coun Inst Marine Bio Sci 1411 Oxford St Halifax NS B3H 3Z1 Can

VANDERMEER, R(OY) A, b Chicago, Ill, Sept 7, 34; m 56; c 2. PHYSICAL METALLURGY. *Educ:* Ill Inst Technol, BS, 56, PhD(phys metall), 61. *Prof Exp:* Res assoc metall, Ill Inst Technol, 58-60; metallurgist, Oak Ridge Nat Lab, 60-80; metallurgist, Nuclear Div, Y-12 Plant, Union Carbide Corp, 80; prof metall eng, Univ Tenn, 81-; AT OAK RIDGE NAT LAB. *Concurrent Pos:* Lectr, Univ Tenn, 63-66 & Ford Found grant prof, 66-80; vis prof mat sci, Univ Rochester, 74-75, vis prof metall eng, Ill Inst Tech, 80. *Mem:* Am Inst Mining, Metall & Petrol Engrs; Am Soc Metals. *Res:* Grain boundary migration in metals; recovery and recrystallization of metals and alloys; uranium alloy; phase transformations. *Mailing Add:* 935 Wooter Dr Oak Ridge TN 37830

VANDER MEER, ROBERT KENNETH, b Chicago, Ill, Nov 29, 42; m 64, 86; c 3. CHEMICAL ECOLOGY, NATURAL PRODUCT CHEMISTRY. *Educ:* Blackburn Col, BA, 64; John Carroll Univ, MS, 66; Pa State Univ, PhD(org chem), 72. *Prof Exp:* Lectr chem, Univ South Pac, Fiji Islands, 72-76; RES CHEMIST, AGR RES SERV, USDA, 77- *Concurrent Pos:* Am Cancer Soc fel, Cornell Univ, 76-77; adj asst prof, Zool Dept & Entom-Nematol Dept, Univ Fla, 83-; ed, Fire Ants & Leaf-Cutting Ants & Biol & Mgt, 86, Appl Myrmecology A World Perspective, 90. *Honors & Awards:* Patent Award, USDA, 85; Res Award, USDA, 86. *Mem:* Am Chem Soc; Int Soc Chem Ecol; Entom Soc Am; Sigma Xi; Int Union Study Social Insects. *Res:* Isolation and identification of insect pheromones and defensive secretions; analytical organic chemistry; insect biochemistry; the interaction of plants and insects; insect and animal behavior and physiology. *Mailing Add:* ARS-USDA PO Box 14565 Gainesville FL 32604

VAN DER MEER, SIMON, b The Hague, Neth, Nov 24, 25. PHYSICS. *Prof Exp:* Delft, Philps Phys Lab, Eindhoven, 52-56; sr engr, Europ Orgn Nuclear Res, 56-90; RETIRED. *Honors & Awards:* Nobel Prize in Physics, 84. *Mem:* Am Acad Arts & Sci; Royal Neth Acad Sci. *Mailing Add:* Four Chemin de Corbillettes 1218 GD Saconnex Switzerland

VANDERMEULEN, JOHN HENRI, b Ryswyk, Neth, Oct 2, 33; Can citizen; m 62; c 4. OIL POLLUTION ECOTOXICOLOGY. *Educ:* Univ Alta, Can, BSc, 58, MSc, 67; Univ Calif, Los Angeles, PhD(biol), 72. *Prof Exp:* Res affil marine biol, Hawaii Inst Marine Biol, 69-71; teaching scholar, dept zool, Duke Univ, 72-73; RES SCIENTIST ENVIRON OCEANOG, MARINE ECOL LAB, BEDFORD INST OCEANOG, 73- *Concurrent Pos:* Mem, steering comt, Nat Acad Sci, 81-85; sr lectr, Int Ocean Inst, Malta, 81-; mem bd dirs, Int Ctr Ocean Develop, Halifax, 85-; hon res assoc, Dalhousie Univ, 85-; adj prof, Sch Resources & Environ Studies, Dalhousie Univ. *Res:* Physical and chemical fate of spilled petroleum hydrocarbons, and biological-ecological effects in marine and freshwater environments; pollution related physiological stress; ecotoxicology; biomineralization in marine invertebrate organisms; calcification in reef-corals, shell and ligament formation in bivalve molluscs; petroleum pollution in tropical ecosystems, mangroves and coral reefs; coastal zone management. *Mailing Add:* Marine Chem Fisheries & Oceans Bedford Inst Oceanog Dartmouth NS B2Y 4A2 Can

VAN DER MEULEN, JOSEPH PIERRE, b Boston, Mass, Aug 22, 29; m 60; c 3. NEUROLOGY, NEUROPHYSIOLOGY. *Educ:* Boston Col, AB, 50; Boston Univ, MD, 54. *Prof Exp:* Intern med, Cornell Med Div, Bellevue Hosp, New York, 54-55, asst resident, 55-56; asst resident neurol, Harvard Neurol Unit, Boston City Hosp, 58-59, resident, 59-60; Nat Inst Neurol Dis & Blindness fel, Nobel Inst Neurophysiol, Karolinska Inst, Sweden, 60-62; instr, Harvard Neurol Unit, Boston City Hosp, 62-66, assoc, 66-67; asst prof, Sch Med, Case Western Reserve Univ, 67-69, assoc prof neurol & biomed eng, 69-71; prof neurol & chmn dept, 71-79, dir dept neurol, 71-79, VPRES HEALTH AFFAIRS, UNIV SOUTHERN CALIF, 77- *Concurrent Pos:* Fel, Harvard Neurol Unit, Boston City Hosp, Mass, 62-66; ed, Arch of Neurol, 76-78. *Mem:* fel Am Acad Neurol; Am Neurol Asn. *Res:* Neurophysiology of abnormalities of posture and movement; computer assisted image analysis muscle biopsies; motor control systems in humans. *Mailing Add:* USC Health Sci Campus 1985 Zonal Ave Ste 100 Los Angeles CA 90033

VANDERPLAATS, GARRET NIEL, b Modesto, Calif, Feb 14, 44; m 79; c 2. AUTOMATED DESIGN OPTIMIZATION. *Educ:* Ariz State Univ, BS, 67, MS, 68; Case Western Reserve Univ, PhD(eng mech), 71. *Prof Exp:* Res scientist aero eng, Ames Res Ctr, NASA, 71-79; assoc prof mech eng, Naval Postgrad Sch, 79-84; prof mech eng, Univ Calif, Santa Barbara, 84-89; PRES, VMA ENG, GOLETA, CA, 89- *Concurrent Pos:* Pres, Eng Design Optimization, Inc, Santa Barbara, 84-89. *Honors & Awards:* Wright Brothers Medal, Soc Automotive Engrs, 77. *Mem:* Am Soc Mech Engrs; assoc fel Am Inst Aeronaut & Astronaut; Am Soc Civil Engrs; Opers Res Soc Am. *Res:* Development of automated optimization techniques for engineering design, with application to structural, aeronautical and mechanical systems. *Mailing Add:* 1275 Camino Rio Verde Santa Barbara CA 93111-1014

VANDERPLOEG, HENRY ALFRED, b Chicago, Ill, Aug 23, 44; m 68. AQUATIC ECOLOGY. *Educ:* Mich Technol Univ, BS, 66; Univ Wis-Madison, MS, 68; Ore State Univ, PhD(biol oceanog), 72. *Prof Exp:* Aquatic ecologist Environ Sci Div, Oak Ridge Nat Lab, 72-74; AQUATIC ECOLOGIST, GREAT LAKES ENVIRON RES LAB, NAT OCEANIC & ATMOSPHERIC ADMIN, 74- *Mem:* AAAS; Am Soc Limnol & Oceanog; Sigma Xi; Int Soc Limnol; Oceanog Soc; World Asn Copepodologists. *Res:* Ecology of selective feeding of zooplankton; dynamics of seasonal succession of Great Lakes plankton; plankton life cycle strategies. *Mailing Add:* 2808 Brockman Ann Arbor MI 48104

VANDERRYN, JACK, b Groningen, Neth, Apr 14, 30; nat US; m 56; c 4. PHYSICAL CHEMISTRY, RESEARCH ADMINISTRATION. *Educ:* Lehigh Univ, BA, 51, MS, 52, PhD(chem), 55. *Prof Exp:* Asst, Lehigh Univ, 51-55; asst prof chem, Va Polytech Inst, 55-58; chemist, Res & Develop Div, US AEC, Tenn, 58-62, tech adv to asst gen mgr res & develop, Washington, DC, 62-67; US Dept State sr sci adv to US Mission, Int Atomic Energy Agency, 67-71; tech asst, Off Gen Mgr, US AEC, Washington, DC, 71-72, tech asst to dir, Div Appl Technol, 72-74, chief, Energy Technol Br, Div Appl Technol, 74-75, actg dir div energy storage, 75, dir, Off Int Res & Develop Progs, US Energy Res & Develop Admin, 75-77, dir, Off Int Res & Develop Progs, US Dept Energy, 77-82; AGENCY DIR, ENERGY & NATURAL RESOURCES, USAID, 82- *Mem:* AAAS. *Res:* Science policy development; energy technology development; research and development administration; international cooperation in energy research and development; science policy. *Mailing Add:* 8112 Whittier Blvd Bethesda MD 20187

VANDERSALL, JOHN HENRY, b Helena, Ohio, July 20, 28; m 63; c 2. ANIMAL NUTRITION. *Educ:* Ohio State Univ, BS, 50, MS, 54, PhD(dairy sci), 59. *Prof Exp:* Instr dairy sci, Agr Exp Sta, Ohio State Univ, 57-59; from asst prof to assoc prof, 59-71, PROF DAIRY SCI, UNIV MD, COLLEGE PARK, 71- *Mem:* Fel AAAS; Am Soc Animal Sci; Am Dairy Sci Asn; Sigma Xi. *Res:* Effects of forages on milk production and growth and physiological bases for differences; effects of feeds upon the composition of milk. *Mailing Add:* 10906 Ashfield Rd Adelphi MD 20783

VANDER SANDE, JOHN BRUCE, b Baltimore, Md, Mar 27, 44; m 72; c 2. MATERIALS SCIENCE. *Educ:* Stevens Inst Technol, BE, 66; Northwestern Univ, Evanston, PhD(mat sci), 70. *Prof Exp:* Fulbright scholar metall, Oxford Univ, 70-71; from asst prof to assoc prof, 71-81, PROF MAT SCI, MASS INST TECHNOL, 81- *Mem:* Am Inst Mining, Metall & Petrol Engrs; Electron Microscopy Soc Am. *Res:* Physical and mechanical behavior of crystalline solids; electron microscopy and electron diffraction. *Mailing Add:* Dept Mat Sci & Eng Mass Inst Technol Rm 5025 77 Massachusetts Ave Cambridge MA 02139

VANDERSLICE, JOSEPH THOMAS, b Philadelphia, Pa, Dec 21, 27; m 54; c 8. PHYSICAL CHEMISTRY. *Educ:* Boston Col, BS, 49; Mass Inst Technol, PhD(phys chem), 53. *Prof Exp:* From instr to asst prof chem, Cath Univ, 52-56; from asst prof to prof chem, Univ Md, College Park, 56-78, dir, Inst Molecular Physics, 67-68, prof molecular physics, 62-76, head, Dept Chem, 68-76; PROF CHEM, NUTRIT INST, USDA, 78-, RES CHEMIST, 78- *Concurrent Pos:* Consult, US Naval Res Lab, 63-68. *Mem:* Inst Food Technol; Am Chem Soc; fel Am Phys Soc. *Res:* Intermolecular forces; thermodynamic temperature scale; interpretations of molecular beam experiments; transport properties of high temperature gases; franck-condon factors and interpretation of spectroscopic data on diatomic molecules; auroral spectroscopy; interpretation of rocket experiments on the ionosphere; vitamin B6; vitamin C, thiamine and folic acid composition in biological materials; theoretical foundations of flow injection analysis. *Mailing Add:* Nutrit Inst Beltsville Agr Res Ctr Beltsville MD 20705

VANDERSLICE, THOMAS AQUINAS, b Philadelphia, Pa, Jan 8, 32; m 56; c 4. PHYSICAL CHEMISTRY. *Educ:* Boston Col, BS, 53; Cath Univ, PhD(phys chem), 56. *Prof Exp:* Asst, Cath Univ, 53-65, Fulbright fel, 56; res assoc, Res & Develop Ctr, Gen Elec Co, 56-62, mgr eng, Vacuum Prod Oper, 62-64, mgr, Vacuum Prod Bus Sect, 64-66, gen mgr, Info Devices Dept, Okla, 66-68, dep div gen mgr, Info Systs Progs Dep Div, Ariz, 68-70, vpres & div gen mgr, Electronic Components Bus Div, 70-72, vpres & group exec, Spec Systs & Prod Group, 72-77, vpres & sect exec, Power Systs Sect, 78-79; pres & chief oper officer, GTE Corp, 79-83; chief exec officer, Apollo Computer Inc, 84-89, chmn, 86-89; CHMN & CHIEF EXEC OFFICER, M/A-COM INC, 89- *Concurrent Pos:* Trustee, Comt Econ Develop; mem, Aspen Inst Humanistic Studies; chmn bd trustees, Boston Col. *Mem:* Nat Acad Sci; Nat Acad Eng; Am Chem Soc; Am Phys Soc; Am Vacuum Soc; Am Inst Physics; Sigma Xi. *Res:* Surface chemistry; mass spectrometry; high vacuum technology; gaseous discharges. *Mailing Add:* 401 Edgewater Pl Suite 560 Wakefield MA 01880

VANDER SLUIS, KENNETH LEROY, b Holland, Mich, Dec 19, 25; m 52; c 3. PHYSICS. *Educ:* Baldwin-Wallace Col, BS, 47; Pa State Univ, MS, 50, PhD(physics), 52. *Prof Exp:* Physicist, Oak Ridge Nat Lab, 52-90; RETIRED. *Concurrent Pos:* Res guest, Spectros Lab, Mass Inst Technol, 60-61; mem comt line spectra of the elements, Nat Res Coun, 61-72. *Mem:* Am Phys Soc; Optical Soc Am; Sigma Xi. *Res:* Experimental atomic spectroscopy, echelle gratings, interferometry, gas laser systems. *Mailing Add:* 954 W Outer Dr Oak Ridge TN 37830

VAN DER SLUYS, WILLIAM, mechanical engineering, for more information see previous edition

VAN DER SPIEGEL, JAN, b Aalst, Belg, Apr 12, 51. NEURAL NETWORK IMPLEMENTATIONS, INTEGRATED SENSOR TECHNOLOGY. *Educ:* Kath Univ Leuven, Belg, BEE, 73, Masters, 74, PhD(elec eng), 79. *Hon Degrees:* MA, Univ Pa, 88. *Prof Exp:* Postdoctoral, 80-81, asst prof, 81-87, ASSOC PROF ELECTRONICS, DEPT ELEC ENG, UNIV PA, 87- *Concurrent Pos:* Consult, I-Stat, Inc, Princeton, NJ, 83- & Corticon, Inc, Philadelphia, 90-; assoc prof, Dept Mat Sci & Eng, Univ Pa, 87-, dir, Ctr Sensor Technologies, 89-; vis prof, Scuola Superioe S Anna, Pisa, Italy, 89; mem adv bd, Ben Franklin Technol Ctr Mfg Processes & Sensor, 90- *Honors & Awards:* NSF Presidential Young Investr, 84. *Mem:* Sr mem Inst Elec & Electronics Engrs; Mat Res Soc; Int Neural Network Soc. *Res:* Integrated and smart sensors; hardware implementation of neural networks for sensory processing; sensor technologies and electronic materials for integrated circuits. *Mailing Add:* Dept Elec Eng Univ Pa 200 S 33rd St Philadelphia PA 19104-6390

VANDERSPURT, THOMAS HENRY, b Lawrence, Mass, Apr 1, 46; m 71; c 2. PHYSICAL INORGANIC CHEMISTRY. *Educ:* Lowell Technol Inst, BS, 67; Princeton Univ, MA & PhD(chem), 72. *Prof Exp:* Fel catalysis chem, Princeton Univ, 72-73; res chemist, 73-77, SR RES CHEMIST CATALYSIS CHEM, CELANESE RES CO, CELANESE INC, 77- *Mem:* NY Acad Sci; Sigma Xi; Am Chem Soc; NAm Catalysis Soc. *Res:* Supported metal alloy catalysts; metal/metal oxide selective oxidation catalysts; homogenous selective dimerization carbonylation catalysis; homogeneous selective aromatic acetoxylation; catalysis; supported hydroformylation catalysis. *Mailing Add:* 46 Upper Creek Rd Stockton NJ 08559

VAN DER VAART, HUBERTUS ROBERT, b Makassar, Celebes, Indonesia, Mar 2, 22; m 50; c 3. MATHEMATICS, STATISTICS. *Educ:* Univ Leiden, Drs, 50, PhD(theoret biol), 53. *Prof Exp:* Sci officer, Univ Leiden, 50-57; vis assoc prof exp statist, NC State Col, 57-58 & statist, Univ Chicago, 58; extraordinary prof theoret biol, Univ Leiden, 58-60, prof, 60-62, dir Inst Theoret Biol, 58-62, dir Cent Comput Inst, 61-62; assoc prof, 62-63, PROF STATIST & MATH, NC STATE UNIV, 63-, DREXEL PROF BIOMATH, 74- *Concurrent Pos:* Co-ed, Acta Biotheoretica, 53-75; Neth Orgn Pure Res fel, 57-58; co-ed, Statistica Neerlandica, 60-74; mem, Panel Life Sci, Comt Undergrad Prog Math, 67-70; mem, Panel Instructional Mats Appl Math, Comt Undergrad Prog Math, Math Asn Am, 74-76. *Mem:* Biomet Soc; Am Math Soc; Math Asn Am; Inst Math Statist; Soc Indust & Appl Math. *Res:* Mathematical statistics; probability; stochastic processes; theoretical biology; principles of scientific method; mathematical models for biosystems. *Mailing Add:* Dept Statist NC State Univ Campus Box 8203 Raleigh NC 27695

VAN DER VEEN, JAMES MORRIS, b Chicago, Ill, Sept 19, 31; m 59; c 3. ORGANIC CHEMISTRY, X-RAY CRYSTALLOGRAPHY. *Educ:* Swarthmore Col, BA, 53; Harvard Univ, AM, 56, PhD(org chem), 59. *Hon Degrees:* MEng, Stevens Inst Technol, 80. *Prof Exp:* Fels, Ga Inst Technol, 58-59 & Argonne Nat Lab, 59-60; res assoc nuclear magnetic resonance spectros, Retina Found, 60-61; from asst prof to assoc prof, 61-76, PROF ORG CHEM, STEVENS INST TECHNOL, 76- *Concurrent Pos:* Consult, Picatinny Arsenal, NJ, 63-68 & Exxon Corp, 74-78. *Mem:* AAAS; Am Crystallog Asn; Am Chem Soc; Sigma Xi; The Chem Soc. *Res:* Physical organic chemistry; x-ray crystallography; computer methods in chemistry. *Mailing Add:* Dept Chem Stevens Inst Technol Hoboken NJ 07030

VANDERVEEN, JOHN EDWARD, b Prospect Park, NJ, May 13, 34; m 67; c 2. NUTRITION, CHEMISTRY. *Educ:* Rutgers Univ, BS, 56; Univ NH, PhD(chem, nutrit), 61. *Prof Exp:* Res chemist, USAF Sch Aerospace Med, 64-75; DIR DIV NUTRIT, CTR FOOD SAFETY & APPL NUTRIT, FOOD & DRUG ADMIN, DEPT HEALTH & HUMAN SERV, 75- *Mem:* Am Inst Nutrit; Am Chem Soc; Am Soc Clin Nutrit; Inst Food Technologists; Am Dairy Sci Asn; Aerospace Med Asn. *Res:* Nutritional requirements of the American population; assessment of the nutritional quality of the national food supply; energy and mineral requirements and effects of excess nutrient intakes. *Mailing Add:* HGG 260 FDA 200 C St SW Washington DC 20204

VANDERVEEN, JOHN WARREN, b Mount Vernon, NY, Dec 2, 33; m 59; c 2. CHEMICAL ENGINEERING. *Educ:* Univ Nebr, BS, 60, MS, 61; Univ Minn, PhD(chem eng), 65. *Prof Exp:* SUPVR CHEM PROCESSES SECT, PHILLIPS RES CTR, 65- *Mem:* Sigma Xi. *Res:* Reactor analysis; combustion; process simulation and optimization; chemical kinetics; fluid mechanics; process control; fermentation; economics; design. *Mailing Add:* 1410 SE Macklyn Lane Bartlesville OK 74003

VANDER VELDE, GEORGE, b Chicago, Ill, June 24, 43; m 66; c 3. ANALYTICAL CHEMISTRY, MASS SPECTROMETRY. *Educ:* Hope Col, BA, 65; Univ Houston, PhD(biophysics), 71. *Prof Exp:* Res scientist assoc, natural prod, Dept Bot, Univ Tex, Austin, 70-73, instr biol, Exten Serv, 72-73; res chemist anal chem, Nat Ctr Toxicol Res, Jefferson, 73-74; appl support chemist & prod mgr mass spectrometry, Finnigan Corp, Sunnyvale, Calif, 74-77; assoc dir, Finnigan Inst, 77-79; dir tech serv, O H Mat Co, Findlay, Ohio, 79-81; vpres develop, Environ Testing & Cert Corp, Edison, NJ, 81-82; VPRES SCI & TECHNOL, CHEM WASTE MGT CORP, OAK BROOK, IL, 82- *Mem:* Am Chem Soc; Am Soc Mass Spectrometry; Am Soc Testing & Mat; Sigma Xi; NY Acad Sci. *Res:* Applications of analytical chemistry to environmental protection; development of new technology to environmental protection and cleanup; waste process technology; analytical chemistry methodology, development; hazardous materials characterization and analysis. *Mailing Add:* 5715 Lawn Dr Western Springs IL 60558-2226

VANDER VELDE, JOHN CHRISTIAN, b Mich, Sept 25, 30; m 53; c 3. PHYSICS. *Educ:* Hope Col, AB, 52; Univ Mich, MA, 53, PhD(physics), 58. *Prof Exp:* From instr to assoc prof physics, 58-67, PROF PHYSICS, UNIV MICH, ANN ARBOR, 67- *Concurrent Pos:* Assoc in res, LePrince-Ringuet Lab, Polytech Sch, Paris, 66-67; mem prog comt, Argonne Nat Lab, 71-74; mem users exec comt, Nat Accelerator Lab, 72-74; vis scientist, Saclay Nuclear Res Ctr, France, 74. *Mem:* Fel Am Phys Soc. *Res:* Elementary particle physics. *Mailing Add:* Randall Lab Univ Mich Ann Arbor MI 48109-1120

VANDER VELDE, W(ALLACE) E(ARL), b Jamestown, Mich, June 4, 29; m 54; c 2. AERONAUTICAL ENGINEERING, NAVIGATION SYSTEMS. *Educ:* Purdue Univ, BSAE, 51; Mass Inst Technol, ScD(instrumentation), 56. *Prof Exp:* Staff engr, Instrumentation Lab, Mass Inst Technol, 53-56; dir applns eng, GPS Instrument Co, Inc, 56-57; asst prof aeronaut eng, 57-61, assoc prof aeronaut & astronaut, 61-65, PROF AERONAUT & ASTRONAUT, MASS INST TECHNOL, 65- *Honors & Awards:* Educ Award, Am Automatic Control Coun, 88. *Mem:* Fel Am Inst Aeronaut & Astronaut; Inst Elec & Electronics Engrs. *Res:* Automatic control systems; navigation and guidance of aerospace vehicles; instrumentation systems. *Mailing Add:* Dept Aeronaut & Astronaut Rm 33-109 Mass Inst Technol Cambridge MA 02139

VANDERVEN, NED STUART, b Ann Arbor, Mich, July 15, 32; m 61; c 2. PHYSICS. *Educ:* Harvard Col, AB, 55; Princeton Univ, PhD(physics), 62. *Prof Exp:* Instr physics, Princeton Univ, 59-61; from instr to assoc prof, 61-79, PROF PHYSICS, CARNEGIE-MELLON UNIV, 79- *Mem:* Am Phys Soc. *Res:* Magnetic resonance; solid state physics; biophysics. *Mailing Add:* Dept Physics Carnegie-Mellon Univ Schenly Park 5000 Forbes Ave Pittsburgh PA 15213

VAN DER VOO, ROB, b Zeist, Netherlands, Aug 4, 40; m 66; c 2. GEOLOGY, GEOPHYSICS. *Educ:* State Univ Utrecht, BSc, 61, Drs, 65 & 69, PhD(geol, geophys), 69. *Prof Exp:* Res assoc paleomagnetism, State Univ Utrecht, 65-70; vis asst prof, 70-72, from asst prof to assoc prof, 72-79, chmn dept, 81-88, PROF GEOPHYS, UNIV MICH, ANN ARBOR, 79-, CHMN DEPT GEOL SCI, 91- *Mem:* Am Geophys Union; Geol Soc Am; Ger Geol Asn; Royal Netherlands Geol & Mining Soc; Royal Acad Sci Neth. *Res:* Paleomagnetism, plate tectonics of the Atlantic Ocean and Mediterranean Sea; stratigraphy and tectonics of Appalachian, Hercynian, Laramide and Pyrenean-Alpine mountain belts. *Mailing Add:* Dept Geol Sci Univ Mich Ann Arbor MI 48109

VAN DER VOORN, PETER C, b Haarlem, Netherlands, Feb 25, 40; US citizen; m 65. ELECTROPHOTOGRAPHY. *Educ:* Wichita Univ, BS, 61; Univ Ill, PhD(inorg chem), 65. *Prof Exp:* Tech assoc, 65-85, MGR QUAL ASSURANCE COPY PRODS, EASTMAN KODAK CO, 85- *Mem:* Am Chem Soc. *Res:* Electrophotographic developers; inorganic photoconductors; solid state chemistry. *Mailing Add:* 26 Tartarian Circle Rochester NY 14612

VANDERVOORT, PETER OLIVER, b Detroit, Mich, Apr 25, 35; m 56; c 2. ASTRONOMY. *Educ:* Univ Chicago, AB, 54, SB, 55, SM, 56, PhD(physics), 60. *Prof Exp:* Vis res assoc, Nat Radio Astron Observ, Assoc Univs, Inc, WVa, 60; NSF fel, Princeton Univ Observ, 60-61; asst prof, 61-65, assoc prof astron, 65-80, PROF ASTRON & ASTROPHYS, UNIV CHICAGO, 80- *Concurrent Pos:* NSF sr fel, Leiden Observ, Neth, 67-68. *Mem:* Int Astron Union; Am Astron Soc; Am Phys Soc; Royal Astron Soc; Sigma Xi. *Res:* Hydrodynamic stability; gas dynamics; interstellar matter; stellar dynamics; galactic structure. *Mailing Add:* 5471 S Ellis Ave Chicago IL 60615

VANDER VORST, ANDRE, b Schaerbeek, Belg, Oct 22, 35; m 59; c 5. MICROWAVE COMMUNICATIONS, BIOMEDICAL MICROWAVES. *Educ:* Cath Univ Louvain, Belg, Engr, 58, PhD(elec eng), 65; Mass Inst Technol, MS, 65. *Prof Exp:* Asst prof, 62-68, PROF, CATH UNIV LOUVAIN, 68- *Concurrent Pos:* Postdoctoral fel, Stanford Univ, 65-66; head, Elec Eng Dept, Cath Univ Louvain, 70-72, dean, 72-75, vpres, 73-75, pres, Open Fac, 73-87. *Mem:* Fel Inst Elec & Electronics Engrs; Electromagnetics Acad; Inst Elec Engrs. *Res:* Microwaves; electrical engineering; design of circuits up to 700 gigahertz; atmospheric transmission up to 300 gigahertz; interaction of electromagnetic fields with the nervous system, using microwave acupuncture as a stimulus; author of several books and more than 100 publications. *Mailing Add:* Microwaves Univ Cath Louvain Batiment-Maxwell Louvain-la-Neuve B-1348 Belgium

VANDER WALL, EUGENE, b Munster, Ind, Feb 8, 31; m 53; c 4. PHYSICAL INORGANIC CHEMISTRY. *Educ:* Calvin Col, BS, 52; Univ Colo, PhD(chem), 57. *Prof Exp:* Res chemist, Atomic Energy Div, Phillips Petrol Corp, Idaho, 56-60, supvr chemist, 60-63; mem sci staff, Aerojet-Gen Corp, Sacramento, 63-64, supvr liquid propellant res, 64-67, tech supvr res & develop, 67-71, supvr, Chem Res Lab, 71-74, mgr chem processes, 74-82, mgr, Eng & Mfg Labs, 83-89, SR MGR ENG LABS, AEROJET PROPULSION DIV, AEROJET LIQUID ROCKET CO, 90- *Concurrent Pos:* Nat Reactor Testing Sta prof, Univ Idaho, 57-58 & 60-61; mem bd trustees, Bethesda Hosp, Denver, Colo. *Mem:* Fel Am Inst Chem; Am Chem Soc. *Res:* Gelation of liquids; characterization of hazardous chemicals; material-fluid compatibility evaluation; liquid propellant research; development of chemical processes for disposal of hazardous wastes. *Mailing Add:* 5552 Wildwood Way Citrus Heights CA 95610

VANDER WENDE, CHRISTINA, b Paterson, NJ, June 12, 30. BIOCHEMISTRY, PHARMACOLOGY. *Educ:* Upsala Col, BS, 52; Rutgers Univ, MS, 56, PhD(biochem), 59. *Prof Exp:* Jr pharmacologist, Wallace & Tiernan, Inc, 52-53 & Schering Corp, 53-55; pharmacologist, Maltbie Labs, 55-56; asst physiol & biochem, Rutgers Univ, 56-59; sr res scientist, E R Squibb & Sons, 59-60; sr res scientist, Vet Admin Hosp, 60-64; assoc prof pharmacol & biochem, 64-69, PROF PHARMACOL, RUTGERS UNIV, NEW BRUNSWICK, 69- *Concurrent Pos:* Nat Cancer Inst res grant, 61-67; Eastern Leukemia Asn Inc res scholar, 67-69; Epilepsy Found grants, 69-70 & 72-73; Pharmaceut Mfrs Asn grant, 70-72; Nat Drug Abuse Inst grant, 73-76; lectr, Exten Serv, Rutgers Univ, 61-63; dir undergrad res, Upsala Col, 61-64; lectr, All Souls Hosp, Morristown, NJ, 62. *Mem:* AAAS; Am Soc Pharmacol & Exp Theraput; Soc Neurosci; Acad Pharmaceut Sci; NY Acad Sci; Sigma Xi. *Res:* Biochemical pharmacology of central nervous system metabolism and enzymology; toxicology. *Mailing Add:* 252 Carol Jean Way Somerville NJ 08876

VAN DER WERFF, TERRY JAY, b Hammond, Ind, May 16, 44; m 68; c 5. EDUCATIONAL ADMINISTRATION. *Educ:* Mass Inst Technol, SB & SM, 68; Oxford Univ, DPhil(eng sci), 72. *Prof Exp:* Staff engr, ARO Inc, 67-68; asst prof mech eng, physiol & biophys & clin sci, Colo State Univ, 70-73; vis asst prof med, Univ Colo Med Ctr, 73-74; head, Dept Biomed Eng, Univ Cape Town & Groote Schuur Hosp, SAfrica, 74-80; dean sci & eng, Seattle Univ, 81-90; EXEC VPRES ACAD AFFAIRS, ST JOSEPH'S UNIV, 90- *Concurrent Pos:* Consult, Rand Corp, 67 & 69-70 & Los Alamos Sci Lab, 72. *Honors & Awards:* Teetor Award, Soc Automotive Engrs, 72. *Mem:* AAAS; Am Phys Soc; Am Soc Mech Engrs; fel Biomed Eng Soc SAfrica; Am Soc Eng Educ; fel Royal Soc SAfrica; Soc Math Biol. *Res:* Nonlinear oscillations; reaction-diffusion systems; cardiovascular fluid dynamics; similarity transformations; science and technology policy. *Mailing Add:* St Joseph's Univ Philadelphia PA 19131-1395

VANDERWERFF, WILLIAM D, b Philadelphia, Pa, Dec 31, 29; m 58; c 4. ORGANIC CHEMISTRY. *Educ:* Univ Pa, BS, 51, PhD(org chem), 60. *Prof Exp:* Prod supvr explosives, E I du Pont de Nemours & Co, 51-53; res chemist, Suntech, Inc, 59-70, chief new prods res, 70-77, staff scientist, 77-82, strategic planning specialist, 82, res sect chief, 82-83, RES MGR, SUN REFINING & MKT CO, 83- *Mem:* Am Chem Soc. *Res:* Auto-oxidation; hydrocarbon chemistry; petrochemicals; polymer technology. *Mailing Add:* 37 Green Tree Dr Westchester PA 19382-8408

VANDERWIEL, CAROLE JEAN, b Cleveland, Ohio, May 6, 50; m 76; c 2. MICROSCOPIC ANATOMY, PHYSIOLOGY. *Educ:* Univ Tex, BS, 72; Baylor Univ Med Ctr, PhD(anat), 76. *Prof Exp:* Technician biochem, Univ Tex Health Sci Ctr, 68-70; technician serol, Harris Hosp, Ft Worth, 70-72; grad asst anat, Baylor Col Dent, Dallas, 73-76; res assoc endocrinol, Dent Res Ctr, Sch Med, Univ NC, 77-78, fel orthop, 78-80, asst prof, Dept Surg, 80-82; dir, Metab Bone Clin, Dallas, Tex, 82-86; EDUC CONSULT, MIDAS REX INST, FT WORTH, TEX, 91- *Concurrent Pos:* Grant, Baylor Col Dent, 74-76 & NSF, 80-83; nat res serv award, NIH, 78-80; Pfizer pharmaceut award, 81, young investr award, 82; res scientist award, Parathyroid Conf, 83. *Mem:* AAAS; Am Soc Bone & Mineral Res; Am Acad Orthop Res. *Res:* Role of hormones, especially parathyroid hormone and calcitonin, on calcium fluxes between bone fluid and blood utilizing histological and physiological techniques; diagnosis of metabolic bone disease by quantitative histomorphometric analysis of human bone biopsies. *Mailing Add:* 141 Vista Dr Willow Park TX 76087

VANDERWIELE, JAMES MILTON, b July 23, 58; US citizen. DIGITAL TESTING, STATISTICAL PROCESS CONTROL. *Educ:* Okla State Univ, BS, 80; Univ Okla, MS, 87. *Prof Exp:* Develop engr, 80-88, SR ENGR, AT&T TECHNOLOGIES, INC, 88- *Mem:* Inst Elec & Electronics Engrs Computer Soc. *Res:* Data communications, specifically carrie sense multiple access protocols; digital circuitry testability. *Mailing Add:* 1712 Whispering Creek Ct Edmond OK 73013

VANDERWIELEN, ADRIANUS JOHANNES, b Germany, Mar 28, 44; US citizen; m 68; c 2. ANALYTICAL CHEMISTRY, PHYSICAL CHEMISTRY. *Educ:* San Diego State Univ, BSc, 70; Univ Calif, PhD(chem), 74. *Prof Exp:* Res fel phys chem, Univ Alta, 74-75; res scientist anal chem, 75-80, mgr, Anal Support, Control Div, 80-86; GROUP MGR, PHARM LAB, UPJOHN CO, 86- *Mem:* Am Chem Soc; Am Pharmaceut Asn. *Res:* Analytical research and development in particle sizing methods; physical characterization of pharmaceutical powders; thermal analyses; chromatography and methods validation. *Mailing Add:* 7700 Portage Rd Upjohn Co Kalamazoo MI 490091

VANDERWOLF, CORNELIUS HENDRIK, b Edmonton, Alta, Dec 13, 35; m 62. BEHAVIOR-ETHOLOGY. *Educ:* Univ Alta, BSc, 58; McGill Univ, MSc, 59, PhD, 62. *Prof Exp:* Res fel biol, Calif Inst Technol, 62-63; Nat Res Coun fel, Brain Res Inst, Switz, 63-64; from asst prof to assoc prof psychol, McMaster Univ, 64-68; assoc prof, 68-73, PROF PSYCHOL, UNIV WESTERN ONT, 73- *Concurrent Pos:* Nat Res Coun res grants, 64- *Mem:* Sigma Xi; Can Col Neuropsychopharmacol; Soc Neurosci. *Res:* Role of forebrain structures in patterning and control of motor activity; role of cholinergic and serotonergic brain systems. *Mailing Add:* Dept Psychol Univ Western Ont London ON N6A 5C2 Can

VANDERZANT, CARL, b Nymegen, Neth, Sept 7, 25; nat US; m 52; c 2. FOOD MICROBIOLOGY. *Educ:* State Agr Univ, Wageningen, BS, 47, MS, 49; Iowa State Univ, MS, 50, PhD(dairy bact), 53. *Prof Exp:* From asst prof to assoc prof dairy sci, 53-62, PROF FOOD MICROBIOL, TEX A&M UNIV, 62- *Mem:* Am Soc Microbiol; Inst Food Technologists. *Res:* Bacteriological problems of foods. *Mailing Add:* Dept Animal Sci Tex A&M Univ College Station TX 77843

VANDERZANT, ERMA SCHUMACHER, b Elwood, Ill, Jan 30, 20; m 52; c 2. BIOCHEMISTRY. *Educ:* Iowa State Univ, BS, 42, PhD(biochem), 53. *Prof Exp:* BIOCHEMIST, COTTON INSECTS BR, ENTOM RES DIV, SCI & EDUC ADMIN-AGR RES, USDA, TEX A&M UNIV, 54- *Honors & Awards:* J Everett Bussart Mem Award, Entom Soc Am, 71. *Mem:* Am Chem Soc; Entom Soc Am; Am Inst Nutrit; Sigma Xi. *Res:* Chemically defined diets for insects; nutrition and metabolism of growth factors, lipides and amino acids. *Mailing Add:* 1303 Broadmoor Bryan TX 77802

VANDERZEE, CECIL EDWARD, b Wetonka, SDak, Apr 26, 12; m 44. PHYSICAL CHEMISTRY. *Educ:* Jamestown Col, BS, 38; Univ Iowa, PhD(chem), 49. *Prof Exp:* Instr, Jamestown Col, 39 & high sch, SDak, 39-42; from instr to assoc prof chem, Univ Nebr, Lincoln, 49-58, vchmn dept, 65-70, prof chem, 58-; RETIRED. *Concurrent Pos:* Mem bd dirs, Calorimetry Conf, 64-66, chmn elec, 67-68, chmn, 68-69, counsellor, 73-76; assoc mem, Comn Thermodyn, Int Union Pure & Appl Chem, 76-83. *Honors & Awards:* Huffman Mem Award, Calorimetry Conf, 75. *Mem:* Am Chem Soc. *Res:* Thermodynamics; calorimetry. *Mailing Add:* Dept Chem Univ Nebr Lincoln NE 65888

VAN DER ZIEL, ALDERT, physics, fluctuation phenomenology; deceased, see previous edition for last biography

VAN DER ZIEL, JAN PETER, b Eindhoven, Neth, Aug 17, 37; US citizen; m 65; c 2. SOLID STATE PHYSICS. *Educ:* Univ Minn, BS, 59; Harvard Univ, MS, 61, PhD(appl physics), 64. *Prof Exp:* Res fel appl physics, Harvard Univ, 64-65; MEM TECH STAFF, BELL LABS, 65- *Mem:* Sr mem Inst Elec & Electronics Engrs; Am Phys Soc; Soc Photo-Optical Instrumentation Engrs. *Res:* Lasers and nonlinear optics; optical spectroscopy; integrated optics; semiconductor lasers. *Mailing Add:* 7A-221 AT&T Bell labs Murray Hill NJ 07974-2070

VANDER ZWAAG, ROGER, b Holland, Mich, Dec 27, 38; m 64; c 3. BIOSTATISTICS. *Educ:* Hope Col, AB, 60; Purdue Univ, MS, 62; Johns Hopkins Univ, PhD(biostat), 68. *Prof Exp:* Asst prof biostat, Sch Med, Vanderbilt Univ, 68-77; ASSOC PROF BIOSTAT, UNIV TENN, 77- *Mem:* Am Statist Asn. *Res:* Epidemiology. *Mailing Add:* 6157 Quince Rd Memphis TN 38119

VAN DER ZWET, TOM, b Borneo, Indonesia, Apr 7, 32; nat US; m 55, 80; c 3. PLANT PATHOLOGY. *Educ:* Col Trop Agr, Deventer, Netherlands, BS, 52; La State Univ, BS, 55, MS, 57, PhD(plant path), 59. *Prof Exp:* Plant pathologist, Tung Res Lab, Bogalusa, La, 59-65, Fruit Lab, Plant Genetics & Germplasm Inst, Agr Res Ctr, Beltsville, Md, 65-79, PLANT PATHOLOGIST, APPALACIAN FRUIT RES STA, KEARNEYSVILLE, WVA, USDA, 79- *Concurrent Pos:* Adj prof, WVa Univ, Morgantown, 79. *Honors & Awards:* Stark Award, Am Soc Hort Sci, 78. *Mem:* Am Phytopath Soc; Int Soc Hort Sci; Int Soc Plant Path. *Res:* Diseases of tropical and subtropical crops; soil and leaf spot fungi; fire blight of pome fruit. *Mailing Add:* US Dept Agr Appalachian Fruit Res Sta Kearneysville WV 25430

VAN DE SANDE, JOHAN HUBERT, b Bersen Op Zoom, Neth, Oct 28, 41; Can citizen; m 65; c 2. NUCLEIC ACIDS CONFORMATION. *Educ:* Univ Leiden, Neth, Candidaats, 63; Univ Alta, Can, PhD(org chem), 68. *Prof Exp:* Fel nucleic acids, Enzyme Inst, Madison, Wis, 68-70; res assoc biol & chem, Mass Inst Technol, 70-72; from asst prof to assoc prof, 72-79, PROF BIOCHEM, UNIV CALGARY, 79-, HEAD, DEPT MED BIOCHEM, 88- *Concurrent Pos:* Vis prof, Med Res Coun Can, 80-81. *Res:* Study of conformational polymorphism in topologically stressed DNA; proximal and distal effects of conformational transitions on gene expression; role of conformational parameters in DNA-protein interaction; enzymology of DNA repair. *Mailing Add:* Dept Med Biochem Fac Med Univ Calgary 2500 University Dr Calgary AB T2N 1N4 Can

VAN DE STEEG, GARET EDWARD, b Minneapolis, Minn, Feb 8, 40; m 65; c 2. RADIOCHEMISTRY, NUCLEAR CHEMISTRY. *Educ:* Marquette Univ, BS, 62; Univ NMex, PhD(chem), 68. *Prof Exp:* Sr res chemist, Kerr-McGee Corp, 68-75, proj res chemist, 75-78, sr proj anal chemist, 78-83, mgr anal chem, 83-89, SR PROJ MGR, KERR-MCGEE CORP, 89- *Concurrent Pos:* Adj prof, Univ Okla, 88- *Mem:* Sigma Xi; Am Chem Soc. *Res:* Environmental chemistry and process design; dilute solution chemistry of actinides, lanthanides, transition metals and fission products; solvent extraction of boron, actinides, lanthanides and transition metals; analytical methods development; radiochemical tracer studies; design, development and install environmental control and remediation processes. *Mailing Add:* 2312 NW 113th Pl Oklahoma City OK 73120

VAN DEUSEN, RICHARD L, b Norwich, NY, Jan 31, 26; m 52; c 3. POLYMER CHEMISTRY. *Educ:* Muhlenberg Col, BS, 50; Univ Miami, MS, 55; Univ Buffalo, PhD(chem), 60. *Prof Exp:* Cytotechnologist, Cancer Inst Miami, 51-54; chemist, Olin-Mathieson Chem Corp, 56; instr chem, Univ Buffalo, 56-59; res chemist, Ansco Div, Gen Aniline & Film Co, 59-60; res chemist, 60-70, CHIEF POLYMER BR, USAF MAT LAB, 70- *Mem:* Am Chem Soc; AAAS; Sigma Xi; Am Inst Chem; Int Soc Heterocyclic Chem. *Res:* Photopolymerization; cyclopolymerization kinetics; solid state polymerization; polycondensation; thermally stable plastics, elastomers and fibrous materials. *Mailing Add:* Two Verano Pl Santa Fe NM 87505-8826

VAN DE VAART, HERMAN, b Arnhem, Neth, Apr 11, 34; m 60; c 2. SOLID STATE ELECTRONICS. *Educ:* Delft Univ Technol, Ing, 58, PhD(tech sci), 69. *Prof Exp:* Res asst elec eng, Delft Univ Technol, 56-58; res engr, Transitron Electronic Corp, Mass, 60-62; res asst, 62-65, res staff mem, 65-73, mgr solid state devices dept, 73-80, mgr signal processing dept, 80-81, DIR APPL PHYSICS LAB, SPERRY RES CTR, SPERRY RAND CORP, 81- *Mem:* Am Phys Soc; sr mem Inst Elec & Electronics Engrs. *Res:* Semiconductor technology; quadrupole and ferro magnetic resonance; microwave magnetics; solid state delay lines; ultrasonics; ferrites. *Mailing Add:* Allied Signal Inc PO Box 1021 Morristown NJ 07962

VAN DE VAN, THEODORUS GERTRUDUS MARIA, b Hertogenbosch, Holland, Feb 22, 46. PHYSICAL CHEMISTRY. *Educ:* State Univ Utrecht, BSc, 69, MSc, 71; McGill Univ, PhD(phys chem), 76. *Prof Exp:* Res fel rheology, Univ Sydney, 76-77; SR SCIENTIST, PHYS CHEM, PULP & PAPER RES INST CAN, MCGILL UNIV, 78-, RES ASSOC, DEPT CHEM, 81- *Concurrent Pos:* Vis prof, Dept Chem, Univ Bristol, Eng, 87-88; C B Purves fel, 73-75; Royal Soc Guest Res fel, 87-88. *Mem:* Can Pulp & Paper Asn; Soc Rheology; Int Asn Colloid & Interface Scientists; Polymer Colloid Group. *Res:* Fundamental research in the areas of microrheology and wetting and spreading. *Mailing Add:* Pulp & Paper Res Ctr McGill Univ 3420 University St Montreal PQ H3A 2A7 Can

VANDEVENDER, JOHN PACE, b Jackson, Miss, Sept 12, 47; m 71; c 3. INTENSE PARTICLE BEAMS, PULSED POWER. *Educ:* Vanderbilt Univ, BA, 69; Dartmouth Col, MA, 71; Imp Col Sci & Tech, Univ London, MPhil, 72, PhD(physics), 74. *Prof Exp:* Mem tech staff res, Sandia Nat Lab, 74-79, div supvr, Pulsed Power Res, 79-82, dept mgr, Fusion Res, 82-84, prog mgr, Internal Fusion, 84-90, DIR PULSED POWER SCI, SANDIA NAT LAB, 84- *Concurrent Pos:* Prog mgr, Inertial Fusion, Sandia Nat Lab, 84-90; comt mem, Strategic Defense Initiative Orgn & dir energy, 85-; res comt mem, Nat Acad Sci, Space Power, 87-88, Los Alamos Nat Lab, Chem & Lasers, 88- & Naval Studies Bd, 89- *Mem:* Fel Am Phys Soc; Inst Elec & Electronics Engrs. *Res:* Driving inertial confinement fusion with intense beams of lithium ions for military applications and economic power production is major initiative; exploring the feasibility of new concepts in directed and kinetic energy and understanding transient radiation effects on electronic systems. *Mailing Add:* 7604 Lamplighter NE Albuquerque NM 87109

VAN DEVENTER, WILLIAM CARLSTEAD, b Salisbury, Mo, Oct 22, 08; m 34; c 1. BIOLOGY. *Educ:* Cent Methodist Col, AB, 30; Univ Ill, AM, 32, PhD(field biol), 35. *Prof Exp:* Asst biol, Cent Methodist Col, 28-30; asst zool, Univ Ill, 30-34; biologist, Monroe County Parks, NY, 34-35; prof biol, St Viator Col, 35-38; prof biol, Stephens Col, 38-53; head dept, 53-63, prof, 53-79, EMER PROF BIOL, WESTERN MICH UNIV, 79- *Mem:* AAAS; Ecol Soc Am; Nat Sci Teachers Asn; Nat Asn Res Sci Teaching (vpres, 54, pres, 55). *Res:* Biology of Crustacea; ecology of birds; field biology; human ecology; science education. *Mailing Add:* 3629 Canterbury Ave Kalamazoo MI 49007

VAN DE WATER, JOSEPH M, b Bonne Terre, Mo, Nov 26, 34. CARDIOVASCULAR & RESPIRATORY PHYSIOLOGY, TRAUMA SURGERY. *Educ:* Stanford Univ, BS, 56, MD, 60. *Prof Exp:* Chief surg, Episcopal Hosp, 85-89; PROF SURG, ALBERT EINSTEIN COL MED, 89-; DIR SURG EDUC, LONG ISLAND JEWISH MED CTR, 89- *Concurrent Pos:* Prof surg & physiol & co-dir trauma, Sch Med, Temple Univ, 85-89. *Mem:* Biomed Eng Soc; AMA; Am Col Surg; Soc Univ Surgeons; Am Asn Surg Trauma; Am Asn Thoracic Surg. *Mailing Add:* Dept Surg Long Island Jewish Med Ctr 270-05 76th Ave New Hyde Park NY 11042

VAN DE WETERING, RICHARD LEE, b Bellingham, Wash, Aug 2, 28; m 60; c 2. MATHEMATICS. *Educ:* Univ Wash, Seattle, BS, 50; Western Wash State Col, EdM, 55; Stanford Univ, PhD(math), 60. *Prof Exp:* From asst prof to assoc prof, 60-67, dept chair, 79-84, PROF MATH, SAN DIEGO STATE UNIV, 67- *Concurrent Pos:* Res grant, Delft Univ Technol, 66-67; res assoc, Math Inst, Univ Groningen, 73-74. *Mem:* Am Math Soc; Math Asn Am. *Res:* Ordinary differential equations and integral transforms. *Mailing Add:* Dept Math San Diego State Univ 5300 Campanile Dr San Diego CA 92182

VANDE WOUDE, GEORGE, b Brooklyn, NY, Dec 25, 35; m 59; c 4. BIOCHEMISTRY, VIROLOGY. *Educ:* Hofstra Col, BA, 59; Rutgers Univ, MS, 62, PhD(biochem), 64. *Prof Exp:* Res assoc fel, USDA, Plum Island, 64-65; res chemist, Plum Island Animal Dis Lab, USDA, 65-72; head, Human Tumor Studies Sect, Viral Biol Br, Nat Cancer Inst, Bethesda, 72-75, head, Virus Tumor Biochem Sect, 75-81, chief, Lab Molecular Biol, 81-83, DIR ABL-BASIC RES PROG, NAT CANCER INST-FREDERICK CANCER RES & DEVELOP CTR, 83- *Honors & Awards:* Robert J & Claire Pasarow Found Award Cancer Res, 89. *Mem:* AAAS; Am Chem Soc; Am Asn Cancer Res; Am Soc Microbiol. *Res:* Molecular mechanisms of carcinogenesis; oncogenes; gene expression in eukaryotes; mammalian retrovirus vectors. *Mailing Add:* ABL-Basic Res Prog Nat Cancer Inst Frederick Cancer Res & Develop Ctr Frederick MD 21702

VAN DIJK, CHRISTIAAN PIETER, b Amsterdam, Neth, Nov 10, 15; US citizen; m 63; c 3. PHYSICAL ORGANIC CHEMISTRY, CHEMICAL ENGINEERING. *Educ:* Amsterdam Munic Univ, BS, 36, MS, 40; Delft Univ Technol, PhD(org chem), 46. *Prof Exp:* Lab mgr org chem, Delft Univ Technol, 40-48; asst dept head, Shell Lab Amsterdam, Royal Dutch Shell Co, Neth, 48-57; sr chemist, Dow Chem Co, 57-60; sr scientist, M W Kellogg Co, NJ, 60-75, sr scientist, Pullman Kellogg Res & Develop Ctr, 75-85; INDEPENDENT CONSULT, 85- *Concurrent Pos:* Chmn, Nat Exam Comt towards Qual Sci Coworkers, 43-56; consult, Chemische Fabriek Rotterdam, Neth, 46-48. *Honors & Awards:* Chem Pioneer Award, Am Inst Chemists, 76. *Mem:* Am Chem Soc; Royal Neth Chem Soc; Sigma Xi. *Res:* Inventing new process forms for conversions in the petroleum, petrochemical and polymer field; combined information from the fields of organic and physical chemistry with design-engineering calculations. *Mailing Add:* 10722 Glenway Houston TX 77070

VAN DILLA, MARVIN ALBERT, b New York, NY, June 18, 19; div; c 4. MOLECULAR BIOLOGY. *Educ:* Mass Inst Technol, PhD(physics), 51. *Prof Exp:* Res asst, Mass Inst Technol, 46-51; asst res prof physics, Radiobiol Lab, Univ Utah, 51-57; mem staff, Biomed Res Group, Los Alamos Nat Lab, 57-72; cytophysics sect leader, 72-82, SR SCIENTIST, GENETICS SECT, BIOMED RES DIV, LAWRENCE LIVERMORE NAT LAB 83- *Mem:* AAAS; Soc Analytical Cytol; Am Soc Human Genetics. *Res:* Cell analysis and sorting by high speed flow methods; flow cytometry; cell cycle analysis; flow cytogenetics; sperm cell analysis; chromosome sorting; gene library construction. *Mailing Add:* Lawrence Livermore Nat Lab Livermore CA 94550

VANDIVER, BRADFORD B, b Orlando, Fla, Mar 7, 27; m 80; c 2. GEOLOGY. *Educ:* Univ Colo, BA, 57, MS, 58; Univ Wash, Seattle, PhD(geol), 64. *Prof Exp:* Explor geologist, Tenn Gas Transmission Co, Bolivia, 58-60; asst prof geol, Univ Idaho, 64 & Univ Ore, 64-65; PROF GEOL, STATE UNIV NY COL POTSDAM, 65- *Concurrent Pos:* Vis prof, Univ Colo, 70 & 71; prof, Univ Munich, Ger, 72. *Mem:* Fel Geol Soc Am; Sigma Xi. *Res:* Metamorphic petrology and structural geology, Cascades, Rocky Mountains, Alps, Adirondacks, Odenwald, (Germany); environmental geology. *Mailing Add:* PO Box 327 Hannawa Falls NY 13647

VANDIVIERE, H MAC, b Dawsonville, Ga, Mar 26, 21; m 41; c 2. PREVENTIVE PEDIATRICS, CHEST DISEASE. *Educ:* Mercer Univ, AB, 43, MA, 44; Univ NC, MD, 60. *Prof Exp:* From instr to asst prof biol, Mercer Univ, 42-48; chief spec serv, Res Lab, State Dept Pub Health, Ga, 48-51; res bacteriologist, NC Sanatorium Syst, 51-53, dir dept res, 53-67; assoc prof community med, 67-72, prof, 72-80, PROF PEDIAT & DIR, DIV PREV & COMMUNITY PEDIAT, COL MED, UNIV KY, 80- *Concurrent Pos:* Am Pub Health Asn fel epidemiol; instr bact, Univ Mich, 44-46; dir labs & res, Gravely Sanatorium, 53-62; med dir, Haitian Am Tuberc Inst, Jeremie, Haiti, 62-; asst prof community health sci, Sch Med, Duke Univ, 65-67; consult, Dept Pub Health & adv, President's Comt Control Tuberc, Repub Haiti, 65-; clin assoc prof, Sch Pub Health, Univ NC, 69-; dir div tuberc & fungal dis, State Health Dept, Ky, 72-73, dir div remedial health servs, 73-76. *Mem:* Am Thoracic Soc; Am Pub Health Asn; AMA; Am Teachers Prev Med; Soc Epidemiol Res. *Res:* Antituberculosis vaccination; purification of tuberculo-proteins, diagnostic skin-testing methods; epidemiology. *Mailing Add:* Dept Pediatrics Col Med Univ Ky 800 Rose St Lexington KY 40536

VANDLEN, RICHARD LEE, b Battle Creek, Mich, Oct 22, 47. BIOCHEMISTRY, NEUROCHEMISTRY. *Educ:* Mich State Univ, BS, 69, PhD(phys chem), 72. *Prof Exp:* Fel neurochem, Calif Inst Technol, 72-75, res fel, 75-76; sr res biochemist, Merck Sharp & Dohme Res Labs, 76-85; sr scientist, 85-89, DIR, PROTEIN CHEM, GENENTECH INC, SOUTH SAN FRANCISCO, 89- *Concurrent Pos:* NIH fel, Calif Inst Technol, 74-75. *Mem:* Sigma Xi; Am Crystallog Asn; NY Acad Sci; AAAS. *Res:* Regulation of hormone synthesis and release; hormone and neurotransmitter receptors; protein structure and function; cell culture. *Mailing Add:* Genentech Inc 460 Pt San Bruno Blvd South San Francisco CA 94080

VAN DOEREN, RICHARD EDGERLY, b Tulsa, Okla, Mar 31, 37; m 60; c 3. ACOUSTICS, ELECTRONICS. *Educ:* Colo Sch Mines, BSc, 60; Ohio State Univ, MSc, 64, PhD(elec eng), 68. *Prof Exp:* Physicist, US Naval Air Develop Ctr, Pa, 60-64; res assoc electromagnetic theory, Electro Sci Lab, Ohio State Univ, 64-69; sr engr, N Star Res & Develop Inst, 69-74; PRES, MIDWEST ACOUST & ELECTRONICS, INC, 74- *Mem:* Inst Elec & Electronics Engrs; Acoust Soc Am; Inst Noise Control Eng; Am Consult Engrs Coun; Audio Eng Soc; Am Soc Testing & Mat. *Res:* Acoustical noise reduction techniques, acoustical response of rooms; acoustic measurement techniques and systems; human response to sound; electroacoustic systems; computer methods in acoustics. *Mailing Add:* Midwest Acoustics & Elec Inc 6950 France Ave S Minneapolis MN 55435

VAN DOLAH, ROBERT FREDERICK, b Portland, Ore, Nov 4, 49; m; c 2. MARINE ECOLOGY, INVERTEBRATE ZOOLOGY. *Educ:* Marietta Col, BS, 71; Univ Md, MS, 75, PhD(zool), 77. *Prof Exp:* SR MARINE SCIENTIST MARINE ECOL & ASST DIR, SC MARINE RESOURCES RES INST, 88- *Concurrent Pos:* Adj fac, Col Charleston. *Mem:* Ecol Soc Am; Estuarine Res Fedn. *Res:* Population and community ecology with particular emphasis on the regulatory processes operating in marine and estuarine systems; environmental research on effects of habitat perturbations and pollution. *Mailing Add:* SC Marine Resources Res Inst PO Box 12559 Charleston SC 29412

VAN DOLAH, ROBERT WAYNE, b Cheyenne, Wyo, Feb 1, 19; m 42; c 3. CHEMISTRY. *Educ:* Whitman Col, AB, 40; Ohio State Univ, PhD(org chem), 43. *Prof Exp:* Asst chem, Ohio State Univ, 40-42; asst to sci dir, William S Merrell Co, Ohio, 43-44, res chemist & group leader, 44-46; actg head org chem br, US Naval Ord Test Sta, Calif, 46-48, head, 48-53, head chem div, 53-54; chief, Explosives Res Lab, Pittsburgh Mining & Safety Res Ctr, US Bur Mines, 54-71, res dir, 71-78; CONSULT, 78- *Honors & Awards:* Distinguished Serv Medal, US Dept Interior, 65; Nitro Nobel Medal, 67; H H Storch Award, Am Chem Soc, 72. *Mem:* Fel AAAS; fel Am Inst Chem; Am Chem Soc; Sigma Xi; Nat Fire Protection Asn. *Res:* Propellants and explosives; combustion; mine safety; industrial safety. *Mailing Add:* 202 Cherokee Rd Pittsburgh PA 15241-1516

VAN DOMELEN, BRUCE HAROLD, b Shelby, Mich, May 27, 33; m 57; c 4. PHYSICS. *Educ:* Kalamazoo Col, BA, 55; Univ Wis, MA, 57, PhD(physics), 60. *Prof Exp:* Staff mem, Sandia Nat labs, 60-62, sect supvr phys metall, 62-65, div supvr anal physics, 65, tech adv syst res, 65-69, div supvr explor power sources, 69-78, Explosives Projs & Tests Div, 78-85, New Hire Projs, 85-88, DIV SUPVR TECHNOL TRANSFER, SANDIA NAT LABS, 88- *Concurrent Pos:* Actg chmn, Governor's Sci Adv Comt, NMex, 66-70, Governor's sci adv, 66-75; NMex mem, Western Interstate Nuclear Bd, 67-77, chmn, 71-73; mem Nat Governors' Coun Sci & Technol, 70-75. *Mem:* Am Phys Soc; Sigma Xi. *Res:* Explosives; physical chemistry. *Mailing Add:* 3204 La Sala Cuadra NE Albuquerque NM 87111

VAN DONGEN, CORNELIS GODEFRIDUS, b Geertruidenberg, Neth, Mar 20, 34; m 69; c 2. ANIMAL HUSBANDRY, REPRODUCTIVE PHYSIOLOGY. *Educ:* Wageningen State Agr Univ, BS, 57, MS, 59; Univ Ill, Urbana, MS, 62, PhD(dairy sci), 64. *Prof Exp:* Res fel pharmacol, Harvard Univ, 64-65; res assoc physiol, Brown Univ, 66; asst prof pharmacol, NY Med Col, 66-67; res assoc, Bio-Res Inst, Inc, 67-76, vpres, 76-85; PRES, BIO BREEDERS INC, 85- *Concurrent Pos:* Grants, NIH; investr, USDA Contract; consult, Bio Res Consults. *Mem:* Am Dairy Sci Asn; Am Soc Animal Sci; Brit Soc Study Fertil; Soc Study Reproduction; Am Asn Lab Animal Sci. *Res:* Management of breeding colony of inbred and hybrid Syrian hamsters; development of animal models of human disease; husbandry and nutritional factors; muscular dystrophy, carcinogenicity, aging and reproductive physiology. *Mailing Add:* 45 Hall Ave Watertown MA 02172

VAN DOORNE, WILLIAM, b Utrecht, Neth, Dec 12, 37; US citizen; m 61; c 3. INORGANIC CHEMISTRY. *Educ:* Calvin Col, BS, 60; Univ Mich, MS, 62, PhD, 65. *Prof Exp:* From asst prof to assoc prof chem, 66-74, PROF CHEM, CALVIN COL, 74-, CHMN DEPT, 77- *Concurrent Pos:* Vis assoc prof, Univ Hawaii, 72-73. *Mem:* Am Chem Soc. *Res:* Phophorus-nitrogen compounds; synthetic inorganic chemistry; crystallography. *Mailing Add:* 3201 Burton St SE Grand Rapids MI 49546-4349

VAN DREAL, PAUL ARTHUR, b Chicago, Ill, Feb 15, 32; m 57; c 2. BIOCHEMISTRY, CYTOLOGY. *Educ:* Calvin Col, BS, 57; Mich State Univ, PhD(bot, biochem, cytol), 61. *Prof Exp:* Clin biochemist, St Lawrence Hosp, 61-63; NIH fel, Biol Div, Oak Ridge Nat Lab, 63-64; asst prof biochem, Med Sch, Univ Ore, 64-66; from asst prof to assoc prof clin chem, Med Sch, Univ Wash, 66-71, dir lab computer div, 69-71; vpres & dir res, Hycel Inc, 71-72; tech mgr radioimmunoassay develop, Corning Glass Works, Inc, 72-73; assoc prof path & clin chem, Med Sch, Univ Ky, 73-74; lab dir & tech dir, Nat Health Labs, Inc, 74-75; vpres & lab dir, Herner Analytics, 75-77; lab dir, Clin Lab Med Serv, 77-80; mgr chem applicator, Diag Div, Abbot Labs, 80-83; ASSOC PROF, PATH DEPT, UNIV TEX SOUTHWESTERN MED SCH, 83-; TECH DIR, NAT HEALTH LABS, DALLAS TEX, 90- *Concurrent Pos:* Lectr biochem, Mich State Univ, 62-63; Nat Cancer Inst fel carcinogenesis, Med Sch, Univ Ore, 64-66; consult, Ortec Div, EG&G, 67-71, Clin Instruments Div, Beckman Instruments, 67-, Pesticide Res Lab, Dept of Health, State of Wash, 68-69, Bausch & Lomb Inc, 69- & Corning Glass Inc,

74-; Bausch & Lomb Grant electrophoresis, Med Sch, Univ Wash, 69-70. *Mem:* AAAS; Am Chem Soc; fel Am Asn Clin Chemists; Am Soc Clin Path; Asn Clin Scientists; Sigma Xi. *Res:* New clinical laboratory diagnostic techniques; aging; patient normals as a function of age and disease onset; biochemistry of the cell cycle. *Mailing Add:* 1010 Pheasant Ridge Dr Grapevine TX 76051

VAN DRESER, MERTON LAWRENCE, b Des Moines, Iowa, June 5, 29; m 52; c 2. CERAMIC ENGINEERING, MATERIALS SCIENCE. *Educ:* Iowa State Univ, BS, 51. *Prof Exp:* Tech supvr fiberglass mfg, Owens-Corning Fiberglas Corp, 54-57; res engr, Kaiser Aluminum & Chem Corp, 57-60, res sect head basic refractory res, 60-63, lab mgr, 63-65, assoc dir res, 65-69, dir, refractories res, 69-72, dir, non-metall mat res, 72-83, vpres & dir res, indust chemicals & Harshaw/Filtrol partnership, 83-85, dir business div, 85-88, consult, Kaiser Aluminum & Chem Corp, 88-89; CONSULT, 89- *Concurrent Pos:* Mem tech adv comn, Refractories Inst, 74-84, chmn, 80-84; mem adv bd, Dept Ceramic Eng, Univ Ill, 75-78; vol exec, Intl Exec Serv Corp, Pakistan, 90. *Mem:* Fel Am Ceramic Soc (vpres, 73-74); Brit Ceramic Soc; Nat Inst Ceramic Eng; Am Inst Mining, Metall & Petrol Eng; hon mem Am Soc Testing & Mat. *Res:* Refractories, sintering of refractory oxides and silicates; chemical and ceramic bonding of refractory powders; development of refractory products for application in iron, steel, glass, non-ferrous metal and petro-chemical industries. *Mailing Add:* 40 Castledown Rd Pleasanton CA 94566

VAN DRIEL, HENRY MARTIN, b Breda, Neth, Dec 27, 46; Can citizen; m 70; c 3. LASER PHYSICS, NONLINEAR OPTICS. *Educ:* Univ Toronto, BSc, 70, MSc, 71 & PhD(physics), 75. *Prof Exp:* fel physics, Univ Ariz, 75-76; PROF PHYSICS, UNIV TORONTO, 76- *Concurrent Pos:* Vis scientist, Harvard Univ, 83 & IBM, 83-85; John Simon Guggenheim Found fel, 86; Nat Res Coun fel, 75. *Mem:* Can Asn Physicists; Am Phys Soc; fel Optical Soc Am. *Res:* Usage of ultrafast laser pulses in investigating picosecond and femtosecond optoelectronic phenomena in semiconductor; nonlinear optical response of solids including metals and superlattice semiconductors. *Mailing Add:* Dept Physics Univ Toronto 60 St George St Toronto ON M5S 1A7 Can

VAN DRIESSCHE, WILLY, b Stekene, Belg, Feb 8, 40; m 65; c 2. GENERAL MEDICAL SCIENCES. *Educ:* K U Leuven, PhD(physics), 68. *Prof Exp:* PROF PHYSIOL, K U LEUVEN, 75- *Mem:* Am Physiol Soc; Ger Physiol Soc; Belg Physiol Soc. *Res:* Electrophysiology of epithelial cells; inlizo cellular ion concentration. *Mailing Add:* Labo voor Fysiologie K U Leuven Gasthuisberg Leuven 3000 Belgium

VANDRUFF, LARRY WAYNE, b Elmira, NY, Apr 28, 42; m 66; c 2. WILDLIFE BIOLOGY, URBAN ECOLOGY. *Educ:* Mansfield State Col, BS, 64; Cornell Univ, MS, 66, PhD(wildlife ecol), 71. *Prof Exp:* Res asst vert ecol & genetics, Cornell Univ, 64-66, res asst wildlife ecol, NY Wildlife Res Unit, 66-70; asst prof vert biol, 70-77, assoc prof, 77-85, PROF, WILDLIFE BIOL, STATE UNIV NY COL ENVIRON SCI & FORESTRY, 85- *Honors & Awards:* Daniel L Leedy Urban Wildlife Conserv Award, Nat Inst Urban Wildlife, 87. *Mem:* Am Soc Mammal; Wildlife Soc; Ecol Soc Am. *Res:* Field studies in the ecology of urban wildlife species; waterfowl biology and wetland ecology; dynamics of homeotherm populations and wildlife; habitat relationships. *Mailing Add:* Dept Biol State Univ NY Col Environ Sci & Forestry Syracuse NY 13210

VAN DUUREN, BENJAMIN LOUIS, b SAfrica, May 5, 27; nat US; div; c 2. ORGANIC CHEMISTRY. *Educ:* Univ SAfrica, BS, 46 & 48, MS, 49; Univ Orange Free State, ScD, 51. *Prof Exp:* Res assoc, Univ Ill, 51-53 & Univ Calif, Los Angeles, 53-54; res chemist, E I du Pont de Nemours & Co, NY, 54-55; from instr chem to assoc prof, 55-69, PROF ENVIRON MED, NY UNIV MED CTR, 69- *Concurrent Pos:* Consult, Environ Protection Agency, 75- & NSF, 79- *Mem:* Am Chem Soc; Am Asn Cancer Res. *Res:* Chemistry of pyrrolizidine alkaloids and benzofuran fish toxic compounds; infrared spectroscopy; fluorescence spectroscopy of aromatic compounds; environmental carcinogens; carcinogenesis and metabolism of carcinogens; tobacco and cancer. *Mailing Add:* Dept Environ Med NY Univ Med Ctr 550 First Ave New York NY 10016

VAN DUYNE, RICHARD PALMER, b Orange, NJ, Oct 28, 45; m. ANALYTICAL CHEMISTRY, CHEMICAL PHYSICS. *Educ:* Rensselaer Polytech Inst, BS, 67; Univ NC, PhD(anal chem), 71. *Prof Exp:* Asst prof, 71-76, assoc prof anal chem, 76-79, prof anal & phys chem, 79-86, MORRISON PROF CHEM, NORTHWESTERN UNIV, EVANSTON, 86- *Concurrent Pos:* Fel, Alfred P Sloan Found, 74-78. *Honors & Awards:* Coblentz Mem Prize, 80; Fresenius Award, 81; Pittsburgh Spectros Award, 91. *Mem:* Am Chem Soc; fel AAAS; fel Am Phys Soc. *Res:* Radical ion chemistry; surface-enhanced Raman spectroscopy; chemical applications of lasers; time-resolved fluorescence spectroscopy; tunable dye laser resonance Raman spectroscopy; laboratory computer systems; microfabrication, scanning electron, tunneling and optical microscopy; theory of electron transfer. *Mailing Add:* Dept Chem Northwestern Univ Evanston IL 60208

VAN DYK, JOHN WILLIAM, b Paterson, NJ, May 2, 28; m 51; c 3. PHYSICAL CHEMISTRY, COMPUTER SOFTWARE. *Educ:* Rutgers Univ, AB, 50; Columbia, AM, 51, PhD(chem), 54. *Prof Exp:* Asst chem, Columbia Univ, 50-52; res chemist, Polychems Dept, 54-64, staff chemist, Fabrics & Finishes Dept, 64-82, res assoc, E I du Pont de Nemours & Co, Inc, 82-85; CONSULT, 85- *Mem:* Am Chem Soc; Sigma Xi. *Res:* Polymerization kinetics; surface chemistry; polymer chemistry; paint chemistry; color science; visual perception; computer science; solubility parameters, solvent selection; scientific computer program. *Mailing Add:* 106 Cambridge Dr Wilmington DE 19803

VAN DYKE, CECIL GERALD, b Effingham, Ill, Feb 4, 41; m 69; c 5. PLANT PATHOLOGY, FUNGUS-HOST ULTRASTRUCTURE. *Educ:* East Ill Univ, BSEd, 63; Univ Ill, Urbana, MS, 66, PhD(plant path), 68. *Prof Exp:* NIH res assoc, Univ Ill, Urbana, 68; res assoc, 68-69, from instr to assoc prof, 69-89, PROF BOT, NC STATE UNIV, 89- *Mem:* Am Phytopath Soc; Mycol Soc Am; Sigma Xi; Creation Res Soc. *Res:* Ultrastructure of fungi and fungus-host pathological interactions and biological control. *Mailing Add:* Dept Bot Box 7612 NC State Univ Raleigh NC 27695-7612

VAN DYKE, CHARLES H, b Rochester, Pa, Sept 19, 37; m 66; c 1. INORGANIC CHEMISTRY, ORGANOMETALLIC CHEMISTRY. *Educ:* Geneva Col, BS, 59; Univ Pa, PhD(inorg chem), 64. *Prof Exp:* Asst prof chem, 63-70, ASSOC PROF CHEM, CARNEGIE-MELLON UNIV, 70- *Mem:* Am Chem Soc; Royal Soc Chem. *Res:* Synthesis and study of volatile hydride derivatives of the Group IV elements. *Mailing Add:* Dept Chem Carnegie-Mellon Univ Dorothy Hall 2121 Pittsburgh PA 15213

VAN DYKE, CRAIG, b Detroit, Mich, Oct 4, 41; m 69; c 2. PSYCHIATRY, PSYCHOPHARMACOLOGY. *Educ:* Univ Wash, BS, 63, MD, 67. *Prof Exp:* Asst prof psychiat, Yale Univ, 74-78; assoc prof, 79-86, PROF PSYCHIAT, UNIV CALIF, SAN FRANCISCO, 86. *Concurrent Pos:* Chief, Psychiat Serv, 87. *Mem:* Am Psychosom Soc; Int Col Psychosom Med; Soc Neurosci; Int Neuropsychol Soc. *Res:* Psychoimmunology; neuropsychiatry. *Mailing Add:* Vet Admin Med Ctr 116A 4150 Clement San Francisco CA 94121

VAN DYKE, HENRY, b Pittsburgh, Pa, Oct 1, 21; m 43; c 4. MICROBIOLOGY. *Educ:* Western Reserve Univ, BS, 47; Univ Mich, MA, 49, PhD(zool), 55. *Prof Exp:* Malariologist, USPHS, 52; instr zool, Univ Mich, 52-53; asst prof biol, Carleton Col, 53-60 & Ore Col Educ, 60-63; ASSOC PROF BIOL, ORE STATE UNIV, 63- *Mem:* AAAS; Soc Protozool; Am Soc Microbiol; Marine Biol Asn UK, Am Soc Limnol & Oceanog. *Res:* Ecology, physiology, culture and photobiology of marine communities. *Mailing Add:* 3300 NW Van Buren Ave Corvallis OR 97331

VAN DYKE, JOHN WILLIAM, JR, b Holland, Mich, Nov 15, 35; m 59; c 3. ORGANIC CHEMISTRY. *Educ:* Hope Col, AB, 58; Univ Ill, PhD(org chem), 62. *Prof Exp:* SR RES CHEMIST, THERAPEUT RES LAB, MILES LABS, INC, 62- *Mem:* Am Chem Soc. *Res:* Organic synthesis of pharmacologically active compounds. *Mailing Add:* 2917 E Jackson Blvd Elkhart IN 46516-5025

VAN DYKE, KNOX, b Chicago, Ill, June 23, 39; m 78; c 5. PHARMACOLOGY, BIOCHEMISTRY. *Educ:* Knox Col, AB, 61; St Louis Univ, PhD(biochem), 66. *Prof Exp:* Res assoc pharmacol, 66-68, sr res pharmacologist, 68-69, from asst prof to assoc prof pharmacol, 69-77, PROF PHARMACOL & TOXICOL, MED CTR, WVA UNIV, 77- *Concurrent Pos:* WHO grant, 70-, training grant malaria; WVa Heart Asn grant, 70-; NIH instnl cancer & gen res WVa rep, Oak Ridge Assoc Univs, 72-75, Merck, Sharp & Dohme, Sandoz, 86-88, Knoll, 88-, FDA US-Bur of Mines. *Mem:* AAAS; Am Chem Soc; Am Soc Pharmacol & Exp Therapeut; Int Soc Biochem Pharmacol; Am Soc Photobiol. *Res:* Malariology and mechanism of drug resistance; automated analysis of enzyme and nucleic acid systems; radioimmunoassay; mechanisms of antimalarial drugs; adrenergic transmitter-energy complexes; adenosine utilization and syntheses; measurement of bioluminescent and chemiluminescent reaction; inflammatory drugs and free radicals; multiple drug resistance. *Mailing Add:* Dept Pharmacol WVa Univ Med Ctr Morgantown WV 26506

VAN DYKE, MILTON D(ENMAN), b Chicago, Ill, Aug 1, 22; m 46, 62; c 6. FLUID MECHANICS. *Educ:* Harvard Univ, BS, 43; Calif Inst Technol, MS, 47, PhD(aeronaut), 49. *Prof Exp:* Aeronaut res scientist, Nat Adv Comt Aeronaut, 43-46 & 50-58; aeronaut engr, Douglas Aircraft Co, 48; consult aerodynamicist, Rand Corp, 49-50; vis prof, Univ Paris, 58-59; prof aeronaut eng, 59-75, PROF APPL MECH, STANFORD UNIV, 75- *Concurrent Pos:* Lectr, Stanford Univ, 50-58; Guggenheim fel, 54-55; Nat Acad Sci exchange vis, USSR, 65. *Mem:* Nat Acad Eng; Am Phys Soc; Am Acad Arts & Sci. *Res:* Compressible flow theory; viscous flow theory. *Mailing Add:* Div Appl Mech Stanford Univ Stanford CA 94305-4040

VAN DYKE, RUSSELL AUSTIN, b Rochester, NY, Feb 8, 30; m 56; c 2. PHARMACOLOGY, TOXICOLOGY. *Educ:* Hope Col, BS, 51; Univ Mich, MS, 53; Univ Ill, PhD(nutrit biochem), 60. *Prof Exp:* Biochemist, Dow Chem Co, Midland, Mich, 61-68; consult, Dept Anesthesiol, Mayo Clin, Rochester, Minn, 68-89, Dept Cell Biol, 80-89, prof biochem, Mayo Med Sch, 79-89, prof pharmacol, 84-89; RES BIOSCIENTIST, HENRY FORD HOSP, DETROIT, MICH, 89- *Concurrent Pos:* Instr biochem, Saginaw Valley Col, Mich, 64-68; prin & co-investr, NIH; lectr, Columbia Univ, 79, Japanese Soc Anesthesiol, Tokyo, 86. *Mem:* AAAS; Am Soc Anesthesiologists; Am Soc Pharmacol & Exp Therapeut; Soc Toxicol; Int Soc Study Xenobiotics. *Res:* Metabolism of xenobiotics; mechanism of organ damage by drugs or metabolites; influence of various altered physiological states on drug metabolism and toxicity. *Mailing Add:* Henry Ford Hosp 2799 W Grand Blvd Detroit MI 48202

VANE, ARTHUR B(AYARD), b Portland, Maine, June 1, 15; m 42; c 3. PHYSICAL CHEMISTRY, ELECTRICAL ENGINEERING. *Educ:* Univ Wash, Seattle, BS, 37; Ore State Col, MS, 41; Stanford Univ, EE, 49. *Prof Exp:* Staff mem, Radiation Lab, Mass Inst Technol, 42-45; physicist, US Naval Ord Test Sta, 45-47; res assoc, Microwave Lab, Stanford Univ, 47-49; sr engr, Varian Assocs, 49-55, mgr systs develop, 55-58, mgr systs dept, Radiation Div, 58-62; mgr microwave dept, Melabs, Inc, 62-65; sr scientist, Cent Res Labs, Varian Assocs, 65-71; vpres, Sonoma Eng & Res, Santa Rosa, 71-75; sr scientist, Addington Labs, Sunnyvale, 75-77; vpres, Westmont Labs, Palo Alto, 77-79; RETIRED. *Concurrent Pos:* Consult, Vane Microwave Consult Co, 78-; comptroller, Cult Systs Res, 79- *Mem:* AAAS; fel Am Inst Chemists; Am Chem Soc; Inst Elec & Electronics Engrs. *Res:* Microwave techniques; solid state microwave devices. *Mailing Add:* 823 Valparaiso Ave Menlo Park CA 94025

VANE, FLOIE MARIE, b Dawson, Minn, Nov 25, 37. DRUG METABOLISM. *Educ:* Gustavus Adolphus Col, BS, 59; Mich State Univ, PhD(org chem), 63. *Prof Exp:* Sr chemist, 64-72, group chief, 73-78, sect head, 79-84, RES INVESTR, HOFFMANN-LA ROCHE, INC, 85- *Concurrent Pos:* NIH fel, Mass Inst Technol, 63-64; vis asst prof, Baylor Col Med, 68-69. *Mem:* Am Chem Soc; Am Soc Pharmacol & Exp Therapeut; Sigma Xi. *Res:* Structure determination of organic compounds by spectroscopic methods such as nuclear magnetic resonance and mass spectroscopy; structure identification of drug metabolites. *Mailing Add:* 770 H Anderson Ave Apt 19A Cliffside Park NJ 07010

VANE, JOHN ROBERT, b Worcestershire, UK, Mar 29, 27; m 48; c 2. ENDOTHELIAL CELLS, PROSTAGLANDINS. *Educ:* Univ Birmingham, BSc, 46; Univ Oxford, BSc, 49, DPhil, 53, DSc, 70. *Hon Degrees:* DMed, Copernicus Acad Med, Carcow, 77; Dr, Rene Descartes Univ, Paris, 78; DSc, City Univ New York, 80, Aberdeen Univ, 83, NY Med Col, 84, Birmingham Univ, 84, Camerino Univ, Italy, 84, Catholic Univ, Belg, 86; Dr Hon Causa, Univ Buenos Aires, Arg, 86. *Prof Exp:* Instr & asst prof pharmacol, Yale Univ, 53-55; sr lectr pharmacol, Inst Basic Med Sci, Royal Col Surgeons Eng, 55-61, reader, London Univ, 61-65, prof exp pharmacol, 66-73; group res & develop dir, Wellcome Found Ltd, 73-85; DIR, WILLIAM HARVEY RES INST, ST. BARTHOLOMEW'S HOSP MED COL, 86-, CHMN RES COMT, 87-; PROF MED, NY MED COL, VALHALLA, 86- *Concurrent Pos:* Lectr, numerous univs & soc, 68-88; Walter C MacKenzie vis prof, Univ Alta, 77, vis prof, Harvard Univ, 79, vis prof pharmacol, NY Med Col, 86-; mem coun, Imp Cancer Res Fund, 85-; chmn, Imp Cancer Res Technol Ltd, 87-; mem sci adv comt, Osaka Biosci Inst, Japan, 87-; mem sch med vis comt, Case Western Reserve, Univ Cleveland, 88- *Honors & Awards:* Nobel Prize physiol, 82; Baly Medallist, Royal Col Physicians, 77; Albert Lasker Basic Med Res Award, Albert & Mary Lasker Found, 77; Joseph J Bunim Medal, Am Rheumatism Asn, 79; Peter Debye Prize, Univ Maastricht, Holland, 80; Nuffield Lectr & Gold Medal, Royal Soc Med, Eng, 80; Dale Medallist, Soc Endocrinol, 81; Galen Medallist, Worshipful Soc Apothecaries, 83; Louis Pasteur Found Prize, 84; Royal Medal, Royal Soc, 89. *Mem:* Foreign assoc Nat Acad Sci; hon mem Brit Pharmacol Soc (meetings secy, 66-71, gen secy, 71-73, foreign secy, 79-84); hon mem Physiol Soc; fel Inst Biol; fel Royal Soc (vpres, 85-87); hon mem Am Physiol Soc; hon fel Am Col Physicians; foreign hon mem, Am Acad Arts & Sci; foreign mem, Nat Acad Med Buenos Aires; hon fel Royal Col Physicians. *Res:* Mode of action of aspirin and similar drugs; discovery prostacyclin and its relationship to other prostaglandins and thromboxanes; underlying mechanism which initiates atherosclerosis, especially in the involvement of endothelial cell. *Mailing Add:* William Harvey Res Inst St Bartholomew's Hosp Med Col Charterhouse Square London ECIM 6BQ England

VAN ECHO, ANDREW, b Barton, Ohio, Jan 27, 18; m 45; c 1. ENGINEERING & MATERIALS SCIENCE. *Educ:* Ohio State Univ, BS, 42. *Prof Exp:* Res asst, Battelle Mem Inst, Columbus, Ohio, 41-43, Manhattan Proj, Univ Chicago, 43-45; chief inspector, Joslyn Mfg & Supply Co, 45-47, asst works mgr, Wm E Pratt Mfg Co Div, 47-49, asst supt wire mill, 49-54, supvr prod control, 54-56, chief metallurgist & mgr processing & qual control, 56-63; asst chief fuels & mat br, US AEC, 63-73; metall engr, US Energy Res & Develop Admin, 73-77; metall engr, US Dept Energy, 77-80, sr mat eng, 80-86, mgr metall absorbers & standards & mat eng, 86-90, MGR INT PROGS DIV, OFF NUCLEAR ENERGY, US DEPT ENERGY, 90- *Concurrent Pos:* Mem, World Metall Cong, 57; Govt liaison rep of Dept Energy, Nat Acad Sci adv bd, Comt on Fatigue Crack Initiation at Elevated Temperatures, 77-78; mem joint US-USSR working group metall, Joint US-USSR Comn Sci & Tech Coop, 73-78 & 78- *Honors & Awards:* Award of Merit, Am Soc Testing & Mat, 73, Hon Award, 75. *Mem:* Am Soc Metals; fel Am Soc Testing & Mat. *Res:* Uranium processing and fabrication; stainless steel melting, processing and fabrication; refractory alloy consolidation, processing and fabrication; silicon vaporphase plating of steels; V-alloy development; engineering properties of structural materials for high temperature design; uranium dioxide, uranium-plutonium dioxide, uranium metal fuels for reactors; dispersion strengthened ferritic steels for advanced liquid metal reactors. *Mailing Add:* US Dept Energy NE-14/GTN Washington DC 20545

VAN ECHO, DAVID ANDREW, b Ft Wayne, Ind, July 19, 47; m 71; c 2. ONCOLOGY, INTERNAL MEDICINE. *Educ:* Xavier Univ, BS, 69; Univ Md, MD, 73. *Prof Exp:* Intern & resident internal med, Univ Hosp, Baltimore, Md, 73-75; fel med oncol, Baltimore Cancer Res Ctr, 75-77; ASSOC PROF MED, UNIV MD SCH MED, 78- *Concurrent Pos:* Investr, Nat Cancer Inst, NIH, 77-78, sr investr, 78- *Mem:* Am Col Physicians. *Res:* Phase I and II studies in solid tumors and acute leukemia; whole-body hyperthermia. *Mailing Add:* Univ Hosp 22 S Greene St Baltimore MD 21201

VAN ECK, EDWARD ARTHUR, b Grand Rapids, Mich, May 26, 16; m 46; c 2. MICROBIOLOGY. *Educ:* Hope Col, BA, 38; Univ Mich, MSc, 41, PhD(bact), 50. *Prof Exp:* Instr bact, Univ Mich, 48-50; asst prof, Univ Kans, 50-53; purchasing agent, Stand Grocer Co, Mich, 53-58; lectr & reader microbiol, Christian Med Col, Vellore, India, 58-63, assoc prof, 62-63; prof, 63-81, EMER PROF BIOL, NORTHWESTERN COL, IOWA, 81- *Concurrent Pos:* Vis prof Univ Kans, 72-73. *Mem:* AAAS; Am Soc Microbiol; Sigma Xi. *Res:* Antigenic constitution of the Salmonellae; serological survey for presence of leptospirosis in South India; tumor immunology. *Mailing Add:* 118 Florida Ave NW Orange City IA 51041

VAN ECK, WILLEM ADOLPH, b Wageningen, Netherlands, July 27, 28; nat US; m 56; c 3. SOIL SCIENCE, HYDROLOGY. *Educ:* Wageningen State Agr Univ, BSc, 51; Mich State Univ, MSc, 54, PhD(soil sci), 58. *Prof Exp:* Asst plant ecol, Wageningen State Agr Univ, 50, soil surv, 51; forester, Gold Coast Govt Surv Team, 52; asst soil fertil, Mich State Univ, 52-53, soil surv & forest soils, 52-56; from asst prof to assoc prof soil sci, WVa Univ, 57-66; sr lectr land planning, Univ EAfrica, 66-72; PROF SOIL SCI & STATE EXTEN SPECIALIST, WVA UNIV, 72- *Concurrent Pos:* Environ scientist, US Environ Protection Agency, Washington, DC, 76-77; pres, Acad Assocs, Econ-Environ Consults. *Mem:* AAAS; Soil Sci Soc Am; Ecol Soc Am; Sigma Xi; fel Soil Conserv Soc Am; Water Pollution Control Fedn. *Res:* Effect of environment on soil and vegetation development, especially as applied to forest and watershed management; relation of soil morphology and pedology to soil and water conservation and to physical land use planning; assessment of soil fertility in agronomy and forestry; water pollution control. *Mailing Add:* 1702 Kilarney Dr Cary NC 27511

VAN EEDEN, CONSTANCE, b Delft, Netherlands, Apr 6, 27; wid; c 1. PROBABILITY. *Educ:* Univ Amsterdam, BSc, 49, MA, 54, PhD, 58. *Prof Exp:* Res assoc, Math Ctr, Univ Amsterdam, 54-60; vis assoc prof, Mich State Univ, 60-61; res assoc, Univ Minn, Minneapolis, 61-64; from assoc prof to prof, 65-89, EMER PROF MATH, UNIV MONTREAL, 89- *Concurrent Pos:* Assoc prof & actg dir statist ctr, Univ Minn, Minneapolis, 64-65; res mem, Math Res Ctr, Univ Wis-Madison, 69; adj prof statist, Univ BC, 89-, adj prof math, Univ Que, Montreal, 89- *Honors & Awards:* Gold Medal, Statist Soc Can, 90. *Mem:* Can Statist Soc; Int Statist Inst; Inst Math Statist; Am Statist Asn; Can Math Soc. *Res:* Mathematical statistics. *Mailing Add:* Moerland 19 Broek in Waterland 1151 BH Netherlands

VANEFFEN, RICHARD MICHAEL, b Milwaukee, Wis, June 24, 53. ELECTROCHEMICAL METHODS. *Educ:* Univ Notre Dame, BS, 75; Univ Wis, Madison, PhD(anal chem), 79. *Prof Exp:* SR RES CHEMIST, DOW CHEM USA, 79- *Mem:* Am Chem Soc; Sigma Xi. *Res:* Electrochemical analysis and fundamental electrochemical research; application of electrochemical techniques to the solution of industrial process problems. *Mailing Add:* Analytical Sci 1897 Bldg Dow Chem Co Midland MI 48667

VAN EIKEREN, PAUL, b Hilversum, Neth, July 6, 46; US citizen; m 70; c 2. MEMBRANE SEPARATIONS, ENZYMOLOGY. *Educ:* Columbia Col, AB, 68; Mass Inst Technol, PhD(org chem), 71. *Prof Exp:* Dreyfus instr chem, Mass Inst Technol, 71-72; prof chem, Harvey Mudd Col Sci & Eng, 72-86; DIR RES, BEND RES, INC, 86- *Concurrent Pos:* Vis prof, Inst Molecular Biol, Univ Ore, 79-81. *Mem:* Am Chem Soc. *Res:* Development of membrane-based systems for fermentation, enzymatic synthesis, natural product separation and controlled-release of pharmaceuticals; drug development; chemical and biochemical process research. *Mailing Add:* 1922 NW Seventh St Bend OR 97701-8599

VAN ELDIK, LINDA JO, CALCIUM BINDING PROTEINS. *Educ:* Duke Univ, PhD(microbiol & immunol), 78. *Prof Exp:* ASSOC PROF PHARMACOL & CELL BIOL, VANDERBILT UNIV, 86- *Mailing Add:* Howard Hughes Med Inst Vanderbilt Univ Nashville TN 37232

VANELLI, RONALD EDWARD, b Quincy, Mass, July 5, 19; m 53; c 2. ORGANIC CHEMISTRY. *Educ:* Harvard Univ, AB, 41, MA & PhD(chem), 50. *Prof Exp:* Sr res chemist, Photo Prods Dept, E I du Pont de Nemours & Co, 50-51; dir, Chem Labs & lectr chem, Harvard Univ, 51-89, dir sci ctr, 72-89; RETIRED. *Mem:* Am Chem Soc. *Res:* Isonorcamphor; color and constitution. *Mailing Add:* 89 Woodridge Rd Wayland MA 01778

VAN ELSWYK, MARINUS, JR, plant breeding, genetics; deceased, see previous edition for last biography

VAN EMDEN, MAARTEN HERMAN, b Rheden, Neth. COMPUTER SCIENCE. *Educ:* Delft Univ Technol, MEng, 66; Univ Amsterdam, DSc(math & natural sci), 71. *Prof Exp:* Res assoc comput, Math Ctr, Amsterdam, 66-71; postdoctoral fel, IBM Thomas J Watson Res Ctr, 71-72; res fel artificial intel, Univ Edinburgh, 72-75; fac comput sci, Univ Waterloo, 75-87; PROF COMPUT SCI, UNIV VICTORIA, 87- *Mem:* Inst Elec & Electronics Engrs; Asn Comput Mach. *Res:* Logic programming; software engineering. *Mailing Add:* Dept Computer Sci Univ Victoria PO Box 3055 Victoria BC V8W 3P6 Can

VAN ENKEVORT, RONALD LEE, b Escanaba, Mich, Dec 20, 39; m 62; c 1. MATHEMATICS. *Educ:* Univ Wash, BS, 62; Ore State Univ, MS, 66, PhD(math), 72. *Prof Exp:* High sch teacher, 62-67; asst prof, 71-77, ASSOC PROF MATH, UNIV PUGET SOUND, 77- *Mem:* Am Math Soc. *Res:* Additive number theory. *Mailing Add:* Dept Math Univ Puget Sound Tacoma WA 98416

VAN EPPS, DENNIS EUGENE, b Rock Island, Ill, Nov 26, 46; m 73; c 2. IMMUNOLOGY. *Educ:* Western Ill Univ, BS, 68; Univ Ill, PhD(microbiol), 72. *Prof Exp:* NIH fel immunol, Univ NMex, 72-74, from asst prof to assoc prof, med & microbiol, 72-85, prof med & path, 85-88; DIR DEPT APPL CELLULAR BIOL, BAXTER HEALTH CARE INC, ILL, 88- *Concurrent Pos:* Arthritis Found fel, 74-; sect ed, J Leukocyte Biol; mem, neurol C study sect, NIH; prin investr grants, Nat Heart, Lung & Blood Inst, Nat Cancer Inst & Nat Inst Neurol & Commun Dis. *Honors & Awards:* Young Investr Pulmonary Res Award, Nat Heart & Lung Inst, 74; Sr Investr Award, Nat Arthritis Found. *Mem:* Am Soc Microbiol; Am Asn Immunologists; Am Fedn Clin Res; Sigma Xi; Am Asn Pathologists; Reticuloendothelial Soc. *Res:* Normal and abnormal phagocytic cell function and humoral factors which may alter this function; basic mechanisms of leukocyte locomotion; neutrophyl activation; lymphokines; interaction of neuropeptides and the immune system; cell surface receptor modulation. *Mailing Add:* Baxter Health Care Inc Wilson Rd Rte 120 Round Lake IL 60073

VAN ESELTINE, WILLIAM PARKER, b Syracuse, NY, Aug 21, 24; m 48; c 2. BACTERIOLOGY. *Educ:* Oberlin Col, AB, 44; Cornell Univ, MS, 47, PhD(bact), 49. *Prof Exp:* Asst bact, NY State Agr Exp Sta, 44-45 & Cornell Univ, 46-48; assoc prof, Clemson Col, 48-52; asst prof vet hyg, 52-59, assoc prof vet microbiol & prev med, 59-67, prof, med microbiol, 67-87, EMER PROF MED MICROBIOL, COL VET MED, UNIV GA, 87- *Mem:* AAAS; Am Soc Microbiol; Am Inst Biol Sci; NY Acad Sci. *Res:* Microbiology of foods; bactericidal and bacteriostatic agents; physiology and taxonomy of bacteria, especially animal pathogens. *Mailing Add:* Dept Med Microbiol Col Vet Med Univ Ga Athens GA 30602

VAN ESSEN, DAVID CLINTON, b Glendale, Calif, Sept 14, 45; m 69; c 2. BIOLOGY, NEUROSCIENCE. *Educ:* Calif Inst Technol, BS, 67; Harvard Univ, PhD(neurobiol), 71. *Prof Exp:* Res fel neurobiol, Harvard Med Sch, 71-73, neurophysiol, Inst Physiol, Univ Oslo, 73-75 & anat, Univ Col London, 75-76; from asst prof to assoc prof, 76-84, PROF BIOL, CALIF INST TECHNOL, 84- *Concurrent Pos:* Fel NIH, 71-73, Helen Hay Whitney Found, 73-76; mem adv panel, Sensory Physiol & Perception Prog, NSF, 78-81; Sloan res fel, 78-80. *Mem:* AAAS; Soc Neurosci; Asn Res Vision & Ophthal. *Res:* Visual cortex; functional organization of extrastriate areas in primates; neuromuscular development; control of synapse formation and elimination. *Mailing Add:* Div Biol Calif Inst Technol Pasadena CA 91125

VAN ETTEN, HANS D, b Peoria, Ill, Sept 16, 41; m 63; c 2. PLANT PATHOLOGY. *Educ:* Wabash Col, BA, 63; Cornell Univ, MS, 66, PhD(plant path), 70. *Prof Exp:* From asst prof to assoc prof plant path, Cornell Univ, 70-77; PROF PLANT PATHOL, UNIV ARIZ, 77- *Concurrent Pos:* Vis prof, Univ Munster, WGer, 78-79. *Mem:* AAAS; Am Phytopath Soc; NAm Photochem Soc; Am Soc Microbiol. *Res:* Physiology of disease. *Mailing Add:* Dept Plant Path Forbes Rm 104 Univ Ariz Tucson AZ 85721-0001

VAN ETTEN, JAMES L, b Cherrydale, Va, Jan 7, 38; m 60; c 3. MICROBIAL PHYSIOLOGY. *Educ:* Carleton Col, BA, 60; Univ Ill, MS, 63, PhD(plant path), 65. *Prof Exp:* NSF fel microbiol, Univ Pavia, 65-66; from asst prof to assoc prof, 66-74, PROF PLANT PATH, UNIV NEBR, LINCOLN, 74- *Mem:* AAAS; Am Phytopath Soc; Soc Gen Microbiol; Am Soc Microbiol; Am Soc Virol. *Res:* Biochemistry of fungal spore germination and bacteriophage and viruses of eukaryotic algae; biochemistry. *Mailing Add:* Dept Plant Path Univ Nebr Lincoln NE 68583-0722

VAN ETTEN, JAMES P(AUL), b Perry, NY, Mar 27, 22; m 47; c 7. ELECTRONICS ENGINEERING. *Educ:* US Coast Guard Acad, BS, 43; Mass Inst Technol, EE, 50. *Prof Exp:* Sr proj engr, ITT Corp, 58-59, exec engr, 59-60, assoc lab dir, 60-62, lab dir, 62-66, dir navig systs, 66-69, chief scientist avionics, 69-70, dir commun, navig & identification systs, labs, 70-73, tech adv to vpres & dir eng, 73-79, dir mkt, 80-84; RETIRED. *Concurrent Pos:* Mem, sci adv comt, US Coast Guard, 70-72; dir, Wild Goose Asn, 72-; consult, 84- *Mem:* Am Inst Navig; fel Inst Elec & Electronics Engrs; Wild Goose Asn (pres, 74-76). *Res:* Hyperbolic, rho-rho and rho-theta radio navigation systems and equipment; integrated airborne navigation systems and equipment; ground transmitting equipment for LORAN and TACAN; radio navigation systems. *Mailing Add:* 230 Rutgers Pl Nutley NJ 07110

VAN ETTEN, ROBERT LEE, b Evergreen Park, Ill, June 11, 37; c 3. ENZYMOLOGY, CLINICAL CHEMISTRY. *Educ:* Univ Chicago, BS, 59; Univ Calif, Davis, MS 64, PhD(chem), 65. *Prof Exp:* Technician, Ben May Labs, Cancer Res, Univ Chicago, 57-59; teaching asst, Univ Calif, Davis, 60-63; NIH fel Northwestern Univ, 65-66; from asst prof to prof chem, 66-83, head biochem div, 83-87, ASSOC HEAD, CHEM DEPT, PURDUE UNIV, WEST LAFAYETTE, 87- *Concurrent Pos:* Res career develop award, NIH, 69-73; Alexander von Humboldt fel, Marburg, Ger, 75-76; Nat Acad Sci Exchange, Poland, 85. *Honors & Awards:* Silver Medal, Polish Acad Sci, Krakow, 82. *Mem:* Am Chem Soc; Am Soc Biol Chemists; NY Acad Sci; Am Asn Clin Chem. *Res:* Mechanisms of enzymatic catalysis; phosphotyrosyl protein phosphatases; clinical chemistry of phosphatases and sulfatases; oxygen-18 isotope effects on carbon-13 and nitrogen-15 nuclear magnetic resonance spectra. *Mailing Add:* Dept Chem Purdue Univ West Lafayette IN 47907-1393

VAN EYS, JAN, b Hilversum, Neth, Jan 25, 29; nat US; m 55; c 2. BIOCHEMISTRY. *Educ:* Vanderbilt Univ, PhD, 55; Univ Wash, MD, 66. *Prof Exp:* Fel biochem, McCollum-Pratt Inst, Johns Hopkins Univ, 55-57; from asst prof to prof biochem, Sch Med, Vanderbilt Univ, 57-73, from asst prof to prof pediat, 68-73; prof pediat, Univ Tex, 73-79, Mosbacher chair pediat, M D Anderson Cancer Syst, 79-90, head dept pediat, 73-90, CHMN, DEPT PEDIAT, UNIV TEX SCH MED, HOUSTON, 90- *Concurrent Pos:* Investr, Howard Hughes Med Inst, 57-66. *Mem:* Am Soc Biol Chemists; Am Inst Nutrit; NY Acad Sci. *Res:* Metabolism and enzymology in glycolysis; pediatric hematology, oncology and nutrition in cancer. *Mailing Add:* Dept Pediat Univ Tex Med Sch PO Box 20708 Houston TX 77225

VAN FAASEN, PAUL, b Holland, Mich, June 6, 34; m 58; c 2. PLANT TAXONOMY. *Educ:* Hope Col, BA, 56; Mich State Univ, MS, 62, PhD(bot), 71. *Prof Exp:* Instr biol, Lake Forest Col, 62-63; PROF BIOL, HOPE COL, 63- *Mem:* Bot Soc Am; Am Soc Plant Taxonomists; Int Asn Plant Taxonomists. *Res:* Biosystematics of Aster, especially those of northeast United States; biology of weeds. *Mailing Add:* Dept Biol Hope Col Holland MI 49423

VAN FLANDERN, THOMAS C(HARLES), b Cleveland, Ohio, June 26, 40; m 63; c 4. CELESTIAL MECHANICS. *Educ:* Xavier Univ, Ohio, BS, 62; Yale Univ, PhD(astron), 69. *Prof Exp:* Astronr, 63-75, chief, celestial mech br, 75-83, PRES, META RES, US NAVAL OBSERV, 90- *Concurrent Pos:* Consult, Jet Propulsion Lab, 71. *Mem:* Am Astron Soc; Int Astron Union; AAAS. *Res:* Lunar motion; asteroids; comets; occulations; cosmology; gravitation; solar system astronomy. *Mailing Add:* 6327 Western Ave NW Washington DC 20015

VANFLEET, HOWARD BAY, b Salt Lake City, Utah, June 5, 31; m 54; c 7. SOLID STATE PHYSICS. *Educ:* Brigham Young Univ, BS, 55; Univ Utah, PhD(physics), 61. *Prof Exp:* Asst physics, Univ Utah, 56-60; from asst prof to assoc prof physics, 60-69, chmn, Dept Physics, 79-88, PROF PHYSICS, BRIGHAM YOUNG UNIV, 69- *Concurrent Pos:* Res grants, USAF Off Sci Res, Brigham Young Univ, 62-66, NSF, 69-77; phys scientist, US Army Electronics Command, 66-67; vis prof physics, Am Univ, Cairo, 73-74; consult, Codevintec Pac, Inc, 75-76. *Mem:* Am Phys Soc; Am Asn Physics Teachers; Sigma Xi. *Res:* Ultra high pressure solid state physics; particular phenomena, such as solid state diffusion, Mossbauer effects, melting and high pressure calibration; high temperature superconductors. *Mailing Add:* Dept Physics Brigham Young Univ Provo UT 84602

VAN FOSSAN, DONALD DUANE, b El Paso, Tex, Jan 5, 29; m 49; c 3. BIOCHEMISTRY. *Educ:* Sul Ross State Col, BS, 49; Univ Tex, MA, 52, PhD(biochem), 54, MD, 61. *Prof Exp:* Asst, Univ Tex, 52-54; res biochemist, Air Force Sch Aviation Med, 54-57, head lab sect, Dept Physiol-Biophys, 56-57; instr clin path & consult clin labs, Hosp, Univ Tex Med Br, 57-61; intern, St Joseph Hosp, Ft Worth, Tex, 61-62, resident path, Univ Tex, 62-66; dir clin path, Med Ctr, Baylor Univ, 66-69; DIR CHEM, ST JOHN'S HOSP, 69-; CLIN PROF PATH, SOUTHERN ILL UNIV, 69-, ASST CHMN DEPT, 73- *Concurrent Pos:* Instr anal chem, Trinity Univ, 56; dir clin path, St John's Hosp, 68-; med dir labs, St John's Hosp, 76- *Mem:* Col Am Pathologists; AMA; Am Soc Clin Pathologists. *Res:* Analytic biochemistry; pathology; endocrinology; toxicology. *Mailing Add:* 701 E Mason St Springfield IL 62708

VAN FOSSEN, DON B, b Des Moines, Iowa, Apr 15, 42; m 62; c 1. APPLIED MECHANICS, AEROSPACE ENGINEERING. *Educ:* Iowa State Univ, BS, 64; Univ Mo, Rolla, MS, 68. *Prof Exp:* Test engr struct, McDonnell Aircraft, 64-69; sr res engr, Babcock & Wilcox Co, 69-72, group supvr appl mech, 72-83; PRES, FINITE ELEMENT TECHNOL CORP, 83- *Concurrent Pos:* mem subcomt shells, Pressure Vessel Res Comt, Welding Res Coun, 77-; tech adv, Struct Anal Prog, User's Group, 77-83; past chmn, Cam Sect, Am Soc Mech Engrs & Pressure Vessel Piping Div, Am Soc Mech Engrs. *Mem:* Am Soc Mech Engrs. *Res:* Applied research in the application of the finite element method to structural analysis and heat transfer for general structures. *Mailing Add:* 10945 Hazelview Ave Alliance OH 44601

VAN FRANK, RICHARD MARK, b Lansing, Mich, Oct 11, 30; m 54; c 2. CELL BIOLOGY, ANALYTICAL BIOCHEMISTRY. *Educ:* Mich State Univ, BS, 52, MS, 56. *Prof Exp:* Officer in-chg biol lab, US Naval Damage Control Training Ctr, Philadelphia, 52-54; sr scientist, Div Molecular & Cell Biol, Lilly Res Labs, Eli Lilly & Co, 57-90; PUB POLICY ANALYST, IND ENVIRON INST, 91- *Mem:* AAAS; NY Acad Sci; Electrophoresis Soc. *Res:* Development of methodology for fractionation of cells and isolation and analysis of subcellular particles and substances; isolation and analysis of products produced using R-DNA technology; microsequencing of proteins; two dimensional gel electrophoresis image analysis. *Mailing Add:* 7620 Brookview Lane Indianapolis IN 46250

VAN FURTH, RALPH, b The Hague, Neth, Apr 30, 29; m; c 3. IMMUNOLOGY, PHARMACOLOGY. *Educ:* Univ Leiden, MD, 55, PhD(cum laude), 64; FRCP(E), 77. *Prof Exp:* Asst psychiat, Univ Utrecht, 55; asst gen practr, 56; resident internal med, St Elisabeth's Hosp, Groote Gasthuis, Haarlem, 57-61; chief resident internal med & res fel, Dept Immuno-Hemat & Bloodbank, 61-63, chief clin, Dept Microbiol Dis, 64-72, HEAD LAB CELLULAR IMMUNOL, DEPT INFECTIOUS DIS, UNIV HOSP, LEIDEN, 67- *Concurrent Pos:* Guest investr, Rockefeller Univ, 64-67 & 68; lectr, Internal Med & Immunol, Univ Leiden, 72-75; vis assoc prof, Rockefeller Univ, 72, vis prof, 74 & 85, Ruitinga Found, Acad Med Ctr, Amsterdam, 86; Lister fel, Royal Col Physicians, Edinburgh, 74, fel, 77; secy, Infectious Dis Soc, Neth & Flanders, 76-82; Macarthur post-grad lectureship, Edinburgh Univ Med Sch, 79; foreign corresp mem, Royal Acad Med, Belgium, 79; adv comt, Med Res, Neth Orgn, Adv Pure Res, 82-, Int Immunol-compromised Host Soc, 88-; chmn, Infectious Dis Soc, Neth & Flanders, 83-; Fulbright fel, 85; mem, Nat Comt AIDS Prev, 90-, Jury Found van Gysel, Med Res, 91- *Honors & Awards:* Alexandre Besredka Price, French Ger Found, Univ Berlin, 87; Friedrich Sasse Award, 88. *Mem:* Am Soc Immunologists; Am Soc Microbiol; Cell Kinetics Soc; Europ Soc Clin Invest; fel Infectious Dis Soc Am; NY Acad Sci; Reticuloendothelial Soc; Soc Gen Microbiol. *Res:* Author of various publications and journals. *Mailing Add:* Dept Infectious Dis Univ Hosp Bldg 1 C5-P PO Box 9600 Leiden RC 2300 Netherlands

VAN GEET, ANTHONY LEENDERT, b Rotterdam, Neth, July 24, 29; US citizen; m 56; c 3. PHYSICAL CHEMISTRY, ANALYTICAL CHEMISTRY. *Educ:* Delft Univ Technol, ChemEng, 55; Univ Southern Calif, PhD(phys chem), 61. *Prof Exp:* Res assoc chem, Mass Inst Technol, 61-63; asst prof, State Univ NY, Buffalo, 63-69; assoc prof, Oakland Univ, 69-70; ASSOC PROF CHEM, STATE UNIV NY COL OSWEGO, 70- *Mem:* AAAS; Am Chem Soc; Royal Neth Chem Soc; Sigma Xi. *Res:* Nuclear magnetic resonance of protons, lithium, sodium and fluorine in solution; hydration, complexation and ion-pairing of monovalent ions; determination of trace amounts of heavy metals in lake water; instrumentation. *Mailing Add:* Dept Chem State Univ NY Oswego NY 13126

VAN GELDER, ARTHUR, b Paterson, NJ, Jan 13, 38; m 60; c 2. ELECTRICAL ENGINEERING. *Educ:* Univ Pa, BSEE, 59; City Col New York, MEE, 64, PhD(elec eng), 68. *Prof Exp:* Asst proj engr, Kearfott Div, Gen Precision, Inc, 60-61; lectr elec eng, City Col New York, 61-68; asst prof, Univ Del, 68-75 & Miami Univ, 75-80; asst prof elec eng, Lafayette Col, 80-87; DIR ELECTRONICS ENG, UNIV SCRANTON, 87- *Mem:* Inst Elec & Electronics Engrs; Am Soc Eng Educ. *Res:* Electronic circuits; control systems; digital systems. *Mailing Add:* Dept Phys & Elec Eng Univ Scranton Scranton PA 18510

VAN GELDER, NICO MICHEL, b Sumatra, Neth E Indies, Dec 24, 33; Can citizen; m 59; c 3. BIOCHEMISTRY, PHYSIOLOGY. *Educ:* McGill Univ, BSc, 55, PhD(biochem), 59. *Prof Exp:* Life Ins Med Res Fund fel, Cambridge Univ, Eng, 59-60; res fel neurophysiol & neuropharmacol, Harvard Med Sch, 60-62; asst prof pharmacol, Sch Med, Tufts Univ, 62-67; assoc prof physiol, Univ Montreal, 67-77, exec, Neurol Sci Res Ctr, 76-82, bd adv, Neurochem Res, 75-88, PROF PHYSIOL, UNIV MONTREAL, 77-, ASST VDEAN RES, 89- *Concurrent Pos:* Res grants, Nat Inst Neurol Dis & Blindness, 62-66, Nat Multiple Sclerosis Soc, 66- & Med Res Coun Can Neurol Sci Group, 67-79; UNESCO prof, grad studies, Latin Am, 73-; assoc ed, Can J Biochem, 76-78; med adv, Savoy Found Epilepsy, 80-, pres med bd, 83-; adj prof, Montreal Neurol Inst, 80-; expert consult, Pan Am Health Orgn, 88-; mem bd, Spinal Cord Res Found, 89- *Mem:* Am Soc Neurochem; Int Soc Neurochem; NY Acad Sci; Int Brain Res Orgn; Europ Soc Neurochem;

Venezuelan Soc Neurosci. *Res:* Biochemistry of epilepsy; structure-activity relationships; biochemistry of brain damage; malnutrition; genetic contribution to epilepsy; function of taurine. *Mailing Add:* Dept Physiol Univ Montreal CP 6128 Succursale A Montreal PQ H3C 3J7 Can

VAN GELDER, RICHARD GEORGE, b New York, NY, Dec 17, 28; m 62; c 3. MAMMALOGY. *Educ:* Colo Agr & Mech Col, BS, 50; Univ Ill, MS, 52, PhD(zool), 58. *Prof Exp:* Asst zool, Colo Agr & Mech Col, 47-50 & Univ Ill, 50-53; from asst to instr mammal, Univ Kans, 54-56; from asst cur to assoc cur mammals, 56-69, chmn dept, 59-74, CUR MAMMALS, AM MUS NATURAL HIST, 69- *Concurrent Pos:* Lectr, Columbia Univ, 58-59, asst prof, 59-63; mem bd dirs, Archbold Exped, Inc, 64-75 & Quincy Bog Natural Area, 76-; prof lectr, State Univ NY Downstate Med Ctr, 70-73. *Mem:* AAAS; Am Soc Mammalogists (vpres, 67-68, pres, 68-70); Wildlife Soc; Soc Syst Zool. *Res:* Mammalian taxonomy, evolution, behavior and ecology; mammalian natural history, classification and history. *Mailing Add:* Seven Florence Rd Harrington Park NJ 07640

VAN GELUWE, JOHN DAVID, b Rochester, NY, Sept 18, 16; m 53; c 3. ENTOMOLOGY. *Educ:* State Univ NY, BS, 39. *Prof Exp:* Asst, Exten Serv, State Univ NY Col Agr, Cornell Univ, 39-44; dir res & develop, Soil Bldg Div, Coop GLF Exchange, Inc, 44-64; mgr prod develop, Agr Chem Div, Ciba-Geigy Corp, 64-67, tech dir, 67-70, asst dir field & farm res, 70-82; RETIRED. *Mem:* Fel Am Soc Hort Sci; fel Entom Soc Am; fel Am Phytopath Soc; Weed Sci Soc Am. *Res:* Formulations, basic laboratory evaluation and field testing of insecticides, fungicides and herbicides. *Mailing Add:* Seven Lakes PO Box 541 West End NC 27376

VAN GEMERT, BARRY, b Attleboro, Mass, Feb 17, 46; m 69; c 3. ORGANIC CHEMISTRY. *Educ:* Univ Mass, BS, 68; Univ RI, MS, 72; Purdue Univ, PhD(org chem), 76. *Prof Exp:* SR RES CHEMIST, PPG INDUSTS, 76- *Mem:* Am Chem Soc. *Res:* Organic synthesis; preparation of novel photochronic compounds for use in sunglass applications. *Mailing Add:* 2004 High Pointe Dr Murrysville PA 15668-8515

VAN GINNEKEN, ANDREAS J, b Wynegem, Belg, Jan 1, 35; m 72; c 1. PHYSICS. *Educ:* Univ Chicago, MSc, 59, PhD(chem), 66. *Prof Exp:* Res assoc physics, McGill Univ, 66-70; PHYSICIST, FERMI NAT LAB, 70- *Mem:* Am Phys Soc. *Res:* Nuclear and particle physics; radiation physics; nuclear chemistry. *Mailing Add:* Fermi Nat Lab Batavia IL 60510

VAN GROENEWOUD, HERMAN, b Breda, Netherlands, May 27, 26; Can citizen; m 50; c 2. FOREST ECOLOGY. *Educ:* Univ Sask, BA, 56, MA, 60; Swiss Fed Inst Technol, ScD(geobot), 65. *Prof Exp:* Res scientist forest path, Lab Saskatoon, 54-65, RES SCIENTIST FOREST ECOL, MARITIMES FOREST RES CTR, GOVT CAN, 65- *Mem:* Can Bot Asn; Brit Ecol Soc; Ecol Soc Am; Can Inst Forestry. *Res:* Forest site studies; multivariate analysis; watershed studies. *Mailing Add:* Maritimes Forest Res Ctr Box 4000 Fredericton NB E3B 5P7 Can

VAN GULICK, NORMAN MARTIN, b Los Angeles, Calif, July 1, 26; m 48; c 2. PHYSICAL MATHEMATICS. *Educ:* Univ Colo, AB, 48; Univ Southern Calif, PhD, 54. *Prof Exp:* Asst prof chem, Univ Ore, 56-60; res chemist, 60-80, res assoc, Polymer Prod Dept, 80-89, CONSULT, E I DU PONT DE NEMOURS & CO, INC, 90- *Concurrent Pos:* Post doctorate, Harvard Univ, 54; chem warfare, US Army, 54-56. *Mem:* Am Chem Soc; Sigma Xi. *Res:* Organic polymer chemistry; organometallic and organic fluorine chemistry; reaction mechanisms; liquid-drop models. *Mailing Add:* 2022 Lower Lane Arden Wilmington DE 19810-4224

VAN GUNDY, SEYMOUR DEAN, b Whitehouse, Ohio, Feb 24, 31; m 54; c 2. PLANT PATHOLOGY, NEMATOLOGY. *Educ:* Bowling Green State Univ, BA, 53; Univ Wis, PhD, 57. *Prof Exp:* Res assoc, Univ Wis, 53-57; from asst nematologist to assoc nematologist, Univ Calif, Riverside, 57-68, assoc dean res, 68-71 & 85-88, asst vchancellor res, 71-72, chmn dept, 72-84, interim dean, 88-90, PROF NEMATOL, UNIV CALIF, RIVERSIDE, 68-, DEAN RES, 90- *Concurrent Pos:* NSF sr fel, Australia, 65-66; ed-in-chief, J Nematol, 67-71. *Mem:* Fel AAAS; fel Am Phytopath Soc; fel Soc Nematol (vpres, 72-73, pres, 73-74); Soc Europ Nematol; Am Inst Biol Sci. *Res:* Biology and control of nematodes. *Mailing Add:* Dept Nematol Univ Calif Riverside CA 92521

VAN HALL, CLAYTON EDWARD, b Grand Rapids, Mich, Apr 24, 24; m 51; c 2. ANALYTICAL CHEMISTRY. *Educ:* Hope Col, AB, 49; Mich State Univ, MSc, 54, PhD(anal chem), 56. *Prof Exp:* Asst, Mich State Univ, 52-56; chemist, 56-58, anal chemist, 58-61, anal specialist, 61-65, anal res specialist, 65-72, ASSOC SCIENTIST, DOW CHEM CO, 72- *Res:* Instrumental methods; trace gas methods; trace element methods; purity of inorganic compounds; primary standards; water analysis. *Mailing Add:* 3712 Wintergreen Dr Midland MI 48640

VAN HANDEL, EMILE, b Rotterdam, Holland, Mar 29, 18; nat US; m 46; c 2. ORGANIC CHEMISTRY. *Educ:* State Univ Leiden, BS, 38, MS, 41; State Inst Technol, Delft, MS, 41; Univ Amsterdam, PhD(biochem), 54. *Prof Exp:* Indust chemist, 45-54; res biochemist, St Anthon's Hosp, Voorburg, Holland, 54-55; asst prof physiol, Univ Tenn, 55-58; biochemist, 58-83, PROF BIOCHEM, FLA MED ENTOM LAB, UNIV FLA, VERO BEACH, 83- *Concurrent Pos:* Consult, study sect trop med & parasitol, NIH, 78-82. *Mem:* Am Heart Asn; Am Soc Biol Chemists. *Res:* Lipid and carbohydrate chemistry and metabolism; insect biochemistry; atherosclerosis. *Mailing Add:* Fla Med Entom Lab Univ Fla 200 Ninth St SE Vero Beach FL 32962

VAN HARN, GORDON L, b Grand Rapids, Mich, Dec 30, 35; m 58; c 3. PHYSIOLOGY. *Educ:* Calvin Col, AB, 57; Univ Ill, MS, 59, PhD(physiol), 61. *Prof Exp:* Assoc prof biol, Calvin Col, 61-68 & Oberlin Col, 68-70; PROF BIOL, CALVIN COL, 70- *Concurrent Pos:* Res assoc, Blodgett Mem Hosp, 70-76. *Mem:* AAAS; Am Sci Affil. *Res:* Smooth muscle contractile and electrical activity; intestinal smooth muscle; cardiac muscle. *Mailing Add:* Provost Calvin Col Grand Rapids MI 49546

VAN HARREVELD, ANTHONIE, physiology, for more information see previous edition

VAN HASSEL, HENRY JOHN, b Paterson, NJ, May 2, 33; m 60. DENTISTRY, PHYSIOLOGY. *Educ:* Maryville Col, BA, 54; Univ Md, DDS, 63; Univ Wash, MSD, 64, PhD(physiol), 69. *Prof Exp:* Asst prof endodont & physiol, Med Sch, Univ Wash, 69-71, res assoc physiol, Regional Primate Res Ctr, 69-76, assoc prof physiol, 71-, assoc prof endodont, Dent Sch, 71-, assoc prof biophys, 76-; dep chief dent serv & dir endodont residency, USPHS Hosp, Seattle, 71-; chmn, dept endodont, Univ Md; DEAN, SCH DENT, ORE HEALTH SCI UNIV, 84- *Concurrent Pos:* Consult, US Army, Ft Lewis, Wash, 69-; chmn, Nat Workshop Pulp Biol, 71-; vchmn, Sect Physiol, Am Asn Dent Schs, 72- *Honors & Awards:* Carl A Schlack Award, Asn Mil Surgeons US, 71. *Mem:* Am Dent Asn; Am Asn Endodont; Int Asn Dent Res. *Res:* Oral physiology; psychophysiology; neurophysiology of pain. *Mailing Add:* Sch Dent Ore Health Sci Univ 611 SW Campus Dr Portland OR 97201-3097

VAN HATTUM, ROLLAND JAMES, speech pathology, audiology, for more information see previous edition

VAN HAVERBEKE, DAVID F, b Eureka, Kans, July 15, 28. TAXONOMY, SILVICULTURE. *Educ:* Kans State Univ, BS, 50; Colo State Univ, MS, 59; Univ Nebr, PhD(bot), 67. *Prof Exp:* Res forester, Rocky Mountain Forest & Range Exp Sta, USDA, 58 & Southeastern Forest Exp Sta, 59-62, res forester, 62-; RETIRED. *Concurrent Pos:* Assoc prof forestry, Univ Nebr, Lincoln. *Mem:* Soc Am Foresters. *Res:* Forest botany and genetics; forest tree improvement; noise abatement; shelterbelt management. *Mailing Add:* 6919 W 101st St Shawnee Mission KS 66212

VAN HECKE, GERALD RAYMOND, b Evanston, Ill, Nov 1, 39. PHYSICAL CHEMISTRY. *Educ:* Harvey Mudd Col, BS, 61; Princeton Univ, AM, 63, PhD(phys chem), 66. *Prof Exp:* Chemist, Shell Develop Co, 66-70; asst prof, 70-74, assoc prof, 74-80, PROF CHEM, HARVEY MUDD COL, 80- *Concurrent Pos:* Vis res assoc fundamental physics, Univ Lille, France, 77; vis res assoc biophys, Boston Univ, 77; Nat Acad Sci exchange scientist, Inst Phys Chem, Polish Acad Sci, Warsaw, 80 & Cent Inst Electron Physics, Ger Dem Repub Acad Sci, EBerlin, 83; fac fel, Jet Propulsion Lab, NASA, 82, 83; univ guest researcher, Chem Thermodyn Lab, Osaka Univ, 84, 89. *Mem:* AAAS; Am Chem Soc; Royal Soc Chem; fel Am Inst Chemists; Sigma Xi; NY Acad Sci. *Res:* Nuclear magnetic resonance studies of paramagnetic transition metal complexes; thermodynamics and physical properties of liquid crystals. *Mailing Add:* Dept Chem Harvey Mudd Col Claremont CA 91711-5990

VAN HEERDEN, PIETER JACOBUS, b Utrecht, Neth, Apr 14, 15; nat US; m 49; c 2. PHYSICS. *Educ:* Univ Utrecht, PhD(physics), 45. *Prof Exp:* Res physicist, Bataafse Petrol Co, Neth, 44-45; vis lectr, Harvard Univ, 48-49, res fel nuclear physics, 49-53; res assoc, Gen Elec Res Lab, NY, 53-62; physicist, Polaroid Res Labs, 62-81; LECTR & CONSULT, FOUND PHYSICS, MATH THEORY INTEL & ARTIFICIAL INTEL, 82- *Mem:* Am Phys Soc; Neth Phys Soc; Am Philos Sci Asn; Inst Elec & Electronics Engrs; Cognitive Sci Soc. *Res:* Experimental nuclear and solid state physics; foundation of physics; foundation of scientific knowledge and intelligence. *Mailing Add:* 18217 145th Ct NE Woodinville WA 98072

VAN HEUVELEN, ALAN, b Buffalo, Wyo, Dec 15, 38; m 62; c 1. PHYSICS. *Educ:* Rutgers Univ, BA, 60; Univ Colo, PhD(physics), 64. *Prof Exp:* Assoc prof physics, 64-74, PROF PHYSICS, NMEX STATE UNIV, 74- *Mem:* Am Phys Soc. *Res:* Biophysics using electron spin resonance to study enzymes. *Mailing Add:* Dept Physics NMex State Univ Las Cruces NM 88003

VAN HEYNINGEN, EARLE MARVIN, b Chicago, Ill, Oct 15, 21; m 51; c 4. ORGANIC CHEMISTRY. *Educ:* Calvin Col, AB, 43; Univ Ill, PhD(org chem), 46. *Prof Exp:* Res chemist, Eli Lilly & Co, 46-65, res scientist, 65-66, res assoc, 66-69, dir agr chem, Greenfield Labs, 69-72, dir chem, 72-83, dir biochem & phys chem, Lilly Res Labs, 83-85; RETIRED. *Mem:* Am Chem Soc. *Res:* Synthesis of barbituric acids; antimalarials, anti-arthritics; cholesterol lowering agents; cephalosporin antibiotics. *Mailing Add:* 2919 S Post Rd Indianapolis IN 46239

VAN HEYNINGEN, ROGER, b Chicago, Ill, Oct 2, 27; m 51; c 3. SOLID STATE PHYSICS. *Educ:* Calvin Col, AB, 51; Univ Ill, MS, 55, PhD, 58. *Prof Exp:* Sr physicist, 58-62, res assoc, 62-66, asst div head, 66-67, PHYSICS DIV DIR, EASTMAN KODAK CO, 67- *Mem:* Am Phys Soc; Optical Soc Am. *Res:* Electronic and optical properties of insulating and semiconducting solids. *Mailing Add:* Res Labs Eastman Kodak Co Rochester NY 14650

VAN HISE, JAMES R, b Tracy, Calif, Aug 11, 37; m 64; c 2. PHYSICAL CHEMISTRY. *Educ:* Walla Walla Col, BS, 59; Univ Ill, PhD(phys chem), 63. *Prof Exp:* Res assoc nuclear chem, Oak Ridge Nat Lab, 63-65; from asst prof to assoc prof chem & physics, Andrews Univ, 65-69; prof physics & chmn dept, Tri-State Col, 69-72; PROF CHEM, PAC UNION COL, 72- *Concurrent Pos:* Consult radiol phys & nuclear chem. *Mem:* Am Chem Soc; Am Phys Soc; Sigma Xi. *Res:* Nuclear photodisintegration; alpha, beta and gamma ray spectroscopy; positron annihilation in organic media. *Mailing Add:* 566 Sunset Dr Angwin CA 94508

VAN HOLDE, KENSAL EDWARD, b Eau Claire, Wis, May 14, 28; m 50; c 4. PHYSICAL CHEMISTRY. *Educ:* Univ Wis, BS, 49, PhD(chem), 52. *Prof Exp:* Res chemist textile fibers dept, E I du Pont de Nemours & Co, 52-55; res assoc, Univ Wis, 55-56, asst prof chem, Univ Wis-Milwaukee, 56-57; from asst prof to prof, Univ Ill, Urbana, 57-67; PROF BIOPHYS, ORE STATE UNIV, 67- *Concurrent Pos:* Guggenheim fel, 73-74; Am Cancer Soc res professorship, 76- *Mem:* Nat Acad Sci; Am Soc Biol Chem. *Res:* Physical chemistry of biological macromolecules; biophysical chemistry. *Mailing Add:* Dept Biochem & Biophys Ore State Univ Corvallis OR 97331

VAN HOOK, ANDREW, physical chemistry; deceased, see previous edition for last biography

VAN HOOK, JAMES PAUL, b Paterson, NJ, Oct 16, 31; m 57; c 5. PHYSICAL CHEMISTRY, FUEL TECHNOLOGY. *Educ:* Col of the Holy Cross, BS, 53; Princeton Univ, PhD(chem), 58. *Prof Exp:* Res chemist, M W Kellogg Co Div, Pullman, Inc, 57-62, supvr, 62-65, sect head process res, 65-74; process develop mgr, Corp Eng Dept, Allied Chem Corp, 74-76; process technol mgr & lab coordr, Foster Wheeler Energy Corp, 77-81, technol mgr, Foster Wheeler Synfuels Corp, 81-86, SR RES ASSOC, FORSTER WHEELER DEVELOP CORP, 86- *Mem:* AAAS; Am Chem Soc; Am Inst Chem Engrs. *Res:* Catalytic oxidation for chlorine production; steam-hydrocarbon reactions for production of synthesis gas, hydrogen or synthetic natural gas; coal gasification; air pollution control; delayed coking and solvent deasphalting; shale oil processing; petroleum engineering. *Mailing Add:* 102 Harrison Brook Dr Basking Ridge NJ 07920

VAN HOOK, ROBERT IRVING, JR, b Rome, Ga, Jan 21, 42; m 64; c 2. ENVIRONMENTAL SCIENCE, ECOLOGY. *Educ:* Clemson Univ, BS, 66, PhD(entom), 69. *Prof Exp:* Assoc res ecologist radiation effects, Ecol Sci Div, 70-72, res ecologist animal ecol, Environ Sci Div, 73-76, tech asst life sci, Dir Staff, 76-77, prog mgr, ecosyst studies, 77-80, SECT HEAD, TERRESTRIAL ECOL SECT, ENVIRON SCI DIV, OAK RIDGE NAT LAB, 80- *Concurrent Pos:* Adj asst prof, Ecol Prog, Univ Tenn, 77- *Mem:* Ecol Soc Am; Sigma Xi; AAAS. *Res:* Management of and participation in basic and applied ecological research concerning productivity, biogeochemical cycling, and pollutant effects associated with fossil and non-fossil energy systems; current emphasis on glorac carbon cycling, acid rain and biomass production. *Mailing Add:* Bldg 1505 Oak Ridge Nat Lab Oak Ridge TN 37830

VAN HOOK, WILLIAM ALEXANDER, b Paterson, NJ, Jan 14, 36; m 62; c 3. PHYSICAL CHEMISTRY. *Educ:* Col of the Holy Cross, BS, 57; Johns Hopkins Univ, MA, 59, PhD(chem), 61. *Prof Exp:* Res assoc phys chem, Brookhaven Nat Lab, 61-62; from asst prof to assoc prof, 62-72, PROF CHEM, UNIV TENN, KNOXVILLE, 72- *Concurrent Pos:* Fulbright res fel, Belg, 67-68; Nat Acad Sci exchange fel, Yugoslavia, 71; vis prof, Univ Beijing, 85, Lunzhou, 87. *Mem:* AAAS; Am Chem Soc. *Res:* Isotope effects on chemical and physical properties of molecular systems; solutions. *Mailing Add:* Dept Chem Univ Tenn Knoxville TN 37996

VAN HOOSIER, GERALD L, JR, b Weatherford, Tex, June 4, 34; m 59; c 2. LABORATORY ANIMAL SCIENCE, ANIMAL VIROLOGY. *Educ:* Agr & Mech Col Tex, DVM, 57. *Prof Exp:* Head animal test sect, Div Biol Standards, NIH, 57-59, in serv training, Viral & Rickettsial Dis Lab, Calif State Dept Health, 59-60, head appl virol sect, Div Biol Standards, 60-62; from instr to assoc prof exp biol, Baylor Col Med, 62-69; from asst prof to assoc prof vet path & dir lab animal resources, Wash State Univ, 69-75; dir, Div Animal Med & Prof Animal Med & Path, 75-88, PROF & CHMN, DEPT COMP MED, UNIV WASH, 89- *Concurrent Pos:* Resident path, Baylor Col Med, 69-70 & Wash State Univ, 70-71; mem animal resources adv comt, Animal Res Bd, NIH, 74-78. *Honors & Awards:* Charles A Griffin Award, 86. *Mem:* AAAS; Am Asn Lab Animal Sci; Am Soc Exp Path; Am Vet Med Asn. *Res:* Laboratory animal disease and medicine; comparative pathology; animal virology. *Mailing Add:* Dept Comp Med Univ Wash Seattle WA 98195

VAN HORN, D(AVID) A(LAN), b Des Moines, Iowa, Apr 4, 30; m 60; c 2. CIVIL ENGINEERING. *Educ:* Iowa State Univ, BS, 51, MS, 56, PhD(struct eng), 59. *Prof Exp:* Hwy engr, Fed Hwy Admin, 51-54; asst, Iowa State Univ, 54-55, instr civil eng, 55 & 56-58, instr theoret & appl mech, 55-56, from asst prof to assoc prof civil eng, 58-62; res assoc prof, 62-66, PROF CIVIL ENG, LEHIGH UNIV, 66-, CHMN DEPT, 67- *Mem:* Am Soc Civil Engrs; Am Concrete Inst; Am Soc Eng Educ; Sigma Xi. *Res:* Structural engineering; behavior of prestressed and reinforced concrete members and structures; behavior of structural materials; structural analysis. *Mailing Add:* Dept Civil Eng Lehigh Univ Bldg 13 Bethlehem PA 18015-3176

VAN HORN, DAVID DOWNING, b Rochester, NY, Apr 23, 21; m 45. METAL PHYSICS, MATHEMATICS. *Educ:* Univ Rochester, BA, 42; Case Inst Technol, PhD(physics), 49. *Prof Exp:* Instr physics, Univ Rochester, 43-44; jr physicist, Clinton Eng Works, 44-46; instr physics, Case Inst Technol, 46-49; res assoc metall, Knolls Atomic Power Lab, 49-57, group leader chem & metall eng, Incandescent Lamp Dept, 57-81, CONSULT PHYSICIST, INCANDESCENT & SPECIALTY LAMP ENG DEPT, GEN ELEC CO, 81- *Mem:* Am Phys Soc; Am Soc Metals; Am Asn Physics Teachers; Math Asn Am; Am Inst Mining, Metall & Petrol Engrs; Math Asn Am; Soc Indust Appl Math; Sigma Xi. *Res:* Solid state diffusion; mechanical properties; heat transfer; tungsten; incandescent lamps; radiation measurements. *Mailing Add:* 15959 Glynn Rd Gen Elec Co East Cleveland OH 44112

VAN HORN, DIANE LILLIAN, b Waukesha, Wis, Aug 21, 39; m 72; c 3. PHYSIOLOGY, ELECTRON MICROSCOPY. *Educ:* Univ Wis-Madison, BS, 61; Marquette Univ, MS, 66, PhD(physiol), 68. *Prof Exp:* Res assoc, Wood Vet Admin Ctr, 66-67, supvry scientist, Electron Micros Lab, 69-75; from instr to assoc prof, 68-77, PROF PHYSIOL & OPHTHAL, MED COL WIS, 77-; CHIEF ELECTRON MICROS SECT, WOOD VET ADMIN CTR, 75- *Concurrent Pos:* Seeing Eye Inc res grant, Med Col Wis, 69-72, Nat Eye Inst res grant, 72-79; Nat Eye Inst Ctr grant, 77-81. *Honors & Awards:* William & Mary Greve Int Res Scholar Award, Res Prevent Blindness, Inc, 81. *Mem:* Am Asn Univ Prof; Am Physiol Soc; Asn Res Vision & Ophthal; Electron Micros Soc Am. *Res:* Corneal physiology and ultrastructure; electron microscopy of ocular and other tissues. *Mailing Add:* 8505 W Rae Ct Greendale WI 53219

VAN HORN, DONALD H, b Hinsdale, Ill, Oct 9, 28; m 59; c 2. ECOLOGY. *Educ:* Kalamazoo Col, BA, 50; Univ Ill, MS, 52; Univ Colo, PhD(zool), 61. *Prof Exp:* Asst prof biol, Lake Forest Col, 61-62 & Utica Col, 62-65; vis asst prof, Univ Colo, Colorado Springs Ctr, 65-71, assoc prof, 71-74, chmn dept, 74-77, prof biol, 74-89; RETIRED. *Mem:* AAAS; Am Ornith Union; Ecol Soc Am; Am Soc Zoologists. *Res:* Terrestrial ecology, especially community and population analysis of mountain animals. *Mailing Add:* 4118 Tumbleweed Dr Colorado Springs CO 80907

VAN HORN, GENE STANLEY, b Oakland, Calif, June 26, 40; m 62; c 3. SYSTEMATIC BOTANY. *Educ:* Humboldt State Univ, AB, 63; Univ Calif, Berkeley, PhD(bot), 70. *Prof Exp:* Vis asst prof biol, Tex Tech Univ, 70-71; asst prof, 71-78, assoc prof, 78-80, PROF BIOL, UNIV TENN, CHATTANOOGA, 80- *Concurrent Pos:* Tex State Inst Funds grant, Tex Tech Univ, 71; Univ Chattanooga Found grant, Univ Tenn, Chattanooga, 72-73 & 86. *Mem:* Bot Soc Am; Inst Asn Plant Taxonomists; Am Soc Plant Taxonomists. *Res:* Biosystematics and evolution of angiosperms, especially asteraceae; floristics; biogeography. *Mailing Add:* Dept Biol Univ Tenn Chattanooga TN 37403

VAN HORN, HAROLD H, JR, b Pomona, Kans, Jan 13, 37; m 58; c 3. DAIRY SCIENCE. *Educ:* Kans State Univ, BS, 58, MS, 59; Iowa State Univ, PhD(dairy nutrit), 62. *Prof Exp:* Assoc prof dairy nutrit & mgt & exten dairyman, Iowa State Univ, 61-70; PROF DAIRY SCI & CHMN DEPT, UNIV FLA, 70-, ANIMAL NUTRITIONIST, 74- *Mem:* Am Dairy Sci Asn; Am Soc Animal Sci. *Res:* Improved dairy feeding and management practices; nutrition research in the use of urea in dairy rations. *Mailing Add:* Dept Dairy Sci Univ Fla Gainesville FL 32611

VAN HORN, HUGH MOODY, b Williamsport, Pa, Mar 5, 38; m 60; c 3. ASTROPHYSICS. *Educ:* Case Inst Technol, BS, 60; Cornell Univ, PhD(astrophys), 66. *Prof Exp:* Res assoc, 65-67, asst prof, 67-72, assoc prof astrophys, 72-77, chmn, Dept Physics & Astron, 80-86, actg assoc dean, 87-89, PROF PHYSICS & ASTRON, UNIV ROCHESTER 77- *Concurrent Pos:* Vis fel, Joint Inst Lab Astrophys, Univ Colo, Boulder, 73-74; invited speaker, Am Astron Soc, 80; vis prof, Univ Texas, Austin, 87. *Mem:* Int Astron Union; Am Astron Soc; AAAS. *Res:* Degenerate dwarfs, brown dwarfs, and neutron stars; structure, evolution, oscillations and atmospheres; nuclear reactions and equation of state in stars; accretion disk structure and oscillations. *Mailing Add:* Dept Physics & Astron Univ Rochester Rochester NY 14627-0011

VAN HORN, KENT R(OBERTSON), metallurgy, for more information see previous edition

VAN HORN, LLOYD DIXON, b Bartlesville, Okla, Mar 25, 38; m 59; c 2. CHEMICAL ENGINEERING. *Educ:* Rice Univ, BA, 59, PhD(chem eng), 66. *Prof Exp:* Res engr, Shell Oil Co, Tex, 66-68, supvr chem eng res & develop, 68-69, asst to mgr mfg res & develop, head off, NY, 69-70, sr engr, head off, Houston, 70-72; CONSULT ENGR, BILES & ASSOCS, HOUSTON, 72- *Mem:* Am Chem Soc. *Res:* Chemical reactor analysis and simulation; multiphase fluid flow in packed beds; applications of advanced process computer control; thermodynamics of hydrocarbon-hydrogen systems. *Mailing Add:* 14838 LaQuinta Lane Houston TX 77079

VAN HORN, RUTH WARNER, b Waterloo, Iowa, Mar 24, 18; m 45. ORGANIC CHEMISTRY. *Educ:* Univ Calif, Los Angeles, BA, 39, MA, 40; Pa State Univ, PhD(org chem), 44. *Prof Exp:* Org chemist, Am Cyanamid Co, 44-48; instr chem, Hunter Col, 48-49; from asst prof to assoc prof, 49-64, PROF CHEM, FRANKLIN & MARSHALL COL, 64- *Mem:* AAAS; Am Chem Soc. *Res:* Synthesis. *Mailing Add:* 1726 Old Philadelphia Pike Lancaster PA 17602

VAN HORN, WENDELL EARL, b Cantril, Iowa, Feb 8, 29; m 59, 80; c 4. CHEMICAL ENGINEERING. *Educ:* Iowa State Col, BS, 52; Univ Chicago, MBA, 74. *Prof Exp:* Develop engr, Minn Mining & Mfg Co, 53-56; process develop engr, Quaker Oats Co, Tex, 56-70, mgr chem eng res & develop, 70-77, sr res assoc, 77-81, sr process assoc, 81-84, ENVIRON & OCCUP SAFETY, QUAKER OATS CHEM INC, TENN, 84- *Mem:* Am Inst Chem Engrs. *Res:* Pollution control; development of waste acetic acid recovery methods; process development of furfural manufacturing methods; liquid phase and vapor phase catalytic processes; fluized bed combustion. *Mailing Add:* Quaker Oats Chem Inc 3324 Chelsea Ave PO Box 8035 Memphis TN 38108-0035

VAN HORNE, ROBERT LOREN, b Malvern, Iowa, Dec 26, 15; m 41, 63; c 5. PHARMACOGNOSY, PHARMACY. *Educ:* Univ Iowa, BS, 41, MS, 47, PhD, 49. *Prof Exp:* Instr pharm, Univ Iowa, 49-51, from asst prof to assoc prof pharmacog, 51-56; dean, Sch Pharm, Univ Mont, 56-75, dir continuing educ, Sch Pharm, 75-80, prof pharm, 56-85; RETIRED. *Concurrent Pos:* Mem fac adv coun, Gov of Mont; secy-treas, Western States Pharm Conf, 66-83; dir, Western Area Alcohol Educ & Training Prog, Nev, 74-; chmn, Mont Adv Coun Alcohol & Drug Dependence, 76-85. *Mem:* Am Asn Cols Pharm; Am Pharmaceut Asn. *Res:* Water soluble embedding materials for microtechnique; polyethylene glycols as substitutes for glycerin and ethanol in pharmaceutical preparations; surfactants in the preparation of coal tar lotions; anionic exchange resins for alkaloid separation; phytochemistry of mistletoe species. *Mailing Add:* 91 Brookside Way Missoula MT 59802-3278

VAN HOUTEN, FRANKLYN BOSWORTH, b New York, NY, July 14, 14; m 43; c 3. GEOLOGY. *Educ:* Rutgers Univ, BS, 36; Princeton Univ, PhD(geol), 41. *Prof Exp:* Instr geol, Williams Col, 39-42; from asst prof to prof, 47-85, EMER PROF GEOL, PRINCETON UNIV, 85- *Concurrent Pos:* Consult, 41-; geologist, US Geol Surv, 48-55 & Geol Surv Can, 53; vis prof, Univ Calif, Los Angeles, 63, State Univ NY, Binghamton, 71 & Univ Basel, 71. *Honors & Awards:* Twenhofel Medal, Soc Econ Paleontologists & Mineralogists. *Mem:* Fel Geol Soc Am; hon mem Soc Econ Paleontologists & Mineralogists; Am Asn Petrol Geologists; Int Asn Sedimentol; hon mem Colombian Geol Soc. *Res:* Sedimentology; clay minerals; zeolites; iron oxides; red beds; Triassic rocks, eastern North America and northwestern Africa, continental drift reconstructions; Cenozoic nonmarine deposits, western United States and northern South America; modern marine sediments; molasse facies in orogenic belts; Phanerozoic oolitic ironstones and glauconitic greensands; Nubian sandstone of northern Africa. *Mailing Add:* Dept Geol & Geophys Sci Princeton Univ Princeton NJ 08540

VAN HOUTEN, ROBERT, b Peoria, Ill, Oct 2, 23; m 47; c 4. MATERIALS ENGINEERING, NUCLEAR ENGINEERING. *Educ:* Washington Univ, St Louis, BS, 47, PhD(chem eng), 50;. *Prof Exp:* Res engr, Victor Div, Radio Corp Am, 50-52; sect head res eng, Metals & Ceramics Div, P R Mallory & Co, 52-56, assoc lab dir chem & metall res, 57-58; prin & lead engr, Aircraft Nuclear Propulsion Dept, Gen Elec Co, Ohio, 58-64, mgr reactor mat develop, Nuclear Mat & Propulsion Oper, 64-70; mem staff, Atomics Int, Canoga Park, 70-73; reactor safety engr, 74-84, sr nuclear engr, 84-89, TECH COORDR, US NUCLEAR REGULATORY COMN, 90- *Concurrent Pos:* Consult, Oak Ridge Nat Lab, 56-59; lectr, Univ Cincinnati, 61-66, adj assoc prof, 66-70. *Mem:* Sigma Xi; Res Soc Am. *Res:* Chemical, metallurgical, electrochemical and nuclear materials and processes for high temperature extreme duty; metal hydrides fabrication; light water reactor nuclear fuels testing and evaluation under accident conditions. *Mailing Add:* 20139 Laurel Hill Way Germantown MD 20874

VAN HOUTEN, RONALD G, b 1944; m 86; c 6. AUTISM, TREATMENT OF LEARNING DISABILITIES. *Educ:* State Univ NY, Stony Brook, BA, 68, Dalhousie Univ, MA, 69, PhD (psychol), 71. *Prof Exp:* PROF PSYCHOL, MT ST VINCENT UNIV, 71- *Concurrent Pos:* Assoc ed, Educ & Treat Children, 80-83, J Appl Behav Anal, 84-87; dir Soc Exp Anal Behav, 84-91, vpres, 88; prin investr, 71-; chair, Asn Behav Anal Right Effective Treat Task Force, 86-88. *Honors & Awards:* Can Crime Prevention Award, 83. *Mem:* Asn Behav Anal; Can Psychol Asn; NY Acad Sci. *Res:* Treatment of severe behavioral problems in developmentally delayed children and adults; treatment of learning disabilities; treatment of bed wetting; variables influencing the effects of social antecedents and contingencies of human behavior; traffic safety, speed control, impaired driving, pedestrian safety. *Mailing Add:* Psychol Dept Mt St Vincent Univ Halifax NS B3M 2J6 Can

VANHOUTTE, JEAN JACQUES, b Courtrai, Belg, Aug 27, 32; US citizen; m 60; c 4. PEDIATRIC RADIOLOGY. *Educ:* Cath Univ Louvain, MD, 59. *Prof Exp:* Instr radiol, Johns Hopkins Univ, 63-65; asst prof, Univ Colo, 65-71; assoc prof, 71-74, PROF & VCHMN, DEPT RADIOL, UNIV OKLA HEALTH SCI CTR, 74- *Concurrent Pos:* Chief radiologist, Okla Childrens Mem Hosp, 71- *Mem:* fel Am Col Radiol; Am Roentgen Ray Soc; Radiol Soc NAm; Soc Pediat Radiol; Asn Univ Radiologists. *Res:* Application of radiological sciences and imaging sciences to pediatrics. *Mailing Add:* State Okla Teaching Hosp PO Box 26307 Oklahoma City OK 73126

VANHOUTTE, PAUL MICHEL, PHARMACOLOGY, PHYSIOLOGY. *Educ:* Univ Ghent, Belg, MD, 65; Univ Antwerp, Belg, PhD(pharmacol), 73. *Prof Exp:* PROF PHYS PHARMACOL, MAYO CLIN, 81- *Res:* Pharmacology and physiology of blood vessel walls; endothelium and autonomic nerves; hypertension and coronary vapospasm. *Mailing Add:* 252 Hedwig Houston TX 77024

VAN HOUWELING, CORNELIUS DONALD, b Mahaska Co, Iowa, July 19, 18; m 42; c 5. VETERINARY MEDICINE. *Educ:* Iowa State Univ, DVM, 42, MS, 66. *Prof Exp:* Vet, Springfield, Ill, 42-43; dir vet med rels, Ill Agr Asn, 46-48; dir prof rels & asst exec secy, Am Vet Med Asn, 48-53; instr, Col Vet Med, Univ Ill, 53-54; dir livestock regulatory progs, Agr Res Serv, USDA, 54-56, asst adminr, 56-61, asst dir regulatory labs, Nat Animal Dis Lab, Iowa, 61-66; dir bur vet med, US Food & Drug Admin, Rockville, MD, 67-78, spec asst to comnr agr matters, 78-79; consult, 80-86; RETIRED. *Concurrent Pos:* Mem comn vet educ, Southern Regional Educ Bd; mem subcomt laws, rules & regulations animal health, Nat Res Coun; chmn adv comt humane slaughter, US Secy Agr; mem & chmn, Coun Pub Health & Regulatory Vet Med & organizing comt, Am Col Vet Preventive Med; chmn, Food & Drug Admin task force on antibiotics in animal feeds, 70-71; consult. *Honors & Awards:* Award, Am Mgt Asn, 57; Karl F Meyer Gold Headed Cane Award, Am Vet Epidemiol Soc, 78. *Mem:* Fel AAAS; Am Asn Food Hyg Veterinarians (pres, 78); US Animal Health Asn; Nat Asn Fed Veterinarians; World Asn Vet Food Hyg. *Res:* Regulatory veterinary medicine. *Mailing Add:* 1200 Peace St Pella IA 50219

VAN HOVEN, GERARD, b Los Angeles, Calif, Nov 23, 32; m 56; c 2. PLASMA PHYSICS, SOLAR PHYSICS. *Educ:* Calif Inst Technol, BS, 54; Stanford Univ, PhD(physics), 63. *Prof Exp:* Mem tech staff, Bell Tel Labs, 54-56; electron physicist, Gen Elec Co, 56-63; res assoc, W W Hansen Labs Physics, Stanford Univ, 63-65, res physicist, Inst Plasma Res, 65-68; from asst prof to assoc prof, 68-79, PROF PHYSICS, PRIN INVESTR, UNIV CALIF, IRVINE, 79- *Concurrent Pos:* Fulbright fel, Vienna Tech Univ, 63-64; consult, Gen Elec Co, 63-65, Varian Assocs, 65-68, Smithsonian Astrophys Observ, 75, Aerospace Corp, 75-77 & NASA, 79-81, 89; Langley-Abbot vis scientist, Ctr Astrophys, Harvard Univ, 75; vis astrophysicist, Oss Astrofisico Arcetri, Univ Florence, 76, vis prof, 81-82 & 88. *Mem:* Fel Am Phys Soc; Am Astron Soc; Int Astron Union; Am Geophys Union. *Res:* Solar-terrestrial activity magnetohydrodynamics, especially magnetic field reconnection, energy-transport instabilities and coronal structure. *Mailing Add:* Dept Physics Univ Calif Irvine CA 92717

VAN HUSS, WAYNE D, effects of exercise on risk factors & aging, for more information see previous edition

VAN HUYSTEE, ROBERT BERNARD, b Amsterdam, Holland, Sept 29, 31; Can citizen; m 58; c 1. BIOCHEMISTRY, BOTANY. *Educ:* Univ Sask, BA, 59, MA, 61; Univ Minn, St Paul, PhD(hort), 64. *Prof Exp:* Fel physiol, Purdue Univ, 64-66; asst prof radiobiol, 66-69, assoc prof plant sci, 69-79, PROF PLANT SCI, UNIV WESTERN ONT, 79- *Concurrent Pos:* Exchange scientist Can-France, 70, 73; vis prof, Univ Paris, France, 87. *Mem:* Am Soc Plant Physiol; Can Soc Plant Physiol. *Res:* Process of cold acclimation in plants; metabolism in cultured plant cells; study of peroxidase, hemo-, glyco-, calcium protein. *Mailing Add:* Dept Plant Sci Univ Western Ont London ON N6A 5B7 Can

VANICEK, C DAVID, b Waterloo, Iowa, Oct 12, 39. FISHERIES MANAGEMENT. *Educ:* Iowa State Univ, BS, 61, MS, 63; Utah State Univ, PhD(fishery biol), 67. *Prof Exp:* Fishery biologist, US Fish & Wildlife Serv, 63-67; from asst prof to assoc prof, 67-79, PROF BIOL, CALIF STATE UNIV, SACRAMENTO, 79- *Mem:* Am Fisheries Soc; Am Inst Biol Sci; Am Inst Fishery Res Biol; Pac Fishery Biologists; Desert Fishes Coun. *Res:* Freshwater fishery biology and management. *Mailing Add:* Dept Biol Sci Calif State Univ 6000 J Street Sacramento CA 95819

VANICEK, PETR, b Susice, Czech, July 18, 35; m 60; c 3. GEODESY, GEOPHYSICS. *Educ:* Prague Tech Univ, Dipl Ing, 59; Czech Acad Sci, PhD(math physics), 68. *Prof Exp:* Div head land surv, Prague Inst Surv & Cartog, 59-63; consult numerical anal & comput prog, Fac Tech & Nuclear Physics, Prague Tech Univ, 63-67; sr res fel & sr sci officer, Inst Coastal Oceanog & Tides, Nat Environ Res Coun Gt Brit, 68-69; Nat Res Coun Can Postdoctoral fel, Dept Energy, Mines & Resources, 69-71; prof surv sci, Univ Toronto, 81-83; assoc prof, 71-76, PROF GEOD, UNIV NB 76-, DIR GRAD STUDIES, DEPT SURV ENG, 83-85 & 91- *Concurrent Pos:* Can rep, Comn Recent Crustal Movements, Int Union Geod & Geophys, 77-; secy subcomt geod, Nat Res Coun Can, 72-74; mem, Can Subcomt Geodynamics, 75-80; vis prof, Nat Res Coun Can, Univ Parana, Brazil, 75, 76, 79, 84 & 87; vis prof, Univ Stuttgart, WGer, 82 & 83; mem, NAS comt Geod, 82-85; adj prof surv sci, Univ Toronto, 83-90; pres spec study group, Int Asn Geod, 83-87 & 89- *Honors & Awards:* Humboldt Distinguished Sr Scientist Award. *Mem:* fel Am Geophys Union; Can Inst Surv; fel Geol Asn Can; Can Geophys Union (pres, 87-89). *Res:* Geodesy; earth tides, crustal movements and mean sea level; applied mathematics, especially spectral analysis and mechanics. *Mailing Add:* Surv Eng Univ NB PO Box 4400 Fredericton NB E3B 5A3 Can

VANIER, JACQUES, b Dorion, Que, Jan 4, 34; m 61; c 2. QUANTUM ELECTRONICS, ATOMIC & MOLLECULAR PHYSICS. *Educ:* Univ Montreal, BA, 55, BSc, 58; McGill Univ, MSc, 60, PhD(physics), 63. *Prof Exp:* Lectr physics, McGill Univ, 61-63; physicist, Quantum Electronics Div, Varian Assocs, Mass, 63-67 & Hewlett-Packard Co, 67; from assoc prof to prof elec eng, Laval Univ, 68-83; prin res, Nat Res Coun, 83-85, asst dir physics div, 85-90, dir, Lab Basic Standards, 86-90, DIR GEN, INST NAT MEASUREMENTS STANDARDS. *Honors & Awards:* Centennial Medal, Inst Elec & Electronics Engrs. *Mem:* Can Asn Physicists; fel Inst Elec & Electronics Engrs; fel Am Phys Soc; fel Royal Soc Can. *Res:* Electron paramagnetic resonance; nuclear magnetic resonance; optical pumping; masers; frequency standards; atomic clocks; atomic and molecular physics; electromagnetism; solid state physics; thermal physics. *Mailing Add:* Nat Res Coun Inst Nat Measurements Standards Montreal Rd Ottawa ON K1A 0R6 Can

VANIER, PETER EUGENE, b St Kitts, Leeward Islands, Feb 26, 46; nat US; m 71; c 1. MATERIALS SCIENCE, SOLID STATE PHYSICS. *Educ:* Cambridge Univ, BA, 67; Syracuse Univ, MS, 69, PhD(physics), 76. *Prof Exp:* Res assoc physics, Yeshiva Univ, 76-78; from asst scientist to assoc scientist, Mat Sci, Brookhaven Nat Lab, 78-83, scientist, 83-87, physicist, NPB Div, 87-90, PHYSICIST, ADV REACTOR DIV, BROOKHAVEN NAT LAB, 91- *Concurrent Pos:* Vis scientist, electrotech lab, Ibarak, Japan, 84. *Mem:* Am Phys Soc; Mat Res Soc. *Res:* High-temperature properties of carbon and carbides; electronic, magnetic, and optical properties of semiconductors; properties of amorphous semiconductors related to solar cell applications. *Mailing Add:* Dept Nuclear Energy Bldg 701 Brookhaven Nat Lab Upton NY 11973

VAN INWEGEN, RICHARD GLEN, b Brooklyn, NY, Nov 24, 44; m 67; c 3. BIOCHEMISTRY, PHYSIOLOGY. *Educ:* State Univ NY, Binghamton, BA, 66, MA, 69; Univ Ill, Urbana-Champaign, PhD(physiol), 72. *Prof Exp:* Res assoc pharmacol, Univ Tex Med Sch, Houston, 72-76; sr res scientist biochem, USV Pharmaceut Corp, 76-79; GROUP LEADER BIOCHEM, REVLON HEALTH CARE GROUP, 79- *Mem:* Sigma Xi; NY Acad Sci; Am Soc Pharmacol & Exp Therapeut. *Res:* Cyclic nucleotides in asthma and hypertension; cyclic nucleotide associated enzymes-cyclases, kinases and phosphodiesterases; receptor biochemistry; biochemistry of leukotrienes. *Mailing Add:* 1720 Concord Ct Blue Bell PA 19422

VAN ITALLIE, THEODORE BERTUS, b Hackensack, NJ, Nov 8, 19; m 48; c 5. MEDICINE. *Educ:* Harvard Univ, SB, 41; Columbia Univ, MD, 45; Am Bd Internal Med, dipl, 54. *Prof Exp:* Intern med, St Luke's Hosp, New York, 45-46, from asst resident to resident, 48-50; res fellow nutrit, Sch Pub Health, Harvard Univ, 50-51, res assoc, 51-52, asst prof clin nutrit, Schs Med & Pub Health, 55-57; instr, 52-55, from assoc clin prof to clin prof, 57-71, PROF MED, COL PHYSICIANS & SURGEONS, COLUMBIA UNIV, 71-, ASSOC DIR INST HUMAN NUTRIT, 67- *Concurrent Pos:* From asst to assoc, Peter Bent Brigham Hosp, Boston, 50-57; dir lab nutrit res, St Luke's Hosp, 52-55, asst attend physician, 53-55, attend physician, 57-, dir med, 57-75; vis lectr, Sch Pub Health, Harvard Univ, 57-60; mem gastroenterol & nutrit training comt, NIH, 69-73; mem food & nutrit bd, Nat Acad Sci, 70-74; Nat Inst Arthritis, Metab & Digestive Dis Adv Coun, 78-81; ed-in-chief, Am J Clin Nutrit, 79-81; special adv, Surgeon Gen Human Nutrit, 80-81. *Honors & Awards:* McCollum Award, Am Soc Clin Nutrit, 85; Goldberger Award, AMA, 85. *Mem:* Soc Exp Biol & Med; Am Clin & Climat Asn; Am Fedn Clin Res; fel Am Col Physicians; Am Soc Clin Nutrit (pres, 75-76); fel Am Inst Nutrit; Am Soc Clin Invest. *Res:* Carbohydrate and lipid physiology and biochemistry; clinical nutrition; metabolism; control of food intake and regulation of body fat; body composition. *Mailing Add:* St Luke's Hosp Ctr Amsterdam Ave at 114th St New York NY 10025

VAN KAMMEN, DANIEL PAUL, b Dordrecht, Neth, Aug 26, 43; US citizen; m 70; c 1. PSYCHOPHARMACOLOGY. *Educ:* Univ Utrecht Med Sch, MD, 66, PhD(pharmacol), 78. *Prof Exp:* Unit chief & staff psychiatrist, NIMH, 73-82; PROF PSYCHIAT & CHIEF STAFF, WESTERN PSYCHIAT INST & CLIN, 82- *Concurrent Pos:* Mem, Behav & Clin Res Rev Comt, 75-78; fac mem, Wash Sch Psychiat, 75-; vis prof, Dept Psychiat, Univ Ala, Birmingham, 78; consult, Vet Admin & Nat Heart, Lung & Blood Inst,

79; mem, RAG Ment Health & Behav Sci, Va. *Mem:* Fel Am Col Neuropsychopharmacol; AAAS; Int Psychoneuroendocrinol; Soc Neurosci; Collegium Int Neuropsychopharmacol; Soc Biol Psychiat. *Res:* Biochemical and pharmacological exploration of schizophrenia; spinal fluid studies; endocrinology. *Mailing Add:* Vet Admin Med Ctr Highland Dr Pittsburgh PA 15206

VAN KAMPEN, KENT RIGBY, b Brigham City, Utah, July 30, 36; m 59; c 3. VETERINARY PATHOLOGY. *Educ:* Utah State Univ, BS, 61; Colo State Univ, DVM, 67; Univ Calif, Davis, PhD(comp path), 67. *Prof Exp:* Vet pathologist, Poisonous Plant Res Lab, Agr Res Serv, USDA, Utah, 67-70; assoc prof vet sci, Utah State Univ, 68-70; dir res, Intermountain Labs, Inc, 69-77; prof & head, Dept Animal, Dairy & Vet Sci, Utah State Univ, Logan, 77-80; ADJ ASSOC PROF, COL MED UNIV UTAH, 69- *Mem:* AAAS; Am Col Vet Path; Am Vet Med Asn; fel Am Col Vet Toxicol. *Res:* Pathogenesis of animal diseases related to similar disorders in man; pathology of natural and man made toxicants in animals; mechanisms of carcinogenesis. *Mailing Add:* 3300 Saratoga Lane Plymouth MN 55441

VANKIN, GEORGE LAWRENCE, b Baltimore, Md, Apr 22, 31; m 56; c 2. EVOLUTIONARY THEORY. *Educ:* NY Univ, BS, 54, PhD(zool), 62; Wesleyan Univ, MA, 56. *Prof Exp:* Res asst genetics, Wesleyan Univ, 56; teaching fel biol, NY Univ, 56-59, res asst embryol, 59-62, lectr biol, 61-62; from asst prof to assoc prof, 62-75, PROF BIOL, WILLIAMS COL, 75- *Concurrent Pos:* Vis asst prof, Med Col, Cornell Univ, 68. *Mem:* Soc Syst Zool; Willi Hennig Soc. *Res:* Evolutionary theory; history and philosophy of evolutionism. *Mailing Add:* Bronfman Sci Ctr Williams Col 18 Hoxsey St Williamstown MA 01267

VAN KLAVEREN, NICO, b Amersfoort, Neth, Feb 2, 34; m 59; c 3. KNOWLEDGE SYSTEMS, COMPUTER AIDED ENGINEERING. *Educ:* Delft Univ Technol, MSc, 59, DSc(chem eng), 66; Pepperdine Univ, MBA, 78. *Prof Exp:* Asst prof chem eng, Delft Univ Technol, 61-66; SR ENG ASSOC, CHEVRON RES & TECHNOL CO, 66- *Concurrent Pos:* Vis prof, Steven's Inst Technol, 64. *Mem:* Am Inst Chem Engrs; Neth Royal Inst Eng; Am Asn Artificial Intel. *Res:* Chemical engineering science; process design; user oriented software for process engineering and business and strategic planning; modeling of technical, economical and social processes; management of complex organizations; organizational development; application of artificial intelligence (knowledge systems); computer aided training. *Mailing Add:* 2635 Mira Vista Dr El Cerrito CA 94530

VAN KLEY, HAROLD, b Chicago, Ill, Mar 7, 32; m 59; c 2. BIOCHEMISTRY, PROTEIN CHEMISTRY. *Educ:* Calvin Col, AB, 53; Univ Wis, MS, 55, PhD(biochem), 58. *Prof Exp:* From instr to sr instr, Sch Med, St Louis Univ, 58-61, asst prof biochem, 61-82; dir biochem res, St Mary's Health Ctr, 67-82; PROF, DEPT CHEM, TRINITY CHRISTIAN COL, 82- *Concurrent Pos:* High Sch Student Res Apprenticeship Prog, Argonne Nat Lab, 83-88; chemist/consult, Chem Waste Mgt Tech Ctr, 89- *Mem:* Am Pancreatic Asn; AAAS; Am Chem Soc; Sigma Xi; Am Soc Biochem & Molecular Biol. *Res:* Protein structure, especially as related to biological function and regulatory mechanisms in metabolism; protein changes in neoplasia; pancreatic enzymes; metals analysis in hazardous waste. *Mailing Add:* Dept Chem Trinity Christian Col 6601 W College Dr Palos Heights IL 60463-0929

VAN KRANENDONK, JAN, b Delft, Neth, Feb 8, 24; m 52; c 3. THEORETICAL PHYSICS. *Educ:* Univ Amsterdam, PhD(physics), 52. *Prof Exp:* Res asst, Univ Amsterdam, 50-54; lectr, State Univ Leiden, 55-58; assoc prof, Univ Toronto, 58-60, prof physics, 60-90; RETIRED. *Concurrent Pos:* Neth Orgn Pure Res fel, Harvard Univ, 53-54. *Honors & Awards:* Steacie Prize, 64. *Mem:* Am Phys Soc; fel Royal Soc Can; Can Asn Physicists; corresp mem Neth Acad Sci. *Res:* Molecular and solid-state physics. *Mailing Add:* 1129 Sunny Side Rd Kelowna BC V1O 2N7 Can

VAN KREY, HARRY P, b Combined Locks, Wis, Oct 2, 31; m 52; c 2. PHYSIOLOGY, AGRICULTURE. *Educ:* Univ Calif, Davis, BS, 60, PhD(animal physiol), 64. *Prof Exp:* Fel, Univ Wis, 64-65; assoc prof, 65-80, PROF AVIAN PHYSIOL, VA POLYTECH INST & STATE UNIV, 80- *Mem:* AAAS; Sigma Xi; Poultry Sci Asn; World Poultry Sci Asn; Soc Study Reproduction. *Res:* Avian reproductive physiology; poultry science. *Mailing Add:* 1837 St Andrews Circle Blacksburg VA 24060

VAN LANCKER, JULIEN L, b Auderghem, Belg, Aug 14, 24; m 49; c 3. PATHOLOGY. *Educ:* Cath Univ Louvain, MD, 50. *Prof Exp:* Asst path, Cath Univ Louvain, 50-53; vis instr, Univ Kans, 53-54; Runyon fel oncol, Univ Wis, 54-55; asst path, Cath Univ Louvain, 55-56; asst prof path, Univ Utah, 56-60; assoc prof path & chief path sect, Primate Res Ctr, Univ Wis, 60-66; prof med sci, Brown Univ, 66-70; PROF PATH & CHMN DEPT, UNIV CALIF, LOS ANGELES, 70- *Mem:* Radiation Res Soc; Am Soc Exp Path; Am Soc Biol Chemists; NY Acad Sci; Int Acad Path. *Res:* Cell biology; chemical pathology; molecular mechanisms in disease. *Mailing Add:* Dept Path 13-327 chs Univ Calif 405 Hilgard Ave Los Angeles CA 90024

VAN LANDINGHAM, HUGH F(OCH), b Greensboro, NC, Apr 12, 35; m 64; c 3. ELECTRICAL ENGINEERING. *Educ:* NC State Univ, BS, 57; NY Univ, MEE, 59; Cornell Univ, PhD(elec eng), 67. *Prof Exp:* Mem tech staff, Bell Tel Labs, NJ, 57-62; from asst prof to assoc prof, 66-80, PROF ELEC ENG, VA POLYTECH INST & STATE UNIV, 80- *Mem:* Inst Elec & Electronics Engrs. *Res:* Communication and control systems. *Mailing Add:* Dept Elec Eng Va Polytech Inst & State Univ Blacksburg VA 24061

VAN LANEN, ROBERT JEROME, b Green Bay, Wis, July 30, 43; m 78; c 1. BIO-ORGANIC CHEMISTRY, PHYSICAL ORGANIC CHEMISTRY. *Educ:* St Norbert Col, BS, 65; Univ Colo, Boulder, PhD(org chem), 71. *Prof Exp:* NIH fel chem, Univ Wis-Madison, 71-73; ASSOC PROF CHEM, ST XAVIER COL, 73- *Mem:* AAAS; Am Chem Soc. *Res:* Mechanisms of enzyme-catalyzed reactions; chemical education; chemistry of small ring compounds. *Mailing Add:* St Xavier Col 103rd & Central Park Ave Chicago IL 60655-3198

VAN LEAR, DAVID HYDE, b Clifton Forge, Va, Dec 1, 40. FORESTRY, SOILS. *Educ:* Va Polytech Inst, BS, 63, MS, 65; Univ Idaho, PhD(forest sci), 69. *Prof Exp:* Fel sch forestry, Univ Fla, 68-69; soil scientist, US Forest Serv, 69-71; assoc prof, 71-77, PROF SILVICULT, CLEMSON UNIV, 77- *Mem:* Soc Am Foresters; Soil Sci Soc Am. *Res:* Hardwood silviculture; environmental forestry; forest fertilization; soil-site relationships; strip mining. *Mailing Add:* Dept Forestry Clemson Univ 201 Sykes Hall Clemson SC 29634

VAN LEER, JOHN CLOUD, b Washington, DC, Feb 14, 40; m 62; c 3. PHYSICAL OCEANOGRAPHY, OCEAN ENGINEERING. *Educ:* Case Inst Technol, BSME, 62; Mass Inst Technol, ScD(phys oceanog), 71. *Prof Exp:* Engr, Draper Lab, Mass Inst Technol, 62-65, from res asst to res assoc phys oceanog, 65-71; asst prof, 71-75, ASSOC PROF PHYS OCEANOG, ROSENSTIEL SCH MARINE & ATMOSPHERIC SCI, UNIV MIAMI, 76- *Mem:* Sigma Xi; Am Geophys Union; Marine Technol Soc. *Res:* Physical oceanographic research on continental shelves and deep ocean; response to wind forcing, surface and bottom boundary layers; development of ocean instruments, notably the Cyclesonde, an automatic oceanographic radiosonde. *Mailing Add:* Oceanog Dept Naval Postgrad Sch Monterey CA 93940

VAN LEEUWEN, GERARD, b Hull, Iowa, July 4, 29; m 52; c 2. PEDIATRICS. *Educ:* Calvin Col, BA, 50; Univ Iowa, MD, 54. *Prof Exp:* Intern, Butterworth Hosp, Mich, 54-55; resident pediat, Univ Mo-Columbia, 57-59, from instr to asst prof, 62-69; prof pediat & chmn dept, Univ Nebr Med Ctr, Omaha, 69-78; assoc med dir, Sect Med Care, Mo Div Health, 78-80; prof & chmn, pediat dept, Med Sch, Univ Kans, Wichita, 80-90; PROF PEDIAT, UNIV OSTEOP MED, DES MOINES, IA, 90- *Concurrent Pos:* NIH trainee, 62-64; Am Thoracic Soc fel, 64-66; Nat Found Birth Defects grant, 66; proj consult, Head Start, 65; dir, Nat Found Treatment Ctr, 66; dir, Emergency Med, Univ Nebr, 76-78. *Mem:* Am Acad Pediat; Am Pediat Soc. *Res:* Neonatal physiology and hypogylcemia; hyaline membrane syndrome; teratology. *Mailing Add:* Pediat Dept Univ Osteop Med 3200 Grand Ave Des Moines IA 50312

VAN LENTE, KENNETH ANTHONY, b Holland, Mich, Mar 29, 03; m 29; c 4. PHYSICAL CHEMISTRY. *Educ:* Hope Col, AB, 25, MS, 26; Univ Mich, PhD, 31. *Prof Exp:* Asst, Univ Mich, 27-31; from asst prof to prof, 31-71, EMER PROF PHYS CHEM, SOUTHERN ILL UNIV, CARBONDALE, 71- *Mem:* Am Chem Soc. *Res:* Liquid junction potentials; constant temperature baths; composition of plating baths; chemical education. *Mailing Add:* 1209 W Chautauqua Carbondale IL 62901-2454

VAN LIER, JAN ANTONIUS, b Ginneken, Neth, Nov 17, 24; m 54; c 4. PHYSICAL CHEMISTRY. *Educ:* Univ Utrecht, BS, 51, MS, 54, PhD(phys chem), 59. *Hon Degrees:* JD, Cleveland State Univ, 77. *Prof Exp:* Res assoc mineral eng, Mass Inst Technol, 55-58; instr colloid chem, Univ Utrecht, 58-59; res chemist, Philips Electronics, Holland, 59-60; sr res chemist, Parma Tech Ctr, Union Carbide Corp, 60-74, staff res chemist, 74-86; TECHNOL ASSOC, WESTLAKE TECHNOL LAB, EVEREADY BATTERY CO, 86- *Mem:* Am Chem Soc; Royal Neth Chem Soc; Sigma Xi. *Res:* Microbalance techniques; solubility of quartz; battery materials; differential and thermogravimetric analysis; solid electrolytes; basic electrochemistry; ternary phase diagrams; surface chemistry; wetting phenomena; plasma polymerization. *Mailing Add:* 9937 Little Mountain Rd Concord OH 44060

VAN LIER, JOHANNES ERNESTINUS, b Amsterdam, Neth, May 26, 42. BIOCHEMISTRY. *Educ:* Delft Univ Technol, Neth, 66; Univ Tex Med Br Galveston, PhD(biochem), 69. *Prof Exp:* Res assoc, Univ Tex Med Br Galveston, 66-69, instr, 69-70; asst prof, 70-75, assoc prof, 75-81, PROF NUCLEAR MED RADIOBIOL, MED CTR, UNIV SHERBROOKE, 81- *Concurrent Pos:* Med Res Coun Can res grant, 70- *Mem:* AAAS; Fedn Am Socs Exp Biol; Am Chem Soc; Can Fedn Biol Socs; Soc Nuclear Med. *Res:* Radiopharmaceuticals for nuclear med; photosensitizers and anticancer agents. *Mailing Add:* Dept Nuclear Med Univ Sherbrooke Med Ctr Sherbrooke PQ J1H 5N4 Can

VAN LIEW, HUGH DAVENPORT, b Spokane, Wash, Jan 28, 30; m 59; c 3. MEDICAL PHYSIOLOGY. *Educ:* State Col Wash, BS, 51; Univ Rochester, MS, 53, PhD(physiol), 56. *Prof Exp:* Res fel, Sch Pub Health, Harvard Univ, 59-61; asst prof physiol, Stanford Univ, 61-63; from asst prof to assoc prof, 63-74, PROF PHYSIOL, STATE UNIV NY, BUFFALO, 74- *Concurrent Pos:* Ed, Undersea Biomed Res; mem, Coun Biol Ed. *Honors & Awards:* Stover-Link Award, 86. *Mem:* AAAS; Am Physiol Soc; Undersea Med Soc. *Res:* Diffusion of gases through tissues; gas tensions in the tissues; subcutaneous gas pockets as models of decompression sickness bubbles; diffusion and convection in the lung; pulmonary function in hyperbaric environments. *Mailing Add:* Dept Physiol State Univ NY Buffalo NY 14214

VAN LIEW, JUDITH BRADFORD, b Boston, Mass, Jan 22, 30; m 59; c 3. PHYSIOLOGY. *Educ:* Bates Col, BS, 51; Univ Wash, MS, 54; Univ Rochester, PhD(physiol), 58. *Prof Exp:* Res assoc biochem, Woman's Med Col Pa, 58-60; res assoc med, 64-70, res asst prof, 70-74, ASST PROF PHYSIOL, SCH MED, STATE UNIV NY, BUFFALO, 74-; ASSOC PROF PHYSIOL & RES PHYSIOLOGIST, VET ADMIN HOSP, BUFFALO, 73- *Mem:* Am Physiol Soc; Am Soc Nephrology. *Res:* Physiology of normal and abnormal proteinuria. *Mailing Add:* Dept Physiol Vet Admin Med Ctr 3495 Bailey Ave Buffalo NY 14215

VAN LIGTEN, RAOUL FREDRICK, b Bandung, Indonesia, Sept 27, 32; US citizen; m 86; c 5. OPTICAL PRODUCTION, PHYSICAL OPTICS. *Educ:* Delft Inst Tech, Neth, MEng, 57; Sorbonne Univ, PhD(physics), 72. *Prof Exp:* Physicist, Res & Develop, Nat Res Coun, Neth, 56-60, Am Optical Corp, 60-61; dept head, Image Eval, Itek Corp, Lexington, MA, 61-63; sr physicist res & develop, Am Optical Corp, USA, 63-73; dir res & develop, Am Optical Corp, Europe, 73-76; exec vpres, Polycore Optical, Singapore, 76-82; prof elec eng, Nat Univ Singapore, 82-86; dir res & develop, Younger Optics,

Los Angeles, 86-89. *Concurrent Pos:* Adv group, USAF Systs Command, 66-67; mem exec comt, Sci Coun Singapore, 81-86; mem, Adv Comm, Sci & Technol, Ministry Trade & Indust, Singapore, 81-86; chmn, Comt Laser Appln, Econ Develop Bd, Singapore, 84-86. *Honors & Awards:* Karl Fairbanks Mem Award, Int Soc Optical Engr, 69. *Mem:* Fel Optical Soc Am; Int Soc Optical Engr; Soc Mgt Engrs. *Res:* Optical physics; holography; electro-optical instrumentation; techniques in lens manufacturing. *Mailing Add:* PO Box 65117 Ft Lauderdale FL 33316

VAN LINT, VICTOR ANTON JACOBUS, b Samarinda, Indonesia, May 10, 28; US citizen; m 50; c 4. PHYSICS. *Educ:* Calif Inst Technol, BS & PhD(physics), 54. *Prof Exp:* Instr physics, Princeton Univ, 54-55; physicist, Gen Atomic Div, Gen Dynamics Corp, 57-65, assoc dir, Spec Nuclear Effects Lab, 65-69, mgr, Defense Sci Dept, Gulf Radiation Technol, 69-70, vpres, Gulf Energy & Environ Systs & mgr, Gulf Radiation Technol Div, 70-73; pres, Intelcom Radiation Technol, 73-74, consult, 74-75; mgr, Elec Appln Div, 75-82, special asst to dep dir sci & tech, Defense Nuclear Agency, 82-83, MGR, EXP PHYSICS DIV, MISSION RES CORP, 83- *Honors & Awards:* NASA Pub Serv Award, 81. *Mem:* Am Phys Soc; fel Inst Elec & Electronics Engrs. *Res:* Radiation effects, including solid state physics, atomic physics, and electronic systems analysis. *Mailing Add:* 1032 Skylark Dr La Jolla CA 92037

VAN LOON, EDWARD JOHN, b Danville, Ill, Dec 3, 11; m 54; c 1. BIOCHEMISTRY. *Educ:* Univ Ill, AB, 36; Rensselaer Polytech Inst, MS, 37, PhD(chem), 39. *Prof Exp:* Res assoc & instr physiol & pharmacol, Albany Med Col, Union NY, 39-43; instr & asst prof biochem, Mich State Univ, 46; asst & assoc prof, Sch Med, Univ Louisville, 46-49; chief, Med Res Lab, Vet Admin Hosp, 49-55; group leader & head biochem sect, Smith Kline & French Labs, 55-67; chief, Spec Pharmacol Animal Lab, Food & Drug Admin, 67-75; RETIRED. *Concurrent Pos:* Mem, Am Bd Clin Chemists. *Mem:* Am Soc Clin Invest; Am Inst Nutrit; Am Soc Pharmacol & Exp Therapeut; Am Chem Soc; Am Soc Biol Chem. *Res:* Clinical biochemistry; intermediary metabolism; biochemical pharmacology and drug metabolism; clinical biochemistry. *Mailing Add:* 1412 Kent St Durham NC 27707-1534

VAN LOON, JON CLEMENT, b Hamilton, Ont, Jan 9, 37; m 61; c 3. GEOLOGY, CHEMISTRY. *Educ:* McMaster Univ, BSc, 59; Univ Toronto, PhD(analytical chem), 64. *Prof Exp:* From asst prof to assoc prof anal geochem, 64-77, PROF GEOL, UNIV TORONTO, 77- *Mem:* Sigma Xi. *Res:* Application of modern analytical methods, particularly ICP source mass spectrometry, to the analysis of natural products, with particular emphasis on environmental samples. *Mailing Add:* Dept Geol Univ Toronto St George Campus 170 College St Toronto ON M5S 1A1 Can

VAN LOPIK, JACK RICHARD, b Holland, Mich, Feb 25, 29; m; c 1. RESEARCH ADMINISTRATION, MARINE SCIENCES. *Educ:* Mich State Univ, BS, 50; La State Univ, MS, 53, PhD(geol), 55. *Prof Exp:* Field investr, Coastal Studies Inst, La State Univ, 51-54, instr geol, 54; geologist, Waterways Exp Sta, Corps Engrs, US Army, 54-57, asst chief & chief geol br, 57-61; res scientist, chief area eval sect & mgr space & environ sci prog, Geosci Opers, Tex Instruments Inc, 61-66, tech requirements dir, 66-68; chmn dept marine sci, 68-74, PROF MARINE SCI & DIR SEA GRANT DEVELOP, LA STATE UNIV, BATON ROUGE, 68-, DEAN, CTR WETLAND RESOURCES, 70- *Concurrent Pos:* Mem, Nat Res Coun Earth Sci Div, Nat Acad Sci-Nat Res Coun, 67-72, chmn panel geog & human & cult resources, Comt Remote Sensing Progs for Earth Resources Surv, 69-77; mem, La Adv Comn Coastal & Marine Resources, 71-73, Nat Adv Comt on Oceans & Atmosphere, 78-84; mem bd dirs, Gulf South Res Inst, 74-84; chmn, Coastal Resources Directorate, US Nat Comt for Man & the Biosphere, US Nat Comn for UN Educ, Sci & Cult Orgn, 75-82; mem, Lower Miss River Waterways Safety Adv Comt, Eighth Coast Guard Dist, 83-; mem adv coun, Nat Coastal Resources Res & Develop Inst, 85-; mem, Chief Engrs Environ Adv Bd, US Army CEngr, 88-, bd dirs, La Partnership Technol & Innovation, 89- *Mem:* Fel Geol Soc Am; Am Mgt Asn; Soc Res Adminr; fel AAAS; sr mem Am Astronaut Soc; Am Geophys Union. *Res:* Photogeology and remote sensing; deltaic and arid zone geomorphology and sedimentation; military and engineering geology; terrain analysis and quantification; lunar and earth-orbiting-satellite exploration; coastal zone management. *Mailing Add:* Ctr Wetland Resources La State Univ Baton Rouge LA 70803-7500

VAN LOVEREN, HENK, b Utrecht, Neth, Dec 1, 51; m 72; c 2. IMMUNOTOXICOLOGY, IMMUNOPHARMACOLOGY. *Educ:* State Univ Utrecht, MSc, 75, PhD(immunol), 81. *Prof Exp:* Postdoctoral fel immunol, Yale Univ, New Haven, 82-83; immunotoxicologist, 84-89, SECT HEAD IMMUNOTOXICOL, NAT INST PUB HEALTH & ENVIRON PROTECTION, 89- *Concurrent Pos:* Prin investr, validation studies immunotoxicol, 87-; coordr, EC Res Prog Develop Immunotoxicol Test Batteries in collab UK, Ger & France, 90- *Mem:* Am Asn Immunologists; Soc Toxicol. *Res:* Test systems for assessing undesired effects of exposure to chemicals, environmental, natural or drugs, to the immune system. *Mailing Add:* Dept Path Nat Inst Pub Health & Environ Protection PO Box 1 Bilthoven 3720 BA Netherlands

VAN MAANEN, EVERT FLORUS, b Harderwyk, Neth, Sept 17, 18. PHARMACOLOGY. *Educ:* State Univ Utrecht, BS, 38, Phil Drs, 45; Harvard Univ, PhD(pharmacol), 49. *Prof Exp:* From instr to assoc prof, 49-66, PROF PHARMACOL, COL MED, UNIV CINCINNATI, 66- *Concurrent Pos:* Dir biol sci, William S Merrell Co Div, Richardson-Merrell, Inc, 55-62. *Mem:* AAAS; Am Soc Pharmacol & Exp Therapeut; Soc Exp Biol & Med. *Res:* Neuromuscular transmission; cholinesterase inhibitors; cardiovascular agents; autonomic drugs; theories of drug action. *Mailing Add:* Dept Pharmacol Univ Cincinnati Col Med 231 Bethesda Ave Cincinnati OH 45267-0575

VANMARCKE, ERIK HECTOR, b Menen, Belg, Aug 6, 41; m 65; c 3. CIVIL ENGINEERING, OPERATIONS RESEARCH. *Educ:* Cath Univ Louvain, Engr, 65; Univ Del, MS, 67; Mass Inst Technol, PhD(eng), 70. *Prof Exp:* From instr to profcivil eng, Mass Inst Technol, 68-85, NSF res grants, 70-85; PROF CIVIL ENG, PRINCETON UNIV, 85- *Concurrent Pos:* Consult eng; consult, Off Sci & Technol Policy, 78-80; ed, Struct Safety, 82-; vis scholar, Harvard Univ, 84-85. *Honors & Awards:* Raymond C Reese Res Prize, Am Soc Civil Engrs, 75,; Walter Huber Res Prize, 84. *Mem:* Am Soc Civil Engrs; Earthquake Eng Res Inst; Am Geophys Union; Seismol Soc Am; Sigma Xi; AAAS. *Res:* Structural and geotechnical engineering; vibrations induced by wind, earthquakes and water waves; structural safety; random fields. *Mailing Add:* Dept Civil Eng & Opers Res E-QUAD E-223 Princeton Univ Princeton NJ 08544

VAN METER, DAVID, b Southampton, NY, Mar 27, 19; m 50. APPLIED PHYSICS, INFORMATION SCIENCES. *Educ:* Mass Inst Technol, BS & SM, 43; Harvard Univ, MA, 53, PhD, 55. *Prof Exp:* Mem tech staff, Bell Tel Labs, 43-46; asst prof elec eng, Pa State Univ, 46-52; lab mgr, Melpar, Inc, 55-60 & Litton Systs, Inc, 60-66; chief comput res lab, Electronics Res Ctr, NASA, 66-70; chief, Info Sci Div, US Dept Transp, 70-77, Systs Develop Div, 78-80, sr tech staff mem, Transp Systs Ctr, 80-84; RETIRED. *Concurrent Pos:* Lectr, Harvard Univ, 55 & 59; ed, Trans Info Theory, 64-67. *Res:* Computer science; information theory. *Mailing Add:* 10 Byron Boston MA 02108

VAN METER, DONALD EUGENE, b Ashtabula, Ohio, Aug 30, 42; m 64; c 1. SOIL CONSERVATION. *Educ:* Purdue Univ, Lafayette, BS, 64; Mich State Univ, MS, 65; Ind Univ, Bloomington, DEduc, 71. *Prof Exp:* County agt agr, Coop Exten Serv, Purdue Univ, 65-68; PROF NATURAL RESOURCES, BALL STATE UNIV, 69-, CHMN DEPT, 79- *Mem:* Soil Conserv Soc Am; Am Soc Agron. *Res:* Natural resource management; agriculture extension education in developing nations. *Mailing Add:* Dept Natural Resources Ball State Univ Muncie IN 47306

VAN METER, WAYNE PAUL, b Fresno, Calif, Feb 16, 26; m 48; c 4. INORGANIC CHEMISTRY. *Educ:* Ore State Col, BS, 50, MS, 52; Univ Wash, PhD(inorg chem), 59. *Prof Exp:* Chemist, Hanford Atomic Prod Oper, Gen Elec Co, Wash, 51-56; from asst prof to assoc prof, 59-71, PROF CHEM, UNIV MONT, 71- *Mem:* Am Chem Soc. *Res:* Measurement of trace concentrations of metals in biological systems; development of sample processing techniques and instrumentation for atomic absorption spectrometry. *Mailing Add:* 2224 1/2 Rattlesnake Dr Missoula MT 59801

VAN METRE, THOMAS EARLE, JR, b Newport, RI, Jan 11, 23; m 47; c 5. MEDICINE, ALLERGY. *Educ:* Harvard Univ, BS, 43, MD, 46; Am Bd Internal Med, dipl, 55. *Prof Exp:* Intern med, Johns Hopkins Hosp, 46-47, asst resident, 47-48, 50 & 51-52, Am Cancer Soc fel, 52-53; asst prof internal med, Sch Med, St Louis Univ, 53-54; from instr to asst prof, 56-70, physician-in-chg Adult Allergy Clin, 66-84, PHYSICIAN, JOHNS HOPKINS HOSP, 56-, ASSOC PROF MED, MED SCH, JOHNS HOPKINS UNIV, 70- *Concurrent Pos:* Pvt pract, 54-; mem attend staff, Baltimore City Hosp, 54- & Union Mem Hosp, 59- *Mem:* AMA; Am Fedn Clin Res; Am Col Physicians; Am Acad Allergy (pres, 78); Am Clin & Climat Asn. *Res:* Allergy; effect of corticosteroids on growth; uveitis; asthma. *Mailing Add:* 11 E Chase St Baltimore MD 21202

VANMIDDLESWORTH, FRANK L, b Memphis, Tenn, Nov 26, 55; m 79. BIOORGANIC CHEMISTRY, MEDICINAL CHEMISTRY. *Educ:* Vanderbilt Univ, BA, 77; Emory Univ, PhD(chem), 82. *Prof Exp:* Res assoc chem, Emory Univ, 77-82; postdoctoral res assoc, Dept Pharm, Univ Wis, 82-84; org chemist, Northern Regional Res Ctr, USDA, 84-87; natural prod chemist, 87-90, SYNTHETIC CHEMIST & RES FEL, MERCK & CO, 90- *Concurrent Pos:* Adj prof, Plant Path Dept, Univ Ill, 86. *Mem:* Am Chem Soc; AAAS; Am Pharmacog Soc; Am Peptide Soc. *Res:* Medicinal, bioorganic, natural product, and immunochemistry; rational design and synthesis of biologically active compounds; chemico-enzymatic systhesis; isolation, structure determination, modification, and biosynthesis of natural products such as enzyme inhibitors, antibiotics and mycotoxins; peptidomimetic vaccines. *Mailing Add:* Merck & Co R50G-146 PO Box 2000 Rahway NJ 07065

VAN MIDDLESWORTH, LESTER, b Washington, DC, Jan 13, 19; m 48; c 4. PHYSIOLOGY, MEDICINE. *Educ:* Univ Va, BS, 40, MS, 42 & 44; Univ Calif, Berkeley, PhD(physiol), 47; Univ Tenn, MD, 51. *Prof Exp:* Chief chemist, Piedmont Apple Prod Corp, Va, 39-44; res assoc, Radiation Lab & teaching asst physiol, Univ Calif, 44-46; from instr to assoc prof, 46-59, PROF PHYSIOL & BIOPHYS, CTR HEALTH SCI, UNIV TENN, MEMPHIS, 59-, PROF MED, 74- *Concurrent Pos:* Res asst physiol, Univ Va, 42-44; intern, John Gaston Hosp, 51-52; USPHS career res award, 61-89. *Mem:* Am Chem Soc; Am Physiol Soc; Endocrine Soc; Am Thyroid Asn; Health Physics Soc. *Res:* Vapor phase catalysis; hormone synthesis; carbohydrate metabolism; aviation medicine; anoxia; metabolism of plutonium, radium, iodide, thiocyanate and thyroxine; thyroid physiology; goiter; audiogenic seizures; radioactive fallout; mycotoxins. *Mailing Add:* Dept Physiol & Biophys Univ Tenn Ctr Health Sci 894 Union Ave Memphis TN 38163

VAN MIEROP, LODEWYK H S, b Surabaya, Java, Mar 31, 27; US citizen; m 54; c 5. MEDICINE. *Educ:* State Univ Leiden, MD, 52; Am Bd Pediat, cert pediat cardiol. *Prof Exp:* Lectr anat, McGill Univ, 61-62; from asst prof to assoc prof pediat, Albany Med Col, 62-66; assoc prof, 66-68, PROF PEDIAT & PATH, COL MED, UNIV FLA, 68-, GRAD RES PROF, 78- *Concurrent Pos:* NIH res career develop award, 64-73; res assoc, Mt Sinai Hosp, NY, 63-66; mem comt nomenclature of heart, NIH, 67-68; mem southern regional res rev comt, Am Heart Asn, 68-72. *Mem:* Am Pediat Soc; Am Heart Asn; Am Asn Anat; fel Am Acad Pediat; fel Am Col Cardiol. *Res:* Pediatric cardiology; pathology and pathogenesis of congenital heart disease; cardiac embryology. *Mailing Add:* Dept Pediat Univ Fla Col Med Gainesville FL 32610

VANN, DOUGLAS CARROLL, b Coronado, Calif, May 3, 39; m 61; c 1. IMMUNOBIOLOGY. *Educ:* Univ Calif, Berkeley, AB, 60; Univ Calif, Santa Barbara, PhD(biol), 66. *Prof Exp:* Jr scientist, Inter-Am Trop Tuna Comn, 60-62; res assoc immunol, Biol Div, Oak Ridge Nat Lab, 66-68; USPHS training grant, Scripps Clin & Res Found, 68-70; asst prof, 70-73, ASSOC PROF GENETICS, UNIV HAWAII, HONOLULU, 73- *Concurrent Pos:* Prin investr, USPHS res grant, 71-77. *Mem:* AAAS. *Res:* Cellular basis of immune responses. *Mailing Add:* 1739 Ala Moana Blvd-C Honolulu HI 96815

VANN, JOSEPH M, b Clinton, NC, Dec 30, 37; m 61; c 1. RADIATION PROTECTION, EMERGENCY MANAGEMENT. *Educ:* NC State Univ, BS, 58; Univ NC, Chapel Hill, MEd, 61; ECarolina Univ, MPhysics, 72; Va Polytech Inst, MS, 73. *Prof Exp:* Prof math, Mt Olive Col, 61-71; safety & licensing engr, Nuclear Eng, Gen Pub Utilities, 73-76; nuclear engr, Nuclear Eng, NJ Radiation Protection, 76-81; sr engr, Nuclear Eng, Ebasco Serv, 81-89; SR ENGR, NUCLEAR ENG, WVALLEY NUCLEAR SERV, 90- *Mem:* Am Phys Soc; Sigma Xi. *Res:* Applications of Green's functions to reactor kinetics equations; transform and asymptotic analysis of wave motion; integral transform theory; radiation protection and emergency planning concerns for nucleus facilities, including high-level radioactive waste facilities. *Mailing Add:* 23 Hillview Ave Madison NJ 07940

VANN, W(ILLIAM) PENNINGTON, b Belton, Tex, Sept 9, 35; m 62; c 3. CIVIL ENGINEERING. *Educ:* Columbia Univ, BA, 58, BS, 59, MS, 60; Rice Univ, PhD(civil eng), 66. *Prof Exp:* Struct designer civil eng, Walter P Moore, Consult Engr, Tex, 60; asst prof, Rice Univ, 66-72; ASSOC PROF CIVIL ENG, TEX TECH UNIV, 72- *Concurrent Pos:* Mem, Earthquake Eng Res Inst, 67- *Mem:* Am Soc Civil Engrs; Am Concrete Inst; Seismol Soc Am; Sigma Xi. *Res:* Dynamic response of inelastic structural systems; static and dynamic behavior of inelastic structure, including buckling, energy absorption and failure; mobile homes; wind engineering. *Mailing Add:* Dept Civil Eng Tex Tech Univ Lubbock TX 79409

VANN, WILLIAM L(ONNIE), b McAlpin, Fla, Sept 21, 23; m 46; c 1. ELECTRICAL ENGINEERING. *Educ:* Univ Fla, BEE, 50. *Prof Exp:* Test engr, Gen Elec Co, 50; assoc engr, Johns Hopkins Univ, 50-55, proj supvr, 58-62, asst group supvr, 62-69, group supvr, 69-72, sr engr, Appl Physics Lab, 55-88, staff engr, 72-88, proj engr, 81-88; RETIRED. *Mem:* Inst Elec & Electronics Engrs. *Res:* Microwave and radar development; data acquisition and tracking radars; systems engineering; engineering management. *Mailing Add:* 20 Southview Ct Silver Spring MD 20905

VAN NESS, HENDRICK C(HARLES), b New York, NY, Jan 18, 24; wid; c 1. VAPOR & LIQUID EQUILIBRIUM. *Educ:* Univ Rochester, BS, 44, MS, 46; Yale Univ, DEng, 53. *Prof Exp:* Instr eng, Univ Rochester, 45-47; chem engr, M W Kellogg Co, Pullman, Inc, 47-49; asst prof chem eng, Purdue Univ, 52-56; from asst prof to prof chem eng, Rensselaer Polytech Inst, 56-83, chmn, Div Fluid, Chem & Thermal Processes, 69-74, inst prof, 83-89, EMER PROF CHEM ENG, RENSSELAER POLYTECH INST, 89- *Concurrent Pos:* Fulbright lectr, King's Col, Univ Durham, 58-59; vis prof, Univ Calif, Berkeley, 66 & Inst for Chem Technol, Denmark Tech Sch, Lyngby, Denmark, 77. *Honors & Awards:* Warren K Lewis Award, Am Inst Chem Engrs, 88. *Mem:* Am Chem Soc; fel Am Inst Chem Engrs. *Res:* Solution thermodynamics; phase equilibria. *Mailing Add:* Dept Chem Eng Rensselaer Polytech Inst Troy NY 12180-3590

VAN NESS, JAMES E(DWARD), b Omaha, Nebr, June 24, 26; m 48; c 4. ELECTRICAL ENGINEERING. *Educ:* Iowa State Col, BS, 49; Northwestern Univ, MS, 51, PhD(elec eng), 54. *Prof Exp:* Asst & res engr, 49-51, lectr elec eng, 52-53, from asst prof to assoc prof, 54-60, dir, Univ Comput Ctr, 62-65, chmn dept elec eng, 69-72, PROF ELEC ENG & COMPUT SCI, NORTHWESTERN UNIV, EVANSTON, 60- *Concurrent Pos:* Vis assoc prof, Univ Calif, 58-59; vis prof, Mass Inst Technol, 73-74. *Mem:* Fel Inst Elec & Electronics Engrs; Asn Comput Mach. *Res:* Use of digital computers in power system problems; numerical analysis; control systems. *Mailing Add:* Dept Elec Eng & Comput Sci Northwestern Univ Evanston IL 60208-3118

VAN NESS, JOHN WINSLOW, b McLean Co, Ill, Aug 16, 36; m 64; c 2. STATISTICS. *Educ:* Northwestern Univ, BS, 59; Brown Univ, PhD(appl math), 64. *Prof Exp:* Vis asst prof statist, Stanford Univ, 64-65, actg asst prof, 65-66; asst prof math, Univ Wash, 66-71; assoc prof statist, Carnegie-Mellon Univ, 71-73; assoc prof, 73-75, head, Prog Math Sci, 73-78, assoc dean, Sch Natural Sci & Math, 78-83, PROF, UNIV TEX, DALLAS, 75-, HEAD, PROG MATH SCI, 80- *Mem:* Inst Math Statist; fel Am Statist Asn; Classification Soc. *Res:* Theoretical and applied statistics; biostatistics; multivariate and time series analysis; classification and discriminant analysis; nonstandard regression analysis. *Mailing Add:* Prog Math Sci Univ Tex-Dallas PO Box 830688 Richardson TX 75083-0688

VAN NESS, KENNETH E, THERMODYNAMICS OF SURFACES. *Educ:* Bucknell Univ, BS, 67; Rutgers Univ, PhD(mat sci), 86. *Prof Exp:* ASST PROF, ENG, WASHINGTON & LEE UNIV, 86- *Mem:* Am Phys Soc. *Res:* Thermodynamics of surfaces; polymer liquids (evaluation of state, surface phenomena). *Mailing Add:* Dept Physics & Eng Washington & Lee Univ Lexington VA 24450

VAN NESTE, ANDRE, b Brussels, Belg, July 15, 38; Can citizen; m 67; c 3. MATERIALS SCIENCE, METALLURGY. *Educ:* Laval Univ, BSc, 60, DSc(metall), 63. *Prof Exp:* Researcher electron micros, Chem Metall Div, Nat Ctr Sci Res, Paris, 63-65; assoc prof metall, 65-77, PROF METALL, DEPT METALL & VDEAN EDUC, FAC SCI & ENG, LAVAL UNIV, 77- *Concurrent Pos:* Asn Foreign Tech Trainees in France fel, 63-64; Nat Res Coun Can fel, 64-65. *Mem:* Am Soc Metals; Can Inst Mining & Metall; Am Soc Eng Educ. *Res:* Mechanical properties of materials, fatigue, fracture, wear; service failure analysis. *Mailing Add:* 1127 Beau-Pre Sainte-Foy PQ G5H 3N7 Can

VANNICE, MERLIN ALBERT, b Broken Bow, Nebr, Jan 11, 43; m 71. CHEMICAL ENGINEERING, CATALYSIS. *Educ:* Mich State Univ, BS, 64; Stanford Univ, MS, 66, PhD(chem eng), 70. *Prof Exp:* Engr, Dow Chem Co, 66; res engr, Corp Res Labs, Exxon Res & Eng Co, 71-75, sr res engr, 75-76; assoc prof, 76-80, distinguished alumni prof eng, 85-91, PROF CHEM ENG, PA STATE UNIV, 80-, DISTINGUISHED PROF ENG, 91- *Concurrent Pos:* Indust fel, Sun Oil Co, Pa, 69-71; chmn, Gordon Conf on Catalysis, 82; Humboldt res award, 90. *Honors & Awards:* Prof Progress Award, Am Inst Chem Engrs, 86; Emmett Award, NA Catalysis Soc, 87; Schuit Lectr, Univ Del, 89. *Mem:* Am Chem Soc; Am Inst Chem Engrs; Catalysis Soc NAm; Sigma Xi; Nat Res Soc. *Res:* Adsorption and heterogeneous catalysis, including hydrogenation reactions of CO, aromatics and oxygenates, kinetics, catalyst preparation and characterization, surface diffusion and metal-support interactions. *Mailing Add:* Dept Chem Eng Pa State Univ University Park PA 16802

VANNIER, WILTON EMILE, b Pasadena, Calif, June 6, 24; m 53; c 2. IMMUNOCHEMISTRY, BIOCHEMISTRY. *Educ:* Univ Calif, San Francisco, MD, 48; Calif Inst Technol, PhD(immunochem), 58. *Prof Exp:* Instr exp med, Sch Pub Health, Univ NC, 51-54; immunochemist, Lab Immunol, Nat Inst Allergy & Infectious Dis, Md, 58-64, head, Immunochem Sect, 64-68; assoc prof biochem, Sch Med, Univ Southern Calif, 68-70; res med officer, Naval Med Res Inst, Nat Naval Med Ctr, 70-84; RETIRED. *Concurrent Pos:* Res fel, Calif Inst Technol, 57-60, vis assoc chem, 85- *Mem:* Fel AAAS; Am Chem Soc; Am Asn Immunol. *Res:* Parasite immunology; chemistry of antibodies and antigen-antibody reactions; liposome cell interactions. *Mailing Add:* 209 Oak Meadow Rd Siera Madre CA 91024

VAN NORMAN, GILDEN RAMON, b Jamestown, NY, Dec 11, 32; m 58; c 2. PHOTOGRAPHIC CHEMISTRY, POLYMER CHEMISTRY. *Educ:* Univ Rochester, BS, 54; Mass Inst Technol, PhD(org chem), 57. *Prof Exp:* TECH ASSOC, EASTMAN KODAK CO, 57- *Mem:* AAAS; Am Chem Soc. *Mailing Add:* 317 Orchard Creek Lane Rochester NY 14612

VAN NORMAN, JOHN DONALD, b Jamestown, NY, Sept 11, 34; m 58; c 1. ANALYTICAL CHEMISTRY. *Educ:* Univ Rochester, BS, 55; Rensselaer Polytech Inst, PhD(anal chem), 59. *Prof Exp:* Res assoc chem, Brookhaven Nat Lab, 59-61, from asst chemist to chemist, 61-69; assoc prof, 69-77, prof chem, Youngstown State Univ, 77-79; AT DEPT CHEM, OLD DOMINION UNIV, 79- *Concurrent Pos:* Fac fel, Am Soc Eng Educ, NASA, 87-88. *Mem:* AAAS; Am Chem Soc; Sigma Xi. *Res:* Spectrophotometry and electroanalytical chemistry of bile pigments; catalysis; gas permeation of polymeric material. *Mailing Add:* Dept Chem Old Dominion Univ Norfolk VA 23508-8501

VAN NORMAN, RICHARD WAYNE, plant physiology, for more information see previous edition

VANNOTE, ROBIN L, b Summit, NJ, Aug 12, 34; m 59; c 4. STREAM & RIVER ECOLOGY. *Educ:* Univ Maine, BS, 57; Mich State Univ, MS, 62, PhD(limnol), 63. *Prof Exp:* Biologist, Water Qual Br, Tenn Valley Authority, 63-65, chief biol sect, 65-66; DIR, STROUD WATER RES CTR, ACAD NATURAL SCI PHILADELPHIA, 66- *Concurrent Pos:* Mem pesticide monitoring subcomt, Fed Comt Pest Control, 64- 66; adj prof entom & appl ecol, Univ Del, 74- *Mem:* AAAS; Am Fisheries Soc; Am Soc Limnol & Oceanog; Ecol Soc Am; Sigma Xi; Am Entom Soc; NAm Benthological Soc. *Res:* Ecology of streams and rivers; interactions with terrestrial landscapes, production ecology; detrital systems, energy flow and nutrient budgets, geomorphology of streams; effects of channel modifications, drainage, and rural runoff on biotic productivity and system stability; ecology of regulated river systems. *Mailing Add:* 315 W Street Rd Kennett Square PA 19348

VAN OERO, WILLEM THEODORUS HENDRICUS, b Amsterdam, Neth, Mar 17, 34; m 60; c 2. NUCLEAR & PARTICLE PHYSICS. *Educ:* Univ Amsterdam, PhD(math, physics), 63. *Prof Exp:* From res asst to res assoc, Inst Nuclear Physics Res, Amsterdam, Neth, 57-64; asst res physicist, Univ Calif, Los Angeles, 64-66; assoc prof, 67-74, PROF PHYSICS, UNIV MAN, 74- *Concurrent Pos:* Vis assoc prof, Univ Calif, Los Angeles, 71-72; vis scientist, CEN de Saclay, Gif-sur-Yvette, France, 75-76, Tri Univ Meson Facil, Vancouver, BC, 81-83, 87-89, Saturne Nat Lab, Gif-sur-Yvette, France, 89-90; vis staff mem, Los Alamos Sci Lab, NMex, 79-80; prog dir NSF, Wash, DC, 86-87; Killam fel, 87-89. *Mem:* Fel Am Phys Soc; Can Asn Physicists; NY Acad Sci; Sigma Xi. *Res:* Nuclear reactions induced by various types of particle beams at low and intermediate energies; phenomenological and theoretical analyses of few-nucleon problems; nuclear optical model; fundamental symmetries; strangeness in nuclear matter. *Mailing Add:* Dept Physics Univ Man Winnipeg MB R3T 2N2 Can

VANONI, VITO A(UGUST), b Camarillo, Calif, Aug 30, 04; m 34. HYDRAULICS. *Educ:* Calif Inst Technol, BS, 26, MSc, 32, PhD(civil eng, hydraul), 40. *Prof Exp:* Jr struct engr, Bethlehem Steel Corp, 26-28 & 30-31; draftsman, Am Bridge Co, 28-30; from proj supvr to hydraul engr & supvr lab, USDA, 35-47, assoc dir, Hydrodyn Lab, Nat Defense Res Comt, 41-46, from asst prof to prof hydraul, 42-74, EMER PROF HYDRAUL, CALIF INST TECHNOL, 74- *Mem:* Nat Acad Eng; hon mem Am Soc Civil Engrs. *Res:* Sediment transportation; hydraulics of open channels; harbor and coastal engineering. *Mailing Add:* Environ Eng Dept Calif Inst Technol Pasadena CA 91125

VAN ORDEN, HARRIS O, b Smithfield, Utah, Oct 6, 17; m 48; c 1. ORGANIC CHEMISTRY. *Educ:* Utah State Agr Col, BS, 38; Wash State Univ, MS, 42; Mass Inst Technol, PhD(org chem), 51. *Prof Exp:* From asst prof to assoc prof chem, Utah State Agr Col, 46-52; NIH spec fel, Univ Utah, 53; from assoc prof to prof, 54-83, actg head dept, 58-59, EMER PROF CHEM, UTAH STATE UNIV, 83- *Mem:* Fel AAAS; Am Chem Soc; Sigma Xi. *Res:* Synthetic organic and bio-organic chemistry; protein sequence studies; synthesis of peptides; enzyme specificity studies. *Mailing Add:* 281 E Eighth N Logan UT 84321

VAN ORDEN, LUCAS SCHUYLER, III, b Chicago, Ill, Nov 3, 28; m 53; c 4. CHEMICAL DEPENDENCY, PSYCHOPHARMACOLOGY. *Educ:* Northwestern Univ, BS, 50, MS, 52, MD, 56; Yale Univ, PhD(pharmacol), 66. *Prof Exp:* Intern, Harper Hosp, Detroit, 56-57; asst resident surg, Med Ctr, Yale Univ, 61-62; Nat Inst Neurol Dis & Blindness spec fel neuroanat, Dept Anat, Harvard Med Sch, 66-67; from asst prof to assoc prof, 67-73, PROF PHARMACOL, COL MED, UNIV IOWA, 73- *Concurrent Pos:* Resident psychiat, Univ of Iowa, 75-76; dir, Chem Dependency Unit, Ment Health Inst, Mt Pleasant & adj prof pharmacol & psychiat, Univ Iowa Col Med, 78-79. *Mem:* AMA; Am Med Soc Alcoholism & Other Drug Dependencies; Am Psychiat Asn. *Res:* Neuropharmacology; autonomic nervous system fine structure and histochemistry of adrenergic transmitter; quantitative cytochemistry, immunocytochemistry; alcohol and drug abuse. *Mailing Add:* 12208 Camino Arbustos Albuquerque NM 87111

VAN ORDER, ROBERT BRUCE, b Glenvale, Ont, Mar 19, 15; nat US; m 42; c 3. ORGANIC CHEMISTRY. *Educ:* Queen's Univ, Can, BA, 38, MA, 39; NY Univ, PhD(org chem), 42. *Prof Exp:* Asst, NY Univ, 39-42; res org chemist, Stamford Res Labs, Am Cyanamid Co, 42-46, plant chemist, Calco Chem Div, 46-53, asst chief chemist, Org Chem Div, Bound Brook Lab, 53-55 & Mkt Develop Dept, 55-62; tech dir, Pearsall Chem Co, 62-63; mgr sales develop, Am Cyanamid Co, Bound Brook, 63-69, mgr org chem res & develop, Wayne, 69-77 & Bound Brook, 77-80; RETIRED. *Mem:* Fel Am Inst Chemists; Am Chem Soc. *Res:* Structure and synthesis of antibiotics; inorganic pigments and chemicals; dyestuffs and pigment dispersions; market development and sales development of new chemical products. *Mailing Add:* 58 Sycamore Ave Berkeley Heights NJ 07922

VAN OSS, CAREL J, b Amsterdam, Neth, Sept 7, 23; m 51; c 3. IMMUNOCHEMISTRY, PHYSICAL BIOCHEMISTRY. *Educ:* Univ Paris, PhD(phys biochem), 55. *Prof Exp:* Fel colloid chem, Van't Hoff Lab, Univ Utrecht, 55-56; fel phys chem, Ctr Electrophoresis, Sorbonne, 56-57; dir lab phys biochem, Nat Vet Col Alfort, 57-63; asst head dept microbiol, Montefiore Hosp, NY, 63-65; assoc prof biol, Marquette Univ, 66-68; assoc prof, 68-72, PROF MICROBIOL, SCH MED, STATE UNIV NY, BUFFALO, 72-, HEAD IMMUNOCHEM LAB, 68- *Concurrent Pos:* French Ministry Agr res fel, 55-57; master of res, French Nat Agron Inst, 62-63; consult, Amicon Corp, 64-67; dir serum & plasma depts, Milwaukee Blood Ctr, 65-68; prin investr, USPHS res grant, 66-75; consult, Gen Elec Co, 67; mem consult comt electrophoresis & other chem separation processes in outer space, NASA, 71-81; exec ed, Preparative Biochem, 71-, Separation & Purification Methods, 72-90, ed, Immunol Commun, 82-84 & Immunol Invest, 85-; consult mem, immunol Panel Diag Prod Adv Comt, Food & Drug Admin, 75-83; adj prof chem eng, State Univ NY, Buffalo, 79- *Mem:* Am Chem Soc; Am Asn Immunologists; Electrophoresis Soc. *Res:* Membrane separation methods; precipitation in immunochemical, organic and inorganic systems; diffusion; sedimentation; physical surface properties of cells; mechanism of phagocytic engulfment; cell separation methods; van der Waals and polar interactions between cells and/or polymers in liquids; opsonins; hydrophobic interactions; colloid and surface science. *Mailing Add:* Dept Microbiol Immunochem Lab State Univ NY Sch Med Buffalo NY 14214

VAN OSTENBURG, DONALD ORA, b East Grand Rapids, Mich, July 19, 29; m 51; c 2. SOLID STATE PHYSICS. *Educ:* Calvin Col, BS, 51; Mich State Univ, MS, 53, PhD(physics), 56. *Prof Exp:* Assoc physicist, Armour Res Found, Ill Inst Technol, 56-59; from asst physicist to assoc physicist, Argonne Nat Lab, 59-70; prof, 70-87, CHMN, PHYSICS DEPT, DE PAUL UNIV, 87- *Concurrent Pos:* Pres, Cent States Univ Inc, 81-82. *Mem:* Am Phys Soc; Am Sci Affiliation. *Res:* Static electrification; electron paramagnetic resonance; lattice dynamics; magnetism; nuclear magnetic resonance; electronic structure of metals and alloys; semiconductor devices; biophysics; catalysts. *Mailing Add:* Dept Physics De Paul Univ 2219 N Kenmore Chicago IL 60614

VAN OVERBEEK, JOHANNES, b Schiedam, Holland, Jan 2, 08; nat US; m 32, 48; c 6. PLANT PHYSIOLOGY, BIOLOGY. *Educ:* State Univ Leiden, BS, 28; Univ Utrecht, MS, 32, PhD(bot), 33. *Hon Degrees:* Dr, Univ Belg, 60. *Prof Exp:* Asst bot, Univ Utrecht, 33-34; asst plant hormones, Calif Inst Technol, 34-37, instr, 37-39, asst prof plant physiol, 39-43; plant physiologist, Inst Trop Agr, PR, 43-44, head dept, 44-46, asst dir, 46-47; chief plant physiologist & head dept plant physiol, Agr Lab, Shell Develop Co, 47-67; head dept, 67-73, prof biol, 67-78, EMER PROF BIOL, TEX A&M UNIV, 78- *Concurrent Pos:* Hon prof, Col Agr, Univ PR, 44-47; ed, Plant Physiol, 67-78; mem, Gov Adv Panel Use of Agr Chem, Tex, 70. *Honors & Awards:* Award, Am Soc Agr Sci, PR, 48. *Mem:* AAAS; Charles Reed Barnes hon mem Am Soc Plant Physiologists; Bot Soc Am; Soc Gen Physiol; Am Inst Biol Sci. *Res:* Plant hormones; physiology of growth; applied plant physiology. *Mailing Add:* 3615 Sunnybrook Lane Bryan TX 77802

VAN OVERSTRAETEN, ROGER JOSEPH, b Vlezenbeek, Belg, Dec 7, 37; m 60; c 2. MICRO-ELECTRONICS, SOLAR CELLS. *Educ:* Katholieke Univ, Leuven, cert nuclear eng, 60; Univ Stanford, PhD(phys electronics), 63. *Hon Degrees:* Dr, INPG, France, 87. *Prof Exp:* Res asst, Univ Stanford, 62-65; asst prof, 65-68, PROF SOLID STATE PHYSICS & ELECTRONICS, KATHOLIEKE UNIV, LEUVEN, BELG, 68-; PRES, IMEC, 84- *Concurrent Pos:* Lectr, Iria, France, 71 & Univ Newcaster Tyne, UK, 73; vis prof, Univ Fla, 74, Stanford Univ, 79 & Cent Electronic Res Inst, India, 80; mem, bd dirs, Imec, Cobrain, EDC, Janssen Pharmaceut & Soltech, 84- *Mem:* Fel Inst Elec & Electronics Engrs; Am Phys Soc. *Res:* Physics and electronics of semiconductor devices related to integrated circuits, sensors and solar cells. *Mailing Add:* Imec Kapeldreef 75-B Pres Imec Vzw Kapeldreef 75-B Leuven 3001 Belgium

VAN PATTER, DOUGLAS MACPHERSON, b Montreal, Que, July 4, 23; m 50; c 4. NUCLEAR PHYSICS. *Educ:* Queen's Univ, Can, BSc, 45; Mass Inst Technol, PhD(physics), 49. *Prof Exp:* Jr physicist, Nat Res Coun Can, 45-46; asst physics, Mass Inst Technol, 46-49, res assoc, 49-52; res assoc, Univ Minn, 52, asst prof, 52-54; physicist, Bartol Res Found, Franklin Inst, 54-76; BARTOL PROF PHYSICS & BARTOL RES FOUND, UNIV DEL, 76- *Concurrent Pos:* Chmn subcomt nuclear constants, Nat Acad Sci-Nat Res Coun, 59-64; vis prof, Inst Nuclear Physics, Univ Frankfurt, 72. *Mem:* Fel Am Phys Soc; AAAS; Meteoritical Soc. *Res:* Nuclear reactions using electrostatic accelerators; proton microprobe; elemental compositions using induced x-rays of tektites and meteorites. *Mailing Add:* 97 Sproul Rd Springfield PA 19064

VANPEE, MARCEL, b Hasselt, Belg, Dec 4, 16; US citizen; m 48; c 3. PHYSICAL CHEMISTRY. *Educ:* Cath Univ Louvain, BS, MS & PhD(phys chem), 40, Agrege de l'Enseignement Superieur, 56. *Prof Exp:* Nat Res fel radio & photochem, Nat Found Sci Res, Belg, 40-45; head lab sci res, Nat Inst Mines, Belg, 46-56; prof physics, Univ Leopoldville, Congo, 56-57; supvry chemist, US Bur Mines, Pa, 57-60; sr scientist, Reaction Motor Div, Thiokol Chem Corp, 60-68; prof chem eng, Univ Mass, Amherst, 68-80; RETIRED. *Concurrent Pos:* Lectr & researcher, Cath Univ Louvain, 40-45; mem, Nat Found Sci Res, Belg, 40-; res fel, Univ Minn, Minneapolis, 48; staff physicist, Edsel B Ford Inst, Mich, 50. *Honors & Awards:* Awards, Prix Jean Stass, 40, Prix Louis Empain, Belg Acad Sci, 43 & Prix Frederic Swartz, 54. *Mem:* Combustion Inst; Am Inst Aeronaut & Astronaut. *Res:* Radiochemistry; photochemistry; kinetics of combustion reactions; cool flames; ignition; flame spectroscopy and structure; high energy fuel and oxidizers; rocket exhaust radiation; hypergolic ignition; atomic and chemiluminescent reactions; reentry observables. *Mailing Add:* 151 Rolling Ridge Rd Amherst MA 01002

VAN PELT, ARNOLD FRANCIS, JR, b Orange, NJ, Sept 24, 24; m 47; c 2. ECOLOGY OF ANTS, ANT COLONY BEHAVIOR. *Educ:* Swarthmore Col, BA, 45; Univ Fla, MS, 47, PhD(biol), 50. *Prof Exp:* Assoc prof biol, Appalachian State Teachers Col, 50-54; prof, Tusculum Col, 54-57, prof biol & chem, 57-63; prof biol, Greensboro Col, 63-80, chmn, Dept Sci & Math, 64-71 & 81-83, dir, Allied Health Progs, 75-89, chmn, Div Natural Sci & Math, 81-83, Moore prof biol, 81-88, EMER DISTINGUISHED MOORE PROF BIOL, GREENSBORO COL, 83- *Concurrent Pos:* Sewell grant, Highlands Biol Sta, 53-54; res grants, 53-78; NSF grants, 55-56 & 62; consult, Univ Ga ecol team, Savannah River Plant, AEC, 60, Oak Ridge Nat Lab, Biol Div, 69-70, 73; mem, conf plant biochem, Inst Paper Chem, 61, conf molecular genetics, SW Ctr Advan Studies, 68; Piedmont Univ Ctr grants, 64-67, 71 & 72; Res Corp Brown-Hazen Fund grant, 65-67; mem radiation biol conf, Oak Ridge Inst Nuclear Studies, 65; courtesy appointment, Sch Med Technol, Bowman Gray Sch Med, Winston-Salem, NC, 66-, physician asst prog, 71-, Sch Radiol Technol, Moses H Cone Mem Hosp, Greensboro, NC, 68-, Sch Med Technol, 85- & Sch Med Technol, Forsyth Mem Hosp, Winston-Salem, NC, 72-; Savannah River Ecol Lab, Ecol of Ants, 74, 76-78; adj prof, continuing educ, Univ NC-Greensboro, 74; vis lectr radiation biol, Moses H Cone Mem Hosp, Greensboro, NC, 75 & 77; Burroughs-Wellcome grant, 79, 86-88; Southern Regional Educ Bd grant, 82; Greensboro Col grants, 82-88; grants, Off Resource Mgt, Big Bend Nat Park, Tex, 82-89, Nat Park Syst grant-in-aid, Tex A&M Univ, 90. *Mem:* AAAS; Sigma Xi; Entom Soc Am. *Res:* Ecology of mountain ants; mouse genetics; guinea pig leukemia; nest relocation in harvester ants. *Mailing Add:* 203 Howell Pl Greensboro NC 27408-1712

VAN PELT, RICHARD H, b St Louis, Mo, Apr 11, 22. METALLURGICAL ENGINEERING. *Educ:* Univ Ill, BS, 43. *Prof Exp:* Mgr, Res Metall Div, Caterpillar Inc, 46-85; RETIRED. *Mem:* Fel Am Soc Metals Int. *Mailing Add:* 915 Birchwood Dr Washington IL 61571

VAN PELT, RICHARD W(ARREN), b Chicago, Ill, Dec 22, 33; c 3. NUMERICAL CONTROL, SERVO SYSTEMS. *Educ:* Stanford Univ, BS, 55; Univ Ill, MS, 60; Case Inst Technol, PhD(mech eng), 65. *Prof Exp:* Sr engr, Systs Develop Div, IBM Corp, 65-78; ADV ENGR, TAPE DIV, STORAGE TECHNOL CORP, 78- *Mem:* Instrument Soc Am; Inst Elec & Electronics Engrs. *Res:* Numerical control, development of special purpose computers for process control and for real time control of fluid and electric servo systems; simulation and analysis of dynamics of electromechanical systems. *Mailing Add:* 665 Meadowbrook Boulder CO 80303

VAN PELT, ROLLO WINSLOW, JR, b Chicago, Ill, Dec 14, 29; m 54; c 2. PATHOLOGY. *Educ:* Wash State Univ, BA, 54, DVM, 56; Mich State Univ, MS, 61, PhD(path), 65; Am Col Vet Pathologists, dipl, 65. *Prof Exp:* Vet, Rose City Vet Hosp, Ore, 56-57, Tigard Vet Hosp, 57 & Willamette Vet Hosp, 57-59; res asst arthrology, Mich State Univ, 59-60, res assoc, 60-62, Nat Inst Arthritis & Metab Dis fel, 62-64, spec fel, 64-65, from asst prof to assoc prof path, 65-70; vis assoc prof zoophysiol & path, Inst Arctic Biol, Univ Alaska, Fairbanks, 70-71, assoc prof zoophysiol & path, 71-78; CHIEF OF STAFF & PATHOLOGIST, ALASKA VET MED CLIN, 78- *Concurrent Pos:* Upjohn Co grant-in-aid, 62-64; All Univ res grant, Mich State Univ, 63-65; Paul Harris fel, Rotary Found, Rotary Int, 85. *Mem:* Am Vet Med Asn; Am Col Vet Pathologists; Am Vet Radiol Soc; Am Soc Vet Clin Path; Sigma Xi; US & Can Acad Path; NY Acad Sci. *Res:* Hereditary and congenital abnormalities of the respiratory tract in Alaskan husky dogs; exhaustrion pneumonia syndrome in racing sled dogs. *Mailing Add:* Alaska Vet Med Clin 410 Trainor Gate Rd Fairbanks AK 99701

VAN PELT, WESLEY RICHARD, b Passaic, NJ, Oct 17, 43; m 65; c 1. HEALTH PHYSICS, INDUSTRIAL HYGIENE. *Educ:* Rutgers Univ, New Brunswick, BA, 65, MS, 66; NY Univ, PhD(nuclear eng), 71; Am Bd Health Physics, cert, 73, Am Bd Indust Hyg, 76. *Prof Exp:* Asst res scientist aerosol physics, Med Ctr, NY Univ, 67-71; environ scientist, Environ Analysts, Inc, NY, 71-72; indust hygienist, Radiation Safety Off, 72-77, asst mgr safety & indust hyg, 77-80, mgr safety & indust hygiene, 80-82, corp safety dir, Hoffmann-La Roche Inc, 82-85; PRES & CONSULT, WESLEY R VAN PELT ASSOC, INC, 85- *Concurrent Pos:* Consult, NJ Comn Radiation Protection, 76-77 & mem, NJ X-ray Technician Bd Examnrs, 78- & NJ Panel Sci Adv, 81-; adj asst prof, Univ Med & Dent, Robert Wood Johnson Med Sch, 86- *Mem:* Sigma Xi; Health Physics Soc; Am Indust Hyg Asn; NY Acad Sci; AAAS; Am Biol Safety Asn; Am Soc Safety Engrs; Air Pollution Control Asn. *Res:* Applied health physics; radiation protection program development;

interaction of airborne radioactivity with natural aerosols; measurement and study of the natural ionizing radiation background; indoor air pollution; laboratory and industrial health and safety. *Mailing Add:* 773 Paramus Rd Paramus NJ 07652

VAN PERNIS, PAUL ANTON, medicine, for more information see previous edition

VAN PILSUM, JOHN FRANKLIN, b Prairie City, Iowa, Jan 28, 22; m 58; c 6. BIOCHEMISTRY. *Educ:* Univ Iowa, BS, 43, PhD, 49. *Prof Exp:* Instr biochem, Long Island Col Med, 49-51; asst prof, Univ Utah, 51-54; from asst prof to assoc prof, 54-63, PROF BIOCHEM, MED SCH, UNIV MINN, MINNEAPOLIS, 71- *Mem:* Am Soc Biol Chemists; Am Inst Nutrit. *Res:* Guanidinium compound metabolism. *Mailing Add:* Dept Biochem 4-225 Millard Hall Univ Minn Col Med Sci Minneapolis MN 55455

VAN POOLEN, LAMBERT JOHN, b Detroit, Mich, Apr 20, 39; m 62; c 2. MECHANICAL ENGINEERING DESIGN, PHILOSOPHY OF TECHNOLOGY. *Educ:* Calvin Col, BS, 64; Ill Inst Technol, BSME, 64, MSME, 65, PhD(mech & aerospace eng), 69. *Prof Exp:* PROF ENG, CALVIN COL, 69- *Concurrent Pos:* Res scientist, Nat Inst Standards & Technol. *Mem:* Am Soc Eng Educ; Soc Philos & Technol; Am Soc Mech Engrs. *Res:* Thermodynamics; heat transfer; initiator and foremost authority on the use of liquid volume fractions to predict critical density of pure fluids mixtures; thermodynamic properties of coexistence data. *Mailing Add:* Dept Eng Calvin Col Grand Rapids MI 49506

VAN POZNAK, ALAN, b Newark, NJ, Dec 30, 27; m 50; c 4. ANESTHESIOLOGY, PHARMACOLOGY. *Educ:* Cornell Univ, AB, 48, MD, 52. *Prof Exp:* Asst instr surg, Cornell Univ Med Col, 55-56, asst resident & resident anesthesiol, 56-58, instr surg, 58-61, from asst prof to assoc prof, 61-72, PROF ANESTHESIOL, CORNELL UNIV, 72- *Concurrent Pos:* Res fel pharmacol, Cornell Univ, 62-64, clin asst prof, 67-70, clin assoc prof, 70-73, assoc prof, 73-74, prof, 74- *Mem:* AMA; Am Soc Anesthesiol; Asn Univ Anesthesiologists. *Res:* New inhalation anesthetics; neuromuscular effects of inhalation anesthetics. *Mailing Add:* 525 E 68th St New York NY 10021

VAN PRAAG, HERMAN M, b Schiedam, Neth, Oct 17, 29; m 56; c 4. BIOLOGICAL PSYCHIATRY. *Educ:* State Univ, Leiden Neth, MD, 56; Univ Utrecht, PhD(neurobiol), 62. *Prof Exp:* Chief-of-staff, dept psychiat, Dijkzigt Hosp, Rotterdam, Neth, 63-66; founder & first head, dept biol psychiat, Psychiat Univ Clin, State Univ, Groningen, 66-77, from assoc prof to prof psychiat & biol psychiat, 68-70; prof psychiat & head dept, Acad Hosp, State Univ, Utrecht, 77-82; PROF PSYCHIAT & HEAD DEPT, ALBERT EINSTEIN COL MED, NY, 82-; PSYCHIATRIST-IN-CHIEF, MONTEFIORE MED CTR, NY, 82- *Concurrent Pos:* Lady Davis vis prof, dept psychiat, Hebrew Univ, Hadassah Univ Hosp, Jerusalem, Israel, 76-77; mem, comt on therapeut, NY State Off Ment Health, 83; comt on ment health, Greater NY Hosp Asn, 83; adv comt, Ment Health Asn, 83, Clin Res Ctr, Albert Einstein Col Med, 83; bd trustees, Asn Nervous & Ment Dis, 83, High Point Hosp, NY, 83; guest lectr, var univs & insts throughout Europe, US, Asia and SAfrica. *Honors & Awards:* Ramaer Medal, Neth Soc Psychiat & Neurol, 65; Anna-Monika Prize, Anna-Monika Found, Switz, 73; Saal van Zwanenberg Prize, Dutch Soc Pharmacol & Physiol, 76; Harold Himwich Mem Lectr, Soc Biol Psychiat, US, 76; Hinck Mem Lectr, Univ Toronto, 76; Maudsley Bequest Lectr, Royal Col Psychiatrists, Eng, 81; Eli Robins Lectr, Wash Univ, Mo, 83; Leonard Cammer Mem Award, Col Physicians & Surgeons, NY, 83; Reynier de Graaf Medal, Dutch Soc Advan Biol & Med Sci, 83; Duplar Award, 86. *Mem:* Hon mem Arg Soc Biol Psychiat; hon mem Interdisciplinary Soc Biol Psychiat; corresp mem Royal Acad Sci, Neth; fel Am Psychopath Asn; Int Col Neurobiol, Biol Psychiat & Psychopharmacol; Int Soc Psychoneuroendocrinol; fel Am Col Neuropsychopharmacol. *Res:* Mental health; biological psychiatry. *Mailing Add:* Dept Psychiat Albert Einstein Col Med Bronx NY 10461

VANPRAAGH, RICHARD, b Ont, Can, Apr 11, 30; m 62; c 3. PEDIATRIC CARDIOLOGY, PEDIATRIC CARDIAC PATHOLOGY. *Educ:* Univ Toronto, MD, 54. *Hon Degrees:* AM, Harvard Univ, 89. *Prof Exp:* Asst prof pediat, Sch Med, Northwestern Univ, 65; clin assoc path, 65-67, asst clin prof, 67-70, assoc prof, 70-73, PROF PATH, HARVARD MED SCH, 74- *Concurrent Pos:* Asst dir, Congenital Heart Dis Res & Training Ctr, Hektoen Inst Med Res, Chicago, 63-65; dir cardiac path & embryol & res assoc, Children's Hosp, 65-; vis prof pediat & cardiol, Univ Ore, 78; Pfizer vis lectr, Children's Mem Hosp, Chicago, 86, 87. *Honors & Awards:* Morgani lectr, Univ Padova, Italy, 73; Haile Selassie lectr, Nat Heart Hosp, London, 73; Hammersmith Cardiac Surg lectr, London, 76; Ann Taran Man lectr, Sch Med, NY Univ, 79; Beth Raby Lectr, Children's Hosp, Montreal, 88; Jerome Liebman Lectr, Rainbow Babies' & Children's Hosp, Cleveland. *Mem:* Fel Am Col Cardiol; corresp mem Brit Cardiac Soc. *Res:* Correlation of clinical, pathologic, embryologic and etiologic findings concerning congenital heart disease, in order to improve diagnostic accuracy and surgical success; description of newly recognized forms of congenital heart disease; new surgical operations; causes of heart disease in infants and children. *Mailing Add:* Children's Hosp 300 Longwood Ave Boston MA 02115

VAN PUTTEN, JAMES D, JR, b Grand Rapids, Mich, Apr 14, 34; m 59; c 2. INDUSTRIAL PROCESS CONROL. *Educ:* Hope Col, AB, 55; Univ Mich, AM, 57, PhD(physics), 60. *Prof Exp:* Instr physics, Univ Mich, 60-61; NATO fel, Europ Orgn Nuclear Res, Geneva, 61-62; asst prof, Calif Inst Technol, 62-67; assoc prof, 67-70, PROF PHYSICS, HOPE COL, 70-, CHMN DEPT, 75-86, 90- *Concurrent Pos:* Consult, Electro-Optical Systs, 63-67, Teledyne Corp, 67-70, Donnelly Inc, 67-, White Westing House, 82- & Gen Motors, 85- *Mem:* AAAS; Am Phys Soc. *Res:* Bubble chambers; counter and spark chamber techniques; satellite borne space physics experiments; nuclear charge structure; use of microcomputers in process control; real time statistical process control. *Mailing Add:* Dept Physics Hope Col Holland MI 49423

VAN RAALTE, JOHN A, b Copenhagen, Denmark, Apr 10, 38; US Citizen; m 63; c 2. RESEARCH ADMINISTRATION. *Educ:* Mass Inst Technol, SB & SM, 60, EE, 62, PhD(solid state physics, elec eng), 64. *Prof Exp:* Res asst lab insulation res, Mass Inst Technol, 60-64; mem tech staff, RCA Res Labs, 64-70, head displays & device concepts, 70-79, head video disc rec & playback res, 79-83, dir video disc syst res, 83-84, dir syst mat & process res, 84-86; dir mat & process technol lab, David Sarnoff Res Ctr, 86-88; AT THOMSON CONSUMER ELECTRONICS, 88- *Mem:* Fel Inst Elec & Electronics Engrs; Am Phys Soc; fel Soc Info Display (secy, 81-82, treas, 81-, vpres, 83-84, pres, 84-86); Sigma Xi. *Res:* Materials; dielectrics; electro-optic materials; lasers; displays; video disc; consumer electronics. *Mailing Add:* Thomson Consumer Electronics 1002 New Holland Ave Lancaster PA 17601

VAN REEN, ROBERT, b Paterson, NJ, June 12, 21. BIOCHEMISTRY. *Educ:* NJ State Teachers Col, Montclair, AB, 43; Rutgers Univ, PhD(biochem), 49. *Prof Exp:* Assoc biochemist, Brookhaven Nat Lab, 49-51; res assoc, McCollum-Pratt Inst, Johns Hopkins Univ, 51-53, asst prof biol, 53-56; supv chemist & assoc head dent div, Naval Med Res Inst, 56-61, head nutrit biochem div, 61-70; prof, 70-85, chmn dept, 70-83, EMER PROF FOOD SCI & HUMAN NUTRIT, UNIV HAWAII, HONOLULU, 85- *Honors & Awards:* McLester Award, Asn Mil Surgeons of US, 59. *Mem:* Am Dietetic Asn; Soc Nutrit Educ; Am Soc Biol Chemists; Inst Food Technologists; Am Inst Nutrit. *Res:* Mammalian and avian nutrition; requirements and functions of vitamins and trace elements; interrelationships between trace elements and enzyme systems; experimental dental caries; metabolism in calcified tissues; nutrition and urolithiasis. *Mailing Add:* Dept Food Sci & Human Nutrit Univ Hawaii-Manoa Honolulu HI 96822

VAN REMOORTERE, EMILE C, cardiovascular physiology, pharmacology, for more information see previous edition

VAN RENSBURG, WILLEM CORNELIUS JANSE, b SAfrica, Nov 28, 38; m 62; c 3. APPLIED GEOLOGY. *Educ:* Univ Pretoria, BSc, 61, MSc, 63; Univ Wis, PhD(geol), 65. *Prof Exp:* Sr geologist, SAfrica Geol Surv, 60-67; dep dir, SAfrica Nat Dept Planning, 67-73; head, Econ & Corting Div, Nat Inst Metall, 73-75; tech dir, SAfrica Minerals Bur, 79-81; assoc dir, Bur Econ Geol, 79-81, PROF GEOL & PETROL ENG, UNIV TEX, 81- *Concurrent Pos:* Commissioner, Pres Comn Conserv SAfrica's Coal Resources, 70-75; B P Prof, Rand Afrikaans Univ, 76-78; dir, Tex Mining & Mineral Resources Res Inst & Coal Res Consortium, Tex Univ, 79-81. *Mem:* Soc Econ Geologists; SAfrican Coal Processing Soc. *Res:* Strategic minerals; evaluation of coal resources; international coal trade. *Mailing Add:* 7619 Rockpoint Dr Austin TX 78731

VAN REUTH, EDWARD C, CERAMICS ENGINEERING. *Educ:* Va Polytechnic Inst & State Univ, BS, MS; Univ Ill, PhD. *Prof Exp:* Asst prof mat eng, Va Polytechnic Inst & State Univ, 57-60; res asst, Univ Ill, 60-63; sr res scientist, David Taylor Res Lab, 63-66; postdoctoral res, Kammerlingh Onnes Lab, Univ Leiden, Neth, 66-67; head, Composites & Speciality Mat Br, David Taylor Res Lab, Annapolis, Md, 67-71, asst to dir, lab, & dir superconductivity, 71-72; staff specialist metall & ceramic sci, Defense Advan Res Proj Agency, 72-79, dir, Mat Sci Div, 79-83; PRES, TECHNOL STRATEGIES, INC, 84- *Concurrent Pos:* Partic adv group, Aeronaut Res & Develop, NATO; invited lectr, Switz, Ger, Gt Brit, France, Norway, Neth, Soviet Union & US; prof lectr, George Washington Univ. *Honors & Awards:* George Kimball Burgess Award; Tech Achievement Award, Am Soc Mech Engrs; Distinguished Pioneer Award, Am Soc Mfg Engrs. *Mem:* Fel Am Soc Metals Int. *Res:* Developing strategy for new or existing technologies, primarily in the areas of technology assessment; marketing strategies; system integration; advanced materials; superconductivity; advanced engines; chemical and metallurgical processing. *Mailing Add:* Technol Strategies Inc 10722 Shingle Oak Ct Burke VA 22015

VAN RHEENEN, VERLAN H, b Oskaloosa, Iowa, Feb 15, 39; m 62; c 2. ORGANIC CHEMICAL PROCESS RESEARCH & DEVELOPMENT, MANAGEMENT. *Educ:* Cent Col, Iowa, BA, 61; Univ Wis, PhD(org chem), 66. *Prof Exp:* Distinguished scientist, Upjohn Co, 66-84; CONSULT, 84- *Mem:* Am Chem Soc; Sigma Xi. *Res:* Total synthesis of natural products; new synthetic methods; steroid chemistry; prostaglandin synthesis. *Mailing Add:* 2112 Vanderbilt Rd Kalamazoo MI 49002

VAN RIJ, WILLEM IDANIEL, b Brielle, Neth, Apr 19, 42; US citizen; m 68; c 2. COMPUTER SCIENCES, PLASMA PHYSICS. *Educ:* Univ Auckland, BS, 64, MS, 66; Fla State Univ, PhD(physics), 70. *Prof Exp:* Res assoc theoret nuclear physics, Brookhaven Nat Lab, 70-72 & Univ Wash, 72-74; COMPUT PHYSICIST, OAK RIDGE NAT LAB, 74- *Mem:* Am Phys Soc. *Res:* Computer simulations of plasmas for controlled thermonuclear research. *Mailing Add:* 7820 Castlecomb Rd Powell TN 37849

VAN RIPER, CHARLES, III, b Mahopac, NY, Sept 24, 43; m 77; c 4. ORNITHOLOGY, EPIDEMIOLOGY. *Educ:* Colo State Univ, BS, 66, MEd, 67; Univ Hawaii, PhD(zool), 78. *Prof Exp:* Instr, Mahopac High Sch, NY, 67-68, Hawaii Prep Acad, Kamuela, Hawaii, 68-72; teaching asst, dept zool, Univ Hawaii, Honolulu, 72-74, res asst, 74-75, asst researcher, 77-79; unit leader, Coop Nat Park Resources Studies Unit, Davis, Calif, 79-87, asst adj prof, Dept Zool, 80-85, res scientist, Nat Park Serv, 87-90, UNIT LEADER, COOP NAT PARK RESOURCES STUDIES UNIT, 90-; PROF, NORTHERN ARIZ UNIV, FLAGSTAFF, 90- *Concurrent Pos:* Numerous res grants & fels, US Dept Interior, Nat Park Ser & Nat Park Res, 72-; Palila recovery team, US Fish & Wildlife Serv, 74, Peregrine Falcon recovery team Western US, 82, asst ed, J Wildlife Dis, 86 US Dept Interior Pub Task Force, 87, Wildlife biologist, dept Interior, US Fish Wildlife Serv. *Honors & Awards:* Spec Merit Award, US Fish Wildlife Serv. *Mem:* Am Ornithologists' Union; Ecol Soc Am; Soc Am Naturalists; Wildlife Dis Asn. *Res:* Seasonal change in bird communities of the chaparal and blue- oak woodlands in Central California; Mill and Deer Creek drainages; evaluation of wildlife habitat relationships database for predicting bird community composition in Central California; over 80 publications of ornithological subjects. *Mailing Add:* PO Box 5614 Northern Ariz Univ Flagstaff AZ 86011-5614

VAN RIPER, GORDON EVERETT, b Flat Rock, Mich, Dec 7, 17; m 43, 75; c 1. AGRONOMY. *Educ:* Mich State Univ, BS, 55; Univ Wis, MS, 57, PhD(agron), 58. *Prof Exp:* From asst prof to assoc prof agron, Univ Nebr, 58-64; mgr, Dept Agron, Deere & Co, 64-69, mgr, Dept Res Coord, 69-73; pres, Jay Dee Equip, Inc, 73-81; consult, 81-88; RETIRED. *Concurrent Pos:* Vpres, Agr Res Inst, 71-72; pres, Am Forage & Grassland Coun, 72. *Mem:* Am Soc Agron; Crop Sci Soc Am. *Res:* Crop physiology. *Mailing Add:* 28 Pine Tree Rd Kewanee IL 61443

VAN RIPER, KENNETH ALAN, b New Brunswick, NJ, Feb 7, 49. ASTROPHYSICS, PHYSICS. *Educ:* Cornell Univ, AB, 70; Univ Pa, PhD(physics), 76. *Prof Exp:* Res assoc, Enrico Fermi Inst, Univ Chicago, 76-78; res assoc astrophys, dept physics, Univ Ill, Urbana, 78-81; WITH LOS ALAMOS NAT LAB, 81- *Mem:* Am Phys Soc; Am Astron Soc; Sigma Xi. *Res:* Stellar collapse and explosion; supernovae; neutrino astrophysics; neutron star formation and evolution; dense, hot matter. *Mailing Add:* Los Alamos Nat Lab X-6 MSB226 PO Box 1663 Los Alamos NM 87545

VAN ROGGEN, AREND, b Nijmegen, Neth, Jan 2, 28; m 52; c 1. ELECTRODYNAMICS, COMPUTER SCIENCE. *Educ:* State Univ Leiden, Drs(phys chem), 53; Duke Univ, PhD(physics), 56. *Prof Exp:* Asst, Lab Phys Chem, State Univ Leiden, 48-54, sci asst, 56; asst physics, Duke Univ, 54-56; from res physicist to sr res physicist, E I du Pont de Nemours & Co, Inc, 56-72, sr res specialist, 72-75, res assoc, 75-81; CONSULT, 82- *Concurrent Pos:* Ed, Inst Elec & Electronics Engrs Elec Insulation Soc Trans Elec Insulation, 77-; secy, Conf Elec Insulation & Dielec Phenomena, 80-81, vchmn-treas, 82-83 & chmn, 84-85; adj prof, molecular electronics, Cath Univ, 90- *Mem:* Am Phys Soc; Am Soc Testing & Mat; fel Inst Elec & Electronics Engrs; Inst Elec & Electronics Engrs Dielectrics & Elec Insulation Soc; NY Acad Sci. *Res:* Electronic and magnetic structure of matter; paramagnetic resonance; dielectrics; physics instrumentation; interaction of electric fields and materials; computational aspects of above. *Mailing Add:* RD 2 Kennett Sq Wilmington PA 19348

VAN ROOSBROECK, WILLY WERNER, b Antwerp, Belg, Aug 10, 13; nat US; m 45. SEMICONDUCTORS. *Educ:* Columbia Univ, AB, 34, MA, 37. *Prof Exp:* Res physicist, 37-78, CONSULT, BELL LABS, 78- *Honors & Awards:* Van Buren Prize Math, 34. *Mem:* Fel Am Phys Soc; NY Acad Sci. *Res:* Mathematical physics of semiconductors; theory of current-carrier injection and transport and of amorphous and relaxation semiconductors; rectification, trapping, recombination, radiation, high-field, space-charge and switching effects. *Mailing Add:* 19 Whittredge Rd Summit NJ 07901

VAN ROSSUM, GEORGE DONALD VICTOR, b London, Eng, Dec 13, 31; m 59; c 4. CELL PHYSIOLOGY, BIOCHEMICAL PHARMACOLOGY. *Educ:* Oxford Univ, MA, 59, DPhil(biochem), 60. *Prof Exp:* NATO fel physiol chem, Univ Amsterdam, 60-62; NIH fel, Johnson Res Found, Univ Pa, 62-63, res assoc phys biochem, Sch Med, 63-64; reader biochem, Christian Med Col, Vellore, India, 65-67; vis asst prof phys biochem, Johnson Res Found, Sch Med, Univ Pa, 67-69; actg chmn dept, 73-75, PROF PHARMACOL, SCH MED, TEMPLE UNIV, 69- *Concurrent Pos:* NATO vis prof, Inst Gen Path, Cath Univ, Rome, 71; contract prof, Univ Rome, 85-86. *Mem:* Brit Biochem Soc; Am Soc Biol Chemists; Am Soc Pharmacol & Exp Therapeut; NY Acad Sci. *Res:* Ion and water transport; tissue electrolytes in cancer; control of energy metabolism; cell toxicity of lead. *Mailing Add:* Dept Pharmacol Temple Univ Sch Med Philadelphia PA 19140

VAN RYZIN, MARTINA, b Appleton, Wis, June 10, 23. HISTORY OF SCIENCE, MATHEMATICS. *Educ:* Silver Lake Col, Wis, BA, 46; Marquette Univ, MS, 56; Univ Wis, PhD(hist of sci), 60. *Prof Exp:* Head dept math, 57-70, from instr to assoc prof math, 60-69, PROF MATH, SILVER LAKE COL, 69-, ACAD DEAN, 70-, CHMN, DEPT MATH & COMPUT SCI. *Mem:* Math Asn Am; Hist Sci Soc. *Res:* History of mathematics, especially Medieval period; Arabic-Latin tradition of Euclid's elements in the 12th century. *Mailing Add:* Dept Introd Comput & Prog Silver Lake Col 2406 S Alverno Rd Manitowoc WI 54220

VAN SAMBEEK, JEROME WILLIAM, b Milbank, SDak, Aug 1, 47; m 72; c 5. MICROPROPAGATION. *Educ:* SDak State Univ, BS, 69; Wash Univ, PhD(plant physiol), 75. *Prof Exp:* Fel plant path, Univ Mo, 75; res plant physiologist, Southern Forest Exp Sta, La, 75-79; RES PLANT PHYSIOLOGIST, NORTH CENT FOREST EXP STA, ILL, 79-, PROJ LEADER PHYSIOL, GENETICS & PROCESSING CENT HARDWOODS, 89- *Concurrent Pos:* Adj asst prof, Southern Ill Univ, 80-; ed, Walnut Coun Bull, 84-88, coun pres, 90-91. *Honors & Awards:* Black Walnut Achievement Award, 88. *Mem:* Am Soc Plant Physiologists; Plant Growth Regulator Soc Am; Sigma Xi; Walnut Coun. *Res:* Physiological research on micropropagation, seedling establishment, ground cover management and fruiting of black walnut and the other fine hardwoods. *Mailing Add:* N Cent Forest Exp Sta Southern Ill Univ Carbondale IL 62901

VAN SANT, JAMES HURLEY, JR, b Ashland, Ohio, Jan 16, 33; m 60; c 2. MECHANICAL ENGINEERING. *Educ:* Univ Idaho, BS, 56, MS, 60; Ore State Univ, PhD(mech eng), 64. *Prof Exp:* Asst flight test engr, Test Div, Douglas Aircraft Co, Calif, 56-57; instr mech eng, Univ Idaho, 57-59; res staff assoc heat transfer & fluid mech res, Gen Atomic Div, Gen Dynamics Corp, Calif, 59-61; res asst, Eng Exp Sta, Ore State Univ, 62-64; mech engr, Heat Transfer Res, Lawrence Livermore Lab, Univ Calif, 64-71; sr researcher, Hydro Que Inst Res, Can, 71-78; RES & DEVELOP ENGR, LAWRENCE LIVERMORE NAT LAB, UNIV CALIF, 78- *Concurrent Pos:* Guest prof, Univ Que. *Mem:* Am Soc Mech Engrs. *Res:* Heat transfer and fluid mechanics; free and forced flow; conduction; evaporation; thermal properties; high temperature and cryogenic systems. *Mailing Add:* Lawrence Livermore Nat Lab L-197 PO Box 808 Livermore CA 94551

VAN SAUN, WILLIAM ARTHUR, b Ashland, Pa, Dec 23, 46; m 69; c 3. ORGANIC CHEMISTRY, PESTICIDE CHEMISTRY. *Educ:* Ursinus Col, BS, 68; Villanova Univ, PhD(chem), 75. *Prof Exp:* Chemist med chem, Merck, Sharp & Dohme Res Labs, 68-72; res chemist, 75-79, sr res chemist, 79-80, mgr org synthesis, 80-85, mgr biol eval, 85-87, dir, biol res, 87-91, DIR DISCOVERY RES, AGR CHEM GROUP, FMC CORP, 91- *Concurrent Pos:* Adj asst prof, Mercer Co Community Col, 78-; vis lectr org chem, Rider Col, 79. *Mem:* Am Chem Soc; AAAS; Soc Chem Indust. *Res:* Organic synthesis, pesticides, quantitative structure-activity relationships; biological testing. *Mailing Add:* FMC Agr Chem Group Box 8 Princeton NJ 08543

VAN SCHAIK, PETER HENDRIK, b Arnhem, Neth, Apr 18, 27; US citizen; m 54; c 5. PLANT BREEDING. *Educ:* Ont Agr Col, Univ Guelph, BSA, 52; Univ Toronto, MSA, 54; Purdue Univ, PhD(plant breeding), 56. *Prof Exp:* Res agronomist cotton, Agr Res Serv, USDA, Brawley, Calif, 57-64, coord res agronomist food legumes, Tehran, Iran & New Delhi, India, 64-70, res agronomist peanut res, Holland, Va, 70-72, asst area dir, Fresno, Calif, 72-80, assoc area dir, 80-86; RETIRED. *Concurrent Pos:* Consult, Ford Found, 70, Rockefeller Found, 70 & Experience Inc, 71, 76. *Mem:* Am Soc Agron. *Res:* Crop breeding; agronomy; foreign development; tropical agriculture. *Mailing Add:* 9210 N Stoneridge Fresno CA 93710

VAN SCHILFGAARDE, JAN, b The Hague, Neth, Feb 7, 29; nat US; m 51; c 3. SOILS & SOIL SCIENCE. *Educ:* Iowa State Col, BS, 49, MS, 50, PhD(agr eng, soil physics), 54. *Prof Exp:* Instr & assoc agr eng, Iowa State Col, 49-54; from asst prof to prof agr eng, NC State Col, 54-64; chief water mgt engr, Soil & Water Conserv Div, Agr Res Serv, 64-67, assoc dir, 67-71, dir, 71-72, dir, US Salinity Lab, USDA, 72-84, dir, Mountain States Area, 84-87, ASSOC DIR, NORTHERN PLAINS AREA, AGR RES SERV, 87- *Concurrent Pos:* Vis prof, Ohio State Univ, 62; adj prof soils, Univ Calif, Riverside, 74-84. *Honors & Awards:* John Deere Medal, Am Soc Agr Engrs, 77; Royce J Tipton Award, Am Soc Civil Engr, 86. *Mem:* Nat Acad Eng; Am Soc Civil Engrs; Soil Sci Soc Am; Soil Conserv Soc Am; Am Soc Agr Engrs. *Res:* Management of water for crop production, especially by agricultural drainage. *Mailing Add:* 2625 Redwing Rd Suite 350 Ft Collins CO 80526

VAN SCHMUS, WILLIAM RANDALL, b Aurora, Ill, Oct 4, 38; m 61; c 3. GEOCHRONOLOGY, PRECAMBRIAN GEOLOGY. *Educ:* Calif Inst Technol, BS, 60; Univ Calif, Los Angeles, PhD(geol), 64. *Prof Exp:* From asst prof to assoc prof, 67-75, PROF GEOL, UNIV KANS, 75- *Concurrent Pos:* Mem adv panels, NASA, NSF & Nat Acad Sci. *Mem:* Am Geophys Union; Geochem Soc; Meteoritical Soc; Geol Soc Am. *Res:* Geochronology and geochemistry of Precambrian continental crust. *Mailing Add:* Dept Geol Univ Kans Lawrence KS 66045

VAN SCIVER, STEVEN W, b Philadelphia, Pa, Mar 13, 48; m 69; c 2. CRYOGENICS, HEAT TRANSFER. *Educ:* Lehigh Univ, BS, 70; Univ Wash, MS, 72, PhD(physics), 76. *Prof Exp:* Proj assoc superconductivity, 76-77, asst scientist, 77-79, asst prof, 79-82, ASSOC PROF, NUCLEAR ENG, UNIV WIS-MADISON, 82- *Concurrent Pos:* Consult, Tex Accelerator Ctr, 80- *Mem:* Am Phys Soc; Inst Elec & Electronics Engrs. *Res:* Cryogenics; heat transfer and transport in superfluid helium; low temperature properties of materials; design of superconducting magnets for fusion and energy storage. *Mailing Add:* Nuclear Eng 921 Eng Res Univ Wis 1500 Johnson Dr Madison WI 53706

VAN SCOTT, EUGENE JOSEPH, b Macedon, NY, May 27, 22; m 48; c 3. DERMATOLOGY. *Educ:* Univ Chicago, BS, 45, MD, 48. *Prof Exp:* Intern, Millard Fillmore Hosp, Buffalo, NY, 48-49; resident physician dermat, Univ Chicago, 49-52; assoc, Univ Pa, 52-53; chief dermat br, Nat Cancer Inst, 53-68, sci dir gen labs & clins, 66-68; PROF DERMAT, HEALTH SCI CTR, TEMPLE UNIV, 68-; CLIN PROF, HAHNEMANN, 89- *Concurrent Pos:* Assoc dir, Skin & Cancer Hosp, Philadelphia, 68-89. *Honors & Awards:* Taub Int Mem Award, 64; Clarke White Award, 65; Albert Lasker Award, 72; Stephen Rothman Award, Soc Invest Dermat, 75; Lila Gruber Cancer Res Award, Am Acad Dermat, 80; Howard Fox lectr, NY Acad Med, 81. *Mem:* Hon mem Can Dermat Asn; Soc Invest Dermat; Am Dermat Asn; Am Asn Cancer Res; Am Soc Clin Invest; Am Acad Dermat. *Res:* Biology and physiology of epithelial growth; differentiation and neoplasia; pathogenesis of psoriasis; biology; immunologic aspects and clinical management of cutaneous lymphomas; dermatopharmacology. *Mailing Add:* Three Hidden Lane Abington PA 19001

VANSELOW, CLARENCE HUGO, b Syracuse, NY, Sept 30, 28; m 51; c 7. PHYSICAL CHEMISTRY. *Educ:* Syracuse Univ, BS, 50, MS, 51, PhD, 58. *Prof Exp:* Instr chem, Colgate Univ, 55-56; prof, Thiel Col, 56-64; ASSOC PROF CHEM, UNIV NC, GREENSBORO, 64- *Mem:* Am Chem Soc. *Res:* Gas phase radiation chemistry; kinetic theory of precipitation processes. *Mailing Add:* Dept Chem 221 Petty Sci Bldg Univ NC 100 Spring Garden St Greensboro NC 27412

VANSELOW, NEAL A, b Milwaukee, Wis, Mar 18, 32; m 58; c 2. MEDICAL ADMINISTRATION, ALLERGY. *Educ:* Univ Mich, AB, 54, MD, 58, MS, 63; Am Bd Internal Med, dipl, 65, cert, 68; Am Bd Allergy & Immunol, dipl, 72. *Prof Exp:* From instr to assoc prof internal med, Univ Mich Med Sch, Ann Arbor, 63-74, from assoc prof to prof postgrad med, 67-74, asst to chmn, Dept Postgrad Med, 68-71, actg chmn, 71-72, chmn, Dept Postgrad Med & Health Professions Educ, 72-74; prof internal med & dean, Col Med, Univ Ariz, 74-77; prof internal med & vpres, Univ Nebr, 77-82, chancellor, Med Ctr, 77-82; vpres health sci & prof internal med, Univ Minn, 82-89; CHANCELLOR & PROF INTERNAL MED, TULANE UNIV MED CTR, 89- *Concurrent Pos:* Staff physician, Univ Hosp, Ann Arbor, Mich, 63-74, Tucson, Ariz, 74-77, Univ Nebr Hosp, Omaha, 77-82, Univ Minn Hosp & Clin, 83-89, Tulane Univ Hosp & Clin, New Orleans, La, 90-; consult, Vet Admin Hosp, Ann Arbor, Mich, 65-74; surv team mem, Continuing Med Educ Accreditation Prog, AMA & Liaison Comt for Continuing Med Educ, 68-80; mem, Task Force Continuing Med Educ, Asn Am Med Cols, 75-76, Mgt Educ Network Adv Comt, 76-78, Task Force Minority Student

Opportunities in Med, 76-78; mem, Adv Comt Governance Study, Consortium for Study of Univ Hosps, 81-82; mem, Nat Adv Panel, Essentials of Univ Educ for Nursing Proj, Am Asn Cols Nursing, 85-86; chairperson, Coun Grad Med Educ, US Dept Health & Human Serv, 86-; chmn bd dirs, Asn Acad Health Ctrs, 87-88; mem, Comt to Study Strategies for Supporting Grad Med Educ in Primary Care, Inst Med-Nat Acad Sci, 89. *Mem:* Inst Med-Nat Acad Sci; fel Am Col Physicians; Sigma Xi; fel Am Col Physician Execs; Soc Med Adminrs; AMA; fel Am Acad Allergy. *Res:* Mechanisms of aspirin sensitivity; immunosuppression and immunologic adjuvants; author of numerous publications and several chapters in books. *Mailing Add:* Tulane Univ Med Ctr 1430 Tulane Ave New Orleans LA 70112

VANSELOW, RALF W, b Berlin, Ger, July 12, 31; m 61. PHYSICAL CHEMISTRY, SURFACE CHEMISTRY. *Educ:* Tech Univ Berlin, BS, 57, Dipl Ing, 62, Dr Ing, 66. *Prof Exp:* Res asst field emission micros, Fritz Haber Inst, Max Planck Soc, 52-66, res assoc, 66-68; asst prof, 68-74, dir, Lab Surface Studies, 76-78, assoc prof chem, 74-80, chmn dept, 78-81, PROF, UNIV WIS-MILWAUKEE, 80- *Mem:* Ger Chem Soc; Am Chem Soc; Ger Vacuum Soc; Am Vacuum Soc. *Res:* Studies of metal surfaces by means of field electron and field ion microscopy; adsorption; surface migration; epitaxial growth. *Mailing Add:* Dept Chem Univ Wis Milwaukee WI 53201

VAN SICKLE, DALE ELBERT, b Ft Collins, Colo, Oct 8, 32; m 62; c 2. PHYSICAL ORGANIC CHEMISTRY. *Educ:* Colo State Univ, BS, 54; Univ Utah, MS, 56; Univ Calif, PhD, 59. *Prof Exp:* Asst chem, Univ Utah, 54-55 & Univ Calif, 56-59; org chemist, Stanford Res Inst, 59-69; PRIN RES CHEMIST, EASTMAN CHEM DIV, EASTMAN KODAK CO, 69- *Mem:* AAAS; Am Chem Soc; Sigma Xi. *Res:* Mechanisms and kinetics of reactions; oxidation of hydrocarbons; free radical reactions. *Mailing Add:* 2113 Sheffield St Kingsport TN 37660

VAN SICKLE, DAVID C, b Des Moines, Iowa, Jan 9, 34; m 56. HISTOLOGY, PATHOBIOLOGY. *Educ:* Iowa State Univ, DVM, 57; Purdue Univ, PhD(develop anat), 66. *Prof Exp:* Gen pract, Ill, 57-58 & 60-61; from instr to assoc prof, 61-75, FROM HISTOL & EMBRYOL, PURDUE UNIV, WEST LAFAYETTE, 75-, HEAD, DEPT ANAT, 82- *Concurrent Pos:* Morris Animal Found fel, 64-66; adj prof anat, Sch Med, Ind Univ, 75-; pres, Confr Res Workers Animal Dis, 81; mem, Col USAF Reserve, 77-82; adj prof grad studies, Wright State Univ, Ohio, 81- *Mem:* Sigma Xi; Am Asn Anat; NY Acad Sci; Orthopedic Res Soc. *Res:* Osteogenesis and abnormalities associated with errors in osteogenesis; orthopedic pathobiology in arthritis, hip dysplasia and osteochondritis; fracture healing and effect of growth factors on bone and cartilage. *Mailing Add:* Dept Vet Anat Purdue Univ West Lafayette IN 47907

VAN SICLEN, DEWITT CLINTON, b Carlisle, Pa, Oct 25, 18; m 49; c 4. TECTONICS. *Educ:* Princeton Univ, AB, 40, MA, 47, PhD(geol), 51; Univ Ill, MS, 41. *Prof Exp:* Jr geologist, Peoples Natural Gas Co, Pa, 41-42 & 46; field geologist, Drilling & Explor Co, Inc, 47-50; exec officer, Off Sci Res, US Dept Air Force, 51-52; res geologist, Pan-Am Prod Co, 52-56; sr geologist, Pan-Am Petrol Corp, 56-59; from assoc prof to prof, 59-82, chmn dept, 60-67, CONSULT GEOLOGIST, UNIV HOUSTON, 66- *Mem:* Am Asn Petrol Geologists; Nat Asn Geol Teachers; Am Inst Prof Geologists; Soc Petrol Engrs; Asn Eng Geologists; Geol Soc Am. *Res:* Behavior of subsurface fluids; petroleum migration and entrapment; surficial geology and active faults of Texas-Louisiana coastal plain; tectonics of Gulf of Mexico and surrounding regions. *Mailing Add:* 4909 Bellaire Blvd Bellaire TX 77401

VAN SLUYTERS, RICHARD CHARLES, b Chicago, Ill, June 12, 45; m 68. PHYSIOLOGICAL OPTICS, NEUROPHYSIOLOGY. *Educ:* Ill Col Optom, BS, 67, OD, 68; Ind Univ, Bloomington, PhD(physiol optics), 72. *Prof Exp:* ASSOC PROF OPTOM-PHYSIOL OPTICS & NEUROBIOL, SCH OPTOM, UNIV CALIF, BERKELEY, 75- *Concurrent Pos:* Am Optom Found Res fel, 68-69; Nat Eye Inst fel, 69-71 & spec res fel, 72-74; NSF fel, 71-72; Miller Inst Basic Res Sci fel, Univ Calif, Berkeley, 74-76; Sloan fel, 78-80. *Mem:* Soc Neurosci; AAAS; Asn Res Vision & Ophthal; fel Am Acad Optom. *Res:* Neurophysiology of developing mammalian visual systems. *Mailing Add:* Sch Optom Univ Calif Berkeley CA 94720

VAN SLYKE, RICHARD M, b Manila, Philippines, Aug 17, 37; US citizen; m 69. OPERATIONS RESEARCH. *Educ:* Stanford Univ, BS, 59; Univ Calif, Berkeley, PhD(opers res), 65. *Prof Exp:* Asst prof elec & indust eng, Univ Calif, Berkeley, 65-69; vpres, Network Anal Corp, 69-80; prof elec eng & comput sci, Stevens Inst Technol, 80-83; dir, Telecommun Ctr, 83-88, PROF POLYTECH UNIV, 88- *Concurrent Pos:* Consult. *Mem:* Soc Indust & Appl Math; Opers Res Soc Am; Am Math Soc; Inst Elec & Electronics Engrs. *Res:* Mathematical techniques for optimization, especially for information network design and analysis. *Mailing Add:* Polytech Univ 333 Jay St Brooklyn NY 11201

VANSOEST, PETER JOHN, b Seattle, Wash, June 30, 29; m 59; c 3. ANIMAL NUTRITION. *Educ:* Wash State Univ, BS, 51, MS, 52; Univ Wis, PhD(nutrit), 55. *Prof Exp:* Biochemist, Agr Res Serv, USDA, 57-68; assoc prof animal nutrit, 68-73, PROF ANIMAL NUTRIT, CORNELL UNIV, 73- *Honors & Awards:* Am Feed Mfrs Award, 67; Hoblitzelle Nat Award Agr, 68. *Mem:* AAAS; Am Dairy Sci Asn; Am Soc Animal Sci; Asn Off Anal Chem. *Res:* Ruminant digestion and metabolism; chemistry of fibrous feedstuffs and methods of analysis; forage chemistry. *Mailing Add:* Dept Animal Sci Cornell Univ Ithaca NY 14853

VANSPEYBROECK, LEON PAUL, b Wichita, Kans, Aug 27, 35; m 59; c 3. X-RAY ASTRONOMY. *Educ:* Mass Inst Technol, BS, 57, PhD, 65. *Prof Exp:* Res assoc high energy physics, Mass Inst Technol, 65-67; staff scientist x-ray astron, Am Sci & Eng, Inc, Mass, 67-74; STAFF SCIENTIST, CTR ASTROPHYS, 74- *Honors & Awards:* Goddard Award, Int Soc Optical Eng, 85. *Mem:* Fel Am Phys Soc; Am Astron Soc; Int Astron Union. *Res:* Solar and stellar astronomy; x-ray optics. *Mailing Add:* Ctr Astrophys 60 Garden St B423 Cambridge MA 02138

VAN STEE, ETHARD WENDEL, b Traverse City, Mich, July 17, 36; m 60; c 2. PHARMACOLOGY, TOXICOLOGY. *Educ:* Mich State Univ, BS, 58, DVM, 60; Ohio State Univ, MS, 66, PhD(vet physiol, pharmacol), 70. *Prof Exp:* Res pharmacologist, Aerospace Med Res Lab, Wright-Patterson AFB, Ohio, 67-75; HEAD, INHALATION TOXICOL SECT, NAT INST ENVIRON HEALTH SCI, 75- *Concurrent Pos:* Adj assoc prof pharmacol, Univ NC, 75- *Mem:* Soc Toxicol; Am Soc Pharmacol & Exp Therapeut; Am Vet Med Asn; AAAS; Am Soc Vet Physiologists & Pharmacologists. *Res:* Inhalation toxicology; halogenated alkanes, anesthetics; cardiovascular pharmacology. *Mailing Add:* Rte 9 Box 432F Chapel Hill NC 27514

VAN STEENBERGEN, ARIE, b Vlaardingen, Neth, Feb 26, 28; m 53; c 4. PHYSICS. *Educ:* Delft Univ Technol, MSc, 52; McGill Univ, PhD(physics, math), 57. *Prof Exp:* Res asst electron beam optics, Delft Univ Technol, 50-53; res physicist, Nat Defense Res Lab, The Hague, Neth, 53-54; res assoc nuclear magnetic resonance, McGill Univ, 54-57; head alternating gradient synchrotron div, 65-74, sr physicist, 75-77, HEAD NAT SYNCHROTRON LIGHT SOURCE, BROOKHAVEN NAT LAB, 77- *Concurrent Pos:* Consult, Radiation Dynamics, Inc, 60- *Mem:* Am Phys Soc; Europ Phys Soc. *Res:* High energy particle accelerators and storage rings; particle beam dynamics; synchrotron radiation sources. *Mailing Add:* Nat Synchrotron Light Source Div Bldg 725B Brookhaven Nat Lab Upton NY 11973

VANSTONE, J R, b Owen Sound, Ont, Aug 12, 33; m 56; c 3. MATHEMATICS. *Educ:* Univ Toronto, BA, 55, MA, 56; Univ Natal, PhD(math), 59. *Prof Exp:* Lectr math, 59-61, from asst prof to assoc prof, 61-76, PROF MATH, UNIV TORONTO, 76- *Mem:* Can Math Cong; Math Asn Am; Am Math Soc; Soc Indust & Appl Math; Sigma Xi. *Res:* Differential geometry. *Mailing Add:* Dept Math Univ Col 15 Kings Coll Circle Toronto ON M5S 1A1 Can

VANSTONE, SCOTT ALEXANDER, b Chatham, Ont, Sept 14, 47; m 70; c 1. MATHEMATICS. *Educ:* Univ Waterloo, BMath, 70, MMath, 71, PhD(math), 74. *Prof Exp:* From asst prof to assoc prof, 74-85, PROF MATH, ST JEROME'S COL, 85- *Concurrent Pos:* Ed-in-chief, Designs, Codes & Cryptography. *Res:* The existence and construction of balanced incomplete block designs and regular pairwise balanced designs which are closely related to finite linear spaces and balanced equidistant codes; coding theory; cryptography and finite fields. *Mailing Add:* Dept Combinatorics & Optimiz Univ Jermes Col Westmount Rd N Waterloo ON N2L 3G3 Can

VAN STRIEN, RICHARD EDWARD, b Battle Creek, Mich, Sept 17, 20; m 42; c 3. ORGANIC CHEMISTRY. *Educ:* Hope Col, AB, 42; Univ Pa, MS, 44, PhD(org chem), 48. *Prof Exp:* Res chemist, Standard Oil Co, 47-60; sect leader, res & develop dept, Amoco Am Chem Corp, 60-61, sect leader in-chg prod appln new chem, 61-67, asst dir polymers & plastic div, 67-69, conensation polymer div, 69-76, div dir, explor res div, 76-82; RETIRE. *Mem:* Am Chem Soc; Sigma Xi. *Res:* Synthetic detergents; gelling agents; surface coatings; high-performance plastics; adhesives. *Mailing Add:* 9148 Southmoor Ave Highland IL 46322-2513

VAN STRYLAND, ERIC WILLIAM, b South Bend, Ind, June 3, 47; m 78. LASER PHYSICS. *Educ:* Humboldt State Univ, BS, 70; Univ Ariz, MS, 75, PhD(physics), 76. *Prof Exp:* Res assoc laser physics, Optical Sci Ctr, Univ Ariz, 72-76; res scientist, Ctr Laser Studies, Univ Southern Calif, 76-78; prof laser physics, dept physics, N Tex State Univ, 78-87; PROF PHYSICS & ELEC ENG, UNIV CENT FLA, 87- *Mem:* Am Phys Soc; Optical Soc Am; Laser Inst Am; Inst Elec & Electronics Engrs. *Res:* Ultrashort light pulses and their generation measurement and uses; optical coherent transient effects; lifetime measurements; nonlinear absorption; laser induced damage; laser material interactions. *Mailing Add:* Creol Res Pavillion 12424 Research Pkwy Suite 400 Orlando FL 32826

VAN SWAAY, MAARTEN, b The Hague, Neth, Aug 1, 30; m 54; c 4. COMPUTER SCIENCES. *Educ:* State Univ Leiden, BS, 53, Princeton Univ, PhD(chem), 56. *Hon Degrees:* Drs, State Univ, Leiden, 56. *Prof Exp:* Sr res asst phys chem, State Univ Leiden, 56-59; res assoc instr anal, Eindhoven Technol Univ, 59-63; from asst prof to assoc prof anal chem, 63-81, ASSOC PROF COMPUT SCI, KANS STATE UNIV, 81- *Concurrent Pos:* Consult, Am Inst Prof Educ. *Mem:* Asn Comput Mach; Sigma Xi; Inst Elec & Electronics Engrs Comput Soc. *Res:* Social and ethical issues; interfacing and control. *Mailing Add:* Dept Comput & Info Sci Kans State Univ Nichols Halls Manhattan KS 66506

VAN TAMELEN, EUGENE EARL, b Zeeland, Mich, July 20, 25; m 51; c 3. CHEMISTRY. *Educ:* Hope Col, AB, 47; Harvard Univ, MA, 49, PhD(chem), 50. *Hon Degrees:* DSc, Hope Col & Bucknell Univ, 71. *Prof Exp:* From instr to prof org chem, Univ Wis, 50-61, Adkins prof chem, 61-62; prof chem, 62-87, chmn dept, 74-78, EMER PROF CHEM, STANFORD UNIV, 87- *Concurrent Pos:* Guggenheim fels, 65 & 73; prof extraordinarius, Neth, 67-74; mem adv bd, Chem & Eng News, 68-70, Synthesis, 69- & Accounts of Chem Res, 70-73; ed, Bioorg Chem, 71-82. *Honors & Awards:* Award in Pure Chem, Am Chem Soc, 61 & Award for Creative Work in Synthetic Org Chem, 70; Baekeland Award, 65. *Mem:* Nat Acad Sci; Am Acad Arts & Sci; Am Chem Soc. *Res:* Chemistry of natural products including structure, synthesis and biosynthesis; new reactions. *Mailing Add:* Dept Chem Stanford Univ Stanford CA 94305

VAN TASSEL, ROGER A, b Orange, NJ, Oct 19, 36. AERONOMY. *Educ:* Wesleyan Univ, AB, 58; Northeastern Univ, MS, 68, PhD, 72. *Prof Exp:* Chemist, Lunar-Planetary Lab, 61-67, res chemist, Aeronomy Lab, 67-82, RES CHEMIST, INFRARED TECHNOL DIV, AIR FORCE GEOPHYS LAB, 83- *Mem:* AAAS; Sigma Xi; Am Geophys Union; Optical Soc Am. *Res:* Atomic spectra in the vacuum ultraviolet; oscillator strengths of atomic transitions; ultraviolet airglow originating in the upper atmosphere; Fourier infrared instrumentation. *Mailing Add:* 11 Toddy Brook Rd Hollis NH 03049

VAN TASSELL, MORGAN HOWARD, b Johnson City, NY, Oct 31, 23; m 53; c 2. MICROBIOLOGY. *Educ:* Univ Wis, BS, 50. *Prof Exp:* Plant bacteriologist, Com Solvents Corp, Ill, 50-52; res microbiologist, Cent Res Dept, Anheuser-Busch, Inc, 53-57, supvr microbiologist, Fermentation Pilot Lab, 57-60; asst plant mgr, Sheffield Chem Div, Nat Dairy Prod Corp, 60-62, plant mgr, Kraftco Corp, 62-75; dir water serv, City of Oneonta, NY, 75-85; RETIRED. *Mem:* Am Soc Microbiol. *Res:* Industrial microbiology; applied and developmental research of fermentations; mutational development and selection of cultures; pilot scale equipment; automated continuous fermentations; enzymatic hydrolysis of proteins. *Mailing Add:* 32 Hudson St Oneonta NY 13820

VAN THIEL, DAVID H, b Cut Bank, Mont, Sept 5, 41. GASTROENTEROLOGY, ENDOCRINOLOGY. *Educ:* Univ Calif, Los Angeles, MD, 67. *Hon Degrees:* LLD, Freed-Hardeman Col, 82. *Prof Exp:* From instr to asst prof, 73-78, ASSOC PROF MED, UNIV PITTSBURGH SCH MED, 78- *Concurrent Pos:* Res fel, Univ Calif Med Ctr, Los Angeles, 64; intern, NY Hosp, 67-68, asst resident, 68-69; clin assoc, Endocrinol Br Nat Cancer Inst, 69-70 & Reprod Res Br, NIH, 70-71; sr asst resident, Univ Hosp, Boston, 71-72, res fel, 72-73; USPHS Career Develop Award, NIH, 77. *Mem:* Am Fedn Clin Res; Am Asn Study Liver Dis; fel Am Inst Chem Engrs; fel Inst Chemists. *Res:* Endocrine alterations associated with liver disease with special interest in alcoholic liver disease. *Mailing Add:* Sch Med 1000J Scaife Hall Univ Pittsburgh 3601 Fifth Ave Pittsburgh PA 15213

VAN THIEL, MATHIAS, b Sitobonda, Java, Sept 18, 30; US citizen; m 59; c 3. PHYSICAL CHEMISTRY, PHYSICS. *Educ:* Cornell Univ, BA, 54, Univ Calif, Berkeley, PhD(phys chem), 58. *Prof Exp:* Fel phys chem & shock tube kinetics, Univ Minn, Minneapolis, 58-59; RES PHYSICIST HIGH PRESSURE EQUATION OF STATE, LAWRENCE LIVERMORE LAB, 59- *Mem:* Am Chem Soc; AAAS; Sigma Xi; Am Physics Soc. *Res:* Infrared spectroscopy-matrix isolation; high temperature gas phase kinetics-shock tube; shock waves and high pressure equations of state; hydrodynamic code calculations; shaped charge design and penetration modeling; computation of explosive mixture properties; compilation of shock wave data. *Mailing Add:* Lawrence Livermore Nat Lab PO Box 808 Livermore CA 94550

VAN'T HOF, JACK, b Grand Rapids, Mich, Apr 11, 32; m 52; c 2. CELL BIOLOGY, PLANT MOLECULAR BIOLOGY. *Educ:* Calvin Col, AB, 57; Mich State Univ, PhD(bot), 61. *Prof Exp:* Biologist, Hanford Labs, Gen Elec Corp, 61-62; res assoc radiobiol & fel, Biol Dept, Brookhaven Nat Lab, 62-64, asst cytologist, 64-65; asst prof cytol, Dept Bot, Univ Minn, 65-66; cytologist, 66-80, SR CYTOLOGIST, BIOL DEPT, BROOKHAVEN NAT LAB, 80- *Mem:* Am Soc Cell Biol; Bot Soc Am; Am Soc Plant Physiol; Genetics Soc Am. *Res:* Cellular and molecular biology of plant cell division and chromosomal DNA replication. *Mailing Add:* Biol Dept Brookhaven Nat Lab Upton NY 11973

VANTHULL, LORIN L, b Sioux, Iowa, June 26, 32; m 55; c 3. DEVELOPMENT OF ALTERNATIVE ENERGY TECHNOLOGY,. *Educ:* Univ Minn, Minneapolis, BS, 54; Univ Calif, Los Angeles, MS, 56; Calif Inst Technol, PhD(physics), 67. *Prof Exp:* Res engr, res lab, Hughes Aircraft Co, 54-58; sr res scientist, cryogenic devices, sci lab, Ford Motor Co, 66-69; AEC grant, Univ Calif, San Diego, 69; assoc prof physics, 69-77, PROF PHYSICS, UNIV HOUSTON, 77- *Concurrent Pos:* Consult, Lawrence Berkeley Lab, Univ Calif, 70-71, Manned Spacecraft Ctr, NASA, 71-72; prin invest, Solar Cent Reciever Analysis & Code Develop, Univ Houston, 73-; prog mgr, Solar Thermal Adv Res Ctr, Univ Houston, 81-; div chief, Solar Thermal Div Energy Lab, Univ Houston, 75- *Mem:* Am Phys Soc; Sigma Xi; AAAS; Int Solar Energy Soc. *Res:* Develop computer techniques for optimization and design of solar central receiver systems represented by Solar One at Barstow, Calif; current effects support attempted commercialization of this new alternative energy technology. *Mailing Add:* Univ Houston Houston TX 77204-6421

VANT-HULL, LORIN LEE, b Sioux Co, Iowa, June 26, 32; m 55; c 3. SOLAR THERMAL POWER, SUPERCONDUCTING QUANTUM DEVICES. *Educ:* Univ Minn, Minneapolis, BS, 54; Univ Calif, Los Angeles, MS, 55; Calif Inst Technol, PhD(physics), 67. *Prof Exp:* Res engr, Res Lab, Hughes Aircraft Co, 54-58; sr res scientist cryogenic devices, Sci Lab, Ford Motor Co, 66-69; assoc prof, 69-77, PROF PHYSICS, UNIV HOUSTON, 77- *Concurrent Pos:* Consult, Lawrence Berkeley Lab, Univ Calif, 70-71 & Manned Spacecraft Ctr, NASA, 71-72; prog mgr, Solar Energy Lab, Univ Houston, 74-; consult, Battelle Pac Northwest Lab, 78-80, Jet Propulsion Lab, 80-81, Solar Energy Res Inst, 84; prin investr, numerous solar central receiver grants & contracts, Dept Energy & var indust firms, 73-; assoc ed, J Solar Energy, 76-; dir radiation div, Solar Energy Soc, 78-82; prog mgr, Solar Thermal Advan Res Ctr, 81- *Mem:* Am Inst Physics; Solar Energy Soc; Sigma Xi; AAAS. *Res:* Superconductivity; quantum interference and its use in instrumentation; solar energy; large scale efficient generation of solar electricity; solar tower central receiver systems optimization and performance. *Mailing Add:* Dept Physics Univ Houston Houston TX 77204

VAN TIENHOVEN, ARI, b The Hague, Neth, Apr 22, 22; nat US; m 50; c 3. ANIMAL PHYSIOLOGY. *Educ:* Univ Ill, MS & PhD(animal sci), 53. *Prof Exp:* Asst prof poultry husb, Miss State Col, 53-55; from asst prof to assoc prof avian physiol, 55-69, prof, 69-87, EMER PROF ANIMAL PHYSIOL, COL AGR & LIFE SCI, CORNELL UNIV, 87- *Concurrent Pos:* NATO fel, 61-62; assoc ed, Biol of Reproduction, 74- *Mem:* Fel AAAS; Am Soc Zoologists; fel Poultry Sci Asn; Am Asn Anat; Soc Study Reproduction. *Res:* Neuroendocrinology; reproductive physiology; temperature regulation of birds. *Mailing Add:* Dept Poultry & Avian Sci Cornell Univ 103 Rice Hall Ithaca NY 14853-5601

VAN TILBORG, ANDRÉ MARCEL, b Delft, Neth, Dec 6, 53; US citizen; m 90. REAL-TIME COMPUTING, PARALLEL COMPUTING. *Educ:* State Univ NY, Buffalo, BA, 75, PhD(computer sci), 82. *Prof Exp:* Sr computer scientist, Calspan Corp, 80-83, prin computer scientist, 84; prin computer scientist, Honeywell Systs & Res Ctr, 83-84; sr scientist, Carnegie Mellon Univ, 84-86; prog mgr computer systs, 87-89, DIR, COMPUTER SCI DIV, OFF NAVAL RES, 89- *Concurrent Pos:* Lectr, State Univ NY, Buffalo, 77-80; chair, Tech Comt Real-Time Systs, Inst Elec & Electronics Engrs Computer Soc, 87-91. *Mem:* Inst Elec & Electronics Engrs Computer Soc; Asn Comput Mach; Sigma Xi. *Res:* Resource management in distributed computing systems; author of numerous technical publications. *Mailing Add:* 13129 New Parkland Dr Herndon VA 22071-2765

VAN TILL, HOWARD JAY, b Ripon, Calif, Nov 28, 38; m 58; c 4. ASTRONOMY. *Educ:* Calvin Col, BS, 60; Mich State Univ, PhD(physics), 65. *Prof Exp:* Res scientist physics, Univ Calif, Riverside, 65-66; asst prof, Univ Redlands, 66-67; PROF PHYSICS, CALVIN COL, 67- *Concurrent Pos:* Res scientist, Dept Astron, Univ Tex, Austin, 74. *Mem:* Am Astron Soc; Am Phys Soc; Am Sci Affil. *Res:* Study of interstellar molecular clouds using millimeter-wave techniques. *Mailing Add:* Dept Physics Calvin Col Grand Rapids MI 49546

VAN'T RIET, BARTHOLOMEUS, b Arnhem, Neth, June 25, 22; nat US; m 55; c 3. ANALYTICAL CHEMISTRY. *Educ:* Vrye Univ, Neth, BSc, 50; Univ Minn, PhD(anal chem), 57. *Prof Exp:* Asst anal chem, Vrye Univ, Neth, 48-51; from asst to instr, Univ Minn, 51-57; instr chem, Univ Va, 58-64; assoc prof 64-85, EMER PROF CHEM, MED COL VA, 86- *Honors & Awards:* Chem Pioneer Award, Am Inst Chemists, 73. *Mem:* Am Chem Soc; Am Inst Chemists; Sigma Xi. *Res:* Application of complexing agents in mammals; dispersion of calculi by surface reactions; drug analysis. *Mailing Add:* Dept Med Chem Med Col Va Box 540 Richmond VA 23298-0540

VAN TRUMP, JAMES EDMOND, b Wood River, Nebr, Mar 1, 43; m 68; c 2. ARAMID FIBERS, BALLISTICS PROTECTION. *Educ:* Univ Wyo, BS, 64; Univ Calif, San Diego, MS, 73, PhD(chem), 75. *Prof Exp:* Res chemist, Am Potash & Chem Corp, 68-70; res chemist, Kingston Plant Fibers, 75-77 & Exp Sta Pigments, 77-79, RES ASSOC, DU PONT EXP STA FIBERS, 79- *Mem:* Am Chem Soc. *Res:* Fiber science; manufacture, structure and use of high strength industrial fibers; aramid fibers, pulps and fibrids, carbon fibers, nylon and polyester fibers, papers, ballistic fibers. *Mailing Add:* 4A Wood Rd Wilmington DE 19806

VAN TUYL, ANDREW HEUER, b Fresno, Calif, July 6, 22; m 55; c 4. MATHEMATICS. *Educ:* Fresno State Col, AB, 43; Stanford Univ, MA, 46, PhD(math), 47. *Prof Exp:* Asst chem, Stanford Univ, 43-44, asst elec eng, 44-45, asst physics, 46, asst math, 46-47; MATHEMATICIAN, NAVAL SURFACE WARFARE CTR, 47- *Concurrent Pos:* Res assoc, Ind Univ, 53. *Mem:* Fel AAAS; Am Inst Aeronaut & Astronaut; Am Math Soc; Soc Indust & Appl Math; NY Acad Sci; Math Asn Am. *Res:* Potential theory; special functions; hydrodynamics; gas dynamics. *Mailing Add:* Naval Surface Warfare Ctr White Oak Silver Spring MD 20903-5000

VAN TUYL, HAROLD HUTCHISON, b Ft Worth, Tex, Oct 13, 27; m 52; c 4. RADIOCHEMISTRY. *Educ:* Agr & Mech Col, Tex, BS, 48. *Prof Exp:* Res chemist, Gen Elec Co, 48-65; res chemist, 65-70, mgr nuclear fuel cycle, 70-84, MGR, CRIT MASS LAB, PAC NORTHWEST LABS, BATTELLE MEM INST, 84- *Mem:* Am Nuclear Soc; Am Chem Soc. *Res:* Fission product recovery; dose rate and shielding calculations; fission product and transuranics generation calculations; transuranic element separations; nuclear chemistry. *Mailing Add:* 2158 Hudson Ave Richland WA 99352-2027

VAN TUYLE, GLENN CHARLES, b Wilkes-Barre, Pa, May 28, 43. BIOCHEMISTRY. *Educ:* Lafayette Col, AB, 65; Thomas Jefferson Univ, PhD(biochem), 71. *Prof Exp:* Chemist org chem, Rohm & Haas Chem Co, 66-68; teaching & res asst biochem, Thomas Jefferson Univ, 68-71; assoc, State Univ NY, Stony Brook, 71-74; ASSOC PROF BIOCHEM, MED COL VA, VA COMMONWEALTH UNIV, 74- *Concurrent Pos:* Assoc prof res grant, Nat Inst Gen Med Sci, 76-82. *Mem:* Am Soc Cell Biol; Am Soc Biol Chemists. *Res:* Mitochondrial biogenesis; DNA structure, packaging and replications. *Mailing Add:* Dept Biochem & Molecular Biophys Va Commonwealth Univ Box 614 MCV Sta Richmond VA 23298-0614

VAN TYLE, WILLIAM KENT, b Frankfort, Ind, Feb 10, 44; m 82; c 2. PHARMACOLOGY. *Educ:* Butler Univ, BS, 67; Ohio State Univ, MSc, 69, PhD(pharmacol), 72. *Prof Exp:* From asst prof to assoc prof, 72-83, PROF PHARMACOL, BUTLER UNIV, 83- *Concurrent Pos:* Secy-treas, Dist 4, Am Asn Cols Pharm-Nat Asn Bds Pharm, 75-88. *Mem:* Am Asn Col Pharm. *Res:* Bioavailability of drugs to central nervous system and drug effects on central neurotransmitters. *Mailing Add:* Butler Univ 4600 Sunset Ave Indianapolis IN 46208

VAN UITERT, LEGRAND G(ERARD), b Salt Lake City, Utah, May 6, 22; m 45; c 3. MATERIALS SCIENCE, INORGANIC CHEMISTRY. *Educ:* George Washington Univ, BS, 49; Pa State Univ, MS, 51, PhD(chem), 52. *Prof Exp:* Mem tech staff, Bell Tel Labs, Inc, 52-88; RETIRED. *Honors & Awards:* W R G Baker Award, Inst Elec & Electronics Engrs, 71; H N Potts Award, Franklin Inst, 75; IRI Award, Indust Res Inst, 76; Creative Invention Award, Am Chem Soc, 78; Am Phys Soc Int Prize, 81. *Mem:* Nat Acad Eng; Am Chem Soc. *Res:* Magnetic oxides; luminesence; lasers, electro-optic and non-linear devices; magnetic bubble domain materials, fiber optics, passive displays and dielectric films. *Mailing Add:* Two Terry Dr Morristown NJ 07960-4713

VAN UMMERSEN, CLAIRE ANN, b Chelsea, Mass, July 28, 35; m 58; c 2. DEVELOPMENTAL BIOLOGY, ANIMAL PHYSIOLOGY. *Educ:* Tufts Univ, BS, 57, MS, 60, PhD(biol), 63. *Prof Exp:* Res asst radiobiol, Tufts Univ, 57-60, res assoc, 60-67, lectr biol, 67-68; asst prof, 68-74, assoc dean acad affairs, Liberal Arts Col, 75-76, assoc vchancellor acad affairs, 76-77, interim chancellor, 78-79, dir, Environ Sci Ctr & Biol Grad Prog, 79-81, assoc vice chancellor acad affairs, Bd Regents, 81-85, ASSOC PROF BIOL, UNIV MASS, BOSTON, 74-, VCHANCELLOR MGT SYSTS & TELECOMMUN, BD REGENTS, 85- *Concurrent Pos:* Fel, Tufts Univ,

63-67; mem teaching fac, Lancaster Courses in Ophthal, Colby Col, 62- *Mem:* AAAS; Am Soc Zoologists; Soc Develop Biol; Sigma Xi. *Res:* Biological effects of microwave radiation on the eye and the developing embryo. *Mailing Add:* 12 Frost Dr Durham NH 03824

VAN VALEN, LEIGH MAIORANA, b Albany, NY, Aug 12, 35; m 74; c 2. EVOLUTIONARY BIOLOGY, PALEONTOLOGY. *Educ:* Miami Univ, BA, 56; Columbia Univ, MA, 57, PhD(zool), 61. *Prof Exp:* Boese fel, Columbia Univ, 61-62; NATO fel, Univ Col, London, 62-63; res fel vert paleont, Am Mus Natural Hist, 63-66; asst prof anat, 67-68, from asst prof to assoc prof evolutionary biol, 68-73, assoc prof to prof biol, 73-88, PROF ECOL & EVOLUTION, UNIV CHICAGO, 88- *Concurrent Pos:* Res assoc geol, Field Mus Natural Hist, 71-; managing ed & ed, Evolutionary Theory, 73- & ed Evolutionary Monographs, 77- *Mem:* Soc Study Evolution (vpres, 73, 80); Soc Vert Paleont; Ecol Soc Am; Philos Sci Asn; Am Soc Naturalists (treas, 69-72, vpres, 74-75); Genetics Soc Am. *Res:* Energy in ecology and evolution; extinction; ecological control of large-scale evolutionary patterns; analytical paleoecology; mammalian evolution; the phenotype; competition; natural selection of plants and animals; evolutionary theory; biological variation; evolution of development; basal radiation of placental mammals; body size. *Mailing Add:* Dept Ecol & Evolution Univ Chicago 1101 E 57th St Chicago IL 60637

VAN VALIN, CHARLES CARROLL, b Wakefield, Nebr, Aug 10, 29; m 54. ATMOSPHERIC CHEMISTRY, ATMOSPHERIC PHYSICS. *Educ:* Nebr State Teachers Col, BA, 51; Univ Colo, Boulder, MS, 58. *Prof Exp:* Teacher sci, Ralston High Sch, Nebr, 53-54; chemist pesticide mfg, Shell Chem Co, Denver, 57-58; res chemist sugar chem, Great Western Sugar Co, Denver, 58-60; res chemist pesticide res, Fish-Pesticide Res Lab, Bur Sport Fisheries & Wildlife, Dept Interior, 61-66; RES CHEMIST ATMOSPHERIC RES, ENVIRON RES LABS, NAT OCEANIC & ATMOSPHERIC ADMIN, US DEPT COM, 66- *Concurrent Pos:* Abstractor, Chem Abstracts Serv, Am Chem Soc, 62-64. *Mem:* Am Chem Soc; Am Geophys Union; Sigma Xi. *Res:* Chemical and physical processes of acidic precipitation formation; reactions in polluted atmospheres; natural sources of acidic precursors; atmospheric transport of aerosols and trace gases. *Mailing Add:* Nat Oceanic & Atmospheric Admin US Dept Com 325 Broadway Boulder CO 80303

VAN VALKENBURG, ERNEST SCOFIELD, applied physics, electrical engineering, for more information see previous edition

VAN VALKENBURG, JEPTHA WADE, JR, b Ann Arbor, Mich, Mar 26, 25; m 49, 76; c 3. PHYSICAL CHEMISTRY, SURFACE CHEMISTRY. *Educ:* Kalamazoo Col, BS, 49; Univ Wis, MS, 51; Univ Mich, PhD(phys chem), 55; William Mitchell Col Law, JD, 79. *Prof Exp:* Chemist, Dow Chem Co, Mich, 54-58, proj leader pesticidal formulations res, 58-60, group leader, 60-66, mgr patent admin, 66-68; supvr phys chem res, 3M Co 68-71, info scientist specialist, 71-73, mem patent liaison staff, 73-80, mgr regulatory affairs & int res & develop, 81-84, mgr, bus ext, Data Rec Prod Div, 3M CO, 84-86; indust serv rep, NASA/ARAC, 86-; dir, comt sci & technol, res & dev, State of Minn, 87-88; CONSULT, ACCESSIBLE TECHNOLOGIES INC, 88- *Concurrent Pos:* Adv Pesticide Formulations, UNIDO, 86-87; chmn, Coun Comt Patents, Am Chem Soc, 85-87, Technol Transfer, Univ Minn Ctr Interfacial Eng, 91- *Mem:* Am Chem Soc; Sigma Xi. *Res:* Pesticidal formulations; surfactants; emulsions; diffusion; kinetics; encapsulation and surface coatings; biological correlations; computer media; federal and state regulations. *Mailing Add:* 494 Curfew St Paul MN 55104

VAN VALKENBURG, M(AC) E(LWYN), b Union, Utah, Oct 5, 21; m 43; c 6. ELECTRICAL ENGINEERING, SYSTEM THEORY. *Educ:* Univ Utah, BS, 43; Mass Inst Technol, SM, 46; Stanford Univ, PhD(elec eng), 52. *Prof Exp:* Mem staff, Radiation Lab, Mass Inst Technol, 43-45, asst, Electronics Lab, 45-46; from instr to assoc prof elec eng, Univ Utah, 46-55; acting instr, Stanford Univ, 49-51; from assoc prof to prof, Univ Ill, Urbana, 55-66, assoc dir, Coord Sci Lab, 59-66; prof & chmn dept, Princeton Univ, 66-74; prof, 74-82, actg dean, 84-85, W W Grainger prof elec eng, 82-88, dean, 85-88, EMER DEAN, COL ENG, UNIV ILL, URBANA, 88- *Concurrent Pos:* Ed, Trans Circuit Theory, Inst Elec & Electronics Engrs, 60-63, ed Proc, 65-68, ed-in-chief, Press, 85-; vis prof, Univ Calif, Berkeley, 62-63 & Univ Hawaii, 78-79; Thompson vis prof, Univ Ill, 72-73, Univ Ariz, 82-83. *Honors & Awards:* George Westinghouse Award, Am Soc Eng Educ, 63, Lamme Award, 78; Educ Medal, Inst Elec & Electronics Engrs, 72. *Mem:* Nat Acad Eng; fel Inst Elec & Electronics Engrs (vpres, 69-71); fel Am Soc Eng Educ; Sigma Xi. *Res:* Circuit theory; analog filter theory; systems theory; energy systems. *Mailing Add:* Everitt Lab Univ Ill 1406 Green St Urbana IL 61801

VAN VECHTEN, DEBORAH, b Washington, DC, Oct 24, 47; m 70. LOW TEMPERATURE PHYSICS. *Educ:* Brown Univ, ScB, 69; Univ Md, College Park, MS, 75, PhD(physics), 79. *Prof Exp:* Vis asst prof physics, Ga Inst Technol, 78-79; asst prof physics, Howard Univ, 81-83; mat scientist, Sachs Freeman Assocs, 85-89; assoc, Nat Res Coun, 79-81, postdoctoral res assoc, 83-85, PHYSICIST, US NAVAL RES LAB, 89- *Concurrent Pos:* Guest scientist, Nat Bur Standards, 81. *Mem:* Am Phys Soc; Am Vacuum Soc; Mat Res Soc. *Res:* Non-equilibrium superconductivity; properties and applications of superconductors; high energy resolution x and gamma ray detectors; role of fluctuations in phase transitions; ion beam assisted thin film deposition; Josephson tunnel junctions; resistive phase transition in 2 dimensions; percolation theory; layered thin film structures including Rugates; quantum measurement theory; theoretical solid state physics; experiments which seek to define physical phenomena and on theoretical models of these phenomena, basic solid state physics. *Mailing Add:* 5107 Edmondson Ave Baltimore MD 21229-2336

VAN VECHTEN, JAMES ALDEN, b Washington, DC, July 29, 42; m 83; c 2. THEORETICAL SOLID STATE PHYSICS. *Educ:* Univ Calif, Berkeley, AB, 65; Univ Chicago, PhD(physics), 69. *Prof Exp:* Infrared res officer semiconductors, US Naval Res Lab, Wash, 69-71; mem tech staff electro optic res, Bell Tel Lab, Murray Hill, 71-74; res staff mem semiconductor physics, Thomas J Watson Res Ctr, IBM, 74-85; AT DEPT ELEC ENG, ORE STATE UNIV, 85- *Concurrent Pos:* Fannie & John Horte Found fel. *Mem:* Fel Am Phys Soc; Electrochem Soc; fel Inst Physics (London); sr mem Inst Elec & Electronics Engrs; Electrochem Soc Am; Mat Res Soc. *Res:* Theoretical study of covalently bonded solids, their electronic, optical, mechanical and thermochemical properties; co-developer of the dielectric scale of electronegativity; semiconductors. *Mailing Add:* Dept Elec & Computer Eng Ore State Univ Corvallis OR 97331

VAN VELDHUIZEN, PHILIP ANDROCLES, b Hospers, Iowa, Nov 6, 30; m 84; c 4. MATHEMATICS, STATISTICS. *Educ:* Cent Col, BA, 52; Univ Iowa, MS, 60. *Prof Exp:* Teacher jr high sch, 54; instr, Exten Ctr, Univ Ga, 54-56; instr math, Cent Col, 56-59; asst prof, Sacramento State Col, 60-63; assoc prof, 63-74, PROF MATH, UNIV ALASKA, FAIRBANKS, 74- *Concurrent Pos:* Spec lectr & resource personnel, Mod Math Prog, Fairbanks, Anchorage & Kodiak, Alaska, 64-67; mem adv bd every pupil eval prog, Northwest Regional Lab, State Dept Educ, 75-81; mem, Fairbanks North Star Borough Sch Bd, 78-81, clerk, 78-80, treas, 80-81; mem, Nat Coun Teachers Math. *Mem:* Math Asn Am; Am Statist Asn. *Res:* Social basis of mathematics teaching and learning; effect of the social setting has on the learning and teaching atmosphere in a secondary classroom. *Mailing Add:* PO Box 82593 Fairbanks AK 99708

VAN VERTH, JAMES EDWARD, b Huntington, WVa, Jan 26, 28; m 65; c 2. ORGANIC SYNTHESIS, ORGANIC MECHANISMS. *Educ:* Xavier Univ, Ohio, BS, 50; Univ Detroit, MS, 52; Ind Univ, PhD(org chem), 57. *Prof Exp:* Sr res chemist, Monsanto Chem Co, 56-61; asst, Yale Univ, 61-63; from asst prof to assoc prof, 63-75, PROF CHEM, CANISIUS COL, 75- *Concurrent Pos:* Vis scientist, State Univ NY, Buffalo, 80-81; prin investr, Petrol Res Fund grant, 68-69 & 81-83. *Mem:* Am Chem Soc. *Res:* Organic synthesis; mechanisms. *Mailing Add:* Dept Chem Canisius Col Buffalo NY 14208

VAN VLACK, LAWRENCE H(ALL), b Atlantic, Iowa, July 21, 20; m 43; c 2. MATERIALS SCIENCE & ENGINEERING. *Educ:* Iowa State Col, BS, 42; Univ Chicago, PhD(geol), 50. *Prof Exp:* Ceramist, US Steel Corp, 42-43, petrogr, 43-52, process metallurgist, 52-53; assoc prof mat & metall eng, 53-58, chmn dept, 67-73, PROF MAT & METALL ENG, UNIV MICH, ANN ARBOR, 58- *Concurrent Pos:* Vis prof, Univ Calif, Berkeley, 61, Univ Melbourne, 67, Univ Kanpur, 69, Monash Univ, 73, & Zhejiang Univ, 84. *Honors & Awards:* A Sauveur Award, Am Soc Metals, 79, Gold Medal, 84, White Award, 85. *Mem:* Fel AAAS; fel Am Ceramic Soc; fel Am Soc Metals; Am Soc Eng Educ; Am Inst Mining, Metall & Petrol Engrs. *Res:* Refractories; slags; nonmetallic inclusions; ceramic materials; process metallurgy; materials science instruction; nickel oxide. *Mailing Add:* Unit 309 2115 Nature Cove Ann Arbor MI 48104

VAN VLECK, FRED SCOTT, b Clearwater, Nebr, Dec 12, 34; m 60; c 5. CONTROL THEORY, MULTIPLE-VALUED FUNCTIONS. *Educ:* Univ Nebr, BSc, 56, MA, 57; Univ Minn, PhD(math), 60. *Prof Exp:* Instr math, Mass Inst Technol, 60-62; from asst prof to assoc prof, 62-68, PROF MATH, UNIV KANS, 68- *Concurrent Pos:* Vis prof, Univ Colo, 71-72. *Mem:* Math Asn Am; Am Math Soc; Soc Indust & Appl Math. *Res:* Control theory; measurable multiplevalued functions; ordinary differential equations; optimization. *Mailing Add:* Dept Math Univ Kans Lawrence KS 66045

VAN VLECK, LLOYD DALE, b Clearwater, Nebr, June 11, 33; m 58; c 2. GENETICS, ANIMAL SCIENCE. *Educ:* Univ Nebr, BS, 54, MS, 55; Cornell Univ, PhD(animal breeding), 60. *Hon Degrees:* DSc, Univ Nebr, 86. *Prof Exp:* Res assoc animal breeding, 59-60, res geneticist, 60-62, from asst prof to assoc prof animal genetics, 62-73, prof animal genetics, Cornell Univ, 73-88; PROF ANIMAL GENETICS, UNIV NEBR, LINCOLN, 88-; RES GENETICIST, USDA, 88- *Concurrent Pos:* Vis prof, Univ Nebr, Lincoln, 73, Scandinavian Grad Prog Animal Breeding, Uppsala, 75, Univ Calif, Davis, 85. *Honors & Awards:* Am Soc Animal Sci Award, 72; Nat Asn Animal Breeders Award, Am Dairy Sci Asn, 74 & 83. *Mem:* Biomet Soc; Am Dairy Sci Asn; Am Soc Animal Sci; Am Genetic Asn. *Res:* Methods of improving genetic value of large animals using genetic theory, statistical technique for unbalanced data, and computer processing. *Mailing Add:* Animal Sci A218 Univ Nebr Lincoln NE 68583-0908

VAN VLEET, JOHN F, b Lodi, NY, Mar 23, 38; m 61; c 2. VETERINARY PATHOLOGY. *Educ:* Cornell Univ, DVM, 62; Univ Ill, MS, 65, PhD(vitamin E deficiency), 67. *Prof Exp:* Asst vet, 62-63; USPHS trainee vet path, Univ Ill, 63-66, instr, 66-67; from asst prof to assoc prof, 67-76, PROF VET PATH, PURDUE UNIV, WEST LAFAYETTE, 76-, ASSOC DEAN ACAD AFFAIRS, 88- *Mem:* Vet Med Asn; Int Acad Path; Am Col Vet Path. *Res:* Ultrastructural and nutritional pathology; myocardial diseases; selenium-vitamin E deficiency; cardiomyopathy; cardiovascular and skeletal muscular pathology. *Mailing Add:* Sch Vet Med Purdue Univ West Lafayette IN 47907

VAN VLIET, ANTONE CORNELIS, b San Francisco, Calif, Jan 11, 30; m 53; c 4. WOOD SCIENCE, COMMUNICATIONS. *Educ:* Ore State Univ, BS, 52, MS, 58; Mich State Univ, PhD, 70. *Prof Exp:* Instr forest prod, Ore State Univ, 55-59; asst to plant mgr plywood prod, Bohemia Lumber Co, 59-60; asst prof wood prod, Extent, 63-71, from asst prof to assoc prof, 63-80, dir, off careers, planning & placement, 71-90, PROF FOREST PROD, ORE STATE UNIV, 80- *Concurrent Pos:* State Rep, Dist 35, Ore Legis, 75, 77, 79, 81, 83, 85, 87, 89, & 91. *Mem:* Forest Prod Res Soc. *Res:* Plywood production; wood anatomy and utilization; company educational programs; behavioral aspect of communications; management science. *Mailing Add:* 1530 NW 13th Corvallis OR 97330

VAN VLIET, CAROLYNE MARINA, b Dordrecht, Neth, Dec 27, 29; US citizen; div; c 4. STATISTICAL MECHANICS. *Educ:* Free Univ, Amsterdam, BS, 49, MA, 53, PhD(physics), 56. *Prof Exp:* Fel elec eng, Univ Minn, Minneapolis, 56-57, asst prof, 57-58; asst dir physics lab, Free Univ, Amsterdam, 58-60; from assoc prof to prof elec eng, Univ Minn, Minneapolis,

60-66, prof elec eng & physics, 66-69; PROF THEORET PHYSICS, CTR MATH RES, UNIV MONTREAL, 69- *Concurrent Pos:* Fulbright scholar, 56-58; vis prof, Univ Fla, 74 & 78-89. *Mem:* Am Phys Soc; fel Am Sci Affil; Europ Phys Soc; Can Asn Physicists; fel Inst Elec & Electronics Engrs; Can Res Inst Advan Women. *Res:* Non-equilibrium statistical mechanics; kinetic equations and electrical transport in solids; solid state electronics; noise and fluctuation phenomena. *Mailing Add:* Ctr Math Res Univ Montreal Montreal PQ H3C 3J7 Can

VAN VORIS, PETER, b Bethlehem, Pa, Mar 10, 48; m 71; c 1. ECOLOGY, ECOSYSTEM ANALYSIS. *Educ:* Kenyon Col, BA, 70; State Univ NY, Buffalo, MS, 73; Univ Tenn, PhD(ecol), 77. *Prof Exp:* Instr biol & ecol, Trocaire Col, 72-74; res asst, Oak Ridge Nat Lab, 74-76; res scientist ecol, Columbus Div, Battelle Mem Inst, 77-82, STAFF SCIENTIST, EARTH SCI DEPT, BATTELLE NORTHWEST, RICHLAND, WA, 83- *Concurrent Pos:* Fed fel, State Univ NY, Buffalo, 71-72; consult, Chem-Trol Pollution Serv Inc, Model City, NY, 71-73; Dept Energy/Oak Ridge Assoc Univs fel, Oak Ridge Nat Lab, 76-77. *Honors & Awards:* Outstanding Col & Univ Educator Award, Asn Am Cols & Univs, 74. *Mem:* Ecol Soc Am; Am Inst Biol Sci; AAAS; Am Soc Naturalists; Audubon Soc; Am Soc Testing & Mat; Soc Environ Toxicol & Chem. *Res:* Phenomena of stability in ecosystems; the relationship of the measure of complexity with ecosystem stability; terrestrial microcosm development; environmental toxicology. *Mailing Add:* Pac NW Lab Batelle Mem Inst PO Box 999 Richland WA 99352

VAN VOROUS, TED, b Billings, Mont, Jan 6, 29; m 51; c 5. ANALYTICAL CHEMISTRY. *Educ:* Mont State Col, BS, 53, MS, 54. *Prof Exp:* Analytical chemist, Dow Chem Co, 54-56, scientist-chemist, 56-59, sr chemist, 59-60, group supvr, 60-62, res group mgr, 62-69; pres, VTA Inc, 69-75; pres, Vac-Tac Systs, 76-81; PRES, VAN VOROUS CONSULTS & E-LINE/USA, 81-; PRES, VACUUM INC, 82- *Mem:* Am Vacuum Soc; Geochem Soc; Sigma Xi. *Res:* High vacuum research; evaporation processes; ionphenomena; electron microsopy-metallurgy; electron microprobe analysis; instrumentation development; x-ray and emission spectroscopy; high temperature materials; epitaxial structures; planar magnetron sputtering. *Mailing Add:* Van Vorous Consults 992 Sycamore Ave Boulder CO 80303

VAN VORST, WILLIAM D, b Cuthbert, Ga, Aug 20, 19; m 49; c 3. CHEMICAL & SYSTEMS ENGINEERING. *Educ:* Rice Inst, BS, 41, ChE, 42; Mass Inst Technol, SM, 43; Univ Calif, Los Angeles, PhD, 53. *Prof Exp:* Res engr, Northrop Aircraft, Inc, 43-46 & NAm Aviation, Inc, 46; from lectr to prof eng, 46-87, chmn eng syst dept, 75-78, PROF ENG, UNIV CALIF, LOS ANGELES, 73- *Concurrent Pos:* Consult, Aerospace Industs, Inc, 49-59; overseas exp & consult int develop, 59-; vpres acad affairs, Robert Col, Istanbul, 69-71, adv foreign fac, Bogazici Univ, 71-72. *Honors & Awards:* Ralph Teetor Award, Soc Automotive Engrs. *Mem:* AAAS; Am Soc Eng Educ; fel Am Inst Chem Engrs; fel Inst Advan Eng; Int Asn Hydrogen Energy. *Res:* Engineering systems; industrial development and role of technology in developing countries; environmental engineering-pollution abatement; hydrogen and alcohols as alternative vehicular fuels; energy conversion-alternative sources. *Mailing Add:* 5531 Boelter Hall Univ Calif 405 Hilgard Ave Los Angeles CA 90024-1592

VAN VUNAKIS, HELEN, b New York, NY, June 15, 24; m 58; c 2. BIOCHEMISTRY. *Educ:* Hunter Col, BA, 46; Columbia Univ, PhD(biochem), 51. *Prof Exp:* USPHS fel & res assoc, Johns Hopkins Univ, 51-54; sr res scientist, State Dept Health, NY, 54-58; from asst prof to assoc prof, 58-74, PROF BIOCHEM, BRANDEIS UNIV, 74- *Concurrent Pos:* NIH career res award; mem, Soc Scholars, Johns Hopkins Univ. *Mem:* Am Soc Biol Chem. *Res:* Structure of proteins and nucleic acids; nicotine metabolism; assay and metabolism of other tobacco compounds. *Mailing Add:* Dept Biochem Brandeis Univ 415 South St Waltham MA 02254

VAN WAGNER, CHARLES EDWARD, b Montreal, Que, Dec 9, 24; m 55, 87; c 3. FOREST FIRE BEHAVIOR, FIRE ECOLOGY. *Educ:* McGill Univ, BEng, 46; Univ Toronto, BScF, 61. *Prof Exp:* Chief chemist, Can Pittsburg Industs, 46-58; RES SCIENTIST, CAN FOREST SERV, 60- *Concurrent Pos:* Assoc ed, Forest Sci, Can J Forest Res. *Mem:* Can Inst Forestry. *Res:* Forest fire; measurement and theory of fire behavior; variation in moisture content of forest fuel with weather; use of prescribed fire in forest management; effects of fire on forest ecology and timber supply. *Mailing Add:* Petawawa Nat Forestry Inst Chalk River ON K0J 1J0 Can

VAN WAGNER, EDWARD M, b Highland Falls, NY, Nov 3, 24; m 52; c 3. ELECTRICAL ENGINEERING. *Educ:* Univ Rochester, BS, 51. *Prof Exp:* Lab technician, Distillation Prod Div, Eastman Kodak Co, 51-53; elec engr, Consol Vacuum Corp, 53-55, physicist, Haloid Co, 55-58 & Haloid Xerox Corp, 58-65, scientist, 65-67, mgt photoreceptor process design, 67-69, SYSTS ANALYST, RES LAB DIV, XEROX CORP, ROCHESTER, 69- *Res:* Vacuum gauges; electrical instrumentation and control of vacuum production systems; graphic arts applications for xerography; design of chemical process equipment; vacuum equipment; welding and forming equipment and electrooptical test equipment. *Mailing Add:* 89 Anytrell Dr Webster NY 14580

VAN WAGTENDONK, JAN WILLEM, b Palo Alto, Calif, Feb 21, 40; m 68; c 2. FOREST ECOLOGY. *Educ:* Ore State Univ, BS, 63; Univ Calif, Berkeley, MS, 68, PhD(wildland res sci), 72. *Prof Exp:* RES SCIENTIST FIRE ECOL, YOSEMITE NAT PARK, NAT PARK SERV, 72- *Mem:* Soc Am Foresters; Ecol Soc Am. *Res:* Ecological role of fire in the Sierra Nevada ecosystems; recreational carrying capacities for wilderness areas. *Mailing Add:* Nat Park Serv Yosemite Nat Park El Portal CA 95318

VAN WAGTENDONK, WILLEM JOHAN, biochemistry; deceased, see previous edition for last biography

VAN WART, HAROLD EDGAR, b Bay Shore, NY, Oct 29, 47; m 74; c 3. BIOCHEMISTRY, PHYSICAL CHEMISTRY. *Educ:* State Univ NY, Binghamton, BA, 69; Cornell Univ, MS, 71, PhD(biophys chem), 74. *Prof Exp:* Temp asst prof chem, Cornell Univ, 74-75; NIH fel biochem & assoc staff med, Harvard Med Sch, 75-78; asst prof, 78-82, assoc prof chem, 82-86, PROF CHEM, FLA STATE UNIV, 86- *Concurrent Pos:* NSF, undergrad res Participation grant, 67-68; res assoc, Univ Vienna, Austria, 69; teaching assistantship, Cornell Univ, 69-71; NIH Predoctoral Traineeship, Cornell Univ, 74-75; postdoctoral fel, Harvard Med Sch, 75-78; NIH Res Career Develop Award, 82-87; dir, Molecular Biophys, PhD Prog, 82-86; dir, Molecular Biophys Laser Spectros Lab, 80-; vis lectr, var schs, co & socs. *Mem:* Am Chem Soc; AAAS; Fedn Am Soc Biol Chemists; Am Soc Biochem & Molecular Biol. *Res:* Structure-function relationships in enzymes, enzyme mechanism, Raman and resonance Raman studies of biomolecules; enzymology, proteolytic enzymes, role of proteolysis in biological control, metalloenzymes. *Mailing Add:* Dept Chem Fla State Univ Tallahassee FL 32306

VAN WAZER, JOHN ROBERT, b Chicago, Ill, Apr 11, 18; m 40; c 1. CHEMISTRY. *Educ:* Northwestern Univ, BS, 40; Harvard Univ, AM, 41, PhD(phys chem), 42. *Prof Exp:* Phys chemist, Eastman Kodak Co, 42-44; res group leader, Clinton Eng Works, Tenn, 44-46; phys chemist, Rumford Chem Works, RI, 46-49; head physics res, Great Lakes Carbon Corp, 49-50; sr scientist, Monsanto Co, 50-68; prof, 68-88, EMER PROF CHEM, VANDERBILT UNIV, 88- *Concurrent Pos:* Asst res dir, Monsanto Co, 51-60, dir chem dynamics res, 64-67. *Mem:* AAAS; Am Chem Soc; Soc Rheol; NY Acad Sci; Royal Soc Chem London. *Res:* Applied quantum mechanics; chemistry of phosphorus compounds; substituent-exchange or redistribution reactions; inorganic chemistry; rheology; applied nuclear-magnetic resonance; photoelectron spectroscopy; nutrition. *Mailing Add:* 6666 Brookmont Terr No 903 Nashville TN 37205

VAN WEERT, GEZINUS, b Rotterdam, Neth, Aug 24, 33; Can citizen; m 58; c 2. EXTRACTIVE METALLURGY, IRON FOUNDRY METALLURGY. *Educ:* Technol Univ Delft, BEng, 56, MEng, 58, DTechnol, 89; Univ Toronto, MASc, 59. *Prof Exp:* Test engr metall, Inco, Copper Cliff, Ont, 59-61; res & develop investr, NJ Zinc, Pa, 61-63; mgr process metall, Falconbridge Nickel Mines, Toronto, 63-78; dir, tech & econ serv, QLT-Fer Et Titane Inc, 78-85; mgr, Metall Technol, Prochem Ltd, 85-89; GEN MGR, HYDROCHEM DEVELOPMENTS LTD, 90-; PROF MINERAL PROCESSING & EXTRACTIVE METALL, TECHNOL UNIV DELFT. *Concurrent Pos:* Chmn, Hydrometall Sect, Can Inst Mining & Metall, 72-77; ed, CIM Bulletin, 78-85. *Honors & Awards:* Technol Award, Metall Soc, Am Inst Mining, Metall & Petrol Engrs, 73; Sherritt Hydrometall Award, 79. *Mem:* Can Inst Mining & Metall; Am Inst Mining, Metall & Petrol Engrs; Am Inst Chem Engrs; Inst Mining & Metall UK. *Res:* Non-ferrous extractive metallurgy, both hydro and pyro, and operational aspects of the ductile iron foundry industry. *Mailing Add:* Mevesteyn 20 Maasland 3155 XK Netherlands

VAN WIJNGAARDEN, ARIE, b Holland, Apr 8, 33; Can citizen; m 57; c 3. PHYSICS. *Educ:* McMaster Univ, PhD(physics), 62. *Prof Exp:* Teacher physics, 62-70, assoc prof, 70-73, PROF PHYSICS, UNIV WINDSOR, 73- *Concurrent Pos:* Nat Res Coun-Ont Res Found res grants, 62- *Res:* Radiative processes and lamb shift. *Mailing Add:* Dept Physics Univ Windsor Windsor ON N9B 3P4 Can

VAN WINGEN, N(ICO), petroleum engineering, for more information see previous edition

VAN WINKLE, MICHAEL GEORGE, b Newark, Ohio, July 25, 39; c 3. IMMUNOLOGY, IMMUNOCHEMISTRY. *Educ:* Ohio State Univ, BSc, 62, MSc, 64, PhD(immunol), 66. *Prof Exp:* Group leader res atopic allergy, Riker Lab, 3M Co, 66-71; head, Dept Immunol, Nucleic Acid Res Inst, Int Chem & Nuclear Corp, 71-73; mgr, Dept Hepatitis, Curtis Labs, Inc, 73-76; mgr hepatitis opers, Nuclear Med Labs, Inc, 76-78; SR RES IMMUNOCHEMIST, CLIN IMMUNOASSAY RES & DEVELOP, BECKMAN INSTRUMENTS, INC, 78- *Mem:* Am Asn Immunol; Am Soc Microbiol; AAAS. *Res:* Research and development on clinical immunoassays. *Mailing Add:* Beckman Instruments Inc 1761 Kaiser Ave Irvine CA 92714

VAN WINKLE, QUENTIN, b Grand Forks, NDak, Mar 10, 19; m 41; c 4. CHEMISTRY. *Educ:* SDak Sch Mines & Technol, BS, 40; Ohio State Univ, PhD(chem), 47. *Prof Exp:* Asst chem, Ohio State Univ, 40-43, res assoc eng, Exp Sta, 44; asst chemist metall lab, Univ Chicago, 44-46; res assoc chem, Ohio State Univ, 46-48, from asst prof to prof,48-80; RETIRED. *Concurrent Pos:* Consult, E I du Pont de Nemours & Co, 57-75. *Mem:* AAAS; Am Chem Soc. *Res:* Physico-chemical properties of proteins, high polymers; nucleic acids; surface chemistry. *Mailing Add:* 271 S Main St Box 91 West Mansfield OH 43358-0091

VAN WINKLE, THOMAS LEO, chemical engineering; deceased, see previous edition for last biography

VAN WINKLE, WEBSTER, JR, b Plainfield, NJ, Nov 18, 38; m 61; c 3. ENVIRONMENTAL SCIENCES. *Educ:* Oberlin Col, BA, 61; Rutgers Univ, New Brunswick, PhD(zool), 67. *Prof Exp:* Res assoc, Shellfish Res Lab, Rutgers Univ, 66-67; asst prof biol, Col William Mary, 67-70; USPHS fels, NC State Univ, 70 & 72; res assoc, 72-75, res staff mem, Environ Sci Div, 75-78, HEAD, AQUATIC ECOL SECT, OAK RIDGE NAT LAB, 79- *Concurrent Pos:* NSF fel, Marine Lab, Duke Univ, 69; NSF sci fac fel, NC State Univ, 71-72. *Mem:* AAAS; Ecol Soc Am; Am Fisheries Soc. *Res:* Assessment of environmental impacts on aquatic ecosystems; fish population modeling; spectral analysis of environmental time series; data analysis. *Mailing Add:* Environ Sci Div Bldg 1505 Oak Ridge Nat Lab Oak Ridge TN 37831-6038

VAN WINTER, CLASINE, b Amsterdam, Neth, Apr 8, 29. MATHEMATICAL PHYSICS. *Educ:* Univ Groningen, BSc, 50, MSc, 54, PhD(physics), 57. *Prof Exp:* Res asst physics, Univ Groningen, 51-58, sci officer, 58-68; PROF MATH & PHYSICS, UNIV KY, 68- *Concurrent Pos:* Fel physics, Univ Birmingham, 57 & Niels Bohr Inst Theoret Physics, Univ Copenhagen, 63; vis assoc prof physics, Ind Univ, Bloomington, 67-68; scientist-in-residence, Math & Comput Sci Div, Argonne Nat Lab, 85. *Mem:* Am Math Soc; Am Phys Soc; Int Asn Math Physics. *Res:* Three-and more-body problem in quantum mechanics; quantum scattering theory; functional analysis; complex variables. *Mailing Add:* Dept Math Univ Ky Lexington KY 40506-0027

VAN WOERT, MELVIN H, b Brooklyn, NY, Nov 3, 29; m 55. INTERNAL MEDICINE. *Educ:* Columbia Univ, BA, 51; NY Med Col, MD, 56. *Prof Exp:* From intern to resident internal med, Univ Chicago, 56-60, res asst gastroenterol, 62-63; from asst scientist to assoc scientist, Brookhaven Nat Lab, 63-67; from asst prof to assoc prof med & pharmacol, Sch Med, Yale Univ, 67-74; prof internal med & head, Sect Clin Pharmacol, 74-78, PROF PHARMACOL, MT SINAI SCH MED, 74-, PROF NEUROL, 78- *Mem:* AAAS; fel Am Col Physicians; Soc Neurosci; Soc Neurochem; Am Soc Pharmacol & Exp Therapeut. *Res:* Neuropharmacological approaches to extrapyramidal disease; serotonin metabolism and myoclonus. *Mailing Add:* Dept Neurol Mt Sinai Sch Med Fifth Ave & 100th St New York NY 10029

VAN WORMER, KENNETH A(UGUSTUS), JR, b Mannsville, NY, Oct 4, 30; m 57; c 4. CHEMICAL ENGINEERING. *Educ:* Clarkson Col Technol, BS, 52, MS, 54; Mass Inst Technol, ScD(chem eng), 61. *Prof Exp:* Engr chem eng, Gen Elec Co, 52-53; from instr to assoc prof, 54-79, chmn dept, 71-81, PROF CHEM ENG, TUFTS UNIV, 79- *Mem:* Am Soc Eng Educ; Am Inst Chem Engrs. *Res:* Solar engineering; energy conservation; waste water treatment; liquid-liquid extraction; heterogeneous catalysis in the direct reduction of iron ore; applied thermodynamics; applications of digital computers to chemical engineering; reaction kinetics; contribution of nucleation to phase transformations; applied mathematics; optimization; electronic materials processing. *Mailing Add:* Dept Chem Eng Tufts Univ Medford MA 02155

VAN WYK, CHRISTOPHER JOHN, b Fairborn, Ohio, Sept 5, 55; m 80; c 2. OBJECT-ORIENTED PROGRAMMING, COMPUTATIONAL GEOMETRY. *Educ:* Swarthmore Col, BA, 77; Stanford Univ, PhD(computer sci), 80. *Prof Exp:* Mem tech staff, AT&T Bell Labs, 80-91; ASSOC PROF COMPUTER SCI, DREW UNIV, 90- *Concurrent Pos:* Instr, NJ Gov Sch Sci, 84 & 89; vis asst prof, Stevens Inst Technol, 84-85; vis lectr, Princeton Univ, 87; assoc ed, J Computer & Syst Sci, 90- *Mem:* Math Asn Am; Asn Comput Mach; Inst Elec & Electronics Engrs; Sigma Xi; Soc Indust & Appl Math. *Res:* Application of techniques from data structures, algorithms and computational geometry to problems in graphics, circuit design and document preparation. *Mailing Add:* Dept Math & Computer Sci Drew Univ Madison NJ 07940

VAN WYK, JUDSON JOHN, b Maurice, Iowa, June 10, 21; m 44; c 4. PEDIATRICS, ENDOCRINOLOGY. *Educ:* Hope Col, AB, 43; Johns Hopkins Univ, MD, 48; Am Bd Pediat, dipl. *Hon Degrees:* ScD, Hope Col, 76. *Prof Exp:* Fel biochem, St Louis Univ, 43-44; Henry Strong Dennison scholar physiol chem, Johns Hopkins Univ, 47, intern & asst resident pediat, Johns Hopkins Hosp, 48-50; investr metab, Nat Heart Inst, 51-53; fel pediat endocrinol, Johns Hopkins Univ, 53-55; from asst prof to assoc prof pediat, 55-62, prof, 62-75, KENAN PROF PEDIAT, SCH MED, UNIV NC, CHAPEL HILL, 75- *Concurrent Pos:* Attend physician, NC Mem Hosp, 55-; Markle scholar med sci, 56-61; USPHS res career award, 62-88; mem training grants comt in diabetes & metab, NIH, 67-71 & endocrine study sect, 71-75; vis scientist, Karolinska Inst, Sweden, 68-69; consult, Womack Army Hosp, Ft Bragg; assoc ed, J Clin Endocrinol & Metab, 84-89. *Honors & Awards:* Fred Konrad Koch Award & Medal, Endocrine Soc, 88. *Mem:* Endocrine Soc; Soc Pediat Res; fel Am Acad Pediat; Am Pediat Soc. *Res:* Human sex differentiation; pituitary function and hormonal control of growth and sexual maturation; isolation and physiologic role of somatomedin; growth factors in cellular proliferation. *Mailing Add:* Univ NC Chapel Hill Div Pediat Endocrinol CB No 7220 509 Burnett-Womack Chapel Hill NC 27599-7220

VAN WYLEN, GORDON J(OHN), b Grant, Mich, Feb 6, 20; m 51; c 5. MECHANICAL ENGINEERING. *Educ:* Calvin Col, AB, 42; Univ Mich, BSE, 42, MSE, 47; Mass Inst Technol, ScD(mech eng), 51. *Hon Degrees:* DLitt, Hope Col, 72. *Prof Exp:* Indust engr, E I du Pont de Nemours & Co, 42-43; instr mech eng, Pa State Univ, 46-48; asst, Mass Inst Technol, 49-51; from asst prof to prof, Univ Mich, Ann Arbor, 51-72, chmn dept, 58-65, dean col eng, 65-72; pres, 72-87, EMER PRES, HOPE COL, 87- *Mem:* Fel AAAS; fel Am Soc Mech Engrs. *Res:* Thermodynamics and cryogenics. *Mailing Add:* 817 Brook Village Dr Holland MI 49423

VANYO, JAMES PATRICK, b Wheeling, WVa, Jan 29, 28. ROTATING FLUIDS, HELIOTROPIC STROMATOLITES. *Educ:* WVa Univ, BSME, 52; Salmon P Chase Col, JD, 59; Univ Calif, Los Angeles, MA, 66, PhD(eng), 69. *Prof Exp:* Asst supvr commun, Am Tel & Tel Co, NY, 52-53; pres, Van Industs, Inc, Ohio, 54-59; asst to pres, Remanco, Inc, Calif, 59-61; proposal specialist, Marquardt Corp, 61-63; planning analyst, Litton Industs, 69-70; from asst prof to assoc prof, 71-84, PROF ENG, UNIV CALIF, SANTA BARBARA, 84- *Concurrent Pos:* Lectr, Sinclair Col, Ohio, 56-57; consult rotating fluids; vis mem staff, Commonwealth Sci Indust Res Orgn, Div Atmos Physics, Melbourne, Australia, 80, Sch Phys, Univ Newcastle, Eng, 83, & Geophys Inst, Fairbanks, Alaska, 87. *Mem:* Am Phys Soc; Am Geophys Union. *Res:* Dynamics of rotating nonrigid bodies and fluids; Earth-Sun-Moon dynamics and heliotropic stromatolites. *Mailing Add:* Dept Mech Eng Univ Calif Santa Barbara CA 93106

VANÝSEK, PETR, b Ostrava, Czech, June 12, 52. ELECTROCHEMISTRY, ELECTROANALYTICAL CHEMISTRY. *Educ:* Charles Univ Prague, MS, 76, RNDr, 77; Czechoslovak Acad Sci, CSc, 82. *Prof Exp:* Postdoctoral fel, Univ NC, 82-84; fac in residence chem, Univ NH, 84-85; ASSOC PROF ANALYTICAL CHEM, NORTHERN ILL UNIV, 85- *Mem:* Electrochem Soc; Sigma Xi; Soc Electroanal Chem. *Res:* Electrochemical properties of interfaces between immiscible solutions, impedance studies of electrochemical systems; electroanalytical chemistry and small domain electrochemistry; editing of electrochemical literature and data; computer interfacing in electrochemistry. *Mailing Add:* Dept Chem Northern Ill Univ DeKalb IL 60115-2862

VAN ZANDT, LONNIE L, b Bound Brook, NJ, Sept 29, 37; m 75; c 3. SOLID STATE PHYSICS. *Educ:* Lafayette Col, BS, 58; Harvard Univ, AM, 59, PhD(physics), 64. *Prof Exp:* Mem staff, Sci Lab, Ford Motor Co, 62-64 & Lincoln Lab, Mass Inst Technol, 64-67; from asst prof to assoc prof, 67-83, PROF PHYSICS, PURDUE UNIV, WEST LAFAYETTE, 83- *Mem:* Sigma Xi; Bioelectromagnetic Soc. *Res:* Physics of biological molecules. *Mailing Add:* Dept Physics Purdue Univ West Lafayette IN 47907

VAN ZANDT, PAUL DOYLE, b Vandalia, Ill, Dec 29, 27; m 54; c 1. PARASITOLOGY. *Educ:* Greenville Col, AB, 52; Univ Ill, MS, 53; Univ NC, MSPH, 55, PhD(parasitol), 60. *Prof Exp:* Asst pub health, Univ NC, 58-61; from asst prof to assoc prof biol, 61-69, PROF BIOL, YOUNGSTOWN STATE UNIV, 69- *Mem:* Fel AAAS; Am Soc Parasitol; Am Soc Trop Med & Hyg; Royal Soc Trop Med & Hyg. *Res:* Immunology of animal parasites; medical parasitology and microbiology. *Mailing Add:* 7222 Pittsburgh Rd Poland OH 44514

VAN ZANDT, THOMAS EDWARD, b Highland Park, Mich, July 10, 29; m 61; c 2. RADAR METEOROLOGY. *Educ:* Duke Univ, BS, 50; Yale Univ, PhD, 55. *Prof Exp:* Physicist, Sandia Corp, 54-57; PHYSICIST, AERONOMY LAB, NAT OCEANIC & ATMOSPHERIC ADMIN, 57- *Concurrent Pos:* Vis lectr, Univ Colo, 61-70, adj prof, 70- *Mem:* Am Geophys Union; Am Meteorol Soc; Int Union Radio Sci. *Res:* Atmospheric internal gravity waves; interpretation of clear-air doppler radar observations. *Mailing Add:* Nat Oceanic & Atmospheric Admin 325 Broadway Boulder CO 80303

VAN ZANT, KENT LEE, b Humboldt, Nebr, July 5, 47; m 71. GEOLOGY, PALYNOLOGY. *Educ:* Earlham Col, AB, 69; Univ Iowa, MS, 73, PhD(geol), 76. *Prof Exp:* Asst prof geol, Beloit Col, 76-78; asst prof geol, Earlham Col, 78-81; STAFF PALEONTOLOGIST, AMOCO PROD CO, 81- *Concurrent Pos:* Nat Acad Sci res grant, USSR, 76-77; Res Corp grants, 77, 80. *Mem:* Geol Soc Am; Am Asn Petrol Geologists; Am Asn Stratig Palynologists. *Res:* Mesozoic and Cenozoic palynology of the Western US. *Mailing Add:* Amoco Prod Co Box 800 Denver CO 80202

VAN ZEE, RICHARD JERRY, b Kalamazoo, Mich, Feb 15, 47; m 77; c 5. ELECTRON SPIN RESONANCE SPECTROSCOPY. *Educ:* Western Ky Univ, BA, 69; Mich State Univ, PhD(phys chem), 76. *Prof Exp:* Grad asst chem, Mich State Univ, 69-75; FEL PHYS CHEM, UNIV FLA, 75- *Res:* The electronic and magnetic characterization of high temperature, high spin molecules; high temperature vaporization techniques. *Mailing Add:* 850 Old Oaks Rd Archer FL 32618-9452

VAN ZWALENBERG, GEORGE, b Neth, Sept 7, 30; US citizen; m 53, 76; c 3. MATHEMATICS. *Educ:* Calvin Col, BS, 53; Univ Fla, MA, 55; Univ Calif, Berkeley, PhD(math), 68. *Prof Exp:* Instr math, Bowling Green State Univ, 59-60; vis lectr, Calvin Col, 60-61, asst prof, 61-63; asst prof, Calif State Univ, Fresno, 63-67; chmn dept, 74-77, PROF MATH, CALVIN COL, 68- *Concurrent Pos:* Math Asn Am vis lectr, High Schs, 65-66; head math dept, Egyptian Air Force Acad, Bilbeis, Egypt, 86-88. *Mem:* Am Math Soc; Math Asn Am. *Res:* Complex variables. *Mailing Add:* Dept Math Calvin Col 3201 Burton St Grand Rapids MI 49546

VAN ZWIETEN, MATTHEW JACOBUS, b Zeist, Netherlands, Apr 6, 45; US citizen; m 66; c 2. VETERINARY PATHOLOGY. *Educ:* Univ Calif, Davis, BS, 67, DVM, 69; Am Col Vet Pathologists, dipl, 74; Univ Utrecht, PhD, 84. *Prof Exp:* Sci investr cell biol & path, Med Res Inst Infectious Dis, US Army, 69-71; res fel, New Eng Regional Primate Res Ctr & Animal Res Ctr, Harvard Med Sch, 71-75, assoc path, 75-76; mem staff, Inst Exp Gerontol, 76-84; assoc dir, 84-86, dir path, 86-89, SR DIR, DEPT SAFETY ASSESSMENT, MERCK SHARP & DOHME RES LABS, 89- *Concurrent Pos:* Res assoc path, Angell Mem Animal Hosp, 71-74, consult, 75-76; res assoc path, Children's Hosp Med Ctr, 73-75; head diag procedures, New Eng Regional Primate Res Ctr & Animal Res Ctr, Harvard Med Sch, 75-76. *Mem:* Am Col Vet Pathologists; Int Acad Path; Am Asn Lab Animal Sci; Am Vet Med Asn; Soc Toxicol Pathologists. *Res:* Pathology of aging in laboratory animals; mechanisms of radiation induced mammary carcinogenesis in rats; identification and development of spontaneous animal diseases as models for their human counterparts. *Mailing Add:* Dept Safety Assessment Merck Sharp & Dohme Res Labs West Point PA 19486

VAN ZYTVELD, JOHN BOS, b Hammond, Ind, Nov 12, 40; m 61; c 3. SOLID STATE PHYSICS. *Educ:* Calvin Col, AB, 62; Mich State Univ, MS, 64, PhD(physics), 67. *Prof Exp:* Fel physics, Univ Sheffield, 67-68; from asst prof to assoc prof, 68-76, PROF PHYSICS, CALVIN COL, 76-, DEPT CHMN, 85- *Concurrent Pos:* Res physicist, Battelle Mem Inst, Ohio, 69; sr fel, dept physics, Univ Leicester, Eng, 74-75; Fulbright-Hays sr lectr physics, Yarmouk Univ, Irbid, Jordan, 80-81; assoc prog dir, Solid State Phys Prog, Div Mat Res, NSF, Washington, DC, 83-84, prog dir, 84-85; mem phys coun, Undergrad Res, 85- *Mem:* AAAS; Am Phys Soc; Am Asn Physics Teachers; fel Am Sci Affiliation. *Res:* Electron transport properties of solid and liquid metals, alloys and semiconductors. *Mailing Add:* Dept Physics Calvin Col Grand Rapids MI 49546

VARADAN, VASUNDARA VENKATRAMAN, b Guntur, India, June 10, 48; m 73; c 3. PHYSICS. *Educ:* Kerala Univ, India, BSc, 67, MSc, 69; Univ Ill, Chicago Circle, PhD(physics), 74. *Prof Exp:* Res assoc mech, Cornell Univ, 74-77; asst prof, 77-81, assoc prof mech, Ohio State Univ, 81-85; PROF, DEPT ENG SCI & MECH, PA STATE UNIV, 85- *Concurrent Pos:* Prin investr, Rockwell Int, 77-81; co-prin investr, Off Naval Res, 78-, & Res Ctr Eng Electronic & Acoust Mat; Nat Oceanic & Atmospheric Asn grant, 78-80, Naval Res Lab, 79 & Ames Lab, Army Res Off & Naval Coastal Systs Ctr, 81-; ed, J Wave-Mat Interaction; chmn bd, HVS Technols Inc, 88-; distinguished prof eng sci & mech & elec eng, Pa State Univ. *Mem:* Fel Acoust Soc Am; Soc Eng Sci; Inst Elec & Electronics Engrs. *Res:* Wave-material interaction; composite materials; radar materials, chiral materials, active control, saw sensors. *Mailing Add:* Dept Eng Sci & Mech Pa State Univ Hammond Bldg University Park PA 16802

VARADAN, VIJAY K, b Madurai, India, Feb 23, 43; US citizen; m 73; c 3. ELECTROMAGNETIC ABSORBING, OPTICAL COATINGS & SENSORS. *Educ:* Univ Madras, BE, 64; Pa State Univ, MS, 69; Northwestern Univ, PhD(eng mech), 74. *Prof Exp:* Asst prof eng, Cornell Univ, 75-77; from asst prof to assoc prof eng, Ohio State Univ, 77-83, dir, Wave Propagation Lab, 79-83; PROF ENG SCI, PA STATE UNIV, 83-, PROF ENG SCI & ELEC ENG, & DIR, CTR ENG ELECTRONIC & ACOUST MAT, 86-, ALUMNI DISTINGUISHED PROF ENG, 88- *Concurrent Pos:* Mem adv bd mat, Serial Digit Input/Output Civil Appln, 86- *Mem:* Am Soc Mech Engrs; fel Acoust Soc Am; Inst Elec & Electronics Engrs; Am Defense Preparedness Asn; Am Ceramics Soc; Mat Res Soc. *Res:* All aspects of wave-material interaction, sonar, radar, microwave and optically absorbing composites and piezoelectric, chiral, ferrite and polymer composites and conducting polymers; involved in the design and development of various electronic, acoustic and structural composites and devices including sensors, transducers, acoustic and ultrasonic wave absorbers and filters; also interested in microwave and ultrasonic experiments to measure the dielectric, magnetic, mechanical and optical properties of composites; microwave welding of ceramics, polymers and composites, electromagnetic interference/radio frequency interference materials, coatings, gaskets, sealants; magnetic shileding for power lines, computers; electromagnetic interference conformal coatings for circuit boards. *Mailing Add:* Ctr Eng Electronic & Acoust Mat Pa State Univ 149 Hammond Bldg University Park PA 16802

VARADARAJAN, KALATHOOR, b Bezwada, India, Apr 13, 35; m 61; c 2. MATHEMATICS, TOPOLOGY. *Educ:* Loyola Col, Madras, India, BA, 55; Columbia Univ, PhD(topology), 60. *Prof Exp:* Res fel math, Tata Inst Fundamental Res, India, 60-61, fel, 61-67; vis assoc prof, Univ Ill, Urbana, 67-69; reader, Tata Inst Fundamental Res, India, 69-71; vis prof, Ramanujan Inst, Madras, 71; assoc prof, 71-73, PROF MATH, UNIV CALGARY, 73- *Mem:* Am Math Soc; Can Math Cong. *Res:* Algebraic and differential topology; homological algebra. *Mailing Add:* Dept Math Statist & Comput Sci Univ Calgary 2920 24th Ave NW Calgary AB T2N 1N4 Can

VARADY, JOHN CARL, b Niagara Falls, NY, Feb 26, 35; m 66. BIOSTATISTICS. *Educ:* Calif Inst Technol, BS, 56; Univ Wash, MA, 58; Univ Calif, Los Angeles, PhD(biostatist), 65. *Prof Exp:* Opers res analyst, Radioplane Div, Northrop Corp, 57-58; sr mathematician, Systs Develop Corp, 58-65; chief biostatist, Calif Dept Ment Hyg, 65-66; dir comput servs, Univ Cincinnati, 66-70; DIR BIOSTATIST, SYNTEX LABS, 70- *Mem:* Am Statist Asn; Asn Comput Mach. *Mailing Add:* Biostatist Syntex Lab 3401 Hillview Palo Alto CA 94303

VARAIYA, PRAVIN PRATAP, b Bombay, India, Oct 29, 40; m 63. ELECTRICAL ENGINEERING, ECONOMICS. *Educ:* Univ Bombay, India, BS, 60; Univ Calif, Berkeley, MS, 62, PhD(elec eng), 66. *Prof Exp:* Mem tech staff commun, Bell Tel Labs, 62-63; from asst prof to assoc prof, 66-70, PROF ELEC ENG, UNIV CALIF, BERKELEY, 70-, PROF ECON, 77- *Concurrent Pos:* Fel Guggenheim Found, 71-72; vis prof elec eng, Mass Inst Technol, 74-75; res prof, Miller Found, 78-79. *Mem:* Inst Elec & Electronics Engrs. *Res:* System theory; urban economics; computer communications. *Mailing Add:* Dept Elec Eng & Econ Univ Calif 2120 Oxford St Berkeley CA 94720

VARAN, CYRUS O, b Hamadan, Iran, Mar 26, 34; m; c 1. CIVIL ENGINEERING. *Educ:* SDak State Univ, BS, 58; Univ Kans, MS, 60; Univ Del, PhD(appl sci), 64. *Prof Exp:* Struct engr, Howard, Needles, Tammen & Bergendoff, Consult Engrs, 58-60; asst prof civil eng, Univ NMex, 64-71, assoc prof, 71-80. *Concurrent Pos:* Res assoc, Univ Del Res Found res grant, 63-64; prin investr, grants, Univ NMex, 65, NSF, 65-67 & Sandia Labs, 70-71 & 71-72; investr, US CEngr grant, 70. *Mem:* Am Soc Civil Engrs; Am Soc Eng Educ. *Res:* Dynamics of structures; discrete and macro mechanics; design of guyed towers and articulated lattice shell structures; dynamics of grid-stiffened and ribbed plates; seismic design of building structures. *Mailing Add:* Dept Civil Eng Univ NMex Albuquerque NM 87131

VARANASI, PRASAD, b Vijayavada, India, Dec 20, 38; m 72. PLANETARY ATMOSPHERES, SPECTROSCOPY. *Educ:* Andhra Univ, India, BSc Hons, 57; Indian Inst Sci, Bangalore, MSc, 61; Mass Inst Technol, SM, 62; Univ Calif, San Diego, PhD(eng physics), 67. *Prof Exp:* From asst prof to assoc prof eng physics, 67-81, PROF ATMOSPHERIC SCI, STATE UNIV NY, STONY BROOK, 81- *Concurrent Pos:* NASA grants; assoc ed, J Quant Spectros & Radiative Transfer, 73; mem, Atmospheric Spectras Applications Working Group, 88- *Mem:* Am Geophys Union; Optical Soc Am; Am Astron Soc. *Res:* Infrared spectroscopy as applied to planetary atmospheres; experimental work on collision broadening of spectral lines and molecular structure; global warming; remote sensing of atmosphere; atmospheric radiation measurements. *Mailing Add:* Inst Terrestrial & Planetary Atmospheres State Univ NY Stony Brook NY 11794-2300

VARANASI, SURYANARAYANA RAO, b Narsapur, India, Oct 4, 39; US citizen; m 65. ENGINEERING MECHANICS. *Educ:* Andhra Univ, BE, 60; Calif Inst Technol, MS, 61; Univ Wash, PhD(aerospace eng), 68. *Prof Exp:* Res engr appl mech, Gas Turbine Div, Boeing Co, 65-66; res engr, Dept Comput, 66-69, SPECIALIST ENGR FATIGUE & FRACTURE RES, STRUCT TECHNOL, BOEING COM AIRPLANE CO, 69- *Concurrent Pos:* Consult, Math Sci Corp, 63. *Mem:* Am Soc Testing & Mat. *Res:* Fracture mechanics; fatigue; structural integrity; damage tolerance; stress analysis; finite element methods; plasticity; viscoelasticity; wave propagation; applied mathematics; computer applications; working with civilian military aircraft air worthiness. *Mailing Add:* Struct Technol Boeing Co Seattle WA 98124

VARANASI, USHA, b Bassien, Burma. ENVIRONMENTAL BIOCHEMISTRY, AQUATIC TOXICOLOGY. *Educ:* Univ Bombay, BSc, 61; Calif Inst Technol, MS, 64; Univ Wash, PhD(chem), 68. *Prof Exp:* Res assoc lipid biochem, Oceanic Inst, Oahu, Hawaii, 69-71; assoc res prof, 71-75; supvr res chemist & task mgr, Northwest Fisheries Ctr, 75-87, DIR, ENVIRON CONSERV DIV, NAT MARINE FISHERIES SERV, NAT OCEANIC & ATMOSPHERIC ADMIN, SEATTLE, WASH, 87-; RES PROF CHEM, SEATTLE UNIV, 75- *Concurrent Pos:* Vis scientist, Pioneer Res Unit, Northwest Fisheries Ctr, Nat Marine Fisheries Serv, Nat Oceanic & Atmospheric Admin, Wash, 69-72, res chemist, 75-80; from affil assoc prof to affil prof chem, Univ Wash, 80-88. *Mem:* Am Soc Biol Chemists; Am Chem Soc; Am Asn Cancer Res; Am Soc Toxicol & Exp Therapeut; Am Women Sci. *Res:* Biochemical effects of xenobiotics in fish; interactions of carcinogens with cellular macromolecules such as DNA and protein; lipid structure and metabolism; chemistry of bioacoustics; reaction kinetics in binary solvent systems; adaptive mechanisms in marine organisms; environmental conservation research in biochemistry. *Mailing Add:* 2725 Montlake Blvd E Nat Oceanic & Atmospheric Admin Seattle WA 98112

VARANI, JAMES, METASTASIS, CELL BIOLOGY. *Educ:* Univ NDak, PhD(microbiol), 74. *Prof Exp:* ASSOC PROF IMMUNOL, MED SCH, UNIV MICH, 80- *Res:* Cancer. *Mailing Add:* Dept Path Med Sch Univ Mich Ann Arbor MI 48109

VARBERG, DALE ELTHON, b Forest City, Iowa, Sept 9, 30; m 55; c 3. MATHEMATICS. *Educ:* Univ Minn, BA, 54, MA, 57, PhD(math), 59. *Prof Exp:* From asst prof to assoc prof, 58-65, PROF MATH, HAMLINE UNIV, 65- *Concurrent Pos:* NSF fel, Inst Advan Study, 64-65; sci fac fel, Univ Wash, 71-72. *Mem:* Am Math Soc; Math Asn Am. *Res:* Stochastic and Gaussian processes; measure theory; convexity theory. *Mailing Add:* 1363 W Roselawn St Paul MN 55113

VARCO-SHEA, THERESA CAMILLE, b Buffalo, NY, Apr 23, 59; m 82; c 1. ELECTROCHEMICAL SENSOR DESIGN. *Educ:* St John Fisher Col, BS, 81; Univ Tex, Austin, PhD(anal chem), 87. *Prof Exp:* Consult, Hydrolab Corp, 86-87; electrochem design engr, Div High Voltage Eng, 87-88, TECH DIR, ANACON CORP, 88- *Mem:* Am Chem Soc. *Res:* Design of process control instrumentation for a wide range of industries; electrochemical sensor design. *Mailing Add:* 15 Cross St Westborough MA 01581

VARDANIS, ALEXANDER, b Athens, Greece, Mar 13, 33; Can citizen; m 59; c 3. BIOCHEMISTRY. *Educ:* Univ Leeds, BSc, 55; McGill Univ, MSc, 58, PhD(biochem), 60. *Prof Exp:* RES OFF BIOCHEM, RES INST, CAN DEPT AGR, UNIV WESTERN ONT, 61- *Concurrent Pos:* Nat Res Coun Can fel, 59-61. *Mem:* Chem Inst Can; Am Soc Biochem & Molecular Biol. *Res:* Intermediary metabolism of carbohydrates, particularly glycogen metabolism; chitin biosynthesis; protein kinases. *Mailing Add:* Res Ctr Can Dept Agr 1400 Western Rd London ON N6G 2V4 Can

VARDARIS, RICHARD MILES, b Lakewood, Ohio, Nov 28, 34; m 70; c 2. NEUROENDOCRINOLOGY, NEUROPSYCHOPHARMACOLOGY. *Educ:* Case Western Reserve Univ, BA, 62; Univ Ore, MS, 67, PhD(med psychol), 68. *Prof Exp:* From asst prof to assoc prof, 67-77, PROF PSYCHOL, KENT STATE UNIV, 78- *Concurrent Pos:* NIH res grants, 72-82; chmn, Biopsychol Prog, Kent State Univ, 76-, dir, Div Biomed Sci, 76-79; res prof neurobiol, Northeastern Col Med, Ohio Univ, 77-; co-prin investr, NSF res grant, 78-80. *Mem:* Soc Neurosci; NY Acad Sci; AAAS; Sigma Xi. *Res:* Effects of steroid hormones and related compounds on excitability of brain tissue; behavioral effects of gonadal steroids and drugs of abuse; neurobiological analysis of linguistic phenomena. *Mailing Add:* 3175 Bird Dr Ravena OH 44266

VARDEMAN, STEPHEN BRUCE, b Louisville, Ky, Aug 27, 49; m 70; c 2. ENGINEERING STATISTICS, STATISTICAL QUALITY CONTROL. *Educ:* Iowa State Univ, BS, 71, MS, 73; Mich State Univ, PhD(statist), 75. *Prof Exp:* Asst prof, Purdue Univ, West Lafayette, 75-81; from asst prof to assoc prof statist, 81-86, assoc prof indust eng, 83-86, PROF STATIST, IOWA STATE UNIV, AMES, 86-, PROF INDUST ENG, 86- *Concurrent Pos:* Assoc ed, Am Statistician, 84-87, Technometrics, 86-; chair, Sect Phys & Eng Sci, Am Statist Asn, 91. *Mem:* Fel Am Statist Asn; Am Soc Qual Control; Inst Math Statist. *Res:* Statistical quality control; general engineering statistics; applied stochastic control. *Mailing Add:* 303 Snedecor Hall Dept Statist Iowa State Univ Ames IA 50011-1210

VARDI, JOSEPH, US citizen. CHEMICAL ENGINEERING. *Educ:* Univ Cincinnati, PhD(chem eng), 64. *Prof Exp:* Engr, 64-67, SR RES ENGR, EXXON RES & ENG CO, 67- *Mem:* Am Inst Chem Engrs; Am Chem Soc; Sigma Xi. *Res:* Air pollution and fuels research; impact of fuels and their combustion products on the environment; application of catalysis and adsorption to new processes to control automotive air pollution; conduction and radiation heat transfer. *Mailing Add:* 85 Orchard Rd Demarest NJ 07627-1717

VARDI, YEHUDA, US citizen. INFERENCE & ESTIMATION FROM BIASED & INCOMPLETE DATA, STATISTICAL METHODS IN TOMOGRAPHY & IMAGE ANALYSIS. *Educ:* Hebrew Univ, BSc, 70; Technion Univ, MSc, 73; Cornell Univ, MSc, 75, PhD(opers res), 77. *Prof Exp:* Scientist math & statist, AT&T Bell Labs, 77-87; PROF STATIST, RUTGERS UNIV, 87- *Mem:* Am Statist Asn; Inst Math Statist; Int Statist Inst. *Res:* Developing statistical methods for real life application, including image reconstruction from projections, inference from sample-selection-biased data. *Mailing Add:* Statist Dept Rutgers Univ Hill Ctr Busch Campus New Brunswick NJ 08903

VARDIMAN, RONALD G, b Louisville, KY, Sept 17, 32; m 69; c 2. MATERIALS CHARACTERIZATION, MECHANICAL PROPERTIES. *Educ:* Univ Notre Dame, BS, 54, MS, 56, PhD(metall), 61. *Prof Exp:* METALLURGIST, MAT SCI & TECH, US NAVAL RES LAB, WASHINGTON, DC, 61- *Mem:* Sigma Xi; Minerals, Metals & Mat Soc; Am Soc Metals. *Res:* Materials characterization and effects of defect structure on properties, including work on ion implantation, superconducting materials, diffusion, dislocation observation, sintering and crystal growth. *Mailing Add:* US Naval Res Lab Code 6320 Washington DC 20375

VARGA, CHARLES E, physical organic chemistry, for more information see previous edition

VARGA, GABRIELLA ANNE, b Budapest, Hungary, Aug 8, 51; US citizen; m 77. ANIMAL NUTRITION, NUTRITION. *Educ:* Duquesne Univ, BS, 73; Univ RI, MS, 75; Univ Md, PhD(animal sci), 78. *Prof Exp:* Res asst ruminant nutrit, Univ RI, 73-75; res asst, Univ Md, 75-78; FEL RUMINANT NUTRIT, WVA UNIV, 79-; DEPT DAIRY & ANIMAL SCI, PA STATE UNIV. *Concurrent Pos:* Res animal scientist, Ruminant Nutrit Lab USDA, Beltsville, MD, 82-85. *Mem:* Am Soc Animal Sci; AAAS; Appl Environ Micros; Am Soc Dairy Soc; Sigma Xi. *Res:* Specializing in utilization of non-protein nitrogen compounds by ruminants; feeding of animal waste to ruminants; utilization and microbiological aspects are also of interest; effect of feed intake in dairy cows in early lactation; utilization of dietary fiber by dairy cows; digesta kinetics; nutrient absorption; amino acid metabolism; forage and fiber utilization by ruminants. *Mailing Add:* Dept Dairy & Animal Sci Pa State Univ 225B Borland Lab University Park PA 16802

VARGA, GIDEON MICHAEL, JR, b Brooklyn, NY, Feb 13, 41. INORGANIC CHEMISTRY. *Educ:* Manhattan Col, BS, 62; Georgetown Univ, PhD(inorg chem), 67; New York Univ, MBA, 74. *Prof Exp:* STAFF SCIENTIST, EXXON RES & ENG CO, 67- *Mem:* Am Chem Soc; Sigma Xi. *Res:* Properties and production of aviation and distillate fuels; characterization of inorganic and organic compounds; nox emission control; pollution monitoring instrumentation; energy conservation; combustion control; electrochemistry; heteropoly electrolytes and blues; polarography; heterogeneous catalysis; fuel cells. *Mailing Add:* 1196 Lake Ave Apt 20 Clark NJ 07066

VARGA, JANOS M, b Nagyoroszi, Hungary, June 19, 35; US citizen; c 3. RADIO-DERIVATIZATION OF POLYMERS, LIGAND-BINDING ASSAYS. *Educ:* Univ Technol, Budapest, Hungary, BS, 59; Eotvos L Univ, Budapest, Hungary, PhD(biochem), 65. *Prof Exp:* Fel biochem, Royal Inst Technol, Stockholm, 67-69; res assoc molecular biol, Dept Molecular Biol & Biochem, Yale Univ, 69-71, res assoc immunochem, Sch Med, 71-74, asst prof, 74-76, assoc prof melanoma, 76-81, prof & sr scientist, 81-84; prog dir molecular immunol, Nat Cancer Inst, NIH, 84-87; PROF IMMUNOCHEM, UNIV INNSBRUCK, AUSTRIA, 88- *Concurrent Pos:* Res career develop award, NIH, 86; consult, Epipharm Co, 88- *Mem:* Am Asn Immunologists; Austrian Biochem Soc. *Res:* Isolation of new antibiotics; in vitro diagnosis for allergies; hormone receptors on cancer cells; drug targeting; multispecificity of antibodies; multispecific allergic reactions; solid-phase assays. *Mailing Add:* Dept Dermat Univ Innsbruck Anichstr 35 Innsbruck 6020 Austria

VARGA, LOUIS P, b Portland, Ore, Mar 25, 22; m 48; c 4. ANALYTICAL CHEMISTRY, RADIOCHEMISTRY. *Educ:* Reed Col, BA, 48; Univ Chicago, MS, 50; Ore State Univ, PhD(anal chem), 60. *Prof Exp:* Chemist, Hanford Labs, 50-53; instr & res assoc, Reed Col, 53-57; res assoc anal chem, Mass Inst Technol, 60-61; asst prof chem, 61-67, ASSOC PROF CHEM, OKLA STATE UNIV, 67- *Concurrent Pos:* Vis staff mem, Los Alamos Sci Lab, 68-78. *Mem:* Am Chem Soc; Sigma Xi. *Res:* Analytical instrumentation; water analysis; rare earth spectra. *Mailing Add:* Dept Chem Okla State Univ Stillwater OK 74078

VARGA, RICHARD S, b US, Oct 9, 28; m 51; c 1. MATHEMATICS. *Educ:* Case Inst Technol, BS, 50; Harvard Univ, AM, 51, PhD(math), 54. *Prof Exp:* Adv mathematician, Bettis Atomic Power Lab, Westinghouse Elec Co, 54-60; prof math, Case Western Reserve Univ, 60-69; UNIV PROF MATH, KENT STATE UNIV, 69- *Concurrent Pos:* Consult, Gulf Res & Develop Co, 60, Argonne Nat Lab, 61 & Los Alamos Sci Lab, 68-; Guggenheim fel, Harvard Univ & Calif Inst Technol, 63; Sherman Fairchild scholar, Calif Inst Technol, 74; Gast prof, Munich Tech Univ, 76. *Honors & Awards:* von Humboldt prize. *Mem:* Am Math Soc; Soc Indust & Appl Math. *Res:* Numerical analysis; approximation theory. *Mailing Add:* Dept Math Kent State Univ Kent OH 44242

VARGAS, FERNANDO FIGUEROA, b Puerto Montt, Chile, Aug 9, 26; US citizen; m 51; c 3. STUDY OF OSMOTIC PHENOMENA. *Educ:* Univ Chile, BSc, 45, DDS(oral surg), 51; Univ Minn, PhD, 63. *Prof Exp:* Asst prof physiol, Inst Physiol, Univ Concepcioń, 52-55; assoc prof, Inst Physiol, Univ Chile, 58-65, prof, Sch Sci, 65-70; lectr physiol, Univ Minn, 74-80; PROF PHYSIOL, UNIV PR, 80- *Concurrent Pos:* Vis prof, dept physiol, Univ Calif, Los Angeles, 67; mem bd, Univ Chile, 70-72; consult, Medtronic Inc, 79-80; chmn, dept biol, Sch Sci, Univ Chile, 67-70, Atomic Energy Comn, Chile, 72-73, Gordon Conf, Microcirculation, 85; prin investr & dir, Univ PR, Health Sci Campus, NIH, Mem Prog, 85- *Honors & Awards:* Winthrop Award, Chilean Biol Soc, 58. *Mem:* Am Physiol Soc; NY Acad Sci; Microcirculation Soc; AAAS. *Res:* Permeability of vascular endothelium to water and non electrolytes; endothelial cell electrophysiology and surface phenomena and their relationship with permeability of endothelium; mechanisms of plasma protein effects on hydraulic and electrical conductivity of large vessel endothelium. *Mailing Add:* Dept Physiol Med Sci Campus Univ PR GPO Box 5067 San Juan PR 00936

VARGAS, JOSEPH MARTIN, JR, b Fall River, Mass, Mar 11, 42; m 63; c 2. PLANT PATHOLOGY. *Educ:* Univ RI, BS, 63; Okla State Univ, MS, 65; Univ Minn, Minneapolis, PhD(plant path), 68. *Prof Exp:* from asst prof to assoc prof, 68-82, PROF BOT & PLANT PATH, MICH STATE UNIV, 82- *Mem:* Am Phytopath Soc; Am Soc Agron; Int Turfgrass Soc. *Res:* Turfgrass pathology; resistance to fungicides; biological control. *Mailing Add:* Dept Bot Mich State Univ East Lansing MI 48823

VARGAS, ROGER I, b Long Beach, Calif. ENTOMOLOGY. *Educ:* Univ Calif, Riverside, BA, 69; San Diego State Univ, MS, 74; Univ Hawaii, PhD(entom), 79. *Prof Exp:* Res assoc entom, Univ Hawaii, 79-80; RES SCIENTIST, TROP FRUIT & VEG RES LAB, HONOLULU, 80- *Concurrent Pos:* Affil fac, Univ Hawaii, 82-91. *Mem:* AAAS; Entom Soc Am. *Res:* Ecology; pest management, mass rearing and eradication of tropical fruit flies. *Mailing Add:* Trop Fruit & Veg Res Lab PO Box 2280 Honolulu HI 96804

VARGHESE, SANKOORIKAL LONAPPAN, b Narakal, Kerala, Mar 13, 43; US citizen; m 71; c 4. EXPERIMENTAL ATOMIC PHYSICS. *Educ:* Kerala Univ, BSc, 63, MSc, 65; Univ Louisville, MS, 67; Yale Univ, PhD(physics), 74. *Prof Exp:* Res asst physics, Yale Univ, 68-74, res staff physicist, 74; res assoc physics, Kans State Univ, 74-76; vis asst prof, Univ Okla, 76-77, E Carolina, 77-80; from asst prof to assoc prof, 80-85, PROF PHYSICS, UNIV S ALA, 85- *Concurrent Pos:* Consult, Burroughs-Wellcome Co, 81-85; vis researcher, Oak Ridge Nat Lab, 81-; US patent, high-pressure chromatographic system. *Mem:* Am Phys Soc; Sigma Xi. *Res:* Positron and positronium research, first observation of the n-2 state of positronium; accelerator based atomic physics, first direct lifetime measurement of x-ray emitters in the pico-second range; Mo06ssbauer studies. *Mailing Add:* Dept Physics Univ S Ala Mobile AL 36688

VARGO, STEVEN WILLIAM, b Whiting, Ind, Sept 9, 31; m 56; c 3. AUDIOLOGY. *Educ:* Ind State Univ, BS, 54; Purdue Univ, MS, 57; Ind Univ, PhD(audiol), 65. *Prof Exp:* Assoc prof audiol, Ill State Univ, 65-71; Nat Inst Neurol Dis & Stroke spec res fel, Auditory Res Lab, Northwestern Univ, Evanston, 71-73; assoc prof surg, Hershey Med Ctr, Pa State Univ, 73-79; PVT PRACT AUDIOL, 79- *Concurrent Pos:* Mem, State Bd Examr Audiol & Speech Path Pa. *Mem:* Am Speech & Hearing Asn; Acoust Soc Am. *Res:* Scientific study of communication behavior with primary emphasis on the auditory mechanism of both normal and abnormal systems. *Mailing Add:* 431 E Chocolate Ave Hershey PA 17033

VARIN, ROBERT ANDRZEJ, b Piastow, Poland, May 1, 46; Can citizen; m 67; c 2. MATERIALS SCIENCE & ENGINEERING. *Educ:* Warsaw Tech Univ, MASc, 72 & PhD(mats sci & eng), 76. *Prof Exp:* Asst prof, Warsaw Tech Univ, 76-78 & 80-82; fel, Univ Man, 78-80, res assoc, 82-83; asst prof, 83-88, ASSOC PROF MAT SCI & ENG, UNIV WATERLOO, 88- *Mem:* Metall Soc; Am Soc Metals Int; NY Acad Sci; Mat Res Soc; Can Asn Composite Struct & Mat. *Res:* Mechanical properties of metallic polycrystalline materials and advanced composites with special emphasis on the role of grain boundaries and interfaces in their behavior; intermetallics and their composites. *Mailing Add:* Dept Mech Eng Univ Waterloo 200 University Ave W Waterloo ON N2L 3G1 Can

VARIN, ROGER ROBERT, b Bern, Switz, Feb 15, 25; nat US; m 51; c 3. PHYSICAL CHEMISTRY. *Educ:* Univ Bern, PhD(chem), 51. *Prof Exp:* Fel phys chem, Harvard Univ, 51-52; res chemist, E I du Pont de Nemours & Co, 52-57, res assoc, 57-62; dir res, Riegel Textile Corp, 62-71; PRES, VARINIT CORP, 71- *Concurrent Pos:* Pres, Technol Assocs, Greenville, SC, 71- & Varinit S A, Carouge, Switz, 74- *Mem:* AAAS; Am Chem Soc; Fiber Soc; Soc Advan Mat & Process Eng; Soc Mfg Engrs. *Res:* Fiber physics and chemistry; textile technology; polymer physics and chemistry; rheology. *Mailing Add:* Four Barksdale Rd Greenville SC 29607

VARINEAU, VERNE JOHN, b Escanaba, Mich, Mar 11, 15; m 45; c 4. MATHEMATICS. *Educ:* Col St Thomas, BS, 36; Univ Wis, AM, 38, PhD(math), 40. *Prof Exp:* Asst, Univ Wis, 36-39; from instr to prof, 40-85, EMER PROF MATH, UNIV WYO, 85- *Concurrent Pos:* NSF sci fac fel, Stanford Univ, 63-64. *Mem:* Am Math Soc; Math Asn Am. *Res:* Matrices with elements in a principal ideal ring. *Mailing Add:* 714 Tenth St Laramie WY 82070

VARKEY, THANKAMMA EAPEN, b Palai, India, Oct 31, 36; m 64; c 1. ORGANIC CHEMISTRY. *Educ:* Kerala Univ, BSc, 56, MSc, 57; Temple Univ, PhD(chem), 74. *Prof Exp:* Lectr chem, Kerala Univ, 57-64; prof, Alphonsa Col, India, 64-68; fel, Drexel Univ, 74-76 & Tex Christian Univ, 76-77; fel chem, Lamar Univ, 77-79; fel chem, Lehigh Univ, 79-80; ASST PROF, WILSON COL, CHAMBERSBURG, PA, 80- *Mem:* Am Chem Soc; Sigma Xi. *Res:* Synthesis and conformational analysis of steroids; synthesis, structure and stereochemistry of nitrogen-sulfur ylides (iminosulfuranes). *Mailing Add:* Box 65 ESU Stroudsburg PA 18301-2999

VARKI, AJIT POTHAN, b Jan 4, 52; US citizen; c 1. CANCER BIOLOGY, GLYCOPROTEIN BIOCHEMISTRY. *Educ:* Christian Med Col, India, MB, BS, 75. *Prof Exp:* Resident med officer, Malankara Mission Hosp, 75; res asst biochem, Univ Nebr, 76, resident, 77-78; resident med, Episcopal Hosp, Temple Univ, 76-77; fel hemat & oncol, Sch Med, Wash Univ, 78-82, instr med, 80-82; asst prof med, 82-86, ASSOC PROF & CO-HEAD, DIV HEMAT-ONCOL, UNIV CALIF, SAN DIEGO, 87- *Concurrent Pos:* Prin investr, Nat Inst Gen Med Sci, 83-, Nat Cancer Inst, 85-; dir, training prog cancer res, Nat Cancer Inst, 85- *Mem:* Am Soc Biol Chem & Molecular Biol; Am Fedn Clin Res; Am Col Physicians; Am Soc Clin Invest. *Res:* Oligosaccharide units of glycoproteins and glycolipids and their biological roles. *Mailing Add:* Cancer Ctr V-111E Univ Calif La Jolla CA 92093

VARLASHKIN, PAUL, b San Antonio, Tex, Aug 28, 31; m 54; c 4. SOLID STATE PHYSICS. *Educ:* Univ Tex, BS, 52, MA, 54, PhD(physics), 63. *Prof Exp:* Asst, Defense Res Lab, Univ Tex, 51-52, physicist, 52-53; from res scientist to chief res & develop, Electro-Mech Co, 53-63; fel & res assoc, Univ NC, 64-66; asst prof physics, La State Univ, Baton Rouge, 66-72; ASSOC PROF PHYSICS, ECAROLINA UNIV, 72- *Concurrent Pos:* Res physicist, White Sands Proving Grounds, 52. *Mem:* Am Phys Soc; Sigma Xi. *Res:* Positron annihilation; liquid metals; positronium formation; solid state physics; chemical physics; metal-ammonia solutions. *Mailing Add:* Dept Physics ECarolina Univ Greenville NC 27858-4353

VARMA, ANDRE A O, b Paramaribo, Surinam, June 10, 27; nat US; m 52; c 3. MEDICINE, BIOSTATISTICS. *Educ:* Sch Med Surinam, Med Doct, 50; Columbia Univ, MSc, 60. *Prof Exp:* Dist health officer med, Govt Surinam, 58-60, head biostatist pub health, 60-67; from asst prof to assoc prof biostatist, Pub Health, Columbia Univ, 67-74; assoc prof, 74-81, PROF COMMUNITY & PREV MED, STATE UNIV NY, STONY BROOK, 81- *Concurrent Pos:* Consult, Neth Govt, 65-66, Pan Am Health Orgn, 66 & WHO, 78; fel, Surinam Govt, 62-64 & US & Surinam Govts, 68-70; chmn, Dept Community Med, State Univ NY, Stony Brook, 78- *Mem:* Am Statist Asn; Biomet Soc; Am Pub Health Asn; NY Acad Sci; Asn Teachers Prev Med. *Res:* Design and analysis of clinical trials; epidemiology of cancer; emergency medical services; complications in post-abortion pregnancies; public health. *Mailing Add:* Dept Prev Med State Univ NY HSC L3-086 Stony Brook NY 11794-8036

VARMA, ARUN KUMAR, b Faizabad, India. MATHEMATICS. *Educ:* Banaras Hindu Univ, BSc, 55; Univ Lucknow, MSc, 58; Univ Alta, PhD(math), 64. *Prof Exp:* Lectr math, Univ Rajasthan, 64-66; fel, Univ Alta, 66-67; from asst prof to assoc prof, 67-77, PROF MATH, UNIV FLA, 77- *Mem:* Am Math Soc. *Res:* Interpolation theory; approximation theory; numerical analysis. *Mailing Add:* Dept Math Univ Fla Gainesville FL 32611

VARMA, ARVIND, b Ferozabad, India, Oct 13, 47; nat US; m 71; c 2. CHEMICAL REACTION ENGINEERING, KINETICS & CATALYSIS. *Educ:* Panjab Univ, India, BS, 66; Univ NB, MS, 68; Univ Minn, PhD(chem eng), 72. *Prof Exp:* Asst prof chem eng, Univ Minn, 72-73; sr res engr, Union Carbide Corp, 73-75; from asst prof to prof chem eng, 75-88, chmn dept, 83-88, ARTHUR J SCHMITT PROF CHEM ENG, UNIV NOTRE DAME, 88- *Concurrent Pos:* Vis prof, Univ Wis-Madison, 81, Chevron vis prof, Calif Inst Technol, 82, vis prof, Indian Inst Technol, Kan pur, 89 & vis chair prof, Univ Caligari, Italy, 89; co-ed, Math Understanding Chem Eng Systs, 80 & Chem Reaction & Reactor Eng, 87; Fulbright scholar award, Indo-Am fel, 88-89. *Mem:* Am Inst Chem Engrs; Am Chem Soc; Sigma Xi; Fel Am Inst Chemists; NY Acad Sci; Am Ceramic Soc. *Res:* Fundamental research in chemical and catalytic reaction engineering and synthesis of advanced materials. *Mailing Add:* Dept Chem Eng Univ Notre Dame Notre Dame IN 46556-5637

VARMA, ASHA, b Bareilly, India, Mar 19, 42; US citizen; m 67; c 2. ANALYTICAL CHEMISTRY. *Educ:* Agra Univ, India, BSc, 58, MSc, 60; Banaras Hindu Univ, India, PhD(chem), 63. *Prof Exp:* Sr res fel chem, Banaras Hindu Univ, India, 63-64, Nat Chem Lab, 64-66; asst dir res chem, Forensic Sci Lab, Sagar, India, 66-68; sci pool officer chem, H B Technol Inst, Kanpur, India, 69-70; res assoc anal chem, Inst Mat Sci, Univ Conn, Storrs, 73-75; res scientist anal chem, Lab Res Structure Matter, Univ Pa, 77-82; chemist, 82-87, DEP DIR, NAVAL AIR DEVELOP CTR, 87- *Concurrent Pos:* Res fel, Banaras Hindu Univ, Varanasi, India, 60-63; fel chem, Univ Conn, Storrs, 66-67 & 73-75; asst dir, Forensic Sci lab, Sagar, India, 67-69. *Mem:* Am Chem Soc; fel Am Inst Chemists; Int Union Pure & Appl Chem; Coblentz. *Res:* Atomic absorption; emission ultraviolet and infrared spectroscopy; electroanalytical techniques; author of 40 publications. *Mailing Add:* Off Sci & Technol Code 01B Naval Air Develop Ctr Warminster PA 18974-5000

VARMA, DAYA RAM, REPRODUCTIVE TOXICITY, ENVIRONMENTAL CHEMICALS. *Educ:* McGill Univ, PhD(pharmacol), 61. *Prof Exp:* ASSOC PROF PHARMACOL, MCGILL UNIV, 61- *Mailing Add:* Dept Pharmacol McGill Univ Sch Med 3655 Drummond St Montreal PQ H3G 1Y6 Can

VARMA, MAN MOHAN, b Patiala, India, July 5, 32; m 66. ENVIRONMENTAL ENGINEERING. *Educ:* Ala Polytech Inst, BS, 57; Iowa State Col, MS, 58; Okla State Univ, MS, 60; Univ Okla, PhD(eng sci), 63. *Prof Exp:* Instr eng sci, Univ Okla, 60-61, res assoc, 61-62, asst prof, 62-63; asst prof, Tufts Univ, 63-65; dir sanit eng, Morgeroth & Assocs, Mass, 65-66; assoc prof, 66-71, PROF BIO-ENVIRON ENG, HOWARD UNIV, 71-, DIR BIO-ENVIRON ENG & SCI, 66- *Concurrent Pos:* Consult, Morgeroth & Assocs, 66-; WHO consult, 72-; environ health consult, Ministry Pub Health, Kuwait, 75-; vis prof, Harvard Univ, 75-76. *Honors & Awards:* World Cult Prize for Lett, Arts & Sci, Calvatore, Italy. *Mem:* Water Pollution Control Fedn; Am Water Works Asn; Nat Environ Health Asn; NY Acad Sci; Am Soc Testing & Mat. *Res:* Health effects caused by certain environmental contaminants; formation of trihalomethanes in water, enumeration of bacteria by ATP and Luciferase; kinetics of removal of halogenated compounds by adsorption and biokinetics. *Mailing Add:* 704 Chichester Lane Silver Spring MD 20904

VARMA, MATESH NARAYAN, b Saugor, India, Sept 9, 43; m 67; c 3. PHYSICS, RADIATION PHYSICS. *Educ:* Univ Jabalpur, India, BSc, 61, MSc, 63; Case Western Reserve Univ, MS, 69; Case Inst of Technol, PhD(physics), 71. *Prof Exp:* Sci officer reactor eng, Bhabha Atomic Res Ctr, India, 63-67; univ fel physics, Case Western Reserve Univ, 67-71, res assoc solid state physics, 71-72; sr scientist radiation physics, Safety & Environ Protection Div, Brookhaven Nat Lab, 72-87; PROG MGR, RADIOL & CHEM PHYS, DOSIMETRY RADON, OFF HEALTH ENVIRON RES, US DEPT ENERGY, 84- *Mem:* Health Physics Soc; Bevalac Users Asn; Radiation Res Soc. *Res:* Microdosimetry; surface physics; thin films; Mossbauer spectroscopy; ultrahigh vacuum technology; reactor physics; health physics; radiological and chemical physics; bio-physics and biophysical modeling. *Mailing Add:* 7220 Deer Lake Lane Derwood MD 20855

VARMA, RAJ NARAYAN, b Oct 11, 28; m; c 2. NUTRITION IN CANCER. *Educ:* Univ Calif, Davis, PhD(biochem), 62. *Prof Exp:* ASSOC PROF NUTRIT, YOUNGSTOWN STATE UNIV, 83- *Mem:* Am Dietetic Asn; Am Soc Parenteral & Enteral Nutrit; Soc Nutrit Educ. *Res:* Nutrition in sickle cell anemia; nutrition and the elderly; nutrition research. *Mailing Add:* Dept Home Econ Youngstown State Univ 410 Wick Ave Youngstown OH 44555

VARMA, RAJENDER S, b New Delhi, July 26, 51; c 2. ORGANIC CHEMISTRY. *Educ:* Punjab Univ, India, BSc, 70; Kurukshetra Univ, MSc, 72; Univ Delhi, PhD(natural prod chem), 76; Norweg Inst Technol, dipl, 78. *Prof Exp:* Asst res, Inst Cellulose Technol, Norway, 77-79; res fel, Dept Org Chem, Univ Liverpool, Eng, 79-82; res assoc chem, Univ Tenn, Knoxville, 82-86; ASST PROF, CTR BIOTECHNOL, BAYLOR COL MED, 86- *Concurrent Pos:* Norweg Agency Int Develop fel, Norweg Inst Technol, Trondheim; group leader, Houston Biotechnol, Inc, 86-90; grantee, Am Cancer Soc, 88-91. *Mem:* Am Chem Soc. *Res:* Reduction of nitroalkenes to useful synthetic precursors; chemiselective reductions; synthesis of novel heterocycles; application of organoboranes in the rapid synthesis of radiopharmaceuticals. *Mailing Add:* Ctr Biotechnol Baylor Col Med 4000 Res Forest Dr The Woodlands TX 77381

VARMA, RAVI KANNADIKOVILAKOM, b Tripunithura, India, Dec 23, 37; m 65; c 2. MEDICINAL CHEMISTRY. *Educ:* Maharaja's Col, Ernakulam, India, BSc, 57, MSc, 59; Univ Poona, PhD(org chem), 65. *Prof Exp:* Sci asst chem, Nat Chem Labs, Poona, India, 59-65; res fel org chem, Purdue Univ, Lafayette, 65-66; staff scientist bio-org chem, Worcester Found, Mass, 66-69; res fel org chem, Harvard Univ, 70-72; RES FEL, E R SQUIBB & SONS, INC, 72- *Mem:* Am Chem Soc; Royal Soc Chem; Sigma Xi. *Res:* Organic synthesis; chemistry of biologically active molecules; mechanism of action of drugs. *Mailing Add:* 7887 Tyson Oaks Circle Vienna VA 22180

VARMA, SHAMBHU D, b Ghazipur, India; US citizen; m; c 3. BIOCHEMISTRY. *Educ:* Univ Allahabad, India, BSc, 55, MSc, 57; Univ Rajasthan, PhD(biochem), 64. *Prof Exp:* Chmn biochem, chem dept, Punjab Agrc Univ, India, 65-69; vis scientist, Nat Eye Inst, NIH, Bethesda, Md, 72-76; PROF & DIR RES EYE BIOCHEM, DEPT OPHTHAL, MED SCH, UNIV MD, 76- *Concurrent Pos:* Consult, Nat Eye Inst, NIH. *Honors & Awards:* William Friedkin Res Award, 77; Alexander Von Humboldt Prize, 87. *Mem:* AAAS; Asn Res Vision & Ophthal. *Res:* Intermediary metabolism, diabetes; sorbitol pathway in lens; aldose reductase; mechanism of action of flavonoids and other drugs in lens; superoxide, its implications in ocular diseases, particularly cataracts; nutrients and ocular manifestations; transport mechanisms. *Mailing Add:* Ophthal Dept Sch Med Univ Md Ten S Pine St Baltimore MD 21201

VARMA, SURENDRA K, b Lucknow, India, Dec 10, 39; m 67; c 2. PEDIATRICS, ENDOCRINOLOGY. *Educ:* King George Med Col, MBBS, 62, MD, 68; SN Med Col, Agra, DCH, 64; Am Bd Pediat, dipl, 76, dipl & cert endocrinol, 78. *Prof Exp:* Res assoc endocrinol, Dept Nutrit & Food Sci, Mass Inst Technol, 72-74; instr pediat, Harvard Med Sch, 73-74; from asst prof to assoc prof pediat, Tex Tech Univ Health Sci Ctr, 74-83, actg assoc chmn, 77-79, assoc chmn, 79-87, interim chmn, 84-86, asst dean, Sch Med, 87-89, PROF PEDIAT, TEX TECH UNIV HEALTH SCI CTR, 83-, ASSOC CHMN PEDIAT, 91- *Concurrent Pos:* Res fel, K G Med Col, 65-66; res fel, Harvard Med Sch, Peter Bent Brigham Hosp, 68-69 & Children's Hosp Med Ctr, 69-71; clin res fel, Mass Gen Hosp, 72-73, clin assoc, 73-74; dir, Endocrine Div, Tex Tech Univ Sch Med, 78-; pres, Am Diabetes Asn, NTex Affil, Inc, 81-82, Tex Affil, 82. *Mem:* Endocrine Soc; Am Thyroid Asn; Lawson Wilkins Pediat Endocrine Soc; Am Diabetes Asn; Am Acad Pediat. *Res:* Perinatal thyroid pathophysiology; growth hormone and factors; carbohydrate metabolism. *Mailing Add:* Dept Pediat Tex Tech Univ Sch Med Lubbock TX 79430

VARMUS, HAROLD ELLIOT, b Oceanside, NY, Dec 18, 39; m 69; c 2. MOLECULAR VIROLOGY, ONCOGENESIS. *Educ:* Amherst Col, BA, 61; Harvard Univ, MA, 62; Columbia Univ, MD, 66. *Prof Exp:* Lectr, 70-72, from asst prof to prof microbiol, 72-82, PROF BIOCHEM & BIOPHYS, UNIV CALIF, SAN FRANCISCO, 82- *Concurrent Pos:* Assoc ed, Cell & Virol, 74-; hon prof molecular virol, Am Cancer Soc. *Honors & Awards:* Nobel Prize in physiol or med, 89. *Mem:* Nat Acad Sci; Inst Med-Nat Acad Sci; Am Soc Microbiol; Am Soc Virol; AAAS. *Res:* Mechanisms of viral replication and oncogenesis, using retroviruses and hepatitis B viruses. *Mailing Add:* Dept Microbiol & Immunol Box 0502 Univ Calif San Francisco CA 94143

VARNELL, THOMAS RAYMOND, b Whiteriver, Ariz, Jan 27, 31; m 57; c 3. BIOCHEMISTRY, PHYSIOLOGY. *Educ:* Univ Ariz, BS, 57, MS, 58, PhD(agr biochem), 60. *Prof Exp:* Instr animal nutrit, Univ Wyo, 60-62; res biochemist, Dept HEW & Food & Drug Admin, 62-63; asst prof animal nutrit, Univ Wyo, 63-65, from asst prof to assoc prof animal physiol, 65-73, prof animal physiol, 73-84; RETIRED. *Concurrent Pos:* Stanford Res Inst grant, 61-64. *Mem:* Animal Nutrit Res Coun. *Res:* Metabolism of vitamin A and carotene; lipid metabolism; intestinal transport and metabolism of amino acids; potassium requirements and availability. *Mailing Add:* 914 Stratton Dr Safford AZ 85546

VARNER, JOSEPH ELMER, b Nashport, Ohio, Oct 7, 21; c 4. PLANT PHYSIOLOGY. *Educ:* Ohio State Univ, BSc, 42, MSc, 43, PhD(biochem), 49. *Hon Degrees:* Dr, L'Universite De Nancy, 77. *Prof Exp:* Chemist, Owens-Corning Fiberglas Corp, 43-44; res engr, Battelle Mem Inst, 46-47; res assoc, Res Found, Ohio State Univ, 49-50, asst prof agr biochem, 50-53; res fel, Calif Inst Technol, 53-54; from assoc prof to prof biochem, Ohio State Univ, 54-61; prof, Res Inst Advan Study, 61-65 & Mich State Univ, 65-73; PROF BIOCHEM, WASH UNIV, 73- *Concurrent Pos:* NSF fel, Cambridge Univ, 59-60 & Univ Wash, 71-72. *Mem:* Nat Acad Sci; Am Soc Biol Chem; Am Soc Plant Physiol; Fel AAAS. *Res:* Plant biochemistry; biochemistry of aging cells; action mechanism of plant hormones; glycoproteins. *Mailing Add:* Dept Biol Wash Univ St Louis MO 63130

VARNER, LARRY WELDON, b San Antonio, Tex, June 25, 44; m 67; c 1. ANIMAL NUTRITION, WILDLIFE RESEARCH. *Educ:* Abilene Christian Col, BS, 66; Univ Nebr, Lincoln, MS, 68, PhD(nutrit), 70. *Prof Exp:* Asst animal sci, Univ Nebr, Lincoln, 66-69, res assoc animal nutrit, 69-70, asst prof, 70-71; res scientist, USDA, 71-74, assoc prof animal nutrit, Tex A&M Univ, 74-88; consult, natural res, 88-89; CONSULT NUTRITIONIST, PURINA MILLS, INC, 89- *Mem:* Wildlife Soc; Am Soc Animal Sci; Soc Range Mgt. *Res:* Ruminant nutrition; nitrogen and energy metabolism; wildlife nutrition. *Mailing Add:* Two Sky Vue Ave New Braunfels TX 78132-4744

VARNERIN, LAWRENCE J(OHN), b Boston, Mass, July 10, 23; m 52; c 8. SOLID STATE PHYSICS, MATERIALS SCIENCE. *Educ:* Mass Inst Technol, PhD(physics), 49. *Prof Exp:* Res engr, Electronics Div, Sylvania Elec Prod, Inc, 49-52; res physicist, Res Labs, Westinghouse Elec Corp, 52-57; mem tech staff solid state devices, AT&T Bell Labs, 57-66, head, Magnetic & Microwave Mat & Device Dept, 66-80, head, microwave device dept, 80-84, head, Heterojunction ICS & Mat Dept, 85-86; CHMN COMPUT SCI & ELEC ENG DEPT, LEHIGH UNIV, 86- *Concurrent Pos:* Assoc ed, J Magnetism & Magnetic Mat. *Mem:* Fel Am Phys Soc; Inst Elec & Electronics Engrs. *Res:* Gallium arsenide integrated circuits, microwave field effect transistors and associated materials. *Mailing Add:* Computer Sci & Elec Dept Lehigh Univ Packard Lab 19 Bethlehem PA 18015

VARNES, DAVID JOSEPH, b Howe, Ind, Apr 5, 19; m 43, 66; c 2. ENGINEERING GEOLOGY, SEISMOLOGY. *Educ:* Calif Inst Technol, BS, 40. *Prof Exp:* Lab instr geol, Northwestern Univ, 40-41; recorder & jr geologist, 41, asst geologist, 43-45, assoc geologist, 45-48, chief br eng geol, 61-64, GEOLOGIST, US GEOL SURV, 48- *Concurrent Pos:* Mem comts, Transp Res Bd, Nat Acad Sci-Nat Res Coun, 53-73; vis lectr, Chinese Univ Develop Proj, 87. *Honors & Awards:* E B Burwell Jr Award, Geol Soc Am, 70 & 76, Distinguished Prof Pract, 87; Hans Cloos Medal, Int Asn Eng Geol, 89. *Mem:* Hon mem Asn Eng Geologists; Geol Soc Am; Int Asn Eng Geol (vpres, 82-86); Geol Soc London. *Res:* Geologic studies of Lake Bonneville; landslides; mechanics of soil and rock deformation; logic of mapping; earthquake prediction. *Mailing Add:* US Geol Surv MS 966 Box 25046 Denver Fed Ctr Denver CO 80225

VARNES, MARIE ELIZABETH, b Cleveland, Ohio, Dec 20, 42; m 68; c 2. BIOCHEMISTRY, RADIATION BIOLOGY. *Educ:* Notre Dame Col, BS, 65; Ind Univ, PhD(biochem), 73. *Prof Exp:* Res asst, 76-77, res assoc, 77-79, instr, 79-81, ASST PROF, DEPT RADIOL, DIV RADIATION BIOL, CASE WESTERN RESERVE UNIV, 81- *Concurrent Pos:* Co-investr, Am Cancer Soc grant, 77-78 & Nat Cancer Inst grant, 78- *Mem:* Radiation Res Soc; Sigma Xi; Am Asn Cancer Res. *Res:* Metabolism of nitro drugs used as radiosensitizers; influence of hormones and thiols on radiation response; modification of repair of radiation damage in tumor cells. *Mailing Add:* 2549 Kingston Rd Cleveland OH 44118

VARNEY, EUGENE HARVEY, b South Egremont, Mass, Dec 25, 23; m 56; c 3. BOTANY, PHYTOPATHOLOGY. *Educ:* Univ Mass, BS, 49; Univ Wis, PhD, 53. *Prof Exp:* Plant pathologist, USDA, 53-56; asst res specialist, 56-59, from assoc prof plant path to prof plant path, 59-88, EMER PROF PLANT PATH, RUTGERS UNIV, NEW BRUNSWICK, 88- *Mem:* AAAS; Am Phytopath Soc; Mycol Soc Am; Sigma Xi. *Res:* Diseases of small fruits; plant virology; mycology. *Mailing Add:* 17 Hadler Dr Somerset NJ 08873

VARNEY, ROBERT NATHAN, b San Francisco, Calif, Nov 7, 10; m 48; c 2. MOLECULAR PHYSICS. *Educ:* Univ Calif, AB, 31, MA, 32, PhD(physics), 35. *Hon Degrees:* DSc, Univ Innsbruck, Austria, 83. *Prof Exp:* Instr physics, Univ Calif, 35-36 & NY Univ, 36-38; asst prof, Univ Wash, 38-41, from assoc prof to prof, 46-64; asst exp labs officer, USNR, 41-45; sr mem & sr consult scientist, Lockhead Palo Alto Res Lab, 64-75; NSF sr fel, US Army Ballistic Res Labs, 75-76; CONSULT, 78- *Concurrent Pos:* Vis mem tech staff, Bell Tel Labs, NJ, 51-52; mem exec comt, Gaseous Electronics Conf, 51-53, 60-62 & 65-68, secy, 67; NSF sr fel, Royal Inst Technol, Sweden, 58-59; mem, Gov Sci Adv Comt, Mo, 61-64; Fulbright lectr, Inst Atomic Physics, Innsbruck Univ, 71-72, 76-77, guest prof atomic physics, 77-78. *Honors & Awards:* Austrian Cross Honor for Sci & Art, 81. *Mem:* Fel AAAS; fel Am Phys Soc; Am Asn Physics Teachers. *Res:* Ion-molecule reactions; collisions of positive ions in gases; spark breakdown; secondary electron emission. *Mailing Add:* 4156 Maybell Way Palo Alto CA 94306

VARNEY, WILLIAM YORK, b Forest Hills, Ky, Apr 1, 17; m 40; c 2. ANIMAL HUSBANDRY. *Educ:* Univ Ky, BS, 51, MS, 52; Mich State Univ, PhD, 60. *Prof Exp:* Prin pub schs, Ky, 39-43; dist mgr, Southern States Coop, Va, 52-53; sales promoter, Swift & Co, Ill, 53-54; from instr to prof animal husb, Univ Ky, 64-77, exten prof animal sci, 77-82; RETIRED. *Mem:* Am Soc Animal Sci; Inst Food Technologists. *Res:* Carcass studies of beef, pork and lamb. *Mailing Add:* 317 Malabu Circle Lexington KY 40502

VARNHORN, MARY CATHERINE, mathematics; deceased, see previous edition for last biography

VARNUM, WILLIAM SLOAN, b St Louis, Mo, Jan 23, 41. PHYSICS. *Educ:* Wash Univ, BS, 63; Fla State Univ, PhD(physics), 67. *Prof Exp:* Instr physics, Fla State Univ, 67-68; assoc scientist missile physics, Radiation Serv Co, 68-69; guest res scientist, Max-Planck-Institut Plasmaphysik, 69-70; res fel ionospheric physics, Max Planck Institut Ionosph-ren-Physik, 70-72; asst prof physics, New Col Sarasota, 72-73; NIH fel med physics, Univ Calif, Los Angeles, 73-74; STAFF MEM PHYSICS, LOS ALAMOS SCI LAB, 74- *Concurrent Pos:* Consult, Rand Corp, 73-75. *Mem:* Am Phys Soc. *Res:* Plasma physics; laser fusion; nuclear weapons design. *Mailing Add:* Los Alamos Sci Lab MSE531 PO Box 1663 Los Alamos NM 87545

VARON, MYRON IZAK, b Chicago, Ill, Aug 20, 30; m 59; c 3. RADIOBIOLOGY, MEDICINE. *Educ:* Univ Chicago, PHB, 50; Northwestern Univ, BSM, 52, MD, 55; Univ Rochester, MS, 63, PhD(radiation biol), 65. *Prof Exp:* Intern, Cook County Hosp, Chicago, 55-56; Med Corps, USN, 56-79, med officer, USS Lenawee, 56-58 & Armed Forces Spec Weapons Proj, 58-59, asst to mgr naval reactor br, Idaho Br Off, AEC, 59-60, sr med officer & radiation safety officer, USS Long Beach, 60-62, assoc surg, Univ Rochester, 62-65, med dir radiation biol, Naval Radiol Defense Lab, 65-67, from asst dep sci dir to dir, Armed Forces Radiobiol Res Inst, 67-75, dep comndg officer, Nav Med Res & Develop command, Med Corps, USN, 75-79; vpres & sci dir, Amyotrophic Lateral Sclerosis Soc Am, 79-; RETIRED. *Concurrent Pos:* Lectr, US Naval Hosp, Oakland, Calif, 66-67. *Mem:* Health Physics Soc; Radiation Res Soc; Asn Mil Surg US; NY Acad Sci; Soc Nuclear Med; Sigma Xi. *Res:* Health cells; molecular mechanism of action of nerve growth factor; tropic factors directed to neurons and glial cell; radiation safety of nuclear reactors; labeled antibody localization for cancer therapy; nuclear weapons effects; radiation behavioral effects; radiation recovery and residual injury; experimental pathology. *Mailing Add:* 3536 Holboro Dr Los Angeles CA 90027

VARON, SILVIO SALOMONE, b Milan, Italy, July 25, 24; c 2. NEUROCHEMISTRY, NEUROBIOLOGY. *Educ:* Univ Lausanne, EngD, 45; Univ Milan, MD, 59. *Prof Exp:* Resident asst prof neurochem, Inst Psychiat, Univ Milan, 60-63; res assoc, Dept Biochem, City of Hope Med Ctr, Duarte, Calif, 61-63; res assoc neurobiol, Dept Biol, Wash Univ, 63-64, assoc prof, 64-65; vis assoc prof, Dept Genetics, Sch Med, Stanford Univ, 65-67; assoc prof, 67-72, PROF NEUROBIOL, DEPT BIOL, MED SCH, UNIV CALIF, SAN DIEGO, 72- *Concurrent Pos:* App mem, Pres Coun Spinal Cord Injury. *Mem:* Int Soc Neurochem; Am Soc Neurochem; Soc Neurosci; Am Soc Cell Biol; Int Soc Develop Neurosci. *Res:* Structure and properties of the nerve growth factor protein; dissociation fractionation and culture of cells from nervous tissues; in vitro study of neuroglial cells; molecular mechanism of action of nerve growth factor; search for trophic factors directed to neurons and glial cells; in vivo models for neural regeneration. *Mailing Add:* Dept Biol Univ Calif San Diego Med Sch Box 109 La Jolla CA 92093

VARRESE, FRANCIS RAYMOND, METALLURGICAL ENGINEERING, MATERIALS SCIENCE. *Educ:* Lehigh Univ, BS, 61. *Prof Exp:* Mat engr, Pratt & Whitney Aircraft, 61-63; metallurgist, Cabot Stellits Div, 63-65; sr metallurgist, SPS Technologies, 65-68, mgr metall & chem eng, 68-70, staff metallurgist, Tool Div, 71-73, corp mgr mat res, 73-78; plant metallurgist, Robert Wooler Co, 70-71; SR PRIN MAT ENGR, HONEYWELL, 78- *Concurrent Pos:* Mem, Adv & Tech Awareness Coun, Am Soc Metals Int, 76-79 & Handbk Comt, 77. *Mem:* Fel Am Soc Metals Int; Am Soc Testing & Mat. *Res:* Application of development of materials, processing, research, and engineering in the fastener and process control industry. *Mailing Add:* Honeywell MS 9-7 1100 Virginia Dr Ft Washington PA 19034-3260

VARRICCHIO, FREDERICK, b Brooklyn, NY, May 18, 38; m 62; c 2. BIOCHEMISTRY, DEVELOPMENTAL BIOLOGY. *Educ:* Univ Maine, Orono, BS, 60; Univ NDak, MS, 64; Univ Md, Baltimore, PhD(biochem), 66; Univ Autonoma de Ciudad Juarez, MD, 86. *Prof Exp:* Asst biochem, Univ Freiburg, WGer, 66-67; researcher molecular biol, Nat Ctr Sci Res, France, 67-69; fel internal med, Yale Univ, 69-72; asst prof, Mem Sloan-Kettering Inst Cancer Res, 72-77; prof exp oncol, Nova Univ, 77-79; prof chem & chmn dept, Nat Col Chiropractic, Lombard, Ill, 80-83; fac res assoc, Oak Ridge Nat Lab, 79-84; RESIDENT PATH, COOK COUNTY HOSP, CHICAGO, 86-89, CONSULT, 89- *Concurrent Pos:* Vis prof, Max Planck Inst Nutrit, Dortmund, Ger, 79; res assoc, Argonne Nat Lab, 80-; adj prof chem, Col DuPage, 80-84. *Mem:* Am Soc Biol Chemists; Am Chem Soc; Am Soc Microbiol; Am Asn Pathologists; Ger Soc Biol Chemists; Sigma Xi. *Res:* Nuclear proteins; transfer ribonucleic in growth, differentiation and tumors. *Mailing Add:* 26 W 285 Blackhawk Dr Wheaton IL 60187

VARSA, EDWARD CHARLES, b Marissa, Ill, Oct 18, 38; m 65; c 2. SOIL FERTILITY. *Educ:* Southern Ill Univ, Carbondale, BS, 61; Univ Ill, Urbana, MS, 65; Mich State Univ, PhD(soil sci), 70. *Prof Exp:* Asst agron, Univ Ill, 64-65; asst soil sci, Mich State Univ, 65-66, instr, 66-68, teaching asst, 68-70; asst prof, 70-84, ASSOC PROF SOILS, SOUTHERN ILL UNIV, CARBONDALE, 84- *Mem:* Am Soc Agron; Soil Sci Soc Am; Sigma Xi. *Res:* Soil fertility research on, and the fate of, applied fertilizer nitrogen in soils. *Mailing Add:* Dept Plant & Soil Sci Southern Ill Univ Carbondale IL 62901

VARSAMIS, IOANNIS, b Alexandria, UAR, July 24, 32; Can citizen; m 58. PSYCHIATRY. *Educ:* Univ Alexandria, MB, ChB, 57; Conjoint Bd, London, Eng, dipl psychol med, 61; Univ Man, dipl psychiat, 64; FRCP(C), 65. *Prof Exp:* Psychiatrist, Winnipeg Psychiat Inst, 64-68, med supt, 68-73; PSYCHIATRIST, GRACE GEN HOSP, 74-; ASSOC PROF PSYCHIAT, UNIV MAN, 67- *Mem:* Can Med Asn; Can Psychiat Asn. *Res:* Phenomenology of schizophrenia; geriatric psychiatry. *Mailing Add:* Man Adolescent Treat Ctr 120 Tecumseh St Winnipeg MB R3E 2A9 Can

VARSEL, CHARLES JOHN, b Fayette City, Pa, Jan 11, 30; m 50; c 6. FOOD CHEMISTRY, FOOD BIOCHEMISTRY. *Educ:* St Vincent Col, BA, 54; Univ Richmond, MS, 58; Med Col Va, PhD, 70. *Prof Exp:* Chemist, Linde Air Prod Co, Union Carbide & Carbon Corp, 54-55 & Res & Develop Dept, Philip Morris, Inc, 55-59; res assoc fuel technol, Pa State Univ, 59-60; from res chemist to sr res chemist, Res & Develop Dept, Philip Morris, Inc, Va, 60-69; prin chemist, Food Div, Citrus Res & Develop, 69-70, mgr chem, 70-73, dir citrus res & develop, 73, DIR RES & DEVELOP, COCA-COLA FOODS, 73- *Concurrent Pos:* Counr, Tex Sect Inst Food Technol; Bd Gov, Food Update. *Mem:* AAAS; Am Chem Soc; Inst Food Technol; NY Acad Sci. *Res:* Instrumental analysis; citrus chemistry; essential oils; mass spectroscopy; chemistry of natural products. *Mailing Add:* Coca-Cola Foods Dept Res & Develop PO Box 2079 Houston TX 77252-1900

VARSHNEY, PRAMOD KUMAR, b Allahabad, India, July 1, 52; m 78; c 2. COMMUNICATIONS, COMPUTER ENGINEERING. *Educ:* Univ Ill, Urbana-Champaign, BS, 72, MS, 74, PhD(elec eng), 76. *Prof Exp:* Teaching asst, elec eng, Univ Ill, Urbana-Champaign, 72-76; from asst prof to assoc prof, 76-86, PROF ELEC & COMPUT ENG, SYRACUSE UNIV, 86- *Concurrent Pos:* Res assoc, Rome Air Develop Ctr, 79; vis prof, Ind Inst Technol, Delhi, India, 84-85. *Honors & Awards:* Am Soc Eng Educ. *Mem:* Sr mem Inst Elec & Electronics Engrs. *Res:* Communication theory; communication networks; distributed algorithms for signal processing; application of information theory to computer algorithms; testing and reliability; parallel processing. *Mailing Add:* 121 Link Hall Syracuse Univ Syracuse NY 13244-1240

VARSHNI, YATENDRA PAL, b Allahabad, India, May 21, 32. ASTROPHYSICS, THEORETICAL PHYSICS. *Educ:* Univ Allahabad, BSc, 50, MSc, 52, PhD(physics), 56. *Prof Exp:* Asst prof physics, Univ Allahabad, 55-60, fel, Nat Res Coun Can, 60-62; from asst prof to assoc prof, 62-69, PROF PHYSICS, UNIV OTTAWA, 69- *Mem:* Am Phys Soc; Can Asn Physicists; Brit Inst Physics; Royal Astron Soc UK; Am Astron Soc; AAAS. *Res:* Molecular and atomic structure; quasi-stellar objects; quantum theory; energy levels of nuclei. *Mailing Add:* Dept Physics Univ Ottawa Ottawa ON K1N 6N5 Can

VARTANIAN, LEO, b Ludlow, Mass, Apr 22, 33; m 68. INTERNATIONAL TECHNICAL LIAISON, TECHNOLOGY LICENSING. *Educ:* Rensselaer Polytech Inst, BChE, 54; Am Int Col, MBA, 59; Stanford Univ, MChE, 67. *Prof Exp:* Mem tech staff, Monsanto Chem Co, 54-66, supvr plant technol, 67-71, opers supt, 71-73, gen supt, 73-81, mgr plastics res & develop, 81-85, mgr int liaison, 85-90, MGR PLASTICS MKT DEVELOP, MONSANTO CHEM CO, 90- *Res:* Polymerization processes and products; coordination of technology exchange including worldwide conferences; technology in and out licensing and alliances. *Mailing Add:* 500 Innerness Lane Longmeadow MA 01106

VARTANIAN, PERRY H(ATCH), JR, b Rochester, NY, June 14, 31; m 58; c 3. ELECTRICAL ENGINEERING. *Educ:* Calif Inst Technol, BS, 53; Stanford Univ, PhD(elec eng), 56. *Prof Exp:* Sect head, Sylvania Elec Prod, Inc, 54-57, vpres & dir, Melabs, Calif, 57-69; dir res & develop, Bus Equip Div, SCM Corp, 69-80; CONSULT, 80- *Mem:* Inst Elec & Electronics Engrs. *Mailing Add:* Vartron Corp 199 Brookwood Rd Woodside CA 94062

VARTERESSIAN, K(EGHAM) A(RSHAVIR), b Bahcecik, Izmit, Turkey, Jan 25, 08; nat US; m 46; c 2. FLUIDIZED BED CALCINATION. *Educ:* Pa State Col, BS, 30, MS, 31, PhD(chem eng), 35. *Prof Exp:* Res asst chem, Pa State Col, 35-37, instr chem eng, 37-38 & 42-43, res assoc, 39-42 & 44-45, instr physics, Army Specialized Training Prog, 43-44; instr chem & head dept, Am Univ Cairo, 45-49; assoc chem engr, Argonne Nat Lab, 49-50, sr chem engr, Chem Eng Div, 50-55 & 65-68, sr chem engr, Reactor Anal & Safety Div, 68-73; RETIRED. *Concurrent Pos:* Instr & chmn Chem Eng Dept, Int Sch Nuclear Sci & Eng, Argonne Nat Lab, 55-65; Standard Oil Develop Co, 38-39. *Mem:* Emer mem Am Chem Soc; emer mem Am Inst Chem Engrs; emer fel Am Inst Chemists; Nat Geog Soc. *Res:* Phase distribution of radioactive decay heat sources in spent nuclear fuel; unit operations in the reprocessing and waste treatment in nuclear fuel cycles; fission product spectra of spent nuclear fuel; liquid-liquid extraction. *Mailing Add:* 4120 George Ave No 2 San Mateo CA 94403-4704

VARTSKY, DAVID, b Bielawa, Poland, July 2, 46; Israel citizen; m 71; c 2. MEDICAL PHYSICS. *Educ:* Technion, Israel, BSc, 71; Birmingham Univ, MSc, 74, PhD(physics), 76. *Prof Exp:* Res assoc, Brookhaven Nat Lab, 76-77, asst scientist, 77-80, assoc scientist med physics, 80-; SOREQ NUCLEAR RES CTR, YAVINE, ISRAEL. *Mem:* Inst Elec & Electronics Engrs; Am Phys Soc. *Res:* Medical applications of nuclear physics; determination of body composition by neutron activation analysis in-vivo; in-vivo determination of toxic metals in the body. *Mailing Add:* Rad Prof Dept Soreq Nuclear Res Ctr Yavine 70600 Israel

VARTY, ISAAC WILLIAM, b Consett, Eng, Feb 9, 24; m 52; c 3. FOREST ENTOMOLOGY. *Educ:* Aberdeen Univ, BSc, 50, PhD(entom), 54. *Prof Exp:* Asst forest zool, Aberdeen Univ, 50-54; dist forest officer, Forestry Comn, Scotland, 54-58; forest res scientist, Maritimes Forest Res Ctr, Can Forestry Serv, 58-88; RETIRED. *Mem:* Fel Entom Soc Can; Can Inst Forestry. *Res:* Environmental impact of forest spraying; introduction of exotic parasites for control of forest pests; insecticide spray efficacy. *Mailing Add:* RRH One Ripples NB E0E 1M0 Can

VARY, JAMES CORYDON, b Feb 24, 39; c 2. SPORE GERMINATION, SPORULATION. *Educ:* Univ Wis, MS, 64, PhD(bacteriol), 67. *Prof Exp:* Postdoctoral fel, Stanford Univ, 67-69; PROF BIOCHEM, UNIV ILL, CHICAGO, 69- *Concurrent Pos:* Nat Bd Dent Examiners, 72-78. *Mem:* Am Soc Microbiol; AAAS; Am Soc Biochem & Molecular Biol. *Res:* Biochemical mechanisms for triggering germination and sporulation. *Mailing Add:* Sch Med Dept Biochem M/C 563 Box 6998 Univ Ill 1853 W Polk Chicago IL 60612

VARY, PATRICIA SUSAN, b Wewoka, Okla, Nov 20, 41; m 67; c 2. MICROBIAL GENETICS, GENE REGULATION. *Educ:* Tex Christian Univ, BS, 63, MS, 65; Univ Wis-Madison, MS, 67; Stanford Univ, PhD(microbiol genetics), 69. *Prof Exp:* Asst prof bact, Northern Ill Univ, 73-76; asst prof chem, NCent Col, 77; from asst prof to assoc prof, 77-88, PROF MICROBIOL GENETICS, NORTHERN ILL UNIV, 88-, PRESIDENTIAL RES PROF, 91- *Concurrent Pos:* Indust consult, 86-; mem, adv bd, Bacillus Stock Cult Colle & study group, Int Comn Taxon Viruses, 88-; vis researcher, Inst Pasteur, Paris, 89-90; sr fel, Fogarty Int Ctr, NIH, 91. *Mem:* Am Soc Microbiol; Genetics Soc Am; Sigma Xi; AAAS. *Res:* Genetic methods for Bacillus megaterium; isolation of the only transducing phages, plasmid analysis, transposition, gene fusion techniques, mapping; cell reaction to stress such as sporulation, nitrogen deprivation; cloning hosts for industrial applications. *Mailing Add:* Dept Biol Sci Northern Ill Univ Dekalb IL 60115-2861

VAS, STEPHEN ISTVAN, b Budapest, Hungary, June 4, 26; Can citizen; m 53. MEDICAL MICROBIOLOGY, IMMUNOLOGY. *Educ:* Pazmany Peter Univ, Budapest, MD, 50, PhD(microbiol), 56. *Prof Exp:* Lectr microbiol, Pazmany Peter Univ, 48-49, asst prof, 49-50; Rockefeller res fel microbiol, McGill Univ, 57-59, asst virologist, 59-60, from asst prof to prof immunol, 60-77, chmn dept, 72-77; assoc prof med, 77-80, PROF MED MICROBIOL, UNIV TORONTO, 77-, PROF MED, 80- *Concurrent Pos:* Microbiologist-in-chief & sr physician, Royal Victoria Hosp, 72-77; microbiologist-in-chief & physician, Toronto Western Hosp, 77- *Mem:* Am Soc Microbiol; Can Soc Microbiol; Can Soc Immunol; Can Med Asn; Can Asn Med Microbiol. *Res:* Antibody synthesis; synthesis of complement; effect of antibodies on bacteria and on tissue cells; peritoneal dialysis. *Mailing Add:* Dept Microbiol Toronto Western Hosp 399 Bathurst St Toronto ON M5T 2S8 Can

VASARHELYI, DESI D, b Hungary, Sept 27, 10; nat US; m 57. CIVIL ENGINEERING. *Educ:* Univ Cluj, BA, 28; Budapest Tech Univ, Dipl, 33, DSc(eng), 44. *Prof Exp:* Asst bridge struct & concrete, Budapest Tech Univ, 32; design & field engr, Palatinus Co, 36; lectr, Inst Higher Tech Educ, Budapest, 44; engr, Brit Army, Austria, 46; res engr, 50-52, from instr to prof, 52-80, EMER PROF CIVIL ENG, UNIV WASH, 80- *Mem:* Am Soc Civil Engrs; Am Welding Soc; Soc Exp Stress Anal; Int Asn Bridge & Struct Engrs. *Res:* Stress analysis; structural theory; engineering materials; steel structures. *Mailing Add:* 4055 NE 57th St Seattle WA 98105

VASAVADA, KASHYAP V, b Ahmedabad, India, July 25, 38; m 69; c 2. THEORETICAL PHYSICS. *Educ:* Univ Baroda, BS, 58; Univ Delhi, MS, 60; Univ Md, PhD(physics), 64. *Prof Exp:* Nat Acad Sci fel physics, Goddard Space Flight Ctr, NASA, 64-66; asst prof, Univ Conn, 66-70; assoc prof, 70-74, PROF PHYSICS, IND UNIV-PURDUE UNIV, INDIANAPOLIS, 74- *Concurrent Pos:* Nat Insts Health Sr Fel, 85-86. *Mem:* Am Phys Soc. *Res:* High energy physics; scattering theory; magnetic resonance; theoretical physics. *Mailing Add:* 1316 Brookton Ct Indianapolis IN 46260-3368

VASCO, DONALD WYMAN, b Oakland, Calif, Jan 22, 58. SEISMOLOGY, GEOPHYSICAL INVERSE THEORY. *Educ:* Univ Tex, Austin, BSc, 81; Univ Calif, Berkeley, PhD(seismol), 87. *Prof Exp:* Postdoctoral seismol, Geophys Lab, 87-89, Seismographic Sta, 89-91, STAFF SCIENTIST, SEISMOL, EARTH SCI DIV, LAWRENCE BERKELEY LAB, UNIV CALIF, 91- *Concurrent Pos:* Vis fel, Australian Nat Univ, 91. *Mem:* Am Geophys Union; Soc Explor Geophys; Soc Indust & Appl Math; Seismol Soc Am. *Res:* Development of a technique to determine fluid intrusion within the earth using observations of surface displacement; analysis of global seismic traveltime observations for the velocity structure of the earth's mantle; development of techniques for the inversion of geophysical data; author of numerous publications. *Mailing Add:* Ctr Computational Seismol Earth Sci Div/Bldg 50E Lawrence Berkeley Lab One Cyclotron Rd Berkeley CA 94720

VASCONCELOS, AUREA C, b Caquas, PR, Dec 25, 35; US citizen; m 68. PLANT PHYSIOLOGY. *Educ:* Univ PR, BS, 57; George Washington Univ, MA, 61; Univ Chicago, PhD(biol), 69. *Prof Exp:* Asst prof biol, Univ PR, 61-65; vis asst prof plant physiol, 70-72, assoc prof bot, 72-90, PROF BOT, RUTGERS UNIV, 90- *Concurrent Pos:* Mem, Int Cell Res Orgn, UNESCO. *Mem:* AAAS; Am Soc Plant Physiol; NY Acad Sci; Am Soc Cell Biol; Int Soc Plant Biol. *Res:* Chloroplast development; import of nuclear coded polypeptides into chloroplasts. *Mailing Add:* Dept Biol Sci Rutgers Univ PO Box 1059 New Brunswick NJ 08903

VASCONCELOS, WOLNER V, b Recife, Brazil, May 17, 37. ALGEBRA, COMPUTER ALGEBRA. *Educ:* Univ Chicago, PhD(math), 66. *Prof Exp:* From asst prof to assoc prof math, 67-75, PROF MATH, RUTGERS UNIV, 75- *Mailing Add:* Dept Math Rutgers Univ New Brunswick NJ 08903

VASEEN, V(ESPER) ALBERT, b Denver, Colo, Sept 13, 17; m 41; c 2. ACCELERATED GROWTH OF CLONES OF PLANT TISSUE BY DUAL HYDROPHONIC MEANS. *Hon Degrees:* Dr Sci, Univ Del Norte, Coquimbo, Chile, 81. *Prof Exp:* Asst state sanit engr, Colo State Health Dept, 41-43; sanit officer, US Army, 43-46; pres, Ripple & Howe Inc, 46-66; proj engr, Stearns Rogers Inc, 66-80; PRES & CONSULT ENVIRON TECHNOL AVASCO, 80- *Concurrent Pos:* Trustee, Water Pollution Control Fedn, 56-58; mem, US Dept Com, 55-78, Int Exec Serv Corps, 68-, adv bd, Chem Week Mgt, 74; consult coal gasification, In Situ Technol Inc, 79-80; pres & consult, Technometrics Inc, 80-87. *Honors & Awards:* Cert Accomplishment, Int Graphoanal Soc Inc, 62, Master Graphoanal, 64. *Mem:* Water Pollution Control Fedn; Inter Am Asn Sanit Engrs. *Res:* Inventor of over 300 disclosures relating to environment and power; 33 US patents; development of continuous processing formenter; author of numerous publications. *Mailing Add:* 9840 W 35th Ave Wheat Ridge CO 80033

VASEK, FRANK CHARLES, b Maple Heights, Ohio, May 9, 27; m 54; c 2. BOTANY. *Educ:* Ohio Univ, BS, 50; Univ Calif, Los Angeles, PhD(bot), 55. *Prof Exp:* Asst bot & teaching asst, Univ Calif, Los Angeles, 50-54; from instr to prof bot, Univ Calif, Riverside, 54-89; RETIRED. *Mem:* Bot Soc Am; Ecol Soc Am; Soc Study Evolution; Am Soc Plant Taxon. *Res:* Plant taxonomy; evolution; population dynamics. *Mailing Add:* 18756 Los Hermanus Ranch Rd Valley Center CA 92082

VASERSTEIN, LEONID, IV, b Kuibyshev, USSR, Sept 15, 44; m 68; c 2. ALGEBRAIC K-THEORY, ARITHMETIC GROUPS. *Educ:* Moscow State Univ, MS, 66, PhD(math), 69. *Prof Exp:* Sr researcher & head, Sect Oper Res & Prog, All-Union Inst Info & Tech Econ Res Elec Indust, 69-78; PROF MATH, PA STATE UNIV, 79- *Concurrent Pos:* Mem jury, All-Union Math Olympiads, USSR, 62-77; vis prof, Univ Bielefeld, Inst des Hautes Etudes Sci, France, 78, Univ Chicago, 79 & Cornell Univ, 79-80; Guggenheim fel, 84. *Mem:* Am Math Soc. *Res:* Operations research, dynamical systems. *Mailing Add:* Dept Math Pa State Univ University Park PA 16802

VASEY, CAREY EDWARD, b Bristol, Pa, Feb 12, 27; m 49; c 2. INSECT MORPHOLOGY, INSECT TAXONOMY. *Educ:* Lycoming Col, BA, 53; Syracuse Univ, MS, 59; State Univ NY Col Environ Sci & Forestry, PhD(forest entom), 75. *Prof Exp:* Indust microbiologist vitamin assay, Publicker Indust, Philadelphia, Pa, 51-53; lab technician histol technol, Med Ctr, Univ Kans, 53-55; teacher biol, James Buchanan Sch, Mercersburg, Pa, 55-58 & Coatesville Area Schs, Coatesville, Pa, 59-60; instr, Philadelphia Col Pharm & Sci, 60-64; asst prof, 64-74, ASSOC PROF BIOL, COL ARTS & SCI, STATE UNIV NY, GENESEO, 74- *Mem:* Am Entom Soc. *Res:* Systematics morphology and ultrastructure of selected families of diptera. *Mailing Add:* Biol Dept Col Arts & Sci State Univ NY Geneseo NY 14454

VASEY, EDFRED H, b Mott, NDak, Aug 29, 33; m 55; c 4. SOIL FERTILITY, SOIL CONSERVATION. *Educ:* NDak State Univ, BS, 55, MS, 57; Purdue Univ, PhD(plant nutrit), 62. *Prof Exp:* From asst prof to assoc prof, 61-67, head Plant Sci Sect, NDak Coop Exten Serv, 71-82, EXTEN SOILS SPECIALIST, EXTEN SERV, NDAK STATE UNIV, 67-, PROF SOILS, 69- *Concurrent Pos:* Assoc ed, Appl Agr Res, Springer-Verleg, New York, NY. *Mem:* Soil Sci Soc Am; Coun Agr Sci & Technol; Sigma Xi; Soil & Water Conserv Soc. *Res:* Fertility needs of crops grown on North Dakota soils; computer software for crop production. *Mailing Add:* 2802 Maple St N Box 5575 Fargo ND 58102

VASEY, FRANK BARNETT, b Webster City, Iowa, Feb 6, 42; m 68; c 1. RHEUMATOLOGY, INTERNAL MEDICINE. *Educ:* Cornell Col, BA, 64; Univ NDak, BS, 66; Univ Pa, MD, 68. *Prof Exp:* Asst prof internal med, McGill Univ & Royal Victoria Hosp, Montreal, 76-77; ASST PROF MED, RHEUMATOLOGY, UNIV SFLA, 77- *Concurrent Pos:* Rheumatology fel, Royal Victoria Hosp, McGill Univ, 74; dir, Rheumatology & Immunol Clin Lab, Univ SFla, 77-78 & Rheumatology & Immunol Res Lab, Vet Admin Hosp, Tampa, 78- *Mem:* Fel Am Col Physicians. *Res:* Immunogenetics basis of psoriatic arthritis. *Mailing Add:* 12901 Bruce B Downs Blvd Tampa FL 33612

VASHISHTA, PRIYA DARSHAN, b Aligarh, India, Aug 24, 44; m 70; c 1. SOLID STATE PHYSICS, MATERIALS SCIENCE. *Educ:* Agra Univ, BS, 60; Aligarh Univ, MS, 62; Indian Inst Technol, PhD(physics), 67. *Prof Exp:* Fel, St Andrews Univ, UK, 66-68; res assoc, McMaster Univ, 68-70; asst prof physics, Northwestern Univ, 70-71; asst prof, Western Mich Univ, 71-72; res physicist, Argonne Nat Lab, 72-90, dir, Solid State Sci Div, 79-90; FLOATING POINT SYST CHAIRED PROF COMPUTATIONAL METHODS, DEPT PHYSICS, LA STATE UNIV, 90- *Concurrent Pos:* Vis scientist, Theoret Physics Div, Atomic Energy Res Estab, UK, 68, Inst Theoret Physics, Sweden, 71; fel appl comput, Thomas J Watson Res Ctr, IBM, 72; mem tech staff, Bell Labs, 76; vis assoc prof, Univ Calif, San Diego, 76-77; co-ed, J Solid State Ionics, Amsterdam, 79-; mem, Solid State Sci Panel, NSF, 79-82. *Mem:* Am Phys Soc; AAAS; Sigma Xi. *Res:* Theoretical physics of condensed matter; electron-phonon interaction and superconductivity in metals and alloys; many-body interactions in metals and semiconductors; molecular dynamics studies of condensed matter; phase transitions on surfaces. *Mailing Add:* Dept Physics & Astron La State Univ Baton Rouge LA 70803-4001

VASICEK, DANIEL J, b Cleveland, Ohio, Nov 6, 42; m 73; c 4. PARALLEL PROGRAMMING, FUNCTIONAL LANGUAGES. *Educ:* Purdue Univ, BS, 64, MS, 65; Univ Colo, PhD(aerospace eng sci), 73. *Prof Exp:* Dir pre-eng prog, Univ Colo, 72-75; sem leader math & res assoc math, physics, eng & statist, 73-89, SEM LEADER COMPUTER SCI, AMOCO PROD CO, 85- *Concurrent Pos:* Instr calculus, Tulsa Jr Col, 76; sem leader kinetic theory, Tulsa Free Univ, 83. *Mem:* Asn Computer Mach; Soc Indust & Appl Math. *Res:* Parallel extensions for programming languages; successful demonstration of a practical parallel program of large size, 100,000 lines; statistical analysis of seismic, well log and core data; numerical representations of the rotation group; interpolation and smoothing. *Mailing Add:* Amoco Prod Co PO Box 3385 Tulsa OK 74102

VASIL, INDRA KUMAR, b Basti, India, Aug 31, 32; m 59; c 2. BOTANY. *Educ:* Banaras Hindu Univ, BSc, 52; Univ Delhi, MSc, 54, PhD(bot), 58. *Prof Exp:* Res asst bot, Univ Delhi, 54-58, asst prof, 59-63; res assoc, Univ Wis, 63-65; scientist, Indian Agr Res Inst, 65-67; from assoc prof to prof, 67-79, GRAD RES PROF, UNIV FLA, 79- *Concurrent Pos:* Res assoc, Univ Ill, 62-63. *Honors & Awards:* Sr US Scientist Award for Res, Fed Repub Ger, 74. *Mem:* AAAS; Int Asn Plant Tissue Cult; Bot Soc Am; Int Soc Plant Morphol; Int Soc Plant Molecular Biol. *Res:* Developmental morphology; physiology of reproduction in flowering plants; morphogenesis and differentiation in higher plants; plant tissue and organ culture, especially of cereals and grasses. *Mailing Add:* Veg Crops Dept Univ Fla Gainesville FL 32611-0514

VASIL, MICHAEL LAWRENCE, b San Diego, Calif, Sept 6, 45; m 71; c 1. PARASITOLOGY, BIOCHEMISTRY. *Educ:* Univ Tex, El Paso, BS, 71, Med Sch Dallas, PhD(microbiol), 75. *Prof Exp:* Chief bacteriologist, Providence Hosp, 67-71; res asst, Univ Tex, 71-75; instr, Univ Ore Med Sch, 75-77; asst prof, Univ Calif Los Angeles Sch Med, 77-78; asst prof microbiol, 78-84, assoc prof, 84-90, PROF MICROBIOL, UNIV COLO MED SCH, 90- *Concurrent Pos:* Prin investr res grants, Procter & Gamble, 80-86, Cystic Fibrosis Found, 79-81, NIH, 79-93. *Mem:* Am Soc Microbiol; Am Soc Clin Pathologists; Sigma Xi. *Res:* Mechanism of bacterial pathogenesis and the development of bacterial vaccines; clinical microbiology-DNA probes; microbial genetics as well as recombinant DNA technology in psendomenas aeruginosa escherichia coli, staphylococcus anreus; industrial microbiology-over expression of gene products by bacteria. *Mailing Add:* Dept Microbiol & Immunol Univ Colo Med Sch 4200 E Ninth Ave Denver CO 80262

VASIL, VIMLA, b New Delhi, India, Dec 11, 32; US citizen; m 59; c 2. CELL & PROTOPLAST CULTURE, TISSUE CULTURE. *Educ:* Univ Delhi, BSc, 53, MSc, 55, PhD(bot), 59. *Prof Exp:* Res assoc bot, Univ Delhi, 59-62; res assoc agron, Univ Ill, 62-63; res assoc plant path, Univ Wis, 63-65; scientist, Coun Sci & Indust Res, New Delhi, 66; assoc bot, 67-69, assoc res scientist, 79-85, RES SCIENTIST, UNIV FLA, 85- *Mem:* Bot Soc Am. *Res:* Developmental morphology and embryology of angiosperms and gymnosperms; cell and tissue culture of higher plants, particularly cereal and grass species. *Mailing Add:* Dept Veg Crops Univ Fla Gainesville FL 32611-0514

VASILAKOS, NICHOLAS PETROU, b Athens, Greece, Jan 1, 54. CHEMICAL ENGINEERING. *Educ:* Nat Tech Univ Athens, BS, 76; Calif Inst Technol, MS, 78, PhD(coal desulfurization), 81. *Prof Exp:* ASST PROF CHEM ENG, UNIV TEX, AUSTIN, 81- *Mem:* Am Inst Chem Engrs; Am Chem Soc; Sigma Xi. *Res:* Coal desulfurization by chemical treatment; coal liquefaction by super critical solvent extraction; production of liquid hydrocarbon fuels from urban, industrial and agricultural wastes by catalytic hydrogenolysis. *Mailing Add:* Dept Chem Eng Univ Tex Austin TX 78712

VASILATOS-YOUNKEN, REGINA, b New York, NY, Nov 30, 54; m 83. GROWTH HORMONE, GROWTH HORMONE BINDING PROTEINS. *Educ:* Univ Maine, Orono, BS, 76; Pa State Univ, PhD(animal nutrit), 82. *Prof Exp:* Res fel nutrit, Roman L Hruska US Meat Animal Res Ctr, 82-83; res assoc molecular & cell biol, 83, asst prof poultry sci, 83-89, ASSOC PROF POULTRY SCI, PA STATE UNIV, 89- *Concurrent Pos:* Assoc ed, Physiol & Reproduction Sect, Poultry Sci, 88-; chair, Physiol Prog, Poultry Sci Asn, 89. *Mem:* AAAS; Am Inst Nutrit; Endocrine Soc; Am Soc Animal Sci; Poultry Sci Asn. *Res:* Endocrine regulation of growth and development with emphasis on factors influencing the biological action of growth hormone and growth hormone binding proteins. *Mailing Add:* 203 Henning Bldg University Park PA 16802

VASILE, MICHAEL JOSEPH, b Newton, NJ, May 11, 40. PHYSICAL CHEMISTRY, SPECTROSCOPY. *Educ:* Rutgers Univ, BS, 62; Princeton Univ, MA, 64, PhD(chem), 66. *Prof Exp:* Res fel chem, Nat Res Coun Can, 66-68; DISTINGUISHED MEM TECH STAFF, BELL LABS, 68- *Mem:* Electrochem Soc; Am Vacuum Soc; Mat Res Soc. *Res:* Plasma chemistry; surface chemistry; electronic materials. *Mailing Add:* AT&T Bell Labs 2C-105 600 Mountain Ave Murray Hill NJ 07974

VASILEVSKIS, STANISLAUS, astronomy, for more information see previous edition

VASILIAUSKAS, EDMUND, b Lithuania, June 18, 38; US citizen; m 70; c 4. ORGANIC CHEMISTRY. *Educ:* Rochester Inst Technol, BS, 63; Loyola Univ Chicago, PhD(org chem), 70. *Prof Exp:* Chemist, Olin Corp, 63-65 & Witco Chem Corp, 70-71; from instr to assoc prof, 71-80, PROF CHEM, MORAINE VALLEY COMMUNITY COL, 80- *Mem:* Am Chem Soc. *Res:* Study of the derivatives of benzonorbornene. *Mailing Add:* Dept Chem Moraine Valley Community Col Palos Hills IL 60465

VASILOS, THOMAS, b New York, NY, Oct 18, 29; m 54; c 2. CERAMICS, CHEMISTRY. *Educ:* Brooklyn Col, BS, 50; Mass Inst Technol, DSc(ceramics), 54. *Prof Exp:* Asst ceramics, Mass Inst Technol, 50-53; res engr, Ford Motor Co, 53; mgr, Ceramics Res Dept, Corning Glass Works, 55-57; sect chief metals & ceramics, Res & Adv Develop Div, 57-66, mgr mat sci dept, Avco Systs Div, 66-77, mgr, mat develop dept, Avco Corp, 77-79, prin scientist, Avco Systs Div, 79-87, PROF CHEM ENG, UNIV LOWELL, 87- *Concurrent Pos:* Consult, Mat Adv Bd, Nat Acad Sci, 61- *Honors & Awards:* Ross Coffin Purdy Award, Am Ceramic Soc, 67. *Mem:* Fel Am Ceramic Soc. *Res:* Thermal and mechanical properties of ceramics; crystal growing; ferroelectric ceramics and crystals; nuclear fuel ceramics; ceramic coatings; metal reinforced ceramics; diffusion in crystals; plastic-ceramic composites; composite formulation and properties; refractory metals. *Mailing Add:* Chem Eng Dept Univ Lowell One University Ave Lowell MA 01854

VASINGTON, FRANK D, b Norwich, Conn, Nov 3, 28; div; c 4. BIOCHEMISTRY. *Educ:* Univ Conn, AB, 50, MS, 52; Univ Md, PhD(biochem), 55. *Prof Exp:* Asst prof biochem, Sch Med, Univ Md, 55-57; Nat Found res fel, McCollum-Pratt Inst, Johns Hopkins Univ, 57-59, asst prof physiol chem, Sch Med, 59-64; assoc prof, Univ Conn, 64-68, head biol sci group, 67-71, head biochem & biophys sect, 67-77, assoc dean, Col Lib Arts & Sci, 76-78, assoc vpres acad affairs, 78-82, interim dean, Col Lib Arts & Sci, 86-88, PROF BIOCHEM, UNIV CONN, 68-, DEAN, COL LIB ARTS & SCI, 88-; INTERIM DEAN, COL LIB ARTS & SCI, 86- *Mem:* AAAS; Am Chem Soc; Am Soc Biol Chem. *Res:* Biosynthesis and turnover of intracellular membranes; secretion; active transport processes in mitochondria and bacteria. *Mailing Add:* Wood Hall Box U-98 Univ Conn Storrs CT 06268

VASKA, LAURI, b Rakvere, Estonia, May 7, 25; nat US; m 54; c 5. INORGANIC CHEMISTRY. *Educ:* Univ Göttingen, BS, 49; Univ Tex, PhD(chem), 56. *Prof Exp:* Fel Magnetism & chemisorption, Northwestern Univ, 56-57; res fel inorg chem, Mellon Inst, 57-64; assoc prof, 64-67, PROF INORG CHEM, CLARKSON UNIV, 67 - *Concurrent Pos:* Fulbright-Hays fel, Univ Helsinki, 72. *Honors & Awards:* Boris Pregel Award, NY Acad Sci, 71. *Mem:* AAAS; Am Chem Soc; fel NY Acad Sci; Royal Soc Chem. *Res:* Coordination chemistry; catalysis; oxygen-carrying complexes; noble metal chemistry. *Mailing Add:* Dept Chem Clarkson Univ Potsdam NY 13676

VASKO, JOHN STEPHEN, b Cleveland, Ohio, Mar 17, 29; m 52; c 6. CARDIOVASCULAR & THORACIC SURGERY. *Educ:* Ohio State Univ, DDS, 54, MD, 58. *Prof Exp:* Instr anat & physiol, Cols Med & Dent, Ohio State Univ, 54-55, instr oper dent & consult prosthetic dent, Col Dent, 54-58; intern surg, Johns Hopkins Univ Hosp, 58-59, asst instr cardiac surg, Sch Med, Johns Hopkins Univ, 59-60; asst resident surgeon, Vanderbilt Univ Hosp, chief resident surgeon & chief labs exp surg, Nat Heart Inst, 64-66; assoc prof, 66-73, PROF THORACIC & CARDIOVASC SURG, COL MED, OHIO STATE UNIV, 73- *Concurrent Pos:* Halsted res fel cardiac surg, Sch Med, Johns Hopkins Univ, 59-60 asst resident surgeon cardiac, 59-61; mem, Tech Rev Comt, Artificial Heart Prog, NIH & mem, Adv Comt, Coun Cardiovasc Surg, Am Heart Asn, 65-; partic, NIH Grad Prog, 65. *Mem:* AAAS; Asn Acad Surg; Soc Thoracic Surg; AMA; Am Col Surgeons. *Res:* Cardiovascular physiology. *Mailing Add:* Dept Surg Ohio State Univ Col Med Columbus OH 43210

VASKO, MICHAEL RICHARD, b Detroit, Mich, Mar 29, 48; m 80; c 2. NEUROCHEMISTRY, NEUROPHARMACOLOGY. *Educ:* Univ Mich, Ann Arbor, BS, 70, PhD(pharmacol), 76. *Prof Exp:* Res fel, Dept Pharmacol, Med Sch, Univ Mich, Ann Arbor, 71-75; instr pharmacol & neurol, Univ Tex Health Sci Ctr Dallas, 75-77, asst prof, 77-80; vis res fel, Dept Pharmacol, Inst Animal Physiol, Eng, 80-81; chief, pharmacol sect, Vet Admin Ctr, Dallas, 81-88; asst prof, Dept Pharmacol, Univ Tex Health Sci Ctr, Dallas, 81-88, mem grad fac biomed sci, 81-88; ASSOC PROF PHARMACOL, IND UNIV SCH MED, 88- *Concurrent Pos:* Chief, Neuropharmacol Lab & staff pharmacologist, Vet Admin Med Ctr, Dallas, 75-80; Nat Inst Drug Abuse res fel, 80. *Mem:* Soc Neurosci; Int Soc Neurochem; Am Soc Pharmacol Exp Therapeut; Am Pain Soc. *Res:* Involvement of neurotransmitters in the antinociceptive effects of narcotic analgesics; regulation of neurotransmitter release from sensory neurons. *Mailing Add:* Dept Pharmacol & Toxicol Ind Univ Sch Med 635 Barnhill Dr Indianapolis IN 46202

VASLOW, DALE FRANKLIN, b Chicago, Ill, Aug 27, 45. PLASMA PHYSICS. *Educ:* Univ Wis, BS, 67, MS, 69, PhD(elec eng), 73. *Prof Exp:* Staff scientist plasma physics, Gen Atomic Co, 74-84; PHYSICIAN, USCD MED CTR 84- *Mem:* Inst Elec & Electronics Engrs; Sigma Xi. *Res:* Solid state laser design; low light level image detection system; optical design; laser Thomson Scattering experiment; solid ablation in a plasma; fast neutron induced chemistry. *Mailing Add:* 4608 Huggins Way San Diego CA 92122

VASOFSKY, RICHARD WILLIAM, b Waukegan, Ill, May 14, 46. PHYSICAL CHEMISTRY, SURFACE SCIENCE. *Educ:* Univ Denver, BS, 68; Ore State Univ, MS, 71, PhD(phys chem), 75. *Prof Exp:* Res asst phys chem, Ore State Univ, 69-74; res assoc physics, Clarkson Col Technol, 75-77, res asst prof, 78; consult surface sci, Rome Air Develop Ctr, 78-80; PVT CONSULT, 84- *Mem:* Am Chem Soc; Am Vacuum Soc; Clay Mineral Soc; Int Soc Hybrid Microelectron; AAAS; Sigma xi. *Res:* Adsorption by clays; gas-solid interactions; chemisorption; thin films; desorption; vacuum and ultramicrobalance techniques and applications. *Mailing Add:* Rte 3 Box 846 Louden TN 37774

VASQUEZ, ALPHONSE THOMAS, b Boston, Mass, Apr 19, 38; m 65. MATHEMATICS AND COMPUTER SCIENCE, ALGORITHMS FOR ALGEBRAIC GEOMETRY. *Educ:* Mass Inst Technol, BS, 59; Univ Calif, Berkeley, PhD(math), 62. *Prof Exp:* Mem, Inst Advan Study, 62-64; res assoc math, Brandeis Univ, 64-65, asst prof, 65-67; assoc prof, 67-77, PROF MATH & COMPUT SCI, GRAD DIV, CITY UNIV NEW YORK, 77- *Concurrent Pos:* Vis prof, Univ Mich, Univ Calif, Berkeley. *Mem:* Am Math Soc. *Res:* Algebraic and differential topology; homological algebra; error-correcting codes; computational complexity of curves defined over finite fields. *Mailing Add:* Grad Div City Univ New York 33 W 42nd St New York NY 10036

VASSALLE, MARIO, b Viareggio, Italy, May 26, 28; US citizen; m 59; c 5. CARDIOVASCULAR & ELECTROPHYSIOLOGY. *Educ:* Liceo-Ginnasio G Carducci, Viareggio, BA, 47; Univ Pisa, MD, 53. *Hon Degrees:* Dr, Univ Ferrara, 91. *Prof Exp:* Actg chief resident med, French Hosp, NY, 58-59; NIH trainee, Cardiovasc Res & Training Prog, Med Col Ga, 59-60; teaching fel, dept physiol, State Univ NY, Downstate Med Ctr, 60-61, NY Heart Asn fel, 61-62, instr, 62; NIH fel, Physiol Inst, Bern, Switz, 62-64; from asst prof to assoc prof, 65-71, PROF PHYSIOL, STATE UNIV NY HEALTH SCI CTR, 71- *Concurrent Pos:* Mem, Coun Basic Sci, Am Heart Asn, 69- & Nat Conf Cardiovasc Dis, 69; assoc ed, Am J Physiol, 76-80; consult, NIH; vis prof, dept gen physiol, Univ Ferrara, Italy, 71, dept physiol & biophys, Univ Vt, Burlington, 78 & dept med, Cath Univ A Gemelli, Rome, Italy, 84-85; mem, NY Health Res Coun, 72-75. *Honors & Awards:* Sinsheimer Fund Award, 66-71. *Mem:* AAAS; Am Physiol Soc; Harvey Soc; NY Acad Sci; Am Heart Asn; Cardiac Muscle Soc; Sigma Xi. *Res:* Cardiac electrophysiology, particularly cardiac automaticity and its control. *Mailing Add:* Dept Physiol State Univ NY Health Sci Ctr Brooklyn NY 11203

VASSALLO, DONALD ARTHUR, b Waterbury, Conn, June 7, 32; m 60; c 4. POLYMER CHEMISTRY. *Educ:* Univ Conn, BA, 54; Univ Ill, MS, 56, PhD(anal chem), 58. *Prof Exp:* res assoc, Plastics Dept, E I du Pont de Nemours & Co, Inc, 58-85, consult, 86-87; RETIRED. *Concurrent Pos:* Consult, Maldermid Chem Co, 55, Naugatuck Chem Co, 56-57. *Mem:* AAAS; Am Chem Soc; Soc Plastics Eng; Sigma Xi. *Res:* Automatic nonaqueous titrations; thermogravimetry; differential thermal analysis, especially polymers; polymer melt rheology; polymer structure-property correlations; polyolefins; polyvinyl alcohol; polymeric gas barriers. *Mailing Add:* Five Aldham Ct Wilmington DE 19803

VASSALLO, FRANKLIN A(LLEN), b Waterbury, Conn, Feb 22, 34; m 66; c 3. MECHANICAL ENGINEERING. *Educ:* Univ Conn, BS, 56; Univ Ill, MS, 57. *Prof Exp:* Asst mech engr, 57-59, assoc mech engr, 59-62, res engr, 62-67, prin res engr, 67-70, sect head, Cornell Aeronaut Lab, 70-78; SECT HEAD, CALSPAN CORP, 78- *Mem:* Nat Res Coun; Nat Mat Adv Bd. *Res:* Heat transfer and erosion in rapid fire weapons; determination of gas enthalpy; heat transfer and flow processes in high heat flux environments; hazardous materials transport; assessment of use of non-metallic composite materials in conventional weapon systems; space refrigeration. *Mailing Add:* 1273 Ransom Rd Lancaster NY 14086

VASSAMILLET, LAWRENCE FRANCOIS, b Elizabethville, Congo, Sept 14, 24; nat US; m 54; c 2. SOLID STATE ANALYSIS. *Educ:* Mass Inst Technol, BSc, 46, MS, 50; Univ Liege, DSc(phys sci), 52; Carnegie Inst Technol, PhD(physics), 57. *Prof Exp:* Jr physicist, Monsanto Chem Co, Ohio, 47-48; physicist, Nat Carbon Co Div, Union Carbide Corp, 53; fel, Carnegie-Mellon Univ, 57-63, sr fel, Inst Sci, 63-67, assoc prof metall & mat sci, 67-80; prin res scientist, Columbus Labs, Battelle Mem Inst, 80-83; mgr anal lab, Varian Spec Metals Div, 84-90; RETIRED. *Mem:* Metall Soc. *Res:* X-ray diffraction; imperfections in crystals; electron probe microanalysis; electron microscopy. *Mailing Add:* PO Box 21293 Columbus OH 43221-0293

VASSEL, BRUNO, b Allahabath, Brit India, Oct 17, 08; nat US; m 36; c 3. BIOCHEMISTRY. *Educ:* Yale Univ, BS, 36; Univ Mich, MS, 37, PhD(biochem), 39. *Prof Exp:* Lab asst biochem, Univ Mich, 37-39; res biochemist, Am Cyanamid Co, 39-43; assoc prof biochem & res agr chemist exp sta, NDak Col, 43-46; supvr org & biochem res, Int Minerals & Chem Corp, 46-55; DIR RES, JOHNSON & JOHNSON BRAZIL, 55-, MEM EXEC COMT, 59- *Honors & Awards:* Robert W Johnson Medal Res & Develop, 64. *Mem:* AAAS; Am Chem Soc; Am Oil Chem Soc; Am Soc Sugar Beet Technol. *Res:* Protein isolations; monosodium glutamate processes; amino acid analyses and syntheses; pharmaceuticals; polarograph; flotation reagents; detergents; starch derivatives; surgical and pharmaceutical products. *Mailing Add:* 109 70 S 700 East St Apt 202 Sandy UT 84070-3454

VASSELL, GREGORY S, b Moscow, Russia, Dec 24, 21; m 57; c 2. ELECTRICAL ENGINEERING. *Educ:* Tech Univ, Berlin, Ger, Dipl Eng, 51; NY Univ, MBA, 54. *Prof Exp:* From asst engr to sr engr elec eng, 51-61, sect head high voltage planning, 62-66, assoc chief syst planning engr, 66-67, chief syst planning engr, 67-68, asst vpres bulk power supply planning, 68-73, vpres syst planning & dir, Am Elec Power Serv Corp, 73-76, sr vpres systs planning, & dir, 76-88; CONSULT, 88- *Concurrent Pos:* Chmn, Syst Reliability Adv Panel, East Cent Area Reliability Group, 67-69; mem, Tech Adv Comt Transmission, Fed Power Comn, 68-70, NCent Reg Task Force, Fed Energy Regulatory Comn, 79-81, US Nat Comt, World Energy Conf; mem, comt rev, Nat Comm Syst Initiatives, Nat Res Coun, 82-84, comt, Elec Energy Systs, 85; mem, Atlantic Coun, USA World Energy Conf Study Group on Energy In Less Develop Countries, 84-86. *Mem:* Nat Acad Eng; Int Conf Large High Voltage Elec Systs; fel Inst Elec & Electronics Engrs. *Res:* Electric power supply planning; energy resource planning. *Mailing Add:* Consult 2247 Pinebrook Rd Columbus OH 43220

VASSELL, MILTON O, b Jamaica, West Indies, May 8, 31; m 58; c 2. THEORETICAL SOLID STATE PHYSICS. *Educ:* NY Univ, BA, 58, PhD(physics), 64. *Prof Exp:* SR RES SCIENTIST, GTE LABS, INC, 64- *Mem:* Am Phys Soc; Sigma Xi; NY Acad Sci. *Res:* Many particle physics; transport theory in semiconductors and metals; nonlinear optics; physics of lasers; acoustic surface wave propagation; acousto-electric effects; electron optics; integrated optics; optical guided wave propagation; heterostructure device physics. *Mailing Add:* GTE Labs Inc 40 Sylvan Rd Waltham MA 02154

VASSILIADES, ANTHONY E, b Chios, Greece, Nov 26, 33; US citizen; m 57; c 2. PHYSICAL CHEMISTRY, POLYMER CHEMISTRY. *Educ:* Wagner Col, BS, 56; Syracuse Univ, MS, 58; Polytech Inst Brooklyn, PhD(phys chem), 62. *Prof Exp:* From asst prof to assoc prof chem, Wagner Col, 61-66; assoc dir res, Champion Papers, Inc, 66-68, dir res, Champion Papers Group, US Plywood-Champion Papers, Inc, 68-70, vpres & dir res & develop, 70-76, vpres & dir sci, Champion Int Corp, 76-78; PRES, EPACOR, 80- *Concurrent Pos:* Consult, Champion Papers, Inc, 64-66. *Mem:* Am Chem Soc; NY Acad Sci; Soc Plastics Eng. *Res:* Coacervation of charged colloidal systems; transport phenomena in liquids; physical chemistry of high polymers; surface phenomena; engineering plastics; carbonless papers. *Mailing Add:* 8738 Tanager Woods Dr Cincinnati OH 45249

VASSILIOU, ANDREAS H, b Ora, Larnaca, Cyprus, Nov 30, 36; m 65; c 2. MINERALOGY. *Educ:* Columbia Univ, BS, 63, MA, 65, PhD(mineral), 69. *Prof Exp:* From asst prof to assoc prof geol, 69-81, PROF GEOL, RUTGERS UNIV, 81- *Concurrent Pos:* Chairperson, Dept Geol Sci, Rutgers Univ, 77-86 & 88- *Mem:* Geol Soc Am; Mineral Soc Am; Soc Mining Eng; Minerals Metals & Mat Soc; Sigma Xi; Nat Asn Geol Teachers. *Res:* Mineralogy of uranium, particularly the nature of urano-organic deposits in the Colorado Plateau and elsewhere; economic ore deposits of manganese, gold, etc. *Mailing Add:* 47 Glenbrook Rd Morris Plains NJ 07950

VASSILIOU, EUSTATHIOS, b Athens, Greece, Aug 22, 34; m 60; c 3. COATINGS CHEMISTRY & PHYSICS. *Educ:* Nat Tech Univ Athens, BScChE, 58; Univ Manchester, PhD(chem), 64. *Prof Exp:* Res chemist, Nuclear Res Ctr, Democritus, Greece, 64-66; res fel solid state physics, Harvard Univ, 66-67; res chemist, 67-73, staff chemist, 73-78, res assoc, Marshall Lab, 78-79, res supvr, 79-80, res assoc, 80-84, SR RES ASSOC, EXP STA, E I DU PONT DE NEMOURS & CO, INC, 84- *Mem:* Asn Harvard Chemists. *Res:* Inorganic physical chemistry; physics and chemistry of glasses; vanadium; thermistors; anodic films; luminescence; ozone physics; semiconductor surface phenomena; magnetic phenomena; lubrication; structural plastics; fluorocarbon and other high temperature resistant coatings; corrosion resistant coatings; electrodeposition of coatings; photoresists; electronic materials for hybrid circuits. *Mailing Add:* 12 S Townview Lane Newark DE 19711

VASSILIOU, MARIUS SIMON, b June 7, 57; US citizen; m 89. COMPUTATIONAL PHYSICS & ENGINEERING, PUBLISHING SYSTEMS. *Educ:* Harvard Univ, AB, 78; Calif Inst Technol, MS, 79, PhD(geophys & elec eng), 83; Univ Southern Calif, MS, 87; Univ Calif, Los Angeles, MBA, 91. *Prof Exp:* Grad res asst, Calif Inst Technol, 79-83, sr scientist, 83; sr res geophysicist, Arco Oil & Gas Co, 83-85; MEM TECH STAFF & PROJ MGR, ROCKWELL INT SCI CTR, 85- *Concurrent Pos:* Consult, Rockwell Int Sci Ctr, 81-82 & TRW, Inc, 82-83; vis prof, Univ Tex, Dallas, 84 & Moorpark Col, 86. *Mem:* Sr mem Inst Elec & Electronics Engrs; Soc Explor Geophysicists; Asn Comput Mach; Am Geophys Union; Sigma Xi; Am Soc Metals. *Res:* New methods to consolidate powdered materials preserving nonequilibrium properties; elastic wave propagation and seismology; computational electromagnetics; digital typography, halftoning and publishing systems. *Mailing Add:* Rockwell Int 1049 Camino Dos Rios PO Box 1085 Thousand Oaks CA 91360

VASTANO, ANDREW CHARLES, b New York, NY, Feb 26, 36; m 58; c 4. PHYSICAL OCEANOGRAPHY, NUMERICAL ANALYSIS. *Educ:* NC State Univ, BS, 56; Univ NC, MS, 60; Tex A&M Univ, PhD(phys oceanog), 67. *Prof Exp:* Anal engr, Pratt & Whitney Aircraft Corp, 56-57; sonar engr,

Western Elec Co, Bell Tel Co, 60-62; instr physics, Tex A&M Univ, 62-63, res scientist, 63-66; assoc prof phys oceanog, Univ Fla, 66-67; asst scientist, Woods Hole Oceanog Inst, 67-69; instr, 64, ASSOC PROF OCEANOG, TEX A&M UNIV, 69- *Mem:* Am Geophys Union. *Res:* Numerical studies of tsunami and storm surges; theory of gravity waves; mesoscale ocean dynamics; topographic interaction of current systems. *Mailing Add:* Dept Oceanog Tex A&M Univ College Station TX 77843

VASTOLA, FRANCIS J, b Buffalo, NY, Feb 22, 28; m 69. LABORATORY INSTRUMENTATION & CONTROL. *Educ:* Univ Buffalo, BA, 50; Pa State Univ, PhD(fuel technol), 59. *Prof Exp:* Chemist, Nat Bur Standards, Washington, DC, 50-52; asst & res assoc, 56-59, from asst prof to assoc prof, 59-72, prof, 72-86, EMER PROF FUEL SCI, PA STATE UNIV, 86- *Mem:* Am Chem Soc; Sigma Xi. *Res:* Mass spectrometry; solid and gaseous combustion; kinetics and instrumentation. *Mailing Add:* 406 Hillcrest Ave State College PA 16801

VASU, BANGALORE SESHACHALAM, b Bangalore, India, May 20, 29; nat US; m 65; c 1. BIOLOGY. *Educ:* Univ Madras, BSc, 49, MSc, 62; Stanford Univ, PhD(biol), 65. *Prof Exp:* Asst prof zool, Pachaiyappa's Col, Madras Univ, 50-59, lectr, Zool Res Lab, 59-62; Fulbright res fel biol, Stanford Univ, 62-65; AEC fel, Univ Notre Dame, 65-67; asst prof zool, Ohio Wesleyan Univ, 67-68; UNESCO specialist biol, Univ Zambia, 68-73; asst prof biol, Calif State Univ, Chico, 74-77; PROF BIOL, MENLO COL, 78- *Concurrent Pos:* Fulbright grant, US Educ Found India, 62. *Mem:* Sigma Xi; Am Inst Biol Sci; AAAS; Radiation Res Soc; Marine Biol Asn UK. *Res:* Age-related changes at the cellular and molecular level. *Mailing Add:* 35686 Nuttman Dr Fremont CA 94536-2546

VATISTAS, GEORGIOS H, b Neapolis, Laconias, Greece, Jan 25, 53; Can citizen; m 82; c 2. VORTEX DYNAMICS, MICROGRAVITY FLUID MECHANICS. *Educ:* Concordia Univ, BEng, 78, MEng, 80, PhD(mech eng), 84. *Prof Exp:* Res asst mech eng, Concordia Univ, 78-82, teacher & lab instr, 82-85, asst prof, 85-89, ASSOC PROF MECH ENG, CONCORDIA UNIV, 89- *Concurrent Pos:* Consult, Dept Surg, Royal Victoria Hosp Montreal, 84-87 & Bendix Avelex Inc, 88; grad prog dir mech eng, Concordia Univ, 90- *Honors & Awards:* Ralph R Teetor Educ Award, Soc Automotive Engrs, 87. *Mem:* Am Inst Aeronaut & Astronaut; Can Aeronaut & Space Inst. *Res:* Flow instabilities; dynamics of concentrated vortices; physics of fluids; microgravity fluid mechanics; dynamics of liquid sloshing; biofluid mechanics; computational fluid dynamics. *Mailing Add:* Dept Mech Eng Concordia Univ 1455 de Maisonneuve Blvd W Montreal PQ H3G 1M8 Can

VATNE, ROBERT DAHLMEIER, b Pipestone, Minn, Oct 21, 34; m 63; c 1. PARASITOLOGY. *Educ:* Augustana Col, SDak, BA, 56; Kans State Univ, MS, 58, PhD(parasitol), 63. *Prof Exp:* Asst prof biol, St Cloud State Col, 63-64; Salsbury Labs, Iowa, 65-69; res specialist, 70-73, Dow Chem Co, USA, 70-73, regist specialist, 74-81, assoc scientist, 86-90, PROD REGIST MGR, DOW CHEM CO, USA, 81-; SR RES SCIENTIST, DOW ELANCO, 91- *Mem:* Am Soc Parasitol. *Res:* Etiology and chemotherapy of histomoniasis, coccidiosis and helminthiasis. *Mailing Add:* Dow Elanco Quad IV 9002 Purdue Rd Indianapolis IN 46268-1189

VATSIS, KOSTAS PETROS, b Patras, Greece, May 6, 45. ENZYMOLOGY. *Educ:* Calif State Univ, Long Beach, BS, 67, MS, 69; Univ Ill Med Ctr, PhD(pharmacol), 75. *Prof Exp:* Res asst pharmacol, Col Med, Univ Ill, 69-75; scholar biol chem, Med Sch, Univ Mich, 75-76, lectr, 76-78; from asst prof to assoc prof pharmacol, Med Sch, Northwestern Univ, 78-87; RES SCIENTIST, UNIV MICH MED SCH, 87- *Concurrent Pos:* Consult, Dept Pharmacol, Col Med, Univ Ill, 77-; prin investr, starter grant, Pharmaceut Mfg Asn Fedn, Inc, 80-81, res grant NIH, 80-83. *Mem:* Am Soc Pharmacol & Exp Therapeut; Soc Toxicol; Am Soc Biol Chemists. *Res:* Physicochemical studies on the mechanism of interaction of components of the nicotinamide-adenine dinucleotide phosphate and nicotinamide dinucleotide linked electron transport chains in mammalian hepatic microsomes: modulation of the cytochrome P-450 containing monooxygenase system by cytochrome b5 and related membrane-bound proteins. *Mailing Add:* Dept Pharmacol Univ Mich Med Sch Med Sci 1 M6322-0626 Ann Arbor MI 48109-0626

VAUCHER, JEAN G, b Can, 42; m; c 2. COMPUTER SCIENCE. *Educ:* Univ Ottawa, BSc, 62; Univ Manchester, MSc, 64, PhD, 68. *Prof Exp:* Researcher, IBM, Can, 68-70; from asst prof to assoc prof, 70-80, chmn, 81-83, PROF COMPUT SCI, UNIV MONTREAL, 80- *Concurrent Pos:* Athlone fel, 62-64. *Mem:* Asn Comput Mach. *Res:* Simulation; intelligent systems; parallelism; structured programming. *Mailing Add:* Dept Informatique Sci & RO PO 6128 Sta A Montreal PQ H3C 3J7 Can

VAUDO, ANTHONY FRANK, b Brooklyn, NY, Jan 21, 46; m 67; c 1. PHYSICAL CHEMISTRY. *Educ:* City Col New York, BS, 66; Mass Inst Technol, PhD(phys chem), 70. *Prof Exp:* NSF fel, Boston Univ, 70-71; sr develop chemist, St Regis Paper Co, 71-74, tech dir, Laminated & Coated Prods Div, 74-80; CONSULT, 80- *Concurrent Pos:* Tech dir, flexible package div, Princeton Pkg Inc, 81-84, dir opers, 84-85, general mgr, 85. *Mem:* Am Chem Soc; Soc Photog Sci & Eng; Tech Asn Pulp & Paper Indust. *Res:* Photochemistry and its applications to imaging materials. *Mailing Add:* 6534 Wickerwood Dr Dallas TX 75248

VAUGHAN, BURTON EUGENE, b Santa Rosa, Calif, May 31, 26; m 49; c 2. PHYSIOLOGY, BIOPHYSICS. *Educ:* Univ Calif, Berkeley, AB, 49, PhD, 55. *Prof Exp:* Vis scientist, White Mountain High Altitude Res Sta, Calif, 53-54; proj leader, Oper Deepfreeze I, US Exped to Antarctic, 55; staff scientist biophys br, US Naval Radiol Defense Lab, 56-61, br head, 62-69; MGR ECOSYSTS DEPT, PAC NORTHWEST LABS, BATTELLE MEM INST, 69- *Concurrent Pos:* Actg instr, Sch Med, Stanford Univ, 57, Lectr, 60-63, res assoc, 63-70; consult, Clin Invest Ctr, Oakland Naval Hosp, 60-61 & US Naval Med Res Unit 2, Taipei, Taiwan, 61, 64 & 65; trustee, Independent Sch Dist, 63, presiding off, 64-65; rep, County Comt Sch Dist Orgn, 65; mem bd educ, Castro Valley Unified Sch Dist, 65, pres, 66-67; consult, Govt Health Facil, Manila, Philippines; sci coun chmn, Pac Sci Ctr Found, Seattle, 75-78, mem exec comn & trustee, 78- *Mem:* Am Soc Plant Physiol; Am Physiol Soc; Biophys Soc; NY Acad Sci; Radiation Res Soc; Sigma Xi. *Res:* Electrolyte and other absorption processes, including permeability, metabolically coupled transport; gastro-intestinal irradiation injury in the mammal; ionic uptake, discrimination and fixation by terrestrial plant and marine algal tissues; radiocontamination processes in organisms and ecological systems; pollution biology and pathways; plant and animal physiology. *Mailing Add:* 2456 Harris Ave Richland WA 99352

VAUGHAN, DAVID ARTHUR, b Mattoon, Wis, Mar 5, 23; m 51; c 1. NUTRITION. *Educ:* Univ Calif, BA, 49; Univ Ill, MA, 54, PhD(animal nutrit), 55. *Prof Exp:* Res physiologist, Arctic Aeromed Lab, USAF, 55-59, supvry chemist, 59-68, res biochemist, USAF Sch Aerospace Med, 68-69; res biochemist, Nutrit Inst, Sci & Educ Admin-Agr Res, USDA, 69-79; RETIRED. *Mem:* Am Physiol Soc; Am Inst Nutrit. *Res:* Survival nutrition in the Arctic; vitamin B requirements and intermediary metabolism during stress; biochemistry of hibernation; protein nutrition. *Mailing Add:* 8533 Pineway Dr Laurel MD 20723-1239

VAUGHAN, DAVID SHERWOOD, mechanical engineering; deceased, see previous edition for last biography

VAUGHAN, DEBORAH WHITTAKER, b Concord, NH, Nov 30, 43; m 66. NEUROANATOMY. *Educ:* Univ Vt, BA, 66; Boston Univ, PhD(biol), 71. *Prof Exp:* USPHS fel, 71-72, res asst prof, 72-78, ASST PROF NEUROANAT, SCH MED, BOSTON UNIV, 78- *Res:* Electron microscopic analysis of neocortex, primarily of rat, in regards to the effects of aging on the brain. *Mailing Add:* Dept Anat Boston Univ Sch Med 80 E Concord St Boston MA 02118

VAUGHAN, DOUGLAS STANWOOD, b Biddeford, Maine, July 12, 46; m 76. POPULATION DYNAMICS, RISK ASSESSMENT. *Educ:* Univ NH, BS, 68; Pa State Univ, MA, 70; Univ RI, PhD(oceanog), 77. *Prof Exp:* Statistician, Environ Protection Agency, 71-73; res assoc, Oak Ridge Nat Lab, 77-82; leader, Stock Dynamics Br, 82-85, LEADER, MENHADEN TEAM, BEAUFORT LAB, SOUTHEASTERN FISHERIES CTR, NAT MARINE FISHERIES, 85- *Concurrent Pos:* Co-prin investr, eval use life hist data to assess impact pollution on fish pop, Ocean Assessment Div, Nat Oceanic & Atmospheric Admin. *Mem:* Biomet Soc; Am Fisheries Soc; Int Oceanog Found; fel Am Inst Fish Res Biol; Sigma Xi. *Res:* Application of statistical methods and population and community modeling approaches to assessing the effects of one or more stresses on fish populations, and the development of approaches for environmental risk assessment; application of statistics to fish stock assessment; use of matrix models for describing population dynamics. *Mailing Add:* Nat Marine Fisheries Serv Beaufort Lab Beaufort NC 28516

VAUGHAN, HERBERT EDWARD, b Ogdensburg, NY, Feb 18, 11; m 47. MATHEMATICS. *Educ:* Univ Mich, BS, 32, AM, 33, PhD(math), 35. *Prof Exp:* Instr math, Brown Univ, 35-36; Lloyd fel, Univ Mich, 36-37; instr, 37-41, assoc, 41-45, from asst prof to assoc prof, 45-60, PROF MATH, UNIV ILL, URBANA, 60-, MEM UNIV COMT ON SCH MATH PROJ EDUC, 56- *Mem:* Am Math Soc; Math Asn Am; Asn Symbolic Logic. *Res:* Topology; abstract spaces. *Mailing Add:* 907 S Vine St Urbana IL 61801

VAUGHAN, J RODNEY M, b Margate, Eng, May 2, 21; US citizen; m 48; c 3. MICROWAVE ENGINEERING, COMPUTER AIDED DESIGN. *Educ:* Cambridge Univ, BA, 48, MA, 57, PhD, 72. *Prof Exp:* Engr, Res Labs, Elec & Musical Indust, Eng, 48-57; sr engr & proj mgr, Tube Dept, Gen Elec Co, NY, 57-68; chief scientist, Tube Div, Litton Industs, Inc, San Carlos, 68-89; PRES, RODNEY VAUGHAN ASSOCS, INC, 89- *Concurrent Pos:* Chmn, MIL-E-I, Microwave Rev Comt, 66-68; lectr math, Col Notre Dame, Calif, 69-70; chmn adv bd, Air Force Thermionic Eng Res Prog, 79-80. *Mem:* Fel Inst Elec & Electronics Engrs; Electron Devices Soc (secy, 81). *Res:* High power microwave tubes; advanced computer methods. *Mailing Add:* Two Sequoia Way Redwood City CA 94061

VAUGHAN, JAMES ROLAND, b Allentown, Pa, June 7, 28; m 50; c 3. MICROBIOLOGY, BIOCHEMISTRY. *Educ:* Muhlenberg Col, BS, 52; Lehigh Univ, MS, 54, PhD(biol), 61. *Prof Exp:* From instr to assoc prof microbiol, Muhlenberg Col, 56-67, head dept, 65-90, PROF MICROBIOL, MUHLENBERG COL, 67-, SR PROF, 90- *Mem:* AAAS; Am Soc Microbiol; Am Chem Soc; Am Soc Cell Biol; Sigma Xi. *Res:* Microbial physiology. *Mailing Add:* Dept Biol Muhlenberg Col Allentown PA 18104

VAUGHAN, JERRY EUGENE, b Gastonia, NC, Oct 30, 39; m 69. MATHEMATICS, TOPOLOGY. *Educ:* Davidson Col, BS, 61; Duke Univ, PhD(math), 65. *Prof Exp:* Teaching asst math, Duke Univ, 62-63; assoc prof, Eve Div, Univ Md, 66-67; asst prof, Univ NC, Chapel Hill, 67-73; assoc prof, 73-76, PROF MATH, UNIV NC, GREENSBORO, 76- *Mem:* Am Math Soc; Math Asn Am. *Res:* General topology; generalized metric spaces; product spaces; cardinal invariant properties. *Mailing Add:* Dept Math 383 Bus Econ Bldg Univ NC 1000 Spring Garden St Greensboro NC 27412

VAUGHAN, JOHN DIXON, b Clarksville, Va, Mar 3, 25; m 62. PHYSICAL CHEMISTRY. *Educ:* Col William & Mary, BS, 50; Univ Ill, PhD(chem), 54. *Prof Exp:* Asst phys chem, Univ Ill, 50-51, AEC, 51-54; res chemist, Chem Dept, Exp Sta, E I du Pont de Nemours & Co, 54-58; asst prof phys chem, Va Polytech Inst, 59-62 & Univ Hawaii, 62-64; assoc prof, 64-71, PROF PHYS CHEM, COLO STATE UNIV, 71- *Mem:* Am Chem Soc; Sigma Xi. *Res:* Computation chemistry, molecular modeling; reaction profile modeling. *Mailing Add:* Dept Chem Colo State Univ Ft Collins CO 80523

VAUGHAN, JOHN HEATH, b Richmond, Va, Nov 11, 21; m 46, 83; c 4. IMMUNOLOGY, MEDICINE. *Educ:* Harvard Univ, AB, 42, MD, 45. *Prof Exp:* Intern med, Peter Bent Brigham Hosp, 45-56, resident, 48-51; Nat Res Coun fel med sci, Col Physicians & Surgeons, Columbia Univ, 51-53; asst prof,

Med Col Va, 53-58; assoc prof med & asst prof bact, Sch Med & Dent, Univ Rochester, 58-63, prof med & head div immunol & infectious dis, 63-70; chmn clin div, Scripps Clin & Res Found, 70-74, chmn dept clin res, 74-77, head div clin immunol, 77-87; PROF MED RESIDENCE, UNIV CALIF, SAN DIEGO, 90- *Concurrent Pos:* Fel, Peter Bent Brigham Hosp, Boston, 48-51; consult, NIH, 56-63, mem bd sci counr, Nat Inst Allergy & Infectious Dis, 68-72; mem allergy clin immunol res comt, NIH, 80-84. *Honors & Awards:* Gold Medal Award, Am Col Rheumatology, 90. *Mem:* Am Soc Clin Invest; Asn Am Physicians; Am Asn Immunol; Am Rheumatism Asn (pres, 70-71); Am Acad Allergy (pres, 66-67). *Res:* Immunological phenomena in internal medicine; the basis of poor anti-viral responses in autoimmune and acquired immunodeficiency diseases; immunity to Epstein-Barr virus; autoimmunity and rheumatoid arthritis. *Mailing Add:* Dept Med Univ Calif San Diego La Jolla CA 92093-0945

VAUGHAN, JOHN THOMAS, b Tuskegee, Ala, Feb 6, 32; m 56; c 3. VETERINARY MEDICINE. *Educ:* Auburn Univ, DVM, 55, MS, 63. *Prof Exp:* From instr to assoc prof large animal surg & med, Auburn Univ, 55-70; prof vet surg & dir, Large Animal Hosp, NY State Vet Col, Cornell Univ, 70-74; chmn, dept large animal surg & med, 74-77, DEAN, COL VET MED, AUBURN UNIV, 77- *Mem:* Am Vet Med Asn; Am Asn Equine Practitioners (pres, 81); Am Asn Vet Clinicians; Am Col Vet Surg (pres, 80); Asn Am Vet Med Cols; Nat Academies Pract. *Res:* Large animal and equine surgery and medicine; general surgery of the equine system with emphasis on urogenital and gastrointestinal surgery. *Mailing Add:* Col Vet Med Auburn Univ 104 Greene Hall Auburn AL 36849

VAUGHAN, LINDA ANN, b Brooklyn, NY, July 26, 50; m 74; c 1. HUMAN NUTRITION. *Educ:* Univ Calif, Davis, BS, 72; Cornell Univ, MNS, 74; Univ Ariz, PhD(agr biochem, nutrit), 77. *Prof Exp:* Nutritionist II pub health nutrit, Maricopa Co Health Dept, 74-75; nutrit consult cardiac rehab, Dr A E Smith, Tempe, Ariz, 77; asst prof, Dept Food & Nutrit, Univ Nebr, 77-79; ASST PROF, DEPT HOME ECON, ARIZ STATE UNIV, 79- *Mem:* Am Dietetic Asn; Inst Food Technologists; Soc Nutrit Educ; Am Inst Nutrit; Am Sch Health Asn. *Res:* Maternal and infant nutrition; composition of human milk. *Mailing Add:* Dept Home Econ Ariz State Univ Tempe AZ 85287

VAUGHAN, LOY OTTIS, JR, b Birmingham, Ala, June 30, 45; m 66; c 1. MATHEMATICS. *Educ:* Fla State Univ, BA, 66; Univ Ala, MA, 67, PhD(math), 70. *Prof Exp:* Asst prof, 69-80, ASSOC PROF MATH, UNIV ALA, BIRMINGHAM, 80- *Mem:* Am Math Soc; Math Asn Am. *Res:* General topology, fixed and almost fixed point theory of continua. *Mailing Add:* Dept Math Univ Ala Univ Sta Birmingham AL 35294

VAUGHAN, MARTHA, b Dodgeville, Wis, Aug 4, 26; wid; c 3. BIOCHEMISTRY. *Educ:* Univ Chicago, PhB, 44; Yale Univ, MD, 49. *Prof Exp:* Asst instr res med, Univ Pa, 51-52; Nat Res Coun fel, NIH, 52-54; sr asst surgeon to med dir, USPHS, 54-89; CHIEF, LAB CELLULAR METAB, NAT HEART LUNG & BLOOD INST, NIH, 74- *Concurrent Pos:* mem metab study sect, Div Res Grants, USPHS, 65-68, head sect metab, 68-74; int fel rev comt, Fogarty Int Ctr, 73-74, 77-78; actg chief, Molecular Dis Br, Nat Heart & Lung Inst, 74-76; bd dir, Found Adv Educ Sci, Inc, 79-; vchmn, Gordon Conf Cyclei Nucleotides, 81, chmn, 82; ed, Biochem Biophys Res Commun, 90-91. *Honors & Awards:* G Burroughs Mider Lectr, NIH, 79; Harvey Soc Lectr, 82. *Mem:* Nat Acad Sci; Am Soc Biochem & Molecular Biol; Am Soc Clin Invest; Asn Am Physicians; Harvey Soc. *Res:* Mechanism of hormone action. *Mailing Add:* NIH Bldg 10 Rm 5N-307 9000 Rockville Pike Bethesda MD 20892

VAUGHAN, MARY KATHLEEN, b Houston, Tex, Sept 7, 43; m 66; c 3. PINEAL, NEUROENDOCRINOLOGY. *Educ:* Univ St Thomas, Houston, BA, 65; Univ Tex, Galveston, PhD(anat), 70. *Prof Exp:* Lectr human anat & fel, Univ Rochester, NY, 70-71; guest worker, NIH, Bethesda, Md, 71-73; fel, 73-75, asst prof, 75-80, ASSOC PROF CELL & STRUCT BIOL, UNIV TEX HEALTH SCI CTR, 80- *Mem:* Endocrine Soc; Soc Neurosci; Am Asn Anatomists; Int Soc Chronobiol; Am Soc Zoologists; Int Soc Psychoneuroendocrinol. *Res:* Interaction of the natural environment and pineal gland on the neuroendocrine-gonadal and neuroendocrine-thyroid axes. *Mailing Add:* 16707 Turkey Point San Antonio TX 78232

VAUGHAN, MICHAEL RAY, b Newport News, Va, Aug 11, 44; m 71; c 3. POPULATION DYNAMICS, HABITAT ECOLOGY. *Educ:* NC State Univ, BS, 71; Ore State Univ, MS, 74; Univ Wis, Madison, PhD(wildlife ecol), 79. *Prof Exp:* Res asst wildlife ecol, Ore State Univ, 71-74, Univ Wis-Madison, 74-79; actg asst leader, Wis Coop Wildlife Res Univ, Va Polytech Inst & State Univ, US Fish & Wildlife Serv, 79-80, asst leader, Va Coop Wildlife Res Unit Pop Dynamics, 80-82, leader, Va Coop Wildlife Res Unit, 82-85; CONSULT, 85- *Mem:* Wildlife Soc; Int Bear Asn. *Res:* Population dynamics and cyclic phenomena especially snowshoe hares, mountain goats, black bears, wild turkeys and white-tailed deer. *Mailing Add:* Va Coop Wildlife Res Unit 148 Cheatham Hall Va Polytech & State Univ Blacksburg VA 24061

VAUGHAN, MICHAEL THOMAS, b Washington, DC, Sept 29, 40; m 78; c 2. MINERALS ELASTICITY, MANTLE MINERALOGY. *Educ:* Shimer Col, BS, 60; Univ Cincinnati, MS, 65; State Univ, NY, MS, 76, PhD(geophysics), 79. *Prof Exp:* Instr physics, Thomas More Col, 68-70; asst prof, WVa State Col, 71-74; teaching asst geol, State Univ NY, 75-76, fel res asst geophys, 76-79; asst geophysicist, Hawaii Inst Geophys, 79-81; asst prof geol, Univ Ill, Chicago, 81-; DEPT EARTH & SPACE SCI, STATE UNIV NY. *Mem:* Am Geophys Union; Geol Soc Am; Mineral Soc Am. *Res:* Relations between crystal structure and elasticity, especially in high-pressure phases of oxides and silicates, by measuring the single-crystal elastic constants of suitable well-characterized crystals with known crystal structures. *Mailing Add:* Dept Earth & Space Sci State Univ NY Stonybrook NY 11794

VAUGHAN, NICK HAMPTON, b Graham, Tex, Feb 11, 23; m 60; c 2. MATHEMATICS. *Educ:* NTex State Univ, BS, 47, MS, 48; La State Univ, PhD(math), 68. *Prof Exp:* Mathematician, US Naval Ord Plant, Ind, 51-53; res assoc math, Statist Lab, Purdue Univ, Lafayette, 53-55; sr aerophysics engr, Gen Dynamics, Tex, 55; instr math, La State Univ, 55-58; asst prof, NTex State Univ, 58-65; instr, La State Univ, 65-68; assoc prof, 68-78, PROF MATH, UNIV NTEX, 78- *Mem:* Am Math Soc; Math Asn Am. *Res:* Commutative rings; ideal theory; algebraic number theory. *Mailing Add:* Dept Math NTex State Univ Denton TX 76203

VAUGHAN, PHILIP ALFRED, physical chemistry, for more information see previous edition

VAUGHAN, TERRY ALFRED, b Los Angeles, Calif, May 5, 28; m 50; c 2. VERTEBRATE ZOOLOGY. *Educ:* Pomona Col, BA, 50; Claremont Cols, MA, 52; Univ Kans, PhD, 58. *Prof Exp:* Asst zool, Univ Kans, 52-54 & 56-58; asst biologist, Colo State Univ, 58-64, assoc biologist, 64-70; PROF ZOOL, NORTHERN ARIZ UNIV, 70- *Mem:* Am Soc Mammalogists; Cooper Ornith Soc. *Res:* Chiropteran and rodent ecology; functional morphology. *Mailing Add:* PO Box 655 Northern Ariz Univ Box 5640 Lake Montezuma AZ 86342

VAUGHAN, THERESA PHILLIPS, b Kearney, Nebr, Oct 13, 41; m 69. ALGEBRA. *Educ:* Antioch Col, BA, 64; Am Univ, MA, 68; Duke Univ, PhD(math), 72. *Prof Exp:* Math programmer, Naval Ship Res & Develop Ctr, 64-69; asst prof math, NC Wesleyan Col, Rocky Mt, 72-73; LECTR MATH, UNIV NC, GREENSBORO, 74- *Mem:* Am Math Soc; Sigma Xi. *Res:* Polynomials and linear structure of finite fields; enumeration of sequences of integers by patterns; 2x2 matrices with positive integer entries; computation of discriminants. *Mailing Add:* 4112 Dogwood Dr Greensboro NC 27410

VAUGHAN, VICTOR CLARENCE, III, b Toledo, Ohio, July 19, 19; m 41; c 3. PEDIATRICS. *Educ:* Harvard Univ, AB, 39, MD, 43; Am Bd Pediat, dipl, 51, cert pediat allergy, 60. *Prof Exp:* Instr pediat, Sch Med, Yale Univ, 45-47 & 49-50, asst prof, 50-52; assoc prof, Sch Med, Temple Univ, 52-57; prof & chmn dept, Med Col Ga, 57-64; chmn dept, 64-76, PROF PEDIAT, SCH MED, TEMPLE UNIV, 64-; SR MED EVAL OFFICER, 81- *Concurrent Pos:* Res fel, Harvard Med Sch, 47-49; mem bd, Am Bd Pediat, 60-65 & 68-73, secy, 63-65, pres, 73; med dir, St Christopher's Hosp Children, 64-76; nat bd med examr, sr fel med eval, 77-81. *Mem:* Fel Am Acad Allergy; Soc Pediat Res (vpres, 64-65); Am Acad Pediat; Am Fedn Clin Res; Am Pediat Soc. *Res:* Hemolytic disease of newborn; human genetics; allergic disorders of children; human growth and development; medical education and evaluation. *Mailing Add:* 656 Junipero Serra Blvd Stanford CA 94305-8444

VAUGHAN, WILLIAM MACE, b Mt Vernon, NY, Aug 26, 42; m 65; c 2. ENVIRONMENTAL SCIENCE, SOFTWARE SYSTEMS. *Educ:* Wittenberg Univ, BS, 64; Univ Ill, Urbana-Champaign, MS, 66, PhD(biophys), 69. *Prof Exp:* Res assoc radiation sensitivity environ & monitoring, Ctr Biol Natural Systs, Wash Univ, St Louis, 69-71, coordr nitrogen proj, fertilizers-environ impact, 71-74; pres software serv, Eng Mgt Info Corp, 81-84; vpres serv, Air Qual Measurements, Environ Measurements Inc, 74-88; mgr midwest opers, Aerovironment, Inc, 85-87; PRES, ENVIRON SOLUTIONS INC, 87- *Concurrent Pos:* Lectr dept technol & human affairs, Wash Univ, 73-77; chmn EM-6 comt, Remote & Moving Measurements Air & Waste Mgt Asn, 85- *Mem:* Air & Waste Mgt Asn; AAAS; Am Asn Radon Scientists & Technologists. *Res:* Air toxic investigations; document progress in environment cleanup; assessment of hazardous waste problems through audits and field measurements; indoor air quality surveys; field determination of important processes in the transport and transformation of air pollutants through ambient measurements. *Mailing Add:* Environ Solutions Inc PO Box 11323 Clayton MO 63105-0123

VAUGHAN, WILLIAM WALTON, b Clearwater, Fla, Sept 7, 30; m 51; c 4. AEROSPACE SCIENCES. *Educ:* Univ Fla, BS, 51; Fla State Univ, cert meteorol, 52; Univ Tenn, PhD, 76. *Prof Exp:* Meteorologist, USAF, 52-55, res & develop meteorologist, Air Force Armament Ctr, 55-57, tech asst meteorol, Army Ballistic Missile Agency, 58-60; chief, Aerospace Environ Off, 60-65, chief, Aerospace Environ Div, 65-76, chief, Atmospheric Sci Div, Marshall Space Flight Ctr, NASA, 76-86; RES PROF & DIR, RES INST, UNIV ALA, HUNTSVILLE, 86- *Honors & Awards:* Exceptional Serv Medal, NASA, 69; Losey Atmospheric Sci Medal, Am Inst Aeronaut & Astronaut, 80. *Mem:* Am Meteorol Soc; Am Inst Aeronaut & Astronaut; Am Geophys Soc; AAAS; Sigma Xi. *Res:* Applied research in aerospace sciences and especially atmospheric science relative to space system and spacecraft experiment development; management of interdisciplinary vomousity research program. *Mailing Add:* 5606 Alta Dena Dr Huntsville AL 35802

VAUGHAN, WORTH E, b New York, NY, Feb 1, 36; m 69; c 4. PHYSICAL CHEMISTRY. *Educ:* Oberlin Col, AB, 57; Princeton Univ, AM, 59, PhD(phys chem), 60. *Prof Exp:* Res assoc phys chem, Princeton Univ, 60-61; from asst prof to assoc prof, 61-78, PROF CHEM, UNIV WIS-MADISON, 78- *Mem:* AAAS; Am Phys Soc; Am Chem Soc. *Res:* Dielectric and nuclear magnetic relaxation in liquids; irreversible statistical mechanics. *Mailing Add:* Dept Chem Univ Wis Madison WI 53706

VAUGHAN, WYMAN RISTINE, b Minneapolis, Minn, Oct 28, 16; m 43; c 2. CHEMISTRY. *Educ:* Dartmouth Col, AB, 39, AM, 41; Harvard Univ, AM, 42, PhD(org chem), 44. *Prof Exp:* Instr chem, Dartmouth Col, 39-41; res assoc, Harvard Univ, 42-44; res assoc, Dartmouth Col, 44-46; res assoc, Univ Mich, 46-47, from instr to prof, 47-66; prof chem & head dept, 66-76, EMER PROF, UNIV CONN, 81- *Mem:* Am Chem Soc; NY Acad Sci. *Res:* Synthetic organic chemistry; Diels-Alder reaction; sterochemistry; reaction mechanisms; potential anticancer agents; molecular rearrangements. *Mailing Add:* Dept Chem Univ Conn Storrs CT 06268

VAUGHEN, VICTOR C(ORNELIUS) A(DOLPH), b Wilmington, Del, Oct 8, 33; div; c 4. CHEMICAL ENGINEERING, CHEMISTRY. *Educ:* Stetson Univ, BS, 57; Mass Inst Technol, SB, 56, SM, 57, PhD(chem eng), 60. *Prof Exp:* Group leader, 63-76, mgr hot cell opers, 76-80, mgr fossil energy assessments, 80-81, prog mgr fossil energy eval, 81, SECT HEAD, ENG COORD & ANALYSIS SECT, CHEM TECHNOL DIV, OAK RIDGE NAT LAB, MARTIN MARIETTA ENERGY SYSTS, INC, 81- *Concurrent Pos:* Exchange scientist, Fed Repub Ger, 72-74; Oak Ridge assoc univ travelling lectr, ethics of technol, 83; chmn session on ethics & technol (nuclear power, biotechnology, etc), Welding & Testing Technol Energy Conf. *Mem:* Am Nuclear Soc; Am Inst Chem Engrs; Nat Soc Prof Engrs; AAAS; Soc Risk Anal; Sigma Xi. *Res:* Nuclear fuel reprocessing; high radioactivity level chemical development; special isotope purification and production by solvent extraction and ion exchange; safety and radiation control and operation and maintenance of hot cells; coal liquefaction technology assessments; ethics of technology; professional and business ethics. *Mailing Add:* 106 Gordon Rd Oak Ridge TN 37830-6233

VAUGHN, CHARLES MELVIN, b Deadwood, SDak, Nov 23, 15; m 41; c 2. PARASITOLOGY, PROTOZOOLOGY. *Educ:* Univ Ill, BA, 39, MA, 40; Univ Wis, PhD(invert zool), 43. *Prof Exp:* Asst zool, Univ Ill, 39-40 & Univ Wis, 40-43; from asst prof to assoc prof, Miami Univ, 46-51; assoc dir, Field Serv Unit, Am Found Trop Med, 52-53; prof zool & chmn dept, Univ SDak, 53-65; prof zool, physiol & chmn dept, 65-71, actg dean res, 69-71, chmn dept, 71-78, prof zool, 71-81, EMER PROF ZOOL & PARASITOLOGIST, MIAMI UNIV, 81- *Concurrent Pos:* Asst prof parasitol & sr parasitologist, Inst Trop Med, Bowman Gray Sch Med, Wake Forest Col, 50-52, assoc dir, 51-52; dir, NSF Acad Year Inst, Univ SDak, 58-64, prog dir, NSF Col & Elem Prog, Res Training & Acad Year Study Prog, 64-65; consult, NSF, 65-69; biologist, Ohio State Univ-USAID Indian Educ Proj, 66 & NSF-USAID India Educ Proj, 67. *Honors & Awards:* Benjamin Harrison Medal, Miami Univ, 79. *Mem:* AAAS; Am Soc Parasitol; Am Micros Soc (pres, 73); Soc Protozool; Am Soc Trop Med & Hyg; Am Soc Zoologists. *Res:* Human and animal parasitology; malaria parasite and vector surveys; intestinal parasite surveys; schistosomiasis survey and control; physiology and ecology of gastropoda; life history and culture methods in protozoa, other invertebrates. *Mailing Add:* Dept Zool 180 Biol Sci Bldg 6295 Devonshire Dr Oxford OH 45056

VAUGHN, CLARENCE BENJAMIN, b Philadelphia, Pa, Dec 14, 28; m 53; c 4. ONCOLOGY, PHYSIOLOGICAL CHEMISTRY. *Educ:* Benedict Col, BS, 51; Howard Univ, MS, 55, MD, 57; Wayne State Univ, PhD(physiol chem), 65. *Prof Exp:* NIH res fel, 62-64, asst prof oncol, 67-81, CLIN ASSOC PROF INTERNAL MED, SCH MED, WAYNE STATE UNIV, 78-; DIR ONCOL, PROVIDENCE HOSP, 73- *Concurrent Pos:* Lab instr, Wayne State Univ, 63-67, assoc dept biochem, 68-; res physician, Milton A Darling Mem Ctr, Mich Cancer Found, 64-70, clin dir, 70-72; mem consult staff, Depts Med, Oakwood Hosp, 68-, Detroit Mem Hosp, 70-, Harper Hosp, 73-, Hutzel Hosp, 75-, Southfield Rehab, 76-, Sinai Hosp, 77-, Detroit Macomb Hosp Corp, 88-, Oakland Gen Hosp, 89-; mem, HEW Pub Adv Comt, 76-, Nat Cancer Inst Educ Rev Comt, 83-, adv comt, Nat Cancer Inst, 85- & bd trustees, Providence Hosp Found, 87-89; med dir oncol, Samaritan Health Ctr, 86-; clin prof, Wayne State Univ, 88-, Oakland Univ, 89-; chmn, Minority Res Subcomt, SWOG, 90- *Mem:* AMA; Am Col Physicians; NY Acad Sci; Am Chem Soc; Am Cancer Soc; AAAS; fel Am Col Clin Pharmacol; Am Soc Clin Oncol; Am Soc Prev Oncol; fel Am Inst Chemists. *Res:* Organic acid metabolism; ferritin; monoclonal antibodies; estrogen; progesterone receptors. *Mailing Add:* Southfield Oncol Inst Inc 27211 Lahser Rd No 200 Southfield MI 48034

VAUGHN, DANNY MACK, b Muskegon, Mich, Aug 19, 48; m 78; c 1. SURFACE GEOLOGY. *Educ:* Ind State Univ, BS, 78, PhD(phys geog), 84. *Prof Exp:* Fel phys geog, Ind State Univ, 79-83; asst prof geog & geol, Lake Superior State Col, 84-85; ASST PROF PHYS GEOG, GEOMORPHOL & AIR PHOTO MAP INTERPRETATION, UNIV MO, KANSAS CITY, 85- *Mem:* Asn Am Geographers; Geol Soc Am; Nat Speleol Soc; Sigma Xi. *Res:* Geologic and geomorphic interpretations from remote sensed imagery and topographic maps; paleohydrology. *Mailing Add:* Dept Geog Weber State Univ Ogden UT 84408-2510

VAUGHN, JACK C, b Burbank, Calif, July 4, 37; m 63; c 3. CELL BIOLOGY. *Educ:* Univ Calif, Los Angeles, BA, 60; Univ Tex, PhD(bot), 64. *Prof Exp:* Asst zool, Univ Calif, Los Angeles, 60-61; USPHS fel, Univ Wis, 64-66; from asst prof to assoc prof, 66-75, PROF ZOOL, MIAMI UNIV, 75- *Concurrent Pos:* NSF res grant, 67-69 & 70-71; investr, Marine Biol Lab, Woods Hole, 70 & 71. *Mem:* AAAS; Am Inst Biol Scientists; Am Soc Cell Biol; Int Fedn Cell Biol; Am Soc Zoologists; Am Phys Soc. *Res:* Cell biology; chromosome structure and function. *Mailing Add:* Dept Zool Miami Univ Oxford OH 45056

VAUGHN, JAMES E, JR, b Kansas City, Mo, Sept 17, 39; m 61; c 2. MOLECULAR NEUROMORPHOLOGY, DEVELOPMENTAL NEUROBIOLOGY. *Educ:* Westminster Col, BA, 61; Univ Calif, Los Angeles, PhD(anat), 65. *Prof Exp:* Fel brain res, Univ Edinburgh, 65-66; asst prof anat, Boston Univ, 66-70; assoc chmn, 83-86, HEAD SECT MOLECULAR NEUROMORPHOL, CITY OF HOPE, 70-, CHMN, DIV NEUROSCI, BECKMAN RES INST, 86- *Concurrent Pos:* Prin investr, NIH & NSF grants, 69-; fel, Neurosci Res Prog, 69; prog dir NIH prog proj grant, 80-; assoc ed J Neurocytol, 78-86, ed Synapse, 86-; pres res staff, Orgn of City of Hope, 86; mem adv comt, Lee Int Ctr Biomed Res, 87-90. *Mem:* Soc Neurosci Am Asn Anat; Am Soc Cell Biol; Int Brain Res Orgn; NY Acad Sci; AAAS. *Res:* Fine structure, molecular morphology and immunocytochemistry of adult and developing central nervous system. *Mailing Add:* Div Neurosci Beckman Res Inst City Hope 1450 E Duarte Rd Duarte CA 91010-0269

VAUGHN, JAMES L, b Marshfield, Wis, Mar 2, 34; div; c 4. INSECT PATHOLOGY. *Educ:* Univ Wis, BS, 57, MS, 59, PhD(bact), 62. *Prof Exp:* Res officer tissue cult virol, Insect Path Res Inst, Can Dept Forestry, 61-65; SUPVRY MICROBIOLOGIST, INSECT PATH LAB, USDA, 65-, RES LEADER, INSECT PATH LAB, 79- *Concurrent Pos:* Invert ed, In Vitro. *Mem:* Am Soc Microbiol; Soc Invert Path; Tissue Cult Asn; AAAS. *Res:* Methods for growth of insect tissue in vitro; study of processes of viral infection and development in in vitro insect systems. *Mailing Add:* Insect Biocontrol Lab Rm 214 Bldg 011A Agr Res Ctr-W USDA Beltsville MD 20705

VAUGHN, JOE WARREN, b Otterbein, Ind, Oct 8, 33; m 55; c 4. INORGANIC CHEMISTRY, PHYSICAL CHEMISTRY. *Educ:* DePauw Univ, BA, 55; Univ Ky, MS, 57, PhD(phys chem), 59. *Prof Exp:* Welch Found fel, Univ Tex, 59-61; from asst prof to assoc prof, 61-70, PROF CHEM, NORTHERN ILL UNIV, 70-, CHMN DEPT, 84- *Mem:* Am Chem Soc. *Res:* Nonaqueous solvent; fluoro complexes of trivalent chromium; stereochemistry of coordination compounds; synthesis of cis-trans isomers. *Mailing Add:* Dept Chem Northern Ill Univ De Kalb IL 60115

VAUGHN, JOHN B, b Birmingham, Ala, Mar 3, 24; m 45; c 6. EPIDEMIOLOGY. *Educ:* Auburn Univ, DVM, 49; Tulane Univ, MPH, 56. *Prof Exp:* From instr to asst prof, 57-60, assoc prof, 70-79, assoc dean, 77-79, PROF EPIDEMIOL & ACTG CHMN, DEPT APPL HEALTH SCI, SCH PUB HEALTH & TROP MED, TULANE UNIV, 79- *Concurrent Pos:* Consult, Epidemiol Sect, La State Dept Health, New Orleans, 57-; epidemic aid coordr, Sch Pub Health & Trop Med with Asn Schs Pub Health & Ctr Dis Control, 80-; liaison person, Sch Pub Health & Trop Med with Asn Schs Pub Health & Ctr Dis Control, 80-; mem contrib fac, Prev Med Residency Training Prog, Sch Pub Health & Trop Med, 80- *Mem:* Am Pub Health Asn; Soc Epidemiol Res. *Res:* Zoonotic epidemiology; author of numerous publications. *Mailing Add:* Tulane Univ 1501 Canal St Rm 613 New Orleans LA 70112

VAUGHN, MICHAEL THAYER, b Chicago, Ill, Aug 6, 36. UNIFIED GAUGE THEORIES. *Educ:* Columbia Univ, AB, 55; Purdue Univ, PhD(physics), 60. *Prof Exp:* Res assoc physics, Univ Pa, 59-62; asst prof, Ind Univ, 62-64; assoc prof, 64-73, PROF, NORTHEASTERN UNIV, 73- *Concurrent Pos:* Vis scientist, Argonne Nat Lab, 67, Deutsches Elektronen Synchrotron, Hamburg, 70, Univ Vienna, 70 & Int Ctr Theoret Physics, Trieste, 71; vis prof, Tex A&M Univ, 75 & Southampton Univ, 79; sr vis fel, Sci Res Coun, UK, 79, 86; vis lectr, Vanderbilt Univ, 85. *Honors & Awards:* Karl Lark Horovitz Prize, Purdue, 59. *Mem:* Am Phys Soc; AAAS. *Res:* Unified gauge theories; renormalization group analysis; group theory; exceptional groups and octonions. *Mailing Add:* Dept Physics Northeastern Univ Boston MA 02115

VAUGHN, MOSES WILLIAM, b Rock Hill, SC, Nov 2, 13; m 42; c 3. FOOD TECHNOLOGY. *Educ:* WVa State Col, BS, 38; Mich State Univ, MS, 42 & 49; Univ Mass, PhD(food technol), 51. *Prof Exp:* Prof food technol, Univ Md, Eastern Shores, 46-69, prof agr, 69-76, prof animal nutrit & food technol, 76-84; RETIRED. *Mem:* AAAS; Inst Food Technologists. *Res:* Combination jellies and jams; chemical methods for detecting meat spoilage; nitrogen partitions in fish meal; bacteriological effects of ionizing radiations in fishery products; bacterial flora of bottom muds; effects of washing shellstock and shucked oysters to bacterial quality, microwave opening of oysters; nutritive value of selected pork products. *Mailing Add:* Rte 1 Box 49 Springhill Rd Hebron MD 21830

VAUGHN, PETER PAUL, b Altoona, Pa, Aug 22, 28; m 48; c 3. VERTEBRATE PALEONTOLOGY. *Educ:* Brooklyn Col, BA, 50; Harvard Univ, MA, 52, PhD(biol), 54. *Prof Exp:* Instr anat, Univ NC, 54-56; asst prof zool & asst cur fossil vert, Univ Kans, 56-57; assoc cur vert paleont, US Nat Mus, 57-58; from asst prof to assoc prof, 59-67, PROF ZOOL, UNIV CALIF, LOS ANGELES, 67- *Concurrent Pos:* Res assoc, Los Angeles County Mus, 61- *Mem:* Soc Vert Paleont; Paleont Soc. *Res:* Comparative vertebrate anatomy and paleontology; Paleozoic tetrapods; late Paleozoic vertebrate faunas and paleobiogeography. *Mailing Add:* Dept Biol 2203 Life Sci Univ Calif 405 Hilgard Ave Los Angeles CA 90024

VAUGHN, REESE HASKELL, food microbiology; deceased, see previous edition for last biography

VAUGHN, RICHARD CLEMENTS, b Ionia, Mich, Jan 17, 25; m 47; c 5. INDUSTRIAL ENGINEERING. *Educ:* Mich State Univ, BA, 48; Toledo Univ, MIE, 55. *Prof Exp:* Res & indust engr, various Ohio Co, 51-57; asst prof indust eng, Univ Fla, 57-62; from assoc prof to prof, Iowa State Univ, 62-86, emer prof indust eng, 87-; RETIRED. *Concurrent Pos:* Consult, Liberty Bell Mfg Co, 59-61; instr math, Univ Toledo, 51-57; sabbatical, Univ Wales, Swansea, 80-81. *Mem:* Sigma Xi; Am Soc Eng Educ; Am Soc Qual Control; Inst Mgt Sci. *Res:* Quality control, product liability, operations research. *Mailing Add:* 1519 Harding Ave Ames IA 50010

VAUGHN, ROBERT DONALD, chemical engineering, for more information see previous edition

VAUGHN, THOMAS HUNT, b Clay, Ky, Nov 11, 09; m 30; c 3. ORGANIC CHEMISTRY. *Educ:* Univ Notre Dame, BS, 31, MS, 32, PhD(org chem), 34. *Prof Exp:* Vpres, Vitox Labs, Ind, 33-34; head dept org res, Union Carbide & Carbon Res Labs, Inc, NY, 34-39; dir org res, Mich Alkali Co, 39-41, asst dir res, 41-43; asst dir, Wyandotte Chem Corp, 43-45, dir, 45-48, vpres, 48-53; vpres res & develop, Colgate-Palmolive Co, 53-57; exec vpres, Pabst Brewing Co, 57-60; PRES, THOMAS H VAUGHN & CO, WARETOWN, 60- *Concurrent Pos:* Consult to Secy Navy & Qm Gen, US Army, 43-53; mem Scand Res & Indust Tour, 46; lectr, Prog Bus Admin for Top Level Ger Indust Leaders, Mutual Security Agency Sem, 52; pres, Tech Sect, World Conf Surface Active Agents, Paris, 54; dir, Indust Res Inst, 54-58, from vpres to pres, 57-58; chmn, Conf Admin Res, 55; mem adv bd, Off Crit Tables, Nat

Acad Sci, 56; mem adv bd, Col Sci, Univ Notre Dame, 56-; ed, Res Mgt, 58-69; pres, Myzon Labs, Chicago, Ill, 60-62; tech consult, UN, 70- *Mem:* AAAS; fel Am Inst Chem; Am Inst Chem Eng; Am Chem Soc; Soc Indust Chem. *Res:* Unsaturated compounds and their derivatives; polyols; detergents; polymers; research and business administration. *Mailing Add:* Potter & Brumfield Inc 200 Richland Creek Dr Princeton IN 47671

VAUGHN, WILLIAM KING, b Denison, Tex, July 28, 38; m 60; c 2. BIOSTATISTICS. *Educ:* Tex Wesleyan Col, BS, 60; Southern Methodist Univ, MS, 65; Tex A&M Univ, PhD(statist), 70. *Prof Exp:* Med technologist, St Joseph Hosp, Ft Worth, Tex, 61-63; res analyst biostatist, Univ Tex M D Anderson Hosp & Tumor Inst, 65-67; asst prof, 70-77, ASSOC PROF BIOSTATIST, SCH MED, VANDERBILT UNIV, 77- *Mem:* Am Statist Asn; Sigma Xi. *Res:* Clinical trials; simulation and Monte Carlo studies; design of experiments; statistical methodology in toxicology; biometrics. *Mailing Add:* 1632 Doug Olson El Paso TX 79936

VAUGHT, JIMMIE BARTON, EPIDEMIOLOGY OCCUPATIONAL CANCER. *Educ:* Med Col Ga, PhD(biochem), 77. *Prof Exp:* MEM STAFF, DEPT OCCUP STUDIES, WESTAT INC. *Mailing Add:* Dept Occup Studies Westat Inc 1650 Research Blvd Rockville MD 20850

VAUGHT, ROBERT L, b Alhambra, Calif, Apr 4, 26; m 55; c 2. MATHEMATICS. *Educ:* Univ Calif, AB, 45, PhD(math), 54. *Prof Exp:* From instr to asst prof math, Univ Wash, 54-58; from asst prof to assoc prof, 58-63, PROF MATH, UNIV CALIF, BERKELEY, 63- *Concurrent Pos:* Fulbright scholar, Univ Amsterdam, 56-57; NSF fel, Univ Calif, Los Angeles, 63-64; Guggenheim fel, 67. *Honors & Awards:* Carol Karp Prize, Int Asn Symbolic Logic, 78. *Mem:* Am Math Soc; Int Asn Symbolic Logic. *Res:* Foundations of mathematics. *Mailing Add:* Dept Math Univ Calif Berkeley CA 94720

VAUN, WILLIAM STRATIN, b Hartford, Conn, Oct 10, 29; m 61; c 1. MEDICINE. *Educ:* Trinity Col, Conn, BS, 51; Univ Pa, MD, 55. *Prof Exp:* Resident internal med, Hartford Hosp, Conn, 55-57 & 59-61; asst dir med, St Luke's Hosp, Cleveland, Ohio, 62-65; assoc prof med, 69-72, actg dean, Sch Continuing Educ, 72-73, PROF MED, SCH MED, HAHNEMANN UNIV, 72-, ASST DEAN, 85-; DIR MED EDUC, MONMOUTH MED CTR, 65- *Concurrent Pos:* Consult, Am Bd Internal Med, 73-77 & Nat Acad Sci, 75-76; chmn adv coun, Off Consumer Health Educ, Rutgers Med Sch, 73-75; dir, Greenwall Found, NY, 75-; trustee, Monmouth Col, West Long Branch, NJ, 79-, Hosp Res & Educ Trust NJ, 80-84; pres, Frank & Louise Groff Found, Red Bank, NJ,79-; vchmn bd trustees, Monmouth Co, 84- *Mem:* Asn Hosp Med Educ (vpres, 72-73); fel Am Col Physicians; Am Soc Internal Med; Am Geriat Soc. *Mailing Add:* Monmouth Med Ctr Long Branch NJ 07740

VAUPEL, DONALD BRUCE, b Hackensack, NJ, Aug 30, 42; m 67; c 2. PHARMACOLOGY. *Educ:* Wittenberg Univ, BA, 64; Univ Ky, MS, 70, PhD(pharmacol), 74. *Prof Exp:* Chemist qual control, Lederle Labs, Am Cyanamid Co, 64-66; PHARMACOLOGIST, ADDICTION RES CTR, NAT INST DRUG ABUSE, 72- *Concurrent Pos:* Asst adj prof pharmacol, Col Med, Univ Ky, 78- *Mem:* Sigma Xi; Am Soc Pharmacol & Exp Therapeut; Soc Neurosci. *Res:* Assess pharmacological equivalence and study mechanisms of action of abused drugs, particularly opioids and hallucinogens; animal physiology and behavior isolated tissue preparations subjective assessments in humans. *Mailing Add:* 900 Fairway Dr Towson MD 21204

VAUPEL, MARTIN ROBERT, b Evansville, Ind, July 24, 28. ANATOMY, EMBRYOLOGY. *Educ:* Ind Univ, AB, 49; Tulane Univ, PhD(anat), 54. *Prof Exp:* Res asst, 49-50, grad asst, 50-54, from instr to assoc prof, 54-86, PROF ANAT, TULANE UNIV, 86- *Mem:* Teratology Soc; Am Asn Anatomists. *Res:* Congenital anomalies of nervous system; teratology; reproductive endocrinology. *Mailing Add:* Dept Anat Tulane Univ Sch Med 1430 Tulane Ave New Orleans LA 70112

VAUSE, EDWIN H(AMILTON), b Chicago, Ill, Mar 30, 23; m 51; c 5. CHEMICAL ENGINEERING, CHEMISTRY. *Educ:* Univ Ill, BS, 47, MS, 48; Univ Chicago, MBA, 52. *Hon Degrees:* DSc, Univ Evansville, 77. *Prof Exp:* Chem engr, Standard Oil Co, 48-49, sr chem engr, 49-50, res projs engr, 50-51, asst gen foreman, Mfg Projs Div, 51-52, Light Oils Div, 52-53 & Hwy Oils Div, 53-57; dir lab admin, Mead Johnson & Co, 57-60; vpres, Charles F Kettering Found, 60-; RETIRED. *Mem:* Am Inst Chem Engrs; NY Acad Sci; Agr Res Inst. *Res:* Management of research on science technology and public policy formulation; science and technology policy issues. *Mailing Add:* 11834 Calle Parral San Diego CA 92128

VAUX, HENRY JAMES, b Bryn Mawr, Pa, Nov 6, 12; m 37; c 2. ECONOMICS POLICY. *Educ:* Haverford Col, BS, 33; Univ Calif, MS, 35, PhD(agr econ), 48. *Hon Degrees:* DSc, Haverford Col, 85. *Prof Exp:* Instr forestry, Ore State Col, 37-42; asst economist, La Agr Exp Sta, 42-43; assoc economist, US Army, 43; assoc economist, US Forest Serv, 46-48; from lectr to assoc prof forestry, Sch Forestry, 48-53, dean, Sch Forestry & assoc dir, Agr Exp Sta, 55-65, prof, 53-78, EMER PROF FORESTRY, SCH FORESTRY, UNIV CALIF, BERKELEY, 78- *Concurrent Pos:* Consult ed, McGraw-Hill Bk Co, 54-76; chmn, Calif Bd Forestry, 76-83. *Honors & Awards:* Gifford Pinchot Medal, Soc Am Foresters, 83. *Mem:* AAAS; fel Soc Am Foresters; Forest Hist Soc; hon mem Soc Foresters Finland. *Res:* Long term timber supply; price behavior and market structures for forest products. *Mailing Add:* 145 Mulford Univ Calif Berkeley CA 94720

VAUX, JAMES EDWARD, JR, b Pittsburgh, Pa, June 13, 32; m 54; c 3. CHEMISTRY, STATISTICS. *Educ:* Carnegie Inst, BS, 52, MS, 64, PhD(chem), 67. *Prof Exp:* Chemist, E I du Pont de Nemours & Co, Inc, 53-57; teacher, Shady Side Acad, 57-64; asst dir, Found Study Cycles, 64-67, exec dir, 67-76; LECTR CHEM, UNIV PITTSBURGH, 65-; ASSOC PROF & CHMN, CARLOW COL, 81- *Mem:* Am Chem Soc; Am Statist Asn. *Res:* Organic and analytical chemistry; statistics. *Mailing Add:* Carlow Col 3333 5th Ave Pittsburgh PA 15213-3165

VAVICH, MITCHELL GEORGE, b Miami, Ariz, Aug 24, 16; m 37; c 1. NUTRITION. *Educ:* Univ Ariz, BS, 38, MS, 40; Pa State Univ, PhD(biochem), 43. *Prof Exp:* Asst physiol chem, Pa State Univ, 40-42, instr biochem, 43-46; from assoc prof to prof agr biochem, 46-75, head dept, 69-75, prof food sci, 75-81, EMER PROF, UNIV ARIZ, 81- *Concurrent Pos:* Spec field staff mem, Rockefeller Found, 65-67; chmn, Comt Agr Biochem & Nutrit, 69- *Mem:* AAAS; Am Chem Soc; Am Inst Nutrit. *Res:* Fluorides, ascorbic acid; vitamins in canned foods; interrelationships of vitamins and other food components; carotenes and vitamin A; nutritional status; biochemistry of cyclopropenoid fatty acids; nutrient value of dietary proteins. *Mailing Add:* 4733 E Cherry Hills Dr Tucson AZ 85718

VAVRA, JAMES JOSEPH, b Boulder, Colo, Aug 30, 29; m 51; c 5. INFECTIOUS DISEASES. *Educ:* Univ Colo, BA, 51; Univ Wis, MS, 53, PhD, 55. *Prof Exp:* Res assoc, dept biochem, Upjohn Co, 55-57, res assoc & proj leader, dept microbiol, 57-62, sr res scientist, dept clin res, 62-64 & dept microbiol, 64-68, assoc dir infectious dis res, 68-89; CONSULT, 89- *Mem:* AAAS; Am Soc Microbiol; NY Acad Sci; Sigma Xi. *Res:* Fermentation biochemistry, especially metabolism; antibiotic-pathogen relationships, especially resistance development. *Mailing Add:* 1505 Royal Oak Ave Portage MI 49002

VAWTER, ALFRED THOMAS, b Los Angeles, Calif, Mar 30, 43; m 66; c 2. ECOLOGY, EVOLUTIONARY BIOLOGY. *Educ:* Univ Calif, Irvine, BS, 70; Cornell Univ, PhD(biol), 77. *Prof Exp:* Lectr ecol, Cornell Univ, 76-77; scholar evolution, Univ Calif, Los Angeles, 77-78; ASST PROF BIOL, WELLS COL, 78- *Concurrent Pos:* Consult, Resource Planning Assocs, 78-80, Glacier Nat Park, 87. *Mem:* Am Soc Naturalists; Soc Study Evolution; Soc Conserv Biol; Lepidoptera Res Found. *Res:* Population biology, especially ecological genetics of Lepidoptera; molecular evolution; conserv biol. *Mailing Add:* Dept Biol Wells Col Aurora NY 13026

VAWTER, SPENCER MAX, b Morgan Co, Ind, Feb 18, 37; m 72; c 3. PHYSICS. *Educ:* Franklin Col, BA, 59; DePaul Univ, MS, 67; Univ Mich, Ann Arbor, cert physiol, 68; Univ Southern Calif, cert comput systs, 70. *Prof Exp:* Student, Defense Projs Div, Western Elec Co at Lincoln Lab, Mass Inst Technol, 59-60, systs planning & develop engr, 60-63; dir ionizing radiation sect, AMA, Chicago, 63-66 & med physics sect, 66-70, assoc dir dept med instrumentation, 70-73; asst to pres, Bio-Dynamics, Inc, 73-76; pres, Bio-Sound, Inc, 76-85, chmn bd, 85-86; pres, Vascular Diagnostics, Inc, 86-87; PRES, LABSONICS, INC, 87- *Mem:* Am Phys Soc; Instrument Soc Am; Asn Advan Med Instrumentation. *Res:* Interaction of electromagnetic energy with human tissue, especially the effect of laser energy on the human eye; application of physical principles to medical practice; application of physical sciences to health care. *Mailing Add:* 3330 Bay Rd South Dr Indianapolis IN 46240

VAYO, HARRIS WESTCOTT, b Chicago, Ill, Nov 15, 35; m 62; c 3. APPLIED MATHEMATICS. *Educ:* Culver-Stockton Col, BA, 57; Univ Ill, MS, 59, PhD(math), 63. *Prof Exp:* Asst math, Univ Ill, 57-62; res fel biomath, Harvard Univ, 63-65; from asst prof to assoc prof, 65-74, PROF MATH, UNIV TOLEDO, 74- *Concurrent Pos:* Mem coun basic sci, Am Heart Asn, 67-; vis prof, McGill Univ, 76, Ill Col, 86-87; vis scientist, Cornell Univ, 89. *Mem:* Math Asn Am; Am Math Soc; Am Acad Mech; Soc Math Biol; Soc Indust & Appl Math. *Res:* Applications of mathematics to biological and medical problems, particularly cardiovascular work and red cell biomechanics. *Mailing Add:* Dept Math Univ Toledo Toledo OH 43606

VAZ, NUNO A, b Carmona, Portugal, Nov 23, 51; US citizen; m 77; c 3. FLAT PANEL DISPLAYS & LIQUID CRYSTAL DEVICES, ELECTRO-OPTICS. *Educ:* Tech Univ Lisbon, Portugal, EE, 75; Kent State Univ, BS, 77, PhD(physics), 80. *Prof Exp:* Res asst, Inst Physics & Math, Lisbon, Portugal, 75-76; res fel, dept physics, Kent State Univ, 81-82, res assoc, 82-83; sr res scientist, 84-86, STAFF RES SCIENTIST, GEN MOTORS RES LABS, 86- *Concurrent Pos:* Fel, Matsumae Int Found, 83. *Honors & Awards:* Campbell Award, 90. *Mem:* Sigma Xi; Am Phys Soc; Int Soc Magnetic Resonance; Soc Automotive Engrs. *Res:* Experimental and applied research on liquid crystals, in particular, phase behavior, phase transitions and microdispersions, using primarily optical techniques, nuclear magnetic resonance and other spectroscopic techniques. *Mailing Add:* Dept Physics Gen Motors Res Labs 30050 Mound Rd Warren MI 48090-9055

VAZIRI, MENOUCHEHR, b Tehran, Iran, June, 3, 51; US citizen. TRANSPORTATION, SYSTEMS ANALYSIS. *Educ:* Pahlavi Univ, Siraz, Iran, BSc, 75; Univ Calif, Davis, MSc, 76, PhD(civil eng), 80; Armstrong Col, Berkeley, Calif, MBA, 79; Golden Gate Univ, San Francisco, Calif, MPA. *Hon Degrees:* Cert, Kyoto Univ, Japan & Int Coop Comput, Japan, 83. *Prof Exp:* Design engr civil eng, Ziband Consult Engrs, 74-75; res asst transp, dept civil engr, Univ Calif, 77-78, teaching asst civil engr, 79-80; asst prof, 80-83, res assoc, 83-84, ASST PROF TRANSP, DEPT CIVIL ENG, UNIV KY, 84- *Concurrent Pos:* Prin investr, Univ res prog, US Dept Transp, Urban Mass Transp Admin, 82-83; consult, Am Engrs, 86 & Spalding Assocs, 85-; co-prin investr, Div Mass Transp, Ky Dept Transp, 85. *Honors & Awards:* Ralph Teetor Award, Soc Automotive Eng, 82; Mobusho, Japanese Govt, 83. *Mem:* Am Soc Civil Engrs; Inst Transp Engrs; Transp Res Bd; Japan Soc Civil Eng; Sigma Xi; Soc Automotive Engrs. *Res:* Transportation engineering and planning; transportation modeling; public transit; analysis of relationships between land-use and transportation; transportation technology and developments and their relationships to national goals and development; systems analysis and operation research; network analysis; land-use and urban planning; water resources planning and operation; business and public administration. *Mailing Add:* Dept Civil Eng Univ Ky Lexington KY 40506-0046

VAZQUEZ, ALFREDO JORGE, b Buenos Aires, Arg, Jan 21, 37; m 62; c 1. NEUROPHARMACOLOGY, ELECTROPHYSIOLOGY. *Educ:* Bernadino Rivadavia Col, Arg, BS, 54; Univ Buenos Aires, MD, 62. *Prof Exp:* From intern to resident med, Tigre Hosp, Arg, 59-61; instr pharmacol,

Chicago Med Sch, 62-65, assoc, 65-67; asst prof, Fac Med, Univ Man, 67-71; ASSOC PROF PHARMACOL, CHICAGO MED SCH, 71-, VCHMN DEPT, 77- & ACTG CHMN DEPT, 85- *Concurrent Pos:* Res assoc, Lab Psychopharmacol, Nat Neuropsychiat Inst, Buenos Aires, 60-62. *Mem:* Am Soc Pharmacol & Exp Therapeut. *Res:* Physiology and pharmacology of the cerebral cortex; epilepsy; drug abuse and hallucinogenic drugs; cardiovascular. *Mailing Add:* Dept Pharmacol Chicago Med Sch 3333 N Greenbay Rd North Chicago IL 60064

VAZQUEZ, JACINTO JOSEPH, pathology, for more information see previous edition

VEACH, ALLEN MARSHALL, b Lancaster, SC, Sept 21, 33. ACCELERATOR PHYSICS. *Educ:* Univ Ala, BS, 56; Univ Akron, MS, 59. *Prof Exp:* Physicist phys testing, B F Goodrich Co, 56-57, microscopist chem micros, 57-59; develop specialist isotope separation, 59-60, physicist plasma physics, 60-70, RES PHYSICIST ION SOURCES & ACCELERATORS, OAK RIDGE NAT LAB, 70- *Mem:* Am Phys Soc. *Res:* Plasma physics; accelerators; ion sources; high vacuum; isotope separation; electron and ion emission; electrical and gas discharges; ion and electron optics; ionization phenomena; mass spectroscopy. *Mailing Add:* RR 3 No 227A Crossville TN 38555

VEAL, BOYD WILLIAM, JR, b Chance, SDak, May 21, 37; m 62; c 2. SOLID STATE PHYSICS. *Educ:* SDak State Univ, BS, 59; Univ Pittsburgh, MS, 62; Univ Wis, PhD(physics), 69. *Prof Exp:* Engr, Westinghouse Res Labs, 59-63; asst physicist, 69-73, PHYSICIST, ARGONNE NAT LAB, 73- *Mem:* Am Phys Soc. *Res:* Electronic properties of solids. *Mailing Add:* Div Med Sci Argonne Nat Lab 9700 Cass Ave Argonne IL 60439

VEAL, DONALD L, b Chance, SDak, Apr 17, 31; m 53; c 2. METEOROLOGY. *Educ:* SDak State Univ, BS, 53; Univ Wyo, MS, 60, PhD, 64. *Prof Exp:* Instr civil eng, SDak State Univ, 57-58; from instr to prof atmospheric sci, Univ Wyo, 58-87, asst dir, Natural Resources Res Inst, 67-70, head dept, 70-77, vpres res, 77-81, pres res, 81-87; PRES & CHIEF EXEC OFFICER, PARTICLE MEASURING SYSTS, INC, 87- *Mem:* Am Meteorol Soc; Royal Meteorol Soc; Sigma Xi; Am Soc Eng Educ; Am Soc Civil Engrs; Nat Soc Prof Engrs; Am Geophys Union. *Res:* Cloud physics and weather modification. *Mailing Add:* 7018 Indian Peaks Trail Boulder CO 80301

VEALE, WARREN LORNE, b Antler, Sask, Mar 13, 43; m 66; c 1. PHYSIOLOGY, NEUROPSYCHOLOGY. *Educ:* Univ Man, BSc, 64; Purdue Univ, West Lafayette, MSc, 68, PhD(neuropsychol), 71. *Prof Exp:* Instr psychol, Brandon Univ, 64-66, lectr, 66-67; from asst prof to assoc prof, 70-76, PROF MED PHYSIOL, FAC MED, UNIV CALGARY, 76-, ASSOC DEAN RES & ADMINR GRAD STUDIES PROG, 74- *Concurrent Pos:* Vis scientist, Nat Inst Med Res, London, Eng, 69; mem sci adv comt non-med use of drugs, Nat Health & Welfare-Med Res Coun, 73-74, chmn comt, 74-77; mem, Med Res Coun, 77-, mem, Studentship Comt, 77-, chmn, 78-, mem, Prog Grants Comt, 78-; consult, Health & Welfare, Health Prevention & Promotion Directorate, 78-; mem res comt, Alta Provincial Cancer Hosps Bd, 78-; regional ed, Pharmacol, Biochem & Behav. *Mem:* Can Psychol Soc; Soc Neurosci; Can Biochem Soc; NY Acad Sci; Am Physiol Soc. *Res:* Central nervous systems' involvement in temperature regulation, fever and action of antipyretics; neurohumoral changes in brain related to alcoholism. *Mailing Add:* Div Med Physiol Univ Calgary Calgary AB T2N 1N4 Can

VEATCH, RALPH WILSON, mathematics; deceased, see previous edition for last biography

VEAZEY, SIDNEY EDWIN, b Wilmington, NC, Sept 18, 37; m 62; c 4. PHYSICS. *Educ:* US Naval Acad, BS, 59; Duke Univ, PhD(physics), 65. *Prof Exp:* Electronics mat officer & main propulsion asst, USS Pollack, 66-68, navigator-opers off, USS Ulysses S Grant, 68-71 & USS James Madison, 71-72, dep dir Trident submarine design develop proj, Naval Ship Eng Ctr, 72, chief Naval mat, Combat Systs Adv Group, Naval Mat Command, 72-73, exec asst to chief Naval develop, 73-74, design mgr nuclear attack submarines, Naval Ship Eng Ctr, 74-76, dep div head, Combat Systs Dept, Naval Surface Weapons Ctr, 76-77, dep dept head, Strategic Systs Dept, 77-78, dept comndr eval & officer-in-chg, Naval Surface Weapons Ctr, White Oak Lab, Silver Spring, Md, 78-80, chmn, Naval Syst Eng Dept, US Naval Acad, Annapolis, 80-82; exec scientist, ORI Inc, 82-84; div dir, ASG, Inc, 84-86; partner, Creative Eng & Construct, 86-89; PRES, S E VENTURES, INC, 89- *Concurrent Pos:* Prof mech eng, Va. *Res:* Microwave spectroscopy of the alkali fluorides; application of lasers to communication from ships; SEAMOD weapons systems for advanced platforms; nuclear power; submarines. *Mailing Add:* Rte 2 Box 497 Fredericksburg VA 22405

VEAZEY, THOMAS MABRY, b Paris, Tenn, Jan 13, 20; m 38; c 3. ORGANIC CHEMISTRY. *Educ:* Murray State Univ, BS, 40; Univ Ill, PhD(org chem), 53. *Prof Exp:* Chemist & job instr supvr, E I du Pont de Nemours & Co, Inc, 41-45; res chemist, Devoe & Raynolds Co, 45-50; res chemist, Chemstrand Corp, 53-55, develop group leader synthetic fibers, 55-58, supvr develop, 58-63; mgr patent liaison, Monsanto Textiles Div, Monsanto Textiles Co, 63-70, sr develop assoc, 71-82; SECY-TREAS, WYOMING ANALYSIS LAB, 79- *Mem:* Am Chem Soc; Sigma Xi. *Res:* Chemical and spinning process development of synthetic textile fibers. *Mailing Add:* 2026 Woodland St SE Decatur AL 35601

VEBER, DANIEL FRANK, b New Brunswick, NJ, Sept 9, 39; m 59; c 2. MEDICINAL CHEMISTRY, PEPTIDE CHEMISTRY. *Educ:* Yale Univ, BA, 61, MS, 62, PhD(org chem), 64. *Prof Exp:* Sr chemist, Merck Sharp & Dohme Res Labs, 64-66, res fel, 66-72, sr res fel, 72-75, assoc dir med chem, 75-79, dir, 79-80, SR DIR MED CHEM, MERCK SHARP & DOHME RES LABS, 80- *Concurrent Pos:* Mem planning comt, Am Peptide Symp, 79-85; Consult, NIH Contraceptive Develop Br, 81-82, 85; Biorg Study Sect, NIH & Pharmacol Sci Rev Comt, 81-85; vis prof chem, Wis Univ, 83; fac, Residential Sch Med Chem, Drew Univ, 87-; invited expert analyst, Biochem & Molecular Biol, Chemtracts, 90-; chmn, Am Peptide Soc, 90- *Mem:* Sigma Xi; NY Acad Sci; AAAS; Am Chem Soc; Am Soc Biol Chemists; Am Peptide Soc. *Res:* Protein and peptide synthesis; new methods and protecting groups in peptide synthesis; chemical and biological properties of enzymes and peptide hormones; chemistry of heterocyclic compounds; conformational analysis; medicinal chemistry. *Mailing Add:* 290 Batleson Rd Ambler PA 19002

VEBLEN, DAVID RODLI, b Minneapolis, Minn, Apr 27, 47; m 76; c 2. MINERALOGY, PETROLOGY. *Educ:* Harvard Univ, BA, 69, MA, 74, PhD(geol), 76. *Prof Exp:* fac res assoc geol, Ariz State Univ, 76-79, asst prof geol, 79-81; from asst prof to assoc prof, 81-84, PROF GEOL, JOHNS HOPKINS UNIV, 84- *Honors & Awards:* Mineral Soc Am Award, 83. *Mem:* Mineral Soc Am; Am Geophys Union; Mineral Asn Can; AAAS; Electron Micros Soc Am; Microbeam Analysis Soc. *Res:* Crystal chemistry and defect structures of silicate minerals, especially chain and sheet silicates; solid state reactions; x-ray diffraction and electron microscopy; applications of mineralogy to igneous and metamorphic petrology. *Mailing Add:* Dept Earth & Planetary Sci Johns Hopkins Univ 3400 N Charles St Baltimore MD 21218

VEDAM, KUPPUSWAMY, b Vedharanyam, India, Jan 15, 26; m 56; c 2. PHYSICS, MATERIALS SCIENCE. *Educ:* Univ Nagpur, BSc, 46, MSc, 47; Univ Saugor, PhD(physics), 51. *Prof Exp:* Lectr physics, Indian Govt Educ Serv, 46-47; lectr, Univ Saugor, 47-48 & 51-53; sr res asst, Indian Inst Sci, 53-56; res assoc, Pa State Univ, 56-57, asst prof, 57-59; sr res officer, Atomic Energy Estab, Bombay, India, 60-62; sr res assoc, 62-64, assoc prof, 63-70, PROF PHYSICS, PA STATE UNIV, 70- *Concurrent Pos:* Mem panel piezoelec transducers, Indian Stand Inst, 61-62. *Mem:* Fel Am Phys Soc; fel Optical Soc Am; Phys Soc Japan; Sigma Xi. *Res:* Crystal physics; optics; ferroelectricity; x-ray and neutron diffraction; high pressure physics, physics of surfaces and materials characterization; spectroscopic ellipsometry. *Mailing Add:* Dept Physics Pa State Univ University Park PA 16802

VEDAMUTHU, EBENEZER RAJKUMAR, b Tamilnadu, India, June 23, 32; m 63; c 2. FOOD SCIENCE, FOOD TECHNOLOGY. *Educ:* Univ Madras, BSc, 53; Nat Dairy Res Inst, India, dipl, 56; Univ Ky, MS, 61; Ore State Univ, PhD(microbiol), 65. *Prof Exp:* Dairy asst, Govt Madras, Animal Husb Serv India, 57-59; assoc microbiol, Ore State Univ, 65-66; Dept HEW trainee dairy microbiol, Iowa State Univ, 66-67, asst prof food technol, 67-71; asst prof microbiol, Ore State Univ, 71-72; sr microbiologist, 72-75, CHIEF RES MICROBIOLOGIST, MICROLIFE TECHNICS, SARASOTA, 75- *Concurrent Pos:* Cheese technol consult, 67-; mem chapter revision comt stand methods examination dairy prod, Am Pub Health Asn, 75- *Mem:* AAAS; Int Asn Milk, Food & Environ Sanit; Am Dairy Sci Asn; Am Soc Microbiol. *Res:* Microbiology of dairy and food products, especially starter and spoilage flora. *Mailing Add:* Microlife Technics Box 3917 Sarasota FL 33578

VEDDER, JAMES FORREST, b Pomona, Calif, June 3, 28; m 70; c 2. PLANETARY SCIENCES, MICROPARTICLE ACCELERATORS. *Educ:* Pomona Col, BA, 49; Univ Calif, PhD(nuclear physics), 58. *Prof Exp:* Asst nuclear physics, Radiation Lab, Univ Calif, 51-58; res scientist, Missiles & Space Co, Lockheed Aircraft Corp, 58-63; res scientist, Ames Res Ctr, NASA, 63-89; RES SCIENTIST, SAN JOSE STATE UNIV FOUND, 90- *Mem:* Am Phys Soc; Am Geophys Union; Sigma Xi. *Res:* Nuclear physics; beta decay; space physics; meteoroids; microparticle accelerators; craters formed by hypervelocity microparticles; remote sensing of soil moisture; measurement of stratospheric halocarbons, methane, nitrous oxide, ozone, etc; polar ozone depletion. *Mailing Add:* 26355 Calle del Sol Los Altos Hills CA 94022-3301

VEDEJS, EDWIN, b Riga, Latvia, Jan 31, 41; US citizen. ORGANIC CHEMISTRY. *Educ:* Univ Mich, Ann Arbor, BS, 62; Univ Wis-Madison, PhD(chem), 66. *Prof Exp:* Nat Acad Sci-Air Force Off Sci Res fel chem, Harvard Univ, 66-67; assoc prof, 67-77, PROF CHEM, UNIV WIS-MADISON, 77- *Concurrent Pos:* A P Sloan fel, 71-73. *Mem:* Am Chem Soc; Sigma Xi. *Res:* Synthetic and organophosphorus chemistry; thermal rearrangements. *Mailing Add:* Dept Chem Univ Wis 1101 University Ave Madison WI 53706-1322

VEDERAS, JOHN CHRISTOPHER, b Detmold, Ger, 47; US citizen. BIOORGANIC CHEMISTRY. *Educ:* Stanford Univ, BSc, 69; Mass Inst Technol, PhD(chem), 73. *Prof Exp:* Res assoc chem, Univ Basel, Switz, 73-76 & Purdue Univ, 76-77; from asst prof to assoc prof chem, 77-87 PROF CHEM, UNIV ALTA, 87- *Concurrent Pos:* Chmn, Biol Chem Div, Chem Inst Can, 83-84, fel, 86. *Honors & Awards:* Merck, Sharp, Dohme Award, Chem Inst Can, 86, Labatt Award, 91. *Mem:* Am Chem Soc; Brit Chem Soc; Chem Inst Can; Sigma Xi. *Res:* Mechanism and stereochemistry of enzymes in amino acid metabolism; radiochemical synthesis; biosynthesis of secondary metabolites. *Mailing Add:* Dept Chem Univ Alta Edmonton AB T6G 2G2 Can

VEDROS, NEYLAN ANTHONY, b New Orleans, La, Oct 6, 29; m 55; c 2. MICROBIOLOGY, IMMUNOLOGY. *Educ:* La State Univ, BSc, 51, MSc, 57; Univ Colo, PhD(microbiol), 60. *Prof Exp:* Nat Inst Allergy & Infectious Dis fel, Med Sch, Univ Ore, 60-62; Bact Div, Naval Med Res Inst, Bethesda, Md, 62-66; res microbiologist, Biol Lab, 66-68, DIR NAVAL BIOMED RES LAB & PROF MED MICROBIOL & IMMUNOL, UNIV CALIF, BERKELEY, 68- *Honors & Awards:* Lab Sect Award, Am Pub Health Asn, 66. *Mem:* Int Asn Aquatic Animal Med; Am Soc Microbiol; Am Asn Immunol; Soc Exp Biol & Med; Asn Mil Surg US. *Res:* Immunochemistry of Neisseria Meningitidis; host-parasite studies in marine pinnipeds; ecology of terrestrial and marine fungi. *Mailing Add:* Dept Biomed & Environ Health Sci Sch Pub Health Univ Calif Berkeley CA 94720

VEDVICK, THOMAS SCOTT, b Tacoma, Wash, June 23, 44; m 67; c 2. HEMOGLOBINOPATHIES, PROTEINS. *Educ:* Univ Puget Sound, BS, 66; Western Wash State Col, MS, 68; Univ Ore, PhD(biochem), 72. *Prof Exp:* Lectr, biol, 76-77, asst res biochemist, 72-87, RES SCIENTIST, AGOURON INST, 88-, STAFF SCIENTIST, 89- *Concurrent Pos:* Prin investr, NIH, 78-80; lectr, chem dept, Univ Calif, San Diego, 81- *Mem:* Sigma Xi; Protein Soc; Asn Biomolecular Res Facil. *Res:* Thalassemia syndromes; primary structure of proteins, development of techniques, and equipment for microsequencing; determination of the microheterogeneity of human fetal hemoglobin gamma chains; protein sequencing. *Mailing Add:* SIBIA 505 Coast Blvd S La Jolla CA 92037

VEECH, RICHARD L, b Decatur, Ill, Sept 19, 35; m 65; c 3. BIOCHEMISTRY, MEDICINE. *Educ:* Harvard Univ, BA, 57, MD, 62; Oxford Univ, PhD(biochem), 69. *Prof Exp:* CHIEF, LAB METAB, NAT INST ALCOHOL ABUSE & ALCOHOLISM, 78- *Mem:* Brit Biochem Soc; Am Soc Biol Chemists; Am Inst Nutrit; Neurochem Soc. *Res:* Control of metabolic processes. *Mailing Add:* Nat Inst Alcohol Abuse & Alcoholism Lab Metab & Molecular Biol 551 DANAC 12501 Washington Ave Rockville MD 20852

VEECH, WILLIAM AUSTIN, b Detroit, Mich, Dec 24, 38; m 65; c 2. MATHEMATICS. *Educ:* Dartmouth Col, AB, 60; Princeton Univ, PhD(math), 63. *Prof Exp:* H B Fine instr math, Princeton Univ, 63-64, Higgins lectr, 64-66; asst prof, Univ Calif, Berkeley, 66-69; assoc prof, 69-72, chmn dept, 82-86, PROF MATH, RICE UNIV, 72- *Concurrent Pos:* Mem math, Inst Advan Study, Princeton Univ, 68-69, 72, 76-77 & 83-84; NSF grant, Rice Univ, 69-, Alfred P Sloan fel, 71-73; adv comt math & comput sci, NSF, 79-82; ed bd, Ergodic Theory & Dynamical Systs, 81-, Annals of Math, 85-; at-large-mem & cour, Am Math Soc, 86- *Mem:* Am Math Soc. *Res:* Topological dynamics; ergodic theory; probability theory; functional analysis; almost periodic functions; number theory. *Mailing Add:* Dept Math Rice Univ PO Box 1892 Houston TX 77251-1892

VEEN-BAIGENT, MARGARET JOAN, b Toronto, Ont, Dec 23, 33; m 69; c 2. NUTRITION. *Educ:* Univ Toronto, BA, 55, MA, 56, PhD(nutrit), 64. *Prof Exp:* From lectr to asst prof, Sch Hyg, 56-75, ASSOC PROF NUTRIT, FAC MED, UNIV TORONTO, 75- *Mem:* Nutrit Soc Can; Brit Nutrit Soc; Am Inst Nutrit; NY Acad Sci. *Res:* Calcium requirements. *Mailing Add:* Dept Nutrit Sci Fac Med Univ Toronto Toronto ON M5S 1A8 Can

VEENEMA, RALPH J, b Prospect Park, NJ, Dec 13, 21; m 44; c 4. UROLOGY. *Educ:* Calvin Col, AB, 42; Jefferson Med Col, MD, 45; Am Bd Urol, dipl, 57. *Prof Exp:* Asst resident, Vet Admin Hosps, Alexandria, La, 46 & Jackson, Miss, 47; asst resident path, Paterson Gen Hosp, 48 & surg path, Col Physicians & Surgeons, Columbia Univ, 49; asst resident urol, Vet Admin Hosp, Bronx, 49; asst resident & resident, Columbia-Presby Med Ctr, 50-52, from asst to assoc, 53-58, asst clin prof, 58-60, from asst prof to assoc prof clin urol, 60-68, PROF CLIN UROL, COL PHYSICIANS & SURGEONS, COLUMBIA UNIV, 68- *Concurrent Pos:* Assoc urologist, St Joseph Hosp, Paterson, NJ, 53-56; chief, Urol Outpatient Clin, Columbia-Presby Med Ctr, 55-60, from asst attend urologist to assoc attend urologist, 55-68, attend urologist, 68-, chief urol, Francis Delafield Hosp, Cancer Res Inst, 60-75; attend urologist & chief urol serv, Valley Hosp, Ridgewood, NJ, 56-60, consult, 60-; consult, USPHS Hosp, Staten Island, NY, 61- & Harlem Hosp, New York, 62- *Honors & Awards:* Am Urol Asn 2nd Prize, 62, 1st Prize, 64. *Mem:* Am Asn Genito-Urinary Surg; Am Urol Asn; fel Am Col Surgeons; fel AMA; NY Acad Med (secy, 60-61). *Res:* Pathophysiology of genitourinary neoplasms. *Mailing Add:* 161 Ft Washington Ave New York NY 10032

VEENING, HANS, b Neth, May 7, 31; nat US; m 57. ANALYTICAL CHEMISTRY. *Educ:* Hope Col, AB, 53; Purdue Univ, MS, 55, PhD, 59. *Prof Exp:* From instr to assoc prof, 58-72, PROF CHEM, BUCKNELL UNIV, 72-, DEPT CHMN, 86- *Concurrent Pos:* NSF fac fel with Dr J F K Huber, Univ Amsterdam, 66-67; NIH spec res fel, Biochem Separations Sect, Oak Ridge Nat Lab, 72-73; NSF grants, 68-72, 76-78 & 84-85; Petrol Res Fund grants, 68-86; prof in charge short course on automated anal, Am Chem Soc, 76-78; NIH grant, 78-81; mem, Anal Chem Deleg Sci Exchange Vis to People's Repub China, 85; NSF res grant, Univ Amsterdam, 84-85. *Mem:* Sigma Xi; Am Chem Soc; Royal Dutch Chem Soc. *Res:* High performance liquid chromatography of biochemically active compounds; mass spectrometry of biochemically active compounds; gas chromatography. *Mailing Add:* Dept Chem Bucknell Univ Lewisburg PA 17837

VEESER, LYNN RAYMOND, b Sturgeon Bay, Wis, Sept 18, 42; m 76. NUCLEAR PHYSICS, FIBER OPTIC SENSORS. *Educ:* Univ Wis-Madison, BS, 64, MS, 65, PhD(physics), 68. *Prof Exp:* STAFF MEM PHYSICS, LOS ALAMOS SCI LAB, 67- *Mem:* Am Phys Soc; Soc Photo-Optical Instrumentation Engrs. *Res:* Pulse power diagnostics. *Mailing Add:* Los Alamos Sci Lab D-14 MS410 PO Box 1663 Los Alamos NM 87545

VEGA, ROBETO, b Mayaguez, PR, Dec 28, 56; m 78; c 1. PARTICLE PHYSICS PHENOMENOLOGY. *Educ:* Univ PR, Mayaquez, BS, 78; Ga Inst Technol, MS, 82; Univ Tex, Austin, PhD(physics), 88. *Prof Exp:* Pres postdoctoral fel, theo-particle physics, Univ Calif, Davis, 88-90; Ford Found postdoctoral fel, theo-particle physics, Nat Res Coun, 90-91; RES ASSOC THEO-PARTICLE PHYSICS, STANFORD LINEAR ACCELERATOR CTR, 91- *Mem:* Am Phys Soc. *Res:* Elementary particle phenomenology with emphasis in the study of Higgs physics and spontaneous/dynamical symmetry breaking at future colliders; high precision physics and Higgs-induced CP violation. *Mailing Add:* Stanford Linear Accelerator Ctr Bin 81 PO Box 4349 Stanford CA 94309

VEGH, EMANUEL, b New York, NY, Nov 20, 36; m 60; c 3. MATHEMATICS. *Educ:* Univ Del, BA, 58, MA, 60; Univ NC, PhD(math), 65. *Prof Exp:* Lectr math, Univ Del, 58-60 & Univ NC, 60-63; RES MATHEMATICIAN, US NAVAL RES LAB, 63- *Concurrent Pos:* Assoc prof lectr, Univ Md, 67-89; assoc prof lectr, George Washington Univ, 65-67, prof lectr, 89- *Honors & Awards:* Res Publ Award, US Naval Res Lab, 69, 86, & 87. *Mem:* Math Asn Am; Am Math Soc; London Math Soc; Sigma Xi. *Res:* Number theory; applied mathematics. *Mailing Add:* US Naval Res Lab 4555 Overlook Ave SW Washington DC 20375

VEGORS, STANLEY H, JR, b Detroit, Mich, Jan 5, 29; m 51; c 3. NUCLEAR PHYSICS. *Educ:* Middlebury Col, BA, 51; Mass Inst Technol, BS, 51; Univ Ill, MS, 52, PhD(physics), 55. *Prof Exp:* Res assoc physics, Univ Ill, 55-56; physicist, Phillips Petrol Co, 56-58; assoc prof, 58-61, head dept, 58-65, PROF PHYSICS, IDAHO STATE UNIV, 61- *Concurrent Pos:* Prof physics, Univ Petrol & Minerals, Saudi Arabia, 82-84. *Mem:* Am Phys Soc; Int Solar Energy Soc. *Res:* Radioactivity, solar energy, nuclear safeguards; nuclear waste disposal. *Mailing Add:* Dept Physics Idaho State Univ Pocatello ID 83209

VEGOTSKY, ALLEN, b New York, NY, Mar 2, 31; m 67; c 1. BIOLOGICAL CHEMISTRY. *Educ:* City Col New York, BS, 52; Fla State Univ, MS, 57, PhD(chem), 61. *Prof Exp:* Asst biochem, NY Univ, 52, US Army Chem Corps Lab, 53-55 & Fla State Univ, 55-60; NIH fel, Purdue Univ, 60-63; asst prof biol & chem, Wheaton Col, Mass, 63-69; assoc prof, Wells Col, 69-74; biosci coordr, Biomed Interdisciplinary Curric Proj, 74-77; asst med dir, Cystic Fibrosis Found, 77-78; res adminr, 78-88, SCI PROG DIR, AM CANCER SOC, 88- *Mem:* Am Asn Cancer Res; Sigma Xi; Am Inst Hist Pharm; Soc Hist Archeol. *Res:* Research administration; cancer research; history of pharmacy; historic archaeology. *Mailing Add:* 2215 Greencrest Dr Atlanta GA 30345

VEHAR, GORDON ALLEN, b Cleveland, Ohio, Apr 26, 48; m 77; c 1. RECOMBINANT THERAPEUTICS. *Educ:* Bowling Green State Univ, BS, 70; Univ Cincinnati, PhD(biol chem), 76. *Prof Exp:* Postdoctoral fel, Dept Biochem, Univ Wash, 75-80; sr scientist, 80-86, dir cardiovasc res, 86-90, STAFF SCIENTIST, GENENTECH INC, 90- *Honors & Awards:* Murray Thelin Award for Outstanding Res, Nat Hemophilia Found, 89. *Mem:* Am Soc Biochem & Molecular Biol; Am Fedn Clin Res. *Res:* Coagulation and fibrinolysis with application to the development of recombinant therapeutics; development of such drugs as recombinant tissue plasminogen activator and recombinant human factor VIII. *Mailing Add:* Genentech Inc 460 Point San Bruno Blvd South San Francisco CA 94080

VEHRENCAMP, SANDRA LEE, b Glendale, Calif, Feb 11, 48; m 73; c 2. ANIMAL BEHAVIOR, ORNITHOLOGY. *Educ:* Univ Calif, Berkeley, BA, 70; Cornell Univ, PhD(animal behav), 76. *Prof Exp:* Lectr, 76-79, asst prof, 79-85, ASSOC PROF BIOL, UNIV CALIF, SAN DIEGO, 85- *Concurrent Pos:* Exped leader, grant, Nat Geog Soc, 78-79; mem, NIMH panel, 81-85. *Mem:* Am Ornith Union. *Res:* Evolution of avian and mammalian social organization and communication; sociobiology, ecological energetics. *Mailing Add:* Dept Biol 0116 Univ Calif San Diego 3500 Gilman Dr La Jolla CA 92093-0116

VEHSE, ROBERT CHASE, b Morgantown, WVa, Sept 9, 36; m 61; c 2. SOLID STATE PHYSICS. *Educ:* WVa Univ, BA, 58; Univ Tenn, Knoxville, PhD(physics), 64. *Prof Exp:* Mem tech staff compound semiconductor mat, Bell Tel Labs, 68-72, SUPVR COMPOUND SEMICONDUCTOR MAT GROUP, BELL LABS, 73- *Mem:* Am Phys Soc; Electrochem Soc; Sigma Xi. *Res:* Development of processes useful for production of epitaxial layers of semiconductor materials. *Mailing Add:* 16 Cardinal Pl Wyomissing PA 19610

VEHSE, WILLIAM E, b Morgantown, WVa, Apr 28, 32; m 56; c 4. PHYSICS. *Educ:* WVa Univ, BA, 55; Carnegie Inst, MS, 59, PhD(physics), 62. *Prof Exp:* From asst prof to assoc prof physics, 61-72, PROF PHYSICS, WVA UNIV, 72-, CHMN DEPT, 75- *Mem:* Am Phys Soc; Am Asn Physics Teachers; Sigma Xi. *Res:* Nuclear and electron resonance in metals; optics. *Mailing Add:* 1173 Cambridge Ave Morgantown WV 26505

VEICSTEINAS, ARSENIO, b Bellano, Italy, Oct 9, 44; m 70; c 3. TEMPERATURE REGULATION, ELECTRICITY. *Educ:* Univ Milan, Italy, Med Dr. *Prof Exp:* From asst prof to assoc prof physiol, Univ Milan, 71-87; PROF PHYSIOL, UNIV BRESCIA, 87- *Concurrent Pos:* Chmn, Sch Specialization Sport Med, 88- *Mem:* Am Physiol Soc; Ital Physiol Soc. *Res:* Cardiorespiratory and metabolic changes during muscular exercise in healthy and disabled persons; temperature regulation in water immersion in humans; physiology of muscle contraction; effects of electromagnetic fields on animals; sport medicine. *Mailing Add:* Inst Physiol Univ Brescia Brescia 25124 Italy

VEIDIS, MIKELIS VALDIS, b Riga, Latvia, Jan 25, 39; m 63; c 3. CHEMISTRY. *Educ:* Univ Queensland, BSc, 63, MSc, 67; Univ Waterloo, PhD(chem), 69. *Prof Exp:* Chemist, Queensland Govt Chem Lab, 63-67; res fel chem, Harvard Univ, 69-71; metallurgist, Wakefield Corp, 71-75, vpres, res & develop, 75-85; STAFF SCIENTIST, UNITRODE CORP, 85- *Concurrent Pos:* Vis res scientist, Chem Dept, Northeastern Univ, 78-85. *Mem:* Am Chem Soc; Royal Australian Chem Inst; Sigma Xi. *Res:* Chemistry of metal surfaces as related to sintering phenomena; crystallographic studies of structures of inorganic complexes and compounds of biological interest. *Mailing Add:* 71 Walnut Hill Rd Newton MA 02161

VEIGEL, JON MICHAEL, b Mankato, Minn, Nov 10, 38; m 62. SCIENCE POLICY, ENERGY POLICY. *Educ:* Univ Wash, BS, 60; Univ Calif, Los Angeles, PhD(phys inorg chem), 65. *Prof Exp:* Res chemist, Jackson Lab, E I du Pont de Nemours & Co, Inc, Del, 65; asst prof phys inorg chem & res chemist, F J Seiler Res Lab, USAF Acad, 65-68; asst prof, Joint Sci Dept, Claremont Cols, 68-73; assoc prof energy & environ, Calif State Col, Domingue Hills, Calif, 73-74; dir energy prog & cong sci fel, Off Technol Assessment, US Cong, 74-75; adminr alternatives div, Energy Comn, Sacramento, Calif, 75-78; chief mkt develop, Solar Energy Res Inst, Golden, Colo, 78- 79, asst dir technol commercialization, 79, div mgr planning applications & impacts, 79-81; pres, Alternative Energy Corp, Res Triangle Park, NC, 81-88; PRES, OAK RIDGE ASSOC UNIVS, OAK RIDGE, TN, 88- *Concurrent Pos:* Mem sci & pub policies comt, AAAS, 75-79; mem synthesis panel, Nat Acad Study Nuclear Power & Alternative Systs, 75-76; mem, NC Energy Develop Authority, 83-88, chmn, 87-88; chmn, Oak Ridge Community Found, 89-; bd mem, Am Coun Energy Efficient Econ, 89-, Alliance Environ Educ, 89-, Energy & Eng bd, Nat Acad Sci & Eng, 89- *Mem:* AAAS. *Res:* National science policy; energy policy; technology assessment. *Mailing Add:* Oak Ridge Assoc Univs Box 117 Oak Ridge TN 37831-0117

VEIGELE, WILLIAM JOHN, b New York, NY, June 18, 25; m 56; c 4. PHYSICS. *Educ:* Hofstra Col, Hempstead, NY, BA, 49, MA, 51; Univ Colo, Boulder, PhD(physics), 60. *Prof Exp:* Testing engr, NY Testing Labs, 49-50; instr physics, Williams Col, Mass, 51-52; instr eng & physics, Hofstra Col, NY, 52-57; instr physics, Univ Colo, Boulder, 57-58; thermodynamicist, Cryogenic Sect, Nat Bur Standards, 58-59; prof & head, Physics Dept, Parsons Col, Iowa, 60-61; sr scientist, Solid State Physics Lab, Martin Marietta Aerospace Corp, 61-64; sr scientist & proj mgr, Kaman Sci Corp, 64-74; founder & pres, Resource Sci Inc, 74-78; LECTR & VIS ASSOC PROF PHYSICS, MECH ENG, ELEC & COMPUTER ENG, NUCLEAR & CHEM ENG, UNIV CALIF, SANTA BARBARA, CALIF, 78- *Concurrent Pos:* Lectr physics, Univ Colo, Colorado Springs, 66-77; prog mgr, sr scientist & prod line mgr, Santa Barbara Res Ctr, 78-84; dept dir & prog mgr, GRC, 84-89; consult, GRC, Santa Barbara, Calif, Raytheon Electromagnetics Systs Div, Santa Barbara, Calif, Santa Barbara Res Ctr, Goleta, Calif & County of Santa Barbara; consult, Med Care & Res Found, Denver, Colo, Colo Dept Hwys, Denver, Colo Fairchild Camera & Instrument Corp, Long Island, NY & Elec & Computer Eng Dept, Univ Calif, Santa Barbara. *Mem:* Sigma Xi. *Res:* Radiation effects; solid state and atomic physics; thermodynamics; electrooptics; infrared technology; Monte Carlo photon transport codes; photon cross sections; three-phase equations of state; author of 75 publications. *Mailing Add:* 333 Old Mill Rd No 324 Santa Barbara CA 93110

VEILLON, CLAUDE, b Church Point, La, Jan 11, 40; m; c 2. ANALYTICAL CHEMISTRY, SPECTROSCOPY. *Educ:* Univ Southwestern La, BS, 62; Univ Fla, MS, 63, PhD(anal chem), 65. *Prof Exp:* Res chemist, Nat Bur Standards, 65-67; from asst prof to assoc prof anal chem, Univ Houston, 67-74; vis scientist, Harvard Med Sch, 74-76; RES CHEMIST, HUMAN NUTRIT RES CTR, 76- *Concurrent Pos:* Nat Acad Sci-Nat Res Coun res assoc, 65-67; res fel, NIH/Nat Cancer Inst, 74-76. *Mem:* Am Chem Soc; Soc Appl Spectros; Optical Soc Am; Am Inst Physics. *Res:* Isotopic analysis; trace metal analysis; analytical instrumentation; trace metal metabolism. *Mailing Add:* Human Nutrit Res Ctr USDA Bldg 307 Rm 226A Beltsville MD 20705

VEINOTT, ARTHUR FALES, JR, b Boston, Mass, Oct 12, 34; m 60, 88; c 2. OPERATIONS RESEARCH. *Educ:* Lehigh Univ, BS & BA, 56; Columbia Univ, EngScD(indust eng), 60. *Prof Exp:* From asst prof to assoc prof indust eng, 62-67, chmn dept, 75-85, PROF OPERS RES, STANFORD UNIV, 67- *Concurrent Pos:* Western Mgt Sci Inst grant, 64-65; Off Naval Res contract, 64-80; consult, Rand Corp, 65- & IBM Res Ctr, 68-69; NSF grant, 67-, mem res initiation grant panel, 71; vis prof, Yale Univ, 72-73; ed, J Math Opers Res, 74-80; Guggenheim fel, 78-79. *Mem:* Nat Acad Eng; Inst Mgt Sci; Opers Res Soc Am; fel Inst Math Statist. *Res:* Development of lattice programming, a qualitative theory of optimization for predicting the direction of change of optimal decisions resulting from alteration of problem parameters; structure and computation of optimal policies for inventory systems and dynamic programs. *Mailing Add:* Dept Opers Res Stanford Univ Stanford CA 94305-4022

VEINOTT, CYRIL G, b Somerville, Mass, Feb 15, 05; m 36; c 1. ACOUSTIC MEASUREMENTS OF MACHINES. *Educ:* Univ Vt, BS, 26. *Hon Degrees:* DEng, Univ Vt, 51. *Prof Exp:* Mgr indust sect, Westinghouse Elec Corp, 26-52; chief, AC Eng, Reliance Electric Co, 53-70; invited prof elec machs, Univ Laval, Que, 70-72; vol exec, Int Exec Serv Corps, 72-79; PVT INDUST CONSULT, 70- *Honors & Awards:* Tesla Medal, Inst Elec & Electronics Engrs, 77, Centennial Medal, 84. *Mem:* Inst Elec & Electronics Engrs (vpres, 49-51). *Res:* Investigation and development of tools and methods for the design of electric machinery; the use of digital computers using thenavaoluble mainframes and personal computers; author of books; holds US patent. *Mailing Add:* 4197 Oakhurst Cir W Sarasota FL 34233

VEIRS, VAL RHODES, b Allegan, Mich, Sept 20, 42; m 64; c 1. INTELLIGENT SYSTEMS. *Educ:* Case Inst Technol, BS, 64; Ill Inst Technol, PhD(physics), 69. *Prof Exp:* Res physicist, Zenith Radio Corp, 64-65; asst prof physics, Ill Inst Technol, 69-71; asst prof, 71-80, ASSOC PROF PHYSICS, COLO COL, 80- *Mem:* AAAS; Am Asn Physics Teachers. *Res:* Intelligent tutoring. *Mailing Add:* Dept Physics Colo Col Colorado Springs CO 80903

VEIS, ARTHUR, b Pittsburgh, Pa, Dec 23, 25; m 51; c 3. BIOCHEMISTRY, PHYSICAL CHEMISTRY. *Educ:* Univ Okla, BS, 47; Northwestern Univ, PhD(phys chem), 51. *Prof Exp:* Instr phys chem, Univ Okla, 51-52; res chemist, Dept Phys Chem, Armour & Co, 52-60, head dept, 59-60; assoc prof biochem, 60-65, asst dean grad affairs, 68-70, assoc dean med & grad schs, 70-76, PROF BIOCHEM, SCH MED, NORTHWESTERN UNIV, CHICAGO, 65-, CHMN DEPT ORAL BIOL, SCH DENT, 77-, PROF MOLECULAR BIOL, 80- *Concurrent Pos:* Spec instr, Crane Jr Col, 55-56 & Loyola Univ, 57-58; Guggenheim fel, 67; fel NIH Fogarty Sr Int Scholar Award, European Molecular Biol Lab, Grenoble & Weizmann Inst Sci, Rehovot, Israel, 77; chmn & mem spec prog adv comt, Nat Inst Dent Res, Dent Res Inst, 74-78; centennial scholar, Case Inst Technol, 80; distinguished vis prof, Univ Adelaide, Australia, 81; mem pathobiochemistry study sect, NIH, 83-87; chmn, Gordon Conf Chem & Biol Bones & Teeth, 85 & Gordon Conf Struct Macromolecules-Collagen, 81. *Honors & Awards:* Biol Mineralization Award, Int Asn Dental Res, 81. *Mem:* Am Chem Soc; Am Soc Biol Chemists; Biophys Soc; NY Acad Sci; Int Asn Dent Res; fel AAAS. *Res:* Physical chemistry and biology of the connective tissue systems; colloid chemistry; biological mineralization; study of the connective tissue macromolecules, particularly the collagens and the phosphorylated proteins of mineralized tissues; mechanism of biomineralization. *Mailing Add:* Dept Oral Biol Northwestern Univ Dent Sch Chicago IL 60611

VEIT, BRUCE CLINTON, b Cleveland, Ohio, Aug 22, 42; c 2. MICROBIOLOGY. *Educ:* Univ Cincinnati, PhD(microbiol), 72. *Prof Exp:* CHIEF SECT IMMUNOL & MICROBIOL, WILLIAM BEAUMONT ARMY MED CTR, 84-; ASSOC PROF IMMUNOL, UNIV TEX, EL PASO, 85- *Concurrent Pos:* Consult, Tobacco Health & Res Inst, Lexington Ky. *Mem:* Am Asn Immunologists. *Res:* Immuno-regulation in allergy; anti-idiotype vaccines. *Mailing Add:* Dept Clin Invest William Beaumont Army Med Ctr El Paso TX 79920

VEIT, JIRI JOSEPH, b Prague, Czech, Apr 15, 34; m 61. OPTICS, NUCLEAR PHYSICS. *Educ:* Univ London, BSc, 55, PhD(nuclear physics), 59; Univ Birmingham, MSc, 56. *Prof Exp:* Instr physics, Univ BC, 59-62; lectr, Univ London, 62-63; from asst prof to assoc prof, 63-71, PROF PHYSICS, WESTERN WASH UNIV, 71- *Mem:* Am Phys Soc; Am Asn Physics Teachers. *Res:* Positronium; stripping and pickup reactions; nuclear reaction mechanisms; nuclear structure. *Mailing Add:* Dept Physics & Astron Western Wash Univ 516 High St Bellingham WA 98225

VEITCH, FLETCHER PEARRE, JR, b College Park, Md, Dec 21, 08; m 39; c 2. ENZYME ISOLATION, ENZYME KINETICS. *Educ:* Univ Md, BS, 31, MS, 33, PhD(org chem), 35. *Prof Exp:* Res chemist, Nat Canners Asn, 35-37; asst prof biochem, Sch Med, Georgetown Univ, 37-47; prof, 47-74, EMER PROF BIOCHEM, UNIV MD, 75- *Mem:* Sigma Xi; Am Chem Soc; NY Acad Sci; AAAS. *Res:* Hormones and enzymes. *Mailing Add:* Box 513 Lexington Park MD 20653

VEITH, DANIEL A, b Metairie, La, Apr 18, 36; m 56; c 2. SOLID STATE PHYSICS. *Educ:* Tulane Univ, BS, 56, PhD(physics), 63; Univ Calif, Los Angeles, MS, 58. *Prof Exp:* Mem tech staff, Hughes Aircraft Co, 56-59; sci specialist space div, Chrysler Corp, 63-67; PROF PHYSICS & DEPT HEAD, NICHOLLS STATE UNIV, 67- *Res:* Nucleation and growth of thin crystalline films. *Mailing Add:* Dept Chem & Physics Nicholls State Univ Thibodaux LA 70310

VEITH, FRANK JAMES, b New York, NY, Aug 29, 31; m 54; c 4. SURGERY, TRANSPLANTATION BIOLOGY. *Educ:* Cornell Univ, AB, 52, MD, 55. *Prof Exp:* NIH fel, Harvard Med Sch, 63-64; asst prof surg, Cornell Univ, 64-67; assoc prof, 67-71, PROF SURG, ALBERT EINSTEIN COL MED, 71-; CO-DIR KIDNEY TRANSPLANT UNIT, MONTEFIORE HOSP, 67-, ATTEND SURG & CHIEF VASCULAR SURG, 72- *Concurrent Pos:* Markle scholar acad med, Cornell Univ, Albert Einstein Col Med & Montefiore Hosp, 64-69; career scientist award, Health Res Coun, City New York & Montefiore Hosp, 65-72; assoc attend surgeon, Montefiore Hosp, 67-71; consult, Heart-Lung Proj Comt, 71- *Mem:* Soc Univ Surgeons; Soc Vascular Surg; Am Asn Thoracic Surg; Am Surg Asn; Transplantation Soc. *Res:* Lung transplantation, pulmonary physiology, kidney transplantation and vascular surgery. *Mailing Add:* Dept Surg Albert Einstein Col Med 111 E 210th St Bronx NY 10467

VEIZER, JÁN, b Pobedim, Czech, June 22, 41; Can citizen; m 66; c 2. EARTH SCIENCE, GEOCHEMISTRY. *Educ:* Comenius Univ, Czech, PG, 64, RNDr, 68; Slovak Acad Sci, Czech, CSc, 68; Australian Nat Univ, PhD(geochem), 71. *Prof Exp:* Lectr, Comenius Univ, Czech, 63-66; res sci geol, Slovak Acad Sci, Czech, 66-71; vis asst prof geol, Univ Calif, Los Angeles, 72; vis res scientist geochem, Univ Göttingen, WGer, 72-73; res scientist geol, Univ Tübingen, WGer, 73; PROF GEOL, UNIV OTTAWA, CAN 73-; PROF & CHAIR GEOL, RUHR UNIV, BOCHUM, GER, 88- *Concurrent Pos:* Vis prof, Univ Tübingen, WGer, 74, Northwestern Univ, 83 & Lady Davis Prof, Hebrew Univ, Israel, 87; vis fel, Australian Nat Univ, 79; consult, NASA, 83-87. *Honors & Awards:* Killam Res Prof, Can Res Coun, 86. *Mem:* Fel Royal Soc Can; fel Geol Asn Can; fel Geol Soc Am; Geochem Soc Am. *Res:* Evolution of sedimentation, atmosphere and life in geologic history. *Mailing Add:* Derry Lab Dept Geol Univ Ottawa Ottawa ON K1N 6N5 Can

VEJVODA, EDWARD, b New York, NY, Apr 18, 24; m 49; c 3. INDUSTRIAL CHEMISTRY. *Educ:* Univ Northern Colo, BA, 49, MA, 51. *Prof Exp:* Anal chemist, Anal Labs, Dow Chem Co, 52-56, res chemist, 56-60, sr res chemist res & develop labs, 60-62, anal supvr, 62-64, anal proj supvr, Dow Chem Int, Ger, 64-65, res staff asst, Chem-Physics Res & Develop Labs, 65-68, res mgr, Chem Res & Develop, 68-75, sr res mgr, 75-81; chem opers dir, Rockwell Int, 75-85, plutonium opers dir, 86-87; CONSULT, LAMB ASSOCS & ACTINIDE PROCESS CHEM & PLUTONIUM RECOVERY, 87-; SR ENGR, LOS ALAMOS TECH ASSOCS, 87- *Mem:* Am Chem Soc; Sigma Xi; Inst Nuclear Mat Mgt. *Res:* Actinide chemistry; development of analytical methods for the assay and impurity analysis of the actinide elements, especially optical emission spectroscopy for the impurity analysis of plutonium and americium compounds; process development for the separation and purification of plutonium compounds; waste management and processing; plutonian fabrication. *Mailing Add:* 2625 Juilliard St Boulder CO 80303

VELA, ADAN RICHARD, b Laredo, Tex, Oct 28, 30; m 55; c 4. PHYSIOLOGY. *Educ:* Baylor Univ, BS, 52; Univ Tenn, PhD(physiol) 62. *Prof Exp:* Res assoc data anal, Comput Ctr, Univ Tenn, 63; asst prof surg, 63-71, ASSOC PROF SURG, SCH MED, LA STATE UNIV MED CTR, 71- *Mem:* Sigma Xi; Am Physiol Soc; Gastrointestinal Res Group. *Res:* Gastrointestinal physiology; esophageal motility; small intestine motility; andrectal manometry endotoxin. *Mailing Add:* 5024 Tartan Dr Metairie LA 70003

VELA, GERARD ROLAND, b Eagle Pass, Tex, Sept 18, 27; m 53; c 4. MICROBIOLOGY, FOOD SCIENCE & TECHNOLOGY. *Educ:* Univ Tex, BA, 50, MA, 51, PhD(microbiol), 63. *Prof Exp:* Res asst, Univ Tex, 50-51; res asst biochem, Southwest Found Res, 52-54; res asst immunol, Sch Pub Health, Harvard Univ, 54-57; head clin chemist, Santa Rosa Hosp, San Antonio, Tex, 57-59; res microbiologist, USAF Sch Aerospace Med, 59-65; from asst prof to prof microbiol, 65-85, ASSOC DEAN, SCI-TECH A&S,

NTEX STATE UNIV, 85- *Concurrent Pos:* Fulbright lectr, Bogota, Colombia, 72; ed, Tex J Sci, 75-83; adj prof microbiol, Univ Monterey, Nuevo Leon, Mex, Univ Granada, Spain, Univ Tex, Austin, Univ Chihuahua, Mex. *Mem:* Am Acad Microbiol; Am Soc Microbiol; Soc Gen Microbiol; Can Soc Microbiol; AAAS. *Res:* Nature of microorganisms in their natural habitat; biochemical interrelationships in mixed cultures of microorganisms; physiology and morphology of azotobacter; microbiology of industrial waste-waters; radiation effects on microorganisms; ATP-ADP-AMP kinetics. *Mailing Add:* Dept Biol Sci Univ NTex Denton TX 76203

VELARDO, JOSEPH THOMAS, b Newark, NJ, Jan 27, 23; wid. PHYSIOLOGY, BIOCHEMISTRY. *Educ:* Northern Colo Univ, AB, 48; Miami Univ, SM, 49; Harvard Univ, PhD(biol, physiol, endocrinol), 52. *Prof Exp:* Asst org & inorg chem, Northern Colo Univ, 47-48; asst & instr zool & human heredity, Miami Univ, 48-49; teaching & res fel, biol & endocrinol & histochem morphology, Harvard Univ, 49-52, res fel biol & endocrinol, 52-53, res assoc path, Sch Med, 53-54, res assoc surg, 54-55; asst prof anat, Sch Med, Yale Univ, 55-61; prof & chmn dept anat, NY Med Col, 61-62; dir, Inst Study Human Reproduction & dir educ prog, 62-67; chmn dept, Stritch Sch Med, Loyola Univ, Chicago, 67-73, prof anat, 67-88; CONSULT BIOSCIENTIST, 88- *Concurrent Pos:* Asst surg, Peter Bent Brigham Hosp, Boston, 54-55; Lederle med fac award, 55-58; prof biol, John Carroll Univ, 62-67; US deleg, Int Cong Reproduction, Vatican, 64; head dept res, St Ann Hosp, Cleveland, 64-67; ed, Endocrinol Reproduction, Essentials Human Reproduction, Biol Reproduction, Uterus & Enzymes Female Gen Systs; hon vpres, res, develop & educ, Universal Res Systs, 73-82; dir biomed, Curric Consult, 83-; mem adv coun, Int Biographical Centre, Cambridge, Eng, 90- *Honors & Awards:* Rubin Award, Am Soc Study Steril, 55. *Mem:* Brit Soc Endocrinol; Endocrine Soc; fel Geront Soc; Am Physiol Soc; fel NY Acad Sci; fel AAAS; Am Asn Anat; Soc Zool; Pan Am Asn Anat. *Res:* Endocrinology of reproduction; anatomy, physiology, biochemistry, molecular biology, histochemistry and cytochemistry of reproductive organs; molecular interactions and steroid-gonadotropic hormone interrelationships-induction in ovulation; zoology; research administration; science education; science policy. *Mailing Add:* 607 E Wilson Rd Old Grove East Lombard IL 60148-4062

VELECKIS, EWALD, b Kybartai, Lithuania, Aug 1, 26; US citizen; m 55; c 1. CHEMISTRY. *Educ:* Univ Ill, BS, 53; Ill Inst Technol, MS, 57, PhD(chem), 60. *Prof Exp:* CHEMIST, ARGONNE NAT LAB, 59- *Mem:* Am Chem Soc; Sigma Xi. *Res:* Phase equilibria in inorganic systems; molecular beams; alloy thermodynamics; metallic solutions and liquid state; fusion reactors; metal hydrides; hydrogen storage. *Mailing Add:* 23 Ruggles Ct Orland Park IL 60462-1926

VELENYI, LOUIS JOSEPH, b Budapest, Hungary, June 17, 34; US citizen; m 57; c 3. CATALYSIS, FUELS TECHNOLOGY. *Educ:* Case Western Reserve Univ, BA, 70, PhD(chem), 75. *Prof Exp:* Res asst clin chemist, Case Western Reserve Univ, Highland View Hosp, 62-72; proj leader, chemist, Standard Oil Co, Ohio, 75-87; RES SCIENTIST, BP AM, 88- *Mem:* Am Chem Soc; Sigma Xi. *Res:* Synthesis and nuclear magnetic resonance study of porphyrins and phthalocyanines; catalysis of various reactions; carbon chemistry; fuels technology. *Mailing Add:* 1266 Roland Rd Lyndhurst OH 44124

VELETSOS, A(NESTIS), b Istanbul, Turkey, Apr 28, 27; nat US; m 66; c 2. CIVIL ENGINEERING, STRUCTURAL ENGINEERING & MECHANICS. *Educ:* Robert Col, Istanbul, BS, 48; Univ Ill, Urbana, MS, 50, PhD(civil eng), 53. *Prof Exp:* From asst prof to prof civil eng, Univ Ill, Urbana, 53-64, assoc mem, Ctr Advan Study, 61-62; prof & chmn dept eng, 64-72, BROWN & ROOT PROF ENG, RICE UNIV, 66- *Concurrent Pos:* Struct designer, F L Ehasz, NY, 50 & Skidmore, Owings & Merrill, Ill, 53; consult, 53-; NSF consult, Indian Inst Technol, Bombay, 69; vis prof, Cath Univ Rio de Janeiro, Univ Calif, Berkeley, 77; mem, Earthquake Eng Res Inst, vpres, 74-77; chmn, US Joint Comt Earthquake Eng, 77-80; mem, Eng Mech Div, Am Soc Civil Engrs. *Honors & Awards:* Norman Medal, Am Soc Civil Engrs, 59 & 90, Res Prize, 61, Newmark Medal, 78, Howard Award, 90. *Mem:* Nat Acad Eng; Am Soc Civil Engrs; Seismol Soc Am; Int Asn Bridge & Struct Engrs; Earthquake Eng Res Inst. *Res:* Structural engineering and mechanics, particularly dynamics of structures and earthquake engineering; offshore structures. *Mailing Add:* Dept Civil Eng Rice Univ PO Box 1892 Houston TX 77251-1892

VELEZ, SAMUEL JOSE, b San Juan, PR, July 19, 45; m 67; c 4. NEUROPHYSIOLOGY. *Educ:* Univ PR, BS, 66, MS, 69; Yale Univ, PhD(neurophysiol), 74. *Prof Exp:* Instr biol, Univ PR, 66; biologist, US Naval Sta, San Juan, PR, 66 & 67-68; res asst pharmacol, Sch Med, Univ PR, 67-68; teaching asst neurophysiol, Yale Univ, 70 & 71; NIH fel zool, Univ Tex, Austin, 74-76; asst prof, 76-82, ASSOC PROF BIOL SCI, DARTMOUTH COL, 82- *Mem:* AAAS; Soc Neurosci; Sigma Xi. *Res:* Patterns of neuronal connections; nerve-muscle trophic interactions; facilitation at the neuromuscular junction; regeneration of neuromuscular connections; developmental neurobiology. *Mailing Add:* Dept Biol Sci Dartmouth Col Hanover NH 03755

VELEZ, WILLIAM YSLAS, b Tucson, Ariz, Jan 15, 47; m 68; c 2. NUMBER THEORY, ALGEBRA. *Educ:* Univ Ariz, BS, 68, MS, 72, PhD(math), 75. *Prof Exp:* Mem tech staff math, Sandia Labs, 75-77; asst prof, 77-80, ASSOC PROF MATH, UNIV ARIZ, 81- *Mem:* Math Asn Am. *Res:* Elementary and algebraic number theory; field theory. *Mailing Add:* Dept Math Univ Ariz Tucson AZ 85721

VELICK, SIDNEY FREDERICK, b Detroit, Mich, May 3, 13; m 41; c 2. BIOCHEMISTRY. *Educ:* Wayne State Univ, BS, 35; Univ Mich, MS, 36, PhD(biol chem), 38. *Prof Exp:* Rockefeller Found fel, Johns Hopkins Univ, 39-40; Int Cancer Res Found fel, Yale Univ, 41-45; from asst prof to prof biol chem, Sch Med, Wash Univ, 45-64; prof & head dept, 64-79, EMER PROF BIOL CHEM, COL MED, UNIV UTAH, 88- *Concurrent Pos:* Mem biochem study sect, NIH, 65-69; assoc ed, Archives Biochem. *Honors & Awards:* Sr Alexander von Humboldt Award, 73. *Mem:* Nat Acad Sci; Am Soc Biol Chemists; Am Chem Soc; AAAS. *Res:* Bacterial lipids; protein chemistry and metabolism; mechanism of enzyme action. *Mailing Add:* Dept Biochem Col Med Univ Utah Salt Lake City UT 84112

VELIKY, IVAN ALOIS, b Zilina, Czech, Mar 23, 29; m 52; c 2. BIOLOGICAL CHEMISTRY. *Educ:* Slovak Tech Univ, Bratislava, EngC, 50, DiplEng, 52; Slovak Acad Sci, PhD, 60. *Prof Exp:* From asst prof to assoc prof biochem, Slovak Tech Univ, Bratislava, 52-65; fel, Prairie Regional Lab, Nat Res Coun Can, 65-67, assoc res officer, 67-75, sr res officer, Div Biol Sci, 75-90; PRES, TIVELCO INT INC, 90- *Concurrent Pos:* Sr sci adv & consult mem, Sci Adv Bd. *Mem:* Chem Inst Can; Can Biochem Soc; Int Asn Plant Tissue Cult; Can Soc Microbiologists. *Res:* Physiology and biochemistry of microorganisms; physiology of cell growth in suspension cultures (fermentors); biosynthesis of secondary metabolites and biotransformation of biologically active compounds by cell cultures; immobilized cells and proteins; synthesis of surfactants, fine chemicals; cosmetics; skin and hair care products; formulations. *Mailing Add:* Tivelco Int Inc 613 Fielding Dr Ottawa ON K1V 7G7 Can

VELKOFF, HENRY RENE, b Cleveland, Ohio, May 14, 21; m 50; c 3. MECHANICAL & AERONAUTICAL ENGINEERING. *Educ:* Purdue Univ, BSME, 42; Ohio State Univ, MSME, 52, PhD, 62. *Prof Exp:* Analyst, Lockheed Aircraft Corp, 42-44; res engr, Aerotor Assocs, 44 & Nat Adv Comt Aeronaut, 44; develop engr, Wright-Patterson AFB, 44-46, sect head rotary wing helicopter rotor res, 47-56, br chief, 56-57, br chief vertical takeoff landing turbine res, 57-59, consult elec propulsion, 59-63; chief scientist, US Army Aviation Res & Develop Lab, 72-74; PROF MECH ENG, OHIO STATE UNIV, 63-72 & 74- *Concurrent Pos:* Consult, Air Br, Off Naval Res, 63-64, Wright-Patterson AFB, 64-66, Aviation Labs, US Army, 66-72, Vertol Div, Boeing Co, 65-67 & Battelle Mem Inst, 68-70; US Army Res Off grant electrofluidmech, 64-72; mem spec subcomt vertical takeoff landing aircraft, NASA, 65, beyond the horizon study comt, USAF, 66, aviation sci adv group, US Army, 70-72, sci adv bd, United Aircraft Corp, 71-72 & 74-, adv bd NASA helicopters, Nat Res Coun, 78 & Heli Air on Frost Prev, 82-85. *Honors & Awards:* Distinguished Res Award, USAF, 62. *Mem:* Am Soc Mech Engrs; hon fel Am Helicopter Soc; Am Inst Aeronaut & Astronaut; Electrostatic Soc Am. *Res:* Electrofluidmechanics; interactions of electro-static fields with gaseous boundary layers, heat transfer, condensation, freezing and flames; helicopter rotor advanced concepts; helicopter rotor flow fields; cavitation detection. *Mailing Add:* 210 Easy St-35 Mountain View CA 94043

VELLA, FRANCIS, b Malta, July 24, 29; m 56; c 5. BIOCHEMISTRY, GENETICS. *Educ:* Royal Univ Malta, BSc, 49, MD, 52; Oxford Univ, BA, 54, MA, 58; Univ Singapore, PhD(biochem), 62. *Hon Degrees:* DSc, Univ Malta, 89. *Prof Exp:* Asst lectr biochem, Univ Singapore, 56-57, lectr, 57-60; sr lectr, Univ Khartoum, 60-64, reader biochem genetics, 64-65; vis assoc prof biochem, 65-66, assoc prof, 66-71, PROF BIOCHEM, UNIV SASK, 71- *Concurrent Pos:* Tutor, WHO Lab Course in Abnormal Hemoglobins, Ibadan, Nigeria, 63; lectr, NATO Advan Course in Pop Genetics, Rome, Italy, 64; vis prof biochem, Univ Cambridge, 73-74; mem, Comt Educ, Int Union Biochem, 79-, chmn, Comt on Educ, 83-; external examr & vis prof, El Fateh Univ, Libya, 80-; external examr & vis prof, Kuwait Univ, 81-; chmn, Comt Educ, Int Union Biochem, 82-91; numerous workshops on biochem educ. *Honors & Awards:* Chevalier, Order of St Sylvester, Vatican City, Italy, 65. *Mem:* fel Chem Inst Can; fel Royal Soc Chem; fel Royal Col Path. *Res:* Molecular genetics; abnormal human hemoglobins; hereditary enzyme deficiencies in man; teaching methods in biochemistry. *Mailing Add:* Dept Biochem Univ Sask Saskatoon SK S7N 0W0 Can

VELLACCIO, FRANK, b New Haven, Conn, Sept 24, 48; m 70; c 1. BIO-ORGANIC CHEMISTRY. *Educ:* Fordham Univ, BS, 70; Mass Inst Technol, PhD(org chem), 74. *Prof Exp:* ASST PROF CHEM, COL HOLY CROSS, 74- *Mem:* Sigma Xi; Am Chem Soc. *Res:* Synthetic methods for peptide synthesis; intromolecular acyl transfers. *Mailing Add:* Dept Chem Col Holy Cross Worcester MA 01610

VELLA-COLEIRO, GEORGE, b Malta, Mar 15, 41. PHYSICS. *Educ:* Royal Univ Malta, BSc, 61; Oxford Univ, MA, 63, DPhil(physics), 67. *Prof Exp:* Mem tech staff physics, 67-80, SUPVR, BELL LABS, 80- *Concurrent Pos:* Rhodes Scholar, 61-64. *Mem:* Am Phys Soc; Inst Elec & Electeonics Engrs. *Res:* Magnetism; semiconductors; Josephson devices. *Mailing Add:* AT&T Bell Labs Rm 7E-505 Murray Hill NJ 07974

VELLEKAMP, GARY JOHN, b Englewood, NJ, May 29, 51. PROTEINS. *Educ:* Hartwick Col, Oneonta, BA, 73; State Univ NY, Binghamton, MA, 78, PhD(biol sci), 82. *Prof Exp:* Postdoctoral, State Univ NY, Binghamton, 82-83 & Univ Conn Health Ctr, 83-87; sr scientist, 87-90, PRIN SCIENTIST, SCHERING-PLOUGH RES, 91- *Mem:* Am Soc Biochem & Molecular Biol; Protein Soc; AAAS; Am Chem Soc. *Res:* Protein purification processes for potential human therapeutics. *Mailing Add:* Schering-Plough Res Bldg U-13-1 1011 Morris Ave Union NJ 07083

VELLETRI, PAUL A, b 1950; m. HEALTH SCIENCE ADMINISTRATION. *Educ:* George Washington Univ Med Ctr, PhD(pharmacol), 81. *Prof Exp:* Prog adminr pharmacol, Nat Inst Gen Med Sci, 85-89, sr staff fel, 83-84, HEALTH SCI ADMINR, NAT HEART LUNG BLOOD INST, NIH, 89- *Mem:* Am Soc Pharmacol & Exp Therapeut. *Res:* Angiotensin-converting enzyme; hypertension; molecular pharmacology; neurobiology. *Mailing Add:* Div Extramural Affairs Westwood Bldg Rm 648 Nat Heart Lung Blood Inst NIH Bethesda MD 20892

VELLTURO, ANTHONY FRANCIS, b Ansonia, Conn, Dec 3, 36. APPLIED CHEMISTRY, ORGANIC CHEMISTRY. *Educ:* Yale Univ, BS, 58, MS, 59, PhD(org chem), 62. *Prof Exp:* NIH fel, Tulane Univ, La, 64-65; sr res chemist, Techni-Chem Co, Conn, 65-70; sr develop chemist, 70-73, group leader, 73-77, mgr, tech dept, 77-85, PROD MGR, CIBA-GEIGY CHEM CO, 85- *Mem:* Am Chem Soc. *Res:* Reaction mechanisms; synthetic organic chemistry. *Mailing Add:* 655 Montana Dr Toms River NJ 08753-2797

VELTRI, ROBERT WILLIAM, b McKeesport, Pa, Dec 1, 41; m 62; c 2. CANCER RESEARCH, IMMUNOPHARMACEUTICALS. *Educ:* Youngstown Univ, BA, 63; WVa Univ, MS, 65, PhD(microbiol), 68. *Prof Exp:* Asst prof microbiol, Med Ctr, WVa Univ, 68-72, assoc prof, 72-75, prof microbiol & otolaryngol, 76-81, dir otolaryngic res, Div Otolaryngol, 68-81; dir res & develop, Cooper Biomed Inc, 81-84; PRES & CO-FOUNDER, AM BIOTECHNOL CO, 84. *Concurrent Pos:* Immunol consult, Dent Sci Inst, Univ Tex, Houston, 72-74; vis prof microbiol, Univ Chile, Santiago, 81; regional dir, Cancer Res Lab, Nat Found Cancer Res, 79-86. *Mem:* AAAS; Sigma Xi; Soc Gen Microbiol; Am Soc Microbiol; Am Asn Cancer Res; Am Asn Immunol. *Res:* Role of tonsils in immunobiology; virology and immunology of herpesvirus infections; microbiology and immunology of otolaryngic infections; isolation and identification of human tumor-associated antigens; Epstein-Barr virus-host relationships; development of synthetic butyvolactone immunopharmaceuticals; patents issued and pending in field of immunopharmaceuticals, anti cancer therapy. *Mailing Add:* 1920 E Second St No 2208 Edmond OK 73037-3376

VELZY, CHARLES O, b Oak Park, Ill, Mar 17, 30; m 57; c 3. MUNICIPAL SOLID WASTE MANAGEMENT & WASTEWATER TREATMENT. *Educ:* Univ Ill, BS, 53, MS, 59, BS, 60. *Prof Exp:* Design & proj engr, Nussbaumer, Clarke & Velzy, 59-66; secy-treas & dir, 66-76, PRES, CHARLES R VELZY ASSOC, INC, 76-; VPRES, ROY F WESTON, INC, 87- *Concurrent Pos:* Vchmn, Comt D-22 Air Qual, Am Soc Testing & Mat, 74-75; mem, bd gov, Am Soc Mech Engrs, 83-84 & Sci Adv Bd, Environ Protection Agcy, 85-; consult, WHO, Geneva, Switz, 85-87. *Honors & Awards:* Centennial Medal, Am Soc Mech Engrs, 80. *Mem:* Fel Am Soc Mech Engrs (pres, 89-90); Am Water Works Asn; Water Pollution Control Fedn; Am Soc Civil Engrs; Am Acad Environ Engrs; Am Soc Testing & Mat; fel Am Consult Engrs Coun. *Res:* Author of numerous publications on environmental engineering. *Mailing Add:* 131 Woodcrest Ave White Plains NY 10604

VEMULA, SUBBA RAO, b Andhra Pradesh, India, Dec 30, 24; m 56; c 2. FRICTION, DESIGNS. *Educ:* Rensselaer Polytech Inst, MS, 68, PhD(mech), 74. *Prof Exp:* Jr civil engr design bridges, 77-79, asst civil engr, 79-81, CIVIL ENGR II, STRUCT ENG RESOURCE & SYST GROUP, 81- *Mem:* Am Soc Mech Engrs. *Res:* Friction. *Mailing Add:* 27 Pinehurst Ave Albany NY 12205

VEMURI, SURYANARAYANA, b Dec 23, 43; m 68; c 2. ELECTRICAL ENGINEERING. *Educ:* Osmania Univ, BE, 65; Indian Inst Sci, ME, 68; Univ NB, MScE, 70, PhD(elec eng), 75. *Prof Exp:* Res assoc, Univ NB, 73-75; asst prof, Univ Alaska, Fairbanks, 75-76; asst prof elec eng, Univ Nebr, 77-79; PRIN ENGR, HARRIS CONTROLS DIV, 79- *Concurrent Pos:* Consult, Lincoln Elec Syst, 78; adj prof, Fla Inst Technol, 79-; NSF grant. *Honors & Awards:* Elec Power Res Inst Award. *Mem:* Inst Elec & Electronics Engrs. *Res:* Computer modeling and analysis of electric power systems; optimum operation of power systems including unit maintenance scheduling, thermal and hydro resource scheduling over long term, short term and daily problems; unit commitment and load forecasting. *Mailing Add:* Harris Controls Div Harris Corp PO Box 430 Melbourne FL 32901

VEMURI, VENKATESWARARAO, b Chodavaram, India, Jan 17, 38; m 65; c 3. DISTRIBUTED PROCESSING, ARTIFICIAL NEURAL NETWORKS. *Educ:* Andhra Univ, India, BE, 58; Univ Detroit, MS, 63; Univ Calif, PhD(eng), 68. *Prof Exp:* Tech asst elec testing, Bhilai Steel Works, India, 58-61; jr engr, Radio Corp Am, 63-64; systs analyst, Environ Dynamics, Inc, 68-69; asst res engr, Univ Calif, Los Angeles, 69-70; asst prof aeronaut, astronaut & eng sci, Purdue Univ, 70-73; assoc prof comput sci, State Univ NY, Binghamton, 73-81; mem tech staff, TRW, Redondo Beach, Calif, 81-85; prof, Dept Appl Sci, Univ Calif, Davis, 85- *Concurrent Pos:* Res engr, Univ Southern Calif, 68-69. *Mem:* Inst Elec & Electronics Engrs. *Res:* Computational methods and computer architecture; distributed processing; modeling and simulation; artificial neural networks. *Mailing Add:* Univ Calif PO Box 808 L 794 Livermore CA 94550

VENA, JOSEPH AUGUSTUS, b Jersey City, NJ, Apr 18, 31; m 56; c 2. CYTOLOGY. *Educ:* St Peter's Col, BS, 52; Fordham Univ, MS, 55, PhD(cytol), 63. *Prof Exp:* Teacher high sch, 53-57; from instr to asst prof biol, Fordham Univ, 57-63; assoc prof, 63-66, PROF BIOL, TRENTON STATE COL, 66- *Concurrent Pos:* Sigma Xi res grant, 66-67; res consult, Univ Calif, Berkeley; consult, USPHS, 73-; NSF grant, 74. *Mem:* AAAS; Sigma Xi; Am Soc Cell Biol. *Res:* Coordinated studies of ultrastructural changes and electrophysiological properties in conduction system of canine heart during pharmacohogically induced alterations. *Mailing Add:* Dept Biol Trenton State Col Pennington Rd Trenton NJ 08625

VENABLE, DOUGLAS, b Charleston, WVa, Aug 17, 20; m 43; c 1. PHYSICS. *Educ:* Hampden-Sydney Col, BS, 42; Univ Va, MS, 47, PhD(physics), 50. *Prof Exp:* Design engr indust electronics div, Westinghouse Elec Corp, 42-46; mem staff, Los Alamos Nat Lab, Univ Calif, 50-57, alt group leader, 57-65, group leader, 65-72, alt div leader, 72-76, dep asst dir, 76-79, prog mgr, 79-80, dep assoc dir, 80-82; RETIRED. *Concurrent Pos:* Adj prof, Los Alamos Grad Ctr, Univ NMex, 57, 58 & 61. *Mem:* Fel AAAS; fel Am Phys Soc; Sigma Xi. *Res:* Crystal physics; electron beam dynamics; electron linear accelerators, gaseous discharges; flash radiography; detonation phenomena; hydrodynamics and shock wave phenomena. *Mailing Add:* 118 Aztec Ave Los Alamos NM 87544

VENABLE, EMERSON, b Cincinnati, Ohio, Dec 3, 11; m 35; c 4. PHYSICAL CHEMISTRY. *Educ:* Univ Pittsburgh, BS, 33. *Prof Exp:* Chemist, Agfa-Ansco Corp, 33-35; res chemist, Mine Safety Appliances Co, 35-37 & 44-46; res engr, Westinghouse Elec & Mfg Co, 37-44; res dir, Freedom Valvoline Oil Co, Freedom, Pa, 46-51; CONSULT, CHEMIST & ENGR, PITTSBURGH, 51- *Concurrent Pos:* Lectr, Univ Pittsburgh, 37-45; consult, Off Sci Res, USN, 44-46, AEC, 55-56. *Mem:* Fel AAAS; Am Chem Soc; hon fel Am Inst Chemists (pres-elect, 67-69 & pres, 69-71); fel Am Pub Health Asn; Nat Soc Prof Engrs; Nat Acad Indust Hyg. *Res:* Physical chemistry of delectrics; gas detection; fire and explosion; industrial hygiene; air pollution; corrosion; forensic science. *Mailing Add:* 6111 Fifth Ave Pittsburgh PA 15232-2807

VENABLE, JOHN HEINZ, JR, b Atlanta, Ga, June 9, 38; m 62. MOLECULAR BIOPHYSICS. *Educ:* Duke Univ, BS, 60; Yale Univ, MS, 63, PhD(biophys), 65. *Prof Exp:* Vis scientist, King's Col, Univ London, 65-67; asst prof molecular biol, 67-72, ASSOC PROF MOLECULAR BIOL, VANDERBILT UNIV, 72-, ASSOC DEAN, 81- *Concurrent Pos:* NSF fel, 65-66; USPHS fel, 66-67. *Mem:* AAAS; Biophys Soc; Sigma Xi. *Res:* Macromolecular structure; biophysical chemistry; transition-metal complexes; x-ray diffraction; electron paramagnetic resonance. *Mailing Add:* Vanderbilt Univ Box 1796 Sta B Nashville TN 37235

VENABLE, PATRICIA LENGEL, b Elyria, Ohio, Aug 6, 30; m 65; c 2. BOTANY. *Educ:* Col Wooster, BA, 52; Ohio State Univ, MSc, 54, PhD(bot), 63. *Prof Exp:* Instr bot, Hanover Col, 54-55; instr biol, Muskingum Col, 55-56 & Col Wooster, 56-60; vis instr bot & zool, Ohio State Univ, Lakewood Br, 62; assoc prof biol, State Univ NY Col Buffalo, 63-65 & Rider Col, 66-70; master biol, Lawrenceville Sch, 74-79; co-adj biol, Trenton State Col, 79-80; prog dir, Stony Brook-Millstone Watersheds Asn, 80-81; BIOL COORDR & TEACHER, PRINCETON DAY SCH, 81- *Mem:* Sigma Xi; Am Inst Biol Sci; Am Orchid Soc. *Res:* Morphological and biosystematic work with vascular plants; orchids; intertidal invertebrates. *Mailing Add:* 10 Monroe Ave Lawrenceville NJ 08648-1606

VENABLE, WALLACE STARR, b Wilkensburg, Pa, Apr 19, 40; m 62. ENGINEERING EDUCATION, ENGINEERING MECHANICS. *Educ:* Cornell Univ, BA, 62; Univ Toledo, MSES, 64; WVa Univ, EdD(eng educ), 72. *Prof Exp:* Instr, 66-72, lectr, 72-74, asst prof mech, 74-80, ASSOC PROF MECH & AEROSPACE ENG, WVA UNIV, 80- *Concurrent Pos:* Eng analyst, Hedenburg & Venable, 62- *Mem:* Fel Am Soc Eng Educ; Am Soc Mech Engrs. *Res:* Development and evaluation of educational methods and materials in engineering; accident analysis. *Mailing Add:* Box 125 Rte 13 Morgantown WV 26505

VENABLES, JOHN ANTHONY, b Leicester, UK, May 19, 36; m 61; c 2. ELECTRON MICROSCOPY MATERIALS, SURFACE PHYSICS. *Educ:* Cambridge Univ, UK, BA, 58, PhD(physics), 61. *Hon Degrees:* MA, Cambridge Univ, UK, 61. *Prof Exp:* Res assoc physics, Univ Ill, Urbana, 61-64; lectr, 64-71, reader, 71-88, PROF PHYSICS, UNIV SUSSEX, BRIGHTON, UK, 88- *Concurrent Pos:* Vis scientist, Max Planck Inst Stuttgart, Ger, 69; fel, Inst Physics, Eng, 72; prof assoc, CRMC2-CNRS, Marseille, 73-82, France, 74-86; fel, Japan Soc Prom Sci, 76; sci adv bd, Fritz-Haber Inst, Berlin, 81-, Lab Maurice lectr, Nancy, France, 89- *Mem:* Inst Physics; Am Inst Physics; Electron Micros Soc Am. *Res:* Electron microscopy and surface science; adsorption and crystal growth mechanisms; development of analytical techniques; auger electron and other spectroscopies; molecular solids and interatomic forces. *Mailing Add:* Dept Physics & Astron Ariz State Univ Tempe AZ 85287-1504

VENABLES, JOHN DUXBURY, b Cleveland, Ohio, Feb 6, 27; m 48; c 3. MATERIALS SCIENCE. *Educ:* Case Inst Technol, BS, 54; Univ Warwick, PhD, 71. *Prof Exp:* Physicist, Parma Res Ctr, Union Carbide Corp, 54-64; corp scientist, Martin Marietta Labs, 64-90; PRES, VENABLES & ASSOCS, 90- *Concurrent Pos:* Lectr, Univ Calif, 87. *Mem:* Am Phys Soc; Electron Micros Soc Am; Mat Res Soc; Sigma Xi. *Res:* Defect structure of solids; radiation effects in solids; ordering effects in transition metal carbides; high temperature ceramics; adhesive bonding; electron microscopy. *Mailing Add:* Venables & Assocs 848 Bosley Ave Baltimore MD 21204

VENARD, CARL ERNEST, b Marion, Ohio, Jan 10, 09; m 34; c 2. ZOOLOGY. *Educ:* Ohio State Univ, BA, 31, MSc, 32; NY Univ, PhD(helminth), 36. *Prof Exp:* Asst zool, Ohio State Univ, 32-34; asst biol, NY Univ, 34-36; from instr to prof zool & entom, 36-73, EMER PROF ZOOL & ENTOM, OHIO STATE UNIV, 73- *Mem:* Am Soc Parasitol; Entom Soc Am; Am Soc Zoologists. *Res:* Parasites of game birds and fishes; taxonomy and distribution of helminths; morphology of linguatulida; biology of fleas and mosquitoes. *Mailing Add:* Dept Entom Ohio State Univ 1735 Neil Ave Columbus OH 43210

VENDITTI, JOHN M, b Baltimore, Md, Feb 19, 27; m 51; c 3. PHARMACOLOGY, BIOCHEMISTRY. *Educ:* Univ Md, BS, 49, MS, 57; George Wash Univ, PhD(pharmacol), 65. *Prof Exp:* Biologist, 51-58, head, Screening Sect, 63-66, CHIEF DRUG EVAL BR, NAT CANCER INST, 66-, PHARMACOLOGIST, 58- *Concurrent Pos:* Vpres & dir, res, MicroBiotest, Inc, Chantilly, Va. *Mem:* AAAS; Am Asn Cancer Res; Soc Exp Biol & Med; Am Soc Pharmacol & Exp Therapeut; NY Acad Sci; Am Soc Microbiol. *Res:* Experimental cancer chemotherapy; biochemical and pharmacological actions of potential antitumor agents; antimicrobiol efficacy. *Mailing Add:* 6222 Stoneham Ct Bethesda VA 22021

VENEMA, GERARD ALAN, b Grand Rapids, Mich, Jan 26, 49; m 69; c 3. TOPOLOGY. *Educ:* Calvin Col, AB, 71; Univ Utah, PhD(math), 75. *Prof Exp:* Instr math, Univ Tex, Austin, 75-77; mem, Inst Advan Study, Princeton, NJ, 77-79; from asst prof to assoc prof, 79-83, PROF MATH, CALVIN COL, 83- *Mem:* Am Math Soc; Math Asn Am. *Res:* Geometric topology and applications to shape theory. *Mailing Add:* Dept Math Calvin Col Grand Rapids MI 49506

VENEMA, HARRY J(AMES), b Grand Rapids, Mich, July 28, 22; m 45; c 5. ELECTRICAL ENGINEERING. *Educ:* Univ Ill, BS, 44, MS, 47, PhD(elec eng), 50. *Prof Exp:* Asst, Elec Eng Lab, Univ Ill, 47-50; elec engr, Missile Guid Sect, Gen Elec Co, 50-53, elec engr, Magnetic Appln Sect, 53-56; mgr mil eng, Electronics Div, Stewart-Warner Corp, 56-59; electronic res, Roy C Ingersoll Res Ctr, Borg-Warner Corp, 59-86; RETIRED. *Mem:* Sigma Xi; Inst Elec & Electronics Engrs. *Mailing Add:* 1908 Driving Park Rd Wheaton IL 60187

VENEMAN, PETER LOURENS MARINUS, b Oudenrijn, The Netherlands, Nov 27, 47; m 73; c 2. SOIL GENESIS, SOIL PHYSICS. *Educ:* State Agr Univ, Wageningen, BS, 72; Univ Wis-Madison, MS, 75, PhD(soils), 77. *Prof Exp:* Res asst soils, Univ Wis-Madison, 72-77; from asst prof to assoc prof, 77-89, PROF SOILS, UNIV MASS, AMHERST, 89- *Mem:* Am Soc Agron; Soil Sci Soc Am; Int Soil Sci Soc; Soil Conserv Soc Am. *Res:* Formation, morphology and classification of spodosol soils, suitability rating of soils for the disposal of liquid wastes; suitability of soil for fruit production; lead and arsenic pesticide residues in soils; hydric soils; wetland identification. *Mailing Add:* Dept Plant & Soil Sci Univ Mass Amherst MA 01003

VENER, KIRT J, b Highland Park, Mich, Feb 1, 43; m 67; c 2. DIGESTIVE DISEASES. *Educ:* Wayne State Univ, Bs, 64, PhD(biol), 74. *Prof Exp:* Prin investr, NSF, 73-74; asst prof, Dept Biol, Layola Univ Chicago, Ill, 74-75; spec asst to the assoc dir for Digestive Dis & Nutrit, Nat Inst Diabetes & Digestive & Kidney Dis, NIH, 79-81; asst vice chancellor Res Affairs, Univ Tenn, Memphis, 87; EXEC SECY, NAT INST ARTHRITIS, MUSCULOSKELETAL & SKIN DISEASE, NIH, 87- *Mem:* Int Soc Chronobiol. *Res:* The time domain in experimental biology and medicine. *Mailing Add:* Res Br Rm 5A07 NIAMS-NIH, Westwood Bldg Bethesda MD 20892

VENERABLE, JAMES THOMAS, organic chemistry, for more information see previous edition

VENETSANOPOULOS, ANASTASIOS NICOLAOS, b Athens, Greece, June 19, 41; c 2. ELECTRICAL ENGINEERING, COMMUNICATIONS. *Educ:* Nat Tech Univ, Athens, dipl elec & mech eng, 65; Yale Univ, MS, 66, MPh, 68, PhD(commun), 69. *Prof Exp:* Res asst, N V Phillips, Neth, 64; asst in instr, Yale Univ, 66-68, res asst, 68-69; lectr elec eng, 68-70, from asst prof to assoc prof, 70-81, chmn commun group, 74-78, assoc chmn dept, 78-79, PROF ELEC ENG & CHMN COMMUN GROUP, UNIV TORONTO, 81- *Concurrent Pos:* Consult, Elec Eng Consociates Ltd, 69-; fel, Nat Sci & Eng Res Coun Can, Univ Toronto, 69-; Defense Res Bd Can fel, 72-75; lectr continuing educ, George Washington Univ, 80-, Northeastern Univ. *Mem:* Fel Inst Elec & Electronics Engrs; fel Eng Inst Can (vpres, 82-85); Sigma Xi; AAAS; Can Soc Elec Engrs (pres, 82-85); NY Acad Sci. *Res:* Digital signal processing; neural networks; channel modeling; signal design; image processing, analysis and computer vision. *Mailing Add:* Dept Elec Eng Univ Toronto Toronto ON M5S 1A4 Can

VENEZIAN, GIULIO, b Torino, Italy, Dec 9, 38; m 68; c 2. HYDRODYNAMICS, APPLIED MECHANICS. *Educ:* McGill Univ, BEng, 60; Calif Inst Technol, PhD(eng sci), 65. *Prof Exp:* Res fel eng sci, Calif Inst Technol, 65-68; from asst prof to assoc prof ocean eng, Univ Hawaii, 68-89; ASSOC PROF, SOUTHEAST MO STATE UNIV, 89- *Res:* Rotating fluid dynamics; magnetohydrodynamics; classical physics; applications to geophysics; water waves; geophysical fluid dynamics. *Mailing Add:* Dept Physics SE Mo State Univ Cape Girardeau MO 63701

VENEZKY, DAVID LESTER, b Washington, DC, Sept 12, 24; m 50; c 2. INORGANIC CHEMISTRY. *Educ:* George Washington Univ, BS, 48; Univ NC, PhD(chem), 62. *Prof Exp:* Phys sci aide trace elements unit, US Geol Surv, 48-49; chemist, US Naval Res Lab, 49-55; instr chem, Univ NC, 58-60; asst prof inorg chem, Auburn Univ, 60-62; res chemist, Naval Res Lab, 62-69, head reaction mechanism sect, Inorg Chem Div, 69-75, head solution chem sect, 75-81, head inorg & electrochem br & assoc supt, Chem Div, 81-84, liaison scientist chem, London Br Off, Off Naval Res, 84-85, sci dir, 85- 87, HEAD, SURFACE CHEM BR, CHEM DIV, NAVAL RES LAB, WASHINGTON, DC, 87- *Mem:* Am Chem Soc; Sigma Xi; Royal Chem Soc. *Res:* Coordination compounds and aggregation of inorganic substances in solutions; studies to elucidate the methods of preparation, structure and properties of inorganic polymers; chemical microsensors. *Mailing Add:* Naval Res Lab Code 6170 Washington DC 20375

VENHAM, LARRY LEE, b Akron, Ohio, June 24, 41; m 63; c 1. PEDODONTICS, PSYCHOLOGY. *Educ:* Ohio State Univ, DDS, 65, MS, 67, PhD(psychol), 72. *Prof Exp:* NIH fel, 67-69; asst prof, 70-78, assoc prof dent educ, Health Ctr Sch Dent Med, Univ Conn, 78-85; PVT PRACT. *Concurrent Pos:* Am Inst Res Creative Talent Award, 72; Nat Inst Dent Res Spec Dent Award, 75- *Mem:* Am Psychol Asn; Int Asn Dent Res; Soc Res Child Develop; Am Soc Dent Children. *Res:* Child development; situational stress, anxiety and coping behavior in response to dental stress; developmental factors in developing stress tolerance. *Mailing Add:* Pediat Dent 390 Broad St Windsor CT 06095

VENIER, CLIFFORD GEORGE, b Trenton, Mich, June 17, 39; m 65; c 3. ORGANIC SULFUR CHEMISTRY, LUBRICATION CHEMISTRY. *Educ:* Univ Mich, BS, 62; Ore State Univ, PhD(org chem), 66. *Prof Exp:* Res assoc chem, Univ Tex, 66-67; asst prof chem, Tex Christian Univ, 67-74, assoc prof, 74-80; sr chemist, Ames Lab, Iowa State Univ, 80-84; SR RES ASSOC, PENNZOIL PROD CO, 84- *Concurrent Pos:* Vis assoc prof, Univ Nijmegen, Neth, 75. *Mem:* Am Chem Soc; Royal Soc Chem; Sigma Xi. *Res:* Organic sulfur chemistry; coal chemistry with chemistry of lubrication; quantum organic chemistry. *Mailing Add:* Nine Splitrock Ct The Woodlands TX 77381

VENIT, STEWART MARK, b New York, NY, Apr 4, 46; m 72. MATHEMATICS. *Educ:* Queens Col, NY, BA, 66; Univ Calif, Berkeley, MA, 69, PhD(math), 71. *Prof Exp:* Asst prof, 71-77, assoc prof, 77-80, PROF MATH, CALIF STATE UNIV, LOS ANGELES, 80- *Mem:* Am Math Soc. *Res:* Numerical solution of partial differential equations. *Mailing Add:* Dept Math Calif State Univ 5151 State University Dr Los Angeles CA 90032

VENKATA, SUBRAHMANYAM SARASWATI, b Nellore, India, June 28, 42; Indian & US citizen; m 71; c 2. ELECTRICAL ENGINEERING. *Educ:* Andhra Univ, BSEE, 63; Indian Inst Technol, MSEE, 65; Univ SC, PhD(eng), 71. *Prof Exp:* Lectr elec eng, Coimbatore Inst Technol, 65-66; asst, Univ SC, 68-71; instr, Univ Lowell, 71-72; asst prof elec eng, WVa Univ, 72-75, assoc prof, 75-79; PROF ELEC ENG, UNIV WASH, 79- *Concurrent Pos:* Consult, SC Elec & Gas Co, 69-70; fel, Univ SC, 71; consult, Union Carbide Corp, 77-78, Puget Sound Power & Light, 79-, UIC, 84-, Scott & Scott Consults, 88-; ser ed, Power Systs. *Mem:* Fel Inst Elec & Electronics Engrs; Am Soc Eng Educ; Sigma Xi; Nat Soc Prof Engrs. *Res:* Six-phase power transmission; mine power system safety; reliability, availability and optimum maintainability; energy conservation; digital and analog simulation of energy systems; electrical power distribution. *Mailing Add:* Dept Elec Eng Univ Wash Seattle WA 98195

VENKATACHALAM, MANJERI A, b Calcutta, India, May 24, 40. ANATOMIC PATHOLOGY. *Educ:* Calcutta Med Col, BS & MS, 62. *Prof Exp:* PROF PATH, UNIV TEX HEALTH SCI CTR, SAN ANTONIO, 79- *Mem:* Am Asn Pathologists; Am Soc Cell Biol; Am Soc Clin Invest; Int Acad Path; AAAS; Am Soc Nephrology. *Mailing Add:* Dept Path Univ Tex Health Sci Ctr 7703 Floyd Curl Dr San Antonio TX 78284-7750

VENKATACHALAM, TARACAD KRISHNAN, b Cochin, India, Apr 28, 37. ORGANIC CHEMISTRY. *Educ:* Univ Bombay, BSc, 58, MSc, 62; Univ Louisville, PhD(chem), 65. *Prof Exp:* Res chemist, 65-69, sr res chemist, 69-85, RES ASSOC, E I DU PONT DE NEMOURS & CO, INC, 85- *Mem:* Am Chem Soc; Indian Chem Soc; Royal Inst Chem. *Res:* Polymer technology; natural and synthetic resins; rubber chemistry; textile fibers; tire cord adhesion and processing; ropes and cables; tires; seat belts. *Mailing Add:* Du Pont Co Fibers Dept Bldg 702 Chestnut Run Plaza Wilmington DE 19805

VENKATARAGHAVAN, R, b Madras, India, June 29, 39. INFORMATION SCIENCE & SYSTEMS. *Educ:* Univ Madras, BSc, 58, MSc, 60; Indian Inst Sci, Bangalore, PhD(chem), 63. *Prof Exp:* Fel spectros, Nat Res Coun Can, 63-65; NIH res assoc mass spectros, Purdue Univ, Lafayette, 65-69; sr res assoc chem, Cornell Univ, 69-77; mem staff, Res Data Processing, 77-80, DIR, RES COMPUT, LEDERLE LABS, AM CYANAMID CO, 84- *Concurrent Pos:* Consult, US Army Labs; mem, bd sci adv, Nat Cancer Inst. *Mem:* AAAS; Am Chem Soc. *Res:* Pharmaceutical chemistry; chemical information systems; computer aided analytical techniques; structure-activity studies; artificial intelligence techniques. *Mailing Add:* Lederle Labs Pearl River NY 10965

VENKATARAMAN, M, b June 10, 43; US citizen; m 75; c 2. T&B LYMPHOCYTES SUBSETS, HUMAN LUNG CANCER. *Educ:* Madras Vet Col, India, BVSc, 67; Inst Med Sci, New Delhi, India, MSc, 72, PhD(immunol), 75. *Prof Exp:* ASST PROF IMMUNOL, MT SINAI HOSP MED CTR, 79- *Mem:* Am Asn Immunologists; Soc Cryobiol. *Res:* Cellular immunology; tumor immunology; cell growth and differentiation factors; cryopreservation effects on immunocompetent cell functions; monoclonal antibodies; hybridomas; bone marrow transplantation; radiation effects. *Mailing Add:* Dept Med Mt Sinai Hosp Med Ctr Calif Ave & 15th St Chicago IL 60608

VENKATARAMANAN, RAMAN, b Kallal, Madras, India, Aug 30, 51; m 84. BIOPHARMACEUTICS, PHARMACOKINETICS. *Educ:* Madras Med Col, BS, 72; Birla Inst Technol & Sci, MS, 74; Univ BC, PhD(pharmacokinetics & biopharmaceut), 79. *Prof Exp:* Fel pharm, Univ Wash, Seattle, 78-80; asst prof, 80-86, ASSOC PROF PHARM, UNIV PITTSBURGH, 86- *Concurrent Pos:* Hosp training, Stanley Hosp, Madras, India, 71-72; mem task force bioequivalence, Dept Health; mem study group cyclosporine, Nat Asn Clin Biochemists; consult transplant teams, Univ Pittsburgh; consult, Magee Women's Hosp Pittsburgh; mem, Am Heart Asn, Western Pa affil, 81-; vis scientist, Alta Heritage Found, 85, Med Res Coun, 90. *Mem:* Fel Am Col Clin Pharm; Am Asn Pharmaceut Sci; Am Asn Col Pharm. *Res:* Pharmacokinetic and pharmacodynamic studies of drugs in organ transplant patients in order to optimize drug therapy in this patient population. *Mailing Add:* 718 Salk Hall Univ Pittsburgh Pittsburgh PA 15261

VENKATARAMIAH, AMARANENI, environmental physiology, marine zoology, for more information see previous edition

VENKATESAM, MALABI M, b Ranchi, India, July 3, 50; US citizen. MICROBIOL PATHOGENESIS, CELL BIOLOGY. *Educ:* Nagpur Univ, BSc, 67, MSc, 69; Univ Pittsburgh, PhD(biochem), 77. *Prof Exp:* Jr res fel, Dept Biochem, All India Inst Med Sci, 69-71; grad teaching asst, Dept Molecular Biol, State Univ NY, Stony Brook, 71-72; grad teaching asst, Dept Biochem, Univ Pittsburgh, Pa, 72-77; vis fel, Lab Biochem & Pharmacol, Nat Inst Arthritis, Diabetes & Digestive & Kidney Dis, NIH, 78-81, staff fel, Lab Cell & Develop Biol, 81-84, sr staff fel, 84-85; sr nat res coun fel, 85-87, RES CHEMIST BACT IMMUNOL, WALTER REED ARMY INST RES, 87- *Concurrent Pos:* Prin investr, Prog Appropriate Technol Health, USAID, Seattle, Wash, 89-91; found lectr, Am Soc Microbiol, 90-91; sponsor, Nat Res Coun, Washington, DC, 90- *Mem:* Am Soc Microbiol. *Res:* Characterization of bacterial and host-associated factors that determine pathogenesis of bacillary dysentery. *Mailing Add:* Bact Immunol Walter Reed Army Inst Res Washington DC 20307-5100

VENKATESAN, DORASWAMY, b Coimbatore, India. SPACE PHYSICS, ASTROPHYSICS. *Educ:* Loyola Col, Madras, India, BSc, 43; Benares Hindu Univ, MSc, 45; Gujarat Univ, India, PhD(cosmic rays), 55. *Prof Exp:* Lectr physics, K P Col, Allahabad, India, 47-48; lectr, Durbar Col, Rewa, 49; sr res asst cosmic rays, Phys Res Lab, Ahmedabad, 49-56; fel, Inst Electron Physics, Royal Inst Technol, Stockholm, 56-57; fel, Nat Res Coun Can, 57-60; res assoc space physics & astrophysics, Univ Iowa, 60-63, asst prof physics, 63-65, consult, High Altitude Balloon Prog, 65; assoc prof physics, Univ Alta, 65-66; assoc prof, 66-69, PROF PHYSICS, UNIV CALGARY, 69- *Mem:* Am Geophys Union; Can Asn Physicists; Can Astron Soc; fel Brit Inst Physics. *Res:* Solar terrestrial relations; astrophysics involving studies of cosmic rays, radiation belts, ionospheric absorption, auroral x-rays, geomagnetism, solar activity, cosmic x-ray sources and interplanetary medium. *Mailing Add:* Dept Physics Univ Calgary 2500 Univ Dr NW Calgary AB T2N 1N4 Can

VENKATESAN, S, MOLECULAR BIOLOGY, AIDS RETRO VIRUS. *Educ:* Guntur Med Col AP Guntur, India, MD, 67. *Prof Exp:* SR SCIENTIST, FREDRICK CANCER CTR, NIH, GOVT RES INST, 77- *Mailing Add:* NIH Bldg 4 Rm 326 Bethesda MD 20892

VENKATESAN, THIRUMALAI, b Madras, India, June 19, 49; m 77. PHYSICS. *Educ:* Indian Inst Technol, Kharagpur, BS, 69, Kanpur, MS, 71; City Univ New York, PhD(physics), 77. *Prof Exp:* Mem tech staff, Optical Commun Res, Bell Labs, 77-79, mem staff, Radiation Physics Res, 79-89; PROF ELEC ENG & PHYSICS, CTR SUPERCONDUCTIVITY RES, DEPT PHYSICS, UNIV MD, COLLEGE PARK, 89- *Concurrent Pos:* Mgr, Mat Modifications Group, Bellcore Dir, Lab Surface Modification. *Mem:* Sigma Xi; Am Inst Physics; fel Am Phys Soc. *Res:* Ion solid interaction, ion beam lithography, germanium selenide resists, metal-insulatory transition; optical properties of semiconductors; optical nonlinear devices; optical communication systems and associated solid state and device physics; laser-solid interaction; epitaxial crystalline metal oxides and high temperature superconducting film; physics and applications. *Mailing Add:* Ctr Superconductivity Dept Physics Univ Md College Park MD 20742

VENKATESWARAN, UMA D, b India, Aug 6, 53. SOLID STATE PHYSICS-EXPERIMENTAL, SEMICONDUCTORS. *Educ:* Madurai Univ, India, BSc, 73, MSc, 75; Univ Mo, Columbia, PhD(physics), 85. *Prof Exp:* Scientist, Mat Sci Lab, Kalpakkam, India, 76-82; guest scientist, Max-Planck-Inst, Stuttgart, Ger, 86-88; postdoctoral res assoc, Dept Physics, State Univ NY, Buffalo, 88-90; ASST PROF SOLID STATE PHYSICS, DEPT PHYSICS, OAKLAND UNIV, 91- *Concurrent Pos:* Lectr, Dept Physics, State Univ NY, Buffalo, 91. *Mem:* Am Phys Soc. *Res:* Study of optical properties of electro-optic materials and high temperature superconductors under high hydrostatic pressure (0-20GPa). *Mailing Add:* Dept Physics Oakland Univ Rochester MI 48309

VENKATU, DOULATABAD A, b Bangalore, India, July 31, 36; m 67; c 2. METALLURGY, CERAMICS. *Educ:* Univ Mysore, BSc, 55; Indian Inst Sci, Bangalore, Dipl, 58; Univ Notre Dame, MS, 61, PhD(metall eng & mat sci), 65. *Prof Exp:* Sr res asst metall, Indian Inst Sci, Bangalore, 58-59; res asst metall eng, Univ Notre Dame, 59-64; asst prof, Clemson Univ, 64-67; res assoc, Rensselaer Polytech Inst, 67-68; asst prof, Clemson Univ, 68-69; sr mat scientist, Owens-Ill Inc, 69-73; SR ENG SPECIALIST, GOODYEAR AEROSPACE CORP, 73- *Mem:* Am Soc Metals; Am Inst Mining, Metall & Petrol Engrs; Am Ceramic Soc. *Res:* Fatigue of metals; fracture mechanics; light metals technology; powder metallurgy; high temperature and technical ceramics. *Mailing Add:* 1278 Goldfinch Trail Stow OH 44224

VENKAYYA, VIPPERLA, b Raghudevapuram, Andhra, India, May 16, 31; m 65; c 3. STRUCTURAL ENGINEERING. *Educ:* Andhra Univ, BSc, 52; Indian Inst Technol, BTech, 56; Univ Mo, MS, 59; Univ Ill, Urbana, PhD(struct eng), 62. *Prof Exp:* Asst engr, Damodar Valley Corp, India, 56-57; bridge engr, State Hwy Dept, Pa, 58-59; asst prof struct eng, State Univ NY, Buffalo, 62-67; AEROSPACE ENGR, AIR FORCE FLIGHT DYNAMICS LAB, WRIGHT-PATTERSON AFB, 67- *Concurrent Pos:* Assoc ed, Am Inst Aeronaut & Astronaut J, 78-80; adj prof, Air Force Inst Technol, 78- *Honors & Awards:* Gen Foulois Award, Air Force Flight Dynamics Lab. *Mem:* Am Soc Civil Engrs; Am Inst Aeronaut & Astronaut; Am Soc Mech Engrs; Am Acad Mech. *Res:* Response of discrete and continuous elastic systems to static and dynamic disturbances, particularly civil and aeronautical structures; stability of elastic systems; dynamics and control of space structures. *Mailing Add:* 5464 Honeyleaf Way Dayton OH 45424

VENKETESWARAN, S, b Alleppey, Kerala, India, July 21, 31; m 75; c 1. CELL BIOLOGY, BOTANY. *Educ:* Univ Madras, BSc, 50; Univ Bombay, MSc, 53; Univ Pittsburgh, PhD(bot), 61. *Prof Exp:* Demonstr biol, Jai Hind Col, Univ Bombay, 52-57; ed asst, Coun Sci & Indust Res, Govt India, 57-58; teaching fel & asst, Univ Pittsburgh, 59-61, res assoc, 61-62, instr biol, 62-63, NASA res assoc, 63-65; asst prof, 65-68, ASSOC PROF BIOL, UNIV HOUSTON, 68- *Concurrent Pos:* Am Cancer Soc instnl grant, Univ Pittsburgh, 65; consult & NASA res grant, Lunar Sample Receiving Lab, 66-75; vis scientist, Argonne Nat Lab, 67, NASA & JSC, 87-88; NSF travel award, India, 80; Bioenergy grant, 68-71, USAID grant, 71-74 & 83-85. *Mem:* AAAS; Bot Soc Am; Am Inst Biol Sci; Tissue Cult Asn. *Res:* Plant morphogenesis using tissue culture techniques; wood biomass as energy sources; forestry; tissue culture biomass. *Mailing Add:* Dept Biol Univ Houston Houston TX 77204

VENNART, GEORGE PIERCY, b Boston, Mass, Apr 1, 26; m 51; c 3. PATHOLOGY. *Educ:* Wesleyan Univ, AB, 48; Univ Rochester, MD, 53. *Prof Exp:* Asst biol, Wesleyan Univ, 47-48; intern & resident path, NC Mem Hosp, 53-56; asst prof, Col Physicians & Surgeons, Columbia Univ, 56-60; assoc prof path, Univ NC, 60-65; PROF PATH & CHMN DIV CLIN PATH, MED COL VA, 65-, CHMN DEPT PATH, 78- *Concurrent Pos:* Instr, Univ NC, 54-56; asst attend pathologist, Preby Hosp, New York, 56-60. *Res:* Experimental liver disease; platelet agglutination; pulmonary morphology and physiology. *Mailing Add:* Dept Pathol Va Commonwealth Univ Box 662 MCV Sta Richmond VA 23298

VENNES, JACK A, b Wheeler, Wis, June 12, 23. GASTROENTEROLOGY, INTERNAL MEDICINE. *Educ:* Univ Minn, BS, 47, MD, 51. *Prof Exp:* Intern med, Hennepin County Med Ctr, Minneapolis, 51-52; residency, Vet Admin Hosp, Minneapolis, 52-55; pvt pract, St Louis Park Med Ctr, Minneapolis, 57-63; staff physician gastroenterol, 64-67, asst chief med, 67-71, STAFF PHYSICIAN GASTROENTEROL, VET ADMIN HOSP, MINNEAPOLIS, 71- *Concurrent Pos:* Instr med, Univ Minn, 55-57, from asst prof to assoc prof, 65-76, prof med, 76- *Mem:* Am Gastroenterol Asn; Am Soc Gastrointestinal Endoscopy; Am Asn Study Liver Dis; Am Soc Clin Invest; Am Fedn Clin Res. *Res:* Development and applications of fiberoptic endoscopy to improved diagnosis in upper gastrointestinal tract, pancreas and biliary tree; treatment of gastrointestinal hemorrhage; non-surgical endoscopic removal of common duct gallstones; improved teaching methods of fiberoptic endoscopy. *Mailing Add:* Vet Admin Hosp Minneapolis MN 55417

VENNES, JOHN WESLEY, b Grenora, NDak, Aug 28, 24; m 48; c 3. BACTERIOLOGY. *Educ:* Univ NDak, BS, 51, MS, 52; Univ Mich, PhD(bact), 57. *Prof Exp:* Instr bact, Univ NDak, 52-54; asst, Univ Mich, 54-56; from instr to assoc prof, 56-66, actg dean, sch med, 73-75, assoc dean acad affairs, 73-77, PROF BACT, UNIV NDAK, 66-, CHMN, DEPT MICROBIOL, 81- *Mem:* Am Soc Microbiol. *Res:* Bacterial physiology and industrial microbiology. *Mailing Add:* Dept Microbiol Univ NDak Sch Med 501 Columbia Rd N Grand Forks ND 58201

VENNESLAND, BIRGIT, b Kristiansand, Norway, Nov 17, 13; US citizen. ENZYMOLOGY. *Educ:* Univ Chicago, BS, 34, PhD(biochem), 38. *Hon Degrees:* DSc, Mt Holyoke Col, 60. *Prof Exp:* Asst biochem, Univ Chicago, 38-39; fel, Harvard Med Sch, 39-41; from instr to prof biochem, Univ Chicago, 41-68; dir, Max Planck Inst Cell Physiol, WBerlin, Ger, 68-70; leader, 70-81, EMER, VENNESLAND RES INST, W BERLIN, GER, 81-; ADJ PROF, BIOCHEM & BIOPHYSICS, UNIV HAWAII, 87- *Honors & Awards:* Hales Award, Am Soc Plant Physiol, 50; Garvan Medal, Am Chem Soc, 64. *Mem:* Am Chem Soc; Am Soc Biol Chemists; fel AAAS; fel NY Acad Sci; Am Soc Plant Physiologists. *Res:* Carboxylation reactions in animals and plants; mechanisms of hydrogen transfer in pyridine nucleotide dehydrogenases; enzymology and mechanism of photosynthesis; mechanism of nitrate reduction. *Mailing Add:* 1206 Mokapu Blvd Kailua HI 96734

VENNOS, MARY SUSANNAH, b Oct 14, 31; Can citizen; m 58; c 4. CHEMISTRY. *Educ:* Univ London, BSc, 53; Univ NB, Fredericton, PhD(chem), 56. *Prof Exp:* Instr chem, Univ NB, 56-59; from asst prof to assoc prof, Russell Sage Col, 59-70; assoc prof, 70-80, PROF CHEM, ESSEX COMMUNITY COL, BALTIMORE COUNTY, MD, 80- *Mem:* Am Chem Soc; Sigma Xi. *Res:* Analytical instrumentation; polarography and chemical kinetics. *Mailing Add:* 4003 Milldale Ct Phoenix MD 21131-2103

VENT, ROBERT JOSEPH, b Ford City, Pa, Feb 13, 40; m; c 2. UNDERWATER ACOUSTICS. *Educ:* San Diego State Univ, BS, 61, MS, 69. *Prof Exp:* Physicist underwater acoust, Navy Electronics Lab, 61-68; supv physicist, Naval Undersea Ctr, 68-77, RES PHYSICIST, ACOUST, NAVAL OCEAN SYSTS CTR, 77- *Concurrent Pos:* Work in musical acoustics. *Res:* Underwater acoustics, especially attenuation, surface, bottom and volume scattering; applied ocean sciences. *Mailing Add:* Naval Ocean Systs Ctr Code 663 San Diego CA 92152

VENTA, PATRICK JOHN, b Rock Springs, Wyo, Nov 6, 51; m 79; c 1. RECOMBINANT DNA, HUMAN GENETICS. *Educ:* Univ Calif, Irvine, BS, 74; Univ Mich, MS, 77, PhD(human genetics), 83. *Prof Exp:* Scholar, 83-84, res assoc human genetics, Univ Mich, 84-90; ASST PROF, MICH STATE UNIV, 90- *Mem:* AAAS; Genetics Soc Am; Sigma Xi; Am Soc Human Genetics. *Res:* Structure and regulation of eukaryotic genes; molecular evolution; mapping mammalian genomes. *Mailing Add:* Small Animal Clin Sci Mich State Univ East Lansing MI 48824-1314

VENTER, J CRAIG, b Salt Lake City, Utah, Oct 14, 46; m 81. PROTEIN CHEMISTRY, HYBRIDOMA. *Educ:* Univ Calif, San Diego, BA, 72, PhD(physiol & pharmacol), 75. *Prof Exp:* Res assoc cardiovasc pharmacol, Univ Calif, San Diego, 75-76; asst prof pharmacol, 76-81, ASSOC PROF BIOCHEM, STATE UNIV NY, BUFFALO, 82- *Concurrent Pos:* Mem, Basic Sci Coun, Am Heart Asn. *Mem:* Am Soc Pharmacol & Exp Therapeut; AAAS. *Res:* Purification and molecular characterization of B-adrenergic and muscarinic acetylcholine receptors; production of monoclonal antibodies to each receptor. *Mailing Add:* Dept Pharmacol-Therapeut State Univ NY Sch Med Buffalo NY 14214

VENTERS, MICHAEL DYAR, NONINVASIVE VASCULAR DIAGNOSTICS. *Educ:* Univ NMex, PhD(physiol), 79. *Prof Exp:* DIR CLIN PHYSIOL & CARDIOL SERV, ST JOSEPH HEALTHCARE CORP, 80- *Res:* Critical care physiological monitoring. *Mailing Add:* 9408 Spain NE Albuquerque NM 87111

VENTRE, FRANCIS THOMAS, b Old Forge, Pa, Sept 16, 37; m 64; c 2. BUILDING SCIENCE, REGULATION OF TECHNOLOGY. *Educ:* Pa State Univ, BArch, 61; Univ Calif, Berkeley, MCP, 66; Mass Inst Technol, PhD(urban studies & planning), 73. *Prof Exp:* Asst prof urban design, Univ Calif, Los Angeles, 66-68; res assoc building technol, Mass Inst Technol, 70-73; asst chief, Off Building Standards & Codes, Ctr Building Technol, Nat Bur Standards, 73-75, asst to dir, Inst Appl Technol, Nat Eng Lab, 76-78, chief, Environ Design Res Div, Ctr Building Technol, 78-83; PROF ENVIRON DESIGN & POLICY, COL ARCHIT & URBAN STUDIES, VA POLYTECH INST & STATE UNIV, 83- *Concurrent Pos:* Guest lectr, Sch Archit, Carnegie Mellon Univ, 80 & Pa State Col Eng, 81; vis prof, Sch Archit, Univ Md, 82-83; prin investr, NSF grant, 85-87; mem comt technol advan buildings, Building Res Bd, Nat Acad Sci-Nat Acad Eng, 85-86. *Mem:* Environ Design Res Asn; Am Econ Asn; Am Soc Testing & Mat; Am Inst Architects; Am Planning Asn. *Res:* Methods for measuring building performance; competitive conditions for international trade in design and construction services; size, structure, deployment and economic impact of the design and construction services industries; measurement theory; design theory. *Mailing Add:* Dept Arch Design VA Polytech Inst Blacksburg VA 24061

VENTRES, CHARLES SAMUEL, b Tucson, Ariz, Oct 21, 42. AERONAUTICAL ENGINEERING. *Educ:* Univ Ariz, BSME, 64; Princeton Univ, MA, 67, PhD(eng), 70. *Prof Exp:* Res assoc aeronaut eng, Princeton Univ, 69-70, mem res staff, 70-74; SR CONSULT, BOLT, BERANEK & NEWMAN, INC, 74- *Mem:* Am Inst Aeronaut & Astronaut; Acoust Soc Am. *Res:* Structural dynamics; aeroelasticity; acoustics; physical acoustics; aerodynamics. *Mailing Add:* 395 Broadway R5G Cambridge MA 02139

VENTRESCA, CAROL, b Columbus, Ohio; c 2. PERFORMANCE IMPROVEMENT IN MANUFACTURING. *Educ:* Ohio State Univ, BS, 78, MS, 81. *Prof Exp:* Tech staff, Metrek Div, Mitre Corp, 78; vpres & chief exec officer, SynGenetics Corp, 88-90; res scientist, 79-85, mgr, 85-88, SR MGR, MFG SOFTWARE SYSTS, BATTELLE MEM INST, 90- *Mem:* Opers Res Soc Am; Mil Opers Res Soc; Am Defense Preparedness Asn; Soc Mfg Engrs. *Res:* Assists manufacturing companies in improving performance in 6 key areas: quality, cost, timeliness, product performance, legal/social impact and risk; solutions include product design engineering, process optimization, automation, computer integrated manufacturing, computer system maintenance and enhancement; applies operations research analysis techniques to a variety of problems for government and industry; chemical warfare defense; nuclear waste isolation; cost analysis and other types of optimization. *Mailing Add:* Battelle Mem Inst 505 King Ave Columbus OH 43201-2693

VENTRICE, CARL ALFRED, b York, Pa, Aug 7, 30; m 60; c 3. PLASMA PHYSICS, NUCLEAR PHYSICS. *Educ:* Pa State Univ, BS, 56, MS, 58, PhD(physics), 62. *Prof Exp:* Sr analyst, Anal Serv Inc, 63-64; assoc prof physics, Tenn Technol Univ, 64-66; assoc prof elec eng, Auburn Univ, 66-68; PROF ELEC ENG, TENN TECHNOL UNIV, 68- *Mem:* AAAS; Inst Elec & Electronics Engrs; Am Phys Soc. *Res:* Interaction of electromagnetic waves in plasmas; plasma stability; lasers. *Mailing Add:* Dept Elec Eng Tenn Technol Univ Cookeville TN 38505

VENTRICE, MARIE BUSCK, b Allentown, Pa, Oct 17, 40; m 60; c 3. THERMAL & FLUID SCIENCES. *Educ:* Tenn Technol Univ, BSES, 66, PhD(mech eng), 74; Auburn Univ, MS, 68. *Prof Exp:* Instr eng sci, 69-70, from asst prof to assoc prof, 74-86, PROF MECH ENG, TENN TECHNOL UNIV, 86- *Concurrent Pos:* Interim dir, Ctr Elec Power, Tenn Technol Univ, 85-88, assoc dean eng, 89- *Mem:* Am Soc Mech Engrs; Am Soc Eng Educ; Nat Soc Prof Engrs; Sigma Xi; Am Inst Aeronaut & Astronaut. *Res:* Experimental studies of liquid propellant rocket combustion instabilities using an analog technique; cogeneration. *Mailing Add:* Col Eng Tenn Technol Univ Cookeville TN 38505

VENTRIGLIA, ANTHONY E, b New York, NY, June 20, 22; m 53; c 2. APPLIED MATHEMATICS. *Educ:* Columbia Univ, AB, 42; Brown Univ, ScM, 43. *Prof Exp:* Instr math, Rutgers Univ, 47; from instr to asst prof, 47-61, ASSOC PROF MATH, MANHATTAN COL, 61- *Concurrent Pos:* Social Sci Res Fel, Stanford Univ, 57, instr, Hunter Col, 61-63; adj asst prof, City Col New York; NSF fel, Inst Math Teachers, Univ Wyo, 59. *Mem:* AAAS; Am Math Soc; Math Asn Am; Am Acad Polit & Soc Sci; Am Asn Univ Prof. *Res:* Linear algebra and analysis; partial differential equations. *Mailing Add:* One Georgia Ave Bronxville NY 10708

VENTURA, JOAQUIN CALVO, b Cadiz, Spain, Mar 22, 29; Can citizen; m 61; c 2. PATHOLOGY. *Educ:* Univ Seville, MD, 52; Univ Montreal, PhD, 58; FRCP(C), 61. *Prof Exp:* SR LECTR PATH, UNIV MONTREAL, 58-; MEM STAFF, SANTA CABRINI HOSP, 71- *Concurrent Pos:* Chief serv & dir res, St Joseph of Rosemont Hosp, Montreal, 62-69; consult pathologist, Louis H LaFontaine Hosp, Montreal, 69- & Maisonneuve Rosemont Hosp, Montreal, 70-; assoc dir labs, Santa Cabrini Hosp, 69-70. *Mem:* Can Asn Path; NY Acad Sci; Can Med Asn. *Res:* Chronic bronchitis, role of sensitization; pathogenesis of bronchiectasis; effects of pollution on chronic experimental bronchitis. *Mailing Add:* Santa Cabrini Hosp 5655 St Zotique E Montreal PQ H1T 1P7 Can

VENTURA, JOSE ANTONIO, b Barcelona, Spain, Nov 3, 54; m 81; c 3. NONLINEAR PROGRAMMING, VISION SYSTEMS. *Educ:* Polytech Univ Barcelona, Spain, BSIE, 79; Univ Fla, ME, 84, PhD (indust eng), 86. *Prof Exp:* Asst prof opers res, Univ Mo, Columbia, 86-89; ASST PROF OPERS RES, PA STATE UNIV, 89- *Concurrent Pos:* Panelist, NSF, 89-91; NSF presidential young investr, 90. *Honors & Awards:* Ralph R Teetor Award, Soc Automotive Engrs, 89. *Mem:* Opers Res Soc Am; sr mem Soc Mfg Engrs; sr mem Inst Indust Engrs; Soc Automotive Engrs; Am Soc Eng Educ. *Res:* Applied optimization; computerized manufacturing and inspection; production management; applied probability. *Mailing Add:* 207 Hammond Bldg Pa State Univ University Park PA 16802

VENTURA, WILLIAM PAUL, b Braddock, Pa, Dec 1, 42; m 69; c 2. ECOLOGY. *Educ:* Duquesne Univ, BS, 64, MS, 66; New York Med Col, PhD(pharmacol), 69, Pace Univ, MBA, 80. *Prof Exp:* Res assoc endocrinol, Duquesne Univ, 66; from instr to asst prof pharmacol, New York Med Col, 69-74; assoc prof, 74-81, PROF PHARMACOL & CHMN, DEPT BIOL SCI, PACE UNIV, 81- *Concurrent Pos:* Lalor Found grant, 70; NSF equip grant, 80, Dorr Found grant, 89; Dyson Col fel, 85. *Mem:* Am Physiol Soc; NY Acad Sci; Int Fertil Asn; Am Chem Soc; Am Col Clin Pharm. *Res:* Reproductive pharmacology, male and female reproductive studies. *Mailing Add:* 368 Elm Rd Briarcliff Manor NY 10510

VENTURELLA, VINCENT STEVEN, b Pittsburgh, Pa, Aug 24, 30; m 54; c 3. HIGH PRESSURE LIQUID CHROMATOGRAPHY. *Educ:* Univ Pittsburgh, BS, 54, MS, 56, PhD(med chem), 61. *Prof Exp:* Asst prof pharm chem, Fordham Univ, 60-63; anal res chemist, Abbott Labs, 63-64; asst prof pharm & pharm chem, Temple Univ Philadelphia, Pa, 64-67; mgr anal res, Hoffman-LaRoche Inc, 67-71, group leader, 72-76, mgr anal develop, 76-79, sr tech fel, 79-85; chief res br, US Bur Customs Lab, New York, 71-72; group leader, Anaquest Div, BOC Group, 85-88, SECT MGR, ANAQUEST INC, 88- *Concurrent Pos:* Adj asst prof, dept chem, Rutgers Univ, 69-70, vis prof, Sch Pharm, 81-; adj assoc prof, Fairleigh Dickinson Univ, Rutherford, NJ, 72-75; consult nuclear magnetic resonance, NF Rev Comt, 72-78. *Mem:* NY Acad Sci; Am Pharmaceut Asn; Am Chem Soc; NAm Thermal Analysis Soc; Int Soc Magnetic Resonance; Soc Appl Spectros. *Res:* Method development for bulk pharmaceutical chemicals, drug products, metabolisms and degradation products; identification of impurities and intermediates by high resolution nuclear magnetic resonance and mass spectroscopy, particularly carbon 13 and phosphorous 31 resonance. *Mailing Add:* Anaquest Inc 100 Mountain Ave Murray Hill NJ 07974

VENUGOPALAN, SRINIVASA I, b Madras, Tamilnadu, India, Dec 19, 44; Indian citizen; m 70; c 2. LASER SPECTROSCOPY, CONDENSED MATTER PHYSICS. *Educ:* Univ Madras, BSc, 63; Purdue Univ, MS, 69, PhD(physics), 73. *Prof Exp:* Sci officer, Bhabha Atomic Res Ctr, India, 63-67; res/teaching asst physics, Purdue Univ, 67-73; scientist, Raman Res Inst, India, 73-79; res physicist, Purdue Univ, 79-81; asst prof, 81-84, ASSOC PROF PHYSICS, STATE UNIV NY, BINGHAMTON, 84- *Concurrent Pos:* Vis scientist, US Naval Res Lab, 83; consult, IBM-Endicott, 83-87; vis assoc prof, Purdue Univ, 84 & Cornell Univ, 87-88. *Mem:* Sigma Xi. *Res:* Experimental solid state physics; raman, infrared and quasielastic light scattering investigations of excitations in semiconductors, liquid crystals, colloidal materials, and magnetically ordered systems. *Mailing Add:* Dept Physics State Univ NY Binghamton NY 13901

VENUTI, WILLIAM J(OSEPH), b Philadelphia, Pa, Aug 16, 24; m 49; c 6. CIVIL ENGINEERING. *Educ:* Univ Pa, AB, 47; Univ Colo, BS, 50, MS, 55; Stanford Univ, PhD(civil eng), 63. *Prof Exp:* Civil engr, US Bur Reclamation, Colo & Mont, 49-52; instr civil eng, Univ Colo, 52-55; assoc prof, 55-63, PROF CIVIL ENG, SAN JOSE STATE UNIV, 63- *Concurrent Pos:* Sr design engr, Food Mach & Chem Corp, 56-; assoc, Boeing Airplane Co, 59; prof struct eng, SEATO Grad Sch Eng, Bangkok, Thailand, 64-66; vis prof, Stanford Univ, 70-71, 78- & Univ Dundee, 72; NSF int travel grant, India, 72. *Mem:* Fel Am Soc Civil Engrs; Am Soc Eng Educ; fel Concrete Inst; Am Rwy Eng Soc. *Res:* Lightweight prestressed concrete; framed structures; effects of impact loads on structures; fatigue of concrete; structural dynamics; blast loading and effects; prestressed concrete connectors; concrete railroad ties. *Mailing Add:* Dept Civil Eng San Jose St Univ Mich Sqg San Jose CA 95192

VENUTO, PAUL B, b Flushing, NY, Feb 8, 33; m 58; c 3. PETROLEUM PROCESSING, PRODUCTION TECHNOLOGY. *Educ:* Univ Pa, AB, 54, PhD(org chem), 62. *Prof Exp:* Res & develop chemist, Columbian Carbon Co, 57-59; sr res chemist, Mobil Oil Corp, 62-66, group leader heterogeneous catalysis, 66-67, group leader appl res & develop div, 67-69, res assoc, Paulsboro Res Lab, 69-75, mgr anal & spec technol, Cent Res Lab, 75-77, mgr, heavy oils & energy minerals, Dallas Res Lab, 77-82, mgr, Enhanced Oil Recovery, Dallas Res Lab, 82-86, MGR CATALYSIS, CTR RES LAB, MOBIL RES & DEVELOP CORP, 86- *Honors & Awards:* Ipatieff Award, Am Chem Soc, 71. *Mem:* Am Chem Soc; Am Inst Chem Eng; Soc Petrol Engrs. *Res:* Organic heterogeneous catalysis; zeolite technology; process scoping and economics; catalysis in petroleum refining; uranium in-situ leaching; heavy oil and tar sands thermal recovery; enhanced oil recovery. *Mailing Add:* Mobil Res & Develop Corp Ctr Res Lab PO Box 1025 Princeton NJ 08543-1025

VENZKE, WALTER GEORGE, b White Lake, SDak, June 18, 12; m 39; c 1. VETERINARY ANATOMY. *Educ:* Iowa State Col, DVM, 35, PhD(vet anat), 42; Univ Wis, MS, 37. *Prof Exp:* Asst genetics, Univ Wis, 35-37; instr vet anat, Iowa State Col, 37-41, asst prof, 41-42, vet physiol, 42; instr zool, Ohio State Univ, 46, asst prof vet prev med, 46-48, assoc prof vet med, 48-53, prof vet anat & head dept, 54-80, asst dean & secy, Col Vet Med, Ohio State Univ, 60-80; RETIRED. *Mem:* Am Vet Med Asn; Am Asn Anat; Conf Res Workers Animal Dis. *Res:* Endocrinology of the thymus and pineal gland. *Mailing Add:* 2535 Andover Rd Columbus OH 43221

VEOMETT, GEORGE ECTOR, b Rochester, NY, Aug 27, 44; m 70; c 4. CELL BIOLOGY, ENDOCRINOLOGY. *Educ:* Univ Rochester, AB, 66; Univ Colo, PhD, 72. *Prof Exp:* Res assoc virol, Univ Colo, Boulder, 72-74, res assoc cell biol, 74-76; asst prof, 77-81, ASSOC PROF CELL BIOL, UNIV NEBR, 81- *Concurrent Pos:* Prin investr, USDA grant, 87- *Mem:* AAAS; Am Soc Cell Biol. *Res:* Somatomedin binding protein; role of somatomedins in cellular gene expression; effects of interferons on cell growth and differentiation. *Mailing Add:* 348 Manter Hall Life Sci Univ Nebr Lincoln NE 68588-0118

VERA, HARRIETTE DRYDEN, b Washington, Pa, Feb 22, 09. MEDICAL MICROBIOLOGY. *Educ:* Mt Holyoke Col, AB, 30; Yale Univ, PhD(bact), 38; Am Bd Microbiol, dipl. *Prof Exp:* Asst zool, Mt Holyoke Col, 30-31; teacher high sch, Conn, 31-37; from instr to asst prof physiol & hyg, Goucher Col, 38-43; res bacteriologist, Baltimore Biol Lab, 43-60; dir qual control lab prod, Becton, Dickinson & Co, 60-62, dir qual control, B-D Labs, Inc, 62-75; consult, 75-76; RETIRED. *Concurrent Pos:* Vis lectr, Goucher Col, 46-51; consult, Becton, Dickinson & Co, 52-60 & US Dept Army, 56-59. *Honors & Awards:* Barnett L Cohen Award, Am Soc Microbiol, 63. *Mem:* Fel AAAS; fel Am Acad Microbiol; fel Am Pub Health Asn; Am Soc Microbiol; NY Acad Sci; Sigma Xi. *Res:* Bacterial morphology and physiology, especially nutrition and quality control methods and specifications. *Mailing Add:* Apt 110 800 Southerly Rd Baltimore MD 21204-8408

VERBANAC, FRANK, b Yugoslavia, Jan 12, 20; nat US; m 45; c 2. ORGANIC CHEMISTRY. *Educ:* Wayne State Univ, BS, 41; Univ Ill, PhD(chem), 49. *Prof Exp:* Chemist, Gelatin Prod Corp, 42-46 & Merck & Co, Inc, 49-57; sr res chemist, A E Staley Mfg Co, 57-60, group leader, 60-70, sr scientist, 70-85; RETIRED. *Mem:* AAAS; Am Chem Soc; Sigma Xi. *Res:* N-arylpyrazolines; antibiotics; natural products; carbohydrates; polymers; proteins. *Mailing Add:* 12 Dakota Dr Decatur IL 62526-2331

VERBEEK, EARL RAYMOND, b Philadelphia, Pa, Mar 4, 48; div. MINERAL LUMINESCENCE. *Educ:* Pa State Univ, BS, 69, PhD(struct geol), 75. *Prof Exp:* GEOLOGIST, STRUCT GEOL, US GEOL SURV, 74- *Mem:* Franklin-Ogdensburg Mineral Soc. *Res:* Minor structures of deformed rocks; mechanics of folding, origin of joints and quantitative characterization of three-dimensional joint networks in rock masses; luminescence spectroscopy of minerals. *Mailing Add:* US Geol Surv MS 913 Fed Ctr Denver CO 80225

VERBEKE, JUDITH ANN, b St Louis, Mo, Jan 27, 48. PLANT ANATOMY, DEVELOPMENTAL PLANT BIOLOGY. *Educ:* Rockhurst Col, BS, 80; Univ Calif, Los Angeles, PhD(biol), 85. *Prof Exp:* asst prof plant biol, Univ Ill, Chicago, 85-90; ASSOC PROF PLANT SCI, UNIV ARIZ, TUCSON, 90- *Mem:* AAAS; Bot Soc Am; Am Soc Plant Physiologists; Am Soc Cell Biol; Electron Micros Soc Am; Soc Develop Biol. *Res:* Cell communication and differentiation in plants from a cellular and organismal point of view; redifferentiation response in the form of diffusible factors which move between cells. *Mailing Add:* Dept Plant Sci Univ Ariz Tucson AZ 85721

VERBER, CARL MICHAEL, b New York, NY, May 20, 35; m 57; c 2. OPTICAL PHYSICS. *Educ:* Yale Univ, BS, 55; Univ Rochester, MA, 58; Univ Colo, PhD(physics), 61. *Prof Exp:* sr physicist, Columbus Lab, Battelle Mem Inst, 61-86; PROF ELEC ENG, GA TECH, 86- *Mem:* AAAS; fel Optical Soc Am; Soc Photo-Optical Instrumentation Engrs; sr mem Inst Elec & Electronics Engrs. *Res:* Optical properties of solids; integrated optics; optical data processing; optical computing. *Mailing Add:* Ga Inst Technol Sch Elec Eng Atlanta GA 30332

VERBINSKI, VICTOR V, b Shickshinny, Pa, May 7, 22; m 58; c 5. NUCLEAR PHYSICS. *Educ:* Mass Inst Technol, SB, 48; Univ Pa, PhD(physics), 57. *Prof Exp:* Physicist, Gen Elec Co, 57-59, Oak Ridge Nat Lab, 59-67 & Gulf Gen Atomic, 67-74; mem staff, IRT Corp, 74-75; MEM STAFF, SCI APPLN, INC, 75- *Mem:* Am Phys Soc; Am Nuclear Soc. *Res:* Low energy nuclear physics; neutron spectroscopy; nuclear structure physics; photonuclear reactions and fission studies; radiation measurements in nondestructive testing. *Mailing Add:* Sci Appln Inc 4161 Campus Point Ct San Diego CA 92121

VERBISCAR, ANTHONY JAMES, b Chicago, Ill, Mar 22, 29; m 59; c 3. SYNTHETIC ORGANIC & NATURAL PRODUCTS CHEMISTRY. *Educ:* DePaul Univ, BS, 51; Univ Notre Dame, PhD(org chem), 55. *Prof Exp:* Res chemist, Hercules, US Army Sci & Prof Personnel Prog, Edgewood Arsenal 54-56; fel, Univ Chicago, 56-67; vpres res, Regis Chem Co, 57-63; fel, Univ Calif, Los Angeles, 64-65; PRES, ANVER BIOSCI DESIGN, 65- *Concurrent Pos:* Chmn subcomt biogenic amines, spcif & criteria biochem compounds, Nat Res Coun, NSF. *Mem:* Am Chem Soc; Am Inst Chemists; Sigma Xi; Am Soc Pharmacog; Orient Healing Arts Inst. *Res:* Organic synthesis; medicinal chemistry; microbial systems; natural products; biogenic amines, indole chemistry; metabolism of foreign compounds in mammals; plant materials, jojoba, guayule and red squill, oleander, antiviral and anticancer plant derived immuno modulators. *Mailing Add:* Anver Biosci Design Inc 160 E Montecito Ave Sierra Madre CA 91024

VERBIT, LAWRENCE, organic chemistry, for more information see previous edition

VERBRUGGE, CALVIN JAMES, b Sioux Falls, SDak, July 26, 37; m 61; c 2. POLYMER CHEMISTRY. *Educ:* Calvin Col, BA, 59; Purdue Univ, PhD(org chem), 63. *Prof Exp:* Sr chemist, 63-69, sr res chemist, 69-80, RES ASSOC, S C JOHNSON & SON, INC, 80- *Mem:* Sigma Xi; Am Chem Soc. *Res:* Polymer emulsion and solution polymerization and coatings therefrom; Alpha olefin maleic anhydride polymers. *Mailing Add:* Polymer Res Dept 1525 Howe St S C Johnson & Son Inc Racine WI 53403-5011

VERBY, JOHN E, b St Paul, Minn, May 24, 23; m 46; c 4. FAMILY MEDICINE, COMMUNITY HEALTH. *Educ:* Carleton Col, BA, 44; Univ Minn, MB, BS, MD, 47. *Prof Exp:* Physician, pvt family pract, Minn, 49-68; PROF FAMILY PRACT & COMMUNITY HEALTH, MED SCH, UNIV MINN, MINNEAPOLIS, 69- *Concurrent Pos:* Sci assoc, Mayo Clin, Rochester, Minn, 67-68. *Mem:* Int Soc Gen Med; Am Asn Family Pract; AMA. *Res:* Thyroid disease. *Mailing Add:* Dept Family Pract Univ Minn Box 81 Mayo Minneapolis MN 55455

VERCELLOTTI, JOHN R, b Joliet, Ill, May 2, 33; m 66; c 2. BIOCHEMISTRY, AGRICULTURAL & FOOD CHEMISTRY. *Educ:* St Bonaventure Univ, BA, 55; Marquette Univ, MS, 60; Ohio State Univ, PhD(chem), 63. *Prof Exp:* Asst chem, Marquette Univ, 58-60; fel Ohio State Univ, 60-63, lectr & vis res assoc, 63-64; asst prof, Marquette Univ, 64-67; asst prof, Univ Tenn, Knoxville, 67-70; assoc prof, Va Polytech Inst & State Univ, 70-74, prof biochem & nutrit, 74-80; SUPRVY CHEMIST & RES LEADER, SOUTHERN REGIONAL RES CTR, AGR RES SERV, USDA, NEW ORLEANS, 85- *Concurrent Pos:* Res chemist, Freeman Chem Corp, Wis, 59 & V-Labs, Inc, Covington, La, 80-85; consult, US Vet Hosp, Wood, Wis, 65-67 & Oak Ridge Nat Lab, 67-71; vis scientist, Ronzoni Inst, Milan Italy, 77-78; indust & res grants, NSF, NIH, USDA & NATO; chmn, Carbohydrate Div, Southern Regional Res Ctr, Agr Res Serv, USDA, 88. *Honors & Awards:* Crinos Fel, Univ Milan, 77-78. *Mem:* Am Chem Soc; The Chem Soc; Am Soc Biol Chem; Inst Food Technologists; Sigma Xi; Am Inst Chem. *Res:* Biosynthesis and reactivity of glycoproteins and mucopolysaccharides; bacterial and fungal carbohydrate metabolism; food flavor quality and sensory evaluation; carbohydrate chemistry and enzymology. *Mailing Add:* 215 E Fourth Ave Covington LA 70433

VERCH, RICHARD LEE, b Wakefield, Mich, Feb 15, 37; m 66; c 1. AQUATIC BIOLOGY. *Educ:* Northland Col, BS, 62; Northern Mich Univ, MA, 66; Univ NDak, DA(biol), 71. *Prof Exp:* Asst prof biol, Bay de Noc Col, 66-69; asst prof, 71-75, ASSOC PROF BIOL & CHMN DIV NATURAL SCI, NORTHLAND COL, 75- *Mem:* Am Inst Biol Scientists; Nat Asn Biol Teachers; Nat Asn Sci Teachers. *Res:* Biology teaching, self study units. *Mailing Add:* Dept Natural Sci Northland Col 1411 Ellis Ave Ashland WI 54806

VERDEAL, KATHEY MARIE, b Denver, Colo, June 29, 49; div. MAMMALIAN & ENVIRONMENTAL TOXICOLOGY. *Educ:* Colo State Univ, BS, 76; Univ Wis-Madison, MS, 78, PhD(environ toxicol), 82. *Prof Exp:* Territorial mgr, Upjohn Pharmaceut Co, 85; toxicologist, Chematox Lab, Inc, 85-87; DIR TOXICOL. *Concurrent Pos:* Assoc scientist, Wis Clin Cancer Ctr. *Honors & Awards:* James Price Cancer Res Award. *Mem:* AAAS. *Res:* Alteration of pituitary hormones as a side effect of the use of antidepressant and antianxiety agents, their resultant carcinogenic potential; oncology. *Mailing Add:* Kathey M Verdeal PhD Inc 4481 Clay Boulder CO 80301

VERDERBER, JOSEPH ANTHONY, mechanical engineering, for more information see previous edition

VERDERBER, NADINE LUCILLE, b St Louis, Mo, Jan 28, 40. MATHEMATICS EDUCATION. *Educ:* Wash Univ, AB, 62; Univ Mo, MA, 65; Ohio State Univ, PhD(math educ), 74. *Prof Exp:* Teaching asst math, Univ Mo, 63-64; from instr to asst prof, 65-80, ASSOC PROF MATH, SOUTHERN ILL UNIV, EDWARDSVILLE, 80- *Concurrent Pos:* Teaching asst, Ohio State Univ, 71-72. *Mem:* Nat Coun Teachers Math; Am Asn Univ Prof; Math Asn Am. *Res:* Mathematics education; readability of mathematics texts; teaching mathematics with a spread sheet; spatial visualization. *Mailing Add:* Dept Math Statist & Comput Sci Southern Ill Univ Edwardsville IL 62026-1653

VERDEYEN, JOSEPH T, b Terre Haute, Ind, Aug 15, 32; m 54; c 4. ELECTRICAL ENGINEERING, PLASMA PHYSICS. *Educ:* Rose Polytech Inst, BS, 54; Rutgers Univ, MS, 58; Univ Ill, Urbana, PhD(elec eng), 62. *Prof Exp:* Mem tech staff, Bell Tel Labs, 54-55 & 57; asst instr elec eng, Rutgers Univ, 57-58; from instr to assoc prof, 62-69, PROF ELEC & NUCLEAR ENG, UNIV ILL, URBANA, 69-, DIR, GASEOUS ELECTRONIC LAB, 72- *Concurrent Pos:* Mem tech staff, Stavid Eng, 57-58; mem awards comt, Nat Electronics Conf, 65; consult, Corps Engrs, 70-, Zenith Radio Corp, Chicago, 71-79, Gen Elec Lamp Div, 76- & Lucitron Inc, 79-81, Sandia Nat Labs, 74-; dir, Ctr Compound Semiconductor Microelectronics, 88-89; chmn, Gaseous Elec Conf, 84-85, sec, 90. *Mem:* Am Phys Soc; Inst Elec & Electronics Engrs; Sigma Xi. *Res:* Studies of laser discharges and plasmas used for semiconductor processing; semiconductor lasers; microwave. *Mailing Add:* Dept Elec Eng Univ Ill 104 Geb Urbana IL 61801

VERDI, JAMES L, b New Haven, Conn, Aug 11, 41; m 63; c 2. PHYSIOLOGY. *Educ:* SConn State Col, BS, 63; Univ Nev, Reno, MS, 65, PhD(biochem & physiol), 71. *Prof Exp:* Instr physiol, New Haven Col, 65-66; assoc prof biochem, Univ Nev, Reno, 66-68; CLIN LAB DIR, VET ADMIN MED CTR, NEV, 68-; ASSOC PROF, MED SCH, UNIV NEV, 78- *Concurrent Pos:* Consult, State Lab Syst, Nev, 78. *Mem:* Am Asn Clin Chem; Am Chem Soc; AAAS. *Res:* Immunocompetence in aging; nutrition in aging. *Mailing Add:* 105 Mansur Rd New Haven CT 06514

VERDIER, PETER HOWARD, b Pasadena, Calif, Feb 16, 31; m 53; c 1. PHYSICAL CHEMISTRY, POLYMER PHYSICS. *Educ:* Calif Inst Technol, BS, 52; Harvard Univ, PhD(phys chem), 57. *Prof Exp:* Res assoc chem, Mass Inst Technol, 57-58; res fel, Harvard Univ, 58-59; res chemist, Union Carbide Res Inst, 59-64, staff consult, 64-65; chemist, Nat Bur Standards, 65-70, chief molecular characterization sect, Polymers Div, 70-75; CHEMIST, NAT INST STANDARDS & TECHNOL, 75- *Mem:* Am Phys Soc. *Res:* Chemical physics, especially molecular structure and dynamics; polymer solution properties; polymer chain dynamics; polymer molecular weight determination. *Mailing Add:* Polymer Div Nat Inst Standards & Technol Gaithersburg MD 20899

VERDINA, JOSEPH, b Palermo, Italy, Dec 7, 21; US citizen; m 60; c 2. MATHEMATICS. *Educ:* Univ Palermo, PhD(math), 49. *Prof Exp:* Prof physics, Harbor Col, Calif, 58-59; PROF MATH, CALIF STATE UNIV, LONG BEACH, 59- *Mem:* Math Asn Am; Ital Math Union. *Res:* Geometrical transformations; computer simulation. *Mailing Add:* Dept Math Calif State Univ 1250 Bellflower Blvd Long Beach CA 90840

VERDON, JOSEPH MICHAEL, b New York, NY, July 4, 41; m 63; c 3. THEORETICAL & COMPUTATIONAL FLUID DYNAMICS, MECHANICAL SYSTEMS. *Educ:* Webb Inst Naval Archit, BS, 63; Univ Notre Dame, MS, 65, PhD(eng sci), 67. *Prof Exp:* Res engr, United Aircraft Res Labs, 67-68; asst prof mech eng, Univ Conn, 68-72; sr res engr, 72-81, prin scientist, 81-88, MGR, THEORET FLUID DYNAMICS, UNITED TECHNOL RES CTR, 88- *Concurrent Pos:* Consult, Pratt & Whitney Aircraft, 70-71; prin investr, res contracts sponsored by NASA, USN & USAF, 77-; assoc ed, Am Inst Aeronaut & Astronaut J, 88 - *Mem:* Am Soc Mech Engrs; Soc Indust & Appl Math; Am Inst Aeronaut & Astronaut; Sigma Xi; Am Acad Mech. *Res:* Unsteady aerodynamics, viscous flows, random vibrations; applied mathematics; author of several survey articles on unsteady aerodynamics for turbomachinery. *Mailing Add:* United Technol Res Ctr Silver Lane East Hartford CT 06108-1049

VERDU, SERGIO, b Barcelona, Spain, Aug 15, 58; m 82; c 1. INFORMATION THEORY, COMMUNICATION THEORY. *Educ:* Polytech Univ Barcelona, Engr, 80; Univ Ill, Urbana, MS, 82, PhD(elec eng), 84. *Prof Exp:* Asst prof elec eng & computer sci, 84-89, ASSOC PROF ELEC ENG, PRINCETON UNIV, 89- *Concurrent Pos:* Fac develop award, Int Bus Mach, 86; NSF presidential young investr award, 87; prin investr, Off Naval Res & Army Res Off, 87-; assoc ed, Inst Elec & Electronics Engrs Trans Automatic Control, 87-90 & Trans Info Theory, 90-; vis fel, Inst Advan Study, Australian Nat Univ, 89; mem bd gov, Inst Elec & Electronics Engrs Info Theory Soc, 89-; consult, Bell Commun Res, 90- *Mem:* Sr mem Inst Elec & Electronics Engrs. *Res:* Communication and information theory of noisy channels; analysis and design of multiuser communication systems; statistical signal processing in decision and communication systems. *Mailing Add:* Dept Elec Eng Princeton Univ Princeton NJ 08544

VERDUIN, JACOB, b Orange City, Iowa, Nov 19, 13; m 42; c 5. PLANT PHYSIOLOGY. *Educ:* Iowa State Col, BS, 39, MS, 41, PhD(plant physiol), 47. *Prof Exp:* Instr bot, Iowa State Col, 41-42, instr plant physiol, 45-46; assoc prof bot & head dept, Univ SDak, 46-48; prof biol & dept chmn, Bowling Green State Univ, 55-64; prof bot, 64-84, EMER PROF, SOUTHERN ILL UNIV, CARBONDALE, 84- *Concurrent Pos:* Consult, Commonwealth Edison, Chicago, 73- & Nat Environ Res Ctr, Environ Protection Agency, Nev, 75-77. *Mem:* AAAS; Ecol Soc Am; Am Soc Limnol & Oceanog; Am Fisheries Soc; Am Inst Biol Scientists. *Res:* Photosynthesis under natural conditions; respiration; diffusion problems; aquatic ecology; impact of electric power on aquatic systems. *Mailing Add:* R Four Box 202 Carbondale IL 62901

VERDY, MAURICE, b Montreal, Que, Sept 14, 33; c 3. OBESITY, DIABETES & THYROID. *Educ:* Univ Montreal, MD, 56. *Prof Exp:* ENDOCRINOLOGIST, HOTEL-DIEU HOSP, 63-; PROF MED, UNIV MONTREAL, 76- *Mailing Add:* Hotel-Dieu Hosp 3840 St Urbain Montreal PQ H2W 1T8 Can

VEREBEY, KARL G, b Budapest, Hungary, Mar 12, 38; US citizen; m 62; c 2. CLINICAL PHARMACOLOGY, ANALYTICAL TOXICOLOGY. *Educ:* City Univ New York Hunter Col, BA, 65, MA, 68; Cornell Univ, Med Col, PhD(pharmacol), 72. *Prof Exp:* Res assoc neurol, Cornell Univ Med Col, 72-73; dir, clin pharmacol, 73-84, ASST DIR, NY STATE DSAS TESTING & RES LABS, 84-; RES PROF PSYCHIAT, NY MED COL, VALHALLA, 76-; ASSOC PROF PSYCHIAT, SUNY HEALTH SCI CTR AT BROOKLYN, 81-; CLIN LAB DIR, PSYCHIAT DIAG LAB AM, 82- *Concurrent Pos:* USPHS fel, 68-73; sr res assoc, Biobehav Res Found, Inc, 81-; candidate, Am Bd Forensic Toxicol. *Mem:* Am Soc Pharmacol & Exp Therapeut; NY Acad Sci; Am Acad Forensic Sci. *Res:* Clinical pharmacology; analytical toxicology; psychopharmacology; drug biotransformation in animals and man; identification of new metabolites and toxic drug interactions; development of sensitive analytical methods for various drugs. *Mailing Add:* Bur Labs New York City Dept Health 455 First Ave New York NY 10016

VEREEN, LARRY EDWIN, b Loris, SC, Mar 24, 40. MICROBIOLOGY. *Educ:* Clemson Univ, BS, 63, MS, 64; Colo State Univ, PhD(microbiol), 68. *Prof Exp:* Asst prof food sci & biochem, Clemson Univ, 68-70; ASSOC PROF BIOL, LANDER COL, 70- *Mem:* Am Soc Microbiol. *Res:* Behavior of Clostridium perfringens in vacuum-sealed foods. *Mailing Add:* Dept Biol Lander Col Stanley Ave Greenwood SC 29646

VERELL, RUTH ANN, b New York, NY, Mar 8, 35; m 66; c 2. ORGANIC CHEMISTRY. *Educ:* Allegheny Col, BS, 57; Univ Ill, MS, 58; Columbia Univ, PhD(chem), 62. *Prof Exp:* Res chemist, Nat Bur Standards, 62-64; prof asst, 64-65, asst prog dir, Instrnl Sci Equip Prog, 65-68, assoc prog dir, Col Sci Improv Progs, 68-69 & 71-73, proj mgr exp projs & develop progs, NSF, 73-76, prof assoc, Energy Res & Develop Admin, 76-77; employee develop specialist, Dept Air Force, 82-88; mem prof staff, 77-82, dep dir, Div Univ & Indust Progs, 88-90, DEP ASSOC DIR, UNIV & SCI EDUC, US DEPT ENERGY, 90- *Mem:* AAAS; Am Chem Soc; Sigma Xi. *Res:* Cyclopropenones; alkaline conversions of labeled sugars. *Mailing Add:* 6215 Thornwood Dr Alexandria VA 22310-2961

VERESS, SANDOR A, b Jaszkiser, Hungary, Mar 13, 27; US citizen; m 51; c 2. PHOTOGRAMMETRY, GEODESY. *Educ:* Univ Forestry & Timber Indust, Hungary, BS, 51; Hungarian Tech Univ, Sopron, MS, 56; Laval Univ, DSc(photogram), 68. *Prof Exp:* From instr to asst prof geod & photogram, Univ Forestry & Timber Indust, Hungary, 51-56; asst prof, Univ BC, 56-59; photogrammetrist, Ohio State Hwy Dept, 60-62; asst prof surv & mapping, Purdue Univ, 62-65; PROF CIVIL ENG, UNIV WASH, 65- *Concurrent Pos:* Consult to several pvt & fed photogram orgns, 61- *Honors & Awards:* Pres Citation, Am Soc Photogram, 77 & 79, Am Cong Surveying & Mapping, 80. *Mem:* Am Soc Civil Engrs; Am Cong Surv & Mapping; Am Soc Photogram; Sigma Xi. *Res:* Determination of structural deformations by photogrammetry; biomedical and x-ray photogrammetry. *Mailing Add:* Dept Civil Eng 121 More Hall Univ Wash Seattle WA 98195

VERGARA, WILLIAM CHARLES, b Far Rockaway, NY, July 6, 23; m 46. ELECTRONICS ENGINEERING. *Educ:* Rensselaer Polytech Inst, BEE, 45. *Prof Exp:* Engr, Cardwell Mfg Corp, 46-48; proj engr, Commun Div, 48-53, prin engr, 53-58, dir advan res, 58-72, dir phys electronics, commun div, 72-76, head microselectronics engr, Bendix Corp, Baltimore, 76-82; RETIRED. *Concurrent Pos:* Mem, Gov Sci Adv Bd, State of Md, 66- *Mem:* Inst Elec & Electronics Engrs. *Res:* Physics of thin films; microelectronics; radiowave propagation; radio receiver design. *Mailing Add:* 910 Dunellen Dr Towson MD 21204

VERGENZ, ROBERT ALLAN, b Milwaukee, Wis, Dec 26, 56; m 82. CHEMISTRY, PHYSICS. *Educ:* Rollins Col, Winter Park, Fla, BA, 78; Rutgers Univ, New Brunswick, NJ, MS, 85, PhD(chem), 89. *Prof Exp:* Anal chemist, Merck Chem Mfg, Albany, Ga, 78-80; ASST PROF CHEM & PHYSICS, UNIV NFLA, JACKSONVILLE, 87- *Mem:* Am Chem Soc. *Res:* Electronic state of matter from theory and experiment; chemical education; interrelations of science and technology. *Mailing Add:* Dept Natural Sci Univ NFla Jacksonville FL 32216

VERGHESE, KURUVILLA, b Kottayam, India, June 29, 36; m 64; c 3. NUCLEAR ENGINEERING. *Educ:* Univ Kerala, BSc, 58; Univ Iowa, MS, 60, PhD(nuclear eng), 63. *Prof Exp:* Asst prof nuclear eng, 63-84, grad adminr, 76-82, PROF NUCLEAR ENG, NC STATE UNIV, 84- *Concurrent Pos:* Consult, various industrial corps. *Honors & Awards:* Glenn Murphy Award, Am Soc Eng Educ. *Mem:* Am Nuclear Soc; Am Soc Eng Educ. *Res:* Reactor physics problems; radiation applications. *Mailing Add:* Dept Nuclear Eng NC State Univ Box 7909 Raleigh NC 27607

VERGHESE, MARGRITH WEHRLI, b Davos, Switz, May 12, 39; m 64; c 3. GENETICS, IMMUNOGENETICS. *Educ:* Iowa State Univ, BS, 61, PhD(poultry breeding), 64. *Prof Exp:* Res asst poultry breeding, Iowa State Univ, 62-64; res assoc quant genetics, NC State Univ, 65-68; researcher biostatist, Univ NC, Chapel Hill, 68-69; res assoc, Duke Univ, 75- *Honors & Awards:* Nat Res Serv Award, NIH, 75. *Mem:* AAAS; Am Asn Immunologists. *Res:* Quantitative genetics; selection theory for quantitative traits; interaction between artificial and natural selection in genetic populations; simulation of genetic populations; effects of murine anti-H-Z sera and human anti-DrW sera on the human mixed lymphocyte culture reaction; Con-A induced suppression in human cellular immune function. *Mailing Add:* 5709 Crutchfield Rd Raleigh NC 27606

VERGONA, KATHLEEN ANNE DOBROSIELSKI, b Pittsburgh, Pa, Dec 6, 48; m 73; c 1. CELLULAR AGING, PRIMARY LIVER CULTURE. *Educ:* Univ Pittsburgh, BS, 70, PhD(cell biol), 76. *Prof Exp:* Fel res assoc, Cancer Res Unit, Allegheny Gen Hosp, 76; fel, Sch Dent Med, Univ Pittsburgh, 77-79, asst prof histol, 76-81, RES ASST PROF, SCH MED, UNIV PITTSBURGH, 79-, ASSOC PROF HISTOL, SCH DENT MED, 81- *Concurrent Pos:* Actg chmn, Dept Anat/Histol, Sch Dent Med, Univ Pittsburgh, 90- *Mem:* Am Soc Cell Biol; Tissue Cult Asn; Sigma Xi; Am Asn Dent Res. *Res:* Hormonal regulation of hepatic microsomal glucose-6-phosphatase, in vivo and in primary cell cultures; age-related changes in salivary gland function. *Mailing Add:* 615-2 Salk Hall Sch Dent Med Univ Pittsburgh Pittsburgh PA 15261

VERHAGE, HAROLD GLENN, b May 18, 37; m; c 3. CELLULAR REPRODUCTIVE BIOLOGY, OVIDUCT PHYSIOLOGY. *Educ:* Colo State Univ, PhD(reproductive biol), 72. *Prof Exp:* PROF, UNIV ILL, CHICAGO, 88- *Mem:* Soc Study Reprod; Soc Gynec Invest. *Res:* Mechanisms of hormone action; steroid induced proteins. *Mailing Add:* Dept Obstet & Gynec Univ Ill PO Box 6998 Chicago IL 60680

VERHALEN, LAVAL, b Knox City, Tex, May 8, 41; m 64, 88; c 2. COTTON BREEDING, COTTON GENETICS. *Educ:* Tex Tech Col, BS, 63; Okla State Univ, PhD(plant breeding, genetics), 68. *Prof Exp:* From instr to assoc prof agron, 67-77, PROF AGRON, OKLA STATE UNIV, 77- *Concurrent Pos:* Prin investr, Cotton Breeding & Genetics, 67-, Eval Cotton Varieties for Okla, 67-; secy, grad fac, genetics, Okla State Univ, 70-72, group I-biol sci grad fac, 85-87, vchair, 87-89, chair, 89-; chmn, 26th Cotton Improv Conf, 73-74, S-77 ann meeting, 80-81; assoc ed, Crop Sci, 81-85; treas, Sigma Xi, Okla State Univ, 82-86; mem, planning comn W Cotton Prod Conf, 84, 89, tech panel, determine minimum distance cotton, Am Seed Trade Asn, 88-; consult, several co, 85, 87-90; deleg, SW Reg Task Force, Nat Cotton Coun, 87; cotton genetics res award, 88. *Mem:* Sigma Xi; Am Soc Agron; Crop Sci Soc Am. *Res:* Cotton breeding; genetics, particularly population genetics; variety testing; cultural practices. *Mailing Add:* Dept Agron Okla State Univ Stillwater OK 74078

VERHANOVITZ, RICHARD FRANK, b Walsenburg, Colo, Sept 20, 44; m 67; c 2. SYSTEMS ENGINEERING. *Educ:* Wilkes Col, BS, 66; Lehigh Univ, MS, 69, PhD(physics), 74. *Prof Exp:* Mathematician, Defense Intel Agency, Dept Defense, 66; engr, Univac, Sperry-Rand Corp, 66-67; SR PROGRAMMER, SPACE DIV, GEN ELEC CO, 75-, MGR SYSTS ENG, 80-, PROG MGR, 85- *Mem:* Am Phys Soc; AAAS; NY Acad Sci; Opers Res Soc. *Res:* Low energy nuclear physics; two and three body interactions; charge symmetry and nuclear coulomb interactions; computer simulation and numerical analysis. *Mailing Add:* Ten Wampenog Circle Royersford PA 19468

VERHEY, ROGER FRANK, b Grand Rapids, Mich, Sept 12, 38; m 60; c 4. MATHEMATICS & COMPUTER EDUCATION. *Educ:* Calvin Col, AB, 60; Univ Mich, MA, 61, PhD(math), 66. *Prof Exp:* Lectr, 65-66, from asst prof to assoc prof, 66-72, chmn dept math & statist, 71-77, PROF MATH, UNIV MICH, DEARBORN, 72-, DIR, CIS PROG, 82- *Concurrent Pos:* Fulbright lectr, Univ Ceylon, 69-70. *Mem:* Math Asn Am; Nat Coun Teachers Math; Int Coun Comput Educ. *Res:* Mathematics education; computers in the classroom. *Mailing Add:* 1454 Crawford Lane Ann Arbor MI 48105

VERHEYDEN, JULIEN P H, b Brussels, Belg, May 22, 33; m 59; c 2. ORGANIC CHEMISTRY, MOLECULAR BIOLOGY. *Educ:* Free Univ Brussels, Lic en sci, 55, PhD(chem), 58. *Prof Exp:* Res assoc, Free Univ Brussels, 58-59; res assoc, Inst Sci Res Indust & Agr, Brussels, Belg, 60-61; fel, 61-63, res chemist, 63-72, head bio-org dept, Syntex Inst Molecular Biol, 70-77, head dept, Syntex Int Org Chem, 77-81, ASST DIR, SYNTEX INST BIO-ORG CHEM, 85-, HEAD DEPT CHEM, 81- *Mem:* Am Chem Soc; Chem Soc Belg; Royal Soc Chem. *Res:* Carbohydrates; nucleosides; nucleotides; genetic engineering. *Mailing Add:* Syntex Inst Bio-Org Chem 3401 Hillview Ave Palo Alto CA 94304-1320

VERHOEK, FRANK HENRY, b Grand Rapids, Mich, Feb 12, 09; m 40; c 3. PHYSICAL CHEMISTRY, CHEMICAL KINETICS. *Educ:* Harvard Univ, SB, 29; Univ Wis, MS, 30, PhD(phys chem), 33; Oxford Univ, DPhil(phys chem), 35. *Prof Exp:* Asst chem, Univ Wis, 29-33; Rhodes scholar, Oxford Univ, 33-35, Copenhagen Univ, 35-36; from instr to assoc prof chem, 36-53, supvr res found, 42-65, vchmn dept chem, 60-64 & 66-68, PROF CHEM, OHIO STATE UNIV, 53- *Concurrent Pos:* Res chemist, Gen Elec Co, 38; res assoc, Stanford Univ, 40; consult, Liberty Mirror Div, Libby-Owens-Ford Glass Co, 43-52; prin chemist, Argonne Nat Lab, 47; sr chemist, Olin Mathieson Chem Corp, 55; consult, US Naval Weapons Ctr, 57-62; vis prof, Univ Fla, 58-59; lectr chem bond approach proj, NSF, 59-68; Rhodes scholar, 33-36. *Mem:* Am Chem Soc. *Res:* Solution kinetics; gas kinetics; complex ion equilibria; strength of acids in nonaqueous solvents; solubility of electrolytes in nonaqueous solvents; hydrocarbon oxidation; oxidation of boron alkanes. *Mailing Add:* Dept Chem Ohio State Univ 120 W 18th Ave Columbus OH 43210

VERHOEK, SUSAN ELIZABETH, b Columbus, Ohio; m; c 1. ECONOMIC BOTANY, BIOSYSTEMATICS & PLANT MATERIALS. *Educ:* Ohio Wesleyan Univ, BA, 64; Ind Univ, MA, 66; Cornell Univ, PhD(bot), 75. *Prof Exp:* Herbarium supvr, Mo Bot Garden, 66-70; from asst prof to assoc prof, 74-85, PROF BIOL, LEBANON VALLEY COL, 85- *Concurrent Pos:* Bot consult, Merrill Publ Co, 87-88. *Mem:* Am Soc Plant Taxon; Bot Soc Am; Int Asn Plant Taxon; Soc Econ Bot (vpres, 84-85, pres, 85-86); Am Asn Bot Gardens & Arboreta. *Res:* Biosystematics, economic botany, hybridization and pollination of tribe Polianthеae (Agavaceae); US spring flora. *Mailing Add:* Dept Biol Lebanon Valley Col Annville PA 17003-0501

VERHOEVEN, JOHN DANIEL, b Monroe, Mich, Aug 26, 34; m 62; c 5. METALLURGY. *Educ:* Univ Mich, BS, 57, MS, 59, PhD(metall eng), 63. *Prof Exp:* From asst prof to assoc prof, 63-69, PROF METALL, IOWA STATE UNIV, 69- *Mem:* Am Soc Metals; Am Inst Mining, Metall & Petrol Engrs; Metals Soc; Am Asn Crystal Growth. *Res:* Solidification in metals and metal alloys; physical metallurgy; superconducting alloys. *Mailing Add:* 2111 Graeber Ames IA 50011

VERHOOGEN, JOHN, b Brussels, Belg, Feb 1, 12; nat US; wid; c 4. GEOPHYSICS. *Educ:* Free Univ Brussels, MinEng, 33; Univ Liege, GeolEng, 34; Stanford Univ, PhD(volcanol), 36. *Prof Exp:* Asst geol, Free Univ Brussels, 36-39; Nat Sci Res Found fel, Belg, 39-40; chief prospecting serv, Kilo-Moto Gold Mines, Congo, 40-43; engr, Belg Congo Govt, 43-46; from assoc prof to prof geol, 47-77, EMER PROF GEOL, UNIV CALIF, BERKELEY, 77- *Concurrent Pos:* Guggenheim fel, 53-54, 61. *Mem:* Nat Acad Sci; fel Geol Soc Am; fel Am Acad Arts & Sci; fel Am Geophys Union. *Res:* Paleomagnetism; thermodynamics of geologic phenomena; volcanology. *Mailing Add:* Dept Geol & Geophys Univ Calif Berkeley CA 94720

VERINK, ELLIS D(ANIEL), JR, b Peking, China, Feb 9, 20; US citizen; m 42; c 2. METALLURGICAL ENGINEERING. *Educ:* Purdue Univ, BS, 41; Ohio State Univ, MS, 63, PhD(metall eng), 65. *Prof Exp:* Engr, Aluminum Co Am, 46-48, mgr chem sect, Develop Div, 48-59, mgr chem & petrol indust sales, 59-62; assoc prof metall, 65-68, asst chmn dept metall & mat eng, 70-73, chmn mat sci & eng dept, 73-86, PROF METALL, UNIV FLA, 68-, DISTINGUISHED SERV PROF, 84- *Concurrent Pos:* Consult, Aluminum Asn, 67-86, Copper Develop Asn, 84-; pres, Mat Consult, Inc; pres, Metall Soc, Am Inst Mining, Metall & Petrol Engrs, 84, fel award, 89. *Honors & Awards:* Sam Tour Award, Am Soc Testing & Mat, 78; Willis Rodney Whitney Award, Nat Asn Corrosion Engrs, 82. *Mem:* Am Inst Mining, Metall & Petrol Engrs; fel Am Soc Metals; Am Welding Soc; Nat Asn Corrosion Engrs; Nat Soc Prof Engrs. *Res:* Corrosion; materials selection. *Mailing Add:* Dept Mat Sci & Eng Univ Fla Gainesville FL 32611

VERITY, MAURICE ANTHONY, b Bradford, Eng, Apr 21, 31; c 3. PATHOLOGY, NEUROPATHOLOGY. *Educ:* Univ London, MB, BS, 56. *Prof Exp:* Intern surg, Paddington Gen Hosp, London, Eng & Portsmouth Group Hosps, 56; intern med, Royal Hosp, Wolverhampton, 57; clin pathologist, United Bristol Hosps, 58-59; vis asst prof pharmacol, 59-60, assoc resident path, Sch Med, 60-61, assoc prof, 68-74, PROF PATH, SCH MED, UNIV CALIF, LOS ANGELES, 74-, MEM, BRAIN RES INST, 67- *Concurrent Pos:* NIH travel award, Int Neuropath-Neurol Cong, Europe , 65 & 70; NIH fel biochem, Med Sch, Bristol Univ, 68-69; Milheim Found grant, Sch Med, Univ Calif, Los Angeles, 70-71, USPHS grant, 71-74. *Mem:* AAAS; Am Soc Exp Path; Am Asn Pathologists & Bacteriologists; foreign mem Royal Soc Med; Brit Biochem Soc. *Res:* Biochemical and histochemical studies of subcellular organelle function in pathologic states, including mercury intoxication, partial hepatectomy; investigations of neurogenic control of vascular smooth muscle; thyroid hormone modulation of brain development. *Mailing Add:* Dept Path Univ Calif Med Ctr 405 Hilgard Ave Los Angeles CA 90024

VERKADE, JOHN GEORGE, b Chicago, Ill, Jan 15, 35; div; c 3. BIOINORGANIC CHEMISTRY, ORGANOMETALLIC CHEMISTRY. *Educ:* Univ Ill, BS, 56, PhD(inorg chem), 60; Harvard Univ, AM, 57. *Prof Exp:* From instr to assoc prof, 60-70, PROF INORG CHEM, IOWA STATE UNIV, 70- *Concurrent Pos:* Grants, NSF, 61-, Petrol Res Found, 63-66 & NIH, 72-78; Sloan fel, 66-68. *Mem:* Am Chem Soc; Sigma Xi. *Res:* Spectroscopic studies of coordination compounds containing phosphorus ligands; catalytic studies of transition metal compounds hypervalent non metallic compounds; stereospecific reactions of phosphorus compounds. *Mailing Add:* Dept Chem A-125 Gilman Iowa State Univ Ames IA 50011-0061

VERKADE, STEPHEN DUNNING, b New London, Conn, July 15, 57. MYCORRHIZAE, PLANT PRODUCTION. *Educ:* Univ RI, BS, 79; Purdue Univ, MS, 81, PhD(plant physiol), 85. *Prof Exp:* Res asst, Purdue Univ, 79-85, teaching asst, 83-85; asst prof, 85-88, ASSOC PROF, UNIV FLA, 88- *Mem:* Am Soc Hort Sci; Am Mycol Soc; Int Plant Propagators Soc. *Res:* Use of mycorrhizal fungi to enhance plant growth in production; development of plant propagation techniques for endangered species. *Mailing Add:* Ft Lauderdale Res & Educ Ctr 3205 College Ave Ft Lauderdale FL 33314

VERLANGIERI, ANTHONY JOSEPH, b Newark, NJ, Aug 2, 45; m 67; c 3. BIOCHEMISTRY, TOXICOLOGY. *Educ:* Rutgers Univ, BS, 68; Pa State Univ, PhD(biochem), 73. *Prof Exp:* Asst biochem, Pa State Univ, 68-72; asst prof toxicol, Cook Col, Rutgers Univ, 72-80; PROF PHARMACOL & TOXICOL, UNIV MISS, 80- *Concurrent Pos:* Consult toxicologist, independent labs & litigation. *Mem:* Am Col Vet Toxicol; Am Chem Soc; Am Inst Ultrasound Med; Tissue Cult Asn; AAAS; Am Col Toxicol; Soc Environ Toxicol & Chem. *Res:* Effects of sulfating agents on atherogenesis and the influence of lead intoxication on the central nervous system, behavior and learning; sulfated glycosaminoglycans, analytical methods, role in atherogenesis; endothelial cell culture model systems; animal models of atherogenesis and diabetes; ultrasound carotid analysis in primates; nutrition; pathology. *Mailing Add:* Dept Pharmacol Univ Miss University MS 38677

VERLEUR, HANS WILLEM, b Hillegom, Holland, July 1, 32; US citizen; m 56, 83; c 4. PHYSICS, MATERIAL SCIENCE. *Educ:* Cooper Union, BS, 63; NY Univ, MS, 64, PhD(physics), 66. *Prof Exp:* Supvr Display Develop Group, Bell Labs, 60-85, DESIGN, DEVELOP & MFG MGR, PLASMA DISPLAY PRODUCTS, AT&T TECHNOL, INC, 85- *Mem:* Soc Info Display. *Res:* Optoelectronic device research. *Mailing Add:* AT&T Microelectronics Inc 2525 N 12th St Reading PA 19612

VERLEY, FRANK A, b Kingston, Jamaica, Dec 18, 33; m 67. GENETICS. *Educ:* Univ Conn, BS, 59; Univ Ill, Urbana, MS, 60, PhD(genetics), 64. *Prof Exp:* Resident res assoc radiation genetics, Argonne Nat Lab, 64-67; from asst prof to assoc prof biol genetics, 67-74, PROF BIOL GENETICS, NORTHERN MICH UNIV, 74- *Mem:* Genetics Soc Am; Am Inst Biol Sci; Am Genetics Asn; Biomet Soc Am. *Res:* Genetic effects of a recessive sex-linked lethal gene on prenatal development in mice, copper metabolism in the mottled mice and biosynthesis of metallothionein in mice; restriction mapping of mouse X-chromosome and mitochondrial DNA; DNA sequence analysis of restriction fragment of X-chromosome and mitochondrial DNA. *Mailing Add:* Dept Biol Northern Mich Univ 33 W Sci Bldg Marquette MI 49855

VERMA, AJIT K, b India, Aug 9, 44; US citizen; m 69; c 4. POLYAMINES IN GROWTH CONTROL, VITAMIN A MECHANISM OF ACTION. *Educ:* Punjab Agr Univ, India, BSc, 66, MSc, 68; Flinders Univ SAustralia, PhD(biochem & cancer bone), 76. *Prof Exp:* Fel cancer res, McArdle Lab, Univ Wis-Madison, 76-79, proj assoc, 79-81, scientist, Human Oncol, Clin Cancer Ctr, 81-84, asst prof, 84-88, ASSOC PROF CANCER RES, HUMAN ONCOL, CLIN CANCER CTR, UNIV WIS-MADISON, 88- *Mem:* Am Asn Cancer Res; AAAS; Am Soc Biol Chemists. *Res:* To analyze the biochemical and molecular mechanisms of the induction of cancer as a rational approach for the choice of agents for cancer prevention. *Mailing Add:* Dept Human Oncol K4/532 CSC Univ Wis 600 Highland Ave Madison WI 53792

VERMA, ANIL KUMAR, b Lucknow, India, Dec 15, 50; US citizen; m 77; c 2. PLASMA MEMBRANE CALCIUM TRANSPORT ATPASE. *Educ:* Lucknow Univ, BSc, 69, MSc, 71; Kanpur Univ, PhD(biochem), 76. *Prof Exp:* Jr res fel biochem, Cent Drug Res Inst, Lucknow, India, 71-75; postdoctoral fel biochem, Univ Ala, 75-76; postdoctoral fel biochem, McGill Univ, 77-79; ASSOC BIOCHEM & MOLECULAR BIOL, MAYO CLIN-FOUND, 79-, INSTR, 87- *Concurrent Pos:* Vis fel cell biol, NIH, 76-77. *Mem:* Am Soc Biochem & Molecular Biol. *Res:* Molecular biology of plasma membrane calcium transport atpase. *Mailing Add:* Mayo Clin-Found Guggenheim 16 Floor Rochester MN 55906

VERMA, DEEPAK KUMAR, b India. METALS PROCESSING, FRACTURE MECHANICS. *Educ:* Bihar Inst Technol, India, BS, 67; Univ Mass, MSME, 70; Ga Inst Technol, PhD(mech eng), 79. *Prof Exp:* Trainee engr, Hindustan Motors Ltd, India, 67-68; res fel mech eng, Univ Mass, 68-71; tech asst, Tata Steel Ltd, India, 71-73; tech exec, Texmaco Ltd, India, 73-74; teaching fel mat eng, Ga Inst Technol, 74-78; res develop engr, Southwire Corp, Ga, 78-80; STAFF ENGR, IBM CORP, 80- *Mem:* Am Soc Mech Engrs. *Res:* Processing of metallic materials, fracture behavior, machining, metal forming, casting, heat treatment, surface finishing and environmental degradation. *Mailing Add:* 2613 Third Pl Rochester MN 55906

VERMA, DEVI C, b Barsalu-Haryana, India, Apr 30, 46. AGRICULTURAL BIOCHEMISTRY. *Educ:* Punjab Agr Univ, BS, 68; Univ Calif, Davis, MS, 70; State Univ NY, Buffalo, PhD(plant biochem), 75. *Prof Exp:* Fel plant tissue cult, W Alton Jones Cell Sci Ctr, 75-78; SR RES FEL, INST PAPER CHEM, 78- *Mem:* Am Soc Plant Physiologists; Int Asn Plant Tissue Cult. *Res:* Biochemistry of cellular differentiation; cell wall formation and development in plant tissue cultures; gymnosperm tissue culture; propagation of forest tree species by tissue culture. *Mailing Add:* 302 Tulip Lane Freehold NJ 07728

VERMA, GHASI RAM, b Sigari, India, Aug 1, 29; m 54; c 3. APPLIED MATHEMATICS. *Educ:* Birla Eng Col, India, BA, 50; Benaras Hindu Univ, MA, 54; Univ Rajasthan, India, PhD(math), 57. *Prof Exp:* Tutor math, Birla Eng Col, India, 54-57, lectr, 57-58; fel, Courant Inst Math Sci, NY Univ, 58-59; asst prof, Fordham Univ, 59-61; reader, Birla Inst Technol & Sci, India, 61-64; assoc prof, 64-80, PROF MATH, UNIV RI, 80- *Concurrent Pos:* Sr sci res fel, Coun Sci & Indust Res, New Delhi, India, 55-57. *Mem:* Am Math Soc; Math Asn Am; Soc Indust & Appl Math. *Res:* Elasticity; fluid mechanics. *Mailing Add:* Dept Math Univ RI Kingston RI 02881

VERMA, PRAMODE KUMAR, b Barauli, India, Sept 1, 41; Can citizen; m 66; c 1. ELECTRICAL ENGINEERING. *Educ:* Patna Univ, BSc hons, 59; Indian Inst Sci, BEng, 62; Sir George Williams Univ, DEng(elec eng), 70, Univ Pa, Wharton Sch, MBA, 84. *Prof Exp:* Sr res asst elec commun eng, Indian Inst Sci, 62-64; asst div engr commun, Indian Posts & Tel, 64-67; asst prof elec eng & comput sci, Sir George Williams Univ, 71; supv engr comput commun, Bell Can, 72-78; mem tech staff comput commun, Bell Labs, 78-80; dist mgr, AT&T, 80-84, SUPVR, AT&T BELL LABS, 84- *Concurrent Pos:* Lectr, Univ Ottawa, 75-78; gov Int Coun Comput Commun, Wash, DC. *Mem:* Sr mem Inst Elec & Electronics Engrs; Commun Soc. *Res:* Computer networks; communication networks. *Mailing Add:* AT&T 200 Laurel Ave Middletown NJ 07748

VERMA, RAM D, b May 31, 29; Can citizen; m 62; c 3. SPECTROSCOPY. *Educ:* Univ Agra, BSc, 52, MSc, 54, PhD(physics), 58. *Prof Exp:* Lectr physics, Dav Col, Aligarh, 54-55; res asst, Aligarh Muslim Univ, India, 55-57, sr res fel, 57-58; res assoc, Univ Chicago, 58-61; fel Nat Res Coun Can, 61-63; from asst prof to assoc prof, 63-71, PROF PHYSICS, UNIV NB, 71- *Concurrent Pos:* Can deleg, Int Conf Spectros, 67; vis prof, Univ Calif, Santa Barbara, Univ Stockholm & Bhabha Atomic Res Ctr, Bombay, India. *Mem:* Am Phys Soc; Can Asn Physicists. *Res:* Molecular structure and spectra of stable and free radicals; investigation covering region from near infrared to far ultraviolet; laser physics and laser spectroscopy. *Mailing Add:* Dept Physics Univ NB Col Hill Box 4400 Fredericton NB E3B 5A3 Can

VERMA, RAM S, b Barabanki, India, Mar 3, 46; nat US; m 62; c 2. CLINICAL CYTOGENETICS, HUMAN GENETICS. *Educ:* Agra Univ, India, BSc, 65, MSc, 67; Univ Western Can, PhD(cytogenetics), 72; Royal Col Pathologists, London, DCC, 84. *Prof Exp:* Res assoc, dept pediat, Med Ctr, Univ Colo, Denver, 73-74, fel, 74-76; chief, Div Cytogenetics, Interfaith Med Ctr, Brooklyn, NY, 80-88; from instr to assoc prof, 76-85, PROF, DEPT MED ANAT & CELL BIOL, STATE UNIV NY HEALTH SCI CTR, BROOKLYN, 85-; CHIEF, DIV GENETICS, LONG ISLAND COL HOSP, 86- *Concurrent Pos:* Consult, Phototake, 82-87; mem cytogenetic adv comt, Dept Health, New York, 78-; mem Genetic Task Force of New York, 76-; consult, Nat Geog Socd, Washington, DC, 82; consult, WHO, Geneva, Switz, 82. *Mem:* Fel AAAS; Am Fedn Clin Res; Am Genetic Asn; Am Soc Cell Biol; Am Soc Human Genetics; Europ Soc Human Genetics; Royal Col Path; Inst Biol. *Res:* Structural organization of human chromosomes with respect to banding techniques; molecular roots of cancer including the role of oncogenes in pathogenesis of cancer; author of three books, over 250 original articles and 200 abstracts and presentations. *Mailing Add:* Div Genetics Long Island Col Hosp 340 Henry St Brooklyn NY 11201

VERMA, SADANAND, b Muzaffarpur, India, Jan 24, 30. ALGEBRA, TOPOLOGY. *Educ:* Patna Univ, BSc, 50; Univ Bihar, MSc, 52; Wayne State Univ, MS & PhD(math), 58. *Prof Exp:* Hon lectr math, L S Col, Univ Bihar, 52-53, lectr univ, 53-55 & 58-60; asst prof, Univ Windsor, 60-65; assoc prof, Western Mich Univ, 65-67; chmn dept 68-90, PROF MATH, UNIV NEV, LAS VEGAS, 67- *Mem:* Math Asn Am; Can Math Cong; Asn Comput Mach. *Res:* Algebraic topology; homotopy theory; elementary number theory; magnetohydrodynamics; general topology; elementary ordinary differential equations. *Mailing Add:* 121 Rancho Vista Dr Las Vegas NV 89106

VERMA, SHASHI BHUSHAN, b Buxar, Bihar, India, July 27, 44; US citizen; m 72; c 2. AGRICULTURAL & FOREST SCIENCES. *Educ:* Ranchi Univ, India, BS, 65; Univ Colo, MS, 67; Colo State Univ, PhD(fluid dynamics), 71. *Prof Exp:* Res asst, Colo State Univ, 67-71, postdoctoral fel, Fluid Dynamics & Diffusion Lab, 71-72; postdoctoral res assoc, Agr Meteorol Sect, Dept Hort & Forestry, Univ Nebr, Lincoln, 72-74, staff meteorologist, Dames & Moore, San Francisco, 74; asst prof, Agr Meteorol Sect, Dept Agr Eng, 74-78, assoc prof, Ctr Agr Meteorol & Climat, 78-84, PROF, DEPT AGR METEOROL, UNIV NEBR, LINCOLN, 84-, CO-DIR, CTR LASER-ANAL STUDIES TRACE GAS DYNAMICS, 88- *Concurrent Pos:* Mem, Field Lab Task Force, Univ Nebr, 83-84 & oversight comt, Agr Res & Develop Ctr, 84-; res grants, NSF, 75-78, 78-79, 79-83, 83-86, 84-85, 86-90 & 90-91, US Dept Agr, 79-83, Nebr Soybean Develop, 80 & 82-83, Nebr Grain Sorghum Develop, 82-84, Nat Aeronaut & Space Admin, 85-86, 86-87, 87-90 & 90-91; lectr, NSF, US-India Exchange Scientists Prog, 81 & Univ Tuscia, Italy, 85; mem, Comt Agr & Forest Meteorol, Am Meteorol Soc, 83-86, Coun Agr Sci & Technol Task Force on Improving Irrig Efficiency, 86-87, adv panel, Nat Ctr Atmospheric Res Field Observing Facil, 86-89; consult, World Meteorol Orgn-UN Develop Prog, Proj India, 89 & 90. *Mem:* Am Meteorol Soc; AAAS; Am Soc Agron; Sigma Xi. *Res:* Micrometeorology; atmosphere-biosphere interactions; surface exchange processes; trace gas fluxes; energy and matter exchanges; water use efficiency; micrometeorological/eddy correlation instrumentation. *Mailing Add:* Dept Agr Meteorol Univ Nebr Lincoln 242 LW Chase Hall Lincoln NE 68583-0728

VERMA, SURENDRA KUMAR, b India, Jan 1, 43; m 75. MECHANICAL & CHEMICAL ENGINEERING. *Educ:* Agra Univ, India, BSc, 61; Indian Inst Technol, Kharagpur, BTech, 65; Univ Louisville, MS, 70, PhD(chem eng), 74. *Prof Exp:* Mech engr, Shriram Fertilizers & Chem, India, 65-68; res assoc mech eng, Univ Louisville, 70; res assoc, Universal Restoration, Washington, DC, 73-74; DEVELOP SCIENTIST TECH CTR, UNION CARBIDE CORP, 74- *Mem:* Am Inst Chem Engrs; Sigma Xi. *Res:* New separation techniques development for energy conservation and process improvement; hydrocarbons and amines technology. *Mailing Add:* 5341 Shadowbrook Dr West Brook Cross Lanes WV 25313

VERMA, SURENDRA P, b Manglore, India, Jan 1, 41; US citizen; m; c 3. CELL FUNCTIONS, LIPOSOMES. *Educ:* Agra Univ, India, BSc, 60; Aligarh Univ, India, MSc, 62; Roorkee Univ, India, PhD(chem), 66. *Prof Exp:* Res assoc biophys, Mich State Univ, East Lansing, 68-70 & Nat Res Coun Can, 70-72; from instr to assoc prof radiobiol, Tufts New Eng Med Ctr, Boston, 72-88; ASSOC PROF COMMUNITY HEALTH, SCH MED, TUFTS UNIV, BOSTON, 88- *Concurrent Pos:* Vis prof, INSERM, Unit 58, Montplier, France, 82-83; consult, Allied Instrumentation Lab, Lexington, Mass, 83-86; prin investr, NIH & Environ Protection Agency grants, 83- *Res:* Biophysics; membrane structure; raman spectroscopy; radiation effects on cell membrane structure; effect of pesticides on membrane structure. *Mailing Add:* Dept Community Health Sch Med Tufts Univ 136 Harrison Ave Boston MA 02111

VERMAAS, WILLEM F J, b Rhoon, Neth, June 3, 59; m 85; c 1. PHOTOSYNTHESIS, DIRECTED MUTAGENESIS. *Educ:* Agr Univ, Wageningen, Neth, Ingenieurs, 82, Doctorate agr sci, 84. *Prof Exp:* Res assoc plant biol, Univ Ill, Urbana-Champaign, 80-81 & Mich State Univ, 81-82; res assoc biophys, Tech Univ, Berlin, Ger, 82-83; scientist, Agr Univ, Wageningen, Neth, 83-84; vis scientist, E I du Pont de Nemours & Co, Inc, 84-86; asst prof, 86-90, ASSOC PROF BOT, ARIZ STATE UNIV, 90- *Concurrent Pos:* Prin investr, NSF, 87, NASA, 88-, Dept Energy & Binat Agr Res & Develop Fund, 89-; NSF presidential young investr award, 90. *Honors & Awards:* Unilever Chem Award, 80. *Mem:* AAAS; Int Soc Plant Molecular Biol. *Res:* Using molecular biological tools, specific cyanobacterial mutants are generated with alterations at targeted sites of photosynthesis related proteins; analysis of mutants in terms of structure and function of the photosynthetic machinery; structure/function relationships of specific proteins and residues. *Mailing Add:* Dept Bot Ariz State Univ Tempe AZ 85287-1601

VERMEIJ, GEERAT JACOBUS, b Sappemeer, Neth, Sept 28, 46; m 72; c 1. EVOLUTIONARY BIOLOGY, BIOGEOGRAPHY. *Educ:* Princeton Univ, AB, 68; Yale Univ, MPhil, 70, PhD(biol), 71. *Prof Exp:* From instr to asst prof 71-74, assoc prof, 74-80, PROF ZOOL, UNIV MD, COLLEGE PARK, 80- *Concurrent Pos:* J S Guggenheim Mem fel, 75. *Mem:* Soc Study Evolution; Ecol Soc Am; Soc Am Naturalists; Paleont Soc; Neth Malacol Soc. *Res:* Comparative ecology and history of shallow-water benthic marine communities; adaptive morphology, especially molluscs and decapods; temporal patterns of adaptation. *Mailing Add:* Dept Zool Univ Md College Park MD 20742

VERMEULEN, CARL WILLIAM, b Chicago, Ill, July 23, 39; div; c 2. MICROBIOLOGY, BIOCHEMISTRY. *Educ:* Hope Col, AB, 61; Univ Ill, Urbana, MS, 63, PhD(microbiol), 66. *Prof Exp:* Asst prof microbiol & biochem, 66-71, ASSOC PROF MICROBIOL & BIOCHEM, COL WILLIAM & MARY, 71- *Concurrent Pos:* Desert soils mapping. *Mem:* Fel Am Inst Chemists. *Res:* Microbial genetics; biochemical contributions to soil genesis; quantification of bacterial capsular polysaccharides. *Mailing Add:* Dept Biol Col William & Mary Williamsburg VA 23185

VERMEULEN, THEODORE (COLE), b Los Angeles, Calif, May 7, 16; m 39; c 2. CHEMICAL ENGINEERING, CATALYSIS WATER TECHNOLOGY. *Educ:* Calif Inst Technol, BS, 36, MS, 37; Univ Calif, Los Angeles, PhD(chem), 42. *Prof Exp:* Jr chem engr, Union Oil Co, 37-39; asst chem, Univ Calif, Los Angeles, 39-41; chem engr, Shell Develop Co, 41-47; assoc prof chem eng, 47-51, chmn div, 52-53, Miller res prof, 59-60, PROF CHEM ENG, UNIV CALIF, BERKELEY, 51-, DIR WATER TECHNOL CTR, 80- *Concurrent Pos:* Consult, Lawrence Berkeley Lab, 47-, Savannah River Lab, E I du Pont de Nemours & Co, 61-67, US Borax Res Corp, 63-67, Upjohn Co, 67-71, Teknekron, Inc, 71-76 & Exxon Res & Eng, 80-; Fulbright prof, Univ Liege & Ghent, 53; vis prof, French Petrol Inst, 54, Nat Univ Mex, 67 & South China Inst Technol, 81; consult & dir, Memorex Corp, 63-81; Guggenheim fel, Cambridge Univ, 64; res assoc, Scripps Inst Oceanog, 70-71. *Honors & Awards:* W H Walker Award, Am Inst Chem Engrs, 71. *Mem:* Am Nuclear Soc; fel Am Inst Chem Engrs; Am Chem Soc; Am Inst Aeronaut & Astronaut; Am Water Works Asn. *Res:* Water purification and recovery; magnetochemistry and coal liquefication; multicomponent thermodynamics and diffusion; homogeneous and heterogeneous catalysis; chemical kinetics and reactor design; liquid extraction; agitation and fluidization; ion exchange; adsorption; interfacial phenomena; atmospheric modeling. *Mailing Add:* 725 Cragmont Ave Berkeley CA 94720

VERMILLION, ROBERT EVERETT, b Kingsport, Tenn, Aug 17, 37; m 63; c 2. TEACHING PHYSICS. *Educ:* King Col, AB, 59; Vanderbilt Univ, MS, 61, PhD(physics), 65. *Prof Exp:* Asst prof physics, 65-70, assoc prof, 70-77, PROF PHYSICS, UNIV NC, CHARLOTTE, 77- *Mem:* Am Phys Soc; Am Asn Physics Teachers; Sigma Xi. *Res:* Electric shock-tube production of plasmas; techniques and apparatus for teaching undergraduate physics; experimental plasma physics. *Mailing Add:* Dept Physics Univ NC Charlotte NC 28223

VERMILYEA, BARRY LYNN, b Dayton, Ohio, Dec 17, 41; m 65; c 2. FOOD SCIENCE, MICROBIOLOGY. *Educ:* Univ Wyo, BS, 64; Iowa State Univ, MS, 67, PhD(food sci), 69. *Prof Exp:* Sr scientist food sci, Gen Mills Inc, 69-70; develop mgr food sci, Cargill Inc, 70-71; tech mgr food sci, Int Multifoods Inc, 71-80; food consult, Kingdom Saudi Arabia, Riyadh, 80-82; DIR, QUAL CONTROL, LAND O'LAKES INC, MINNEAPOLIS, MINN, 82- *Mem:* Inst Food Technol. *Res:* New foods products, processes and packaging concepts for retail, industrial and institutional areas. *Mailing Add:* 7417 N Douglas Dr Brooklyn Park MN 55443

VERMILYEA, D(AVID) A(UGUSTUS), b Troy, NY, Oct 11, 23; m 50; c 3. METALLURGY. *Educ:* Rensselaer Polytech Inst, PhD(metall), 53. *Prof Exp:* Instr, Rensselaer Polytech Inst, 48-51; chemist, Gen Elec Co, 51-58, mgr, Chem Metall Sect, Res Lab, 58-59, metallurgist, Phys Chem Lab, 59-74, energy analyst, Energy Sci & Eng Sector, 74-78, consult res & develop strategic anal, Gen Elec Res & Develop Ctr, 78-84; RETIRED. *Honors & Awards:* Whitney Award, Nat Asn Corrosion Engrs, 75; Acheson Award. *Res:* Oxidation of metals; crystal growth; electrochemistry; corrosion. *Mailing Add:* 2505 Whamer Lane Schenectady NY 12309

VERMUND, HALVOR, b Norway, Aug 8, 16; nat US; m 43; c 2. RADIOLOGY. *Educ:* Univ Oslo, MD, 43; Univ Minn, PhD(radiol), 51. *Prof Exp:* Fel, Halden Munic & Vestfold County Hosps, Norway, 44-48; Picker Found fel radiol, Univ Minn, 51-53, res assoc radiol, 53, from asst prof to assoc prof, 54-57; prof radiol & dir radiation ther, Univ Wis-Madison Hosps, 57-68; prof radiol & dir radiother res & develop, 68-78, EMER PROF RADIOL, UNIV CALIF, IRVINE, 78- *Concurrent Pos:* Mem radiation study sect, NIH, 62-66, 65-69; mem, Comt Diag & Ther Cancer, Am Cancer Soc Adv Comt on Ther Cancer, 60-62, 72-76; mem staff, Norweg Radiumhosp, Oslo, Norway, 78-85; vis prof radiation oncol, Univ Wis-Madison, 85-87; consult, Wendt Regional Cancer Ctr, Dubuque, Iowa, 87-89; vis prof, ECarolina Sch Med, 89- *Mem:* Fel Am Col Radiol; Soc Exp Biol & Med; Am Radium Soc; Radiol Soc NAm; Am Roentgen Ray Soc; hon mem Soc Oncol & Therapeut Radiol Norway, 87. *Res:* Radiation therapy; medical radiology; radioactive isotopes. *Mailing Add:* 1210 Sand Point Way Corona Del Mar CA 92625

VERMUND, STEN HALVOR, b Minneapolis, Minn, Jan 31, 54; m 78; c 2. INFECTIOUS DISEASE EPIDEMIOLOGY, PEDIATRICS. *Educ:* Stanford Univ, BA, 74; Albert Einstein Col Med, MD, 77; London Sch Hyg & Trop Med, MSc, 81; Columbia Univ, PhD(epidemiol), 90, Royal Inst Pub Health, UK, DPH, 81. *Prof Exp:* Resident pediat, Presby Hosp, 77-80; fel epidemiol, Columbia Univ, 81-83, asst prof epidemiol & pediat, 83-85; asst prof epidemiol & pediat, Albert Einstein Col Med, 85-88; CHIEF, EPIDEMIOL BR, DIV AIDS, NAT INST ALLERGY & INFECTIOUS DIS, NIH, 88- *Concurrent Pos:* Asst attend pediatrician, Presby Hosp, 82-85 & Montefiore Med Ctr, 85-89; adj asst prof, Columbia Univ, 85-89 & Cornell

Univ, 86-89; assoc clin prof, Albert Einstein Col Med, 89- *Mem:* Am Soc Trop Med & Hyg; Soc Epidemiol Res; Am Acad Pediat; Am Col Epidemiol; Soc Adolescent Med. *Res:* Infectious disease epidemiology; parasitic diseases; tropical public health; sexually transmitted diseases; HIV/AIDS; HPV and HIV, relationship to cervix disease. *Mailing Add:* Epidemiol Br Div AIDS Nat Inst Allergy & Infectious Dis NIH 6003 Executive Blvd Bethesda MD 20892

VERNADAKIS, ANTONIA, b Canea, Crete, Greece, May 11, 30; m 61. DEVELOPMENTAL NEUROBIOLOGY. *Educ:* Univ Utah, BA, 55, MS, 57, PhD(anat, pharmacol), 61. *Prof Exp:* Interdisciplinary training prog fel pharmacol, Univ Calif Sch Med, San Francisco Med Ctr, 64-65; asst res physiologist, Univ Calif, Berkeley, 65-67; from asst prof to assoc prof, 67-78, PROF PSYCHIAT & PHARMACOL, UNIV COLO SCH MED, 78- *Concurrent Pos:* Res scientist develop award, NIMH, 69-79; vis prof, Univ Athens, Greece, 81 & Univ Zimbabwe Sch Med, Harare, 83. *Mem:* Am Soc Pharmacol & Exp Therapeut; Am Physiol Soc; Am Neurochem Soc; Int Soc Neurochem; Int Soc Psychoneuroendocrinol; Int Soc Develop Neurosci (pres, 83-85). *Res:* Regulatory mechanisms in brain maturation; drugs and hormones; neurotransmission maturation; neural cell growth and differentiation using neural cell culture. *Mailing Add:* Dept Psychiat & Pharmacol Univ Colo Sch Med 4200 E Ninth Ave Denver CO 80220

VERNAZZA, JORGE ENRIQUE, b Buenos Aires, Arg, Jan 16, 43; US citizen; m 73; c 2. SOLAR PHYSICS, PLASMA PHYSICS. *Educ:* Univ Buenos Aires, Licenciate, 67; Harvard Univ, PhD(astron), 72. *Prof Exp:* Res fel astron, Harvard Univ, 72-74, res assoc, 74-80; MEM STAFF, LAWRENCE LIVERMORE NAT LAB, 80- *Mem:* Am Astron Soc; Am Phys Soc. *Res:* Radiative transfers; plasma physics. *Mailing Add:* Lawrence Livermore Nat Lab L-95 PO Box 408 Livermore CA 94550

VERNBERG, FRANK JOHN, b Fenton, Mich, Nov 6, 25; m 45; c 3. MARINE BIOLOGY, PHYSIOLOGICAL ECOLOGY. *Educ:* DePauw Univ, AB, 49, MA, 50; Purdue Univ, PhD(zool), 51. *Prof Exp:* From instr to prof zool, Duke Univ, 51-69, asst dir res, 58-63, asst dir marine lab, 63-69; BARUCH PROF MARINE BIOL & DIR BELLE W BARUCH COASTAL RES INST, UNIV SC, 69- *Concurrent Pos:* Guggenheim fel, 57-58; Fulbright-Hayes res award, Brazil, 65 & Fulbright-Hayes fel, 65; lectr, Univs Kiel & Sao Paulo, Chulalongkorn Univ, Thailand; mem comt manned orbital res lab, Am Inst Biol Sci, 66-; dir int biol prog-prog exp anal, Biogeog of the Sea, Nat Acad Sci, 67-69; consult, Environ Protection Agency, 74-, Nat Sci Found, 80-; managing ed, J Exp Marine Biol & Ecol; chmn, Comn Physiol Ecol, Int Asn Ecol. *Honors & Awards:* Russell Award, 77; W S Proctor Award, Sigma Xi, 83. *Mem:* AAAS; Estuarine Res Fedn (pres, 75-77); Am Soc Zoologists (pres, 82); Ecol Soc Am; Int Asn Ecol. *Res:* Physiological ecology of marine animals; distribution of decapod crustacea; tissue metabolism; mechanisms of temperature acclimation; physiological diversity of latitudinally separated populations. *Mailing Add:* Belle W Baruch Inst Univ SC Columbia SC 29208

VERNBERG, WINONA B, b Kans, Nov 6, 25; c 3. ESTUARINE SYSTEMS, ENVIRONMENTAL PHYSIOLOGY. *Educ:* Kans State Col, BS, 44; DePauw Univ, MA, 47; Purdue Univ, PhD, 51. *Prof Exp:* Instr DePauw Univ, 47-49; res assoc, Duke Univ, 51-73; res prof biol, Univ SC, 69-75, prof pub health & prog dir, Environ Health Staff, 75-77, actg dean, Sch Pub Health, 77-78, dean, 78-79, DEAN COL HEALTH, UNIV SC, 79- *Concurrent Pos:* Mem, Nat Adv Comt Oceans & Atmosphere, 74-76, exec comt sci adv bd, Environ Protection Agency, 78-82, Nat Adv Coun, NASA, 79-80, educ comt, Asn Sch Pub Health, 81, peer rev panel, Off Explor Res, Environ Protection Agency, 87-90; chairperson, Sci Adv Comt Off Toxic Substances, Environmental Protection Agency, 78-82; pres, Coun Educ Pub Health, 85 & 86; bd dir, Univ SC Educ Found, 86 & 90. *Honors & Awards:* Proctor Prize, Sigma Xi, 83. *Mem:* Am Pub Health Asn; Sigma Xi; Royal Soc Trop Med & Hyg; Am Soc Zoologists (treas, 73-76). *Res:* Effects of urbanization; various sources of pollution; other potential disturbances on the coastal estuarine system. *Mailing Add:* Sch Pub Health Univ SC Columbia SC 29208

VERNEKAR, ANANDU DEVARAO, b Hosali, India, July 5, 32; m 59; c 3. METEOROLOGY. *Educ:* Univ Poona, BSc, 55, BSc, 56, MSc, 59; Univ Mich, MS, 63, PhD(meteorol), 66. *Prof Exp:* Sci asst meteorol, Upper Air Sect, India Meteorol Dept, 56-61; res scientist, Travelers Res Ctr, Inc, 67-69; from asst prof to assoc prof, 69-79, PROF METEOROL, UNIV MD, COLLEGE PARK, 79- *Mem:* Am Geophys Union; Am Meteorol Soc. *Res:* Dynamical meteorology; general circulation; theory of climate and statistical meteorology. *Mailing Add:* Dept Meteorol Univ Md College Park MD 20742

VERNER, JAMES HAMILTON, b Hitchin, Eng, Feb 22, 40; Can citizen; m 64; c 3. MATHEMATICS. *Educ:* Queen's Univ, Ont, BSc, 62, MSc, 65; Univ Edinburgh, PhD(comput sci), 69. *Prof Exp:* Lectr math, Royal Mil Col Can, 63-64; External Aids Off teaching adv, Umuahia, Eastern Nigeria, 64-66; res assoc, 69-72, asst prof, 72-77, ASSOC PROF MATH, QUEENS UNIV, ONT, 77- *Concurrent Pos:* Hon lectr math, Univ Auckland, NZ, 75-76. *Mem:* Soc Indust & Appl Math. *Res:* Numerical analysis, especially numerical solution of initial value problems for ordinary differential equations. *Mailing Add:* Dept Math & Statist Queen's Univ Kingston ON K7L 3N6 Can

VERNER, JARED, b Baltimore, Md, Aug 16, 34; m 58; c 3. ANIMAL ECOLOGY. *Educ:* Wash State Univ, BS, 57; La State Univ, MS, 59; Univ Wash, PhD(zool), 63. *Prof Exp:* Res assoc zool, Univ Calif, Berkeley, 63-65; from asst prof to prof biol, Cent Wash State Col, 65-73; prof biol, Ill State Univ, 73-, adj prof ecol, 77-; AT FORESTRY SCI LAB, FRESNO. *Concurrent Pos:* NSF fel, 63-65, res grant, 66-71. *Mem:* Am Ornith Union; Cooper Ornith Soc; Wilson Ornith Soc; Soc Study Evolution; Ecol Soc Am. *Res:* Evolution and natural selection; avian social organization and communication systems; avian population ecology. *Mailing Add:* 936 E Locust Ave Fresno CA 93710

VERNIER, ROBERT L, b El Paso, Tex, July 29, 24; m 45; c 5. PEDIATRICS, NEPHROLOGY. *Educ:* Univ Dayton, BS, 48; Univ Cincinnati, MD, 52. *Prof Exp:* Clin fel pediat, Univ Ark, 52-54; clin fel pediat, Univ Minn, 54-55, USPHS res fel, 55-57, Am Heart Asn res fel, 57-59; asst prof pediat & Am Heart Asn estab investr, Med Sch, Univ Minn, 59-65; prof, Sch Med, Univ Calif, Los Angeles, 65-68; PROF PEDIAT, MED SCH, UNIV MINN, 68- *Concurrent Pos:* Estab investr, Am Heart Asn, 59-60; Guggenheim fel, Dept Biophys, State Serum Inst, Copenhagen, Denmark, 60-61; sr fel, Fogarty Inst Ctr, Dept Path, Grotingen, Neth, 75-76. *Honors & Awards:* Mead Johnson Award Pediat Res, 62. *Mem:* AAAS; Am Soc Clin Invest; Am Soc Exp Path; Soc Pediat Res; Am Soc Nephrology (pres, 79-80); Am Soc Pediat Nephrology (pres, 76-77). *Res:* Clinical pediatrics; renal disease in childhood; electron microscopy in the kidney. *Mailing Add:* Dept Pediat Univ Minn Med Sch Minneapolis MN 55455

VERNIER, VERNON GEORGE, b Norwalk, Conn, Nov 14, 24; m 55; c 4. PHARMACOLOGY. *Educ:* Univ Ill, BS, 47, MD, 49. *Prof Exp:* Intern, Res & Educ Hosp, Univ Ill, 49-50, res assoc pharmacol, Med Col, 50-51, instr, 51-52; res assoc, Sharpe & Dohme, Inc, 52-54; res assoc physiol, Merck Inst Therapeut Res, 56-63; mgr, Pharmacol Sect, Stine Lab, 63-74, dir pharmacol, 74-82, assoc dir prof serv, Med Res, E I du Pont de Nemours Co, 82-85; CONSULT, PHARMACEUT INDUST. *Concurrent Pos:* Lectr, Sch Med, Temple Univ, 56-66, vis prof, 66-87. *Mem:* AAAS; Am Soc Pharmacol & Exp Therapeut; Am Col Neuropsychopharmacol; NY Acad Sci. *Res:* Neuropsychopharmacology; psychopharmacology; toxicology. *Mailing Add:* 303 Lark Dr Newark DE 19713

VERNIKOS, JOAN, b Alexandria, Egypt, May 9, 34; m 78; c 2. ENDOCRINE PHARMACOLOGY, STRESS & COPING. *Educ:* Univ Alexandria, BPharm, 55; Univ London, PhD(pharmacol), 60. *Prof Exp:* Muelhaupt scholar, Ohio State Univ, 60-61, asst prof pharmacol, 61-64; Nat Acad Sci-Nat Res Coun res assoc, 64-66, chief human studies br, 72-76, actg dep dir, life sci, 76, RES SCIENTIST, AMES RES CTR, 66- ACTG ASSOC DIR, SPACE RES, 86-, ACTG CHIEF, LIFE SCI DIV, NASA, 88- *Concurrent Pos:* HEW/NIH, Pharmacol Study Sect, 74-78; Endocrinol, 73-77, assoc ed, Pharmacol Rev, 77-81; hon clin prof pharmacol, Sch Med, Wright State Univ, Ohio, 75-81, ASPET subcomt, Women in Pharmacol, 79-83; Life Sci & Systs Tech Comt, Am Inst Aeronaut & Astronaut, 90-; Const & By-laws comt, Aerospace Med Asn, 90- *Honors & Awards:* NASA Medal for Except Sci Achievement, 73; Hubertus Strughold Award Space Med, 90. *Mem:* Endocrine Soc; Aerospace Med Asn; Int Neuroendocrine Soc; Am Soc Pharmacol & Exp Therapeut; Int Brain Res Orgn; AIAA. *Res:* Stress and the environmental, behavioral and physiological factors that affect the stress response including weightlessness and inactivity; mechanisms regulating pituitary-adrenal function, fluids, electrolytes and drug/stress and pain/stress interactions. *Mailing Add:* Ames Res Ctr Space Res Dir NASA Moffett Field CA 94035

VERNON, C(ARL) WAYNE, b Alamosa, Colo, Oct 29, 39; m 58; c 2. ELEMENTARY PARTICLE PHYSICS. *Educ:* Univ Wash, BPhys, 61; Princeton Univ, PhD(physics), 66. *Prof Exp:* Univ Wash, BPhys, 61; asst prof, 66-74, assoc prof, 74-80, PROF PHYSICS, UNIV CALIF, SAN DIEGO, 80- *Res:* Cosmic ray and deep inelastic scattering muons; electron-positron colliding beam physics; x-ray detectors for crystallography; free-electron lasers and particle accelerators. *Mailing Add:* Dept Physics Univ Calif San Diego La Jolla CA 92093-0319

VERNON, EUGENE HAWORTH, animal breeding; deceased, see previous edition for last biography

VERNON, FRANK LEE, JR, b Dallas, Tex, Sept 16, 27; m 50; c 3. LOW TEMPERATURE PHYSICS, QUANTUM PHYSICS. *Educ:* Southern Methodist Univ, BS, 49; Univ Calif, Berkeley, MS, 52; Calif Inst Technol, PhD(elec eng, physics), 59. *Prof Exp:* Asst recorder geophys prospecting, Tex Co, 49-50; head, Sect Microwave Physics, Hughes Aircraft Co, 51-61; SR STAFF SCIENTIST LOW TEMPERATURE & QUANTUM PHYSICS, AEROSPACE CORP, 61- *Concurrent Pos:* Res fel physics, Calif Inst Technol, 59-60. *Mem:* Am Phys Soc; Inst Elec & Electronics Engrs; Sigma Xi; AAAS. *Res:* Experimental and theoretical investigations in the fields of low temperature, microwave and quantum physics including superconductors, lasers and the interaction of electron tunneling mechanisms with high frequency radiation. *Mailing Add:* 1560 Knollwood Terr Pasadena CA 91103

VERNON, GREGORY ALLEN, b Akron, Ohio, July 27, 47. PHYSICAL CHEMISTRY. *Educ:* Pa State Univ, BS, 69; Univ Ill, MS, 71, PhD(chem), 75. *Prof Exp:* Res anal chemist, Atomic Int Div, Rockwell Int, 75-80; RES CHEMIST, CHINA LAKE NAVAL WEAPONS CTR, 80- *Mem:* Am Chem Soc; Sigma Xi. *Res:* Detonation physics; thermal hazards; decomposition kinetics. *Mailing Add:* 436 N Warner St Ridgecrest CA 93555-3646

VERNON, JOHN ASHBRIDGE, b Camden, NJ, Jan 19, 40; m 62; c 3. ORGANIC CHEMISTRY. *Educ:* Rutgers Univ, BS, 61; Univ Md, PhD(org chem), 65. *Prof Exp:* Asst chem, Univ Md, 61-63, Gillette Harris res fel org chem, 63-64; res chemist, E I du Pont de Nemours & Co, Inc, 65-70; plant mgr, Pioneer Labs, Chesebrough-Ponds, Inc, 71-73; mgr labs, Vick Mfg Div, Richardson-Vicks, Inc, 73-77, qual assurance mgr, 77-85, group dir tech serv/qual assurance, 85-90, GROUP MGR TECH SERV, VICKS HEALTH CARE DIV, RICHARDSON-VICKS, INC, 90- *Mem:* Am Chem Soc. *Res:* Reactions of organic compounds over alumina; synthesis and reaction of azides; heterocyclic compounds; synthesis of liquid crystals. *Mailing Add:* 330 Warminster Rd Hatboro PA 19040

VERNON, LEO PRESTON, b Roosevelt, Utah, Oct 10, 25; m 46; c 5. BIOCHEMISTRY. *Educ:* Brigham Young Univ, BA, 48; Iowa State Col, PhD, 51. *Prof Exp:* Fel, Enzyme Inst, Univ Wis, 51-52; res assoc, Washington Univ, 52-54; assoc prof chem, Brigham Young Univ, 54-61; dir, C F Kettering Res Lab, Kettering Found, Ohio, 61-70; dir res, Brigham Young Univ, 70-74, asst acad vpres res, 74-81; vpres, Billings Energy Corp, 81-; PROF CHEM, BRIGHAM YOUNG UNIV. *Concurrent Pos:* Researcher, Nobel Inst, Stockholm, Sweden, 60-61. *Honors & Awards:* Utah Award, Am Chem Soc, 85. *Mem:* Am Soc Biol Chem; Am Soc Plant Physiol. *Res:* Photosynthesis; cytochrome chemistry; respiratory enzymes; peptide toxins. *Mailing Add:* Dept Chem Brigham Young Univ 675 WIDB Provo UT 84602

VERNON, LONNIE WILLIAM, b Dallas, Tex, Mar 16, 22; m 49; c 3. PHYSICAL CHEMISTRY, FUEL SCIENCE. *Educ:* Rice Univ, BA, 48, MA, 50, PhD(chem), 52. *Prof Exp:* SR RES ASSOC, EXXON RES & ENG CO, 52- *Mem:* Am Chem Soc; Sigma Xi. *Res:* Coal conversion processes; surface chemistry; catalysis. *Mailing Add:* 5017 Ashwood Dr Baytown TX 77521

VERNON, RALPH JACKSON, b Greenville, SC, Apr 6, 20; m 47; c 3. OCCUPATIONAL HEALTH, SAFETY ENGINEERING. *Educ:* Clemson Univ, BS, 50; Tex A&M Univ, MEd, 51; Univ Iowa, PhD(prev med, environ hyg), 68. *Prof Exp:* Instr safety, Tex A&M Univ, 51-53; safety engr, Liberty Mutual Ins Co, 53-60, loss prev mgr, 60-66; prof eng technol & indust eng, 68-77, PROF INDUST ENG, TEX A&M UNIV, 77-, HEAD INDUST HYG & SAFETY ENG DIV, 68- *Concurrent Pos:* Vpres, Hill, Stocker, Vernon & Assocs, 70-; mem, Bd Cert Safety Prof. *Mem:* Human Factors Soc; Am Inst Indust Engrs; Am Indust Hyg Asn; Am Conf Govt Indust Hyg; Am Pub Health Asn; Sigma Xi. *Res:* Occupational diseases; accident causation; noise; biodynamics and resulting effects on performance decrement. *Mailing Add:* 1003 Hereford St College Station TX 77840

VERNON, ROBERT CAREY, b Wilmington, Ohio, Feb 7, 23; m 49; c 2. SOLID STATE PHYSICS. *Educ:* Bates Col, BS, 47; Wesleyan Univ, MA, 49; Pa State Univ, PhD(physics), 52. *Prof Exp:* Asst physics, Pa State Univ, 49-52; instr, Williams Col, 52-54, lectr, 54-55, asst prof, 55-58; asst prof, Clarkson Col Technol, 58-60, assoc prof, 60-61; chmn dept physics, 61-72, PROF PHYSICS, SIMMONS COL, 61- *Mem:* Am Phys Soc; Am Asn Physics Teachers; Geol Soc Am. *Res:* Imperfections in nearly perfect crystals; optical properties of semiconductors. *Mailing Add:* 189 Pond Westwood MA 02090

VERNON, RONALD J, b Chicago, Ill, June 3, 36; m 71. ELECTRICAL ENGINEERING. *Educ:* Northwestern Univ, BS, 59, MS, 61, PhD(elec eng), 65. *Prof Exp:* From asst prof to assoc prof, 65-77, PROF ELEC ENG, UNIV WIS-MADISON, 77- *Concurrent Pos:* NSF res grants, 66 & 68. *Mem:* Inst Elec & Electronics Engrs. *Res:* Electromagnetic field theory; microwave engineering; microwave interaction with solids; semiconductor physics. *Mailing Add:* Dept Elec Eng 3546a Eng Bldg Univ Wis 1415 Johnson Dr Madison WI 53706

VERNON, RUSSEL, strategic management software, unix system applications, for more information see previous edition

VERNON, WILLIAM W, b Concord, NH, Nov 1, 25; m 51; c 1. GEOLOGY, ARCHAEOLOGY. *Educ:* Univ NH, BA, 52; Lehigh Univ, MS, 55, PhD, 64; Univ Pa, MS, 84. *Prof Exp:* Civil engr, Dept Pub Rds & Hwys, NH, 52-53; geologist, US Geol Surv, 56-57; from asst prof to assoc prof geol, 57-71, chmn dept, 65-74, PROF GEOL & ANTHROP, DICKINSON COL, 71- *Concurrent Pos:* Res assoc, Univ Mus, Univ Pa, Philadelphia. *Mem:* AAAS; Sigma Xi; Nat Asn Geol Teachers; Archaeol Inst Am; Geol Soc Am. *Res:* Mineralogy, petrology and structure of igneous and metamorphic rocks in south-central New Hampshire; archaeological investigations of Early Man in New York State; archaeoceramic and archaeometallurgical investigations in Thailand. *Mailing Add:* Dept Geol Dickinson Col Carlisle PA 17013

VERONIS, GEORGE, b New Brunswick, NJ, June 3, 26; m 63; c 2. OCEANOGRAPHY. *Educ:* Lafayette Col, AB, 50; Brown Univ, PhD(appl math), 54. *Hon Degrees:* MA, Yale Univ, 66. *Prof Exp:* Staff meteorologist, Inst Advan Study, 53-56; staff mathematician, Woods Hole Oceanog Inst, 56-63; assoc prof oceanog, Mass Inst Technol, 61-63, res oceanogr, 64-66; PROF GEOPHYS & APPL SCI, YALE UNIV, 66- *Concurrent Pos:* Guggenheim fel, Stockholm, Sweden, 60-61 & 66-67; sr queen's fel, Australia, 81; ed, J Marine Res, 73- *Honors & Awards:* Alexander von Humboldt Award, 86; Robert L & Bettie P Cody Award in Ocean Sci, 89. *Mem:* AAAS; fel Am Geophys Union; Am Acad Arts & Sci; Norweg Acad Arts & Lett. *Res:* Ocean circulations; rotating and stratified fluids. *Mailing Add:* Dept Geol & Geophys Yale Univ New Haven CT 06520

VEROSUB, KENNETH LEE, b New York, NY, July 10, 44; m 67; c 2. PALEOMAGNETISM, TECTONICS. *Educ:* Univ Mich, BA, 66; Stanford Univ, MS, 71, PhD(physics), 73. *Prof Exp:* Asst prof geophys, Amherst Col, 72-75; from asst prof to assoc prof, 75-84, PROF GEOPHYS, UNIV CALIF, DAVIS, 84- *Concurrent Pos:* Vis researcher, Ctr Faibles Radioactivities, Gif-Sur-Yvette, France, 82-83; vis prof, Ctr Geol & Geophys, Univ Sci Tech Languedoc, Montpellier, France; Fulbright fel, France, 89-90. *Mem:* Am Geophys Union; Geol Soc Am; Sigma Xi; Earthquake Eng Res Inst; Nat Asn Geol Teachers. *Res:* Paleomagnetism of sediments; history of the earth's magnetic field; geomagnetic polarity transitions and excursions; tephrostratigraphy; California tectonics; gravity and magnetic modelling; geologic hazards; satellite remote sensing. *Mailing Add:* Dept Geol Univ Calif Davis CA 95616

VERPOORTE, JACOB A, b Utrecht, Neth, Oct 17, 36; m 61; c 2. BIOPHYSICAL CHEMISTRY. *Educ:* Univ Utrecht, MSc, 60; Univ Pretoria, PhD(biochem), 64. *Prof Exp:* Fel phys protein chem, Univ Alta, 64-65 & Harvard Univ, 65-67; from asst prof to assoc prof, 67-84, PROF BIOCHEM, DALHOUSIE UNIV, 84- *Concurrent Pos:* Muscular Dystrophy Asn Can fel, 64-66. *Res:* Isolation and characterization of biologically active proteins; physico-chemical studies, including studies on the structure of proteins. *Mailing Add:* Dept Biochem Dalhousie Univ Halifax NS B3H 4H7 Can

VERRALL, RONALD ERNEST, b Ottawa, Ont, Feb 26, 37; m 61; c 2. PHYSICAL CHEMISTRY. *Educ:* Univ Ottawa, Ont, BSc, 62, PhD(phys chem), 66. *Prof Exp:* Nat Res Coun Can-NATO fel, Mellon Inst, Carnegie-Mellon Univ, 66-68; from asst prof to assoc prof, 68-77, PROF CHEM, UNIV SASK, 77- *Concurrent Pos:* Vis fel, Mellon Inst, Carnegie-Mellon Univ, 68. *Mem:* Sigma Xi; AAAS; Chem Inst Can. *Res:* Thermodynamic studies of electrolyte and non-electrolyte solutions; fluorescence studies of protein dynamics; ultrasonic studies of liquid phase, particularly microemulsions; effect of high intensity ultrasound on biological systems. *Mailing Add:* Dept Chem Univ Sask Saskatoon SK S7N 0W0 Can

VERRIER, RICHARD LEONARD, BEHAVIORAL BIOLOGY, NERVOUS SYSTEM. *Educ:* Univ Va, PhD(cardiovasc physiol), 69. *Prof Exp:* ASSOC PROF CARDIOVASC DIS, SCH PUB HEALTH, HARVARD UNIV, 83- *Res:* Neural factors and cardiac arrhythmias. *Mailing Add:* Dept Pharmacol Georgetown Univ Med Sch 3900 Reservoir Rd NW Washington DC 20007

VERRILLO, RONALD THOMAS, b Hartford, Conn, July 31, 27; m 50; c 3. PSYCHOPHYSICS, NEUROSCIENCES. *Educ:* Syracuse Univ, BA, 52; Univ Rochester, PhD(psychol), 58. *Prof Exp:* Asst prof spec educ, 57-62, res assoc, Bioacoust Lab, 59-63, res fel, Lab Sensory Commun, 63-67, assoc prof sensory commun, 67-74, assoc dir, Inst Sensory Res, 80-84, prof sensory sci, 74-83, PROF NEUROSCI, SYRACUSE UNIV, 83-, DIR, INST SENSORY RES, 84- *Concurrent Pos:* NATO sr fel, Oxford Univ, 70-71; vis prof, Karolinska Hosp, Stockholm, Sweden, 77. *Honors & Awards:* Res Award, Am Personnel & Guid Asn, 62. *Mem:* Fel Acoust Soc Am; Psychonomic Soc; Int Asn Study Pain; Soc Neurosci. *Res:* Cutaneous sensitivity; effects of the physical parameters of vibratory stimuli on threshold and suprathreshold responses in humans; sensory characteristics of pain. *Mailing Add:* Inst Sensory Res Syracuse Univ Merrill Lane Syracuse NY 13244-5290

VERSCHINGEL, ROGER H C, b Jan 19, 28; Can citizen; m 59. CHEMICAL INSTRUMENTATION. *Educ:* Sir George Williams Univ, BSc, 49; McGill Univ, PhD(chem), 55. *Prof Exp:* Lectr chem, Sir George Williams Univ, 54-56, from asst prof to assoc prof, 56-67, PROF CHEM, CONCORDIA UNIV, 67-, CHMN DEPT, 68-, DEAN, SIR GEORGE WILLIAMS FAC SCI, 73- *Mem:* Am Chem Soc; Chem Inst Can; Sigma Xi. *Res:* Chemical spectroscopy. *Mailing Add:* 200 Gaspe Nun's Island Montreal PQ H3E 1E6 Can

VERSCHOOR, J(ACK) D(AHLSTROM), b Grand Rapids, Mich, Nov 11, 23; m 47; c 3. HEAT TRANSFER, BUILDING CONSTRUCTION TECHNOLOGY. *Educ:* Univ Mich, BS, 45; Calif Inst Technol, MS, 47. *Prof Exp:* Physicist heat transfer, Johns-Manville Res Ctr, 47-59, proj mgr, Johns-Manville Corp, 59-69, mgr corp planning, 69-72, mgr venture anal & develop, 72-75, res assoc testing & contracts, Johns-Manville Res & Develop Ctr, 75-82; CONSULT, VERSCHOOR ASSOCS, 82- *Concurrent Pos:* Teaching fel, Calif Inst Technol, 45-47; instr, Rutgers Univ, 47-58; adv comt mem, US Dept Energy, 79-81. *Mem:* Am Soc Heating Refrig & Air Conditioning Engrs; Am Soc Testing & Mat; Sigma Xi. *Res:* Energy conservation for industrial and building applications; air leakage; moisture and sound control in building construction; physics of heat transfer in thermal insulations; development of heat transfer test apparatus. *Mailing Add:* 179 Gail Lane Bailey CO 80421-1820

VERSCHUREN, JACOBUS PETRUS, b Delft, Neth, Apr 13, 30; nat Can; m 54; c 2. ENGINEERING, HYDROLOGY. *Educ:* Delft Univ Technol, CE, 52; Univ Alta, MSc, 60; Colo State Univ, PhD, 68. *Prof Exp:* Asst civil eng, Lehigh Univ, 52-53; engr, Howard Needles Tammen & Bergendoff, NY, 53; soil mech, Soil Mech Labs, Delft Univ Technol, 53-54; resident bridge engr, Prov Dept of Hwy, Man, 54-55; lectr, 55-57, from asst prof to assoc prof, 57-70, PROF CIVIL ENG, UNIV ALTA, 70- *Concurrent Pos:* Dir, T Blench & Assocs, Ltd, 59-; spec consult, Northwest Hydraul Consult, Ltd. *Mem:* Can Water Resources Asn. *Res:* Engineering and environmental aspects of hydrology. *Mailing Add:* Dept Civil Eng Univ Alta Edmonton AB T6G 2E2 Can

VERSCHUUR, GERRIT L, b Capetown, SAfrica, June 5, 37; m 66; c 1. ASTRONOMY. *Educ:* Rhodes Univ, SAfrica, BSc, 57, MSc, 60; Univ Manchester, PhD(radio astron), 65. *Prof Exp:* Jr lectr physics, Rhodes Univ, SAfrica, 60; lectr, Univ Manchester, 64-67; res assoc, Nat Radio Astron Observ, 67-69, asst scientist, 69-72, assoc scientist, 72-73; prof astrogeophys & dir, Fiske Planetarium, Univ Colo, Boulder, 73-80; vpres mkt & sales, AET, Boulder, 82-84; WRITER & CONSULT, 88- *Concurrent Pos:* Vis scientist, Nat Radio Astron Observ, 84-85, Nat Radio Astron Observ, 84-85; lectr, Univ Md, 88. *Mem:* Am Astron Soc; Hist Sci Soc; Astron Soc Pac. *Res:* Interstellar neutral hydrogen studies; interstellar magnetic field measurements. *Mailing Add:* 4802 Brookstone Terr Bowie MD 20720

VERSEPUT, HERMAN WARD, b Grand Rapids, Mich, Sept 25, 21; m 51. PAPER TECHNOLOGY. *Educ:* Yale Univ, BE, 42; Lawrence Univ, MS, 48, PhD(pulp & paper technol), 51. *Prof Exp:* Res chemist, Robert Gair Co, Inc, 51-56, chief appl res sect, Gair Paper Prod Group, Continental Can Co, Inc, 56-61; dir res & develop, Folding Carton Div, Riegel Paper Corp, 61-66; sr develop engr, Beloit Corp, 67-71; RES MGR PAPERBOARD, BOXBOARD RES & DEVELOP ASN, 71- *Mem:* Tech Asn Pulp & Paper Indust; Am Chem Soc. *Res:* Product and process development in the manufacture of packaging materials from recycled fibers. *Mailing Add:* Boxboard Res & Develop Asn 350 S Burdick Mall Rm 207 Kalamazoo MI 49007-4820

VERSES, CHRIST JAMES, b Stamford, Conn, Apr 12, 39; div; c 1. PHYSIOLOGY, MOLECULAR BIOLOGY. *Educ:* Valparaiso Univ, BS, 61; Univ Conn, PhD(bact), 67. *Prof Exp:* Fel microbiol, Univ Colo Med Ctr, 66-68; asst prof biol, Moorhead State Col, 68-69; res microbiologist, Pollution Control Industs Inc, 69-72; dir pub health lab, Water & Sewage Anal, State of Conn, 72-73; asst prof, 73-75, dir honors prog, 78-81, ASSOC PROF BIOL, SACRED HEART UNIV, 75-, DIR ALLIED HEALTH SCI, 81- *Concurrent Pos:* Owner anal lab; Lilly vis fac fel, Yale Univ. *Mem:* Fel Am

Inst Chemists; Am Soc Microbiologists; assoc Sigma Xi. *Res:* Pollution survey of municipal harbor systems; mechanism of attachment of phage to a host cell; biochemical reactions; viral genetics; new sterilization device and indicators; heart metabolism; microbiology; monoamine oxidase inhibition. *Mailing Add:* 2600 Park Ave No 1-T Bridgeport CT 06604

VERSIC, RONALD JAMES, b Dayton, Ohio, Oct 19, 42; m 66; c 2. PRODUCT DEVELOPMENT. *Educ:* Univ Dayton, BS, 64; Johns Hopkins Univ, MA, 68; Ohio State Univ, PhD(mat eng), 69. *Prof Exp:* Sr scientist, Standard Register Co, 71-76; dir res & develop, Monarch Marking Systs, Subsidiary of Pitney Bowes, 76-79; vpres, 79-88, PRES, RONALD T DODGE CO, 88- *Concurrent Pos:* Adj asst prof, Univ Cincinnati, 89- *Mem:* Sigma Xi; AAAS; Am Chem Soc; Am Asn Physics Teachers; Soc Imaging Sci & Technol. *Res:* Engineering management related to commercial product development in the chemical and paper converting areas. *Mailing Add:* Ronald T Dodge Co PO Box 630 Dayton OH 45459

VERSTEEGH, LARRY ROBERT, b Minneapolis, Minn, Feb 13, 49; m 70; c 2. DRUG REGULATORY AFFAIRS. *Educ:* Cent Col, Iowa, BA, 71; Iowa State Univ, PhD(biochem), 75. *Prof Exp:* Staff scientist, Procter & Gamble Co, 75-79; assoc dir drug regulatory affairs, Mead Johnson Pharmaceut Co, 79-82; dir regulatory affairs, NAm, G D Searle, 82-87; VPRES REGULATORY AFFAIRS, ERBAMONT NEV, 87- *Concurrent Pos:* Lectr biol sci, Univ Cincinnati, 76-77. *Mem:* AAAS; Regulatory Affairs Profs Soc; Fedn Am Scientists; Drug Info Asn. *Res:* Drug regulatory affairs; drug development; quality assurance; quality control. *Mailing Add:* 4240 Woodhall Rd Columbus OH 43220

VER STRATE, GARY WILLIAM, b Metuchen, NJ, Jan 29, 40; m 61; c 2. POLYMER CHEMISTRY, POLYMER PHYSICS. *Educ:* Hope Col, BS, 63; Univ Del, PhD(chem), 67. *Prof Exp:* Sr res chemist, Exxon Chem Co, Linden, 66-75, res assoc, 75-82; SR RES, EXXON CHEM CO, 82- *Mem:* Am Phys Soc; Am Chem Soc; Soc Rheolology. *Res:* Crystallinity; rheological properties; characterization by physical methods; light scattering; kinetics and molecular weight distribution; branching; cationic polymerization; chemical modification of polymers; polymer networks; liquid rubbers. *Mailing Add:* 28 Bayside Dr Atlantic Highlands NJ 07716

VERTER, HERBERT SIGMUND, b New York, NY, Jan 30, 36; m 60, 81; c 1. ORGANIC CHEMISTRY. *Educ:* City Col New York, BS, 56; Harvard Univ, MA, 57, PhD(chem), 60. *Prof Exp:* NATO fel, Imp Col, Univ London, 60-61; asst prof chem, Cent Mich Univ, 61-66; assoc prof, Inter-Am Univ, PR, 66-67, dean acad affairs, 73, prof chem, 67-77, chmn dept, 66-72 & 74-77; admin dir, Dept Chem, Brandeis Univ, 77-79; CHMN, SCI DEPT, WOODMERE ACAD, 79- *Mem:* Am Chem Soc; Sigma Xi. *Res:* Chemistry of natural products; organic synthesis; mechanism; carbon oxides. *Mailing Add:* 150 Corbin Pl Brooklyn NY 11235

VERTER, JOEL I, b Brooklyn, NY, June 14, 42. BIOSTATISTICS. *Educ:* City Col NY, BS, 65; Univ Mass, MS, 67; Univ NC, PhD(biostatist), 79. *Prof Exp:* Statistician, Biostatist Res Br, Nat Heart, Lung & Blood Inst, NIH, 67-90; SR STATISTICIAN, HENRY FORD HOSP, 90- *Mem:* Am Statist Asn; Biomet Soc; Soc Clin Trials. *Mailing Add:* Henry Ford Hosp 23725 Northwestern Southfield MI 48075

VERTES, VICTOR, b Cleveland, Ohio, Sept 10, 27. INTERNAL MEDICINE. *Educ:* Western Reserve Univ, BS, 49, MD, 53; Am Bd Internal Med, dipl, 62. *Prof Exp:* Intern med, Univ Hosps, Cleveland, 53-54; resident, Mt Sinai Hosp, Cleveland, 54-56; fel metab & endocrinol, NY Hosp-Cornell Med Ctr, 56-57; demonstr & instr med, 58-62 & 64, from asst clin prof to assoc clin prof, 64-72, PROF MED, SCH MED, CASE WESTERN RESERVE UNIV, 72-, DIR DEPT MED, MT SINAI MED CTR, 64- *Concurrent Pos:* Assoc vis physician, Mt Sinai Hosp, 58-64, dir metab & endocrine lab, 59-65, proj dir chronic dialysis ctr, 65-68; investr, USPHS, 65-; mem, Coun Hemoglobin-Blood Pressure Res. *Mem:* Fel Am Col Physicians; Am Soc Nephrology; fel Am Col Angiol; dipl mem Pan-Am Med Asn; Europ Dialysis & Transplant Asn. *Res:* Treatment and mechanisms of uremia; role of the kidney in hypertension. *Mailing Add:* One Mt Sinai Dr Cleveland OH 44106

VERTREES, ROBERT LAYMAN, b Louisville, Ky, Nov 1, 39; m 66; c 3. NATURAL RESOURCES POLICY. *Educ:* Purdue Univ, BS, 61; Mich State Univ, MS, 67, PhD(resource develop), 74. *Prof Exp:* Instr resource econ, dept agr & food econ, Univ Mass, 69-72; asst prof, dept econ, SDak State Univ, 73-76; ASST PROF LAND & WATER RESOURCES PLANNING & POLICY, OHIO STATE UNIV, 76- *Mem:* Am Water Resources Asn; Soil Conserv Soc Am. *Res:* Economic and land-use impacts of alternative water and land resource programs and projects; formulation and implementation of public water and land resource projects, policies and programs; natural resources information systems. *Mailing Add:* Sch Natural Resources Ohio State Univ Columbus OH 43210-1085

VERVOORT, GERARDUS, b Utrecht, Neth, July 17, 33; Can citizen; m 64; c 2. MATHEMATICS EDUCATION, COMMUNICATIONS SCIENCE. *Educ:* Loras Col, BA, 60; Univ Iowa, MSc, 64, PhD(math), 70. *Prof Exp:* Teacher elem sch, Neth, 53-55 & Indian Sch, Ont, Can, 55-58; itinerant teacher, Dept Northern Affairs, NW Territories, 59-60; teacher sec sch, Dept Indian Affairs, Alta, 60-62; PROF MATH & EDUC, LAKEHEAD UNIV, 70- *Concurrent Pos:* IBM Corp grant, Lakehead Univ, 71-72; consult, Ont Educ Commun Authority, 74-75; dept commun res contract, 74-75 & 75-76. *Mem:* Math Asn Am; Can Asn Prof Educ. *Res:* Instructor effectiveness in the teaching of university and college level mathematics courses; satellite communication for delivery of higher education in remote areas. *Mailing Add:* Dept Math Sci Lakehead Univ Oliver Rd Thunder Bay ON P7B 5E1 Can

VERWOERDT, ADRIAN, b Voorburg, Neth, July 5, 27; US citizen; m 65; c 2. PSYCHIATRY, PSYCHOANALYSIS. *Educ:* Univ Amsterdam, MD, 52; Am Bd Psychiat & Neurol, dipl, 62. *Prof Exp:* Intern, Touro Infirmary, New Orleans, 53-54; resident, Med Ctr, Duke Univ, 54-55 & 58-60, chief resident, 59-60, from instr to assoc prof, 60-71, dir gero-psychiat training, 66-80; dir residency training psychiat, 68-80, dir, Geropsychiat Inst, John Umstead Hosp, Butner, 80-86; PROF PSYCHIAT, MED CTR, DUKE UNIV, 71- *Concurrent Pos:* Consult, Vet Admin Regional Off, Winston-Salem, 61-63, Serv to Aging, NC Dept Pub Welfare, 66-68, Dorothea Dix Hosp, Raleigh, 69-72 & Cherry Hosp, Goldsboro, 72-74; fel psychiat res, Duke Univ Med Ctr, 60-62, NIMH career teacher training award, 64-66; mem, NC Multiversity Comt, 68-75; corresp ed, J Geriat Psychiat, 70-86; instr psychoanal, Duke-Univ NC Psychoanal Inst, 72. *Mem:* Fel Am Psychiat Asn; Am Asn Geriat Psychiat; Am Psychoanal Asn; Pan-Am Med Asn. *Res:* Physical illness and depressive symptomatology; psychological reactions in fatal illness; depression in the aged; sexual behavior in senescence; training in geriatric psychiatry; psychiatric education. *Mailing Add:* Dept Psychiat Duke Univ Med Ctr Durham NC 27710

VESELL, ELLIOT S, b New York, NY, Dec 24, 33. PHARMACOGENETICS, BIOCHEMICAL PHARMACOLOGY. *Educ:* Harvard Univ, MD, 59. *Prof Exp:* Intern pediat, Mass Gen Hosp, 59-60; res assoc human genet & asst physician, Rockefeller Inst, 60-62; asst resident med, Peter Bent Brigham Hosp, 62-63; clin assoc, Nat Inst Arthritis & Metab Dis, 63-65, head sect pharmacogenet, Lab Chem Pharmacol, Nat Heart Inst, 65-68; EVAN PUGH PROF PHARMACOL, GENETICS & MED & CHMN DEPT PHARMACOL, COL MED, HERSHEY MED CTR, PA STATE UNIV, 68-, ASST DEAN GRAD EDUC, 73- *Concurrent Pos:* William N Creasy vis prof clin pharmacol, Sch Med, George Washington Univ, 75, Sch Med, Univ Conn, 79; Pfizer lectr clin pharmacol, Univ Iowa, Med Col Pa, 76, Univ Conn, 77, Med Col Wis, 78, Georgetown Univ, 79, State Univ NY, Buffalo, 81, Morehouse Sch Med, Atlanta, Ga & Col Med, Howard Univ, Washington, DC, 85; Wellcome vis prof basic med sci, Col Physicians & Surgeons, Columbia Univ, NY, 82. *Honors & Awards:* Samuel James Meltzer Award, 67; Am Soc Pharmacol & Exp Therapeut Award, 71; Julius W Sturmer Mem Lectr; Allan D Bass Lectr, dept pharmacol, Sch Med, Vanderbilt Univ, Nashville, Tenn, 85. *Mem:* Harvey Soc; Am Soc Human Genetics; Soc Exp Biol & Med; Am Soc Clin Invest; Am Fedn Clin Res; Am Asn Physicians; Am Soc Clin Pharmacol & Therapeut; Asn Med Sch Pharmacol. *Res:* Multiple molecular forms of enzymes; effect of heredity and environmental factors on disposition of drugs; biochemical pharmacology. *Mailing Add:* Hershey Med Ctr Pa State Univ Col Med Hershey PA 17033

VESELY, DAVID LYNN, b Omaha, Nebr, Mar 6, 43; c 4. ENDOCRINOLOGY & METABOLISM. *Educ:* Creighton Univ, BS, 67; Univ Ariz, MD & PhD(Physiol), 72. *Prof Exp:* Fel, NIH, Univ Ariz, 68-72; intern med, Univ Miami Affil Hosp, 72-73, resident internal med, 73-74, fel endocrinol, 74-76; asst prof med, Univ Miami Med Sch, 76-78; from asst prof to assoc prof, 78-83, PROF MED, UNIV ARK MED SCI, 83- *Concurrent Pos:* Fel, NIH, Univ Miami, 74-76; actg chief endocrinol, Univ Ark Med Sci, 78-85; sr int scholar, Fogarty Int Ctr, Nat Inst Health, Nice, France, 84-85. *Honors & Awards:* Lange Award, Lange Publ Co, 72; Mosby Award, Mosby Publ Co, 72. *Mem:* Endocrine Soc; NY Acad Sci; Am Asn Cancer Res; Am Diabetes Asn; Am Fedn Clin Res; Southern Soc Clin Invest. *Res:* Atriol natriuretic factor and prohormone; atriol natriuretic factor peptides in normal and pathophysiologic states with respect to their circulating concentrations, physiologic effects and cellular mechanism of action. *Mailing Add:* Div Endocrinol & Metab Univ Ark Med Sci 4301 W Markham St Little Rock AR 72205

VESLEY, DONALD, b Astoria, NY, Nov 7, 32; m 62; c 1. ENVIRONMENTAL HEALTH. *Educ:* Cornell Univ, BS, 55; Univ Minn, MS, 58, PhD(environ health), 68. *Prof Exp:* From instr to assoc prof, 60-78, PROF PUB HEALTH, UNIV MINN, MINNEAPOLIS, 78-, DIR ENVIRON HEALTH & SAFETY, 80- *Mem:* Am Soc Microbiol; Am Pub Health Asn; Am Biol Safety Asn. *Res:* Environmental microbiology. *Mailing Add:* Univ Minn Rm W-136 Boynton Health Serv Minneapolis MN 55455

VESSEL, EUGENE DAVID, b Mt Olive, Ill, Dec 1, 27; m 46; c 2. ORGANIC POLYMER CHEMISTRY. *Educ:* Univ Ill, BS, 57; Univ Iowa, MS, 59, PhD(org chem), 60. *Prof Exp:* Res chemist, Chevron Res Corp, Standard Oil Calif, 60-61; sr res chemist, United Tech Ctr, United Aircraft Corp, 61-67; RES SPECIALIST NONMETALLIC MAT, COM AIRPLANE GROUP, BOEING CO, 67- *Mem:* Am Chem Soc; Sigma Xi. *Res:* Adhesives; elastomers; plastics; composite materials; materials research; fire research and technology. *Mailing Add:* 12521 163rd Ave SE Snohomish WA 98290-8802

VESSELINOVITCH, STAN DUSHAN, b Zagreb, Yugoslavia, Feb 20, 22; Can citizen; m 47; c 3. PHYSIOLOGY, ONCOLOGY. *Educ:* Univ Belgrade, DVM, 49; Univ Toronto, MVSc, 57, DVSc, 58. *Prof Exp:* Lectr physiol, Univ Belgrade, 49-51; res assoc oncol, Univ Toronto, 52-53, asst prof, 54-58; res assoc, Michael Reese Hosp, Chicago, Ill, 59-61, asst dir oncol, 62-64; assoc prof, Chicago Med Sch, 64-69; assoc prof radiol, 69-72, PROF RADIOL & PATH, FRANKLIN MCLEAN MEM RES INST, UNIV CHICAGO, 72-, DIR, SECT RADIATION BIOL & EXP ONCOL, 80- *Concurrent Pos:* Res assoc fel, Nat Cancer Inst Can, 55-57, prin investr res grant, 57-58; co-investr res grant, Nat Cancer Inst, 59-64, prin investr, 64-69, res contracts, 69- *Mem:* AAAS; Am Vet Med Asn; Am Asn Cancer Res; NY Acad Sci. *Res:* Animal physiology; environmental carcinogenesis; factors and mechanisms modifying carcinogenesis. *Mailing Add:* Dept Radiol & Pathol Univ Chicago Pritzker Sch Med 5841 Maryland Ave Chicago IL 60637

VESSEY, ADELE RUTH, b South Charleston, WVa, 47. IMMUNOLOGY, VIROLOGY. *Educ:* Ohio Univ, BS, 69, MS, 72; Case Western Reserve Univ, PhD(microbiol), 76. *Prof Exp:* Fel immunol, Cleveland Clin Found, 76-80. *Mem:* Am Soc Microbiol; Int Asn Comp Res Leukemia & Related Dis; Sigma Xi. *Res:* Interaction of leukemia viruses with the immunological system; autoaggressive and immunosuppressive properties of lymphomas. *Mailing Add:* RD 3 Box 194B Elverston PA 19520

VESSEY, STEPHEN H, b Stamford, Conn, Mar 7, 39; m 62; c 2. ANIMAL BEHAVIOR, ECOLOGY. *Educ:* Swarthmore Col, BA, 61; Pa State Univ, MS, 63, PhD(zool), 65. *Prof Exp:* Biologist, NIH, 65-69; from asst prof to assoc prof, 69-80, PROF BIOL SCI, BOWLING GREEN STATE UNIV, 80- *Mem:* Am Soc Primatologists; Animal Behav Soc; Am Soc Mammal; AAAS; Int Primatol Soc; Sigma Xi. *Res:* Social behavior and population dynamics; field studies of non-human primates and small mammals. *Mailing Add:* Dept Biol Bowling Green State Univ Bowling Green OH 43403

VESSEY, THEODORE ALAN, b St Paul, Minn, June 16, 38; m 65; c 2. MATHEMATICAL ANALYSIS. *Educ:* Univ Minn, BA, 60, PhD(math), 66. *Prof Exp:* Assoc res engr, Honeywell, Inc, 62-63; asst prof math, Univ Wis-Milwaukee, 66-70; asst prof, 70-72, ASSOC PROF MATH, ST OLAF COL, 72- *Concurrent Pos:* Adj prof, Naval Grad Sch, 77; vis scholar, Stanford Univ, 77-78; vis prof, Univ Lund, Sweden, 86-87. *Mem:* Am Math Soc; Math Asn Am. *Res:* Cluster set theory in determination of boundary behavior of complex functions. *Mailing Add:* Dept Math St Olaf Col Northfield MN 55057

VESSOT, ROBERT F C, b Montreal, Que, Apr 16, 30; c 3. PHYSICS. *Educ:* McGill Univ, BA, 51, MSc, 54, PhD(physics), 57. *Prof Exp:* Mem, Div Sponsored Res Staff, Mass Inst Technol, 56-60; mgr maser res & develop, Varian Assocs, 60-67 & Hewlett-Packard Co, 67-69; physicist, 69-72, SR PHYSICIST, SMITHSONIAN ASTROPHYS OBSERV, CAMBRIDGE, 72- *Honors & Awards:* Exceptional Sci Achievement Medal, NASA, 78. *Mem:* Am Phys Soc; Sigma Xi. *Res:* Physical electronics; noise in electron beams; atomic beams; atomic resonance physics; atomic hydrogen maser frequency standard; tests of gravitation and relativity. *Mailing Add:* 334 Ocean Ave Marblehead MA 01945

VEST, CHARLES MARSTILLER, b Morgantown, WVa, Sept 9, 41; m 63; c 2. MECHANICAL ENGINEERING, OPTICS. *Educ:* Univ WVa, BSME, 63; Univ Mich, MSE, 64, PhD(mech eng), 67. *Prof Exp:* From asst prof to assoc prof, Univ Mich, Ann Arbor, 67-72, assoc dean acad affairs, 81-87, prof mech eng & dean, Col Eng, 87-89, provost, 89-90; PRES, MASS INST TECHNOL, CAMBRIDGE, 90- *Concurrent Pos:* Vis assoc prof aero & elec eng, Stanford Univ, 74-75; assoc ed, J Optical Soc Am, 82-83; trustee, Environ Res, Inst Mich, WGBH Educ Found & New Eng Aquarium; coop mem, Woods Hole Oceanog Inst. *Mem:* Soc Mech Engrs; fel Optical Soc Am; fel Am Acad Arts & Sci; AAAS; Sigma Xi. *Res:* Heat transfer and fluid mechanics; hydrodynamic stability; optical holography and coherent optical measurement techniques; computer tomography. *Mailing Add:* 111 Memorial Dr Cambridge MA 02142-1348

VEST, FLOYD RUSSELL, b Orland, Calif, Feb 12, 34; m 55; c 2. MATHEMATICS EDUCATION. *Educ:* ECent State Col, BSEd, 56; Univ Okla, MA, 59; NTex State Univ, EdD(math), 68. *Prof Exp:* Instr math, ETex State Univ, 59-61; asst prof, 61-74, ASSOC PROF MATH, NTEX STATE UNIV, 74- *Mem:* Math Asn Am. *Res:* Learning theory; curriculum. *Mailing Add:* 1103 Brighton Terr Denton TX 76201

VEST, HYRUM GRANT, JR, b Salt Lake City, Utah, Sept 23, 35; m 58; c 5. PLANT PATHOLOGY, PLANT GENETICS. *Educ:* Utah State Univ, BS, 60, MS, 65; Univ Minn, PhD(plant path), 67. *Prof Exp:* Res plant pathologist, Crops Res Div, Agr Res Serv, USDA, Md, 67-70; assoc prof hort, Mich State Univ, 70-76; prof hort & head dept, Okla State Univ, 76-83; head hort sci, Tex A&M Univ, 83-89; HEAD, PLANTS, SOILS & BIOMETEOROL DEPT, UTAH STATE UNIV, 89- *Concurrent Pos:* Mem, Nat Plant Genetic Resources Bd, USDA, 82-88. *Mem:* fel Am Soc Hort Sci; Am Soc Agron. *Res:* Genetics of nodulation and nitrogen fixation in soybean; breeding and genetics of onions, lettuce and asparagus. *Mailing Add:* Plants Soils & Biometeorol Logan Sta UT 84321-4820

VEST, ROBERT W(ILSON), b Lawrenceburg, Ind, Oct 17, 30; m 52, 82; c 3. PHYSICAL CHEMISTRY, MATERIALS ENGINEERING. *Educ:* Purdue Univ, BS, 52; Iowa State Univ, PhD, 57. *Prof Exp:* Chemist, Nat Lead Co Ohio, 52-53; asst, Ames Lab & Inst Atomic Res, Iowa State Univ, 53-57; res chemist, Monsanto Chem Co, 57-61; sr scientist, Systs Res Labs, Inc, 61-66; prof, 66-72, BASIL S TURNER PROF ENG, PURDUE UNIV, WEST LAFAYETTE, 72- *Concurrent Pos:* Consult, CTS Microelectronics, Inc, 66 & Gen Motors Corp, 72- *Honors & Awards:* Ross Coffin Purdy Award, Am Ceramic Soc, 67; Tech Achievement Award, Int Soc Hybrid Microelectronics, 81. *Mem:* Am Soc Metals; fel Am Ceramic Soc; Nat Inst Ceramic Engrs; Int Soc Hybrid Microelectronics. *Res:* Electrical properties of metal oxides; electroceramics; transport properties on non-metallic materials; hybrid microelectronics. *Mailing Add:* MSEE Bldg Purdue Univ West Lafayette IN 47907

VESTAL, BEDFORD MATHER, b Gainesville, Tex, Mar 8, 43; m 65; c 3. ANIMAL BEHAVIOR, BEHAVIORAL ECOLOGY. *Educ:* Austin Col, BA, 65; Mich State Univ, MS, 67, PhD(zool), 70. *Prof Exp:* Instr biol, Univ Mo, St Louis, 69-70, asst prof, 70-73; asst prof, 73-80, ASSOC PROF ZOOL, UNIV OKLA, 80- *Concurrent Pos:* Res assoc, Mich State Univ, 70-71; res cur, Oklahoma City Zoo, 73-76; vis scientist zool, Monash Univ, 83-84; fac admin fel, Col Lib Studies, Univ Okla, 89- *Mem:* AAAS; Animal Behav Soc; Sigma Xi; Am Soc Mammal. *Res:* Comparative social behavior of mammals; behavioral ecology of rodents. *Mailing Add:* Sutton Hall Dept Zool Univ Okla Norman OK 73019

VESTAL, CHARLES RUSSELL, b Moran, Kans, Feb 21, 40; m 62; c 2. CHEMICAL ENGINEERING. *Educ:* Colo Sch Mines, BS, 62, MS, 69, PhD(chem eng), 73. *Prof Exp:* Process design engr, Continental Oil Co, 62-63; assoc engr, Marathon Oil Co, 65-69, engr, 69-72, advan engr, 72-74, res engr chem eng, 74-77, adv res engr, 77-80, res engr, 80-82, sr res engr, Denver Res Ctr, 82-84, ADVAN SR ENGR, PETROL TECHNOL CTR, MARATHON OIL CO, 84- *Concurrent Pos:* Adj assoc prof, Colo Sch Mines, 75-78; lectr, Univ Colo, Boulder, 75-84. *Mem:* Am Inst Chem Engrs; Soc Petrol Engrs. *Res:* Geostatistics, supercomputing and petroleum reservoir simulation. *Mailing Add:* Petrol Technol Ctr Marathon Oil Co PO Box 269 Littleton CO 80160-0269

VESTAL, CLAUDE KENDRICK, b High Point, NC, Mar 11, 16; m 46. METEOROLOGY, CLIMATOLOGY. *Educ:* Guilford Col, AB, 46. *Prof Exp:* Observer, Weather Bur, NC, 37-39, observer & forecaster, DC, 39-42, proj head climat, NY, 42-43, sect head, DC, 43-50; foreign serv staff officer, Dept State, Monrovia, Liberia, 50-52; sect head, Weather Bur, DC, 52-56, regional climatologist, Tex, 57-71; PVT CONSULT, 71- *Res:* Application of modern statistical methods to climatological data analysis, for design purposes and evaluation of operational risks. *Mailing Add:* 1720 Gun Wood Pl Crofton MD 21114

VESTAL, J ROBIE, b Orlando, Fla, Oct 16, 42; m 67, 85. MICROBIOLOGY. *Educ:* Hanover Col, BA, 64; Miami Univ, MS, 66; NC State Univ, PhD(microbiol), 69. *Prof Exp:* Fed Water Pollution Control Admin res assoc microbiol, Syracuse Univ, 69-71; from asst prof to assoc prof, 71-83, PROF BIOL SCI, UNIV CINCINNATI, 83- *Concurrent Pos:* Vis prof, Fla State Univ, 80-81. *Mem:* Sigma Xi; Arctic Inst NAm; Am Soc Microbiol; AAAS; Am Soc Limnol & Oceanog; fel Am Acad Microbiol. *Res:* Functional microbial ecology of aquatic, marine and terrestrial habitats; microbial metabolism; effects of pollutants on microbial habitats. *Mailing Add:* Dept Biol Sci Univ Cincinnati Cincinnati OH 45221-0006

VESTAL, ROBERT ELDEN, b Auburn, Calif, Oct 25, 45; m 68; c 2. CLINICAL PHARMACOLOGY, GERONTOLOGY. *Educ:* Stanford Univ, AB, 67; Univ Calif, San Francisco, MD, 71. *Prof Exp:* Intern med, Univ Col Med Ctr, 71-72, resident, 72-73; clin assoc, Geront Res Ctr, Nat Inst Aging, NIH, 73-75; fel clin pharmacol, Vanderbilt Univ, 75-77; staff physician, 77-85, CHIEF CLIN PHARMACOL & GERONTOL UNIT & VET ADMIN MED CTR, 78-, ASSOC CHIEF STAFF RES & DEVELOP, 85- *Concurrent Pos:* Asst med, Johns Hopkins Univ, 73-75; vis physician, Baltimore City Hosp, 73-75; instr pharmacol & med, Vanderbilt Univ, 76-77; asst & prof med, Univ Wash, 77-81, assoc prof med, 81-; mem, Adv Panel Geriat, US Pharmacopeial Conv, 81-; mem, Res Adv Group, Vet Admin, 83-87; mem, Comt Chem Toxicity & Aging, Nat Res Coun, Nat Acad Sci, 86-87. *Honors & Awards:* Arthur S Flemming Award, 82. *Mem:* Am Soc Pharmacol & Exp Therapeut; Am Soc Clin Pharmacol & Therapeut; fel Am Col Physicians; Brit Pharmacol Soc; fel Gerontol Soc Am; Am Fedn Clin Res. *Res:* Effects of aging on drug metabolism and drug response; the clinical pharmacology of methylxanthines; drug interactions; stable isotope methodology in clinical pharmacology research; research administration. *Mailing Add:* Res Serv 151 Vet Admin Med Ctr 500 W Fort St Boise ID 83702

VESTER, JOHN WILLIAM, b Cincinnati, Ohio, June 5, 24; c 5. BIOCHEMISTRY, INTERNAL MEDICINE. *Educ:* Univ Cincinnati, MD, 47. *Prof Exp:* Porter fel res med, Hosp Univ Pa, 54-56; from asst prof to assoc prof biochem & nutrit, Grad Sch Pub Health, Univ Pittsburgh, 56-61, from asst prof to assoc prof med, Sch Med, 56-67, asst prof biochem, 61-67; assoc prof biochem, 67-74, assoc prof med, 67-71, asst dean, 81-85, PROF MED, COL MED, UNIV CINNINATI, 71-, PROF BIOCHEM, 74-, ASSOC DEAN, 85-; dir res, Good Samaritan Hosp, 67- *Concurrent Pos:* Asst ward chief, Hosp Univ Pa, 54-56; chief sect isotopes & metab, Vet Admin Hosp, Pittsburgh, Pa, 61-67 & assoc chief of staff, 62-67. *Mem:* Endocrine Soc; Am Fedn Clin Res; Am Diabetes Asn; fel Am Col Physicians; fel Am Col Cardiol. *Res:* Diabetes and mechanism of insulin actions; alcoholism; obesity; muscle diseases. *Mailing Add:* Good Samaritan Hosp Clifton-Dixmyth Ave Cincinnati OH 45220

VESTLING, CARL SWENSSON, b Northfield, Minn, May 6, 13; m 38; c 3. BIOCHEMISTRY. *Educ:* Carleton Col, BA, 34; Johns Hopkins Univ, PhD(biochem), 38. *Prof Exp:* From instr to prof chem, Univ Ill, Urbana, 38-63; prof biochem, Univ Iowa, 63-81, head dept, 63-76; VIS PROF BIOCHEM, UNIV ARIZ, 82- *Concurrent Pos:* Guggenheim fel, Nobel Inst, Sweden, 54. *Mem:* AAAS; Am Chem Soc; Am Soc Biol Chemists; Soc Exp Biol & Med; Brit Biochem Soc. *Res:* Isolation, structure, mechanism, lactate and malate dehydrogenases and certain other liver and hepatoma enzymes. *Mailing Add:* 7981 N Sendero Uno Tucson AZ 85704

VESTLING, MARTHA MEREDITH, b Urbana, Ill, Sept 4, 41. ORGANIC CHEMISTRY. *Educ:* Oberlin Col, AB, 62; Northwestern Univ, Evanston, PhD(chem), 67. *Prof Exp:* Fel, Univ Fla, 67-68; res assoc, Vanderbilt Univ, 68-69; asst prof chem, 70-75, ASSOC PROF CHEM, STATE UNIV NY, BROCKPORT, 75- *Mem:* Am Chem Soc; AAAS; Am Soc Mass Spectrometry; Asn Women Sci; Sigma Xi. *Res:* Environmental fate of organic chemicals; mass spectrometry; chromatography. *Mailing Add:* Dept Chem & Biochem Univ Md Baltimore MD 21228

VETELINO, JOHN FRANK, b Westerly, RI, Oct 17, 42; m 67; c 2. ELECTRICAL ENGINEERING, PHYSICS. *Educ:* Univ RI, BS, 64, MS, 66, PhD(elec eng), 69. *Prof Exp:* From asst prof to assoc prof, 69-79, PROF ELEC ENG, UNIV MAINE, ORONO, 79- *Concurrent Pos:* Consult, Navy Underwater Systs Ctr, Allied Chem Corp. *Mem:* Am Phys Soc; Acoust Soc Am; sr mem Inst Elec & Electronics Engrs. *Res:* Microwave acoustics; surface acoustic waves; solid state phenomena; lattice dynamics and related thermodynamic optical and electronic properties of solids; phase transitions and impurity studies; electromagnetic and acoustic wave propagation; switched capacitor filters; sonar signal processing. *Mailing Add:* Dept Elec Eng Univ Maine Orono ME 04469

VETHAMANY, VICTOR GLADSTONE, b Servaikaramadam, India, Feb 7, 35; Can citizen; m 62; c 1. BIOLOGY, ANATOMY. *Educ:* Univ Madras, BA, 54, MA, 57; Univ Toronto, PhD(zool), 65. *Prof Exp:* Asst lectr biol, Univ Madras, 57-59; high sch teacher, Ethiopia, 60-61; demonstr zool, Univ Toronto, 61-64; res assoc path, Isaac Albert Res Inst, Kingsbrook Med Ctr, 65-67; asst prof, 67-73, ASSOC PROF ANAT, MED SCH, DALHOUSIE UNIV, 73- *Concurrent Pos:* Fel, Isaac Albert Res Inst, Kingsbrook Med Ctr, NY, 65-67; Med Res Coun vis scientist, Inst Cellular Path, Paris, 73. *Honors & Awards:* Nat Sci Prize, 54; Bourne Prize, 67. *Mem:* AAAS; Am Asn Anat; Can Asn Anat; NY Acad Sci; Can Soc Cell Biol. *Res:* Ultrastructure and histochemistry of blood cells and blood forming organs; ultrastructural studies on Niemann-Pick disease; comparative ultrastructural studies of blood in vertebrates and invertebrates; chemotaxis and cell injury. *Mailing Add:* Dept of Anat Dalhousie Univ Med Sch Halifax NS B3H 4H7 Can

VETTE, JAMES IRA, b Evanston, Ill, Mar 4, 27; m 51; c 4. PHYSICS. *Educ:* Rice Univ, BS, 52; Calif Inst Technol, PhD(physics), 58. *Prof Exp:* Jr geophysicist, Humble Oil Co, 52; asst physics, Calif Inst Technol, 52-54; staff scientist, Sci Res Lab, Convair Div, Gen Dynamics Corp, 58-62, sr staff scientist, Sci Res Lab, Astronaut Div, 62; mgr nuclear physics, Vela Satellite Prog, Aerospace Corp, 62-63, staff scientist, Space Physics Lab, 63-67; dir, 67-84, SR STAFF SCIENTIST, NAT SPACE SCI DATA CTR, NASA GODDARD SPACE FLIGHT CTR, 84- *Honors & Awards:* Except Achievement Award, NASA, 69 & Goddard Group Achievement Award, 79; Scostep Award, 84. *Mem:* AAAS; Am Phys Soc; Am Geophys Union; Inst Elec & Electronics Engrs; Sigma Xi. *Res:* Synchrotron; meson physics; high altitude radiation with balloons; solar physics; high energy physics; cosmic rays; magnetospheric physics; satellite measurements; space physics. *Mailing Add:* 12502 White Dr Silver Spring MD 20904

VETTER, ARTHUR FREDERICK, b De Witt, Iowa, July 26, 18; m 41; c 4. CHEMICAL & NUCLEAR ENGINEERING. *Educ:* Coe Col, BA, 39; NC State Col, MS, 56. *Prof Exp:* Meteorologist, US Air Force, 47-54, res adminr, 56-57, asst prof physics & nuclear eng, Air Force Inst Technol, 57-61, mem staff, Hq, US Air Force, 61-62; ASSOC PROF CHEM ENG, UNIV IOWA, 62- *Concurrent Pos:* With dept nuclear eng, Univ Ill, Urbana, 69-71. *Mem:* Am Inst Chem Engrs; Am Nuclear Soc; Fine Particle Soc; Sigma Xi. *Res:* Radiation protection; particle morphology; abrasive wear. *Mailing Add:* Dept of Chem Eng Univ of Iowa Iowa City IA 52240

VETTER, BETTY M, b Center, Colo, Oct 25, 24; m 51; c 3. SCIENCE & ENGINEERING MANPOWER. *Educ:* Univ Colo, Boulder, BA, 44; Stanford Univ, MA, 48. *Prof Exp:* Analytical chemist, Shell Develop Co, 44-45; meteorologist, US Naval Women's Reserve, 45-46; instr speech, Fresno State Col, Calif, 48-50; instr, Far East Div, Univ Calif, 50-51; adj prof, Am Univ, 52-64; EXEC DIR, SCI MANPOWER COMN, COMN PROFESSIONALS SCI & TECHNOL, 63- *Concurrent Pos:* Prin investr res projs, NSF, 76-90, Off Technol Assessment, 85-87. *Res:* All aspects of supply and demand for scientists and engineers, including participation of women and minorities. *Mailing Add:* Comn Professionals Sci & Technol 1500 Massachusetts Ave NW Suite 831 Washington DC 20005

VETTER, JAMES LOUIS, b St Louis, Mo, Jan 26, 33; m 54; c 2. FOOD TECHNOLOGY. *Educ:* Washington Univ, AB, 54; Univ Ill, MS, 55, PhD(food technol), 58. *Prof Exp:* Food technologist, Monsanto Chem Co, 58-63; mgr res & develop labs, Keebler Co, Ill, 63-67, dir res & develop, 67-72; corp dir res & develop, Confectionery, Nut & Snack Prod Lab, Planters/Curtiss Div, Standard Brands, Inc, 72-74, vpres & tech dir res & develop lab, 74-75; vpres food sci & technol, Triticale Industs, Inc, Amarillo, Tex, 75-77; dir res, 77-80, VPRES TECH, AM INST BAKING, 81- *Mem:* Am Asn Cereal Chem (secy, 83-85, pres-elect, 85-86, pres, 86-87); Inst Food Technologists. *Res:* Basic, nutrition and applied research on cereal grains and their utilization in processed, grain-based foods. *Mailing Add:* Am Inst of Baking 1213 Bakers Way Manhattan KS 66502

VETTER, RICHARD J, b Castlewood, SDak, July 17, 43; m 65; c 2. HEALTH PHYSICS, RADIOBIOLOGY. *Educ:* SDak State Univ, BS, 65, MS, 67; Purdue Univ, PhD(bionucleonics), 69. *Prof Exp:* Asst prof biol, Point Park Col, Pittsburgh, 69-70; from asst prof to prof bionucleonics, 70-80, PROF BIOPHYSICS, PURDUE UNIV, 80-; HEAD OCCUP SAFETY SECT, MAYO CLIN, 90- *Concurrent Pos:* Asst radiol control officer, Purdue Univ, 70-80; asst radiation safety officer, Mayo Clin, 80-83, radiation safety officer, 83-; ed-in-chief, Health Physics, 88- *Mem:* AAAS; Health Physics Soc; Int Radiation Protection Asn; Soc Nuclear Med; Sigma Xi; Am Asn Physicists Med. *Res:* Biological effects and dosimetry of ionizing and nonionizing radiation including radioecology. *Mailing Add:* Radiation Safety Off Mayo Clin Rochester MN 55905

VETTER, RICHARD L, b Henry Co, Ill, Dec 28, 30; m 50; c 2. ANIMAL SCIENCE, ENVIRONMENTAL SCIENCES. *Educ:* Univ Ill, BS, 53, MS, 57; Univ Wis, PhD(nutrit, biochem), 60. *Prof Exp:* Asst, Univ Ill, 55-57; asst biochem, Univ Wis, 57-60, fel, 60-61; res nutritionist, Hess & Clark Co Div, Richardson-Merrill, Inc, 61-62; from asst prof to prof animal sci, Iowa State Univ, 62-78; DIR RES, A O SMITH HARVESTORE PROD, INC, 78- *Concurrent Pos:* Researcher, Inst Animal Physiol, Cambridge, Eng, 68-69; mem Coun Agr Sci & Technol. *Mem:* Am Soc Animal Sci; Am Soc Dairy Sci; Am Inst Nutrit; Fedn Am Soc Exp Biol; Am Regist Prof Animal Scientists. *Res:* Nutrition and metabolic disorders; animal production; chemistry, nutrition and utilization of plant, animal and by-product wastes; bioenergy production and nutrient conservation. *Mailing Add:* A O Smith Harvestore Prod Inc 345 Harvestore Dr De Kalb IL 60115

VETTER, WILLIAM J, b Grenfell, Sask. ELECTRICAL ENGINEERING. *Educ:* Univ Toronto, BASc, 59; Univ Waterloo, MASc, 61, PhD(elec eng), 65. *Prof Exp:* Lectr elec eng, Univ Waterloo, 61-65, from asst prof to assoc prof, 65-71; prof, 71-77, PROF ENG & APPL SCI, MEM UNIV NFLD, 77- *Mem:* Inst Elec & Electronics Engrs. *Res:* Theoretical and practical aspects of dynamical systems, particularly modeling, analysis, control and simulation; instrumentation and computer control of industrial systems and processes. *Mailing Add:* Dept Eng Mem Univ Nfld St John's NF A1B 3X5 Can

VETTERLING, JOHN MARTIN, b Fitzsimons, Colo, July 21, 34; m 57; c 3. PROTOZOOLOGY, PARASITOLOGY. *Educ:* Colo State Univ, BS, 56, MS, 62; Univ Ill, PhD(vet med sci), 65, Univ Calif, Davis, MPVM, 85. *Prof Exp:* Sr res parasitologist, Beltsville Parasitol Lab, Sci & Educ Admin-Agr Res, 65-76, asstt area dir, Sci & Educ Admin-Agr Res, 76-81, dir, Rocky Mountain Area, Agr Red Serv, USDA, 81-86; PVT CONSULT, PARASITOLOGIC SERV, FT COLLINS, 86- *Mem:* Am Soc Parasitol; World Asn Advan Vet Parasitol; Sigma Xi; Am Asn Vet Parasitol. *Res:* Biology and taxonomy of coccidia; patholgenesis of intestinal parasites. *Mailing Add:* PO Box 475 Ft Collins CO 80522-0475

VETTERLING, WILLIAM THOMAS, b Greenfield, Mass, July 16, 48; m. SOLID STATE PHYSICS. *Educ:* Amherst Col, BA, 70; Harvard Univ, MA, 71, PhD(physics), 76. *Prof Exp:* Asst prof, 76-81, assoc prof physics, Harvard Univ, 81-84; sr scientist, 84-86, PRIN SCIENTIST, POLAROID CORP, 86- *Mem:* Am Phys Soc; Inst Elec & Electronic Eng; Soc Photo-Optical Instrument Engrs. *Res:* Mossbauer effect, solid state physics. *Mailing Add:* 35 Turning Mill Rd Lexington MA 02173

VEUM, TRYGVE LAURITZ, b Virogua, Wis, Mar 16, 40; m 67; c 2. ANIMAL NUTRITION, VETERINARY PHYSIOLOGY. *Educ:* Univ Wis, BS, 62; Cornell Univ, MS, 65, PhD(animal nutrit, vet physiol & path), 68. *Prof Exp:* From asst prof to assoc prof, 67-80, PROF ANIMAL NUTRIT, UNIV MO-COLUMBIA, 80- *Concurrent Pos:* Mem, Animal Nutrit Res Coun. *Honors & Awards:* Young Researcher Award, Am Soc Animal Sci, 80. *Mem:* Am Soc Animal Sci; Am Inst Nutrit. *Res:* Swine nutrition. *Mailing Add:* Animal Sci Res Ctr Univ Mo Columbia MO 65211

VEVERKA, JOSEPH F, b Pelrimov, Czech, June 8, 41; Can citizen; m 69; c 1. PLANETARY SCIENCE, ASTRONOMY. *Educ:* Queen's Univ, Kingston, Ont, BSc, 64, MSc, 65; Harvard Univ, MA, PhD(astron), 70. *Prof Exp:* Res assoc, 70-72, sr res assoc space sci, 72-74, asst prof, 74-77, ASSOC PROF ASTRON, CORNELL UNIV, 77- *Mem:* Am Astron Soc; Am Geophys Union; Meteoritical Soc; Royal Astron Soc Can. *Res:* Spacecraft investigation of planetary and satellite surfaces; evolution of planets and satellites. *Mailing Add:* Dept Astron Cornell Univ 426 Space Sci Bldg Ithaca NY 14853

VEZERIDIS, MICHAEL PANAGIOTIS, b Thessaloniki, Greece, Dec 16, 43; m 78; c 2. SURGICAL ONCOLOGY. *Educ:* Nat Univ Athens Med Sch, MD, 67. *Hon Degrees:* MA, Brown Univ, 89. *Prof Exp:* Attending surgeon, Roswell Park Mem Inst, 81-82, staff surgeon, Vet Admin Med Ctr, Providence, RI, 82-84, asst prof surg, Brown Univ, 82-88; CHIEF SURGEON ONCOL, VET ADMIN MED CTR, BROWN UNIV, RI, 84-, ASSOC CHIEF SURG, 86- *Concurrent Pos:* Prin investr, Heterogeneity & Metastasis in Human Pancreatic Cancer, Vet Admin Merit Rev Grant, 83-; consult, Roger Williams Gen Hosp, RI, 86-, RI Hosp, Providence, 87-; vis prof, Univ Patras Med Sch, Greece, 88; assoc prof, Brown Univ, 88- *Mem:* NY Acad Sci; Am Col Surgeons; Asn Acad Surg; Am Soc Clin Oncol; Soc Surg Oncol; Am Asn Cancer Res. *Res:* Heterogeneity and metastasis of human pancreatic cancer and other solid human tumors; human tumor metastasis. *Mailing Add:* Surgical Serv 112 Vet Admin Med Ctr Providence RI 02908

VEZINA, CLAUDE, b Oka, Que, Feb 19, 26; m 52; c 3. MICROBIAL GENETICS & BIOCHEMISTRY. *Educ:* Univ Montreal, BA, 46, BSA, 50, MSc, 52; Univ Wis, PhD(bact), 56. *Prof Exp:* Prof bact & biochem, Oka Agr Inst, Univ Montreal, 50-60; head gen microbiol, Res Labs, Ayerst, McKenna & Harrison, Ltd, 60-64, assoc dir res-microbiol, 64-80, dir microbiol, Ayerst Labs, Am Home Prod Corp, 80-83; asst dir, Inst Armand-Frappier, Univ Quebec, 83-88; INST ROSELL INC, 88- *Concurrent Pos:* Nat Res Coun Can res assoc, 57; asst prof, Fac Med, Univ Montreal, 64- *Mem:* Am Soc Microbiol; Mycol Soc Am; Soc Indust Microbiol; Am Chem Soc; Soc Gen Microbiol; Can Soc Microbiol; fel Chem Inst Can. *Res:* Physiology; transformation of steroids; antibiotics; heterokaryosis in fungi; recombination in streptomyces and nocardia; nocardiophages; antitumor agents; molecular biology of streptomycetes; fermentation; lactic starters. *Mailing Add:* 11 St-Sulpice St Oka PQ J0N 1E0 Can

VEZIROGLU, T NEJAT, b Istanbul, Turkey, Jan 24, 24; m 61; c 2. HEAT TRANSFER, NUCLEAR ENGINEERING. *Educ:* Univ London, BSc, 46, PhD(heat transfer), 51, Imp Col, dipl, 47. *Prof Exp:* Sci adv, Off Soil Prod, Ankara, Turkey, 53-55, assoc dir steel silos, 55-57; eng consult, 57-59; tech dir, M K Veziroglu Construct Co Ltd, Istanbul, 59-62; assoc prof mech eng, 62-66, chmn dept, 71-75, assoc dean res, 75-79, PROF MECH ENG, UNIV MIAMI, 66-, DIR, CLEAN ENERGY RES INST, 74- *Concurrent Pos:* Ed, Int J Hydrogen Energy. *Mem:* Fel AAAS; fel Am Soc Mech Engrs; Am Inst Aeronaut & Astronaut; fel Brit Inst Mech Engrs; Int Asn Hydrogen Energy (pres). *Res:* Thermal conductance of metal surfaces in contact; two-phase flow instabilities; solar and hydrogen energy. *Mailing Add:* Col Eng Univ Miami PO Box 248294 Coral Gables FL 33124

VIA, FRANCIS ANTHONY, b Frostburg, Md; m 70; c 2. CHEMISTRY, SYNTHETIC INORGANIC & ORGANOMETTALLI CHEMISTRY. *Educ:* WVa Univ, BS, 65; Ohio State Univ, MS, 67, PhD(phys org chem), 70. *Prof Exp:* Teaching asst chem, Ohio State Univ, 65-66, res fel phys org chem, 66-70; from res chemist to sr res chemist, 70-75, supvr org res, 75-76, asst to res dir, 76-78, mgr inorg res, Stauffer Chem Co, 78-82; MGR CATALYSTS & MATS RES, AKZO CHEM, INC, 82- *Mem:* Am Chem Soc; Sigma Xi; AAAS; Am Ceramic Soc; Mats Res Soc. *Res:* Zeigler-Natta catalysis; materials research-electronic and structural; homogenous and heterogenous catalysis; main group inorganic chemical sulfur, phosphorus; sol-gel synthesis-oxide powders and films; specialty chemicals and agricultural chemicals. *Mailing Add:* AKZO Chem Inc Dobbs Ferry Res Ctr Dobbs Ferry NY 10522-1697

VIA, GIORGIO G, physics, for more information see previous edition

VIA, WILLIAM FREDRICK, JR, b Ironton, Ohio, Dec 27, 20; m 47; c 2. PEDODONTICS. *Educ:* Ohio State Univ, DDS, 45; Univ Mich, MS, 53. *Prof Exp:* Instr oper dent, Col Dent, Ohio State Univ, 48-51; instr, Col Dent, Univ Calif, 51-52; mem staff, Henry Ford Hosp, Detroit, Mich, 53-68; chmn dept oral radiol, Sch Dent Med, Univ Conn, 68-69; prof oral diag & chmn dept, Sch Dent, Univ NC, Chapel Hill, 69-80; RETIRED. *Mem:* AAAS; fel Am Col Dentists; fel Am Acad Dent Radiol; Am Acad Pedodontics; Int Asn Dent Res. *Res:* Prenatal, neonatal and post-natal influences upon dental enamel development; healing of dental pulp following bacterial, chemical or mechanical trauma; oral roentgenographic technique. *Mailing Add:* 810 Indian Spring Rd Chapel Hill NC 27514

VIAL, JAMES LESLIE, b Taft, Calif, Dec 19, 24; div; c 1. VERTEBRATE BIOLOGY, POPULATION ECOLOGY. *Educ:* Calif State Univ, Long Beach, BA, 52, MA, 54; Univ Southern Calif, PhD, 65. *Prof Exp:* From instr to assoc prof biol, Los Angeles Valley Col, 55-61; vis prof zool, Univ Costa Rica, 61-62, Ford Found Prof ecol, 62-64; vis assoc prof biol, Western Mich Univ, 64-66; from assoc prof to prof, Univ Mo-Kansas City, 66-75, assoc dean res & dir res admin, 66-68; prof biol & chmn fac biol sci, 75-82, PROF BIOL, UNIV TULSA, 75- *Concurrent Pos:* Assoc dir, Orgn Trop Studies, Inst Trop Ecol, Costa Rica, 63, dir, 64; herpet ed, Am Soc Ichthyologists & Herpetologists, 72-78; managing ed, J Herpetol, 83-90; cert sr ecol, Ecol Soc Am, 84. *Mem:* AAAS; Ecol Soc Am; Am Soc Mammal; Soc Study Amphibians & Reptiles (pres, 77); Am Soc Icthyologists & Herpetologists; Sigma Xi; Am Inst Biol Sci; Herpetologists League. *Res:* Vertebrate ecology and population dynamics, especially amphibians and reptiles; environmental inventories and impact reports. *Mailing Add:* Fac Biol Sci Univ Tulsa Tulsa OK 74104

VIAL, LESTER JOSEPH, JR, b New Orleans, La, Mar 19, 44; m 71. MEDICINE, PATHOLOGY. *Educ:* La State Univ, New Orleans, BS, 66; Med Sch, La State Univ, MD, 70; Am Bd Path, dipl, 74. *Prof Exp:* Intern path, Charity Hosp, New Orleans, 70-71, resident, 71-74; instr, 74-75, ASST PROF PATH, MED SCH, LA STATE UNIV, 75- *Concurrent Pos:* Vis staff, Charity Hosp, New Orleans, 74- *Mem:* AOA; Am Soc Clin Path; Am Soc Cytol. *Mailing Add:* 1901 Perdido St New Orleans LA 70118

VIAL, THEODORE MERRIAM, b Ware, Iowa, Feb 27, 21; m 49; c 5. RUBBER CHEMISTRY. *Educ:* Univ Md, BS, 42; Univ Ill, PhD(chem), 49. *Prof Exp:* Res chemist, Chas Pfizer & Co, 48-50; res chemist, 50-51, tech rep, 51-58, tech mgr rubber chem dept, 58-60, com develop rubber chemicals & elastomers, 60-66, group leader, Chem Res Div, Am Cyanamid Co, 66-86; RETIRED. *Mem:* AAAS; Am Chem Soc; Sigma Xi. *Res:* Rubber and elastomer compounding; theory and application of vulcanization and protective agents; polyacrylate and other specialty elastomers. *Mailing Add:* 35 Woodside Lane Princeton NJ 08540

VIAMONTES, GEORGE IGNACIO, IMMUNOREGULATION. *Educ:* Wash Univ, PhD(cell & develop biol), 79. *Prof Exp:* SR SCIENTIST, ORTHO PHARMACEUT CORP, 82- *Res:* Development of the immune system. *Mailing Add:* 298 Kinderhook Dr Chesterfield MO 63017-2920

VIANNA, NICHOLAS JOSEPH, b New York, NY, Dec 20, 41; m 67; c 2. EPIDEMIOLOGY. *Educ:* St Peter's Col, BS, 63; Cornell Univ Med Col, 67; Albany Med Col, MSPH, 71. *Prof Exp:* Intern & resident med, Montefiore Hosp & Med Ctr, New York, 68-70; clin instr, Albany Med Col, 70-71; fel infectious dis, NJ Col Med & Sloan Kettering Mem Inst, 71-72; DIR BUR OCCUP HEALTH & CHRONIC DIS RES, NY STATE DEPT HEALTH, ALBANY, 72-; ASST PROF, DEPT MED & PREV MED, ALBANY MED COL, 72- *Concurrent Pos:* Asst to the dir, Bur Epidemiol & Cancer, NY State Health Dept, Albany, 69-71; officer epidemiol, Ctr Dis Control, Atlanta, 69-71; consult, Kettering Inst Cancer Res, NY, 72-73 & Cancer Epidemiol Sect, WHO, Geneva, Switz, 72-73; travel fel, WHO, Oxford Univ, 72; dir, Coeymans Med Clin, Albany, NY, 72-; consult infectious dis, Ellis Hosp, Schenectady, NY, 73- *Res:* Epidemiology of lymphoreticular malignancies with major emphasis on etiology. *Mailing Add:* Eight Victoria Dr Guilderland NY 12084

VIANO, DAVID CHARLES, b San Mateo, Calif, May 7, 46. INJURY BIOMECHANICS, AUTOMOTIVE SAFETY ENGINEERING. *Educ:* Santa Clara Univ, BS, 68; Calif Inst Technol, MS, 69, PhD(appl mech), 72. *Prof Exp:* Postdoctoral biomed sci, Univ & Swiss Fed Inst Technol, Zurich, Switz, 72-74; sr res engr, Biomed Sci Dept, Gen Motors Res Labs, 74-76, staff res engr, 76-78, asst dept head, 78-87, PRIN RES SCIENTIST, BIOMED SCI DEPT, GEN MOTORS RES LABS, 87-, PROG LEADER SAFETY RES, 88- *Concurrent Pos:* Chmn, Passenger Protection Comt, Soc Automotive Engrs, 80-; from adj asst prof to adj assoc prof eng, Wayne State Univ, Detroit, 81-89, adj prof, 89-; mem, Comt Fed Trauma Res, Nat Acad Sci, Nat Res Coun, 84-85, Comt Sch Bus Safety, Transp Res Bd, 87-89, Comt Occupant Restraint Res Needs, 88-90, Adv Comt Injury Prev & Control, Ctr Dis Control, 89-92 & bd dirs, Nat Head Injury Found, 87-; assoc ed, Accident Anal & Prev, 88- *Honors & Awards:* R H Isbrandt Medal, Soc Automotive Engrs, 81, 85 & 86, Colwell Award, 82, 88, 89 & 90; Award for Safety Eng Excellence, Nat Hwy Traffic Safety Admin, 89. *Mem:* Asn Advan Automotive Med (pres, 89); fel Am Soc Mech Engrs; fel Soc Automotive Engrs; Nat Head Injury Found; Am Trauma Soc; Am Soc Biomech. *Res:* Injury biomechanics, human tolerance, trauma mechanisms, injury disability; automotive safety, crash injury prevention, occupant protection by safety belts, airbags; public health programs for injury control; safety research and policy; author of numerous publications. *Mailing Add:* Biomed Sci Dept Gen Motors Res Labs Warren MI 48090-9055

VIAVANT, WILLIAM JOSEPH, computer science, for more information see previous edition

VICE, JOHN LEONARD, b Evergreen Park, Ill, Jan 12, 42; m 63. MICROBIOLOGIST, BIOCHEMISTRY. *Educ:* Loyola Univ Chicago, BS, 63; Univ Ill, Chicago, MS, 65, PhD(microbiol), 69. *Prof Exp:* Res microbiologist, Nat Cancer Inst, 65-66; dir clin microbiol, Alexian Bros Med Ctr, Ill, 67-68; dir clin microbiol, Med Ctr, Loyola Univ, Chicago, 68-77; sect head, Clin Microbiol, City of Hope, Nat Med Ctr, Duarte, Calif, 77-88; MICROBIOLOGIST, SPECIALTY LABS, 88- *Concurrent Pos:* Consult, Community Gen Hosp, Sterling, Ill, 72- *Mem:* Am Soc Microbiol; Sigma Xi. *Res:* Immunogenetics of rabbit immunoglobulins; rabbit lymphocyte antigens; immunochemical characterization of gram negative nonfermentative bacteria. *Mailing Add:* 1500 E Duarte Rd Duarte CA 91010

VICEPS-MADORE, DACE I, b Esslingen, Ger, Feb 22, 47; US citizen; m 75. CELL BIOLOGY. *Educ:* Univ Rochester, AB, 69; Temple Univ, MA, 71, PhD(biol), 74. *Prof Exp:* Res investr cell biol, Wistar, Inst, 73-75; fel cell biol, Inst Cancer Res, 75-77; fel pharmacol, Yale Univ, 77-80; MEM FAC, DEPT BIOCHEM, UNIV VT, 80- *Mem:* AAAS; Asn Women Sci; Tissue Cult Asn; Sigma Xi. *Res:* The regulation of enzyme activity in eukaryotic cells. *Mailing Add:* One Schoen Rd Pittsford NY 14534

VICHICH, THOMAS E, b Calumet, Mich, Oct 5, 17; m 45; c 2. MATHEMATICS. *Educ:* Mich Col Mining & Technol, BS, 39; Univ Mich, MS, 52. *Prof Exp:* From instr to assoc prof, Mich Technol Univ, 41-61, prof math, 61-80; RETIRED. *Res:* Differential and integral calculus; differential equations. *Mailing Add:* 41 Agate St Houghton MI 49931

VICK, CHARLES BOOKER, b Seaboard, NC, Sept 15, 32; m 63; c 2. WOOD ADHESIVES SCIENCES & TECHNOLOGY. *Educ:* Duke Univ, AB, 54, MF, 58. *Prof Exp:* Forest prod technologist, 58-60, wood scientist, 60-74, prin wood scientist, 74-86, RES SCIENTIST, FOREST PROD LAB, SOUTHEASTERN FOREST EXP STA, 86- *Concurrent Pos:* Chmn, Subcomt Construct Adhesives, Am Soc Testing & Mat, 75. *Mem:* Am Soc Testing & Mat; Forest Prod Res Soc; Adhesion Soc. *Res:* Develop the technology to bond wood that has been chemically treated; with preservatives, and dimensional stabilizers. *Mailing Add:* Forest Prod Lab One Gifford Pinchot Dr Madison WI 53705

VICK, GEORGE R, b New Waverly, Tex, Dec 18, 20; m 43; c 2. ALGEBRA, GEOMETRY. *Educ:* Sam Houston State Teachers Col, BA, 41, MA, 42; Univ Tex, PhD(math), 64. *Prof Exp:* Teacher high sch, Tex, 41-42; from asst prof to assoc prof math, 46-51, assoc prof, 55-56, actg dir dept, 56-64, dir dept, 64-67, PROF MATH, SAM HOUSTON STATE UNIV, 64- *Mem:* Math Asn Am; Nat Coun Teachers Math. *Res:* Foundations of geometry; mathematical pedagogy at all levels, especially training and re-training teachers. *Mailing Add:* 3101 Ave Q Huntsville TX 77340

VICK, GERALD KIETH, b Dixon, Ill, Mar 6, 30; m 50; c 3. TEACHING. *Educ:* Univ Ill, BS, 52; Univ Rochester, PhD(org chem), 56. *Prof Exp:* Res chemist, Esso Res & Eng Co, 55-58, proj leader engine oils, 58-62, sect head motor fuels & lubricants, 62-65, sr staff adv petrol fuels & lubricants, 65-67, dir lubricants & specialties lab, 67-75, sr planning adv, Corp Planning Dept, Exxon Corp, 75-76, sr staff adv, Petrol & Synthetic Fuels Res, Exxon Res & Eng Co, 76-86; ADJ INSTR, BUCKS CO COMMUNITY COL, 88- *Mem:* AAAS; Am Chem Soc; Soc Automotive Eng; fel Am Inst Chemists. *Res:* Organic chemistry; fuel technology and petroleum engineering. *Mailing Add:* Three Devon Dr New Hope PA 18938

VICK, JAMES, b Crookston, Minn, Sept 6, 31. HERPETOLOGY, MARINE BIOLOGY. *Educ:* Univ NDak, MS, 55, PhD(pharmacol), 58. *Prof Exp:* Chief CU Labs, Walter Reed, Army, 69-73 & neurophysiol, Aberdeen Proving Grounds, 74-78; NATO prin deleg, Surgeon Gen Off, Pentagon, 78-80; DIV DIR (A), FOOD & DRUG ADMIN, WASHINGTON, DC, 81- *Concurrent Pos:* Assoc prof pharmacol, Univ Md, 79-81; assoc prof pathophysiol, Walter Reed Army Inst Res, Washington, DC, 85-; prof pharmacol, American Univ, 89-91; chmn, Inst Animal Care & Use Comt, Food & Drug Admin, 90- *Honors & Awards:* Cert Recognition, Sigma Xi, 91. *Mem:* Underseas Med Soc; Sigma Xi; Soc Exp Biol Med. *Res:* Shock; venoms; toxins; treatment of cyanide poisoning; nerve agent toxicity; author of various publications. *Mailing Add:* Food & Drug Admin 200 C St SW Washington DC 20204

VICK, JAMES WHITFIELD, b Hope, Ark, Mar 8, 42; m 64; c 2. MATHEMATICS. *Educ:* La State Univ, Baton Rouge, BS, 64; Univ Va, MA, 66, PhD(math), 68. *Prof Exp:* Instr math, Princeton Univ, 68-70; from asst prof to assoc prof, 70-82, asst dean, 80-83, PROF MATH, UNIV TEX, AUSTIN, 82-, ASSOC DEAN, COL NATURAL SCI, 83- *Mem:* Math Asn Am; Am Math Soc; AAAS. *Res:* Algebraic and differential topology, K-theory and transformation groups. *Mailing Add:* Dept Math Univ Tex Austin TX 78712

VICK, ROBERT LORE, b Courtland, Miss, Sept 1, 29; m 53; c 1. PHARMACOLOGY, PHYSIOLOGY. *Educ:* Univ Miss, BS, 52, MS, 54; Univ Cincinnati, PhD(pharmacol), 57. *Prof Exp:* Actg asst prof pharm, Southwestern Okla State Col, 53; instr, Univ Miss, 53-54; res assoc pharmacol, Univ Cincinnati, 57-58; instr physiol, State Univ NY Upstate Med Ctr, 58-61; from asst prof to prof physiol, 61-85, PROF MOLECULAR PHYSIOL & BIOPHYS, BAYLOR COL MED, 85- *Concurrent Pos:* USPHS Career Develop Award 66-71. *Mem:* AAAS; Am Physiol Soc; Am Soc Pharmacol & Exp Therapeut; Soc Exp Biol & Med. *Res:* Heart and circulation; ion movements; electrophysiology; autonomic nervous system. *Mailing Add:* Dept Molecular Physiol & Biophys Baylor Col Med Houston TX 77030

VICKERS, DAVID HYLE, b Sturgis, Miss, Jan 14, 40; m 70; c 2. ENTOMOLOGY, BIOCHEMISTRY. *Educ:* Miss State Univ, BS, 61, MS, 64; La State Univ, PhD(entom), 69. *Prof Exp:* Instr entom, Southeastern La Col, 64-66; asst prof physiol, Fla Technol Univ, 69-72, actg chmn dept, 74-75, chmn, 75-81, ASSOC PROF PHYSIOL, UNIV CENT FLA, 72- *Mem:* AAAS. *Res:* Genetics; molecular biology. *Mailing Add:* Dept Biol Univ Cent Fla Orlando FL 32816

VICKERS, FLORENCE FOSTER, b Philadelphia, Pa; m 76. PHARMACOLOGY. *Educ:* Pa State Univ, BS, 69; Hahnemann Med Col, PhD(cardiovasc physiol & pharmacol), 81. *Prof Exp:* Res biologist, Merck Sharp & Dohme Res Labs, 69-74, med writer, 74-78; doctoral res fel, Hahnemann Med Col, 78-81, instr, 79-80; clin res assoc, Cardiovasc Div, Merck Sharp & Dohme, 82-84; assoc dir clin res, Wyeth Int Ltd, 84-87; VIS ASSOC PROF, COL MED, PA STATE UNIV, 87-; ASSOC DIR CLIN RES, CIBA-GEIGY PHARMACEUT, 88- *Mem:* AAAS; Am Soc Clin Pharmacol & Therapeut; fel Am Col Clin Pharmacol; Drug Info Asn; Sigma Xi. *Res:* Cardiovascular physiology and pharmacology; renin-angiotension system; anti hypertensive drugs; beta adrenoceptor antagonists; catecholamines; drug metabolism. *Mailing Add:* Box 243 Ctr Sch & Slotter Rds Bedminster PA 18944

VICKERS, J(OHN) M(ICHAEL) F(RANK), electrical engineering, for more information see previous edition

VICKERS, JAMES HUDSON, b Columbus, Ohio, Apr 21, 30; m 64; c 1. VETERINARY MEDICINE, PATHOLOGY. *Educ:* Ohio State Univ, BSc, 52, DVM, 58; Univ Conn, MS, 66; Am Col Vet Pathologists, dipl, 75. *Prof Exp:* Vet, Columbus Zoo, Ohio, 58-60; vet, Lab Animal Colony, Lederle Labs, Am Cyanamid Co, NY, 60-64, pathologist, 64-66, head dept vet path, 66-68, head dept exp path, 68-70; vpres & dir res, Primelabs, Inc, 70-73; DIR PATH & PRIMATOL BR, CTR DRUGS & BIOL, FOOD & DRUG ADMIN, 73- *Concurrent Pos:* Lectureship, State Univ NY Downstate Med Sch; Food & Drug Admin rep, Interagency Animal Res Comt, 74-, proj officer, Primate Breeding Colony Contracts. *Honors & Awards:* Presidential Citation, 79; Commissioner's Spec Citation, Food & Drug Admin, 88. *Mem:* Am Vet Med Asn; Am Asn Lab Animal Sci; Am Col Vet Pathologists; Soc Toxicol; Int Acad Pathol. *Res:* Diseases and pathology of primates, laboratory animals and exotic zoological species; testing and quality control of vaccines, especially polio vaccine, toxicological testing and pathology of pharmaceuticals. *Mailing Add:* 2324 Oak Dr Ijamsville MD 21754

VICKERS, ROGER SPENCER, b Hitchin, Eng, Nov 13, 37; m 67; c 3. REMOTE SENSING, MEASUREMENT TECHNOLOGY. *Educ:* Univ Southampton, BSc, 59, PhD(physics), 63. *Prof Exp:* Res physicist, IIT Res Inst, 63-66 & Stanford Univ, 66-68; res assoc remote sensing, Colo State Univ, 68-69; vpres, Environ Res Assocs, 69-70; assoc prof remote sensing, Colo State Univ, 70-73; SR PHYSICIST, SRI INT, 73- *Concurrent Pos:* Consult, commercial & industrial, 71-81, NSF, 74-79. *Honors & Awards:* IR-100 Award, 76. *Mem:* Soc Explor Geophysicists. *Res:* Remote sensing techniques; infrared absorption and Raman spectroscopy; high resolution radars; subsurface profiling; radar sounding of ice; development of electromagnetic geophysical techniques; airborne ground-penetrating radar. *Mailing Add:* SRI Int Menlo Park CA 94025

VICKERS, STANLEY, b Blackpool, Eng, Sept 27, 39; US citizen. DRUG METABOLISM. *Educ:* Univ London, BSc, 62; State Univ NY Buffalo, PhD(biochem pharm), 67. *Prof Exp:* Fel, Univ Kans, 66-69; res fel, 69-80, SR RES FEL DRUG METAB, MERCK INST THERAPEUT RES, 80- *Mem:* Am Soc Pharmacol & Exp Therapeut; Am Chem Soc; AAAS; NY Acad Sci. *Res:* Detoxification mechanisms and metabolic transformations which control the fate of foreign compounds. *Mailing Add:* RR 2 Box 243 Perkasie PA 18944

VICKERS, THOMAS J, b Miami, Fla, Mar 29, 39; m 63; c 4. ANALYTICAL CHEMISTRY. *Educ:* Spring Hill Col, BS, 61; Univ Fla, PhD(chem), 64. *Prof Exp:* From asst prof to assoc prof, 66-76, PROF ANALYTICAL CHEM, FLA STATE UNIV, 76- *Mem:* Am Chem Soc; Soc Appl Spectros. *Res:* Spectroscopic methods of analysis; Raman spectroscopy. *Mailing Add:* Dept Chem Fla State Univ Tallahassee FL 32306-3006

VICKERS, WILLIAM W, b San Francisco, Calif, June 21, 23; m 54; c 3. ATMOSPHERIC PHYSICS. *Educ:* Univ Calif, BA, 54, MA, 56; McGill Univ, PhD(hydrol, meteorol), 65. *Prof Exp:* Res assoc, Inst Polar Studies, Ohio State Univ, 57-61; head geophys res group, Tech Opers, Inc, 61-66; sr sci exec, EG&G, Inc & Environ Sensor Systs Div, Mitre Corp, Bedford, 66-85; consult, 85-87; RETIRED. *Concurrent Pos:* Dir, Explorers Club, 69-72. *Mem:* Am Geophys Union; Am Meteorol Soc; NY Acad Sci; Cosmos Club. *Res:* Studies of snow and ice in Antarctica during the geophysical year; Air Force investigations of weather modification; atmospheric transport of toxic materials; atmospheric effects on radio-radar propagation; analyses of electronic warfare scenarios. *Mailing Add:* Box 471 Fair Haven NY 13064

VICKERS, ZATA MARIE, b Salem, Ore, Oct 13, 50. FOOD SCIENCE. *Educ:* Ore State Univ, BS, 72; Cornell Univ, PhD(food sci), 75. *Prof Exp:* Asst prof, 75-80, ASSOC PROF FOOD SCI, UNIV MINN, ST PAUL, 80- *Mem:* Inst Food Technol. *Res:* Relationships between the physical, acoustical and sensory properties of foods. *Mailing Add:* Dept Food Sci & Nutrit Univ Minn 1334 Eckles Ave St Paul MN 55108

VICKERS-RICH, PATRICIA, b Exter, Calif, July 11, 44; m 66; c 2. VERTEBRATE PALEONTOLOGY, BIOGEOGRAPHY. *Educ:* Univ Calif, Berkeley, AB, 66; Columbia Univ, MA, 69, PhD(geol), 73. *Prof Exp:* Asst geol, Columbia Univ, 67-68; asst prof geosci geol, Dept Geosci, Tex Tech Univ & assoc cur vert paleont, Mus, 73-76; LECTR-READER, DEPT EARTH SCI & DEPT ECOL/EVOLUTIONARY BIOL, MONASH UNIV, 76- *Concurrent Pos:* Fulbright fel, Australian Am Educ Found, Australia, 73-74; hon res assoc, Mus Victoria, 78 & 80-; Nat Geog Soc grants, 78, 80-; Australian Res grants comn grantee, 78-; Australian China Coun grantee, 80-81; Australian Acad Sci Chinese exchange scientist, 79; Australian Nat Parks Wildlife grantee, 81; Sunshine Found, Utah Mining, Ingram Trust grants, 76- *Mem:* Am Ornith Union; Geol Soc Australia; Royal Australian Ornith Union; Soc Vert Paleont; Orgn Trop Studies; Planetary Soc. *Res:* Evaluation of cenozoic fossil birds from Australasia; paleogene birds from Asia and South America; Neogene avifaunas from SAfrica; development of bird bone key for archaeological use; investigation into phylogenetic value of avian quadrate; development of a machine translation system for Chinese scientific literature; avian systematics; polar dinosaurs and other biota from late mesozic polar environment in Southeastern Australia. *Mailing Add:* Dept Earth Sci Monash Univ Clayton Victoria 3168 Australia

VICKERY, LARRY EDWARD, b Atlanta, Ga, Nov 26, 45; c 2. ENZYMOLOGY, STEROID CHEMISTRTY. *Educ:* Univ Calif, Santa Barbara, BA, 67, PhD(biol), 71. *Prof Exp:* Res assoc biochem, Western Regional Res Lab, USDA, 71; res assoc biophys chem, Lawrence Berkeley Lab, 72; fel, Univ Calif, Berkeley, 73-74, res assoc biophys chem, Dept Chem, 75-76; asst prof biophys, 77-82, assoc prof, physiol, biophys & biol chem, 82-88, PROF PHYSIOL, BIOPHYS & BIOL CHEM, UNIV CALIF, IRVINE, 88- *Concurrent Pos:* Res assoc, Nat Res Coun, Nat Acad Sci, Nat Acad Engr, 71; fel, Nat Inst Gen Med Sci, USPHS, 73-74. *Mem:* Biophys Soc; AAAS; Am Soc Biochem & Molecular Biol; Protein Soc. *Res:* Investigations on the molecular mechanisms and regulation of steroid hormone biosynthesis; characterization of enzymes and reaction mechanisms involved; development of inhibitors of steroid synthesis; protein chemistry, engineering and mutagenesis. *Mailing Add:* Dept Physiol & Biophys Univ Calif Irvine CA 92717

VICKERY, ROBERT KINGSTON, JR, b Saratoga, Calif, Sept 18, 22; m 51; c 2. PLANT EVOLUTION. *Educ:* Stanford Univ, AB, 44, AM, 48, PhD(biol, bot), 52. *Prof Exp:* Instr bot, Pomona Col, 50-51; from instr to assoc prof biol, 52-64, head dept genetics & cytol, 62-65, PROF BIOL, UNIV UTAH, 64- *Concurrent Pos:* Researcher, Carnegie Inst Wash, 48-52; res fel, Calif Inst Technol, 55; vis assoc prof, Harvard Univ, 63; mem cellular biol & genetics fel panel, NIH, 65-69, 7th, 12th & 13th Int Bot Cong & 10th-13th Int Gen Cong; assoc ed, Evolution, Soc Study Evolution, 68-72; mem, Int Orgn Plant Biosysts Coun, 75-81. *Mem:* AAAS; Soc Study Evolution (vpres, 77-78); Am Soc Nat; Ecol Soc Am; Genetics Soc Am. *Res:* Cytogenetics, ecologic, numerical and classical taxonomic, and biochemical approaches to problems of the evolutionary mechanisms and patterns of the genus Mimulus, particularly sections Simiolus and Erythranthe. *Mailing Add:* Dept Biol Univ Utah Salt Lake City UT 84112

VICKREY, HERTA MILLER, b San Gregorio, Calif; div; c 4. IMMUNOLOGY, MEDICAL MICROBIOLOGY. *Educ:* Jan Jose State Col, BA, 57; Univ Calif, Berkeley, MA, 63, PhD(bact & immunol), 70. *Prof Exp:* Microbiologist, Viral & Rickettsial Dis Lab, Calif State Dept Pub Health, 57-60, 61-62; res bacteriologist, Univ Calif, Berkeley, 63-64; asst prof immunol, virol & microbiol, Univ Victoria, BC, 70-72; res assoc cancer immunol, dept res & educ, Wayne County Gen Hosp, 72-83; lab supvr, dept med admin, Univ Mich, Ann Harbor, 77-83; lab dir pub health, Shasta County Pub Health Lab, 83-84; PUB HEALTH MICROBIOLOGIST, TULARE COUNTY PUB HEALTH LAB, 84- *Concurrent Pos:* Bacteriologist, Children's Hosp Med Ctr Northern Calif, Oakland, 58-70; res grants, Univ Victoria, BC, 70-72; med staff res & educ grants, 73-83; vis scientist, dept nutrit & food sci, Mass Inst Technol, Cambridge, 82; mycol trainer, Calif State Dept Health Serv, Berkeley, 85-86. *Mem:* Am Soc Microbiol; Clin Ligand Assay Soc; Am Soc Clin Path; NY Acad Sci. *Res:* Cellular immunological resistance: tuberculosis and oncogenesis; cell culture assays to autoallergies, oncogenesis and hypersensitivities; tissue culture studies: lymphatic leukemia immunotherapy; growth and metabolic responses of liver cells (canine, rabbit, murine-diabetic versus normal) to hormones, drugs, lectins, medium additives; hybridoma immunological technology; general medical sciences. *Mailing Add:* 3505 Campus Dr No 5 Visalia CA 93277

VICKROY, DAVID GILL, b San Antonio, Tex, July 5, 41; m 64; c 2. ANALYTICAL CHEMISTRY. *Educ:* Vanderbilt Univ, BA, 63; Rice Univ, MA, 66; Univ Tenn, PhD(inorg chem), 69. *Prof Exp:* Res chemist, Celanese Fibers Co, 69-72, sr res chemist, 72-75, res assoc anal chem, 75-76, group leader, 76-82, MGR, HOECHST CELANESE RES DIV, 82- *Mem:* Sigma Xi; Am Chem Soc; NY Acad Sci. *Res:* Analytical characterization of complex mixtures such as tobacco smoke and environmental samples; development of synthetic alternatives to natural products; analytical testing of advanced materials. *Mailing Add:* Hoechst Celanese Res Div 86 Morris Ave Summit NJ 07901

VICK ROY, THOMAS ROGERS, b Denver, Colo, Oct 17, 22; m 48; c 3. AUTOMATIC CONTROL OF CHEMICAL PROCESSES, COMPUTER CONTROL OF INDUSTRIAL PROCESSES. *Educ:* Univ Colo, BS, 44; Mass Inst Technol, MS, 48. *Prof Exp:* Res engr, E I du Pont de Nemours & Co, Inc, 48-50, design engr, 50-54, consult mgr automatic process control, 54-69 & polymer processing, 69-73, res consult, 73-77, prin consult nuclear eng, 77-86; PRIN CONSULT PROCESS CONTROL, CONDUX, INC, 86- *Concurrent Pos:* Consult ed, Control Eng, 58-62. *Mem:* Fel Instrument Soc Am; Inst Elec & Electronics Engrs; Am Inst Chem Engrs; Am Chem Soc; Soc Plastics Engrs. *Res:* Direct digital computer control of chemical processes and nuclear reactors; polymer synthesis and processing. *Mailing Add:* 404 Concord Ave Talleyville DE 19803-2316

VICKROY, VIRGIL VESTER, JR, b San Antonio, Tex, Aug 8, 31; m 55; c 2. POLYMER CHEMISTRY, PHYSICAL CHEMISTRY. *Educ:* Auburn Univ, BS, 52; Univ Akron, MS, 62, PhD(polymer sci), 65. *Prof Exp:* Jr chemist, B F Goodrich Co, 55-61; res chemist, Harrison-Morton Labs, Ohio, 61-63; res chemist, Univ Akron, 63-65; sr res chemist, Monsanto Co, 65-73; lab mgr, Marathon-Morco Co, 73-74; prin chemist, Dart Industs, 74-75; RES CHEMIST, ALLIED CORP, 75- *Mem:* Am Chem Soc; Soc Rheology; NAm Thermal Anal Soc; Soc Plastics Engrs. *Res:* Effect of thermal and thermo-oxidative history on morphological, mechanical and molecular properties of polyolefins; physical chemistry of Ziegler and Phillips catalyst systems. *Mailing Add:* PO Box 53006 Baton Rouge LA 70892-3006

VICORY, WILLIAM ANTHONY, b Beauford, SC, June 7, 58; m 82; c 3. HIGH STRENGTH FIBER OPTIC SPLICING, MECHANICAL FIBER OPTIC SPLICING. *Educ:* Southern Col Technol, BMET, 82. *Prof Exp:* Sr tech assoc, 82-90, MEM TECH STAFF, AT&T BELL LABS, 90- *Res:* Fiber optic splicing design and development; testing; customer interfacing; documentation; granted two patents. *Mailing Add:* 2000 Northeast Expressway Rm 1B01 Norcross GA 30071

VICTERY, WINONA WHITWELL, b Abilene, Tex, Apr 15, 41; m 63. RENAL PHYSIOLOGY, METAL TOXICOLOGY. *Educ:* Rice Univ, BA, 63; Univ Wis-Madison, MS, 67; Univ Mich, Ann Arbor, PhD(physiol), 78; Am Bd Toxicol, dipl, 86. *Prof Exp:* Res biologist nuclear med, Univ Calif, Los Angeles, 67-74; scholar physiol, Univ Mich, 78-81; scholar pharm, Nat Inst Environ Health Sci, 81-83, expert, Biomet & Risk Assessment Prog, 83-86; HEALTH SCIENTIST, ENVIRON CRITERIA & ASSESSMENT OFF, US ENVIRON PROTECTION AGENCY, 86- *Mem:* Am Physiol Soc; Soc Risk Anal. *Res:* Physiological relationships between essential and toxic trace metals; risk assessment and risk communication. *Mailing Add:* US Environ Protection Agency Reg 1X Off Reg Adminr 1235 Mission St San Francisco CA 94103

VICTOR, ANDREW C, b New York, NY, Nov 4, 34; m 55; c 3. ENGINEERING PHYSICS. *Educ:* Swarthmore Col, BA, 56; Univ Md, MS, 61. *Prof Exp:* Physicist, Nat Bur Standards, 56-62; head, Standards Lab, US Naval Ord Test Sta, US Naval Weapons Ctr, 62-64, physicist systs anal & rocket plume technol, 64-68, head, Anal Br, 68-73, head Propulsion Anal Br, 73-76, head appl propulsion res, 76-79, coordr independent explor develop, 79-80, proj mgr explosives advan develop, 80-81, head, Thermal/Struct Br, 81-84, prog mgr, Navy Insensitive Munitions Advan Develop Propulsion Prog, 84-89; CONSULT PHYSICIST, VICTOR TECHNOL, 90- *Concurrent Pos:* Chmn, plume technology subcomt, JANNAF, 78-80, mem exhaust plume technol subcomt, 64-89, propulsion systs hazards subcomt, 84-89; TTcp lead focus officer, propulsion hazards, 85-89; mem, plume technol working group, 87-89; US focus officer plume technol, 67-78. *Mem:* Assoc fel Am Inst Aeronaut & Astronaut; Sigma Xi; Syst Safety Soc. *Res:* Missile propulsion analysis; rocket exhaust plume; ramjet engine cost analysis; weapon energy utilization; energetic material hazards; insensitive munitions. *Mailing Add:* Victor Technol 712 N Peg St Ridgecrest CA 93555

VICTOR, GEORGE A, b Ridgway, Pa, Nov 15, 36; m 63; c 2. ATOMIC AND MOLECULAR PROCESSES AND APPLICATIONS. *Educ:* Rensselaer Polytech Inst, BS, 58; Queen's Univ, Belfast, PhD(appl math-physics), 66. *Prof Exp:* Staff scientist physics, GCA Corp, 61-71; PHYSICIST PHYSICS, SMITHSONIAN ASTROPHYS OBSERV, 71- *Concurrent Pos:* Sr res fel, Queen's Univ, Belfast, 66-67; lectr, dept astron, Harvard Univ, 71-; vis fel, Joint Inst Lab Astrophys, Univ Colo Nat Bur Standards, 78-79; prog comt, Am Phys Soc, 83-85. *Mem:* Am Asn Physics Teachers; Am Geophys Union; Sigma Xi; fel Am Phys Soc. *Res:* Quantum theory of atomic and molecular structure and scattering processes; interaction of radiation with matter; applications of cross section data to problems in atmospheric physics, astrophysics, lasers and plasmas. *Mailing Add:* Dept Astron Harvard Univ Cambridge MA 02138

VICTOR, JOE MAYER, b Houston, Tex, Sept 20, 39; m 81; c 4. ELECTRICAL ENGINEERING. *Educ:* Univ Tex, Austin, BS, 62, MS, 64, PhD, 67. *Prof Exp:* Teaching asst elec eng, Univ Tex, Austin, 61-66; sr res engr, Southwest Res Inst, 66-75; ENG MGR, PETROLITE INSTRUMENTS, 75-; AT BIO QUANTUM. *Mem:* Laser Inst Am; Inst Elec & Electronics Engrs. *Mailing Add:* 4939 Valkeith Houston TX 77096

VICTOR, JONATHAN DAVID, b Yonkers, NY, Nov 21, 54. VISUAL NEUROPHYSIOLOGY, MATHEMATICAL MODELLING. *Educ:* Harvard Col, BA, 73; Rockefeller Univ, PhD(neurophysiol), 79; Cornell Univ, MD, 80. *Prof Exp:* Asst prof biophys, Rockefeller Univ, 84-86; from asst prof to assoc prof, 86-91, PROF NEUROL & NEUROSCI, MED COL & GRAD SCH MED SCI, CORNELL UNIV, 91- *Concurrent Pos:* Consult, Bell Tel Labs, 83; adj prof, Rockefeller Univ, 86-; assoc attend, Rockefeller Univ Hosp, 86-, NY Hosp, 87- & Hosp Spec Surg, 90- *Mem:* Soc Neurosci; Asn Res Vision & Opthal; Sigma Xi; Am Neurol Asn; Asn Res Nerv & Ment Dis; Biomed Eng Soc. *Res:* Neural computations underlying the process of visual information by the mammalian brain at retinal, thalamic, and cortical levels; new techniques of nonlinear systems analysis and their application to biological systems. *Mailing Add:* Med Col Cornell Univ 1300 York Ave New York NY 10021

VICTOR, LEONARD BAKER, b Schenectady, NY, Aug 3, 34; m 66; c 2. CLINICAL PATHOLOGY, LEGAL MEDICINE. *Educ:* NY Univ, AB, 53; Univ Brussels, MD, 60; Royal Col Trop Med, TMD, 60. *Prof Exp:* Intern & resident path, Strong Hosp, Univ Rochester, 61-65, sr instr, Univ, 65-67; assoc prof path & lab med, Meharry Med Col, 68-72, assoc prof path, Grad Sch & dir Meharry Multiphasic Lab, 68-72; prof path, Univ Tenn, Memphis & dir clin labs, City of Memphis Hosps, 72-78; prof & chmn, Dept Path, Sch Med, Marshall Univ, 78-80; DEAN, CLIN SCI, ROSS UNIV SCH MED, 84- *Concurrent Pos:* Dep med examr, Monroe County, NY, 65-67; consult, State Hosp, Rochester, NY, 65-67; assoc prof biomed eng, Sch Eng, Vanderbilt Univ, 69-72; chmn, Comt Health Fitness Sci, Nashville, 70-71; chmn, Nat Adv Task Force for Regional Med Prog Eval of Automated Multiphasic Health Testing, 72; med dir & dir, Mid-South Comprehensive Home Health Serv Agency, 73-77; mem bd dirs, Mid-South Regional Blood Ctr, 73-77. *Mem:* Fel Col Am Path; fel Am Soc Clin Path; fel Soc Advan Med Systs; fel Royal Soc Health; AMA. *Res:* Administrative and legal medicine and pathology including curriculum development and interdisciplinary functions; lab medicine and prospective medicine, including automation, computerization and management techniques. *Mailing Add:* 1402 Washington Ave Suite 1 Huntington WV 25704

VICTOR, WALTER K, b Bronx, NY, Dec 18, 22. RADIO SYSTEMS FOR SPACE COMMUNICATION. *Prof Exp:* RETIRED. *Mem:* Nat Acad Eng; fel Inst Elec & Electronics Engrs. *Mailing Add:* 1630 Pegfair Estates Dr Pasadena CA 91103-1934

VICTORIA, EDWARD JESS, JR, b San Diego, Calif, Sept 11, 41. BIOCHEMISTRY, CELL BIOLOGY. *Educ:* Univ Calif, Los Angeles, AB, 63, MA, 65, PhD(molecular & cell biol), 68. *Prof Exp:* Asst res biologist, Univ Calif, Los Angeles, 68-69; fel, Univ Utrecht, 70-71; spec res fel, Lab Biochem, NIH, 71-73; ASSOC RES BIOCHEMIST, DEPT PATH, UNIV CALIF, SAN DIEGO, 73- *Concurrent Pos:* Res fel, Am Cancer Soc, Biochem Lab, Utrecht, Neth, 70-71; USPHS, NIH fel, 71-73; prin investr, NIH res grant, 78- *Mem:* Am Chem Soc; Am Soc Cell Biol; AAAS; Biochem & Molecular Biol. *Res:* Membrane biochemistry. *Mailing Add:* 3503 Pershing Ave San Diego CA 92104-3413

VICTORICA, BENJAMIN (EDUARDO), b Mendoza, Arg, June 9, 36; m 63; c 3. PEDIATRICS. *Educ:* Nat Univ Cuyo, MD, 62; Educ Coun For Med Grad, cert, 63; Am Bd Pediat, dipl, 68, cert pediat cardiol, 71. *Prof Exp:* Intern, MedSch, Nat Univ Cuyo, 63-63 & St Benedict's Hosp, Ogden, Utah, 63-64; from resident pediat to chief resident, 64-66, instr & spec trainee pediat cardiol, 67-70, asst prof, 70-74, assoc prof, 74-79, PROF PEDIAT CARDIOL, COL MED, UNIV FLA, 80- *Mem:* Fel Am Acad Pediat; Am Col Cardiol. *Res:* Pediatric cardiology. *Mailing Add:* Dept Pediat & Cardiol Univ Fla Col Med 3 Hillis Miller Health Gainesville FL 32610

VICTORIUS, CLAUS, b Hamburg, Ger, Aug 24, 23; nat US; m 52; c 2. ORGANIC COATINGS. *Educ:* Guilford Col, BS, 43; Univ NC, MA, 46. *Prof Exp:* Res assoc, Exp Sta, E I du Pont de Nemours & Co, Inc, Wilmington, Del, 46-83; RETIRED. *Concurrent Pos:* Consult, 85- *Mem:* Am Chem Soc; Sigma Xi. *Res:* Organic coatings; automotive finishes; high solids, powder and water-based coatings; gel treatments for petroleum reserviors. *Mailing Add:* 21 Paxon Hollow Rd Media PA 19063

VIDA, JULIUS, b Losonc, Czech, May 30, 28; US citizen; m; c 4. MEDICINAL CHEMISTRY. *Educ:* Pazmany Peter Univ, Budapest, Dipl, 50; Carnegie Inst Technol, MS, 59, PhD(org chem), 60; Columbia Univ, MBA, 81. *Prof Exp:* Chemist, EGIS Co, (Wander Co), Hungary, 50-56 & Merck & Co, Inc, NJ, 57-58; res fel, Harvard Univ, 61-62; chemist, Worcester Found Exp Biol, Mass, 62-67; group leader, T Clark Lab, Kendall Co, Lexington, Mass, 67-72, sect head, 72-75; asst dir res planning & licensing, Bristol Lab, 75-76; dir chem, 76-79, DIR CHEM RES, DEVELOP & LICENSING, BRISTOL-MYERS CO, INT DIV, 80-, VPRES LICENSING, 85- *Concurrent Pos:* Adj prof med chem, Grad Sch Pharmaceut Sci, Northeastern Univ, 73-75; lectr ophthal, Columbia Univ, 77-84. *Mem:* Sigma Xi; Am Chem Soc; AAAS; NY Acad Sci. *Res:* Drugs acting on the central nervous systems; heterocyclic compounds; antibiotics; anticonvulsants; anabolic agents; antiosteoporotics. *Mailing Add:* Bristol-Myers Squibb Co Rte 206 & Province Line Rd Prınceton NJ 08543-4000

VIDAL, JACQUES J, b Liege, Belg, Apr 18, 30; div; c 2. COMPUTER SCIENCE, NEUROSCIENCES. *Educ:* Univ Liege, MS, 54; Saclay Nuclear Res Ctr, France, nuclear engr, 58; Univ Paris, PhD(elec eng), 61. *Prof Exp:* Lectr, Univ Liege, 56-63; from asst prof to assoc prof eng, 63-70, PROF ENG, UNIV CALIF, LOS ANGELES, 70- *Concurrent Pos:* Orgn Econ Coop & Develop & NATO res fels, 62-63; consult comput educ, USAID mission to Tunisia, 70-85; mem, Brain Res Inst; res fel, Inst Nat Sante & Rech Med, France, 78-81; prof, Univ Paris VII, France, 89. *Mem:* Inst Elec & Electronics Engrs; Soc Neurosci; Asn Comput Mach; NY Acad Sci. *Res:* Neuroscience; machine intelligence; biocybernetics; system identification. *Mailing Add:* Dept Comput Sci 3731 Boelter Univ Calif 405 Hilgard Ave Los Angeles CA 90024-1596

VIDALE, RICHARD F(RANCIS), b Rochester, NY, Apr 22, 36; m 65; c 2. SYSTEMS ENGINEERING. *Educ:* Univ Rochester, BS, 58; Univ Wis, PhD(mech), 64. *Prof Exp:* From asst prof to assoc prof, 64-70, prof syst eng & chmn dept, 70-81, PROF ELEC, COMPUT & SYST ENG, BOSTON UNIV, 84- *Concurrent Pos:* Consult, Raytheon Co, Space & Info Systs Div, Sudbury, 66-67, Sanders Assocs, Inc, Bedford, 67-70, Chas T Main, Int, 77-, GTE, Needham Heights, 82, The MITRE Corp, 84-, Data Gen Corp, Westboro, 85-, Charles Stark Draper Lab, Cambridge, 85-, Kollsman, Inc, Merrimack, NH, 87, The Anal Scis Corp, Reading, 87- *Mem:* Inst Elec & Electronics Engrs; Am Soc Eng Educ. *Res:* Software engineering; systems engineering methodology. *Mailing Add:* Dept Elec Comput & Syst Eng Col Eng Boston Univ Boston MA 02215

VIDALI, GIAN FRANCO, b Trieste, Italy. SURFACE PHYSICS, LOW TEMPERATURE PHYSICS. *Educ:* Univ Genoa, DSc, 77; Pa State Univ, PhD(physics), 82. *Prof Exp:* Res fel chem eng, Calif Inst Technol, 82-83, res fel low temperature physics, 83-84; asst prof, 84-90, ASSOC PROF PHYSICS, SYRACUSE UNIV, 90- *Concurrent Pos:* Alfred P Sloan fel, 86. *Mem:* Am Phys Soc; Am Asn Physics Teachers; Sigma Xi; Mat Res Soc. *Res:* Experimental surface and low temperature physics; interaction of atoms-molecules with surfaces using atom beam scattering and other probes; adsorption-desorption phenomena at low temperature. *Mailing Add:* Syracuse Univ 201 Physics Bldg Syracuse NY 13244

VIDAURRETA, LUIS E, b Havana, Cuba, Dec 15, 20; US citizen; m 43; c 2. ANALYTICAL CHEMISTRY. *Educ:* Univ Havana, PhD(chem), 43. *Prof Exp:* Prof analytical chem, Univ Havana, 43-65; ASSOC PROF CHEM, LA STATE UNIV, BATON ROUGE, 66- *Mem:* Am Chem Soc; Asn Off Analytical Chemists; Am Soc Sugar Cane Technologists. *Res:* Instrumental analysis; gas chromatography; sugar and sugar by-products analysis. *Mailing Add:* 1246 Seyburn Dr La State Univ Baton Rouge LA 70808

VIDAURRI, FERNANDO C, JR, b Laredo, Tex, Feb 23, 39; m 67; c 3. CHEMICAL ENGINEERING. *Educ:* Tex Tech Univ, BSChE, 62, MSChE, 65, PhD, 68. *Prof Exp:* Develop engr, Phillips Petrol Co, 62-63; teaching asst, Tex Tech Univ, 64-68; sr develop engr, Phillips Petrol Co, Okla, 68- 78, tech mgr, Phillips Chem Co, Tex, 78-81, opers mgr, Petrochem Plant, 81-82, opers mgr engr, Plastics Plant, 82-85, mgr engr, Plastics Plant, 85-, PHILLIPS PETROL CO, OKLA. *Mem:* Am Inst Chem Engrs. *Res:* Process development; kinetics; heat transfer; high temperature polymers; experimental and theoretical thermodynamics; acid gas treating; economic evaluation; mixing and reactor design. *Mailing Add:* Phillips Petrol Co Phillips Res Ctr Bartlesville OH 74004

VIDAVER, ANNE MARIE KOPECKY, b Vienna, Austria, Mar 29, 38; US citizen; wid; c 2. BACTERIOLOGY. *Educ:* Russell Sage Col, BA, 60; Ind Univ, Bloomington, MA, 62, PhD(bact), 65. *Prof Exp:* Instr bact, 65-66, res assoc plant path, 66-72, from asst prof to assoc prof, 72-79, PROF, DEPT PLANT PATH, UNIV NEBR, LINCOLN, 79-, HEAD DEPT, 84- *Concurrent Pos:* Interim dir, Ctr Biotechnol, 88-89. *Mem:* Am Soc Microbiol; fel Am Phytopath Soc (secy, 80-83, vpres, 84-85, pres elect, 85-86, pres, 86-88); fel AAAS. *Res:* Phytopathogenic and beneficial bacteria; bacteriophages; bacteriocins. *Mailing Add:* Dept Plant Path Univ Nebr Lincoln NE 68508

VIDAVER, GEORGE ALEXANDER, biochemistry; deceased, see previous edition for last biography

VIDAVER, WILLIAM ELLIOTT, b San Francisco, Calif, Feb 2, 21; m 51; c 3. PLANT PHYSIOLOGY. *Educ:* San Francisco State Univ, AB, 58; Stanford Univ, PhD(biol), 64. *Prof Exp:* Fel plant biol, Carnegie Inst Dept Plant Biol, 63-65; assoc prof biol, 65-69, PROF BIOL, SIMON FRASER UNIV, 69- *Concurrent Pos:* Nat Res Coun Can operating grants, 65- *Mem:* AAAS; Am Soc Plant Physiol; Can Soc Plant Physiol; Am Inst Biol Sci; Sigma Xi; Am Soc Photobiol. *Res:* Mechanisms of photosynthesis; physiological investigations related to vigor, survivorship phenotypic expression and general development of forest tree, horticultural and crop plants undergoing in vitro clonal micropropagation. *Mailing Add:* Dept Biol Sci Simon Fraser Univ Burnaby BC V5A 1S6 Can

VIDEON, FRED F(RANCIS), b Hayden, Colo, Oct 4, 34; m 57; c 3. CIVIL ENGINEERING. *Educ:* Colo State Univ, BS, 58, MCE, 60; Univ Ill, PhD(civil eng), 65. *Prof Exp:* From asst prof to prof Civil Eng & Eng Mech, 65-88, EMER PROF, MONT STATE UNIV, 90-; SR STRUCT ENGR, HKM ASSOC, 88- *Honors & Awards:* Western Elec Award, Am Soc Eng Educ. *Mem:* Am Soc Civil Engrs; Am Soc Eng Educ; Nat Soc Prof Engrs. *Res:* Feasibility of using nuclear explosives for peaceful purposes; brittle fracture of mild steel; state of stress in solids containing cracks; structural mechanics and design; structural design and loading; plate girder behavior; post tensioned concrete systems; design of steel, concrete and segmental bridges. *Mailing Add:* HKM Assoc PO Box 1090 Bozeman MT 59715

VIDMAR, PAUL JOSEPH, b Vallejo, Calif, May 22, 44. UNDERWATER ACOUSTICS. *Educ:* Univ Notre Dame, BS, 66; Univ Calif, San Diego, MS, 72, PhD(physics), 75. *Prof Exp:* Res assoc physics, Fusion Res Ctr, 75-78, RES ASSOC, APPL RES LAB, UNIV TEX, AUSTIN, 78- *Mem:* Acoust Soc Am; Am Phys Soc; Sigma Xi. *Res:* Theoretical studies of long-range propagation and bottom interaction in underwater acoustics. *Mailing Add:* 6220 Cellini St Coral Gables FL 33146-3442

VIDOLI, VIVIAN ANN, b Bridgeport, Conn, Nov 2, 41. PHYSIOLOGY, NEUROPHYSIOLOGY. *Educ:* Southern Conn State Col, BS, 63; Ariz State Univ, MS, 66, PhD(zool & physiol), 69. *Prof Exp:* From asst prof to assoc prof, 70-78, PROF BIOL, CALIF STATE UNIV, FRESNO, 78-, ASST DIR, DIV HEALTH PROFESSIONS, 78-, DEAN, DIV GRAD STUDIES & RES, 80- *Concurrent Pos:* Consult, Area Health Educ Consortium, San Joaquin Valley, Calif, 74- *Mem:* Am Physiol Soc; AAAS; Sigma Xi. *Res:* Anatomical and physiological correlates of sensory mechanisms. *Mailing Add:* 1888 E Ness Fresno CA 93710

VIDONE, ROMEO ALBERT, b Greenwich, Conn, July 1, 30; m 55; c 3. PATHOLOGY. *Educ:* Davis & Elkins Col, BS, 52; Yale Univ, MD, 57. *Prof Exp:* From instr to assoc prof, 59-68, ASSOC CLIN PROF PATH, YALE UNIV, 68-; CHMN DEPT PATH, HOSP ST RAPHAEL, 77- *Concurrent Pos:* Asst clin prof, Health Ctr, Univ Conn, 72-; pvt pract; dir lab, Charlotte Hungerford Hosp, 68-77. *Mem:* Am Soc Clin Path; Col Am Path; AMA; Int Acad Path; Am Cancer Soc. *Res:* Cardiopulmonary physiology and pathology; cancer, clinicopathologic correlation of tumors. *Mailing Add:* Dept Path 1450 Chapel St New Haven CT 06511

VIDOSIC, J(OSEPH) P(AUL), b Lovran, Austria, June 10, 09; US citizen; m 35; c 2. ENGINEERING DESIGN. *Educ:* Stevens Inst Technol, ME, 32, MS, 34; Purdue Univ, PhD(mech), 51. *Prof Exp:* Instr elec, Stevens Inst Technol, 32-34; res engr, Keuffel & Esser Co, 35-36; plant engr, Whitlock Cordage Co, 36-37; from instr to assoc prof theoret mech, Ga Inst Technol, 38-49; instr mech eng, Purdue Univ, 49-51; prof, Ga Inst Technol, 51-59, regents' prof, 60-68; dean admin, 68-73, EMER DEAN, MID GA COL, 73-; EMER REGENTS' PROF, GA INST TECHNOL, 73- *Concurrent Pos:* Vis prof, Univ Baghdad, 64-65, Tuskegee Inst Technol, 67-68 & 74, Korea Advan Inst Sci & Eng, 79. *Honors & Awards:* Am Defense Medal, 45. *Mem:* Fel Am Soc Mech Engrs; Am Soc Eng Educ; Soc Exp Stress Anal; Sigma Xi. *Res:* Stress analysis; lubrication; bearings; vibration; photoelasticity; plastics; design; mechanics; mechanisms; materials science. *Mailing Add:* 38 Walnut St Cooperstown NY 13326

VIDRINE, MALCOLM FRANCIS, b Eunice, La, June 23, 49; m 83; c 2. ARTHROPOD-VECTORS, AQUATIC MOLLUSKS. *Educ:* La State Univ, BS, 70, MS, 74; Univ Southwestern La, PhD(biol), 80. *Prof Exp:* Biologist, Gulf S Res Inst, 73-76; instr zool & physiol, Univ Southwestern La, 77-80; asst dir, Jefferson Davis Parish Mosquito Abatement Dist, 80-84; from instr to asst prof, 84-90, ASSOC PROF, LA STATE UNIV, EUNICE, 90- *Concurrent Pos:* Jessup fel, Acad Nat Sci Philadelphia, 78. *Mem:* Sigma Xi; Am Malacol Union. *Res:* Aquatic ecology; invertebrates in parasitic ecosystems; parasitology; evolution and systematics of fresh-water mollusks and their parasites; evolution and systematics of unionicolid water-mites of the world; biodiversity. *Mailing Add:* Rte 2 Box 489J Eunice LA 70535-9410

VIDT, EDWARD JAMES, b Pittsburgh, Pa, June 16, 27; m 53; c 2. COAL CONVERSION, COST ENGINEERING. *Educ:* Carnegie Inst Technol, BSc, 48. *Prof Exp:* Sr develop engr chem eng, Air Reduction Co, 59-60; process proj engr, Chem Plants Div, Blaw-Knox Co, 60-65, asst mgr synthetic fuels, 65-72; sr res scientist, 72-74, fel engr, 74-78, adv engr chem eng, Process Eng Dept, Westinghouse Elec Corp, 78-87; CHEM ENG CONSULT, 87- *Concurrent Pos:* Mem, Ad Hoc Comt Data Coal Conversion, Nat Bur Standards, 72-74; mem, Tech Adv Comt, US Off Coal Res, 74-75. *Mem:* Am Inst Chem Engrs; Carnegie Inst. *Res:* Coal conversion to synthetic fuels for clean, efficient production of heat and power. *Mailing Add:* Consult Engr 2510 Hollywood Dr Pittsburgh PA 15235

VIDULICH, GEORGE A, environmental chemistry; deceased, see previous edition for last biography

VIDYASAGAR, MATHUKUMALLI, b Guntur, Andhra Pradesh, India, Sept 29, 47. ELECTRICAL ENGINEERING, APPLIED MATHEMATICS. *Educ:* Univ Wis-Madison, BS, 65, MS, 67, PhD(elec eng), 69. *Prof Exp:* Asst prof elec eng, Marquette Univ, 69-70; asst prof & Nat Res Coun Can grants, 70-73, assoc prof, 73-77, PROF ELEC ENG, CONCORDIA UNIV, 77-; PROF ELEC ENG, UNIV WATERLOO. *Concurrent Pos:* Consult, Alcan Smelters & Chem, Ltd, 75-78; mem grant selection comt for elec engrs, Natural Sci & Eng Res Coun, 78- *Honors & Awards:* Prix George Montefiore, Asn Elec Engrs Belg, 76. *Mem:* Inst Elec & Electronics Engrs; Soc Indust & Appl Math. *Res:* Control and system theory; large-scale systems. *Mailing Add:* Dept Elec Eng Univ Waterloo Waterloo ON N2L 3G1 Can

VIEBROCK, FREDERICK WILLIAM, b Staten Island, NY, Nov 23, 35; m 56; c 2. BIOCHEMISTRY. *Educ:* Wagner Col, BS, 57; Polytech Inst Brooklyn, MS, 68; Va Polytech Inst & State Univ, PhD(biochem), 74. *Prof Exp:* Res scientist enzymol, Wallerstein Lab, Div Travenol Lab, 58-69; SR SCIENTIST BIOCHEM, JOHNSON & JOHNSON RES CTR, 72- *Mem:* AAAS; Am Chem Soc; Sigma Xi. *Res:* Purification and kinetic analysis of enzymes of the purine metabolic pathways; wound healing; eczematous skin diseases; local anesthetic; interactions of flouride with tooth enamel. *Mailing Add:* 1779 W Circle Rd Martinsville NJ 08836

VIECHNICKI, DENNIS J, b Passaic, NJ, Dec 25, 40; m 65; c 4. CERAMICS, MATERIAL SCIENCE. *Educ:* Rutgers Univ, BS, 62; Pa State Univ, PhD(ceramics), 66. *Prof Exp:* Sr scientist, Westinghouse Res & Develop Labs, 66; br chief, Processing & Applications Br, 68-87, BR CHIEF, CERAMICS RES BR, US ARMY MAT TECHNOL LAB, 87- *Concurrent Pos:* Nat Sci Res Ctr-NSF France-US exchange of scientists grant, Lab Appl Solid State Chem, Metall Chem Res Ctr, Vitry-sur-Seine, France, 72-73. *Mem:* Fel Am Ceramic Soc. *Res:* Reactions in oxides above 1000 C; eutectoid decomposition; high temperature growth using heat exchanger method of sapphire, spinel, Nd:YAG and eutectics; ceramic-metal seals; radome materials; high energy laser-ceramic interactions; ceramic processing; ceramic properties at high strain rates; armor materials. *Mailing Add:* Five Poplar Rd Wellesley MA 02181

VIEHLAND, LARRY ALAN, b St Louis, Mo, Apr 30, 47; m 69; c 2. CHEMICAL PHYSICS. *Educ:* Mass Inst Technol, BS, 69; Univ Wis-Madison, PhD(chem), 73. *Prof Exp:* Res assoc chem, Brown Univ, 73-76, asst prof (res) chem, 76-77; from asst prof to assoc prof, 77-82, PROF CHEM, PARKS COL, 82- *Concurrent Pos:* Fulbright scholar, Australia, 88. *Mem:* Sigma Xi; Am Phys Soc. *Res:* Theoretical chemistry and atomic physics, specifically kinetic theory and nonequilibrium statistical mechanics as a tool for understanding intermolecular potentials and other microscopic properties. *Mailing Add:* Dept Chem Parks Col St Louis Univ Cahokia IL 62206

VIEIRA, DAVID JOHN, b Oakland, Calif, May 5, 50; m 72; c 3. NUCLEAR CHEMISTRY. *Educ:* Ore State Univ, BS, 72; Univ Calif, Berkeley, PhD(nuclear chem), 78. *Prof Exp:* Res fel, Nat Sci Found, Ore State Univ, 71; res asst, Los Alamos Sci Lab, 72; res & teaching asst nuclear chem, Univ Calif, Berkeley & Lawrence Berkeley Lab, 72-78; dir fel, Los Alamos Nat Lab, 78-79, staff scientist, 79-85, sect leader nuclear chem, 85-89, TEAM LEADER NUCLEAR CHEM, LOS ALAMOS NAT LAB, 89- *Concurrent Pos:* Alexander von Humboldt fel, GSI, Darmstadt, Ger, 90-91. *Mem:* Am Chem Soc; Am Phys Soc. *Res:* Nuclear mass measurements; radioactive beam research; decay and reaction studies of light nuclei far from B-stability; magnetic spectrometers; fast-timing devices and beams optics; author of numerous papers in various professional journals. *Mailing Add:* INC-11/LAMPF MS No H 824 Los Alamos Nat Lab PO Box 1663 Los Alamos NM 87545-0001

VIELE, GEORGE WASHINGTON, b Wausau, Wis; m 58; c 2. GEOLOGY, TECTONICS. *Educ:* Yale Univ, BS, 51; Univ Utah, PhD(geol), 60. *Prof Exp:* Geologist, US Geol Surv, 51-56 & Standard Oil Co Calif, 57-59; from asst prof to assoc prof, 59-72, chmn dept, 74-77, PROF GEOL, UNIV MO, COLUMBIA, 72- *Mem:* Geol Soc Am; Am Geophys Union. *Res:* Structural geology; regional tectonics; Northern Rocky and Ouachita Mountains. *Mailing Add:* Dept Geol Univ Mo Columbia MO 65211

VIEN, STEVE HUNG, b Sept 13, 57; m; c 1. INORGANIC & ORGANIC ANALYSES. *Educ:* Iowa State Univ, BS, 81; Kans State Univ, MS, 86, PhD(anal chem), 88. *Prof Exp:* Anal chemist, Sterling Drug Inc, 81-83; SR RES CHEMIST, DOW CHEM, 88- *Mem:* Am Chem Soc; Appl Spectros Soc. *Res:* Developing methods for inorganic and organic analyses; solving problems related to chemical production. *Mailing Add:* Dow Chem A-915 Freeport TX 77541

VIER, DWAYNE TROWBRIDGE, b Washington, DC, Sept 17, 14; m 51; c 2. THERMODYNAMICS & MATERIAL PROPERTIES. *Educ:* Univ NH, BS, 37, MS, 39; Columbia Univ, PhD(chem phys), 43. *Prof Exp:* Asst, Univ NH, 37-39, asst Columbia Univ, 39-40, asst chem, 40-42, res chemist, S A M Labs, 43-45; assoc scientist, Manhattan Dist, 45-46, group leader, Los Alamos Sci Lab, 46-70, staff mem, Los Alamos Nat Lab, Univ Calif, 70-82, consult, Los Alamos Nat Lab, NMex, 82-90; RETIRED. *Mem:* Am Chem Soc; AAAS. *Res:* Fields in physical chemistry; inorganic chemistry of rare radioactive elements; high temperature chemistry. *Mailing Add:* 764 43rd St Los Alamos NM 87544

VIERCK, CHARLES JOHN, JR, b Columbus, Ohio, July 6, 36; m 60; c 2. NEUROSCIENCE, SOMATOSENSATION. *Educ:* Univ Fla, BS, 59, MS, 61, PhD(psychol), 63. *Prof Exp:* Fel neurosci, Inst Neurol Sci, Univ Pa, 63-65; asst prof, 65-71, assoc prof, 71-76, PROF NEUROSCI, COL MED, UNIV FLA, 76-; ADJ RES PROF PHYSIOL, SCH MED, UNIV NC, 75- *Concurrent Pos:* Nat Inst Neurol Dis & Stroke res grant, 67-, mem neurol B study sect, 72-76; assoc ed, J Neurosci, 80-83; mem, Animal Resources Rev Comt, 81-85; ed, Somatosensory Res, 83- *Mem:* Am Psychol Asn; Psychonomic Soc; Soc Neurosci; Int Neuropsychol Soc; Int Asn Study Pain. *Res:* Central nervous system mechanisms relating to somesthetic discrimination; discrimination and perception of pain; recovery of function after nervous system damage. *Mailing Add:* Dept Neurosci J Hillis Miller Health Ctr Univ Fla Col Med Gainesville FL 32610

VIERCK, ROBERT K, b Avoca, Iowa, Jan 5, 08; m 33; c 3. ENGINEERING MECHANICS. *Educ:* Univ Iowa, BS, 32, MS, 33. *Prof Exp:* Engr, State of Iowa, 33-34; jr engr, US Bur Reclamation, 34-36; instr eng, Univ Ill, 36-39; from asst engr to assoc engr, Fed Power Comn, 39-43; from asst prof to prof eng mech, 43-73, actg head dept, 65-67, EMER PROF ENG MECH, PA STATE UNIV, 73- *Concurrent Pos:* Consult, Boeing Co, Wash, 54 & NAm Aviation, Inc, Calif, 55. *Mem:* Am Soc Eng Educ; Am Acad Mech. *Res:* Mechanical vibrations; mechanical properties of materials. *Mailing Add:* 299 Nimitz Ave State College PA 16801-6308

VIERECK, LESLIE A, b New Bedford, Mass, Feb 20, 30; m 55; c 3. PLANT ECOLOGY, PLANT TAXONOMY. *Educ:* Dartmouth Col, BA, 51; Univ Colo, MA, 57, PhD(plant ecol), 62. *Prof Exp:* Asst bot, McGill Subarctic Res Sta, 54-55; asst, Herbarium, Univ Colo, 55-57, actg cur, 56-57, res assoc ecol, Inst Arctic & Alpine Res, 55-59; res assoc ecol, Univ Alaska, 59-60, asst prof bot, 60-61; res biologist, Alaska Dept Fish & Game, 61-63; PRIN PLANT ECOLOGIST, INST NORTHERN FORESTRY, 63- *Concurrent Pos:* Affil prof div life sci, Univ Alaska, 75- *Mem:* Fel AAAS; fel Arctic Inst NAm; Ecol Soc Am; Soc Am Foresters; Sigma Xi. *Res:* Plant ecology and plant taxonomy of arctic, subarctic and alpine regions. *Mailing Add:* 308 Tanana Dr Inst Northern Forestry Fairbanks AK 99775-5500

VIERNSTEIN, LAWRENCE J, b New York, NY, Feb 20, 19; m 69; c 2. ELECTRONICS ENGINEERING. *Educ:* Okla State Univ, BS, 50, MS, 51; Johns Hopkins Univ, PhD, 70. *Prof Exp:* Mem assoc staff, Appl Physics Lab, Johns Hopkins Univ, 52-57, sr physicist, 57-59, mem prof staff, Appl Physics Lab, 59-81, Dept Physiol, 61-66, Wilmer Inst, 66-76, MEM PROF STAFF, NEUROSURG, JOHNS HOPKINS UNIV, 76- *Mem:* Sigma Xi. *Res:* Theoretical biology and biomedical engineering; neurophysiology; artificial intelligence in medicine. *Mailing Add:* 12 W 96th St 5D New York NY 10025

VIERS, JIMMY WAYNE, b Grundy, Va, Feb 26, 43; m 65; c 2. PHYSICAL CHEMISTRY. *Educ:* Berea Col, AB, 65; Wake Forest Univ, MA, 67; Stanford Univ, PhD(chem), 71. *Prof Exp:* ASSOC PROF CHEM, VA POLYTECH INST & STATE UNIV, 71- *Mem:* Am Chem Soc. *Res:* Quantum chemistry. *Mailing Add:* Dept Chem Va Polytech Inst & State Univ Blacksburg VA 24060

VIERTL, JOHN RUEDIGER MADER, b New York, NY, Sept 25, 41; m 69; c 2. ELECTROMAGNETISM, SOLID STATE PHYSICS. *Educ:* Fordham Univ, BS, 63; Rutgers Univ, MS, 65; Cornell Univ, PhD(appl physics), 73. *Prof Exp:* Res training prog physics & geophys fission tracks & semiconductors, 67-69, RES PHYSICIST ULTRASONIC PHENOMENA & LASER INTERACTIONS, GEN ELEC TURBIN TECHNOL LAB, 73- *Honors & Awards:* Managerial Award, Gen Elec Turbine, 85. *Mem:* Fel Am Phys Soc; Soc Photo-Optical Instrumentation Engrs; Am Soc Nondestructive Testing. *Res:* Optical properties of thin films; point defects in solids; ultrasonic imaging, scattering theory, phenomena, ultrasonic non destructive testing; laser target interactions; ultrasonic transducer design; electromagnetism, eddy currats, coatings metallic and non metallic automotive testing systems. *Mailing Add:* 1403 Clifton Park Rd Schenectady NY 12309

VIESSMAN, WARREN, JR, b Baltimore, Md, Nov 9, 30; m 53; c 9. HYDROLOGY, WATER RESOURCES. *Educ:* Johns Hopkins Univ, BE, 52, MSE, 58, DEng(water resources), 61. *Prof Exp:* Proj engr, Johns Hopkins Univ, 56-61; from asst prof to assoc prof civil eng, Univ NMex, 61-66; prof civil eng & dir water resources ctr, Univ Maine, 66-68; prof civil eng & dir water resources res inst, Univ Nebr, Lincoln, 68-75; sr specialist eng & pub words, Cong Res Serv, Libr of Cong, 75-83; chmn & prof environ eng sci, 83-90, ASSOC DEAN RES & GRAD STUDY, UNIV FLA, GAINESVILLE, 90- *Honors & Awards:* Icko Iben Award, 83. *Mem:* Am Soc Civil Engrs; Am Water Resouces Asn. *Res:* Water resources systems and policy. *Mailing Add:* Col Eng Univ Fla Gainesville FL 32611

VIEST, IVAN M, b Slovakia, Oct 10, 22; nat US; m 53. STRUCTURAL ENGINEERING, ENGINEERING PROMOTION. *Educ:* Slovak Tech Univ, Slovakia, CE, 46; Ga Inst Technol, MS, 48; Univ Ill, PhD(eng), 51. *Prof Exp:* Asst, Univ Ill, 48-51, from res assoc to res assoc prof, 51-57; bridge res eng, Am Asn State Hwy Off Rd Test, Nat Acad Sci, 57-61; struct engr, Bethlehem, Steel Corp, 61-67, sr struct consult, 67-70, asst mgr sales eng, 70-82; CONSULT STRUCT ENGR, 83- *Concurrent Pos:* Consult, Nelson Student Welding, Ohio, 54-61; mem, Transp Res Bd, Nat Acad Sci-Nat Res Coun; mem bd, Eng Found, 75-87, Am Soc Civil Engrs, 69-71 & 74-75. *Honors & Awards:* Wason Medal, Am Concrete Inst, 55; Res Prize, Am Soc Civil Engrs, 58; Construct Award, Eng News Rec, 62. *Mem:* Nat Acad Eng; hon mem Am Soc Civil Engrs (vpres, 74-75); fel Am Concrete Inst; Int Asn Bridge & Struct Eng; fel AAAS; prof mem, Am Inst Steel Construct. *Res:* Steel structures; composite construction; reinforced concrete structures. *Mailing Add:* PO Box 1428 Bethlehem PA 18016

VIETH, JOACHIM, b Hamburg, Ger, Oct 26, 25; m 53; c 2. PLANT MORPHOLOGY. *Educ:* Univ Saarbruecken, Lic natural sci, 53, Dr rer nat, 57; Univ Dijon, DSc(bot), 65. *Prof Exp:* Asst bot, Univ Saarbruecken, 53-57; res fel, Nat Ctr Sci Res, Univ Dijon, 57-65; vis prof, 66-67, assoc prof, 67-77, PROF BOT, UNIV MONTREAL, 77- *Mem:* Can Bot Asn; Int Asn Plant Tissue Cult. *Res:* Anatomy of flowers and inflorescences, both normal and anomalous; relationship between vegetative and inflorescential regions, between normal and anomalous forms experimentally induced; plant propagation by tissue culture. *Mailing Add:* Dept Biol Sci Univ Montreal CP 6128 Succursale A Montreal PQ H3C 3J1 Can

VIETH, WOLF R(ANDOLPH), b St Louis, Mo, May 5, 34; m 57; c 4. CHEMICAL ENGINEERING. *Educ:* Mass Inst Technol, SB, 56, ScD(chem eng), 61; Ohio State Univ, MSc, 58. *Prof Exp:* Res engr, NAm Aviation, Inc, 56-57; Ford fel eng, 61-62; dir practice sch sta, Mass Inst Technol Sta-Am Cyanamid Co, NJ, 62-64; from asst prof to assoc prof chem eng, Mass Inst Technol, 62-68, overall dir sch chem eng pract, 65-68; chmn chem & biochem eng dept, 68-78, PROF CHEM & BIOCHEM ENG, RUTGERS UNIV, NEW BRUNSWICK, 68- *Concurrent Pos:* Consult, Am Cyanamid Co, 63, Ashland Oil Co, 63-, Carter's Ink, 64- & US Army Natick Labs, 65- *Mem:* Am Chem Soc; Am Inst Chem Engrs. *Mailing Add:* Dept of Chem & Biochem Eng Rutgers Univ New Brunswick NJ 08903

VIETMEYER, NOEL DUNCAN, b Wellington, NZ, Nov 9, 40; m 65; c 3. ECONOMIC BIOLOGY, SCIENCE WRITING. *Educ:* Univ Otago, NZ, BSc, 63; Univ Calif, Berkeley, PhD, 67. *Prof Exp:* Lectr org chem, Univ Calif, Berkeley, 67-68; NIH fel, Stanford Univ, 68-69, fel, 68-70; PROF ASSOC, NAT ACAD SCI, 70- *Res:* Innovative technology for developing countries; development of neglected plants and animals with promising economic potential. *Mailing Add:* Nat Acad Sci 2101 Constitution Ave Washington DC 20418

VIETOR, DONALD MELVIN, b Urbana, Ill, Sept 29, 45; m 71; c 2. CROP PHYSIOLOGY. *Educ:* Univ Minn, BS, 67, MS, 69; Cornell Univ, PhD(crop sci), 75. *Prof Exp:* Biol sci asst soil sci, US Army Cold Regions Res & Engr Lab, 69-71; asst prof agron, Univ Mass, 74-76; asst prof, 76-82, ASSOC PROF AGRON, DEPT SOIL & CROP SCI, TEX A&M UNIV, 82- *Concurrent Pos:* Chmn, Student Activities, Am Soc Agron, 84; res, Curric Proj, Nat Agr, 84-89. *Mem:* Am Soc Agron; Crop Sci Soc Am; Am Soc Plant Physiol. *Res:* Physical and enzymatic regulation of carbon export from leaves, and of carbon partitioning within plant during growth and development; photosynthate partitioning during seeding development and after seed maturation of grasses and legumes; systems approach to research planning. *Mailing Add:* Dept Soil & Crop Sci Tex A&M Univ College Station TX 77843

VIETS, FRANK GARFIELD, JR, b Stanberry, Mo, Apr 3, 16; m 38; c 3. SOIL SCIENCE. *Educ:* Colo Agr & Mech Col, BS, 37; Univ Calif, MS, 39, PhD(plant physiol), 42. *Prof Exp:* Agent div cereal crops, USDA, Calif, 37-39; asst div plant nutrit, Univ Calif, 39-42; supv chemist, Cutter Labs, Calif, 42-44; assoc agr chemist, Exp Sta, SDak State Col, 44-45; agronomist div soil mgt & irrig, USDA, 45-49, soil scientist, Agr Res Serv, 49-53, soil & water conserv res div, 53-74; CONSULT AGR, 80- *Concurrent Pos:* Vis prof, Univ Ill, 59 & Iowa State Univ, 64, Col State Univ, 74 & Univ Saskatchewan, 75; ed in chief, Soil Sci Soc Am, 63-65; agr consult, 74- *Mem:* Fel AAAS; fel Soil Sci Soc Am (vpres, 66, pres, 67); fel Am Soc Agron; Int Soc Soil Sci. *Res:* Mineral nutrition of plants; zinc deficiency in soils and plants; water pollution by animal wastes, fertilizers and agriculture; soil fertility and productivity; tropical soils. *Mailing Add:* 102 Yale Way Ft Collins CO 80525-1718

VIETS, HERMANN, b Quedlinburg, Ger, Jan 28, 43; US citizen; m 68; c 4. AEROSPACE ENGINEERING, MECHANICAL ENGINEERING. *Educ:* Polytech Inst Brooklyn, BS, 65, MS, 66, PhD(astronaut), 70. *Prof Exp:* Res asst fluid mech res, Polytech Inst Brooklyn, 68-69; group leader, Aerospace Res Labs, US Air Force, 70-75; assoc prof mech eng, Wright State Univ, 76-80, prof, 80-81; prof mech eng & assoc dean, WVa Univ, 81-83; PROF MECH ENG & DEAN, UNIV RI, 83- *Concurrent Pos:* NATO res grant, 69-70; fel, von Karman Inst, Brussels, Belg, 69-70; USAF Off Aerospace res grant, 77-; consult, USAF Aero Propulsion Lab, 76-80; chmn bd, Precision Stampings Inc, Beaumont, Calif, 78-; mem & bd dir, Astro-Med Inc, West Warwick, RI, Promptus Commun, Portsmouth, RI. *Honors & Awards:* Gov Award for Sci & Technol, 87. *Mem:* Assoc fel Am Inst Aeronaut & Astronaut; Am Helicopter Soc; Soc Mfg Engrs; Ger Soc Air & Space Travel. *Res:* Positive aspects of time dependent flows; fluidically and mechanically generated unsteadiness; advanced ramjet combustors; vortex dynamics; jets and wakes; computational methods; nozzles, diffusers and thrust augmentors. *Mailing Add:* Col Eng Univ RI Kingston RI 02881

VIETTE, MICHAEL ANTHONY, b Pittsburg, Kans, Mar 27, 41; m 63; c 1. PHYSICS. *Educ:* Kans State Col Pittsburg, BA, 64, MS, 66; Univ Mo, Rolla, PhD(physics), 72. *Prof Exp:* Res asst cloud physics, Grad Ctr Cloud Physics Res, Univ Mo, Rolla, 66-70; asst prof, 71-80, ASSOC PROF PHYSICS, UNIV MAINE, ORONO, 80- *Concurrent Pos:* NSF res grant, 72- *Mem:* Am Phys Soc; Am Geophys Union; Am Meteorol Soc. *Res:* Condensation and growth of micron sized water droplets. *Mailing Add:* Box 443 25 Fed St Bar Harbor ME 04609

VIETTI, TERESA JANE, b Ft Worth, Tex, Nov 5, 27. PEDIATRICS, HEMATOLOGY. *Educ:* Rice Inst, AB, 49; Baylor Univ, MD, 53; Am Bd Pediat, dipl, 59; Bd Pediat Hemat & Oncol, dipl, 74. *Prof Exp:* Instr pediat, Wayne State Univ, 58 & Southwestern Med Sch, Univ Tex, 58-60; vis pediatrician, Hacettepe Children's Hosp, Ankara, Turkey, 60-61; from asst prof to assoc prof pediat, 61-72, PROF PEDIAT, SCH MED, WASH UNIV, 72-, PROF PEDIAT IN RADIOL, 80- *Concurrent Pos:* Dir hemat labs, attend pediatrician & consult, Tex Children's Hosp, Dallas, 58-60; attend pediatrician & consult, Parkland Mem Hosp, 58-60; Am Cancer Soc fel, 58-59; USPHS trainee, 59-60, grant, 61-; vchmn, Southwest Oncol Group, 61-; asst pediatrician, St Louis Children's Hosp, 61-65, assoc pediatrician, 65-, dir div hemat & oncol, 70-; from asst pediatrician to assoc pediatrician, Barnes & Allied Hosps, 61-65; consult, St Louis County Hosp; assoc in pediat, Mo Crippled Children's Serv; mem, Cancer Clin Invest Rev Comt, 74-78; pediat consult high risk maternity & child care prog, Mo Div Health, 75; chmn Pediat Oncol Group, 81-85. *Mem:* Am Acad Pediat; Am Hemat Soc; Int Soc Hemat; Am Asn Cancer Res; Am Pediat Soc. *Res:* Oncology; cancer chemotherapy. *Mailing Add:* St Louis Childrens Hosp 400 S Kingshighway St Louis MO 63110

VIG, BALDEV K, b India, Oct 1, 35; c 2. CYTOGENETICS. *Educ:* Khalsa Col, India, BSAgr, 57; Panjab Univ, India, MS, 61; Ohio State Univ, PhD(genetics), 67. *Prof Exp:* Demonstr agr, Khalsa Col, India, 58-61; assoc prof bot, Rajasthan Col Agr, India, 61-64; res cytogeneticist, Dept Pediat, Children's Hosp, Ohio State Univ, 67-68; from asst prof to assoc prof biol, 68-78, PROF GENETICS, UNIV NEV, RENO, 78- *Concurrent Pos:* Consult, Western Environ Res Ctr, Environ Protection Agency; res grant, Environ Protection Agency, NIH & Univ Nev, 78-; human & med geneticist, Nev Ment Health Inst, 76-81; Humboldt fel & Jones fel, Ger Cancer Res Ctr fel. *Mem:* Am Genetics Asn; Genetic Soc Am; Genetics Soc Can; Environ

Mutagen Soc; Sigma Xi; Am Soc Human Genetics. *Res:* Action of antileukemic drugs on chromosomes; sequence of centromere separation; kinetochore/centromere function. *Mailing Add:* Dept Biol Univ Nev Reno NV 89557-0015

VIGDOR, STEVEN ELLIOT, b New York, NY, July 23, 47; m 70; c 2. EXPERIMENTAL NUCLEAR PHYSICS. *Educ:* City Col NY, BS, 67; Univ Wis-Madison, MS, 69, PhD(physics), 73. *Prof Exp:* Res assoc, Dept Physics, Univ Wis, 73-74; fel appointee, Argonne Nat Lat, 74-76, res assoc, 76; asst prof, 76-79, ASSOC PROF PHYSICS, IND UNIV, 79- *Mem:* Am Phys Soc. *Res:* Nuclear structure and nuclear reactions at intermediate energies; interplay of microscopic and macroscopic aspects of heavy-ion-induced reactions; polarization measurements in nuclear structure studies and tests of fundamental symmetry principles. *Mailing Add:* Dept Physics Ind Univ Bloomington IN 47401

VIGEE, GERALD S, b Crowley, La, Mar 4, 31; c 3. PHYSICAL INORGANIC CHEMISTRY. *Educ:* US Mil Acad, BS, 54, La State Univ, Baton Rouge, BS, 60, PhD(chem), 68. *Prof Exp:* Prof engr, NASA, Ala, 60-61; propulsion design engr, Chrysler Corp, 61-64; asst prof chem, Univ Miss, 68-69; ASSOC PROF CHEM, UNIV ALA, BIRMINGHAM, 69- *Mem:* Sigma Xi. *Res:* Synthesis of coordination complexes, investigation of the magneto chemistry and spectroscopic energy levels of these complexes. *Mailing Add:* 2304 Pine Crest Dr Birmingham AL 35216-2112

VIGFUSSON, NORMAN V, b Ashern, Man, July 1, 30; m 54; c 5. GENETICS, MEDICAL GENETICS. *Educ:* Univ Man, BSA, 51; Univ Alta, PhD(fungal genetics), 69. *Prof Exp:* Asst prof genetics, 69-72, assoc prof, 72-76, PROF BIOL, EASTERN WASH UNIV, 76- *Concurrent Pos:* Genetic consult, Sacred Heart Med Ctr, Spokane, Wash, 75- *Mem:* Genetics Soc Am; AAAS; Sigma Xi. *Res:* Sexuality in Neorospora crassa with respect to stages and control of the sexual cycle and attempts to arrive at elucidation of incompatibility control mechanism; human cytogenetics. *Mailing Add:* Sacred Heart Med Ctr Cyto-Genetic Lab 101 W Eighth Spokane WA 99204

VIGGERS, ROBERT F, b Tacoma, Wash, Jan 18, 23; m 45; c 3. MECHANICAL ENGINEERING. *Educ:* Univ Wash, BS, 44; Ore State Col, MS, 50. *Prof Exp:* Instr eng, Univ Wash, 46-47 & Ore State Col, 47-49; from instr to prof eng, Seattle Univ, 49-70, chmn dept, 78-82, prof mech eng, 70-89, EMER PROF MECH ENG, SEATTLE UNIV, 90- *Concurrent Pos:* NSF fac fel, 60-61; head hydraul sect, Reconstruct Cardiovasc Res Lab, Providence Hosp, 61- *Res:* Automatic control; stability problems; cardiovascular hydraulics; machine design. *Mailing Add:* Seattle Univ 900 Broadway Seattle WA 98122

VIGGIANO, ALBERT, b Derby, Conn, Feb 28, 54. ATMOSPHERIC ION CHEMISTRY, STATE SELECTED REACTIVITY. *Educ:* Univ Calif, Berkeley, BS, 76; Univ Colo, Boulder, PhD(chem physics), 80. *Prof Exp:* Res chemist, Max Planck Inst, 80-82; RES CHEMIST, GEOPHYS DIRECTORATE, PHILLIPS LAB, 83- *Mem:* Am Chem Soc; Am Soc Mass Spectrometry; Am Geophys Union. *Res:* Kinetic energy, temperature, rotational and vibrational energy dependencies on the rate constants and branching ratios of ion molecule reactions in the gas phase. *Mailing Add:* Geophys Directorate-LID Phillips Lab Hanscom AFB MA 01731-5000

VIGIL, EUGENE LEON, b Chicago, Ill, Mar 14, 41; div; c 3. CELL BIOLOGY, PLANT PHYSIOLOGY. *Educ:* Loyola Univ Chicago, BS, 63; Univ Iowa, MS, 65, PhD(bot), 67. *Prof Exp:* NIH fel, Univ Wis-Madison, 67-69; trainee cell biol, Univ Chicago, 69-71; asst prof cell biol, Marquette Univ, 71-79; plant cell biologist, Univ Md, 79-81; res assoc, 81-88, PLANT PHYSIOLOGIST, BELTSVILLE AGR RES CTR, USDA, 88- *Concurrent Pos:* Distinguished vis scientist, Dept Physiol & Biophys, Colo State Univ, 74-75; prog chmn, Histochem Soc, 76-79, counr, 79-82. *Mem:* NY Acad Sci; Am Soc Cell Biol; Am Soc Plant Physiol; Histochem Soc. *Res:* Biogenesis and turnover of microbodies in cotyledons of fatty seeds during germination; effects of drought stress on development and utilization of protein bodies in radicles of oil seeds; effects of drought stress on cotton fiber development and maturity. *Mailing Add:* B001 BARC-W Climate Stress Lab Beltsville MD 20705-2350

VIGIL, JOHN CARLOS, b Espanola, NMex, Mar 28, 39; m 58; c 4. NUCLEAR ENGINEERING. *Educ:* NMex Inst Mining & Technol, BS, 61; Univ NMex, MS, 63, PhD(nuclear eng), 66. *Prof Exp:* Staff mem reactor physics, Los Alamos Nat Lab, 63-77, group leader, Thermal Reactor Safety group, 77-80, asst div leader, Energy Div, 80-81, asst to assoc dir, 81-84, dep div leader, Personnel Admin, 84-86, dir staff, 86-89, DIV LEADER, HUMAN RESOURCES DEVELOP, LOS ALAMOS NAT LAB, 89- *Concurrent Pos:* Mem, Bd Regents, NMex State Univ, 89- *Mem:* Am Nuclear Soc; NY Acad Sci; Sigma Xi. *Res:* Reactor safety, physics, codes and computations; energy technology. *Mailing Add:* 215 Kimberly Los Alamos NM 87544

VIGLIERCHIO, DAVID RICHARD, b Madera, Calif, Nov 25, 25; m 67; c 1. NEMATOLOGY. *Educ:* Calif Inst Technol, BS, 50, PhD(bio-org chem), 55. *Prof Exp:* Jr res nematologist, 55-57, asst res nematologist, 57-63, assoc nematologist, 63-69, chmn nematol, 78-85, NEMATOLOGIST, UNIV CALIF, DAVIS, 69- *Concurrent Pos:* Fulbright fel, 64-65, 76-77; J S Guggenheim fel, 65; partic, US Antarctic Prog, 69-70; Nat Acad Sci exchange USSR, 70-71; Minister Agr & Forestry fel, WGer, 84. *Honors & Awards:* Minister Agr Invitation Lectr, PRC, 81. *Mem:* Am Chem Soc; Soc Nematol; Soc Europ Nematol; AAAS. *Res:* Chemistry and physiology of plant parasitic and free-living nematodes; host-parasite relationships; physiological methods of nematode control; behavioral properties of nematodes. *Mailing Add:* 710 Miller Dr Davis CA 95616-3621

VIGLIONE, SAM S, b Erie, Pa, July 12, 29; m 57; c 4. ELECTRICAL ENGINEERING. *Educ:* Carnegie Inst Technol, BS, 54; Univ Southern Calif, MS, 56. *Prof Exp:* Electronic engr, Hughes Aircraft Co, 54-58; sr design engr, Astronaut Div, Convair Corp, 58-59; sr res scientist, Aeronutronic Div, Ford Motor Co, 59-61; mgr, Pattern Recognition Systs Dept, McDonnell Douglas Corp, Huntington Beach, 61-76, dir, Res & Develop Directorate, 76-78; Sr Dir, S S Viglione & Assocs, 78-; PRES, INTERSTATE VOICE PRODS. *Mem:* Sr mem Inst Elec & Electronics Engrs. *Res:* Mathematical procedures for the simulation of pattern recognition systems; pattern recognition systems for classification of photographic and physiologic data; investigation of biological neural networks and their electronic replication; development of speech recognition systems. *Mailing Add:* 13301 Prospect Santa Ana CA 92705

VIGNERY, AGNES M C, b Poitieres, France, Mar 2, 49. BONE CELL BIOLOGY. *Educ:* Paris Univ, DDS, 72, PhD, 75. *Prof Exp:* ASST PROF ORTHOP SURG, SCH MED, YALE UNIV, 77- *Mem:* Am Soc Cell Biol; Am Soc Bone & Mineral Res; Calcified Tissue Soc. *Mailing Add:* Dept Cell Biol Yale Univ 333 Cedar St New Haven CT 06510

VIGNOS, JAMES HENRY, b Cleveland, Ohio, July 27, 33; m 62; c 2. MATHEMATICS. *Educ:* Case Inst, BS, 55; Yale Univ, MS, 57, PhD(physics), 62. *Prof Exp:* Vis res scientist, Low Temperature Inst, Bavarian Acad Sci, Ger, 62-64; resident res assoc, Chem Div, Argonne Nat Lab, 64-66; asst prof physics, Dartmouth Col, 66-72; sr res scientist, 74-78, prin res scientist, 78-84, CONSULT SCIENTIST, CORP RES CTR, FOXBORO CO, 84- *Concurrent Pos:* Fulbright res scholar, 62-63; von Humboldt fel, 63-64; asst ed, Am J Physics, 70-73, ed, Am J Physics Ten-Year Cumulative Index, 63-72 & 72-73; mem exec comt, Am Phys Soc, New Eng, 75-77; subcomt, MFC-SC9, Ultrasonic Flowmeters, Coun Codes & Standards, Am Soc Mech Engrs, 76-, subcomt, MFC-SC15, Installation Effects on Primary Devices, 79-, chmn subcomt, MFC-SC5, electromagnetic flowmeters, 88-, mem main comt, MFC, 88- *Mem:* Am Phys Soc; Sigma Xi. *Res:* Acoustic, electromagnetic, thermal and fluid mechanic investigations relating to advanced approaches to fluid flow measurement; liquid and solid helium; superconductivity; ultrasonics. *Mailing Add:* 129 Manning St Needham Heights MA 02194

VIGNOS, PAUL JOSEPH, JR, b Canton, Ohio, Nov 10, 19; m 46; c 3. MYOLOGY, RHEUMATOLOGY. *Educ:* Univ Notre Dame, BS, 41; Western Reserve Univ, MD, 44. *Prof Exp:* From intern to resident, Univ Hosps, Cleveland, 44-46; resident, Presby Hosp, New York, 48-49; Am Cancer Soc fel, Univ Hosps, Cleveland, 49-50; Rees fel med, Sch Med, 50-51, USPHS fel pharmacol, 51-52, from instr to prof, 52-85, EMER PROF MED, SCH MED, CASE WESTERN RESERVE UNIV, 85. *Honors & Awards:* Medal, Univ Marseille. *Mem:* Am Rheumatism Asn; Am Cong Med Rehab; Cent Soc Clin Res; Royal Soc Health; AAAS. *Res:* Bioclinical effect of myopathic disease of the locomotor system on skeletal muscle and ambulation; biochemistry of normal and diseased muscle. *Mailing Add:* 2875 River Rd Chagrin Falls OH 44022

VIGO, TYRONE LAWRENCE, b New Orleans, La, Feb 1, 39; m 63; c 2. POLYMER CHEMISTRY, TEXTILE CHEMISTRY. *Educ:* Loyola Univ, La, BS, 60; Tulane Univ La, MS, 63, PhD(org chem), 69. *Prof Exp:* Res chemist, Southern Regional Res Ctr, Agr Res Serv, USDA, 63-66, proj leader, textile & polymer chem, 68-76, dir & res leader, textiles & clothing lab, 76-85; LEAD SCIENTIST, SOUTHERN REGIONAL RES CTR, AGR RES SERV, USDA, 85- *Concurrent Pos:* Vis prof chem dept, Tulane Univ, La, 70-75, 87-; adj prof, Textile Dept, Univ Ga, 90- *Mem:* Fiber Soc; Am Chem Soc; Am Asn Textile Chemists & Colorists; Textile Inst. *Res:* Chemical modification of polymers and textiles by application of new synthetic techniques; synthetic organic chemistry; polymer chemistry and physics of textiles and polymers; industrial microbiology of polymeric materials; thermal analysis. *Mailing Add:* Southern Regional Res Ctr USDA Agr Res Serv PO Box 19687 New Orleans LA 70179

VIGRASS, LAURENCE WILLIAM, b Melfort, Sask, May 9, 29; m 54; c 3. GEOLOGY, ENGINEERING. *Educ:* Univ Sask, BE, 51, MSc, 52; Stanford Univ, PhD(geol), 61. *Prof Exp:* Geologist, Calif Stand Co, 52-55; res geologist, Imp Oil Ltd, 58-65; consult geologist, Western Resources Consult Ltd, 65-68; assoc prof geol, Univ Sask, Regina, 68-73, actg chmn dept, 72-73, dir energy res, 76-87; head dept, 88-91, PROF GEOL, UNIV REGINA, 73- *Mem:* Geol Asn Can; Can Inst Mining & Metall; Am Asn Petrol Geol; Can Soc Petrol Geologists; Can Geothermal Energy Asn. *Res:* Sedimentary geology; occurence of petroleum and natural gas; geothermal energy in sedimentary basins; water movement and occurrence in the subsurface. *Mailing Add:* Dept Geol Univ Regina Regina SK S4S 0A2 Can

VIJAY, HARI MOHAN, Can citizen; m 68; c 2. MECHANISMS OF HYPERSENSITIVITY. *Educ:* Univ Bombay India BSc, 56, MSc, 58; Univ Manchester Eng, MSc, 63; Univ Birmingham, PhD(org chem), 66. *Prof Exp:* Fel indole alkaloids, Univ Man, 68-70, res assoc immunol, 70-74; RES SCIENTIST, NAT HEALTH & WELFARE CAN, 74- *Concurrent Pos:* Assoc prof, Univ Montreal; adj prof, Univ Ottawa. *Mem:* Can Soc Immunol; Int Asn Aerobiol; Am Asn Immunologists; NY Acad Sci; fel Am Acad Allergy & Immunol. *Res:* Standardization of allergens. *Mailing Add:* Drug Toxicol Div Health Protection Br NHW Sir F Banting Bldg Tunney's Pasture Ottawa ON K1A 0L2 Can

VIJAY, INDER KRISHAN, b Lahore, India, Dec 25, 40. BIOCHEMISTRY, FOOD SCIENCE. *Educ:* Panjab Univ, BS, 61; Univ Sask, MS, 66; Univ Calif, Davis, PhD(biochem), 71. *Prof Exp:* Prod supvr food prod, Nestle Int, 61-63; fel biochem, Sch Med, Univ Mich, 71-72; NIH trainee & fel, Sch Med, Univ Calif, Davis, 72-75; asst prof, 75-80, ASSOC PROF DAIRY SCI, UNIV MD, 80- *Concurrent Pos:* Multiple grants, 75-; NIH res career develop award, 78-83. *Mem:* AAAS; Am Dairy Sci Asn; Inst Food Technologists; Sigma Xi. *Res:* Biochemistry of glycoproteins; enzyme activities in sterilized milk. *Mailing Add:* Dept Dairy Sci Univ Md College Park MD 20742

VIJAY, MOHAN MADAN, b Hospet, Karnataka, India, may 18, 37; Can citizen; m 68; c 2. HIGH PRESSURE WATER JET TECHNOLOGY. *Educ:* Univ Bombay, India, BS, 59; Univ Manchester, UK, BS, 63; Univ Birmingham, UK, MS, 65; Univ Man, PhD(nuclear eng), 78. *Prof Exp:* Lectr mech eng, Univ Man, Winnipeg, Can, 66-74; SR RES OFFICER, INST MECH ENG, NAT RES COUN CAN, 75- *Concurrent Pos:* Ed, Int J Water Jet Technol, 90-; vis prof, Colo Sch Mines, Golden, 90- *Mem:* Sigma Xi; Water Jet Technol Asn (vpres, 89-); Int Soc Water Jet Technol (pres, 90-). *Res:* High speed water jets; field of water jet technology including manufacturing, mining, medical applications and worldwide transfer of technology. *Mailing Add:* Nat Res Coun Can Ottawa ON K1A 0R6 Can

VIJAYAGOPAL, PARAKAT, US citizen; m; c 2. BIOCHEMISTRY. *Educ:* Univ Kerala, India, BS, 66, PhD(biochem), 73; Banaras Hindu Univ, MS, 68. *Prof Exp:* Res asst biochem, Univ Kerala, India, 68-70, lectr, 73-75; teaching fel, Australian Nat Univ, 75-77; res assoc, 78-79, instr, 79-83, asst prof med, 83-87, ASSOC PROF MED, LA STATE UNIV MED CTR, 87- *Mem:* Am Soc Biol Chemists; Am Heart Asn; Soc Complex Carbohydrates; Sigma Xi. *Res:* Arterial wall proteoglycan metabolism in normal and atherosclerotic state; lipoprotein-proteoglycan interactions in cultured cells. *Mailing Add:* Dept Med La State Univ Med Ctr 1542 Tulane Ave New Orleans LA 70112

VIJAYAN, SIVARAMAN, b Thuckalay, Madras, June 14, 45. CHEMICAL ENGINEERING, SURFACE SCIENCE. *Educ:* Madras Univ, BSc, 64; Indian Inst Technol, Madras, B Tech, 67, M Tech, 69; Univ NB, MSc, 71; Swiss Fed Inst Technol, Lausanne, DSc(chem eng), 74. *Prof Exp:* First res asst & lectr, Swiss Fed Inst Technol, Lausanne, 72-76; ASST PROF CHEM ENG, MCMASTER UNIV, 78- *Concurrent Pos:* Fel McMaster Univ, 76-77; adj asst eng, Univ Fla, 77-78; grant, Swiss Nat Found Advan Sci Res, 73-76; consult, Biazzi, SA, Vevey, Switz, 75-76; ed, J Chem Eng 73-76. *Mem:* Can Soc Chem Eng; Can Asn Physicists; Am Inst Chem Eng; Brit Inst Chem Eng; Indian Inst Chem Eng. *Res:* Interfacial phenomena in chemical engineering transport processes; stability of macroemulsions; surface chemistry; surfactants microstructure; dispersion phase separation. *Mailing Add:* Atomic Energy Can Ltd Chalk River Lab Chalk River ON K0J 1J0 Can

VIJAYAN, VIJAYA KUMARI, b Trivandrum, India, Feb 25, 42; m 66; c 2. HUMAN ANATOMY, NEUROANATOMY. *Educ:* Univ Kerala, MBBS, 65; Univ Calif, Davis, PhD(anat), 72. *Prof Exp:* Tutor human anat, Med Col, Trivandrum, India, 65-68; ASST PROF HUMAN ANAT, MED SCH, UNIV CALIF, DAVIS, 73- *Res:* Biochemistry and ultrastructure of developing and aging nervous system; neuroglial reaction to injury; neurotransmitters. *Mailing Add:* Dept Anat Vet Admin Med Ctr 150 Muir Rd Martinez CA 94553

VIJAYENDRAN, BHEEMA R, b Bangalore, India, 1941; m 70; c 2. COLLOID & SURFACE CHEMISTRY. *Educ:* Univ Madras, BTech, 63, MTech, 65; Univ Southern Calif, PhD(chem), 69; Univ New Haven, MBA, 77. *Prof Exp:* Lectr, Cent Leather Res Inst, India, 65-66; indust fel surface chem, R J Reynolds Indust, NC, 69-70; mgr res, Copier Prod Div, Pitney Bowes, Inc, 74-75, chemist, 70-77; res assoc, Celanese Res Co, 76-78, proj mgr surface & colloid chem, Celanese Polymer Specialty Co, 78-83; VENTURE MGR, AIR PROD & CHEM, NEW POLYMER TECHNOL, 83- *Concurrent Pos:* Teaching asst, Univ Southern Calif, 66-68; adj fac, Ind Univ, 81-83. *Mem:* Am Chem Soc; Soc Petrol Engrs; Sigma Xi; Soc Plastic Inst; Tech Asn Paper & Pulp. *Res:* Physical chemistry of surfaces; colloidal systems; emulsions; biopolymers and synthetic polymers; interfacial phenomena and their application in graphic arts such as printing, photography, xerography and other reprographic techniques; emulsion and water borne polymers for coatings, adhesives and sealants; extrusion technology for food and industrial packaging; water soluble polymers in oil field chemicals, paper and water treatment. *Mailing Add:* Millbrook Farms 4702 Parkview Dr N Emmaus PA 18049

VIJH, ASHOK KUMAR, b Multan, India, Mar 15, 38; Can citizen; div; c 1. INTERFACIAL ELECTROCHEMISTRY. *Educ:* Punjab Univ, India, BSc, 60, MSc, 61; Ottawa Univ, PhD(electrochem), 66. *Hon Degrees:* LLD, Concordia Univ, 88. *Prof Exp:* MASTER-IN-RES, RES INST, HYDRO-QUE, 69- *Concurrent Pos:* Invited prof, INRS-Energy, Univ Que, 70- *Honors & Awards:* Lash Miller Award, Can Sect, Electrochem Soc, 73; Noranda Award, Chem Inst Can, 79; Archambault Prize, French-Can Asn Advan Sci, 84; Killam Mem Prize Eng, Can Coun, 87; Thomas Eadie Medal, Royal Soc Can, 89; Palladium Medal, Chem Inst Can, 90; Officer, Order of Can, 90. *Mem:* Fel Nat Acad Sci India; fel Royal Soc Can; fel Royal Soc Chem UK; fel Am Phys Soc; fel Inst Elec & Electronics Engrs; fel Inst Physics UK; fel Chem Inst Can; assoc fel Third World Acad Sci; Europ Acad Arts Sci & Humanities. *Res:* Mechanisms of electrochemical reactions; interfacial electrochemistry in relation to the principles of solid state physics; electrochemical physics. *Mailing Add:* IREQ CP 1000 Varennes PQ J0L 2P0 Can

VIKIS, ANDREAS CHARALAMBOUS, b Moni, Cyprus, July 8, 42; Can citizen; m 71; c 2. PHYSICAL CHEMISTRY. *Educ:* Col Emporia, BSc, 64; Kans State Univ, PhD(phys chem), 69. *Prof Exp:* Fel, Univ Toronto, 69-70, lectr & res assoc, 70-74, asst prof & res assoc, 74-75; asst res officer chem, Nat Res Coun Can, 75-79; res officer, 79-85, head res chem, 85-90, DIR CHEM DIV, ATOMIC ENERGY CAN LTD RES, 90. *Mem:* Am Chem Soc; fel Can Inst Chem; Can Nuclear Soc. *Res:* Gas phase kinetics; photochemistry; isotope enrichment; reactor safety; nuclear waste management; air pollution. *Mailing Add:* Chem Div Whiteshell Labs Atomic Energy Can Ltd Res Pinawa MB R0E 1L0 Can

VIKRAM, CHANDRA SHEKHAR, b Payagpur, India, Oct 31, 50; m 75; c 2. OPTICS. *Educ:* Indian Inst Technol, Delhi, MTech, 70, PhD(optics), 73. *Prof Exp:* Sr res fel holography, Indian Inst Technol, Delhi, 70-75, sci pool officer, 75-77; res assoc holography, Pa State Univ, University Park, 77-82, sr res assoc optics, 82-89; SR RES SCIENTIST, UNIV ALA, HUNTSVILLE, 89- *Mem:* fel Optical Soc Am; Soc Photo-Optical Instrumentation Engrs. *Res:* Holography; speckle metrology; particle analysis; ultra-low thermal expansion measurements. *Mailing Add:* Ctr Appl Optics Univ Ala Huntsville AL 35899

VIKSNE, ANDY, b Jan 27, 34. GEOPHYSICS. *Educ:* Harvard Univ, AB, 56; Univ Utah, MS, 58. *Prof Exp:* Geophysicist, Texaco, Inc, 59-65, Systs Sci Corp, 65-66; scientist, Raytheon Co, 67-68; geophysicist, US Bur Mines, 68-72; CHIEF, GEOPHYS SECT, US BUR RECLAMATION, 72- *Mem:* Soc Explor Geophys; Europ Asn Explor Geophysicists; Earthquake Eng Res Inst. *Res:* Application of geophysical exploration methods in solving geotechnical engineering problems; in situ determination of elastic moduli for earth dams and foundation sites; strong motion earthquake instrumentation and site characteristics. *Mailing Add:* 350 Dover Ct Broomfield CO 80020-1547

VILA, SAMUEL CAMPDERROS, b Rubi, Spain, May 7, 30. ASTROPHYSICS. *Educ:* Univ Barcelona, Lic physics, 52; Univ Rochester, PhD(astron), 65. *Prof Exp:* Res assoc astron, Ind Univ, 65-67 & Inst Space Studies, NASA, NY, 67-69; asst prof, 69-74, ASSOC PROF ASTRON, UNIV PA, 74- *Mem:* Am Astron Soc; Int Astron Union. *Res:* Stellar evolution. *Mailing Add:* Dept Astron Univ Pa Philadelphia PA 19104

VILCEK, JAN TOMAS, b Bratislava, Czech, June 17, 33; m 62. CYTOKINES, GROWTH FACTORS. *Educ:* Univ Bratislava, MD, 57; Czech Acad Sci, CSc(virol), 62. *Prof Exp:* Res assoc virol, Inst Virol Czech Acad Sci, Bratislava, 57-59, head lab, 62-64; from asst prof to assoc prof, 65-72, PROF MICROBIOL, SCH MED, NY UNIV, 72-, HEAD, CYTOKINE RES UNIT, 84- DIR, MICROBIOL GRAD TRAINING PROG, 84- *Concurrent Pos:* Am Cancer Soc grant, 65-66; USPHS grants, 65-88, career develop award, 68-73 & contract, 70-81; Irwin Strasburger Mem Med Found grant, 69-73; ed, Arch Virol 72-74, ed in chief, 75- & assoc ed, 84-; assoc ed, Virol, 77-79, Interferon Monographs, 79-88, J Interferon Res, 80-85, Appl Biochem Biotechnol, 81-86 & Infect Immunity, 83-85; chmn, adv comt microbiol & virol, Am Cancer Soc, 84; assoc ed, J Immunol Methods, 86- & J Immunol, 87-89, J Biol Chem, 88-90; adv ed, ISI Atlas Sci Immunol, 87-89; mem, Sci Adv Bd, Max-Planck-Inst Biochem, Munich, WGer, 87- *Mem:* AAAS; Am Asn Immunol; Am Soc Microbiol; Brit Soc Gen Microbiol; Am Soc Virol; Int Soc Interferon Res. *Res:* Interferons; regulations of immune responses by lymphokines and other cytokines; regulation of cell growth and gene expression. *Mailing Add:* Dept Microbiol NY Univ Sch Med 550 First Ave New York NY 10016

VILCHES, OSCAR EDGARDO, b Mercedes, Arg, Feb 20, 36. LOW TEMPERATURE, SURFACE PHYSICS. *Educ:* Nat Univ Cuyo, lic physics, 59, Dr en Fisica, 66. *Prof Exp:* Investr physics, Cent Atomico Bariloche, Arg, 60-64; res asst, Univ Ill, Urbana, 64-65, res assoc, 65-67; res assoc, Univ Calif, San Diego, 67-68; asst prof, 68-73, assoc prof, 73-80, PROF PHYSICS, UNIV WASH, 80- *Concurrent Pos:* NSF res grant, 70-; vis prof, Unicamp, Brazil, 75, Univ Lumny, France, 81; vis scientist, Weizmann Inst, Israel, 86. *Mem:* Am Phys Soc. *Res:* Properties of liquid and solid helium and helium films; experimental physical adsorption; measurement of thermodynamic properties of very thin films, with emphasis on quartum monolayers of helium and hydrogen. *Mailing Add:* Dept Physics Univ Wash Seattle WA 98195

VILCINS, GUNARS, b Riga, Latvia, May 8, 30; m 61; c 2. ANALYTICAL CHEMISTRY, SPECTROSCOPY. *Educ:* Univ Richmond, BS, 54, MS, 62. *Prof Exp:* Assoc chemist, 57-63, res chemist, 63-77, SR SCIENTIST, PHILIP MORRIS, INC, 77- *Mem:* Am Chem Soc; Soc Appl Spectros; Coblentz Soc. *Res:* Infrared and Raman spectroscopy; cigarette smoke; tobacco; low temperature studies; microanalysis; tunable diode laser spectroscopy; air quality analysis; ion chromatography; GC-FTIR. *Mailing Add:* 2504 Haviland Dr Richmond VA 23229

VILENKIN, ALEXANDER, b Kharkov, USSR, May 13, 49; m 73; c 1. THEORETICAL PHYSICS. *Educ:* Kharkov State Univ, USSR, MS, 71; State Univ NY, Buffalo, PhD(physics), 78. *Prof Exp:* Res assoc physics, Case Western Reserve Univ, 77-78; vis asst prof, 78-79, from asst prof to assoc prof, 79-87, PROF PHYSICS, TUFTS UNIV, 87- *Concurrent Pos:* Presidential young investr award, 84- *Mem:* Am Phys Soc. *Res:* General relativity; quantum field theory; cosmology. *Mailing Add:* Dept Physics Tufts Univ Medford MA 02155

VILKER, VINCENT LEE, b Beaver Dam, Wis, Jan 17, 43; m 81; c 1. COLLOID SCIENCE, BIOENGINEERING. *Educ:* Univ Wis, Madison, BS, 67; Mass Inst Technol, PhD(chem eng), 76. *Prof Exp:* Res engr, Exxon Res & Eng, 67-70; from asst prof to assoc prof, 75-86, PROF CHEM ENG, UNIV CALIF, LOS ANGELES, 86- *Concurrent Pos:* Prin investr, Nat Ctr Intermedia Transp Res, Univ Calif, Los Angeles, 80-; vchmn, Dept Chem Eng, Univ Calif, Los Angeles, 86-; Fulbright Fel, 84-85. *Mem:* AAAS; Am Chem Soc; Am Inst Chem Engrs. *Res:* Physical chemistry of solutions of biological macromolecules; membrane transport phenomena; bioelectrochemistry of redox enzymes; movement and fate of toxic materials (viruses, volatile organics) in soils. *Mailing Add:* 5531 Boelter Hall Univ Calif Los Angeles CA 90024-6900

VILKITIS, JAMES RICHARD, b Rush, Pa. REGIONAL RESOURCE PLANNING & MANAGEMENT, TERRESTRIAL ECOLOGY. *Educ:* Mich State Univ, BS, 65; Univ Idaho, MS, 68; Univ Mass, PhD(wildlife biol), 70. *Prof Exp:* Res asst, Water Resource Res Ctr, 68-70; spec big game wildlife leader, Dept Inland Fisheries & Game, 70-71; prin partner, Carlozzi, Sinton & Vilkitis Inc, 71-79; owner & mgr, TLC Leather, 76-81; PROF NATURAL RESOURCES MGT, CALIF POLYTECH STATE UNIV, 80- *Concurrent Pos:* Biostatistician, Regional Plannning & Design Assocs, 69-71; lectr, Univ Mass, 71-79; res assoc ecol, Inst Man & Environ, 73-74; asst prof biol sci, Mt Holyoke Col, 79-80; assoc, R Stollars Assocs, 86-; staff officer, Land Mgt Planning, US Forest Serv, 87-88. *Honors & Awards:* Cert Merit, USDA, 88. *Mem:* Wildlife Soc; Am Forestry Asn; Asn Environ Prof. *Res:* Developing regional resource planning and management strategies for terrestrial, aquatic, coastal and riparian systems; wastewater recycling and use. *Mailing Add:* Dept Natural Resource Mgt Calif Polytech State Univ San Luis Obispo CA 93407

VILKKI, ERKKI UUNO, b Soanlahti, Finland, July 5, 22; m 52; c 1. PHOTOGRAPHIC ASTROMETRY, DOUBLE STARS. *Educ:* Navig Sch Rauma, Finland, dipl, 48 & 52; Navig Sch Kotka, Finland, dipl, 59. *Prof Exp:* Collabr, Pic du Midi Observ, France, 64-66; observer & res technician, 67-73, from asst astronr to assoc astronr, 73-85, ASTRONR, YERKES OBSERV, 85- *Concurrent Pos:* Merchant Marine, 45-62. *Mem:* Am Astron Soc; Int Astron Union. *Res:* Trigonometric stellar parallaxes; photographic observations of double stars. *Mailing Add:* Yerkes Observ Williams Bay WI 53191

VILKS, GUSTAVS, b Riga, Latvia, May 7, 29; Can citizen; wid; c 2. MICROPALEONTOLOGY. *Educ:* McMaster Univ, BSc, 61; Dalhousie Univ, MSc, 66, PhD, 73. *Prof Exp:* MICROPALEONTOLOGIST, BEDFORD INST OCEANOG, 62- *Mem:* Geol Asn Can. *Res:* Ecology and paleoecology of Recent Foraminifera in the Canadian Arctic and Labrador Shelf; ecology of planktonic Foraminifera in the North Atlantic; glacial limits off eastern Canada; sedimentary processes on Labrador shelf and Abyssal Plains. *Mailing Add:* Atlantic Geosci Ctr Bedford Inst Oceanog Dartmouth NS B2Y 4A2 Can

VILKS, PETER, b Hamilton, Ont, Mar 30, 56; m 78; c 2. ENVIRONMENTAL RADIOCOLLOIDS. *Educ:* Dalhousie Univ, BSc, 78; McMaster Univ, MSc, 81, PhD(geol), 85. *Prof Exp:* RES SCIENTIST, ATOMIC ENERGY CAN LTD, 85- *Res:* Likelihood of radiocolloid formation and its impact on the concept of nuclear fuel waste disposal; kinetic approach to the study of metal sorption mechanisms on clay. *Mailing Add:* Nine Dalhouise Dr Pinawa MB R0E 1L0 Can

VILLA, JUAN FRANCISCO, b Matanzas, Cuba, Sept 23, 41; US citizen; m 67; c 4. INORGANIC CHEMISTRY. *Educ:* Univ Miami, BS, 65, MS, 67, PhD(inorg chem), 69. *Prof Exp:* Teaching asst chem, Univ Miami, 65-69; res assoc inorg chem, Univ NC, Chapel Hill, 69-71; from asst prof to assoc prof, 71-78, actg dean natural & social sci, 80-81, PROF CHEM, LEHMAN COL, 78- *Concurrent Pos:* George N Shuster fel, Lehman Col, 71-72 & 74-77; Petrol Res Fund fel, 71-73; Fulbright-Hays sr lectr, Colombia, SAm, 76; adj prof, Sarah Lawrence Col, 77 & 78. *Mem:* Am Chem Soc; The Chem Soc; Sigma Xi. *Res:* Study of transition metal coordination compounds of biological importance including synthesis, electron paramagnetic resonance spectroscopy, magnetic susceptibility measurements, electronic and infrared spectra; ligand field and molecular orbital calculations. *Mailing Add:* Dept Chem Herbert H Lehman Col Bronx NY 10468

VILLA, VICENTE DOMINGO, b Laredo, Tex, Dec 1, 40; m 62; c 2. MICROBIAL PHYSIOLOGY. *Educ:* Univ Tex, Austin, BA, 64; Rice Univ, PhD(microbiol), 70. *Prof Exp:* Fel molecular biol, Molecular Biol Lab, Univ Wis, 69-71; res assoc, Rosenstiel Res Ctr, Brandeis Univ, 71-72; from asst prof to prof, NMex State Univ, 72-84; PROF BIOL & DISHMAN CHMN SCI, SOUTHWESTERN UNIV, 85- *Concurrent Pos:* NIH fel, 70-71; ad hoc consult, Minority Biomed Support Prog, NIH, 72-75, mem gen res support prog adv comt, Div Res Resources, 76-80; panelist, NSF Rev Panel-Res Initiation & Support Prog, 76. *Mem:* Am Soc Microbiol. *Res:* Cell wall polyuronides and morphogenesis in fungi; chracterizing the polyuronides found in the cell wall of Mvcor and relating their characteristics to the morphogenetic development of the organism. *Mailing Add:* 507 Meadowbrooke Dr Georgetown TX 78628

VILLABLANCA, JAIME ROLANDO, b Chillan, Chile, Feb 28, 29, US citizen; m 55; c 5. NEUROPHYSIOLOGY, EXPERIMENTAL NEUROLOGY. *Educ:* Univ Chile, BA, 46, Lic Med, 53, Dr(med), 54; Univ Calif, Los Angeles, cert neurophysiol, 68. *Prof Exp:* From instr to prof pathophysiol, Sch Med, Univ Chile, 54-71; assoc res anat & psychiat, 71-72, PROF PSYCHIAT, UNIV CALIF, LOS ANGELES, 72-, PROF ANAT, 77- *Concurrent Pos:* Rockefeller Found fel physiol, Johns Hopkins Univ, 59-61; fel, Neurol Unit, Harvard Med Sch, 61; USAF Off Sci res grant, 62-65; NIH fogarty res fel anat, Univ Calif, Los Angeles, 66-68; Found Fund Res in Psychiat grant, 69-72; Nat Inst Child Health & Human Develop prog proj grant, 71-94; Nat Inst Drug Abuse Grant, 81-85; mem, Ment Retardation Res Ctr, Univ Calif, Los Angeles; chmn, Brain Res Inst, Univ Calif, Los Angeles; sci adv coun, Int Inst Res & Advice Ment Deficiency, Madrid, Spain; Nat Inst Neurol Commun Dis & Stroke grant, 88-92; Sci Coun Avepane, Caracas, Venezuela; chief ed, J Brain Dysfunction. *Mem:* Am Asn Anatomists; Am Physiol Soc; Soc Neurosci; Sigma Xi; Int Brain Res Orgn. *Res:* Neurophysiology of sleep-wakefulness; neurological, behavioral and electrophysiological effects of lesions upon the mature and upon the developing brain; physiology and pathophysiology of the basal ganglia; role of the basal ganglia on the effects of opiates; recovery of function and anatomical reorganization following lesions of the mature and the developing brain. *Mailing Add:* Dept Psychiat Univ Calif Los Angeles CA 90024

VILLACORTE, GUILLERMO VILAR, b Rizal NE, Philippines, Dec 25, 33; m 61; c 2. ALLERGY, IMMUNOLOGY. *Educ:* Univ St Tomas, Manila, AA, 52, MD, 57; Am Bd Allergy & Immunol, dipl, 74, recert, 80. *Prof Exp:* Asst prof pediat, Sch Med, Creighton Univ, 69-79; mem staff, allergy-immunol serv, Wilford Hall Med Ctr, Lackland AFB, Tex, 79-84; mem staff, Allergy Immunol Serv, USAFR, 84-90; CONSULT, 90- *Concurrent Pos:* Res pediat, Good Samaritan Hosp, Cincinnati, Ohio, 62-65; fels pediat allergy & immunol, Med Ctr, Univ Cincinnati, 65-67; fel immunol, Med Ctr, Univ Cincinnati, 67-69. *Mem:* Fel Am Acad Allergy. *Res:* Etiopathogenesis of allergic and immunodeficiency diseases. *Mailing Add:* 8306 Brixton San Antonio TX 78250-2402

VILLAFANA, THEODORE, b New York, NY, Sept 23, 36; m 61; c 4. NUCLEAR MEDICAL PHYSICS, MEDICAL IMAGING PHYSICS. *Educ:* Hunter Col, BA, 59; Univ Pittsburgh, MSc, 65; Johns Hopkins Univ, PhD(radiol physics), 69. *Prof Exp:* Physicist, Columbia-Presby Med Ctr, 59-63 & Montefiore Hosp, Pittsburgh, 63-65; asst prof radiol physics, PR Nuclear Ctr, Univ PR, 69-72; assoc prof, George Washington Univ Med Sch, 72-75; PROF RADIOL PHYSICS, MED CTR, TEMPLE UNIV, 75- *Concurrent Pos:* Vis prof, Univ PR Med Sch, 77-; consult, Med Sch, Hahnemann Univ, 80-; mem, radiographics panel, Radiol Soc NAm, 83-91, bd dirs, Am Asn Physicists Med, 86-89 & Task Force Educ, Am Col Radiol, 88-91. *Mem:* Am Asn Physicists Med; Health Physics Soc; Radiol Soc NAm; Am Col Radiol. *Res:* Mammographic medical imaging; magnetic resonance imaging; publications on radiological physics. *Mailing Add:* Radiol Dept Temple Univ Hosp Philadelphia PA 19140

VILLAFRANCA, JOSEPH JOHN, b Silver Creek, NY, Mar 23, 44; m 90; c 2. BIOCHEMISTRY, BIO-ORGANIC CHEMISTRY. *Educ:* State Univ NY Col Fredonia, BS, 65; Purdue Univ, Lafayette, PhD(biochem), 69. *Prof Exp:* USPHS fel, Inst Cancer Res, 69-71; from asst prof to prof, 71-76, EVAN PUGH PROF CHEM, PA STATE UNIV, UNIVERSITY PARK, 86- *Concurrent Pos:* NSF grant, 72- & USPHS grant, 74-; estab investr, Am Heart Asn, 78-83, mem brant rev comt, 84- 86; mem, biochem study sect, NIH, 82-86, chmn, 85-86; prog chmn, Am Chem Soc, 82-84; mem adv comt, Oak Ridge Nat Lab, 88-91; mem, major chem instrument rev panel, NJF, 90-93. *Mem:* Am Chem Soc; Biophys Soc; Am Soc Biol Chemists. *Res:* Mechanism of enzyme action studied by magnetic resonance techniques; biophysics. *Mailing Add:* Dept Chem Pa State Univ University Park PA 16802

VILLA-KOMAROFF, LYDIA, b Las Vegas, NMex, Aug 7, 47; m 70. MOLECULAR BIOLOGY, DEVELOPMENTAL BIOLOGY. *Educ:* Goucher Col, AB, 70; Mass Inst Technol, PhD(cell biol), 75. *Prof Exp:* Res fel biol, Harvard Univ, 75-78; from asst prof to assoc prof microbiol, Med Sch, Univ Mass, 78-85; ASSOC PROF, DEPT NEUROL, HARVARD MED SCH, 85-; SR RES ASSOC, DEPT NEUROL, DIV NEUROSCI, 85-, ASSOC DIR, MENT RETARDATION CTR, CHILDREN'S HOSP, BOSTON. *Concurrent Pos:* Vis fel, Cold Spring Harbor Lab, 76-77; fel, Helen Hay Whitney Found, 75-78; mem, Mammalian Genetics Study Sect, NIH, 82-84, Neurol Dis Prog Proj Rev Comt, 89-; chair, Neurol Res Steering Comt, Children's Hosp, Boston, 88- *Mem:* Am Soc Microbiol; Am Soc Cell Biol; AAAS; Fedn Am Sci; Sigma Xi; Soc Neurosci. *Res:* Growth factors in brain development; structure & function of insulin-like growth factors; structure and function of genes expressed in central and peripheral nervous system; role of amino acid sequence in regulated secretion of insulin. *Mailing Add:* Neurol Res Enders 250 Children's Hosp 300 Longwood Ave Boston MA 02115

VILLANI, FRANK JOHN, SR, b Brooklyn, NY, May 9, 21; m 51; c 4. MEDICINAL CHEMISTRY. *Educ:* Brooklyn Col, BA, 41; Fordham Univ, MS, 43, PhD(chem), 46. *Prof Exp:* Org chemist, Schering Corp, Bloomfield, 46-64, fel med chem, 64-83; RETIRED. *Mem:* Am Chem Soc. *Res:* Aldehyde condensations; synthetic medicinals; heterocyclic chemistry. *Mailing Add:* 22 Oakland Terr Fairfield NJ 07004-3827

VILLANUEVA, GERMAN BAID, ANTITHROMBIN, THROMBIN-HEPARIN REACTION. *Educ:* Fordham Univ, PhD(biochem), 71. *Prof Exp:* PROF BIOCHEM, DEPT BIOCHEM, NY MED COL, 76- *Mailing Add:* Dept Biochem Elmwood Hall NY Med Col Valhalla NY 10595-1690

VILLANUEVA, JOSE, b Santiago de Cuba, Cuba, Mar 31, 37; US citizen; m 61; c 2. MECHANICAL ENGINEERING. *Educ:* Ga Inst Technol, BS, 59, MS, 61, PhD(eng mech), 65. *Prof Exp:* Prof docent mech eng, Univ Oriente, Cuba, 61; asst prof eng mech, Ga Inst Technol, 61-68; assoc prof mech eng, 68-77, actg chmn dept, 71-77, PROF MECH ENG, FLA ATLANTIC UNIV, 77- *Concurrent Pos:* Eng consult, Am Art Metals, Ga, 61 & Lockheed-Ga Co, 66- *Honors & Awards:* Ralph R Teetor Award, Soc Automotive Engrs, 67. *Mem:* Am Soc Mech Engrs. *Res:* Continuum mechanics; dynamics and vibrations; sloshing of liquids and design of mechanical models to describe the phenomena; energy efficiency in buildings. *Mailing Add:* Dept Mech Eng Fla Atlantic Univ Boca Raton FL 33431

VILLAR, JAMES WALTER, b New York, NY, July 25, 30; m 58; c 3. METALLURGY, GEOLOGY. *Educ:* Mich State Univ, BS, 52, MS, 56, PhD(geol), 64. *Prof Exp:* Geologist, Cleveland-Cliffs Iron Co, 56-58 & 60-63, from sr metallurgist to asst chief metallurgist, 63-65, asst mgr res & develop, 65-68, mgr, 68-78, gen mgr, 78-79, gen mgr res & eng, 79-81, vpres, 81-86; SR VPRES OPERS & SR VPRES TECH, CLEVELAND CLIFFS INC, 86- *Mem:* Fel Geol Soc Am; Am Inst Mining, Metall & Petrol Engrs; Am Iron & Steel Inst; Am Iron Ore Asn. *Res:* Development of beneficiation methods for low grade iron deposits, especially fine grained hematites. *Mailing Add:* Cliffs Mining Co 1100 Superior Ave Cleveland OH 44114-2589

VILLARD, OSWALD G(ARRISON), JR, b Dobbs Ferry, NY, Sept 17, 16; m 42; c 3. ELECTRONICS, DEFENSE RESEARCH. *Educ:* Yale Univ, AB, 38; Stanford Univ, EE, 43, PhD(radio eng), 49. *Prof Exp:* Actg instr elec eng, Stanford Univ, 41-42; spec res assoc, Radio Res Lab, Harvard Univ, 42-43, mem sr staff, 43-46; actg asst prof elec eng, 46-50, from asst prof to assoc prof, 50-55, dir, Radiosci Lab, 60-73, dir ionospheric dynamics lab, 70-72, PROF ELEC ENG, STANFORD UNIV, 55-, SR SCI ADV, SYST TECHNOL DIV, SRI INT, 72- *Concurrent Pos:* Emer mem, Air Force Studies Bd, Nat Res Coun, 62-; mem geophys panel, USAF Sci Adv Bd, 66-75; mem, Naval Res Adv Comt, 69-75, chmn, 72-75; mem adv bd, Nat Security Agency, 76-86. *Honors & Awards:* Morris Liebmann Mem Award, Inst Elec & Electronics Engrs, 57, Centennial Award, 85; SRI Int Presidential Fel Award for Excellence, 88. *Mem:* Nat Acad Sci; Nat Acad Eng; fel AAAS; fel Inst Elec & Electronics Engrs; Am Geophys Union; fel Am Acad Arts & Sci; Int Sci Radio Union. *Res:* Ionospheric radio propagation; upper atmosphere research; radar techniques; defense electronic systems. *Mailing Add:* SRI Int 333 Ravenswood Ave Menlo Park CA 94025

VILLAREAL, RAMIRO, b Monterrey, Nuevo Leon, Mex, Sept 24, 50. RESEARCH ADMINISTRATION. *Educ:* Monterrey Inst Technol, BS, 72. *Prof Exp:* SOFTWARE ENGR MGR, ARMOR ELEVATOR DIV, KONE CORP, 83- *Res:* High speed coin and paper currency recognition and counting systems; intelligent vertical transportation systems; high speed dynamic weighing systems. *Mailing Add:* 9409 Walhampton Dr Louisville KY 40242

VILLAREJO, MERNA, b New York, NY, June 19, 39; m 59; c 2. BIOCHEMISTRY. *Educ:* Univ Chicago, BS, 59, PhD(biochem), 63. *Prof Exp:* Res assoc & asst prof biochem, Univ Chicago, 63-68; res assoc biol chem, Sch Med, Univ Calif, Los Angeles, 68-75; asst prof, 75-82, assoc prof, 82-88, ASSOC DEAN BIOL SCI, UNIV CALIF, DAVIS, 87-, PROF BIOCHEM, 88- *Concurrent Pos:* USPHS fel, Univ Chicago, 63-65; prin investr, NIH, 75-; Howard Hughes Med Inst Grants, 89-94. *Mem:* Am Soc Biochem & Molecular Biol; Am Soc Microbiol. *Res:* Regulation of gene expression in response to osmotic stress. *Mailing Add:* Dept Biochem & Biophys Univ Calif Davis CA 95616

VILLAREJOS, VICTOR MOISES, b La Paz, Bolivia, Sept 4, 18; US citizen; m 41; c 4. EPIDEMIOLOGY, TROPICAL MEDICINE. *Educ:* Univ Heidelberg, MD, 41; Tulane Univ, MPH & TM, 59, DrPH, 61. *Prof Exp:* Chief serv med, Miraflores Gen Hosp, La Paz, 45-58; from assoc prof to prof trop med, Med Sch, Univ La Paz, 47-58; assoc prof trop med, Sch Med, La State Univ, 61-66, chief epidemiol sect, La State Univ Int Ctr Med Res & Training, San Jose, Costa Rica, 62-66, prog coordr, 66-69, prof trop med, Sch Med, La State Univ, New Orleans, 66-88, dir, La State Univ-Int Ctr Med Res & Training, San Jose, Costa Rica, 69-88, EMER PROF, LA STATE UNIV, 88- *Concurrent Pos:* Alexander von Humboldt fel, Ger, 41-42; Pan Am Health Orgn fel, 58-59; USPHS & Armed Forces Epidemiol Bd res grants. *Mem:* Am Soc Trop Med & Hyg; Am Soc Parasitol; Am Pub Health Asn; Sigma Xi. *Res:* Epidemiology of diseases prevalent in tropical areas; pathogenesis of E histolytica; diarrheal diseases; infectious hepatitis. *Mailing Add:* 4241 Briarwood Dr Independence KY 41051

VILLAR-PALASI, CARLOS, b Valencia, Spain, Mar 3, 28; m 57; c 4. PHARMACOLOGY, BIOCHEMISTRY. *Educ:* Univ Valencia, MS, 51; Univ Madrid, PhD(biochem), 55; Univ Barcelona, MS, 62. *Prof Exp:* NIH fel, Case Western Reserve Univ, 57-60, res assoc pharmacol, 63-64; res assoc enzymol, Univ Madrid, 60-63, asst prof, 62-63; res assoc biochem, Univ Minn, Minneapolis, 64-65, asst prof, 65-69; assoc prof, 69-73, PROF PHARMACOL, UNIV VA, 73- *Concurrent Pos:* Span Res Coun fel, Univ Hamburg, 53-54; NIH grant, Univ Va, 70- *Honors & Awards:* AAAS Res Award, 60; Span Soc Biochem Res Award, 72. *Mem:* AAAS; Brit Biochem Soc; Am Soc Biol Chemists; Span Biochem Soc; Am Soc Pharmacol & Exp Therapeut. *Res:* Control of metabolic and muscle function by phosphorylation of proteins; protein kinases and phosphatases; glycogen metabolism; mechanism of action of insulin, glucagon and adrenergic hormones; contractile protein structure and function; carbohydrate metabolism; hormonal effects on energy metabolism. *Mailing Add:* Dept Pharmacol Box 448 Sch Med Univ Va 1300 Jefferson Park Ave Charlottesville VA 22908

VILLARREAL, JESSE JAMES, b San Antonio, Tex, Oct 22, 13; m 35; c 2. SPEECH PATHOLOGY. *Educ:* Univ Tex, BA, 35, MA, 37; Northwestern Univ, PhD(speech path, audiol), 47. *Prof Exp:* Dir, Speech & Hearing Clin, Univ Tex, Austin, 39-62, prof, 52-65, chmn, Dept Speech, 62-68, prof speech commun & educ, 65-80. *Mem:* Fel Am Speech & Hearing Asn; Speech Commun Asn. *Res:* English as a second language. *Mailing Add:* 5104 Crest Way Austin TX 78731

VILLARREAL, LUIS PEREZ, b Los Angeles, Calif, July 6, 49; m 82. MOLECULAR VIROLOGY. *Educ:* Los Angeles State Univ, BS, 71; Univ Calif, San Diego, PhD(biol), 76. *Prof Exp:* Fel molecular biol, Biochem Dept, Stanford Univ, 76-78; asst prof microbiol & virol, Health Sci Ctr, Univ Colo, 78-; PROF MOLECULAR BIOL & BIOCHEM, UNIV CALIF, IRVINE. *Concurrent Pos:* Res fel, Jane Coffin Childs Mem Res Fund, 76. *Mem:* Am Soc Microbiol. *Res:* Control of gene expression in animal viruses. *Mailing Add:* Dept Molecular Biol & Chem Univ Calif Irvine CA 92717

VILLARS, FELIX MARC HERMANN, b Biel, Switz, Jan 6, 21; nat US; m 49; c 4. THEORETICAL PHYSICS. *Educ:* Swiss Fed Inst Technol, Dipl, 45, DSc, 46. *Prof Exp:* Res asst physics, Swiss Fed Inst Technol, 46-49; vis mem, Inst Adv Study, 49-50; res assoc 50-52, from asst prof to assoc prof, 52-60, PROF PHYSICS, MASS INST TECHNOL, 60- *Concurrent Pos:* Consult, Lincoln Lab, 54-64, 68-70; Guggenheim fel, 56-57; lectr physics, Harvard Med Sch, 74- *Mem:* Am Phys Soc; Am Acad Arts & Sci. *Res:* Nuclear physics, mainly nuclear models and reactions; quantum field theory; physics of upper atmosphere; turbulence; plasma probes; biophysics. *Mailing Add:* Dept Physics Mass Inst Technol Cambridge MA 02139

VILLEE, CLAUDE ALVIN, JR, b Lancaster, Pa, Feb 9, 17; m 52; c 4. BIOCHEMISTRY. *Educ:* Franklin & Marshall Col, BS, 37; Univ Calif, PhD(physiol genetics), 41. *Hon Degrees:* AM, Harvard Univ, 57; ScD, Franklin & Marshall, 91. *Prof Exp:* Res assoc zool, Univ Calif, 41-42; from instr to asst prof, Univ NC, 42-45; from instr to assoc prof biol chem, 46-63, PROF BIOL CHEM, HARVARD UNIV, 63-, ANDELOT PROF, 64-, TUTOR PRECLIN SCI, MED SCH, 47- *Concurrent Pos:* Asst prof, Armstrong Col, 41-42; tech aide, Comt Growth, Nat Res Coun, 46; Lalor fel, Marine Biol Lab, Woods Hole, 47 & 48; Guggenheim fel, Denmark, 49-50; res assoc, Boston Lying-in-Hosp, 50-; consult, Mass Gen Hosp, 50- & NSF, 59-; consult, NIH, 58-, mem nat adv child health & human develop coun, 63-65; dir lab reproductive biol, Boston Hosp Women, 66-; mem sci adv comt, Ore Regional Primate Ctr, 70-; distinguished vis prof, Univ Belgrade & Mahidol Univ, Bangkok, 74; consult, March Dimes Found, 78- *Honors & Awards:* Ciba Award, Endocrine Soc, 56; Rubin Award, Am Soc Study Steril, 57. *Mem:* Hon mem Soc Gynec Invest; hon fel Am Col Obstet & Gynec; hon fel Am Gynec Soc; Am Soc Biol Chemists; Genetic Soc Am. *Res:* Nucleic acid chemistry and metabolism; carbohydrate metabolism; effects of hormones on intermediary metabolism; function of the placenta; biochemical genetics; metabolism of fetal tissues. *Mailing Add:* Dept Biol Chem Harvard Med Sch 25 Shattuck St Boston MA 02115

VILLEE, DOROTHY BALZER, b Charleston, SC, Nov 25, 27; m 52; c 4. PEDIATRIC ENDOCRINOLOGY, AGING. *Educ:* Barnard Col, Columbia Univ, BA, 50; Harvard Med Sch, MD, 55. *Prof Exp:* Res assoc pediat, Children's Hosp, 62-68; ASST PROF PEDIAT, HARVARD MED SCH, 68- *Concurrent Pos:* Fel, Mass Gen Hosp, 56-62; consult, March of Dimes/Birth Defects Found, 75-85; mem, nat adv coun, NIH, 76-81. *Mem:* Endocrine Soc. *Res:* Nature and causes of progeria; endocrine functions of tumors; phemomenon of aging; enzymology of steroid hormones; biosynthesis and catabolism; studies of growth in humans. *Mailing Add:* Pediat Dept Harvard Med Sch 25 Shattuck St Boston MA 02115

VILLEGAS, CESAREO, b Montevideo, Uruguay, April 5, 21; Can citizen; m 50; c 4. MATHEMATICAL STATISTICS. *Educ:* Univ de La Repub, Uruguay, Ing Ind, 53. *Prof Exp:* Prof math, Univ de La Repub, Uruguay, 58-68; vis assoc prof, Univ Rochester, 68-70; assoc prof, 70-79, PROF STATIST, SIMON FRASER UNIV, 79- *Mem:* Int Statist Inst; Inst Math Statist; Can Statist Soc. *Res:* Developing a new approach to statistics called structural bayesian inference, and its applications, especially to linear functional models. *Mailing Add:* Dept Math & Statist Simon Fraser Univ Burnaby BC V5A 1S6 Can

VILLELLA, JOHN BAPTIST, zoology, for more information see previous edition

VILLEMEZ, CLARENCE LOUIS, JR, b Port Arthur, Tex, Sept 6, 38; m 64; c 4. BIOCHEMISTRY. *Educ:* Harvard Univ, AB, 58; Purdue Univ, MS, 61, PhD(biochem), 63. *Prof Exp:* Fel biochem, Purdue Univ, 63-65; asst res biochemist, Univ Calif, Berkeley, 65-66; res assoc biochem, Univ Colo, 66-67; from asst prof to assoc prof, Ohio Univ, 67-72; assoc prof, 72-74, PROF BIOCHEM, UNIV WYO, 74- *Concurrent Pos:* Vis prof microbiol, Univ Tex Health Sci Ctr, Dallas, 79-80. *Mem:* Am Soc Cell Biol; AAAS; Am Soc Biol Chemists; Am Soc Plant Physiologists; NY Acad Sci. *Res:* Polysaccharide biosynthesis; cell wall formation; protein glycosylation; hybridomas; specific cytotoxic reagents. *Mailing Add:* Dept Molecular Biol Univ Wyo Laramie WY 82071

VILLEMURE, M PAUL JAMES, b Newberry, Mich, Nov 28, 28. MATHEMATICS. *Educ:* Siena Heights Col, BS, 50; Univ Notre Dame, PhD(math), 58. *Prof Exp:* Teacher high sch, 58-59; instr math & sci, Col San Antonio, 51-54; from instr math to asst prof math, 59-69, PROF MATH, BARRY COL, 69- *Mem:* Math Asn Am. *Mailing Add:* Dept Math Barry Col 11300 NE Second Ave Miami Shores FL 33161

VILLENEUVE, A(LFRED) T(HOMAS), b Syracuse, NY, Mar 14, 30; m 56; c 5. ANTENNAS, MICROWAVES. *Educ:* Manhattan Col, BEE, 52; Syracuse Univ, MEE, 55, PhD(elec eng), 59. *Prof Exp:* Res assoc elec eng, Syracuse Univ, 52-56, instr, 56-59, asst prof, 59; mem tech staff, Hughes Aircraft Co, 59-60, staff engr, 60-63, sr staff engr, 63-72, sr scientist, 72-84; chief scientist, Antenna Systs Lab, 84-89; RETIRED. *Concurrent Pos:* Lectr, Univ Southern Calif, 59, Loyola Univ, Calif, 62 & Univ Calif, Los Angeles, 69; mem comn B, Int Union Radio Sci. *Mem:* Inst Elec & Electronics Engrs; Sigma Xi. *Res:* Ultra-high-frequency and microwave antennas; microwave filters; large aperture antennas; electromagnetic fields in anisotropic media; advanced antenna techniques; satellite antennas; space communications; phase scanned arrays. *Mailing Add:* 8123 Kenyon Ave Los Angeles CA 90045

VILLENEUVE, ANDRE, b Chicoutimi, Que, Sept 17, 32; m 58; c 1. NEUROPSYCHOPHARMACOLOGY, FORENSIC PSYCHIATRY. *Educ:* Laval Univ, BA, 52, MD, 58; McGill Univ, MSc, 66; FRCPS(C); FRCPsychiat. *Prof Exp:* Residency psychiat, Cent Islip State Hosp & NY Sch Psychiat, 61-64; researcher psychiat, Fac Grad Studies, McGill Univ, Montreal & clin fel, Royal Victoria Hosp, 64-66, researcher psychopharmacol, Hosp St Anne & Fac Med, Paris, 66-67; lectr, 68-70, assoc prof, 70-76, PROF, DEPT PSYCHIAT, LAVAL UNIV, QUE, 77- *Concurrent Pos:* Res psychiatrist neuropsychopharmacol, Univ Ctr Hosp Robert Giffard, Beauport, Que, 67-86, chief, 67-74 & 79-85; dir res, Dept Psychiat, Hosp de l'Enfant-Jesus, Que, 73-82; pres, Examining Bd, French Sect, Psychiat Specialty, Royal Col Physicians & Surgeons, Can, 78-84; chief, Neuropsychopharmacol Sect, Clinique Roy-Rousseau, Beauport, Que, 82-, chief, Dept Psychiat, 83-; hon mem, Foreign Sci Adv Bd, Vinohrady Sch Med, Charles Univ, Prague, Czech, 90- *Mem:* Fel Am Psychiat Asn; fel Am Col Physicians; fel Am Col Forensic Psychiat; fel Col Int Neuropsychopharmacol; found fel & pres Can Col Neuropsychopharmacol; fel Royal Col Psychiatrists (Eng). *Res:* Methodology; clinical trials; pharmacokinetics; drug interactions; extrapyramidal system (side effects, tardive dyskinesia, endorphins, estrogens); neuropsychoendocrinology; forensic psychiatry. *Mailing Add:* Clinique Roy-Roussau Box CP 850 Limoilou Quebec PQ G1L 4Y9 Can

VILLERE, KAREN R, b Teaneck, NJ, Apr 9, 44; div. STELLAR EVOLUTION, STAR FORMATION. *Educ:* Univ Pa, BA, 65; Univ Calif, Berkeley, MA, 68; Univ Calif, Santa Cruz, PhD(astron), 76. *Prof Exp:* Nat Res Coun resident res assoc, Ames Res Ctr, NASA, 77-79; res astronr, Univ Calif, Santa Cruz & Ames Res Ctr, NASA, 79-86; assoc prof physics, San Francisco State Univ, 86-87; res scientist, 74-76, SCIENTIST, SCI APPLNS INT CORP, 87- *Concurrent Pos:* Vis scientist, Tata Inst Fundamental Res, Bombay, India, 84; adj prof, San Francisco State Univ, 87- *Mem:* Am Astron Soc. *Res:* Theoretical astrophysics; numerical modeling of radiative and hydrodynamic processes in stars and the interstellar medium. *Mailing Add:* 5150 El Camino Real Suite B 31 Los Altos CA 94022

VILLET, RUXTON HERRER, b June 26, 33; US citizen; m 60; c 2. BIOCHEMISTRY, BIOCHEMICAL ENGINEERING. *Educ:* Univ Cape Town, BSc, 53; Princeton Univ, MS, 59; Univ Oxford, DPhil(biochem), 68. *Prof Exp:* Chem engr petrol, Mobil Oil Refinery, SAfrica, 53-57; chem engr res, Shell Develop Co, Calif, 59-60; assoc prof chem eng, Univ Witwatersrand, 61-65; res biochem, Inst Nat de la Sante et de la Rech Medicale, France, 73-75; reader, biochem eng, Massey Univ, NZ, 76-78; br chief, Biotechnol Br, Solar Energy Res Inst, Colo, 78-82; NAT PROG LEADER, AGR RES SERV, USDA, WASHINGTON, DC, 86- *Concurrent Pos:* Res fel, Calif Inst Technol, 68-70 & Max-Planck Inst, WGer, 71-73; dir res biotechnol, Elf Aquitaine, France, 82-83, consult, 83-85; dir biotechnol consortium, ARBS,

France, 83-84; mgr, Dept Biotechnol, Stauffer Chem Co, Calif, 85-86. *Mem:* Am Inst Chem Engrs. *Res:* Biochemical, molecular genetic and chemical engineer research directed toward developing biotechnological processes. *Mailing Add:* USDA/ARS/NPS Rm 334 Bldg 005 Barc-West Beltsville MD 20705

VILMS, JAAK, b Viljandi, Estonia, Jan 30, 37; US citizen; m 62; c 2. MATHEMATICS. *Educ:* Dickinson Col, AB, 59; Columbia Univ, MS, 61, PhD(math), 67. *Prof Exp:* Instr math, Purdue Univ, 65-67, asst prof, 67-71; ASSOC PROF MATH, COLO STATE UNIV, 71- *Mem:* Am Math Soc; Math Asn Am. *Res:* Differential geometry. *Mailing Add:* Dept Math Colo State Univ Ft Collins CO 80523

VILTER, RICHARD WILLIAM, b Cincinnati, Ohio, Mar 21, 11; m 35; c 1. MEDICINE. *Educ:* Harvard Univ, BA, 33, MD, 37; Am Bd Internal Med, dipl; Am Bd Nutrit, dipl, MA, Am Col Physicians, 80. *Prof Exp:* Intern, Cincinnati Gen Hosp, Ohio, 37-38, sr asst resident, 40-41, chief resident, Med Serv, 41-42; from asst prof to assoc prof med, Col Med, Univ Cincinnati, 42-56, asst to dean, Col Med, 43-45, asst dean, 45-52, asst dir dept, 52-56, dir, Lab Hemat & Nutrit, 45-56, Taylor prof med & dir, Dept Internal Med, 56-78; prof on spec assign, 78-81, EMER PROF, 81- *Concurrent Pos:* Fel nutrit, Hillman Hosp, Ala, 39-40; attend physician, Cincinnati Gen Hosp, 45-; Musser lectr, Sch Med, Tulane Univ, 65; chmn hemat study sect, NIH, 65-69, Comt Invest & Ther Cancer, Am Cancer Soc, 63-64. *Honors & Awards:* Goldberger Award, AMA, 60, Tehan Award, 78; Daniel Drake Medal Award, 85. *Mem:* Am Soc Clin Invest; Am Clin & Climat Asn (vpres, 64-65, 83-84); Asn Am Physicians; fel Am Col Physicians (secy-gen, 74-77, pres-elect, 78-79, pres, 79-80, emer pres, 84); Am Soc Hemat; Am Soc Clin Nutrit (pres, 60-61). *Res:* Hematology; nutrition; refractory and aplastic anemias; megaloblastic anemias; nutritional anemias. *Mailing Add:* 6067 Col Med Bldg 231 Bethesda Ave Cincinnati OH 45267

VIMMERSTEDT, JOHN P, b Jamestown, NY, June 5, 31; m 53; c 4. SOIL SCIENCE, FORESTRY. *Educ:* State Univ NY Col Forestry, Syracuse, BS, 53; Yale Univ, MS, 58, DF, 65. *Prof Exp:* Res forester, Southeastern Forest Exp Sta, 55-58; asst prof, 63-68, ASSOC PROF FOREST SOILS, SCH NATURAL RESOURCES, OHIO STATE UNIV, 68- *Mem:* Soil Sci Soc Am; fel Soc Am Foresters; AAAS; Sigma Xi. *Res:* Cation exchange properties of plant roots; mineral nutrition of trees; reclamation of spoil banks from coal mining; ecological impacts of forest recreation; soil fauna; use of legumes for nitrogen enrichment of forest planting sites; tree ring chemistry as indicator of past chemical environment; soil forming factors and soil formation. *Mailing Add:* Dept Forestry Ohio Agr Res & Develop Ctr Wooster OH 44691-4096

VINA, JUAN R, b Valencia, Spain, Feb 11, 56; m 87; c 1. BIOCHEMISTRY. *Educ:* Univ Valencia, MD, 79, PhD(biochem), 81. *Prof Exp:* Asst prof biochem, Fac Med, Valencia, Spain, 81-82; asst prof metab res, Dept Anesthesia, Hershey Med Ctr, 83 & 87-88; titular prof, 89-90, PROF BIOCHEM, DEPT BIOCHEM, UNIV VALENCIA, 91- *Mem:* Am Physiol Soc; Biochem Soc; Nutrit Soc. *Res:* Role of the gamma-glutamyl cycle in mammalian cells. *Mailing Add:* Dept Biochem Fac Med Ave Blasco Ibanez 17 Valencia 46010 Spain

VINAL, RICHARD S, b Worcester, Mass, May 5, 38; m 61; c 2. INORGANIC CHEMISTRY. *Educ:* Bates Col, BS, 60; Cornell Univ, PhD(inorg chem), 65. *Prof Exp:* Sr res chemist, 65-71, RES ASSOC, EASTMAN KODAK CO, 71- *Mem:* Am Chem Soc; Soc Photog Scientists & Engrs; Sigma Xi. *Res:* Transition metal; coordination chemistry; structural studies; photographic science; redox reactions; non-silver photographic systems development. *Mailing Add:* 350 Mt Airy Dr Rochester NY 14617-2126

VINATIERI, JAMES EDWARD, b Yankton, SDak, June 27, 47; m 69; c 2. INTERACTIVE COMPUTER GRAPHICS. *Educ:* Univ SDak, BA, 69; Univ Nebr, PhD(chem), 74. *Prof Exp:* Sr res chemist, 73-81, syst specialist, 82-84, STAFF DIR, PHILLIPS PETROL CO, 84- *Mem:* Am Chem Soc. *Res:* Chemical and physical properties of surfactant systems for enhanced oil recovery, especially phase behavior and interfacial phenomena. *Mailing Add:* 4627 Rolling Meadows Dr Bartlesville OK 74006-5529

VINCE, ROBERT, b Auburn, NY, Nov 20, 40; m 61; c 2. MEDICINAL CHEMISTRY. *Educ:* Univ Buffalo, BS, 62; State Univ NY Buffalo, PhD(med chem), 66. *Prof Exp:* Asst prof med chem, Col Pharm, Univ Miss, 66-67; from asst prof to assoc prof, 67-73, PROF MED CHEM, COL PHARM, UNIV MINN, 76- *Concurrent Pos:* Vis scientist, Roche Inst Molecular Biol, 74-75; res career develop award, Nat Cancer Inst, 72-76. *Honors & Awards:* Lunsford Richardson Grad Res Award, Richarson-Merrell Inc, 66. *Mem:* Am Chem Soc; Am Pharmaceut Asn; Am Soc Biol Chemists; Am Asn Cancer Res; Sigma Xi; Am Soc Microbiol; AAAS. *Res:* Design and synthesis of inhibitors of protein biosynthesis; nucleoside analogs as cancer chemotherapy agents; antiviral drug design. *Mailing Add:* Col Pharm Health Sci Unit F Univ Minn Minneapolis MN 55455

VINCENT, DAYTON GEORGE, b Hornell, NY, Apr 23, 36; m 59, 75, 86; c 4. METEOROLOGY. *Educ:* Univ Rochester, AB, 58; St Louis Univ, dipl meteorol, 59; Univ Okla, MS, 64; Mass Inst Technol, PhD(meteorol), 70. *Prof Exp:* Weather officer meteorol, Air Weather Serv, USAF, 59-62; res asst, Univ Okla Res Inst, 64; res meteorologist, Naval Weapons Lab, Dahlgren, Va, 64-65; res assoc, Mass Inst Technol, 69-70; from asst prof to assoc prof, 70-82, PROF ATMOSPHERIC SCI, DEPT GEOSCI, PURDUE UNIV, WEST LAFAYETTE, 82- *Mem:* Fel Am Meteorol Soc; Sigma Xi; Royal Meterol Soc. *Res:* Impact of convection on large-scale circulations in the tropics and mid-latitudes. *Mailing Add:* Dept Earth & Atmospheric Sci Purdue Univ West Lafayette IN 47907

VINCENT, DIETRICH H(ERMANN), b Leszno, Poland, May 11, 25; m 54; c 2. SURFACE STUDIES, GASES IN METALS & AMORPHOUS MATERIALS. *Educ:* Univ Goettingen, dipl phys, 50, Dr rer nat, 56. *Prof Exp:* Sci asst, Isotope Lab, Max Planck Inst Med Res, 51-58; resident res assoc, Argonne Nat Lab, 58-60; from assoc prof to prof, 60-89, EMER PROF NUCLEAR ENG, UNIV MICH, ANN ARBOR, 89- *Concurrent Pos:* Vis prof, Taiwan Univ & Nat Taiwan Normal Univ, Taipei, Taiwan, Repub China, 69; Int Atomic Energy Agency expert, Nat Tsing-Hua Univ, Hsinchu, Taiwan, Repub China, 68-69; Inter-Agency Personnel Agreement assignee, Ctr Anal Chem, Nat Bur Standards, Wash, DC, 82-83. *Mem:* AAAS; Am Phys Soc; Am Nuclear Soc; Sigma Xi. *Res:* Radiation effects in solids; Mossbauer spectroscopy; ion beam analysis; gases in metals; gases in amorphous semiconductors. *Mailing Add:* 3024 Phoenix Lab Univ Mich Ann Arbor MI 48109-2100

VINCENT, DONALD LESLIE, b St John, NB, Can, July 23, 21; m 49; c 3. ORGANIC CHEMISTRY. *Educ:* Acadia Univ, BSc, 42; McGill Univ, PhD(chem), 53. *Prof Exp:* Instr chem, Acadia Univ, 46-49; asst res officer, Nat Res Coun Can, 53-60; org group leader res lab, Coal Tar Prod Div Dom Tar & Chem Co, Ltd, 60-63; group leader organic chem, Domtar Res Ctr, 63-79, sr res scientist, 79-83; RETIRED. *Mem:* Chem Inst Can. *Res:* Wood chemistry; lignin; carbohydrates; pulp and paper; natural products; organic synthesis. *Mailing Add:* 153 Douglas Shand Ave Pointe Claire PQ H9R 2E2 Can

VINCENT, GEORGE PAUL, b Cleveland, Ohio, Oct 20, 01; m 68; c 1. CHEMISTRY. *Educ:* Hiram Col, AB, 23; Cornell Univ, MS, 24, PhD(phys chem), 27. *Prof Exp:* Asst chem, Cornell Univ, 23-25; res chemist, Eastman Kodak Co, 27-30; res chemist, Olin Mathieson Chem Corp, NY, 30-33, mgr res lab, 33-34, asst to res dir, 35-38, mgr sales develop, 49-52, mgr sales, Hydrocarbon Div, 52-53, govt serv, 53-67; CONSULT, 67- *Concurrent Pos:* Fed Govt liaison in food additive fields. *Mem:* Am Chem Soc; Am Inst Chem Eng. *Res:* Electrochemistry; sodium chlorite manufacturing methods and uses; inorganic chemistry in alkali-chlorine industry. *Mailing Add:* 865 First Ave New York NY 10017

VINCENT, GERALD GLENN, b Winnipeg, Man, Apr 13, 34; m 56; c 4. RESEARCH ADMINISTRATION, POLYMER CHEMISTRY. *Educ:* Univ Man, BS, 58, MS, 60, PhD(phys chem), 63. *Prof Exp:* Res chemist adhesion, Dow Chem Co, 63-65, proj mgr adhesives, 65-67; sect leader resins, DeSoto Inc, Des Plaines, 67-69, tech mgr aerospace, 69-72, mgr resin res, 72-73; asst dir res, 73-74, DIR RES, CENT RES, CROWN ZELLERBACH, 74- *Mem:* Am Chem Soc; Sigma Xi; Tech Asn Pulp & Paper Indust; Indust Res Inst. *Res:* Adhesion chemistry. *Mailing Add:* 15 Storrs Ct Mahwah NJ 07430-1581

VINCENT, HAROLD ARTHUR, b Lake City, Iowa, Jan 19, 30; m 57; c 2. ANALYTICAL CHEMISTRY. *Educ:* Univ Iowa, BS, 53; Univ Nev, MS, 60; Univ Ariz, PhD(chem), 64. *Prof Exp:* Chemist, Mining Anal Lab, Univ Nev, Reno, 56-68; chief chemist, Anaconda Co, 68-77, dir, Geol Labs, 77-81, mgr geol res, 81-84; consult, 85-86; CHEMIST, US ENVIRON PROTECTION AGENCY, LAS VEGAS, 87- *Mem:* AAAS; Am Chem Soc; Soc Appl Spectros; Sigma Xi. *Res:* Electroanalytical chemistry; fast neutron activation analysis; flame emission and absorption spectroscopy; x-ray emission spectroscopy; thermal methods of analysis. *Mailing Add:* US Environ Protection Agency PO Box 93478 Las Vegas NV 89193-3478

VINCENT, JAMES SIDNEY, b Redlands, Calif, Sept 19, 35; m 69; c 2. PHYSICAL CHEMISTRY. *Educ:* Univ Redlands, BS, 57; Harvard Univ, PhD(chem), 63. *Prof Exp:* Fel, Harvard Univ, 63-64 & Calif Inst Technol, 64-65; asst prof chem, Univ Calif, Davis, 65-71; ASSOC PROF CHEM, UNIV MD, BALTIMORE COUNTY, 71- *Mem:* Am Phys Soc; Soc Appl Spectroscopists. *Res:* Raman and infrared spectroscopic investigations of proteins and model membrane systems. *Mailing Add:* Dept Chem Univ Md Baltimore County Baltimore MD 21228

VINCENT, JERRY WILLIAM, b Chicago, Ill, June 24, 35; m 59; c 3. PALEONTOLOGY, PALEOECOLOGY. *Educ:* Univ Maine, Orono, BA, 58; Tex A&M Univ, MEd, 66, PhD(geol), 71. *Prof Exp:* Teacher geol & biol, N Yarmouth Acad, 58-61; teacher high sch, Maine, 61-67; assoc prof, 69-80, PROF GEOL, STEPHEN F AUSTIN STATE UNIV, 80- *Mem:* Soc Econ Paleont & Mineral; Nat Asn Geol Teachers. *Res:* Paleoecology of carbonate rocks, numerical taxonomy of various fossil taxa; biostratigraphy of lower Cretaceous rocks of Texas; invertebrate paleontology; earth science education in elementary and secondary schools. *Mailing Add:* Box 13011 SFA Sta Nacogdoches TX 75962

VINCENT, LEONARD STUART, b Cleveland, Ohio, July 27, 47. ARANEOLOGY. *Educ:* Calif State Univ, Northridge, BA, 70; Univ Calif, Davis, MS, 72; Univ Calif, Berkeley, PhD(entom), 80. *Prof Exp:* Lectr biol & arachnol, Univ Calif, Berkeley, 80-81; ASST PROF BIOL, GA SOUTHERN COL, 81- *Mem:* Sigma Xi; Am Arachnol Soc; Brit Arachnol Soc; Entom Soc Am. *Res:* Ecology and natural history of spiders. *Mailing Add:* Div Biol Sci Fullerton Col 321 E Chapman Ave Fullerton GA 92634

VINCENT, LLOYD DREXELL, b DeQuincy, La, Jan 7, 24; m 51; c 2. NUCLEAR PHYSICS. *Educ:* Univ Tex, BS, 52, MA, 53, PhD(physics), 60. *Prof Exp:* From asst prof to assoc prof, Univ Southwestern La, 53-58; instr, Tex A&M, 55-56; res scientist, Tex Nuclear Corp, 59-60, prof & dir physics dept, Sam Houston State Col, 60-65, asst to pres, 65-67; PRES, ANGELO STATE UNIV, 67- *Concurrent Pos:* Bd dir, Asn Tex Col & Univ, 81-85; dir, WTex Utilities Co, 78-; bd dir, WTex Rehab Ctr, 77- *Mem:* Am Phys Soc; Am Asn Physics Teachers; Am Asn State Col & Univ; Sigma Xi. *Mailing Add:* President's Off Angelo State Univ San Angelo TX 76909

VINCENT, MONROE MORTIMER, b Cleveland, Ohio, July 28, 12; m 41; c 1. CELL BIOLOGY, IMMUNOBIOLOGY. *Educ:* Adelbert Col, Western Res Univ, BA, 34; Edinburgh Univ, BA, 35. *Prof Exp:* Res asst parasitol, Univ Chicago, 36-41; parasitologist, US Army, Maj, SNC, Cmndg Officer, 15th Malaria Surv Unit, 42-46, vpres, North-Strong Corp, 48-50; vpres, Microbiol Assocs, 50-74; CONSULT, 74-; SR RES ASSOC, DEPT PEDIAT, SCH MED, UNIFORMED SERV UNIV OF THE HEALTH SCI, BETHESDA, MD, 78- *Concurrent Pos:* Assoc ed, In Vitro, 70-79; mem bd dir, Am Found Biol Res, 74-82; exec ed, Tissue Cult manual, 74-79; exec ed, Index Tissue Cult, 71-80. *Mem:* Tissue Cult Asn; Soc Cryobiol (treas, 73-74); Am Soc Cell Biol; Am Soc Trop Med; NY Acad Sci; Am Asn Tissue Banks. *Res:* Cell tissue and organ culture, virology, cell-mediated immunity, cryobiology. *Mailing Add:* 3905 Jones Bridge Rd Chevy Chase MD 20815-6721

VINCENT, MURIEL C, b Spokane, Wash, Sept 8, 22. PHARMACY, PHARMACEUTICAL CHEMISTRY. *Educ:* Ore State Col, BS, 44; Univ Wash, MS, 51, PhD, 55. *Prof Exp:* Instr pharm, Univ Wash, 53-54 & Ore State Col, 54-56; from asst prof to assoc prof, 56-58, chmn dept, 58-74, asst dean, Col Pharm, 65-82, PROF PHARM PRACT, COL PHARM, NDAK STATE UNIV, 58- *Concurrent Pos:* Consult, Vet Admin Hosp, 60-74. *Mem:* AAAS; Am Chem Soc; Am Pharmaceut Asn; Acad Pharm Sci. *Res:* Application of ion exchange resins and chromatography to the analysis of pharmaceutical products; drug absorption. *Mailing Add:* 515 30th Ave N Apt 13 Fargo ND 58102

VINCENT, PHILLIP G, b East Machias, Maine, July 18, 41; m 72; c 4. BIOCHEMISTRY, MICROBIOLOGY. *Educ:* Univ Md, BS, 64, PhD(fungus physiol), 67. *Prof Exp:* Res asst bot, Univ Md, 64, res fel fungus physiol & biochem, 65-67; res phytopathologist, 67-68, res microbiologist, 68-78, res plant physiologist, 78-85, PROJ LEADER & SR SCIENTIST, AGR ENVIRON INST, USDA, 85- *Concurrent Pos:* Consult, UN Div Narcotics, secy, working group on Papaver bracteatum, UN Secretariat; chmn, peer rev comt, Nat Inst Drug Abuse; peer reviewer, drug abuse biomed res rev comt, Alcohol, Drug Abuse, Ment Health & Human Serv, Dept HEW; mem, plant derived narcotics comt, Weed Sci Soc Am. *Mem:* AAAS; Am Soc Microbiol; Am Phytopath Soc; Scand Soc Plant Physiol; Sigma Xi; NY Acad Sci; Am Inst Chemists; Am Chem Soc; Am Inst Biol Sci. *Res:* Fungus physiology and biochemistry; mechanism of action of toxicants; phytopathology; plant physiology; biochemical characteristics and mechanisms of quality deterioration of meat; narcotic plant research, including marijuana, opium, poppy and cocaine. *Mailing Add:* Ctr Drug Eval & Res Rm 14B-45 HFD 102 5600 Fishers Rockville MD 20857

VINCENT, STEVEN ROBERT, b New Westminster, BC, Oct 12, 54; c 1. NEUROSCIENCES. *Educ:* Carleton Univ, BSc, 76; Univ BC, PhD(interdisciplinary), 80. *Prof Exp:* Asst histol, Karolinska Inst, 80-82; fel physiol, 82-83, asst prof, 83-88, ASSOC PROF PSYCHIAT, UNIV BC, 88- *Concurrent Pos:* Scholar, 83-88, scientist, Med Res Coun Can, 88- *Mem:* Soc Neurosci (chap pres, 87-88); Can Asn Neurosci. *Res:* Biochemical and neuroanatomical organization of the mammalian brain; localization and function of neurotransmitters, their receptors and second messengers and involvement of these systems in human neuro-degenerative disease. *Mailing Add:* Dept Psychiat Div Neurol Sci Univ BC Vancouver BC V6T 1W5 Can

VINCENT, THOMAS LANGE, b Portland, Ore, Sept 16, 35; m; c 2. AEROSPACE ENGINEERING. *Educ:* Ore State Univ, BS, 58, MS, 60; Univ Ariz, PhD(aerospace eng), 63. *Prof Exp:* Res engr, Boeing Airplane Co, Wash, 59-60; from asst prof to assoc prof aerospace & mech eng, 63-68, PROF AEROSPACE & MECH ENG, UNIV ARIZ, 68- *Concurrent Pos:* Guest prof, Tech Univ Munich, 67; NSF sci fac fel, Univ Calif, Berkeley, 69-70; guest lectr, Inst Mech & Appl Mach, Univ Genoa, 70; recipient, US-Australia Coop Sci Prog Award, NSF, 76-77, 83-84, 85 & 89; Hill vis professorship, Univ Minn, 90. *Res:* Optimal control and game theory with applications in control design, management programs for dynamical systems and evolution and adaption of biological systems. *Mailing Add:* Dept Aerospace & Mech Eng Univ Ariz Tucson AZ 85721

VINCENT, WALTER SAMPSON, b Veneta, Ore, Aug 6, 21; div; c 3. BACTERIAL ADHESION. *Educ:* Ore State Col, BS, 46, MS, 48; Univ Pa, PhD(zool), 52. *Prof Exp:* Asst zool, Ore State Col, 46; asst instr, Univ Pa, 48; res assoc genetics, Iowa State Col, 51-52; from instr to asst prof anat, Col Med, State Univ NY Upstate Med Ctr, 52-61; assoc prof, Sch Med, Univ Pittsburgh, 61-69; vis prof, Dept Biol, Brooklyn Col, 69-70; prof biol sci & chmn dept, 71-76, PROF CELL & MOLECULAR BIOL, SCH LIFE & HEALTH SCI, UNIV DEL, 76- *Concurrent Pos:* USPHS sr res fel, 59-61, career develop award, 61-63; Lalor fel, Marine Biol Lab, Woods Hole, 55, trustee, 66- & mem corp; res prof, Univ Edinburgh, 64, Sci Res Coun sr vis res fel, 67; mem exec comt, Marine Biol Lab, 71-75. *Mem:* AAAS; Am Soc Cell Biol; Genetics Soc Am; Int Inst Embryol; Am Soc Microbiol; Am Inst Biol Sci. *Res:* Evolution of ribosomal genes; adhesion of bacteria. *Mailing Add:* Zool Dept Ariz State Univ Tempe AZ 85287

VINCENTI, WALTER G(UIDO), b Baltimore, Md, Apr 20, 17; m 47; c 2. HISTORY OF TECHNOLOGY, GAS DYNAMICS. *Educ:* Stanford Univ, AB, 38. *Prof Exp:* Aeronaut res scientist, Ames Aeronaut Lab, Nat Adv Comt Aeronaut, Calif, 40-57; prof, 57-83, EMER PROF AERONAUT ENG, STANFORD UNIV, 83- *Concurrent Pos:* Lectr, Stanford Univ, 46-47, 52-54; Guggenheim fel, 63; co-ed, Ann Rev Fluid Mech, 70-77. *Honors & Awards:* Rockefeller Pub Serv Award, 56; Usher Prize, 84. *Mem:* Nat Acad Eng; Newcomen Soc; Soc Hist Technol; corresp mem Int Acad Astronaut; Hist Sci Soc; fel Am Inst Aeronaut & Astronaut. *Res:* History and philosophy of technology, especially regarding nature and sources of engineering knowledge. *Mailing Add:* Dept Aeronaut & Astronaut Stanford Univ Stanford CA 94305-4035

VINCENZ, STANISLAW ALEKSANDER, b Oskrzesince, Poland, Feb 4, 15; m 49; c 1. EXPLORATION GEOPHYSICS, GEOMAGNETISM & PALEOMAGNETISM. *Educ:* Univ London, ARCS & BSc, 37, DIC, 39, PhD(geophys), 52. *Prof Exp:* Demonstr geophys, Imp Col, London, 48-49, asst lectr, 49-51, res asst, 51-53; geophysicist & head geophys div, Jamaica Indust Develop, Corp, Jamaica, WI, 53-61; assoc prof geophys & geophys eng, 61-67, prof, 67-85, EMER PROF GEOPHYS, ST LOUIS UNIV, 85- *Concurrent Pos:* NSF prin investr grants, St Louis Univ, 62-67, 67-68, 69-70, 71-73 & 74-85 & US Geol Surv, 73-76. *Mem:* Soc Explor Geophys; Am Geophys Union; fel Royal Astron Soc; Europ Asn Explor Geophys; Sigma Xi. *Res:* Exploration geophysics; rock magnetism and paleomagnetism; geomagnetism; paleomagnetic investigations of Paleozoic sediments of Spitsbergen and of Central North America; interpretation of aeromagnetic anomalies of Northern Mississippi Embayment, specifically New Madrid earthquake zone. *Mailing Add:* Dept Earth & Atmospheric Sci St Louis Univ 3507 Laclede Ave St Louis MO 63103

VINCENZI, FRANK FOSTER, b Seattle, Wash, Mar 14, 38; m 60; c 3. PHARMACOLOGY. *Educ:* Univ Wash, BS, 60, MS, 62, PhD(pharmacol), 65. *Prof Exp:* NSF fel, Berne, 65-67; from asst prof to assoc prof, 67-80, PROF PHARMACOL, SCH MED, UNIV WASH, 80-, VCHMN DEPT, 77- *Mem:* AAAS; NY Acad Sci; Am Soc Pharmacol & Exp Therapeut; Biophys Soc; Cardiac Muscle Soc. *Res:* Autonomic transmitters; mechanisms of cardioactive drugs; membrane transport; red blood cell physiology and pathology; calmodulin and anti-calmodulin drugs; microcomputers in research and teaching. *Mailing Add:* Dept Pharmacol Univ Wash Seattle WA 98195

VINCETT, PAUL STAMFORD, b Southend, Eng, Jan 23, 44; m 65; c 3. THIN FILM PHYSICS, PHOTOGRAPHIC SCIENCE. *Educ:* Univ Cambridge, BA, 65, PhD(physics), 68. *Prof Exp:* Fel physics, Simon Fraser Univ, 68-70; scientist, Corp Lab, ICI Ltd, Runcorn, Eng, 70-74; scientist physics, Xerox Res Ctr Can, 74-80, mgr thin film sci & memory, 80-88, MGR, ADVAN TECHNOL, XEROX CAN INC, 88- *Concurrent Pos:* Tutor, Open Univ, NW Region, Manchester, Eng, 71-74; mem ed bd, Thin Solid Films, 80-85; chmn, Div Appl Physics, Can Asn Physicists, 89-90, dir, Corp Members, 90- *Honors & Awards:* Kosar Mem Award, Soc Photog Scientists Engrs, 87- *Mem:* Can Asn Physicists; Soc Photog Scientists & Engrs. *Res:* Solid state and chemical physics, particularly thin film physics and conduction processes in insulating solids and liquids; novel electrophotographic and photographic processes; high density information recording; business development of new technologies; author of article in encyclopedia of physical science. *Mailing Add:* Xerox Can Inc 2660 Speakman Dr Mississauga ON L5K 2L1 Can

VINCIGUERRA, MICHAEL JOSEPH, b New York, NY, Mar 19, 45; m 70; c 1. PHYSICAL CHEMISTRY. *Educ:* Iona Col, BS, 66; Adelphi Univ, MS, 69, PhD(phys chem), 71. *Prof Exp:* From asst prof to assoc prof chem, 70-78, asst vpres acad affairs, 80-83, vpres, 83-87, PROF CHEM, STATE UNIV NY AGR & TECH COL, FARMINGDALE, 78-, CHMN, DIV ARTS & SCI, 76-, PROVOST, 87- *Concurrent Pos:* Res asst, Adelphi Univ, 71-72; adj asst prof, St John's Univ, NY, 72; res consult, Unichem Res Assoc, 75- *Mem:* Am Chem Soc; NY Acad Sci; Am Asn Higher Educ; AAAS. *Res:* Physical properties of bio-polymers; light scattering by biological gels; nuclear magnetic relaxation of polymer solutions. *Mailing Add:* Provost State Univ NY Agr & Tech Col Farmingdale NY 11735

VINCOW, GERSHON, b New York, NY, Feb 27, 35; m 64; c 2. PHYSICAL CHEMISTRY. *Educ:* Columbia Univ, AB, 56, MA, 57, PhD(chem), 59. *Prof Exp:* Fel, Hebrew Univ, Israel, 60; NSF fel, Calif Inst Technol, 60-61; from asst prof to prof chem, Univ Wash, 61-71; chmn, Dept Chem, 71-77, vpres res & grad affairs, 77-78, actg dean, 79-80, dean, col arts & sci, 80-85, VCHANCELLOR ACAD AFFAIRS, SYRACUSE UNIV, 85- *Concurrent Pos:* Sloan Found res fel, 64-67; NSF fel, Harvard Univ, 70-71. *Mem:* Am Chem Soc; Am Phys Soc; AAAS. *Res:* Electron paramagnetic resonance spectroscopy. *Mailing Add:* Dept Chem Syracuse Univ Syracuse NY 13210

VINE, ALLYN COLLINS, b Garrettsville, Ohio, June 1, 14; m 40; c 3. OCEAN ENGINEERING, GEOPHYSICS. *Educ:* Hiram Col, BA, 36; Lehigh Univ, MS, 38. *Hon Degrees:* PhD, Lehigh Univ, 73. *Prof Exp:* Physicist, 40-50, phys oceanogr, 50-63, sr scientist, 63-79, EMER SR SCIENTIST, WOODS HOLE OCEANOG INST, 79- *Concurrent Pos:* Oceanogr, USN Bur Ships, 46-49. *Honors & Awards:* Compass Award, Marine Technol Soc, 68; David Stone Award, New Eng Aquarium, 77; Lockheed Award, Marine Technol Soc, 87. *Mem:* Nat Acad Eng; Am Geophys Soc; Marine Technol Soc; AAAS; Acoust Soc Am. *Res:* Underwater acoustics; design of deep submersibles and research ships; improved measurement techniques for the floor of the ocean; the ocean as a stabilizing influence in world affairs. *Mailing Add:* Woods Hole Oceanog Inst Box 416 Woods Hole MA 02543

VINE, JAMES DAVID, b Detroit, Mich, Dec 14, 21; m 48; c 2. GEOCHEMISTRY. *Educ:* Univ Mich, BS, 43. *Prof Exp:* Geologist, US Geol Surv, 46-80; consult, 80-86; RETIRED. *Concurrent Pos:* Mem, Nat Battery Adv Comt. *Res:* Exploration for lithium brines and clays. *Mailing Add:* 21736 Panorama Dr Golden CO 80401

VINES, DARRELL LEE, b Crane, Tex; m; c 3. ELECTRICAL ENGINEERING. *Educ:* McMurry Col, BA, 59; Tex Tech Univ, BS, 59, MS, 60; Tex A&M Univ, PhD(elec eng), 67. *Prof Exp:* Engr, Tex Instruments, 60-62; instr elec eng, Tex Tech Univ, 62-63 & Tex A&M Univ, 63-66; assoc prof, 66-76, PROF ELEC ENG & COMPUT SCI, TEX TECH UNIV, 76- *Concurrent Pos:* Distinguished vis prof , USAF Acad, 81-82. *Honors & Awards:* Meritorious Serv, Inst Elec & Electronics Engrs Educ Soc, 90. *Mem:* Inst Elec & Electronics Engrs; Am Soc Eng Educ. *Res:* Computer applications and logic circuit design; digital signal processing. *Mailing Add:* Dept Elec Eng & Comput Sci Tex Tech Univ Lubbock TX 79409-3103

VINES, HERBERT MAX, b Ala, Feb 4, 18; m 42; c 2. PLANT PHYSIOLOGY. *Educ:* Ala Polytech Inst, BS, 40; Univ Calif, MS, 49; Univ Calif, Los Angeles, PhD, 59. *Prof Exp:* Specialist postharvest physiol, Univ Calif, 49-53; tech rep nutrit, Shell Chem Co, 53-56; technician plant biochem, Univ Calif, 57-59, fel, 59-60; assoc biochemist, Citrus Exp Sta, Univ Fla, 61-67; prof plant physiol, Univ Ga, 67-88; RETIRED. *Mem:* Am Soc Hort Sci; Am Soc Plant Physiol. *Res:* Post-harvest physiology; plant metabolism, especially metabolic blocks in electron transport system. *Mailing Add:* 494 W Cloverhurst Athens GA 30606

VINEYARD, BILLY DALE, b Clarkton, Mo, Sept 7, 31; m 56; c 1. BIOORGANIC & PEPTIDE CHEMISTRY. *Educ:* Southeast Mo State Col, BS, 53; Univ Mo, PhD(org chem), 59. *Prof Exp:* Res chemist, Celanese Corp, 59-60; sr res chemist, Org Div, 60-64, res specialist, 64-68, sci fel, 68-76, sr fel, 76-86, DISTINGUISHED FEL, MONSANTO CO, 86- *Mem:* Am Chem Soc; AAAS. *Res:* Catalytic homogeneous asymmetric hydrogenation; food, fine and feed chemicals. *Mailing Add:* 754 Ambois Dr St Louis MO 63141

VINGIELLO, FRANK ANTHONY, b New York, NY, Aug 20, 21; m 42; c 3. CHEMISTRY. *Educ:* Polytech Inst Brooklyn, BS, 42; Duke Univ, PhD(org chem), 47. *Prof Exp:* Lab instr org chem, Polytech Inst Brooklyn, 43-44; lab instr, Duke Univ, 44-47; instr, Univ Pittsburgh, 47; res assoc, Northwestern Univ, 47-48; from asst prof to assoc prof chem, Va Polytech Inst & State Univ, 48-57, prof org chem, 57-68; PROF ORG CHEM, NORTHEAST LA UNIV, 68- *Concurrent Pos:* Chemist, WVa Ord Works, 42-43; consult chem indust. *Honors & Awards:* J Shelton Horsley Award, Va Acad Sci, 66. *Mem:* Am Chem Soc; NY Acad Sci. *Res:* Organic synthesis in steroids and aromatic molecules; mechanisms of organic reactions; cyclization of o-benzylphenones; synthesis of aromatic hydrocarbons, research in air pollution. *Mailing Add:* Dept Chem Northeast La Univ 700 University Ave Monroe LA 71209

VINH, NGUYEN XUAN, b Yenbay, Viet Nam, Jan 3, 30; m 55; c 4. ANALYSIS & FUNCTIONAL ANALYSIS. *Educ:* Air Inst, France, BS, 53; Univ Marseille, MS, 54; Univ Colo, MS, 63, PhD(aerospace), 65; Univ Paris, DSc(math), 72. *Prof Exp:* Asst prof aerospace eng, Univ Colo, 65-68; assoc prof, 68-72, PROF AEROSPACE ENG, UNIV MICH, ANN ARBOR, 72- *Concurrent Pos:* Vis lectr, Univ Calif, Berkeley, 67; vis prof ecol nat sup aero, France, 74; assoc ed, Acta Astronautica, 79-; chair prof, Nat Tsing Hua Univ, Taiwan, 82. *Mem:* Math Asn Am; Int Acad Astronaut; French Nat Acad Air & Space. *Res:* Ordinary differential equations; astrodynamics and optimization of space flight trajectories; theory of non linear oscillations. *Mailing Add:* Dept Aerospace Eng Univ Mich Ann Arbor MI 48104

VINICK, FREDRIC JAMES, b Amsterdam, NY, June 18, 47; m 70; c 3. SYNTHETIC ORGANIC CHEMISTRY. *Educ:* Williams Col, BA, 69; Yale Univ, PhD(chem), 73. *Prof Exp:* NIH res fel, Columbia Univ, 73-75; sr scientist org chem, Pharmaceut Div, Ciba-Geigy Corp, 75-78; ASST DIR MED CHEM, CENT RES, PFIZER, INC, 78- *Concurrent Pos:* NSF fel. *Mem:* Am Chem Soc; NY Acad Sci. *Res:* Synthesis of biologically and/or medicinally important compounds; discovery of new drug leads. *Mailing Add:* Pfizer Inc Eastern Point Rd Groton CT 06340

VINING, LEO CHARLES, b Whangarei, NZ, Mar 28, 25; m 53, 89; c 4. BIO-ORGANIC CHEMISTRY, MICROBIOLOGY. *Educ:* Univ NZ, BSc, 48, MSc, 49; Cambridge Univ, PhD, 51. *Prof Exp:* Scholar, Univ Kiel, Ger, 51-53; fel, Rutgers Univ, 53-54, instr, Inst Microbiol, 54-55; asst res off, Prairie Regional Lab, Nat Res Coun Can, 55-58, assoc res off, 58-62, sr res off, Atlantic Regional Lab, 62-69, prin res off, 69-71; prof biol, 71-86, Killam Prof, 86-90, PROF BIOL PALHOUSIE UNIV, 90- *Concurrent Pos:* vis scientist, Mass Inst Technol, 77-78 & Univ Alta, 84-85. *Honors & Awards:* Harrison Prize, Royal Soc Can, 72; Can Soc Microbiologists Award, 76; Merck, Sharpe & Dohme lectr, 65; John Labatt Award, 85; Charles Thom Award, Soc Indust Microbiol, 85. *Mem:* Fel Royal Soc Chem; fel Can Inst Chem; Can Soc Microbiol; Am Soc Microbiol; fel Royal Soc Can. *Res:* Chemistry of antibiotics; fungal metabolites; biosynthesis of natural products; control of secondary metabolism. *Mailing Add:* Dept Biol Dalhousie Univ Halifax NS B3H 4J1 Can

VINJE, MARY M (TAYLOR), b Madison, Wis, Sept 8, 13; m 38. BIOLOGY. *Educ:* Univ Wis, PhD(bot), 38. *Hon Degrees:* Dr Humanities, St Ambrose Col, Davenport, Iowa, 80. *Prof Exp:* Asst bot, Univ Wis, 35-38; teacher high schs, 45-48; from asst prof to prof biol, St Ambrose Col, 48-77, head dept, 74-77, emer prof, 77-79, Hauber chair, 80-81; RETIRED. *Concurrent Pos:* NSF vis scientist, Iowa High Schs, 60-66. *Mem:* AAAS; Nat Asn Biol Teachers; Nat Sci Teachers Asn; Bot Soc Am; Sigma Xi. *Res:* Aerobiology, particularly the investigation of microbiota in atmosphere; isolation of a fungus from an ozone meter; origin of blue rain. *Mailing Add:* 786 Carlita Circle Rohnert Park CA 94928

VINOCUR, MYRON, medicine, pathology, for more information see previous edition

VINOGRADE, BERNARD, b Chicago, Ill, May 7, 15; m 42; c 4. MATHEMATICS. *Educ:* City Col, BS, 37; Univ Mich, MA, 40, PhD(math), 42. *Prof Exp:* Instr math, Univ Wis, 42-44 & Tulane Univ, 44-45; staff mem, Radiation Lab, Mass Inst Technol, 45; from asst prof to assoc prof math, Iowa State Univ, 45-55, prof, 55-74, distinguished prof, Sci & Humanities, 74-80 chmn dept, 61-64, actg head, 60; RETIRED. *Concurrent Pos:* Opers analyst, Standby Unit, USAF, 50-; vis prof, San Diego State Col, 59-60 & City Col New York, 64-65. *Mem:* Am Math Soc; Math Asn Am. *Res:* Abstract and linear algebra; field theory; forest resouce management. *Mailing Add:* PO Box 1089 Santa Clara UT 84765-1089

VINOGRADOFF, ANNA PATRICIA, b Essex, Eng; div. PHARMACEUTICAL CHEMISTRY & AGRICULTURAL PRODUCTS CHEMISTRY. *Educ:* Univ Calif, Los Angeles, BSc, 76, PhD(chem), 81. *Prof Exp:* Teaching asst org chem, Univ Calif, Los Angeles, 76-77, res assoc, 78-80; sr res chemist, 80-84, PROJ LEADER, DOW CHEM USA, 84- *Mem:* Am Chem Soc. *Res:* Organic synthesis with emphasis on molecules of biological interest and natural products. *Mailing Add:* Agr Prod Group Dow Chem USA PO Box 9002 Walnut Creek CA 94598-0997

VINOGRADOV, SERGE, b Beirut, Lebanon, Aug 27, 33; US citizen; c 2. BIOCHEMISTRY. *Educ:* Am Univ Beirut, BA, 52, MA, 54; Ill Inst Technol, PhD(phys chem), 59. *Prof Exp:* Fel, Univ Alta, 59-62; res assoc, Yale Univ, 62-66; asst prof biochem, 66-68, assoc prof, 68-71, PROF BIOCHEM & ADJ ASSOC PROF BIOL, WAYNE STATE UNIV, 71- *Mem:* Am Soc Biol Chem; The Chem Soc; Am Chem Soc; Biophys Soc. *Mailing Add:* Dept Biochem 4320 Scott Wayne State Univ 5950 Cass Ave Detroit MI 48201

VINOKUR, MARCEL, b Moravska-Ostrava, Czech, Feb 16, 29; US citizen; m 54, 63; c 1. FLUID MECHANICS, NUMERICAL ANALYSIS. *Educ:* Cornell Univ, BEngPhys, 51; Princeton Univ, PhD(aeronaut eng), 57. *Prof Exp:* Assoc res scientist, Lockheed Palo Alto Res Lab, 55-59, res scientist, Lockheed Missiles & Space Co, Calif, 59-61, staff scientist, 61-71; lectr mech, Univ Santa Clara, 65-75, res assoc mech eng, 75-84; RES SPECIALIST, STERLING SOFTWARE, 84- *Mem:* Assoc fel Am Inst Aeronaut & Astronaut. *Res:* Inviscid flow; radiation gas dynamics; non-equilibrium flow; numerical methods. *Mailing Add:* 919 Channing Ave Palo Alto CA 94301

VINORES, STANLEY ANTHONY, b Pottsville, Pa, July 7, 50; m 77; c 4. OPHTHALMOLOGY, DEVELOPMENTAL NEURO-ONCOLOGY. *Educ:* Pa State Univ, BS, 72; Univ Tex, PhD(zool), 76. *Prof Exp:* Teaching asst biol & zool, Univ Tex, 72-74, trainee carcinogenesis & genetics, 74-76; fel, Ohio State Univ, 77-78; staff fel, NIH, 78-81; res assoc, Inst Cancer Res, 81-82; RES ASST PROF, MED SCH, UNIV VA, 82- *Mem:* Asn Res Vision & Ophthal; NY Acad Sci; Soc Neurosci; Am Asn Pathologists; Fedn Am Soc Exp Biol; AAAS; Histochem Soc. *Res:* Immunocytochemistry and electron microscopy of diabetic retina and nervous system tumors. *Mailing Add:* Dept Ophthal Univ Va Sch Med Charlotteville VA 22908

VINSON, DAVID BERWICK, b Houston, Tex, Oct 7, 17; m 40; c 2. PSYCHOPHYSIOLOGY. *Educ:* Univ Calif, Los Angeles, BA, 41; Univ London, PhD, 52. *Prof Exp:* Dir, Tex Acad Advan Life Sci, 60-87, pres, Microset Inc,78-87; PRES, ASSESSMENT SYSTS INC, 67-, PRES, FACTOR, INC, 86- *Concurrent Pos:* Consult, Life Sci, 54- *Mem:* Aerospace Med Asn; Am Inst Aeronaut & Astronaut; Am Psychol Asn; Inst Elec & Electronics Engrs; Soc Biol Psychiat; Sigma Xi. *Res:* Man-machine systems; neuropsychology. *Mailing Add:* 106 Schattenbaum Fredericksburg TX 78624

VINSON, JAMES S, b Chambersburg, Pa, May 17, 41; m 67. PHYSICS. *Educ:* Gettysburg Col, BA, 63; Univ Va, MS, 65, PhD(physics), 67. *Prof Exp:* Res asst low temperature physics, Univ Va, 64-67; asst prof physics, MacMurray Col, 69-71; assoc prof, Univ NC, Asheville, 71-75, prof physics, 75-78, chmn dept, 71-78, dir comput ctr, 74-78; DEAN COL ARTS & SCI, UNIV HARTFORD, 78- *Mem:* AAAS; Am Phys Soc; Am Asn Physics Teachers; Nat Sci Teachers Asn; World Future Soc; Sigma Xi. *Res:* Low temperature physics; scintillations in liquid helium; quantum mechanics; computer based instructions; future studies. *Mailing Add:* Pres Univ Evansville 1800 Lincoln Ave Evansville IN 47722

VINSON, JOE ALLEN, b Ft Smith, Ark, Nov 16, 41; m 66; c 2. ANALYTICAL CHEMISTRY. *Educ:* Univ Calif, Berkeley, BS, 63; Iowa State Univ, MS, 66, PhD(org & analytical chem), 67. *Prof Exp:* Res asst, Anal Chem Sect, Ames Lab, AEC, 66-67; asst prof chem, Shippensburg State Col, 67-68 & Washington & Jefferson Col, 68-72; mem staff, J T Baker Chem Co, 72, prod develop chemist, 72-74; assoc prof, 74-90, PROF CHEM, UNIV SCRANTON, 90- *Concurrent Pos:* Res Corp Cottrell grant, 69-70; Law Enforcement Assistance Admin grant, 71-72; NSF Instruct Sci Equip grant, 75-77; Int Copper Res Asn grant, 79-82, Ben Franklin grant, 88-90. *Mem:* Am Chem Soc; NY Acad Sci. *Res:* Clinical, drug and pollution analysis; thin layer chromatography; analysis of marijuana in biological fluids; vitamins and minerals in health and disease. *Mailing Add:* Dept Chem Univ Scranton Scranton PA 18072

VINSON, LEONARD J, biochemistry; deceased, see previous edition for last biography

VINSON, RICHARD G, b Prattville, Ala, Nov 18, 31; m 55; c 2. MATHEMATICS. *Educ:* Huntingdon Col, BA, 54; Fla State Univ, MA, 56; Univ Ala, PhD(math), 62. *Prof Exp:* Instr math, Fla State Univ, 55-56, Univ Tenn, 56-58 & Univ Ala, 58-61; prof, Huntingdon Col, 61-69; PROF MATH, UNIV S ALA, 69- *Concurrent Pos:* Instr state-wide educ TV network, 63-67; dir NSF two-yr col prog. *Mem:* AAAS; Am Math Soc; Math Asn Am; Am Asn Univ Professors; Nat Coun Teachers Math. *Res:* Non-Euclidean geometry; statistics; calculus. *Mailing Add:* Dept Math Univ S Ala Mobile AL 36688

VINSON, S BRADLEIGH, b Mansfield, Ohio, Apr 8, 38; m 60; c 2. CHEMICAL ECOLOGY. *Educ:* Ohio State Univ, BS, 61; Miss State Univ, MS, 63, PhD(entom), 65. *Prof Exp:* Res asst entom, Miss State Univ, 64-65, asst prof, 65-69; assoc prof, 69-75, PROF ENTOM, TEX A&M UNIV, 75- *Concurrent Pos:* Prog chmn, Entom Soc Am, 73, chmn physiol sect, 79; subj ed, Entom Sci, 84-; chmn, Imported Fire Ant Rev Comt, Environ Protection Agency, 80. *Honors & Awards:* Outstanding Res Award, Am Registry Prof Entomologists, 79, Ital Silvestry Award Biol Control, 86. *Mem:* AAAS; Entom Soc Am; Am Inst Biol Sci; Am Soc Zool; Am Chem Soc. *Res:* Vertebrate insecticide resistance; mechanisms of arthropod resistance; insect physiology; parasite-host and predator-host relationships; biology of social insects. *Mailing Add:* Dept Entom Tex A&M Univ College Station TX 77843

VINSON, WILLIAM ELLIS, b Greensboro, NC, Apr 4, 43; m 63; c 2. POPULATION GENETICS, DAIRY SCIENCE. *Educ:* NC State Univ, BS, 65, MS, 68; Iowa State Univ, PhD(pop genetics), 71. *Prof Exp:* asst prof dairy cattle genetics, 71-76, assoc prof, 76-82, PROF DAIRY SCI, VA

POLYTECH INST & STATE UNIV, 82-, DEPT HEAD, 87- *Honors & Awards:* J L Lush Award, Am Dairy Sci Asn. *Mem:* Am Dairy Sci Asn; Am Genetic Asn; Am Soc Animal Sci; Biomet Soc; Sigma Xi. *Res:* Direct and correlated responses to selection; pedigree evaluation of genetic merit; inheritance of discontinuous characters; genetic evaluations from field data; computer simulation of genetic populations. *Mailing Add:* Dept Dairy Sci Va Polytech Inst & State Univ Blacksburg VA 24061

VINSONHALER, CHARLES I, b Winfield, Kans, Mar 29, 42. MATHEMATICS. *Educ:* Calif Inst Technol, BS, 64; Univ Wash, PhD(math), 68. *Prof Exp:* Asst prof, 68-76, ASSOC PROF MATH, UNIV CONN, 76- *Mem:* Am Math Soc. *Res:* Ring theory and Abelian groups. *Mailing Add:* Dept Math Univ Conn Storrs CT 06269

VINT, LARRY FRANCIS, b Davenport, Iowa, May 12, 41; m 85; c 6. ANIMAL BREEDING, POULTRY BREEDING. *Educ:* Iowa State Univ, BS, 63, MS, 69, PhD(animal breeding), 71. *Prof Exp:* Data processing mgr biomet, Pilch-De Kalb, De Kalb AgRes, 71-72, dir res poultry breeding, 72-73, res investr corp develop, 73-74; geneticist animal breeding, USDA, 74; asst dir, Res Poultry Breeding, De Kalb AgRes, 74-88, dir Vet & Tech Serv, 88-91, DIR RES POULTRY BREEDING, DE KALB POULTRY RES, 91- *Mem:* Am Soc Animal Sci; Poultry Sci Asn; Coun Agr Sci & Technol; AAAS; World's Poultry Asn. *Res:* Genetic improvement in poultry populations; poultry management systems, growth and nutrition. *Mailing Add:* De Kalb Poultry Res Inc 3100 Sycamore Rd De Kalb IL 60115

VINTERS, HARRY VALDIS, b Port Arthur, Ont, Can, Dec 8, 50. NEUROPATHOLOGY, CLINICAL NEUROSCIENCES. *Educ:* Univ Col, Univ Toronto, BSc, 72, MD, 76; FRCPS(C), 81. *Prof Exp:* Fel neuropath, Med Ctr, Univ Calif, Los Angeles, 82-83; asst prof path & clin neurol sci, Univ Western Ont, 84-85; ASST PROF PATH & MEM, BRAIN RES INST, MED CTR, UNIV CALIF, LOS ANGELES, 85- *Concurrent Pos:* Staff neuropathologist, Univ Hosp, London, Can, 84-85 & Med Ctr, Univ Calif, Los Angeles, 85-; prin investr, NIH First Award, Nat Inst Neurol Commun Dis & Stroke; Can Heart Found fel, 84-85; John D French Found & Wilson Found fel, 88-89. *Honors & Awards:* Desmond Magner Award, Nat Cancer Inst, Can, 80. *Mem:* Can Asn Neuropathologists; Am Asn Neuropathologists; Am Soc Cell Biol; AAAS. *Res:* Micro-vascular changes in brain related to aging and Alzheimer's disease; neurologic and neuropathologic complications of AIDS; miscellaneous basic and clinical aspects of stroke and cerebro-vascular disease. *Mailing Add:* Dept Path CHS18-170 Med Ctr Univ Calif Los Angeles CA 90024-1732

VINTI, JOHN PASCAL, b Newport, RI, Jan 16, 07. CELESTIAL MECHANICS. *Educ:* Mass Inst Technol, SB, 27, ScD(physics), 32. *Prof Exp:* Harrison res fel physics, Univ Pa, 32-34; res asst, Mass Inst Technol, 34-35; instr, Brown Univ, 36-37; asst prof, The Citadel, 37-38; instr, Worcester Polytech Inst, 39-41; physicist ballistic res labs, Aberdeen Proving Ground, 41-57 & Nat Bur Standards, 57-65; prof appl math, NC State Univ, 66; consult, Exp Astron Lab, 67-70, vis assoc prof aeronaut & astronaut, Inst, 69-70, LECTR AERONAUT & ASTRONAUT, MASS INST TECHNOL, 71-, CONSULT, MEASUREMENT SYSTS LAB, 70- *Concurrent Pos:* Lectr, Univ Del, 48-49 & 54-55 & Univ Md, 50-52; consult, Army Chem Ctr, 52; prof lectr, Georgetown Univ, 63-64; adj prof, Cath Univ Am, 65-66; mem, Comn Celestial Mech, Int Astron Union. *Mem:* Fel AAAS; fel Am Phys Soc; assoc fel Am Inst Aeronaut & Astronaut; Am Astron Soc; Am Geophys Union; fel Royal Astron Soc. *Res:* Possible effects of variation of the gravitational constant, both in dynamical astronomy and in cosmology. *Mailing Add:* 273 Clarendon St Boston MA 02116-1404

VINYARD, GARY LEE, b Harrisburg, Ill, Mar 13, 49. AQUATIC ECOLOGY, BEHAVIORAL ECOLOGY. *Educ:* Univ Kans, BA, 71, PhD(biol), 77. *Prof Exp:* Vis asst prof, Sch Biol Sci, Okla State Univ, 76-77; asst prof, Dept Zool, Univ Mont, 77-78; asst prof, 78-84, ASSOC PROF DEPT BIOL, UNIV NEV, 84- *Mem:* Am Soc Limnol & Oceanog; AAAS; Ecol Soc Am; Sigma Xi; Am Fish Soc. *Res:* Predatory interactions of aquatic organisms, both fish and invertebrates; behavioral ecology of aquatic organisms. *Mailing Add:* Dept Biol Univ Nev Reno NV 89557

VINYARD, WILLIAM CORWIN, b McArthur, Calif, Apr 30, 22; m 60. BOTANY. *Educ:* Chico State Col, BA, 42; Mich State Univ, MS, 51, PhD(bot), 58. *Prof Exp:* Instr bot, Univ Okla, 53-54; instr, Mich State Univ, 55-56; instr, Univ Mont, 56-57; asst prof, Univ Kans, 57-58; from asst prof to assoc prof, 58-72, PROF BOT, HUMBOLDT STATE UNIV, 72- *Concurrent Pos:* Algological consult, Calif State Dept Water Resources, 62-64 & Klamath Basin Study, US Dept Interior, 67-68 & Nat Park Serv, 71-72. Phycol Soc Am; Int Asn Plant Taxon; Int Phycol Soc. *Mem:* Sigma Xi. *Res:* Taxonomy and ecology of freshwater algae; synopsis of the desmids of North America; algae of western North America; relation of algae to water pollution; algal food of herbivorous fishes; eipzoic algae. *Mailing Add:* Dept Biol Humboldt State Univ Arcata CA 95521

VIOLA, ALFRED, b Vienna, Austria, July 8, 28; nat US; m 63. ORGANIC CHEMISTRY. *Educ:* Johns Hopkins Univ, BA, 49, MA, 50; Univ Md, PhD(chem), 55. *Prof Exp:* Asst instr chem, Johns Hopkins Univ, 49-50; teaching asst, Univ Md, 50-54; res assoc, Boston Univ, 55-57; from asst prof to assoc prof chem, 57-68, PROF CHEM, NORTHEASTERN UNIV, 68- *Concurrent Pos:* Vis prof, Univ Munich, Ger, 77, Monash Univ, Melbourne, Australia, 84. *Mem:* Am Chem Soc; Sigma Xi. *Res:* Preparation and properties of highly unsaturated organic compounds; thermal rearrangements; stereochemistry of transition states; pericyclic reactions of acetylenes and allenes. *Mailing Add:* Dept Chem Northeastern Univ Boston MA 02115

VIOLA, JOHN THOMAS, b Haverhill, Mass, Mar 6, 38; m 60; c 2. TECHNICAL MANAGEMENT. *Educ:* Univ NH, BS, 60; Pa State Univ, MS, 61; Mass Inst Technol, PhD(chem), 67. *Prof Exp:* US Air Force, 60-84, nuclear effects res officer, thermodyn, Air Force Weapons Lab, Albuquerque, NMex, 61-64, physicist, 67-71; instr & asst prof chem, USAF Acad, 71-74, assoc prof & dir advan courses chem, 74-75, prog mgr chem, Directorate Chem Sci, Air Force Off Sci Res, 75-78, dir bus mgt, E-4 Prog Off, 78-80, prog dir, Joint Surveillance Systs, 80-82, dep space transp systs & testing, Air Force Space Div, El Segundo, Calif, 82-84, PROG MGR SPACE PROGS, ROCKWELL INT SCI CTR, 84- *Res:* Infrared detectors materials and devices, development of infrared focal plane arrays; microgravity materials processing and crystal growth. *Mailing Add:* 1007 Brookview Ave Westlake Village CA 91361

VIOLA, RONALD EDWARD, b Brooklyn, NY, July 28, 46; m 71; c 2. ENZYMOLOGY, MAGNETIC RESONANCE SPECTROSCOPY. *Educ:* Fordham Univ, BS, 67; Pa State Univ, MS, 73, PhD(biochem), 76. *Prof Exp:* Asst biochem, Univ Wis, 76-79; asst prof, Southern Ill Univ, 79-84; assoc prof, 84-89, PROF BIOCHEM, UNIV AKRON, 89- *Mem:* Am Chem Soc; Am Soc Biochem & Molecular Biol; Int Soc Magnetic Resonance. *Res:* Magnetic resonance studies of metal ion complexes; determination of enzyme mechanisms by kinetic analysis; studies of aluminium and platinum coordination chemistry and their role in biological systems. *Mailing Add:* Dept Chem Univ Akron 190 E Buchtel Ave Akron OH 44325-3601

VIOLA, VICTOR E, JR, b Abilene, Kans, Apr 8, 35; m 62; c 3. PHYSICAL CHEMISTRY. *Educ:* Univ Kans, AB, 57; Univ Calif, Berkeley, PhD(nuclear chem), 61. *Prof Exp:* Instr & res fel nuclear chem, Univ Calif, Berkeley, 61-62; NSF fel, Europ Orgn Nuclear Res, 63, Ford fel, 64; res assoc chem, Argonne Nat Lab, 64-66; from asst prof to assoc prof chem, Univ Md, College Park, 66-74, prof, 74-80; dir, Cyclotron Facil, 86-87, prof, 80-90, DISTINGUISHED PROF CHEM, IND UNIV, 90- *Concurrent Pos:* Consult, Argonne Nat Lab, 66-73; vis prof, Univ Calif, Berkeley, 73-74; Guggenheim fel, 80-81; consult, Lawrence Livermore Lab, Lawrence Berkeley Lab. *Honors & Awards:* Am Chem Soc Award in Nuclear Chem, 86. *Mem:* Fel AAAS; Am Asn Univ Professors; Am Chem Soc; fel Am Phys Soc; Sigma Xi. *Res:* Reaction mechanism studies in heavy ion and intermediate-energy collisions; nuclear astrophysics; nuclear fission at moderate. *Mailing Add:* Dept Chem Ind Univ Bloomington IN 47405

VIOLANTE, MICHAEL ROBERT, b Buffalo, NY, Mar 6, 44; div; c 2. COLLOID CHEMISTRY, PHYSICAL CHEMISTRY. *Educ:* State Univ NY, Buffalo, BA, 66; Fla State Univ, PhD(inorg chem), 70. *Prof Exp:* res scientist pulp & paper, Res & Develop Div, Union-Camp Corp, 70-74; asst prof, 74-82, ASSOC PROF RADIOL, SCH MED & DENT, UNIV ROCHESTER, 82-; VPRES RES & DEVELOP, STERILIZATION TECH SERV, INC, 83- *Mem:* Am Chem Soc; NY Acad Sci; AAAS. *Res:* Development and formulation of particulate drug delivery system; water insoluble drugs can be administered intravenously, orally or by any other route of administration; membrane transport phenomena; colloid and solution chemistry. *Mailing Add:* Dept Radiol Univ Rochester Med Ctr Rochester NY 14642

VIOLET, CHARLES EARL, b Des Moines, Iowa, May 1, 24; m 51; c 5. PHYSICS. *Educ:* Univ Chicago, BS, 48; Univ Calif, AB, 49, PhD(physics), 53. *Prof Exp:* Physicist, 50-57, test group dir, Oper Plumbbob, 57-58, dep test mgr, Oper Hardtack, 58-59, test div leader, 59-61, PHYSICIST, LAWRENCE LIVERMORE LAB, UNIV CALIF, 61- *Honors & Awards:* Fel, Am Phys Soc. *Mem:* Am Phys Soc. *Res:* Mössbauer spectroscopy; magnetism of metals and alloys; physical metallurgy; high Tc superconductors. *Mailing Add:* Lawrence Livermore Nat Lab Univ Calif Box 808 Livermore CA 94550

VIOLETT, THEODORE DEAN, b Great Bend, Kans, Apr, 27, 32; m 53; c 3. PHYSICS. *Educ:* Univ Mo, BS, 53, MA, 54; Univ Colo, PhD(physics), 59. *Prof Exp:* PROF PHYSICS, WESTERN STATE COL COLO, 59- *Mem:* Am Phys Soc; Am Asn Physics Teachers. *Res:* Vacuum ultraviolet radiation and solar spectroscopy. *Mailing Add:* Dept Physics Western State Col 75E01 Heights Gunnison CO 81230

VIOLETTE, JOSEPH LAWRENCE NORMAN, b Winslow, Maine, Aug 24, 32; m 57; c 7. ELECTRICAL ENGINEERING. *Educ:* Rensselaer Polytech Inst, BEE, 56; NC State Univ, PhD(elec eng), 71; Auburn Univ, Montgomery, MBA, 72. *Prof Exp:* USAF, 56-77, from instr to asst prof elec eng, USAF Acad, 62-67, proj mgr tactical air commun systs, Electronic Systs Div, L G Hanscom Field, Mass, 69-71, proj mgr, Advan Res Projs Agency, Washington, DC, 72-73; C3I proj engr, Washington Off, TRW, 77-79; PRIN ENGR, DON WHITE CONSULTS, EMC ENG, 80-; PRES & FOUNDER, EMC & SYSTS ENG, JLN VIOLETTE & ASSOC. *Mem:* Inst Elec & Electronics Engrs. *Res:* Integral equation solution for nonplanar obstacles in coaxial waveguides obtained from dyadic Green's function formulation of boundary value problem; computer solution of singular integral equations so derived. *Mailing Add:* Violette Eng Corp 120 E Borad St PO Box 639 Falls Church VA 22046

VIRARAGHAVAN, THIRUVENKATACHARI, b Madras, India, July 15, 34; m 67; c 2. WATER TREATMENT, WASTEWATER TREATMENT. *Educ:* Univ Madras, India, BE, 55, MSc, 63; Univ Ottawa, PhD(civil eng), 75. *Prof Exp:* Jr public health engr, Dept Pub Health Eng & Munic Works, Govt Madras, India, 55-61, asst pub health engr, 61-65; asst adv pub health eng, Ministry Health, Govt India, New Delhi, 65-70; res asst fluid mech, Dept Civil Eng, Univ Ottawa, 70-75; sr environ engr, ADI Ltd Consult Engrs, NB, Can, 75-82; assoc prof, 82-83, PROF, FAC ENG, UNIV REGINA, 83- *Concurrent Pos:* Hon res assoc, Univ NB, 78-86. *Honors & Awards:* Nawab Zain Yar Jung Bahadur Mem Gold Medal, Inst Engrs (India), 77-78. *Mem:* Fel Inst Engrs; fel Am Soc Civil Engrs; Can Soc Civil Eng; Am Water Works Asn; Water Pollution Control Fedn. *Res:* On-site wastewater treatment and disposal; biological treatment of wastewaters, especially anaerobic treatment of wastewaters; use of peat in pollution control; groundwater pollution. *Mailing Add:* Fac Eng Univ Regina Regina SK S4S 0A2 Can

VIRELLA, GABRIEL T, b Vilanova ila Geltiu, Barcelona, Spain, May 23, 43; US citizen; m 67; c 2. CLINICAL & DIAGNOSTIC IMMUNOLOGY. *Educ:* Med Sch Univ Lisbon, Portugal, MD, 67, PhD(microbiol), 74; Am Bd Med Lab Immunol, cert, 79. *Prof Exp:* Res immunol, Gulbenkian Inst Sci, Portugal, 74-75; from asst prof to assoc prof immunol, 76-80, PROF IMMUNOL & MICROBIOL, MED UNIV SC, 80-, PROF PATH & LAB MED, 82- *Concurrent Pos:* Consult, Dept Path Lab Med, Med Sch Univ SC, 76; prin investr, Kroc Found Grant, 82-84, juvenile Diabetes Found Grant, 82-84. *Honors & Awards:* Pfizer Award, Pfizer Inc, 73. *Mem:* Brit Soc Immunol; Am Asn Immunologist; Am Acad Microbiol; Asn Med Lab Immunologist; Clin Immunol Soc. *Mailing Add:* Dept Microbiol & Immunol Med Univ SC 171 Ashley Ave Charleston SC 29425

VIRGILI, LUCIANO, b Carassai, Italy, Mar 15, 48; m 72; c 2. THERMAL ANALYSIS, MICROSCOPY. *Educ:* Univ Firenze, Italy, Dr, 75. *Prof Exp:* ASST MGR, E R SQUIBB & SONS, INC, 78- *Mem:* Am Chem Soc. *Res:* Methods development in pharmaceuticals and related raw materials; problem solving; non routine complaint analysis; materials characterization and properties particle analysis. *Mailing Add:* 426 Wheeler Rd New Brunswick NJ 08902-2710

VIRGO, BRUCE BARTON, b Vancouver, BC, Mar 18, 43; Can citizen; m 69; c 2. TOXICOLOGY, DRUG & STEROID METABOLISM. *Educ:* Univ BC, BSc, 65, MSc, 70, PhD(pharmacol, toxicol), 74. *Prof Exp:* Res biologist econ ornith, Can Wildlife Serv, Govt Can, 65-66, contractee ecol, 67-68; Nat Res Coun fel toxicol, McGill Univ, 74-75; Nat Res Coun fel pharmacol, Univ Montreal, 75; asst prof, 75-81, assoc prof physiol & pharmacol, 81-88, Univ Windsor, 81-88; PROF TOXICOL/PHARMACOL & DIR, CTR TOXICOL, MEMORIAL UNIV, 89- *Concurrent Pos:* Consult, Indust Res Inst, Windsor, 79-87 & Govt New Brunswick, 80; assoc ed, Can J Physiol Pharmacol, 84-; bd mem, Can Fed Biol Soc, 89-; mem, Pharmacol Grant Panel, MRCC, 90- *Mem:* Soc Toxicol Can (sec, 89-); Pharmacol Soc Can; Soc Toxicol. *Res:* Pharmacology of the hepatic drug and steroid metabolizing enzymes; physiology, pharmacology and toxicology of reproduction processes at all organizational levels; toxicology of environmental chemicals with emphasis on the effects of chronic, low-dose exposure. *Mailing Add:* Ctr Toxicol Fac Pharm Memorial Univ St John's NF A1B 3V6 Can

VIRK, KASHMIR SINGH, b India; US citizen; m 58; c 2. COMBUSTION CHAMBER DEPOSITS EFFECTS ON OCTANE REQUIREMENT. *Educ:* U P Agr Univ, India, BSc, 66; Univ Ill, MS, 68. *Prof Exp:* Sr proj engr, 83-88, TECHNOLOGIST, TEXACO RES CTR, 89- *Mem:* Soc Automotive Eng. *Res:* Diesel exhaust filters; diesel fuel composition effect on exhaust emissions; diesel fuel additives effect on exhaust emissions. *Mailing Add:* 12 Ladue Rd Hopewell Junction NY 12533

VIRKAR, RAGHUNATH ATMARAM, b Vir, Maharashtra, India, Dec 12, 30; US citizen; m 53; c 1. PHYSIOLOGY, INVERTEBRATE ZOOLOGY. *Educ:* Univ Bombay, BS, 50, MS, 52; Univ Minn, PhD(zool), 64. *Prof Exp:* Demonstr zool, Wilson Col, Bombay, 52-53; lectr, Vithalbhai Patel Col, 53-61; instr, Univ Minn, 64; asst res biologist, Univ Calif, Irvine, 65-66 & Univ Calif, Riverside, 66; asst prof biol, Wis State Univ-Superior, 66-68; assoc prof, Newark State Col, 68-73; chmn dept biol sci, 77-83, PROF BIOL, KEAN COL NJ, 73-, CHMN BIOL SCI, 88- *Mem:* AAAS; Am Soc Zool; Am Inst Biol Sci. *Res:* Nutritional role of dissolved organic matter in invertebrates; role of free amino acids in osmoregulation. *Mailing Add:* Dept Biol Kean Col NJ Union NJ 07083

VIRKKI, NIILO, b Vuoksela, Finland, May 7, 24; m 49; c 3. INSECT CYTOGENETICS. *Educ:* Univ Helsinki, Lic phil, 51, PhD(genetics), 52. *Prof Exp:* Inspector pest animals, Vet Sect, Health Bd, Helsinki, Finland, 49-53; lab keeper, Univ Helsinki, 53-61, asst prof genetics, 55-61; from asst cytogeneticist to assoc cytogeneticist, 61-64, CYTOGENETICIST, AGR EXP STA, UNIV PR, 64- *Concurrent Pos:* Fel fores insect lab, Can Dept Agr, Ont, 55-56; NSF grants, 65-68, Fundaçao Amparo de Pesquisa de Sao Paulo grant, 80; vis prof, Univ Cuzco, Perú, 66, Univ Estadual Paulista, Brazil, 81. *Res:* Problems concerning evolution of karyotypes in the beetle sub-family Alticinae, especially formation of giant asynaptic sex chromosomes and their mode of orientation and segregation in meiosis; beetle cytogenetics. *Mailing Add:* Crop Protection Dept Agr Exp Sta Rio Piedras PR 00928-1360

VIRNIG, MICHAEL JOSEPH, b Rochester, Minn, Mar 31, 46; m 68; c 3. ORGANIC CHEMISTRY. *Educ:* St Mary's Col, BA, 68; Iowa State Univ, PhD(org chem), 74. *Prof Exp:* Sr res chemist, Henkel Corp, 75-80, group leader, 80-84, res assoc, 84-88, tech mgr, 88-91; TECH MGR, COGNIS INT, 91- *Mem:* Am Chem Soc; Am Inst Mining Engrs; Sigma Xi. *Res:* Solvent extraction, process development organic synthesis, natural products. *Mailing Add:* 1105 Lanewood Way Santa Rosa CA 95404-2720

VIRNSTEIN, ROBERT W, b Washington, DC, Mar 19, 43; m 69; c 2. MARINE ECOLOGY. *Educ:* Johns Hopkins Univ, BA, 66; Univ SFla, MA, 72; Col William & Mary, PhD(marine sci), 76. *Prof Exp:* Marine scientist, Va Inst Marine Sci, 75-76; fel, Harbor Br Inst, 76-77, asst res scientist, Harbor Br Found, 77-85; ENVIRON SPECIALIST, ST JOHNS RIVER WATER MGT DIST, 87- *Concurrent Pos:* Consult, Seagrass Ecosysts Analysts, 85- *Mem:* Ecol Soc Am; Int Asn Meiobenthologists; Southeastern Estuarine Res Soc. *Res:* Estuarine benthic ecology; trophic relationships; experimental field ecology; role of predation; benthic invertebrate taxonomy; seagrass growth rates; seagrass communities. *Mailing Add:* St Johns River Water Mgt Dist PO Box 1429 Palatka FL 32178-1429

VIRTUE, ROBERT, b Denver, Colo, Sept 12, 04. ANESTHESIOLOGY. *Educ:* Univ Mich, PhD(biochem), 33; Univ Colo, MD, 46. *Prof Exp:* EMER PROF ANESTHESIOL, UNIV COLO, 69- *Mem:* Am Chem Soc; Am Soc Biol Chemists; AMA; Sigma Xi. *Mailing Add:* 727 Birch St Denver CO 80220-4953

VISCOMI, B(RUNO) VINCENT, b Philadelphia, Pa, Sept 21, 33; m 58; c 3. MECHANICAL ENGINEERING. *Educ:* Drexel Inst, BS, 56; Lehigh Univ, MS, 57; Univ Colo, PhD(civil eng), 69. *Prof Exp:* Engr, Westinghouse Elec Co, 57; res engr, Philadelphia Elec Co, 58-62, nuclear engr, 62-64; from asst prof to assoc prof eng, 64-75, PROF ENG, LAFAYETTE COL, 75-, HEAD DEPT CIVIL ENG, 72-; vis fac assoc, Lehigh Univ, 87-88. *Concurrent Pos:* Roy T & Laura Jones fac lectureship, Lafayette Col, 69; NSF fel, Nat Ctr Resource Recovery, 74; consult, Resource Recovery Serv, Woodbridge, NJ & Naval Sea Systs Command, Washington, DC, 81; res engr & cluster group leader, Nat Found ATLSS Ctr, Lehigh Univ, 88. *Mem:* Am Soc Eng Educ; NY Acad Sci; Am Soc Civil Engrs; Sigma Xi; Am Soc Testing & Mat. *Res:* Dynamic response of elastic mechanisms; structural dynamics. *Mailing Add:* Dept Civil Eng Lafayette Col Easton PA 18042

VISCONTI, JAMES ANDREW, b St Louis, Mo, Apr 13, 39; m 63; c 2. PHARMACY. *Educ:* St Louis Col Pharm, BS, 61, MS, 63; Univ Miss, PhD(pharm), 69. *Prof Exp:* Resident, John Cochran Vet Admin Hosp, St Louis, Mo, 63, staff pharmacist, 64-66; res pharmacist, Vet Admin Hosp, Long Beach, Calif, 63-64; asst prof pharm, 68-72, ASSOC PROF PHARM, COL PHARM, OHIO STATE UNIV, 72-, DIR DRUG INFO CTR, UNIV HOSPS, 68- *Concurrent Pos:* Lehn & Fink Pharm gold medal award, 61-62; Robert Lincoln McNeil citation fel award, 66-67. *Mem:* Am Soc Hosp Pharmacists; Am Pharmaceut Asn; assoc mem AMA; Drug Info Asn. *Res:* Epidemiology and econmics of adverse drug reactions; pharmacology of drug-drug, drug-laboratory tests and drug-food interactions; computerized drug information services; health and disease economics. *Mailing Add:* 217 Lloyd Parks Hall Ohio St Univ Col Med Columbus OH 43210

VISEK, WILLARD JAMES, b Sargent, Nebr, Sept 19, 22; m 49; c 3. NUTRITION, TOXICOLOGY. *Educ:* Univ Nebr, BSc, 47; Cornell Univ, MSc, 49, PhD(nutrit biochem), 51; Univ Chicago, MD, 57. *Hon Degrees:* DSc, Univ Nebr, 80. *Prof Exp:* Asst animal nutrit, Cornell Univ, 48-51; AEC fel, Univ-Atomic Energy Agr Res Prog, Tenn, 51-52, res assoc, 52-53; res asst pharmacol, Univ Chicago, 53-57, from asst prof to assoc prof, 57-64; prof nutrit & comp metab, Cornell Univ, 64-75; PROF CLIN SCI, COL MED, UNIV ILL, URBANA, 75-, PROF NUTRIT & METAB, DEPT NUTRIT SCI, 78-, PROF, DEPT INTERNAL MED, 84- *Concurrent Pos:* Intern univ hosp & clins, Univ Chicago, 57-59; mem teratol subcomt, Comn Drug Safety, 63 & subcomt animal nutrit, Nat Res Coun-Nat Acad Sci, 65-72, adv coun, Inst Lab Animal Resources, 66-69, subcomt animal care facilities surv, 68-70; consult sect health related facilities, USPHS, 67; assoc ed, Nutrit Rev, 68-71; George Henry Durgin lectr, Bridgewater State Col, 70; Nat Cancer Inst-USPHS spec fel, Mass Inst Technol, 70-71; res fel, Mass Gen Hosp, 70-71; grad fac rep nutrit, Cornell Univ, 74-; mem adv comt, Diet, Nutrit & Cancer Prog, Nat Cancer Inst, 76-; mem inst rev bd, Univ Ill, 76-; mem exec comt, Sch Basic Med Sci & Clin Med, Univ Ill, Urbana-Champaign; consult & mem dean's comt, Danville Vet Admin Hosp, 77-; mem study sect nutrit, NIH, 80-84; Brittingham vis prof, Univ Wis-Madison, 82-83; comt technol options to inprove health, 86-88; mem, Nutrit Res Sci Adv Comt, Nat Dairy Coun, 87-90; mem, Coun Sci Adv Human Nutrit, USDA, 88-; mem bd dirs, Am Bd Nutrit, 88; sr univ scholar, Univ Ill, 88-91; ed-in-chief, J Nutrit, 90- *Honors & Awards:* Osborne Mendel Award, 85; Hogan Mem Lectr, Univ Mo, 87. *Mem:* Fel AAAS; Am Soc Pharmacol & Exp Therapeut; Soc Exp Biol & Med; fel Am Soc Animal Sci; Am Soc Clin Nutrit; fel Am Inst Nutrit; Am Asn Cancer Res; Am Gastroenterol Asn; Cent Soc Clin Res. *Res:* Effects of ammonia on energy metabolism, area cycle activity and nucleic acid synthesis; interactions of amino acids; enzyme immunity; influence of diet on cancer incidence and influence of dietary fat level on gene expression. *Mailing Add:* 190 Med Sci Bldg Univ Ill 506 S Mathews Urbana IL 61801

VISELTEAR, ARTHUR JACK, b New York, NY, Mar 19, 38; m 66; c 2. HISTORY OF MEDICINE, PUBLIC HEALTH. *Educ:* Tulane Univ, BA, 59; Univ Calif, Los Angeles, MPH, 63, PhD(hist), 65. *Prof Exp:* Lectr pub health, Univ Calif, Los Angeles, 65-69; from asst prof to assoc prof pub health, 69-79, res assoc hist of sci & med, 74-79, ASSOC PROF HIST MED & PUB HEALTH, SCH MED, YALE UNIV, 79- *Concurrent Pos:* Robert Wood Johnson Health Policy fel, Inst Med, Nat Acad Sci, 74-75. *Mem:* AAAS; Am Asn Hist Med; Am Pub Health Asn. *Res:* History and health policy; social medicine; health services research. *Mailing Add:* Sect Hist Med Yale Univ Sch Med 333 Cedar St New Haven CT 06510

VISHER, FRANK N, b Twin Falls, Idaho, Mar 10, 23; m 48; c 4. HYDROLOGY, GEOLOGY. *Educ:* Tex Tech Col, BS, 46 & 47. *Prof Exp:* Apprentice engr, Tex Hwy Dept, 47-48; geologist, US Geol Surv, 48-56, engr, 56-66, hydrologist, Fla, 66-67, res hydrologist, 67-80, CONSULT HYDROLOGIST, US GEOL SURV, 80- *Mem:* AAAS. *Res:* Hydrologic studies, especially the principals of occurrence of ground water, water budget studies, fresh-salt water interrelationships, geochemistry of water and relation of geomorphology to ground water; environmental studies. *Mailing Add:* 3351 Vivian Ct Wheat Ridge CO 80033

VISHER, GLENN S, b May 20, 30; US citizen; m 53; c 3. GEOLOGY. *Educ:* Univ Cincinnati, BS, 52; Northwestern Univ, MS, 56, PhD(geol), 60. *Prof Exp:* Explorationist, Shell Oil Co, 58-60; res geologist, Sinclair Res, Inc, 60-66; from adj asst prof to prof geol, Univ Tulsa, 64-80; PRES, GEOL SERV & VENTURES INC, 80-; PRES, BNJ OIL PROPERTIES INC, 81- *Concurrent Pos:* Lectr training courses, Domestic & Int Petrol Co Personnel; consult, Int Petrol Co; adj res prof, Wichita State Univ, 86-, Univ Okla, 88- *Mem:* Am Asn Petrol Geol; fel Geol Soc Am; Soc Econ Paleont & Mineral; Int Asn Sedimentol; fel AAAS. *Res:* Stratigraphic models; depostional processes; physical characteristics of sandstone units; texture of sandstones; petrology of shales and sandstones. *Mailing Add:* Geol Serv & Ventures Inc 2920 E 73rd St Tulsa OK 74136

VISHNIAC, HELEN SIMPSON, b New Haven, Conn, Dec 22, 23; m 51; c 3. MICROBIAL ECOLOGY, MYCOLOGY. *Educ:* Univ Mich, BA, 45; Radcliffe Col, MA, 47; Columbia Univ, PhD(bot), 50. *Prof Exp:* Tutor biol, Queens Col, NY, 48-51, instr, 51-52; lectr microbiol, Sch Med, Yale Univ,

53-61; res assoc biol, Univ Rochester, 74-78; asst prof microbiol, Dept Cell, Molecular & Develop Biol, 78-83, ASSOC PROF BOT & MICROBIOL, OKLA STATE UNIV, 83- *Concurrent Pos:* Lectr, Nazareth Col Rochester, 75-76; fel, Am Asn Univ Women, 75-76; vis assoc prof, State Univ NY Brockport, 76-78. *Mem:* AAAS; Mycol Soc Am; Am Soc Microbiol. *Res:* Antarctic yeasts, aquatic fungi; molecular systematics. *Mailing Add:* Dept Bot & Microbiol Okla State Univ Stillwater OK 74078

VISHNUBHOTLA, SARMA RAGHUNADHA, b Masuli Patam, India, July 7, 46; m 74. COMPUTER SCIENCE. *Educ:* Madras Inst Technol, India, DMIT, 68; Wash Univ, MS, 72, DSc(comput sci), 73. *Prof Exp:* Asst prof comput sci, Cent Mich Univ, 73-78, assoc prof, 78-; AT DEPT ENG, OAKLAND UNIV, ROCHESTER, MICH 78- *Mem:* Asn Comput Mach; Inst Elec & Electronics Engrs. *Res:* Computer science education; fault diagnosis in computer hardware and software systems; data structures and data bases. *Mailing Add:* Dept Comput Sci Oakland Univ 108 Dodge Hall Rochester MI 48063

VISICH, MARIAN, JR, b Brooklyn, NY, Jan 8, 30; m 59; c 3. AEROSPACE ENGINEERING. *Educ:* Polytech Inst Brooklyn, BAeE, 51, MAeE, 53, PhD, 56. *Prof Exp:* From asst to res assoc aeronaut eng, Polytech Inst Brooklyn, 51-56, from res asst prof to prof, 56-77; ASSOC DEAN, COL ENG & APPL SCI, STATE UNIV NY, STONY BROOK, 77- *Concurrent Pos:* Consult, Curtiss-Wright Corp, 55, Gen Elec Corp, 56, Gen Appl Sci Lab, 56-68, Advan Technol Labs, 69-75 & US Army Res Labs, 69-75. *Mem:* AAAS; Am Inst Aeronaut & Astronaut; Am Soc Eng Educ; Sigma Xi. *Res:* Physics of fluids; high-speed aerodynamics; aircraft and missile propulsion. *Mailing Add:* Eight Whitehall Dr Huntington NY 11743-3916

VISKANTA, RAYMOND, b Lithuania, July 16, 31; US citizen; m 56; c 3. HEAT TRANSFER. *Educ:* Univ Ill, BS, 55; Purdue Univ, MS, 56, PhD(heat transfer), 60. *Prof Exp:* Asst mech engr heat transfer, Argonne Nat Lab, 56-58, assoc mech engr, 60-62; assoc prof, 62-66, prof mech eng, 66-86, GOSS DISTINGUISHED PROF ENG, PURDUE UNIV, 86- *Concurrent Pos:* Vis prof mech eng, Univ Calif, Berkeley, 68-69; consult to various pvt & govt orgn, 72-; Alexander von Humboldt Found award, 76-78; guest prof mech eng, Tech Univ Munich, 76-77; vis prof, Tokyo Inst Technol, 83. *Honors & Awards:* US Sr Scientist Award, Alexander von Humboldt Found, 75; Heat Transfer Mem Award, Am Soc Mech Engrs, 76; Thermophys Award, Am Inst Aeronaut & Astronaut, 79; Sr Res Award, Am Soc Eng Educ, 84; Max Jakob Mem Award, Am Soc Mech Engrs-Am Inst Chem Engrs, 86; Melville Medal, Am Soc Mech Engrs, 88. *Mem:* Nat Acad Eng; fel Am Inst Aeronaut & Astronaut; fel Am Soc Mech Engrs. *Res:* Radiation transfer in gases and solids; applied thermodynamics; heat transfer in combustion systems; solar energy utilization; solid/liquid phase change heat transfer. *Mailing Add:* Sch Mech Eng Purdue Univ West Lafayette IN 47907

VISNER, SIDNEY, b New York, NY, Dec 10, 17; wid; c 2. PHYSICS, NUCLEAR REACTOR SAFETY. *Educ:* City Col New York, BS, 37, MS, 38; Univ Tenn, PhD(physics), 51. *Prof Exp:* Res asst physics, Columbia Univ, 41-43, res scientist & sect leader, S A M Labs, 43-45; sr res physicist & dept head, Gaseous Diffusion Plant, Carbide & Carbon Chems Corp, 45-51; group leader, Exp Reactor Physics, Oak Ridge Nat Lab, 51-55; mgr physics dept, Nuclear Div, Combustion Eng, Inc, 55-69, dir physics & comput anal, Nuclear Power Dept, 69-82; STAFF CONSULT, INST NUCLEAR POWER OPERS, 82- *Concurrent Pos:* Reactor Physics Adv Comt, US Energy Res & Develop Admin, 75-77. *Mem:* Sigma Xi; Am Phys Soc; fel Am Nuclear Soc. *Res:* Hydrodynamic studies in isotope separation by gaseous diffusion; transport properties of gases at low pressures; neutron reactor physics and analysis; experimental physics; nuclear reactor development. *Mailing Add:* Inst Nuclear Power Opers 1100 Circle 75 Pkwy Suite 1500 Atlanta GA 30339

VISOTSKY, HAROLD M, b Chicago, Ill, May 25, 24; m 55; c 2. MEDICINE, PSYCHIATRY. *Educ:* Univ Ill, BS, 48, MD, 51. *Prof Exp:* Coordr psychiat residency training, Med Sch, Univ Ill, Chicago, from asst prof to assoc prof psychiat, 59-69; prof psychiat & chmn dept, Med Sch, 69-91, DIR, INST PSYCHIAT, NORTHWESTERN & MEM HOSP, 75- *Concurrent Pos:* Nat Found Infantile Paralysis res fel, 55-56; chief of serv, Chicago State Psychiat Hosp, 57-59; dir ment health sect, City Bd Health, Chicago, 59-62; dir, Ill Dept Ment Health, 62-69; chmn task force, Joint Comn Ment Health of Children, 66-68; mem adv comt, Secy Dept HEW, 66-67; mem, Eval Ment Health Prog, First Mission, USSR, 67, NIMH Stud Mission, USSR, 89 & IOM Study Vet Admin Resources, 88-; mem task force, President's Comn Ment Health; Sr consult Ctr Ment Health & Psychiat Serv, Am Hosp Asn, 79-; chmn, Int Comn Abuse Psychiat & Psychiatrists, Am Psychiat Asn, 80-84, Coun Int Affairs, 84-; mem, Mission to Japan Ment Health Syst, Int Comn Jurists, 85-87. *Honors & Awards:* Edward A Strecker Award, Inst Pa Hosp, 69; Bowis Award, Am Col Psychiat, 81, Distinguished Serv Award, 88; Presidential Commendation Award, Am Psychiat Asn, 88. *Mem:* Am Asn Social Psychiat (vpres, 76-77, pres, 88-91); Am Hosp Asn; fel Am Psychiat Asn (secy, 71-73, vpres, 73-74); Am Col Psychoanalysts; Am Orthopsychiat Asn (pres, 76-77); Am Col Psychiatrists (pres, 83-84). *Res:* Social and milieu psychiatry; effects of hallucinogenic drugs in understanding mental illness; stress: coping and adaptation. *Mailing Add:* Dept Psychiat & Behav Sci Northwestern Univ Med Sch 303 E Superior St Chicago IL 60611

VISSCHER, PIETER BERNARD, b Minneapolis, Minn, Dec 11, 45; m 72; c 2. NON-EQUILIBRIUM STATISTICAL MECHANICS. *Educ:* Harvard Univ, BA, 67; Univ Calif, Berkeley, MA, 68, PhD(solid state physics), 71. *Prof Exp:* Res assoc, Univ Ill, Urbana, 71-73; res physicist, Univ Calif, San Diego, 73-75; asst prof physics, Univ Ore, 75-78; asst prof, 78-80, assoc prof, 80-84, PROF PHYSICS, UNIV ALA, 84- *Concurrent Pos:* Prin investr, NSF res grant, 79-84; consult, Los Alamos Nat Lab, 85-86. *Mem:* Am Phys Soc; Am Asn Univ Professors. *Res:* Developed exactly renormalizable theory of transport; applications to calculation of liquid viscosity, disordered systems and viscoelastic hydrodynamics; solid state and surface physics. *Mailing Add:* Dept Physics & Astron Univ Ala University AL 35486

VISSCHER, SARALEE NEUMANN, b Lewistown, Mont, Jan 9, 29; m 69; c 4. ENTOMOLOGY. *Educ:* Univ Mont, BA, 49; Mont State Univ, MS, 58, PhD(entom), 63. *Prof Exp:* Asst prof entom, Mont State Univ, 62-65; NIH res fel insect develop, Univ Va, 65-66; assoc prof, 67-71, PROF ENTOM, MONT STATE UNIV, 72- *Concurrent Pos:* Co-investr, NIH res grant, 65-69; US-Japan grant, 77-79; AID grant, 79-80, 85-87; Rockefeller grant, 80-81; Dow Chem grant, 80-81; NSF grant, 80-82; Agr & Resources Inventory Surv through Aerospace Remote Sensing grant, 80-82; hon guest, Arthropodan Embryol Soc Japan. *Mem:* AAAS; Entom Soc Am; Am Soc Zoologists; Soc Develop Biol; Sigma Xi; Pan-Am Acridological Soc; Int Soc Chem Ecol. *Res:* Maternal/embryonic interrelationships; physiology of embryonic diapause of insects; role of plant growth hormones in regulation of insect growth, reproduction and longevity; host plant effects on grasshopper population dynamics. *Mailing Add:* Dept Entom Mont State Univ Bozeman MT 59715

VISSCHER, WILLIAM M, b Memphis, Tenn, May 16, 28; m 51; c 4. THEORETICAL PHYSICS, THERMODYNAMICS & MATERIAL PROPERTIES. *Educ:* Univ Minn, BA, 49; Cornell Univ, PhD(theoret physics), 53. *Prof Exp:* Res assoc physics, Univ Md, 53-56; STAFF MEM THEORET PHYSICS, LOS ALAMOS SCI LAB, 56- *Concurrent Pos:* Vis prof, Univ Wash, 67. *Mem:* Fel Am Phys Soc; Sigma Xi. *Res:* Meson theory; theory of nuclear structure and spectra; lattice dynamics and Mössbauer effect; particle accelerator physics; solid state physics; statistical mechanics; random packing; transport processes; scattering theory; materials science; acoustics. *Mailing Add:* 102 Loma Del Escolar Los Alamos NM 87544

VISSER, CORNELIS, engineering mechanics, civil engineering, for more information see previous edition

VISTE, ARLEN E, b Austin, Minn, Aug 13, 36; m 59; c 3. INORGANIC CHEMISTRY. *Educ:* St Olaf Col, BA, 58; Univ Chicago, PhD(inorg chem), 62. *Prof Exp:* Asst prof chem, St Olaf Col, 62-63; NSF fel, Columbia Univ, 63-64; asst prof, 64-68, assoc prof, 68-73, PROF CHEM, AUGUSTANA COL, SDAK, 73- *Concurrent Pos:* Partic fac res participation prog, Argonne Nat Lab, Ill, 70-71; vis scientist, Dept Phys Chem, Abo Akademi, Turku, Finland, 81-82. *Mem:* Am Chem Soc; Royal Soc Chem; Sigma Xi. *Res:* Reaction mechanisms; relativistic quantum chemistry; spectroscopy. *Mailing Add:* 1500 W 30th St Sioux Falls SD 57105

VISVANATHAN, T R, b India, June 4, 22; US citizen. EARTH SCIENCE, STATISTICS. *Educ:* Univ Madras, BS, 41; Fla State Univ, MS, 67; Univ SC, PhD(geol), 73. *Prof Exp:* Asst meteorologist, India Meteorol Serv, 43-65; instr geog, Benedict Col, 67-70; asst prof geol & geog, 71-78, ASSOC PROF GEOL & GEOG, UNIV SC, UNION, 78- *Mem:* Am Geophys Union. *Res:* Seismology; earthquake prediction; meteorology; prediction of atmospheric phenomena; solar terrestrial relationship planetary influences on terrestrial phenomena. *Mailing Add:* Univ Campuses & Dept Lifelong Learning & Continuing Educ Univ SC Columbia SC 29208

VISWANADHAM, RAMAMURTHY K, b Tiruvur, India, July 16, 46; US citizen; m 70; c 2. CEMENTED CARBIDES & COMPOSITES, STRUCTURE PROPERTY RELATIONSHIPS. *Educ:* Osmania Univ, Hyderabad, India, BE, 68; Univ Ill, Urbana- Champaign, MS, 70, PhD(metall eng), 73. *Prof Exp:* Sr scientist res & develop, Martin Marietta Labs, Baltimore, Md, 75-79, staff scientist, 86-88; mgr res & develop, Reed Tool Co/Baker Int, Houston, Tex, 79-86; DIR RES & DEVELOP, MULTI-METALS/VT-AM, LOUISVILLE, KY, 88- *Concurrent Pos:* Chmn, First Int Conf Sci of Hard Mat, 80-81; mem, ed bd, Metall Trans, Metall Soc, Am Inst Mining, Metall & Petrol Engrs, 80-86; int liaison, Second Int Conf Sci of Hard Mat, 83-84. *Honors & Awards:* Robert Glen Lye Mem Award, 77 & 81. *Mem:* Am Soc Mat; Mat Res Soc; Am Powder Metall Inst; Am Inst Mining & Metall Engrs. *Res:* Injection molding of cemented carbide composites; process development and process optimization; structure-property relationships in a variety of different composites. *Mailing Add:* 4806 Clipping Ct Louisville KY 40241

VISWANATHA, THAMMAIAH, b Channapatna, India, Sept 22, 26. BIOCHEMISTRY. *Educ:* Univ Mysore, PhD, 55. *Prof Exp:* Rask-Orsted fel, Carlsberg Lab, Denmark, 56-57; res assoc, Univ Minn, 57-58; vis scientist, Nat Inst Arthritis & Metab Dis, 58-62 & Inst Molecular Biol & Dept Chem, Univ Ore, 62-64; PROF CHEM, UNIV WATERLOO, 64- *Mem:* Sigma Xi. *Res:* Enzymes; proteins; nucleic acids. *Mailing Add:* Dept Chem Univ Waterloo Waterloo ON N2L 3G1 Can

VISWANATHAN, C T, b Madras, India; US citizen. BIOPHARMACEUTICS, PHARMACOKINETICS. *Educ:* Presidency Col, BSc, 62, MSc, 64; Marquette Univ, MS, 72; Univ Wis-Madison, MS 74, PhD(pharmacokinetics), 77. *Prof Exp:* Res assoc, Sch Pharm, Univ Ga, 77-78 & Univ Wash, 78-79; REVIEWING SCIENTIST, DIV BIOPHARMACEUT, FOOD & DRUG ADMIN, 79- *Res:* Drug absorption, distribution, metabolism and elimination in humans; phenobarbital; valproic acid; drug interactions; protein binding. *Mailing Add:* HFD 400 Rm 13B19 5600 Fishers Lane Rockville MD 20855

VISWANATHAN, CHAND R, b Madras, India, Oct 23, 29. SOLID STATE PHYSICS, ELECTRONICS. *Educ:* Univ Madras, BSc, 48, MA, 49; Univ Calif, Los Angeles, MS, 59, PhD(solid state physics), 64. *Prof Exp:* Engr, Res Dept, All India Radio, 49-57; asst prof eng, 62-68, assoc prof eng & appl sci, 68-74, asst dean, Sch Eng & Appl Sci, 74-77, chmn, Elec Eng Dept, 79-85, PROF ENG & APPL SCI, UNIV CALIF, LOS ANGELES, 74- *Honors & Awards:* Centennial Medal, Inst Elec & Electronics Engrs. *Mem:* Fel Inst Elec & Electronics Engrs; Am Phys Soc. *Res:* Solid state electronics; magnetic properties of materials; electron energy levels in solids; electron emission from solids; semiconductor device physics; integrated electronics. *Mailing Add:* Dept Elec Eng Univ Calif 7731 Boelter Hall Los Angeles CA 90024

VISWANATHAN, KADAYAM SANKARAN, b Madras, India, Apr 25, 37; m 67. PHYSICS, FIELD THEORY. *Educ:* Univ Madras, BSc, 57; Univ Calif, Riverside, MA, 64, PhD(physics), 65. *Prof Exp:* Res officer crystallog, Atomic Energy Estab, India, 57-60; from asst prof to assoc prof, 65-81, PROF THEORET PHYSICS, SIMON FRASER UNIV, 81- *Mem:* Am Phys Soc; Can Asn Physicists. *Res:* Theoretical high energy physics; study of 2-d gravity using extrinsic geometry of string world sheet and its relation to W-algebras; mechanisms for CP violation in particle physics; q-deformation of Lie algebras and superalgebras. *Mailing Add:* Dept Physics Simon Fraser Univ Burnaby BC V5A 1S6 Can

VISWANATHAN, R, b Tenkasi, India, Dec 17, 38; US citizen; m 65; c 2. LOW TEMPERATURE PHYSICS. *Educ:* Univ Madras, India, MA, 60, MSc, 60; Indian Inst Sci, PhD(physics), 64. *Prof Exp:* Res assoc, Univ Ill, 65-66; res physicist, Battelle Mem Inst, 66-67; res assoc, Univ Cincinnati, 67-69; asst scientist, Univ Calif, San Diego, 69-74; assoc scientist, Brookhaven Nat Lab, 74-78; DEPT MGR, HUGHES AIRCRAFT CO, 78-; AT INT MINERALS & CHEM CO. *Concurrent Pos:* Vis scientist, Inst Solid State Physics, Ger, 73; prin investr, Hughes Aircraft Co, 78-79. *Mem:* Am Phys Soc; Am Inst Mining & Metal Engrs; Am Vacuum Soc; AAAS; Sigma Xi. *Res:* Low temperature calorimetry; radiation damage in superconductors and devices; physical property measurements; author or coauthor of over 50 puplications in international journals. *Mailing Add:* Hughes Aircraft Co MS B364 Bldg S41 PO Box 92919 Los Angeles CA 90009

VISWANATHAN, RAMASWAMI, b Ranchi, India. LASER-SURFACE INTERACTIONS, SPECTROELECTROCHEMISTRY. *Educ:* Bombay Univ, BSc, 73; Indian Inst Technol, MSc, 75; Univ Ore, Eugene, PhD(phys chem), 80. *Prof Exp:* Fel dept chem, Northwestern Univ, 80-83; asst prof, 83-89, ASSOC PROF CHEM & COMPUT EDUC, BELOIT COL, WIS, 89- *Concurrent Pos:* Vis scientist, IBM Almaden Res Ctr, San Jose, 86-87; vis assoc prof, Dept Chem, Northwestern Univ, 89-90. *Mem:* Am Chem Soc. *Res:* Chemical physics and physical chemistry; molecular beams, laser-surface interactions and spectroelectrochemistry; parallel processing computer hardware. *Mailing Add:* Dept Chem Beloit Col Beloit WI 53511

VITAGLIANO, VINCENT J, b New York, NY, Oct 29, 27; m 51; c 9. COMPUTER AIDED DESIGN, INTERACTIVE COMPUTER GRAPHICS. *Educ:* Manhattan Col, BCE, 49; Va Polytech Inst, MS, 51; NY Univ, EngScD, 60. *Prof Exp:* Instr appl mech, Va Polytech Inst, 49-50; struct designer, Praeger-Maguire, Consult Engrs, 50-52 & M W Kellogg Co, Pullman, Inc, NY, 52-54; from instr to assoc prof civil eng, Manhattan Col, 54-63; univ consult, IBM Corp, 63-90; CONSULT, 90- *Concurrent Pos:* Consult, 55-; Smith-Mundt vis lectr, AlHikma Univ Baghdad, 61-62. *Mem:* Am Soc Civil Engrs; Am Soc Eng Educ. *Res:* Structural engineering; electronic computers. *Mailing Add:* 20532 Sausalito Dr Boca Raton FL 33498

VITALE, JOSEPH JOHN, b Boston, Mass, Dec 14, 24; m 49; c 1. NUTRITION, BIOCHEMISTRY. *Educ:* Northeastern Univ, BS, 47; NY Univ, MS, 49; Harvard Univ, DSc(nutrit biochem), 51; Antioquia Univ, Colombia, MD, 66. *Prof Exp:* Res assoc nutrit, Sch Pub Health, Harvard Univ, 51-54, assoc, 54-55, asst prof, 55-66; prof food, nutrit & med, Univ Wis, 66-67; dir nutrit progs, Sch Med, Tufts Univ, 67-72; prof path & community med, Sch Med, Boston Univ, 72-; AT MALLORY INST-PATH, BOSTON CITY HOSP. *Concurrent Pos:* Res assoc path, Sch Med, Boston Univ, 52-66; spec consult, Interdept Comt Nutrit for Nat Defense, 59-; Claude Bernard prof, Med Sch, Univ Montreal, 60; vis prof, Univ del Valle, Colombia, 60-62. *Mem:* Am Inst Nutrit; Am Soc Clin Nutrit; Brit Nutrit Soc; Sigma Xi. *Res:* Atherosclerosis, gastrointestinal metabolism, nutritional anemias and public health. *Mailing Add:* Mallory Inst-Path Boston City Hosp 734 Mass Ave Boston MA 02118

VITALE, RICHARD ALBERT, b New Haven, Conn, Sept 7, 44. MATHEMATICS, STATISTICS. *Educ:* Harvard Univ, AB, 66; Brown Univ, PhD(appl math), 70. *Prof Exp:* Asst prof appl math, Brown Univ, 70-76; asst scientist, Math Res Ctr, Madison, 76-77; assoc prof, 77-84, prof math, Claremont Grad Sch, 84-87, PROF STATIST, UNIV CONN, 87. *Mem:* Sigma Xi; Am Math Soc; Bernoulli Soc Math Statist & Probability; Soc Indust & Appl Math; Inst Math Statist; Math Asn Am. *Res:* Probability and statistics. *Mailing Add:* Dept Statist Univ Conn Storrs CT 06268

VITALIANO, CHARLES JOSEPH, b New York, NY, Apr 2, 10; m 40; c 2. GEOLOGY. *Educ:* City Col New York, BS, 36; Columbia Univ, AM, 38, PhD(mineral), 44. *Prof Exp:* Asst mineral, Columbia Univ, 37-39, lab instr gems & precious stones, Exten, 39-40; instr ceramic petrog, Rutgers Univ, 40-42; from asst geologist to assoc geologist, US Geol Surv, 42-46; assoc prof, 47-57, PROF GEOL, IND UNIV, BLOOMINGTON, 57- *Concurrent Pos:* Geologist, US Geol Surv, 46-59; Fulbright scholar, NZ Geol Surv, 54-55; NSF grant, 57-60 & 62-67; mem consortium crystalline basement rocks cent US, 80-82; mem sci comt, Indiana Univ, 79-83; panelist, Archaeol Implications of 1500 BC Minoan Eruption of Santorini Volcano, Oxford Univ, Eng; chmn elect, Archaeol Geol Div, Geol Soc Am, 85-86. *Mem:* Fel Mineral Soc Am; Soc Econ Geologists; fel Geol Soc Am; fel AAAS. *Res:* Geology and ore deposits of the Paradise Peak Quadrangle, Nevada; igneous and metamorphic petrography of western Nevada, southern New Zealand and southwest Montana; volcanic rocks of western United States; archaeological geology of Mediterranean regions. *Mailing Add:* Dept Geol Ind Univ Bloomington IN 47401

VITALIANO, DOROTHY BRAUNECK, b New York, NY, Feb 10, 16; m 40; c 2. GEOLOGY. *Educ:* Barnard Col, AB, 36; Columbia Univ, AM, 38, MPhil, 73. *Prof Exp:* Teaching asst geol, Barnard Col, 36-39; field asst, US Geol Surv, 42-43, geologist, 53-86; FREE-LANCE GEOL TRANSLR, 86- *Concurrent Pos:* Mem-at-large, Sect Comt E Geol & Geog, AAAS, 78-82; nat lectr, Sigma Xi, 81-83; adj prof geol, Ind Univ, 83- *Mem:* Fel Geol Soc Am; Geosci Info Soc; fel AAAS. *Res:* Bronze Age eruption of Santorini Volcano; scientific basis of Atlantis; geomythology; technical translation; tephrochronology; archeological geology. *Mailing Add:* 1114 Brooks Dr Bloomington IN 47401

VITELLO, PETER A, b Glendale, Calif, Sept 15, 50; m 76. THEORETICAL ASTROPHYSICS. *Educ:* Univ Southern Calif, BS, 72; Cornell Univ, PhD(theoret phys), 77. *Prof Exp:* Fel theoret astrophys, Ctr Astrophys, Harvard Col Observ, 77-80; mem staff, Sci Applications, Inc. 80-89; MEM STAFF, LAWRENCE LIVERMORE NAT LABS, 89- *Mem:* Sigma Xi; Am Astron Soc. *Res:* Theoretical studies of radiation-driven stellar winds in binary x-ray source systems, and of accretion onto black holes. *Mailing Add:* Lawrence Livermore Nat Lab 7000 East Ave L-296 Livermore CA 94550

VITERBI, ANDREW J, b Bergamo, Italy, Mar 9, 35; US citizen; m 58; c 3. COMMUNICATIONS. *Educ:* Mass Inst Technol, BS & MS, 57; Univ Southern Calif, PhD(elec eng), 62. *Hon Degrees:* Dr, Univ Waterloo, 90. *Prof Exp:* Res engr, Commun Res Sect, Jet Propulsion Lab, Calif Inst Technol, 57-62, res group supvr, 62-63; from asst prof to prof eng, Univ Calif, Los Angeles, 63-73; exec vpres, Linkabit Corp, 73-82, pres, 82-84; VCHMN & CHIEF TECH OFFICER, QUALCOMM, INC, 85-; PROF ELEC & COMPUTER ENG, UNIV CALIF, SAN DIEGO, 85- *Concurrent Pos:* Chmn vis comt, Elec Eng Dept, Technion Israel Inst Technol; mem, Mass Inst Technol Corp vis comt elec eng & comput sci Army Sci Bd; chmn US Comn Signal Processing, Int Radio Sci Union; distinguished lectr, Univ Ill Coord Sci Lab; Marconi Int fel award, 90. *Honors & Awards:* Inst Elec & Electronics Engrs Award, 62, Ann Award, 68, Alexander Graham Bell Medal, 84; Columbus Int Commun Award, Ital Nat Res Coun, 75. *Mem:* Nat Acad Eng; fel Inst Elec & Electronics Engrs. *Res:* Communication and information theory; coding; detection; modulation; signal processing; new spread spectrum processing techniques for jam resistant communications and for digital cellular radio; author of various publications. *Mailing Add:* Qualcomm Inc 10555 Sorrento Valley Rd San Diego CA 92121

VITKAUSKAS, GRACE, ANTI-TUMOR DRUGS, METABOLIC COOPERATION. *Educ:* Univ Conn, PhD(molecular biol), 79. *Prof Exp:* ASSOC RES SCIENTIST, DEPT PHARMACOL, SCH MED, YALE UNIV, 80- *Res:* Cell-to-cell interactions. *Mailing Add:* PO Box 854 Middlebury CT 06762

VITKOVITS, JOHN A(NDREW), b Cleveland, Ohio, Apr 7, 21; m 48; c 2. MECHANICAL ENGINEERING. *Educ:* Southern Methodist Univ, BS, 48. *Prof Exp:* Proj engr, Lubrizol Corp, Ohio, 48-49; mech engr, Southwest Res Inst, 49-55, asst sect mgr, 55-56, mgr, Standardized Tests Sect, 56-59, dir, Dept Engines, Fuels & Lubricants Eval, 59-72, vpres, Div Engines, Fuels & Lubricant Eval, 72-86; RETIRED. *Concurrent Pos:* Mem diesel rating panel & comt automatic transmission fluids, Coord Res Coun, Inc, 56-; mem automatic transmission fluid panel, Gen Motors Corp. *Mem:* Soc Automotive Engrs. *Res:* Automotive engines; torque converters; high speed and high torque hypoid gear lubricant evaluation; copper-lead, tin overlay and babbitt bearing endurance testing; design of engine research labs. *Mailing Add:* 131 Mountridge Dr San Antonio TX 78228-1717

VITOLS, VISVALDIS ALBERTS, b Riga, Latvia, Aug 24, 36; US citizen; m 63; c 2. ELECTRICAL ENGINEERING, INFORMATION SCIENCE. *Educ:* Iowa State Univ, BS, 58, MS, 59, PhD(elec eng), 62. *Prof Exp:* Asst prof elec eng, Iowa State Univ, 62-63; group scientist, NAm Aviation, Inc, 63-65; mem tech staff comput technol, IBM Corp, 65-68; mgr pattern recognition, Rockwell Int, 68-75; MGR, INFO SCI, 75- *Concurrent Pos:* Consult, Nat Acad Sci, 69-72. *Mem:* Inst Elec & Electronics Engrs; Sigma Xi; Int Asn Identification. *Res:* Pattern recognition and signal processing techniques utilizing general and special purpose digital processors; processor architectures for information classification and retrieval systems; machine recognition of unconstrained speech and image processing; imaged based scene and target recognition. *Mailing Add:* 505 S Aberdeen St Anaheim Hills CA 92807-4656

VITOSH, MAURICE LEE, b Odell, Nebr, Jan 16, 39; m 63; c 1. AGRONOMY, SOIL SCIENCE. *Educ:* Univ Nebr, BS, 62, MS, 64; NC State Univ, PhD(soils), 68. *Prof Exp:* Agronomist, NC Dept Agr, 65-68; EXTEN SPECIALIST SOIL FERTIL PROF CROP & SOIL SCI, MICH STATE UNIV, 68- *Mem:* Am Soc Agron; Soil Sci Soc Am; Potato Asn Am; Am Soybean Asn. *Res:* Soil fertility with potato, corn, soybeans and field beans. *Mailing Add:* Dept Crop & Soil Sci Mich State Univ East Lansing MI 48824-1325

VITOUSEK, MARTIN J, b Honolulu, Hawaii, July 30, 24; m 65; c 4. GEOPHYSICS. *Educ:* Stanford Univ, BS, 49, PhD(math), 54. *Prof Exp:* Radar lab worker, Pearl Harbor, 43-46; asst prof math, Univ Hawaii, 53-55; sr engr, Scripps Inst, Calif, 56-59; from assoc geophysicist to geophysicist, 61-74, SPECIALIST OCEANOG INSTRUMENT, HAWAII INST GEOPHYS, UNIV HAWAII AT MANOA, 74- *Mem:* Marine Technol Soc; Solar Energy Soc. *Res:* Applied mathematics; solid earth geophysics and oceanography; long period ocean waves, instrumentation and analysis. *Mailing Add:* Jimar PO Box 150 Kealakekua HI 96750

VITOUSEK, PETER MORRISON, b Honolulu, Hawaii, Jan 24, 49. ECOLOGY. *Educ:* Amherst Col, BA, 71; Dartmouth Col, PhD(biol sci), 75. *Prof Exp:* Asst prof biol, Ind Univ, Bloomington, 75-79; ASSOC PROF BOT, UNIV NC, CHAPEL HILL, 80-; AT BIOL DEPT, STANFORD UNIV. *Mem:* Ecol Soc Am; AAAS; Soil Sci Soc Am. *Res:* Regulation of nutrient cycling in terrestrial ecosystems; land-water interactions. *Mailing Add:* Dept Biol Herrin Labs Rm 445 Stanford Univ Stanford CA 94305

VITOVEC, FRANZ H, b Vienna, Austria, June 7, 21; nat Can; m 47; c 2. PHYSICAL METALLURGY, ENGINEERING MECHANICS. *Educ:* Vienna Tech Univ, dipl, 46, Dr tech sci, 47. *Prof Exp:* Docent, Vienna Tech Univ, 51; from asst prof to assoc prof mech & mat, Univ Minn, 52-58; from assoc prof to prof metall eng, Univ, Wis, 58-65; chmn, Dept Mineral Eng, Univ Alta, 71-80, prof metall eng, 65-81, prof mech eng, 81-; RETIRED. *Mem:* Fel Am Soc Metals; Am Soc Testing & Mat. *Res:* Relationship between the mechanical behavior of metals and alloys, their microstructure, and the environment. *Mailing Add:* 8942 Forest Park Dr Sydney BC V8C 4E9 Can

VITT, DALE HADLEY, b Washington, Mo, Feb 9, 44; c 2. BOTANY, BRYOLOGY. *Educ:* Southeast Mo State Col, BS, 67; Univ Mich, Ann Arbor, MS, 68, PhD(bot), 70. *Prof Exp:* From asst prof to prof, 70-88, MC CALLA PROF BOT, UNIV ALTA, 88- *Mem:* Int Asn Plant Taxon; Am Bryol & Lichenological Soc; Can Bot Soc; Danish Bryol Soc; Brit Bryol Soc; Japanese Bryol Soc. *Res:* Taxonomic, phylogenetic and ecological studies of bryophytes; monographic treatment of arctic, antarctic and tropical mosses; ecological analyses and productivity of bryophytes in arctic and alpine tundras and evolution and development of boreal peatlands; biogeochemistry of wetlands; peatland ecology. *Mailing Add:* Dept Bot Univ Alta Edmonton AB T6G 2M7 Can

VITT, LAURIE JOSEPH, b Bremerton, Wash, Aug 20, 45; m. ECOLOGY, HERPETOLOGY. *Educ:* Western Wash State Col, BA, 67, MS, 71; Ariz State Univ, PhD(zool), 76. *Prof Exp:* Res fel ecol, Academia Brasileira de Ciencas, Brazil, 77-78; scholar ecol, Mus Zool, Univ Mich, 78-79; res assoc fel, Mus Natural Hist, Univ Ga, 79-81, res assoc, 81; asst prof, 81-85, ASSOC PROF BIOL, UNIV CALIF, LOS ANGELES, 85- *Concurrent Pos:* Res ecologist, Ariz State Univ grant, 73-75; res assoc, Ariz State Univ, 75-76; consult, Desert Plan Off, Bur Land Mgt, 78 & Athene Wildlife Found, 79; adj asst prof, Savannah River Ecol Lab & Univ Ga, 81- *Mem:* Am Soc Ichthyol & Herpet; Soc Study Evolution; Herpet League; Soc Study Amphibians & Reptiles; Ecol Soc Am; Sigma Xi. *Res:* Community structure, competition, predation, reproductive effort, parental investment, demographies, life histories and reproduction of vertebrates, tail autotomy of lizards. *Mailing Add:* Biol Dept Univ Calif Los Angeles CA 90024

VITTER, JEFFREY SCOTT, b New Orleans, La, Nov 13, 55; m 82; c 2. ANALYSIS OF ALGORITHMS, COMPUTATIONAL COMPLEXITY. *Educ:* Univ Notre Dame, BS, 77; Stanford Univ, PhD(comput sci), 80. *Hon Degrees:* AM, Brown Univ, 86. *Prof Exp:* from asst prof to assoc prof, Comput Sci, Brown Univ, 80-88; PROF COMPUT SCI, BROWN UNIV, 88- *Concurrent Pos:* Grad fel, NSF, 77-80; Teaching fel, Stanford Univ, 79; consult, IBM, Xerox, Inst Defense Anal, Knowledge Eng, Univ Space Res Asn, 81-; prin investr, NSF & IBM, 81- & Defense Advan Res Proj Agency-Off Naval Res, 83-85; guest ed, Inst Elec & Electronics Engrs Trans on Comput, 85 & J Algorithmica, 88 & 92; vis scientist, Inst Nat Res Informatics & Automation, 86-87; mem, Math Sci Res Inst, 86; mem-at-large, Spec Interest Group Automata & Computability Theory, 87-91; ed, Inst Elec & Electronics Engrs Trans on Comput, 87-91, Commun Asn Comput Mach, 88-, Soc Indust & Appl Math J Comput, 89-; Guggenheim fel, 86- *Honors & Awards:* Fac Develop Award, IBM, 84; Presidential Young Investr Award, NSF, 85-91. *Mem:* Asn Comput Mach; Inst Elec & Electronics Engrs; Sigma Xi. *Res:* Software design and optimization; order statistics; concept learning; parallel optimization; computational geometry; parallel processing; machine learning. *Mailing Add:* Dept Comput Sci Brown Univ Box 1910 Providence RI 02912

VITTETOE, MARIE CLARE, b Keota, Iowa, May 19, 27. CLINICAL LABORATORY SCIENCES. *Educ:* Marycrest Col, BS, 50; WVa Univ, Morgantown, MS, 71, EdD(higher educ admin), 73. *Prof Exp:* Staff technician, St Joseph Hosp, Ottumwa, Iowa, 50-67, lab supvr, 67-70; asst prof health occup, Univ Ill, Urbana-Champaign, 73-78; PROF MED TECHNOL, UNIV KY, LEXINGTON, 78- *Concurrent Pos:* Instr microbiol & chem, Sch Nursing, St Joseph Hosp, 50-70; instr, didactic & clin, Ottumwa Sch Med Technol, 57-70. *Mem:* Am Soc Med Technol; Am Soc Allied Health Professions; affil mem Am Soc Clin Pathologists. *Res:* Administration in science education: program development, cost effectiveness and student learning patterns. *Mailing Add:* CAHP MCAnnex No Two Univ Ky Lexington KY 40536-0080

VITTI, TRIESTE GUIDO, b Detroit, Mich, May 22, 25; m 53; c 4. BIOPHARMACEUTICS, PHARMACOKINETICS. *Educ:* Univ Detroit, BS, 49, MS, 51; Wayne State Univ, PhD(biochem), 61. *Prof Exp:* Lectr pharmacol, Fac Med, Univ Man, 64-67; chief bioavailability, Upjohn Co, Mich, 67-71; dir clin res, Bur Drugs, Food & Drug Admin, 71-72; PROF BIOPHARMACEUT, FAC PHARM, UNIV MAN, 72- *Concurrent Pos:* USPHS fel, Univ Man, 64-67; consult, Biodecision Labs, Pittsburgh, Pa, 72-76; mem permanent adv expert comt bioavailability, Health & Welfare, Health Progs Bd, Can, 74-; mem bd, Alcohol & Drug Educ Serv, Man, 78-; mem adv res comt, Col Family Physicians, Family Med Ctr, Winnipeg, 78-80. *Mem:* Am Chem Soc; Pharmacol Soc Can; Can Pharmaceut Asn; Sigma Xi. *Res:* Biochemical pharmacology. *Mailing Add:* Fac Pharm Univ Man Winnipeg MB R3T 2N2 Can

VITTITOE, CHARLES NORMAN, b Louisville, Ky, Oct 3, 34; m 58; c 2. ELECTROMAGNETIC PULSE, NUMERICAL MODELING. *Educ:* Univ Ky, BS, 56; Univ Wis, MS, 58; Univ Ky, PhD(physics), 63. *Prof Exp:* Instr physics, Univ Ky, 59-60, from res asst to res assoc, 62-63; asst prof, Univ Ohio, 63-66, univ res comt, grant, 66; staff mem, Radiation Phenomena Div, Sandia Labs, 66-70, staff mem, 70-84, distinguished mem tech staff, Electromagnetic Applns Dir, 84-90, DISTINGUISHED MEM TECH STAFF, RADIATION & ELECTROMAGNETIC ANALYSIS DIV, 90- *Concurrent Pos:* Guest ed, J Radiation Effects Res & Eng, 84. *Mem:* Am Phys Soc. *Res:* Nuclear Weapon effects; optical transport theory; central receiver solar energy collection; geological probing by electrical methods; time-domain finite-difference solutions; to boundary-valve problems of electrodynamics; non-linear effects; computerized tomography subject models. *Mailing Add:* 3304 Ocotillo Ct NE Albuquerque NM 87111

VITTORIA, CARMINE, b Avella, Italy, May 15, 41; US citizen; m 67; c 3. MAGNETISM. *Educ:* Toledo Univ, BS, 62; Yale Univ, MS, 67, PhD(physics), 70. *Prof Exp:* Elec engr bionics, Naval Ord Lab, 62-63; teacher elec eng, Toledo Univ, 63-64; PHYSICIST, NAVAL RES LAB, 70- *Concurrent Pos:* Naval Res Coun adv, Naval Res Lab, 74, res award, 76; mem tech prog comt, Nat Magnetism Conf, 78; consult, Navy Electronic Syst Agencies. *Honors & Awards:* Outstanding Achievement Award, Naval Res Lab, 72 & 74. *Mem:* Am Phys Soc; Inst Elec & Electronics Engrs; Sigma Xi. *Res:* Electromagnetic wave propagation in magnetic materials. *Mailing Add:* Dept Elec & Comput Eng N Eastern Univ 360 Huntington Ave Boston MA 02115

VITTUM, MORRILL THAYER, b Haverhill, Mass, May 4, 19; m 41; c 3. HORTICULTURE, AGRONOMY. *Educ:* Univ Mass, BS, 39; Univ Conn, MS, 41; Purdue Univ, PhD(soil sci), 44. *Prof Exp:* Asst agron, Univ Conn, 39-41; asst agron, Purdue Univ, 41-42, tech asst soils, 42-45; from asst prof to prof, 46-83, head dept, 60-69 & 71-83, EMER PROF VEG CROPS, NY STATE AGR EXP STA, CORNELL UNIV, 83- *Concurrent Pos:* Actg asst olericulturist, Univ Calif, Davis, 56-57; vis prof hort, Ore State Univ, 64; proj leader, Univ Philippines-Cornell Grad Educ Prog, Col Agr, Univ Philippines, 69-71; actg horticulturist, Coop State Res Serv, USDA, 73-74 & 80-81. *Mem:* Am Soc Agron; Soil Sci Soc Am; fel Am Soc Hort Sci; Int Soc Hort Sci; Int Soc Soil Sci. *Res:* Effects of fertilizers, irrigation, rotation and cultural practices on the yield and quality of processing vegetables; evapotranspiration and soil-plant-water relationships. *Mailing Add:* NY State Agr Exp Sta Cornell Univ Geneva NY 14456-0462

VITULLO, VICTOR PATRICK, b Chicago, Ill, Oct 18, 39; m 62; c 1. ORGANIC CHEMISTRY. *Educ:* Loyola Univ, Chicago, BS, 61; Ill Inst Technol, PhD(chem), 65. *Prof Exp:* NSF fel, Mass Inst Technol, 65-66; res chemist, E I du Pont de Nemours & Co, Inc, 66-68; instr, Univ Kans, 68-69; asst prof, 69-74, ASSOC PROF CHEM, UNIV MD, BALTIMORE COUNTY, 69- *Mem:* AAAS; Am Chem Soc. *Res:* Physical organic chemistry; organic reaction mechanisms; acid-base catalysis; transition state structure; isotope effects. *Mailing Add:* Dept Chem Univ Md Baltimore County E 401 Wilkens Ave Catonsville MD 21228

VIVIAN, J(OHNSON) EDWARD, b Montreal, Que, July 6, 13; nat US; m 40; c 3. CHEMICAL ENGINEERING. *Educ:* McGill Univ, BEng, 36; Mass Inst Technol, SM, 39, ScD(chem eng), 45. *Prof Exp:* Asst, Can Pulp & Paper Asn, Que, 34-36; asst chem eng, 37-38, asst dir, Bangor Pract Sch, 38-41, from asst prof to assoc prof chem eng, 42-56, dir, Buffalo Pract Sch, 41-43, adminr, Metall Proj, 44-45, res group leader, 45-46, dir, Sch Chem Eng Pract, 46-57 & 73-80, dir, Eng Pract, Sch, Oak Ridge, 48-57, dir, AEC Proj Separation, 54, prof chem eng, 56-80, Exec Off Dept Chem Eng, 73-79, sr lectr, 80-84, EMER PROF CHEM ENG, MASS INST TECHNOL, 80- *Concurrent Pos:* Vis prof, Birla Inst Technol & Sci, Pilani, India, 72. *Honors & Awards:* Colburn Award, Am Inst Chem Engrs, 48. *Mem:* AAAS; Am Soc Eng Educ; Am Chem Soc; Tech Asn Pulp & Paper Indust; fel Am Inst Chem Engrs. *Res:* Unit operations; mass transfer; chemical engineering design. *Mailing Add:* 80 Austin Dr No 203 Burlington VT 05401

VIVIAN, VIRGINIA M, b Barneveld, Wis, July 1, 23. NUTRITION. *Educ:* Univ Wis, BS, 45; Columbia Univ, MS, 47; Univ Wis, PhD(home econ, biochem), 59. *Prof Exp:* Instr foods, nutrit & dietetics, Sch Nursing, Presby Hosp, 48-49; instr foods, nutrit & dietetics, Sch Nursing, Univ Mich, 49-51, asst dir dietary dept, Univ Hosp, 51-55; from asst prof to prof home econ, Ohio State Univ & Ohio Agr Res & Develop Ctr, 59-77, chairperson dept human nutrit & food mgt, 77-82, Carol S Kennedy distinguished prof nutrit, 82-88; RETIRED. *Concurrent Pos:* Pvt consult, 88- *Mem:* AAAS; Am Home Econ Asn; Am Dietetic Asn; Am Inst Nutrit; NY Acad Sci; Sigma Xi. *Res:* Amino acid-lipid metabolism with humans, adolescent and elderly dietary adequacy and nutrition status studies; exercise, diet and metabolism. *Mailing Add:* 1298 La Rochelle Dr Columbus OH 43221

VIVONA, STEFANO, b St Louis, Mo, Mar 25, 19; m 50; c 5. PREVENTIVE MEDICINE, PUBLIC HEALTH. *Educ:* St Louis Univ, MD, 43; Harvard Univ, MPH, 52. *Prof Exp:* Chief prev med, 5th Army Corps, US Army, Ger, 52-54, chief prev med, 7th Army, 54-55, actg chief biostatist, Walter Reed Army Inst Res, DC, 55-58, chief prev med res Br, Med Res & Develop Command, 58-60, chief prev med, 8th Army, Korea, 60-61, dir div commun dis & immunol, 62-64, chief med res team, Vietnam, 64-65, dir med component, SEATO, Thailand, 65-67, dir div prev med, Walter Reed Army Inst Res, 67-69; res adminr, 69-70, vpres res grant awards, 70-74, VPRES RES, RES DEPT, AM CANCER SOC, INC, 74- *Concurrent Pos:* Fel biologics res, Walter Reed Army Inst Res, 61-62. *Mem:* AMA; Am Pub Health Asn; Am Asn Cancer Res; Am Col Prev Med. *Res:* Infectious diseases; epidemiologic aspects of biostatistics; cancer. *Mailing Add:* 25 Fanshaw Ave Yonkers NY 10705

VIZY, KALMAN NICHOLAS, b Gyor, Hungary, July 7, 40; US citizen; m 68; c 2. APPLIED PHYSICS. *Educ:* Cleveland State Univ, BES, 63, BS, 64; John Carroll Univ, MS, 67; Walden Inst Advan Studies, PhD, 90. *Prof Exp:* Dept head sci, Byzantine Educ Ctr, 64-67; sr physicist, res labs, 67-80, CORP CONSULT, EASTMAN KODAK CO, 80- *Concurrent Pos:* Adj lectr modern physics, Rochester Inst Technol, 67-; adj asst prof radiol, Univ Rochester, 90- *Honors & Awards:* Autometric Award, Am Soc Photogram & Am Soc Mech Engrs, 75. *Mem:* Am Soc Mech Engrs; Am Asn Physics Teachers; Am Asn Physicists Med; Am Phys Soc; Soc Photog Scientists & Engrs; Am Soc Prof Engrs. *Res:* Research and development on the application of imaging for reconnaissance, micrographics, and medical applications; systems designer and analyst. *Mailing Add:* Health Sci Div Eastman Kodak Co Riverwood Rochester NY 14650

VLACH, JIRI, b Praha, Czech, Oct 5, 22; m 49; c 1. ELECTRICAL ENGINEERING. *Educ:* Prague Tech Univ, Dipl Eng, 47, PhD(elec eng), 57. *Prof Exp:* Mem res staff, Res Inst Radiocommun, Czech, 48-67; vis prof elec eng, Univ Ill, 67-69; PROF ELEC ENG, UNIV WATERLOO, 69- *Mem:* Fel Inst Elec & Electronics Engrs. *Res:* Network theory; computer aided design; VLSI. *Mailing Add:* Dept Elec Eng Univ Waterloo Waterloo ON N2L 3G1 Can

VLACHOPOULOS, JOHN A(POSTOLOS), b Volos, Greece, Aug 11, 42. CHEMICAL ENGINEERING. *Educ:* Athens Tech Univ, dipl, 65; Wash Univ, St Louis, MS, 68, DSc, 69. *Prof Exp:* From asst prof to assoc prof, 68-79, chmn, 85-88, PROF CHEM ENG, MCMASTER UNIV, 79. *Concurrent Pos:* Sabbatical, Univ Stuttgart Ger, 75 & Ecole Des Mines, Paris, France, 81-82 & 88-89. *Mem:* Am Inst Chem Engrs; Soc Plastics Engrs; Soc Rheol; fel Chem Inst Can; Can Soc Chem Eng; Sigma Xi; Polymer Proc Soc. *Res:* Polymer processing and polymer rheology; finite difference and finite element methods for numerical simulations; fluid mechanics; polymer extrusion; calendering; injection molding; thermoforming; compression molding. *Mailing Add:* Dept Chem Eng McMaster Univ Hamilton ON L8S 4L8 Can

VLADECK, BRUCE C, b New York, NY, Sept 13, 49; m; c 3. HEALTH ADMINISTRATION. *Educ:* Harvard Univ, BA, 70; Univ Mich, MA, 72, PhD, 73. *Prof Exp:* From asst prof to assoc prof pub health & health admin, Ctr Community Health Systs, Columbia Univ, 74-79; asst comnr, Div Health Planning & Resources Develop, NJ State Dept Health, 79-82; asst vpres, Robert Wood Johnson Found, 82-83; PRES, UNITED HOSP FUND NY, 83- *Concurrent Pos:* Adj prof pub admin, NY Univ, 88-; vis lectr, Princeton Univ, 81-83; mem, Comt Health Planning, Inst Med-Nat Acad Sci, 79-80; consult, Off Secy, US Dept HEW, 78. *Mem:* Inst Med-Nat Acad Sci; assoc fel NY Acad Med. *Mailing Add:* United Hosp Fund NY 55 Fifth Ave New York NY 10003

VLADUTIU, ADRIAN O, b Bucharest, Romania, Aug 5, 40; US citizen; m 71; c 2. CLINICAL PATHOLOGY, IMMUNOPATHOLOGY. *Educ:* Spiru Haret, Romania, BS, 56; Bucharest Univ, MD, 62; Jassy Univ, PhD(physiopath), 68. *Prof Exp:* Asst prof physiopath, Sch Med, Univ Bucharest, 68-71; res asst prof microbiol, State Univ NY, Buffalo, 69-71, clin assoc prof path, 77-81, res assoc prof, 79-86, PROF PATH, SCH MED, STATE UNIV NY, BUFFALO, 81-, PROF MICROBIOL, 82-, RES PROF MED, 86- *Concurrent Pos:* Intern, G Marinesco Hosp, Bucharest, 62-65, Millard Fillmore Hosp, Buffalo, 71-72; resident, State Univ NY, Buffalo, E J Meyer & Buffalo Gen Hosp, 72-74; dir, Immunopath Lab, Buffalo Gen Hosp, 74-, assoc pathologist, 79-, dir, Chem Lab, 81-, dir, Clin Lab, 82-, pathologist, 82-; consult, path, Niagara Falls Mem Hosp, 76-82; mem, E Witebsky Ctr Immunol, 81-, prof microbiol, State Univ NY, Buffalo, 82-; prin investr, NIH, 85-89. *Mem:* Am Asn Immunologists; Am Asn Pathologists; Soc Exp Biol & Med; AAAS; NY Acad Sci. *Res:* Autoimmunity in animals and man; diagnosis, pathogenesis and particularly its genetic control; laboratory medicine particularly isoenzymes, immune complexes, receptor assays and differential diagnosis of pleural effusions; tumor markers. *Mailing Add:* Dept Path & Microbiol State Univ NY Buffalo Gen Hosp 100 High St Buffalo NY 14203

VLADUTIU, GEORGIRENE DIETRICH, b Bremerton, Wash, Dec 21, 44; m 71; c 2. MICROBIOLOGY. *Educ:* Syracuse Univ, BS, 66; State Univ NY, Buffalo, MA, 70, PhD(microbiol), 73. *Prof Exp:* Res instr, 76-77, res asst prof, 77-81, RES ASSOC PROF PEDIAT, CHILDREN'S HOSP, BUFFALO, 81- *Concurrent Pos:* Prin investr, NSF, 77-79, NIH, 79-85 & Cystic Fibrosis Found, 81-82. *Mem:* Am Soc Biol Chemists; Am Soc Human Genetics; Soc Pediat Res; Asn Women Sci. *Res:* Transport of lysosomal enzymes in fibroblasts via specific glycosylated recognition markers; hereditary lysosomal storage diseases such as I-cell disease (mucolipidosis II) in which lysosomal enzymes are abnormally glycosylated and excreted; neonatal screening tests for inborn errors of metabolism such as familial hypercholesterolemia and cystic fibrosis. *Mailing Add:* Children's Hosp 219 Bryant St Buffalo NY 14222

VLADUTZ, GEORGE E, b Oradea, Rumania, Apr 6, 28; US citizen; m 54; c 1. CHEMICAL ACTION RETRIEVAL, KNOWLEDGE REPRESENTATION FOR INFORMATION RETRIEVAL. *Educ:* Lensovet Technol Inst, MSc, 52; Mendelevev Inst Chem Technol, Candidate of Sci, 56; Inst Elementorganic Chem, DSc, 68. *Prof Exp:* Sr res assoc, head dept chem info, head dept semiotics, All -Union Inst Sci & Tech Info, USSR Acad Sci, 56-74; vis res fel, Postgrad Libr & Info Sci Sch, Univ Sheffield, UK, 75-76; MGR BASIC RES, INST SCI INFO, PHILADELPHIA, PA, 76- *Concurrent Pos:* Prof info sci, Inst Continued Educ for Info Officers, Moscow, 72-74; res & develop proj leader, All-Union Inst Classification & Encoding, State Comt Standards, Moscow, 68-70; head, Sect Info Systs, Coun Sci Tech Info, USSR State Comt Sci & Technol, 70-74. *Mem:* Am Soc Info Sci; Am Chem Soc; Asn Comput Ling. *Res:* Development of methods for encoding chemical reactions and compounds for computerized retrieval purposes; theoretical foundations for information science; computer representation of documents and knowledge for information retrieval purposes; phrase based methods of automatic indexing, in particular of the key word phrase subject indexes; bibliographic coupling based associative methods of retrieval. *Mailing Add:* Inst Sci Info 3501 Market St Philadelphia PA 19104-3302

VLAHAKIS, GEORGE, b New York, NY, Oct 12, 23; m 49; c 2. GENETICS. *Educ:* Johns Hopkins Univ, AB, 51; Univ Tex, MA, 53. *Prof Exp:* Biologist, NIH, 52-53; chemist, US Testing Co, 54-55; biologist, Nat Cancer Inst, 55-86; RETIRED. *Mem:* AAAS. *Res:* Role of genes and their relationship to non-genetic factors in the development of tumors in mice. *Mailing Add:* 1720 Evelyn Dr Rockville MD 20852

VLAOVIC, MILAN STEPHEN, b Novi Sad, Yugoslavia, Feb 1, 36; m 69; c 3. VETERINARY PATHOLOGY. *Educ:* Univ Belgrade, DVM, 61; Univ Sask, MSc, 70; Univ Mo, Columbia, PhD(vet med), 74. *Prof Exp:* Gen practr, WGer, 65-67; tech officer, Can Dept Agr, 67-68; res asst vet microbiol, Univ Sask, 68-70; res asst, Wash State Univ, 70-71; res assoc, Univ Mo, 71-74; vet pathologist, Frederick Cancer Res Ctr, 74-78; PATHOLOGIST, EASTMAN KODAK CO, 78- *Mem:* Am Vet Med Asn; Am Asn Lab Animal Sci; Soc Pharmacol & Environ Pathologists; Tissue Cult Asn. *Res:* Immunopathology. *Mailing Add:* Eastman Kodak Seven Dixon Woods Honeoye Falls NY 14472

VLASES, GEORGE CHARPENTIER, b New York, NY, Oct 22, 36; m 58; c 4. PLASMA PHYSICS. *Educ:* Johns Hopkins Univ, BES, 58; Calif Inst Technol, MS, 59, PhD(aeronaut), 63. *Prof Exp:* Res fel aeronaut, Calif Inst Technol, 63; from asst prof to assoc prof aerospace eng sci, Univ Colo, Boulder, 63-69; res assoc prof, 69-73, PROF NUCLEAR ENG, AEROSPACE RES LAB, UNIV WASH, 73- *Concurrent Pos:* Consult, Aerospace Corp, Calif, 63 & Spectra Technol, Inc; vis scientist, Max Planck Inst Plasma Phys, Munich, 81 & 85. *Mem:* Am Phys Soc. *Res:* Plasma physics and controlled thermonuclear fusion. *Mailing Add:* Dept Nuclear Eng Univ Wash Seattle WA 98195

VLASUK, GEORGE P, b Miami, Fla, Oct 9, 55; m 81; c 2. PROTEOLYTIC ENZYMES, INHIBITION OF PROTEOLYTIC ENZYMES. *Educ:* Miami Dade Community Col, AA, 75; Univ SFla, BS, 77; Kent State Univ, PhD(biochem), 81. *Prof Exp:* Postdoctoral molecular biol, State Univ NY, Stony Brook, 82-83; staff scientist molecular biol, Calif Biotechnol Inc, 83-85; sr res biochemist, 85-87, res fel biol chem, 87-90, ASSOC DIR PHARMACOL, MERCK SHARP & DOHME, 90- *Concurrent Pos:* Mem Thrombosis Coun, Am Heart Asn. *Mem:* Am Soc Biochem & Molecular Biol; AAAS; Am Heart Asn. *Res:* Identification and utilization of specific inhibition of blood coagulation factors to understand the mechanism of vascular thrombosis in vivo and in vitro. *Mailing Add:* WP26-431 Merck Sharp & Dohme West Point PA 19486

VLATTAS, ISIDOROS, b Chios, Greece, Apr 28, 35; US citizen; m 67; c 3. ORGANIC CHEMISTRY. *Educ:* Nat Univ Athens, BS, 59; Univ BC, MsD, 63, PhD(chem), 66. *Prof Exp:* Fel, Harvard Univ, 67-68; res chemist, 68-80, SR STAFF SCIENTIST ORG CHEM, CIBA-GEIGY CORP, 80- *Mem:* Am Chem Soc. *Res:* Natural products; medicinal organic chemistry research. *Mailing Add:* Chem Corp Ciba Geigy Morris Ave Summit NJ 07901

VLAY, GEORGE JOHN, b Buffalo, NY, Dec 1, 27; m 49; c 3. MATHEMATICS, SYSTEMS DESIGN & SYSTEMS SCIENCE. *Educ:* Univ Buffalo, BS, 53. *Prof Exp:* Mgr advan req satellite prog, Ford Aerospace Corp, W Develop Labs, 68-73, mgr advan space systs satellite design, 74-75, mgr commun systs satellite commun systs, 76-77, dir bus develop & planning, plans & mkt, 78-82, dir tech affairs IR&D, 82-85, dir prod assurance qual, reliability & metrol, Space Systs Div, 85- 88, dir systs mgt, 88-90; PRES SYSTS MGT, SYSTS MGT ASSOCS, 90- *Concurrent Pos:* Mem, Coun Defense Space Indust Asn-Streamling Acquisition, 87- & USAF/AFSC Indust RFP Crit Process Team, 89-; distinguished lectr, Am Inst Aeronaut & Astronaut, 89-91. *Honors & Awards:* Eng Award of Excellence, Electronic Industs Asn, 90. *Mem:* Assoc fel Am Inst Aeronaut & Astronaut; sr mem Inst Elec & Electronics Engrs; Am Soc Eng Mgt. *Res:* Program management; systems engineering management; risk management. *Mailing Add:* 32 Yerba Buena Ave Los Altos CA 94022-2208

VLCEK, DONALD HENRY, b Holyrood, Kans, Nov 17, 18; m 44; c 4. ELECTRONICS, RESEARCH ADMINISTRATION. *Educ:* US Mil Acad, BA, 43; Stanford Univ, ME, 49. *Prof Exp:* Chief electronic requirements br, Hq, Air Defense Command, USAF, 49-51, electronics staff officer to Asst Secy Defense for Res & Develop, 52-54, semi-atomic ground environ proj officer, Air Defense Systs Proj Off, Hq, Air Res & Develop Command, 55-56, chief track test div, Missile Develop Ctr, 56-61, test instrumentation develop div, Hq, Air Force Systs Command, 61-64, dir plans & requirements, Hq, Nat Range Div, 64-67, dir eng, Hq, Ground Electronics Eng Installation Agency, 67-69, comdr, Ballistic Missile Early Warning Site, Eng, 69-72; COORDR, CANCER RES CTR, MED SCH, OHIO STATE UNIV, 72-, COORDR & BUS MGR, COMP CANCER CTR, 72- *Mem:* Inst Elec & Electronics Engrs. *Res:* Captive testing and the influence of the rate of change of acceleration upon the behavior of components and systems; design criteria for containment laboratories. *Mailing Add:* 2588 Edgevale Upper Arlington Columbus OH 43221

VLIET, DANIEL H(ENDRICKS), b New Orleans, La, May 30, 21; m 43; c 3. ELECTRICAL ENGINEERING. *Educ:* Tulane Univ, BSEE, 49; Univ Mich, MSEE, 52; Univ Wis, PhD, 65. *Prof Exp:* Assoc prof, 58-65, PROF ELEC ENG, TULANE UNIV LA, 65- *Mem:* Am Soc Eng Educ; Inst Elec & Electronics Engrs. *Res:* Power system analysis; electrical machinery; automation and control. *Mailing Add:* 1036 Jefferson Ave New Orleans LA 70115

VLIET, GARY CLARK, b Bassano, Alta, June 3, 33; US citizen; m 62; c 3. HEAT TRANSFER, SOLAR ENERGY. *Educ:* Univ Alta, BSc, 55; Stanford Univ, MS, 57, PhD(mech eng), 62. *Prof Exp:* Res scientist, Lockheed Missiles & Space Co, 61-71; assoc prof, 71-79, PROF MECH ENG, UNIV TEX, AUSTIN, 79-, W R WOOLRICH PROF, 85- *Concurrent Pos:* Consult, Various Cos; Fluor fel, Stanford, 56. *Mem:* Assoc mem Am Soc Mech Engrs; Int Solar Energy Soc; Am Soc Heating, Refrig & Air-Conditioning Engrs. *Res:* Thermal energy systems; solar energy; energy conversion. *Mailing Add:* Dept Mech Eng Univ Tex Austin TX 78712

VLITOS, AUGUST JOHN, plant physiology, for more information see previous edition

VOBACH, ARNOLD R, b Chicago, Ill, Nov 20, 32; m 57; c 2. MATHEMATICS. *Educ:* Harvard Univ, AB, 54, SB, 56; Ill Inst Technol, MS, 59; La State Univ, PhD(math), 63. *Prof Exp:* From instr to asst prof, Univ Ga, 62-68; ASSOC PROF MATH, UNIV HOUSTON, 68- *Mem:* Am Math Soc; Math Asn Am. *Res:* Topology, logic. *Mailing Add:* Dept Math Univ Houston Cullen Blvd Houston TX 77204-3476

VOBECKY, JOSEF, b Brno, Czech, Sept 29, 23; m 55; c 3. EPIDEMIOLOGY OF CHRONIC DISEASE, NUTRITION EPIDEMIOLOGY. *Educ:* Masaryk Univ, Brno, MD, 50; Postgrad Med Sch, Prague, DPH, 56, dipl epidemiol, 60; CSPQ, 74. *Prof Exp:* Epidemiologist, Czech Pub Health Serv, Prague, 50-55, head dept epidemiol, Brno, 56-62; dir dept epidemiol, Inst Epidemiol & Microbiol, Prague, 63-68; from asst prof to assoc prof epidemiol, 69-78, chmn dept community health sci, 78-85, PROF EPIDEMIOL, MED FAC, UNIV SHERBROOKE, 78- *Concurrent Pos:* Vis prof, Fac Med, Charles Univ, Prague, 60-62; consult field proj, WHO, Mongolia, 63-65, Iraq, 66, lectr, 66-69; sr lectr, Postgrad Med Sch, Prague, 66-69. *Mem:* Int Epidemiol Asn; Soc Epidemiol Res; fel Am Col Epidemiol; NY Acad Sci. *Res:* Epidemiology of nutrition; epidemiological surveillance; chronic disease epidemiology; environmental factors; cancer epidemiology. *Mailing Add:* Dept Community Med Fac Med Univ Sherbrooke Sherbrooke PQ J1H 5N4 Can

VOCCI, FRANK JOSEPH, b Baltimore, Md, Aug 13, 24; wid. TOXICOLOGY. *Educ:* Loyola Col, Md, BS, 49. *Prof Exp:* Group leader & gen chemist, Aerosol Br, 49-61, chief, Basic Toxicol Br, Toxicol Div, Edgewood Arsenal, 61-76, chief, Whole Animal Toxicol Br, Toxicol Div, 76-80, CONSULT, TOXICOL, FOOD & DRUG ADMIN, DOD, CDC, ENVRON, EPA(DYNAMAC), 80- *Mem:* Sigma Xi; Am Chem Soc; Am Indust Hyg Asn. *Mailing Add:* 6009 Winthrope Ave Baltimore MD 21206

VOCKE, MERLYN C, b Milwaukee, Wis, Nov 17, 33; m 61; c 3. ELECTRICAL ENGINEERING. *Educ:* Valparaiso Univ, BS, 55; Univ Notre Dame, MS, 57; Univ Iowa, PhD, 71. *Prof Exp:* From instr to assoc prof, 55-76, PROF ELEC ENG, VALPARAISO UNIV, 76- *Concurrent Pos:* Consult, Naval Weapons Support Ctr. *Honors & Awards:* Centennial Medal, Inst Elec & Electronics Engrs. *Mem:* Inst Elec & Electronics Engrs; Am Soc Eng Educ. *Res:* Microprocessor applications; digital and analog system design. *Mailing Add:* Dept Elec & Comput Eng Valparaiso Univ Valparaiso IN 46383

VODICNIK, MARY JO, b Milwaukee, Wis, Dec 23, 51; m 75. METABOLISM & PHARMACOKINETICS, REPRODUCTIVE TOXICOLOGY. *Educ:* Marquette Univ, BS, 74, PhD(biol & comp physiol), 78. *Prof Exp:* Asst toxicol, 78-80, asst prof, 80-84, ASSOC PROF PHARMACOL & TOXICOL, MED COL WIS, 84- *Concurrent Pos:* Mem, spec study sect, NIH, 83-85 & metab path study sect, 85-, chmn, physiol chem ad hoc study sect, 86; prin investr, Nat Inst Environ Health Sci, NIH, 81-90. *Mem:* Am Soc Pharmacol & Exp Therapeut; Int Soc Study Xenobiotics; Soc Toxicol. *Res:* Mechanisms responsible for transplacental and milk transfer of environmental chemicals; effects of pregnancy and lactation on pharmacokinetics; effects of pregancy on efficacy and toxicity of drugs and environmental chemicals. *Mailing Add:* Lilly Res Labs Eli Lilly & Co PO Box 708 Greenfield IN 46140

VO-DINH, TUAN, b Nhatrang, Vietnam; US citizen. ENVIRONMENTAL HEALTH EFFECTS MONITORING, ADVANCED INSTRUMENTATION. *Educ:* Swiss Fed Inst Tech, Lausanne, Switz, BS, 70, PhD(phys chem), 75. *Prof Exp:* Postdoctoral fel chem, Univ Fla, 75-77; res scientist, 77-84, GROUP LEADER, OAK RIDGE NAT LAB, 84- *Concurrent Pos:* Chmn, Dept Energy, Sci Panel Monitoring Instrumentation, Off Health & Environ Res, 84; co-chmn, Int Comt Polycyclic Aromatic Compounds, 86-; adj prof, Univ Tenn, 88-; tech ed, Polycyclic Aromatic Compounds J, 89-; chmn, Subcomt E13-09, Asn Standards & Testing Mat. *Honors & Awards:* IR 100 Award, Res & Develop Mag, 81 & 82; Tech Event Award, Martin Marietta Energy Systs, 86; Excellence Award, Fed Lab Consortium, 86; Gold Medal Award, Soc Appl Spectros, 88; Languedoc-Rousillon Medal, Univ Perpignan, France, 89. *Mem:* Fel Am Inst Chemists; Am Chem Soc; Int Union Pure & Appl Chem; Soc Appl Chem; Soc Advan Sci; Asn Standards & Testing Mat. *Res:* Development of advanced methods and instrumentation for the detection of environmental pollutants and health effects associated with industrial processes and technology development. *Mailing Add:* Oak Ridge Nat Lab Bldg 4500 S PO Box 2008 MS-6101 Oak Ridge TN 37831-6101

VODKIN, LILA OTT, b Laurens, SC, Dec 21, 50; m 75. PLANT MOLECULAR BIOLOGY. *Educ:* Univ SC, BS, 73, MS, 75; NC State Univ, PhD, 78. *Prof Exp:* Res geneticist, USDA, Beltsville, Md, 78-87; ASSOC PROF AGRON, UNIV ILL, URBANA, 88- *Mem:* Genetics Soc Am; Am Soc Plant Physiologists. *Res:* Gene expression in soybeans. *Mailing Add:* Agron Dept Turner Hall Univ Ill Urbana IL 61801

VODKIN, MICHAEL HAROLD, b Boston, Mass, Dec 4, 42; m 75. GENETICS, MOLECULAR BIOLOGY. *Educ:* Boston Col, BS, 64, MS, 66; Univ Ariz, PhD(genetics), 71. *Prof Exp:* Fel genetics, Cornell Univ, 71-73; asst prof biol, Univ SC, 73-79; staff fel, NIH, 79-81; mem staff, Med Res Inst Infectious Dis, US Army, 81-88; RES SPECIALIST PATHOL DEPT, UNIV ILL, 88- *Mem:* Sigma Xi; Genetics Soc Am; Am Soc Microbiol. *Res:* Genetics and biochemistry of of hemoparasites; cloning of antigens for use as vaccines. *Mailing Add:* Div Path Col Vet Med Univ Ill Urbana IL 61801

VOEDISCH, ROBERT W, b Ft Eustis, Va, Nov 5, 24; m 52, 84; c 3. ORGANIC CHEMISTRY, ENVIRONMENTAL SCIENCES. *Educ:* Beloit Col, BS, 48. *Prof Exp:* Res & develop chemist, Lawter Chem Inc, Chicago, 50-55, group leader, 55-56, chief chemist, 56-60, tech dir, 60-67, vpres res & develop, 67-85. *Concurrent Pos:* Consult, 85- *Mem:* AAAS; Am Chem Soc; Am Inst Chemists; Am Asn Textile Chemists & Colorists. *Res:* Luminescent compounds; ink vehicles; synthetic resins; alkyds, phenolics, maleics, ketone and polyamide resins; environmental science. *Mailing Add:* 722 N Hadow St Arlington Heights IL 60004-5616

VOEKS, ROBERT ALLEN, b Seattle, Wash, Nov 10, 50; m 86; c 2. TROPICAL RAIN FORESTS, BIOGEOGRAPHY. *Educ:* Portland State Univ, BS, 75, MS, 80; Univ Calif, Berkeley, PhD(geog), 87. *Prof Exp:* ASSOC PROF GEOG, CALIF STATE UNIV, FULLERTON, 87- *Concurrent Pos:* Prin investr, Siuslaw Nat Forest, US Forest Serv, 89; vis prof, Fed Univ Bahia, Brazil, 90-91. *Mem:* Asn Am Geographers. *Res:* Ecology, biogeography and human use of tropical forests; African ethnobotany in the New World; palm ecology; tropical folk medicine. *Mailing Add:* Geog Dept Calif State Univ Fullerton CA 92634

VOELCKER, HERBERT B(ERNHARD), b Tonawanda, NY, Jan 7, 30; m 54; c 2. MECHANICAL ENGINEERING, INDUSTRIAL & MECHANICAL ENGINEERING. *Educ:* Mass Inst Technol, BS, 51, MS, 54; Imp Col, Univ London, PhD(eng), 61. *Prof Exp:* Lectr elec eng, Imp Col, Univ London, 60-61; from asst prof to prof elec eng, Univ Rochester, 71-85, dir, Prod Automation Proj, 72-85; dep dir, DMCE Div, NSF, 85-86; CHARLES LAKE PROF MECH ENGR, CORNELL UNIV, 86- *Concurrent Pos:* US Army Signal Corps, 51-58; consult indust, 58-; NATO fel, 67-68; sr vis scientist, UK Sci Res Coun, 81-83; dir, Cornell Mfg Eng & Productivity Prog, 87-91. *Honors & Awards:* Inst Elec & Electronics Engrs Award, 67. *Mem:* Fel Inst Elec & Electronics Engrs; Asn Comput Mach; Am Soc Mech Engrs; Soc Mfg Engrs. *Res:* Automation engineering for design and production in the mechanical industries; national science and technology policy. *Mailing Add:* Sibley Sch Mech Eng Cornell Univ Ithaca NY 14853-7501

VOELKER, ALAN MORRIS, b Eau Claire, Wis, Aug 12, 38; m 60; c 2. SCIENCE EDUCATION. *Educ:* Wis State Univ-River Falls, BS, 59; Syracuse Univ, MS, 63; Univ Wis-Madison, PhD(sci educ), 67. *Prof Exp:* Teacher chem, physics, gen sci & math & chmn dept, High Schs, Wis, 59-64; asst prof sci educ, Ohio State Univ, 67-69; asst prof sci educ, Univ Wis-Madison & prin investr cognitive learning, Res & Develop Ctr, 69-73; ASSOC PROF & PROF SCI EDUC, COL EDUC, NORTHERN ILL UNIV, 73- *Concurrent Pos:* Mem, adv bd sci educ sect, Educ Resources Info Ctr, Info Anal Ctr Sci, Math & Environ Educ, 72-76; dist dir VIII, Nat Sci Teachers Asn, 84-86. *Mem:* Fel AAAS; Nat Sci Teachers Asn; Nat Asn Res Sci Teaching; Am Educ Res Asn; Asn Educ Teachers Sci. *Res:* Science concept learning; science teacher education; attitudes toward science; scientific literacy; attentive publics for organized science. *Mailing Add:* Curric & Instr Northern Ill Univ De Kalb IL 60115

VOELKER, C(LARENCE) E(LMER), b Two Rivers, Wis, July 6, 23; m 47; c 3. CHEMICAL ENGINEERING. *Educ:* Univ Wis, BS, 49, MS, 50. *Prof Exp:* Process control engr, Food Mach & Chem Corp, 50-52; process engr, 52-56, group leader, Process Eng Dept, 56-57, proj leader polychems, Res Dept, 57-59, from res engr to sr res engr, Process Fundamentals Res Lab, 59-64, sr process engr, Comput Res Lab, 64-65, sr process engr, Process Eng Dept, 65-68, tech expert, 68-70, SR PROCESS SPECIALIST, PROCESS ENG DEPT, DOW CHEM CO, 70- *Mem:* AAAS; Am Chem Soc; Sigma Xi; Am Inst Chem Engrs. *Res:* Heat transfer; fluid dynamics; mathematics; crystallization. *Mailing Add:* 300 Sinclair St Midland MI 48640

VOELKER, RICHARD WILLIAM, b Stanton, Nebr, July 16, 36; m 61; c 4. VETERINARY PATHOLOGY. *Educ:* Kans State Univ, BS & DVM, 59; Purdue Univ, MS, 64, PhD(vet path), 69; Am Col Vet Pathologists, cert, 70. *Prof Exp:* Vet food inspector, US Army Vet Corps, 59-61; med lab officer, US Armed Forces, Europe, 61-64; instr vet path, Purdue Univ, West Lafayette, 64-68; staff pathologist, Hazleton Labs, 68-71; sect head toxicol, William S Merrell Co, 71-73; DIR PATH, HAZLETON LABS AM, 73- *Concurrent Pos:* Adj asst prof path, Med Sch, Univ Cincinnati, 71-73. *Mem:* Am Vet Med Asn; Int Acad Path; Am Col Vet Pathologists; Soc Pharmacol & Environ Pathologists; Indust Vet Asn. *Res:* Toxocologic pathology in the investigation and description of various tissue responses caused by a wide variety of chemical and pharmaceutical compounds; tumor induction in laboratory animals by a wide variety of environmental. *Mailing Add:* 10409 Huntrace Way Vienna VA 22180

VOELKER, ROBERT ALLEN, b Palmer, Kans, Jan 24, 43; m 65; c 1. MOLECULAR BIOLOGY. *Educ:* Concordia Teachers Col, Nebr, BSEd, 65; Univ Nebr, Lincoln, MS, 67; Univ Tex, Austin, PhD(zool), 70. *Prof Exp:* Asst, Univ Nebr, Lincoln, 67; NIH fel, Univ Tex, Austin, 67-70, Univ Ore, 70-71; res assoc genetics, NC State Univ, 71-73; sr staff fel, Lab Environ Mutagenesis, 76-80, RES GENETICIST LAB MOLECULAR GENETICS, NAT INST ENVIRON HEALTH SCI, 80- *Concurrent Pos:* Vis asst prof, Dept Genetics, NC State Univ, Raleigh, 73-76. *Mem:* Genetics Soc Am. *Res:* Drosophila population genetics; Drosophila salivary gland chromosome cytogenetics; meiotic drive in Drosophila; genetic control of the structure and function of RNA ploymerase II in Drosophila; genetic and molecular analysis of suppressor of sable in drosophila. *Mailing Add:* Lab Genetics Nat Inst Environ Health Sci Research Triangle Park NC 27709

VOELKER, ROBERT HETH, US citizen. INTEGRATED CIRCUIT DESIGN, COMPUTATIONAL ELECTROMAGNETICS. *Educ:* Univ Mich, BSE, 82, MSE, 83, PhD(elec eng), 89. *Prof Exp:* ASST PROF ELEC ENG, UNIV NEBR, LINCOLN, 90- *Mem:* Inst Elec & Electronics Engrs. *Res:* Electromagnetic field analysis of signal propagation in high-speed digital and microwave integrated circuits; computer-aided design of integrated circuits; high-speed electro-optical measurements; supercomputing. *Mailing Add:* 243 N Walter Scott Eng Ctr Univ Nebr Lincoln NE 68588-0511

VOELLY, RICHARD WALTER, RECOMBINANT DNA, GENETICS. *Educ:* Swiss Fed Inst Technol, Zurich, PhD(biochem), 75. *Prof Exp:* from asst prof to PROF BIOCHEM, SCH MED, UNIV MIAMI, 82- *Mailing Add:* Sch Med Dept Biochem PO Box 016129 Univ Miami 1011 NW 15th St Miami FL 33101

VOELZ, FREDERICK, b Wheaton, Ill, May 22, 27; m 50; c 2. ENVIRONMENTAL ENGINEERING. *Educ:* Ill Inst Technol, BS, 51, MS, 53, PhD(physics, math), 55. *Prof Exp:* Asst, Ill Inst Technol, 51-55; proj chemist, Sinclair Res, Inc, 55-62, sr res physicist, 62-69, sr proj engr spec projs, 69-78, RES ASSOC NEW TECHNOL APPL, ATLANTIC RICHFIELD CO, 78- *Mem:* Soc Automotive Engrs; Air Pollution Control Asn; Sigma Xi. *Res:* Raman and infrared spectroscopy; molecular structure; automotive exhaust emissions instrumentation and testing; ambient air and environmental technology. *Mailing Add:* 1338 Macarthur Munster IN 46321

VOELZ, GEORGE LEO, b Wittenberg, Wis, Oct 13, 26; m 50; c 4. OCCUPATIONAL MEDICINE. *Educ:* Univ Wis, BS, 48, MD, 50. *Prof Exp:* AEC fel indust med, 51-52; indust physician, Los Alamos Sci Lab, Univ Calif, 52-57; chief med br, Idaho Opers Off, AEC, 57-63, asst dir, Health Serv Lab, 63-67, dir, 67-70; health div leader, Los Alamos Sci Lab, 70-82, asst div leader, Health, Safety & Environ Div, 82-87, SECT LEADER EPIDEMIOL, LOS ALAMOS NAT LAB, 87- *Concurrent Pos:* Mem, Nat Coun Radiation Protection & Measurements, 75 -; ed adv, Occup Health & Safety J; mem, Int Comn Radiol Protection, 84- *Mem:* Am Col Occup Med; Am Acad Occup Med; Am Col Prev Med; Am Indust Hyg Asn; Health Physics Soc. *Res:* Occupational health problems, especially in the atomic energy industries; radiological health problems; radiobiological research and radiation dosimetry; epidemiology. *Mailing Add:* 117 La Vista Dr Los Alamos NM 87544

VOELZ, MICHAEL H, b Columbus, Ind, March 5, 56; m 81; c 2. INSTRUMENT DESIGN, PHYSIOLOGIC HUMAN MONITORING. *Educ:* Purdue Univ, BS, 79. *Prof Exp:* Res assoc psychol, Purdue Univ Biomed Ctr, 79-81; design asst, blood oxygenation, Bio-Tek Inc, Ohmeda, 81-82; develop engr blood pheresis, Cobe Labs Inc, 82-83, proj engr, blood oxygenation, 83-85; mgr res & develop, physiologic instrument, Lafayette Instruments, 85-87; DIR RES & DEVELOP, BISSELL HEALTH CORE, 88- *Concurrent Pos:* Consult, Water Pure Inc, 86. *Mem:* Asn Advan Med Instrumentation. *Res:* Research and products development research in physiologic monitoring, especially non-invasive monitoring; blood pressure measurement, analog and digital circuit design and human physiology. *Mailing Add:* 6914 Baldwin Dr Battle Ground IN 47920

VOET, DONALD HERMAN, b Amsterdam, Neth, Nov 29, 38; US citizen; m 65; c 2. CRYSTALLOGRAPHY, BIOCHEMISTRY. *Educ:* Calif Inst Technol, BS, 60; Harvard Univ, PhD(chem), 67. *Prof Exp:* Res assoc biol, Mass Inst Technol, 66-69; asst prof, 69-74, ASSOC PROF CHEM, UNIV PA, 74- *Mem:* Am Chem Soc; Am Crystallog Asn; Sigma Xi. *Res:* X-ray structural determination of molecules of biological interest, particularly proteins, and nucleic acids; author of one textbook. *Mailing Add:* Dept Chem Univ Pa Philadelphia PA 19104

VOET, JUDITH GREENWALD, b New York, NY, Mar 10, 41; m 65; c 2. BIOCHEMISTRY. *Educ:* Antioch Col, BS, 63; Brandeis Univ, PhD(biochem), 69. *Prof Exp:* Res assoc, Haverford Col, 72-74 & Inst Cancer Res, 75; vis scientist, Univ Oxford, 76; lectr chem, Univ Pa, 76-77; vis asst prof, Univ Del, 77-78; asst prof, 78-85, ASSOC PROF CHEM, SWARTHMORE COL, 85-, CHEM DEPT CHAIR, 89- *Concurrent Pos:* NIH fel, 70-72. *Mem:* Am Chem Soc; AAAS; Am Soc Plant Physiologists; Sigma Xi. *Res:* Mechanisms of enzyme action; chemical modification of enzymes; membrane transport mechanisms. *Mailing Add:* Dept Chem Swarthmore Col Swarthmore PA 19081

VOGAN, DAVID A, JR, b Mercer, Pa, Sept 8, 54. GROUP REPRESENTATIONS. *Educ:* Mass Inst Technol, PhD(math), 76. *Prof Exp:* From asst prof to assoc prof math, 79-84, PROF MATH, MASS INST TECHNOL, 84- *Mailing Add:* Dept Math Mass Inst Tech Cambridge MA 02139

VOGAN, ERIC LLOYD, b London, Ont, Sept 3, 24; m 51; c 3. PHYSICS. *Educ:* Univ Western Ont, BSc, 46, MSc, 47; McGill Univ, PhD(physics), 52. *Prof Exp:* Sci officer, Defense Res Telecommun Estab, Can, 52-57; Can liaison officer, Lincoln Lab, Mass Inst Technol & Air Force Cambridge Res Ctr, 57-60; sci officer, Defense Res Telecommun Estab, Can, 60-64; assoc prof physics, 64-70, PROF PHYSICS, UNIV WESTERN ONT, 70- *Mem:* Am Asn Physics Teachers; Am Geophys Union; Can Asn Physicists. *Res:* Aeronomy; physics of the upper atmosphere. *Mailing Add:* Dept Physics Univ Western Ont London ON N6A 5B9 Can

VOGEL, ALFRED MORRIS, b New York, NY, Mar 11, 15; m 40; c 2. CHEMISTRY, INSTRUMENTATION. *Educ:* City Col New York, BS, 34; NY Univ, MS, 48, PhD(chem), 50. *Prof Exp:* Jr chemist, USN, 36-41; chemist, New York City Bd Transp, 41-47; asst, NY Univ, 47-49; instr, Sch Indust Technol, 49-51; from asst prof chem to prof chem, Adelphi Univ, 50-67, chmn dept, 53-67; PROF CHEM, C W POST COL, LONG ISLAND UNIV, 67-, CHMN DEPT, 74- *Mem:* Am Chem Soc; Sigma Xi. *Res:* Analytical and chemical instrumentation; methods of assay of pharmaceutical products; coordination compounds. *Mailing Add:* 209-20 18th Ave Bayside NY 11360

VOGEL, ARTHUR MARK, TUMOR DIAGNOSIS, MELANOMA. *Educ:* NY Univ, PhD(path) & MD, 75. *Prof Exp:* Asst prof path, Univ Wash, 79-87. *Res:* Cancer research. *Mailing Add:* Dept Path St Louis Univ Med Ctr 1402 S Grand Blvd St Louis MO 63104

VOGEL, CARL-WILHELM E, b Hamburg, Ger, Mar 9, 51; m 89. ANTIBODY CONJUGATES, COMPLEMENT SYSTEM. *Educ:* Univ Hamburg, WGer, MD, 76, PhD(biochem), 85. *Prof Exp:* Intern, Dept Med, Univ Hosp, Christian-Aalbrechts Univ, Kiel, WGer & Dept Surg, Gen Hosp, Geesthacht, WGer, 76-77; res fel, Dept Enzym, Inst Physiol Chem, Univ Hamburg, WGer, 78; res fel, Dept Molecular Immunol, Res Inst Scripps Clin, La Jolla, Calif, 79-82, res assoc, Depts Molecular Immunol & Immunol, 82; asst prof, 82-87, ASSOC PROF BIOCHEM & MED, GEORGETOWN UNIV SCH MED & DENT & GRAD SCH ARTS & SCI, WASHINGTON DC, 87- *Concurrent Pos:* Mem, Int Ctr Interdisciplinary Studies Immunol, Georgetown Univ Med Ctr, Washington, DC, 82-, sci dir, 87-; mem, Vincent T Lombardi Cancer Res Ctr, Georgetown Univ Med Ctr, 82-,; mem spec rev comt, Nat Cancer Inst, 87, 89 & 90; mem ad hoc study sect exp immunol, NIH, 87. *Honors & Awards:* Alan Beuman Award, 88. *Mem:* Ger Soc Biol Chem; Am Soc Microbiol; Am Asn Immunologists; Am Soc Biol Chemists; Int Soc Develop & Comp Immunol; AAAS; Am Asn Cancer Res; Am Soc Trop Med & Hyg; Am Fedn Clin Res; Ger Soc Immunol; AMA. *Res:* Molecular mechanisms of complement killing of tumor cells; biochemistry of complement protein. *Mailing Add:* Dept Biochem Georgetown Univ Sch Med 3900 Reservoir Rd NW Washington DC 20007

VOGEL, F(ERDINAND), L(INCOLN), electronic materials, for more information see previous edition

VOGEL, FRANCIS STEPHEN, b Middletown, Del, Sept 29, 19; m 49; c 5. PATHOLOGY. *Educ:* Villanova Col, AB, 41; Western Reserve Univ, MD, 44; Am Bd Path, dipl, 51. *Prof Exp:* From asst prof to assoc prof path, Med Col, Cornell Univ, 50-61, asst prof path in surg, 50-61; PROF PATH, MED CTR, DUKE UNIV, 61- *Concurrent Pos:* Consult, Vet Admin Hosp, New York. *Mem:* Am Soc Exp Path; Am Asn Pathologists & Bacteriologists. *Res:* Neuropathology; metabolic function of mitochondrial nucleic acids. *Mailing Add:* Dept Path Duke Univ Med Ctr Box 3712 Durham NC 27710

VOGEL, GEORGE, b Prague, Czech, May 28, 24; nat US; m 50; c 2. ORGANIC CHEMISTRY. *Educ:* Prague Inst Technol, DSc(chem), 50. *Prof Exp:* Lectr chem, Univ Col, Ethiopia, 51-54; res chemist, Monsanto Chem, Ltd, Eng, 54-55; res fel, Ohio State Univ, 55-56; from asst prof to assoc prof, 56-69, PROF CHEM, BOSTON COL, 69- *Mem:* Am Chem Soc. *Res:* Heterocyclic chemistry; steric effects in conjugated systems; mass spectrometry. *Mailing Add:* Dept Chem Boston Col 140 Commonwealth Ave Chestnut Hill MA 02167-3800

VOGEL, GERALD LEE, b Janesville, Wis, Feb 6, 43. ANALYTICAL CHEMISTRY. *Educ:* Univ Wis-Madison, BS, 65; Georgetown Univ, MS, 70, PhD(chem), 73. *Prof Exp:* Supvr, Washington Ref Lab, 71-73, Pharmacopathics Res Lab, 73-74; PROJ LEADER, AM DENT ASN HEALTH FOUND, 74- *Concurrent Pos:* NIH fel. *Mem:* Int Asn Dent Res. *Res:* Dissolution and precipitation of biological calcium phosphates as it relates to dental caries. *Mailing Add:* Am Dent Asn Health Found Passenbarger Res Ctr Nat Inst Sci & Technol Gaithersburg ND 20899

VOGEL, GLENN CHARLES, b Columbia, Pa, Mar 7, 43; m 69. INORGANIC CHEMISTRY. *Educ:* Pa State Univ, University Park, BS, 65; Univ Ill, Urbana, MS, 67, PhD(inorg chem), 70. *Prof Exp:* Asst prof, 70-74, assoc prof, 74-82, PROF CHEM, ITHACA COL, 82- *Concurrent Pos:* Grants, Am Chem Soc-Petrol Res Fund, 73, NATO, 79 & Res Corp, 80, W R Grace, 89-91; fac res opportunity award, NSF, 83 & 85; Dana teaching fel. *Mem:* Sigma Xi; Am Chem Soc. *Res:* Complex formation of metalloporphyrins; transition metal complexes of oxocarbon ligands; ternary copper catechol complexes; zeolites. *Mailing Add:* Dept Chem Ithaca Col Danby Rd Ithaca NY 14850

VOGEL, HENRY, b New York, NY, Sept 2, 16; m 47; c 1. MICROBIOLOGY, SEROLOGY. *Educ:* La State Univ, BS, 40; NY Univ, MS, 49, PhD(biol), 56. *Prof Exp:* Bacteriologist, Jewish Mem Hosp, New York, 45-49; sr bacteriologist, Willard Parker Hosp, New York, 49-52; sr bacteriologist, Bur Labs, New York City Dept Health, 52-66, sr res scientist, 66-79. *Mem:* NY Acad Sci; Brit Soc Appl Bact; Brit Soc Gen Microbiol; Sigma Xi. *Res:* Enteric microbiology; metabolism of cold-blooded acid fast organisms; metabolism of Leptospira; serologic studies of genetic relationships and diagnosis. *Mailing Add:* Dept Pathol Columbia Univ Col Physicians & Surgeons New York NY 10032

VOGEL, HENRY ELLIOTT, b Greenville, SC, Sept 16, 25; m 53; c 4. SOLID STATE PHYSICS. *Educ:* Furman Univ, BS, 48; Univ NC, MS, 50, PhD(physics), 62. *Prof Exp:* From instr to assoc prof, Clemson Univ, 50-65, head dept physics, 67-71, dean, Col Sci, 71-87, prof physics, 65-90, EMER PROF PHYSICS, CLEMSON UNIV, 90- *Mem:* Am Phys Soc; Am Asn Physics Teachers. *Res:* Superconductivity; thin vacuum-deposited films; tunneling between films. *Mailing Add:* 222 Wyatt Ave Clemson SC 29631-3003

VOGEL, HOWARD H, JR, b New York, NY, Nov 30, 14; m 40, 61; c 4. ZOOLOGY. *Educ:* Bowdoin Col, AB, 36; Harvard Univ, MA, 37, PhD(biol), 40. *Prof Exp:* From asst prof to assoc prof, Wabash Col, 41-47, actg head dept, 46; chmn col biol sci, Univ Chicago, 47-50; assoc biol & group leader neutron radiobiol, Argonne Nat Lab, 50-67; prof radiol, physiol & biophys & head radiation biol, dept radiation oncol, 67-80, prof radiation oncol, 73-80, EMER PROF RADIOL, CTR HEALTH SCI, UNIV TENN, MEMPHIS, 80- *Concurrent Pos:* Ornithologist, Bowdoin-MacMillan Arctic Exped, 34; assoc, Roscoe B Jackson Mem Lab; adj prof biol, Memphis State Univ, 83-91; mem nat lectr comt, Sigma Xi, 89-; Austin teaching fel, Harvard & Radcliffe Col. *Mem:* AAAS; Am Soc Zool; assoc Am Ornith Union; assoc Arctic Int NAm; Transplantation Soc. *Res:* History of arctic aviation; social behavior of birds and mammals; skin transplantation; radiobiology; biological effects of neutrons; radiation carcinogenesis. *Mailing Add:* 208 Ben Avon Way Memphis TN 38111

VOGEL, JAMES ALAN, b Snohomish, Wash, Dec 22, 35; m 59; c 3. OCCUPATIONAL PHYSIOLOGY, BODY COMPOSITION. *Educ:* Wash State Univ, BS, 57; Rutgers Univ, PhD(physiol), 61. *Prof Exp:* Res physiologist, US Army Med Res & Nutrit Lab, Fitzsimons Gen Hosp, US Army Res Inst Environ Med, 61-67, dir exercise physiol, 73-90, RES PHYSIOLOGIST, US ARMY RES INST ENVIRON MED, 67-, DIR OCCUP HEALTH & PERFORMANCE DIRECTORATE. *Mem:* Am Physiol Soc; Am Col Sports Med. *Res:* Cardiac output physiology; exercise and physical fitness training; cardiopulmonary physiology of high altitude; body composition. *Mailing Add:* US Army Res Inst Environ Med Natick MA 01760-5007

VOGEL, JAMES JOHN, b Longmont, Colo, June 16, 35; m 60; c 2. BIOCHEMISTRY, NUTRITION. *Educ:* William Jewell Col, AB, 57; Univ Wis, MS, 59, PhD(biochem), 61. *Prof Exp:* Res fel dent biochem, Forsyth Dent Ctr, Harvard Univ, 61-63; res assoc biochem, Med Sch, Univ Minn, 63-67; asst mem, 67-71, assoc prof, 71-86, PROF, UNIV TEX DENT BR, 86- *Mem:* Int Asn Dent Res; Am Chem Soc. *Res:* Dietary factors involved in dental caries; nutritional and metabolic aspects of magnesium, phosphorus and fluorine with respect to skeletal tissues; microbiologic calcification; phospholipid-protein interactions in calcification. *Mailing Add:* Univ Tex Dent Br PO Box 20068 Houston TX 77225

VOGEL, KATHRYN GIEBLER, b Los Angeles, Calif, May 19, 42; m 64; c 3. EXTRACELLULAR MATRIX BIOCHEMISTRY, PROTEOGLYCANS. *Educ:* Pomona Col, BA, 63; Univ Calif, Los Angeles, MA, 66, PhD(zool & chem), 68. *Prof Exp:* Asst cell physiol, Boston Biomed Res Inst, 72-74; asst prof, Simmons Col, Boston, 74-75; res assoc, 75-77, asst prof, 77-83, ASSOC PROF BIOL, UNIV NMEX, 83- *Concurrent Pos:* Vis res prof, Univ Lund, Sweden, 82-83; chmn, Women in Cell Biol. *Mem:* Am Soc Cell Biol; Soc Complex Carbohydrates. *Res:* Structure, metabolism and role of proteoglycans in fibrous tissues; how the type and amount of proteoglycan content is regulated, how these interact with collagen in forming matrix and how that matrix suits the mechanical forces to which it is subjected. *Mailing Add:* Dept Biol Univ NMex Albuquerque NM 87131

VOGEL, MARTIN, b Los Angeles, Calif, Mar 7, 35; m 63; c 2. ORGANIC CHEMISTRY, POLYMER CHEMISTRY. *Educ:* Calif Inst Technol, BS, 55, PhD(org chem), 61. *Prof Exp:* Asst prof org chem, Rutgers Univ, 60-65; sr res chemist, 65-83, res fel, 83-87, SR RES FEL, ROHM & HAAS CO, 87- *Mem:* Am Chem Soc. *Res:* Organic coatings; emulsion polymerization; latex paints. *Mailing Add:* 550 Pine Tree Rd Jenkintown PA 19046

VOGEL, NORMAN WILLIAM, b Brooklyn, NY, May 17, 17; m 47; c 4. ZOOLOGY. *Educ:* Univ Mich, AB, 40, MS, 43; Univ Ind, PhD(zool), 56. *Hon Degrees:* ScD, Col Washington & Jefferson, 86. *Prof Exp:* Asst zool, Univ Mich, 42-43; asst physiol, Vanderbilt Univ, 43-44; Lawrason Brown res fel, Saranac Lab, NY, 48-49; instr biol, Champlain Col, 49-52; asst zool, Univ Ind, 52-55; from asst prof to prof, 56-85, EMER PROF BIOL, WASHINGTON & JEFFERSON COL, 85- *Concurrent Pos:* USPHS res grant, Nat Cancer Inst, 58-59; Fulbright-Hays lectr physiol, Fac Med, Univ Nangrahar, Afghanistan, 68-69. *Mem:* AAAS; Am Inst Biol Sci. *Res:* Role of the pituitary gland in chick growth and development, prior to hatching by accomplishing hypophysectomy through ablation of the free-head region of the early embryo. *Mailing Add:* RD 1 Box 258A West Alexander PA 15376

VOGEL, PAUL WILLIAM, b Swayzee, Ind, Oct 5, 19; m 48; c 2. PHYSICAL CHEMISTRY. *Educ:* DePauw Univ, AB, 41; Ind Univ, PhD(org chem), 46. *Prof Exp:* Asst, Ind Univ, 41-44; res chemist, Lubrizol Corp, 45-46, res supvr, 47-50, tech asst to dir res & develop, 50-53, supvr org & anal res, 53-59, dir testing, 59-62, dir chem res, 62-78; RETIRED. *Mem:* AAAS; Am Chem Soc; Sigma Xi. *Res:* Synthetic organic chemistry; pyroxonium and pyrylium salts; organic phosphorous compounds; lubricant additives. *Mailing Add:* 348 Thistle Trail Cleveland OH 44124-4182

VOGEL, PETER, b Prague, Czech, Aug 12, 37. NUCLEAR PHYSICS. *Educ:* Czech Inst Technol, Prague, EngrTechPhysics, 60; Acad Sci USSR, CandSci(physics), 66. *Prof Exp:* Res fel, Joint Inst Nuclear Res, Dubna, USSR, 62-66, Nuclear Res Inst, Rez, Czech, 66-68 & Niels Bohr Inst, Copenhagen, Denmark, 68-69; res assoc, Nordic Inst Theoret Atomic Physics, Univ Bergen, 69-70; sr res fel physics, 70-75, res assoc, 75-81, SR RES ASSOC PHYSICS, CALIF INST TECHNOL, 81- *Mem:* Am Phys Soc. *Res:* Nuclear structure theory; vibrations, rotations and deformations of nuclei; intermediate energy physics, mesonic atoms, neutrinos. *Mailing Add:* Phys 34 Norman Bridge Lab Physics Calif Inst Technol Pasadena CA 91125

VOGEL, PHILIP CHRISTIAN, b Fargo, NDak, Nov 28, 41; m 66; c 2. PHYSICAL ORGANIC CHEMISTRY. *Educ:* Lawrence Univ, AB & BS, 63; Ind Univ, PhD(chem), 67. *Prof Exp:* From res assoc to sr res assoc chem, Yeshiva Univ, 67-70; asst prof chem, Col Pharmaceut Sci, Columbia Univ, 70-73; guest scientist, Max Planck Inst Chem, 73-75; res staff scientist, BASF Corp, 75-76, deleg Dyestuffs Div, BASF Ag, 76-78, mgr tech develop, Colors & Auxiliaries Div, 78-82, mgr tech develop & qual control, Dyestuffs Div, 82-86, MGR PROD SERVS, RENSSELAER WORKS, BASF CHEMICALS, BASF CORP, 86- *Mem:* Am Chem Soc. *Res:* Theory of isotope effects in organic reaction mechanisms; theory of structure of aqueous solutions. *Mailing Add:* BASF Corp Riverside Ave Rensselaer NY 12144

VOGEL, RALPH A, b Brooklyn, NY, June 13, 23; m 52; c 3. MICROBIOLOGY. *Educ:* Wagner Col, BS, 46; Univ Buffalo, MS, 49; Duke Univ, PhD(microbiol), 52. *Prof Exp:* Instr bact, Sch Med, Duke Univ, 52-53; from instr to asst prof, 54-76, ADJ ASSOC PROF MICROBIOL, SCH MED, EMORY 76-; MICROBIOLOGIST, VET ADMIN HOSP, 54- *Concurrent Pos:* Fel, Yale Univ, 52; instr, Ga State Col, 53- *Mem:* Fel Am Acad Microbiol; Am Soc Microbiol; Sigma Xi; NY Acad Med. *Res:* Immunology of the mycosis; pathogenesis of bacterial and mycotic diseases. *Mailing Add:* 3862 Gunnin Rd Norcross GA 30092

VOGEL, RICHARD CLARK, b Ames, Iowa, Jan 28, 18; m 44; c 3. PHYSICAL CHEMISTRY. *Educ:* Iowa State Univ, BS, 39; Pa State Univ, MS, 41; Harvard Univ, AM, 43, PhD(chem), 46. *Prof Exp:* Asst prof chem, Ill Inst Technol, 46-49; sr chemist, Argonne Nat Lab, 49-54, assoc dir, Chem Eng Div, 54-63, div dir, 63-73; mem staff, Exxon Nuclear Co, 73-83; mem staff, Elec Power Res Inst, 83-89; RETIRED. *Honors & Awards:* Robert E Wilson Award, Am Inst Chem Eng. *Mem:* Am Chem Soc; fel Am Nuclear Soc; fel Am Inst Chem Engrs. *Res:* Physical inorganic chemistry; fluorine chemistry; pyrochemical processes; separations processes; nuclear reactor safety. *Mailing Add:* 2081 Robin Hood Lane Los Altos CA 94024

VOGEL, RICHARD E, b Chicago, Ill, July 7, 30; m 53; c 2. COMPUTER SCIENCE. *Educ:* Colo State Univ, BS, 53; Univ NMex, MS, 60. *Prof Exp:* Staff scientist, Los Alamos Sci Lab, 57-59; consult statist & comput, Corp Econ & Indust Res, 59-60; dir comput facil, Kaman Nuclear Div, Kaman Aircraft Corp, 60-69; PRES, DATA MGT ASSOCS INC, 69- *Concurrent Pos:* Lectr, Univ Colo, 61-; adj prof, Colo Col, 65- *Mem:* Am Meteorol Soc. *Res:* Statistics; meteorology; mathematics. *Mailing Add:* 2614 Pavo Pl Colorado Springs CO 80906

VOGEL, ROGER FREDERICK, b Pittsburgh, Pa, Nov 1, 42; m 69; c 9. CATALYSIS, SYNTHESIS. *Educ:* Valparaiso Univ, BS, 64. *Prof Exp:* Chemist, Sherwin Williams Co, 64-66; sr res chemist, Gulf Oil Corp, 66-85; RES ASSOC, CHEVRON CORP, 85- *Res:* Synthesize new catalytic materials for process applications; controlling catalyst pore properties; author of publications on affects of anions upon pore properties of modified aluminas. *Mailing Add:* 1814 Salisbury Dr Fairfield CA 94533

VOGEL, STEFANIE N, MACROPHAGES, ENDOTOXINS. *Educ:* Univ Md, PhD(microbiol), 77. *Prof Exp:* ASSOC PROF MICROBIOL & IMMUNOL, UNIFORMED SERV UNIV HEALTH SCI, 80- *Mailing Add:* 10061 Cape Ann Dr Columbia MD 21046

VOGEL, STEVEN, b Beacon, NY, Apr 7, 40; m 63, 74; c 1. ZOOLOGY, BIOFLUIDMECHANICS. *Educ:* Tufts Univ, BS, 61; Harvard Univ, AM, 63, PhD(biol), 66. *Prof Exp:* From asst prof to assoc prof, 66-79, PROF ZOOL, DUKE UNIV, 79- *Concurrent Pos:* Jr fel, Harvard Univ, 64-66; vis fac, Marine Biol Lab, 72, Univ Wash Marine Lab, 79-83, Tjärnö, Sweden Marine Lab, 89. *Mem:* Fel AAAS; Sigma Xi; Am Soc Zoologists. *Res:* Fluid flow through and around organisms of all phyla in both air and water; convective cooling; general writing in comparative biomechanics. *Mailing Add:* Dept Zool Duke Univ Durham NC 27706

VOGEL, THOMAS A, b Janesville, Wis, July 5, 37; m 60; c 3. PETROLOGY, VOLCANOLOGY. *Educ:* Univ Wis, BS, 59, MS, 61, PhD(geol), 63. *Prof Exp:* Asst prof geol, Rutgers Univ, New Brunswick, 63-68; assoc prof, 68-74, PROF GEOL, MICH STATE UNIV, 74- *Concurrent Pos:* Vis prof, Univ SC, 74-75; vis scientist, Lawrence Livermore Nat Labs, 81-82 & 88-89, consult, 82-91. *Mem:* Geol Soc Am; Am Geophys Union. *Res:* Evolution of high-level silicic magma bodies; origin of batholiths; origin of studies of zoned ash-flow sheets; volatiles in magmas. *Mailing Add:* Dept Geol Mich State Univ East Lansing MI 48824

VOGEL, THOMAS TIMOTHY, b Columbus, Ohio, Feb 1, 34; m 65; c 4. HISTORY & PHILOSOPHY OF SCIENCE. *Educ:* Col Holy Cross, AB, 55; Ohio State Univ, MS, 60, PhD(physiol), 62; Georgetown Med Sch, MD, 65. *Prof Exp:* Teaching asst physiol, 59-61, instr surg, 69-70, CLIN ASST PROF SURG, OHIO STATE UNIV, 73- *Concurrent Pos:* Consult, Bur Voc Rehab, Ohio, 70-; adv, Peer Rev Orgn, 85- *Mem:* Am Physiol Soc; Fed Am Socs Exp Biol; Soc Acad Surg; Am Col Surgeons. *Res:* Metabolism in critically ill patients; tumors producing hormones; blood flow in activity; insulin relationships to muscle metabolism. *Mailing Add:* 621 S Cassingham Rd Columbus OH 43209

VOGEL, VERONICA LEE, b New York, NY, Mar 9, 43; m 69; c 1. TOXICOLOGY, CLINICAL CHEMISTRY. *Educ:* Univ Mich, BS, 64; NY Univ, MS, 67, PhD(phys chem), 72. *Prof Exp:* Res fel chem, NSF, Feltman Res Lab, 72-74; asst prof chem, County Col Morris, 74-75; lectr chem, Rutgers Univ, 75-79; anal chemist, Forensic Toxicol Lab, NJ State Med Examr's Off, 79-81; lab dir, Spec Chem, Toxicol & Environ Lab, MetPath Clin Labs, Teterboro, NJ, 81-83; dir, Lifechem Lab, Nat Med Care, 83-84; CONSULT, 84- *Concurrent Pos:* Consult, Energetics Mat Lab, 74-81 & Batelle Res Labs; postdoctoral fel, NSF. *Mem:* Am Chem Soc; Am Asn Clin Chem. *Res:* Theoretical quantum chemistry, approximate molecular orbital calculations; drug analysis. *Mailing Add:* Three Ihnen Ct Montvale NJ 07645

VOGEL, WILLIS GENE, b Seward, Nebr, Nov 27, 30; m 54; c 4. SURFACE MINE REVEGETATION, RANGE SCIENCE. *Educ:* Univ Nebr, BS, 52; Mont State Col, MS, 61. *Prof Exp:* Range conservationist, Soil Conserv Serv, USDA, Idaho, 59-60, range conservationist, Forest Serv, Mo, 60-63, range scientist, Forest Serv, Northeastern Forest Exp Sta, 63-86. *Mem:* Sigma Xi. *Res:* Revegetation of coal strip mine spoils. *Mailing Add:* 116 Cumberland St Berea KY 40403

VOGEL, WOLFGANG HELLMUT, b Dresden, Ger, Aug 4, 30; m 61; c 2. BIOCHEMISTRY, PHARMACOLOGY. *Educ:* Dresden Tech Univ, BS, 49; Stuttgart Tech Univ, MS, 56, PhD(chem), 58. *Prof Exp:* Postdoctoral fel, State Univ NY Upstate Med Ctr, 58-59; chemist, Farbwerke Hoechst, Ger, 59-61; res assoc biochem pharmacol, Col Med, Univ Ill, 61-63, instr, 63-64; vis scientist, NIH, 64-65; asst prof pharmacol, Col Med, Univ Ill, 65-67; assoc prof, 67-74, PROF PHARMACOL, 74-, JEFFERSON MED COL, 74-, PROF PSYCHIAT & HUMAN BEHAV, 76-, VCHMN, DEPT PSYCHIAT, 84-, ACTG CHMN, 87- *Concurrent Pos:* Med res assoc, L B Mendel Res Lab, Elgin State Hosp, 65-67. *Honors & Awards:* Humboldt Prize; Lindback Award. *Res:* Biochemistry of mental disorders; biochemical pharmacology; stress research; development of drug assays; neurochemical correlates of behavior. *Mailing Add:* 14 Darien Dr Cherry Hill NJ 08034

VOGELBERGER, PETER JOHN, JR, b Youngstown, Ohio, Apr 14, 32; m 54; c 4. NUCLEAR ENGINEERING. *Educ:* US Naval Acad, BS, 54. *Prof Exp:* Assoc nuclear engr, Argonne Nat Lab, 63-65; mgr tech liaison, Isotopes, Inc, 65-67, vpres & gen mgr, Energy Systs Div, Teledyne Isotopes, Inc, 68-76, PRES, TELEDYNE ENERGY SYSTS, TIMONIUM, 76- *Concurrent Pos:* Mem, Md Adv Comn Atomic Energy, 68-78. *Res:* Design, development and production of nuclear and fossil-fuel thermoelectric power systems for space and terrestrial use; electrochemical gas generators and fuel cell. *Mailing Add:* 225 Tunbridge Rd Baltimore MD 21212

VOGELFANGER, ELLIOT AARON, b New York, NY, Apr 5, 37; m 58; c 2. POLYMER SCIENCE, ORGANIC CHEMISTRY. *Educ:* Columbia Univ, BA, 58; Univ Calif, Los Angeles, PhD(phys org chem), 63. *Prof Exp:* Res chemist, Esso Res & Eng Co, 63-66; group leader polymer sci, Celanese Corp, 66-74; MGR RES & DEVELOP, SOLTEX POLYMER CORP, 74- *Concurrent Pos:* Lectr, Hunter Col, 65-71. *Mem:* Am Chem Soc; Soc Plastics Engrs; Am Soc Testing & Mat; Sigma Xi. *Res:* Polymer rheology; high temperature polymers; physical organic chemistry; polyolefins research, application, development, and catalysis. *Mailing Add:* Seven Hampton Ct Houston TX 77024-5447

VOGELHUT, PAUL OTTO, b Vienna, Austria, Dec 2, 35; m 60; c 3. BIOPHYSICS. *Educ:* Univ Calif, Berkeley, AB, 57, PhD(biophys), 62. *Prof Exp:* Asst prof bioelectronics, Univ Calif, Berkeley, 62-70; sr res physicist, Ames Co, Div Miles Labs Inc, 70-80, prin res & develop engr, 80-90; CONSULT BIOPHYS, 91- *Concurrent Pos:* Consult, Electro-Neutronics, 65-68; Ames Co div, Miles Labs Inc, 69-70. *Mem:* AAAS. *Res:* Molecular sensing approaches using LB techniques and monolayer polymers including receptors and antibody molecules; biosensor development, cond polymers, radio frequency techniques and biosystems; immunoassay systems. *Mailing Add:* 12497 Dragoon Trail Mishawaka IN 46544

VOGELI, BRUCE R, b Alliance, Ohio, Nov 25, 29; m 56; c 2. MATHEMATICS. *Educ:* Mt Union Col, BS, 51; Kent State Univ, MA, 57; Univ Mich, PhD(math educ), 60. *Prof Exp:* Assoc prof math, Bowling Green State Univ, 59-65; prof math, Teachers Col, 65-76, CLIFFORD BREWSTER UPTON PROF, COLUMBIA UNIV, 76- *Concurrent Pos:* Vis prof, Lenin Inst, Moscow, USSR, 64 & Kurukshetra Univ, India, 65; consult, Ministry Educ, Chile, 66-67 & Silver Burdett Co, 60-; NSF fel; sr Fulbright lectr. *Honors & Awards:* Harold Benjamin Prize. *Mem:* Math Asn Am; Nat Coun Teachers Math. *Res:* Mathematics education; international mathematical activities. *Mailing Add:* Dept Math & Sci Educ Columbia Univ-Teachers Col Box 210 New York NY 10027

VOGELMAN, JOSEPH H(ERBERT), b New York, NY, Aug 18, 20; m 46; c 3. ELECTRONICS, BIOMEDICAL ENGINEERING. *Educ:* City Col New York, BS, 40; Polytech Inst Brooklyn, MEE, 48, DEE, 57. *Prof Exp:* Res analyst, Signal Corps Radar Lab, US Dept Army, 42, proj engr, 42-43, proj engr, Signal Corps Eng Labs, 43-44; chief test equip sect, Watson Labs, USAF, 44-47, chief develop br, 47-51; chief scientist gen eng, Rome Air Develop Ctr, NY, 51-52, chief electronic warfare lab, 52-55, dir commun & electronics, 55-59; vpres res & develop, Capehart Corp, 59-64; dir electronics, Chromalloy Am Corp, 64-67, vpres, 67-73, gen mgr, Pocket Fone Div, 65-67, vpres, 67-73; vpres & dir, Cro-Med Bionics Corp, 68-73; vchmn bd & sr vpres, Laser Link Corp, 71-73; PRES, VOGELMAN DEVELOPMENT CO, 73-; CHIEF SCIENTIST, ORENTREICH FOUND ADVAN SCI, 73- *Concurrent Pos:* Mem, Army-Navy Radio Frequency Cable Coord Comt, 44-48; mem test equip comt, Res & Develop Bd, 48-52, chmn waveguide comt, 48-52; chmn commun tech adv comt, Air Res & Develop Command, 58-59, award, 58, consult, 59-; consult, Dept of Defense, 59-67; vpres, ACR Electronics Corp, 66-68; dir & consult, Orentreich Found Advan Sci, 63-73. *Mem:* Fel AAAS; fel Inst Elec & Electronics Engrs; Sigma Xi. *Res:* Radio frequency instrumentation; microwave theory and techniques; bio-medical instrumentation; communications; computers for medicine; biochemistry; radio immune assays; chromatography. *Mailing Add:* 48 Green Dr Roslyn NY 11576-3221

VOGELMANN, HUBERT WALTER, b Buffalo, NY, Nov 13, 28; m 51; c 2. BOTANY. *Educ:* Heidelberg Col, BS, 51; Univ Mich, MA, 52, PhD(bot), 55. *Prof Exp:* Asst prof taxon bot, 59-62, assoc prof, 62-70, PROF BOT, UNIV VT, 70- *Mem:* Ecol Soc Am. *Res:* Ecology of mountain forests; natural areas protection. *Mailing Add:* Dept Bot Univ Vt Agr Col 85 S Prospect St Burlington VT 05405

VOGH, BETTY POHL, b Georgetown, Ohio, Apr 19, 27; m 47; c 4. PHARMACOLOGY, PHYSIOLOGY. *Educ:* Tex Woman's Univ, BA, 46; Univ Fla, PhD(physiol), 64. *Prof Exp:* Res assoc physiol, 65-66, res assoc pharmacol, 66-67, asst prof, 68-74, ASSOC PROF PHARMACOL, COL MED, UNIV FLA, 74- *Mem:* Sigma Xi. *Res:* Physiology and pharmacology of body fluids; regulation of cerebrospinal fluid. *Mailing Add:* 1119 SW 11th Ave Gainesville FL 32601

VOGL, OTTO, b Traiskirchen, Austria, Nov 6, 27; nat US; m 55; c 2. POLYMER SCIENCE & ENGINEERING. *Educ:* Univ Vienna, PhD, 50. *Hon Degrees:* Dr, Univ Jena. *Prof Exp:* Instr chem, Univ Vienna, 48-53; res assoc, Univ Mich, 53-55 & Princeton Univ, 55-56; chemist, E I du Pont de Nemours & Co, Del, 56-70; prof, 70-83, EMER PROF POLYMER SCI ENG, UNIV MASS, AMHERST, 83-; HERMAN F MARK PROF, POLYTECH UNIV, 83- *Concurrent Pos:* Vis prof, Kyoto Univ & Osaka Univ, 68 & 80, Royal Inst Technol, Stockholm, 71 & 87, Univ Freiburg, 73, Univ Strasburg, 76, Univ Berlin, 77 & Tech Univ Dresden, 82; chmn, Am Chem So Div Polymer Chem & Conn Valley Sect, 74; comt mem macromolecular chem, Nat Res Coun-Nat Acad Sci, 75-78, chmn, 78-80; pres, Pac Polymer Fedn, 87-90; Fulbright fel, 76; sr scientist fel, Japan Soc Promotion Sci, 80. *Honors & Awards:* Humboldt Award, Fed Repub Ger, 77; W H Rauscher Mem lectr, 84; Mobay lectr, Univ Pittsburgh, 85; Chem Pioneer Award, Am Inst Chemists, 85; Exner Medal, 87; Chem & Phys Medal, Austrian Res Inst, 89; Appl Polymer Chem Award, Am Chem Soc, 90. *Mem:* AAAS; Am Chem Soc; Austrian Chem Soc; Japanese Soc Polymer Sci; NY Acad Sci; foreign mem Austrian Acad Sci; Sigma Xi. *Res:* Ionic and stereoselective polymerization; polyaldehydes; ring opening polymerization; regular copolyamides; reactions on polymers; functional polymers; biologically and ultraviolet active polymers; head to head polymers; optically active polymers. *Mailing Add:* 333 Jay St Polytech Univ Brooklyn NY 11201

VOGL, RICHARD J, b Milwaukee, Wis, Jan 19, 32; m 61; c 3. BOTANY, ECOLOGY. *Educ:* Marquette Univ, BS, 53, MS, 55; Univ Wis, PhD(ecol), 61. *Prof Exp:* Instr bot, Marquette Univ, 55-56; res asst, Univ Wis, 58-61; PROF BOT, CALIF STATE UNIV, LOS ANGELES, 61- *Concurrent Pos:* Ed, Ecol Soc Am, 72-75. *Mem:* Ecol Soc Am; Wildlife Soc. *Res:* Plant and fire ecology. *Mailing Add:* Dept of Biol Calif State Univ 5151 State University Dr Los Angeles CA 90032

VOGL, THOMAS PAUL, b Vienna, Austria, July 10, 29; nat US; c 3. BIOMEDICAL SCIENCE, SCIENCE POLICY & ADMINISTRATION. *Educ:* Columbia Univ, BA, 52; Univ Pittsburgh, MS, 57; Carnegie-Mellon Univ, PhD(syst sci), 69. *Prof Exp:* Sr res physicist, Res Lab, Westinghouse Elec Corp, 52-60; head infrared sect, Res Labs, Hughes Aircraft Co, 60-61; mgr optical physics, Res Labs, Westinghouse Elec Corp, 61-69, mgr optics, 69-74; prin staff officer, Assembly Life Sci, Nat Acad Sci, 74-77; exec secy, Nat Comn Digestive Dis, NIH, 77-79, nutrit coordr comt, off dir, 79-86; SR RES SCIENTIST & MGR, ADVAN CONCEPTS DEPT, ENVIRON RES INST MICH, 86- *Concurrent Pos:* Lectr, Univ Calif, Los Angeles, 59-74; mem comt photother in newborn, Div Med, Nat Acad Sci-Nat Res Coun & chmn subcomt bioeng aspects; adj prof radiation biophysics, Dept Radiol & Pediat, Col Physicians & Surgeons, Columbia Univ, 73-79, mem, Bioeng Inst, 75-79. *Mem:* Optical Soc Am; AAAS; Am Soc Photobiol. *Res:* Optical imaging and illuminating systems; non-linear optimization; clinical applications of light; phototherapy of hyperbilirubinemia; biological and artificial neural networks. *Mailing Add:* 8130 Old Georgetown Rd Bethesda MD 20814-1450

VOGLER, LARRY B, b Houston, Tex, Feb 5, 47. B LYMPHOCYTE GROWTH & DIFFERENTIATION. *Educ:* Baylor Col Med, MD, 73. *Prof Exp:* ASST PROF IMMUNOL, SCH MED, VANDERBILT UNIV, 81- *Mem:* Soc Pediat Res; Am Asn Immunol. *Mailing Add:* Dept Pediat Southeastern Health Serv Inc 3280 Howell Mill Rd Terr Level Atlanta GA 30327

VOGT, ALBERT R, b St Louis, Mo, Apr 6, 38; c 2. PLANT PHYSIOLOGY. *Educ:* Univ Mo, BS, 61, MS, 62, PhD(forestry), 66. *Prof Exp:* Instr forestry, Univ Mo, 65-66; from asst prof to prof tree physiol, Ohio Agr Res & Develop Ctr, Ohio State Univ, 66-85, from actg assoc chmn to chmn, Div Forestry, 69-85; DIR, SCH NATURAL RESOURCES, UNIV MO, COLUMBIA, 85- *Mem:* Am Soc Plant Physiol; Soc Am Foresters. *Res:* Physiology of tree growth and development; bud dormancy in oak; flowering of trees. *Mailing Add:* Sch Natural Resources Univ Mo Columbia MO 65211

VOGT, ERICH WOLFGANG, b Steinbach, Man, Nov 12, 29; m 52; c 5. NUCLEAR REACTIONS, INTERMEDIATE ENERGY PHYSICS. *Educ:* Univ Man, BSc, 51, MSc, 52; Princeton Univ, PhD(physics), 55. *Hon Degrees:* DSc, Univ Man, 82, Queens Univ, 84, Carleton Univ, 88; LLD, Univ Regina, 86. *Prof Exp:* Nat Res Coun Can fel, Univ Birmingham, 55-56; res officer physics, Atomic Energy Can Ltd, 56-65; PROF PHYSICS, UNIV BC, 65-; DIR, TRIUMF PROJ, CAN NAT MESON LAB, 81- *Concurrent Pos:* Vis assoc prof, Univ Rochester, 58-59; Nat Res Coun Can sr travelling fel, Oxford Univ, 71-72; vpres fac & student affairs, Univ BC, 75-81; chmn, Sci Coun BC, 78-80. *Honors & Awards:* Centennial Medal Can, 67. *Mem:* Am Phys Soc; Royal Soc Can; Can Asn Physicists (pres, 70-71). *Res:* Theory of nuclear reactions, nuclear structure and intermediate energy physics; physics of pion-nucleus interactions. *Mailing Add:* Triumf 4004 Wesbrook Mall Vancouver BC V6T 2A3 Can

VOGT, HERWART CURT, b Elizabeth, NJ, Sept 14, 29; m 58; c 2. POLYMER CHEMISTRY. *Educ:* Northwestern Univ, 52; Univ Del, MS, 54, PhD(chem), 57; Wayne State Univ, MS, 77. *Prof Exp:* Res chemist, Hercules Inc, 57-59; sr res chemist, 59-66, res assoc, 67-73, supvr, 73-77, corp toxicologist 77-79, MGR ADMIN, BASF WYANDOTTE CORP, 79- *Concurrent Pos:* Vis lectr, Oakland Univ, 66-69 & Wayne State Univ, 69-71; exchange chemist, BASF-AG, WGer, 71-73; adj prof, Upsala Col, 81- *Mem:* Am Chem Soc; AAAS; Sigma Xi. *Res:* Organic phosphorus compounds pertaining to polymers; novel halogen containing unsaturated polyesters; isocyanate and urethane chemistry; chlorine containing elastomers; research and development in noncellular urethane plastics. *Mailing Add:* Administration BASF Wyandotte Corp 100 Cherry Hill Rd Parsippany NJ 07054

VOGT, KRISTIINA ANN, b Turku, Finland, Mar 3, 49; US citizen; m 73. MICROBIAL ECOLOGY, FORESTRY. *Educ:* Univ Tex, El Paso, BS, 71; NMex State Univ, MS, 74, PhD(biol), 75. *Prof Exp:* Res assoc ecosyst, 76-80, RES ASST PROF, COL FOREST RESOURCES, UNIV WASH, 80- *Mem:* Soc Indust Microbiol; AAAS; Sigma Xi. *Res:* Physiology; decomposition and nutrient cycling; mycorrhizae and below ground root dynamics. *Mailing Add:* Dept Forest Resources Univ Wash Seattle WA 98195

VOGT, MOLLY THOMAS, b Lyndhurst, Eng, Apr 15, 39; div; c 2. BIOCHEMISTRY. *Educ:* Bristol Univ, BSc, 60; Univ Pittsburgh, PhD(biochem), 67. *Prof Exp:* Jr res officer, Toxicol Unit, Med Res Coun, Eng, 60-62; res asst biochem path, Sch Med, 62-63, from asst prof biochem to assoc prof, 70-78, chmn div health related prof interdisciplinary progs, 72-74, assoc dean, 77-83, PROF BIOCHEM, SCH HEALTH RELATED PROF, UNIV PITTSBURGH, 78-, DIR CONTINUING MED EDUC, 84- *Concurrent Pos:* NIH fel biochem, Sch Med, Univ Pittsburgh, 67-70, Health Res Serv Found grant, 70-71; Am Coun Educ Admin intern, 74-75. *Honors & Awards:* Pres's Award, Am Soc Allied Health Professions, 81. *Mem:* AAAS; Am Inst Biol Scientists; Sigma Xi; fel Am Soc Allied Health Professions. *Res:* Cell metabolism in health and disease; phagocytic process; mitochondrial metabolism; biochemical effects of typical air pollutants. *Mailing Add:* 711 Field Club Dr Pittsburgh PA 15238

VOGT, PETER KLAUS, b Broumov, Czech, Mar 10, 32; US citizen. GENETICS, MOLECULAR BIOLOGY. *Educ:* Univ Wurzburg, Ger, BS, 55; Univ Tubingen, Ger, PhD(biol), 59. *Prof Exp:* From asst prof to assoc prof path, Sch Med, Univ Colo, 62-67; from assoc prof to prof microbiol, Sch Med, Univ Wash, 67-71; Hastings prof, 71-78, Hastings distinguished prof, 78-80, HASTINGS DISTINGUISHED PROF MICROBIOL & CHMN DEPT, SCH MED, UNIV SOUTHERN CALIF, LOS ANGELES, 80- *Concurrent Pos:* Damon Runyon cancer fel, Virus Lab, Univ Calif, Berkeley, 59-62; res grants, USPHS, 62- & Am Cancer Soc, 63-68; mem virol study sect, NIH, 67-71; mem cell biol & virol adv comt, Am Cancer Soc, 72-76; bd sci consults, Sloan Kettering Inst Cancer Res, 72-80, chmn, 80-81; mem, Coun Res & Clin Invest, Am Cancer Soc, 79-82, Cancer Spec Prog Rev Comt, Nat Cancer Inst, 81-85, sci adv bd, Coun Tobacco Res, 87- & comt sci adv, Ger Cancer Ctr, 89- *Honors & Awards:* Vogeler Prize, Max-Planck-Soc, 76; Alexander von Humboldt Award, Fed Repub Ger, 84; Ernst Jung Prize for Med, 85; Waterford Biomed Sci Award, 86; Robert J & Claire Pasarow Award, 87; Paul-Ehrlich & Ludwig-Darmstaedter Prize, 88; Bristol Myers Award, 89; ICN Int Prize in Virol, 89. *Mem:* Nat Acad Sci; hon mem Japanese Cancer Asn; Am Soc Microbiol; AAAS; Genetics Soc Am; Am Soc Virol; Am Asn Cancer Res. *Res:* Mechanism of neoplastic cellular transformation induced by viruses; virology; genetics; cellular differentiation; neoplastic transformation; immunology. *Mailing Add:* Dept Microbiol Rm HMR 401 Univ Southern Calif Sch Med 2011 Zonal Ave Los Angeles CA 90033-1054

VOGT, PETER RICHARD, b Hamburg, Ger, June 8, 39; US citizen; m 67; c 2. MARINE GEOPHYSICS. *Educ:* Calif Inst Technol, BS, 61; Univ Wis, MA, 65, PhD(oceanog), 68. *Prof Exp:* Geophysicist, US Naval Oceanog Off, 67-75, GEOPHYSICIST, US NAVAL RES LAB, 76-77 & 78- *Concurrent Pos:* Mem staff, Univ Oslo, Norway, 77-78. *Honors & Awards:* Henry A Kaminski Award, Sci Res Soc Am. *Mem:* Am Geophys Union; fel Geol Soc

Am; fel Explorers Club. *Res:* Geophysical research on the constitution and history of the crust beneath the sea, especially the analysis of marine magnetic field, gravity, altimetric, acoustic backscatter, and topographic anomalies and mantle hot spot phenomena as related to ocean floor (plate) movement, crustal composition and continental drift. *Mailing Add:* Code 5110 US Naval Res Lab Washington DC 20375-5000

VOGT, ROCHUS E, b Neckarelz, Ger, Dec 21, 29; US citizen; m 58; c 2. PHYSICS. *Educ:* Univ Chicago, SM, 57, PhD(physics), 61. *Prof Exp:* Res assoc cosmic rays, Univ Chicago, 61-62; from asst prof to assoc prof physics, Calif Inst Technol, 62-70, chmn fac, 75-77, chief scientist, Jet Propulsion Lab, 77-78, chmn div physics, math & astron, 78-83, vpres & provost, 83-87, PROF PHYSICS, CALIF INST TECHNOL, 70-, R STANTON AVERY DISTINGUISHED SERV PROF, 82-, DIR MASS INST TECHNOL LASER INTERFEROMETER GRAVITATIONAL WAVE OBSERV PROJ, 87- *Concurrent Pos:* Actg dir, Owens Valley Radio Observ, Calif, 80-81; vis prof, Mass Inst Technol, 88- *Mem:* Fel Am Phys Soc; AAAS. *Res:* Cosmic rays; astrophysics; experimental gravitation. *Mailing Add:* 102-33 Calif Inst Technol Pasadena CA 91125

VOGT, STEVEN SCOTT, b Rock Island, Ill, Dec 20, 49; m; c 3. ASTRONOMY. *Educ:* Univ Calif, Berkeley, AB(physics) & AB(astron), 72; Univ Tex, MA, 75, PhD(astron), 78. *Prof Exp:* Res asst astron, Univ Tex, 72-78; asst prof & asst astronr, 78-84, assoc prof & astronr, 84-87, PROF & ASTRONR, LICK OBSERV, UNIV CALIF, SANTA CRUZ, 87- *Mem:* Am Astron Soc; Optical Soc Am; Soc Photo-Optical Instrumentation Engrs. *Res:* Astronomical instrumentation; solid state imaging detectors, stellar spectroscopy. *Mailing Add:* Lick Observ Univ Calif Santa Cruz CA 95064

VOGT, THOMAS CLARENCE, JR, b San Antonio, Tex, Sept 21, 32; m 63; c 6. PHYSICAL CHEMISTRY, PETROLEUM ENGINEERING. *Educ:* St Mary's Univ, BS, 54, PhD(phys chem), 61. *Prof Exp:* Res chemist, Mobil Oil Corp, 57; asst, Univ Notre Dame, 57-58, fel diffusion kinetics, Radiation Proj, 58-61; assoc chemist, Field Res Lab, Mobil Res & Develop Corp, 61-80, res assoc, 80-82; uranium field opers mgr, Mobil Alternative Energy, Inc, 82-84; eng supvr, 85-87, compl eng adv, 88-89, acquisitions trades & sales, 90, COMPLETIONS ADV, MOBIL OIL CORP, DENVER-WEST, 91- *Mem:* Sigma Xi; Am Chem Soc; Soc Petrol Eng. *Res:* Diffusion and recombination of free radicals in liquid systems; effects of high pressure on reaction rates; chemical stimulation of petroleum production wells; in-situ uranium leaching; hydraulic fracturing of subsurface formations; economics of well treatments; in situ coal gasification. *Mailing Add:* Mobil Oil Corp PO Box 5444 Denver CO 80217

VOGT, WILLIAM G(EORGE), b McKeesport, Pa, June 1, 31; div; c 2. COMPUTER CONTROL, ACQUISITION & POINTING & TRACKING. *Educ:* Univ Pittsburgh, BS, 53, MS, 57, PhD(elec eng), 62. *Prof Exp:* Engr, Adv Systs & Eng Div, Westinghouse Elec Corp, 61; res engr, Eng Res Div, 53-60, assoc prof, 62-68, PROF ELEC ENG, UNIV PITTSBURGH, 68- *Concurrent Pos:* NASA res assoc, Univ Pittsburgh, 63-64; consult, Astrionics Labs, Marshall Space Flight Ctr, NASA, Ala, 64-67, Fecker Systs Div, Owens-Ill, Pa, 68-74, Contraves-Goerz 78-, TASC, 78-, NASA Indust Appln Ctr, 82-, Compunetics, Inc, 82- & Tex Instruments, Inc, 83-; NSF fel, 61-62 & NSF sci fac fel, 75-76; mem bd dirs, Univ Res & Develop Assocs, Inc. *Mem:* Inst Elec & Electronics Engrs; Am Soc Mech Engrs; Instrument Soc Am; Sigma Xi; Soc Comput Simulation. *Res:* Computer control; robotics; high speed real time digital signal processing; acquisition, pointing and tracking systems; visual inspection systems; parallel processing; microcomputer and personal computer applications; modeling and simulation. *Mailing Add:* Dept Elec Eng Univ Pittsburgh 348 BEH Pittsburgh PA 15261

VOHR, JOHN H, b Laconia, NH, Nov 27, 34; m 56; c 4. MECHANICAL ENGINEERING. *Educ:* Harvard Univ, AB, 56; Columbia Univ, MS, 58, PhD(mech eng), 64. *Prof Exp:* Res engr, Mech Tech Inc, NY, 62-68, supvr anal mech, 68-69; from asst prof to assoc prof mech eng, Rensselaer Polytech Inst, 69-73, chmn mech eng curric, 70-73; SR ENGR, GEN ELEC CO, 73- *Concurrent Pos:* Consult, Gen Elec Co, 69- *Honors & Awards:* Best Paper of Year Award, Lubrication Div, Am Soc Mech Engrs, 67. *Mem:* Am Soc Mech Engrs; Am Soc Lubrication Engrs. *Res:* Analytical studies of hydrodynamic and hydrostatic bearings; heat transfer analysis and two-phase flow studies; experimental studies of flow stability phenomena. *Mailing Add:* 1400 Dean St Schenectady NY 12309

VOHRA, PRAN NATH, b Gwaliar, India, June 11, 19. NUTRITION, BIOCHEMISTRY. *Educ:* Univ Panjab, India, MSc, 42; Wash State Univ, MS, 54; Univ Calif, Davis, PhD(nutrit), 59. *Prof Exp:* Res asst chem, Sci & Indust Res Orgn, India, 42-49; int trainee fermentations, Joseph E Seagram & Sons, Ky, 49-50; specialist nutrit, Dept Poultry Husb, Univ Calif, Davis, 58-59; asst, BO&C Mills, Eng, 59-60; specialist poultry, US AID India, 61-62; asst res nutritionist, Dept Poultry Husb, 62-70, from assoc prof to prof, 70-89, EMER PROF AVIAN SCI, UNIV CALIF, DAVIS, 89- *Concurrent Pos:* Consult. *Honors & Awards:* Am Feed Mfrs Asn Award. *Mem:* Poultry Sci Asn; Am Inst Nutrit; Brit Biochem Soc; Brit Nutrit Soc. *Res:* Trace elements in nutrition; improvement of nutrition in developing countries; comparative nutrition of avian species; nutrition evaluation of cereals and legumes; poultry husbandry. *Mailing Add:* Dept Avian Sci Univ Calif Davis CA 95616

VOHS, JAMES A, b Idaho Falls, Idaho, Sept 26, 28; m 53; c 4. MEDICINE. *Educ:* Univ Calif, Berkeley, BA, 52. *Prof Exp:* CHMN & CHIEF EXEC OFFICER, KAISER FOUND HEALTH PLAN & HOSP, 80- *Honors & Awards:* Justin Ford Kimball Award, Am Hosp Asn. *Mem:* Inst Med-Nat Acad Sci. *Mailing Add:* Kaiser Found Health Plan & Hosp One Kaiser Plaza Ordway Bldg 27th Floor Oakland CA 94612

VOHS, PAUL ANTHONY, JR, b Kansas City, Kans, Jan 19, 31; m 53; c 5. ZOOLOGY. *Educ:* Kans State Univ, BS, 55; Southern Ill Univ, MA, 58; Iowa State Univ, PhD, 64. *Prof Exp:* Proj leader, Coop Proj, Ill Dept Conserv, Ill Natural Hist Surv & Southern Ill Univ, 55-58; res assoc, Coop Wildlife Res Proj, Southern Ill Univ, 59-61; res asst, Coop Wildlife Res Unit, Iowa State Univ, 61-62, instr zool & entom, 63-64, asst prof zool, 64-68, assoc prof wildlife biol, 68; assoc prof wildlife ecol, Ore State Univ, 68-73, prof & exten wildlife specialist, Coop Exten Serv & Dept Fisheries & Wildlife, 73-74; prof wildlife & fisheries sci & head dept, SDak State Univ, Brookings, 74-76; leader, Okla Coop, Wildlife Res Unit, Okla State Univ, Fish & Wildlife Serv, Stillwater, 76-79; supvr, Coopr Wildlife Res Units, US Dept Interior-Fish & Wildlife Serv, Washington, DC, 79-81; dir, Denver Wildlife Res Ctr, Denver Co, 82-87; LEADER, IOWA COOP FISH & WILDLIFE RES UNIT, IOWA STATE UNIV FISH & WILDLIFE SERV, AMES, IOWA, 87- *Mem:* Wilson Ornith Soc; Wildlife Soc; Am Soc Mammal; Ecol Soc Am; Am Ornith Union. *Res:* Vertebrate ecology; response of birds and mammals to manipulations of habitat; landscape ecology. *Mailing Add:* Iowa Coop Fish & Wildlife Res Unit Iowa State Univ 11 Sci Hall II Ames IA 50011

VOICHICK, MICHAEL, b Yonkers, NY, May 28, 34; m 60; c 3. MATHEMATICS. *Educ:* Oberlin Col, BA, 57; Brown Univ, PhD(math), 62. *Prof Exp:* Res instr math, Dartmouth Col, 62-64; from asst prof to assoc prof, 64-73, PROF MATH, UNIV WIS-MADISON, 73- *Mem:* Am Math Soc; Math Asn Am. *Res:* Function theory. *Mailing Add:* Dept Math Univ Wis 405 Van Vleck Hall Madison WI 53706

VOIGE, WILLIAM HUNTLEY, b Pittsburgh, Pa, Sept 15, 47. BIOCHEMISTRY. *Educ:* Mich State Univ, BS, 69; Case Western Reserve Univ, PhD(biochem), 75. *Prof Exp:* Instr chem, St Olaf Col, 74-75; asst prof, 75-83, ASSOC PROF CHEM, JAMES MADISON UNIV, 83- *Concurrent Pos:* Vis fel biochem, Princeton Univ, 82-83. *Mem:* AAAS. *Res:* Computer-aided instruction in biochemistry; nucleotide-polyamine complexation; enzymes of poly-B-hydroxybutyrate pathway. *Mailing Add:* Dept Chem James Madison Univ Harrisonburg VA 22807

VOIGHT, BARRY, b Yonkers, NY, Dec 17, 37; m 59; c 2. GEOLOGY. *Educ:* Univ Notre Dame, BS, 59 & 60, MS, 61; Columbia Univ, PhD(struct geol), 65. *Prof Exp:* Asst prof eng geol, 64-70, assoc prof, 70-78, PROF GEOL, PA STATE UNIV, 78- *Concurrent Pos:* Vis prof, Delft Technol Inst, 72, Univ Toronto, 73 & Univ Calif, Santa Barbara, 81; adj geologist, US Geol Survey, 80-; consult, US Geol Surv Rev Volcano Prog, 86 & Nat Res Coun Comt Ground Failure Hazard, 82- *Honors & Awards:* Res award, Nat Res Coun, 84; George Stephenson Medal, Inst Civil Engrs London, 84. *Mem:* Geol Soc Am; Int Soc Rock Mech. *Res:* Stress measurements; residual stresses in rocks; fault mechanics; engineering geology; rock mechanics; mechanics of landslides; geology of Iceland; geology of Mt St Helens and Cascade volcanos; disaster prevention of landslides and volcano eruptions; failure predictions; methods to predict volcano eruptions; applied volcanology world wide. *Mailing Add:* Dept Geosci Pa State Univ University Park PA 16802

VOIGT, ADOLF F, b Upland, Calif, Jan 31, 14; m 41; c 2. NUCLEAR CHEMISTRY. *Educ:* Pomona Col, BA, 35; Claremont Col, MA, 36; Univ Mich, PhD, 41. *Prof Exp:* From asst prof to prof, 46-55, EMER PROF CHEM, IOWA STATE UNIV, 82- *Concurrent Pos:* Asst dir, Ames Lab, 65-80. *Mem:* Am Chem Soc; Am Phys Soc; Am Nuclear Soc; AAAS; Sigma Xi. *Mailing Add:* Ames Lab USDOE Iowa State Univ Ames IA 50011

VOIGT, CHARLES FREDERICK, b Woodside, NY, Dec 17, 42; m 62; c 3. ORGANIC CHEMISTRY. *Educ:* Univ SFla, Tampa, BA, 65; Duke Univ, PhD(org chem), 70. *Prof Exp:* Assoc indexer org chem, Chem Abstr Serv, 70-71, sr assoc ed macromolecular chem, 71-73, sr ed appl chem, 73-79, asst mgr chem technol, 79-85, Patent Serv, 85-87, doc anal mgr, patent serv, 87-90, DEPT MGR, CHEM TECHNOL, CHEM ABSTR SERV, 90- *Mem:* Am Chem Soc. *Res:* Heterocyclic chemistry; polymers; applied chemistry. *Mailing Add:* Chem Abstr Serv Dept 60 PO Box 3012 Columbus OH 43210

VOIGT, EVA-MARIA, b Dortmund, WGer, Feb 2, 28; Can citizen. PHYSICAL CHEMISTRY, MOLECULAR PHYSICS. *Educ:* McMaster Univ, BSc, 53, MSc, 54; Univ BC, PhD(phys chem), 63. *Prof Exp:* Head res sect, Aylmer Foods, Inc, 55-56; lectr chem, Mt Allison Univ, 56-57; fel phys chem, Univ Calif, Berkeley, 63-65; asst prof, 66-69, assoc prof, 69-79, PROF CHEM, SIMON FRASER UNIV, 79- *Mem:* AAAS; Am Chem Soc; Am Phys Soc; Chem Inst Can; Can Inst Phys. *Res:* Molecular spectroscopy; charge-transfer interactions; energy transfer. *Mailing Add:* Dept Chem Simon Fraser Univ Burnaby BC V5A 1S6 Can

VOIGT, GARTH KENNETH, b Merrill, Wis, Jan 17, 23; m 46; c 3. SOILS, PLANT NUTRITION. *Educ:* Univ Wis, BS, 48, MS, 49, PhD(soils), 51. *Prof Exp:* From instr to asst prof soils, Univ Wis, 51-55; from asst prof to prof, 55-67, actg dean, Sch Forestry, 70- 71, 75-76 & 86-87, dir admis, 70-75, dir grad studies, Dept Forestry, 71-75, MARGARET K MUSSER PROF FOREST SOILS, YALE UNIV, 67- *Concurrent Pos:* Collabr, Lake States Forest Exp Sta, US Forest Serv, 54-60. *Mem:* AAAS; Soil Sci Soc Am; Am Soc Agron; Am Soc Plant Physiol. *Res:* Relationships between soil and the growth of plants. *Mailing Add:* 280 Kenwood Ave Mt Carmel CT 06518

VOIGT, GERD-HANNES, b Ger; m 75. MAGNETO-HYDRODYNAMICS, SPACE PHYSICS. *Educ:* Brunswick Tech Univ, Dipl, 70, PhD(physics), 75. *Prof Exp:* Sci asst teach & res, Tech Univ Brunsick, 70-75, Univ Darmstadt, 75-80; SR RES SCIENTIST, RICE UNIV, 80- *Concurrent Pos:* Mem magnetospheric working group, Int Asn Geomagnetism & Aeronomy, 78- *Mem:* Am Geophys Union; Ger Geophys Soc; Europ Geophys Soc. *Res:* Theory of planetary magnetospheres; development of computer codes for earth's magnetosphere; development of numerical methods for solving non linear equations in magneto-hydrodynamics; space plasma physics. *Mailing Add:* Space Phys & Astron Rice Univ Houston TX 77251

VOIGT, HERBERT FREDERICK, b New York, NY, Oct 27, 52; m 75; c 2. AUDITORY NEUROSCIENCE, NEUROPHYSIOLOGY. *Educ:* City Col New York, BE, 74; Johns Hopkins Univ, PhD(biomed eng), 80. *Prof Exp:* Fel, Neural Encoding Lab, 79-80, ASSOC PROF BIOMED ENG, COL ENG & ASSOC PROF OTOLARYNGOL, SCH MED, BOSTON UNIV, 81-

Concurrent Pos: Panelist, NSF fel prog, 85-87. *Mem:* Acoust Soc Am; AAAS; Inst Elec & Electronics Engrs; Int Brain Res Orgn; Soc Neurosci; World Fedn Neuroscientists. *Res:* Auditory neuroscience; physiology of the auditory nerve and cochlear nucleus; correlation of physiological responses with anatomical properties of neurons in cochlear nucleus; multi-unit recording and analysis; mathematical modeling of the neuronal circuitry of the cochlear nucleus. *Mailing Add:* Dept Biomed Eng Boston Univ 44 Cummington St Boston MA 02215

VOIGT, JOHN WILBUR, b Sullivan, Ind, July 6, 20; m 43; c 2. PLANT ECOLOGY. *Educ:* Univ Nebr, PhD, 50. *Prof Exp:* From asst prof to assoc prof, 50-60, dean gen studies div, 62-75, assoc dean col sci, 80-84, PROF BOT, SOUTHERN ILL UNIV, CARBONDALE, 60- *Concurrent Pos:* Dir, Biol Sci Progs, 84. *Mem:* Soc Range Mgt; Sigma Xi. *Res:* Geography of southern Illinois vascular plants; vegetation of southern Illinois; prairie and pasture research. *Mailing Add:* Dept Bot Southern Ill Univ Carbondale IL 62903

VOIGT, PAUL WARREN, b Ann Arbor, Mich, Mar 20, 40; m 63; c 3. PLANT BREEDING. *Educ:* Iowa State Univ, BS, 62; Univ Wis, MS, 64, PhD(agron), 67. *Prof Exp:* Res geneticist, Southern Great Plains Field Sta, 67-74, res geneticist, 74-80, SUPVR RES GENETICIST & RES LEADER, GRASSLAND, SOIL & WATER RES LAB, AGR RES SERV, USDA, 80- *Concurrent Pos:* Assoc ed, Crop Sci, 80-82, tech ed, 85-87. *Mem:* Fel Am Soc Agron; fel Crop Sci Soc Am; Soc Range Mgt; Am Forest & Grassland Coun. *Res:* Forage grass breeding and genetics. *Mailing Add:* Grassland Soil & Water Res Lab 808 E Blackland Rd Temple TX 76502

VOIGT, ROBERT GARY, b Olney, Ill, Dec 21, 39; m 62; c 2. NUMERICAL ANALYSIS. *Educ:* Wabash Col, BA, 61; Purdue Univ, West Lafayette, MS, 63; Univ Md, College Park, PhD(math), 69. *Prof Exp:* Res assoc, Comput Sci Dept, Univ Md, 69-70, vis asst prof, 70-71; mathematician, Naval Ship Res & Develop Ctr, Washington, DC, 71-73; asst dir, 73-83, ASSOC DIR, INST COMPUT APPLNS SCI & ENG, 83- *Mem:* Soc Indust & Appl Math; Asn Comput Mach; Am Math Soc; AAAS; Inst Elec & Electronics Engrs. *Res:* Numerical analysis for parallel and vector computers and the development of parallel computing systems. *Mailing Add:* ICASE MS-132 C NASA-Langley Res Ctr Hampton VA 23665

VOIGT, ROBERT LEE, b Hebron, Nebr, Nov 23, 24; m 51; c 4. PLANT BREEDING. *Educ:* Univ Nebr, BS, 49, MS, 55; Iowa State Univ, PhD(crop breeding), 59. *Prof Exp:* Instr soybeans, Iowa State Univ, 55-59; from asst prof & asst plant breeder to assoc prof & assoc plant breeder, 59-69, PROF PLANT SCI & PLANT BREEDER, AGR EXP STA, UNIV ARIZ, 69- *Concurrent Pos:* Ed, Sorghum Newslett, Sorghum Improvement Conf NAm, 72- *Mem:* Am Soc Agron; Crop Sci Soc Am; Nat Asn Cols & Teachers Agr; Coun Agr Sci & Technol. *Res:* Crop breeding; forage and grain sorghum; soybeans. *Mailing Add:* Dept Plant Sci Univ Ariz Tucson AZ 85721

VOIGT, WALTER, b Havana, Cuba, Feb 26, 38; US citizen; m 61; c 3. BIOCHEMISTRY, ENDOCRINOLOGY. *Educ:* Univ Villanueva, Cuba, MS, 60; Univ Miami, PhD(biochem), 68. *Prof Exp:* Res assoc skin biochem, 69-70, asst prof dermat, 70-74, ASSOC PROF PATHOL & ONCOL, SCH MED, UNIV MIAMI, 74- *Concurrent Pos:* Fel bile acid metab, Sch Med, Univ Miami, 68-69, Am Cancer Soc grant, 70-71, Nat Cancer Inst grant, 72-75. *Mem:* AAAS; Brit Biochem Soc; Am Chem Soc; Endocrine Soc; Am Fedn Clin Res; Sigma Xi. *Res:* Mechanism of androgen action and prostatic neoplasia; enzymes of bile acids and steroid metabolism; biochemistry of the skin; membrane electron transport; cancer. *Mailing Add:* Dept Pathol Univ Miami Sch Med PO Box 016960 Miami FL 33101

VOISARD, WALTER BRYAN, b Dayton, Ohio, Dec 14, 25; m 46; c 4. MECHANICAL ENGINEERING. *Educ:* Univ Cincinnati, BSME, 50. *Prof Exp:* Proj engr, Haines Designed Prod Co, 50-53, proj engr & chief eng, McCauley Industrial Corp, 53-60, chief engr, 60-82, Gen mgr, McCauley Assessory Div, Cessna Aircraft Co, 82-90; RETIRED. *Concurrent Pos:* Consult, 68- *Mem:* Soc Automotive Engrs; Am Inst Aeronaut & Astronaut; Soc Air Safety Invest. *Res:* Aircrcraft propellers, propeller deicing, governors, synchrophasers and aircraft wheels and brakes. *Mailing Add:* 1072 Grange Hall Rd Dayton OH 45340

VOIT, EBERHARD OTTO, b Dortmund, WGer, Feb 8, 53; m 80; c 3. MATHEMATICAL MODELLING, NONLINEAR SYSTEMS ANALYSIS. *Educ:* Univ Koln. *Prof Exp:* Res asst, Dept Ecol, Univ Koln; asst, Univ Mich, 81-82, sr res assoc, 85, asst res scientist, Dept Microbiol & Immunol, 85-86; ASSOC PROF BIOMET, MED UNIV SC, 86- *Mem:* NY Acad Sci; Soc Indust & Appl Math; Soc Math Biol; Sigma Xi. *Res:* Nonlinear formalism for description, analysis and understanding of organizationally complex systems; biomedical phenomena. *Mailing Add:* Dept Biomet Med Univ SC 909 Summerall Ctr Charleston SC 29425-2501

VOITLE, ROBERT ALLEN, b Parkersburg, WVa, May 12, 38; m 75; c 4. POULTRY PHYSIOLOGY. *Educ:* Univ WVa, BS, 62, MS, 64; Univ Tenn, PhD(physiol), 69. *Prof Exp:* From asst prof physiol to assoc prof, Univ Fla & from asst poultry physiologist to assoc poultry physiologist, 69-79; prof physiol & head, Dept Poultry Indust, Calif Polytech State Univ, 79-81; ASSOC DEAN, COL AGR, AUBURN UNIV, 81- *Concurrent Pos:* Dir, mgr & pres, Alachia County Fair Asn, 69-79. *Mem:* Poultry Sci Asn; Sigma Xi. *Res:* Environmental and reproductive physiology with special emphasis on the effect of nutrition and photoperiod; breeding and genetics, especially radiation effects. *Mailing Add:* Sch Agr Forestry & Biol Sci Auburn Univ Auburn AL 36849

VOJNOVICH, THEODORE, b Weirton, WVa, Oct 7, 32; m 58; c 2. CERAMIC ENGINEERING. *Educ:* Iowa State Univ, BS, 59, MS, 61, PhD(ceramic eng), 67. *Prof Exp:* Mgt trainee, Weirton Steel Co Div, Nat Steel Corp, 59-60; planning engr, Western Elec Co, 61-64; SR ENGR, RES LABS, WESTINGHOUSE ELEC CO, 67- *Mem:* Am Ceramic Soc; Nat Inst Ceramic Engrs; Am Soc Metals. *Res:* Electronic and magnetic properties of materials; high temperature properties of oxides and alloys; materials processing including sintering, chemical vapor deposition and sputtering. *Mailing Add:* 500 Colony Ct NW Vienna VA 22180

VOKES, EMILY HOSKINS, b Monroe, La, May 21, 30; m 59. INVERTEBRATE PALEONTOLOGY, MALACOLOGY. *Educ:* Tulane Univ, La, BS, 60, MS, 62, PhD(paleont), 67. *Prof Exp:* Cur paleont, Dept Geol, 57-74, assoc prof, 73-81, chmn dept, 74-82, PROF GEOL, TULANE UNIV, 81- *Concurrent Pos:* Lectr geog, Tulane Univ, 69-, assoc ed, Tulane Studies Geol & Paleont, 70-; vis prof, Univ Rio Grande do Sul, Brazil, 71; vis cur, Australian Mus, 80; actg dean, Newcomb Col, Tulane Univ, 87-88. *Mem:* Am Malacol Union; Paleont Soc; Paleont Res Inst; Sigma Xi. *Res:* Systematic paleontology and zoology of Cenozoic Gastropoda, including both fossil and recent members. *Mailing Add:* Dept Geol Tulane Univ New Orleans LA 70118-5698

VOKES, HAROLD ERNEST, b Windsor, Ont, June 27, 08; nat US; m 32, 59; c 4. STRATIGRAPHY, INVERTEBRATE PALEONTOLOGY. *Educ:* Occidental Col, BA, 31; Univ Calif, PhD(paleont), 35. *Prof Exp:* Hon fel paleont, Yale Univ, 35-36; asst geologist, State Geol Surv, Ill, 37; from asst cur to assoc cur invert paleont, Am Mus Natural Hist, 37-43, actg chmn dept invert, 43; geologist, US Geol Surv, 43-45; from assoc prof to prof geol, Johns Hopkins Univ, 45-56; prof, 56-72, chmn dept, 57-67 & 70-71, W R Irby prof, 72-77, EMER PROF GEOL, TULANE UNIV, 77- *Concurrent Pos:* Guggenheim fel, Am Univ Beirut, 40; geologist, US Geol Surv, PI, 52-53; vis prof, Univ Rio Grande do Sul, Brazil, 71; mem, Int Comn Zool Nomenclature, 44-83; trustee, Paleont Res Inst, 72-77, pres, 76-77. *Mem:* Fel Geol Soc Am (vpres, 52); Am Asn Petrol Geologists; Paleont Soc (secy, 40-49, pres, 51); Am Malacol Union; Sigma Xi. *Res:* Cretaceous and Tertiary stratigraphy and molluscan paleontology; fossil and recent pelecypoda. *Mailing Add:* Dept Geol Tulane Univ New Orleans LA 70118

VOLANTE, RALPH PAUL, b Nelson, Pa, Aug 10, 49; m 75. ORGANIC CHEMISTRY. *Educ:* Pa State Univ, BS, 71; Harvard Univ, MA, 73, PhD(org chem), 76. *Prof Exp:* Fel org chem, Cornell Univ, 76-77; SR RES CHEMIST ORG CHEM, MERCK & CO, 77- *Mem:* Am Chem Soc; Sigma Xi. *Res:* Synthetic organic chemistry, reaction processes and mechanisms; new synthetic methods. *Mailing Add:* 22 Hawthorne Lane East Windsor NJ 08520-2220

VOLAVKA, JAN, b Prague, Czech, Dec 29, 34; m 86; c 1. PSYCHIATRY, ELECTROPHYSIOLOGY. *Educ:* Charles Univ, Prague, BA & MD, 59; Czech Acad Sci, PhD(med sci), 65. *Prof Exp:* Intern internal med, Psychiat Hosp, Horni Berkovice, Czech, 59-60, resident psychiat, 60-63; resident psychiatrist, Psychiat Res Inst, Prague, 63-66; electroencephalographer, London Hosp, Eng, 66-67; resident psychiatrist, Psychiat Res Inst, Prague, 67-68; fel neurophysiol, Max Planck Inst Psychiat, 68-69; asst prof, New York Med Col, 69-73, assoc prof psychiat, 73-76; prof psychiat, Inst Psychiat, Univ Mo, 76-79. *Concurrent Pos:* Prin investr, NIMH grant, 87-90. *Mem:* Soc Biol Psychiat; Am EEG Soc. *Res:* Psychopharmacology; EEG. *Mailing Add:* Nathan S Kline Inst Psych Res Orangeburg NY 10962

VOLBORTH, ALEXIS, b Viipuri, Finland, July 11, 24; nat US; m 47; c 7. MINERALOGY, GEOLOGICAL ENGINEERING. *Educ:* Univ Helsinki, PhC, 50, PhLic & PhD(geol, mineral), 54. *Prof Exp:* Res asst, Geol Surv, Finland, 50, field asst, 52; asst, Inst Technol, Finland, 50-51; field geologist, Finnish Mineral Co, 53; sr asst geol, Univ Helsinki, 53-54; traveling res fel, Outokumpu Found, Univ Vienna, Univ Heidelbera, 54-55, Calif Inst Technol, Hoover fel, 55-56; from asst mineralogist to mineralogist, Nev Mining Anal Lab, Univ Nev, Reno, 56-68, res assoc & consult, Desert Res Inst, 61-62, assoc prof, Univ, 63, prof, 64-68, mem radioactiv safety bd, 64-66; Killam vis prof geol, Dalhousie Univ, 68-71, Killam res prof, 71-72; vis prof, Lunar Sci Inst, NASA, Univ Houston, 72-73; vis res chemist, Univ Calif, Irvine, 73-76; prof geol & chem, NDak State Univ, 75-78; prof geol & scientist, Nuclear Radiation Ctr, Wash State Univ, 78-79; dir accelerator lab, 83-86, sr radiation safety officer, 83-86, PROF GEOCHEM & CHEM, MONT TECH, BUTTE, 79-, PROF GEOL ENG, 87- *Concurrent Pos:* Consult, first seimic undergound atomic test, US AEC, Proj SHOAL, 61-63, geochem King Abdul Aziz Univ, Jeddah, Saudi Arabia, 75; Australian Acad Sci sr fel, 65; J S Guggenheim Mem Found fel, 65; adj prof geol, Mackay Sch Mines, Univ Nev, Reno, 69-73; prin investr, Stoichiometry Study of Lunar Rocks, NASA, 72-73; consult, US AEC, 61-63, NASA, 65-73, Anaconda Co, 68 & Johns Manville Corp, Chevron 80-83, Pegasus Gold Inc, 87-; US Rep, Second Conf Natural Reactors, OKLO Phenomenon, Int Atomic Energy Agency, Paris, 77, rep, Int Geol Correlation Prog, Prog 315, 90-; interpreter & translr, Soviet Siberia, Major US & Can Mining Co, 90-; Placer Dome Inc, Echo Bay, Inc, 90. *Honors & Awards:* White Cross, Finnish Chem Soc, 55. *Mem:* Fel Mineral Soc Am; fel Am Inst Chemists; Am Chem Soc; Am Nuclear Soc; Soc Econ Geologists; Int Precious Metals Inst. *Res:* Geochemistry and analytical chemistry of complex systems, mainly nondestructive instrumental neutron activation and x-ray, fluorescence analysis of major and trace elements; oxygen stoichiometry in rocks, minerals, chemicals and industrial products; mineralogy of and deficiency of oxygen in lunar rocks and fines; nondestructive analysis of coal and lignite; mineralogy of platinum group elements; gold deposits; tin deposits. *Mailing Add:* Dept Chem & Geochem Mont Tech Butte MT 59701

VOLBRECHT, STANLEY GORDON, b Lodi, Calif, Sept 12, 23; m 45; c 4. MINING GEOLOGY. *Educ:* Col of the Pac, BA, 53; Stanford Univ, MS, 62. *Prof Exp:* Explor geologist, Am Copper Co, 54-55; instr geol, Stockton Col, 56-61; from asst prof to assoc prof, 61-70, PROF GEOL, UNIV OF THE PAC, 70-, CHMN DEPT GEOL & GEOG, 66- *Mem:* Nat Asn Geol Teachers; Geol Soc Am. *Res:* Economics. *Mailing Add:* Dept Geol & Geog Univ of the Pac 3601 Pacific Ave Stockton CA 95211

VOLCANI, BENJAMIN ELAZARI, b Ben-Shemen, Israel, Jan 4, 15 ; US citizen; m 48; c 1. MICROBIOLOGY, BIOCHEMISTRY. *Educ:* Hebrew Univ, MSc, 36, PhD, 41. *Prof Exp:* Vis scientist microbiol, Inst Tech, Delft Univ, 37-38 & chem, State Univ Utrecht, 38-39; mem staff, Sieff Res Inst, Weizmann Inst, 39-58; prof microbiol, Univ Calif, 59-85, EMER PROF SCRIPPS INST OCEANOG, 85- *Concurrent Pos:* Res fel, Univ Calif, Berkeley, 45-46, res assoc biochem, 56-59; res fel microbiol, Hopkins Marine Sta, Stanford Univ, 46-47 & Calif Inst Technol, 47; res fel biochem, Univ Wis, 48; res assoc, Pasteur Inst, Paris, 51; vis prof, Univ Col, Welsh Nat Sch Med, Cardiff, Wales, UK, 73-74. *Mem:* Am Soc Cell Biol; Am Soc Microbiol; Soc Gen Microbiol; AAAS. *Res:* Microbial metabolism and ecology; antimetabolites; bacterial pigments; halophilic microorganisms; biochemistry and ultra-fine structure of the diatoms; siliceous organisms and dinoflagellates; silicon metabolism; mineralization in biological systems; role of silicon in life processes and pathogenicity; molecular biology of silicon. *Mailing Add:* Scripps Inst Oceanog Univ Calif La Jolla CA 92093

VOLD, BARBARA SCHNEIDER, b Oakland, Calif, Jan 3, 42. BIOCHEMISTRY, MOLECULAR BIOLOGY. *Educ:* Univ Calif, Berkeley, BA, 63; Univ Ill, MS, 64, PhD(cell biol), 67. *Prof Exp:* NIH fel biol, Mass Inst Technol, 67-69; assoc microbiol, Scripps Clin & Res Found, 69-76; mem staff Biomed Res Dept, 77-85, MEM, MOLECULAR BIOL DEPT, SRI INT, 85- *Concurrent Pos:* Nat Inst Gen Med Sci career develop award, 71-76; consult, Physiol Chem Study Sect, NIH, 73-77 & NSF Study Sect, 88; mem ed bd, J Biol Chem, 86- *Mem:* Am Soc Microbiol; Am Soc Biochem & Molecular Biol. *Res:* Structure and function of transfer ribonucleic acids; changes in nucleic acids during development; antibodies to modified nucleosides. *Mailing Add:* Syva Co 2-218 PO Box 10058 Palo Alto CA 94303-0847

VOLD, CARL LEROY, b McVille, NDak, Dec 9, 32; m 61; c 2. SOLID STATE PHYSICS, METALLURGY. *Educ:* Concordia Col, Moorhead, Minn, BA, 54; Iowa State Univ, MS, 59. *Prof Exp:* Jr scientist, Ames Lab, Iowa State Univ, 54-56; RES PHYSICIST, US NAVAL RES LAB, 59- *Mem:* Am Soc Metals; Am Crystallog Asn; Sigma Xi. *Res:* Application of x-ray diffraction techniques to the study of defects in metals; energetics of solid-liquid and solid-solid interfaces in pure metals; crystalline anisotrophy; dislocation energetics. *Mailing Add:* Code 6320 Mat Sci & Technol Div Naval Res Lab Washington DC 20375

VOLD, MARJORIE JEAN, b Ottawa, Ont, Oct 25, 13; US citizen; wid; c 3. COLLOID CHEMISTRY. *Educ:* Univ Calif, BS, 34, PhD(chem), 36. *Prof Exp:* Jr res assoc chem, Stanford Univ, 37-41; res assoc & lectr, 41-58, adj prof, 58-73, EMER PROF CHEM, UNIV SOUTHERN CALIF, 73- *Concurrent Pos:* Res chemist, Union Oil Co Calif, 42-46; Guggenheim Mem fel, State Univ Utrecht, 53-54. *Honors & Awards:* Garvan Medal, Am Chem Soc, 67. *Mem:* Am Chem Soc; Royal Soc Chem; Int Asn Colloid-Interface Scientists; AAAS; Sigma Xi. *Res:* Association colloids; mesomorphic phases; gels and other colloidal solids; stability of emulsions, foams, films and suspensions; adsorption; rheology; computer simulation of colloidal processes. *Mailing Add:* 15632 Pomcrado Rd Poway CA 92064-2456

VOLD, REGITZE ROSENORN, b Copenhagen, Denmark, July 2, 37; US citizen; m 72. NUCLEAR MAGNETIC RESONANCE, MOLECULAR DYNAMICS. *Educ:* Tech Univ Denmark, MS, 60, PhD(org chem), 62. *Prof Exp:* Lectr org chem, Tech Univ Denmark, 62; fel chem, Univ NMex, 62-64; staff fel magnetic resonance, NIH, 65-71; res chemist, 71-82, PROF, UNIV CALIF, SAN DIEGO, 82- *Concurrent Pos:* Fel, NIH, 68-69; guest worker, Nat Bur Standards, 69-71; lectr chem, Univ Calif, San Diego, 72-82; mem bd trustees, Exp Nuclear Magnetic Resonance Conf, 74-78 & 81-84, treas, 75-78, chairwoman, 83; assoc ed, J Am Chem Soc, 77-79. *Mem:* Am Chem Soc; fel AAAS. *Res:* Nuclear magnetic resonance as used in study of molecular dynamics in condensed phases. *Mailing Add:* Dept Chem Univ Calif San Diego 9500 Gilman Dr La Jolla CA 92093-0342

VOLD, ROBERT LAWRENCE, b Los Angeles, Calif, Sept 20, 42; m 71. CHEMICAL PHYSICS. *Educ:* Univ Calif, Berkeley, BS, 63; Univ Ill, Urbana, MS, 65, PhD(chem), 66. *Prof Exp:* From asst prof to assoc prof, 68-80, PROF CHEM, UNIV CALIF, SAN DIEGO, 80- *Concurrent Pos:* A P Sloan fel, 72-74. *Mem:* Am Inst Physics; Am Chem Soc. *Res:* Nuclear magnetic resonance; relaxation mechanisms; theory and applications of pulsed nuclear magnetic resonance techniques. *Mailing Add:* Dept Physics Col William & Mary Williamsburg VA 23185

VOLDENG, ALBERT NELSON, medicinal chemistry; deceased, see previous edition for last biography

VOLESKY, BOHUMIL, b Prague, Czech, Oct 29, 39; m 67. BIOCHEMICAL ENGINEERING, WATER POLLUTION CONTROL. *Educ:* Prague Tech Univ, MESc, 62; Univ Western Ont, PhD(biochem eng), 71. *Prof Exp:* Proj engr, Cent Res Inst Food Indust, 63-66; res assoc food eng, Prague Tech Univ, 66-67; res asst biochem eng, Univ Western Ont, 67-70, teaching & res fel, Fanshawe Col Arts & Technol, 70-72, lectr, 72-73; ASSOC PROF BIOCHEM ENG, MCGILL UNIV, 73- *Concurrent Pos:* Consult, var assignments, 70-; vis prof chem eng, Univ PR, Mayaquez, 81, Ecole Polytech Fed, ETH, Lausanne, Switz, 88-89; exchange scientist, Acad Sci, Prague, Czech, 81, 83 & 87, Nankai Univ, Tianjin, China, 88, Ecole Polytech, Lorraine, Nancy, France, 88, Swiss Nat Sci Found, 88-89; pres, BV Sorbex, Inc, 88- *Mem:* Am Chem Soc; Chem Inst Can; Can Soc Chem Engrs; Eng Inst Can. *Res:* Biotechnology; fermentation process engineering and optimization; microbial product/process development; biosorbent detoxification of industrial effluents; biosorbent recovery of nuclear fuel and metallic elements; development of new biosorbent materials; industrial water pollution control; environmental studies. *Mailing Add:* Dept Chem Eng McGill Univ 3480 Univ St Montreal PQ H3A 2A7 Can

VOLGENAU, LEWIS, b Buffalo, NY, Nov 30, 40; m 69; c 2. PAPER CHEMISTRY. *Educ:* Syracuse Univ, BE, 61; Inst Paper Chem, MS, 65, PhD(paper chem), 69. *Prof Exp:* Engr res & develop, Riegel Paper, 61-63; engr pulp & paper, Champion Int, 69-72; MGR RES & DEVELOP, BETZ LABS, INC, 72- *Res:* Corrosion, scale and foam control in aqueous systems; development of biocides and specialty chemical formulations. *Mailing Add:* 9669 Grogans Mill Rd PO Box 4300 The Woodlands TX 77380

VOLICER, LADISLAV, b Prague, Czech, May 21, 35; m 72; c 5. PHARMACOLOGY, AGING. *Educ:* Charles Univ, Prague, MD, 59; Czech Acad Sci, PhD(pharmacol), 64. *Prof Exp:* Resident med, Hosp Jindr Hradec, 59-61; instr pharmacol, Sch Pediat, Charles Univ, Prague, 61-65; vis assoc, Nat Heart Inst, Md, 65-66; res assoc & lectr, Inst Pharmacol, Czech Acad Sci, 66-68; res asst prof, Sch Med, Univ Munich, 68-69; from asst prof to assoc prof, 69-77, PROF PHARMACOL, SCH MED & GRAD SCH, BOSTON UNIV, 77-, ASST PROF MED, 75-, PROF PSYCHIAT, 85- *Concurrent Pos:* Asst vis physician, Boston City Hosp, 75-86; clin pharmacologist, Geriat Res Educ Clin Ctr, Vet Admin Hosp, Bedford, 80-86, dep dir, 86- *Mem:* Soc Neurosci; Am Soc Pharmacol & Exp Therapeut; Geront Soc. *Res:* Pharmacology of aging; Alzheimer's disease; neurochemical and clinical research. *Mailing Add:* Dept Pharmacol Boston Univ Boston MA 02118

VOLIN, RAYMOND BRADFORD, b Kalispell, Mont, May 22, 43; m 64; c 1. HORTICULTURE, AGRONOMY. *Educ:* Mont State Univ, BS, 66, MS, 68, PhD(plant path), 71. *Prof Exp:* Trainee agron, Mont State Univ, 66-68, res asst plant path, 68-71; from asst prof to assoc prof, Agr Res & Educ Ctr, Univ Fla, 71-84; RES SCIENTIST/PLANT BREEDER, ROGERS NK SEED CO, 84- *Concurrent Pos:* Vis scientist, Int Inst Trop Agr, Nigeria, WAfrica; Sabbitic fel, 79-80; rural credit assoc, Queensland, Australia, 80; attend, NSF conf on underexploited crops, Taiwan & Japan, 82. *Mem:* Am Soc Hort Sci; Am Phytopath Soc; Tomato Genetics Coop. *Res:* Genetic improvement of field and vegetable crops; physiological relationship and genetic interaction between plant hosts and plant disease organisms. *Mailing Add:* 10270 Greenway Rd Naples FL 33961

VOLK, BOB G, b Auburn, Ala, July 13, 43; m 66; c 4. SOIL CHEMISTRY. *Educ:* Ohio State Univ, BS, 65, MS, 67; Mich State Univ, PhD(soil sci), 70. *Prof Exp:* Asst prof soil sci, Agr Res & Educ Ctr, Belle Glade, Fla, 70-73; from asst prof to assoc prof, Univ Fla, 73-84; Dept Agron, Univ Mo, Columbia, 85-90; WATER CTR, UNIV NE, 90- *Concurrent Pos:* Vis prof, Environ Protection Agency, Ore State Univ, Corvallis, 76-77. *Mem:* Am Soc Agron; Soil Sci Soc Am; Int Soil Sci Soc; Int Humic Acid Soc. *Res:* Chemistry of soil organic matter and the movement of nutrients through the soil profile; effects of acid rain on soils. *Mailing Add:* Water Ctr 103 Natural Resources Hall Univ NE Lincoln NE 68583-0844

VOLK, BRUNO W, b Vienna, Austria, 09; m; c 1. DIABETES, TAY-SACHS DISEASE. *Educ:* Univ Vienna, Austria, BS, 28, MD, 34. *Prof Exp:* Instr path, Univ Vienna, Austria, 34-38, Northwestern Univ, Chicago, 40-44; res fel, Hektoen Inst, Chicago, 46-49; vis assoc prof, Albert Einstein Col Med, NY, 56-60; from clin assoc prof to prof, Downstate Med Ctr, State Univ NY, 60-77; prof-in-residence, 77-82, EMER PROF-IN-RESIDENCE PATH, UNIV CALIF, IRVINE, 82- *Concurrent Pos:* Past pres, NY Path Soc. *Mem:* Am Diabetes Asn; Endocrine Soc; AAAS; Am Asn Path; Am Soc Clin Path; Europ Asn Study Diabetes. *Res:* Endocrine pancreas and morphology of insulin secretion; lipidoses, primarily Tay-Sachs Disease, involving the brain and genetics; enzyme chemical and biochemical studies in lipidoses. *Mailing Add:* 2324 S Beverly Glen Blvd Los Angeles CA 90064

VOLK, MURRAY EDWARD, b Cleveland, Ohio, Aug 23, 22; m 49; c 3. ORGANIC CHEMISTRY, RADIOCHEMISTRY. *Educ:* Oberlin Col, BA, 43; Univ Chicago, MS, 48; Temple Univ, PhD(chem), 53. *Prof Exp:* Assoc chemist, Nuclear Instrument & Chem Corp, 53-55; pres, Volk Radiochem Co, 55-65; mkt mgr res prod, Miles Labs, 66-69; PRES, ISOLAB INC, 69- *Mem:* AAAS; Am Chem Soc; Am Asn Clin Chem; Soc Nuclear Med; Sigma Xi. *Res:* Application of isotopes to biological and chemical research; preparation of radioactive pharmaceuticals; liquid chromatography applied to clinical diagnostic problems. *Mailing Add:* Isolab Inc Drawer 4350 Akron OH 44321-0350

VOLK, RICHARD JAMES, b Tela, Honduras, Nov 5, 28; m 51; c 3. PLANT NUTRITION. *Educ:* Purdue Univ, BS, 50, MS, 51; NC State Univ, PhD(soil chem), 54. *Prof Exp:* Res specialist, Crops Div, Biol Warfare Labs, Ft Detrick, Md, 54-56; from asst prof to assoc prof, 56-66, PROF SOIL SCI, NC STATE UNIV, 66- *Concurrent Pos:* Grants, NSF, 62-64, 78-80 & 82-84, Am Potash Inst, 63-66; res contract, USDA, 65-68, Westraco, 82-85 & Enichem Americas, Inc, 88-90; vis scientist, Australian Nat Univ, 81, Nat Inst Agr Res, France, 88. *Honors & Awards:* Co-recipient Campbell Award, Am Inst Biol Sci, 65. *Mem:* Am Soc Plant Physiol; Soil Sci Soc Am; Crop Sci Soc Am; fel Am Soc Agron. *Res:* Application of mass spectrometry and stable isotopes to plant nutrition and biochemistry; absorption and metabolism of ammonium and nitrate nitrogen by plants; regulatory role of mineral nutrition in photosynthesis and respiration. *Mailing Add:* Dept Soil Sci NC State Univ Box 7619 Raleigh NC 27695

VOLK, THOMAS LEWIS, b Dayton, Ohio, Nov 4, 33; m 61; c 4. HUMAN PATHOLOGY, ENDOCRINOLOGY. *Educ:* Univ Dayton, BS, 55; Marquette Univ, MD, 59. *Prof Exp:* Instr path, Ohio State Univ, 65-66; asst prof, Univ Kans, 66-67; asst prof path, Univ Calif, Davis, 68-72; DIR LABS, KAWEAH DELTA DIST HOSP, 72- *Concurrent Pos:* NIH path training grant, 63-65; Am Cancer Soc adv clin fel, 65-66; consult, Vet Admin Hosp, Kansas City, Mo, 67-68 & Sacramento County Hosp, Calif, 68-72. *Mem:* AMA; Int Acad Path; Am Asn Path & Bact. *Res:* Ultrastructural-functional relationships of steroidogenesis in the adrenal cortex and placenta, ovary and testis; ultrastructural changes in the adrenal cortex and placenta, produced by drugs inhibiting steroidogenesis. *Mailing Add:* Kaweah Delta Dist Hosp 400 W Mineral King Visalia CA 93291

VOLK, VERIL VAN, b Montgomery, Ala, Nov 18, 38; m 68; c 2. SOIL CHEMISTRY. *Educ:* Ohio State Univ, BS, 60, MS, 61; Univ Wis, PhD(soils), 66. *Prof Exp:* Proj assoc soils, Univ Wis, 66; from asst prof to assoc prof, 66-80, PROF SOILS, ORE STATE UNIV, 80-, ASSOC DIR, ORE AGR EXP STA. *Concurrent Pos:* Coop State Res Serv, Washington, DC, 84-85. *Mem:* Am Soc Agron; Soil Sci Soc; Coun Agr Sci & Technol. *Res:* Ion exchange and soil acidity interactions; waste utilization on soils. *Mailing Add:* 2280 NW Huntington Dr Corvallis OR 97330

VOLK, WESLEY AARON, b Mankato, Minn, Nov 23, 24; m 45; c 2. MICROBIOLOGY. *Educ:* Univ Wash, BS & BS(food technol), 48, MS, 49, PhD, 51. *Prof Exp:* From asst prof to assoc prof, 51-64, PROF MICROBIOL, SCH MED, UNIV VA, 64- *Concurrent Pos:* NIH spec fel, 62-63; spec fel, Max Planck Inst Immunobiol, 69-70; fel, Pasteur Inst, 83-84. *Mem:* Am Soc Microbiol; Am Soc Biol Chemists. *Res:* Carbohydrate metabolism; monoclonal antibodies; virus neutralization; tetanus toxin. *Mailing Add:* Dept Microbiol Box 441 Univ Va Sch Med Charlottesville VA 22908

VOLKAN, VAMIK, b Nicosia, Cyprus, Dec 13, 32; US citizen; c 4. PSYCHIATRY, PSYCHOANALYSIS. *Educ:* Univ Ankara, MD, 56; Wash Psychoanal Inst, grad, 71; bd cert psychoanal, 73. *Prof Exp:* Staff physician, NC State Hosp, 61-63; from instr to assoc prof, 63-72, PROF PSYCHIAT, SCH MED, 72-, DIR, CTR PSYCHO-POLIT STUDIES, UNIV VA, 86- *Concurrent Pos:* Vis prof psychiat, Univ Ankara, Turkey, 74-75; dir Blue Ridge Hosp Div, Univ Va, 78- & dir Div Psychoanalytic Studies, 79- *Mem:* Am Psychoanal Asn; fel Am Psychiat Asn; Int Psychoanal Asn; Int Soc Polit Psychol (pres, 83-84); Group Adv Psychiat. *Res:* Psychotherapy of schizophrenia; pathological grief reactions; psycho-history; psycho-politics. *Mailing Add:* Dept Psychiat Univ Va Sch Med Box 395 Med Ctr Charlottesville VA 22908

VOLKER, EUGENE JENO, b Sopron, Hungary, May 13, 42; US citizen; m 79; c 2. ORGANIC CHEMISTRY. *Educ:* Univ Md, BS, 64; Mass Inst Technol, MS, 67; Univ Del, PhD(chem), 70. *Prof Exp:* From asst prof to assoc prof, 69-79, PROF CHEM, SHEPHERD COL, WVA, 79- *Concurrent Pos:* Vis sci, Frederick Cancer Res Facil, 86-90. *Mem:* Am Chem Soc. *Res:* Chemical education; coal chemistry; peptide chemistry. *Mailing Add:* Dept Chem Shepherd Col Shepherdstown WV 25443

VOLKER, JOSEPH FRANCIS, biochemistry, dentistry; deceased, see previous edition for last biography

VOLKERT, WYNN ARTHUR, b St Louis, Mo, Apr 6, 41; m 67; c 3. RADIOCHEMISTRY, RADIOBIOLOGY. *Educ:* St Louis Univ, BS, 63; Univ Mo, Columbia, PhD(chem), 68. *Prof Exp:* NASA fel, 67-69, from asst prof to assoc prof, 69-80, PROF RADIOL SCI & BIOCHEM, UNIV MO, COLUMBIA, 80- *Concurrent Pos:* NIH grant, Univ Mo-Columbia, 73-; consult, Vet Admin Hosp, Columbia, 73- *Mem:* Radiation Res Soc; Biophys Soc; Soc Exp Biol & Med; Am Chem Soc; Soc Nuclear Med; Radiophys Sci Coun (pres-elect, 87, pres, 88). *Res:* Radiopharmaceutical chemistry; chemistry of Tc-99m nitrogen ligands; radionuclidic therapy. *Mailing Add:* Dept Radiol Univ Mo Med Ctr Columbia MO 65212

VOLKIN, ELLIOT, b Mt Pleasant, Pa, Apr 23, 19; m 47; c 2. BIOCHEMISTRY. *Educ:* Pa State Col, BS, 42; Duke Univ, MA, 45, PhD(biochem), 47. *Prof Exp:* Res assoc biochem, Duke Univ, 47-48; sci dir biochem, Oak Ridge Nat Lab, 48-65, sr res scientist, 65-84; CONSULT, 85- *Concurrent Pos:* Prof, Univ Tenn, 77-84. *Mem:* Fel AAAS; Am Chem Soc; Am Soc Biol Chem; Am Soc Microbiol; Sigma Xi. *Res:* Biochemical and biophysical studies of nucleic acids and nucleoproteins. *Mailing Add:* 899 W Outer Dr Oak Ridge TN 37830

VOLKMAN, ALVIN, b Brooklyn, NY, June 10, 26; m 47, 67, 73; c 7. PATHOLOGY, IMMUNOLOGY. *Educ:* Union Col, BS, 47; Univ Buffalo, MD, 51; Oxford Univ, DPhil, 63. *Prof Exp:* Asst prof path, Columbia Univ, 60-66; from asst mem to assoc mem, Trudeau Inst, 66-77; actg chmn, 89-90, PROF PATH, SCH MED, ECAROLINA UNIV, 77-, ASSOC DEAN RES & GRAD STUDIES, 90- *Concurrent Pos:* Arthritis & Rheumatism Found fel, 52-54; resident to sr resident path, Peter Bent Brigham Hosp; Am Cancer Soc scholar, 61-62; adj assoc prof path, Sch Med, NY Univ, 69-79; mem, IMS study sect, NIH, 75-79, chmn, 77-79. *Mem:* Am Asn Immunol; AAAS; Am Soc Hemat; Am Soc Microbiologists; NY Acad Sci; Reticuloendothelial Soc; Am Asn Pathol; Sigma Xi. *Res:* Origin and differentiation of mononuclear phagocytes. *Mailing Add:* Off Res & Grad Studies ECarolina Univ Sch Med Greenville NC 27858-4354

VOLKMANN, FRANCES COOPER, b Harlingen, Tex, May 4, 35; m 58; c 2. EXPERIMENTAL PSYCHOLOGY, NEUROSCIENCE. *Educ:* Mt Holyoke Col, AB, 57; Brown Univ, MA, 59, PhD(psychol), 61. *Hon Degrees:* DSc, Mt Holyoke Col, 87. *Prof Exp:* USPHS fel, Brown Univ, 61-62; res assoc psychol, Mt Holyoke Col, 64-65; lectr, Smith Col, 66-67, from asst prof to assoc prof, 67-78, dean fac, 83-88, PROF PSYCHOL, SMITH COL, 78- *Concurrent Pos:* Lectr, Univ Mass, 64-65; vis assoc prof, Brown Univ, 74, vis prof, 78-82; NSF res grant, Smith Col & Brown Univ, 74-78; Nat Eye Inst res grant, Brown Univ, 78-82; actg pres, Smith Col, 91. *Mem:* Fel Am Psychol Asn; fel AAAS; fel Optical Soc Am; Soc Neurosci; Asn Res Vision & Ophthal; Psychonomic Soc. *Res:* Vision and visual perception; vision during eye movements; visual development. *Mailing Add:* Col Hall 27 Smith Col Northampton MA 01063

VOLKMANN, KEITH ROBERT, b Milwaukee, Wis, May 30, 42; m 65; c 1. MICROBIOLOGY. *Educ:* Univ Rochester, BA, 65, PhD(microbiol), 73; Ga Sch Dent, DMD, 76. *Prof Exp:* Res asst virol, Med Ctr, Univ Rochester, 68-73; ASST PROF ORAL BIOL, MED COL, GA SCH DENT, 76- *Mem:* Int Asn Dent Res. *Res:* Penetration of oral antigens into oral mucosa and connective tissue; effect of endotoxins on the human periodontium. *Mailing Add:* Dept Oral Biol/Microbiol Med Col 1120 15th St Augusta GA 30912

VOLKMANN, ROBERT ALFRED, b Pittsburgh, Pa, Aug 21, 45; m 79; c 1. ORGANIC CHEMISTRY. *Educ:* Lafayette Col, AB, 67; Univ Pittsburgh, PhD(org chem), 72. *Prof Exp:* Res assoc, Stanford Univ, 72-74; RES SCIENTIST ORG CHEM, PFIZER INC, 74- *Concurrent Pos:* Instr, Conn Col, 76- *Mem:* Am Chem Soc. *Res:* Development of novel methodology for the design and or synthesis of biologically active molecules. *Mailing Add:* 135 Dogwood Lane Mystic CT 06355-1040

VOLKOFF, GEORGE MICHAEL, b Moscow, Russia, Feb 23, 14; Can Citizen; m 40; c 3. THEORETICAL PHYSICS. *Educ:* Univ BC, BA, 34, MA, 36; Univ Calif, PhD(theoret physics), 40. *Hon Degrees:* DSc, Univ BC, 45. *Prof Exp:* Asst prof physics, Univ BC, 40-43; assoc res physicist, Montreal Lab, Nat Res Coun, Can, 43-45, res physicist & head theoret physics br, Atomic Energy Proj, Que & Ont, 45-46; head, Dept Physics, Univ BC, 61-72, prof, 46-79, dean, 72-79, EMER DEAN, UNIV BC, 79- *Concurrent Pos:* Ed, Can J Physics, 50-56; mem, Nat Res Coun Can, 69-75; ed, Soviet Physics Usp, 79-; mem, Tech Adv Comt, Nuclear Fuel Waste Mgt Progr, Atomic Energy Can Ltd, 79-89. *Honors & Awards:* Centennial Medal Can, 67. *Mem:* Fel AAAS; fel Am Phys Soc; Am Asn Physics Teachers; fel Royal Soc Can; Can Asn Physicists (vpres, 61-62, pres, 62-63). *Res:* Theoretical nuclear physics; neutron diffusion; nuclear magnetic and quadrupole resonance. *Mailing Add:* 1776 Western Parkway Vancouver BC V6T 1V3 Can

VOLKOV, ANATOLE BORIS, b San Francisco, Calif, Oct 29, 24; m 50; c 2. NUCLEAR PHYSICS. *Educ:* Univ NC, BS, 48; Univ Wis, MS, 50, PhD(physics), 53. *Prof Exp:* Longwood fel, Univ Del, 53-55; asst prof physics, Univ Miami, 58-59; sr lectr, Israel Inst Technol, 59-62; res intermediate scientist, Weizmann Inst, 62-63; Ford Found fel, Niels Bohr Inst, Copenhagen, Denmark, 63-64; from asst prof to assoc prof, 64-68, PROF PHYSICS, MCMASTER UNIV, 68- *Mem:* Fel Am Phys Soc; Can Asn Physicists. *Res:* Theoretical physics, especially low energy nuclear physics and nuclear deformations. *Mailing Add:* Dept Physics & Math McMaster Univ 1280 Main St W Hamilton ON L8S 4M1 Can

VOLKSEN, WILLI, b Gitter, Ger, Mar 9, 50; m 72; c 2. POLYMER CHEMISTRY, MATERIALS SCIENCE. *Educ:* NMex Inst Mining & Technol, BS, 72; Univ Lowell, PhD(polymer chem), 75. *Prof Exp:* Fel polymer chem, Calif Inst Technol, 75-76; sr scientist, Jet Propulsion Lab, 76-77; MEM RES STAFF POLYMER CHEM, IBM CORP, 77- *Mem:* Am Chem Soc. *Res:* Synthesis of new and novel polymeric materials including polyelectrolytes and polymers of high temperature stability. *Mailing Add:* 372 El Portal Way San Jose CA 95123

VOLL, MARY JANE, b Baltimore, Md, June 29, 33. BACTERIAL GENETICS. *Educ:* Loyola Col, BA, 55; Johns Hopkins Univ, MSc, 61; Univ Pa, PhD(microbiol), 64. *Prof Exp:* Staff fel microbiol, NIH, 64-66, USPHS fel, 66-67, microbiologist, 67-69; res assoc biol, Johns Hopkins Univ, 69-71; asst prof microbiol, 71-76, ASSOC PROF MICROBIOL, UNIV MD, COLLEGE PARK, 76- *Mem:* Am Soc Microbiol. *Res:* Molecular basis of phenotypic variation in bacteria; genetics of vibrio species; bacterial degradation of organic compounds. *Mailing Add:* Dept Microbiol Univ Md College Park MD 20742

VOLLAND, LEONARD ALLAN, b Cleveland, Ohio, Apr 26, 37; m 63; c 2. PLANT ECOLOGY. *Educ:* Univ Idaho, BS, 59; Ore State Univ, MS, 63; Colo State Univ, PhD(quant ecol), 74. *Prof Exp:* Forester natural resources, 59-66, plant ecologist, 66-73, QUANT ECOLOGIST, US FOREST SERV, 73- *Mem:* Soc Am Foresters; Soc Range Mgt. *Res:* Plant community ecology and its application to natural resource management. *Mailing Add:* US Forest Serv PO Box 3623 Portland OR 97208

VOLLE, ROBERT LEON, b Houston, Pa, June 2, 30; m 52; c 5. PHARMACOLOGY. *Educ:* WVa Wesleyan Col, BS, 53; Univ Kans, PhD(pharmacol), 59. *Prof Exp:* From instr to assoc prof pharmacol, Sch Med, Univ Pa, 60-65; prof, Sch Med, Tulane Univ, 65-68; prof pharmacol, chmn & assoc dean, Sch Med, Univ Conn, 68-83; prof & assoc vpres, res & grad studies, 83-84, vpres acad affairs & res, WVa Univ, 84-85; prof & assoc dean of res & basic sci, Univ Ky, 85-86; PRES, NAT BD MED EXAMR, NAT BD MED, 86- *Concurrent Pos:* Marsh fel pharmacol, Sch Med, Univ Pa, 59-60, Pa Plan scholar, 60-63; USPHS career develop award, 63-65. *Mem:* Fel AAAS; Am Soc Pharmacol & Exp Therapeut; Soc Neurosci; Sigma Xi. *Res:* Neuropharmacology. *Mailing Add:* 3930 Chestnut St Nat Bd Med Examr Philadelphia PA 19104-3194

VOLLENWEIDER, RICHARD A, b Zurich, Switz, June 27, 22; m 65. LIMNOLOGY. *Educ:* Univ Zurich, dipl biol, 46, PhD(biol), 51. *Hon Degrees:* DSc, McGill Univ, Montreal, 86. *Prof Exp:* Teacher undergrad schs, Lucern, Switz, 49-54; fel limnol, Ital Hydrobiol Inst, Palanza, Italy, 54-55 & Swiss Swed Res Coun, Uppsala, 55-56; field expert limnol & fisheries, UNESCO Dept Agr, Egypt, 57-59; res assoc limnol, Ital Hydrobiol Inst, Pallanza, 59-66; consult water pollution, Orgn Econ Coop Develop, Paris, France, 66-68; chief limnologist & head fisheries res bd, 68-70, chief, Lakes Res Div, 70-73, SR SCIENTIST, CAN CENTRE INLAND WATERS, 73- *Concurrent Pos:* Prof biol, McMaster Univ, Hamilton, Ont, 78; consult, Pan Am Health Orgn, Venezuela, 77-80, Italy, 77-, Arg, 80-, Ecuador, 82-, Brazil, 83, Mex, 83-, Int Lake Environ Comn, 85 & WHO, 85- *Honors & Awards:* Int Award, Premio Cervia/Ambiente, 78; Int Tyler Prize Environ Achievement, 86; Naumann-T Lienemann Medal, 87. *Mem:* Ital Asn Ecol; Int Asn Theoret & Appl Limnol; Royal Soc Can. *Res:* Inland and marine water research; biological communities; water chemistry and physics; eutrophication; water pollution. *Mailing Add:* Can Ctr Inland Waters Box 5050 Burlington ON L7R 4A6 Can

VOLLHARDT, K PETER C, b Madrid, Spain, Mar 7, 46; Ger citizen. ORGANIC SYNTHESIS, MECHANISTIC CHEMISTRY. *Educ:* Univ Munich, Ger, dipl, 67; Univ Col, London, Eng, PhD(chem), 72. *Prof Exp:* Fel, Cal Inst Technol, 72-74; from asst prof to assoc prof, 74-82, PROF CHEM, UNIV CALIF, BERKELEY, 82- *Concurrent Pos:* Fel, Regents' Summer Fac,

75, Alfred P Sloan Found, 76-80; prin investr, Mat & Molecular Res Div, Lawrence Berkeley Lab, Calif, 75-; vis prof, Univ Paris-Orsay, 79, Univ Bordeaux, 85, Univ Lyon, 87; ad hoc reviewer, Med Chem Study Sect, NIH, 84; assoc ed, Synthesis, 84-89; mem, Comt Org Chem, Int Union Pure & Appl Chem, 87-; ed, Synlett, 89- *Honors & Awards:* Chevron Res Award, 77; Adolf Windhaus Award, Ger Chem Soc, 83; Organometallic Chem Award, Am Chem Soc, 87, Arthur C Cope Scholar Award, 91. *Mem:* Am Chem Soc; The Chem Soc, London; Ger Chem Soc; Sigma Xi. *Res:* Transition metals in organic synthesis; organometallic clusters in catalysis; new synthetic methods; antiaromatics; strained systems; natural products; gas phase pyrolyses; pyrolyses. *Mailing Add:* Dept Chem Univ Calif Berkeley CA 94720

VOLLMAR, ARNULF R, b Pluderhausen, Ger, Apr 15, 28. ORGANIC CHEMISTRY. *Educ:* Univ Heidelberg, dipl chem, 55, PhD(org chem), 57. *Prof Exp:* Res assoc, Univ Heidelberg, 57-58; fel, Univ Calif, Los Angeles, 58-60; res chemist, Chevron Res Corp, Calif, 60-64; assoc prof, 65-74, PROF CHEM, CALIF STATE POLYTECH UNIV, POMONA, 74- *Concurrent Pos:* NSF res grant, 69. *Mem:* Am Chem Soc; Sigma Xi. *Res:* Chemistry of nitronium acetate adducts of alkylbenzenes and furan derivatives. *Mailing Add:* Dept Chem Calif State Polytech Univ Pomona CA 91768

VOLLMER, ERWIN PAUL, b New York, NY, Jan 16, 06; m 34; c 2. PHYSIOLOGY. *Educ:* Dartmouth Col, AB, 29; NY Univ, MS, 39, PhD(physiol), 41. *Prof Exp:* Bacteriologist, Calco Chem Co Div, Am Cyanamid Co, 37; asst instr biol, NY Univ, 41-42; tutor, Brooklyn Col, 42-43; physiologist, US Naval Med Inst, 47-56; chief endocrinol, Cancer Chemother Nat Serv Ctr, 56-66; chief endocrine eval br, Gen Labs & Clins & exec secy breast cancer task force, Nat Cancer Inst, Bethesda, 66-74; pres, DC Inst Ment Hyg, 85-87; RETIRED. *Mem:* AAAS; Endocrine Soc; NY Acad Sci. *Res:* Physiology of resistance to infection; endocrine factors in hemopoiesis; endocrine etiology and chemotherapy in cancer. *Mailing Add:* 7202 44th St Chevy Chase MD 20815

VOLLMER, FREDERICK WOLFER, b Corning, NY, Apr 18, 56; m 84; c 2. STRUCTURAL GEOLOGY. *Educ:* Univ Calif, Davis, BS, 78; State Univ NY, Albany, MS, 81; Univ Minn, PhD(geol), 85. *Prof Exp:* Teaching asst geol, State Univ NY, Albany, 78-80, res asst, 79-81; teaching asst, Univ Minn, Minneapolis, 81-84, res asst, 82-83; from instr to asst prof, 84-90, ASSOC PROF GEOL, STATE UNIV NY, NEW PALTZ, 90- *Concurrent Pos:* Prin investr, Res Found State Univ NY, 90-92. *Mem:* Am Geophys Union; Geol Soc Am; Int Asn Struct Tectonic Geologists. *Res:* Structural geology; ductile fold-nappes; computer-aided analysis of complexly deformed regions; stress and strain theory; melange fabric. *Mailing Add:* Dept Geol Sci State Univ NY New Paltz NY 12561

VOLLMER, JAMES, b Philadelphia, Pa, Apr 19, 24; m 46; c 3. PHYSICS. *Educ:* Union Col, BS, 45; Temple Univ, MA, 51, PhD(physics), 56; Harvard Univ, advan mgt prog, 71. *Prof Exp:* Instr physics, Temple Univ, 46-51; res engr, Indust Div, Honeywell Inc, 51-59; engr appl res, RCA Corp, 59, group leader appl plasma physics, 59-63, mgr appl physics, 63-66, mgr dir, Advan Technol Labs, 68-72, gen mgr, Palm Beach Div, 72-74, div vpres & gen mgr, 74-75, div vpres & gen mgr, Govt Commun Systs Div, 75-76 div vpres & gen mgr, Govt Systs Div, 76-79, group vpres, 79-83, sr vpres, RCA Corp, 83-84; PRES, JAMES VOLLMER ASSOC, 84- *Concurrent Pos:* Lectr, Temple Univ, 57-59; adj prof, Drexel Inst Technol, 64-66; chmn session on low noise technol, Int Conf Commun, 66; chmn, Bd Dir, Bartol Res Inst, Bd Gov, Franklin Inst. *Honors & Awards:* Centennial Award, Inst Elec & Electronics Engrs, 84. *Mem:* Fel AAAS; fel Inst Elec & Electronics Engrs; Am Phys Soc. *Res:* Infrared properties of materials; plasma physics; quantum electronics; microsonics; lasers; photosensors; radiometry. *Mailing Add:* 186 Shelter Lane J1C Jupiter FL 33469

VOLLMER, REGIS ROBERT, b Wilkinsburg, Pa, Aug 20, 46; m 67; c 2. CARDIOVASCULAR PHARMACOLOGY. *Educ:* St Vincent Col, Latrobe, Pa, BA, 68; Univ Houston, PhD(pharmacol), 75. *Prof Exp:* Res scientist, Squibb Inst Med Res, 75-77; ASST PROF PHARMACOL, SCH PHARM, UNIV PITTSBURGH, 77- *Concurrent Pos:* Prin investr, NIH grant, 81-; mem, Am Heart Asn. *Mem:* Am Soc Pharmacol & Exp Therapeut; AAAS; Sigma Xi. *Res:* Cardiovascular pharmacology with specific focus upon the role of dietary sodium and its influence on the sympathetic nervous system control of cardiovascular function relating to hypertensive disease; antihypertensive drugs. *Mailing Add:* Dept Pharmacol Univ Pittsburgh Salk Hall 908 4200 Fifth Ave Pittsburgh PA 15261

VOLMAN, DAVID H, b Los Angeles, Calif, July 10, 16; m 44; c 3. PHYSICAL CHEMISTRY. *Educ:* Univ Calif, Los Angeles, AB, 37, AM, 38; Stanford Univ, PhD(chem), 40. *Prof Exp:* Asst chem, Univ Calif, Los Angeles, 37-38, res chemist, Nat Defense Res Comt Proj, 41-42; asst chem, Stanford Univ, 38-39; instr chem & jr chemist, Exp Sta, Univ Calif, 40-41; res chemist, Off Sci Res & Develop, Northwestern Univ, 41-45 & Univ Ill, 45-46; from asst prof & asst chemist to assoc prof & assoc chemist, Exp Sta, 46-56, chmn dept, 74-80, prof, 56-87, EMER PROF CHEM, UNIV CALIF, DAVIS, 87- *Concurrent Pos:* Guggenheim fel, Harvard Univ, 49-50; Standard Oil Co Calif fel, Stanford Univ, 39-40; vis prof, Univ Wash, 58; ed bd, J Photochem, 72-; ed, Advan Photochem, 83- *Mem:* Am Chem Soc; Int-Am Photochem Soc. *Res:* Photochemistry; kinetics; chemistry of the atmosphere. *Mailing Add:* Dept Chem Univ Calif Davis CA 95616

VOLOSHIN, ARKADY S, b Kishinev, USSR, Aug 7, 46; c 2. ENGINEERING MECHANICS. *Educ:* Leningrad Polytech Inst, USSR, dipl, 69; Tel-Aviv Univ, Israel, PhD(mech), 78. *Prof Exp:* Sr res officer, Kishinev Inst Non-Destructive Testing, 69-70; asst exp stress anal, Tel-Aviv Univ, 73-78, fel biomech, 78-79; asst prof exp stress anal & biomech, Iowa State Univ, 79-84; ASSOC PROF MECH ENG & MECH, LEHIGH UNIV, 84- *Mem:* Soc Exp Mech; Am Soc Biomech. *Res:* Biomechanics of gait and impulse wave propagation through human locomotor system; photoelasticity through digital image analysis-application to fracture mechanics and nondestructive evaluation; fractional fringe moire; composite materials. *Mailing Add:* 354 Packard Lab Dept Mech Eng & Mech Lehigh Univ Bethlehem PA 18015

VOLP, ROBERT FRANCIS, b Elkhorn, Wis, Oct 20, 52; m 77; c 4. METABOLISM OF FOREIGN COMPOUNDS, BIOCHEMICAL TOXICOLOGY. *Educ:* Univ Wis-Stevens Pt, BS, 75; Univ Wis-Madison, MS, 77, PhD(pharmacol & toxicol), 79. *Prof Exp:* Res assoc, Inst Pharmacol & Toxicol, Univ Goettingen, Fed Repub Ger, 79-80, Univ Ariz, 80-82; res asst prof, Primate Res Inst, NMex State Univ, 82-83; asst prof, 83-88, ASSOC PROF CHEM, MURRAY STATE UNIV, 88- *Concurrent Pos:* Prin investr, NIH, 85-87 & 90-92. *Mem:* Am Chem Soc; Soc Toxicol. *Res:* Metabolism of halogenated aliphatic compounds; biochemical mechanisms of toxicity of halogenated aliphatic compounds; regulation of cellular UDPGA metabolism. *Mailing Add:* Dept Chem Murray State Univ Murray KY 42071-3306

VOLPE, ANGELO ANTHONY, b New York, NY, Nov 8, 38; m 65. ORGANIC CHEMISTRY, POLYMER CHEMISTRY. *Educ:* Brooklyn Col, BS, 59; Univ Md, MS, 62, PhD(org chem), 66. *Hon Degrees:* ME, Stevens Inst Technol, 75. *Prof Exp:* Res chemist, US Naval Ord Lab, 61-66; from asst prof to prof chem, Stevens Inst Technol, 66-77, actg head dept chem & chem eng, 74-75; chmn dept, dean col arts & sci, 80-83, PROF CHEM, EAST CAROLINA UNIV, 77-, VCHANCELLOR ACAD AFFAIRS, 83- *Mem:* Am Chem Soc; Sigma Xi. *Res:* Correlation of polymer properties to molecular structure; synthesis and mechanisms of formation and degradation of thermally stable polymers; monomer synthesis; synthesis and study of biopolymers; educational administration. *Mailing Add:* Walton House Tenn Tech Univ PO Box 5007 Cookeville TN 38505

VOLPE, ERMINIO PETER, b New York, NY, Apr 7, 27; m 55; c 3. ZOOLOGY. *Educ:* City Col New York, BS, 48; Columbia Univ, MA, 49, PhD(zool), 52. *Prof Exp:* Asst zool, Columbia Univ, 48-51; instr biol, City Col New York, 51-52; from asst prof to prof zool, Newcomb Col, Tulane Univ, 52-81, chmn dept, Col, 54-64, chmn dept, Univ, 64-66, assoc dean, Grad Sch, 67-69, chmn dept, 69-79; PROF BASIC MED SCI, SCH MED, MERCER UNIV, 81- *Concurrent Pos:* Mem steering comt, Biol Sci Curric Study, 66-69; consult, Comn Undergrad Educ Biol Sci, NSF, 67-70; US Nat comnr, UNESCO, 68-72; mem exam comt, Col Entrance Exam Bd, Princeton Univ, 69-72; chmn advan placement test biol, Educ Testing Serv, 75-81; ed, Am Zoologist, Am Soc Zoologists, 76-81; mem grad record exam comt biol, Educ Testing Serv, 81-90. *Mem:* Fel AAAS; Genetics Soc Am; Soc Study Evolution; Am Soc Zoologists (pres, 81); Soc Syst Zool; Am Soc Naturalists; Am Soc Human Genetics. *Res:* Embryology, genetics and evolution of amphibians; transplantation immunity and tolerance in anurans; writer in medical genetics. *Mailing Add:* Sch Med Mercer Univ Macon GA 31207

VOLPE, GERALD T, b New York, NY, Feb 15, 35; m 57; c 2. ELECTRICAL ENGINEERING. *Educ:* City Col New York, BEE, 57, MEE, 61; NY Univ, EngScD, 64. *Prof Exp:* Jr engr, Bendix Res Labs, Mich, 57-59; design engr, Loral Electronics Div, Loral Corp, NY, 59-61; instr elec eng, NY Univ, 61-64; sr eng, CBS Labs, Columbia Broadcasting Syst, 64-66, Marchand Electronic Labs, 66-67 & Perkin-Elmer Corp, Conn, 67-68; Assoc Prof Elec Eng, Cooper Union, 68-; AT UNIV BRIDGEPORT. *Concurrent Pos:* Consult, Perkin-Elmer Corp, 69-78 & Gen Instrument Corp, NY, 73-77. *Mem:* Inst Elec & Electronics Engrs; Optical Soc Am. *Res:* Feedback controls; communication theory; electro-optics and acoustics; circuit theory. *Mailing Add:* Dept Elec Eng Univ Bridgeport 380 University Ave Bridgeport CT 06601

VOLPE, JOSEPH J, b Salem, Mass, Dec 17, 38. PEDIATRICS, NEUROLOGY. *Educ:* Harvard Univ, MD, 64. *Prof Exp:* PROF PEDIAT, NEUROL & BIOL CHEM, WASH UNIV, ST LOUIS, MO, 80- *Mailing Add:* Neurologist-in-Chief Harvard Med Sch Childrens Hosp 300 Longwood Ave Boston MA 02115

VOLPE, P(ETER) J, JR, b New York, NY, Mar 13, 34; m 57; c 4. CHEMICAL ENGINEERING, ECONOMICS. *Educ:* Rice Inst, BA, 56, BS, 57. *Prof Exp:* Jr prod engr, 57-58, from jr engr to engr, 58-60, from res engr to sr res engr, 60-65, group leader chem eng econ & design, 65-66, sect head process develop, 66-71, mgr chem eng, 71-74, tech mgr, 74-76, opers mgr, Bishop Tex Plant, 76-80, mgr facil & admin, Tech Ctr, 80-81, DIR FAC, TECH CTR, CORPUS CHRISTI PLANT, CELANESE CHEM CO, 81- *Mem:* Am Inst Chem Engrs. *Res:* Laboratory and pilot plant process development for bulk organic chemicals, especially nylon salt; economics and process design for new bulk organic chemical products and processes. *Mailing Add:* 318 Cape Hatteras Corpus Christi TX 78412

VOLPE, ROBERT, b Toronto, Ont, Mar 6, 26; m 49; c 5. IMMUNOLOGY. *Educ:* Univ Toronto, MD, 50; FRCP(C), 56; FACP, 65. *Prof Exp:* Dept Vet Affairs med res fel, Clin Invest Unit, Sunnybrook Hosp, Toronto, 52-53; Med Res Coun Can fel, Toronto Gen Hosp, 55-57; sr res fel endocrinol, Fac Med, Univ Toronto, 57-65; from clin teacher to assoc prof, 65-71, PROF MED, FAC MED, UNIV TORONTO, 71-; DIR ENDOCRINE RES LAB, WELLESLEY HOSP, 67- *Concurrent Pos:* Physician-in-chief, Dept Med, Wellesley Hosp, Toronto, 74-87; gov, Am Col Physicians, 80-84; dir, Univ Toronto Div Endocrinol, Metab, 87-; mem coun, Royal Col Physicians, 88- *Honors & Awards:* Jamieson Prize, Can Soc Nuclear Med, 80; Baxter Prize lectr, Toronto Soc Clin Res, 84; Sandoz Prize lectr, Can Soc Endocrinol Metab, 85. *Mem:* Fel Am Col Physicians; Am Thyroid Asn (pres, 80-81); Am Fedn Clin Res; Endocrine Soc; Can Soc Endocrinol & Metab (pres, 72); Assoc Am Physicians; fel Royal Col Physicians Can. *Res:* Immune mechanisms in thyroid diseases; autoimmunity in endocrine system; immunoregulatory abnormalities in autoimmune endocrine diseases; thyroid diseases. *Mailing Add:* Three Daleberry Pl Don Mills ON M3B 2A5 Can

VOLPE, ROSALIND ANN, b New Haven, Conn, May 1, 54; m 87. HEALTH EFFECTS OF LEAD & CADMIUM & ZINC. *Educ:* Barnard Col, BA, 75; Columbia Univ, MS, 80, PhD(environ sci), 88. *Prof Exp:* Res asst virol & oncol, Rockefeller Univ, 75-77; intern, New York Attorney Gen Office Environ Proj Bur, 79; asst mgr environ health, Int Lead Zinc Res Org, 80-86; asst dir environ health, Lead Ind Assoc, 84-86; MGR ENVIRON HEALTH, INT LEAD ZINC RES ORG, 86- *Concurrent Pos:* Counr, Soc Geochem & Health, 88-; prin invest, Impart Eng Control, 88. *Mem:* Am Inst Health Hosp;

Soc Environ Geochem Health. *Res:* Managed program covering the heavy metals lead zinc and cadmium; most studies directed at human target organs and range from occupational, general population, epidemiology and toxicology projects, conferences and information exchanges. *Mailing Add:* ILZRO 2525 Meridian Pkwy PO Box 12036 Research Triangle Park NC 27709-2036

VOLPP, GERT PAUL JUSTUS, b Loerrach, Ger, July 30, 30; nat US; m 62; c 4. AGRICULTURAL CHEMISTRY. *Educ:* Univ Basel, PhD(chem), 58. *Prof Exp:* Res fel org chem, Harvard Univ, 58-63; interdisciplinary scientist, 63-65, mgr explor org res, 65-72, mgr prod res, 72-73, tech dir alkali chem, 73-75, asst venture mgr, pyrethroids, 75-76, asst dir res, 76-77, acquisition mgr, 77-79, DIR COM DEVELOP, FMC CORP, 80- *Mem:* Am Chem Soc; NY Acad Sci. *Res:* Synthetic organic chemistry; intermediates for dyestuffs; additives for plastics; antioxidants; detergent chemistry; agricultural chemistry; manufacturing technology for soda ash, caustic, chlorine, glycerine, allyl alcohol, barium and strontium chemicals; commercial development of pesticides. *Mailing Add:* 116 Poe Rd Princeton NJ 08540-4122

VOLTZ, STERLING ERNEST, b Philadelphia, Pa, Apr 17, 21; m 43; c 2. PHYSICAL CHEMISTRY, RESEARCH ADMINISTRATION. *Educ:* Temple Univ, AB, 43, MA, 47, PhD(phys chem), 52. *Prof Exp:* Lab asst, Temple Univ, 46-47; res fel, Univ Pa, 47-48; instr, Temple Univ, 48-51; res chemist, Houdry Process Corp, 51-58; group leader, Sun Oil Co, 58-60; supv chemist, Missile & Space Div, Gen Elec Co, 60-62, consult liaison scientist, 62-68; res assoc, Mobil Res & Develop Corp, Paulsboro, NJ, 68-80, admin adv, 80-82, admin mgr, 82-86; PVT CONSULT, 86- *Mem:* AAAS; Am Chem Soc; Catalysis Soc; Sigma Xi. *Res:* Catalysis; surface and solid state chemistry; chemical kinetics; electrochemistry; fuel cells; petroleum and petrochemical processes; synthetic fuels; automotive emission control systems; program management; research administration and planning; research contracts. *Mailing Add:* Six E Glen Circle Media PA 19063

VOLTZOW, JANICE, b New Haven, Conn, May 7, 58; m 86; c 1. INVERTEBRATE ZOOLOGY, FUNCTIONAL MORPHOLOGY. *Educ:* Yale Univ, BS, 80; Duke Univ, PhD(zool), 85. *Prof Exp:* Teaching asst zool, Duke Univ, 80-82, trainee, Cocos Found, 82-85; fel, Friday Harbor Labs, Univ Wash, 85-86; ASST PROF ZOOL & MARINE BIOL, UNIV PR, 87- *Concurrent Pos:* Chairperson, Pub Affairs Comt, Am Soc Zoologists, 87-89; counr-at-large, Am Malacol Union, 90-92. *Mem:* Am Soc Zoologists; Am Malacol Union; Sigma Xi; AAAS. *Res:* Development, organization, biomechanics and evolution of invertebrate muscle systems, especially prosobranch gastropods. *Mailing Add:* Dept Biol Univ PR Rio Piedras PR 00931

VOLWILER, WADE, b Grand Forks, NDak, Sept 16, 17; m 43; c 3. MEDICINE. *Educ:* Oberlin Col, AB, 39; Harvard Med Sch, MD, 43; Am Bd Internal Med, dipl, 50, recert, 77; Am Bd Gastroenterol, dipl, 54. *Prof Exp:* From intern to resident med, Mass Gen Hosp, Boston, 43-45, asst, 45-48; asst, Harvard Med Sch, 46-48; from instr to prof, 49-82, head, Div Gastroenterol, 50-81, EMER PROF MED, SCH MED, UNIV WASH, 82- *Concurrent Pos:* Teaching fel med, Harvard Med Sch, 45-46; res fel gastroenterol, Mass Gen Hosp, Boston, 45-48; Am Gastroenterol Asn res fel, 47; Nat Res Coun fel, 48-49; Markle scholar, 50-55; res assoc, Mayo Found, Univ Minn, 48-49; attend physician, King County Hosp Syst, Seattle, 50-82 & Vet Admin Hosp, 51-82; consult, USPHS Hosp, 55-82 & Univ Wash Hosp, 60-82; mem subspecialty bd gastroenterol, Am Bd Internal Med, 70-76. *Honors & Awards:* Friedenwald Medal, Am Gastroenterol Asn, 81. *Mem:* Am Soc Clin Invest; Am Gastroenterol Asn (secy, 59-62, pres, 67); Asn Am Physicians; Am Asn Study Liver Dis (pres, 56). *Res:* Liver diseases; gastroenterology; plasma proteins. *Mailing Add:* Dept Med RG-24 Univ Wash Sch Med Seattle WA 98195

VOLZ, FREDERIC ERNST, b Singen, Ger, Oct 29, 22; m 57; c 3. ATMOSPHERIC PHYSICS. *Educ:* Univ Frankfurt, dipl, 50, PhD(meteorol), 54. *Prof Exp:* Res asst, Lichtklimat Observ Arosa, 50-52; res asst meteorol, Univ Mainz, 52-57; res fel atmospheric physics, Harvard Univ, 57-61 & Astron Inst, Univ Tübingen, 62-67; RES PHYSICIST, AIR FORCE GEOPHYS LAB, 67- *Mem:* Am Meteorol Soc; Am Geophys Union; Optical Soc Am; Ger Meteorol Soc. *Res:* Atmospheric optics, optical constants of aerosol, twilight, stratospheric aerosol; infrared. *Mailing Add:* 24 Tyler Rd Lexington MA 02173

VOLZ, JOHN EDWARD, b Baltimore, Md, March 23, 40; m 61; c 4. ADIPOSE TISSUE, TEMPOROMANDIBULAR JOINT. *Educ:* Towson State Univ, BS, 69; Univ Md, PhD(anat), 75. *Prof Exp:* ASSOC PROF GROSS ANAT, MED SCH, TEMPLE UNIV, 73- *Concurrent Pos:* Consult, Vet Admin Hosp, Dent Br, Del, 86. *Mem:* Am Asn Anatomists; Am Asn Dent Sch. *Res:* Adipose tissue, especially the effects of diet and exercise on the growth and development of fat stores in rats and mice; temporomandibular joint disfunction. *Mailing Add:* Dept Anat Sch Med Temple Univ 3223 N Broad St Philadelphia PA 19140-5096

VOLZ, MICHAEL GEORGE, b Long Beach, Calif, Nov 30, 45; m 68; c 2. ENVIRONMENTAL BIOCHEMISTRY & MICROBIOLOGY. *Educ:* Univ Calif, Berkeley, BS, 67, PhD(soil sci, plant physiol), 72. *Prof Exp:* Asst res biochemist, Univ Calif, Berkeley, 72-74, res biochemist, 74-75; asst plant physiologist, Conn Agr Exp Sta, 75-77; assoc sanit microbiologist, Sanit & Radiation Lab, Calif State Dept Health, 77-82, pub health chemist, Hazardous Mat Lab, 82-84, environ biochemist, Sanit & Radiation Lab, 85-89, res scientist, 89, CHIEF, DIV LABS, CALIF STATE DEPT HEALTH, 89- *Concurrent Pos:* Chairperson, Occup Health & Environ Health, LIFT 7000; mem, Comts on Training & Human Retrovirus Testing, Asn State & Territorial Pub Health Labs Dirs. *Mem:* Am Soc Agron; Soil Sci Soc Am; Am Water Works Asn; AAAS; Water Pollution Control Fedn; Soc Environ Geochem Health; Asn State & Territorial Pub Health Lab Dirs. *Res:* Assessing the significance of environmental microbiol/chemical contamination, attendant environmental data quality and subsequent influence on regulatory decisions and activities impacting upon the public health. *Mailing Add:* Div Labs State Calif Dept Health Serv 2151 Berkeley Way Berkeley CA 94704-9980

VOLZ, PAUL ALBERT, b Ann Arbor, Mich, Mar 26, 36. MYCOLOGY, BOTANY. *Educ:* Heidelberg Col, BA, 58; Mich State Univ, MS, 62, PhD(mycol), 66; Century Univ, PhD(bus admin), 91. *Prof Exp:* Instr bot, Univ Wis-Milwaukee, 62-63; USPHS postdoctoral res grant, Med Ctr, Ind Univ, 67-68; assoc prof bot & mycol, Purdue Univ, 68-69; asst prof & mycologist, 69-72, assoc prof, 72-80, PROF BOT & MYCOL, EASTERN MICH UNIV, 80- *Concurrent Pos:* Sr res assoc, Nat Res Coun, 71-73; res contractor, NASA Manned Spacecraft Ctr, 71-74; vis prof mycol, Nat Taiwan Univ, 74-75, Wayne State Univ, 80-81 & NATO Advan Study Inst, 82- *Honors & Awards:* Kholodny Medal, Inst Bot, Acad Sci, USSR. *Mem:* AAAS, Fel, Explorers Club; Am Inst Biol Sci; NY Acad Sci; Am Fern Soc; Asn Trop Biol; Am Soc Eng Educ. *Res:* Fern anatomy; marine and soil fungi of the Bahamas and The Republic of China; keratinophilic fungi; drug sensitivity and nutritional requirements of fungi; effects of space flight parameters on select fungal species; fungal cytogenetics and morphology; medical mycology; microbial ecology. *Mailing Add:* 1805 Jackson Ave Ann Arbor MI 48103

VOLZ, RICHARD A, b Woodstock, Ill, July 10, 37; m 61; c 3. ROBOTICS, REAL TIME COMPUTING. *Educ:* Northwestern Univ, BS, 60, MS, 61, PhD(elec eng), 64. *Prof Exp:* Assoc prof, 64-77, assoc chmn elec & comput eng dept, 78-79, assoc dir comput ctr, 79-82, PROF ELEC & COMPUT ENG, UNIV MICH, ANN ARBOR, 77-, DIR ROBOTICS LAB, 81- *Concurrent Pos:* NSF grants, 65-68, 70-72 & 85; Air Force Off Sci Res contract, 81-; assoc ed Robotics, Trans Aerospace & Electronics Systs, Inst Elec & Electronics Engrs, 83-; mem, Automation & Robotics Panel, NASA, 84-85. *Mem:* Inst Elec & Electronics Engrs; Asn Comput Mach; Soc Mfg Engrs; Int Inst Robotics. *Res:* Robotics; task planning techniques for robotics; real-time computing techniques, particularly real-time software; distributing languages for embedded real-time systems; machine vision. *Mailing Add:* Dept Computer Sci Tex A&M Univ College Station TX 77843

VOLZ, WILLIAM BECKHAM, electrooptics, for more information see previous edition

VOLZ, WILLIAM K(URT), b Mannheim, Ger, Feb 28, 19; nat US; m 42; c 4. CHEMICAL ENGINEERING. *Educ:* Columbia Univ, BSc, 41. *Prof Exp:* Mem eng staff, Monsanto Co, 41-48, proj engr, 48-55, sect mgr eng, 55-60, asst dir, 60-62, eng dir, Spain, 64-65, mgr int eng, 65-77, mgr eng, 77-79; asst prof chem eng, 79-81, CONSULT & AFFIL PROF, WASHINGTON UNIV, ST LOUIS, 81- *Mem:* Nat Soc Prof Engrs; Instrument Soc Am; Sigma Xi; Am Inst Chem Engrs. *Res:* Management; design and construction of new chemical plant facilities. *Mailing Add:* 46 Morwood Lane St Louis MO 63141

VOMACHKA, ARCHIE JOEL, b Duluth, Minn, Sept 28, 46; m 73. REPRODUCTIVE BIOLOGY, ENDOCRINOLOGY. *Educ:* Univ Minn, Duluth, BA, 68; Mich State Univ, PhD(zool), 76. *Prof Exp:* Fel reprod physiol, Univ Kans Med Ctr, 76-79; lectr exp vert biol, Princeton Univ, 79-81; ASST PROF HUMAN PHYSIOL, MARQUETTE UNIV, 81- *Concurrent Pos:* Res assoc, Princeton Univ, 79-81. *Mem:* Animal Behav Soc; AAAS; Sigma Xi. *Res:* Neural and hormonal regulation of sexual behavior in rodents, including the sexual differentiation of behavior and endocrine physiology in hamsters. *Mailing Add:* Dept Biol Marquette Univ Milwaukee WI 53233-2274

VOMHOF, DANIEL WILLIAM, b Grant, Nebr, Apr 19, 38; m 60, 78; c 3. FORENSIC SCIENCE, CHEMISTRY. *Educ:* Augsburg Col, BA, 62; Univ Ariz, MS, 66, PhD(plant physiol), 67; Am Inst Chemists, cert, 69. *Prof Exp:* Chemist, Ariz Agr Exp Sta, Univ Ariz, 63-67; res chemist, Corn Refiners Asn, 67-69; dir, Region IX Lab, US Bur Customs, Ill, 69-72, forensic scientist, Region VII, 72-74, dir, Lab Div, US Customs Serv, Region IX, Chicago, 74; PRES, EXPERT WITNESS SERV, 74- *Concurrent Pos:* Res assoc, Nat Bur Stand, 67-69; adj prof occup safety, Nat Univ, 84-; chmn gen educ, Coleman Col, La Mesa Ca, 86-90. *Mem:* AAAS; Am Chem Soc; fel Am Inst Chemists; Sigma Xi; Am Soc Testing & Mat. *Res:* Accident dynamics; document identification; biomechanics; analytical chemistry; driver behavior; safety engineering, human factor; intelligent systems. *Mailing Add:* Expert Witness Serv 8387 University Ave La Mesa CA 91941

VOMOCIL, JAMES ARTHUR, b Jacumba, Calif, Sept 12, 26; m 46; c 3. SOIL SCIENCE, AGRONOMY. *Educ:* Univ Ariz, BS, 50; Mich State Univ, MS, 52; Rutgers Univ, PhD, 55. *Prof Exp:* Asst soil sci, Mich State Univ, 50-52, instr, 52; asst, Rutgers Univ, 52-55; instr soil physics, Univ Calif, Davis, 55, from asst prof to assoc prof & assoc exp sta, 55-67; EXTEN SOILS SPECIALIST & PROF SOILS, ORE STATE UNIV, 67- *Mem:* Am Soc Agron; Soil Sci Soc Am; Int Soc Soil Sci. *Res:* Soil physical condition and plant growth; soil strength and deformation. *Mailing Add:* Dept Soils Ore State Univ Corvallis OR 97331

VON, ISAIAH, b Philadelphia, Pa, Dec 28, 18; m 45; c 3. INDUSTRIAL ORGANIC CHEMISTRY. *Educ:* Univ Buffalo, BA, 40; Univ Pa, MS, 41, PhD(org chem), 43. *Prof Exp:* Res assoc, Nat Defense Res Comt Proj, Univ Pa, 43-45, mem comt on med res proj, 45-46; res chemist, Am Cyanamid Co, 46-53, develop chemist, 53-54, group leader, 54-56, sect chief chemist, 56-64, dep chief chemist, 65-81; CONSULT, 81- *Mem:* Am Chem Soc; Am Asn Textile Chemists & Colorists. *Res:* Dyestuffs; pigments; organic intermediates. *Mailing Add:* Apt 16E 1050 George St New Brunswick NJ 08901

VONA, JOSEPH ALBERT, b Brooklyn, NY, Aug 15, 20; m 46; c 2. ORGANIC CHEMISTRY. *Educ:* Brooklyn Col, BA, 41, MA, 44; Polytech Inst Brooklyn, PhD, 54. *Prof Exp:* Head lab sect plastics res, Barrett Chem Co, 45-46; res & develop chemist, Nat Lead Co, 46-50; asst to tech dir, Baker

Castor Oil Co, 50-55; mgr, Tech Serv Lab, 55-69, DIR, MTD LAB, CELANESE CHEM CO, 69- *Mem:* Am Chem Soc; Com Develop Asn; Chem Indust Asn; Asn Res Dirs; NY Acad Sci. *Res:* Research and development in radiation technology; new types of coatings; emulsion solution and bulk polymerization of monomers; new compounds which can produce durable coatings. *Mailing Add:* Celanese Chem Co 108 Marlboro St Westfield NJ 07090

VON ALMEN, WILLIAM FREDERICK I, b Olney, Ill, May 6, 28; m 50; c 4. PALYNOLOGY, GEOLOGY. *Educ:* Southern Ill Univ, Carbondale, BA, 57; Univ Mo, Columbia, MA, 59; Mich State Univ, PhD(geol), 70. *Prof Exp:* Geologist, Pure Oil Co, 59-60; geologist, Stand Oil Co Tex, 60-64, geologist-palynologist, 66-68, palynologist, Chevron Oil Field Res Co, 68-69, from lead palynologist to div paleontologist, 69-71, SR PALEONTOLOGIST, CHEVRON USA, INC, 71- *Mem:* Am Asn Stratig Palynol. *Res:* Palynology of Devonian-Mississippian Boundary; Mesozoic palynostratigraphy. *Mailing Add:* 110 Holly Dr Metairie LA 70005

VON AULOCK, WILHELM HEINRICH, b Pirna, Ger, Jan 24, 15; US citizen; m 56; c 4. ELECTRICAL & SYSTEMS ENGINEERING. *Educ:* Tech Univ, Berlin, Dipl Eng, 37; Stuttgart Tech Univ, DEng, 53. *Prof Exp:* Div head engr, Torpedoversuchsanstalt Eckernfoerde, Ger, 42-45; physicist, Bur Ships, Navy Dept, Washington, DC, 47-53; mem tech staff, 54-62, dept head nuclear effects, 62-71, dir tech support, Am Bell Int, 77-79, dept head installation studies, 71-79, DEPT HEAD DEVELOP, PLANNING & ANALYSIS, BELL LABS, 79- *Concurrent Pos:* Instr, Postgrad Sch, Univ Md, 51-52. *Mem:* Fel Inst Elec & Electronics Engrs. *Res:* Guided acoustic torpedoes; electromagnetic fields in sea water; microwave ferrite materials and devices; phased arrays for radar applications; systems engineering for communication facilities. *Mailing Add:* PO Box 84 Fairview Dr Bedminster NJ 07921

VON BACHO, PAUL STEPHAN, JR, photographic chemistry; deceased, see previous edition for last biography

VON BAEYER, HANS CHRISTIAN, b Berlin, Ger, Apr 6, 38; US citizen; m 83; c 4. THEORETICAL PHYSICS, PHYSICS POPULARIZATION. *Educ:* Columbia Univ, AB, 58; Univ Miami, MSc, 61; Vanderbilt Univ, PhD(physics), 64. *Prof Exp:* Res assoc physics, McGill Univ, 64-65, asst prof, 65-68; from asst prof to assoc prof, 68-75, chmn dept, 72-78, PROF PHYSICS, COL WILLIAM & MARY, 75- *Concurrent Pos:* Vis prof, Tri-Univ Meson Facil & Simon Fraser Univ, 78-79; dir, Va Asn Res Campus, 79-84; secy, Southeastern Univ Res Asn, 80-85. *Mem:* Fel Am Phys Soc; Fedn Am Sci; Am Asn Univ Prof; Am Asn Physics Teachers. *Res:* Theory of elementary particles; public understanding of science. *Mailing Add:* Dept Physics Col William & Mary Williamsburg VA 23185

VON BARGEN, KENNETH LOUIS, b Alliance, Nebr, Apr 6, 31; m 82; c 3. TRACTOR & EQUIPMENT TESTING. *Educ:* Univ Nebr, Lincoln, BS, 52, MS, 62; Purdue Univ, Lafayette, PhD(agr eng), 70. *Prof Exp:* Design engr, Lockwood Grader Corp, Nebr, 55-56; from instr to asst prof agr eng, 56-69, assoc prof systs eng, 69-77, PROF AGR SYSTS ENG, UNIV NEBR, LINCOLN, 77- *Concurrent Pos:* Mem bd dirs, Am Forge & Grasslands Coun, 77-80; mem, Nebr Bd Tractor Testing Engrs, 76-80, chmn, 81- *Mem:* Sigma Xi; Am Soc Agr Engrs; Am Forage & Grassland Coun; Soc Automotive Engrs. *Res:* Forage systems, particle size analysis, systems engineering of agricultural production systems emphasizing weather and equiptment sizing; man-machine performance; agriculture tractor performance and energy requirements; automatic control of agriculture machines. *Mailing Add:* Univ Nebr 205 Chase Hall Lincoln NE 68583-0726

VON BECKH, HARALD JOHANNES, aerospace medicine, for more information see previous edition

VON BERG, ROBERT L(EE), b Wheeling, WVa, June 14, 18; m 47; c 4. PROCESS DESIGN, REACTOR DESIGN. *Educ:* WVa Univ, BSChE, 40, MS, 41; Mass Inst Technol, ScD(chem eng), 44. *Prof Exp:* Instr chem eng, WVa Univ, 39-40; indust engr, E I du Pont de Nemours & Co, Del, 44-46; from asst prof to prof, 46-88, EMER PROF CHEM ENG, CORNELL UNIV, 88- *Concurrent Pos:* Consult, Atomic Energy Comn, 50 & Dow Chem Co, Mich, 53-54; NATO res fel, Delft Univ Technol, 60-61; vis prof, Univ New Castle, Australia, 74-75, Univ Canterbury, NZ, 82, 88. *Mem:* Am Chem Soc; Am Inst Chem Engrs. *Res:* Chemical processes; liquid-liquid extraction; reaction kinetics and reactor design; nuclear processing; radiation chemistry. *Mailing Add:* Sch Chem Eng Cornell Univ Ithaca NY 14853

VON BODUNGEN, GEORGE ANTHONY, b New Orleans, La, Oct 12, 40; m 69; c 3. PHYSICAL CHEMISTRY. *Educ:* Loyola Univ, New Orleans, 62; Tulane Univ La, PhD(phys chem), 66. *Prof Exp:* sr res chemist, 66-80, PROCESS DEVELOP GROUP LEADER, COPOLYMER RUBBER & CHEM CORP, 80- *Mem:* Soc Plastics Engrs. *Res:* Synthesis and rheology of impact resistant plastics and thermo plastic elastomers; computer simulations and mathematical models of process and products; applied mathematics; development and scale-up of emulsion products and processes. *Mailing Add:* 3836 Partridge Lane Baton Rouge LA 70809

VON BORSTEL, ROBERT CARSTEN, b Kent, Ore, Jan 24, 25; m 48. GENETICS. *Educ:* Ore State Col, BA, 47, MS, 49; Univ Pa, PhD(zool), 53. *Prof Exp:* Fel, Carnegie Inst, NY, 52-53; biologist, Oak Ridge Nat Lab, 53-71; chmn dept, 71-81, McCalla res prof, 84-85, PROF GENETICS, UNIV ALTA, 71- *Concurrent Pos:* NSF fel, Univ Pavia, 59-60; mem, Basil Inst Immunol, 82-83. *Mem:* Genetics Soc Am; Am Soc Naturalists; Genetic Soc Can (pres, 79-80); Int Asn Environ Mutagen Socs (secy, 73-78); Am Environ Mutagen Soc. *Res:* Dominant lethality and cell-killing by radiation; microorganism genetics; spontaneous mutation rates; mutator genes in yeast. *Mailing Add:* 12312 Grandview Dr Edmonton AB T6H 4K4 Can

VON BUN, FRIEDRICH OTTO, b Vienna, Austria, June 22, 25; US citizen; m 52; c 2. PHYSICS, MATHEMATICS. *Educ:* Vienna Tech Univ, MS, 52; Graz Tech Univ, PhD(physics, math), 56. *Prof Exp:* Physicist, US Army Signal Corps, 53-57, chief molecular beam sect, Atomic Resonance Br, 57-59, sr scientist & dir frequency control div, 59-60; consult, Tracking & Data Systs Directorate, NASA, 60-61, head plans off, 61-63, chief systs anal off, 63-65, chief mission anal off, 65-67, chief mission & trajectory anal div, 67-71, chief trajectory anal & geodyn div, Mission & Data Opers Directorate, 71-72, chief geodyn prog div, Goddard Space Flight Ctr, 72-82, asst dir appl sci, Appln Directorate, 74-79; RETIRED. *Concurrent Pos:* Mem panel tracking & data anal, Nat Acad Sci, 63-65; spec adv, Range Tech Adv Group, 66-; mem, Steering Comt, Working Group 1, COSPAR, 72-; chmn, Working Group Earth & Ocean Dynamics, Int Astronaut Fedn, 75 & Working Group Global Data Collection, 76. *Mem:* Assoc fel Am Inst Aeronaut & Astronaut; Am Geophys Union; Int Astronaut Fedn. *Res:* Space systems analysis; navigation; geodynamics; ocean dynamics; gravity and magnetic field studies; active and passive microwave observations of the Earth's surface from space; application of space science and technology toward solutions of practical problems. *Mailing Add:* 12506 White Dr Silver Spring MD 20904

VONDERHAAR, BARBARA KAY, b Des Moines, Iowa, July 4, 43; m 74; c 2. ENDOCRINOLOGY. *Educ:* Clarke Col, BA, 65; Univ Wis-Madison, PhD(oncol), 70. *Prof Exp:* Fel, McArdle Lab Cancer Res, 70-71; staff fel, Nat Inst Arthritis, Metab & Digestive Dis, 71-73, sr staff fel, 73-76; cancer expert, 76-80, sr res chemist, Lab Pathophysiol, 80-85, SR RES CHEMIST, LAB TUMOR IMMUNOL BIOL, NAT CANCER INST, 85- *Concurrent Pos:* Assoc ed, Cancer Res, 85-89; mem coun, Gordon Res Conf, 85-88; mem comt, Mammary Gland Biol & Lactation, 85-; Buuroughs Wellcome vis prof, Clarke Col, Dubuque, Iowa, 84-85. *Mem:* AAAS; Endocrine Soc; Am Soc Cell Biol; Am Soc Biochem & Molecular Biol; Int Asn Breast Cancer Res; Am Asn Cancer Res. *Res:* Multiple hormone interactions in mammary gland development and milk protein production; the effects of neonatal hormone treatment on development of mammary glands and tumor formation; thyroid hormone and prolactin interactions in breast development and tumorigenesis. *Mailing Add:* NIH Bldg 10 Rm 5B56 Bethesda MD 20892

VONDER HAAR, RAYMOND A, b St Louis, Mo, Nov 18, 46; m 68; c 2. BIOCHEMISTRY, CELL BIOLOGY. *Educ:* Univ Mo, Columbia, AB, 68; Purdue Univ, PhD(molecular biol), 73. *Prof Exp:* Fel virol & cell biol, Univ Utah, 74-78; ASST PROF MED BIOCHEM, TEX A&M UNIV, 78- *Res:* Frameshift suppression in mammalian cells; mutagenesis in mammalian cells and animal viruses. *Mailing Add:* 2801 Jennifer Dr College Station TX 77843

VONDER HAAR, THOMAS HENRY, b Quincy, Ill, Dec 28, 42; m 80; c 3. METEOROLOGY, SPACE SCIENCE. *Educ:* St Louis Univ, BS, 63; Univ Wis-Madison, MS, 64, PhD(meteorol), 68. *Prof Exp:* Assoc scientist meteorol, Space Sci & Eng Ctr, Univ Wis, 68-70; assoc prof, 70-77, PROF ATMOSPHERIC SCI, COLO STATE UNIV, 77-, HEAD DEPT, 74-, DIR, COOP INST RES, 80-, PROF & DIR, CIRA. *Concurrent Pos:* Consult, US Army, McDonnell-Douglas Corp, Ball Bros & Res Corp, 69-, NASA; mem int radiation comn, Int Union Geod & Geophys, 75; mem climate res comt, Nat Acad Sci. *Honors & Awards:* Second Half Century Award, Am Meteorol Soc, 81. *Mem:* Am Meteorol Soc. *Res:* Application of measurements from meteorological satellites to problems of atmospheric and environmental science; radiation measurement; air pollution; weather forecasting. *Mailing Add:* 515 S Howes St Ft Collins CO 80521

VONDRA, CARL FRANK, b Seward, Nebr, June 3, 34; m 55; c 4. GEOLOGY. *Educ:* Univ Nebr, BS, 56, MS, 58, PhD(geol), 63. *Prof Exp:* Develop geologist, Calif Co, 61-62; geologist, Pan Am Petrol Corp, 62-63; from asst prof to assoc prof, 63-71, PROF GEOL, IOWA STATE UNIV, 71- *Mem:* Geol Soc Am; Am Asn Petrol Geol; Paleont Soc; Soc Vert Paleont; Soc Econ Paleont & Mineral. *Res:* Stratigraphy of the Eocene deposits of the Big Horn Basin, Wyoming; stratigraphy of the upper Eocene and Oligocene deposits of Egypt; stratigraphy of the Siwalik deposits in northern India; stratigraphy and sedimentation of the Plio-Pleistocene deposits in the East Rudolf Basin, Kenya. *Mailing Add:* Dept Earth Sci 253 Sci Iowa State Univ Ames IA 50011

VONDRAK, EDWARD ANDREW, b Chicago, Ill, Nov 12, 38; m 61; c 3. MATHEMATICS. *Educ:* Knox Col, AB, 60; Vanderbilt Univ, MA, 63, PhD(physics), 65. *Prof Exp:* Teaching fel physics, Vanderbilt Univ, 61-64; from asst prof to assoc prof, 67-72, PROF PHYSICS & MATH, UNIV INDIANAPOLIS, 72- *Mem:* Am Asn Physics Teachers; Math Asn Am. *Mailing Add:* 8219 Burn Ct Indianapolis IN 46217

VON DREELE, ROBERT BRUCE, b Minneapolis, Minn, Dec 10, 43; m 78; c 1. SOLID STATE CHEMISTRY, CRYSTALLOGRAPHY. *Educ:* Cornell Univ, BS, 66, PhD(chem), 71. *Prof Exp:* From asst prof to prof chem, Ariz State Univ, 71-86; STAFF MEM, LOS ALAMOS NAT LAB, 86- *Concurrent Pos:* NSF fel, Dept Inorg Chem, Oxford Univ, 72-73; Fulbright fel, Rutherford Lab, 86. *Mem:* Am Crystallog Asn; Sigma Xi. *Res:* Neutron scattering x-ray crystal structure analysis; solid state chemistry; powder diffraction. *Mailing Add:* Lansce MS H805 Los Alamos Nat Lab Los Alamos NM 87545

VONEIDA, THEODORE J, b Auburn, NY, Aug 26, 30; m 56; c 3. NEUROBIOLOGY. *Educ:* Ithaca Col, BS, 53; Cornell Univ, MEd, 54, PhD(zool), 60. *Prof Exp:* Res assoc neuroanat, Walter Reed Army Inst Res, 54-56; asst comp neurol, Cornell Univ, 56-59; from asst prof to prof, Sch Med, Case Western Reserve Univ, 62-76; PROF & CHMN NEUROBIOL DEPT, COL MED, NORTHEASTERN OHIO UNIV, 78- *Concurrent Pos:* USPHS res fel neurobiol, Calif Inst Technol, 60-62; adj prof anat & biol, Case Western Reserve Univ, 76- *Honors & Awards:* Environ Qual Award, Environ Protection Agency, 74. *Mem:* AAAS; Am Asn Anatomists; Soc Neurosci. *Res:* Utilization of neuroanatomical and behavioral techniques to investigate the central nervous system. *Mailing Add:* NE Ohio Univ Col Med State Rte 44 Rootstown OH 44272

VON ESCHEN, GARVIN L(EONARD), b Morristown, Minn, July 22, 13; m 38; c 3. AERONAUTICAL ENGINEERING. *Educ:* Univ Minn, BAeroE, 36, MS, 39. *Prof Exp:* Instr, Univ Minn, 36-40, instr aeronaut eng, 40-42, from asst prof to assoc prof, 42-46, consult & proj supvr, Res Found, 46-76, prof & chmn dept, 46-80, EMER PROF AERONAUT & ASTRONAUT ENG, OHIO STATE UNIV, 80- *Concurrent Pos:* Consult, Minneapolis-Honeywell Regulator Corp, 44-46 & Denison Res Found, 57-60; engr, Lockheed Aircraft, 39-40; res specialist, NAm Aviation, 55, 57; mem, Am Soc Eng Educ, 71-75. *Mem:* Am Inst Aeronaut & Astronaut; Am Phys Soc; Am Soc Eng Educ. *Res:* Compressible flow; experimental aerodynamics. *Mailing Add:* 3758 Chevington Rd Columbus OH 43220

VON ESSEN, CARL FRANCOIS, b Tokyo, Japan, May 17, 26; nat US; m 79; c 3. RADIOTHERAPY & ONCOLOGY. *Educ:* Stanford Univ, AB, 48, MD, 52; Am Bd Radiol, dipl, 58. *Prof Exp:* Res fel cancer, Stanford Univ, 57-59; from instr to assoc prof radiol, Yale Univ, 59-69; prof radiol & oncol & dir radiation ther, Univ Calif, San Diego, 69-77; res prof radiol, Univ NMex, 77-78; leader, Pion Ther Proj, Swiss Inst Nuclear Res, 78-84; PROF, MED FAC, UNIV BASEL, SWITZ, 82-; ADJ PROF, RADIATION MED, BROWN UNIV, PROVIDENCE, RI, 85- *Concurrent Pos:* Vis prof, Christian Med Col, Vellore, India, 65-66; mem rev comt, Radiation Study Sect, NIH, 67-69 & Cancer Res Ctr, 70-74; mem staff, Ludwig Inst Cancer Res, 75-76; WHO consult, Sri Lanka, 84-85, Zimbabwe & Jordan, 88; clin assoc radiation med, Mass Gen Hosp, 85-; dir, Dept Radiation Oncol, Southwood Community Hosp, Norfolk, Ma, 85- *Honors & Awards:* Swiss Cancer Prize, 81. *Mem:* Am Soc Ther Radiologists; Europ Soc Therapeut Radiol Oncol; Am Col Radiol. *Res:* Organization of postgraduate training programs in oncology and radiotherapy in developing countries. *Mailing Add:* Dept Radiation Oncol Southwood Community Hosp Norfolk MA 02056

VON EULER, LEO HANS, b Stockholm, Sweden, Jan 31, 31; US citizen; m 55; c 2. PATHOLOGY. *Educ:* Williams Col, BA, 52; Yale Univ, MD, 59. *Prof Exp:* Trainee path, Sch Med, Yale Univ, 59-61, trainee pharmacol, 61-63; fel hemat, Dept Clin Path, NIH, 65-66; scientist, Sect Nutrit Biochem, Nat Inst Arthritis & Metab Dis, 66-67, prog adminr path res training progs, 67-72, spec asst to dir, 72-74, DEP DIR, NAT INST GEN MED SCI, 74-, ACTG DIR, PHYSIOL & BIOMED ENG PROG, 78- *Mem:* Am Asn Path. *Res:* Purine and pyrimidine metabolism; orotic acid induced fatty liver in the rat; biochemical and histological changes. *Mailing Add:* NIH 9000 Rockville Pike Bethesda MD 20892

VON FISCHER, WILLIAM, chemistry, for more information see previous edition

VON GIERKE, HENNING EDGAR, b Karlsruhe, Ger, May 22, 17; m 50; c 2. BIOACOUSTICS, BIOMECHANICS. *Educ:* Karlsruhe Tech, Dipl Ing, 43, DrEng, 44. *Prof Exp:* Asst acoust, Inst Theoret Elec Eng & commun techniques, Karlsruhe Tech, 44-47, lectr, 46; consult, Wright-Patterson AFB, 47-54, chief bioacoust br, 54-63, dir, Biodynamics & Bionics Div, 63-88, EMER DIR, BIODYNAMICS & BIONICS DIV, WRIGHT PATTERSON AFB, 88- *Concurrent Pos:* Mem comt hearing bioacoust & biomech, Nat Res Coun, 53-, mem bioastronaut comt, 59-61; mem adv comt flight med & biol, NASA, 60-61; assoc prof, Ohio State Univ, 63-; mem, White House Ad Hoc Panel Jet Aircraft Noise, 66; clin prof, Wright State Univ, 80- *Honors & Awards:* Eric LiljenKrantz Award, Aerospace Med Asn, 66 & Arnold D Tuttle Award, 74; Hubertus Strughold Medal, 80; Silver Medal, Acoust Soc Am, 81; H R Lissner Award, Am Soc Mech Engrs, 83; Rayleigh Medal, UK Inst Acoust, 89. *Mem:* Nat Acad Eng; fel Aerospace Med Asn (vpres, 66-67); hon fel Inst Environ Sci; Int Acad Astronaut; fel Acoust Soc Am (pres, 78); Int Acad Aviation & Space Med; Inst Noise Control Eng; Biomed Eng Soc. *Res:* Physical, physiological and psychological acoustics; biodynamics; effects of noise, vibration and impact on man; communication biophysics; bionics; bioengineering; author of over 160 technical publications, book chapters. *Mailing Add:* 1325 Meadow Lane Yellow Springs OH 45387

VON GOELER, EBERHARD, b Berlin, Ger, Feb 22, 30; m 60; c 3. HIGH ENERGY PHYSICS INSTRUMENTATION. *Educ:* Univ Ill, MS, 55, PhD(physics), 61. *Prof Exp:* Res assoc physics, Univ Ill, 60-61; res scientist, Deutsches Elektronen Synchrotron, Hamburg, Ger, 61-63; from asst prof to assoc prof, 63-73, PROF PHYSICS, NORTHEASTERN UNIV, 73- *Concurrent Pos:* Vis prof, Univ Hamburg, 67-68; vis scientist, Nat Accelerator Lab, Ill, 71-72 & Stanford Linear Accelerator, Calif, 78-79; exec officer, Physics Dept, Northeastern Univ, 80-81; vis prof, Univ Houston, 86-87. *Mem:* Am Phys Soc. *Res:* Surface physics; tests of quantum electrodynamics; photoproduction of vector mesons, antibaryons; meson spectroscopy; high mass bosons; colliding electron beam experiments; non-accelerator elementary particle physics; nucleon spin-structure functions; counter techniques in high energy physics. *Mailing Add:* Dept Physics Northeastern Univ 360 Huntington Ave Boston MA 02115

VON GONTEN, (WILLIAM) DOUGLAS, b Rockdale, Tex, Feb 15, 34; m 60; c 2. PETROLEUM ENGINEERING. *Educ:* Tex A&M Univ, BS(geol) & BS(petrol eng), 57, MS, 65, PhD(petrol eng), 66. *Prof Exp:* Asst res engr, Tex Eng Exp Sta, 66-67; from asst prof to assoc prof, 67-76, PROF PETROL ENG, TEX A&M UNIV, 76-, HEAD DEPT, 76- *Concurrent Pos:* Petrol prod engr, Magnolia Petroleum Co, 57-58 & Mobil Oil Co, 60-62; res eng lectr, Exxon Corp, Houston, 74- & Soc Petrol Engrs, 75-; reservoir eng consult, Hunt Oil, Getty Oil, Expolsives Corp Am, Tenneco Oil & Gulf; mem bd dir, Soc Petrol Engrs, 81-84. *Mem:* Soc Petrol Engrs; Am Asn Petrol Geol; Soc Prof Well Log Analysts; Am Soc Eng Educ. *Res:* Subsurface formation evaluation; reservoir engineering. *Mailing Add:* 2506 Memorial Dr Bryan TX 77801

VON GUTFELD, ROBERT J, b Berlin, Ger, Mar 5, 34; US citizen; m 77; c 2. SOLID STATE PHYSICS. *Educ:* Queens Col, BS, 54; Columbia Univ, MA, 57; NY Univ, PhD(physics), 65. *Prof Exp:* Substitute instr physics, Queens Col, 54-55; engr, Sperry Gyroscope Co, 57-60; RES STAFF MEM, T J WATSON RES CTR, IBM CORP, 60- *Honors & Awards:* Res Award, Electrochem Soc, 84. *Mem:* Fel Am Phys Soc; Electrochem Soc. *Res:* Thermal transport in solids using heat pulse techniques; amorphous semiconductor and dye laser research and applications to optical memories; transverse thermoelectric effects in metallic thin films; laser thermoelastic waves; laser enhanced plating and etching; electrochemical techniques for repair of microcircuits. *Mailing Add:* T J Watson Res Ctr IBM Corp Yorktown Heights NY 10598

VON HAGEN, D STANLEY, b Nashville, Tenn, Dec 21, 37; m 59; c 2. PHARMACOLOGY. *Educ:* Carson-Newman Col, BS, 59; Vanderbilt Univ, PhD(pharmacol), 65. *Prof Exp:* Res assoc pharmacol, Vanderbilt Univ, 65-66, instr, 66; instr, 67-69, ASST PROF PHARMACOL, NJ MED SCH, COL MED & DENT, NJ, 69- *Mem:* AAAS; NY Acad Sci; Sigma Xi. *Res:* Smooth muscle physiology and pharmacology; physiological role of calcium ion in smooth muscle function. *Mailing Add:* Dept Pharmacol New Jersey Col Med 100 Bergen St NJ Col Med 185 S Orange Ave NJ 07103

VON HERZEN, RICHARD P, b Los Angeles, Calif, May 21, 30; m 58; c 2. MARINE GEOPHYSICS. *Educ:* Calif Inst Technol, BS, 52; Harvard Univ, AM, 56; Univ Calif, PhD(oceanog), 60. *Prof Exp:* Lab asst oceanog, Scripps Inst Oceanog, 52-53, geophysicist, 58-60, asst res geophysicist, 60-64; dep dir off oceanog, UNESCO, 64-66; assoc scientist, 66-73, chmn, Dept Geol & Geophys, 82-85, SR SCIENTIST, WOODS HOLE OCEANOG INST, 73- *Concurrent Pos:* Assoc ed, J Geophys Res, Am Geophys Union, 69-71; vis res geophysicist & lectr, Scripps Inst Oceanog, 74-75; vis prof & lectr, Mass Inst Technol, 82. *Mem:* Fel Am Geophys Union. *Res:* Structure and dynamics of the earth beneath the ocean floor especially as evidenced from geothermal studies. *Mailing Add:* Woods Hole Oceanog Inst Woods Hole MA 02543

VON HIPPEL, ARTHUR R, b Rockstock MecKlenburg, EGer, Nov 19, 1898; wid; c 5. MOLECULAR ENGINEERING. *Educ:* Univ Gottingen, PhD, 24. *Prof Exp:* Prof molecular sci, Göttingen Univ, 26-34; prof, Nils Bohr Lab, Copenhagen, 35-36; prof molecular sci, Lab Insulation Res, Mass Inst Technol, 36-85, emer prof, 85-; RETIRED. *Mem:* Nat Acad Eng; Am Phys Soc; AAAS. *Mailing Add:* 265 Glen Rd Weston MA 02193

VON HIPPEL, FRANK, b Cambridge, Mass, Dec 26, 37; m 87; c 1. THEORETICAL PHYSICS. *Educ:* Mass Inst Technol, SB, 59; Oxford Univ, DPhil(physics), 62. *Prof Exp:* Res assoc physics, Univ Chicago, 62-64; res assoc, Cornell Univ, 64-66; asst prof, Stanford Univ, 66-70; mem staff theory group, High Energy Physics Div, Argonne Nat Lab, 70-73; resident fel, Nat Acad Sci, 73-74; res scientist, 74-78, sr res physicist, Ctr Environ Studies, 78-83, PROF PUB & INT AFFAIRS, PRINCETON UNIV, 83- *Concurrent Pos:* Sloan Found fel, 67-70; consult nuclear energy policy, Off Technol Assessment, Gen Acct Off, House Interior Comt, US Cong, Nuclear Regulatory Comn, Dept Energy, 75. *Honors & Awards:* Am Phys Soc Award, 77; Pub Interest Award, Fedn Am Scientists, 90. *Mem:* Am Phys Soc; AAAS; Fedn Am Scientists. *Res:* Nuclear energy policy; energy general policy; nuclear arms control and disarmament. *Mailing Add:* Dept Pub Int Affairs Princeton Univ Princeton NJ 08544

VON HIPPEL, PETER HANS, b Gottingen, Ger, Mar 13, 31; nat US; m 54; c 3. BIOPHYSICAL CHEMISTRY, MOLECULAR BIOLOGY. *Educ:* Mass Inst Technol, BS, 52, MS, 53, PhD(biophys), 55. *Prof Exp:* Asst phys biochem, Mass Inst Technol, 53, NIH fel, 55-56; phys biochemist, US Naval Med Res Inst, Md, 56-59; from asst prof to assoc prof biochem, Dartmouth Med Sch, 59-67; res assoc, Univ Ore, 67-69, dir, Inst Molecular Biol, 69-80, chmn dept, 80-86, PROF CHEM, UNIV ORE, 67- *Concurrent Pos:* Sr fel, NIH, 59-67; chmn, Gordon Res Conf Physics & Phys Chem of Biopolymers, 68; mem corp vis comn, Dept Biol, Mass Inst Technol, 73-77; Guggenheim found fel, 73-74; mem bd sci counr, Nat Inst Arthritis, Metab & Digestive Dis, NIH, 74-78; mem coun, Inst Gen Med Sci, NIH, 81-86; mem adv comt to dir, NIH, 88-; res prof, Am Cancer Soc, 88- *Mem:* Nat Acad Sci; AAAS; Biophys Soc (pres, 73-74); Am Soc Biol Chem; Am Chem Soc; Soc Gen Physiol; fel Am Acad Arts & Sci; Sigma Xi. *Res:* Physical biochemistry of macromolecules; structure, function and interactions of proteins and nucleic acids; molecular aspects of control of genetic expression. *Mailing Add:* Inst Molecular Biol Univ Ore Eugene OR 97403

VON HOERNER, SEBASTIAN, b Goerlitz, Ger, Apr 15, 19; m 42; c 3. ASTRONOMY, ENGINEERING. *Educ:* Univ Goettingen, Ger, dipl, 49, PhD(physics), 51. *Prof Exp:* Scientist astrophys, Max Planck Inst Physics, Goettingen, 49-57; Fulbright fel, Mt Wilson & Palomar Observ, Pasadena, Calif, 55-56; scientist, Astron Rechen-Inst, Heidelberg, Ger, 57-62; scientist astrosphys & eng, Nat Radio Astron Observ, Green Bank, 62-85; RETIRED. *Concurrent Pos:* Vis prof, Univ Switz, 62, Univ Calif, Los Angeles, 69, Nat Univ, Mexico City, 71, Max Planck Inst Radioastron, Bonn, 72 & 75, Cornell Univ, Ithaca, NY, 74 & Univ Okla, Norman, 77; eng consult, 85- *Mem:* Deut Astron Gesellschaft; Max-Planck-Gesellschaft; Ver Deut Wiss; Int Astron Union; Alexander von Humboldt award, 84. *Res:* Astrophysics, star formation, stellar dynamics; radio astronomy, cosmology; antenna design, structural optimization; life in space; interstellar communication. *Mailing Add:* Krummenacker-Str 186 Esslingen 7300 Germany

VON HUENE, ROLAND, b Los Angeles, Calif, Jan 30, 29; m 53; c 3. GEOLOGY. *Educ:* Univ Calif, Los Angeles, AB, 53, PhD, 60. *Prof Exp:* Gen geologist, US Naval Ord Test Sta, 53-67; geophysicist, US Geol Surv, 67-88; GEOPHYSICIST, UNIV KIEL, 88- *Concurrent Pos:* Asst, Univ Calif, Los Angeles, 55-57; Fulbright grant, Innsbruck, 57-58; mem comt Alaska earthquake, Nat Acad Sci; Pac site panel, Joint Ocean Insts for Deep Earth Sampling; dep chief off marine geol, US Geol Surv, 73-75; panel mem active margin & site surv, Int Prog Ocean Drilling, 74-; mem nat comt, US Geodynamics Comt, 75- *Mem:* AAAS; Am Geophys Union; Sigma Xi; Geol Soc Am; Ger Geol Asn. *Res:* Marine gravity; magnetics; seismic profiling; tectonics; structural geology; gravimetry. *Mailing Add:* Geomar Wischhofstr 1-4 Kiel 2300 14 Germany

VON HUNGEN, KERN, b Modesto, Calif, May 2, 40; m 67; c 2. NEUROCHEMISTRY, PSYCHOPHARMACOLOGY. *Educ:* Reed Col, BA, 62; Ind Univ, PhD(biol chem), 68. *Prof Exp:* Fel neurobiochem, Chem Biodynamics Lab, Univ Calif, Berkeley, 68-70; fel neurobiochem, Dept Biol Chem, Univ Calif, Los Angeles, 70-71; res chemist, Neurochem Lab, 76-79, CHIEF, BRAIN BIOCHEM LAB, VET ADMIN MED CTR, SEPULVEDA, CALIF, 79- *Concurrent Pos:* Asst res biochemist, Dept Biol Chem, Univ Calif, Los Angeles, 71- *Mem:* Am Soc Neurochem. *Res:* Functional neurochemistry: regulatory mechanisms involving brain membranes, proteins, neurotransmitters and cyclic nucleotides and the effects of alcohol and age on these systems; biochemical aspects of behavior. *Mailing Add:* Brain Biochem Mail Code 151B2 Vet Admin Med Ctr Sepulveda CA 91343

VON KESZYCKI, CARL HEINRICH, physics, for more information see previous edition

VON KLITZING, KLAUS, b Schroda, Ger, June 28, 43. SOLID STATE PHYSICS. *Educ:* Technische Univ Braunschweig, Univ Wurzburg. *Hon Degrees:* Dr, chemnitz, Maryland, Antwerp, Monpellier. *Prof Exp:* Prof physics, Technische Univ, Munich, 80-84; DIR, MAX-PLANCK INST SOLID RES, STUTTGARD, 84- *Honors & Awards:* Nobel Prize in Physics, 85. *Mailing Add:* Max-Planck Inst fur Festkorperforschung Heisenbergstra 1 Postfach 800665 Stuttgart 80 7000 Germany

VON KORFF, RICHARD WALTER, b Davenport, Iowa, Jan 6, 16; m 43; c 3. BIOCHEMISTRY, ENZYMOLOGY. *Educ:* Univ Minn, BA, 47, PhD(physiol chem), 51; Am Bd Clin Chem, dipl, 74. *Prof Exp:* Anal chemist, Testing & Res Lab, Deere & Co, Ill, 37-41; asst & sr sci aide, Anal & Phys Chem Div, Northern Regional Res Lab, USDA, 41-43, jr chemist, Agr Residues Div, 43-45; asst prof pediat & physiol chem, Univ Minn, 55-66; dir biochem res, Friends of Psychiat Res, Spring Grove State Hosp, Baltimore, 66-68; dir biochem res, Md Psychiat Res Ctr, 68-77; res prof biochem, Mich Molecular Inst, Midland, Mich, 77-85; RETIRED. *Concurrent Pos:* Fel, Inst Enzyme Res, Univ Wis, 51-52; Whitney Found fel biochem, Dept Pediat, Heart Hosp, Univ Minn, 52-53, Am Heart Asn fel, 53-55, USPHS sr res fel biochem, 60-66; chmn subcomt enzymes, Comt Biol Chem, Nat Acad Sci-Nat Res Coun, 61-67 & 69-76, adj prof, Dept Med Chem, Sch Pharm, 71-73; adj assoc prof, Dept Macromolecular Sci, Case-Western Univ, 79-; adj prof, Dept Biochem, Dent Sch, Univ Md, 75-77; adj prof dept chem, Cent Mich Univ, 80-85. *Mem:* Am Soc Biol Chemists; Am Chem Soc; Nat Acad Clin Biochem; Int Soc Neurochem; Am Soc Neurochem. *Res:* Monoamine oxidase, reaction mechanism, nature of action of reversible inhibitors; enzymic control mechanisms; monoamine oxidase. *Mailing Add:* 15 Rosemary Ct Midland MI 48640

VON LEDEN, HANS VICTOR, b Ger, Nov 20, 18; nat US; m 48; c 2. LARYNGOLOGY. *Educ:* Loyola Univ, Ill, MD, 42; Am Bd Otolaryngol, dipl, 45. *Prof Exp:* Intern, Mercy Hosp-Loyola Univ Clins, 41-42; resident, Presby Hosp, Chicago, 42-43; fel otolaryngol & plastic surg, Mayo Found, Univ Minn, 43-45, first asst, Mayo Clin, 45; clin assoc otolaryngol, Stritch Sch Med, Loyola Univ, Ill, 47-51; from asst prof to assoc prof, Sch Med, Northwestern Univ, 52-61; assoc prof surg, Sch Med, Univ Calif, Los Angeles, 61-66; PROF BIO-COMMUN, UNIV SOUTHERN CALIF, 66- *Concurrent Pos:* Consult, USN, 47-; assoc prof, Cook County Grad Sch Med, 49-58; med dir, William & Harriet Gould Found, 55-59; pres, Inst Laryngol & Voice Dis, 59-65, med dir, 65-; vis prof, US & 25 for countries. *Honors & Awards:* Bucranio, Univ Padua, 58; Gold Medal, Ital Res Cross, 59; Hektoen Medal, AMA, 60; Sci Awards, Am Speech & Hearing Asn, 60, 62 & 65; Casselberry Award, Am Laryngol Asn, 62; Manuel Garcia Prize, Int Asn Logoped & Phoniatrics, 68; Gutzmann Medal, Ger Soc Otolaryngol, 80; Sci Achievement Award, AMA, 80. *Mem:* Fel AAAS; fel Am Acad Otolaryngol; fel Am Col Surgeons; fel Int Col Surgeons (pres, 72); fel Am Speech & Hearing Asn; Sigma Xi. *Res:* Voice and speech; laryngology. *Mailing Add:* Westwood Med Plaza 10921 Wilshire Blvd Los Angeles CA 90024

VON LICHTENBERG, FRANZ, b Miskolc, Hungary, Nov 29, 19; nat US; m 49; c 6. PATHOLOGY. *Educ:* Nat Univ Mex, MD, 45; Am Bd Path, dipl, 51. *Hon Degrees:* Dr, Nat Univ Nicaragua, 59; MA, Harvard Univ, 68. *Prof Exp:* Pathologist, Hosp Exp Nutrit, Mex, 47; prof path, Nat Univ Mex, 48-52; from asst prof to assoc prof, Univ PR, 53-58; instr, 58-59, assoc, 59-62, from asst prof to assoc prof, 62-74, PROF PATH, HARVARD MED SCH, 74- *Concurrent Pos:* Fel, Mex Dept Health, 46; Kellogg Found, Am Col Physicians & Latin Am fels, 50-51; pathologist, Clin Hosp, Bahia, Brazil, 51-52, Gen Hosp, Mex, 52-53 & San Juan City Hosp, PR, 53-58; assoc pathologist, Peter Bent Brigham Hosp, 58-62, sr assoc, 62-68, pathologist, 68-81, sr pathologist, 82; assoc mem comn parasitol, Armed Forces Epidemiol Bd, 59, mem, 64-71; consult, Div Parasitic Dis, WHO, 65; mem study sect trop med & parasitol, NIH, 68-73; mem steering comn Schistosomiasis, WHO-TDR, 77- (chmn, 83); James W McLaughlin vis prof, Univ Tex Med Br, 80; Theobald Smith Lectr, NY Soc Trop Med; Fogarty Scholar-in-Residence, NIHFIC, 83, 85; Wellcome lectr, London, 86. *Mem:* Am Soc Trop Med & Hyg (vpres, 76, pres, 85); Am Asn Path; Fedn Am Socs Exp Biol. *Res:* Tropical and parasitic diseases; schistosomiasis; filariasis; liver pathology; immunopathology. *Mailing Add:* Brigham & Women's Hosp 75 Francis St Boston MA 02115

VON MALTZAHN, WOLF W, b Ger, Oct 3, 46; m 73; c 3. MEDICAL INSTRUMENTATION & CLINICAL ENGINEERING, SIGNAL PROCESSING. *Educ:* Ohio State Univ, MS, 71; Univ Stuttgart, Ger, dipl elec eng, 74; Univ Hannover, Ger, Dr(biomed eng), 79. *Prof Exp:* Res assoc med instrumentation, Inst Biomed Eng, Univ Stuttgart, 71-74; asst prof physiol, Inst Physiol, Univ Essen, Ger, 74-79; asst prof, 79-84, ASSOC PROF BIOMED ENG, UNIV TEX, ARLINGTON, 84- *Concurrent Pos:* Consult investr, St Elizabeth Hosp, Beaumont, 87; mem, forensic testing staff, St Michael's Hosp, 88 & P Chamblin, Weber & Mahaffy, Attorneys, 90. *Mem:* Inst Elec & Electronics Engrs Eng Med & Biol Soc; Am Soc Eng Educ; Biomed Eng Soc. *Res:* Measurement of blood pressure and blood flow; infiltration detection; sensors; medical devices for patient monitoring. *Mailing Add:* PO Box 19138 Arlington TX 76019

VON MEERWALL, ERNST DIETER, b Vienna, Austria, Dec 29, 40. PHYSICS. *Educ:* Northern Ill Univ, BS, 63, MS, 65; Northwestern Univ, Evanston, PhD(physics), 69. *Prof Exp:* Res assoc, Dept Metall & Mat Res Lab, Univ Ill, Urbana, 69-71; asst prof, 71-74, assoc prof, 74-80, PROF PHYSICS, UNIV AKRON, 80- *Mem:* Am Phys Soc; Sigma Xi. *Res:* Solid state experiment; nuclear magnetic resonance, M-ssbauer effect and magnetic susceptibility; alloys; nuclear quadrupole effect; polymers; numerical methods. *Mailing Add:* Dept Physics Univ Akron Akron OH 44325

VON MOLNÁR, STEPHAN, b Leipzig, Ger, June 26, 35; US citizen; m 56; c 2. SEMICONDUCTORS, MAGNETISM. *Educ:* Trinity Col (Conn), BS, 57; Univ Maine, MS, 59; Univ Calif, Riverside, PhD(physics), 65. *Prof Exp:* Mem res staff physics, Polychem Div, Exp Sta, E I du Pont de Nemours & Co, 59-60; mgr, Coop Phenomena Group, 68-89, MEM RES STAFF PHYSICS, THOMAS J WATSON RES CTR, IBM CORP, 65-, SR MGR NOVEL STRUCT PHYSICS, 89- *Concurrent Pos:* Vis scientist, Cavendish Lab, Cambridge, 69; Sr res fel, Imp Col, Univ London, 73-74; vis scientist, Nat Ctr Soc Republicans, Grenoble, 82, 83; mem, Prog Adv Comt, Meson Physics Facil, Los Alamos Nat Lab, 84-87; mem, Pane Lon Diluted Magnetism Semiconductors, Comn Phys Sci, Nat Res Coun 89-92. *Honors & Awards:* Alexander von Humboldt Sr US Scientist Award, 86. *Mem:* Fel Am Phys Soc; Mat Res Soc. *Res:* Paramagnetic and ferromagnetic resonance; transport, optical and magneto-optical properties of magnetic semiconductors; tunneling spectroscopy of superconductors and semiconductors; low temperature specific heat. *Mailing Add:* Thomas J Watson Res Ctr IBM Corp PO Box 218 Yorktown Heights NY 10598

VONNEGUT, BERNARD, b Indianapolis, Ind, Aug 29, 14; wid; c 5. PHYSICAL CHEMISTRY. *Educ:* Mass Inst Technol, BS, 36, PhD(phys chem), 39. *Prof Exp:* Res assoc, Preston Labs, Pa, 39-40 & Hartford Empire Co, Conn, 40-41; chem eng, Mass Inst Technol, 41-42, meteorol, 42-45; res labs, Gen Elec Co, 45-52; mem staff, Arthur D Little Inc, 52-67; PROF ATMOSPHERIC SCI, STATE UNIV NY ALBANY, 67-, SR RES SCIENTIST ATMOSPHERIC SCI RES CTR, 67- *Mem:* AAAS; Am Meteorol Soc; Am Geophys Union; Meteorol Soc Japan; Royal Meteorol Soc. *Res:* Nucleation phenomena; cloud seeding; surface chemistry; aerosols; atmospheric electricity. *Mailing Add:* 35 Norwood St Albany NY 12203

VON NEIDA, ALLYN ROBERT, b West Reading, Pa, May 7, 32; m 55; c 3. METALLURGY, MATERIALS SCIENCE. *Educ:* Lehigh Univ, BS(elec eng) & BS(metall eng), 55; Yale Univ, PhD(metall), 60. *Prof Exp:* Res metallurgist, Olin Mathieson Chem Corp, 55-57; res asst metall, Yale Univ, 60-61; MEM TECH STAFF, BELL TEL LABS, MURRAY HILL, 61- *Res:* Magnetics; crystal growth. *Mailing Add:* 133 Ashland Rd Summit NJ 07901

VON NOORDEN, GUNTER KONSTANTIN, b Frankfurt, Ger, Mar 19, 28; US citizen; c 1. OPHTHALMOLOGY. *Educ:* Univ Frankfurt, MD, 54; State Univ Iowa, MS, 60. *Prof Exp:* Rotating intern, St Vincent Infirmary, Little Rock, Ark, 54-56; fel ophthal, Cleveland Clin, 56-57; resident, Med Ctr, State Univ Iowa, 57-60, asst prof, Sch Med, 61-63; from assoc prof to prof, Johns Hopkins Univ, 63-72; PROF OPHTHAL, BAYLOR COL MED, 72- *Concurrent Pos:* Nat Inst Neurol Dis & Blindness spec trainee, Univ Tubingen, 60-61; Nat Inst Neurol Dis & Blindness spec fel, Univ Iowa, 61-62; mem, Armed Forces Nat Res Coun Vision, 64-68; Int Strabismological Asn Bielschowsky lectr, 70; adj prof neurol sci, Sch Biol Sci, Univ Tex, 72-; pres, Am Orthoptic Coun, 73-74. *Honors & Awards:* Hectoen Gold Medal, AMA, 60; Honor Award, Am Acad Ophthal & Otolaryngol, 70. *Mem:* Am Ophthal Soc; fel Am Acad Ophthal & Otolaryngol; Asn Res Strabismus (secy, 72-73); Int Strabismological Asn (secy-treas, 68-74); Pan Am Ophthal Asn. *Res:* Investigation of clinical and laboratory aspects of neuromuscular anomalies of the eyes, especially amblyopia; improvement of our knowledge of strabismus and the basic morphological and neurophysiological aspects of different forms of amblyopia. *Mailing Add:* Tex Childrens Hosp PO Box 20269 Houston TX 77025

VON OHAIN, HANS JOACHIM, b Dessau, Ger, Dec 14, 11; US citizen; m 49; c 4. ENERGY CONVERSION, GAS TURBINE TECHNOLOGY. *Educ:* Univ Goettingen, PhD(physics), 35. *Prof Exp:* Inventor & developer, Heinkel A/C Co, Ger, 36-47; res scientist, Aero Res Labs, 47-63; chief scientist, Aero Space Res Lab, USAF, 63-75, Propulsion Lab, Wright-Patterson AFB, 75-79; SR RES ENGR, RES INST, UNIV DAYTON, 80-, PROF MECH ENG, 84- *Concurrent Pos:* Adj prof aero propulsion, Res Inst, Univ Dayton, 79-84; adj prof, Univ Fla; consult aero propulsion, 80-; Charles Lindbergh prof, Nat Air & Space Mus, Smithsonian Inst, 84-85. *Honors & Awards:* Goddard Award, Am Inst Aeronaut & Astronaut, 66; Aachen & Munich Prize Technol & Appl Natural Sci, 85; Charles Stark Draper Prize, Nat Acad Eng, 91. *Mem:* Nat Acad Eng; hon fel Am Inst Aeronaut & Astronaut; distinguished mem Jet Pioneer's Asn USA. *Res:* Advanced ejector application to aircraft; advanced methods for particle separation in multicomponent flows; feasibility of a vapor injection heat pipe; multicomponent flow compressor concept; ejector heat pump system concept. *Mailing Add:* Aerospace Mech Div Res Inst Univ Dayton 300 Col Park Dayton OH 45469-0110

VON OSTWALDEN, PETER WEBER, b Reichenberg, Czech, June 1, 23; m 46; c 1. ORGANIC CHEMISTRY. *Educ:* Univ Graz, Doctorandum, 50; Columbia Univ, MA, 54, PhD(pyridine chem), 58. *Prof Exp:* Process develop chemist, Merck & Co, Inc, Cherokee Plant, Pa, 57-63; from asst prof to assoc prof, 63-77, PROF CHEM, YOUNGSTOWN STATE UNIV, 77- *Concurrent Pos:* Vis assoc, Calif Inst Technol, 70-71; vis fel, Princeton Univ, 79-80. *Mem:* Am Chem Soc; Sigma Xi. *Res:* Pyridine and steroid chemistry; heterocyclic nitrogen oxides; chemistry of heterocyclic compounds; spectroscopy; organic applications. *Mailing Add:* Dept Chem Youngstown State Univ Youngstown OH 44503

VON-RECKLINGHAUSEN, DANIEL R, b New York, NY, Jan 22, 25; m 60; c 2. ACOUSTICS, ELECTRONICS ENGINEERING. *Educ:* Mass Inst Technol, SB, 51. *Prof Exp:* Chief engr, HH Scott, 51-73; staff consult, Electro-Audio Dynamics, 73-84; CONSULT, DR VON-RECKLINGHAUSEN CONSULTS, 84- *Honors & Awards:* Gold Medal, Audio Eng Soc, 78. *Mem:* Fel Inst Elec & Electronics Engrs; Audio Eng Soc; Acoust Soc Am. *Mailing Add:* Dr Von-Recklinghausen Consults 17 Glen Dr Hudson NH 03051

VON RIESEMANN, WALTER ARTHUR, b Brooklyn, NY, Feb 12, 30; c 2. STRUCTURAL MECHANICS, SEISMIC ENGINEERING. *Educ:* Polytech Inst Brooklyn, BCE, 58; Univ Ill, Urbana, MSCE, 59; Stanford Univ, PhD(civil eng), 68. *Prof Exp:* Res engr, Alcoa Res Labs, 59-60; staff mem, 60-77, SUPVR STRUCT MECH, SANDIA NAT LABS, 77- *Concurrent Pos:* Adj prof, Univ NMex, 74-77; pres, NMex Sect, Am Soc Civil Engrs, 75-76; chmn, NMex Sect, Am Soc Mech Engrs, 80-81; mem, Sr Seismic Rev & Adv Panel, Nuclear Regulatory Comn, 84- *Honors & Awards:* Robert Ridgway Award, Am Soc Civil Engrs, 58. *Mem:* Am Nuclear Soc. *Res:* Analytical (finite element method) and experimental investigations of mechanical components and nuclear power plant containments; high-speed aircraft impact tests. *Mailing Add:* 7928 Woodhaven Dr NE Albuquerque NM 87109-5261

VON RIESEN, DANIEL DEAN, b Beatrice, Nebr, Nov 20, 43. ORGANIC CHEMISTRY. *Educ:* Hastings Col, BA, 65; Univ Nebr, PhD(chem), 71. *Prof Exp:* Instr chem, Hastings Col, 70-71; asst prof, Hamilton Col, 71-72; prof chem, Roger Williams Col, 72-91; CONSULT, 91- *Concurrent Pos:* Vis prof, Univ Nebr, 84; high performance liquid chromatography specialist, Isco Inc, 85-86. *Mem:* Am Chem Soc. *Res:* Cycloaddition reactions of heterocumulenes; chemical education; biopolymer separations; high performance liquid chromatography; microcomputer applications. *Mailing Add:* Chem Dept Roger Williams Col Bristol RI 02809

VON ROSENBERG, DALE URSINI, b Austin, Tex, Sept 5, 28; m 53; c 4. KINETIC MODELS FOR GENERATION OF PETROLEUM FROM SOURCE ROCKS, PETROLEUM EXPLORATION & PRODUCTION. *Educ:* Univ Tex, BS, 49; Mass Inst Technol, ScD(chem eng), 53. *Prof Exp:* Sr res engr, Humble Oil & Refining Co, 53-57; assoc prof chem eng, La State Univ, 57-63; prof, Tulane Univ, 63-76 & Univ Tulsa, 76-79; SR RES ASSOC, MOBIL RES & DEVELOP CORP, 79- *Concurrent Pos:* Consult, Esso Res Labs, 55-62; prin investr, NASA, 65-70; vis prof, Univ Tex, 75-79. *Mem:* Soc Petrol Engrs. *Res:* Development of methods for numerical solution of partial differential equations which describe problems in engineering; new applications of mathematics to solve real world problems. *Mailing Add:* 6036 Del Norte Lane Dallas TX 75225

VON ROSENBERG, H(ERMANN) E(UGENE), b Austin, Tex, Mar 6, 26; m 57. CHEMICAL ENGINEERING. *Educ:* Univ Tex, BS, 49, MS, 51; Univ Del, PhD(chem eng), 55. *Prof Exp:* Res chem engr, 54-60, sr res chem engr, 60-65, res specialist, 65-72, res assoc, 72-77, HEAD M&C, EXXON RES & ENG CO, 77- *Mem:* Am Chem Soc; Am Inst Chem Engrs; Sigma Xi. *Res:* Chemical reactor dynamics applied chiefly to the petroleum refining processes; feed, catalyst and process studies in fluid catalytic cracking; coal utilization and gasification. *Mailing Add:* 105 Crow Rd Baytown TX 77520-1809

VON ROSENBERG, JOSEPH LESLIE, JR, b Lockhart, Tex, Aug 22, 32; m 58; c 3. ORGANIC CHEMISTRY. *Educ:* Univ Tex, Austin, BA, 54, PhD(chem), 63. *Prof Exp:* Res chemist, Ethyl Corp, 57-58; res chemist, Celanese Chem Co, 62-64; Robert A Welch fel, Univ Tex, Austin, 64-65; asst prof, 65-69, assoc prof, 69-80, PROF CHEM, CLEMSON UNIV, 80- *Mem:* Am Chem Soc; The Chem Soc; Sigma Xi. *Res:* Physical organic and organometallic chemistry. *Mailing Add:* Dept Chem Clemson Univ 201 Sykes Hall Clemson SC 29634

VON ROSENVINGE, TYCHO TOR, b Beverly, Mass, Apr 18, 42; m 66; c 3. SPACE PHYSICS. *Educ:* Amherst Col, AB, 63; Univ Minn, PhD(physics), 70. *Prof Exp:* Nat Acad Sci fel, 69-71, ASTROPHYSICIST, HIGH ENERGY ASTROPHYS DIV, GODDARD SPACE FLIGHT CTR, NASA, 71-, PROJ SCIENTIST, INT COMETARY EXPLORER, 72- *Mem:* Am Phys Soc; Am Geophys Union. *Res:* Charge composition of solar energetic particles; charge composition, origin and propagation of galactic cosmic rays. *Mailing Add:* Code 661 NASA Goddard Space Flight Ctr Greenbelt MD 20771

VON RUMKER, ROSMARIE, b Halberstadt, Ger, July 30, 26; nat US. PESTICIDE EFFICACY. *Educ:* Univ Bonn, dipl & DAgr(plant path, entom, agr econ), 50. *Prof Exp:* Farm adminr seed breeding, Ger, 50-51; agr res biologist, Farbenfabriken Bayer, Ag, 51-54; dir res, Chemagro Corp, NY, 54-58, vpres res & develop, Kansas City, Mo, 59-71; managing partner, RVR Consults, 71-89; RETIRED. *Concurrent Pos:* Mem state & nat sci & environ adv comt, Sci Adv Bd, US Environ Protection Agency & Nat Acad Sci. *Mem:* Entom Soc Am; Am Chem Soc; Am Soc Agr Eng; Weed Sci Soc Am; Plant Growth Regulator Soc Am. *Res:* Crop protection; benefits, costs and environmental effects of pesticides; pest management problems and opportunities; pesticide research, development, marketing and economics, market research and forecasting. *Mailing Add:* 6400 Hodges Dr Shawnee Mission KS 66208

VON SCHONFELDT, HILMAR ARMIN, b Delitzsch, Ger, May 3, 37. PETROLEUM ENGINEERING, ROCK MECHANICS. *Educ:* Clausthal Tech Univ, Dipl Ing, 64; Univ Minn, Minneapolis, PhD(mineral resources), 70. *Prof Exp:* Res engr, Shell Develop Co, 66 & Continental Oil Co, 67; asst prof petrol eng, Univ Tex, Austin, 69-77; MGR MINING RES, OCCIDENTAL RES CORP, 77- *Mem:* Soc Petrol Engrs. *Res:* In situ stress measurement; hydraulic fracturing; drilling; underground caverns; solution mining. *Mailing Add:* Island Creek Corp PO Box 11430 Lexington KY 40575

VON STRANDTMANN, MAXIMILLIAN, b Grodno, Poland, April 17, 27; US citizen. CHEMICAL RESEARCH & DEVELOPMENT, CHEMICAL MANUFACTURING. *Educ:* Univ Bamberg, BS, 52, MS, 53; Univ Erlagen, PhD(org chem), 55. *Prof Exp:* Scientist med chem res, Chem Factory, Bamberg, Ger, 55-57; scientist med chem res, Warner Lambert Res Inst, 57-60, sr scientist, 61-67, sr res scientist, 68-74, assoc dir org chem & med chem res, 75-76; prin chem, Imp Chem Indust, 77-79; vpres res & develop, Custom Chem Labs, 80-84; PRES, MEDEA RES LABS, 85- *Mem:* Am Chem Soc; NY Acad Sci. *Res:* Organic chemical research in reactive intermediates, heterocyclic chemistry, new synthetic pathways and building blocks; medicinal chemistry research in immune system diseases, antimicrobials and central nervous system. *Mailing Add:* 200 Wilson St Bldg D6 Port Jefferson NY 11776

VON STRYK, FREDERICK GEORGE, b Pollenhof, Estonia, Sept 6, 12; Can citizen; m 44; c 1. ORGANIC CHEMISTRY. *Educ:* Tartu State Univ, Chem, 34; Univ Leipzig, dipl chem, 38, Dr rer nat(chem), 40. *Prof Exp:* Res asst org chem, Univ Leipzig, 40-41; res chemist, Badische Soda & Anilin Fabrik, Ger, 41-48, Tex Co, Que, 48-51 & Dom Rubber Co, Ont, 51-65; res scientist, Can Dept Agr, 65-77; RETIRED. *Concurrent Pos:* Consult. *Mem:* Chem Inst Can. *Res:* Plant protection by chemicals, such as insecticides, fungicides and herbicides; translocation of these chemicals in plants and changes in their structure due to plant metabolism. *Mailing Add:* Apartado Postal 228 Palma de Mallorca Spain

VON TERSCH, LAWRENCE W, b Waverly, Iowa, Mar 17, 23; m 48. ELECTRICAL ENGINEERING. *Educ:* Iowa State Univ, BS, 43, MS, 48, PhD(elec eng), 53. *Prof Exp:* From asst prof to prof elec eng, Iowa State Univ, 46-56; chmn dept elec eng, 58-65, assoc dean eng, 65-67, actg dean, 67-68, PROF ELEC ENG & DIR COMPUT LAB, MICH STATE UNIV, 56-, DEAN COL ENG, 68- *Mem:* Inst Elec & Electronics Engrs. *Res:* Computer applications. *Mailing Add:* Dean Eng Mich State Univ 107 Engr Bldg East Lansing MI 48824

VON TURKOVICH, BRANIMIR F(RANCIS), b Zagreb, Croatia, Dec 23, 24; US citizen; m 51; c 5. MATERIALS SCIENCE, MECHANICAL ENGINEERING. *Educ:* Univ Naples, BSc, 47; Univ Madrid, MSc & DNav Eng, 51; Univ Ill, Urbana, PhD(mech eng, physics), 62. *Prof Exp:* Naval architect, Forgas & Font SA, Madrid, 50-51; sr res engr, Kearney & Trecker Corp, Wis, 52-57; lectr mech eng, Univ Ill, Urbana, 57-62, assoc prof mech eng & physics, 62-69, prof mech & indust eng, 69-70; PROF MECH ENG & CHMN DEPT, UNIV VT, 70-, DIV DIR, NSF, 88- *Concurrent Pos:* Lectr, Marquette Univ, 55-57; vis prof, Torino Polytech, Italy, 67-68; consult, Vermont Am Corp, Louisville, Ky, 71-; NATO sr prof, Italy, 76, 84, 87. *Honors & Awards:* Res Medal, Soc Mfg Engrs, 76. *Mem:* Fel Am Soc Mech Engrs; Am Phys Soc; Int Inst Prod Res; fel Soc Mfg Engrs; Sigma Xi; Metall Soc. *Res:* Mechanical and physical metallurgy; production engineering; theoretical and applied mechanics; metal cutting and forming. *Mailing Add:* Dept Mech Eng Univ Vt Votey Bldg Burlington VT 05401

VONVOIGTLANDER, PHILIP FRIEDRICH, b Jackson, Mich, Feb 3, 46; m 68; c 3. NEUROPHARMACOLOGY. *Educ:* Mich State Univ, BS, 68, DVM, 69, MS, 71, PhD(pharmacol), 72. *Prof Exp:* NIH trainee cent nerv syst pharmacol, Mich State Univ, 69-72; from res scientist to sr res scientist, 72-82, sr scientist, 82-89, DISTINGUISHED SCIENTIST & DIR, CENT NERV SYST RES, UPJOHN CO, 91- *Concurrent Pos:* Adj prof pharmacol, La State Univ, 85-; Mich State Univ, 86- *Mem:* AAAS; Soc Neurosci; Am Soc Pharmacol & Exp Therapeut. *Res:* Development of animal models of psychiatric and neurological diseases for the purpose of studying the mechanisms of action of centrally acting drugs and identification of new therapeutic agents. *Mailing Add:* Upjohn Co Kalamazoo MI 49001

VON WINBUSH, SAMUEL, b Henderson, NC, Aug 2, 32; m 62; c 1. INORGANIC CHEMISTRY, PHYSICAL CHEMISTRY. *Educ:* Tenn State Univ, AB, 53; Iowa State Univ, MS, 56; Univ Kans, PhD(inorg chem), 60. *Prof Exp:* Asst prof chem & chmn dept, Tenn State Univ, 60-62; prof, NC A&T State Univ, 62-65; prof, Fisk Univ, 65-71; prof, 71-80, DISTINGUISHED PROF CHEM, STATE UNIV NY COL OLD WESTBURY, 80- *Concurrent Pos:* Consult metals & ceramics div, Oak Ridge Nat Lab, 66-; vis prof, Wesleyan Univ, 69-70; consult, State Univ NY Col Old Westbury, 70-71. *Mem:* AAAS; Am Chem Soc; Sigma Xi. *Res:* Coordination chemistry; ligand field and charge transfer spectra; inorganic polymers; unfamiliar oxidation states of metals in molten salts and other nonaqueous solvents. *Mailing Add:* Dept Chem & Physics State Univ NY Box 210 Old Westbury NY 11568

VON WINKLE, WILLIAM A, b Bridgeport, Conn, Nov 29, 28; m 40; c 8. HYDRODYNAMICS, UNDERWATER ACOUSTICS. *Educ:* Yale Univ, BSEE, 50-52, MSEE, 52; Univ Calif, Berkeley, PhD(eng sci), 61. *Prof Exp:* Asst, Yale Univ, 50-52; electronic engr, USN Underwater Sound Lab, 52-61, electronic engr, mgr & dir, Bur Ships Trident Lab, Bermuda, 61-63, head signal processing br, 63-66, assoc tech dir res, 66-70, ASSOC TECH DIR TECHNOL, NAVAL UNDERWATER SYSTS CTR, 70- *Concurrent Pos:* Instr, YMCA Jr Col, 51-52 & Mitchell Jr Col, 54-57; vis lectr, Univ Conn, 54-81 & lectr, 81-, distiguished lectr, Univ New Haven, 81-; assoc ed, Acoust Signal Processing, Acoust Soc Am, 69-81 & 85-; adj asst prof, Rensselaer Polytech Inst, 55-65; instr, Univ Md, 62; US Nat Rep to Sci Comn of Nat Rep, NATO, Saclant Res Ctr, 71- *Mem:* Fel Inst Elec & Electronics Engrs; fel Acoust Soc Am; Sigma Xi. *Res:* Acoustics, sonar systems, target characteristics, transducer array design, fire control, submarine weaponry, hydrodynamics, optical signal processing, transient detection and classification, adaptive filtering techniques and holography; boundary layer hydrodynamics with application to naval warfare problems. *Mailing Add:* 105 Gardner Ave New London CT 06320

VON ZELLEN, BRUCE WALFRED, b Ann Arbor, Mich, Feb 14, 22; m 49; c 1. PARASITOLOGY, PROTOZOOLOGY. *Educ:* Northern Mich Col, AB, 47; Univ Mich, MS, 49; Duke Univ, PhD(zool), 59. *Prof Exp:* Lab asst physiol & zool, Univ Mich, 49-50; assoc prof biol sci, Ky Wesleyan Col, 50-57; instr zool, Duke Univ, 57-59; ASSOC PROF BIOL SCI, NORTHERN ILL UNIV, 59- *Concurrent Pos:* Consult, panels on equip grants, NSF, Washington, DC & Chicago. *Mem:* AAAS; Am Soc Trop Med & Hyg; Soc Protozool; Am Soc Parasitol. *Res:* Parasitic protozoa; life history of coccidiosis; blood parasite infection; cell culture. *Mailing Add:* Dept Biol Sci Northern Ill Univ De Kalb IL 60115

VOOGT, JAMES LEONARD, b Grand Rapids, Mich, Feb 8, 44; m 66; c 3. PHYSIOLOGY, NEUROENDOCRINOLOGY. *Educ:* Mich Technol Univ, BS, 66; Mich State Univ, MS, 68, PhD(physiol), 70. *Prof Exp:* NIH fel, Med Ctr, Univ Calif, San Francisco, 70-71; asst prof physiol, Univ Louisville, 71-77, NIH res grant, 72-81; assoc prof, 77-82, PROF PHYSIOL, UNIV KANS, 82- *Mem:* Am Physiol Soc; Int Soc Neuroendocrinol; Endocrine Soc; Soc Study Reproduction; Soc Neurosci; Sigma Xi. *Res:* Control of anterior pituitary function by hypothalamus; feedback systems; hormone analysis; reproduction control. *Mailing Add:* Dept Physiol Univ Kans Kansas City KS 66103

VOOK, FREDERICK LUDWIG, b Milwaukee, Wis, Jan 17, 31; m 58; c 2. PHYSICS. *Educ:* Univ Chicago, BA, 51, BS, 52; Univ Ill, MS, 54, PhD(physics), 58. *Prof Exp:* Mem staff, 58-62, div supvr, 62-71, dept mgr, 71-78, DIR, SANDIA LABS, 78- *Concurrent Pos:* Chmn, Am Phys Soc, Oliver & Buckley Prize Selection Comt, 87; mem policy bd, Nanofabrication Facil, Cornell Univ, Col Eng Adv Bd, Univ Ill. *Mem:* Fel Am Phys Soc; Mat Res Soc; Bohmishe Phys Soc; Sigma Xi. *Res:* Defects in solids, primarily semiconductors; defects investigated by means of radiation damage at low temperatures; infrared absorption; ion implantation in semiconductors; ion backscattering and channeling studies of solids. *Mailing Add:* Sandia Labs Org 1100 Albuquerque NM 87185

VOOK, RICHARD WERNER, b Milwaukee, Wis, Aug 2, 29; m 57; c 4. ENGINEERING PHYSICS. *Educ:* Carleton Col, BA, 51; Univ Ill, MS, 52, PhD(physics), 57. *Prof Exp:* Mem res staff, Res Ctr, Int Bus Mach Corp, NY, 57-61; res labs, Franklin Inst, Pa, 61-65; assoc prof metall, Syracuse Univ, 65-70, prof mat sci, 70-84, chmn, Solid State Sci & Technol Prog, 84-87 & 90-91, PROF PHYSICS, SYRACUSE UNIV, 84-, DIR ELECTRON MICROS LAB, 68- *Concurrent Pos:* Consult, Alcoa, 90, Amperex Corp, 72, Carrier Corp, 72-73, Revere Copper & Brass, 77 & 84-86, Inficon Leybold-Heraeus, 81 & 84-90, Litton Panelvision, 87, Forensic Sci, 67-90; prin investr, AEC, 66-72, NSF, 66-76 & 87-91, Energy Res Develop Agency, 77-78, Dept Energy, 78-89, Off Naval Res, 79-82 & 86-90, IBM Corp, 82-84 & 88-89, Niagara Mohawk Power Corp, 83-86, Sandia Nat Lab, 84-86, Alcoa Found, 83-89, Rome Air Develop Command, USAF, 89-91; res contract, Westinghouse, 77-79. *Honors & Awards:* L P Pfeil Medal & Prize, Metals Soc Gt Brit, 83. *Mem:* Am Phys Soc; Am Vacuum Soc; Electron Micros Soc Am; Mat Res Soc. *Res:* Electron microscopy and diffraction; x-ray diffraction; thin films; epitaxial growth; surface physics and chemistry; imperfections in solids; electrical contact phenomena; electromigration. *Mailing Add:* 201 Physics Bldg Syracuse Univ Syracuse NY 13244-1130

VOORHEES, BURTON HAMILTON, b Tucson, Ariz, Dec 3, 42. BIOMATHEMATICS. *Educ:* Univ Calif, Berkeley, AB, 64; Univ Ariz, MS, 66; Univ Tex, Austin, PhD(physics), 71. *Prof Exp:* Asst prof math & physics, Pars Col, Iran, 71-73; res assoc math, Univ Alta, 73-82; ASSOC PROF MATH, ATHABASCA UNIV, 82- *Res:* Stochastic geometry; quantization of gravitation; black hole physics; mathematical models of evolutive processes; general systems theory. *Mailing Add:* Dept Math Athabasca Univ Box 10000 Athabasca AB T0G 2R0 Can

VOORHEES, FRANK RAY, b Pekin, Ill, Dec 8, 35; m 58; c 2. PHYSIOLOGY. *Educ:* Univ Fla, BS, 58; Univ Ill, MS, 68, PhD(entom), 69. *Prof Exp:* PROF BIOL, CENT MO STATE UNIV, 75- *Mem:* Entom Soc Am; Soc Develop Biol; Am Mosquito Control Asn. *Res:* Gender-determining mechanisms in mosquitos. *Mailing Add:* Dept Biol WCM 306 Cent Mo State Univ Warrensburg MO 64093-5053

VOORHEES, HOWARD R(OBERT), b Eatontown, NJ, Feb 13, 21; m 46; c 3. CHEMICAL ENGINEERING, MATERIALS SCIENCE. *Educ:* Rutgers Univ, BS, 42; Mass Inst Technol, MS, 47; Univ Mich, PhD(chem & metall eng), 56. *Prof Exp:* Asst prof chem eng, Univ Toledo, 48-50; instr chem & metall eng, Univ Mich, 50-55, assoc res engr, 55-63; TECH DIR, MAT TECHNOL CORP, 64- *Mem:* Am Soc Testing & Mat; Am Soc Metals. *Res:* Creep-rupture of alloys under variable and complex stresses. *Mailing Add:* 2646 Park Ridge Dr Ann Arbor MI 48103

VOORHEES, JOHN E, b Lima, Ohio, Aug 20, 29; m 53; c 2. MACHINE DESIGN & VIBRATION CONTROL. *Educ:* Univ Toledo, BSME, 51; Ohio State Univ, MSME, 52. *Prof Exp:* Prin mech engr, Battelle Mem Inst, 52-58, asst chief mech res, 58-60, group dir mech dynamics, 60-66, chief mech dynamics, 66-70; mgr res & develop, Minster Mach Co, 70-77; mgr res, Hobart Corp, 77-86; ENG CONSULT, 86- *Concurrent Pos:* Westinghouse fel, 52-53. *Mem:* Am Soc Mech Engrs. *Res:* Vibrations analysis; design of balancing equip; control system and machine design; development of high speed mechanisms and heavy machines; dynamic analysis of heavy machinery; design of machine foundations and mounts. *Mailing Add:* 1429 Fox Dale Pl Sidney OH 45365

VOORHEES, JOHN JAMES, b Cleveland, Ohio, Dec 5, 38; m 61; c 4. DERMATOLOGY, MEDICAL RESEARCH. *Educ:* Bowling Green State Univ, BS, 60; Univ Mich, Ann Arbor, MD, 63. *Prof Exp:* Intern internal med, 63-64, trainee clin internal med, 66, NIH trainee biochem, 67-68, trainee clin dermat, 67-69, Carl Herzog scholar biochem, 68-70, from instr to assoc prof dermat, 69-74, PROF DERMAT, MED SCH, UNIV MICH, ANN ARBOR, 74-, CHMN DEPT, 75-, CHIEF DERMAT SERV, UNIV HOSP, 75- *Concurrent Pos:* Consult dermat, Dept Med, Wayne County Gen Hosp, 69-84, Vet Admin Hosp, Ann Arbor, 71-, St Joseph Mercy Hosp, 78- & Chelsea Community Hosp, 80-; assoc ed, J Cutaneous Path, 72-; mem med & sci adv bd, Nat Psoriasis Found, 71-; mem revision panel, 1980 Ed, US Pharmacopeia, 75-80; contrib ed, Int Psoriasis Bull; chmn, Dermat Found. *Honors & Awards:* Taub Int Mem Award Psoriasis Res, 73; Henry Russell Award Distinguished Res, Univ Mich, 73; Outstanding Serv Award, Nat Psoriasis Found, 73. *Mem:* Am Soc Clin Invest; Am Soc Pharmacol & Exp Therapeut; Am Soc Exp Path; Am Soc Cell Biol; Soc Exp Biol & Med; AMA; Soc Investigative Dermat; Skin Pharmacol Soc; Am Asn Cancer Res; Endocrine Soc. *Res:* Role of cyclic nucleotides, glucocorticoids, the arachidonate, HETE, thromboxane, prostaglandin cascade and immunology in the molecular pathophysiology and pharmacology of skin diseases with inflammation, induced proliferation and reduced differentiation. *Mailing Add:* Dept Dermat Out-Patient Bldg Rm C-2064 Med Sch Univ Mich Ann Arbor MI 48109

VOORHEES, KENT JAY, b Provo, Utah, Sept 7, 43; m 66; c 2. MATERIAL SCIENCE, CHEMOMETRICS. *Educ:* Utah State Univ, BS, 65, MS, 68, PhD(org chem), 70. *Prof Exp:* Res fel phys chem, Mich State Univ, 70-71; instr org chem, Univ Utah, 71-73, asst res prof anal polymer chem, 73-76, assoc res prof, 76-79; assoc prof, 79-85, PROF CHEM, COLO SCH MINES, 85- *Concurrent Pos:* Mem, Mat Flammability Comt, Nat Acad Sci & Mertings & Expositions Comt, Am Chem Soc. *Mem:* Am Chem Soc; Am Soc Mass Spectrometry. *Res:* Formulation of complex structures and/or degradation mechanisms by the application of gas chromatography and/or mass spectrometry in the analysis of thermal decomposition products of synthetic and natural polymers; flammability of materials. *Mailing Add:* Dept Chem Colo Sch Mines Golden CO 80401

VOORHEES, LARRY DONALD, b Benson, Minn, Dec 23, 46; m 70; c 2. ZOOLOGY, ECOLOGY. *Educ:* Univ Minn, Morris, BS, 70; NDak State Univ, MS, 72, PhD(zool), 76. *Prof Exp:* Instr zool, NDak State Univ, 74-75, res assoc ecol, 75-76; res assoc, 76-83, MEM RES STAFF, ECOL, OAK RIDGE LAB-MARTIN MARIETTA, 83- *Concurrent Pos:* Comt mem, Roadside Maint Transp Res Bd, Nat Res Coun, 78-85, Task Force, Wildlife & Fisheries Issues, 81. *Mem:* Wildlife Soc; AAAS; Sigma Xi; Ecol Soc Am. *Res:* Applied problems in terrestrial community ecology; management of habitat for wildlife; management and analysis of environmental monitoring data. *Mailing Add:* Environ Sci Div Oak Ridge Nat Lab Oak Ridge TN 37831

VOORHEES, PETER WILLIS, b Staten Island, NY, Dec 24, 55; m 77; c 2. PHASE TRANSFORMATIONS, THERMODYNAMICS. *Educ:* Rensselaer Polytech Inst, BS, 77, PhD(mat eng), 82. *Prof Exp:* Instr mat eng, Rensselaer Polytech Inst, 80-81; Nat Res Coun res fel, Nat Bur Standards, 82-84, metallurgist, 84-88; ASSOC PROF, DEPT MAT SCI & ENG, NORTHWESTERN UNIV. *Concurrent Pos:* Vis scientist, Inst Theoret Physics, Univ Calif, Santa Barbara; vis prof, Groupe de Physique des Solides, Universite Paris VII. *Honors & Awards:* Presidential Young Investr Award, NSF. *Mem:* Am Soc Metals; AAAS; Am Inst Mining Metall & Petrol Engrs; Am Phys Soc; Sigma Xi. *Res:* First order solid-solid and solid-liquid phase transformations, specifically the kinetics and morphological development accompanying the phase transformation process. *Mailing Add:* Dept Mat Sci & Eng Northwestern Univ Evanston IL 60203

VOORHESS, MARY LOUISE, b Livingston Manor, NY, June 2, 26. PEDIATRICS, ENDOCRINOLOGY. *Educ:* Univ Tex, BA, 52; Baylor Univ, MD, 56. *Prof Exp:* From intern to resident pediat, Albany Med Ctr, New York, 56-59; from asst prof to prof pediat, State Univ NY Upstate Med Ctr, 61-76; PROF PEDIAT, STATE UNIV NY, BUFFALO, 76- *Concurrent Pos:* Res fel pediat endocrinol & genetics, State Univ NY Upstate Med Ctr, 59-61; Nat Cancer Inst res grant, 62-69, career develop award, 61-71; consult pediat, Roswell Park Mem Inst, Buffalo, 78-; mem, Nat Adv Environ Health Sci Coun, 80-83. *Mem:* Endocrine Soc; Am Fedn Clin Res; Am Acad Pediat; Lawson Wilkins Pediat Endocrine Soc; Am Pediat Soc; fel AAAS. *Res:* Pediatric endocrinology; catecholamine metabolism in children; growth disorders in children. *Mailing Add:* Children's Hosp 219 Bryant St Buffalo NY 14222

VOORHIES, ALEXIS, JR, b New Iberia, La, Sept 8, 99; m 24; c 3. CHEMICAL ENGINEERING, CATALYSIS. *Educ:* St Charles Col, La, AB, 17; La State Univ, BS, 22; Loyola Univ, La, MS, 26. *Hon Degrees:* DSc, Loyola Univ, La, 64. *Prof Exp:* Prof chem, physics & math, Northwestern State Col, La, 22-24 & Loyola Univ, La, 24-30; from chem engr to assoc dir, Baton Rouge Refinery, Esso Res Labs, Humble Oil & Refining Co, 30-47, dir, Esso Res Labs, 47-64; VIS PROF CHEM ENG, LA STATE UNIV, BATON ROUGE, 64- *Concurrent Pos:* Consult, Esso Res & Eng Co, 64- *Honors & Awards:* E V Murphree Award Indust & Eng Chem, Am Chem Soc, 77. *Mem:* Am Chem Soc; Am Inst Chem Engrs; Soc Chem Industs. *Res:* Catalysis in petroleum refining; catalytic reforming; hydrocracking; hydroisomerization. *Mailing Add:* 2569 E Lake Shore Dr Baton Rouge LA 70808-2145

VOORHIES, JOHN DAVIDSON, b Hartford, Conn, Nov 26, 33; m 59. SURFACE CHEMISTRY. *Educ:* Princeton Univ, AB, 55, MA, 57, PhD(chem), 58. *Prof Exp:* Res chemist, Am Cyanamid Co, 58-62, sr res chemist, 62-64, group leader, 64-71, sr res scientist, 71-82, specialist, safety & environ control, 78-81; DIR & CHIEF EXEC OFFICER, ENVIRON ANALYSIS CORP, 82- *Concurrent Pos:* Vis scholar, Dept Chem Eng, Stanford Univ, 70-71. *Mem:* AAAS; Am Chem Soc; Electrochem Soc. *Res:* Electrochemistry; electrochemical power sources; electrochemistry of organic compounds; electroanalytical techniques; chronopotentiometry; polarography; coulometry; interfacial chemistry; heterogeneous catalysis; hydrotreating catalysts; water analysis. *Mailing Add:* 14 Harrison Ave New Canaan CT 06840

VOORHIES, MICHAEL REGINALD, b Orchard, Nebr, June 17, 41; m 68; c 2. VERTEBRATE PALEONTOLOGY. *Educ:* Univ Nebr, BS, 62; Univ Wyo, PhD(geol), 66. *Prof Exp:* From asst prof to assoc prof geol, Univ Ga, 66-75; assoc cur, 75-80, CUR FOSSIL VERT, UNIV NEBR STATE MUS, 80-; PROF GEOL, UNIV NEBR, 77- *Mem:* Paleont Soc; Soc Vert Paleont; Soc Study Evolution; Soc Syst Zool; Sigma Xi. *Res:* Taphonomy and population dynamics of Cenozoic mammals; community evolution; neogene stratigraphy of the Great Plains. *Mailing Add:* 2736 Ossa Wintna Dr Orchard NE 68764

VOORHOEVE, RUDOLF JOHANNES HERMAN, b Sentang, Sumatra, Indonesia, Oct 4, 38; m 62, 68, 75; c 5. SURFACE CHEMISTRY, PHYSICAL INORGANIC CHEMISTRY. *Educ:* Delft Univ Technol, Ing, 61, Dr(organosilicon & catalytic chem), 64. *Prof Exp:* Instr organosilicon & catalysis chem, Delft Univ Technol, 61-64; res chemist, Nat Defense Res Orgn, 64-66; res chemist, Koninklyke/Shell Lab, Amsterdam, 66-68; mem tech staff, Bell Labs, 68-80; tech dir, Res & Develop Ctr, Hoechst Celanese Advan Technol Group, Corpus Christi, Tex, 80-90; CORP VPRES TECHNOL, GREAT LAKES CHEM CORP, 90- *Mem:* Am Chem Soc; Am Phys Soc; Mat Res Soc (pres, 79). *Res:* Organosilicon chemistry, especially direct synthesis of organohalosilanes; heterogeneous catalysis, studies by gas-solid kinetics, mechanistic studies; gas-solid reactions for chemical vapors deposition, molecular beams; process studies; commodity and intermediate chemicals; product research; flame retardant specialty chemicals; pharmaceutical intermediates. *Mailing Add:* c/o Great Lakes Chem Corp PO Box 2200 West Lafayette IN 47906

VOOS, JANE RHEIN, b Nuremberg, Ger, Oct 2, 27; m 50; c 3. MICROBIOLOGY. *Educ:* Hunter Col, BA, 52; Columbia Univ, PhD(biol sci), 68. *Prof Exp:* Res bacteriologist, Bellevue Hosp, New York, 53-54; instr biol, Stern Col Women, 56-58; from asst prof to assoc prof biol sci, 68-75, asst to dean, 71-72, prof, 75-80, chmn dept, 72-84, ASST TO VPRES ACAD AFFAIRS, WILLIAM PATTERSON COL, 84- *Mem:* AAAS; Bot Soc Am; Mycol Soc Am. *Res:* Electron microscopy of spores and modern pollens. *Mailing Add:* 28 Wenonak Ave Rockaway NJ 07866

VOOTS, RICHARD JOSEPH, b Quincy, Ill, Feb 18, 21; m 45; c 3. ACOUSTICS. *Educ:* NY Univ, BS, 44; Univ Iowa, PhD, 55. *Prof Exp:* Engr, Boeing Aircraft Co, Kans, 46; instr, Benton High Sch, 47-48; physics lab, Univ Wichita, 48; audio technician, Bennett Music House, 49-51; asst psychologist, Univ Iowa, Col Med, 51-52, res assoc speech path & audiol, 51-61, res asst prof speech path, audiol & music, 61-89, asst prof, 77-89, EMER ASST PROF OTOLARYNGOL & MAXILLOFACIAL SURG, UNIV IOWA COL MED, 89- *Concurrent Pos:* Res assoc otolaryngol, Univ Hosps, 55-89. *Mem:* AAAS; Acoust Soc Am; Am Auditory Soc. *Res:* Otological and musical acoustics; auditory stimulus-response relationship, particularly the frequency-pitch dimension; automated apparatus for sensitivity threshold and differential pitch threshold audiometry; physical acoustics of musical instruments. *Mailing Add:* Dept Otolaryngol Univ Iowa Iowa City IA 52242

VOPAT, WILLIAM A, b New York, NY, Mar 8, 10; m 35; c 2. MECHANICAL ENGINEERING. *Educ:* Cooper Union, BS, 31; Univ Mich, MSE, 37. *Hon Degrees:* ME, Cooper Union, 39. *Prof Exp:* Asst sales refrig engr, Lynbrook, NY, 31-32; from instr to prof, 32-76, head dept mech eng, 48-69, dean sch eng & sci, 69-76, EMER PROF MECH ENG & EMER DEAN SCH ENG & SCI, 76- *Concurrent Pos:* Consult ed, McGraw-Hill Pub Co, Inc, 38-45; asst engr, Kennedy Van Saun Mfg Co, 42; consult, NY Civil Serv Comn. *Mem:* Am Soc Mech Engrs; Am Soc Eng Educ. *Res:* Applied energy conversion; steam and gas turbines; power station engineering and economy. *Mailing Add:* 150 Park Blvd Malverne NY 11565-1717

VORCHHEIMER, NORMAN, b Thungen, Ger, Sept 10, 35; US citizen; m 67; c 3. POLYMER CHEMISTRY. *Educ:* Brooklyn Col, BS, 57; Polytech Inst Brooklyn, PhD(org chem), 62. *Prof Exp:* Res chemist textile fibers, E I du Pont de Nemours & Co, 62-67; sr res chemist polymer synthesis, Betz Labs Inc, 67-70; exec vpres, Shasta Fund Inc, 70-71; sr res chemist, 71-73, group leader, 73-77, supvr polymer synthesis, 77-80, sect head synthesis-polymer appln, 80-82, mgr, synthesis-pilot plant-water-wastewater res, 82-86, RES FEL, BETZ LABS INC, TREVOSE, PA, 86- *Mem:* Am Chem Soc; Royal Soc Chem. *Res:* Water-soluble polymers. *Mailing Add:* PO Box 403 Buckingham PA 18912

VORE, MARY EDITH, b Guatemala City, Guatemala, June 27, 47; US citizen; m 76; c 2. TOXICOLOGY, PHARMACOLOGY. *Educ:* Asbury Col, BA, 68; Vanderbilt Univ, PhD(pharmacol), 72. *Prof Exp:* Fel, Dept Biochem & Drug Metab, Hoffmann-LaRoche Inc, 72-74; asst prof toxicol, Dept Pharmacol, Univ Calif Med Ctr, San Francisco, 74-78; from asst prof to assoc prof, 78-86, PROF PHARMACOL, COL MED, UNIV KY, 86- *Concurrent Pos:* Pharmacol study sect, NIH, 83-87, prin investr, 79-95; coun, Nat Inst Environ Health Sci, 91-95. *Mem:* Am Soc Pharmacol & Exp Therapeut (secy-treas, 87-88); Soc Toxicol; Am Asn Study Liver Dis. *Res:* Mechanisms of organic anion transport across the liver; regulation by estrogens and environmental pollutants; hepatic drug elimination in pregnancy. *Mailing Add:* Dept Pharmacol Univ Ky Med Sch Lexington KY 40536

VOREADES, DEMETRIOS, b Athens, Greece, Dec 8, 42; m 76; c 2. APPLIED PHYSICS, SEMICONDUCTOR DEVICE CHARACTERIZATION. *Educ:* Univ Athens, dipl physics, 66; Univ Chicago, MS, 70, PhD(physics), 75. *Prof Exp:* Res asst physics, Univ Athens, 67-68; teaching asst, Univ Chicago, 68-70, res asst, Enrico Fermi Inst, 70-75; res fel biol, Calif Inst Technol, 75-77; syst mgr cytol, Obstet & Gynec Lying-In Hosp, Chicago, 77-78; assoc biophysicist biol, Brookhaven Nat Lab, 78-79; sr engr, Burroughs Corp, 79-81; SR STAFF PHYSICIST, HUGHES AIRCRAFT, 81- *Mem:* Am Phys Soc; Inst Elec & Electronics Engrs. *Res:* Scanning transmission electron microscopy as analytical tool in biology, physics and materials science; ion implantation; semiconductor devices. *Mailing Add:* 2330 Bancroft St San Diego CA 92104

VORHAUS, JAMES LOUIS, b St Louis, Mo, Aug 2, 50; m 75; c 1. SOLID STATE PHYSICS. *Educ:* Lehigh Univ, BS, 72; Univ Ill, Champaign-Urbana, MS, 74, PhD(physics), 76. *Prof Exp:* Res asst low temp physics, Dept Physics, Univ Ill, Champaign-Urbana, 72-76; sr scientist device physics, Res Div, Raytheon Co, 76-85; dir opers, Solid State Microwave Group, Epsco, Inc, 85-87; PROD LINE MGR, MICROWAVE POWER PROD, AVANTEK INC, 87- *Mem:* Am Phys Soc; Inst Elec & Electronics Engrs. *Res:* Design, fabrication and evaluation of galium arsenide monolithic microwave integrated circuits. *Mailing Add:* MS 2-G Avantek Inc 3175 Bowers Ave Santa Clara CA 95054

VORHEES, CHARLES V, b Columbus, Ohio, Oct 9, 48; m 82; c 4. PSYCHOTERATOLOGY. *Educ:* Univ Cincinnati, BA, 71; Vanderbilt Univ, MA, 73, PhD(psychopharmacol), 77. *Prof Exp:* Fel res scholar psychoteratology, 76-78, asst prof, 78-82, assoc prof, Dept Pediat, 82-88, DIR, PSYCHOTERATOLOGY LAB, CHILDRENS HOSP RES FOUND, INST DEVELOP RES, CINCINNATI, 80-, PROF DEVELOP BIOL & ENVIRON HEALTH, 88- *Concurrent Pos:* Assoc prof, Social Sci Prog, Evening Col, Univ Cincinnati, 77- *Mem:* Behav Teratology Soc (pres, 84-85); Teratology Soc; Int Soc Develop Psychobiol; AAAS; Neurobehav Toxicol Soc; Soc Neurosci; Am Psychol Asn. *Res:* Behavioral birth defects; psychoactive drugs and food additives as possible causes of mental retardation or other learning or emotional problems caused by early (usually prenatal) exposure to these agents; developmental psychopharmacology. *Mailing Add:* Inst Develop Res Childrens Hosp Res Found Cincinnati OH 45229

VORHERR, HELMUTH WILHELM, b Alzey, WGer, Feb 6, 28; m 55; c 2. OBSTETRICS & GYNECOLOGY, PHARMACOLOGY. *Educ:* Univ Mainz, MD, 55, specialist obstet & gynec, 62. *Prof Exp:* Mem staff obstet & gynec, Univ Frankfurt, 62-65; res pharmacologist, Cedars-Sinai Med Ctr, Los Angeles, Calif, 65-68; assoc prof, 68-71, PROF OBSTET, GYNEC & PHARMACOL, 71-, DIR BREAST CLIN, SCH MED, UNIV NMEX, 79- *Concurrent Pos:* Damon Runyon Mem Fund res fel, Cedars-Sinai Med Ctr & Univ Calif, Los Angeles, 65-66, NIH spec fel, 66-67; asst prof, Sch Med, Univ Calif, Los Angeles, 66-68. *Mem:* Am Fedn Clin Res; Am Soc Pharmacol & Exp Therapeut; Soc Gynec Invest. *Res:* Effects of sex steroid hormones on reproductive tissues; factors influencing embryonic/fetal growth; lactation; pathobiology of breast cancer. *Mailing Add:* Dept Obstet & Gynec Univ NMex Albuquerque NM 87131

VORHIS, ROBERT C, b Covington, OH, Aug 31, 17; m 47; c 2. HYDROGEOLOGY, LARGER AMINIFERA. *Educ:* Ohio Wesleyan Univ, BA, 39; Univ Iowa, MS, 41. *Prof Exp:* Geologist, US Geol Surv, 46-74; geologist, Near East Found, Tanzania, 76-79; geologist Dimpex Co, Upper Volta, 80-83; CONSULT GEOL, 83- *Concurrent Pos:* Mem, Hydrol Panel, Comt Alaskan Earthquake, Nat Acad Sci, 64-70. *Mem:* Geol Soc Am. *Res:* Worldwide hydrologic effects of the Alaskan earthquake of 1964; synonomy of the American larger foraminifera. *Mailing Add:* 1560 Stoneleigh Hills Rd Lithonia GA 30058-5631

VORIS, HAROLD K, b Chicago, Ill, Oct 5, 40. HERPETOLOGY. *Educ:* Hanover Col, BA, 62; Univ Chicago, PhD(biol), 69. *Prof Exp:* Instr biol, Yale Univ, 67-69; asst prof, Dickinson Col, 69-73; asst to dir, 83-84, head div ,84, chmn, Dept Zool, 85, vpres collections & res, 85-89, CUR, FIELD MUS NATURAL HIST, 73- *Concurrent Pos:* Assoc prog dir, Syst Biol Prog, NSF, 81-82. *Mem:* Am Soc Ichthyol & Herpet; Soc Syst Zool; Ecol Soc Am; Soc Study Evolution. *Res:* Evolution and systematics; ecology; sea snake ecology and systematics; rain forest ecosystems; numerical taxonomy. *Mailing Add:* Div Reptiles Field Mus Natural Hist Chicago IL 60605

VOROSMARTI, JAMES, JR, b Palmerton, Pa, Oct 18, 35; c 3. OCCUPATIONAL MEDICINE, DIVING PHYSIOLOGY. *Educ:* Lafayette Col, AB, 57; Jefferson Med Col, MD, 61. *Prof Exp:* Post doc fel, State Univ NY, Buffalo, 70-72; asst to undersecy defense, Res & Eng, 83-86; CONSULT, TECHNOL TRANSFER & OCCUP MED, 86- *Concurrent Pos:* Exchange med officer, Royal Naval Physiol Lab, Royal Naval Inst Naval Med, Gosport, Eng & Off Naval Res, London, 72-75; dep dir, Naval Med Res Inst, Bethesda, 75-78, prog mgr diving med, Naval Med Res & Develop Command, 78-80, cmndg officer, 80-83. *Honors & Awards:* Shilling Award, Undersea & Hyperbaric Med Soc, 87. *Mem:* Fel Am Col Physicians; fel Am Col Prev Med; fel Am Acad Family Med; Am Physiol Soc; fel Am Acad Occup Med; Undersea & Hyperbaric Med Soc (pres, 77-78). *Res:* Hyperbaric physiology. *Mailing Add:* 16 Orchard Way S Rockville MD 20854

VORST, JAMES J, b Cloverdale, Ohio, Mar 20, 42; m 66; c 3. AGRONOMY, CROP ECOLOGY. *Educ:* Ohio State Univ, BS, 64, MS, 66; Univ Nebr, PhD(agron), 69. *Prof Exp:* Teaching asst agron, Ohio State Univ, 64-66; instr, Univ Nebr, 66-69; MEM FAC, DEPT AGRON, PURDUE UNIV, LAFAYETTE, 69- *Mem:* Am Soc Agron; Crop Sci Soc Am; Sigma Xi; Ctr Appln Sci & Technol. *Res:* Crop production and physiology; cropping systems; teaching methods in agronomy. *Mailing Add:* Dept Agron Purdue Univ West Lafayette IN 47906

VORTMAN, L(UKE) J(EROME), b Springfield, Ill, Apr 18, 20; m 46. PHYSICS, ENGINEERING. *Educ:* Univ Ill, BS, 47, MS, 49. *Prof Exp:* Mem tech staff, Sandia Labs, 49-83, distinguished mem tech staff, 83-87; RETIRED. *Concurrent Pos:* Mem adv comt civil defense, Nat Acad Sci, 61-73, chmn, Protective Struct Subcomt, 65-66& blast& thermal effects subcomt, 66-70, mem, phys effects subcomt Supersonic Transport, 66-71; consult, Boeing Co, 64. *Mem:* AAAS; Am Nuclear Soc. *Res:* Effects of nuclear and chemical explosions in air and underground; peaceful uses of nuclear explosives; protective construction. *Mailing Add:* 933 McDuffie Circle NE Albuquerque NM 87110

VORTUBA, JAN, b Nov 14, 38; US citizen. PHYSICS. *Educ:* Tech Univ Prague, Czech, Dipl Ing, 61; Czech Acad Sci, PhD(physics), 68. *Prof Exp:* Physicist, Inst Physics, Prague, 63-69, 70-79 & 80-82 & Brookhaven Nat Lab, 79-80; PHYSICIST, FONAR CORP, 82- *Concurrent Pos:* Dir, Radio

Frequency Technol, dept head, & sr systs engr, Fonar Corp, 82- *Res:* Microwave electronics; accelerator physics; medical physics; high frequency superconductivity; vacuum technology; experimental physics; designed and developed 25 automatically shaped high performance receiver coils for nuclear magnetic resonance imaging. *Mailing Add:* Fonar Corp 110 Marcus Dr Melville NY 11747

VOS, BERT JOHN, TOXICOLOGY. *Educ:* Univ Chicago, PhD(phys chem & pharmacol), 34. *Prof Exp:* Dep dir, Div Toxicol, US Food & Drug Admin, 69-70; CONSULT, SELF-EMPLOYED, 70- *Mailing Add:* PO Box 569 McLean VA 22101

VOS, KENNETH DEAN, b Oskaloosa, Iowa, Nov 13, 35; m 60; c 2. PHYSICAL CHEMISTRY. *Educ:* Cent Col, Iowa, BA, 57; Mich State Univ, PhD(phys chem), 63. *Prof Exp:* Res asst chem, Los Alamos Sci Lab, 60; staff assoc, John Jay Hopkins Lab, Gen Atomic Div, Gen Dynamics Corp, Calif, 63-68; sr res chemist, S C Johnson & Son, Inc, 68-72, supvr pressurized prod res sect, 72-76, prod safety dir, 76-78, phys sci mgr & prod safety dir, 78-82, dir regulatory affairs, 82-87; VPRES, PALMER HOUSTON, INC, 88- *Honors & Awards:* First Charles E Allderdice Jr Award, Chem Specialties Mfrs Asn. *Mem:* AAAS; Am Chem Soc; Am Phys Soc; NY Acad Sci. *Res:* Physical chemistry, electron paramagnetic presonance of metalamines and free radicals; semiconductor and high polymer chemistry biomedical research; fine particle and aerosol research; toxicology; product safety; Environ Protection Agency, Food & Drug Admin, USDA, Consumer Prod Safety Comn, Bur Alcohol, Tobacco & Firearms regulations. *Mailing Add:* Palmer Houston Inc 8226 Kerr St Houston TX 77029-3908

VOSBURG, DAVID LEE, b Enid, Okla, Dec 24, 30; m 60; c 2. STRATIGRAPHY. *Educ:* Phillips Univ, BS, 52; Univ Okla, MS, 54, PhD(geol), 63. *Prof Exp:* From instr to asst prof geol, Univ RI, 60-65; asst prof & chmn dept, Phillips Univ, 65-66; asst prof, 66-67, ASSOC PROF GEOL, ARK STATE UNIV, 67- *Mem:* Am Asn Petrol Geol. *Res:* Occurrence and distribution of subsurface evaporites within shelf sediments related to the Permian Basin, especially economic potential and stratigraphic relations. *Mailing Add:* 4121 Oak Hill Lane Jonesboro AR 72401

VOSBURGH, KIRBY GANNETT, b Pasadena, Calif, May 27, 44; m 67; c 2. APPLIED PHYSICS, EXPERIMENTAL PHYSICS. *Educ:* Cornell Univ, BS, 65, MS, 67; Rutgers Univ, PhD(physics), 71. *Prof Exp:* Res asst applied physics, Cornell Univ, 65-67; mem tech staff accelerator physics, Princeton-Penn Accelerator, Princeton Univ, 67-68; res fel physics, Rutgers Univ, 68-71; mem tech staff & asst to dir particle physics, Princeton Particle Accelerator, Princeton Univ, 71-72; physicist, Gen Elec Co, 72-77, mgr, signal electronic systs, Corp Res & Develop, 77-79, mgr electronic mat, 79-80, mgr, silicon processing, 80-84, mgr, Very-Large-Scale Integration Technol Lab, 84-87, mgr, Electronic Mat Lab, 88-89, MGR, APPL PHYSICS LAB, GEN ELEC CO, 89- *Mem:* Am Phys Soc; Inst Elec & Electronics Engrs. *Res:* Solid state electronics; very-large-scale integration devices; silicon process technology; electronic materials; photovoltaics; power generation and transmission; electronic systems, devices and peripherals computer memory systems; medical diagnostic imaging (NMR, xray, nuclear); radiation therapy; electron optics, lamps & lighting systems. *Mailing Add:* Corp Res & Develop Gen Elec Co PO Box 8 Schenectady NY 12301

VOSHALL, ROY EDWARD, b Beacon, NY, May 29, 33; m 56; c 2. ELECTRICAL ENGINEERING. *Educ:* Carnegie Inst Technol, BS, 56, MS, 57, PhD, 61. *Prof Exp:* From instr to asst prof elec eng, Carnegie Inst Technol, 57-63, lectr, 63-69; sr res scientist, Westinghouse Elec Corp, 63-76, fel engr, Res Labs, 76-89; PROF-ELECT ENG, GANNON UNIV, 89- *Mem:* Inst Elec & Electronics Engrs; Am Phys Soc; Sigma Xi. *Res:* Electrical gaseous discharges; plasma physics; magnetohydrodynamics; vacuum arcs; vacuum breakdown. *Mailing Add:* 106 Walten Pointe Rd Erie PA 16511

VOSKO, SEYMOUR H, b Montreal, Que, Sept 9, 29; m 55; c 2. THEORETICAL PHYSICS. *Educ:* McGill Univ, BEng Phys, 51, MSc, 52; Carnegie Inst Technol, PhD(theoret physics), 57. *Prof Exp:* Instr physics, Carnegie Inst Technol, 56-58, vis asst prof, 58-60; from asst prof to assoc prof, McMaster Univ, 60-64; fel scientist, Westinghouse Res Labs, Pa, 64-70; PROF PHYSICS, UNIV TORONTO, 70- *Concurrent Pos:* Chmn, Theoret Physics Div, Can Asn Physicists, 72-73; vis prof physics, Univ Calif, Irvine, 78-79 & Rutgers Univ, 86; vis scientist, IBM, Thomas J Watson Res Ctr, 79 & Solid State Theory Group, Argonne Nat Lab, 85; Hooker distinguished vis prof, McMaster Univ, 85-86. *Mem:* Am Phys Soc; Can Asn Physicists. *Res:* Theoretical condensed matter physics; density functional theory; electronic structure of many electron systems; atomic electron affinities; phonons in metals; magnetism in metals. *Mailing Add:* Dept Physics Univ Toronto Toronto ON M5S 1A7 Can

VOSS, ANNE COBLE, b Richmond, Ind, Aug 22, 46; m 69; c 3. FOOD SCIENCE & NUTRITION. *Educ:* Ohio State Univ, BS, 68, PhD(food sci & nutrit), 84. *Prof Exp:* Clin dietician, Johns Hopkins Hosp, 68-69; regist dietitian, US Army Clins, Ger, 70-75 & pvt pract, 75-78; clin res assoc, Ross Labs, Div Abbott Labs, 78; proj dir, Nutrit Educ, Ohio Dept Educ, 79-80; therapeut dietician & clin instr dietetics, 69-70, grad res assoc, Dept Food Sci & Technol, 80-84, RES ASSOC NUTRIT, DEPT MED BIOCHEM, OHIO STATE UNIV, 85- *Concurrent Pos:* Consult, Ohio Dent Asn, 76- & Am Dent Asn Dent Health, 88-; lectr, Dept Food Sci & Technol, Col Dent & Pharm, 87-; adj asst prof, Col Human Ecol, Ohio State Univ, 87- & Col Nursing, Otterbein Col, 90- *Mem:* Sigma Xi; Am Dietetic Asn; Am Diabetes Asn; NY Acad Sci. *Res:* Fatty acid metabolism particularly desaturation and chain elongation using diabetic model. *Mailing Add:* Dept Med Biochem Ohio State Univ 337 Hamilton Hall Columbus OH 43210

VOSS, CHARLES HENRY, JR, b Kiangyen, China, Sept 28, 26; US citizen; m 54; c 2. ELECTRICAL ENGINEERING, BIOENGINEERING. *Educ:* La State Univ, BS, 49, MS, 56; NC State Univ, PhD(elec eng), 63. *Prof Exp:* Engr, WJBO-WBRL-FM, 51-53; div transmission engr, Southern Bell Tel Co, 53-54; instr elec eng, La State Univ, 54-56 & NC State Univ, 58-61; assoc prof, 62-67, PROF, LA STATE UNIV, 67-, UNDERGRAD COORDR, ELEC & COMPUT ENG DEPT, 84- *Concurrent Pos:* Res Coun fac fel, 65; Delta Regional Primate Ctr consult, 66-70; FJ Haydol Jr Kiaser Aluminum prof, 90. *Mem:* Sigma Xi. *Mailing Add:* Dept Elec Eng La State Univ Baton Rouge LA 70803

VOSS, EDWARD GROESBECK, b Delaware, Ohio, Feb 22, 29. TAXONOMIC BOTANY, NOMENCLATURE. *Educ:* Denison Univ, BA, 50; Univ Mich, MA, 51, PhD(bot), 54. *Prof Exp:* Asst syst bot, Biol Sta, 49, bot, uUiv, 50-51, Biol Sta, 51-53, res assoc, bot gardens, 54, res asst, Metab Res Lab, Univ Hosp, 54-56, res assoc herbarium, 56-61, from asst prof to assoc prof bot, 60-69, CUR VASCULAR PLANTS, HERBARIUM, UNIV MICH, ANN ARBOR, 61-, PROF BOT, 69- *Concurrent Pos:* Ed, Mich Botanist, 62-76; secy gen comt on bot nomenclature, 69-87, secy ed comt, Int Code of Bot Nomenclature, 69-81, chmn ed comt, 81-87; vice rapporteur, Bur of Nomenclature, Int Bot Congresses, 69 & 75, rapporteur, 81. *Honors & Awards:* Gleason Award, NY Bot Garden, 86. *Mem:* Am Soc Plant Taxon; Soc Syst Zool; Lepidop Soc; Int Asn Plant Taxon; Soc Bibliog Natural Hist; Linnean Soc London. *Res:* Floristics; vascular flora and vegetational history of Great Lakes region; history of biology; nomenclature; Lepidoptera of Michigan; natural areas in Michigan. *Mailing Add:* Herbarium North Univ Bldg Univ Mich Ann Arbor MI 48109-1057

VOSS, EDWARD WILLIAM, JR, b Chicago, Ill, Dec 2, 33; m 58, 74; c 2. IMMUNOCHEMISTRY, MICROBIOLOGY. *Educ:* Cornell Col, AB, 55; Univ Ind, Indianapolis, MS, 64, PhD(immunol), 66. *Prof Exp:* USPHS fel, Sch Med, Wash Univ, 66-67; from asst prof to assoc prof immunochem, 67-74, PROF MICROBIOL, UNIV ILL, URBANA, 74-, DIR, CELL SCI CTR, 88-, JUBILEE PROF, 90- *Concurrent Pos:* Fac fel sci, NSF, 75; vis prof microbiol, Ore State Univ, 77, dept biochem, molecular biol & cell biol, Northwestern Univ, 85; assoc appointment, Ctr Advan Study, 81-88; vis scholar, Univ Utah, Salt Lake City, 84-85; hon bd mem, Cent Ill Chap, Am Lupus Soc, 86-; mem comt rev, USP, 90- *Honors & Awards:* Nat Lupus Hall of Fame, Am Lupus Soc. *Mem:* AAAS; Am Asn Immunol; Sigma Xi; Protein Soc; Fedn Am Scientists; Am Soc Biochemists; Reticuloendothelial Soc; Am Chem Soc; NY Acad Sci; Am Soc Biochem & Molecular Biol. *Res:* Structure function and biosynthesis of immunoglobulins; lymphocyte receptors and autoimmune diseases. *Mailing Add:* Dept Microbiol Univ Ill 407 S Goodwin 131 Burrill Hall Urbana IL 61801

VOSS, GILBERT LINCOLN, biological oceanography, systematic zoology; deceased, see previous edition for last biography

VOSS, HENRY DAVID, b Evergreen Park, Ill, Jan 5, 50; m 73; c 4. ATMOSPHERIC PHYSICS, SPACE SCIENCES. *Educ:* Ill Inst Technol, BS, 72; Univ Ill, Urbana-Champaign, MS, 74, PhD(elec eng), 77. *Prof Exp:* res assoc, Aeronomy Lab, Univ Ill, 77-79; RES SCIENTIST, LOCKHEED PALO ALTO RES LAB, 79- *Concurrent Pos:* Consult prof elec, Urbana, 76-78. *Mem:* Am Geophys Union; Inst Elec & Electronics Engrs. *Res:* Upper atmosphere and ionosphere investigations; plasma physics; energetic particles in the magnetosphere and their global precipitation patterns on the atmosphere; rocket-borne instrumentation development; thermal physics; satellite instrumentation development. *Mailing Add:* Lockheed Palo Alto Res Lab 3251 91-20 B255 3251 Hanover St Palo Alto CA 94304

VOSS, JAMES LEO, b Grand Junction, Colo, Apr 7, 34; m 54; c 3. ANIMAL PHYSIOLOGY. *Educ:* Colo State Univ, BS, 56, DVM, 58, MS, 65. *Prof Exp:* From instr to assoc prof med & surg, 58-71, head dept, 75- 86, PROF CLIN SCI, COLO STATE UNIV, 72-, DEAN, COL VET MED & BIOMED SCI, 86- *Mem:* Am Vet Med Asn; Soc Study Reproduction; Am Soc Animal Sci; Am Asn Equine Practioners. *Res:* Equine reproduction; sexual behavior; artificial insemination; spermatogenesis; female reproductive cycle; ova transfer; pregnancy maintenance and control of ovulation. *Mailing Add:* 1217 Southridge Dr Ft Collins CO 80521

VOSS, KENNETH EDWIN, b Hastings, Nebr, Nov 12, 46; m 79; c 2. COLLOID SCIENCE. *Educ:* Univ Nebr, Lincoln, BSc, 69; Kans Univ, PhD(inorg chem), 75. *Prof Exp:* Fel res staff, Iowa State Univ, 74-76; from res chemist to sr 79-81, GROUP LEADER, NEW BUS RES, ENGELHARD CORP, 81- *Mem:* Am Chem Soc; Am Ceramic Soc; Mat Res Soc. *Res:* Advanced materials, particularly ceramics, ceramic composites and coatings; inorganic compounds, industrial minerals and chemicals; characterization of physical, chemical and colloidal properties for applications as catalysts, pigments and sorbents. *Mailing Add:* Engelhard Corp Menlo Park CN 28 Edison NJ 08818-2415

VOSS, PAUL JOSEPH, b Chicago, Ill, March 10, 43; div; c 2. OPERATIONS RESEARCH, SYSTEMS DESIGN. *Educ:* Syracuse Univ, BS, 69; Johns Hopkins Univ, MS, 72. *Prof Exp:* Assoc physicist, 69-75, SR PHYSICIST, APPL PHYSICS LAB, JOHNS HOPKINS UNIV, 75- *Concurrent Pos:* Facil mgr, Guidance Syst Eval Lab, 83-; chmn, Aegis Scenario Cert Comt, 85- *Res:* Analysis and simulation of missile and radar processing systems. *Mailing Add:* Appl Physics Lab Johns Hopkins Univ Johns Hopkins Rd Laurel MD 20707

VOSS, REGIS D, b Cedar Rapids, Iowa, Jan 4, 31; m 56; c 3. SOIL FERTILITY. *Educ:* Iowa State Univ, BS, 52, MS, 60, PhD(soil fertil), 62. *Prof Exp:* Res asst soil fertil, Iowa State Univ, 57-62; agriculturist, Test Demonstration Br, Tenn Valley Auth, 62-64; from asst prof to assoc prof, 64-66, PROF AGRON, IOWA STATE UNIV, 66- *Concurrent Pos:* Vis prof, Univ Ill, 70-71; consult, Int Maize & Wheat Improvement Ctr, Arg, 71-; mem bd dirs, Am Soc Agron, 77-78 & Soil Sci Soc Am, 80-83; leader, Agron Exten, Iowa State Univ, 83- *Honors & Awards:* Agron Exten Educ Award, Am Soc Agron, 84; Agron Achievement Award-Soils, Am Soc Agron, 89. *Mem:* Fel AAAS; fel Am Soc Agron; fel Soil Sci Soc Am. *Res:* Effect of uncontrolled factors on the response of field crops to applied fertilizers by using biological statistical methods. *Mailing Add:* Dept Agron Iowa State Univ Ames IA 50011

VOSS, RICHARD FREDERICK, b St Paul, Minn, Aug 27, 48. NOISE, JOSEPHSON JUNCTIONS. *Educ:* Mass Inst Technol, BS, 70; Univ Calif, Berkeley, PhD(physics), 75. *Prof Exp:* RES STAFF MEM, THOMAS J WATSON RES LAB, IBM, 75- *Honors & Awards:* Siefert Mem Lectr. *Mem:* Am Phys Soc; Sigma Xi. *Res:* Thermal and quantum mechanical limitations to physical devices, particularly Josephson Junctions; one-over-F noise and connection with music; computer generation and display of fractal objects; fractal characterization and analysis. *Mailing Add:* IBM Res PO Box 218 Yorktown Heights NY 10598

VOSSEN, JOHN LOUIS, b Philadelphia, Pa, Apr 4, 37; m 63; c 2. PHYSICS. *Educ:* St Joseph's Col, Pa, 58. *Prof Exp:* Engr, RCA Semiconductor & Mat Div, 58-62, group leader thin-film physics, RCA Advan Commun Lab, 62-65, mem tech staff, Process Res Lab, David Sarnoff Res Ctr, 65-78, HEAD THIN FILM TECHNOL, RCA LABS, 78- *Honors & Awards:* Achievement Awards, RCA Labs, 68, 69 & 71; Nerken Award, Am Vacuum Soc, 85. *Mem:* Am Vacuum Soc (pres, 80); Am Phys Soc; Electrochem Soc; Am Inst Physics. *Res:* Study of the methods by which the properties of thin films may be controlled or etched; principally, sputtering, ion plating, plasma anodization, evaporation, plasma deposition and etching; thin film technology. *Mailing Add:* RCA Labs Princeton NJ 08540

VOSTAL, JAROSLAV JOSEPH, b Prague, Czech, Mar 17, 27; m 52; c 4. PHARMACOLOGY, TOXICOLOGY. *Educ:* Charles Univ, Prague, MD, 51; Czech Acad Sci, PhD(med sci), 61. *Prof Exp:* Physician, Regional Inst Nat Health, Jihlava, Czech, 51-55; vis scientist, Nat Inst Pub Health, Stockholm, Sweden, 67-68; assoc prof pharmacol & toxicol, Sch Med, Univ Rochester, 68-77, assoc prof prev med & commun health, 69-77; HEAD BIOMED SCI DEPT, GMC RES LABS, 77- *Concurrent Pos:* Mem, Permanent Comn & Int Asn Occup Health, 66-, mem, Int Subcomt Toxicol Metals, 69-; chmn, Panel Fluorides, Nat Acad Sci-Nat Res Coun, 70-71, mem, Comt Biol Effects Atmospheric Pollutants, 70- *Mem:* AAAS; Am Soc Pharmacol & Exp Therapeut; Fedn Am Soc Exp Biol; Soc Toxicol. *Res:* Pharmacology of organomercurial compounds; toxicology of heavy metals and inorganic poisons, inter-species differences in pharmacokinetics and biotransformation of toxic substances. *Mailing Add:* Biomed Sci Dept GMC Res Labs GM Tech Ctr Warren MI 48090

VOSTI, DONALD CURTIS, b Modesto, Calif, Aug 26, 27; m 51; c 6. ORGANIC CHEMISTRY, FOOD TECHNOLOGY. *Educ:* Univ Calif, BS, 47, PhD, 52. *Prof Exp:* Chemist, E & J Gallo Winery, Calif, 47-52; sr chemist, 52-61, supvr, Container Specif Group, Western Area Lab, 61-62, supvr, Eval & Inspection Group, 62-64, supvr, Packaging Technol Group, 64-66, mgr, Customer Rels Sect, 66-69, proj mgr, Prod Technol Sect, 69-72, assoc dir, Gen Technol Sect, 72-73, proj mgr plastic hot fill foods, 73-76, PROJ MGR PLASTIC TECHNOL, AM CAN CO, 76- *Mem:* Inst Food Technologists. *Res:* Commercialization of new rigid containers developed for foods, non-foods and beverages; container utilization technology involving food processing, microbiological spoilage of food, container corrosion and related problems; container materials and manufacturing process improvments; development of plastic container for hot filled foods. *Mailing Add:* 433 N Northwest Hwy Barrington IL 60010

VOTAW, CHARLES ISAC, b Farris, Okla, Jan 3, 33; m 56; c 2. MATHEMATICS. *Educ:* Okla State Univ, BS, 57; NTex State Univ, MS, 67; Univ Kans, PhD(math), 71. *Prof Exp:* Staff mem, Sandia Corp, 57-60; mkt engr, Raytheon Co, 60-62; prod engr, Tex Instruments, 62-65; prod & test engr, Hunt Electronics, 65-66; PROF MATH, FT HAYS STATE UNIV, 71- *Mem:* Math Asn Am; Soc Indust & Appl Math; Am Math Soc. *Res:* Topology; applied mathematics. *Mailing Add:* Dept Math Ft Hays State Univ 600 Park St Hays KS 67601

VOTAW, ROBERT BARNETT, b Cincinnati, Ohio, Oct 19, 39; m 84; c 4. BIOSTRATIGRAPHY, PALEOECOLOGY. *Educ:* Ind Univ, BS, 62, MA, 64; Ohio State Univ, PhD(geol), 71. *Prof Exp:* Geologist, Standard Oil Co, Calif, 64-68; asst prof, 71-77, dept chmn, 73-76, asst dean facs, 76-77, dept chmn, 82-88, ASSOC PROF GEOL, IND UNIV NORTHWEST, 77- *Mem:* Geol Soc Am; Paleontol Soc. *Res:* Conodont biostratigraphy of the Ordovician rocks of the North American midcontinent; paleoecology of the middle and upper Ordovician rocks of the North American midcontinent. *Mailing Add:* Dept Geosci Ind Univ Northwest 3400 Broadway Gary IN 46408

VOTAW, ROBERT GRIMM, b St Louis, Mo, Sept 13, 38; m 61; c 3. BIOCHEMISTRY, MEDICAL EDUCATION. *Educ:* Wesleyan Univ, BA, 60; Case Western Reserve Univ, PhD(microbiol), 66. *Prof Exp:* Instr microbiol, Sch Med, Case Western Reserve Univ, 66-67; instr biochem, 67-70, asst prof, 70-77, assoc dean med educ, 74-77, ASST PROF RES HEALTH EDUC & DIR COMPUT BASED EDUC, SCH MED, UNIV CONN, 77- *Mem:* AAAS; Am Soc Microbiol; Am Chem Soc; Am Asn Med Cols. *Res:* Medical education; computer based education; use of simulation of teaching and assessing complex behavior, clinical problem solving. *Mailing Add:* 390 Middle Rd Farmington CT 06032

VOTH, DAVID RICHARD, b St Cloud, Minn, July 11, 41; m 69; c 2. PARASITOLOGY, INVERTEBRATE ZOOLOGY. *Educ:* Univ NDak, BS, 63, MS, 65; Ore State Univ, PhD(zool), 71. *Prof Exp:* Asst prof biol, Haile Selassie I Univ, 66-68; instr zool, Ore State Univ, 69-71; assoc prof, 71-75, PROF BIOL, METROP STATE COL, 75- *Mem:* Am Soc Parasitologists; Nat Educ Asn. *Res:* Helminth taxonomy, ecology and natural history. *Mailing Add:* Dept Biol Box 53 1006 11th St Denver CO 80204

VOTH, HAROLD MOSER, b Newton, Kans, Dec 29, 22; m 46; c 3. PSYCHIATRY. *Educ:* Washburn Univ, BS, 43; Univ Kans, MD, 47; Menninger Sch Psychiat, MD(psychiat), 52. *Prof Exp:* Asst sect chief, Vet Admin Hosp, Topeka, Kans, 52-53, asst chief, Acute Intensive Treatment Sect, 53-54, chief women's neuropsychiat serv, 54-57; STAFF PSYCHIATRIST, MENNINGER FOUND, 57-; CHIEF OF STAFF, VET ADMIN HOSP, TOPEKA, 80- *Concurrent Pos:* Mem fac, Menninger Sch Psychiat, 55-; NIMH res grant, Menninger Found, 63-72; examr, Am Bd Psychiat & Neurol, 70-71; consult, Walter Reed Army Med Ctr, 72-; assoc chief psychiat for educ, Vet Admin Hosp, Topeka, Kans, 75-; rear admiral, Med Corps, USN. *Honors & Awards:* William Porter Award, Am Asn Mil Surgeons, 79. *Mem:* Fel AAAS; fel Am Psychiat Asn; Am Col Psychoanal. *Res:* Personality organization; psychotherapy; the study of autokinesis as a research and clinical instrument. *Mailing Add:* Chief Staff Vet Admin Hosp Topeka KS 66622

VOTTA, FERDINAND, JR, b Providence, RI, June 8, 16; m 47; c 2. CHEMICAL ENGINEERING. *Educ:* Univ RI, BS, 39, MS, 41; Yale Univ, DEng, 58. *Prof Exp:* Develop engr, Allied Chem & Dye Corp, 41-43; res engr, Manhattan Proj, Columbia, 44-46; from instr to prof chem eng, Univ RI, 46-81; RETIRED. *Mem:* Am Inst Chem Engrs; Sigma Xi. *Res:* Thermodynamics; heat and mass transfer. *Mailing Add:* 2294 Kingstown Rd Box 215 Kingston RI 02881

VOUGHT, ELDON JON, b Chicago, Ill, May 21, 35; m 59; c 4. TOPOLOGY. *Educ:* Manchester Col, AB, 57; Univ Mich, Ann Arbor, MA, 58; Univ Calif, Riverside, PhD(math), 67. *Prof Exp:* Instr math, Pomona Col, 60-61; assoc prof math, Calif State Polytech Univ, Pomona, 61-70; PROF MATH, CALIF STATE UNIV, CHICO, 70- *Concurrent Pos:* NSF res fel, 71-73; vis prof math, Ariz State Univ, 75-76, 84; mathematician, Lockheed Aircraft Co, 77-88; vis prof math, Univ Richmond, 89-90. *Mem:* Am Math Soc; Math Asn Am. *Res:* The study of the invariance of various topological properties of compact, connected metric spaces under certain types of continuous functions, for example, local homeomorphisms, confluent and refinable functions. *Mailing Add:* Dept Math Calif State Univ Chico CA 95926

VOUGHT, ROBERT HOWARD, b Ridgway, Pa, Jan 30, 20; m 49; c 3. PHYSICS. *Educ:* Allegheny Col, BA, 41; Univ Pa, PhD(physics), 46. *Prof Exp:* Res physicist, Univ Pa, 42-46, instr phsyics, 44-46; res assoc, Gen Elec Co, 46-48; from asst prof to assoc prof, 48-56; Solid State physicist, Gen Elec Co, 56-85; RETIRED. *Concurrent Pos:* Exchange prof, St Andrews Univ, 53-54. *Mem:* Am Phys Soc; Am Asn Physics Teachers; AAAS; Am Nuclear Soc; Inst Elec & Electronics Engrs. *Res:* Semiconductors; mass spectrometry; surface physics; corrosion; energy conversion materials and devices; optoelectron devices; reactor technology; radiation damage. *Mailing Add:* 1465 Myron St Schenectady NY 12309

VOULGAROPOULOS, EMMANUEL, b Lowell, Mass, Apr 16, 31; m 59; c 2. PUBLIC HEALTH. *Educ:* Tufts Col, BSc, 52; Cath Univ Louvain, MD, 57; Johns Hopkins Univ, MPH, 62. *Prof Exp:* Med dir, Med Int Corp, Cambodia, 58-60, exec field dir for Asia, Cent Am & Africa, 60-61; dep, Health Serv Develop Proj, Vietnam AID, Saigon, 62-64, chief pub health div, 64-65; from assoc prof to prof pub health, Sch Pub Health, Univ Hawaii, Manoa, 65-85, head, Int Health Prog, 65-70, head, Int Health-Pop & Family Planning Studies Prog, 70-71, assoc dean, 74-79; chief, Off Pop & Health, Indonesia, 85-90, CHIEF, OFF POP, HEALTH & NUTRIT, USAID/MANILA, 90-; CONSULT, INT ASN SCHS SOCIAL WORK, NY, 73- *Concurrent Pos:* Consult, AID, 62, US Civil Admin Ryuku Islands, 65-67, US Trust Territory Pac Islands, 65-71, Govt of Guam & Govt of Am Samoa, 65-71, Peace Corps, 66-69, PPac Comn, Noumea New Caledonia, 67; vis prof pub health & adv, Udayana Community Health Prog, Udayana State Univ, 71-72; vis prof pub health & adv to fac pub health, Univ Indonesia, 72-73; consult, USPHS Global Community, Arlington, Va, 69,Mininstry Health, Govt Indonesia, 71-72, Ministry Educ & Cult, 71-73 & Ministry Social Welfare, 73; dir, Lampang Health Develop Proj, Thailand, 75-79; vis prof pub health & adv, Univ Indonesia, 79-81; adv, Pop & World Bank, USAID, Indonesia, 77-81. *Res:* Health and medical education systems; health delivery systems; international health. *Mailing Add:* Off Pop Health & Nutrit APO USAID San Francisco CA 96528

VOURNAKIS, JOHN NICHOLAS, b Cambridge, Ohio, Dec 1, 39; m 61; c 1. BIOPHYSICS, MOLECULAR BIOLOGY. *Educ:* Albion Col, BA, 61; Cornell Univ, PhD(chem), 68. *Prof Exp:* Nat Acad Sci exchange fel, Inst Org Chem & Biochem, Prague, Czech, fall 68; NIH fel biol, Mass Inst Technol, 69-71, res assoc, 71-72; res assoc, Harvard Univ, 72-73; prof biol, Syracuse Univ, 73-85; dir, Molec Genetics Corp, Dartmouth Univ, 85-87; dir, Sci Affairs, Molecular Therapeut, Inc, 86-88; vpres sci, Venax Corp, 88-90; SR VPRES, GENMAN INC, 90- *Concurrent Pos:* Vis assoc prof, Amherst Col, 70-71; NIH res grant gen med, 75; dir, Genetic Eng, Bristol Myers Co, 82-83; consult, Bolt, Beranek & Newman, 81, Bristol Myers Co, 81, Nat Itellenic Biotech Co, Biohellas, 83-84; chmn, sci adv bd, Chitin Co, Inc, 85-88. *Mem:* Biophys Soc; Soc Develop Biol; AAAS. *Res:* Secondary and tertiary structure of eukaryotic mRNA; structure of mapping techniques using enzymes as probes; secondary structure of rRNA; anti-caner drug-DNA interactions; human genome analysis; gene finding technology. *Mailing Add:* RR 1 Six Carriage Lane Lyme NH 03763

VOUROS, PAUL, b Thessaloniki, Greece, Apr 1, 38; US citizen; m 65; c 2. ANALYTICAL CHEMISTRY, ORGANIC CHEMISTRY. *Educ:* Wesleyan Univ, BA, 61; Mass Inst Technol, PhD(chem), 65. *Prof Exp:* Staff scientist, Tech Opers, Inc, 66-67, proj mgr, 67-68; asst prof chem, Inst Lipid Res, Col Med, Baylor Univ, 68-74; sr scientist, Inst Chem Anal, 74-79, assoc prof, 79-85, PROF CHEM, NORTHEASTERN UNIV, 85- *Mem:* Am Chem Soc; Sigma Xi; Am Soc Mass Spectrometry. *Res:* Organic mass spectrometry; mass spectrometry of biological compounds; gas chromatography-mass spectrometry; photographic ion detection; applications of chromatography and mass spectrometry to forensic problems; high performance liquid chromatography-mass spectrometry. *Mailing Add:* Dept Chem Northeastern Univ Boston MA 02115

VOURVOPOULOS, GEORGE, b June 11, 36; US citizen; m 63; c 3. NUCLEAR PHYSICS, APPLICATIONS OF NUCLEAR METHODS. *Educ:* Nat Univ Athens, BS, 58; Fla State Univ, MS, 65, PhD(physics), 67. *Prof Exp:* From assoc prof to prof physics, Fla A&M Univ, 67-76, chmn dept,

74-76; dir, Tandem Accelerator Lab, Greek Atomic Energy Comn, 76-83; vis prof, Vanderbilt Univ, 83-84; PROF PHYSICS, WESTERN KY UNIV, 84- *Concurrent Pos:* Res fel, Israel Inst Technol, 69-70; Res Corp grant, Fla A&M Univ, 71-76, NSF grant, 72-74, 84-88. *Mem:* Am Phys Soc; Sigma Xi. *Res:* Nuclear reactions; applications of nuclear techniques. *Mailing Add:* Dept Physics & Astron Western Ky Univ Bowling Green KY 42101

VOUTSAS, ALEXANDER MATTHEW (VOUTSADAKIS), b New York, NY, Mar 26, 23; m 57; c 2. TECHNOECONOMIC MANAGEMENT. *Educ:* Rensselaer Polytech Inst, BAeE, 44; Northwestern Univ, MS, 45; Harvard Grad Bus Sch, AMP, 69; Century Univ, Calif, PhD(eng), 85. *Prof Exp:* Aerodynamicist transonic windtunnel, Ames Labs, Nat Aeronaut & Space Admin, 45-46, XP-92, F102 delta wing, Gen Dynamics-Convair, San Diego, Calif, 47-48; proj aerodynamicist, Hermes Intercontinental Missile, Gen Elec Co, Schenectady, NY, 48-51; aerodynamicist rigel missile (pre polaris), Grumman Corp, Bethpage, NY, 51-52; proj design engr terrapin NASA iono rocket, Republic Aviation Corp, Hicksville, NY, 52-56; res & develop syst mgr atlas-titan intercontinental ballistic missile inertial guidance systems, Am Bosch Arma Corp, Garden City, NY, 56-61; mgr mkt commun & aerospace syst, Int Tel & Tel Corp, NY, 62-67; pres adv aerospace, Hellenic Aerospace Indust, Ltd, Greece, 80-91; RETIRED. *Concurrent Pos:* Guest comt mem, US Naval Bur Ordnance Comt Aerobalastics, 49-; lectr, Am Mgr Asn NY, 68-69; mkt consult, ITT Fed Elec Corp NY, 67-73; indust consult & pres adv, Motor Oil Hellas Ltd, NY & Greece, 69-73; pres, A Voutsas Assoc, aerospace & indust tech consult, 73-; bd dirs, Am Acad Greece, 76-78. *Mem:* Am Inst Aeronaut & Astronaut; fel Brit Interplanetary Soc. *Res:* Inventor of ultrasonic c-scan theory and equipment for diagnosis of materials fatigue and cancer growths and its cure; vibrating string accelerometer, Apollo 17 lunar mission gravimeter theory and equipment; author or coauthor of a number of publications. *Mailing Add:* Kalliga 21 Athens 114-73 Greece

VO-VAN, TRUONG, b Saigon, Viet-Nam, Dec 3, 48; m 71; c 2. THIN FILMS PHYSICS. *Educ:* Polytech Col Montreal, BScA, 70; Univ Moncton, MSc, 72; Univ Toronto, PhD(physics), 76. *Prof Exp:* PROF PHYSICS, UNIV MONCTON, 76- *Concurrent Pos:* Nat Res Coun Can grant, 76-, Regional Develop Prog grant, 76- & Coun Res, Univ Moncton, 76- *Mem:* Optical Soc Am; Can Asn Physicists; Asn Can-Fr pour L'Advan Sci. *Res:* Experimental solid state physics; specialization in thin films; optical properties of solids, spectroscopy and optics; solar energy applications. *Mailing Add:* Dept Physics Univ Moncton Moncton NB E1A 3E9 Can

VOXMAN, WILLIAM L, b Iowa City, Iowa, Feb 1, 39; m 63; c 2. TOPOLOGY. *Educ:* Univ Iowa, BA, 60, MS, 63, PhD(math), 68. *Prof Exp:* Latin Am teaching fel & prof math, Univ Chile, 68-69; prof, Concepcion Univ, 69; Fulbright travel grant & prof, State Tech Univ, Chile, 69-70; asst prof, 70-72, assoc prof, 72-76, PROF MATH, UNIV IDAHO, 76- *Concurrent Pos:* Latin Am teaching fel & prof math, Nat Polytech Sch, Quito, Ecuador, 74-76. *Mem:* Am Math Soc; Sigma Xi. *Res:* General topology, especially upper semicontinuous decompositions of topological spaces. *Mailing Add:* 1118 King Rd Moscow ID 83843

VOYTUK, JAMES A, b Pittsburg, Pa, July 5, 36; m. APPLIED MATHEMATICS, MATHEMATICS GENERAL. *Educ:* Carnegie Inst Technol, BS, 58, MA, 59, PhD(math), 63. *Prof Exp:* Asst prof math, Western Reserve Univ, 62-66; from asst prof to assoc prof math, Rensselaer Polytech, 66-85; assoc exec dir, Am Math Soc, 85-89; SR PROJ DIR, NAT RES COUN, 89- *Concurrent Pos:* Exec officer, Dept Math, Rensselaer Polytech, 73-85. *Mem:* Am Math Soc; Math Asn Am; AAAS; Sigma Xi; Soc Indust & Appl Math. *Mailing Add:* Nat Res Coun 2101 Constitution Ave NW Washington DC 20418

VOYVODIC, LOUIS, b Yugoslavia, Oct 22, 21; US citizen; m 51; c 5. HIGH ENERGY PHYSICS. *Educ:* McGill Univ, BSc, 43, PhD(physics), 48. *Prof Exp:* Physicist cosmic rays, Nat Res Coun Can, 48-56; tech proj dir radiation physics, Isotope Prod Ltd Can, 56-58; physicist, Armour Res Found, Ill Inst Technol, 58-60; physicist, Argonne Nat Lab, 62-72; PHYSICIST, FERMI NAT ACCELERATOR LAB, 72- *Mem:* Am Inst Physics; Am Phys Soc; Fedn Am Scientists. *Res:* Interactions of elementary particles at high energies, particularly as studied by optical track chamber techniques, and development of improved detection techniques. *Mailing Add:* 707 S Monroe Hinsdale IL 60521

VOZOFF, KEEVA, b Minneapolis, Minn, Jan 26, 28; m 57; c 4. EXPLORATION GEOPHYSICS, ENGINEERING GEOPHYSICS. *Educ:* Univ Minn, BPhys, 49; Pa State Univ, MSc, 51; Mass Inst Technol, PhD(geophys), 56. *Prof Exp:* Geophysicist, Nucom-McPhar Geophys, Ltd, 55-58; assoc prof geophys, Univ Alta, 58-64; vpres, Geosci Inc, Mass, 64-69; consult, 69-72; PROF GEOPHYS, MACQUARIE UNIV, AUSTRALIA, 72-, DIR, CTR GEOPHYS EXPLOR RES, 81-; DIR, HARBOUR DOM GMBH. *Concurrent Pos:* Mem earth sci ad hoc comt, Soviet-Australian Coop, 74-75; Hearst vis prof, Univ Calif, Berkeley, 78-80; prof geophys, Inst Geophys & Meteorol, Univ Cologne, 89- *Mem:* Hon mem Australian Soc Explor Geophysicists; Am Geophys Union; Europ Asn Explor Geophys; Petrol Explor Soc Australia; fel Australian Acad Technol Sci & Eng; hon mem Soc Explor Geophysicists; hon fel Asn Explor Geophysicists India; Am Asn Petrol Geologists; Australian Soc Explor Geophysicists (pres, 76-77). *Res:* Electrical and electromagnetic methods of determining earth structure; natural electromagnetic fields; tomographic imaging. *Mailing Add:* Sch Earth Sci Macquarie Univ Sydney NSW 2109 Australia

VRANA, NORMAN M, b Hudson Heights, NJ, Feb 16, 20; m 42; c 4. ELECTRICAL ENGINEERING. *Educ:* NY Univ, BEE, 47; Cornell Univ, MEE, 51. *Prof Exp:* Sr engr, ADT Co, NY, 40-49; assoc prof elec eng, Cornell Univ, 49-57; sr res engr, Autonetics Co, Calif, 57-58; assoc prof elec eng, Cornell Univ, 58-65; res & develop engr, Hewlett Packard Co, Colo, 65-66; assoc prof, 66-75, PROF ELEC ENG, CORNELL UNIV, 75- *Concurrent Pos:* Commun engr & seminar consult, NY Tel Co, 62-65; consult, Frankford Arsenal, Pa, 68-71 & Ballistics Res Lab, Md, 70- *Mem:* Inst Elec & Electronics Engrs. *Res:* Hybrid computer simulation and computation; communication and instrumentation; electronic design and development; computer and digital systems. *Mailing Add:* Dept Elec Eng Phillips Hall Cornell Univ Ithaca NY 14853

VRANIC, MLADEN, b Zagreb, Yugoslavia, Apr 3, 30; Can citizen; m 83; c 3. PHYSIOLOGY, ENDOCRINOLOGY. *Educ:* Univ Zagreb, MD, 55, DSc (physiol), 62; FRCP(C), 86. *Prof Exp:* Fel, 63-65, from asst prof to assoc prof, 65-72, PROF PHYSIOL, FAC MED, UNIV TORONTO, 72-, PROF MED, 78- *Concurrent Pos:* Mem, Inst Med Sci, Univ Toronto, 73-, Inst Biomed Elec & Eng, 73-78, sen res comt, 80-82; fac scholar, Josia Macy Fdn, Univ Geneva, 76; guest prof, Univ Geneva, 76-77; assoc ed, Am J Physiol, 82-89; vis res fel, Merton Col, Univ Oxford, Eng, 86; hon vis prof, Fac Med, Univ Zagreb, 87; fel, Merton Col, Oxford, 86; Killam res fel, 88, 89. *Honors & Awards:* Pfizer Lectr, Clin Res Inst, Univ Montreal, 85; Inaugural Banting & Best Mem Lectr, Madrid, Spain, 85; Upjohn lectr, Univ Ottawa, Can, 80; Vuk Vrhovac mem lectr, Univ Zagreb, Yugoslavia, 77. *Mem:* Endocrine Soc; Can Diabetes Asn; Am Diabetes Asn; Can Physiol Soc; Am Physiol Soc; fel Royal Col Physicians & Surg; Int Diabetes Fed; NY Acad Sci; Europ Asn Study Diabetes; Can Soc Clin Res. *Res:* Metabolic roles and interactions of insulin, glucagon, epinephrine, growth hormone, B-endorphims and glucocorticoid hormones in health and disease-diabetes and obesity; origin, structure and secretion of nonpancreatic glucagon; endocrine responses and effects during exercise and stress; tracer methodology-hepatic futile cycles in health and disease; regulation of glucose transporters in the muscle. *Mailing Add:* Dept Physiol Med Sci Bldg Univ Toronto Fac Med Toronto ON M5S 1A8 Can

VRATSANOS, SPYROS M, b Athens, Greece, Apr 10, 20; US citizen; m 58; c 2. BIOCHEMISTRY, ORGANIC CHEMISTRY. *Educ:* Univ Athens, dipl chem, 50; Adelphi Univ, MS, 56; Fordham Univ, PhD(enzymol, org chem), 61. *Prof Exp:* Asst prof biochem, Adelphi Univ, 61-63; res assoc, 63-65, ASST PROF MICROBIOL, COL PHYSICIANS & SURGEONS, COLUMBIA UNIV, 65- *Mem:* Am Chem Soc; Neuberg Socl Harvey Soc; Sigma Xi. *Res:* Organophosphorous compounds; origin of life on the earth; proteins; active sites of enzymes; conversion of light energy to chemical signals; chemistry of vision; immunochemistry. *Mailing Add:* 11 Chadwick Rd Syosset NY 11791-6508

VRBA, FREDERICK JOHN, b Cedar Rapids, Iowa, May 25, 49; m 71; c 2. ASTRONOMY. *Educ:* Univ Iowa, BA, 71; Univ Ariz, PhD(astron), 76. *Prof Exp:* STAFF ASTRONR, FLAGSTAFF STA, NAVAL OBSERV, 76- *Honors & Awards:* Simon Newcomb Award, US Naval Observ, 87. *Mem:* Sigma Xi; Am Astron Soc; Int Astron Union. *Res:* Infrared, optical, and polarimetric observations of young stars, dark nebulae, and the general interstellar medium; charge coupled device photometry and polarimetry; trigonometric parallaxes of nearby stars. *Mailing Add:* Naval Observ Flagstaff Sta Box 1149 Flagstaff AZ 86002

VRBANAC, JOHN JAMES, b Grand Rapids, Mich, Jan 30, 49; m 83; c 1. MASS SPECTROMETRY, PHARMACOLOGY. *Educ:* Mich State Univ, BS, 72, PhD(pharmacol), 84. *Prof Exp:* ASST PROF PHARMACOL, MED UNIV SC, 84- *Concurrent Pos:* Consult, Ciba-Geigy Corp, 88. *Mem:* Am Soc Mass Spectrometry; Am Chem Soc. *Res:* The use of mass spectrometry to solve qualitative and quantitative problems in medicine and biology. *Mailing Add:* Upjohn Co 7256-126-222 Kalamazoo MI 49001-3298

VREBALOVICH, THOMAS, b Los Angeles, Calif, July 10, 26; m 51; c 2. SPACE PHYSICS, FLUID MECHANICS. *Educ:* Calif Inst Technol, BS, 48, MS, 49, PhD(aeronaut eng), 54. *Prof Exp:* From res scientist to sr res scientist, Jet Propulsion Lab, Calif Inst Technol, 52-61, res specialist, 61-62, Ranger proj scientist, 63-65, group supvr photosci, 65-66, Surveyor assoc proj scientist, 65-67, div rep space sci, 66-67, Voyager landed capsule syst scientist, 67-68, on leave, 68-70, Sci Recommendation Chief, Mariner Mars 1971 proj, 70-73, mission sci coordr, Mariner Jupiter Saturn Proj, 73-74, mgr res, 74-75; mem coun sci & technol affairs, Am Embassy, New Delhi, India, 75-80, MEM COUN SCI & TECHNOL AFFAIR, AM EMBASSY, CAIRO, EGYPT, 80- *Concurrent Pos:* Instr, Univ Southern Calif, 56 & Univ Calif, Los Angeles, 57; consult, Flow Corp, 66-; vis prof aeronaut, Indian Inst Technol, Kanpur, 68-70; mem bd gov, Photog Art & Sci Found, 72-; mem bd dir, US Educ Found in India, 75. *Honors & Awards:* Fairbanks Mem Award, Soc Photog Instrumentation Engrs, 66; NASA Group Achievement Award for Mariner Mars 71 proj. *Mem:* Am Inst Aeronaut & Astronaut; Sigma Xi; Am Phys Soc; Explorers Club. *Res:* Supersonic aerodynamics; space photography and science. *Mailing Add:* Apt B202 Tossa De Mar Costa Brava Spain

VREDEVELD, NICHOLAS GENE, b Hudsonville, Mich, May 5, 29; m 53; c 2. PLANT PATHOLOGY, MICROBIOLOGY. *Educ:* Calvin Col, AB, 51; Mich State Univ, MS, 55, PhD(plant path), 65. *Prof Exp:* Biochemist, St Lawrence Hosp, Lansing, Mich, 62-64; from asst prof to assoc prof, 64-84, PROF BIOL, UNIV TENN, CHATTANOOGA, 84- *Mem:* Am Phytopath Soc. *Res:* Fungicides and fungus physiology, environmental effect on plant disease distribution; effect of microbiol toxins on plants, whole and in tissue culture; animal cell culture (lymphocytes). *Mailing Add:* Dept Biol Univ Tenn Chattanooga TN 37401-2598

VREDEVOE, DONNA LOU, b Ann Arbor, Mich, Jan 11, 38; m 62; c 1. IMMUNOTOXICOLOGY, TUMOR IMMUNOLOGY. *Educ:* Univ Calif, Los Angeles, BA, 59, PhD(microbiol), 63. *Prof Exp:* Instr bact, 63, asst res immunologist, 64-67, asst prof nursing res, 67-70, assoc prof, 70-76, consult lab nuclear med & radiation biol, 67-80, assoc dean, 76-78, chmn, Allied Sci Sect, 79-88, actg assoc dean, Sch Nursing, 85-86, Dir Space Planning, Cancer Ctr, 74-90, PROF NURSING RES, UNIV CALIF, LOS ANGELES, 76- *Concurrent Pos:* USPHS fel microbiol, Stanford Univ, 63-64; res grants, Calif Inst Cancer Res, Univ Calif, Cancer Res Coord Comt, Am Cancer Soc, Calif Div & Nat Am Cancer Soc, US Dept Energy & USPHS. *Mem:* Am Asn Immunol; Am Soc Microbiol; Am Asn Cancer Res; Sigma Xi; Nat League Nursing (vpres, 79-81). *Res:* Serotyping for human kidney transplantation; delayed hypersensitivity; immunosuppression; immunotherapy; tumor immunology; mouse lymphoma; effects of metallic ions on the immune response; carcinogenesis. *Mailing Add:* Sch Nursing Univ Calif Ctr Health Sci 10833 Le Conte Ave Los Angeles CA 90024-6918

VREDEVOE, LAWRENCE A, b Ann Arbor, Mich, Aug 2, 40; m 66; c 2. ANESTHESIOLOGY. *Educ:* Univ Calif, Los Angeles, BA, 62, MA, 64, PhD(physics), 66; Univ Calif, San Francisco, MD, 75. *Prof Exp:* Mem tech staff theoret physics, Sci Ctr, NAm Rockwell Corp, Calif, 66-70; assoc prof physics, Ind Univ Bloomington, 70-72; mem staff med training, Sch Med, Univ Calif, San Francisco, 72-75, surg resident, 75-77, head & neck surg resident, 77-78, anesthesiol resident, 78-80, ASST PROF ANESTHESIOL, CTR HEALTH SCI, UNIV CALIF, LOS ANGELES, 80-; CHIEF, DEPT ANESTHESIOL, SAINT JOHN'S HOSP, SANTA MONICA, CALIF, 84- *Mem:* AMA; Am Soc Anesthesiologists; Am Phys Soc. *Mailing Add:* 2210 Wilshire Blvd Suite 230 Santa Monica CA 90402

VREELAND, JOHN ALLEN, b Orlando, Fla, Jan 6, 25; m 52; c 2. PHYSICS. *Educ:* Presby Col, BS, 49; Univ Wis, MA, 51, PhD(physics), 56. *Prof Exp:* Asst physicist, Univ Wis, 50-55; fel scientist, Atomic Power Div, Westinghouse Elec Corp, 55-60; sr nuclear specialist, Rocketdyne Div, NAm Aviation, 60-62; mgr nuclear anal dept, nuclear rocket opers, Aerojet-Gen Corp, 62-69; PROF NUCLEAR ENG, SCH ENG, CALIF STATE UNIV, SACRAMENTO, 69- *Concurrent Pos:* Lectr, Univ Calif, 60-; consult, Univ Fla, 62; mem Atomic Indust Forum. *Mem:* Am Physics Soc; Am Nuclear Soc; Am Inst Aeronaut & Astronaut; Sigma Xi. *Res:* Reactor physics; nuclear structure; analysis and detection of nuclear transport phenomena; systems analysis related to nuclear power plant design. *Mailing Add:* 4901 Shamrock Dr Fair Oaks CA 95628

VREELAND, THAD, JR, b Portland, Ore, Oct 20, 24; m 48; c 3. MATERIALS SCIENCE. *Educ:* Calif Inst Technol, BS, 49, MS, 50, PhD(mech eng), 52. *Prof Exp:* Res fel mech eng, 52-54, from asst prof to assoc prof, 54-63, assoc prof mat sci, 63-67, PROF MAT SCI, CALIF INST TECHNOL, 68- *Concurrent Pos:* Consult, indust & govt labs. *Res:* Dynamic powder consolidation; crystal analysis by x-ray rocking curves; plastic deformation of crystals; dislocation dynamics. *Mailing Add:* Eng & Appl Sci 128-95 Calif Inst Technol 1201 E Calif Blvd Pasadena CA 91125

VREELAND, VALERIE JANE, b Brooklyn, NY, Oct 15, 43; m 73; c 2. CELL BIOLOGY, CARBOHYDRATE BIOCHEMISTRY. *Educ:* Lake Forest Col, BA, 65; Stanford Univ, MA, 67, PhD(biol sci), 71. *Prof Exp:* Fel carbohydrate immunochem, Univ Trondheim, Norway, 71; fel marine pollution, Woods Hole Oceanog Inst, 72-73; fel carbohydrate immunocytochem, 74-78, RES BOTANIST DEVELOP BIOL, DEPT BOT, UNIV CALIF, BERKELEY, 78- *Concurrent Pos:* Lectr, dept biol, Univ Calif, Berkeley, 75, res assoc, Los Angeles, 75-78; res assoc, biol dept, Univ Calif, Los Angeles, 75-78. *Mem:* Am Soc Cell Biol; Am Soc Plant Physiol; Phycol Soc Am; Int Phycol Soc. *Res:* Cell wall carbohydrate production and developmental modifications investigated in a brown algal embryo system; monoclonal antibodies and carbohydrate hybridization probes utilized as molecular markers for important carbohydrate structures. *Mailing Add:* Dept Plant Biol Univ Calif Berkeley CA 94720

VREMAN, HENDRIK JAN, b Soest, Netherlands, Jan 22, 39; US citizen; m 64; c 2. BIOLOGICAL CHEMISTRY. *Educ:* Univ NC, Chapel Hill, BA, 68; Univ Wis-Madison, PhD(bot), 73. *Prof Exp:* Lab asst dairy chem & bact, United Gooi Dairies, Hilversum, Netherlands, 56-57; anal chemist vet med, Lab Medical Vet Medicine, Univ Utrecht, 57-60; anal chemist pub health, Pharmaceut & Toxicol Lab, Nat Inst Pub Health, Utrecht, Netherlands, 60-62; res technician biochem, E R Johnson Found, Univ Pa, 62-64; res technician physiol & pharmacol, Duke Univ, 64-68; res asst bot, Inst Plant Develop, Univ Wis-Madison, 68-73; Nat Res Coun res assoc, Western Regional Res Ctr, Agr Res Serv, USDA, Calif, 73-75; res assoc med, Stanford Univ Serv, Vet Admin Hosp, 75-80, lab dir, Gen Clin Res Ctr, 80-82, LAB DIR, NEONATAL METAB LAB, SCH MED, STANFORD UNIV, 82- *Mem:* AAAS; Am Soc Plant Physiologists; Am Chem Soc; Am Soc Photobiol; Oxygen Soc; Int Soc Free Radical Res. *Res:* Role of cytokinins in plant growth and development; isolation, separation, identification of naturally occurring cytokinins; plant cell and tissue cultures; acetate metabolism in humans with chronic renal failure; zinc nutrition; atherosclerosis; carbohydrate metabolism; diabetes mellitus in human subjects; neonatal jaundice; metabolism of heme; biological production of carbon monoxide; photosensitizers in medicine; lipiol peroxidation; metalloporphyrins in medicine. *Mailing Add:* Dept Pediat Stanford Univ Med Ctr S214 Stanford CA 94305

VRENTAS, CHRISTINE MARY, b Chicago, Ill, June 16, 53; m 75; c 2. POLYMER RHEOLOGY, DIFFUSION IN POLYMERS. *Educ:* Ill Inst Technol, BS, 75; Northwestern Univ, MS, 77, PhD(chem eng), 81. *Prof Exp:* Asst prof, 81-83, from adj asst prof to adj assoc prof, 83-90, ADJ PROF CHEM ENG, PA STATE UNIV, 90- *Mem:* Am Inst Chem Engrs; Soc Rheology. *Res:* Transport phenomena in fluids, primarily polymeric materials; polymer rheology (development of constitutive equations and material properties measurement); theoretical studies of polymer-solvent diffusion and sorption. *Mailing Add:* Dept Chem Eng Pa State Univ University Park PA 16802

VRENTAS, JAMES SPIRO, b Danville, Ill, Apr 14, 36; m 75; c 2. CHEMICAL ENGINEERING. *Educ:* Univ Ill, BS, 58; Univ Del, MChE, 61, PhD(chem eng), 63. *Prof Exp:* Res engr, Dow Chem Co Mich, 63-72; from asst prof to prof, 72-85, DOW PROF CHEM ENG, PA STATE UNIV, 85- *Honors & Awards:* William H Walker Award, Am Inst Chem Engrs, 81, Charles M A Stine Mat Eng & Sci Award, 89. *Mem:* Am Inst Chem Engrs; Am Chem Soc; Sigma Xi. *Res:* Transport phenomena; fluid mechanics; diffusion; applied mathematics; polymer science. *Mailing Add:* Dept Chem Eng Pa State Univ University Park PA 16802

VRIELAND, GAIL EDWIN, b Grand Rapids, Mich, Jan 4, 38; m 63; c 3. INDUSTRIAL CHEMISTRY. *Educ:* Calvin Col, AB, 59; Northwestern Univ, PhD(chem), 63. *Prof Exp:* Res specialist, 63-80, RES ASSOC CHEM, CENT RES LAB, DOW CHEM CO, 80- *Mem:* Am Chem Soc; Sigma Xi. *Res:* Heterogeneous catalysis and high temperature vapor phase reaction including oxidations and hydrocyanation. *Mailing Add:* 1420 Crescent Dr Midland MD 48640-3315

VRIENS, GERARD N(ICHOLS), b New York, NY, July 4, 24; m 47; c 2. CHEMICAL ENGINEERING. *Educ:* Purdue Univ, BSChE, 44, PhD(chem eng), 49. *Prof Exp:* Sr chemist, Process Develop Dept, Am Cyanamid Co, 49-58, chief chemist, Rubber Chem Dept, 58-65, group leader dyes res & develop, 65-69, mgr process develop, Agr Div, 69-79, prin engr, Agr Res Div, 79-87; RETIRED. *Res:* Decision and risk analysis. *Mailing Add:* 1024A Thornbury Lane Lakehurst NJ 08733

VRIESEN, CALVIN W, b Elkhart Lake, Wis, Aug 31, 16; m 41; c 2. ORGANIC POLYMER CHEMISTRY, ORGANIC CHEMISTRY. *Educ:* Univ Minn, BS, 39, MS, 47; Purdue Univ, PhD(org chem), 52. *Prof Exp:* Assoc prof chem, Ill Col, 47-49; res chemist, Chattanooga Nylon Plant, E I du Pont de Nemours & Co, 52-56, Chambers Works, 56-58; res chemist, 58-62, staff chemist, 62-68, sr scientist, 68-74, GROUP SUPVR, THIOKOL CORP, ELKTON, MD, 74- *Honors & Awards:* Aerospace Scientist of Year Award, Am Inst Aeronaut & Astronaut, 67. *Mem:* Am Chem Soc; Sigma Xi. *Res:* Condensation, cationic, anionic polymerization; synthesis of new binders, oxidizers, coolants for solid rocket propellants. *Mailing Add:* 12 Mitchell Circle Brookside Newark DE 19713-2522

VRIJENHOEK, ROBERT CHARLES, b Rotterdam, Netherlands, Mar 13, 46; US citizen; m 68; c 2. EVOLUTIONARY BIOLOGY. *Educ:* Univ Mass, BA, 68; Univ Conn, PhD(zool), 72. *Prof Exp:* Asst prof biol, Southern Methodist Univ, 72-74; from asst prof to assoc prof zool, 74-84, prof biol, 84-, PROF GENETICS & DIR CTR THEORET GENETICS, RUTGERS UNIV, NEW BRUNSWICK. *Concurrent Pos:* NSF grants, 74, 76, 77, 79, 82 & 85; mem, Pop Biol & Physiol Ecol Panel, NSF, 78-81; mem, Conf Biol Diversity, US Dept State, 81; mem coun, Am Genetics Asn, 81-83, 85-, assoc ed, Evolution; ed, Evolution, 87-90; assoc ed, Conservation Biol, 87- *Mem:* Soc Study Evolution; Am Soc Ichthyol & Herpetol; Genetics Soc Am; Am Genetic Asn; Am Soc Naturalists; Am Soc Persatologist; Soc Conserv Biol. *Res:* Population genetic studies of evolutionary relationships and genetic variation in fishes, snails and parasitic helminths; the effects of various sexual and asexual mating systems on the genetic structure and evolutionary potential of populations. *Mailing Add:* Ctr Theoret & Appl Genetics Paul Rutgers Univ PO Box 231 New Brunswick NJ 08903

VROMAN, HUGH EGMONT, b Detroit, Mich, Apr 18, 28; m 59. BIOCHEMISTRY. *Educ:* Univ Md, BS, 50, PhD(zool), 62. *Prof Exp:* Biologist, Nat Heart Inst, 57-58, biochemist, 58-61; res biologist, Insect Physiol Lab, Agr Res Serv, USDA, 61-66; biochemist, Dept Dermat, Sch Med, Univ Miami, 66-69; actg chmn dept biol, Claflin Col, 71-73, prof biol, 69-76; from asst prof to prof biol, Cleveland State Community Col, 76-90; RETIRED. *Mem:* AAAS; Entom Soc Am; Am Soc Zool; Brit Biochem Soc; Am Inst Biol Sci. *Res:* Cholesterol metabolism; lipid biosynthesis by insects; insect hormones; sterol metabolism by insects; lipid biosynthesis and metabolism in skin. *Mailing Add:* 3840 Sycamore Dr NW Cleveland TN 37312

VROMAN, LEO, b Gouda, Holland, Apr 10, 15; nat US; m 47; c 2. PHYSIOLOGY, BIOPHYSICS. *Educ:* Jakarta Med Col, Indonesia, Drs, 41; Univ Utrecht, PhD(animal physiol), 58. *Prof Exp:* Asst zool, anat & physiol, Jakarta Med Col, 41; res assoc, St Peter's Gen Hosp, New Brunswick, NJ, 46-55; asst, Mt Sinai Hosp, New York, 56-58; sr physiologist, Stress-Tension Proj, Dept Animal Behav, Am Mus Natural Hist, 58-61; biochemist, Vet Admin Hosp, Brooklyn, 61-78, res career scientist, 78-86; ASSOC PROF BIOPHYS DEPT, STATE UNIV NY, DOWNSTATE MED CTR, 77-; SR RES SCIENTIST, CHEM ENG DEPT, COLUMBIA UNIV, 87- *Honors & Awards:* Clemson Award for Basic Sci in Biomater Res, 86; Silver Medal, Neth Royal Acad Sci, 87. *Mem:* Fel NY Acad Sci; fel NY Acad Med. *Res:* Behavior of blood at interfaces; biomaterials. *Mailing Add:* 2365 E 13th St Apt 6U Brooklyn NY 11229

VROOM, ALAN HEARD, b Montreal, Que, Can, Oct 5, 20; m 43, 84; c 2. INSTRUMENTATION, MATERIALS SCIENCE. *Educ:* McGill Univ, BSc, 42, PhD(phys chem), 45. *Prof Exp:* Nat Res Coun Can, 44-46; asst dir res pulp & paper, Fraser Co, Ltd, 46-49; Hibbert Mem fel & hon lectr, McGill Univ, 50; res fel bark chem, Pulp & Paper Res Inst Can, 50-51; asst chief appl chem sect, Weyerhaeuser Timber Co, 51-52, chief appl physics sect, 52-54, chief appl chem sect, 54; asst dir res, Consol Paper Corp, Ltd, 55-56, dir res, 56-67, dir res & develop Consol-Bathurst Ltd, 67-71; spec consult, Nat Res Coun Can, 71-73; PRES, SULFURCRETE PRODS, INC, 73- *Concurrent Pos:* Consult, Pulp & Paper Res Inst, Can, 51-54. *Mem:* Am Concrete Inst; fel Chem Inst Can; Can Soc Chem Eng. *Res:* Bark chemistry; pulp and paper; wood and fiber technology; sulfur utilization; development of new sulfur-based construction materials, primarily sulfur concrete. *Mailing Add:* 10728 Willowfern Dr SE Calgary AB T2J 1R4 Can

VROOM, DAVID ARCHIE, b Vancouver, BC, Sept 12, 41; m 69; c 1. POLYMER ENGINEERING, RADIATION TECHNOLOGY. *Educ:* Univ BC, BSc, 63, PhD(phys chem), 67. *Prof Exp:* Nat Res Coun Can overseas fel, 67-68; staff chemist, Atomic Physics Br, Gulf Radiation Technol Div, Gulf Energy & Environ Systs, 68-73; prin scientist, IRT Corp, 73-77, mgr atomic physics dept, 77-81; tech dir corp res & develop process eng, 81-84, mfg mgr, HTG Group, 85-87, DIR RADIATION SERV, RAYCHEM CORP, 87- *Mem:* Am Phys Soc; Am Chem Soc. *Res:* Photoionization and photoelectron spectroscopy; electron impact studies of excitation; dissociation and ionization; low energy ion neutral reactions and pulse radiolysis studies; polymer processing; polymer physics; radiation technology; accelerator technology. *Mailing Add:* 300 Constitution Dr Raychem Corp Menlo Park CA 94025

VUCHIC, VUKAN R, b Belgrade, Yugoslavia, Jan 14, 35; US citizen; m 60; c 4. TRANSPORTATION ENGINEERING. *Educ:* Univ Belgrade, dipl transp eng, 60; Univ Calif, Berkeley, MEng, 65, PhD(civil eng, transp), 66. *Hon Degrees:* MA, Univ Pa, 71. *Prof Exp:* Planning engr, Hamburger Hochbahn AG, Ger, 60-61; asst & prin engr, Wilbur Smith & Assocs, Conn, 61-63; asst prof civil eng-transp, 67-70, assoc prof, 70-75, PROF TRANSP

ENG, DEPT SYST, UNIV PA, 75- *Concurrent Pos:* Mem, Transp Res Bd, Nat Acad Sci-Nat Res Coun; consult, Urban Mass Transp Admin, US Dept Transp & Off Technol Assessment, US Cong, 75. *Honors & Awards:* Friedrich Lehner Medal, Munich, Ger, 82. *Mem:* Am Soc Civil Engrs; assoc mem Inst Transp Engrs; assoc mem Int Union Pub Transp; Asn Transp & Commun Eng & Technicians Yugoslavia. *Res:* Transportation policy, analysis, economics, systems definition, operation, evaluation and facilities; public transportation systems, rail transit; city planning and urban development; urban transportation planning, design and operations; traffic engineering; highway, street and transit network design; transportation in developing countries. *Mailing Add:* Dept Syst Sci 113 Towne Bldg Univ Pa Philadelphia PA 19104-6315

VUCICH, M(ICHAEL) G(EORGE), b Bower Hill, Pa, Oct 30, 26; m 57; c 4. ELECTROCHEMISTRY, LUBRICATION. *Educ:* US Merchant Marine Acad, BS, 48; Carnegie Inst Technol, BS, 52. *Prof Exp:* From chem engr to res chem engr, Weirton Steel Co Div, Nat Steel Corp, 52-58, res engr, 58-59, sr res engr, 59-73, res assoc, 73-74, supvr corrosion & lubrication res, Res & Develop Dept, 74-85, res & develop, 85-91; RETIRED. *Mem:* Electrochem Soc; Nat Asn Corrosion Engrs. *Res:* Electrodeposition; surface chemistry and corrosion; lubrication; cold reduction. *Mailing Add:* 115 Forest Rd Wierton WV 26062

VUCKOVIC, VLADETA, b Aleksinac, Yugoslavia, Mar 30, 23; m 54; c 2. MATHEMATICS. *Educ:* Univ Belgrade, MS, 49; PhD(math), Serbian Acad Sci, 53. *Prof Exp:* Instr math, Univ Belgrade, 49-52; sci collabr, Math Inst, Serbian Acad Sci, 52-54; prof math, Teacher Inst, Zrenjanin, Yugoslavia, 54-60; from asst prof to assoc prof, Univ Belgrade, 60-63; asst prof, 63-66, ASSOC PROF MATH, UNIV NOTRE DAME, 66- *Mem:* Math Asn Am; Asn Symbolic Logic. *Res:* Foundations of mathematics; mathematical analysis; summability of divergent series and integrals; theory of recursive functions. *Mailing Add:* 2754 Southridge Dr Univ Notre Dame South Bend IN 46614

VUILLEMIN, JOSEPH J, b Waco, Tex, July 22, 34; m 57; c 3. GENERAL PHYSICS. *Educ:* Univ Tex, BS, 56; Baylor Univ, MS, 57; Univ Chicago, PhD(physics), 65. *Prof Exp:* NSF fel, Cambridge Univ, 65-66; asst prof, 66-70, assoc prof, 70-81, PROF PHYSICS, UNIV ARIZ, 81- *Concurrent Pos:* Sci res coun fel, Univ Bristol, 74-75. *Mem:* Am Phys Soc. *Res:* Electronic structure of metals and low temperature physics. *Mailing Add:* Dept Physics Univ Ariz Tucson AZ 85721

VUILLEUMIER, FRANCOIS, b Berne, Switz, Nov 26, 38; m 64, 72, 83; c 3. POPULATION BIOLOGY, BIOGEOGRAPHY. *Educ:* Univ Geneva, Lic nat sci, 61; Harvard Univ, PhD(biol), 67. *Prof Exp:* From instr to asst prof biol, Univ Mass, Boston, 66-71; prof zool & dir inst animal ecol, Univ Lausanne, 71-72; res fel marine biol, Biol Sta Roscoff, France, 72-73; vis prof, Lab Ecol, Ecole Normale Superieure, Univ Paris, 73-74; CUR ORNITH, AM MUS NATURAL HIST, 74-, CHMN ORNITH, 87- *Concurrent Pos:* Am Mus Natural Hist Chapman fel, 67-68; vis prof, Univ Andes, Merida, Venezuela, 81. *Mem:* AAAS; fel Am Ornith Union; Soc Study Evolution; Soc Syst Zool; Am Soc Naturalists; Ecol Soc Am; corresp mem Soc Ornith France. *Res:* Avian migration; avian speciation in South American Andes; biogeography; patterns of species diversity in continental habitats; ecological genetics of marine molluscs; community ecology; evolution ornithology; southern hemisphere biography. *Mailing Add:* Am Mus Natural Hist Central Park W at 79th St New York NY 10024

VUKASOVICH, MARK SAMUEL, b Detroit, Mich, Dec 6, 27; m 53, 70; c 5. CORROSION PROTECTION. *Educ:* Wayne State Univ, BS, 51, MS, 53. *Prof Exp:* Res scientist friction/wear/high temperature mat, Chrysler Corp, 53-60; dept head fiber optics, pigments & electron mat, Ceramics Dept, Horizons Res Inc, 60-64; res mgr microencapsulated pigments & fluorescent pigments, Sherwin Williams Co, 64-67; dir res & develop, Prof Dent Mat & Prod, Kerr Mfg Co, 67-69; suprv chem res, 70-88, CONSULT CORROSION PROTECTION, LUBRICATION & CORROSION INHIBITION, CLIMAX MOLYBDENUM CO, 88- *Concurrent Pos:* Part-time fac, Corrosion Protection, Chem Eng Dept, Wayne State Univ, 90- *Honors & Awards:* Wilbur Deutsch Mem Award, Soc Tribologists & Lubrication Engrs, 86. *Mem:* Nat Asn Corrosion Engrs; Soc Tribologists & Lubrication Engrs; Am Chem Soc; Fedn Soc Coating Technol. *Res:* Corrosion inhibition; corrosion protection; pigment synthesis; fiber reinforced composities; fiber optics; microencapsulated products; electronic materials; lubrication. *Mailing Add:* 1457 Woodland Dr Ann Arbor MI 48103

VUKOV, RASTKO, b Belgrade, Yugoslavia, June 23, 42; Can citizen; m 66; c 2. ORGANIC & POLYMER CHEMISTRY. *Educ:* Univ Belgrade, BSc, 65; Univ Alta, PhD(chem), 72. *Prof Exp:* Res scientist, Raylo Chem Ltd, 72-76, proj mgr, 76-78, dir indust res, 78-80, mgr contract res, 80-82; SCI ADV, POLYSAR LTD, CAN, 82- *Mem:* Can Inst Chem; Am Chem Soc; Sigma Xi. *Res:* Investigation of the micro-structure of polymers; studies of polymer reactions; synthesis of novel monomers, antioxidants and cross-linking agents, preparation and characterization of specialty polymers; development of structure-properties correlations for polymers. *Mailing Add:* 569 Sayre Dr Princeton NJ 08540

VUKOVICH, FRED MATTHEW, b Chicago, Ill, July 13, 39; m 66; c 4. DYNAMIC METEOROLOGY, PHYSICAL OCEANOGRAPHY. *Educ:* Parks Col Aeronaut Technol, St Louis, BS, 60; St Louis Univ, MS, 63, PhD(meteorol), 66. *Prof Exp:* Res meteorologist, Meteorol Res Inc, 63-64; res meteorologist, Res Triangle Inst, 66-68, supvr, 68-71, sr scientist, 71-74, mgr, Geosci Dept, 74-85, dir, 85-90, PRIN SCIENTIST, ATMOSPHERIC SCI DEPT, RES TRIANGLE INST, 90- *Concurrent Pos:* Assoc prof, Duke Univ, 67-77. *Mem:* Am Meteorol Soc. *Res:* Dynamic meteorology of urban atmosphere; physical oceanographic studies on continental shelf, Gulf Stream, and Gulf of Mexico; dynamics of synoptic-scale pollution; satellite meteorology and oceanography; tropical meteorology; global climate studies; cloud modelling. *Mailing Add:* Res Triangle Inst PO Box 12194 Research Triangle Park NC 27709

VUKOVICH, ROBERT ANTHONY, b Hoboken, NJ, Aug 6, 43; m 65; c 3. CLINICAL PHARMACOLOGY, CLINICAL RESEARCH. *Educ:* Allegheny Col, BS, 65; Jefferson Med Col, PhD(pharm), 69. *Prof Exp:* Clin res scientist clin pharmacol, Warner-Lambert Res Inst, 69-70; assoc dir med res, USV Labs, 70-71; dir clin pharmacol, The Sqiubb Inst Med Res, 72-79; dir clin res, Div Develop Therapeut, Res & Develop Div, Revlon Health Care Group, 79-83; CHIEF EXEC OFFICER & PRES, ROBERTS PHARMACEUT CORP, 83- *Honors & Awards:* Fel, Am Col Clin Pharmacol. *Mem:* Am Soc Clin Pharmacol & Therapeut; Am Soc Pharmacol & Exp Therapeut; NY Acad Sci; fel Am Col Clin Pharmacol; AMA; AAAS; Parental Drug Asn; Sigma Xi; Royal Soc Med. *Res:* Clinical pharmacology; clinical research; new drug development; cardiovascular; inflammatory, infectious diseases; oncology; endocrinology; urology; pain management. *Mailing Add:* Seven Taylor Run Holmdel NJ 07733

VULLIEMOZ, YVONNE, b Switz; US citizen. BIOCHEMICAL PHARMACOLOGY. *Educ:* Univ Lausanne, Switz, PharmD, 58; Univ Paris, France, PhD(pharmacol), 69. *Prof Exp:* Res assoc, Col Physicians & Surgeons, Columbia Univ, 69-77, asst prof, 77-84, assoc res scientist, 84-89, RES SCIENTIST, COL PHYSICIANS & SURGEONS, COLUMBIA UNIV, 90- *Mem:* Am Soc Pharmacol & Exp Therapeut. *Res:* Pharmacology of the autonomic nervous system. *Mailing Add:* Dept Anesthesiol Col Physicians & Surgeons Box 46 Columbia Univ 630 W 168th St New York NY 10032

VULLO, WILLIAM JOSEPH, b Buffalo, NY, June 14, 33; m 59; c 2. ORGANIC CHEMISTRY. *Educ:* Univ Buffalo, BA, 55; Northwestern Univ, PhD(org chem), 59. *Prof Exp:* Instr gen chem, Northwestern Univ, 59; sr chemist, Hooker Chem Corp, 59-69; mgr textile res, Mohasco Industs, 69-79; mem staff, environ protection oper, 80, MGR ENVIRON & LAB SERV, GEN ELEC CO, 85- *Mem:* Am Chem Soc; Am Asn Textile Chem & Colorists; Sigma Xi. *Res:* Organometallic chemistry; organic reaction mechanisms; metal conversion coatings and treatments; cellulose reactive chemicals; organic phosphorus and fluorine chemistry; fire retardants; textile chemicals and finishes; environmental sciences. *Mailing Add:* 24 Velina Dr Burnt Hills NY 12027

VUREK, GERALD G, b San Francisco, Calif, May 7, 35; m 59; c 1. BIOMEDICAL ENGINEERING, ELECTRICAL ENGINEERING. *Educ:* Calif Inst Technol, BS, 56; Stanford Univ, MS, 57, Engr & MS, 61, PhD(physiol), 64. *Prof Exp:* Elec engr, Instrument Sect, NIH, 57-59, develop engr, Instrument Eng & Develop Br, 61-63, sr investr, Lab Tech Develop, 63-83, SR SCIENTIST, NAT HEART INST, 83- *Concurrent Pos:* Vpres, Joint Comt Eng in Med & Biol, 66, treas, 67-68, gen chmn ann conf, 70. *Mem:* AAAS; Inst Elec & Electronics Engrs; Biomed Eng Soc; Int Soc Optical Eng; Optical Soc Am. *Res:* Instrument and methods development for biochemical analysis; microchemical analysis instrumentation; cardiac anaphylaxis; microimmunochemistry; artificial circulatory support devices; in vivochemical sensors. *Mailing Add:* Abbott Labs 1212 Terra Bella Mountain View CA 94043

VUSKOVIC01, LEPOSAVA, b Lesnica, Yugoslavia, Apr 23, 41; m 87; c 3. ATOMIC INTERACTIONS WITH ELECTRONS & FIELDS IN GROUND & EXCITED STATES. *Educ:* Univ Belgrade, Yugoslavia, dipl physics/chem, 63, MS, 68, PhD(physics), 72. *Prof Exp:* Res fel physics, Inst Physics, Univ Belgrade, Yugoslavia, 64-73, res scientist, 73-78, head, Atomic Physics Lab, 75-78, assoc prof physics, 80-85, dir, Atomic Laser & High Energy Physics Div, 81-85; assoc prof, Univ Arts, Belgrade, Yugoslavia, 73-85; ASSOC PROF PHYSICS, NY UNIV, 85- *Concurrent Pos:* Mem, Gen Comn, Int Conf Physics, Int Conf Physics of Electronic & Atomic Collisions, 77-81; sr res assoc, Nat Res Coun/NASA Jet Propulsion Lab, Pasadena, Calif, 78-80. *Mem:* Am Phys Soc; Europ Phys Soc; Yugaslav Phys Soc. *Res:* Experimental research in atomic beam collision physics; scattering studies, particularly measurements that describe fundamental interactions between electrons and atomic and molecular species and clusters, including collision in the presence of laser fields. *Mailing Add:* Physics Dept NY Univ Four Washington Pl New York NY 10003

VYAS, BRIJESH, b Feb 7, 48. INDUSTRIAL & MANUFACTURING ENGINEERING, METALLURGY & PHYSICAL METALLURICAL ENGINEERING. *Educ:* Indian Inst Technol, Bombay, India, BTech, 71; State Univ NY, Stony Brook PhD(mat sci), 75. *Prof Exp:* Assoc mem tech staff, Brookhaven Nat Lab, Upton, NY, 75-80; mem tech staff, 80-89, SUPVR, AT&T BELL LABS, 89- *Concurrent Pos:* Vis prof, Tech Univ Denmark, Lyngby, 82. *Honors & Awards:* Sam Tour Award, Am Soc Testing & Mat, 82. *Mem:* Electrochem Soc. *Res:* Energy conversion technologies primarily advanced batteries; erosion and corrosion of materials specially localized corrosion. *Mailing Add:* AT&T Bell Labs Rm 1E-241 600 Mountain Ave Murray Hill NJ 07974-2070

VYAS, GIRISH NARMADASHANKAR, b Aglod, India, June 11, 33; m 62; c 2. IMMUNOLOGY, GENETICS. *Educ:* Univ Bombay, BSc, 54, MSc, 57, PhD(microbiol), 64. *Prof Exp:* Asst res officer, Blood Group Ref Ctr, Indian Coun Med Res, Bombay, 57-64; officer-in-chg, Bombay Munic Blood Ctr, King Edward Mem Hosp, India, 64-65; lectr immunol, Univ Calif, San Francisco, 67-69, asst prof path, 69-73, assoc prof, 73-77, dir blood bank, 69-87, PROF LAB MED & DIR TRANSFUSION RES, SCH MED, UNIV CALIF, SAN FRANCISCO, 77- *Concurrent Pos:* Jr res fel hemat, J J Hosp, Bombay, India, 56-57; fel genetics, Western Reserve Univ, 65-67; Fulbright scholar, Pasteur Inst, 80. *Honors & Awards:* Julliard Prize, Int Soc Blood Transfusion, 69. *Mem:* AAAS; Am Asn Immunol; Am Soc Microbiol; Am Soc Hemat. *Res:* Microbiology; blood group serology; immunogenetics; blood banking; viral hepatitis; transfusion and circulatory physiology; genetics of gamma globulin and its structure; molecular biology of transfusion transmitted viral infections causing aids and hepatitis. *Mailing Add:* Univ Calif Med Ctr San Francisco CA 94143-0134

VYAS, RAVINDRA KANTILAL, structural engineering, applied mechanics; deceased, see previous edition for last biography

VYAS, REETA, b Gujarat, India, May 25, 53; m 82; c 1. QUANTUM OPTICS, NUCLEAR REACTION. *Educ:* Banaras Hindu Univ, BSc, 73 & MSc, 75; State Univ NY, Buffalo, PhD(physics), 84. *Prof Exp:* Jr res fel, Coun Sci & Indust Res, India, 76-78; teaching asst, State Univ NY, Buffalo, 79-81, Univ grad fel, 81-83, postdoctoral res assoc physics, 85; vis asst prof, 84-88, res asst prof physics, 88-89, ASST PROF PHYSICS, UNIV ARK, 89- *Concurrent Pos:* Hon lectr, Banaras Hindu Univ, India, 76. *Mem:* Am Phys Soc; Optical Soc Am. *Res:* quantum statistics of light matter interactions and meson exchange effects in the nuclei; author of 26 scientific articles. *Mailing Add:* Phys Dept Univ Ark Fayetteville AR 72701

VYBORNY, CARL JOSEPH, b Oak Park, Ill, Nov 23, 50; m 75; c 1. RADIOLOGY, MEDICAL PHYSICS. *Educ:* Univ Ill, Chicago, BS, 72, Urbana, MS, 73; Univ Chicago, PhD(radiol), 76, MD, 80. *Prof Exp:* Res assoc radiol physics, 78-80, resident physician radiol, 80-84, asst prof, 84-85, CLIN ASSOC PROF RADIOL, UNIV CHICAGO, 85- *Honors & Awards:* Eastman Kodak Sci Award, 76; Itek Award, 79; Andrew W Mellon Found Fel Award, 84. *Mem:* Am Asn Phys Med; Radiol Soc NAm; Am Col Radiol; AMA. *Res:* Applications of computer methods to radiologic diagnosis; chest radiography; mammography. *Mailing Add:* Dept Radiol Box 429 Univ Chicago Chicago IL 60637

VYDELINGUM, NADARAJEN AMEERDANADEN, b Mauritius Plaines Wilhems Curepipe, June 1, 45; US citizen; m 71; c 2. METABOLISM, CELL BIOLOGY. *Educ:* Univ London, BSc, 72, MSc, 74, PhD(clin biochem), 79; Inst Biol, MIBiol, 90. *Prof Exp:* Res assoc endocrinol & metab, St Mary's Med Sch, Univ London, UK, 74-77; from instr to asst prof med & pharmacol, Med Col Wis, Milwaukee, 77-86; ASST ATTEND BIOCHEM, MEM SLOAN-KETTERING CANCER CTR, NEW YORK, 86-, DIR RES METAB & CANCER, 86- *Concurrent Pos:* Prin investr, Am Heart Asn, 79-81, Am Diabetes Asn, 82-84, NIH, 85-88; adj asst prof biochem & cell biol, Univ Wis-Milwaukee, 79- 82, lectr, 79-82; ad hoc reviewer, sci jour, 80-; peer reviewer, Health Sci Consortium, 84-, Am Diabetes Asn, 85-90; asst lab mem metab & cancer, Mem Sloan-Kettering Cancer Ctr, New York, 86- *Mem:* Am Diabetes Asn; Am Soc Biochem & Molecular Biol; Am Inst Nutrit; Biochem Soc UK; Am Asn Cancer Res; Inst Biol UK. *Res:* Biochemical and molecular mechanisms responsible for the alteration of lipid and carbohydrate metabolism in cancer cachexia. *Mailing Add:* 2416 Westwood Bldg NIH 5333 Westbard Ave Bethesda MD 20892

VYE, MALCOLM VINCENT, b Gary, Ind, Feb 17, 36; m 60; c 1. HEMATOLOGY, PATHOLOGY. *Educ:* Marquette Univ, MD, 61. *Prof Exp:* Asst instr, 62-64, from instr to asst prof path, Univ Ill Col Med, 64-66 & 68-71; ASST PROF PATH, MED SCH, NORTHWESTERN UNIV, EVANSTON, 71-; ASSOC PATHOLOGIST, EVANSTON HOSP, 71- *Concurrent Pos:* Mem staff, Armed Forces Inst Path. *Mem:* AAAS; Am Soc Clin Path; Am Asn Path & Bact; Col Am Pathologists. *Res:* Cell differentiation embryonic muscle; ultrastructure of glycogen; hematology laboratory methodology. *Mailing Add:* 145 N Sheridan Rd Hubbard Woods IL 60093

W

WAACK, RICHARD, b Syracuse, NY, May 18, 31; m 60. PHYSICAL CHEMISTRY, POLYMER CHEMISTRY. *Educ:* State Univ NY, BS, 53, MS, 54, PhD, 58. *Prof Exp:* Tech serv rep, Dow Chem Co, Mich, 54-56; chemist, Solvay Process Div, Allied Chem Co, NY, 58-59; res chemist, Eastern Res Lab, Dow Chem Co, Mich, 59-60, res chemist, Phys Res Lab, 67-69; MGR DEVELOP LAB, POLAROID CORP, WALTHAM, MD 69- *Mem:* Am Chem Soc; Soc Photog Sci & Eng. *Res:* Polymer synthesis; spectroscopy; ionic polymerization mechanisms; photographic science; silver halide emulsion technology; colloidal processes; water soluble polymers; diffusion processes. *Mailing Add:* 19 Morrill Dr Wayland MA 01778-4709

WAAG, CHARLES JOSEPH, b Oct 25, 31; US citizen; m 56; c 1. GEOLOGY. *Educ:* Univ Pittsburgh, BS, 56, MS, 58; Univ Ariz, PhD(geol), 68. *Prof Exp:* Sr geologist, Orinoco Mining Co, US Steel Corp, 58-63; sr geologist, Va Div Mineral Resources, 63-64; asst & lectr, Univ Ariz, 64-68; asst prof, 68-71, assoc prof geol, Ga State Univ, 71-; AT BOISE STATE UNIV. *Concurrent Pos:* Consult hydrogeol, 69- *Mem:* Geol Soc Am. *Res:* Glaciers as models in structural geology; gravity tectonics attendant to mantled gneiss domes in the Basin and Range Province. *Mailing Add:* Dept Geol & Geophys Boise State Univ 1910 University Dr Boise ID 83725

WAAG, ROBERT CHARLES, b Upper Darby, Pa, Oct 8, 38; m 61; c 2. ELECTRICAL ENGINEERING, ACOUSTICS. *Educ:* Cornell Univ, BEE, 61, MS, 63, PhD(commun), 65. *Prof Exp:* Mem tech staff, Sandia Sci Labs, 65-66; proj officer commun, Rome Air Develop Ctr, USAF, 66-69; ASSOC PROF ELEC ENG & RADIOL, UNIV ROCHESTER, 69- *Concurrent Pos:* NSF grant, 75-; Career Develop Award, NIH, 76- *Mem:* Inst Elec & Electronics Engrs; Acoust Soc Am; Am Inst Ultrasound Med. *Res:* Apply principles of physics and signal processing along with computer-based technology to improve imaging in diagnostic ultrasound and to characterize materials from measurements of ultrasonic scattering. *Mailing Add:* Dept Radiol Univ Rochester Med Ctr Box 648 Rochester NY 14642

WAAGE, JONATHAN KING, b Pueblo, Colo, July 27, 44; m 88. BEHAVIORAL ECOLOGY, EVOLUTIONARY BIOLOGY. *Educ:* Princeton Univ, AB, 66; Univ Mich, PhD(zool), 71. *Prof Exp:* from instr to assoc prof, 72-90, PROF BIOL, BROWN UNIV, 90- *Mem:* Am Soc Naturalists; Soc Study Evolution; Animal Behav Soc; Ecol Soc Am; Societas Internationalis Odonatologics. *Res:* Evolutionary and ecological determinants of reproductive behavior in odonates and other insects; manipultive and comparative field studies of sexual selection, including sperm competition and mating systems. *Mailing Add:* Box G-W208 Div Biol & Med Brown Univ Providence RI 02912

WAAGE, KARL MENSCH, b Philadelphia, Pa, Dec 17, 15; m 42; c 2. GEOLOGY. *Educ:* Princeton Univ, AB, 39, MA, 42, PhD(geol), 46. *Prof Exp:* From instr to assoc prof, 46-67, chmn dept geol-geophys, 73-76, dir, Peabody Mus, 79-82, PROF GEOL, YALE UNIV, 67-, CUR INVERT PALEONT, PEABODY MUS, 46- *Concurrent Pos:* Geologist, US Geol Surv, Washington, DC, 42-59. *Mem:* Fel Geol Soc Am; Paleont Soc; fel AAAS. *Res:* Field exploration for non-metalliferous deposits; stratigraphic geology and paleontology of cretaceous of western interior. *Mailing Add:* Peabody Mus Yale Univ New Haven CT 06520

WAALAND, IRVING T, b Brooklyn, NY, July 2, 27. AERONAUTICAL DESIGN. *Educ:* NY Univ, BAe, 53. *Prof Exp:* Proj aerodynamicist & proj engr, Grumman Aerospace Corp, 53-74; proj dir, Northrop Corp, 74-75, eng mgr var classified progs, 75-78, prog mgr concept develop, 79-80, dep prog mgr, 81-83, prog mgr, 86-88, VPRES, NORTHROP CORP, 81-, CHIEF DESIGNER, 88- *Concurrent Pos:* Ad hoc mem, Naval Studies Bd, Nat Acad Sci & Aeronaut Facil Bd, NASA. *Honors & Awards:* Aircraft Design Award, Am Inst Aeronaut & Astronaut, 89 & Aircraft Design Cert Merit, 89; Leslie E Simon Award, Am Defense Preparedness Asn, 90. *Mem:* Nat Acad Eng; fel Am Inst Aeronaut & Astronaut. *Res:* Advanced concept development and system synthesis; multi-discipline team building; program or engineering management; low observables; aeropropulsion; stability and control; avionics; sensors; author of various publications; granted one patent. *Mailing Add:* Northrop Corp 8900 E Washington Blvd I001/AP Pico Rivera CA 90660-3783

WAALAND, JOSEPH ROBERT, b San Mateo, Calif, Feb 22, 43; m 69; c 1. ALGOLOGY, CYTOLOGY. *Educ:* Univ Calif, Berkeley, BA, 66, PhD(bot), 69. *Prof Exp:* From asst prof to assoc prof, 69-80, PROF BOT, UNIV WASH, 83-, ASSOC CHMN, 87- *Mem:* Brit Phycol Soc; Phycol Soc Am; Int Phycol Soc; Marine Biol Asn UK. *Res:* Development, cytology and ecology of algae; aquaculture of marine algae. *Mailing Add:* Dept Bot Univ Wash Seattle WA 98195

WAALKES, T PHILLIP, b Belmond, Iowa, Oct 30, 19; m 45; c 6. PUBLIC HEALTH. *Educ:* Hope Col, AB, 41; Ohio State Univ, PhD(org chem), 45; George Washington Univ, MD, 51. *Prof Exp:* Asst chem, Ohio State Univ, 41-43, res assoc, Res Found, 44, Am Petrol Inst res assoc, Univ, 45, instr chem, 46-47; intern, USPHS Hosp, 51-52, res med, 52-55, Nat Heart Inst, 55-58, asst chief in chg clin activ, Cancer Chemother Nat Serv Ctr, 58-63, assoc dir, Nat Cancer Inst, 63-68, sr investr, 68-75; PROF ONCOL, MED SCH, JOHNS HOPKINS UNIV, 75- *Mem:* Am Asn Cancer Res; Am Soc Clin Oncol; AAAS. *Res:* Organic fluorine compounds; biochemistry; amino acids; fluorinated derivatives of propane and propylene; addition of fluorine to double bonds; cancer chemotherapy; biological markers. *Mailing Add:* Dept Oncol Johns Hopkins Univ 720 Rutland Ave Baltimore MD 21205

WABECK, CHARLES J, b Montague, Mass, July 16, 38; m 64; c 2. FOOD SCIENCE. *Educ:* Univ Mass, BS, 62; Univ NH, MS, 64; Purdue Univ, PhD(food sci), 66. *Prof Exp:* Res assoc poultry & frozen foods, Armour & Co, 66-69; dir res frozen foods, Ocoma Foods Co, Nebr, 69; prof poultry prod, 69-76, PROF POULTRY SCI, UNIV MD, COLLEGE PARK, 76- *Mem:* Inst Food Technologists; Poultry Sci Asn. *Res:* Research and development; quality control; frozen foods; poultry and meat products. *Mailing Add:* Dept Poultry Sci Lower Eastern Shore Res & Educ Ctr Princess Anne MD 21853

WABER, JAMES THOMAS, b Chicago, Ill, Apr 8, 20; m 51; c 3. ATOMIC PHYSICS, SOLID STATE PHYSICS. *Educ:* Ill Inst Technol, BS, 41, MS, 43, PhD(metall), 46. *Prof Exp:* Res assoc, Ill Inst Technol, 46, asst prof chem, 46-47; assoc metallurgist, Los Alamos Sci Lab, 47-49, staff mem, 49-66; PROF MAT SCI, NORTHWESTERN UNIV, EVANSTON, 67-; PROF PHYSICS, MECH TECH UNIV, 85- *Concurrent Pos:* NSF sr fel, Univ Birmingham, 60-61; chmn comt alloy phases, past chmn nuclear metall comt & mem exec comt, Inst Metals Div, NY; partic, Robert A Welch Found Conf on Chem Res XIII Mendeleef Centennial-The Transuranium Elements; Alexander von Humboldt Found sr scientist award, Bonn, WGermany; prof physics, Mich Technol Univ, Houghton. *Honors & Awards:* Turner Prize, Electrochem Soc, 47; Whitney Prize, Nat Asn Corrosion Eng, 63. *Mem:* Am Soc Metals; Electrochem Soc; Am Inst Mining, Metall & Petrol Engrs; Nat Asn Corrosion Engrs; fel Am Phys Soc. *Res:* Corrosion and oxidation of metals; relativistic self-consistent field Dirac-Slater and Hartree-Fock calculations for atoms and ions; energy band calculations; chemistry and physics of superheavy elements; positron annhilation in metals. *Mailing Add:* Dept Physics MI Tech Univ Houghton MI 49931

WACHHOLZ, BRUCE WILLIAM, b Chicago, Ill, Aug 16, 36; m 63; c 1. RADIATION BIOLOGY. *Educ:* Valparaiso Univ, BA, 58; Univ Rochester, MS, 59, PhD(radiation biol), 67. *Prof Exp:* Sr res scientist, Pac Northwest Labs, Battelle Mem Inst, 66-71; radiation biologist, Off Health & Environ, Energy Res & Develop Admin, 71-; AT NAT CANCER INST, BETHESDA, MD. *Mem:* AAAS; Am Phys Soc; Radiation Res Soc; NY Acad Sci; Geront Soc. *Res:* Pathological, physiological and endocrinological effects of radiation; metabolism and toxicity of radionuclides; gerontology. *Mailing Add:* Div Cancer Etiology Nat Cancer Inst Exec Plaza N Rm 630 9000 Rockville Pike Bethesda MD 20892

WACHMAN, HAROLD YEHUDA, b Tel Aviv, Israel, Dec 2, 27; nat US; m 54; c 3. SURFACE PHYSICS, COMBUSTION AT ZERO-G. *Educ:* City Col New York, BS, 49; Univ Mo, MA, 52, PhD(phys chem), 57. *Prof Exp:* Specialist chem physics, Aerosci Lab, Gen Elec Co, 57-63; vis prof, 63-64, assoc prof, 64-69, PROF AERONAUT & ASTRONAUT, MASS INST TECHNOL, 69-, CHMN, GRAD DIV DEPT AERONAUT & ASTRONAUT, 76- *Concurrent Pos:* Consult, Space Sci Lab, Gen Elec Co; vis sr res fel, Jesus Col, Oxford Univ, 72-73. *Mem:* Sigma Xi; Am Inst Aeronaut & Astronaut. *Res:* Rarefied gas phenomena; adsorption; high temperature chemical equilibrium studies; gas surface interactions; nucleation phenomena; spacecraft-environment interactions; molecular dynamics computations. *Mailing Add:* Dept Aeronaut & Astronaut Mass Inst Technol Cambridge MA 02139

WACHMAN, MURRAY, b Tel Aviv, Israel, Feb 1, 31; US citizen; m 58; c 3. MATHEMATICS. *Educ:* Brooklyn Col, BA, 53; NY Univ, MS, 56, PhD(math), 61. *Prof Exp:* Math analyst, Repub Aviation Corp, 57-59; appl mathematician, Gen Elec Co, 59-63, consult mathematician, 63-65, group leader appl math, 65-67; assoc prof, 67-73, PROF MATH, UNIV CONN, 73- *Concurrent Pos:* Consult, Missile & Space Div, Gen Elec Co, 67-69, Naval Underwater Systs Ctr, 84- *Mem:* Am Math Soc; Soc Indust & Appl Math. *Res:* fluid mechanics; biological and economic models. *Mailing Add:* Dept Math Univ Conn Storrs CT 06268

WACHOWSKI, HILLARD M(ARION), electrodynamics, applied mathematics, for more information see previous edition

WACHS, ALAN LEONARD, b Kalamazoo, Mich, Oct 25, 59. MATERIALS DEVELOPMENT & CHARACTERIZATION, SURFACE & INTERFACE STUDIES. *Educ:* Cornell Univ, AB, 81; Univ Ill Urbana-Champaign, MS, 82, PhD(physics), 87. *Prof Exp:* Teaching asst, Univ Ill Urbana-Champaign, 81-82, assoc res asst physics, 82-86; postdoc res assoc, Lawrence Livermore Nat Lab, 86-89; MEM, RES STAFF, OAK RIDGE NAT LAB, 89- *Concurrent Pos:* Summer res assoc, Brookhaven Nat Lab, 80. *Mem:* Mat Res Soc; Am Vacuum Soc; Am Phys Soc. *Res:* Structure - property - process relationships in thin-film materials; experimental study of electronic structure; surface and interface science; condensed matter physics. *Mailing Add:* Oak Ridge Nat Lab MS6030 PO Box 2008 Oak Ridge TN 37831

WACHS, GERALD N, b Chicago, Ill, Nov 5, 37; m 62; c 4. DERMATOLOGY. *Educ:* Univ Ill, BS, 58, MD, 62; Am Bd Dermat, dipl, 68. *Prof Exp:* Intern med, Michael Reese Hosp, Ill, 62-63; resident dermat, Univ Calif, 63-65, chief resident, 65-66; mem dept clin invest, Schering-Plough Corp, Kenilworth, NJ, 66-67, from asst med dir to assoc med dir, 67-74, dir new prod planning, 74-78, sr assoc med dir, 78-82; pres, bd health, Millburn, NJ, 84-90; PVT PRACT, DERMAT & COSMETIC SURG, 82- *Concurrent Pos:* Clin asst dermatologist, St Vincent's Hosp, NY, 67-85; attend staff dermatologist, Mary Manning Walsh Home, New York, 71-81; attend staff, St Barnabas Med Ctr, Livingston, NJ; team dermatologist, NJ Nets, NJ Devils. *Mem:* Fel Am Col Physicians; fel Am Acad Dermat; fel Am Col Allergists; Int Soc Trop Dermat; Am Acad Allergy. *Res:* Clinical investigation of drugs in dermatology, allergy and ophthamology; development of new concepts in approaching the therapy of difficult clinical diseases in dermatology, allergy and ophthamology. *Mailing Add:* 116 Millburn Ave Millburn NJ 07041

WACHS, MELVIN WALTER, b Detroit, Mich, 1933; m; c 3. ENERGY CONSERVATION & STUDIES, APPLIED ENVIRONMENTAL SCIENCES. *Educ:* Univ Mich, BA, 52, MA, 54; Indust Col Armed Forces, MA & cert, 60; Am Univ, PhD(polymer sci & admin), 68; Nat Defense Univ, MA & cert, 84. *Prof Exp:* Asst prof Polit Sci & chmn, Asia Prog, Western Mich Univ, 59-62; assoc dir, Exec Inst, Off Career Develop, US Civil Comn, 62-64, Educ Resources, 64-66; planning-prog-budgeting & Mgt Info Systs coord & chief, Planning Br, NIMH, USPHS, HEW, 66-68; dir, Community Develop Training Div, 68-72, SR PROG OFFICER, COMMUNITY PLANNING & DEVELOP, HUD, 72-; PROF PUB, ADMIN, ADV PROG GOVT STUDIES, UNIV OKLA, 65- *Concurrent Pos:* Consult, Off Sci & Technol, Exec Off of Pres, 63-72, staff dir & mem, President's Comt Fed Labs, 63-72; mem, Fed Interagency Comn Sci & Eng, 64-78, Fed Interagency Coun Educ, 64- & var other groups & councils, 64-; vis prof pub admin, Maritime Col, State Univ NY, 67-68 & vis prof, Cent Mich Univ, 69-; Coastal Zone mgt coordr, Off Environ & Energy, 78-82. *Honors & Awards:* Int Medal of Honor, Museu de Baleen (Pres), Portugal, 83. *Mem:* AAAS; NY Acad Sci; Am Acad Polit & Social Sci. *Res:* Technology transfer; Science and public policy; energy conservation and efficiency research and policy. *Mailing Add:* 4832 Drummond Ave Chevy Chase MD 20815

WACHSBERGER, PHYLLIS RACHELLE, b New York, NY; m 67; c 2. CELL BIOLOGY. *Educ:* City Univ New York, BS, 64; Med Col Pa, PhD(physiol & biophys), 71. *Prof Exp:* Res instr muscle biophys, Dept Physiol & Biophys, Med Col Pa, 70-71; res assoc muscle biophys, Dept Anat, Univ Pa, 73-82; lectr biol, Bryn Mawr Col, 83-86; INSTR, DEPT RADIATION ONCOL, THOMAS JEFFERSON UNIV, 86- *Concurrent Pos:* NIH fel, Dept Anat, Univ Pa, 71-73; asst prof physiol & biophys, Hahnemann Med Col & Hosp, Philadelphia, 77-81; vis asst prof, Haverford Col, 84. *Mem:* Biophys Soc; AAAS; Am Inst Biol Sci; Sigma Xi; NAm Hypothermia Group; Am Soc Cell Biol. *Res:* Studies of the self assembly of synthetic vertebrate smooth muscle myosin filaments; comparative studies of molecular substructure of myosin filaments from various muscle types; immunochemical studies of the Myosin Crossbridge; identification of monsclonal antibodies raised to myosin and other myofibrillar proteins; effect of hyperthermia on cytoskeleton and nuclear matrix in mammalian cells. *Mailing Add:* 418 Newton Rd Berwyn PA 19312

WACHSMAN, JOSEPH T, b New York, NY, July 25, 27; m 60; c 1. MICROBIOLOGY, BIOCHEMISTRY. *Educ:* NY Univ, AB, 48; Univ Calif, PhD(microbiol), 55. *Prof Exp:* USPHS fel, Brussels, Belg, 55-57; from asst prof to assoc prof microbiol, Univ Ill, Urbana- Champaign, 57-85; dir res, Touch Sci Inc, 89-91; CONSULT, 91- *Concurrent Pos:* USPHS res career develop award, 62- *Mem:* Am Soc Microbiol. *Res:* Biochemistry and molecular biology of mammalian cells; unique properties of malignant cells-emphasis on the plasminogen activator associated with transformed cells-nucleic acid synthesis and repair. *Mailing Add:* 425 Ridgecrest Dr Chapel Hill NC 27514

WACHSPRESS, EUGENE LEON, b New York, NY, Apr 17, 29; m 52; c 3. NUMERICAL ANALYSIS, ITERATIVE SOLUTION OF LARGE SYSTEMS. *Educ:* The Cooper Union, BME, 50; Union Col, BS, 56; Rensselaer Polytech Inst, PhD(math), 68. *Hon Degrees:* Dr, Univ Libre de Bruxelles, Belg, 85. *Prof Exp:* Nuclear engr, Knolls Atomic Power Lab, Gen Elec, 52-65, mathematician, 65-70, consult mathematician, 71-85; PROF MATH, UNIV TENN, KNOXVILLE, 83- *Concurrent Pos:* Vis fel numerical anal, Univ Dundee, Scotland, 70-71; ed, Comput & Math with Appln, 75- & J Applied Math Letters, 87-; numerical analyst, Res Inst Tallahassee, 85-86; vis fel, Electromechanics Br, GE, 86; consult, Supercomputer Computations. *Mem:* Fel Am Nuclear Soc; Soc Indust & Appl Math; Math Asn Am. *Res:* Numerical solution of large linear systems; finite element basis function construction for complex geometry using algebraic-geometry foundations; numerical linear algebra; alternating direction implicit iteration theory and application. *Mailing Add:* Math Dept Univ Tenn Ayres 18B Knoxville TN 37996

WACHTEL, ALLEN W, b New York, NY, Aug 13, 25; m 46, 61; c 2. CELL BIOLOGY, CYTOLOGY. *Educ:* Columbia Univ, BS, 53, MA, 54, PhD(zool), 62. *Prof Exp:* Res asst cell biol, Cell Res Lab, Mt Sinai Hosp, New York, 56-63; from asst prof to assoc prof zool, 63-72, PROF BIOL, UNIV CONN, 72- *Mem:* Am Soc Cell Biol; Electron Micros Soc Am; Histochem Soc. *Res:* Histochemistry; cytology of electric organs; receptor structure. *Mailing Add:* Dept Physiol Univ Conn U-42 75 N Eaglevill Storrs CT 06268

WACHTEL, STEPHEN SHOEL, b Philadelphia, Pa, June 17, 37; m 62; c 2. EMBRYO TRANSFER. *Educ:* Kenyon Col, Ohio, AB, 59; Univ Pa, PhD(biol), 71. *Prof Exp:* Assoc immunogenetics, Sloan-Kettering Inst, 75-78, assoc mem, 78-81; assoc prof immunol, Cornell Univ Med Col, 77-81, assoc res prof pediat, 82-84; dir res, Ctr Reprod & Biol, 84-87, prof physiol & biophys, 85-86, PROF OBSTET & GYNEC & DIR, IMMUNOGENETICS LAB, DEPT OBSTET & GYNEC, UNIV TENN MEMPHIS, 87-, CHIEF RES REPRODUCTIVE GENETICS LABS, 88- *Concurrent Pos:* NIH res career develop award, 75-80; prin investr, NIH res grants, 76-; co-investr, Pop Coun grant, 77-78; ad hoc consult, maternal, child health res comt, NIH, 82- *Honors & Awards:* Zurkow lectr, Univ Pa, 81; John Lattimer lectr, Am Acad Pediat, 78; Pres Lectr, Obstet Soc Philadelphia, 83. *Mem:* Int Transplantation Soc; Am Asn Immunologists; Int Embryo Transfer Soc; Am Genetic Asn; Am Soc Immunol Reproduction; Am Fertil Soc; Soc Gynec Invest; Am Soc Human Genetics. *Res:* H-Y antigen, biology of sex determination; immunogenetics of tissue transplantation; sex predetermination in bovine embryo transfer; serological analysis of epidermal cell antigens; fetal cells in maternal circulation. *Mailing Add:* Dept Obstet & Gynec Univ Tenn Memphis 853 Jefferson Memphis TN 38163

WACHTELL, GEORGE PETER, b New York, NY, Mar 18, 23; div; c 2. PHYSICS. *Educ:* Princeton Univ, PhD(physics), 51. *Prof Exp:* Mem staff, Radiation Lab, Mass Inst Technol, 43-45; asst, Princeton Univ, 45-51; prin scientist, Energy Eng Lab, Franklin Res Ctr, Philadelphia, 51-88; RETIRED. *Mem:* Sigma Xi. *Res:* Optics; supersonics; heat transfer; fluid dynamics. *Mailing Add:* Kings Hwy Towers Apt D-209 Maple Shade NJ 08052

WACHTELL, RICHARD L(LOYD), b New York, NY, Feb 18, 20; m 41; c 1. ENGINEERING. *Educ:* Columbia Univ, BS, 41. *Prof Exp:* Asst metallurgist, Repub Aviation Corp, NY, 42-46; metall engr, Tech Serv Sect, Res Lab, Air Reduction Co, 46-48; supvr metallog lab, Am Electro-Metals Corp, 48-49, proj engr, 49-51, asst tech dir, 51-52; chief metallurgist, Chromalloy Corp, 52-54, tech dir, 54-60, vpres & gen mgr, Chromalloy Div, 60-63, vpres, 63-68, pres, 63-76, pres, Turbine Support Div, 66-76, pres res & tech div & mem bd dirs, 68-71, exec vpres technol, 71-76, div pres & res tech div & mem bd dirs, 68-71, exec vpres Chromalloy Metal Tectonics, 77-80, chmn bd, 80-85; PRES, METALOGOS, 85- *Concurrent Pos:* Consult metall, 85- *Mem:* Am Soc Metals; Am Ord Asn; Am Inst Mining, Metall & Petrol Engrs; Am Iron & Steel Inst. *Res:* High temperature metallurgy, especially refractory metals and super-alloys and their protection from oxidation damage; coatings by diffusion; metallurgy techniques; intermetallic systems for high temperature service. *Mailing Add:* Chromalloy Am Corp Tuxedo Park NY 10987

WACHTER, RALPH FRANKLIN, b Frederick, Md, Mar 6, 18; m 47; c 3. BIOCHEMISTRY, VIROLOGY. *Educ:* Univ Notre Dame, BS, 39; Catholic Univ, MS, 41; Purdue Univ, PhD(biochem), 50. *Prof Exp:* City chemist, Frederick, Md, 41-43; biochemist, US Army Biol Labs, 50-72; res chemist, Rickettsial Div, US Army Med Res Inst Infectious Dis, 72-83; RETIRED. *Mem:* Sigma Xi; Am Soc Microbiol. *Res:* Biochemical aspects of virology; virus stabilization and inactivation; biochemical and biological characterization of rickettsiae; rickettsial vaccines. *Mailing Add:* 310 N College Pkwy Frederick MD 21701

WACHTL, CARL, organic chemistry; deceased, see previous edition for last biography

WACHTMAN, JOHN BRYAN, JR, b Conway, SC, Feb 6, 28; m 55. SOLID STATE SCIENCE. *Educ:* Carnegie Inst Technol, BS, 48, MS, 49; Univ Md, PhD(physics), 61. *Prof Exp:* Physicist, Nat Bur Standards, 51-62, chief, Phys Properties Sect, 62-68, chief, Inorg Mat Div, 68-78, dir, Ctr Mat Sci, 78-83; dir, 83-88, SUSMAN PROF CERAMICS RES, CTR CERAMICS RES, RUTGERS UNIV, 88- *Concurrent Pos:* Ed, Ceramics & Glass, 68- & Sci & Technol, 68-; trustee, Edward Orton, Jr Ceramic Found, 70-82; mem ceramic eng adv bd, Univ Ill Urbana, 73-76, Alfred Univ, 74-82, Pa State Univ, 76-83, Northwestern Univ, 77, & Mass Inst Technol,78-82; prog mgr mat, Off Technol Assessment, US Cong, 74-75; mem adv coun ceramics, Univ NY Alfred, 74-80; mem mat dept adv comt, State Univ NY, Stonybrook, 79; lectr, Johns Hopkins Univ, 81; mem, Policy Comt, Nat Acad Eng, 82-85 & chmn, Comt Mem, 83; mem, Comt High Technol Ceramics Japan, Nat Res Coun, 83-84, Comt Sci & Technol Implications Processing Strategic Mat 84-85, Subcomt Advan Mat, Comt Future Chem Eng, 85, Comt Mat Eng Res Bd, 85 & Comt res Opportunities Mat Res, 87; ed, Am Ceramic Soc, 88- *Honors & Awards:* Silver Medal, Dept of Com, 60, Gold Medal, 71; Sosman Mem Lectr Award, Am Ceramic Soc, 74; Stratton Award, Nat Bur Standards, 75; Hobart N Kramer Award, 78; Orton Mem Lectr, 81; Dorn Mem Lectr, 81; Nat mat Achievement Award, Int Acad Ceramics, 86; Arthur L Fredburg Lectr, Int Acad Ceramics, 89. *Mem:* Nat Acad Eng; fel Am Ceramic Soc (pres, 78-79); fel Am Phys Soc; Nat Inst Ceramic Engrs; Fedn Mat Socs (secy/treas, 73, pres-elect, 74, pres, 75); Am Soc Metals Inc; Am Asn Crystal

Growth; Am Chem Soc; AAAS; Mat Res Soc. *Res:* Mechanical properties and effective utilization of inorganic materials; composites; thin films; microporous ceramic composites; tribology; superconducting ceramics; national materials policy; author of various publications. *Mailing Add:* Ctr Ceramics Res Rutgers Univ Box 909 Piscataway NJ 08855

WACK, PAUL EDWARD, b Council Bluffs, Iowa, Apr 28, 19; m 52; c 4. NUCLEAR PHYSICS. *Educ:* Creighton Univ, AB, 41; Univ Notre Dame, MS, 42, PhD(physics), 47. *Prof Exp:* Asst physics, Univ Notre Dame, 41-43, instr, 43-46; res assoc, Off Naval Res, 46-47; dir dept physics, Creighton Univ, 47-49; from asst prof to prof, 49-86, head dept, 66-73, EMER PROF PHYSICS, UNIV PORTLAND, 86- *Concurrent Pos:* Res assoc, Off Rubber Reserve, 43-45; res assoc, Gen Tire Co, 45-46. *Honors & Awards:* Culligan Award, 61. *Mem:* Am Phys Soc; Am Asn Physics Teachers. *Res:* Electron optics; stress relaxation, low temperature behavior, equation of state and electrical conductivity of natural and synthetic rubbers; nuclear spectroscopy. *Mailing Add:* Dept Physics Univ Portland Portland OR 97203

WACKER, GEORGE ADOLF, b New York, NY, Aug 18, 39; m 63; c 2. METALLURGY, CORROSION. *Educ:* Polytech Inst, Brooklyn, BSMetE, 62, MSMetE, 64. *Prof Exp:* Foundry metallurgist, Naval Appl Sci Lab, 62, phys metallurgist, 63-65 & Marine Eng Lab, 65-67; metallurgist, 68-71, head metal physics br, 71-77, head, high temperature alloys br, 77-78, HEAD, METALS DIV, NAVAL SHIP RES & DEVELOP CTR, 79- *Mem:* Am Soc Metals; Am Soc Testing & Mat; Nat Asn Corrosion Engrs; Sigma Xi. *Res:* Effects of the marine environment on metals and alloys; physical metallurgy of engineering materials used in saline environments; effects of elevated temperature corrosive environments on engineering materials. *Mailing Add:* 1128 Rutlandview Dr Davidsonville MD 21035

WACKER, WALDON BURDETTE, b Garrison, NDak, Aug 13, 23; m 55; c 4. IMMUNOLOGY, MICROBIOLOGY. *Educ:* Washington Univ, AB, 49; Univ Mich, MS, 51; Ohio State Univ, PhD(bact), 57. *Prof Exp:* Res assoc virol, Ohio State Univ, 58-59; from asst prof to assoc prof microbiol, 59-80, prof ophthal res, 80-85, EMER PROF, UNIV LOUISVILLE, 85- *Concurrent Pos:* NIH career develop award, Univ Louisville, 62-69, NIH res grant, 62-85. *Honors & Awards:* Am Uveitis Soc Award, 88; Proctor Medal, 91. *Mem:* Asn Res Vision & Ophthal; Sigma Xi; Am Uveitis Soc. *Res:* Autoimmune disease; immunopathology; uveitis. *Mailing Add:* Eye Res Inst Dept Ophthal Univ Louisville Sch Med Louisville KY 40202

WACKER, WARREN ERNEST CLYDE, b Brooklyn, NY, Feb 29, 24; m 48; c 2. MEDICINE, BIOCHEMISTRY. *Educ:* George Washington Univ, MD, 51; Harvard Univ, MS, 68. *Prof Exp:* Intern, George Washington Univ Hosp, 51-52, resident, 52-53; resident, Peter Bent Brigham Hosp, 53-55; res fel biophys, Harvard Med Sch & Peter Bent Brigham Hosp, 55-57; from instr to assoc prof med, Harvard Med Sch, 57-71; PROF HYG, HARVARD UNIV, 71- *Concurrent Pos:* Nat Found Infantile Paralysis fel, 55-57; investr, Howard Hughes Med Inst, 57-68. *Mem:* Am Soc Biol Chemists; Am Soc Clin Invest; Biochem Soc; Am Chem Soc; Am Col Physicians. *Res:* Biochemistry of metals; studies of metalloenzymes; use of enzymatic methods in diagnoses; analytical chemistry of metals in biological material. *Mailing Add:* 91 Glen Rd Apt NC-1 Brookline MA 02146-7764

WACKER, WILLIAM DENNIS, b St Louis, Mo, Dec 5, 41; m 73; c 2. STATISTICS. *Educ:* Wash Univ, BS, 64, MS, 67, DSc(probability), 71. *Prof Exp:* ASSOC PROF MATH, PARKS COL, ST LOUIS UNIV, 68- *Mem:* Am Statist Asn; Inst Mgt Sci. *Res:* Application of optimization theory and statistics to real world problems. *Mailing Add:* Dept Math Parks Col St Louis Univ Cahokia IL 62206

WACKERLE, JERRY DONALD, b Edna, Kans, May 21, 30; m 49; c 4. PHYSICS, MATHEMATICS. *Educ:* Univ Kans, BS, 51, MS, 54, PhD(physics), 56. *Prof Exp:* Staff mem, 56-64, ASST GROUP LEADER, UNIV CALIF, LOS ALAMOS SCI LAB, 64- *Concurrent Pos:* Adj prof physics, Univ NMex, 58-68. *Mem:* Am Phys Soc; AAAS. *Res:* Shock wave physics and equation of state; initiation and detonation of chemical explosives. *Mailing Add:* Group M9 MS P952 Los Alamos Nat Lab Los Alamos NM 87545

WACKERNAGEL, HANS BEAT, b Basel, Switz, Aug 31, 31; US citizen; m 74; c 4. ASTRODYNAMICS, DATA PROCESSING. *Educ:* Univ Basel, PhD(astron), 58. *Prof Exp:* Observer, Observ Neuchatel, 53-54; res asst, Observ Basel, 55-58; astronr, Proj Spacetrack, Air Force Cambridge Res Labs, Mass, 58-59, 496L Syst Proj Off, 60-61, First Aerospace Control Squadron, Ent AFB, Colo, 61-62, Ninth Aerospace Defense Div, 62-68 & Fourteenth Aerospace Force, 68-73, comput specialist, Hq NAm Air Defense, 73-75, opers res analyst, Second Commun Squadron, Buckley Air Nat Guard Base, 75, mathematician, 75-79, physicist, Hq NAm Air Defense, Peterson AFB, Colo, 79, opers res analyst, GS-14, Hq Air Force Space Com, 83-86 & sr mathematician, 86-89, sr mathematician, GS-15, Hq Air Force Space Com, Colo Springs, 89-90, CONSULT, AEROSPACE ENG, 90- *Concurrent Pos:* Lectr, Dept Math, Univ Colo, Colo Springs, 64-72, Dept Eng, Navig Space, 90-; astron, GS-14, Hq 14th Aerospace Force, Colo Springs, 66-73. *Honors & Awards:* Meritorious Civil Serv Award, 88; Outstanding Civilian Career Serv Award, 90. *Mem:* Am Astron Soc; fel Brit Interplanetary Soc; Swiss Astron Soc; Int Astron Union. *Res:* Design and evaluation of advanced space defense systems; applied celestial mechanics. *Mailing Add:* 51 Broadmoor Hills Dr Colorado Springs CO 80906-4355

WACKMAN, PETER HUSTING, b Cleveland, Ohio, June 16, 28; m 51; c 3. THEORETICAL PHYSICS. *Educ:* Univ Wis, BS, 51, MS, 53; Univ Pittsburgh, PhD(physics), 60. *Prof Exp:* Sr scientist, Bettis Atomic Power Lab, Westinghouse Elec Corp, 53-60; staff scientist, A-C Spark Plug Div, Gen Motors Corp, 60-61; ASSOC PROF MECH ENG, MARQUETTE UNIV, 61- *Concurrent Pos:* Mem staff, McGraw Edison Power Syst, 69- *Mem:* Am Phys Soc; Am Nuclear Soc. *Res:* Nuclear structure; low energy nuclear physics; nuclear reactor physics; inertial guidance systems; radiation effects; materials science. *Mailing Add:* 2622 N 91st St Milwaukee WI 53226

WADA, GEORGE, b Lomita, Calif, Oct 18, 27; m 58; c 2. ELECTRICAL ENGINEERING. *Educ:* Calif Inst Technol, BS, 54; Stanford Univ, MS, 55, PhD(elec eng), 58. *Prof Exp:* Res assoc elec eng, Stanford Univ, 55-58; mem tech staff, Devices Dept, Watkins-Johnson Co, 58-80; STAFF SCIENTIST, HUGHES AIRCRAFT CORP, 80- *Mem:* Inst Elec & Electronics Engrs. *Res:* Microwave electron tubes and devices. *Mailing Add:* Hughes Aircraft Corp 3100 Lomita Blvd Torrance CA 90501

WADA, JAMES YASUO, b Lomita, Calif, May 15, 34; m 57; c 4. LASERS, PLASMA PHYSICS. *Educ:* Univ Calif, Los Angeles, BS, 56; Univ Southern Calif, MS, 58, PhD(elec eng), 63. *Prof Exp:* Sect head, Elec-Gasdyn Lasers, Hughes Res Labs, Malibu, 56-74, SR SCIENTIST & DEPT MGR, HUGHES SPACE-COMMUN GROUP, LOS ANGELES, 74- *Concurrent Pos:* Lectr, Univ Southern Calif, 64, asst prof, 64-66. *Mem:* Inst Elec & Electronics Engrs; Sigma Xi. *Res:* High power lasers and optics; physical optics; microwave tubes; electromagnetic theory; space communications. *Mailing Add:* 7401 Asman Ave Canoga Park CA 91307

WADA, JUHN A, b Tokyo, Japan, Mar 28, 24; nat Can; m 56; c 2. MEDICINE, NEUROLOGY. *Educ:* Hokkaido Imp Univ, Japan, MD, 45, DMedSci, 51; FRCPS(C), 72. *Prof Exp:* Asst prof neurol & psychiat, dir labs exp neurol & brain surgeon-in-chief, Univ Hosps, Hokkaido Imp Univ, Japan, 52-57; res assoc neurol, 57-59, asst prof neurol res & psychiat & chief labs EEG & neurophysiol, 60-63, assoc prof med neurol & dir EEG labs, 63-70, PROF NEUROL SCI, UNIV BC & DIR NEUROL & EEG DEPT, HEALTH SCI CTR HOSP, 70- *Concurrent Pos:* Fel, Univ Minn, 54-55 & Montreal Neurol Inst, McGill Univ, 55-56; Can Med Res Coun assoc, 66; attend neurologist & assoc dir EEG dept, Vancouver Gen Hosp. *Honors & Awards:* William G Lennox Lectr, Am Epilepsy Soc, 81; Wilder Penfield Award, Can League Against Epilepsy, 88. *Mem:* Am Electroencephalog Soc; Am Epilepsy Soc; fel Am Acad Neurol; Can Neurol Soc; Can Soc Electroencephalog; Epilepsy Int Cong Vancouver (pres, 78); Can League Against Epilepsy (pres, 77-79); Am & E G Soc (pres, 85-88); Am Epilepsy Soc (pres, 88-89). *Res:* Neurological mechanism of human behavior; epilepsy; electrical activity of brain; cerebral speech function. *Mailing Add:* Dept Neurol Sci Univ BC Vancouver BC V6T 1A2 Can

WADA, WALTER W, b Loomis, Calif, Feb 26, 19; m 46; c 4. PHYSICS. *Educ:* Univ Utah, BA, 43; Univ Mich, MA, 46, PhD, 51. *Prof Exp:* Physicist nucleonics div, US Naval Res Lab, 51-62; PROF PHYSICS, OHIO STATE UNIV, 64- *Concurrent Pos:* Lectr, Univ Md, 51-62; vis prof, Northwestern Univ, 62-64. *Mem:* Fel Am Phys Soc. *Res:* Quantum theory of fields and applications in electrodynamics; theoretical high energy physics. *Mailing Add:* Dept Physics Ohio State Univ Columbus OH 43210

WADDELL, CHARLES NOEL, b Omaha, Nebr, Nov 11, 22; m 45; c 6. PHYSICS. *Educ:* Univ Calif, BA, 50, PhD(physics), 58. *Prof Exp:* Assoc physics, Univ Calif, 50-52, physicist, Radiation Lab, 52-58; asst prof, 58-65, ASSOC PROF PHYSICS, UNIV SOUTHERN CALIF, 65- *Mem:* Am Phys Soc. *Res:* Nuclear reaction mechanisms; semiconductor physics, ion implantation and optical properties. *Mailing Add:* Dept Physics Univ Southern Calif University Park Los Angeles CA 90089

WADDELL, HENRY THOMAS, b Wilson, Ark, Apr 19, 18; m 45; c 2. BOTANY. *Educ:* Peabody Col, BS, 49, MA, 51; Univ Fla, PhD(plant path), 59. *Prof Exp:* Asst prof biol, Martin Br, Univ Tenn, 49-56; assoc prof, Peabody Col, 59-63; PROF BIOL, LAMAR UNIV, 63- *Mem:* AAAS; Am Phytopath Soc; Mycol Soc Am; Sigma Xi. *Res:* Plant pathology; mycology. *Mailing Add:* 965 Dowlen Rd Beaumont TX 77706

WADDELL, KIDD M, b Roby, Tex, June 23, 37; m 83; c 3. LIMNOLOGY, GEOCHEMISTRY. *Educ:* Univ Tex, El Paso, BS, 61, BS, 62; Univ Utah, MS, 71. *Prof Exp:* Anal chemist, Qual Water Br, 62-66, hydrologist, Water Res Div, 66-84, SUPVR HYDROLOGIST, WATER RES DIV, US GEOL SURV, 64- *Res:* Propose, plan and manage hydrologic and limnologic investigations; remedial investigations at contaminated sites; modeling of ground and surface water including water quality. *Mailing Add:* 6544 S Rothmoor Dr Salt Lake City UT 84121

WADDELL, ROBERT CLINTON, b Mattoon, Ill, Aug 15, 21; m 60; c 3. PHYSICS. *Educ:* Eastern Ill Univ, BS, 47; Univ Ill, MS, 48; Iowa State Univ, PhD(physics), 55. *Prof Exp:* From instr to prof, 48-81, EMER PROF PHYSICS, EASTERN ILL UNIV, 81- *Concurrent Pos:* Res asst, Iowa State Univ, 53-55. *Mem:* Int Res Group Physics Teaching; Sigma Xi. *Res:* Physics education. *Mailing Add:* Dept Physics Eastern Ill Univ Charleston IL 61920

WADDELL, THOMAS GROTH, b Madison, Wis, July 29, 44. BIO-ORGANIC CHEMISTRY. *Educ:* Univ Wis-Madison, BS, 66; Univ Calif, Los Angeles, PhD(org chem), 69. *Prof Exp:* Scholar org chem, Univ Calif, Los Angeles, 69; NIH res fel, Univ Calif, Berkeley, 70-71; from asst prof to assoc prof, 71-81, PROF CHEM, UNIV TENN, CHATTANOOGA, 81- *Concurrent Pos:* Vis prof, Univ Denver, 80. *Honors & Awards:* Irvine W Grote Professorship. *Mem:* Am Chem Soc; Am Soc Pharmacog. *Res:* Chemical evolution; chemical constituents of medicinal plants; organic reactions. *Mailing Add:* Dept Chem Univ Tenn Chattanooga TN 37402

WADDELL, WALTER HARVEY, b Chicago, Ill, Sept 26, 47. SURFACE SPECTROSCOPY. *Educ:* Univ Ill, Chicago, BS, 69; Univ Houston, PhD(chem), 73. *Prof Exp:* Res assoc chem, Columbia Univ, 73-75; from asst prof to assoc prof chem, Carnegie-Mellon Univ, 79-83; sect head, Goodyear Tire & Rubber, 83-90; SCIENTIST, PPG INDUSTS, 90- *Concurrent Pos:* NIH res fel, Nat Eye Inst, 75. *Mem:* Am Chem Soc. *Res:* Spectroscopic and mechanistic investigations of tire compound chemical interactions; compounding with silica; surface characterizations of polymers; surface modifications; polymer filler interactions; adhesion. *Mailing Add:* PPG Industries Inc 440 College Park Dr Monroeville PA 15146

WADDELL, WILLIAM JOSEPH, b Commerce, Ga, Mar 16, 29; m 74; c 4. PHARMACOLOGY. *Educ:* Univ NC, AB, 51, MD, 55. *Prof Exp:* From asst prof to assoc prof pharmacol, Univ NC, Chapel Hill, 58-71, assoc prof oral biol, 67-69, prof oral biol & assoc dir dent res ctr, 69-72, assoc div dir, Ctr Res Pharmacol & Toxicol, 66-67; prof pharmacol, Univ Ky, 72-77; PROF PHARMACOL & TOXICOL & CHMN DEPT, UNIV LOUISVILLE, 77- *Concurrent Pos:* USPHS res fel, Univ NC, Chapel Hill, 55-58; NIH spec fel, Royal Vet Col, Sweden, 65-66. *Mem:* Teratology Soc; Soc Toxicol; Am Soc Pharmacol & Exp Therapeut; Soc Exp Biol & Med; Am Physiol Soc. *Res:* Intracellular pH; teratogenic agents, carcinogenic agents. *Mailing Add:* Dept Pharmacol & Toxicol Univ Louisville Health Sci Ctr Louisville KY 40292

WADDEN, RICHARD ALBERT, b Sioux City, Iowa. ENVIRONMENTAL HEALTH, ENVIRONMENTAL ENGINEERING. *Educ:* Iowa State Univ, BS, 59; NC State Univ, MS, 62; Northwestern Univ, PhD(chem, environ eng), 72. *Prof Exp:* Develop engr, Linde Co, Tonawanda, NY, 59-60; engr, Humble Oil & Refining Co, Houston, 62-65; instr chem & mech eng, Pahlavi Univ, Iran, 65-67; tech adv, Ill Pollution Control Bd, Chicago, 71-72; asst dir, Environ Health Resource Ctr, Univ Ill, 72-74, from asst prof to assoc prof, 72-79, PROF ENVIRON HEALTH SCI, SCH PUB HEALTH, UNIV ILL, 79-, DIR OFF TECHNOL TRANSFER, CTR SOLID WASTE MGT & RES, 87- *Concurrent Pos:* Adv, Northeastern Ill Planning Comn, 73-76; lectr, Nat Safety Coun, 74-76 & Nat Inst Safety & Health, 74-76; mem task force hazardous mat in environ, Am Pub Health Asn, 75-77; vis scientist, Japanese Nat Inst Environ Studies, 78-79, invited scientist, 83, 84 & 88; mem, prog rev comt indoor air pollution, personal exposure, atmospheric chemistry & physics progs, US Environ Protection Agency; consult, Beijer Inst Energy & Environ, Swed Acad Sci, Off Toxic Substances, Environ Protection Agency; dir environ & occup health sci, 84-86 & 88-; reviewer, NSF Japan Prog; dipl, Am Acad Environ Eng & Am Acad Indust Hyg. *Mem:* Am Inst Chem Engrs; Am Chem Soc; Air Pollution Control Asn; Am Acad Environ Engrs; Am Acad Indust Hyg; Am Indust Hyg Asn. *Res:* Characterization and modeling of air pollution in inside and outside environments; fine particle modeling and measurements; ozone episode detection; methodologies for predicting pollution source impacts on human health; air pollution source-receptor modeling; engineering control of workplace hazards. *Mailing Add:* Univ Ill Sch Pub Health Box 6998 Chicago IL 60680

WADDEN, THOMAS ANTONY, b Richmond, Va, Sept 3, 52; m 84; c 2. PSYCHIATRY, NUTRITION. *Educ:* Brown Univ, AB, 75; Univ NC, Chapel Hill, PhD(psychol), 81. *Prof Exp:* NIMH traineeship, Univ NC, 76-78; from instr to assoc prof psychol in psychiat, Sch Med, Univ Pa, 81-91; PROF PSYCHOL, SYRACUSE UNIV, 92- *Concurrent Pos:* Prin investr, Res Scientist Award, Behav Treatment Obese Children & Adults, NIMH, 87-; assoc ed, Ann Behavior Med, 91- *Honors & Awards:* New Res Award, Asn Advan Behav Ther, 86. *Mem:* Am Psychol Asn; Soc Behav Med; Asn Advan Behav Ther; NAm Asn Study Obesity. *Res:* Causes and treatment of obesity; short- and long-term effects of caloric restriction and weight loss on metabolic rate; long-term use of behavior therapy to improve the maintenance of weight loss. *Mailing Add:* Dept Psychol Syracuse Univ Syracuse NY 13244

WADDILL, VAN HULEN, b Brady, Tex, Aug 24, 47; m 69. ENTOMOLOGY. *Educ:* Tex A&M Univ, BS, 70, MS, 71; Clemson Univ, PhD(entom), 74. *Prof Exp:* From asst prof to assoc prof, 75-88, PROF ENTOM, AGR RES & EDUC CTR, INST FOOD & AGR SCI, UNIV FLA, 88-, CTR DIR, 88- *Concurrent Pos:* Coun, Arg Sci & Technol. *Mem:* Entom Soc Am. *Res:* Management of insect pests of vegetables. *Mailing Add:* 148 Parkwood Dr West Palm Beach FL 33411

WADDINGTON, CECIL JACOB, b Cambridge, Eng, July 6, 29; m 56. PHYSICS, ASTROPHYSICS. *Educ:* Bristol Univ, BSc, 52, PhD(physics), 55. *Prof Exp:* Royal Soc McKinnon res studentship physics, Bristol Univ, 56-59, lectr, 59-62; assoc prof, 62-68, PROF, SCH PHYSICS & ASTRON, UNIV MINN, MINNEAPOLIS, 68- *Concurrent Pos:* Res assoc & lectr, Univ Minn, 57-58; Nat Acad Sci sr fel, Goddard Space Flight Ctr, Md, 61; sr vis fel, Imp Col, Univ London, 72-73; mem, Cosmic Ray Comn, Int Union Pure & Appl Physics, 72-78. *Honors & Awards:* Except Sci Achievement Medal, NASA. *Mem:* AAAS; fel Am Phys Soc; Am Astron Soc; Int Astron Union. *Res:* Physics and astrophysics of relativistic heavy ions, using detectors exposed on balloons and satellites to the cosmic radiation and to beams of ions accelerated by machines. *Mailing Add:* Sch Physics & Astron Univ Minn 116 Church St SE Minneapolis MN 55455

WADDINGTON, DONALD VAN PELT, b Norristown, Pa, Dec 31, 31; m 55; c 6. NITROGEN FERTILIZERS, TURFGRASS MANAGEMENT. *Educ:* Pa State Univ, BS, 53; Rutgers Univ, MS, 60; Univ Mass, PhD(agron), 64. *Prof Exp:* Asst chemist, Eastern States Farmer's Exchange, Inc, 56-57; instr agron, Univ Mass, 60-65; asst prof soil technol, 65-68, assoc prof soil sci, 68-75, PROF SOIL SCI, PA STATE UNIV, UNIVERSITY PARK, 75- *Concurrent Pos:* Assoc ed, Crop Sci, Crop Sci Soc Am, 86-88. *Mem:* Fel Am Soc Agron; Soil Sci Soc Am; Int Soil Sci Soc; Soil & Water Conserv Soc; Int Turfgrass Soc; Crop Sci Soc Am. *Res:* Soil physical properties, especially soil modification for turfgrass; turfgrass nutrition; controlled-release fertilizers; impact attenuation and traction on athletic field surfaces. *Mailing Add:* 115 Pine Tree Ave Boalsburg PA 16827

WADDINGTON, JOHN, b Manchester, Eng, Mar 15, 38; Can citizen; m 66; c 1. WEEDS, FORAGE CROPS. *Educ:* Univ Leeds, Eng, BSc, 60; Univ Man, MSc, 62, PhD(agron), 68. *Prof Exp:* RES SCIENTIST FORAGE CROPS, RES BR, AGR CAN, 68- *Mem:* Agr Inst Can. *Res:* Forage crops management, particularly establishment, weed control, plant competition and effects of weather on yield and quality of hay and seed. *Mailing Add:* Agr Can Res Sta PO Box 1030 Swift Current SK S9H 3X2 Can

WADDINGTON, RAYMOND, b Eng, June 16, 62. CASE TECHNOLOGY, HUMAN-COMPUTER INTERACTION. *Educ:* Univ Keele, BSc, 84; Univ Nottingham, PhD(psychol), 89. *Prof Exp:* Sr res assoc computer sci, Univ London, 87-89; ASST PROF COMPUTER SCI, UNIV GUELPH, 90- *Concurrent Pos:* Mem, Human-Computer Interaction Specialist Group, Brit Computer Soc, London, 89- *Mem:* Brit Computer Soc; Asn Comput Mach. *Res:* Human-computer interaction; psychology of programming; psychology of software engineering; case technology; reverse engineering; educational software. *Mailing Add:* Dept Computer Sci Univ Guelph Guelph ON N1G 2W1

WADDLE, BRADFORD AVON, b Tex, Jan 26, 20; m 45. AGRONOMY, PLANT BREEDING. *Educ:* Agr & Mech Col, Tex, BS, 42, MS, 50; Purdue Univ, PhD(plant breeding), 54. *Prof Exp:* Instr, Hunt County Voc Schs, Tex, 46-47; jr agronomist, Greenville Cotton Sta, USDA, 48; instr cotton breeding, Agr Exp Sta, Univ Tex A&M, 50; asst agron & Altheimer chair cotton, Agr Exp Sta, Univ Ark, 51-56, assoc, 56-59, prof agron & Altheimer chair cotton res, 59-74, distinguished prof & Altheimer chair cotton, 74-86, EMER DISTINGUISHED PROF & ALTHEIMER CHAIR COTTON, UNIV ARK, FAYETTEVILLE, 86- *Mem:* Am Soc Agron; Am Genetic Asn. *Res:* Cotton breeding and genetics, especially breeding for resistance to disease and insects. *Mailing Add:* Dept Agron Univ Ark Fayetteville AR 72701

WADE, ADELBERT ELTON, b Hilliard, Fla, Apr 29, 26; m 50; c 2. PHARMACOLOGY, BIOCHEMISTRY. *Educ:* Univ Fla, BS, 54, MS, 56, PhD(pharmacol), 59. *Prof Exp:* Asst chemother, Univ Fla, 54-56, asst biochem, 56-57, asst chemother, 57-59; from asst prof to assoc prof, 59-67, PROF PHARMACOL, UNIV GA, 67-, HEAD DEPT, 68- *Mem:* Am Asn Cols Pharm; Soc Exp Biol & Med; Am Soc Pharmacol & Exp Therapeut; Int Soc Biochem Pharmacol; Sigma Xi. *Res:* Influence of diet and drugs on mixed-function oxidases; effects of diet on drug and carcinogen metabolism. *Mailing Add:* Dept Pharm Univ Ga Athens GA 30602

WADE, ADRIAN PAUL, b Amersham, UK, Apr 7, 60; m 82; c 3. FLOW INJECTION ANALYSIS, CHEMICAL ACOUSTIC EMISSION. *Educ:* Southampton Univ, UK, BSc, 81; Univ Wales, UK, PhD(anal chem), 85. *Prof Exp:* Extra-mural res assoc expert systs, Brit Petrol Res Ctr, Univ Col Swansea, 84-85; res chemist & computer scientist expert systs, Brit Petrol Res Ctr, Sunbury-on-Thames, UK, 85-87; ASST PROF CHEM, UNIV BC, 87- *Concurrent Pos:* Vis res assoc anal chem, Mich State Univ, 85-87; contrib ed, Trends in Anal Chem, 89-; corresp ed, J Automatic Chem, 90-; fac assoc, Pulp & Paper Res Inst Can, 90-; affil, Pulp & Paper Ctr, Univ BC, 90- *Mem:* Royal Soc Chem; fel Inst Analysts & Programmers; Asn Anal Chemists; Chem Inst Can; Can Soc Chem. *Res:* Flow injection analysis; chemical acoustic emission; chemometrics; expert systems; author of several publications. *Mailing Add:* Chem Dept Univ BC 2036 Main Mall Vancouver BC V6T 1Y6 Can

WADE, CAMPBELL MARION, b Elizabethtown, Ky, Nov 25, 30; m 56, 78; c 4. ASTRONOMY. *Educ:* Harvard Univ, AB, 54, AM, 55, PhD(astron), 57. *Prof Exp:* Res officer, Div Radiophysics, Commonwealth Sci & Indust Res Orgn, Australia, 57-59; res assoc, 60-66, SCIENTIST, NAT RADIO ASTRON OBSERV, 66- *Concurrent Pos:* Adv ed, Soviet Astron, Am Inst Physics, 69- *Mem:* Am Astron Soc; Int Astron Union. *Res:* Galactic and extragalactic radio astronomy. *Mailing Add:* 1224 Apache Dr NW Socorro NM 87801

WADE, CHARLES GARY, b Spring City, Pa, Dec 8, 38; m 64; c 2. ORGANIC & INORGANIC CHEMISTRY. *Educ:* Ursinus Col, BS, 60; Univ Del, MS, 68. *Prof Exp:* Chemist, Abex Corp, 66-68; sr chemist, 68-74, res chemist, 74-76, supvr, 76-83, sr res scientist & team leader, 83-85, TECH MGR & LICENSING, ATLAS POWDER CO, 85- *Concurrent Pos:* Chmn, Am Chem Soc, 85. *Honors & Awards:* IR 100 Award, Ind Res Mag, 77. *Mem:* Am Chem Soc. *Res:* Water based explosives based on emulsion technology. *Mailing Add:* Atlas Powder Co PO Box 577 Tamaqua PA 18252

WADE, CHARLES GORDON, b Griggsville, Ill, Apr 5, 37; m 64. POLYMER CHEMISTRY, PHYSICAL CHEMISTRY. *Educ:* Southern Ill Univ, BA, 60; Mass Inst Technol, PhD(phys chem), 65. *Prof Exp:* Res assoc chem, Enrico Fermi Inst Nuclear Studies, Univ Chicago, 65-67; from asst prof to assoc prof chem, Univ Tex, Austin, 67-80; appln scientist, IBM Instruments, Inc, 80-82, mgr nuclear magnetic resonance, 82-83, mgr magnetics, 83-85, mgr WCoast Opers, 85-87; mgr mat anal, 87-88, mgr poly char, 88-90, MGR TECH STAFF, IBM ALMADEN RES CTR, 90- *Mem:* Am Phys Soc; Am Chem Soc; Sigma Xi; AAAS. *Res:* Nuclear magnetic resonance in biological systems; nuclear magnetic resonance in polymers; computer controlled instrumentation; properties of liquid crystals; fluorescence of carcinogens; chemical carcinogenesis; structure and diffusion in membrane systems; spectroscopic studies of biological systems. *Mailing Add:* IBM Almaden Res Ctr 650 Harry Rd San Jose CA 95120

WADE, CLARENCE W R, b Laurinburg, NC, Mar 31, 27; m 55; c 1. ORGANIC CHEMISTRY. *Educ:* J C Smith Univ, BS, 48; Tuskegee Inst, MS, 50; Georgetown Univ, PhD(org chem), 65. *Prof Exp:* From instr to asst prof chem, St Augustine's Col, 50-57; from chemist to res chemist, Nat Bur Stand, 57-66; res chemist, US Army Med & Biomech Lab, Walter Reed Army Med Ctr, 66-68, chief, Synthesis Br, 68-70, chief mat & applns div, 70-72, chief, Mat & Applns Div, 72-76, sr res chemist, 76-, proj area mgr chem systs, 80-; AT DEPT PHARMACOL, HOWARD UNIV, WASHINGTON, DC. *Concurrent Pos:* Consult, Nat Heart Inst, 68-70; adj prof chem, Univ DC, 75-; prof orthop res, Howard Univ, 77-79. *Mem:* AAAS; Am Chem Soc; NY Acad Sci; Asn Off Anal Chemists. *Res:* Development of inert or degradable implant materials, tissue and bone adhesives, sutures, tendons, vascular tubes, wound and burn dressings, bone repair polymers; mechanisms of implant degradation. *Mailing Add:* 1736 Buchanan St NE Washington DC 20017-3123

WADE, DALE A, b Buffalo, SDak, May 23, 28; m 53; c 5. WILDLIFE MANAGEMENT. *Educ:* SDak State Univ, BS, 69, PhD(animal sci), 72. *Prof Exp:* Mem staff mammal control, US Fish & Wildlife Serv, 62-65; wildlife specialist, Colo State Univ, 72-74, Univ Calif, Davis, 74-78; wildlife specialist, Tex A&M Univ, San Angelo, 78-86; dir Nat Tech Support Staff, ADC-APHIS-USDA, 86-90; RETIRED. *Concurrent Pos:* Consult, US Environ

Protection Agency & US Dept Agr, Various State Agencies, 70-85. *Mem:* Sigma Xi; AAAS; Am Inst Biol Sci; Soc Range Mgt; Wildlife Soc; Am Soc Testing & Mat; Nat Parks & Conserv Asn; Coun Agr Sci & Tech. *Res:* Evaluation of biological, economic conflicts and possible solutions in human, wildlife and agricultural relationships. *Mailing Add:* 8610 Bonita Pl Cheyenne WY 82009

WADE, DAVID ROBERT, b London, Eng, May 25, 39; m 62; c 3. BIOCHEMISTRY. *Educ:* Univ Cambridge, BA & MA, 63, PhD(biochem), 67. *Prof Exp:* Res assoc physiol, Col Med, Pa State Univ, 67-69; Bank Am Giannini fel biochem, Sch Med, Univ Calif, Davis, 69-71; USPHS grant metab regulation & asst prof physiol, Col Med, Pa State Univ, 71-74; ASSOC PROF PHYSIOL, SCH MED, SOUTHERN ILL UNIV, 74- *Res:* Metabolic regulation. *Mailing Add:* Dept Physiol Southern Ill Univ Sch Med Carbondale IL 62901

WADE, EARL KENNETH, b Toledo, Iowa, July 13, 14; m 47; c 3. PLANT PATHOLOGY. *Educ:* Univ Wis, BS, 38, MS, 50. *Prof Exp:* Instr high sch, Wis, 38-42; asst potato cert serv, 46-50, prof plant path, 69-79, EXTEN PLANT PATHOLOGIST, UNIV WIS-MADISON, 50-, EMER PROF PLANT PATH, 79- *Mem:* Am Phytopath Soc; Am Potato Asn. *Res:* Vegetable and fruit diseases. *Mailing Add:* 5007 Marathon Dr Madison WI 53705

WADE, GLEN, b Ogden, Utah, Mar 19, 21; m 45; c 4. ENGINEERING PHYSICS. *Educ:* Univ Utah, BS, 48, MS, 49; Stanford Univ, PhD, 54. *Prof Exp:* Electronic scientist, US Naval Res Labs, Washington, DC, 49-50; res assoc, Microwave Lab, Gen Elec Co, 55; mem sr staff, Electronics Labs, Stanford Univ, 55-60, assoc prof elec eng, 58-60; asst gen mgr, Res Div, Raytheon Co, Mass, 60-63; dir sch elec eng & J Preston Levis prof eng, Cornell Univ, 63-66; PROF ELEC ENG, UNIV CALIF, SANTA BARBARA, 66- *Concurrent Pos:* Consult, Gen Elec Co, 55-58, Diamond Ord Fuze Labs, 57-60, Zenith Radio Corp, 57-60, 63-73 & EG&G, Inc, 66-79, Innovision, 87-, Wiley Publ Co, 84-; ed, Trans on Electron Devices, Inst Elec & Electronics Engrs, 61-71 & J Quantum Electronics, 65-67; ser ed, Harcourt, Brace & World, Inc, 64-70; consult mem adv group electron devices, Comt of Dept Defense, 66-74; Japan Soc Promotion of Sci vis prof award, Univ Tokyo, 71; Fulbright-Hays fel, Spain, 72-73; ed, Proc of Inst Elec & Electronics Engrs, 77-80; Taiwanese Nat Sci Coun spec chair award, Nat Taiwan Univ, 80-81; UN vis prof, Nanjing Inst Technol, 86; lectr, Taiwanese Nat Sci Coun, 88; UN vis prof, Southeast Univ China, 89. *Honors & Awards:* Nat Electronics Conf Ann Award, 59. *Mem:* Am Phys Soc; Inst Elec & Electronics Engrs; Sigma Xi. *Res:* Physical and quantum electronics; ultrasonics; optical systems. *Mailing Add:* Dept Elec Eng Univ Calif Santa Barbara CA 93106

WADE, JAMES B, PHYSIOLOGY, MEMBRANE STRUCTURE. *Educ:* Princeton Univ, PhD(biol), 72. *Prof Exp:* PROF PHYSIOL, SCH MED, UNIV MD, 83- *Mailing Add:* Dept Physiol Univ Md Sch Med 655 N Baltimore St Baltimore MD 21201

WADE, JAMES JOSEPH, b St Paul, Minn, Jan 7, 46; m 70; c 3. MEDICINAL CHEMISTRY. *Educ:* Col St Thomas, BA, 68; Univ Minn, PhD(org chem), 72. *Prof Exp:* NIH fel org chem, Univ Rochester, 72-73; sr med chemist, 73-76, RES SPECIALIST, RIKER LABS, 3M CO, 76- *Mem:* Am Chem Soc; Sigma Xi. *Res:* Design and synthesis of organic compounds for possible medicinal use, particularly in the antiallergy and antithrombotic areas. *Mailing Add:* 1385 N Hallmark Ave St Paul MN 55119-1830

WADE, LEROY GROVER, JR, b Jacksonville, Fla, Oct 8, 47; m 74; c 2. ORGANIC SYNTHETIC METHODS. *Educ:* Rice Univ, BA, 69; Harvard Univ, AM, 70, PhD(chem), 74. *Prof Exp:* From asst prof to assoc prof chem, Colo State Univ, 74-89; PROF CHEM, WHITMAN COL, 89- *Mem:* Am Chem Soc; AAAS; Sigma Xi; Am Acad Forensic Sci. *Res:* Organic chemistry; organic synthesis; chemical education; author of organic chemistry textbooks. *Mailing Add:* Chem Dept Whitman Col Walla Walla WA 99362

WADE, MICHAEL GEORGE, b Watford, Eng, Nov 5, 41; c 4. KINISIOLOGY, HUMAN FACTORS ENGINEERING. *Educ:* Loughborough Col, DLC, 63; Univ Ill, MS, 68, PhD(Kinesiology & human factors), 70. *Prof Exp:* Lectr phys educ, Univ Guelph, Can, 65-66; asst prof, Inst Child Behav & Develop, Univ Ill, 66-70, asst prof leisure studies & spec educ, 70-75, assoc prof, 75-81; prof phys educ, Southern Ill Univ, 81-86; DIR & PROF, SCH PHYS EDUC, RECREATION & SCH HEALTH, PROF, CTR RES ON LEARNING, PERCEPTION & COGNITION, UNIV MINN, 86- *Mem:* NAm Soc Psychol Sport & Phys Activ; Am Asn Ment Deficiency; fel Am Acad Phys Educ; fel Am Asn Ment Retardation. *Res:* Motor behavior and developmental disabilities and problems of control and coordination; mental retardation; play behavior of children; biorhythms. *Mailing Add:* Sch Phys Educ Recreation & Sch Health Col Educ Univ Minn 110 Cooke Hall 1900 University Ave Minneapolis MN 55455

WADE, MICHAEL JAMES, b Salt Lake City, UT, May 18, 42; m 71. TOXICOLOGY, NUTRITIONAL TOXICOLOGY. *Educ:* Univ Utah, BS, 64, MS, 67; Wash Univ, PhD(molecular biol), 71. *Prof Exp:* Res assoc biophysics, Max-Planck Inst Med Res, Heidelberg, Ger, 72-73, chem, Boston Univ, 73-75; fel toxicol, Univ Calif, San Francisco, 75-76; staff scientist, Life Sci Res Off, Fedn Am Soc Exp Biol, 76-78; REV SCIENTIST TOXICOL, US FOOD & DRUG ADMIN, 78- *Mem:* Am Col Toxicol; Soc Toxicol & Environ Chem; Am Chem Soc. *Mailing Add:* 5716 Thames Way Carmichael CA 95608-5557

WADE, MICHAEL JOHN, b Evanston, Ill, Oct 21, 49; m 72. POPULATION BIOLOGY, POPULATION GENETICS. *Educ:* Boston Col, BA, 71; Univ Chicago, PhD(theoret biol), 75. *Prof Exp:* Asst prof, 75-82, ASSOC PROF BIOL, UNIV CHICAGO, 82- *Concurrent Pos:* Res career develop award, NIH, 81-86. *Mem:* Soc Am Naturalists; Soc Study Evolution. *Res:* Role of population structure in evolution; evolution of social behaviors. *Mailing Add:* Dept Ecol & Evolution Univ Chicago 940 E 57th St Chicago IL 60637

WADE, PETER ALLEN, b Taunton, Mass, Nov 12, 46; m 67; c 2. ORGANIC CHEMISTRY. *Educ:* Lowell Technol Inst, BS, 68; Purdue Univ, PhD(org chem), 73. *Prof Exp:* Fel org chem, Univ Groningen, Neth, 73-74; IBM fel, Harvard Univ, 74-76; asst prof, 76-87, ASSOC PROF CHEM, DREXEL UNIV, 87- *Mem:* Am Chem Soc. *Res:* New synthetic methods; cycloaddition reactions; reactive intermediates; strained rings; carbohydrates. *Mailing Add:* Dept Chem Drexel Univ Philadelphia PA 19104

WADE, PETER CAWTHORN, b Washington, DC, Feb 15, 44; m 66; c 1. MEDICINAL CHEMISTRY, ORGANIC CHEMISTRY. *Educ:* Middlebury Col, AB, 66; Univ Wash, PhD(org chem), 71. *Prof Exp:* res scientist, Squibb Inst Med Res, 71-80; res assoc, Diamond Shamrock Corp, 80-; AT ORSYNEX, INC, COLUMBUS, OHIO. *Mem:* Am Chem Soc. *Res:* Anxiolytic, antidepressive, neuroleptic and anti-inflammatory, anti-hypertensive and anthelmintic agents; heterocyclic chemistry. *Mailing Add:* 372 Nutt Rd Spring Valley OH 45370-9690

WADE, RICHARD ARCHER, b Fitchburg, Mass, Aug 16, 30; m 70; c 2. BIOLOGICAL OCEANOGRAPHY. *Educ:* Univ Miami, BS, 56, MS, 62, PhD(biol oceanog), 68. *Prof Exp:* Marine scientist, Ayerst Labs, Div Am Home Prod Corp, 66-68; head dept ecol & pollution, Va Inst Marine Sci, 68-69; chief lab, Environ Protection Agency, Fed Water Qual Admin, 68-69 & 70-71; exec secy, Sport Fishing Inst, 71-72; exec dir, Am Fisheries Soc, 72-75; MARINE ECOLOGIST, US FISH & WILDLIFE SERV, 75- *Concurrent Pos:* Consult, NIH Pesticide Proj, Univ Miami, 66-68; mem, Water Qual Mgt Comt, US Govt Interagency Group, 68-69; mem res subcomt, Fed Comt Pest Control, 68; clin res assoc, Med Univ SC, 70-71; mem, Subcomt Marine Water Qual Criteria, Nat Acad Sci, 71; treas, Sport Fishing Res Found, 71-72. *Mem:* Am Fisheries Soc; Am Soc Ichthyol & Herpet; Gulf & Caribbean Fisheries Inst; Marine Technol Soc. *Res:* Coastal ecosystems of the United States, including dredge disposal, offshore oil and gas development, development of deepwater ports, power plant construction and operation; marine and estuarine water quality problems. *Mailing Add:* 2310 15th St Alamogordo NM 88310-4803

WADE, ROBERT HAROLD, b Opportunity, Wash, Sept 16, 20; m 44; c 2. ORGANIC CHEMISTRY, POLYMER CHEMISTRY. *Educ:* Univ Wash, BS, 46, PhD(chem), 51. *Prof Exp:* Res chemist, M W Kellog Co Div, Pullman, Inc, 51-57; org chemist & proj leader, Stanford Res Inst, 57-63; sr res scientist, US Naval Undersea Ctr, 63-83; RETIRED. *Res:* Synthesis of polynuclear aromatic compounds; high temperature metal-chelate polymers; physical and chemical fate of fluoride in plants; synthesis and properties of water soluble and friction reducing polymers; marine natural products. *Mailing Add:* 7810 Golfcrest Dr San Diego CA 92119

WADE, ROBERT SIMSON, b Gorrie, Ont, Aug 20, 20; m 43; c 3. CHEMISTRY, RESEARCH ADMINISTRATION. *Educ:* Univ Western Ont, BA, 42, MA, 43. *Prof Exp:* Res chemist, Imp Oil, Ltd, 43-47; chief chemist, Imp Tobacco Co Can, Ltd, 47-53, mgr lab, 53-57, mgr res develop & tech serv, 57-69, mgr res & develop, 69-76, mgr res & lab serv, 76-80, presidential sci adv, 80-84; RETIRED. *Mem:* Chem Inst Can; Can Res Mgt Asn; Am Chem Soc. *Res:* Growing, processing and manufacturing of tobacco and tobacco products; development of processes and products; technology of tobacco and tobacco smoke. *Mailing Add:* 97 Hillsmount Crescent Imp Tobacco Ltd 3810 St Antoine London ON N6K 1V6 Can

WADE, THOMAS EDWARD, b Jacksonville, Fla, Sept 14, 43; m 66; c 3. SOLID STATE MICROELECTRONICS, VLSI MULTILEVEL INTERCONNECTIONS SYSTEMS. *Educ:* Univ Fla, BSEE, 66, MSEE, 69, PhD(microelectronics), 74. *Prof Exp:* Asst prof elec eng, Univ Fla, 74-76; prof elec eng & dir microelectron res lab, Miss State Univ, 76-85; actg dir, Ctr Microelectronics Design & Test, 86-87, ASSOC DEAN RES, COL ENG, UNIV SFLA, 85-, EXEC DIR, CTR ENG DEVELOP & RES, 85- *Concurrent Pos:* Solid State Circuit Specialist, Appl Micro Circuit Corp, San Diego, Calif, 81; res scientist, NASA, Marshall Space Flight Ctr, Huntsville, 83 & 84; consult, 82-; dir, Eng Indust Exp Sta, Univ SFla, 85-; bd dirs, Eng Res Coun, Am Soc Eng Educ; chmn, External Relations Comt, Soc Res Adminr, 87-90; educ comt, Int Soc Hybrid microelectronics, 80-84. *Mem:* Am Soc Eng Educ; Soc Res Adminr; Am Vacuum Soc; Nat Coun Univ Res Adminr; Nat Soc Prof Engrs; Int Soc Hybrid Microelectronics. *Res:* Technical specialization in solid state microelectronics and related topics; VSLI multilevel interconnection systems; fluctuation phenomena (electronic noise) in semiconductors and solid state devices; solid state device and related materials fabrications and characterizations; novel research administration techniques. *Mailing Add:* Eng Deans Off Univ SFla 4202 Fowler Ave Tampa FL 33620

WADE, WILLIAM H, b San Antonio, Tex, Nov 3, 30; wid. PHYSICAL CHEMISTRY, PETROLEUM ENGINEERING. *Educ:* St Mary's Univ, Tex, BS, 51; Univ Tex, PhD(chem), 55. *Prof Exp:* Res scientist, Univ Calif, Berkeley, 55-58; res scientist, 58-61, from asst prof to assoc prof, 61-72, chmn dept, 74-80, PROF CHEM, UNIV TEX, AUSTIN, 72- *Mem:* Am Chem Soc; Soc Petrol Engrs. *Res:* Surface chemistry; emulsions; surfactants for enhanced oil recovery. *Mailing Add:* Dept Chem Univ Tex Austin TX 78712-1104

WADE, WILLIAM HOWARD, b Stoughton, Wis, Apr 18, 23; m 43; c 1. ENTOMOLOGY, PHYTOPATHOLOGY. *Educ:* Univ Calif, BS, 50, PhD(entom), 56. *Prof Exp:* Res & teaching asst, Univ Calif, 50-53; mgr tech serv & prod prom, Agr Chem Div, 53-72, mgr develop, 72-75, MGR TECH SERV, AGR CHEM GROUP, FMC CORP, 75- *Mem:* Entom Soc Am; Sigma Xi. *Res:* Insect biology; field evaluation of pesticides; investigation of pesticide related problems. *Mailing Add:* 214 W Andrews Fresno CA 93705

WADE, WILLIAM RAYMOND, II, b Los Angeles, Calif, Oct 28, 43; m 65; c 2. MATHEMATICS. *Educ:* Univ Calif, Riverside, BA, 65, MA, 66, PhD(math), 68. *Prof Exp:* From asst prof to assoc prof, 68-78, PROF MATH, UNIV TENN, KNOXVILLE, 78- *Concurrent Pos:* Consult, Oak Ridge Nat Lab, 69-76; vis assoc prof math, Univ Southern Calif, 77; Fulbright prof math,

Moscow State Univ, 77-78, Indian Statist Inst, Bangalore, India, 83; vis prof math, Eotvos Univ, Hungary, 85; Res Grants, NSF, 78-79, 84-86 & 88-91. *Mem:* Am Math Soc; Math Asn Am. *Res:* Fourier analysis on groups; Haar and Walsh series; sets of uniqueness; harmonic analysis of p-series fields; transform theory. *Mailing Add:* Dept Math Univ Tenn Knoxville TN 37916

WADELIN, COE WILLIAM, b Dover, Ohio, Aug 18, 27; m 50; c 1. ANALYTICAL CHEMISTRY. *Educ:* Mt Union Col, BS, 50; Purdue Univ, MS, 51, PhD, 53. *Prof Exp:* Res chemist, Goodyear Tire & Rubber Co, Akron, Ohio, 53-65, sect head anal chem, 65-75, sect head spectros, 75-77, mgr anal sci & technol, Fiber & Polymer Res & Develop Div, 77-87; RETIRED. *Concurrent Pos:* Fel, Ctr Advan Eng Study, Mass Inst Technol, 68-69. *Mem:* Am Chem Soc. *Res:* Analysis of polymers and organic chemicals; absorption spectroscopy. *Mailing Add:* 1195 Inverness Lane Stow OH 44224-2275

WADELL, LYLE H, b Elsie, Mich, Mar 7, 34; div; c 5. ANIMAL BREEDING, DATA PROCESSING MANAGEMENT. *Educ:* Mich State Univ, BS, 55, MS, 57; Iowa State Univ, PhD(animal breeding, statist, genetics), 59. *Prof Exp:* Res assoc animal breeding res, 59-60, res animal geneticist, 60-61, admin supvr comput ctr mgt, 61-66, DIR COMPUT CTR MGT, CORNELL UNIV, 66 - *Res:* Computing center management; data processing techniques; statistics. *Mailing Add:* B-24 Morrison Hall Cornell Univ Ithaca NY 14853

WADEY, WALTER GEOFFREY, b Whangarei, NZ, Sept 9, 18; nat US; m 45; c 3. PHYSICS. *Educ:* Univ Mich, BSc, 41, MA, 42, PhD(physics), 47. *Prof Exp:* Res assoc, Radio Res Lab, Harvard Univ, 43-45; instr physics, Yale Univ, 47-50, asst prof, 50-56; prof, Southern Ill Univ, 56-57; mgr sci prog, Remington Rand Univac Div, Sperry Rand Corp, 57-58, tech coordr, 59; physicist, Hughes Aircraft Co, 59-60; mgr advan electromech develop dept, Univac Div, Sperry Rand Corp, 60-62; chief scientist, Bowles Eng Corp, 62-63 & Wash Tech Assocs, 63-64; sr scientist, Opers Res, Inc, Silver Spring, MD, 64-84; RETIRED. *Mem:* AAAS; Am Phys Soc; Asn Comput Mach; Opers Res Soc Am; Marine Technol Soc. *Res:* Experimental nuclear physics; nuclear spectroscopy; linear electron accelerators; electronics; computer programming and arithmetics; fluid mechanics; electromechanical design; fluid-amplifier technology; operations research; systems analysis; anti-submarine warfare; information systems. *Mailing Add:* Six Greenfield Lane Scituate MA 02066

WADKE, DEODATT ANANT, b July 7, 38; US citizen; m 67; c 2. PHYSICAL PHARMACY. *Educ:* Banaras Hindu Univ, BPharm, 61; Ohio State Univ, MS, 63; State Univ NY, Buffalo, PhD(pharmaceut), 67. *Prof Exp:* Res assoc formulations, Merck Sharpe & Dohme Res Labs, 66-69; res investr pharmaceut res & develop, 69-71, sr res investr, 72-73, head, preformulation studies sect, 73-76, head, solid formulation develop sect, 76-79, asst dir pharm res & develop, 79-85, DIR, PHARMACEUT RES & DEVELOP, BRISTOL-MYERS SQUIBB PHARMACEUT RES INST, NEW BRUNSWICK, 85- *Mem:* Am Pharmaceut Asn; Acad Pharmaceut Sci. *Res:* Thermodynamics of dissolution, solubilization and absorption; dissolution of polyphase systems; drug stability; pharmacokinetics. *Mailing Add:* Bristol-Myers Squibb Pharmaceut Res Inst New Brunswick NJ 08903

WADKINS, CHARLES L, b Joplin, Mo, May 8, 29; m 52; c 2. BIOCHEMISTRY. *Educ:* Univ Kans, AB, 51, PhD(biochem), 56. *Prof Exp:* Instr biochem, Univ Kans, 55-56; from instr to assoc prof, Sch Med, Johns Hopkins Univ, 57-66; chmn dept, 66-80, PROF BIOCHEM, MED CTR, UNIV ARK, LITTLE ROCK, 66- *Concurrent Pos:* Fel biochem, Sch Med, Johns Hopkins Univ, 56-57; USPHS sr res fel, 59-64; mem adv panel, NSF; planning officer, Nat Inst Aging, NIH, 80-82. *Mem:* Am Chem Soc; Am Soc Biol Chemists; Brit Biochem Soc. *Res:* Biological oxidation reactions; oxidative phosphorylation; mechanism and control of biological calcification reactions. *Mailing Add:* Dept Biochem Univ Ark Col Med Markham & Hooper Dr Little Rock AR 72205

WADLEIGH, CECIL HERBERT, b Gilbertville, Mass, Oct 1, 07; m 30; c 4. PLANT PHYSIOLOGY, RESOURCE CONSERVATION. *Educ:* Mass Col, BS, 30; Ohio State Univ, MS, 32; Rutgers Univ, PhD(plant physiol), 35. *Hon Degrees:* DSc, Univ Mass, 74. *Prof Exp:* Asst, Rutgers Univ, 33-36; asst prof agron, Univ Ark, 36-41; sr chemist, Regional Salinity Lab, Bur Plant Indust, USDA, 41-42 & Bur Plant Indust, Soils & Agr Eng, 42-48, prin plant physiologist, 48-51, head physiologist in chg div sugar plant invests, 51-54, head sect soils & plant relationships, Soil & Water Conserv Res Br, 54-55, dir, Soil & Water Conserv Res Div, 55-74; RETIRED. *Concurrent Pos:* Mem, Comn Watershed Hydrol, White House Off Sci & Technol, 61-62; mem, White House Panel on Indus Basin Pakistan, 61-63; Off deleg USDA, White House Conf Conserv, 63; mem, Comn Water Resources Res, 63-69; mem, Comn Environ Quality, 67-69; consult, Lake Verrett Watershed Planning Group, La, 71-76; Off deleg, USDA, Bilateral Conf Environ Protection, Pilsen, Czechoslovakia, 73. *Mem:* Nat Acad Sci; fel Soil Conserv Soc Am; Am Soc Agron; Soil Sci Soc Am; Am Soc Sugar Beet Technol; Am Soc Plant Physiol (pres, 51-52); Am Soc Hort Sci. *Res:* Mineral nutrition of plants; carbohydrate and nitrogen metabolism of plants; salt tolerance of plants; soil moisture stress. *Mailing Add:* 5621 Whitefield Chapel Rd Lanham MD 20706-2515

WADLEIGH, KENNETH R(OBERT), b Passaic, NJ, March 27, 21; m 48; c 2. MECHANICAL ENGINEERING. *Educ:* Mass Inst Technol, SM & SB, 43, ScD, 53. *Hon Degrees:* MA, Univ Cambridge, Eng, 54. *Prof Exp:* From instr to assoc prof, 46-61, dean student affairs, 61-69, PROF MECH ENG, MASS INST TECHNOL, 61-, VPRES, 69-, DEAN GRAD SCH, 75- *Concurrent Pos:* Lectr, Univ Cambridge, Eng, 53-54. *Honors & Awards:* Goodwin Medal, 52; Bronze Beaver Award, Mass Inst Technol, 69. *Res:* Applied thermodynamics; fluid mechanics; two-phase flows. *Mailing Add:* Mass Inst Technol Rm 3-136 77 Massachusetts Ave Cambridge MA 02139

WADLEY, MARGIL WARREN, b Cisco, Tex, Dec 4, 31; m 66; c 1. ENVIRONMENTAL SCIENCES, ATMOSPHERIC SCIENCES. *Educ:* Southern Nazarene Univ, BS, 53; Okla State Univ, MS, 60; Purdue Univ, PhD(inorg chem), 63. *Prof Exp:* Sr chemist, Autonetics Div, NAm Aviation, Inc, 63-64 & Korad Dept, Union Carbide Corp, 64-65; sr res engr, Autonetics Div, NAm Aviation, Inc, 65 & 66-69 & Space Systems Div, Rockwell Int, Inc, 65-66; environ scientist, 69-71, prin chemist, 71-75, supv chemist, Southern Calif Air Pollution Control Dis, 75-77; prin chemist, 77-85, DIR LAB SERVS, SOUTH COAST AIR QUAL MGT DIST, 85- *Concurrent Pos:* Mem bd, Henry George Sch Soc Sci, Los Angeles. *Mem:* Am Chem Soc; Sigma Xi; Am Sci Affil; Air & Waste Mgt Asn. *Res:* Size and mass distribution of airborne particulate matter and associated visibility relationships. *Mailing Add:* 520 E Riverdale Ave Orange CA 92665

WADLINGER, ROBERT LOUIS PETER, b Philadelphia, Pa, Mar 20, 32; m 55; c 3. CHEMISTRY. *Educ:* LaSalle Col, AB, 53; Cath Univ, PhD(photochem kinetics, phys chem), 61. *Prof Exp:* Res asst, Benjamin Franklin Inst Labs, Pa, 53-55; sr res chemist, Mobil Oil Co, Inc, Labs, NJ, 60-62; assoc prof chem, State Univ NY Col Oneonta, 62-65; assoc prof chem, Niagara Univ, 65-76; sales rep, Equitable Life Assurance Soc, USA, 76-77; sr scientist, SCA Serv, Inc, Model City, NY, 77-78; with Recra Res, Inc, Tonawanda, NY, 78-79; assoc prof chem, Elmira Col, Elmira, NY, 80-81; ASST PROF CHEM, PA STATE UNIV, MIDDDLETOWN, PA, 81- *Concurrent Pos:* Assoc prof chem, Daemen Col, Amherst, NY, 80; consult environ chem, Hooker Chem Corp, Grand Island, NY, 80. *Mem:* Am Chem Soc. *Res:* Zeolite syntheses. *Mailing Add:* 4963 Creek Rd Lewiston NY 14092-1838

WADLOW, DAVID, b London, Eng, 1950. BIOMEDICAL SENSORS, GAS HANDLING. *Educ:* Univ Manchester, UK, BSc, 71; Univ Leicester, UK, MSc, 72; Univ Reading, UK, PhD(electrogas dynamics), 84. *Prof Exp:* Res assoc, Dept Eng & Cybernet, Univ Reading, UK, 73-76 & Cryog Ctr, Stevens Inst Technol, 76-78; sr engr, 78-88, LEAD ENGR, BOC GROUP TECH CTR, 88- *Mem:* Inst Physics. *Res:* Sensors, including new technologies in gas flow measurement and gas composition analysis; mass spectrometry and visible emission spectroscopy; biomedical instrumentation. *Mailing Add:* Boc Group Tech Ctr 100 Mountain Ave Murray HIll NJ 07974

WADMAN, W HUGH, b Marlborough, Eng, Sept 18, 26; m 52; c 2. CHEMISTRY, BIOCHEMISTRY. *Educ:* Bristol Univ, BSc, 47, PhD, 51. *Prof Exp:* Res assoc plant biochem, Univ Calif, 51-53; group leader cellulose chem, Rayonier Inc, Wash, 53-55; prof org chem, Univ Pac, 55-89; RETIRED. *Mem:* Am Chem Soc. *Res:* Carbohydrate chemistry and chromatographic techniques; origin of life; primitive biochemical systems. *Mailing Add:* 1880 River Dr Stockton CA 95204

WADSWORTH, DALLAS FREMONT, b Arcadia, Okla, Mar 2, 22; c 1. PLANT PATHOLOGY. *Educ:* Okla State Univ, BS, 48, MS, 49; Univ Calif, PhD(plant path), 66. *Prof Exp:* From asst prof to prof bot & plant path, Okla State Univ, 71-85; RETIRED. *Mem:* Am Phytopath Soc. *Res:* Diseases of peanuts; plant virology. *Mailing Add:* PO Box 877 Rockport TX 78382

WADSWORTH, FRANK H, b Chicago, Ill, Nov 26, 15; m 41, 84; c 5. FORESTRY. *Educ:* Univ Mich, BSF & MF, 37, PhD(forestry), 50. *Prof Exp:* STAFF MEM, INST TROP FORESTRY, US FOREST SERV, 42- *Concurrent Pos:* Consult 20 trop countries, 49-80. *Honors & Awards:* Fernow Award, Am Forestry Asn, 73; Gulf Cons Award, 85. *Mem:* Soc Am Foresters; Am Forestry Asn; Int Soc Trop Foresters. *Res:* Multiple forest land use and management, silviculture, growth and yield of naturally regenerated forests and timber plantations in the humid tropics. *Mailing Add:* Sacarello 1016 Urbanizacion San Martin Rio Piedras PR 00924

WADSWORTH, HARRISON M(ORTON), b Duluth, Minn, Aug 20, 24; m 50; c 2. INDUSTRIAL ENGINEERING, STATISTICS. *Educ:* Ga Inst Technol, BInd Eng, 50, MS, 55; Case Western Reserve Univ, PhD(statist), 60. *Prof Exp:* Indust engr, Steel Heddle Mfg Co, SC, 52-54; qual control engr, Nat Carbon Co, 54-56; asst prof mech eng, Mich State Univ, 56-57; instr & res assoc statist, Case Western Reserve Univ, 57-60; from assoc prof to prof indust eng, Ga Inst Technol, 64-91; CONSULT, 91- *Concurrent Pos:* Consult statist, Lockheed-Ga Co, 66-70; NSF grant, exp design course for eng profs, Univ Wis, 66; Orgn Econ Coop & Develop consult, Mid East Tech Univ, Ankara, 67-68; ed, J Quality Technol, 79-82; consult, qual control & statist. *Honors & Awards:* Brumbaugh Award, Am Soc Qual Control, 71, Austin Bonis Award, 85, Howard P Jones Medal, 86; Shewhart Medal, 89. *Mem:* Fel Am Soc Qual Control; sr mem Inst Indust Eng; Am Statist Asn; Sigma Xi; AAAS. *Res:* Economics of statistical sampling plans; design of experiments; quality control; reliability. *Mailing Add:* 660 Valley Green Dr NE Atlanta GA 30342

WADSWORTH, JEFFREY, b Hamburg, WGer, May 12, 50; US citizen; m 76; c 2. METALLURGY. *Educ:* Sheffield Univ, Eng, BMetall, 72, PhD(metall), 75. *Hon Degrees:* DMet, Sheffield Univ, 90. *Prof Exp:* Fel metall, Stanford Univ, 76-78, res assoc, 78-81; MGR, METALL DEPT, LOCKHEED RES & DEVELOP DIV, 81- *Concurrent Pos:* Ed, Sheffield Univ Metall Soc J, 74; consult prof, Dept Mat Sci, Stanford Univ, 81- *Honors & Awards:* Metallurgica Aparecida Medal; Brunton Medal. *Mem:* Am Inst Mining, Metall & Petrol Engrs; Fel Am Soc Metals; Am Ceramic Soc; Mat Res Soc. *Res:* Research into the physical and mechanical properties of metals and non-metals with an emphasis on materials of significant technological importance. *Mailing Add:* Lockheed Res & Develop Div 3251 Hanover St B204 0/93-10 Palo Alto CA 94304

WADSWORTH, MILTON E(LLIOT), b Salt Lake City, Utah, Feb 9, 22; m 43; c 6. ELECTROCHEMISTRY, EXTRACTIVE METALLURGY. *Educ:* Univ Utah, BS, 48, PhD(metall), 51. *Hon Degrees:* Dr, Univ Liege Belgium 79; Deng, Colo Sch Mines, 91. *Prof Exp:* From instr to prof, Univ Utah, 48-83, chmn dept, 55-66 & 74-76, assoc dean, Col Mines & Mineral Industs, 76-83, dean, Col Mines & Earth Sci, 83-91, DISTINGUISHED PROF METALL

ENG, UNIV UTAH, 83- *Concurrent Pos:* Milton E Wadsworth fel, Univ Utah, 65-, dir, Utah Mining & Minerals Resources Res Inst, 79-80; vis prof, dept metall, Univ BC, 67-68 & US Steel Res Lab, Pa; lectr, Extractive Metall Div, Am Inst Mining, Metall & Petrol Engrs, 69; consult, numerous insts; sci adv, US Bur Mines Metall Res Ctr, Salt Lake City Sta, 72-79; bd dirs, Am Chemet Corp, 83-, Tech Res Assocs, 85-, Utah Biores Inc, 85-, Utah Geol & Mineral Surv, 88-91; mem, Comt Mineral Resources, Nat Asn State Univs & Land-Grant Cols, 82-83, Future Iron & Steel Prod Adv Comt, Geneva Steel, 89. *Honors & Awards:* Warren Lectr, Sch Mines & Metall, Univ Minn, 62; James Douglas Gold Medal Award, Am Inst Mining, Metall & Petrol Engrs, 78, Mineral Indust Educ Award, 81; Henry Crumb Lectr, 79; Antoine M Gaudin Award, Soc Mining Engrs, Am Inst Mining, Metall, & Petrol Engrs, 84 & Educ Award, Metall Soc, 89. *Mem:* Nat Acad Eng; Am Inst Mining, Metall & Petrol Engrs (pres, 91); Am Soc Eng Educ; Can Metals Soc; fel Am Soc Metals Int; Am Chem Soc; Sigma Xi; Electrochem Soc; Can Inst Mining & Metall. *Res:* Surface chemistry of mineral systems in mineral dressing and extractive metallurgy processes; hydrometallurgy; intermediate temperature processing such as roasting, decomposition and reduction; application of reaction rate kinetics in determination of mechanisms in extractive metallurgy processes and electrochemistry as applied to metals extraction; author of numerous publications. *Mailing Add:* Dept Metall & Metall Eng Univ Utah 412 WBB Salt Lake City UT 84112

WADSWORTH, WILLIAM BINGHAM, b Cortland, NY, Dec 4, 34; m 62; c 2. PETROLOGY OF GRANITES, STATISTICS IN GEOLOGY. *Educ:* Brown Univ, AB, 57; Northwestern Univ, MS, 62, PhD(geol), 66. *Prof Exp:* Asst prof, Univ SDak, 63-66; from asst prof to assoc prof, Idaho State Univ, 66-72; assoc prof, 72-78, PROF GEOL, WHITTIER COL, 78- *Concurrent Pos:* NSF Sci fac fel, Pomona Col, 71-72; chmn, Geol Dept, Whittier Col, 73-85 & 87-90, interim dean fac, 85-86. *Mem:* Am Asn Geol Teachers; Am Geophys Union; Am Mineral Soc; fel Geol Soc Am; Int Asn Math Geol; Soc Econ Paleontologists & Mineralogists. *Res:* Petrology of the Cornelia porphyry-copper stock at Ajo, Arizona; methods to quantify textures in granitic plutons; x-ray diffraction methods in modal analysis of granites; computer applications in geology. *Mailing Add:* Dept Geol Whittier Col 1306 Philadelphia St Whittier CA 90608

WADSWORTH, WILLIAM STEELE, JR, b Hartford, Conn, May 6, 27; m 56; c 4. ORGANIC CHEMISTRY. *Educ:* Trinity Col, Conn, BS, 50, MS, 52; Pa State Univ, PhD(chem), 56. *Prof Exp:* Res chemist, Rohm & Haas Co, 56-63; assoc prof, 63-68, PROF CHEM, SDAK STATE UNIV, 68- *Mem:* Am Chem Soc; Sigma Xi. *Res:* New reactions and mechanisms in organic chemistry; heterocyclic and organophosphorus chemistry. *Mailing Add:* 1938 Victory St Brookings SD 57006-3546

WADT, WILLARD ROGERS, b Bayonne, NJ, Jan 6, 49. THEORETICAL CHEMISTRY. *Educ:* Williams Col, BA, 70; Calif Inst Technol, PhD(chem), 75. *Prof Exp:* Sr res chemist, Mound Lab, Monsanto Res Corp, 74-76; mem staff, Los Alamos Nat Lab, 76-81, dep group leader & proj mgr, 81-85; vpres, BioDesign, Inc, 86; DEP DIV LEADER, LOS ALAMOS NAT LAB, 87- *Mem:* Am Chem Soc; AAAS. *Res:* Ab initio electronic structure theory of molecules; electronic transition lasers; molecular photochemistry; mesic molecules; atmospheric chemistry. *Mailing Add:* CLS-DO/MS-J 563 Los Alamos Nat Lab Los Alamos NM 87545

WAEHNER, KENNETH ARTHUR, analytical chemistry; deceased, see previous edition for last biography

WAELSCH, SALOME GLUECKSOHN, b Ger, Oct 6, 07; nat US; m 43; c 2. GENETICS, DEVELOPMENTAL BIOLOGY. *Educ:* Univ Freiburg, PhD(zool), 32. *Prof Exp:* Res assoc & lectr zool, Columbia Univ, 36-55; from assoc prof to prof anat, Albert Einstein Col Med, 55- 63, chmn dept genetics, 63-76, prof, 58-88, DISTINGUISHED EMER PROF GENETICS, ALBERT EINSTEIN COL MED, 88- *Mem:* Nat Acad Sci; Am Acad Arts & Sci; Am Asn Anatomists; Genetics Soc Am; Soc Develop Biol; Am Soc Zoologists. *Res:* Developmental and mammalian genetics; role and control of genes in differentiation. *Mailing Add:* Dept Molecular Genetics Albert Einstein Col Med 1300 Mars Park Ave Bronx NY 10461

WAESCHE, R(ICHARD) H(ENLEY) WOODWARD, b Baltimore, Md, Dec 20, 30; m 57; c 2. AEROSPACE SCIENCE. *Educ:* Williams Col, BA, 52; Princeton Univ, MA, 62, PhD(aerospace & mech sci), 65. *Prof Exp:* Intermediate scientist, Redstone Arsenal Res Div, Rohm and Haas Co, Ala, 54-59, scientist, 64-66; asst in res aerospace sci, Princeton Univ, 61-64, res aide, 64; sr res engr, Propulsion Applications, United Technol Res Ctr, 66-81; PRIN SCIENTIST TECHNOL, ATLANTIC RES CORP, 81- *Concurrent Pos:* Consult, Goodyear Aircraft Corp, 59, Princeton Univ, 64-65, Nat Res Coun, 85-86, Defense Advan Res Projs Agency, 88- & Directed Technologies, Inc, 89-; assoc ed, J Spacecraft & Rockets, Am Inst Aeronaut & Astronaut, 75-80, ed-in-chief, 80-86, J Propulsion & Power, 86- *Mem:* Am Phys Soc; fel Am Inst Aeronaut & Astronaut; Combustion Inst. *Res:* Combustion related to chemical propulsion, rockets, ramjets ducted rockets; management of propulsion-related programs; optical spectroscopy; ablation heat transfer; fuel-spray atomization and combustion in a ramjet environment. *Mailing Add:* Atlantic Res Corp 5945 Wellington Rd Gainesville VA 22065-1699

WAFFLE, ELIZABETH LENORA, b Marion, Iowa, Feb 14, 38. PARASITOLOGY, INVERTEBRATE ZOOLOGY. *Educ:* Cornell Col, BA, 60; Univ Iowa, MS, 63; Iowa State Univ, PhD(parasitol), 67. *Prof Exp:* Assoc prof biol, Armstrong State Col, 66-67; asst prof, Iowa Wesleyan Col, 67-68; asst prof, 68-77, ASSOC PROF BIOL, EASTERN MICH UNIV, 77- *Concurrent Pos:* Consult, Parasitol Prog, Ann Arbor Biol Ctr, 70-82. *Mem:* Am Soc Parasitol; Am Mosquito Control Asn; Wildlife Dis Asn. *Res:* Dog heartworm and other parasites of dogs; parasites of fish; mosquitoes feeding habits in relation to disease transmission; marine biology; entomology. *Mailing Add:* Dept Biol Eastern Mich Univ Ypsilanti MI 48197

WAGENAAR, EMILE B, b Poerwokerto, Indonesia, Apr 7, 23; Can citizen; m 54; c 3. CELL BIOLOGY, GENETICS. *Educ:* St Agr Univ, Wageningen, Ir, 54; Univ Alta, PhD(genetics), 58. *Prof Exp:* Nat Res Coun Can res fel, 58-60; res scientist, Genetics & Plant Breeding Res Inst, Can Dept Agr, 60-65; res zoologist, Univ Calif, Berkeley, 65-67; assoc prof, 67-69, PROF BIOL, UNIV LETHBRIDGE, 69- *Concurrent Pos:* Vis assoc & sessional lectr, Ottawa, Ont, 64-65. *Mem:* Genetics Soc Am; Genetics Soc Can. *Res:* Cytogenetics and evolution of species in Triticum and Hordeum; chemistry, ultrastructure and behavior of chromosomes in nucleus and during cell division. *Mailing Add:* Dept Biol Sci Univ Lethbridge 4401 Univ Dr Lethbridge AB T1K 3M4 Can

WAGENAAR, RAPHAEL OMER, b Spokane, Wash, Jan 9, 16; wid; c 2. DAIRY BACTERIOLOGY. *Educ:* Wash State Univ, BS, 42, MS, 47; Univ Minn, PhD(dairy bact), 51. *Prof Exp:* Asst dairy bact, Univ Minn, 49-51; res assoc, Food Res Inst, Univ Chicago, 51-56; sect leader microbiol, Food Develop Dept, 56-62, res assoc microbiol, food develop activ, Tech Ctr, 62-83, CONSULT, GEN MILLS, INC, 84- *Mem:* Am Soc Microbiol; Am Dairy Sci Asn; Inst Food Technologists; Sigma Xi. *Res:* Bacterial food poisoning; lactic acid bacteria; effect of irradiation on bacterial spores and toxins; psychrophilic bacteria causing food spoilage. *Mailing Add:* 1493 Fulham St St Paul MN 55108-1438

WAGENBACH, GARY EDWARD, b Barron, Wis, Mar 24, 40; m 60; c 3. PARASITOLOGY, ZOOLOGY. *Educ:* Univ Wis-River Falls, BS, 62; Univ Wis-Madison, MS, 64, PhD(zool), 68. *Prof Exp:* NIH proj assoc, Univ Wis-Madison, 68-69; from asst prof to assoc prof, 69-82, PROF BIOL, CARLETON COL, 82- *Concurrent Pos:* Vis prof, Stanford Univ, 80; prin investr, Marine Biol Lab, Woods Hole, 83-88; Carleton Foreign Study Prog, Australia; prin investr, Univ Wash, Seattle, 89- *Mem:* AAAS; Sigma Xi; Am Soc Zoologists. *Res:* Biology of parasites, especially digenetic trematodes and gregarines; analysis of gut structure and function. *Mailing Add:* Dept Biol Carleton Col Northfield MN 55057

WAGENET, ROBERT JEFFREY, b Pittsburg, Calif, Aug 10, 50. SOIL PHYSICS, SOIL CHEMISTRY. *Educ:* Univ Calif, Davis, BS, 71, PhD(soil sci), 75; Univ Okla, MS, 72. *Prof Exp:* Staff res assoc, Univ Calif, Davis, 73-74, water scientist, 74-75; from asst prof to assoc prof soil sci, Utah State Univ, 76-82; PROF SOIL, CROP & ATMOSPHERIC SCI, CORNELL UNIV, ITHACA, 82-, CHMN DEPT, 87- *Honors & Awards:* Honor Award, Soil Conserv Soc Am, 89. *Mem:* Am Soc Agron; Am Geophys Union; Sigma Xi; fel Soil Sci Soc Am; Int Soil Sci Soc. *Res:* Simulation modeling of soil water and solutes including description of transient nitrogen and pesticide fluxes under field conditions; utilization and improvement of salt-affected soils and saline waters. *Mailing Add:* Dept Soil Crop & Atmospheric Sci Cornell Univ 235 Emerson Hall Ithaca NY 14853

WAGENKNECHT, BURDETTE LEWIS, b Cotter, Iowa, Sept 9, 25; m 51; c 4. BOTANY. *Educ:* Univ Iowa, BA, 48, MS, 54; Univ Kans, PhD(bot), 58. *Prof Exp:* Instr biol & phys sci, Franklin Col, 54-55; asst bot, Univ Kans, 57-58; hort taxonomist, Arnold Arboretum, 58-61; from asst prof to assoc prof biol, Norwich Univ, 61-68; PROF BIOL & HEAD DEPT, WILLIAM JEWELL COL, 68- *Mem:* AAAS; Am Inst Biol Sci; Am Boxwood Soc. *Res:* Floristics of Washington County, Iowa; Heterotheca; taxonomy of cultivated wood plants; registration of cultivars in the genus Buxus. *Mailing Add:* Dept Biol William Jewell Col Liberty MO 64068

WAGENKNECHT, JOHN HENRY, b Washington, Iowa, Jan 30, 39; m 60; c 3. ORGANIC CHEMISTRY, ELECTROCHEMISTRY. *Educ:* Monmouth Col, AB, 60; Univ Iowa, PhD(chem), 64. *Prof Exp:* Sr res chemist, 64-70, res specialist, 70-74, SCI FEL, MONSANTO CO, 74- *Concurrent Pos:* Div Ed, J Electrochem Soc, 84; div secy-treas, Electrochem Soc, 77-79, vchmn, 79-81, chmn, 81-83. *Mem:* Electrochem Soc; Int Soc Electrochem; Am Chem Soc. *Res:* Synthesis of organic chemicals by electrochemistry; electroanalytical chemistry; electrical discharge chemistry; scale-up of electro-organic processes; sensors. *Mailing Add:* Monsanto Co 800 N Lindbergh Blvd St Louis MO 63167

WAGER, JOHN FISHER, b Glendale, Calif, Jan 22, 53; m 75; c 1. ELECTRICAL ENGINEERING. *Educ:* Ore State Univ, BS, 77; Colo State Univ, MS, 78, PhD(elec eng), 81. *Prof Exp:* Mem tech staff, Hughes Res Lab, 82-84; asst prof, 84-89, ASSOC PROF ELEC ENG, ORE STATE UNIV, 89- *Mem:* Am Vacuum Soc; Inst Elec & Electronics Engrs. *Res:* Electrical characterization, modeling and exploratory materials development of alternating current thin film electroluminescent devices for flat panel displays; electrical characterization and atomistic thermodynamic modeling of compound semiconductor defects. *Mailing Add:* Dept Elec & Computer Eng Ore State Univ Corvallis OR 97331

WAGGENER, ROBERT GLENN, b Benton, Ky, June 12, 32; m 59; c 2. MEDICAL PHYSICS, BIOPHYSICS. *Educ:* Univ Tex, Austin, BA, 54, MA, 63; Univ Tex M D Anderson Hosp & Tumor Inst Houston, PhD(biophys), 67; Am Bd Radiol, cert, 72. *Prof Exp:* Res asst physics, Nuclear Physics Lab, Balcones Res Ctr, Univ Tex, Austin, 60-61; pres, Nucleonics Res & Develop Corp, Tex, 61-63; radiol health specialist, Tex State Health Dept, 63-64; Nat Cancer Inst fel physics, Univ Tex M D Anderson Hosp & Tumor Inst Houston, 67-68; asst prof, 68-72, ASSOC PROF RADIOL, UNIV TEX MED SCH SAN ANTONIO, 72-, ASSOC PROF DIAG & ROENTGENOL, 77- *Concurrent Pos:* Consult, Brooke Army Med Ctr, Ft Sam Houston, Tex, 71- *Mem:* Am Asn Physicists in Med; Biophys Soc; Am Col Radiol; Soc Nuclear Med; Radiol Soc NAm. *Res:* Measurement of x-ray spectra; calculation of information content in diagnostic x-rays; dosimetry and measurement of ionizing radiation; computerized tomography. *Mailing Add:* Dept Radiol Univ Tex Med Sch Radiol Med Phys Div San Antonio TX 78284

WAGGENER, RONALD E, b Green River, Wyo, Oct 6, 26; m 48; c 4. RADIOLOGY. *Educ:* Univ Nebr, BS, 49, MS, 53, MD, 54, PhD, 57; Am Bd Radiol, dipl, 59. *Prof Exp:* ASSOC PROF RADIOL, UNIV NEBR MED CTR, OMAHA, 58- *Concurrent Pos:* Radiotherapist, Methodist Hosp, 59- *Mem:* Am Asn Cancer Res; Radiol Soc NAm; Royal Soc Med; Brit Inst Radiol; fel Am Col Radiol. *Res:* Biological effects of radiation stressing the hematological effects of ionizing rays. *Mailing Add:* 1227 S 109th St Omaha NE 68144

WAGGENER, THOMAS BARROW, b Alvin, Tex, Apr 15, 51; c 2. PEDIATRICS, BIOENGINEERING. *Educ:* Harvard Univ, PhD(bioeng), 79. *Prof Exp:* Fel, Cardiovasc Res Inst, Univ Calif San Francisco, 79-81; instr pediat & eng, Harvard Med Sch, 81-85; ASST PROF PEDIAT, NEW ENG MED CTR, 85- *Mem:* Am Physiol Soc; Inst Elec & Electronics Engrs. *Res:* Physiological control system analysis; respiratory control in infants and adults; apnea of prematurity and sudden infant death syndrome; physiological effects of hypoxia. *Mailing Add:* Dept Pediat Box 45 New Eng Med Ctr 750 Washington St Boston MA 02111

WAGGENER, THOMAS RUNYAN, b Indianapolis, Ind, July 20, 38; m 64; c 2. FOREST ECONOMICS, POLICY ANALYSIS. *Educ:* Purdue Univ, BSF, 62; Univ Wash, MF, 63, MA, 65, PhD(forest econ), 66. *Prof Exp:* Asst prof, 67-71, chmn mgt & soc sci div, 72-75, assoc prof, 71-78, PROF FOREST ECON, COL FOREST RESOURCES, UNIV WASH, 78-, ASSOC DEAN INSTR, 78- *Concurrent Pos:* Assoc coordr course in trop forestry, Orgn Trop Studies, Inc, Costa Rica & Honduras, 68 & 71; economist & analyst, Pub Land Law Rev Comn, DC, 68-69. *Mem:* Am Econ Asn; Soc Am Foresters; Sigma Xi. *Res:* Natural resources economics and analysis of economic impact of resource management policies; regional economic analysis; industrial organization and market structure; forest policy. *Mailing Add:* Col Forest Resources Univ Wash Seattle WA 98195

WAGGENER, WILLIAM COLE, b Princeton, Ky, Feb 5, 17; m 51; c 2. PHYSICAL CHEMISTRY. *Educ:* Centre Col, Ky, AB, 39; Univ Buffalo, PhD(inorg chem), 49. *Prof Exp:* Asst chem, Univ Buffalo, 39-41; supvr, Ammonia Oxidation Plant, Lake Ont Ord Works, 41-42; chemist, Res Lab, Nat Carbon Co, 43-45; MEM STAFF, OAK RIDGE NAT LAB, 49- *Mem:* AAAS; Am Chem Soc; Sigma Xi. *Res:* Chemistry of thorium; inorganic complexes; coordination properties of thiocyanates; solution spectrophotometry over wide ranges of temperature and pressure; near infrared spectroscopy of pure substances in condensed states; liquid effluents from light water-cooled nuclear reactors; heterogeneous calatysis. *Mailing Add:* 309 Camelot Ct Knoxville TN 37922

WAGGLE, DOYLE H, b Osborne, Kans, Aug 11, 39; m 60; c 2. CEREAL CHEMISTRY. *Educ:* Ft Hays Kans State Col, BS, 61; Kans State Univ, MS, 63, PhD(milling indust), 66. *Prof Exp:* Res asst feed technol, Kans State Univ, 65-66, res assoc, 66-67; process res chemist, 67-68, mgr process res, 68-72, DIR RES & DEVELOP, VENTURE MGT, RALSTON PURINA CO, 72-, DIV VPRES, 78- *Mem:* Am Asn Cereal Chemists; Inst Food Technologists; Am Chem Soc. *Res:* Chemistry of processes related to foods and feeds. *Mailing Add:* 348 Rieth Terr St Louis MO 63122

WAGGONER, EUGENE B, GEOLOGY ENGINEERING. *Educ:* Univ Calif, Los Angeles, BS, 37, MS, 39. *Prof Exp:* Petrol engr & geologist, Tidewater-Assocs Oil Co Calif & Super Oil Co Calif, 40-45; eng geologist, US Bur Reclamation, 45-54; eng geologist consult, 54-60; exec vpres & co-mgr, Woodward-Clyde & Assocs, Denver, Colo, 60-67, chief exec officer, 68-71, pres, 68-73; RETIRED. *Concurrent Pos:* Chmn, Comt Fels, Am Consult Engrs Coun, 77-78; consult eng geol, 73- *Mem:* Nat Acad Eng; Am Consult Engrs Coun (pres, 66-67); hon mem Asn Eng Geologists; fel Am Soc Civil Engrs; fel Asn Soil & Found Engrs; Sigma Xi. *Res:* Author of various publications. *Mailing Add:* 336 Seawind Dr Vallejo CA 94590

WAGGONER, JACK HOLMES, JR, b Pittsburgh, Pa, Sept 4, 27; m 61. THEORETICAL PHYSICS. *Educ:* Ohio State Univ, BS, 49, PhD(physics), 57. *Prof Exp:* Asst photochem, Res Found, Ohio State Univ, 49-53, asst physics, 53-55 from inst to asst prof & res assoc 58-59, from prof assoc supvr to prof supvr, 59; asst prof, Univ Calif, Riverside, 59-61; asst prof, 61-65, ASSOC PROF PHYSICS, HARVEY MUDD COL, 65- *Concurrent Pos:* Vis assoc, Calif Inst Technol, 67-68. *Mem:* AAAS; Am Phys Soc; Am Inst Physics; Am Asn Physics Teachers. *Res:* Methods of theoretical physics; theory of molecular spectroscopy. *Mailing Add:* Dept Physics Harvey Mudd Col Claremont CA 91711

WAGGONER, JAMES ARTHUR, b West Lafayette, Ind, Dec 31, 31; m 53; c 4. NUCLEAR PHYSICS. *Educ:* Univ Ill, BS, 53; Cornell Univ, PhD(exp physics), 60. *Prof Exp:* Physicist, Lawrence Radiation Lab, Univ Calif, 60-70, Physics Int Co, Calif, 71 & Maxwell Labs, 71; PHYSICIST, SCHLUMBERGER WELL SERVS, 71- *Concurrent Pos:* Consult, NASA. *Mem:* Am Phys Soc. *Res:* Lunar and planetary surface composition analysis using neutron inelastic scattering; Van Allen zone charged particles; geophysical instrumentation. *Mailing Add:* Schlumberger Well Servs PO Box 2175 Houston TX 77252-2175

WAGGONER, PAUL EDWARD, b Appanoose Co, Iowa, Mar 29, 23; m 45; c 2. CLIMATOLOGY. *Educ:* Univ Chicago, SB, 46; Iowa State Col, MS, 49, PhD, 51. *Prof Exp:* From asst to assoc plant pathologist, 51-56, chief dept soils & climat, 56-69, vdir, 69-71, dir, 72-87, DISTINGUISHED SCIENTIST, CONN AGR EXP STA, 87- *Concurrent Pos:* Guggenheim fel, 63; lectr, Yale Univ, 62- *Honors & Awards:* Am Meteorol Soc Award, 67. *Mem:* Nat Acad Sci; fel Am Phytopath Soc; Am Meteorol Soc; Ecol Soc Am; Am Soc Plant Physiologists. *Res:* Agriculture; plant pathology; effect of environment on plants, especially plant diseases; water resources. *Mailing Add:* Conn Agr Exp Sta PO Box 1106 New Haven CT 06504-1106

WAGGONER, PHILLIP RAY, b Parkersburg, WVa, Apr 4, 43; m 67; c 2. DEVELOPMENTAL BIOLOGY. *Educ:* WVa Univ, BS, 65, MS, 68, PhD(genetics & develop biol), 72. *Prof Exp:* Instr biol, Fairmont State Col, 68 & WVa Univ, 68-69; ASST PROF ANAT, WAYNE STATE UNIV, 72- *Mem:* Am Asn Anat; AAAS; Sigma Xi. *Res:* Development of the vertebrate eye. *Mailing Add:* Anat Med Sch Univ Miami Sch Med PO Box 016960 Miami FL 33101

WAGGONER, RAYMOND C, b Louisville, Ky, Feb 13, 30; m 54; c 3. CHEMICAL ENGINEERING. *Educ:* Univ Louisville, BChE, 52; Tex A&M Univ, MEng, 61, PhD(chem eng), 64. *Prof Exp:* Prod control engr & sr prod control engr, Dow Chem Co, 56-61; res asst chem engr, Tex A&M Univ, 61-64; sr engr, Humble Oil & Refining Co, 64-65; assoc prof, 65-79, PROF CHEM ENG, UNIV MO, ROLLA, 79- *Concurrent Pos:* Res engr, Savannah River Lab, E I du Pont de Nemours & Co Inc, 81-82. *Mem:* Am Inst Chem Engrs; Am Chem Soc; Instrument Soc Am. *Res:* Distributed control micro processor based instruments for laboratory pilot scale and process plant operations; advanced algorithms for on line control applications particularly stage wise separation processes such as distillation and air stripping. *Mailing Add:* Dept Chem Eng Univ Mo Rolla MO 65401

WAGGONER, RAYMOND WALTER, b Carson City, Mich, Aug 2, 01; m 30; c 2. PSYCHIATRY, NEUROLOGY. *Educ:* Univ Mich, MD, 24; Univ Pa, ScD, 30. *Prof Exp:* Intern, Harper Hosp, Detroit, 24-25; resident, Philadelphia Orthop Hosp & Infirmary Nerv Dis, 25-26; lab intern, Pa Hosp, Philadelphia, 26; from asst prof to assoc prof neurol, Med Sch, 29-36, from asst neurologist to neurologist, Univ Hosp, 29-36, chmn dept psychiat, Med Sch, 37-70, prof psychiat & dir neuropsychiat inst, 37-70, EMER PROF PSYCHIAT & EMER DIR NEUROPSYCHIAT INST, MED SCH, UNIV MICH, ANN ARBOR, 70- *Concurrent Pos:* Consult, spec comt rights ment ill, Am Bar Found, 59-66, indust personnel security, Dept Defense, 48-60, Surgeon-Gen, US Army, Selective Serv Syst, Peace Corps, Vet Admin & Social Security Admin, 43-50; mem med adv bd, Social Security Admin, 65-; adv comt, Nat Paraplegic Found; mem test comt psychiat, Nat Bd Med Examr, 65-70; mem, Res Socs Coun, 70-73; vpres bd trustees, Mich Inst Pastoral Care; consult & bd mem, Reproductive Biol Res Found, 70-; consult, Mich State Dept Ment Health, 74-; distinguished vis prof psychiat, Univ Louisville, 74- *Honors & Awards:* E B Bowis Award, Am Col Psychiat, 68. *Mem:* Fel AAAS; Am Acad Psychoanal; fel Am Col Psychiat (vpres, 64-65, pres elect, 65-66, pres, 66-67); Am Geriat Soc; fel Am Psychiat Asn (vpres, 60-61, pres elect, 68-69, pres, 69-70). *Res:* Personality studies in chorea; the convulsive state; myopathies; psychotherapy. *Mailing Add:* 3333 Geddes Rd Ann Arbor MI 48105

WAGGONER, WILBUR J, b Sutherland, Iowa, May 28, 24; m 46; c 3. MATHEMATICS. *Educ:* Buena Vista Col, BA, 47; Drake Univ, MSE, 50; Univ Wyo, EdD, 56. *Prof Exp:* Prin, coach & teacher high sch, 47-51; supt twp sch, 51-55; asst, Univ Wyo, 55-56; from asst prof to assoc prof, 56-62, actg dean, Sch Grad Studies, 73-75, PROF MATH, CENT MICH UNIV, 62- *Mem:* Am Statist Asn; Nat Coun Teachers Math. *Res:* Statistics. *Mailing Add:* Dept Math Cent Mich Univ Mt Pleasant MI 48858

WAGGONER, WILLIAM CHARLES, b Alma, Mich, Jan 18, 36; m 80; c 8. PHYSIOLOGY, TOXICOLOGY. *Educ:* Hope Col, AB, 58; Mich State Univ, MS, 61, PhD(physiol), 63. *Prof Exp:* Teaching fel physiol & pharmacol, Med Sch, Mich State Univ, 60-63; res physiologist, Colgate-Palmolive Res Ctr, 63-64, res projs coordr oral health, 64-66; asst dir med res, Unimed, Inc, 66-70; assoc dir med serv & govt affairs, Wallace Pharmaceut, NJ, 70-74; mgr med & regulatory affairs, 74-80, dir govt affairs, Baby Prod Co, Johnson & Johnson, 80-; AT R & D THOMPSON MED COL, NEW YORK, NY. *Concurrent Pos:* Consult fac mem, Inst Clin Toxicol, 74- *Mem:* Am Physiol Soc; fel Am Acad Clin Toxicol; Soc Toxicol. *Res:* Clinical toxicology and pharmacology. *Mailing Add:* 60F Apgar Way Lebanon NJ 08833

WAGGONER, WILLIAM HORACE, b Ravenna, Ohio, June 8, 24; m 46; c 1. INORGANIC CHEMISTRY. *Educ:* Hiram Col, AB, 49; Western Reserve Univ, MS, 51, PhD(chem), 53. *Prof Exp:* Asst chem, Hiram Col, 48-49; asst prof, 52-59, ASSOC PROF INORG CHEM, UNIV GA, 59- *Mem:* Am Chem Soc. *Res:* Solubilities in non-aqueous systems; history of chemistry; spectral properties of inorganic materials. *Mailing Add:* 160 University Dr Athens GA 30605-1436

WAGH, MEGHANAD D, b Bombay, India, Sept 23, 48; m 75; c 2. PARALLEL PROCESSING, DIGITAL SYSTEM DESIGN. *Educ:* Indian Inst Technol, Bombay, BTech, 71, PhD(elec eng), 77. *Prof Exp:* Asst prof elec eng, Old Dominion Univ, 80-84; ASSOC PROF COMPUTER ENG, LEHIGH UNIV, 84- *Mem:* Inst Elec & Electronics Engrs Computer Soc; Inst Elec & Electronics Engrs Speech & Signal Processing Soc. *Res:* Interdependence between parallel algorithms and parallel architectures; optimal parallel algorithms for realistic architectural overheads; new architectures matching algorithmic skeletons. *Mailing Add:* Dept Computer Sci & Elec Eng Lehigh Univ Bethleham PA 18015

WAGH, PREMANAND VINAYAK, b Sadashivgad, India, July 9, 34; US citizen; c 3. BIOCHEMISTRY. *Educ:* Univ Bombay, BSc Hons, 54, MSc, 56; Southern Ill Univ, Carbondale, MS, 62; Univ Minn, Minneapolis, PhD(animal sci, biochem), 65. *Prof Exp:* Res assoc biochem, Univ Ill Med Ctr, 65; res assoc, State Univ NY Buffalo, 65-68, instr, 68-69; res biochemist, Vet Admin Hosp, 69-73, chief connective tissue res, 75-77; asst prof biochem, Med Ctr, Univ Ark, Little Rock, 69-76; res assoc prof, Div Cell & Molecular Biol, Dept Biol Sci, 79-87, RES ASSOC PROF, DEPT BIOCHEM PHARMACOL, STATE UNIV NY, AMHERST, 88- *Mem:* Soc Complex Carbohydrates; Am Soc Biochem & Molecular Biol. *Res:* Cardiovascular glycoproteins in Atherosclerosis; structure and function of connective tissue glycoproteins; glycosidases. *Mailing Add:* Dept Biochem Pharmacol 308 Hochstetter Hall State Univ NY Amherst NY 14260

WAGLE, GILMOUR LAWRENCE, b Staten Island, NY, Nov 17, 22; m 49; c 3. PHARMACOLOGY. *Educ:* Wagner Col, BS, 50; Rutgers Univ, MS, 56; Princeton Univ, MA, 59, PhD(biol), 60. *Prof Exp:* Pharmacologist, Res Dept, Ciba Pharmaceut Prod, Inc, NJ, 48- 61; sr pharmacologist, Chas Pfizer & Co, 61-64; asst to dir res admin, Toxicol Res Sect, 64-72, asst dir, Med Controls Br, Off Govt Controls, 72-77, dir, Sci Compliance Qual Assurance Dept, 77-79, res toxicologist, Environ Serv Div, 80, mgr med Regulatory Surveillance, Med Res Div, Lederle Labs, Am Cyanamid Co, 81-85,; RETIRED. *Concurrent Pos:* Drug Regulatory Affairs, Qual Assurance Drug Res & Develop; consult. *Mem:* Soc Toxicol; Fedn Am Scientists. *Res:* Cardiovascular, renal and central nervous system pharmacology; government regulatory affairs; toxicology; quality assurance. *Mailing Add:* 20 Starboard Tack Keowee Key Salem SC 29676

WAGLE, ROBERT FAY, b Jamestown, NDak, Sept 3, 16. FORESTRY, BOTANY. *Educ:* Univ Minn, BS, 40; Univ Wash, MF, 55; Univ Calif, PhD(bot), 58. *Prof Exp:* Asst forestry, Univ Wash, 47-48; logging engr, Shasta Plywood Co, Calif, 48-49; sr lab asst, Univ Calif, 49-54; asst, Calif Forest & Range Exp Sta, US Forest Serv, 54-57; assoc prof watershed mgt, 57-69, PROF WATERSHED MGT & WATERSHED SPECIALIST, UNIV ARIZ, 69- *Concurrent Pos:* Consult fire & silvicult, Univ Nev, 64, wood technol, Tucson Power, Indust Res Inst, World Bank, forest res, Yeman, Arabia, pulp wood availability, SW Wilburt Assoc, fire & forest regeneration & forestry educ, Ft Apache Indian Reservation, Bur Indian Affairs, Az,. *Mem:* Fel AAAS; Soc Am Foresters; Ecol Soc Am. *Res:* Silvics, genetics, ecology and silviculture; plant variation and its relationship to environment; nutrient and water relationships of wildland plants; effects of fire on plants and their environments; growth and variation in containerized pine seedling; effect of nutrients and mycorrhizae on containerized pine seedling growth, and root/shoot ratios and field survival; tree species for ornamentals and Christmas trees in the southwest. *Mailing Add:* Sch Renew Nat Resource Univ Ariz Tucson AZ 85721

WAGLE, SHREEPAD R, b Bombay, India, Jan 1, 31; nat US; m 62; c 3. PHARMACOLOGY. *Educ:* Univ Bombay, BS, 52, MS, 55; Univ Ill, PhD(biochem, nutrit), 59. *Prof Exp:* Res chemist, Haffkine Inst, India, 52-55; asst, Univ Ill, 55-59; res chemist, Sigma Lab, India, 60; from res assoc to asst prof, 60-65, assoc prof, 65-68, dir grad progs, 68-80 PROF PHARMACOL, SCH MED, IND UNIV, INDIANAPOLIS, 68- *Concurrent Pos:* Fulbright Award. *Mem:* Am Cancer Soc; Am Inst Nutrit; Soc Exp Biol & Med; Am Diabetes Asn; Am Soc Pharmacol & Exp Therapeut; Am Soc Biochem & Molecular Biologists. *Res:* Cofactors in protein and RNA biosynthesis; role of hormones and nutritional factors in protein synthesis and cancer cells; protein kinases; lipases and cyclic adenosine monophosphates; metabolism of isolated liver parenchymal and sinusoidal cells; hepatic carcinogens and eicosanoids. *Mailing Add:* Dept Pharmacol Ind Univ Sch Med 1100 W Michigan St Indianapolis IN 46203

WAGLEY, PHILIP FRANKLIN, b Mineral Wells, Tex, Feb 5, 17; m 53. MEDICINE. *Educ:* Southern Methodist Univ, BS, 38; Johns Hopkins Univ, MD, 43; Am Bd Internal Med, dipl. *Prof Exp:* Instr med, 45-47 & 49-64, asst prof, 64-74, ASSOC PROF MED, JOHNS HOPKINS UNIV, 74- *Concurrent Pos:* Am Col Physicians res fel, Harvard Med Sch, 47-48; Nat Res Coun fel med sci, Mass Inst Technol, 48-49. *Mem:* Am Clin & Climat Asn; fel Am Col Physicians; Am Thoracic Soc; Am Soc Hemat. *Res:* Chest disease. *Mailing Add:* Nine E Chase St Baltimore MD 21202

WAGMAN, GERALD HOWARD, b Newark, NJ, Mar 4, 26; m 48; c 2. MICROBIAL BIOCHEMISTRY, TECHNICAL MANAGEMENT. *Educ:* Lehigh Univ, BS, 46; Va Polytech Inst, MS, 47. *Prof Exp:* Tech asst antibiotics, Squibb Inst Med Res, 47-49, electronics, 49-54, microbial biochemist, 54-57; from assoc biochemist to sr biochemist Schering Corp, 57-69, Sect leader, 69-70, mgr antibiotics dept, 70-74, assoc dir microbiol sci/antibiotics, 74-77, assoc dir, Microbiol Sci/Screening Lab, 77-78, dir, Microbiol Strain Lab, 79-85, prin scientist, 85-89, MGR LIBR INFO CTR, SCHERING-PLOUGH RES, 89- *Concurrent Pos:* Mem adv bd, Nat Cert Comn in Chem & Chem Eng, 85-88. *Mem:* AAAS; Am Chem Soc; Am Soc Microbiol; fel Am Inst Chemists; Sigma Xi; Royal Soc Chem. *Res:* Antibiotics, especially isolation, identification and evaluation; strain development; fermentation biosynthesis and development; isolation and identification of natural products; author and editor of four books on isolation, separation and purification of natural products. *Mailing Add:* Schering-Plough Res 60 Orange St Bloomfield NJ 07003

WAGNER, ALAN R, b Columbus, Ohio, Aug 20, 23; m 52; c 3. PATHOLOGY. *Educ:* Ohio State Univ, DVM, 46, MSc, 52. *Prof Exp:* Sr veterinarian, UNRRA, 46-47; field veterinarian, USDA, 47; veterinarian, Columbus Health Dept, 47-48; pathologist, Path Serv Labs, Ohio State Dept Agr, 48-56 & Lederle Labs Div, Am Cyanamid Co, 56-58; dir vet med res, Warren-Teed Pharmaceut Inc, 58-69; pres, Arlington Res Labs, Inc, 69-72; dir res & develop, Pet Chem, Inc, 75-78; ASST DIR & PATHOLOGIST, LAB ANIMAL CTR, OHIO STATE UNIV, 72-75, 78- *Mem:* Am Vet Med Asn; Am Asn Lab Animal Sci; NY Acad Sci; Soc Toxicol; Sigma Xi. *Res:* Pathology in fields of human and animal investigations and clinical as well as histopathological and bacteriological determination; new pharmaceutical and biological products. *Mailing Add:* Lab Animal Ctr Ohio State Univ 6089 Godown Rd Columbus OH 43235

WAGNER, ALBERT FORDYCE, b Rochester, NY, Feb 3, 45; m 69; c 2. CHEMICAL PHYSICS. *Educ:* Boston Col, BS, 66; Calif Inst Technol, PhD(chem), 72. *Prof Exp:* Presidential intern, 72-74, asst chemist, 74-77, CHEMIST, CHEM DIV, ARGONNE NAT LAB, 77- *Mem:* Am Inst Physics; Am Phys Soc; Am Chem Soc. *Res:* Theory and modeling of chemical reactions. *Mailing Add:* Chem Div Argonne Nat Lab Argonne IL 60439

WAGNER, ANDREW JAMES, b Greenwich, Conn, Apr 12, 34; m 69; c 2. METEOROLOGY, CLIMATOLOGY. *Educ:* Wesleyan Univ, BA, 56; Mass Inst Technol, MS, 58. *Prof Exp:* Res meteorologist, Extended Forecast Div, Weather Bur, Climate Anal Ctr, Nat Weather Serv, Nat Oceanic & Atmospheric Admin, 65-69, res meteorologist forecasting & interpretation, 69-72, meteorologist long range prediction group, 72-78, meteorologist, Dept Com, 78-89, SR FORECASTER, PREDICTION BR, CLIMATE ANALYSIS CTR, NAT WEATHER SERV, NAT OCEANIC & ATMOSPHERIC ADMIN, 90- *Mem:* Am Meteorol Soc; Royal Meteorol Soc; Am Geophys Union; Am Sci Affil; Nat Weather Asn; Asn Am Weather Observers. *Res:* Understanding and predicting statistical relationships of monthly and seasonal weather patterns over the northern hemisphere; authored articles in professional journals. *Mailing Add:* Rm 604 World Weather Bldg Climate Analysis Ctr W/NMC51 Washington DC 20233

WAGNER, ARTHUR FRANKLIN, b Jersey City, NJ, Oct 25, 22; m 45; c 3. ORGANIC CHEMISTRY. *Educ:* Princeton Univ, AB, 48, MA, 49, PhD(org chem), 51. *Prof Exp:* SR RES FEL ORG CHEM, MERCK SHARP & DOHME RES LABS, 65- *Mem:* Am Chem Soc. *Res:* Synthetic organic chemistry in natural products, isolation, structure determination and synthesis of vitamins and cofactors; synthesis of benzimidazoles; synthesis of peptides, biopolymers and immobilized biopolymers; B-lactam antibiotic synthesis; nucleoside and nucleotide synthesis. *Mailing Add:* 24 Sturges Way Princeton NJ 08540-5335

WAGNER, AUBREY JOSEPH, energy applications; deceased, see previous edition for last biography

WAGNER, BERNARD MEYER, b Philadelphia, Pa, Jan 17, 28; m 51; c 3. PATHOLOGY. *Educ:* Hahnemann Med Col, MD. *Prof Exp:* Dir exp path labs, Hahnemann Med Col, 54-55; asst prof path, Med Sch & Grad Sch Med, Univ Pa, 56-58; assoc prof path & Robert L King chair cardiovasc res, Sch Med, Univ Wash, 58-60; prof path & chmn dept, New York Med Col, 60-67, clin prof, 67-68; CLIN PROF PATH, COLUMBIA UNIV, 68-; dir labs, Beekman Downtown Hosp, NY, 71-; DEP DIR, NATHAN KLINE RES INST, RES PROF, UNIV SCH MED, NY, 86- *Concurrent Pos:* Dazian Found Med Res fel, Hahnemann Med Col, 53, Am Heart Asn Southeast Pa fels, 54-55; asst vis chief serv, Philadelphia Gen Hosp, 54-58; pathologist & dir path, Children's Hosp, Philadelphia, 55-58; lectr, Philadelphia Col Pharm, 56-58; attend pathologist, Vet Admin Hosp, 58-; Burroughs Wellcome Fund travel grant & spec investr, Hosp for Sick Children, London, Eng, 59; vpres, Warner Lambert Res Inst, 67-; dir labs, Francis Delafield Hosp, 68-71; ed-in-chief, Human Path, 73-88; assoc ed, Conn Tissue Res, 74-80, J Environ Path & Toxicol, 76-80; dir labs, Overlook Hosp, Summit, NJ, 76-86. *Honors & Awards:* Hon mem, Am Col Vet Pathologists, 85; fel, Royal Col Pathologists, London, 86. *Mem:* Am Soc Exp Path; Soc Pediat Res; Fedn Am Soc Exp Biol; Am Asn Path & Bact; Am Rheumatism Asn; fel Royal Col Pathologists Eng. *Res:* Diseases of connective tissue; rheumatic heart disease; toxicology; comparative pathology. *Mailing Add:* Nathan Kline Inst Orangeburg NY 10962

WAGNER, C(HRISTIAN) N(IKOLAUS) J(OHANN), b Saarbruecken-Dudweiler, Ger, Mar 6, 27; US citizen; m 52; c 3. PHYSICAL METALLURGY, MATERIALS SCIENCE. *Educ:* Univ Saarland, Lic es sc, 51, Dipl Ing, 54, Dr rer nat, 57. *Prof Exp:* Asst x-ray metall, Inst Metall Res, Univ Saarland, 53-55, 57-58, res asst phys metall, 59; from asst prof to assoc prof eng & appl sci, Yale Univ, 59-70; chmn, Mat Dept, 74-79, asst dean, Undergrad Studies, 82-85, PROF ENG & APPL SCI, UNIV CALIF, LOS ANGELES, 70- *Concurrent Pos:* Vis prof, Univ Saarbrucken, 79-80. *Honors & Awards:* Humboldt Award, 89. *Mem:* Am Soc Metals; Am Phys Soc; Am Crystallog Asn; Am Inst Mining, Metall & Petrol Engrs; Mat Res Soc. *Res:* Diffraction studies including x-rays, neutrons, electrons of amorphous materials and liquids, plastic deformation and transformation in alloys, thin films, and biomaterials. *Mailing Add:* Mat Dept 5731-Boelter Hall Univ Calif Los Angeles CA 90024-1595

WAGNER, CARL E, b NJ, July 10, 40; m 79; c 2. ARTIFICIAL INTELLIGENCE, PLASMA ENGINEERING. *Educ:* Mass Inst Technol, SB, 61, ScD(plasma physics), 70. *Prof Exp:* Physicist, US Naval Res Lab, 70-75; sr physicist, Sci Appl Inc, 75-81; head physicist, Int Nuclear Energy Systs Co, 81-84; sr physicist, TRW Inc, 84-87; PRIN SCIENTIST, JAYCOR INC, 87- *Concurrent Pos:* Consult, Sci Appl Inc, 81 & Energy Appl & Systs Inc, 84-85; adj prof, Calif State Univ, 86. *Mem:* Am Phys Soc; Am Asn Artificial Intel; Sigma Xi. *Res:* Development of expert system simulations; tokamak fusion reactor engineering; computational and theoretical plasma physics. *Mailing Add:* PO Box 9103 Rancho Santa Fe CA 92067

WAGNER, CARL GEORGE, b Newark, NJ, Sept 26, 43. MATHEMATICS. *Educ:* Princeton Univ, AB, 65; Duke Univ, PhD(math), 69. *Prof Exp:* From asst prof to assoc prof, 69-81, PROF MATH, UNIV TENN, KNOXVILLE, 81- *Concurrent Pos:* Fel, Ctr Advan Study Behav Sci, Stanford, 78-79. *Mem:* Soc Indust & Appl Math. *Res:* Combinatorics; decision theory. *Mailing Add:* Dept Math Univ Tenn Knoxville TN 37996-1300

WAGNER, CHARLES EUGENE, b Memphis, Tenn, June 21, 23; m 49; c 3. MORPHOLOGY. *Educ:* Princeton Univ, AB, 47; Ind Univ, PhD(zool), 54. *Prof Exp:* Asst zool, Ind Univ, 48-50; from instr to prof anat, Sch Med, Univ Louisville, 52-88, from asst dean to assoc dean, 61-74; RETIRED. *Concurrent Pos:* Vis prof, Univ Nottingham, 74-75. *Mem:* Sigma Xi; Am Asn Clin Anatomists. *Res:* Experimental morphology; regeneration; movements at synovial joints. *Mailing Add:* 506 Fairlawn Ave Louisville KY 40207-3658

WAGNER, CHARLES KENYON, b Cleveland, Ohio, Mar 11, 43; m 66; c 2. ECOLOGY, BIOLOGY. *Educ:* Emory Univ, BA, 65; Univ Ga, MS, 68, PhD(zool), 73. *Prof Exp:* Lectr zool, Univ Ga, 70-72; asst prof biol, Southwestern at Memphis, 72-77; asst prof, 77-81, ASSOC PROF BIOL, CLEMSON UNIV, 81- *Concurrent Pos:* Consult, Community Develop Task Force, 75-77. *Honors & Awards:* Am Soc Mammalogists Award, 72. *Mem:* Ecol Soc Am; Am Soc Mammalogists; Am Inst Biol Sci; Sigma Xi. *Res:* Bioenergetics of terrestrial populations; microecosystem investigation of population growth. *Mailing Add:* 121 Clemson St Clemson SC 29631

WAGNER, CHARLES ROE, b Olivet, SDak, Dec 3, 25; m 50; c 2. ORGANIC CHEMISTRY. *Educ:* SDak Sch Mines & Technol, BSc, 50; Mich State Col, PhD, 55. *Prof Exp:* Res chemist, 55-62, res supvr, Specialty Chem Div, 62-69, mgr process res, 69-71, dir develop, 71-74, DIR, BUFFALO RES LABS, SPECIALTY CHEM DIV, ALLIED CHEM CORP, 74- *Mem:* Am Chem Soc. *Res:* Organic isocyanates; organic acids. *Mailing Add:* 4302 MacDonnell Dr Murrysville PA 15668-1358

WAGNER, CLIFFORD HENRY, b Cincinnati, Ohio. COMPUTER GRAPHICS. *Educ:* Univ Cincinnati, AB, 67; Univ Mich, AM, 68; State Univ NY, Albany, PhD(math), 73. *Prof Exp:* Asst prof math, Fitchburg State Col, 73-78; asst prof math, 78-82, ASSOC PROF MATH & COMPUTER SCI, PA STATE UNIV, HARRISBURG, 83- *Honors & Awards:* Allendoerfer Award, Math Asn Am, 83. *Mem:* Am Math Soc; Math Asn Am; Nat Coun Teachers Math; Asn Comput Mach; Inst Elec & Electronics Engrs. *Res:* Applied mathematics; statistics; computer graphics; iterative methods; probability theory; medical applications. *Mailing Add:* Pa State Univ-Harrisburg Middletown PA 17057

WAGNER, CONRAD, b Brooklyn, NY, Nov 1, 29; m 53; c 2. BIOCHEMISTRY, NUTRITION. *Educ:* City Col New York, BA, 51; Univ Mich, MS, 52, PhD(biochem), 56. *Prof Exp:* USPHS fel biochem, NIH, 59-61; from asst prof to assoc prof biochem, 61-75, PROF BIOCHEM, SCH MED, VANDERBILT UNIV, 75- *Concurrent Pos:* Res biochemist, Vet Admin Hosp, 61-68, chief biochem res, 68-, assoc chief of staff res, 74- *Honors & Awards:* Borden Award, Am Inst Nutrit, 83. *Mem:* Am Soc Biol Chemists; Am Inst Nutrit. *Res:* Gluconeogenesis from lipid in Tetrahymena pyriformis; regulation of tryptophan-niacin relation in animals and microorganisms; sulfonium compounds and one carbon metabolism in bacteria; role and function of natural folate coenzymes; characterization of folate binding proteins; cellular transport of folate and other cofactors; nutritional biochemistry; nutrition and cancer. *Mailing Add:* Res Serv Vet Admin Med Ctr 24th Ave S Nashville TN 37212

WAGNER, DANIEL HOBSON, b Jersey Shore, Pa, Aug 24, 25; m 49; c 4. MATHEMATICS. *Educ:* Haverford Col, BS, 47; Brown Univ, PhD(math), 51. *Prof Exp:* Mem sci staff, Opers Eval Group, Mass Inst Technol, 51-56; supvr math anal, Burroughs Corp, 56-58; partner, Kettelle & Wagner, 58-63; PRES, DANIEL H WAGNER ASSOCS, 63- *Concurrent Pos:* Chmn electronics reliability task group, Off Asst Secy Defense Res & Eng, 56; lectr, Swarthmore Col, 58 & Univ Pa, 58 & 61-62; comt appl math training, Nat Acad Sci- Nat Res Coun, 77-78. *Mem:* Am Math Soc; Opers Res Soc Am; Soc Indust & Appl Math; Math Asn Am; Inst Mgt Sci. *Res:* Operations research; constrained optimization; measurable set-valued functions. *Mailing Add:* US Naval Acad Annapolis MD 21402

WAGNER, DAVID DARLEY, b Ft Riley, Kans, Sept 27, 44; m 71; c 2. INTESTINAL ECOLOGY, INTESTINE FUNCTION. *Educ:* Univ Md, BSc, 72, MSc, 74, PhD(nutrit), 77. *Prof Exp:* Res assoc microbiol, Dept Food Sci, NC State Univ, 77; reviewing staff scientist metab drugs, Div Drugs Avian Species, 77-79, biores monitoring prog mgr, toxicol, Off Sci Eval, 79-80, CHIEF, DIV VET RES, ANIMAL NUTRIT & BIOL BR, CTR VET MED, FOOD & DRUG ADMIN, 80- *Mem:* Sigma Xi; Am Soc Animal Sci; Poultry Sci Asn. *Res:* Effects of diet composition on intestine microecology; intestine digestive and absorptive function; intestine mucosal integrity and diet plus drug interactions. *Mailing Add:* Food & Drug Admin Div Vet Res Bldg 328-A Agr Res Ctr Beltsville MD 20705

WAGNER, DAVID HENRY, b Detroit, Mich, Aug 18, 45. TAXONOMY, COLLECTIONS MANAGEMENT. *Educ:* Univ Puget Sound, BA, 68; Wash State Univ, MS, 74, PhD(bot), 76. *Prof Exp:* Asst prof, 76-82, DIR & CUR, HERBARIUM, UNIV ORE, 76-, ASSOC PROF BIOL, 82- *Concurrent Pos:* Pres, Mount Pisgah Arboretum, 79-81 & 83-85, vpres, 81-83; pres, Eugene Natural Hist Soc, 88-89 & 90- *Mem:* Int Asn Plant Taxon; Bot Soc Am; Am Bryological & Lichenological Soc; Am Soc Plant Taxonomists; Am Fern Soc; Brit Pteridological Soc. *Res:* Floristics of Pacific Northwest (especially Pteridophyta & Hepaticae); reproductive biology of ferns and flowering plants; cytology of fern hybrids; community phenology; endangered species management. *Mailing Add:* Dept Biol Univ Ore Eugene OR 97403

WAGNER, DAVID LOREN, b Erie, Pa, Nov 19, 42; m 64; c 1. ENGINEERING PHYSICS. *Educ:* Case Western Reserve Univ, BS, 64, MS, 66, PhD(physics), 70. *Prof Exp:* From asst prof to assoc prof, 70-75, CHMN, DEPT PHYSICS, EDINBORO STATE COL, 72-, PROF, 75- *Concurrent Pos:* NSF acad year exten grant, 71-73, student sci training grant, 75, teacher grant, 78, robotics training, 86, research grant, 86. *Mem:* Am Asn Physics Teachers; AAAS. *Res:* Fermi surface of metals and semi-metals. *Mailing Add:* Dept Physics Edinboro Univ Edinboro PA 16444

WAGNER, EDWARD D, b Eureka, SDak, June 28, 19; m 42; c 3. PARASITOLOGY. *Educ:* Walla Walla Col, BA, 42; Wash State Univ, MS, 45; Univ Southern Calif, PhD, 53. *Prof Exp:* Instr biol, Atlantic Union Col, 45-47 & Andrews Univ, 47-49; assoc, Univ Southern Calif, 50-52; head dept parasitol, Sch Trop & Prev Med, Loma Linda Univ, 50-59, from instr to prof microbiol, Sch Med, 53-88; RETIRED. *Res:* Schistosomiasis; parasite therapy; helminths. *Mailing Add:* 10961 Desert Lane Dr No 164 Calimesa CA 92320

WAGNER, EDWARD KNAPP, b Akron, Ohio, May 4, 40; m 61; c 2. ANIMAL VIROLOGY, BIOCHEMISTRY. *Educ:* Univ Calif, Berkeley, BA, 62; Mass Inst Technol, PhD(biochem), 67. *Prof Exp:* Helen Hay Whitney Found fel, Univ Chicago, 67-70; asst prof, 70-75, assoc prof, 75-80, PROF VIROL, UNIV CALIF, IRVINE, 80- *Concurrent Pos:* Nat Cancer Inst res grant, 70- *Mem:* AAAS; Am Soc Microbiol; Tissue Cult Asn; Am Soc Biol Chemists; Am Soc Cell Biol. *Res:* Control of gene action in animal virus infection; mechanism of viral carcinogenesis; control of information transfer between nucleus and cytoplasm in eucaryotic cells. *Mailing Add:* Dept Molecular Biol & Biochem Univ Calif Irvine CA 92717

WAGNER, ERIC G, b Ossining, NY, Oct 1, 31; m 60; c 3. MATHEMATICS. *Educ:* Harvard Univ, BA, 53; Columbia Univ, MA, 59, PhD(math), 63. *Prof Exp:* Tech engr, IBM Corp, 53-54, assoc engr switching theory, 56-58, RES STAFF MEM, T J WATSON RES CTR, IBM CORP, 58- *Concurrent Pos:* Lectr, NY Univ, 64-65, adj asst prof, 65-66; sr vis res fel, Queen Mary Col, Univ London, 73-74. *Mem:* Am Math Soc; Asn Comput Mach; Asn Symbolic Logic; NY Acad Sci; Europ Asn Theoret Comput Sci. *Res:* Computability theory and category theory with emphasis on their relationship to computer science; theory of programming languages. *Mailing Add:* IBM Watson Res Ctr PO Box 218 Yorktown Heights NY 10598

WAGNER, EUGENE ROSS, b Monroe, Wis, Nov 21, 37; m 58; c 2. RADIOCHEMISTRY. *Educ:* Univ Wis, BS, 59, PhD(org chem), 64. *Prof Exp:* Chemist, Spec Assignment Prog, Dow Chem Co, 63-64, Dow Human Res & Develop Labs, Pitman-Moore Div, 64-67, sr res chemist, 67-72, res specialist, Chem Biol Res, 72-75, sr res specialist, Midland, Mich, 75-78, sr res specialist, pharmaceut chem, 78-80; sr res specialist, Merrell Dow, 81-88; ASSOC SCIENTIST, MARION MERRELL DOW RES INST, 89- *Mem:* Am Chem Soc; Int Isotope Soc. *Res:* Radiolabeling of new drug candidates. *Mailing Add:* Marion Merrell Dow Res Inst PO Box 68470 Indianapolis IN 46268

WAGNER, EUGENE STEPHEN, b Gary, Ind, Mar 30, 34; m 62; c 3. BIOLOGICAL CHEMISTRY, PHYSICAL CHEMISTRY. *Educ:* Ind Univ, BS, 59; Purdue Univ, PhD(chem), 64. *Prof Exp:* Instr chem, Purdue Univ, 62-64; sr phys chemist, Eli Lily & Co, Ind, 64-71; from asst prof to assoc prof, 71-79, PROF MED EDUC & CHEM, MUNCIE CTR MED EDUC, BALL STATE UNIV, 79- *Mem:* Am Chem Soc. *Res:* Interactions of antibiotics and ascorbic acid with constituents of blood. *Mailing Add:* Ctr Med Educ Ball State Univ Muncie IN 47306-1099

WAGNER, FLORENCE SIGNAIGO, b Birmingham, Mich, Feb 18, 19; m 48; c 2. BOTANY. *Educ:* Univ Mich, Ann Arbor, AB, 41, MA, 43; Univ Calif, Berkeley, PhD(bot), 52. *Prof Exp:* Res asst soc sci, Off Coordr Inter-Am Affairs, 43-45 & Off Strategic Serv, 45; from res asst to sr res assoc bot, 61-76, ASSOC RES SCIENTIST, UNIV MICH, ANN ARBOR, 76- *Concurrent Pos:* Lectr, Univ Ctr Adult Educ, Ann Arbor, 71-78; adj lectr, Univ Mich, 86. *Honors & Awards:* Investigatora Asociada ad hon en Citologia, Museo Nacional de Costa Rica. *Mem:* Am Fern Soc (vpres, 84-85, pres, 86-87); Brit Pteridological Soc; Bot Soc Am (chair, Pteridological sect, 83-84); Am Soc Plant Taxonomists; Int Asn Plant Taxon; Int Orgn Plant Biosystematists. *Res:* Analysis of chromosomal behavior in ferns and fern hybrids and comparative studies of their morphology. *Mailing Add:* Dept Biol Univ Mich Ann Arbor MI 48109-1048

WAGNER, FRANK A, JR, b New Haven, Conn, Apr 19, 32; m 61; c 4. ORGANIC CHEMISTRY. *Educ:* Yale Univ, BS, 58; Rutgers Univ, NB, PhD(org chem), 68. *Prof Exp:* From chemist to sr res chemist, 58-77, group leader, 77-85, SR GROUP LEADER, AM CYANAMID CO, 85- *Mem:* Am Chem Soc; AAAS. *Res:* Preparation of compounds as agricultural pesticides; design of procedures for large-scale syntheses. *Mailing Add:* Am Cyanamid Co PO Box 400 Princeton NJ 08543-0400

WAGNER, FRANK S, JR, b Temple, Tex, Aug 26, 25; m 53; c 5. ORGANIC CHEMISTRY. *Educ:* Southwest Tex State Col, BA & MA, 47. *Prof Exp:* Assoc prof chem, Schreiner Inst, Tex, 48-50; analyst, Celanese Corp, 50-52, group leader, 52-53, librn, 53-65, head info ctr, Tech Ctr, Celanese Chem Co, 65-83; RETIRED. *Concurrent Pos:* Organizer, Nandina Corp, 84- *Mem:* AAAS; Am Chem Soc; Spec Libr Asn; Egypt Explor Soc. *Res:* Application of machine methods to critical literature reviews and commerical intelligence activities; writing of encyclopedic reviews. *Mailing Add:* 834 Oak Park Ave Corpus Christi TX 78408

WAGNER, FREDERIC HAMILTON, b Corpus Christi, Tex, Sept 26, 26; m 49; c 2. BIOLOGY. *Educ:* Southern Methodist Univ, BS, 49; Univ Wis, MS, 53, PhD(wildlife mgt, zool), 61. *Prof Exp:* Refuge asst, US Fish & Wildlife Serv, 45; asst zool & bot, Southern Methodist Univ, 46-49; asst wildlife mgt, Univ Wis, 49-51; res fel, Wildlife Mgt Inst, 51; res biologist, Wis Conserv Dept, 52-58; asst prof wildlife resources, Utah State Univ, 58-59; res biologist, Wis Conserv Dept, 59-61; assoc prof, 61-66, assoc dean, Col Natural Resources, 70-77, PROF WILDLIFE RESOURCES, UTAH STATE UNIV, 66-, ASSOC DEAN, COL NATURAL RESOURCES, 77- *Concurrent Pos:* Dir Desert Biomed, US-Int Biol Prog & mem US Exec Comt, Int Biol Prog, 71-74; mem comt predator control, President's Coun Environ Qual, 71. *Honors & Awards:* Award, Wildlife Soc, 68. *Mem:* Ecol Soc Am; Am Soc Mammal; Wildlife Soc; Am Inst Biol Sci; Cooper Ornith Soc; Sigma Xi. *Res:* Vertebrate population ecology, especially population dynamics, limiting factors and homeostatic mechanisms; wildlife management; conservation of natural resources; systems ecology. *Mailing Add:* Dept Wildlife & Fisheries Utah State Univ Logan UT 84322-5200

WAGNER, FREDERICK WILLIAM, b Erie, Pa, Feb 4, 40; m 65; c 2. BIOCHEMISTRY. *Educ:* Southwest Tex State Col, BSc, 62; Tex A&M Univ, PhD(biochem), 66. *Prof Exp:* US Air Force res assoc, Brooks AFB, 66-67; res fel biochem & biophys, Tex A&M Univ, 67-68; asst prof biochem & nutrit, 68-73, assoc prof, 73-80, PROF, DEPT AGR BIOCHEM, UNIV NEBR, LINCOLN, 80- *Concurrent Pos:* Vis prof, Biophys Res Lab, Med Sch, Harvard Univ, Boston, 81-82. *Mem:* Sigma Xi. *Res:* Structure and function of proteins with special emphasis on proteolytic enzymes. *Mailing Add:* Univ Nebr Biochem Hall E Campus Lincoln NE 68583-0718

WAGNER, GEORGE HOYT, b Mulberry, Ark, Dec 28, 14; m 39; c 3. GEOLOGY. *Educ:* Univ Ark, BS, 37, MS, 74; Univ Iowa, MS, 39, PhD(phys chem), 41. *Prof Exp:* Chemist, Univ Ark, 35-37; asst chem, Univ Iowa, 37-41; res chemist, Linde Div, Union Carbide Corp, 41-47, head div phys chem, 47-51, asst to supt, 51-53, res supvr, 53-55, mgr res, 55-59, dir, 59-64, mgr develop, 64-65, dir res, Mining & Metals Div, 65-66, vpres, 66-70, ferroalloy div, 70-71; CONSULT RES & DEVELOP, 71- *Concurrent Pos:* Adj prof,

Univ Ark, 74- *Honors & Awards:* Schoellkopf Medal, Am Chem Soc, 60. *Mem:* Am Chem Soc; Geol Soc Am; AAAS. *Res:* Synthetic lubricants; corrosion inhibition; organometallics; geochemistry, economic geology and atmospheric chemistry. *Mailing Add:* Box 144 Fayetteville AR 72702

WAGNER, GEORGE JOSEPH, b Buffalo, NY, Sept 15, 43; m 70; c 3. PLANT PHYSIOLOGY, ENVIRONMENTAL SCIENCES. *Educ:* State Univ NY Buffalo, BA, 70, MA, 71, PhD(biol), 74. *Prof Exp:* Res assoc, Brookhaven Nat Lab, 74-77, assoc scientist plant biochem, 79-83; assoc prof, 83-87, PROF, DEPT AGRON, UNIV KY, 87- *Mem:* Am Soc Plant Physiologists; AAAS; Phytochem Soc NAm; Soc Environ Geochem & Health. *Res:* Study of the physiology and biochemistry of the mature plant cell vacuole, the mechanisms of solute accumulation and the fate of heavy metals in plants; mechanisms of secretion in plants. *Mailing Add:* Agron Dept Univ Ky Lexington KY 11973

WAGNER, GEORGE RICHARD, b Chicago, Ill, Nov 12, 33; m 54; c 3. SOLID STATE PHYSICS. *Educ:* Univ Ill, Urbana, BS, 60; Carnegie-Mellon Univ, MS, 62, PhD(physics), 65. *Prof Exp:* Sr engr, 65-73, fel scientist physics, 74-89, MGR SUPERCONDUCTOR MAT & ELECTRONICS, SCI & TECHNOL CTR, WESTINGHOUSE ELEC CORP, 89- *Mem:* Am Phys Soc. *Res:* Superconductor electronics, materials and devices. *Mailing Add:* Westinghouse Sci & Technol Ctr 1310 Beulah Rd Pittsburgh PA 15235

WAGNER, GERALD C, BIOCHEMISTRY, PROTEIN STRUCTURE & FUNCTION. *Educ:* Univ Ill, Chicago, PhD(chem), 75. *Prof Exp:* RES ASST PROF PHYSICS & CHEM, UNIV ILL, URBANA, 84- *Mem:* Am Chem Soc; Am Soc Biochem & Molecular Biol; Biophys Soc. *Res:* Biophysics. *Mailing Add:* 308 N Prairie St Apt 402 Champaign IL 61820-3449

WAGNER, GERALD GALE, b Plainview, Tex, June 3, 41; m 62; c 2. IMMUNOLOGY, IMMUNOPARASITOLOGY. *Educ:* Tex Tech Col, BS, 63; Univ Kans, MA, 65, PhD, 68. *Prof Exp:* Microbiologist, Immunol Div, Plum Island Animal Dis Lab, 68- 71, microbiologist, Coop Res Div, EAfrican Vet Res Orgn, USDA, 71-77; assoc prof, 77-86, PROF VET MICROBIOL & PARASITOL, TEX A&M UNIV, 86- *Concurrent Pos:* Nat Acad Sci-Agr Res Serv res fel, Plum Island Animal Dis Lab, 68-70; tech adv, Interam Inst for Coop in Agr for Animal Health Projs in Mex, 80-84; subcomt foreign animal dis & ectoparasite diag & res, Bd Agr, Nat Acad Sci-Nat Res Coun, 82-84; consult, rev trypanosomiasis res & training, UN Develop Prog, 84; coordr, int progs, Col Vet Med, Tex A&M Univ, 89- *Mem:* Fedn Am Soc Exp Biol; Am Soc Microbiol; Am Asn Immunol; Am Soc Parasitol. *Res:* Pathogenesis of protozoal infections in domestic animals; cellular effector mechanisms of immunity. *Mailing Add:* Dept Vet Pathobiol Tex A&M Univ College Station TX 77843-4467

WAGNER, GERALD ROY, b Evansville, Ind, Feb 14, 28; m 50; c 2. ORGANIC CHEMISTRY, GEOLOGY. *Educ:* Mt Union Col, BS, 50; Univ Ark, MS, 52. *Prof Exp:* Res chemist, Com Solvents Corp, NY, 52-53, Olin Mathieson Chem Corp, 53-55 & Nat Aniline Div, Allied Chem Corp, 55-61; res chemist, 61-66, asst prof, 66, head dept, 66-72, PROF CHEM, ERIE COMMUNITY COL, 66- *Mem:* Am Chem Soc; Nat Asn Geol Teachers; Am Inst Chemists. *Res:* Geological education; chemical education. *Mailing Add:* PO Box 834 Erie Community Col North Camp Main Youngs Rd Williamsville NY 14221-0834

WAGNER, HANS, b July 19, 32; US citizen. ORGANIC CHEMISTRY. *Educ:* Univ Iowa, BS, 55; Pa State Univ, PhD(chem), 59. *Prof Exp:* Sr res investr, 59-70, GROUP LEADER CHEM, G D SEARLE & CO, 70- *Mem:* Am Chem Soc. *Res:* Dipolar cycloadditions; mesoionic compounds; heterocyclic azido compounds. *Mailing Add:* Hercules Inc Res Ctr Bldg 8134-357 Wilmington DE 19894-0002

WAGNER, HARRY HENRY, b San Diego, Calif, Jan 10, 33; m 56; c 3. FISH BIOLOGY, ECOLOGY. *Educ:* Humboldt State Col, BS, 55; Ore State Univ, MS, 59, PhD(fisheries), 70. *Prof Exp:* Fishery res biologist & physiol ecologist, Ore Dept Fish & Wildlife, 59-68, fishery res coordr, 69-73, res supvr, 73-78, asst chief, 79-82, chief, Fisheries Div, 83-89; PLANNING ASSOC, NORTHWEST POWER PLANNING COUN, 90- *Concurrent Pos:* Courtesy assoc prof, Ore State Univ, 59- *Mem:* Am Fisheries Soc. *Res:* Parr-smolt transformation of anadromous salmonids. *Mailing Add:* Northwest Power Planning Coun 851 SW Sixth Ave Suite 1100 Portland OR 97204-1348

WAGNER, HARRY MAHLON, b Iola, Kans, June 1, 24; m 44; c 5. MATHEMATICS. *Educ:* Naval Postgrad Sch, BS, 54; Kans State Teachers Col, MS, 64; Univ Ark, EdD(higher educ), 69. *Prof Exp:* Instr math, John Brown Univ, 64-67; res grad asst psychol, Univ Ark, 67-69; from asst prof to assoc prof math, Cameron Univ, 69-90; RETIRED. *Mem:* Math Asn Am. *Mailing Add:* 307 Tanglewood Lane Lawton OK 73505-5317

WAGNER, HARVEY ARTHUR, b Ann Arbor, Mich, Jan 2, 05; m 29. ENGINEERING. *Educ:* Univ Mich, BS, 27; Lawrence Inst Technol, DEng, 69. *Prof Exp:* Mem staff, Procter & Gamble Co, 27-28; mem staff, Detroit Edison Co, 28-69, exec vpres, 69-70; VCHMN & DIR, OVERSEAS ADV ASSOCS, INC, 74- *Concurrent Pos:* Trustee, Nat Sanit Found, 65-84; consult engr, 70-; chmn, Comt Nuclear Fuels & Comt Advan Projs, Edison Elec Inst; mem, Tech & Eng Comt, Power Reactor Develop Co & Atomic Power Develop Assocs, Inc; vchmn, Econ Comt, Atomic Power Develop Assocs. *Honors & Awards:* Cert Pub Serv, Fed Power Comn, 64; Sesquicentennial Award as Outstanding Exec & Nuclear Power Consult, Univ Mich, 67. *Mem:* Nat Acad Eng; fel Am Soc Mech Engrs; fel Am Nuclear Soc. *Res:* Power engineering and management; author of over 60 publications. *Mailing Add:* Overseas Adv Assocs Inc 3000 Book Bldg 1249 Washington Blvd Detroit MI 48226

WAGNER, HENRY GEORGE, b Washington, DC, Sept 13, 17; m 45; c 3. NEUROSCIENCES, AEROSPACE MEDICINE. *Educ:* George Washington Univ, AB, 39, MD, 42; Univ Pa, cert ophthal, 49. *Prof Exp:* Intern, Naval Hosp, Brooklyn, NY, 42-43, med officer, Naval Med Sch, Md, 43, flight surgeon, Sch Aviation Med, Naval Air Sta, Fla, 43-44, flight surgeon, Naval Air Base, Guam, 44-45, flight surgeon, Naval Air Sta, Tex, 45-46, asst supt aeromed equip lab, Naval Air Exp Sta, Pa, 46-48, med res investr, Naval Med Res Inst, Md, 51-54, sr med officer, USS Valley Forge, 54-56, head physiol div, Naval Med Res Inst, 56-60, cmndg officer, 60-61, exec officer, 61-64, actg dir physiol sci dept, 61-64, dir aerospace crew equip lab, Naval Air Eng Ctr, Pa, 64-66; dir intramural res, 66-74, actg chief lab neurophysiol, 76-80, CHIEF SECT NEURONAL INTERACTIONS, LAB NEUROPATH & NEUROANAT SCI, NAT INST NEUROL, COMMUN DIS & STROKE, 74- *Concurrent Pos:* Fel biophys, Johns Hopkins Univ, 49-51, hon prof, 58-64; mem vision comt, hearing & bioacoust, Nat Acad Sci-Nat Res Coun; vis prof ophthal, Duke Univ, 76-85. *Mem:* AAAS; Am Physiol Soc; Soc Neurosci; Am Col Prev Med; AMA; Int Brain Res Orgn. *Res:* Neuroscience of the visual system; pathophysiology of stroke. *Mailing Add:* 3319 P St NW Washington DC 20007

WAGNER, HENRY N, JR, b Baltimore, Md, May 12, 27; m 51; c 4. INTERNAL MEDICINE, NUCLEAR MEDICINE. *Educ:* Johns Hopkins Univ, AB, 48, MD, 52. *Hon Degrees:* DSc, Washington Col. *Prof Exp:* From asst prof med & radiol, Sch Med to assoc prof med, radiol & radiol sci, 59-67, assoc prof med, 67-68, PROF RADIOL SCI & RADIOL, SCH MED & SCH HYG & PUB HEALTH, JOHNS HOPKINS UNIV, 67-, PROF MED, 68-, DIR, DIV NUCLEAR MED, 65-, DIR, DIV RADIATION HEALTH SCI, 77- *Honors & Awards:* George Hevesy Medal, 76; Nuclear Pioneer, Soc Nuclear Med, 83. *Mem:* Am Fedn Clin Res (past pres); Asn Am Physicians; Am Soc Clin Invest; Soc Nuclear Med (past pres); World Fedn Nuclear Med & Biol (past pres); Inst Med; AMA; Am Fedn Clin Res (past pres). *Mailing Add:* Div Radiation Health Sci Sch Hyg & Pub Health Baltimore MD 21205

WAGNER, HERMAN LEON, b New York, NY, Mar 21, 21; m; c 1. POLYMER CHEMISTRY. *Educ:* City Col, New York, BS, 42; Polytech Inst Brooklyn, MS, 46; Cornell Univ, PhD(chem), 50. *Prof Exp:* Chemist, SAM Labs, Manhattan Proj, Columbia Univ, 42-46; res assoc, Cornell Univ, 50-51; phys chemist, E I du Pont de Nemours & Co, 51-55, M W Kellog Co, 55-57 & Cleanese Corp Am, 57-68; res chemist, Nat Bur Standards, 68-87; SR SCIENTIST, ATLANTIC RES CORP, 87- *Mem:* Am Chem Soc. *Res:* Physical chemistry of high polymers; dilute solution properties; characterization of high polymers; thermal analysis; composites; melt rheology; fibers; gel permeation chromatography; correlation of molecular structure with physical properties; low shear viscosity of ultra high molecular weight polyethylene solutions. *Mailing Add:* 12038 Gatewater Dr Potomac MD 20854

WAGNER, J ROBERT, b Philadelphia, Pa, Jan 8, 32; m 57; c 2. TEXTILES & NONWOVENS, PAPER. *Educ:* Philadelphia Col Textiles & Sci, BS, 57; NC State Univ, Raleigh, MS, 66; Univ Leeds, MPhil, 78. *Prof Exp:* Mgr res, Formex Co, Huyck Corp, 57-65; PROF & DIR TEXTILE & APPAREL RES, PHILADELPHIA COL TEXTILES & SCI, 66- *Concurrent Pos:* Consult, J Robert Wagner Co, 66-; div chmn, Tech Asn Pulp & Paper Indust, 83-85. *Honors & Awards:* Leadership & Serv Award, Tech Asn Pulp & Paper Indust, 90. *Mem:* Fel Tech Asn Pulp & Paper Indust; Int Nonwovens & Disposables Asn; Am Asn Textile Chemists & Colorists. *Res:* Development of first synthetic Fourdrinier fabric, inside press fabric, open mesh dryer fabric, thermally bounded polypropylene nonwoven diaper cover stock and nonwoven fabric for space shuttle astronaut spacesuits. *Mailing Add:* 2996 Runnymede Dr Norristown PA 19401

WAGNER, JAMES BRUCE, JR, b Hampton, Va, July 28, 27; m 51; c 3. MATERIALS SCIENCE. *Educ:* Univ Va, BS, 50, PhD(chem), 55. *Prof Exp:* Fel metall, Mass Inst Technol, 54-56; asst prof, Pa State Univ, 56-58 & Yale Univ, 58-62; assoc prof, 62-65, prof mat sci, Northwestern Univ, 65-77, dir, Mat Res Ctr, 72-76; dir, 80-84, PROF, CTR SOLID STATE SCI, ARIZ STATE UNIV, 77- *Concurrent Pos:* Ford Found resident engr pract, Semiconductor Prod Div, Motorola, Inc, Ariz, 68-69; lectr, Univ Nacional del Sur, Argentina, 81; fel Japan Soc Promotion Sci, Nagoya Univ, 83; regents prof, Ariz State Univ, 89. *Honors & Awards:* Found Res Award, Aluminum Co Am, 86; Outstanding Achievement Award of High Temperature Mat Div, Electrochem Soc, 86. *Mem:* Am Phys Soc; Am Inst Mining, Metall & Petrol Eng; hon men Electrochem Soc (pres, 83-84); Solid State Ionics Soc; Asian Solid State Ionics. *Res:* Oxidation of metals; thermodynamics and transport properties of compound semiconductors; solid electrolytes. *Mailing Add:* Ctr Solid State Sci Tempe AZ 85287-1704

WAGNER, JEAMES ARTHUR, b New Praque, Minn, Sept 5, 44; div; c 1. ENVIRONMENTAL PHYSIOLOGY. *Educ:* St Johns Univ, BSc, 66; Univ SDak, MA, 67; Univ Western Ont, 70. *Prof Exp:* Asst prof physiol, Ind Univ, Bloomington, 69-71; from asst res to assoc res, 71-84, RES PHYSIOL, UNIV CALIF, SANTA BARBARA, 84- *Concurrent Pos:* Prog chmn, Int Symposium, Univ Calif, Santa Barbara, 77; lectr, Westmont Col, 77, Univ Calif, Santa Barbara, 71-; prin investr, NIH, 79-; co-prin investr, Calif ARB, 83-85, HEI, 84- *Mem:* AAAS; Am Col Sports Med; Am Physiol Soc; Fedn Am Scientists; NY Acad Sci. *Res:* Environmental and cardiorespiratory physiology, specifically physiological response differences to environmental stressors that are related to age and gender; physiological responses to heat, cold, altitude, air pollution and exercise stressors. *Mailing Add:* Inst Environ Stress Univ Calif Santa Barbara CA 93106

WAGNER, JOHN ALEXANDER, b Kansas City, Mo, Feb 9, 35; m 63, 86; c 2. ENTOMOLOGY. *Educ:* Northwestern Univ, BS, 57, MS, 59, PhD(biol), 62. *Prof Exp:* Mem fac, 62-88, PROF BIOL & SCI, KENDALL COL, 80- *Concurrent Pos:* Collabr & consult, Encycl Britannica Films, 62-75; res assoc, Dept Biol Sci, Northwestern Univ, 63-78; mem, Environ Studies Group, Alfred Benesch & Co, Consult Engrs, 74-78; res assoc, Dept of Zool, Div Insects, Field Mus Natural Hist. *Mem:* Am Inst Biol Sci; Soc Syst Zool;

Coleopterists Soc; Nature Conservancy; AAAS; Sigma Xi. *Res:* Coleoptera, family Pselaphidae, especially nearctic and neotropics. *Mailing Add:* Dept Educ Field Mus Nat Hist Roosevelt Rd at Lake Shore Dr Chicago IL 60605-2496

WAGNER, JOHN EDWARD, b Springfield, Mo, Oct 11, 27; m 50; c 2. CIVIL ENGINEERING, EARTH SCIENCES. *Educ:* US Mil Acad, BS, 50; Univ Ill, MS, 59, PhD(civil eng), 61; George Washington Univ, MBA, 80. *Prof Exp:* Instr, US Army Eng Sch, Ft Belvoir, Va, 52-53, asst proj engr, Eng Dist, Little Rock, Ark, 53-54, co comdr & staff engr, VII Corps, Europe, 55-58, res engr soil mech, Waterways Exp Sta, Miss, 60-63, staff engr adv, Army Repub Vietnam, 63, dep dir, Nuclear Cratering Group, Lawrence Radiation Labs, Calif, 64-65, chief engr br, Test Command, Defense Atomic Support Agency, NMex, 65-67, comdr & dir, Cold Regions Res & Eng Lab, NH, 67-70, staff engr, First Field Force, Vietnam, 70-71, comdr & dir, Engr Topographic Labs, Ft Belvoir, Va, 71-74, dep dir res, Off Dep Chief Staff, Res Develop & Acquisitions, 74-77, asst for conserv, Off Secy Defense, 77-79, dep div eng, NAtlantic Div, Corps Engrs, 79-81; exec secy, UN Nat Comt Rock Mech & Tunneling Technol, 81-84, SR STAFF OFFICER BD ARMY SCI & TECHNOL, NAT ACAD SCI, WASHINGTON, DC, 84-; SR CIVIL ENGR, ANALYTICAL SERV, INC, ARLINGTON, VA, 84- *Concurrent Pos:* Consult Deep Underground Construct, Anal Serv, Inc, Va, 82-84. *Mem:* Sigma Xi; fel Am Soc Civil Engrs; Nat Soc Prof Engrs; Soc Am Mil Engrs. *Res:* Soil mechanics especially arching of soils and slope stability; all areas of Army funded research; cold regions research and topographic sciences. *Mailing Add:* 3229 First Pl N Arlington VA 22201-1038

WAGNER, JOHN GARNET, b Weston, Ont, Mar 28, 21; m 46; c 2. PHARMACEUTICAL CHEMISTRY, ORGANIC CHEMISTRY. *Educ:* Univ Toronto, PhmB, 47; Univ Sask, BSP, 48, BA, 49; Ohio State Univ, PhD(pharmaceut chem), 52. *Hon Degrees:* DSc, Ohio State Univ, 80. *Prof Exp:* Instr pharmaceut chem, Ohio State Univ, 51-52, asst prof, 52-53; res scientist, Upjohn Co, Mich, 53-56, sect head, 56-63, sr res scientist, 63-68; asst dir res & develop, Pharm Serv, Univ Hosp, 68-72, Albert G Prescott prof, Col Pharm, 82-86, PROF PHARM, UNIV MICH, ANN ARBOR, 68-, JOHN G FEARLE PROF PHARMACEUT, 86-, PROF PHARMACOL, MED SCH, 86- *Honors & Awards:* Ebert Prize, Am Pharmaceut Asn, 61; Host-Madsen Medal, Int Pharmaceut Fedn, 72; Propter Merita Medal, Czech Med Soc, 74; Res Award,Am Pharmaceut Asn Acad Scientists, 83; Volwiler Award, Am Asn Col Pharm, 83. *Mem:* AAAS; Am Pharmaceut Asn; Am Fedn Clin Res; Am Soc Clin Pharmacol & Therapeut; Am Soc Pharmacol & Exp Therapeut; NY Acad Sci; Soc Study Xenobiotics. *Res:* Pharmacokinetics and biopharmaceutics; absorption, metabolism and excretion of drugs in man. *Mailing Add:* Upjohn Ctr Clin Pharmacol Univ Mich Hosp UJ 3705/0509 Ann Arbor MI 48109-0504

WAGNER, JOHN GEORGE, b Bowmansville, NY, July 9, 42; m 67; c 3. EXPERIMENTAL MECHANICS. *Educ:* State Univ NY Buffalo, BS, 65; Univ Calif, Berkeley, MS, 67; Brown Univ, PhD(solid mech), 69. *Prof Exp:* Asst prof mech eng, Univ Pittsburgh, 69-75; res engr optical recording, 76-80, PROG LEADER POWDER TECHNOL, PHILIPS LAB, NAM PHILIPS CORP, 80- *Concurrent Pos:* Consult, various law firms, 72-75; prin investr conveyor belt transport anal, US Bur Mines, 75; vis engr, Res Lab, US Steel, 75; vis scientist, Natuurkundig Lab, Neth, 80-81. *Mem:* Am Soc Mech Engrs; Am Soc Metals; Am Powder Metall Inst; Sigma Xi; Soc Exp Stress Anal. *Res:* Theoretical and experimental mechanics of flow and deformation of solid and powdered materials with particular emphasis on constitutive and void behavior. *Mailing Add:* NAm Philips Corp 100 E 42nd St New York NY 10017

WAGNER, JOHN PHILIP, b Trenton, NJ, Feb 29, 40; m 69; c 2. SEPARATIONS SCIENCES, COMBUSTION-FLAMMABILITY. *Educ:* St Joseph's Univ, BS, 61; Johns Hopkins Univ, MS, 64, PhD(chem eng), 66. *Prof Exp:* Res asst, Dept Chem, Johns Hopkins Univ, 61-62, assoc chemist, Appl Physics Lab, 62, res fel, Dept Chem Eng, 62-66, sr engr, Appl Physics Lab, 66-72; assoc dir & res engr, Food Protein Res & Develop Ctr, Tex Eng Exp Sta, 83-90, assoc prof, Dept Indust Eng, 85-89, ASSOC PROF, DEPT NUCLEAR ENG, TEX A&M UNIV, 89-, ASSOC DIR & RES ENGR, ENG BIOSCI RES CTR, 90- *Concurrent Pos:* Res asst, RCA Labs, Princeton, NJ, 61; sr res engr, Factory Mutual Res Corp, Norwood, Mass, 72-73; res supv, Gillette Res Inst, 73-78, group leader, 78; staff engr, Exxon Res & Eng Co, 78-83, sr staff engr, 83; consult, Lawrence Livermore Nat Lab, Exxon Co, Englehard Industs, Gillette Res Inst, Liberty Mutual & Champion Int; mem, Environ Comt, Am Oil Chemists Soc, 85-86; co-guest ed, spec ed Bioresource Technol, 91. *Mem:* Am Inst Chem Engrs; Am Chem Soc; Am Soc Eng Educ; Asn Advan Indust Crops. *Res:* Advanced electrically enhanced separations in two and three- phase systems; process engineering involving biomaterials, petrochemicals and oily sludges; combustion/ flammability characteristics of plastics and wood derived products; electrostatics; fire and gas sensor technology; author of numerous publications in science journals; awarded three US patents. *Mailing Add:* Dept Nuclear Eng Tex A&M Univ College Station TX 77843-3133

WAGNER, JOSEPH EDWARD, b Dubuque, Iowa, July 29, 38; m 59; c 4. LABORATORY ANIMAL SCIENCE. *Educ:* Iowa State Univ, DVM, 63; Tulane Univ, MPH, 64; Univ Ill, Urbana, PhD, 67. *Prof Exp:* PROF VET MED, COL VET MED, UNIV MO, COLUMBIA, 69- *Concurrent Pos:* Mem, Animal Resources Adv Comt, Div Res Resources, NIH, 80- *Mem:* Am Vet Med Asn; Am Asn Lab Animal Sci; Am Col Lab Animal Med; Am Soc Lab Animal Practitioners; NY Acad Sci. *Res:* Pathogenesis and etiology of naturally occurring diseases of animals used in human health related research. *Mailing Add:* W213 Vet Path Univ Mo Columbia MO 65211

WAGNER, KENNETH, botany, documentary photography, for more information see previous edition

WAGNER, KIT KERN, b Chickasha, Okla, Nov 13, 47; m 70. DYNAMIC METEOROLOGY, AIR POLLUTION METEOROLOGY. *Educ:* Univ Okla, BS, 70, MS, 71, PhD(meteorol), 75. *Prof Exp:* Res asst meteorol, Univ Okla Res Inst, 70-75; meteorologist, Nat Severe Storm Lab, Nat Oceanic & Atmospheric Admin, 75; asst prof meteorol, Univ Calif, Davis, 75-80; air pollution res specialist, 81-88, SR AIR POLLUTION SPEC, CALIF AIR RESOURCES BD, 88- *Concurrent Pos:* Cert Meteorol Consult. *Mem:* Am Meteorol Soc; Air Pollution Control Asn. *Res:* Mesoscale dynamic meteorology and air pollution modeling. *Mailing Add:* Calif Air Resources Bd PO Box 2815 Sacramento CA 95812

WAGNER, LAWRENCE CARL, b Campbellsport, Wis, Dec 28, 46. HIGH TEMPERATURE CHEMISTRY. *Educ:* Marquette Univ, BS, 68; Purdue Univ, PhD(chem), 74. *Prof Exp:* Appointee chem div, Argonne Nat Lab, 74-76; sr mem tech staff, 76-79, MGR, DEVICE ANALYSIS LAB, TEX INSTRUMENTS, 79- *Mem:* Am Chem Soc; Am Soc Mass Spectrometry; Sigma Xi. *Res:* High temperature mass spectrometry; photoelectron spectrometry; diffusion controlled processes; scanning electron microscopy; semiconductor failure analysis. *Mailing Add:* 4405 Brigade Ct Plano TX 75024-5431

WAGNER, MARTIN GERALD, b New York, NY, Mar 19, 42; m 65; c 3. POLYMER SCIENCE, RHEOLOGY. *Educ:* Cooper Union, BChE, 62; Northwestern Univ, Evanston, MS, 64, PhD(chem eng), 67. *Prof Exp:* SR ENG ASSOC, RES & DEVELOP LAB, DUPONT CHEMICALS, E I DU PONT DE NEMOURS & CO, INC, 66- *Mem:* AAAS; Am Inst Chem Engrs; Soc Rheology; Am Chem Soc. *Res:* Polymer processing; polymerization technology; formulation of high explosive compositions. *Mailing Add:* DuPont Chemicals Du Pont Exp Sta PO Box 80353 Wilmington DE 19880-0353

WAGNER, MARTIN JAMES, b Independence, Kans, Oct 4, 31; m 53; c 3. BIOCHEMISTRY, SCIENCE EDUCATION. *Educ:* Pittsburg (KS) State Univ, BS, 54; Ind Univ, PhD(biochem), 58. *Prof Exp:* Asst chem, Ind Univ, 54-58; from asst prof to assoc prof, 58-63, chmn dept, 58-87, PROF BIOCHEM, BAYLOR COL DENT, 63- *Mem:* Am Chem Soc; Int Asn Dent Res; Sigma Xi; Am Asn Dent Schs. *Res:* Intermediary metabolism of inorganic fluoride ion; trace element nutrition; toxicology of trace elements; microcomputer applications in education. *Mailing Add:* Dept Biochem Baylor Col Dent Dallas TX 75246

WAGNER, MELVIN PETER, b Nebr, Nov 16, 26; m 53; c 5. ORGANIC CHEMISTRY, POLYMER CHEMISTRY. *Educ:* Creighton Univ, BS, 49, MS, 52; Univ Akron, PhD(polymer chem), 60. *Prof Exp:* Res chemist, 52-60, sr res chemist, 60-64, supvr rubber chem res, 64-78, RES SCIENTIST, BARBERTON LAB, CHEM DIV, PPG INDUSTS, 78- *Mem:* Am Chem Soc; Am Soc Testing & Mat; Soc Plastic Engrs. *Res:* High polymers; rubber reinforcement; vulcanization. *Mailing Add:* 416 Forest Lane Wadsworth OH 44281-2329

WAGNER, MORRIS, b Chicago, Ill, Aug 6, 17; m 47; c 4. BACTERIOLOGY & GNOTOBIOLOGY, IMMUNOLOGY. *Educ:* Cornell Univ, BS, 41; Univ Notre Dame, MS, 46; Purdue Univ, PhD, 66. *Prof Exp:* Grad asst, 41-43, bacteriologist, Lobund Labs, 43-46, from instr to prof microbiol, 46-84, asst chmn dept, 69-84, EMER PROF MICROBIOL, UNIV NOTRE DAME, 84-, RES SCIENTIST, LOBUND LAB, 41- *Concurrent Pos:* NSF sci fac fel, 63; mem, Subcomt Stand Gnotobiotics, Nat Acad Sci; travel awards, Am Soc Microbiol, 66 & 70, NATO, 65 & 69 & NSF, 67; pres, Ind Br, Am Soc Microbiol, 67; adj prof microbiol, Sch Med, Ind Univ, 79-85. *Mem:* Am Soc Microbiol; Asn Gnotobiotics; Am Asn Lab Animal Sci; Am Soc Dent Children; Acad Pedodontics. *Res:* Gnotobiotics; experimental dental caries; defined intestinal flora. *Mailing Add:* Lobund Lab Univ Notre Dame Notre Dame IN 46556

WAGNER, NEAL RICHARD, b Topeka, Kans, May 4, 40; m 71. TOPOLOGY. *Educ:* Univ Kans, AB, 62; Univ Ill, Urbana-Champaign, AM, 64, PhD(math), 70. *Prof Exp:* From asst prof to assoc prof math, El Paso, 69-79, PROF COMPUTER SCI, UNIV TEX, SAN ANTONIO, 79- *Concurrent Pos:* Vis assoc prof comput sci, Univ Houston, Tex, 79-81 & Drexel Univ, Philadelphia, 81-; consult modern cryptography, 81- *Mem:* Am Math Soc; Math Asn Am. *Res:* Computer sciences in general; software systems; theory; real-time simulation of NASA space shuttle. *Mailing Add:* Univ Tex San Antonio TX 78285-0664

WAGNER, NORMAN KEITH, b Longview, Wash, Oct 3, 32; m 54; c 2. MICROMETEOROLOGY. *Educ:* Univ Wash, BS, 54, MS, 56; Univ Hawaii, PhD(meteorol), 66. *Prof Exp:* Instr meteorol, Univ Tex, 56-57 & 58-63, res meteorologist, 57-58; asst prof meteorol, Univ Hawaii, 65, asst researcher, 65-66; asst prof, 66-70, dir, Atmospheric Sci Group, 72-76, ASSOC PROF METEOROL, UNIV TEX, AUSTIN, 70- *Concurrent Pos:* Vis prof, Univ Okla, 73 & 75. *Mem:* Am Meteorol Soc. *Res:* Micrometeorology; atmospheric boundary layer. *Mailing Add:* Dept Civil Eng Univ Tex Austin TX 78712

WAGNER, ORVIN EDSON, b Los Angeles, Calif, Jan 23, 30; m 77; c 3. COHERENT WAVE BEHAVIOR IN PLANTS. *Educ:* Walla Walla Col, BA, 53, BS, 57; Ariz State Univ, MS, 63; Univ Tenn, PhD(physics), 68. *Prof Exp:* Scientist, Lockheed Res Lab, Palo Alto, Calif, 61-62; instr physics, Walla Walla Col, College Place, Wash, 62-64; fel, Oak Ridge Nat Lab, 68-69; asst prof physics, Calif State Polytech Univ, San Luis Obispo, 69-74; PRES, WAGNER RES LAB, ROGUE RIVER, ORE, 74- *Concurrent Pos:* Consult, Wagner Electronic Prods, 69-; lectr, Rogue Community Col, Grant Pass, Ore, 74-79. *Mem:* Sigma Xi; Am Phys Soc. *Res:* Wave behavior in plants and application of this theory to the solar system; author of five articles and one book on wave behavior in plants. *Mailing Add:* Wagner Res Lab 2645 Sykes Creek Rd Rogue River OR 97537

WAGNER, PATRICIA ANTHONY, b Kirksville, Mo, Nov 26, 37; m 59; c 2. NUTRITIONAL SCIENCES. *Educ:* Northeast Mo State Teachers Col, BS, 59; Univ Wis-Madison, MS, 73, PhD(nutrit sci, biochem), 75. *Prof Exp:* Teacher home econ, Mo Pub Sch Syst, 59-69; res asst & NIH trainee nutrit sci, Univ Wis-Madison, 69-75; asst prof, 75-80, ASSOC PROF HUMAN NUTRIT, UNIV FLA, 80- *Concurrent Pos:* Prin investr, NIH-Nat Inst Aging res grant, 77-80 & Area Agency Aging, Older Am Nutrit Proj, 81; co-prin investr, USDA-Sci & Educ Admin grant, 78-81. *Mem:* Sigma Xi; Am Inst Nutrit; Am Soc Clin Nutrit; Proc Soc Exp Biol & Med. *Res:* Human nutrition; trace element requirements and metabolism; community nutritional assessment; nutrition and aging; nutrition in international development particularly in subsaharan Africa. *Mailing Add:* 1510 NW 35th Terr Gainesville FL 32605

WAGNER, PETER EWING, b Ann Arbor, Mich, July 4, 29; m 51; c 2. PHYSICS, ENVIRONMENTAL ENGINEERING. *Educ:* Univ Calif, AB, 50, PhD(physics), 56. *Prof Exp:* Asst, Univ Calif, 50-56; physicist, Westinghouse Res Labs, 56-59; from assoc prof to prof elec eng, Johns Hopkins Univ, 59-73; prof & dir, Ctr Environ & Estuarine Studies, Univ Md, 73-80; prof physics, Univ Ala, 80-81; acad vchancellor & prof physics, Univ Miss, 81-84; provost & prof physics & elec eng, Utah State Univ, 84-89; PROVOST & PROF PHYSICS & ELEC ENG, STATE UNIV NY, BINGHAMTON, 89- *Concurrent Pos:* Consult, Radiation Lab & Carlyle Barton Lab, Johns Hopkins Univ, 59-65, Westinghouse Elec Corp, 59-66, Am Cyanamid Co, 64-70 & US Army, 69-70; Guggenheim fel, 66-67; physicist, Appl Physics Lab, Johns Hopkins Univ, 71; spec proj engr, State of Md, 71-72; exec secy, Md Power Plant Siting Adv Comt, 72-73. *Mem:* AAAS; Am Phys Soc; Sigma Xi. *Res:* Solid state physics; paramagnetic resonance; microwave acoustics; static electrification; environmental measurements. *Mailing Add:* Off Provost State Univ NY Binghamton NY 13902-6000

WAGNER, PETER J, b Chicago, Ill, Dec 25, 38; m 63; c 6. PHYSICAL ORGANIC CHEMISTRY, PHOTOCHEMISTRY. *Educ:* Loyola Univ, Ill, BS, 60; Columbia Univ, MA, 61, PhD(chem), 63. *Prof Exp:* Res assoc chem, Columbia Univ, 63-64; NSF fel, Calif Inst Technol, 64-65; from asst prof to assoc prof, 65-70, PROF CHEM, MICH STATE UNIV, 70- *Concurrent Pos:* Sloan fel, 68-70; NSF sr fel, Univ Calif, Los Angeles, 71-72; consult, Hercules, Inc, 73-77; assoc ed, J Am Chem Soc, 75-86; Guggenheim fel, 83-84. *Mem:* Sigma Xi; Am Chem Soc. *Res:* Mechanisms of free radical and photochemical reactions; electronic energy transfer; photoexcited polyfunctional molecules. *Mailing Add:* Dept Chem Col Natural Sci Mich State Univ East Lansing MI 48824

WAGNER, RAYMOND LEE, b Kansas City, Mo, Aug 21, 46; m 69; c 2. ASTRODYNAMICS. *Educ:* Rice Univ, BA, 68; Univ Tex, Austin, PhD(astron), 72. *Prof Exp:* Asst prof astron, Univ Wash, 72-74; asst prof astron & physics, La State Univ, Baton Rouge, 74-79; software engr, Ford Aerospace, 79-82, prin software engr, 82-83, supvr, 83-86, prog mgr, 86-89, dept mgr, 89-90; MGR, ADVAN SYSTS, ROCKWELL INT, 90- *Concurrent Pos:* Fac assoc, Univ Colo, 83-84. *Mem:* fel Am Inst Aeronaut & Astronaut; Sigma Xi; Int Astron Union; Nat Mil Intel Asn; Armed Forces Commun & Electronics Asn. *Res:* Formal methods; system security and integrity; astrodynamics; orbit theory, command and control. *Mailing Add:* PO Box 19000 San Bernardino CA 92423-9000

WAGNER, RICHARD CARL, b Orange, NJ, Sept 25, 41. MATHEMATICS. *Educ:* Rutgers Univ, AB, 63; Univ Chicago, MS, 64, PhD(math), 68. *Prof Exp:* From asst prof to assoc prof, 68-79, PROF MATH, FAIRLEIGH DICKINSON UNIV, 79- *Mem:* Asn Comput Mach; Am Math Soc; Math Asn Am; London Math Soc. *Res:* Quadratic forms; algebraic k-theory. *Mailing Add:* Dept Math Fairleigh Dickinson Univ Madison NJ 07940

WAGNER, RICHARD JOHN, b Barnesville, Minn, Jan 13, 36; m 58; c 2. SOLID STATE PHYSICS. *Educ:* St John's Univ, Minn, BS, 58; Univ Calif, Los Angeles, MS, 60, PhD(physics), 66. *Prof Exp:* Mem tech staff eng, Hughes Aircraft Co, 58-68, staff physicist, 68-69, sr tech staff asst physics, 69-71, sr staff physicist, 71-75, sr scientist, 75-78, dept mgr, Receiver Dept, 78-80, asst lab mgr, Microwave Systs Lab, 80-90, SR DIV STAFF MGR, MICROWAVE DIV, HUGHES AIRCRAFT CO, 90- *Concurrent Pos:* Teaching asst, Univ Calif, Los Angeles, 61-63, res asst, 63-66, asst res physicist, 66-67. *Mem:* Am Phys Soc. *Res:* Solid state physics, particularly as applicable to solid state microwave devices. *Mailing Add:* Div Microwave Hughes Aircraft Mail Sta Re/R/9M30 Box 92426 Los Angeles CA 90009

WAGNER, RICHARD JOHN, b New Ulm, Minn, Dec 3, 32; m 53; c 4. MATHEMATICAL PHYSICS. *Educ:* Univ Minn, BA, 53, MS, 55; Rice Univ, PhD(physics), 58. *Prof Exp:* Mem tech staff theoret physics, Ramo-Wooldridge Corp, 58-59 & TRW Space Tech Labs, 59-61, sect head quantum theory, 62-65, sect head, TRW Systs, 65-66, sect head wave propagation, 66-74, SR SCIENTIST, TRW, 75- *Concurrent Pos:* Lectr, Dept Math, Univ Southern Calif, 60-62. *Honors & Awards:* Cedric K Ferguson Medal, Am Inst Mining, Metall & Petrol Engrs, 58. *Mem:* Am Phys Soc. *Res:* Scattering theory; missile and space vehicle re-entry physics; radio propagation; electromagnetic diffraction theory; radar concealment; underwater acoustics; statistical scattering theory; remote sensing; laser propagation; non-linear optics. *Mailing Add:* TRW One Space Park Bldg 01-1070 Redondo Beach CA 90278

WAGNER, RICHARD LLOYD, b Manitowoc, Wis, May 30, 34; m 56; c 5. POLYMER CHEMISTRY, PHOTOCHEMISTRY. *Educ:* Univ Wis-Madison, BS, 60. *Prof Exp:* Chemist, Hercules Res Ctr, 60-65, res chemist & proj leader mat sci & appl res, 66-71, sr venture analyst, New Enterprise Dept, 71-73, supvr mkt serv, 73-75, mgr eng & develop, Graphic Systs Div, Org Dept, 75-78, MGR, DEVELOP DEPT, HERCULES INC, 78- *Mem:* Am Chem Soc. *Res:* Applications research and product development work related to uses of company products in graphic arts areas and other commercially important areas; product and market development with photochemical systems in graphic arts uses; chemical and equipment systems for graphic arts products involving photopolymers. *Mailing Add:* One Wellington West Dr RD 1 Hockessin DE 19707-1014

WAGNER, RICHARD LORRAINE, JR, b Oklahoma City, Okla, July 7, 36; m 58; c 3. PHYSICS, MATHEMATICS. *Educ:* Williams Col, BA, 58; Univ Utah, PhD(physics), 63. *Prof Exp:* Student teacher & res asst physics, Univ Utah, 58-63; physics & defense res, Lawrence Livermore Lab, 63-75, assoc dir, 76-81; ASST SECY DEFENSE, ATOMIC ENERGY, DEPT DEFENSE, 81- *Concurrent Pos:* Mem, Advan Res Projs Agency-Defense Nuclear Agency Long Range Res & Develop Panel, 73-74, Joint Strategic Target Planning Staff Sci Adv Group, Joint Chiefs of Staff, Offutt AFB, Nebr, 73-79, US Army Sci Adv Panel, 76-78 & Defense Sci Bd, Off Under Secy Defense, 79- *Mem:* Am Phys Soc. *Res:* Cosmic rays; high energy physics; nuclear explosive design; weapons effects; antiballistic missile system studies. *Mailing Add:* 1003 Congress Lane McLean VA 22101

WAGNER, RICHARD S(IEGFRIED), b Wels, Austria, July 3, 25; m 57; c 4. MATERIALS SCIENCE, SOLID STATE PHYSICS. *Educ:* Vienna Tech Univ, MS, 51; Harvard Univ, MS, 57, PhD(appl physics), 59. *Prof Exp:* MEM TECH STAFF MAT SCI, AT&T BELL TEL LABS, 59-, DEPT HEAD, 70- *Honors & Awards:* C Mathewson Gold Medal, Am Inst Mining, Metall & Petrol Engrs, 66. *Mem:* AAAS; Am Inst Mining, Metall & Petrol Engrs; Am Phys Soc; Electrochem Soc; Am Vacuum Soc; Sigma Xi. *Res:* Physical metallurgy; solidification; crystal growth; defects in solids; vapor phase reactions; semiconductor device and magnetic bubble device technology development; technology development Silicon VLS1 and VLS1 devices. *Mailing Add:* Bell Tel Labs Rm 2A-325 Murray Hill NJ 07974

WAGNER, ROBERT ALAN, b Philadelphia, Pa, Mar 25, 41; m 68. COMPUTER SCIENCE. *Educ:* Mass Inst Technol, BS, 62; Carnegie-Mellon Univ, PhD(comput sci), 69. *Prof Exp:* Programmer, Rand Corp, 62-65; asst prof comput sci, Cornell Univ, 68-71; assoc prof systs & info sci, Vanderbilt Univ, 71-78; ASSOC PROF COMPUT SCI, DUKE UNIV, 78- *Mem:* Soc Indust & Appl Math; Asn Comput Mach; Sigma Xi. *Res:* Algorithms, especially techniques for constructing optimal algorithms; application of dynamic programming to computer-suggested problems; programming languages; operating systems. *Mailing Add:* 4800 University Dr Apt 4B Durham NC 27707

WAGNER, ROBERT E(ARL), b Baltimore, Md, July 30, 20; m 49; c 3. CHEMICAL ENGINEERING. *Educ:* Drexel Inst, BS, 46; Princeton Univ, MS, 48, PhD(chem eng), 55. *Prof Exp:* Lab asst chem, E I du Pont de Nemours & Co, 39-41 & Lever Bros Co, 41-42; thermodynamist, Glenn L Martin Co, 46; from asst prof to assoc prof, 49-62, PROF CHEM ENG, WORCESTER POLYTECH INST, 62- *Concurrent Pos:* Consult, New Eng Gas & Elec Asn. *Mem:* Am Chem Soc; Am Inst Chem Engrs; Sigma Xi. *Res:* Distillation; vapor liquid equilibria. *Mailing Add:* Dept Chem Eng Worcester Polytech Inst 100 Institute Rd Worcester MA 01609

WAGNER, ROBERT EDWIN, b Akron, Ohio, May 5, 20; m 52; c 2. ORGANIC CHEMISTRY. *Educ:* Mass Inst Technol, SB, 42; Princeton Univ, MA, 49, PhD(chem), 51. *Prof Exp:* Jr technologist, Shell Oil Co, 43-47; res chemist, 53-60, TECH SERV SUPVR, EXP STA LAB, E I DU PONT DE NEMOURS & CO, INC, 60- *Concurrent Pos:* Mem, Nat Defense Res Comt, 42. *Res:* Cellulose and polymer chemistry. *Mailing Add:* Burnt Mill Rd Chadds Ford PA 19317

WAGNER, ROBERT G, b Kansas City, Mo, Apr 2, 34; m 57; c 3. SOLID STATE PHYSICS, DIGITAL IMAGE ANALYSIS. *Educ:* Grinnell Col, AB, 56; Univ Mo, MS, 60, PhD(physics), 66. *Prof Exp:* Res engr, NAm Aviation, Inc, 56-58; res asst solid state physics, 60-66, from res scientist to assoc scientist, 66-71, sr group engr electronics, 71-76, sr tech specialist, 76-81, PRIN TECH SPECIALIST, MCDONNELL DOUGLAS CORP, 81- *Concurrent Pos:* Asst prof, Univ Mo, St Louis, 71- *Mem:* AAAS; Inst Elec & Electronics Engrs; Am Phys Soc; Sigma Xi. *Res:* Thins films; device physics. *Mailing Add:* McDonnell Douglas Corp Mail Code 1063413 PO Box B-516 St Louis MO 63166

WAGNER, ROBERT G, b Freeport, Ill, Sept 14, 50; c 2. PARTICLE PHYSICS DETECTORS, ELECTROWEAK INTERACTIONS. *Educ:* Univ Ill, BS, 72, MS, 74, PhD(physics), 78. *Prof Exp:* Postdoctoral, 77-80, asst physicist, 80-82, PHYSICIST, ARGONNE NAT LAB, 82- *Mem:* Am Phys Soc. *Res:* Hadron collider; detector physics. *Mailing Add:* Bldg 362 Rm E-285 Argonne Nat Lab 9700 S Cass Ave Argonne IL 60439-4815

WAGNER, ROBERT H, b Peru, Ind, Aug 11, 21; m 45; c 6. BIOCHEMISTRY. *Educ:* DePauw Univ, AB, 43; Univ Cincinnati, PhD(biochem), 50. *Prof Exp:* Asst, DePauw Univ, 43-44; asst, Res Found, Children's Hosp, 46-50; res assoc path & instr biochem, 53-56, asst prof path & biochem, 57-61, assoc prof, 61-67, PROF PATH, SCH MED, UNIV NC, CHAPEL HILL, 67-, PROF BIOCHEM, 72- *Concurrent Pos:* USPHS sr fel, 59-63, res career develop fel, Univ NC, Chapel Hill, 64-69. *Honors & Awards:* Muray Thelin Hemophilia Award; Int Prize French Asn Hemophilia. *Mem:* AAAS; Am Chem Soc; Soc Exp Biol & Med; Am Inst Chemists; Am Soc Exp Path; Sigma Xi. *Res:* Plasma proteins; enzymes; blood clotting; antihemophilic factors. *Mailing Add:* 311 Burlage Circle Chapel Hill NC 27514

WAGNER, ROBERT PHILIP, b New York, NY, May 11, 18; m 47; c 3. GENETICS. *Educ:* City Col New York, BS, 40; Univ Tex, PhD(genetics), 43. *Prof Exp:* Instr zool, Univ Tex, 43-44; res biologist, Nat Cotton Coun, Dallas, 44-45; from asst prof to prof, 45-77, EMER PROF ZOOL, UNIV TEX, AUSTIN, 77- *Concurrent Pos:* Nat Res Coun fel, Calif Inst Technol, 46; Guggenheim fel, 57; mem genetics panel, NSF, 61-64 & prog projs comt, 64-68 & Genetics Training Grant Comt, Nat Inst Gen Med Sci, 70-73; consult, Los Alamos Nat Lab, 78-; vis prof, Ind Univ, 62- & Univ Tex, Austin, 88- *Mem:* AAAS; Soc Study Evolution; Genetics Soc Am (secy, 65-66, vpres, 70, pres, 71); Am Soc Naturalists; Am Soc Biochem & Molecular Biol; Am Soc Human Genetics. *Res:* Chromosome structure and function. *Mailing Add:* 313 Los Arboles Dr Santa Fe NM 87501

WAGNER, ROBERT RODERICK, b New York, NY, Jan 5, 23. VIROLOGY, MICROBIOLOGY. *Educ:* Yale Univ, MD, 46. *Prof Exp:* Intern med, New Haven Hosp, 46-47, asst resident, 49-50; instr med, Yale Univ, 51-53, asst prof, 53-55; from asst prof to prof microbiol, Johns Hopkins Univ, 56-67, from asst dean to assoc dean med fac, 57-63; PROF MICROBIOL & CHMN DEPT, UNIV VA, 67-, MARION MCNULTY WEAVER PROF ONCOL, 84- *Concurrent Pos:* USPHS fel, Nat Inst Med Res, London, 50-51; vis fel, All Souls Col, Oxford Univ, 67; vis scientist, Dept Path, Oxford Univ, 67, Chinese Acad Med Sci, 82; ed-in-chief, J Virol, 66-82; consult, NIH, NSF & Am Cancer Soc; Josiah Macy Jr Found Fac Scholar, Oxford Univ, 75-76; Distinguished US scientist award, Alexander von Humboldt Found, 83; vis prof, Univ Giessen & Univ Würzburg, 83; dir Cancer Ctr, 84- *Mem:* Am Soc Microbiol; Am Soc Clin Invest; Asn Am Med Cols; Asn Am Physicians; Am Soc Biol Chemists; Am Soc Virol (pres, 84-85). *Res:* Biochemistry of viruses. *Mailing Add:* Dept Microbiol Univ Va Sch Med Box 441 Charlottesville VA 22908

WAGNER, ROBERT THOMAS, b Winona, Minn, July 15, 23; c 3. NUCLEAR PHYSICS. *Educ:* US Mil Acad, BS, 46; Univ Va, PhD(nuclear physics), 55. *Prof Exp:* Staff mem, Los Alamos Sci Lab, 55-64; prof physics, St Mary's Col, Minn, 64-66; chief tech develop div, Nike-X Syst Off, US Army, 66-67; prof & head physics dept, Northern Mich Univ, 67-75; dean fac, Sch Sci & Math, Univ Southern Colo, 75-78; acad dean, NMex Military Inst, 78-86; RADIOLOGIC PHYSICIST, RADIOL DEPT, EASTERN NMEX MED CTR, ROSWELL, NMEX, 88- *Mem:* Am Asn Physics Teachers; Am Phys Soc; Sigma Xi. *Res:* Ferroelectric ceramics; particle accelerators; fission physics; nuclear decay schemes; neutron and x-ray transport and diffusion; theoretical mechanics; scintillation radiation detectors. *Mailing Add:* 2609 Sherrill Lane Roswell NM 88201

WAGNER, ROBERT WANNER, mathematics; deceased, see previous edition for last biography

WAGNER, ROGER CURTIS, b Aitkin, Minn, May 14, 43; m 68; c 2. CELL BIOLOGY. *Educ:* Hamline Univ, BS, 65; Ohio Univ, MS, 67; Univ Minn, Minneapolis, PhD(cell biol), 71. *Prof Exp:* Teaching asst zool, anat & hist, Ohio Univ, 65-67; teaching asst biol & physiol, Univ Minn, Minneapolis, 67-68, res asst electrophysiol, St Paul; fel cell biol, Med Sch, Yale Univ, 71-74; asst prof biol sci, 74-77, PROF LIFE & HEALTH SCI, UNIV DEL, 77- *Concurrent Pos:* Nat Heart Lung & Blood Inst, 76-88. *Honors & Awards:* Res Career Develop Award. *Mem:* AAAS; Am Soc Cell Biol; Biophys Soc; Am Inst Biol Sci; Microcirculatory Soc Am. *Res:* Mechanism and function of macropinocytosis and micropinocytosis in mammalian cells; biomembranes; cell and molecular biology; histology; electron microscopy. *Mailing Add:* Sch Life & Health Sci Univ Del Newark DE 19711

WAGNER, ROSS IRVING, b Los Angeles, Calif, Apr 8, 25; m 49; c 3. SYNTHETIC INORGANIC & ORGANOMETALLIC CHEMISTRY. *Educ:* Univ Calif, Los Angeles, BS, 47; Univ Southern Calif, MS, 50, PhD(chem), 53. *Prof Exp:* Asst chem, Univ Southern Calif, 49-53; sr res chemist, Am Potash & Chem Corp, 53-63; mem tech staff, Rocketdyne Div, Rockwell Int, 63-90; RETIRED. *Mem:* Am Chem Soc. *Res:* Chemistry of boron and phosphorus sulfur and fluorine. *Mailing Add:* 4943 Queen Victoria Rd Woodland Hills CA 91364-4755

WAGNER, SIGURD, b Gaenserndorf, Austria, Nov 13, 41. SEMICONDUCTORS APPLICATION. *Educ:* Univ Vienna, Austria, PhD(phys chem), 68. *Prof Exp:* Fel metall eng, Ohio State Univ, 69-70; mem tech staff semiconductor res, Bell Labs, 70-78; chief, Photovoltaic Res Br, Solar Energy Res Inst, 78-80; PROF ELEC ENG, PRINCETON UNIV, 80- *Res:* Preparation and properties of new semiconductors and their application to devices; new semiconductors for solar cells. *Mailing Add:* Dept Elec Eng Princeton Univ Princeton NJ 08544

WAGNER, THOMAS CHARLES GORDON, b Pittsburgh, Pa, Jan 9, 16; m 42; c 3. ELECTRICAL ENGINEERING. *Educ:* Harvard Univ, SB, 37; Univ Md, MA, 40, PhD(math), 43. *Prof Exp:* With W Jett Lauck, DC, 37-38; asst mathematician, Univ Md, College Park, 38-40, instr, 40-45, prof elec eng, 46-76; PRES, TCG INC, 73- *Concurrent Pos:* Consult, Wash Inst Technol, Md, 40-46, Minneapolis-Honeywell Regulator Co, 47-59, Litton Industs, Inc, 59-62, Keltec Industs, Inc, 62-67 & Aero Geo Astro Div, Aiken Industs, Inc, 67-80. *Mem:* Inst Elec & Electronics Engrs; Sigma Xi. *Res:* Circuit analysis; timing devices; topology of networks. *Mailing Add:* 201 W Montgomery Ave Rockville MD 20850

WAGNER, THOMAS EDWARDS, b Cleveland, Ohio, Nov 29, 42; m 66, 90; c 3. BIOCHEMISTRY, ENDOCRINOLOGY. *Educ:* Princeton Univ, AB, 64; Northwestern Univ, PhD(biochem), 66. *Prof Exp:* Asst prof chem, Wellesley Col, 66-67; asst prof biochem, Med Col, Cornell Univ, 67-70; from asst prof to assoc prof, 70-78, PROF DEPT CHEM & BIOCHEM, OHIO UNIV, 78-, PROF DEPT ZOOL & BIOMED SCI, 82-, DIR MAMMALIAN RECOMBINANT GENETICS INST, 82- *Concurrent Pos:* Petrol Res Fund grant, 67-70; assoc endocrinol, Sloan-Kettering Inst Cancer Res, 68-70; prof & chmn, Grad Prog Molecular & Cell Biol, Ohio Univ, 82-86, dir, Edison Animal Biotechnol Ctr, 84-89, sci dir, 89-; sr vis fel, Animal Res Ctr, Cambridge Univ, 82; sci rev panelist, Nat Res Coun, 83-; distinguished lectr, Univ Tenn, 84 & Boyce Thompson Inst Plant Res, Cornell Univ, 87; mem, Comt Maintaining Global Genetic Resources, Nat Acad Sci, 86. *Mem:* Am Chem Soc. *Res:* Author of 158 publications. *Mailing Add:* Dept Chem Ohio Univ Athens OH 45701

WAGNER, TIMOTHY KNIGHT, b Pearl River, NY, July 5, 39; m 65; c 3. SOLID STATE PHYSICS. *Educ:* Univ Rochester, BS, 61; Univ Md, PhD(physics), 68. *Prof Exp:* Assoc solid state physics, Ames Lab, AEC, Iowa, 67-68, asst physicist, 68-70; from instr to asst prof, Iowa State Univ, 68-70; assoc prof, 70-74, chmn dept, 74-79, PROF PHYSICS, E STROUDSBURG STATE COL, 74- *Mem:* Sigma Xi; Am Phys Soc; Am Asn Physics Teachers. *Res:* Fermi surface measurements using radio-frequency size effect; ferromagnetic resonance in rare earth metals. *Mailing Add:* Dept Physics East Stroudsburg Univ East Stroudsburg PA 18301-2999

WAGNER, VAUGHN EDWIN, environmental health, medical entomology, for more information see previous edition

WAGNER, WARREN HERBERT, JR, b Washington, DC, Aug 29, 20; m 48; c 2. BOTANY. *Educ:* Univ Pa, AB, 42; Univ Calif, PhD, 50. *Prof Exp:* Res fel, Harvard Univ, 50-51; from instr to assoc prof, 51-61, dir bot gardens, 66-71, chmn dept, 75-78, PROF BOT & NATURAL RESOURCES, UNIV MICH, ANN ARBOR, 61-, CUR HERBARIUM, 62- *Concurrent Pos:* Mem, Ad Hoc Comt Plant Taxon, Nat Acad Sci, 56-57, Plant Sci Planning Comt, 64-65, dep for bot, Subcomt Syst Biogeog, US Nat Comt, Int Biol Prog, 65-68; trustee, Cranbrook Inst Sci, 63-78; mem, Fairchild Trop Garden Res Comt, 66-69; mem, Smithsonian Inst Coun, 67-72 & Panel Syst Biol, NSF; consult mem, Int Union Conserv Nature & Natural Resources, 72-; hon mem, Smithsonian Coun, Curado Asn ad hon Herbario Nacional de Costa Rica; vis prof, Univ Hawaii, 87, Harvard Univ, 91. *Honors & Awards:* Merit Award, Bot Soc Am; Asa Gray Award, Am Soc Plant Taxon. *Mem:* Nat Acad Sci; fel AAAS (secy, 63-67, vpres bot sci sect, 68); Soc Study Evolution (vpres, 66, pres, 72); Am Soc Plant Taxon (pres, 66); Am Fern Soc (secy, 51-53, cur, 57-, pres, 70-71); Bot Soc Am (vpres, 72, pres, 77); Am Acad Arts & Sci; Int Asn Pteridologists (vpres, 83-87, pres, 88-93). *Res:* Morphology, life cycles, evolution and systematics of vascular plants, especially pteridophytes; science education; biology of higher plants, especially ferns; phylogenetic theory. *Mailing Add:* Dept Bot Univ Mich Ann Arbor MI 48109

WAGNER, WILLIAM CHARLES, b Elma, NY, Nov 12, 32; m 54; c 4. ENDOCRINOLOGY, EDUCATION ADMINISTRATION. *Educ:* Cornell Univ, DVM, 56, PhD(physiol), 68. *Prof Exp:* Res vet, Col Vet Med, Cornell Univ, 57-65, fel physiol, Dept Animal Sci, 65-68; from asst prof to prof physiol, Col Vet Med Iowa State Univ, 68-77; prof & head, Dept Vet Biosci, 77-90, ASSOC DEAN RES, COL VET MED, UNIV ILL, 90- *Concurrent Pos:* Vis prof, Inst Physiol, Tech Univ Munich, WGer, 73-74; consult, Res Adv Bd, Morris Animal Found, 78-81; Alexander von Humboldt scientist award, 73; Fulbright sr res prof, Inst Tierzucht, WGer, 84-85; mem, exec comt, Int Cong Animal Reprod & Artificial Insemination, 84-, PRES, 88-; Am Vet Med Assoc, Coun Educ, 87-93; prog vet sci, USDA-CSRS, 90-91. *Mem:* Am Col Theriogenology (pres, 77-78); Am Physiol Soc; Am Vet Med Asn; Am Soc Animal Sci; Soc Study Reproduction. *Res:* Physiology of parturition and the postpartum female, especially in regard to ruminants; placental hormone synthesis and leucocyte function. *Mailing Add:* Dept Vet Biosci Col Vet Med Univ Ill 2001 S Lincoln Urbana IL 61801

WAGNER, WILLIAM EDWARD, JR, b New York, NY, June 17, 25; m 63; c 2. CLINICAL PHARMACOLOGY. *Educ:* Princeton Univ, BA, 45; Columbia Univ, MD, 50; Am Bd Family Pract, dipl, 70, 77 & 83. *Prof Exp:* Instr clin med, Med Sch, NY Univ, 60-77; SR FEL CLIN PHARMACOL, PHARMACEUT DIV, CIBA-GEIGY CORP, 51-; ASSOC CLIN PROF MED, MED SCH, COLUMBIA UNIV COL PHYSICIANS & SURGEONS, 77- *Concurrent Pos:* Attending staff, Overlook Hosp, 62; secy, FPD, 76- *Mem:* AMA; Am Soc Clin Pharmacol & Therapeut; fel Am Acad Family Pract. *Res:* Drug metabolism; pharmacokinetics; biopharmaceuticals. *Mailing Add:* 3301 Valley Rd Basking Ridge NJ 07920

WAGNER, WILLIAM FREDERICK, b Canton, Mo, Sept 13, 16. CHEMISTRY. *Educ:* Culver-Stockton Col, AB, 38; Univ Chicago, SM, 40; Univ Ill, PhD(anal chem), 47. *Prof Exp:* Asst chemist, State Geol Surv, Ill, 40-45; asst chem, Univ Ill, 45-47; asst prof, Hanover Col, 47-49; from instr to prof, 49-83, chmn dept, 65-68 & 76-82, EMER PROF CHEM, UNIV KY, 83- *Mem:* AAAS; Am Chem Soc. *Res:* X-ray applied to chemical analysis; solvent extraction of metal chelates; thermal methods of analysis. *Mailing Add:* Dept Chem Univ Ky Lexington KY 40506

WAGNER, WILLIAM GERARD, b St Cloud, Minn, Aug 22, 36; m 68; c 4. THEORETICAL PHYSICS, QUANTUM ELECTRONICS. *Educ:* Calif Inst Technol, BS, 58, PhD(physics), 62. *Prof Exp:* Mem tech staff, Res Labs, Hughes Aircraft Co, 62-65, sr staff physicist, 65-70; assoc prof, 66-69, PROF PHYSICS & ELEC ENG, UNIV SOUTHERN CALIF, 69-, DEAN, 73- *Concurrent Pos:* Consult, Rand Corp, 60-65; Tolman res fel theoret physics, Calif Inst Technol, 62-65, lectr, Caltech, 63-65; asst prof, Univ Calif, Irvine, 65-66; consult, Janus Mgt Corp, 70-71 & Croesus Capital Corp, 71-74; dean, Div Natural Sci & Math, Col Lett, Arts & Sci, Univ Southern Calif, 73-87, spec asst, Acad Record Serv, 75-81; dean, Interdisciplinary Progs, 87-; NSF fel, Hughes fel. *Mem:* Am Phys Soc; Sigma Xi. *Res:* Neuroscience; informatics; Computer applications. *Mailing Add:* 2828 Patricia Ave Los Angeles CA 90064-4425

WAGNER, WILLIAM JOHN, b Gary, Ind, Mar 29, 38; m 62; c 5. SOLAR PHYSICS, SOLAR-TERRESTRIAL PHYSICS. *Educ:* John Carroll Univ, BS, 60, MS, 62; Univ Colo, PhD(astro-geophys), 69. *Prof Exp:* Sr physicist, 62-64, sci eng fel, Rocketdyne Div, NAm Aviation, Inc, 64-69; astrophysicist & Big Dome facil sect chief, Sacramento Peak Observ, Air Force Geophys Lab, 69-76; SMM coronagraph-polarimeter exp scientist, Nat Oceanic & Atmospheric Admin, 76-81, staff scientist, High Altitude observ, Nat Ctr Atmospheric Res, 81-84, physicist, Space Environ Lab, 84- 90; SOLAR PHYSICS DISCIPLINE SCIENTIST, SPACE PHYSICS DIV, OFF SPACE SCI & APPLN, NASA HQ, 90- *Concurrent Pos:* Mem sub comn solar physics, Comt on Space Res, 81-; responsible scientist GOES Solar X-ray Imager, 85-90, Solar X- ray Physics, proj leader, 86-90; prog scientist SMM, Standards for Occup Safety & Health Admin, Solar-A, Orbiting Space Lab, 90- *Mem:* Fel AAAS; Int Astron Union; Am Astron Soc; Am Phys Soc; Am Geophys Union. *Res:* Observational research concerning solar physics, solar activity, the corona and solar wind; solar-terrestrial physics; spectroscopy. *Mailing Add:* 1612 Sherwood Rd Silver Spring MD 20902

WAGNER, WILLIAM S, b Cincinnati, Ohio, Apr 8, 36; m 65. ELECTRICAL ENGINEERING. *Educ:* Univ Ky, BS, 59; Case Western Reserve Univ, MS, 61; Univ Cincinnati, PhD, 67. *Prof Exp:* Instr, 61-67, ASST PROF ELEC ENG, UNIV CINCINNATI, 67- *Concurrent Pos:* Ford Found yr in residency prog grant, 68-69. *Res:* Nonlinear system analysis; network synthesis; electronics; control systems. *Mailing Add:* Dept Phys Sci Northern Ky Univ Univ Dr Highland Heights KY 41076

WAGNER, WILLIAM SHERWOOD, b Mora, Minn, Sept 21, 28; m 62. ORGANIC CHEMISTRY. *Educ:* Univ Minn, BChem, 49; Univ Mo, PhD(org chem), 52. *Prof Exp:* Sr res chemist, Chemstrand Corp, Ala, 52-59; mgr org res paper prod, Fiber Prod Res Ctr, Inc, 59-62, asst dir paper prod, 62-63; res assoc, Celanese Res Co, 63-65, head spinning res sect, 65-70, group mgr, Hoechst Celanese Corp, 70-71, mgr res & labs, 71-72, dir develop, 72-81, dir fiber technol, Hoechst Fibers Indust Div, 81-88, sr tech assoc, 88-90; RETIRED. *Mem:* Am Chem Soc. *Res:* Synthetic fibers; polymers; organic synthesis; polymer synthesis and properties, processes for forming and treating fibers, fiber properties. *Mailing Add:* 696 Perrin Dr Spartanburg SC 29302-2458

WAGNER, WILTZ WALKER, JR, b New Orleans, La, July 7, 39; m 67; c 1. PULMONARY PHYSIOLOGY. *Educ:* Colo State Univ, PhD(physiol), 74. *Prof Exp:* Res fel physiol, Univ Colo Med Ctr, Denver, 60-67, res assoc, 67-74, instr, 74-80, asst prof med, 80-85; assoc prof physiol, biophys & anesthesia, 85-90, V K STOELTING PROF ANESTHESIOL, IND UNIV SCH MED, 90- *Concurrent Pos:* Site vis, NIH, 74; consult, Med Sch, Univ Calif, Los Angeles, 74 & Univ Calif, La Jolla, 75, Med Sch, Univ Calif, San Francisco, 70 & 79, Harvard Med Sch, Boston, 79 & 84. *Mem:* Fel Royal Micros Soc; Sigma Xi; Am Physiol Soc; Microcirc Soc; Am Heart Asn. *Res:* Pulmonary microcirculation using methods for direct visualization of capillary perfusion in vivo; capillary control mechanisms and functional implications in health and disease; collateral ventilation and relation to ventilation-perfusion balance; athletic amenorrhea. *Mailing Add:* Dept Physiol & Biophys Ind Univ Sch Med Indianapolis IN 46223

WAGNER-BARTAK, CLAUS GUNTER, b Munich, Ger, Sept 9, 37: Can citizen; m 69; c 3. MOLECULAR PHYSICS, AEROSPACE ENGINEERING. *Educ:* Ludwig Maximilian Univ, Munich, BSc, 62, MSc, 66, DrRerNat, 69. *Prof Exp:* Res scientist, Univ Munich, 66-69; proj scientist, aerospace technol, Messerschmitt-Boelkow-Blohm, 69-74, eng mgr & proj mgr, 69-74; proj mgr, Spar Aerospace Ltd, 74-80, vpres, 80-83; vpres, Energy Dynamics Inc, 83-91; PRES, DIASYN TECHNOLOGIES LTD, 88- *Honors & Awards:* Pub Serv Medal, NASA, 82; Joseph F Engelberger Award, Robotics Int, 86. *Mem:* Am Inst Aeronaut & Astronaut. *Res:* Applied and theoretical physics; science; biology; biotechnology; material science and applications in medicine and biology; science and society technology assessments; new technology developments; robotics; radiology. *Mailing Add:* 4092 Lee Hwy Arlington VA 22207

WAGONER, DALE E, b Niagara Falls, NY, Oct 12, 36; m 65, 74; c 3. GENETICS, BIOLOGY. *Educ:* Ind Univ, AB(music) & AB(zool), 59, MA, 64, PhD(genetics), 65. *Prof Exp:* Res asst Drosophila genetics, H J Muller Lab, Ind Univ, 60; res geneticist, Metab & Radiation Res Lab, USDA, 64-75; prof biol, Maharishi Int Univ, 75-83; SR TEACHING LAB SPECIALIST, NEW COL UNIV S FLA, 88- *Concurrent Pos:* From asst prof to assoc prof entom, Grad Fac, NDak State Univ, 68-75. *Mem:* Sigma Xi; Am Genetic Asn. *Res:* Basic formal genetics of house flies; karyotype-linkage group relationship; insect control by the use of genetic mechanisms, such as chromosomal translocation, meiotic drive, hybrid sterility, cytoplasmic incompatability, compound chromosomes and conditional lethal mutations; aging and oxygen consumption research in practitioners of the transcendental meditation and TM Sidhis Program. *Mailing Add:* 4427 Violet Ave Sarasota FL 34233

WAGONER, DAVID EUGENE, b Clarinda, Iowa, July 8, 49; m 82. EXPERIMENTAL HIGH ENERGY PHYSICS, HADRO-PRODUCTION RESEARCH. *Educ:* Iowa State Univ, BS, 71; Cornell Univ, MS, 78, PhD(exp physics), 81. *Prof Exp:* Teaching asst, Dept Physics, Cornell Univ, 75-76, res asst, Lab Nuclear Studies, 76-81; res assoc, Dept Physics, Fermi Nat Accelerator Lab, 81-85; assoc res prof, Dept Physics, Fla A&M Univ, 85-88; ASSOC PROF, PRAIRIE VIEW A&M UNIV, 88- *Concurrent Pos:* Co-prin investr, Fla A&M Univ, 86-88; co-prin investr, Prairie View A&M Univ, 88- *Mem:* Am Phys Soc; AAAS; Inst Elec & Electronics Engrs. *Res:* Hadronic production of charmonium states and direct phaton processes; Monte Carlo computer simulations; operation and analysis of large electromagnetic calorimeter; hadronic production of beauty mesons; study of massive dimuon states. *Mailing Add:* Dept Physics Box 488 Prairie View A&M Univ Prairie View TX 77446-0488

WAGONER, GLEN, b Terreton, Idaho, July 28, 27; m 52; c 1. PHYSICS, INSTRUMENT DESIGN. *Educ:* Idaho State Col, BS, 49; Univ Chicago, MS, 52; Univ Calif, PhD(physics), 57. *Prof Exp:* Physicist, Res Labs, Union Carbide Corp, 57-85; RETIRED. *Concurrent Pos:* Prof lectr, Case Inst Technol, 59-61; staff lectr, Baldwin Wallace Col, 84- *Mem:* Am Phys Soc. *Res:* Solid state physics; magnetic resonance; electronics; electric arcs. *Mailing Add:* 26564 Lake Rd Bay Village OH 44140

WAGONER, ROBERT H, b Columbus, Ohio, Jan 8, 52; m 80; c 2. SHEET METAL FORMING, MECHANICAL METALLURGY. *Educ:* Ohio State Univ, BS, 74, MS, 75, PhD(metall eng), 76. *Prof Exp:* Res scientist, Physics Dept, Gen Motors Res Labs, 77-80, staff res scientist, 80-83; PROF, DEPT METALL ENG, OHIO STATE UNIV, COLUMBUS, 83- *Concurrent Pos:* NSF fel, Univ Oxford, 76-77; Maitre de Rechercle, Ecole des Mines de Paris, 90-91. *Honors & Awards:* Rossiter W Raymond Mem Award & Robert Lansing Hardy Gold Medal, Am Inst Mining, Metall & Petrol Engrs, 81; H Mathewson Gold Medal, Metall Soc, 81, 83 & 88, Hardy Gold Medal, 81; Raymond Mem Award, Am Inst Mining, Metall & Petrol Engrs, 81, 83; Pres Young Investr, NSF, 84. *Mem:* Am Inst Mining, Metall & Petrol Engrs; Am Soc Mech Engrs; Soc Automotive Engrs; fel Am Soc Metals Int; Mat Res Soc; Metall Soc; Am Acad Mech. *Res:* Metal elasticity and plasticity; sheet metal forming; deformation testing, dislocation modelling; mechanical equation of state studies; finite element modeling. *Mailing Add:* Dept Mat Sci Eng 116 W 119th Ave Columbus OH 43210-1179

WAGONER, ROBERT VERNON, b Teaneck, NJ, Aug 6, 38; m 63, 87; c 2. THEORETICAL ASTROPHYSICS, COSMOLOGY. *Educ:* Cornell Univ, BME, 61; Stanford Univ, MS, 62, PhD(physics), 65. *Prof Exp:* Res fel physics, Calif Inst Technol, 65-68, Sherman Fairchild distinguished scholar, 76; from asst prof to assoc prof astron, Cornell Univ, 68-73; assoc prof, 73-77, PROF PHYSICS, STANFORD UNIV, 77- *Concurrent Pos:* Sloan res fel, 69-71; Guggenheim fel, 79; George Ellery Hale distinguished vis prof, Univ Chicago, 78; mem comt space astron & astrophys, NAS, 79-82, physics surv comt & theory study panel, Space Sci Bd, Nat Res Coun, 80-86, prog adv comt, Ctr Particle Astrophys, 90, Nat Sci & Eng Res Coun grant selection comt, Can, 90-93; prin investr, NSF & NASA. *Mem:* Fel Am Phys Soc; Int Astron Union; Am Astron Soc. *Res:* Astrophysics of compact objects and supernovae; cosmology; gravitation theory. *Mailing Add:* Dept Physics Stanford Univ Stanford CA 94305-4060

WAGONER, RONALD LEWIS, b Fairfield, Calif, Aug 4, 42; m 60; c 2. MATHEMATICS. *Educ:* Fresno State Col, BA, 65, MA, 66; Univ Ore, PhD(math), 69. *Prof Exp:* Asst prof, 69-74, assoc prof, 74-77, PROF MATH, CALIF STATE UNIV, FRESNO, 77- *Concurrent Pos:* Math specialist, Fresno City Unified Sch Dist, 70-71. *Res:* Ring theory: associative rings with identity. *Mailing Add:* Dept Math Calif State Univ 6241 N Maple Ave Fresno CA 93740

WAGREICH, PHILIP DONALD, b New York, NY, July 25, 41; m 62; c 3. PURE MATHEMATICS. *Educ:* Brandeis Univ, BA, 62; Columbia Univ, PhD(math), 66. *Prof Exp:* Lectr math, Brandeis Univ, 66-68; asst prof, Univ Pa, 68-73; assoc prof, 73-80, PROF, UNIV ILL, CHICAGO, 80- *Concurrent Pos:* Off Naval Res fel, 68-69; mem, Inst Advan Study, 68-70; NSF res grants, 74-; mem, Nat Coun Teachers Math. *Mem:* Am Math Soc; Math Asn Am. *Res:* Algebraic geometry; topology; transformation groups; development of curricula and programs to improve the quality of pre-college education. *Mailing Add:* Dept Math Statist & Comput Sci Univ Ill Chicago M/C 249 Box 4348 Chicago IL 60680

WAGSTAFF, DAVID JESSE, b Lehi, Utah, Feb 22, 35; m 63; c 3. TOXICOLOGY. *Educ:* Utah State Univ, BS, 59, PhD(toxicol), 70; Cornell Univ, DVM, 62. *Prof Exp:* Vet epidemiologist, USPHS, 62-64; vet meat inspector, USDA, 64-65; vet epidemiologist, USPHS, 65-66; asst prof toxicol, Univ Mo, Columbia, 69-73; toxicologist, 73-77, EPIDEMIOLOGIST, FOOD & DRUG ADMIN, 77- *Concurrent Pos:* NIH fel toxicol, Utah State Univ, 66-69; mem, Am Bd Vet Toxicol. *Mem:* Soc Toxicol; Am Col Vet Toxicol; Am Vet Med Asn. *Res:* Induction of liver microsomal enzymes; drug toxicity; environmental contaminants; toxicants in natural foods; interaction of toxicology with other fields; toxicant interactions; poisonous plants; food safety epidemiology. *Mailing Add:* 200 C St SW Washington DC 20204

WAGSTAFF, SAMUEL STANDFIELD, JR, b New Bedford, Mass, Feb 21, 45. COMPUTATIONAL NUMBER THEORY. *Educ:* Mass Inst Technol, BS, 66; Cornell Univ, PhD(math), 70. *Prof Exp:* Instr math, Univ Rochester, 70-71; vis mem, Inst Advan Study, 71-72; vis lectr, Univ Ill, 72-75, asst prof math, 75-81; assoc prof statist & comput sci, Univ Ga, 81-83; ASSOC PROF COMPUT SCI, PURDUE UNIV, 83- *Mem:* Am Math Soc; Math Asn Am; Soc Actuaries. *Res:* Factoring; primality testing; diophantine equations; computational complexity. *Mailing Add:* Dept Comput Sci Purdue Univ West Lafayette IN 47907

WAH, THEIN, b Rangoon, Burma, Apr 11, 19; nat US; m 52; c 3. MECHANICS, STRUCTURAL ENGINEERING. *Educ:* Univ Rangoon, BS, 41; Univ Utah, MS, 48; Harvard Univ, MS, 49; Univ Ill, PhD(eng), 53. *Prof Exp:* Asst engr, Burma Rwy, 41-47; bridge designer, State Div, Hwys, Ill, 52-53; asst prof civil eng & mech, Lehigh Univ, 53-54 & Univ Conn, 54-57; sr res engr, Southwest Res Inst, Tex, 57-61, staff scientist, 62-71; prof civil & mech eng, Tex A&I Univ, 71-84; RETIRED. *Concurrent Pos:* Vis prof, Indian Inst Technol, Kharagpur, 61-62; eng consult, San Antonio, Tex, 84- *Mem:* Am Soc Mech Engrs. *Res:* Elasticity; plasticity; vibrations; creep; thermoelasticity. *Mailing Add:* 6821 Stonykirk San Antonio TX 78240

WAHAB, JAMES HATTON, b Bridgeton, NC, Aug 29, 20; m 47; c 2. MATHEMATICS. *Educ:* Col William & Mary, BS, 40; Univ NC, AM, 50, PhD(math), 51. *Prof Exp:* Instr math & eng, Norfolk Div, Col William & Mary, 40-42 & 46-47, asst prof, 47-48; instr math, Univ NC, 50-51; from asst prof to assoc prof, Ga Inst Technol, 51-58; prof, La State Univ, 58-61, chmn dept, 60-61; prof, NC State Col, 61-63; prof & chmn dept, Univ NC, Charlotte, 63-68, actg acad dean, 64-68; head dept, 68-73, dir undergrad studies math, 77-80, prof math, 68-84, VIS PROF MATH, UNIV SC, 84- *Mem:* Am Math Soc; Math Asn Am. *Res:* Irreducibility of legendre polynomials; algebra; statistics; numerical analysis. *Mailing Add:* Rollins Col Box 2743 Winter Park FL 32789

WAHBA, ALBERT J, b Alexandria, Egypt, Feb 27, 28; US citizen; m 65; c 3. BIOCHEMISTRY. *Educ:* Univ Calif, Berkeley, AB, 51; Univ Tex, MA, 54; Tufts Univ, PhD(biochem & pharamacol), 61. *Prof Exp:* Instr, Dept Biochem, Sch Med, NY Univ, 63, asst prof, 63-65, assoc prof, 66-69; prof & dir biochem, Lab Molecular Biol, Can, 70-77; PROF & CHMN, DEPT BIOCHEM, UNIV MISS MED CTR, 77- *Concurrent Pos:* Jane Coffin Childs Fund med res fel, Dept Biochem, Sch Med, NY Univ, 62; vis scientist, Salk Inst Biol Studies, 66; Med Res Coun Assoc, Can, 70-77. *Mem:* Am Soc Biol Chemists; AAAS; Sigma Xi; Am Chem Soc; Am Soc Microbiol. *Res:* Nucleic acids and protein synthesis; molecular mechanisms and regulation during early embryonic development; transcriptional and translated control of gene expression during development of brine shrimp Artemia embryos; author or coauthor of over 70 publications. *Mailing Add:* Dept Biochem Univ Miss Med Ctr 2500 N State St Jackson MS 39216-4505

WAHBA, GRACE, b Washington, DC. MATHEMATICAL STATISTICS. *Educ:* Cornell Univ, BA, 56; Univ Md, College Park, MA, 62; Stanford Univ, PhD(math statist), 66. *Prof Exp:* Res mathematician, Opers Res, Inc, 57-61; systs analyst, Int Bus Mach Corp, 61-66; res assoc math statist, Stanford Univ,

66-67; from asst prof to assoc prof, 67-74, JOHN BASCOM PROF STATIST, UNIV WIS-MADISON, 75- *Concurrent Pos:* Assoc ed, Annals Statist, 74-80; mem int ed bd, Commun in Statist, 71-; fel St Cross Col; sr vis fel, Oxford Univ, 74-75; Lady Davis fel, tech, 80; mem, Coun Soc Indust & Appl Math, 80-82, Coun Inst Math Statist, 78-80, adv panel, math sci, NSF, 80-82, Peer Rev Panel, 81. *Mem:* Fel Inst Math Statist; fel Am Statist Asn; Am Math Soc; Soc Indust & Appl Math; fel AAAS. *Res:* Statistical model building; inverse problems; multivariate function estimation; experimental design. *Mailing Add:* Statist Univ Wis 1210 W Dayton St Madison WI 53706

WAHL, A(RTHUR) J, b Saxman, Kans, Feb 5, 20; m 50; c 2. ELECTRICAL ENGINEERING. *Educ:* Univ Kans, BS, 42; Princeton Univ, PhD(elec eng), 50. *Prof Exp:* Mem tech staff, Bell Tel Labs, 53-56, supvr semiconductor device develop, 56-83; RETIRED. *Mailing Add:* 1618 Meadowlark Rd Wyomissing PA 19610

WAHL, ARTHUR CHARLES, b Des Moines, Iowa, Sept 8, 17; m 43; c 1. NUCLEAR FISSION. *Educ:* Iowa State Univ, BS, 39; Univ Calif, PhD(chem), 42. *Prof Exp:* Res assoc, Manhattan Proj, Univ Calif, 42-43, group leader, Los Alamos Sci Lab, 43-46; assoc prof, 49-53, Farr prof radiochem, 53-83, EMER PROF CHEM, WASH UNIV, 83- *Concurrent Pos:* Consult, Los Alamos Sci Lab, 50-; NSF fel, 67. *Honors & Awards:* Am Chem Soc Award, 66; Humboldt Award, 77. *Mem:* Am Chem Soc. *Res:* Nuclear-charge distribution in fission; rapid electron-transfer reactions. *Mailing Add:* 1550 Los Pueblos Los Alamos NM 87544

WAHL, EBERHARD WILHELM, meteorology, space sciences; deceased, see previous edition for last biography

WAHL, FLOYD MICHAEL, b Hebron, Ind, July 7, 31; m 53; c 4. GEOCHEMISTRY. *Educ:* DePauw Univ, AB, 53; Univ Ill, MA, 57, PhD(mineral & geochem), 58. *Prof Exp:* Instr geol, Univ Ill, Urbana, 58-59, res asst prof, 59-60, from asst prof to assoc prof, 60-69; prof & chmn dept, Univ Fla, 69-73, prof geol, 70-82, dir div phys sci & math, 71-73, assoc dean, grad sch & assoc dir res, 73-79 & 80-81, actg dean grad study & res, 79-80; EXEC DIR, GEOL SOC AM, 82- *Mem:* Fel Geol Soc Am; Mineral Soc Am; Geochem Soc; Clay Minerals Soc; Am Inst Prof Geologists; Sigma Xi; Soc Econ Paleont & Mineral. *Res:* Clay mineralogy and sedimentary geochemistry; development of mineral resources; chemical alteration and those factors that lead to and control element concentration; phase changes in minerals at elevated temperatures. *Mailing Add:* Geol Soc Am PO Box 9140 Boulder CO 80301

WAHL, GEORGE HENRY, JR, b New York, NY, Sept 17, 36; m 58; c 3. STRUCTURAL CHEMISTRY. *Educ:* Fordham Univ, BS, 58; NY Univ, MS, 61, PhD(org chem), 63. *Prof Exp:* Res chemist, Pittsburgh Plate Glass Chem Co, Ohio, 58-59; NIH res fel org chem, Cornell Univ, 63-64; from asst prof to assoc prof, 64-75, PROF ORG CHEM, NC STATE UNIV, 75- *Concurrent Pos:* Guest prof, Swiss Fed Inst, Zurich, 73-74; consult, Environ Protection Agency, 78- *Honors & Awards:* Sigma Xi Res Award, 74. *Mem:* Am Chem Soc; Royal Soc Chem. *Res:* Organic stereochemistry; nuclear magnetic resonance spectroscopy; mass spectrometry; synthesis of unusual structures for physical investigation; synthesis and structure of adamantane and biphenyl derivatives. *Mailing Add:* Dept Chem NC State Univ Raleigh NC 27695-8204

WAHL, JONATHAN MICHAEL, b Washington, DC, Jan 29, 45; m 70; c 1. MATHEMATICS. *Educ:* Yale Univ, BS & MA, 65; Harvard Univ, PhD(math), 71. *Prof Exp:* Instr math, Univ Calif, Berkeley, 70-72; vis, Inst Advan Study, Princeton Univ, 72-73, 79; from asst prof to assoc prof, 73-81, PROF MATH, UNIV NC, CHAPEL HILL, 81- *Concurrent Pos:* NSF res grant, 71- *Mem:* Am Math Soc. *Res:* Singularities; deformation theory; algebraic geometry. *Mailing Add:* Dept Math Univ NC Chapel Hill NC 27599-3250

WAHL, PATRICIA WALKER, b La Grande, Ore, Dec 6, 38; m 63; c 1. BIOSTATISTICS. *Educ:* San Jose State Univ, BA, 60; Univ Wash, PhD(biostatist), 71. *Prof Exp:* Res analyst comput programming, Lockheed Missiles, Lockheed Aircraft Corp, 60-62; systs analyst, Control Data Corp, 63-64; head programmer, 64-66, instr biostatist, 71-73, asst prof, 74-85, PROF BIOSTATIST, UNIV WASH, 85-; ASSOC DEAN, SCH PUB HEALTH, 85- *Mem:* Am Statist Asn; Biomet Soc. *Res:* Use of regression analysis and other multivariate statistical techniques for exploratory data analysis; effect on classification by discriminant analysis when model assumptions fail. *Mailing Add:* Dept Biostatist Univ Wash Seattle WA 98195

WAHL, SHARON KNUDSON, b Mt Vernon, Wash, Mar 16, 45; m 71; c 2. IMMUNOLOGY. *Educ:* Pac Lutheran Univ, BS, 67; Univ Wash, PhD(biol struct), 71. *Prof Exp:* Fel path, Sch Med, Univ Wash, 71-72; fel cellular immunol, 72-74, staff fel humoral immunity, 74-75, sr staff fel humoral immunity, 75-76, res microbiologist, 76-83, CHIEF, CELLULAR IMMUNOL, NAT INST DENT RES, 83- *Concurrent Pos:* Preceptor, PRAT fel; lectr, grad course, FAES; vis prof, Cleveland Clin, 85; mem, Subcomt for Joint Adv-Comt Clin Hyperbaric Med Res. *Honors & Awards:* NIH Director's Award, 85; Howard & Martha Holley Res Prize Rheumatology, 90. *Mem:* Am Asn Immunol; Am Fedn Clin Res; AAAS. *Res:* Mechanisms of activation and characterization of T and B lymphocyte participation in cellular immune reactions and effect of immunosuppressive agents on these responses; influence of immune system on connective tissue metabolism; monocyte phenotype and function in inflammation and disease; immunomodulation by transforming growth factor beta; polypeptide growth factors. *Mailing Add:* Nat Inst Dent Res Bldg 30 Rm 326 9000 Rockville Pike Bethesda MD 20892

WAHL, WERNER HENRY, b Buffalo, NY, Oct 1, 30; m 51; c 2. NUCLEAR CHEMISTRY, RADIOCHEMISTRY. *Educ:* Univ Buffalo, BA, 54; Purdue Univ, MS, 56, PhD(phys inorg chem), 57. *Prof Exp:* Chem operator, Pathfinder Chem Corp, 49; asst, Linde Co Div, Union Carbide Co, 51-53; asst, Durez Plastics, Inc, 53; asst chem, Purdue Univ, 54-57; res chemist, Union Carbide Nuclear Corp, 57-61, group leader, 61-65, asst mgr res, 65-66, dir radiopharmaceut, Neisler Labs, Inc, Union Carbide Corp, NY, 66-69; dir opers, Mallinckrodt/Nuclear, Mo, 69-70; vpres & gen mgr, Amersham-Searle Corp, Ill, 70, exec vpres, 71, pres, 71-75; vpres new bus develop, Searle Diag, Inc, 75-78; PRES, NUCLEAR DIAGNOSTICS, INC, 78- *Concurrent Pos:* Asst chem, Univ Buffalo, 53. *Mem:* Fel AAAS; Am Chem Soc; Am Asn Physicists Med; Clin Radio Assay Soc; fel Am Inst Chem. *Mailing Add:* Nuclear Diag Inc 575 Robbins Dr Troy MI 48084

WAHL, WILLIAM G, b Winnepeg, Can, Sept 11, 30. MINERALOGY. *Educ:* Univ Manitoba, BSc, 53, MSc, 56. *Prof Exp:* PROF GEOL, UNIV CALIF, 65- *Mem:* Geol Asn Can; Mineral Asn Can; Minerol Soc Am; Geol Soc Am; Mineral Soc UK. *Mailing Add:* RR 1 Corbyville ON K0K 1V0 Can

WAHLBECK, PHILLIP GLENN, b Kankakee, Ill, Mar 29, 33; m 56; c 3. HIGH TEMPERATURE CHEMISTRY & SUPERCONDUCTIVITY, SURFACE CHEMISTRY. *Educ:* Univ Ill, BS, 54, PhD(chem), 58. *Prof Exp:* Asst chem, Univ Ill, 54-58; res assoc, Univ Kans, 58-60; from instr to assoc prof, Ill Inst Technol, 60-72; chmn dept, 72-78, PROF CHEM, WICHITA STATE UNIV, 72- *Concurrent Pos:* Vis prof, Tech Univ Norway, 70, 78 & 82; vis scientist, Los Alamos Nat Lab, 91. *Mem:* Am Sci Affil; Metall Soc; Am Chem Soc; Mat Res Soc. *Res:* Molecular beams; thermodynamics at high temperatures; transition metal hydrides, oxides, selenides and tellurides; vapor pressure measurements; effusion of gases; gas-surface interactions adsorption phenomena; mean residence times; surface diffusion; spatial distributions of restituted molecules; high temperature superconductors. *Mailing Add:* Dept Chem Wichita State Univ Wichita KS 67208

WAHLERT, JOHN HOWARD, b New York, NY, May 12, 43; m 69; c 3. VERTEBRATE PALEONTOLOGY. *Educ:* Amherst Col, BA, 65; Harvard Univ, MA, 66, PhD(geol), 72. *Prof Exp:* Curatorial asst vert paleont, 72-77, assoc, 76-81, RES ASSOC, DEPT VERT PALEONT, AM MUS NATURAL HISTORY, 81- *Concurrent Pos:* Vis asst prof biol, Franklin & Marshall Col, 77-78, cur mammal, North Mus, 78-81; asst prof biol, Millersville State Col, 80-81; asst prof biol, Baruch Col, 81-86, prof, 89; mem doctoral fac, Grad Sch & Univ Ctrs Prog Biol, City Univ New York, 87- *Mem:* Soc Vert Paleont; Am Soc Mammal. *Res:* Cenozoic rodents and their anatomy, taxonomy and phylogeny. *Mailing Add:* Dept Vert Paleont Am Mus Natural Hist New York NY 10024

WAHLGREN, MORRIS A, b Wildrose, NDak, May 31, 29; m 55; c 3. ENVIRONMENTAL CHEMISTRY. *Educ:* Jamestown Col, BS, 51; Univ Mich, PhD(chem), 61. *Prof Exp:* Radiochemist, Atomic Energy Div, Phillips Petrol Co, Idaho, 53-56; asst chemist, Chem Div, Argonne Nat Lab, 61-66, assoc chemist, 66-72, chemist, Radiol & Environ Res Div, 72-80, chemist, Chem Tech Div, 80-87, CHEMIST, ENVIRON, SAFETY & HEALTH DIV, ARGONNE NAT LAB, 87- *Mem:* AAAS; Am Chem Soc; fel Am Inst Chemists. *Res:* Nuclear and analytical chemistry; radiochemical separations; chemical limnology; behavior of artificial radionuclides in the Great Lakes; radiochemical bioassay. *Mailing Add:* 1278 Fellows St St Charles IL 60174

WAHLIG, MICHAEL ALEXANDER, b New York, NY, Oct 21, 34; m 56; c 4. ENERGY CONVERSION. *Educ:* Manhattan Col, BS, 55; Mass Inst Technol, PhD(physics), 62. *Prof Exp:* Res assoc physics, Mass Inst Technol, 62-66; res staff physics, 66-72, MEM RES STAFF ENERGY CONVERSION & ENERGY CONSERV, APPL SCI DIV, LAWRENCE BERKELEY LAB, UNIV CALIF, BERKELEY, 72- *Mem:* Am Phys Soc; Int Solar Energy Soc; AAAS. *Res:* Research, development and analysis of conversion and use of thermal energy for providing cooling and heating; energy conservation in building energy systems. *Mailing Add:* Lawrence Berkeley Lab Univ Calif Berkeley CA 94720

WAHLS, HARVEY E(DWARD), b Evanston, Ill, Aug 8, 31; m 60; c 2. FOUNDATION ENGINEERING, SOIL MECHANICS. *Educ:* Northwestern Univ, BS, 54, MS, 55, PhD(civil eng), 61. *Prof Exp:* From instr to asst prof civil eng, Worcester Polytech Inst, 55-60; from asst prof to assoc prof, 60-69, PROF CIVIL ENG, NC STATE UNIV, 69-, ASSOC HEAD DEPT, 84- *Concurrent Pos:* Instr, Northwestern Univ, 57-59; consult, Transp Res Bd, 70-71 & 80-83; chmn, Am Soc Civil Engers, Geotech Eng Div, 82-83. *Mem:* Am Soc Civil Engrs; Am Soc Testing & Mat; Am Soc Eng Educ; Int Soc Soil Mech & Found Engrs; Transp Res Bd; US Nat Soc Soil Mech & Found Eng. *Res:* Consolidation theory for cohesive soils; settlement analysis; compaction process and the behavior of compacted soils; soil dynamics. *Mailing Add:* Dept Civil Eng NC State Univ Box 7908 Raleigh NC 27695-7908

WAHLSTROM, ERNEST E, b Boulder, Co, Dec 30, 09. GEOLOGY. *Educ:* Univ Colo, BS, 31; Harvard Univ, MS, 36, PhD, 39. *Prof Exp:* Prof geol, Univ Colo, 36-78; RETIRED. *Mem:* Geol Soc Am; Mineral Soc Am; Soc Econ Geologists; Am Inst Petrol Geologists. *Mailing Add:* 174 Ave NE Redmond WA 98052

WAHLSTROM, LAWRENCE F, b Aurora, Wis, Feb 4, 15; m 38; c 2. MATHEMATICS. *Educ:* Lawrence Col, BA, 36; Univ Wis, MA, 37, PhD(math educ), 50. *Prof Exp:* Pub sch teacher, Ill, 37-41, chmn dept math, jr high sch, 41-45; chmn dept, Elgin Acad, 45-47; asst, Univ Wis, 47-48; PROF MATH & CHMN DEPT, UNIV WIS-EAU CLAIRE, 48- *Concurrent Pos:* NSF fac sci grant, 57-58. *Mem:* Math Asn Am. *Res:* Geometry. *Mailing Add:* 110 Skyline Dr Eau Claire WI 54703

WAHLSTROM, RICHARD CARL, b Craig, Nebr, Feb 13, 23; m 47; c 3. ANIMAL SCIENCE. *Educ:* Univ Nebr, BS, 48; Univ Ill, MS, 50, PhD(animal nutrit), 52. *Prof Exp:* Asst animal husb, Univ Ill, 48-51; res assoc nutrit, Merck Inst Therapeut Res, 51-52; assoc prof animal husb, 52-59, head dept, 60-67, PROF ANIMAL HUSB, SDAK STATE UNIV, 59- *Concurrent Pos:* Vis prof, Univ Nottingham & Nat Inst Res, Dairying, Eng, 74-75. *Honors & Awards:* Animal Mgt Award, Am Soc Animal Sci, 76. *Mem:* Am Soc

Animal Sci; Am Inst Nutrit. *Res:* Swine nutrition; antibiotics; selenium poisoning; protein levels and amino acid requirements; high protein cereals; mineral nutrition; by-product feeds. *Mailing Add:* 1817 Garden Sq Brookings SD 57006

WAHNSIEDLER, WALTER EDWARD, b Ind, Jan 23, 47; m 69; c 1. CHEMICAL PHYSICS. *Educ:* Purdue Univ, BS, 67, PhD(chem physics), 75. *Prof Exp:* Vis scholar mat sci, Northwestern Univ, 74; TECH SPECIALIST, CHEM SYST DIV, ALUMINUM CO AM, 75- *Mem:* Am Chem Soc; Sigma Xi; Inst Elec & Electronics Engrs. *Res:* Theoretical solid state studies; numerical modelling of chemical processes; properties of oxides; aluminum smelting; environmental impact of industry; fluid state modeling. *Mailing Add:* 16 Oakwood Terr Oakmont PA 15139

WAHR, JOHN CANNON, b Ann Arbor, Mich, Apr 2, 26; m 49; c 2. PHYSICS. *Educ:* Univ Mich, BSE, 48, MS, 49, PhD(physics), 53. *Prof Exp:* Asst, Univ Mich, 48-49; physicist, Cent Res, Dow Chem Co, 53-80. *Mem:* AAAS; Am Phys Soc; Optical Soc Am. *Res:* Quantum electronics; holography; atomic and molecular physics; surface physics. *Mailing Add:* PO Box 3558 Boulder CO 80307

WAHRHAFTIG, AUSTIN LEVY, b Sacramento, Calif, May 5, 17; m 57. MASS SPECTROMETRY. *Educ:* Univ Calif, AB, 38; Calif Inst Technol, PhD(phys chem), 41. *Prof Exp:* Fel, Calif Inst Technol, 41-45; res chemist, Dr W E Williams, 45-46; univ fel, Ohio State Univ, 46-47; from asst prof to prof, 47-87, EMER PROF CHEM, UNIV UTAH, 87- *Concurrent Pos:* Vis prof, Latrobe Univ, Australia, 72 & 80. *Mem:* AAAS; Am Chem Soc; Am Phys Soc; Am Soc Mass Spectrometry. *Res:* Molecular spectra; mass spectrometry; kinetics of gas-phase ion reactions; dense (supercritical) gas chromatography. *Mailing Add:* Dept Chem Univ Utah Salt Lake City UT 84112

WAHRHAFTIG, CLYDE (ADOLPH), b Fresno, Calif, Dec 1, 19. GEOLOGY. *Educ:* Calif Inst Technol, BS, 41; Harvard Univ, MA, 47, PhD, 53. *Prof Exp:* Jr geologist, US Geol Surv, 41 & 42-43, asst geologist, 43-45; assoc prof, 60-67, PROF GEOL, UNIV CALIF, BERKELEY, 67-; GEOLOGIST, US GEOL SURV, 45- *Concurrent Pos:* Am Geol Inst vis geoscientist, 69; mem comt geol sci, Nat Acad Sci, 70-72; consult, Conserv Found, 71-72. *Honors & Awards:* Kirk Bryan Award, Geol Soc Am, 67. *Mem:* AAAS; fel Geol Soc Am; Am Geophys Union; Sigma Xi. *Res:* Geomorphology; igneous petrology; stratigraphy and sedimentation; geology applied to land use; geology of California and Alaska. *Mailing Add:* 554 Valley St San Francisco CA 94131

WAI, CHIEN MOO, b China, Aug 8, 37; m 65; c 2. GEOCHEMISTRY, ENVIRONMENTAL CHEMISTRY. *Educ:* Nat Taiwan Univ, BS, 60; Univ Calif, Irvine, PhD(chem), 67. *Prof Exp:* Fel, Univ Calif, Los Angeles, 66-69; from asst prof to assoc prof chem & geol, 69-78, PROF CHEM, UNIV IDAHO, 78- *Concurrent Pos:* Vis assoc prof, Inst Geophys & Planetary Physics, Univ Calif, Los Angeles, 75-76; scientist in residence, Argonne Nat Lab, 82-83. *Mem:* AAAS; Am Chem Soc; Geochem Soc. *Res:* Chemical effects of nuclear transformation; origin of meteorites; heavy metal pollution. *Mailing Add:* Dept Chem Univ Idaho Moscow ID 83843

WAIBEL, PAUL EDWARD, b Hawthorne, NJ, June 22, 27; m 71; c 3. POULTRY NUTRITION. *Educ:* Rutgers Univ, BS, 48; Univ Wis, MS, 51, PhD(poultry nutrit, biochem), 53. *Prof Exp:* Teaching asst poultry husb, Univ Wis, 49-53; res assoc poultry nutrit, Cornell Univ, 53-54; res assoc, 54-55, from asst prof to assoc prof, 55-64, PROF POULTRY NUTRIT, UNIV MINN, ST PAUL, 64- *Honors & Awards:* Nat Turkey Fedn Res Award; Am Feed Manufacturers Poultry Nutrit Res Award. *Mem:* Am Inst Nutrit; Poultry Sci Asn; NY Acad Sci. *Res:* Nutrition of turkeys. *Mailing Add:* Dept Animal Sci Univ Minn 1404 Gortner Ave St Paul MN 55108

WAID, MARGARET COWSAR, b Baton Rouge, La, Feb 21, 41; m 63; c 2. APPLIED MATHEMATICS. *Educ:* La State Univ, Baton Rouge, BS, 61, MS, 63; Tex Tech Univ, PhD(math), 71. *Prof Exp:* Asst prof math, DC Teachers Col, 71-72; assoc prof math, Univ Del, 72-81; sr develop engr, Schlumberger Well Serv, 81; supvr, s/w & anal, NL Sperry Sun, 84, mgr prod serv, 84-87; mgt consult, Waid Consult Serv, 87-90; SUPVR ELECTRO-MECH RES, HALLIBURTON LOGGING SERV, 90- *Concurrent Pos:* Vis assoc prof math, Univ Tex, 79-80. *Mem:* Soc Indust & Appl Math; Soc Petrol Engrs; Soc Prof Well Log Analysts; Nat Asn Corrosion Engrs. *Res:* Partial differential equations, including applications to fluid flow through porous media analysis; applications to well log analysis, especially pressure measurements and production logging. *Mailing Add:* 16603 Ben Nevis Dr Houston TX 77084

WAID, REX A(DNEY), b Dardanelle, Ark, Jan 14, 33; m 59; c 3. ELECTRICAL ENGINEERING. *Educ:* William Jewell Col, BA, 54; Univ Mo-Columbia, BS, 58, MS, 59; Univ Wis-Madison, PhD(elec eng), 68. *Prof Exp:* From instr to asst prof elec eng, Univ Mo, Columbia, 59-63; asst prof, Univ Wis-Madison, 65-66; asst prof, 66-76, PROF ELEC ENG, UNIV MO-COLUMBIA, 76- *Concurrent Pos:* Consult, Univ Mo Network Analyzer, 62-63. *Mem:* Inst Elec & Electronics Engrs; Am Soc Eng Educ; Simulation Coun. *Res:* Pattern recognition; data acquisition and processing; computer design and development. *Mailing Add:* 207 Elec Eng Univ Mo Columbia MO 65211

WAID, TED HENRY, b Warsaw, Poland, Mar 28, 25; Can citizen; m 58; c 2. CHEMICAL ENGINEERING, ORGANIC CHEMISTRY. *Educ:* Univ Caen, BSc, 50, BEng, 51; McGill Univ, PhD(org chem), 57. *Prof Exp:* Chemist, Sherwin-Williams Co, Can, 52-53; sr res chemist, Monsanto Can Ltd, 57-60, res group leader surface finishes, 60-64, develop specialist, 64-65; PRES, CHEMOR INC, 65- *Concurrent Pos:* Mem Can Govt Specifications Bd, 63- *Mem:* Am Chem Soc; Sigma Xi; Am Concrete Inst. *Res:* Syntheses of nitrogen containing steroids, chloromethylated aromatic hydrocarbons, aromatic polyamides and sulphur containing heterocyclic compounds; chemical coatings and adhesives; polymer chemistry. *Mailing Add:* 6160 Bernard Mergler Montreal PQ H3X 4A5 Can

WAIDELICH, D(ONALD) L(ONG), b Allentown, Pa, May 3, 15; m 39; c 1. ELECTROMAGNETIC FIELDS. *Educ:* Lehigh Univ, BS, 36, MS, 38; Iowa State Univ, PhD(elec eng), 46. *Prof Exp:* Asst, Lehigh Univ, 36-38; from instr to prof, 38-85, chmn dept, 60-61, assoc dir eng exp sta, 54-58, EMER PROF ELEC ENG, UNIV MO-COLUMBIA, 85- *Concurrent Pos:* Elec engr, US Naval Ord Lab, 44-45; Fulbright grant & vis prof, Univ Cairo, 51-52; Fulbright res grant, Univ Australia, 61-62; vis prof, Univ NSW, 61-62; consult, USN Electronics Lab, 49-52, UNESCO, Egypt, 52, Argonne Nat Lab, 53-56, 65-72, Bendix Aviation Corp, 57, Int Tel & Tel Co Labs, 58-59, Midwest Res Inst, 60-61, Goddard Space Flight Ctr, NASA, 62-67, US Naval Underwater Systs Ctr, 71, Hughes Aircraft Co, 72-88 & McDonnell Douglas Corp, 83-85. *Honors & Awards:* Res Award, Sigma Xi, 77; Excellence Award, Inst Elec & Electronics Engrs, 85. *Mem:* Am Soc Eng Educ; Nat Soc Prof Engrs; fel Inst Elec & Electronics Engrs; Am Soc Nondestructive Testing; Am Soc Testing & Mat; Electrostatic Soc Am. *Res:* Electromagnetic fields; nondestructive testing; pulsed electromagnetic waves in metals and dielectrics; antennae for communication satellites; mathematical transforms. *Mailing Add:* Dept Elec & Comp Engr Univ Mo Columbia MO 65211

WAIFE, SHOLOM OMI, b New York, NY, Feb 20, 19; m 42; c 2. INTERNAL MEDICINE, MEDICAL EDUCATION. *Educ:* Johns Hopkins Univ, AB, 40; NY Univ, MD, 43; Am Bd Internal Med, dipl, 51. *Prof Exp:* Res assoc med, Sch Med, Yale Univ, 45; resen resident physician, Long Island Col Hosp, 45-46; asst med, Med Sch, Johns Hopkins Univ, 46-48; instr, Sch Med, Univ Pa, 48-52; assoc, Sch Med, Ind Univ, Indianapolis, 52-60, asst prof, 60-68, assoc prof med, 68-81; dir med serv div, Res Lab, Eli Lilly & Co, 64-81; RETIRED. *Concurrent Pos:* Ed-in-chief, Am J Clin Nutrit, 52-62; head med educ dept, Eli Lilly & Co, 52-64; co-ed, Perspectives Biol & Med, 58-64. *Mem:* Am Diabetes Asn; Am Med Writers' Asn; fel Am Col Physicians; Am Fedn Clin Res. *Res:* Metabolism; diabetes; obesity; vitamins. *Mailing Add:* Seaview F 9150 SE Riverfront Terr Tequesta FL 33469

WAILES, JOHN LEONARD, b Loveland, Colo, Oct 9, 23; m 47; c 3. PHARMACY. *Educ:* Univ Colo, BS, 47, MS, 50, PhD(pharm), 54. *Prof Exp:* Chemist, US Food & Drug Admin, 47-48; pharmacist, Park-Hill Drug Co, Colo, 48-50; instr pharm, Univ Colo, 50-54; from asst prof to assoc prof, 43-61, PROF PHARM, UNIV MONT, 61- *Concurrent Pos:* USPHS grant, 59; with Merck Sharp & Dohme Div, Merck & Co, Colo, 51-54. *Mem:* Am Pharmaceut Asn; Asn Cols Pharm. *Res:* Respiration of mold and yeast in the presence and absence of inhibitors and antagonists using the Warburg apparatus; preservation of pharmaceutical products; synergism and antagonism of various preservatives and their possible inactivation; complexing of macromolecules. *Mailing Add:* 525 34 St Missoula MT 59801

WAINBERG, MARK ARNOLD, b Montreal, Que, Apr 21, 45; m 69; c 2. AIDS, CANCER. *Educ:* McGill Univ, BSc, 66; Columbia Univ, PhD(microbiol), 72. *Prof Exp:* Lectr immunol, Hebrew Univ-Hadassah Med Sch, 72-74; STAFF INVESTR AIDS RES, LADY DAVIS INST MED RES, JEWISH GEN HOSP, MONTREAL, 74-; PROF MED, MCGILL UNIV. *Concurrent Pos:* Europ Molecular Biol Orgn res fel, 72-74; Que Med Res Coun res scholar, 75-; Nat Cancer Inst Can res grant, 75-; researcher, Dept Microbiol & Immunol, Univ Montreal, 75- *Mem:* Am Soc Microbiol; Sigma Xi; Can Soc Immunol; NY Acad Sci; Can Oncol Soc. *Res:* Virus-induced immune suppression; transmission and growth of HIV virus in aquired immune deficiency syndrome (AIDS); viral gene expression at different stages of retrovirus-induced tumor growth. *Mailing Add:* Jewish Gen Hosp McGill Univ 3755 Cote Ste Catherine Rd Montreal PQ H3T 1E2 Can

WAINE, MARTIN, b Berlin, Ger, Apr 8, 33; US citizen; m 63; c 2. PHYSICS. *Educ:* Columbia Univ, BS, 58; Yale Univ, MS, 59, PhD(physics), 65. *Prof Exp:* Asst prof physics, Mt Holyoke Col, 64-70; prin engr, MRC Corp, Md, 71-72; chief engr, Diamondex Enterprises Inc, 72-74; vpres mfg, Evershield Prod Inc, 74-75; prin engr, MRC Corp, 75-79; dir eng, 79-80, vpres, 81-82, PRES, GENERAL CLUTCH CORP, 82- *Res:* Dynamic nuclear orientation; nuclear magnetic resonance; instrumentation and control theory. *Mailing Add:* 29 Coventry Lane Riverside CT 06878

WAINER, ARTHUR, b Cincinnati, Ohio, Jan 28, 38; m 57; c 3. BIOCHEMISTRY. *Educ:* Univ Miami, BS, 57; Univ Fla, PhD(biochem), 61. *Prof Exp:* Instr biochem, Univ Fla, 61-62; from instr to assoc prof, Bowman Gray Sch Med, 62-70; PROF CHEM, EDINBORO STATE COL, 70-, CHMN DEPT, 80- *Mem:* AAAS; Am Chem Soc; Am Soc Biol Chemists; Am Asn Clin Chemists. *Res:* Sulfur amino acid metabolism, ion exchange column chromatography. *Mailing Add:* Dept Biol & Health Servs Edinboro Univ Edinboro PA 16444-0001

WAINER, IRVING WILLIAM, stereochemical separations, pharmacokinetics, for more information see previous edition

WAINERDI, RICHARD E(LLIOTT), b New York, NY, Nov 27, 31; m 56; c 2. NUCLEAR ACTIVATION ANALYSIS. *Educ:* Univ Okla, BS, 52; Pa State Univ, MS, 55, PhD, 58. *Prof Exp:* Jr exploitation engr, Shell Oil Co, 52; asst petrol eng, Pa State Univ, 53-55; coordr nuclear activities, Dresser Industs, Inc, 56-57; assoc prof petrol & nuclear eng, Tex A&M Univ, 57-61, supvr training reactor facil & radiol safety off, 57-58, head nuclear sci ctr, Eng Exp Sta, 57-59, head activation anal res lab, 58-77, asst to dean eng, 59-62, prof chem eng, 61-77, assoc dean eng, 62-72, assoc vpres acad affairs, 72-77; sr vpres & dir of spec proj, 3D/Int, Inc, 77-82; pres, Gulf Res & Develop Co, 82-84; PRES, TEX MED CTR, 84- *Honors & Awards:* George Henesy Medal, 77. *Mem:* Am Nuclear Soc; Nat Soc Prof Engrs. *Res:* Activation analysis and isotope utilization. *Mailing Add:* 406 Jesse Jones Library Texas Med Ctr Houston TX 77030-5382

WAINFAN, ELSIE, b New York, NY, Aug 2, 26; m 47; c 2. BIOCHEMISTRY. *Educ:* City Col New York, BS, 47; Univ Southern Calif, PhD(biochem), 54. *Prof Exp:* Res technician, NY Psychiat Inst, 47-49; USPHS fel, Med Sch, Univ Ore, 55-56; res assoc biochem, Cornell Univ, 56-59; res assoc, Col Physicians & Surgeons, Columbia Univ, 59-67; asst prof,

Univ Southern Calif, 67-68; assoc investr, 68-80, INVESTR, NY BLOOD CTR, 80- *Concurrent Pos:* Assoc scientist, Sloan-Kettering Inst Cancer Res, 77- *Mem:* Am Chem Soc; Am Soc Biol Chemists; Am Soc Microbiologists; Am Asn Cancer Res. *Res:* Carcinogenesis, nucleic acids, enzymes; metabolic inhibitors. *Mailing Add:* NY Blood Ctr 310 E 67th St New York NY 10021-6204

WAINGER, STEPHEN, HARMONIC ANALYSIS. *Educ:* Univ Chicago, PhD(math), 61. *Prof Exp:* PROF MATH, UNIV WIS, 67- *Mailing Add:* Dept Math Univ Wis Madison WI 53706

WAINIO, WALTER W, enzymology, biochemistry; deceased, see previous edition for last biography

WAINWRIGHT, JOHN, b Sheffield, Eng, Jan 23, 43. APPLIED MATHEMATICS. *Educ:* Univ SAfrica, PhD(math), 67. *Prof Exp:* From asst prof to assoc prof, 53-80, PROF MATH, UNIV WATERLOO, 81- *Mem:* Am Math Soc; Soc Indust & Appl Math. *Mailing Add:* Dept Math Univ Waterloo Waterloo ON N2L 3G1 Can

WAINWRIGHT, LILLIAN K (SCHNEIDER), b Brooklyn, NY, June 30, 23; m 52; c 2. GENETICS. *Educ:* Brooklyn Col, BA, 43; Columbia Univ, MA, 51, PhD(zool), 56. *Prof Exp:* Res asst zool, Columbia Univ, 43-52; from asst prof to assoc prof biol, 57-70, PROF BIOL, MT ST VINCENT UNIV, 70-, CHMN DEPT, 79- *Mem:* Genetics Soc Am; Can Soc Cell Biol; Sigma Xi; Can Fedn Biol Soc. *Res:* Organ cultures as models of tissues in vivo; control of the diurnal cycle of NAT activity in the chick pineal gland. *Mailing Add:* Dept Biol Mt St Vincent Univ Halifax NS B3M 2J6 Can

WAINWRIGHT, RAY M, b Deep River, Iowa, July 24, 13; m 52; c 3. ELECTRICAL ENGINEERING. *Educ:* Mont State Univ, BSEE, 36; Univ Ill, MSEE, 49. *Prof Exp:* Asst elec engr, Mont Power Co, 36-38; elec engr, Mont-Dakota Utilities Co, 38-42, res engr, 45-46; res engr, US Signal Corps, 42-45; from asst prof to assoc prof elec eng, Univ Ill, 46-56; dir eng, Good-All Elec Mfg Co, 56-61; mgr res, Capacitor Div, TRW, Inc, 61-63; prof elec eng, Colo State Univ, 63-66; prof elec eng & coordr continuing educ, Univ Denver, 66-74; STAFF ELEC ENGR, STEARNS ROGER , 74- *Mem:* Nat Asn Corrosion Engrs; sr mem Inst Elec & Electronics Engrs. *Res:* Cathodic protection; reliability engineering; engineering economy. *Mailing Add:* Stearns Roger PO Box 5888 Denver CO 80217

WAINWRIGHT, RICHARD ADOLPH, electronics engineering, for more information see previous edition

WAINWRIGHT, STANLEY D, b Hull, Eng, Apr 15, 27; Can citizen; m 52; c 2. NEUROCHEMISTRY, BIOLOGICAL CLOCKS. *Educ:* Cambridge Univ, BA, 47; Univ London, PhD(biochem), 50. *Prof Exp:* Brit Med Res Coun exchange scholar biochem, Physiol Microbiol Serv, Pasteur Inst, Paris, 50-51; res assoc microbial genetics, Columbia Univ, 51-52; Nat Res Coun Can & Atomic Energy Can, Ltd fel, Biol Div, Atomic Energy Can, Ltd, 52-55; res assoc microbial physiol, Yale, 55-56; res asst prof & Med Res Coun assoc biochem, 56-58, res assoc prof & Med Res Coun assoc, 58-64, PROF BIOCHEM, DALHOUSIE UNIV, 65- *Concurrent Pos:* Career investr, Med Res Coun, 65- *Mem:* AAAS; Genetics Soc Am; Am Soc Cell Biologists; Can Biochem Soc; Can Soc Cell Biol (pres, 75-76). *Res:* Biochemical neuroendocrinology of the developing chick pineal gland. *Mailing Add:* Dept Biochem Fac Med Dalhousie Univ Halifax NS B3H 3J5 Can

WAINWRIGHT, STEPHEN ANDREW, b Indianapolis, Ind, Oct 9, 31; m 56; c 4. BIOMECHANICS. *Educ:* Duke Univ, BS, 53; Univ Cambridge, BA, 58, MA, 63; Univ Calif, Berkeley, PhD(zool), 62. *Prof Exp:* NSF fel med physics, Karolinska Inst, Sweden, 62-63; NSF fel biol, Woods Hole Oceanog Inst, 63-64; from assoc prof to prof zool, 76-85, JAMES B DUKE PROF ZOOL, DUKE UNIV, 85-; ADJ PROF, SCH DESIGN, NC STATE UNIV, 83- *Mem:* Soc Exp Biol UK; Marine Biol Asn UK; Sigma Xi; Am Soc Biomech (pres, 81); Am Soc Zoologists (pres, 88). *Res:* Functional morphology and mechanics of supportive systems of animals and plants from the macromolecular through the organism levels of organization. *Mailing Add:* Dept Zool Duke Univ Durham NC 27706

WAINWRIGHT, THOMAS EVERETT, b Seattle, Wash, Sept 22, 27; m 59; c 5. PHYSICS. *Educ:* Mont State Col, BS, 50; Univ Notre Dame, PhD(physics), 54. *Prof Exp:* Instr physics, Univ Notre Dame, 53-54; STAFF PHYSICIST, LAWRENCE LIVERMORE LAB, UNIV CALIF, 54- *Honors & Awards:* Lawrence Award, US AEC, 73. *Mem:* Fel Am Phys Soc; Am Geophys Union; Am Phys Soc. *Res:* Statistical mechanics; applied physics. *Mailing Add:* 220 Grover Lane Walnut Creek CA 94596

WAINWRIGHT, WILLIAM LLOYD, b Fostoria, Ohio; m 51; c 4. ENGINEERING. *Educ:* Purdue Univ, West Lafayette, BS, 51, MS, 54; Univ Mich, Ann Arbor, PhD(eng mech), 58. *Prof Exp:* Assoc prof mech, US Naval Postgrad Sch, 58-61; res asst, Univ Calif, Berkeley, 61-63, asst prof, 63-64; ASSOC PROF MECH, UNIV COLO, BOULDER, 64- *Mem:* AAAS; Soc Eng Sci; Sigma Xi. *Res:* Continuum mechanics. *Mailing Add:* 4305 Chippewa Dr Boulder CO 80302

WAISMAN, JERRY, b Borger, Tex, Sept 14, 34; m 58; c 3. PATHOLOGY. *Educ:* Univ Tex, BA, 56, MD, 60. *Prof Exp:* Pathologist & chief lab div path, US Air Force Hosp, Sheppard AFB, Tex, 62-64; fel path, Univ Utah, 64-65, instr path, 65-68; from asst prof to assoc prof, Univ Calif, Los Angeles, 68-76, prof path, 76-81; DIR LAB, UNIV HOSP, NEW YORK UNIV MED CTR, 81- *Concurrent Pos:* Attend physician, Ft Douglas Vet Admin Hosp, Salt Lake City, 67, part-time sr physician, 67-68; consult, Sepulveda Vet Admin Hosp, 76-81, NY Vet Admin Med Ctr, 82- *Mem:* Int Acad Path; Am Asn Path; Electron Micros Soc Am; Am Soc Cytol. *Res:* Ultrastructure of benign and malignant neoplasms; fine needle aspiration of tumors. *Mailing Add:* Dept Path NY Univ Med Ctr 560 First Ave Rm H461 New York NY 10016

WAISMAN, JOSEPH L, b Racine, Wis, Mar 10, 19; m 40; c 1. METALLURGICAL ENGINEERING, MATERIALS SCIENCE. *Educ:* Univ Ill, MetE, 40; Univ Calif, Los Angeles, PhD(eng), 69. *Prof Exp:* Asst chief metallurgist, Douglas Aircraft Co, Calif, 45-57; western mgr, Tatnall Measuring Systs Div, Budd Co, 57-59; chief metallurgist, Douglas Aircraft Co, Calif, 59-60, chief mat res & prod methods, 60-62, asst chief engr, Missiles & Space Systs, 62-64, asst dir res & develop, 64-66, dir, 66-67, dir res & develop, McDonnell Douglas Astronaut Co, Huntington Beach, 67-68, dir res & develop, 68-73, dir advan prod applns, 73-80, dir cryogenic insulation prog, 80-81, dir energy prog, 81-84, consult, McDonnell Douglas Astronaut Co West, 84-90; PVT CONSULT, 90- *Concurrent Pos:* Consult, Metals Adv Bd, NSF, 63-66. *Mem:* Fel Am Soc Metals; assoc fel Am Inst Aeronaut & Astronaut; Soc Exp Stress Anal; Am Inst Mining, Metall & Petrol Engrs; Am Soc Testing & Mat. *Res:* Fatigue of metals; stress corrosion cracking; residual stresses. *Mailing Add:* 25 Redwood Tree Lane Irvine CA 92715

WAISS, ANTHONY C, JR, b China, Sept 30, 36; US citizen; m 58; c 3. ORGANIC CHEMISTRY, NATURAL PRODUCTS. *Educ:* Univ Calif, Berkeley, BS, 58; Univ Calif, Los Angeles, PhD(chem), 62. *Prof Exp:* Res fel chem, Harvard Univ, 62-63; RES LEADER, WESTERN REGION RES CTR, AGR RES SERV, USDA, 63- *Mem:* Am Chem Soc; Royal Soc Chem; Phytochem Soc NAm; Entom Soc Am. *Res:* Isolation and structural determination of biologically active compounds; chemical basis of host plant resistance to insects and diseases; crop protection. *Mailing Add:* 505 Grandview Ct Richmond CA 94801

WAIT, DAVID FRANCIS, b Sidney, Nebr, Sept 28, 33; div; c 4. METROLOGY. *Educ:* Colo State Univ, BS, 55, MS, 57; Univ Mich, PhD(physics), 63. *Prof Exp:* Instr & res asst physics, Univ Mich, 62-63; sr scientist, Laser Systs Ctr, Lear Siegler, Inc, 63; PHYSICIST, NAT BUR STAND, 63- *Mem:* Sigma Xi; Inst Elec & Electronics Engrs; Microwave Theory & Tech Soc. *Res:* Noise in communications; radiometers; microwave cryogenic noise standards. *Mailing Add:* 2795 Iliff St Boulder CO 80303-7019

WAIT, JAMES RICHARD, b Ottawa, Can, Jan 23, 24; US citizen; m 51; c 2. GEOENVIRONMENTAL SCIENCE. *Educ:* Univ Toronto, BASc, 48, MASc, 49, PhD(elec eng), 51. *Prof Exp:* Radio Physics Lab, Ottawa, Can, 52-55; consult appl physics, US Dept Com, Boulder, Co, 55-80; prof elec eng, 80-88, REGENTS PROF, UNIV ARIZ, 88- *Concurrent Pos:* Mem nat comt, Int Union Radio Sci, 58-61 & 65-68, secy, US Nat Comt, 75-78; adj prof elec eng, Univ Colo, Boulder, 61-83, fel, Coop Inst Res Environ Sci, 68-80; mem, Electromagnetics Lab, Univ Ariz, 83- *Honors & Awards:* Flemming Award, US Chamber Com, 64; Harry Diamond Award, Inst Elec & Electronics Engrs, 64; Res & Achievement Award, Nat Oceanic & Atmospheric Admin, 73; Van der Pol Gold Medal, Int Union Radio Sci, 78; Centennial Award, Inst Elec & Electronics Engrs, 84; Geoscience & Remote Sensing Soc Award, Inst Elec & Electronics Engrs, 85; Antennas & Propagation Soc Distinguished Achievement Award, Inst Elec & Electronics Engrs, 90. *Mem:* Nat Acad Eng; fel Inst Elec Engrs; Am Geophys Union; fel Inst Elec & Electronics Engrs; Int Union Radio Sci. *Res:* Applications of electromagnetic theory to problems in geophysics and telecommunications. *Mailing Add:* 2210 E Waverly Tucson AZ 85719

WAIT, JOHN V, b Chicago, Ill, Oct 1, 32; m 61; c 2. ELECTRONICS ENGINEERING. *Educ:* Univ Iowa, BSEE, 55; Univ NMex, MSEE, 59; Univ Ariz, PhD, 63. *Prof Exp:* Res engr, RCA Labs, 55; instr, Univ NMex, 57-59; res engr & instr, Univ Ariz, 59-63; asst prof, Univ Calif, Santa Barbara, 63-64; assoc prof grad eng educ syst, Univ Fla, 64-66; assoc prof, 66-71, PROF ELEC ENG, UNIV ARIZ, 71- *Mem:* Inst Elec & Electronics Engrs; Soc Comput Sci. *Res:* Electronics; computers; signal processing. *Mailing Add:* Dept Elec & Comput Eng Univ Ariz Tucson AZ 85721

WAIT, SAMUEL CHARLES, JR, b Albany, NY, Jan 26, 32; m 57; c 2. PHYSICAL CHEMISTRY. *Educ:* Rensselaer Polytech Inst, BS, 53, MS, 55, PhD(chem), 56. *Prof Exp:* Fulbright fel, Univ Col, London, 56-57, asst lectr chem, 57-58; res fel, Univ Minn, 58-59; asst prof, Carnegie Inst Technol, 59-60; chemist, Nat Bur Standards, 60-61; from asst prof to assoc prof chem, Rensselaer Polytech Inst, 61-71, asst dean, Sch Sci, 72-74, actg dean, 78-80 & 88-89, PROF CHEM, RENSSELAER POLYTECH INST, 71-, ASSOC DEAN SCH SCI, 74- *Concurrent Pos:* Mem adv coun sci & math, Schenectady County Community Col, 75-, chmn, 76-78; mem, Schenectady County Fire Adv Bd, 76-81, vchmn, 78; mem bd fire comnrs, Niskayuna Dist Two, 78-84; mem bd trustees, Dudley Observ, 80-, pres trustees, 91- *Mem:* Am Chem Soc; Optical Soc Am; Coblentz Soc. *Res:* High resolution ultraviolet, infrared and Raman spectroscopy; asymmetric rotor theory and calculation; quantum biology; molecular orbital theory; vibrational and fine structural analyses; theoretical methods; simple and polyatomic systems. *Mailing Add:* Sch Sci Rensselaer Polytech Inst Troy NY 12180-3590

WAITE, ALBERT B, reproductive physiology, genetics, for more information see previous edition

WAITE, DANIEL ELMER, b Grand Rapids, Mich, Feb 19, 26; m 48; c 4. ORAL SURGERY, MAXILLOFACIAL SURGERY. *Educ:* Univ Iowa, DDS, 53, MS, 55; Am Bd Oral Surg, dipl, 59. *Prof Exp:* Resident oral surg, Univ Hosp, Univ Iowa, 53-55, from instr to prof & head dept, Col Dent, 55-59, assoc prof, Hosp Dent Dept, Univ Hosps, 57-63; asst prof dent, Mayo Grad Sch Med, 63-68, PROF ORAL SURG, CHMN DIV & HEAD HOSP DENT, SCH DENT, UNIV MINN, MINNEAPOLIS, 68- *Concurrent Pos:* Mem staff, Proj Hope, Peru, Ceylon & Haiti; trustee, Park Col, Mo. *Mem:* Am Soc Oral Surg; Am Dent Asn; Am Col Dent; Int Asn Dent Res; Sigma Xi. *Mailing Add:* 3302 Gaston Ave Dallas TX 75246

WAITE, LEONARD CHARLES, b Reynoldsville, Pa, Sept 10, 41; m 60; c 3. PHARMACOLOGY. *Educ:* Alderson-Braddus Col, BS, 65; WVa Univ, MS, 67; Univ Mo, Columbia, PhD(pharmacol), 69. *Prof Exp:* From asst prof to assoc prof, 70-79, PROF PHARMACOL, SCH MED, UNIV LOUISVILLE, 79- *Mem:* Endocrine Soc; Am Soc Pharmacol & Exp Therapeut. *Res:* Endocrinology; physiology; calcium metabolism. *Mailing Add:* Dept Pharmacol Univ Louisville Sch Med Louisville KY 40292

WAITE, MOSELEY, b Durham, NC, Oct 22, 36; m 59; c 3. BIOCHEMISTRY, ORGANIC CHEMISTRY. *Educ:* Rollins Col, BS, 58; Duke Univ, PhD(biochem), 63. *Prof Exp:* From asst prof to assoc prof, 67-76, PROF BIOCHEM, BOWMAN GRAY SCH MED, 76-, CHMN, 78-, ASSOC MED, PULMONARY SECT, 89- *Concurrent Pos:* Am Cancer Soc fel biochem, Duke Univ, 62-65; Am Heart Asn advan fel, Univ Utrecht, 65-67; grants, Am Heart Asn, 66-69, USPHS, 67-81 & NC Heart Asn, 70-73; USPHS res career develop award, 73-78 & Environ Protection Agency, 80-82. *Mem:* AAAS; Am Chem Soc; Am Soc Biol Chemists. *Res:* Phospholipid and fatty acid metabolism; enzyme purification and characterization; relation of metabolism of lipids to certain morphological changes, especially mitochondrion; prostaglandin synthesis; phospholipid; prostaglandins. *Mailing Add:* Dept Biochem Bowman Gray Sch Med Medical Center Blvd Winston-Salem NC 27157-1016

WAITE, PAUL J, b New Salem, Ill, June 21, 18; m 43; c 2. CLIMATOLOGY. *Educ:* Western State Univ, BEd, 40; Univ Mich, MS, 66. *Prof Exp:* Meteorologist, USAF, 42-46; meteorologist & climatologist, Nat Weather Bur, 48-74; dep mgr Lacie Proj, Environ Data Serv, 74-76; climatologist, 76-88, CLIMAT CONSULT, IOWA DEPT AGR, 88- *Concurrent Pos:* Teacher & coach, Ill schs, 38-39 & 40-42; coach & instr, Ill High Schs, 46-48; adj prof geol & geog, Drake Univ, 70-75 & 77-; pres, Iowa Sci Acad, 86-87. *Honors & Awards:* Group Achievement Award, NASA, 79. *Mem:* Am Asn State Climatologists (secy, 76-77; pres, 77-78); Am Meteorol Soc; Nat Weather Asn. *Res:* Iowa storm climatology; agricultural climatology; climatology for decision making and descriptive climatology; approximately 100 professional and popular articles, reviews and books published. *Mailing Add:* 6657 NW Timberline Dr Des Moines IA 50313-5436

WAITE, WILLIAM MCCASTLINE, b New York, NY, Jan 14, 39; m 60; c 1. PROGRAMMING LANGUAGES, SOFTWARE. *Educ:* Oberlin Col, AB, 60; Columbia Univ, MS, 62, PhD(elec eng), 65. *Prof Exp:* Res asst, Electronics Res Labs, Columbia Univ, 62-65; NSF fel, 65-66; from asst prof to assoc prof, 66-75, PROF ELEC ENG, UNIV COLO, BOULDER, 75- *Concurrent Pos:* Vis lectr, Dept Info Sci, Monash Univ, Australia, 70-71; temp res assoc, Culham Lab, UK Atomic Energy Auth, Eng, 71; ed, Spec Interest Group on Operating Systs, Asn Comput Mach, 72-; mem staff, Inst Informatik, Univ Karlsruhe, Ger, 80; vis lectr, Melbourne Univ, Australia, 77, 82 & 88; chmn, Int Fed Info Processing Working Group 2.4, 83- *Mem:* Asn Comput Mach; Brit Comput Soc. *Res:* Programming languages and software system design. *Mailing Add:* Elec Eng Univ Colo Campus Box 425 Boulder CO 80309

WAITER, SERGE-ALBERT, b Paris, France, Feb 8, 30; US citizen; m 53; c 1. MATHEMATICS, GAS DYNAMICS. *Educ:* Univs Lille & Paris, Lic es Sci, 51; Univ Paris, Dr es Sci, Univ Paris, 54. *Prof Exp:* Res engr, Off Aeronaut & Astronaut Res, France, 49-51; flight test engr, Fouga Aircraft, 51-53; mgr prototype dept, Sud Aviation, 53-59; res scientist plasma physics, Eng Ctr, Univ Southern Calif, 59-62; PRIN SR SCIENTIST, SPACE DIV, N AM ROCKWELL CORP, 62- *Mem:* Assoc fel Am Inst Aeronaut & Astronaut; Sigma Xi; Soc Civil Engrs France. *Res:* Fluid dynamics; plasma physics; space sciences. *Mailing Add:* 801 S Crest Vista Dr Monterey Park CA 91754

WAITES, ROBERT ELLSWORTH, b Middletown, Ohio, Apr 28, 16; m 45; c 2. ENTOMOLOGY. *Educ:* Otterbein Col, BS, 41; Ohio State Univ, MS, 46, PhD(entom), 49. *Prof Exp:* Asst entomologist, 51-68, ASSOC ENTOMOLOGIST, AGR EXP STA, UNIV FLA, 68-, ASSOC PROF ENTOM, 74- *Mem:* Am Registry Prof Entomologists; Coleopterists Soc; Sigma Xi; Entom Soc Am. *Res:* Chemical control of insects on vegetable crops; systematics and biology of the Coccinellidae; insects on peanuts and field corn. *Mailing Add:* 1061 NE 20th Ave Gainesville FL 32609

WAITHE, WILLIAM IRWIN, b New York, NY, May 3, 37; m 58, 74; c 6. CELL PHYSIOLOGY, CARCINOGEN METABOLISM. *Educ:* St Francis Col, BS, 58; NY Univ, MS, 63, PhD(cell biol), 69. *Prof Exp:* Res asst med genetics, Sch Med, NY Univ, 58-66; res assoc immunol & cell biol, Mt Sinai Sch Med, City Univ NY, 66-68, instr genetics, 68-71; asst prof, 71-77, ASSOC PROF MED, FAC MED, UNIV LAVAL, 77- *Concurrent Pos:* Scholar, Med Res Coun Can, 71-76. *Mem:* Can Biochem Soc; Can Soc Cell Biol. *Res:* Regulation of growth by nuclear and cytoplasmic proteins; biochemical mechanisms controlling lymphocyte activation in vitro; carcinogen metabolism by human lymphocytes; cytochrome P-450, enzymatic activity and gene expression in lymphocytes; Epstein-Barr virus transformed lymphocyte cell lines. *Mailing Add:* Hotel Dieu de Quebec 11 Cote du Palais Quebec PQ G1R 2J6 Can

WAITKINS, GEORGE RAYMOND, b Glasgow, Scotland, Feb 28, 11; nat US; m 37; c 2. PHYSICAL CHEMISTRY. *Educ:* Syracuse Univ, 33, MS, 34, PhD(chem), 38. *Prof Exp:* Res engr, Battelle Mem Inst, 38-43; chemist, Mutual Chem Co Am, 43; chem supvr, Can Copper Refiners, Ltd, Que, 44-45; res chemist, Calco Div, Am Cyanamid Co, 45-52; asst mgr res dept, Am Zinc Lead & Smelting Co, 52-62; phys chemist, Mattin Labs, Mearl Corp, Ossining, 62-76; RETIRED. *Mem:* Am Chem Soc; Am Inst Chemists; AAAS; Sigma Xi. *Res:* Inorganic, organic and nacreous pigments; crystal growth. *Mailing Add:* One Hughes St Croton-on-Hudson NY 10520

WAITS, BERT KERR, b New Orleans, La, Dec 21, 40; m 63; c 2. MATHEMATICS. *Educ:* Ohio State Univ, BSc, 62, MSc, 64, PhD(math educ), 69. *Prof Exp:* Asst to chmn dept math, 65-69, from asst prof to assoc prof 69-88, PROF MATH, OHIO STATE UNIV, 88- *Concurrent Pos:* Mem, Nat Coun Teachers Math, Curric & Eval Standards for Sch Math; dir, Ohio Early Col Math Placement Testing Prog, 78- *Mem:* Math Asn Am; Nat Coun Teachers Math. *Res:* Mathematics education; individualized instruction at the college level; micro-compter applications of pre-calculus and calculus; curriculum revision and graphing software. *Mailing Add:* Dept Math Ohio State Univ 231 W 18th Ave Columbus OH 43210

WAITZ, JAY ALLAN, b Elizabeth, NJ, Nov 26, 35; m 60; c 2. CHEMOTHERAPY. *Educ:* Univ Idaho, BS, 57, MS, 59; Univ Ill, PhD(parasitol), 62. *Prof Exp:* Assoc res parasitologist, Parke, Davis & Co, 62-65, res parasitologist, 65-66; sr microbiologist, Schering Corp, 66-68, sect head, 69-70, mgr chemother dept, 70-73, assoc dir microbiol, 73-77, dir antibiotic res, 77-81, vpres microbiol res, 81-82; PRES, DNAX RES INST, 82- *Mem:* Am Soc Microbiol. *Res:* Chemotherapy of parasitic, bacterial and fungal diseases. *Mailing Add:* DNAX Research Inst Palo Alto CA 94304-1104

WAITZMAN, MORTON BENJAMIN, b Chicago, Ill, Nov 8, 23; m 49; c 3. PHYSIOLOGY, BIOCHEMISTRY. *Educ:* Univ Miami, BS, 48; Univ Ill, MS, 50, PhD(physiol), 53. *Prof Exp:* Res asst physiol, Univ Ill, 51-54; res assoc, Dept Pharmacol & Lab Res Ophthal, Sch Med, Western Reserve Univ, 54-56, instr, 56-59, asst prof ophthal res & pharmacol & dir lab res ophthal, 59-62; assoc prof & dir, Lab Ophthal Res, 62-68, asst prof, Dept Physiol, 62-67, PROF & DIR, LAB OPHTHAL RES, EMORY UNIV, 68-; SCI & CLIN INVEST DIR, GRADY EYE CLINS, 81- *Concurrent Pos:* Chair, tech adv comt on Clean Air, 70-71; extensive teaching at undergrad, grad & prof levels. *Honors & Awards:* Numerous Hon Invited Lectureships. *Mem:* AAAS; Asn Res Vision & Ophthal; Am Asn Univ Prof; Am Physiol Soc; NY Acad Sci; Sigma Xi. *Res:* Ophthalmic research; metabolic and hormonal aspects of aqueous humor and cerebrospinal fluid production; glaucoma; metabolic aspects of diabetes mellitus; neuro-chemistry; autonomic nature of ocular extracts; numerous major publications and chapters. *Mailing Add:* 1137 Mason Woods Dr NE Atlanta GA 30329

WAIWOOD, KENNETH GEORGE, b St Boniface, Man, Feb 12, 47; m 72. MARINE FISH AQUACULTURE. *Educ:* Sir George Williams Univ, BSc, 68; Queen's Univ, MSc, 72; Univ Guelph, PhD(zool), 77. *Prof Exp:* RES SCIENTIST MARINE FISH AQUACULT, FISHERIES & OCEANS, 77- *Mem:* Can Soc Zoologists; Aquacult Asn Can. *Res:* Eco-physiology of marine fishes in support of aquaculture; bioenergetics; physiology of reproduction, feeding and growth. *Mailing Add:* Fisheries & Oceans Biol Sci Br Aquacult & Invert Fisheries Div St Andrews NB E0G 2X0 Can

WAJDA, EDWARD STANLEY, b Schenectady, NY, Oct 31, 24; m 50; c 2. PHYSICS. *Educ:* Union Col, NY, BS, 45; Cornell Univ, MS, 48; Rensselaer Polytech Inst, PhD(physics), 53. *Prof Exp:* Instr physics, Amherst Col, 46-47; Col, instr, Union Col, NY, 48-49, asst prof, 53-55; SR PHYSICIST, IBM CORP, 55- *Mem:* Inst Elec & Electronics Engrs; Sigma Xi. *Res:* Solid state physics; semiconductors. *Mailing Add:* 39 Spy Hill Rd Poughkeepsie NY 12603

WAJDA, ISABEL, b Cracow, Poland, Apr 3, 13; US citizen; m 34. PHARMACOLOGY, NEUROCHEMISTRY. *Educ:* Jagiellonian Univ, BSc, 36; Univ Birmingham, PhD(pharmacol), 51. *Prof Exp:* Res pharmacologist, Oxford Univ, 43-46; res pharmacologist, Med Sch, Univ Birmingham, 46-51; head dept pharmacol, Med Sch, Univ Mendoza, Arg, 53-55; instr, New York Med Col, Flower & Fifth Ave Hosps, 55, asst prof, 56-59; sr res scientist, NY State Psychiat Inst, 59-66, assoc res scientist, 66-68; assoc res scientist, NY State Ctr Neurochem, 68-90; RETIRED. *Concurrent Pos:* Assoc prof pharmacol & chmn dept, Sch Dent, Fairleigh Dickinson Univ, 68-74. *Mem:* Am Soc Pharmacol & Exp Therapeut; Am Soc Neurochem; Brit Pharmacol Soc; Int Soc Neurochem. *Res:* Biological standardization; biochemistry and pharmacology of the central nervous system related to neurochemical transmission; enzyme metabolism in experimental autoimmune diseases; metabolism of neurotransmitters in drug addiction; receptor binding and effects of lithium. *Mailing Add:* NY State Ctr Neurochem N S Kline Inst Orangeburg NY 10962

WAKADE, ARUN RAMCHANDRA, pharmacology, for more information see previous edition

WAKAYAMA, YOSHIHIRO, b Oogaki City, Japan, Apr 30, 45; m 71; c 2. NEUROLOGY, MYOLOGY. *Educ:* Nagoya Univ, MD, 76, PhD(med sci), 81. *Prof Exp:* Myology fel, Muscular Dystrophy Asn Am, 76-79; asst prof res, Sch Med, Univ Pa, 79; assoc prof neurol, 80-89, PROF NEUROL, SCH MED, SHOWA UNIV, 89- *Mem:* Japanese Soc Neurol; Am Acad Neurol; Am Soc Cell Biol; Am Asn Neuropathologists. *Res:* Ultrastructural investigations of muscle and plasma membrane associated cytoskeletons of dystrophic skeletal muscles by using cytochemical, immunoelectron microscopic and freeze etching electron microscopic techniques. *Mailing Add:* Div Neurol Dept Med Showa Univ Fujigaoka Hosp 1-30 Fujigaoka Midori-ku Yokohama 227 Japan

WAKE, DAVID BURTON, b Webster, SDak, June 8, 36; m 62; c 1. EVOLUTIONARY BIOLOGY. *Educ:* Pac Lutheran Univ, BA, 58; Univ Southern Calif, MSc, 60, PhD(biol), 64. *Prof Exp:* Asst biol, Univ Southern Calif, 58-59, head lab assoc, 62-63, instr, 63-64; instr anat & biol, Univ Chicago, 64-66, asst prof, 66-69; assoc prof zool, Univ Calif, Berkeley, 69-73, assoc cur, Mus Vert Zool, 69-71, prof zool, 73-89, DIR, MUS VERT ZOOL, UNIV CALIF, BERKELEY, 71-, CUR HERPET, 73-, PROF INTEGRATIVE BIOL, 89- *Concurrent Pos:* Fel, John Simon Guggenheim Mem Found, 81-82; Bd Biol, Nat Res Coun, 86- *Honors & Awards:* Quantrell Award, Univ Chicago, 67. *Mem:* AAAS; Am Soc Ichthyologists & Herpetologists; Soc Syst Zool; Am Soc Zoologists; Soc Study Evolution (pres, 83); Am Soc Naturalists (pres, 89). *Res:* Functional, developmental and evolutionary morphology; evolution, systematics and zoogeography of modern Amphibia, with emphasis on salamanders; evolutionary theory. *Mailing Add:* Mus Vert Zool Univ Cailf Berkeley CA 94720

WAKE, MARVALEE H, b Orange, Calif, July 31, 39; m 62; c 1. VERTEBRATE BIOLOGY. *Educ:* Univ Southern Calif, BA, 61, MS, 64, PhD(biol), 68. *Prof Exp:* Teaching asst biol, Univ Ill, Chicago, 64-66, instr, 66-68, asst prof, 68-69; lectr, 69-73, from asst prof to assoc prof zool, 73-80, assoc dean, Col Lett & Sci, 75-78, chmn dept zool, 85-89, PROF INTEGRATIVE BIOL, UNIV CALIF, BERKELEY, 89- *Concurrent Pos:*

Prin investr, NSF, 77-; vis prof, Univ Bremen, 82, Univ Paris VII, 89. *Honors & Awards:* Guggenheim Fel, 88-89. *Mem:* AAAS; Am Soc Zoologists; Am Soc Ichthyologists & Herpetologists (pres, 84); Soc Study Evolution; Soc Study Amphibians & Reptiles. *Res:* Evolution of vertebrates; morphology; reproductive biology. *Mailing Add:* Dept Integrative Biol Univ Calif Berkeley CA 94720

WAKEFIELD, CAROLINE LEONE, neuroanatomy, for more information see previous edition

WAKEFIELD, ERNEST HENRY, b Vermilion, Ohio, Feb 11, 15; m 39; c 2. ELECTRICAL ENGINEERING. *Educ:* Univ Mich, BS, 38, MS, 39, PhD, 52. *Prof Exp:* Instr elec eng, Univ Tenn, 39-41; assoc physicist, Mass Inst Technol, 42; assoc physicist, Manhattan Proj, Chicago, 43-46; pres, RCL, Inc, 46-62 & RCL Calif, Inc, 59-62; PRES, LINEAR ALPHA, INC, 62-; CHMN, THIRD WORLD ENERGY INST INT, 78-; CHMN, INT INST MGT & APPROPRIATE TECHNOL FOR EMERGING NATIONS, 81- *Concurrent Pos:* Hon prof, Cent Univ Ecuador, 58; dir, Atomic Indust Forum, 60-62; pres, Evanston Bd Educ, 66-67. *Res:* Data processing systems; electronic controls; technology for third world; electric vehicle design. *Mailing Add:* 2300 Noyes Ct Evanston IL 60201

WAKEFIELD, LUCILLE MARION, b Dayville, Conn, June 13, 25. NUTRITION. *Educ:* Univ Conn, BS, 49, MS, 56; Ohio State Univ, PhD(nutrit), 65. *Prof Exp:* Intern dietetics, Mt Auburn Hosp, 49-50; therapeut dietitian, New Brit Gen Hosp, 50-52, admin dietitian, 52-53; dir dietetics, Auburn Mem Hosp, 53-57; asst prof food & nutrit, Univ Vt, 57-65, head, Dept Nutrit & Inst Mgt, 62-65; prof foods & nutrit & head dept, Kans State Univ, 65-75 & Fla State Univ, 75-79; CHMN, DEPT FOOD & NUTRIT, UNIV NC, GREENSBORO, 79- *Concurrent Pos:* Int work in India, Mexico & Morroco. *Mem:* AAAS; fel Am Inst Chemists; Am Dietetic Asn; Am Pub Health Asn; Am Inst Nutrit; Am Col Nutrit. *Res:* Nutritional, sociological, psychological aspects of humans and their body composition as it relates to population groups and nutritional status; clinical nutrition and community health problems; nutrition in aging; food patterning. *Mailing Add:* Dept Foods & Nutrit Univ NC Greensboro NC 27412

WAKEFIELD, ROBERT CHESTER, b Providence, RI, Sept 14, 25; m 49; c 3. AGRONOMY. *Educ:* Univ RI, BS, 50; Rutgers Univ, MS, 51, PhD, 54. *Prof Exp:* Res assoc farm crops, Rutgers Univ, 51-54; from asst prof to assoc prof agron, 54-65, chmn dept, 61-70, PROF AGRON, UNIV RI, 65- *Res:* Crop ecology; landscape ecology. *Mailing Add:* 850 Usquepaugh Rd West Kingston RI 02892

WAKEFIELD, SHIRLEY LORRAINE, b Milwaukee, Wis, Nov 20, 34; m 57; c 2. OTHER ENGINEERING. *Educ:* Univ Wis-Madison, BS, 57; Univ Cincinnati, PhD(phys chem), 69; Xavier Univ, MBA, 76. *Prof Exp:* Chemist, Forest Prod Lab, US Forest Serv, 57-58; lab technician, Pulmonary Dis Res Lab, Vet Admin Hosp, 59-62; staff engr, Avco Electronics, 69-73; spec progs mgr, Measurements & Sensors Develop, Gen Elec Co, 73-75, mgr composites, Mat & Processes Technol Progs, 75-81, mgr, Survivability & Exhaust Syst Mat, Advan Eng & Technol Prog Dept, 81-82, mgr visible signature reduction technol, Aircraft Engine Group, 82-85, MGR ELECTROOPTICS DESIGN TECHNOL, GEN ELEC CO, 85- *Mem:* Am Chem Soc; Am Phys Soc; Am Soc Nondestructive Testing; Soc Advan Mat & Process Eng. *Res:* Infrared detectors; nuclear quadrupole resonance spectroscopy; pulmonary disease; gas chromatography; high energy x-ray, Raman spectroscopy; ceramics; high temperature alloy development; countermeasure materials; countermeasure technology. *Mailing Add:* Gen Elec Aircraft Engine Group MD J185 175 & Bypass 50 Cincinnati OH 45215

WAKEHAM, HELMUT, b Hamburg, Ger, Apr 15, 16; US citizen; m 39; c 3. PHYSICAL CHEMISTRY. *Educ:* Univ Nebr, BA, 36, MA, 37; Univ Calif, PhD(phys chem), 39. *Prof Exp:* Asst chem, Univ Nebr, 36-37; asst, Univ Calif, 37-39; res chemist, Stand Oil Co, Calif, 39-41; res chemist, Southern Regional Res Lab, USDA, 41-47; from res assoc to proj head & dir res, Chem Physics Sect, Textile Res Inst, NJ, 49-56; dir, Ahmedabad Textile Industs Res Asn, India, 56-58; staff asst to vpres & chief opers & subsidiaries, Res Ctr, Philip Morris, Inc, 58-60, dir, Res Ctr, 60-61, vpres res & develop, 61-82; RETIRED. *Concurrent Pos:* Mem, Tobacco Working Group, 68-76; mem, Nat Cancer Plan, 71; mem bd dir, Indust Res Inst, Inc, 71-72; Va Laureat Technol. *Mem:* Fel AAAS; Am Chem Soc; Fiber Soc; Soc Rheol; fel Am Inst Chemists. *Res:* Agricultural and food chemistry; technical management. *Mailing Add:* 8905 Norwick Rd Richmond VA 23229

WAKELAND, WILLIAM RICHARD, b Mound City, Ill, Nov 14, 21; m 45; c 4. ELECTRICAL ENGINEERING, CONTROL SYSTEMS. *Educ:* US Naval Acad, BS, 43; US Naval Postgrad Sch, BS, 50, MS, 51; Univ Houston, PhD(elec eng), 68. *Prof Exp:* Dir astronaut directorate, Naval Missile Ctr, Point Mugu, Calif, 61-62; mgr Gemini-agena develop, Manned Spacecraft Ctr, NASA, Houston, 62-64; dir aeronaut elec dept, Naval Air Develop Ctr, Pa, 64-65; instr, Univ Houston, 65-68; from asst prof to assoc prof, Trinity Univ, Tex, 68-78; PROF & HEAD, ELEC ENG DEPT, LAMAR UNIV, 78- *Mem:* Inst Elec & Electronics Engrs; Nat Soc Prof Engrs; Am Soc Eng Educ. *Res:* Control system design; optimal design; weighting factor for quadratic performance index. *Mailing Add:* 1065 Monterey Beaumont TX 77706

WAKELEY, JAMES STUART, b Raleigh, NC, July 8, 50; m 73; c 1. WILDLIFE ECOLOGY, WETLAND DELINEATION. *Educ:* Univ Calif, Santa Barbara, BA, 71; Univ Maine, MS, 73; Utah State Univ, PhD(wildlife ecol), 76. *Prof Exp:* From asst prof to assoc prof wildlife ecology, Pa State Univ, 76-86; WILDLIFE BIOLOGIST, US ARMY ENGR WATERWAYS EXP STA, 86- *Mem:* Wildlife Soc; Am Ornithologists' Union; Ecol Soc Am; Raptor Res Found; Soc Wetland Scientists. *Res:* Ecology and management of game and nongame birds; applied population biology; wetland ecology. *Mailing Add:* US Army Engrs Waterways Exp Sta 3909 Halls Ferry Rd Vicksburg MS 39180-6199

WAKELIN, DAVID HERBERT, b Southampton, Eng, Dec 8, 40; US citizen; m 66; c 2. METALLURGICAL ENGINEERING. *Educ:* Univ London, BSc, 62, ARSM, 62, PhD(eng metall) & DIC, 66. *Prof Exp:* Res engr, Graham Res Lab, Jones & Laughlin Steel Corp, 66-68, sr res engr, 68-70, res assoc metall, 70-74, develop engr ironmaking, 74-79, supervr ironmaking eng, Graham Res Lab, 79-84, MGR, DEVELOP ENG, PRIMARY, LTV STEEL CO, 84- *Mem:* Am Inst Mining, Metall & Petrol Engrs; Iron & Steel Soc; Asn Iron & Steel Engrs. *Res:* Primary steelmaking processes and energy requirements. *Mailing Add:* LTV Steel Co Technol Ctr 6801 Brecksville Rd Independence OH 44131

WAKELIN, JAMES HENRY, JR, physics; deceased, see previous edition for last biography

WAKELYN, PHILLIP JEFFREY, b Akron, Ohio, Apr 29, 40. TEXTILE CHEMISTRY. *Educ:* Emory Univ, BS, 63; Ga Inst Technol, MS, 68; Univ Leeds, PhD, 71. *Prof Exp:* Res chemist, Fibers Div, Dow Chem Co, 63-66 & Dow-Badische Co, 66-67; res assoc textile chem, Textile Res Ctr, Tex Tech Univ, 71-73, head chem res, 73; MGR ENVIRON HEALTH & SAFETY, NAT COTTON COUN, 73- *Concurrent Pos:* Lectr textile chem, Tex Tech Univ, 72-73, adj prof chem eng, 74-78. *Mem:* Am Chem Soc; Sigma Xi; Am Asn Textile Chemists & Colorists; NY Acad Sci; assoc Brit Textile Inst; Am Oil Chem Soc. *Res:* Physical and chemical properties of textile fibers; dyeing and finishing of textile materials; fire retardants and flammability of textiles; chemistry of wool and cotton; environmental, industrial health and safety, consumer problems; cotton dust and occupational diseases, formaldehyde, toxic chemicals, air toxics. *Mailing Add:* Nat Cotton Coun 1521 New Hampshire Ave NW Washington DC 20036

WAKEMAN, CHARLES B, b New Haven, Conn, Aug 4, 27; m 48; c 3. ELECTRONICS. *Educ:* Yale Univ, BE, 50, ME, 52, PhD(elec eng), 55. *Prof Exp:* Res engr, Magnetics Inc, 55-57, dir res & develop, 57-61, vpres res & develop, 61-62; dir electronics res, Corning Glass Works, 62-66, dir phys res, 66-72, dir corp develop, 72-74, dir res & develop Europe, 74-78, dir tech & admin serv, 78-80; vpres res & develop, Siecor Corp, 80-85, pres, 85-86; RETIRED. *Mem:* AAAS; Inst Elec & Electronics Engrs; Sigma Xi. *Res:* Magnetic domain effects in highly rectangular hysteresis loop magnetic materials of interest in magnetic switching, memory and similar devices. *Mailing Add:* PO Box 300 Hickory NC 28603-0300

WAKEMAN, DONALD LEE, b Lebanon, Mo, Nov 17, 29; m 50; c 3. ANIMAL HUSBANDRY. *Educ:* Okla State Univ, BSA, 51; Univ Fla, MSA, 55. *Prof Exp:* Instr animal sci, Univ Tenn, 51; instr, 55-57, asst prof, Univ & asst animal husbandman, Agr Exp Sta, 57-67, assoc prof, Univ, 67-75, PROF ANIMAL SCI, UNIV FLA, 75-, ASSOC ANIMAL HUSBANDMAN, AGR EXP STA, 67- *Mem:* Am Soc Animal Sci. *Res:* Animal production and nutrition; beef cattle production. *Mailing Add:* Dept Animal Sci Univ Fla Gainesville FL 32611

WAKEMAN, JOHN MARSHALL, b Victoria, Australia, June 12, 37; m 70; c 2. FISH & WILDLIFE SCIENCES. *Educ:* Southern Ill Univ, BS, 73; Univ Ala, MS, 75; Univ Tex, Austin, PhD(zool), 78. *Prof Exp:* Res scientist, Maricult Ctr, Univ Tex, 78-79; PROF ANIMAL PHYSIOL, LA TECH UNIV, 79- *Concurrent Pos:* Res scientist, Marine Consortium, La Tech Univ, 80, Marine Sci Inst, Univ Tex, Austin, 81. *Mem:* Sigma Xi; Am Fisheries Soc; Gulf Estuarine Res Soc. *Res:* Swimming energetics and physiological responses of fishes with respect to salinity variations and to environmental pollutants; spawning and culture of marine fishes; closed-system aquaculture and mariculture. *Mailing Add:* Dept Biol Sci La Tech Univ PO Box 3187 Ruston LA 71272

WAKIL, SALIH J, b Kerballa, Iraq; nat US; m 52; c 4. BIOCHEMISTRY. *Educ:* Am Univ Beirut, BSc, 48; Univ Wash, PhD(biochem), 52. *Prof Exp:* Res assoc, Inst Enzyme Res, Univ Wis, 52-56, asst prof, Univ, 56-59; from asst prof to prof biochem, Sch Med, Duke Univ, 59-71; PROF BIOCHEM & CHMN DEPT, BAYLOR COL MED, 71- *Concurrent Pos:* Vis prof, Pasteur Inst Paris, 68-69; John Simon Guggenheim fel, 68-69; ad hoc mem physiol study sect, NIH, 71; mem metab biol panel, NSF, 71-74; pres, Asn Med Sch Departments Biochem, 88-89. *Honors & Awards:* Paul Lewis Award, Am Chem Soc, 67; Kuwait Prize, Kuwait Found Advan Sci, 88; Distinguished Serv Award, Arab Am Med Asn, 90- *Mem:* Nat Acad Sci; Am Soc Biochem & Molecular Biol; Sigma Xi; Am Soc Neurochem; Am Soc Microbiol; Am Chem Soc. *Res:* Genetic and metabolic control of fatty acid metabolism. *Mailing Add:* Dept Biochem Baylor Col Med One Baylor Plaza Houston TX 77030

WAKIM, KHALIL GEORGES, internal medicine; deceased, see previous edition for last biography

WAKIMOTO, BARBARA TOSHIKO, b Phoenix, Ariz, July 22, 54. DEVELOPMENTAL GENETICS, CYTOGENETICS. *Educ:* Ariz State Univ, BS, 76; Ind Univ, PhD(genetics), 81. *Prof Exp:* Res asst, Carnegie Inst Wash, 81-84; ASST PROF DEVELOP BIOL, UNIV WASH, 85- *Concurrent Pos:* Prin investr, Dept Zool, Univ Wash, 85- *Mem:* Genetics Soc Am; Develop Biol Soc. *Res:* Importance of chromosome structure for the expression of genes during development of Drosophila. *Mailing Add:* Dept Zool Univ Wash Seattle WA 98195

WAKSBERG, ARMAND L, b Paris, France; Can citizen; m 60; c 1. LASERS, COMMUNICATIONS. *Educ:* McGill Univ, BS, 56, MS, 60. *Prof Exp:* Scientist, Canadair, Ltd, 56-58 & Can Aviation Electronics, 60-63; sr scientist res dept, RCA Ltd, 63-77; DIR, LASER & ELECTRO-OPTICS, MPB TECHNOLOGIES, 77- *Concurrent Pos:* Chmn, Div Optics, Can Asn Physicists. *Mem:* Inst Elec & Electronics Engrs; Can Asn Physicists. *Res:* Lasers, including sidelight spectroscopy, laser noise and phase locking phenomena; laser communications and propagation; laser systems; laser receivers. *Mailing Add:* MPB Technol 1725 Transcanada Hwy Dorval PQ H9P 1J1 Can

WAKSBERG, JOSEPH, b Kielce, Poland, Sept 20, 15; US citizen; m 41; c 2. APPLIED STATISTICS. *Educ:* City Col New York, BS, 36. *Prof Exp:* Jr mathematician, USN Dept, 37-38; asst proj dir math, US Works Proj Admin, 38-40; asst proj dir, US Bur Census, 40-59, asst chief construction statist div, 59-63, chief statist methods div, 63-71, assoc dir statist, Bur, 72-73; VPRES, WESTAT INC, 73- *Concurrent Pos:* Instr statist, USDA Grad Sch, 63-73; consult, CBS News, 66- *Mem:* Fel Am Statist Asn; Int Asn Surv Statist. *Res:* Sample design for surveys; research in survey methodology, especially sampling and response errors. *Mailing Add:* 6302 Tone Dr Bethesda MD 20817-5814

WAKSMAN, BYRON HALSTEAD, b New York, NY, Sept 15, 19; m 44; c 2. IMMUNOLOGY. *Educ:* Swarthmore Col, BA, 40; Univ Pa, MD, 43. *Prof Exp:* Intern, Michael Reese Hosp, Ill, 44; res assoc neuropath, Harvard Med Sch, 49-52, assoc bact & immunol, 52-57, asst prof, 57-63; prof microbiol, 63-74, prof path, 74-78, ADJ PROF PATH, YALE UNIV, 79-; VPRES RES, NAT MULTIPLE SCLEROSIS SOC, 80- *Concurrent Pos:* Fel, Mayo Clin, 46-48; NIH fel, Columbia Univ, 48-49; res fel neuropath, Mass Gen Hosp, 49-52; assoc bacteriologist, Mass Gen Hosp, 52-63; consult assoc bacteriologist, Mass Eye & Ear Infirmary, 57-63; mem microbiol fels panel, NIH, 61-64, mem study sect B on allergy & immunol, 65-69; mem res rev panel, Nat Multiple Sclerosis Soc, 61-66; mem expert adv panel immunol, WHO, 63-68; chmn dept microbiol, Yale Univ, 64-70 & 72-74. *Mem:* Fel AAAS; Am Asn Immunol (secy-treas, 61-64, pres, 70-71); Am Soc Microbiol; Am Neurol Asn; Am Acad Neurol. *Res:* Role of thymus and lymphocytes in immune responses; immunologic tolerance; delayed hypersensitivity; mechanism of action of suppressor T-cells and lymphokines; immunologic and pathologic character of experimental autoallergic diseases; neuroimmunology; demyelinative diseases. *Mailing Add:* 300 E 54th St Apt 5K New York NY 10022

WALASZEK, EDWARD JOSEPH, b Chicago, Ill, July 4, 27; m 55; c 2. PHARMACOLOGY. *Educ:* Univ Ill, BSc, 49; Univ Chicago, PhD(pharmacol), 53. *Hon Degrees:* MD, Univ Helsinki, 90. *Prof Exp:* Asst prof neurophysiol & biochem, Univ Ill, 55-56; from asst prof to assoc prof, 57-62, PROF PHARMACOL, UNIV KANS MED CTR, KANSAS CITY, 62-, CHMN DEPT, 64- *Concurrent Pos:* Res fel, Univ Edinburgh, 53-55; USPHS spec res fel, 56-61, res career develop award, 61-63, res career award, 63-64; mem health study sect med chem, NIH, 62-66, mem res career award study sect, Nat Inst Gen Med Sci, 66-70; mem adv coun, Int Union Pharmacol, 72-76; mem, Health Study Sect Pharmacol-Toxicol, 74-78; mem comt teaching of sci, Int Coun Sci Unions, 72-78; foreign mem, Finland Acad Sci, 79; UN consult, China Develop Prog; consult, USSR through Ministry Health, Comput Med Educ; chmn, teaching comn, Int Union Pharmacol, 75-84; consult, Cutter Labs, 62-70, Alza Pharmaceut, 68-80, Inter X Corp, 73-81. *Honors & Awards:* Bela Issekutz medal, Hungarian Acad Sci; Pharmaceut Award, Polish Acad Sci; Recognition Award, Nat Drug Inst, Poland. *Mem:* AAAS; Am Chem Soc; Soc Neurosci; Am Soc Pharmacol & Exp Therapeut; fel Am Col Clin Pharmacol. *Res:* Pharmacologically active polypeptides; neurohumoral substances; naturally-occurring biogenic amines; pharmacology and physiology of the central nervous system. *Mailing Add:* Dept Pharmacol Univ Kans Med Ctr Kansas City KS 66103

WALAWENDER, MICHAEL JOHN, b Auburn, NY, Dec 16, 39; m 67. PETROLOGY, GEOLOGY. *Educ:* Syracuse Univ, BS, 65; SDak Sch Mines & Technol, MS, 67; Pa State Univ, University Park, PhD(petrol), 72. *Prof Exp:* Res asst mineral, SDak Sch Mines & Technol, 65-67; asst petrol, Pa State Univ, University Park, 67-72; asst prof geol, 72-77, ASSOC PROF GEOL, SAN DIEGO STATE UNIV, 77- *Mem:* Geol Soc Am. *Res:* Igneous and metamorphic petrology; mineralogy; planetology. *Mailing Add:* Dept Geol San Diego State Univ 5300 Campanile Dr San Diego CA 92182

WALBA, DAVID MARK, b Oakland, Calif, June 29, 49; m; c 1. ORGANIC CHEMISTRY. *Educ:* Univ Calif, Berkeley, BS, 71; Calif Inst Technol, PhD(chem), 75. *Prof Exp:* Fel, Univ Calif, Los Angeles, 75-77; from asst prof to assoc prof, 77-87, PROF CHEM, UNIV COLO, BOULDER, 87- *Concurrent Pos:* Fel, A P Sloan Found; Camille & Henry Dreyfus Teacher-Scholar. *Mem:* Am Chem Soc; Sigma Xi. *Res:* Host-guest chemistry; liquid crystals; topological stereochemistry. *Mailing Add:* Dept Chem Univ Colo Boulder CO 80309-0215

WALBA, HAROLD, b Chelsea, Mass, Mar 10, 21; m 46; c 2. ORGANIC CHEMISTRY. *Educ:* Univ Mass, BS, 46; Univ Calif, PhD(chem), 49. *Prof Exp:* From instr to assoc prof, 49-58, chmn dept, 61-64, PROF CHEM, SAN DIEGO STATE UNIV, 58- *Mem:* AAAS; Am Chem Soc; The Chem Soc; Sigma Xi. *Res:* Substituent effects and their transmission in organic molecules; tautomerism; acid-base strengths. *Mailing Add:* 3870 Carancho St La Mesa CA 91941-7606

WALBERG, CLIFFORD BENNETT, b Watkins, Minn, Feb 24, 15; m 46; c 4. CLINICAL CHEMISTRY. *Educ:* Univ Sask, BS, 39; Univ Southern Calif, AB, 43, MS, 45, PhD(biochem), 57. *Prof Exp:* Asst prof path, 69-76, CLIN CHEMIST, TOXICOL LAB, LOS ANGELES COUNTY-UNIV SOUTHERN CALIF MED CTR, 57-, ASSOC PROF PATH, SCH MED, UNIV SOUTHERN CALIF, 76- *Mem:* Am Asn Clin Chem. *Res:* Clinical biochemistry; toxicology. *Mailing Add:* 364 W Spazier Ave Burbank CA 91506

WALBORG, EARL FREDRICK, JR, b Chicago, Ill, Nov 13, 35; m 58; c 3. BIOCHEMISTRY. *Educ:* Austin Col, BA, 58; Baylor Univ, PhD(biochem), 62. *Prof Exp:* Asst biochemist & asst prof biochem, M D Anderson Hosp & Tumor Inst, Houston, 65-70, assoc prof biochem, 70-73, assoc biochemist & chief sect protein struct, 70-77, prof biochem, 73-77; prof biochem, Univ Tex Syst Cancer Ctr, Sci Park Res Div, Smithville, 77-85; mem grad fac, Univ Tex Grad Sch Biomed Sci, Houston, 70-85; CONSULT, 85- *Concurrent Pos:* USPHS res fel physiol chem, Univ Lund, 62-65; Eleanor Roosevelt Int Cancer fel biochem, Neth Cancer Inst, Amsterdam, 74; consult, 85- *Mem:* AAAS; Am Chem Soc; fel Am Inst Chem; Am Asn Cancer Res; Am Soc Biol Chem. *Res:* Chemistry of the cell-surface, glyoproteins, hepatocarcinogenesis. *Mailing Add:* PO Box 728 Smithville TX 78957-0728

WALBORN, NOLAN REVERE, b Bloomsburg, Pa, Sept 30, 44; m 75; c 1. ASTRONOMY. *Educ:* Gettysburg Col, BA, 66; Univ Chicago, PhD(astron, astrophys), 70. *Prof Exp:* Fel, Yerkes Observ, Univ Chicago, 71 & David Dunlap Observ, Univ Toronto, 71-73; staff astronr, Cerro Tololo Inter-Am Observ, 73-81; nat res coun sr res assoc, Goddard Space Flight Ctr, NASA, 82-84; STAFF ASTRONR, SPACE TELESCOPE SCI INST, 84- *Mem:* Am Astron Soc; Can Astron Soc; Int Astron Union; Astron Soc Pac. *Res:* Stellar spectroscopy; spectral classification; early-type stars; galactic structure; interstellar lines; Magellanic Clouds. *Mailing Add:* Space Telescope Sci Inst 3700 San Martin Dr Baltimore MD 21218

WALBORSKY, HARRY M, b Lodz, Poland, Dec 25, 23; nat US; m 53; c 4. ORGANIC CHEMISTRY. *Educ:* City Col New York, BS, 45; Ohio State Univ, PhD(chem), 49. *Prof Exp:* Res assoc, Calif Inst Technol, 48; res assoc, Atomic Energy Proj, Univ Calif, Los Angeles, 49-50; from asst prof to assoc prof chem, 50-59, PROF CHEM, FLA STATE UNIV, 59- *Concurrent Pos:* USPHS fel, Basel, Switz, 52-53; R O Lawton distinguished prof, 80. *Honors & Awards:* Alexander von Humboldt Sr Scientist Award, 88. *Mem:* Am Chem Soc; Royal Soc Chem. *Res:* Small ring compounds; organometallics; asymmetric synthesis; electrolytic and dissolving metal reductions; synthetic methods. *Mailing Add:* Dept Chem Fla State Univ Tallahassee FL 32306

WALBOT, VIRGINIA ELIZABETH, US citizen. PLANT GENETICS. *Educ:* Stanford Univ, AB, 67; Yale Univ, MPhil, 69, PhD(biol), 72. *Prof Exp:* NIH fel, Univ Ga, 72-75; asst prof biol, Washington Univ, 75-80; PROF BIOL SCI, STANFORD UNIV, 80- *Mem:* AAAS; Bot Soc Am; Soc Develop Biol; Am Soc Plant Physiol; Sigma Xi. *Res:* Plant molecular biology and development; genetics; botany. *Mailing Add:* Dept Biol Sci Stanford Univ Stanford CA 94305-5020

WALBRICK, JOHNNY MAC, b Wichita Falls, Tex, Sept 14, 41; m 64; c 2. INDUSTRIAL ORGANIC CHEMISTRY. *Educ:* Midwestern Univ, BS, 63; Univ Fla, PhD(chem), 67. *Prof Exp:* NSF fel, Univ Fla, 67-68; res chemist, Res Labs, Merichem Co, 70-74, mgr res, 73-77, dir res, 77-89, VPRES, RES & DEVELOP, MERICHEM CO, 89- *Mem:* Am Chem Soc; Chem Marketing Res Asn; Indust Res Inst; Com Develop Asn. *Res:* Mechanism of electroorganic reaction processes; new industrial processes for organic chemicals. *Mailing Add:* 1503 Central Houston TX 77012

WALBURG, H E, b Newark, NJ, Feb 6, 32; m 54; c 4. VETERINARY MEDICINE. *Educ:* Dartmouth Col, AB, 53; Va Polytech Inst, MS, 58; Univ Ga, DVM, 58; Univ Ill, PhD(radiobiol), 61. *Prof Exp:* Biologist, 61-73, dir, Comp Animal Res Lab, Oak Ridge Nat Lab, 73-83; CONSULT, 83- *Honors & Awards:* Animal Care Panel Res Award, 65. *Mem:* Radiation Res Soc; Geront Soc; Am Asn Cancer Res; Am Vet Med Asn. *Res:* Radiation carcinogenesis and radiation induced life-shortening and aging. *Mailing Add:* PO Box 9 Seymour TN 37865

WALBURN, FREDERICK J, b Cumberland, Md, Feb 7, 51. BIOMEDICAL ENGINEERING. *Educ:* Va Polytech Inst, BS, 73, MS, 75, PhD(eng sci & mech), 79. *Prof Exp:* Res bioengr, 82-90, dir corp res develop, 90-91, PRIN INVESTR, RES DIV, MIAMI HEART INST, 91- *Concurrent Pos:* Adj prof bioeng, Univ Miami, 82- *Mem:* Biomed Eng Soc; Am Soc Mech Engr; Am Heart Asn; Inst Elec & Electronics Engrs; Soc Clin Trials. *Mailing Add:* Miami Heart Inst 4701 N Meridian Ave Miami Beach FL 33140

WALCH, HENRY ANDREW, JR, b Minneapolis, Minn, June 3, 22; m 53; c 2. MYCOLOGY. *Educ:* Univ Calif, Los Angeles, BA, 50, PhD(microbiol), 54. *Prof Exp:* Mycol technician, Univ Calif, Los Angeles, 50-54, res asst, 54-55; from instr to assoc prof microbiol, 55-64, chmn dept, 60-64 & 72-75, PROF MICROBIOL, SAN DIEGO STATE UNIV, 64- *Concurrent Pos:* Res grants, San Diego Imp Counties Tuberc & Respiratory Health Asn, 57-72 & Respiratory Dis Asn Calif, 72-73; consult & lectr, Sharp Mem Hosp, San Diego, 58- & US Naval Hosp, 64-; consult, Palomar Mem Hosp, Escondido, Calif; NIH spec fel, Mycol Unit, Commun Dis Ctr, Atlanta, Ga. *Mem:* Am Soc Microbiol; Mycol Soc Am; Am Inst Biol Sci; AAAS; Sigma Xi. *Res:* Human and animal pathogenic fungi, particularly virulence factors, immunology and ecology. *Mailing Add:* 4605 El Cerrito Dr San Diego CA 92115

WALCHLI, HAROLD E(DWARD), b Warren, Pa, Nov 13, 22; m 45; c 2. ENGINEERING, PHYSICS. *Educ:* Pa State Univ, BS, 44; Univ Tenn, MS, 54. *Prof Exp:* Field engr, Bell Tel Co, Pa, 41; instr preradar, Pa State Univ, 42-43, staff asst elec eng, 43, asst instr physics, 43-44; asst engr, Tenn Eastman Corp, 44-47; engr, Carbide & Carbon Chem Co, 47-55; asst to tech dir, Westinghouse Elec Corp, 56-57, mgr eng serv, Atomic Power Dept, 57-58, asst proj mgr, Yankee Atomic Plant Proj, 58-60, fuel serv supvr, 60-64, fuel serv mgr, Nuclear Fuel Div, 64-71, mgr & adv engr, pressurized water reactor systs, Nuclear Servs Dept, Nuclear Energy Systs, 71-77, mgr strategic progs, Nuclear Serv Div, 77-80, facil & financial planning, fel engr data & commun serv, 80-86, Info Ctr Consult, microcomput specialist, Water Reactor Div, 83-86; PRES, HAL-COM ASSOCS, ENG COMPUTER CONSULTS, 80- *Concurrent Pos:* Chmn, Adv Comt Radioactive Mat, Hazardous Substances Transp Bd, Commonwealth Pa, 72-77; Am Soc Mech Engrs Y-32 Drafting Standards, 61-88; Am Nat Sci Inst N14 Transp Stds, 60- *Mem:* Am Soc Mech Engrs; Inst Elec & Electronics Engrs; Inst Nuclear Mat Mgt. *Res:* Nuclear magnetic resonance spectroscopy; electronic circuit design; radioactive materials transport; nuclear materials management; engineering management; computerized data communications; atomic plant design and construction; atomic fuel cycle services; microprocessor equipment applications; engineering and business applications software. *Mailing Add:* 1329 Foxboro Dr Monroeville PA 15146

WALCOTT, BENJAMIN, b Boston, Mass, May 31, 41; m 72. COMPARATIVE PHYSIOLOGY. *Educ:* Harvard Univ, BA, 63; Univ Ore, PhD(biol), 68. *Prof Exp:* USPHS physiol trainee, Univ Ore, 64-67, instr biol, 67-68; vis res fel biol, Res Sch Biol Sci, Australian Nat Univ, 69-71, fel biol, 71-72; asst prof, 72-79, ASSOC PROF, NEUROBIOL & BEHAVIOR,

STATE UNIV NY, STONY BROOK, 79-, ASSOC PROVOST, 87- *Mem:* Am Soc Cell Biol; foreign mem Brit Soc Exp Biol; Soc Gen Physiol; Biophys Soc; Soc Neurosci. *Res:* neural control of tear glands and of the cells of the immune system within the tear glands. *Mailing Add:* Dept Neurobiol & Behavior State Univ NY Stony Brook NY 11794

WALCOTT, CHARLES, b Boston, Mass, July 19, 34; m 76; c 2. BEHAVIORAL PHYSIOLOGY. *Educ:* Harvard Univ, AB, 56; Cornell Univ, PhD, 59. *Prof Exp:* Asst, Cornell Univ, 56-58; res fel biol, Harvard Univ, 59-60, asst prof appl biol, Div Eng & Appl Physics, 60-65; asst prof biol, Tufts Univ, 65-67; assoc prof biol, 67-74, actg dir, Ctr Curriculum Develop, 67-71, chmn dept cellular & comp biol, 71-76, PROF BIOL, STATE UNIV NY, STONY BROOK, 74-; PROF BIOL & EXEC DIR, LAB ORNITH, CORNELL UNIV, 81- *Concurrent Pos:* Dir, Natural Sci TV Proj, 59-60; dir elem sci study, Educ Develop Ctr, Inc, 65-67; dir, Content Res 3-2-1 Contact, Children's Television Workshop, 78-80. *Mem:* AAAS; Am Soc Zoologists; Animal Behav Soc; Sigma Xi. *Res:* Neurophysiology; animal behavior and orientation. *Mailing Add:* Cornell Lab Ornith 159 Sapsucker Woods Rd Ithaca NY 14850

WALCZAK, HUBERT R, b South Saint Paul, Minn, Jan 21, 34; m 61; c 3. MATHEMATICS. *Educ:* Col St Thomas, BA, 55; Univ Minn, PhD(math), 63. *Prof Exp:* Asst prof, 63-72, assoc prof, 72-77, PROF MATH, COL ST THOMAS, 77- *Mem:* Math Asn Am. *Res:* Analysis and quasiconformal mappings. *Mailing Add:* Dept Math Col St Thomas 2115 Summit Ave St Paul MN 55105

WALD, ALVIN STANLEY, b New York, NY, May 17, 34; m 77; c 2. HOSPITAL DESIGN, HOSPITAL SYSTEMS MANAGEMENT. *Educ:* Cooper Union, BEE, 55; Polytech Inst Brooklyn, MEE, 61; NY Univ, PhD(biomed eng), 74. *Prof Exp:* Sr engr, Bulova Res Develop Labs, 61-64; biomed eng res scientist & asst prof exp neurosurg, dept neurosurg, Med Ctr, NY Univ, 64-78; SR RES SCIENTIST, COLUMBIA UNIV COL PHYSICIANS & SURGEONS, 78-; TECH DIR, ANESTHESIOL BIOMED ENG SERV, COLUMBIA PRESBY MED CTR, 78- *Concurrent Pos:* Instr, Sch Respiratory Ther, NY Univ-Bellevue Hosp, 68-81; NY State Health Res Coun grant, 76-77; mem, comt construct exam for cert clin engrs, Am Bd Clin Eng, 77; mem neurol device panel, Bur Med Devices, Food & Drug Admin, 78-81; rep, med devices standards mgt bd & med device comt, Am Nat Standards Inst, 79-83; tech specialist, physiol monitoring comt, Univ Hosp, State Univ NY, Stony Brook, 80; mem hosp comt, long-range planning comt, Columbia Presby Med Ctr, 82-; ed, Eng in Med & Biol Mag, Inst Elec & Electronics Engrs, 84- *Mem:* AAAS; Am Physiol Soc; NY Acad Sci; Asn Advan Med Instrumentation; Inst Elec & Electronics Engrs. *Res:* Development and application of modern technology to health care needs. *Mailing Add:* Dept Anesthesiol Columbia Univ Col P&S 630 W 168th St New York NY 10032

WALD, ARNOLD, b New York, NY, June 10, 42; m 66; c 2. GASTROENTEROLOGY, INTERNAL MEDICINE. *Educ:* Colgate Univ, BA, 64; State Univ NY, Downstate Med Ctr, MD, 68. *Prof Exp:* From instr to asst prof med, Sch Med, Johns Hopkins Univ, 75-78; asst prof, 78-83, ASSOC PROF MED, SCH MED, UNIV PITTSBURGH, 83- *Concurrent Pos:* Head, gastroenterol unit, Montefiore Univ Hosp, Pittsburgh, 83- *Mem:* Am Gastroenterol Asn; Am Fedn Clin Res; Am Motility Soc; fel Am Col Physicians; fel Am Col Gastroenterol. *Res:* Disorders of gastrointestinal motor function (motility); behavioral modification of gastrointestinal function. *Mailing Add:* Dept Med Montefiore Univ Hosp Pittsburgh PA 15213

WALD, FRANCINE JOY WEINTRAUB, b Brooklyn, NY, Jan 13, 38; m 64; c 2. PHYSICS, SCIENCE EDUCATION. *Educ:* City Col New York, BEE, 60; Polytech Inst Brooklyn, MS, 62, PhD(chem physics), 69. *Prof Exp:* Engr solid state physics, Remington Rand Univac Div, 60; instr physics, Polytech Inst Brooklyn, 62-64, adj res assoc, 69-70; sci consult physics & biol, 72-75, SCI INSTR, FRIENDS SEM, 75-, CHAIRPERSON DEPT SCI, 76- *Concurrent Pos:* Lectr phys sci, New York Community Col, 69 & 70. *Mem:* Am Phys Soc; Sigma Xi; Am Asn Physics Teachers; Nat Sci Teachers Asn; Asn Teachers Independent Schs; NY Acad Sci; AAAS. *Res:* Investigating how children of various ages respond to science, particularly physics, how they assimilate the concepts and language encountered; gender issues in science education. *Mailing Add:* 520 La Guardia Pl New York NY 10012

WALD, FRITZ VEIT, b Dieringhausen, WGer, Apr 28, 33; m 59; c 3. SOLID STATE CHEMISTRY, MATERIALS SCIENCE. *Educ:* Sch Tech Chem, Cologne, Ger, BS, 55. *Prof Exp:* Tech chemist, Ed Doerrenberg Soehne, Steelworks Ruenderoth, WGer, 55-57; res asst metall & solid state chem, Philips Cent Lab Aachen, 57-61; res metallurgist, Frigistors Ltd Que, Can, 61-63; sr scientist, Tyco Labs Inc, 63-69, head, Mat Sci Dept, Corp Technol Ctr, 69-75, assoc tech dir, 75-80, dir res, 81-87, SR SCI ADV, MOBIL SOLAR ENERGY CORP, 88- *Concurrent Pos:* Mem bd dirs, Radiation Monitoring Devices Inc, 76- *Honors & Awards:* Indust Res Award, 70. *Mem:* Metall Soc; Am Inst Mining, Metall & Petrol Engrs; fel Am Inst Chemists. *Res:* Solid state chemistry; electronic materials; photovoltaic solar energy conversion; crystal growth. *Mailing Add:* Mobil Solar Corp Four Suburban Park Dr Billerica MA 01821-3980

WALD, GEORGE, b New York, NY, Nov 18, 06. BIOCHEMISTRY, MOLECULAR BIOLOGY. *Educ:* NY Univ, BS, 27; Columbia Univ, MA, 28, PhD(zool), 32. *Prof Exp:* Prof, 44-68, Higgins prof, 68-77, EMER PROF BIOL, HARVARD UNIV, 77- *Honors & Awards:* Nobel Prize in Physiol, 67. *Mem:* Nat Acad Sci; AAAS; Am Philos Soc. *Res:* Cosmology of life and mind. *Mailing Add:* 21 Lakeview Ave Cambridge MA 02138

WALD, MILTON M, b San Francisco, Calif, Oct 29, 25; m 56; c 2. CHEMISTRY. *Educ:* Univ Calif, Los Angeles, BS, 49; Univ Southern Calif, PhD(chem), 54. *Prof Exp:* Res assoc, Brookhaven Nat Lab, 54-56; chemist, Shell Develop Co, 56-88; RETIRED. *Mem:* Am Chem Soc. *Res:* Organic and catalytic chemistry; petroleum chemistry. *Mailing Add:* 10706 Holly Springs Houston TX 77042

WALD, NIEL, b New York, NY, Oct 1, 25; m 53; c 2. PUBLIC HEALTH, RADIATION MEDICINE. *Educ:* Columbia Univ, AB, 45; NY Univ, MD, 48. *Prof Exp:* Intern & resident med affiliated hosps, NY Univ, 48-49, 50-52; sr hematologist & head radioisotope lab, Atomic Bomb Casualty Comn, Japan, 54-57; head biologist, Health Physics Div, Oak Ridge Nat Lab, 57-58; assoc res prof, Univ Pittsburgh, 58-60, assoc prof, 60-62, chmn, Dept Radiation Health, 69-76 & 77-89, Dept Occup Health, 75-76 & Dept Indust Environ Health Sci, 76-77, PROF RADIATION HEALTH, GRAD SCH PUB HEALTH, UNIV PITTSBURGH, 62-, PROF RADIOL, SCH MED, 65- & PROF HUMAN GENETICS, GRAD SCH PUB HEALTH, 91- *Concurrent Pos:* Fel immunohemat, NY Univ affiliated hosps, 49-50; asst prof med, Sch Med, Univ Pittsburgh, 58-65; mem Pa Governor's Adv Comt Atomic Energy Develop & Radiation Control, 66-84, chmn, 74-76; consult, Div Oper Safety, US AEC, 68-75, Div Compliance, 69-75, US Navy Submarine & Radiation Med Div, 73-, US Energy Res & Develop Admin, 75-77, US Nuclear Regulatory Comn, 75- & US Dept of Energy, 78-80; mem, Nat Coun Radiation Protection, 70-82, consociate mem, 82-, US Nuclear Regulatory Comn Adv Panel Decontamination, Three Mile Island-2 Reactor Facil, 83-; mem, Pa Adv Comt on Low Level Radioactive Waste Disposal, 85-; vis Am prof, Royal Soc Med, UK, 86; US mem working group Health Effects, US-USSR Joint Coordr Comt, Civilian Nuclear Reactor Safety, 89- *Mem:* Health Physics Soc (pres, 73-74); Radiation Res Soc; Environ Mutagen Soc; Soc Human Genetics; AMA. *Res:* Diagnosis and treatment of radiation injury; health physics; cytogenetics. *Mailing Add:* Grad Sch Pub Health Univ Pittsburgh Pittsburgh PA 15261

WALD, ROBERT MANUEL, b New York, NY, June 29, 47. THEORETICAL PHYSICS. *Educ:* Columbia Univ, AB, 68; Princeton Univ, PhD(physics), 72. *Prof Exp:* Res assoc physics, Univ Md, 72-74; res assoc, 74-76, from asst prof to assoc prof, 76-85, PROF PHYSICS, UNIV CHICAGO, 85- *Mem:* Am Phys Soc. *Res:* General relativity and gravitation; black holes; quantum field theory in curved spacetime. *Mailing Add:* Enrico Fermi Inst Univ Chicago Chicago IL 60637

WALD, SAMUEL STANLEY, b New York, NY, Feb 5, 07; m 32; c 2. ROENTGENOLOGY. *Educ:* NY Univ, DDS, 28; Am Bd Oral & Maxillofacial Radiol, dipl, 80. *Prof Exp:* CLIN PROF RADIOL, DIAG & ORAL CANCER, COL DENT, NY UNIV, 28-, ASSOC PROF ROENTGENOL, 54-, CLIN PROF RADIOL, SCH MED & POSTGRAD MED SCH, 30- *Concurrent Pos:* Head dept roentgenol & diag, Guggenheim Dent Clin & Sch Dent Hyg, 30-42; consult, Dent Clins, Community Serv Soc, NY, 30-; spec lectr, US Naval Hosp, St Albans, 48- & Vet Admin, Brooklyn, 51-; chmn adv comt radiol, State Dept Health, NY & mem, Mayor's Adv Comt Radiation & Dent, New York, 56-; vis prof, NJ Col Med & Dent, 58-; consult radiol, diag & oral med, Mem Hosp Cancer & Allied Dis, New York; rear admiral, US Naval Reserves, DC, 60. *Honors & Awards:* Jarvy Burkhart Laureate Gold Medal Award, NY Dent Soc, 81. *Mem:* Fel Royal Soc Health; Sci Res Soc Am; Am Dent Asn; Asn Mil Surg US; fel Am Col Dent; Sigma Xi. *Res:* Radiology and diagnosis; oral surgery and oral cancer; dental medicine. *Mailing Add:* 420 E 72nd St New York NY 10021

WALDBAUER, EUGENE CHARLES, b Philadelphia, Pa, July 4, 26; m 56; c 5. NATURAL HISTORY. *Educ:* East Stroudsburg State Col, BS, 52; Pa State Univ, MS, 56; Cornell Univ, PhD(wildlife biol, natural hist & parasitol), 66. *Prof Exp:* From asst prof to assoc prof, 56-69, PROF BIOL, STATE UNIV NY COL CORTLAND, 69- *Mem:* Wilderness Soc. Sigma Xi. *Res:* Flora of Cortland County, New York; pollen analysis of central New York bogs; ferns and lycopodiums of central New York. *Mailing Add:* Dept Biol State Univ NY Col Cortland NY 13045

WALDBAUER, GILBERT PETER, b Bridgeport, Conn, Apr 18, 28; m 55; c 2. ENTOMOLOGY. *Educ:* Univ Mass, BS, 53; Univ Ill, MS, 56, PhD, 60. *Prof Exp:* Asst, 53-58, from instr to assoc prof, 58-71, PROF ENTOM, UNIV ILL, URBANA, 71- *Concurrent Pos:* Sr scientist, Int Rice Res Inst, Los Ban13s, Phillipines, 78-79; USAID consult, Pakistan Agr Res Coun, 85. *Mem:* Sigma Xi; Entom Soc Am; Ecol Soc Am. *Res:* Ecology, behavior and physiology of insects; mimicry. *Mailing Add:* Dept Entom 320 Morrill Hall Univ Ill Urbana IL 618014

WALDBILLIG, RONALD CHARLES, b Iron Mountain, Mich, Mar 17, 43; m 61; c 2. MEMBRANE BIOPHYSICS, ELECTRO-OPTICAL DIELECTRICS. *Educ:* Northern Mich Univ, BS, 67; Univ Rochester, PhD(neurobiol), 73. *Prof Exp:* Res fel, Anat Dept, Duke Univ, 74-76; ASST PROF BIOPHYS, PHYSIOL DEPT, MED BR, UNIV TEX, 76- *Mem:* Biophys Soc; AAAS. *Res:* Molecular and physical aspects of membrane function and structure; electrical and optical analysis of the insulating characteristics of single lipid bilayer membranes. *Mailing Add:* Physiol & Biophys Dept 411 Basic Sci Bldg F41 Univ Tex Med Sch 301 University Blvd Galveston TX 77550

WALDE, RALPH ELDON, b Perham, Minn, Mar 8, 43; div; c 2. MATHEMATICS. *Educ:* Univ Minn, Minneapolis, BA, 64; Univ Calif, Berkeley, PhD(math), 67. *Prof Exp:* Asst prof math, Univ Minn, Minneapolis, 67-72; ASST PROF MATH, TRINITY COL, CONN, 72- *Mem:* Am Math Soc; Math Asn Am; Asn Comput Mach. *Res:* Lie algebras; non-associative algebras. *Mailing Add:* Dept Math Trinity Col 300 Summit St Hartford CT 06106

WALDEN, C(ECIL) CRAIG, b Guernsey, Sask, Apr 19, 21; m 41; c 4. POLLUTION CONTROL, MICROBIOLOGY. *Educ:* Univ Sask, BA, 40, MA, 41; Univ Minn, PhD(agr biochem), 54. *Prof Exp:* Chemist & supvr munitions, Defence Industs, Ltd, 41-44; chemist in charge cereal chem, Quaker Oats Co Can, Ltd, 44-50; assoc res chemist, BC Res Coun, 50-51; asst prof cereal chem, Univ Sask, 51-54; res chemist, 54-61, head div appl biol, 61-70, assoc dir, BC Res, 70-81; dir, Forintek Can Corp, 81-84; CONSULT, 84- *Concurrent Pos:* Hon prof chem eng, Univ BC. *Mem:* Fed Asns Can Environ; Can Pulp & Paper Asn. *Res:* Water quality and effluent treatment; mineral microbiology. *Mailing Add:* 4008 Quesnel Vancouver BC V6L 2X2 Can

WALDEN, CLYDE HARRISON, b Kansas City, Mo, Dec 19, 21; m 46; c 3. SCIENCE ADMINISTRATION. *Educ:* William Jewell Col, BA, 42; Univ Colo, MS, 46, PhD(phys chem), 49. *Prof Exp:* Mgr uranium accountability, Mallinckrodt Chem Co, 48-51; sr scientist sec recovery oil, Phillips Petrol Co, 51; mgr qual control, Nat Lead Co, Ohio, 51-57; dir qual control, Gen Tire & Rubber Co, 57-64; DIR PROCESS TECHNOL, KAISER ALUMINUM & CHEM CORP, 64- *Mem:* Am Chem Soc; fel Am Soc Qual Control. *Res:* Administration of technical and engineering functions; heats of chemical reactions. *Mailing Add:* 6106 Bullard Dr Oakland CA 94611-3109

WALDEN, DAVID BURTON, b New Haven, Conn, Mar 29, 32; c 2. PLANT GENETICS, CYTOGENETICS. *Educ:* Wesleyan Univ, BA, 54; Cornell Univ, MSc, 58, PhD(genetics), 59. *Prof Exp:* Fel bot, Ind Univ, 59-61; from asst prof to assoc prof, 61-71, actg chmn dept, 71-73, PROF PLANT SCI, UNIV WESTERN ONT, 71- *Concurrent Pos:* Vis prof, Dept Genetics, Univ Birmingham, 73-74 & 81 & Dept Genetics & Develop & Dept Agron, Univ Ill, 74; pres, Biol Coun Can, 74-75; assoc ed, Can J Genetics & Cytol, 78-81; secy-gen, 16th Int Genetics Congress. *Mem:* AAAS; Crop Sci Soc Am; Genetics Soc Am; Genetics Soc Can (pres, 81-83); Int Genetics Fedn (pres, 88-). *Res:* Pollen biology; corn genetics; plant and human cytogenetics; somatic cell genetics; heat shock and stress proteins. *Mailing Add:* Dept Plant Sci Univ Western Ont Fac Sci Western Sci Ctr London ON N6A 5B8 Can

WALDEN, JACK M, b Sheridan, Wyo, July 1, 22; m 46; c 5. ELECTRICAL ENGINEERING, COMPUTER SCIENCE. *Educ:* SDak Sch Mines & Technol, BS, 44; Okla State Univ, MS, 62, PhD(eng), 65. *Prof Exp:* Chief engr, Midnight Sun Broadcasting, Alaska, 48-53; vpres & tech dir, Northern TV, Inc, 53-60; from instr to assoc prof elec eng, Okla State Univ, 60-69; eng group leader, Calculator Prod Div, Hewlett-Packard Co, 69-78; SECT TECH DIR, FISCAL INFO COLO, INC, 78- *Mem:* Sr mem Inst Elec & Electronics Engrs; Asn Comput Mach. *Res:* Computer logic design; computer programming of operating systems and compilers; engineering applications of computers; computer-aided instruction. *Mailing Add:* 2507 Lake Dr Loveland CO 80538

WALDEN, ROBERT HENRY, b New York, NY, May 24, 39; div; c 2. SEMICONDUCTORS, OPTOELECTRONIC INTEGRATED CIRCUITS. *Educ:* NY Univ, BES, 62, MEE, 63, PhD(eng sci), 66. *Prof Exp:* Mem tech staff, Bell Tel Labs Inc, 66-78; sr proj engr, 78-80, sect head, 80-81, mgr tech dept, Hughes Aircraft Co, 81-83, sr scientist, 83-85, SR STAFF ENGR, HUGHES RES LABS, 85- *Honors & Awards:* Hughes Res Lab outstanding achievement award, 87. *Mem:* Inst Elec & Electronics Engrs. *Res:* Optoelectronic integrated circuits, InP-based heterostructures; analog-to-digital conversion, sigma delta modulators; sub-micrometer III-V integrated circuits and device modelling; MOS, H8T device physics. *Mailing Add:* Hughes Res Labs 3011 Malibu Canyon Rd Malibu CA 90265

WALDERN, DONALD E, b Lacombe, Alta, June 8, 28; m 53; c 4. ANIMAL NUTRITION, BIOCHEMISTRY. *Educ:* Univ BC, BSA, 51, MSA, 54; Wash State Univ, PhD(nutrit, biochem), 62. *Prof Exp:* Res scientist, Exp Sta, Can Dept Agr, 53-57 & 61-62; assoc prof dairy nutrit, Wash State Univ, 62-67; res scientist, West Region Res Br, Agr Can, 67-73, dir, Res Sta, 73-78, prog specialist, 78-80, dir, Lacombe Res Sta, 80-89; RETIRED. *Concurrent Pos:* Coordr, agr res progs, WCan. *Mem:* Am Dairy Sci Asn; Am Soc Animal Sci; Agr Inst Can; Can Soc Animal Sci. *Res:* Nutritive value of forages and cereal grains for dairy and beef cattle; complete feeds for dairy cows and early weaned calves; relationship of blood biochemical parameters to performance factors in dairy and beef cows; nutritional management systems for beef cows grazing grassland and forested rangeland. *Mailing Add:* Box 14 Site 75 RR 2 Summerland BC V0H 1Z0 Can

WALDHAUER, F D, b Brooklyn, NY, Dec 6, 27; m 55; c 5. ENGINEERING. *Educ:* Cornell Univ, BEE, 48; Columbia Univ, MSEE, 60. *Prof Exp:* Engr, Radio Corp Am, 48-53, patent agent, 53-55; engr, Bell Tel Labs, 56-63; supvr, AT&T Bell Labs, 63-87; DIR, RESOUND CORP, 87- *Mem:* Fel Inst Elec & Electronics Engrs. *Res:* Physiological processes of hearing to understand how build hearing prostheses; mathematical and physical feedback processes. *Mailing Add:* Resound Corp 220 Saginaw Dr Redwood City CA 94063

WALDHAUSEN, JOHN ANTON, b New York, NY, May 22, 29; m 57; c 3. SURGERY. *Educ:* Col Great Falls, BS, 50; St Louis Univ, MD, 54. *Prof Exp:* Intern, Johns Hopkins Hosp, Baltimore, 54-55, resident, 56-57; surgeon, Nat Heart Inst, Md, 57-59: resident, Hosp, Univ Pa, 59-60; resident, Med Ctr, Ind Univ, Indianapolis, 60-62, instr surg, Sch Med, 62-63, asst prof, 63-66; assoc prof, Sch Med, Univ Pa, 66-70; prof surg, 70-84, interim provost & dean, Col Med, 72-73, prof, 70-84, JOHN W OSWALD PROF SURG, HERSHEY MED CTR, PA STATE UNIV, 84-, CHMN DEPT, 70- *Concurrent Pos:* Surg res fel, Johns Hopkins Hosp, Baltimore, Md, 55-56; NIH career develop award, Sch Med, Ind Univ, Indianapolis, 63-66; assoc surgeon, Children's Hosp, Philadelphia, 66-70 & Hosp Univ Pa, 66-70; mem surg study sect B, NIH, 74-78; mem, Anesthesia Grant Comt, Nat Inst Gen Med Sci, NIH, 70-73; chmn, Coronary Artery Dis Opers Comt, Vet Admin Coop Study, 75-; mem, Pvt Doctors Am Collab Study, Harvard Sch Pub Health, 79-82; dir, Am Bd Surg, 85. *Honors & Awards:* Fel, Am Asn Adv Sci, 86. *Mem:* Soc Univ Surg; Am Col Cardiol (secy, 81-82); Am Asn Thoracic Surg; Am Col Surgeons; Am Surg Asn (first vpres, 85-86); Am Physiol Soc; fel AAAS. *Res:* Effects of operative repair of congenital heart defects on pulmonary circulation; newer methods in repair of congenital heart defects; effects of cardiac surgery on ventricular function. *Mailing Add:* Hershey Med Ctr Pa State Univ PO Box 850 Hershey PA 17033

WALDICHUK, MICHAEL, b Mitkau, Roumania, Oct 23, 23; nat Can; m 55; c 3. OCEANOGRAPHY. *Educ:* Univ BC, BA, 48, MA, 50; Univ Wash, PhD, 55. *Prof Exp:* Assoc scientist, Pac Biol Sta, Fisheries Res Bd Can, Nanaimo, BC, 54-58, sr scientist, 58-63, prin scientist, 63-66, oceanogr-in-chg, 66-69; oceanog consult & secy, Can Comt Oceanog, Ottawa, 69-70; prog head, 70-77, SR SCIENTIST, WEST VANCOUVER LAB, FISHERIES & OCEANS, CAN, 77- *Concurrent Pos:* Mem, Intergovt Maritime Consult, Orgn-Food & Agr Orgn-UNESCO-World Meteorol Orgn-Int Atomic Energy Agency-UN Joint Group Experts Sci Aspects of Marine Pollution, 69-77, chmn, 70-73; partic coastal wastes mgt study session, Nat Acad Sci, Wyo, 69, workshop on marine environ qual, 71; mem panel marine aquatic life & wildlife, Comt Water Qual Criteria, Nat Acad Sci, Washington, DC, 71-73. *Mem:* AAAS; Can Meteorol & Oceanog Soc; Am Geophys Union; Am Soc Limnol & Oceanog; fel Chem Inst Can; Sigma Xi. *Res:* Chemical and physical oceanography; industrial wastes and marine pollution. *Mailing Add:* West Vancouver Lab 4160 Marine Dr West Vancouver BC V7V 1N6 Can

WALDINGER, HERMANN V, b Vienna, Austria, June 17, 23; US citizen; m 48; c 2. MATHEMATICS. *Educ:* Pomona Col, BA, 43; Brown Univ, MS, 44; Columbia Univ, PhD(math), 51. *Prof Exp:* Res engr, Repub Aviation Corp, 45-46; appl mathematician, M W Kellogg Co, 46-53; sr mathematician, Nuclear Develop Corp Am, 53-59; sr scientist, Repub Aviation Corp, 59-61; ASSOC PROF MATH, POLYTECH UNIV, 61- *Mem:* AAAS; Am Math Soc; Math Asn Am; Sigma Xi. *Res:* Group theory. *Mailing Add:* 600 Maitland Ave Teaneck NJ 07666

WALDINGER, RICHARD J, b Brooklyn, NY, Mar 1, 44; c 2. COMPUTER SCIENCE. *Educ:* Columbia Univ, AB, 64; Carnegie-Mellon Univ, PhD(comput sci), 69. *Prof Exp:* Res mathematician, 69-76, sr comput scientist, 76-81, STAFF SCIENTIST, ARTIFICIAL INTEL CTR, SRI INT, 81- *Concurrent Pos:* Vis instr, Stanford Univ, 73, vis scholar, 79-85, consult prof, 85-; NSF grants. *Mem:* Asn Automatic Reasoning. *Res:* Artificial intelligence; automatic program synthesis; mechanical theorem proving; planning. *Mailing Add:* Artificial Intel Ctr SRI Int Menlo Park CA 94025

WALDMAN, ALAN S, b Bronx, NY, Aug 15, 59; m 85. MAMMALIAN CELL TRANSFECTION. *Educ:* State Univ NY, Albany, BS, 80; Johns Hopkins Univ, PhD(biochem), 85. *Prof Exp:* Postdoctoral fel, Sch Med, Yale Univ, 85-89; ASST PROF BIOCHEM & MOLECULAR BIOL, SCH MED, IND UNIV, 89- *Concurrent Pos:* Ad hoc reviewer, Molecular & Cellular Biol, 89-, J Molecular Biol, 89-, Somatic Cell & Molecular Genetics, 89-, Biotech, 90-, Proc Nat Acad Sci, 91- *Mem:* Am Soc Biochem & Molecular Biol; Sigma Xi. *Res:* Basic mechanisms of homologous recombination, genetic rearrangements, in mammalian cells, using molecular genetics as well as biochemical approaches. *Mailing Add:* Walther Oncol Ctr Sch Med Ind Univ 975 W Walnut St Indianapolis IN 46202-5121

WALDMAN, BARBARA CRISCUOLO, b Bethesda, Md, Jan 24, 56; m 85. GLYCOSYLATION, CELL BIOLOGY. *Educ:* Va Polytech Inst & State Univ, BS, 78; Johns Hopkins Univ, PhD(biochem), 85. *Prof Exp:* Postdoctoral fel pharmacol, Yale Univ Sch Med, 85-89, postdoctoral assoc therapeut radiol, 89; ASST SCIENTIST, IND UNIV SCH MED, 89- *Mem:* Am Soc Cell Biol; Am Soc Biochem & Molecular Biol; Sigma Xi; Am Chem Soc. *Res:* Regulation of glycosylation in mammalian cells; biosynthesis of asparagine-linked glycoproteins and nucleotide-sugar transport into the Golgi apparatus; protein trafficking. *Mailing Add:* Ind Sch Med Riley Hosp Rm A593 702 Barnhill Dr Indianapolis IN 46202

WALDMAN, BERNARD, b New York, NY, Oct 12, 13; m 42, 64; c 4. NUCLEAR PHYSICS. *Educ:* NY Univ, AB, 34, PhD(physics), 39. *Prof Exp:* Asst physics, NY Univ, 35-38; res assoc, Univ Notre Dame, 38-40, instr, 40-42, asst prof, 42-43; staff mem & group leader, Los Alamos Sci Lab, 43-45; assoc prof physics, 45-51, assoc dean, Col Sci, 64-67, PROF PHYSICS, UNIV NOTRE DAME, 51-, DEAN, COL SCI, 79-; PROF, MICH STATE UNIV, 79- *Concurrent Pos:* Mem staff, Midwestern Univs Res Asn, 58-59, vpres, 59-60, 65-, lab dir, 60-65; mem physics adv panel, NSF, 65-68, chmn, 68; trustee, Univs Res Asn, Inc, 65-71; assoc dir, Nat Superconducting Cyclotron Lab, 79- *Mem:* Fel Am Phys Soc; Am Asn Physics Teachers; Sigma Xi. *Res:* Medium energy electrons and x-rays; electrostatic and high energy accelerators. *Mailing Add:* 1909 Wedgewood Dr Carolina Trace Sanford NC 27330

WALDMAN, GEORGE D(EWEY), b Hartford, Conn, Aug 5, 32. AERONAUTICAL ENGINEERING. *Educ:* Trinity Col, BS, 54; Brown Univ, MS, 57, PhD(appl math), 59. *Prof Exp:* Staff scientist, Res & Adv Develop Div, 59-65, SR STAFF SCIENTIST, SYSTS DIV, AVCO CORP, 65- *Mem:* Am Inst Aeronaut & Astronaut. *Res:* Theory of flows at high speeds; pressure and heat transfer distributions over reentering vehicles; aerodynamic stability and control; multiphase flows and pollution. *Mailing Add:* 511 West St Reading MA 01867

WALDMAN, JEFFREY, b Philadelphia, Pa, Jan 10, 41; m 62; c 3. ALUMINUM ALLOY RESEARCH, THERMAL MECHANICAL PROCESSING. *Educ:* Drexel Univ, BS, 63; Mass Inst Technol, ScD(metall), 67. *Prof Exp:* Mat engr, Us Army Armament Res & Develop, 77-83; BR CHIEF, NAVAL AIR DEVELOP CTR, 83- *Concurrent Pos:* Vis scientist, Light Metals Res Inst, Novara, Italy, 71; adj prof, Drexed Univ, 75-; mem, Mfg Tech Adv Group Metals Subcomt, Dept Defense, 76-85, Nat Mat Adv Bd Panel Aluminum Powder Metall Alloys, 82-83, Agard Struct & Mat Panel, NATO, 87- & Aerospace Mat Conf, Am Soc Metals, 90-; res adv, Nat Res Coun, 78- & Off Naval Technol, 85- *Mem:* Fel Am Soc Metals; Metall Soc; Sigma Xi. *Res:* Advanced metals and alloys for Navy aircraft, such as aluminum-lithium alloys, thermal mechanical processing, as well as on metal matrix and ceramic matrix composites. *Mailing Add:* 3894 Donna Dr Huntingdon Valley PA 19006

WALDMAN, JOSEPH, b Philadelphia, Pa, May 12, 06; c 3. OPHTHALMOLOGY. *Educ:* Jefferson Med Col, MD, 30; Am Bd Ophthal, dipl, 35. *Prof Exp:* Prof, 30-80, EMER PROF OPHTHAL, JEFFERSON MED COL, 30- *Mem:* AMA. *Mailing Add:* 404 Meadowbrook Lane Erdenheim PA 19118

WALDMAN, L(OUIS) A(BRAHAM), b Toledo, Ohio, Oct 13, 29; m 61; c 2. NUCLEAR REACTOR TECHNOLOGY. *Educ:* Univ Toledo, BS, 51; Carnegie Inst Technol, MS, 56; Univ Pittsburgh, PhD(chem eng), 64. *Prof Exp:* Student-employee, Oak Ridge Sch Reactor Tech, Oak Ridge Nat Lab, Union Carbide & Carbon Chem Corp, 51-52; FEL ENGR, BETTIS ATOMIC POWER LAB, WESTINGHOUSE ELEC CORP, 52- *Concurrent Pos:* Lectr, Carnegie Inst Technol, 65-66. *Mem:* Am Inst Chem Engrs; Am Nuclear Soc. *Res:* Heat and mass transfer; corrosion; erosion; fretting wear; release, transport and deposition of fission products from nuclear fuels; material and fuel element development for nuclear reactors. *Mailing Add:* 6550 Lilac St Pittsburgh PA 15217

WALDMAN, ROBERT H, b Dallas, Tex, Dec 21, 38; m 63; c 4. MICROBIOLOGY. *Educ:* Rice Inst, BA, 59; Wash Univ, MD, 63. *Prof Exp:* Intern, Johns Hopkins Hosp, 64, resident, 65; clin assoc, Nat Inst Allergy & Infectious Dis, 65-67; prof & actg chmn dept med, Col Med, Univ Fla, 67-76; prof & chmn, Dept Med, Sch Med, WVa Univ, 76-85; DEAN, COL MED, UNIV NEBR MED CTR, 85- *Concurrent Pos:* Consult, cholera, WHO, 69, vis scientist, Int Res & Training Ctr Immunol, 69-70; mem cholera adv comt, NIH. *Honors & Awards:* Alexandre Besredka Prize, Ger Immunol Soc, 77. *Mem:* Am Soc Clin Invest; Am Fedn Clin Res; Am Asn Immunol; Am Soc Microbiol; fel Am Col Physicians; Am Clin Climat Asn. *Res:* Immunology of viral respiratory infections; study of the secretory immunologic system; immunization by application of antigen to mucous surfaces; host defense mechanisms resident on mucosal surfaces. *Mailing Add:* Dean's Off Univ Nebr Col Med Omaha NE 68198-6545

WALDMANN, THOMAS A, b New York, NY, Sept 21, 30; m 58; c 3. MEDICINE, IMMUNOLOGY. *Educ:* Univ Chicago, AB, 51; Harvard Univ, MD, 55. *Hon Degrees:* DSc, Debrecen Hungary, 90. *Prof Exp:* Intern, Mass Gen Hosp, 55-56; clin assoc, 56-58, sr investr, Metab Br, 59-65, head, Immunophysiol Sect, 65-73, CHIEF METAB BR, NAT CANCER INST, 71- *Concurrent Pos:* Am Heart Asn fel, Nat Cancer Inst, 58-59; mem, Nat Cancer Plan Comt, 72; consult, Fed Trade Comn & WHO; assoc ed, J Immunol. *Honors & Awards:* Michael Heidelberger lectr, Columbia Univ; Irvin Strasberger lectr, Cornell Univ; Lucy Klein Mem lectr, Northwestern Univ; Merril lectr, Thomas Jefferson Univ; Phillips McMaster Mem lectr, Rockefeller Univ; Bela Schick Award, Am Col Allergists; Henry M Stratton Medal, Am Soc Hemat; Larry S Bernton Award, Allergy Soc; Kroc Honor Award, Am Asn Physicians; G Burroughs Mider Award, NIH; Ciba-Geigy Drew Award; Lila Gruber Prize; Artois-Baillet LaTour Health Prize. *Mem:* Nat Acad Sci; Am Asn Immunologists; Am Fedn Clin Res; Am Soc Clin Invest; Asn Am Physicians; Am Physiol Soc; fel Am Soc Microbiologists; fel Am Acad Allergy & Immunol; AAAS; Sigma Xi; Am Acad Arts & Sci. *Res:* Factors controlling the human immune responses; discovery of the diseases intestinal lymphangiectasia, allergic enteropathy and familial hypercatabolic hypoproteinemia; new mechanisms of human disease including abnormalities of suppressor and helper T cells in the pathogenesis of primary immune deficiency disease, immunodeficiency associated with cancer and autoimmune disease; molecular analysis of immunoglobulen and T cell receptor rearrangement; interleukin-2 receptor. *Mailing Add:* Metab Br Nat Cancer Inst Bethesda MD 20892

WALDNER, MICHAEL, b St Louis, Mo, Mar 21, 24; m 55; c 5. ENGINEERING PHYSICS. *Educ:* Wash Univ, St Louis, BS, 44; Cornell Univ, MS, 48, PhD(eng physics), 54. *Prof Exp:* Test engr, McQuay Norris Co, Mo, 44-45; res engr, Northrop Aircraft Corp, Calif, 48-50; physicist, Gen Elec Co, NY, 53-59 & Hughes Aircraft Co, Calif, 59-61; mem tech staff, NAm Aviation Sci Ctr, 61-68; sr staff physicist, 68-80, SR SCIENTIST, HUGHES RES LABS, 80- *Mem:* Inst Elec & Electronics Engrs. *Res:* Semiconductor devices and surfaces; optical properties of semiconductors; thin film insulators; acousto-electric devices; semiconductor integrated circuits. *Mailing Add:* Hughes Aircraft Co 3011 Malibu Canyon Rd RI S5 Malibu CA 90265

WALDO, GEORGE VAN PELT, JR, b Montgomery, Ala, July 20, 40; m 66. ENGINEERING PHYSICS. *Educ:* Johns Hopkins Univ, AB, 61; Univ Md, PhD(physics), 72. *Prof Exp:* PHYSICIST, DAVID TAYLOR RES CTR, MD, 62-, PRIN INVESTR, 70- *Mem:* Am Phys Soc; Sigma Xi. *Res:* Statistical mechanics of phase transitions; cavitation induced by shock waves; interaction of shock waves with structures; monte-carlo simulation of ship vulnerability; perturbation theory of phase transitions; underwater explosions. *Mailing Add:* David Taylor Res Ctr Code 1750-2 Bethesda MD 20084

WALDO, WILLIS HENRY, b Detroit, Mich, Sept 27, 20; m 49; c 5. INORGANIC CHEMISTRY. *Educ:* Washington & Jefferson Col, BS, 42; Univ Md, MS, 50. *Prof Exp:* Chemist, E I du Pont de Nemours & Co, 42-45; chemist, Socony Vacuum Oil Co, 45-46; asst chem, Univ Md, 46-49; tech ed, Monsanto Chem Co, 49-60, admin mgr agr div, Monsanto Co, 60-77, ed, Monsanto Tech Rev, 56-77; opers mgr, Indust Res Inst, Res Corp, St Louis, 77-80; RETIRED. *Concurrent Pos:* Mem, Nat Acad Sci-Nat Res Coun Comt on Mod Methods Handling Chem Info, 62-66; lectr, Southern Ill Univ, Edwardsville, 72-74. *Mem:* Soc Tech Commun; Am Chem Soc; Sigma Xi; Am Soc Info Sci. *Res:* Chromium complexes; sulfur; machine documentation. *Mailing Add:* 3057 So Higuera No 211 San Luis Obispo CA 93401

WALDREN, CHARLES ALLEN, b Syracuse, Kans, June 2, 34; m 61; c 3. BIOPHYSICS. *Educ:* Univ Colo, Boulder, BA, 59; Univ Colo Med Ctr, Denver, MS, 65, PhD(biophys), 72. *Prof Exp:* Chemist, 60-61, fel, 61-65, res tech III, 65-67, instr, 67-75, ASST PROF BIOPHYS & GENETICS, UNIV COLO MED CTR, 75-; SR FEL, ELEANOR ROOSEVELT INST CANCER RES, 74-; ASSOC PROF, DEPT RADIOL, UNIV COLO HEALTH SCI CTR, 80- *Concurrent Pos:* Vis scientist & fel, CRC DNA Repair Unit, Univ Cambridge, 72-73, 75-76, 81 & 85. *Mem:* Sigma Xi; Am Soc Cell Biol; Cancer Res Soc; Radiation Res Soc; Tissue Cult Asn; AAAS. *Res:* Genetic-biochemical-molecular biological analysis of mutagenesis and repair mechanisms in somatic mammalian cells; relationship to human developmental disease. *Mailing Add:* Radiol & Radiot Biol MRB Colo State Univ Ft Collins CO 80523

WALDREP, ALFRED CARSON, JR, b Orange, Tex, Apr 17, 23; m; c 4. ORAL SURGERY, DENTISTRY. *Educ:* Loyola Univ, La, DDS, 46; Baylor Univ, BS, 59, MS, 61; Am Bd Oral & Maxillofacial Surg, dipl, 63. *Prof Exp:* Oral surgeon, Valley Forge Gen Hosp, Valley Forge Pa Hq, US Army, San Antonio, Tex, 46-54, dent surgeon, Task Force 7, Cent Pac, 54-55, chief hosp surg dent, US Army Hosp, Ft Polk, La, 55-58, resident oral surg, Hosp, Baylor Univ, 58-59, resident, Brooks Army Med Ctr, 59-61, consult, US Army Med Area, Stuttgart, Ger, 61-64, chief host dent, US Army Hosp, Ft Polk, La, 64-68; asst dean extramural affairs, Col Dent Med, Univ SC, 76-77, asst dean curric & extramural affairs, 77-85, prof oral surg, 68-85; RETIRED. *Concurrent Pos:* Fel, Hosp, Baylor Univ, 58-59 & Brooks Army Med Ctr, 59-61; consult, Vet Admin Hosp, Charleston, SC, 68-, US Navy Hosp, 69-85 & SC Dept Corrections, 69-85; coordr dent activ, SC Area Health Educ Ctr, 72-85. *Mem:* Am Dent Asn; Am Soc Oral Surg; assoc Brit Asn Oral Surg; Am Soc Maxillofacial Surg. *Res:* Dental education; precautions for patients on drug therapy; oral surgery for patients on anticoagulant therapy. *Mailing Add:* 1042 Ft Sumter Dr Charleston SC 29412

WALDREP, THOMAS WILLIAM, b Madison, Fla, Feb 14, 34; m 55; c 2. PLANT PHYSIOLOGY. *Educ:* Univ Fla, BSA, 61; Univ Ky, MSA, 63; NC State Univ, PhD(crop sci), 67. *Prof Exp:* Instr crop sci, NC State Univ, 66-67, asst prof, 67-69; RES SCIENTIST, LILLY RES LABS, ELI LILLY & CO, 69- *Mem:* Weed Sci Soc Am. *Res:* Basic and applied research in plant physiology; growth regulators and herbicides; herbicide research. *Mailing Add:* 1501 Chapman Dr Greenfield IN 46140

WALDRON, ACIE CHANDLER, b Malad, Idaho, Feb 4, 30; m 57; c 5. AGRONOMY, ENTOMOLOGY. *Educ:* Brigham Young Univ, BSc, 57; Ohio State Univ, MSc, 59, PhD(agron), 61. *Prof Exp:* Res asst agron, Agr Exp Sta, Ohio State Univ, 57-61; develop chemist, Agr Div, Am Cyanamid Co, NJ, 61-66; exten specialist pesticide chem & state coordr agr chem, Ohio Coop Exten Serv, 66-77, COORDR, N CENT REGION PESTICIDE IMPACT ASSESSMENT PROG, OHIO AGR RES & DEVELOP CTR, OHIO STATE UNIV, 77-; PESTICIDE COORDINATOR & COORDINATOR, PESTICIDE APPLICATOR TRAINING, OHIO COOP EXTEN SERV, 89- *Concurrent Pos:* State IR-4 rep, Ohio State Univ & Ohio Agr Res & Develop Ctr, 73-; pesticide appln educ training coordr, Ohio Coop Exten Serv, 75-77. *Mem:* Am Chem Soc; Am Soc Agron; Coun Agr Sci & Technol; Soil Sci Soc Am; Entom Soc Am. *Res:* Pesticide residue chemistry; pesticide residues in plant and animal crops, in soil and water; pesticide safety; administration of regional research for pesticide impact assessment; chemistry of organic nitrogen and phosphorous in soil organic matter. *Mailing Add:* Pesticide Impact Assessment Prog Ohio State Univ 1991 Kenny Rd Columbus OH 43210

WALDRON, CHARLES A, b Minneapolis, Minn, July 16, 22; m 43; c 2. ORAL PATHOLOGY. *Educ:* Univ Minn, DDS, 45, MSD, 51; Am Bd Oral Path, dipl, 52. *Prof Exp:* From asst prof to prof path, Sch Dent Wash Univ, 50-57; prof path, Sch Dent, Emory Univ, 57-82; RETIRED. *Concurrent Pos:* Consult, Vet Admin Hosp, Ga, 57 & Dent Intern Prog, Ft Benning, 59; mem med bd dirs, Am Bd Oral Path, 59-70; sci adv bd consult, Armed Forces Inst Path, 70-75. *Mem:* Am Dent Asn; fel Am Col Dent; Am Acad Oral Path (pres, 59, ed, 70-76). *Res:* Oral tumors; diagnostic oral pathology. *Mailing Add:* Dept Oral Path Sch Dent 1462 Clifton Rd Atlanta GA 30322

WALDRON, HAROLD FRANCIS, b Manchester, Ohio, Sept 24, 29; m 49; c 1. APPLIED CHEMISTRY. *Educ:* Capital Univ, BS, 52; Purdue Univ, West Lafayette, MS, 54. *Prof Exp:* Chemist, Uranium Div, Mallinckrodt Chem Works, Mallinckrodt, Inc, 54-62, supvr anal methods develop, 62-66, supvr anal res, Opers Div, 66-67, res group leader, Indust Chem Div, 67-69, res mgr chem group, 69-73, res & develop mgr, 73-75, res assoc, Food Prod Div, 75-78, res fel, 78-89, CONSULT, PROCESS & CORROSION CHEM, MALLINCKRODT, INC, 89- *Mem:* Am Chem Soc; Nat Asn Corrosion Eng. *Res:* Chemical analytical methods, especially the use of vacuum techniques and application of complex ion formation; design and application of electronic instrumentation; general inorganic and uranium chemistry; process research and development. *Mailing Add:* Six Garden Lane Kirkwood MO 63122

WALDRON, HOWARD HAMILTON (HANK), b Nampa, Idaho, Nov 6, 17; m 43; c 3. GEOLOGY. *Educ:* Univ Wash, BS, 40. *Prof Exp:* Photogrammetrist, US Hydrographic Off, Washington, DC, 42-46; geologist, US Geol Surv, 46-73; STAFF CONSULT GEOL, SHANNON & WILSON, INC, 73- *Concurrent Pos:* Tech adv, Geol Surv Indonesia, 60-62, Costa Rica, 64 & Colo Eng Coun, 65-73; eng geol consult & adv, AEC, 67-72 & Vet Admin, 71-73; mem, Earthquake Eng Res Inst; consult geol, 80- *Mem:* Asn Eng Geol; fel Geol Soc Am; Am Soc Civil Eng. *Res:* Engineering geology; areal and glacial geology of Pacific Northwest; urban and environmental geology; geologic volcanic and earthquake hazards evaluations. *Mailing Add:* Box C-30313 Seattle WA 98103-8067

WALDRON, INGRID LORE, b Nyack, NY, Dec 8, 39; c 2. PSYCHOSOMATIC MEDICINE. *Educ:* Radcliffe Col, AB, 61; Univ Calif, Berkeley, PhD(biol), 67. *Prof Exp:* NSF fel, Univ Cambridge, 67-68; ASSOC PROF BIOL, UNIV PA, 68- *Concurrent Pos:* NIH grant, Univ Pa, 69-72; social security grant, Univ Pa, 79-80; res assoc, Pop Studies Ctr, & adj assoc prof psychol, Univ Pa, 82-; regional ed, 83-88, adv ed, Social Sci & Med, 88- *Mem:* Pop Asn Am. *Res:* Human biology; sex differences; social and psychological origins of disease; employment and women's health. *Mailing Add:* Dept Biol Univ Pa Philadelphia PA 19104-6018

WALDRON, KENNETH JOHN, b Sydney, Australia, Feb 11, 43; m 68; c 3. ROBOTICS, MACHINE DESIGN. *Educ:* Univ Sydney, BE, 64, MEngSci, 65; Stanford Univ, PhD(mech eng), 69. *Prof Exp:* Res asst mech eng, Stanford Univ, 65-68, actg asst prof, 68-69; from lectr to sr lectr, Univ NSW, 69-74; assoc prof, Univ Houston, 74-80; MEM FAC, DEPT MECH ENG, OHIO STATE UNIV, 80-, NORDHOLT PROF, 84- *Concurrent Pos:* Tech ed, Trans J Mechanisms, Transmissions & Automation in Design, Am Soc Mech Engrs, 88- *Honors & Awards:* Ralph R Teetor Award, Soc Automotive Engrs,

77; Leonardo Da Vinci Award, Am Soc Mechanical Engrs, 88. *Mem:* Fel Am Soc Mech Engrs; Soc Automotive Engrs; Sigma Xi; Soc Mfg Engrs; Am Soc Eng Educ. *Res:* Mechanism kinematics; manipulator design and control; design of mobile robotic systems; computer-aided mechanism design. *Mailing Add:* Dept Mech Eng Ohio State Univ Columbus OH 43210

WALDROP, ANN LYNEVE, b Winters, Tex, Oct 9, 39; m 64. CRYSTALLOGRAPHY. *Educ:* McMurry Col, BA, 61; Vanderbilt Univ, MA, 65; Mass Inst Technol, PhD(crystallog), 70. *Prof Exp:* Teacher high sch, Mass, 64-65; ASST PROF CHEM, SIENA COL, NY, 70- *Mem:* Am Crystallog Asn; Am Chem Soc. *Res:* Determination of crystal and molecular structure by x-ray crystallography; relationship between structure and properties; polymorphism. *Mailing Add:* 20 Michael Dr Saratoga Springs NY 12866

WALDROP, FRANCIS N, b Asheville, NC, Oct 5, 26; m 50; c 2. MEDICINE. *Educ:* Univ Minn, AB, 46; George Washington Univ, MD, 50. *Prof Exp:* Intern med, Univ Hosp, George Washington Univ, 50-51; resident psychiat, US Dept HEW, 51-54, med officer, 54-59, assoc dir res, 59-65, dir prof training psychiat, 63-71, dir, Clin & Behav Studies Res Ctr, St Elizabeth's Hosp, 65-79, dep adminr, Alcohol, Drug Abuse & Ment Health Admin, 75-79; RETIRED. *Concurrent Pos:* Clin asst prof, George Washington Univ, 62-65, clin assoc prof, 65-; spec asst res & training, Nat Inst Ment Health, 66-68, dep dir, Nat Ctr Ment Health Serv, Training & Res, 68-71, from assoc dir to dir, Div Manpower & Training Progs, 71-75. *Honors & Awards:* Vestermark Award, Am Psychiat Asn, 80. *Mem:* AAAS; AMA; fel Am Psychiat Asn. *Res:* Clinical psychiatry; psychopharmacology; drug dependence; basic biological sciences in relation to psychiatric disorders. *Mailing Add:* 1775 Elton Rd Silver Spring MD 20903

WALDROP, MORGAN A, b Ft Worth, Tex, Jan 8, 37. SOLID STATE PHYSICS, ATOMIC PHYSICS. *Educ:* Rice Inst, BA, 59, MA, 62, PhD(physics), 64. *Prof Exp:* SR RES PHYSICIST, PHILLIPS PETROL CO, 63- *Mem:* Am Phys Soc. *Res:* Nuclear magnetic and electron paramagnetic resonance in solids; atomic aspects of heterogeneous catalysis. *Mailing Add:* 1330 Melmart Dr SE Bartlesville OK 74006

WALDROUP, PARK WILLIAM, b Maryville, Tenn, Oct 17, 37; m 61; c 3. NUTRITION, BIOCHEMISTRY. *Educ:* Univ Tenn, Knoxville, BS, 59; Univ Fla, MS, 62, PhD(nutrit, biochem), 65. *Prof Exp:* Res assoc poultry nutrit, Univ Fla, 64-65, asst prof, 65-66; from asst prof to assoc prof, 66-75, PROF POULTRY NUTRIT, UNIV ARK, FAYETTEVILLE, 75- *Honors & Awards:* Nat Broiler Coun Award, 80. *Mem:* Poultry Sci Asn; Am Inst Nutrit; Animal Nutrit Res Coun. *Res:* Studies concerned with nutrient requirements of poultry in terms of nutrient balance and interrelationships of nutrients; effects of processing on nutritive value of feeds. *Mailing Add:* Dept Animal Sci Univ Ark Fayetteville AR 72701

WALDSTEIN, SHELDON SAUL, b Chicago, Ill, June 23, 24; m 52; c 3. INTERNAL MEDICINE, ENDOCRINOLOGY & METABOLISM. *Educ:* Northwestern Univ, BS, 46, MD, 47, MS, 51; Am Bd Internal Med, dipl. *Prof Exp:* Intern & resident internal med, Cook County Hosp, 47-51; from clin asst to assoc prof, 51-66, PROF INTERNAL MED, MED SCH, NORTHWESTERN UNIV, CHICAGO, 66- *Concurrent Pos:* Res assoc, Hektoen Inst & assoc attend physician, Cook County Hosp, 54-57, attend physician, 57-69, chief, Northwestern Med Div, 59-62, exec dir dept med, 62-64, chmn dept med, 64-69; exec dir, NSuburban Asn Health Resources, Northbrook, 69-72; dir, Northwestern Univ Med Asn, 74-77; exec dir, Cook County Grad Sch Med, 77- *Mem:* Endocrine Soc; AMA; Am Col Physicians; Am Fedn Clin Res; Cent Soc Clin Res. *Res:* Endocrinology. *Mailing Add:* 222 E Superior St Chicago IL 60611

WALECKA, JERROLD ALBERTS, b Highland Park, Ill, Aug 30, 30; m 53; c 2. ORGANIC CHEMISTRY. *Educ:* Lawrence Col, BS, 51, MS, 53, PhD(pulp & paper), 56. *Prof Exp:* Asst tech dir, WVa Pulp & Paper Co, 55-59; res chemist, Mead Corp, 59-63, assoc res dir new prod develop, 63-70; gen mgr res & develop, Forest Prod Group, Continental Can Co, 70-77; MGR BROWNBOARD RES & DEVELOP, WEYERHAEUSER CO, 77- *Mem:* Tech Asn Pulp & Paper Indust. *Res:* Pulp and paper industry; new products; cellulose and plastic chemistry; photochemistry; reproduction processes; graphic arts; information handling; electronic data processing; packaging. *Mailing Add:* Weyerhaeuser Co Tacoma WA 98477

WALECKA, JOHN DIRK, b Milwaukee, Wis, Mar 11, 32; m 54; c 3. THEORETICAL PHYSICS. *Educ:* Harvard Col, BA, 54; Mass Inst Technol, PhD(physics), 58. *Prof Exp:* NSF fel, Europ Orgn Nuclear Res, Switz, 58-59; NSF fel, Stanford, 59-60, from asst prof to assoc prof physics, 60-66, assoc dean humanities & sci, 70-72, chmn acad senate, 73-74, chmn, Dept Physics, 77-82, PROF PHYSICS, STANFORD UNIV, 66- *Concurrent Pos:* A P Sloan Found fel, 62-66; mem sci prog adv comts, Bates Linac, Mass Inst Technol, 71-76, Nevis Cyclotron, Columbia Univ, 71-76, & Los Alamos Meson Physics Facil, 74-76; mem ad hoc panel on future nuclear sci, Nat Res Coun-Nat Acad Sci, 75-77; mem Nuclear Sci Adv Comt, 77; mem vis comt, Lab Nuclear Sci, Nat Inst Technol, 80- *Mem:* Fel Am Phys Soc. *Res:* Nuclear structure; high energy physics. *Mailing Add:* 12000 Jefferson Ave Newport News VA 23606

WALENGA, RONALD W, b New York, NY, Sept 25, 46. ARACHIDONIC ACID METABOLISM. *Educ:* Antioch Col, BS, 68; Univ Mich, PhD(biochem), 74. *Prof Exp:* ASST PROF PEDIAT, STATE UNIV NY HEALTH SCI CTR, SYRACUSE, 81- *Mem:* Am Asn Pathologists; NY Acad Sci. *Res:* Prostaglandins hydroxy-acids. *Mailing Add:* Rainbow Baby & Children's Hosp 2074 Abington Cleveland OH 44106

WALERIAN, SZYSZKOWSKI, b Ludwigsburg, Ger, Aug 10, 45; Can citizen; m 72; c 2. FINITE ELEMENT METHODS IN APPLIED MECHANICS, MECHANICS OF TIME-DEPENDENT MATERIALS. *Educ:* Univ Warsaw, Poland, MSc, 69, PhD(struct stability), 74. *Prof Exp:* From asst prof to assoc prof struct mech, Aeronaut Technol & Appl Mech Dept, Univ Warsaw, 74-82; vis prof ice mech, Univ Calgary, 82-86, assoc prof time-dependent mat, 86-91; PROF, DEPT MECH ENG, UNIV SASK, 91- *Mem:* Am Acad Mech; Am Soc Mech Engrs; Soc Eng Sci; Int Asn Shell & Spatial Struct. *Res:* Computer aided structural analysis; mechanics of composite thin-walled structures; structural optimization; modeling of visco-elasto-plastic and brittle materials; stability and dynamics of large flexible structures; holographic interferometry in strain-stress measurements. *Mailing Add:* Dept Mech Eng Univ Sask Saskatoon SK S7J 0W0 Can

WALES, CHARLES E, b Chicago, Ill, Dec 20, 28; m 53; c 3. EDUCATIONAL ENGINEERING. *Educ:* Wayne State Univ, BS, 53; Univ Mich, MS, 54; Purdue Univ, PhD, 65. *Prof Exp:* From instr to assoc prof chem eng, Wayne State Univ, 54-65; asst prof, Purdue Univ, 65-67; assoc prof eng, Wright State Univ, 67-69; dir freshman eng, 69-81, PROF ENG, WVA UNIV, 72-, DIR, CTR GUIDED DESIGN, 81- *Concurrent Pos:* Am Oil fel, 64. *Honors & Awards:* George Westinghouse Teaching Award, Am Soc Eng Educ, 71. *Mem:* Int Cong Individualized Instruction; Am Soc Eng Educ. *Res:* Engineering education; programmed learning; guided and educational systems design. *Mailing Add:* WVa Univ Eng Sci Bldg PO Box 6101 Morgantown WV 26506

WALES, DAVID BERTRAM, b Vancouver, BC, July 31, 39; m 61; c 2. COMBINATORICS, FINITE MATHEMATICS. *Educ:* Univ BC, BSc, 61, MA, 62; Harvard Univ, PhD(math), 67. *Prof Exp:* Bateman res fel, 67-68, from asst prof to assoc prof, 68-77, assoc dean, 76-80, dean, 80-84, PROF MATH, CALIF INST TECHNOL, 77-, EXEC OFFICER MATH, 85- *Mem:* Am Math Soc. *Res:* Representation theory of finite groups; combinatorics. *Mailing Add:* Dept Math Calif Inst Technol 1201 E California Blvd Pasadena CA 91125

WALES, WALTER D, b Oneonta, NY, Aug 2, 33; m 55; c 2. PHYSICS. *Educ:* Carleton Col, BA, 54; Calif Inst Technol, MS, 55, PhD(physics), 60. *Prof Exp:* From instr to assoc prof, 59-72, chmn dept, 73-82, PROF PHYSICS, UNIV PA, 72- *Concurrent Pos:* Assoc dir, Princeton Pa Accelerator, 68-71; physicist, AEC, 72-73. *Mem:* Am Phys Soc. *Res:* Particle physics. *Mailing Add:* Dept Physics Univ Pa Philadelphia PA 19174

WALFORD, ROY LEE, JR, b San Diego, Calif, June 29, 24; m 50; c 3. PATHOLOGY. *Educ:* Univ Chicago, BS, 46, MD, 48. *Prof Exp:* Intern, Gorgas Hosp, CZ, 50-51; resident path, Vet Admin Hosp, Los Angeles, 51-52; chief lab, Chanute AFB Hosp, Ill, 52-54; from asst prof to prof path, 54-70, PROF PATH & HEMATOPATH, SCH MED, UNIV CALIF, LOS ANGELES, 70- *Concurrent Pos:* Attend physician, Brentwood Vet Admin Hosp, 55-56; consult, Los Angeles Harbor Gen Hosp, 59-70, Space Biopheres Ventures, 82- *Honors & Awards:* Kleemeier Award, Gerontol Soc Am, 80; Res Award, Am Aging Asn, 80; Henderson Award, Am Geriat Soc, 82. *Mem:* Am Soc Exp Path; Am Asn Path & Bact; Geront Soc Am. *Res:* Hematologic pathology; immunology of the white blood cell; space biology; gerontology. *Mailing Add:* Dept Path Univ Calif Sch Med Los Angeles CA 90024

WALGENBACH, DAVID D, b Marshall, Minn, Sept 4, 37; m 61; c 4. ENTOMOLOGY, AGRONOMY. *Educ:* Iowa State Univ, BS, 59; Univ Wis-Madison, MS, 62, PhD(entom), 65. *Prof Exp:* Asst prof biol, Stout State Univ, 64-65; field tech specialist entom, Chevron Chem Co, 65-66, crop specialist, Agr Chem Res, 66-67, sr res specialist, 67-73; MEM FAC, DEPT ENTOM & ZOOL, SDAK STATE UNIV, 73- *Mem:* AAAS; Entom Soc Am; Am Soc Agron. *Res:* Foreward agricultural chemical research; plant physiology. *Mailing Add:* Dept Plant Sci SDak State Univ Box 2207a Brookings SD 57007

WALGENBACH-TELFORD, SUSAN CAROL, b Sacramento, Calif, June 29, 52; m 83; c 2. NEURAL CONTROL OF CIRCULATION, SMALL INTESTINAL TRANSPLANTATION. *Educ:* Univ Calif, BS, 74, PhD(physiol), 79. *Prof Exp:* Postgrad res physiol, Univ Calif, Davis, 76-79; NIH postdoctoral fel physiol, Mayo Grad Sch Med, Minn, 79-82, instr, 81-83; res assoc physiol, Mayo Clin, Rochester, Minn, 82-83; asst prof anesthesiol, Med Col Wis, Milwaukee, 83-88, asst prof physiol, 84-88; RES PHYSIOLOGIST, VET ADMIN MED CTR, MILWAUKEE, WIS, 84-; ASST CLIN PROF ANESTHESIOL & PHYSIOL, MED COL WIS, 88- *Concurrent Pos:* Prin investr, new fac award, Med Col Wis, 83-84, Am Heart Asn, 84-85, NIH new investr award, 84-87 & Vet Admin merit award, 87-90; mem, Coun Circulation, Am Heart Asn. *Mem:* Am Physiol Soc; AAAS. *Res:* Neural control of the circulation, specifically the role of baroreflexes and chemoreflexes in the regulation of circulation during exercise and environmental stress; secondary interest in physiology of small intestinal transplantation. *Mailing Add:* Dept Anesthesiol Med Col Wis Vet Admin Med Ctr Res Serv 151 Milwaukee WI 53295

WALHOUT, JUSTINE ISABEL SIMON, b Aberdeen, SD, Dec 11, 30; m 58; c 4. SCIENCE EDUCATION, EDUCATIONAL POLICY. *Educ:* Wheaton Col, BS, 52; Northwestern Univ, PhD(org chem), 56. *Prof Exp:* Instr chem, Wright Br, Chicago City Jr Col, 55-56; from asst prof to assoc prof, 56-89, PROF CHEM & CHAIR, ROCKFORD COL, 89- *Concurrent Pos:* Consult, Pierce Chem Co, 68-69; mem, Ill State Bd Educ, Gubernatorial Appointee, 74-81. *Mem:* Am Chem Soc; Sigma Xi; Midwest Asn Chem Teachers Lib Arts Cols. *Res:* High temperature studies of nitrogen heterocyclic compounds; educational research and papers in nurses education in chemistry; literature research and writings of trimethylsilylation reactions. *Mailing Add:* Dept Chem Rockford Col Rockford IL 61108-2393

WALI, KAMESHWAR C, b Bijapur, India, Oct 15, 27; m 52; c 3. THEORETICAL PHYSICS. *Educ:* Univ Bombay, BSc, 48; Benares Hindu Univ, MSc, 52, MA, 54; Univ Wis, PhD(theoret physics), 59. *Prof Exp:* Res assoc theoret physics, Univ Wis, 59-60 & Johns Hopkins Univ, 60-62; from asst physicist to sr physicist, Argonne Nat Lab, 62-69; chmn dept, 86-89, PROF PHYSICS, SYRACUSE UNIV, 69- *Concurrent Pos:* Co-ed, Int Conf Weak Interactions, Argonne Nat Lab, 65; vis mem, Inst Advan Sci Study, Bures-sur-Yvette, France, 71-72, 75-76, 79-80, 83-84, 90; co-ed, Int Symp

Nucleon-Nucleon Annihilations, Syracuse, 75; vis scientist, Int Inst Theoret Physics, Trieste, Italy, 67; ed, Proceedings of eight workshop on grand unification, Syracuse Univ, NY, 87. *Mem:* Am Phys Soc; Sigma Xi. *Res:* Elementary particles; high energy physics; higher symmetries; grand unified theories. *Mailing Add:* Dept Physics/201 Physics Syracuse Univ Syracuse NY 13244-1130

WALI, MOHAN KISHEN, b Srinagar, India, Mar 1, 37; m 60; c 2. PLANT ECOLOGY, ENVIRONMENTAL BIOLOGY. *Educ:* Univ Jammu & Kashmir, India, BSc, 57; Allahabad Univ, MSc, 60; Univ BC, PhD(plant ecol), 69. *Prof Exp:* Demonstr bot, S P Col, Srinagar, India, 61-63, lectr, 63-65; teaching asst biol, Univ BC, 66-67, teaching asst plant ecol, 68-69; asst prof, Univ NDak, 69-73, assoc prof, 73-79, prof biol, 79-, spec asst univ pres, 77-; DIR GRAD PROG ENVIRON SCI, COL ENVIRON SCI, STATE UNIV NY. *Concurrent Pos:* Res asst, Nat Res Coun Can, 67-69; dir, Proj Reclamation, 75-; sr ed, Reclamation Rev, 76-81, chief ed, Reclamation & Revegetation res, 82- *Honors & Awards:* Sigma Xi Award, Univ NDak, 75; B C Gamble Award, 77. *Mem:* Ecol Soc Am; Can Bot Asn; Am Inst Biol Sci; Brit Ecol Soc; Int Asn Ecol. *Res:* Environmental ecology; influence of water, nutrients, temperature and light on plant populations and communities; nutrient cycling; ecosystem model building; soil-plant relationship; pollution; phytosociology; systems approach to the reclamation of strip mined areas. *Mailing Add:* Dept Environ Sci State Univ NY Syracuse NY 13210

WALIA, AMRIK SINGH, b Punjab, India, Aug 6, 47; m 74; c 2. IMMUNOLOGY, CHEMISTRY. *Educ:* Punjab Univ, BS, 65; Meerut Univ, MS, 68; Loyola Univ, PhD(chem), 75. *Prof Exp:* Lectr chem, GMN Col, Ambala India, 68-70; instr, Loyola Univ, 74-75; tumor immunologist, 75-77, ASST RES PROF, DEPT SURG, SCH MED, UNIV ALA, 80-; CHIEF, ALCOHOLISM RES, VET ADMIN MED CTR, 82- *Concurrent Pos:* Fel, Univ Ala Cancer Ctr, 77-78; NIH fel, Nat Cancer Inst, 78-80. *Honors & Awards:* Inst Cancer Res Technol Trans Award, Int Union Against Cancer, 78. *Mem:* AAAS; Am Chem Soc; Sigma Xi; Am Asn Immunologists; NY Acad Sci; Int Soc Immunopharmacology; Res Soc Alcholism. *Res:* Immune response to polyoma induced tumors, IgG, IgM, and C3 receptors for T cells; biological and biochemical characterization of C3 receptors; effect of alcohol on immune functions. *Mailing Add:* Surg Dept LHR 746 Univ Ala Sch Med Birmingham AL 35294

WALIA, JASJIT SINGH, b Lahore, India, Mar 19, 34; m 66; c 2. ORGANIC CHEMISTRY. *Educ:* Univ Punjab, India, BS(hons), 55, MS(hons), 56; Univ Southern Calif, PhD(org chem), 60. *Prof Exp:* Res assoc org chem, Univ Southern Calif, 60; res assoc, Mass Inst Technol, 60-61; lectr, Benaras Hindu Univ, 62-66; assoc prof, 66-73, PROF CHEM, LOYOLA UNIV, LA, 73- *Honors & Awards:* Prof Prem Singh Medal, 55; Dux Academicus Award, 82. *Mem:* Am Chem Soc; Royal Soc Chem. *Res:* New reactions of organic nitrogen compounds; new synthetic reactions; novel heterocyclization reactions; carbanion chemistry; cyanide ion-catalyzed reactions; mechanism of reactions; new compounds of potential therapeutic and/or other commercial importance. *Mailing Add:* Dept Chem Loyola Univ New Orleans LA 70118

WALING, J(OSEPH) L(EE), b Brook, Ind, Mar 17, 16; m 42; c 4. STRUCTURAL ENGINEERING. *Educ:* Purdue Univ, BS, 38, MSE, 40; Univ Ill, PhD, 52. *Prof Exp:* Asst appl mech, Purdue Univ, 38-40, instr, 40-41; naval architect, Norfolk Navy Yard, 41-45; from asst prof to assoc prof eng mech, 45-54, prof eng sci, 54-55, prof struct eng, 55-83, from asst dean to assoc dean grad sch, 60-82, dir div sponsored progs, 66-83, EMER PROF CIVIL ENG, PURDUE UNIV, 83-, EMER DIR DIV SPONSORED PROGS, 83- *Concurrent Pos:* Mem Nat Coun Univ Res Adminr. *Mem:* AAAS; Am Soc Civil Engrs; Am Soc Eng Educ; Am Concrete Inst; Int Asn Shell Struct. *Res:* Photoelasticity; structural mechanics; reinforced concrete; shell structures. *Mailing Add:* 3731 Capilano Dr West Lafayette IN 47906

WALKENSTEIN, SIDNEY S, b Philadelphia, Pa, Dec 21, 20; m 46; c 1. DRUG METABOLISM. *Educ:* Temple Univ, BS, 42, AM, 50, PhD(biochem), 53. *Prof Exp:* Biochemist, Mold Metab, Pitman-Dunn Labs, 52-53; chief radiochemist drug metab, Wyeth Inst Med Res, 53-58; pharmaceut specialist, Union of Burma Appl Res Inst, 58-60; sr res scientist, Wyeth Labs, Inc, 60-62, mgr radiochem sect, 62-67; ASSOC DIR BIOL RES, SMITH KLINE & FRENCH LABS, 67- *Concurrent Pos:* Consult, Burma pharmaceut indust, 58-60. *Mem:* Am Soc Pharmacol & Exp Therapeut; Am Pharmaceut Asn; Int Soc Biochem Pharmacol; NY Acad Sci. *Res:* Metabolism of aldehydes and fatty acids; effects of toxins on yeast respiration; pantothenate-deficient yeast metabolism; utilization of hydrocarbons by molds; biotransformation and physiological disposition of isotopically-labeled drugs; medicinal plants; mechanism of drug action; trace drug analysis; pharmacokinetics; biopharmaceutics. *Mailing Add:* 804 N 29th St Philadelphia PA 19130

WALKER, A EARL, b Winnipeg, Man, Mar 12, 07; nat US; m. MEDICINE, NEUROLOGICAL SURGERY. *Educ:* Univ Alta, BA, 26, MD, 30. *Hon Degrees:* LLD, Univ Alta, 52; DHL, Johns Hopkins Univ. *Prof Exp:* Intern, Toronto Western Hosp, 30-31; Smith fel, Univ Chicago, 31, res neurol & neurosurg, 31-34, instr neurol, 37-38, prof neursurg, 37-47, from instr to assoc prof neurol surg, 38-45, prof & chief div, 46-47; instr neurosurg, Univ Iowa, 34; Rockefeller fel, Yale Univ, Univ Amsterdam & Brussels, 35-37; prof neurol surg, 47-72, EMER PROF NEUROL SURG, JOHNS HOPKINS UNIV, 72- *Concurrent Pos:* Vis prof neurol neurosurg, Sch Med, Univ NMex, 71-83, emer, 83- *Mem:* Soc Neurol Surg (pres, 66); Am Asn Neurol Surg (pres, 69); Am EEG Soc (pres, 54); Am Neurol Asn (pres, 65-); fel Am Col Surg. *Res:* Neurophysiological basis of epilepsy; anatomy and physiology of thalamus; experimental physiology of cerebral cortex; cerebello-cerebral relationships; visual mechanisms; neurosurgical therapy of pain; physiology of cerebral injuries; cerebral death; intracranial pressure; epidemiology of brain tumors; history of neurosurgery. *Mailing Add:* 1445 Wagontrain Dr SE Albuquerque NM 87123

WALKER, ALAN, b Bridlington, Eng, Apr 30, 37; m 62; c 2. INORGANIC CHEMISTRY. *Educ:* Univ Nottingham, BSc, 59, PhD(inorg chem), 62. *Prof Exp:* Dept Indust & Sci Res res fel inorg chem, Univ Nottingham, 62-63; resident res assoc fel chem, Argonne Nat Lab, 63-65; ASSOC PROF CHEM, UNIV TORONTO, 65- *Concurrent Pos:* Natural Sci & Eng Res Coun Can res grant, 65- *Mem:* Fel Royal Soc Chem. *Res:* Nonaqueous solvent chemistry, particularly of nitrates; inorganic spectroscopy; reactions of coordinated ligands to platinum group metals. *Mailing Add:* Dept Chem Univ Toronto-Scarborough Scarborough ON M1Z 1A4 Can

WALKER, ALAN KENT, b Albert Lea, Minn, Jan 23, 50; m 73; c 2. AGRONOMY. *Educ:* Univ Minn, BS, 72; Univ Md, MS, 74; Iowa State Univ, PhD(plant breeding), 77. *Prof Exp:* Res asst soybean breeding, Univ Md, 72-74; res assoc, Iowa State Univ, 75-77; asst prof soybean breeding, Ohio Agr Res & Develop Ctr, 77-; PROF CELL BIOL & ANAT, JOHNS HOPKINS UNIV. *Mem:* Am Soc Agron; Crop Soc Agron. *Res:* Breeding for Phytophthora stem and root rot resistance and tolerance in soybeans; development of superior cultivars of soybeans. *Mailing Add:* Dept Cell Biol & Anat Wood Basic Sci Rm Baltimore MD 21205

WALKER, ALMA TOEVS, b Charlson, NDak, Aug 6, 11; m 41. BOTANY. *Educ:* Iowa State Col, BS, 40, PhD, 52; Tex State Col Women, MA, 43. *Prof Exp:* Asst prof home econ, Col Idaho, 40-42; anal chemist, Armour & Co, Inc, Tex, 44; asst prof home econ, Utah, 45-46; asst, Ohio State Univ, 46-48; microbiologist, Tuberc Res Lab, Vet Admin Hosp, Atlanta, Ga, 57-58; assoc investr zool, Univ Ga, 59-62, res physiologist, 62-77; RETIRED. *Concurrent Pos:* Grants in aid, Am Acad Arts & Sci, 62 & Sigma Xi, 65. *Mem:* AAAS; Sigma Xi. *Res:* Fungus physiology; catalase of mycobacteria; lipid deposition in migratory birds; lichens; medicinal plants; conservation and environmental concerns. *Mailing Add:* 182 Ebony Lounge Rd Milledgeville GA 31061

WALKER, ALTA SHARON, b Ogdensburg, NY, Apr 28, 42. PLANETOLOGY, GEOMORPHOLOGY. *Educ:* Syracuse Univ, BA, 64; Univ Minn, MS, 71; Rice Univ, PhD(geol), 77. *Prof Exp:* Geologist, Tetra Tech, Inc, 77-78, Nat Air & Space Mus, Smithsonian Inst, 78-79; RES GEOLOGIST, US GEOL SURV, 79- *Mem:* Am Geophys Union; Sigma Xi; AAAS; Am Soc Photogram; Geol Soc Am. *Res:* Use of remotely sensed data to evaluate mineral resources, to analyze eolian and fluvial regimes and to assess paleoclimate indicators in arid environments; photogeology of the Galilean satellites. *Mailing Add:* US Geol Surv Mail Stop 927 Reston VA 22092

WALKER, AMEAE M, b Epsom, Eng, Oct 29, 51; m 82; c 4. PROLACTIN MODIFICATION, CELL BIOLOGY. *Educ:* Univ Liverpool, PhD(cell biol), 76. *Prof Exp:* Asst prof, 79-86, ASSOC PROF BIOMED SCI, UNIV CALIF, RIVERSIDE, 86- *Concurrent Pos:* Ad hoc reviewer, NSF & USDA grants; prin investr, var res grants; teacher, med histol. *Mem:* Am Soc Cell Biol; Endocrine Soc; Brit Soc Cell Biol. *Res:* Protein secretion; cell endocrinology. *Mailing Add:* Div Biomed Sci Univ Calif 900 University Ave Riverside CA 92521

WALKER, ARTHUR BERTRAM CUTHBERT, JR, b Cleveland, Ohio, Aug 24, 36; m 59; c 1. SPACE PHYSICS, ASTRONOMY. *Educ:* Case Inst Technol, BS, 57; Univ Ill, Urbana, MS, 58, PhD(physics), 62. *Prof Exp:* Mem tech staff, Space Physics Lab, Aerospace Corp, 65-68, staff scientist, 68-70, sr staff scientist, 70-72, dir, Space Astron Proj, 72-73, ASSOC PROF APPL PHYSICS, STANFORD UNIV, 74-, ASSOC DEAN GRAD STUDIES, 75- *Concurrent Pos:* Mem exec comt, Inst Plasma Res, Stanford Univ, 75-, chmn, Astron Course Prog, 76- *Mem:* Sigma Xi; Am Phys Soc; Am Geophys Union; Am Astron Soc; Int Astron Union. *Res:* Solar physics; solar coronal structure; solar x-rays; solar abundances; high energy astrophysics; stellar x-ray sources; interstellar medium; physics of the upper atmosphere. *Mailing Add:* Stanford Univ CSSA ERL 310 Stanford CA 94305

WALKER, AUGUSTUS CHAPMAN, b Brooklyn, NY, Oct 2, 23; m 47; c 4. TECHNICAL MANAGEMENT, EDUCATION ADMINISTRATION. *Educ:* Harvard Univ, BS, 48. *Prof Exp:* Asst biochemist, Thanhauser Lab, New Eng Med Ctr, 48-51; res chemist high polymers, Cryovac Div, W R Grace Co, 52-57; res assoc, Plastics Lab, Mass Inst Technol, 57-58; lectr, Lowell Technol Inst, 57-59, asst prof chem, 58-59; consult, group leader, sect chief & asst to gen mgr, Res & Adv Develop Div, Avco Corp, 59-65; dir res, Polymer Corp, Pa, 65-70; dir res, Resin Products Div, PPG Indust, Inc, 70-73; consult & dir, Off Post-Col Prof Educ, 75-80, SR LECTR ENG MGT, CARNEGIE-MELLON UNIV, 74-; PRES, EFFECTIVE RES, PITTSBURGH, 73- *Honors & Awards:* Award, Am Inst Chem Engrs, 63. *Mem:* Am Chem Soc; Inst Elec & Electronics Engrs; Am Soc Eng Educ; Sigma Xi. *Res:* Writer and lecturer subjects include methods of managing technical activities; training scientists and engineers; problem solving in science and technology. *Mailing Add:* 2297 Colony Ct Pittsburgh PA 15237

WALKER, BAILUS S, PUBLIC HEALTH. *Educ:* Univ Mich, MPH, 59; Univ Minn, PhD(occup & environ health), 75. *Prof Exp:* Environ health scientist & adminr, Environ Health Serv, Washington, DC, 72-79; dir, Occup Health Standards, US Dept Labor, 79-81; comnr pub health, Commonwealth Mass, 83-87 & toxicol, Sch Pub Health, State Univ NY, Albany, 87-90; PROF, DEPT OCCUP & ENVIRON HEALTH & DEAN, COL PUB HEALTH, UNIV OKLA HEALTH SCI CTR, 90- *Concurrent Pos:* Head, US Exchange Mission to Japan, Collab US-Japanese efforts in occup med, 80; mem, Physicians Human Rights Mission, SKorea, 87; mem, Mozambique Health Assessment Mission, 88; mem, Secy's Coun Health Prom & Dis Prev, US Dept Health & Human Serv; mem, Comn Study Future Pub Health US, Nat Acad Sci. *Honors & Awards:* Hildrus Poindexter Asn, Am Pub Health Asn, 88. *Mem:* Fel Royal Soc Health. *Res:* Physical, chemical and biological hazards in macroenvironments. *Mailing Add:* Col Pub Health Univ Okla Health Sci Ctr 1000 Stanton L Young Blvd Oklahoma City OK 73190

WALKER, BENNIE FRANK, b Mt Pleasant, Tex, Sept 19, 37. PHYSICAL CHEMISTRY. *Educ:* Sam Houston State Col, BS, 59, MA, 62; Univ Tex, Austin, PhD(phys chem), 70. *Prof Exp:* Instr chem, Sam Houston State Col, 59-62; from asst prof to assoc prof, 68-79, PROF CHEM, STEPHEN F AUSTIN STATE UNIV, 79- *Mem:* Am Chem Soc. *Res:* Kinetics; hydrogen bonding; applications of computers in chemistry. *Mailing Add:* Dept Chem Stephen F Austin State Univ Nacogdoches TX 75962

WALKER, BILLY KENNETH, b Canyon, Tex, June 17, 46; m 80. LARGE SOFTWARE SYSTEMS, PROGRAM CORRECTNESS. *Educ:* WTex State Univ, BS, 68; Tex Tech Univ, MS, 70, PhD(math), 74. *Prof Exp:* Asst prof computer sci, WTex State Univ, 75-76, Amarillo Col, 76-79, Univ Okla, 79-83; assoc prof, 83-87, PROF & CHMN, DEPT COMPUTER SCI, E CENT UNIV, 87- *Mem:* Asn Comput Mach; sr mem Inst Elec & Electronics Engrs; Am Math Soc; Math Asn Am. *Res:* Author of three textbooks; author of more than 50 papers. *Mailing Add:* PO Box 2107 Ada OK 74820

WALKER, BRUCE DAVID, b Champaign, Ill, Apr 18, 52; m 82; c 2. IMMUNOLOGY. *Educ:* Univ Colo, BS, 76; Case Western Res Univ, MD, 80. *Prof Exp:* ASST PROF MED, HARVARD MED SCH, 88- *Mem:* Am Asn Immunologists. *Res:* HIV-1-specific cytoxic 7 lymphocytes; hepatitis c virus immunology. *Mailing Add:* Infectious Dis Unit Mass Gen Hosp Boston MA 02114

WALKER, BRUCE EDWARD, b Montreal, Que, June 17, 26; m 48; c 4. ANATOMY. *Educ:* McGill Univ BSc, 47, MSc, 52, PhD(genetics), 54; Univ Tex, MD, 66. *Prof Exp:* Lectr anat, McGill Univ, 54-57; from asst prof to assoc prof, Univ Tex Med Bd, Galveston, 57-67; prof anat & chmn dept, 67-75; PROF ANAT, MICH STATE UNIV, 75- *Mem:* Am Asn Anat; Teratology Soc; Am Asn Cancer Res. *Res:* Experimental teratology; carcinogenesis. *Mailing Add:* Dept Anat Mich State Univ East Lansing MI 48824-1316

WALKER, CAROL L, b Martinez, Calif, Aug 19, 35; m 62; c 4. MATHEMATICS. *Educ:* Univ Colo, BME, 57; NMex State Univ, MS, 61, PhD(math), 63. *Prof Exp:* Mem math, Inst Advan Study, 63-64; from asst prof to assoc prof, 64-72, PROF MATH, NMEX STATE UNIV, 72-, DEPT HEAD, 79- *Concurrent Pos:* NSF fel, 63-64, NSF grants, 64-72. *Mem:* Am Math Soc. *Res:* Algebra, primarily Abelian group theory and homological algebra. *Mailing Add:* Dept Math NMex State Univ Las Cruces NM 88003

WALKER, CEDRIC FRANK, b Los Angeles, Calif, Jan 26, 50; m 85; c 2. NEUROPROSTHESES, MICROPROCESSORS. *Educ:* Stanford Univ, BS, 72, MS, 72; Duke Univ, PhD(biomed eng), 78. *Prof Exp:* Res assoc, Div Neurosurg, Duke Med Ctr, 77; asst prof, 77-82, ASSOC PROF BIOMED ENG, TULANE UNIV, 82- *Concurrent Pos:* Adj prof, Dept Orthop Surg & Dept Psychiat & Neurol, Tulane Univ, 78-; prin investr, NSF, 78-80 & Nat Inst Neurol & Comn Dis & Stroke, NIH, 82-85. *Mem:* Am Bd Clin Eng; Inst Elec & Electronics Engrs; Biomed Eng Soc. *Res:* Neuroprosthetic devices, particularly for cerebellar stimulation; bioeffects of low frequency electromagnetic fields; electrical injuries. *Mailing Add:* Dept Biomed Eng Tulane Univ New Orleans LA 70118

WALKER, CHARLES A, b Foreman, Ark, Dec 14, 35; m 57; c 3. NEUROPHARMACOLOGY. *Educ:* Ark Agr, Mech & Normal Col, BS, 57; Wash State Univ, MS, 59; Loyola Univ, PhD(pharmacol), 69. *Prof Exp:* Res asst, Wash State Univ, 57-59; asst prof biol, Ft Valley State Col, 59-63; asst prof physiol, Tuskegee Inst, 63-65, assoc prof pharmacol, 68-71, prof vet pharmacol & chmn dept pharmacol, 71-74; prof pharm & dean, Sch Pharm, 74-86, DEPT PHARMACOL, FLA A&M UNIV, 86-; CHANCELLOR, UNIV ARK PINE BLUFF. *Mem:* AAAS; Am Asn Clin Chem; NY Acad Sci; Am Soc Pharmacol & Exp Therapeut; Int Soc Chronobiol. *Res:* Circadian rhythms of biogenic amines in the central nervous system. *Mailing Add:* Univ Ark Pine Bluff PO Box 4008 1501 N University Dr Pine Bluff AR 71601

WALKER, CHARLES A(LLEN), b Wise Co, Tex, June 18, 14; m 42; c 3. CHEMICAL ENGINEERING. *Educ:* Univ Tex, BSChE, 38, MSChE, 40; Yale Univ, DEng, 48. *Prof Exp:* Asst prof chem, Tex Col Arts & Indust, 40-41 & Univ Ark, 41-42; from instr to prof, 42-84, master, Berkeley Col, 59-69, dept chmn, 74-76 & 81-84, EMER PROF CHEM ENG, YALE UNIV, 84- *Concurrent Pos:* Mem adv bd, Petrol Res Fund, 70-72, chmn, 72-81. *Mem:* AAAS; Am Chem Soc; Sigma Xi; Am Soc Eng Educ; Am Inst Chem Engrs. *Res:* Water quality control; technology and the social sciences; role of the social sciences in energy policy; nuclear waste management. *Mailing Add:* 313-A Mason Lab Yale Univ Box 2159 New Haven CT 06520-2159

WALKER, CHARLES R, b Chicago, Ill, Dec 18, 28; m 50; c 4. BIOCHEMISTRY, FISH BIOLOGY. *Educ:* Southern Ill Univ, BA, 51, MA, 52. *Prof Exp:* Biochemist & fishery biologist, Mo Conserv Comn, 52-61; biochemist, Fish Control Lab, US Fish & Wildlife Serv, 61-67, chief br pest control res, Div Fishery Res, 67-72, chief off environ assistance, 72-75, sr environ scientist, 75-81; CONSULT ENVIRON CONTAMINANTS & TOXICOL, 81- *Concurrent Pos:* Consult fishery biol, hazardous mat, pesticides & pond cult, 54-82; lectr, Viterbo Col, 64-65; instr, USDA Grad Sch, 69-; mem adj fac environ systs mgt, Am Univ, 75- *Mem:* Am Chem Soc; Am Fisheries Soc; Weed Sci Soc Am; Am Soc Testing & Mat; Am Soc Limnol & Oceanog. *Res:* Fishery research; aquatic ecology, fish-pesticide relationships; pollution biology; pond culture; aquatic herbicides; toxicity; efficacy residues of drugs and pest control agents for fisheries; analytical chemistry; limnology; soil science; environment impact statements control; environmental impact assessments; biological testing methods and hazard assessment of toxic substances; monitoring environmental contaminants. *Mailing Add:* 4613 Dixie Hill Rd Fairfax VA 22030

WALKER, CHARLES THOMAS, b Chicago, Ill, Sept 5, 32; m 73; c 3. SOLID STATE PHYSICS. *Educ:* Univ Louisville, AB, 56, MS, 58; Brown Univ, PhD(physics), 61. *Prof Exp:* Res asst physics, Brown Univ, 58-60; res assoc, Cornell Univ, 61-63; asst prof, Northwestern Univ, Evanston, 63-67, assoc prof, 67-71; PROF PHYSICS, ARIZ STATE UNIV, 71-, CHMN DEPT, 81- *Concurrent Pos:* Guggenheim fel, Oxford Univ, 67-68; vis prof, Munich Tech Univ, 71; consult, Motorola, Inc, 74-; vis prof, Univ Sao Paulo, 76 & Univ Regensburg, 77; dir, Ctr Solid State Sci, Ariz State Univ, 76-78; Alexander von Humboldt fel, Max-Planck Inst, Stuttgart, 78-79. *Mem:* AAAS; fel Am Phys Soc; Am Asn Physics Teachers. *Res:* Light scattering; lattice dynamics and impurity studies in solids; magnetism. *Mailing Add:* Dept Physics Ariz State Univ Tempe AZ 85287

WALKER, CHARLES WAYNE, b Oberlin, Ohio, Mar 27, 47; m 69. DEVELOPMENTAL BIOLOGY. *Educ:* Miami Univ, BA, 69; Cornell Univ, MS, 73, PhD(invertebrate zool), 76. *Prof Exp:* Lectr embryol, Cornell Univ, 75-76; asst prof, 76-80, ASSOC PROF EMBRYOL OF INVERTEBRATES, UNIV NH, 80- *Concurrent Pos:* Hubbard Fund grant, 78-80, NSF grant, 80-81, NATO grant, 80-81. *Mem:* Am Soc Zoologists; Sigma Xi; Soc Develop Biol; Int Soc Invertebrate Reproduction. *Res:* Physiological and ultrastructural aspects of cellular interaction during spermatogenesis and regeneration; comparative aspects of invertebrate development. *Mailing Add:* Dept Zool Univ NH Durham NH 03824

WALKER, CHERYL LYN, b Portland, Ore, July 7, 55; m; c 2. CARCINOGENESIS, CELL BIOLOGY. *Educ:* Univ Colo, BA, 77; Univ Tex Health Sci Ctr, PhD(cell biol), 84. *Prof Exp:* Staff fel, Lab Pulmonary Pathobiol, Nat Inst Environ Health Sci, NIH, 84-87, sr staff fel, 87-88; SCIENTIST I, DEPT CELLULAR & MOLECULAR TOXICOL, CHEM INDUST INST TOXICOL, 88- *Concurrent Pos:* Consult, BioSearch Labs Inc, Arlington, Tex, 73-77; coordr, Chem Indust Inst Toxicol summer student intern prog, 89-90; adj asst prof, Path, Univ NC, 89-, Toxicol, NC State Univ, 91-; mem Pub Commun Comt, Soc Toxicol, 91-93, Small Bus Innovation Res Concept Rev Panel, NIH, Biol Models & Mat Resources Prog Study Sect, Sci Rev Panel, Nat Inst Environ Health Sci, prog comt, Am Asn Cancer Res, 91-92. *Mem:* Am Soc Cell Biol; Am Asn Cancer Res; Soc Toxicol; AAAS. *Res:* Molecular mechanisms of chemical carcinogenesis with emphasis on oncogenes and tumor suppressor genes involved in renal cell carcinoma and mesothelioma; current investigations utilize rodent models for renal carcinogenesis and in vitro and in vivo studies on mesothelial cell transformation; numerous publications. *Mailing Add:* PO Box 12137 Research Triangle Park NC 27709

WALKER, CHRISTOPHER BLAND, b Lakeland, Fla, July 25, 25; m 61; c 3. PHYSICS. *Educ:* Davidson Col, BS, 48; Mass Inst Technol, PhD(physics), 51. *Prof Exp:* Fulbright scholar, France, 51-52; instr physics, Mass Inst Technol, 52-53; from asst prof to assoc prof, Inst Metals, Chicago, 53-63; RES PHYSICIST, ARMY MAT & MECH RES CTR, 63- *Concurrent Pos:* Guggenheim fel, 63-64. *Mem:* Am Phys Soc; Am Crystallog Asn; Fr Soc Mineral & Crystallog. *Res:* X-ray diffraction; imperfections in crystals; thermal vibrations; neutron inelastic scattering. *Mailing Add:* 22 Baskin Rd Lexington MA 02173

WALKER, DAN B, b Connersville, Ind, Apr 18, 45. PLANT ANATOMY & DEVELOPMENT. *Educ:* Ind Univ, Bloomington, AB, 68; Univ Calif, Berkeley, PhD(bot), 74. *Prof Exp:* Lectr bot, Univ Calif, Berkeley, 73-74; asst prof bot, Univ Ga, 74-78; asst prof biol, 78-85, dir sci educ, 85-86, RES SCIENTIST, CTR STUDY EVOLUTION & ORIGIN OF LIFE, UNIV CALIF, LOS ANGELES, 85-; PROF & ASSOC DEAN, SAN JOSE STATE UNIV, 86- *Mem:* Sigma Xi; AAAS; Bot Soc Am; Am Soc Plant Physiologists; Soc Develop Biol. *Res:* Investigations of structure-function and developmental problems at the cellular level in higher plants, especially on the mechanisms of intercellular communication and pattern formation in plants. *Mailing Add:* Dept Biol Sci San Jose State Univ San Jose CA 95192

WALKER, DANIEL ALVIN, b Cleveland, Ohio, Dec 18, 40; m 70; c 4. SEISMOLOGY. *Educ:* John Carroll Univ, BS, 63; Univ Hawaii, MS, 65, PhD(geophys), 71. *Prof Exp:* Res asst seismol, 63-68, jr seismologist, 69-71, asst geophysicist, 72-76, ASSOC GEOPHYSICIST, HAWAII INST GEOPHYS, UNIV HAWAII, 76- *Mem:* Seismol Soc Am; Am Geophys Union; AAAS. *Mailing Add:* Hawaii Inst Geophys Univ Hawaii 2525 Correa Rd Honolulu HI 96822

WALKER, DAVID, b Troy, New York, NY, Aug 9, 46; m 80; c 3. PETROLOGY. *Educ:* Oberlin Col, AB, 68; Harvard Univ, AM, 70, PhD(geol), 72. *Prof Exp:* Lectr geol, Harvard Univ, 73-74, res fel geophysics, 72-77, sr res assoc, 78-82; LAMONT-DOHERTY GEOL OBSERV, 82- *Honors & Awards:* F W Clarke Medal, Geochem Soc, 75; Guggenheim Fel, 88. *Mem:* Am Geophys Union; Geochem Soc; Mineral Soc Am; AAAS. *Res:* General geology with specialty in petrology, particularly experimental petrology. *Mailing Add:* Lamont-Doherty Geol Observ Columbia Univ Palisades NY 10964

WALKER, DAVID CROSBY, b York, Eng, June 16, 34; m 78; c 3. PHYSICAL CHEMISTRY, NUCLEAR CHEMISTRY. *Educ:* Univ St Andrews, BSc, 55, Hons, 56; Univ Leeds, PhD(chem), 59. *Hon Degrees:* DSc, Univ St Andrews, 74. *Prof Exp:* Fel chem, Nat Res Coun Can, 59-61; res lectr, Univ Leeds, 61-64; from asst prof to assoc prof, 64-75, PROF CHEM, UNIV BC, 75- *Concurrent Pos:* Vis prof, Univ Leeds, Paris & JSPS, 81. *Honors & Awards:* Miller Prize, 56; Irvine Chem Medal, 56. *Mem:* Fel Chem Inst Can; fel Royal Soc Chem; Am Chem Soc; Am Inst Physics; Radiation Res Soc. *Res:* Radiation chemistry of water and organic liquids; muonium chemistry; solvated electron studies in polar liquids; orgins of optical activity in nature. *Mailing Add:* Dept Chem Univ BC Vancouver BC V6T 1Y6 Can

WALKER, DAVID HUGHES, b Nashville, Tenn, May 31, 43; m 68; c 2. RICKETTSIOLOGY, MICROBIAL PATHOGENESIS. *Educ:* Davidson Col, BA, 65; Vanderbilt Univ, MD, 69. *Prof Exp:* Resident path, Peter Bent Brigham Hosp, 69-73; USPHS surgeon virol, Ctr Dis Control, 73-75; from asst prof to prof path, Univ NC, 75-87; PROF & CHMN DEPT PATH, UNIV TEX MED BR, 87- *Concurrent Pos:* Fel, Med Sch, Harvard Univ, 71-73; clin asst prof, Med Sch, Emory Univ, 74-75; prin investr, NIH grants, 79-82, 84-

& 85-88, NIH contract, 79-82 & US Army Med Res & Develop Command contract, 83-86. *Mem:* Infectious Dis Soc Am; Am Soc Rickettsiology & Rickettsial Dis (vpres, 88); Am Asn Pathologists; Am Soc Trop Med & Hyg; Int Acad Pathologists; Soc Pediat Path. *Res:* Molecular rickettsiology, immunity to rickettsiae, rickettsial pathogenesis and diagnosis; role of proteases in infectious disease pathogenesis, and respiratory syncytial virus pathogenesis and antiviral treatment; tropical medicine, arenaviruses and hemorrhagic fevers, particularly Lassa fever; ecology of diseases; treatment and pathogenesis of Pheumocystis infections. *Mailing Add:* 101 Keiller Bldg F09 Galveston TX 77550

WALKER, DAVID KENNETH, b Youngstown, Ohio, Apr 4, 43; m 67. COMPUTER INTERFACING, INTERACTIVE VIDEO LEARNING. *Educ:* Pa State Univ, University Park, BS, 65; WVa Univ, MS, 68, PhD(physics), 71. *Prof Exp:* Instr physics, WVa Univ, 69-71; from asst prof to assoc prof, 71-80, PROF PHYSICS, WAYNESBURG COL, 81- *Mem:* Asn Comput Mach; Math Asn Am; Am Asn Physics Teachers. *Res:* Computer applications in science; laboratory interfacing; interactive videodisc learning; hard disk organization and management systems. *Mailing Add:* Dept Math/Comput Sci Waynesburg Col Waynesburg PA 15370

WALKER, DAVID N(ORTON), b Yuba City, Calif, Aug 17, 43; m 68; c 2. PLASMA PHYSICS, SPACE PLASMA RESEARCH. *Educ:* Univ Md, BS, 65; Univ NH, MA, 72, PhD(physics), 75. *Prof Exp:* Engr, McDonnell-Douglas Corp, 66-67; assoc engr, Elec Assoc Inc, 68-70; instr physics, Univ NH, 70-75; analyst, Anal Servs Inc, 75-77; PHYSICIST, NAVAL RES LAB, 77- *Mem:* Am Phys Soc; Am Geophys Union; Am Acad Sci. *Res:* Plasma physics as related to space plasma. *Mailing Add:* 336 N Edison ST Arlington VA 22203

WALKER, DAVID RUDGER, b Ames, Iowa, Sept 15, 29; m 48; c 10. POMOLOGY. *Educ:* Utah State Univ, BS, 51, MS, 52; Cornell Univ, PhD, 55. *Prof Exp:* From asst prof to assoc prof hort, NC State Univ, 55-60; assoc prof, 60-65, PROF PLANT SCI, UTAH STATE UNIV, 65- *Honors & Awards:* Shepard Award, Am Pomol Soc; Stark Award, Am Soc Hort Sci. *Mem:* Fel Am Soc Hort Sci; Am Pomol Soc; fel AAAS. *Res:* Plant hardiness; mineral nutrition; growth substances; rootstocks. *Mailing Add:* Dept Plant Sci Utah State Univ Logan UT 84322-4820

WALKER, DAVID TUTHERLY, b Huntington, WVa, July 10, 22; m 57; c 1. MATHEMATICS. *Educ:* Wofford Col, BS, 49; Univ Ga, MS, 51, PhD, 55. *Prof Exp:* Asst math, Univ Ga, 51-53; instr, Univ SC, 53-54; asst, Univ Ga, 54-55; from asst prof to prof math, Memphis State Univ, 67-87; RETIRED. *Mem:* Sigma Xi. *Res:* Mathematical analysis; modern algebra; theory of numbers; geometry. *Mailing Add:* 4344 Tuckahoe Rd Memphis TN 38117

WALKER, DENNIS KENDON, b Sacramento, Calif, Aug 1, 38; m 60; c 4. BOTANY. *Educ:* Humboldt State Col, BA, 60; Univ Calif, Davis, MS, 64, PhD(bot), 66. *Prof Exp:* Asst prof, 65-70, assoc prof, 70-76, PROF BOT, HUMBOLDT STATE UNIV, 76- *Mem:* Electron Micros Soc Am. *Res:* Plant morphology; developmental plant anatomy and plant ultrastructure, specifically the ultrastructure of differentiating elements of vascular tissues. *Mailing Add:* Dept Bot Humboldt State Univ Arcata CA 95521

WALKER, DON WESLEY, b Ft Worth, Tex, July 30, 42; m 64; c 2. NEUROSCIENCE, NEUROPHARMACOLOGY. *Educ:* Univ Tex, Arlington, BA, 64; Tex Christian Univ, MA & PhD(psychol), 68. *Prof Exp:* Asst prof, 70-75, assoc prof, 75-80, PROF NEUROSCI & PSYCHOL, UNIV FLA, 80-; RES CAREER SCIENTIST, VET ADMIN HOSP, GAINESVILLE, 83- *Concurrent Pos:* Nat Inst Ment Health training grant, Col Med, Univ Fla, 68-70, NIH res grant neurosci, 72-; Vet Admin res fund grant, Vet Admin Hosp, Gainesville, 70-; res investr, 70-83. *Mem:* AAAS; Soc Neurosci; Res Soc Alcoholism. *Res:* Neurobiology of alcoholism; chronic effects of ethanol on the brain. *Mailing Add:* Dept Neurosci Univ Fla Col Med Gainesville FL 32610

WALKER, DONALD F, b Brush, Colo, July 16, 23; m 44; c 1. VETERINARY MEDICINE. *Educ:* Colo State Univ, DVM, 44. *Prof Exp:* Pvt pract, Grassland Hosp, 45-58; assoc prof, 58-66, PROF LARGE ANIMAL SURG & MED, AUBURN UNIV, 66-, DEPT HEAD, 78- *Mem:* Am Col Theriogenology; Sigma Xi. *Res:* Urogenital surgery; ultrasonic therapy. *Mailing Add:* Dept Large Animal Surg & Med Auburn Univ Auburn AL 36830

WALKER, DONALD I, b Lombard, Ill, Jan 13, 22; m 44; c 3. ANALYTICAL CHEMISTRY, CHEMICAL MICROSCOPY. *Educ:* Univ Ill, BS, 48; Univ Colo, PhD(chem), 56. *Prof Exp:* Asst chem, Univ Colo, 48-50, asst instr, 53-56; res chemist, Los Alamos Sci Lab, Univ Calif, 50-53; dep dir health & safety div, Idaho Opers Off, AEC, 56-57, dir licensee compliance div, 57-60, dir region VIII, Div Compliance, 60-62, dir region IV, 62-70, dir, Health Serv Lab, Idaho Opers Off, Energy Res & Develop Admin, 70-76; exec dir, Assoc Western Univs, Inc, 76-85. *Concurrent Pos:* Consult, Rocky Flats Div, Dow Chem Co, 54. *Mem:* AAAS; Am Chem Soc; Health Physics Soc. *Res:* Administration of environmental monitoring, radiation dosimetry, ecology. *Mailing Add:* 2706 E Creek Rd Sandy UT 84093

WALKER, DUARD LEE, b Bishop, Calif, June 2, 21; m 45; c 4. VIROLOGY, MICROBIOLOGY. *Educ:* Univ Calif, Berkeley, AB, 43, MA, 47; Univ Calif, San Francisco, MD, 45; Am Bd Med Microbiol, dipl. *Prof Exp:* Asst resident physician internal med, Stanford Univ Serv, San Francisco Hosp, 50-52; assoc prof, Univ Wis-Madison, 52-59, chmn dept, 70-76 & 81-88, prof med microbiol, Med Sch, 59-88, EMER PAUL F CLARK PROF, UNIV WIS-MADISON, 88- *Concurrent Pos:* Nat Res Coun fel, Rockefeller Inst, NY, 47-49; USPHS fel, George Williams Hooper Found, Univ Calif, San Francisco, 49-50, res assoc, 50-51; consult, Naval Med Res Unit 4, Great Lakes, Ill, 58-74; mem microbiol training comt, Nat Inst Gen Med Sci, 66-70; mem, Nat Adv Allergy & Infectious Dis Coun, 70-74, study group papovaviridae, Int Comt Taxon Viruses, 76-87, adv comt blood prog res, Am Red Cross, 78-79 & bd sci adv, Delta Regional Primate Ctr, Tulane Univ, 80-83; mem vaccines & related biol prod adv comt, US Food & Drug Admin, 85- 89; mem, Behring diagnostic award comt, Am Soc Microbiol, 86-89, Am Type Cult Adv Comt, 86- *Mem:* Nat Acad Sci; Am Soc Microbiol; fel Am Acad Microbiol; Am Soc Virol; fel Infectious Dis Soc Am; Am Asn Immunol. *Res:* Persistent and chronic viral infections; host response to viral infection. *Mailing Add:* Dept Microbiol & Immunol Univ Wis Med Sch Madison WI 53706

WALKER, EDWARD BELL MAR, b Ogden, Utah, Mar 19, 52; m 75; c 2. PHOTOCHEMISTRY, PHOTOBIOLOGY. *Educ:* Weber State Col, BA, 76; Tex Tech Univ, PhD(chem), 80. *Prof Exp:* Grad student chem, Tex Tech Univ, 76-80; scholar biochem pharm, Stanford Univ Med Ctr, 80-81; asst prof, 81-85, ASSOC PROF BIOCHEM, WEBER STATE COL, 85- *Mem:* Am Soc Photobiol; Am Chem Soc; Am Soc Plant Physiologists; Sigma Xi. *Res:* Photochemistry and photobiology, particularly the mechanisms of photoreception in both plants and animals. *Mailing Add:* Dept Chem Weber State Col Ogden UT 84408

WALKER, EDWARD JOHN, b Detroit, Mich, Apr 16, 27; m 60; c 1. SOLID STATE PHYSICS. *Educ:* Univ Mich, BSE, 49; Yale Univ, PhD(physics), 60. *Prof Exp:* Asst electronics, Tube Lab, Nat Bur Stand, 49-53; physicist, Res Ctr, IBM Corp, 60-; RETIRED. *Res:* Semiconductor physics. *Mailing Add:* Spring Valley Rd Ossining NY 10562

WALKER, EDWARD ROBERT, b Winnipeg, Man, July 29, 22; m 54; c 3. METEOROLOGY. *Educ:* Univ Man, BSc, 43; Univ Toronto, MA, 49; McGill Univ, PhD(Meteorol), 61. *Prof Exp:* Meteorologist, Meteorol Serv Can, 43-59; res asst meteorol, McGill Univ, 59-60; res micrometeorologist, Defence Res Bd, Can, 61-67; res Arctic meteorologist, Can Dept Environ, 67-78; RETIRED. *Concurrent Pos:* Environ consult, 78-85. *Mem:* Am Meteorol Soc; Royal Astron Soc Can; fel Royal Meteorol Soc. *Res:* Arctic meteorology; oceanography. *Mailing Add:* 3350 Woodburn Ave Victoria BC V8P 5C1 Can

WALKER, ELBERT ABNER, b Huntsville, Tex, Mar 11, 30; m 51; c 4. MATHEMATICS. *Educ:* Sam Houston State Col, BA, 50, MA, 52; Univ Kans, PhD(math), 55; Colo State Univ, MS, 78. *Prof Exp:* High sch teacher, Tex, 50-52; mathematician, US Dept Defense, Washington, DC, 55-56; asst prof math, Univ Kans, 56-57; from asst prof to prof math, NMex State Univ, 57-87; prog officer, NSF, 87-89; CONSULT, 89- *Mem:* Am Math Soc; Math Asn Am; Am Statist Asn; Sigma Xi. *Res:* Abelian group theory; category theory; ring theory; statistics. *Mailing Add:* Dept Math Sci NMex State Univ Las Cruces NM 88003

WALKER, ELIZABETH REED, b Rochester, Pa, July 2, 41; m 67; c 1. HUMAN ANATOMY. *Educ:* Mich State Univ, BA, 63; WVa Univ, MS, 71, PhD(human anat), 75. *Prof Exp:* Res asst microbiol, Rockefeller Univ, 64-66; technologist electron micros, 67-71, lectr human anat, 74-75, INSTR HUMAN ANAT, WVA UNIV, 75- *Res:* Investigation of rheumatology by transmission and scanning electron microscopy, particularly pathogenesis of rheumatoid arthritis and other connective tissue diseases, and pulmonary research, with emphasis on macrophage uptake of respirable mineral particulates. *Mailing Add:* Dept Anat WVa Univ Sch Med Morgantown WV 26506

WALKER, ERIC A(RTHUR), b Long Eaton, Eng, Apr 29, 10; nat US; m 37; c 2. ACOUSTICAL ENGINEERING. *Educ:* Harvard Univ, BS, 32, SM, 33, ScD(elec eng), 35. *Hon Degrees:* LLD, Temple Univ & Lehigh Univ, 57, Hofstra Col, Lafayette Col & Univ Pa, 60, Univ RI, 62; LHD, Elizabethtown Col, 58; LittD, Jefferson Med Col, 60; DL, St Vincent Col, 68; ScD, Wayne State Univ, 65, Thiel Col, 66, Univ Notre Dame, 68, Univ Pittsburgh, 70 & Univ Bridgeport, 72. *Prof Exp:* Instr elec eng & math, Tufts Col, 34-38, asst prof math, 38-39, chmn dept elec eng, 39-40; assoc prof elec eng & head dept, Univ Conn, 40-42; assoc dir, underwater sound lab, Harvard Univ, 42-45; prof elec eng & head dept, Pa State Univ 45-51, dir Ord Res Lab, 45-52, dean col eng & archit, 51-56, vpres, 56, pres, 56-70; vpres, Aluminum Co Am, 70-75; EMER PRES, PA STATE UNIV, 70- *Concurrent Pos:* Mem & former chmn, Comt Undersea Warfare, Nat Res Coun; chmn comt eng, NSF, 51-53, mem, Nat Sci Bd; exec secy, Res & Develop Bd, US Dept Defense, 50-51; mem sci adv panel, US Dept Army, 56-58; chmn res adv comt, USN, 71-; vchmn, President's Comt Scientists & Engrs, 56-58; gen chmn, President's Conf Technol & Distribution Res for Benefit of Small Bus, 57; mem exec comt, Am Asn Land Grant Cols & State Univs; bd vis, US Naval Acad, 58-60; chmn, Inst Defense Anal, 58-87; mem, adv comt, Memphis State Univ. *Honors & Awards:* Bliss Award, Am Soc Mil Engrs, 59; Horatio Alger Award; Lamme Award, Am Soc Eng Educ. *Mem:* Nat Acad Eng (pres, 66-70); fel Am Phys Soc; fel Acoust Soc Am; Am Soc Eng Educ (vpres, 52-54, pres, 60-61); fel Inst Elec & Electronics Engrs. *Res:* Acoustic properties of liquids; high voltage insulation; electrostatics precipitation. *Mailing Add:* Pa State Univ 222-A Hammond Bldg Univ Park PA 16802

WALKER, FRANCIS EDWIN, b Morris, Ill, Nov 29, 31; m 51; c 3. AGRICULTURAL ECONOMICS. *Educ:* Univ Ill, BS, 54, MS, 58, PhD(agr econ), 60. *Prof Exp:* Asst prof agr econ, Purdue Univ, 60-61; from asst prof to assoc prof, 61-68, PROF AGR ECON, OHIO STATE UNIV, 68- *Mem:* Am Statist Asn; Am Agr Econ Asn. *Res:* International trade policy; interregional competition. *Mailing Add:* 632 Wedgewood Dr Apt 3 Columbus OH 43220

WALKER, FRANCIS H, b San Francisco, Calif, Jan 15, 36; m 66; c 2. ORGANIC CHEMISTRY. *Educ:* Stanford Univ, BS, 58, MS, 60. *Prof Exp:* Chemist, 60-76, SR RES CHEMIST, ICI AMERICAS, 76- *Res:* Organic synthesis of agricultural chemicals; metabolite synthesis. *Mailing Add:* ICI Americas 1200 S 47th St Richmond CA 94804

WALKER, FREDERICK, b Woodbury, NJ, Jan 1, 54; m 79; c 2. POLYMER CHEMISTRY. *Educ:* Bloomfield Col, BA, 76; Yale Univ, MS, 78, PhD(org chem), 82. *Prof Exp:* Sr scientist, Rohm & Haas, 82-88; RES DIR, AKZO COATINGS, INC, 88- *Concurrent Pos:* Chmn, Great Lakes Polymer Conf, 91- *Mem:* Am Chem Soc; Fedn Socs Coatings Technol. *Res:* Synthesis of solution polymers, nonaqueous dispersions, and acrylic latexes for use in coatings and related applications. *Mailing Add:* Akzo Coatings Inc PO Box 7062 Troy MI 48007-7062

WALKER, GENE B(ERT), b Gladewater, Tex, Feb 24, 32; m 56; c 4. ELECTRICAL ENGINEERING. *Educ:* Univ Tex, BS, 59, MS, 62, PhD(elec eng), 64. *Prof Exp:* Res engr, Elec Eng Res Lab, Univ Tex, 59-64; sr res engr, Southwest Res Inst, 64-67; from asst prof to assoc prof, 67-77, PROF ELEC ENG, UNIV OKLA, 77-, ASSOC DEAN ENG, 79- *Concurrent Pos:* Consult, Nat Severe Storms Lab, Nat Oceanic & Atmospheric Admin, 69-78. *Mem:* Inst Elec & Electronics Engrs; Sigma Xi. *Res:* Radio wave propagation; radio direction finding; radar sounding of troposphere; acoustic sounding of the troposphere. *Mailing Add:* 2408 Cypress Ave Norman OK 73069

WALKER, GEORGE EDWARD, b Chillicothe, Ohio, Nov 5, 40; m 64; c 3. THEORETICAL NUCLEAR PHYSICS. *Educ:* Wesleyan Univ, BA, 62; Case Western Reserve Univ, MS, 64, PhD(physics), 66. *Prof Exp:* Res assoc physics, Los Alamos Sci Lab, 66-68; res assoc, Stanford Univ, 68-70; from asst prof to assoc prof, 70-76, PROF PHYSICS, IND UNIV, BLOOMINGTON, 76-, CHAIRPERSON, 86- *Concurrent Pos:* Vis staff mem, Los Alamos Sci Lab, 68- *Mem:* fel Am Phys Soc; Am Asn Physics Teachers. *Res:* Nuclear theory; electron scattering; meson-nucleus interactions; nucleon-nucleus interactions; heavy ion scattering. *Mailing Add:* Dept Physics Ind Univ Bloomington IN 47401

WALKER, GLENN KENNETH, b South Weymouth, Mass, May 15, 48. CELL BIOLOGY, PROTOZOOLOGY. *Educ:* Univ Mass, Amherst, BS, 70; Northern Ariz Univ, MS, 72; Univ Md, Col Park, PhD(cell biol), 75. *Prof Exp:* Teaching asst biol, Northern Ariz Univ, 70-72; teaching asst zool, cell biol & protozool, Univ Md, College Park, 72-75; NIH res fel, Cell Chem Lab, Univ Mich, Ann Arbor, 75-76; from asst prof to assoc prof, 76-85, PROF BIOL, EASTERN MICH UNIV, 85- *Concurrent Pos:* Adj prof, Med Sch, Univ Mich, 82-; consult, Occup Safety & Health Admin, 82- *Mem:* Am Micros Soc; Sigma Xi. *Res:* Examination of the molecular mechanisms associated with pathologies which may represent abnormalities in skeletal muscle differentiation; ultrastructural and cytochemical examination of protozoan encystment and differentiation. *Mailing Add:* Dept Biol Eastern Mich Univ Ypsilanti MI 48197

WALKER, GORDON ARTHUR HUNTER, b Kinghorn, Scotland, Jan 30, 36; m 62; c 2. ASTROPHYSICS. *Educ:* Univ Edinburgh, BSc, 58; Univ Cambridge, PhD(astrophys), 62. *Prof Exp:* Nat Res Coun fel astrophys, Dept Mines & Technol Surv, Dom Astrophys Observ, 62-63, res scientist II, 63-69; assoc prof, 69-74, dir inst astron & space sci, 72-78, PROF, UNIV BC, 74- *Mem:* Can Astron Soc; fel Royal Soc Can; Am Astron Soc. *Res:* Interstellar materials, particularly interstellar dust; early type stars, their distance, luminosity and rotational velocities; telescope auxilliary instrumentation; low light level multichannel detection systems. *Mailing Add:* Dept Geophys & Astron Univ BC 2075 Westbrook Pl Vancouver BC V6T 1W5 Can

WALKER, GRAYSON HOWARD, b North Wilkesboro, NC, Dec 9, 38; m 71. STATISTICAL MECHANICS, CHEMICAL PHYSICS. *Educ:* Univ NC, BS, 61; Univ Ill, MS, 62; Ga Inst Technol, PhD(physics), 69. *Prof Exp:* From instr to prof physics, Clark Col, 67-77; prof, 77-81, UNIV CHATTANOOGA FOUND PROF PHYSICS & DIR ENVIRON STUDIES PROG, UNIV TENN, CHATTANOOGA, 81- *Mem:* Am Phys Soc; Am Meteorol Soc; Am Asn Physics Teachers. *Res:* Applications of the methods of statistical physics to problems in chemical physics, atomospheric research, planetary atmospheres; nonlinear systems. *Mailing Add:* Dept Physics & Astron Univ Tenn 615 McCallie Ave Chattanooga TN 37402

WALKER, GRAYSON WATKINS, b Norfolk, Va, Feb 10, 44; m 70; c 1. ELECTROCHEMISTRY, MINERAL PROCESSING. *Educ:* Va Polytech Inst, BS, 67; Am Univ, MS, 79, PhD(phys chem), 81. *Prof Exp:* Res chemist, Mobility Equip Res & Develop Ctr, US Army, Ft Belvoir, Va, 67-80; res chemist, 80-83, SUPVRY RES CHEMIST, US BUR MINES, AVONDALE, MD, 83- *Concurrent Pos:* Adj prof, chem dept, Am Univ, Washington, DC, 84-85. *Mem:* Am Inst Mining Engrs; Electrochem Soc; Sigma Xi; Am Chem Soc. *Res:* Surface chemistry of mineral flotation systems; electrochemistry of sulfide minerals; kinetics of fuel cell reactions on platinum electrodes; new electrolytes for acid fuel cell systems. *Mailing Add:* 1407 Key Dr Alexandria VA 22302

WALKER, GUSTAV ADOLPHUS, b Locust Grove, Ga, Dec 5, 44; m 70; c 2. BIOCHEMISTRY, MICROBIOLOGY. *Educ:* Clark Col, BS, 67; Purdue Univ, PhD(biochem), 74. *Prof Exp:* NIH fel, 74-76 & Med Sch, St Louis Univ, 76-77; RES SCIENTIST, UPJOHN CO, 77- *Mem:* Am Chem Soc; Am Soc Microbiologists; AAAS. *Res:* Protein chemistry; assay design and development of pharmaceutical products, including liquid chromatography, electrophoresis, and spectrophotometric techniques. *Mailing Add:* 5125 Midfield Kalamazoo MI 49001-3298

WALKER, HARLEY JESSE, b Bushnell, Mich, July 4, 21; m 53; c 3. COASTAL MORPHOLOGY, COASTAL ENGINEERING. *Educ:* Univ Calif, Berkeley, BA, 47, MA, 54; La State Univ, Baton Rouge, PhD(geog), 60. *Hon Degrees:* Doctorate, Univ Uppsala, Sweden, 86. *Prof Exp:* From asst prof to assoc prof geog, Ga State Univ, Atlanta, 50-59, chmn dept 53-59; res geogr, Off Naval Res, Washington, DC, 59-60; from asst prof to prof, 60-77, chmn dept, 62-71, Boyd prof, 77-84, EMER BOYD PROF GEOG, LA STATE UNIV, BATON ROUGE, 84- *Concurrent Pos:* Vis prof, Univ Calif, Berkeley, 67, Univ Hawaii, Manoa, 80, 85 & 88 & Tsukuba Univ, Japan, 81-; liaison scientist, Off Naval Res, London, 68-69; vchmn, Comn Coastal Environ, 76-84; mem geog sci bd, Nat Acad Sci, 78-82; grant, Japan Found, 78-77, fel, 79; chmn, US Nat Comt, Int Geog Union-Nat Acad Sci, 80-84; consult, N Slope Bor; first hon fel, Int Asn Geomorphologists, 89. *Honors & Awards:* Honor Award, Asn Am Geogrs, 77, First Distinguished Career Award in Geomorphol, 89. *Mem:* Fel Arctic Inst NAm; Asn Am Geogrs; Coastal Soc; fel AAAS; Brit Geomorphol Res Group; Japanese Geomorphol Union; Am Geophys Union; Am Quaternary Asn. *Res:* Morphologic, hydrologic and nearshore oceanographic research on Arctic coasts; morphologic and human modification studies of Oriental shorelines. *Mailing Add:* Dept Geog La State Univ Baton Rouge LA 70803-4105

WALKER, HARRELL LYNN, b Minden, La, May 14, 45; wid; c 2. PLANT PATHOLOGY, BIOLOGICAL CONTROL. *Educ:* La Tech Univ, BS, 66; Univ Ky, MS, 69, PhD(plant path), 70. *Prof Exp:* Plant pathologist, Plant Indust Div, Ala Dept Agr, 74-75, asst dir, 75-76; res plant pathologist, Southern Weed Sci Lab, Sci & Educ Admin- USDA, 76-84; dir, La Res Sta, Mycogen Corp, 84-87; PROF BOT, LA TECH UNIV, 87- *Mem:* Am Phytopath Soc; Weed Sci Soc Am; Sigma Xi. *Res:* Biological control of weeds. *Mailing Add:* Dept Biol Sci La Tech Univ PO Box 3158 TS Ruston LA 71272-0001

WALKER, HOMER FRANKLIN, b Beaumont, Tex, Sept 7, 43; m 84; c 2. NUMERICAL ANALYSIS, SCIENTIFIC & STATISTICAL COMPUTING. *Educ:* Rice Univ, BA, 66; NY Univ, MS, 68, PhD(math), 70. *Prof Exp:* From asst prof to assoc prof math, Tex Tech Univ, 70-74; vis assoc prof, 74-75, assoc prof, 75-80; prof math, Univ Houston, 80-85; PROF MATH, UTAH STATE UNIV, 85- *Concurrent Pos:* Vis assoc prof, Univ Denver, 73-74; vis assoc prof comput sci, Cornell Univ, 78; vis prof math, Univ NMex, 81-82; vis prof, computer sci, Yale Univ, 89- *Mem:* Soc Indust & Appl Math; Am Math Soc; Math Asn. *Res:* Numerical analysis; partial differential equations; statistical pattern recognition. *Mailing Add:* Dept Math & Stat Utah State Univ Logan UT 84322-3900

WALKER, HOMER WAYNE, b Saxonburg, Pa, May 22, 25; m 63; c 1. FOOD MICROBIOLOGY. *Educ:* Pa State Univ, BS, 51; Univ Wis, MS, 53, PhD(bact), 55. *Prof Exp:* Asst bact, Univ Wis, 51-55; from asst prof to prof, 55-90, EMER PROF FOOD TECHNOL, IOWA STATE UNIV, 90- *Concurrent Pos:* Fulbright fel, Denmark, 77. *Mem:* Am Soc Microbiol; fel Inst Food Technologists; AAAS; Int Asn Milk, Food & Environ Sanit; Brit Soc Appl Bact. *Res:* Resistance of bacterial spores to heat and chemicals; bacterial toxins; antibiotics in foods and use as preservatives; microbiology of processed poultry and meats; sanitary bacteriology of food and water; mycotoxins. *Mailing Add:* 1513 Harding Ames IA 50010

WALKER, HOWARD DAVID, b New York, NY, May 7, 25; m 47; c 3. BIOCHEMISTRY. *Educ:* NY Univ, BA, 47, MS, 48; Univ Calif, Los Angeles, PhD(biochem), 55. *Prof Exp:* USPHS fel biochem, Am Meat Inst Found, Chicago, 55-56; instr, Northwestern Univ, 56-57; from asst prof to assoc prof chem, 57-68, PROF CHEM, CALIF POLYTECH STATE UNIV, SAN LUIS OBISPO, 68- *Concurrent Pos:* Group leader, Vet Admin Hosp, Downey, Ill, 56-57. *Mem:* Am Chem Soc. *Res:* Chemistry of pesticides and foods. *Mailing Add:* Dept Chem Calif Polytech State Univ San Luis Obispo CA 93407

WALKER, HOWARD GEORGE, JR, agricultural chemistry, for more information see previous edition

WALKER, HUGH S(ANDERS), b Mooringsport, La, July 31, 35; m 58; c 4. MECHANICAL ENGINEERING, APPLIED MECHANICS. *Educ:* La State Univ, BS, 57, MS, 60; Kans State Univ, PhD(mech eng), 65. *Prof Exp:* Instr eng mech, La State Univ, 57-60; res asst, 60-64, from instr to assoc prof, 64-76, PROF MECH ENG, KANS STATE UNIV, 76-, ASSOC DIR INST COMPUT RES IN ENG, 69- *Concurrent Pos:* NSF res grant, 66-67. *Mem:* Am Soc Mech Engrs; Am Inst Aeronaut & Astronaut; Soc Exp Stress Anal; Am Soc Eng Educ. *Res:* Analytical and experimental investigations in stress analysis, vibrations and acoustics; numerical analysis and computer techniques. *Mailing Add:* Dept Mech Eng Kans State Univ Manhattan KS 66506

WALKER, IAN GARDNER, b Saskatoon, Sask, Apr 20, 28; m 52, 80; c 4. BIOCHEMISTRY, CELL BIOLOGY. *Educ:* Univ Sask, BA, 48; Univ Toronto, MA, 51, PhD(biochem), 54. *Prof Exp:* Defense sci serv officer, Defence Res Med Labs, Toronto, 54-60, spec lectr, Fac Pharm, Univ Toronto, 54-62; from asst prof to assoc prof, 66-74, PROF BIOCHEM, CANCER RES LAB & DEPT BIOCHEM, UNIV WESTERN ONT, 74- *Concurrent Pos:* Nat Cancer Inst Can fel, Ont Cancer Inst, 60-62; Eleanor Roosevelt Int Cancer fel, 68-69; Stanford Univ, 81; Erasmus Univ, Rotterdam, 87-88. *Mem:* Am Asn Cancer Res; Can Biochem Soc; Can Soc Cell Biol. *Res:* Biochemistry of nucleic acids, cell division, anticancer agents; biochemistry and toxicology of omega-fluorinated compounds; toxicity of oxygen at high pressures; excision repair. *Mailing Add:* Dept Biochem Univ Western Ont London ON N6A 5C1 Can

WALKER, IAN MUNRO, b Toronto, Ont, Aug 18, 40; US citizen; m 66. INORGANIC CHEMISTRY. *Educ:* Bowdoin Col, BA, 62; Brown Univ, PhD(chem), 67. *Prof Exp:* NIH fel chem, Univ Ill, Urbana, 67-68; asst prof, 68-73, ASSOC PROF CHEM, YORK UNIV, 73- *Mem:* Am Chem Soc. *Res:* Structure of ion-aggregates in solution; single crystal near infrared spectroscopy. *Mailing Add:* Dept Chem York Univ Downsview ON M3J 1P3 Can

WALKER, J CALVIN, b Mooresville, NC, Jan 16, 35; m 58; c 3. NUCLEAR PHYSICS, SOLID STATE PHYSICS. *Educ:* Harvard Univ, AB, 56; Princeton Univ, PhD(physics), 61. *Prof Exp:* Instr physics, Princeton Univ, 61-62; fel, Atomic Energy Res Estab, Harwell, Eng, 62-63; from asst prof to assoc prof, 63-70, PROF PHYSICS, JOHNS HOPKINS UNIV, 70-, CHMN, 87- *Concurrent Pos:* Alfred P Sloan Found fel, 66-68; Shaw Travelling fel, Harvard, 57. *Mem:* Fel Am Phys Soc; NY Acad Sci. *Res:* Atomic beam studies of radioactive nuclei; solid state and nuclear studies using gamma resonance techniques. *Mailing Add:* Dept Physics & Astron Johns Hopkins Univ Baltimore MD 21218

WALKER, J KNOX, b Bryan, Tex, Nov 16, 27; m 51; c 5. INSECT MANAGEMENT IN COTTON. *Educ:* Agr & Mech Col, Tex, BS, 50, MS, 56. *Prof Exp:* From instr to assoc prof, 53-81, PROF ENTOM, TEX A&M UNIV, 81- *Mem:* Entom Soc Am. *Res:* Management systems for insects in cotton. *Mailing Add:* Dept Entom Tex A&M Univ College Station TX 77843

WALKER, JAMES BENJAMIN, b Dallas, Tex, May 15, 22; m 56; c 3. BIOCHEMISTRY. *Educ:* Rice Inst, BS, 43; Univ Tex, MA, 49, PhD(biochem), 52. *Prof Exp:* Res scientist, Biochem Inst, Univ Tex, 52-55; Nat Cancer Inst fel biochem, Univ Wis, 55-56; from asst prof to assoc prof, Baylor Col Med, 56-64; PROF BIOCHEM, RICE UNIV, 64- *Concurrent Pos:* USPHS sr res fel, 57-64. *Mem:* Am Soc Biol Chemists; Am Chem Soc; Am Soc Microbiol. *Res:* Enzymes involved in biosynthesis of creatine and certain antibiotics especially gentamicin, spectinomycin and streptomycin and their regulation; physiological effects of introduction of synthetic phosphagens into brain, heart, muscle and tumor cells; feedback repression during embryonic development. *Mailing Add:* Dept Biochem William Marsh Rice Univ PO Box 1892 Houston TX 77251-1892

WALKER, JAMES CALLAN GRAY, b Johannesburg, SAfrica, Jan 31, 39; m 59, 82; c 4. ATMOSPHERIC CHEMISTRY & EVOLUTION. *Educ:* Yale Univ, BS, 60; Columbia Univ, PhD(geophys), 64. *Prof Exp:* Res assoc aeronomy, Inst Space Studies, NY, 64-65; res fel, Queen's Univ, Belfast, 65-66; res assoc, Goddard Space Flight Ctr, NASA, 66-67; asst prof geol, Yale Univ, 67-70, assoc prof geophys, 70-74; sr res assoc, Nat Astron & Ionosphere Ctr, 74-80; assoc dir, Space Physics Res Lab, 80-85, PROF ATMOSPHERIC SCI, UNIV MICH, 80- *Concurrent Pos:* Adj asst prof, NY Univ, 64-65; mem, comt solar terrestrial res, Geophys Res Bd, Nat Acad Sci, 71-76 & comt on planetary & lunar exploration, Space Sci Bd, 77-78; assoc ed, J Geophys Res, 74-76; mem, Int Comn Planetary Atmospheres & Evolution, 78-; mem, Comt Planetary Biol & Chem Evolution, Space Sci Bd, 79-82; mem, Atmospheric Sci Adv Comt, NSF, 80-83; mem Planetary Atmospheres Mgt Oper Working Group, NASA, 81-85; assoc ed, EOS, 88- *Mem:* AAAS; fel Am Geophys Union; Sigma Xi; Am Asn Univ Professors; Fedn Am Scientists; Int Soc Study Origin Life; Geol Soc Am. *Res:* Aeronomy; atmospheric physics; ionospheric physics; evolution of the atmosphere. *Mailing Add:* Space Physics Res Lab Univ Mich Ann Arbor MI 48109

WALKER, JAMES ELLIOT CABOT, b Bryn Mawr, Pa, Sept 28, 26; m 65; c 1. INTERNAL MEDICINE. *Educ:* Williams Col, BA, 49; Univ Pa, MD, 53; Harvard Univ, MS, 66. *Prof Exp:* Intern, Univ Wis Hosp, 53-54; resident med, Univ Mich Hosp, 54-55; res fel, Harvard Med Sch, 57-60; sr resident, Peter Bent Brigham Hosp, 59-60, asst to assoc dir ambulatory serv, 60-65; prof med & soc, Univ Conn, 65-67, prof clin med & health care & chmn dept, 67-71, prof med & chmn, Dept Community Med & Health Care, 71-86, PROF MED & ASSSOC DIR TRAVELERS CTR AGING, SCH MED, UNIV CONN, 87- *Concurrent Pos:* Mass Heart Asn fel, 58-59; Commonwealth Fund traveling fel, 65-66; from instr to lectr, Harvard Med Sch, 60-66; dir div med care res, Dept Med, Peter Bent Brigham Hosp, 63-66; consult, Univ Wis Hosp & Univ NB Hosp, 65; chief med serv, Univ Conn Health Ctr, McCook Div & actg chief med serv, Vet Admin Hosps, Newington, Univ Conn, 69-71; pres, Can Am Health Coun, 79-87; dir, Ctr Int Community Health Studies, 81-86; dir, Geriat Assessment Clin, 89. *Mem:* AAAS; Asn Am Med Cols; fel Am Col Physicians; Am Fedn Clin Res; AMA. *Res:* Pulmonary physiology; airway temperatures; delivery of health care services; responsibilities of medical education and the university to medical care and society. *Mailing Add:* Travelers Ctr Aging Univ Conn 263 Farmington Ave Farmington CT 06032

WALKER, JAMES FREDERICK, b Riverton, Ala, July 30, 04; m 27; c 1. PHYSIOLOGY, HISTOLOGY. *Educ:* Univ Miss, AB, 27, MS, 31; Univ Iowa, PhD(zool), 35. *Prof Exp:* Instr sci, 26-30, actg head dept, 29-30, assoc prof, Miss Southern Col, 30-31; asst zool, Univ Iowa, 31-32; assoc prof sci, Miss Southern Col, 32-33 & 35-41; instr naval aviation ground sch, Boston Naval Aviation Base, 41-43; instr naval preflight sch, Univ Iowa, 43-44; from assoc prof to prof anat, Univ Calif-Calif Col Med, 44-45; prof biol, 45-72, head div biol sci, 46-57, chmn dept biol, 57-68, assoc dean arts & sci, 68-70, distinguished univ prof biol, 72-80, EMER DISTINGUISHED UNIV PROF BIOL, UNIV SOUTHERN MISS, 80- *Concurrent Pos:* Researcher zool, Univ Calif, Los Angeles, 62-63. *Mem:* AAAS; assoc Am Physiol Soc; Sigma Xi. *Res:* Experimental histology; marine biology; histology and cytology of invertebrates; physiology of vertebrates and invertebrates; histology and histochemistry of the commercial shrimp integument and the phenomena of black spotting in shrimp integument; transverse fission of two types in hydra; correlation of excessive bud formation and excessive tentacles in individual hydra. *Mailing Add:* Eight Woodland Sq Petal MS 39465

WALKER, JAMES FREDERICK, JR, b Minneapolis, Minn, July 22, 37; m 59; c 3. NUCLEAR THEORY, INTERMEDIATE ENERGY REACTIONS. *Educ:* Univ Minn, BPhys, 59, MS, 61, PhD(physics), 64. *Prof Exp:* Asst res scientist, NY Univ, 64-66; mem res staff, Mass Inst Technol, 66-68; asst prof, 68-74, ASSOC PROF PHYSICS, UNIV MASS, AMHERST, 74- *Mem:* Am Phys Soc. *Res:* Pion interactions with nuclei; nuclear reaction theories. *Mailing Add:* Dept Physics & Astron Univ Mass Amherst MA 01003

WALKER, JAMES HARRIS, b Washington, DC, Oct 13, 44; m 69; c 2. SPECTRORADIOMETRY, RADIOMETRIC PHYSICS. *Educ:* Univ Md, BS, 70. *Prof Exp:* Physics technician optical pyrometry, Nat Bur Standards, 65-69, PHYSICIST RADIOMETRY, NAT INST STANDARDS & TECHNOL, 70- *Concurrent Pos:* Consult. *Mem:* Optical Soc Am. *Res:* Metal freezing point blackbodies. *Mailing Add:* Nat Inst Standards & Technol Bldg 221 Rm A221 Gaithersburg MO 20899

WALKER, JAMES JOSEPH, b Philadelphia, Pa, Dec 29, 33; m 57; c 3. THEORETICAL PHYSICS. *Educ:* Univ NMex, BS, 59; Univ SC, PhD(physics), 65. *Prof Exp:* Gen mgr, EG&G, Inc, NMex, 65-75; GROUP LEADER NEUTRON PHYSICS, LOS ALAMOS SCI LAB, J-16, 75- *Mem:* Sigma Xi. *Res:* Integral equations; holography; nuclear physics. *Mailing Add:* PO Box 8129 Santa Fe NM 87504-8129

WALKER, JAMES KING, b Greenock, Scotland, Oct 9, 35; m 60; c 2. PARTICLE PHYSICS. *Educ:* Glasgow Univ, BSc, 57, PhD(physics), 60. *Prof Exp:* Res scientist, Advan Training Sch, Paris, 60-62; res assoc physics, Harvard Univ, 62-64, from asst prof to assoc prof, 64-69; scientist, Nat Accelerator Lab, 69-84; PROF, UNIV FLA, 85- *Concurrent Pos:* Consult, Pilot Chem Co, 67-69. *Honors & Awards:* Kelvin Award, Exp Physics, 60. *Res:* Electromagnetic and weak properties and structure of elementary particles; elementary particle physics. *Mailing Add:* Physics Dept Univ Fla Gainesville FL 32611

WALKER, JAMES MARTIN, b Jonesboro, La, Oct 21, 38; m 61; c 2. HERPETOLOGY. *Educ:* La Polytech Inst, BS, 60, MS, 61; Univ Colo, PhD(zool), 66. *Prof Exp:* Assoc prof, 65-76, PROF ZOOL, UNIV ARK, FAYETTEVILLE, 76- *Mem:* Am Soc Ichthyologists & Herpetologists; Herpetologists League. *Res:* Reptiles and amphibians of North America, with special interest in the ecology and systematics of lizards of the genus Cnemidophorus of the family Teiidae. *Mailing Add:* Dept Zool Univ Ark Fayetteville AR 72701

WALKER, JAMES RICHARD, b Boise, Idaho, Feb 26, 33; m 61; c 3. PHYSIOLOGY. *Educ:* Ariz State Univ, BS, 56; Univ Miss, PhD(physiol), 65. *Prof Exp:* Instr, 65-66, ASST PROF PHYSIOL, UNIV TEX MED BR, GALVESTON, 66-, ASST DIR INTEGRATED FUNCTIONAL LAB, 74- *Concurrent Pos:* Mem staff, Commun Sci Lab, Univ Fla, 72-73. *Mem:* Acoust Soc Am; Am Inst Physics. *Res:* Mathematical modelling of physiological systems. *Mailing Add:* Integrated Functional Lab F-39 Univ Tex Med Br Galveston TX 77550

WALKER, JAMES ROY, b Chestnut, La, Nov 8, 37; m 59; c 2. MICROBIOLOGY. *Educ:* Northwestern State Col, La, BS, 60; Univ Tex, PhD(microbiol), 63. *Prof Exp:* Nat Cancer Inst fel biochem sci, Princeton Univ, 65-67; from asst prof to assoc prof, 67-78, PROF MICROBIOL, UNIV TEX, AUSTIN, 78-, CHMN DEPT, 81- *Concurrent Pos:* Res assoc dept chem, Harvard Univ, 72-73. *Mem:* Genetics Soc Am; Am Soc Microbiol. *Res:* Microbial genetics; regulation of cell division; mechanism of DNA replication. *Mailing Add:* Dept Microbiol Univ Tex Austin TX 78712

WALKER, JAMES WILLARD, b Taylor, Tex, Mar 23, 43; m 73; c 2. EVOLUTIONARY BIOLOGY. *Educ:* Univ Tex, Austin, BA, 64; Harvard Univ, PhD(biol), 70. *Prof Exp:* From asst prof to assoc prof, 69-83, PROF BOT, UNIV MASS, AMHERST, 83-, HEAD DEPT, 86- *Honors & Awards:* George R Cooley Award, Am Soc Plant Taxon, 72. *Mem:* AAAS; Bot Soc Am; Am Soc Plant Taxon; Am Inst Biol Sci; Linnean Soc London. *Res:* Angiosperm systematics; morphology, phylogeny and evolution of primitive angiosperms; pollen morphology of primitive dicots. *Mailing Add:* Dept Bot Univ Mass Amherst MA 01003

WALKER, JAMES WILSON, b NC, July 17, 22; m 45; c 2. MATHEMATICAL STATISTICS. *Educ:* Univ NC, PhD(math statist), 57. *Prof Exp:* Intel specialist, US Dept Air Force, 50-55; asst statist, Univ NC, 55-56; from asst prof to assoc prof math, 56-64, res assoc eng exp sta, 58-65, PROF MATH, GA INST TECHNOL, 64- *Concurrent Pos:* Consult, WVa Pulp & Paper Co, 66-68 & Union Camp, Inc, 74-75. *Mem:* Am Statist Asn; Math Asn Am. *Res:* Statistical inference from grouped data; optimal grouping of statistical data; inefficiency of certain estimates based on grouped data. *Mailing Add:* Univ SC Columbia SC 29208

WALKER, JEAN TWEEDY, b Dublin, Ireland, Mar 3, 44; m 72. MICROBIAL GENETICS, ELECTRON MICROSCOPY. *Educ:* Trinity Col, Ireland, BA, 65; Univ Reading, Eng, PhD(microbiol), 71. *Prof Exp:* Res demonstr, Univ Reading, 65-67; from asst prof to assoc prof, Trinity Col, 67-72; asst res scientist, 72-75, ASSOC RES SCIENTIST, UNIV IOWA, 75- *Mem:* Soc Gen Microbiol; Am Soc Microbiol; Royal Micros; Genetics Soc of Am. *Res:* Bacterial and phage genetics and molecular biology; morphogenesis of phage; plasmids. *Mailing Add:* Dept Microbiol Univ Iowa Iowa City IA 52240

WALKER, JEARL DALTON, b Pensacola, Fla, Jan 20, 45; m 84; c 4. OPTICS. *Educ:* Mass Inst Technol, BS, 67; Univ Md, PhD(physics), 73. *Prof Exp:* Chmn, 85-89, from asst prof to assoc prof, 73-81, PROF PHYSICS, CLEVELAND STATE UNIV, 81- *Concurrent Pos:* Mem staff, Sci Am, 77-90. *Mem:* Am Asn Physics Teachers. *Res:* General physics. *Mailing Add:* Dept Physics Cleveland State Univ Cleveland OH 44115

WALKER, JERRY ARNOLD, b Olney, Ill, Mar 4, 48; m 76. SYNTHETIC ORGANIC CHEMISTRY. *Educ:* Univ Ill, BS, 69; Mass Inst Technol, PhD(org chem), 73. *Prof Exp:* Fel org chem, Univ Calif, Los Angeles, 73-74 & Calif Inst Technol, 74-75; res chemist, 75-83, RES MGR, UPJOHN CO, 83- *Mem:* Am Chem Soc; Royal Soc Chem. *Res:* Research and development of methods for the synthesis of biologically active compounds. *Mailing Add:* Upjohn Co 1510-91-1 Kalamazoo MI 49001

WALKER, JERRY C, b El Paso, Tex, Feb 20, 38; m 55; c 3. AGRONOMY, TOXICOLOGY. *Educ:* Pan Am Univ, BA, 64; Clemson Univ, PhD(agron), 69. *Prof Exp:* Plant sci rep, Lilly Res Labs, 69-73, regional res rep, 73-75, regional res mgr, 75-79, res dir, 79-82, head, 82-83, dir, 83-86, DIR TOXICOL, LILLY RES LABS, 86- *Mem:* Am Soc Agron; Soil Sci Soc Am; Weed Sci Soc Am; Coun Agr Sci & Technol; Am Chem Soc. *Res:* Agrichemical product development. *Mailing Add:* Lilly Res Labs Greenfield Labs PO Box 708 Greenfield IN 46140

WALKER, JERRY TYLER, b Cincinnati, Ohio, Sept 7, 30; m 53; c 2. PLANT PATHOLOGY. *Educ:* Miami Univ, Ohio, BA, 52; Ohio State Univ, MSc, 57, PhD, 60. *Prof Exp:* Asst, Ohio State Univ, 55-59, asst agr exp sta, 59-61; plant pathologist, Brooklyn Bot Garden, 61-69; assoc prof, 69-79, PROF PLANT PATH & HEAD DEPT, AGR EXP STA, GA STA, UNIV GA, 79- *Concurrent Pos:* NSF grant, 63-66; actg chmn res, Kitchawan Lab, NY, 67-69. *Mem:* Am Phytopath Soc; Soc Nematol; Int Soc Arboriculture. *Res:* Phytonematology, including control; diseases of ornamentals; air pollution effects on plants. *Mailing Add:* Dept of Plant Path Ga Sta Agr Exp Sta Univ Ga Griffin GA 30223-1797

WALKER, JIMMY NEWTON, b Eldorado, Okla, Mar 6, 24; m 49; c 2. CHEMICAL ENGINEERING, CHEMISTRY. *Educ:* Okla State Univ, BS, 49; Univ Chicago, MBA, 63. *Prof Exp:* Engr, US Gypsum Co, 49-50; prod foreman paint plant, Tex, 50-51; res chemist, 57-58, sect mgr joint compounds, 58-61, div mgr, Formulated Prod, 61-63 & Fiber & Formulated Prod, 63-66, dir res & develop, 66-72, VPRES RES & DEVELOP, US GYPSUM CO, LIBERTYVILLE, 72- *Concurrent Pos:* Indust Res Inst seminar comt, 69- *Mem:* Am Chem Soc; Indust Res Inst. *Res:* Development of flat latex paints; dissociation of gypsum; industrial resin tooling products; advanced joint compounds and systems; research and project evaluation. *Mailing Add:* 13033 Decant Dr Poway CA 92064-1118

WALKER, JOAN MARION, b New Plymouth, NZ, May 21, 37; NZ & Can citizen. MORPHOMETRIC ARTICULATIONS. *Educ:* Univ Man, BPT, 71, MA, 73; McMaster Univ, PhD(med sci), 77. *Prof Exp:* Lectr phys ther, Univ Toronto, 63-66, Univ Witwatersrand, 66-69, Univ Man, 70-71; lectr anat, McMaster Univ, 73-76; from asst prof to assoc prof, Growth & Develop Arthrology, Dept Phys Ther, Univ Southern Calif, 78-86; DIR PHYS THER DEPT, DALHOUSIE UNIV, 86- *Mem:* Am Phys Ther Asn; Can Physiol Ther Asn; Can Asn Phys Arthropology; AAAS; Am Asn Anatomists; Can Asn Anatomists. *Res:* Relationships between growth of the human fetal and infant hip joint and congenital hip disease; aging mechanisms in synovial joints including range of motion studies. *Mailing Add:* Sch Phys Ther Forrest Bldg 5869 University Dr Halifax NS B3H 3J5 Can

WALKER, JOHN J, b Alma, Nebr, July 4, 35; m 60. ORGANIC CHEMISTRY. *Educ:* Univ Nebr, Lincoln, BS, 58; Atlanta Univ, MS, 68; Ga Inst Technol, PhD(org chem), 73. *Prof Exp:* Instr, Ga Inst Technol, 70-73; tech dir, Dettelbach Chem Corp, 73-81; DIR RES & DEVELOP, I SCHNEID, 81- *Concurrent Pos:* Adj prof, DeKalb Community Col, 73- *Mem:* Am Chem Soc; fel Am Inst Chemists; AAAS. *Res:* Synthesis of physiologically active barbiturates. *Mailing Add:* 2952 Greenrock Trail Doraville GA 30340

WALKER, JOHN LAWRENCE, JR, b Whitewater, Wis, Dec 12, 31; m 56; c 2. PHYSIOLOGY. *Educ:* Univ Wis, BS, 56; Duke Univ, MA, 58; Univ Minn, Minneapolis, PhD(physiol), 63. *Prof Exp:* Instr physiol, Univ Minn, Minneapolis, 62-64, asst prof, 64-65; asst prof, 66-71, assoc prof, 75-76, PROF PHYSIOL, UNIV UTAH, 76- *Concurrent Pos:* USPHS fel, 64-66, res grant, 66. *Mem:* Am Physiol Soc. *Res:* Mechanism of movement of ions and molecules through membranes, especially permeation of ions and electrical properties of membranes; ion selective microelectrodes. *Mailing Add:* Dept Physiol Univ Utah Sch Med 410 Chipeta Way Salt Lake City UT 84108

WALKER, JOHN MARTIN, b Norfolk, Va, July 6, 35; m 55; c 2. WATER POLLUTION, SOIL SCIENCE. *Educ:* Rutgers Univ, BS, 57, MS, 59, PhD(agron), 61. *Prof Exp:* Asst soil fertility and plant nutrit, Purdue Univ, 57-60; NATO fel soil chem, Rothamsted Exp Sta, Eng, 61-62; res soil scientist soils lab plant indust sta, Soil & Water Conserv Res Div, Agr Res Serv, USDA, 63-72, soil scientist biol waste mgt lab, 72-74, actg chief, 75; regional sci adv wastewater, sludge & soil, Off Res & Develop, 75-77, PHYS SCIENTIST, OFF WATER PROGS OPERS, US ENVIRON PROTECTION AGENCY, 77- *Concurrent Pos:* Adj prof dept crop & soil sci, Mich State Univ, 75-77. *Mem:* Fel AAAS; Am Soc Agron; Soil Sci Soc Am; Int Soc Soil Sci; Water Pollution Control Fedn. *Res:* Utilization of sewage sludge and wastewater treatment and use on land; soil temperature effects on movement and uptake of water and ions; plant response to controlled environments. *Mailing Add:* 1419 Monroe St Washington DC 20010

WALKER, JOHN NEAL, b Erie, Pa, Feb 19, 30; m 54; c 2. AGRICULTURAL & ENVIRONMENTAL ENGINEERING. *Educ:* Pa State Univ, BS, 51, MS, 58; Purdue Univ, PhD(agr eng), 61. *Prof Exp:* Exten agr engr, Pa State Univ, 54-58; asst agr eng, Purdue Univ, 58-60; from asst prof to assoc prof, 60-66, dept chmn agr eng, 74-81, PROF AGR ENG, UNIV KY, 66-, ACTG DIR, INST MINING & MINERALS RES, 81- *Concurrent Pos:* Vis scientist, Nat Inst Agr Eng, Eng, 70. *Mem:* Fel Am Soc Agr Eng; Am Soc Eng Educ. *Res:* Environmental and structural problems associated with plant and animal structures, especially greenhouse problems. *Mailing Add:* 3220 Tates Creek Rd Lexington KY 40502

WALKER, JOHN ROBERT, b Newbern, Tenn, Nov 27, 31; m 55; c 1. ENTOMOLOGY, RESEARCH ADMINISTRATION. *Educ:* La State Univ, BS, 55, MS, 59; Iowa State Univ, PhD(entom), 62. *Prof Exp:* Asst prof, 62-65, asst to vpres res, 66-68, asst to vpres instr & res, 66-80, asst vpres acad affairs, univ syst, 80-85, ASSOC PROF ENTOM, LA STATE UNIV, BATON ROUGE, 65-, ACTG VPRES ACAD AFFAIRS, UNIV SYST, 85- *Mem:* AAAS; Entom Soc Am; Nat Conf Advan Res; Nat Coun Univ Res Adminr; Sigma Xi. *Res:* Effects of ionizing radiations on reproductive system of insects. *Mailing Add:* PO Box 16070 La State Univ Syst Baton Rouge LA 70893

WALKER, JOHN SCOTT, b Washington, DC, May 25, 44; m 70. MAGNETOHYDRODYNAMICS, MATERIALS PROCESSING. *Educ:* Webb Inst Naval Archit, BS, 66; Cornell Univ, PhD(fluid mech), 70. *Prof Exp:* Res assoc, Cornell Univ, 70-71; from asst prof to assoc prof, Univ Ill, 71-78, asst dean, Col Eng, 80-81, prof Dept Theoret & Appl Mech, 78-88, PROF, DEPT MECH & INDUST ENG, UNIV ILL, 88- *Concurrent Pos:* NSF grants, 73-91; consult, Oak Ridge Nat Lab, 78-80, IBM Res Lab, 81, Westinghouse Res & Develop Ctr, 81-83 & Monsanto Elec Mat Co, 83-85; Dept Energy contract, 84-91; Dept Defense contract, 86-91; Nat Ctr Composite Mat Res, 86-91. *Mem:* Am Acad Mech; Am Soc Mech Engrs. *Res:* Processing thick-section composite materials; growth of crystals for electronics; design of homopolar machines; fusion reactor thermal hydraulics. *Mailing Add:* Dept Mech & Indust Eng 138 Mech Eng Bldg 1206 W Green St Urbana IL 61801

WALKER, JOSEPH, b Rockford, Ill, Dec 28, 22; m 44; c 4. ANALYTICAL CHEMISTRY, ORGANIC CHEMISTRY. *Educ:* Beloit Col, BS, 43; Univ Wis, MS, 48, PhD(chem), 50. *Prof Exp:* Sr res chemist res ctr, Pure Oil Co, Ill, 50-51, proj technologist, 51-56, sect supvr phys chem, 56-58, dir anal res & serv div, 58-64, res coordr, 64-65, dir res, 65-66, assoc dir res, 66-78, vpres, Chem Res Dept, Union Oil Co Calif, Brea, 78-85; RETIRED. *Mem:* Am Chem Soc; Sigma Xi. *Res:* Petroleum technology; analysis of petroleum products; petrochemicals research; fuels development; petroleum research administration. *Mailing Add:* 406-A Pasadena Court San Clemente CA 92672-5478

WALKER, KEITH GERALD, b Carthage, Mo, Aug 22, 41; m 63; c 2. ATOMIC PHYSICS, MOLECULAR PHYSICS. *Educ:* Bethany Nazarene Col, BS, 63; Ohio State Univ, MS, 66; Univ Okla, PhD(physics), 71. *Prof Exp:* Technician, State of Ohio, summer 64; instr physics, 65-67, from asst prof to assoc prof, 67-72, PROF PHYSICS, BETHANY NAZARENE COL, 72- *Mem:* Am Asn Physics Teachers; Optical Soc Am; Am Phys Soc. *Res:* Electron-atom impact and resulting cross-sections. *Mailing Add:* Point Loma Nazarene Col 3900 Lomaland Dr San Diego CA 92106

WALKER, KELSEY, JR, b Columbus, Tex, Nov 16, 25; m 45; c 5. THEORETICAL GAS DYNAMICS, APPLIED MATHEMATICS. *Educ:* Rensselaer Polytech Inst, BAE, 50; Mass Inst Technol, SM, 52. *Prof Exp:* Res scientist, Douglas Aircraft Co, 52-54; sr scientist, Lockheed Missile Systs Div, 54-55 & Aeronutronic Systs, Inc, 56-57; sect head syst anal, Space Tech Labs, 57-61; dept mgr systs eng, Aerospace Corp, 61-66; mgr sr staff, Los Angeles Opers, 66-72, asst prog mgr site defense, Ballistic Missile Defense Bus Area, 72-75, asst mgr adv defense systs, 75, PROG MGR, BALLISTIC MISSILE DEFENSE BUS AREA, TRW SYSTS, REDONDO BEACH, 75- *Mem:* Am Inst Aeronaut & Astronaut; Sigma Xi. *Res:* Transonic gas dynamics; supersonic wing body interference; hypersonic gas dynamics; reentry body ablation theory; flight mechanics of reentry vehicles and ballistic missiles; systems analysis and design of ballistic missiles and space craft. *Mailing Add:* 5011 Casa Dr Tarzana CA 91356

WALKER, KENNETH MERRIAM, b Blaine, Ore, Mar 30, 21; m 41; c 4. BIOLOGY. *Educ:* Ore State Col, BS, 42, MS, 49, PhD(zool), 55. *Prof Exp:* From instr to asst prof biol, Univ Puget Sound, 51-57; from asst prof to prof, 57-86, EMER PROF BIOL, WESTERN ORE STATE COL, 86- *Mem:* Am Soc Mammalogists. *Res:* Vertebrate taxonomy and ecology. *Mailing Add:* 36 Walnut Dr Monmouth OR 97361

WALKER, KENNETH RUSSELL, b Spartanburg, SC, June 21, 37; div; c 3. STRATIGRAPHY, PALEOECOLOGY. *Educ:* Univ NC, Chapel Hill, BS, 59, MS, 64; Yale Univ, MPh, 67, PhD(paleoecol), 69. *Prof Exp:* From asst prof to prof, 68-82, head dept, 77-87, CARDEN PROF GEOL, UNIV TENN, KNOXVILLE, 82- *Concurrent Pos:* Spec publ co-ed Paleont Soc, 82; pres SE Sect Paleont Soc 84-85; vchmn SE Sect Paleont Soc, 82, Geol Soc Am, 85; NSF panel mem, 90-93. *Mem:* Paleont Soc; Am Asn Petrol Geol; Soc Econ Paleont & Mineral; Geol Soc Am; Int Asn Sedimentologist. *Res:* Cambro-Ordovician problems; Holocene and ancient carbonate environments; carbonate geochemistry; ancient marine organic communities; lower paleozoic paleoenvironments; trophic relationships in organic communities; invertebrate paleontology; sedimentology. *Mailing Add:* Dept Geol Sci Univ Tenn Knoxville TN 37996-1410

WALKER, LAURENCE COLTON, b Washington, DC, Sept 8, 24; m 48; c 4. SILVICULTURE, NATURAL RESOURCE POLICY. *Educ:* Pa State Univ, BS, 48; Yale Univ, MF, 49; State Univ NY, PhD(silvicult, soils), 53. *Prof Exp:* Forester, US Forest Serv, 48-51; asst, State Univ NY Col Forestry, Syracuse, 51-53; res forester, US Forest Serv, 53-54; prof silvicult res, Univ Ga, 54-63; prof forestry & dean sch, Stephen F Austin State Univ, 63-76, Hunt prof, 76-88; RETIRED. *Concurrent Pos:* Consult, Nat Plant Food Inst, & USAID; consult, forest indust, surface-mining indust, educ insts & Int Exec Serv Corps. *Honors & Awards:* William T Hornaday Gold Medal for distinguished serv in conserv, 86. *Mem:* Fel AAAS; fel Soc Am Foresters; fel Am Sci Affil. *Res:* Silvicides for hardwood control; soil-water relationships in forests; forest fertilization; natural resource policy; technology transfer; transferring technical information for ready use by professional foresters and laymen; international forestry relationships; author of one book. *Mailing Add:* Sch Forestry Stephen F Austin State Univ Nacogdoches TX 75962-6109

WALKER, LELAND J, b Fallon, Nev, Apr 18, 23; m; c 3. CIVIL ENGINEERING. *Educ:* Iowa State Univ, BS, 44. *Hon Degrees:* PhD, Mont State Univ, 83. *Prof Exp:* Lieutenant, USN, 44-46, lieutenant comdr, Civil Eng Res Lab, 51-53; mat engr, US Bur Reclamation, 46-51, construct engr, 53-55; vpres, Wenzel & Co Consult Eng, 55-58; chmn bd, Northern Eng & Testing Co, 58-88; RETIRED. *Concurrent Pos:* Dir, Mont Power Co, Entech Cos, Sletten Construct Co, Advan Technol Inst; pres, McLaughlin Res Inst, 89- *Honors & Awards:* Grinter Award, Accrediting Bd Eng & Technol, 84; Truesdail Award, Am Coun Independent Labs, 85. *Mem:* Nat Acad Eng; fel AAAS; fel Am Consult Engrs Coun; Am Soc Civil Engrs (pres, 76-77); Accrediting Bd Eng & Technol (pres, 80-83). *Mailing Add:* Consult Engr PO Box 7425 Great Falls MT 59406

WALKER, LEON BRYAN, JR, b Gulfport, Miss, June 9, 25; m 47; c 1. ANATOMY. *Educ:* Univ Houston, BS, 50, MS, 52; Duke Univ, PhD(anat), 55. *Prof Exp:* Asst biol, Univ Houston, 50-52; asst anat, Duke Univ, 53-55; instr, Sch Med, Temple Univ, 55-59; from asst prof to assoc prof, 59-71, PROF ANAT, SCH MED, TULANE UNIV, 71- *Mem:* Am Asn Anatomists. *Res:* Muscle innervation; morphology of neuromuscular spindles; stress-strain studies in tendon. *Mailing Add:* Dept Anat Tulane Univ Med Sch 1430 Tulane Ave New Orleans LA 70112

WALKER, LEROY HAROLD, b Union, Utah, Sept 24, 33; m 63; c 3. MATHEMATICS, COMPUTER SCIENCE. *Educ:* Univ Utah, BS, 55; Mass Inst Technol, SM, 57, EE, 58; Univ Calif, Los Angeles, PhD(math), 68. *Prof Exp:* Res assoc opers res ctr, Mass Inst Technol, 58-60; asst prof math,

Brigham Young Univ, 68-73; sr programmer & analyst, Univ Utah, 73-81; SYST ANALYST, INTERMOUNTAIN CONSUMER POWER ASN, 81- *Concurrent Pos:* Fac res fel, Brigham Young Univ, 69-70. *Mem:* Inst Math Statist; Math Asn Am; Am Math Soc; Asn Comput Mach. *Res:* Stopping rules for stochastic processes. *Mailing Add:* Intermountain Consumer Power Asn 8722 S 300 W Sandy UT 84070-1419

WALKER, LOREN HAINES, b Bartow, Fla, Sept 25, 36; m 61; c 3. HIGH POWER ELECTRONICS, AC MOTOR DRIVES. *Educ:* Univ Fla, BEE, 58; Mass Inst Technol, MS, 61. *Prof Exp:* Design engr, Spec Control Dept, Gen Elec Co, 58-60, sr design engr, 61-70; develop engr, Exide Power Systs, Div ESB Inc, 70-72; elec engr Gen Elec Corp Res & Develop, 72-76, CONSULT ENG, DRIVE SYSTS DEPT, GEN ELEC CORP, 76- *Honors & Awards:* IR 100 Award, Indust Res Inc, 74. *Mem:* Fel Inst Elec & Electronics Engrs; Indust Appln Soc; Power Eng Soc. *Res:* Development, design and research of solid state power conversion equipment including variable frequency drives, cycloconverters, un-interruptible power supplies, power conversion for utility energy storage and reactive power controllers. *Mailing Add:* Gen Elec Drive Systs - Rm 500 2823 Titleist Dr Salem VA 24153

WALKER, M LUCIUS, JR, b Washington, DC, Dec 16, 36; m 60; c 2. MECHANICAL ENGINEERING. *Educ:* Howard Univ, BSME, 57; Carnegie Inst Technol, MSME, 58, PhD(mech eng), 66. *Prof Exp:* Teaching asst, Carnegie Inst Technol, 57-58; instr eng, Howard Univ, 58-59 & Carnegie Inst Technol, 61-63; from asst prof to assoc prof eng, 63-68, asst dean, 65-66, actg head dept, 66-67, head, 67-68, assoc dean, Sch Eng, 73-74, actg dean, 77-78, PROF MECH ENG, HOWARD UNIV, 70-, DEAN, SCH ENG, 78- *Concurrent Pos:* Vis sr staff mem, Int Res & Technol Corp, Washington, DC, 69-70; Ford teaching fel, Carnegie Inst Technol; mem eng manpower comn, Accrediting Bd Eng & Technol & Biotechnol Resources Rev Comt, Nat Inst Health, 72-; mem bd trustees, Carnegie-Mellon Univ. *Mem:* Sigma Xi; NY Acad Sci; Am Soc Mech Engrs; Am Soc Eng Educ. *Res:* New transportation systems planning and economics; cardiovascular mechanics. *Mailing Add:* Sch Eng Howard Univ 2400 6th St NW Washington DC 20059

WALKER, MARSHALL JOHN, optics; deceased, see previous edition for last biography

WALKER, MARY CLARE, b San Francisco, Calif. HISTOCOMPATIBILITY TESTING, HUMAN MONOCLONAL ANTIBODIES. *Educ:* Univ Tex, El Paso, BS, 67; NY Univ, PhD(med sci), 75. *Prof Exp:* Res asst biochem, Res Inst, Hosp Joint Dis, Mt Sinai Sch Med, 68-69; adj lectr biol, Bernard Baruch Col, City Univ New York, 74-75; fel immunol, dept path, Sch Med, NY Univ, 75-80, res asst prof, 80-81; ASSOC PROF IMMUNOL & DIR HISTOCOMPATIBILITY LAB, IMMUNOL RES CTR, INST ARMAND-FRAPPIER, UNIV QUE, 81- *Concurrent Pos:* Vis scientist trainee, tissue typing lab, Mem Sloan-Kettering Cancer Ctr, 81-82; mem, Nat Health Res & Develop Prog Rev Comt, Health & Welfare Can, 84- *Mem:* Am Asn Immunologists; Am Soc Histocompatibility & Immunogenetics; Can Soc Immunol; NY Acad Sci. *Res:* Production of human monoclonal antibodies against histocompatibility locus antigens; primary and secondary in vitro immunization of human B lymphocytes; chemotherapy of AIDS. *Mailing Add:* Inst Armand Frappier 531 Blvd des Prairies CP100 Laval PQ H7N 4Z3 Can

WALKER, MERLE F, b Pasadena, Calif, Mar 3, 26; m 59; c 1. ASTRONOMY. *Educ:* Univ Calif, AB, 49, PhD(astron), 52. *Prof Exp:* Asst astron, Univ Calif, 49-52, jr res astronr, 55-56; Carnegie fel, Mt Wilson & Palomar Observ, 52-54; res assoc, Yerkes Observ, Univ Chicago, 54-55; instr, Warner & Swasey Observ, Case Inst Technol, 56-57; from asst astronr to assoc astronr, 57-71, PROF ASTRON & ASTRONR, LICK OBSERV, UNIV CALIF, SANTA CRUZ, 71- *Concurrent Pos:* Sr resident astronr, Cerro Tololo Interam Observ, 68-69. *Honors & Awards:* Helen B Warner Prize Astron, Am Astron Soc, 58. *Mem:* Int Astron Union; Am Astron Soc; Sigma Xi. *Res:* Photoelectric photometry of short period variable stars; photoelectric magnitudes and colors of stars; stellar spectra and radial velocities; electronic image intensification; astronomical seeing and observatory sites. *Mailing Add:* Lick Observ Univ Calif Santa Cruz CA 95064

WALKER, MICHAEL BARRY, b Regina, Sask. PHYSICS. *Educ:* McGill Univ, BEng, 61; Oxford Univ, PhD, 65. *Prof Exp:* Fel, 66-68, assoc prof, 68-77, PROF PHYSICS, UNIV TORONTO, 77- *Honors & Awards:* Herzberg Medal, 77; Rhodes Scholar. *Mem:* Can Asn Physicists. *Res:* Theoretical solid state physics. *Mailing Add:* Dept Physics Univ Toronto Toronto ON M5S 1A7 Can

WALKER, MICHAEL DIRCK, b New York, NY, Jan 24, 31; m 53; c 3. NEUROSURGERY. *Educ:* Yale Univ, BA, 56; Boston Univ, MD, 60. *Prof Exp:* Intern surg, Mass Mem Hosps, 60-61; resident neurosurg, Boston City Hosp & Lahey Clin, 61-65; sr investr pharmacol, Nat Cancer Inst, 65-67; chief sect neurosurg, Baltimore Cancer Res Ctr, 67-80, chief ctr, 71-80; DIR STROKE & TRAUMA PROG, NAT INST NEUROL & COMMUN DIS & STROKE, NIH, 80- *Concurrent Pos:* Chmn, Brain Tumor Study Group, 67; assoc dir, Div Cancer Treatment, Nat Cancer Inst, 73-; assoc prof neurosurg, Sch Med, Univ Md, 73-; asst prof neurol surg, Sch Med, Johns Hopkins Univ, 74- *Mem:* Am Asn Cancer Res; Am Acad Neruol; Am Soc Clin Oncol; Cong Neurol Surg; NY Acad Sci; Sigma Xi. *Res:* Neurological surgery; analysis and treatment of human brain tumors with cytotoxic agents able to penetrate the blood-brain barrier. *Mailing Add:* Div Stroke & Trauma Nat Inst Neurol Dis & Stroke Fed Bldg Rm 812 Bethesda MD 20892

WALKER, MICHAEL STEPHEN, b Hull, Eng, Sept 13, 40; m 66; c 2. PHOTOPHYSICS. *Educ:* Univ Sheffield, BS, 62, PhD(chem), 65. *Prof Exp:* AEC fel dept chem, Univ Minn, Minneapolis, 65-67; scientist res labs, 67-71, Mgr Labs, Xerox Corp, 71- *Mem:* Inst Elec & Electronics Engrs. *Res:* Photophysics of organic molecules including semiconductors and photoconductors; materials characterization. *Mailing Add:* 12/2 Kandaogawamachi Chiyoba/ku Tokyo 101 Japan

WALKER, MICHAEL STEPHEN, b Detroit, Mich, Dec 16, 39; m 70; c 2. ENGINEERING & SOLID STATE PHYSICS. *Educ:* Mass Inst Technol, BS, 61; Carnegie Inst Technol, MS, 64; Carnegie Mellon Univ, PhD(physics), 71. *Prof Exp:* Sr engr, Westinghouse Elec Corp, 61-75; TECH DIR, MAGSTREAM, INTERMAGNETICS GEN CORP, 76- *Mem:* Am Phys Soc; Inst Elec & Electronics Engrs; Soc Mining Engrs. *Res:* Research and development on superconducting materials and devices, including the development of niobium-tin multifilament conductors and superconducting magnets for machinery, energy storage and for plasma confinement for fusion devices; mineral separation with magnetic fluids. *Mailing Add:* Intermagnetics Gen Corp PO Box 566 Guilderland NY 12084

WALKER, NATHANIEL, b Cincinnati, Ohio, Apr 9, 09; m 34; c 2. FOREST MANAGEMENT, FOREST ECONOMICS. *Educ:* Colo Col, BS, 33; Pa State Univ, MS, 55; NC State Univ, PhD, 70. *Prof Exp:* Dist forest ranger, US Forest Serv, 33-44; asst exten forester, Exten Serv, 44-47, from assoc prof to prof forestry, 47-74, EMER PROF FORESTRY, OKLA STATE UNIV, 74- *Mem:* Fel Soc Am Foresters. *Res:* Frequency analyses and measurements; development and yield of cottonwood and red cedar; stand structures and valuation in several species of forest trees; forest soil-site evaluations. *Mailing Add:* 5505 W 19th St Apt 249 Stillwater OK 74074-1323

WALKER, NEIL ALLAN, b Flint, Mich, May 4, 24; m 53; c 2. ACAROLOGY. *Educ:* Southern Methodist Univ, BS, 47; Univ Mich, MA, 48; Univ Calif, Berkeley, PhD(entom), 64. *Prof Exp:* Fisheries res technician inst fisheries res, Mich Dept Conserv, 49-50; from asst prof to assoc prof zool, 58-67, chmn dept biol sci & agr, 70-73, prof biol, 67-80, prof biol, FT Hays State Univ, 80-83; RETIRED. *Concurrent Pos:* Consult, Dept Sci & Indust Res, NZ, 65 & 72. *Mem:* Entom Soc Am; Acarol Soc Am. *Res:* Taxonomy; biogeography; ecology of ptyctimous Oribatei; biology of Opiliones. *Mailing Add:* Box No 4 Galesburg KS 66740-0004

WALKER, PETER ROY, b Batley, Eng, Nov 24, 45; m 81; c 1. BIOCHEMISTRY, MOLECULAR BIOLOGY. *Educ:* Univ Sheffield, BSc Hons, 67, PhD(biochem), 70. *Prof Exp:* Damon Runyon Mem Fund fel oncol, McArdle Lab, Univ Wis, 70-73; hon lectr biochem, Univ Sheffield, 73-75; RES OFFICER BIOL, NAT RES COUN, 75- *Concurrent Pos:* Med Res Coun grant, Brit Empire Cancer Campaign grant & Wellcome Found grant, Univ Sheffield, 73-75. *Res:* Mechanisms of gene expulsion; changes in chromosome structure and action of transcription factors; molecular and cell biological aspect of cell growth and cell death. *Mailing Add:* Molecular Cell Biol Group Bldg M54 Montreal Rd Ottawa ON K1A 0R6 Can

WALKER, PHILIP CALEB, b Pittsburgh, Pa, Nov 26, 11; m 40; c 4. BIOLOGY, PALYNOLOGY. *Educ:* Univ Pittsburgh, BS, 34, PhD, 58. *Prof Exp:* Asst bot & biol, Univ Pittsburgh, 34-40; park naturalist, Bur Parks, Pittsburgh, Pa, 40-43; asst prof biol, WVa Inst Technol, 46-51; prof 51-82, EMER PROF BIOL, STATE UNIV NY COL PLATTSBURGH, 82- *Concurrent Pos:* Instr bot, Geneva Col, 37-38; panelist instrnl sci equip prog, NSF. *Mem:* AAAS. *Res:* Forest ecology and sequence; palynology and forest sequence studies in Pleistocene and post-Wisconsin glacial bogs in New York, Pennsylvania and New Jersey; forest sequence of Hartstown bog area in Pennsylvania; palynology of Adirondack bogs and anemophilous and zoophilous pollen transport in the atmosphere. *Mailing Add:* RD 1 Box 516 Plattsburgh NY 12902

WALKER, PHILIP L(EROY), JR, b Baltimore, Md, Jan 10, 24; m 49; c 3. MATERIALS SCIENCE. *Educ:* Johns Hopkins Univ, BS, 47, MS, 48; Pa State Univ, PhD(fuel technol), 52. *Prof Exp:* Control chemist, Lever Bros Co, 48-49; from asst prof to assoc prof, Pa State Univ, University Park, 52-55, head dept fuel technol, 54-59, chmn div mineral technol, 59-65, head dept mat sci, 67-78, prof fuel technol, 55-74, Evan Pugh prof mat sci, 74-83, EMER PROF, PA STATE UNIV, UNIVERSITY PARK, 83- *Concurrent Pos:* Chmn, Am Carbon Comt, 62-70; ed, Chem Physics of Carbon, 62-81; assoc ed, Carbon, 64-81. *Honors & Awards:* Henry H Storch Award, Am Chem Soc, 69; George Skakel Mem Award, 71. *Mem:* Am Chem Soc; Am Phys Soc; Am Carbon Soc; Sigma Xi. *Res:* Catalysis; adsorption; kinetics; crystal growth; solid state chemistry; heterogeneous reactions; pollution control; carbon and coal science. *Mailing Add:* 223 Acad Projs Bldg Pa State Univ University Park PA 16802

WALKER, RAYMOND JOHN, b Los Angeles, Calif, Oct 26, 42. SPACE PHYSICS. *Educ:* San Diego State Univ, BA, 64; Univ Calif, Los Angeles, MS, 69, PhD(planetary & space physics), 73. *Prof Exp:* Res assoc physics, Univ Minn, Minneapolis, 73-77; from asst researcher to assoc researcher, 77-84, RESEARCHER GEOPHYS, UNIV CALIF, LOS ANGELES, 84- *Concurrent Pos:* Mem, Comt Data Mgt & Computation, Space Sci Bd, Nat Res Coun, 81-86, Comt Geophys Data, Comn Phys Sci, Math & Res, 84-87, Comt NASA Info Systs, Bd Telecommun & Computer Appln, Comn Eng & Tech Systs, 86, Comt Solar & Space Physics, Space Studies Bd, Nat Res Coun, 90-; mgr, Planetary Plasma Interactions Node, NASA Planetary Data Syst, 89- *Mem:* Am Geophys Union; AAAS. *Res:* Magnetospheric physics; the magnetospheres of the earth and Jupiter; the dynamics of charged particles in the magnetosphere; numerical studies of magnetospheric convection; quantitative modeling of magnetospheric magnetic fields; organization and analysis of multi-parameter satellite data sets; magnetohydrodynamic simulation of magnetospheric processes. *Mailing Add:* Inst Geophys Univ Calif Los Angeles CA 90024-1567

WALKER, RICHARD BATTSON, b Tennessee, Ill, Oct 24, 16; m 40; c 3. BOTANY. *Educ:* Univ Ill, BS, 38; Univ Calif, PhD(bot), 48. *Prof Exp:* From instr to assoc prof, 48-60, chmn dept, 62-71, dir biol educ, 75-82, prof, 60-87, EMER PROF BOT, UNIV WASH, 87- *Mem:* AAAS; Ecol Soc Am; Bot Soc Am; Am Soc Plant Physiologists. *Res:* Mineral nutrition and water relations of conifers; comparative calcium-magnesium nutrition; iron nutrition. *Mailing Add:* Dept Bot Univ Wash Seattle WA 98195

WALKER, RICHARD DAVID, b Washington, DC, Feb 19, 31; m 53; c 3. CIVIL ENGINEERING. *Educ:* Univ Md, BS, 53; Purdue Univ, MSCE, 55, PhD(civil eng), 61. *Prof Exp:* Asst civil eng, Purdue Univ, 53-55; from instr to assoc prof, 57-68, actg head dept, 69-70, head dept, 70-83, PROF CIVIL ENG, VA POLYTECH INST & STATE UNIV, 68- *Concurrent Pos:* Mem comt A2-E1 & A2-H03 & chmn comt A3-E5, Transp Res Bd, Nat Acad Sci-Nat Res Coun, 70-76. *Mem:* Am Soc Civil Engrs; Am Soc Eng Educ; Am Soc Testing & Mat. *Res:* Highway materials; durability of concrete; design of flexible pavements; lime-stabilization; identification of aggregates causing poor concrete performance when frozen. *Mailing Add:* 701 Broce Dr NW Blacksburg VA 24060

WALKER, RICHARD E, b Cincinnati, Ohio, Dec 24, 23; m 46; c 2. CHEMICAL ENGINEERING, RHEOLOGY. *Educ:* Purdue Univ, BS, 45; Bucknell Univ, MS, 48; Iowa State Col, PhD(chem eng), 52. *Prof Exp:* Res chem engr, Stand Register Co, 46-47; from asst to instr, Bucknell Univ, 47-48; res assoc, Iowa Eng Exp Sta, 48-52; res engr, Jersey Prod Res Co, Stand Oil Co NJ, 52-63; PROF CHEM ENG, LAMAR UNIV, 63- *Mem:* Am Inst Chem Engrs; Am Inst Mining, Metall & Petrol Engrs; Soc Rheology. *Res:* Means of drilling for and producing oil; rheology of elastic and non-elastic liquids; non-Newtonian flow applications. *Mailing Add:* Dept Chem Eng Lamar Univ Beaumont TX 77710

WALKER, RICHARD FRANCIS, b South Amboy, NJ, Dec 14, 39. GERONTOLOGY, REPRODUCTIVE ENDOCRINOLOGY. *Educ:* Rutgers Univ, BS, 61, PhD(endocrinol), 72; NMex State Univ, MS, 68. *Prof Exp:* Asst prof biol, Clemson Univ, 72-76; trainee geront, Duke Univ, 77-78; fel neuroendocrinol, Univ Calif, Berkeley, 78-80; ASST PROF ANAT, MED CTR, UNIV KY, 80- *Mem:* Endocrine Soc; Am Asn Anatomists; Geront Soc; Int Soc Develop Neurosci. *Res:* Role of hypothalamic monoamines, specifically in suprachismatic nucleus, in aging of the female reproductive system. *Mailing Add:* Dept Biol Sterling Col Sterling KS 67579

WALKER, RICHARD IVES, b Portsmouth, Va, Nov 2, 42; m 67; c 2. IMMUNOMODULATION, NUCOSAL IMMUNITY. *Educ:* Tex Christian Univ, BS, 69, MS, 68; Univ NH, PhD, 73. *Prof Exp:* Microbiologist, Naval Radiol Defense Lab, 68-69; head, Bacteriol Div, Naval Med Res Unit Two, 69-71; microbiologist, Armed Forces Radiobiol Res Inst, 73-78; microbiologist, 78-79, head Enteric Diseases div, 80, head, Med Microbiol br, 80-82, dep dir, Infectious Dis, Naval Med Res Inst, 82-84; DEP DIR, ARMED FORCES RADIOBIOL RES INST, 84- *Mem:* Am Soc Microbiol; Soc Intestinal Microbiol Ecol & Dis; Sigma Xi. *Res:* Management of opportunistic infections and protection against enteric infections; understanding mechanisms and use of immunomodulators and means to enhance nucosal immunity. *Mailing Add:* 19404 Faber Ct Gaithersburg MD 20879

WALKER, RICHARD V, b Pueblo, Colo, Mar 8, 18; m 45; c 3. MEDICAL BACTERIOLOGY, IMMUNOLOGY. *Educ:* Univ Calif, Berkeley, BS, 49, MPH, 52, PhD(bact), 60. *Prof Exp:* Assoc & instr, Pub Health Lab, Sch Pub Health, Univ Calif, Berkeley, 49-57; from grad res immunologist to asst res immunologist, George Williams Hooper Found, Med Ctr, Univ Calif, San Francisco, 57-65; asst res immunologist, Nat Ctr Primate Biol, Univ Calif, Davis, 65-67; ASSOC PROF ZOOL, OHIO UNIV, 67- *Mem:* Am Soc Microbiol. *Res:* Bacterial toxins; immunochemistry; fluorescent antibody; Salmonella-Shigella diagnosis. *Mailing Add:* Dept Zool & Microbiol Ohio Univ Athens OH 45701

WALKER, ROBERT BRIDGES, b Houston, Tex, Sept 24, 46; m 71. THEORETICAL CHEMISTRY, CHEMICAL PHYSICS. *Educ:* La State Univ, New Orleans, BS, 68; Univ Tex, Austin, PhD(chem), 73. *Prof Exp:* Res assoc, James Franck Inst, Univ Chicago, 73-76; appointee, 76-77, STAFF MEM, LOS ALAMOS NAT LAB, 77- *Mem:* AAAS; Am Phys Soc. *Res:* Quantum reactive scattering of light atom-diatom systems; quantum and classical description of infrared multiple photon excitation dynamics of polyatomic molecules. *Mailing Add:* Group T-12 MS B268 PO Box 1663 Los Alamos Nat Lab Los Alamos NM 87545

WALKER, ROBERT D(IXON), JR, b Atlanta, Ga, Mar 6, 12; m 35; c 3. CHEMICAL ENGINEERING. *Educ:* Ga Inst Technol, BS, 35; Univ Fla, MS, 51. *Prof Exp:* Res chemist, Eastman Kodak Co, NY, 35-44; prof, 44-82, EMER PROF, CHEM ENG, UNIV FLA, 82- *Honors & Awards:* Sigma Xi, 46. *Mem:* Am Chem Soc; Electrochem Soc; Soc Petrol Engrs; Am Inst Chem Engrs. *Res:* Adsorption fractionation; electrochemistry and thermodynamics of fused salt systems; transport phenomena; electrochemical engineering; solubility and diffusion in biological systems; fuel cells; thermal batteries; enhanced oil recovery. *Mailing Add:* 4740 NW 20th Pl Gainesville FL 32605

WALKER, ROBERT HUGH, b O'Donnell, Tex, Sept 8, 35; m 55; c 4. PHYSICS. *Educ:* Tex Christian Univ, BS, 57, MS, 59; Mass Inst Technol, PhD(physics), 62. *Prof Exp:* Asst prof physics, 64-67, assoc dean Col Arts & Sci, 73-74, dean Col Natural Sci & Math, 74-82, interim chancellor, 82-83, ASSOC PROF PHYSICS, UNIV HOUSTON, 67-, VPRES ACAD AFFAIRS, 84- *Concurrent Pos:* Lectr, Baylor Col Med, 66-; consult, Int Inst Educ, 67-69. *Mem:* Am Phys Soc; Am Asn Physics Teachers. *Res:* Theoretical physics; solid state physics; atomic physics. *Mailing Add:* 10930 Chimney Rock Houston TX 77096

WALKER, ROBERT LEE, b St Louis, Mo, June 29, 19; m 46. PHYSICS. *Educ:* Univ Chicago, BS, 41; Cornell Univ, PhD(exp physics), 48. *Prof Exp:* Asst metall lab, Univ Chicago, 42-43; scientist, Los Alamos Sci Lab, 43-46; res assoc, Cornell Univ, 48-49; from asst prof to assoc prof, 49-59, PROF PHYSICS, CALIF INST TECHNOL, 59- *Mem:* Am Phys Soc; Sigma Xi. *Res:* Photoproduction experiments and analyses; interaction of gamma rays with matter; high energy physics. *Mailing Add:* 200 Barbara St Frederick MD 21701

WALKER, ROBERT MOWBRAY, b Philadelphia, Pa, Feb 6, 29; m 51, 73; c 2. SPACE PHYSICS. *Educ:* Union Univ, NY, BS, 50; Yale Univ, MS, 51, PhD(physics), 54. *Hon Degrees:* DSc, Union Univ, NY, 67; Dr, Univ Clermont-Ferrand, 75. *Prof Exp:* Res assoc, Gen Elec Co, 54-66; dir lab space physics, 66-75, MCDONNELL PROF PHYSICS, WASH UNIV, 66-, DIR, MCDONNELL CTR SPACE SCI, 75- *Concurrent Pos:* NSF sr fel, 62; vis prof, Univ Paris, 62-63 & Calif Inst Technol, 72; adj prof, Rensselaer Polytech Inst, 65-66; mem, Lunar Sample Anal Planning Team, 68-70 & Lunar Sample Rev Bd, 70-72; mem bd dirs, Vols Tech Assistance & Univs Space Res Asn, 69-71; mem, Lunar Sci Inst Adv Comt, 72-76, Space Sci Bd, Nat Sci, bd sci & tech for int develop, 74-77, comt lunar & planetary explor, 77-80; vis phys res lab, Ahmedabad, India & Inst Astron, Paris, 81; mem, task force on sci uses of space sta, 85-, meteorite working group, 85-88, chmn, 90-92, NASA Planetary Geosci Strategy Comn, 86; mem, Org Comt Soc, Europ Sci Found, 89. *Honors & Awards:* Am Nuclear Soc Award, 64; Yale Eng Asn Award, 66; NASA Medal Except Sci Achievement, 70; E O Lawrence Award, AEC, 71; Lawrence Smith Medal, Nat Acad Sci, 91. *Mem:* Nat Acad Sci; fel Am Phys Soc; fel Meteoritical Soc; fel Am Geophys Union; Am Astron Soc; fel AAAS. *Res:* Radiation effects in solids; development of dielectric nuclear track detectors and their application to nuclear science; geochronology; space science; cosmic rays; meteorites; astrophysics; planetary surfaces; archeometry; laboratory studies of interplanetary dust. *Mailing Add:* Dept Physics Wash Univ One Brookings Dr Box 1105 St Louis MO 63130

WALKER, ROBERT PAUL, b Washington, DC, Mar 15, 43; m 65; c 2. MATHEMATICS. *Educ:* Univ Md, BS, 65; Mass Inst Technol, PhD(math), 68. *Prof Exp:* Asst prof math, Univ NC, Chapel Hill, 68-73; prof & chmn dept, Talladega Col, 73-76; assoc prof & dir, Bowie State Col, 76-79; assoc dir, Syst Planning Corp, 79-86; DESIGNER, DEWBERRY DAVIS CO, 86- *Mem:* Am Math Soc; Math Asn Am; Nat Asn Mathematicians. *Res:* Systems engineering, structured analysis and software systems design and development; operations research; strategic systems analysis. *Mailing Add:* Dewberry Davis 8401 Arlington Blvd Fairfax VA 22031

WALKER, ROBERT W, b Arlington, Mass, Mar 15, 33; m 59; c 1. ENVIRONMENTAL SCIENCES. *Educ:* Univ Mass, BS, 55, MS, 59; Mich State Univ, PhD(microbiol), 63. *Prof Exp:* Hatch fel, 63-64; instr, 64-65, asst prof, 65-74, ASSOC PROF ENVIRON SCI, UNIV MASS, AMHERST, 74- *Concurrent Pos:* Vis prof, res Univ Toulouse, France, 72-73, Univ Otago, Dunedin, NZ, 80, INSA Univ Toulouse, 90, Univ El Salvador, San Salvador, 91. *Honors & Awards:* co-recipient DIFCO Lab Award, Am Pub Health Asn, 79. *Res:* Environmental microbiology; bioremediation. *Mailing Add:* Environ Sci Skinner Hall Univ Mass Amherst MA 01003

WALKER, ROBERT WINN, b Montgomery, Ala, Jan 5, 25; m 49; c 3. PHYSICAL CHEMISTRY. *Educ:* Auburn Univ, BS, 48; Mass Inst Technol, PhD(phys chem), 52. *Prof Exp:* Res chemist, Redstone Labs, Rohm & Haas Co, 52-53, group leader propellant res, 53-59, sect head phys & polymer chem, 59-65, lab head ion exchange appln res div, 65-73, mgr ion exchange res dept, 73-76, proj leader fluid process chem res, 76-83; LECTR, PA STATE UNIV, 84- *Mem:* Am Chem Soc; Am Inst Chemists. *Res:* Molecular structure; chemical thermodynamics; rocket propulsion; ion exchange resins; adsorbents; flocculants. *Mailing Add:* 6008 Cannon Hill Rd Ft Washington PA 19034-1802

WALKER, ROGER GEOFFREY, b London, Eng, Mar 26, 39; m 65; c 2. SEDIMENTOLOGY. *Educ:* Oxford Univ, BA, 61, DPhil(geol), 64. *Prof Exp:* NATO fel geol, Johns Hopkins Univ, 64-66; from asst prof to assoc prof, 66-73, PROF GEOL, MCMASTER UNIV, 73- *Concurrent Pos:* Vis scientist, Denver Res Ctr, Marathon Oil Co, 73-74, Amoco Can Petrol Co, 82; distinguished lectr, Am Asn Petrol Geologists, 79-80; vis fel, Australian Nat Univ, 81; vis prof, Fed Univ Ouro Preto, Brazil, 87, 89 & 90. *Honors & Awards:* Past Pres' Medal, Geol Asn Can, 75; Link Award, Can Soc Petrol Geologists, 83, R J W Douglas Mem Medal, 90; Judd A & Cynthia S Oualline Centennial lectr, Univ Tex, Austin, 86. *Mem:* Soc Econ Paleont & Mineral; Int Asn Sedimentol; Am Asn Petrol Geologists; Geol Asn Can; fel Royal Soc Can. *Res:* Sedimentary facies analysis; sedimentology of turbidites; quantitative basin analysis; sedimentology of Western Canadian Cretaceous clastic wedge. *Mailing Add:* Dept Geol McMaster Univ 1280 Maint St W Hamilton ON L8S 4M1 Can

WALKER, ROLAND, b Stellenbosch, SAfrica, Feb 8, 07; US citizen; wid; c 2. TOXICOLOGY. *Educ:* Oberlin Col, AB, 28, AM, 29; Yale Univ, PhD(zool), 34. *Prof Exp:* Asst biol, Yale Univ, 29-31; instr physiol, Oberlin Col, 31-32; instr biol, 34-42, from asst prof to assoc prof, 42-54, prof, 54-72, EMER PROF BIOL, RENSSELAER POLYTECH INST, 72- *Mem:* AAAS; Wildlife Dis Asn. *Res:* Neurology of fish and crustacea; fish parasitology; ultrastructure of fish blood and tumor cells with virus. *Mailing Add:* Dept Biol Rensselaer Polytech Inst Troy NY 12180-3590

WALKER, RONALD ELLIOT, physics, physical chemistry, for more information see previous edition

WALKER, RUSSELL GLENN, b Cincinnati, Ohio, May 3, 31; m; c 2. ASTRONOMY, INFRARED PHYSICS. *Educ:* Ohio State Univ, BSc, 53, MSc, 54; Harvard Univ, PhD(astron), 67. *Prof Exp:* Staff scientist Fourier spectros, Block Assocs, 60-61; physicist infrared physics, Air Force Cambridge Res Labs, 61-66, chief, Infrared Physics Br, 67-75; astronr, Int Sci Sta, Jungfraujoch, Switz, 66-67; staff scientist astrophys, Ames Res Ctr, NASA, 75-81; ASSOC SCIENTIST, JAMIESON SCI & ENG, INC, 81 - *Concurrent Pos:* Consult, Smithsonian Astrophys Observ, 65-66 & TOM subgroup, Dir of Defense Res & Eng Reentry Progs, 68-69; mem, Infrared Panel Astron Study Group, Nat Acad Sci, 71-72; mem, Space Sci Bd Infrared & Submillimeter Astron, 74-75. *Honors & Awards:* Medal for Except Sci Achievement, NASA. *Mem:* Am Astron Soc; fel Optical Soc Am. *Res:* Infrared astronomy; cryogenically cooled telescopes and instruments for space research; infrared sky surveys, atmospheric infrared phenomena. *Mailing Add:* PO Box F1 Felton CA 95018

WALKER, RUSSELL WAGNER, b Fredericktown, Ohio, Nov 14, 24; m 46; c 3. ORGANIC CHEMISTRY, PHYSICAL CHEMISTRY. *Educ:* Ohio Wesleyan Univ, BS, 47; Ohio State Univ, PhD(org chem), 52. *Prof Exp:* Res assoc, Am Petrol Inst, 48-52; res chemist, Sinclair Res, Inc, 53-57, group leader, 57-60, div dir, 60-66, res dir, Sinclair Petrochem, Inc, 66-67, tech mgr, Sinclair Res, Inc, 67-68, vpres & dir assoc opers, Sinclair Petrochem, 68-69, mgr res & develop, Sinclair-Koppers Co, 69-74; DIR RES & DEVELOP, ARCO POLYMERS INC, 74- *Mem:* AAAS; Am Chem Soc; Indust Res Inst; Sigma Xi. *Res:* Relation of hydrocarbon structure to combustion characteristics; biodegradation and environmental pollution. *Mailing Add:* Dept Neurol Mem Hosp 1275 York Ave New York NY 10021

WALKER, RUTH ANGELINA, b New York, NY, July 11, 20. ORGANIC CHEMISTRY. *Educ:* Vassar Col, BA, 42; Yale Univ, PhD(org chem), 45. *Prof Exp:* Asst chem, Chas Pfizer & Co, NY, 45; asst chemist, Col Med, NY Univ, 45-50; sr res chemist, Celanese Corp Am, 50-56; sr res chemist, Johnson & Johnson, 57; instr chem, Hunter Col, Bronx, 57-60, asst prof, 61-68, assoc prof, Lehman Col, 69-71, dir, Health Prof Inst, 77-78, prof chem, 71-84, assoc dean health professions, 79-84, EMER PROF, LEHMAN COL, CITY UNIV NEW YORK, 85- *Concurrent Pos:* Sigma Delta Epsilon grant-in-aid metal complexes hydroxyanthraquinones, Hunter Col, 63, Sigma Xi grant-in-aid res, 65, George N Shuster fel grant, 65, City Univ New York res grant, 65; chmn, Lehman Col Comt Curric, 68-77. *Honors & Awards:* Award, Am Asn Textile Chem & Colorists, 60. *Mem:* Sigma Xi; Am Chem Soc; fel NY Acad Sci; Sci Res Soc Am. *Res:* Synthesis of medicinal products; dyestuff synthesis; organometallic complexes of 1,4-dihydroxynathraquinones; methods of teaching; development of interdisciplinary programs to teach the team delivery of health care. *Mailing Add:* 3300 Darby Rd-Pine 7310 Quadrang Haverford PA 19041-1095

WALKER, SHARON LESLIE, b Orange, NJ, May 7, 58. SUSPENDED PARTICLE DETECTION, SEDIMENT TRANSPORT. *Educ:* Univ Wash, BS, 81. *Prof Exp:* OCEANOGR, PAC MARINE ENVIRON LAB, NAT OCEANIC & ATMOSPHERIC ADMIN, DEPT COM, 79- *Mem:* Am Geophys Union. *Res:* Analysis of suspended particle size distributions; chemistry and transport of suspended particles in marine estuarine environments and in hydrothermal vent plumes. *Mailing Add:* Nat Oceanic & Atmospheric Admin PMEL-R-E-PM 7600 Sandpoint Way NE Bldg No 3-Bin C15700 Seattle WA 98115

WALKER, SHEPPARD MATTHEW, b Perkinston, Miss, Feb 2, 09; m 32; c 1. PHYSIOLOGY, BIOPHYSICS. *Educ:* Western Ky Univ, BS, 32, AM, 33; La State Univ, PhD(physiol), 41. *Prof Exp:* Instr biol, Perkinston Jr Col, 33-38; asst prof sci, Delta State Teachers Col, Miss, 41-42; from instr to asst prof physiol, Sch Med, Wash Univ, 42-49; assoc prof, 49-62, prof, 62-74, EMER PROF PHYSIOL, SCH MED, UNIV LOUISVILLE, 74- *Concurrent Pos:* Mem, Spec Rev Muscle Contraction, 60 & 67; actg chmn dept physiol & biophys, Sch Med, Univ Louisville, 65-67. *Mem:* Soc Exp Biol & Med; Am Physiol Soc; Biophys Soc. *Res:* Muscle structure and function; development of fine structures in muscle fibers; neurophysiology. *Mailing Add:* Dept Physiol & Biophys Sch Med Univ Louisville Health Sci Ctr 115A Louisville KY 40292

WALKER, TERRY M, b Chicago, Ill, Dec 15, 38; m 60; c 2. COMPUTER SCIENCE. *Educ:* Fla State Univ, BS, 61; Univ Ala, PhD(statist), 66. *Prof Exp:* Asst prof comput sci & economet, Ga State Col, 65-67; assoc prof comput sci, Univ Houston, 67-72; prof comput sci & head dept, Univ Southwestern La, 72-87; RETIRED. *Concurrent Pos:* Lectr & Ford Found consult, Atlanta, 66-67. *Mem:* Asn Comput Mach. *Res:* Simulation of industrial processes; computer programming languages. *Mailing Add:* 1060 Kupalu Dr Kihei Maui HI 96753-9234

WALKER, THEODORE ROSCOE, b Madison, Wis, Feb 8, 21; m 49; c 4. GEOLOGY, SEDIMENTARY PETROLOGY. *Educ:* Univ Wis, BS, 47, PhD, 52. *Prof Exp:* Asst geologist, State Geol Surv, Ill, 52-53; from asst prof to assoc prof geol, 53-65, fac res lectr, 72-73, chmn dept geol sci, 72-75, prof, 65-86, EMER PROF GEOL, UNIV COLO, BOULDER, 86- *Concurrent Pos:* NSF sr fel, 62-63; Am Asn Petrol Geol distinguished lectr, 65. *Mem:* Geol Soc Am; hon mem Soc Econ Paleont & Mineral (pres, 82-83); Am Asn Petrol Geologists. *Res:* Sedimentation; sedimentary petrology. *Mailing Add:* Dept Geol Sci Univ Colo Box 250 Boulder CO 80309

WALKER, THERESA ANNE, b Rantoul, Ill, Mar 26, 52. SCIENCE & TECHNOLOGY FOR ECONOMIC DEVELOPMENT. *Educ:* Eastern Ill Univ, BS, 73; Yale Univ, MPhil, 77, PhD(biochem), 80. *Prof Exp:* Sci & technol res asst, Ill Legis Coun, 80-81; dir res, Dept Obstet-Gynec, Quillen-Dishner Col Med, E Tenn State Univ, 81-82; staff scientist, Ill Legis Res Unit, 82-85; dir, NY State Legis Comn Sci & Technol, 85-88; prin sci assoc, 88-90, MGR UNIV-INDUST PROGS, NY STATE SCI TECHNOL FOUND, 90- *Concurrent Pos:* Teaching asst, Dept Biol, Yale Univ, 74-76, res asst, 78-80; teaching asst, Dept Biochem, Quillen-Dishner Col Med, 80; guest lectr, E Tenn State Univ, 80; adj asst prof, Dept Pharmacol, Sch Med, Southern Ill Univ, 84-85. *Mem:* Sigma Xi; AAAS; Am Chem Soc; Scientists' Inst Pub Info; Am Inst Biol Sci. *Res:* Science and technology for economic development. *Mailing Add:* 99 Washington Ave Albany NY 12210

WALKER, THOMAS CARL, biomaterials, instrumentation, for more information see previous edition

WALKER, THOMAS EUGENE, b Glendale, Calif, Feb 1, 48; m 73. BIOCHEMISTRY. *Educ:* Westmar Col, BA, 69; Univ Iowa, PhD(biochem), 74. *Prof Exp:* Res assoc, Mich State Univ, 74-75; fel, 75-77, STAFF MEM, LOS ALAMOS NAT LAB, 77- *Mem:* Am Chem Soc. *Res:* The synthesis by chemical and biosynthetic techniques and nuclear magnetic resonance and mass spectral analysis of carbon 13, nitrogen 15, oxygen 17 and oxygen 18 enriched compounds of biological interest. *Mailing Add:* Isotec 3858 Benner Rd Miamisburg OH 45342-4304

WALKER, THOMAS JEFFERSON, b Dyer Co, Tenn, July 24, 31; m 59; c 2. BEHAVIORAL ECOLOGY, SYSTEMATICS. *Educ:* Univ Tenn, BA, 53; Ohio State Univ, MSc, 54, PhD(entom), 57. *Prof Exp:* From asst prof to assoc prof biol sci & entom, 57-68, prof biol sci, 68-71, PROF ENTOM, UNIV FLA, 71- *Concurrent Pos:* Res assoc dept tropic res, NY Zool Soc, 66; res assoc, Fla State Collection Arthropods, 63-; ed, Fla Entomologist, 64-66. *Mem:* Fel AAAS; Entom Soc Am; Soc Study Evolution; Orthopterists Soc; Sigma Xi; Lepidopterist's Soc. *Res:* Acoustical behavior of insects; systematics, behavior, ecology and evolution of Gryllidae and Tettigoniidae; migratory behavior of butterflies. *Mailing Add:* Dept Entom & Nematol Univ Fla Gainesville FL 32611-0740

WALKER, WALDO SYLVESTER, b Fayette, Iowa, June 12, 31; m 52; c 2. BOTANY. *Educ:* Upper Iowa Univ, BS, 53; Univ Iowa, MS, 57, PhD(bot), 59. *Prof Exp:* From asst prof to assoc prof biol, 58-68, assoc dean, 63-65, dean admin, 69-73, dean col, 73-80, PROF BIOL, GRINNEL COL, 68, EXEC VPRES, 80- *Res:* Ultrastructure of plants; experimental morphology; plant physiology. *Mailing Add:* Off Exec Vpres Grinnell Col 1202 Park St Grinnell IA 50112

WALKER, WALTER W(YRICK), b Winslow, Ariz, Jan 14, 24; m 52. PHYSICAL METALLURGY, PHYSICAL CHEMISTRY. *Educ:* Univ Ariz, BSc, 50, MSc, 62, PhD(metall), 68. *Hon Degrees:* PhD, Univ Phys Sci, 57. *Prof Exp:* Asst metallurgist, Buick Motor Div, Gen Motors Corp, 50-51; res engr, Oak Ridge Nat Lab, 51-52; tech supvr metall, Livermore Res Lab, AEC, 52-53; sr metallurgist, Tucson Div, Hughes Aircraft Co, 53-56, tech supvr metall engr, 56-59; lectr metall, Univ Ariz, 59-62; group leader metall eng, Tucson Div, Hughes Aircraft Co, 62-63, group head mat eng, 63-66; assoc prof metall eng, Univ Ariz, 67-73; sr process engr, Tucson Div, Hughes Aircraft, 73-78, staff engr, Tucson Eng Labs, 78-84, sr staff engr, 84-86, sr scientist, 86-89; RETIRED. *Concurrent Pos:* Assoc instr, Pima Community Col, 74-86. *Mem:* Am Soc Metals; Am Inst Mining, Metall & Petrol Engrs. *Res:* Nucleation and solidification of metals; crystal growth phenomena; thermodynamics and kinetics of condensed systems; metallurgy of meteorites; plasticity of ionic crystals; surface chemistry and physics of solid interfaces; novel optical materials; microindentation hardness testing. *Mailing Add:* 5643 E Seventh St Tucson AZ 85711

WALKER, WARREN ELLIOTT, b New York, NY, Apr 7, 42; div; c 3. OPERATIONS RESEARCH, PUBLIC POLICY ANALYSIS. *Educ:* Cornell Univ, BA, 63, MS, 64, PhD(opers res), 68. *Prof Exp:* Pres, Compuvisor, Inc, 68-70; sr opers res analyst & proj dir, NY City-Rand Inst, Rand Corp, 70-75; asst vpres, Chem Bank, 75-76; dep dir, Urban Acad, 76-77; SR POLICY ANALYST, RAND CORP, 77- *Concurrent Pos:* Consult, US Environ Protection Agency Off Solid Waste Mgt Progs, 68-72; adj prof opers res, Columbia Univ, 71-77; pres, Urbatronics Inc, 75-80; chmn, Los Angeles Prod Adv Comt, 85-88; vis prof, Delft Univ Technol, 88-89. *Honors & Awards:* Lanchester Prize, Opers Res Soc Am, 74; Edelman Award, Inst Mgt Sci, 74 & 84; NATO Systs Sci Prize, 76. *Mem:* Inst Mgt Sci; Opers Res Soc Am. *Res:* Applying quantitative methods to the analysis of problems in governmental management and operations; water management, fire department deployment, the criminal justice system and military manpower. *Mailing Add:* Rand 1700 Main St Santa Monica CA 90406-2138

WALKER, WARREN FRANKLIN, JR, b Malden, Mass, Sept 27, 18; m 44; c 4. ZOOLOGY. *Educ:* Harvard Univ, SB, 41, PhD(zool), 46. *Prof Exp:* Instr anat sch med, Boston Univ, 45-47; instr zool, Oberlin Col, 47-48, from asst prof to prof biol, 49-85, actg provost, 74-75, chmn dept, 67-74; RETIRED. *Mem:* AAAS; Am Asn Anat; Am Soc Zool; Soc Syst Zool; Am Soc Ichthyol & Herpet. *Res:* Herpetology of South America; vertebrate anatomy and evolution; myology; vertebrate locomotion. *Mailing Add:* PO Box 436 Ossipee NH 03864

WALKER, WELLINGTON EPLER, industrial chemistry, organic chemistry; deceased, see previous edition for last biography

WALKER, WILBUR GORDON, b Lena, La, Sept 18, 26; m 47; c 4. INTERNAL MEDICINE, NEPHROLOGY. *Educ:* Tulane Univ, MD, 51. *Prof Exp:* Intern med, Osler Med Serv, Johns Hopkins Hosp, 51-52; resident, Tulane Serv, Charity Hosp, La, 52-53; asst resident, Osler Med Serv, Johns Hopkins Hosp, 53-54, Am Heart Asn fel, 54-56, resident physician, 56-57, from asst prof to assoc prof, 58-68, dir, Renal Div, 58-88, dir clin res ctr, 60-89, PROF MED, SCH MED, JOHNS HOPKINS UNIV, 68-, PHYSICIAN, JOHNS HOPKINS HOSP, 57- *Concurrent Pos:* Estab investr, Am Heart Asn, 57-60; consult, Vet Admin Hosp, Loch Raven; dir renal div, Johns Hopkins & Good Samaritan Hosps; physician-in-chg, Dept Res Med, Good Samaritan Hosp; mem urol & renal dis training grant comt, Nat Inst Arthritis, Metab & Digestive Dis, 68-72; mem gen clin res ctrs comt, Div Res Resources, NIH, 72-76, chmn, 75-76; mem, Kidney Comn Md, 72-79 & 81-90, chmn, 75-78; chmn, Qual Control Comt, modification & diet in renal dis, multictr clin trial Nat Inst Diabetes Digestive Kidney, NIH, 87-; chmn, AIDS Nephropathy Study Sect Nat Inst Diabetes Digestive Kidney, NIH, 88; mem, Working Party Nat Heart, Lung & Blood Inst, Working Group Report on Hypertension & Chronic Renal Dis, Nat High Blood Pressure Educ Prog. *Mem:* Am Physiol Soc; Am Soc Nephrology; Am Soc Clin Invest; Am Fedn Clin Res. *Res:* Renal function and renal diseases; electrolyte metabolism; renal ion exchange mechanisms; metabolic investigations in renal disease; energy metabolism and protein synthesis in isolated glomeruli; capillary and glomerular permeability to protein; renin-angiotensin-aldosterone physiology; factors influencing biological calcification; pathogenesis of diabetic nephropathy; nutritional disturbances in hypertension; nutritional aspects of renal failure; nature of renal damage in hypertension; role of K deficiency in hypertension. *Mailing Add:* Johns Hopkins Hosp Baltimore MD 21205

WALKER, WILLIAM CHARLES, b Santa Barbara, Calif, Aug 22, 28; m 51; c 3. SOLID STATE PHYSICS. *Educ:* Univ Calif, AB, 50; Univ Southern Calif, MS, 53, PhD(physics), 55. *Prof Exp:* Asst physics, Univ Southern Calif, 50-52, res assoc, 52-55; instr, 55-57, from asst prof to assoc prof, 57-68, chmn dept, 72-75, PROF PHYSICS, UNIV CALIF, SANTA BARBARA, 68- *Concurrent Pos:* Consult, Servomechanisms, Inc, 59-60; Nat Acad Sci-NASA fel, Goddard Space Flight Ctr, 63-64; consult, Sloan Technol, 68-74. *Mem:* AAAS; Am Phys Soc; Sigma Xi. *Res:* Solid state and ultraviolet spectroscopy; optical properties of solids; conducting polymers. *Mailing Add:* Dept Physics Univ Calif Santa Barbara CA 93106

WALKER, WILLIAM COMSTOCK, b Milwaukee, Wis, July 6, 21; m 45; c 2. PAPER CHEMISTRY, PHYSICAL CHEMISTRY. *Educ:* Lehigh Univ, BS, 43, MS, 44, PhD(phys chem), 46. *Prof Exp:* Inst res fel, Lehigh Univ, 46-47; dir, Nat Printing Ink Res Inst, 47-55, res asst prof chem, 53-55; res dir, Westvaco Corp, New York, NY, 55-65, tech asst to corp vpres res, 64-86; RETIRED. *Concurrent Pos:* Instr, Muhlenberg Col, 46-47. *Honors & Awards:* Robert F Reid Award, Graphic Arts Tech Found, 77; Silver Medal Award & Charles W Englehard Medallion, Tech Asn Pulp & Paper Indust, 82. *Mem:* Am Chem Soc; Tech Asn Graphic Arts; Tech Asn Pulp & Paper Indust; Can Pulp & Paper Asn; Sigma Xi. *Res:* Printability of paper; novel crop systems; paper production technology; printing inks; adsorption of gases on solids; removal of sulfur oxides from flue gases. *Mailing Add:* One Springfield Pl Savannah GA 31411

WALKER, WILLIAM DELANY, b Dallas, Tex, Nov 23, 23; m 46, 75; c 3. PARTICLE PHYSICS. *Educ:* Rice Inst, BA, 44; Cornell Univ, PhD(cosmic ray physics), 49. *Prof Exp:* Physicist, US Naval Res Lab, 44-45; asst prof physics, Rice Inst, 49-51; lectr, Univ Calif, 51-52; asst prof, Univ Rochester, 52-54; from asst prof to prof, 54-67, Max Mason prof, 67-71, chmn dept, Univ Wis-Madison, 64-66; chmn dept, 75-81, prof, 71-90, J B DUKE PROF PHYSICS, DUKE UNIV, 90- *Concurrent Pos:* Mem physics panel, NSF, 64-67; mem high energy surv comt, Nat Acad Sci, 64-65; chmn, Argonne User's Group, 64-66; mem user's exec comt, Fermilab, 72-75; chmn, Fermilab User's Exec Comt, 73-74; mem bd dir, Oak Ridge Assoc Univ, 80-85; secy, Region 5, Univ Res Asn. *Mem:* Fel Am Phys Soc; Sigma Xi. *Res:* Strong interaction physics; technology of bubble chambers; technology of particle detection. *Mailing Add:* Dept Physics Duke Univ Durham NC 27706

WALKER, WILLIAM F(RED), b Sherman, Tex, Dec 1, 37; m 60; c 2. MECHANICAL ENGINEERING, BIOMEDICAL ENGINEERING. *Educ:* Univ Tex, BS, 60, MS, 61; Okla State Univ, PhD(mech eng), 66. *Prof Exp:* Aerodyn engr, Ling-Temco-Vought, Inc, 61-62; from asst prof to assoc prof, 65-75, PROF AEROSPACE ENG, RICE UNIV, 75-, CHMN, DEPT MECH ENG & MAT SCI, 76- *Mem:* Am Inst Aeronaut & Astronaut; Am Soc Mech Engrs; Am Soc Artificial Internal Organs. *Res:* Compressible turbulent boundary layers; separated and reattached jet flows; transpiration and ablation cooling. *Mailing Add:* Col Eng Auburn Univ 108 Ramsey Hall Auburn AL 36849-5330

WALKER, WILLIAM HAMILTON, b Brookline, Mass, Dec 28, 34; m 62; c 2. CIVIL ENGINEERING, STRUCTURAL DYNAMICS. *Educ:* Univ Mass, BS, 56; Univ Ill, Urbana, MS, 58, PhD(civil eng), 63. *Prof Exp:* Res asst civil eng, Univ Ill, 56-61, from instr to assoc prof, 61-90, asst head dept, 73-76, PROF CIVIL ENG, UNIV ILL, URBANA, 90-, ASSOC HEAD DEPT, 86- *Concurrent Pos:* Mem comt bridge dynamics, proj panels, Transp Res Bd, Nat Acad Sci, 65- *Mem:* Am Soc Civil Engrs; Am Soc Eng Educ; Am Asn Univ Professors; Sigma Xi. *Res:* Structural mechanics with emphasis on dynamics; dynamic response of bridges by means of field tests and computer analysis; earthquake engineering; fatigue and fracture metal structures. *Mailing Add:* 1116 Newmark Lab Univ Ill 205 N Mathews Ave Urbana IL 61801

WALKER, WILLIAM J, JR, RADIATION SAFETY. *Educ:* Va Mil Inst, BS, 58; Univ Kans, MS, 64; Univ Fla, PhD(environ eng), 71; Am Bd Health Physics, cert. *Prof Exp:* Base sanitary & indust hyg eng, Castle AFB, Calif, 58-60; staff environ engr, High Wycombe, Eng, 60-62; res health physicist, Kirtland AFB, Albuquerque, NMex, 64-68; pres & prin consult, Physics Control Inc, 73-78; chief med physics, Malcolm Grow USAF Med Ctr, Andrew AFB, Washington, DC, 71-78; sect leader, Med & Acad Licensing Sect, US Nuclear Regulatory Comn, Washington, DC, 78-83; vpres, Med Div, RSO Inc, 83-84; CHIEF, RADIATION SAFETY BR, NIH, 89-; sr vpres & chief scientist, Health Physics Serv, Inc, 85-88. *Concurrent Pos:* Consult radiol physicist, Sacred Heart Hosp, 84-, sr consult, Inst Radiol Imaging Sci Inc, 88-89, consult, US Nuclear Regulatory Comn, 89; pres & chief operating officer, Radiopharmaceut Mgt Serv Inc, 88; mem, Nuclear Med Sci Comt, Am Col Nuclear Physicians, Standardization Nuclear Med Instrumentation Comt, Subcomt Nuclear Med Technol. *Mem:* Health Physics Soc; Sigma Xi; Am Col Nuclear Physicians; Am Asn Physicists Med; Am Col Med Physics; Soc Nuclear Med. *Mailing Add:* NIH Radiation Safety Br Bldg 21 Rm 112 Bethesda MD 20892

WALKER, WILLIAM M, b Savannah, Tenn, Sept 17, 28; m 51; c 4. SOIL FERTILITY, BIOMETRICS. *Educ:* Florence State Col, BS, 50; Univ Tenn, MS, 57; Iowa State Univ, PhD(soil fertil), 61. *Prof Exp:* Res assoc agron, Iowa State Univ, 57-61; asst agronomist, Univ Tenn, 61-66; from asst prof to prof, 66-88, EMER PROF BIOMET, UNIV ILL, URBANA, 88- *Mem:* Am Soc Agron; Coun Soil Testing & Plant Anal. *Res:* Application of biomathematics to soil-plant relationships. *Mailing Add:* Dept Agron Univ Ill 1102 S Goodwin Urbana IL 61801

WALKER, WILLIAM R, b Lincoln, Nebr, June 8, 25; m 49; c 3. CIVIL ENGINEERING, LAW. *Educ:* Univ Nebr, Lincoln, BSCE, 49, JD, 52; Univ NC, Chapel Hill, MSSE, 64. *Prof Exp:* DIR WATER RESOURCES RES CTR, VA POLYTECH INST & STATE UNIV, 65- *Concurrent Pos:* Vis scholar, Bd Rivers & Harbors, CEngr, 70-71; mem exec bd, Univ Coun on Water Resources. *Mem:* AAAS; Am Soc Civil Engrs; Am Soc Eng Educ; Water Pollution Control Fedn; Am Water Works Asn. *Res:* Legal and institutional arrangements for water resources. *Mailing Add:* 1502 Greenwood Dr Blacksburg VA 24060

WALKER, WILLIAM STANLEY, b Glendale, Calif, Apr 21, 40; m 64; c 2. IMMUNOLOGY, MICROBIOLOGY. *Educ:* Univ Southern Calif, AB, 63, PhD(microbiol), 68. *Prof Exp:* Lectr bact, Univ Southern Calif, 64-65; sci officer first class, Dept Immunohaemat, Acad Hosp, State Univ Leiden, 71; asst mem labs virol & immunol, 72-74, assoc mem div immunol, 75-84, MEM DEPT IMMUNOL, ST JUDE CHILDREN'S RES HOSP, 84- *Concurrent Pos:* Fel immunol, Pub Health Res Inst City New York, Inc, 68-71. *Mem:* Soc Leukocyte Biol; Am Asn Immunologists; Am Soc Cell Biol. *Res:* Immunobiology of macrophages. *Mailing Add:* Immunol Saint Jude Res Hosp PO Box 318 Memphis TN 38101

WALKER, WILLIAM WALDRUM, b Alexander City, Ala, Jan 16, 33; m 58; c 3. PHYSICS. *Educ:* Auburn Univ, BS, 55; Univ Va, MA, 57, PhD(physics), 59. *Prof Exp:* Asst prof physics, Col William & Mary, 59-60; physicist, Signal Res & Develop Lab, US Army, NJ, 60; from asst prof to assoc prof, 67-71, PROF PHYSICS, UNIV ALA, 71- *Concurrent Pos:* Radiation Safety Officer, Univ Ala, 84- *Mem:* Am Phys Soc; Am Asn Physics Teachers. *Res:* Positron life-times; low energy nuclear physics. *Mailing Add:* Dept Physics Univ Ala Tuscaloosa AL 35487

WALKER-NASIR, EVELYNE, carbohydrate chemistry, for more information see previous edition

WALKIEWICZ, THOMAS ADAM, b Erie, Pa, Dec 25, 39; m 62; c 3. NUCLEAR PHYSICS, SOLID STATE PHYSICS. *Educ:* Xavier Univ, Ohio, BS, 62; Pa State Univ, PhD(physics), 69. *Prof Exp:* Asst prof physics, East Stroudsburg State Col, 69; physicist, Picatinny Arsenal, NJ, 69-70; ASSOC PROF PHYSICS, EDINBORO STATE COL, 70- *Concurrent Pos:* AEC collab researcher, Oak Ridge Nat Lab, 71-77, consult, 75-76. *Mem:* Am Phys Soc; Am Asn Physics Teachers. *Res:* Experimental low-energy nuclear physics, especially gamma ray spectroscopy; neutron and charged particle reactions; experimental solid state physics, especially electrooptical and radiation effects in amorphous semiconductors. *Mailing Add:* Dept Physics Edinboro Univ Edinboro PA 16444

WALKINGTON, DAVID L, b Waukegan, Ill, July 20, 30; m 55; c 1. BOTANY. *Educ:* Ariz State Univ, BA, 57, MS, 59; Claremont Grad Sch, PhD(bot), 65. *Prof Exp:* Instr bot, Ariz State Univ, 59-60; res assoc, Calif State Univ, 60-63, lectr, 62-63, from asst prof to assoc prof biol, 63-72, actg assoc dean, Sch Math, Sci & Eng, 73-74, actg dean, 74-75 & 77-78, assoc dean, 75-77, actg assoc vpres, 78-80, assoc vpres, 80-84, PROF BIOL, CALIF STATE UNIV, FULLERTON, 72-, DIR, FULLERTON ARBORETUM, 85- *Concurrent Pos:* NSF res grant, 65-66; pres bd trustees, Mus Asn NOrange County, 75- *Mem:* Nat Asn Biol Teachers; Cactus & Succulent Soc Am; Am Inst Biol Sci. *Res:* Morphology and chemistry of pollen; chemotaxonomy of cacti; naturally occurring antibiotics in plants especially bryophytes and cacti. *Mailing Add:* Fullerton Arboretum Calif State Univ Fullerton CA 92634

WALKINSHAW, CHARLES HOWARD, JR, b Blairsville, Pa, Nov 14, 35; m 57, 83; c 5. PHYTOPATHOLOGY, MICROBIOLOGY. *Educ:* Univ Fla, BSA, 57; Univ Wis, PhD(plant path), 60. *Prof Exp:* Asst plant path, Univ Wis, 57-60, proj assoc, 60-61, trainee biochem & path sch med, 61-63; plant pathologist, Southern Forest Exp Sta, US Forest Serv, Miss, 63-65; asst prof microbiol sch med, Univ Miss, 65-68; plant pathologist med support br, NASA-Manned Spacecraft Ctr, 68-73; PRIN PLANT PATHOLOGIST, SOUTHERN FOREST EXP STA, US FOREST SERV, 73- *Honors & Awards:* Super Serv Award, USDA, 72. *Mem:* AAAS; Am Phytopath Soc; Am Soc Cell Biol; Sigma Xi. *Res:* Plant diseases; plant tissue culture; fungus diseases of pines; biochemistry of plant diseases; agriculture in Lunar and Mars bases. *Mailing Add:* USDA Forest Serv PO Box 8937 Asheville NC 28814

WALKLING, ROBERT ADOLPH, b Philadelphia, Pa, Sept 11, 31; m 59; c 2. ACOUSTICS. *Educ:* Swarthmore Col, BA, 53; Harvard Univ, SM, 54, PhD(acoustics), 62. *Prof Exp:* Res fel appl physics, Harvard Univ, 62-63; asst prof physics, Bowdoin Col, 63-69; ASSOC PROF PHYSICS, UNIV SOUTHERN MAINE, 69- *Mem:* AAAS; Acoust Soc Am; Audio Eng Soc; Am Sci Affiliation; Am Asn Physics Teachers; Sigma Xi. *Res:* Electroacoustics; noise and vibration; architectural and musical acoustics. *Mailing Add:* 34 Boody St Brunswick ME 04011

WALKLING, WALTER DOUGLAS, b Baltimore, Md, Feb 27, 39; m 61; c 2. PHARMACY. *Educ:* Univ Md, BS, 61, MS, 63, PhD(pharm), 66. *Prof Exp:* Pharmacist, Yager Drug Co, 64-66; Nat Inst Gen Med Sci fel, Swiss Fed Inst Technol, 66-67; sr pharmaceut chemist, Eli Lilly & Co, 67-70; sr scientist, McNeil Pharmaceut, 70-72, group leader, 72-76, sect head, Pharm Res Dept, 76-86, dir Pharmaceut Technol Dept, McNeil Pharmaceut, 86-90, DIR PHARMACEUT TECHNOL DEPT, RW JOHNSON, 90- *Mem:* Fel Am Asn Pharmaceut Scientists; Controlled Release Soc; Am Chem Soc. *Res:* Design and evaluation of pharmaceutical dosage forms. *Mailing Add:* Pharmaceut Technol Dept RW Johnson Spring House PA 19477

WALKOWIAK, EDMUND FRANCIS, b Webster, Mass, June 12, 35; m 60; c 2. PHYSIOLOGY. *Educ:* Boston Univ, AB, 58; Univ Conn, PhD(zool), 67. *Prof Exp:* Instr biol, Salem State Col, 59-62; from asst prof to assoc prof, Augusta Col, 66-69; assoc prof, 70-74, PROF PHYSIOL, NEW ENG COL OPTOM, 74-, DIR, INSTNL AFFAIRS, 74- *Concurrent Pos:* Mem Biol Curric Comt, Univ Syst Ga, 69. *Mem:* Am Inst Biol Sci; Am Soc Mammal. *Res:* Application of serological and electrophoretic techniques to the study of tear films. *Mailing Add:* NE Col Optom 424 Beacon St Boston MA 02115

WALKUP, JOHN FRANK, b Oakland, Calif, Feb 7, 41; m 65; c 3. OPTICAL COMPUTING, DIGITAL IMAGE PROCESSING. *Educ:* Dartmouth Col, BA, 62, BEE, 63; Stanford Univ, MS, 65, PhD(elec eng), 71. *Prof Exp:* Res asst, Stanford Univ, 63-71; from asst prof to prof elec eng, 71-85, PW HORN PROF ELEC ENG, TEX TECH UNIV, 85- *Concurrent Pos:* Assoc, Gen Motors Res Labs, 62; assoc engr, Aerojet - Gen Corp, 63; consult, ESL, Inc,

74-76, Optics Technol, 74 & NSF, 75; vis scholar, Optical Sci Ctr, Univ Ariz, 82; assoc dean eng, Tex Tech Univ, 82-83; chmn, Educ Coun, Optical Soc Am, 87-88 & Gordon Res Conf Holography & Optical Info Processing, 91. *Honors & Awards:* Halliburton Award, Tex Tech Univ, 80; AT&T Found Award, Am Soc Eng Educ, 85. *Mem:* Fel Inst Elec & Electronics Engrs; fel Optical Soc Am; Int Soc Optical Eng; Am Soc Eng Educ; Sigma Xi. *Res:* Basic and applied research in optical information processing, optical computing, statistical optics, digital image processing, neural networks and signal processing. *Mailing Add:* Optical Systs Lab Dept Elec Eng Tex Tech Univ Lubbock TX 79409-3102

WALL, CONRAD, III, b Boston, Mass, June 13, 39; m 61; c 2. BIOENGINEERING. *Educ:* Tulane Univ, BS, 62, MS, 68; Carnegie-Mellon Univ, PhD(bioeng), 75. *Prof Exp:* Proj officer elec eng, US Army AV Labs, 63-65; mem tech staff appl physics, Boeing Co, 65-70; NIH res assoc sensory physiol, Dept Otolaryngol, Med Sch, 75-77, SCI DIR, HUMAN VESTIBULAR LAB, UNIV PITTSBURGH, 76- *Mem:* Soc Neurosci; Inst Elec & Electronics Engrs; Barany Soc; Sigma Xi. *Res:* Information processing in the vestibular and visual systems; digital signal processing of clinical and experimental sensory systems data. *Mailing Add:* 5873 Hobart Pittsburgh PA 15217

WALL, DONALD DINES, b Kansas City, Mo, Aug 13, 21; m 43; c 3. COMPUTER SCIENCES. *Educ:* Univ Calif, PhD(math), 49. *Prof Exp:* Instr, Santa Barbara Col, 49-51; appl sci rep, IBM Corp, 51-74, mkt analyst, 74-81; PRES, DATAMAPS CONSULT CO, 81- *Mem:* Math Asn Am. *Res:* Number theory; computing machines. *Mailing Add:* 1505 Briar Lake Ct Atlanta GA 30345

WALL, EDWARD THOMAS, b Brooklyn, NY, May 16, 20; m 47; c 4. ELECTRICAL ENGINEERING, CONTROL SYSTEMS. *Educ:* Purdue Univ, West Lafayette, BS, 47; Lehigh Univ, MS, 49; Univ Denver, PhD(elec eng), 67. *Prof Exp:* Instr elec eng, Univ Maine, 49; engr, Pac Gas & Elec Co, 49-51; asst prof, Calif State Polytech Univ, 51-54; analytical engr, Gen Elec Co, 54-57; staff engr, Martin-Marietta Corp, 57-64; res elec engr, US Bur Reclamation, 64-66; assoc prof, 66-73, PROF ELEC ENG, UNIV COLO, DENVER, 66- *Concurrent Pos:* NSF grants, Univ Colo, Denver, 68-71. *Mem:* AAAS; Inst Elec & Electronics Engrs; Sigma Xi. *Res:* Adaptive control system design, digital control systems with application to aerospace systems; nonlinear control systems; power system analysis. *Mailing Add:* Dept Elec Eng & Comput Sci Univ Colo Denver PO Box 173364 Denver CO 80217-3364

WALL, FRANCIS JOSEPH, b Moss Point, Miss, Mar 22, 27; m 50; c 3. BIOSTATISTICS, APPLIED STATISTICS. *Educ:* Sul Ross State Col, BS, 47; Univ Colo, MS, 56; Univ Minn, PhD, 61. *Prof Exp:* Sr statistician, Dow Chem Co, 52-57; statistician, Remington Rand Univac Div, Sperry Rand Corp, 57-61; sr res mathematician, Dikewood Corp, 61-69; biostatistician, Lovelace Found Med Educ & Res, 69-71; CONSULT BIOSTATIST & STATIST ANALYSIS, 72- *Concurrent Pos:* Consult, Shell Oil Co, Colo, 56; vis lectr, Univ NMex, 67, adj asst prof sch med, 70-74, clin assoc, 68-69. *Mem:* Biomet Soc; Am Statist Asn; Sigma Xi. *Res:* Application of statistical methods to biomedical research; design and analysis of clinical trials. *Mailing Add:* 3016 Waverly Dr No 103 Los Angeles CA 90039-2035

WALL, FREDERICK THEODORE, b Chisholm, Minn, Dec 14, 12; m 40; c 2. STATISTICAL MECHANICS. *Educ:* Univ Minn, BCh, 33, PhD(chem), 37. *Prof Exp:* Univ Ill, Urbana, 37-64, dean grad col, 55-63; prof chem & chmn dept, Univ Calif, Santa Barbara, 64-66, vchancellor res, 65-66; vchancellor grad studies & res & prof chem, Univ Calif, San Diego, 66-69; exec dir, Am Chem Soc, 69-72; prof, Rice Univ, 72-78; prof chem, San Diego State Univ, 79-82; ADJ PROF CHEM, UNIV CALIF, SAN DIEGO, 82- *Honors & Awards:* Am Chem Soc Award, 45. *Mem:* Nat Acad Sci; AAAS; Am Acad Arts & Sci; Am Chem Soc; Am Phys Soc; Finnish Chem Soc. *Res:* Physical chemistry of macromolecular configurations; discrete wave mechanics. *Mailing Add:* 2468 Via Viesta La Jolla CA 92037

WALL, GREGORY JOHN, b Toronto, Ont, Aug 16, 44; m 68; c 3. SOIL SCIENCE. *Educ:* Univ Guelph, BSA, 67, MSc, 69; Ohio State Univ, PhD(soil sci), 73. *Prof Exp:* Res officer soil sci, Agr Can, 67-70; teaching asst agron, Ohio State Univ, 70-71, res assoc soil mineral, 71-73; SOIL SCIENTIST, AGR CAN, 73- *Mem:* Can Soc Soil Sci; Am Soc Agron; Int Asn Great Lakes Res; Int Soc Soil Sci; Soil Conserv Soc Am. *Res:* Sources and magnitude of water pollution by sediment in agricultural regions; mineralogy and exchange properties of fluvial sediments; variability of soil physical and engineering properties in the mineralogy of soils; interpretation of soil survey data. *Mailing Add:* 12 Princeton Pl Guelph ON N1G 3S4 Can

WALL, JOHN HALLETT, b St Stephen, NB, Aug 10, 24; m 61; c 1. MICROPALEONTOLOGY. *Educ:* Univ NB, BSc, 45; Univ Alta, MSc, 51; Univ Mo, PhD(geol), 58. *Prof Exp:* Asst geologist, NB Dept Mines, 43; asst geologist, Geol Surv Can, 44; jr geologist, Imp Oil Ltd, 45-46, subsurface geologist & micropaleontologist, 47-51; subsurface geologist, J C Sproule & Assocs, Explor Consults, 52; micropaleontologist, Calif Standard Co, 53; res officer, Res Coun Alta, 57-74; RES SCIENTIST, GEOL SURV CAN, 74- *Concurrent Pos:* Lectr geol, Univ Alta, 60-72. *Mem:* Fel Geol Soc Am; Soc Econ Paleont & Mineral; Paleont Soc; fel Geol Asn Can. *Res:* Mesozoic microfossils of western and arctic Canada. *Mailing Add:* 3303 33rd St NW Geol Surv Can Calgary AB T2L 2A7 Can

WALL, JOSEPH S, b Madison, Wis, Nov 17, 42. BIOPHYSICS. *Educ:* Univ Wis-Madison, BS, 64; Univ Chicago, PhD(biophys), 71. *Prof Exp:* Fel biophys, Univ Chicago, 71-73; assoc biophysicist, 73-78, BIOPHYSICIST DEPT BIOL, BROOKHAVEN NAT LAB, 78- *Honors & Awards:* Lawrence Award, US Dept Energy, 88. *Mem:* AAAS; NY Acad Sci; Electron Micros Soc Am. *Res:* Development and biological application of the high resolution scanning transmission electron microscope. *Mailing Add:* Dept Biol Brookhaven Nat Lab Upton NY 11973

WALL, JOSEPH SENNEN, b Chicago, Ill, June 2, 23; m 50; c 3. BIOCHEMISTRY, PROTEIN CHEMISTRY. *Educ:* Univ Chicago, BS, 46, MS, 49; Univ Wis, PhD(biochem), 52. *Prof Exp:* Instr chem, Lincoln Col, 47-49; asst biochem, Univ Wis, 50-52; instr pharmacol sch med, NY Univ, 52-56; head chem reactions & structure invests, 56-72, res leader, 72-83, RES CHEMIST CEREAL PROTEINS RES UNIT, CEREAL SCI & FOODS LAB, NORTHERN REGIONAL RES CTR, USDA, PEORIA, ILL, 83- *Mem:* AAAS; Am Chem Soc; Am Soc Biol Chemists; Am Asn Cereal Chemists; Inst Food Technologists. *Res:* Mechanism of nitrogen fixation by bacteria; isolation of natural products; hormonal regulation of carbohydrate metabolism in mammals; protein chemistry; cereal chemistry; nutrition; enzymology. *Mailing Add:* 8606 N Servite Dr Milwaukee WI 53223-2514

WALL, LEONARD WONG, b Tallulah, La, Nov 7, 41; m 68; c 3. PHYSICS. *Educ:* La Tech Univ, BS, 63; Iowa State Univ, PhD(physics), 69. *Prof Exp:* Vis asst prof, Univ Kans, 68-69; PROF PHYSICS, CALIF POLYTECH STATE UNIV, 69- *Mem:* Am Phys Soc; Am Asn Physics Teachers; AAAS; Sigma Xi. *Res:* Physics of energy; energy usage in buildings; passive solar. *Mailing Add:* Dept Physics Calif Polytech State Univ San Luis Obispo CA 93401

WALL, MALCOLM JEFFERSON, JR, b Meridian, Miss, Jan 25, 41; m 65; c 2. PHYSIOLOGY. *Educ:* Lamar Univ, BS, 63; Univ Tex Med Br, Galveston, PhD(physiol), 70; NTex State Univ, MA, 67. *Prof Exp:* Asst biol, NTex State Univ, 63-66; from instr to asst prof physiol & med, Med Col Wis, 70-74; res assoc physiol & biophys, 74-76, surg internship, 76-77, RESIDENT PHYSICIAN, DIV ORTHOP SURG, UNIV TEX MED BR, GALVESTON, 74- *Concurrent Pos:* USPHS grant, Med Col Wis, 72-74; lectr, Univ Wis-Milwaukee, 72. *Mem:* AAAS; Am Inst Biol Sci; Biophys Soc; Assoc Am Physiol Soc. *Res:* Intestinal transport mechanisms for electrolytes, orgainc solutes and water. *Mailing Add:* 6701 Heritage Pkwy No 17 Rockwall TX 75087

WALL, MONROE ELIOT, b Newark, NJ, July 25, 16; m 41; c 2. BIOCHEMISTRY. *Educ:* Rutgers Univ, BS, 36, MS, 38, PhD(biochem), 39. *Prof Exp:* Chemist, NJ Exp Sta, 39-40; res chemist, Wallerstein Labs, 40; res assoc, Barrett Co, 41; from asst chemist to supvr plant steroid units eastern regional res lab, Bur Agr & Indust Chem, USDA, 41-53, supvr eastern utilization res br, Agr Res Serv, 53-60; head natural prod lab, 60-66, dir chem & life sci lab, 66-71, vpres phys & life sci div, 71-83, CHIEF SCIENTIST, RES TRIANGLE INST, 83- *Concurrent Pos:* Adj prof, NC State Univ, 62- & Univ NC, Chapel Hill, 66-; consult, NIH, 62- *Honors & Awards:* Walter Harrung Mem lect, Univ NC, Col Pharm, 77; Eber Lect Award, Univ Ill, Col Pharm, 83. *Mem:* AAAS; Am Chem Soc; fel Am Asn Pharmaceut Scientists; Am Pharmacol Soc; Soc Econ Bot (pres, 75). *Res:* Plant chemistry; steroids; cancer chemotherapy; drug metabolism; chemistry and metabolism of cannabinoids. *Mailing Add:* PO Box 12194 Research Triangle Park NC 27709-2194

WALL, ROBERT ECKI, b Aurora, Ill, Aug 1, 35; m 63; c 3. GEOPHYSICS. *Educ:* Carleton Col, AB, 57; Columbia Univ, PhD(geophys), 65. *Prof Exp:* Res assoc marine geophys, Lamont Geol Observ, Columbia Univ, 65-66; sci officer marine geol & geophys, Off Naval Res, 66-70; prog dir submarine geol & geophys, 70-75, head oceanog sect, 75-81, HEAD OCEAN SCI RES SECT, NSF, 81- *Mem:* AAAS; Am Geophys Union; Geol Soc Am; Soc Explor Geophys. *Res:* Marine geophysics; geological oceanography. *Mailing Add:* Coburn Hall Univ Maine Orono ME 04469

WALL, ROBERT GENE, b Mo, Nov 17, 37; m 64; c 3. ORGANIC CHEMISTRY. *Educ:* Ore State Univ, BS, 61, MS, 63; Univ Wis-Madison, PhD(org chem), 66. *Prof Exp:* NIH fel, Univ Mich, Ann Arbor, 66-67; res chemist, 67-72, sr res chemist, 72-81, SR RES ASSOC CHEMIST, CHEVRON RES CO, RICHMOND, CALIF, 81- *Mem:* AAAS; Am Chem Soc; Soc Petrol Engrs. *Res:* Physical organic chemistry; catalysis and surfactants for enhanced oil recovery. *Mailing Add:* 2826 Wright Ave Pinole CA 94564-1040

WALL, ROBERT LEROY, b Doylestown, Pa, July 7, 21; m 52; c 4. MEDICINE. *Educ:* Oberlin Col, AB, 43; Temple Univ, MD, 46; Am Bd Internal Med, dipl, 55. *Prof Exp:* Intern, Bryn Mawr Hosp, 46-47, resident clin path, 47-48; mem staff med res, Univ Hosp, 50-52, asst prof med & dir lymphoma clin, Col Med, 52-57, assoc prof med, 57-65, asst dean res & secy col med, 71-73, PROF MED, COL MED, OHIO STATE UNIV, 65- *Concurrent Pos:* Consult, Dayton Vet Hosp, 54-; mem plasma proteins comt, Nat Res Coun, 63-71. *Mem:* AAAS; Am Soc Hemat; AMA; Am Asn Cancer Res; Am Fedn Clin Res. *Res:* Blood preservation; plasma protein fractionation; qualitative changes in globulins of diseased plasma. *Mailing Add:* 3650 Olentangy River Rd Columbus OH 43214

WALL, RONALD EUGENE, b Dryden, Ont, Feb 19, 36; m 57; c 4. PLANT PATHOLOGY. *Educ:* Ont Agr Col, BSA, 58; Univ Wis, PhD(plant path), 62. *Prof Exp:* Plant pathologist, Can Dept Agr, 62-66; plant pathologist, 66-73, RES SCIENTIST, DEPT ENVIRON, CAN FORESTRY SERV, 73- *Concurrent Pos:* Teacher, forest path. *Mem:* Am Phytopath Soc; Can Phytopath Soc. *Res:* Decays of maturing plants; trunk rots of forest trees; diseases of conifer seedlings; diseases of forest weeds. *Mailing Add:* Pac Forestry Ctr 506 W Barnside Rd Victoria BC V8Z 1M5 Can

WALL, THOMAS RANDOLPH, b Lakeland, Fla, Mar 23, 43; m; c 2. MOLECULAR BIOLOGY, IMMUNOLOGY. *Educ:* Univ SFla, AB, 65; Ind Univ, PhD(microbiol), 70. *Prof Exp:* Res assoc cell & molecular biol, Columbia Univ, 70-72; asst prof, 72-78, assoc prof, 78-79, PROF MICROBIOL & IMMUNOL, SCH MED, UNIV CALIF, LOS ANGELES, 79- *Concurrent Pos:* Damon Runyon Mem Fund Cancer Res fel, Columbia Univ, 70-72; mem, Molecular Biol Inst, Univ Calif, Los Angeles, 72-; founder & dir, INGENE Inc, Santa Monica, Calif, 80-; sci adv bd, FMC Bio Products; mem sci adv bd, XOMA Corp. *Mem:* Am Soc Microbiol; Am Asn Immunologists. *Res:* Expression and regulation of genes in eukaryotic cells; molecular analysis of the expression and control of genes in lymphocytes; developmental control of the immune response. *Mailing Add:* Molecular Biol Inst Univ Calif Los Angeles Los Angeles CA 90024

WALL, WILLIAM JAMES, b Northampton, Mass, June 25, 21; m 46; c 1. ENTOMOLOGY. *Educ:* Univ Mass, BS, 42, MS, 49; Univ Calif, PhD(entom), 52. *Prof Exp:* Instr biol, Univ Mass, 46-49; asst, Univ Calif, 49-52; instr biol, State Univ NY Albany, 52-54, asst prof, 54-56; PROF BIOL, BRIDGEWATER STATE COL, 56- *Concurrent Pos:* NIH-USPHS-Nat Commun Dis Ctr res grant, Cape Cod, Mass, 67-70; entomologist, Cape Cod Mosquito Control Proj, 58-; mem sci adv panel, Volta River Basin Area, WAfrica, WHO, 74-79. *Mem:* AAAS; Entom Soc Am; Am Mosquito Control Asn; Am Asn Univ Professors; Sigma Xi. *Res:* Biology, life history, control and taxonomy of Thysanura, Tabanidae, Ceratopogonidae and Culicidae. *Mailing Add:* Dept Biol Bridgewater State Col Bridgewater MA 02324

WALLACE, AARON, health systems, physics; deceased, see previous edition for last biography

WALLACE, ALEXANDER CAMERON, b St Thomas, Ont, Aug 27, 21; m 48; c 4. MEDICINE, PATHOLOGY. *Educ:* Univ Western Ont, BA, 47, MD, 48; FRCP(C), 72. *Prof Exp:* Clin intern, Victoria Hosp, London, Ont, 48-49; intern path, New Haven Hosp, 49-51; resident, 51-52; lectr med res, Univ Western Ont, 52-53, asst prof, 53-55; assoc prof path, Univ Man, 55-61; dir cancer res lab, 61-65, head dept path, 65-74, PROF PATH, FAC MED, UNIV WESTERN ONT, 65- *Concurrent Pos:* Markle Found scholar, 52-57; pathologist, Winnipeg Munic Hosp, 55-61; mem res adv comt, Nat Cancer Inst Can, 60-68, dir, 69-80; consult, Westminster Hosp, London, 63-78; pathologist, Univ Hosp, London, 72- *Mem:* Fel Am Col Physicians; Can Asn Path; Int Acad Path. *Res:* Cancer; biology of neoplasia; metastases; renal pathology; immunopathology. *Mailing Add:* 19 King St Unit 1001 340 Huron St London ON N6A 5N8 Can

WALLACE, ALFRED THOMAS, b Cranford, NJ, Nov 26, 35; m 58; c 3. ENVIRONMENTAL ENGINEERING. *Educ:* Rutgers Univ, BS, 59, Univ Wis, MS, 60, PhD(sanit eng), 65. *Prof Exp:* Asst engr, Triangle Conduit & Cable Co, 57-58 & Am Oil Co, 60-62; instr sanit eng, Univ Wis, 62-65; asst prof environ eng, Clemson Univ, 65-67; assoc prof, 67-72, PROF CIVIL ENG, UNIV IDAHO, 72- *Concurrent Pos:* Lectr, Calumet Ctr, Purdue Univ, 60-62. *Mem:* Am Water Works Asn; Water Pollution Control Fedn; Am Soc Civil Engrs. *Res:* Unit operations of environmental engineering; industrial wastes; ground water pollution. *Mailing Add:* Dept Civil Eng Univ Idaho Moscow ID 83843

WALLACE, ALTON SMITH, b New Bern, NC, Jan 3, 44; m 69; c 2. OPERATIONS RESEARCH. *Educ:* NC A&T State Univ, BS, 66; Pa State Univ, MS, 68; Univ Md, PhD(math), 74. *Prof Exp:* Eng officer, US Army Corps Engrs, 68-70; sr scientist, BDM Corp, Va, 73-75; assoc dir, 75-87, DIR, SYST PLANNING CORP, VA, 87- *Mem:* Asn US Army; Armed Forces Commun Electronics Asn. *Res:* Development of military weapons systems and surveillance systems; test and evaluation of systems; survivability assesments; weapons effectiveness. *Mailing Add:* 11803 Maher Dr Ft Washington MD 20744

WALLACE, ANDREW GROVER, b Columbus, Ohio, Mar 22, 35; m 57; c 3. INTERNAL MEDICINE, CARDIOVASCULAR PHYSIOLOGY. *Educ:* Duke Univ, BS, 58, MD, 59. *Prof Exp:* Intern med, Med Ctr, Duke Univ, 59-60, asst resident, 60-61; investr cardiovasc physiol, Nat Heart Inst, 61-63; chief resident med, 63-64, assoc, 64-65, from asst prof to assoc prof, 65-70, dir cardiac intensive care unit, 65-70, PROF MED, ASST PROF PHYSIOL, ASST DIR GRAD MED EDUC & CHIEF CARDIOL DIV, MED CTR, DUKE UNIV, 70-, ASSOC VPRES & CHIEF EXEC OFFICER, MED CTR, 81- *Concurrent Pos:* Fel cardiol, Duke Univ, 61; Markle scholar acad med, 65-70; USPHS career develop award, 65-70. *Mem:* Am Fedn Clin Res; Am Heart Asn. *Res:* Cardiology; electrocardiology; electrophysiology of the heart. *Mailing Add:* Med Ctr Med Box 3708 Duke Univ Durham NC 27710

WALLACE, ANDREW HUGH, b Glasgow, Scotland, June 14, 26. MATHEMATICS. *Educ:* Univ Edinburgh, MA, 46; St Andrews Univ, PhD(math), 49. *Prof Exp:* Asst lectr math, Univ Col, Dundee, 46-49, lectr, 49-50; Commonwealth Fund fel, Univ Chicago, 50-52; lectr, Univ Col, Dundee, 52-53; from lectr math to sr lectr, Univ Col, NStaffordshire, 53-57; asst prof, Univ Toronto, 57-59; asst prof, Ind Univ, 59-61, prof, 61-64; grant in aid, Inst Advan Study, 64-65; prof math, Univ Pa, 65-88; RETIRED. *Concurrent Pos:* Vis prof & assoc dir, Univ Pa Group, Pahlavi Univ, Iran, 71-72. *Mem:* Am Math Soc; Can Math Cong; London Math Soc. *Res:* Algebra; algebraic geometry and topology. *Mailing Add:* Parados Nikolaou Skoula Chania Crete 73100 Greece

WALLACE, ARTHUR, b Bear River, Utah, Jan 4, 19; m 43; c 4. PLANT NUTRITION, PLANT PHYSIOLOGY. *Educ:* Utah State Univ, BS, 43; Rutgers Univ, PhD, 49. *Prof Exp:* Chief div environ biol, 66-72, asst chief div environ biol, 72-79, PROF PLANT NUTRIT, UNIV CALIF, LOS ANGELES, 49- *Mem:* Am Soc Plant Physiol; Am Chem Soc; Am Soc Hort Sci; fel Am Soc Agron; fel Soil Sci Soc Am. *Res:* Inorganic plant nutrition and related physiology; major cations, nitrogen and micronutrient elements including their supply to plants by synthetic chelating agents; comparative mineral nutrition of plants; ecophysiology; trace element toxicity. *Mailing Add:* 10215 Clematis Ct Los Angeles CA 90077

WALLACE, BRUCE, b McKean, Pa, May 18, 20; m 45; c 2. GENETICS. *Educ:* Columbia Univ, AB, 41, PhD(genetics), 49. *Prof Exp:* Res assoc dept genetics, Carnegie Inst, 47-49; res assoc, LI Biol Asn, 49-58; assoc prof genetics, Cornell Univ, 58-61, prof genetics, 61-81; prof biol, 81-82, DISTINGUISHED PROF, VA POLYTECH INST & STATE UNIV, 83- *Concurrent Pos:* Humboldt preisträger, 86-87. *Mem:* Nat Acad Sci; Genetics Soc Am (secy, 68-70, pres, 74); Am Acad Arts & Sci; Am Soc Naturalists (secy, 56-58, pres, 70); AAAS; Soc Study Evolution (pres, 74); Am Genetics Asn (pres, 90). *Res:* Population dynamics and speciation of Drosophila. *Mailing Add:* Dept Biol Va Polytech Inst & State Univ Blacksburg VA 24061

WALLACE, C(HARLES) E(DWARD), b Portland, Ore, Apr 19, 29; m 65; c 2. ENGINEERING MECHANICS, PHYSICS. *Educ:* Lewis & Clark Col, BS, 51; Ore State Univ, MS, 54; Stanford Univ, PhD(eng mech), 59. *Prof Exp:* Engr, Douglas Aircraft Co, 53-55; asst prof eng sci, 58-59, from assoc prof to prof & chmn dept, 59-67, prof eng mech & mat & chmn dept, 67-74, prof & chmn aerospace eng & eng sci, 74-82, prof mech & aerospace, 82-87, PROF & ASST DEAN, COL ENG & APPL SCI, ARIZ STATE UNIV, 87- *Concurrent Pos:* Consult, Gen Elec Co, 59 & NAm Aviation, Inc, 60-; Nat Res Coun sr resident res associateship, NASA-Langley Res Ctr, 68-69. *Mem:* Am Inst Aeronaut & Astronaut; Acoust Soc Am; Am Soc Eng Educ; Am Acad Mech. *Res:* Plasticity of anisotropic materials; thermoelasticity of plates; vibration of aircraft structure; acoustic fatigue and damping; community noise control; radiation resistance. *Mailing Add:* Col Eng Ariz State Univ Tempe AZ 85287-5506

WALLACE, CARL J, b Lane's Prairie, Mo, Feb 11, 38; m 61; c 3. CHEMICAL ENGINEERING. *Educ:* Univ Mo, Rolla, BS, 61, MS, 62, PhD(chem eng), 66. *Prof Exp:* Teaching asst chem, Sch Mines, Univ Mo, Rolla, 61-62, 63-64 & 65-66; from asst prof to assoc prof chem eng, SDak Sch Mines & Technol, 66-75; group leader, Jet Propulsion Lab, Calif Inst Technol, 75-80; detailee, US Dept Energy, 80-83; tech staff, Argonne Nat Lab, 84-88; TECH PROG INTEGRATOR, SOLAR ENERGY RES INST, 88- *Concurrent Pos:* Lectr, NSF Summer Inst, 67; NSF res initiation grant, 69-70. *Mem:* Am Inst Chem Engrs; Am Chem Soc; Am Soc Eng Educ; AAAS; Sigma Xi. *Res:* Biochemical engineering; chemical processing by ultraviolet and ultrasonic energy; bioconversion; energy recovery from wastes; environmental engineering. *Mailing Add:* 907 Sixth St SW Apt 303 C Washington DC 20024

WALLACE, CHESTER ALAN, b Rockville Centre, NY, Oct 1, 42; m 68; c 2. STRATIGRAPHY, SEDIMENTARY PETROLOGY. *Educ:* Antioch Col, BA, 65; Univ Calif, Santa Barbara, PhD(geol), 72. *Prof Exp:* Asst prof geol, Ga Southwestern Col, 69-74; GEOLOGIST, US GEOL SURV, 74- *Mem:* Geol Soc Am; Soc Econ Paleontologists & Mineralogists. *Res:* Stratigraphy, sedimentology, paleocurrent analysis, and diagenesis of clastic sedimentary rocks; tectonic history and sedimentary basin analysis of Precambrian sedimentary rocks in the western United States. *Mailing Add:* US Geol Surv Mail Stop 913 Federal Ctr Box 25046 Denver CO 80225

WALLACE, CRAIG KESTING, b Woodbury, NJ, Dec 4, 28; m 60; c 3. MEDICINE, RESEARCH ADMINISTRATION. *Educ:* Princeton Univ, AB, 50; NY Med Col, MD, 55. *Prof Exp:* Clin investr, US Naval Med Res Unit, Taiwan, 60-63; instr med, Jefferson Med Col Hosp, 60-64; from asst prof to assoc prof, Sch Med & Sch Hyg & Pub Health, Johns Hopkins Univ, 64-72 & Ctr Med Res & Training Calcutta India, 64-72; cmndg officer, US Naval Med Res Unit Ethiopia, 72-76, prog mgr infectious dis, Naval Med Res & Develop Command, 76-78, chief internal med, Camp Pendleton, 78-80, dir clin serv, Naval Regional Med Ctr, Jacksonville, Fla, 80-82, cmdg officer US Naval Med Res Unit, Egypt, 80-84; dir, Fogarty Int Ctr, 84-87, ASSOC DIR INT RES, NIH, 84- *Concurrent Pos:* Adv & consult, WHO , 62-; consult, Magee Mem Hosp Philadelphia, Pa, 63-64; physician & consult, Vet Admin Hosp, Perry Point, Md, 67-72; mem Bact & Mycol Study Sect, NIH, 68-72, chmn, 71-72; physician Good Samaritan Hosp, Baltimore, Md, 69-72; ed bd & correp ed, Ethiopian Med J, 74-; clin assoc prof med, Uniformed Servs Univ Sci, 77-; mem, sci adv bd, Leonard Wood Mem Liploser Found, 84-88; physician, Good Samaritan Hosp, Baltimore, 69-72; dir, John E Fogarty Int Ctr Advan Study Health Sci, NIH, 84-87. *Mem:* AAAS; fel Am Col Physicians; Infectious Dis Soc Am; Am Soc Microbiol; Royal & Am Soc Trop Med & Hyg; Royal Soc Trop Med & Hyg. *Res:* Pathophysiology and treatment of infectious diseases, especially in the areas of tropical enteric infections; physician training and research in international medicine. *Mailing Add:* Dept Int Res NIH Bldg 1 Rm 338 Bethesda MD 20892

WALLACE, DAVID H, b Cambridge, Mass, May 15, 37; m 63. MICROORGANISMS. *Educ:* Harvard Univ, AB, 58; Mass Inst Technol, SM, 63, ScD(food sci & eng), 66. *Prof Exp:* Dep dir, New Eng Enzyme Ctr, Tufts Univ, 66-80; MICROBIOLOGIST, NEW ENG BIOLABS, 80- *Mem:* Sigma Xi. *Res:* Carry out large scale culture of microorganisms, optimize growth conditions and improve methods for large scale purification. *Mailing Add:* 61 Donazette St Wellesley MA 02181

WALLACE, DAVID LEE, b Homestead, Pa, Dec 24, 28; m 55; c 3. STATISTICS, DATA ANALYSIS. *Educ:* Carnegie Inst Technol, BS, 48, MS, 49; Princeton Univ, PhD(math), 53. *Prof Exp:* Moore instr math, Mass Inst Technol, 53-54; from asst prof to assoc prof statist, 54-67, PROF STATIST, UNIV CHICAGO, 67- *Concurrent Pos:* Fel, Ctr Advan Study Behav Sci, 60-61; mem comput & biomath sci study sect, NIH, 70-74. *Mem:* Fel Am Statist Asn; Inst Math Statist; Biometric Soc; fel Royal Statist Soc. *Res:* Theoretical statistics; computer methods. *Mailing Add:* Dept Statist Univ Chicago 5734 S University Ave Chicago IL 60637

WALLACE, DONALD HOWARD, b Driggs, Idaho, June 27, 26; m 49; c 5. PLANT GENETICS. *Educ:* Utah State Agr Col, BS, 53; Cornell Univ, PhD(plant breeding), 58. *Prof Exp:* Asst plant breeding, 53-55 & 57, actg asst prof plant breeding & veg crops, 55-57, from asst prof to assoc prof, 58-71, PROF PLANT BREEDING & VEG CROPS, CORNELL UNIV, 71- *Concurrent Pos:* Vis prof, Univ Philippines, Los Bonos, 68-69, Mich State Univ, 78-79. *Honors & Awards:* Campbell Award, 70; Asgrow Award, 81. *Mem:* Am Soc Hort Sci; Crop Sci Soc Am; Am Soc Plant Physiol; fel AAAS. *Res:* Development of hybrid and improved varieties of vegetables; physiological genetics of crop yield. *Mailing Add:* Dept Plant Breeding Cornell Univ Ithaca NY 14850

WALLACE, DONALD MACPHERSON, JR, b Montclair, NJ, June 24, 34. MECHANICAL ENGINEERING. *Educ:* Univ Vt, BSME, 60; Univ Ill, Urbana, MS, 62; Columbia Univ, EngScD(mech eng), 68. *Prof Exp:* Instr mech eng, Univ Ill, 60-62; PROF MECH ENG, NORWICH UNIV, 62- *Concurrent Pos:* Engr, Allis-Chalmers Corp, 73-74. *Mem:* Am Soc Mech Engrs; Am Soc Eng Educ; Sigma Xi. *Res:* Kinematics of spatial mechanisms. *Mailing Add:* Dept Mech Eng Norwich Univ Northfield VT 05663

WALLACE, DOUGLAS CECIL, b Cumberland, Md, Nov 6, 46. MITOCHONDRIAL GENETICS. *Educ:* Cornell Univ, BS, 68; Yale Univ, MPh, 72, PhD(somatic cell genetics), 75. *Prof Exp:* Res microbiologist, USPHS Northwestern Water Hyg Lab, Gig Harbor, Wash, 68-70; NIH fel, Sch Med, Yale Univ, 75-76; asst prof, Sch Med, Stanford Univ, 76-83; prof biochem, 83-90, DIR, CTR GENETICS & MOLECULAR MED, EMORY UNIV, 90- *Mem:* Sigma Xi; Am Soc Microbiol; AAAS; Am Soc Human Genetics. *Res:* Study of the genetics, biogenesis and phylogenetic relationships of mammalian mitochondria. *Mailing Add:* Emory Univ Sch Med 1510 Clifton Rd 3031 Atlanta GA 30322

WALLACE, EDITH WINCHELL, b Jersey City, NJ, Oct 3, 35; m; c 3. SPERMATOGENESIS, ENDOCRINOLOGY. *Educ:* Montclair State Col, BA, 56, MA, 61; Rutgers Univ, PhD(zool), 69. *Prof Exp:* Instr biol, Westwood Bd Educ, NJ, 56-61; instr sci, Englewood Hosp Sch Nursing, 65-68; PROF BIOL, WILLIAM PATERSON COL, NJ, 68- *Mem:* AAAS; NY Acad Sci; Soc Study Reproduction. *Res:* Morphological, histological, transmission electron microscope and physiological investigations in selenium and zinc deficient rodents of the roles of these trace elements in spermatogenesis. *Mailing Add:* 19 Iris Circle Glen Rock NJ 07452

WALLACE, EDWIN GARFIELD, b Akron, Ohio, Jan 27, 17; m 44; c 6. ORGANIC CHEMISTRY. *Educ:* Univ Miami, Ohio, AB, 38; Ohio State Univ, PhD(chem), 42. *Prof Exp:* Res chemist, Eastman Kodak Co, NY, 42-47; res chemist, US Naval Ord Test Sta, Calif, 47-48; res chemist & group leader, Shell Chem Corp, 48-54; from asst to vpres res to mgr chem res, Western Res Ctr, Stauffer Chem Co, 54-65, lab dir, 65-67, sr res assoc, 67-82; RETIRED. *Concurrent Pos:* Chmn Calif sect, Am Chem Soc, 79. *Mem:* Am Chem Soc. *Res:* Agricultural chemicals; industrial organic chemicals and products; polymers; fluorine chemicals. *Mailing Add:* 133 Sleepy Hollow Lane Orinda CA 94563

WALLACE, FRANKLIN GERHARD, b Deer River, Minn, Apr 3, 09; m 35, 75; c 4. PARASITOLOGY. *Educ:* Carleton Col, BA, 28; Univ Minn, MA, 30, PhD(zool), 33. *Prof Exp:* Asst prof biol, Lingnam Univ, 33-37; from inst to prof, 37-77, EMER PROF ZOOL, UNIV MINN, MINNEAPOLIS, 77- *Concurrent Pos:* Consult, Vet Admin Hosp, 46-73; mem trop med & parasitol study asst, NIH, 70-74; consult, Brazil, 77, 79, 81 & 87. *Mem:* Am Micros Soc (vpres, 66); Am Soc Parasitol (vpres, 74); Am Soc Trop Med & Hyg; Soc Protozool (pres, 77). *Res:* Trypanosomatid flagellates and parasites of insects. *Mailing Add:* 2603 Cohansey St St Paul MN 55113

WALLACE, FREDERIC ANDREW, b Boston, Mass, Apr 28, 33; m 54; c 2. PHYSICAL CHEMISTRY, ANALYTICAL CHEMISTRY. *Educ:* Harvard Univ, AB, 58; Calif Inst Technol, MS, 60; Tufts Univ, PhD(chem), 67. *Prof Exp:* GROUP LEADER PHYS & ANALYTICAL CHEM RES, CENT ANALYSIS LAB, RES DIV, POLAROID CORP, 67- *Concurrent Pos:* Chemist, Monsanto Chem Co, 70-74. *Mem:* Am Chem Soc. *Res:* Investigations of structure and bonding of organo-silver complexes and salts; measurement of stability constants, and solubility products of same; chemical analysis and method development for photographic chemicals. *Mailing Add:* 53 Eaton Rd Framingham MA 01701-2727

WALLACE, GARY DEAN, b Pasadena, Calif, Jan 4, 46; m 77; c 2. PLANT SYSTEMATICS, PLANT ANATOMY. *Educ:* Calif State Univ, BA, 67, MA, 72; Claremont Grad Sch, PhD(bot), 75. *Prof Exp:* Biologist, Los Angeles Count Aboretum, 75-81; ASSOC CUR & BOTANIST, NATURAL HIST MUS, LOS ANGELES, 82- *Concurrent Pos:* Exten Instr, Univ Calif, Los Angeles, 76- *Mem:* Am Soc Plant Taxonomists; Int Asn Plant Taxonomy; Bot Soc Am; AAAS; Asn Trop Biol; Sigma Xi. *Res:* Taxonomy of the monotropoidese (ericpase); checklist of the flora of the offshore islands of Southern California; ecological word anatomy of arctostaphylos (ericpase). *Mailing Add:* Bot Sect Nat Hist Mus 900 Exposition Blvd Los Angeles CA 90007

WALLACE, GARY OREN, b Stewart Co, Tenn, Apr 2, 40; m 62; c 2. ECOLOGY. *Educ:* Austin Peay State Univ, BS, 62; Univ Tenn, MS, 64, PhD(zool), 70. *Prof Exp:* Teacher biol, Maryville Col, 64-65; ASST PROF BIOL, MILLIGAN COL, 67-68 & 71- *Concurrent Pos:* Ed, The Migrant, 71-81. *Mem:* Wilson Ornith Soc; Nat Audubon Soc; Sigma Xi. *Res:* Abundance and distribution of certain bird populations. *Mailing Add:* Rte 7 Box 3430 Sunrise Dr Elizabethton TN 37643

WALLACE, GERALD WAYNE, b Sault Ste Marie, Mich, July 20, 33; m 54; c 3. ANALYTICAL CHEMISTRY, PHYSICAL CHEMISTRY. *Educ:* Univ Mich, BS, 55; Purdue Univ, MS, 57, PhD(anal chem), 59. *Prof Exp:* Sr chemist, Esso Res & Eng Co, 59-60; sr anal chemist, 60-65, res scientist, 65-67, head anal develop phys, 67-71, dir, Phys Chem Res Div, 71-79, dir pharmaceut res, 79-80, DIR ANALYTICAL DEVELOP, ELI LILLY & CO, 80- *Concurrent Pos:* USP Comt Rev. *Mem:* Am Chem Soc; AAAS. *Res:* Spectroscopy; fluorescence; photochemistry. *Mailing Add:* Lilly Res Labs Eli Lilly Corp Ctr Bldg 58/1 Indianapolis IN 46285

WALLACE, GORDON DEAN, b Los Angeles, Calif, Dec 17, 27; m 52; c 2. MICROBIOLOGY. *Educ:* Colo State Univ, BS, 52, DVM, 54; Univ Calif, MPH, 62. *Prof Exp:* Epidemiologist, Commun Dis Ctr, USPHS, 54-61, med researcher, 61-79, asst sci dir, 78-83, sr policy analyst, exec Off Pres, off Sci & Technol Policy, 83-84, assoc sci dir, actg dep dir & sci dir, Nat Inst Allergy & Infectious Dis, 84-86; sr assoc, Ling Technol, 86-87; pres, WBA, 87-91; PRES, BIO-BRITE, 89- *Concurrent Pos:* Consult infectious dis, Tripler Us Army Hosp, Childrens Hosp & Queens Med Ctr, Honolulu, Hawaii, 61-78; assoc clin prof trop med & med microbiol, Sch Med, Univ Hawaii, 70-78; counr, Am Soc Trop Med & Hyg, 77-81, Sci prog chmn, 81-83; guest instr, Georgetown Univ, 83-84; mem competetive grants policy adv comt, USDA, 83-85; pres, NIH Alumni Asn, 90-91. *Honors & Awards:* McCallam Award, Asn Mil Surgeons US, 83. *Mem:* fel AAAS; Am Soc Trop Med & Hyg. *Res:* Epidemiology of infectious diseases, including eosinophilic meningitis, influenza and toxoplasmosis. *Mailing Add:* 10233 Holly Hill Pl Potomic MD 00854

WALLACE, GORDON THOMAS, b Chicago, Ill, Sept 22, 42; m 69; c 2. CHEMICAL OCEANOGRAPHY, ATMOSPHERIC CHEMISTRY. *Educ:* Antioch Col, BS, 65; Univ RI, PhD(oceanog), 76. *Prof Exp:* Anal chemist, Geigy Chem Corp, 62-63; chemist, Naval Res Lab, 63-69; res asst, Univ RI, 72-76; res assoc, Skidaway Inst Oceanog, 75-76, asst prof, 76-82; ASSOC PROF ENVIRON SCI, UNIV MASS, 82-, DIR, ENVIRON SCI PROG, 90- *Concurrent Pos:* Consult, Amos T Shaler Inc, 73-74, Knoll Atomic Power Labs; adj asst prof, Ga Inst Technol, 78-88. *Mem:* Am Soc Limnol & Oceanog; AAAS; Am Geophys Union; Estaurin Res Fed; Am Chem Soc. *Res:* Biogeochemistry of metals in the marine environment; trace metal-organic interactions; chemical fractionation at the sea-air interface; waste disposal in coast-marine environment. *Mailing Add:* Environ Sci Prog Univ Mass Boston Harbor Campus Boston MA 02125

WALLACE, GRAHAM FRANKLIN, b Santa Rosa, Calif, Mar 27, 35; m 59; c 2. SOFTWARE ENGINEERING, SOFTWARE DEVELOPMENT. *Educ:* Pomona Col, BA, 57; Univ Calif, Berkeley, MA, 60. *Prof Exp:* Mathematician, Math Sci Dept, 60-64, res mathematician, 64-66, asst mgr admin, 66-70, sr systs programmer, Info Sci Lab, 70-74, sr res engr, Systs Tech Lab, Stanford Res Inst, 74-84, SR RES ENGR, SPEC COMMUN SYSTS LAB, SRI INT, 84- *Concurrent Pos:* Chmn seventh symposium gaming & chmn steering comt, Nat Gaming Coun, 68; vchmn tech prog eight int conf, Inst Elec & Electronics Engrs Comput Soc, 74. *Mem:* AAAS; Sigma Xi. *Res:* Development of computer software; design, analysis and evaluation of computer-based information systems. *Mailing Add:* 1331 Hillview Dr Menlo Park CA 94025

WALLACE, HAROLD DEAN, b Walnut, Ill, June 8, 22; m 45; c 4. ANIMAL NUTRITION. *Educ:* Univ Ill, BS, 45, MS, 47; Cornell Univ, PhD(animal nutrit), 50. *Prof Exp:* From asst prof to assoc prof, 50-61, PROF ANIMAL NUTRIT, UNIV FLA, 61-, CHMN ANIMAL SCI, 76- *Honors & Awards:* Am Feed Mfrs Award, 62. *Mem:* AAAS; Am Soc Animal Sci; Am Dairy Sci Asn; Am Inst Nutrit. *Res:* Swine nutrition; vitamins; antibiotics; amino acids; carcass quality. *Mailing Add:* 1812 SW 36th Pl Gainesville FL 32608

WALLACE, HELEN M, b Hoosick Falls, NY, Feb 18, 13. PUBLIC HEALTH. *Educ:* Wellesley Col, AB, 33; Columbia Univ, MD, 37; Harvard Univ, MPH, 43. *Prof Exp:* NY Dept Health, 43-55; prof prev med, New York Med Col, 55-56; prof maternal & child health, Univ Minn, 56-59; chief child health studies, US Children's Bur, 59-62; prof maternal, child & family health, Sch Pub Health, Univ Calif, Berkeley, 62-80; PROF, GRAD SCH PUB HEALTH, SAN DIEGO STATE UNIV, 80- *Concurrent Pos:* WHO traveling fel, 57; consult, WHO, Uganda, 61, Philippines, 66, India, 68 & 69, Turkey, 69, Geneva, 70 & 74 & Iran, 72; Ford Found consult, Sch Pub Health, Univ Antioquaia, Colombia, 71; consult, Health Bur, Panama Canal Co, 72, India, Thailand, Burma & Ceylon, 75; dir Uganda Prog, Univ Calif, India & Thailand, 81, Nepal, 83 & Burma, 85, China, 87, Zimbabwe, 84-87. *Honors & Awards:* Am Pub Health Asn; Am Acad Pediat. *Mem:* Asn Teachers Maternal & Child Health (pres); Am Pub Health Asn; Am Acad Pediat. *Res:* Maternal and child health. *Mailing Add:* San Diego State Univ Grad Sch Pub Health San Diego CA 92182

WALLACE, HERBERT WILLIAM, b Brooklyn, NY, Dec 11, 30; m 54; c 3. BIOCHEMISTRY, SURGERY. *Educ:* Harvard Univ, AB, 52; Tufts Univ, MD, 56, MS, 60; Wharton Sch Univ Pa, MBA, 81. *Prof Exp:* Asst attend surgeon, Elmhurst Gen Hosp Div, Mt Sinai Hosp, 65-66; assoc, 66-70, from asst prof to assoc prof surg & assoc prof physiol, Sch Med, 70-79, assoc prof bioeng, Col Eng & Appl Sci, 74-76, PROF SURG & PHYSIOL, SCH MED & PROF BIOENG, COL ENG & APPL SCI, UNIV PA, 76-; RES ASSOC, DIV CARDIOL, PHILADELPHIA GEN HOSP. *Concurrent Pos:* Res fel, Nat Heart Inst, 58-59; teaching fel surg, Sch Med, Univ Pittsburgh, 61-62; Am Thoracic Soc fel, 62-65; John Polachek Found Med Res fel, Div Cardio-Thoracic Surg, Mt Sinai Hosp, New York, 65-66; asst surgeon, Grad Hosp, Univ Pa, 66-74, assoc surg, 74-, assoc dir, Gen Clin Res Ctr, 67-70, actg dir, 70-73. *Mem:* AAAS; Asn Acad Surg; Am Chem Soc; Am Col Surg; Am Fedn Clin Res; Sigma Xi. *Res:* Biochemistry and physiology of extracorporeal circulation and respiration; lung metabolism; hemoglobin; artificial red cell; cardiac metabolism; cancer immunology; management and marketing of medical services. *Mailing Add:* 255 Harrogate Rd Wynnewood PA 19096

WALLACE, JACK E, b Harrisburg, Ill, Jan 5, 34; m 55; c 2. ANALYTICAL BIOCHEMISTRY, BIOCHEMICAL PATHOLOGY. *Educ:* Univ Southern Ill, BA, 55, MA, 57; Purdue Univ, PhD(biochem), 62. *Prof Exp:* Instr chem, Univ Southern Ill, 55-57; instr biochem, Purdue Univ, 59-61; chemist anal lab, State Chemist's Lab, Ind, 57-59; chief forensic toxicol br, USAF Sch Aerospace Med, 61-72; from asst prof to assoc prof, 72-79, PROF PATH, UNIV TEX HEALTH SCI CTR, SAN ANTONIO, 79- *Concurrent Pos:* Consult, Army Med Lab, Ft Sam Houston, Tex, Audie Murphy Vet Admin Hosp, San Antonio, Tex, Wilford Hall USAF Med Ctr, Lackland AFB, Tex, & Harris Med Labs, Ft Worth, Tex, SW Res Inst, San Antonio, Tex, Abuse Testing Prog, Nat Inst Drug Abuse; inspector, Matrix Technologies, Houston, Tex; dir, Precision Anal Labs, San Antonio, Tex, 90-; consult, Nichols Inst N Tepous, 90- *Honors & Awards:* Alexander O Gettler Award in Forensic Toxicol, 91. *Mem:* Sr mem Am Chem Soc; fel Am Inst Chem; fel Am Acad Forensic Sci; Am Acad Clin Toxicol; Am Asn Clin Chemists. *Res:* Forensic toxicology; drug metabolism; drug analysis in biological specimens; pharmacology; clinical chemistry; biochemical pathology; over 150 publications in various scientific journals. *Mailing Add:* Dept Path Univ Tex Health Sci Ctr 7703 Floyd Curl Dr San Antonio TX 78284

WALLACE, JAMES, b New Brunswick, NJ, Oct 6, 32; m 58; c 5. OPTICS, FLUID MECHANICS. *Educ:* US Merchant Marine Acad, BS, 54; Univ Notre Dame, MS, 59; Brown Univ, PhD(fluid mech), 63. *Prof Exp:* Res engr, Bendix Corp, 58-59; prin res scientist optics, Fluid Mech, Avco Everett Res Lab & Avco Systs Div, 63-73; PRES, FAR FIELD, INC, 73- *Mem:* Optical Soc Am; Inst Elec & Electronics Engrs. *Res:* Theoretical optics; lasers; fluid mechanics. *Mailing Add:* Six Thoreau Way Sudbury MA 01776

WALLACE, JAMES BRUCE, b Williamsburg Co, SC, Mar 2, 39; m 62; c 1. ENTOMOLOGY, HYDROBIOLOGY. *Educ:* Clemson Univ, BS, 61; Va Polytech Inst, MS, 63, PhD(entom), 67. *Prof Exp:* Res asst entom, Va Polytech Inst, 61-66; asst prof, 67-71, assoc prof, 71-77, PROF ENTOM, UNIV GA, 77-, MEM STAFF, INST ECOL, 69- *Concurrent Pos:* Environ Protection Agency res grant, Univ Ga, 68-72; NSF res grant, 74-88; vis scientist, Univ Lund, Sweden, 80. *Mem:* AAAS; Ecol Soc Am; Am Entom Soc; NAm Benthol Soc. *Res:* Stream ecology; aquatic entomology; secondary production of invertebrates. *Mailing Add:* Dept Entom Univ Ga Athens GA 30602

WALLACE, JAMES D, b Miss, Mar 6, 04. ELECTRICAL ENGINEERING. *Educ:* Univ Miss, BA, 25, MA, 27. *Prof Exp:* Electronic scientist, radio physicist, consult, Naval Res Lab, 28-66; RETIRED. *Mem:* Fel Inst Elec & Electronics Engrs. *Res:* Military communication with voice communication; teletype and data transmission. *Mailing Add:* 3501 Bayshore Blvd Apt 803 Tampa FL 33629

WALLACE, JAMES M, b Augusta, Ga, Aug 11, 39; m 84; c 1. TURBULENT FLOW. *Educ:* Ga Inst Technol, BCE, 62; MSc, 64; Oxford Univ, PhD(eng sci), 69. *Prof Exp:* Hydraul engr, Harza Eng Co, 64-65 & Sir William Halcrow & Partners, London, 65-66; res scientist, Max-Planck Inst, 69-75; from asst prof to assoc prof, 75-83, PROF MECH ENG, UNIV MD, COLLEGE PARK, 83- *Concurrent Pos:* Vis asst prof, Ohio State Univ, 71-73; asst provost, Univ Md, Col Park, 85-86, asst dean, 86-87. *Mem:* Fel Am Phys Soc. *Res:* Interaction of science and technology with society; physics of turbulent flow. *Mailing Add:* Dept Mech Eng Univ Md College Park MD 20742

WALLACE, JAMES ROBERT, b Magnolia, Ark, Nov 9, 38; m 63; c 3. HYDROLOGY, FLUID MECHANICS. *Educ:* Ga Inst Technol, BCivE, 61, MS, 63; Mass Inst Technol, ScD, 66. *Prof Exp:* Res asst civil eng, Mass Inst Technol, 62-66; asst prof, 66-69, ASSOC PROF CIVIL ENG, GA INST TECHNOL, 69- *Mem:* Am Geophys Union; Am Soc Civil Engrs. *Res:* Flow of fluids in open channels; design and operation of water resource systems; computer simulation of hydrologic processes. *Mailing Add:* 2680 Peppermint Dr Tucker GA 30084

WALLACE, JAMES WILLIAM, JR, b Cincinnati, Ohio, July 31, 40; m 62; c 2. PLANT PHYSIOLOGY BIOCHEMISTRY, BIOLOGICAL PHOTOGRAPHY. *Educ:* Miami Univ, BS, 62, MS, 64; Univ Tex, Austin, PhD(plant biochem), 67. *Prof Exp:* Technician, Kimberly-Clarke Corp, 62; from asst to assoc prof, 67-77, PROF BIOL, WESTERN CAROLINA UNIV, 78- *Concurrent Pos:* Secy, Phytochem Soc NAm, 71-74, ed-in-chief, 74-76; fac exchange prog, People's Repub China, 84; sr res fel, NZ Nat Res Adv Coun, 76-77; chmn, Southeastern Sect, Bot Soc Am, 82-85, Phytochem Sect, 85-87; biol photogr, Lobdell & Assoc, Anchorage, Alaska; chmn fac, Western Carolina Univ, 88-92. *Mem:* Sigma Xi; Bot Soc Am; Phytochem Soc NAm (secy, 71-74); Explorers Club; Am Fern Soc. *Res:* Flavonoids, biosynthesis, physiology and distribution; biological photography. *Mailing Add:* Dept Biol Western Carolina Univ Cullowhee NC 28723

WALLACE, JOAN M, b Rochester, NY, Mar 7, 28. PLANT BIOCHEMISTRY, TOXICOLOGY. *Educ:* Cornell Univ, BS, 51; Rutgers Univ, MS, 54, PhD(plant physiol), 57. *Prof Exp:* Res fel biol, Calif Inst Technol, 58-63; res chemist, Western Regional Res Lab, USDA, 63-87; RETIRED. *Mem:* Am Inst Biol Sci; Sigma Xi. *Res:* Toxicology of natural plant constituents; effects of fumigants on food composition. *Mailing Add:* 5951 Dam Road El Sobrante CA 94803

WALLACE, JOHN F(RANCIS), b Boston, Mass, Oct 26, 19; wid; c 3. METALLURGY, FOUNDRY TECHNOLOGY. *Educ:* Mass Inst Technol, BS, 41, MS, 53. *Prof Exp:* Asst metallurgist, Watertown Arsenal, Mass, 41-42, metallurgist, 46-47, sr metallurgist, 47-52, prin metallurgist, 52-54; from assoc prof to prof, 54-80, dir NASA-CCDS, 87-90, REP STEEL PROF METALL, CASE WESTERN RESERVE UNIV, 80-, CHMN DEPT METALL & MAT SCI, 74-, LTV STEEL PROF METALL, 87-, CO-DIR, NASA-CCDS, 91- *Concurrent Pos:* Consult, 55; Hoyt Mem lectr, Am Foundrymen's Soc, 75. *Honors & Awards:* Pangborn Gold Medal, Foundrymen's Soc, 62; Nyselius Award, Am Die Casting Inst, 67; Gold Medal, Gray & Ductile Iron Founders' Soc, 70; Howard Taylor Award, Am Foundry Men's Soc, 81; Doehler Award, Am Die Casting Inst, 84. *Mem:* Foundrymen's Soc; fel Am Soc Metals; Am Inst Mining, Metall & Petrol Engrs; Soc Die Casting Engrs; Sigma Xi. *Res:* Cast metals; control of casting processes, including solidification behavior, gating, rising, mold selection and behavior; die casting; heat treatment of metals; welding; metal forming and mechanical behavior of metals. *Mailing Add:* DMSE 10900 Euclid Case Western Reserve Univ Cleveland OH 44106

WALLACE, JOHN HOWARD, b Cincinnati, Ohio, Mar 8, 25; m 45, 79; c 2. MICROBIOLOGY, IMMUNOLOGY. *Educ:* Howard Univ, BS, 47; Ohio State Univ, MS, 49, PhD(bact), 53. *Prof Exp:* Asst virol, Children's Hosp Res Found, Cincinnati, 47 & 49-51; asst bact, Ohio State Univ, 51-53, asst instr, 53; asst bacteriologist, Leonard Wood Mem Lab & res assoc bact & immunol, Harvard Univ Med Sch, 55-59; from asst prof to prof microbiol, Meharry Med Col, 59-66; from assoc prof to prof, Sch Med, Tulane Univ, 66-70; prof, Ohio State Univ, 70-72; PROF MICROBIOL & IMMUNOL & CHMN DEPT MICROBIOL, SCH MED, UNIV LOUISVILLE, 72- *Concurrent Pos:* USPHS fel, Ohio State Univ, 54-55; USPHS sr res fel, 59-61; NIH career res develop award, 61-66; mem bd sci counrs, Nat Inst Allergy & Infectious Dis, 72-75, chmn, 75-76; prof microbiol & immunol & chmn dept & assoc dean for acad affairs, Sch Med, Morehouse Col, 79-80; mem cancer res manpower rev comn, Nat Cancer Inst, 77-81; mem microbiol & infectious dis adv comn, Nat Inst Allergy & Infectious Dis, 82- *Mem:* Am Asn Immunol; Transplantation Soc; fel Am Acad Microbiol; Am Asn Cancer Res; Soc Exp Biol & Med. *Res:* Virus modified erythrocytes; in vitro cultivation of murine leprosy bacilli; tissue culture in agar systems; delayed hypersensitivity; immunology of leprosy; tissue transplantation; tumor immunology; effects of tobacco smoke on immune responses; regulation of the immune response. *Mailing Add:* Dept Microbiol Immunol Univ La Sch Med 319 Abraham Frexner Way Louisville KY 40292

WALLACE, JOHN M, JR, b Cincinnati, Ohio, Dec 29, 24; m 66. ELECTRICAL ENGINEERING. *Educ:* Rensselaer Polytech Inst, BEE, 51; Ga Inst Technol, MS, 55. *Prof Exp:* From instr to asst prof, 52-65, assoc prof, 65-80, PROF ELEC ENG, GA INST TECHNOL, 80- *Mem:* Am Soc Eng Educ; Am Asn Univ Professors. *Res:* Engineering education; instrumentation; digital hardware. *Mailing Add:* Eight Ardmore Sq Atlanta GA 30309

WALLACE, JOHN MICHAEL, b Flushing, NY, Oct 28, 40; c 3. METEOROLOGY. *Educ:* Webb Inst Naval Archit, BS, 62; Mass Inst Technol, PhD(meteorol), 66. *Prof Exp:* From asst prof to assoc prof, 66-77, PROF ATMOSPHERIC SCI, UNIV WASH, 77- *Concurrent Pos:* Adj assoc prof environ studies, Univ Wash, 73-; dir, Joint Inst Study Atmosphere & Ocean, 80- *Honors & Awards:* Macelwane Award, Am Geophys Union, 72; Meisinger Award, Am Meteorol Soc, 75. *Mem:* Am Meteorol Soc; Am Geophys Union. *Res:* General circulation; tropical meteorology. *Mailing Add:* Dept Atmospheric Sci Univ Wash Seattle WA 98195

WALLACE, JON MARQUES, b Pensacola, Fla, Dec 21, 43; m 70; c 2. PLASMA PHYSICS. *Educ:* Univ Calif, Berkeley, AB, 66; Harvard Univ, PhD(physics), 71. *Prof Exp:* Fel physics, Univ Md, 71-73; STAFF MEM, LOS ALAMOS NAT LAB, 73- *Mem:* Am Phys Soc. *Res:* Nuclear scattering theory; transport theory; hydrodynamics; inertial confinement fusion; cosmic ray theory; laser-plasma interactions. *Mailing Add:* Los Alamos Nat Lab MS F645 PO Box 1663 Los Alamos NM 87545

WALLACE, KENDALL B, Alpena, Mich, May 3, 53. BIOCHEMICAL MECHANISMS, COMPARATIVE TOXICOLOGY. *Educ:* Mich State Univ, PhD(physiol), 79. *Prof Exp:* ASST PROF PHARMACOL & TOXICOL, UNIV MINN, 81-, DIR, CHEM TOXICOL RES CTR, DULUTH SCH MED, 85- *Mem:* Int Soc Xenobiotics; Soc Environ Toxicol & Chem. *Mailing Add:* Dept Pharmacol Univ Minn Ten University Dr Duluth MN 55812-2487

WALLACE, KYLE DAVID, b Nancy, Ky, Apr 3, 43; m 64; c 2. MATHEMATICS. *Educ:* Eastern Ky State Col, BS, 63; Vanderbilt Univ, MS, 65, PhD(math), 70. *Prof Exp:* Instr math, Easten Ky Univ, 65-67; asst prof, 70-75, assoc prof, 75-80, PROF MATH, WESTERN KY UNIV, 80- *Mem:* Math Asn Am. *Res:* Infinite Abelian groups; structure and classification of groups. *Mailing Add:* Dept Math Western Ky Univ Bowling Green KY 42101

WALLACE, LANCE ARTHUR, b San Francisco, Calif, Dec 29, 38; m 64, 75; c 2. HUMAN EXPOSURE ASSESSMENT, RELATION OF EXPOSURE TO BODY BURDEN. *Educ:* Univ Wash, BA, 59; City Univ NY, PhD(physics), 73. *Prof Exp:* Asst prof physics & astron, Rose-Hulman Inst Technol, 73-75; staff assoc, Nat Acad Sci, 75-77; MEM STAFF ENVIRON SCI, US ENVIRON PROTECTION AGENCY, 77- *Concurrent Pos:* Vis prof, Harvard Univ Sch Pub Health, 84-86. *Mem:* AAAS; Am Cancer Soc. *Res:* Measurement of human exposure to environmental pollutants and associated body burden; development of personal air quality monitors for volatile organic compounds, carbon monoxide, respirable particles and pesticides; large scale (100-1000 persons) field studies of exposure. *Mailing Add:* 11568 Woodhollow Ct Reston VA 22091

WALLACE, LARRY J, b South Lyon, Mich, Sept 17, 37; m 58; c 3. ORTHOPEDIC SURGERY. *Educ:* Mich State Univ, BS, 60, DVM, 62, MS, 64; Am Col Vet Surgeons, dipl. *Prof Exp:* Asst instr surg & med, Mich State Univ, 62-63; asst prof exp path, Lab Animal Med, Univ Fla, 63-66; asst prof surg, Kans State Univ, 66-69, assoc prof, 69-72; assoc prof, 72-74, PROF ORTHOP SURG, UNIV MINN, 74- *Concurrent Pos:* Dir surg residencies, Col Vet Med, Univ Minn, 72-, head, Div Small Animal Surg, 73-78; consult vet, Cardiac Pacemakers, Inc, St Paul, 78-; lectr orthop sur, Ohio State Univ, 82. *Mem:* Am Vet Med Asn; Am Animal Hosp Asn; Vet Orthop Soc; Sigma Xi. *Res:* Joint transplantation; bone grafting; bone healing; methods for the fixation of fractures; arthritis; reconstructive surgery; joint prostheses; growth deformities of bone; hip dysplasia; bone banking; bone infections; experimental surgery. *Mailing Add:* 233 17th Ave NW New Brighton MN 55112

WALLACE, MARION BROOKS, b Mankato, Minn, Dec 17, 17; m 36, 75; c 1. ENTOMOLOGY. *Educ:* Univ Minn, BA, 47, MS, 50, PhD(zool), 54. *Prof Exp:* Res fel, 54-56, from instr to assoc prof, 56-71, prof entom, Univ Minn, St Paul, 71-86; RETIRED. *Mem:* NY Acad Sci; Entom Soc Am; Soc Invert Path; Am Soc Zoologists; Tissue Cult Asn. *Res:* Insect tissue culture; physiology of intracellular symbiotes; infectious diseases; nutrition and development of insects. *Mailing Add:* 2603 Cohansey St Roseville MN 55113

WALLACE, MICHAEL DWIGHT, b Columbia, SC, Feb 15, 47; m 68; c 2. PROCESS CONTROL ENGINEERING, CHEMICAL ENGINEERING. *Educ:* Ga Inst Technol, BChE, 68; Univ Fla, ME, 69; Ohio State Univ, MBA, 82. *Prof Exp:* Asst chem eng, Univ Fla, 68-69; develop engr, Celanese Plastics Co, 70-73; process control engr, Brunswick Pulp & Paper Co, 73-76, group leader, 76-78, mgr, 78; mgr systs-process eng, 78-81, mgr opers technol, 81-83, mgr fac anal, 83-87, MGR MFG SUPPORT, MEAD CENT RES, 87- *Mem:* Tech Asn Pulp & Paper Indust; Am Mgt Asn. *Mailing Add:* Mead World Hq Ct House Plaza Northeast Dayton OH 45463

WALLACE, PAUL FRANCIS, b Tyrone, Pa, June 11, 27; m 52; c 3. SURFACE CHEMISTRY. *Educ:* Pa State Univ, BS, 50, MS, 52. *Prof Exp:* Res asst ceramics, Pa State Univ, 50-52; res engr, Res Labs, Carborundum Co, 52-54; res engr, Alloy Tech Div, 54-70, sr scientist, 70-80, staff scientist, 80-81, TECH SPECIALIST, PROCESS CHEM & PHYSICS DIV, ALCOA RES LABS, ALUMINUM CO AM, 81- *Mem:* Sigma Xi; Am Chem Soc. *Res:* Surface chemistry of aluminum and aluminum alloys as related to fabricating processes and end uses of fabricated products. *Mailing Add:* 200 McLaughlin Dr New Kensington PA 15068

WALLACE, PAUL WILLIAM, b Cork, Ireland, Dec 27, 36, US citizen; m 64; c 4. MECHANICAL ENGINEERING, MANUFACTURING ENGINEERING. *Educ:* Col Technol, Dublin, BS, 58; Univ Salford, MS, 62; Univ Bristol, PhD(mech eng), 65, Univ Dublin, MA, 81. *Prof Exp:* Engr, Unidare Ltd, Dublin, 58-60; res asst, Univ Salford, 60-62; res engr, Univ Bristol, 62-65; res metallurgist, IIT Res Inst, Chicago, 65-68; mgr, Mat Develop, SPS Technol, 68-69, mgr, Irish Lab, 70-76, dir Res & Develop, 76-80, vpres eng, 83-85, PRES, ASSEMBLY SYSTS DIV, SPS TECHNOL, 85- *Concurrent Pos:* Prof & head, Dept Mech Eng, Trinity Col, Dublin, Ireland, 80-82. *Honors & Awards:* James Clayton Prize, Inst Mech Eng, 85. *Mem:* Inst Mech Engrs, London; Am Soc Mech Engrs; Soc Mech Eng. *Res:* Mechanics; metallurgy; machine tools; metal working; automatic control; adhesive and mechanical fastening. *Mailing Add:* SPS Technol Jenkintown PA 19046

WALLACE, PHILIP RUSSELL, b Toronto, Ont, Apr 19, 15; m 40; c 3. THEORETICAL PHYSICS. *Educ:* Univ Toronto, BA, 37, MA, 38, PhD, 40. *Prof Exp:* Instr math, Univ Cincinnati, 40-42; instr, Mass Inst Technol, 42; assoc res physicist, Atomic Energy Div, Nat Res Coun, 43-46; from asst prof to prof appl math, McGill Univ, 46-63, prof physics, Rutherford Physics Bldg, 63-72, dir inst theoret physics, 66-70; MacDonald prof physics, Univ Paul Sabatier, Toulouse, France, 72-82, prof solid physics lab, 81-82, vis prof, 82-83, emer prof, 82-; prin, sci col, Concordia Univ, 84-87; RETIRED. *Concurrent Pos:* Mem comn higher educ, Super Coun Educ, Que, 70-73; mem grant selection comt physics, Nat Res Coun Can, 71-74; ed, Can J Physics, 73-80; mem, Int Adv Comt, Int Conf Semiconductors Edinburgh, 78, Kyoto, 80, Montpellier, 82; prin, sci Col, Concordia Univ, 84-87. *Mem:* Am Phys Soc; Am Asn Physics Teachers; Can Asn Physicists; Europ Phys Soc; fel Royal Soc Can. *Res:* Theoretical and solid state physics; physics of semiconductors and semimetals in intense magnetic fields. *Mailing Add:* Dept Physics Rutherford Physics Bldg McGill Univ Montreal PQ H3A 2T8 Can

WALLACE, RAYMOND HOWARD, JR, b Columbus, Ga, July 29, 36; m 58; c 2. GEOPRESSURED-GEOTHERMAL ENERGY RESEARCH. *Educ:* Fla State Univ, BS, 60; La State Univ, MS, 66. *Prof Exp:* Res asst, La Water Resources Res Inst, 65-66; geologist, US Geol Surv, 66-69, hydrologist, 69-74, res proj chief, 75-83, geothermal liaison, 84-87, LIAISON, HIGH-LEVEL RADIOACTIVE WASTE MGT, US GEOL SURV, 87- *Concurrent Pos:* Dep chief, Gulf Coast Hydrosci Ctr, US Geol Surv, 74-83; geothermal tech adv, US Dept Energy, 75-, geothermal prog mgr, 84-87; mem, Task Force Non-Conventional Gas, Fed Energy Regulatory Comn, 75-78 & Hydrocarbon Geosci Res Coord Comt, Dept Energy, 88-89; sci tech adv & mgr, Nat Continental Sci Drilling, US Geol Surv & Dept Energy, 84- *Honors & Awards:* Super Serv Award, Dept Interior, 82. *Mem:* Geol Soc Am; Am Asn Petrol Geologists; Sigma Xi. *Res:* Deep-basin hydrogeologic research and geothermal resources; scientific drilling activities related to geothermal resources and geologic disposal of radioactive wastes; author of 28 publications. *Mailing Add:* 6663 Tennyson Dr McLean VA 22101-5716

WALLACE, RICHARD KENT, b Washington, DC, Jan 29, 54; m 76; c 1. NUCLEAR ASTROPHYSICS, RADIATION HYDRODYNAMICS. *Educ:* La State Univ, Baton Rouge, BS, 75; Univ Calif, Santa Cruz, MS, 77, PhD(astrophysics), 81. *Prof Exp:* MEM STAFF, LOS ALAMOS NAT LAB, 81- *Mem:* Am Astron Soc. *Res:* Basic thermonuclear burn physics; laser fusion target physics; applications to astrophyscics includes nucleosynthesis and nuclear sources for novae, supernovae and x- and gamma-ray bursts. *Mailing Add:* Los Alamos Nat Lab PO Box 1663 MS B-257 Los Alamos NM 87545

WALLACE, ROBERT ALLAN, b Chicago, Ill, Aug 23, 30. BIOCHEMISTRY, ORGANIC CHEMISTRY. *Educ:* Northern Ill Univ, BS, 53; Univ Bonn, MS, 57, PhD(org chem), 59. *Prof Exp:* Fel, Calif Inst Technol, 60-62; assoc prof biochem, 62-71, PROF BIOCHEM, HUMBOLDT STATE UNIV, 71- *Mem:* AAAS; Am Chem Soc. *Res:* Enzyme and pteridine chemistry. *Mailing Add:* Dept Chem Humboldt State Univ Arcata CA 95521

WALLACE, ROBERT B, b Stoneham, Mass, Jan 16, 37. PSYCHOBIOLOGY, NEUROANATOMY. *Educ:* Boston Univ, AB, 60, AM, 61, PhD(psychol), 66. *Prof Exp:* Instr psychol, 66-67, lectr, 67-68; from asst prof to assoc prof psychol, 72-80, PROF PSYCHOL & BIOL, UNIV HARTFORD, 80-, CHAIR DEPT PSYCHOL, 86- *Concurrent Pos:* Res assoc, Mass Inst Technol, 66-68 & Inst Living, 71-; vis assoc prof, Univ Conn Health Ctr, 74-77, res assoc, 78. *Mem:* AAAS; Am Asn Anat; Soc Neurosci; Psychonomic Soc; NY Acad Sci. *Res:* Relation of central nervous system structure to behavior; plasticity of mammalian nervous system; postnatal neurogenesis; neuraltransplantation. *Mailing Add:* Dept Psychol Univ Hartford 200 Bloomfield Ave West Hartford CT 06117

WALLACE, ROBERT BRUCE, b Vancouver, BC, Jan 28, 50; m 71; c 2. RECOMBINANT DNA. *Educ:* Simon Fraser Univ, BSc, 72; McMaster Univ, PhD(biochem),75. *Prof Exp:* Res asst biol & chem, Simon Fraser Univ, 70; student biochem, McMaster Univ, 71-75, fel, 75; fel biol, Calif Inst Technol, 75-77 & 77-78; ASSOC RES SCIENTIST, DEPT MOLECULAR GENETICS, CITY OF HOPE RES INST, 78- *Concurrent Pos:* Prin investr, NIH, 79- *Mem:* AAAS. *Res:* Molecular basis of human genetic disease; genes of the murine major histocompatibility complex and other multigene families; gene evolution; application of synthetic DNA. *Mailing Add:* Dept Molecular Genetics City of Hope Res Inst Duarte CA 91010

WALLACE, ROBERT EARL, b New York, NY, July 16, 16; m 45; c 1. GEOLOGY. *Educ:* Northwestern Univ, BS, 38; Calif Inst Technol, MS, 40, PhD(struct geol, vert paleont), 46. *Prof Exp:* Geologist, US Geol Surv, 42-70, chief Southwestern Br, 60-65, regional geologist, 70-73, chief scientist, Off Earthquake Studies, 73-87; RETIRED. *Concurrent Pos:* From asst prof to assoc prof, Wash State Univ, 46-51; vis lectr, Stanford Univ, 60; mem comt seismol, Nat Acad Sci-Nat Res Coun; chmn, US/USSR Environ Agreement, US Working Group Earthquake Prediction; mem eng criteria rev bd, San Francisco Bay Conserv & Develop Comn, 78-, chmn, 81-90. *Honors & Awards:* Medal, Seismol Soc Am, 89. *Mem:* Fel Geol Soc Am; Soc Econ Geol; Seismol Soc Am; Earthquake Eng Res Inst; fel AAAS. *Res:* Active faults; tectonics; earthquakes; engineering geology and environment; mineral deposits. *Mailing Add:* 240 Cervantes Rd Portola Valley CA 94028

WALLACE, ROBERT HENRY, b Cleveland, Ohio. MEDICAL PHYSICS. *Educ:* Transylvania Col, BA, 64; Purdue Univ, West Lafayette, MS, 68, PhD(bionucleonics), 70. *Prof Exp:* Assoc radiol health physicist, Ky State Dept Health, 65-66; ASST PROF RADIOL, ALBANY MED COL, 70-, RADIATION PHYSICIST, ALBANY MED CTR HOSP, 70- *Mem:* Am Asn Physicists in Med; Health Physics Soc. *Res:* Radiation dosimetry and protection; nuclear medicine instrumentation. *Mailing Add:* Radiol Oncol Dept Mem Med Ctr PO Box 23089 Savannah GA 31403

WALLACE, ROBERT WILLIAM, b Central Falls, RI, Jan 1, 43; m 65; c 2. SCIENCE EDUCATION, ENVIRONMENTAL CHEMISTRY. *Educ:* Providence Col, BS, 64; Niagara Univ, MS, 66; Boston Univ, PhD(chem), 73. *Prof Exp:* ASST PROF CHEM, BENTLEY COL, 72- *Mem:* Am Chem Soc; AAAS. *Res:* Air and water quality as well as the quality of consumer products, including chemical composition versus the list of ingredients. *Mailing Add:* 19 Coolidge Ave Westford MA 01886-1807

WALLACE, ROBIN A, b Chicago, Ill, Nov 11, 33; m 55, 84; c 2. REPRODUCTIVE BIOLOGY. *Educ:* Columbia Univ, BA, 55, PhD(zool), 61. *Prof Exp:* Res assoc, Sloan-Kettering Inst, NY, 57; consult, Oak Ridge Nat Lab, 60-61, USPHS fel, 61-63, staff mem biol div, 63-81; vis prof, anat dept, Col Med, Univ Fla, Gainesville, 81-84; PROF ANAT & CELL BIOL, WHITNEY LAB, ST AUGUSTINE, 84- *Concurrent Pos:* Vis investr, Nat Res Coun Can, 63-64; mem, Marine Biol Lab Corp; dir reprod biol prog, Marine Biol Lab, Woods Hole, 74 & 75; co-ed, Develop Biol, 72-74; mem develop biol panel, NSF, 75-76. *Mem:* AAAS; Am Soc Biochem Molecular Biol; Soc Develop Biol; Fedn Am Sci; Am Soc Ichthyol Herpetol; Am Soc Zool; Int Soc Develop Biol. *Res:* Comparative biochemical studies on yolk proteins and the mechanisms of oocyte growth in vertebrates. *Mailing Add:* Whitney Lab 9505 Ocean Shore Blvd St Augustine FL 32086-8623

WALLACE, RONALD GARY, b Cadiz, Ohio, July 6, 38; m 65; c 2. GEOMORPHOLOGY, PETROLEUM GEOLOGY. *Educ:* Kent State Univ, BS, 61; Ohio State Univ, MS, 64, PhD(geol), 67. *Prof Exp:* Petroleum geologist, Standard Oil Co, Tex, 67-69; asst prof geol, Ohio State Univ, 69-70; PROF GEOL & CHMN DEPT, EASTERN ILL UNIV, 70- *Mem:* Am Asn Petrol Geol. *Res:* Alpine mass movement; strip mine erosion. *Mailing Add:* Dept Geog & Geol Eastern Ill Univ Charleston IL 61920

WALLACE, SIDNEY, b Philadelphia, Pa, Feb 26, 29; c 3. RADIOLOGY. *Educ:* Temple Univ, BA, 49, MD, 54; Am Bd Radiol, dipl, 62. *Prof Exp:* Intern, Philadelphia Gen Hosp, 54-55; resident radiol, Hosp, Jefferson Med Col, 59-62, from instr to asst prof, Col, 62-66; assoc prof, 66-69, prof, 69-80, ASHBEL SMITH PROF RADIOL, UNIV TEX M D ANDERSON HOSP & TUMOR INST HOUSTON, 80- *Concurrent Pos:* Fel radiol, Univ Lund, 63-64. *Mem:* AMA; Am Col Radiol; Int Soc Lymphology. *Res:* Lymphangiography and angiography. *Mailing Add:* Univ Tex Anderson Cancer Ctr 1515 Holcomb Blvd Houston TX 77030

WALLACE, STEPHEN JOSEPH, b Youngstown, Ohio, May 10, 39; m 61; c 2. THEORETICAL NUCLEAR PHYSICS. *Educ:* Case Inst Technol, BS, 61; Univ Wash, MS, 69, PhD(physics), 71. *Prof Exp:* Res engr, Boeing Co, 61-68; res asst physics, Univ Wash, 68-71; res assoc, Univ Fla, 71-72 & Harvard Univ, 72-74; from asst prof to assoc prof, 74-82, PROF PHYSICS, UNIV MD, COLLEGE PARK, 82- *Mem:* Am Phys Soc. *Res:* Intermediate to high energy nuclear multiple scattering theory; relativistic dynamics of nuclei. *Mailing Add:* Dept Physics & Astron Univ Md College Park MD 20742

WALLACE, STEWART RAYNOR, b Freeport, NY, Mar 31, 19; m 46; c 2. GEOLOGY. *Educ:* Dartmouth Col, BA, 41; Univ Mich, MS, 48, PhD(geol), 53. *Prof Exp:* Geologist, US Geol Surv, 48-55; resident geologist, Climax Molybdenum Co, 55-58, chief geologist, 58-64, chief geol & explor, 64-69; pres & dir explor, Mine Finders, Inc, 70-75; CONSULT, 76- *Honors & Awards:* D C Jackling Award, Am Inst Mining, Metal & Petrol Engrs, 74; Distinguished Mem, Soc Mining Engrs, 84. *Mem:* Geol Soc Am; Am Inst Mining, Metall & Petrol Engrs; Soc Econ Geol; Asn Explor Geochemists. *Res:* Genesis of metallic ore deposits; igneous and metamorphic structure and petrology related to ore deposits. *Mailing Add:* 8700 W 14th Ave Lakewood CO 80215

WALLACE, SUSAN SCHOLES, b Brooklyn, NY, Jan 10, 38; c 3. MOLECULAR BIOLOGY, BIOPHYSICS. *Educ:* Marymount Col, NY, BS, 59; Univ Calif, Berkeley, MS, 61; Cornell Univ, PhD(biophys), 65. *Prof Exp:* USPHS fel, Columbia Univ, 65-67; instr biol sci, Lehman Col, 67-68, asst prof, 69-73, assoc prof, 73-76; from assoc prof to prof microbiol, NY Med Col, 76-88; PROF & CHAIRPERSON MICROBIOL & MOLECULAR GENETICS, UNIV VT, 88- *Concurrent Pos:* City Univ New York res grant, Lehman Col, 68-76, NIH res grant, 72-; vis prof, Albert Einstein Col Med, 74-75; scholar, Am Cancer Soc, 74-75; mem, NIH Radiation Study Sect, 77-81; Dept Energy res grant, 77- *Honors & Awards:* Aaron Bendich Award, 84. *Mem:* AAAS; Biophys Soc; Radiation Res Soc (pres, 91-92); Am Soc Microbiologists; Am Soc Biol Chemists; Am Soc Photobiol. *Res:* Quantitation and repair of DNA damage in bacteria and bacteriophages; in vivo and in vitro processing and repair of ionizing radiation and oxidative DNA damage. *Mailing Add:* Dept Microbiol & Molecular Genetics Univ Vt Burlington VT 05405

WALLACE, SUSAN ULMER, b Demopolis, Ala, Jan 31, 52; m 81. CROP PHYSIOLOGY. *Educ:* Univ Ala, BS, 73, MS, 75; Iowa State Univ, PhD(agron), 79. *Prof Exp:* From asst prof to assoc prof, 80-90, PROF AGRON, CLEMSON UNIV, 90- *Mem:* Am Soc Agron; Crop Sci Soc Am; Am Soc Plant Physiologists. *Res:* Production-oriented physiology of soybeans and other field crops; crop plant tolerance to environmental stress. *Mailing Add:* Dept Agron & Soils Clemson Univ Clemson SC 29634-0359

WALLACE, TERRY CHARLES, b Phoenix, Ariz, May 18, 33; m 55; c 5. MATERIALS CHEMISTRY, CHEMICAL TECHNOLOGY. *Educ:* Ariz State Univ, BS, 55; Iowa State Univ, PhD(phys chem), 58. *Prof Exp:* Mem staff, Los Alamos Nat Lab, Univ Calif, 58-70, alt group leader, 70-79, group leader, 79-83, assoc div leader, 83-87, prog mgr, 87-88, TECH PROG COORDR, LOS ALAMOS NAT LAB, UNIV CALIF, 88- *Concurrent Pos:* Chief tech projs br, Environ Test Div, Dugway Proving Ground. *Honors & Awards:* Fed Lab Consortium Award, 88. *Mem:* AAAS; Am Chem Soc; fel Am Inst Chem. *Res:* Structural properties and mass transport of materials at high temperature; materials synthesis; materials characterization; high temperature thermodynamics; chemical vapor deposition; modeling of high temperature chemical processes. *Mailing Add:* 146 Monte Rey S Pajarito Acres Los Alamos NM 87544

WALLACE, TERRY CHARLES, JR, b Ames, Iowa, June 30, 56; m; c 1. SEISMOLOGY, CRUSTAL DYNAMICS. *Educ:* NMex Inst Mining Technol, BS(geophys) & BS(math), 78; Calif Inst Technol, MS, 80, PhD(geophys), 83. *Prof Exp:* Asst Prof, 83-88, ASSOC PROF, SEISMOG, DEPT GEOSCI, UNIV ARIZ, 88- *Concurrent Pos:* Curator, Univ Ariz, Mineral Mus, 85- *Mem:* Seismog Soc Am; Am Geophys Union; AAAS. *Res:* Computational Seismology; global crustal structure; explosion source physics; tectonics. *Mailing Add:* Dept Geosci Univ Ariz Bldg 77 Tucson AZ 85721

WALLACE, THOMAS PATRICK, b Washington, DC, Apr 11, 35; m 59; c 3. PHYSICAL CHEMISTRY, POLYMER SCIENCE. *Educ:* State Univ NY Col Potsdam, BS, 58; Syracuse Univ, MS, 61; St Lawrence Univ, MS, 64; Clarkson Col Technol, PhD(phys chem), 68. *Prof Exp:* Asst prof chem, State Univ NY Col Potsdam, 61-67; from asst prof to prof chem, Rochester Inst Technol, 68-78, head, Dept Chem, 70-72, assoc dean, Col Sci, 72-73, dean, Col Sci, 73-78; PROF CHEM SCI & DEAN, SCH SCI & HEALTH PROF, OLD DOMINION UNIV, NORFOLK, VA, 78- *Mem:* AAAS; Am Chem Soc; NY Acad Sci; Sigma Xi. *Res:* Light scattering; polymer latex systems and characterization; size distribution analysis; statistical thermodynamics of dilute polymer solutions. *Mailing Add:* 500 Pacific S 1103 Virginia Beach VA 23451

WALLACE, TRACY I, b Irvine, Ky, Nov 20, 24; m 51; c 3. INTERNAL MEDICINE. *Educ:* Univ Ky, BS, 46; Univ Cincinnati, MD, 49. *Prof Exp:* Staff physician, Vet Admin Hosp, McKinney, Tex, 55-57, asst chief med serv, Temple Vet Admin Hosp, 57-59, chief med serv, 61-; AT DEPT INTERNAL MED, TEX A&M UNIV. *Mem:* AMA. *Res:* Adrenal function in patients with pulmonary emphysema and other hypoxic states. *Mailing Add:* Col Med Tex Univ Vet Admin Hosp Temple TX 76501

WALLACE, VICTOR LEW, b Brooklyn, NY, Mar 20, 33; m 62; c 2. COMPUTER SCIENCE. *Educ:* Polytech Inst Brooklyn, BS, 55; Univ Mich, PhD(elec eng), 69. *Prof Exp:* Mem tech staff syst eng, Bell Tel Labs, 55-56; mathematician-programmer, IBM Corp, 56-57; instr elec eng, Univ Mich, 57-62; assoc res scientist, 62-69; assoc prof comput sci, Univ NC, 69-76; PROF COMPUT SCI, UNIV KANS, 76- *Concurrent Pos:* Vis scientist, Imp Col, Univ London, 70. *Mem:* Asn Comput Mach; Inst Elec & Electronics Engrs; Inst Mgt Sci; Am Asn Univ Professors; Sigma Xi. *Res:* Computer system modeling; operating system theory; computer graphics software; man-machine interface in computer-aided design. *Mailing Add:* 1509 Massachusetts St Lawrence KS 66044

WALLACE, VOLNEY, b Idaho Falls, Idaho, Oct 9, 25; m 44; c 5. GEOCHEMISTRY. *Educ:* Univ Idaho, BS, 46; Purdue Univ, MS, 49, PhD(agr chem), 53. *Prof Exp:* Asst agr chem, Purdue Univ, 47-53; res assoc, Wash State Univ, 53-55; asst prof biochem, SDak State Col, 55-61; res chemist, Dugway Proving Ground, 61-90; RETIRED. *Concurrent Pos:* Vpres, Terracopia. *Res:* Geo-'agricultural and analytical chemistry; environmental energy and gardening. *Mailing Add:* 72 W 6100 S Murray UT 84107

WALLACE, WILLIAM, b Manchester, Eng, Oct 15, 40, Can citizen; m 64; c 3. MATERIALS ENGINEERING, METALLURGY. *Educ:* Manchester Univ, BSc, 63; Victoria Univ Manchester, PhD(metall), 66. *Prof Exp:* Lectr, John Dalton Col Technol, 66-67; asst res officer, 67-70, assoc res officer, 70-76, head, Mat Sect, 76- 81, HEAD, STRUCT & MAT LAB, NAT AERONAUT ESTAB CAN, NAT RES COUN CAN, 81- *Concurrent Pos:* Hon adj prof, Dept mech & Aeronaut Eng, Carleton Univ, Ottawa, Dept Mech Eng, Univ Toronto, Ont; mem adv group aerospace res & develop, NATO, mat panel, chmn, 83-84. *Mem:* Inst Metals; Inst Metallurgists; Am Soc Metals; Can Aero Space Inst. *Res:* Metal processing; powder metallurgy; isothermal and superplastic forging; mechanical behavior, creep, fatigue, and fracture; airframe and engine structural technology. *Mailing Add:* Struct & Mat Lab Montreal Rd Ottawa ON K1A 0R6 Can

WALLACE, WILLIAM DONALD, b Detroit, Mich, Sept 19, 33; m; c 2. SOLID STATE PHYSICS. *Educ:* Eastern Mich Univ, BA, 55; Univ Md, College Park, MS, 60; Wayne State Univ, PhD(physics), 66. *Prof Exp:* Asst prof physics, Eastern Mich Univ, 59-62; res asst, Cornell Univ, 67-70; asst prof, 70-74, ASSOC PROF PHYSICS, OAKLAND UNIV, 74- *Concurrent Pos:* Leverhulme vis fel, Univ Essex, 66-67. *Mem:* Am Phys Soc; Am Asn Physics Teachers. *Res:* Ultrasonic properties and magnetic properties of solids; amorphous materials and low temperature physics. *Mailing Add:* Dept Physics Oakland Univ Rochester MI 48309-4401

WALLACE, WILLIAM EDWARD, b Fayette, Miss, Mar 11, 17; m 47; c 3. MATERIALS SCIENCE. *Educ:* Miss Col, BA, 36; Univ Pittsburgh, PhD(chem), 41. *Prof Exp:* Asst, Univ Pittsburgh, 36-40; Carnegie Found fel, 40-42, sr res fel, 42-44; res assoc, Ohio State Univ, 44-45; from asst res prof to assoc res prof chem, 45-53, chmn dept, 63-77, distinguished serv prof, 77-83, prof chem, Univ Pittsburgh, 53-; PROF APPL SCI & ENG, CARNEGIE-MELLON UNIV, 83-; PRES, ADV MAT CORP, 84- *Concurrent Pos:* Guggenheim fel, 54; consult numerous industs and govt agencies. *Honors & Awards:* Morley Award, 83. *Mem:* Fel AAAS; Am Chem Soc. *Res:* Magnetic behavior and hydrogen absorption of intermetallic compounds containing lanthanides; high energy permanent magnets. *Mailing Add:* Carnegie Mellon Res Inst Carnegie Mellon Univ Pittsburgh PA 15213

WALLACE, WILLIAM EDWARD, JR, b Charleston, WVa, Sept 25, 42; m 70; c 2. ENVIRON MENTAL HEALTH, CHEMICAL ENGINEERING. *Educ:* WVa Univ, BS, 63, MS, 67, PhD(physics), 69. *Prof Exp:* Nat Res Coun-US Bur Mines res assoc, Morgantown Energy Res Ctr, US Bur Mines, 70-71, res physicist, 71-74,; res physicist-asst dir, Morgantown Energy Technol Ctr, US Dept Energy, 75-80; RES PHYSICIST & AEROSOL RES TEAM LEADER, DIV RESPIRATORY DIS STUDIES, NAT INST OCCUP SAFETY & HEALTH, 80- *Concurrent Pos:* Adj prof, Dept Chem Eng, WVa Univ, 82-; comt mem, Nat Res Coun, 82; monograph workshop mem, WHO-IARC, 83. *Mem:* Sigma Xi. *Res:* Spectroscopic methods development and bioassay methods development for analysis of respirable particle toxicity an pulmonary response. *Mailing Add:* Nat Inst Occup Safety & Health 944 Chestnut Ridge Rd Morgantown WV 26504

WALLACE, WILLIAM J, b Knoxville, Tenn, July 27, 35; m 58; c 2. PHYSICAL CHEMISTRY. *Educ:* Carson-Newman Col, BS, 56; Purdue Univ, PhD(inorg chem), 61. *Prof Exp:* Asst prof inorg chem, Univ Miss, 60-63; from asst prof to assoc prof, Muskingum Col, 63-72, actg chmn dep, 68-69, chmn dept, 79-80 & 85-89, coordr, Sci Div, 72-76, PROF INORG & PHYS CHEM, MUSKINGUM COL, 72- *Concurrent Pos:* Res grants, Res Corp, 61-63 & Am Acad Arts & Sci, 62-63; dir, NSF Res Partic High Sch Teachers, 66, 67 & 68; res fel with P S Braterman, Univ Glasgow, 70-71; fac fel & res grants, Lewis Res Ctr, NASA, 78-80; vis fac fel, Fla State Univ, R J Clark, 87, 88 & 90. *Mem:* Am Chem Soc; Sigma Xi. *Res:* Study of systems using polyether solvents with inorganic compounds, especially solubility, reaction and spectral phenomena; iron carbonyl reactions with Lewis bases; electrocatalysis of chronium; reduction in flow cell; super conductor synthesis. *Mailing Add:* Dept Chem Muskingum Col New Concord OH 43762

WALLACE, WILLIAM JAMES LORD, b Salisbury, NC, Jan 13, 08; m 29; c 1. PHYSICAL CHEMISTRY. *Educ:* Univ Pittsburgh, BS, 27; Columbia Univ, AM, 31; Cornell Univ, PhD(phys chem), 37; Livingstone Col, LLD, 59; Concord Col, LHD, 70; Alderson Broaddus Col, DSc, 71. *Prof Exp:* Instr chem, Livingston Col, 27-32; instr, Lincoln Univ, Mo, 32-33; from instr to prof, WVa State Col, 33-75, actg admin asst to pres, 44-45, admin asst, 45-50, actg pres, 52-53, pres, 53-73, emer pres, 73-; RETIRED. *Concurrent Pos:* Mem, Kanawha County Pub Libr Bd, 68- *Honors & Awards:* Outstanding Civilian Serv Medal, Dept Army, 72; Annual Award, Educ Comn States, 75. *Mem:* Am Chem Soc; Sigma Xi. *Res:* Freezing points of aqueous solutions of alpha amino acids; teaching problems in general chemistry. *Mailing Add:* PO Box 417 Institute WV 25112

WALLACH, EDWARD E, b Brooklyn, NY, Oct 8, 33; m 56; c 2. OBSTETRICS & GYNECOLOGY. *Educ:* Swarthmore Col, BA, 54; Cornell Univ, MD, 58; Nat Bd Med Examr, dipl, 59; Am Bd Obstet & Gynec, dipl, 66. *Prof Exp:* Intern internal med, Cornell Med Div, Bellevue Hosp, NY, 58-59; resident obstet & gynec, Kings County Hosp, 59-63; assoc, Sch Med, Univ Pa, 65-66, from asst prof to prof obstet & gynec, 71-84; dir obstet & gynec, Pa Hosp, 71-84; PROF & CHMN, DEPT GYNEC & OBSTET, SCH MED, JOHNS HOPKINS UNIV, 84-, PHYSICIAN-IN-CHIEF, DEPT GYNEC & OBSTET, JOHNS HOPKINS HOSP, 84- *Concurrent Pos:* Res fel reproductive physiol, Worcester Found Exp Biol, 61-62; Josiah Macy Jr Found fel, 65-, Lalor Found fel, 67-69; consult, US Naval Hosp, Philadelphia, 74-84; vis prof, Dept Obstet & Gynec, Kyoto Univ Med Sch, Japan, 81-82. *Honors & Awards:* Serono lectr, Am Fertil Soc, 84. *Mem:* Am Col Obstetricians & Gynecologists; Soc Gynec Invest (pres, 86-87); Am Fertil Soc (pres, 85-86); Soc Study Reproduction; Endocrine Soc; AAAS; Am Asn Hist Med; Am Col Surgeons; Asn Planned Parenthood Physicians. *Res:* Reproductive biology; ovarian physiology; gynecologic endocrinology; infertility; family planning. *Mailing Add:* Johns Hopkins Hosp 600 N Wolfe St Baltimore MD 21205

WALLACH, JACQUES BURTON, b New York, NY, Jan 25, 26; m 53; c 3. MEDICINE, PATHOLOGY. *Educ:* Long Island Col Med, MD, 47; Am Bd Path, dipl, 55. *Prof Exp:* From instr to asst prof, 54-59, VIS ASST PROF PATH, ALBERT EINSTEIN COL MED, 59-; CLIN PROF PATH, STATE UNIV NY HEALTH SCI CTR, BROOKLYN, 79- *Concurrent Pos:* Resident fel & asst pathologist, Queens Gen Hosp Ctr, New York, 48-55; asst vis pathologist, Bronx Munic Hosp Ctr, 54-59; consult, NY Zool Park, 54-84; vis asst prof, Rutgers Med Sch, Col Med & Dent NJ, 69-72, vis assoc prof, 72-; Fel, Nat Libr Med, Computers Med, Mt Sinai Sch Med, New York, NY, 78-79. *Mem:* Fel Am Soc Clin Path; fel Col Am Path; fel Am Col Physicians; NY Acad Sci. *Res:* Rheumatic heart disease; clinical pathology; comparative pathology; computers in interpretative medical laboratory reporting. *Mailing Add:* 18 Dartmouth Rd Cranford NJ 07016

WALLACH, MARSHALL BEN, b Buffalo, NY, June 15, 40; m 63; c 3. NEUROPHARMACOLOGY. *Educ:* Columbia Univ, BS, 62; Univ Minn, PhD(pharmacol), 67. *Prof Exp:* Assoc res scientist neuropharmacol, Dept Psychiat, Med Ctr, NY Univ, 68-70, res scientist, 70-71, instr, 71-72; staff researcher II, Syntex res, 73-76, sr staff researcher, 76-79, prin scientist, 79-83, head dept neuropharmacol, 83-86, MGR BIOL INFO, SYNTEX RES, 87- *Concurrent Pos:* Consult, New York City Rand Inst, 71-72. *Mem:* AAAS; Am Soc Exp Pharmacol & Therapeut; Western Pharmacology Soc; Soc Neurosci. *Res:* Models of psychiatric and neurological diseases; electrophysiology of sleep; neurohumoral mechanisms; anorexigens; antidepressants; neuroleptics; analgesics; antiparkinsonian agents; anticonvulsants; antitussives and psychotomimetic agents. *Mailing Add:* Sci Info Dept Syntex Res 3401 Hillview Ave Palo Alto CA 94304

WALLACH, STANLEY, b Brooklyn, NY, Dec 10, 28; m 54, 73; c 8. MEDICINE. *Educ:* Cornell Univ, AB, 48; Columbia Univ, MA, 49; State Univ NY Downstate Med Ctr, MD, 53. *Prof Exp:* Intern med, Univ Med Serv, Kings County Hosp, Brooklyn, 53-54; resident, Col Med, Univ Utah, 54-56; clin & res fel, Mass Gen Hosp, 56-57; from instr to prof, State Univ NY Downstate Med Ctr, 57-73, prog dir, USPHS Clin Res Ctr, 66-73; prof med & assoc chmn dept, Albany Med Col, 73-83; chief med, Albany Vet

Admin Med Ctr, 73-83; chief med, Bay Pines Vet Admin Med Ctr, 83-90, PROF MED & ASSOC CHMN DEPT, COL MED, UNIV SFLA, 83- *Concurrent Pos:* Vis physician, Kings County Hosp, Brooklyn, 57-73; career scientist, Health Res Coun City of New York, 61-71; consult, St Johns Episcopal Hosp, Brooklyn, 65-73; attend physician, State Univ Hosp, Brooklyn, 66-73; consult & lectr, US Naval Hosp, St Albans, NY, 66-73; res collabr, Med Dept, Brookhaven Nat Lab, 69-81; attend physician, Albany Med Ctr, 73-83; chmn, Med Adv Panel, Paget's Dis Found, mem bd dirs, 90- *Honors & Awards:* Hektoen Silver Award, AMA, 59; John B Johnson Award, Paget's Dis Found, 89. *Mem:* Asn Am Phys; Am Soc Clin Invest; fel Am Col Pharmacol; Am Soc Bone & Mineral Res; fel Am Col Nutrit (vpres, 83-85, pres-elect, 85-87, pres, 87-); Am Soc Magnesium Res. *Res:* Endocrine and metabolic diseases; radioisotopes; calcium, magnesium, trace element and mineral metabolism; metabolic bone disease. *Mailing Add:* Dept Internal Med Col Med Univ SFla 12901 Bruce B Downs Blvd Box 19 Tampa FL 33612-4799

WALLACH, SYLVAN, b San Antonio, Tex, Jan 9, 14; m 38; c 5. MATHEMATICS. *Educ:* Rutgers Univ, BS, 34; Johns Hopkins Univ, PhD(math), 48. *Prof Exp:* Chemist Mass & Waldstein Co, NJ, 36-38; patent examr, US Patent Off, Washington, DC, 38-42; indust analyst, War Prod Bd, 42-45; instr math, Johns Hopkins Univ, 48-49; sr scientist atomic power div, Westinghouse Elec Corp, 49-52; sr scientist, Walter Kidde Nuclear Labs, Inc, 52-57; asst proj mgr, Gibbs & Cox, Inc, 57-61; chmn dept math, 62-72, PROF MATH, C W POST COL, LONG ISLAND UNIV, 62- *Mem:* AAAS; Am Math Soc; Math Asn Am. *Mailing Add:* 101 Central Park W New York NY 10023

WALLACK, PAUL MARK, b Girard, Kans, Aug 3, 27; m 50, 80; c 6. INDUSTRIAL ENGINEERING. *Educ:* Univ Tulsa, BS, 50; Okla State Univ, MS, 56, PhD(indust eng), 67. *Prof Exp:* Div engr, Pure Oil Co, 50-55; supvr, Sandia Labs, 56-59; assoc prof, Kans State Univ, 62-64; supvr, Space Div, Rockwell Corp, 64-67; staff specialist, Autonetics Div, 67-69, mgr prod eng, 69-74, mgr indust eng, Admiral Div, 74-75, staff specialist, Strategic Systs Div, 76-78, mgr advan mfg, Autonetics Strategic Systs Div, 78-84, eng staff specialist, Strategic Defense & Electro-Optical Systs Div, 84-87, prod oper specialist, Satellite & Space Electronics Systs Div, Rockwell Int, 87-88; PRES, WALCO CONSULT, INC, 88- *Concurrent Pos:* consult, Sandia Corp, 63; lectr, Calif State Univ, Fullerton, 72-74 & 79-80; lectr, Prod Opers Mgmt, Univ Calif, Irvine, 88. *Mem:* Inst Indust Engrs; Nat Soc Prof Engrs. *Res:* Production design economics and forecasting; sources and statistics associated with inspector error; organization behavior and the management process. *Mailing Add:* 2306 Arbutus St Newport Beach CA 92660

WALLANDER, JEROME F, b Cato, Wis, Aug 29, 39; m 65; c 3. FOOD SCIENCE, BIOCHEMISTRY. *Educ:* Univ Wis, BS, 62, MS, 65, PhD(food sci), 68. *Prof Exp:* Sr scientist, Mead Johnson Nutritionals, 67-71, assoc dir food prod develop, Mead Johnson & Co, 71-89, DIR, NUTRIT PROD DEVELOP, MEAD JOHNSON NUTRITIONALS, 90- *Mem:* Sigma Xi; Am Dairy Sci Asn; Inst Food Technol; Am Chem Soc. *Res:* Milk lipase and milk protein studies. *Mailing Add:* Nutrit Prod Develop Mead Johnson Nutritionals Evansville IN 47721

WALLBANK, ALFRED MILLS, b Farmington, Mich, May 13, 25; wid. VIROLOGY. *Educ:* Mich State Univ, BS, 48, MS, 53, PhD, 57. *Prof Exp:* Bacteriologist, Barry Labs, Inc, Mich, 48-50; bacteriologist, Blood Sterilization Proj, Henry Ford Hosp, Detroit, 50-52; asst microbiol & pub health, Mich State Univ, 53-56; res assoc & microbiologist, Duke Univ, 56-59; res asst prof microbiol & virologist, Sch Vet Med, Univ Pa, 59-66; head dept virol, Va Inst Sci Res, 66-67; ASSOC PROF MICROBIOL, MED COL, UNIV MAN, 67- *Mem:* Am Soc Microbiol; Am Asn Cancer Res; Can Col Microbiologists; Tissue Cult Asn; Am Acad Microbiol; Sigma Xi. *Res:* Virucides, disinfectants and biohazards. *Mailing Add:* Dept Med Microbiol Univ Man Med Sch Winnipeg MB R3E 0W3 Can

WALLBRUNN, HENRY MAURICE, b Chicago, Ill, Apr 30, 18; m 53; c 4. ZOOLOGY. *Educ:* Univ Chicago, BS, 40, PhD, 51. *Prof Exp:* Instr zool, Univ Chicago, 50; asst prof biol, 51-62, ASSOC PROF ZOOL, UNIV FLA, 63- *Concurrent Pos:* Fel statist & zool, Univ Chicago, 53-54. *Mem:* Genetics Soc Am; Soc Study Evolution. *Res:* Genetics; genetics and evolution of orchids. *Mailing Add:* 7016 NW 20th Pl Gainesville FL 32605

WALLCAVE, LAWRENCE, b Schenectady, NY, Apr 21, 26; m 65. BIOCHEMISTRY. *Educ:* Univ Calif, Berkeley, BS, 48; Calif Inst Technol, PhD, 53. *Prof Exp:* Assoc prof biochem, Eppley Inst Med Ctr, Univ Nebr, Omaha, 68-74, res prof, 74-81; CONSULT, CALSEC CONSULTS, INC, BERKELEY, CALIF, 81- *Res:* Chemical carcinogenesis; analytical biochemistry. *Mailing Add:* 6578 Birch Dr Santa Rosa CA 95404

WALLEIGH, ROBERT S(HULER), b Washington, DC, Mar 31, 15; m 38; c 2. ELECTRICAL ENGINEERING. *Educ:* George Washington Univ, BS, 36. *Prof Exp:* Test engr, Gen Elec Co, 36-37; lighting & power apparatus specialist, Gen Elec Supply Corp, DC, 37-38; rating exam, US Civil Serv Comn, 38-39, secy, Bd US Civil Serv Exam, US Pub Rd Admin, 39-43; elec engr, Lab, Nat Bur Standards, 43, phys sci adminr, 43-53; phys sci adminr, Diamond Ord Fuze Labs, Ord Corps, US Dept Army, 53-55; assoc dir admin, Nat Bur Standards, 55-74, actg dep dir, 75-78, sr adv int affairs, 78-79; RETIRED. *Concurrent Pos:* Consult, Inst Elec & Electronics Engrs, 79- *Honors & Awards:* Gold Medal, Dept Com, 67. *Mem:* AAAS; Inst Elec & Electronics Engrs. *Res:* Radio; illumination engineering; electric apparatus and machinery; electronics management. *Mailing Add:* 5701 Springfield Dr Bethesda MD 20816-1237

WALLEN, CLARENCE JOSEPH, b Phoenix, Ariz, Oct 17, 16. MATHEMATICS. *Educ:* St Louis Univ, BA, 45, MS, 46, PhD(math), 56; Alma Col, STL, 52. *Prof Exp:* Instr math, 46-48 & 56-57, from asst prof to assoc prof, 57-70, chmn dept, 70-73, PROF MATH, LOYOLA MARYMOUNT UNIV, LOS ANGELES, 70-, MEM BD TRUSTEES, 67- *Mem:* AAAS; Am Math Soc; Math Asn Am; Sigma Xi. *Res:* Limit theory in mathematics. *Mailing Add:* Dept Math Loyola Univ Los Angeles CA 90045

WALLEN, CYNTHIA ANNE, b Asheville, NC, June 27, 52; m 80; c 2. EXPERIMENTAL THERAPEUTICS, CELL KINETICS. *Educ:* Univ NC, Greensboro, BA, 74; Univ Rochester, MS, 78, PhD(biophys), 80. *Prof Exp:* Res assoc exp therapeut, dept radiol, Univ Utah, 80-82, res asst prof, 82-84; asst scientist radiation biophys, Univ Kans, 84-; DEPT RADIOL, BOWMAN GRAY SCH MED, NC. *Concurrent Pos:* Fel, Univ Utah, 80-81. *Mem:* Radiation Res Soc; Sigma Xi; Soc Anal Cytologists; Cell Kinetics Soc; Am Asn Cancer Res; NAm Hyperthermia Group. *Res:* The regulation of cell proliferation within tumors and the influence a cell's proliferative status has on determining its response to various treatment modalities such as x-ray, heat and nitro sources and their combinations. *Mailing Add:* Dept Radiol Bowman Gray Sch Med 300 S Hawthorne Winston Salem NC 27103

WALLEN, DONALD GEORGE, b Ont, Can, July 6, 33; m 65; c 2. ECOLOGY, ALGAL PHYSIOLOGICAL ECOLOGY. *Educ:* Dalhousie Univ, BS, 61, MEd, 62; Simon Fraser Univ, MS, 67, PhD(algal physiol & ecol), 70. *Prof Exp:* Asst prof, 70-75, ASSOC PROF BIOL, UNIV WINDSOR, 75- *Mem:* Am Soc Limnol & Oceanog; Phycol Soc Am; Can Soc Plant Physiol; Int Asn Theoret Appl Limnol. *Res:* Interrelationship between light quality, intensity and temperature on growth, photosynthesis and metabolism of algae; toxic substances, chromium and other heavy metal effects on growth and photosynthesis of diatoms, and phytoplankton assemblages in lakes Erie and St Clair; heterotrophic utiization of amino acids by algae. *Mailing Add:* Dept Biol Sci Univ Windsor Windsor ON N9B 3P4 Can

WALLEN, LOWELL LAWRENCE, b Rockford, Ill, May 22, 21; m 49; c 2. BIO-ORGANIC CHEMISTRY. *Educ:* Wheaton Col, Ill, BS, 44; Univ Ark, MS, 48; Iowa State Col, PhD, 54. *Prof Exp:* Asst chemist, Goodyear Tire & Rubber Co, Ohio, 44-46; asst, Univ Ark, 46-48; asst, Iowa State Col, 51-54; biochemist, Northern Regional Res Ctr, USDA, 54-83; RETIRED. *Mem:* Am Chem Soc; Coblentz Soc; Am Oil Chem Soc. *Res:* Chemistry of fermentation products; microbiological type reactions; oxidation; reduction; fermentations; chemical structure elucidation; bio-organic chemistry; infrared spectroscopy; fatty acid chemistry; mycotoxin chemistry. *Mailing Add:* 2929 N Gale Ave Peoria IL 61604

WALLEN, STANLEY EUGENE, b Lincoln, Nebr, Jan 31, 48; m 70; c 2. DAIRY SCIENCE, FOOD SAFETY. *Educ:* Univ Nebr, BS, 69; Iowa State Univ, PhD(food technol statist), 76. *Prof Exp:* Vet food inspector, US Army, 69-72; grad res & teaching asst, Dept Food Technol, Iowa State Univ, 72-76; food technologist, Res & Develop Command, US Army, 76-77; EXTEN FOOD SCIENTIST, DEPT FOOD SCI TECHNOL, UNIV NEBR, LINCOLN, 77- *Mem:* Inst Food Technologists; Int Asn Milk, Food & Environ Sanitarians; Am Dairy Sci Asn. *Res:* Milk quality: factors affecting iodine concentration, beta-carotene concentration, somatic cell count, bacteria count and flavor of milk; mastitis control in dairy cattle. *Mailing Add:* 2610 Surrey Ct Lincoln NE 68512

WALLENBERGER, FREDERICK THEODORE, b St Peter, Austria, Aug 28, 30; nat US; m; c 4. ADVANCED CERAMIC FIBERS. *Educ:* Fordham Univ, MS, 56, PhD(chem), 58. *Prof Exp:* Instr chem, Fordham Univ, 57-58; res fel, Harvard Univ, 58-59; res chemist pioneerins res div, 59-63, res supvr, Carothers Res Lab, Chestnut Run & Christina Lab, 63-69, res assoc, Chestnut Run, 78-85; SR RES ASSOC, EXP STA, FIBERS DEPT, E I DU PONT DE NEMOURS & CO, INC, 85- *Concurrent Pos:* Gordon Conf lectr, 64, 75 & 91; gen chmn, 5th Mid Atlantic regional meeting, Am Chem Soc, 70, chmn, 1st Conf Chem & Environ, 70; lectr, Fiber Soc, 80-81. *Mem:* AAAS; Am Chem Soc; NY Acad Sci. *Res:* Organic polymer synthesis; advanced ceramic fibers; textile technology; advanced composites; consumer research; ozonization; high performance fibers; structural foams. *Mailing Add:* Du Pont Co Wilmington DE 19898

WALLENDER, WESLEY WILLIAM, b Bismark, NDak, Feb 16, 54; m 78; c 2. IRRIGATION ENGINEERING, WATER SCIENCE. *Educ:* Oregon State Univ, BS, 76; Univ Calif, Davis, MS, 78; Utah State Univ, BS, 81, PhD(agr & irrigational eng), 82. *Prof Exp:* Consult eng, irrig, Keller Eng, 80-82; PROF WATER SCI & IRRIG ENG, UNIV CALIF, DAVIS, 82- *Concurrent Pos:* Consult engr, 80- *Mem:* Am Soc Agr Engrs; Am Geophys Union; Sigma Xi. *Res:* Surface irrigation hydraulic modeling with stochastic inputs; spatial variability of infiltration characteristics related to optimization within environmental constraints; sprinkle irrigation hydraulics and optimization; volume variance relations on sampling. *Mailing Add:* Dept Land Air & Water Resources Univ Calif Davis CA 95616

WALLENFELDT, EVERT, b Stanton, Iowa, June 26, 04; m 28; c 2. FOOD SCIENCE. *Educ:* Iowa State Col, BS, 26; Cornell Univ, MS, 29. *Prof Exp:* Instr high sch, Wis, 26-28; dairy fieldman, Borden Farm Prod Co, Ill, 29-31, supvr spec prod & tech probs, 31-34; supvr, Borden-Wieland Co, 34-37; res bacteriologist, Borden Co, 37-38; prof dairy indust, 38-70, EMER PROF FOOD SCI, UNIV WIS-MADISON, 70- *Mem:* AAAS; Am Dairy Sci Asn; Nat Environ Health Asn; Int Asn Milk, Food & Environ Sanitarians. *Res:* Market milk; butter manufacturing; concentrated milks and related products. *Mailing Add:* 2238 Hollister Ave Univ Wis Madison WI 53705

WALLENFELS, MIKLOS, b Budapest, Hungary, July 14, 34; US citizen; m 57; c 2. MECHANICAL ENGINEERING. *Educ:* Budapest Tech Univ, BS, 56; Univ Buffalo, MS, 62. *Prof Exp:* Engr, 59-62, res engr, 62-69, ASSOC ENGR, YERKES RES & DEVELOP, E I DU PONT DE NEMOURS & CO, INC, 69- *Mem:* Soc Plastics Engrs. *Res:* Process development for manufacturing and coating thermoplastic films. *Mailing Add:* 424 Countryside Lane Williamsville NY 14221

WALLENMEYER, WILLIAM ANTON, b Evansville, Ind, Feb 3, 26; m 52; c 4. HIGH ENERGY PHYSICS. *Educ:* Purdue Univ, BS, 50, MS, 54, PhD(physics), 57. *Hon Degrees:* DSc, Purdue Univ, 89. *Prof Exp:* Asst physics, Purdue Univ, 50-54 & 56; jr res assoc high energy particle interactions, Brookhaven Nat Lab, 54-55; asst prof physics, Wabash Col,

55-56; physicist, Midwestern Univs Res Asn, Wis, 56-60, div dir particle accelerators, 60-62; physicist res div, US AEC, 62-64, dir high energy physics, 64-75; dir, high energy physics, US Energy Res & Develop Admin, 75-77; dir high energy physics, Dept Energy, 77-87, assoc dir high energy & nuclear physics, Dept Energy, 85-86; PRES, SOUTHEASTERN UNIV RES ASN, 87- *Mem:* Fel AAAS; fel Am Phys Soc. *Res:* Elementary particle physics; accelerator physics; science management and administration. *Mailing Add:* 1709 New York Ave NW Suite 320 Washington DC 20006

WALLENSTEIN, MARTIN CECIL, EPILEPSY, THERMOREGULATION. *Educ:* Univ Pa, PhD(physiol), 74. *Prof Exp:* ASSOC PROF PHYSIOL, NY UNIV, 78- *Mailing Add:* Dept Physiol NY Univ 421 First Ave New York NY 10010

WALLENTINE, MAX V, b Paris, Idaho, Apr 19, 31; m 53; c 9. FARM & RANCH MANAGEMENT, DIARY MANAGEMENT. *Educ:* Utah State Univ, BS, 55; Cornell Univ, MS, 56, PhD(animal sci & physiol), 60. *Prof Exp:* Asst animal sci, Cornell Univ, 55-56 & 58-60, res assoc meat sci, 60-61; asst prof meat & animal sci, 58-60, res assoc meat sci, Purdue, 61-62; col biol & agr, 71-82, chmn, Animal Sci, 85-88, PROF, MEAT & ANIMAL SCI, BRIGHAM YOUNG UNIV, 67-, DIR AGR STA, 68- *Concurrent Pos:* Consult, Algerian Govt, 68-70. *Mem:* Am Meat Sci Asn; Sigma Xi; Coun Agr Sci & Technol. *Res:* Ultrasonic evaluation of live meat animals; carcass effects from stilbestrol and pelleted roughages; early breeding of ewe lambs; effects of nutritional level; altatta silage (plastic bags) quality and effect on milk production. *Mailing Add:* 301 Widstoe Bldg Brigham Young Univ Provo UT 84602

WALLER, BRUCE FRANK, b Austin, Minn, Oct 18, 47; m; c 4. CARDIOVASCULAR PATHOLOGY, CARDIOLOGY. *Educ:* Luther Col, Decorah, Iowa, BA, 69; Univ Minn, Minneapolis, MD, 73, MS, 76. *Prof Exp:* Staff assoc path, path br, Nat Heart, Lung & Blood Inst, NIH, Bethesda, Md, 78-82; clin asst & prof med, Sch Med, Georgetown Univ, 78-82; prof, 82-89, CLIN PROF PATH & MED, IND UNIV SCH MED, INDIANAPOLIS, 89- *Concurrent Pos:* Res assoc, Krannert Inst Cardiol, 82-89. *Mem:* Am Col Cardiol; Am Heart Asn; Am Col Physicians; Am Col Chest Physicians; Int Acad Path; Am Col Sports Med. *Res:* Cardiovascular research; angioplasty and valvular heart disease; sudden death in athletes. *Mailing Add:* 8402 Harcourt Rd Suite 400 Indianapolis IN 46260

WALLER, COY WEBSTER, b Dover, NC, Feb 25, 14; m 45; c 4. PHARMACEUTICAL CHEMISTRY. *Educ:* Univ NC, BS, 37; Univ Buffalo, MS, 39; Univ Minn, PhD(pharmaceut chem), 42. *Prof Exp:* Instr pharm, State Col Wash, 42-44; dir org chem, Lederle Labs, Am Cyanamid Co, 44-57; dir chem res, Mead Johnson & Co, 57-61; vpres res, Res Ctr, 61-68; from assoc dir to dir, Res Inst Pharmaceut Sci & prof pharm, 68-79, RES PROF, UNIV MISS, 79- *Mem:* Am Chem Soc; Am Pharmaceut Asn; Royal Soc Chem; Sigma Xi. *Res:* Structural determination of natural organic compounds. *Mailing Add:* Res Inst Pharmaceut Sci Univ Miss Sch Pharm University MS 38677

WALLER, DAVID PERCIVAL, b Buffalo, NY, Jan 18, 43; m 78; c 3. ORGANIC CHEMISTRY, PHOTOGRAPHIC CHEMISTRY. *Educ:* Tex Christian Univ, BS, 64; Northeastern Univ, MS, 75. *Prof Exp:* Res asst chem, Res Found, Tex Christian Univ, 64-65; from asst scientist to assoc scientist chem, 65-74, scientist, 74-80, sr scientist chem, 80-87, RES ASSOC, POLAROID CORP, 87- *Mem:* Am Chem Soc; Royal Soc Chem. *Res:* Organic synthesis; novel photographic systems; dye chemistry. *Mailing Add:* 31 Coolidge Ave Lexington MA 02173

WALLER, DONALD MACGREGOR, b Northampton, Mass, Oct 15, 51. PLANT ECOLOGY, POPULATION BIOLOGY. *Educ:* Amherst Col, AB, 73; Princeton Univ, PhD(biol), 78. *Prof Exp:* Asst instr biol, Princeton Univ, 74-75; from asst prof to assoc prof, 78-89, PROF BOT, UNIV WIS-MADISON, 90- *Concurrent Pos:* Res fel bot, Gray Herbarium, Harvard Univ, 77-78; exec vpres, Soc Study Evolution, 91- *Mem:* Bot Soc Am; Soc Study Evolution; Sigma Xi; Soc Conserv Biol; Ecol Soc Am. *Res:* Ecology; competitive and reproductive strategies of plants; evolution; adaptive significance of various breeding systems of plants; demography and genetics of rare plants. *Mailing Add:* Dept Bot Birge Hall Univ Wis Madison WI 53706-1381

WALLER, FRANCIS JOSEPH, b Rome, NY, Mar 12, 43; m 69. ORGANIC CHEMISTRY, ORGANOMETALLIC CHEMISTRY. *Educ:* Niagara Univ, BS, 65; Univ Vt, PhD(org chem), 70. *Prof Exp:* Res investr low temperature kinetics, St Louis Univ, 70-71; vis asst prof org chem, St Lawrence Univ, 71-72; asst prof, Simmons Col, 72-74; res chemist, Polymer Prod Dept & Petrochem Dept, 74-79, staff scientist, 79-82, proj leader, Cent Res & Develop Dept, E I du Pont de Nemours & Co Inc, 82-88; RES ASSOC, CORP SCI & TECH CTR, AIR PROD & CHEM INC, 88- *Mem:* AAAS; Am Chem Soc; NY Acad Sci; Catalysis Soc. *Res:* Homogeneous catalysis; chemical process assessment; perfluorinated ion-exchange polymers; carbonylation technology; organic chemical synthesis; photochemistry. *Mailing Add:* 7201 Hamilton Blvd Air Prod & Chem Inc Allentown PA 18195-1501

WALLER, GEORGE ROZIER, JR, b Clinton, NC, July 14, 27; m 47; c 3. BIOCHEMISTRY. *Educ:* NC State Univ, BS, 50; Univ Del, MS, 52; Okla State Univ, PhD(biochem), 61. *Prof Exp:* Instr agr & biol chem, Univ Del, 50-53; res chemist, Imp Paper & Color Div, Hercules Powder Co, NY, 53-56; from asst prof to prof, 56-68, asst dir, Agr Exp Sta, 69-76, EMER PROF BIOCHEM, OKLA STATE UNIV, 88- *Concurrent Pos:* NIH fel, Nobel Med Inst & Karolinska Inst, Sweden, 63-64; Seydell-Woolley lectr, Ga Inst Technol, 68; mem, Okla Environ Qual Task Force & Gov Task Force on Recommending Sci Policy Struct for State of Okla, 70; pres, Midcontinent Environ Ctr Asn, 69-75; sabbatical leave, Swed Food Inst, Univ Zurich & Univ London; ed, Mass Spectrometry Rev, 81-83, founding ed, 84; head, USA team for USA & Taiwan sem, NSF, 82, bilateral sci mtg, 88; mgr nat meeting, Phytochem Soc NAm & Am Soc Pharmacog, 78. *Mem:* AAAS; Am Soc Biol Chemists; Am Soc Mass Spectrometry; Am Chem Soc; Phytochem Soc NAm (pres, 78-79); Sigma Xi. *Res:* Plant biochemistry, autotoxicity and allelopathy in plants; Maillard reaction products; biochemical applications of mass spectrometry; alkaloid metabolism; author of numerous publications; cholesterol reducing ability of saponins. *Mailing Add:* Dept Biochem Okla State Univ Stillwater OK 74074

WALLER, GORDON DAVID, b Gale, Wis, July 19, 35; m 59; c 2. APICULTURE. *Educ:* Wis State Univ-River Falls, BS, 59; Utah State Univ, MS, 67, PhD(entom, statist), 73. *Prof Exp:* Sci teacher, Pub Schs, Wis, 62-64; res entomologist, Carl Hayden Bee Res Ctr, Agr Res Serv, USDA, 67-89; adj prof entom, Univ Ariz, 69-89; RETIRED. *Mem:* Int Bee Res Asn; Entom Soc Am; Int Comn Bee Bot; Int Union Study Social Insects; Am Soc Agron. *Res:* Applied pollination ecology with special emphasis on honey bee responses to olfactory and gustatory stimuli and the evaluation of foraging behavior of honey bees subjected to different management practices; protecting honey bees from insecticides. *Mailing Add:* 2421 E First St Tucson AZ 85719

WALLER, HARDRESS JOCELYN, b San Diego, Calif, Aug 27, 28; m 53; c 3. GENERAL PHYSIOLOGY. *Educ:* San Diego State Col, AB, 50; Univ Wash, PhD(physiol), 57. *Prof Exp:* From instr to assoc prof physiol, Albert Einstein Col Med, Yeshiva Univ, 58-73; ASSOC PROF NEUROSCI, MED COL OHIO, 73- *Mem:* Soc Neurosci; AAAS; Am Physiol Soc; Sigma Xi. *Res:* Electrical activity of central nervous system. *Mailing Add:* Med Col Ohio Neurol Surg PO Box 10008 Toledo OH 43699

WALLER, JAMES R, b Eureka, Mont, Dec 8, 31; m 55; c 8. MICROBIOLOGY, BIOCHEMISTRY. *Educ:* St John's Univ, Minn, BA, 57; Univ Minn, MS, 60, PhD(microbiol), 64. *Prof Exp:* Res assoc microbial physiol, Univ Cincinnati, 63-66; asst prof, 66-70, ASSOC PROF MICROBIOL, UNIV NDAK, 70- *Mem:* Am Soc Microbiol; Sigma Xi. *Res:* Clinical microbiology; physiology and mechanisms of vitamin transport in microorganisms and tissue cultures. *Mailing Add:* 2514 S Tenth St Grand Forks ND 58201

WALLER, JOHN WAYNE, b Johnson City, Tenn, Dec 15, 37; m; c 3. ELECTRICAL ENGINEERING, ELECTRONICS. *Educ:* Univ Tenn, PhD(eng sci). *Prof Exp:* From asst prof to assoc prof elec eng, 64-88, ASSOC DEPT HEAD, UNIV TENN, KNOXVILLE, 83- *Concurrent Pos:* Vis prof, Ga Tech, 74-75. *Res:* Circuits; non-majors courses. *Mailing Add:* Dept Elec Eng Ferris Hall 402 Univ Tenn Knoxville TN 37996-2100

WALLER, JULIAN ARNOLD, b New York, NY, Apr 17, 32; m 56; c 2. MEDICINE, PUBLIC HEALTH. *Educ:* Columbia Univ, AB, 53, Boston Univ, MD, 57; Harvard Univ, MPH, 60; Am Bd Prev Med, dipl, 65. *Prof Exp:* Intern, Mary Fletcher Hosp, Burlington, Vt, 57-58; resident, Contra Costa Health Dept, Calif, 58-59; regional consult chronic dis & voc rehab, USPHS, 60-62; resident, Calif Dept Pub Health, 62, coordr accident prev, 62-63, med officer occup health, 64-66, med officer chronic dis, 66-67, chief emergency health serv, 68; prof community med, 68-72, prof & chmn, Dept Epidemiol & Environ Med, 72-79, PROF MED, UNIV VT, 79- *Concurrent Pos:* Consult, US Dept Transp, 67-80, Vt State Health Dept, 68-78 & Vt Dept Ment Health, 70-76; mem safety & occup health study sect, Dept HEW, 68-71; mem, Nat Hwy Safety Adv Comt, 69-72 & Nat Motor Vehicle Safety Adv Coun, 74-77; mem, Nat Res Coun Comt Trauma Res, 84-85, expert panel accident prev, WHO, 85-; bd dir, Consumers Union US, 67-70 & 86-89. *Honors & Awards:* Int Asn Accident & Traffic Med Award, 78; A J Mirkin Award, Asn Advan Automotive Med, 87; Injury Control Special Interest Group Award, Am Pub Health Asn, 89. *Mem:* AAAS; fel Am Pub Health Asn; fel Am Col Epidemiol; fel Am Col Prev Med; Asn Advan Automotive Med (pres, 74). *Res:* Epidemiology and control of highway and non-highway injury; epidemiology and control of problem drinking; health hazards to visual artists. *Mailing Add:* Dept Med Mansfield House Univ Vt Burlington VT 05405

WALLER, RAY ALBERT, b Grenola, Kans, Mar 4, 37; m 60; c 2. MATHEMATICAL STATISTICS. *Educ:* Southwestern Col, BA, 59; Kans State Univ, MS, 63; Johns Hopkins Univ, PhD(math statist), 67. *Prof Exp:* Instr math, St John's Sch, PR, 60-61; asst prof math, Towson State Col, 66-67; from asst prof to assoc prof statist, Kans State Univ, 67-74; dep div leader, 80-87, STAFF MEM, LOS ALAMOS NAT LAB, 74-, STAFF ASST, 87- *Concurrent Pos:* Consult, White Sands Missile Range, 68-72. *Mem:* Am Statist Asn; Soc Risk Analysis. *Res:* Bayesian inference and reliability estimation. *Mailing Add:* 525 Navajo St Los Alamos NM 87544

WALLER, RICHARD CONRAD, b Victory, Wis, Sept 24, 15; m 36, 48; c 5. CHEMICAL ENGINEERING, PHYSICAL CHEMISTRY. *Educ:* Wash State Univ, BS, 38; Iowa State Univ, PhD(phys chem), 42. *Prof Exp:* Res chemist, E I du Pont de Nemours & Co, Inc, 42-48; res scientist fibers, Goodyear Tire & Rubber Co, 48-53, sect head polymer, 53-58, mgr, 58-65, dir res & develop polymer, 65-76, vpres res, 76-80; CONSULT, 80- *Res:* Polymer; fiber. *Mailing Add:* 123 Leland St SE Port Charlotte FL 33952

WALLER, ROGER MILTON, b Taylor, Wis, Dec 30, 26; m 55; c 3. GROUNDWATER GEOLOGY. *Educ:* Univ Wis, BS, 50; Univ Ariz, MS, 69. *Prof Exp:* Jr computer seismic explor, NMex-Tex, Nat Geophys Co, 50-51; groundwater geologist, US Geol Surv, Calif Dist, 51-54, admin hydrologist, Alaska Dist, 55-63, res hydrologist, Alaska Earthquake Effects, 64-65, hydrologist, NY Dist, 66-68, Great Lakes Basin groundwater coordr, Madison, Wis, 68-71, assoc dist chief, Ohio Dist, 72-73, hydrologist, 73-86; RETIRED. *Concurrent Pos:* Hydrol panel mem, Nat Acad Sci Comt Alaska Earthquake, 64-74; consult, 87- *Mem:* Geol Soc Am; Am Geophys Union; Nat Well Water Asn. *Res:* Groundwater occurence in glacial deposits. *Mailing Add:* RD 1 Box 184 East Greenbush NY 12061

WALLER, STEVEN SCOBEE, b Indianapolis, Ind, Aug 29, 47; m 70; c 3. RANGE SCIENCE, PLANT ECOLOGY. *Educ:* Purdue Univ, BSc, 70; Tex A&M Univ, PhD(range sci), 75. *Prof Exp:* Res asst, Tex A&M Univ, 71-73, teaching asst, 73-74, res fel, 74-75; asst prof, SDak State Univ, 75-77, asst to dir resident instr & asst prof, 77-78; assoc prof, 78-84, PROF RANGE SCI, UNIV NEBR, 84- *Concurrent Pos:* Assoc ed, J Range Mgt, 82-84; chmn, Range Sci Educ Coun, 85; regional coordr, NCent Region, LISA Prog, 89- *Honors & Awards:* Serv Award, Soc Range Mgt, 88. *Mem:* Soc Range Mgt; Sigma Xi; Nat Asn Cols & Teachers Agr. *Res:* Development of successful range improvement practices within the field of range science. *Mailing Add:* Keim Hall 347 E Campus Lincoln NE 68583-0914

WALLER, THOMAS RICHARD, b Chicago, Ill, July 18, 37; m 59; c 2. MALACOLOGY, EVOLUTION. *Educ:* Univ Wis, BS, 59, MS, 61; Columbia Univ, PhD(geol), 66. *Prof Exp:* Assoc cur, 66-74, CUR CENOZOIC MOLLUSCA, SMITHSONIAN INST, 74- *Mem:* AAAS; Paleont Soc; Soc Syst Zool; Am Malacological Union; Nat Shellfisheries Asn; Paleont Res Inst; Unitas Malacologia. *Res:* Cenozoic and living mollusca; evolution; zoogeography; bivalve morphology, development and phylogeny; upper Cenozoic biostratigraphy of eastern North America and Caribbean. *Mailing Add:* Dept Paleobiol Smithsonian Inst Washington DC 20560

WALLER, WILLIAM T, b Girard, Kans, Apr 29, 41; m 61; c 2. AQUATIC BIOLOGY, LIMNOLOGY. *Educ:* Kans State Col, BS, 65, MS, 67; Va Polytech Inst & State Univ, PhD(biol), 71. *Prof Exp:* Res asst, Kans State Col, 65-67, Univ Kans, 68, Va Polytech Inst & State Univ, 68-71 & Region VII, Environ Protection Agency, 72; assoc res scientist environ med, Med Ctr, NY Univ, 72-75; ASSOC PROF, GRAD SCH ENVIRON SCI, UNIV TEX, DALLAS, 75- *Mem:* Am Fisheries Soc. *Res:* Organismic, population and community structure and function as it relates to acute and chronic stresses. *Mailing Add:* Univ Tex PO Box 688 Richardson TX 75080

WALLERSTEIN, DAVID VANDERMERE, b New York, NY, Nov 19, 37; m 72; c 1. APPLIED MECHANICS, SYSTEMS ANALYSIS. *Educ:* Mich Technol Univ, BS, 60, MS, 61, PhD(eng mech), 69. *Prof Exp:* Design specialist, Lockheed Calif Co, 69-71; asst prof res, Va Polytech Inst, 71-72; sr opers res specialist, Lockheed Calif Co, 72-81; SR STAFF ENGR, MACNEAL-SCHWENDLER CORP, 82- *Concurrent Pos:* Adj assoc prof aerospace eng, Univ Southern Calif, 82- *Mem:* Am Inst Mining, Metall & Petrol Engrs. *Res:* Structuring of engineering systems into mathematical models; development of solutions that yield optimal values of system measures of desirability; finite element analysis. *Mailing Add:* Dept Aerospace Eng Univ Southern Calif Univ Park Los Angeles CA 90089

WALLERSTEIN, EDWARD PERRY, b New York, NY, May 23, 28; m 51; c 3. OPTICAL SYSTEM DESIGN. *Prof Exp:* Engr, Perkin-Elmer Corp, 51-55; res asst, Phys Res Labs, Boston Univ, 55-58; engr, Itek Corp, 58-60; founder & exec, Diffraction Limited, Inc, 60-69; mgr systs group, Valtec Corp, 70-74; group leader/proj engr, Lawrence Livermore Nat Lab, 74-90, dep assoc prog leader/group leader, 90-91; CONSULT, 91- *Concurrent Pos:* Consult design optical systs, 55- *Honors & Awards:* IR 100 Award, Indust Res & Develop, 79, 87 & 88. *Mem:* Optical Soc Am; Soc Photo-Optical Instrumentation Engrs; Int Soc Optical Eng. *Res:* Design and manufacture of advanced high precision optical systems for reconnaisance, astronomy and laser fusion; large operture frequency conversion with nonlinear materials; numerous articles in optical industry journals. *Mailing Add:* 1742 Beachwood Way Pleasanton CA 94566

WALLERSTEIN, GEORGE, b New York, NY, Jan 13, 30. ASTRONOMY. *Educ:* Brown Univ, AB, 51; Calif Inst Technol, PhD, 58. *Prof Exp:* From instr to assoc prof astron, Univ Calif, Berkeley, 58-64; chmn, Astron Dept, 65-80, PROF ASTRON, UNIV WASH, 65- *Concurrent Pos:* Bd trustees, Brown Univ, 75-80; Bd Dir, Coun Econ Priorities, 84- *Mem:* Am Astron Soc; Royal Astron Soc; Arctic Inst NAm; Astron Soc Pac; AAAS. *Res:* Spectra of variable stars; abundances of the elements in stellar atmospheres; interstellar absorption lines. *Mailing Add:* Dept Astron Univ Wash Seattle WA 98195

WALLERSTEIN, RALPH OLIVER, JR, b San Francisco, Calif, July 31, 53; m 76; c 2. HEMATOLOGY, ONCOLOGY. *Educ:* Mass Inst Technol, BS, 75; Univ Calif, San Francisco, MD, 79; Am Bd Internal med, cert, 82. *Prof Exp:* Med intern, Peter Bent Brigham Hosp & Med Sch, Harvard Univ, 79-80, jr med resident, 80-81, sr med resident, 81-82; pvt pract internal med, Queen Valley Hosp, 82-84; fel hemat & oncol, 84-86, ASST CLIN PROF MED, UNIV CALIF, SAN FRANCISCO, 86- *Concurrent Pos:* Clin instr med, Univ Calif, San Francisco, 83-84, attend physician, Cancer Res Inst & Hemat & Oncol Div, Vet Admin Hosp, 86-87, instr & lectr, Hemat Sect, Sch Med, 86-88; attend physician, Children's Hosp San Francisco, 86-, Med Ctr, Stanford Univ & Oncol Div, Palo Alto Vet Admin, 87-88. *Mem:* Inst Med-Nat Acad Sci; fel Am Col Physicians; Am Soc Clin Oncol. *Res:* Hematology; molecular biology of chronic myelogenous leukemia; metastatic melanoma; author of 12 technical publications. *Mailing Add:* Dept Lab Med Univ Calif San Francisco CA 94143

WALLERSTEIN, ROBERT SOLOMON, b Berlin, Ger, Jan 28, 21; US citizen; m 47; c 3. PSYCHIATRY, PSYCHOANALYSIS. *Educ:* Columbia Univ, BA, 41, MD, 44. *Prof Exp:* Intern med, Mt Sinai Hosp, New York, 44-45, asst resident, 45-46, resident, 48, resident psychiat, 49; resident, Vet Admin Hosp, Topeka, Kans, 49-51, chief psychosom sect, 51-53; assoc dir dept res, Menninger Found, Kans, 54-65, dir dept, 65-66; chief, Dept Psychiat, Mt Zion Hosp, San Francisco, 66-78; clin prof psychiat, Sch Med & Langley-Porter Neuropsychiat Inst, 67-75, prof psychiat, Sch Med & Dir, Langley Porter Inst, 75-85, PROF PSYCHIAT, UNIV CALIF, SAN FRANCISCO, 85- *Concurrent Pos:* Lectr psychiat, Menninger Sch Psychiat, Kans, 51-66; lectr, Topeka Inst Psychoanal, 59-66, training & supv analyst, 65-66; fel, Ctr Advan Studies Behav Sci, Stanford, Calif, 64-65 & 81-82; training & supv analyst, San Francisco Psychoanal Inst, 66-; mem res sci career develop comn, NIMH, 66-70, chmn, 68-70; vis prof psychiat, Sch Med, La State Univ & New Orleans Psychoanal Inst, 72-73 & Sch Med, Pahlavi Univ, Iran, 77. *Honors & Awards:* J Elliott Royer Award, Univ Calif, 73. *Mem:* Am Psychoanal Asn (pres, 71-72); Int Psychoanal Asn (vpres, 77-85, pres, 85-89); fel Am Psychiat Asn; fel Am Col Physicians; fel Am Orthopsychiat Asn. *Res:* Psychotherapy research, especially the processes and outcomes of psychoanalytic therapy; supervision processes; alcoholism; psychosomatic medicine. *Mailing Add:* Dept Psychiat Univ Calif San Francisco CA 94143

WALLES, WILHELM EGBERT, b Enschede, Neth, May 25, 25; nat US; m 51; c 5. ORGANIC CHEMISTRY, POLYMER CHEMISTRY. *Educ:* Univ Amsterdam, MSc, 48, PhD(org chem, physics), 51, DSc, 53. *Prof Exp:* Asst indust res, Univ Amsterdam, 49-51; Nat Res Coun Can fel, 51-53; dir res, N V Neth Refining Co, 53-55; assoc scientist, Dow Chem Co, 55-67, sr assoc res scientist plastics, Cent Res Lab, 67-89; RETIRED. *Honors & Awards:* IR 100 Award, 74. *Mem:* Am Chem Soc. *Res:* Synthesis and physical properties of high polymers; surface reactions of polymers; polymolecular complexes; magneto-organic chemistry; aerosol physics; surface chemistry of plastic films, fibers and articles. *Mailing Add:* 6648 N River Rd Freeland MI 48623-9202

WALLEY, WILLIS WAYNE, b Brooklyn, Miss, July 26, 34; m 66. ZOOLOGY, ECOLOGY. *Educ:* Southern Miss Univ, BS, 56; Miss State Univ, MS, 61, PhD(zool)j 65. *Prof Exp:* Pub sch teacher, Miss, 56-61; asst prof biol, Southeastern La Col, 64-68; prof biol & chmn dept & chmn div III, Belhaven Col, 68-79; PROF BIOL & CHMN DEPT, DELTA STATE UNIV, 79- *Mem:* AAAS; Am Inst Biol Sci; Nat Audubon Soc; Sigma Xi. *Res:* Absorption, metabolism and excretion of chlorinated hydrocarbons in birds; in vitro metabolism of dichloro-diphenyl-trichloro-ethane by various tissues of the common grackle; biogeochemical cycling of boron; site fertility and primary production in hydrosoils. *Mailing Add:* Dept Biol Delta State Univ Cleveland MS 38733

WALLICK, EARL TAYLOR, b Monticello, Ark, Jan 11, 38; m 62; c 2. BIOLOGICAL CHEMISTRY. *Educ:* Miss State Univ, BS, 60, MS, 62; Rice Univ, PhD(org chem), 66. *Prof Exp:* Res chemist, Dacron Res Lab, E I du Pont de Nemours & Co, Inc, 66-67; assoc prof chem, King Col, 67-71; trainee myocardial biol, Baylor Col Med, 71-73, instr cell biophys, 73-76, asst prof, 76-77; from asst prof to assoc prof, 77-85, PROF PHARMACOL & CELL BIOPHYS, COL MED, UNIV CINCINNATI, 86- *Concurrent Pos:* Mem, Pharmacol Study Sect, NIH. *Mem:* Am Chem Soc; Am Soc Pharmacol & Exp Therapeut. *Res:* Isotope effects; cardiac glycosides, adenosine triphosphatase. *Mailing Add:* Dept Pharmacol & Cell Biophys Col Med Univ Cincinnati 231 Bethesda Ave Cincinnati OH 45267-0575

WALLICK, GEORGE CASTOR, b Grand Rapids, Mich, July 2, 23; m 45; c 2. PHYSICS. *Educ:* Univ Mich, BS, 43, MS, 46, PhD(physics), 52. *Prof Exp:* Radio engr radar res & develop, US Naval Res Lab, 44-45; sr res technologist appl math, 51-59, RES ASSOC, FIELD RES LAB, MOBIL RES & DEVELOP CORP, DALLAS, 59- *Mem:* Am Phys Soc; Am Asn Physics Teachers. *Res:* Petroleum production; flow of fluids through porous media; non-Newtonian flow; heat transfer; numerical analysis; computer programming and utilization; geophysics; underground coal gasification. *Mailing Add:* 518 Towne Pl Duncanville TX 75116

WALLIN, JACK ROBB, b Omaha, Nebr, Nov 21, 15; m 37; c 2. PLANT PATHOLOGY. *Educ:* Iowa State Univ, BS, 39, PhD(plant path), 44. *Prof Exp:* Asst bot, Univ Mo, 39-41 & N Iowa Agr Exp Asn, 41-44; collabr, E May Seed Co, 44-45; res asst prof, Iowa State Univ, 45-47; plant pathologist, Agr Res Serv, USDA, 47-58 & res plant pathologist, 75-86; sr plant pathologist, Iowa State Univ, 59-75; prof, 75-86, EMER PROF PLANT PATH, UNIV MO, COLUMBIA, 88- *Concurrent Pos:* Mem, working group, Comn Instruments & Methods Observation, World Meteorol Orgn; mem, aerobiol comt, Nat Acad Sci, 76-80. *Honors & Awards:* William F Peterson Award, Int Soc Biometeorol, 67. *Mem:* Am Phytopath Soc; Am Meteorol Soc; Int Soc Biometeorol. *Res:* Investigating the cause of epidemics of corn diseases and experimental forecasts of these diseases especially Aspergillus flavus and genetic resistance to the fungus and levels of aflatoxin; epidemiology; disease forecasting; aerobiology; corn diseases. *Mailing Add:* Rte 5 Box 5208 Fulton MO 65251-9805

WALLIN, JOHN DAVID, b Pasadena, Calif, June 30, 37; m 59; c 2. NEPHROLOGY, RENAL PHYSIOLOGY. *Educ:* Stanford Univ, BS, 58; Sch Med, Yale Univ, MD, 62. *Prof Exp:* Intern & resident med, Naval Regional Med Ctr, PR, 62-66, chief med, 66-70; res fel nephrology, Southwest Med Sch, Univ Tex, 70-72; dir clin invest, Naval Regional Med Ctr, Calif, 72-78; chief nephrology & prof internal med, Sch Med, Tulane Univ, 78-90; CHIEF NEPHROLOLGY & PROF INTERNAL MED, SCH MED, LA STATE UNIV, 90- *Concurrent Pos:* Instr internal med, Univ PR, 67-70; chief nephrology, Vet Admin Hosp, New Orleans, 78-81. *Mem:* Fel Am Col Physicians; Am Soc Nephrology; Am Fedn Clin Res; Int Soc Nephrol. *Res:* Examination of homeostasis of water metabolism with specific reference to control of vasopresion release from hypophysis and its end organ effects in renal tubule. *Mailing Add:* Dept Med Sch Med La State Univ 1542 Tulane Ave New Orleans LA 70112

WALLIN, RICHARD FRANKLIN, b Chicago, Ill, Jan 31, 39; m 61; c 2. TOXICOLOGY, MEDICAL DEVICE EVALUATION. *Educ:* Univ Ill, BS, 61, DVM, 63, MS, 64, PhD(vet med sci), 66. *Prof Exp:* Physiologist, McDonnell Aircraft Corp, 66-67; sr res pharmacologist, Baxter Labs, Inc, Morton Grove, 67-70, dir res admin, 70-72, actg dir pharm & microbiol res, 72-73, assoc dir pharmacol res, 73-77; sci dir, 77-78, vpres & sci dir, 78-81, PRES, NAM SCI ASSOCS, INC, 81- *Concurrent Pos:* NIH fel, 63-66; adj assoc prof, Med Col Ohio; lectr, & course dir, Ctr Prof Advan, East Brunswick, NJ; mem USP adv comt. *Mem:* Am Vet Med Asn; Parenteral Drug Asn; Am Asn Lab Animal Sci; Soc Biomat. *Res:* Biomaterials; medical devices; biocompatibility of materials. *Mailing Add:* 29969 St Andrews Dr Perrysburg OH 43551

WALLING, CHEVES (THOMSON), b Evanston, Ill, Feb 28, 16; m 40; c 5. PHYSICAL ORGANIC CHEMISTRY, FREE RADICAL REACTIONS. *Educ:* Harvard Univ, BA, 37; Univ Chicago, PhD(org chem), 39. *Prof Exp:* Res chemist, Jackson Lab, E I DuPont de Nemours & Co, 39-42 & Gen Labs, US Rubber Co, 43-49; res assoc, Lever Bros Co, 49-52; prof chem, Columbia Univ, 52-70, chmn dept, 63-66; DISTINGUISHED PROF, UNIV UTAH, 70- *Concurrent Pos:* Tech aide, Comt Med Res, Off Sci Res & Develop, 45-46; chmn div chem & chem technol, Nat Res Coun, 72-73; ed, J Am Chem Soc, 75-81. *Honors & Awards:* James Flack Norris Award, Am Chem Soc, 70, Lubrizol Award in Petrol Chem, 84. *Mem:* Nat Acad Sci; Am Acad Arts & Sci; AAAS; Am Chem Soc. *Res:* Organic reaction mechanisms; free radical reactions; polymerization; peroxides and autoxidation. *Mailing Add:* Dept Chem Univ Utah Salt Lake City UT 84112

WALLING, DERALD DEE, b Granger, Iowa, Feb 14, 37; m 58; c 4. MATHEMATICAL STATISTICS. *Educ:* Iowa State Univ, BS, 58, MS, 61, PhD(math), 63. *Prof Exp:* Mathematician, Ames Lab, US AEC, 62-63; asst prof math, Univ Ariz, 63-66; assoc prof, 66-87, PROF MATH, TEX TECH UNIV, 87- *Res:* Psychophysics; mathematics sociology; statistics; probability. *Mailing Add:* Dept Math Tex Tech Univ Lubbock TX 79409-1042

WALLINGFORD, ERROL E(LWOOD), b Ottawa, Ont, Jan 25, 28; m 54; c 3. COMMUNICATIONS. *Educ:* Carleton Col, BSc, 53; Univ Ottawa, MSc, 61. *Prof Exp:* Inspector, Dept Nat Defense Inspection Servs, 52-57; lectr physics, Waterloo Col, 57-58; from lectr to asst prof, Royal Mil Col Can, 61-68, assoc prof elec eng, 68-88; RETIRED. *Concurrent Pos:* Defence Res Bd Can grants, 65- *Mem:* Inst Elec & Electronics Engrs. *Res:* Primary source redundancy identification; removal and replacement using secondary source data to maintain a block code structure; method applicable to both first and second order redundancies. *Mailing Add:* RR 1 Sydenham ON K0H 2T0 Can

WALLINGFORD, JOHN STUART, b El Paso, Tex, Apr 13, 35; m 70; c 2. PHYSICS. *Educ:* Univ Minn, Minneapolis, BPhys, 61; Fla State Univ, MS, 66, PhD(physics), 67. *Prof Exp:* Instr physics, Fullerton Jr Col, 61-62 & Cerritos Col, 62-63; asst, Fla State Univ, 63-66; from asst prof to assoc prof, Fla A&M Univ, 66-69; vis assoc prof, Temple Univ, 69-70; assoc prof, 70-75, PROF PHYSICS, PEMBROKE STATE UNIV, 75- *Mem:* Am Asn Physics Teachers; Am Inst Physics. *Res:* Consequences of general relativity theory; making science interesting and accessible to all; hydrolysis of organic wastes; computer assisted tomography (reconstruction techniques); linear programming (simplex algorithm Kachiyan's algorithm); personal (micro-) computer programming and interfacing. *Mailing Add:* Dept Phys Sci Pembroke State Univ Pembroke NC 28372

WALLIS, CLIFFORD MERRILL, b Waitsfield, Vt, Mar 7, 04; m 28; c 2. ELECTRONICS, RADIO ENGINEERING. *Educ:* Univ Vt, BS, 26; Mass Inst Technol, MS, 28; Harvard Univ, ScD(eng), 41. *Prof Exp:* Mem staff eng training course, Gen Elec Co, Mass, 26-28; from instr to prof elec eng, 28-70, chmn dept, 47-68, EMER PROF ELEC ENG, UNIV MO, COLUMBIA, 70- *Concurrent Pos:* Res assoc, Underwater Sound Lab, Harvard Univ, 44-45; mem & vchmn, Nat Res Comt, Am Inst Elec Engrs, 58-59; Fulbright lectr, Ankara, 60-61, award, Taiwan, 67-68; dir prof develop, USN Underwater Systs Ctr, Conn, 70-72. *Mem:* Am Soc Eng Educ; fel Inst Elec & Electronics Engrs. *Res:* Rectifier analysis; half wave gas rectifier circuits; single phase full wave rectifier circuits; current division in tetrode and pentode power tubes; low frequency constant time delay lines; solid state theory of semiconductors. *Mailing Add:* Moretown VT 05660

WALLIS, DONALD DOUGLAS JAMES H, b Brandon, Man, Apr 20, 43. IONOSPHERIC PHYSICS. *Educ:* Univ Alta, Calgary, BSc, 65; Univ Calgary, MSc, 68; Univ Alaska, PhD(geophys), 74. *Prof Exp:* Fel geophys, Univ Alta, 73-75; res assoc geophys, Univ Calgary, 75-76; res assoc, 76-81, RES OFFICER SPACE PHYS, NAT RES COUN CAN, 81- *Mem:* Can Asn Physicists; Am Geophys Union. *Res:* Auroral spectroscopy; ionospheric winds and currents; atmospheric changes; magnetospheric physics. *Mailing Add:* 499 Blair St Ottawa ON K1G 0J3 Can

WALLIS, GRAHAM B, b Rugby, Eng, Apr 1, 36; m 59; c 4. ENGINEERING. *Educ:* Cambridge Univ, BA, 57, MA, 61, PhD(eng), 61; Mass Inst Technol, SM, 59. *Prof Exp:* Res fel, UK Atomic Energy Authority, 59-62; from asst prof to assoc prof, 62-72, PROF ENG, DARTMOUTH COL, 72- *Concurrent Pos:* Sr vis res fel, Heriot-Watt Univ, 64; vis reader, Univ Warwick, 70-71. *Mem:* Am Soc Mech Engrs. *Res:* Two-phase and multicomponent flow; heat and mass transfer; boiling and condensation. *Mailing Add:* Dept Eng Sci Dartmouth Col Hanover NH 03755

WALLIS, PETER MALCOLM, b London, Eng, May 9, 52; Can citizen; m 76; c 3. BIOGEOCHEMISTRY. *Educ:* Univ Toronto, BSc, 74; Univ Waterloo, MSc, 75, PhD(biol), 78. *Prof Exp:* Fel, 78-79, res assoc, 79-80, PROF ASSOC, KANANASKIS CTR, UNIV CALGARY, 80- *Honors & Awards:* E E Ballantyne Award Environ Res, 86. *Mem:* NAm Benthological Soc; Soc Int Limnol; Freshwater Biol Asn; Sigma Xi; Wildlife Dis Asn; Am Soc Microbiol. *Res:* Microbiology; biogeochemistry; chemical evolution of groundwater; host reserviors and transmission of giardiasis; survival of pathogens in the environment. *Mailing Add:* Biosci Bldg 42 Univ Calgary Kananaskis Ctr Calgary AB T2N 1N4 Can

WALLIS, RICHARD FISHER, b Washington, DC, May 14, 24; m 55; c 2. SOLID STATE PHYSICS. *Educ:* George Washington Univ, BS, 45, MS, 48; Cath Univ, PhD(chem), 52. *Prof Exp:* Fel, Inst Fluid Dynamics, Univ Md, 51-53; chemist, Appl Physics Lab, Johns Hopkins Univ, 53-56; physicist, US Naval Res Lab, 56-58, from actg head to head,Semiconductors Br, 58-66; prof physics, Univ Calif, Irvine, 66-67; head Semiconductors Br, US Naval Res Lab, 67-69; chmn dept, 72-75 & 80-83, PROF PHYSICS, UNIV CALIF, IRVINE, 69- *Concurrent Pos:* Consult, Res Labs, Gen Motors Corp, 58-79, US Naval Res Lab, 69-79. *Honors & Awards:* Pure Sci Award, Naval Res Lab, 64. *Mem:* Fel Am Phys Soc. *Res:* Quantum and statistical mechanics; solid state theory. *Mailing Add:* Dept Physics Univ Calif Irvine CA 92717

WALLIS, ROBERT CHARLES, medical entomology, parasitology; deceased, see previous edition for last biography

WALLIS, ROBERT L, b Sheridan, Wyo, Sept 22, 34; m 60; c 2. PHYSICS. *Educ:* Univ Colo, BA, 56, MA, 58, PhD(physics), 62. *Prof Exp:* From asst prof to assoc prof, 62-75, PROF PHYSICS, BALDWIN-WALLACE COL, 75-, CHMN DEPT, 70- *Mem:* AAAS; Am Phys Soc; Am Asn Physics Teachers; Sigma Xi. *Res:* Educational techniques. *Mailing Add:* Dept Physics Baldwin-Wallace Col Berea OH 44017

WALLIS, THOMAS GARY, b Paducah, Ky, 48; m 69; c 2. PHOTOGRAPHIC SCIENCE. *Educ:* Murray State Univ, BA, 70; Duke Univ, PhD(org chem), 74. *Prof Exp:* Assoc chemist, Ohio State Univ, 74-76; res chemist, 76-86, TECH MGR, EASTMAN KODAK CO, 86- *Mem:* Soc Motion Picture & TV Engrs. *Res:* Synthesis of photographically active compounds; image structure improvement in color films; new motion picture color films. *Mailing Add:* Eastman Kodak Co Res Lab Rochester NY 14650

WALLIS, W ALLEN, b Philadelphia, Pa, Nov 5, 12; m 35; c 2. ECONOMICS. *Educ:* Univ Minn, AB, 32. *Hon Degrees:* DSc, Hobart & William Smith Cols, 73; LLD, Roberts Wesleyan Col, 73 & Univ Rochester, 84; LHD, Grove City Col, 75. *Prof Exp:* Instr econ, dept econ, Yale Univ, 37-38; from asst prof to prof, Stanford Univ, 38-46; prof statist & econ, Grad Sch Bus, Univ Chicago, 46-62, chmn, dept statist, 49-57, dean grad sch, 56-62; prof econ & statist, pres & trustee, Univ Rochester, 62-78, emer prof, chancellor & hon trustee, 78-82; under secy econ affairs, US State Dept, 82-89; RESIDENCE SCHOLAR, AM ENTERPRISE INST, 89- *Concurrent Pos:* Consult, Rand Corp, 46-66; pres, Nat Comn Study Nursing & Nursing Educ, 67-70; mem, adv coun on higher educ, NY State Dept Educ, 70-78; chmn, adv comt social security studies, Am Enterprise Inst, 76-82; subcomt postperformance eval res, Nat Acad Sci, 81-82; W Allen Wallis Prof, Univ Chicago & Fel, Univ Rochester, 82. *Honors & Awards:* Wilks Mem Award, Am Statist Asn, 80. *Mem:* Fel AAAS; fel Am Acad Arts & Sci; Am Econ Asn; fel Am Statist Asn (pres, 65). *Res:* Author or co-author of ten books and monographs and 55 articles on statistics, economics and higher education. *Mailing Add:* Am Enterprise Inst 1150 17th St NW Washington DC 20036

WALLIS, WALTER DENIS, b Sidney, NSW, Australia, June 26, 41. COMBINATORIAL DESIGNS, GRAPH THEORY. *Educ:* Univ Sydney, BSc, 63, PhD(pure math), 68. *Prof Exp:* Lectr math, La Trobe Univ, 67-70; from lectr to assoc prof math, Univ Newcastle, 70-85; PROF MATH, SOUTHERN ILL UNIV, 85- *Concurrent Pos:* Vis prof, Univ Waterloo, 72 & 78, Univ Man, 73, Univ Surrey, 76 & Simon Fraser Univ, 84-85; ed, J Combinatorial Math & Comput, 87-; vis prof, Curtin Univ Technol, 88. *Mem:* Am Math Soc; Math Asn Am; Soc Indust Appl Math; Inst Combinatorics Appln. *Res:* Combinatorial mathematics including graph theory and experimental designs. *Mailing Add:* Math Dept Southern Ill Univ Carbondale IL 62901-4408

WALLMAN, JOSHUA, b New York, NY, July 4, 43. NEUROBIOLOGY. *Educ:* Harvard Univ, AB, 65; Tufts Univ, PhD(biol), 72. *Prof Exp:* Fel neurophysiol, Inst Animal Behav, Rutgers Univ, 72-74; from asst prof to assoc prof, 74-85, PROF BIOL, CITY COL NEW YORK, 86- *Concurrent Pos:* NIH fel, 72-73; res grants, Nat Eye Inst, NIH, 78- 81 & 81-92, NSF, 85-86; mem, Working Group on Myopia, Nat Res Coun, Integrative Neural Sci Panel, NSF, 84-87. *Mem:* Asn Res Vision Ophthal; Soc Neurosci; Sigma Xi. *Res:* Neurophysiology of the accessory optic system; physiological and behavioral studies of the development of the oculomotor and visual systems in birds; studies on experimentally-induced myopia; regulation of growth of the eye. *Mailing Add:* Dept Biol City Col New York New York NY 10031

WALLMARK, J(OHN) TORKEL, b Stockholm, Sweden, June 4, 19; m 49; c 2. ELECTRONICS. *Educ:* Royal Inst Technol, Sweden, Civilingenjor, 44, Techn lic, 47, Techn dr, 53. *Prof Exp:* Asst electronics, Royal Inst Technol, Sweden, 44; engr, A B Standard Radio Mfg Co, Stockholm, 44-45; asst electronics, Royal Inst Technol, Sweden, 45-47; trainee, RCA Labs, 47-48; asst electronics, Royal Inst Technol, 49-53; res engr, RCA Labs, 53-68; prof electronics, Chalmers Univ Technol, Sweden, 68-90; RETIRED. *Concurrent Pos:* Secy, State Tech Res Coun, Stockholm, 50-51; res engr, Elektrovarme Inst, Sweden, 52-53; consult, Royal Swedish Air Force, 52-53; prof, Chalmers Univ Technol, 64-66. *Honors & Awards:* L J Wallmark Award, Royal Acad Sci Sweden, 54; Polhem Award, Sweden Eng Soc, 82; Cedergren Medal, Royal Inst Tech, 85. *Mem:* Am Phys Soc; fel Inst Elec & Electronics Engrs; Royal Swed Acad Eng Sci; Royal Swed Acad Sci; fel AAAS. *Res:* Integrated circuits; solid state devices. *Mailing Add:* Chalmers Univ Technol Gothenburg 41296 Sweden

WALLNER, STEPHEN JOHN, b Sioux Falls, SDak, Mar 22, 45; m 69; c 2. PLANT PHYSIOLOGY. *Educ:* SDak State Univ, BS, 67, MS, 69; Iowa State Univ, PhD(plant physiol), 73. *Prof Exp:* Plant physiologist, US Army Natick Labs, 73-75; asst prof hort physiol, Pa State Univ, University Park, 75-80; ASSOC PROF PLANT PHYSIOL, COLO STATE UNIV, 80- *Mem:* Am Soc Plant Physiologists; Am Inst Biol Sci; Sigma Xi; AAAS; Int Asn Plant Tissue Cult. *Res:* Postharvest physiology, especially involving cell wall changes during fruit ripening. *Mailing Add:* Hort Dept Pa State Univ Main Campus University Park PA 16802

WALLNER, WILLIAM E, b Greenfield, Mass, Nov 25, 36; m 64; c 2. ENTOMOLOGY, PLANT PATHOLOGY. *Educ:* Univ Conn, BS, 59; Cornell Univ, PhD, 65. *Prof Exp:* Res asst entom, Cornell Univ, 59-65; exten entomologist, Mich State Univ, 65-76, assoc prof entom, 69-74, prof, 74-76; RES PROJ LEADER, FOREST INSECT & DIS LAB, US FOREST SERV, 76- *Concurrent Pos:* Vis scientist, USSR 81, 86 & 88, China, 88. *Mem:* Entom Soc Am; Soc Am Foresters; Sigma Xi. *Res:* Biology and control of forest ornamental and plantation insects with emphasis on population suppression and pest management of insects for forest-recreational areas; gypsy moth; biolcology of forest insects with emphasis on biological control and population dynamics; recent research centers on gypsy moth and insects associated with acid rain deposition; outbreak insects especially the lepidoptera. *Mailing Add:* Forest Insect & Dis Lab US Forest Serv Hamden CT 06514

WALLRAFF, EVELYN BARTELS, b Chicago, Ill, Oct 21, 20. MICROBIOLOGY, IMMUNOLOGY. *Educ:* Rosary Col, BS, 40; Univ Chicago, MS, 42; Univ Ariz, PhD(virol, immunol), 61. *Prof Exp:* Res technician bact & immunol, Univ Chicago & Zoller Dent Clin, 41-43; res microbiologist, Vet Admin Hosp, 61-71; res assoc microbiol & med technol, Univ Ariz, 70-87; prof microbiol & Life Sci, Pima Col, 72-87; RETIRED. *Concurrent Pos:* Consult, Vet Admin Hosp, Tucson, 71- *Mem:* AAAS; Am Soc Microbiol; Am Asn Immunol; Reticuloendothelial Soc; Am Thoracic Soc. *Res:* Coccidioidinn hypersensitivity used for study of mechanisms of delayed hypersensitivity in man; cellular and transplantation immunology; anti-macrophage serum. *Mailing Add:* 2708 E Mabel St Tucson AZ 85716

WALLS, HUGH A(LAN), b Wellsboro, Pa, Feb 14, 34; m 71; c 3. CHEMICAL ENGINEERING, MECHANICAL ENGINEERING. *Educ:* Univ Okla, BS, 57, MChE, 58, PhD(chem eng), 63. *Prof Exp:* Res asst chem & metall eng, Res Inst, Univ Okla, 58-62, proj dir inst & asst prof metall eng, 62-63; res engr, Bur Eng Res, 63-65, asst prof, 64-68, assoc prof, 68-80, PROF MECH ENG, UNIV TEX, AUSTIN, 68-, ASSOC DIR, 77- *Concurrent Pos:* NSF res initiation grant, 65-67. *Mem:* AAAS; Am Soc Mech Engrs; NY Acad Sci; Am Inst Chemists; Am Chem Soc. *Res:* Liquid structure and transport phenomena, fluid mechanics; heat transfer, materials science; thermodynamics; engineering design; numerical methods. *Mailing Add:* Dept Mech Eng Univ Tex Austin TX 78712

WALLS, KENNETH W, b Ft Lauderdale, Fla, Dec 4, 28; m 56. IMMUNOLOGY, PARASITOLOGY. *Educ:* Ind Univ, AB, 49 & 50; Univ Mich, MS, 52, PhD(bact), 55. *Prof Exp:* Lab chief parasitol & mycol serol, 55-59, lab chief toxoplasmosis, 59-66, CHIEF PARASITOL SEROL UNIT, CTR DIS CONTROL, 66- *Mem:* AAAS; Am Soc Trop Med & Hyg. *Res:* Immunology, serology and epidemiology of parasitic diseases with special emphasis on toxoplasmosis and related diseases. *Mailing Add:* 4006 Northlake Creek Ct Tucker GA 30084

WALLS, NANCY WILLIAMS, b Johnstown, Pa, Sept 19, 30; div. BACTERIOLOGY. *Educ:* Univ Mich, BS, 52, MS, 53, PhD(bact), 59. *Prof Exp:* Instr bact, Emory Univ, 58-59; asst res biologist, 59-61, res asst prof, 62-67, sr res biologist, 67-69, actg dir sch biol, 69-70, ASSOC PROF BIOL, GA INST TECHNOL, 69- *Mem:* AAAS; Am Soc Microbiol; Radiation Res Soc; Am Inst Biol Sci; NY Acad Sci; Sigma Xi. *Res:* Physiology of Clostridium botulinum; marine microbial ecology; anaerobic bacterial spores; mechanisms of bacterial toxin formation; behavioral mechanisms of sea turtles. *Mailing Add:* Sch Biol Sci Ga Inst Technol Atlanta GA 30332

WALLS, ROBERT CLARENCE, b Batesville, Ark, Mar 9, 34; m 66; c 3. BIOMETRICS & BIOSTATISTICS, STATISTICS. *Educ:* Harding Univ, BS, 59; Univ Ark, MS, 61; Okla State Univ, PhD(statist), 67. *Prof Exp:* Mathematician, Res & Technol Dept, Texaco Inc, 61-64; from asst prof to assoc prof, 66-77, head biomet, 77-82, PROF BIOMET, UNIV ARK FOR MED SCI, LITTLE ROCK, 82- *Mem:* Am Statist Asn; Soc for Clin Trials; Biomet Soc; Sigma Xi. *Res:* Health services research; mathematical models in biology and medicine. *Mailing Add:* Div Biomet No 585 Univ Ark for Med Sci Little Rock AR 72205

WALMSLEY, FRANK, b New Bedford, Mass, June 26, 35; m 59; c 2. INORGANIC CHEMISTRY. *Educ:* Univ NH, BS, 57; Univ NC, Chapel Hill, PhD(chem), 62. *Prof Exp:* From asst prof to prof, 62-87, EMER PROF CHEM, UNIV TOLEDO, 87- *Concurrent Pos:* Vis prof, Mich State Univ, 79-80; lectr, Univ Tex, San Antonio, 87-89 & 91, Trinity Univ, 89-91. *Mem:* Am Chem Soc; Royal Soc Chem; Am Sci Affil; Sigma Xi. *Res:* Spectral and magnetic properties of coordination compounds; heteropolyions; author, general and inorganic chemistry. *Mailing Add:* Dept Chem Univ Toledo Toledo OH 43606

WALMSLEY, IAN ALEXANDER, b Manchester, Eng, Jan 13, 60. NONLINEAR OPTICS, QUANTUM OPTICS. *Educ:* Univ London, BSc, 80; Univ Rochster, PhD(optics), 86. *Prof Exp:* Res assoc, Cornell Univ, 86-87; ASST PROF OPTICS, UNIV ROCHSTER, 88- *Concurrent Pos:* NSF presidential young investr, 90. *Mem:* Optical Soc Am; Am Phys Soc. *Res:* Nonlinear and quantum optics, especially ultrafast phenomena in these areas. *Mailing Add:* Inst Optics Univ Rochester Rochester NY 14627

WALMSLEY, JUDITH ABRAMS, b Oak Park, Ill, Feb 6, 36; m 59; c 2. INORGANIC CHEMISTRY. *Educ:* Fla State Univ, BA, 58; Univ NC, Chapel Hill, PhD(chem), 62. *Prof Exp:* Res scientist chem, Owens-Ill, Inc. 63-66; vis res assoc, 73-75, asst prof chem, 81-82, sr res assoc chem, Univ Toledo, 74-87; ASST PROF CHEM, UNIV TEX, SAN ANTONIO, 87- *Mem:* Am Chem Soc; AAAS; Sigma Xi; Asn Women Sci. *Res:* Chemistry of metal complexes of biological significance; transition metal complexes of organophosphorus ligands; self-association of hydrogen-bonding molecules in nonpolar solvents; solute-solvent interactions; properties of nucleotides in solution. *Mailing Add:* Div Earth & Phys Sci Univ Tex San Antonio TX 78249-0603

WALMSLEY, PETER N(EWTON), b Oldham, Eng, Mar 11, 36; m 56; c 3. CHEMICAL ENGINEERING. *Educ:* Univ Manchester, BScTech, 57, PhD(chem eng), 60. *Prof Exp:* Engr, Sabine River Works, E I du Pont de Nemours & Co, Inc, 60-63, engr, Plastics Dept, Exp Sta, 63-67, sr res engr, 67-70, res supvr, 70-74, sr res supvr, 74-75, prin consult, Corp Plans Dept, 75-79, mgr acquisitions & divestitures, Corp Plans Dept, 79-89; PRES, AMT MGT, INC, 89- *Mem:* Am Inst Chem Engrs; Soc Advan Mat & Process Eng. *Mailing Add:* 3216 Fordham Rd Westmorland Ave Wilmington DE 19807

WALNE, PATRICIA LEE, b Newark, NJ, Nov 27, 32. CELL BIOLOGY, PHYCOLOGY. *Educ:* Hanover Col, BS, 54; Ind Univ, MS, 59; Univ Tex, PhD(phycol, cell biol), 65. *Prof Exp:* Res fel, Cell Res Inst, Univ Tex, 65-66; from asst prof to assoc prof, 66-73, PROF BOT, UNIV TENN, KNOXVILLE, 73-, BENWOOD DISTINGUISHED PROF, 85- *Concurrent Pos:* Ed, Phycol Soc Am Newslett, 66-69; consult, Biol Div, Oak Ridge Nat Lab, 66-74; Fulbright sr res scholar Denmark, Inter-country Exchange, Turkey & WGer; prin investr res grants, NSF, 68-, adv panel, 76-78; Am Asn Univ Women sr fel, 74-75; vis res prof, Univ Copenhagen, 76, 77, 82 & Nencki Inst Exp Biol, Warsaw, 81 & 87; Nat Acad Sci US Exchange Scientist Poland, 83; Adv panel life sci, 78-80, chair, 81, area adv panel, WEurop/Scand, 86-88, coun int exchange scholars; int org comt, 2nd Int Phycol Congr, Copenhagen, Denmark, 85. *Honors & Awards:* Darbaker Prize, Bot Soc Am, 78. *Mem:* Am Soc Cell Biol; Int Soc Evolutionary Protistology; Phycol Soc Am (secy, 69-72, vpres, 73, pres, 74); Electron Micros Soc Am; Int Phycol Soc; Brit Phycol Soc; Soc Protozoologists; fel AAAS. *Res:* Cell biology and experimental phycology; ultrastructure and development of algae; photoresponse and sensory transduction in algal flagellates; biomineralization, especially of extracellular matrices; systematic and evolutionary biology. *Mailing Add:* Dept Bot Univ Tenn Knoxville TN 37996

WALNUT, THOMAS HENRY, JR, b Philadelphia, Pa, May 22, 24; m 70; c 2. QUANTUM CHEMISTRY. *Educ:* Harvard Univ, AB, 47; Brown Univ, PhD(chem), 51. *Prof Exp:* Instr, Inst Study Metals, Univ Chicago, 50-52; from asst prof tc assoc prof, 52-63, PROF CHEM, SYRACUSE UNIV, 63- *Mem:* Am Chem Soc; Am Phys Soc. *Res:* Magnetic susceptibility of molecules; quantum chemistry; magnetic vibrational circular dichroism. *Mailing Add:* Dept Chem Syracuse Univ Syracuse NY 13244

WALOGA, GERALDINE, b Warrenton, NC, June 17, 46. NEUROBIOLOGY. *Educ:* Northwestern Univ, BA, 68; Purdue Univ, PhD(biol sci), 75. *Prof Exp:* Res assoc, Harvard Univ, 75-77 & State Univ NY, Stony Brook, 77-79; res assoc, 79-80, asst prof, 80-87, ASSOC PROF PHYSIOL, SCH MED, BOSTON UNIV, 87- *Mem:* Asn Res Vision & Ophthal; Biophys Soc. *Res:* Interactions of cyclic nucleotides, calcium ions and the inositol polyphosphates in excitation and adaption of vertebrate photoreceptors; electrophysiology of human retinoblastoma cells. *Mailing Add:* Dept Physiol Sch Med Boston Univ 80 E Concord St Boston MA 02118

WALPER, JACK LOUIS, b Excel, Alta, Nov 29, 16; nat US; m 43. GEOLOGY. *Educ:* Univ Okla, BS, 47, MS, 49; Univ Tex, PhD(geol), 58. *Prof Exp:* Asst geol, Univ Okla, 47-48; asst prof, Univ Tulsa, 48-54; instr, Univ Tex, 55-58; assoc prof, Univ Tulsa, 58-63; prof geol, Tex Christian Univ, 63-81; RETIRED. *Concurrent Pos:* Consult & mem bd dirs, Tex Archit Aggregate Co & Empresa Centro Americana, 74- *Mem:* Am Geophys Union; Asn Eng Geologists; Am Asn Petrol Geologists; Geol Soc Am; Nat Asn Geol Teachers. *Res:* Field exploration; tectonics, especially Central American tectonics. *Mailing Add:* Dept Geol Tex Christian Univ Ft Worth TX 76129

WALPERT, GEORGE W, b Monett, Mo, Dec 10, 24; m 48; c 6. CHEMICAL ENGINEERING. *Educ:* Mo Sch Mines, BS, 47; Univ Colo, MS, 51, PhD(chem eng), 54. *Prof Exp:* Chem engr, Koppers Co, Inc, 47-50; sr engr, Monsanto Chem Co, 53-57; group leader atomic fuel recovery, Phillips Petrol Co, 57-58; dir res, Wigton Develop Lab, 58-59; mgr process develop, Kordite Co Div, Nat Distillers & Chem Corp, 59-62, USI Film Prod Div, 62-63 & Kordite Co Div, Socony Mobil Oil Co, Inc, 63-64; mgr mat eng, 64-68, MGR OPERS REV STAFF, XEROX CORP, ROCHESTER, 68- *Mem:* Am Chem Soc; Soc Plastics Engrs. *Res:* Process and materials development in fields of coal byproducts, plastics and intermediates, pure metals, packaging materials and consumables for office machinery. *Mailing Add:* 3471 Rutgers Rd Behtlehem PA 18017

WALPOLE, RONALD EDGAR, b Wiarton, Ont, June 19, 31; m 64; c 1. MATHEMATICS, STATISTICS. *Educ:* McMaster Univ, BA, 54, MSc, 55; Va Polytech Inst, PhD, 58. *Prof Exp:* From asst prof to assoc prof, 57-61, PROF MATH & STATIST, ROANOKE COL, 61- *Concurrent Pos:* Consult, White Sands Missile Range, 50-63. *Mem:* Inst Math Statist; Am Statist Asn; Biomet Soc. *Res:* Statistical design of experiments. *Mailing Add:* Dept Math Roanoke Col Salem VA 24153

WALRADT, JOHN PIERCE, b Caldwell, Idaho, Feb 12, 42; m 64; c 4. FLAVOR & FRAGRANCE CHEMISTRY, LABORATORY AUTOMATION. *Educ:* Univ Idaho, BS, 65; Ore State Univ, MS, 67, PhD(food sci), 69. *Prof Exp:* Sr chemist, Flavor Res, Int Flavors & Fragrances, Inc, 69-71, proj leader, 71-73, group leader, Flavor Res, Anal Chem, 73-76, sr group leader, Instrumental Anal, 76-85, mgr, 85-87, DIR, RES & DEVELOP ADMIN, INT FLAVORS & FLAGRANCES, INC, 87- *Mem:* Am Chem Soc; Inst Food Technologists. *Res:* Analytical chemistry; laboratory automation; robotics; flavor chemistry, gas chromatography, natural component identification; pioneer in laboratory robotics. *Mailing Add:* Int Flavors & Fragrances Inc 1515 Hwy 36 Union Beach NJ 07735

WALRAFEN, GEORGE EDOUARD, b Topeka, Kans, May 18, 29; m 60. STRUCTURE WATER, STRUCTURE GLASSES. *Educ:* Univ Kans, BS, 51, MS, 57, PhD (chem), 59. *Prof Exp:* Mem tech staff phys chem, AT & T Bell Labs, 60-75; PROF CHEM, HOWARD UNIV, 75- *Concurrent Pos:* Prof phys chem, Univ Marburg, 72-73. *Mem:* Sigma Xi; Am Phys Soc. *Res:* Roman and infrared spectroscopy of water; aqueous solutions, glasses, optical fibers. *Mailing Add:* Chem Dept Howard Univ 525 Col NW Washington DC 20059

WALSBERG, GLENN ERIC, b Long Beach, Calif, June 25, 49; m 74; c 2. ENVIRONMENTAL PHYSIOLOGY, BIOPHYSICAL ECOLOGY. *Educ:* Calif State Univ, BS, 71; Univ Calif, Los Angeles, PhD(biol), 75. *Prof Exp:* NIH fel, Wash State Univ, 76-78; from asst prof to assoc prof zool, 78-89, PROF ZOOL, ARIZ STATE UNIV, 89- *Concurrent Pos:* Ed, The Condor, 90- *Mem:* Am Ornithologists' Union; Ecol Soc Am; Cooper Ornith Soc; Am Soc Zoologists; Sci Res Soc NAm; AAAS. *Res:* Avian ecological energetics; physiological and biophysical ecology of birds and mammals; desert ecology; avian physiology. *Mailing Add:* Dept Zool Ariz State Univ Tempe AZ 85287-1501

WALSER, ARMIN, b Walzenhausen, Switz, Apr 6, 37; m 57; c 3. MEDICINAL CHEMISTRY. *Educ:* Swiss Fed Inst Technol, Dipl Ing Chem, 60, PhD(org chem), 63. *Prof Exp:* Fel org chem, Stanford Univ, 63-64; sr chemist, Nutley, NJ, 64-66, Basel, Switz, 66-69 & Nutley, NJ, 69-72, res fel med chem, 72-74, group chief med chem, 74-79, SR RES FEL, HOFFMANN-LA ROCHE, INC, 79- *Mem:* Am Chem Soc. *Res:* Synthesis of new compounds of potential pharmaceutical interest, in particular compounds acting on central nervous system. *Mailing Add:* 19 Crane Ave West Caldwell NJ 07006-7903

WALSER, MACKENZIE, b New York, NY, Sept 19, 24; m; c 4. MEDICINE. *Educ:* Yale Univ, BA, 44; Columbia Univ, MD, 48; Am Bd Internal Med, dipl, 56. *Prof Exp:* Intern med, Mass Gen Hosp, 48-49, asst resident, 49-50; from instr to asst prof, Univ Tex Southwestern Med Sch, Dallas, 50-52; investr, Nat Heart Inst, 54-57; from asst prof to assoc prof, 57-70, PROF MED & PHARMACOL, SCH MED, JOHNS HOPKINS UNIV, 70-, PHYSICIAN, JOHNS HOPKINS HOSP, 60- *Concurrent Pos:* Resident, City-County Hosp, Dallas, Tex, 50-52. *Honors & Awards:* Exp Therapeut Award, Am Soc Pharmacol & Exp Therapeut, 75; Herman Award, Am Soc Clin Nutrit, 88. *Mem:* Am Physiol Soc; Am Soc Clin Invest; Am Soc Pharmacol & Exp Therapeut; Am Inst Nutrit; Asn Am Physicians. *Res:* Medical, physiological, and pharmacological aspects of electrolyte and amino acid metabolism and renal function. *Mailing Add:* Dept Pharmacol Sch Med Johns Hopkins Univ Baltimore MD 21205

WALSER, RONALD HERMAN, b Juarez, Mex, Jan 24, 44; m 67; c 5. HORTICULTURE, PLANT PHYSIOLOGY. *Educ:* Brigham Young Univ, BS, 68; Utah State Univ, PhD(crop physiol), 75. *Prof Exp:* Mgr res & develop, Hyponex Co, 75-76; asst prof, Univ Ky, 76-78; asst prof hort, Tex Tech Univ, 78-80; assoc prof agron, 80-91, PROF AGRON, BRIGHAM YOUNG UNIV, 91- *Mem:* Am Soc Hort Sci. *Res:* Environmental physiology of fruits and vegetables. *Mailing Add:* Dept Agron Brigham Young Univ 275 Widtsoe Bldg Provo UT 84062

WALSH, ARTHUR CAMPBELL, b Vancouver, BC, Dec 21, 19; US citizen; m 44; c 3. GERIATRIC PSYCHIATRY, ALZHEIMERS DISEASE. *Educ:* Univ Alta, MD, 43. *Prof Exp:* Family med pract, 45-64; fel psychiat, Western Psychiat Inst & Clin, Univ Pittsburgh, 64-67, clin asst prof, 67-89; PRES, CTR SENILITY STUDIES, 80- *Concurrent Pos:* Pvt pract adult psychiat, 68-; psychiat consult, Vet Admin Hosp, Pittsburgh, 70-88; staff psychiatrist, Woodville State Hosp, 75-86. *Mem:* Am Psychiat Asn; AMA; Am Asn Geriat Psychiat; Am Geriat Soc. *Res:* Treatment program for dementia, including Alzheimer's disease; combining psychiatric therapy with medicine to improve the blood flow to the brain; legal aspects of dementia, such as competency to make a will and abnormal behavior; author of over 20 publications and two books. *Mailing Add:* Ctr Senility Studies 161 N Dithridge St Pittsburgh PA 15213

WALSH, BERTRAM (JOHN), b Lansing, Mich, May 7, 38; m 60. MATHEMATICAL ANALYSIS. *Educ:* Univ Mich, PhD(math), 63. *Prof Exp:* Lectr math, Univ Mich, 63; from asst prof to assoc prof, Univ Calif, Los Angeles, 63-70; assoc prof, 70-71, PROF MATH, RUTGERS UNIV, NEW BRUNSWICK, 71- *Concurrent Pos:* Vis asst prof, Univ Wash, 66-67. *Mem:* Am Math Soc. *Res:* Functional analysis, locally convex spaces, linear transformations and spectral theory; measure theory; potential theory. *Mailing Add:* Dept Math Rutgers Univ New Brunswick NJ 08903

WALSH, CHARLES JOSEPH, b July 30, 40; m 62; c 2. CELL BIOLOGY, MOLECULAR BIOLOGY. *Educ:* Univ Calif, Riverside, PhD(cell biol), 68. *Prof Exp:* ASSOC PROF BIOL SCI, UNIV PITTSBURGH, 79- *Mem:* Am Soc Cell Biol. *Res:* Cell differentiation; regulation of gene expression; biology of cilia and flagella, naegleria biology. *Mailing Add:* Dept Biol Sci Univ Pittsburgh Pittsburgh PA 15260

WALSH, CHRISTOPHER THOMAS, b Boston, Mass, Feb 16, 44; m 66; c 1. BIOCHEMISTRY. *Educ:* Harvard Univ, BA, 65; Rockefeller Univ, PhD(life sci), 70. *Prof Exp:* Helen Hay Whitney Found fel, Brandeis Univ, 70-72; from asst prof to prof chem & biol, Mass Inst Technol, 72- 87, assoc dir, Whitaker Col Mgt, 79-82, Uncas & Helen Whitaker prof, 80-85, Karl Taylor Compton prof, 85-87, chmn, Chem Dept, 82-87; David Wesley Gaiser prof, 87-91, CHMN, DEPT BIOL CHEM & MOLECULAR PHARMACOL, MED SCH, HARVARD UNIV, 87-, HAMILTON KUHN PROF, 91- *Concurrent Pos:* Alfred P Sloan Found fel, 75-77; Camille & Henry Dreyfus teacher-scholar grant, 76-80; consult, Merck, Sharp & Dohme Res Labs, 75-81, Monsanto Corp Res Labs, 80-81, Johnson & Johnson, 82-83, Hoffman LaRoche, 82-, Genzyme & Bioinfo Assocs, 83-, Firmenich, S A, 86-90, Enzymatics, 88- & Biotage, 89-; mem, Panel Res Grants Study, NSF, 77-79, Panel Study Sect Biochem, NIH, 78-82, Gen Med Coun, NIH, 83-85 & Chemal Rev Group, WHO, 84-86; biol sect ed, Ann Reports Med Chem, 78-80; co-chmn, Gordon Res Conf Enzymes, Coenzymes & Molecular Biol, 78 & Conf Methanogenesis, 84; chmn, Study Sect Biochem, NIH, 82-; assoc ed, Ann Rev Biochem, 90- *Honors & Awards:* Eli Lilly Award, 79; Baker Lectr, Univ Calif, Santa Barbara, 83; Lutz Lectr, Univ Va, 85; Troy C Daniels Lectr, Univ Calif, San Francisco, 86; Edward E Smissman Lectr, Univ Kans, 86; Guthikonda Lectr, Columbia Univ, 86; Ida Beam Lectr, Univ Iowa, 87; Nelson Leonard Lectr, Univ Ill, 89; Melvin Calvin Lectr, Univ Calif, Berkeley, 89; Dauben Lectr, Univ Wash, 90; David Green Lectr, Univ Wis, 90. *Mem:* Nat Acad Sci; Inst Med-Nat Acad Sci; Am Acad Arts & Sci; Am Soc Biol Chemists; Am Chem Soc; Am Soc Microbiol. *Res:* Enzymatic reaction mechanisms, phosphoryl and pyrophosphoryl transfers; flavin-dependent enzymes; membrane biochemistry and mechanism of active transport. *Mailing Add:* Dept Biol Chem & Molecular Pharmacol Harvard Med Sch 240 Longwood Ave C1-213 Boston MA 02115

WALSH, DAVID ALLAN, b Schenectady, NY, Aug 3, 45; m 67; c 2. ORGANIC CHEMISTRY, MEDICINAL CHEMISTRY. *Educ:* Clarkson Col Tech, NY, BS, 67; Univ NH, MS, 70, PhD(org chem), 73. *Prof Exp:* Sr res chemist, 74-79, GROUP MGR, A H ROBINS CO INC, 79- *Concurrent Pos:* NIH fel, Dept Med Chem, Sch Pharm, Univ Kans, 73-74. *Mem:* Am Chem Soc; Int Soc Heterocydic Chem. *Res:* Synthesis of new nonsteroidal anti-inflammatory agents; synthesis of various nitrogen-containing heterocydes including piperidines, pyrrolidines, isoquinolines, and quinolines; synthesis of novel antiallergy agents. *Mailing Add:* Nutra Sweet Co Box 2387 Augusta GA 30903

WALSH, DAVID ERVIN, b DeGraff, Minn, Aug 7, 39; m 63; c 2. CEREAL CHEMISTRY, BIOCHEMISTRY. *Educ:* St Cloud State Col, BS, 61; NDak State Univ, MS, 63, PhD(cereal chem), 69. *Prof Exp:* Proj leader cereal prod, Squibb Beech-Nut Corp, 63-65; instr cereal chem, NDak State Univ, 65-69, assoc prof cereal chem & technol, 69-77; STAFF MEM, GEN NUTRIT CORP, 77- *Mem:* Am Asn Cereal Chemists; Inst Food Technologists. *Res:* Macaroni products; industrial engineering; computer applications to food processes; protein compositional studies of wheat. *Mailing Add:* 920 Sixth Ave No 203 Fargo ND 58102

WALSH, DON, b Berkeley, Calif, Nov 2, 31; m 62; c 2. PHYSICAL OCEANOGRAPHY, OCEAN ENGINEERING. *Educ:* US Naval Acad, BS, 54; Tex A&M Univ, MS, 67, PhD(oceanog), 68; San Diego State Col, MA, 68. *Prof Exp:* Officer-in-chg bathyscaphe Trieste, Navy Electronics Lab, USN, San Diego, Calif, 58-62, prin investr remote sensor oceanog proj, Tex A&M Univ, 65-68, sci liaison officer ocean eng, Submarine Develop Group One, San Diego, 69-70, spec asst to Asst Secy Navy Res & Develop, Navy Dept, Washington, DC, 70-73, dep dir, Navy Labs, Hq Naval Mat Command, 74-75; prof ocean eng & dir, Inst Marine & Coastal Studies, Univ Southern Calif, 75-83; PRES, INT MARITIME INC, 76- *Concurrent Pos:* Partic, Deep Freeze, Antarctic, 71; fel, Woodrow Wilson Int Ctr Scholars, Smithsonian Inst, 72-74; mem US adv comt, Eng Comt Ocean Resources, Nat Acad Eng, 72-82; dir, US Naval Inst, 74-75; ed, Marine Technol Soc J, 75-80; chmn comt aquacult, Nat Res Coun, 76-78; pres & mem bd dirs, Int Maritime, Inc, 76-; mem, State Dept Law of Sea Adv Group, 79-83, Nat Adv Comt Oceans & Atmosphere, 79-86; mem comt maritime indust opportunities & requirements for develop ocean resources, Nat Res Coun, Nat Acad Sci, 78-80; mem, Space Appln Adv Comt, NASA, 83-86, bd gov, Calif Marine Acad, 85 & Marine bd, Nat Res Coun, 90. *Mem:* AAAS; Am Geophys Union; fel Marine Technol Soc (vpres, 75-79); Soc Naval Architects & Marine Engrs; fel Explorers Club. *Res:* Application of deep submersibles to ocean sciences; deep ocean engineering research and development; application of remote sensors to oceanography; ocean resource planning and policy; author of numerous publications. *Mailing Add:* Int Maritime Inc 839 S Beacon St 217 San Pedro CA 90731-3739

WALSH, EDWARD JOHN, b Brooklyn, NY, Aug 29, 42; m 64; c 3. INORGANIC CHEMISTRY, POLYMER CHEMISTRY. *Educ:* Franklin & Marshall Col, BA, 64; Middlebury Col, MS, 66; Pa State Univ, University Park, PhD(inorg chem), 70. *Prof Exp:* Asst prof chem, Pa State Univ, Shenango Valley Campus, 70-; MGR, MAT DEVELOP, WESTINGHOUSE & MGR, ENVIRON CONTROL, ABB POWER T&D CO. *Mem:* Am Chem Soc; The Chem Soc; sr mem Inst Elec & Electronic Engrs. *Res:* Phosphazene derivatives; germazanes; trace elements in water aseneazenes; pcb's and fire products. *Mailing Add:* 4712 Cedarfield Dr Raleigh NC 27606

WALSH, EDWARD JOSEPH, b Woonsocket, RI, June 13, 41; m 73; c 1. ELECTRICAL ENGINEERING, RADIO OCEANOGRAPHY. *Educ:* Northeastern Univ, BS, 63, PhD(elec eng), 67. *Prof Exp:* Instr elec eng, Northeastern Univ, 66-67; aerospace technologist, Electronics Res Ctr, 67-70, AEROSPACE TECHNOLOGIST, NASA & GODDARD SPACE FLIGHT CTR, WALLOPS FLIGHT FACIL, 70- *Concurrent Pos:* Mem, Comn F, Union Radio Sci Int. *Mem:* Inst Elec & Electronics Engrs. *Res:* Electromagnetic theory; radio wave propagation and scattering; radio oceanography; radar altimetry. *Mailing Add:* NASA/Goddard Space Flight Ctr Wallops Flight Facil E 106 Wallops Island VA 23337

WALSH, EDWARD JOSEPH, JR, b Philadelphia, Pa, Aug 2, 35; m 59; c 2. ORGANIC CHEMISTRY. *Educ:* State Univ NY Albany, BS, 60; Univ NH, PhD(chem), 64. *Prof Exp:* Teaching asst, State Univ NY Albany, 60-61; asst prof, 64-86, assoc prof, 68-70, PROF CHEM & DEPT CHMN, ALLEGHENY COL, 70- *Concurrent Pos:* NSF sci fac grant, Mass Inst Technol, 69-70. *Mem:* Am Chem Soc. *Res:* Free radical reactions involving the acyl radical; free radical reactions of certain organotin hydrides. *Mailing Add:* Dept Chem Allegheny Col Meadville PA 16335

WALSH, EDWARD KYRAN, b Philadelphia, Pa, Feb 19, 31; m 52; c 6. ENGINEERING SCIENCE, APPLIED MECHANICS. *Educ:* Union Col, BME, 63; Brown Univ, PhD(appl math), 67. *Prof Exp:* Engr, Mech Technol Inc, 62-63; res fel, Mellon Inst Sci, 66-68; asst prof civil eng, Carnegie-Mellon Univ, 67-70; assoc prof, 70-74, PROF ENG SCI, UNIV FLA, 74- *Concurrent Pos:* Consult, Gen Eng & Consult Lab, Gen Elec Co, 64-65, Sandia Labs, Albuquerque, 69-82, Vet Admin Hosp, Richmond, 80- *Mem:* Soc Natural Philos. *Res:* Continuum mechanics; dynamic material response; wave propagation; bioengineering. *Mailing Add:* Dept Eng Sci Univ Fla Gainesville FL 32611

WALSH, EDWARD NELSON, b Chicago, Ill, Nov 22, 25; m 50; c 2. SYNTHETIC INORGANIC & ORGANOMETALLIC CHEMISTRY. *Educ:* Ill Inst Technol, BS, 48, PhD, 65; DePaul Univ, MS, 52. *Prof Exp:* Chemist, Swift & Co, Ill, 48-51 & Victor Chem Works, 51-59; supvr org res, 59-63, mgr chem res, 63-65, mgr chem prod develop sect, Dobbs Ferry, 65-69, sr sect mgr org res, 69-75, mgr chem dept, 75-83, sr scientist, Stauffer Chem Co, Dobbs Ferry, 83-85, adj lectr, 85-86; asst prof, St Peters Col, Jersey City, NJ, 86-89; CONSULT, 85- *Mem:* AAAS; Am Chem Soc; Sigma Xi. *Res:* Organophosphorus compounds; agricultural chemicals; solvents; surfactants; flame retardants; synthetic lubricants; pharmaceutical intermediates; organometallics; photochemistry; catalysts; inorganic phosphorus compounds. *Mailing Add:* 33 Concord Dr New City NY 10956

WALSH, GARY LYNN, b Fremont, Nebr, June 30, 40; m 64. AIR POLLUTION, ENVIRONMENTAL SCIENCES. *Educ:* Midland Lutheran Col, BS, 62; Univ Nebr, MS, 65; Univ SDak, PhD(zool), 69. *Prof Exp:* Asst prof zool, Ind Univ Northwest, 69-70; air pollution control chief, Michigan City, Ind, 70-72; admin asst to air pollution control officer, 72-74, supvr, Air Pollution Control Sect, 74-79, asst chief, 79-84, CHIEF, DIV ENVIRON HEALTH, LINCOLN LANCASTER COUNTY HEALTH DEPT, 84- *Mem:* Air Pollution Control Asn. *Res:* Role of vitamin B12, biotin and thiamine on seasonal fluctuations of euglenophyte populations; taxonomy of Antarctic freshwater and soil amoeba. *Mailing Add:* 3160 Woodsdale Blvd Lincoln NE 68502

WALSH, GERALD MICHAEL, b Portland, Ore, Sept, 1, 44; m 70; c 5. PHARMACOLOGY. *Educ:* Univ Santa Clara, BS, 66; Ore State Univ, PhD(pharmacol), 71; John Marshall Law Sch, JD, 84. *Prof Exp:* Res asst pharmacol, Ore State Univ, 67-69; asst prof, Univ Ga, 70-74; asst prof res med, Sch Med, Univ Okla, 74-76; staff scientist, Alton Ochsner Med Found, 76-78; mgr cardiovasc pharmacol, Baxter-Travenol, 78-81; GROUP LEADER, DEPT PHARMACOL, GD SEARLE & CO, 81- *Concurrent Pos:* NIH instnl res grant, 81-; Ochsner Found, 76-79. *Mem:* Am Soc Pharmacol & Exp Therapeut; Soc Exp Biol Med. *Res:* Cardiovascular pharmacology and toxicology; hypertension. *Mailing Add:* Dept Pharmacol GD Searle Co Box 5110 Chicago IL 60680

WALSH, JAMES ALOYSIUS, b Brooklyn, NY, Dec 15, 33; m 60. ORGANOSULFUR CHEMISTRY. *Educ:* Fordham Univ, BS, 55; Purdue Univ, MS, 58, PhD(org chem), 63. *Prof Exp:* Instr chem, Purdue Univ, 60-63; from asst prof to assoc prof, 63-73, chmn dept, 69-72, 77-81, PROF CHEM, JOHN CARROLL UNIV, 73- *Concurrent Pos:* Res assoc, Univ Calif, Santa Cruz, 73; summer res consult, Diamond Shamrock Corp, 80; vis prof, Ohio State Univ, 84. *Mem:* Am Chem Soc; Sigma Xi. *Res:* Chemistry of organosulfur compounds, especially sulfoxides and derivatives of sulfurtrioxide; phthalocyanines. *Mailing Add:* Dept Chem John Carroll Univ Cleveland OH 44118

WALSH, JAMES PAUL, b Fall River, Mass, Apr 3, 17; m 42; c 3. ENGINEERING. *Educ:* Stevens Inst Technol, ME, 38; Univ Md, MS, 50. *Hon Degrees:* EngD, Stevens Inst Technol, 58. *Prof Exp:* Sect head, US Naval Res Lab, 42-55, Bikini Bomb Test, 46, dep dir Vanguard Earth Satellite Proj, 55-59; vpres, C-E-I-R, Inc, 60-63; pres & chmn bd, Matrix Corp, 63-69; supt ocean technol, Naval Res Lab, Washington, DC, 69-78; CONSULT OCEAN ENG, SHOCK, VIBRATION & DYNAMICS & PRES, J PAUL WALSH & ASSOC INC, 78- *Concurrent Pos:* Consult, US Naval Res Lab, directed shock design of first nuclear powered submarine, mechanics of deep ocean power cables. *Mem:* Am Soc Mech Engrs; Sigma Xi. *Res:* Shock vibration; dynamics of structures; instrumentation; systems analysis; ocean technology; ocean operations; mechanics of deep sea power cables. *Mailing Add:* Box 946 Sherwood Forest MD 21405

WALSH, JOHN BREFFNI, b Brooklyn, NY, Aug 20, 27; m 55; c 3. RADAR SYSTEMS, OPERATION ANALYSIS. *Educ:* Manhattan Col, BEE, 48; Columbia Univ, MS, 50. *Prof Exp:* Instr elec eng, Columbia Univ, 49-51; asst chief radar systs, Rome Air Develop Ctr, NY, 51-52, asst chief air defense systs & lab, 52-53, tech dir intel & reconnaissance div, 53; asst prof elec eng & asst dir electronic res lab, Columbia Univ, 53-65; dep res to asst secy, USAF, Washington, DC, 66-71; sr staff mem, Nat Security Coun & asst to President's sci adv, White House, 71-72; dep dir defense res & eng, Off Secy Defense, Washington, DC, 72-77, asst secy gen defense support, NATO, Off Secy Defense, Brussels, Belg, 77-80; prof & dean, Exec Inst, 80-81, EMER PROF SYSTS ACQUISITION MGT, DEFENSE SYSTS MGT COL, 82; VPRES & CHIEF SCIENTIST, BOEING MIL AIRPLANES, WICHITA, KANS, 82- *Concurrent Pos:* Consult, Defense Sci Bd, 83-, Defense Nuclear Agency & NASA, 84-; mem, Cong Adv Comt Aeronaut, 84-85; vpres tech activ, Am Inst Aeronaut & Astronaut, 87- *Mem:* Fel Inst Elec & Electronics Engrs; fel Am Inst Aeronaut & Astronaut; NY Acad Sci. *Res:* Radar systems; operation analysis; aircraft and missile guidance and navigation; nuclear weapons effects. *Mailing Add:* Boeing Aerospace & Electronics Div PO Box 3999 MS 82-55 Seattle WA 98124-2499

WALSH, JOHN EDMOND, b New York, NY, Aug 20, 39; m 66; c 3. PLASMA PHYSICS. *Educ:* NS Tech Col, BSc, 62; Columbia Univ, MSc, 65, DSc, 68. *Prof Exp:* Res engr, US Army Signal Res & Develop Lab, Ft Monmouth, 62-65; from asst prof to assoc prof, 68-79, PROF PHYSICS, DARTMOUTH COL, 79- *Mem:* Am Phys Soc; Sigma Xi; Optical Soc Am. *Res:* Diffusion of turbulent plasmas; electron scattering in turbulent plasmas; nonlinear interactions in plasmas; millimeter and submillimeter radiation sources. *Mailing Add:* Dept Physics Dartmouth Col Hanover NH 03755

WALSH, JOHN H, b Jackson, Miss, Aug 22, 38. INTERNAL MEDICINE, GASTROENTEROLOGY. *Educ:* Vanderbilt Univ, MD, 63. *Prof Exp:* PROF MED, SCH MED, UNIV CALIF, LOS ANGELES, 78- *Mailing Add:* Dept Med Univ Calif Los Angeles Sch Med Los Angeles CA 90024

WALSH, JOHN HERITAGE, b Montreal, Que, Jan 3, 29; m 58; c 2. ENERGY & ENVIRONMENT INTERFACE, CARBON DIOXIDE & CLIMATE CHANGE. *Educ:* McGill Univ, BEng, 50, MEng, 51; Mass Inst Technol, ScD, 55. *Prof Exp:* Res scientist, Dept Energy, Mines & Resources, 55-80, sr adv, Coal, 80-85; PVT PRACT, ENERGY ADV, 85- *Concurrent Pos:* French grant fel, study French steel indust, 58; ed/publ, Can Metall Quart, 62-68; Imp Oil Ltd lectr, Univ Western Ont, 83. *Honors & Awards:* Joseph Beck Award, Int Steel Soc, 78. *Mem:* Fel Am Soc Metals; Can Inst Mining & Metall; Am Inst Mining, Metall & Petrol Engrs; Int Asn Energy Economists. *Res:* Application of fossil fuels under conditions of a limit of carbon dioxide emissions; technical and policy aspects; thermodynamics; material properties. *Mailing Add:* 19 Lambton Ave Ottawa ON K1M 0Z6 Can

WALSH, JOHN JOSEPH, b Cambridge, Mass, Sept 11, 42; m 69. ECOLOGY, OCEANOGRAPHY. *Educ:* Harvard Univ, AB, 64; Univ Miami, MS, 68, PHD(marine sci), 69. *Prof Exp:* Fel, Univ Wash, 69-70, res asst prof oceanog, 70-75; head div oceanog sci, Brookhaven Nat Lab, 75-84; grad res prof, 84-91, DISTINGUISHED RES PROF MARINE SCI, UNIV SFLA, 91- *Honors & Awards:* Gold Medal Sci, Univ de Liege, 80. *Mem:* AAAS; Am Soc Limnol & Oceanog. *Res:* Shelf ecosystems; systems analysis; statistics; phytoplankton ecology; mathematical models; theoretical ecology. *Mailing Add:* Univ SFla Marine Sci 140 Seventh Ave S St Petersburg FL 33701

WALSH, JOHN JOSEPH, b New York, NY, July 31, 24; wid; c 3. CARDIOLOGY. *Educ:* Long Island Col Med, MD, 48; Am Bd Internal Med, dipl, 58. *Prof Exp:* Intern USPHS Hosp, NY, 48-49, resident med, Seattle, Wash, 51-54, asst chief med, New Orleans, La, 54-56, dep chief, 56; from instr med to asst prof clin med, 57-60, dean sch med & coordr health serv, 68-69, PROF MED, SCH MED, TULANE UNIV, 60-, VPRES HEALTH AFFAIRS, 69-, CHANCELLOR MED CTR, 72- *Concurrent Pos:* Fel cardiol, Sch Med, Tulane Univ, 57-58, instr, 55-; vis physician, Charity Hosp, 55-; chief res activities, USPHS, 58-64, chief med, 63-64, med officer in charge, 64-66, dir div direct health serv, 66-68. *Mem:* Am Thoracic Soc; AMA; fel Am Col Cardiol; fel Am Col Physicians; fel Am Col Chest Physicians. *Res:* Cardiopulmonary diseases. *Mailing Add:* Tulane Univ Med Ctr 1430 Tulane Ave New Orleans LA 70112

WALSH, JOHN M, b Wichita Falls, Tex, Nov 6, 23; m 52; c 3. PHYSICS, CONTINUUM DYNAMICS. *Educ:* Univ Tex, BS, 47, PhD(physics), 50. *Prof Exp:* Staff mem, Los Alamos Sci Lab, 50-60; staff mem, Gen Atomic Div, Gen Dynamics Corp, 60-67; mgr continuum mech div, Systs Sci & Software, 67-74; mem staff, 74-88, LAB ASSOC, LOS ALAMOS NAT LAB, 88 - *Honors & Awards:* Shock Compression Award, Am Phys Soc, 87. *Mem:* Am Phys Soc. *Res:* Shock hydrodynamics; shock wave physics; properties of materials at extreme pressures; experimental, theoretical and numerical work in these areas and supervision of groups so involved; fluid dynamics. *Mailing Add:* Los Alamos Nat Lab MS J960 PO Box 1663 Los Alamos NM 87545

WALSH, JOHN PAUL, b Rochester, NY, Dec 29, 42; c 2. ORGANIC CHEMISTRY. *Educ:* Purdue Univ, Lafayette, BS, 64; Univ Wis-Madison, MS, 66; Univ Tex, Austin, PhD(org chem), 70. *Prof Exp:* Sr res chemist, Org Res Dept, Pennwalt Corp, Pa, 69-74; res chemist, Para-Chem Inc, 74-76; sr appl chemist, Celanese Chem Co, Inc, 76-86; at BASF CORP, 86- *Mem:* Am Chem Soc. *Res:* Organic synthesis; organosulfur, nitrogen and phosphorus; alkyds; polyesters; urea-formaldehyde resins; rosins. *Mailing Add:* 295 Whippany Rd Whippany NJ 07981

WALSH, JOHN RICHARD, b San Francisco, Calif, Aug 22, 20; m 44; c 5. INTERNAL MEDICINE. *Educ:* Creighton Univ, BS, 43, MD, 45, MSc, 51; Am Bd Internal Med, dipl, 53. *Prof Exp:* Actg asst med, Sch Med, Creighton Univ, 51-52, asst, 52-53, from instr to assoc prof, 53-57, prof & dir dept, 57-60; PROF MED, MED SCH, UNIV ORE, 60- *Concurrent Pos:* Ward physician, Vet Admin Hosp, Omaha, Nebr, 51-52, asst chief med serv, 52-53, actg chief, 53-54, chief, 54-56, chief, Portland, Ore, 60-, actg chief radioisotope serv, 62-70; assoc, Col Med, Univ Nebr, 53-55, asst prof, 55-56. *Honors & Awards:* Milo D Leavitt Mem Lectr Award, Am Geriat Soc, 90. *Mem:* Am Soc Hemat; Am Fedn Clin Res; fel Am Col Physicians; Int Soc Hemat. *Res:* Hematology. *Mailing Add:* Health Sci Ctr Sch Med Univ Ore 3181 SW Jackson Park Rd Portland OR 97201

WALSH, JOHN THOMAS, b Lincoln, RI, Dec 23, 27; m 52; c 3. ANALYTICAL CHEMISTRY. *Educ:* Providence Col, BS, 50; Univ RI, MS, 52, PhD(chem), 65. *Prof Exp:* Res chemist, Rumford Chem Works, RI, 52-54; res chemist, US Army Natick Labs, 54-86; RES CHEMIST, STATE RI DEPT TRANSP LABS, 86- *Mem:* Am Chem Soc; Am Soc Testing & Mat. *Res:* Analytical instrumentation research in the areas of gas chromatography and mass spectrometry; applications have been the composition study of natural products such as foods and biologicals; investigation of toxic pollutants in air, water and solid wastes by analytical methods of gas and liquid chromatography and mass spectrometry; fourier transform infrared spectroscopy as an area of analytical instrumentation research; bacterial degradation products of military materials such as foods, clothing and munitions; investigation of tricothecene mycotoxins by analytical methods of gas and liquid chromatography and mass spectronmetry; investigation of protective coating materials via fourier transform IR spectroscopy and gas chromatography. *Mailing Add:* 15 Ridgeland Dr Cumberland RI 02864

WALSH, JOHN V, b Okla, Dec 11, 42. MEMBRANE PHYSIOLOGY. *Educ:* Harvard Univ, MD, 70. *Prof Exp:* Assoc prof, 81-86, PROF PHYSIOL, SCH MED, UNIV MASS, 86- *Mem:* Soc Neurosci; Am Physiol Soc; Biophys Soc; AAAS. *Mailing Add:* Dept Physiol Univ Mass Sch Med 55 Lake Ave N Worcester MA 01605

WALSH, JOSEPH BROUGHTON, b Utica, NY, Sept 5, 30; m 62; c 1. GEOLOGY, GEOPHYSICS. *Educ:* Mass Inst Technol, SB, 52, SM, 54, ME, 56, ScD(mech eng), 58. *Prof Exp:* Engr, Foster Miller Assocs, Inc, 57-59, A-b DeLaval Ljungstrom Angturbin, 59-60 & Woods Hole Oceanog Inst, 60-63; part-time vis prof, 63-64; res assoc geol & geophys, 63-72, SR RES SCIENTIST, DEPT EARTH & PLANETARY SCI, MASS INST TECHNOL, 72- *Res:* Theoretical analysis of various properties of rock, especially strength, elastic moduli and seismic attenuation, and analysis of how these properties should affect behavior in situ. *Mailing Add:* Rm 54-720 Bldg 722 Mass Inst Technol 77 Massachusetts Ave Cambridge MA 02139

WALSH, KENNETH ALBERT, b Yankton, SDak, May 23, 22; m 44; c 5. CHEMICAL METALLURGY, CERAMICS. *Educ:* Yankton Col, BA, 42; Iowa State Univ, PhD(chem), 50. *Prof Exp:* Asst prof chem, Iowa State Univ, 50-51; mem staff, Los Alamos Sci Lab, 51-57; supvr inorg chem res, Int Minerals & Chem Corp, 57-60; assoc dir technol, Brush Wellman Inc, Elmore, 60-86; RETIRED. *Concurrent Pos:* Consult, 86- *Mem:* Soc Mining Engrs; Am Chem Soc; Am Soc Metals. *Res:* Beryllium metal extraction; role of trace elements in properties of beryllium; beryllium chemicals, ecology, and electronic materials. *Mailing Add:* 2106 Kensington Dr Tyler TX 75703-2232

WALSH, KENNETH ANDREW, b Sherbrooke, Que, Aug 7, 31; m 53; c 3. BIOCHEMISTRY. *Educ:* McGill Univ, BSc, 51; Purdue Univ, MS, 53; Univ Toronto, PhD, 59. *Prof Exp:* Jr res officer, Nat Res Coun Can, 53-55; res instr, 59-62, from asst prof to assoc prof, 62-68, PROF BIOCHEM, UNIV WASH, 68- *Mem:* Am Soc Biol Chemists; Protein Soc. *Res:* Structure and function of proteins; mechanisms of zymogen activation and protease action; amino acid sequence and protein conformation; molecular evolution; domain structure of regulated proteins. *Mailing Add:* Dept Biochem 5370 Univ Wash Seattle WA 98195

WALSH, LEO MARCELLUS, b Moorland, Iowa, Jan 16, 31; m 58. SOIL FERTILITY, SOIL SCIENCE. *Educ:* Iowa State Univ, BS, 52; Univ Wis, MS, 57, PhD(soils), 59. *Prof Exp:* Asst prof & exten specialist, 59-64, assoc prof, 64-68, chmn, 72-79, PROF SOILS, UNIV WIS-MADISON, 68-, DEAN, COL AGR & LIFE SCI, 79- *Concurrent Pos:* Mem, comn consumer needs & opportunities, Bd Agr, Nat Res Coun; bd dirs, Nat Nonpoint Source Inst, secy & treas, 85. *Mem:* Fel Soil Sci Soc Am (pres, 79); fel AAAS; fel Am Soc Agron; Soil Conserv Soc Am. *Res:* Use of nitrogen and sulfur fertilizers; use of zinc, manganese and other micronutrients; soil fertility, especially for corn and other cash crops; disposal of wastes on agricultural land; soil conservation; water quality as influenced by agricultural practices. *Mailing Add:* Univ Wis 1450 Linden Dr Madison WI 53706

WALSH, MICHAEL PATRICK, b Liverpool, Eng, Feb 20, 51; Brit & Can citizenship; m 72; c 2. MUSCLE BIOCHEMISTRY, CELL REGULATION. *Educ:* Univ Col, Dublin, BSc, 74; Univ Manitoba, PhD(biochem), 78. *Prof Exp:* Teaching fel, Nat Ctr Sci Res, Montpellier, France, 78-80; res assoc, Dept Nutrit & Food Sci, Univ Ariz, 80-82; from asst prof to assoc prof, 82-90, PROF BIOCHEM, DEPT MED BIOCHEM, UNIV CALGARY, 90- *Honors & Awards:* Ayerst Award, Can Biochem Soc, 90; EWR Steacie Prize, 90. *Mem:* Am Soc Biochem & Molecular Biol; Biophys Soc; Biochem Soc UK; Can Biochem Soc. *Res:* Study of the biochemical mechanisms involved in the regulation of smooth muscle contraction, particularly protein phosphorylations; study of calcium-binding proteins and their involvement in the physiological regulation of the activities of various enzymes. *Mailing Add:* Dept Med Biochem Univ Calgary 3330 Calgary AB T2N 4N1 Can

WALSH, PATRICK NOEL, b New York, NY, Dec 7, 30; m 62; c 5. HIGH TEMPERATURE CHEMISTRY, METAL & CERAMIC COATINGS. *Educ:* Fordham Univ, BS, 51, MS, 52, PhD(chem), 56. *Prof Exp:* Res assoc high temperature chem, Ohio State Univ, 56-60; mem staff, Union Carbide Res Inst, NY, 60-66 & Space Sci & Eng Lab, 66-68, res assoc, Linde Div, 68-81, CONSULT, COATINGS SERV DEPT, UNION CARBIDE CORP, 82- *Mem:* Am Soc Metals; Electrochem Soc; Am Chem Soc. *Res:* Thermodynamics and kinetics of high temperature chemical processes; formulation, deposition and analysis of wear-resistant coatings. *Mailing Add:* Union Carbide Corp 1500 Polco St Indianapolis IN 46224

WALSH, PETER, b New York, NY, Aug 21, 29; m 52; c 5. PHYSICS. *Educ:* Fordham Univ, BS, 51; NY Univ, MS, 53, PhD(physics), 60. *Prof Exp:* Sr scientist, Westinghouse Lamp Div, NY, 51-61; instr physics & math, eve sch, Wagner Col, 57-63; supvr physics res, Am Stand Res Lab, NJ, 61-63; PROF PHYSICS & ELEC ENG, FAIRLEIGH DICKINSON UNIV, 63- *Concurrent Pos:* Dir, NSF Undergrad Partic Prog, Fairleigh Dickinson Univ, 63-; consult, S-F-D Labs, 63-66, Am Standard, 63, Belock Instr, 63-64, Nuclear Res Assocs, 64-65, Thiokol Chem, 64, US Army Res Off, 64-71, 77, 80, Curtiss Wright Corp, 65, Singer Corp, 68, Picatinny Arsenal, 70-74, Corning Glass, 71, Columbus Labs, Battelle Mem Inst, 74-75, Duro Test Corp, 76-89, Xerox Corp, 79-83, Mass Inst Technol, 80, Polaroid, 81 & 84, Peak Systs & Vet Admin, 85, Valore McAllister, 87, Qwest, 88-; vis res scientist, Mass Inst Technol, 77; vis prof, Univ Sheffield, 78 & 79, Univ Genova, 84 & Stanford Univ, 84 & 85; NASA fel, 80, 83 & 87; Nat Res Lab fel, 81, 82, 86, 88 & 90; invited speaker for various asn, 70-90; Res Award, DOA, 75. *Honors & Awards:* Outstanding Res Award, Picatinny Arsenal, 72 & 73. *Mem:* AAAS; Am Phys Soc; Optical Soc Am; Mat Res Soc. *Res:* Optics; amorphous semiconductors; lasers; quantum physics; plasmas; artificial intelligence; super conductivity. *Mailing Add:* Dept Elec Eng Fairleigh Dickinson Univ Teaneck NJ 07666

WALSH, PETER NEWTON, b Chicago, Ill, Apr 16, 35; m 58; c 3. HEMATOLOGY. *Educ:* Amherst Col, BA, 57; Washington Univ, MD, 61; Oxford Univ, DPhil(med), 72; Am Bd Internal Med, dipl, 68. *Prof Exp:* Intern internal med, Barnes Hosp, St Louis, 61-62, resident, 62-63; fel hemat, Sch Med, Wash Univ, 63-69; res fel blood coagulation, Oxford Haemophilia Ctr, Churchill Hosp, 69-72; asst prof, 72-74, ASSOC PROF INTERNAL MED, HEALTH SCI CTR, TEMPLE UNIV, 75- *Concurrent Pos:* Sr resident, Palo Alto Stanford Hosp, 64-65; chief resident, Sch Med, Wash Univ, 65-66; asst physician, Barnes Hosp, 65-66; med liaison officer, Nat Heart Inst, 66-69; NIH res fel, 69-72; hon sr registr med, United Oxford Hosps, 69-72; assoc ed, Thrombosis et Diathesis Haemorrhagica, 76-; mem Exec Comt Coun Thrombosis, Am Heart Asn. *Honors & Awards:* First Int Prize, Viviana Luckhaus Found, Arg, 72; Jane Nugent Cochems Prize, Univ Colo Sch Med, 74. *Mem:* Int Soc Thrombosis & Haemostasis; Am Physiol Soc; Soc Exp Med & Biol; Am Soc Clin Invest; Am Soc Biochem Molecular Biol. *Res:* Role of blood platelets in blood coagulation, hemostasis and thrombosis; coagulation factor biochemistry; mechanisms of binding of coagulation factors to platelets; role of platelet coagulant activities in thrombosis. *Mailing Add:* Thrombosis Res Ctr Temple Univ Sch Med 3400 N Broad St Philadelphia PA 19140

WALSH, RAYMOND ROBERT, b Denver, Colo, Apr 9, 25; m 52; c 6. PHYSIOLOGY. *Educ:* Cornell Univ, AB, 50, PhD(zool), 53. *Prof Exp:* Assoc res physiologist, Brookhaven Nat Lab, 53-55; from instr to assoc prof physiol, Sch Med, Univ Colo, Denver, 55-71; prof, Sch Dent, Southern Ill Univ, 71-72; PROF BIOL & CHMN DEPT, ST LOUIS UNIV, 72- *Mem:* Am Physiol Soc; Am Soc Zoologists; Soc Exp Biol & Med; Sigma Xi. *Res:* Neurobiology; comparative physiology. *Mailing Add:* Dept Biol St Louis Univ St Louis MO 63103

WALSH, ROBERT JEROME, b Chicago, Ill, Jan 12, 29; div; c 3. SEMICONDUCTOR MATERIALS, CHEMO-MECHANICAL POLISHING. *Educ:* Univ Wis, BS, 50. *Prof Exp:* Res engr, Monsanto Co, 51-55, sr res engr, 55-58, res group leader, 58-62, sr res group leader, 62-70, fel, 70-81, sr fel, 81-85; RETIRED. *Res:* Semiconductor materials technology including growth of single crystals, epitaxial deposition, damage free polished surfaces and ultracleaning of surfaces, ultraflat sicicon wafers; ultraflat silicon wafers. *Mailing Add:* 356 Sudbury Lane Ballwin MO 63011

WALSH, ROBERT MICHAEL, b Wilmington, Del, Jan 28, 38; m 61; c 4. PHYSICAL CHEMISTRY. *Educ:* Univ Del, BS, 60; Univ Calif, Berkeley, PhD(chem), 65. *Prof Exp:* Res chemist, 65-72, sr res chemist, 72-80, res scientist, 80-86, res assoc, 86-90, CORP MGR, TECHNOL, HERCULES INC, 90- *Concurrent Pos:* Tech dir, Esgraph, 85-86. *Mem:* Licensing Exec Soc; Am Chem Soc; Sigma Xi; Technol Transfer Soc. *Res:* Photoprocesses in materials; photographic systems; polymer systems. *Mailing Add:* 1800 Mount Salem Ln Wilmington DE 19806

WALSH, ROBERT R(EDDINGTON), b Wilmington, Del, Nov 4, 27. ELECTRICAL ENGINEERING, PHYSICS. *Educ:* St Mary's Col, Md, AB, 53. *Prof Exp:* Engr, E I du Pont de Nemours & Co, 54-59, res engr, 59-60; asst to dir appl physics, All Am Eng Co, 60-61, dir mkt & prod develop div, 61-63, dir res, 63-66; exec vpres & gen mgr, Technidyne, Inc, 67-71; pub rels asst, 71-72; electronics consult, Advan Technol Prod, Inc, Newark, 73-76; tech planning staff mem, 76-79, ASST VPRES TECH PLANNING, WILMINGTON TRUST CO, 80- *Concurrent Pos:* Chief engr, Reynolds Broadcasting Co, 57-60; teacher, Adult Prog, Wilmington Pub Schs, 56-60. *Mem:* Inst Elec & Electronics Engrs; AAAS; NY Acad Sci. *Res:* Circuit logic; systems; automata; memory and computing circuits; design of cybernetic systems; instruments for medical applications and devices and systems for instrumentation, control and automation of industrial processes; engineering applications of lasers; data and communication networks; energy management. *Mailing Add:* Wilmington Trust Ctr Rodney Sq N Wilmington DE 19890

WALSH, SCOTT WESLEY, b Wauwatosa, Wis, July 23, 47. PERINATAL PHYSIOLOGY, REPRODUCTIVE ENDOCRINOLOGY. *Educ:* Univ Wis-Milwaukee, BS, 70; Univ Wis-Madison, MS, 72, PhD(endocrinol, reproductive physiol), 75. *Prof Exp:* Asst prof physiol, Sch Med Univ NDak, 75-76; asst scientist perinatal physiol, Ore Regional Primate Res Ctr, 76-80; asst prof dept physiol, Health Sci Ctr, Sch Med, Univ Ore, 78-80; asst prof dept physiol, Mich State Univ, 80-85; ASSOC PROF, DEPT OBSTET & GYNEC, UNIV TEX MED SCH, 85- *Mem:* Soc Study Reproduction; Sigma Xi; Endocrine Soc; Am Physiol Soc; Soc Gynec Invest; Perinatal Res Soc; Int Soc for the Study of Hypertension in Pregnancy. *Res:* Endocrine functions of the primate placenta as they relate to hypertension in pregnancy (preeclampsia), regulation of fetoplacental blood flow, and the onset of labor; placental production rates of eicosanoids (prostacyclin, thromboxane, prostaglandins, HETEs, leukotrienes) and their effects on fetoplacental blood flow and maternal systemic blood pressure; human placental tissues are studied in vitro, and pregnant rhesus monkeys and pregnant sheep are used as in vivo animal models. *Mailing Add:* Dept Obstet & Gynec Reprod Sci Univ Tex Med Sch 6431 Fannin Suite 3204 Houston TX 77030

WALSH, STEPHEN G, b Brooklyn, NY, Apr 23, 47. BIOCHEMISTRY. *Educ:* Cath Univ Am, BA, 69; State Univ NY Buffalo, MA, 80, PhD(biochem), 81. *Prof Exp:* Teacher sci & math, St Thomas Community Sch, 69-73; teacher chem, La Salle Acad, 73-77; asst prof chem, Col Mt St Vincent, 81-87; MEM STAFF, BUR PATTERN & TRADEMARKS, 87- *Mem:* AAAS; Am Chem Soc; Sigma Xi. *Res:* Protein chemistry and immunochemistry of venom allergens investigated by biochemical and immunochemical methods. *Mailing Add:* 1400 20th St NW Apt 918 Washington DC 20036

WALSH, TERESA MARIE, b Philadelphia, Pa, May 1, 62; m 86. RHEOLOGY. *Educ:* Villanova Univ, BS, 84. *Prof Exp:* Patent examr, US Patent & Trademark Off, 84-86; APPLN ENG, NAMETRE CO, 86- *Mem:* Inst Elec & Electronics Engrs; Soc Woman Eng. *Res:* Reviewing, evaluating and improving electronic designs for custom application. *Mailing Add:* 101 Forrest St Metuchen NJ 08840

WALSH, THOMAS DAVID, b Chicago, Ill, Oct 30, 36; m 87; c 2. CHEMISTRY. *Educ:* Univ Notre Dame, AB, 58; Univ Calif, PhD(chem), 62. *Prof Exp:* Asst prof chem, Univ Ga, 62-67; vis asst prof, Ohio State Univ, 67-68; assoc prof, Univ SDak, 68-70; asst prof, 70-72, ASSOC PROF CHEM, UNIV NC, CHARLOTTE, 72- *Concurrent Pos:* NSF postdoctoral fel, Calif Inst Technol, 61-62; Danforth Found assoc, 64-; res assoc, Univ NC, Chapel Hill, 69-70. *Mem:* Am Chem Soc. *Res:* Organic reaction mechanisms; computer- aided instruction. *Mailing Add:* Dept Chem Univ NC Charlotte NC 28223

WALSH, WALTER MICHAEL, JR, b Los Angeles, Calif, July 28, 31; m 56; c 2. SOLID STATE PHYSICS. *Educ:* Harvard Univ, AB, 54, AM, 55, PhD(physics), 58. *Prof Exp:* Res fel solid state physics, Harvard Univ, 58-59; mem tech staff, Bell Labs, 59-67 & 77, dept head, 67-77, head, Solid State & Physics Metals Res Dept, 67-77. *Mem:* Fel Am Phys Soc; Am Phys Soc. *Res:* Experimental physics of solids using microwave resonance techniques; effects of pressure and temperature on solids; resonance and wave propagation phenomena in metals and organic conductors. *Mailing Add:* 28 Sherbrook Dr Berkeley Heights NJ 07922

WALSH, WILLIAM ARTHUR, b Manchester, Conn, June 28, 54. ICHTHYOLOGY. *Educ:* Fairfield Univ, BS, 77; Univ Conn, MS, 83, PhD(ecol), 86. *Prof Exp:* RES BIOLOGIST, OCEANIC INST, WALMANALO, HAWAII, 86- *Mem:* Am Fisheries Soc; Am Soc Ichthyologists & Herpetologists. *Res:* Effects of abiotic factors on the ecology and physiology of early life stages of marine fishes. *Mailing Add:* 1524 Aalapapa Dr Kailua HI 96734

WALSH, WILLIAM J, b Saginaw, Mich, Oct 2, 36; m 62; c 5. CHEMICAL & NUCLEAR ENGINEERING. *Educ:* Notre Dame Univ, BS, 58; Univ Mich, MS, 60, MS, 61; Iowa State Univ, PhD(chem eng), 64. *Prof Exp:* Res asst, Univ Mich Res Inst, 58-61 & Inst Atomic Res, Ames, Iowa, 61-64; assoc engr, Argonne Nat Lab, 64-77; MEM STAFF, DIAMOND SHAMROCK CORP, 77- *Mem:* Am Inst Chem Engrs; Inst Elec & Electronics Engrs. *Res:* High temperature battery development; nuclear fuels reprocessing; liquid metal distillation, nuclear criticality; radiotracer experiments; mass transfer; heat transfer; high vacuum experiments; fast breeder reactor design and economics. *Mailing Add:* PO Box 889 Alamo CA 94507

WALSH, WILLIAM K, b Columbus, Ohio, Sept 29, 32; m; c 1. TEXTILE CHEMISTRY, CHEMICAL ENGINEERING. *Educ:* Univ SC, BS, 54; NC State Univ, PhD(chem eng), 67. *Prof Exp:* Engr, Celanese Corp Am, SC, 59-60; res asst, 60-67, asst prof, 67-72, assoc prof, 72-77, prof textile chem, 77-80, asst dean res, Sch Textiles, 80-81, ASSOC DEAN, RES & GRAD EDUC, SCH TEXTILES, NC STATE UNIV, 81- *Mem:* AAAS; Am Chem Soc; Am Asn Textile Chemists & Colorists; Fiber Soc. *Res:* Applications of ionizing radiation to textile chemistry; radiation graft copolymerization, cross-linking, mechanical properties of textiles; physical and surface chemistry of polymers. *Mailing Add:* Head Textile Eng Dept Auburn Univ Auburn AL 36849-5327

WALSKE, MAX CARL, b Seattle, Wash, June 2, 22; m 46; c 3. PHYSICS. *Educ:* Univ Wash, BS, 44; Cornell Univ, PhD(physics), 51. *Prof Exp:* Mem staff, Los Alamos Sci Lab, 51-55, asst leader theoret div, 55-56; dep res dir, Atomics Int Div, NAm Aviation, Inc, 56-59; mem US deleg, Conf Suspension Nuclear Tests, Geneva, Switz, 59-61; sci rep, AEC, London, Eng, 61-62; theoret physicist, Rand Corp, 62-63; sci attache, US Missions to NATO & Orgn Econ Coop & Develop, Paris, France, 63-65; staff mem, Los Alamos Sci Lab, 65-66; asst to secy defense for atomic energy & chmn mil liaison comt, US Dept Defense, 66-73; pres, Atomic Indust Forum Inc, 73-87; RETIRED. *Concurrent Pos:* Consult, Los Alamos Sci Lab, 56-59 & 62-63. *Mem:* Am Phys Soc; fel Am Nuclear Soc; fel Explorers Club; Sigma Xi. *Res:* Nuclear and reactor physics. *Mailing Add:* PO Box 370 Silverdale WA 98383-0370

WALSTAD, JOHN DANIEL, b Minneapolis, Minn, Aug 22, 44; m 66; c 2. FOREST PROTECTION. *Educ:* Col William & Mary, BS, 66; Duke Univ, MF, 68; Cornell Univ, PhD(entom), 71. *Prof Exp:* Forest scientist, Weyerhaeuser Co, 71-76, admin asst res, 76-77, res mgr forestry, 77-80; ASSOC PROF FOREST VEG MGT, ORE STATE UNIV, 80- *Mem:* Soc Am Foresters; Weed Sci Soc Am; Entom Soc Am; AAAS. *Res:* Forest vegetation management. *Mailing Add:* Forest Sci Dept Ore State Univ Corvallis OR 97331

WALSTEDT, RUSSELL E, b Minneapolis, Minn, June 12, 36; m 64; c 2. SOLID STATE PHYSICS. *Educ:* Mass Inst Technol, BS, 58; Univ Calif, Berkeley, PhD(physics), 62. *Prof Exp:* NSF fel, Clarendon Lab, Eng, 61-62; asst res physicist, Univ Calif, Berkeley, 62-65; MEM TECH STAFF SOLID STATE MAGNETISM, BELL LABS, 65- *Mem:* Am Phys Soc. *Res:* Nuclear magnetic resonance and its application to the study of magnetism and atomic motion in the solid state; computer modeling of amorphous magnetic systems. *Mailing Add:* Room 1D362 Box 261 Bell Labs Murray Hill NJ 07974

WALSTON, DALE EDOUARD, b Woodsboro, Tex, Dec 1, 30. MATHEMATICS. *Educ:* Tex A&M Univ, BA, 52; Univ Tex, MA, 59, PhD(math), 61. *Prof Exp:* Asst prof, 61-72, ASSOC PROF MATH, UNIV TEX, AUSTIN, 72- *Concurrent Pos:* Consult, Manned Space Ctr, NASA, 66. *Mem:* Am Math Soc; Math Asn Am. *Res:* Numerical solution of differential equations. *Mailing Add:* Dept Math Univ Tex Austin TX 78712

WALSTON, WILLIAM H(OWARD), JR, b Salisbury, Md, Apr 13, 37; m 62; c 2. MECHANICAL ENGINEERING. *Educ:* Univ Del, BME, 59, MME, 61, PhD(appl sci), 64. *Prof Exp:* Asst prof, 65-67, ASSOC PROF MECH ENG, UNIV MD, COLLEGE PARK, 67- *Mem:* Am Soc Mech Engrs. *Res:* Signal propagation; shock and vibrations analysis; applied mathematics; design; automotive drag reduction, acoustics, noise control. *Mailing Add:* Dept Mech Eng Univ Md College Park MD 20742

WALSTROM, ROBERT JOHN, b Omaha, Nebr, Apr 24, 22; m 44; c 2. ENTOMOLOGY. *Educ:* Univ Nebr, BS, 47, MS, 49; Iowa State Univ, PhD, 55. *Prof Exp:* State entomologist, State Dept Agr & Inspection, Nebr, 48-50; exten entomologist, Iowa State Univ, 50-55; PROF ENTOM, SDAK STATE UNIV, 55- *Mem:* AAAS; Entom Soc Am. *Res:* Control of beneficial and injurious legume insects; apiculture. *Mailing Add:* 1409 First St Brookings SD 57006

WALT, ALEXANDER JEFFREY, b Cape Town, SAfrica, June 13, 23; US citizen; c 3. SURGERY. *Educ:* Univ Cape Town, MB & ChB, 48; FRCS(C), 55; Univ Minn, MS, 56; FRCS, 56; Am Bd Surg, dipl, 62. *Prof Exp:* Lectr path, Univ Cape Town, 49-50; registr surg, St Martin's Hosp, Bath, Eng, 56-57; asst surgeon, Groote Schuur Hosp, Cape Town, 57-61; asst chief surg, Vet Admin Hosp, Allen Park, Mich, 61-62; from asst prof to assoc prof, 61-66, from asst dean to assoc dean med, 64-70, prof surg & chmn dept, 66-88, PROF SURG, SCH MED, WAYNE STATE UNIV, 88- *Concurrent Pos:* Consult div physician manpower & clin cancer training comt, NIH, 72-73. *Mem:* Am Surg Asn; Int Soc Surg; Soc Surg Alimentary Tract; Am Asn Surg of Trauma. *Res:* Studies of the effects and clinical management of severe trauma of the liver and stomach of humans. *Mailing Add:* Dept Surg Wayne State Univ Sch Med Detroit MI 48201

WALT, MARTIN, b West Plains, Mo, June 1, 26; m 50; c 4. SPACE PLASMA PHYSICS, GEOPHYSICS. *Educ:* Calif Inst Technol, BS, 50; Univ Wis, MS, 51, PhD(physics), 53. *Prof Exp:* Mem staff, Los Alamos Sci Lab, 53-56; mem sci staff, 56-65, mgr physics, 65-71, dir phys sci, 71-84, DIR RES, LOCKHEED MISSILES & SPACE CO, 84- *Concurrent Pos:* Mem, Panel Nuclear Physics, 70-72, comt solar terrestrial res, Nat Acad Sci, 83-89; mem adv comt, Space Sci Lab, Univ Calif, Berkeley, 72-76, mem sci & educ adv comt, Lawrence Berkeley Lab, 82-; mem space & earth sci adv comt, NASA, 84-88; consult prof, Stanford Univ, 85-; mem exec comt, 87-89, mem governing bd, 86- , Am Inst Phyics; bd overseers, Superconducting Supercollider, 89-; mem adv comt, Ctr Particle Astrophysics, Univ Calif, Berkeley, 89- *Mem:* Fel Am Phys Soc; fel Am Geophys Union; Am Inst Aeronaut & Astronaut; AAAS. *Res:* Experiments and theory of interaction of fast neutrons with nuclei; space research, including measurements and theory on geomagnetically trapped radiation belts, aurora and cosmic rays; diffusion of ions and electrons in plasmas. *Mailing Add:* 12650 Viscaino Ct Los Altos Hills CA 94022

WALTAR, ALAN EDWARD, b Chehalis, Wash, July 10, 39; m 61; c 4. FAST REACTORS, SAFETY. *Educ:* Univ Wash, BS, 61; MIT, MS, 62; Univ Calif Berkeley, PhD(eng sci), 66. *Prof Exp:* Sr res sci, fast reactor modeling, Battelle Northwest, 66-72; mgr Reactor Dynamics, Westinghouse Hanford Co, 72-76; vis prof heat transfer fast reactors, Univ Va, 76-77; adv eng, 77-79, MGR, REACTOR & PHYSICS APPLIED PHYSICS & SAFETY, WESTINGHOUSE HANFORD CO, 79- *Concurrent Pos:* Chmn, Prog Comt, Nuclear Reactor Safety Div, Am Nuclear Soc, 78-80, Richland Sect, 82-83, Tech Prog Comt, 82-85, Bylaws & Rules Comt, ANS, 84-86, Nuclear Reactor Safety Div, 86-87, Tech Prog Comt, 86-88; instr, Power Reactors Short Course on Fast Breeder Reactors, Joint Ctr Grad Study, 81-82 & 83, Short Course Fast Breeder Reactors, Los Alamos, Nat Lab, 83-84. *Mem:* Am Nuclear Soc (chair, Nuclear Reactor Safety Div, 86-87, Bylaws & Rules Comt, 84-86); Am Assoc Adv Sci. *Res:* Development of computational models to describe the intrinsic safety response of fast breeder reactors during off normal conditions; wide range energy transformation systems. *Mailing Add:* 1617 Sunset St Richland WA 99352

WALTCHER, AZELLE BROWN, b New York, NY, Mar 27, 25; m 55; c 2. MATHEMATICS. *Educ:* Columbia Univ, BA, 45, MA, 46; NY Univ, PhD(math, educ), 54. *Prof Exp:* Asst math, Barnard Col, Columbia Univ, 45-46; instr, Hollins Col, 46-48; teacher, Calhoun Sch, NY, 48-52; from instr to assoc prof, Univ, 52-72, teaching fel, New Col, 61-72, PROF MATH, HOFSTRA UNIV, 72- *Concurrent Pos:* Mem fac, Sarah Laurence Col, 53-55. *Mem:* Am Math Soc; Math Asn Am. *Res:* Logic and foundations of mathematics; number theory; group theory. *Mailing Add:* 84-19 Kent St Jamaica NY 11432

WALTCHER, IRVING, b Newport, RI, Mar 6, 17. POLYMER CHEMISTRY. *Educ:* Univ RI, BS, 38; Duke Univ, MA, 40; Ohio State Univ, PhD(org chem), 47. *Prof Exp:* Chemist, War Ord Dept, Ala, 41-42; res chemist, B F Goodrich Co, Ohio, 42-44; asst org chem, Res Found, Ohio State Univ, 44-47; res assoc chem, Polytech Inst Brooklyn, 47-48; assoc prof, State Univ NY Col Forestry, Syracuse, 48-55; asst prof, 55-58, ASSOC PROF CHEM, CITY COL NEW YORK, 58, DEP CHMN DEPT, 74- *Mem:* Am Chem Soc. *Res:* Selective hydrogenation of acetylenes; preparation and characterization of graft copolymers. *Mailing Add:* 8419 Kent St Jamaica NY 11432

WALTENBAUGH, CARL, b Canton, Ohio, July 17, 48; m 73; c 3. IMMUNOGENETICS, REGULATION OF IMMUNE RESPONSE. *Educ:* Baldwin-Wallace Col, Berea, Ohio, BS, 70; Univ Ill Med Ctr, Chicago, MS, 73, PhD(immunol), 75. *Prof Exp:* Res fel, 75-77, instr path, Sch Med, Harvard Univ, 77-79; asst prof, 79-84, ASSOC PROF MICROBIOL & IMMUNOL, SCH MED, NORTHWESTERN UNIV, 84- *Mem:* Soc Develop Biol; Sigma Xi; Am Asn Immunologists; Reticuloendothelial Soc. *Res:* Immunogenetic regulation of the immune response by suppressor T cells and their soluble factors; development of monoclonal antibodies and cell lines. *Mailing Add:* Dept Microbiol-Immunol Northwestern Univ Sch Med 303 E Chicago Ave Chicago IL 60611

WALTER, CARL, b Cleveland, Ohio, Nov 30, 05; m 29; c 6. SURGERY. *Educ:* Harvard Col, AB, 28; Harvard Med Sch, MD, 32. *Prof Exp:* EMER PROF SURG, MED SCH, HARVARD UNIV, 72-; EMER SURGEON, PETER BENT BRIGHAM HOSP, 72- *Mem:* Am Chem Soc; Am Surg Asn; Surg Infection Soc; AMA. *Res:* Nosocomia infections; parenteral fluids. *Mailing Add:* Ten Shattuck St Boston MA 02115

WALTER, CARLTON H(ARRY), b Willard, Ohio, July 22, 24; m 48; c 2. ELECTRICAL ENGINEERING, PHYSICS. *Educ:* Ohio State Univ, BEE, 48, MS, 51, PhD(elec eng), 57. *Prof Exp:* Res assoc, Ohio State Univ, 48-54, from instr to assoc prof, 54-65, asst supvr lab, 54-57, assoc supvr, 57-69, prof elec eng, Electro Sci Lab, 65-83, tech area dir antennas, 69-83; MGR, ANTENNA SYSTS PROD, TRW, 83- *Concurrent Pos:* Mem bd dirs, Ladar Systs, Inc, 64-71. *Mem:* Fel Inst Elec & Electronics Engrs. *Res:* Microwave, traveling wave, Luneberg lens and electrically small antennas. *Mailing Add:* TRW MS 505 1550A 13208 Tining Dr Poway CA 92064

WALTER, CHARLES FRANK, b Sarasota, Fla, June 19, 36. BIOCHEMISTRY, TECHNICAL AND PATENT LAW. *Educ:* Ga Inst Technol, BS, 57; Fla State Univ, MS, 59, PhD(chem), 62; Univ Houston, JD, 79. *Prof Exp:* NIH fel, Med Sch, Univ Calif, San Francisco, 62-64; from asst prof to assoc prof biochem, Med Sch, Univ Tenn, Memphis, 64-70; assoc prof biomath & biochem, M D Anderson Hosp & Tumor Inst, Univ Tex, Houston, 70-74; prof chem eng, Univ Houston, 74-77; PATENT & TRADEMARK ATTY, 79-; DIR, PROG ON LAW & TECHNOL, UNIV PARK LAW CTR, UNIV HOUSTON, 83- *Concurrent Pos:* NIH career develop award, Med Sch, Univ Tenn, Memphis, 65-70. *Mem:* Biophys Soc; Soc Math Biol; Am Soc Biol; Chem Fed Soc Exp Biol; Am Bar Asn; Tex Bar Asn. *Res:* Information science; biological control; enzyme mechanisms and kinetics; data bases; communications; real-time computer applications; cognitive processes and models; artificial intelligence and expert systems; cybernetics of law and science; impact of technological developments; societal regulation of science and technology. *Mailing Add:* 9131 Timberside Houston TX 77025

WALTER, CHARLES ROBERT, JR, b Charlottesville, Va, Oct 31, 22; m 50; c 3. ORGANIC CHEMISTRY. *Educ:* Univ Va, BA, 43, PhD(chem), 49. *Prof Exp:* Res asst chem, Univ Ill, 49-50; asst prof, Univ NC, 50-52; sr res chemist, Nitrogen Div, Allied Chem Corp, 52-58, supvry res chemist, 58-60, mgr res, 60-66; chmn dept, 66-83, prof chem, 66-87, EMER PROF CHEM, GEORGE MASON UNIV, 88- *Mem:* Am Chem Soc. *Res:* Synthetic organic chemistry; Diels-Alder reaction of quinoneimides; industrial process development; organic nitrogen chemicals; vapor phase catalysis; chlorination of olefins. *Mailing Add:* 4221 SE Eighth Ave Cape Coral FL 33904

WALTER, CHARLTON M, b Altoona, Pa, July 1, 23; m 47; c 2. INFORMATION SCIENCE, SIGNAL DATA ANALYSIS. *Educ:* Columbia Univ, BA, 49; Harvard Univ, MA, 51. *Prof Exp:* Mathematician, Commun Lab, USAF Cambridge Res Ctr, 51-54, chief simulation & eval br, Comput & Math Sci Lab, 54-63, chief dynamic processes br, Data Sci Lab, 63-70, chief multisensor processing br, data sci lab, USAF Cambridge Res Labs, 70-73, chief anal & simulation br, Comput Ctr, 73-76; sr systs anal, dir res servs, USAF Geophys Lab, 76-79; CONSULT, APPL RES CONSULTS, INC, 80- *Mem:* AAAS; Inst Elec & Electronics Engrs; Soc Gen Syst Res. *Res:* Development of interactive, computer-based, display-oriented signal processing systems, with applications to environmental sensor data collection; statistical data reduction; dynamic modelling; simulation and systems evaluation. *Mailing Add:* 58 Conant Rd Lincoln MA 01773

WALTER, DONALD K, b Philadelphia, Pa, May 28, 31; m 63; c 1. ENERGY CONSERVATION. *Educ:* Drexel Inst Technol, BS, 53; Drexel Univ, MS, 66. *Prof Exp:* Chief opers, facil engr, 1st Cav Div, US Army, 59-60, resident engr, Area Off, Warren AFB, 60-61, real estate & facil engr, HQ Area Command, Vietnam, 66-67, chief supply, Engr Command, 67-69, facil engr, Wuertzberg, Ger, 69-71 & Ft Detrick, Md, 71-73; asst prof mil sci & tactics, Drexel Univ, 61-66; city engr, Annapolis, MD, 73-75; DIR, DEPT ENERGY, 75- *Mem:* Am Soc Mech Engrs; Nat Soc Prof Engrs; Soc Am Mil Engrs; Am Soc Testing & Mats. *Res:* Management of programs related to productive use of municipal solid waste and energy conservation in municipal functions; conservation technologies including mechanical processing for solid fuels and recyclable materials, thermochemical conversion for steam, gaseous or liquid fuels and biochemical conversion for gaseous or liquid fuels; energy from municipal waste; management of programs on industrial waste materials utilization, conversion and reduction; 06094259xxxof programs on solar industrial applications to include hazardous wastes detoxification, solar industrial heat and application of high photon and heat fluxes to industrial processes. *Mailing Add:* 289 Marlinspike Dr Severna Park MD 21146

WALTER, EDWARD JOSEPH, b St Louis, Mo, Dec 6, 14; m 48; c 9. GEOPHYSICS. *Educ:* St Louis Univ, BS, 37, MS, 40, PhD(geophys), 44. *Prof Exp:* Seismic computer, Root Petrol Co, Ark, 37; asst seismologist, Shell Oil Co, 38, seismologist, 45-46; assoc prof math & asst dir seismol observ, 46-50, dir dept, 57-59, dir seismol observ, 62, PROF MATH, JOHN CARROLL UNIV, 50- *Mem:* AAAS; Seismol Soc Am; Soc Explor Geophys; Am Geophys Union; Geol Soc Am; Sigma Xi. *Res:* Seismology; local earthquakes; crustal structure; wave motion; attenuation coefficients; volcanology; engineering seismology; environmental acoustics and vibration. *Mailing Add:* 4174 Carroll Blvd Cleveland OH 44118-4529

WALTER, EUGENE LEROY, JR, b St Thomas, VI, Aug 14, 22; m 64; c 4. MEDICAL MICROBIOLOGY, VIROLOGY. *Educ:* Univ Calif, Los Angeles, BA, 47; Univ Southern Calif, MS, 58; Univ Wis, PhD(med microbiol), 64. *Prof Exp:* Eng asst, Los Angeles Bur Standard, 47-51; microbiologist, Epidemic Dis Control Unit, Pearl Harbor, 51-56; med microbiologist med res units, Berkeley, Calif, Great Lakes, Ill, Cairo, Egypt, Washington, DC, 56-69; health sci adminr, Nat Heart Blood & Lung Inst, NIH, 71-87; RETIRED. *Concurrent Pos:* Instr bact, US Navy Hosp Corps Sch, 53-54. *Mem:* AAAS; Am Soc Microbiol. *Res:* Bacteriophage; papilloma virus; infectious hepatitis diagnosed by fluorescent antibody; latent herpes simplex virus; leptospirosis; environmental engineering; health science administration. *Mailing Add:* 8410 Post-Oak Rd Potomac MD 20854-3480

WALTER, EVERETT L, b Rensselaer, Ind, July 1, 29; m 51; c 4. MATHEMATICS. *Educ:* Ariz State Univ, BS, 51; NMex State Univ, MS, 57, PhD(math), 61. *Prof Exp:* Dir comput, Army Field Forces, Ft Bliss, Tex, 54-56; instr math, NMex State Univ, 56-60; res mathematician, White Sands Missile Range, 61-62; assoc prof, 62-68, PROF MATH, NORTHERN ARIZ UNIV, 68- *Mem:* Math Asn Am; Am Math Soc. *Res:* Functional analysis. *Mailing Add:* Dept Math NAriz Univ Box 5717 Flagstaff AZ 86011

WALTER, F JOHN, b 1931; c 4. LOW TEMPERATURE NUCLEAR ALIGNMENT & POLARIZATION, NUCLEAR PHYSICS. *Educ:* Kans State Univ, BS, 53; Univ Tenn, MS, 58, PhD(physics), 65. *Prof Exp:* Develop engr, Reactor Exp Eng Div, ORNL, 53-57, res assoc, Physics Div, 57-64; chief physicst Div Nuclear, RIDL, Chief, 64-65; dir semiconductor res & develop, ORTEC Inc, 66-71, tech dir, 71-73, vpres & tech dir, 73-76; vpres & gen mgr, Phys Sci Dir, Egg Ortec, 76-77, vpres, asst gen mgr & dir, Detection Div, 77-80; pres, Waltec Inc, 80-82; mgr, Semiconductor Div, Tennelec Inc, 82-85; PRES, INSTRAPEC INC, 85- *Concurrent Pos:* Develop dir, portable gold analyzer, SAfrica Chamber Mines; admin comnr, Nuclear & Plasma Sci Soc, 80-, chmn, Nuclear Instruments & Detectors Comn, proj engr, Nuclear Instruments Standards, asst prog chmn, Semiconductor & Scintillation Counter & Nuclear Sci Symp; prog chmn, Nuclear Sci Symp, Inst Elec & Electronics Engrs & Nuclear & Plasma sci Soc, 80, fel review comt, 82-, proj leader video teaching pilot prog, Educ & Continued Prof Develop, mem, Environ Instructing & Montoring; Nuclear & Plasma Sci Soc rep, Coun Ocean Eng, Inst Elec & Electronics Engrs-Coun Ocean Eng; delegate, nuclear instruments & detection comt, Int Electro Tech Comn, 80-; tech dir, Life Sci Prog, ORTEC. *Mem:* Fel, Inst Elec & Electronics Engrs; Nuclear & Plasma Sci Soc; Sigma Xi; Am Phys Soc. *Res:* Author and co-author of more than 40 papers on semiconductor radiation detectors, nuclear instrumentation, cryogenics, kinetics and kinematics of heavy nuclei decay and fission. *Mailing Add:* PO Box 847 Oak Ridge TN 37830

WALTER, GILBERT G, b Ottawa, Ill, Nov 24, 30; m 58; c 3. ANALYSIS & FUNCTIONAL ANALYSIS, BIOMATHEMATICS. *Educ:* Gen Motors Inst, BIE, 53; NMex State Univ, BSEE, 56; Univ Wis, MS, 59, PhD(math), 62. *Prof Exp:* Proj engr, AC Electronics Div, Gen Motors Corp, 56-57; from instr to PROF MATH, UNIV WIS-MILWAUKEE, 61- *Concurrent Pos:* Vis prof, Univ Calif, San Diego, 65-66, Univ Agraria, Lima, Peru, 68-69, Univ Costa Rica, 78, Imp Col, London, 80, Univ Nac Auto Mex, Mexico City, 81, Univ Calif, Davis, 82, Calif Polytech Univ, San Luis Obispo, 85, Univ Del, 88. *Mem:* Am Math Soc; Math Asn Am; Inst Math Statist; Soc Indust & Appl Math. *Res:* Mathematical analysis: generalized functions, sampling theorems; statistics: density estimation, empiric Bayes estimation; biomathematics; compartmental models and fisheries models. *Mailing Add:* Math Dept Univ Wis-Milwaukee Box 413 Milwaukee WI 53201

WALTER, GORDON H, METALLURGY. *Educ:* Ill Inst Technol, BS. *Prof Exp:* MGR, MAT TECH & STANDARDS, AGR & COMPONENTS ENG, JI CASE CO. *Concurrent Pos:* Staff mem mat spec develop, Int Harvester, metal res, mat engr, Agr Group, mgr, metals res. *Mem:* Am Soc Metals; Am Soc Agr Engrs; Soc Automotive Engrs. *Res:* Materials specifications development; agricultural and components engineering. *Mailing Add:* Agr Equip & Component Eng J I Case Seven S 600 County Line Rd Hinsdale IL 60521

WALTER, HARRY, b Vienna, Austria, May 15, 30; m 56; c 3. BIOCHEMISTRY, CELL BIOLOGY. *Educ:* City Col New York, BS, 51; Ind Univ, MS, 53, PhD(biochem), 55. *Prof Exp:* Teaching asst chem, Ind Univ, 51-52, res asst biochem, 52-55, res assoc, 55-57; prin scientist, Vet Admin Hosp, Brooklyn, NY, 57-62; RES CHEMIST, LAB CHEM BIOL, VET AFFAIRS MED CTR, LONG BEACH, CALIF, 62-, RES CAREER SCIENTIST, 78- *Concurrent Pos:* Asst clin prof, Dept Biol Chem, Sch Med, Univ Calif, Los Angeles, 62-75; clin prof, Dept Physiol, Col Med, Univ Calif, Irvine, 75-79, prof in residence, Dept Physiol & Biophys, 79-89. *Mem:* Am Soc Biochem & Molecular Biol; Am Soc Cell Biol; Biophys Soc; Swed Biochem Soc. *Res:* Characterization of membrane surface properties by cell partitioning in two-polymer aqueous phase systems; factors in cell partitioning. *Mailing Add:* Lab Chem Biol Vet Affairs Med Ctr Long Beach CA 90822

WALTER, HARTMUT, b Stettin, Ger, July 13, 40; m 69. BIOGEOGRAPHY, ORNITHOLOGY. *Educ:* Univ Bonn, Dr rer nat(bird ecol), 67. *Prof Exp:* Harkness fel geog, Univ Calif, Berkeley, 67-68 & Univ Chicago, 68; assoc regional expert ecol & conserv for Africa, UNESCO Field Sci Off, Nairobi, Kenya, 70-72; actg asst prof, 72-73, asst prof, 73-74, assoc prof biogeog, 74-80, PROF GEOG, UNIV CALIF, LOS ANGELES, 80- *Honors & Awards:* Hoerlein Prize, Ger Biol Asn, 60. *Mem:* AAAS; Am Ornith Union; Asn Am Geogr; Cooper Ornith Soc. *Res:* Island biogeography; evolutionary ecology; raptor ecology; wildlife conservation; design and management of nature reserves and national parks; Mediterranean and African environments. *Mailing Add:* Dept Geog 1255 Univ Calif 405 Hilgard Ave Los Angeles CA 90024

WALTER, HENRY ALEXANDER, b Muehlhausen, Ger, Jan 8, 12; nat US; m 39; c 4. ORGANIC CHEMISTRY. *Educ:* Univ Heidelberg, dipl, 39. *Prof Exp:* Res chemist, Plaskon Co, 39-42; asst prof chem, Univ Mo, 42-44; res specialist, Monsanto Chem Co, 44-62; sr scientist, Plastic Coating Corp, Scott Paper Co, 62-71, consult, Scott Graphics, Inc, 71-76; CONSULT CHEMIST, CATAUMET, MASS, 76- *Concurrent Pos:* Assoc mem, Woodshole Oceanog Inst. *Mem:* AAAS; Am Chem Soc; NY Acad Sci. *Res:* Polymer chemistry; technical information services; patent liaison. *Mailing Add:* PO Box 86 Cataumet MA 02534

WALTER, HENRY CLEMENT, b Boston, Mass, Sept 12, 19; m 54; c 6. ORGANIC CHEMISTRY. *Educ:* Mass Inst Technol, SB, 41, PhD(org chem), 46. *Prof Exp:* Asst, Mass Inst Technol, 42-43 & 44-45; RES CHEMIST, EXP STA, E I DU PONT DE NEMOURS & CO, INC, 46- *Mem:* Am Chem Soc. *Res:* Elastomers; adhesives. *Mailing Add:* 310 Hampton Rd Sharpley Wilmington DE 19803-2420

WALTER, JOHN FITLER, b Philadelphia, Pa, Mar 19, 43; m 68; c 3. APPLIED PHYSICS, ELECTRO-OPTICS. *Educ:* Drexel Univ, BSEE, 66, MS, 68, PhD(physics), 70. *Prof Exp:* PHYSICIST, APPL PHYSICS LAB, JOHNS HOPKINS UNIV, 70- *Mem:* Am Defense Prep Asn; Am Inst Aeronaut & Astronaut. *Res:* Electro-optics applications to missle navigation and control; laser physics with applications to missle guidance. *Mailing Add:* Appl Physics Lab Johns Hopkins Rd Laurel MD 20723-6099

WALTER, JOHN HARRIS, b Los Angeles, Calif, Dec 14, 27; m 55; c 3. ALGEBRA. *Educ:* Calif Inst Technol, BS, 51; Univ Mich, MS, 53, PhD, 54. *Prof Exp:* From instr to asst prof math, Univ Wash, 54-61; assoc prof, 61-66, PROF MATH, UNIV ILL, URBANA, 66- *Concurrent Pos:* NSF fel, 57-58; vis asst prof, Univ Chicago, 60-61, vis assoc prof, 65-66; res assoc, Harvard Univ, 67-68 & Cambridge Univ, 72-73. *Mem:* Am Math Soc. *Res:* Finite groups; classical groups; representation theory. *Mailing Add:* Dept Math St Altgeld Hall Univ Ill 1409 W Green St Urbana IL 61801

WALTER, JOSEPH DAVID, b Merchantville, NJ, July 6, 39; m 62; c 3. ENGINEERING MECHANICS, MECHANICAL ENGINEERING. *Educ:* Va Polytech Inst, BS, 62, MS, 64, PhD(eng mech), 66; Univ Akron, MBA, 85. *Prof Exp:* Asst, Va Polytech Inst, 65-66; res physicist, Cent Res Labs, Firestone Tire & Rubber Co, 66-69, mgr physics & math res, 69-74, asst dir, 74-89; dir res, 89-90, DIR RES & ENG, BRIDGESTONE/FIRESTONE INC, 90- *Concurrent Pos:* Adj prof, Dept Mech Eng, Univ Akron, 75- *Mem:* Am Chem Soc; Accreditation Bd Eng & Technol; Am Soc Mech Engrs; Soc Automotive Engrs. *Res:* Composite materials; polymer physics; stress analysis; tire mechanics. *Mailing Add:* 343 Barnstable Rd Akron OH 44313

WALTER, JOSEPH L, b Braddock, Pa, Jan 23, 30. INORGANIC CHEMISTRY. *Educ:* Duquesne Univ, BS, 51; Univ Pittsburgh, PhD(chem), 55. *Prof Exp:* ASSOC PROF INORG CHEM, UNIV NOTRE DAME, 60- *Concurrent Pos:* NIH fels, 62-70; AEC fel, 63-67. *Mem:* Am Chem Soc; Soc Appl Spectros; Am Asn Med Col. *Res:* Normal coordinate analysis of inorganic coordination compounds using the Urey-Bradley Force Field calculations and the thermodynamic studies of metal chelate formation. *Mailing Add:* Dept Chem Univ Notre Dame Notre Dame IN 46556

WALTER, LOUIS S, b New York, NY, Aug 11, 33; m 57; c 2. GEOCHEMISTRY. *Educ:* City Col New York, BS, 54; Univ Tenn, MS, 55; Pa State Univ, PhD(geochem), 60. *Prof Exp:* Res fel geochem, Pa State Univ, 60-62; res assoc, Nat Acad Sci, 62-63; geochemist, 63-73, asst div chief, Earth Sci Div, 73-74, CHIEF EARTH SURV APPLN DIV, GODDARD SPACE FLIGHT CTR, NASA, 74- *Mem:* Am Geophys Union; Geochem Soc; Mineral Soc Am; Meteoritical Soc; Sigma Xi. *Res:* Experimental petrology and mineralogy; crystal chemistry; phase equilibria; petrography; electron microprobe analyses; planetology; theoretical petrology; application of remote sensing to agriculture, geology; use of geophysical measurements from space for tectonic studies and oil and mineral exploration. *Mailing Add:* 2111 Appletree Lane Silver Spring MD 20904

WALTER, MARTIN EDWARD, b Lone Pine, Calif, Jan 26, 45; m 67; c 1. DUALITY BETWEEN GEOMETRY & ALGEBRA. *Educ:* Univ Redlands, Calif, BS, 66; Univ Calif, Irvine, MA, 68, PhD(math), 71. *Prof Exp:* Fel math, Univ Calif, Los Angeles, 70-71; res assoc, Queens Univ, Kingston, Can, 71-73; from asst prof to assoc prof, 73-84, PROF MATH, UNIV COLO, BOULDER, 84- *Concurrent Pos:* Prin investr, NSF, 73-85; fel, Univ Pa, 77; fel, Univ Calif, Berkeley, 78; vis prof, math Univ Trondheim, Norway, 82; res fel math sci, Res Inst, Berkeley, 84; NSF fel; Woodrow Wilson fel; Alfred P Sloan fel. *Mem:* Am Math Soc; Math Asn Am; AAAS. *Res:* Duality between geometry and algebra with application to physics and other branches of mathematics. *Mailing Add:* Campus Box 426 Univ Colo Boulder CO 80309

WALTER, PAUL HERMANN LAWRENCE, b Jersey City, NJ, Sept 22, 34; m 56; c 2. INORGANIC CHEMISTRY. *Educ:* Mass Inst Technol, SB, 56; Univ Kans, PhD(inorg chem), 60. *Prof Exp:* Res chemist, Cent Res Dept, E I du Pont de Nemours & Co, 60-66, col rels rep, Employee Rels Dept, 66-67; from asst prof to assoc prof, 67-78, chmn dept, 75-85, PROF CHEM, SKIDMORE COL, 78- *Concurrent Pos:* Guest, Univ Stuttgart, 64-65; mem, bd dirs, Am Chem Soc, 91-93. *Mem:* AAAS; Am Chem Soc; fel Chem Inst Can; fel Am Inst Chemists; Am Asn Univ Profs (pres, 84-86). *Res:* Solid state inorganic chemistry; chemical education; rhenium chemistry; inorganic analytical chemistry; environmental chemistry. *Mailing Add:* Dept Chem Skidmore Col Saratoga Springs NY 12866

WALTER, REGINALD HENRY, b St John's, Antigua, Feb 13, 33; m 65. FOOD CHEMISTRY. *Educ:* Tuskegee Inst, BS, 60, MS, 63; Univ Mass, PhD(food sci), 67. *Prof Exp:* Asst food technologist, Univ PR, 67-68; sr chemist, Int Flavors & Fragrances, Inc, 68-71; asst prof, 71-77, ASSOC PROF FOOD SCI, CORNELL UNIV, 77- *Mem:* Am Chem Soc; Inst Food Technologists. *Res:* Food waste; carbohydrate polymers. *Mailing Add:* Dept Foods Sci Cornell Univ NY Agr Sta PO Box 462 Geneva NY 14456-9311

WALTER, RICHARD D, b Alameda, Calif, Aug 16, 21; m 47. MEDICINE. *Educ:* St Louis Univ, MD, 46; Am Bd Psychiat & Neurol, cert psychiat, 55, cert neurol, 60. *Prof Exp:* From instr to assoc prof, 55-70, PROF NEUROL & CHMN DEPT, MED CTR, UNIV CALIF, LOS ANGELES, 70- *Concurrent Pos:* Res fel psychiat, Langley Porter Clin, Univ Calif, Los Angeles, 53-55; dir, Reed Neurol Res Ctr. *Mem:* Am Acad Neurol; fel Am Electronencephalogram; Am Psychiat Asn; Am Neural Asn. *Res:* Clinical neurophysiology and electroencephalography as it relates to the convulsive disorders. *Mailing Add:* Box 27A02 Los Angeles CA 90019

WALTER, RICHARD L, b Chicago, Ill, Nov 1, 33; m 58; c 3. NUCLEAR PHYSICS. *Educ:* St Procopius Col, BS, 55; Univ Notre Dame, PhD(physics), 60. *Prof Exp:* Res assoc nuclear physics, Univ Wis, 59-61, instr physics, 61-62; from asst prof to assoc prof, 62-74, PROF PHYSICS, DUKE UNIV, 74- *Concurrent Pos:* Vis prof, Max Planck Inst Nuclear Physics, Heidelberg, Ger, 70-71; Fulbright res fel, 70-71; vis scientist, Los Alamos Sci Lab, 75. *Mem:* Am Phys Soc; Sigma Xi. *Res:* Neutron physics; polarization of nucleons produced in reactions; scattering of polarized nucleons; low energy accelerator physics; studies involving trace metals in the environment. *Mailing Add:* Dept Physics Duke Univ Durham NC 27706

WALTER, RICHARD WEBB, JR, b West Chester, Pa, Oct 5, 44; m 67; c 2. MICROBIAL BIOCHEMISTRY, FERMENTATION. *Educ:* Pa State Univ, BS, 66; Mich State Univ, PhD(biochem), 72. *Prof Exp:* Fel biochem, Univ Colo Med Ctr, Denver, 72-74; sr res biochemist, Dow Chem Co, 74-80, res leader, 80-85, develop leader, 85-89, DEVELOP ASSOC, DOW CHEM CO, 89- *Mem:* Am Soc Microbiol; Am Chem Soc; Am Soc Testing & Mats; Soc Indust Microbiol. *Res:* Biochemical transformation and synthesis of sterospecific molecules that are of interest to both the chemical and pharmaceutical industries and are difficult to synthesize by normal chemical means; fermentation process design and optimization; structure activity, relationship for antimicrobials; development of new industrial antimicrobials; mechanism of action studies. *Mailing Add:* Larkin Lab Dow Chem Co Midland MI 48674

WALTER, ROBERT IRVING, b Johnstown, Pa, Mar 12, 20. PHYSICAL ORGANIC CHEMISTRY. *Educ:* Swarthmore Col, AB, 41; Johns Hopkins Univ, MA, 42; Univ Chicago, PhD(chem), 49. *Prof Exp:* Asst, Swarthmore Col, 40-41 & Johns Hopkins Univ, 41-42; res chemist, Wyeth, Inc, 42-44; instr chem, Univ Colo, 49-51; from res asst prof to res assoc prof, Rutgers Univ, 51-53; instr chem, Univ Conn, 53-55; assoc physicist, Brookhaven Nat Lab, 55-56; from asst prof to prof chem, Haverford Col, 56-68; prof, 68-90, EMER PROF CHEM, UNIV ILL, CHICAGO, 90- *Concurrent Pos:* NSF fac fel, 60-61; mem, Adv Coun Col Chem, 66-70, sr staff assoc, 67; vis prof, Stanford Univ, 67; acad guest, Inst Phys Chem, Univ Zurich, 75-76; US Nat Acad Sci exchange vis, Romania, 82 & 88. *Mem:* Fel AAAS; Am Chem Soc; Sigma Xi. *Res:* Equilibria in porphyrin systems; preparation and properties of stable organic free radicals; mechanisms in heterogeneous catalysis. *Mailing Add:* Dept Chem Univ Ill Box 4348 Chicago IL 60680

WALTER, ROBERT JOHN, b Cleveland, Ohio, Nov 15, 50. CELL MOTILITY, TUMOR IMMUNOLOGY. *Educ:* Case Western Reserve Univ, PhD(anat & cell biol), 78. *Prof Exp:* Asst prof cell biol, Univ Ill, 80-87; ADJ ASSOC PROF ANAT, LOYOLA UNIV, 87-; SR SCIENTIFIC OFF, HEKTOEN INST, 88- *Mem:* Am Soc Cell Biol; AAAS; Soc Leukocyte Biol. *Res:* Cell motility research with specific interests in leukocyte chemotaxis receptors, the role of cytoskeleton in cell locomotion, and abnormal chemotaxis as in Kartagener's syndrome and in tumor patients; hormonal effects on the immune system and tumors. *Mailing Add:* Div Surg Res Hektoen Inst 625 S Wood St Chicago IL 60612

WALTER, RONALD BRUCE, b South Bend, Ind, July 15, 57; m 87. NUCLEIC ACID CHEMISTRY, MOLECULAR GENETIC. *Educ:* Fla State Univ, BS, 79, MS, 81, PhD(molecular genetic), 85. *Prof Exp:* Res assoc, Nat Res Coun, Environ Protection Agency, Environ Res Lab, 85-87; fel, syst cancer ctr, 87-88, res assoc, Univ Tex, 88; ASST PROF, GENETIC & MOLECULAR BIOL, SOUTHWEST TEX STATE UNIV, 88- *Concurrent Pos:* Adj res assoc, Cancer Ctr Sci Park Res Div, Univ Tex, 88-; nat res coun res assoc award, Nat. *Honors & Awards:* Arnold Ravin Award, 84. *Mem:* Am Soc Microbiol. *Res:* Molecular mechanisms of DNA repair in prokaryotes particularly with the bacterium Haemophilus influenzae; evolution of DNA repair enzymes from lower vertabrates to mammals by genetic mapping of homologous genes in fish. *Mailing Add:* Dept Biol Sw Tex State Univ San Marcos TX 78666-4616

WALTER, THOMAS JAMES, b Dodgeville, Wis, Aug 20, 39; m 65; c 3. ORGANIC CHEMISTRY. *Educ:* Univ Wis-Madison, BS, 61; Cornell Univ, MST, 69; Univ Ga, PhD(chem), 73. *Prof Exp:* Pub sch teacher, Wis, 65-68; res chemist, 73-78, sr res chemist, 78-80, supvr, 80-82, BUS RES, ETHYL CORP, 82- *Mem:* Am Chem Soc; Am Asn Crystal Growth; Chem Mkt Res Asn. *Res:* Paracyclophanes, synthesis of heterocycles; synthesis of drugs and drug intermediates, chemical intermediates, metal organic chemical vapor deposition. *Mailing Add:* Ethyl Corp 451 Florida St Baton Rouge LA 70801

WALTER, TREVOR JOHN, b West Bromwich, Eng, June 9, 44; US citizen. BIOCHEMISTRY. *Educ:* Univ Hull, Eng, BS, 65; Univ Manchester, Eng, PhD(biochem), 68. *Prof Exp:* Res assoc biochem, Dept Bot, Univ Md, 68-69 & Dept Biochem, Univ Fla, 69-70; chemist, A Guinness Son & Co Ltd, 70-74 & Dept Biochem, Univ Fla, 74-76; res assoc, Dept Biochem, Univ Minn, 76-79; GROUP LEADER FOOD BIOCHEM, KRAFT RES & DEVELOP, 79- *Mem:* Sigma Xi. *Res:* Food chemistry. *Mailing Add:* Kraft Res & Develop 801 Waukegan Rd Glenview IL 60025

WALTER, WILBERT GEORGE, b Lingle, Wyo, Nov 16, 33; m 55; c 2. NATURAL PRODUCTS, MEDICINAL CHEMISTRY. *Educ:* Univ Colo, BA, 55, BS & MS, 58; Univ Conn, PhD(pharmaceut chem), 62. *Prof Exp:* Asst, Univ Colo, 55-58; asst, Univ Conn, 58-61, spec res technologist, 60; res assoc pharmacog & asst prof pharmaceut chem, Sch Pharm, Univ Tenn, 61-63; assoc prof, Sch Pharm, Univ Miss, 63-68; chmn dept, 68-77, PROF MED CHEM, COL PHARM, MED UNIV SC, 68- *Mem:* Am Pharmaceut Asn; Am Soc Pharmacog; Am Chem Soc; Soc Econ Bot. *Res:* Organic medicinal chemistry; natural product chemistry. *Mailing Add:* Dept Pharm Sci Med Univ SC 171 Ashley Ave Charleston SC 29425

WALTER, WILLIAM ARNOLD, JR, b Pittsburgh, Pa, May 17, 22; m 46; c 2. EPIDEMIOLOGY. *Educ:* Ind Univ, AB, 43, MD, 45; Johns Hopkins Univ, MPH, 51. *Prof Exp:* Med officer, Ky State Bd Health, Louisville, 48-50; dir, Venereal Dis Control Div, Fla State Bd Health, Jacksonville, 51-53, assoc dir, Bur Prev Dis, 53-55; epidemiologist, Epidemiol Sect, Nat Cancer Inst, 55-57; med officer in-chg, Houston Pulmonary Cytol Proj, Univ Tex M D Anderson Hosp & Tumor Inst, 57-59; med officer in-chg uterine cancer cytol proj, Women's Med Col Pa, 59-60; res grants adminr, Grants & Training, 60-66, chief, Spec Prog Br, Extramural Activ, 66-77, dep dir, Div Cancer Res Resource & Centers, 77-80, actg dir & dep dir, Div Extramural Activ, Nat Cancer Inst, NIH, 80-84; PVT CONSULT RES ADMIN, 84- *Mem:* Am Pub Health Asn; AMA. *Res:* Chronic disease epidemiology with special interest in geographic pathology of leukemia. *Mailing Add:* 6310 Wilson Lane Bethesda MD 20817

WALTER, WILLIAM MOOD, JR, b Sumter, SC, Nov 20, 36; m 59; c 1. FOOD SCIENCE. *Educ:* The Citadel, BS, 58; Univ Ga, PhD(org chem), 63. *Prof Exp:* Res asst, Univ Ga, 60-63; asst prof, 65-70, assoc prof, 70-77 PROF FOOD SCI, NC STATE UNIV, 77- RES CHEMIST, AGR RES SERV, USDA, 65- *Mem:* Am Chem Soc; Inst Food Technologists; Sigma Xi. *Res:* Effect of processing and storage on organic constituents of food; emphasis on quality and nutritional value of processed foods. *Mailing Add:* 2128 Cowper Dr Raleigh NC 27608

WALTER, WILLIAM TRUMP, b Jamaica, NY, Dec 28, 31; m 60; c 4. ELECTROPHYSICS. *Educ:* Middlebury Col, AB, 53; Mass Inst Technol, PhD(physics), 62. *Prof Exp:* Res asst, Res Lab Electronics, Mass Inst Technol, 59-62; sr scientist, TRG, Inc, 62-67; res scientist, Polytech Inst NY, 67-79, res assoc prof, 79-84; CONSULT, AIL DIV, EATON CORP, 84- *Concurrent Pos:* Guest, Res Lab Electronics, Mass Inst Technol, 62-63; pres, Laser Consults, Inc, 68- *Mem:* Am Phys Soc; Optical Soc Am; NY Acad Sci; Soc Photo-Optical Instrumentation Engrs; Am Asn Physics Teachers; AAAS. *Res:* Wave-matter interactions; metal vapor lasers; laser research development and applications; microwaves; gas discharges; atomic physics; resonance phenomena in dilute gases, including optical pumping, orientation and nuclear magnetic resonance; optical and radiofrequency spectroscopy; infrared and electro-optics. *Mailing Add:* 344 W Hills Rd Huntington NY 11743

WALTERS, CARL JOHN, b Albuquerque, NMex, Sept 14, 44. SYSTEMS ECOLOGY. *Educ:* Humboldt State Col, BS, 65; Colo State Univ, MS, 67, PhD(fisheries), 69. *Prof Exp:* Res asst fisheries biol, Colo Coop Wildlife Unit, Colo State Univ, 66-67, NSF fel, 67-69, consult, 68-70; assoc prof, 69-82, PROF ZOOL & ANIMAL RESOURCE ECOL, UNIV BC, 82- *Concurrent Pos:* Can Depts Environ & Fisheries consult, 71-; res scholar, Int Inst Appl Systs Anal, Austria, 74-75, 82-83. *Res:* Dynamics of ecological communities; application of mathematical models and computer simulation techniques to problems in resource ecology; adaptive management of renewable resources. *Mailing Add:* Dept Zool Univ BC 2075 Westbrook Mall Vancouver BC V6T 1Z2 Can

WALTERS, CAROL PRICE, b Lansing, Mich, Oct 15, 41; m 88; c 1. PROTEIN CHEMISTRY. *Educ:* Albion Col, AB, 63; Univ Vt, PhD(physiol), 72. *Prof Exp:* From res assoc med to res assoc prof med, 72-85, RES ASSOC PROF PED, COL MED, UNIV VT, 85- *Concurrent Pos:* Dir, Vt Alpha-Fetoprotein Prenatal Screening Prog, 81-; coordr, Vt Newborn Screening Prog, 89- *Mem:* Sigma Xi; Int Soc Oncodevelop Biol & Med; Am Asn Cancer Res. *Res:* Investigation of the physicochemistry and metabolism of the carcino-embryonic protein alpha-fetoprotein and its molecular variants in sera of fetal and hepatoma-bearing rats and also in human maternal serum and amniotic fluid in the presence of fetal malformations. *Mailing Add:* Dept Pediat Univ Vt Med Alumni Bldg Burlington VT 05405

WALTERS, CHARLES PHILIP, b Kansas City, Mo, May 1, 15; m 36; c 3. ASTROGEOLOGY. *Educ:* Kans State Univ, BS, 36, MS, 38; Cornell Univ, PhD, 57. *Prof Exp:* Asst geologist, Continental Oil Co, 44-48; from asst prof to assoc prof, 45-72, PROF GEOL, KANS STATE UNIV, 72- *Concurrent Pos:* Ford Found fac fel, 51-52; mem comt exam natural sci test & geol subj matter test, Educ Testing Serv, 62-70. *Mem:* AAAS; Am Asn Petrol Geologists; Am Quaternary Asn; Meteoritical Soc; Sigma Xi. *Res:* Structural and tectonic geology; geophysics; planetology; environmental geology; deterioration of Kansas salt beds; tektites from cryptovolcanic eruptions; paleoclimatology; climate change. *Mailing Add:* 920 Vattier Manhattan KS 66502

WALTERS, CHARLES SEBASTIAN, b Detroit, Mich, Aug 18, 13; m 39; c 2. FORESTRY. *Educ:* Purdue Univ, BS, 38; Yale Univ, DFor, 57. *Prof Exp:* Asst, Tenn Valley Authority, 40 & Univ Ill, 40-41; proj forester, Timber Prod War Proj, Ill, 41-45; asst chief wood technol & utilization in forestry, Univ Ill, Urbana, 45-47, from asst prof to assoc prof, 47-57, prof, 57-79; consult, 79-81; RETIRED. *Concurrent Pos:* Consult, Indonesia, 71; adv, Nat Bur Standards, 71-78; mem standing comt hardboard, US Bur Standards, 74-78. *Mem:* AAAS; Sigma Xi; Forest Prod Res Soc; Soc Wood Sci & Technol; Am Soc Testing & Mat. *Res:* Technology of wood and its use; wood preservation. *Mailing Add:* 101 W Windsor Rd Univ Ill Urbana IL 61801

WALTERS, CRAIG THOMPSON, b Columbus, Ohio, July 23, 40; m 62; c 1. LASER EFFECTS, MATERIALS PROCESSING. *Educ:* Ohio State Univ, BS & MS, 63, PhD(physics), 71. *Prof Exp:* From res physicist to sr physicist, 63-71, assoc fel plasma physics, 71-73, sr researcher laser effects, 73-75, assoc sect mgr laser effects & electromagnetics, 75-87, RES LEADER, LASER EFFECTS CTR, BATTELLE COLUMBUS LABS, BATTELLE MEM INST, 87- *Concurrent Pos:* Mem adv panel laser-supported absorption waves, Defense Advan Proj Agency, 74; mem adv group sensor susceptibility, Forum Military Appln Directed Energy, 89-90. *Honors & Awards:* NASA Tech Brief Award, 69. *Mem:* Am Phys Soc; Am Inst Aeronaut & Astronaut. *Res:* Interaction of intense laser beams with materials; study of plasma produced in laser beam interaction with condensed matter; simulation of free-electron laser effects; study of effects of laser generated shocks; laser processing of materials. *Mailing Add:* Battelle Columbus Labs 505 King Ave Columbus OH 43201

WALTERS, CURLA SYBIL, b Jamaica, June 3, 29; c 1. IMMUNOLOGY, MICROBIOLOGY. *Educ:* Andrews Univ, BA, 61; Howard Univ, MSc, 64; Georgetown Univ, PhD(microbiol & immunol), 69. *Prof Exp:* From instr to asst prof immunol, Med Ctr, Univ Colo, Denver, 71-74; assoc prof dept med, Med Ctr, Howard Univ, 74-77; MEM STAFF, DEPT MED, HOWARD UNIV HOSP, 77- *Concurrent Pos:* Am Asn Univ Women fel, Karolinska Inst, Sweden, 69-70; NIH training grant, Med Ctr, Univ Colo, 71; mem study sect, NIH, 81-85 & Nat Kidney Found. *Mem:* Am Soc Microbiologists; Am Asn Immunol; Sigma Xi. *Res:* Basic and tumor immunology. *Mailing Add:* 8404 11th Ave Silver Springs MD 20903

WALTERS, DEBORAH K W, b Baltimore, Md, June 14, 51. COMPUTER VISION, VISUAL PSYCHOPHYSICS. *Educ:* Guilford Col, AB, 73; Univ Brimingham, Eng, MSc, 78, PhD(neurocommun), 80. *Prof Exp:* Physicist, med image processing, King's Col Hosp & Med Sch, Eng, 73-76; teaching fel exp neurol, Med Sch, Univ Birmingham, Eng, 79-81; vis asst prof visual psycho-physics, Dept Psychol, 81-83, ASSOC PROF COMPUT SCI & ADJ PROF PSYCHOL, STATE UNIV NY, BUFFALO, 83- *Mem:* Inst Elec & Electronics Engrs; Cognitive Sci Soc; Am Assoc Artificial Intel; Int Soc Optical Eng; Am Asn Comput Mach; Int Soc Neural Networks. *Res:* Computer vision; the early stages of visual processing; development of algorithms based on human psychophysics and neurophysiology; geometrical considerations; inferences about the regularities in the physical world; parallel computations. *Mailing Add:* Dept Comput Sci State Univ NY 226 Bell Hall Buffalo NY 14260

WALTERS, DOUGLAS BRUCE, b Brooklyn, NY, Apr 6, 42; m; c 1. CHEMICAL HEALTH & SAFETY, INDUSTRIAL HYGIENE. *Educ:* Long Island Univ, BS, 63, MS, 65; Univ Ga, PhD(chem), 71. *Prof Exp:* Food chemist, A&P, New York, 62-64; res chemist, Farbewerke Hoechst A G, Frankfurt, Ger, 65, US Environ Protection Agency, Athens, Ga, 69-71 & USDA, 71-77; tech progs mgr chem, NIEHS, 77-80, HEAD, LAB HEALTH & SAFETY, NAT TOXICOL PROG, NAT INST ENVIRON HEALTH SCI, NIH HEALTH HUMAN SERV, 80- *Honors & Awards:* Nat Award, Outstanding Contrib Sci & Technol Chem Health & Safety, Am Chem Soc, 89. *Mem:* Am Chem Soc; Am Indust Hyg Asn. *Res:* Chemical health and safety; human factors; ergonomics; laboratory health and safety; safety engineering; laboratory and equipment design; safety and health; health and safety aspects of toxicology testing; safe handling chemical carcinogens; waste disposal of hazardous chemicals; industrial hygiene. *Mailing Add:* Nat Inst Environ Health Sci PO Box 12233 Research Triangle Park NC 27709

WALTERS, EDWARD ALBERT, b Whitefish, Mont, Jan 2, 40; m 64; c 3. CHEMICAL DYNAMICS, QUANTUM CHEMISTRY. *Educ:* Pac Lutheran Univ, BS, 62; Univ Minn, Minneapolis, PhD(org chem), 66. *Prof Exp:* Res assoc chem, Cornell Univ, 66-68; from asst prof to assoc prof, 68-74, PROF CHEM, UNIV NMEX, 85- *Concurrent Pos:* Res Corp grant, Univ NMex, 69-70, NSF grant, 69-72; sabbatical, Univ Kent, Canterbury, 75, MPI für Strömungsforshung, Göttingen, 76, Brookhaven Nat Lab, 84. *Honors & Awards:* Lee Irving Smith Award, 66. *Mem:* AAAS; Am Vacuum Soc; Am Chem Soc. *Res:* Vacuum ultra violet spectroscopy; kinetic isotope effects; potential energy surfaces for reactive collisions; photoionization mass spectrometry of cluster molecules. *Mailing Add:* Dept Chem Univ NMex Albuquerque NM 87131

WALTERS, FRED HENRY, b Owen Sound, Ont, Aug 8, 47. ANALYTICAL CHEMISTRY. *Educ:* Univ Waterloo, BSc, 71; Univ Mass, PhD(chem), 75. *Prof Exp:* Res asst, Toronto Gen Hosp, Kitchener Waterloo Hosp & Ashland Oil Can, 66-71; teaching asst chem, Univ Mass, 71-75; res assoc chem, Univ Windsor, 75-76; asst prof, Quinnipiac Col, Hamden, Conn, 76-79; ASSOC PROF, UNIV SOUTHWESTERN LA, 79- *Mem:* Am Statist Asn; Am Chem Soc; Sigma Xi. *Res:* Chemometrics and statistics; high pressure liquid chromatography. *Mailing Add:* Dept Chem No 44370 Univ Southwestern La Lafayette LA 70504

WALTERS, GEOFFREY KING, b Baton Rouge, La, Aug 23, 31; m 54; c 3. ATOMIC PHYSICS. *Educ:* Rice Univ, BA, 53; Duke Univ, PhD(physics), 56. *Prof Exp:* NSF fel, Duke, 56-57; br mgr & physicist, Tex Instruments, Inc, 57-62, corp res assoc, 62-63; prof physics, 63-64, actg dean sci & eng, 68-69 & 72-73, chmn dept physics, 73-77, dean nat sci, 80-87, PROF PHYSICS & SPACE SCI, RICE UNIV, 64- *Concurrent Pos:* Actg chief fire technol div, Nat Bur Standards, 71-72; Guggenheim Found fel, Stanford Univ, 77-78; vis prof, College de France, 87. *Mem:* Fel AAAS; fel Am Phys Soc; Am Geophys Union. *Res:* Magnetic resonance; low temperature, solid state, atomic collisions and reactions, surface physics; optical pumping and dynamic nuclear orientation; solar-terrestrial relationships; radio-astronomy. *Mailing Add:* Dept Physics Rice Univ Box 1892 Houston TX 77251

WALTERS, HUBERT JACK, b Spring, Tex, Nov 2, 15; m 47; c 3. PLANT PATHOLOGY. *Educ:* NMex Col, BS, 41; Univ Ill, MS, 48; Univ Nebr, PhD(plant path), 51. *Prof Exp:* Asst prof plant path & asst plant pathologist, Univ Wyo, 50-54; from asst prof & asst plant pathologist to assoc prof & assoc plant pathologist, 54-62, PROF PLANT PATH & PLANT PATHOLOGIST, UNIV ARK, FAYETTEVILLE, 62-; EMER PROF, 88- *Mem:* Am Phytopath Soc. *Res:* Virus diseases; beetle transmission of plant viruses; diseases of soybeans. *Mailing Add:* Rte 6 PO Box 466 Fayetteville AR 72703

WALTERS, JACK HENRY, b Toronto, Ont, Apr 2, 25; m 49; c 3. OBSTETRICS, GYNECOLOGY. *Educ:* Univ Western Ont, BA, 46, MD, 51; FRCPS(C), 57; FRCOG, 67. *Prof Exp:* Chief, Dept Obstet & Gynec & dir cytol, St Joseph's Hosp, London, Ont, 58-73; prof obstet & gynec, Univ Western Ont, 66-73; prof obstet & gynec & chmn dept, Med Col Ohio, 73-78; prof & chmn, Dept Obstet & Gynec, Univ Ottawa, 78-; RETIRED. *Concurrent Pos:* Can Cancer Soc McEachern traveling fel, 56-57; chief obstet & gynec, Ottawa Gen Hosp, 78- *Mem:* Can Med Asn; Soc Obstet & Gynec Can; Am Soc Cytol; fel Am Col Obstet & Gynec. *Res:* Gynecological, particularly hormonal cytology; screening programs; perinatal mortality, particularly statistical research in computer programs; manpower studies; obstetrics-gynecology health care; delivery systems. *Mailing Add:* c/o David Job 1471 Corley Dr S London ON N6G 2K5 Can

WALTERS, JAMES CARTER, b Zeeland, Mich, June 27, 48; m 71; c 2. GLACIAL & PERIGLACIAL GEOMORPHOLOGY. *Educ:* Grand Valley State Col, Mich, BA, 70; Rutgers Univ, MPhil, 73, PhD(geol), 75. *Prof Exp:* From asst prof to assoc prof, 75-89, PROF GEOL, UNIV NORTHERN IOWA, 89- *Concurrent Pos:* Res assoc, Ctr Northern Studies, 75-87; co-prin investr, Nat Park Serv, Interior & Western Alaska, 78-79; vis prof, Middleburg Col, 80, Univ Vt, 81-82; consult, Northern Tech Serv, Anchorage, Alaska, 83, Cold Regions Res & Eng Lab, US Army Corps Engrs, 89. *Mem:* Fel Geol Soc Am; Am Quaternary Asn; Nat Asn Geol Teachers; Sigma Xi. *Res:* Quaternary geology; glacial and periglacial geomorphology; arctic studies. *Mailing Add:* Dept Earth Sci Univ Northern Iowa Cedar Falls IA 50614-0335

WALTERS, JAMES VERNON, b Dublin, Ga, May 13, 33; m 55; c 2. CIVIL ENGINEERING, SANITARY ENGINEERING. *Educ:* Ga Inst Technol, BCE, 55, MS, 58; Univ Fla, PhD(sanit eng), 63. *Prof Exp:* Res asst, State Hwy Dept Ga, 55-56; jr asst & asst sanit engr, Atlanta Regional Off, USPHS, 56-59; from asst prof to assoc prof, 59-70, PROF CIVIL ENG, UNIV ALA, TUSCALOOSA, 70- *Concurrent Pos:* Ford Found eng resident, Southern Kraft Div, Int Paper Co, 66-67; mem, Univ Ala Environ Inst for Waste Mgt Studies, 84- *Mem:* Am Chem Soc; Am Soc Civil Engrs; Am Water Works Asn; Tech Asn Pulp & Paper Indust; Water Pollution Control Asn; Air Pollution Control Asn; Sigma Xi. *Res:* Water supply and sewerage; water, sewage and industrial waste treatment; hydrology; sanitary nematology; hazardous waste management. *Mailing Add:* PO Box AB University AL 35486

WALTERS, JOHN PHILIP, b Elgin, Ill, July 4, 38; m 61; c 2. ANALYTICAL CHEMISTRY, SPECTROSCOPY. *Educ:* Purdue Univ, BS, 60; Univ Ill, Urbana, PhD(chem), 64. *Prof Exp:* Res assoc spectros, Univ Ill, Urbana, 64-65; asst prof, 65-72, PROF ANALYTICAL CHEM, UNIV WIS-MADISON, 72- *Honors & Awards:* Am Chem Soc Award Chem

Instrumentation; Meggers Award; Lester W Strock Award. *Mem:* Soc Appl Spectros; Am Soc Testing & Mat; Am Chem Soc; fel AAAS. *Res:* Time-resolved emission spectroscopy; mechanisms of spectroscopic discharges; spectrochemical methods and instrumentation; computers and lab information management. *Mailing Add:* 32910 Ensley Northfield MN 55057-5246

WALTERS, JOHN PHILIP, b Manhattan, Kans, Sept 26, 41; m 63; c 4. PHYSICAL CHEMISTRY. *Educ:* Kans State Univ, BS, 63; Iowa State Univ, PhD(phys chem), 68. *Prof Exp:* Res chemist, Phillips Petrol Co, Okla, 68-70; res chemist, Phillips Petrol Co, 76-78, sr res chemist, 78-86; res chemist, 70-76, SR RES CHEMIST, PHILLIPS FIBERS CORP, GREENVILLE, SC, 86- *Mem:* Am Chem Soc; Sigma Xi. *Res:* Polymer stabilization. *Mailing Add:* 307 Bloomfield Ct Greer SC 29650-3804

WALTERS, JUDITH R, NEUROPHARMACOLOGY, NEUROPHYSIOLOGY. *Educ:* Yale Univ, PhD(pharmacol), 72. *Prof Exp:* SECT CHIEF, NAT INST COMMUN DIS & STROKE, NIH, 81- *Res:* Basal ganglia function. *Mailing Add:* Physiol & Neuropharmacol Sec Exp Therapeut Br Nat Inst Commun Dis & Stroke NIH Bldg 10 Rm 5C106 Bethesda MD 20892

WALTERS, LEE RUDYARD, b New York, NY, Jan 20, 28; m 50; c 2. ORGANIC CHEMISTRY. *Educ:* Bucknell Univ, BS, 54; Univ Kans, PhD(chem), 58. *Prof Exp:* Asst org chem, Univ Kans, 55-58; res chemist, Atlas Powder Co, Del, 58-59; ASST PROF CHEM, LAFAYETTE COL, 59- *Mem:* Am Chem Soc. *Res:* Nitrogen heterocyclic and organometallic compounds; chemistry of natural products. *Mailing Add:* Dept Chem Lafayette Col Easton PA 18042

WALTERS, LEON C, b Butte, Mont, Jan 6, 40; m 61; c 3. MATERIALS SCIENCE, METALLURGICAL ENGINEERING. *Educ:* Purdue Univ, BS, 61, MS, 63, PhD(mat sci, metall eng), 66. *Prof Exp:* Staff mem, Sandia Labs, 66-69; assoc metall eng, 69-73, mgr reactor & mfg mat support sect & metall engr, Fuels & Mat Dept, 73-78, assoc dir, 78-85, ASSOC DIR INTEGRAL FAST REACTOR PROG, EBR-II PROJ, ARGONNE NAT LAB, IDAHO, 85- *Mem:* Am Inst Mining, Metall & Petrol Engrs; Am Soc Metals; Am Nuclear Soc. *Res:* Formation and diffusion of point defects in compounds; effects of irradiation on the mechanical properties of metals; fabrication and performance of fast reactor metallic fuels. *Mailing Add:* Argonne Nat Lab Fuels & Processes PO Box 2528 Idaho Falls ID 83401

WALTERS, LESTER JAMES, JR, b Tulsa, Okla, June 3, 40; m 67. GEOCHEMISTRY. *Educ:* Univ Tulsa, BS, 62; Mass Inst Technol, PhD(geochem), 67. *Prof Exp:* Res scientist, Marathon Oil Co, 67-69; tech assistance expert, Int Atomic Energy Agency, 69; from asst prof to prof geol, Bowling Green State Univ, 70-81; prin res scientist, 81-85, SR PRIN RES GEOLOGIST, ARCO OIL & GAS CO, 86- *Mem:* Am Chem Soc; Am Asn Petrol Geologists; Geochem Soc; Soc Explor Paleontologists & Mineralogists. *Res:* Geochemical cycle of iodine, bromine, and chlorine in sediments; neutron activation analysis; geobotany; heavy metal pollution in Lake Erie; exploration geochemistry. *Mailing Add:* 2804 Glencliff Dr Plano TX 75075

WALTERS, LOWELL EUGENE, b Freedom, Okla, Jan 13, 19; m 42; c 2. ANIMAL SCIENCE. *Educ:* Okla State Univ, BS, 40; Univ Mass, MS, 42; Okla State Univ, PhD(animal nutrit), 53. *Prof Exp:* Instr animal husb, La State Univ, 42-44; asst prof, Univ Mass, 44-46; from asst prof to assoc prof, 46-57, PROF ANIMAL HUSB, OKLA STATE UNIV, 58- *Mem:* Am Meat Sci Asn. *Res:* Meats; beef quality; carcass composition of beef, pork and lamb; potassium 40 techniques in live animal and carcass evaluation; growth and performance in slaughter livestock; systems analysis methods as applied to efficient beef production. *Mailing Add:* 2128 W3 Stillwater OK 74074

WALTERS, MARK DAVID, b Wash, DC, July 8, 59; m 83; c 1. ELECTRONICS ENGINEERING. *Educ:* Univ NC, Chapel Hill, BA, 81; Univ Va, Charlottesville, MS, 86; NC State Univ, Raleigh, PhD(mat sci), 89. *Prof Exp:* Teaching asst physics, Dept Physics, Univ Va, 83-84, res asst, Dept Mat Sci, 84-86; MEM TECH STAFF, CTR MICROELECTRONICS, MCNC, 86- *Mem:* Mat Res Soc; Minerals, Metals and Mat Soc; Sigma Xi; Electrochem Soc. *Res:* Study of radiation-induced defects in the gate insulators of field effect transistors; developed models of defect distributions and generation mechanisms, and performed experimental studies in association with these models. *Mailing Add:* MCNC PO Box 12889 Research Triangle Park NC 27709-2889

WALTERS, MARTHA I, b Logan, Ohio, Oct 21, 25. CLINICAL CHEMISTRY. *Educ:* Ohio State Univ, BSc, 47, MSc, 59, PhD(clin path), 70. *Prof Exp:* Supvr biochem labs, Ohio State Univ Hosp, 49-59, instr clin path, Univ, 55-59; sr res asst, Bellevue Med Ctr, NY Univ, 59-62; dir biochem, St Francis Hosp, Bronx, NY, 63-67 & Ohio Valley Hosp, Steubenville, 71-73; supvr endocrinol & toxicol, Consol Biomed Labs, 73-80; DIR THERAPEUT DRUG MONITORING, OHIO STATE UNIV HOSP, 80- *Mem:* Am Chem Soc; Am Asn Clin Chemists; Sigma Xi. *Res:* Relationship between metabolism and function in erythrocytes and leukocytes. *Mailing Add:* Ohio State Univ Col Med 952 Chelsea Ave Columbus OH 43209

WALTERS, RANDALL KEITH, b Cheyenne, Wyo, Aug 15, 43; m 63; c 2. MATHEMATICS, COMPUTER SCIENCE. *Educ:* Univ Wyo, BS, 65; Univ Tex, El Paso, MS, 69; NMex State Univ, PhD(math), 75. *Prof Exp:* Res mathematician, White Sands Missle Range, 65-70; math economist, El Paso Natural Gas Co, 75-77; prof math, Col Redwoods, 77-84; asst prof math & comput sci, Univ Nebr, Omaha, 84-88; SR SCIENTIST, LOGICON INC, 88- *Mem:* Asn Comput Mach; Am Math Soc; Math Asn Am; SIAM; Inst Elec & Electronics Engrs Computer Soc. *Res:* Torsion-free abelian groups; alqebraic automata theory; algebraic coding theory; computational geometry. *Mailing Add:* 4814 Webster St Omaha NE 68182

WALTERS, RICHARD FRANCIS, b Teleajen, Romania, Aug 30, 30; US citizen; m 52; c 2. INFORMATION SCIENCE, MEDICAL EDUCATION. *Educ:* Williams Col, BA, 52, Univ Wyo, MA, 53; Univ Bordeaux, dipl natural sci, 55; Stanford Univ, PhD(geol), 57. *Prof Exp:* Res geologist, Humble Oil & Ref Co, 56-67; from asst prof to prof community health, 67-83, prof elec & comput eng, 80-83, PROF & CHMN DIV COMPUT SCI, UNIV CALIF, DAVIS, 83- *Concurrent Pos:* Fel, Col Med Informatics. *Mem:* Asn Am Med Sys & Informatics; Biomed Eng Soc; Asn Comput Mach; sr mem Inst Elec & Electronics Engrs. *Res:* High level language support of networked distributed data bases on heterogeneous systems; machine-independent implementation of high-level languages; computer support of medical records. *Mailing Add:* Div Comput Sci Univ Calif Davis CA 95616

WALTERS, ROBERT F, b Rochester, NY, July 15, 14; m 48; c 4. HYDROLOGY & WATER RESOURCES. *Educ:* Univ Rochester, BSc, 36, MSc, 38; Johns Hopkins Univ, PhD(geol), 46. *Prof Exp:* Res geol, Gulf Oil Corp, Tulsa, 40-48; geologist, Heathman Drilling Co, 48-51; PRES, WALTERS DRILLING CO, INC, 51- *Concurrent Pos:* Assoc ed, Am Asn Petrol Geologists, 65; consult drilling & coring, Nuclear Div Oak Ridge Nat Lab, Union Carbide Corp, 70-79 & Off Nuclear Waste Isolation, Battelle Mem Inst, 78-82; consult salt & subsidence ind, Vulcan Mat Co, 76- *Honors & Awards:* President's Award, Am Asn Petrol Geologists, 46. *Mem:* Sigma Xi; Am Geophys Union; hon mem Am Asn Petrol Geologists; Geol Soc Am; Soc Econ Geologists. *Res:* Movement of fluids through rocks; migration of oil and gas; dissolution of carbonate rocks(karst, paleo karst) and salt beds; paleohydrology of Cambro-Ordovician and Precambrian rocks underlying Kansas. *Mailing Add:* 100 S Main St Suite 420 Wichita KS 67202

WALTERS, ROLAND DICK, b Portland, Ore, Jan 30, 20; m 45; c 3. ORTHODONTICS. *Educ:* Walla Walla Col, BA, 47; Ore State, MS, 49; Loma Linda Univ, DDS, 57, MS, 67. *Prof Exp:* Instr biol, Walla Walla Col, 47-48; asst prof biol, 49-53, instr oral med, 65-67, asst prof anat, 68-69, PROF ORTHOD & CHMN DEPT, SCH DENT, LOMA LINDA UNIV, 69- *Concurrent Pos:* Ed, SDA Dentist, 63; mem, Bd Trustees, Loma Linda Univ, 63-65; vchmn, Bd Dir, Kern Acad, 64; chmn, Tri-County Dent Soc, 71-73; consult, Jerry L Pettis Mem Vet Hosp, 78- *Mem:* Found Orthod Res; Am Dent Asn; Sigma Xi. *Res:* Facial changes and bone growth; transplantation of human teeth; retraction of the maxillae using orthopedic force; mandibular dental arch; surgical orthodontic expansion; primates; techniques to expand the jaws to correct malocclusion of jaws and/or teeth; method to surgically and mechanically expand the lower jaw. *Mailing Add:* 1715 Highland Ave Redlands CA 92374

WALTERS, RONALD ARLEN, b Greeley, Colo, Apr 25, 40; m 69; c 2. BIOCHEMISTRY, RADIOBIOLOGY. *Educ:* Colo State Univ, BS, 62, MS, 64, PhD(radiation biol), 67. *Prof Exp:* Engr radiation biol, Gen Elec Co, 62-63; dep group leader, Los Alamos Nat Lab, Univ Calif, 80-81, group leader, 82-84, from asst to assoc dir chem, Earth & Life Sci, 85-86, asst to assoc dir res, 86-89, STAFF MEM BIOCHEM, LOS ALAMOS NAT LAB, UNIV CALIF, 67-, PROG DIR, BIOL & ENVIRON RES, 89- *Mem:* AAAS; Am Chem Soc; Am Soc Biochem & Molecular Biol; Radiation Res Soc; Am Soc Cell Biol. *Res:* Cellular biology; gene structure and function; trace metal metabolism; radiation biology. *Mailing Add:* ADR MS-A114 Los Alamos Nat Lab Los Alamos NM 87545

WALTERS, STEPHEN MILO, b Caledonia, Minn, Nov 7, 40; m 73; c 1. HIGH PERFORMANCE LIQUID CHROMATOGRAPHY. *Educ:* Univ Wis, La Crosse, BS, 62. *Prof Exp:* Chemist, Cincinnati, 62-68, res coordr, Detroit, 68-81, DIR PESTICIDE & INDUST CHEM RES CTR, US FOOD & DRUG ADMIN, DETROIT, 81- *Mem:* Fel Asn Anal Chemists; Am Chem Soc; Asn Off Anal Chemists; fel Am Inst Chemists. *Res:* Chemical analysis of foods and drugs for purity, potency and safety; method development for residues of pesticides and related chemicals in food and feeds; gas chromatography; chromatographic separations. *Mailing Add:* US Food & Drug Admin 1560 E Jefferson Ave Detroit MI 48207-3114

WALTERS, THOMAS RICHARD, b Milwaukee, Wis, May 9, 29; m 84. PEDIATRICS, HEMATOLOGY. *Educ:* Marquette Univ, MD, 54. *Prof Exp:* Instr pediat, Stanford Univ, 61-62; asst prof, Univ Kans Med Ctr, Kansas City, 62-67; assoc prof, Univ Tenn, Memphis, 67-71; assoc prof, NJ Med Sch, 71-76, prof pediat & dir div hemat-oncol, 76-90; RETIRED. *Concurrent Pos:* Assoc mem hemat, St Jude Children's Res Hosp, Memphis, 67-71. *Mem:* Am Soc Hemat; Am Acad Pediat. *Res:* Mechanisms of tumor growth. *Mailing Add:* Dept Pediat NJ Med Sch Newark NJ 07103-2757

WALTERS, VIRGINIA F, b New York, NY, May 26, 25; m 45; c 2. PHYSICS. *Educ:* Smith Col, AB, 47; Western Reserve Univ, MA, 58, PhD(physics), 65. *Prof Exp:* Physicist, DeMornay Budd, Inc, 47-49; res asst microwave components, Radiation Lab, Columbia Univ, 49-50; lectr elem physics, Adelphi Col, 50-51; physicist, Servo Corp Am, 51; asst physics, Western Reserve Univ, 54-65, fel, 65-66; asst prof, Cleveland State Univ, 66-67; res physicist, Carnegie-Mellon Univ, 67-68; asst prof phys sci, Point Park Col, 68-69; teacher physics & math, Western Reserve Acad, 69-74; lectr physics, Cleveland State Univ, 75-80; RETIRED. *Concurrent Pos:* Vis asst prof, Dept Physics, Cleveland State Univ, 77-81. *Mem:* Am Phys Soc; Am Asn Physics Teachers. *Res:* Positron annihilation; nuclear instrumentation. *Mailing Add:* 105 S Beach Rd South Burlington VT 05403

WALTERS, WILLIAM BEN, b Highland, Kans, Apr 26, 38; m 62; c 2. NUCLEAR CHEMISTRY, PHYSICAL CHEMISTRY. *Educ:* Kans State Univ, BS, 60; Univ Ill, PhD(chem), 64. *Prof Exp:* Res assoc chem, Mass Inst Technol, 64-65, asst prof, 65-70; assoc prof, 70-77, PROF CHEM, UNIV MD, COLLEGE PARK, 77- *Concurrent Pos:* Vis prof physics, Katolieke Univ Leuven, Belg, 78; Guggenheim fel, Clarendon Lab, Oxford, 86-87. *Mem:* Am Chem Soc; Am Phys Soc. *Res:* Radioactive decay; nuclear spectroscopy; new isotopes and isomers; nuclear reactions; nuclear orientation; nuclear structure; neutron-capture gamma-ray spectroscopy; gamma-gamma angular correlations; in-beam gamma ray spectroscopy. *Mailing Add:* Dept Chem Univ Md College Park MD 20742

WALTERS, WILLIAM LE ROY, b Racine, Wis, Mar 30, 32; m 55; c 4. PHYSICS. *Educ:* Univ Wis, BS, 54, MS, 58, PhD(physics), 61. *Prof Exp:* From asst prof to prof physics, Univ Wis-Milwaukee, 61-68, assoc dean sci, Col Lett & Sci, 65-68, exec asst chancellor, 68-70, actg dean col appl sci & eng, 70, vchancellor, 71-81. *Concurrent Pos:* Exec comt coun for acad affairs, Nat Asn State Univs & Land-grant col, 76-79, chmn Coun, 78. *Mem:* AAAS; Am Asn Physics Teachers; Am Phys Soc. *Res:* Science education; science communications. *Mailing Add:* Univ Wis Milwaukee WI 53201

WALTHER, ADRIAAN, b The Hague, Holland, Apr 22, 34; m 60; c 2. OPTICS. *Educ:* Delft Univ Technol, PhD(physics), 59. *Prof Exp:* Mem res staff, Diffraction Ltd Inc, 60-72; PROF PHYSICS, WORCESTER POLYTECH INST, 72- *Concurrent Pos:* Am Optical vis prof, Worcester Polytech Inst, 68-69. *Mem:* Optical Soc Am; Neth Phys Soc. *Res:* Geometrical and physical optics. *Mailing Add:* 20 Whittier Dr Acton MA 01720

WALTHER, ALINA, b Rosenthal, USSR, Aug 15, 23; Can citizen. PLANT PHYSIOLOGY, PLANT ECOLOGY. *Educ:* Sir George Williams Univ, BCom, 53, BSc, 61; McGill Univ, MSc, 63; Univ Toronto, PhD(plant physiol), 68. *Prof Exp:* Spec lectr, 67-68, asst prof, 68-72, ASSOC PROF BIOL, UNIV REGINA, 72- *Mem:* AAAS; Phytochem Soc NAm; Am Soc Plant Physiologists; Can Soc Plant Physiologists; Ecol Soc Am. *Res:* Plant senescence, especially metabolic changes in developing and senescing sunflower cotyledons and leaves; seed production and germination in native prairie plants (opuntia polyacantha and glycyrrhiza lepiolola); betalaines in emergent cactus seedlings. *Mailing Add:* Dept Biol Sci Univ Regina Regina SK S4S 0A2 Can

WALTHER, CARL H(UGO), engineering mechanics, civil engineering, for more information see previous edition

WALTHER, FRANK H, b Williamsport, Pa, Aug 4, 30; m 54; c 3. MINERALOGY. *Educ:* Franklin & Marshall Col, BSc, 52. *Prof Exp:* Mgr mineral res, Harbison-Walker Refractories, 56-72, mgr res, 72-78, dir res, 78; RETIRED. *Mem:* Fel Am Ceramic Soc. *Res:* Mineralogical aspects of refractory technology. *Mailing Add:* HC 64 Box 397 Trout Run PA 17771

WALTHER, FRITZ R, b Chemnitz, Ger, Sept 8, 21. ANIMAL BEHAVIOR. *Educ:* Univ Frankfurt, BS, 44, MS, 56, PhD(zool), 63. *Prof Exp:* Teacher, Fed Ministry Educ, Ger, 51-59; sci dir res & admin, Opel Zoo, 63-63; res scientist, Zurich Zool, Switz, 64 & Serengeti Nat Park, Tanzania, 65-67; assoc prof zool, Univ Mo, Columbia, 67-70; prof wildlife, 70-81, EMER PROF, TEX A&M UNIV, 81- *Concurrent Pos:* Grants, Ger Res Soc, 63, Gertrud Ruegg Found, 64 & 67, Fritz Thyssen Found, 65-67 & 74-75, Res Coun Univ Mo, 69-70, Smithsonian Foreign Currency, 70-72 & Caesar Kleberg Found, 74-75, 78. *Mem:* Am Soc Mammalogists; Ger Soc Mammalogists; Animal Behav Soc. *Res:* Ethology of game animals, especially horned ungulates. *Mailing Add:* Dorsstrasse 22 D-5419 Diedorf-Wienau Germany

WALTHER, JAMES EUGENE, b Spokane, Wash, May 16, 32; m 54; c 2. CHEMICAL ENGINEERING. *Educ:* Univ Wash, BS, 54; Univ Ill, Urbana, MS, 56, PhD(chem eng), 57. *Prof Exp:* Res engr, Stand Oil Co Calif, 57-62; res engr, 62-70, SR RES ENGR, CROWN ZELLERBACH CORP, CAMAS, 70- *Mem:* Am Inst Chem Engrs; Am Chem Soc. *Res:* Air pollution control from pulp and paper industry; instrumentation development for air pollution control. *Mailing Add:* James River Corp PO Box 2218 Richmond VA 23217-2218

WALTKING, ARTHUR ERNEST, b New York, NY, Nov 7, 37; m 61; c 2. ANALYTICAL CHEMISTRY, FOOD CHEMISTRY. *Educ:* Lehigh Univ, BA, 59. *Prof Exp:* Asst chemist, 59-64, chemist, 64-66, group leader anal res, 66-67, sect head anal serv, 67-72, mgr anal serv, 72-79, assoc res scientist, Best Foods Div, 79-84, PRIN MAT SCIENTIST, CPC INT INC, 84- *Concurrent Pos:* Assoc referee oxidized oils, Asn Off Anal Chemists, 71-83 & 85-86; rep, Am Oil Chemists Soc, Joint Am Oil Chemists Soc-Asn Off Anal Chemists-Am Asn Cereal Chemists, Comt Mycotoxins, 75-90; US rep, comn oils, fats & derivatives, Int Union Pure & Appl Chem, 80-; chmn AOCS Mycotoxin Comt, 78-89, Foods I Comt, AOAC, 80-88. *Honors & Awards:* Golden Peanut Award, Nat Peanut Coun, 71. *Mem:* Am Oil Chemists Soc; Am Chem Soc; Asn Off Anal Chemists; Soc Rheol; NAm Thermal Anal Soc. *Res:* The development of analytical methodology for analysis of food products for mycotoxins, flavor volatiles, texture, rheological properties, essential fatty acids and polymers derived from oxidation or heat abuse of vegetable oils. *Mailing Add:* Best Foods Res & Eng Ctr 1120 Commerce Ave Box 1534 Union NJ 07083

WALTMAN, PAUL ELVIS, b St Louis, Mo, Oct 17, 31; m 53; c 3. MATHEMATICAL ANALYSIS, BIOMATHEMATICS. *Educ:* St Louis Univ, BS, 52; Baylor Univ, MA, 54; Univ Mo, MA, 60, PhD(math), 62. *Prof Exp:* Staff mem, Mitre Corp, 62-63 & Sandia Corp, 63-65; from asst prof to prof, Univ Iowa, 65-83; PROF MATH, UNIV EMORY, ATLANTA, 83- *Mem:* Am Math Soc; Soc Indust & Appl Math. *Res:* Ordinary differential equations; modeling biological phenomena. *Mailing Add:* Dept Math & Comput Sci Emory Univ Atlanta GA 30322

WALTMANN, WILLIAM LEE, b Cedar Falls, Iowa, July 5, 34; m 58; c 3. MATHEMATICS. *Educ:* Wartburg Col, BA, 56; Iowa State Univ, MS, 58, PhD(math), 64. *Prof Exp:* Instr math, Wartburg Col, 58-61 & Iowa State Univ, 63-64; assoc prof, 64-72, PROF MATH, WARTBURG COL, 72-, CHMN DEPT, 71- *Mem:* Am Math Soc; Math Asn Am; Soc Indust & Appl Math. *Res:* Inversion of matrices; non-associative rings; tridiagonalization of matrices. *Mailing Add:* Dept Math & Comput Sci Wartburg Col Waverly IA 50677

WALTNER, ARTHUR, nuclear physics; deceased, see previous edition for last biography

WALTON, ALAN, chemical oceanography, geochemistry, for more information see previous edition

WALTON, ALAN GEORGE, b Birmingham, Eng, Apr 3, 36; m 77; c 4. BIOPHYSICAL CHEMISTRY. *Educ:* Univ Nottingham, BSc, 57, PhD(chem), 60, DSc(biophys chem), 73. *Prof Exp:* Res assoc chem, Ind Univ, 60-62; from asst prof to assoc prof macromolecular sci, Case Western Reserve Univ, 62-71, prof, 71-81; PRES, UNIV GENETICS CO, 81- *Concurrent Pos:* Vis lectr, Harvard Med Sch, 71; mem, Pres Task Force Sci & Technol Policy, 75-76. *Mem:* NY Acad Sci; Am Chem Soc; Biophys Soc. *Res:* Conformation and structure of synthetic biopolymers and fibrous proteins; molecular hematology; cell adhesion; genetic engineering; cartilage research; drug release. *Mailing Add:* 17 Walnut Lane Westport CT 06880

WALTON, ANTHONY WARRICK, b Mt Holly, NJ, Apr 10, 43; m 69; c 2. PHYSICAL SEDIMENTOLOGY, DIAGENESIS. *Educ:* Lafayette Col, BA, 65; Univ Tex Austin, MA, 68, PhD(geol), 72. *Prof Exp:* asst prof geol, Vanderbilt Univ, 72-75; asst prof geol, 75-78, ASSOC PROF GEOL, UNIV KANS, 78-, CHMN GEOL, 87- *Concurrent Pos:* Res assoc, Bureau Econ Geol, Univ Tex, Austin, 73 & 75 & 76; geologist, Conoco, 74; res assoc, Kans Geol Surv, 83 & 86. *Mem:* Geol Soc Am; Soc Econ Paleontologists & Mineralogists; Int Assoc Sedimentologists; Am Assoc Petrol Geol. *Res:* Sedimentology, especially deposition and diagenesis of siliciclestic sediments; oil reservoir characterization; deposition of volcaniclastic sediments by lahars and fluvial processes; diagenesis of volcaniclastics. *Mailing Add:* Dept Geol Univ Kans Lawrence KS 66045-2124

WALTON, BARBARA ANN, b Baltimore, Md, Mar 30, 40; m 70. VERTEBRATE EMBRYOLOGY, HISTOLOGY. *Educ:* Ind Univ Pa, BSEd, 62; Univ Okla, MNatSci, 66, PhD(zool), 70. *Prof Exp:* Teacher, Franklin Twp Sch Dist, Pa, 62-63 & Bethel Park Jr High Sch, 63-65; ASST PROF BIOL, UNIV TENN, CHATTANOOGA, 70- *Mem:* Am Inst Biol Sci; Am Soc Zoologists. *Res:* Development of the chicken embryo following x-irradiation and application of other teratogens; histology and histochemistry of embryonic development. *Mailing Add:* Dept Biol Univ Tenn 615 McCallie Ave Chattanooga TN 37403

WALTON, BRYCE CALVIN, b Lead, SDak, June 5, 23; m 46; c 4. PARASITOLOGY. *Educ:* Univ Southern Calif, MS, 50; Univ Md, PhD(zool, parasitol), 56. *Prof Exp:* US Army, 50-, res parasitologist, Walter Reed Army Inst Res, 52-56, chief dept med zool, 406th Med Gen Lab, 56-59, parasitologist, Third Army Med Lab, 59-62, chief parasitic dis sect, Middle Am Res Unit, 62-65, cmndg officer, Med Res Unit, Panama, CZ, 65-69 & US Army Res & Develop Group Far East, 69-72, cmndg officer, Army Med Res Unit-Panama, 72-76; regional adv parasitic dis, Pan Am Health Orgn-WHO, 76-80; CONSULT, 80-; Am Micros Soc. *Mem:* Am Soc Trop Med & Hyg; fel Royal Soc Trop Med & Hyg; Am Micros Soc. *Res:* American trypanosomiasis and leishmaniasis; toxoplasmosis; immuno-diagnosis of parasitic disease; systematics. *Mailing Add:* 771 Barlow Dr Lake Heritage Gettysburg PA 17325-8968

WALTON, CHARLES ANTHONY, b Auburn, Ala, Apr 3, 26; m 46; c 2. PHARMACY, PHARMACOLOGY. *Educ:* Auburn Univ, BS, 49; Purdue Univ, MS, 50, PhD, 56. *Prof Exp:* Instr, Ala Polytech Inst, 49; from asst prof to prof mat med, Col Pharm, Univ Ky, 50-72, head dept, 56-66, dir drug info ctr, 66-72, prof oral biol, Col Dent, 72-73; PROF & ASSOC DEAN, COL PHARM, UNIV TEX AUSTIN, 73- *Concurrent Pos:* Fulbright prof, Univ Cairo, 64-65; prof pharmacol, Univ Tex Health Sci Ctr, San Antonio, 73- *Mem:* Am Soc Hosp Pharmacists; fel Am Col Clin Pharmacol; Am Coun Pharmaceut Educ. *Res:* Clinical pharmacy and drug information services. *Mailing Add:* Dept Pharmacol Univ Tex Health Sci Ctr 7703 Floyd Curl Dr San Antonio TX 78284-7764

WALTON, CHARLES MICHAEL, b Hickory, NC, June 28, 41; m 63; c 4. TRANSPORTATION ENGINEERING, POLICY PLANNING. *Educ:* Va Mil Inst, BS, 63; NC State Univ, MCE, 69, PhD(civil eng), 71. *Prof Exp:* Asst civil eng, NC State Univ, 67-71; from asst prof to assoc prof, 71-83, prof civil eng, 83-87, BESS HARRIS JONES CENTENNIAL PROF NATURAL RESOURCE POLICY STUDIES, UNIV TEX, AUSTIN, 87-, CHMN CIVIL ENG, 88- *Concurrent Pos:* Chmn exec comt, Transp Res Bd, Nat Acad Sci-Nat Res Coun; mem, Governor's Interagency Transp Coun, Tex, 72-; transp consult, Parsons, Brinckerhoff, Quade, Douglas, Inc, NY, 73-; mem exec comn, Urban Transp Div, Am Soc Civil Engrs, 81-; consult, Asn Am Railroads, Washington, DC, 81-; assoc dir, Ctr Transp Res, Univ Tex, Austin, 80-88. *Honors & Awards:* Harland Bartholomew Award, Am Soc Civil Engrs, 87; Frank M Masters Transp Eng Award, Am Soc Civil Engrs, 87. *Mem:* Am Soc Civil Engrs; Inst Transp Engrs; Soc Am Mil Engrs; Opers Res Soc Am; Transp Res Bd; Urban Land Inst. *Res:* Transportation and land use planning; traffic safety; application of light rail transit; truck use of highways; truck size and weight issues; public/private participation in transportation. *Mailing Add:* ECJ Hall Suite 2 606 Univ Tex Austin TX 78712

WALTON, CHARLES WILLIAM, b Carlinville, Ill, Apr 3, 08; m 33; c 3. CHEMISTRY. *Educ:* Univ Ill, BS, 30; Univ Mich, MS, 31, PhD(chem), 33. *Hon Degrees:* DSc, Blackburn Col, 65. *Prof Exp:* Lectr & lab asst, Univ Mich, 30-31; res chemist, Goodyear Tire & Rubber Co, 33-39, supvr phys chem, 40-41, mgr & tech coordr, Synthetic Rubber Div, 41-44, mgr, Chem Prod Develop Div, 44-46; asst to exec vpres, Minn Mining & Mfg Co, 47, mgr, New Prod Div, 48-51, gen mgr, Adhesives & Coatings Div, 52-59, div vpres & gen mgr, Adhesives, Coatings & Sealers Div, 59-61, vpres res, 61-62, vpres res & develop, 62-69, dir, 62-70; RETIRED. *Concurrent Pos:* Trustee, Blackburn Col, 67-74; consult, 69-73. *Mem:* Am Chem Soc; Soc Chem Indust; Commercial Develop Asn. *Res:* Synthetic rubber; high polymers; surface chemistry and solids. *Mailing Add:* PO Box 136 Sugar Loaf Shores FL 33044

WALTON, DANIEL C, b Philadelphia, Pa, May 16, 34; m 60; c 2. PLANT BIOCHEMISTRY, PLANT PHYSIOLOGY. *Educ:* Univ Del, BS, 55; State Univ NY Col Forestry, Syracuse Univ, PhD(plant physiol), 62. *Prof Exp:* Chem engr, E I du Pont de Nemours & Co, 55-58; fel, Univ Tex, 62-63; from asst prof to assoc prof, 63-74, PROF BIOCHEM, STATE UNIV NY COL

FORESTRY, SYRACUSE UNIV, 74- *Concurrent Pos:* Vis mem, Dept Bot, Univ Col Wales, 71-72 & 78-79. *Mem:* AAAS; Am Chem Soc; Plant Growth Regulator Soc; Am Soc Plant Physiologists. *Res:* Seed germination; plant growth regulation. *Mailing Add:* Dept Biol State Univ NY Col Environ Sci & Forestry Syracuse NY 13210

WALTON, DEREK, b Sao Paulo, Brazil, Mar 1, 31; nat US; m 54; c 4. PHYSICS. *Educ:* Univ Toronto, MSc, 54; Harvard Univ, PhD(appl physics), 58. *Prof Exp:* Asst res engr, Univ Calif, 57-58; sr physicist, Convair Div, Gen Dynamics Corp, 58-60; res assoc eng physics, Cornell Univ, 60-62; physicist, Oak Ridge Nat Lab, 62-68; assoc prof, 68-74, PROF PHYSICS, MCMASTER UNIV, 74- *Mem:* Am Phys Soc. *Res:* Solidification, nucleation and growth; phonon-defect interactions; light scattering; spin-phonon interactions; phase transformations; amorphous materials; glasses archaeomagnetism; physics in archaeology; rock magnetism. *Mailing Add:* Dept Physics McMaster Univ 1280 Main St W Hamilton ON L8S 4L8 Can

WALTON, GEORGE, b Edmunton, Eng, Aug 14, 14; nat US; m 49; c 1. PHYSICAL CHEMISTRY, FORENSIC SCIENCE. *Educ:* San Diego State Col, AB, 36; Columbia Univ, MA, 39, PhD(phys chem), 41. *Prof Exp:* Asst, Columbia Univ, 40-41, Manhattan proj, 41; asst prof chem, Col Pharm, Univ Cincinnati, 41-44, assoc prof, 46-47; consult chemist, 47-50; sr res chemist, Drackett Co, 50-52, tech adminr, 52-57, sr scientist, 57-62; geochem prospecting, 62-66; asst dir, Southwestern NMex Media Ctr, Western NMex Univ, 66-68, from asst prof to assoc prof phys sci, 68-80; DIR, TURQUITE MINERALS ASSAY LAB, 80- *Concurrent Pos:* Consult, precious metals recovery and analysis. *Mem:* Am Acad Forensic Sci. *Res:* X-ray crystallography applied to chemical problems; paper chromatography of metal ions; silver in ores by atomic absorption spectroscopy; precious metals recovery and analysis; recovery of precious metals from scrap; research in solvent-bleed forgery; barite ore processing. *Mailing Add:* 1312 S Silver Ave Deming NM 88030

WALTON, GERALD STEVEN, b Kansas City, Kans, July 23, 35; m 56; c 5. PLANT PATHOLOGY. *Educ:* Wabash Col, AB, 57; Rutgers Univ, PhD(plant path), 61. *Prof Exp:* From asst plant pathologist to assoc plant pathologist, 61-77, PLANT PATHOLOGIST, CONN AGR EXP STA, 77- *Mem:* Am Phytopath Soc. *Res:* Nature of resistance to plant diseases, methods of disease control and determination of causal factor of a disease when unknown. *Mailing Add:* 285 Hillfield Rd New Haven CT 06518

WALTON, HAROLD FREDERIC, b Tregony, Eng, Aug 25, 12; nat US; m 38; c 3. ANALYTICAL CHEMISTRY. *Educ:* Oxford Univ, BA, 34, PhD(chem), 37. *Prof Exp:* Procter vis fel, Princeton Univ, 37-38; res chemist, Permutit Co, 38-40; from instr to asst prof chem, Northwestern Univ, 40-46; from asst prof to prof, 47-82, chmn dept, 62-66, EMER PROF CHEM, UNIV COLO, BOULDER, 83-, SR RES ASSOC, COOP INST RES ENVIRON SCI, 83- *Concurrent Pos:* Vis Fulbright-Hays lectr, Trujillo, 66-67 & 70 & Lima, 66-67; vis prof, Pedag Inst, Caracas, Venezuela, 72; hon prof, Univ San Marcos & Univ Trujillo, Peru. *Honors & Awards:* Dal Nogare Award in Chromatography, 88. *Mem:* AAAS; Am Chem Soc; Royal Soc Chem; corresp mem Chem Soc Peru. *Res:* Ion exchange; chromatography. *Mailing Add:* Coop Inst Res Environ Sci Univ Colo Boulder CO 80309-0216

WALTON, HAROLD V(INCENT), b Christiana, Pa, June 17, 21; m 46; c 3. AGRICULTURAL ENGINEERING. *Educ:* Pa State Univ, BS, 42, MS, 50; Purdue Univ, PhD, 61. *Prof Exp:* Engr, Gen Elec Co, 43-45; prof agr eng, Pa State Univ, 47-62; prof, Univ Mo, Columbia, 62-76, chmn dept, 62-69, prof & chief party, India Contract, 69-71; prof agr eng & head dept, Pa State Univ, University Park, 76-85; RETIRED. *Mem:* fel Am Soc Agr Engrs. *Res:* Physical properties of poultry egg and meat products. *Mailing Add:* 291 E McCormick Ave State College PA 16801

WALTON, HARRIETT J, b Claxton, Ga, Sept 19, 33; m 58; c 4. MATHEMATICS. *Educ:* Clark Col, AB, 52; Howard Univ, MS, 54; Syracuse Univ, MA, 57; Ga State Univ, PhD(math educ), 79. *Prof Exp:* Instr math, Hampton Inst, 54-55, asst prof, 57-58; from asst prof to assoc prof, 54-81, PROF MATH, MOREHOUSE COL, 81- *Concurrent Pos:* Proj dir, Atlanta Univ, 70-73 & Atlanta Univ Ctr, Inc, 81-82. *Mem:* Math Asn Am; Nat Asn Math (secy/treas, 80-). *Res:* Remediation in mathematics at the college level. *Mailing Add:* Dept Math Morehouse Col PO Box 76 Atlanta GA 30314

WALTON, HENRY MILLER, b Frankfurt am Main, Ger, May 7, 12; nat US; m 50; c 2. ORGANIC CHEMISTRY. *Educ:* Univ Frankfurt, PhD(philos), 34; Univ Chicago, PhD(org chem), 38. *Prof Exp:* Res chemist, Continental Carbon Co, NY, 39; Chas Pfizer & Co fel, Columbia Univ, 40-42; sr res chemist, Warner Inst Therapeut Res, 43-47; group leader fundamental res lab, Nat Dairy Res Labs, Inc, 47-57; sr res chemist, A E Staley Mfg Co, 57-61, res assoc, 61-69, patent chemist, 69-72; patent consult, 72-74; prof philos, Richland Community Col, 74-75; RETIRED. *Mem:* Am Chem Soc; Am Philos Asn; Philos Sci Asn. *Res:* Vitamins A and E; unsaturated aliphatics; condensation polymers; free radical polymerization reactions; starches; sugars; enzymes; immobilized enzymes; philosophy of science. *Mailing Add:* 115 S Stevens Ave Decatur IL 62522

WALTON, JAY R, b Gary, Ind, Aug 24, 46; m 84; c 2. SOLID MECHANICS, VISCOELASTICITY. *Educ:* DePauw Univ, BA, 68; Ind Univ, MA, 70, PhD(math), 73. *Prof Exp:* From asst prof to assoc prof, 73-86, PROF MATH, TEX A&M UNIV, 86-, PROF AEROSPACE ENG, 90- *Concurrent Pos:* Vis assoc prof, Math Res Ctr, Univ Wis, 79-80; vis Sci Eng Res Coun res fel, Inst Computational Math, Brunel Univ, Eng, 87-88. *Mem:* Am Acad Mech; Am Math Soc; Am Soc Mech Engrs; Math Asn Am; Soc Indust & Appl Math. *Res:* Solid mechanics, especially viscoelasticity and fractures; applied mathematics, especially integral equations, partial differential equations numerical methods. *Mailing Add:* Math Dept Tex A&M Univ College Station TX 77843

WALTON, JOHN JOSEPH, b Sterling, Ill, Aug 25, 34; m 59; c 2. ATMOSPHERIC PHYSICS. *Educ:* Northwestern Univ, BS, 56; Univ Kans, PhD(physics), 61. *Prof Exp:* Physicist, Lawrence Radiation Lab, 61-72, PHYSICIST, LAWRENCE LIVERMORE LAB, 72- *Concurrent Pos:* Lectr appl sci, Univ Calif, 67-71. *Res:* Regional and global atmospheric modeling with emphasis on transport processes. *Mailing Add:* Atmospheric & Geophys Sci Div Lawrence Livermore Lab L-142 Livermore CA 94550

WALTON, KENNETH NELSON, b Winnipeg, Man, May 1, 35; US citizen; c 4. MEDICINE. *Educ:* Univ Man, MD, 59; Am Bd Urol, dipl, 68. *Prof Exp:* Intern, Winnipeg Gen Hosp, 59-60, resident surg & path, 60-61; jr asst resident urol, Johns Hopkins Hosp, 61-62, res fel, 62-63, sr asst resident, 63-64, co-head resident, 64-65; from instr to asst prof urol, Dept Surg, Med Ctr, Univ Ky, 65-69, chmn div urol, 68-69; chmn, Dept Urol, 69-80, LOUIS MCDONALD ORR PROF SURG, SCH MED, EMORY UNIV, 69- *Concurrent Pos:* Am Cancer Soc res fel, 63-64. *Mem:* Am Col Surgeons; AMA; Am Soc Nephrology; Am Urol Asn; Pan-Am Med Asn; Sigma Xi. *Res:* Kidney transplants; factors influencing renal oxygen consumption; effect of hypothermia and the inhibition of the tubular transport of sodium; difference in carbohydrate metabolism between prostatic cancers which are endocrine sensitive and those which are not. *Mailing Add:* 1365 Clifton Rd NE Atlanta GA 30322

WALTON, MATT SAVAGE, b Lexington, Ky, Sept 16, 15; m 39, 69, 70; c 5. GEOLOGY. *Educ:* Univ Chicago, BA, 36; Columbia Univ, MA, 46, PhD(geol), 51. *Prof Exp:* Geologist, US Geol Surv, 41-47; from instr to assoc prof geol, Yale Univ, 48-64; independent consult, 64-73; prof, 73-86, EMER PROF GEOL, UNIV MINN, 86-; CONSULT GEOL, 86- *Concurrent Pos:* Geologist, NY State Geol & Natural Hist Surv, 48-58; regents lectr, Univ Calif, Los Angeles, 71-73; dir, Minn Geol Surv, 73-86. *Mem:* Geol Soc Am. *Res:* Petrology and tectonic development of ultrabasic granite and high grade unitamorphic rocks, especially in the Adirondack mountains of New York and in Minnesota; aquifer thermal energy storage and environmental geology; Geological survey program, heavy construction site investigations and underground construction. *Mailing Add:* 30 Crocus Pl St Paul MN 55102

WALTON, PETER DAWSON, b Leeds, Eng, Oct 18, 24; Can citizen; m 49; c 4. PLANT BREEDING. *Educ:* Univ Durham, BSc, 49, MSc, 53; Univ Lancaster, PhD(pop genetics), 61. *Prof Exp:* Plant breeder, Res Div, Ministry Agr, Sudan, 50-55; sr plant breeder, Empire Cotton Growing Corp, 55-63; lectr agr & bot, Ahmadu Bello Univ, Nigeria, 63-67; assoc prof plant sci, Univ Sask, 67-69; chmn dept, 75-80, PROF PLANT SCI, UNIV ALTA, 69- *Mem:* Brit Inst Biol; Am Soc Agron; Genetics Soc Can; Am Forage & Grassland Coun; fel Royal Soc Arts. *Res:* Forage crop breeding with special reference to the study of genotype by environment interaction. *Mailing Add:* Dept Plant Sci Univ Alberta Edmonton AB T6G 2M7 Can

WALTON, RAY DANIEL, JR, b Ogden, Utah, Jan 26, 21; wid; c 6. CHEMICAL & NUCLEAR ENGINEERING, OPERATIONS RESEARCH. *Educ:* Ore State Univ, BS, 43, MS, 48. *Prof Exp:* Chem engr, Hanford Labs, Gen Elec Co, 47-56; chem engr, Idaho Opers Off, US Atomic Energy Comn, 56-60, tech analyst, Div Oper Anal & Forecasting, 60-64; chem engr, Int Atomic Energy Agency, 64-66; chief eng br, Div Oper Anal & Forecasting, US AEC, 66-70, mat & process control engr, Div Waste & Scrap Mgt, 70-71, chem engr, Develop Br, Div Waste Mgt & Transp, 71-74; proj engr, US Energy Res & Develop Admin, 74-77; actg chief technol br, Div Waste Prod, US Dept Energy, 77-78, eng prog mgr, Technol Div, Off Defense Waste & Transportation, 79-86; PROG MGR, ARGONNE NAT LAB, 86- *Concurrent Pos:* Leader, Int Atomic Energy Agency nuclear power mission, Turkey; vchmn, Aiche Nuclear Engr Div, 76, chmn, 77, prog chmn, 78. *Mem:* Am Sci Affiliation; Am Inst Chem Engrs; AAAS. *Res:* Technical and economic aspects of the nuclear fuel cycle, irradiated reactor fuel processing and disposal of radioactive waste; initiation, funding, direction, evaluation and management of nuclear waste management research and development projects. *Mailing Add:* 19205 Germantown Rd Germantown MD 20874

WALTON, ROBERT BRUCE, b Jersey City, NJ, Nov 30, 15; m 39; c 4. MICROBIOLOGY. *Educ:* Rutgers Univ, BSc, 48, PhD(microbiol), 53. *Prof Exp:* From asst res microbiologist to assoc res microbiologist, Merck Sharp & Dohme Res Labs, 40-53, sr res microbiologist, 53-78, res fel, 78-81; RETIRED. *Res:* Antibiotics; vaccines; immunology; microbial nutrition and physiology; fermentations; actinophage. *Mailing Add:* 798 Central Ave Rahway NJ 07065

WALTON, ROBERT EUGENE, b Shattuck, Okla, Jan 15, 31; m 59; c 3. ANIMAL BREEDING, ANIMAL GENETICS. *Educ:* Okla State Univ, BS, 52, MS, 56; Iowa State Univ, PhD, 61. *Prof Exp:* Farm mgr, Westhide Farms, Eng, 53-54; asst prof dairy sci, Univ Ky, 58-62; geneticist, Am Breeders Serv, 68, pres & chief exec officer, 68-88; PRES & CHIEF EXEC OFFICER, GRACE ANIMAL SERV, 89-; CHMN, AGRACETUS, 90- *Concurrent Pos:* Mem prog mgt develop, Harvard Univ, 70. *Honors & Awards:* Nat Agribus Award, Nat Agri-Mkt Asn, 85. *Mem:* Am Soc Animal Sci; Am Dairy Sci Asn; Biomet Soc; Nat Asn Animal Breeders (pres, 72-74). *Res:* Application of genetical and statistical tools to problems of animal breeding and improvement; estimation of parameters of domestic large animal populations; genetic evaluation of dairy sires. *Mailing Add:* 4066 Vinburn Rd De Forest WI 53532

WALTON, RODDY BURKE, b Goldthwaite, Tex, Dec 9, 31; m 61; c 2. NUCLEAR PHYSICS. *Educ:* Tex A&M Col, BS, 52; Univ Wis, MS, 54, PhD(nuclear physics), 57. *Prof Exp:* Nuclear res officer, Air Force Weapons Lab, 57-59; staff physicist, Gen Atomic Div, Gen Dynamics Corp, 59-67; STAFF MEM, LOS ALAMOS SCI LAB, 67- *Mem:* Am Phys Soc; Inst Nuclear Mat Mgt; fel Am Nuclear Soc. *Res:* Neutron physics; photonuclear research; positron production with an electron linear accelerator; delayed gamma rays and delayed neutrons from fission; non-destructive assay applications; neutral particle beam applications. *Mailing Add:* Group P15 MS D406 Los Alamos Sci Lab Los Alamos NM 87545

WALTON, THEODORE ROSS, b Takoma Park, Md, Feb 26, 31; m 56; c 4. CONDUCTIVE POLYMERS, THERMAL STABLE POLYMERS. *Educ:* Univ Md, BS, 55; Ohio State Univ, PhD(org chem), 60. *Prof Exp:* Proj dir chem, Atlantic Res Corp, 60-63; res chemist, Author W Sloan Found Va, 63-64; res chemist, 64-90, ACTG SECT HEAD, COATING SECT, US NAVAL RES LAB, WASHINGTON, DC, 90- *Mem:* Am Chem Soc. *Res:* Organic and polymer synthesis and chemistry; thermally stable organic polymers; electrical conducting organic polymers; organic non-linear optical materials; rocket motor case thermal insulation and material compatibility; fire retardant coating systems; water based paints; adhesive bonding. *Mailing Add:* US Naval Res Lab 4555 Overlook Ave SW Washington DC 20375-5000

WALTON, THOMAS EDWARD, b McKeesport, Pa, Dec 2, 40; m 87; c 3. VIROLOGY, ARBOVIROLOGY. *Educ:* Purdue Univ, DVM, 64; Cornell Univ, PhD(microbiol), 68. *Prof Exp:* Res vet microbiol, NIH, Nat Inst Allergy & Infectious Dis, USPHS, Maru, 68-72; res leader, Denver, Colo, 64-85, vet med officer, 72-74, RES LEADER, AGR RES SERV, ARTHROPOD-BORNE ANIMAL DIS RES LAB, USDA, LARAMIE, WYO, 86- *Concurrent Pos:* Affil fac, Dept Microbiol, Colo State Univ, 75-; adj prof, Dept Vet Sci, Univ Wyo, 86-; chair, Exec Coun, Am Comn Arthropod-Borne Viruses. *Mem:* Am Vet Med Asn; Am Soc Trop Med Hyg; Am Soc Trop Vet Med (secy, 91-); Soc Vector Ecologists; Am Biol Safety Asn. *Res:* Diagnosis and control of arthropod borne virus diseases of livestock; author or co-author of over 85 publications. *Mailing Add:* USDA ARS ABADRL PO Box 3965 Univ Sta Laramie WY 82071-3965

WALTON, THOMAS PEYTON, III, b Archer, Fla, Dec 7, 22; m 64; c 3. SURGERY. *Educ:* Tulane Univ, MD, 50; Am Bd Surg, dipl, 64. *Prof Exp:* Intern, Charity Hosp La, New Orleans, 50-51; physician for Seminole Indian Nation, Fla, 51-52; resident surg, Baptist Hosp, Nashville, Tenn, 52-55; chief resident, Nashville Gen Hosp, 55-56; pvt pract surg, Tampa, Fla, 57-67; asst prof, La State Univ Med Ctr New Orleans, 68-69; vchief surg, Vet Admin Hosp, Big Spring, Tex, 69-70; clin dir & chief surg, Lafayette Charity Hosp, 70-77; assoc prof, 77-80, prof surg, La State Univ, 80-85; RETIRED. *Concurrent Pos:* VChief surg, Tampa Gen Hosp, Fla, 61-62; chief of staff, Clara Frye Hosp, Tampa, 62-63; asst prof surg, Med Sch, Univ New Orleans, 70-73; guest examr, Am Bd Surg, 71. *Mem:* Fel Am Col Surgeons. *Mailing Add:* LaFayette Charity Hosp LaFayette LA 70501

WALTON, VINCENT MICHAEL, b Spokane, Wash, Oct 23, 49. SPACECRAFT & POINTING PAYLOAD, ALTITUDE CONTROL. *Educ:* Univ Wash, BS, 73, MS, 75. *Prof Exp:* Mem tech staff, Boeing Aerospace Co, 73-75; mem tech staff, 75-80, MEM STAFF, TRW SYSTS GROUP, 80- *Mem:* Am Inst Aeronaut & Astronaut; Inst Elec & Electronics Engrs. *Res:* Systems engineering; design, analysis and simulation of automatic control sytems for spacecraft; orbitor gimballed payloads; gimballed optical pointer trackers; laser hot antoalignment. *Mailing Add:* VMW Systems Dynamics 302 SW 325 Pl Federal Way WA 98023

WALTON, WARREN LEWIS, b La, Dec 13, 14; m 43; c 3. ORGANIC CHEMISTRY. *Educ:* Millsaps Col, BS, 35; La State Univ, MS, 37; Univ Ill, PhD, 41. *Prof Exp:* Br chemist, Coca-Cola Co, 37-38; res & develop chemist, Hercules Co, 41-46; res & develop chemist silicone synthesis, Gen Elec Co, NY, 46-50, head anal unit, Insulating Mat Dept, 50-69, instrumental anal chemist, 50-71; CONSULT, 79- *Concurrent Pos:* Consult, Schenectady Chem Inc, 78-79. *Mem:* Am Chem Soc. *Res:* Infrared spectroscopy; microscopy, gas and gelpermeation chromatography, nuclear magnetic resonance spectroscopy. *Mailing Add:* PO Box 1564 Hammond LA 70404

WALTON, WAYNE J A, JR, b Chariton, Iowa, Apr 6, 41; m 70; c 2. LUNAR GEOLOGY. *Educ:* Drury Col, AB, 64; Univ Ariz, MS, 66; Ohio State Univ, Columbus, PhD(mineral), 72. *Prof Exp:* Teaching asst geol, Univ Ariz, 64-66; teaching assoc mineral, Ohio State Univ, 69-71; instr geol, Capital Univ, 71-72, asst prof, 72-74; instr, Univ Houston, 76-79; asst prof geol, Midwestern State Univ, Ft Worth, Tex, 79-84; CONSULT, 84- *Concurrent Pos:* Geologist, US Geol Surv, Ariz, 64, Am Metal Climax, 65; photo chemist, Kitt Peak Nat Observ, 64-65; res assoc, Univ Ariz, 65; res chemist, Minn Mining & Mfg Co, 66-69; vis instr geol Ohio State Univ, 72; sr res analysis, Northrop Serv, Inc, 74-78. *Mem:* Meteoritical Soc; Can Mineral Soc; Mineral Soc Gt Brit. *Res:* High temperature ceramic bodies. *Mailing Add:* 01-BP 480 Ouagadougou-01 Burkina Faso West Africa

WALTON, WILLIAM RALPH, b Ft Worth, Tex, Apr 11, 23; m 49; c 2. GEOLOGICAL OCEANOGRAPHY. *Educ:* Amherst Col, BA, 49; Univ Calif, MS, 52, PhD(oceanog), 54. *Prof Exp:* Paleoecologist, Gulf Res & Develop Co, 53-57; paleoecologist & sedimentologist, Pan Am Petrol Corp, 57-60, div consult geologist, 60-63, geol & geochem res dir, Amoco Prod Co, Okla, 63-73, chief geologist, 73-75, chief geologist, Amoco Int Oil Co, Ill, 75-81, explor mgr, 77-81; PROF, NORTHWESTERN UNIV, 85- *Concurrent Pos:* Distinguished lectr, Am Asn Petrol Geologists, 72-73; mem offshore explor & prod task group, Nat Petrol Coun Comt Ocean Petrol Resources, 74. *Honors & Awards:* Pres Award, Am Asn Petrol Geol, 57. *Mem:* Soc Econ Paleontologists & Mineralogists; Geol Soc Am; Paleont Soc; Am Asn Petrol Geologists. *Res:* Biostratigraphy of gulf coast tertiary; paleoecology, marine geology and sedimentology. *Mailing Add:* 434-G Elmwood Evanston IL 60202

WALTRUP, PAUL JOHN, b Baltimore, Md, June 12, 45; c 2. ENGINEERING, MATERIALS SCIENCE. *Educ:* Univ Md, BS, 67, MS, 68; Va Polytech Inst, PhD(aeronaut eng), 71. *Prof Exp:* Fel, Johns Hopkins Univ, 71-72, sr staff engr, 72-74, sect supvr, 74-82, GROUP SUPVR, APPL PHYSICS LAB, JOHNS HOPKINS UNIV, 83- *Concurrent Pos:* Consult, McGraw Hill Info Systs, 77-82, Off Naval Res, 84-88; instr, Univ Md, 78-89; adj prof, Va Polytech Inst, 82-89. *Honors & Awards:* Young Engr Award, Am Inst Aeronaut & Astronaut, 74. *Mem:* Assoc fel Am Inst Aeronaut & Astronaut; Combustion Inst; AAAS. *Res:* Ramjet-Scramjet propulsion. *Mailing Add:* Appl Physics Lab Johns Hopkins Rd Laurel MD 20723

WALTZ, ARTHUR G, b Irwin, Pa, Feb 14, 32; m; c 3. NEUROLOGY, CEREBROVASCULAR DISEASES. *Educ:* Univ Mich, BS, 52, MD, 55; Am Bd Psychiat & Neurol, dipl neurol, 62. *Prof Exp:* Rotating intern, Hosp Univ Pa, 55-56; asst resident neurol, Neurol Unit, Boston City Hosp, Mass, 56-57, sr resident, 57-58; res assoc, Sch Med, Wayne State Univ, 58-59, instr, 59; from instr to assoc prof, Mayo Grad Sch Med, Univ Minn, 62-71; prof neurol, Med Sch, Univ Minn, Minneapolis, 71-74; chmn, Dept Neurol, Pac Med Ctr, 75-81; clin prof neurol, Univ Calif, 81-90; RETIRED. *Concurrent Pos:* Teaching fel neurol, Harvard Med Sch, 56-58; asst to staff neurol, Mayo Clin, 61, consult, 62-71; adj prof neurol, Univ of the Pac, 75-80; fel stroke coun, Am Heart Asn. *Mem:* Fel Am Acad Neurol; Am Neurol Asn; Asn Res Nerv & Ment Dis; Sigma Xi. *Res:* Cerebral circulation, including blood flow, microcirculation and fluid balance, in normal and ischemic brain. *Mailing Add:* 66 Celery Ct No 1305 San Francisco CA 94109

WALTZ, RONALD EDWARD, b Indianapolis, Ind, Nov 21, 43; m 67; c 2. PHYSICS. *Educ:* Purdue Univ, BS, 66; Univ Chicago, PhD(physics), 70. *Prof Exp:* Fel, NSF, CERN, Geneva, Switz, 71-72; res assoc, Ctr Theoret Physics, Mass Inst Technol, 72-73; sr scientist, Visidyne Inc, Burlington, Ma, 73-74; SR STAFF SCIENTIST, GEN ATOMICS, SAN DIEGO, CA, 75- *Concurrent Pos:* Adj prof physics, Univ Calif, San Diego, 86; vis prof, Joint Inst Fusion Theory, Nagoya Univ, Japan, 86. *Mem:* Fel Am Phys Soc. *Res:* Theory and simulation of turbulence in fusion plasmas and the phenomenology of tokamak confinement experiments. *Mailing Add:* Gen Atomics 13-303 PO Box 85608 San Diego CA 92138

WALTZ, WILLIAM LEE, b Berkeley, Calif, June 3, 40; m 63; c 2. PHOTOCHEMISTRY, RADIATION CHEMISTRY. *Educ:* Miami Univ, BS, 62; Northwestern Univ, PhD(phys inorg chem), 67. *Prof Exp:* Res assoc chem, Univ Southern Calif, 66-68; sr chemist, Cent Res Lab, 3M Co, Minn, 68-69; from asst prof to assoc prof, 69-80, PROF CHEM, UNIV SASK, 80-, HEAD, DEPT CHEM, 89- *Concurrent Pos:* Vis assoc prof chem, Ohio State Univ, 77-78; guest scientist, Hahn-Meitner Inst Nuclear Res, West Berlin, Ger, 78-82, 84-85 & 87-90. *Mem:* Sigma Xi; Am Chem Soc; Can Inst Chem. *Res:* Inorganic materials. *Mailing Add:* Dept Chem Univ Sask Saskatoon SK S7N 0W0 Can

WALTZER, WAYNE C, b Brooklyn, NY, Apr 18, 48; m 86; c 1. RENAL TRANSPLANTATION, UROLOGICAL SURGERY. *Educ:* Pa State Univ, BA, 69; Univ Pittsburgh Sch Med, MD, 73. *Prof Exp:* Intern, internal med, Presby Univ Hosp, Pa, 73-74, resident, 74-75, resident, Urol Surg, 75-78; fel renal transplant, Mayo Clin Rochester, Minn, 78-79; from instr surg to asst prof surg & urol, 79-84, ASSOC PROF SURG & UROL, STATE UNIV NY, STONY BROOK, 84- *Concurrent Pos:* Asst ed, Transplantation Proceedings, 80-; consult, J Urol, 87- *Mem:* Sigma Xi; Am Soc Transplant Surgeons; Am Col Surgeons; Am Urol Asn; Soc Univ Urologists; Asn Acad Surg; Nat Kidney Found. *Res:* Immunological monitoring of renal transplant recipient with specific emphasis on the lymphocyte subpopulations in blood and renal allograft. *Mailing Add:* Transplantation Serv Health Sci Ctr T-19 State Univ NY Stony Brook NY 11794-8192

WALUM, HERBERT, b Bremerton, Wash, Aug 14, 36; m 59; c 2. MATHEMATICS. *Educ:* Reed Col, BA, 58; Univ Colo, PhD(math), 62. *Prof Exp:* Asst prof math, Harvey Mudd Col, 62-64; asst prof, 64-71, ASSOC PROF MATH, OHIO STATE UNIV, 71- *Mem:* Am Math Soc; Math Asn Am; Sigma Xi. *Res:* Number theory. *Mailing Add:* 234 Math Bldg Campus 050 Ohio State Univ 231 N 18th Ave Columbus OH 43210

WALVEKAR, ARUN GOVIND, b Belgaum, India, May 7, 42. APPLIED MATHEMATICS, OPERATIONS RESEARCH. *Educ:* Univ Bombay, BE, 63 & 64; Ill Inst Technol, MS, 66, PhD(opers res), 67. *Prof Exp:* Instr math, Northeastern Ill State Col, 67-68; asst prof indust eng, Tex Tech Univ, 68-71, assoc prof Indust Eng & Statist, 71-; AT DEPT INDUST ENG, NORTHERN NMEX COMMUNITY COL, EL RITO. *Mem:* Opers Res Soc Am; Am Inst Indust Engrs. *Res:* Multistage decision processes; calculus of variations; complex variables. *Mailing Add:* Dept Indust Eng Northern NW Mex State Univ Las Cruces NM 88003

WALWICK, EARLE RICHARD, b San Diego, Calif, Oct 9, 29; m 51; c 2. CLINICAL CHEMISTRY. *Educ:* Univ Calif, Berkeley, AB, 52, PhD(biochem), 58. *Prof Exp:* Biochemist, US Naval Sch Aviation Med, Fla, 52-53; aviation physiologist, Naval Air Sta, 53-55; biochemist, Naval Hosp, Oakland, Calif, 57-58, biochemist, Naval Radiol Defense Lab, San Francisco, 58-61; supvr res, Aeronutronic Div, Philco-Ford Corp, 61-71; mem staff, US Govt, 71-72; clin chemist & hosp-pathologist, Cent Lab of Orange County, Inc, 74-77; CHIEF CLIN CHEMIST, JERRY L PETTIS MEM VET HOSP, 77- *Concurrent Pos:* NIH fel path, Univ Calif, Irvine, 72-74. *Mem:* Am Asn Clin Chem. *Mailing Add:* 1317 Olive Ave Redlands CA 92373

WALZ, ALVIN EUGENE, b Hot Springs, SDak, Jan 12, 19. PHYSICAL CHEMISTRY, ANALYTICAL CHEMISTRY. *Educ:* Northern State Teachers Col, BS, 43; Univ Iowa, MS, 45, PhD(phys chem), 50. *Prof Exp:* Teacher, high sch, Iowa, 43-48; from asst prof to prof chem, Mankato State Col, 50-63; chmn dept, 66-86, PROF CHEM, CALIF LUTHERAN COL, 63- *Mem:* AAAS; Am Chem Soc; Sigma Xi. *Res:* Reaction rates; electron affinity; methyl stibine. *Mailing Add:* 119 Sirius Circle Thousand Oaks CA 91360

WALZ, DANIEL ALBERT, b Rochester, NY, July 30, 44; m 66; c 3. PHYSIOLOGY, BIOCHEMISTRY. *Educ:* St John Fisher Col, BS, 66; Wayne State Univ, PhD(physiol), 73. *Prof Exp:* Instr, 73-74, from asst prof to assoc prof, 74-79, PROF PHYSIOL, WAYNE STATE UNIV, 83- *Mem:* Am Chem Soc; Int Soc Thrombosis & Hemostasis; Am Heart Asn; AAAS; Am Soc Biol Chem. *Res:* Properties of macromolecules; protein structure; coagulation biochemistry. *Mailing Add:* Sch Med Wayne State Univ 540 E Canfield Detroit MI 48201-1100

WALZ, DONALD THOMAS, b Newark, NJ, Oct 25, 24; m 59; c 3. PHARMACOLOGY. *Educ:* Upsala Col, BS, 50; Rutgers Univ, MS, 51; Georgetown Univ, PhD(pharmacol), 59. *Prof Exp:* Jr pharmacologist, Hoffmann-La Roche, Inc, 51-55; sr investr pharmacol, 60-68, from asst dir to assoc dir pharmacol, 68-77, ASSOC DIR RES, SMITHKLINE & FRENCH LABS, 77- *Concurrent Pos:* Nat Inst Neurol Dis & Blindness fel, Sch Med, Georgetown Univ, 59-60. *Mem:* AAAS; Am Diabetes Asn; Am Soc Pharmacol & Exp Therapeut; Acad Pharmaceut Sci; Fedn Am Socs Exp Biol; Sigma Xi. *Res:* Biochemical neuropharmacology metabolism; carbohydrate metabolism; cardiovascular-autonomic pharmacology. *Mailing Add:* 14 Pilgrim Lane Drexel Hill PA 19026

WALZ, FREDERICK GEORGE, b Brooklyn, NY, May 11, 40; m 62; c 7. BIOCHEMISTRY. *Educ:* Manhattan Col, BS, 62; State Univ NY Downstate Med Ctr, PhD(biochem), 66. *Prof Exp:* NIH res fel biochem, Cornell Univ, 66-68; asst prof, State Univ NY Albany, 68-75; from asst prof to assoc prof chem, Kent State Univ, 75-82; res assoc prof, 77-83, RES PROF MOLECULAR PATH, COL MED, NORTHEAST OHIO UNIV, 83-; PROF CHEM, KENT STATE UNIV, 82- *Concurrent Pos:* NSF res grant, State Univ NY Albany, 69-75 & Kent State Univ, 76-80, 81-84. *Mem:* Am Soc Biochem; Am Chem Soc; Biophys Soc. *Res:* Genetics, regulation and evolution of cytochromes P-450, liver endoplasmic reticulum proteins, albumin secretion; ribonuclease mechanisms and substrate recognition. *Mailing Add:* Dept Chem Kent State Univ Kent OH 44242

WAMBOLD, JAMES CHARLES, b Emmaus, Pa, Nov 24, 32; m 60; c 4. MECHANICAL ENGINEERING, ENGINEERING MECHANICS. *Educ:* Pa State Univ, BS, 59; Carnegie Inst Technol, MS, 60; Univ NMex, PhD(mech eng), 67. *Prof Exp:* Staff mem, Sandia Corp, 58-62; instr mech eng, Univ NMex, 62-67; from asst prof to assoc prof, 67-81, PROF MECH ENG & DIR AUTOMOTIVE SAFETY, PA STATE UNIV, 81- *Concurrent Pos:* Aircraft mech consult to indust & lawyers; US expert, Vehicle-Surface Interaction Comt 1, Permanent Int Asn Road Congresses. *Honors & Awards:* Kummer Lect Award, Am Soc Testing & Mat, 82, Outstanding Achievement Award, 89. *Mem:* Fel Am Soc Mech Engrs; Am Soc Eng Educ; Am Soc Testing & Mat. *Res:* Random signal processing; roughness effects on vehicle performances; application of solid mechanics to engineering design; modeling of physical systems for computer solution and control system design. *Mailing Add:* Rm 301 Dept Mech Eng Pa State Univ University Park PA 16802

WAMPLER, D EUGENE, b Shanxi, China, July 24, 35; US citizen; m 61; c 3. PROTEIN CHEMISTRY. *Educ:* Bridgewater Col, BA, 59; Mich State Univ, MS, 61 & PhD(biochem), 65. *Prof Exp:* Asst prof, dept biochem, Univ Conn Sch Med, 68-75; sr scientist, Bio Gant Corp, 75-76; res specialist, dept biochem, Mich State Univ, 76-77; SR RES FEL, MERCK RES LABS, 77- *Mem:* Am Soc Biochem & Molecular Biol; Am Chem Soc; AAAS. *Res:* Protein isolation from natural sources and recombinant-DNA microorganisms; vaccine formulation; biochemical process research and development. *Mailing Add:* Merck Res Lab W38T-1 West Point PA 19486

WAMPLER, E JOSEPH, b Taiku, China, Jan 27, 33; US citizen; m 56; c 3. ASTRONOMY. *Educ:* Univ Va, BA, 58; Univ Chicago, MS, 59, PhD(astron), 62. *Prof Exp:* Res asst astron, Univ Chicago, 62-63; fel in residence, Miller Inst Basic Res Sci, Univ Calif, 63-65, asst astronomer, Observ, 65-66, from asst astronomer to astronomer, Lick Observ, 66-85; STAFF ASTRONOMER, EUROP SOUTHERN OBSERV, 85- *Concurrent Pos:* Dir, Anglo-Australian Observ, 74-76. *Mem:* Am Astron Soc; Int Astron Union. *Res:* Quasi stellar sources; extra-galactic astronomy; instrumentation. *Mailing Add:* Europ Southern Observ Karl-Schwarzschild-Str 2 D-8046 Garching bei Munchen Germany

WAMPLER, FRED BENNY, b Kingsport, Tenn, Apr 2, 43; m 84; c 3. PHYSICAL CHEMISTRY, PHOTOCHEMISTRY. *Educ:* Univ Tenn, Knoxville, BS, 65; Univ Mo, Columbia, PhD(phys chem), 70. *Prof Exp:* Fel phys chem, Ohio State Univ, 70-72; sr scientist, Allison Div, Gen Motors Corp, 72-74; staff mem, 74-79, asst group leader, 79-81, dep group leader, 81-82, GROUP LEADER, LOS ALAMOS NAT LAB, 82- *Mem:* Am Chem Soc; Inter-Am Photochem Soc. *Res:* Kinetics; application of lasers to chemical problems; energy transfer; laser spectroscopy; atmospheric chemistry; optical instrumentation. *Mailing Add:* 366 Garver Lane Los Alamos NM 87544

WAMPLER, JESSE MARION, b Harrisonburg, Va, Oct 31, 36; m 62; c 2. NUCLEAR GEOCHEMISTRY, GEOCHRONOLOGY. *Educ:* Bridgewater Col, BA, 57; Columbia Univ, PhD(geochem), 63. *Prof Exp:* Res asst geochem, Lamont Geol Observ, NY, 60-63; res assoc, Brookhaven Nat Lab, 63-65; from asst prof to prof, geophys sci, 65-89, assoc dir, 84- 89, ASSOC PROF, SCH EARTH & ATMOSPHERIC SCI, GA INST TECHNOL, 90- *Mem:* Geochem Soc; Am Geophys Union; Sigma Xi; Nat Asn Geol Teachers; Int Asn Geochem & Cosmochem. *Res:* Nuclear geochemistry; geochemistry of argon and potassium; potassium-argon geochronology; lead isotope geochemistry; geochemistry of natural radionuclides. *Mailing Add:* Sch Earth & Atmospheric Sci Ga Inst Technol Atlanta GA 30332-0340

WAMPLER, JOE FORREST, b Chanute, Kans, Dec 13, 26; m 49; c 2. MATHEMATICS. *Educ:* Univ Kans, AB, 50, MA, 52; Univ Nebr, PhD, 67. *Prof Exp:* Instr math, York Col, 51-54, registrar, 53-54; assoc prof math, 54-66, head dept, 54-83, chmn div natural sci, 71-76, actg chair, 88-89, PROF MATH, NEBR WESLEYAN UNIV, 66- *Concurrent Pos:* Woods Found grant, 60-61; NSF coop teacher develop grant, 66. *Mem:* Math Asn Am (secy-treas, 91-94); Nat Coun Teachers Math; Am Asn Univ Professors. *Res:* Liouville's methods; use of various measures of aptitude to predict achievement in college mathematics; statistics; number theory; author of three publications. *Mailing Add:* Dept Math Nebr Wesleyan Univ 5000 St Paul Ave Lincoln NE 68504-2796

WAMPLER, JOHN E, COMPUTER AUTOMATION, ANALYTICAL SPECTROSCOPY. *Educ:* Univ Tenn, PhD(biochem), 69. *Prof Exp:* PROF BIOCHEM, UNIV GA, 82- *Concurrent Pos:* Ed, Anal Instrumentation. *Res:* Stimulus response coupling in single living animal cells; studies of computation approaches to biochemical problems. *Mailing Add:* Dept Biochem Univ Ga 622 Grad Athens GA 30602

WAMSER, CARL CHRISTIAN, b New York, NY, Aug 10, 44; c 2. ORGANIC CHEMISTRY, PHOTOCHEMISTRY. *Educ:* Brown Univ, ScB, 66; Calif Inst Technol, PhD(chem), 69. *Prof Exp:* US Air Force Off Sci Res-Nat Res Coun fel, Harvard Univ, 69-70; from asst prof to assoc prof chem, 70-77, PROF CHEM, CALIF STATE UNIV, FULLERTON, 77- *Concurrent Pos:* Vis assoc prof, Univ Southern Calif, 75-76; vis prof, Univ Hawaii, 80. *Mem:* Am Chem Soc. *Mailing Add:* Dept Chem Portland State Univ Portland OR 97207-0751

WAMSER, CHRISTIAN ALBERT, b Long Island City, NY, July 15, 13; m 40; c 2. INORGANIC CHEMISTRY. *Educ:* Cooper Union, BS, 34. *Prof Exp:* Anal chemist, J F Jelenko & Co, Inc, NY, 34-41; supvr anal group, Gen Chem Div, Allied Chem Corp, 41-48; res chemist, Vitro Corp Am, NJ, 53-62; res chemist, Gen Chem Div, Allied Chem Corp, 62-65, res scientist, 65-69, res scientist, Syracuse Tech Ctr, 69-74, res assoc, Indust Chem Div, 74-78; RETIRED. *Concurrent Pos:* Consult Allied Chem Corp, 78-86. *Mem:* Am Chem Soc. *Res:* Industrial inorganic chemistry, uranium, fluorine, chromium and aluminum compounds. *Mailing Add:* 207 Rebhahn Dr Camillus NY 13031-1919

WAMSLEY, W(ELCOME) W(ILLARD), b Leavenworth, Wash, July 18, 25; m 47; c 3. CHEMICAL ENGINEERING. *Educ:* Univ Wash, BS, 49, PhD(chem eng), 53. *Prof Exp:* Instr, Univ Wash, 51-53; chem engr, E I du Pont de Nemours & Co Inc, 53-60, tech supvr, 60-61, sr tech supvr, 61-64, spec asst, 64-65, supt process control, 65-75, asst plant mgr, 75-80, plant mgr, 80-83, raw mats mgr, 83-85; RETIRED. *Mem:* Am Chem Soc; assoc Am Inst Chem Engrs; Sigma Xi. *Res:* Heat transfer; gas-solid fluidization; coating and drying of photographic emulsions. *Mailing Add:* 114 Marlbrooke Way Kennett Square PA 19348

WAN, ABRAHAM TAI-HSIN, b Tsingtao, China, Oct 14, 28; US citizen; m 59; c 2. CLINICAL CHEMISTRY, ENDOCRINOLOGY. *Educ:* Nat Taiwan Univ, BS, 55; Univ Minn, Minneapolis, MS, 60; Univ Nebr, Omaha, PhD(med biochem), 64. *Prof Exp:* Biochemist, Dept Pub Health, Sask, Can, 63-64 & Sunland Hosp, Orlando, 64-67; assoc dir clin chem, Clin Lab, Dept Path, Norfolk Gen Hosp, 67-74; asst prof, Eastern Va Med Sch, 74; DEP DIR, DEPT BIOCHEM, ALFRED HOSP, 74- *Concurrent Pos:* NIH grant, Sunland Hosp, Orlando, Fla, 64-67 & Norfolk Gen Hosp, 72-74; adj prof, Old Dominion06095501xxx4. *Mem:* Nat Acad Clin Biochem; Am Asn Clin Chemists; Am Soc Clin Pathologists; Australia Asn Clin Biochemists; NY Acad Sci; Am Chem Soc. *Res:* Bismuth retention and excretion in urine and bile juice in mormal and renal patients after oral intake of Bi-subcitrate. *Mailing Add:* Dept Biochem Alfred Hosp Prahran Victoria 3181 Australia

WAN, FREDERIC YUI-MING, b Shanghai, China, Jan 7, 36; US citizen; m 60. SOLID MECHANICS. *Educ:* Mass Inst Technol, SB, 59, SM, 63, PhD(math), 65. *Prof Exp:* Staff mem struct mech, Lincoln Lab, Mass Inst Technol, 59-62, staff assoc, 62-65, from instr to assoc prof appl math, 65-74; prof math & dir Inst Appl Math & Statist, Univ BC, 74-85; PROF, UNIV WASH, 83-, CHMN, APPL MATH DEPT, 84-, ASSOC DEAN, COL ARTS & SCI, 88- *Concurrent Pos:* Sloan Found fel, 73; Killam sr fel, Univ BC, 79; vis fel Econ, Mass Inst Technol; vis assoc Appl Mech, Calif Technol; consult, indust & govt agencies. *Mem:* Fel Am Soc Mech Eng; Soc Indust & Appl Math; Can Math Soc; Can Appl Math Soc (pres, 83-84); fel Am Acad Mech. *Res:* Classical elasticity; shell theory; random vibrations; stochastic ordinary and partial differential equations; natural resource economics; bio-mathematics. *Mailing Add:* Dept Appl Math Univ Wash Seattle WA 98195

WAN, JEFFREY KWOK-SING, b Hong Kong, June 4, 34; m 62; c 2. PHYSICAL CHEMISTRY. *Educ:* McGill Univ, BSc, 58; Univ Alta, PhD(phys chem), 62. *Prof Exp:* Fel, Univ Alta, 62-63; asst res chemist, Univ Calif, 63-65; Nat Res Coun Can fel chem, 65-66; from asst prof to assoc prof, 66-74, PROF CHEM, QUEEN'S UNIV, 74- *Concurrent Pos:* Hon prof, Univ Lanzhou, China, 81-; pres, W & Y Consults Kingston Ltd, Can, 82- *Mem:* Chem Inst Can. *Res:* Photochemistry and electron paramagnetic resonance spectroscopy; microwave induced catalysis. *Mailing Add:* Dept Chem Queen's Univ Kingston ON K7L 3N6 Can

WAN, PETER J, b Shantong, China, Jan 1, 29, 43; US citizen; m 70; c 2. PROCESS DEVELOPMENT, METHODS DEVELOPMENT. *Educ:* Cheng-Kung Univ, China, BSE, 65; Ill Inst Technol, MS, 70; Tex A&M Univ, PhD(phys chem), 73. *Prof Exp:* Fel, Chem Dept, Tex A&M Univ, 74-75, proj leader & asst res chemist, Res & Develop Ctr Food Protein, 75-79; sr res chemist, Best Foods, CPC Int, 79-80; res assoc, Anderson Clayton Foods, 80-83, mgr, 83-84, dir, 84-87; TECHNOL MGR, KRAFT INC, 87- *Mem:* Inst Food Technologists; Am Oil Chemists Soc; Am Chem Soc; Soc Rheology. *Res:* Improve and assure the quality of product; improve process efficiency; develop new methods, new processes and new products. *Mailing Add:* USDA ARS S Reg Res Ctr Box 19687 1100 Robert Lee Blvd New Orleans LA 70179-0687

WAN, SUK HAN, pharmacodynamics, clinical pharmacology, for more information see previous edition

WAN, YIEH-HEI, b China, Feb 17, 47; m 75. TOPOLOGY, APPLIED MATHEMATICS. *Educ:* Nat Taiwan Univ, BS, 68; Univ Calif, Berkeley, PhD(math), 73. *Prof Exp:* From asst prof to assoc prof, 73-84, PROF MATH, STATE UNIV NY, BUFFALO, 84- *Res:* Application of global analysis to the study of general competitive equilibrium theory in mathematical economics; bifurcation theory for dynamical systems. *Mailing Add:* Dept Math State Univ NY Buffalo Diefendorf Hall Buffalo NY 14214

WANAT, STANLEY FRANK, b Nanticoke, Pa, Dec 31, 39; m 64; c 2. ORGANIC CHEMISTRY. *Educ:* Rutgers Univ, AB, 63; Seton Hall Univ, MS, 69, PhD(org chem), 71. *Prof Exp:* Develop chemist agr chem, Shell Chem Co, 64-65; instr chem & math, Union County Col Syst, NJ, 65-67; process develop chemist agr chem, Ciba-Geigy Corp, 67-70; asst, Seton Hall Univ, 70-71; instr org chem, Upsala Col, 71-73; RES & DEVELOP MGR GRAPHIC ARTS CHEM, PPD DIV, HOECHST CELANESE CORP, 73- *Concurrent Pos:* Teacher chem, Union County Schs, 71-73. *Mem:* Am Chem Soc; Soc Photo Sci & Eng. *Res:* Development of photosensitive lithographic products, printing plates, color proofing systems and chemicals; study of surface chemistry of substrates suitable for coating lithographic materials; coating technology; polymer modification for lithographic use. *Mailing Add:* Three Frances Lane Scotch Plains NJ 07076

WAND, MITCHELL, b Philadelphia, Pa, Nov 6, 48; m 70; c 3. COMPUTER SCIENCES. *Educ:* Mass Inst Technol, SB, 69, PhD(math), 73. *Prof Exp:* Prof computer sci, Ind Univ, 73-85; PROF & ASSOC DEAN COMPUTER SCI, COL COMPUT SCI, NORTHEASTERN UNIV, BOSTON, 85- *Concurrent Pos:* Vis prof, Brandeis Univ, 84-85; vis scientist, Lab Computer Sci, Mass Inst Technol, 84- *Mem:* Asn Comput Mach; Europ Asn Theoret Comput Sci. *Res:* Semantics of programming languages; programming theory; logic; algebra. *Mailing Add:* Col Comput Sci Northeastern Univ 360 Huntington Ave 161CN Boston MA 02115

WAND, RONALD HERBERT, ionospheric physics, for more information see previous edition

WANDASS, JOSEPH HENRY, b Buffalo, NY, Feb 16, 60; m 87. SPECTROSCOPY, MATERIALS SCIENCE ENGINEERING. *Educ:* State Univ NY Buffalo, BS, 82, PhD(chem), 86. *Prof Exp:* Postdoctoral fel surface chem, Naval Res Lab, 86-88; SR RES CHEMIST SURFACE CHEM, AKZO CHEM INC, 88- *Mem:* Am Chem Soc. *Res:* Surface chemistry and analysis using ion, electron, and photon probes; sol-gel synthesis of thin films; computer methods in spectroscopy. *Mailing Add:* Akzo Chem Inc One Livingstone Ave Dobbs Ferry NY 10547-3401

WANDER, JOSEPH DAY, b Columbus, Ohio, July 20, 41; m 67; c 3. ORGANIC CHEMISTRY. *Educ:* Case Inst Technol, BS, 63; Ohio State Univ, PhD(chem), 70. *Prof Exp:* Res fel chem, Tulane Univ, 70-71, La State Univ, Baton Rouge, 71-72 & Ohio State Univ, 72-74; dir, Charles B Stout Neurosci Lab, Univ Tenn Ctr Health Sci, Memphis, 74-77; asst prof chem, Univ Ga, 78-84 & Columbus Col, Ga, 85-86; FUELS CHEMIST, USAF, TYNDALL AFB, 86- *Mem:* Am Chem Soc; Sigma Xi; Soc Am Mil Engrs. *Mailing Add:* HQ AFESC/RDVS Tyndall AFB FL 324036001

WANDERER, PETER JOHN, JR, b Monroe, La, Aug 5, 43; m 72; c 3. HIGH ENERGY PHYSICS. *Educ:* Univ Notre Dame, BS, 65; Yale Univ, PhD(physics), 70. *Prof Exp:* Res assoc high energy physics, Lab Nuclear Studies, Cornell Univ, 70-73 & Univ Wis-Madison, 73-75; assoc scientist, 75-78, SCIENTIST HIGH ENERGY PHYSICS, BROOKHAVEN NAT LAB, 78- *Mem:* Am Phys Soc. *Res:* Experimental research in neutrino-nucleon interactions; construction of high field superconducting magnets for SSC and RHIC accelerators. *Mailing Add:* Bldg 902B Brookhaven Nat Lab Upton NY 11973

WANDMACHER, CORNELIUS, b Brooklyn, NY, Sept 1, 11; m 36; c 3. CIVIL ENGINEERING. *Educ:* Polytech Inst Brooklyn, BCE, 33, MCE, 35. *Hon Degrees:* DEng, Polytech Inst Brooklyn, 69 & Rose-Hulman Inst, 75. *Prof Exp:* Civil engr, Plant Eng Dept, WVa Pulp & Paper Co, NY, 36-37; from instr to assoc prof civil eng, Polytech Inst Brooklyn, 37-51, dir eve session & asst to pres, 43-51; William Thomas prof & head dept, 51-58, from assoc dean to dean, Col Eng, 57-74, prof civil eng, 58-81, EMER DEAN, UNIV CINCINNATI, 81- *Concurrent Pos:* Designer & detailer, Off Struct Engr, NY Cent RR, 39; eng examr, Munic Civil Serv Comn, New York, 39-51; struct designer, Phelps Dodge Corp, 40; mem, Bd Regist Prof Engrs, Ohio, 62-72. *Mem:* Hon mem Am Soc Civil Engrs; hon mem Am Soc Eng Educ; Nat Soc Prof Engrs; Am Nat Metric Coun (dir, 76-82); fel Am Soc Test Mat. *Res:* Professional engineering education. *Mailing Add:* 1722 Larch Ave Cincinnati OH 45224

WANDS, RALPH CLINTON, b Norwich, NY, May 12, 19; m 42; c 3. TOXICOLOGY, INDUSTRIAL HYGIENE. *Educ:* Kent State Univ, BS, 41; Univ Minn, MS, 48; Am Bd Indust Hyg, dipl. *Prof Exp:* Chemist, Firestone Tire & Rubber Co, 41-42, develop engr, 42-43; res chemist, Minn Mining & Mfg Co, 51-61 & Indust Hyg & Toxicol, 61-64; prof assoc, Nat Acad Sci, 64-66, dir, Adv Ctr Toxicol, 66-77; dir cosmetic ingredient rev, Cosmetic Toiletry & Fragrance Asn, 77-79; head, Off Info, Chem Indust Inst Toxicol, 79-80; group leader hazardous substances eval, Mitre Corp, 80-84; SELF EMPLOYED TOXICOL & INDUST HYG, 84- *Honors & Awards:* Herbert E Stokinger Award. *Mem:* AAAS; Am Chem Soc; hon mem Am Indust Hyg Asn; Soc Toxicol; Am Conf Govt Indust Hygienists; Sigma Xi. *Res:* Toxicology data evaluation and risk assessment for occupational and public environments and development of acceptable exposure limits; chemistry. *Mailing Add:* 5445 Mt Corcoran Pl Burke VA 22015

WANE, MALCOLM T(RAFFORD), b Pittston, Pa, Jan 2, 21; m 46; c 1. MINING ENGINEERING. *Educ:* Lehigh Univ, BS, 50, Columbia Univ, MS, 54. *Prof Exp:* Res technician & mineral engr, US Steel Corp, 50-53; assoc, 53-59, from asst prof to assoc prof, 59-69, PROF MINING, COLUMBIA UNIV, 69- *Concurrent Pos:* NSF fac fel, Royal Sch Mines, 65-66; consult geologist, New York City Transit Authority & New York City Bd Water Resource Develop. *Mem:* Am Inst Mining, Metall & Petrol Engrs. *Res:* Rock mechanics; failure of brittle solids; earth vibration and blasting problems. *Mailing Add:* 38 Mile Rd Suffern NY 10901

WANEBO, HAROLD, SURGICAL ONCOLOGY, IMMUNOLOGY. *Educ:* Univ Colo, MD, 61. *Prof Exp:* Prof surg & chief, Div Surg Oncol, Univ Va, 77-87; CHMN SURG, ROGER WILLIAM HOSP, PROVIDENCE, 87- *Mailing Add:* Roger William Hosp 825 Chalkstone Ave Providence RI 02908

WANG, ALBERT SHOW-DWO, b Chifoo, China, July 13, 37; m 67; c 2. APPLIED MECHANICS, COMPOSITE MATERIALS. *Educ:* Univ Taiwan, BS, 59; Univ Nev, MS, 63; Univ Del, PhD(aerospace eng), 67. *Prof Exp:* Fel, Univ Del, 67; from asst prof to assoc prof, 67-77, PROF APPL MECH, DREXEL UNIV, 77- *Concurrent Pos:* Consult, Lockheed Corp, Lawrence Livermore Lab & Gen Motors Corp; res fel, Acad Sci, 83; vis scientist, Air Force Off Sci Res, 86-87. *Mem:* Am Soc Mech Engrs; Am Soc Testing & Mat; Soc Advan Mat & Process Eng. *Res:* Stress analysis of composite materials, including failure, fracture, fatigue and reliability. *Mailing Add:* Dept Mech Eng Drexel Univ 32nd & Chestnut Sts Philadelphia PA 19104

WANG, AN, physics, engineering; deceased, see previous edition for last biography

WANG, AN-CHUAN, b Tsing Tao City, China, Dec 28, 36; m 65; c 2. IMMUNOGENETICS. *Educ:* Nat Taiwan Univ, BS, 59; Univ Tex, PhD(genetics), 66. *Prof Exp:* Sec teacher, Taiwan, 61; res assoc genetics, Univ Tex, 66-67; asst prof genetics, Med Ctr, Univ Calif, San Francisco, 70-72, assoc prof microbiol & assoc researcher med, 72-75; assoc prof immunol, 75-76, PROF IMMUNOL, MED UNIV SC, 76-; CONSULT, 86- *Concurrent Pos:* Fel, Med Ctr, Univ Calif, San Francisco, 67-70; USPHS & NSF res grants, 70-; USPHS career development award, 74; Am Cancer Soc fac res award, 74-79; adv ed, Immunochem, 78- *Mem:* AAAS; Am Soc Human Genetics; Genetics Soc Am; Am Asn Immunol. *Res:* Biochemical and genetical analyses of human plasma proteins, with special emphasis on immunoglobulins; immunology. *Mailing Add:* Med Univ SC 855 Robert E Lee Blvd Charleston SC 29412

WANG, ANDREW H-J, b Taipei, Taiwan, Nov 29, 45; US citizen; m 68; c 3. BIOPHYSICAL CHEMISTRY, STRUCTURAL BIOLOGY. *Educ:* Nat Taiwan Univ, BS, 67, MS, 70; Univ Ill, Urbana, PhD(chem), 74. *Prof Exp:* Res assoc, Dept Biol, Mass Inst Technol, 74-80, res scientist, 80-82, prin res scientist, 82-85, sr res scientist, Dept Biol, Mass Inst Technol, 85-88, PROF, DEPT PHYSIOL & BIOPHYS, UNIV ILL, URBANA-CHAMPAIGN, 88- *Mem:* Am Chem Soc; Am Crystallog Asn; AAAS; Am Soc Biochem Molecular Biol; Am Inst Chem. *Res:* Molecular structure of biological macromolecules; structure-function relationship of proteins and nucleic acids; x-ray crystallography. *Mailing Add:* Dept Physiol & Biophys Univ Ill Urbana IL 61801

WANG, AUGUSTINE WEISHENG, microbiology, for more information see previous edition

WANG, BIN, WAVE & INSTABILITY, GENERAL CIRCULATION & WEATHER SYSTEMS. *Educ:* Qingdao Ocean Univ, BS, 66; Univ Sci & Technol, China, MS, 82; Fla State Univ, PhD(geophys fluid dynamics), 84. *Prof Exp:* Vis scientist geophys fluid dynamics, Geophys Fluid Dynamics Prog, Princeton Univ, 84-86; asst prof, 87-89, ASSOC PROF METEOROL, DEPT METEOROL, UNIV HAWAII, 89- *Concurrent Pos:* Sr fel, Joint Inst Marine & Atmospheric Res, Univ Hawaii, Nat Oceanic & Atmospheric Admin, 88-; guest prof, Ocean Univ Qingdao, Peoples Repub China, 90- & First Inst Oceanog, State Oceanic Admin, Peoples Repub China, 89- *Mem:* Am Meteorol Soc; Am Geophys Soc. *Res:* Atmospheric, oceanic wave and instability; climate dynamics. *Mailing Add:* 7268 Waiopua St Honolulu HI 96825

WANG, BIN CHING, b Taipei, Taiwan, July 11, 41; m 65; c 2. PHYSIOLOGY, BIOLOGY. *Educ:* Nat Taiwan Normal Univ, BEd, 63; Northwestern State Univ La, MS, 71; Univ Kans, PhD(physiol), 75. *Prof Exp:* Health educator, Nat Tuberc Asn, Taiwan, 63-66; res asst, Vet Gen Hosp, Taiwan, 66-70; res physiologist, St Luke's Hosp, 75-78; asst prof physiol, Univ Tenn Med Ctr, 79-81; sr scientist, St Luke's Hosp, 81-; AT BIOCRYSTALLOG LAB, VET ADMIN MED CTR, PITTSBURGH, PA; SR SCI, ST LUKE'S HOSP. *Concurrent Pos:* Adj instr physiol, Univ Kans Med Ctr, 75, adj asst prof, 78. *Mem:* AAAS; assoc mem Am Physiol Soc. *Res:* Cardiovascular physiology, especially control of blood pressure. *Mailing Add:* Div Exp Med St Luke's Hosp 44th St & Wornall Rd PO Box 119000 Kansas City MO 64111-9000

WANG, BOSCO SHANG, b Shang-hai, China, Aug 30, 47; US citizen; m 72; c 2. TUMOR IMMUNOLOGY, IMMUNOENDOCRINOLOGY. *Educ:* Fu-Jen Cath Univ, BS, 69; Mich State Univ, MS, 73; Boston Univ, PhD(microbiol), 76. *Prof Exp:* Postdoctoral path, Harvard Med Sch, 76-77, instr, 77-79, asst prof, 79-81; PRIN RES SCIENTIST, AM CYANAMID CO, 81- *Honors & Awards:* Wilson Stone Mem Award, M D Anderson Hosp & Tumor Inst, Tex Cancer Ctr, 77. *Mem:* Am Asn Immunologists. *Res:* Immunologic regulation of endocrine systems. *Mailing Add:* Am Cyanamid Co PO Box 400 Princeton NJ 08543-0400

WANG, C(HIAO) J(EN), b China, Mar 24, 18; nat US; m 45; c 1. AERONAUTICS, MECHANICAL ENGINEERING. *Educ:* Chiao Tung Univ, China, BS, 42; Mass Inst Technol, MSME, 46; Johns Hopkins Univ, PhD, 53. *Prof Exp:* Instr, Johns Hopkins Univ, 49-53; res & eng specialist, NAm Aviation, Inc, 53-56; assoc mgr aerophys dept & mgr propulsion systs dept, Space Technol Labs, Inc, 56-60; head propulsion dept & dir advan studies, Aerospace Corp, 60-65; dep dir combined arms res off, Booz-Allen Appl Res, Inc, Kans, 65-66; DIR OFF ADVAN ENG, ADVAN RES PROJS AGENCY, US DEPT DEFENSE, DC, 66- *Res:* Aeronautical and space science and technology; systems analysis; operations research. *Mailing Add:* 1300 Army-Navy Dr Arlington VA 22202

WANG, CARL C T, b Hankow, China, Dec 2, 35; m 63; c 3. ELECTRICAL ENGINEERING. *Educ:* Univ Ill, BS, 58, MS, 59, PhD(elec eng), 64. *Prof Exp:* Res asst plasma physics, Univ Ill, 60, res asst microwave, 60-64; mem res staff electron-optics, T J Watson Res Ctr, Int Bus Mach Corp, 64-67; res assoc biomed eng & elec engr, Columbia Univ, 67-69; sr engr, Micro-Bit Corp, 69-75; vpres eng, Berkeley Bio-Eng, Inc, 75-79; vpres eng, Cooper Med

Device Corp, 79-81; pres, Med Instrument Develop Labs Inc, 81-85; vpres res, Alcon Surg Instrument, 85-90; PRES, MED INSTRUMENTATION DEVELOP LABS INC, 91- *Concurrent Pos:* Tech consult, NIMH, 68-75, NEI, 84-85, Univ Calif, Los Angeles, 83-84, Univ Southern Calif, 82-83. *Mem:* Inst Elec & Electronics Engrs; Am Acad Ophthal; Sigma Xi; Asn Res in Vision & Ophthal; Chinese-Am Ophthal Soc; World Eye Found. *Res:* Laser; ultra-microwave; electromagnetohydrodynamics; electron-optics; display systems; man-machine systems; physical electronics; physiological system simulation; bio-medical instrumentation. *Mailing Add:* Med Instrumentation Develop Labs Inc 1240 S Loop Rd Alameda CA 94501

WANG, CHANG-YI, b Kweichow, China, Aug 26, 39; US citizen; m 66; c 3. APPLIED MATHEMATICS. *Educ:* Nat Taiwan Univ, BS, 60; Mass Inst Technol, MS, 63, PhD, 66. *Prof Exp:* Fel appl math, Calif Inst Technol, 66-67, assoc appl math, Jet Propulsion Lab, 67-68; asst prof math, Univ Calif, Los Angeles, 68-69; from asst prof to assoc prof math, 69-77, PROF MATH, PHYSIOL & MECH ENG, MICH STATE UNIV, 77- *Concurrent Pos:* Vis prof, Nat Taiwan Univ, 71-72 & 84-85, Nat Tsing Hua Univ, 84-85. *Mem:* Soc Indust & Appl Math; Am Soc Mech Engrs. *Res:* Fluid mechanics; elasticity; biomathematics. *Mailing Add:* Dept Math Mich State Univ East Lansing MI 48824

WANG, CHAO CHEN, b Changchow, Kiangsu, China, Oct 20, 14; US citizen; m 47; c 1. ENGINEERING PHYSICS. *Educ:* Chiao Tung Univ, Shanghai, China, 36; Harvard Univ, SM, 38, ScD, 40. *Prof Exp:* Proj engr microwave, Westinghouse Elec, 41-44; res engr, head dept eng electronics & chief scientist, div beams & laser, Sperry Corp, 45-73; PRES & DIR INDUST RES, INDUST TECHNOL RES INST, 73- *Concurrent Pos:* Vis prof, Cornell Univ, 60-61. *Mem:* Am Phys Soc; Inst Elec & Electronics Engrs; AAAS. *Res:* Microwave tubes for radar during World War II; high density electron beam control and focusing pioneered laser ring gyro development, electromagnetic wave propagation and interaction with electron beams; high power klystron development for radar and for linear accelerator. *Mailing Add:* Indust Tech Res Inst 12 Chestnut PO Box 54 Ridge Rd Holmdel NJ 07733

WANG, CHAO-CHENG, b China, July 20, 38; US citizen; m 63; c 2. MECHANICS. *Educ:* Nat Taiwan Univ, BS, 59; Johns Hopkins Univ, PhD(mech), 65. *Prof Exp:* From asst prof to assoc prof mech, Mech Dept, Johns Hopkins Univ, 66-69; prof, 69-79, chmn, Math Sci Dept, 83-89, NOAH HARDING PROF MATH SCI & PROF, MECH ENG DEPT, RICE UNIV, 79-, CHMN, MECH ENG & MAT SCI DEPT, 91- *Concurrent Pos:* Prof, Mech Eng Dept, Rice Univ, 79- *Mem:* Soc Natural Philos; Am Acad Mech. *Res:* Continuum mechanics; applied mathematics; mechanical engineering. *Mailing Add:* Mech Eng & Mat Sci Dept Rice Univ Houston TX 77251-1892

WANG, CHARLES C(HEN-DING), b Hankow, China, Sept 4, 33; m 61; c 3. ELECTRICAL ENGINEERING, PHYSICS. *Educ:* Taiwan Col Eng, BS, Brown Univ, MS, 58; Stanford Univ, PhD, 60. *Prof Exp:* Asst prof elec eng, Univ Wash, 60-63; res specialist, Philco Res Lab, 63-65; prin res assoc scientist, Ford Sci Lab, 66-68, staff scientist, 68-77, prin res scientist, 77-87; PRES, PENINSULA TECHNOL INC, 87- *Concurrent Pos:* Adj prof elec eng, Wayne State Univ, 76-87; adj prof physics, Univ Mich, 81- *Mem:* Fel Am Phys Soc; Inst Elec & Electronics Engrs. *Res:* Nonlinear optics, quantum electronics; laser spectroscopy; atomic and molecular physics; detection of oxygen hydrogen in the atmosphere. *Mailing Add:* Box 463 Franklin MI 48025

WANG, CHARLES P, b Shanghai, China, Apr 26, 37; m 63. OPTICS, AUTOMATION & ROBOTICS. *Educ:* Nat Taiwan Univ, Taiwan, BS, 59; Tsinghua Univ, Taiwan, MS, 61; Calif Inst Technol, PhD(aeronaut), 67. *Prof Exp:* Lectr, appl math, Nat Taiwan Univ, 61-62; mem tech staff, aronaut, Bellcom Inc, 67-69; sr scientist, quantum elec, Aerospace Corp, 74-86; PRES, OPTODYNE INC, 86- *Concurrent Pos:* Adj prof, eng physics, Univ Calif, San Diego, 74-88; assoc ed, Am Inst Aeronaut & Astronaut, 80-84. *Mem:* Fel Optical Soc Am; Am Inst Aeronaut & Astronaut; Soc Optical & Quantum Electronics (prog chmn, 85); Chinese Am Engr & Scientists. *Res:* Research on fluid mechanics, laser applications in fluid mechanics, and quantum electronics, has successfully developed a CW 150 W Argon-Ion Laser, discharge excimer laser, stable chemical laser and optical phase arrays for higher power chemical lasers. *Mailing Add:* Optodyne Inc 1180 Mahalo Pl Compton CA 90220

WANG, CHARLES T P, b Shantung, China, May 6, 30; US citizen; m 60; c 2. PHYSICS. *Educ:* Taiwan Norm Univ, BS, 55; Southern Ill Univ, Carbondale, MS, 59; Wash Univ, PhD(physics), 66. *Prof Exp:* Lectr physics, Southern Ill Univ, Edwardsville, 59-62; from asst prof to assoc prof, Parks Col, St Louis Univ, 65-67; PROF PHYSICS, STATE UNIV NY COL ONEONTA, 67- *Concurrent Pos:* NSF res participation fel, La State Univ, Baton Rouge, 69 & 71. *Mem:* Am Asn Physics Teachers; Sigma Xi; Am Phys Soc. *Res:* Strong dynamical correlations in nuclear matter. *Mailing Add:* State Univ NY Col Oneonta NY 13820

WANG, CHEN-SHOW, b Taiwan; US citizen; c 3. SOLID STATE PHYSICS, QUANTUM ELECTRONICS. *Educ:* Nat Taiwan Univ, BS, 59; Univ Iowa, MS, 64; Univ Calif, San Diego, PhD(physics), 68. *Prof Exp:* Fel physics, Univ Calif, San Diego, 68-69; res fel, Harvard Univ, 69-72; from asst prof to assoc prof physics, Bartol Res Found, Univ Del, 72-79; mgr res & develop, 79-82, dir, 82-84, VPRES RES & DEVELOP, GEN OPTRONICS CORP, 84- *Mem:* Am Phys Soc; Sigma Xi; NY Acad Sci. *Res:* Solid state physics and quantum electronics, surface physics, lattice dynamics, nonlinear optics laser light scattering, solid state lasers, gas lasers and semiconductor lasers. *Mailing Add:* 58 Mt Horeb Rd Warren NJ 07060

WANG, CHIA PING, b Philippines; US citizen. NUCLEAR PARTICLE & RADIATION PHYSICS, THERMAL PHYSICS. *Educ:* Univ London, BSc, 50; Univ Malaya, MSc, 51; Univs Malaya & Cambridge, PhD(physics), 53. *Hon Degrees:* DSc, Univ Singapore, 72. *Prof Exp:* Asst lectr, Univ Malaya, 51-53; assoc prof physics, Nankai Univ, 54-56, head electron physics & electronics div, 55-58, prof, 56-58; head electrophysics div, Lanchow Proj, 58; sr lectr, prof & actg head, Depts Physics & Math, Hong Kong Univ & Chinese Univ Hong Kong, 58-63; res assoc, Lab Nuclear Studies, Cornell Univ, 63-64; assoc prof space sci & physics, Cath Univ, 64-66; assoc prof physics, Case Inst Technol & Case Western Reserve Univ, 66-70; vis scientist & vis prof, Univs Cambridge & Louvain, US Naval Res Lab, Univ Md & Mass Inst Technol, 70-75; RES PHYSICIST, SCI & ADVAN TECHNOL DIR, US ARMY NATICK RES & DEVELOP CTR, 75- *Mem:* Am Phys Soc; Inst Physics, London; AAAS; Sigma Xi; NY Acad Sci. *Res:* Nuclear particle physics; cosmic rays; neutrinos; space physics; nucleon sub-structure; ultra-high energy particle production; quantum electrodynamics and quantum electronics; dosimetry; laser physics; nonlinear optics; thermal physics; microwaves. *Mailing Add:* 28 Hallett Hill Rd Weston MA 02193

WANG, CHIA-LIN JEFFREY, b China, June 24, 49; m 73. CHEMISTRY. *Educ:* Nat Taiwan Univ, BS, 71; Univ Pittsburgh, PhD(chem), 77. *Prof Exp:* Res assoc, Dept Chem, Univ Pittsburgh, 77, Harvard Univ, 78-79; res chemist, cent res & develop dept, E I du Pont de Nemours & Co, Inc, 79-85, sr res chemist, Med Dept, 85-90; SR RES CHEMIST, MED DEPT, DU PONT MERCK CO, 91- *Mem:* Am Chem Soc. *Res:* Synthesis of medicinally interesting compounds. *Mailing Add:* Du Pont Merck Co Exp Sta PO Box 80353 Wilmington DE 19880-0353

WANG, CHIEN BANG, b China, Oct 10, 41; m 70; c 2. POLYMER MORPHOLOGY, SYNTHETIC FIBERS. *Educ:* Nat Taiwan Univ, BS, 62; Kans State Univ, MS, 65; Univ Wis, PhD(chem eng), 69. *Prof Exp:* Res chem engr, 69-71, sr res chem engr, 72-85, RES ASSOC, EASTMAN CHEM DIV, EASTMAN KODAK CO, 85- *Res:* Polymer morphology; process and product development of synthetic fibers; textile processing. *Mailing Add:* 2313 Oxford Ct Kingsport TN 37660

WANG, CHIEN YI, b Fengshan, Taiwan, Nov 22, 42; US citizen; m 20; c 2. POSTHARVEST PHYSIOLOGY, HORTICULTURE. *Educ:* Nat Taiwan Univ, BS, 64; Ore State Univ, PhD(hort), 69. *Prof Exp:* Res assoc fruits, Ore State Univ, 69-76; RES HORTICULTURIST, USDA, 76- *Mem:* Am Soc Hort Sci; Am Soc Plant Physiol. *Res:* Postharvest physiology of horticultural crops; basic and applied problems concerning physiological and pathological deterioration and quality maintenance of fruits, vegetables and flowers after harvest. *Mailing Add:* Bldg 002 Agr Res Ctr-West Beltsville MD 20705

WANG, CHIH CHUN, b Peking, China, Oct 9, 32; m 59; c 3. MATERIALS SCIENCE, PHYSICAL CHEMISTRY. *Educ:* Nat Taiwan Univ, BSc, 55; Kans State Univ, MSc, 59; Colo State Univ, PhD(phys chem), 62. *Prof Exp:* Res assoc, High Temp Phys Chem Res Lab, Univ Kans, 62-63; mem tech staff solid state mat res, 63-73, FEL TECH STAFF, RCA LABS, DAVID SARNOFF RES CTR, RCA CORP, PRINCETON, 73- *Mem:* Am Chem Soc; Electrochem Soc; Am Phys Soc. *Res:* Electronic materials; thin films; crystal growth; chemical vapor deposition; high pressure and high temperature chemistry; thermodynamics; x-ray crystallography; vidicon materials and devices; tribology. *Mailing Add:* Am Lumi Cranbury Plaza Bldg B 2525 Rte 130 PO Box 515 Cranbury NJ 08512-0515

WANG, CHIH HSING, b Shanghai, China, Sept 20, 17; nat US; m 58; c 1. RADIOCHEMISTRY. *Educ:* Shantung Univ, China, BS, 37; Ore State Univ, MS, 47, PhD(chem), 50. *Prof Exp:* From asst prof to prof chem, 51-85, dir, Radiation Ctr, 62-85, dir, Inst Nuclear Sci & Eng, 64-85, head, Dept Nuclear Eng, 74-85, EMER PROF & DIR, RADIATION CTR, ORE STATE UNIV, 85- *Concurrent Pos:* Consult, NSF, 65-69; chmn, Ore Nuclear & Thermal Energy Coun, 72-73. *Mem:* Fel AAAS; Am Chem Soc; Am Soc Biol Chem; Am Soc Plant Physiol; FEL Am Nuclear Soc. *Res:* Nuclear education; radiotracer methodology. *Mailing Add:* Radiation Ctr Ore State Univ Corvallis OR 97330

WANG, CHIH-CHUNG, b Wusih, China, Mar 8, 22; US citizen; m 53; c 2. METALLURGY. *Educ:* Chiao-Tung Univ, BS, 45; Ill Inst Technol, MS, 50; Mass Inst Technol, DSc, 53. *Prof Exp:* Sr engr, Sylvania Elec Prod, Inc, 53-55; dir, Mat & Metall Dept, Clevite Transistor Prod, Inc, 55-63; staff scientist, Ledgemont Lab, Kennecott Copper Corp, 63-78, mgr metal prod, Lexington Develop Ctr, 78-80; CHIEF METALLURGIST, DURACELL INC, 81- *Mem:* Am Inst Mining, Metall & Petrol Engrs; Am Soc Metals; Electrochem Soc. *Res:* Solidification of metals; crystal growing; materials research; extractive metallurgy; electroplating; recycling of metals. *Mailing Add:* Nine Gould Rd Lexington MA 02173

WANG, CHIH-LUEH ALBERT, b Chia-yi, Taiwan, China, June 22, 50; US citizen; m; c 2. CONTRACTILE PROTEINS, FLUORESCENCE SPECTROSCOPY. *Educ:* Nat Taiwan Univ, BS, 71; Ohio State Univ, PhD(chem), 78. *Prof Exp:* Postdoctoral fel, Boston Biomed Res Inst, 79-82, res assoc, 82-84, staff scientist, 84-88, PRIN SCIENTIST, BOSTON BIOMED RES INST, 88- *Mem:* Am Chem Soc; Biophys Soc; Am Soc Biochem & Molecular Biol; Soc Chinese Bioscientists Am. *Res:* Structure and function of regulatory proteins in smooth muscle; calcium-binding proteins; application of rare earth ions in biological systems. *Mailing Add:* 20 Staniford St Boston MA 02114

WANG, CHI-HUA, b Peking, China, Apr 18, 23; m 49; c 1. ORGANIC CHEMISTRY. *Educ:* St John's Univ, China, BS, 45; Cath Univ, China, MS, 47; St Louis Univ, PhD(org chem), 51. *Prof Exp:* Fel, Brandeis Univ, 51-53, from instr to assoc prof chem, 53-62; sr chemist, Arthur D Little, Inc, 62-64; assoc prof chem, Wellesley Col, 64-68; assoc prof, 68-70, PROF CHEM, UNIV MASS, HARBOR CAMPUS, 70- *Concurrent Pos:* Vis prof chem, Chinese Acad Sci, 83, 85 & 87. *Mem:* Am Chem Soc; Sigma Xi. *Res:* Chemistry of free radicals in solution; mechanism of organic reactions. *Mailing Add:* Dept Chem Univ Mass Boston MA 02125

WANG, CHIN HSIEN, b Taiwan, Sept 4, 39; US citizen; m 63; c 4. CHEMISTRY. *Educ:* Nat Taiwan Univ, BS, 61; Utah State Univ, MS, 64; Mass Inst Technol, PhD(phys chem), 67. *Prof Exp:* Mem tech staff, Bell Tel Labs, 67-69; from asst prof to assoc prof phys chem, Univ Utah, 69-76, prof chem, 76-90; PROF CHEM, UNIV NEBR, 90- *Concurrent Pos:* Petrol Res Fund grant, 70-78; Res Corp grant, 72-73; Alfred P Sloan Found fel, 73; Off Naval Res grants, 74-; NSF grants, 76-91; adj prof mat sci, Univ Utah, 69-90. *Honors & Awards:* David P Gardner Award, 81. *Mem:* Am Phys Soc; Am Chem Soc. *Res:* Light scattering and Raman spectroscopy; polymer physics; statistical mechanics; relaxation and orientation behavior of polymer chairs in solution and in bulk; studies of the glass transition phenomena in supercooled liquids and solids using light scattering and non-equilibrium statistical mechanics. *Mailing Add:* Dept Chem Univ Nebr Lincoln Lincoln NE 68588-3004

WANG, CHING CHUNG, b Peking, China, Feb 10, 36; m 63; c 2. BIOCHEMISTRY, PARASITOLOGY. *Educ:* Nat Taiwan Univ, BS, 58; Univ Calif, Berkeley, PhD(biochem), 66. *Prof Exp:* Fel biochem, Col Physicians & Surgeons, Columbia Univ, 66-67; res assoc, Princeton Univ, 67-69; sr res biochemist, Merck Inst Therapeut Res, 69-72, from res fel to sr res fel, 72-78, sr investr, 78-81; PROF CHEM & PHARMACEUT CHEM, SCH PHARM, UNIV CALIF, SAN FRANCISCO, 81- *Honors & Awards:* Burroughs Wellcome Molecular Parasitol Award, 83. *Mem:* Am Soc Biol Chemists; fel AAAS. *Res:* Biochemistry and development of protozoan parasites; invertebrate neurobiology; antiparasitic chemotherapy. *Mailing Add:* 22 Miraloma Dr San Francisco CA 94127

WANG, CHING-PING SHIH, b Shanghai, China, Feb 16, 47; m 71; c 1. THEORETICAL SOLID STATE PHYSICS. *Educ:* Tung-Hai Univ, Taiwan, BS, 69; La State Univ, Baton Rouge, MS, 71, PhD(physics), 74. *Prof Exp:* Res assoc physics, Dept Physics & Astron, La State Univ, Baton Rouge, 74-76; res assoc physics, Dept Physics & Astron, Northwestern Univ, Evanston, 76-79; asst prof, 79-85, ASSOC PROF DEPT PHYS & ASTRON, UNIV MD, 85- *Concurrent Pos:* Off Naval Res contract, 79-88; NSF grant, 85-92; consult, Naval Res Lab, 85- *Mem:* Am Phys Soc. *Res:* Electronic structure and other properties of high temperature superconductors, of heavy fermion superconductors, of ferromagnetic metals, of magnetic metal surfaces with absorbed atoms and of semiconductors. *Mailing Add:* Dept Physics & Astron Univ Md College Park MD 20742

WANG, CHI-SUN, b Shanghai, China, Oct 8, 42; m 73; c 1. BIOCHEMISTRY. *Educ:* Nat Taiwan Univ, BS, 66; Univ Okla, PhD(biochem), 71. *Prof Exp:* Staff scienist, 74-75, asst mem, 75-82, ASSOC MEM, OKLA MED RES FOUND, 82- *Concurrent Pos:* Adj assoc prof, Dept Biochem & Molecular Biol, Univ Okla Sch Med, 85- *Honors & Awards:* Eason Award, 77; Merrick Award, 81. *Mem:* Sigma Xi; Am Chem Soc; AAAS; Am Soc Biochemistry & Molecular Biol; Am Oil Chemists Soc. *Res:* Lipoprotein lipase; bile salt-activated lipase; tissue lipases. *Mailing Add:* 825 NE 13th St Oklahoma City OK 73104

WANG, CHIU-CHEN, b Canton, China, Nov 5, 22; US citizen. RADIATION ONCOLOGY. *Educ:* Nat Kwei-Yang Med Col, China, MD, 48; Am Bd Radiol, dipl, 53. *Hon Degrees:* MA, Harvard, 90. *Prof Exp:* Rotation intern, Canton Hosp, China, 47-48, asst resident med, 48-49; intern, Univ Hosp, Syracuse, NY, 49-50; asst resident radiol, Mass Gen Hosp, Boston, 50-51, resident, 52, clin fel, 53-56; asst radiol, 58-60, instr, 60-61, clin assoc, 62-67, asst clin prof, 68-69, asst prof, 69-70, from asst prof to assoc prof radiation ther, 70-75, PROF RADIATION THER, HARVARD MED SCH, 75-; RADIATION THERAPIST & HEAD CLIN SERV, MASS GEN HOSP, 73- *Concurrent Pos:* Damon Runyon res grant, Donner Lab, Univ Calif, Berkeley, 61-62; consult radiologist, Lawrence Berkeley Lab, Univ Calif, Berkeley, 62-65 & Mass Eye & Ear Infirmary, Waltham Hosp & Emerson Hosp, Mass, 62-; guest examr, Am Bd Radiol, 66. *Mem:* Fel Am Col Radiol; Am Radium Soc; Radiol Soc NAm; Am Soc Therapeut Radiol & Oncol; AMA. *Res:* Clinical radiation oncology. *Mailing Add:* Dept Radiation Med Mass Gen Hosp Boston MA 02114

WANG, CHIU-SEN, b Taichung, Formosa, Dec 3, 37; m 67; c 3. CHEMICAL & ENVIRONMENTAL HEALTH ENGINEERING. *Educ:* Taiwan Univ, BS, 60; Kans State Univ, MS, 63; Calif Inst Technol, PhD(chem eng), 66. *Prof Exp:* Res fel chem eng, Calif Inst Technol, 66-68; asst prof environ med, Med Ctr, NY Univ, 68-69; assoc prof chem eng, 69-74, PROF CHEM ENG, SYRACUSE UNIV, 74-,. *Concurrent Pos:* Assoc res scientist, Med Ctr, NY Univ, 66; vis prof, Kyoto Univ, Japan, 78. *Mem:* Soc Powder Technol Japan; Air Pollution Control Asn; Am Inst Chem Engrs; Am Indust Hyg Asn; Am Asn Aerosol Res. *Res:* Systems optimization; aerosol physics and physiology; gas-particle separations. *Mailing Add:* 29115 Firthridge Rd Rancho CA 90274

WANG, CHRISTINE A, b Providence, RI, Sept 20, 55; m 80; c 2. ELECTRONIC MATERIALS. *Educ:* Mass Inst Technol, SB, 77, SM, 78, PhD(electronic mat), 84. *Prof Exp:* STAFF SCIENTIST, LINCOLN LAB, MASS INST TECHNOL, 84- *Mem:* Am Asn Crystal Growth; Mat Res Soc; Optical Soc Am. *Res:* Development of III-V semiconductors for diode lasers; uniformity, controllability, and reproducibility for an epitaxial process for optoelectronic devices; diode lasers and short wavelength lasers; author of over 30 publications. *Mailing Add:* Lincoln Lab Mass Inst Technol Lexington MA 02173

WANG, CHU PING, b China, Mar 25, 31; m 61; c 3. COMPUTER SYSTEMS, INFORMATION SCIENCE. *Educ:* Taiwan Univ, BSc, 54; Univ Toronto, MASc, 56; Stanford Univ, PhD(microwave electronics), 61. *Prof Exp:* Asst prof elec eng, San Jose State Col, 60-61; res staff mem, Thomas J Watson Res Ctr, IBM Corp, 61-67; vis assoc prof, Wash Univ, 67-68; res staff mem, San Jose Res Lab, 68-76, RES STAFF MEM, THOMAS J WATSON RES CTR, IBM CORP, 76- *Mem:* Inst Elec & Electronics Engrs. *Res:* Information system design and evaluation methodology; computer performance evaluation; data base organization. *Mailing Add:* Thomas J Watson Res Ctr PO Box 704 Yorktown Heights NY 10598

WANG, CHUN-JUAN KAO, b Mukden, China, Jan 10, 28; m 55; c 3. MYCOLOGY. *Educ:* Nat Taiwan Univ, BS, 50; Vassar Col, MS, 52; Univ Iowa, PhD(mycol), 55. *Prof Exp:* Asst, Univ Iowa, 52-55; res assoc, Clin Labs, Jewish Hosp, Cincinnati, Ohio, 55-58; from asst prof to assoc prof, 59-72, PROF BOT & MYCOL, COL ENVIRON SCI & FORESTRY, STATE UNIV NY, 72- *Concurrent Pos:* Instr, Sch Med, Univ Cincinnati, 57-58. *Honors & Awards:* W H Weston Award for Excellence in Teaching Mycol, Mycol Soc Am, 90. *Mem:* Mycol Soc Am; Brit Mycol Soc; Sigma Xi. *Res:* Medical mycology; ecology, ultrastructure and systematics of imperfect fungi (Hyphomycetes). *Mailing Add:* Col Environ Sci & Forestry State Univ NY Syracuse NY 13210-2788

WANG, DALTON T, PROTEIN CHEMISTRY. *Educ:* McGill Univ, PhD(biochem), 57. *Prof Exp:* ASSOC PROF BIOCHEM, DOWNSTATE MED CTR, STATE UNIV NY, BROOKLYN, 80- *Res:* Chemical modification of proteins; mechanisms of the interactions of proteases and protease inhibitors. *Mailing Add:* Dept Biochem Downstate Med Ctr Brooklyn NY 11203

WANG, DANIEL I-CHYAU, b Nanking, China, Mar 12, 36; US citizen; m 66; c 1. CHEMICAL & BIOCHEMICAL ENGINEERING. *Educ:* Mass Inst Technol, BS, 59, SM, 61; Univ Pa, PhD(chem eng), 63. *Prof Exp:* Process engr, US Army Biol Labs, 63-65; asst prof, 65-76, PROF BIOCHEM, MASS INST TECHNOL, 76-, CHEVRON PROF CHEM ENG, 85- *Concurrent Pos:* Consult, Environ Protection Agency. *Honors & Awards:* Food, Pharmaceut & Bioeng Award, Am Inst Chem Engrs; M J Johnson Award, Am Chem Soc; Inst Lectr, Am Inst Chem Engrs. *Mem:* Nat Acad Eng; Am Inst Chem Engrs; Inst Food Technol; Am Soc Microbiol; Sigma Xi; Am Chem Soc; Am Acad Arts & Scis. *Res:* Kinetics of biological systems; mass transfer in fermentation process; membrane processes; microbial sterilization; hydrocarbon fermentation; fermentation recovery. *Mailing Add:* 17 Pequosset Rd Belmont MA 02178

WANG, DAZONG, b Jingdezhen, China, May 15, 54; m 80; c 2. ADVANCED VEHICLE SYSTEMS, COMPUTER AIDED ENGINEERING DYNAMICS & CONTROL. *Educ:* Zhong Yuan Tech Sch, China, dipl mech eng, 76; Huazhong Univ Sci & Technol, China, ME, 80; Cornell Univ, MS, 83, PhD(mech eng), 85. *Prof Exp:* Res asst, Cornell Univ, 81-85; sr res engr, Res Labs, 85-89, STAFF RES ENGR, SYSTS ENG CTR, GEN MOTOR CORP, 89- *Concurrent Pos:* Lectr, Huazhong Univ Sci & Technol, China, 81; chmn, Computer Simulation Comt, Am Soc Mech Engrs, 90- *Mem:* Am Soc Mech Engrs; Soc Automotive Engrs. *Res:* Advanced vehicle systems; dynamics and control; computer aided engineering; optimum design. *Mailing Add:* GM Systs Eng Ctr 1151 Crooks Rd Troy MI 48084

WANG, EDWARD YEONG, b Nantung, China, July 30, 33; m 60; c 3. SOLID STATE ELECTRONICS. *Educ:* Morningside Col, BS, 54; Purdue Univ, MS, 59; Tufts Univ, PhD(physics), 66. *Prof Exp:* Jr physicist, Nat Semiconductor Co, Ill, 54-56; assoc staff mem, Res Div, Raytheon Co, Mass, 59-61; sr res Electronics Corp Am, 61-63 & Gen Motors Res Lab, Mich, 66-70; assoc prof elec eng, Wayne State Univ, 70-77, prof elec & comput eng, 77-79; PROF ELEC & COMPUT ENG, ARIZ STATE UNIV, TEMPE, 79- *Mem:* Am Phys Soc; Inst Elec & Electronics Engrs. *Res:* Electrical and optical properties of solids; optoelectronics. *Mailing Add:* Dept Elec Eng Ariz State Univ Tempe AZ 85287

WANG, EUGENIA, b Chungking, China, Feb 26, 45; US citizen; m 76; c 1. MOLECULAR BIOLOGY, BIOCHEMISTRY. *Educ:* Nat Taiwan Univ, BSc, 66; Northern Mich Univ, MA, 69; Case Western Reserve Univ, PhD(cell biol), 74. *Prof Exp:* Res asst, Inst Zool, Acad Sinica, 66-67; teaching asst, Northern Mich Univ, 67-69; teaching fel, Case Western Reserve Univ, 69-74; postdoctoral fel, Virol Lab, Rockefeller Univ, 74-76, res assoc, 76-78, asst prof, 78-86; assoc prof, Dept Anat & Dept Med, McGill Univ, 87-90; AT LADY DAVIS INST, JEWISH GEN HOSP, 90- *Concurrent Pos:* Scientist award, Med Res Coun Can, 88-; mem Study Sect Aging Rev Comt, Nat Inst Aging, NIH, 90-; chairperson biol sci, Can Asn Geront, 90-91; ad hoc reviewer, Nat Cancer Inst, NIH, Med Res Coun Can & NSF; vis specialist, Inst Biomed Sci, Acad Sinica. *Mem:* Am Soc Cell Biol; NY Acad Sci; Geront Soc Am; Can Asn Geront. *Mailing Add:* Lady Davis Inst Jewish Gen Hosp 3755 Cote St Catherine Montreal PQ H3T 1E2 Can

WANG, FRANCIS WEI-YU, b Peikang, Taiwan, July 21, 36; US citizen; m 66; c 3. POLYMER SCIENCE. *Educ:* Calif Inst Technol, BS, 61, MS, 62; Univ Calif, San Diego, PhD(chem), 71. *Prof Exp:* Chemist, Pac Soap Co, 62-66; USPHS fel, 71-72; SUPVRY RES CHEMIST, POLYMERS DIV, NAT BUR STANDARDS, 72- *Honors & Awards:* Bronze Medal Award, Dept Com, 85. *Mem:* Am Chem Soc; Am Phys Soc. *Res:* Thermodynamic and frictional properties of polymer solutions; photophysical processes in polymer molecules; ultracentrifugal analysis of macromolecules; diffusion in polymers. *Mailing Add:* Rm B320 Bldg 224 Polymers Div Nat Inst Standards & Technol Gaithersburg MD 20899

WANG, FRANK FENG HUI, b Hopeh, China, Mar 21, 24; nat US; m 58; c 3. MARINE GEOLOGY. *Educ:* Nat Southwestern Assoc Univ, China, BS, 45; Univ Wash, PhD(geol), 55. *Prof Exp:* Asst geol, Nat Southwestern Assoc Univ, China, 45-46; instr, Nat Peking Univ, 46-48; asst, Univ Wash, 50-54; sedimentologist-stratigrapher, Western Gulf Oil Co, 54-57; res geologist, Gulf Res & Develop Co, 57-63; marine geologist, Int Minerals & Chem Corp, 64-67; MARINE GEOLOGIST, US GEOL SURV, 67- *Concurrent Pos:* Spec consult, Chinese Petrol Corp, 58; vis scholar, Northwestern Univ, 64; tech adv, UN, 67-, spec adv & sr marine geologist, UN Develop Prog on Regional Offshore Prospecting in EAsia, 72-74, prin marine geologist, 74-; vchmn, Marine Geol Panel, US-Japan Coop Prog Natural Resources, 70- *Mem:* Am Asn Petrol Geol; Am Geophys Union; Marine Technol Soc. *Res:* Marine mineral resources; ocean mining; regional climate changes in the Western Pacific; regional marine geology of eastern Asia. *Mailing Add:* US Geol Surv Off Marine Geol 345 Middlefield Rd Menlo Park CA 94025

WANG, FRANKLIN FU-YEN, b China, Sept 19, 28; nat US; m 56; c 2. MATERIALS SCIENCE. *Educ:* Pomona Col, BA, 51; Univ Toledo, MS, 53; Univ Ill, PhD(ceramics), 56. *Prof Exp:* Asst dir res, Glascote Prod, Inc, 56-58; res scientist, A O Smith Corp, 58-61; res staff mem, Sperry Rand Res Ctr, 61-66; assoc prof mat sci, 66-72, chmn dept, 72-74, PROF MAT SCI, STATE UNIV NY STONY BROOK, 72- *Mem:* Am Phys Soc; fel Am Ceramic Soc; Am Chem Soc; Am Soc Metals; Inst Elec & Electronics Engrs. *Res:* Magnetic materials; transport properties; dielectric materials; physical ceramics; glass systems; high temperature materials; engineering education; semiconductor processing; crystal growth. *Mailing Add:* Dept Mat Sci & Eng State Univ NY Stony Brook NY 11794-2275

WANG, FREDERICK E, b She-Tou, Formosa, Aug 1, 32; US citizen; m 61; c 2. NITINOL TECHNOLOGY, SUPERCONDUCTIVITY. *Educ:* Memphis State Univ, BS, 56; Univ Ill, MS, 57; Syracuse Univ, PhD(phys chem), 60. *Prof Exp:* Fel, Harvard Univ, 60-61; res assoc metal alloys, Syracuse Univ, 61-63; chemist, US Naval Surface Weapons Ctr, 63-80; PRES, INNOVATIVE TECHNOL INT, INC, 80- *Concurrent Pos:* Fulbright exchange lectr, 67-68. *Mem:* Am Phys Soc; Am Soc Metals; AAAS. *Res:* Metal and alloy physics; order-disorder phenomena; superconductivity; memory effect in alloy. *Mailing Add:* Innovative Technol Int Inc 10747-3 Tucker St Beltsville MD 20705

WANG, GARY T, b Gansu, China, Mar 7, 63; m; c 1. BIOORGANIC CHEMISTRY, BIOCHEMICAL PROBES. *Educ:* Lanzhou Univ, China, BS, 82; Northwestern Univ, Evanston, MS, 84, PhD(chem), 87. *Prof Exp:* Res asst, Chem Dept, Northwestern Univ, 83-87; postdoctoral fel, Chem Dept, Syracuse Univ, 88; postdoctoral fel, 88-90, RES SCIENTIST, DIAG DIV, ABBOTT LABS, 90- *Mem:* Am Chem Soc; AAAS; NY Acad Sci. *Res:* Organic/bioorganic chemistry and biochemistry; elucidation of mechanisms of biological processes, particularly processes involving enzymes, antibodies etc; design and synthesis of tailored molecular entities for biomedical application as diagnostic probes or therapeutical agents. *Mailing Add:* 8602 Osceola Ave Niles IL 60648

WANG, GUANG TSAN, b Taiwan, China, Mar 6, 35; US citizen; m 62; c 2. VETERINARY MEDICINE, PARASITOLOGY. *Educ:* Nat Taiwan Univ, DVM, 58; Univ Ill, Urbana, MS, 64, PhD(vet med sci), 68. *Prof Exp:* Vet, Taiwan Serum Vaccine Labs, 60-62; res asst parasitol, Univ Ill, 62-68; res vet, Am Cyanamid Co, 68-74, group leader parasitol discovery, 74-77, prog mgr, Agr Div, 77-81, dir Animal Indust Res & Develop, 81-90, SR PROD DEVELOP MGR, AMERICAS/FAR EAST, AGR RES DIV, AM CYANAMID CO, PRINCETON, 90- *Mem:* Am Soc Parasitol; Am Vet Med Asn. *Res:* Toxicity and efficacy of anthelmintics, anticoccidials and antibiotics in domestic animals; industrial parasitic chemotherapy; research administration. *Mailing Add:* 41 Slayback Dr Princeton Junction NJ 08550

WANG, GWO-CHING, b Hu-Pei Prov, China, Oct 10, 46; m; c 1. SURFACE PHYSICS. *Educ:* Cheng Kung Univ, Taiwan, BS, 68; Northern Ill Univ, MS, 73; Univ Wis-Madison, PhD(mat sci), 78. *Prof Exp:* Teaching asst, Fu-Jen Univ, Taiwan, 68-69 & Univ Wyo, 69-71; res asst, Northern Ill Univ, 71-73, Univ Wis-Madison, 73-78; physicist, Nat Bur Standards, 78-80; PHYSICIST, SOLID STATE PHYSICS DIV, OAK RIDGE NAT LAB, 80- *Honors & Awards:* Nottingham Prize, 78. *Mem:* Sigma Xi; Am Phys Soc; Mat Res Soc. *Res:* Geometric properties of surfaces and interfaces; chemisorption, kinetics, and phase transitions using high resolution low energy electron diffraction; growth modes and magnetic properties of ultra thin ferromagnetic films. *Mailing Add:* Physics Dept Rensselaer Polytech Inst Troy NY 12180-3590

WANG, H E FRANK, b China, Oct 23, 29; US citizen; m 55; c 2. FLUID PHYSICS, SYSTEMS ENGINEERING. *Educ:* Nat Taiwan Univ, BS, 52; Bucknell Univ, MS, 54; Brown Univ, PhD(gas dynamics), 59. *Prof Exp:* Res engr gas dynamics, Boeing Co, 58-60; mem tech staff & prog mgr aerophys & systs eng, 60-77, dir, 76-81, PRIN DIR, SPACE TEST PROGS, AEROSPACE CORP, 81- *Mem:* Assoc fel Am Inst Aeronaut & Astronaut. *Mailing Add:* 27241 Sunnyridge Rd Rolling Hills CA 90274

WANG, HAO, b Tsinan, China, May 20, 21; m 48, 77; c 3. MATHEMATICS. *Educ:* Nat Southwestern Assoc Univ, China, BS, 43; Tsing Hua Univ, MA, 45; Harvard Univ, PhD, 48; Oxford Univ, MA, 56. *Prof Exp:* Soc Fels fel, Harvard Univ, 48-51, asst prof philos, 51-56; reader philos of math, Oxford Univ, 56-61; Gordon McKay prof math logic & appl math, Harvard Univ, 61-67; PROF LOGIC, ROCKEFELLER UNIV, 67- *Concurrent Pos:* Res engr, Burroughs Corp, 53-54; fel, Rockefeller Found, 54-55; John Locke lectr philos, Oxford Univ, 55; mem tech staff, Bell Tel Labs, 59-60; res scientist, IBM Res Ctr, 73-74; vis, Inst Advan Study, Princeton, 75-76; hon prof, Beijing Univ, 85- & Tsinghua Univ, 86- *Honors & Awards:* Milestone Award, 83. *Mem:* Asn Symbolic Logic; fel Am Acad Arts & Sci; foreign fel Brit Acad; Kurt Gödel Soc, Vienna (pres, 87-89). *Res:* Mathematical logic; epistemology; philosophy of mathematics; general philosophy; contemporary China; author of over 100 articles and seven books. *Mailing Add:* Rockefeller Univ New York NY 10021-6399

WANG, HENRY, b Shanghai, China, July 22, 51; m 84; c 1. BIOCHEMICAL ENGINEERING, INDUSTRIAL BIOLOGY. *Educ:* Iowa State Univ, BS, 69; Mass Inst Technol, SM, 72, PhD(biochem eng), 77. *Prof Exp:* Asst prof, 79-84, ASSOC PROF BIOENG & CHEM ENG, UNIV MICH, 84- *Concurrent Pos:* Res asst, Mass Inst Technol, 72-77; eng assoc, Merck & Co, 77; sr scientist, Schering Plough, 78-79; distinguished sr scientist, Mich Biotech Inst, 87-; counr, Am Chem Soc, 84-87. *Honors & Awards:* W M Peterson Award, Am Chem Soc, 74. *Mem:* Am Inst Chem Eng; Am Chem Soc; Am Soc Microbiol; Soc Indust Microbiol; AAAS. *Res:* Cell cultivation technology; cellular sensing and screening; bioseparation and downstream processing; sterilization technology and computer applications in biological systems. *Mailing Add:* Dept Chem Eng Univ Mich Ann Arbor MI 48109-2136

WANG, HERBERT FAN, b Shanghai, China, Sept 14, 46; US citizen; m 68; c 4. GEOPHYSICS. *Educ:* Univ Wis-Madison, BA, 66; Harvard Univ, AM, 68; Mass Inst Technol, PhD(geophys), 71. *Prof Exp:* Res assoc geophys, Mass Inst Technol, 71-72; from asst prof to assoc prof, 72-82, PROF GEOPHYS, UNIV WIS-MADISON, 82- *Concurrent Pos:* Geoscientist, Dept Energy, 80-81; physicist, Lawrence Livermore Nat Lab, 86-87. *Mem:* Am Geophys Union. *Res:* Rock mechanics; groundwater, thermal and diffusion modeling. *Mailing Add:* Dept Geol & Geophys Univ Wis 1215 W Dayton St Madison WI 53706

WANG, HOWARD HAO, b Shanghai, China, Jan 24, 42; m 63; c 2. NEUROSCIENCES, MOLECULAR PHARMACOLOGY. *Educ:* Calif Inst Technol, BS, 63; Univ Calif, Los Angeles, PhD(neurophysiol), 68. *Prof Exp:* USPHS fel, Univ Calif, Berkeley, 68-69; resident scientist, Neurosci Res Prog, Mass Inst Technol, 69-70; from asst prof to assoc prof, 70-82, PROF BIOL, STEVENSON COL, UNIV CALIF, SANTA CRUZ, 82- *Mem:* AAAS; Am Asn Anat; Biophys Soc; Soc Neurosci. *Res:* Mechanism of anesthetic action and drug-receptor interaction; effect of environmental chemicals on membrane structure and function; molecular and cellular mechanisms of brain function. *Mailing Add:* Sinsheimer Labs Univ Calif Santa Cruz CA 95064

WANG, HSIANG, b China, Jan 20, 36; m 64; c 3. FLUID MECHANICS. *Educ:* Taiwan Univ, BS, 58; Univ Mass, MS, 63; Univ Iowa, PhD(fluid mech), 65. *Prof Exp:* Res asst fluid mech, Iowa Inst Hydraul Res, 62-63, res assoc, 63-65; res engr ocean, US Naval Civil Eng Lab, 65-67; mem tech staff, Nat Eng Sci Div, Fluor Corp Int, 67, mem sr staff, 67-69; sr engr, Tatra-Technol Inc, 69-70; from assoc prof civil eng to prof, Univ Del, 70-81, actg chmn, 78-79; VPRES, COASTAL & OFFSHORE ENG & RES, INC, 77-; CHMN & PROF COASTAL & OCEAN ENG, UNIV FLA, 82- *Concurrent Pos:* Vis prof, Tech Univ Braunschweig. *Mem:* Am Soc Civil Engrs; submarine wake study; offshore structure and offshore oil exploration; coastal and estuarine research; fluid mechanics and ocean engineering. *Mailing Add:* Dept Coastal & Ocean Eng Univ Fla Gainesville FL 32611

WANG, HSIN-PANG, b Nanking, China, Apr 11, 46; US citizen; m 73; c 2. COMPUTER ENGINEERING, COMPUTATIONAL MECHANICS. *Educ:* Cheng-Kung Univ, BS, 69; Univ Fla, MS, 72; Univ RI, PhD(mech eng), 76. *Prof Exp:* RES STAFF, GEN ELEC RES & DEVELOP CTR, 76- *Concurrent Pos:* Prin investr, Intelligent Processing Mat, 88- *Honors & Awards:* Indust Res 100 Award, 83. *Mem:* Am Soc Mech Engrs; Am Physical Soc. *Res:* Computer simulation of manufacturing processes; integration of CAE, CAD and CAM to improve the productivity and producibility; interrelationship between product design and process development; intelligent processing of material. *Mailing Add:* Gen Elec Res & Develop Ctr PO Box 43 Schenectady NY 12301

WANG, HSIOH-SHAN, b Shanghai, China, Sept 1, 28; US citizen; m 64; c 2. NEUROLOGY, NEUROPHYSIOLOGY. *Educ:* Taiwan Univ, Taipei, MB, 53; Am Bd Psychiat & Neurol, dipl, 66. *Prof Exp:* Assoc prof psychiat, Nat Defense Med Col, 61-63; from asst prof to assoc prof, 66-75, PROF PSYCHIAT, MED SCH, DUKE UNIV, 75-, CHIEF PSYCHIAT DAY UNIT, DUKE UNIV HOSP, 71- *Concurrent Pos:* Chief neuropsychiat, Taiwan Vet Gen Hosp, 61-63; consult, var ment health ctr & hosps, 68-79; sr fel, Ctr Study Aging Human Develop, 75. *Mem:* Am Med Asn; fel Am Psychiat Asn; Am Geriat Soc; fel Geront Soc Am. *Res:* Mental health and mental disorders in the elderly; relationship between changes of brain function and behaviors associated with aging; noninvasive Xenon inhalation method for the determination of regional cerebral blood flow. *Mailing Add:* Dept Psych Duke Med Ctr Durham NC 27710

WANG, HSUEH-HWA, b Peiping, China, July 10, 23; US citizen; m 48; c 3. PHARMACOLOGY, PHYSIOLOGY. *Educ:* Nat Cent Univ, Nanking, China, MB, 46. *Prof Exp:* From assoc prof to prof, 70-90, EMER PROF PHARMACOL, COL PHYSICIANS & SURGEONS, COLUMBIA UNIV, 90- *Concurrent Pos:* NY Heart fel, 53-54. *Mem:* AAAS; Am Physiol Soc; Am Soc Pharmacol & Exp Therapeut. *Res:* Coronary circulation; effects of endogenous mediators (prostaglandins, antidiuretic hormone, angiotensin) on peripheral circulation and blood pressure control. *Mailing Add:* Dept Pharmacol Columbia Univ 630 W 168th St New York NY 10032

WANG, HWA LIH, b Chekiang, China, Nov 29, 21; m 49; c 2. BIOCHEMISTRY. *Educ:* Nat Cent Univ, China, BS, 45; Univ Wis, MS, 50, PhD(biochem), 52. *Prof Exp:* Asst biochem, Med Sch, Nat Cent Univ, China, 45; res assoc, Med Sch, Marquette Univ, 53-55, Med Sch, Univ Wis, 56-61 & Wash Univ, 61-62; RES CHEMIST, NORTHERN REGIONAL LAB, AGR RES SERV, USDA, 63- *Mem:* Am Inst Nutrit; Am Chem Soc; Inst Food Technologists. *Res:* Biochemistry and physiology of molds used in soybean and cereal food fermentation; nutritional value of fermented food products; mixed microbial cultures. *Mailing Add:* AR SEA USDA Northern Regional Res Lab Peoria IL 61604

WANG, JAMES C, b China, Nov 18, 36; m 61; c 2. BIOCHEMISTRY, MOLECULAR BIOLOGY. *Educ:* Nat Taiwan Univ, BS, 59; Univ SDak, MA, 61; Univ Mo, PhD(chem), 64. *Prof Exp:* Res fel chem, Calif Inst Technol, 64-66; from asst prof to prof chem, Univ Calif, Berkeley, 66-77; PROF BIOCHEM & MOLECULAR BIOL, HARVARD UNIV, 77- *Concurrent Pos:* Mem biophys & biochem study sect, NIH, 72-76; mem adv comt physiol, cellular & molecular biol, NSF, 80- *Mem:* Nat Acad Sci; Biophys Soc; Am Acad Arts & Sci; Am Soc Biol Chem. *Res:* Structures and functions of DNAs. *Mailing Add:* Dept Biochem & Microbiol Harvard Univ Cambridge MA 02138

WANG, JAMES LI-MING, b Nan King, China, June 25, 46. ANALYSIS. *Educ:* Brown Univ, PhD(math), 74. *Prof Exp:* From asst prof to assoc prof, 76-85, PROF MATH, UNIV ALA, 85- *Mem:* Am Math Soc. *Mailing Add:* Dept Math Univ Ala PO Box 870350 Tuscaloosa AL 35487-0350

WANG, JAMES TING-SHUN, b Nanking, China, Feb 8, 31; US citizen; m 63; c 1. ENGINEERING MECHANICS. *Educ:* Nat Taiwan Univ, BS, 54; Univ Kans, MS, 58; Purdue Univ, PhD(civil eng), 61. *Prof Exp:* Instr eng mech, Univ Kans, 56-58; from asst prof to assoc prof, 61-69, PROF ENG MECH, GA INST TECHNOL, 69- *Concurrent Pos:* Spec lectr, George Washington Univ, 63; consult, Lockheed-Ga Co, 65-66, aircraft develop engr specialist, 66-67. *Mem:* Am Soc Civil Engrs; Am Soc Eng Educ; Soc Rheology; Sigma Xi. *Res:* Structural mechanics; plate and sheil theory. *Mailing Add:* Dept Eng Mech Ga Inst Technol Atlanta GA 30332

WANG, JAW-KAI, b Nanjing, China, Mar 4, 32; div; c 3. AGRICULTURAL ENGINEERING. *Educ:* Nat Taiwan Univ, BSAE, 53; Mich State Univ, MSAE, 56, PhD, 58. *Prof Exp:* Lectr farm mach, Prov Taoyaun Agr Inst, 54-55; asst farm processing, Mich State Univ, 55-58; from asst prof to assoc prof agr eng, 59-68, chmn dept, 64-75, PROF AGR ENG, UNIV HAWAII, 68- *Concurrent Pos:* Consult, US Army, Okinawa, 65, Taiwan Sugar Co, 67, Int Rice Res Inst, 71, Pac Concrete & Rock Co, 74, USAID, 73, The World Bank, 81 & 82, ABA Int, 81-85, Univ Tankship, Del, 80 & 81, Food & Agr Orgn/UN, 83, County of Maui, 84 & US Dept of State, 85; sr fel, Food Inst, East-West Ctr, 73-74; co dir, Int Sci & Educ Coun, 79; vis assoc dir, Int Prog & Studies Off, Nat Asn State Univ & land-grant col, 79; vis prof, Nat Taiwan Univ, 64, Univ Calif, Davis, 80; mem, expert panel on agr mechanization, Food & Agr Orgn, UN. *Honors & Awards:* Engr of the Yr Award, ASAE, Pac Region, 76; Kishida Int Award, Am Soc Agr Engrs, 91. *Mem:* Nat Soc Prof Engrs; fel Am Soc Agr Engrs; Chinese Soc Agr Engrs; World Mariculture Soc; Sigma Xi. *Res:* Irrigated rice production system design; aquacultural engineering, especially fresh water prawn and oyster; agricultural production systems design and analysis. *Mailing Add:* Dept Agr Eng Univ Hawaii 3050 Maile Way Honolulu HI 96822

WANG, JEN YU, b Foochow, China, Mar 3, 15; c 1. METEOROLOGY. *Educ:* Fukien Christian Univ, China, BS, 38; Univ Chicago, cert, 54; Univ Wis, MS, 55, PhD(meteorol), 58. *Prof Exp:* Instr math & physics, Cols, China & Hong Kong, 38-42; prin meteorologist, Weather Bur, China, 42-47; assoc prof physics, Fukien Christian Univ, 47-50; asst meteorol, Weather Forecasting Res Ctr, Univ Chicago, 53-54; asst, Univ Wis, 54-57, res assoc, 57-60, asst prof, 60-64; assoc prof, 64-68, PROF METEOROL & DIR ENVIRON SCI INST, SAN JOSE STATE UNIV, 68- *Concurrent Pos:* Fel, United Bd Higher Educ Asia, US, 50-54; consult, 10th Weather Squadron, USAF, China, 45, US Weather Bur, Washington, DC, 58, AEC Proj, 65 & Stanford Res Inst, 66-67; pres, Milieu Info Serv, 71-, Blackwell Land Mgt Co, 74- & Sierra-Misco, Inc, 81- *Mem:* Am Meteorol Soc; Am Soc Agron; Int Soc Biometeorol; fel Am Geog Soc. *Res:* New techniques in the investigation of environmental relationships between animals and plants; agricultural meteorology; ecology; phenology; phytoclimatology; environmental assessment studies. *Mailing Add:* 1863 Shulman Ave San Jose CA 95124

WANG, JERRY HSUEH-CHING, b Nanking, China, Mar 12, 37; m 62; c 2. BIOCHEMISTRY. *Educ:* Nat Taiwan Univ, BSc, 58; Iowa State Univ, PhD(biochem), 65. *Prof Exp:* From asst prof to assoc prof, 66-78, Prof Biochem, Fac Med, Univ Man, 78-; AT DEPT BIOCHEM, UNIV CALGARY. *Concurrent Pos:* Nat Res Coun Can fel biochem, 65-66; Med Res Coun scholar, 66- *Mem:* Can Biochem Soc; Am Soc Biol Chemists. *Res:* Quaternary structure and regulatory property of enzymes. *Mailing Add:* Dept Med Biochem Univ Calgary Fac Med 3380 Hospital Dr NW Calgary AB T2N 4N1 Can

WANG, JI CHING, b Kobe, Japan, Nov 29, 38; m 71; c 2. ENGINEERING SCIENCE. *Educ:* Osaka Inst Technol, BS, 61; Univ Calif, Berkeley, MS, 65, PhD(mech eng), 69. *Prof Exp:* Design engr, Shinippon Koki, Japan, 61-62; res asst mech eng, Univ Calif, Berkeley, 64-66; res engr, Kaiser Eng, 66-67; from asst prof to assoc prof, 69-82, PROF MECH ENG, SAN JOSE STATE UNIV, 82- *Concurrent Pos:* NSF grants, 70-71; consult, Ames Res Ctr, NASA, 70-71 & 74-91, res grants, 77-91; res engr, Ford Aerospace, 75-76; res grant, CONTECT, 84-91, Westinghouse, 89-91. *Mem:* Inst Elec & Electronics Engrs; Am Soc Mech Engrs; Am Soc Mfg Engrs. *Res:* System control engineering: theory and application of control theory in aircraft; robotics; process control. *Mailing Add:* Dept Mech Eng San Jose State Univ Sch Eng Wash Sq San Jose CA 95192-0087

WANG, JIA-CHAO, b China, Mar 17, 39; m 66; c 2. SOLID STATE PHYSICS, THEORETICAL PHYSICS. *Educ:* Tunghai Univ, BS, 62; Nat Chiao Tung Univ, Taiwan, MS, 65; Univ NC, PhD(physics), 73. *Prof Exp:* Res assoc, Dept Physics & Astron, Univ NC, 73-75 & Wright-Patterson AFB, 75-77; STAFF MEM, SOLID STATE DIV, OAK RIDGE NAT LAB, 77- *Concurrent Pos:* Res assoc, Nat Res Coun, 75-77. *Mem:* Am Phys Soc; Electrochem Soc; Sigma Xi; Sci Res Soc. *Res:* Solid electrolytes or superionic conductors; laser annealing of ion-implanted semiconductors; electronic density of states of disordered systems. *Mailing Add:* 1075 W Outer Dr Oak Ridge TN 37830

WANG, JIN TSAI, b Inchon, Korea, Apr 7, 31; US citizen; m 58; c 2. INORGANIC CHEMISTRY, ANALYTICAL CHEMISTRY. *Educ:* Ore State Univ, BS, 57; Carnegie Inst Technol, PhD(chem), 68. *Prof Exp:* Instr, Pa State Univ, 68; from asst prof to assoc prof chem, Duquesne Univ, 77-87; RES CHEMIST, BIOPHARMACEUT RES BR, FOOD & DRUG ADMIN, WASHINGTON, DC, 87- *Concurrent Pos:* Vis scientist, NIH, 74, 79 & 85. *Mem:* Am Chem Soc; Soc Asian Comp Philos. *Res:* Infrared and polarographic studies of metal complexes. *Mailing Add:* Food & Drug Admin-HFD 424 200 C St SW Washington DC 20204

WANG, JIN-LIANG, b Chu-Nan, Taiwan, Aug 18, 37; US citizen; c 3. PYROPHOSPHATES, ENZYME-CATALYZED POLYMERIZATION. *Educ:* Taipei Inst Technol, dipl, 58; Kent State Univ, MS, 66; Univ Akron, PhD(polymer chem), 71. *Prof Exp:* Sr chem engr, Hua Min Paper Mill, Taiwan, 60-61; res chem engr, Taiwan Prov Tobacco & Wine Monopoly Bur, 61-63; sr res chemist, Res Div, Goodyear Tire & Rubber Co, 66-87; mgr, Wang Enterprise, 87-88; SR RES CHEMIST, GREAT LAKE CHEM CORP, 88- *Concurrent Pos:* Part-time teaching, Univ Akron, 71-86. *Mem:* Am Chem Soc; Sigma Xi. *Res:* Polymer synthesis compounding, characterization and applications; polymerization mechanism and kinetics; chemical modification of polymers; organic synthesis; isoprenoid oligomer synthesis. *Mailing Add:* Great Lakes Chem Corp PO Box 2200 West Lafayette IN 47906

WANG, JOHN L, b Hunan, China, Oct 1, 46. BIOCHEMISTRY. *Educ:* Rockefeller Univ, PhD(biochem), 73. *Prof Exp:* assoc prof, 81-84, PROF BIOCHEM, MICH STATE UNIV, 85- *Mailing Add:* Dept Biochem Mich State Univ E Lansing MI 48824

WANG, JOHN LING-FAI, b Amoy, China, Sept 21, 42; US citizen; m 70. PHYSICAL CHEMISTRY, CHEMICAL METALLURGY. *Educ:* Hope Col, BA, 65; Univ Calif, Berkeley, PhD(chem), 69. *Prof Exp:* Fel ceramics, Lawrence Radiation Lab, 69-70; fel high temperature chem, Rice Univ, 70-74; staff scientist high field superconductor mat, Lawrence Berkeley Lab, 74-80; sr res chemist, Polymer Res Ctr, Gulf Oil Chem Co, 80-83; MGR, ANALYSIS LAB, BECKMAN INDUST CORP, 83- *Mem:* Am Chem Soc; Am Phys Soc; Am Soc Metals; NY Acad Sci; Sigma Xi; Am Ceramic Soc. *Res:* application of fundamental principles of materials science and high temperature chemistry to failure analysis and material development. *Mailing Add:* Beckman Indust 4141 Palm St Fullerton CA 92635

WANG, JOHNSON JENN-HWA, b Hunan, China, Oct 24, 38; US citizen; m 68; c 2. COMPUTER SOLUTION OF ELECTROMAGNETIC PROBLEMS. *Educ:* Nat Taiwan Univ, BS, 62; Fla State Univ, MS, 65; Ohio State Univ, PhD(elec eng), 68. *Prof Exp:* Prof staff mem, TRW Systs, 69-70; sr engr, Motorola, 70-73; mem tech staff, Tex Instruments, 73-75; sr res engr, 75-80, br head, 85-88, PRIN RES ENGR, GA INST TECHNOL, 80-, ADJ PROF, 88- *Mem:* Inst Elec & Electronic Engrs; Sigma Xi; Electromagnetics Acad. *Res:* Electromagnetic theory; numerical analysis using digital computers; antennas; scattering; bioelectromagnetics; electromagnetic radiation hazard; electromagnetic interference; electromagnetic compatibility; microwave imaging. *Mailing Add:* 445 Cove Dr Marietta GA 30067

WANG, JON Y, b Tainan, Taiwan, June 22, 43; US citizen; m 71; c 1. LASER RADAR TECHNOLOGY, INFRARED SYSTEM ENGINEERING. *Educ:* Nat Taiwan Univ, BS, 65; Mass Inst Technol, MS, 68; Purdue Univ, PhD(aerophys & astrophys), 71. *Prof Exp:* Sr res engr electro optics, Gen Dynamics Convair Div, 71-73; staff scientist, Sci Applns, Inc. 73-75; ENG STAFF SPECIALIST ELECTRO OPTICS, GEN DYNAMICS CONVAIR DIV, 75- *Mem:* Optical Soc Am; Am Inst Physics. *Res:* Electro-optics; atmospheric propagation; laser radar applications; sensor system modelings; publications in science journals. *Mailing Add:* 2665 San Clemente Terr San Diego CA 92122

WANG, JOSEPH, b Haifa, Israel; US citizen; c 1. CHEMISTRY. *Educ:* Israel Inst Technol, PhD(chem), 78. *Prof Exp:* From asst prof to assoc prof, 80-88, PROF CHEM, NMEX STATE UNIV, 88- *Mem:* Am Chem Soc; Electrochemical Soc. *Res:* Design and development of electrochemical sensors; author of 200 papers and 2 books. *Mailing Add:* Dept Chem NMex State Univ Las Cruces NM 88003

WANG, JUI HSIN, b Peking, China, Mar 16, 21; nat US; m 49; c 2. BIOCHEMISTRY. *Educ:* Nat Southwest Assoc Univ, China, BSc, 45; Wash Univ, PhD(chem), 49. *Hon Degrees:* MA, Yale Univ, 60. *Prof Exp:* Fel radiochem, Wash Univ, 49-51; res fel chem, Yale Univ, 51-52, res asst, 52-53, from instr to prof, 53-62, Eugene Higgins 62-72; EINSTEIN PROF, UNIV NY BUFFALO, 72- *Concurrent Pos:* Guggenheim fel, Cambridge Univ, 60-61, Yale Univ, 71-72; mem biophys & biochem study sect, NIH, 65-69; Kennedy lectr, Wash Univ, 72; distinguished vis prof, Mich State Univ, 60, 69 & 80. *Mem:* AAAS; Am Chem Soc; Am Soc Biol Chemists; Am Acad Arts & Sci; Am Soc Photobiol; Biophys Soc; Electrochem Soc. *Res:* Diffusion in liquids; hemoglobin; mechanisms of enzyme action, particularly those related to oxidative phosphorylation, photosynthesis and ion-transport through biological membranes; electrochemistry; superconductivity. *Mailing Add:* Acheson Hall State Univ NY Buffalo NY 14214

WANG, KANG-LUNG, b Taiwan, China, July 3, 41; m 68; c 3. ELECTRICAL ENGINEERING, SEMICONDUCTOR PHYSICS & DEVICES. *Educ:* Cheng Kung Univ, Taiwan, BS, 64; Mass Inst Technol, MS, 66, PhD(elec eng), 70. *Prof Exp:* Res assoc, Div Sponsored Res, Mass Inst Technol, 70-71, asst prof elec eng, 71-72; scientist, Res & Develop Ctr, Gen Elec Co, 72-79; assoc prof, 79-82, PROF, ENG DEPT, UNIV CALIF, LOS ANGELES, 82- *Concurrent Pos:* Adj prof, Physics Dept, State Univ NY Albany; consult, Xerox Corp, El Segundo, Calif, Rockwell Int, Thousand Oaks, Calif, Jet Propulsion Lab, Calif Inst Technol & Hughes Aircraft, Areospace, Inc; Guggenheim Fel & Hon Prof, Xi'an Jiaotong Univ, Peoples Repub China. *Mem:* Inst Elec & Electronics Engrs; Am Phys Soc; Am Vacuum Soc; Sigma Xi. *Res:* Semiconductor physics and devices; quantum wells and superlattices; molecular beam epitaxy of SiGe and III-V; quantum effects in semiconductors. *Mailing Add:* 7619A Boelter Hall Elec Eng Dept Univ Calif Los Angeles Los Angeles CA 90024

WANG, KE-CHIN, b Chekiang, China, Dec 8, 30; US citizen; m 55; c 2. HIGH TEMPERATURE CHEMISTRY, CERAMICS. *Educ:* Univ Wis-Madison, BA, 58, MS, 60; Ill Inst Technol, PhD(chem), 66. *Prof Exp:* Res assoc high temperature chem, Okla State Univ, 67-69 & Ill Inst Technol, 69-70; dir res, Ceramtec Industs, Inc, 70-71, vpres, 71-72; RES MGR INDUST CERAMICS, HARBISON-WALKER REFRACTORIES CO, DIV DRESSER INDUSTS, INC, 73- *Mem:* Am Chem Soc; Am Ceramic Soc. *Res:* Advanced ceramic materials and forming processes. *Mailing Add:* 1821 Tyburn Lane Upper St Clair TWP Pittsburgh PA 15241

WANG, KEN HSI, b Shanghai, China, July 4, 34; m 62; c 3. NUCLEAR PHYSICS. *Educ:* Int Christian Univ, Tokyo, BA, 58; Yale Univ, PhD(physics), 63. *Prof Exp:* Res fel physics, Harvard Univ, 63-66; asst prof, 66-69, ASSOC PROF PHYSICS, BAYLOR UNIV, 69- *Concurrent Pos:* Assoc res scientist, Tex A&M Univ, 80-81. *Mem:* Am Phys Soc; Am Asn Physics Teachers. *Res:* Nuclear reaction and scattering. *Mailing Add:* Dept Physics Baylor Univ Waco TX 76798

WANG, KIA K, b Soochow, China, Oct 10, 24; nat US; m 43; c 3. GEOLOGY, STRATIGRAPHY. *Educ:* Nat Southwestern Assoc Univ, China, BS, 43; La State Univ, MS, 47, PhD, 51. *Prof Exp:* Geologist, Calif Oil Co, La, 46; res geologist, State Geol Surv, La, 47-50; asst prof, 50-70, ASSOC PROF GEOL, BROOKLYN COL, 70- *Concurrent Pos:* Asst, La State Univ, 46-50. *Mem:* Geol Soc Am; Soc Econ Paleontologists & Mineralogists; Am Asn Petrol Geologists. *Res:* Petroleum geology; micropaleontology. *Mailing Add:* 4-816 209th St Flushing NY 11364

WANG, KUAN, BIOCHEMISTRY, CELL BIOLOGY. *Educ:* Yale Univ, PhD(biochem), 74. *Prof Exp:* ASSOC PROF CHEM, UNIV TEX, AUSTIN, 77- *Res:* Contracto proteins; muscle structure and function; cell motility. *Mailing Add:* Dept Chem Univ Tex Welsh 4230B Austin TX 78712

WANG, KUNG-PING, b China, Mar 11, 19; nat US; m; c 1. MINING ENGINEERING, METALLURGICAL ENGINEERING. *Educ:* Yenching Univ, China, BS, 40; Mo Sch Mines, BS, 42; Columbia Univ, MS, 43, PhD(mining), 46. *Hon Degrees:* Prof Mining Engr, Mo Sch Mines, 61. *Prof Exp:* Jr engr, Wah Chong Corp, 42 & Hudson Coal Co, 43-44; surveyor, NJ Zinc Co, 44-45; engr, Warren Pipe & Foundry Corp, 46; chief engr & asst gen mgr, Ping-Hsing Coal Co, 46-48; prof, Peiyang Univ, 48; chief specialist int activities, US Bur Mines, 60-69, supvry phys scientist nonmetallic minerals, 70-75, supvry phys scientist Asia minerals, 75-80; PRES, K P WANG ASSOCS, 80- *Concurrent Pos:* Part-time asst to sci adv, Dept Interior, 64-67; adj assoc prof, Krumb Sch Mines, Columbia Univ, 67-70; consult to UN, 69-71, Pennzoil, 80-84, Mitsubishi, 87- *Mem:* Am Inst Mining, Metall & Petrol Engrs. *Res:* Mineral economics; international natural resources, particularly in developing countries. *Mailing Add:* 14330 Cartwright Way North Potomac MD 20878

WANG, KUO KING, b Wutsin, China, Oct 8, 23; c 3. ENGINEERING. *Educ:* Nat Cent Univ, Nanking, BS, 47; Univ Wis-Madison, MS, 62, PhD(mech eng), 68. *Prof Exp:* Engr, Taiwan Shipbuilding Corp, 47-57; mgr eng, Ingalls-Taiwan Shipbuilding Corp, 57-60; supvr shipbuilding, United Tanker Corp, 60-61; proj engr, Walker Mfg Co, 62-66; asst prof mech eng, Univ Wis-Madison, 68-70; assoc prof, 70-77, PROF MECH ENG, CORNELL UNIV, 77- *Concurrent Pos:* Consult, Xomox Corp, Gen Elec Co, TRW Inc, Polaroid Corp & Gillette Co; TRW fel mfg eng, TRW Found, 77; mem, Int Inst Prod Eng Res, 85- *Honors & Awards:* Blackall Award, Am Soc Mech Engrs, 68, William T Ennor Mfg Technol Award, 91; Adams Mem Award, Am Welding Soc, 76; Frederick W Taylor Res Medal, Soc Mfg Engrs, 87. *Mem:* Nat Acad Eng; fel Am Welding Soc; fel Soc Mfg Engrs; fel Am Soc Mech Engrs; Am Soc Metals Int; Soc Plastics Engrs; Polymer Processing Soc. *Res:* Materials processing; numerical control; author of over 100 publicants and several book chapters; recipient of one patent. *Mailing Add:* Sch Mech Eng-Aero Eng Cornell Univ Upson Hall Ithaca NY 14853

WANG, LAWRENCE CHIA-HUANG, b Wusih, China, Apr 5, 40; m 66. PHYSIOLOGY, ZOOLOGY. *Educ:* Taiwan Norm Univ, BSc, 63; Rice Univ, MA, 67; Cornell Univ, PhD(physiol), 70. *Prof Exp:* Vis asst prof biol, Univ Ore, 69-70; from asst prof to assoc prof, 70-80, McCalla prof, 83-84, prof-in-residence, 84-85, PROF ZOOL, UNIV ALTA, 80- *Concurrent Pos:* Operating grant, Nat Res Coun Can, 71-, Nat Defense Res Bd Can, 73-76 & Nat Defense Can, 83- *Mem:* AAAS; Can Soc Zool; Can Physiol Soc; Soc Cryobiol; Am Soc Zoologists; Int Hibernation Soc. *Res:* Physiology of temperature regulation; hypothermia and hibernation in mammals. *Mailing Add:* Dept Zool Univ Alberta Edmonton AB T6G 2E9 Can

WANG, LAWRENCE K, b China, Nov 20, 40; c 3. WATER & WASTE WATER TREATMENT, HAZARDOUS WASTE DISPOSAL. *Educ:* Nat Cheng Kung Univ, BE, 62; Univ Mo, Rolla, ME, 65; Univ RI, MS, 67; Rutgers Univ, PhD (environ eng), 72. *Prof Exp:* Res asst civil eng, Univ Mo, 65-66, environ eng, Univ RI, 66-67; res fel, Rutgers Univ, 67-70; civil engr, Hackensack Water Co, 70-71; proj mgr consult eng, Calspan Corp, 71-74; asst prof chem eng, Rennselaer PolyTech Inst, 74-77, mech eng, Stevens Inst Technol, 77-81; dir environ eng, Lenox Inst Res, 81-88; VPRES, ZOREX CORP, 88- *Concurrent Pos:* Sr sanitary engr, NY State Dept Environ Conserv, 74-77; examr, Nat Coun Eng Examrs, 74-81; assoc ed, Pergamon Press, 78-81; eng consult, US Environ Protection Agency, 77-81, Krofta Eng Corp, 80-; vis prof, Nat Cheng Kung Univ, Taiwan, 82-86, Chekiang Univ, China, 86-; chmn, standard method subcomt, Am Water Works Asn, 80-85; task group, Nat Sanitation Found, 85-87; water oper training prog, New Eng Water Works Asn, 84-87. *Honors & Awards:* Drinking Water Award, Taiwan, China, 86. *Mem:* Am Acad Environ Engrs; Am Water Works Asn; Am Inst Chem Engrs (treas, 86-87); Nat Sanitation Found; Overseas Environ Engrs & Scientists Asn (pres, 88-90). *Res:* Conducting research areas of drinking water, industrial wastes, domestic sewage, hazardous wastes, acid rain and cooling tower; 5 books, 3 US patents, 600 journals and reports in environmental engineering. *Mailing Add:* One Dawn Dr Latham NY 12110-5305

WANG, LEON RU-LIANG, b Canton, China, June 15, 32; US citizen; m 61; c 3. STRUCTURAL ENGINEERING. *Educ:* Cheng Kung Univ, Taiwan, BS, 57; Univ Ill, MS, 61; Mass Inst Technol, ScD(struct), 65. *Prof Exp:* Res asst struct eng, Mass Inst Technol, 61-65, res engr, 65; asst prof struct, Rensselaer Polytech Inst, 65-69, assoc prof, 69-80; prof civil eng, Univ Okla, 80-84; chmn dept, 84-90, PROF CIVIL ENG, OLD DOMINION UNIV, 84- *Concurrent Pos:* Tech consult, Watervliet Arsenal, 66-80 & Nat Sci Coun, Taiwan, 75-; lectr, Am Inst Steel Construct, 69 & 71; NSF vis scientist, Cheng Kung Univ, Taiwan, 72-73; tech rev, NSF, 76-; vis scholar, Pub Works Res Inst, Japan, 81, Toyohashi Univ Technol, Japan, 86. *Mem:* AAAS; Am Concrete Inst; Am Soc Civil Engrs; Am Soc Eng Educ; Chinese Inst Eng; Earthquake Eng Res Inst. *Res:* Structural mechanics; dynamics; stability; buckling; computer applications; model analysis; plate and shell theories; steel and reinforced concrete construction; earthquake engineering; lifeline earthquake engineering. *Mailing Add:* Civil Eng Dept Old Dominion Univ Norfolk VA 23529-0241

WANG, LI CHUAN, b Kaiyuan, China; US citizen; m; c 2. BIOCHEMISTRY, SOIL FERTILITY. *Educ:* Nat Cent Univ, Nanking, China, BS, 45; Univ Wis-Madison, PhD, 52, MS, 58. *Prof Exp:* Teaching asst, Nat Cent Univ, Nanking, 45-48; fel biochem, Univ Wis-Madison, 58-60; res chemist, Monsanto Chem Co, 60-62; assoc prof chem, Univ Alaska, 62-63; PRIN CHEMIST, NORTHERN REGIONAL RES LAB, USDA, 63- *Mem:* Am Chem Soc; Am Soc Plant Physiologists; Inst Food Technologists. *Res:* Soybean proteins, their utilization, flavor and functionalities. *Mailing Add:* 130 Fernwood Lane Bloomingdale IL 60108-2523

WANG, LIN-SHU, b Shanghai, China, Feb 14, 38; US citizen; m 65; c 2. INTERNAL COMBUSTION ENGINES, ENERGY CONVERSION. *Educ:* Univ Calif, Berkeley, PhD(eng), 66. *Prof Exp:* From asst prof to assoc prof eng, 65-72, ASSOC PROF MECH ENG, STATE UNIV NY STONY BROOK, 72- *Concurrent Pos:* Vis mem, Ctr Earth & Planetary Physics, Harvard Univ, 72; founder & pres, Intercool Develop, Ltd, 87- *Mem:* Sigma Xi; Soc Automotive Engrs; Am Soc Mech Engrs. *Res:* Internal combustion engines; intercooled-supercharged power cycles, gas generator engines; thermodynamics; process principle and the concept of steady states; turbulent fluid flow and atmospheric dynamics. *Mailing Add:* Dept Mech Eng State Univ NY Stony Brook Stony Brook NY 11794

WANG, MARIAN M, b Funkien, China, June 20, 28. NUTRITION. *Educ:* Pa State Univ, PhD(nutrit), 60. *Prof Exp:* ASSOC PROF NUTRIT, UNIV GA, 79- *Mem:* Am Inst Nutrit; Inst Food Technologists; Am Dietary Asn. *Mailing Add:* Dept Food Sci & Nutrit Univ Ga Dawson Hall Athens GA 30602

WANG, MAW SHIU, b Chang Hwa, Formosa, Nov 1, 25; m 51; c 3. AGRONOMY, SPECTROCHEMISTRY. *Educ:* Prov Agr Col, Formosa, BS, 51; Okla State Univ, MS, 56; Univ Ill, PhD, 59. *Prof Exp:* Jr engr, Chem Lab, Pingtong Sugar Exp Sta, Formosa, 51-54; asst soil chem, Univ Ill, Urbana, 55-59, res assoc, 59-61; sr res chemist, Cent Res Dept, 61-68, res specialist, 68-77, FEL, MONSANTO CO, 78- *Concurrent Pos:* Adj prof, St Louis Univ, 70- *Honors & Awards:* Megger's Award, Soc Appl Spectros, 73. *Mem:* Am Soc Testing & Mat; Soc Appl Spectros; Sigma Xi. *Res:* Emission spectroscopy; spectrochemical analysis of major and minor elements in agricultural material; traces in semiconductors and related materials; surface analysis, cleaning, and packaging of semiconductors. *Mailing Add:* Monsanto Co 800 N Lindbergh Blvd St Louis MO 63167

WANG, MUHAO S, b China, Jan 29, 42; US citizen; m 68; c 3. ENVIRONMENTAL MANAGEMENT ENGINEERING. *Educ:* Nat Cheng Kung Univ, China, BE, 65; Univ RI, MS, 68, Rutgers Univ, New Brunswick, PhD(environ eng), 72. *Prof Exp:* Civil engr, Sino-Am Eng Corp, 65-66; res asst environ eng, Univ RI, 66-68; res fel, Rutgers Univ, 68-72; SR SANIT ENG, NY STATE DEPT ENVIRON CONSERV, 72- *Concurrent Pos:* Consult, Calsban Corp, Buffalo, 71-74; Dipl, Am Acad Environ Engrs; vis prof, Nat Cheng Kung Univ, Taiwan, 81-86 & Lenox Inst Res, Mass, 81-; ed, Humana Press, Clifton, NJ, 86- *Mem:* Overseas Chinese Environ Engrs & Scientists Asn; Asn Environ Eng Prof; Water Pollution Control Fedn; Chinese Inst Engrs. *Res:* Environmental management; stream pollution; industrial wastes; solid wastes; harzardous wastes. *Mailing Add:* One Dawn Dr Latham NY 12110-5305

WANG, NAI-SAN, b Changhua, Taiwan, Jan 20, 36; Can citizen; m 63; c 4. PATHOLOGY. *Educ:* Nat Taiwan Univ, MD, 60; McGill Univ, MS, 69, PhD(path), 71. *Prof Exp:* Teaching & res fel path, McGill Univ, 67-71, from asst prof to prof, 71-90; PROF PATH, UNIV CALIF, IRVINE, 90- *Concurrent Pos:* Consult, Mesothelioma Ref Panel Can, Tumor Ref Ctr, 74-81. *Mem:* Am Asn Pathologists; Am Thoracic Soc; Can Thoracic Soc; Int Asn Pathologists. *Res:* Ultrastructural studies of the lung and pleura in normal and diseased conditions. *Mailing Add:* Dept Path Long Beach VA Med Ctr 5901 E Seventh St Long Beach CA 90822

WANG, NANCY YANG, b Peiping, China, Jan 20, 26; m 49; c 1. ORGANIC CHEMISTRY. *Educ:* Cath Univ, Peiping, BS, 45; St Louis Univ, MS, 51; Boston Univ, PhD(org chem), 65. *Prof Exp:* Fel biol, Mass Inst Technol, 64-65, fel nutrit, 65-66, res assoc chem, 66-68; res assoc, Retina Found, Boston Univ, 68-71; res assoc, Univ Mass, Boston, 71-73, lectr, 73-75, res assoc chem, 75-82; RETIRED. *Mem:* Sigma Xi. *Res:* Biochemistry and organic reaction mechanisms. *Mailing Add:* Dept Chem Univ Mass Boston MA 02125

WANG, P(AUL) K(ENG) C(HIEH), b Nanking, China, July 23, 34. ELECTRICAL ENGINEERING. *Educ:* Calif Inst Technol, BS, 55, MS, 56; Univ Calif, Berkeley, PhD(elec eng), 60. *Prof Exp:* Mem res staff, Res Lab, Int Bus Mach Corp, 59-64; asst prof elec eng, Univ Southern Calif, 64-67; assoc prof eng, 67-73, PROF ENG & APPL SCI, UNIV CALIF, LOS ANGELES, 73- *Concurrent Pos:* Vis mem res staff, Dept Physique du Plasma et Fusion Controlee, Centre d'Etudes Nucleaires, France, 74-75; consult, Jet Propulsion Lab, Calif Inst Technol, 81- *Mem:* Inst Elec & Electronics Engrs; Soc Indust & Appl Math; Am Phys Soc; Am Math Soc. *Res:* Automatic control system theory; plasma physics; robotics. *Mailing Add:* Elec Eng Rm 66-147L Engr IV Univ Calif Los Angeles CA 90024-1594

WANG, PAO-KUAN, b Tainan, Taiwan, Dec 1, 49; m 76; c 2. CLOUD PHYSICS, AEROSOL PHYSICS & TECHNOLOGY. *Educ:* Nat Taiwan Univ, BS, 71; Univ Calif, Los Angeles, MS, 75, PhD(atmospheric sci), 78. *Prof Exp:* Res atmospheric physicist, Univ Calif, Los Angeles, 78-80, adj asst prof, 80; from asst prof to assoc prof, 80-88, PROF METEOROL, UNIV WIS-MADISON, 88- *Concurrent Pos:* Prin investr, res projs, Environ

Protection Agency, 81- & NSF, 82-; mem, sci rev panel, atmospheric physics & chem, Environ Protection Agency, 82- *Mem:* Am Meteorol Soc; AAAS; Am Geophys Union; Am Asn Aerosol Res; Royal Meteorol Soc. *Res:* Aerosol physics; cloud and precipitation physics; atmospheric electricity; particle technology; cloud dynamics; atmospheric chemistry; historical climatology; aerosol filtration. *Mailing Add:* Dept Meteorol Univ Wis 1225 W Dayton St Madison WI 53706

WANG, PAUL KENG CHIEH, b Nanking, China, July 23, 34; US citizen. CONTROL OF DISTRIBUTED-PARAMETER SYSTEMS, MICROROBOTICS. *Educ:* Calif Inst Technol, BS, 55, MS, 56; Univ Calif, Berkeley, PhD(elec eng), 60. *Prof Exp:* Mem res staff, Int Bus Mach Res Lab, San Jose, Calif, 59-64; asst prof elec eng, Univ Southern Calif, 64-67; PROF ELEC ENG, UNIV CALIF, LOS ANGELES, 67- *Concurrent Pos:* Vis res staff mem, Dept de Physique du Plasma et de la Fusion Controlée, Centre d'Etude Nucleaires, Fontenay-aux-Roses, France, 74; consult, Jet Propulsion Lab, Pasadena, Calif, 81-; hon vis prof, Dept Computer & Syst Sci, Nankai Univ, Tianjin, China, 84- *Mem:* Am Phys Soc; Am Math Soc; Inst Elec & Electronics Engrs; Soc Indust & Appl Math. *Res:* Theoretical and experimental studies on the control of distributed-parameter and nonlinear dynamical systems with applications to large space structures, microrobotics and hydrodynamics. *Mailing Add:* Dept Elec Eng Rm 56-125B Engr IV Univ Calif Los Angeles CA 90024

WANG, PAUL SHYH-HORNG, b Shensi, China, Jan 11, 44; US citizen; m 84; c 2. SYMBOLIC MATHEMATICAL COMPUTATION, NON-NUMERICAL ALGORITHMS. *Educ:* Taiwan Nat Chung-hsing Univ, BS, 66; Mass Inst Technol, PhD(comput sci), 71. *Prof Exp:* From lectr to asst prof computer sci, Mass Inst Technol, 71-77; asst prof, 77-81, PROF COMPUTER SCI, KENT STATE UNIV 81-, DIR RES, INST COMPUTER MATH, 85. *Concurrent Pos:* Consult, Hewlett-Packard Labs, 86-; prin investr, NSF grants, 87-; chmn spec interest group on symbolic & algebraic manipulation, Asn Comput Mach, 87- *Mem:* Asn Comput Mach. *Res:* Making the digital computer a useful tool for scientific computations. *Mailing Add:* Dept Math Sci Kent State Univ Kent OH 44242

WANG, PAUL WEILY, b Kao-Hsiung, Taiwan, Nov 4, 51; m; c 3. SURFACES OF OPTICAL MATERIALS, PROCESSES TO MANUFACTURE OPTICAL MATERIALS. *Educ:* Nat Taiwan Normal Univ, BS, 74; State Univ NY, Albany, MS, 81, PhD(physics), 86. *Prof Exp:* Res assoc, Inst Studies Defects in Solids, 86; res asst prof, Vanderbilt Univ, 86-90; ASST PROF TEACHING & RES, DEPT PHYSICS, UNIV TEX, EL PASO, 90- *Concurrent Pos:* Consult, EOTec Inc, 87-88. *Mem:* Am Optics Soc; Am Physics Soc; Inst Studies Defects in Solids; Photonics Soc Chinese-Am; Int Soc Optical Eng. *Res:* Stimulated luminescence in glasses by particle bombardments; adsorption and desorption of surfaces by electron spectroscopy and sims; sol-gel thin silicate film. *Mailing Add:* Dept Physics Univ Tex El Paso TX 79968

WANG, PETER CHENG-CHAO, b Shantung, China, Jan 11, 37; US citizen; m 60; c 2. STATISTICS. *Educ:* Pac Lutheran Univ, BA, 60; Wayne State Univ, MA, 62, PhD(math), 66. *Prof Exp:* Instr math, Wayne State Univ, 64-66; asst prof statist, Mich State Univ, 66-67; from asst prof to assoc prof, Univ Iowa, 67-70; ASSOC PROF STATIST, NAVAL POSTGRAD SCH, 70- *Concurrent Pos:* Vis assoc prof statist, Stanford Univ, 69-70; consult, B D M Serv Co, Tex, 71- *Mem:* Am Math Soc; Inst Math Statist; London Math Soc; Sigma Xi. *Res:* Stochastic models; combinatorics; forecast models for military systems; sound propagation models. *Mailing Add:* PO Box 234 Pebble Beach CA 93953

WANG, PIE-YI, b Chanhua, Taiwan, Feb 28, 40; nat US; m 68; c 2. FOOD ENGINEERING, MEAT SCIENCE. *Educ:* Nat Taiwan Univ, BS, 63; Univ Hawaii, MS, 67; Michigan State Univ, PhD(agr eng), 70. *Prof Exp:* Res eng, Peter Eckroch & Sons, Inc, 70-74, sr eng, 74-80; mgr operational eng, Armour Foods, 80-81; mgr basic res, Eckrich & Sons Inc, 81-85; DIR PROCESS RES, SWIFT-ECKRICH INC, 85- *Mem:* Am Soc Agr Eng; Inst Food Technologists. *Res:* Developing, planning and managing research and development activities on meat processing technologies, control and instrumentation, equipment development, physical and chemical properties of meats and transfer new technologies to operations. *Mailing Add:* Beatrice Co Inc Rech Dev Ctr 1919 Swift Dr Oak Brook IL 60522

WANG, PING CHUN, b Kiangsu, China, Mar 10, 20; US citizen; m 55; c 2. ENGINEERING. *Educ:* Nat Cent Univ China, BS, 43; Univ Ill, MS, 48, PhD(eng), 51. *Prof Exp:* Engr, China Bridge Co, 43-47; struct designer, Ammann & Whitney Consult Engrs, 51-52; struct supvr, Seelye, Stevenson, Value & Knecht Consult Engrs, 52-60; assoc prof eng, Stevens Inst Technol, 60-63; PROF CIVIL ENG, POLYTECH INST NEW YORK, 63- *Concurrent Pos:* Dir & vpres, Omnidata Serv Inc, 71-72. *Mem:* Fel Am Soc Civil Engrs; Am Concrete Inst; Am Soc Eng Educ. *Res:* Discrete systems approach in structural mechanics. *Mailing Add:* Dept Civil Eng Polytech Univ 333 Jay St Brooklyn NY 11201

WANG, PING-LIEH THOMAS, b China, Nov 16, 46; m 75; c 2. POULTRY PRODUCT, FEED MILL PRODUCT. *Educ:* Nat Chung-Hsing Univ, BS, 70; Miss State Univ, MS, 75; Tex A&M Univ, PhD(food technol), 82. *Prof Exp:* DIR QUAL ASSURANCE, B C ROGERS POULTRY INC, 83- *Mem:* Poultry Sci Asn; Inst Food Technologists. *Mailing Add:* B C Rogers Sons Inc PO Box A Morton MS 39117

WANG, RICHARD HSU-SHIEN, b Qianshan, Anhui, China, Jan 2, 32; m 62; c 3. POLYMER DEGRADATION & STABILIZATION, PLASTICS ADDITIVES. *Educ:* Nat Taiwan Univ, BS, 56; Univ Ill, Urbana, MS, 61; Univ Kans, PhD(chem), 68. *Prof Exp:* Res asst soil chem, Nat Taiwan Univ, 58-59; res asst entomol, Nat Hist Surv Ill, 61-63; res chemist, 68-74, PRIN RES CHEMIST, EASTMAN CHEM CO RES LABS, EASTMAN KODAK CO, 74- *Mem:* Am Chem Soc; Am Inst Chemists. *Res:* Polymer stabilization and synthesis of new antioxidants and light inhibitors for polymeric compositions; development of new polymer intermediates and improved synthetic methods for fine organic chemicals; synthesis and development of plastics additives. *Mailing Add:* 1414 Fairidge Dr Kingsport TN 37664

WANG, RICHARD I H, b Shanghai, China, Oct 12, 24; US citizen; m 58; c 2. PHARMACOLOGY, INTERNAL MEDICINE. *Educ:* St John's Univ, BS, 45; Utah State Univ, MS, 49; Univ Ill, PhD(pharmacol), 52; Northwestern Univ, MD, 55. *Prof Exp:* Intern, Presby Hosp, Chicago, Ill, 55-56; resident, Indianapolis Gen Hosp, Ind, 56-58; prin scientist, Roswell Park Mem Inst, 61-63; assoc prof pharmacol & med, 63-70, PROF CLIN PHARMACOL, MED COL WIS, 70-, ASSOC PROF MED, 77-; CHIEF DRUG TREATMENT CTR, WOOD VET'S ADMIN CTR, 77- *Concurrent Pos:* Chief clin pharmacol serv, Wood Vet Admin Hosp, 63-, dir drug abuse treatment & rehab prog, 71-; consult physician, Milwaukee County Gen Hosp, 64-; attend physician, Milwaukee County Ment Health Ctr, 65- *Mem:* Am Soc Pharmacol & Exp Therapeut; Fedn Am Socs Exp Biol; Radiation Res Soc; Am Soc Clin Pharmacol & Therapeut. *Res:* Clinical pharmacology; radiation biology. *Mailing Add:* Chief Alcohol & Drug Dependence Drug Treat Med Col Wis Vet Admin Med Ctr Milwaukee WI 53295

WANG, RICHARD J, b Chungking, China, Oct 23, 41; US citizen; m 66; c 3. CELL BIOLOGY, AUTOIMMUNITY. *Educ:* Harvard Univ, BA, 64; Univ Colo, PhD(biophys), 68. *Prof Exp:* Fel cell biol, NY Univ, 68-70; res assoc biol, Mass Inst Technol, 70-71; from asst to assoc prof, 71-83, PROF BIOL, 83-, DIR, HUMAN MONOCLONAL ANTIBODY FAC, UNIV MO, COLUMBIA, 87- *Concurrent Pos:* NIH res career develop award, 72-77; biochem consult, Cancer Res Ctr, Columbia, Mo, 73-; res investr, Dalton Res Ctr, Univ Mo, Columbia, 74- *Mem:* Am Soc Cell Biol; Tissue Cult Asn; AAAS. *Res:* Biochemical genetics of human and mammalian cells in culture; regulatory mechanisms in autoimmunity. *Mailing Add:* Dalton Res Ctr Univ Mo Columbia MO 65211

WANG, RICHARD JINCHI, electrical engineering, for more information see previous edition

WANG, ROBERT T, b Chung King, China, Aug 3, 41; US citizen; m 68; c 2. INORGANIC CHEMISTRY, PHYSICAL CHEMISTRY. *Educ:* Taiwan Cheng Kung Univ, BS, 62; Johns Hopkins Univ, PhD(chem), 68. *Prof Exp:* Fel inorg chem, Iowa State Univ, 68-70; asst prof, 70-76, ASSOC PROF CHEM, SALEM STATE COL, 76- *Concurrent Pos:* Consult, US Summit Corp, New York, NY, 72- *Mem:* Am Chem Soc. *Res:* The studies of kinetics and mechanisms of inorganic transition metal complexes reactions; synthesis of bioinorganic compounds. *Mailing Add:* Dept Chem & Physics Salem State Col 352 Lafayette St Salem MA 01970

WANG, RU-TSANG, b Changhua, China, Sept 13, 28; US citizen; m 55; c 2. PHYSICS OF LIGHT SCATTERING, MICROWAVE ELECTRONICS. *Educ:* Nat Taiwan Univ, BSc, 50; Rensselaer Polytech Inst, PhD(physics), 68. *Prof Exp:* Res asst physics, Rensselaer Polytech Inst, 68-70; res assoc, Dudley Observ, 70-76; res assoc physics, Space Astron Lab, State Univ NY Albany, 77-81; ASSOC RES SCIENTIST, SPACE ASTRON LAB, UNIV FLA, GAINESVILLE, 81- *Mem:* Sigma Xi; Optical Soc Am. *Res:* Scattering of light and other electromagnetic radiations; associated microwave technique. *Mailing Add:* Space Astron Lab 1810 NW Sixth St Gainesville FL 32609

WANG, SAM S(HU) Y(I), b Chungking, China, Sept 21, 36; m 66; c 2. MECHANICAL ENGINEERING. *Educ:* Cheng Kung Univ, Taiwan, BSc, 59; Univ Rochester, MSc, 65, PhD(mech & aerospace sci), 68. *Prof Exp:* Design engr, Yue Loong Motor Co Ltd, 60-61; asst mech, Cheng Kung Univ, 61-63; asst fluid mech, Univ Rochester, 63-64, asst aerospace sci, 64-67; from asst prof to assoc prof eng, 67-81, actg chmn, Dept Mech Eng, 82-83, PROF ENG, UNIV MISS, 81-, DIR, CTR COMPUTATIONAL HYDROSCI & ENG, 83-, FREDERICK A P BARNAROL DISTINGUISHED PROF, 88- *Concurrent Pos:* Res fel, Stanford Univ, 68; vis scientist, Johnson Space Ctr, NASA, 73-74 & Aero Propulsion Lab, USAF, 75-; gen chmn & organizer, 3rd Int Conf on Finite Elements in Water Resources, Univ Miss, 80; exec coun, Int Soc Computational Methods Eng, 80-84; tech prog co-chair, Fluid Dynamics, Plasmadynamics & Lasers Conf, Am Inst Aeronaut & Astronaut, 84; gen chmn, 3rd Int Symp River Sedimentation, 86 & Int Symp Sediment Transp Modeling, Am Soc Civil Engrs, 89. *Honors & Awards:* Ralph R Teetor Award, Soc Automotive Engrs, 75; Hydraul Res Achievement Award, Am Soc Civil Engrs. *Mem:* Assoc fel Am Inst Aeronaut & Astronaut; Am Soc Civil Engrs; Am Soc Eng Educ; Am Soc Mech Engrs; Soc Natural Philos; Int Soc Computational Methods Eng; Int Asn Hydraul Res; Nat Soc Prof Engrs; Nat Geog Soc. *Res:* Fluid mechanics including hypersonic flow theory, boundary layer theory, hydrodynamic stability, magnetohydrodynamics, plasma physics and physics of fluids; computer simulation of river and coastal flows and sediment transports. *Mailing Add:* Sch Eng CCHE Univ Miss University MS 38677

WANG, SAN-PIN, b Taiwan, Nov 7, 20; m 46; c 5. MEDICAL MICROBIOLOGY. *Educ:* Keio Univ, Japan, MD, 44; Univ Mich, MPH, 52. *Hon Degrees:* Dr Med Sci, Keio Univ, Japan, 59. *Prof Exp:* Asst bact, Sch Med, Keio Univ, Japan, 44-46; chief dept bact, Taiwan Prov Hyg Lab, 46-51; chief dept virol, Taiwan Serum Vaccine Lab, 52-58; med officer virus immunol, US Naval Med Res Unit, 58-64; vis assoc prof prev med, Sch Med, 64-66, assoc prof, 66-70, PROF PATHOBIOL, SCH PUB HEALTH & COMMUNITY MED, UNIV WASH, 70- *Mem:* Am Asn Path. *Res:* Biological products, rabies and smallpox vaccines; research on tropical diseases, rabies, influenza, encephalitis and trachoma. *Mailing Add:* Sch Pub Health & Community Med Univ Wash F161 Health Sci Bldg Seattle WA 98195

WANG, SHAO-FU, b Dairen, China, Aug 26, 22; m 44; c 3. THEORETICAL PHYSICS, CONDENSED MATTER PHYSICS. *Educ:* Nagoya Univ, PhD(physics), 64. *Prof Exp:* Lectr physics, Ching Kung Univ, Taiwan, 48-52, assoc prof, 52-56; from assoc prof to prof, Tunghai Univ, Taiwan, 56-65; assoc prof, 65-68, PROF PHYSICS, UNIV WATERLOO, 68- *Concurrent Pos:* Res assoc, Univ Ill, 61-63, vis assoc prof, 67. *Mem:* Am Phys Soc; Can Asn Physicists. *Res:* Theoretical study of disordered metals. *Mailing Add:* Dept Physics Univ Waterloo Waterloo ON N2L 3G1 Can

WANG, SHIEN TSUN, b Changsha, China, Aug 24, 38; m 69; c 2. STRUCTURAL ENGINEERING, ENGINEERING MECHANICS. *Educ:* Nat Taiwan Univ, BS, 60; Mich State Univ, MS, 64; Cornell Univ, PhD(struct eng), 69. *Prof Exp:* Civil eng, Nat Taiwan Univ, 61-62 & Mich State Univ, 62-64; struct engr, Cornell Univ, 64-68, res assoc & instr, 68-69; from asst prof to assoc prof, 69-83, dir grad studies, 80-84, PROF CIVIL ENG, UNIV KY, 83-, DIR GRAD STUDIES, 87- *Concurrent Pos:* Fac res & teaching fels, Univ Ky, 70-72; NSF fel, Syracuse Univ, 71 & Ohio State Univ, 80; various consulting work; chm & mem, several Am Soc Civil Engrs Nat Comts; ed bd, J Thin Walled Struct. *Honors & Awards:* Hon Soc. *Mem:* Am Soc Civil Engrs; Structural Stability Res Coun; Am Soc Eng Educ; Sigma Xi; Am Acad Mech. *Res:* Structural stability and post-buckling analysis; thin-walled and cold formed structures; computer aided design and computer graphics; damage and failure analysis; computer modeling and analysis of space and special structures; nonlinear structural analysis; finite element methods; metal structures. *Mailing Add:* Dept Civil Eng Univ Ky Lexington KY 40506

WANG, SHIH CHUN, b Jan 25, 10. NEUROPHYSIOLOGY, NEUROPHARMACOLOGY. *Educ:* Northwestern Univ, PhD(neurol), 40; Peking Union Med Col, MD. *Prof Exp:* PROF PHARMACOL, COL PHYSICIANS & SURGEONS, COLUMBIA UNIV, 56-, PFEIFFER PROF, 79- *Res:* Vomiting center in the medulla; central mechanisms of respiration. *Mailing Add:* 18 Kent Rd Tenafly NJ 07670

WANG, SHOU-LING, b Shanghai, China, Oct 17, 24; nat US; m 86; c 3. SHIP PROTECTION. *Educ:* St John's Univ, China, BS, 46; Yale Univ, ME, 48; Univ Ill, PhD(theoret & appl mech), 52. *Prof Exp:* Designer, D B Steinman, 52-55; asst prof civil eng, Clarkson Univ, 55-57; assoc prof, Univ Mo, 57-60; vis assoc prof eng mech, NC State Univ, 60-67; struct engr, 67-78, HEAD, UNDERWATER PROTECTION GROUP, DAVID TAYLOR RES CTR, 79- *Concurrent Pos:* Lectr weapon effects and ship protection, Mass Inst Technol, 76- *Honors & Awards:* Meritorious Civilian Serv, Dept Navy, 85. *Mem:* Am Acad Mech. *Res:* Structural dynamics; weapon effects; ship protection. *Mailing Add:* 9132 Kirkdale Rd Bethesda MD 20817

WANG, SHU LUNG, b Sichuan, China, May 2, 25; m 47; c 2. COMPUTER APPLICATIONS. *Educ:* Wash Univ, St Louis, BS, 49, MS, 50, DSc(chem eng), 53. *Prof Exp:* From asst prof to assoc prof chem eng, Kans State Univ, 52-57; sect head eng systs develop & dir comput lab, Linden Div, Tarrytown Tech Ctr, 57-66, mgr, Niagara Frontier Regional Comput Ctr, 66-69, mgr comput applns, sci, 69-83; consult, Advan Comput Technol, Union Carbide Corp, 84-85; ASSOC PROF INFO SYSTS, PACE UNIV, 86- *Mem:* Am Inst Chem Eng; Am Chem Soc; Inst Elec & Electronics Engrs Comput Soc; Asn Comput Mach; Am Asn Artificial Intel. *Res:* Applied artificial intelligence, information system development and data processing management; process control systems engineering; adsorption and cryogenic gas separation processes; vapor-liquid equilibrium data; knowledge based systems. *Mailing Add:* 21 Sunset Dr Ossining NY 10562-2101

WANG, SHYH, b China, June 15, 25. ELECTRONICS, PHYSICS. *Educ:* Chiao Tung Univ, BS, 45; Harvard Univ, MA, 49, PhD(appl physics), 51. *Prof Exp:* Res fel appl physics, Harvard Univ, 51-53; eng specialist, Semiconductor Div, Sylvania Elec Prod Co, 53-58; assoc prof elec eng, 58-64, PROF ELEC ENG & COMPUT SCI, UNIV CALIF, BERKELEY, 64- *Concurrent Pos:* vis scientist, IBM Thomas Watson Res Ctr, Yorktown Heights, NY, 74-75; vis prof, Dept Electronic Eng, Univ Tokyo, 84; distinguished vis researcher, Univ Calif, Santa Barbara, 84-85. *Mem:* Am Phys Soc; fel Inst Elec & Electronics Engrs; Optical Soc Am. *Res:* Solid state physics and quantum electronics; semiconducting devices and materials; magnetic properties of solids at microwave frequencies. *Mailing Add:* Dept Elec Eng & Comput Sci Univ Calif 2200 University Ave Berkeley CA 94720

WANG, SIMON S, electrochemical engineering, triboelectrochemistry, for more information see previous edition

WANG, SOO RAY, b Tainan, Taiwan, Aug 2, 40; m 69; c 3. ALLERGY, IMMUNOLOGY. *Educ:* Kaohsiung Med Col, Taiwan, MD, 67; Univ Ill Med Ctr, PhD(immunol), 73. *Prof Exp:* Intern, Shadyside Hosp, Pittsburgh, 68-69; jr residency, Hahnemann Med Col Hosp, Philadelphia, 73-74; residency, Atlantic City Hosp, NJ, 74-75; clin fel, Hosp Univ Pa, 75-77, res fel, 77-78; dir, Div Allergy & Immunol, Dept Internal Med, Chang Gung Mem Hosp, Taiwan, 78-83; PROF MED & MICROBIOL, KAOHSIUNG MED COL, KAOHSIUNG, TAIWAN, 80- & YANG MING MED COL, TAIPEI, 83-; DIR ALLERGY, IMMUNOL & RHEUMATOLOGY SECT, DEPT MED, VET GEN HOSP, TAIWAN, 83- *Mem:* Sigma Xi; Am Acad Allergy & Immunol; Am Asn Cert Allergists. *Res:* Histamine and corticosteroid suppression on human lymphocyte proliferation; slow-reacting substance of anaphylaxis in asthmatics; liver arginase. *Mailing Add:* Sect Allergy Immunol & Rheumatology Dept Med Vet Gen Hosp Taipei Taiwan

WANG, TAITZER, b Taiwan, Feb 2, 39; m 68; c 3. BIOCHEMISTRY. *Educ:* Nat Univ Taiwan, BS, 61; Rice Univ, PhD(org chem), 67. *Prof Exp:* Res asst prof cell biophysics, Baylor Col Med, 75-77; asst prof, 77-85, ASSOC PROF PHARMACOL & CELL BIOPHYSICS, COL MED, UNIV CINCINNATI, 85- *Concurrent Pos:* Nat Inst Arthritis & Metab Dis res fel org chem, Fla State Univ, 67-69; US Dept Defense res fel inorg chem, Univ Ky, 69-71; Welch fel biochem, Baylor Col Med, 71-73; Welch fel, Rice Univ, 73-74; NIH spec res fel cell biophys, Baylor Col Med, 74-75. *Mem:* Am Chem Soc; AAAS; NY Acad Sci; Am Soc Biol Chemists & Molecular Biologists; Am Heart Asn; Biophys Soc. *Res:* Enzyme kinetics and synthetic chemistry; cell biophysics. *Mailing Add:* Dept Pharmacol & Cell Biophys Univ Cincinnati Col Med Cincinnati OH 45267

WANG, THEODORE JOSEPH, b Chicago, Ill, Dec 8, 06; m 36; c 2. PHYSICS. *Educ:* Univ Ill, BS, 32, PhD(physics), 39. *Prof Exp:* Asst physics, Univ Ill, 35-39; physicist, Oakes Prod Corp, Ill, 39-40; fel, Univ Minn, 41; from instr to asst prof elec eng, Ohio State Univ, 42-48; physicist, Nat Bur Standards, 48-49; biophysicist, Nat Cancer Inst, 49-50; physicist, George Washington Univ, 50-52; asst prof physics, Univ Mass, 52-55; prof & head dept, SDak Sch Mines & Technol, 55-56; prof, Howard Univ, 56-59; analyst, Opers Res Off, Res Anal Corp, 59-62; prin scientist, Booz, Allen Appl Res, 62-66; dir, Inst Creative Studies, 67-88; RETIRED. *Concurrent Pos:* Consult physicist, 50-52; analyst, Opers Res Off, Johns Hopkins Univ, 57; analyst, Res Anal Corp, 57-58; prof, Howard Univ, 60-64, lectr, 64-88. *Res:* Radiation physics. *Mailing Add:* 4700 Essex Ave Chevy Chase MD 20815

WANG, THEODORE SHENG-TAO, b Oct 18, 30; US citizen; c 2. RADIOPHARMACOLOGY, PHARMACEUTICAL CHEMISTRY. *Educ:* Nat Mukden Med Col, China, BS, 52; Univ Nebr, Lincoln, MS, 58; Univ Md, Phd(med chem & pharmacol), 64. *Prof Exp:* Asst prof nuclear med, dept nuclear med, Health Ctr, Univ Conn, Farmington, 74-76; chief radiopharmacist & sr staff assoc, Columbia-Presbyterian Med Ctr, 76-78; asst prof clin radiol, 78-84, ASSOC PROF CLIN RADIOL & PUB HEALTH, COL PHYSICIANS & SURGEONS, COLUMBIA UNIV, 84- *Mem:* Soc Nuclear Med; Am Chem Soc; NY Acad Sci; Am Pharmaceut Asn; AAAS; Int Asn Radiopharmacol; Am Col Nuclear Physicians. *Res:* Radiotracer labelling of monoclonal antibodies for tumor diagnostic imaging and therapy. *Mailing Add:* Columbia-Presby Med Ctr 622 W 168th St New York NY 10032

WANG, THOMAS NIE-CHIN, b Shanghai, China, Feb 17, 38; m 67; c 2. ELECTRICAL ENGINEERING, APPLIED MATHEMATICS. *Educ:* Cheng Kung Univ, Taiwan, BSEE, 60; Univ NMex, MSEE, 64; Stanford Univ, PhD(elec eng), 70. *Prof Exp:* RES ENGR, RADIO PHYSICS LAB, STANFORD RES INST, 67- *Concurrent Pos:* Vis prof, Cheng Kung Univ, Taiwan, 71-72; mem US nat comt, Comn 6, Int Union Radio Sci, Washington, DC, 72. *Mem:* Inst Elec & Electronics Engrs; Int Union Radio Sci. *Res:* Radiation and waves in scattering and diffraction of electromagnetic waves; propagation in magnetosphere and ionosphere. *Mailing Add:* Yamada Int Corp Pan-Am Bldg Suite 5110 200 Park Ave New York NY 10166

WANG, TING CHUNG, MYOCARDIAL METABOLISM, ORGAN TRANSPLANTATION. *Educ:* Univ Minn, PhD(cell biol), 74. *Prof Exp:* ASSOC PROF SURG, SCH MED, TULANE UNIV, 82- *Mailing Add:* Dept Surg Tulane Univ Sch Med 1430 Tulane Ave New Orleans LA 70112

WANG, TING-I, b Chekiang, China, Jan 11, 44; m 71; c 1. WEATHER INSTRUMENTATION, ATMOSPHERIC REMOTE SENSING. *Educ:* Nat Taiwan Univ, BA, 66; Dartmouth Col, MA, 70, PhD(radiophys), 73. *Prof Exp:* Vis fel & res assoc optical remote sensing, Coop Inst Res Environ Sci, Univ Colo, 73-75; physicist optical remote sensing, Wave Propagation Lab, Nat Oceanic & Atmospheric Admin, 75-82; staff scientist, Propagation Studies Satellite Commun, Comsat Labs, 82-84; div mgr, opto-electronics instrumentation, Dynamics Technol Inc, 84-85; PRES, SCI TECHNOL, INC, 85- *Mem:* Fel Optical Soc Am; Am Inst Physics; Am Geophys Union; Sigma Xi; sr mem Inst Elec & Electronics Engrs. *Res:* The use of optical effects to develop novel techniques of remote sensors to probe the atmosphere, including turbulence, wind and precipitation; propagation effects on satellite and terrestrial communication systems. *Mailing Add:* Sci Technol Inc Two Research Pl Rockville MD 20850

WANG, TING-TAI HELEN, b Taipei, Taiwan, May 31, 48; m; c 2. MOTOR OIL RESEARCH, CRUDE OIL RESEARCH. *Educ:* Nat Taiwan Univ, BS, 70; Vanderbilt Univ, MS, 73; Univ Houston, PhD(anal chem), 82. *Prof Exp:* Res chemist, Gulf Oil Chem Co, 81-82; res assoc anal & environ chem, Dept Environ Sci & Eng, Rice Univ, 83-84; SR RES CHEMIST, PENNZOIL PROD CO, 84- *Concurrent Pos:* Chmn intrelations, Southeastern Tex Sect, Am Chem Soc, 81- *Mem:* Am Chem Soc; Am Soc Mass Spectrometry. *Res:* Conduct motor oils, crude oils, base stocks and environmental related research using gas chromatography and mass spectrometry; liquid chromatography and mass spectrometry; high pressure liquid chromatography and other analytical separation and spectroscopy techniques. *Mailing Add:* Pennzoil Prod Co PO Box 7569 The Woodlands TX 77380

WANG, TSUEY TANG, b Tainan, Taiwan, Nov 12, 32; US citizen; m 65; c 3. POLYMER SCIENCE, MATERIALS SCIENCE. *Educ:* Cheng Kung Univ, Taiwan, BSc, 55; Brown Univ, MSc, 61, PhD(appl mech), 65. *Prof Exp:* Asst prof, Polytech Univ NY, 65-67; mem tech staff, Bell Tel Labs, 67-88; MEM, DEPT MAT SCI & ENG, RUTGERS UNIV, BUSCH CAMPUS, 89- *Mem:* Fel Am Phys Soc; Soc Rheology; Am Acad Mech; NY Acad Sci. *Res:* Mechanical behavior of polymers; structures and properties; piezoelectricity in polymers; high temperature polymers; high performance composites; polymer blends. *Mailing Add:* Dept Mat Sci & Eng Rutgers Univ Busch Campus PO Box 909 Piscataway NJ 08855-0909

WANG, TUNG YUE, b Peking, China, Oct 27, 21; US citizen; m 48. BIOCHEMISTRY. *Educ:* Nat CheKiang Univ, BSc, 42; Univ Mo, MA, 49, PhD(biochem), 51. *Prof Exp:* Res fel, Jewish Hosp, St Louis, 51-53; res assoc, Washington Univ, St Louis, 53-57, asst prof, 57-59, assoc prof, 59-63; PROF, DEPT BIOL SCI, STATE UNIV NY, BUFFALO, 63- *Mem:* Am Soc Biol Chemists; Am Soc Cell Biol; Am Chem Soc; AAAS. *Res:* Mechanism of androgen action. *Mailing Add:* 65 Autumnview Rd Williamsville NY 14221-6695

WANG, VICTOR KAI-KUO, b Quei-Chow, China, Mar 18, 44; m 69; c 2. PHYSICAL CHEMISTRY, INDUSTRIAL CHEMISTRY. *Educ:* Chung Yuan Col Sci & Eng, Taiwan, BS, 65; State Univ NY Binghamton, MS, 68; Univ Minn, PhD(phys chem), 73. *Prof Exp:* Res asst phys chem, State Univ NY Binghamton, 66-68; teaching assoc, Univ Minn, 68-73; res chemist, 73-80, SR RES CHEMIST, E I DU PONT DE NEMOURS & CO, INC, 81- *Mem:* Am Chem Soc. *Res:* Process research, kinetics and catalysis. *Mailing Add:* Electronics Dept E I du Pont de Nemours & Co Inc Barley Mill Plaza Reynolds Mill Bldg Wilmington DE 19898

WANG, VIRGINIA LI, b Canton, China, Apr 2, 33; US citizen; c 3. PUBLIC HEALTH. *Educ:* NY Univ, MA, 56; Univ NC, MPH, 65, PhD, 68. *Prof Exp:* Res dietitian, Montefiore Hosp, 56; instr nutrit, Univ NC, 62-64; health educ specialist, Univ Md, 68-74; from assoc prof to prof pub health educ, Johns Hopkins Univ, 82- *Concurrent Pos:* Temp consult, Pan Am Health Orgn, 74; consult, WHO, UN, 74, 76 & 82, Nat Cancer Inst, NIH, Dept HEW, 74-77. *Mem:* Am Dietetic Asn; Am Pub Health Asn; Soc Pub Health Educ. *Res:* Evaluation of nutrition and health education programs in developed and developing societies; health care delivery and health education in the Peoples Republic of China; planning, development and evaluation of continuing education for the health professions; smoking cessation research; rural development. *Mailing Add:* 1432 S Camden Ave No 101 Los Angeles CA 90025

WANG, WEI-YEH, b Sian, China, Oct 10, 44; m 69; c 2. BIOCHEMICAL GENETICS, PLANT PHYSIOLOGY. *Educ:* Nat Taiwan Univ, BS, 66; Univ Mo, Columbia, PhD(genetics), 72. *Prof Exp:* Fel genetics, Duke Univ, 72-75; from asst prof to assoc prof, 75-87, PROF BOT, UNIV IOWA, 87- *Concurrent Pos:* Vis prof, Carlsberg Lab, Copenhagen, 80-81. *Mem:* Genetics Soc Am; Am Soc Plant Physiologists; Sigma Xi. *Res:* Genetics, biochemistry and molecular biology of chlorophyll and heme biosynthesis. *Mailing Add:* Dept Bot Univ Iowa Iowa City IA 52242

WANG, WEN I, b Taiwan, June 11, 53. ELECTRICAL ENGINEERING, APPLIED PHYSICS. *Educ:* Nat Taiwan Univ, BS, 75; Cornell Univ, ME, 79, PhD(elec eng), 81. *Prof Exp:* Res staff mem, Rockwell Int Sci Ctr, 81-82; res staff mem, IBM Res Ctr, 82-87; PROF ELEC ENG & APPL PHYSICS, COLUMBIA UNIV, 87- *Mem:* Am Phys Soc; Inst Elec & Electronics Engrs. *Res:* Solid state physics; molecular beam epitaxy. *Mailing Add:* Elec Eng Dept Columbia Univ 1320 Mudd Bldg New York NY 10027

WANG, WUN-CHENG W(OODROW), b Taichung, China, Mar 10, 36; m 62; c 3. AQUATIC TOXICOLOGY. *Educ:* Nat Taiwan Univ, BS, 58, MS, 61; Univ Wis-Madison, PhD(water chem), 68. *Prof Exp:* Asst prof scientist, 67-70, assoc prof scientist, 70-80, PROF SCIENTIST, WATER QUAL SECT, ILL STATE WATER SURV, 81- *Mem:* Soc Environ Toxicol & Chem; Am Soc Testing & Mat. *Res:* Bioassay; aquatic toxicology; ecotoxicology; toxicity tests. *Mailing Add:* Ill State Water Surv PO Box 697 Peoria IL 61652

WANG, XINGWU, b Hangzhou, China, Feb 19, 53; m 87; c 1. SUPERCONDUCTIVE COMPONENTS, THIN FILMS. *Educ:* Harbin Naval Eng Inst, BS, 78; Hangzhou Univ, MS, 81; State Univ NY, Buffalo, PhD(physics), 87. *Prof Exp:* Teacher elec eng, Hangzhou Naval Eng Sch, 78-81, teacher physics, Hangzhou Univ, 81-84; teaching asst physics, State Univ NY, Buffalo, 82-84, res asst, 84-87, postdoctoral res assoc elec eng, 87-88; ASST PROF ELEC ENG, ALFRED UNIV, 88- *Concurrent Pos:* Prin investr, Ctr Advan Ceramic Technol, 88-, NY State Inst Superconductivity, 89-90 & NSF Glass Res Ctr, 91; session chair, US-Japan workshop superconductivity, 89; abstractor, Am Soc Metal Int, 90-; reviewer, NY State Sci & Technol Found, 90. *Mem:* Inst Elec & Electronics Engrs; Am Phys Soc; Am Ceramic Soc; Mat Res Soc; AAAS; Am Soc Metal Int. *Res:* Superconductivity, superconductive electronics, bulk superconductors, physical properties of super conductors, low temperature physics, superfluidity; thin films by laser and plasma deposition technique, physisorption of films; bimodal switching in lasers, mean switching time; phase transitions, critical phenomena, mathematical physics; author of more than 50 publications, three patents issued. *Mailing Add:* Elec Eng Dept Alfred Univ Alfred NY 14802

WANG, YANG, b Tangshan, China, May 12, 23; US citizen; m 66; c 4. CARDIOLOGY. *Educ:* Nat Med Col Shanghai, MB; Harvard Univ, MD, 52; Am Bd Internal Med, dipl, Am Bd Cardiovasc Dis, dipl. *Prof Exp:* Intern & resident med, Mass Gen Hosp, 52-54 & 56-57; dir cardiac catherization Labs, Univ Hosps, 60-85, from instr to assoc prof, 59-64, PROF MED, MED SCH, UNIV MINN, MINNEAPOLIS, 74- *Concurrent Pos:* P D White fel cardiol, Mass Gen Hosp, Boston, 57-58; fel physiol, Mayo Grad Sch Med, Univ Minn, 58-59; consult, Vet Admin Hosp, Minneapolis, 67; fel coun clin cardiol & circulation, Am Heart Asn; attend physician, Univ Minn Hosps, Minneapolis, 59- *Mem:* Fel AAAS; Am Fedn Clin Res; fel Am Col Physicians; Soc Exp Biol & Med; Asn Univ Cardiol. *Res:* Cardiovascular and exercise physiology; cardiac catheterization in humans. *Mailing Add:* Dept Med Univ Minn Hosps Box 83 Minneapolis MN 55455

WANG, YAR-MING, b Taitung, Taiwan, Sept 22, 47; US citizen; m 75; c 2. ELECTROCHEMISTRY. *Educ:* Nat Cheng-Kong Univ, Taiwan, BS, 68; Univ Mo, Rolla, PhD(metall eng), 75. *Prof Exp:* Plant engr, Taiwan Metal & Mining Corp, 69-70; teaching fel, Univ Mo-Rolla, 75-76, res asst prof metall eng, 76-77; res scientist, Sprague Elec Co, 78-79; sr res engr, 79-84, STAFF RES ENGR, GEN MOTORS RES LABS, 84- *Mem:* Metall Soc Am Inst Mining, Metall & Petrol Engrs; Electrochem Soc; Nat Soc Corrosion Engrs. *Res:* Nonferrous chemical metallurgy; electrolytic processes; corrosion mechanism studies; battery research. *Mailing Add:* Phys Chem Gen Motors Res Labs 12 Mile & Mound Warren MI 48090

WANG, YA-YEN LEE, b Peking, China, Mar 1, 30; US citizen; m 56; c 3. MATHEMATICS, COMPUTER SCIENCE. *Educ:* Villa Maria Col, BS, 56; Univ Fla, MS, 58; Univ Idaho, PhD(math), 65. *Prof Exp:* Instr math, 60-62, asst prof, 65-72, acting dir comput ctr, 67, ASSOC PROF MATH, UNIV IDAHO, 72-, PROF COMPUT SCI, 85- *Mem:* Am Math Soc; Math Asn Am; Asn Comput Mach; Sigma Xi; Am Asn Univ Prof. *Res:* Differential geometry and equations; calculus of variations; numerical analysis; computer languages. *Mailing Add:* Dept Comput Sci Univ Idaho Moscow ID 83843

WANG, YEN, b Dairen, China, Oct 21, 28; m 62. RADIOLOGY, NUCLEAR MEDICINE. *Educ:* Nat Taiwan Univ, MD, 53; Univ Pa, DSc(med), 62. *Prof Exp:* Asst prof, 63-65, assoc prof, 65-75, CLIN PROF RADIOL, UNIV PITTSBURGH, 75-; DIR RADIOL, HOMESTEAD HOSP, 66-; DIR NUCLEAR MED, MAGEE-WOMENS HOSP, PITTSBURGH, 67- *Concurrent Pos:* Res fel, Picker Found Acad Sci, 61-63; vis scientist, Protein Found, 62-64; ed, Critical Rev Clin Radiol & Nuclear Med. *Mem:* AMA; Soc Nuclear Med; Am Roentgen Ray Soc; Am Physiol Soc; Am Radium Soc; Radiol Soc NAm. *Res:* Physiology; protein chemistry. *Mailing Add:* Dept Radiol Thomas Jefferson Univ Hosp Philadelphia PA 19107

WANG, YEN CHU, b China, Nov 25, 38. APPLIED SUPERCONDUCTIVITY. *Educ:* Cheng Kung Univ, Taiwan, BS, 60; Nat Chiao Tung Univ, MS, 62; NY Univ, PhD(elec eng), 69. *Prof Exp:* Mem tech staff, Hughs Aircraft Co, Torrance, CA, 69-70; proj engr, ADT Co, NY, 71-72; proj engr, Gen Microwave Corp, Farmingdale, Long Island, 72-74; PROF ELEC ENG, HOWARD UNIV, 74- *Mem:* Am Phys Soc; AAAS; Inst Electronic Commun Engrs Japan; Am Asn Physics Teachers; Math Asn Am. *Res:* Electromagnetic properties of high Tc superconductors; superconducting transistors and EM scattering. *Mailing Add:* Dept Elec Eng Howard Univ Washington DC 20059

WANG, YEU-MING ALEXANDER, biochemistry, hematology; deceased, see previous edition for last biography

WANG, YI-MING, b Hong Kong, Aug 3, 50; US citizen. SOLAR PHYSICS, HIGH-ENERGY ASTROPHYSICS. *Educ:* Mass Inst Technol, ScD, 76. *Prof Exp:* Res fel, Astron Ctr, Univ Sussex, 76-79 & Astronomische Inst, Univ Bonn, 79-86; assoc scientist, Appl Res Corp, 86-88; ASTROPHYSICIST, NAVAL RES LAB, 88- *Concurrent Pos:* Vis res fel, Max Planck Inst Astrophys, 83-84. *Mem:* Am Astron Soc; Int Astron Union. *Res:* Solar and interplanetary physics; high-energy astrophysics; astrophysical fluids; theoretical and numerical modelling; data analysis. *Mailing Add:* Code 4172W Naval Res Lab Washington DC 20375

WANG, YU, crystallography, inorganic chemistry, for more information see previous edition

WANG, YUAN R, b Wuhan, China, June 3, 34; US citizen; m 61; c 4. ENGINEERING APPLICATIONS OF COMPUTERS, PATTERN RECOGNITION & DATA STRUCTURES. *Educ:* Nat Taiwan Univ, BS, 55; Univ Iowa, MS, 60; Northwestern Univ, PhD(computer sci), 67. *Prof Exp:* Develop engr, Western Elec Co, 60-63; asst prof indust eng, Univ Pittsburgh, 67-70; assoc prof computer sci, Univ Nebr, Lincoln, 70-77; assoc prof, 77-82, PROF ELEC ENG & COMPUTER SCI, TEX A&I UNIV, 82-, CHMN DEPT, 85- *Concurrent Pos:* Res fel, NASA Manned Space Ctr, Houston, 70 & NASA Mountain View Res Ctr, 85; sr lang researcher, US Army Computer Systs Command, 76-77; session chmn, Computer Software & Appln Conf, Inst Elec & Electronics Engrs, 77; res fac fel, Lawrence Livermore Nat Lab, 80. *Mem:* Inst Elec & Electronics Engrs; Asn Comput Mach; Am Soc Eng Educ. *Res:* Formal language and automation; fault tolerant computing and reliable computer architecture; pattern recognition and artificial intelligence; numerical methods and simulations; data structure and database systems. *Mailing Add:* Tex A&I Univ Campus PO Box 192 Kingsville TX 78363

WANG, YU-LI, CYTO-SKELETON, MYOGENESIS. *Educ:* Harvard Univ, PhD(biophys), 80. *Prof Exp:* Staff scientist, Immunol & Respiratory Med, Nat Jewish Ctr, 82-87; SR SCIENTIST, WORCESTER FOUND EXP BIOL, 87- *Concurrent Pos:* Assoc prof, dept cell biol, Univ Mass, 87- *Mailing Add:* 222 Maple Ave Shrewsburg MA 01545

WANG, YUNG-LI, b Canton, China, Jan 8, 37; m 63; c 1. SOLID STATE PHYSICS. *Educ:* Nat Taiwan Univ, BS, 59; Nat Tsing Hua Univ, Taiwan, MS, 61; Univ Pa, PhD(physics), 66. *Prof Exp:* Res assoc physics, Univ Pa, 66-67 & Univ Pittsburgh, 67-68; asst prof, 68-72, assoc prof, 72-77, PROF PHYSICS, FLA STATE UNIV, 77- *Mem:* Am Phys Soc. *Res:* Many body theory of spin systems, magnetic phase transitions; crystal-field effects and impurities in magnetic systems. *Mailing Add:* Dept Physics Fla State Univ Tallahassee FL 32306

WANGAARD, FREDERICK FIELD, b Minneapolis, Minn, Jan 3, 11; m 36; c 3. FOREST PRODUCTS. *Educ:* Univ Minn, BS, 33; State Univ NY, MS, 35, PhD(wood technol), 39. *Hon Degrees:* MA, Yale Univ, 52. *Prof Exp:* Instr forestry, Univ Wash, 36-39, asst prof wood technol, 39-42; technologist, Forest Prod Lab, US Forest Serv, 42-45; from asst prof to prof forest prod, Yale Univ, 45-67; head dept forest & wood sci, Colo State Univ, 68-76; CONSULT, 76- *Concurrent Pos:* Adv, Food & Agr Orgn, Philippines, 57; Fulbright res scholar, Norway, 58. *Honors & Awards:* Borden Chem Award, 73; Distinguished Serv Award, Soc Wood Sci & Technol, 83; Golden Mem Award, Soc Am Foresters, 84. *Mem:* Forest Prod Res Soc (pres, 75); Soc Am Foresters; Soc Wood Sci & Technol (pres, 64); Int Acad Wood Sci. *Res:* Thermal conductivity of wood; properties of wood in relation to growth; properties of tropical woods; plywood and laminated wood; technology of wood fibers; wood structure and properties. *Mailing Add:* 1609 Hillside Dr Ft Collins CO 80524

WANGBERG, JAMES KEITH, b Oakland, Calif, Sept 6, 46. ENTOMOLOGY. *Educ:* Humboldt State Col, Calif, BA, 69; Calif State Univ, Humboldt, MA, 73; Univ Idaho, PhD(entom), 76. *Prof Exp:* Vis instr, 75-76, asst prof, 76-81, ASSOC PROF ENTOM, TEX TECH UNIV, 81- *Concurrent Pos:* Co-prin investr, Smithsonian Inst res grant, 78- *Mem:* AAAS; Entom Soc Am; Soc Range Mgt. *Res:* Rangeland entomology with emphasis on insects affecting native shrubs; gall insect biology. *Mailing Add:* Plant Soil Insect Sci Dept Univ Wyo Laramie WY 82071

WANGEMANN, ROBERT THEODORE, b Rhinelander, Wis, Apr 27, 33; m 54; c 2. BIOPHYSICS. *Educ:* Univ Wis-Madison, BS, 55; Univ Rochester, MS, 64; Med Col Va, PhD(biophys), 74. *Prof Exp:* Pharmacist, Northgate Drugs & Southside Drugs, 55-57; med supply officer, US Army, 57-62, health physicist, 64-67, instr nuclear sci, 67-70, chief, Laser-Microwave Div, 73-77, dir radiation & environ sci, 77-78, consult, Off Surgeon Gen, 78-81, comdr, Environ Hyg Agency, 81-84; assoc prof & dir laser biophys, Uniformed Serv Univ, 84-86; EXEC DIR, INST ELEC & ELECTRONICS ENGRS

LASERS & ELECTRO-OPTICS SOC, 86- *Concurrent Pos:* Mem, Phys Agents Threshold Limit Value Comt, Am Conf Govt Indust Hygienists; Merit Rev Panel, Med Res & Develop Command, US Army; assoc ed, Health Physics Jour. *Mem:* AAAS; Bioelectromagnetics Soc; Health Physics Soc; Inst Elec & Electronics Engrs; Am Soc Asn Execs. *Res:* Biological effects of electromagnetic radiation; interactions of radio frequency and optical energy at the biomolecular level and the photochemical aspects of vision and optical radiation effects. *Mailing Add:* Lasers & Electro-optics Soc Inst Elec & Electronics Engrs PO Box 1331 Piscataway NJ 08855-1331

WANGENSTEEN, OVE DOUGLAS, b St Paul, Minn, Mar 15, 42; m 65; c 2. RESPIRATORY PHYSIOLOGY, MICROVASCULAR EXCHANGE. *Educ:* Univ Minn, Minneapolis, BS, 64, PhD(physiol), 68. *Prof Exp:* Res assoc physiol, Sch Med & Dent, State Univ NY, Buffalo, 68-70; asst prof, 70-76, ASSOC PROF PHYSIOL, MED SCH, UNIV MINN, MINNEAPOLIS, 76- *Concurrent Pos:* Vis prof, Anat Inst, Univ Bern, Switz, 80; consult, 3M Co, 86- *Mem:* Am Physiol Soc; Am Thoracic Soc; Microcirculatory Soc. *Res:* Respiratory and cardiovascular physiology; transcapillary exchange; lung fluid balance. *Mailing Add:* Dept Physiol 6-255 Millard Hall Univ Minn 435 Delaware St SE Minneapolis MN 55455

WANGENSTEEN, STEPHEN LIGHTNER, b Minneapolis, Minn, Aug 30, 33; m 56; c 4. SURGERY. *Educ:* Univ Minn, BA, 54, BS, 55; Harvard Univ, MD, 58. *Prof Exp:* Instr surg, Columbia-Presby Med Ctr, 64-65; from asst prof to prof surg, Univ Va, 67-76; PROF & HEAD DEPT SURG, UNIV ARIZ, 76- *Concurrent Pos:* Vascular fel, Columbia-Presby Med Ctr, 58-59, USPHS fel, 60-63. *Mem:* Asn Acad Surg (vpres, 67-68); Soc Univ Surg; Am Surg Asn. *Res:* Gastrointestinal pathophysiology; circulatory shock; vascular surgery. *Mailing Add:* Dept Surg Univ SFla 12901 Bruce B Downs Blvd Box 16 Tampa FL 33612

WANGERSKY, PETER JOHN, b Woonsocket, RI, Aug 26, 27; m 59; c 3. ECOLOGY. *Educ:* Brown Univ, ScB, 49; Yale Univ, PhD(zool), 58. *Prof Exp:* Marine chem technician, Scripps Inst, Univ Calif, 49-50; chemist, Chem Corps, US Dept Army, 50-51; chem oceanogr, US Fish & Wildlife Serv, 51-54; res asst prof marine sci, Marine Lab, Univ Miami, 58-61; res assoc, Bingham Oceanog Lab, Yale Univ, 61-65; assoc prof chem, 65-68, chmn dept oceanog, 77-80, PROF OCEANOG, DALHOUSIE UNIV, 68- *Concurrent Pos:* Guggenheim fel, John Simon Guggenheim Found, 71-72; ed-in-chief, Marine Chem, 74- *Mem:* AAAS; Am Soc Limnol & Oceanog; Am Chem Soc; Ecol Soc Am; Am Nat Soc. *Res:* Mechanisms of marine sedimentation; chemical oceanography; organic metabolites in sea water; population dynamics. *Mailing Add:* Dept Oceanog Dalhousie Univ Halifax NS B3H 4J1 Can

WANG-IVERSON, PATSY, b Shanghai, China, Jan 6, 47; m 76; c 1. CELL BIOLOGY, LIPID BIOCHEMISTRY. *Educ:* Mt Holyoke Col, BA, 68; Bowman Gray Sch Med, 70, PhD(biochem), 75. *Prof Exp:* Res asst enzyme, Burroughs Wellcome & Co, 70-71; fel bact & immunol, Sch Med, Univ NC, 75-77, fel cellular physiol & immunol, Rockefeller Univ, 77-79; ASST PROF BIOCHEM, MT SINAI SCH MED, 79- *Concurrent Pos:* Mem, Coun Arteriosclerosis. *Mem:* Sigma Xi; NY Acad Sci; Am Soc Biol Chemists. *Res:* Human monocyte; macrophage function, specifically its contribution to lipoprotein metabolism. *Mailing Add:* RD 1 Box 301 Wagner Rd Stockton NJ 08559

WANGLER, ROGER DEAN, b Webster City, Iowa, June 3, 50; m 72; c 2. CELL BIOLOGY, CARDIOVASCULAR PHYSIOLOGY. *Educ:* Iowa State Univ, BS, 72; Univ Iowa, PhD(anat), 81. *Prof Exp:* res assoc & fel, Mich State Univ, 85-87; RES SCIENTIST, EISENHOWER MED CTR, 87- *Mem:* Am Physiol Soc. *Res:* Regulation of coronary blood flow; solute exchange across the coronary microvascular bed. *Mailing Add:* Wallis Res Facil Eisenhower Med Ctr 39000 Bob Hope Dr Rancho Mirage CA 92270

WANGLER, THOMAS P, b Bay City, Mich, Aug 2, 37. PHYSICS. *Educ:* Mich State Univ, BS, 58; Univ Wis, PhD(physics), 64. *Prof Exp:* Res assoc physics, Univ Wis, 64-65 & Brookhaven Nat Lab, 65-66; asst physicist, Argonne Nat Lab, 66-80; MEM STAFF, LOS ALAMOS NAT LAB, 80- *Mem:* Am Phys Soc. *Res:* Accelerator physics; cosmic rays; environmental science; nuclear physics; Experimental high energy physics. *Mailing Add:* Los Alamos Nat Lab AT-1 MS-817 PO Box 1663 Los Alamos NM 87545

WANGSNESS, PAUL JEROME, b Madison, Wis, Mar 27, 44; m 67; c 2. ANIMAL NUTRITION. *Educ:* Univ Wis-Madison, BS, 66; Iowa State Univ, PhD(nutrit & physiol), 71. *Prof Exp:* NDEA fel nutrit, Iowa State Univ, 66-69, NSF fel, 69-71; from asst prof to prof nutrit, 72-89, head, dept dairy & animal sci, 81-89, PROF & DIR, COOP EXTEN, PA STATE UNIV, 89- *Concurrent Pos:* Dir, Am Soc Animal Sci Bd, 88-90. *Honors & Awards:* Young Scientist Award, NE Am Soc Animal Sci, 81. *Mem:* Am Dairy Sci Asn; Am Soc Animal Sci; Am Inst Nutrit. *Res:* Regulatory mechanisms involved in the control of food intake and the regulation of energy balance in lean and obese animals. *Mailing Add:* 401 Agr Admin Bldg Pa State Univ University Park PA 16802

WANGSNESS, ROALD KLINKENBERG, b Sleepy Eye, Minn, July 24, 22; m 44; c 2. PHYSICS. *Educ:* Univ Minn, BA, 44; Stanford Univ, PhD(physics), 50. *Prof Exp:* Asst physics, Univ Minn, 42-44; jr scientist, Los Alamos Sci Lab, 44-45; asst physics, Univ Minn, 45-46; asst physics, Stanford Univ, 46-48; asst prof, Univ Md, 50-51, prof, 53-59; acting head dept, 83-85, prof, 59-89, EMER PROF PHYSICS, UNIV ARIZ, 89- *Concurrent Pos:* Physicist, Naval Ord Lab, Md, 51-59; assoc ed, Am Jour Physics, 83-85. *Mem:* Fel AAAS; fel Am Phys Soc; Sigma Xi; Am Asn Physics Teachers. *Res:* Nuclear induction; nuclear moments; ferrimagnetic resonance; anti-ferromagnetism; atomic spectra; electromagnetism. *Mailing Add:* 5035 E Scarlett St Tucson AZ 85711-4340

WANI, JAGANNATH K, b Maharashtra, India, Sept 10, 34; m 59; c 3. STATISTICS. *Educ:* Univ Poona, BSc, 58, Hons, 59, MSc, 60; McGill Univ, PhD(math statist), 67. *Prof Exp:* Lectr math, Col Agr, Dhulia, India, 60-61; res asst statist, Gokhale Inst Econ, Poona, India, 61-62; res asst math statist, McGill Univ, 62-65; asst prof math, Univ Lethbridge, 65-66; from asst prof to assoc prof, St Mary's Univ, NS, 66-69; assoc prof, 69-85, PROF STATIST, UNIV CALGARY, 85-, CHMN DEPT, 82- *Concurrent Pos:* Nat Res Coun Can fel, St Mary's Univ & Univ Calgary, 67-72; Can Math Cong fel, McGill Univ, Queen's Univ & Univ Alta, 69-71. *Mem:* Am Statist Asn; Statist Soc Can. *Res:* Distribution theory and statistical inference. *Mailing Add:* Dept Statist Univ Calgary 2500 University Dr Calgary AB T2N 1N4 Can

WANI, MANSUKHLAL CHHAGANLAL, b Nandubar, India, Feb 20, 25; US citizen; m 54; c 1. STEROID SYNTHESIS, SYNTHESIS OF HETEROCYCLIC COMPOUNDS. *Educ:* Bombay Univ, BSc, 47, MSc, 50; Ind Univ, PhD(chem), 62. *Prof Exp:* Lectr chem, Bhavan's Col, Bombay, India, 51-58; res asst chem, Ind Univ, Bloomington, 58-61; res assoc chem, Univ Wis-Madison, 61-62; PRIN SCIENTIST, RES TRIANGLE INST, NC, 62- *Honors & Awards:* Prof Develop Award, Res Triangle Inst, 83. *Mem:* Am Chem Soc; Am Soc Pharmacog; Sigma Xi; AAAS. *Res:* Isolation and characterization of biologically active compounds from natural sources; synthesis of anticancer, antifertility and iron chelating agents. *Mailing Add:* Res Triangle Inst PO Box 12194 Research Triangle Park NC 27709-2194

WANIEK, RALPH WALTER, b Milan, Italy, June 1, 25; nat US; m 53; c 1. PHYSICS. *Educ:* Univ Vienna, PhD(physics), 50. *Prof Exp:* Res assoc nuclear physics, Inst Radium Res, Univ Vienna, 48-50; asst prof physics & math, Newton Col, 50-56; sr physicist, Cambridge Electron Accelerator, Harvard Univ & Mass Inst Technol, 50-58; dir res, Plasmadyne Corp, 58-60; pres & dir res, Advan Kinetics, Inc, 60-89; CONSULT, 89- *Concurrent Pos:* Res fel, Synchrocyclotron Lab, Harvard Univ, 50-55; consult, Transistor Prod, Inc, 52-55 & Allied Res Assocs, 56-57; lectr, Boston Col, 56-58 & Exten, Univ Calif, Los Angeles, 59-67. *Mem:* Am Phys Soc; Am Inst Aeronaut & Astronaut; Inst Elec & Electronics Engrs. *Res:* Plasma, laser, space and nuclear physics; solid state; production of very intense magnetic fields; problems of space propulsion. *Mailing Add:* 1388 Pacific Ave Laguna Beach CA 92651

WANIELISTA, MARTIN PAUL, b Taylor, Pa, Dec 7, 41; m 66; c 1. ENVIRONMENTAL ENGINEERING, WATER RESOURCES. *Educ:* Univ Detroit, BS, 64; Manhattan Col, MS, 65; Cornell Univ, PhD(environ eng), 71. *Prof Exp:* From asst prof to assoc prof, 70-76, PROF ENG & GORDON J BARNETT PROF ENVIRON SYSTS & MGT, FLA TECHNOL UNIV, 76-, ACTG CHMN DEPT CIVIL & ENVIRON ENG, 78- *Concurrent Pos:* Dir, Environ Systs Eng Inst, 71-74; NSF grant, 72-73; pres, STE Inc, 73- *Mem:* Am Soc Civil Engrs; Am Soc Eng Educ; Asn Environ Eng Prof; Am Water Works Asn; Am Water Resources Asn. *Res:* Optimization models for water and solid waste systems; lake restoration; mathematical models of surface water systems; atmospheric pollution measurements and control methods. *Mailing Add:* Dept Civil & Environ Eng Univ Cent Fla Box 25000 Orlando FL 32816

WANKAT, PHILLIP CHARLES, b Oak Park, Ill, July 11, 44; m 80; c 2. SEPARATIONS. *Educ:* Purdue Univ, West Lafayette, BS, 66; Princeton Univ, PhD(chem eng), 70; Purdue Univ, MS Ed, 82. *Prof Exp:* Assoc prof, 74-78, PROF CHEM ENG, PURDUE UNIV, WEST LAFAYETTE, 78-, HEAD FRESHMAN ENG, 87- *Concurrent Pos:* NSF eng grants, 72-; vis prof, ENSIC, Nancy, France, 83-84. *Honors & Awards:* Westinghouse Award, Am Soc Eng Educ, 84, Carlson Award, 90. *Mem:* Am Inst Chem Engrs; Am Chem Soc; Am Soc Eng Educ. *Res:* Separation techniques; cascade theory; pressure swing adsorption; chromatography; biochemical separations. *Mailing Add:* Sch Chem Eng Purdue Univ West Lafayette IN 47907-1283

WANKE, SIEGHARD ERNST, b Herrenstein, Ger, July 31, 42; Can citizen; m 69; c 3. CHEMICAL ENGINEERING. *Educ:* Univ Alta, BSc, 64, MSc, 66; Univ Calif, Davis, PhD(chem eng), 69. *Prof Exp:* Engr, Chemcell Ltd, Alta, 65-66; res engr, Celanese Res Co, 69-70; from asst prof to assoc prof, 70-78, chmn dept, 85-90, PROF CHEM ENG, UNIV ALTA, 78- *Mem:* Chem Inst Can; Can Soc Chem Eng; Am Inst Chem Engrs; Electron Micros Soc Am; Can. *Res:* Heterogeneous catalysis; supported metal catalysts; chemical kinetics; olefin polymerization. *Mailing Add:* Dept Chem Eng Univ Alta Edmonton AB T6G 2E2 Can

WANLESS, HAROLD ROGERS, b Champaign, Ill, Feb 14, 42; m 65; c 3. SEDIMENTOLOGY, MARINE GEOLOGY. *Educ:* Princeton Univ, AB, 64; Univ Miami, MS, 68; Johns Hopkins Univ, PhD(geol), 73. *Prof Exp:* Res scientist, 71-73, res asst prof, 73-75, asst prof, 75-81, ASSOC PROF MARINE GEOL, SCH MARINE & ATMOSPHERIC SCI, UNIV MIAMI, 81- *Mem:* Soc Econ Paleontologists & Mineralogists. *Res:* Environments and processes of modern coastal and shelf sediments; petrology and paleoenvironmental reconstruction of ancient sedimentary rocks; fine-grained sediment dynamics; biotic influences on sediments; economic and environmental application. *Mailing Add:* Dept Marine Sci Univ Miami Coral Gables FL 33124

WANN, DONALD FREDERICK, electrical engineering; deceased, see previous edition for last biography

WANN, ELBERT VAN, b Grange, Ark, Dec 29, 30; m 50; c 1. GENETICS, PLANT BREEDING. *Educ:* Univ Ark, BS, 59, MS, 60; Purdue Univ, PhD(genetics), 62. *Prof Exp:* Res assoc veg crops, Univ Ill, 62-63; GENETICIST, USDA, 63-, LAB DIR, 72- *Honors & Awards:* Asgrow Award, Am Soc Hort Sci, 72. *Mem:* Am Soc Hort Sci; Am Soc Agron; Crop Sci Soc Am. *Res:* Genetics and breeding of sweet corn and tomatoes, as related to disease and insect resistance and the improvement of consumer quality. *Mailing Add:* USDA-ARS-SCARL PO Box 159 Hwy 3W Lane OK 74555

WANNEMACHER, ROBERT, JR, b Hackensack, NJ, Jan 12, 29; m 71. BIOCHEMISTRY, NUTRITION. *Educ:* Wagner Col, BS, 50; Rutgers Univ, MS, 51, PhD(biochem, nutrit), 60. *Prof Exp:* From res asst to res assoc nutrit & biochem, Bur Biol Res, Rutgers Univ, 51-60, from asst res prof to assoc res prof, 60-69; SR BIOCHEMIST, PHYS SCI DIV, US ARMY MED RES INST INFECTIOUS DIS, 69- *Mem:* AAAS; Am Inst Nutrit; Am Chem Soc; Biophys Soc; Soc Exp Biol & Med; Am Soc Biol Chem; Soc Toxicol; Int Soc Toxinol. *Res:* Protein and RNA metabolism; infectious diseases; regulatory mechanisms; endocrinology, cancer and radiation; toxicinology. *Mailing Add:* Pathophysiol Div Med Res Inst Infectious Dis US Army Ft Detrick Frederick MD 21702-5011

WANNER, ADAM, PULMONARY DISEASES, ASTHMA. *Educ:* Univ Basil, Switz, MD, 66. *Prof Exp:* PROF MED & CHIEF, PULMONARY DIV, UNIV MIAMI, 83- *Mailing Add:* Dept Med Univ Miami Sch Med 4300 Alton Rd Miami Beach FL 33140

WANNIER, PETER GREGORY, b Iowa City, Iowa, Sept 14, 46; m 78; c 4. RADIO ASTRONOMY. *Educ:* Stanford Univ, BS, 68; Princeton Univ, PhD(astron), 75. *Prof Exp:* Tech consult radio astron, Bell Tel Labs, 74-75; asst prof astron & elec eng, Univ Mass, 75-76; from asst prof to assoc prof radio astron, Calif Inst Technol, 78-83; MEM TECH STAFF, SPACE PHYSICS & ASTROPHYS SECT, JET PROPULSION LAB, PASADENA, 83- *Concurrent Pos:* Mem comn J, Int Union Radio Sci, 77-88; vis prof, Univ Gothenburg, Sweden, 86-87; mem comn 34, Int Astron Union, 86-88. *Mem:* Am Astron Soc; Int Astron Union. *Res:* Millimeter-wave and centimeter-wave studies of the interstellar medium, especially of dense clouds, with special interest in nuclear processing of material in the galaxy. *Mailing Add:* 1531 Gould Ave La Canada Flintridge CA 91011

WANTLAND, EVELYN KENDRICK, b Suffolk, Va, June 22, 17; m 39, 64; c 1. MATHEMATICS. *Educ:* Univ Ill, BA, 48, MA, 49, PhD(math), 58. *Prof Exp:* Asst, Univ Ill, 48-49; prof math, Ferrum Jr Col, 49-51; asst prof, Ill Wesleyan Univ, 51-57; asst prof, Kans State Univ, 57-62; assoc prof, Univ Miss, 62-64; prof math & head dept, Ill Wesleyan Univ, 64-76; RETIRED. *Mem:* Am Math Soc; Math Asn Am; Sigma Xi. *Res:* Complex variables. *Mailing Add:* 110 E Beecher St Bloomington IL 61701

WAPLES, DOUGLAS WENDLE, organic geochemistry, for more information see previous edition

WAPNIR, RAUL A, b Buenos Aires, Arg, Jan 6, 30; US citizen; m 52; c 2. BIOCHEMICAL NUTRITION, INTESTINAL PHYSIOLOGY. *Educ:* Univ Buenos Aires, Arg, MS, 53, PhD(chem), 54; Johns Hopkins Univ, Baltimore, MPH, 70. *Prof Exp:* Sr biochemist, Res Dept, Rosewood State Hosp, 63-70, co-dir, 70-73; assoc prof biochem pediat, 73-80, HEAD LAB, PEDIAT RES, NORTH SHORE UNIV HOSP, CORNELL UNIV MED COL, 73-, PROF BIOCHEM PEDIAT, 80- *Concurrent Pos:* From asst to assoc prof pediat res, Univ Md Sch Med, 66-73; grants, NIH, 87-90, USDA, 88-91. *Mem:* Fel Am Col Nutrit; Am Inst Nutrit; Soc Exp Biol & Med; NY Acad Sci. *Res:* Intestinal absorption research, as related to the transport of trace elements and electrolytes, using animal models of gastrointestinal diseases prevalent in childhood. *Mailing Add:* North Shore Univ Hosp Cornell Univ Med Col 300 Community Dr Manhasset NY 11030

WAPPNER, REBECCA SUE, b Mansfield, Ohio, Feb 25, 44. PEDIATRICS, BIOCHEMICAL GENETICS. *Educ:* Ohio Univ, BS, 66; Ohio State Univ, MD, 70; Am Bd Pediat, dipl, 75; Am Bd Med Genetics, dipl, 82. *Prof Exp:* Intern pediat, Children's Hosp, Columbus, Ohio, 70-71, resident, 71-72, asst chief resident, 72-73; fel pediat metab & genetics, 73-75, asst prof, 75-77, ASSOC PROF PEDIAT, SCH MED, IND UNIV, INDIANAPOLIS, 78- *Mem:* Am Acad Pediat; Sigma Xi; Am Soc Human Genetics; Am Med Women's Asn; Soc Study Inborn Errors Metab; Soc Inherited Metab Dis. *Res:* Inborn errors of metabolism. *Mailing Add:* Dept Pediat Riley A-36 Ind Univ Sch Med Indianapolis IN 46202-5225

WARAVDEKAR, VAMAN SHIVRAM, b Varavda, India, May 11, 14; nat US; m 60; c 2. BIOCHEMISTRY. *Educ:* Univ Bombay, MSc, 40, PhD(chem), 42. *Prof Exp:* Res assoc, V J Tech Inst, India, 42-45; res officer carcinogenesis, Tata Mem Hosp, Bombay, 45-48; prin investr, Georgetown Univ, 50-52; vis scientist, Nat Cancer Inst, 52-57; prof biochem, All-India Inst Med Sci, 57-58; chief biochem br, Armed Forces Inst Path, 58-65; dir cancer chemother dept, Microbial Assocs, Inc, DC, 65-72; res chemist, Off of Assoc Dir Drug Res & Develop, Div Cancer Treatment, 72-73, res planning officer, Nat Cancer Inst, 73-Off Assoc Dir Prof Prog Planning & Anal 83; RETIRED. *Concurrent Pos:* Res fel chemother of cancer, Nat Cancer Inst, 48-50; lectr grad sch, Georgetown Univ, 50-55. *Mem:* AAAS; Soc Exp Biol & Med; Am Soc Pharmacol & Exp Therapeut; Am Chem Soc; Am Asn Cancer Res. *Res:* Cellular chemistry; metabolism; enzymology; fatty acid oxidation; protein synthesis; chemotherapy; cancer research. *Mailing Add:* 9479 Reichs Ford Rd Ijamsville MD 21754

WARBURTON, CHARLES E, JR, b Holyoke, Mass, Nov 20, 41; m 66; c 4. POLYMER CHEMISTRY & PHYSICS. *Educ:* Univ Mass, BChE, 63; Princeton Univ, MA, 65, PhD(chem eng), 67. *Prof Exp:* fel, Textile Res Inst, 63-67; RES SCIENTIST, ROHM & HAAS CO, 67- *Mem:* Sigma Xi. *Res:* Polymer chemistry and physics, adhesion, textile finishes, nonwoven binders, organic coatings, radiation cure coatings. *Mailing Add:* 915 Denston Dr Ambler PA 19002

WARBURTON, DAVID LEWIS, b Hackensack, NJ, Aug 10, 47; m 75. GEOCHEMISTRY, WATER RESOURCES. *Educ:* Univ Calif, San Diego, BA, 69; Univ Chicago, PhD(geochem), 78. *Prof Exp:* ASST PROF GEOL, FLA ATLANTIC UNIV, 75- *Mem:* Am Geophys Union; AAAS; Mineral Soc Am; Sigma Xi; Geol Soc Am; Nat Asn Geol Teachers. *Res:* Inorganic pollution of water. *Mailing Add:* Dept Geol Fla Atlantic Univ PO Box 3091 Boca Raton FL 33431-0991

WARBURTON, DOROTHY, b Toronto, Ont, Jan 12, 36; m 57; c 4. HUMAN GENETICS, CYTOGENETICS. *Educ:* McGill Univ, BSc, 57, PhD(genetics), 61. *Prof Exp:* From res asst to res assoc human genetics, Montreal Children's Hosp & McGill Univ, 58-63; res assoc obstet & gynec, Col Physicians & Surgeons, Columbia Univ, 64-67; dir genetics serv, St Luke's Hosp Ctr, 67-68; instr obstet & gynec, Columbia Univ, 68-69, assoc human genetics & develop, 69-71, asst prof human genetics & develop, 71-75, asst prof pediat, 74-75, COL PHYSICIANS & SURGEONS, COLUMBIA UNIV, 75-; DIR GENETICS DIAG LAB, PRESBY HOSP, 69-; PROF CLIN GENETICS PEDIAT, COL PHYSICIANS & SURGEONS, COLUMBIA UNIV. *Mem:* Am Soc Human Genetics. *Res:* Cytogenetics; congenital malformations; human embryonic and fetal death; human gene mapping. *Mailing Add:* Col Physicians & Surgeons Columbia Univ New York NY 10032

WARBURTON, ERNEST KEELING, b Worcester, Mass, Apr 26, 28; m 47; c 3. NUCLEAR PHYSICS. *Educ:* Miami Univ, BA, 49; Mass Inst Technol, SB, 51; Univ Pittsburgh, PhD(physics), 57. *Prof Exp:* From instr to asst prof physics, Princeton Univ, 58-61; assoc physicist, 61-63, physicist, 63-68, SR PHYSICIST, BROOKHAVEN NAT LAB, 68- *Concurrent Pos:* NSF fel, Oxford Univ, 63-64 & 68-69; consult, Lawrence Livermore Nat Lab; assoc ed, Phys Rev Letters, 82-87; Alexander von Humboldt Award, 88-89. *Mem:* Fel Am Phys Soc. *Res:* Theoretical and experimental investigations of nuclear structure. *Mailing Add:* Dept Physics Brookhaven Nat Lab Upton NY 11973

WARBURTON, WILLIAM KURTZ, b Pasadena, CA, July 21, 42. X-RAY DIFFRACTIONS FROM LIQUIDS, X-RAY DETECTOR & OPTICS. *Educ:* Cornell Univ, BSE, 64, MSE, 66; Harvard Univ, MA, 65, PhD(physics) 72. *Prof Exp:* Res assoc nat sci, Div Eng & Appl Sci, 72-78 & Dept Mat Sci, 79, sr res assoc x-ray phys, Stanford Syrochotron Radiation Lab 79-84; from asst res prof x-ray detectors to assoc res prof x-ray detecetors, 84-85, RES ASSOC PROF RADIOL, UNIV SOUTHERN CALIF SCH MED, 87-; PRES X-RAY INSTRUMENT RES CONSULT, X-RAY INSTRUMENT ASN, 88- *Concurrent Pos:* Sr scientist x-ray instrument, Adv Res & Appln Corp, 87-88. *Mem:* Am Phys Soc; AAAS; Int Soc Optic Eng; Int Radiation Physics Soc. *Res:* Advanced x-ray instrumentation for x-ray research using synchrotron or laboratory sources, including x-ray optics, array detectors, and beamline components; x-ray scattering studies of structure of liquids, amorphous materials, and x-ray optical elements. *Mailing Add:* X-Ray Instrumentation Asn 1300 Mills St Menlo Park CA 94025-3210

WARD, ALFORD L, b Rockville, Md, Aug 14, 19. PHYSICS. *Educ:* Univ Md, BS, 49, PhD, 54. *Prof Exp:* physicist, Harry Diamond Labs, US Army Lab Command, Adelphi, 54-90; RETIRED. *Mem:* Fel AAAS; fel Am Phys Soc. *Res:* Gaseous electronics; semiconductors. *Mailing Add:* 4450 S Park Ave Apt 1106 Chevy Chase MD 20815-3647

WARD, ANTHONY THOMAS, b London, Eng, Mar 9, 41; m 66; c 2. PHYSICAL CHEMISTRY. *Educ:* Univ London, BSc, 62; Rensselaer Polytech Inst, MS, 64, PhD(phys chem), 66. *Prof Exp:* MEM RES STAFF, XEROX CORP, 66- *Mem:* Soc Photog Scientists & Engrs; Sigma Xi. *Res:* Design approaches; fabrication methods and spectroscopic characterization techniques for xerographic photoconductors and optical disk devices. *Mailing Add:* Xerox Corp 934 Little Pond Way Webster NY 14580

WARD, ARTHUR ALLEN, JR, b Manipay, Ceylon, Feb 4, 16; US citizen; m 41. NEUROSURGERY. *Educ:* Yale Univ, BA, 38, MD, 42. *Prof Exp:* Demonstr path, McGill Univ, 43, demonstr neurol & neurosurg, 44-45; asst physiol, Yale Univ, 45; asst, Ill Neuropsychiat Inst, 46; instr neurosurg, Univ Louisville, 46-48; from asst prof to prof surg, 48-65, chmn dept, 65-81, PROF NEUROSURG, MED SCH, UNIV WASH, 65- *Concurrent Pos:* Fel, McGill Univ, 43. *Honors & Awards:* Lennox Award, 76. *Mem:* Soc Neurol Surg (pres, 74); Am Acad Neurol Surg (pres, 77-78); Am Epilepsy Soc (vpres, 49, pres, 72); Am Physiol Soc; Am EEG Soc (pres, 59-60). *Res:* Epilepsy; function of animal and human cerebral cortex; reticular formation of the midbrain. *Mailing Add:* Dept Neurosurg RI-20 Univ Wash Med Sch RR 744 Health Sci Bldg Seattle WA 98195

WARD, BENJAMIN F, JR, b Yazoo City, Miss, Apr 23, 43; m 66; c 2. NEW CHEMICAL PRODUCT DEVELOPMENT, RESEARCH MANAGEMENT. *Educ:* Rhodes Col, BS, 65; Univ NC-Chapel Hill, PhD(chem), 69. *Prof Exp:* Res chemist, 69-72, res group leader, 72-74, prod develop mgr, 74-75, tech dir, 75-79, sr prod mgr, 79-83, RES DIR, CHARLESTON RES CTR, WESTVACO CORP, 83- *Mem:* Am Chem Soc; Tech Asn Pulp & Paper Indust; Can Pulp & Paper Asn; Int Acad Wood Sci. *Res:* Pulping and bleaching technology; derivitization and utilization of the by-products of the paper industry; tall oil fatty acids, rosin, lignin and carbon. *Mailing Add:* Westvaco Corp PO Box 2941105 North Charleston SC 29411-2905

WARD, BENNIE FRANKLIN LEON, b Millen, Ga, Oct 19, 48. THEORETICAL PHYSICS. *Educ:* Mass Inst Technol, BS(physics) & BS(math), 70; Princeton Univ, MA, 71, PhD(physics), 73. *Prof Exp:* Instr physics, Princeton Univ, 73; res assoc physics, Stanford Linear Accelerator Ctr, Stanford Univ, 73-75; asst prof physics, Purdue Univ, West Lafayette, 75-78; staff engr, Intel Corp, 78-79; res specialist, 80-82, staff engr, Lockheed Missiles & Space Co, 82-84; assoc prof, 86-90, PROF PHYSICS, UNIV TENN, KNOXVILLE, 90- *Concurrent Pos:* Vis scientist, Stanford Linear Accelerator Ctr, 78-, res assoc, 78. *Honors & Awards:* Sigma Xi. *Mem:* Am Phys Soc; Inst Elec & Electronics Engrs; NY Acad Sci. *Res:* Pursues primarily renormalization-group-improved Yennie-Frantschi-Suura approach to high precision SU2L x U1; radiative corrections at SLC and LEP and new heavy particle production and decay at SLC and LEP type energies. *Mailing Add:* Physics/Astron Univ Tenn Knoxville TN 37996-1200

WARD, CALVIN HERBERT, b Strawberry, Ark, Mar 1, 33; m 54; c 3. PLANT PATHOLOGY, PHYSIOLOGY. *Educ:* NMex State Univ, BS, 55; Cornell Univ, MS, 58, PhD(plant path), 60; Univ Tex, MPH, 78. *Prof Exp:* Res biologist, USAF Sch Aerospace Med, 60-63, plant physiologist, 63-65; assoc prof biol, 66-70, PROF BIOL & ENVIRON SCI & CHMN DEPT ENVIRON SCI & ENG, RICE UNIV, 70-, CO-DIR, NAT CTR GROUNDWATER RES, 79- *Concurrent Pos:* Grants, NASA, 63-66 & 70-, Environ Protection Agency, 66- & US Dept Air Force, 68-70; mem environ biol adv panel, Am Inst Biol Sci, 66-71, chmn, 69-71, mem comt space shuttle impact eval, 74-; mem bd dirs, Southwest Ctr Urban Res, 69-, chmn, 77-78; mem life sci comt, NASA, 71-78; vis prof, Univ Tex Sch Pub Health Houston, 73-74, adj prof environ health, 74-; vpres, US Nat Comn, Int Water Resource Asn, 77-; ed-in-chief, Environ Toxicol & Chem, 81-; sr ed, J Indust Microbiol & Develop Indust Microbiol, 85-; mem, Expert Panel on Res Needs & Opportunities at Federally- Supervised Hazardous Waste Sites, Coun on Environ Qual, Exec Off President, 85-86; mem, Comt on Multimedia Approaches to Pollution Control, Nat Acad Sci, 86-87, Environ Eng Comt, Sci Adv Bd, US Environ Protection Agency & Adv Comt on Multiagency Hazardous Wastes Res, Nat Acad Sci, 87; mem bd dirs, Am Type Culture Collection, 87. *Honors & Awards:* Group Achievement Award, STS-1 Shuttle Environ Effects Team, NASA, 81; corresp gravitational physiol, Int Union Physiol Sci Comn on Gravitational Physiol, 85; Charles Porter Award, Soc Indust Microbiol, 86; Distinguished Serv Award, Soc Environ Toxicol & Chem, 90. *Mem:* AAAS; Am Phytopath Soc; Am Soc Microbiol; Soc Indust Microbiol (pres, 83-84); Int Water Resource Asn; Am Inst Biol Sci (pres, 84-85); Soc Environ Toxicol & Chem; Asn Ground Water Sci & Eng. *Res:* Algal and plant physiology; bioregeneration; environmental microbiology; ground water contamination and pollution control. *Mailing Add:* Depts Biol & Environ Sci & Eng Rice Univ Houston TX 77251

WARD, CALVIN LUCIAN, b Yancey, Tex, Jan 30, 28; m 66; c 2. GENETICS. *Educ:* Univ Tex, BA, 47, MA, 49, PhD(zool), 51. *Prof Exp:* AEC fel, Oak Ridge Nat Lab, 51-52; from instr to assoc prof zool, 52-78, PROF ZOOL, DUKE UNIV, 78- *Mem:* Genetics Soc Am; Soc Study Evolution. *Res:* Cytology and genetics of Drosophila; speciation. *Mailing Add:* Dean/Dir Summer Sch Duke Univ Durham NC 27706

WARD, CHARLES ALBERT, b Bailey, Tex, May 28, 39; m 59. MECHANICAL ENGINEERING, BIOMEDICAL ENGINEERING. *Educ:* Univ Tex, Arlington, BSc, 62; Northwestern Univ, PhD(mech eng), 67. *Prof Exp:* From asst prof to assoc prof, 67-72, PROF MECH ENG, UNIV TORONTO, 77- *Concurrent Pos:* Res assoc, Res Inst, Hosp Sick Children, 72- *Honors & Awards:* Robert W Angus Medal, Can Soc Mech Eng & Eng Inst Can, 88. *Res:* Surface and chemical kinetics; phase change and stability; biocompatibility of synthetic materials; complement activation. *Mailing Add:* Dept Mech Eng Univ Toronto Toronto ON M5S 1A4 Can

WARD, CHARLES EUGENE WILLOUGHBY, b Madison, Wis, Sept 8, 38; m 61; c 2. COMPUTER SOFTWARE DESIGN. *Educ:* Northwestern Univ, BS, 61; Mass Inst Technol, PhD(physics), 67. *Prof Exp:* Res assoc, Lab Nuclear Sci, Mass Inst Technol, 67-68; post doctoral appointee, High Energy Physics Div, Argonne Nat Lab, 68-70, asst physicist, 70-73, physicist, 74-79; mem tech staff, 79-84, SUPVR MGR RES & DEVELOP, AT&T BELL LABS, 84- *Res:* Design of computer software systems for switching systems and other telecommunications applications. *Mailing Add:* Rm IH-6N-312 AT&T Bell Labs Warrenville & Naperville Rd Naperville IL 60566

WARD, CHARLES RICHARD, b Tahoka, Tex, Mar 25, 40; m 61; c 2. ENTOMOLOGY, AGRICULTURE. *Educ:* Tex Tech Col, BS, 62, MS, 64; Cornell Univ, PhD(med entom), 68. *Prof Exp:* From asst prof to assoc prof entom, Tex Tech Univ, 67-76; assoc prof, Tex Agr Exp Sta, 76; entom specialist, Consortium for Int Develop, Bolivia, 76-78; res assoc, 78-80, assoc prof, 78-85, PROF ENTOM, NMEX STATE UNIV, 85-, EXTEN ENTOMOLOGIST PEST MGT SPECIALIST, 80- *Concurrent Pos:* Consult, Consortium for Int Crop Protection, Indonesia, 81, Dominican Repub, 88, Guatemala, 89, Ecuador, 90; chief, Consortium for Int Develop, Honduras, 83-84, Trop Res & Develop, El Salvador, 89 & 90. *Honors & Awards:* Outstanding Contrib Award, Am Registry Prof Entomologists, 81. *Mem:* Entom Soc Am; Am Registry Prof Entomologists. *Res:* Pest management research and extension; biology and control of ornamental, turf, cotton, alfalfa, pecan and range land pests; ecology of desert and grasslands insects; biological control; insects resistance to pesticides. *Mailing Add:* Entom Plant Path & Weed Sci Dept 9301 Indian School Rd NE-201 Albuquerque NM 87112

WARD, CHARLOTTE REED, b Lexington, Ky, Feb 19, 29; m 51; c 4. PHYSICAL CHEMISTRY. *Educ:* Univ Ky, BS, 49; Purdue Univ, MS, 51, PhD(phys chem), 56. *Prof Exp:* Instr gen sci, Ala Educ TV, 58-60, 61-62 & 63-72, from instr to asst prof, 61-75, ASSOC PROF PHYSICS, AUBURN UNIV, 75- *Concurrent Pos:* Abstractor, Chem Abstr, 58-82; auth; pres state conf, Am Asn Univ Professors, 89-91. *Mem:* Am Asn Physics Teachers; Sigma Xi; Am Asn Univ Professors. *Res:* Molecular spectroscopy; development of physical science courses and textbooks, "hands-on" science for children. *Mailing Add:* 134 Norwood Ave Auburn AL 36830

WARD, COLEMAN YOUNGER, b Millican, Tex, Sept 20, 28; m 47; c 3. AGRONOMY, PHYSIOLOGY. *Educ:* Tex Tech Univ, BS, 50, MS, 54; Va Polytech Inst & State Univ, PhD(agron), 62. *Prof Exp:* Instr agricult, Eastern NMex Univ, 50-51; soil scientist, Soil Conserv Serv, USDA, 51-52; instr agron, Tex Tech Univ, 52-54; asst agronomist, Univ Fla, 54-55 & Va Agr Exp Sta, 55-61; from assoc prof to prof crop sci, Miss State Univ, 61-74; prof agron & chmn dept, Univ Fla, 74-80; prof agron & dept head, 79-83, PROF & TURF SPECIALIST, AUBURN UNIV, 83- *Concurrent Pos:* Biomass Res. *Mem:* Fel Am Soc Agron; Sigma Xi; Crop Sci Soc Am. *Res:* Physiology and ecology of turfgrasses and forage crops. *Mailing Add:* 3809 Heritage Pl Opelika AL 36801

WARD, CURTIS HOWARD, b Round Bottom, Ohio, June 21, 27; m 51; c 4. PHYSICAL CHEMISTRY. *Educ:* Ind State Teachers Col, BS, 47; Univ Ky, MS, 50; Purdue Univ, PhD(phys chem), 54. *Prof Exp:* Res chemist, Linde Co Div, Union Carbide Corp, 53-57; assoc prof chem, Auburn Univ, 57-60; sr staff scientist, Avco Corp, 60-61; assoc prof, 61-65, PROF CHEM, AUBURN UNIV, 65- *Mem:* Am Chem Soc; Sigma Xi. *Res:* Thermodynamics; molecular spectroscopy; organometallic chemistry. *Mailing Add:* 134 Norwood Ave Auburn AL 36830

WARD, DANIEL BERTRAM, b Crawfordsville, Ind, Mar 20, 28; m 56; c 4. PLANT TAXONOMY. *Educ:* Wabash Col, AB, 50; Cornell Univ, MS, 53, PhD(plant taxon), 59. *Prof Exp:* From asst prof to assoc prof, 58-75, PROF BOT, UNIV FLA, 75- *Mem:* Int Asn Plant Taxon; Am Soc Plant Taxonomists. *Res:* Vascular flora of Florida; methods of population analysis; preservation of endangered species. *Mailing Add:* Dept Bot Univ Fla 3165 Mcc Bldg Gainesville FL 32611

WARD, DARRELL N, b Logan, Utah, Jan 22, 24; m 46; c 7. BIOCHEMISTRY. *Educ:* Utah State Univ, BS, 49; Stanford Univ, MS, 51; PhD(biochem), 53. *Prof Exp:* Res assoc & instr biochem, Med Col, Cornell Univ, 52-55; from asst biochemist to biochemist, 55-82, head dept, 61-82, from chmn to pres grad fac, 78-80, PROF BIOCHEM & ANIS J SORRELL PROF, DEPT BIOCHEM & MOLECULAR BIOL, M D ANDERSON HOSP & TUMOR INST, UNIV TEX, HOUSTON, 83-, MEM GRAD FAC, GRAD SCH BIOMED SCI, 61- *Concurrent Pos:* Asst prof, Univ Tex Dent Br, Houston, 56-60; from asst clin prof to assoc clin prof, Baylor Col Med, 56-62; mem reproductive biol study sect, NIH, 67-71, consult, Ctr Pop Res. 69-71; chmn, Biochem Endocrinol Study, NIH, 79-81; mem, gonadotropin subcomt, Nat Hormone & Pituitary Prog, 82- *Honors & Awards:* Ayerst Award, Endocrine Soc, 78. *Mem:* AAAS; Am Chem Soc; Am Asn Cancer Res; Am Soc Biol Chem; Endocrine Soc. *Res:* Protein purification; protein and peptide hormones; gonadotropins amino acid sequence; functional group substitution and effects of biological activity; biochemistry of glycoprotein hormones. *Mailing Add:* Dept Biochem & Molecular Biol Univ Tex M D Anderson Cancer Ctr Houston TX 77030

WARD, DAVID, b Wakefield, Eng, Aug 5, 40; Can citizen. EXPERIMENTAL NUCLEAR PHYSICS. *Educ:* Univ Birmingham, BSc, 61; Univ Manchester, PhD(nuclear physics), 65. *Prof Exp:* Fel nuclear physics, Univ Manchester, 65-66 & Univ Calif, Berkeley, 66-68; PHYSICIST, CHALK RIVER NUCLEAR LABS, ATOMIC ENERGY CAN, 68- *Concurrent Pos:* Vis scientist nuclear physics, Univ Calif, Berkeley, 74-75; vis fel nuclear physics, Australian Nat Univ, 81-82. *Mem:* Am Phys Soc. *Res:* Reactions and coulomb excitation with heavy ions; lifetimes and magnetic moments of short-lived nuclear states; atomic phenomena in nuclear physics; stopping powers for heavy ions. *Mailing Add:* 177 Ridge Rd PO Box 581 Deep River ON K0J 0P0 Can

WARD, DAVID ALOYSIUS, b Joliet, Ill, Nov 13, 30; m 53; c 9. MECHANICAL & NUCLEAR ENGINEERING. *Educ:* Univ Ill, BSME, 53. *Prof Exp:* Res engr, Savannah River Lab, 53-65, engr, Savannah River Plant, 65-67, sr supvr, 67-69, area supvr, 69-72, chief tech supvr, 72-75, supt, Reactor Technol Dept, 75-78, supt, Reactor & Reactor Mat Technol Dept, 78-80, RES MGR NUCLEAR ENG, SAVANNAH RIVER PLANT, E I DU PONT DE NEMOURS & CO, INC, 80- *Concurrent Pos:* Mem adv comt reactor safeguards, US Nuclear Regulatory Comn. *Mem:* Am Nuclear Soc. *Res:* Nuclear reactor heat transfer; hydraulics; safety analysis; engineering management. *Mailing Add:* 822 Audobun Circle North Augusta SC 29841

WARD, DAVID CHRISTIAN, b Sackville, NB, May 22, 41; m 62; c 1. VIROLOGY, BIOCHEMISTRY. *Educ:* Mem Univ Nfld, BSc, 61; Univ BC, MSc, 63; Rockefeller Univ, PhD(biochem), 69. *Prof Exp:* ASST PROF MOLECULAR BIOPHYS & BIOCHEM, SCH MED, YALE UNIV, 71- *Concurrent Pos:* Leukemia Soc Am fel, Imp Cancer Res Fund, Eng, 69-71. *Mem:* AAAS; Am Soc Microbiol. *Res:* Replication and genetic analysis of animal viruses; molecular cytogenetics and gene mapping; 3-D topography of DNA in interphase nuclei. *Mailing Add:* Dept Human Genetics Yale Univ Sch Med New Haven CT 06510

WARD, DAVID GENE, b Modesto, Calif, Feb 19, 49; m 81; c 2. NEUROPHYSIOLOGY, HYPERTENSION. *Educ:* Calif State Univ, Stanislaus, BA, 71; Univ Okla, PhD(neurosci), 74. *Prof Exp:* Fel biomed eng, Sch Med, Johns Hopkins Univ, 74-76, res assoc, 76-77, asst prof, 77-79; asst prof physiol, Sch Med, Univ Va, 79-82; DIR PHYSIOL, H M WARD MEM LAB, 83- *Concurrent Pos:* Res career develop award, USPHS, 81; vis lectr, Calif State Univ, Stanislaus, 83- *Mem:* AAAS; Am Physiol Soc; NY Acad Sci; Soc Neurosci; Fedn Am Socs Exp Biol. *Res:* Central neural mechanisms responsible for control of circulation, blood pressure and fluid balance; neurophysiology; neuroendocrinology. *Mailing Add:* H M Ward Mem Lab PO Box 207 Valley Home CA 95384

WARD, DONALD THOMAS, b Sidney, Tex, Mar 18, 36; m 64; c 2. FLIGHT TESTING, STABILITY & CONTROL. *Educ:* Univ Tex, Austin, BS, 58; Air Force Inst Technol, MS, 65; Miss State Univ, PhD(aerospace eng), 74. *Prof Exp:* Tactical fighter pilot, 474th Tactical Fighter Wing, Cannon AFB, USAF, NMex, 59-63, student test pilot, Empire Test Pilots Sch, Farnborough, Eng, 66, tactical fighter pilot, Air Force Adv Group, DaNang Air Base, Vietnam, 67-68, student, Air Command & Staff Col, Maxwell AFB, Calif, 69-71, & test pilot, Res & Develop Off, Air Force Flight Test Ctr, 75-76, staff officer, HQ Air Force Test Pilot Sch, Edwards AFB, Calif, 78-79, Comdr, 4950th Test Wing, Wright-Patterson AFB, Ohio, 79-81; ASSOC PROF AEROSPACE ENG, TEX A&M UNIV, 81- *Concurrent Pos:* Lectr, Calif State Univ, Fresno, 78-79; consult, E-Systs, Inc, 84-87 & Lockheed-Ga Co, 85- *Mem:* Am Inst Aeronaut & Astronaut; Am Soc Eng Educ; Soc Exp Test Pilots; Am Astronaut Soc; Soc Flight Test Engrs. *Res:* Flight mechanics; flight test methods; control of aircraft and spacecraft; orbital mechanics. *Mailing Add:* Dept Aeromech Tex A&M Univ College Station TX 77843

WARD, DOUGLAS ERIC, b Wooster, Ohio, Aug 23, 57; m 82; c 3. OPTIMIZATION THEORY, NONSMOOTH ANALYSIS. *Educ:* Haverford Col, BA, 79; Carnegie-Mellon Univ, MS, 81; Dalhousie Univ, PhD(math), 85. *Prof Exp:* Instr, 84-85, asst prof math, 85-89, ASSOC PROF MATH, MIAMI UNIV, 89- *Mem:* Math Asn Am; Am Math Soc. *Res:* Optimization problems involving nondifferentiable functions; tangent cones; the calculus of generalized gradients; the application of these concepts in the formulation of optimality conditions. *Mailing Add:* Dept Math Miami Univ Oxford OH 45056-1641

WARD, EDMUND WILLIAM BESWICK, b Stockport, Eng, May 30, 30; m 53; c 3. PLANT PATHOLOGY, MICROBIOLOGY. *Educ:* Univ Wales, BSc, 52; Univ Alta, MSc, 54, PhD, 58. *Prof Exp:* Res officer, Alta, 57-61, PRIN PLANT PATHOLOGIST, LONDON RES CTR & HEAD PLANT PATH LAB, CAN DEPT AGR, 61- *Mem:* Am Phytopath Soc; Can Soc Phytopath; Brit Soc Plant Path. *Res:* Physiology of fungi; disease resistance in plants. *Mailing Add:* 448 Regal Dr London ON N5Y 1K1 Can

WARD, EDWARD HILSON, b Milton, Fla, Sept 15, 30; m 54; c 2. INORGANIC CHEMISTRY, ANALYTICAL CHEMISTRY. *Educ:* Troy State Col, BS, 58; Univ Miss, PhD(chem), 63. *Prof Exp:* Fel chem, Fla State Univ, 63-65; assoc prof, 65-68, PROF CHEM, TROY STATE UNIV, 68-, CHMN DEPT PHYS SCI, 70- *Mem:* Am Chem Soc. *Res:* Inorganic complexes; non-aqueous solvent systems; co-precipitation. *Mailing Add:* 129 Glenwood Ave Troy AL 36081-4513

WARD, FRANCES ELLEN, b Freedom, Maine, Mar 21, 39. IMMUNOGENETICS. *Educ:* Clark Univ, AB, 61; Brown Univ, PhD(biol), 65. *Prof Exp:* Instr, 67-69, asst prof, 69-73, assoc prof immunol, 73-79, PROF IMMUNOL, DUKE UNIV MED CTR, 79- *Concurrent Pos:* NIH fel statist, Iowa State Univ, 66-67; dir, Transplant Lab, Durham Vet Admin Hosp, 69-79; assoc scientist, The Wistar Inst, 80-81. *Mem:* Genetics Soc Am; AAAS; Transplantation Soc; Am Asn Clin Histocompatibility Testing; Am Asn Immunologists; Sigma Xi. *Res:* Genetics of the major human histocompatibility complex; immunogenicity of gene products of the major histocompatibility complex as measured by organ and tissue rejection. *Mailing Add:* Dept Microbiol Duke Univ Med Ctr Box 3010 Durham NC 27710

WARD, FRANK KERNAN, b Brockton, Mass, Jan 19, 31; m 56; c 3. ORGANIC CHEMISTRY, POLYMER CHEMISTRY. *Educ:* Boston Col, BS, 54; Mass Inst Technol, PhD(org chem), 58. *Prof Exp:* Res asst chem, Mass Inst Technol, 54-57; res chemist, Celanese Corp, 57-59; scientist, Avco Corp, 59-60, sr scientist, 60-61; from sr chemist to res chemist, 61-67, group leader, 67-71, environ technologist, 71-78, from coordr to prog coordr, 78-89, SR PROG COORDR, TEXACO INC, BEACON, 89- *Mem:* Sigma Xi; Am Chem Soc. *Res:* Synthetic polymer chemistry; organometallics; environmental affairs; fuel and lubricant additives; petrochemicals. *Mailing Add:* 17 Deerwood Dr Hopewell Junction NY 12533

WARD, FRASER PRESCOTT, b Lancaster, Pa, Aug 22, 40; m 83; c 3. ECOLOGY, VETERINARY MEDICINE. *Educ:* Univ Pa, VMD, 65; Johns Hopkins Univ, PhD(pathobiol), 79. *Prof Exp:* Vet practr, Tredyffrin Vet Hosp, Paoli, Pa, 65-66; Captain, Vet Corps, 66-69, chief, Ecol Br, 69-81, chief, Toxicol Br, 81-82, chief scientist, Res Div, US Army Chem Systs Lab, 82-83, CHIEF, BIOTECHNOL DIV & RES DIR, US ARMY CHEM RD&E CTR, ABERDEEN PROVING GROUND, MD, 83- *Concurrent Pos:* Mem path comt, Raptor Res Found, 70-; comnr, Md Chesapeake Bay & Coastal Zone Adv Bd, 73-; proj sci coordr, US/USSR Agreement on Coop Environ Protection, 73- *Mem:* Raptor Res Found; Wildlife Soc; Wildlife Dis Asn; Sigma Xi. *Res:* Global population dynamics of peregrine falcons; diseases of birds of prey; ecology of turtles; toxicology with emphasis on pesticides; assessment of environmental impacts; endangered species; biotechnology. *Mailing Add:* 137 E Broadway Bel Air MD 21014

WARD, FREDERICK ROGER, b Cleveland, Miss, Oct 30, 40; m 64; c 2. MATHEMATICS. *Educ:* Col William & Mary, BS, 62; Univ Colo, MS, 65; Va Polytech Inst & State Univ, PhD(math), 69. *Prof Exp:* Instr math, Va Polytech Inst & State Univ, 65-69; asst prof, 69-72, ASSOC PROF MATH, BOISE STATE UNIV, 72- *Mailing Add:* Dept Math Boise State Univ 1910 Univ Dr Boise ID 83725

WARD, FREDRICK JAMES, b Alert Bay, BC, Jan 22, 28; m 56; c 3. LIMNOLOGY, FISH BIOLOGY. *Educ:* Univ BC, BA, 52, MA, 57; Cornell Univ, PhD(conserv), 62. *Prof Exp:* Biologist, Int Pac Salmon Fisheries Comn, 52-57, proj supvr, 57-64; from asst prof to assoc prof, 64-73, PROF LIMNOL & INVERT ZOOL, UNIV MAN, 73- *Mem:* Am Fisheries Soc; Am Soc Limnol & Oceanog; Can Soc Wildlife & Fishery Biol; Int Asn Theoret & Appl Limnol. *Res:* Limnology, particularly secondary production; dynamics of Pacific salmon populations. *Mailing Add:* 51 MacAlester Bay Ft Garry MB R3T 2X6 Can

WARD, GEORGE A, b Chicago, Ill, Feb 17, 36; m 57; c 4. ANALYTICAL CHEMISTRY. *Educ:* Univ Ill, BS, 57; Northwestern Univ, PhD(anal chem), 61. *Prof Exp:* Sr res chemist, Hercules, Inc, 60-75, mgr, Anal Div, 75-78, dir Anal & Tech Servs, 78-80, dir, chem lab, res ctr, 80-85, dir corp res, 85-90, MGR ADVAN MAT RES, HERCULES, INC, 90- *Mem:* Am Chem Soc; Soc Advan Mat & Process. *Res:* Electrochemistry; nuclear magnetic resonance; materials science. *Mailing Add:* Hercules Res Ctr Hercules Plaza Wilmington DE 19894

WARD, GEORGE HENRY, b Withrow, Wash, Nov 28, 16; m 46; c 2. SYSTEMATIC BOTANY. *Educ:* State Col Wash, BS, 40, MS, 48; Stanford Univ, PhD(biol), 52. *Prof Exp:* Teacher, Wash High Sch, 40-42; asst bot & taxon, State Col Wash, 42 & 46-48; asst bot & biol, Stanford Univ, 48-52, instr biol, 52-54, curatorial asst, Dudley Herbarium, 49; from asst prof to prof, 54-80. Cornila H Dudley, 80-82, EMER PROF BIOL, KNOX COLL, ILL, 82- *Concurrent Pos:* Vis prof, Wash State Univ, 62, 64, 66 & 70. *Mem:* AAAS; Sigma Xi. *Res:* Cyto-taxonomy of Artemisia; arctic flora; stripmine spoilbank ecology. *Mailing Add:* Box 195 Husum Washington WA 96823

WARD, GEORGE MERRILL, dairying, nutrition, for more information see previous edition

WARD, GERALD MADISON, b Thorndike, Maine, Nov 2, 21; m 48; c 2. ANIMAL SCIENCE. *Educ:* Univ Maine, BS, 47; Univ Wis, MS, 48; Wash State Univ, PhD(animal sci), 51. *Prof Exp:* Asst prof dairy sci, Univ Maine, 48-49; exten specialist, Kans State Univ, 51-52 & Wash State Univ, 52-53; from asst prof to assoc prof, Colo State Univ, 53-60; scientist, Los Alamos Sci Lab, Univ Calif, 60-61; prof animal sci, 61-88, EMER PROF, COLO STATE UNIV, 88- *Concurrent Pos:* Head animal prod & health sect, Joint Div Food & Agr Orgn-Int Atomic Energy Agency, UN, Vienna, Austria, 68-70. *Mem:* AAAS; Am Dairy Sci Asn; Am Inst Nutrit; Am Soc Animal Sci. *Res:* Applications of radioisotope technique to animal nutrition; environmental problems of animal agriculture; nutrition of ruminant animals; systems analysis of livestock production. *Mailing Add:* Dept Animal Sci Colo State Univ Ft Collins CO 80523

WARD, GERALD T(EMPLETON), b Harrow, Eng, Apr 15, 26; Can citizen; m 63; c 3. RENEWABLE ENERGY RESOURCES, ENERGY IN AGRICULTURE. *Educ:* Glasgow Univ, BSc, 46; Univ Durham, PhD(agr eng), 52. *Prof Exp:* Sci officer animal physiol, Rowett Res Inst, Agr Res Coun, Aberdeen, Scotland, 51-52; proj engr, Colonial Develop Corp, Tawau, Brit NBorneo, 52-53; lectr physics, Univ Malaya, 53-55 & eng, 55-57, sr lectr, 58-60; dir res, Brace Res Inst & prof eng, McGill Univ, 60-71; head dept, 71-77, PROF AGR ENG, LINCOLN COL, UNIV CANTERBURY & DIR NZ AGR ENG INST, 71- *Concurrent Pos:* Vis prof, Univ Calif, Berkeley, 57-58, Univ SPac, WSamoa, 78 & Beijing Agr Eng Univ, 88; Carnegie Corp & Asia Found grants, 58; consult, UN Food & Agr Orgn, 58-60; mem adv comt, Univ WIndies, 63-68; mem water control bd, Soil Conserv Bd & Smoke Control Bd, Govt Barbados, WI, 63-68; rep for Malaya, Nat Inst Agr Eng, England, 50; vis fel, Int Inst Appl Syst Anal, Laxenburg, Austria, 80-81; vis scientist, Ger Acad Exchange Serv, Nuclear Res Ctr Juelich, WGer, 84, INRA Montfavet, France, 89, USSR Acad Sci, 89, Univ Hort, Budapest, 90. *Honors & Awards:* Clayton Prize, Brit Inst Mech Eng, 57. *Mem:* Am Soc Mech Engrs; Am Soc Agr Engrs; Am Soc Heat, Refrig & Air-Conditioning Engrs; Solar Energy Soc (vpres); fel Brit Inst Mech Eng; Int Inst Hort Soc. *Res:* Rural development, especially arid areas; desalination; solar energy utilization; irrigation; drainage; design of water pumps; vapor compressors; airscrew windmills; environmental control; drying; processing rural products; development of solar greenhouses; horticultural engineering; wine technology. *Mailing Add:* Dept Nat Resources Eng Lincoln Univ Canterbury New Zealand

WARD, GERTRUDE LUCKHARDT, b Mt Vernon, NY, May 27, 23. ENTOMOLOGY. *Educ:* Mt Holyoke Col, AB, 44; Univ Mich, MS, 48; Purdue Univ, PhD(entom), 70. *Prof Exp:* Teaching asst zool, Iowa State Univ, 44-46; reporter-ed, Palladium-Item, Ind, 46-47; teaching asst zool, Univ Mich, 48-49; instr biol, 49-61, lectr, 61-67, from asst prof to assoc prof, 67-75, PROF BIOL, EARLHAM COL, 75-, ASST DIR MUS, 52- *Mem:* Am Micros Soc; Entom Soc Am. *Res:* Wasp-spider relationships; Chalybion zimmermanni; parasitism of the bagworm; bat populations in Indiana. *Mailing Add:* Earlham Col Box E 68 Richmond IN 47374

WARD, HAROLD NATHANIEL, b Evanston, Ill, Apr 29, 36; m 59; c 3. MATHEMATICS. *Educ:* Swarthmore Col, BA, 58; Harvard Univ, MA, 59, PhD(math), 62. *Prof Exp:* From instr to asst prof math, Brown Univ, 62-67; from asst prof to assoc prof, 67-82, PROF MATH, UNIV VA, 82- *Mem:* Am Math Soc; Math Asn Am; Soc Indust Appl Math. *Res:* Finite groups; representations of groups; coding theory. *Mailing Add:* Dept Math Math-Astro Bldg Univ Va Charlottesville VA 22903-3199

WARD, HAROLD RICHARD, b Lancaster, NY, July 31, 31; m 60; c 3. ELECTRICAL ENGINEERING. *Educ:* Clarkson Col Technol, BEE, 53; Univ Southern Calif, MSEE, 57. *Prof Exp:* Engr electronics, Hughes Aircraft Co, 53-54 & 56-57; res engr radar, Sylvania Elec Prod, 57-64; CONSULT SCIENTIST RADAR, RAYTHEON CO, 64- *Mem:* Am Inst Elec & Electronics Engrs. *Res:* Radar system design. *Mailing Add:* Raytheon Co Equip Boston Post Rd Wayland MA 01778

WARD, HAROLD ROY, b Mt Vernon, Ill, Nov 3, 35. ENVIRONMENTAL CHEMISTRY. *Educ:* Southern Ill Univ, AB, 57; Mass Inst Technol, PhD(org chem), 61; Harvard Univ, JD, 75. *Prof Exp:* NSF fel, 61-62; NATO fel, 62-63; from asst prof to assoc prof, 63-71, assoc dean, 76-80, PROF CHEM, BROWN UNIV, 71-, DIR, CTR ENVIRON STUDIES, 76- *Concurrent Pos:* Spec fel, Environ Protection Agency, 72-75. *Res:* Hazardous and solid waste management; environmental law. *Mailing Add:* Environ Studies Brown Univ Box 1943 Providence RI 02912

WARD, HERBERT BAILEY, b Texarkana, Tex; c 1. APPLIED & ENVIRONMENTAL MICROBIOLOGY, FOSSIL FUEL BIOPROCESSING. *Educ:* NTex State Univ, BS, 65, MS, 66; Univ Tex, Austin, PhD(bot), 71. *Prof Exp:* Asst biol, NTex State Univ, 63-66; instr, Univ Tex, Arlington, 66-67; res scientist II bot, Lab Algal Physiol, Univ Tex, Austin, 68-71; from asst prof to assoc prof, 71-86, PROF BIOL, UNIV MISS, 86- *Concurrent Pos:* Oak Ridge Assoc Univs fac res appointee, Oak Ridge Nat Lab, 84, consult, 85-; consult, US Dept Energy, 85-, EG&G Idaho, Inc, 87- *Mem:* Sigma Xi; Am Chem Soc; Am Soc Microbiol. *Res:* Fossil fuel bioprocessing; microbial bioconversions for fuels and chemicals; environmental microbiology. *Mailing Add:* Dept Biol Univ Miss University MS 38677

WARD, INGEBORG L, b R-tha, Ger, Aug 14, 40; US citizen; m 63; c 2. PHYSIOLOGICAL PSYCHOLOGY. *Educ:* Westhampton Col, BS, 60; Tulane Univ, MS, 65, PhD(psychol), 67. *Prof Exp:* PROF PSYCHOL, VILLANOVA UNIV, 66- *Concurrent Pos:* NSF res grant, 68-69; Nat Inst Child Health & Human Develop res grant, 70-85; mem ment health small grant comt, NIMH, res career develop award, 75-85; mem, Bio-Psychol Study Sect, NIH, 78-82; consult ed, Behav Neurosci, Physiol Psychol. *Mem:* Fel Am Psychol Asn; Soc Study Reproduction; AAAS. *Res:* Hormonal and environmental determinants of reproductive behavior, neural and pharmacological bases of sexual behavior. *Mailing Add:* Dept Psychol Villanova Univ Villanova PA 19085

WARD, JAMES, b Belleville, NJ, July 11, 64. RADAR, SIGNAL PROCESSING. *Educ:* Univ Dayton, BEE, 85; Ohio State Univ, MScEE, 87, PhD(elec eng), 90. *Prof Exp:* Res assoc elec eng, Ohio State Univ, 85-90; STAFF MEM ELEC ENG, LINCOLN LAB, MASS INST TECHNOL, 90- *Mem:* Inst Elec & Electronics Engrs; Sigma Xi. *Res:* Analysis of adaptive antenna array architectures and algorithms for radar and communication systems; signal processing; radar systems analysis; packet radio networks. *Mailing Add:* 301 Great Rd Apt 136 Acton MA 01720

WARD, JAMES ANDREW, b Pittsburgh, Pa, May 11, 38; m 66; c 6. PHYSICAL CHEMISTRY, POLYMER CHEMISTRY. *Educ:* St Vincent Col, BS, 60; Univ Notre Dame, PhD(phys chem), 64. *Prof Exp:* Res scientist radiation chem, Babcock & Wilcox Co, 64-65; res scientist radiation chem, Union Carbide Corp, 65-79; CHIEF POLYMER CHEMIST, BUCKMAN LABS, 79- *Concurrent Pos:* Adj prof chem, Memphis State Univ. *Mem:* Am Chem Soc. *Res:* Effects of radiation on polymers in solution, polymer degradation and stabilization; hydrogels; water soluble polymers, water treatment, biocidal polymers. *Mailing Add:* Buckman Labs 1256 N McLean Blvd Memphis TN 38108

WARD, JAMES AUDLEY, b Timmonsville, SC, May 19, 10; m 35; c 3. MATHEMATICS, COMPUTER SCIENCE. *Educ:* Davidson Col, BA, 31; La State Univ, MS, 34; Univ Wis, PhD(math), 39. *Prof Exp:* Asst prof math, Davidson Col, 37-38 & Tenn Polytech Inst, 39-40; prof, Delta State Col, 40-42; from assoc prof to prof, Univ Ga, 46-49; prof, Univ Ky, 49-55; mathematician, USAF Missile Develop Ctr, 55-57, chief digital comput br, 57-59; spec asst to vpres, Univac Div, Sperry Rand Corp, 59-62; staff specialist electronic comput, Off Dir Defense Res & Eng, Dept Defense, 62-69; staff asst comput progs, Naval Sea Systs Command, 69-76; RETIRED. *Concurrent Pos:* Consult, Proj Scamp, Univ Calif, Los Angeles, 52 & Tube Turns, Ky, 53-55; consult, Appl Physics Lab, 76-78. *Mem:* Math Asn Am; Am Math Soc; Asn Comput Mach; sr mem Inst Elec & Electronics Engrs. *Res:* Digital computers; research in hardware and software; mathematical research in linear algebra and in numerical analysis. *Mailing Add:* 2726 Timbertrail Circle Tallahassee FL 32308

WARD, JAMES B, poultry nutrition, biochemistry; deceased, see previous edition for last biography

WARD, JAMES EDWARD, III, b Greenville, SC, Sept 20, 39; m 62; c 3. MATHEMATICS. *Educ:* Vanderbilt Univ, BA, 61; Univ Va, MA, 64, PhD(math), 68. *Prof Exp:* Asst to pres, George Peabody Col, 61-62; teaching asst math, Univ Va, 62-64; jr instr, 66-68, from asst prof to assoc prof, 68-79, dir sr ctr, 71-76, chmn dept, 78-80, PROF MATH, BOWDOIN COL, 79-, CHMN DEPT, 86- *Mem:* Am Math Soc; Math Asn Am. *Res:* Jordan algebras of characteristic two; structure of two-groups. *Mailing Add:* Dept Math Bowdoin Col Brunswick ME 04011

WARD, JAMES VERNON, b Minneapolis, Minn, Mar 27, 40; m 63. STREAM ECOLOGY, LIMNOLOGY. *Educ:* Univ Minn, Minneapolis, BS, 63; Univ Denver, MS, 67; Univ Colo, PhD(limnol), 73. *Prof Exp:* PROF, DEPT BIOL, COLO STATE UNIV, 73- *Mem:* Am Soc Limnol & Oceanog; Soc Int Limnol, Ecol Soc Am; NAm Benthological Soc (pres, 87-88). *Res:* Stream ecology and limnology, especially aquatic macroinvertebrates and factors influencing their distribution and community structure; gound water ecology. *Mailing Add:* Dept Biol Colo State Univ Ft Collins CO 80523

WARD, JERROLD MICHAEL, b New York, NY, Oct 29, 42; m 71. VETERINARY PATHOLOGY. *Educ:* Cornell Univ, DVM, 66; Univ Calif, Davis, PhD(comp path), 70. *Prof Exp:* Res pathologist, Univ Calif, Davis, 66-68; vet pathologist, Environ Protection Agency, 70-72; vet pathologist, 72-78, chief, tumor pathol, Nat Toxicol Prog, 79-81, ACTG CHIEF, TUMOR PATH BR, CARCINOGENESIS TESTING PROG, NAT CANCER INST, 78- *Mem:* Am Vet Med Asn; Int Acad Path. *Res:* Pathology; cancer research; hematopoietic pathology; rodent pathology. *Mailing Add:* 10513 Wayridge Dr Gaithersburg MD 20879

WARD, JOHN C(LAYTON), sanitary engineering, for more information see previous edition

WARD, JOHN E(RWIN), electrical engineering, for more information see previous edition

WARD, JOHN EDWARD, b Chicago, Ill, Feb 7, 23; m 46; c 6. ORGANIC CHEMISTRY. *Educ:* Wabash Col, AB, 44; Lawrence Col, MS, 48, PhD, 51. *Prof Exp:* Res chemist, P H Glatfelter Co, Pa, 50-51; from chief chemist paper chem lab to tech mgr foreign dept, NopCo Chem Co, NJ, 51-59; MANAGING DIR, HENKEL-NOPCO CHIMIE, SA, 59-, CHMN BD, NOPCO ITALIA SPA, 81- *Concurrent Pos:* Henkel-Nopco Prods AG, 82-; Mgr, Danlink Sa, 83- *Mem:* Am Chem Soc; Tech Asn Pulp & Paper Indust. *Res:* Pulp and paper technology, especially the production of pigment coated papers; market research, product development and international marketing of industrial chemical specialties, especially for the particular requirements of various European markets. *Mailing Add:* 24 Route de la Veveyse CH-1700 Fribourg Switzerland

WARD, JOHN EVERETT, JR, b Mocksville, NC, Feb 23, 41; m 64; c 2. MYCOLOGY, ECOLOGY. *Educ:* High Point Col, BS, 63; Wake Forest Univ, MA, 65; Univ SC, PhD(biol), 70. *Prof Exp:* Asst prof biol, Gaston Col, NC, 65-67; from asst prof to assoc prof, 70-81, PROF BIOL, HIGH POINT COL, 81-, DEPT CHAIR, 88- *Mem:* Mycol Soc Am. *Res:* Ecology of soil fungi. *Mailing Add:* Dept Biol High Point Col HP-2 High Point NC 27261-1949

WARD, JOHN F, b Blyth, Northumberland, Eng, Aug 26, 35; m 75; c 2. RADIATION CHEMISTRY & BIOCHEMISTRY, RADIATION BIOLOGY. *Educ:* Durham Univ, BSc, 56, PhD(radiation chem), 60. *Prof Exp:* Demonstr chem, Kings Col, Durham Univ, 60-62; asst res biophysicist, Lab Nuclear Med & Radiation Biol, Univ Calif, Los Angeles, 62-69; assoc res chemist, 69-79, assoc prof path, Sch Med, 74-78, PROF RADIOL, SCH MED, UNIV CALIF, SAN DIEGO, 78-, CHIEF RADIOBIOL, 80- *Concurrent Pos:* Mem radiation study sect, NIH, 74-78; counr chem, Radiation Res Soc, 78-81; assoc ed, Radiation Res, 81- *Mem:* Royal Soc Chem; Biophys Soc; Radiation Res Soc (pres, 85-86); Asn Radiation Res; Soc Free Radical Res. *Res:* Studies of radiation chemical destruction of biologically significant molecules; molecular mechanisms of cell killing; measurement of repair of DNA damage in mammalian cells. *Mailing Add:* Dept Radiol M-010 Univ Calif San Diego La Jolla CA 92093

WARD, JOHN FRANK, b London, Eng, May 14, 34; m 60; c 2. PHYSICS. *Educ:* Oxford Univ, BA, 57, MA & DPhil(physics), 61. *Prof Exp:* From lectr to asst prof physics, Univ Mich, 61-64; Assoc Elec Industs res fel, Oxford Univ & lectr, Wadham Col, Univ Oxford, 64-67; assoc prof, 67-74, PROF PHYSICS, UNIV MICH, ANN ARBOR, 74- *Concurrent Pos:* Consult, Lear-Siegler Laser Systs Ctr, 62-64, Royal Radar Estab, 66, Photon Sources, 67- & KMS Fusion Inc, 81- *Mem:* Am Phys Soc; Sigma Xi. *Res:* Nonlinear optics; lasers. *Mailing Add:* Randall Physics Lab Univ Mich Ann Arbor MI 48109

WARD, JOHN HENRY, b Springfield, Mass, Oct 10, 50; c 2. ATMOSPHERIC SCIENCE. *Educ:* Worcester Polytech Inst, BS, 73; Purdue Univ, MS, 75, PhD(atmospheric sci), 78. *Prof Exp:* Asst atmospheric sci, Purdue Univ, 73-78; FEL NUMERICAL WEATHER PREDICTION, NAT WEATHER SERV, 78- *Concurrent Pos:* Ed, Nat Meteorol Ctr Monthly Performance Summary. *Mem:* Am Meteorol Soc; Sigma Xi. *Res:* Numerical weather prediction, particularly diagnostic evaluation of regional forecast systems and tropical hurricane forecasting. *Mailing Add:* Nat Meteorol Ctr W/NMC 22 WWB Rm 204 Washington DC 20233

WARD, JOHN K, b Litchfield, Nebr, July 1, 27; m 53; c 3. ANIMAL NUTRITION. *Educ:* McPherson Col, BS, 50; Kans State Univ, BS, 54, PhD, 61. *Prof Exp:* Asst animal husb, Okla State Univ, 54-56; asst prof agr, McPherson Col, 56-66; assoc prof, 67-74, PROF ANIMAL SCI, UNIV NEBR, LINCOLN, 74- *Res:* Beef cattle management. *Mailing Add:* Dept Animal Sci Univ Nebr East Campus Lincoln NE 68583-0908

WARD, JOHN ROBERT, b Boston, Mass, Jan 10, 29; m 55; c 2. ELECTRICAL ENGINEERING. *Educ:* Univ Sydney, BSc, 49, BE, 52, PhD(aeronaut eng), 58. *Prof Exp:* Aerodynamicist, Vickers-Armstrongs, Ltd, Eng, 51-52; lectr, Univ Sydney, 57; asst prof elec eng, Brown Univ, 58-62; assoc prof, 62-68, PROF ELEC ENG, NAVAL POSTGRAD SCH, 68- *Mem:* Inst Elec & Electronics Engrs; Am Soc Eng Educ. *Res:* Educational technology, programmed instruction; systems dynamics, circuits and control. *Mailing Add:* Dept Elec Eng US Naval Postgrad Sch Monterey CA 93940

WARD, JOHN ROBERT, b Salt Lake City, Utah, Nov 23, 23; m 48; c 4. MEDICINE. *Educ:* Univ Utah, BS, 44, MD, 46; Univ Calif, Berkeley, MPH, 67. *Prof Exp:* Resident internal med, Salt Lake County Gen Hosp, 49-51; from instr to assoc prof med, 57-70, chmn dept prev med, 66-70, PROF MED & CHIEF RHEUMATOLOGY DIV, COL MED, UNIV UTAH, 70- *Concurrent Pos:* Res fel physiol, Col Med, Univ Utah, 48-49; res fel med, Salt Lake County Gen Hosp, 53-54; fel med & rheumatic dis, Mass Gen Hosp & res fel rheumatic dis, Harvard Med Sch, 55-57; consult, Vet Admin Hosp, 57-; sr investr, Arthritis & Rheumatism Found, 59; mem arthritis training grants comt, Nat Inst Arthritis & Metab Dis. *Mem:* AAAS; AMA; Am Rheumatism Asn; Am Col Physicians; NY Acad Sci. *Res:* Biology of mycoplasmatales; experimental arthritis; immunology; epidemiology; drug studies & evaluation. *Mailing Add:* Rheumatology Div Univ Utah Med Ctr Salt Lake City UT 84112

WARD, JOHN WESLEY, b Martin, Tenn, Apr 8, 25; m 47; c 4. PHARMACOLOGY. *Educ:* George Washington Univ, BS & MS, 55; Georgetown Univ, PhD(pharmacol), 59. *Prof Exp:* Res assoc pharmacol, Hazleton Labs, Inc, 50-56, head dept pharmacol, 56-58, chief dept pharmacol & biochem, 58, res appln specialist, 59; prin pharmacologist, A H Robins Co, 59-60, dir pharmacol, 60-71, pharmacol develop, 71-72, toxicol, 73-77, dir good lab practs, 77-78, biol res, 78-80, res, 80-81, vpres res, 82-89, vpres & gen mgr, Res & Develop Div, 89-90; RETIRED. *Concurrent Pos:* Lectr, Med Col Va, 60-65, affil assoc prof pharmacol, 82- *Mem:* AAAS; Am Chem Soc; Soc Toxicol; NY Acad Sci; Am Soc Pharmacol & Exp Therapeut; Sigma Xi; Int Soc Regulatory Toxicol & Pharmacol. *Res:* Structure-activity relationships; general pharmacodynamics; autonomics; toxicology. *Mailing Add:* 6767 Forest Hill Ave Suite 204 Richmond VA 23225

WARD, JOHN WILLIAM, b Moline, Ill, Oct 16, 29; m 52; c 4. ACTINIDE CHEMISTRY, METAL HYDRIDES. *Educ:* Augustana Col, Ill, BA, 52; Wash Univ, MA, 55; Univ NMex, PhD(phys chem), 66. *Prof Exp:* STAFF MEM, LOS ALAMOS NAT LAB, 56- *Concurrent Pos:* Consult, US Army Nuclear Defense Lab, Edgewood Arsenal, 66-; sr fel, Alexander von Humboldt Found, Inst Transuranium Elements, Ger, 72-73. *Honors & Awards:* Fel, Los Alamos Nat Lab, 83. *Mem:* Am Chem Soc; Am Vacuum Soc; fel Am Inst Chem. *Res:* Vapor pressure theory; Monte Carlo computer simulation of experiment; metal hydrides; gas-surface reactions; chemistry of actinides. *Mailing Add:* 2501 Calle Melecio Santa Fe NM 87505

WARD, JOHN WILLIAM, b Wigan, Eng, Aug 4, 37; m 69; c 6. PHYSICAL CHEMISTRY. *Educ:* Univ Manchester, BSc, 59, MSc, 60; Cambridge Univ, PhD(phys chem), 62. *Prof Exp:* Res coun Alta fel, 62-63; res scientist, 63-66, sr res scientist, 66-70, res assoc, 70-77, sr res assoc, 77-79, STAFF CONSULT, UNION OIL CO CALIF, 79- *Honors & Awards:* Eugene Howdry Award Appl Catalysis, 85. *Mem:* Am Chem Soc; Catalysis Soc; Royal Soc Chem; assoc Royal Inst Chem; Int Zeolite Asn. *Res:* Application of spectroscopic techniques to the study of surface chemistry and catalysis; heterogeneous catalysis; hydrocarbon conversions; petroleum processing, especially hydrotreating, hydrocracking and reforming petrochemical processing; molecular sieves. *Mailing Add:* Union Oil Co Calif Res Dept 376 S Valencia Ave Brea CA 92621-6345

WARD, JONATHAN BISHOP, JR, b Tacoma, Wash, Oct 13, 43; m 69; c 3. GENETICS, TOXICOLOGY. *Educ:* Whitman Col, AB, 65; Univ Idaho, MS, 68; Cornell Univ, PhD(microbiol), 72. *Prof Exp:* Fel somatic cell genetics, Mass Gen Hosp, 72-74; sr res assoc human genetics, 74-78, asst prof, 78-87, ASSOC PROF GENETIC TOXICOL, UNIV TEX MED BR, 87- *Concurrent Pos:* Prin investr, NIH grants. *Mem:* Am Soc Cell Biol; Sigma Xi; Environ Mutagen Soc; Soc Toxicol. *Res:* Mutagenicity of environmental chemicals in cultured mammalian cells; evaluation of chemical mutagens in animals; human population monitoring for genetic damage from environmental agents. *Mailing Add:* Div Environ Toxicol Univ Tex Med Br Galveston TX 77550

WARD, JOSEPH J, b Boston, Mass, Dec 11, 46. ANALYSIS. *Educ:* Purdue Univ, PhD(math), 73. *Prof Exp:* From asst prof to assoc prof, 74-85, PROF MATH, TEX A&M UNIV, 85- *Concurrent Pos:* Vis prof math, Univ Bonn, Ger, 76. *Mem:* Am Math Soc. *Res:* Approximation theory. *Mailing Add:* Dept Math Tex A&M Univ College Station TX 77843

WARD, JOSEPH RICHARD, b Salt Lake City, Utah, Dec 7, 42. CHEMISTRY. *Educ:* Univ Del, BS, 64; State Univ NY Stony Brook, PhD(chem), 69. *Prof Exp:* RES CHEMIST, US ARMY BALLISTIC RES LABS, ABERDEEN PROVING GROUND, 71- *Mem:* Am Chem Soc. *Res:* Erosion of gun barrels, combustion of solid propellants; effect of propellant combustion in the near-wake of supersonic projectiles. *Mailing Add:* 1332 Sweetbriar Lane Bel Air MD 21014

WARD, KEITH BOLEN, JR, b Paducah, Tex, Feb 20, 43. BIOPHYSICS. *Educ:* Tex A&M Univ, BS, 65; Johns Hopkins Univ, PhD(biophys), 74. *Prof Exp:* Res assoc physics, Appl Res Lab, Gen Dynamics Corp, 65-66; Nat Res Coun res assoc, Lab Struct Matter, Naval Res Lab, 74-76; asst prof, 76-81, assoc prof chem, Univ Wis-Parkside, 82-84; AT PROTEIN CRYSTALLOG LAB, NAVAL RES LAB. *Mem:* AAAS; Am Crystallog Asn. *Res:* Study of the relationship between the structure and function of proteins by using x-ray diffraction analysis; oxygen transport pigments; phospholipases; bioluminescent. *Mailing Add:* Naval Res Lab Washington DC 20375

WARD, KYLE, JR, b Beaumont, Tex, Sept 2, 02; wid. CELLULOSE CHEMISTRY, PULP & PAPER TECHNOLOGY. *Educ:* Univ Tex, BA & BS, 23; George Washington Univ, MS, 26; Univ Berlin, PhD(chem), 32. *Hon Degrees:* MS, Lawrence Univ, 68. *Prof Exp:* Instr, Univ Tex, 23-24; jr chemist, Bur Chem & Soils, USDA, 24-28, collabr, 36-38, sr chemist, Southern Regional Res Lab, Bur Agr & Chem Eng, 38-41, prin chemist, Bur Agr & Indust Chem, 41-51; res chemist, Hercules Powder Co, 28-36; res assoc, 51-59, leader cellulose group, 59-66, chmn dept chem, 59-68, leader carbohydrate group & chmn sect org chem, 66-68, EMER PROF, INST PAPER CHEM, 68- *Concurrent Pos:* Res chemist, Chem Found, 36-38; True Mem lectr, 63; consult, Joint Chiefs of Staff, 45 & Am Can Co, 73-75; mem Am-Egyptian chem workshop, 77; Fulbright prof, Helsinki, 70. *Honors & Awards:* Anselmo Payon Award. *Mem:* AAAS; fel Am Chem Soc; hon mem Fiber Soc; fel Tech Asn Pulp & Paper Indust; Am Soc Testing & Mat. *Res:* Cellulose and derivatives; terpenes and related fields; chlorination and oxidation; cotton fiber properties; textiles; high polymers; wood, pulp and paper. *Mailing Add:* 2600 Heritage Woods Dr No A224 Appleton WI 54915-1409

WARD, LAIRD GORDON LINDSAY, b Wellington, NZ, Dec 6, 31; US citizen. INORGANIC CHEMISTRY. *Educ:* Univ NZ, BSc, 56, MSc, 57; Univ Pa, PhD(inorg chem), 61. *Prof Exp:* Res chemist, Fabrics & Finishes Dept, E I du Pont de Nemours & Co, 61-63; res fel, Mellon Inst Sci, 63-64; res chemist, Int Nickel Co, Inc, NY, 64-71; res assoc chem, Univ Ga, 72; sr chemist, Colonial Metals, Inc, 72-74; res assoc & mem chem fac, Cend, Univ Del, 75-82; GROUP LEADER RES & DEVELOP, JOHNSON MATTHEY INC, 75- *Mem:* Am Chem Soc; Royal Soc Chem; fel Am Inst Chemists. *Res:* Hydrometallurgical and pyrometallurgical processes related to refining and recovery of platinum group metals; kinetics of release of carbon-14 and tritium labeled molecules from biosorbable polymers; synthesis of inorganic complexes, especially platinum group metals; organometallic chemistry with group five elements; inorganic pigment applications of non-stoichiometric transition metal oxides. *Mailing Add:* Johnson Matthey 2001 Nolte Dr West Deptford NJ 08066-1727

WARD, LAWRENCE MCCUE, b Canton, Ohio, Dec 11, 44; div; c 3. PSYCHOPHYSICS, ATTENTION. *Educ:* Harvard Univ, AB, 66; Duke Univ, PhD(exp psychol), 71. *Prof Exp:* Asst prof psychol, Rutgers Univ, 70-73; vis asst prof, 73-74, asst prof, 74-77, assoc prof, 77-88, PROF PSYCHOL, UNIV BC, 88- *Concurrent Pos:* Consult, Dept Hwys, State NJ, 72; assoc, Acoust Eng, Vancouver, BC, 75-83; res fel, Harvard Univ, 78-79; consult, Transport Can, 89-91. *Mem:* AAAS; fel Am Psychol Asn; Psychonomic Soc; Int Soc Psychophysics; Can Psychol Asn; fel Am Psychol Soc. *Res:* Psychophysical scaling; psychophysical judgement; general systems theory; decision theory. *Mailing Add:* Dept Psychol Univ BC Vancouver BC V6T 1Y7 Can

WARD, LAWRENCE W(ATERMAN), b Flushing, NY, Feb 21, 26; m 55; c 3. APPLIED & FLUID MECHANICS. *Educ:* Univ Mich, BS, 48; Stevens Inst Technol, MS, 51, DSc(appl mech), 62. *Prof Exp:* Technician, Gibbs & Cox, Inc, NY, 48-50, res engr, 51-55; res engr, Stevens Inst Technol, 55-58; prof eng, 58-87, ASST DEAN, WEBB INST NAVAL ARCHIT, 87- *Concurrent Pos:* NSF sci fac fel, 65-66. *Mem:* Soc Naval Architects & Marine Engrs; Am Soc Naval Engrs; Am Soc Eng Educ; Ger Ship Bldg Soc; Sigma Xi. *Res:* Ship wave pattern and spectra; experimental determination of wave resistance from wave pattern; ship anti-rolling tanks; ship maneuvering; hull impact; wind effect on ships. *Mailing Add:* Dept Eng Webb Inst Naval Archit Glen Cove NY 11542-1398

WARD, LEONARD GEORGE, b Tupper Lake, NY, Mar 23, 30; m 50; c 2. SYSTEMS DESIGN & SYSTEMS SCIENCE, POLYMER ENGINEERING. *Prof Exp:* Design engr, Westinghouse Elec Corp, 50-54, Flight Refueling Inc, 54-59, Lockheed Aircraft Co, 59-60, Manovox Corp, 60-61 & Int Bus Mach, 61-62; sr mech engr, Sororan Eng Inc, 62-69; prin engr, GDI Inc, 69-72; ADV DEVELOP ENGR, STORAGE TECHNOL CORP, 72- *Concurrent Pos:* Pres, Ocean Edge Publ, 86- *Res:* Non impact and impact high speed printer design; forms transport; fusing systems; magnetic ink character recognition systems; print band technology; print hammer design; design for manufacturing/design for assembly; awarded nine US patents. *Mailing Add:* 315 Pine Tree Dr Indialantic FL 32903

WARD, LEWIS EDES, JR, b Arlington, Mass, July 20, 25; m 49; c 3. CONTINUUM THEORY, ORDERED SPACES. *Educ:* Univ Calif, AB, 49; Tulane Univ, MS, 51, PhD(math), 53. *Prof Exp:* Instr math, Univ Nev, 53-54; asst prof, Univ Utah, 54-56; mathematician, US Naval Ord Test Sta, Univ Calif, 56-59; assoc prof, 59-65, PROF MATH, UNIV ORE, 65- *Mem:* Am Math Soc; Math Asn Am; Mex Soc Math. *Res:* Topology, especially ordered spaces, fixed point theory and continuum theory. *Mailing Add:* Dept Math Univ Ore Eugene OR 97403

WARD, LOUIS EMMERSON, b Mt Vernon, Ill, Jan 19, 18; m 42; c 4. MEDICINE, RHEUMATOLOGY. *Educ:* Univ Ill, AB, 39; Harvard Univ, MD, 43; Univ Minn, MS, 49. *Prof Exp:* Consult sect med, Mayo Clin, 50-83, chmn bd gov, 64-75; from instr med to prof clin med, Mayo Grad Sch Med, Univ Minn, 51-73, prof med, Mayo Med Sch, 73-83; RETIRED. *Mem:* Inst Med-Nat Acad Sci; Am Rheumatism Asn; AMA; Nat Soc Clin Rheumatology; Am Col Phys; Cent Soc Clin Res. *Res:* Rheumatic diseases. *Mailing Add:* 30 Raeburn Ct Port Ludlow WA 98365

WARD, MELVIN A, b Tonkawa, Okla, June 4, 40; m 69; c 2. CANCER ETIOLOGY & PREVENTION. *Educ:* Okla State Univ, BA, 62; Univ Hawaii, MS, 64, PhD(bot), 69. *Prof Exp:* Postdoctoral biochem, Univ Hawaii, 69-70; postdoctoral microbiol, Purdue Univ, 70-71; asst prof, 71-77, ASSOC PROF BIOL, HAWAII LOA COL, 77- *Concurrent Pos:* Marc fac fel, Nat Cancer Inst, Cancer Ctr Hawaii, 78-79. *Mem:* Assoc mem Sigma Xi. *Res:* Polymerase chain reaction induced DNA amplification of HTLV-I: induction of adult T-cell leukemia; cancer etiology and prevention: mutagen testing and detection in human bodily fluids. *Mailing Add:* 1415 Kapau Pl Kailua HI 96734

WARD, OSCAR GARDIEN, b Denver, Colo, Feb 16, 32; m 55; c 2. GENETICS, CYTOGENETICS. *Educ:* Univ Ariz, BS, 58, MS, 60; Purdue Univ, PhD(genetics), 66. *Prof Exp:* Instr biol, Purdue Univ, 60-64; asst prof, 66-74, lectr, 74-77, ASSOC PROF, BIOL SCI, UNIV ARIZ, 77- *Concurrent Pos:* Fogarty sr int fel, Mex, 80. *Mem:* AAAS; Genetics Soc Am; Am Soc Human Genetics; Am Soc Mammalogists; Am Genetic Asn. *Res:* Plant and animal cytogenetics with emphasis on mammalian systems including man; chromosome identification by banding patterns with application to karyotype evolution; role of chromosomes during development; human cytogenetics. *Mailing Add:* Dept Ecol & Evolutionary Biol Univ Ariz Tucson AZ 85721

WARD, PAUL H, b Lawrence, Ind, Apr 24, 28; m 52; c 2. OTOLARYNGOLOGY. *Educ:* Anderson Col, AB, 53; Johns Hopkins Univ, MD, 57; Am Bd Otolaryngol, dipl, 62. *Prof Exp:* Intern, Henry Ford Hosp, Detroit, 57-58; resident otolaryngol, Univ Chicago, 58-61, NIH spec res fel, 61-62, asst prof, 62-64; assoc prof surg & chief div otolaryngol, Sch Med, Vanderbilt Univ, 64-68; PROF SURG & CHIEF HEAD & NECK SURG, CTR HEALTH SCI, UNIV CALIF, LOS ANGELES, 68- *Concurrent Pos:* USPHS res grant, 63-69; Deafness Res Found res grant, 65-67; NIH res grant, 66-70; attend otolaryngologist, Nashville Metrop Gen Hosp, Tenn, 64-68; consult, Thayer Vet Admin Hosp, 64-68 & Surgeon Gen, USN, 74-; mem bd dirs, Bill Wilkerson Hearing & Speech Ctr, 64-68. *Mem:* AAAS; Am Otol Soc; Am Acad Ophthal & Otolaryngol; Am Laryngol Soc; Am Laryngol, Rhinol & Otol Soc; Sigma Xi. *Res:* Cochlear, vestibular and laryngeal physiology; temporal bone pathology; velopharyngeal corrective techniques; laryngeal and palatal reconstruction. *Mailing Add:* Dept Surg Sch Med Univ Calif Los Angeles CA 90024

WARD, PETER A, b Winsted, Conn, Nov 1, 34. PATHOLOGY, IMMUNOLOGY. *Educ:* Univ Mich, BS, 58, MD, 60. *Prof Exp:* Intern med, Third Div, Bellevue Hosp, New York, 60-61; resident path, Hosp, Univ Mich, Ann Arbor, 61-63; res fel immunopath, Div Exp Path, Scripps Clin & Res Found, La Jolla, Calif, 63-65; chief immunobiol, Armed Forces Inst Path, Washington, DC, 65-71; prof path, Sch Med, Univ Conn, 71-80, chmn dept, 73-80; interim dean, 82-85, PROF & CHMN DEPT PATH, SCH MED, UNIV MICH, 80- *Honors & Awards:* Borden Award, 60; Parke Davis Award Except Path, 71. *Mem:* Nat Acad Sci-Inst Med; Am Asn Immunologists; Am Soc Clin Invest; Am Asn Path. *Res:* Immunopathology; inflammation; biological role of complement; antibody formation; immune complexes; oxygen radicals. *Mailing Add:* Dept Path Univ Mich Med Sch 1301 Catherine Rd Box 0602 Ann Arbor MI 48109-0602

WARD, PETER LANGDON, b Washington, DC, Aug 10, 43; m 65, 78; c 4. SEISMOLOGY, VOLCANOLOGY. *Educ:* Dartmouth Col, BA, 65; Columbia Univ, MA, 67, PhD(geophys), 70. *Prof Exp:* Asst seismol, 65-70, res scientist, Columbia Univ, 70-71; geophysicist, 71-75, chief br seismol, 75, chief br earthquake mech & prediction, 75-77, coordr earthquake prediction prog, 77-78, GEOPHYSICIST SEISMOL, US GEOL SURV, 78- *Concurrent Pos:* Mem adv panel magma energy res, Sandia Labs, 74-82; mem, Earth Dynamics Adv Subcomt, NASA, 76-77; mem geophys prediction panel, Nat Acad Sci, 77-78. *Mem:* Int Asn Volcanol & Chem of Earth's Interiors; AAAS; Am Geophys Union; Seismol Soc Am; Geol Soc Am. *Res:* Analysis of earthquakes related to volcanoes, geothermal areas and tectonic features; earthquake seismology; earthquakes and ground deformation near volcanoes; geothermal exploration with seismic techniques; seismic instrumentation; computer techniques; earthquake prediction and hazard reduction. *Mailing Add:* US Geol Surv NCER 345 Middlefield Rd Menlo Park CA 94025

WARD, PHILLIP WAYNE, b Warren, Ark, June 10, 35; m 58; c 4. NAVIGATION SATELLITE RECEIVERS, DIGITAL SIGNAL PROCESSING. *Educ:* Univ Tex, El Paso, BS; Southern Methodist Univ, MS, 65. *Prof Exp:* Lt, US Coast & Geod Surv, 58-60; mem tech staff, Mass Inst Technol Instrumentation Lab, 67-69; design engr, 60-67, SR MEM TECH STAFF, TEX INSTRUMENTS INC, 69- *Honors & Awards:* Colonel Thomas L Thurlow Award, Inst Navig, 89. *Mem:* Inst Navig; sr mem Inst Elec & Electronics Engrs. *Res:* Development of advanced Navstar global positioning systems receivers for navigation using global positioning systems satellites used on a variety of Department of Defense platforms including weapons systems, manpack/vehicular systems, avionics systems and low earth orbit satellites; real time software; integrated circuit technology. *Mailing Add:* 9629 Covemeadow Dr Dallas TX 75238

WARD, RAYMOND LELAND, b San Pedro, Calif, Feb 12, 32; m 58; c 4. PHYSICAL CHEMISTRY. *Educ:* Univ Calif, BSc, 53; Washington Univ, St Louis, PhD(chem), 56. *Prof Exp:* Chemist, Lawrence Radiation Lab, Univ Calif, 56-64; NSF sr fel, Harvard Univ, 64-65; CHEMIST, LAWRENCE LIVERMORE NAT LAB, UNIV CALIF, 65- *Mem:* Am Chem Soc. *Res:* Magnetic resonance studies of molecular interactions. *Mailing Add:* Lawrence Livermore Nat Lab Univ Calif PO Box 808 Mail Code L 310 Livermore CA 94550-0622

WARD, RICHARD BERNARD, organic chemistry, for more information see previous edition

WARD, RICHARD FLOYD, b New York, NY, July 5, 27; m 49; c 3. GEOLOGY. *Educ:* Bradley Univ, BS, 50; NY Univ, MS, 56; Bryn Mawr Col, PhD, 58. *Prof Exp:* Geologist, Del State Geol Surv, 54-58; assoc prof geol, Wayne State Univ, 59-87; RETIRED. *Mem:* Geol Soc Am; Geochem Soc. *Res:* Metamorphic and igneous petrology; evolution of crystalline terrains. *Mailing Add:* 22434 Melrose Ct East Detroit MI 48021

WARD, RICHARD JOHN, b Seattle, Wash, Aug 7, 25; wid; c 3. ANESTHESIOLOGY. *Educ:* Gonzaga Univ, BSc, 46; St Louis Univ, MD, 49; Seattle Univ, MEd, 72; Am Bd Anesthesiol, dipl. *Prof Exp:* Chief anesthesiol serv, Air Force Hosp, Weisbaden, Ger, 54-57; asst chief anesthesiol, Lackland AFB, 57-60, chief, 60-61; chief anesthesiol, Ballard Gen Hosp, 62-63; from instr to prof anesthesiol, Sch Med, Univ Wash, 63-90, chief staff, Univ Hosp, 77-79; RETIRED. *Concurrent Pos:* NIH res fel, 64-65; consult, Surgeon Gen, USAF, Europe, 54-57; chief surg res lab, Lackland AFB, 58-61; admin officer, Sch Med, Univ Wash, 66; consult, Madigan Gen Hosp, 72- *Mem:* AAAS; Am Soc Anesthesiol; AMA; Asn Mil Surg US; fel Am Col Anesthesiol. *Res:* Pharmacology and physiology of anesthetized man. *Mailing Add:* Dept Anesthesiol Med Sch Univ Wash Seattle WA 98195

WARD, RICHARD LEO, b Bozeman, Mont, Nov 16, 42; m 66; c 2. VIROLOGY. *Educ:* Mont State Univ, BS, 65; Univ Calif, PhD(biochem), 69. *Prof Exp:* Fel molecular biol, Max Planck Inst Biochem, 69-70; fel animal virol, Roche Inst Molecular Biol, 70-72; asst res virologist, Sch Med, Univ Calif, Los Angeles, 72-74; mem tech staff, Sandia Labs, Albuquerque, 74-80; chief viral disease, Health & Environ Res Lab, US Environ Protection Agency, Cincinnati, 80-81; assoc mem, Christ Hosp Inst Med Red, Cincinnati, 81-84; MEM & DIR CLIN VIROL, J N GAMBLE INST MED RES, 84- *Concurrent Pos:* Adj assoc prof, NMex State Univ, 77-80; prin investr, US Environ Protection Agency, 82-, US AID, 87- *Mem:* Am Soc Microbiol; AAAS; Am Soc Virol. *Res:* Enteric viral diseases; environmental virology; herpes viruses; chemical carcinogenesis; radiation biology; enteric virus, environmental virology, clinical and molecular studies on rotaviruses, antiviral drug studies and vaccine development. *Mailing Add:* 2282 Spinningwheel Lane Cincinnati OH 45244

WARD, RICHARD S, b Beirut, Lebanon, Oct 9, 20; m 60; c 3. PSYCHIATRY, PEDIATRICS. *Educ:* Amherst Col, BA, 42; Columbia Univ, MD, 45, cert, 57. *Prof Exp:* Clin dir, Child Guid Inst, Jewish Bd Guardians, NY, 56-61; assoc prof, 60-63, PROF PSYCHIAT, SCH MED, EMORY UNIV, 63-, ASSOC PROF PEDIAT, 76- *Concurrent Pos:* Rockefeller fel child psychiat, Babies Hosp, Columbia Univ, 48-50. *Mem:* Am Psychoanal Asn; Am Psychiat Asn; fel Am Orthopsychiat Asn; Am Acad Child Psychiat. *Res:* Child development; psychoanalysis. *Mailing Add:* 27 Lenox Pointe NE Atlanta GA 30324

WARD, RICHARD THEODORE, plant ecology, for more information see previous edition

WARD, ROBERT C, b Mt Clemens, Mich, Mar 9, 32; c 4. OSTEOPATHY, BIOMECHANICS. *Educ:* Kansas City Col Osteop Med, DO, 57. *Prof Exp:* Intern, Mt Clemens Gen Hosp, 57-58, staff mem family med, 58-71; preceptor, Col Osteop Med & asst prof family med, 70-71, prof family med & chmn dept, 72-74, prof med educ res & develop, 74-81, PROF BIOMECH, MICH STATE UNIV, 81-; STAFF MEM, LANSING GEN HOSP, 72- *Concurrent Pos:* Off Med Educ fel, Mich State Univ, 72-73. *Mem:* Am Osteop Asn; Am Asn Study Headache; Am Acad Osteop; NAm Acad Musculoskeletal Med; Int Asn Study Pain. *Res:* Family medicine curriculum design; osteopathic therapeutics; community based medical education; stress management education; spinal radiography soft tissue mechanics. *Mailing Add:* Dept Family Med E Fee Hall Mich State Univ Col Osteopath Med East Lansing MI 48824

WARD, ROBERT CARL, b Swansea, Wales, July 4, 44; US citizen; m 66; c 3. ENVIRONMENTAL ENGINEERING, AGRICULTURAL ENGINEERING. *Educ:* Miss State Univ, BS, 66; NC State Univ, MS, 68, PhD(agr eng), 70. *Prof Exp:* Asst prof agr eng, 70-75, assoc prof, 75-80, actg head dept, 82-83, PROF AGR & CHEM ENG, COLO STATE UNIV, 80-, ASSOC DEAN, COL ENG, 86 - *Concurrent Pos:* Guest researcher, Water Qual Inst, Denmark, 76; systs engr, US Environ Protection Agency, 77; consult, Water Qual Ctr, Hamilton, NZ, 83-84. *Honors & Awards:* Durrell Award, 74; Gunlogson Award, Am Soc Agr Engrs, 76. *Mem:* Nat Water Well Asn; Water Pollution Control Fedn; Am Soc Agr Engrs; Am Water Resources Asn; Am Geophys Union; Am Soc Engr Educ. *Res:* Water quality management; design of water quality monitoring systems; on-site home sewage disposal; data use for regulatory water quality management. *Mailing Add:* Dept Agr & Chem Eng Colo State Univ Ft Collins CO 80523

WARD, ROBERT CLEVELAND, b Sparta, Tenn, Dec 7, 44; m 65; c 2. NUMERICAL ANALYSIS. *Educ:* Tenn Technol Univ, BS, 66; Col William & Mary, MS, 69; Univ Va, PhD(appl math), 74. *Prof Exp:* Mathematician, Langley Res Ctr, NASA, 66-74; res staff mem math, 74-77, HEAD MATH RES SECT, COMPUT SCI DIV, NUCLEAR DIV, UNION CARBIDE CORP, 77- *Mem:* Sigma Xi; AAAS; Soc Indust & Appl Math; Asn Comput Mach. *Res:* Developing, analyzing and improving numerical techniques in the areas of numerical linear algebra and computational statistics. *Mailing Add:* Math Sci Oak Ridge Nat Lab PO Box Y Bldg 9207A Oak Ridge TN 37830

WARD, ROBERT T, b Jersey City, NJ, Feb 7, 20; m 60. ZOOLOGY, CYTOLOGY. *Educ:* NJ State Teachers Col, AB, 42; Columbia Univ, MA, 52, PhD(zool), 60. *Prof Exp:* Res assoc zool, Columbia Univ, 52-57; asst prof 60-79, ASSOC PROF ANAT, STATE UNIV NY DOWNSTATE MED CTR, 79- *Concurrent Pos:* NSF res grant, 62-66. *Mem:* AAAS; Electron Micros Soc Am; Am Soc Cell Biol; Am Asn Anat. *Res:* Electron microscopy; histochemistry. *Mailing Add:* Dept Anat State Univ NY Downstate Med Ctr 450 Clarkson Ave Brooklyn NY 11203

WARD, ROGER WILSON, b Paris, Tex, Dec 2, 44; m 79; c 2. QUARTZ CRYSTAL RESONATORS & TRANSDUCERS. *Educ:* McMurry Col, BA, 67; Purdue Univ, MS, 69. *Prof Exp:* Mem tech staff, Hewlett-Packard, Palo Alto, Calif, 69-75; prod mgr, Litronix, Cupertino, Calif, 75-77; engr, Statek Corp, Orange, Calif, 77-79; vpres eng, Colo Crystal Corp, Loveland, Colo, 79-81; eng mgr, Motorola Inc, Ft Lauderdale, Fla, 81-83; vpres eng, Quartztronics Inc, 83-90, PRES, QUARTZDYNE INC, SALT LAKE CITY, UTAH, 90- *Mem:* Sr mem Inst Elec & Electronics Engrs; sr mem Inst Soc Am. *Mailing Add:* Quartzdyne Inc 1020 Atherton Dr C Salt Lake City UT 84123

WARD, RONALD ANTHONY, b New York, NY, Jan 25, 29; m 50; c 2. MEDICAL ENTOMOLOGY. *Educ:* Cornell Univ, BSc, 50; Univ Chicago, PhD(zool), 55; Univ London, MSc, 67. *Prof Exp:* Instr biol, Gonzaga Univ, 55-58; MED ENTOMOLOGIST, WALTER REED ARMY INST RES, 58- *Concurrent Pos:* US Secy of Army fel, London Sch Hyg & Trop Med, 66-67; res assoc, Smithsonian Inst, Washington, DC, 77-; ed, J Am Mosquito Control Asn, 81- *Mem:* Am Soc Trop Med & Hyg; Am Mosquito Control Asn; Royal Soc Trop Med & Hyg; Entom Soc Am. *Res:* Genetic and ecologic factors affecting susceptibility and resistance of arthropods to infectious agents; host adaptation of malaria parasites; mosquito biosystematics; mosquito control. *Mailing Add:* Dept Entom Walter Reed Army Inst Res Washington DC 20307-5100

WARD, RONALD WAYNE, b Johnson City, Tenn, Dec 17, 43; m 66. AGRICULTURAL ECONOMICS, ECONOMETRICS. *Educ:* Univ Tenn, BS, 65; Iowa State Univ, MS, 67, PhD(econ & statist), 70. *Prof Exp:* Res asst agr econ, Univ Tenn, 65; res asst, Iowa State Univ, 65-69; coop agent, USDA, 69-70; assoc prof, 70-80, PROF ECON, DEPT FOOD & RESOURCE ECON, UNIV FLA, 80- *Concurrent Pos:* Res economist, Fla Dept Citrus, 70-; USDA mkt struct res grant, Univ Fla, 71-72. *Mem:* Am Econ Asn; Am Agr Econ Asn. *Res:* Price analysis; marketing; advertising; market structures. *Mailing Add:* Univ Fla 1099 McCarty Hall Gainesville FL 32611

WARD, RONALD WAYNE, b Burbank, Calif, Sept 20, 44; m 76; c 1. APPLIED SEISMOLOGY, EARTHQUAKE SEISMOLOGY. *Educ:* Mass Inst Technol, BS, 66, PhD(geophysics), 71. *Prof Exp:* Res scientist, Tex Instruments, Inc, 65; res geophysicist, Ray Geophys, Inc, Houston, Tex, 66, Geosci, Inc, 67 & Lincoln Lab, 68; sr res scientist geophysics, Amoco Prod Co, Tulsa, Okla, 71-74; ASSOC PROF GEOPHYS, UNIV TEX, DALLAS, 74-, DIR, CTR LITHOSPHERIC STUDIES, 81- *Concurrent Pos:* Chmn, Tech Adv Comt Seismic Methods, Dept Energy, 77-; adv, External Proposal Rev Panel, US Geol Surv, 78-79, co-covener, Workshop Seismic Model for Geysers-Clear Lake Geothermal Region, Pajaro Dunes, Calif, 79; distinguished lectr, Phillips Petrol Seminars, 80. *Mem:* Am Geophys Union; Soc Explor Geophysicists; Seismol Soc Am; Europ Asn Explor Geophysicists; AAAS. *Res:* Three-dimensional and long-range seismic reflection/refracion surveys of the earth's crust; development of new signal processing techniques; seismic modeling techniques, and seismic inversion techniques; seismic studies of geothermal areas, especially the attenuation of seismic waves; seismic velocity studies in partially melted rock. *Mailing Add:* SOhio Tech Ctr 5400 LBJ Pkwy Suite 1200 Dallas TX 75240

WARD, ROSCOE FREDRICK, b Boise, Idaho, Dec 5, 30; m 63; c 2. CIVIL ENGINEERING. *Educ:* Col Idaho, BA, 53; Ore State Col, BS, 59; Wash State Univ, MS, 61; Wash Univ, DSc(environ & sanit eng), 64; Environ Engrs Intersoc, dipl, 69. *Prof Exp:* Design engr, Standard Oil Co Calif, 59-60; asst prof civil eng, Univ Mo-Columbia, 63-65 & Robert Col, Istanbul, 65-67; assoc prof, Asian Inst Technol, Bangkok, 67-68; assoc prof civil eng & assoc dean sch eng, Univ Mass, Amherst, 68-75; br chief, Fuels From Biomass Systs Br, Dept Energy, 75-79; with UN/World Bank, 79-83; PROF PAPER SCI & ENG, MIAMI UNIV, 83- *Concurrent Pos:* Prof, Istanbul Tech Univ, 66-67; consult, Democ, 66-67; prog mgr undergrad instrnl progs & res appl to nat needs, NSF, 72-73; prof civil eng, Bogazici Univ, Istanbul, 74-75; Fulbright prof, Univ Sao Paulo, Sao Carlos, Brazil, 86; scientist, CSIR - Div Forest Technol, Victoria, SAfrica, 90-91; Fulbright Lectr, Agricult Res Inst, Nicosia, Cyprus, 90. *Mem:* Am Acad Environ Eng; Am Soc Civil Engrs; Water Pollution Control Fedn. *Res:* Water supply, wastewater treatment, solid waste disposal and clean energy production from biomass. *Mailing Add:* Sch Appl Sci Miami Univ Oxford OH 45056

WARD, SAMUEL, b Los Angeles, Calif, Sept 29, 44; m 66; c 2. CELL BIOLOGY. *Educ:* Princeton Univ, AB, 65; Calif Inst Technol, PhD(biochem), 71. *Prof Exp:* Tutor biochem sci, Harvard Col, 73-74; asst prof biol, Med Sch, Harvard Univ, 72-77; from assoc prof to prof biol, Johns Hopkins Univ, 77-89; mem staff, Dept Embryol, Carnegie Inst Washington, 77-89; DEPT HEAD & PROF MOLECULAR & CELLULAR BIOL, UNIV ARIZ, 89- *Concurrent Pos:* NSF fel, Med Res Coun Lab Molecular Biol, Cambridge, Eng, 70-71, NIH spec fel, 71-72. *Mem:* Am Soc Genetics; AAAS; Am Soc Cell Biol; Am Soc Develop Biol; Soc Nematologists. *Res:* Genetic control of cell structure and morphology; nematode sperm development; parasite molecular biology. *Mailing Add:* Dept Molecular & Cellular Biol Univ Ariz Tucson AZ 85721

WARD, SAMUEL ABNER, b Binghamton, NY, Apr 27, 23; m 46; c 2. MATERIAL SCIENCE ENGINEERING. *Educ:* Cornell Univ, BEE, 44, MEE, 46, PhD(electronics, physics), 53. *Prof Exp:* Elec radio engr, US Naval Res Lab, 44-45; asst physics, Cornell Univ, 46-50; res engr, Electron Tubes, Nat Union Radio Corp, 50-51, A B Du Mont Labs, Inc, 51-52 & Electron Tubes & Devices, RCA Labs, Inc, 52-56; physicist, Gen Elec Co, 56-57; chief physicist, Machlett Labs, Inc, 57-61; sr eng physicist, Perkin-Elmer Corp, 61-64; sr scientist, CBS Labs, Stamford, Conn, 64-75; sr engr, Machlett Labs, Div Raytheon, 75-78; sr physicist, Coulter Systs Corp, Bedford, 78-80; prin scientist, res div, Raytheon Co, 80-81; SR MEM TECH STAFF, CHARLES STARK DRAPER LAB, CAMBRIDGE, MA, 84- *Mem:* Am Phys Soc; Inst Elec & Electronics Engrs; Sigma Xi. *Res:* Solid state physics; physical electronics; electron devices; sputtering of insulating films; photoelectric emission from semiconductors and metals; photoeffects in solids; ion implantation doping of semiconductors; electron and ion optics; processing of semiconductor devices and materials. *Mailing Add:* Talmadge Hill Rd S RD 2 Box 279 Waverly NY 14892

WARD, STANLEY HARRY, geophysics, for more information see previous edition

WARD, SUSAN A, b Norbury, Eng, Oct 1, 47. RESPIRATORY PHYSIOLOGY. *Educ:* Oxford Univ, PhD(physiol), 74. *Prof Exp:* ASSOC PROF ANESTHESIOL-PHYSIOL, UNIV CALIF, LOS ANGELES, 78- *Mem:* Am Physiol Soc; Am Col Sports Med; Am Thoracic Soc; Brit Physiol Soc; Europ Soc Clin Respiratory Physiol; Am Soc Gravitational & Space Biol. *Mailing Add:* Dept Anesthesiol & Physiol Univ Calif Ctr Health Sci Los Angeles CA 90024

WARD, THOMAS EDMUND, b Los Angeles, Calif, Nov 10, 44; div; c 1. NUCLEAR CHEMISTRY, NUCLEAR PHYSICS. *Educ:* Northeastern State Col, BSEd, 65; Univ Ark, Fayetteville, MS, 69, PhD(nuclear chem), 71. *Prof Exp:* Res assoc nuclear chem, Brookhaven Nat Lab, 70-72; staff chemist, Dept Physics, Ind Univ, Bloomington, 72-85; assoc physicist, 85-88, PHYSICIST, DEPT NUCLEAR ENERGY, BROOKHAVEN NAT LAB, 88- *Concurrent Pos:* Vis assoc chemist, Brookhaven Nat Lab, 72-75; vis fac mem, Dept Chem, Ind Univ, 81-85. *Mem:* Am Chem Soc; Am Phys Soc; NY Acad Sci; Sigma Xi; Am Inst Chemists. *Res:* Nuclear spectroscopy and radioactive decay; intermediate energy nuclear reactions; pion production; cosmic rays; environmental nuclear chemistry; radiation effects. *Mailing Add:* Dept Nuclear Energy Brookhaven Nat Lab Upton NY 11973

WARD, THOMAS J(ULIAN), b Amsterdam, NY, Aug 14, 30. CHEMICAL ENGINEERING, PROCESS CONTROL. *Educ:* Clarkson Univ BChE, 52; Univ Tex, MS, 56; Rensselaer Polytech Inst, PhD(chem eng), 59. *Prof Exp:* Reactor engr, Carbide & Carbon Chem, Oak Ridge, Tenn, 52-53; design engr, E I du Pont de Nemours & Co, Wilmington, Del, 53; asst to Dr E J Weiss, Austin, Tex, 53-56; mem staff, Rensselaer Polytech Inst, 56-59; from asst prof to assoc prof, 59-82, PROF CHEM ENG, CLARKSON UNIV, 82- *Mem:* Am Inst Chem Engrs; Am Nuclear Soc; sr mem Instrument Soc Am. *Res:* Process control; optimization; design; ceramic materials; nuclear engineering. *Mailing Add:* Dept Chem Eng Clarkson Univ Potsdam NY 13699-5705

WARD, TRUMAN L, b Ft Worth, Tex, Oct 21, 25; m 45; c 3. PHYSICS, PHYSICAL CHEMISTRY. *Educ:* Tulane Univ, BS, 48. *Prof Exp:* Assoc physicist, 48-63, RES PHYSICIST, SOUTHERN REGIONAL LAB, AGR RES SERV, USDA, 63- *Mem:* Am Chem Soc; Sigma Xi; fel Am Inst Chem; Am Asn Textile Chem & Colorists. *Res:* Physical properties of vegetable fats and oils; reaction mechanisms and kinetics; synthesis and reactions of thiorane and epoxy compounds; development of new instrumental procedures; glassification of cellulose; low temperature plasmas; polymers; ion exchanges. *Mailing Add:* 500 Walker St New Orleans LA 70124

WARD, WALLACE DIXON, b Pierre, SDak, June 30, 24; m 49; c 4. PSYCHOACOUSTICS. *Educ:* SDak Sch Mines & Technol, BS, 44; Harvard Univ, PhD(exp psychol), 53. *Hon Degrees:* ScD, SDak Sch Mines & Technol, 71. *Prof Exp:* Asst scientist, Rosemount Res Ctr, Univ Minn, 49; asst, Harvard Univ, 49-53; res engr, Baldwin Piano Co, 53-54; res scientist, Cent Inst Deaf, 54-57; res assoc subcomt noise, Comt Conserv Hearing, Am Acad Ophthal & Otolaryngol, 57-62; PROF OTOLARYNGOL & COMMUN DIS, UNIV MINN, MINNEAPOLIS, 62- & PROF ENVIRON HEALTH & PSYCHOL, 72- *Concurrent Pos:* Fels, Acoust Soc Am, 61 & Am Speech & Hearing Asn, 66; chmn exec coun, Comt Hearing, Bioacoust & Biomech, Nat Acad Sci-Nat Res Coun, 71-73; consult, US Army, 71-, Off Noise Abatement & Control, Environ Protection Agency, 72-73 & Air Transport Asn Am, 73-84; assoc ed, J Acoust Soc Am, 84-86. *Mem:* Am Otol Soc; Acoust Soc Am (vpres, 86-87, pres, 88-89); Int Soc Audiol (vpres, 76-78, pres, 78-80); Am Audiol Soc (vpres, 73-75, pres, 76-77); Int Comt Biol Effects Noise; Soc Res Psychol Music & Music Educ. *Res:* Auditory fatigue and noise-induced hearing loss; musical perception; musical psychoacoustics. *Mailing Add:* Hearing Res Lab 2630 University Ave SE Minneapolis MN 55414

WARD, WALTER FREDERICK, b Darlington, Wis, June 23, 40; m 59; c 3. PHYSIOLOGY, ENDOCRINOLOGY. *Educ:* Univ Wis-Platteville, BSc, 64; Marquette Univ, PhD(physiol), 70. *Prof Exp:* Asst prof physiol, Col Med, Pa State Univ, 73-78; asst prof, 78-80, ASSOC PROF PHYSIOL, HEALTH SCI CTR, UNIV TEX, SAN ANTONIO, 80- *Concurrent Pos:* USPHS fel, Brown Univ, 71-73. *Mem:* AAAS; Am Physiol Soc; Am Geront Soc. *Res:* Regulation of protein metabolism; mechanisms of hormone action; aging and protein metabolism. *Mailing Add:* Dept Physiol Univ Tex Health Sci Ctr San Antonio TX 78284

WARD, WILLIAM CRUSE, b Waco, Tex, Apr 26, 33; m 57; c 3. SEDIMENTARY PETROLOGY. *Educ:* Univ Tex, Austin, BS, 55, MA, 57; Rice Univ, PhD(geol), 70. *Prof Exp:* Geologist, Humble Oil & Refining Co, 57-66; asst prof, 70-73, assoc prof, 73-78, PROF GEOL, UNIV NEW ORLEANS, 78- *Mem:* Soc Econ Paleontologists & Mineralogists; Am Asn Petrol Geologists. *Res:* Petrology and diagenesis of Quaternary limestones of eastern Yucatan; sandstone petrology and diagenesis. *Mailing Add:* 5542 Chatham Dr New Orleans LA 70122

WARD, WILLIAM FRANCIS, b Erie, Pa, June 19, 28. PARASITOLOGY. *Educ:* Gannon Col, BS, 50; Univ Notre Dame, MS, 55. *Prof Exp:* Instr biol, Col St Mary, Utah, 55-57; asst prof, Rosemont Col, 57-66, from actg chmn dept to chmn dept, 64-72, chmn div natural sci & math, 76-79 & 87-90, ASSOC PROF BIOL, ROSEMONT COL, 66- *Mem:* Am Soc Microbiol; Sigma Xi; AAAS; Am Soc Parasitol; Am Inst Biol Sci. *Res:* Interrelationships between intestinal parasites and the bacterial flora. *Mailing Add:* Dept Biol Rosemont Col Rosemont PA 19010

WARD, WILLIAM J, III, b Paterson, NJ, Oct 4, 39; m 62; c 2. CHEMICAL ENGINEERING. *Educ:* Pa State Univ, BS, 61; Univ Ill, MS, 63, PhD(chem eng), 65. *Prof Exp:* RES ENGR, GEN ELEC CO, 65- *Mem:* Nat Acad Eng; AAAS; Am Inst Chem Engrs. *Res:* Research and development of membrane separation processes; catalysis. *Mailing Add:* 1924 Hexam Rd Schenectady NY 12309

WARDE, CARDINAL, b Barbados, July 14, 45. SOLID STATE PHYSICS, OPTICAL ENGINEERING. *Educ:* Stevens Inst Technol, BSc, 69; Yale Univ, MPhil, 71, PhD(physics), 74. *Prof Exp:* Asst prof, 74-79, ASSOC PROF ELEC ENG, MASS INST TECHNOL, 79- *Concurrent Pos:* Vinton Hayes fel, Mass Inst Technol, 75-76; prin investr, NSF grants, 76- & Air Force Off Sci Res grant, 77-; consult, Lincoln Lab, Mass Inst Technol, 77-, Rome Air Develop Ctr, 81- & Hamamatsu TV Co, Japan, 80-; State Univ NY res grant, 80-81. *Mem:* Optical Soc Am; Inst Elec & Electronics Engrs; Soc Photo-Optical Instrumentation Engrs. *Res:* Optical signal processing and storage devices; adaptive optical systems; optical properties of electron-beam-addressed materials. *Mailing Add:* Dept Elec Eng Mass Inst Technol Cambridge MA 02139

WARDELL, JOE RUSSELL, JR, b Omaha, Nebr, Nov 11, 29; m 52; c 3. PHARMACOLOGY, PHYSIOLOGY. *Educ:* Creighton Univ, BS, 51; Univ Nebr, MSc, 59, PhD(pharmacol, physiol), 62. *Prof Exp:* Sr pharmacologist, Smith Kline & French Labs, 62-65, group leader pharmacol, 64-68, asst dir pharmacol, 68-71, assoc dir pharmacol & mission dir cardiopulmonary res area, 71-75, assoc dir biol res & mission dir cardiovasc res area, 75-78, sci dir new prod eval, Res & Mission dir cardiovasc res area, 78-81, dir new compound eval, 81-85, dept dir, Res & Develop Compound Acquisitions, 85-86; PRES, WARDELL ASSOCS, 86- *Mem:* AAAS; Am Chem Soc; NY Acad Sci; Am Soc Pharmacol & Exp Therapeut; Am Acad Allergy. *Res:* Cardiovascular and respiratory pharmacology; immunopharmacology; autonomic pharmacology; regulation of biosynthesis and secretion of respiratory mucus. *Mailing Add:* Wardell Assocs 105 Oak St Beaver PA 15009

WARDELL, WILLIAM MICHAEL, b Christchurch, NZ, Nov 15, 38; m 65; c 2. CLINICAL PHARMACOLOGY. *Educ:* Oxford Univ, BA, 61, DPhil(pharmacol), 64, BM & BCh, 67, DM, 73. *Prof Exp:* Intern med, Radcliffe Infirmary, Oxford, 67; intern med & surg, Dunedin Hosp, Univ Otago, NZ, 68; med res officer clin pharmacol & toxicol, NZ Med Res Coun, 69; lectr clin pharmacol, Med Sch, Univ Otago, NZ, 70; instr clin pharmacol, Univ Rochester, 71-73, from asst prof to assoc prof pharmacol, 73-83; VPRES & MED DIR, BOEHRINGER INGELHEIM PHARMACEUTICALS, INC, 83- *Concurrent Pos:* Hon clin asst, Otago Hosp Bd, Dunedin, NZ, 69-70; co-founder & dir, Ctr Study Drug Develop, Univ Rochester Med Ctr, 75-83. *Honors & Awards:* Christopher Welch Prize Biol. *Mem:* Am Soc Pharmacol & Exp Therapeut; Am Soc Clin Pharmacol & Therapeut; Am Col Clin Pharmacol; Australasian Soc Clin & Exp Pharmacol; AMA. *Res:* Design, methodology and analysis of drug studies in man; analgesic and hypnotic drugs in man; regulation and drug development; adverse drug reactions. *Mailing Add:* Boehringer Ingelheim Pharmaceut 90 East Ridge PO Box 368 Ridgefield CT 06877

WARDEN, HERBERT EDGAR, b Cleveland, Ohio, Aug 30, 20; m 58; c 4. SURGERY, THORACIC SURGERY. *Educ:* Washington & Jefferson Col, BS, 42; Univ Chicago, MD, 46; Am Bd Surg, dipl, 58; Bd Thoracic Surg, dipl, 63. *Prof Exp:* Intern, Clin, Univ Chicago, 46-47; asst resident surg, Hosps, Univ Minn, 51-56, res asst, 53-55, res asst physiol, 55-56, clin instr surg, 55-57, chief resident, 56-57, instr surg, 57-60; assoc prof, 60-62, PROF SURG, MED CTR, WVA UNIV, 62- *Concurrent Pos:* Coordr, USPHS Cardiovasc Surg Training Prog, Univ Minn, 56-60; consult, Anoka State Hosp, Minn, 58-59. *Honors & Awards:* Cert of Merit, AMA, 55 & 58, Hektoen Gold Medal, 57; Lasker Award, Am Pub Health Asn, 55. *Mem:* Soc Univ Surgeons; Am Asn Thoracic Surgeons; Am Surg Asn; Soc Thoracic Surgeons; fel Am Col Surgeons; Am Med Asn; fel Am Col Cardiol. *Res:* Cardiovascular surgery and physiology. *Mailing Add:* Dept Surg WVa Univ Med Ctr Morgantown WV 26506

WARDEN, JOSEPH TALLMAN, b Huntington, WVa, Aug 7, 46; c 1. BIOPHYSICAL CHEMISTRY. *Educ:* Furman Univ, BS, 68; Univ Minn, PhD(phys chem), 72. *Prof Exp:* Vis scientist biophys, State Univ Leiden, 72-73; chemist, Univ Calif, Berkeley, 73-75; from asst prof to assoc prof, 75-88, PROF CHEM, RENSSELAER POLYTECH INST, 88- *Concurrent Pos:* Vis prof, Univ Col, London, 81; Carnegie fel, Int Conf Educ Chem, 86. *Mem:* AAAS; Am Chem Soc; Am Soc Photobiol; Biophys Soc. *Res:* Electron spin resonance investigations of electron transfer components and mechanisms in photosynthesis and mitochondrial respiration; artificial intelligence applications in chemistry; photochemistry; solid state chemistry; microelectronics. *Mailing Add:* Dept Chem Rensselaer Polytech Inst Troy NY 12180-3590

WARDER, DAVID LEE, b Akron, Ohio, June 17, 40; m 63; c 2. CIVIL ENGINEERING. *Educ:* Univ Akron, BSCE, 63; Mich State Univ, MS, 65, PhD(civil eng), 69. *Prof Exp:* Assoc prof civil eng, Tri-State Col, 69-73; SR ENGR, ATEC ASSOCS, INC, 76- *Mem:* Am Soc Civil Engrs; Int Soc Soil Mech & Found Engr. *Res:* Use of artificially frozen soil for temporary support; structure-foundation interaction. *Mailing Add:* 11344 Fieldstone Ct Carmel IN 46032

WARDER, RICHARD C, JR, b Nitro, WVa, Sept 30, 36; m 81; c 2. ASTRONAUTICAL SCIENCES, MECHANICAL ENGINEERING. *Educ:* SDak Sch Mines & Technol, BS, 58; Northwestern Univ, MS, 59, PhD(mech eng, astronaut sci), 63. *Prof Exp:* Asst prof mech eng & astronaut sci, Northwestern Univ, 63-65; mgr energy processes res, Space Sci Labs, Litton Industs, Inc, Calif, 65-68; assoc prof, 68-72, PROF MECH & AEROSPACE ENG, UNIV MO, COLUMBIA, 72-, CHMN DEPT, 88- *Concurrent Pos:* Prof staff mem, NSF, 74-76. *Mem:* AAAS; Am Soc Eng Educ; Am Phys Soc; Am Inst Aeronaut & Astronaut; Am Soc Mech Engrs; Am Asn Aerosol Res. *Res:* Aerosol and particulate mechanics; gas dynamics; plasma physics. *Mailing Add:* Dept Mech & Aerospace Eng Univ Mo Columbia MO 65211

WARDESKA, JEFFREY GWYNN, b Irondale, Ohio, June 13, 41; m 65; c 2. INORGANIC CHEMISTRY, BIOINORGANIC CHEMISTRY. *Educ:* Mt Union Col, BSc, 63; Ohio Univ, PhD(inorg chem), 67. *Prof Exp:* From asst prof to prof, 67-79, PROF CHEM, E TENN STATE UNIV, 79- *Concurrent Pos:* Vis prof, Univ NH, 84-85. *Mem:* Am Chem Soc; Sigma Xi. *Res:* Coordination chemistry; reactions of coordination compounds of transition metals; aminoalcohol complexes; metal ion binding to metalloproteins. *Mailing Add:* Dept Chem E Tenn State Univ Johnson City TN 37614

WARDLAW, JANET MELVILLE, b Toronto, Ont, June 20, 24. NUTRITION, HOME ECONOMICS. *Educ:* Univ Toronto, BA, 46; Univ Tenn, MS, 50; Pa State Univ, PhD(nutrit), 63. *Prof Exp:* Dietition, Can Red Cross Soc, Toronto, 47-49; nutritionist, Mich Dept Health, 50-53 & Toronto Dept Pub Health, 53-56; asst prof nutrit, Fac Household Sci, Univ Toronto, 56-60 & 63-64, assoc prof, Fac Food Sci, 64-66; assoc dean-dean designate, Univ Guelph, 67-68, dean, 69-83, prof nutrit, Col Family & Consumer Studies, 66-87, assoc vpres acad, 84-87; RETIRED. *Concurrent Pos:* Chmn bd gov, Int Devel Res Ctr, Ottawan, Can, 85-; hon fel, Univ Guelph, 89. *Honors & Awards:* Stuart's Branded Foods Ltd Award, Can Dietetic Asn, 71. *Mem:* AAAS; Nutrit Soc Can; Can Dietetic Asn (treas, 65-67). *Res:* Sodium regulation during pregnancy; body composition and feeding frequency; community nutrition. *Mailing Add:* 20 Suffolk St W Guelph ON N1H 2H8 Can

WARDLAW, NORMAN CLAUDE, b Trinidad, Brit WI, Nov 22, 35. GEOLOGY. *Educ:* Univ Manchester, BSc, 57; Univ Glasgow, PhD(geol), 60. *Prof Exp:* Spec lectr sedimentation, Univ Sask, 60-62, from asst prof to assoc prof geol, 62-72; from assoc prof to prof geol, 72-85, HEAD DEPT GEOL & GEOPHYS, UNIV CALGARY, 85- *Concurrent Pos:* Res fel, dept geol, Univ Manchester, 70-71, res assoc, Soc Nat Elf Aquitaine, France. *Honors & Awards:* Link Award, Can Petrol Geologists. *Mem:* Soc Econ Paleontologists & Mineralogists; Geol Soc Am; Am Asn Petrol Geologists; Soc Petrol Engrs. *Res:* Petrophysics of oil and gas reservoirs and petroleum recovery. *Mailing Add:* Dept Geol & Geophys Univ Calgary 2500 University Dr Calgary AB T2N 1N4 Can

WARDLAW, WILLIAM PATTERSON, b Los Angeles, Calif, Mar 3, 36; m 63; c 4. MATHEMATICS. *Educ:* Rice Inst, BA, 58; Univ Calif, Los Angeles, MA, 64, PhD(math), 66. *Prof Exp:* Asst prof, Univ Ga, 66-72; asst prof, 72-84, ASSOC PROF MATH, US NAVAL ACAD, 84- *Mem:* Am Math Soc; Math Asn Am. *Res:* Lie algebras and Chevalley groups; universal algebra. *Mailing Add:* Dept Math US Naval Acad Annapolis MD 21402

WARDLE, JOHN FRANCIS CARLETON, b Hemel Hempstead, Eng, May 8, 45; m 71. RADIO ASTRONOMY. *Educ:* Univ Cambridge, BA, 66; Univ Manchester, MSc, 68, PhD(radio astron), 69. *Prof Exp:* Res asst radio astron, Nat Radio Astron Observ, 69-71; from instr to assoc prof, 71-86, PROF ASTROPHYS, BRANDEIS UNIV, 86- *Mem:* Am Astron Soc; Royal Astron Soc. *Res:* Extragalactic radio astronomy; cosmology. *Mailing Add:* Dept Astrophys Brandeis Univ 415 South St Waltham MA 02254

WARD-MCLEMORE, ETHEL, b Sylvarena, Miss, Jan 22, 08; m 35; c 1. FORTRAN PROGRAMMING, BIBLIOGRAPHIES OF SEDIMENTARY BASINS OF THE APPALACHIAN-OUACHITA OROGEN & CHINA. *Educ:* Miss Woman's Col, BA, 28; Univ NC, MA, 29. *Prof Exp:* Head math, E Miss Jr Col, 29-30; instr chem & math, Miss State Col for Women, 30-33; mathematician, math & geophys, Humble Oil & Refining Co, 33-37; tutorial math, Southern Methodist Univ, 39-41; res geophysicist, United Geophys Co, 42-46; CONSULT GEOPHYS, 48- *Concurrent Pos:* Dir, Geol Info Libr Dallas, 78-88 & Tex Acad Sci, 83-86. *Mem:* Am Math Soc; Am Chem Soc; Am Geophys Union; hon mem Soc Explor Geophysicists; Geol Soc Am; Math Asn Am; assoc mem Sigma Xi. *Res:* System ferrous sulphate, sulfuric acid and water; interpretation of probable structure from analysis of geophysical data; application of Hughen's Principle to the reflection of seismic waves at a free surface; history of the Academies of Science to Texas. *Mailing Add:* 8600 Skyline Dr No 1107 Dallas TX 75243

WARDNER, CARL ARTHUR, organic chemistry; deceased, see previous edition for last biography

WARDOWSKI, WILFRED FRANCIS, II, b Pontiac, Mich, May 23, 37; m 74; c 1. HORTICULTURE. *Educ:* Mich State Univ, BS, 59, MS, 61, PhD(pomol), 66. *Prof Exp:* Foreman fruit prod, Blossom Orchard, Leslie, Mich, 61-63; midwest rep tech exten agr chem, Agr Div, Upjohn Co, 66-69; assoc prof, 69-80, PROF EXTEN SERV, CITRUS HARVESTING & HANDLING, AGR RES & EDUC CTR, UNIV FLA, 80- *Concurrent Pos:* Consult, UN Food & Agr Orgn develop prog, Bhutan, 80. *Mem:* Am Soc Hort Sci. *Res:* Nutrition and histology of apples; agricultural chemicals research and development; harvesting and handling of fresh market citrus. *Mailing Add:* 1293 Mirror Terr Winter Haven FL 33881

WARE, ALAN ALFRED, b Portsmouth, Eng, Dec 4, 24; US citizen; m 52; c 4. PLASMA PHYSICS. *Educ:* Imp Col, Univ London, BSc & ARCS, 44, PhD(physics) & DIC, 49. *Prof Exp:* Res asst plasma physics, Imp Col, Univ London, 47-51; sect leader, Res Lab, Assoc Elec Industs, UK, 51-63; consult, Gen Atomic, San Diego, 60-61; group leader plasma physics, Culham Lab, UK Atomic Energy Authority, 63-65; asst mgr res oper plasma physics, Aerojet Gen Corp, 65-69; SR RES SCIENTIST PLASMA PHYSICS, UNIV TEX, AUSTIN, 69- *Concurrent Pos:* Consult, Los Alamos Sci Lab, 71- *Mem:* Am Phys Soc. *Res:* Theory of Tokamak plasmas in research aimed at controlled nuclear fusion power. *Mailing Add:* Dept Physics Univ Tex Austin TX 78712

WARE, BRENDAN J, b Dublin, Ireland, Aug 27, 32. ELECTRICAL POWER TRANSMISSION, SYSTEM ENGINEERING. *Educ:* Nat Univ Ireland, BE, 54; NJ Inst Technol, MSEE, 67. *Prof Exp:* Engr, eng & res, 60-76, MGR, ELEC RES & DEVELOP DIV, AM ELEC POWER & SERV, COLUMBUS, OHIO, 76- *Mem:* Fel Inst Elec & Electronics Engrs. *Mailing Add:* Am Elec Power Serv One Riverside Plaza Columbus OH 43216

WARE, CARL F, b Fullerton, Calif, May 23, 51. T-CELL MEDIATED IMMUNOLOGY. *Educ:* Univ Calif, Irvine, BS, 74, PhD(molecular biol & biochem), 79. *Prof Exp:* ASST PROF IMMUNOL, UNIV CALIF, RIVERSIDE, 82- *Mem:* AAAS; Am Asn Immunologists. *Mailing Add:* Div Biomed Sci Univ Calif Riverside CA 92521-0121

WARE, CAROLYN BOGARDUS, b Baltimore, Md, Oct 15, 30; div; c 1. NEUROANATOMY, PHYSIOLOGICAL PSYCHOLOGY. *Educ:* Western Reserve Univ, BA, 52; Columbia Univ, cert phys ther, 53; Univ Buffalo, MEd, 63; Duke Univ, PhD(psychol), 71. *Prof Exp:* Instr phys ther, Univ NC, 63-66; instr anat, State Univ NY Downstate Med Ctr, 70-75, asst prof & asst dean, Sch Health Related Professions, State Univ NY, Buffalo, 75-78; asst vpres for acad affairs, State Univ NY, Binghamton, 78-; DEAN ACAD AFFAIRS, CAZENOVIA COL. *Mem:* Soc Neurosci; Am Asn Univ Professors; Sigma Xi. *Res:* Comparative neuroanatomy, visual system and cerebellum; central nervous system and behavior, especially visual discrimination. *Mailing Add:* Acad Affairs State Univ NY Binghamton NY 13901

WARE, CHARLES HARVEY, JR, b New York, NY, July 8, 27; m 52; c 2. CHEMICAL ENGINEERING. *Educ:* Princeton Univ, BSE, 49; Univ Pa, MS, 57, PhD(chem eng), 59. *Prof Exp:* Chem engr, Texaco Res Ctr, 59-60, sr chem engr, 60-61, group leader exp lube process res, 61-65, sr res chem engr, 65-70, group leader process anal, 70-74; CONSULT, COMMERCIALIZATION INSIGHTS, 74- *Concurrent Pos:* Adj assoc prof, Columbia Univ, 68-69; adj prof, Manhattan Col, 78-80. *Mem:* Fel Am Inst Chem Engrs; Am Chem Soc; Am Statist Asn. *Res:* Chemical reaction engineering; research methodology; mathematical modelling; design methods; petroleum processing. *Mailing Add:* 1507 Quorier St Charleston WV 25311-2407

WARE, CHESTER DAWSON, mechanical engineering, heat transfer; deceased, see previous edition for last biography

WARE, DONNA MARIE EGGERS, b Springfield, Mo, Oct 1, 42; m 68. PLANT TAXONOMY. *Educ:* Southwest Mo State Col, BA, 64; Vanderbilt Univ, PhD(biol), 69. *Prof Exp:* HERBARIUM CUR VASCULAR PLANTS, COL WILLIAM & MARY, 69-, CUR, HERBARIUM. *Concurrent Pos:* NSF grant-in-aid, Highlands Biol Sta, NC, 70. *Mem:* Am Soc Plant Taxon. *Res:* Floristics; revisional and biosystematic taxonomy. *Mailing Add:* Herbarium Dept Biol Col William & Mary Williamsburg VA 23185

WARE, FREDERICK, b Omaha, Nebr, June 16, 28; m 50, 78; c 5. PHYSIOLOGY, INTERNAL MEDICINE. *Educ:* Univ Nebr, BS, 49, MS, 53, PhD & MD, 56. *Prof Exp:* Instr physiol, 53-56, instr internal med, 55-56, asst prof physiol & pharmacol, 60-62, assoc prof physiol & asst prof internal med, 62-70, PROF PHYSIOL, BIOPHYS & INTERNAL MED, COL MED, UNIV NEBR, OMAHA, 70- *Mem:* Am Physiol Soc; Am Soc Nephrology; Int Soc Nephrology; Am Col Physicians; Am Soc Artificial Internal Organs. *Res:* Membrane electrophysiology of skeletal muscle and heart; principles of electrocardiography; biophysics of renal function. *Mailing Add:* Dept Physiol-Biophys Univ Nebr Col Med 650 Doctors Bldg Omaha NE 68105

WARE, GEORGE HENRY, b Avery, Okla, Apr 27, 24; m 55; c 4. PLANT ECOLOGY. *Educ:* Univ Okla, BS, 45, MS, 48; Univ Wis, PhD(bot), 55. *Prof Exp:* From asst prof to prof bot, Northwestern State Univ, 48-68; DENDROLOGIST, MORTON ARBORETUM, 68-, RES GROUP ADMINR, 76-; ADMINR, URBAN VEGETATION LAB, 85- *Concurrent Pos:* Res group adminr, Morton Arboretum, 76- *Mem:* Ecol Soc Am; Int Soc Arboricult; Am Inst Biol Sci. *Res:* ecology of swamp and floodplain forests; ecology of urban trees. *Mailing Add:* 573 59th Lisle IL 60532-3102

WARE, GEORGE WHITAKER, JR, b Pine Bluff, Ark, Aug 27, 27; m 52; c 3. ENTOMOLOGY. *Educ:* Univ Ark, BS, 51, MS, 52; Kans State Univ, PhD(entom), 56. *Prof Exp:* Assoc prof entom, Ohio State Univ, 56-66; prof entom & head dept, 67-83, ASSOC DIR, AGR EXP STA, UNIV ARIZ, 83- *Concurrent Pos:* Consult, Environ Protection Agency, Off Pesticide Progs Washington, USAID, Univ Calif, Berkeley, 78- & Nat Agr Chem Asn, 80-82; Coop States Res Serv, USDA; expert witness pesticide drift litigation, PR, Fla & Mass, 84-88; ed, Rev Environ Contamination & Toxicol. *Mem:* Entom Soc Am; Am Chem Soc; Soc Toxicol. *Res:* Insecticide toxicology; pesticide chemistry, metabolism and residues; application of pesticides; pesticide drift reduction. *Mailing Add:* Col Agr Univ Ariz Tucson AZ 85721

WARE, GLENN OREN, b Athens, Ga, Dec 8, 41; m 67; c 2. APPLIED STATISTICS, OPERATIONS RESEARCH. *Educ:* Univ Ga, BSF, 63, PhD(forest biomet), 68; Yale Univ, MF, 64. *Prof Exp:* Res forester, Hudson Pulp & Paper Corp, 64-65; asst prof statist, 66-74, ASSOC PROF FORESTRY, UNIV GA, 77-, STA STATISTICIAN, EXP STA, 74- *Mem:* Soc Am Foresters. *Res:* Application of mathematical and statistical techniques in the physical and biological sciences. *Mailing Add:* Sch Forest Resources Univ Ga Athens GA 30602

WARE, JAMES GARETH, b Baltimore, Md, Aug 19, 29; m 55; c 1. MATHEMATICS. *Educ:* Duke Univ, BS, 50; George Peabody Col, MA, 51, PhD(math), 62. *Prof Exp:* Teacher, McCallie Sch, Tenn, 52-54 & 59-60, dept chmn, 60-65; assoc prof, 67-73, chmn dept, 68-90, PROF MATH, UNIV TENN, CHATTANOOGA, 73- *Mem:* Nat Coun Teachers Math; Am Asn Univ Professors; Math Asn Am. *Res:* Mathematics education; geometry. *Mailing Add:* Dept Math Univ Tenn Chattanooga TN 37403

WARE, JAMES H, b Detroit, Mich, Oct 27, 41; m 72; c 1. MATHEMATICAL STATISTICS. *Educ:* Yale Univ, BA, 63; Stanford Univ, MA, 65, PhD(statist), 69. *Prof Exp:* Instr statist, Calif State Col, Hayward, 69-70; math statistician, NIH, 71-80; MEM FAC, DEPT BIOSTATIST, HARVARD SCH PUB HEALTH, 80- *Concurrent Pos:* Assoc ed, J Am Statist Asn, 73-75; adj prof statist, George Washington Univ, 75- *Mem:* Am Statist Asn; Biomet Soc. *Res:* Interest in the areas of nonparametric methods and sequential analysis; survival data analysis and methods for data analysis in clinical trials of chronic disease. *Mailing Add:* Dept Biostatist Harvard Sch Pub Health 677 Huntington Ave Boston MA 02115

WARE, KENNETH DALE, b Webster Springs, WVa, Aug 30, 35; m 57; c 2. FOREST BIOMETRICS, SURVEY SAMPLING. *Educ:* WVa Univ, BS, 56; Yale Univ, MS, 57, PhD(biomet), 60. *Prof Exp:* Res forester, Northeastern Forest Exp Sta, US Forest Serv, 58-61; asst prof forestry, Iowa State Univ, 61-64, assoc prof, 65-68, prof, 69-71; chief mensurationist, Southeastern Forest Exp Sta, US Forest Serv, 71-88; CONSULT, 88- *Concurrent Pos:* Vis scientist, NSF & Soc Am Forests; adj prof, Sch Forest Resources, Univ Ga, 71-; Tasman fel, Univ Canterbury, NZ, 84; vis fel, Autralian Nat Univ, Canberra, 84. *Mem:* Soc Am Forests; Int Union Forestry Res Orgn. *Res:* Concepts and techniques for forest sampling, including surveys on successive occasions; special unequal probability procedures; multi-phase sampling structures and sampling for predicting forest growth and yield under timber management and for monitoring responses to management. *Mailing Add:* 195 Deerfield Rd Bogart GA 30622

WARE, LAWRENCE LESLIE, JR, b Montgomery, WVa, Sept 12, 20; m 46; c 3. MEDICAL MICROBIOLOGY. *Educ:* Roosevelt Univ, BS, 50. *Prof Exp:* Res asst bact, Univ Chicago, 46-49; bacteriologist, Med Bact Div, US Army Chem Corp, Ft Detrick, Md, 51-53, Process Res Div, 53-57, bio-eng br, Pilot Plant, 59; lab dir, Div Indian Health, USPHS, Ariz, 59-64, area lab dir, 63-65; microbiologist & sci info specialist, 65-72, chief, Sci & Tech Info Div, US Army Med Res & Develop Command, 72-86; RETIRED. *Concurrent Pos:* Ed, newsletter, Nat Registry Microbiol, 65-72. *Mem:* Am Soc Microbiol; NY Acad Sci; Asn Mil Surg US; Soc Indust Microbiol; Sigma Xi; fel Am Acad Microbiol. *Res:* Medical microbiology and immunology; microbial fermentations; medical information and documentation storage and retrieval. *Mailing Add:* 9224 Kristin Lane Fairfax VA 22032

WARE, ROGER PERRY, b San Francisco, Calif, Apr 2, 42; m 65; c 2. ALGEBRA. *Educ:* Univ Calif, Berkeley, AB, 65; Univ Calif, Santa Barbara, MA, 68, PhD(math), 70. *Prof Exp:* Asst prof math, Northwestern Univ, Evanston, 70-72 & Univ Kans, Lawrence, 72-74; assoc prof, 74-80, PROF MATH, PA STATE UNIV, 80- *Concurrent Pos:* NSF grants, 71-72, 73, 74 & 76-82; vis prof math, Univ Calif, Berkely, 80-81; prin invest NSA grant, 88-89. *Mem:* Am Math Soc; Math Asn Am. *Res:* Quadratic forms; field theory. *Mailing Add:* Dept Math Pa State Univ University Park PA 16802

WARE, STEWART ALEXANDER, b Stringer, Miss, Aug 20, 42; m 68. PLANT ECOLOGY. *Educ:* Millsaps Col, BA, 64; Vanderbilt Univ, PhD(biol), 68. *Prof Exp:* From asst prof to assoc prof, 67-82, chmn dept, 76-82, PROF BIOL, COL WILLIAM & MARY, 82- *Concurrent Pos:* Ed, Jeffersonia, Va Bot Newslett, 69-74 & Va J Sci, 79-84; vis prof bot, Univ Ark, 84. *Mem:* Ecol Soc Am; Sigma Xi; Bot Soc Am; Torrey Bot Club; Int Asn Veg Sci. *Res:* Vegetation of the southeastern United States; Quercus systematics and ecology; physiological ecology of rock outcrop plants; ecology and distribution of Talinum. *Mailing Add:* Dept Biol Col William & Mary Williamsburg VA 23185

WARE, VASSIE C, RIBOSOMAL RNA, RNA PROCESSING. *Educ:* Yale Univ, PhD(biol), 81. *Prof Exp:* ASST PROF BIOL, LEHIGH UNIV, 85- *Mailing Add:* Dept Biol & Med Brown Univ Box G 02912 Lab Providence RI 02906

WARE, W(ILLIS) H(OWARD), b Atlantic City, NJ, Aug 31, 20; m 43; c 3. COMPUTER SCIENCE, HARDWARE SYSTEMS. *Educ:* Univ Pa, BS, 41; Mass Inst Technol, SM, 42; Princeton Univ, PhD(elec eng), 51. *Prof Exp:* Res engr, Hazeltine Electronics Corp, 42-46, Princeton Inst Advan Study, 46-51; head comput sci dept, 51-71, dep vpres, 71-73, CORP RES STAFF, RAND CORP, 73- *Concurrent Pos:* chmn, HEW, Secy's Adv Comn Automated Personal Data Systs, 71-73; mem & vchmn, Privacy Protection Study Comn, 75-77. *Honors & Awards:* Except Civilian Serv Medal, USAF, 79. *Mem:* Nat Acad Eng; AAAS; fel Inst Elec & Electronics Engrs; Asn Comput Mach. *Res:* Electronic digital computers; applications of computers to military and civil information processing problems; computer system research; societal impact of information technology. *Mailing Add:* RAND Corp 1700 Main St Santa Monica CA 90407-2138

WARE, WALTER ELISHA, b Jacksonville, Fla, June 1, 33; m 55; c 3. PHYSICS. *Educ:* US Naval Acad, BS, 55; Univ Colo, PhD(physics), 62. *Prof Exp:* From instr to assoc prof physics, US Air Force Acad, 62-65, tenure assoc prof, 65-66; res scientist, Nuclear Technol Lab, Kaman Sci Corp, 66-80; DIV MGR, MISSION RES CORP, 80- *Concurrent Pos:* Proj consult, Kaman Nuclear, 62-66. *Mem:* Am Asn Physics Teachers; Inst Elec & Electronics Engrs. *Res:* Nuclear structure theory; effects of nuclear weapons; electromagnetic theory; quantum theory. *Mailing Add:* Mission Res Corp 4935 N 30 St Colorado Springs CO 80919

WARE, WILLIAM ROMAINE, b Portland, Ore, June 13, 31; m 54. PHYSICAL CHEMISTRY. *Educ:* Reed Col, BA, 53; Univ Rochester, PhD, 58. *Prof Exp:* Res chemist, Parma Res Ctr, Union Carbide Corp, 57-60; res assoc chem, Univ Minn, Minneapolis, 61; from asst prof to assoc prof, San Diego State Col, 62-66; from assoc prof to prof, Univ Minn, Minneapolis, 67-71; PROF CHEM, UNIV WESTERN ONT, 71- *Mem:* fel Chem Inst Can. *Res:* Molecular photochemistry and photophysics. *Mailing Add:* 14 Metamora Crescent London ON N6G 1R3 Can

WAREHAM, ELLSWORTH EDWIN, b Avinger, Tex, Oct 3, 14; m 50; c 5. THORACIC SURGERY. *Educ:* Col Med Evangelists, MD, 42; Am Bd Surg, dipl, 54; Bd Thoracic Surg, dipl, 55. *Hon Degrees:* LLD, Andrews Univ, 72. *Prof Exp:* From asst prof to prof, 58-, EMER PROF SURG, SCH MED, LOMA LINDA UNIV. *Mem:* Am Asn Thoracic Surgeons; Am Col Surgeons; Am Col Cardiol; Soc Thoracic Surg; AMA. *Res:* Cardiovascular surgery; open heart surgery; use of heart-lung machine. *Mailing Add:* 38985 Harris Rd Yucaipa CA 92399

WAREN, ALLAN D(AVID), b Toronto, Ont, Nov 23, 35; m 62; c 4. OPERATIONS RESEARCH, COMPUTER SCIENCE. *Educ:* Univ Toronto, BASc, 60; Case Inst Technol, MS, 62, PhD(eng), 64. *Prof Exp:* Electronics engr, Electronic Res Div, Clevite Corp, 63-64, sr electronics engr, 64-66, staff engr, 66; from asst prof to assoc prof elec eng, Cleveland State Univ, 66-69; founder & pres, Com-Share Ltd, 69-71; chmn dept, 71-76, PROF COMPUT & INFO SCI, CLEVELAND STATE UNIV, 71-, INTERIM DEAN, COL BUS ADMIN, 90- *Concurrent Pos:* Consult, Cleveland Court Mgt Proj, Environ Econ, Gould & Cleveland legal firms; consult, Gould Inc Ocean Systs Div, 77-, Cleveland Pub Utilities, 80, Sci Systs Inc, 80-84, Gould Elastomer Prods Div, 79-80, The World Bank, 79 & PPG, 88-; res grants, NSF, 75-78, Off Naval Res, 73-82, NASA, 80-83, Sci Assoc Inc, 85-87, State of Ohio, 88-89. *Mem:* Asn Comput Mach; Inst Elec & Electronics Engrs; Math Programming Soc. *Res:* Computer-aided design of engineering systems; optimization methods and mathematical programming; micro-computer software development; image processing. *Mailing Add:* Dept Comput & Info Sci Cleveland State Univ Cleveland OH 44115

WARF, JAMES CURREN, b Nashville, Tenn, Sept 1, 17; m 65; c 3. INORGANIC CHEMISTRY. *Educ:* Univ Tulsa, BS, 39; Iowa State Univ, PhD(inorg chem), 46. *Prof Exp:* Jr chemist, Phillips Petrol Co, Okla, 40-41; instr chem, Univ Tulsa, 41-42; group leader, Manhattan Proj, Iowa State Univ, 42-47; Guggenheim fel, Univ Berne, 47-48; from asst prof to prof, 48-84, EMER PROF CHEM, UNIV SOUTHERN CALIF, 84- *Concurrent Pos:* Vis prof, Univ Indonesia, 57-59, Airlangga Univ, Indonesia, 62-64, Tech Univ Vienna, 69-70 & Nat Univ Malaysia, Kuala Lumpur, 74-75; consult, Jet Propulsion Lab, Calif Inst Technol, Hasanuddin Univ & Andalas Univ, Indonesia, 78-79, Nat Univ Malaysia, Sabah, 82-83. *Mem:* Am Chem Soc; Fedn Am Sci. *Res:* Hydrides of heavy metals; chemistry in liquid ammonia; chemistry of europium and ytterbium. *Mailing Add:* Dept Chem Univ Southern Calif Los Angeles CA 90089-1062

WARFEL, DAVID ROSS, b Pana, Ill, Sept 25, 42; m 65; c 2. ORGANIC CHEMISTRY, POLYMER CHEMISTRY. *Educ:* Carthage Col, BA, 64; Univ Tenn, Knoxville, PhD(chem), 70; Univ Pittsburgh, MBA, 76. *Prof Exp:* Scientist polymer chem, Koppers Co, Inc, Monroeville, 69-74; sr scientist polymer chem, Arco/Polymers, Inc, Monroeville, 74, prin scientist, Arco Chem Co, 74-86; NEW PROD SCIENTIST, WASHINGTON PENN PLASTIC CO, 86- *Mem:* Am Chem Soc. *Res:* Influence of catalysis on the reactivity of organometallic compounds. *Mailing Add:* PO Box 355 MeadowLands PA 15347

WARFEL, JOHN HIATT, b Marion, Ind, Mar 3, 16; m 42; c 3. ANATOMY. *Educ:* Capital Univ, BSc, 38; Ohio State Univ, MSc, 41; Western Reserve Univ, PhD, 48. *Prof Exp:* Asst, Western Reserve Univ, 46-48; instr, State Univ NY Buffalo, 49-54, assoc, 54-56, from asst prof to assoc prof anat, Sch Med, 56-86; RETIRED. *Mem:* Am Asn Anatomists; Sigma Xi. *Res:* Gross anatomy. *Mailing Add:* 153 Walton Dr Buffalo NY 14226

WARFIELD, CAROL LARSON, b Oldham, SDak, June 6, 41; m 62; c 3. TEXTILE CHEMISTRY. *Educ:* SDak State Univ, BS, 62; Univ Ill, Urbana, MS, 67, PhD(family econ), 77. *Prof Exp:* Teacher home econ, Bridgewater Pub Sch, SDak, 63, Rock Island Pub Sch, Ill, 63-65; instr textiles, Univ Ill, Urbana, 69-77; asst prof textile sci, 77-82, assoc prof & head consumer affairs, 82-86, COORDR, INDUST RELS, AUBURN UNIV, 86- *Mem:* Sigma Xi; Am Asn Textile Chemists & Colorists; Am Soc Testing & Mat; Am Home Econ Asn; Asn Col Professors Textiles & Clothing. *Res:* Consumer attitudes relating to textiles and textile regulation; end-use performance characteristics of textiles and the effect of maintenance on these characteristics; economic aspects of regulation, selection, use and care; upholstery fabric performance aspects and test method development. *Mailing Add:* Auburn Univ 1108 Felton Lane Auburn AL 36830

WARFIELD, GEORGE, b Piombino, Italy, Apr 21, 19; nat US; m 45; c 3. SOLID STATE ELECTRONICS, PHOTO VOLTAICS. *Educ:* Franklin & Marshall Col, BS, 40; Cornell Univ, PhD(physics), 49. *Prof Exp:* From asst prof to prof elec eng, Princeton Univ, 49-74; prof elec eng & exec dir, Inst Energy Conversion, Univ Del, 74-78; assoc dir technol dissemination, Solar Energy Res Inst, 78-80; CONSULT PHOTOVOLTAICS, 80- *Mem:* Am Phys Soc; Inst Elec & Electronics Engrs. *Res:* Photovoltaic cells; solid state device physics; insulator electronics; behavior of electrons in insulators. *Mailing Add:* Box 2678 Hallock Rd 2 Vergennes VT 05491

WARFIELD, J(OHN) N(ELSON), b Sullivan, Mo, Nov 21, 25; m 48; c 3. SYSTEMS ENGINEERING, APPLIED MATHEMATICS. *Educ:* Univ Mo, AB & BSEE, 48, MSEE, 49; Purdue Univ, PhD, 52. *Prof Exp:* Instr elec eng, Univ Mo, 48; from instr to assoc prof, Pa State Univ, 49-55; from asst prof to assoc prof, Univ Ill, 55-57; assoc prof, Purdue Univ, 57-58; from assoc prof to prof, Univ Kans, 58-66; sr adv, Battelle Mem Inst, 66-74; chmn, Dept Elec Eng, Univ Va, 75-78, Harry Douglas Forsyth prof, 75-83, dir, Ctr Interactive Mgt, 81-83; sr mgr, Burroughs Corp, 83-84; dir, Inst Info Technol, 84-86, DIR, INST ADVAN STUDY INTEGRATIVE SCI, GEORGE MASON UNIV, 87- *Concurrent Pos:* Consult, Ramo-Wooldridge Corp, 56-57, Sylvania Data Systs, 59-60, Wilcox Elec Co, 62-66, IBM Corp, 79-82, NSF, 79-; ed, off jour Inst Fedn Systs Res. *Mem:* fel Inst Elec & Electronics Engrs; Asn Integrative Studies. *Res:* Interdisciplinary research methodology & modeling; science education; systems planning; environmental education; bureaucracy. *Mailing Add:* Inst Advan Study Integrative Sci George Mason Univ Fairfax VA 22030-4444

WARFIELD, PETER FOSTER, b Rye, NY, Aug 4, 18; m 42; c 3. POLYMER CHEMISTRY. *Educ:* Hamilton Col, BS, 40; Univ Ill, MS, 41, PhD(org chem), 44. *Prof Exp:* Asst chem, Univ Ill, 42-44; res chemist, Bakelite Corp, NJ, 44-45; from res chemist to sr res chemist, Ansco Div, Gen Aniline & Film Corp, 45-49, develop specialist, 49-52; chemist, E I du Pont de Nemours & Co, Inc, 52-60, tech serv rep, 60-63, res chemist, 63-68, sr res chemist, 69-78, res assoc, Photo Prods Dept, 78-82; RETIRED. *Mem:* Am Chem Soc; Soc Photog Sci & Eng; Sigma Xi. *Res:* Hindered Grignard reactions; aliphatic polyamines; phenolic resins; synthetic peptides; restrainers for photographic gelatin; photographic emulsions; cyanine dyes; color photography; color formers; color processing; photopolymer printing plates. *Mailing Add:* 508 Birch Ave Westfield NJ 07090

WARFIELD, ROBERT BRECKINRIDGE, JR, algebra; deceased, see previous edition for last biography

WARFIELD, ROBERT WELMORE, b Asbury Park, NJ, Oct 11, 26; m 55; c 2. POLYMER CHEMISTRY, THERMODYNAMICS & MATERIAL PROPERTIES. *Educ:* Univ Va, BS, 50. *Prof Exp:* Chemist, US Bur Mines, Md, 50-55; chemist, 55-64, SR SCIENTIST, NAVAL SURFACE WEAPONS CTR, WHITE OAK, 64- *Mem:* Am Chem Soc. *Res:* Patentee in field; chemistry and physics of the solid state of polymers; compressibility and electrical properties of polymers; transitions of polymers; polymerization kinetics; acoustic properties of polymers. *Mailing Add:* 22712 Ward Ave Hereford Hills Germantown MD 20874

WARFIELD, VIRGINIA MCSHANE, b Charlottesville, Va, Sept 30, 42; m 64; c 3. MATHEMATICAL ANALYSIS. *Educ:* Bryn Mawr Col, AB, 63; Brown Univ, MA, 65, PhD(math), 71. *Prof Exp:* Seattle dir math, Spec Elem Educ Disadvantaged Proj, 70-73; LECTR MATH, 73-, DIR OF REMEDIAL MATH, UNIV WASH, 80- *Mem:* Sigma Xi. *Res:* Stochastic integrals and stochastic control theory. *Mailing Add:* Dept Math Univ Wash Seattle WA 98195

WARGA, JACK, b Warsaw, Poland, Dec 5, 22; nat US; m 49; c 2. MATHEMATICS. *Educ:* NY Univ, PhD(math), 50. *Prof Exp:* Assoc mathematician, Reeves Instrument Corp, 51-52; prin engr, Repub Aviation Corp, 52-53; sr mathematician & head math dept, Electrodata Div, Burroughs Corp, 54-56; fel, Weizmann Inst Sci, Israel, 56-57; sr staff mathematician & mgr math dept, Res & Adv Develop Div, Avco Corp, 57-66; PROF MATH, NORTHEASTERN UNIV, 66- *Concurrent Pos:* Ed, J Control & Optimization, Soc Indust & Appl Math, 62-89. *Mem:* AAAS; Am Math Soc; Soc Indust & Appl Math. *Res:* Mathematical control theory; nonsmooth analysis. *Mailing Add:* Dept Math Northeastern Univ Boston MA 02115

WARGEL, ROBERT JOSEPH, b Evansville, Ind, Dec 24, 40; m 70. BIOCHEMISTRY, FOOD SCIENCE & TECHNOLOGY. *Educ:* Univ Evansville, BA, 66; Northwestern Univ, PhD(chem), 70. *Prof Exp:* Chemist, City of Evansville, Ind, 64-66 & Mead Johnson & Co, 66; group leader, 70-80, SR GROUP LEADER, KRAFT INC, 80- *Mem:* Am Chem Soc; Am Soc Microbiol; Am Dairy Sci Asn; Inst Food Technologists. *Res:* Enzyme use in dairy products; biochemistry of cheese; metabolism of dairy culture; new product development and implementation; technical management; research administration. *Mailing Add:* Kraft Inc 801 Waukegan Rd Glenview IL 60025

WARGO, PHILIP MATTHEW, b Danville, Pa, Mar 1, 40. STRESS PHYSIOLOGY, ROOT PATHOLOGY & PHYSIOLOGY. *Educ:* Gettysburg Col, BA, 62; Iowa State Univ, MS, 64, PhD(plant path), 66. *Prof Exp:* Res scientist, US Army, Ft Detrich, Md, 66-68; res plant pathologist, 68-89, PROJ LEADER & PRIN PLANT PATHOLOGIST, USDA FOREST SERV, HAMDEN, CONN, 89- *Honors & Awards:* Res Award, Int Soc Arboricult, 85. *Mem:* Am Phytopath Soc; Int Soc Arboricult; Soc Am Foresters. *Res:* Effects of stress that predispose trees to pathogens that are normally of minor consequence; host-pathogen interaction of stress-induced disease; fungal physiology; armillaria root disease and its effect on forest management; effects of stress on root physiology. *Mailing Add:* 51 Mill Pond Rd Hamden CT 06514

WARHOL, MICHAEL J, SURGICAL PATHOLOGY, IMMUNOCYTOCHEMISTRY. *Educ:* Univ Pittsburgh, MD, 69. *Prof Exp:* ASSOC PROF PATH, SCH MED, HARVARD UNIV, 85- *Mailing Add:* Brigham & Women's Hosp 75 Francis St Boston MA 02115

WARING, DEREK MORRIS HOLT, b Northern Ireland, June 16, 25; nat US; m 50; c 2. ORGANIC CHEMISTRY. *Educ:* Queen's Univ, Belfast, BSCh, 46, MSc, 47, PhD(chem), 49. *Prof Exp:* Res chemist, Albright & Wilson, Eng, 50-52; from res chemist to res assoc, 52-67, SR SUPVR RES & DEVELOP, E I DU PONT DE NEMOURS & CO, INC, 67-; CONSULT, 88- *Res:* Stereochemistry; polynuclear aromatic hydrocarbons; polymer intermediates and chemistry. *Mailing Add:* 2125 Shipley Rd Wilmington DE 19803

WARING, GAIL L, MOLECULAR BIOLOGY, GENETICS. *Educ:* Univ Ore, PhD(biol), 74. *Prof Exp:* ASSOC PROF CELL BIOL, MARQUETTE UNIV, 84- *Mailing Add:* Dept Biol Marquette Univ 530 N 15th Milwaukee WI 53233

WARING, GEORGE HOUSTOUN, IV, b Denver, Colo, July 15, 39; m 62; c 3. ANIMAL BEHAVIOR, VERTEBRATE ZOOLOGY. *Educ:* Colo State Univ, BS, 62, PhD(zool), 66; Univ Colo, MA, 64. *Prof Exp:* From asst prof to assoc prof, 66-83, PROF ANIMAL BEHAV, SOUTHERN ILL UNIV, CARBONDALE, 83- *Concurrent Pos:* Guest prof, Univ Munich, 72-73; res prog dir, US Marine Mammal Comn, 74-75; chair, pub affairs comt, 83-89, Animal Behav Soc Orgn & Bylaws Comm, 89-, parliamentarian, 89; Equine Pract, 87-90. *Mem:* Fel, AAAS, 89; Animal Behav Soc; Am Soc Mammal; Wildlife Soc; Am Ornith Union; Soc Vet Ethology. *Res:* Communicative behaviors of vertebrates; equine behavior; vertebrate natural history; wildlife and behavioral ecology; marine mammal conservation and ethology; applied ethology of wildlife, pest species, and domestic animals. *Mailing Add:* Dept Zool Southern Ill Univ Carbondale IL 62901-6501

WARING, GEORGE O, III, b Buffalo, NY, Feb 21, 41; m 65; c 3. OPHTHALMOLOGY. *Educ:* Wheaton Col, BS, 63; Baylor Med Col, MD, 67; Am Bd Ophthal, dipl, 78. *Prof Exp:* Sr asst surgeon, USPHS Indian Hosp, Winnebago, NB, 68-70; resident, Wills Eye Hosp, Philadelphia, Pa, 70-73, Heed fel, 73-74; asst prof ophthal, Univ Calif, Davis, 74-79; ASSOC PROF OPHTHAL, EMORY UNIV, 79- *Concurrent Pos:* Attend physician ophthal, Ship of Hope, Natal, Brazil, 72; consult, Vet Admin Hosp, Martinez, Calif, 74-79 & Travis AFB, Fairfield, Calif, 74-78; surg dir, Sacramento Valley Eye Bank, 75-79; res consult, Calif Comn Peace Officer Standards & Training, Sacramento, 76-78; chmn, Corneal & External Dis Sect, Found Systs Postgrad Educ in Ophthal, 76- *Honors & Awards:* Physician's Recognition Award, AMA, 76. *Mem:* Am Acad Ophthal; Am Soc Contemporary Ophthal; Asn Res Vision & Ophthal; AMA. *Res:* Clinico-pathologic correlations in corneal diseases; corneal basement membrane; corneal lipid metabolism. *Mailing Add:* Emory Univ Clin 1327 Clifton Rd NE Atlanta GA 30322

WARING, RICHARD C, b Excelsior Springs, Mo, Mar 25, 36; m 62; c 3. PHYSICS. *Educ:* William Jewell Col, BA, 58; Univ Ark, MS, 61. *Prof Exp:* From instr to asst prof, 60-72, ASSOC PROF PHYSICS, UNIV MO, KANSAS CITY, 72- *Concurrent Pos:* Nat coun mem, Soc Physics Students, 76-82; dir, Math & Physics Inst, 84. *Mem:* Sigma Xi; Am Asn Physics Teachers. *Res:* Infrared reflectance spectroscopy. *Mailing Add:* Dept Physics Univ Mo Kansas City MO 64110

WARING, RICHARD H, b Chicago, Ill, May 17, 35; m 57; c 2. PLANT ECOLOGY. *Educ:* Univ Minn, St Paul, BS, 57, MS, 59; Univ Calif, Berkeley, PhD(bot), 63. *Prof Exp:* Asst prof, 63-72, PROF FOREST ECOL, ORE STATE UNIV, 76- *Concurrent Pos:* Dep dir, Coniferous Forest Biome, 73- *Mem:* AAAS; Ecol Soc Am. *Res:* Ecosystem analysis of watersheds; environmental classification; physiological ecology; plant-water relationships. *Mailing Add:* Dept Forestry Ore State Univ Corvallis OR 97331

WARING, ROBERT KERR, JR, b Palmerton, Pa, Aug 18, 28; m 54; c 4. PIGMENT OPTICS, COLOR IMAGING. *Educ:* Va Mil Inst, BS, 50; Yale Univ, PhD, 55. *Prof Exp:* Res physicist, E I du Pont de Nemours & Co, Inc, 55-87; CONSULT, 87- *Mem:* Am Phys Soc. *Res:* Measurement and interpretation of the optical properties of colored pigments; general color imaging; measurement and interpretation of the magnetic properties of assemblies of single domain ferromagnetic particles. *Mailing Add:* 14 Davenport Dr Downington PA 19335

WARING, WILLIAM WINBURN, b Savannah, Ga, July 20, 23; m 52; c 5. PEDIATRICS. *Educ:* Harvard Univ, MD, 47. *Prof Exp:* Intern pediat, Children's Hosp, Boston, Mass, 47-48; intern, Johns Hopkins Hosp, Md, 48-49, asst res, 49-50, chief res outpatient dept, 50-51, chief res, Hosp, 51-52; from instr to assoc prof, 57-65, lectr physiol, 66-80, PROF PEDIAT, SCH MED, TULANE UNIV, 65-, JANE B ARON PROF PEDIAT, 86- *Concurrent Pos:* Pulmonary dis adv comt, Nat Heart & Lung Inst, 71-73; vchmn gen med & sci adv coun, Cystic Fibrosis Found, 72-73; assoc ed, Am J Dis Children. *Mem:* Am Pediat Soc; Am Acad Pediat; Am Col Chest Physicians; Am Thoracic Soc (vpres, 72-73). *Res:* Respiratory disease and physiology in infants and children; cystic fibrosis. *Mailing Add:* Dept Pediat Tulane Univ Sch Med New Orleans LA 70112

WARING, WORDEN, b Washington, DC, Jan 8, 15; m 49; c 1. BIOMEDICAL ENGINEERING. *Educ:* Cornell Univ, BChem, 36; Mass Inst Technol, PhD(phys chem), 40. *Prof Exp:* Instr chem, Tulane Univ, 40-42, asst prof, 42-43; engr, Shell Develop Co, 43-53; engr opers res group, Arthur D Little, Inc, 53-54; chemist, Semiconductor Div, Raytheon Mfg Co, 54-58; head chem sect, Fairchild Semiconductor Corp, 58-64; prin investr human systs design ctr, Rancho Los Amigos Hosp, Downey, Calif, 64-69; assoc prof biomed eng, 69-72, prof biomed eng, Schs Med & Eng, 72-82, EMER PROF, UNIV CALIF, DAVIS, 82- *Mem:* AAAS; Biomed Eng Soc; Am Chem Soc; Electrochem Soc; Inst Elec & Electronics Engrs. *Res:* Thermodynamics; phase rule; industrial operations research; surface chemistry; diffusion; electrochemistry; biomedical engineering. *Mailing Add:* 27083 Patwin Rd Davis CA 95616

WARINNER, DOUGLAS KEITH, b Little Falls, Minn, Jan 20, 41; m 67; c 2. NUCLEAR ENGINEERING. *Educ:* Univ Calif, Davis, BS, 65, MS, 68; Purdue Univ, PhD(mech eng), 73. *Prof Exp:* Teaching asst fluid mech & aerodyne rocket propulsion, Univ Calif, Davis, 65-67; mech engr aerodyn, Naval Weapons Ctr, 67-68; teaching asst heat transfer & measurements, Purdue Univ, 69-71; asst prof thermosci, Fla Atlantic Univ, 71-75; prof & chairperson, Mech Eng, Manketo State Univ, 85-89; MECH ENGR THERMOHYDRAUL, ARGONNE NAT LAB, 75-85 & 89- *Concurrent Pos:* Instr, Ill Inst Technol, 77-; mem safety comn, US Dept Energy, 78-80. *Mem:* AAAS; Sigma Xi; Am Soc Mech Engrs; Am Soc Eng Educ. *Res:* Analytical study of the in-core behavior and atmospheric consequences of a nuclear research reactor meltdown or loss of coolant accident: prediction of the in core thermohydraulics (natural circulation and natural conversion) and the atmospheric dispersion and decay of nuclid. *Mailing Add:* Div RA 9700 S Cass Ave Bldg 208 Rm C-114 Argonne IL 60439

WARITZ, RICHARD STEFEN, b Portland, Ore, Apr 1, 29; m 50; c 4. TOXICOLOGY. *Educ:* Reed Col, BA, 51; Stanford Univ, PhD(chem), 57; Am Bd Toxicol, dipl, 80; Acad Toxicol Sci, dipl, 83. *Prof Exp:* Actg instr gen chem & biochem, Wash State Univ, 54-55; sr res chemist, E I du Pont de Nemours & Co, Inc, 56-62, sr res scientist, 62-64, sect chief inhalation toxicol, 64-70, res mgr, Biosci Group, 70-75; sr toxicologist, 75-77, MGR TOXICOL, HERCULES INC, 78- *Concurrent Pos:* counr, Int Union Toxicol Soc, 84-85. *Mem:* Sigma Xi; Am Chem Soc; Am Indust Hyg Asn; Int Union Toxicol Soc; Soc Toxicol (treas, 81-85). *Res:* Pulmonary toxicology and pharmacology; biochemical measures of chemical exposure; mechanisms of toxic actions of chemicals. *Mailing Add:* 2613 Turnstone Dr Brookmeade 2 Wilmington DE 19808

WARK, KENNETH, JR, b Indianapolis, Ind, Jan 2, 27; m 55; c 3. THERMODYNAMICS. *Educ:* Purdue Univ, BS, 50, PhD(mech eng), 55; Univ Ill, MS, 51. *Prof Exp:* Res engr, Atlantic Refining Co, Tex, 51-53; ASSOC PROF MECH ENG, PURDUE UNIV, WEST LAFAYETTE, 55- *Concurrent Pos:* Consult, US Steel Corp, 57-58 & Rovac Corp, 76-78; NSF sci fac fel, Stanford Univ, 62-63. *Mem:* Am Chem Soc; Am Soc Eng Educ; Combustion Inst. *Res:* Alternative and innovative energy conversion systems. *Mailing Add:* Dept Mech Eng Purdue Univ West Lafayette IN 47907

WARKANY, JOSEF, b Vienna, Austria, Mar 25, 02; US citizen; m 37; c 2. PEDIATRICS. *Educ:* Univ Vienna, MD, 26. *Hon Degrees:* DSc, Thomas Jefferson Med Col, 74; DSc, Univ Ill, 75, Univ Cincinnati, 86. *Prof Exp:* Intern, Univ Pediat Clin, Vienna, 26-27; asst, Fed Inst Mothers & Children, 27-31; dir ment retardation res, Inst Develop Res, Children's Hosp Res Found, 66-76; from asst prof to prof, 32-72, EMER PROF RES PEDIAT, UNIV CINCINNATI, 72- *Concurrent Pos:* Scholar, Children's Hosp Res Found, 31-34, asst attend pediatrician, 34-35, attend pediatrician & fel, 35-; prog consult, Nat Inst Child Health & Human Develop. *Honors & Awards:* Howland Award, Am Pediat Soc, 70; Child Health Award, Charles H Hood Found, 72. *Mem:* Am Pediat Soc; Soc Pediat Res; corresp mem French Soc Pediat; hon mem Europ Teratology Soc. *Res:* Nutritional deficiencies; congenital malformations in children; experimental teratology; mental retardation; experimental oncology. *Mailing Add:* Children's Hosp Res Found Elland Ave Cincinnati OH 45229

WARKENTIN, BENNO PETER, b Man, Can, June 21, 29; m 56; c 3. ENVIRONMENTAL MANAGEMENT. *Educ:* Univ BC, BSA, 51; Wash State Univ, MS, 53; Cornell Univ, PhD(soils), 56. *Prof Exp:* Nat Res Coun Can Overseas fel, Oxford Univ, 56-57; asst prof agr physics, 57-62, assoc prof soil sci, 62-70, dir environ studies, 72-75, MacDonald Col, McGil Univ, 70-78; PROF & HEAD DEPT SOIL SCI, ORE STATE UNIV, CORVALLIS, 78- *Mem:* Am Soc Agron; Can Soc Soil Sci (pres, 65-66); Int Soc Soil Sci; Sigma Xi. *Res:* Physical and chemical properties of clay minerals; physical properties of soils; water in clay soils; solid waste disposal. *Mailing Add:* Dept Soil Sci Ore State Univ Corvallis OR 97331

WARKENTIN, JOHN, b Grunthal, Man, Aug 18, 31; m 57; c 3. ORGANIC CHEMISTRY. *Educ:* Univ Man, BS, 54, MS, 55; Iowa State Univ, PhD(chem), 59. *Prof Exp:* Fel chem, Calif Inst Technol, 59 & Harvard Univ, 59-60; from asst prof to prof, 60-71, PROF CHEM, MCMASTER UNIV, 71- *Concurrent Pos:* Assoc ed, Can J Chem, 76-81. *Honors & Awards:* Syntex Award, 89. *Mem:* Am Chem Soc; Chem Inst Can. *Res:* Synthetic and mechanistic investigations in organic chemistry including carbenes, free radicals, and ylides. *Mailing Add:* Dept Chem McMaster Univ Hamilton ON L8S 4M1 Can

WARLICK, CHARLES HENRY, b Hickory, NC, May 08, 30; m 58; c 1. MATHEMATICS, COMPUTER SCIENCE. *Educ:* Duke Univ, BS, 52; Univ Md, MA, 55; Univ Cincinnati, PhD(math), 64. *Prof Exp:* Mathematician, US Dept Army, 52-53; programmer, Int Bus Mach Corp, 54; appl mathematician, Gen Elec Co, 55-57, supvr appl math, 57-62, supvr appl math & comput software develop, 63-65; lectr comput sci & dir, Comput Ctr, Univ Tex, Austin, 65-80. *Concurrent Pos:* Vpres, VIM Users Orgn Control Data Corp 6000 Series Comput, 68-70, pres, 70-71. *Mem:* Asn Comput Mach. *Res:* Numerical solution of partial differential equations; fundamental solutions of finite difference equations; computer executive operating systems; algorithmic languages. *Mailing Add:* 4509 Edgemont Dr Austin TX 78731

WARLTIER, DAVID CHARLES, b Hartford, Conn, Mar 28, 47; div; c 4. ANESTHESIOLOGY, CARDIOLOGY. *Educ:* Carroll Col, BS, 69; Med Col Wis, PhD(pharmacol), 76, MD, 82. *Hon Degrees:* DSc, Carroll Col, 91. *Prof Exp:* Asst prof pharmacol, 78-82, assoc prof pharmacol & med, 82-88, PROF ANESTHESIOL, MED & PHARMACOL, VCHMN RES ANESTHESIOL, MED COL WIS, 88- *Mem:* Am Physiol Soc; AMA; Am Heart Asn; Am Soc Pharmacol & Exp Therapeut; Am Soc Anesthesiol; Asn Univ Anesthesiologists. *Res:* Physiology, pathophysiology and pharmacology of the coronary circulation; myocardial ischemia; influence of pharmacological agents on the coronary collateral circulation. *Mailing Add:* Dept Anesthesiol MFRC Med Col Wis 8701 Watertown Plank Rd Milwaukee WI 53226

WARMAN, JAMES CLARK, b Morgantown, WVa, May 27, 27; m 53; c 3. HYDROGEOLOGY. *Educ:* WVa Univ, BA, 50, MS, 52. *Prof Exp:* Geologist, US Geol Surv, 52-65; ASSOC PROF CIVIL ENG, AUBURN UNIV, 70-, DIR WATER RESOURCES RES INST, 65- *Concurrent Pos:* Grants, Auburn Univ, Water Resources Planning, US Water Resources Coun, 68-, Econ of Pollution Abatement, US Dept Interior, 71-73 & Res Mgt, 72-74; mem work group hydrol maps, US Nat Comt for Int Hydrol Decade, 70-73; consult, Study of Nat Water Res Probs & Priorities, Univs Coun Water Resources for US Dept Interior, 71-72 & Harmon Eng, 75-; chmn tech div & vpres, Nat Water Well Asn, 71-72. *Honors & Awards:* Ross L Oliver Award, Nat Water Well Asn, 74. *Mem:* Fel Am Water Resources Asn (pres, 76); Am Geophys Union; fel Geol Soc Am. *Res:* Occurrence and availability of ground water; water resources planning; research management; waste heat storage by injection into a confined aquifer. *Mailing Add:* 470 Cary Dr Auburn AL 36830

WARMAN, PHILIP ROBERT, b Jersey City, NJ, Aug 10, 46; Can citizen; m 88; c 1. SOIL BIOCHEMISTRY, SOIL-PLANT RELATIONSHIPS. *Educ:* Rutgers Univ, BSc, 68; Univ Guelph, MSc, 72, PhD(soil biochem), 77. *Prof Exp:* Lectr, Univ Guelph, 76-77; lectr, Macdonald Col, McGill Univ, 77-78, from auxiliary prof to asst prof, 78-81; from asst prof to assoc prof, 81-87, PROF SOIL SCI, NS AGR COL, 87- *Concurrent Pos:* Consult & pres, Coastal BioAgresearch, Ltd, 84- *Honors & Awards:* Plant Food Inst Award, 68. *Mem:* Am Soc Agron; Soil Sci Soc Am; Agr Inst Can; Can Soil Sci Soc; Soil & Water Conserv Soc Am. *Res:* Effects of organic amendments and alternative fertilizers on soil chemistry, soil fertility, crop nutrition and crop production; effects of pesticides and heavy metals on soils and crops; land reclamation and revegation; extraction and identification of soil organic, Nitrogen, Phosphorus and Sulfur compounds. *Mailing Add:* NS Agr Col PO Box 550 Truro NS B2N 5E3 Can

WARMBRODT, ROBERT DALE, b Boonville, Mo, Aug 30, 47. PLANT ANATOMY, PHLOEM TRANSPORT. *Educ:* Univ Mo, Columbia, AB, 70; Univ Wis-Madison, MS, 73, PhD(bot & hort), 78. *Prof Exp:* Res assoc, Dept Bot, Univ Wis-Madison, 78-79; asst prof plant anat & bot, Ohio State Univ, 80-84; res plant physiologist, Climate Stress Lab, USDA, Agr Res Serv, Beltsville, MD, 84-90; COORDR, BIOTECHNOL INFO CTR, NAT AGR LIB, BELTSVILLE, MD, 90- *Concurrent Pos:* Alexander von Humboldt fel, Dept Forest Bot, Univ Gottingen, 79-80; prin investr grant, Develop Biol Prog, NSF, 81-84, co-prin investr, Multi-use Equip Prog, 81; USDA & Univ Md Coop res award, 85-86; Davis fel, Univ Wis, 77; Humboldt fel, Ger, 79-81. *Mem:* AAAS; Bot Soc Am; Am Inst Biol Sci; Am Fern Soc; Sigma Xi. *Res:* Science communication and education; agricultural biotechnology. *Mailing Add:* Biotechnol Info Ctr Rm 1402 Nat Agr Libr Beltsville MD 20705

WARME, JOHN EDWARD, b Los Angeles, Calif, Jan 16, 37; div; c 2. PALEOECOLOGY, BASIN ANALYSIS. *Educ:* Augustana Col, Ill, BA, 59; Univ Calif, Los Angeles, PhD(geol), 66. *Prof Exp:* Fulbright scholar, Scotland, 66-67; W Maurice Ewing prof oceanog, Rice Univ, 67-79; PROF GEOL, COLO SCH MINES, 79- *Concurrent Pos:* Consult, NSF geol prog, 79-83. *Mem:* Fel AAAS; Am Asn Petrol Geol; Int Asn Sedimentologists; Paleont Asn; fel Geol Soc Am; Soc Econ Paleontologists & Mineralogists (pres, 83-84). *Res:* Depositional systems; basin analysis and history; submarine bio-erosion by invertebrates; burrowing marine invertebrates and trace fossils; shelled invertebrate community identification and analysis; modern and fossil reef ecology; carbonate rock environments and paleoenvironments; lagoonal and deep marine ecology and sedimentation. *Mailing Add:* Dept Geol Colo Sch Mines Golden CO 80401

WARME, PAUL KENNETH, b Westbrook, Minn, Jan 23, 42; m 62; c 1. BIOCHEMISTRY. *Educ:* Univ Minn, Minneapolis, BCh, 64; Univ Ill, Urbana, PhD(biochem), 69. *Prof Exp:* Fel protein chem, Cornell Univ, 69-72; asst prof biochem, Pa State Univ, University Park, 72-79; RES DIR, INTERACTIVE MICROWARE, INC, 79- *Concurrent Pos:* NIH fel, Cornell Univ, 69-71. *Mem:* Inst Elec & Electronics Engrs. *Res:* Micro-computer software and hardware for laboratory use; conformational energy calculations on proteins; computer applications in biochemistry; chemical synthesis of biologically active polypeptides; structure-function relationships of proteins; heme proteins and peptides. *Mailing Add:* Interactive Microware Inc 444 E College Ave No 360 State College PA 16801

WARMKE, HARRY EARL, b Twin Falls, Idaho, Aug 29, 07; m 33, 46; c 3. PLANT CYTOLOGY, PLANT GENETICS. *Educ:* Stanford Univ, AB, 31, PhD(plant cytogenetics), 35. *Prof Exp:* Prof & head dept biol, Seton Hall Col, 35-38; asst genetics, Carnegie Inst, 38-40, cytologist, 41-45; head dept cytogenetics, Inst Trop Agr, Univ PR, 45-47; plant breeder, Fed Exp Sta, USDA, 47-53, officer chg, 53-63, plant geneticist, 64-77; prof, 64-77, EMER PROF AGRON, UNIV FLA, 77- *Mem:* Am Soc NAm; Am Genetic Asn. *Res:* Experimental polyploidy; cytogenetics of sex determination; tropical agriculture; mechanism of cytoplasmic male sterility in plants; electron microscopy of plant viruses and viral inclusions. *Mailing Add:* 3100 S AJO Hwy St Augustine FL 32084

WARNE, RONSON JOSEPH, b East Orange, NJ, June 14, 32; m 50. MATHEMATICS. *Educ:* Columbia Univ, AB, 53; NY Univ, MS, 55; Univ Tenn, PhD(math), 59. *Prof Exp:* Asst & instr math, Univ Tenn, 55-59; asst prof, La State Univ, 59-63; assoc prof, Va Polytech Inst, 63-64; prof, WVa Univ, 64-69; PROF MATH, UNIV ALA, BIRMINGHAM, 69- *Res:* Algebraic theory of semigroups. *Mailing Add:* King Fahd Univ Petrol & Mining KFUPM No 1564 Dhahran 31261 Saudi Arabia

WARNE, THOMAS MARTIN, b Chicago, Ill, Sept 27, 39; m 61; c 3. ORGANIC CHEMISTRY, PETROLEUM CHEMISTRY. *Educ:* Yale Univ, BA, 61; Univ Ill, MS, 63; Northwestern Univ, PhD(org chem), 70. *Prof Exp:* Asst prof org chem, Alliance Col, 63-66; res chemist, 70-80, SR RES CHEMIST, AMOCO OIL CO, DIV STANDARDS OIL CO IND, 80- *Mem:* Am Soc Lubrication Engrs; Am Chem Soc; Am Soc Testing & Mat; Sigma Xi. *Res:* Development of industrial lubricants, particularly hydraulic oils and additives; development of products based upon waxes and petrolatums; product toxicity. *Mailing Add:* PO Box 3011 Naperville IL 60566

WARNER, ALDEN HOWARD, b Central Falls, RI, July 2, 37; c 4. DEVELOPMENTAL BIOLOGY. *Educ:* Univ Maine, BA, 59; Univ Southern Ill, MA, 61, PhD(physiol), 64. *Prof Exp:* USPHS fel biol & biochem, Biol Div, Oak Ridge Nat Lab, 64-65; from asst prof to assoc prof, 65-72, head dept, 79-85, PROF BIOL, UNIV WINDSOR, 72- *Concurrent Pos:* Consult, Biol Div, Oak Ridge Nat Lab, 74-75; res prof, Univ Windsor, 90-91. *Mem:* Am Soc Biochem & Molecular Biol; Soc Develop Biol; Can Soc Cell & Molecular Biol. *Res:* Proteases and protease inhibitors and their control in developmment and the onset of muscular dystrophy; proteases and their control in development and the onset of muscular dystrophy. *Mailing Add:* Dept Biol Univ Windsor Windsor ON N9B 3P4 Can

WARNER, ANN MARIE, b Denver, Colo, Mar 31, 44; m 68; c 1. THERAPEUTIC DRUG MONITORING, TOXICOLOGY. *Educ:* Marymount Col, Kans, BS, 66; Univ Kans, PhD(med chem), 70; Am Bd Clin Chem, dipl, 77. *Prof Exp:* NIH fel, Northeastern Univ, 71-73; assoc dir clin labs, Lahey Clin Med Ctr, 73-81; assoc prof path, Univ PR Med Ctr, 81-85; ASSOC PROF & ASSOC DIR, TOXICOL LAB, UNIV CINCINNATI MED CTR, 85- *Concurrent Pos:* Lectr, Cardinal Cushing Col, 71-72. *Honors & Awards:* Outstanding Speaker Award, Am Asn Clin Chem, 87. *Mem:* Clin Ligand Assay Soc; Am Asn Clin Chem. *Res:* Evaluation of biological versus chronological age in neonates; prodrugs. *Mailing Add:* 10868 Bromwell Lane Cincinnati OH 45249

WARNER, BERT JOSEPH, b Ardmore, Okla, June 15, 25; m 52; c 2. CHEMICAL ENGINEERING. *Educ:* Rice Univ, BS, 49. *Prof Exp:* Chem engr, Colombian Petrol Co, 50-53, petrol eng lab mgr, 53-55; res technologist, 55-58, sr res technologist, 58-64, eng assoc, 64-77, MGR RECOVERY PROCESSES, FIELD RES LAB, MOBIL OIL CORP, 78- *Mem:* Soc Petrol Engrs; Sigma Xi. *Res:* Hydrocarbon phase behavior, processing and production. *Mailing Add:* 504 Town Creek Dr Dallas TX 75232

WARNER, BRENT A, b Charleston, WVa, Sept 17, 53. SOLID STATE PHYSICS, CRYOGENICS ENGINEERING. *Educ:* Col Wooster, BA, 75; Ohio State Univ, MS, 85. *Prof Exp:* AEROSPACE ENGR, GODDARD SPACE CTR, 85- *Mem:* Am Phys Soc; Sigma Xi. *Res:* Magnetic shielding for a magnetic cooling system for satellite use. *Mailing Add:* Code 713-4 Goddard Space Flight Ctr Greenbelt MD 20771

WARNER, CAROL MILLER, b New York, NY, Sept 26, 46; c 2. IMMUNOBIOLOGY, GENETICS. *Educ:* Queens Col, NY, BA, 66; Univ Calif, Los Angeles, PhD(biochem), 70. *Prof Exp:* Fel, Yale Univ, 70-71; from asst prof to assoc prof biochem, Iowas State Univ, 71-88; PROF BIOL, NORTHEASTERN UNIV, 88- *Mem:* Soc Develop Biol; Fedn Am Scientists; Sigma Xi; Am Soc Biol Chemists; Am Asn Immunologists. *Res:* Preimplantation mouse embryo development; major histocompatibility complex in aging; gene mapping. *Mailing Add:* Dept Biol Northeastern Univ 360 Huntington Ave Boston MA 02115

WARNER, CECIL F(RANCIS), b Parker, Ind, June 13, 15; m 39; c 2. MECHANICAL ENGINEERING. *Educ:* Purdue Univ, BS, 39, PhD(heat transfer), 45; Lehigh Univ, MS, 41. *Prof Exp:* Instr mech eng, Lehigh Univ, 40-42; from instr to assoc prof, 42-55, PROF MECH ENG, PURDUE UNIV, WEST LAFAYETTE, 55- *Concurrent Pos:* Staff mem, Aerojet Gen Corp. *Mem:* Am Soc Mech Engrs; Air Pollution Control Asn. *Res:* Problems in air pollution; heat transfer; jet and rocket propulsion; heat transfer characteristics of liquid film cooling. *Mailing Add:* No 2 Westminster West Lafayette IN 47906

WARNER, CHARLES D, b Mt Hope, WVa, Mar 25, 45; m 66; c 1. ORGANIC CHEMISTRY. *Educ:* Univ Mo, Columbia, BS, 67, PhD(org chem), 71. *Prof Exp:* Asst prof chem, Mo Valley Col, 71-74; from asst prof to assoc prof chem, 74-87, HEAD DEPT, HASTINGS COL, 78-, PROF CHEM, 87- *Concurrent Pos:* Vis assoc prof, Eppley Inst Cancer Res, 83- 84, Univ Md, Columbia, 86. *Mem:* Am Chem Soc; Sigma Xi; Am Soc Mass Spectrometry. *Res:* Organic chemistry mass spectrometry, gas-phase ion chemistry. *Mailing Add:* Dept Chem Hastings Col Hastings NE 68901

WARNER, CHARLES ROBERT, b Aug 24, 31; Can citizen. MATHEMATICS. *Educ:* Univ Toronto, BA, 55; Rochester Univ, MS, 57, PhD(math), 62. *Prof Exp:* Instr math, Univ Conn, 58-59; asst prof, Mich State Univ, 62-64; asst prof, 64-69, ASSOC PROF MATH, UNIV MD, COLLEGE PARK, 69- *Concurrent Pos:* Vis assoc prof, Univ Calif, Irvine, 70-71, vis prof, Mittag-Leffler Inst, Djursholm, Sweden, 77, Univ Paris, Orsay, 77-78 & Univ Lausanne, Switz, 78. *Mem:* Am Math Soc; Math Asn Am. *Res:* Banach algebras and harmonic analysis. *Mailing Add:* Dept Math Univ Md College Park MD 20742

WARNER, CHARLES Y, b Pocatello, Idaho, Sept 4, 34. MECHANICAL ENGINEERING. *Educ:* Brigham Young Univ, BES, 57, MS, 63; Univ Mich, PhD(mech eng), 66. *Prof Exp:* Design engr, Hewlett-Packard, Calif, 57; test prog engr, Nat Reactor Test Sta, Gen Elec Co, Idaho, 58; from instr to assoc prof, 61-77, PROF MECH ENG, BRIGHAM YOUNG UNIV, 77- *Concurrent Pos:* Consult, Eimco Corp, Utah & Rich's Soft Cushion Bumper Co, 67- *Mem:* Am Soc Mech Engrs; Am Soc Eng Educ; Sigma Xi. *Res:* Automotive safety; heat transfer; engineering design. *Mailing Add:* 150 S Mountain Way Dr Orem UT 84058

WARNER, DANIEL DOUGLAS, b Mobile, Ala, Sept 1, 42; m 68; c 3. NUMERICAL ANALYSIS. *Educ:* Ariz State Univ, Bs, 65; Univ Calif, San Diego, PhD(math), 74. *Prof Exp:* Programmer, Process Comput Sect, Gen Elec Co, 63-65; comput analyst, Airesearch Corp, 66; mem tech staff, Bell Tel Labs, 74-80; PROF MATH, CLEMSON UNIV, 80- *Mem:* Am Math Soc; Soc Indust & Appl Math; Asn Comput Mach. *Res:* Computational mathematics with emphasis on the numerical solution of ordinary and partial differential equations; design of algorithms for parallel computing. *Mailing Add:* Dept Math Sci Clemson Univ Clemson SC 29634-1907

WARNER, DAVID CHARLES, b Granite City, Ill, Apr 27, 42; m 64. MATHEMATICAL PHYSICS, PLASMA PHYSICS. *Educ:* Univ Mo-Columbia, BS, 65, MS, 67, PhD(physics), 70. *Prof Exp:* Asst prof physics, Lincoln Univ, 70-78; ASST PROF PHYSICS, COLUMBIA COL, MO, 78- *Mem:* Am Asn Physics Teachers; Am Phys Soc. *Res:* Quantum kinetic equations of plasma physics; density matrix formalism applied to damping of plasma waves. *Mailing Add:* Dept Math/Physics Ore Inst Technol Klameth Falls OR 97601

WARNER, DON LEE, b Norfolk, Nebr, Jan 4, 34; m 57; c 2. HYDROGEOLOGY, GEOLOGICAL ENGINEERING. *Educ:* Colo Sch Mines, Geol Engr, 56, MSc, 61; Univ Calif, Berkeley, PhD(eng sci), 64. *Prof Exp:* Res geologist, R A Taft Sanit Eng Ctr, Ohio, 64-69; assoc prof, 69-72, prof geol eng, 72-81, DEAN SCH MINES & METALL, UNIV MO, ROLLA, 81- *Concurrent Pos:* Consult; ed, Ground Water & Ground Water Monitoring Rev, 80-85. *Honors & Awards:* Sci Award, Nat Water Well Asn, 84. *Mem:* Am Inst Mining, Metall & Petrol Eng; Geol Soc Am; Am Asn Petrol Geol; Asn Prof Geol Scientists; Nat Water Well Asn; Nat Soc Prof Engrs. *Res:* Mineral exploration and exploitation, areal geology; stratigraphic geology and geophysics; geohydrology; water pollution control; engineering geology. *Mailing Add:* 305 McNutt Hall Univ Mo Rolla MO 65401

WARNER, DONALD R, b Winston, Mo, July 6, 18; m 60; c 3. ANIMAL HUSBANDRY, ANIMAL NUTRITION. *Educ:* Univ Mo, BS, 42, MS, 49, PhD(animal nutrit & ed), 60. *Hon Degrees:* Hon Am Farmer Org Degree, Nat Future Farmers Am, 82. *Prof Exp:* Mem staff, Mo Agr Exten Serv, 45-47; instr animal husb, Univ Mo, 47-49; asst prof animal sci, Univ Nebr, 49-56; from asst prof to prof animal sci, Iowa State Univ, 60-87, emer prof, 87-88; RETIRED. *Concurrent Pos:* Livestock judging team coach, Iowa State Univ. *Mem:* Am Soc Animal Sci. *Res:* Swine feeding investigations; antibiotics; protein supplements for pigs on pasture and in dry lot; methods of feeding and effects on carcass value; bloat studies of sheep; breeding, feeding and management of sheep. *Mailing Add:* 1309 Glendale Ames IA 50011

WARNER, DONALD THEODORE, b Holland, Mich, Apr 7, 18; m 45; c 2. BIOCHEMISTRY. *Educ:* Hope Col, AB, 39; Univ Ill, PhD(biochem), 43. *Prof Exp:* Res chemist, Gen Mills, Inc, 43-52 & Upjohn Co, 52-82; RETIRED. *Honors & Awards:* Bond Award, Am Oil Chem Soc, 66. *Mem:* Am Chem Soc. *Res:* Amino acids; isolation and synthesis; amino acid diets; isolation and properties of threonine; organic synthesis, 1, 4 addition reactions; polysaccharides; peptide synthesis; protein conformation studies; antiasthmatic drugs. *Mailing Add:* 2723 Winchell Ave Kalamazoo MI 49008-2173

WARNER, DWAIN WILLARD, b Cottonwood Co, Minn, Sept 1, 17; m 40, 66, 85; c 5. ZOOLOGY. *Educ:* Carleton Col, BA, 39; Cornell Univ, PhD(ornith), 47. *Prof Exp:* Lab asst bot, Carleton Col, 38-39; asst zoologist, Cornell Univ, 41, asst, 42-43 & 46, instr, 46-47; cur birds, Bell Mus Natural Hist, 47-87, environ dir, Belwin Outdoor Educ Lab, 83-89, from asst prof to assoc prof, 47-67, PROF ZOOL, UNIV MINN, MINNEAPOLIS, 67- *Mem:* Assoc Am Ornith Union; Cooper Ornith Soc; Wilson Ornith Soc; AAAS. *Res:* Birds of Mexico and New Caledonia; zoogeography; biology and ecology of avifauna; bird migration and habitat selection; population levels in declining habitats in tropical and temperate regions. *Mailing Add:* 2122 Novak Ave N Stillwater MN 55082

WARNER, ELDON DEZELLE, b Whitewater, Wis, Oct 5, 11; m 41; c 3. ENDOCRINOLOGY. *Educ:* Wis State Teachers Col, BEd, 32; Univ Wis, PhM, 38, PhD(zool), 41. *Prof Exp:* Asst prof sci, Adams State Teachers Col, 41-43; from instr to assoc prof, 46-58, chmn dept, 56-58, 59-61 & 66-69, prof zool, 58-78, EMER PROF ZOOL, UNIV WIS-MILWAUKEE, 78- *Concurrent Pos:* NSF fel, 58-59; vis prof, Cornell Univ, Ithaca, NY, 64-65; vis prof, Univ Wis- Madison, 67-69. *Mem:* AAAS; Am Soc Zool. *Res:* Lower vertebrate endocrinology. *Mailing Add:* 7921 W Clarke Wawatosa WI 53213

WARNER, FRANCIS JAMES, b Chicago, Ill, Oct 3, 1897. ANATOMY. *Educ:* Loyola Univ Chicago, MD, 19; Univ Iowa, BA, 22; Univ Mich, MA, 25; Am Bd Path, dipl, 52. *Prof Exp:* Instr neurol, Univ Mich, 32-33; instr med sci, Univ Md, 38-39; mem res staff neuroanat, Med Sch, Univ Calif, 42-44; mem res staff neuropath & neuroanat, Med Sch, Columbia Univ, 44-45; lectr neuroanat, Univ Utah, 46-49; guest investr, Sch Med, Yale Univ, 49-50; fel neurol, Henry Ford Hosp, Detroit, Mich, 51-52; guest investr, Cornell Univ, 53-55 & Univ Col, Univ London, 56-57; instr clin neurol, 57-75, asst prof neurol, Sch Med, Temple Univ, 75-81; ASSOC PROF COMP ANAT, SCH VET SCI, UNIV PA, 82- *Mem:* Am Psychiat Asn; Am Micros Soc; Am Soc Ichthyologists & Herpetologists; Am Soc Zoologists; Am Asn Anatomists. *Res:* Comparative neuroembryology in human and lower mammals and marsupials; neuropathology of the malformation in the human brain in adult and fetal material and nervous system; teratological brain malformations in the human form. *Mailing Add:* 1600 Church Rd Apt D107 Wyncote PA 19095-1924

WARNER, FRANK WILSON, III, b Pittsfield, Mass, Mar 2, 38; m 58; c 2. DIFFERENTIAL GEOMETRY, RIEMANNIAN GEOMETRY. *Educ:* Pa State Univ, BS, 59; Mass Inst Technol, PhD(math), 63. *Prof Exp:* Instr math, Mass Inst Technol, 63-64; actg asst prof math, Univ Calif, Berkeley, 64-65, asst prof, 65-68; assoc prof, 68-73, PROF MATH, UNIV PA, 73- *Concurrent Pos:* Guggenheim Fel, 76-77. *Mem:* Am Math Soc; Math Asn Am; AAAS; Sigma Xi. *Res:* Differential geometry. *Mailing Add:* Dept Math Univ Pa Philadelphia PA 19104-6395

WARNER, FREDERIC COOPER, b Whitesboro, NY, Jan 28, 15; m 40; c 2. GEOMETRY. *Educ:* Col Wooster, BA, 37; Univ Buffalo, MA, 50, PhD, 53. *Prof Exp:* Teacher high schs, NY, 37-46; instr math, Univ Buffalo, 46-53; from asst prof to prof math, St Lawrence Univ, 53-80; RETIRED. *Mem:* Am Math Soc; Sigma Xi; Math Asn Am. *Res:* Operations research. *Mailing Add:* Ten Jay St Canton NY 13617

WARNER, H JACK, US citizen. FOOD TECHNOLOGY. *Educ:* Ohio State Univ, BS, 65. *Prof Exp:* Qual control mgr, Chung King Corp, 65-67; food inspector, US Army, 67-69; food technologist develop, 69-72, dir tech serv, Spec Foods Div, 72-78, asst gen mgr, 79-83, GEN MGR, BEATREME FOODS CO, 83-, PRES, BEATREME FOODS, INC, 85- *Mem:* Inst Food Technol; Am Asn Cereal Chemists. *Res:* Spray dehydration of foods. *Mailing Add:* Beatreme Foods 352 E Grand Ave Beloit WI 53511

WARNER, HARLOW LESTER, b Greenport, NY, Aug 26, 42; m 63; c 3. PLANT PHYSIOLOGY, PLANT BREEDING. *Educ:* Cornell Univ, BS, 64; Univ Idaho, MS, 66; Purdue Univ, PhD(plant physiol), 70. *Prof Exp:* SR SCIENTIST PLANT PHYSIOL, ROHM & HAAS CO, 69- *Mem:* Am Soc Hort Sci; Am Soc Plant Physiologists. *Res:* Agricultural chemicals, specifically plant growth regulators and herbicides; plant growth regulators for developing hybrid seeds. *Mailing Add:* 111 Quince Dr Hatboro PA 19040

WARNER, HAROLD, b Philadelphia, Pa, June 6, 17; m 43; c 2. BIOMEDICAL ENGINEERING, NUTRITION. *Prof Exp:* Chief engr, Microwave Div, Frankford Arsenal, 48-56, Teledynamics, Inc, 56-58; consult engr, Missile & Space Div, Gen Elec Co, 58-65; clin asst prof psychiat, res prof biomed eng & chief, Biomed Eng Lab, 65-82, EMR PROF BIOMED ENG & CONSULT, EMORY UNIV, 82- *Mem:* Sigma Xi; NAm Nutrit & Prev Med Asn. *Res:* Application of electronics, physics and mechanics to the fields of reproductive biology, the neurosciences and psychology; role of nutrition in aging and disease. *Mailing Add:* 1305 Bernadette Lanes Atlanta GA 30329

WARNER, HOMER R, b Salt Lake City, Utah, Apr 18, 22; m 46; c 6. PHYSIOLOGY, BIOENGINEERING. *Educ:* Univ Utah, BA, 46, MD, 49; Univ Minn, PhD(physiol), 53. *Hon Degrees:* Dr, Brigham Young Univ, 71, Univ Linkoping, Sweden, 90. *Prof Exp:* Intern, Parkland Hosp, Dallas, 49-50; resident med, Univ Minn Hosp, 50-51; fels, Mayo Clin, 51-52 & Univ Minn, 52-53; res instr, Dept Internal Med, Univ Utah, 53-54, asst res prof, Dept Physiol, 57-64, prof & chmn, Dept Biophys & Bioeng, 64-73, RES PROF, DEPT SURG, SCH MED, UNIV UTAH, 66-, PROF & CHMN, DEPT MED INFORMATICS, 73-, SPEC ASST INFO MGT TO VPRES HEALTH SCI, 83- *Concurrent Pos:* Dir, Cardiovasc Lab, Latter-Day Saints Hosp, 54-70; estab investr, Am Heart Asn, 59-64; mem adv comt comput res, NIH, 61-63, chmn comput res study sect, 63-66; res career awardee, Nat Heart Inst, NIH, 62-83, mem, Heart Prog Proj Study Sect, 66-70; ed, Comput & Biomed Res, 66-; vis prof, Univ Hawaii, 68 & Univ Southern Calif, 72; mem, President's Adv Comt Deploying Scientists & Engrs into Health Ctrs, 71; mem, Health Care Technol Study Sect, Nat Ctr Health Servs Res & Develop, 72-76; mem, Biomed Libr Rev Comt, Nat Libr Med, 82-86, chmn, Grant Rev Study Sect, 85-86; prin investr, Utah IAIMS Develop Proj, 85-89. *Honors & Awards:* James E Talmage Sci Achievement Award, 68. *Mem:* Sr mem Inst Med-Nat Acad Sci; Am Physiol Soc; Am Col Med Informatics (pres, 89). *Res:* Control of cardiovascular system; application of computers to medicine; author of over 200 scientific journals. *Mailing Add:* Dept Med Infomatics Univ Utah Sch Med AB193 Med Ctr Salt Lake City UT 84132

WARNER, HUBER RICHARD, b Glendale, Ohio, May 16, 36; m 85; c 2. AGING. *Educ:* Ohio Wesleyan Univ, BA, 58; Mass Inst Technol, BS, 58; Univ Mich, PhD(biochem), 62. *Prof Exp:* NSF fel biochem, Mass Inst Technol, 62-64; from asst prof to assoc prof, Univ Minn, St Paul, 64-72, prof biochem, 72-84; DEP ASSOC DIR, BIOL AGING PROG, NAT INST AGING, BETHESDA, MD, 84- *Concurrent Pos:* Vis prof, Karolinska Inst, Stockholm, 71-72, Univ Calif, Berkeley, 79. *Mem:* Am Soc Biochem & Molecular Biol; Geront Soc Am; AAAS. *Res:* Biochemistry of bacteriophage infection and replication; DNA repair; nucleotide metabolism. *Mailing Add:* Nat Inst Aging NIH Bldg 31 Rm 5C21 9000 Rockville Pike Bethesda MD 20892

WARNER, ISIAH MANUEL, b DeQuincy, La, July 20, 46; m 68; c 3. FLUORESCENCE SPECTROSCOPY, CHEMOMETRICS. *Educ:* Southern Univ, BS, 68; Univ Wash, PhD(anal chem), 77. *Prof Exp:* Res chemist, Battelle Northwest, 68-73; teaching asst chem, Univ Wash, 73-75, res asst, 75-77; asst prof chem, Tex A&M Univ, 77-82; from assoc prof to prof chem, 82-87, SAMUEL CANDLER DOBBS PROF ANALYTICAL CHEM, EMORY UNIV, 87- *Concurrent Pos:* Adv & consult, Perkin-Elmer Corp, 83-88, Packard Instruments, 84-85, Fla A&M Univ, 83-88 & Eli-Lilly, 88-89. *Honors & Awards:* Pres Young Investr Award, 84; Percy Julian Award Outstanding & Significant Contrib in Res sponsored by Nat Orgn Black Chemists & Chem Engrs, 88. *Mem:* Am Chem Soc; AAAS; Nat Orgn Black Chemists & Chem Engrs; Soc Appl Spectros; Sigma Xi (secy, 84-86, 89); AAAS. *Res:* Develop and apply new and improved methods (chemical, mathematical and instrumental) for analysis of complex systems; development of novel fluorescence instrumentation and new data reduction strategies; improved multicomponent fluorescence analysis. *Mailing Add:* Dept Chem Emory Univ Atlanta GA 30322

WARNER, JAMES HOWARD, b Angola, Ind, Dec 24, 38. BOTANY, PLANT ECOLOGY. *Educ:* Manchester Col, BS, 61; Univ Wis, Madison, MS, 63; Univ Utah, PhD(biol), 71. *Prof Exp:* Teaching asst bot, Univ Wis-Madison, 61-63; ASSOC PROF BIOL, UNIV WIS-LA CROSSE, 63- *Mem:* AAAS; Am Inst Biol Sci; Ecol Soc Am; Sigma Xi. *Res:* Development of indices of site quality in forest management and plant association. *Mailing Add:* 4071 Terrace Dr La Cross WI 54601

WARNER, JOHN CHARLES, b Quincy, Mass, Oct 25, 62; m 85; c 2. MATERIALS DESIGN, HETEROCYCLIC SYNTHESIS. *Educ:* Univ Mass, Boston, BS, 84; Princeton Univ, MA, 86, PhD(org chem), 88. *Prof Exp:* SCIENTIST, POLAROID CORP, 88- *Concurrent Pos:* Lectr/instr, Univ Mass, Boston, 89- *Mem:* Am Chem Soc; Soc Imaging Sci & Technol. *Res:* Organic synthesis; heterocyclic chemistry; photographic chemistry; photochemistry; molecular recognition and self assembly; construction of multi-component organic materials based on non-covalent interactions; computational chemistry. *Mailing Add:* Polaroid Corp 730 Main St 4A Cambridge MA 02139

WARNER, JOHN CHRISTIAN, chemical dynamics; deceased, see previous edition for last biography

WARNER, JOHN NORTHRUP, plant breeding, agronomy, for more information see previous edition

WARNER, JOHN SCOTT, b Woodstown, NJ, Oct 25, 28; m 61; c 4. CHROMATOGRAPHIC METHODS, ENVIRONMENTAL ANALYSIS. *Educ:* Rensselaer Polytech Inst, BS, 49; Cornell Univ, PhD(org chem), 52. *Prof Exp:* Anal chemist, Socony Vacuum Oil Co, 49; asst, Cornell Univ, 49-52; prin chemist, 52-57, proj leader, 57-62, sr res chemist, 62-81, RES LEADER, ORG CHEM DIV, COLUMBUS LABS, BATTELLE MEM INST, 62- *Mem:* AAAS; Am Chem Soc. *Res:* Development of methods for determining organic compounds in water, sludge and solid wastes; application of gas and liquid chromatography; determination of petroleum components and pesticides in environemntal samples. *Mailing Add:* 3605 219th St SW Brier WA 98036

WARNER, JONATHAN ROBERT, b New York, NY, Feb 19, 37; m 58; c 2. MOLECULAR BIOLOGY, CELL BIOLOGY. *Educ:* Yale Univ, BS, 58; Mass Inst Technol, PhD(biophys), 63. *Prof Exp:* Res assoc biophys, Mass Inst Technol, 63-64; res assoc biochem, 64-65, from asst prof to assoc prof, 65-74, dir Sue Golding Grad Sch, 72-83, prof biochem & cell biol, 74-83, PROF & CHMN, DEPT CELL BIOL, ALBERT EINSTEIN COL MED, 83- *Concurrent Pos:* NSF fel, 64-65; career scientist, Health Res Coun, City New York, 65-72; Guggenheim fel, 71-72; Am Cancer Soc fac res award, 72-77; mem sci adv bd, Damon Runyon-Walter Winchell Cancer Fund, 73-77; mem molecular cytol study sect, NIH, 75-77, mem genetic basis of dis rev comt, 85-90; mem sci adv comt, Am Cancer Soc, 81-84, 89-92; mem bd trustees, Cold Spring Harbor Lab, 87- *Mem:* Am Soc Biol Chemists; Am Soc Cell Biol; Am Soc Microbiol; NY Acad Sci; Harvey Soc (vpres, 87-88, pres, 88-89). *Res:* Synthesis and assembly of ribosomes and its regulation in eucaryotic cells; ribosomal protein genes; transcription and processing of RNA nuclear-cytoplasmic interactions. *Mailing Add:* Dept Cell Biol Albert Einstein Col Med Bronx NY 10461

WARNER, KENDALL, b Westfield, Mass, Oct 2, 27; m 70; c 2. FISHERIES. *Educ:* Univ Maine, BS, 50; Cornell Univ, MS, 52. *Prof Exp:* Fishery aide, US Fish & Wildlife Serv, 50; asst fishery biol, Cornell Univ, 51-52; regional fishery biologist, 52-68, chief res biologist, Fishery Div, 68-84, RES MGT SUPRV, MAINE DEPT INLAND FISHERIES & WILDLIFE, 84- *Honors & Awards:* Prof Award of Merit, Northeast Div, Am Fisheries Soc, 87. *Mem:* Am Fisheries Soc; fel Am Inst Fishery Res Biologists. *Res:* Fresh water fisheries; landlocked salmon; brook trout. *Mailing Add:* Fisheries Div Inland Fisheries & Wildlife PO Box 1298 Bangor ME 04402-1298

WARNER, LAURANCE BLISS, b Brooklyn, NY, Dec 29, 31; m 52; c 3. NUCLEAR PHYSICS. *Educ:* Rensselaer Polytech Inst, BS, 53; Johns Hopkins Univ, MS, 58; Fla State Univ, PhD(physics), 62. *Prof Exp:* US Navy, 53-, instr physics & electronics, US Naval Acad, 56-58, terrier battery officer, USS Long Beach CGN-9, 62-64, exec officer, USS Benjamin Stoddert DDG-22, 64-66, physicist, Lawrence Radiation Lab, 66-69, weapons officer, USS Galveston CLG-3, 69-70, dir, Atomic Energy Div, Chief Naval Opers, Washington, DC, 70-77; PROG MGR, LOS ALAMOS NAT LAB, 77- *Concurrent Pos:* Instr physics, Fed City Col, Washington, DC, 71- *Mem:* Am Phys Soc. *Res:* Nuclear decay and reaction spectroscopy; deformed nuclei; fast electronics instrumentation; computer applications; neutron transport; radiative transfer; missile systems analysis; free electron lasers; advanced microwave sources. *Mailing Add:* 465 Camino Cereza Los Alamos NM 87544

WARNER, LAWRENCE ALLEN, b Monroe, Ohio, Apr 20, 14; m 42; c 2. GEOLOGY. *Educ:* Miami Univ, Ohio, AB, 37; Johns Hopkins Univ, PhD(geol), 42. *Prof Exp:* Geologist, US Antarctic Exped, Marie Byrd Land, 39-41 & Alaskan Br, US Geol Surv, 42-46; from asst prof to prof, Univ Colo, Boulder, 46-82; CONSULT, DENVER WATER BD, 55- *Concurrent Pos:* Geologist, US Geol Surv, 46-52; res consult, Univ Boston, 52-54. *Honors & Awards:* Congressional Medal for Sci & Explor. *Mem:* AAAS; Geol Soc Am; Am Geophys Union; Asn Eng Geol; Mineral Soc Am. *Res:* Antarctic and Alaskan geology; structure and tectonics of eastern Rocky Mountains; environmental and engineering geology. *Mailing Add:* Dept Geol Sci Univ Colo Boulder CO 80309

WARNER, LOUISE, anatomy, for more information see previous edition

WARNER, MARK CLAYSON, b Nephi, Utah, Aug 30, 39; m 62; c 6. AQUATIC BIOLOGY. *Educ:* Utah State Univ, BS, 66, MS, 69; Okla State Univ, PhD(zool), 72. *Prof Exp:* Aquatic biologist, Tenn Valley Authority, 70-74; aquatic biologist, Med Bioeng Res & Develop Lab, Environ Protection Res Div, US Army, Ft Detrick, 74-80; AQUATIC BIOLOGIST, SPORT FISH DIV, ALASKA FISH & GAME 80- *Mem:* Am Fisheries Soc; Wildlife Dis Asn; Am Soc Testing & Mat. *Res:* Aquatic bioassy toxicity studies; review and evaluate research proposals in areas of aquatic biology. *Mailing Add:* 800 F St No Fl Juneau AK 99801

WARNER, MARLENE RYAN, b Philadelphia, Pa, Apr 30, 38; m 69; c 2. ENDOCRINOLOGY, CANCER RESEARCH. *Educ:* Univ Calif, Berkeley, AB, 61, MA, 66; Univ Calif, Davis, PhD(anat), 72. *Prof Exp:* Res asst anat, Univ Calif, Davis, 68-69; res assoc cell biol, 72-75, instr cell biol, 75-77, ASST PROF OBSTET & GYNEC & CELL BIOL, BAYLOR COL MED, 77-; PROF, UNION GRAD SCH, 87- *Concurrent Pos:* Prin investr, Nat Cancer Inst grant, 80-86 & Am Cancer Soc grant, 84-85; reviewer, regulatory biol, NSF, 80-83. *Mem:* Am Asn Cancer Res; Tissue Cult Asn; Endocrine Soc; Soc Study Reproduction; Am Soc Zoologists; Am Asn Aging; Am Asn Bone & Mineral Res. *Res:* Tumor biology; reproductive physiology; differentiation; tumor-endocrinology; perinatal; gene activation; oncogene. *Mailing Add:* Dept Interdisciplinary Studies Union Inst Union Grad Sch 3419 Gannett St Houston TX 77025

WARNER, MONT MARCELLUS, b Fillmore, Utah, Oct 9, 19; m 47. GEOLOGY. *Educ:* Brigham Youn Univ, AB, 47, MA, 49; Univ Iowa, PhD(geol), 63. *Prof Exp:* Geologist, Shell Oil Co, 49-55; consult petrol geol, La & Utah, 55-59; instr geol, Brigham Young Univ, 59-61; asst prof, Ariz State Univ, 63-67, res grant, 64-66; chmn, Dept Ecol, Boise State Col, 67-70, prof geol, 67-84; RETIRED. *Concurrent Pos:* Idaho state rep in geothermal matters, 70-72; res & legis work in geothermal resources, 70-72; mem exec comt & comt info educ, Geothermal Resources Coun, 70-72. *Mem:* Am Asn Petrol Geologists; Nat Asn Geol Teachers; Ecol Soc Am; Nat Inst Prof Geol. *Res:* Sedimentation; structural and petroleum geology. *Mailing Add:* 8916 Brookview Dr Boise ID 83709

WARNER, NANCY ELIZABETH, b Dixon, Ill, July 8, 23. PATHOLOGY. *Educ:* Univ Chicago, SB, 44, MD, 49; Am Bd Path, dipl, 54. *Prof Exp:* From intern to asst resident path, Univ Chicago Clins, 49-50; resident, Cedars of Lebanon Hosp, Los Angeles, Calif, 53-54, asst pathologist, 54-58; from asst prof to assoc prof path, Univ Chicago, 58-65, dir, Lab Surg Path, Univ Clins, 59-65; assoc clin prof path, Univ Southern Calif, 65-66; assoc prof, Sch Med, Univ Wash, 66-67; assoc prof, 67-69, chmn dept, 72-83, PROF PATH UNIV SOUTHERN CALIF, 69-, ASSOC DEAN ACAD AFFAIRS, 77- *Concurrent Pos:* Assoc dir labs, Cedars-Sinai Med Ctr, Los Angeles, 65-66; chief pathologist, Women's Hosp, Los Angeles County-Univ Southern Calif Med Ctr, 68-72, dir labs & path, 72-83; surgical pathologist, Univ Southern Calif-Norris Cancer Hosp, 83- *Mem:* Fel Col Am Pathologists; Endocrine Soc; Am Asn Anatomists; Microcirculatory Soc; fel Am Soc Clin Pathologists; Sigma Xi. *Res:* Pathology of endocrine glands; comparative pathology of tumors of the gonads; pathology of preneoplastic lesions of the breast; effect of toxins on the microcirculation. *Mailing Add:* Dept Path Univ Southern Calif 1441 Eastlake Ave Los Angeles CA 90033-0804

WARNER, PAUL LONGSTREET, JR, b New York, NY, June 20, 40; c 2. MEDICINAL CHEMISTRY. *Educ:* Ursinus Col, BS, 62; Pa State Univ, MS, 64; State Univ NY Buffalo, PhD(med chem), 69. *Prof Exp:* Dir patents & prod liaison, 68-77, DIR CHEM RES, WESTWOOD PHARMACEUT INC, 77- *Mem:* Am Pharmaceut Asn; AAAS; Am Chem Soc; NY Acad Sci. *Res:* Antimicrobial agents; sunscreen agents; design and synthesis of antipsoriatics; antimetabolites; design and synthesis of anti-inflammatory agents. *Mailing Add:* 926 Highland Ave Buffalo NY 14223

WARNER, PETER, b Winnipeg, Man, Apr 22, 20; m 52; c 6. MICROBIOLOGY. *Educ:* Univ London, MB & BS, 44, MD, 48, PhD(path), 51. *Prof Exp:* Asst pathologist, Bland-Sutton Inst Path, Middlesex Hosp, London, Eng, 43-44 & 46-52; pathologist, Winnipeg Gen Hosp, 53-54; assoc prof path, Fac Med, Univ Man, 54-55; head med res div, Inst Med Sci, Australia, 55-58; assoc prof med bact, Fac Med, Univ Man, 58-74; asst regional dir, Man Med Servs Br, Health & Welfare, Can, 74-81; dir, Environ Health, Man Govt, 81-86; RETIRED. *Concurrent Pos:* Brit Empire Cancer Campaign traveling scholar, Walter Reed Army Med Ctr, DC, 48-49; asst dep minister health, Man Govt, 67-71; chmn, Man Clean Environ Comn, 71-72; consult, Man Inst Tech; ed, Man Med Rev. *Mem:* Fel Am Acad Microbiol; fel Am Soc Clin Path; Can Soc Microbiol; Can Asn Path; Path Soc Gt Brit & Ireland. *Res:* Medical microbiology; experimental pathology; health administration. *Mailing Add:* 3202-55 Nassau Winnipeg MB R3L 2G8 Can

WARNER, PHILIP MARK, b New York, NY, Nov 5, 46. ORGANIC CHEMISTRY. *Educ:* Columbia Univ, BA, 66; Univ Calif, Los Angeles, PhD(chem), 70. *Prof Exp:* NSF fel, Yale Univ, 70-71; from instr to asst prof, 71-77, ASSOC PROF ORG CHEM, IOWA STATE UNIV, 77- *Mem:* Am Chem Soc; Royal Soc Chem. *Res:* Strained ring compounds; carbonium ions; synthesis. *Mailing Add:* Dept Chem Northeastern Univ Boston MA 02115-5096

WARNER, R(ICHARD) E(LMORE), b Canton, Ohio, May 7, 21; m 43. CHEMICAL ENGINEERING. *Educ:* Miami Univ, AB, 42; Ohio State Univ, MSc, 48, PhD(chem eng), 51. *Prof Exp:* Asst chem eng, Ohio State Univ, 47-49, res assoc, Res Found, 50-51; res chem engr, Olin Mathieson Chem Corp, 51-53, sr chem engr, 53-56, process develop sect chief, 56-58, tech rep, 58-59, asst to vpres & tech dir energy div, 59-61; consult, 65-66; PROF MECH ENG, UNIV IDAHO, 66- *Concurrent Pos:* Vpres, Techno Fund Inc, 61-65; dir, Wallace Expanding Mach, Inc, Ind, 62-70; assoc dir eng, Exp Sta, Univ Idaho, 66-74. *Mem:* Nat Soc Prof Engrs; Int Solar Energy Soc; Sigma Xi. *Res:* Corrosion; phosphate fertilizer chemistry; organic and inorganic process development; market development; process economics; production engineering; solar energy utilization. *Mailing Add:* 605 Ridge Rd Moscow ID 83843

WARNER, RAY ALLEN, b Davis, Calif, May 5, 38; m 65. EXPERIMENTAL NUCLEAR PHYSICS. *Educ:* Univ Calif, Berkeley, BS, 61; Univ Calif, Davis, PhD(physics), 69. *Prof Exp:* Res assoc, 69-71, asst prof chem & physics, Cyclotron Lab, Mich State Univ, 72-77; sr res scientist, anal & nuclear res, 77-90, PROG MGR, OFF NAT SECURITY TECHNOL, PAC NORTHWEST LAB, 90- *Concurrent Pos:* Tech adv, US Dept Energy Off Arms Control, 88-90. *Mem:* Am Phys Soc; AAAS. *Res:* Experiments in collective and intrinsic nuclear structure, parity mixing, beta strength functions, fission yields and properties of delayed neutron emitters; instrumentation for mass spectrometry radio-analytical chemistry; research into technology applicable to verification of compliance with arms control treaties. *Mailing Add:* Pac Northwest Lab PO Box 999 Richland WA 99352

WARNER, RAYMOND M, JR, b Barberton, Ohio, Mar 21, 22; m 48, 74; c 3. PHYSICS. *Educ:* Carnegie Inst Technol, BS, 47; Case Univ, MS, 50, PhD(physics), 52. *Prof Exp:* Lab asst, Pittsburgh Plate Glass Co, Ohio, 41 & 42; lab instr physics, Carnegie Inst Technol, 43; jr physicist, Res Lab, Corning Glass Works, NY, 47-48; instr physics, Case Univ, 51-52; mem tech staff semiconductor device develop, Bell Tel Labs, Inc, NJ, 52-59; chief engr diode develop, Semiconductor Prod Div, Motorola, Inc, Ariz, 59-61, mgr mat res, 61-63, dir eng, 63-65; mgr metal-oxide-semiconductor devices prog, Semiconductor Components Div, Tex Instruments Inc, 65-67; US tech dir, Semiconductor Div, Int Tel & Telegraph Corp, Fla, 67-69; dir technol, Semiconductor Dept, Union Carbide Corp, Calif, 69-70; prof elec eng, 70-, chmn res microelectronics comt, Univ Minn, Minneapolis, 70-80; RETIRED. *Mem:* AAAS; fel Inst Elec & Electronics Engrs; Sigma Xi. *Res:* Semiconductor device physics and engineering; solid-state electronics; intergrated circuits. *Mailing Add:* 6136 Sherman Circle Edina MN 55436-1954

WARNER, RICHARD CHARLES, b Chicago, Ill, Sept 22, 47. LANDFILL DESIGN, SEDIMENT CONTROL. *Educ:* Univ Ill, BS, 70; Clemson Univ, MS, 72, PhD(environ syst eng), 82. *Prof Exp:* Oper officer oceanog, USN, Lewes, Del, 74-75; res asst polit sci, Polit Sci Dept, Clemson Univ, 75-76, res asst pollution control, Environ Systs Eng, 76-79; hydrologist, US Geol Surv, 79-80; exten specialist irrig, 80-81, asst prof, 81-86, ASSOC PROF POLLUTION CONTROL, AGR ENG DEPT, UNIV KY, 86- *Concurrent Pos:* Spec intel, USN, Keflavik, Iceland, 72-73, comput systs dir, 73-74, oceanog res officer, 73-74; consult, Nat Acad Sci, 84- *Mem:* Am Soc Agr Engrs (secy, 85). *Res:* Technology transfer in the water resources areas; hydrology; sediment control; landfill design, monitoring and modelling; irrigation; computer aided design. *Mailing Add:* Agr Eng Dept Rm 108 Univ Ky Lexington KY 40506

WARNER, RICHARD DUDLEY, b Pittsfield, Mass, Aug 3, 44; m 74; c 2. PLANETARY SCIENCE. *Educ:* Mass Inst Technol, SB, 66; Stanford Univ, PhD(geol), 71. *Prof Exp:* Res assoc geol, Univ Md, 73-74; res assoc geol, Univ NMex, 74-80; asst prof, 80-85, ASSOC PROF GEOL, CLEMSON UNIV, 85-, ACTG DEPT HEAD, 86- *Concurrent Pos:* Resident res assoc, Goddard Space Flight Ctr, NASA, 71-73; NASA/Am Soc Eng Ed summer fac fel, 81, 82. *Mem:* Mineral Soc Am. *Res:* Petrology of lunar samples; petrology of terrestrial mafic and ultramafic rocks; experimental petrology; magnetic petrology. *Mailing Add:* Dept Geol Clemson Univ Clemson SC 29634-1908

WARNER, RICHARD G, b Washington, DC, Nov 1, 22; m 49; c 4. ANIMAL NUTRITION. *Educ:* Ohio State Univ, BSc, 47, MS, 48; Cornell Univ, PhD(animal nutrit), 51. *Prof Exp:* PROF ANIMAL NUTRIT, CORNELL UNIV, 51- *Concurrent Pos:* Consult dairy nutrit, Empresa Brasileira Pesquisa Agropecuaria, Brazil. *Mem:* Am Soc Animal Sci; Am Dairy Sci Asn; Am Inst Nutrit. *Res:* Calf and ruminant nutrition; laboratory animal nutrition; food intake physiology. *Mailing Add:* Dept Animal Sci Cornell Univ 149 Morrison Hall Ithaca NY 14853

WARNER, ROBERT, b Buffalo, NY, Feb 16, 12; m 39; c 2. PEDIATRICS. *Educ:* Harvard Univ, AB, 35; Univ Chicago, MD, 39; Am Bd Pediat, dipl, 49. *Prof Exp:* Intern, Buffalo Gen Hosp, 39-40; spec intern pediat, Buffalo Children's Hosp, 40-41, res, 46-47; asst, Cincinnati Children's Hosp & Res Found, 47-48; pvt pract, 48-55; assoc prof pediat, Sch Med, State Univ NY Buffalo, 56-82; clin assoc prof rehab med, 76-82; med dir children's rehab ctr, Buffalo Children's Hosp, 56-82; RETIRED. *Concurrent Pos:* Vis teacher, Buffalo Gen Hosp; attend, Buffalo Children's Hosp; ed, Acta Geneticae Medecae et Genellologiae. *Mem:* Am Soc Human Genetics; fel Am Pub Health Asn; fel Am Acad Pediat; fel Am Acad Cerebral Palsy; Am Acad Neurol. *Res:* Etiologics and relationships for prevention and prognostication of congenital anomalies and mental retardation, especially Down's Syndrome; endocrinology; human genetics; rehabilitation of children with neuromuscular disease cerebral palsy; juvenile amputee; reading and learning problems; treatment of phenylketonuria. *Mailing Add:* 106 Soldiers Pl Buffalo NY 14222

WARNER, ROBERT COLLETT, b Denver, Colo, Aug 31, 13; m 36, 69; c 3. BIOCHEMISTRY. *Educ:* Calif Inst Technol, BS, 35; NY Univ, MS, 37, PhD(biochem), 41. *Prof Exp:* Chemist protein chem, Eastern Regional Lab, USDA, Philadelphia, 41-46; from asst prof to prof biochem, Sch Med, NY Univ, 46-69; chmn dept, 69-77, prof, 69-84, EMER PROF MOLECULAR BIOL & BIOCHEM, UNIV CALIF, IRVINE, 84- *Concurrent Pos:* Mem panel plasma, Nat Res Coun, 52-57; Guggenheim Mem Found fel, Carlsberg Lab, Copenhagen, Denmark, 58; mem study sect biophys & biophys chem, NIH, 60-64; chmn study sect biochem, 72-74; assoc ed, J Biol Chem, 68-72. *Mem:* Am Soc Biol Chemists; Biophys Soc; Am Chem Soc. *Res:* Physical biochemistry of nucleic acids and proteins; mechanism of genetic recombination in small DNA containing bacteriophages; properties of circular DNA; bacterial plasmids. *Mailing Add:* Dept Molecular Biol & Biochem Univ Calif Irvine CA 92717

WARNER, ROBERT EDSON, b Pomeroy, Ohio, Apr 11, 31; m 86; c 4. PARTICLE-PARTICLE CORRELATIONS, TOTAL REACTION CROSS SECTIONS. *Educ:* Antioch Col, BS, 54; Univ Rochester, PhD(physics), 59. *Prof Exp:* From instr to asst prof physics, Univ Rochester, 59-61; asst prof, Antioch Col, 61-63; from asst prof to assoc prof, Univ Man, 63-65; assoc prof, 65-72, PROF PHYSICS, OBERLIN COL, 72-, DEPT CHAIR, 90- *Concurrent Pos:* NSF sci fac fel, Oxford Univ, 71-72; vis prof, Mich State Univ, 80-81, Univ Notre Dame, 87-88. *Mem:* Am Phys Soc; Am Asn Physics Teachers. *Res:* Nuclear scattering and reactions; nucleon-nucleon scattering; final-state interactions; angular correlations; reaction cross sections. *Mailing Add:* Dept Physics Oberlin Col Oberlin OH 44074

WARNER, ROBERT LEWIS, b Redwood Falls, Minn, June 16, 37; m 64; c 2. AGRONOMY, PLANT PHYSIOLOGY. *Educ:* Univ Minn, BS, 62, MS, 64; Univ Ill, PhD(plant physiol & agron), 68. *Prof Exp:* Res asst agron, Univ Minn, 62-64 & Univ Ill, 64-68; from asst prof to assoc prof, 68-80, PROF AGRON, WASH STATE UNIV, 80- *Mem:* Am Soc Agron; Am Soc Plant Physiol; Sigma Xi. *Res:* Factors involved in cold resistance of alfalfa; inheritance and physiological studies of nitrate reductase. *Mailing Add:* Dept Agron & Soils Wash State Univ Pullman WA 99164-6420

WARNER, ROBERT RONALD, b Long Beach, Calif, Oct 28, 46; c 3. MARINE ECOLOGY. *Educ:* Univ Calif, Berkeley, AB, 68, Univ Calif, San Diego, PhD(marine ecol), 73. *Prof Exp:* Fel biol, Smithsonian Trop Res Inst, 73-75; PROF BIOL, UNIV CALIF, SANTA BARBARA, 75- *Mem:* Soc Study Evolution; Ecol Soc Am; Am Soc Naturalists; Animal Behav Soc; Am Soc Ichthyologists & Herpetologists. *Res:* Reproductive strategies of marine organisms. *Mailing Add:* Dept Biol Sci Univ Calif Santa Barbara CA 93106

WARNER, SETH L, b Muskegon, Mich, July 11, 27; m 62; c 3. ALGEBRA. *Educ:* Yale Univ, BS, 50; Harvard Univ, MA, 51, PhD(math), 55. *Prof Exp:* From res instr to assoc prof, 55-65, PROF MATH, DUKE UNIV, 65- *Concurrent Pos:* NSF fel, Inst Advan Study, 59-60; vis prof, Univ Paris, 64-65, Univ Oslo, 82-83; vis res prof, Reed Col, 70-71. *Mem:* Am Math Soc; Math Asn Am. *Res:* Topological algebra; abstract analysis. *Mailing Add:* Dept Math Duke Univ Durham NC 27706

WARNER, THOMAS CLARK, JR, b Waterbury, Conn, Nov 20, 19; m 42, 84; c 2. MECHANICAL ENGINEERING, ELECTRICAL ENGINEERING. *Educ:* Yale Univ, BE, 42; Mass Inst Technol, MS, 47. *Prof Exp:* Res & develop officer, USAF, Wright Patterson AFB, Ohio, 42-45 & Hq, Washington, DC, 47-52; chief develop engr, MB Electronics Div, Textron Inc, 52-62; chmn dept math, Univ New Haven, 62-64 & dept mech eng, 62-65, dean, Sch Eng, 65-77, prof mech eng, 65-84; RETIRED. *Concurrent Pos:* Consult, MB Electronics Div, Textron Inc, 62-70. *Mem:* Am Soc Mech Engrs; Am Soc Eng Educ; Inst Environ Sci; Nat Soc Prof Engrs; Am Inst Aeronaut & Astronaut. *Res:* Vibration analysis and instrumentation; electromechanical design. *Mailing Add:* 53 Nolin Rd Westbrook CT 06498

WARNER, THOMAS GARRIE, b El Paso, Tex, July 7, 48. BIOCHEMISTRY. *Educ:* Univ Tex, El Paso, BS, 70; Univ Calif, San Diego, PhD(chem), 75. *Prof Exp:* Res chemist, Scripps Inst Oceanog, 75-77; US Dept Pub Health fel, Neurosci Dept, Univ Calif, 77-78; asst res neuroscientist, 78-79, ASST PROF IN RESIDENCE, DEPT NEUROSCI, UNIV CALIF, 79- *Mem:* Am Chem Soc; Am Oil Chemists Soc; AAAS. *Res:* Human biochemical genetics; lysosomal storage diseases; complex carbohydrates structure and function analysis and relation to disease states; chemical anthesis of lipids and phospholipids; hydrophobic enzymes. *Mailing Add:* Genetech Inc 460 Pt San Bruno Blvd Bldg 1 Rm 1301 South San Francisco CA 94080-4918

WARNER, VICTOR DUANE, b Coulee Dam, Wash, Sept 9, 43; m 68; c 1. MEDICINAL CHEMISTRY. *Educ:* Univ Wash, BS, 66; Univ Kans, PhD(med chem), 70. *Prof Exp:* Asst prof med chem, Col Pharm & Allied Health Sci, Northeastern Univ, 70-74, assoc prof, 74-80, prof, 80-81, actg chmn, Dept Med Chem & Pharmacol, 75-76, chmn, 76-79, actg dean, Col Pharm & Allied Health, 77-78, actg assoc provost, 78-79, assoc provost, 79-81; prof med chem & dean, Col Pharm, Univ PR, 81-85; PROF MED CHEM & DEAN, COL PHARM, UNIV CINCINNATI, 85- *Mem:* Am Chem Soc; Am Pharmaceut Asn; Am Asn Cols Pharm; Am Soc Allied Health Prof. *Res:* Antibacterial agents for inhibition of dental plaque. *Mailing Add:* Col Pharm Univ Cincinnati Cincinnati OH 45267

WARNER, WALTER CHARLES, b Barberton, Ohio, June 2, 20; m 47; c 2. POLYMER CHEMISTRY. *Educ:* Oberlin Col, AB, 41; Case Western Reserve Univ, MS, 43, PhD(org chem), 50. *Prof Exp:* Chem engr, Firestone Tire & Rubber Co, 43-47; sr res chemist, 49-55, group leader anal & testing, 55-62, sect head, Tech Serv, 62-71, sect head, Phys Testing Res & Serv, 71-77, res ctr adminr, 78-80, HEAD, INFO CTR, GEN TIRE & RUBBER CO, 80- *Mem:* Am Chem Soc; Soc Rheology; Am Soc Testing & Mat; Spec Libraries Asn; Am Soc Info Sci; Sigma Xi. *Mailing Add:* 156 Colony Dr Hudson OH 44236-3314

WARNER, WILLIAM, OPIATE PHARMACOLOGY, PESTICIDE TOXICOLOGY. *Educ:* State Univ NY, PhD(pharmacol), 75. *Prof Exp:* ASSOC PROF PHARMACOL, NY UNIV, COL DENT, 83- *Mailing Add:* 80-15 Grand Central Pkwy Jackson Heights NY 11370

WARNER, WILLIAM HAMER, b Pittsburgh, Pa, Oct 6, 29; m 57; c 1. APPLIED MATHEMATICS. *Educ:* Carnegie Inst Technol, BS, 50, MS, 51, PhD(math), 53. *Prof Exp:* Asst math, Carnegie Inst Technol, 50-53; res assoc appl math, Brown Univ, 53-55; from asst prof to assoc prof mech, 55-68, PROF AEROSPACE ENG & MECH, UNIV MINN, MINNEAPOLIS, 68- *Mem:* Soc Indust & Appl Math; Am Math Soc; Math Asn Am; Soc Natural Philos. *Res:* Continuum mechanics; dynamic stability; energy methods; nonlinear systems; optimization of structures. *Mailing Add:* Dept Aerospace Eng & Mech Univ Minn 107 Akerman Hall 110 Union St SE Minneapolis MN 55455

WARNER, WILLIS L, b Endicott, NY, Jan 28, 30. MEDICINE. *Educ:* Syracuse Univ, BA, 50; State Univ NY, MD, 60. *Prof Exp:* Res investr biochem, US Naval Radiol Defense Lab, 52-55; intern, San Francisco Hosp, 60-61; resident obstet, St Mary's Hosp, San Francisco, 61-63; assoc clin res, Baxter Labs, Ill, 63-68, assoc dir clin res, 68-71; dir clin res-biol, Hoechst Pharmaceut, Inc, 71-75; dir clin res, 75-76, dir med opers, Cutter Labs, Inc, 77-79; dir & pres, Consults for Health Care, 79-90; OWNER, GLB PUBLISHERS, 90- *Concurrent Pos:* Ed, Plasma Forum, 79-81. *Mem:* Am Soc Pharmacol Therapeut; Am Soc Hemat; Am Asn Blood Banks; Am Heart Asn. *Res:* Clinical pharmaceutical. *Mailing Add:* 935 Howard St Suite B San Francisco CA 94103

WARNES, DENNIS DANIEL, b Stephen, Minn, June 14, 33; m 56; c 3. AGRONOMY, WEED SCIENCE. *Educ:* NDak State Univ, BSc, 55; Univ Minn, St Paul, MSc, 60; Univ Nebr, Lincoln, PhD(plant breeding), 69. *Prof Exp:* Technician agron, NDak State Univ, 51-55; teaching asst agron, Univ Minn, Minneapolis, 57-60; instr agron & outstate testing, Univ Nebr, Lincoln, 60-66, supvr, Mead Field Lab, 66-69; AGRONOMIST, WCENT EXP STA, UNIV MINN, MINNEAPOLIS, 69- *Mem:* Am Soc Agron; Weed Sci Soc Am. *Res:* Variety testing, weed control, row spacing, plant population, disease and insect control in corn, soybeans, field beans, sunflowers, small grains and forage crops; principles of weed control with specific weeds. *Mailing Add:* 11 S Court St Morris MN 56267

WARNES, RICHARD HARRY, b Chicago, Ill, Apr 23, 33; m 55; c 2. OPTICS, INSTRUMENTATION. *Educ:* DePauw Univ, BA, 55; Stanford Univ, MS, 57. *Prof Exp:* STAFF MEM PHYSICS, LOS ALAMOS NAT LAB, 57- *Mem:* Am Phys Soc. *Res:* Shock wave phenomena; dynamic equation of state; constitutive relations; optical instrumentation; instrumentation. *Mailing Add:* 299 La Cueva Los Alamos NM 87544

WARNHOFF, EDGAR WILLIAM, b Knoxville, Tenn, May 5, 29; m 56; c 3. ORGANIC CHEMISTRY. *Educ:* Wash Univ, St Louis, AB, 49; Univ Wis, PhD(org chem), 53. *Prof Exp:* Nat Res Coun fel, Birkbeck Col, Eng, 53-54; asst scientist, Nat Heart Inst, 54-57; NSF fel, Fac Pharm, Univ Paris, 57-58; res assoc chem, Mass Inst Technol, 58-59; asst prof, Univ Southern Calif, 59-62; from asst prof to assoc prof, 62-66, PROF CHEM, UNIV WESTERN ONT, 66- *Concurrent Pos:* Mem ed adv bd, J Org Chem, 70-74 & Can J Chem, 75-77; ed, Can J Chem, 82-88. *Honors & Awards:* Merck, Sharp & Dohme Lect Award, Chem Inst Can, 69. *Mem:* Chem Inst Can; Am Chem Soc; Royal Soc Chem. *Res:* Organic nitrogen chemistry; mechanisms of organic reactions, especially alpha-substituted carbonyl compounds. *Mailing Add:* Dept Chem Univ Western Ont London ON N6A 5B7 Can

WARNICK, ALVIN CROPPER, b Hinckley, Utah, Nov 15, 20; m 47; c 3. PHYSIOLOGY. *Educ:* Utah State Agr Col, BS, 42; Univ Wis, MS, 47, PhD(physiol of reprod), 50. *Prof Exp:* Asst animal husb & genetics, Univ Wis, 46-50; asst prof, Ore State Col, 50-53; from asst physiologist to assoc physiologist, 53-62, PHYSIOLOGIST, UNIV FLA, 62- *Concurrent Pos:* Physiologist, UNFAO, Balcarce, Arg, 63. *Mem:* Am Soc Animal Sci; Am Soc Study Reproduction; Genetic Asn; Sigma Xi; Int Embryo Transfer Soc. *Res:* Physiology of reproduction; effect of nutrition on fertility; estrous synchronization; genetics of reproduction. *Mailing Add:* Dept Animal Sci Univ Fla 106 Animal Sci Bldg Gainesville FL 32611

WARNICK, EDWARD GEORGE, radiology, for more information see previous edition

WARNICK, JORDAN EDWARD, b Boston, Mass, Mar 21, 42; m 70; c 1. NEUROPHARMACOLOGY. *Educ:* Mass Col Pharm, BS, 63; Purdue Univ, Lafayette, PhD(pharmacol), 68. *Prof Exp:* USPHS trainee pharmacol, Sch Med, State Univ NY Buffalo, 68-70 & spec awardee, 70-71, asst prof biochem pharmacol, Sch Pharm, 71-74; asst prof, 74-80, ASSOC PROF PHARMACOL & EXP THERAPEUT, SCH MED, UNIV MD, BALTIMORE CITY, 80- *Concurrent Pos:* Vis assoc prof physiol, Armed Forces Radiobiol Res Inst, Bethesda, MD, 85-86; vis prof neurosurg, Nat Taiwan Med Ctr, Taipei, Taiwan, 88. *Mem:* AAAS; NY Acad Sci; Am Soc Pharmacol & Exp Therapeut; Soc Neurosci; Soc Biophys; Soc Toxicol. *Res:* Physiology and pharmacology of muscular dystrophy and allied neuromuscular disorders; trophic influence of nerve on muscle; degeneration and regeneration in the peripheral and central nervous systems; pharmacology of neurotoxins; effects of psychoactive drugs on nicotinic receptor-channel complex, hypocampus and spinal cord; pharmacology of anticholinesterases, TRH and related peptides in the central nervous system disorders. *Mailing Add:* Dept Pharmacol & Exp Therapeut Univ Md Sch Med Baltimore MD 21201

WARNICK, ROBERT ELDREDGE, b Pleasant Grove, Utah, July 10, 29; m 53; c 6. NUTRITION. *Educ:* Brigham Young Univ, BS, 55; Utah State Univ, MS, 63, PhD(poultry nutrit), 70. *Prof Exp:* Agr inspector, Utah State Dept Agr, 58-59; lab technician chem anal, 59-60, res asst poultry res, 60-76, RES ASST PROF TURKEY RES, UTAH STATE UNIV, 76- *Mem:* Poultry Sci. *Res:* Nutrition and management of growing turkeys. *Mailing Add:* 1565 N University Ave No 17 Provo UT 84604

WARNICK, WALTER LEE, b Baltimore, Md, May 31, 47; m 70; c 2. ENVIRONMENTAL EFFECTS OF ACID RAIN. *Educ:* Johns Hopkins Univ, BS, 69; Univ Md, MS, 72, PhD(mech eng), 77. *Prof Exp:* Engr, Westinghouse Aerospace, 69-71 & US Naval Res Lab, 71- 77; engr, 77-85, DIV DIR, PROG INTEGRATION ANALYSIS DIV, OFF ENERGY RES, US DEPT ENERGY, 85- *Concurrent Pos:* Mem, Aquatics Task Group, Nat Acid Precipitation Assessment Prog, 81- *Mem:* Am Soc Mech Engrs. *Res:* Critical examination of the causal chain linking human activities to environmental effects; environmental model testing. *Mailing Add:* Prog Integration Analysis Div US Dept Energy Off Energy Res Washington DC 20585

WARNKE, DETLEF ANDREAS, b Berlin, Ger, Jan 29, 28; m 64; c 2. EARTH SCIENCES. *Educ:* Freiburg Univ, dipl, 53; Univ Southern Calif, PhD(geol), 65. *Prof Exp:* Res asst geol, Aachen Tech Univ, 55-56; jr exploitation engr, Shell Oil Co, 57-58; res asst oceanog, Alan Hancock Found, 59-61; from res assoc to asst prof oceanog & geol, Fla State Univ, 65-71; asst prof, 71-73, actg chmn dept, 73-74, assoc prof, 73-77, prof earth sci, 77-80, PROF GEOL SCI, CALIF STATE UNIV, HAYWARD, 80- *Concurrent Pos:* Exchange prof, Free Univ Berlin, 80-81; fac res partic, US Geol Surv, 85; Fulbright Scholar, Free Univ, Berlin, 87-88. *Mem:* Am Geophys Union; Soc Econ Paleont & Mineral; Ger Geol Union; Geol Soc Am. *Res:* Geochemistry; marine geology; geomorphology; beach erosion. *Mailing Add:* Dept Geol Sci Calif State Univ Hayward CA 94542

WARNOCK, DAVID GENE, b Parkes, Ariz, Mar 5, 45. RENAL PHYSIOLOGY & TRANSPORT SYSTEMS. *Educ:* Univ Calif, San Francisco, MD, 70. *Prof Exp:* ASSOC PROF MED & PHARMACOL, UNIV CALIF, SAN FRANCISCO, 82-; CHIEF, NEPHROLOGY SECT, VET ADMIN MED CTR, 83- *Mem:* AAAS; Am Physiol Soc; Am Soc Clin Invest; Am Soc Nephrology. *Mailing Add:* Dept Med 111J Univ Calif 4150 Clement St San Francisco CA 94121

WARNOCK, JOHN EDWARD, b Freeport, Ill, Aug 20, 32; m 55; c 2. ZOOLOGY. *Educ:* Univ Ill, BS, 54; Univ Wis, MS, 58, PhD(zool), 63. *Prof Exp:* Field asst wildlife res, Ill Natural Hist Surv, 54 & Southern Ill Univ & Ill Natural Hist Surv, 54-55; res assoc, Ill Natural Hist Surv & Univ Ill, 62-64; from asst prof to assoc prof, 64-72, prof biol, 72-87, dir, Inst Environ Mgt, 79-87, EMER PROF SCI, WESTERN ILL UNIV, 87- *Mem:* Am Soc Mammal; Wildlife Soc; Sigma Xi. *Res:* Mammalogy; animal behavior; vertebrate ecology; ornithology; ecology of small mammal populations, especially the relationships of behavior, physiological condition, density and physical factors of the environment; biology and management of small game animals. *Mailing Add:* Dept Biol Sci Western Ill Univ Macomb IL 61455

WARNOCK, LAKEN GUINN, b Newton Falls, Ohio, Apr 19, 28; m 53; c 2. BIOCHEMISTRY. *Educ:* Milligan Col, BS, 57; Vanderbilt Univ, PhD(biochem), 62. *Prof Exp:* Instr biochem, Okla State Univ, 62-64; ASST PROF BIOCHEM, VANDERBILT UNIV, 64-; BIOCHEMIST, VET ADMIN HOSP, 64- *Concurrent Pos:* Consult, Interdept Comt Nutrit, Nat Defense Nutrit Surv, Lebanon, 61. *Mem:* Am Inst Nutrit. *Res:* Carbohydrate metabolism in vitamin deficiencies; vitamin nutriture in hemodialysis. *Mailing Add:* Lab Dept Vet Admin Hosp Nashville TN 37240

WARNOCK, MARTHA L, b Detroit, Mich, July 19, 34; m 59; c 2. PATHOLOGY. *Educ:* Oberlin Col, AB, 56; Harvard Univ, MD, 60. *Prof Exp:* From instr to prof path, Univ Chicago, 65-78; prof path, 78-89, EMER PROF PATH, UNIV CALIF, SAN FRANCISCO, 89- *Mem:* AAAS; Am Thoracic Soc; Int Acad Path; Am Asn Pathologists. *Mailing Add:* HSW501 Univ Calif San Francisco CA 94143-0506

WARNOCK, ROBERT G, b Salt Lake City, Utah, Mar 28, 25; m 48; c 3. PARASITOLOGY. *Educ:* Univ Utah, BS, 49, MS, 51, PhD(parasitol), 62. *Hon Degrees:* DSc, Westminster Col, 90. *Prof Exp:* Instr biol, Univ Utah, 62-63; from asst prof to prof, 63-90, EMER PROF BIOL, WESTMINSTER COL, UTAH, 90- *Mailing Add:* Dept Biol Westminster Col Salt Lake City UT 84105

WARNOCK, ROBERT LEE, b Portland, Ore, Feb 20, 30; m 59; c 2. MATHEMATICAL PHYSICS, ACCELERATOR THEORY. *Educ:* Reed Col, BA, 52; Harvard Univ, AM, 55, PhD(physics), 59. *Prof Exp:* Res assoc, Boston Univ, 59-60 & Univ Wash, Seattle, 60-62; from asst prof to prof, 62-79, ADJ PROF, THEORET PHYSICS, ILL INST TECHNOL, 79-; PHYSICIST, STANFORD LINEAR ACCELERATOR CTR, STANFORD UNIV, 87- *Concurrent Pos:* Asst physicist, Argonne Nat Lab, 64-67, assoc physicist, 67-71; vis scientist, Int Ctr Theoret Physics, Italy, 67; vis prof, Imp Col, Univ London & Univ Bonn, 72; sci assoc, Inst Theoret Phys, Univ Groningen, 76; participating guest, Lawrence Berkeley Lab, 78-87. *Mem:* Am Phys Soc; Soc Indust & Appl Math; Am Math Soc. *Res:* Accelerator theory; nonlinear mechanics; electromagnetic theory; nonlinear mathematical physics; particles and fields. *Mailing Add:* SLAC-Bin 26 PO Box 4349 Stanford CA 94309

WARPEHOSKI, MARTHA ANNA, b Wausau, Wis, Feb 22, 49; div. ORGANIC CHEMISTRY, POLYMER SCIENCE. *Educ:* Mass Inst Technol, SB, 71, PhD(org chem), 77; Johns Hopkins Univ, MA, 75. *Prof Exp:* Res assoc biomat, Dept Mech Eng, Mass Inst Technol, 77-79; SCIENTIST, UPJOHN CO, 81- *Mem:* Am Chem Soc; Sigma Xi. *Res:* Biological responses to polymeric materials; properties and modifications of biopolymers; oxidation of organic molecules; chemiluminescence of organic molecules; organic synthesis; medicinal chemistry. *Mailing Add:* Upjohn Co 7252-209-6 301 Henrietta St Kalamazoo MI 49001-0199

WARR, WILLIAM BRUCE, b Providence, RI, June 24, 33; wid; c 2. NEUROANATOMY. *Educ:* Brown Univ, BA, 57, MA, 58, PhD(physiol psychol), 63. *Prof Exp:* NIH fel neurophysiol & neuroanat, Eaton-Peabody Lab Auditory Physiol, Mass Eye & Ear Infirmary, 63-64, res assoc neurophysiol & neuroanat, 64-67; assoc prof anat, Sch Med, Boston Univ, 67-78; RES ASSOC & DIR, NEUROANAT LAB, BOYS TOWN INST COMMUN DISORDERS CHILDREN, 78-; PROF HUMAN COMMUN, SCH MED, CREIGHTON UNIV, 78- *Concurrent Pos:* Asst, Harvard Med Sch, 67. *Mem:* AAAS; Am Asn Anatomists. *Res:* Neuroanatomy of the auditory system; efferent innervation of the cochlea. *Mailing Add:* 3512 S 133rd Avenue Circle Omaha NE 68144

WARREN, ALAN, b Philadelphia, Pa, Dec 20, 36; m 73. LABORATORY SAFETY. *Educ:* Univ Pa, BA, 58. *Prof Exp:* Res & develop chemist, Philadelphia Quartz Co, 58-63, tech serv rep, 63-66, res & develop serv mgr, 66-70, tech serv specialist, 70-72; int mkt specialist, 72-75, mkt develop specialist, 76-77, RES & DEVELOP OPERS MGR, PQ CORP, 77- *Concurrent Pos:* Secy-treas, Synthetic Amorphous Silica & Silicate Indust Asn, 87-88, vchmn, 89-90, chmn, 91. *Mem:* Am Inst Chemists; Am Chem Soc; AAAS; Soc Res Adminr; Soc Hist Alchemy & Chem. *Res:* Properties and applications of soluble silicates and derivatives. *Mailing Add:* Box 17124 Philadelphia PA 19105

WARREN, BRUCE ALBERT, b Sydney, Australia, Nov 2, 34; div; c 2. PATHOLOGY. *Educ:* Univ Sydney, BSc, 57, MB, BS, 59; Oxford Univ, PhD(path), 64, MA, 67, DSc, 84; FRCP Australasia, 72; MRCPath, UK, 73, FRCPath, UK, 80. *Prof Exp:* Brit Commonwealth scholar, Sir William Dunn Sch Path, Oxford Univ, 62-64; res fel, Div Oncol, Inst Med Res, Chicago Med Sch, 64-65; lectr, Nuffield Dept Surg, Oxford Univ, 66-68, tutor path, Oxford Med Sch, 67-68; vis asst prof anat, 68, from asst prof to prof path, Univ Western Ont, 68-80; clin prof path, 80-83, PROF PATH, UNIV NSW, 83- *Concurrent Pos:* Consult, Westminster Hosp, London, Ont, 68-77; consult, Univ Hosp, London, Ont, 72-80; dir cytopath serv, Univ Hosp, 76-80; assoc ed, Biosci Commun & Companion Life Sci; head anat path, Prince Henry Hosp, 80- *Mem:* AAAS; fel Royal Micros Soc; fel Royal Soc Med; NY Acad Sci. *Res:* Thrombosis and athero-embolism; platelet aggregation; fibrinolysis; function of endothelium; microvasculature of tumors; transplantation and growth of tumors; tumor emboli in bloodstream; asbestosis; mesothelioma. *Mailing Add:* Dept Path Prince Henry Hosp Sydney NSW 2036 Australia

WARREN, BRUCE ALFRED, b Waltham, Mass, May 14, 37. PHYSICAL OCEANOGRAPHY. *Educ:* Amherst Col, BA, 58; Mass Inst Technol, PhD(phys oceanog), 62. *Prof Exp:* Res asst phys oceanog, 62-63, from asst scientist to assoc scientist, 63-78, SR SCIENTIST, WOODS HOLE OCEANOG INST, 78- *Concurrent Pos:* Co-ed, J Phys Oceanog, 80-85. *Mem:* Am Geophys Union; Sigma Xi; Am Meteorol Soc. *Res:* Dynamics of ocean currents; water-mass structures; general ocean circulation. *Mailing Add:* Dept Phys Oceanog Woods Hole Oceanog Inst Woods Hole MA 02543

WARREN, CHARLES EDWARD, b Portland, Ore, Oct 26, 26; m 48; c 3. ECOLOGY, FISH BIOLOGY. *Educ:* Ore State Col, BS, 49, MS, 51; Univ Calif, PhD(zool), 61. *Prof Exp:* Asst prof, 53-59, assoc prof, 59-65, PROF FISHERIES, ORE STATE UNIV, 65-; GEN COORDR BIOL, DEPT FISHERIES & WILDLIFE, OAK CREEK LAB, 57- *Concurrent Pos:* Actg head, Dept Fisheries & Wildlife, Ore State Univ, 70-74. *Mem:* Sigma Xi. *Res:* Water pollution biology; autecology; population ecology; community ecology; theoretical ecology; resource management; philosophy of science. *Mailing Add:* Dept Fisheries & Wildlife Ore State Univ Corvallis OR 97331

WARREN, CHARLES REYNOLDS, b Kyoto, Japan, Sept 24, 13; US citizen; m 45. PLEISTOCENE GEOLOGY. *Educ:* Yale Univ, BS, 35, PhD(geol), 39. *Prof Exp:* Jr geologist, Socony-Vacuum Oil Co, Venezuela, 39-40; jr geologist, US Geol Surv, Calif & Ore, 40-41, asst geologist, NY, 46; instr geol, Yale Univ, 46-47; asst prof, Washington & Lee Univ, 47-52; geologist, US Geol Surv, 52-80; CONSULT, 80- *Mem:* Fel AAAS; fel Geol Soc Am. *Res:* Geomorphology; glacial geology. *Mailing Add:* 10450 Lottsford Rd Mitchellville MD 20716

WARREN, CHRISTOPHER DAVID, b Luton, Eng, Apr 24, 38; m 66; c 2. CARBOHYDRATE CHEMISTRY. *Educ:* Univ Sheffield, BS, 60, PhD(carbohydrate chem), 63; Royal Inst Chem, ARIC, 64. *Prof Exp:* Mem sci staff lipid & carbohydrate chem, Med Res Coun, Nat Inst Med Res, London, 63-69; res fel biol chem, 69-70, assoc, 70-73, prin assoc biol chem, 73-84, ASSOC PROF BIOL CHEM, HARVARD MED SCH, 84- *Concurrent Pos:* Res fel biochem, Mass Gen Hosp, 69-72; asst biochemist, 72-81, Assoc Biochemist, Mass Gen Hosp, 82. *Mem:* Fel Royal Soc Chem; Asn Biol Chemists; AAAS; Soc Complex Carbohydrates; Am Chem Soc. *Res:* Structure, function, biosynthesis, and catabolism of glycoprotein glycans; role of envelope glycoproteins in the pathogenicity of HIV infection; genetic diseases of complex carbohydrate metabolism; synthetic and natural inhibitors of glycoprotein processing; biochemistry of locoweed toxicosis; biosynthesis of glycosylphosphatidylinositol membrane anchors. *Mailing Add:* Two Hillcrest Dr Acton MA 01720

WARREN, CLAUDE EARL, b Columbus, Ohio, Jan 11, 14; m 39; c 2. ELECTRICAL ENGINEERING. *Educ:* Ohio State Univ, BEE, 38; Mass Inst Technol, MS, 40. *Prof Exp:* Meter tester, Ohio Power Co, 38; asst, Mass Inst Technol, 38-40; engr cent sta, Westinghouse Elec Corp, Pa, 40-45; from asst prof to prof, 45-81, supvr res found, 46, EMER PROF ELEC ENG, OHIO STATE UNIV, 81- *Mem:* Sr mem Inst Elec & Electronics Engrs. *Res:* Analog computers; circuit theory. *Mailing Add:* 221 Northmoore Pl Columbus OH 43214

WARREN, CLIFFORD A, b Plainfield, NJ, Nov 6, 13. ELECTRICAL ENGINEERING. *Educ:* Cooper Union, BSEE, 36; Stevens Inst Technol, MSEE, 49. *Prof Exp:* Exec dir, Bell Tel Labs, Whippany, NJ, 31-76; dir bd, Plantronic, Santa Cruz, Calif, 77-87; RETIRED. *Mem:* Fel Inst Elec & Electronics Engrs. *Mailing Add:* 31 Evergreen Lane Watchung NJ 07060

WARREN, CRAIG BISHOP, b Philadelphia, Pa, Oct 21, 39; m 64; c 2. PHYSICAL ORGANIC CHEMISTRY. *Educ:* Franklin & Marshall Col, AB, 61; Villanova Univ, BS, 63; Cornell Univ, PhD(org chem), 70. *Prof Exp:* Fel prebiol evolution, Corp Res Dept, Monsanto Co, 68-69, sr res chemist, 69-73, res specialist, 73-75; group leader, 75-77, sr group leader, 77-80, dir, 80-82, VPRES, INT FLAVORS & FRAGRANCES RES & DEVELOP, 82- *Mem:* Am Chem Soc; Sigma Xi; Soc Cosmetic Chem; Am Soc Testing & Mat; Indust Res Inst. *Res:* Structure-activity relationships of fragrance molecules; quantitative sensory evaluation of flavors and frangrances; neurochemistry of the olfactory system. *Mailing Add:* Int Flavors & Fragrances 1515 Hwy 36 Union Beach NJ 07735

WARREN, DAVID HENRY, b Ithaca, NY, June 9, 30; m 62; c 4. GEOPHYSICS, SEISMOLOGY. *Educ:* Rensselaer Polytech Inst, BS, 51; Columbia Univ, AM, 56. *Prof Exp:* Res asst marine geophys, Lamont Geol Observ, Columbia Univ, 51-53; geophysicist, Stand Oil Soc Calif, 53-61 & Alpine Geophys Assocs, Inc, 61-62; GEOPHYSICIST, US GEOL SURV, 63- *Mem:* Am Geophys Union; Soc Explor Geophys; Seismol Soc Am. *Res:* Explosion seismology studies of the earth's crust and upper mantle; seismic data processing and display techniques; geologic interpretation of geophysical data. *Mailing Add:* 202 W Arbor Ave Sunnyvale CA 94086

WARREN, DON CAMERON, b Saratoga, Ind, July 16, 90; m 10; c 1. GENETICS. *Educ:* Univ Ind, AB, 14, AM, 17; Columbia Univ, PhD(genetics), 23. *Hon Degrees:* Hon Dr, Univ Ind, 72. *Prof Exp:* Sci asst, Carnegie Inst, 14-15; field agent entom, Exp Sta, Ala Polytech Inst, 17-19; asst state entomologist, State Bd Entom, Ga, 19-21; from asst prof to prof poultry genetics, Kans State Col, 23-48; nat coord poultry breeding, USDA, 48-56; geneticist, 56-68, EMER GENETICIST, KIMBER FARMS, INC, CALIF, 68- *Concurrent Pos:* Consult, US Dept of State, India, 55. *Honors & Awards:* Award, Poultry Sci Asn, 33; Borden Award, 40. *Mem:* AAAS; Am Soc Nat; fel Poultry Sci Asn; Am Genetic Asn. *Res:* Genetics of Drosophila and fowl; physiology of reproduction in the fowl. *Mailing Add:* 2860 County Dr No 136 Fremont CA 94536-5364

WARREN, DONALD W, b Brooklyn, NY, Mar 22, 35; m 56; c 2. DENTISTRY. *Educ:* Univ NC, BS, 56, DDS, 59; Univ Pa, MS, 61, PhD(physiol), 63. *Hon Degrees:* Dr(odontol), Univ Kuopio, Finland, 91. *Prof Exp:* From asst prof to assoc prof dent, Sch Med, Univ NC, Chapel Hill, 63-69, prof dent ecol & chmn dept, 70-85, prof dent surg, 69-80, DIR, ORAL & COMMUN DIS PROB, UNIV NC, CHAPEL HILL, 63-, KENAN PROF, 80-, RES PROF OTOLARYNGOL, 88- *Concurrent Pos:* Asst secy gen, Int Cong Cleft Palate, 66-69; consult, Joint Comt Dent & Speech Path-Audiol, Am Dent Asn & Am Speech & Hearing Asn, 67-71; pres, Am Cleft Palate Educ Found, 76-77. *Mem:* Am Dent Asn; Int Asn Dent Res; fel Am Speech & Hearing Asn; Am Cleft Palate Asn (vpres, 67-68, pres, 81-82). *Res:* Physiology of speech; effects of oral-facial disorders on the speech process; effects of breathing on facial morphology; effect of breathing on olfaction. *Mailing Add:* Dent Res Ctr Univ NC Sch Dent Chapel Hill NC 27599

WARREN, DOUGLAS ROBSON, b Fenelon Falls, Ont, July 16, 16; m 42; c 1. OCCUPATIONAL MEDICINE, ENVIRONMENTAL MEDICINE. *Educ:* Univ Toronto, MD, 41, dipl pub health, 47; Can Bd Occup Med, cert, 81. *Prof Exp:* Indust physician & consult, Can, 47-67; dir & partner, Indust Med Consult, Ltd, 67-85; RETIRED. *Concurrent Pos:* Assoc prof indust health, Sch Hyg, Univ Toronto, 67-75, spec lectr, Fac Med, 68-71 & 82-, mem staff hearing conserv course, Div Exten, 69-71; mem assoc comt sci criteria for environ qual, Nat Res Coun Can, 71-78; Can med dir, Occidental Life Calif, 71-81; chmn, Continued Educ Can Coun Occup Med, 75-; mem, Environ Qual Comt, Metrop Toronto Bd Trade, 75-81; exec secy, Occup Med Asn Can & Can Bd Occup Med, 82-90. *Res:* Industrial medicine, related occupational and environmental subjects; hearing conservation and noise control; administration studies related to sickness absence control; metals in the environment and health factors. *Mailing Add:* Lankin Lane Fenelon Falls ON K0M 1N0 Can

WARREN, DWIGHT WILLIAM, III, b Los Angeles, Calif, Dec 21, 42; m 65; c 1. ENDOCRINOLOGY, REPRODUCTION. *Educ:* Univ Calif, AB, 64; Univ Southern Calif, PhD(physiol), 72. *Prof Exp:* from instr to assoc prof, 72-88, PROF PHYSIOL & BIOPHYS, UNIV SOUTHERN CALIF, 88- *Concurrent Pos:* Consult, Amvac Chem Corp, 79-81; Nat Res Serv sr fel, NIH, 80-81; Fulbright scholar, 90. *Mem:* Endocrine Soc; Soc for Study Reproduction; Am Soc Andrology; AAAS; NY Acad Sci. *Res:* Development of the fetal testis and regulation of androgen production by this organ by both internal and external modulators. *Mailing Add:* Dept Physiol & Biophys Sch Med Univ Southern Calif 2025 Zonal Ave Los Angeles CA 90033

WARREN, FRANCIS A(LBERT), b Ashfield, Mass, May 16, 17; m 55. CHEMICAL ENGINEERING. *Educ:* Mass State Col, BS, 39; Ohio State Univ, MS, 49. *Prof Exp:* Lab supvr, Ciba Co, Inc, NY, 40-42; physicist, Hercules Powder Co, 42-45; res engr chem dept, Battelle Mem Inst, 45-49; supvry chem engr & head internal ballistics br, US Naval Ord Test Sta, Calif, 49-54; propellant technologist, Southwest Res Inst, 54-57, mgr spec projs sect, Dept Chem & Chem Eng, 57-58 & propellants sect, 58-60; sr tech specialist, Eng Dept, Rocketdyne Div, NAm Rockwell Corp, 60-72; RETIRED. *Mem:* AAAS; Am Inst Aeronaut & Astronaut; Sigma Xi; Am Chem Soc; Am Ord Asn. *Res:* Solid propellants; internal ballistics of solid-fuel rockets; gas generators; author of a book on rocket propellants. *Mailing Add:* 2133 Lake James Waco TX 76710

WARREN, FRANCIS SHIRLEY, b Winnipeg, Man, Oct 26, 20; m 43; c 6. AGRONOMY. *Educ:* Ont Agr Col, BSA, 46; Univ Minn, MSc, 48, PhD(plant genetics & path), 49. *Prof Exp:* Asst corn breeding, Univ Minn, 47-49; res officer, Exp Farm Can Dept Agr, Ont, 49-53, head forage & cereal sect, NS, 53-66, RES SCIENTIST, FORAGE CROPS SECT, CENT EXP FARM, CAN DEPT AGR, 66- *Mem:* Can Soc Agron; Agr Inst Can. *Res:* Cereal and forage crop production. *Mailing Add:* 45 Rockfield Crest Ottawa ON K2E 5L6 Can

WARREN, GEORGE FREDERICK, b Ithaca, NY, Sept 23, 13; m 44; c 3. WEED SCIENCE. *Educ:* Cornell Univ, BS, 35, PhD(veg crops), 45. *Prof Exp:* Dist county agr agent, Maine, 35-38; asst nutrit veg crops, Cornell Univ, 38-42; asst prof hort, Univ Wis, 45-48; from assoc prof to prof hort, Purdue Univ, 49-79; CONSULT AGR, 79- *Concurrent Pos:* Exec comt mem, Coun Agr Sci & Technol, 75-76, pres, 77-78; hon mem, N Cent Weed Sci Soc. *Honors & Awards:* Campbell Award, Am Inst Biol Sci, 66. *Mem:* Fel AAAS; fel Weed Sci Soc Am (pres, 64-66); fel Am Soc Hort Sci; Weed Sci Soc Am. *Res:* Basis of selective action of herbicides; fate of herbicides in soil; control of weeds in crops. *Mailing Add:* 1130 Cherry Lane West Lafayette IN 47906

WARREN, GEORGE HARRY, b Morrisville, Pa, Sept 7, 16; m 46; c 1. MICROBIOLOGY, CHEMOTHERAPY. *Educ:* Temple Univ, BA, 39, MA, 40; Princeton Univ, PhD(microbiol), 44. *Prof Exp:* Lab instr microbiol, Princeton Univ, 41-42; instr bact-immunol, Jefferson Med Col, 44-46; head dept antibiotics-bact, Wyeth Inst Med Res, 47-59, dir dept microbiol, 60-79, DIR CLIN MICROBIOL, WYETH LABS, 81-; PROF MICROBIOL & IMMUNOL, JEFFERSON MED COL, THOMAS JEFFERSON UNIV, 66- *Concurrent Pos:* Consult, Wyeth-Ayerst Labs, 81- *Mem:* Fel NY Acad Sci; fel Infectious Dis Soc Am. *Res:* Antimicrobial agents and chemotherapy; tumor research; enzymes; mucopolysaccharides; immunostimulatory agents; host response mechanisms. *Mailing Add:* Dept Microbiol Med Col Thomas Jefferson Univ 1020 Locust St Philadelphia PA 19107

WARREN, GUYLYN REA, b Butte, Mont, Aug 16, 41. MOLECULAR BIOLOGY, ANIMAL GENOME MAPPING. *Educ:* Mont State Col, BS, 63; Mont State Univ, PhD(genetics), 67. *Prof Exp:* NIH fel radiation repair, Palo Alto Med Res Found, Stanford Univ, 68-70, res assoc, 70-72; res assoc vet res, Mont State Univ, 73-74, asst prof, 74-76, adj asst prof, 76-80, adj assoc prof chem, 80-89, ADJ ASSOC PROF ANIMAL & RANGE SCI, MONT STATE UNIV, 90- *Concurrent Pos:* Prin investr, USDA, 74-76 & 79-81, Mont Air Pollution Study, 78-80, Smelter Environ Res Assoc, 76-77, Proctor & Gamble, 79-81 & NIH Cancer Inst, 80-82; sci consult, Dept Energy, 77-78, NIH, 82-, Nat Inst Environ Health Sci, 83-86, Mont Agr Exp Sta, 90-93. *Mem:* Environ Mutagen Soc; AAAS; Int Soc Animal Genetics. *Res:* Pollution assessment by rapid microbial bioassays; natural products as mutagens or antimutagens; molecular mode of action of metals as mutagens; ovine genome mapping. *Mailing Add:* Willow Creek MT 59760

WARREN, H(ERBERT) DALE, b Houston, Tex, Apr 8, 32; m 56, 82; c 1. ANALYTICAL CHEMISTRY, INORGANIC CHEMISTRY. *Educ:* Rice Univ, BA, 54; Univ Idaho, MS, 59; Ore State Univ, PhD(anal chem), 66. *Prof Exp:* Tech grad chem, Hanford Atomic Prod Oper, Gen Elec Co, Wash, 56-58, chemist II, 58, tech librn, 59; tech asst chem, Los Alamos Sci Lab, 61; res asst, Union Oil Res Ctr, Calif, 62; from instr to asst prof, 63-74, ASSOC PROF CHEM, WESTERN MICH UNIV, 74- *Mem:* Am Chem Soc; Hist Sci Soc. *Res:* Organic reagents for spectrophotometric analysis; equilibrium constants of coordination compounds; extraction chromatography of inorganic systems; history of chemistry. *Mailing Add:* Dept Chem Western Mich Univ Kalamazoo MI 49008

WARREN, HALLECK BURKETT, JR, b St Louis, Mo, Sept 3, 22; m 51; c 2. BACTERIOLOGY. *Educ:* Univ St Louis, BS, 43; Univ Ill, MS, 49, PhD(bact), 51. *Prof Exp:* Asst dairy bact, Univ Ill, 49-50; res microbiologist, Res Div, Abbott Labs, 51-57; head bact, Res Labs, Pet Inc, 57-68, mgr food sci & eng, 68-70; mgr food sci, Fairmont Foods Co, 70-72; mgr food sci, 72-77, dir food sci & qual assurance, 77-78, VPRES & DIR TECH SERV, INTERSTATE BRANDS CORP, 78- *Mem:* Am Asn Cereal Chem; Am Soc Microbiol; Sigma Xi. *Res:* Microbial physiology of flavor components; bacteriology of foods and dairy products; antibiotic action, production and assay; food preservation. *Mailing Add:* 9929 Mastin Overland Park KS 66212

WARREN, HAROLD HUBBARD, b Derry, NH, July 5, 22. ORGANIC CHEMISTRY. *Educ:* Univ NH, BS, 44, MS, 47; Princeton Univ, MA, 49, PhD(chem), 50. *Prof Exp:* From instr to prof chem, 50-72, Halford R Clark prof, 72-84, EMER PROF CHEM, WILLIAMS COL, 84- *Mem:* Fel AAAS; Am Chem Soc. *Res:* Determination of structure of natural products and synthesis and evaluation of structural variants. *Mailing Add:* 5807 Tidewood Ave Sarasota FL 34231

WARREN, HARRY VERNEY, b Anacortes, Wash, Aug 27, 04; m 34; c 2. GEOLOGY, MINERALOGY. *Educ:* Univ BC, BA, 26, BASc, 27; Oxford Univ, MSc, 28, DPhil, 29. *Hon Degrees:* DSc, Univ Waterloo, 75, Univ BC, 78; FRCGP, Gt Brit, 73. *Prof Exp:* Commonwealth Fund fel, Calif Inst Technol, 29-32; geochem adv, Placer Develop Ltd, Vancouver, 71-74; from lectr to prof, 32-71, PROF MINERAL & PETROL, UNIV BC, 71- *Concurrent Pos:* Exec mem, BC & Yukon Chamber Mines, 39-, from vpres to pres, 39-54; exec mem, UN Asn Can, 48-, pres, 55-58. *Mem:* Fel Geol Soc Am; Am Inst Mining, Metall & Petrol Eng; fel Royal Soc Can; fel Geol Asn Can; Can Inst Mining & Metall; fel Inst Mining & Metall. *Res:* Lead and zinc deposits in southwestern Europe; precious and base metal relationships in western and North America; rarer metals; precious and base metal deposits of British Columbia; trace elements in relation to mineral exploration, epidemiology and biogeochemistry; relationship existing between geology and health. *Mailing Add:* 1816 Western Pkwy Vancouver BC V6T 1V4 Can

WARREN, HERMAN LECIL, b Tyler, Tex, Nov 13, 32; m 63; c 3. PLANT PATHOLOGY. *Educ:* Prairie View Agr & Mech Col, BS, 53; Mich State Univ, MS, 62; Univ Minn, St Paul, PhD(plant path), 69. *Prof Exp:* Res scientist plant path, Olin Mathieson Chem Corp, 62-67; PLANT PATHOLOGIST, AGR RES SERV, USDA, 69- *Concurrent Pos:* From asst prof to prof, Purdue Univ, West Lafayette, 71-81; Commonwealth vis prof, Dept Plant Path Phys & Weed Sci, Va Polytech Inst & State Univ, Blacksburg, Va, 87-88. *Mem:* Am Phytopath Soc; Mycol Soc Am. *Res:* Relationship of soilborne diseases to stalk rot of corn; survival mechanism of soilborne pathogens; effects of light and temperature on spore germination, growth and production of fungi; physiology of host parasites. *Mailing Add:* Dept Bot & Plant Path Purdue Univ West Lafayette IN 47907

WARREN, HOLLAND DOUGLAS, b Wilkes Co, NC, July 31, 32; m 55; c 3. PHYSICS. *Educ:* Wake Forest Col, BS, 59; Univ Va, MS, 61, PhD(nuclear physics), 63. *Prof Exp:* Develop physicist, Celanese Corp Am, 63-64; sr physicist, 64-70, res specialist, 70-87, ADV ENGR, BABCOCK & WILCOX CORP, 87- *Honors & Awards:* I R Award, Res & Develop Magazine, 84. *Mem:* Am Phys Soc; Am Nuclear Soc. *Res:* Neutron spectroscopy; nuclear physics; nuclear instrumentation; neutron radiography; reactor instrumentation. *Mailing Add:* Babcock & Wilcox Corp PO Box 10935 Lynchburg VA 24506-0935

WARREN, J(OSEPH) E(MMET), b Chicago, Ill, Aug 19, 26; wid; c 5. PETROLEUM ENGINEERING. *Educ:* Univ Pittsburgh, BS, 51; Univ Pa, MS, 54, PhD(petrol eng), 60. *Prof Exp:* Res engr, Stanolind Oil & Gas Co, 51-52; res assoc petrol eng, Pa State Univ, 55-56; sect head reservoir eng appln, Gulf Res & Develop Co, 56-63; gen supt reservoirs, Kuwait Oil Co, 63-66; div dir prod, Gulf Res & Develop Co, Pa, 66-67, dept dir explor & prod, 67-70; vpres opers, Santa Fe Int Corp, 70-76; planning adv, Gulf Oil Corp, 76-83; CONSULT PETROL, 83- *Honors & Awards:* Lucas Medal, Soc Petrol Engrs. *Mem:* Soc Petrol Engrs; Inst Mgt Sci; Brit Inst Petrol. *Res:* Energy management; economics; production systems; computer applications; optimization methods; transportation. *Mailing Add:* 107 Nantucket Dr Pittsburgh PA 15238

WARREN, JAMES C, b Oklahoma City, Okla, May 13, 30; m 51; c 4. ENDOCRINOLOGY, BIOCHEMISTRY. *Educ:* Univ Wichita, AB, 50; Univ Kans, MD, 54; Univ Nebr, PhD(biochem), 61. *Prof Exp:* Fel, Nat Inst Child Health & Human Develop, 59-61; from asst prof to prof obstet & gynec, Univ Kans, 61-70, from instr to assoc prof biochem, 61-70; PROF BIOL CHEM, PROF OBSTET & GYNEC & HEAD DEPT, SCH MED, WASHINGTON UNIV, 70- *Concurrent Pos:* Markle scholar med sci, 61. *Mem:* Am Soc Cell Biol. *Res:* Biosynthesis; metabolism and mechanism of action of free and conjugated steroids; kinetics of steroid interconverting enzymes; biochemistry of menstruation. *Mailing Add:* Dept Obstet/Gynec & Biochem Washington Univ Sch Med 4911 Barnes Hospital Plaza St Louis MO 63110

WARREN, JAMES DONALD, b Ludlow, Mass, June 10, 48; m 74. MEDICINAL CHEMISTRY, SYNTHETIC ORGANIC CHEMISTRY. *Educ:* Western New Eng Col, BS, 70; Brown Univ, PhD(chem), 74. *Prof Exp:* Nat Cancer Inst fel, Temple Univ, 73-74; SR RES CHEMIST, LEDERLE LABS, AM CYANAMID CO, 74- *Mem:* Am Chem Soc. *Res:* Organic synthesis and evaluation of antitumor drug candidates. *Mailing Add:* 40 W Nauraushaun Ave Pearl River NY 10965

WARREN, JAMES VAUGHN, internal medicine, cardiology; deceased, see previous edition for last biography

WARREN, JOEL, b New York, NY, 1914; m 42; c 2. CANCER RESEARCH. *Educ:* Yale Univ, AB, 36; Columbia Univ, AM, 38, PhD(bact), 40. *Prof Exp:* DIR, GOODWIN INST CANCER, 67- *Concurrent Pos:* Fel, Nat Res Coun; scientist, NIH. *Mem:* Soc Exp Biol & Med; Am Asn Immunol; Am Acad Microbiol; Am Asn Cancer Res. *Res:* Viruses and viral diseases; biophysical methodology; toxoplasmosis; instrumentation; immunology of viral infections; cancer chemotherapy and immunology. *Mailing Add:* Goodwin Inst Cancer Res 1850 NW 69th Ave Plantation FL 33313

WARREN, JOHN LUCIUS, b Chicago, Ill, Dec 17, 32; m 58; c 2. ACCELERATOR PHYSICS. *Educ:* Univ Chicago, BA, 53; Univ Md, PhD(physics), 59. *Prof Exp:* Asst prof physics, De Pauw Univ, 59-61; mem staff, Los Alamos Nat Lab, Univ Calif, 61-73, asst to assoc dir res, 73-75, asst div leader, Ctr Div, 75-77, staff mem, 77-84, asst group leader, 84-87; SR ENGR, BOEING AEROSPACE CORP, 87- *Mem:* Am Phys Soc. *Res:* Lattice dynamics using inelastic neutron scattering; group theory applied to lattice dynamics; closed orbit distortions in proton storate rings; design of 30 gey proton synchrotron; accelerator code documentation; design FEL electron beam injectors. *Mailing Add:* 15808 SE 160th Pl Renton WA 98058

WARREN, JOHN STANLEY, b Ithaca, NY, Dec 19, 37; div; c 1. GEOLOGY. *Educ:* Cornell Univ, BA, 60; Stanford Univ, PhD(geol), 67. *Prof Exp:* Asst prof geol, Univ Cincinnati, 65-72; assoc prof geol, Thomas Jefferson Col, Grand Valley State Col, 72-79; vis fac, Evergreen State Col, 79-80; MEM FAC, MARLBORO COL, 80- *Mem:* Paleont Soc; Sigma Xi; AAAS. *Res:* Invertebrate paleontology; micropaleontology; palynology. *Mailing Add:* Marlboro Col Marlboro VT 05344

WARREN, KENNETH S, b New York, NY, June 11, 29; m 59; c 2. TROPICAL MEDICINE. *Educ:* Harvard Univ, AB, 51, MD, 55; Univ London, dipl, 59; FRCP, 86. *Hon Degrees:* DSc, Mahidol Univ, 89. *Prof Exp:* Intern med, Boston City Hosp, Mass, 55-56; mem staff med res, Lab Parasitic Dis, NIH, 56-63; from asst prof to assoc prof prev med, Sch Med, Case Western Res Univ, 63-70, from asst prof to prof med, 63-77, adj prof lib sci, 72-75, assoc prof geog med, 70-76, dir div geog med, 73-76, prof lib sci, 75-77; dir health sci, Rockefeller Found, 77-87, assoc vpres molecular biol & info sci, 87-88; DIR SCI, MAXWELL COMMUN CORP, 89- *Concurrent Pos:* Vis prof, Hosp das Clinicas, Univ Bahia, Brazil, 62-63 & London Sch Hyg & Trop Med, 70-71; asst physician, Univ Hosps Cleveland, 63-76, assoc physician, 76-77; consult parasitol, Cleveland Metrop Gen Hosp, 63-77; physician in charge parasitol, Inst Path, Case Western Reserve Univ, 63-77; NIH res career develop award, 66-71; assoc ed, J Immunol, 74-79; mem, numerous sci adv comts & panels, govt agencies & nat socs, 74-; adj prof, Rockefeller Univ, 77-; prof med, NY Univ Med Ctr, 77-; mem, Sci Adv Bd, Global Health Partners, 88- & Int Med Activ Subcomt, Am Col Physicians, 89-; dir, Am Comt, Weizmann Inst Sci. *Honors & Awards:* Bailey K Ashford Medal, Am Soc Trop Med & Hyg, 74; Squibb Award, 75; Heath-Clark Lectr, Univ London, 88; Frohlich Award, NY Acad Sci, 88; Van Thiel Medal, Dutch Soc Parasitol, 89. *Mem:* Inst Med Nat Acad Sci; Am Soc Clin Invest; Am Asn Immunologists; Infectious Dis Soc Am; Asn Am Physicians; fel Am Col Physicians; Am Fedn Clin Res; Am Soc Trop Med & Hyg; Int Epidemiol Asn; Royal Soc Trop Med & Hyg; NY Acad Sci. *Res:* Schistosomiasis; pathophysiology; immunology; control; published numerous articles in various journals. *Mailing Add:* Maxwell Commun Corp 866 Third Ave New York NY 10022

WARREN, KENNETH WAYNE, b Dallas, Tex, Mar 23, 40; c 3. WATER TREATMENT, CRUDE OIL PROCESSING. *Educ:* Baylor Univ, BS, 62, PhD(chem), 68. *Prof Exp:* Sr chemist, Texaco Inc, 68-71; sr engr, 71-80, sr staff engr, 80-82, DIR RES & DEVELOP, COMBUSTION ENG, NATCO, 82- *Mem:* Am Chem Soc; fel Am Inst Chemists; Am Inst Chem Engrs. *Res:* Water purification for both process use & disposal with emphasis on oil/water separation; crude oil processing for oil/water separation especially by electrostatic processes & removal of salt from crude oil; solvent/water separation in hydrometallurgical processes. *Mailing Add:* C-E NATCO PO Box 1710 Tulsa OK 74101-1710

WARREN, LEONARD, b Toronto, Can, Sept 23, 24; US citizen; m 47; c 3. BIOCHEMISTRY. *Educ:* Univ Toronto, BA, 47, MD, 51; Mass Inst Technol, PhD(biochem), 57. *Hon Degrees:* Dr, Univ Reims, France, 88. *Prof Exp:* Vis scientist biochem, NIH, 57-63; PROF, DEPT ANAT, UNIV PA, 63-; INST PROF, WISTAR INST, PHILADELPHIA PA, 75- *Concurrent Pos:* Vis scientist, Pasteur Inst, Paris, France, 63-64 & 77-78 & Imp Cancer Res Fund, Eng, 70-71; prof, Am Cancer Soc, 64- *Mem:* Am Soc Biol Chem; Soc Gen Physiologists. *Res:* Glycoproteins of membrane, in normal and pathological cells (malignancy, metabolic and other diseases); changes in the structure of the carbohydrate groups of glycoproteins with changing conditions; multidrug resistance; lysomal enzymes secretion. *Mailing Add:* Wistar Inst 36th & Spruce St Philadelphia PA 19104

WARREN, LIONEL GUSTAVE, b New York, NY, May 5, 26; m 52; c 3. PARASITOLOGY, MOLECULAR BIOLOGY. *Educ:* Syracuse Univ, AB, 48, MA, 53; Johns Hopkins Univ, ScD(parasitol), 57. *Prof Exp:* Res assoc biol, Rice Univ, 57-60; vis Int Atomic Energy Agency prof parasitol, Sci Res Inst, Caracas, Venezuela, 60-63; assoc prof med parasitol, 63-79, assoc prof trop med, 77-79, PROF MED PARASITOL & TROP MED, LA STATE UNIV MED CTR, NEW ORLEANS, 79- *Concurrent Pos:* USPHS grant, 64-70; scientist, Charity Hosp La, New Orleans, 67-; coordr grad studies, Dept Trop Med, La State Univ Med Ctr, 80-; Pfizer Latin Am grant, 81-82. *Mem:* Am Soc Parasitologists; Am Soc Trop Med & Hyg; Am Soc Cell Biol; Sigma Xi. *Res:* Carbohydrate and oxidative metabolism of endoparasitic animals; immunology of endoparasites; biochemistry of endemic amoebae and related organisms. *Mailing Add:* Microbiol-Immunol & Parasitol La State Univ Med Ctr 1901 Perdido St New Orleans LA 70112

WARREN, LLOYD OLIVER, b Fayetteville, Ark, Dec 27, 15; m 42; c 3. ENTOMOLOGY. *Educ:* Univ Ark, BS, 47, MS, 48; Kans State Col, PhD, 54. *Prof Exp:* Instr & jr entomologist, Univ Ark, 47-51; instr entom, Kans State Univ, 53-54; from asst prof & asst entomologist to prof entom & entomologist, Univ Ark, Fayetteville, 54-73, dir, Ark Agr Exp Sta, 73-83; RETIRED. *Concurrent Pos:* Fulbright Scholar-Yugoslavia, 72; Off Int Cooperation & Develop Rev Panel, Greece, 81; NSF Sem Panel, Seoul, Korea, 82. *Mem:* Entom Soc Am; Rice Working Tech Group. *Res:* Forest insects; apiculture. *Mailing Add:* 4333 Bridgewater Lane Rte 9 Fayetteville AR 72703

WARREN, MCWILSON, b Wayne Co, NC, Aug 29, 29; m 75; c 1. MALARIOLOGY, TROPICAL MEDICINE. *Educ:* Univ NC, BA, 51, MSPH, 52; Rice Univ, PhD(parasitol), 57. *Prof Exp:* From asst prof to assoc prof prev med & pub health, Sch Med, Univ Okla, 57-61 vchmn dept, 59-61; scientist, Lab Parasite Chemother, NIH, USPHS, 61-69, scientist, Far East Res Proj, Kuala Lumpur, Malaysia, 61-64, officer-in-chg, 63-64, officer-in-chg, Sect Cytol, Chamblee, Ga, 64-65, head sect chemother, Nat Inst Allergy & Infectious Dis, 66-69, parasitologist, Cent Am Malaria Res Sta, Ctr Dis Control, 69-74, scientist dir, Ctr Dis Control, 72-79, parasitologist, Vector Biol & Control Div, 74-79, dir, Cent Am Res Sta, El Salvador, 79-81, dir, WHO secretariat for malaria training & appl res, Bur Trop Dis, Kuala Lumpur, Malaysia, 82-85; DIR, SCI RESOURCE PROG, CTR INFECTIOUS DIS, CTR DIS CONTROL, 85- *Concurrent Pos:* China Med Bd fel trop med, Cent Am, 57; consult various orgns, 66-; res assoc, Sch Trop Med & Hyg, Univ London, 67-68; adj prof, Sch Med, Tulane Univ, 78-; mem, Comt Malaria, Inst Med, 90-91; ed, Am J Trop Med & Hyg, 91- *Mem:* Am Soc Trop Med & Hyg; Am Soc Parasitol; Soc Protozool; Royal Soc Trop Med & Hyg; Am Mosquito Control Asn; Sigma Xi. *Res:* Ecology and immunity of the primate malarias; parasite physiology; pathophysiology of infectious disease agents; global and institutional epidemiology and human ecology; field studies on sero-epidemiology of malaria; genetics of malaria vectors; field problems in chemotherapy of malaria; biology of malaria parasites; technology for community participation in malaria control; training methodologies; management issues in Health Delivery Systems. *Mailing Add:* Sci Resource Prog Ctr Infectious Dis Ctr Disease Control Atlanta GA 30333

WARREN, MASHURI LAIRD, b Findlay, Ohio, Jan 12, 40; m 84; c 2. SOLAR ENERGY, ENERGY EFFECIENT BUILDINGS & BUILDING CONTROLS. *Educ:* Ohio Wesleyan Univ, BA, 61; Univ Calif, Berkely, MA, 63, PhD(plasma physics), 68. *Prof Exp:* Asst prof physics, Calif State Univ, Hayward, 68-74; lectr, San Francisco State Univ, 76-78; science writing, 74-78; staff scientist, appl sci div, Lawrence Berkeley Lab, 78-88; PROD MGR, ASI CONTROLS, 88- *Mem:* Am Soc Heating, Refrig & Air Conditioning Engrs; Int Solar Energy Soc; Am Solar Energy Soc. *Res:* Solar energy physics; application computer for building energy conservation; control theory and application; computer simulation and analysis; building monitoring. *Mailing Add:* 1163 Santa Fe Ave Albany CA 94706

WARREN, MITCHUM ELLISON, JR, b Paris, Tenn, Nov 10 34; m 61; c 2. ORGANIC CHEMISTRY. *Educ:* Vanderbilt Univ, BA, 56, PhD(org chem), 63. *Prof Exp:* NIH fel, 63-66; asst prof chem, George Peabody Col, 66-71, assoc prof, 71-79; CONSULT, WARREN ENTERPRISES, 79- *Mem:* AAAS; Am Chem Soc. *Res:* Stereochemistry; optically active compounds; alkaloids. *Mailing Add:* 803 Kendall Dr Nashville TN 37209

WARREN, PAUL HORTON, b Bay Shore, NY, July 19, 53; m 83. PLANETOLOGY, ASTROGEOLOGY. *Educ:* State Univ NY, Oswego, BS, 75; Univ Calif, Los Angeles, PhD(geochem), 79. *Prof Exp:* Fel, Inst Geophys & Planetary Physics, Univ Calif, Los Angeles, 79; fel, Inst Meteoritics, Univ NMex, 79-83; asst res, 83-85, ASSOC RES GEOCHEM, UNIV CALIF, LOS ANGELES, 85- *Mem:* Fel Meteorit Soc; Mineral Soc Am; Planetary Soc; Geochem Soc; Geol Soc Am; Am Geophys Union. *Res:* Early igneous differentiation of the moon and planets; lunar samples; igneous meteorites; gabbroic-basaltic igneous rocks. *Mailing Add:* Inst Geophys Univ Calif Los Angeles CA 90024

WARREN, PETER, b New York, NY, Sept 30, 38; m 84. MATHEMATICS, COMPUTER SCIENCE. *Educ:* Univ Calif, Berkeley, BA, 60; Univ Wis-Madison, MA, 65, PhD(math), 70. *Prof Exp:* Mem tech staff, IBM Nordic Labs, 61-63; invited fel theory of traffic control, Thomas J Watson Res Labs, 64; lectr math, Med Sch, Univ Wis-Madison, 65-66; dir, div data processing, 85-86, asst prof math, 70-74, ASSOC PROF MATH, UNIV DENVER, 74-, DEAN, UNIV COL, UNIV DENVER, 86- *Concurrent Pos:* Statist consult, 73-; dir res, Energy Rec Inst, Colo, 78-82. *Mem:* Am Math Soc; Math Asn Am. *Res:* Probability theory; epidemiology; probability theory in Banach spaces; computer graphics. *Mailing Add:* Dean Univ Col Univ Denver Denver CO 80210

WARREN, REED PARLEY, b Price, Utah, May 6, 42; m 64; c 5. AUTOIMMUNITY, BIOLOGICAL RESPONSE MODIFIERS-IMMUNE-MODULATORS. *Educ:* Univ Utah, BS, 68, PhD(immunol), 73. *Prof Exp:* Res asst, Immunol Lab, Univ Utah, Salt Lake City, 69-73; postdoctorate & res assoc immunol, Sch Med, Univ Wash, Seattle, 73-79, res asst prof, 79-82; res scientist, Fred Hutchinson Cancer Res Ctr, Seattle, Wash, 73-82; res assoc prof immunol, 82-89, PROF IMMUNOL, BIOL/DEVELOP CTR HANDICAPPED PERSONS, UTAH STATE UNIV, LOGAN, 89- *Concurrent Pos:* Prin investr, many grants from NIH & pvt sources, 73-; vis prof microbiol, Brigham Young Univ, Provo, Utah, 78; consult, Cell Technol Inc, Boulder, Colo, 83-, HyClone Labs, Logan, Utah, 83-, Pennwalt Corp, Rochester, NY, 84-88, Murdock Int, Springville, Utah, 89-90; mem, Rev Panel, NIMH, 88-; chmn, Inst Rev Bd, Utah State Univ, 88-90. *Mem:* Am Asn Immunol; Nat Autism Soc; AAAS; Inter-Am Soc Chemother. *Res:* Immunologic, viral and genetic studies in infantile autism; development of immune modulators for the treatment of viral infections, immune deficiencies and autoimmunity; more than 150 full-length articles and presentations at national meetings. *Mailing Add:* Biol-Develop Ctr Handicapped Persons Utah State Univ UMC 6895 Logan UT 84322

WARREN, RICHARD HAWKS, b Binghamton, NY, Feb 16, 34; m 61; c 2. MATHEMATICS. *Educ:* US Naval Acad, BS, 56; Univ Mich, Ann Arbor, MS, 64; Univ Colo, Boulder, PhD(math), 71. *Prof Exp:* Maintenance officer, USAF, 56-62, from instr to assoc prof math, USAF Acad, 64-69, dep dir, Appl Math Res Lab, Aerospace Res Lab, 72-75, res mathematician, Aerospace Med Res Lab, Wright-Patterson AFB, 76-77, chief, Appl Math Group, Air Force Flight Dynamics Lab, 75-76; assoc prof math, Univ Nebr, Omaha, 77-80; SR SYSTS ANALYSIS, GEN ELEC CO, 80- *Mem:* Math Asn Am; Am Math Soc. *Res:* Scheduling theory; operations research; finite mathematics; combinatorics; applied mathematics; complexity theory. *Mailing Add:* 1528 Green Hill Circle Berwyn PA 19312

WARREN, RICHARD JOSEPH, b Lowell, Mass, Dec 25, 31; m 58; c 4. ANALYTICAL CHEMISTRY. *Educ:* Merrimack Col, BS, 53; Univ Pa, MS, 58. *Prof Exp:* Anal chemist, 56-61, sr anal chemist, 61-73, sr investr, 73-80, asst dir, 80-83, ASSOC DIR, SCI ADMIN, SMITHKLINE & FRENCH LABS, 83- *Mem:* Am Chem Soc; Soc Appl Spectros; AAAS. *Res:* Infrared, ultra violet and nuclear magnetic resonance spectroscopy; mass spectroscopy; x-ray diffraction. *Mailing Add:* 552 Walker Rd Wayne PA 19087-1419

WARREN, RICHARD JOSEPH, b Oklahoma City, Okla, June 30, 33; div; c 7. HUMAN GENETICS, CYTOGENETICS. *Educ:* Okla City Univ, AB, 58; St Louis Univ, PhD(microbiol), 67. *Prof Exp:* NIH fel, Harvard Med Sch-Mass Gen Hosp, 67; NIH fel, Wash Univ, 67-69, asst prof pediat & med, Med Sch, 67-70; asst prof pediat, Univ Miami, 70-74; pres & dir biochem testing labs, Genetics Assocs, 74-90; VPRES, SOUTHEAST IG LABS, INC, 90- *Concurrent Pos:* Dir, Palm Beach Genetics Clin, 73-76; chmn ad hoc comt, NIH Contract-Cytogenetics Registries, 73-81; genetics consult, Bur Health & Rehab, State of Fla, 74-81. *Mem:* Am Soc Human Genetics; AAAS; Mammalian Cell Genetics Soc; Am Tissue Cult Asn; Am Soc Cell Biol. *Res:* Human cytogenetics; molecular biology; cellular regulation. *Mailing Add:* 8217 SW 81 Ct Miami FL 33143

WARREN, RICHARD SCOTT, b Malden, Mass, Oct 21, 42; m 67. PLANT PHYSIOLOGY. *Educ:* Defiance Col, BA, 65; Univ NH, MS, 68, PhD(plant sci), 70. *Prof Exp:* Sigma Xi grant-in-aid res, 70-71, ASST PROF BOT, CONN COL, 70- *Mem:* AAAS; Bot Soc Am; Am Inst Biol Sci; Sigma Xi. *Res:* Physiological ecology of Halophytes; physiology of disease resistance. *Mailing Add:* Dept Bot Conn Col New London CT 06320

WARREN, ROBERT HOLMES, b Austin, Tex, Feb 20, 41; div. CELL BIOLOGY. *Educ:* Rice Univ, BA, 62, MA, 63; Harvard Univ, PhD(cell biol), 69. *Prof Exp:* NIH fels, Cambridge Univ, 69-70 & Univ Tex, Austin, 70-71; asst prof biol struct, 71-75, ASSOC PROF ANAT, MED SCH, UNIV MIAMI, 75- *Mem:* Am Soc Cell Biol; Soc Develop Biol. *Res:* Ultrastructural and biochemical basis of cell motility. *Mailing Add:* Anat Rm 124 Univ Miami PO Box 016960 Miami FL 33101

WARREN, S REID, JR, b Philadelphia, Pa, Jan 31, 08; m 30; c 2. ELECTRICAL ENGINEERING, RADIOLOGIC PHYSICS. *Educ:* Univ Pa, BS, 28, MS, 29, ScD, 37; Am Bd Radiol, dipl, 47. *Prof Exp:* From instr to prof elec eng, Univ Pa, 33-76, asst vpres eng, 54-73, prof radiol physics, Sch Med, 58-76, EMER PROF ELEC ENG & RADIOL, UNIV PA, 76- *Concurrent Pos:* Consult radiol physicist, Vet Admin, Pub Health Serv, Hosps, 36-70. *Mem:* Assoc fel Am Col Radiol; fel Inst Elec & Electronics Engrs; fel AAAS; Soc Hist Technol; Health Physics Soc; Hist Sci Soc; Am Asn Physicists Med. *Res:* Electric circuits and fields; radiologic physics. *Mailing Add:* 3300 Darby Rd Elm 3114 Haverford PA 19041

WARREN, STEPHEN THEODORE, b Mich, Nov 30, 53; m 78; c 1. HUMAN GENETICS, SOMATIC CELL GENETICS. *Educ:* Mich State Univ, BS, 76, PhD (human genetics), 81. *Prof Exp:* Res assoc genetics, Univ Ill Col Med, 81-83, instr, 83-85; asst prof biochem, 85-91, ASST PROF PEDIAT, EMORY UNIV SCH MED, 85-, ASSOC PROF BIOCHEM, 91- *Concurrent Pos:* Consult, Centers Dis Control, 88-; collabr, Centre D'Etude de Polymorphisme Humain, 89- *Honors & Awards:* Basil O'Conner Award, March of Dimes Nat Found, 86, Albert E Levy Award, 87. *Mem:* Am Soc Human Genetics; Am Soc Biochem & Molecular Biol; Am Soc Microbiol; Genetics Soc Am; AAAS. *Res:* Human molecular genetics; molecular analysis of the fragile X syndrome; X-linked genetic diseases particulary those mapping to Xq 28. *Mailing Add:* Dept Biochem Emory Univ Sch Med Atlanta GA 30322

WARREN, WALTER R(AYMOND), JR, b New York, NY, Nov 25, 29; m 54; c 10. CHEMICAL LASERS, EXPERIMENTAL FLUID DYNAMICS. *Educ:* NY Univ, BSE, 50; Princeton Univ, MSE, 52, PhD(aeronaut eng), 57. *Prof Exp:* Asst aeronaut eng, Princeton Univ, 50-55; assoc res scientist, Missiles & Space Div, Lockheed Aircraft Corp, 55-56; mgr aeromech & mat lab, Missile & Space Div, Gen Elec Co, Pa, 56-68; dir aerophys lab, Aerospace Corp, Los Angeles, 68-81; PRES, PACIFIC APPLIED RES, LOS ANGELES, 81- *Concurrent Pos:* Guggenheim fel, Princeton, 52-53; lectr, Univ Pa, 62-68; mem sci adv bd comt, USAF, 69-75; mem fluid mech subcomt, NASA, 70; chmn, plasmadynamics tech comt, Am Inst Aeronaut & Astronaut, 71-72, assoc ed, Am Inst Aeronaut & Astronaut J, 72-74. *Mem:* Fel Am Inst Aeronaut & Astronaut. *Res:* High energy lasers; fluid dynamics; plasma dynamics; atmospheric entry; shock tube/tunnel and plasma jet development and application; lasers. *Mailing Add:* Six Crestwind Dr Rancho Palos Verdes CA 90274

WARREN, WAYNE HUTCHINSON, JR, b Newark, NJ, Dec 11, 40; m 67; c 3. ASTRONOMICAL & ASTROPHYSICAL DATA, COMPUTER APPLICATIONS. *Educ:* Fairleigh Dickinson Univ, BA, 68; Ind Univ, MS, 70, PhD(astron), 75. *Prof Exp:* Res assoc, Nat Acad Sci, 76-77, astronr, Sigma Data Serv Corp, 77-86, PRIN SCIENTIST, ST SYSTS CORP, NAT SPACE SCI DATA CTR, GODDARD SPACE FLIGHT CTR, 86- *Concurrent Pos:* Ed, Astron Data Ctr Bull, 80; mem, PACS Oversight Comt, Am Inst Physics, 88. *Mem:* Int Astron Union; Am Astron Soc; Astron Soc Pac; Royal Astron Soc; Int Amateur Prof Photoelec Photom Asn; Planetary Soc. *Res:* Computerized astronomical catalogs and retrieval systems; data archive management and dissemination; data center operations; astronomical photometry and spectrophotometry; astronomical documentation and literature. *Mailing Add:* Nat Space Sci Data Ctr Code 933 NASA Goddard Space Flight Ctr Greenbelt MD 20771

WARREN, WILLIAM A, b Findlay, Ohio, Mar 29, 36; m 59; c 3. BIOCHEMISTRY. *Educ:* Amherst Col, AB, 58; Western Reserve Univ, MD, 62; Univ Mass, PhD, 68. *Prof Exp:* Staff assoc biochem, Nat Inst Arthritis & Metab Dis, Md, 66-68; assoc res physician, Mary Imogene Bassett Hosp, Cooperstown, NY, 68-77; MEM STAFF, DIV LABS & RES, NY STATE DEPT HEALTH, 77- *Concurrent Pos:* USPHS fel biochem, Dartmouth Med Sch, 64-65; fel, Amherst Col, 65-66. *Mem:* Am Chem Soc; Am Asn Clin Chemists; Am Soc Biol Chemists. *Res:* Protein structure and function; chemistry of pyridine nucleotides and glyoxylate; mast cell tumor biochemistry. *Mailing Add:* Seven Kilmer Ct Delmar NY 12054

WARREN, WILLIAM ERNEST, b Rochester, NY, Aug 11, 30; m 55; c 2. APPLIED MATHEMATICS. *Educ:* Univ Rochester, BS, 56, MS, 59; Cornell Univ, PhD(eng mech), 62. *Prof Exp:* From instr to asst prof mech & mat, Cornell Univ, 57-62; STAFF MEM, SOLID DYNAMICS RES DEPT, SANDIA LAB, 62- *Concurrent Pos:* Mem, NMex House Rep, 71-78; chmn, Legis Sch Study Comt, 75-76. *Mem:* Am Inst Aeronaut & Astronaut; Am Math Soc; fel Am Soc Mech Engrs; Soc Indust & Appl Math; Am Acad Mech. *Res:* Plane elastic systems; thermal stress concentrations; electric field effects on solid dielectrics, particularly dielectric breakdown; wave propagation; solid-fluid interacting systems. *Mailing Add:* Sandia Nat Labs Orgn 1813 Bldg 806 Rm 284 Albuquerque NM 87185

WARREN, WILLIAM WILLARD, JR, b Seattle, Wash, Nov 7, 38; m 65; c 2. NUCLEAR MAGNETIC RESONANCE. *Educ:* Stanford Univ, BS, 60; Wash Univ, PhD(physics), 65. *Prof Exp:* Asst res physicist, Univ Calif, Los Angeles, 65-68; mem tech staff, AT&T Bell Labs, 68-90; PROF PHYSICS, ORE STATE UNIV, 91- *Concurrent Pos:* Alexander von Humbold Found, 74. *Honors & Awards:* US Sr Scientist Award. *Mem:* AAAS; fel Am Phys Soc; NY Acad Sci. *Res:* Application of nuclear magnetic resonance to the study of electronic structure and atomic dynamics of liquids and solids, especially metals and semiconductors under extreme temperature/pressure conditions; metal-nonmetal transitions in liquids; electronic transport properties of liquids; high Tc superconductivity. *Mailing Add:* Dept Physics Ore State Univ Weniger Hall 301 Corvallis OR 97331-6507

WARRICK, ARTHUR W, b Kellerton, Iowa, Dec 4, 40; m 62; c 2. SOIL PHYSICS, MATHEMATICS. *Educ:* Iowa State Univ, BS, 62, MS, 64, PhD(soil physics), 67. *Prof Exp:* Res assoc soil physics, Iowa State Univ, 66-67; asst prof, 67-71, assoc prof, 71-81, PROF SOIL PHYSICS, UNIV ARIZ, 81-, RES SCIENTIST, AGR EXP STA, 71- *Mem:* Soil Sci Soc Am; Am Soc Agron; Am Geophys Union. *Res:* Drainage; soil water flow; porous media flow; potential theory. *Mailing Add:* Dept Soil Sci Univ Ariz Tucson AZ 85721

WARRICK, EARL LEATHEN, b Butler, Pa, Sept 23, 11; m 40; c 2. PHYSICAL CHEMISTRY. *Educ:* Carnegie Inst Technol, BS, 33, MS, 34, DSc(phys chem), 43. *Hon Degrees:* DH, Saginaw Valley State Col, 84. *Prof Exp:* Asst, Mellon Inst Sci, 35-37, fel organosilicon chem, 37-46, sr fel, 46-56; asst dir res, Dow Corning Corp, 57-59, mgr hyper-pure silicon div, 59-62, gen mgr electronic prod div, 62-68, mgr new proj bus, 68-72, sr mgt consult, 72-76; RETIRED. *Concurrent Pos:* Lectr, Univ Pittsburgh, 47-48; interim dean sci & eng technol, Saginaw Valley State Col, 79-80 & 83-84. *Honors & Awards:* Charles Goodyear Medal & Award, 76. *Mem:* Am Chem Soc; Sigma Xi. *Res:* Glass composition; chemical kinetics; gas phase; organosilicon and radiation chemistry; physical chemistry of polymers. *Mailing Add:* 508 Crescent Dr Midland MI 48640

WARRICK, PERCY, JR, b South Bend, Ind, Aug 6, 35; m 61; c 2. PHYSICAL ORGANIC CHEMISTRY. *Educ:* Wabash Col, 57; Univ Rochester, PhD(org chem), 61. *Prof Exp:* Fel phys org chem, Univ Minn, 60-62; res assoc, Mass Inst Technol, 62-63; from asst prof to assoc prof, 63-77, PROF CHEM, WESTMINSTER COL, PA, 77- *Concurrent Pos:* Fel, Univ Utah, 70-71; vis prof, Univ Kent, Canterbury, Eng, 78-79. *Mem:* Am Chem Soc; Royal Soc Chem; Sigma Xi. *Res:* Mechanisms of reactions between metals and solutions; general-acid catalysis; relaxation kinetics; acid-base reactions in mixed solvents. *Mailing Add:* Dept Chem Westminster Col New Wilmington PA 16172

WARRINGTON, PATRICK DOUGLAS, b Winnipeg, Man, Mar 21, 42; m 65; c 2. AQUATIC PLANTS DISTRIBUTION & ECOLOGY. *Educ:* Univ BC, BSc, 64, PhD(bot), 70. *Prof Exp:* Consult bot, 72-73; res officer aerial satellite photog, 73-75, BIOLOGIST, BC GOVT, 75- *Mem:* Asn Prof Biologists BC. *Res:* All aspects of the biology of aquatic plants. *Mailing Add:* Water Qual Br Parliament Bldg Victoria BC V8L 3S1 Can

WARRINGTON, TERRELL L, b Baltimore, Md, June 5, 40; m 64. PHYSICAL CHEMISTRY, BIOCHEMISTRY. *Educ:* Yale Univ, BA, 61; Purdue Univ, PhD(phys chem), 66. *Prof Exp:* ASST PROF CHEM, MICH TECHNOL UNIV, 67- *Mem:* Am Chem Soc. *Res:* Physical chemistry of biological macromolecules concentrating mainly on conformational studies. *Mailing Add:* Dept Chem Mich Technol Univ Houghton MI 49931-1295

WARSH, CATHERINE EVELYN, b San Diego, Calif, Apr 19, 43; m 70. OCEANOGRAPHY. *Educ:* Old Dominion Col, BA, 65; Fla State Univ, MS, 71. *Prof Exp:* Teacher math, Kempsville Jr High Sch, Virginia Beach, Va, 65-66; researcher marine biol, Duke Univ, NC, 66-67; teacher marine biol & math, First Colonial High Sch, Virginia Beach, 67-68; researcher limnol, Fla State Univ, Tallahassee, 71-72, res programmer biol oceanog, 72, environ specialist pollution & limnol, Dept Pollution Control, 73; admin oceanogr, Nat Oceanic & Atmospheric Admin-Nat Marine Fisheries Serv, 74-75, RES OCEANOGR, DEPT COM, NAT OCEANIC & ATMOSPHERIC

ADMIN-NAT OCEAN SURV, WASHINGTON, DC, 75- *Concurrent Pos:* Teacher math & phys sci, Griffin Middle Sch, Tallahassee, Fla, 71; phys oceanogr researcher, Dept Com, Nat Oceanic & Atmospheric Admin-Nat Marine Fisheries Serv, Washington, DC, 74; task team oil spill trajectory models, Environ Res Labs-Nat Oceanic & Atmospheric Admin, Boulder, Colo, 75- *Res:* Study of surface circulation, near-shore dynamics over the continental shelf in the middle Atlantic bight; study of circulation, water properties and transport in Gulf of California; pollution monitoring water chemistry and physics for the northeast monitoring program. *Mailing Add:* 9702 Montauk Ave Bethesda MD 20817

WARSHAUER, STEVEN MICHAEL, b New York, NY, May 20, 45; c 2. PALEOECOLOGY. *Educ:* Queens Col, NY, BA, 67; Univ Cincinnati, MS, 69, PhD(geol), 73. *Prof Exp:* Asst prof geol, WVa Univ, 72-77, assoc prof, 77-81; explor geologist, Tenneco Oil, 81-83, sr explor geologist, 83-89; sr geologist, Brit Gas, 89- *Concurrent Pos:* Consult, WVa Geol Econ Surv, Champlin Petrol, Dept Energy, Amoco Oil & US Geol Surv. *Mem:* Am Asn Petrol Geologists; Soc Econ Paleontologists & Minerologists; Sigma Xi. *Res:* Carbonate depositional models; multivariate statistical methods in analyzing geologic data. *Mailing Add:* Brit Gas 1100 Louisiana Suite 2500 Houston TX 77002

WARSHAW, ISRAEL, b Brooklyn, NY, Nov 30, 25; m 48. MATERIALS SCIENCE. *Educ:* Alfred Univ, BA, 51; Pa State Univ, MS, 53, PhD, 61. *Prof Exp:* Sci aide & chem lab asst, Water Resources Br, US Geol Surv, 42-43, chem lab asst, Geochem & Petrol Br, 46-51, chemist, 51-53; asst geochem, Pa State Univ, 51-53 & 57-58; group leader, Glass Res Labs, Pittsburgh Plate Glass Co, 53-57; assoc res dir, Tem-Pres Res, Inc, Pa, 59-60; assoc prog dir, NSF, 60-63, dir eng mat prog, 63-71, actg dir, Eng Div, 71-74, dep dir, 72-76; phys sci adminr, Dept Energy, 77-90; CONSULT, 90- *Concurrent Pos:* Lectr, Univ Pittsburgh, 56-57. *Res:* Phase equilibrium studies of silicate, aluminate and oxide systems at high temperatures and pressures; synthesis of refractory compounds, particularly those of the rare earths; silicate glasses; materials research. *Mailing Add:* 3703 Stewart Dr Chevy Chase MD 20815

WARSHAW, JOSEPH B, b Miami, Fla. PEDIATRICS. *Educ:* Duke Univ, MD, 61. *Prof Exp:* Fac, Harvard Med Sch, 67-73; dir, Div Perinatal Med & prof pediat, obstet & gynec, Yale Med Sch, 73-82; prof & chmn, Dept Pediat, Univ Tex Health Sci Ctr, Dallas, 82-87; PROF & CHMN, DEPT PEDIAT, YALE UNIV, 87-, CHIEF PEDIAT, YALE NEW HAVEN HOSP, 87- *Concurrent Pos:* Josiah Macy scholar, Univ Oxford, 79-80; mem coun, Am Pediat Soc, 87-; mem adv coun, Nat Inst Child Health & Human Develop, NIH, 87-91; mem sci adv bd, St Jude's Children's Res Hosp; trustee, Int Pediat Res Found. *Mem:* Inst Med-Nat Acad Sci; Am Soc Clin Invest; Asn Am Physicians; Soc Pediat Res (pres, 82); Am Pediat Soc; Am Soc Biol Chemists; Am Soc Cell Biol. *Res:* Developmental biology; neonatal and perinatal medicine. *Mailing Add:* Dept Pediat Sch Med Yale Univ New Haven CT 06510

WARSHAW, STANLEY I(RVING), b Boston, Mass, Nov 5, 31; div; c 1. PRODUCT ENGINEERING, MATERIAL SCIENCE & STANDARDS. *Educ:* Ga Inst Technol, BCerE, 57; Mass Inst Technol, ScD(ceramics), 61. *Prof Exp:* Res asst, Ga Inst Technol, 56-57 & Mass Inst Technol, 57-61; sr res scientist, Raytheon Mfg Co, 61-64; res supvr, Ceramics & Metall Sect, 64-68, mgr ceramic technol, Res & Develop Ctr, 68-69, mgr mat & chem dept, 69-72, gen mgr, Prod Develop & Eng Lab, Am Standard Inc, 72-75; dir, Ctr for Consumer Prod Technol, Nat Bur Standards, 75-80, dir, Off Standards Policy, 81-87, assoc dir indust & standards, 87-89; DIR STANDARDS, NAT INST STANDARDS & TECHNOL, 90- *Mem:* Sigma Xi; fel NY Acad Sci. *Res:* Product design; standards and conformity assessment. *Mailing Add:* Nat Inst Standards & Technol Rm A603 Admin Gaithersburg MD 20899

WARSHAW, STEPHEN I, b New York, NY, Mar 26, 39; m 64; c 2. AEROACOUSTICS, WAVE PROPAGATION & NUCLEAR REACTIONS. *Educ:* Polytech Inst Brooklyn, BSc, 60; Johns Hopkins Univ, PhD(physics), 66. *Prof Exp:* Jr instr physics, Johns Hopkins Univ, 60-63, res asst, 63-66; res assoc, Univ Ill, Urbana-Champaign, 66-68; SR PHYSICIST, LAWRENCE LIVERMORE NAT LAB, UNIV CALIF, 68- *Mem:* Am Phys Soc; Acoust Soc Am; Am Geophys Union. *Res:* Experimental low energy nuclear physics; computer modeling of hydrodynamic phenomena; theoretical atmospheric acoustics; computer code solutions to mathematical and physical problems; interpolation methods; ionospheric radio propagation theory and simulation; nuclear reaction applications. *Mailing Add:* Lawrence Livermore Nat Lab L-298 Univ Calif PO Box 808 Livermore CA 94550

WARSHAWSKY, HERSHEY, b Montreal, Que, Feb 6, 38; m 60; c 3. HISTOLOGY. *Educ:* Sir George Williams Univ, BSc, 59; McGill Univ, MSc, 61, PhD(anat), 66. *Prof Exp:* Lectr anat, McGill Univ, 63-66; res fel orthop res, Harvard Univ, 66-67; from asst prof to assoc prof, 67-77, PROF ANAT, MCGILL UNIV, 77- *Concurrent Pos:* Vis prof, Univ Sao Paulo, Royal Dent Col, Aarhus, Denmark, State Univ Compinas, Brazil, Univ Queensland, Brisbane, Australia. *Mem:* Am Asn Anat; Can Asn Anat. *Res:* Use of the enamel organ in the rat incisor as a model system for structural and radioautographic studies of secretory processes, cell renewal, mineralization and growth factor receptors. *Mailing Add:* Dept Anat McGill Univ 3640 University St Montreal PQ H3A 2B2 Can

WARSHAWSKY, JAY, b Chicago, Ill, Mar 27, 27; m 54; c 3. ELECTRICAL ENGINEERING. *Educ:* Ill Inst Technol, BS, 48; Purdue Univ, MS, 50; Northwestern Univ, PhD(elec eng, biomed eng), 63. *Prof Exp:* Dir servomech sect, Cook Res Labs, 51-56; tech dir mech res div, Am Mach & Foundry Co & Gen Am Transp Corp, 56-65; dir res & develop dept, 65-77, VPRES RES & DEVELOP, FULLER CO, GATX CORP, 77- *Concurrent Pos:* Lectr, Grad Sch, Ill Inst Technol, 54, lectr, Med Sch, 63, res asst, 63. *Mem:* AAAS; Inst Elec & Electronics Engrs; Air Pollution Control Asn; Water Pollution Control Fedn; Instrument Soc Am; Sigma Xi. *Res:* Automatic control systems; analog and digital computers; electronic instrumentation; accommodation in the human eye; automatic focusing devices; physiological optics. *Mailing Add:* 700 Bellefonte St Pittsburgh PA 15232

WARSHAY, MARVIN, b Tel Aviv, Israel, Jan 12, 34; US citizen; m 62; c 3. ELECTROCHEMICAL SYSTEMS, ENERGY TECHNOLOGY. *Educ:* Rensselaer Polytech Inst, BChE, 55; Ill Inst Technol, MS, 57, PhD(chem eng), 60. *Prof Exp:* Res engr, Esso Res & Eng Co, 60-62; res engr, Lewis Res Ctr, NASA, 62-78, mgr fuel cell proj, Dept Energy, 76-87, chief Solar Dynamics & Thermal Mgt Br, 87-89, CHIEF ELECTROCHEM TECHNOL BR, NASA, 89- *Concurrent Pos:* Lectr, Cleveland State Univ, 67-, adj prof chem eng, 74- *Mem:* Am Chem Soc; Am Inst Chem Engrs; Electrochem Soc; Sigma Xi. *Res:* Drop motion; chemical reactions in nozzles; chemical reactions and colloidal phenomena in oil additive formation and in muffler corrosion; shock tubes; high temperature chemical kinetics of gases; electrochemical systems; batteries; fuel cells. *Mailing Add:* Lewis Res Ctr NASA MS 301-5 21000 Brookpark Rd Cleveland OH 44135

WARSHEL, ARIEH, b Sde-Nahom, Israel, Nov 20, 40; m 66; c 2. CHEMICAL PHYSICS, MOLECULAR BIOLOGY. *Educ:* Israel Inst Technol, BSc, 66; Wiezmann Inst Sci, MSc, 67, PhD(chem), 69. *Prof Exp:* Res assoc chem, Harvard Univ, 70-72; res assoc, Wiezmann Inst, 72-73, sr scientist, 73-74; vis scientist, Med Res Coun Lab Molecular Biol, Cambridge, Eng, 74-76; from asst prof to assoc prof, 76-84, PROF CHEM, UNIV SOUTHERN CALIF, 84- *Res:* Theoretical study of the early steps of the visual process; resonance Raman of large molecules; simulation of protein folding; simulation of enzymatic reactions; simulation of electron transfer reactions. *Mailing Add:* Dept Chem Univ Southern Calif Los Angeles CA 90089

WARSHOWSKY, BENJAMIN, b New York, NY, Jan 21, 19; m 57; c 3. ANALYTICAL CHEMISTRY. *Educ:* City Col New York, BS, 40; Univ Minn, MS, 42, PhD(biochem), 45. *Prof Exp:* Anal res chemist, Publicker Industs, Inc, Pa, 45-46; chief anal chem sect, Biol Labs, US Army Chem Corps, Ft Detrick, 47-54, prog coord officer, 54-56, decontamination br, 56-57, spec asst to dir med res, 57-60, chief phys detection br, 60-64, chief rapid warning officer, 64-72, chief biol defense br, Edgewood Arsenal, 72-75; PROF, FREDERICK COMMUNITY COL, FREDERICK, MD, 73- *Mem:* Am Chem Soc; Sci Res Soc Am. *Res:* Microbiological detection techniques; instrument development; administration of chemical and biological research; administration of research and development program on rapid detection of microbiological aerosols. *Mailing Add:* 315 W College Terr Frederick MD 21701

WARSI, NAZIR AHMED, b Sheopur, Uttar Pradesh, India, June 30, 39; m 66. MATHEMATICAL PHYSICS. *Educ:* St Andrew's Col, Gorakhpur, India, BSc, 57; Gorakhpur Univ, MSc, 59, PhD(shock wave), 61. *Prof Exp:* Asst prof math, Gorakhpur Univ, India, 59-63; assoc prof physics & math, Savannah State Col, 63-64; prof physics & math, 64-66, PROF MATH, ATLANTA UNIV, 66-, ACTG CHMN DEPT, 70- *Mem:* Am Math Soc; Tensor Soc. *Res:* Shock waves in ideal and magneto-gas-dynamic flows; nonlinear functional analysis; optimization. *Mailing Add:* Dept Comput Sci Atlanta Univ Atlanta GA 30314

WARSI, ZAHIR U A, b Uttar Pradesh, India, July 7, 36; m 58; c 5. APPLIED MATHEMATICS, AERONAUTICS. *Educ:* Univ Lucknow, BSc, 54, MSc, 56, PhD(math), 65. *Prof Exp:* Scientist, Cent Bldg Res Inst, Roorkee, India, 62-67; fel aerodyn 67-70, from asst prof to prof, 70-89, HEARIN-HESS PROF AEROSPACE ENG, MISS STATE UNIV, 89- *Concurrent Pos:* Minna-James-Heineman Found. *Honors & Awards:* Am Soc Eng Educ Res Award, 75. *Mem:* Sigma Xi; Am Inst Aeronaut & Astronaut; Am Acad Mech. *Res:* Numerical fluid dynamics; solutions of compressible and turbulent incompressible flow equations in general coordinates. *Mailing Add:* Dept Aerospace Eng Miss State Univ Drawer A Mississippi State MS 39762

WARTELL, ROGER MARTIN, b New York, NY, Feb 24, 45; m 68; c 2. BIOPHYSICS. *Educ:* Stevens Inst Technol, BSc, 66; Univ Rochester, PhD(physics), 71. *Prof Exp:* NIH fel & res assoc biochem, Univ Wis-Madison, 71-73; from asst prof to assoc prof, 74-83, PROF PHYSICS & BIOL & ASST DIR, SCH PHYSICS, GA INST TECHNOL, 84- *Concurrent Pos:* Vis prof, Univ Wis-Madison, 78-79; NIH career develop award, 79-84; vis scholar, NIH, 86-87. *Mem:* Biophys Soc; AAAS; Am Phys Soc. *Res:* Conformational properties of DNA gene regulatory sites; influence of cooperative interactions along DNA on genetic processes; Raman spectroscopy. *Mailing Add:* Sch Biol Ga Inst Technol Atlanta GA 30332

WARTEN, RALPH MARTIN, mathematics; deceased, see previous edition for last biography

WARTER, JANET KIRCHNER, b Greensburg, Pa, July 27, 33; m 62. PALYNOLOGY, SCIENCE EDUCATION. *Educ:* Pa State Univ, BS, 55, MEd, 60; La State Univ, PhD(bot), 65. *Prof Exp:* Lectr bot, Calif State Col, Fullerton, 65-66; LECTR GEOL, CALIF STATE UNIV, LONG BEACH, 66-68, & 70- *Concurrent Pos:* Res assoc, Los Angeles County Mus Natural Hist. *Mem:* Am Asn Stratig Palynologists. *Res:* Tertiary palynology; Pleistocene seeds and pollen; archeological macro-plant analysis. *Mailing Add:* 17841 Still Harbor Lane Huntington Beach CA 92647

WARTER, STUART L, b New York, NY, Apr 9, 34; m 62. ORNITHOLOGY, VERTEBRATE PALEONTOLOGY. *Educ:* Univ Miami, Fla, BS, 56, MS, 58; La State Univ, PhD(zool), 65. *Prof Exp:* Instr zool, La State Univ, 64-65; from asst prof to assoc prof, 65-75, PROF BIOL, CALIF STATE UNIV, LONG BEACH, 75- *Concurrent Pos:* Res assoc vert paleont, Los Angeles County Mus Natural Hist, 66- *Mem:* Am Ornith Union; Cooper Ornith Soc; Wilson Ornith Soc; Soc Vert Paleont. *Res:* Avian paleontology, morphology and systematics; osteology and relationships of suboscine passerine birds. *Mailing Add:* Dept Biol Calif State Univ Long Beach 1250 Bellflower Blvd Long Beach CA 90840

WARTERS, MARY, b Rome, Ga, Oct 18, 02. GENETICS. *Educ:* Shorter Col, AB, 23; Ohio State Univ, MA, 25; Univ Tex, PhD(cytogenetics), 43. *Prof Exp:* Asst zool, Ohio State Univ, 23-25; instr biol, Winthrop Col, 25-27; from instr to emer prof zool, Centenary Col, 27-71, head dept, 47-69; RETIRED. *Concurrent Pos:* Vis scientist, Jackson Mem Lab, 51; res partic biol div, Oak Ridge Nat Lab, 59-61. *Mem:* Fel AAAS. *Res:* Bryozoa; cytogenetics of Drosophila; chromosomal aberrations in wild populations of Drosophila; x-autosomal translocations. *Mailing Add:* The Glen 403 E Flournoy Lucas Rd Shreveport LA 71115

WARTERS, RAYMOND LEON, b Atlanta, Ga, Nov 22, 45; m 78. MOLECULAR BIOLOGY, RADIATION BIOLOGY. *Educ:* Emory Univ, BA, 67; Fla State Univ, MS, 72, PhD(molecular biol), 76. *Prof Exp:* Nat Cancer Inst fel, 76-78, from res asst prof to res assoc prof, 78-86, ASSOC PROF RADIOL, UNIV UTAH, 86- *Concurrent Pos:* Assoc ed, Radiation Res, 90- *Mem:* Sigma Xi; Radiation Res Soc; Am Soc Cell Biol; AAAS. *Res:* Effect of enviromental agents, including radiations and chemicals, on molecular and cellular biology and biochemistry, especially with respect to the eukaryotic genetic apparatus. *Mailing Add:* Dept Radiol Univ Utah Col Med Salt Lake City UT 84112

WARTERS, WILLIAM DENNIS, b Des Moines, Iowa, Mar 22, 28; m 52; c 2. MICROWAVE ELECTRONICS. *Educ:* Harvard Univ, AB, 49; Calif Inst Technol, MS, 50, PhD(physics), 53. *Hon Degrees:* LLD, Monmouth Col, 86. *Prof Exp:* Asst, Calif Inst Technol, 50-52; mem tech staff guided wave res, Bell Tell Labs, Inc, 53-61, head repeater res dept, 61-67, dir transmission systs res ctr, 67-69, exec dir tech, Staff Employment, Educ & Salary Admin Div, 69-70, dir, Millimeter Wave Syst Lab, 70-76, dir, Toll Transmission Lab, 76-80, dir, Satellite Transmission Lab, 80-83; ASST VPRES, NETWORK TECHNOL RES LAB, BELL COMMUN RES, INC, 84- *Mem:* Am Phys Soc; fel Inst Elec & Electronics Engrs. *Res:* Multi-mode wave guides; millimeter waves; microwaves; transmission systems; communications satellites, fiber optics. *Mailing Add:* 514 Sunnyside Rd Lincroft NJ 07738

WARTHEN, JOHN DAVID, JR, b Baltimore, Md, Mar 8, 39; m 69; c 2. NATURAL PRODUCT CHEMISTRY. *Educ:* Univ Md, BS, 60, PhD(pharmaceut chem), 66. *Prof Exp:* RES CHEMIST, AGR RES SERV, USDA, 65- *Mem:* Am Chem Soc. *Res:* Isolation and identification of phytochemicals; insect attractants, antifeedants, repellants and insecticides. *Mailing Add:* 4413 Rendale Ct Olney MD 20832

WARTIK, THOMAS, b Cincinnati, Ohio, Oct 1, 21; m 52; c 2. INORGANIC CHEMISTRY. *Educ:* Univ Cincinnati, AB, 43; Univ Chicago, PhD(chem), 49. *Prof Exp:* From asst prof to prof chem & head dept, Pa State Univ, 50-71, dean col sci, 71-87; RETIRED. *Concurrent Pos:* Vis scientist, Radiation Lab, Univ Calif, 57, 59 & 61; mem, Fulbright selection comt chem, Nat Acad Sci-Nat Res Coun, 66-72, chmn, 70-72; mem adv bd, Am Chem Soc-Petrol Res Fund, 68-71; consult, Radiation Lab, Callery Chem Co, Koppers Co, Inc & NY Bd Regents, 73-74. *Mem:* Fel AAAS; Am Chem Soc. *Res:* Chemistry of boron and aluminum compounds; light metal hydrides; organometallic chemistry. *Mailing Add:* 120 Davey Lab Pa State Univ University Park PA 16802

WARTOFSKY, LEONARD, b New York, NY, July 14, 37; m 59; c 1. INTERNAL MEDICINE, ENDOCRINOLOGY. *Educ:* George Washington Univ, BS, 59, MS, 61, MD, 64. *Prof Exp:* CHIEF ENDOCRINOL METAB SERV, WALTER REED ARMY MED CTR, WASHINGTON, DC, 76-; PROF MED & COORDR, ENDOCRINOL DIV, UNIFORMED SERV UNIV HEALTH SCI, 80- *Concurrent Pos:* Mem, endocrinol study sect, NIH, 77-85; gov, Am Col Physicians, 82-87; clin prof med, Georgetown Univ Sch Med, 84. *Mem:* Am Thyroid Asn (vpres); Am Col Physicians; Endocrine Soc; Am Soc Clin Invest; Am Fedn Clin Res. *Res:* Physiology of thyroid gland; pathophysiology of disorders of the thyroid. *Mailing Add:* Endocrinol-Metab Serv 7D Walter Reed Army Med Ctr Washington DC 20307-5001

WARTZOK, DOUGLAS, b Lansing, Mich, May 10, 42; m 66. MARINE MAMMAL BEHAVIOR & ECOLOGY. *Educ:* Andrews Univ, BA, 63; Univ Ill, MS, 65; Johns Hopkins Univ, PhD(biophys), 71. *Prof Exp:* Asst prof pathobiol & pop biol, Johns Hopkins Univ, 72-78, assoc prof immunol & infectious dis, dis ecol, 78-84, dir, Div Ecol, 79-82; prof & chmn dept biol sci, Purdue Univ, Ind, 83-91; ASSOC VCHANCELLOR & DEAN GRAD SCH, UNIV MO, ST LOUIS, 91- *Concurrent Pos:* Prin investr, NSF, Off Naval Res, NASA & var other orgns, 75-; invited expert, Int Coun Exploration of the Sea, 81; vis prof, Inst Marine Sci, Univ Alaska, 83; ed-in-chief, Marine Mammal Sci, 87- *Mem:* Soc Marine Mammal. *Res:* Physiological ecology, behavior, population structure and sensory physiology of marine mammals; sensory components of under-ice navigation of ringed seals; behaviors and movements of bowhead whales in response to industrial activities. *Mailing Add:* Grad Sch Univ Mo St Louis MO 63121-4499

WARWICK, JAMES WALTER, b Toledo, Ohio, May 22, 24; m 47, 66; c 6. RADIO ASTRONOMY. *Educ:* Harvard Univ, AB, 47, AM, 48, PhD(astron), 51. *Prof Exp:* Asst prof astron, Wellesley Col, 50-52; mem res staff, Sacramento Peak Observ, 52-55; mem sr sci staff, High Altitude Observ, 55-61, PROF ASTRO-GEOPHYS, UNIV COLO, BOULDER, 61- *Concurrent Pos:* Prin investr, Voyager Missions Planetary Radio Astron Exp, NASA, 73-88. *Mem:* Am Astron Soc; AAAS; Am Geophys Union. *Res:* Theoretical astrophysics; stellar and planetary magnetism; solar physics; solar-terrestrial physics. *Mailing Add:* 3845 E Northbrook Dr Boulder CO 80304-1440

WARWICK, SUZANNE IRENE, b Winnipeg Beach, Man, Dec 10, 52. PLANT TAXONOMY. *Educ:* Univ Man, BSc, 74; Univ Cambridge, Eng, PhD(bot), 77. *Prof Exp:* RES SCIENTIST, BIOSYSTS RES CTR, AGR CAN, 77- *Concurrent Pos:* Adj prof, Univ Ottawa, 85- & MacDonald Col, Univ McGill, 85- 87. *Mem:* Can Bot Asn; Soc Evolution; Am Bot Soc. *Res:* Weed genecology; weed taxonomy; electrophoresis; crop and weed with molecular systematics; DNA variation. *Mailing Add:* Biosyst Res Ctr Agr Can K W Neatby Bldg CEF Ottawa ON K1A 0C6 Can

WARWICK, WARREN J, b Racine, Wis, Jan 27, 28; m 52; c 2. PEDIATRICS. *Educ:* St Olaf Col, BA, 50; Univ Minn, MD, 54. *Prof Exp:* Med fel pediat, Univ Minn, Minneapolis, 55-57, med fel specialist, 59-60, from instr to assoc prof pediat, 60-78, PROF PEDIAT, UNIV MINN, MINNEAPOLIS, 78- *Concurrent Pos:* Alpha Omega Phi fel cardiovasc res, 55-57; Am Heart Asn res fel, 59-60; USPHS res career develop award, 61-66; mem ctr prog comt, Nat Cystic Fibrosis Res Found, 64-66, chmn med care comt, 66-71, coop study comt, 71-72; mem exec bd, Sci-Med Comt, Int Cystic Fibrosis (Mucoviscidosis) Asn, 70-80; mem, Nat Data Registry Comt, Cystic Fibrosis Found, 72-86 & Annalisa Marzotto chair Cystic fibrosis. *Res:* Pulmonary diseases; experimental pathology; immunology; cystic fibrosis. *Mailing Add:* Dept Pediat Box 184 Univ Minn Minneapolis MN 55455

WARZEL, L(AWRENCE) A(LFRED), b Ft Scott, Kans, Sept 26, 25; m 50; c 3. CHEMICAL ENGINEERING. *Educ:* Univ Tulsa, BS, 47; Univ Mich, MS, 52, PhD(chem eng), 55. *Prof Exp:* Res proj engr, Ethyl Corp, 47-51; res assoc, Eng Res Inst, Univ Mich, 53-55; theoret develop engr, Phillips Petrol Co, NY, 55-60, sect mgr, 60-65, tech rep proj develop, Int Dept, 65-68, sr proj mgr, NJ, 68-76, sr proj mgr, Okla, 76-; AT DEPT PETROL & GEOL, UNIV OKLA, NORMAN. *Mem:* AAAS; Am Chem Soc; Am Inst Chem Engrs. *Res:* Mass transfer; separations; systems engineering; phase equilibria; thermodynamics; mathematical analysis of engineering data and scale-up problems; international projects development. *Mailing Add:* PO Box 1264 Norman OK 73070

WASA, KIYOTAKA, b Osaka, Japan, Feb 24, 37; m 66; c 1. CATHODIC SPUTTERING TECHNOLOGY. *Educ:* Osaka Univ, Bachelor, 60, Dr(elec eng), 68. *Prof Exp:* Researcher mat, Wireless Res Labs, Matsushita Elec, 60-75, sr researcher, Mat Res Labs, 76-81, chief researcher, Cent Res Labs, 81-83, res dir, 83-88, VPRES, MATSUSHITA ELEC CO LTD, 89-; DEP RES DIR, RES INST INNOVATIVE TECHNOL EARTH, 90- *Concurrent Pos:* Comt mem, Agency Sci & Technol, Japan, 81-86 & Ministry Int Trade & Indust, Japan, 89; vis prof, Osaka Univ, 85-89 & 90-91, Nagoya Univ, 89-90 & Osaka Prefecture Univ, 89-90. *Honors & Awards:* IR 100 Indust Award, 84. *Mem:* Fel Inst Elec & Electronics Engrs; Am Vacuum Soc; Mat Res Soc. *Res:* Developments of a cathodic sputtering technology as a material processing; production of thin film electronic devices, including saw devices and superconducting devices. *Mailing Add:* Central Research Lab Matsushita Elec Ind Co Ltd 3-15 Yagumo Nakamachi Moriguchi 570 Japan

WASACZ, JOHN PETER, b Brooklyn, NY, Sept 11, 44; m 70; c 4. ORGANIC CHEMISTRY. *Educ:* St John's Univ, NY, BS, 65; Univ Pa, PhD(org chem), 69. *Prof Exp:* From asst prof to assoc prof, 69-82, chmn dept, 84-88, PROF CHEM, MANHATTAN COL, 82- *Concurrent Pos:* Asst mgr NY sect, Am Chem Soc, 73-77, mgr, 77-84; NSF fac fel, Columbia Univ, 75-77; vis scholar, NY Univ, 80, 88-89; dir-at-large, Sigma Xi, 83-88. *Mem:* Am Chem Soc; AAAS; Sigma Xi. *Res:* Organic synthesis; photochemistry of heterocyclic molecules; synthesis of natural products. *Mailing Add:* Dept Chem Manhattan Col Bronx NY 10471

WASAN, DARSH T, b Sarai Salah Hazara, India, July 15, 38; m 66; c 2. CHEMICAL ENGINEERING. *Educ:* Univ Ill, Urbana, BSChE, 60; Univ Calif, Berkeley, PhD(chem eng), 64. *Prof Exp:* From asst prof to assoc prof, chmn dept, 71-87, actg dean eng, 87-88, PROF CHEM ENG, ILL INST TECHNOL, 70-, VPRES RES & TECHNOL, 88- *Concurrent Pos:* Consult, Inst Gas Technol, 65-, res inst, Ill Inst Technol, 66-, Chicago Bridge & Iron Co, 67-71, Ill Environ Protection Agency, 71-, Continental Can Co, 72-, Exxon Res & Eng Co, 77-, Stauffer Chem Co, 80-87, ICI, Americas, 88-, & Nelson Indust, 76- *Honors & Awards:* Western Elec Fund Award, NSF, 88; E W Treile Award, Am Soc Chem Engrs, 89; 3M Lecturership Award, Am Soc Eng, Educ Chem Eng Div, 91. *Mem:* AAAS; Am Inst Chem Engrs; Am Chem Soc; Am Soc Eng Educ; Am Inst Physics; Fine Particle Soc. *Res:* Interfacial phenomena; particle science and technology; enhanced oil recovery; interfacial rheology. *Mailing Add:* Res & Technol PH Rm 228 Chicago IL 60616

WASAN, MADANLAL T, b Saraisaleh, WPakistan, July 13, 30; m 60; c 4. STATISTICS. *Educ:* Univ Bombay, BA, 52, MA, 54; Univ Ill, PhD(statist), 60. *Prof Exp:* From asst prof to assoc prof, 59-68, PROF MATH, QUEEN'S UNIV, ONT, 68- *Concurrent Pos:* Vis assoc prof, Stanford Univ, 65 & Univ Bombay, 65-66; statist consult, Du Pont of Can, 62-65. *Mem:* Inst Math Statist. *Res:* Sequential estimation; stochastic approximation; stochastic processes and applied probability. *Mailing Add:* Dept Math/Statist Queen's Univ Kingston ON K7L 3N6 Can

WASBAUER, MARIUS SHERIDAN, b Rockford, Ill, Sept 29, 28; m 69; c 3. INSECT TAXONOMY. *Educ:* Univ Calif, Berkeley, BS, 50, PhD(entom), 61. *Prof Exp:* SR INSECT BIOSYSTEMATIST, CALIF DEPT FOOD & AGR, 59- *Concurrent Pos:* Collabr, Animal & Plant Health Inspection Serv, USDA, 59-; res assoc, Univ Calif, Berkeley, 73-, mem, Robert Gordon Sproul Assocs, 77-; fel, Calif Acad Sci, 76- *Mem:* Sigma Xi; Int Orgn Biosystematists. *Res:* Biosystematics of New World Pompilidae Mutillidae (Myrmosinae) and of North American Tiphiidae; larval systematics and biology of North American Tephritidae. *Mailing Add:* Anal & Ident Calif Dept Food & Agr 1220 N St Sacramento CA 95814

WASCOM, EARL RAY, ecology; deceased, see previous edition for last biography

WASE, ARTHUR WILLIAM, b Jersey City, NJ, Nov 16, 15; m 40; c 2. BIOCHEMISTRY. *Educ:* Columbia Col, AB, 47; Rutgers Univ, PhD, 51. *Prof Exp:* From instr to assoc prof, Hahnemann Med Col, 51-68; sr res fel biochem, Merck Inst Therapeut Res, 75-80, chief health physicist, 80; PROF BIOCHEM, HAHNEMANN MED COL, 68- *Concurrent Pos:* Fulbright prof, Univ Brussels & Inst Jules Bordet, Belgium, 56-57; consult, Colgate Biol Res Labs, 58- & pub sect, Radio Corp Am; res fel biochem endocrinol, Merck Inst Therapeut Res, 63-75, head dept health physics, 76-86. *Mem:* Am Cancer Soc; Am Heart Asn Coun Thrombosis. *Res:* Radiobiochemistry as applied to cancer research and neuropsychopharmacology; bone metabolism; biochemistry and endocrinology of atherosclerosis. *Mailing Add:* Four Medford Ct Shadow Lake Village Red Bank NJ 07701

WASER, NICKOLAS MERRITT, b Pasadena, Calif, June 28, 48; m 77. POLLINATION ECOLOGY, POPULATION BIOLOGY. *Educ:* Stanford Univ, AB, 70; Univ Ariz, PhD(biol), 77. *Prof Exp:* Teaching fel, Univ Utah, 77-79; from asst prof to assoc prof, 79-90, PROF BIOL, UNIV CALIF, RIVERSIDE, 90- *Concurrent Pos:* Secy, Rocky Mountain Biol Lab, 82-90. *Mem:* Soc Study Evolution; Ecol Soc Am; Bot Soc Am. *Res:* Plant sexual characteristics; behavior of animal pollinators, and their influence on the ecology, genetics, and evolution of plant populations. *Mailing Add:* Dept Biol Univ Calif Riverside CA 92521-0427

WASER, PETER MERRITT, b Pasadena, Calif, Dec 12, 45. BEHAVIORAL ECOLOGY. *Educ:* Stanford Univ, BS, 68; Rockefeller Univ, PhD(biol), 74. *Prof Exp:* NIH fel, Rockefeller Univ, 74-75; asst prof, 75-80, ASSOC PROF BIOL, PURDUE UNIV, 80- *Concurrent Pos:* Grants, EAfrican Wildlife Soc, 74-75, Am Philos Soc, 76-77, NIMH, 77-78, NSF, 78- & NIH, 80-82. *Mem:* Animal Behav Soc; Int Primate Soc; Ecol Soc Am; AAAS; Am Soc Mammalogists; Sigma Xi. *Res:* Adaptive aspects of animal social behavior; animal communication; territoriality and resource use. *Mailing Add:* Dept Biol Sci Purdue Univ West Lafayette IN 47907

WASFI, SADIQ HASSAN, b Basrah, Iraq, July 1, 37; m 68; c 3. INORGANIC CHEMISTRY, ANALYTICAL CHEMISTRY. *Educ:* Univ Baghdad, BS, 61; Georgetown Univ, MS, 66, PhD(inorg chem), 71. *Prof Exp:* From lectr to asst prof chem, Col Sci, Basrah Univ, Iraq, 71-77; res assoc chem, Univ Hawaii, Manoa, 75-76; res assoc, Georgetown Univ, 77-78; assoc prof chem, Montgomery Col, Md, 78-79; assoc prof, 79-84, PROF CHEM, DEL STATE COL, 84- *Concurrent Pos:* Prin Investigator, NIH, MBRS Grant, 87-; vis assoc prof, Georgetown Univ, 80-81. *Mem:* Am Chem Soc; Iraqi Chem Soc; Sigma Xi. *Res:* Transition metal complexes of organic thiols and disulfides; organic derivatives of heteropoly tungstates and molybdates; heteropoly tungstate and molybdate anions containing several transition metal ions; heteropoly oxometalate anions of biological importance. *Mailing Add:* Dept Chem Del State Col Dover DE 19901

WASHA, GEORGE WILLIAM, b Milwaukee, Wis, May 6, 09; m 34; c 3. MECHANICS. *Educ:* Univ Wis, BS, 30, MS, 32, PhD(mech), 38. *Prof Exp:* Instr, 30-40, chmn dept, 53-75, prof, 40-78, EMER PROF MECH, UNIV WIS-MADISON, 78- *Honors & Awards:* Wason Medal, Am Concrete Inst, 41 & 76. *Mem:* fel Am Concrete Inst; hon mem, Am Con Inst, 90. *Res:* Durability, permeability and plastic flow of concrete; light weight agregates and concrete; vibrated concrete; masonry cements; properties of ferrous metals; concrete block. *Mailing Add:* 202 N Eau Claire Ave Apt 205 Madison WI 53705

WASHBURN, ALBERT LINCOLN, b New York, NY, June 15, 11; m 35; c 3. GEOMORPHOLOGY, QUATERNARY GEOLOGY. *Educ:* Dartmouth Col, AB, 35; Yale Univ, PhD(geol), 42. *Hon Degrees:* DSc, Univ Alaska, 81. *Prof Exp:* Mem, Nat Geog Soc exped, Mt McKinley, 36; asst geologist, Boyd E Greenland exped, 37; geol invests, Can Arctic, 38-41, 49 & 81-; exec dir, Arctic Inst NAm, 45-51; dir snow, ice & permafrost res estab, Corps Engrs, US Army, 51-52; prof northern geol, Dartmouth Col, 53-59; prof geol, Yale Univ, 60-66; prof, 67-76, EMER PROF GEOL, UNIV WASH, 76- *Concurrent Pos:* Consult, Res & Develop Bd, 47-53 & CEngrs, US Army, 52-61; hon lectr, McGill Univ, 48-51; mem, US Nat Comt, Int Geophys Year, 53-59, Comt Polar Res, 58-59 & 63-73, Panel Glaciol, Nat Acad Sci, 59-65 & 67-71; geomorphol investrs, Greenland, 54-58, 60 & 64, Antarctica, 57-58; dir, Quaternary Res Ctr, Univ Wash, 67-76; vpres, Int Quaternary Union, 73-82; chmn bd, Polar Res, 78-81; mem, Geophys Inst Adv Bd, Univ Alaska, Fairbanks, 82-; comn, US Arctic Res Comn, 85-88. *Honors & Awards:* Kirk Bryan Award, Geol Soc Am, 71; Medaile Andre H Dumont, Geol Soc Belg, 73; Medal, Univ Liege, Belg, 71; Distinguished Career Award, Geol Soc Am, 88. *Mem:* Fel Am Geog Soc; hon mem Int Glaciol Soc; fel Geol Soc Am; Am Geophys Union; hon mem Arctic Inst NAm; fel Geol Asn Can; hon mem Int Quaternary Union; Am Quaternary Asn (pres, 70); corresp mem Geog Soc Finland; corres mem Geol Soc Belg. *Res:* Geocryology; periglacial studies. *Mailing Add:* Quaternary Res Ctr AK-60 Univ Wash Seattle WA 98195

WASHBURN, HAROLD W(ILLIAMS), b Jacksonville, Ore, June 23, 02; m 56; c 2. PHYSICS, ELECTRICAL ENGINEERING. *Educ:* Univ Calif, BS, 24, PhD(elec eng), 32; Mass Inst Technol, MS, 27. *Prof Exp:* From instr to asst prof elec eng, Univ Calif, 28-31; instr, Mass Inst Technol, 32-33; chief physicist in charge res lab, Western Geophys Co, Calif, 33-37; vpres in charge res, Consol Electrodyne Corp, 37-60; from chief exp space sci to staff scientist, Jet Propulsion Lab, Calif Inst Technol, 60-65; prof, 65-76, EMER PROF ELEC ENG, CALIF STATE UNIV, LONG BEACH, 77- *Concurrent Pos:* Consult, Jet Propulsion Lab, Calif Inst Technol. *Honors & Awards:* Beckman Award, Am Chem Soc, 56. *Mem:* AAAS; Am Soc Testing & Mat; Am Geophys Union. *Res:* Mass spectroscopy; lunar atmosphere and surface characteristics; electric circuits. *Mailing Add:* 102 Rivo Alto Canal Long Beach CA 90803

WASHBURN, JACK, b Mt Vernon, NY, Apr 30, 21; m 47; c 3. MATERIAL SCIENCE OF SEMICONDUCTOR MATERIALS. *Educ:* Univ Calif, BS, 49, MS, 50, PhD, 54. *Prof Exp:* Res engr, Inst Eng Res, 49-52, from instr to assoc prof metall, 52-61, chmn dept, 67-70, prof mat sci & eng, 61-81, EMER PROF DEPT MAT SCI & MINERAL ENG & SR FAC SCIENTIST, LAWRENCE BERKELEY LAB, UNIV CALIF, BERKELEY, 81- *Concurrent Pos:* Sr fel, Univ Cambridge, 59-60; res prof, Miller Inst Basic Res Sci, 61-62; sr fel, Univ Paris, 65. *Honors & Awards:* Mathewson Gold Medal, Am Inst Mining, Metall & Petrol Engrs, 56. *Mem:* Am Soc Metals; Am Inst Mining, Metall & Petrol Engrs; Mat Res Soc. *Res:* Relation between properties and structure of electronic materials, particularly the characterization and effects of crystal imperfections in semiconductors and at metal-semiconductor interfaces. *Mailing Add:* Dept Mat Sci & Mineral Eng Univ Calif Berkeley CA 94720

WASHBURN, KENNETH W, b Martinsville, Va, June 21, 37; m 59; c 2. POULTRY GENETICS. *Educ:* Va Polytech Inst, BS, 59, MS, 62; Univ Mass, PhD(poultry), 65. *Prof Exp:* Res asst poultry genetics, Va Polytech Inst, 60-62; instr, Univ Mass, 62-65; assoc prof, 65-75, PROF POULTRY GENETICS, UNIV GA, 75- *Honors & Awards:* Poultry Sci Jr Res Award, Poultry Sci Asn, 75. *Mem:* Poultry Sci Asn. *Res:* Genetic-nutrition interrelationships; compensatory growth; feed efficiency; egg shell strength; egg cholesterol; hemoglobins. *Mailing Add:* Dept Poultry Univ Ga Athens GA 30601

WASHBURN, LEE CROSS, b Paducah, Ky, Jan 10, 47; m 69; c 3. ORGANIC CHEMISTRY, RADIOPHARMACEUTICAL CHEMISTRY. *Educ:* Murray State Univ, BA, 68; Vanderbilt Univ, PhD(org chem), 72. *Prof Exp:* Assoc scientist, Oak Ridge Assoc Univs, 72-77, scientist Radiopharmaceut Develop & Preclin Nuclear Med, Med & Health Sci Div, 77-90, actg dir, Nuclear Med Prog, Med Sci Div, 90-91; DIR RADIOPHARMACEUT DEVELOP, EUGENE L SAENGER RADIOISOTOPE LAB, UNIV CINCINNATI MED CTR, 91- *Mem:* Soc Nuclear Med. *Res:* Radiopharmaceutical development; radiolabeled monoclonal antibodies for cancer diagnosis and therapy. *Mailing Add:* Eugene L Saenger Radioisotope Lab Univ Cincinnati Med Ctr 234 Goodman St Cincinnati OH 45267-0577

WASHBURN, ROBERT HENRY, b Lincoln, Nebr, Nov 27, 36; m 66. STRATIGRAPHY, STRUCTURAL GEOLOGY. *Educ:* Univ Nebr, BS, 59, MS, 61; Columbia Univ, PhD(geol), 66. *Prof Exp:* Instr geol, Brooklyn Col, 64-66; from asst prof to assoc prof, 66-77, PROF GEOL, JUNIATA COL, 77- *Mem:* AAAS; Geol Soc Am; Am Asn Petrol Geologists; Soc Econ Paleont & Mineral. *Res:* Paleozoic stratigraphy; structural geology of central Nevada and Pennsylvania. *Mailing Add:* Dept Geol Juniata Col Huntingdon PA 16653

WASHBURN, ROBERT LATHAM, b Malone, NY, June 22, 21; m 74; c 4. MECHANICAL ENGINEERING, POLYMER ENGINEERING. *Educ:* Clarkson Col Technol, BE, 47; Mass Inst Technol, MS, 48. *Prof Exp:* Chief engr design develop, Sklenar Furnace & Mfg Co, 46-47; mgr, Eng Tech Sect, E I du Pont de Nemours & Co, Inc, 48-53, asst supt, 53-55, supt, 55-58, sr supvr res & develop, 58-68 & Res Plastics Prod Div, 68- 72, supt, Res Lab, 68-72 & Polymer Prod Div, 72-78, sr supvr, Admin & Eng Res, 78-79, tech serv mgr, Plastics Mkt Molding Compounding & Extrusion, 79-82, Adhesives & Sealants, 82-83, sr tech consult, Warp Size Develop-Textiles, 83-90; RETIRED. *Concurrent Pos:* Independent consult, 90- *Mem:* Sigma Xi; Am Soc Mech Engrs; Tech Asn Pulp & Paper Indust. *Res:* Concept development, resultant program derivation, and subsequent engineering new polymer systems encompassing process, product and equipment; synthesis, rheology, morphology and conformational alterations such as orientation. *Mailing Add:* 3250 Landsdowne Dr Cardiff Wilmington DE 19810

WASHBURN, SHERWOOD L, b Cambridge, Mass, Nov 26, 11. PRIMATE BEHAVIOR. *Educ:* Harvard Univ, BA, 35, PhD(anthrop), 40. *Prof Exp:* Prof anat, Sch Med, Columbia Univ, 39-47; prof anthrop, Univ Chicago, 47-58; prof, 58-78, EMER UNIV PROF ANTHROP, UNIV CALIF, BERKELEY, 78- *Mem:* Nat Acad Sci; Am Anthrop Soc (pres, 62); AAAS. *Res:* Experimental analysis of behavior. *Mailing Add:* Dept Anthrop Univ Calif Kroeber Hall Rm 232 Berkeley CA 94720

WASHBURN, WILLIAM H, b Milwaukee, Wis, Oct 14, 20; m 42; c 3. ANALYTICAL CHEMISTRY. *Educ:* Univ Wis, BA, 41. *Prof Exp:* Chemist, Abbott Labs, 46-85; RETIRED. *Mem:* Soc Appl Spectros; Am Chem Soc; Coblentz Soc; Sigma Xi. *Res:* Infrared spectroscopy, materials purity and chemical structure analysis. *Mailing Add:* 120 Hawthorne Ct Lake Bluff IL 60044

WASHBURNE, STEPHEN SHEPARD, b Hartford, Conn, Sept 6, 42; m 83; c 1. ORGANIC CHEMISTRY. *Educ:* Trinity Col, Conn, BS, 63; Mass Inst Technol, PhD(org chem), 67. *Prof Exp:* Asst prof, 67-73, asst chmn dept, 74-78, ASSOC PROF CHEM, TEMPLE UNIV, 73- *Concurrent Pos:* NSF fel, 64-66; NIH fel, 67; chief consult, Petrarch Systs, Inc, 70-; Fulbright sr lectr, Portugal, 80. *Mem:* Am Chem Soc. *Res:* Organosilicon chemistry; cancer chemotherapy; organometallic chemistry of the elements of group IV. *Mailing Add:* Dept Chem Temple Univ Philadelphia PA 19122

WASHCHECK, PAUL HOWARD, industrial organic chemistry, for more information see previous edition

WASHINGTON, ARTHUR CLOVER, b Tallulah, La, Aug, 19, 39; m 62; c 3. DEVELOPMENTAL BIOLOGY. *Educ:* Tex Col Tyler, BS, 61; Tuskegee Inst, Ala, MS, 63; Ill Inst Technol, PhD(biol), 71. *Prof Exp:* Res scientist plant path, Wash State Univ, 63-64; instr biol, Talladega Col, 65-67; assoc prof, Amundsen-Mayfair Col, 67-72; assoc prof, Langston Univ, Okla, 72-74; PROF BIOCHEM, PRAIRIE VIEW A&M UNIV, 74-, DEAN, GRAD SCH, 82- *Concurrent Pos:* Res scientist, Pfizer Chem Co, Conn, 66; univ admin trainee, Univ Wis-Extension, Madison, 73; prin investr, NSF res initiation award, 73-74, USDA Tri-Co Nutrit, Prairrie View, 74-79, Robert A Welch Found award, 76-, NIH awards, 76-; sci develop consult, Paul Quinn Col, Tex, 76- *Mem:* Nat Inst Sci; Am Soc Microbiologists; AAAS; Sigma Xi. *Res:* Mechanisms which seem to regulate morphogenesis in the cellular slime mold, Dictyostelium discoideum; DNA, RNA folic acid and mitochondrial enzyme metabolism. *Mailing Add:* 10303 Green Creek Dr Houston TX 77070

WASHINGTON, ELMER L, b Houston, Tex, Oct 18, 35; m 60; c 2. PHYSICAL CHEMISTRY. *Educ:* Tex Southern Univ, BS, 57, MS, 58; Ill Inst Technol, PhD(thermodyn), 66. *Prof Exp:* Asst proj engr, Pratt & Whitney Aircraft Div, United Aircraft Corp, Conn, 65-67, res assoc advan mat res & develop lab, 67-69; asst prof phys sci, 69-72, dean natural sci & math, 72-74, assoc prof, 72-77, prof chem & vpres res & develop, 76-80, dean Col Arts & Sci, 74-76, prof chem, 80-90, ACTG VPRES STUDENT AFFAIRS,

CHICAGO STATE UNIV, 90- *Mem:* Electrochem Soc; Am Chem Soc. *Res:* Thermodynamics of non-electrolytes; electrochemistry as related to fuel cell technology. *Mailing Add:* Dept Chem Chicago State Univ 95th St & King Dr Chicago IL 60628

WASHINGTON, JAMES M(ACKNIGHT), b Hackensack, NJ, Dec 1, 38; m 56, 88; c 5. CHEMICAL ENGINEERING. *Educ:* Clemson Univ, BS, 61; Va Polytech Inst, MS, 64, PhD, 69. *Prof Exp:* Asst prof chem eng, Univ NB, 65-66; res engr, E I du Pont de Nemours & Co, 66-68; sr scientist, Allied Chem Corp, 68-72; res engr, 72-80, ASSOC SR ENGR, PHILIP MORRIS RES CTR, 80- *Mem:* Am Inst Chem Engrs; Sigma Xi. *Res:* Chemical reactor engineering, particularly non-ideal mixing of fluids. *Mailing Add:* 2400 Stuts Lane Richmond VA 23236-1368

WASHINGTON, JOHN A, II, b Istanbul, Turkey, May 29, 36; US citizen; m 59; c 3. CLINICAL MICROBIOLOGY, CLINICAL PATHOLOGY. *Educ:* Univ Va, BA, 57; Johns Hopkins Univ, MD, 61. *Prof Exp:* From intern to asst resident surg, Med Ctr, Duke Univ, 61-63; Nat Cancer Inst fel, 63-65; resident clin path, Clin Ctr, NIH, 65-67; assoc consult, Mayo Clin, 67-68, consult, 68-85, assoc prof microbiol & lab med, Mayo Med Sch, 72-76, head sect clin microbiol, 71-85, prof microbiol & lab med, 76-85; CHMN DEPT MICROBIOL, CLEVELAND CLIN FOUND, 86- *Concurrent Pos:* Asst prof microbiol, Mayo Grad Sch Med, 70-72; ed, J Clin Microbiol, 74-75; ed, Antimicrobiol Agents & Chemother, 81-; trustee, Am Bd Path, 89- *Mem:* Fel Am Soc Clin Path; Am Soc Microbiol; fel Am Col Physicians; fel Am Acad Microbiol; fel Infectious Dis Soc Am; fel Col Am Pathologists; fel Am Col Chest Physicians. *Res:* Antimicrobial agents; antimicrobial susceptibility tests; methodology in clinical bacteriology. *Mailing Add:* Dept Microbiol Cleveland Clin Found Clin Ctr Cleveland OH 44195-5140

WASHINGTON, LAWRENCE C, b Middlebury, Vt, June 14, 51. MATHEMATICS, NUMBER THEORY. *Educ:* Johns Hopkins Univ, BA & MA, 71; Princeton Univ, PhD(math), 74. *Prof Exp:* Assoc prof, 81-86, PROF MATH, UNIV MD, 86- *Concurrent Pos:* Alfred P Sloan fel, 79-81. *Mem:* Am Math Soc; Math Asn Am. *Mailing Add:* Dept Math Univ Md College Park MD 20742

WASHINGTON, WARREN MORTON, b Portland, Ore, Aug 28, 36; wid; c 6. METEOROLOGY. *Educ:* Ore State Univ, BS, 58, MS, 60; Pa State Univ, PhD(meteorol), 64. *Prof Exp:* Res asst meteorol, Pa State Univ, 61-63; SR SCIENTIST, NAT CTR ATMOSPHERIC RES, 63-, DIR, CLIMATE & GLOBAL DYNAMICS. *Concurrent Pos:* Adj prof meteorol & oceanog, Univ Mich, 69-71; mem var panels, Nat Acad Sci & NSF, 69-; mem gov sci adv comt, State Colo, 75-78; Pres appointment, Nat Adv Comt Oceans & Atmosphere, 78-83. *Mem:* AAAS; Am Meteorol Soc; Am Geophys Union. *Res:* Numerical modeling of the atmosphere; climate modeling of climate. *Mailing Add:* Nat Ctr Atmospheric Res PO Box 3000 Boulder CO 80307

WASHINGTON, WILLIE JAMES, b Madison, Fla, Dec 26, 42; m 70. PLANT GENETICS, CYTOGENETICS. *Educ:* Fla A&M Univ, BS, 64; Univ Ariz, MS, 66; Univ Mo-Columbia, PhD(plant genetics, cytogenetics), 70. *Prof Exp:* Asst prof biol, Tougaloo Col, 70-71 & Cent State Univ, Ohio, 71-72; D F Jones fel agron, NDak State Univ, 72-73; from asst prof to assoc prof, 73-82, PROF BIOL, CENT STATE UNIV, OHIO, 82- CHMN DEPT, 85- *Concurrent Pos:* Prin investr, Mutagenic-Teratological Potential Lab Solvents, Minority Biomed Support Prog, NIH, 80-83 & Minority Access to Res Careers Prog, 82-87. *Mem:* Orgn Black Scientists; Nat Inst Sci; Cent Asn Adv Health Prof; AAAS. *Res:* Genetic toxicology; assessment of genetic risks of exposure to environmental pollutants; application of genetical and cytogenetical analysis to the improvement of economic crops; mammalian tissue culture; environmental mutagenesis; genetics and cytogenetics of higher plants and animals. *Mailing Add:* Chief Acad Dean Cent State Univ Wilberforce OH 45384

WASHINO, ROBERT K, b Sacramento, Calif, Mar 14, 32; m 56; c 3. ENTOMOLOGY, PUBLIC HEALTH. *Educ:* Univ Calif, Berkeley, BS, 54; Univ Calif, Davis, MS, 56, PhD(entom), 67. *Prof Exp:* Assoc-sr specialist, Calif State Dept Pub Health, 59-65; asst specialist, 65-67, assoc prof, 67-81, PROF ENTOM, UNIV CALIF, DAVIS, 81-, LECTR & ASST ENTOMOLOGIST, 67- *Concurrent Pos:* NIH grant, 71-73. *Mem:* Entom Soc Am; Am Soc Trop Med & Hyg; Am Mosquito Control Asn (vpres). *Res:* Studies regarding the various aspects of insect biology which affect their role as vectors of human and animal pathogens; sociological as well as entomological studies of mosquito pest problems. *Mailing Add:* Dept Entom Univ Calif Davis CA 95616

WASHKO, FLOYD VICTOR, b New Brunswick, NJ, Oct 17, 22; m 47; c 2. VETERINARY PATHOLOGY. *Educ:* Mich State Univ, DVM, 44; Purdue Univ, MS, 48, PhD, 50. *Prof Exp:* Vet pract, NJ, 44-45; assoc prof vet sci, Purdue Univ, 46-53; mem staff, Plum Island Animal Dis Lab, USDA, 53-54; VET PATHOLOGIST, MERCK, SHARP & DOHME RES LABS, 54-, SR INVESTR, 74- *Mem:* Am Vet Med Asn; NY Acad Sci; Am Asn Avian Pathologists; Am Asn Vet Lab Diagnosticians; US Animal Health Asn. *Res:* Brucellosis of swine and cattle; virus diseases of the bovine; veterinary therapeutics. *Mailing Add:* 523 Colonia Blvd Colonia NJ 07067

WASHKO, WALTER WILLIAM, b New Brunswick, NJ, July 29, 20; m 55; c 4. AGRONOMY. *Educ:* Rutgers Univ, BS, 41, MS, 47, Univ Wis, PhD(agron, bot), 58. *Prof Exp:* Agronomist, Tex Res Found, 47-53 & Eastern States Farmers Exchange, 53-64; ext agronomist & prof agron, Univ Conn, 64-88; RETIRED. *Mem:* Am Soc Agron; Am Inst Biol Sci; Sigma Xi. *Res:* Crop production. *Mailing Add:* RR 3 Mansfield Center CT 06250

WASHTON, NATHAN SEYMOUR, b New York, NY, Nov 9, 16; m 44; c 3. SCIENCE EDUCATION & COMMUNICATIONS. *Educ:* NY Univ, BS, 39, EdD, 50; Columbia Univ, MA, 41. *Prof Exp:* Chmn sci, Newark Jr Col, 39-42; dir tech training, USAF, 42-45; teacher biol, chem & math, Rhodes Sch, 45-46; chmn sci, Rutgers Univ, 46-50; coordr sci educ & teaching sci, 50-80, EMER PROF, QUEENS COL, CITY UNIV NEW YORK, 80- *Concurrent Pos:* Consult, NSF; vis prof, Univ PR, 48, Upsala Col, 48, Univ Hawaii, 62-63, Hebrew Univ, Jerusalem, 67 & Ben Gurion Univ, Israel, 79; educ consult, 80- *Honors & Awards:* Res Award, Libr Sci, 61. *Mem:* Fel AAAS; Nat Asn Res Sci Teaching (vpres, 57, pres, 58); Nat Sci Teachers Asn; Am Educ Res Asn; Am Environ Sci Acad. *Res:* Seminars and workshops for faculty development in teaching of sciences for effectiveness; evaluation of college and medical faculty. *Mailing Add:* 30 Oaktree Lane Manhasset NY 11030

WASIELEWSKI, MICHAEL ROMAN, b Chicago, Ill, June 7, 49; m 75; c 2. PHOTOSYNTHESIS. *Educ:* Univ Chicago, BS, 71, MS, 72, PhD(chem), 75. *Prof Exp:* Fel chem, Columbia Univ, 74-75; res assoc, 75-76, asst chemist, 76-81, CHEMIST, ARGONNE NAT LAB, 81- *Mem:* Am Chem Soc; Biophys Soc; Am Soc Photobiol. *Res:* Mechanism of the primary events of photosynthesis; light induced electron transfer reactions; biomimetic modelling of natural biophysical chemistry. *Mailing Add:* Chem Div Argonne Nat lab Argonne IL 60439

WASIELEWSKI, PAUL FRANCIS, b Bay Shore, NY, Oct 21, 41; m 67; c 3. VEHICLE-BORNE COMPUTER APPLICATIONS. *Educ:* Georgetown Univ, BS, 63; Yale Univ, MS, 65, PhD(physics), 69. *Prof Exp:* SCIENTIST, DEPT TRANSP RES, GEN MOTORS RES LAB, 69- *Mem:* Am Phys Soc. *Res:* Microcomputer-based automotive navigation and map display systems; traffic accident research; mathematical models of traffic flow. *Mailing Add:* 25474 Wareham Dr Huntington Woods MI 48070

WASILIK, JOHN H(UBER), b Franklin, NC, Feb 9, 25; m 51; c 6. SOLID STATE PHYSICS, MICROWAVE ENGINEERING. *Educ:* Manhattan Col, BS, 47; Cath Univ Am, PhD(physics), 57. *Prof Exp:* Physicist, Nat Bur Standards, 50-67; physicist, Harry Diamond Labs, 67-; PRIN, WASILIK ASSOC, SILVER SPRING, MD. *Mem:* Am Phys Soc; Inst Elec & Electronics Engrs. *Res:* Elastic and dielectric properties, their pressure and temperature dependence; associated loss mechanisms; excitons; microwave acoustics; lattice attenuation; theory and experiment; piezoelectricity; electrostriction; bulk and surface wave microwave acoustic delay lines; acousto-optics; fiber optics; operations research; strategic defense initiative; nuclear weapons effects. *Mailing Add:* c/o Wasilik Assoc 1307 Sarah Dr Silver Spring MD 20904

WASLEY, RICHARD J(UNIOR), b Oakland, Calif, June 24, 31; m 60; c 3. CIVIL ENGINEERING. *Educ:* Univ Calif, Berkeley, BS, 54; Stanford Univ, MS, 58, PhD(civil eng), 60. *Prof Exp:* Sanit engr, Bur Sanit Eng, Calif Dept Pub Health, 55-56; civil engr, Alameda County Surveyor Off, 57; ENGR, LAWRENCE LIVERMORE LAB, UNIV CALIF, 60- *Honors & Awards:* Alfred Noble Prize, 62. *Mem:* Am Soc Civil Engrs; Nat Soc Prof Engrs; Am Soc Mech Engrs. *Res:* Dynamic mechanical properties of materials; elastic and viscoelastic waves; fluid mechanics; shock wave phenomena. *Mailing Add:* 4290 Colgate Way Livermore CA 94550

WASLIEN, CAROL IRENE, b Mayville, NDak, Sept 24, 40; m 80. NUTRITION EDUCATION, NUTRITION ASSESSMENT. *Educ:* Univ Calif, Santa Barbara, BA, 61; Cornell Univ, MS, 63; Univ Calif, Berkeley, PhD(nutrit), 68. *Prof Exp:* NIH res training fel, Vanderbilt Univ, Naval Med Res Unit-3, Egypt, 68-69; res assoc nutrit, Vanderbilt Univ, 69-72; assoc prof & head dept nutrit & foods, Auburn Univ, 72-77; chief nutrit planning proj, USAID, 77-79; exec dir, League Int Food Educ, 79-81; prof nutrit & food sci prog, Hunter Col, City Univ New York, 81-90; PROF, PUB HEALTH NUTRIT, UNIV HAWAII, 90- *Concurrent Pos:* Res consult biochem dept, Vanderbilt Univ, 72-74; res consult & mem bd dir, Universal Foods Corp, 80-; adj prof, Mt Sinai Sch Med, 80- *Mem:* Am Inst Nutrit; Am Dietetic Asn; Inst Food Tech; Am Pub Health Asn. *Res:* Human requirements for protein, vitamins and trace minerals; use of micro-organisms as food sources for man; nutrition assessment; nutrition program evaluation; international nutrition; geriative nutrition. *Mailing Add:* Pub Health Nutrit Sch Pub Health Univ Hawaii Honolulu HI 96822

WASON, SATISH KUMAR, b Lyallpur, India, Feb 24, 40; m 70; c 3. PHYSICAL INORGANIC CHEMISTRY. *Educ:* Univ Delhi, BSc, 59, MSc, 61; Cornell Univ, PhD(phys chem), 65. *Prof Exp:* Scientist, Coun Sci & Indust Res, New Delhi, India, 65-66; res assoc phys chem, Boston Univ, 66-67; res chemist, E I du Pont de Nemours & Co, Inc, Del, 67-69; tech dir & asst vpres, 85-87, vpres, 87-88, PRES, CHEM DIV, J M HUBER CORP, 88- *Mem:* Am Chem Soc; Soc Cosmetic Chemists; Sigma Xi; Soc Plastic Indust; Soc Plastic Engrs. *Res:* High temperature thermodynamic and spectroscopic studies; photochemistry and mercury photosensitized reactions; chemistry of synthetic silicas and silicates; surface chemistry, structure and applications of fine-particle synthetic silicas and silicates in paper, rubber, paints, plastics, dentifrices, cosmetics, pharmaceutical and specialty industries; structure and properties of kaolin products, zeolites, micas, hectorites, specialty minerals, fillers and extenders. *Mailing Add:* J M Huber Corp PO Box 310 Havre de Grace MD 21078-3718

WASOW, WOLFGANG RICHARD, b Vevey, Switz, July 25, 09; nat US; m 39, 59; c 3. MATHEMATICS. *Educ:* NY Univ, PhD(math), 42. *Prof Exp:* Instr math, Goddard Col, 39-40, Conn Col for Women, 41-42 & NY Univ, 42-46; asst prof, Swarthmore Col, 46-49; mathematician numerical anal res, Univ Calif, Los Angeles, 49-55; math res ctr, US Dept Army, Wis, 56-57; prof, 57-80, EMER PROF MATH, UNIV WIS-MADISON, 80- *Concurrent Pos:* Fulbright fel, Rome, Italy, 54-55 & Haifa, Israel, 62. *Mem:* Am Math Soc; Soc Indust & Appl Math. *Res:* Asymptotic theory and numerical solution of differential equations. *Mailing Add:* Dept Math Univ Wis Van Vleck Hall Madison WI 53706

WASS, JOHN ALFRED, b Kyoto, Japan, Apr 14, 50; US citizen. TUMOR CELL BIOLOGY, LYMPHOCYTE BIOLOGY. *Educ:* Lake Forest Col, BA, 69; Northern Ill Univ, MS, 70; Southern Ill Univ, PhD(physiol), 74; Univ Cincinnati, BS, 77; Lake Forest Grad Sch, MBA, 87. *Prof Exp:* Clin microbiologist, N Chicago Vet Admin Hosp, 74-75; criteria mgr pharmacol & toxicol, Nat Inst Occup Safety & Health, 75-77; postdoctoral res assoc pharmacol & toxicol, Med Col Wis, 77-79; postdoctoral res assoc path, Univ Conn, 79-80; postdoctoral fel path, Univ Mich, 80-83; mgr oncol & immunol, 83-86, DIR RES & DEVELOP ONCOL & IMMUNOL, AM INT HOSP, 86- *Concurrent Pos:* Treas, Vet Lab Asn, 89- *Mem:* Am Physiol Soc; Am Asn Cancer Res; AAAS; Tissue Cult Asn. *Res:* Investigations deal with in-vitro interactions of tumor cells with lymphocytes and host humoral factors; indicators of immune status in cancer patients. *Mailing Add:* Oncol Res & Develop Lab Am Int Hospital Shiloh Blvd Zion IL 60099

WASS, MARVIN LEROY, zoology, botany; deceased, see previous edition for last biography

WASS, WALLACE M, b Lake Park, Iowa, Nov 19, 29; m 53; c 4. VETERINARY MEDICINE, VETERINARY SURGERY. *Educ:* Univ Minn, BS, 51, DVM, 53, PhD(vet med), 61. *Hon Degrees:* Vet Med, Nat Univ Colombia, 63. *Prof Exp:* From instr to asst prof vet med, Univ Minn, 58-63; res assoc lab animal med, Brookhaven Nat Lab, 63-64; prof vet clin sci & head dept, 64-83, PROF, VET CLIN SCI, IOWA STATE UNIV, 83- *Mem:* Am Vet Med Asn. *Res:* Large animal medicine and surgery; metabolic diseases of domestic animals; bovine porphyria. *Mailing Add:* Dept Vet Clin Sci Iowa State Univ Ames IA 50011

WASSARMAN, PAUL MICHAEL, b Milford, Mass, Mar 26, 40; m 60; c 4. DEVELOPMENTAL BIOLOGY. *Educ:* Univ Mass, BS, 61, MS, 64; Brandeis Univ, PhD(biochem), 68. *Prof Exp:* Helen Hay Whitney Found fel, Molecular Res Coun Lab, Cambridge, Eng, 67-69; asst prof, dept biol sci, Purdue Univ, 69-72; lectr & spec res fel, Harvard Med Sch & Rockefeller Found fel, 72-73; from asst prof to assoc prof, dept biol chem, Harvard Med Sch, 73-85, full mem, 85-86; CHMN, DEPT CELL DEVELOP BIOL, ROCHE INST MOLECULAR BIOL, 86- *Concurrent Pos:* Mem exec comt, Prog Cell & Develop Biol, Harvard Med Sch, 75-85; bd tutors biochem sci, Harvard Univ, 75-82; molecular cytol study sect, NIH, 78-82; ad hoc rev, spec study sect, Nat Inst Child Health & Human Develop, 85-; adj prof, Dept Cell Biol, NY Univ Sch Med. *Mem:* Am Soc Biol Chem; Am Soc Cell Biol; Soc Develop Biol; AAAS. *Res:* Cellular and molecular mechanisms of early mammalian development. *Mailing Add:* Dept Cell & Develop Biol Roche Inst Molecular Biol Nutley NJ 07110

WASSEF, NABILA M, b Cairo, Egypt, Jan 1, 43; US citizen; m 73; c 1. LIPID BIOCHEMISTRY & IMMUNOLOGY, LIPSOMOLOGY. *Educ:* Ain-Shams Univ, Cairo, Egypt, BSc, 63, MSc, 67, PhD(biochem), 74. *Prof Exp:* Instr biochem, Fac Sci, Ain-Shams Univ, 63-68; res assoc, US Naval Med Res Unit No 3, 68-73 & Dept Path, Univ Ky Sch Med, 74-76; asst prof pharmacol & biochem, NY Med Col, 77-80; sr investr, 81-89, DEP CHIEF, MEMBRANE BIOCHEM BR, WALTER REED ARMY INST RES, 90- *Mem:* Am Asn Immunologists; Soc Leukocyte Biol; AAAS. *Res:* Lipid biochemistry; prostaglandins and leukotrienes; lipid immunology; presentation of liposomal antigens by macrophages; development of liposomes as carriers of drugs and vaccines. *Mailing Add:* Membrane Biochem Br Walter Reed Army Inst Res Washington DC 20307-5100

WASSER, CLINTON HOWARD, b Phoenix, Ariz, Nov 11, 15; m 39; c 3. RANGE ECOLOGY. *Educ:* Univ Ariz, BS, 37; Univ Nebr, MS, 47; Colo State Univ, MF, 48. *Prof Exp:* Res asst southwestern forest & range exp sta, US Forest Serv, 37-38; instr range sci & asst range mgt, Colo State Univ, 38-43, from asst prof to assoc prof, 43-52, prof, 52-80; CONSULT, 80- *Concurrent Pos:* Asst range conservationist, Colo State Univ, 43-47, actg head dept range mgt, 47-50, head, 50-57, chief range conservationist & chief forestry and range mgt sect, Agr Exp Sta, 47-52, dean col forestry & natural resources, 52-69, pres, Res Found, 57-59; collabr, State Prod & Mkt Admin, Colo, 43-44 & Rocky Mt Forest & Range Exp Sta, US Forest Serv, 54-60; consult, Bowes & Hart, Inc, 47; admin tech rep, McIntire Stennis Coop State Forestry Prog, 63-69. *Honors & Awards:* Frederic Renner Award, Soc Range Mgt, 89. *Mem:* Soc Am Foresters; Soc Range Mgt (pres-elect, 64, pres, 65); AAAS; Ecol Soc Am; Sigma Xi. *Res:* Range management, ecology and seeding; alpine plant ecology. *Mailing Add:* Range Sci Colo State Univ Ft Collins CO 80523

WASSER, RICHARD BARKMAN, b Oshkosh, Wis, Sept 26, 36; m 68; c 2. PAPER CHEMISTRY, PHYSICAL CHEMISTRY. *Educ:* Univ Wis, BS, 59; Inst Paper Chem, MS, 61, PhD(paper chem), 64; Polytech Inst NY, MS, 85. *Prof Exp:* From res chemist to sr res chemist, 64-73, proj leader paper chem, 73-78, mgr paper chem res & develop, 78-80, PRIN RES SCIENTIST, AM CYANAMID CO, 80- *Mem:* Tech Asn Pulp & Paper Indust. *Res:* Physical chemistry of paper and its modification with chemical additives. *Mailing Add:* Seven Dock Rd SN Norwalk CT 06854

WASSERBURG, GERALD JOSEPH, b New Brunswick, NJ, Mar 25, 27; m 51; c 2. GEOLOGY, GEOPHYSICS. *Educ:* Univ Chicago, BSc, 51, MS, 52, PhD(geol), 54. *Hon Degrees:* Dr, Free Univ Brussels, 85. *Prof Exp:* from asst prof to assoc prof, 55-62, prof, 62-87, JOHN D MACARTHUR PROF GEOL & GEOPHYS, DIV GEOL & PLANETARY SCI, CALIF INST TECHNOL, 87-, CHMN DEPT. *Concurrent Pos:* Vis prof, Univ Kiel, 60, Harvard Univ, 62, Univ Berne, 66 & Swiss Fed Inst Technol, 67; Vinton Hayes sr fel, Harvard, 80; Smithsonian Regents' fel, 82; mem Lunar Base Steering Comt, NASA, 84- *Honors & Awards:* Arthur L Day Medal, Geol Soc Am, 70; Except Sci Achievement Award, NASA 70; J F Kemp Medal, Columbia Univ, 73; Leonard Medal, Meteoritical Soc, 75; V M Goldschmidt Medal, Geochem Soc, 78; Jaeger-Hales Lectr, Australian Nat Univ, 80; Harold Jeffreys Lectr, Royal Astron Soc, 81; Arthur L Day Prize & Lectr, Nat Acad Sci, 81, J Lawrence Smith Medal, 85; Ernst Cloos Lectr, Johns Hopkins Univ, 84; Wollaston Medal, Geol Soc London, 85; Sr US Scientist Award, Alexander von Humboldt Found, 85; Harry H Hess Medal, Am Geophys Union, 85; Crafoord Medal, Royal Swed Acad, 86; Holmes Medal, Europ Union Geol, 87; Gold Medal, Royal Astron Soc, 91. *Mem:* Nat Acad Sci; fel Am Geophys Union; Geol Soc Am; fel Am Acad Arts & Sci; Meteoritical Soc (vpres, 85, pres, 87, 88); Am Philos Soc; hon foreign fel Europ Union Geosci. *Res:* Application of methods of chemical physics to geologic problems; measurement of absolute geologic time, solar system and planetary time scales; nucleosynthesis and variations in isotopic abundances due to long and short lived natural radioactivities and cosmic ray interactions in nature. *Mailing Add:* Arms Lab 170-25 Calif Inst Technol Pasadena CA 91125

WASSERHEIT, JUDITH NINA, b Sept 2, 54; m. MEDICAL RESEARCH. *Educ:* Princeton Univ, BA, 74; Harvard Med Sch, MD, 78; Johns Hopkins Univ, MPH, 89; Am Bd Internal Med, dipl, 83. *Prof Exp:* Co-dir, SE Asian Refugee Clin, Harborview Med Ctr, Univ Wash, Seattle & res fel infectious dis, 82-84; res physician infectious dis, Diarrhoeal Dis Res, Bangladesh, 84-86; asst prof, Dept Med, Div Infectious Dis, Johns Hopkins Univ, 86-89; CHIEF, SEXUALLY TRANSMITTED DIS BR, NAT INST ALLERGY & INFECTIOUS DIS, NIH, 89- *Concurrent Pos:* Asst chief, STD Clin Serv, Baltimore City Health Dept, 86-89, med dir, Druid STD Clin, 86-89 & fac STD training ctr, 86-89; mem steering comt, Task Force Prev & Mgt Infertil Human Reproduction Prog, WHO, 89-, Subcomt AIDS/STD Task Force, 90, res working group, Sexually Transmitted Dis Prog, 91, tech adv group, Women & AIDS Prog, Int Ctr Res Women, 91-; consult, Java & Bali Indonesia, Int Women's Health Coalition, 87, 88, 90, Cairo, Egypt, 89, Lusaka, Zambia, 89 & 91. *Mem:* Am Col Physicians; Am Soc Trop Med & Hyg; Am Med Women's Asn; Am Venereal Dis Asn; Nat Coun Int Health; Int AIDS Soc. *Res:* Prevention and management of infertility; sexually transmitted diseases. *Mailing Add:* NIH Nat Inst Allergy & Infectious Dis Sexually Transmitted Dis Br Westwood Bldg Rm 750 5333 Westbard Ave Bethesda MD 20892

WASSERMAN, AARON E, b Philadelphia, Pa, Dec, 28, 21; m 44; c 2. FOOD CHEMISTRY. *Educ:* Philadelphia Col Pharm, BSc, 42, ScD(bact), 48; Mass Inst Technol, MSc, 47. *Prof Exp:* Instr bact, Pa State Col Optom, 47; res assoc, Sharp & Dohme, Inc, 48-54; head, USDA, Meat Composition & Qual Invest, Meat Lab, 63-77, chemist & biochemist, Eastern Regional Res Ctr, 54-81, chief, Meat Lab, 77-81; RETIRED. *Concurrent Pos:* Distinguished res award, Am Meat Sci Asn. *Honors & Awards:* Distinguished Serv Award, Inst Food Technologists. *Mem:* Am Meat Sci Asn; Am Chem Soc; Inst Food Technol. *Res:* Flavor chemistry; organoleptic and sensory evaluation of food products; isolation and identification techniques; food processing; meat technology; bacterial physiology and metabolism; food safety. *Mailing Add:* 2156 Conwell Ave Philadelphia PA 19115

WASSERMAN, AARON OSIAS, b New York, NY, Oct 15, 27; m 69; c 1. VERTEBRATE ZOOLOGY. *Educ:* City Col New York, BS, 51; Univ Tex, PhD(zool), 56. *Prof Exp:* From asst prof to assoc prof, 63-74, PROF BIOL, CITY COL NEW YORK, 74- *Mem:* Soc Study Evolution; Am Soc Ichthyol & Herpet; Am Soc Zool; Sigma Xi. *Res:* Speciation problems in anuran amphibians; cytogenetics of anurans. *Mailing Add:* 200 Winston Dr Apt 3118 Cliffside Park NJ 07010

WASSERMAN, AARON REUBEN, b Philadelphia, Pa, Apr 14, 32; m 63; c 2. BIOCHEMISTRY. *Educ:* Univ Pa, AB, 53; Univ Wis, MS, 57, PhD(biochem), 60. *Prof Exp:* Asst Univ Wis, 57-59; fel plant biochem, Johnson Found Med Physics, Univ Pa, 60-61, res assoc, Dept Chem, 61-62; fel, Enzyme Inst, Univ Wis-Madison & Dept Molecular Biol, Vanderbilt Univ, 63-65; asst prof, 65-72, ASSOC PROF BIOCHEM, MCGILL UNIV, 72- *Mem:* Can Biochem Soc. *Res:* Biomembranes; membrane proteins; cytochromes; bioenergetics; photosynthesis. *Mailing Add:* Dept Biochem McGill Univ 853 Sherbrooke St W Montreal PQ H3A 2M5 Can

WASSERMAN, ALBERT J, b Richmond, Va, Jan 25, 28; m 48; c 3. CLINICAL PHARMACOLOGY, INTERNAL MEDICINE. *Educ:* Univ Va, BA, 47; Med Col Va, MD, 51. *Prof Exp:* Instr, Med Col Va, 56-57, assoc, 57-60, from asst prof to assoc prof, 60-67, asst dean curric, 78-80, PROF MED & PHARMACOL, MED COL VA, 67-, ASSOC DEAN CURRIC, 80- *Concurrent Pos:* Chief, Med Serv, Vet Admin Hosp, Richmond Va, 60-64. *Mem:* Am Col Physicians; Am Col Chest Physicians; Am Soc Clin Pharmacol & Therapeut; Am Fedn Clin Res. *Res:* Cardiovascular pharmacology; human pharmacology and drug trials. *Mailing Add:* Med Col Va Richmond VA 23298

WASSERMAN, ALLEN LOWELL, b New York, NY, Dec 7, 34; m 58; c 2. SOLID STATE PHYSICS. *Educ:* Carnegie Inst Technol, BS, 56; Iowa State Univ, PhD(physics), 63. *Prof Exp:* Res assoc, Princeton Univ, 63-65; from asst prof to assoc prof, 65-83, PROF PHYSICS, ORE STATE UNIV, 84- *Concurrent Pos:* Vis prof, Univ Sussex, Eng, 80, 87 & Univ Bristol, Eng, 90; vis res fel, Royal Soc, 90-91. *Mem:* Am Phys Soc; Sigma Xi. *Res:* Optical properties of solids; heavy fermians. *Mailing Add:* Dept Physics Ore State Univ Corvallis OR 97331

WASSERMAN, ARTHUR GABRIEL, b Bayonne, NJ, Nov 10, 38; m 64, 74; c 2. MATHEMATICS. *Educ:* Mass Inst Technol, BS, 60; Brandeis Univ, PhD(math), 65. *Prof Exp:* Pierce instr math, Harvard Univ, 65-68; asst prof, 68-70, assoc prof, 70-80, PROF MATH, UNIV MICH, ANN ARBOR, 80- *Concurrent Pos:* Mem, Inst Advan Study, Princeton Univ, 71-72; vis prof, Unicamp, Campinas, Brazil, 80, Univ Oporto, Portugal, 81, Aarhus Univ, 84-85; managing partner, Diamond Venture Mgt, 88- *Mem:* Am Math Soc. *Res:* Topology; transformation groups; applied mathematics. *Mailing Add:* Dept Math 3001 Angell Hall Univ Mich Ann Arbor MI 48109-1003

WASSERMAN, BRUCE P, b Brooklyn, NY, Aug 19, 53; m 84; c 1. ENZYME TECHNOLOGY, BIOLOGICAL MEMBRANES. *Educ:* Rutgers Univ, BA, 75; Univ Mass, MS, 78, PhD(food sci), 79. *Prof Exp:* From asst prof to assoc prof, 81-90, PROF FOOD SCI, RUTGERS UNIV, 90- *Concurrent Pos:* Chair, Biotech Div, Inst Food Technologists, 89-90. *Honors & Awards:* Distinguished Scientist Award, Inst Food Technologists, 89. *Mem:* Am Soc Plant Physiologists; Inst Food Technologists; Am Chem Soc. *Res:* Biosynthesis of polysaccharides (B-glucans and starches) in plants and fungi. *Mailing Add:* Dept Food Sci Cook Col Rutgers Univ New Brunswick NJ 08903-0231

WASSERMAN, DAVID, b New York, NY, Apr 6, 17; m 46; c 2. POLYMER CHEMISTRY. *Educ:* Brooklyn Col, AB, 37; Columbia Univ, PhD(org chem), 43. *Prof Exp:* Irvington Varnish & Insulator Co fel chem, Columbia Univ, 43-46; sr chemist, Irvinton Varnish & Insulator Div, Minn Mining & Mfg Co, 46-58, res assoc, Merck, Sharp & Dohme Res Div, 58-64; sr res scientist, 64-68, PROJ LEADER ETHICON, INC, 68- *Honors & Awards:* P B Hoffman Award for Res, Johnson & Johnson, Inc, 74. *Mem:* AAAS; NY Acad Sci; Am Chem Soc. *Res:* Synthetic polymers; organic synthesis; natural products; pharmaceutical chemistry; developed new synthetic absorbable suture. *Mailing Add:* Polymer Dept Ethicon Inc Div Johnson & Johnson Inc Somerville NJ 08876

WASSERMAN, DONALD EUGENE, b New Haven, Conn, Apr 1, 39; m 67; c 2. ERGONOMICS, OCCUPATIONAL-HUMAN VIBRATION. *Educ:* Univ Conn, Storrs, BA, 62; NY Univ, MSEE, 71; Xavier Univ, MBA, 84. *Prof Exp:* Biomed proj engr, Mnemotron Div, Tech Measurements Corp, 62-63; sr biomed engr, Cambridge Instruments Co, 64; sr biomed proj & res engr, Perkin-Elmer Corp, 65-71; chief, Bioacoust & Occup Vibration Group, USPHS, Nat Inst Occup Safety & Health, Cincinnati, 71-84; dir, Eng & Opers & sr res investr, Nat Ctr Rehab Eng, Wright-State Univ, Dayton, 84-86; dir, Human Vibration Eng, Anatrol Corp, Cincinnati, 86-88; CONSULT BIODYNAMICS, HUMAN VIBRATION & BIOMED ENG, 88- *Concurrent Pos:* Mem, Comt Hearing & Bioacoust, Nat Res Coun; consult, Am Conf Govt Indust Hygienists; reviewer, J Acoust Soc Am & J Occup Med. *Mem:* Int Standards Orgn; Am Nat Standards Inst. *Res:* Biomedical engineering; cardiovascular, peripheralvascular, neurological, ocular, auditory, muscle-skeletal, clinical chemistry, and rehabilitation engineering; biodynamics, biomechanics, and the effects of mechanical vibration on humans; author of 70 technical publications; holder of two US patents. *Mailing Add:* 7910 Mitchell Farm Lane Cincinnati OH 45242

WASSERMAN, EDEL, b New York, NY, July 29, 32; m 55; c 2. THEORETICAL CHEMISTRY, ORGANIC CHEMISTRY. *Educ:* Cornell Univ, BA, 53; Harvard Univ, AM, 54, PhD(chem), 59. *Prof Exp:* Mem tech staff, Bell Tel Labs, Inc, 57-76; prof chem, Rutgers Univ, 67-76; dir, Chem Res Ctr, Allied Chem, 76-78, & Corp Res Ctr, 78-80; CHEM SCIENTIST, E I DU PONT DE NEMOURS & CO, INC, 80-, ASSOC DIR TECHNOL, 83- *Concurrent Pos:* Vis prof, Cornell Univ, 62-63; adv ed, Chem Phys Lett, 68-79, J Am Chem Soc, 71-76 & Chem Rev, 82-; regent's lectr, Univ Calif, Irvine, 73. *Mem:* Am Chem Soc; Am Phys Soc. *Res:* Complex chemical systems. *Mailing Add:* 1904 Acad Pl Wilmington DE 19806-2135

WASSERMAN, EDWARD, b New York, NY, Jan 13, 21. PEDIATRICS. *Educ:* Johns Hopkins Univ, BA, 41; NY Med Col, MD, 46. *Prof Exp:* From clin asst to instr, 51-54, asst, 54-55, clin assoc, 55-56, from asst prof to assoc prof, 56-64, PROF PEDIAT, NEW YORK MED COL, FLOWER & FIFTH AVE HOSPS, 64-, CHMN DEPT, 66- *Concurrent Pos:* Mem pediat adv comt, Dept Health, New York, 64- *Mem:* Fel Am Fedn Clin Res; Am Acad Pediat. *Res:* Renal diseases. *Mailing Add:* Munger Pavilion Valhalla NY 10595

WASSERMAN, FREDERICK E, b New York, NY, Sept 19, 48; m 78; c 2. AVIAN BEHAVIORAL ECOLOGY. *Educ:* State Univ NY, Stony Brook, BS, 71; Univ Md, MS, 74, PhD(zool), 77. *Prof Exp:* Res asst, Smithsonian Inst, 75; ASST PROF BIOL, BOSTON UNIV, 77- *Concurrent Pos:* NSF grant, Orgn Trop Studies, 72; Harris Found grant, 74; Sigma Xi grant-in-aid res, 77; sr scientist, dept energy grant, 79-82; NIMH grant, 79-81 & 86-87; consult, Arthur D Little, Inc, 81-82. *Mem:* AAAS; Am Ornith Union; Animal Behav Soc; Am Soc Naturalists; Sigma Xi; Cooper Ornith Soc; Wilson Ornith Soc. *Res:* Functions of territoriality; ecological sources of selection; behavior of the white-throated sparrow and the rufous-sided towhee; effects of microwaves on birds; predatory relationship between birds and Lepidoptera; neuroendocrine correlates of aggression in birds. *Mailing Add:* Dept Biol Boston Univ Five Cummington St Boston MA 02215

WASSERMAN, GERALD STEWARD, b Brooklyn, NY, Nov 22, 37; m 62; c 2. PSYCHOBIOLOGY, SENSORY CODING. *Educ:* NY Univ, BA, 61; Mass Inst Technol, PhD(psychol), 65. *Prof Exp:* Fel electrophysiol, NIH, 65-67; from asst prof to assoc prof psychol, Univ Wis-Madison, 67-75; PROF PSYCHOL, PURDUE UNIV, 75- *Concurrent Pos:* Adv ed, Contemp Psychol, 81-87. *Mem:* Acoust Soc Am; Int Brain Res Orgn; Int Soc Psychophysics; fel Optical Soc Am; Soc Neurosci; Psychonomic Soc. *Res:* Psychobiology of sensory coding. *Mailing Add:* Dept Psychol Sci Purdue Univ West Lafayette IN 47907-1364

WASSERMAN, HARRY H, b Boston, Mass, Dec 1, 20; m 47; c 3. ORGANIC CHEMISTRY. *Educ:* Mass Inst Technol, BS, 41; Harvard Univ, MS, 42, PhD(org chem), 49. *Prof Exp:* Res asst, Off Sci Res & Develop, 45; from instr to prof chem, Yale Univ, 48-62, dir grad studies, 60-62, chmn, Dept Chem, 62-65, dir, Div Phys Sci, 72-75, EUGENE HIGGINS PROF, DEPT CHEM, YALE UNIV, 82- *Concurrent Pos:* Guggenheim fel, 59-60; Am ed, Tetrahedron & Tetrahedron Lett, 60-; mem study sect med chem, NIH, 62-66; postdoctoral rev panel, NSF, 63-65; vis lectr, Japanese Soc, Promotion Sci, 78; vis distinguished prof, Tex A&M Univ, 82; Arthur C Cope scholar, Am Chem Soc, 90. *Honors & Awards:* Catalyst Award, Chem Mfg Asn, 85; Aldrich Award, Am Chem Soc, 87. *Mem:* Nat Acad Sci; Am Acad Arts & Sci. *Res:* Natural products; reactions of organic systems with oxygen; cyclopropanones. *Mailing Add:* Sterling Chem Lab Yale Univ New Haven CT 06520

WASSERMAN, JACK F, b Dayton, Ohio, July 29, 41; m 75; c 3. BIOMECHANICS, BIOACOUSTICS. *Educ:* Purdue Univ, BS, 64; Univ Cincinnati, MS, 71, PhD(mech eng & biomed acoust), 75. *Prof Exp:* Design engr, Gen Elec Co, 65-71; assoc dir, Stroke Res Lab, Univ Cincinnati, 75-79; assoc prof, 79-85, PROF BIO MECH SIGNAL PROCESSING, DEPT ENG SCI & MECH, UNIV TENN, 85- *Concurrent Pos:* Adj assoc prof, Mech Eng Dept, Univ Cincinnati, 75-79,Vet Med, Univ Tenn; consult, Surg Appliances, Inc, 78-81, Patricia Neal Rehab Ctr, 80-81; prin investr, IBM, 82-83, Moore Mach Tool, 84, Electro-Optic, 84. *Mem:* Am Soc Mech Eng; Orthop Res Soc; Am Soc Biomech. *Res:* Finite element analysis, human vibration analysis. *Mailing Add:* Dept Eng Sci & Mech Univ Tenn Knoxville TN 37996-2030

WASSERMAN, JERRY, b Brooklyn, NY, Sept 22, 31; m 52; c 3. INTERNATIONAL TRADE ISSUES. *Educ:* City Col New York, BS, 51, MA, 53; Newark Col Eng, MSEE, 57. *Prof Exp:* Sales mgr, Lambda Electronics Corp, 61-67; gen mgr, Amerex Trading Corp, 67-69; pres, Intertrade Technol, Inc, 69-70; VPRES, ARTHUR D LITTLE, INC, 70- *Concurrent Pos:* Mem, Int Bus Coun, Electronics Industs Asn, 70-; dir, China Fund, 87- & Rosh Intel Systs, 88- *Res:* International trade issues, especially Far East; business issues faced by electronics and information industries. *Mailing Add:* 11 Winthrop Rd Lexington MA 02173

WASSERMAN, KARLMAN, b Brooklyn, NY, Mar 12, 27; m 53; c 4. PHYSIOLOGY, MEDICINE. *Educ:* Upsala Col, BA, 48; Tulane Univ, PhD(physiol), 51, MD, 58. *Prof Exp:* Intern, Johns Hopkins Univ, 58-59; sr res fel med, Univ Calif, San Francisco, 59-61; asst prof med, Stanford Univ, 61-67, dir respiratory function lab & chest clin, 61-67; assoc prof, 67-72, PROF MED, MED SCH, UNIV CALIF, LOS ANGELES, 72-; CHIEF DIV RESPIRATORY DIS, LOS ANGELES HARBOR GEN HOSP, TORRANCE, 67- *Mem:* Am Physiol Soc; Am Fedn Clin Res. *Res:* Respiratory, circulatory and renal physiology; respiratory disease. *Mailing Add:* Dept Med Pulmonary Dis UCLA Med Ctr 1000 W Carson St Torrance CA 90509

WASSERMAN, LAWRENCE HARVEY, b Bronx, NY, Oct 19, 45; m 70; c 3. PLANETARY ASTRONOMY. *Educ:* Rensselaer Polytech Inst, BS, 67; Cornell Univ, MS, 71, PhD(astron), 73. *Prof Exp:* Res assoc astron, Lab Planetary Studies, Cornell Univ, 73-74; fel, 74-77, ASTRONR, LOWELL OBSERV, 77- *Mem:* Am Astron Soc; Sigma Xi. *Res:* Studies of planets, satellites and asteroids by occultation techniques; astronomical image processing; photographic astrometry. *Mailing Add:* Lowell Observ 1400 W Mars Hill Rd Flagstaff AZ 86001

WASSERMAN, LOUIS ROBERT, b New York, NY, July 11, 10; m 57. HEMATOLOGY. *Educ:* Harvard Univ, AB, 31; Rush Med Col, MD, 35; Am Bd Internal Med, dipl, 46. *Prof Exp:* Intern, Michael Reese Hosp, 35-37; res fel, Mt Sinai Hosp, 37-39, clin asst, 40-42, adj physician physiol hemat, 47-50, assoc physician, 50-54, hematologist & dir dept hemat, 54-72; prof med, 66-72, chmn, Dept Clin Sci, 68-72, distinguished serv prof, 72-79, Albert A & Vera G List prof med chem, 77-79, EMER PROF, MT SINAI SCH MED, 79- *Concurrent Pos:* Res fel, Donner Lab Med Physics, Univ Calif, 46-49, consult, Radiation Lab, 48-51; asst clin prof, Col Physicians & Surgeons, Columbia Univ, 51-60, assoc prof, 60-66; res collabr, Brookhaven Nat Lab, 60-72; chmn polycythemia vera study group, Nat Cancer Inst, 67-, mem nat cancer planning comt, 72-73, mem cancer control comt, 72-73; diag rev adv group, 72-75, mem cancer treatment adv comt, 72-74 & chmn, 74-76. *Mem:* Fel Am Col Physicians; Int Soc Hemat (vpres, 70-74); Am Soc Hemat (vpres, 67-68, pres, 68-69); Asn Am Physicians; Am Asn Cancer Res. *Res:* Diseases of the blood. *Mailing Add:* Mt Sinai Sch Med 19 E 98th St Apt 5-B Box 1275 New York NY 10029

WASSERMAN, MARTIN ALLAN, b Newark, NJ, Nov 20, 41; m 66; c 2. PHARMACOLOGY, PHYSIOLOGY. *Educ:* Rutgers Univ, BS, 63; Univ Tex Med Br Galveston, MA, 71, PhD(pharmacol), 72. *Prof Exp:* Pharmacist, Shor's Med Serv Ctr, 63-68; res assoc pharmacol, Upjohn Co, 72-81; instr, Kalamazoo Valley Community Col, 78-81; ASSOC DIR PHARMACOL, SMITH KLINE & FRENCH LAB, 81- *Concurrent Pos:* Res asst, Univ Tex Med Br, 68-70, McLaughlin fel, 70-72; mem, Am Heart Asn. *Honors & Awards:* Mead-Johnson Excellence in Res Award, 72; Sigma Xi Award, 73. *Mem:* Am Soc Pharmacol & Exp Therapeut; Am Thoracic Soc. *Res:* Prostaglandins and pulmonary function and disease; pharmacology of asthma; leukotriene biology; calcium channel blockers. *Mailing Add:* 1004 Davis Rd Ambler PA 19002

WASSERMAN, MARTIN ALLAN, b Newark, NJ, Nov 20, 41; m 66; c 2. PHARMACOLOGY. *Educ:* Rutgers Univ, BS, 63; Univ Tex, Galveston, MA, 71, PhD(pharmacol), 72. *Prof Exp:* Pharmacist, Shor's Med Serv Ctr, 63-68; res scientist pharmacol, Upjohn Co, 72-81; dir pharmacol, Smithkline & French Labs, 81-88; DIR BIOMED EVAL, BRISTOL-MYERS SQUIBB, 89- *Concurrent Pos:* Adj prof pharmacol, Philadelphia Col Pharm & Sci, 84-; vis prof, Dept Med, Vanderbilt Univ, 86-87; assoc ed, J Pulmonary Pharmacol, 87- *Mem:* Am Soc Pharmacol & Exp Therapeut; Am Thoracic Soc; Soc Exp Biol & Med; NY Acad Sci; Drug Info Asn; Sigma Xi. *Res:* Bronchopulmonary pharmacology of drugs for allergy and inflammation; prostaglandins, thromboxane and leukotrienes. *Mailing Add:* Bristol-Myers Squibb Pharmaceut Res Inst PO Box 4000 Princeton NJ 08543-4000

WASSERMAN, MARTIN S, b New York, NY, Jan 19, 38; m 63, 76; c 2. CHILD PSYCHIATRY. *Educ:* Columbia Col, AB, 59; State Univ NY Downstate Med Ctr, MD, 63. *Prof Exp:* Intern med, Albany Med Ctr, Union Univ, 64; resident psychiat, Kings County Gen Hosp, 64-67; instr, State Univ NY Downstate Med Ctr, 66-68; fel child psychiat, Univ Mich-Childrens Psychiat Hosp, 69-70; asst prof & assoc dir grad educ, Med Sch, 70-75, ASSOC CLIN PROF PSYCHIAT, UNIV SOUTHERN CALIF, 75- *Concurrent Pos:* Asst adj prof soc sci, Long Island Univ, 66-69; staff psychiatrist, Cath Charities, New York, 66-69; consult, Oper Headstart, Off Econ Opportunity, San Diego, Calif, 67-69, Ctr Forensic Psychiat, Mich, 69-70 & Bur Prisons, USPHS, 69-70; assoc, Los Angeles Psychoanal Soc & Inst, 70-75; dir & fac mem, Los Angeles Psychoanal Soc & Inst, 76-78, secy, 77-78; sr fac, LA Psychoanal Soc & Inst, 76-; psychiat consult, MGM, Columbia & 20th Century Fox Film Studios, 77-81. *Mem:* Fel Am Psychiat Asn; Am Col Psychiat; assoc Am Psychoanal Asn; Am Acad Child Psychiat. *Res:* College health; psychiatric education; socio-cultural factors in human development; film as developmental organizer. *Mailing Add:* 510 E Channel Rd Santa Monica CA 90402

WASSERMAN, MARVIN, b New York, NY, Feb 2, 29; m 64; c 2. EVOLUTIONARY BIOLOGY. *Educ:* Cornell Univ, BS, 50; Univ Tex, MA, 52, PhD(biol), 54. *Prof Exp:* Res assoc biol, Univ Tex, 56-60; sr lectr zool, Univ Melbourne, 60-62; from asst prof to assoc prof, 62-67, PROF BIOL,

QUEENS COL, NY, 67- *Mem:* Am Genetic Asn; Genetics Soc Am; Am Soc Naturalists; Soc Study Evolution. *Res:* Evolution and genetics of Drosophila; chromosomal polymorphism. *Mailing Add:* 31-65 138th St Apt 6C Flushing NY 11354

WASSERMAN, ROBERT H, b Chicago, Ill, Jan 3, 23; m 48; c 1. APPLIED MATHEMATICS. *Educ:* Univ Chicago, BS, 43; Univ Mich, MS, 46, PhD(math), 57. *Prof Exp:* Aeronaut res scientist, Nat Adv Comt Aeronaut, 43-57; from asst prof to assoc prof, 57-70, PROF MATH, MICH STATE UNIV, 70- *Mem:* Am Math Soc; Math Asn Am. *Res:* Fluid mechanics. *Mailing Add:* Dept Math Mich State Univ East Lansing MI 48824

WASSERMAN, ROBERT HAROLD, b Schenectady, NY, Feb 11, 26; m 50; c 3. PHYSIOLOGY, BIOCHEMISTRY. *Educ:* Cornell Univ, BS, 49, PhD(nutrit), 53; Mich State Univ, MS, 51. *Prof Exp:* Asst bact, Mich State Univ, 49-50; asst animal nutrit, Cornell Univ, 51-53; res assoc biochem, Univ Tenn-AEC agr res prog, 53-55; sr scientist, Oak Ridge Inst Nuclear Studies, 55-57; assoc prof phys biol, NY State Col Vet Med, Cornell Univ, 57-63, actg chmn dept physiol, 64, 71 & 75-76, prof physiol, 63-90, chmn dept/sect physiol, Div Biol Sci, 83-87, JAMES LAW PROF PHYSIOL, NY STATE COL VET MED, CORNELL UNIV, 90- *Concurrent Pos:* Lectr, Oak Ridge Inst Nuclear Studies, 57-59 & Int Atomic Energy Agency, Buenos Aires, 61; chmn, Conf Calcium & Strontium Transport, Cornell Univ, 62; Guggenheim fel, 64-65 & 72; vis fel, Inst Biol Chem, Denmark, 64-65; Orgn Econ Coop & Develop-NSF fel, 64-65; mem gen med B study sect, NIH, 68-72; consult, Dent Res Inst, NC Med & Dent Sch, Chapel Hill, 69; assoc ed, Calcified Tissue Res, 76-79; mem nutrit educ subcomt, Nat Nutrit Consortium, 77-; mem adv bd, Am Soc Bone & Mineral Res, 78-80; mem bd dirs, Cornell Vet, 81-90; mem, food & nutrit bd, Nat Acad Sci-Nat Res Coun, 84- *Honors & Awards:* Mead-Johnson Award Nutrit, Am Inst Nutrit, 69; Wise & Helen Burroughs lectr, Iona State Univ, 74; Andre Lichtwit Prize, Nat Inst Health & Med Res, Paris, 82; Merit Award, NIH, 88; Neuman Award, Am Soc Bone & Mineral Res, 90. *Mem:* Nat Acad Sci; Soc Exp Biol & Med; Am Physiol Soc; Am Inst Nutrit; NY Acad Sci; Am Soc Bone & Mineral Res; Am Soc Vet Physiologists & Pharmacologists; AAAS. *Res:* Mechanisms by which various ions move across epithelial membranes particularly intestinal membranes; study of calcium, phosphorous, cadmium, zinc, arsenate and lead; macromolecules in intestinal mucosa that might be involved in ion movement. *Mailing Add:* Dept Physiol Cornell Univ 717 Vet Res Tower Ithaca NY 14853

WASSERMAN, STANLEY, b Louisville, Ky, Aug 29, 51; m 74; c 2. APPLIED STATISTICS, SOCIAL SCIENCE. *Educ:* Univ Pa, BS & MA, 73; Harvard Univ, AM, 74, PhD(statist), 77. *Prof Exp:* Instr psychol, Harvard Univ, 76; vis instr urban & pub affairs, Carnegie-Mellon Univ, 76-77; asst prof statics, Univ Minn, 77-82; assoc prof, 82-88, PROF PSYCHOL, STATIST & SOCIOL, UNIV ILL, 88- *Concurrent Pos:* Res assoc social sci, Columbia Univ, 78; fel, Social Sci Res Coun, 78-79; co prin investr grants, NSF, 79, prin investr, 80-81 & 85-89; assoc ed, Sociol Methodology, 80-82, J Am Statist Asn, 87-, Psychometrika, 88-; dir, math & statist consult comt, 85-86. *Mem:* Am Sociol Asn; Am Statist Asn; Inst Math Statist; Royal Statist Soc; Sigma Xi; Psychomet Soc; AAAS. *Res:* Categorical data analysis; mathematical sociology; applied stochastic processes; mathematical psychology. *Mailing Add:* 603 E Daniel St Univ Ill Champaign IL 61820

WASSERMAN, STEPHEN I, INTERNAL MEDICINE. *Educ:* Univ Calif, Los Angeles, MD, 68. *Prof Exp:* PROF MED & CHMN DEPT, MED CTR, UNIV CALIF, SAN DIEGO, 88- *Mem:* Am Soc Clin Invest; Am Acad Allergy & Immunol. *Res:* Allergy; immunology. *Mailing Add:* Dept Med/H-811-W Univ Calif San Diego Med Ctr 402 W Dickinson St San Diego CA 92103

WASSERMAN, WILLIAM JACK, b New York, NY, Apr 27, 25; m 59; c 2. ORGANIC POLYMER CHEMISTRY. *Educ:* Univ Calif, Los Angeles, BS, 47; Univ Southern Calif, MS, 50; Univ Wash, PhD(org chem), 54. *Prof Exp:* Asst prof chem, Humboldt State Univ, 54-57; sr res chemist, Martin-Marietta Corp, Wash, 57-62 & Truesdail Labs, Los Angeles, Calif, 62-63; asst prof chem, San Jose State Univ, 63-67; INSTR CHEM, SEATTLE CENT COMMUNITY COL, 67- *Concurrent Pos:* Dir, NSF Inst Polymer Chem for Col Teachers, San Jose State Univ, 67 & NSF 'CAUSE' proj, Ninety-Seattle Cent Community Col, 80-83; mem writing team, Am Chem Soc-NSF Chemtec Proj, 70-72; chmn, Puget Sound Sect, Am Chem Soc, 81, Nat Spring Meeting, 83; vis lectr, Chem Dept, Western Wash Univ, 89 & Univ Wash, 90; prog co-chair, Capilano Conf, Vancouver, 90, Two Year Col Chem Conf & Col Chem Can. *Mem:* Am Chem Soc; Sigma Xi. *Res:* Synthesis and evaluation of condensation polymers; emphasis on epoxies, polyesters, polyamides, polyurethanes; cross-linking agent synthesis and evaluation; aldol condensations with unsaturated ketones; furanoid ring-opening reactions. *Mailing Add:* Div Sci & Math Seattle Cent Community Col Seattle WA 98122

WASSERMAN, WILLIAM JOHN, b Toronto, Ont, Feb 27, 47. CELL BIOLOGY, DEVELOPMENTAL BIOLOGY. *Educ:* Univ Toronto, BSc, 69, MSc, 72, PhD(biol), 76. *Prof Exp:* Fel, Purdue Univ, 76-80; ASST PROF BIOL, UNIV ROCHESTER, 80- *Concurrent Pos:* Prin investr, Univ Rochester, 81- *Mem:* AAAS; Am Soc Cell Biologists; Soc Develop Biol. *Res:* Determining the molecular mechanisms involved in controlling meiosis and mitosis, in oocytes and embryos. *Mailing Add:* Dept Biol Univ Rochester Wilson Blvd Rochester NY 14627

WASSERMAN, ZELDA RAKOWITZ, b New York, NY, July 19, 35; m 55; c 2. COMPUTATIONAL CHEMISTRY, PROTEIN DESIGN. *Educ:* Radcliffe Col, AB, 56; Rutgers Univ, MS, 65. *Prof Exp:* Asst math, Mass Inst Technol, 56-57; mem tech staff, Bell Labs, 65-81; prin investr, E I du Pont de Nemours & Co, 81-90; PRIN INVESTR, DU PONT MERCK PHARMACEUT CO, 91- *Mem:* Am Chem Soc; Asn Comput Mach. *Res:* Computational methods applied to design of peptides, proteins and biomaterials using computer graphics, energy minimization and computer simulation of molecular dynamics. *Mailing Add:* DuPont Merck Pharmaceut Co Exp Sta E228/316E PO Box 80228 Wilmington DE 19880-0228

WASSERMANN, FELIX EMIL, b Bamberg, Ger, Aug 7, 24; US citizen; m 53; c 3. VIROLOGY, EPIDEMIOLOGY. *Educ:* Univ Wis, BS, 49, MS, 50; NY Univ, PhD(microbiol), 57. *Prof Exp:* Res assoc microbiol, Univ Chicago, 58; asst virol, Pub Health Res Inst City New York, Inc, 58-60, from asst to assoc epidemiol, 60-65; from asst prof to assoc prof, 65-77, actg chmn dept, 70-76, PROF VIROL, NY MED COL, 77- *Mem:* Am Soc Microbiol; AAAS; NY Acad Med. *Res:* Virus-host cell interaction; viral genetics; epidemiology; biomedical ethics. *Mailing Add:* Dept Microbiol NY Med Col Valhalla NY 10595

WASSERSUG, RICHARD JOEL, b Boston, Mass, Apr 13, 46; m; c 2. EVOLUTIONARY BIOLOGY. *Educ:* Tufts Univ, BS, 67; Univ Chicago, PhD(evolutionary biol), 73. *Prof Exp:* Asst prof syst & ecol & asst cur, Mus Natural Hist, Univ Kans, 73-74; asst prof anat & comt evolutionary biol, Univ Chicago, 74-81; assoc prod, 81-86, PROF ANAT, DALHOUSIE UNIV, 86- *Mem:* Soc Study Evolution; Am Soc Ichthyologists & Herpetologists; AAAS; Ecol Soc Am; Am Soc Zoologists. *Res:* Adaptations of amphibian larvae, and of fish; studies on the evolution of complex life cycles and on the evolution of diversity. *Mailing Add:* Dept Anat Dalhousie Univ Halifax NS B3H 4H6 Can

WASSHAUSEN, DIETER CARL, b Jena, Ger, Apr 15, 38; US citizen; m 61; c 2. SYSTEMATIC BOTANY. *Educ:* George Washington Univ, BS, 63, MS, 66, PhD(bot), 72. *Prof Exp:* From asst cur to assoc cur dept bot, 69-76, CUR & CHMN DEPT BOT, MUS NATURAL HIST, SMITHSONIAN INST, 76- *Concurrent Pos:* Smithsonian Res Found awards, 74 & 75. *Honors & Awards:* Willdenow Medal, 79. *Mem:* Int Asn Plant Taxon; Am Soc Plant Taxon; Sigma Xi; AAAS; Am Inst Biol Sci. *Res:* Taxonomy of Acanthaceae and flowering plants of the New World tropics. *Mailing Add:* Dept Bot Mus Natural Hist Smithsonian Inst Washington DC 20560

WASSMUNDT, FREDERICK WILLIAM, b Oak Park, Ill, Aug 6, 32; div; c 1. ORGANIC CHEMISTRY. *Educ:* DePauw Univ, BA, 53; Univ Ill, PhD(chem), 56. *Prof Exp:* Instr chem, Univ Calif, 56-58; from instr to asst prof, 58-69, ASSOC PROF CHEM, UNIV CONN, 69- *Concurrent Pos:* Guest prof, Univ Heidelberg, 66-67; invited lectr, Chem Soc Heidelberg, 74; treas, 8th Biennial Conf Chem Educ, 82-85; nat treas, Phi Lambda Upsilon Hon Chem Soc, 84-90, nat pres, 90-; consult, SmithKline Beecham, 90- *Mem:* NY Acad Sci; Am Chem Soc; Royal Soc Chem; Sigma Xi. *Res:* Exploratory synthesis; reactions of organic nitrogen compounds; diazonium salts; bridged bicyclic compounds. *Mailing Add:* Dept Chem Univ Conn Storrs CT 06269-3060

WASSOM, CLYDE E, b Osceola, Iowa, Feb, 6, 24; m 45; c 3. AGRONOMY. *Educ:* Iowa State Col, BS, 49, MS, 51, PhD(crop breeding), 53. *Prof Exp:* Technician agron, Iowa State Col, 47-51, res assoc, 51-54; asst prof, 54-62, assoc prof, 62-76, PROF, KANS STATE UNIV, 76- *Concurrent Pos:* Temp staff mem, Int Ctr Improv Corn & Wheat, Mexico City, 67. *Mem:* Am Soc Agron. *Res:* Forage breeding; research and breeding for improvement of red clover and other legumes; corn breeding and genetics research. *Mailing Add:* Dept Agron Throckmorton Hall Kans State Univ Manhattan KS 66506

WASSON, JAMES A(LLEN), b Tyrone, Pa, July 5, 26; m 83; c 2. PETROLEUM ENGINEERING. *Educ:* Pa State Univ, BS, 51 & 52, MS, 57. *Prof Exp:* Jr petrol engr, Humble Oil & Refining Co, Exxon, 56-58; asst prof petrol eng, La Polytech Inst, 58-60; from asst prof to assoc prof, 60-80, chmn dept, 77-82, PROF PETROL ENG, COL MINERAL & ENERGY RESOURCES, WVA UNIV, 80- *Concurrent Pos:* Staff res adv, US Dept Energy, Morgantown, WVa, 60-85. *Mem:* Am Inst Mining, Metall & Petrol Engrs. *Res:* Petroleum reservoir engineering; secondary recovery of oil; enhanced oil recovery. *Mailing Add:* Col Mineral & Energy Resources WVa Univ PO Box 6070 Morgantown WV 26506-6070

WASSON, JAMES WALTER, b Pittsburgh, Pa, Dec 9, 51; m 74; c 2. PROGRAM MANAGEMENT & PROJECT ENGINEERING, AVIONICS INTEGRATION. *Educ:* Northrop Univ, BS, 81; Univ Phoenix, MBA, 89. *Prof Exp:* Electronics supvr, Ostgaard Industs, 74-75; sr design engr, Airesearch Aviation Co, 75-81; proj engr, Northrop Aircraft Co, 81-84; RES MGR, MCDONNELL DOUGLAS HELICOPTER CO, 84- *Concurrent Pos:* Sr vpres, Avionics Eng Serv, Inc, 80-82; chmn/consult, Indust Adv Avionics Bd, Northrop Univ & chmn, SAE-AE9B High Speed Data Bus Comt, 82-84, Avionics Comt, Am Helicopter Soc, 90-; adj prof, Univ Phoenix Mgt Develop Ctr, 90- *Mem:* Am Helicopter Soc; Army Aviation Asn Am; Am Defense Preparedness Asn; Nat Soc Prof Engrs; Inst Elec & Electronics Engrs. *Res:* Strategic planning, business development, and program management of contractual and independent research and development programs for new product definition; engineering and business management functions; proposal development, contract negotiations and cost/schedule performance; aeronautical and astronautical engineering; author of ten publications. *Mailing Add:* McDonnell Douglas Helicopter Co 5213 E Fairfield Circle Mesa AZ 85205

WASSON, JOHN R, b St Louis, Mo, Aug 22, 41; m 63; c 2. PHYSICAL CHEMISTRY, INORGANIC CHEMISTRY. *Educ:* Univ Mo-Columbia, BS, 63, MA, 66; Ill Inst Technol, PhD(phys chem), 70. *Prof Exp:* Instr chem, Southern State Col, 65-66; asst prof chem, Univ Ky, 69-75; vis sr scientist, Univ NC, Chapel Hill, 75-78; dir, Chem Div, res dept, Lithium Corp Am, 78-81; pres, Kings Mountain Specialties, Inc, 81-82; pres, Syntheco Inc, 82-89; PRES, ADVAN MAT, 89- *Mem:* Am Chem Soc; Royal Soc Chem; Am Asn Textile Chemists & Colorists; Electrochem Soc; AAAS. *Res:* Electron spin resonance; semi-empirical molecular orbital theory; change-transfer complexation; transition metal complexes; higher oxidation states of silver; polymers; electrolytes; specialty ceramics; inorganic synthesis. *Mailing Add:* Rte 1 Box 197 B New Hill NC 27562

WASSON, JOHN TAYLOR, b Springtown, Ark, July 4, 34; m 60; c 2. COSMOCHEMISTRY, PLANETOLOGY. *Educ:* Univ Ark, BS, 55; Mass Inst Technol, PhD(nuclear chem), 58. *Prof Exp:* NSF fel tech physics lab, Munich Technol Univ, 58-59; res chemist geophys res directorate, Air Force Cambridge Res Labs, 59-63; NIH spec fel phys inst, Univ Berne, 63-64; from asst prof to assoc prof, 64-72, PROF, UNIV CALIF, LOS ANGELES, 72- *Concurrent Pos:* Guggenheim fel, Max Planck Inst Chem, 70-71. *Honors & Awards:* Fel, Am Geophys Union. *Mem:* Am Geophys Union; Geochem Soc; Meteoritical Soc (pres, 79-80); Sigma Xi. *Res:* Composition and origin of the meteorites; major accretionary events on earth; geochemistry; solar nebula formation and evolution. *Mailing Add:* Inst Geophys Univ Calif Los Angeles CA 90024

WASSON, L(OERWOOD) C(HARLES), b Denning, Ark, Feb 27, 09; m 36, 90; c 1. CORROSION SCIENCE, BIO-MEDICAL. *Educ:* Univ Ark, BSEE, 33. *Prof Exp:* Shop foreman, Hydraul Lab, Tenn Valley Authority, 35-36; develop engr, Milwaukee Gas Specialty Co, 36-42, prod engr, 42-46, res engr, 46-48; asst chief engr, Durant Mfg Co, 48-49; elec res engr, A O Smith Corp, 49-52, dir, Electrochem Lab, 52-60, mat engr, 60-65, res scientist, 65-74; PRES, L C WASSON CO, 74- *Concurrent Pos:* Mem, Eng Med & Biol Group, Inst Elec & Electronics Engrs; consult, Aqua Dynamics Group Corp, Adamsville, Tenn, 87- *Mem:* Nat Asn Corrosion Engrs; Inst Elec & Electronics Engrs. *Res:* Cathodic protection theory and application; corrosion mechanisms and corrosion prevention techniques; electrochemistry of corrosion. *Mailing Add:* 8711 W Beloit Rd 431 West Allis WI 53227

WASSON, OREN A, b Wooster, Ohio, Mar 27, 35; m 59; c 2. NUCLEAR PHYSICS. *Educ:* Col Wooster, BA, 57; Yale Univ, MS, 59, PhD(physics), 64. *Prof Exp:* Res staff physicist, Yale Univ, 63-65; from res assoc physics to physicist, Brookhaven Nat Lab, 65-73; physicist, Nat Bur Standards, 73-78, PHYSICIST, NAT INST STANDARDS & TECHNOL, 78- *Concurrent Pos:* Vis mem, Oak Ridge Nat Lab, 71-72; tech expert, Int Atomic Energy Agency, Greece, 75. *Mem:* AAAS; Am Phys Soc; Sigma Xi. *Res:* Experimental nuclear physics; neutron and photon reaction mechanisms; neutron cross sections and dosimetry. *Mailing Add:* Nat Inst Standards & Technol Gaithersburg MD 20899

WASSON, RICHARD LEE, b Farmington, Ill, May 19, 32; m 55; c 2. ORGANIC PROCESS CHEMISTRY, TECHNICAL MANAGEMENT. *Educ:* Univ Ill, BS, 53; Mass Inst Technol, PhD, 56. *Prof Exp:* Asst org chem, Mass Inst Technol, 53-56; sr res chemist, Monsanto Co, 56-63, res specialist, 63-66, res group leader, 66-71, mgr flavor-fragrance res, 71-76, mgr com develop, 76-77, mgr new prod/process res, 77-78, mgr technol planning, 78-80, dir technol support, 80-83; dir, Technol Planning & Eval, Indust Chem, G D Searle, 83-84, dir, chem develop, Health Care Div, 84-86, dir, develop, 86-89; PVT SCI CONSULT, 90- *Mem:* Am Chem Soc; Am Inst Chem; NY Acad Sci. *Res:* Preparation and rearrangements of epoxides; halogenation and carboxylation of aromatic compounds; aromatic nitro compounds; basic condensations; organophosphate compounds; isolation and identification of natural products; food chemistry; organic analysis; environmental hazard assessment; process chemistry and scale up, biological research, technical management; peptide chemistry. *Mailing Add:* 8821 Hemingway Dr Crestwood MO 63126

WASSON, W(ALTER) DANA, b NB, Apr 2, 34; m 59; c 2. COMPUTER SCIENCE, ELECTRICAL ENGINEERING. *Educ:* Univ NB, BSc, 56; Mass Inst Technol, SM, 58; Univ Waterloo, PhD, 72. *Prof Exp:* From asst prof to assoc prof elec eng, Univ NB, 58-70, head, comput sci dept, 69-74, dir, Comput Ctr, 64-74, PROF ELEC ENG, UNIV NB, FREDERICTON, 70- *Concurrent Pos:* Dir, Sch Comput Sci, Univ NB, Frederiction, 74-91, dean, fac computer sci, 91- *Honors & Awards:* Bryden Jack Prize; EIC Prize. *Mem:* Inst Elec & Electronics Engrs; Asn Prof Eng; Asn Comput Mach; Can Info Processing Soc; Pattern Recognition Soc. *Res:* Computer hardware and software; pattern recognition and active circuit analysis. *Mailing Add:* Sch Comput Sci PO Box 4400 Fredericton NB E3B 5A3 Can

WASTERLAIN, CLAUDE GUY, b Courcelles, Belg, Apr 15, 35; US citizen; m 57; c 1. EPILEPTOLOGY, CEREBRAL ISCHEMIA. *Educ:* Univ Liege, CSc, 57, MD, 61; Free Univ Brussels, LSc, 69. *Prof Exp:* Asst physiol, Univ Liege, Belg, 61-63; Intern, Middlesex Mem Hosp, 63-64; resident neurol, NY Hosp, Cornell Univ Med Col, 64-67, from instr to assoc prof neurol, 69-76; assoc prof, 76-79, PROF NEUROL, UNIV CALIF, LOS ANGELES SCH MED, 79-, VCHMN DEPT, 76- *Concurrent Pos:* Chief neurol serv, Vet Admin Med Ctr, Sepulveda, Calif, 76- & Olive View Med Ctr, 88-89; attend neurologist, Univ Calif-Los Angeles Hosp, 79-; mem, Brain Res Inst, Univ Calif-Los Angeles, 77-; chief epilepsy res, Vet Admin Med Ctr, 76-; mem prof adv bd, Epilepsy Found Am, 79-, Nat Ataxia Found, 80-; chair Antiepileptic Drug Develop Comt, NIH, 86-88, res career develop award, 73-76. *Mem:* Am Neurol Asn; fel Am Acad Neurol; Am Soc Neurochem; Am Epilepsy Soc; Int Soc Neurochem. *Res:* Use biochemical and pharmacological methods to investigate how epilepsy is acquired, and how epileptic seizures or ischemia can damage the brain and impair its development. *Mailing Add:* Neurol Serv 127 Vet Admin Med Ctr Sepulveda CA 91343

WASTI, KHIZAR, b Jan 27, 48; US citizen; m 76; c 1. TOXICOLOGY. *Educ:* Univ Punjab, Pakistan, BSc, 66; Univ Peshawar, MSc, 69; Marshall Univ, MS, 72; Univ Pa, PhD(chem), 76. *Prof Exp:* Sr lectr chem, Edwardes Col, Pakistan, 69-71; teaching asst, Marshall Univ, 71-72; teaching fel, Univ Pa, 73-75; res chemist, McNeil Labs, Ft Washington, Pa, 74, Campbell Soup Co, Camden, NJ, 74-75; proj mgr & sr info analyst, Toxicol Dept, Res Labs, Franklin Inst, 75-78; TOXICOLOGIST, VA STATE DEPT HEALTH, RICHMOND, 78- *Mem:* Am Col Toxicol; fel Am Inst Chemists; fel Royal Soc Chem. *Mailing Add:* Bur Toxic Substances Info Va State Dept Health PO Box 2448 Richmond VA 23218

WASYLISHEN, RODERICK ERNEST, b Elk Point, Alta, July 6, 44; m 68; c 2. CHEMISTRY. *Educ:* Univ Waterloo, BSc, 68; Univ Man, MSc, 70, PhD(phys chem), 72. *Prof Exp:* Res fel, NIH, Bethesda, Md, 72-74; from asst prof to assoc prof chem, Univ Winnipeg, 74-82; assoc prof, 82-86, PROF CHEM, DALHOUSIE UNIV, 86- *Concurrent Pos:* Vis prof, Univ Guelph, 80-81; Killam sr fel, Dalhousia Univ, 86-87. *Mem:* Chem Inst Can; Am Chem Soc; Sigma Xi; Int Soc Magnetic Resonance; Spectros Soc Can. *Res:* Nuclear magnetic resonance studies of molecular rotations in orientationally disordered solids; isotope effects on nuclear magnetic resonance parameters; rotations of molecules solubilized in surfactants (Micelles) and various liquid crystal systems including model membranes; dynamics of simple anions in solution; rigid solids. *Mailing Add:* Dept Chem Dalhousie Univ Halifax NS B3H 4H6 Can

WASYLKIWSKYJ, WASYL, b Kowel, Ukraine, Feb 12, 35; US citizen; m 60; c 2. ELECTROMAGNETICS. *Educ:* City Col New York, BEE, 57; Polytech Inst Brooklyn, MS, 65, PhD(elec eng), 68. *Prof Exp:* Microwave component design engr, Missiltron Inc, NY, 59-60; Consult & Designers, NY, 60-62; sr tech specialist, ITT Defense Commun Div, Int Tel & Tel Corp, NJ, 62-69; mem tech staff, Inst Defense Anal, 69-; AT DEPT ELEC ENG, GEORGE WASHINGTON UNIV. *Concurrent Pos:* Consult, ITT Fed Labs, NJ, 63-65; prof lectr, George Washington Univ, 70-; mem comns B & C, Int Union Radio Sci. *Honors & Awards:* Spec Commendation Award, Inst Elec & Electronics Engrs Antennas & Propagation Soc, 72. *Mem:* Inst Elec & Electronics Engrs. *Res:* Electromagnetic radiation and diffraction, guided wave and cavity theory; antenna theory, particularly antenna arrays; microwave technology, particularly parametric and solid state microwave devices. *Mailing Add:* 5506 Westbard Ave Bethesda MD 20816

WASYLYK, JOHN STANLEY, b Passaic, NJ, Feb 15, 42; m 69; c 2. GLASS TECHNOLOGY, CERAMIC SCIENCE. *Educ:* Rutgers Univ, BS, BA, 64, PhD(ceramics sci), 70. *Prof Exp:* Pres & tech dir, Glass Container Indust Res Corp, 69-76; tech dir, Am Glass Res Inc, 76-86, dir res, 86-88, dir corp res, 88-89; DIR INT CONSULT SERV, AGR INT, INC, 89- *Mem:* Am Ceramic Soc; Am Soc Testing & Mat; Soc Glass Technol; Nat Inst Ceramic Engrs; Ger Glass Technol Soc. *Res:* Glass technology; heat transfer during forming and relation to workability; glass microanalysis; lubricating coatings on glass surfaces; glass strength; glass fracture analysis; statistical sampling methods; container production process improvements. *Mailing Add:* Agr Int Inc PO Box 149 Butler PA 16003-0149

WAT, BO YING, b Honolulu, Hawaii, Feb 15, 25; m 48; c 4. MEDICINE. *Educ:* Col Med Evangelists, MD, 49; Am Bd Path, dipl. *Prof Exp:* From asst to assoc prof, 58-62, PROF PATH, SCH MED, LOMA LINDA UNIV, 62- *Mem:* Am Soc Clin Pathologists; AMA. *Mailing Add:* Dept Path Loma Linda Univ Loma Linda CA 92354

WAT, EDWARD KOON WAH, b Honolulu, Hawaii, Aug 27, 40; m 69; c 3. ORGANIC CHEMISTRY. *Educ:* Univ Hawaii, BA, 62; Stanford Univ, PhD(chem), 66. *Prof Exp:* NIH fel chem, Harvard Univ, 65-66; res chemist, Cent Res Dept, 66-77, RES CHEMIST, BIOCHEM DEPT, EXP STA, E I DU PONT DE NEMOURS & CO, INC, 77- *Mem:* Am Chem Soc. *Res:* Synthetic organic chemistry and process development for agricultural and pharmaceutical applications. *Mailing Add:* 216 W Pembrey Dr Wilmington DE 19803-2008

WATABE, NORIMITSU, b Kure City, Japan, Nov 29, 22; m 52; c 2. ELECTRON MICROSCOPY. *Educ:* Tohoku Univ, Japan, MS, 48, DSc(biocrystallog), 60. *Prof Exp:* Res investr biocrystallog pearl cult, Fuji Pearl Res Lab, Japan, 48-52; asst fac fisheries, Mie Prefectural Univ, Japan, 52-55, lectr, 55-59, consult, Fisheries Exp Sta, 53-59; res assoc calcification electron micros, Duke Univ, 57-70; assoc prof, 70-72, PROF BIOL & MARINE SCI, UNIV SC, 72-, DIR, ELECTRON MICROS CTR, 70- *Concurrent Pos:* Asst, Tohoku Univ, Japan, 48-49; consult, Ford Found Off Latin Am & Caribbean, 68 & SC State Develop Bd, 75-; Russel res award, Univ SC, 80; Nat Inst Dent Res & NSF grants, Electron Micros Ctr, Univ SC, 71- *Honors & Awards:* Elmer Elseworth Award, 52; Humboldt Award, Germany, 76; Fel, AAAS. *Mem:* AAAS; Electron Micros Soc Am; Am Soc Zool; Royal Micros Soc; Am Micros Soc; Am Malac Union. *Res:* Microanatomy; ultrastructural and physiological aspects of mechanism of calcification in invertebrates, algae and fish. *Mailing Add:* 3510 Greenway Dr Columbia SC 29206

WATADA, ALLEY E, b Platteville, Colo, July 20, 30; m 56; c 2. HORTICULTURE, PLANT PHYSIOLOGY. *Educ:* Colo State Univ, BS, 52, MS, 53; Univ Calif, Davis, PhD(plant physiol), 65. *Prof Exp:* Lab technician, Univ Calif, Davis, 56-65; from asst prof to assoc prof hort, WVa Univ, 65-71; invests leader, 71-72, proj leader, 72-81, res food technologist, 71-81, CHIEF HORT CROPS QUAL LAB, HORT SCI INST, AGR RES SERV, USDA, 81- *Mem:* Am Soc Hort Sci; Am Soc Plant Physiol; Inst Food Technol; Am Inst Biol Sci. *Res:* Postharvest physiology of fruits and vegetables; development of methods for maintaining quality of intact and partially processed fruits and vegetables and determining the mechanisms of regulation of senescence. *Mailing Add:* Bldg 002 Rm 113 Agr Res Serv USDA Beltsville MD 20705

WATANABE, AKIRA, b Vancouver, BC, Sept 17, 35; m 57; c 5. COMMUNICATIONS SCIENCE. *Educ:* McMaster Univ, BSc, 57; Univ Toronto, MA, 62, PhD(molecular physics), 64. *Prof Exp:* Sci off, 57-69, res scientist, dept commun, Defence Res Telecommun Estab, 69-; ASST PROF, UNIV TOKYO, JAPAN. *Concurrent Pos:* Sci counr, Can Embassy, Tokyo, 76-79. *Mem:* Can Asn Physicists; Optical Soc Am; Inst Elec & Electronics Engrs. *Res:* Optical communications, thin-film waveguides and spectroscopy; laser physics. *Mailing Add:* Univ Tokyo Fac Med Hongo Bunkyo-Ku Tokyo 113 Japan

WATANABE, DANIEL SEISHI, b Honolulu, Hawaii, Oct 30, 40; div; c 1. COMPUTER SCIENCE. *Educ:* Harvard Univ, AB, 62, AM, 67, PhD(appl math), 70. *Prof Exp:* Mathematician, Baird-Atomic, Inc, 63-64; asst prof to assoc prof comput sci, Univ Ill, Urbana, 70-83; PROF & CHMN, DEPT INFO & COMPUTER SCI, UNIV HAWAII AT MANOA, HONOLULU, 84-, ASSOC DEAN, COL NATURAL SCI, 90- *Concurrent Pos:* Vis scholar, Tokyo Univ, 79. *Mem:* Sigma Xi; Asn Comput Mach. *Res:* Numerical software; simulation of semiconductor devices. *Mailing Add:* 60 N Beretania St Suite 3402 Honolulu HI 96817

WATANABE, ITARU S, b Sapporo, Japan, June 20, 33; US citizen; wid; c 3. NEUROPATHOLOGY. *Educ:* Keio Univ Sch Med, MD, 58, DMSc, 63. *Prof Exp:* Fel, Sch Med, Ind Univ, 64-69, asst prof, 69-72; assoc prof, 72-77, PROF PATH, SCH MED UNIV KANS, 77; UNIT DIR, PATH, VET ADMIN MED CTR, 72- *Mem:* Am Asn Neuropathologists; Soc Neurosci; Int Congress Neuropath; Am Asn Pathologists; AMA. *Res:* Neuropathologic studies of various brain disorders, particularly their ultrastructural aspects. *Mailing Add:* Dept Path Vet Admin Hosp Kansas City MO 64128

WATANABE, KYOICHI A, b Amagasaki, Japan, Feb 28, 35; m 62; c 6. ORGANIC CHEMISTRY, BIOCHEMISTRY. *Educ:* Hokkaido Univ, BA, 58, MA, 60, PhD(chem), 63. *Prof Exp:* Lectr chem, Sophia Univ, Japan, 63; res assoc, Sloan-Kettering Inst, 63-66; res fel, Univ Alta, 66-68; from assoc to assoc mem, 68-80, MEM CHEM, SLOAN-KETTERING INST, 80-, PROF, SLOAN-KETTERING DIV, GRAD SCH MED SCI, CORNELL UNIV, 81- *Concurrent Pos:* Assoc prof, Sloan-Kettering Div, Grad Sch Med Sci, Cornell Univ, 72-80; mem, Med Chem A Study Sect, NIH, 81-84. *Honors & Awards:* Boleslawa Szarecky Medal, 88. *Mem:* Pharm Soc Japan; Am Chem Soc; Am Asn Cancer Res; Int Asn Heterocyclic Chem; NY Acad Sci; Korean Pharm Soc; AAAS. *Res:* Structure, syntheses, reactions and stereochemistry of nitrogen heterocyclics, carbohydrates, nucleosides and antibiotics of potential biological activities; medicinal chemistry; antitumor and antiviral compounds; chemistry of nucleic acids. *Mailing Add:* Sloan Kettering Inst 1275 York Ave New York NY 10021-6007

WATANABE, MAMORU, b Vancouver, BC, Mar 15, 33; m 74; c 1. MEDICINE, ENDOCRINOLOGY. *Educ:* McGill Univ, BSc, 55, MD, CM, 57, PhD, 63; FRCPS(C), 63. *Prof Exp:* Assoc molecular biol, Albert Einstein Col Med, 65-66, asst prof, 66-67; from assoc prof med & biochem to prof med, Univ Alta, 67-73; head div med, 74-76, assoc dean educ, 76-80, assoc dean res, 80-81, actg dean, 81, PROF MED, UNIV CALGARY, 73-, DEAN, 82- *Concurrent Pos:* Ayerst fel, Endocrine Soc, 63-64; res fel, Am Col Physicians, 64-67; dir dept med, Foothills Hosp, Calgary, Alta. *Mem:* Am Soc Microbiol; Can Med Asn. *Res:* Secretion rate of aldosterone in normal and abnormal pregnancy; replication of RNA bacteriophages; transport of steroids across cell membranes, using Pseudomonas testosteroni as a model system. *Mailing Add:* Dean Med Univ Calgary 3330 Hospital Dr NW Calgary AB T2N 4N1 Can

WATANABE, MICHAEL SATOSI, theoretical physics, information science, for more information see previous edition

WATANABE, MICHIKO, b Hakodate, Japan, Dec 5, 52; US citizen. CELL-CELL ADHESION, CARDIOGENESIS. *Educ:* Univ Calif, Berkeley, BA, 75; Wesleyan Univ, PhD(develop biol), 81. *Prof Exp:* Postdoctoral, Dept Biol, Ind Univ, 81-83; sr res assoc, Dept Genetics, 84-88, ASST PROF, DEPT PEDIAT, SCH MED, CASE WESTERN RESERVE UNIV, 88- *Concurrent Pos:* Assoc lectr, Ohio Col Podiat Med, 83-84; Young anatomist, Am Asn Anatomist, 91-93. *Mem:* Am Soc Cell Biol; Soc Develop Biol; Asn Res Vision & Opthal; Am Heart Asn; Am Asn Anatomist. *Res:* Cell adhesion molecules in embryogenesis; cardiogenesis using immunological reagents and enzymes to purify, localize and probe function in vitro and in vivo. *Mailing Add:* Dept Pediat Sch Med Case Western Reserve Univ 2109 Adelbert Rd Cleveland OH 44106-4901

WATANABE, PHILIP GLEN, b Inglewood, Calif, Mar 23, 47. TOXICOLOGY, PHARMACOLOGY. *Educ:* Univ Calif, BS, 69; Utah State Univ, PhD(toxicol), 74; Am Bd Toxicol, dipl, 80. *Prof Exp:* assoc scientist, Toxicol Lab, Health & Environ Sci, 74-82, DIR TOXICOL, DOW CHEM CO, 82- *Concurrent Pos:* Vis lectr, Sch Pub Health, Univ Mich, 68-79; toxicol comt, Nat Acad Sci-Nat Res Coun, 78-82. *Honors & Awards:* F R Blood Award, Soc Toxicol, 78 & Achievement Award, 80. *Mem:* Soc Toxicol; Am Soc Pharmacol & Exp Therapeut; AAAS; Sigma Xi. *Res:* Molecular toxicology; interaction of chemicals with intracellular macromolecules to facilitate the assessment of hazards of chemical exposure. *Mailing Add:* 3425 Lawndale Dr Midland MI 48640

WATANABE, RONALD S, biochemistry, for more information see previous edition

WATANABE, TAKESHI, b Osaka, Japan, July 15, 40; m 69; c 3. IMMUNOLOGY. *Educ:* Osaka Univ, MD, 66, PhD(immunol), 77. *Prof Exp:* Physician internal med, Osaka Univ Hosp, 67-69; res assoc immunol, Roswell Park Mem Inst, 69-72; asst prof internal med, Sch Med, Osaka Univ, 72-80; prof immunol, Saga Med Sch, Japan, 80-85; PROF IMMUNOL, MED INST BIOREGULATION, KYUSHU UNIV, 85- *Concurrent Pos:* Researcher immunol, Basel Inst Immunol, 75-77. *Mem:* Am Asn Immunologists. *Res:* Investigation on the allergic and immunological disorders from molecular as well as clinical aspects. *Mailing Add:* Med Inst Bioregulation Kyushu Univ Maidashi 3-1-1 Higashi-ku Fukuoka 812 Japan

WATANABE, TOMIYA, b Koriyama, Japan, Aug 3, 27; m 59; c 2. AERONOMY, PLASMA PHYSICS. *Educ:* Tohoku Univ, Japan, BSc, 53, PhD(geophys), 61. *Prof Exp:* Res asst physics, Univ Md, 60-61; lectr, 61-63, from asst prof to assoc prof, 63-72, PROF GEOPHYS, UNIV BC, 72- *Concurrent Pos:* Mem, working group of magnetospheric field variations, comt IV, Int Asn Geomag & Aeronomy, 67-; vis assoc res physicist, Space Sci Lab, Univ Calif, Berkeley, 69; sr res fel, Nat Res Coun Can, 69-70, mem assoc comt space physics, 78-82; vis prof, Japan Soc Prom Sci, 75; sr Killam fel, Univ BC, 84. *Mem:* Am Geophys Union; Soc Terrestrial Magnetism & Elec Japan; Am Phys Soc; Can Asn Physicists. *Res:* Geomagnetism; generation and propagation of hydromagnetic waves in the terrestrial magnetosphere; electrostatic waves in the ionosphere. *Mailing Add:* Dept Geophys & Astron Univ BC 2075 Westbrook Pl Vancouver BC V6T 1W5 Can

WATANABE, YOSHIO, b Tokyo, Japan, Nov 8, 25; m 58; c 2. CARDIAC ELECTROPHYSIOLOGY, ELECTROCARDIOGRAPHY. *Educ:* Keio Gijuku Univ, Tokyo, MD, 51, DMSc, 60. *Prof Exp:* Asst instr internal med, Sch Med, Keio Gijuku Univ, 52-60; res fel med, Hahnemann Med Col & Hosp, Philadelphia, 61-62, from instr to sr instr, 62-66, res asst prof med, physiol & biophys, 66-70, assoc prof, 70-72; PROF MED & DIR, CARDIOVASC INST, SCH MED, FUJITA HEALTH UNIV, 72- *Concurrent Pos:* Mem, Joint Comt Cardiac Pacemakers Japanese Circulation Soc & Japanese Soc Med Engrs, 73-85; chmn, Sci & Proc Comt, 5th Int Symp Cardiac Pacing, 75-76; distinguished lectr, Tel Aviv Univ, Israel, 83; vis prof, Nagoya City Univ Sch Med, Japan, 84-; ed, Heart & Vessels, 85-; int fel, Coun Clin Cardiol, Am Heart Asn, 85- *Honors & Awards:* Kato Mem Prize for Med & Physiol, 81. *Mem:* Am Physiol Soc; fel Am Col Cardiol; Am Heart Asn; corresp mem Brit Cardiac Soc. *Res:* Cardiac electrophysiology and cardiac arrhythmias; impulse formation and conduction in the antrioventricular node and their ionic mechanisms; genesis of ventricular tachycardia and fibrillation; electropharmacology of antiarrhythmic agents. *Mailing Add:* Cardiovasc Inst Fujita Health Univ Toyoake Aichi 470-11 Japan

WATENPAUGH, KEITH DONALD, b Amarillo, Tex, Sept 3, 39; m 63; c 3. PHYSICAL CHEMISTRY, BIOCHEMISTRY. *Educ:* Univ Idaho, BS, 62; Mont State Univ, PhD(chem), 67. *Prof Exp:* Sr fel, Univ Wash, 66-69, res assoc, 69, 72, from asst prof to assoc prof biochem & molecular struct, 72-87; CONSULT, 87- *Concurrent Pos:* Sr scientist IV, Phys & Anal Chem Res Unit, Upjohn Co, 84-; chmn, Struct Biol Synchrotron Users Asn. *Mem:* Am Chem Soc; Am Crystallog Asn (vpres, 91, pres, 92); AAAS. *Res:* Structure and mechanisms of electron transport proteins, protein-substrate interactions; computer modelling of molecular structure and interactions. *Mailing Add:* Phys & Analytical Chem Upjohn Co Kalamazoo MI 49007

WATERBURY, LOWELL DAVID, b Lansing, Mich, Jan 8, 42. PHARMACOLOGY. *Educ:* Univ Mich, BS, 64; Univ Vt, PhD(pharmacol), 68. *Prof Exp:* Technician, Univ Mich, 62-64, asst lab instr bot, 62; from instr to asst prof biochem, Inst Lipid Res, Col Med, Baylor Univ, 67-69; asst prof pharmacol, Bowman Gray Sch Med, Wake Forest Univ, 69-74; staff researcher, 74-80, SR STAFF RESEARCHER, DEPT EXP PHARMACOL, SYNTEX RES, 80- *Mem:* AAAS; Am Soc Pharmacol & Exp Therapeut; Am Chem Soc; Am Fedn Clin Res; Asn Res Vision & Ophthal. *Res:* Biochemical pharmacology; ocular pharmacology; drugs affecting diabetic retinopathy, cataracts, and ocular inflammation; drugs affecting gastric mucus production and plasma renin activity; determination of mechanism of action of newly synthesized therapeutic agents. *Mailing Add:* Dept Pharmacol Syntex Res 3401 Hillview Ave Palo Alto CA 94304-1320

WATERHOUSE, BARRY D, b Oct 14, 49; m 73; c 2. NEUROPHYSIOLOGY, NEUROPHARMACOLOGY. *Educ:* Temple Univ, PhD(pharmacol), 78. *Prof Exp:* Asst prof cell biol & anat, Univ Tex Health Sci Ctr, 81-87; ASSOC PROF PHYSIOL & BIOPHYS, HAHNEMANN UNIV, PHILADELPHIA, 87- *Concurrent Pos:* Klingenstein fel Neurosci. *Mem:* Soc Neurosci; Sigma Xi; Am Soc Pharmacol & Exp Therapeut. *Res:* Neurobiology. *Mailing Add:* Dept Physiol & Biophys MS 409 Hahnemann Univ Broad & Vine Philadelphia PA 19102

WATERHOUSE, HOWARD N, b Bethel, Maine, Apr 19, 32; m 60; c 2. ANIMAL NUTRITION. *Educ:* Univ Maine, BS, 54; Univ Ill, Urbana, MS, 58, PhD(animal sci), 60. *Prof Exp:* Group leader animal nutrit, Gen Mills, Inc, 60-62, proj leader, 62-67; mgr poultry nutrit res & tech serv, Allied Mills Inc, 67-70; dir nutrit & res, Robin Hood Multifoods Ltd, Can, 70-74; sr poultry scientist, Cent Soya, Inc, Decatur, Ind, 74-78; VPRES NUTRIT, BELL GRAIN & MILLING, 78- *Mem:* Poultry Sci Asn. *Res:* Amino acid studies with chicks; dog and cat management and nutrition; applied animal husbandry; poultry nutrition; sales and dealer training; egg quality studies. *Mailing Add:* 2837 Sandberg St Riverside CA 92506

WATERHOUSE, JOHN P, b Kent, Eng, Dec 4, 20; m 44; c 3. HISTOPATHOLOGY, ELECTRON MICROSCOPY. *Educ:* Univ London, MB, BS, 51, BDS, 56, MD, 63. *Prof Exp:* From jr lectr to sr lectr dent path, London Hosp Med Col, Univ London, 56-66, univ reader oral path, 66-67; actg head dept histol, Col Dent, 68-69, prof oral path & head dept 67-83, PROF PATH, COL MED, UNIV ILL MED CTR, 68-, DIR ADVAN PROGS, COL DENT, 83- *Concurrent Pos:* Assoc ed, J Oral Path, 76-; consult, VA, Chicago, 82-88; consult, Med Res Coun Can, 83, 86, Med Res Coun UK, 84, 86; vis prof, Royal Col Physicians & Surgeons of Glasgow, 87- *Honors & Awards:* Robert Caldwell mem lectr, Univ Glasgow, Scotland, 76. *Mem:* Am Acad Oral Pathologists; Brit Soc Periodont (hon ed, 64, hon secy, 66); Int Asn Dent Res (asst secy-treas, 73-77). *Res:* Electron microscopy; cytochemistry; lysosomes; stereology; normal and damaged oral mucosa; inflammatory arthritis. *Mailing Add:* Dept Path Col Med Univ Ill Box 6998 Chicago IL 60680

WATERHOUSE, JOSEPH STALLARD, b Toronto, Ont, Apr 28, 29; m 54; c 3. HUMAN ANATOMY, HUMAN PHYSIOLOGY. *Educ:* Univ Guelph, BSc, 54; Wash State Univ, MSc, 57, PhD(entom), 62. *Prof Exp:* Aide entom & zool, Wash State Univ, 54-60; asst prof biol, State Univ NY Col Arts & Sci Plattsburgh, 60-64; Ford of Can res fel zool, Carleton Univ, 64-65; assoc prof biol, 65-72, chmn dept biol sci, 74-75, PROF BIOL, STATE UNIV NY COL PLATTSBURGH, 72- *Concurrent Pos:* Pres United Univ Profs, State Univ NY, 87-90. *Mem:* AAAS; Am Inst Biol Sci; Entom Soc Am; Entom Soc Can; Sigma Xi. *Res:* Biology of the garden symphylan, Scutigerella immaculata; ecology and taxonomy of the carabidae (ground beetles) in upper New York state; incidence of hypertension in upper New York State; entomology. *Mailing Add:* Dept Biol Sci State Univ Col Arts & Sci A Beaumont Hall Rm 301 Plattsburgh NY 12901

WATERHOUSE, KEITH R, b Derby, Eng, May 10, 29; US citizen; m 55; c 5. UROLOGY, SURGERY. *Educ:* Cambridge Univ, BA, 50, MA, 57; Oxford Univ, MB & BChir, 53. *Prof Exp:* Instr urol, Kings County Hosp Ctr, 59-61; from asst prof to prof, 61-83, EMER PROF UROL SURG, STATE UNIV NY DOWNSTATE MED CTR, 83- *Concurrent Pos:* Consult, Vet Admin Hosp, Brooklyn, NY, St Mary's Hosp, Passaic, NJ & Paterson Gen Hosp, 65-, Samaritan Hosp, Brooklyn, NY, 66- & St Charles Hosp, Port Jefferson, St Francis Hosp, Poughkeepsie, Arden Hill Hosp, Goshen & Brookhaven Mem Hosp, Patchogue, 67- *Mem:* Am Urol Asn; Am Col Surgeons; Am Acad Pediat; Am Fertil Soc; Int Soc Urol; Fel Royal Col Surgeons Eng. *Res:* Investigation, diagnosis and treatment of congenital anomalies of the urinary tract in children and the subsequent treatment of these patients when they develop chronic renal failure. *Mailing Add:* PO Box 69 Bonita Springs FL 33959

WATERHOUSE, RICHARD (VALENTINE), b Eng, Feb 28, 24; nat US; div; c 4. PHYSICS. *Educ:* Oxford Univ, MA, 49; DSc, 83; Cath Univ, PhD, 59. *Prof Exp:* Res physicist, Royal Navy Torpedo Exp Estab, Scotland, 44-46, Paint Res Asn, London, 46-49 & Nat Bur Stand, 51-59; prof, 61-86, EMER PROF PHYSICS, AM UNIV, 86- *Concurrent Pos:* Vis prof, Univ Calif, Berkeley, 69-70 & Univ Delft, Neth, 75-76; consult, US Navy. *Honors & Awards:* Sabine Medal, Acoust Soc Am, 90. *Mem:* Fel Acoust Soc Am. *Res:* Waves and vibrations; theoretical physics; acoustics. *Mailing Add:* 4000 Massachusetts Ave NW No 1512 Washington DC 20016

WATERHOUSE, WILLIAM CHARLES, b Galveston, Tex, Dec 31, 41; m 80. MATHEMATICS. *Educ:* Harvard Univ, AB, 63, AM, 64, PhD(math), 68. *Prof Exp:* Res assoc math, Off Naval Res & Cornell Univ, 68-69; asst prof, Cornell Univ, 69-75; assoc prof, 75-80, PROF MATH, PA STATE UNIV, UNIVERSITY PARK, 80- *Honors & Awards:* Ford Award, Math Asn Am, 84. *Mem:* Am Math Soc; Math Asn Am. *Res:* Affine group schemes and descent theory; linear algebra; algebraic number theory; history of mathematics. *Mailing Add:* Dept Math Pa State Univ Univ Park PA 16802

WATERLAND, LARRY R, b St Louis, Mo, Aug 11, 48; m 70; c 2. HAZARDOUS WASTE INCINERATION. *Educ:* Calif Inst Technol, BS, 70; Stanford Univ, MS, 72, PhD(chem eng), 75. *Prof Exp:* Staff engr, 75-78, sect leader, 78-79, prog mgr, 75-79, DEPT MGR, ACUREX CORP, 79- *Concurrent Pos:* Mem adv panel to State Calif on hazardous waste incinerator permitting. *Mem:* Am Inst Chem Engrs; Air Pollution Control Asn; Combustion Inst. *Res:* Hazardous waste treatment via incineration or other thermal destruction processes; evaluation and control of combustion source air pollutants emissions. *Mailing Add:* 555 Clyde Ave Mountain View CA 94039-7044

WATERMAN, ALAN T(OWER), JR, b Northampton, Mass, July 8, 18; m 46; c 4. ELECTRICAL ENGINEERING. *Educ:* Princeton Univ, BA, 39; Calif Inst Technol, BS, 40; Harvard Univ, MA, 49, PhD(eng sci, appl physics), 52. *Prof Exp:* Meteorologist, Am Airlines, 40-41; instr, Univ Minn, 41-42; res assoc, Calif Inst Technol, 42-45; res scientist, Columbia Univ, 45; chief meteorologist, Univ Tex, 45-46; asst, Harvard Univ, 46-52; res assoc, 52-57, from assoc prof to prof, 58-83, EMER PROF ELEC ENG, STANFORD UNIV, 83- *Concurrent Pos:* Consult, Weapons Systs Eval Group, US Dept Defense, 56-61, Nat Security Agency, 58-68, Stanford Res Inst, 58-, Inst Sci & Technol, Univ Mich, 58-61, Missile Systs Div, Lockheed Aircraft Corp, 58-59, Sylvania Elec Prods Co, Inc, 58-59 & Norair Res Coun, 63-68; chmn comn II, Int Sci Radio Union, 61-64; secy, US Nat Comt, 64-67, vchmn, 67-69, chmn, 69-72; ed, Radio Sci, 87- *Mem:* Am Meteorol Soc; fel Inst Elec & Electronics Engrs; Am Geophys Union. *Res:* Radio wave and optical propagation; physics of the atmosphere. *Mailing Add:* Star Lab Stanford Univ Stanford CA 94305-4055

WATERMAN, DANIEL, b New York, NY, Oct 24, 27; m 60; c 3. MATHEMATICS. *Educ:* Brooklyn Col, BA, 47; Johns Hopkins Univ, MA, 48; Univ Chicago, PhD, 54. *Prof Exp:* Res assoc, Cowles Comn Res in Econ, 51-52; Fulbright fel, Univ Vienna, 52-53; from instr to asst prof math, Purdue Univ, 53-59; asst prof, Univ Wis-Milwaukee, 59-61; prof, Wayne State Univ, 61-69; PROF MATH, SYRACUSE UNIV, 69-, CHMN, 88- *Mem:* Am Math Soc; Math Asn Am. *Res:* Fourier analysis; real variables; functional analysis; orthogonal series. *Mailing Add:* 116 Donridge Dr Dewitt NY 13214

WATERMAN, FRANK MELVIN, b Delhi, NY, Nov 3, 38; m 61; c 4. MEDICAL PHYSICS. *Educ:* Hartwick Col, BA, 60; Clarkson Technol Col, MS, 69, PhD(physics), 73. *Prof Exp:* Res assoc nuclear physics, Kent State Univ, 72-75; res assoc, dept radiol, Univ Chicago, 75-79; PROF, DEPT RADIATION, ONCOL & NUCLEAR MED, THOMAS JEFFERSON UNIV MED COL, 79- *Mem:* Radiation Res Soc; Am Asn Physicists Med; NAm Hyperthermia Group. *Res:* Thermometry for clinical hyperthermia, response of human tumor blood flow to hyperthermia, heating dynamics and hyperthermia equipment development. *Mailing Add:* Dept Radiation Oncol & Nuclear Med Thomas Jefferson Univ Med Col Philadelphia PA 19107

WATERMAN, MICHAEL ROBERTS, b Tacoma, Wash, Nov 23, 39; m 66; c 2. MOLECULAR BIOLOGY, PROTEIN CHEMISTRY. *Educ:* Willamette Univ, BA, 61; Univ Ore, PhD(biochem), 69. *Prof Exp:* Fel biochem, Johnson Res Found, Sch Med, Univ Pa, 68-70; from asst prof to assoc prof, 70-82, PROF BIOCHEM, HEALTH SCI CTR, UNIV TEX, 82- *Mem:* Am Chem Soc; Am Soc Biol Chem; Endocrine Soc. *Res:* Synthesis of cytochrome P-450; regulation of steroidogenesis in the adrenal cortex; cytochrome P-450 structure and function relationship. *Mailing Add:* Univ Tex Southwestern Med Ctr 5323 Harry Hine Blvd Dallas TX 75235-7200

WATERMAN, MICHAEL S, b Coquille, Ore, June 28, 42; div; c 1. MATHEMATICS, STATISTICS. *Educ:* Ore State Univ, BS, 64, MS, 66; Mich State Univ, MA, 68, PhD(probability, statist), 69. *Prof Exp:* Teaching asst math, Ore State Univ, 64-66; from asst prof to assoc prof, Idaho State Univ, 69-75, NSF res grant, 71-73; staff mem, Los Alamos Sci Lab, 75-81; PROF, UNIV SOUTHERN CALIF, 82- *Concurrent Pos:* Vis prof, Univ Hawaii, 79-80 & Med Sch, Univ Calif, San Francisco, 82. *Mem:* Am Statist Asn; fel AAAS; Math Asn Am; Inst Math Statist. *Res:* Ergodic theory; probabilistic and computational number theory; mathematical biology; combinations and finite mathematics; biological science. *Mailing Add:* Univ Southern Calif University Park Los Angeles CA 90089-1113

WATERMAN, PETER LEWIS, b Bedford, Eng, Dec 28, 55. FUCHSIAN & KLEINIAN GROUPS, RIEMANN SURFACES. *Educ:* Univ Southampton, BSc, 77; Univ Aberdeen, PhD(math), 83. *Prof Exp:* Vis asst prof math, Univ Md, 80-82; Lawton lectr math, Temple Univ, 82-84; asst prof, 84-90, ASSOC PROF MATH, NORTHERN ILL UNIV, 90- *Mem:* London Math Soc; Am Math Soc. *Res:* Discrete groups of mobius transformations. *Mailing Add:* Dept Math Northern Ill Univ DeKalb IL 60115-2888

WATERMAN, TALBOT H(OWE), b East Orange, NJ, July 3, 14. COMPARATIVE PHYSIOLOGY. *Educ:* Harvard Univ, AB, 36, MA, 38, PhD(zool), 43; Yale Univ, MA Privatim, 58. *Prof Exp:* Jr fel, Harvard Univ Soc of Fels, 38-40, res assoc, Psychoacoust Lab, Off Sci Res & Develop, Harvard Univ, 41-43; staff mem, Radiation Lab, Mass Inst Technol, 43-45; sci consult, Off Sci Res & Develop, Off Field Serv, USN & USAAF, 45; secy comt res, Sigma Xi, 46; instr biol, 46-47, from asst prof to prof zool, 47-85, EMER PROF, SR RES ASSOC, YALE UNIV, 85- *Concurrent Pos:* Exec fel, Trumbull Col, 46-56; instr invert zool, Marine Biol Lab, Woods Hole, 47-52, mem corp, 48-; secy corp, Bermuda Biol Sta, 51-61, trustee, 62-75 & 76-80; mem Am Inst Biol Sci Adv Comt Hydrobiol, Off Naval Res, 59-65; Guggenheim fel, 62-63; Sigma Xi nat lectr, 64-65; vis prof, Sch Med, Keio Univ, Japan, 68; assoc ed, J Exp Zool, 71-75; vis prof, Japan Soc Prom Sci, 74; vis lectr & investr, Woods Hole Oceanog Inst, 76-; vis prof, Sophia Univ, Japan, 77-78 & 80; lectr, Star Island Conf, 90. *Honors & Awards:* Excellence in Res Award, Crustacean Soc. *Mem:* AAAS; Am Physiol Soc; Am Soc Zool. *Res:* Visual physiology and spatial orientation; deepsea plankton and vertical migrations; compound eye fine structure and information processing; photoreceptor membrane turnover in compound eyes; polarized light behavior in the sea; animal navigation. *Mailing Add:* Dept Biol 912 KBT Yale Univ PO Box 6666 New Haven CT 06511

WATERS, AARON CLEMENT, volcanology, petrology; deceased, see previous edition for last biography

WATERS, DEAN ALLISON, b Jersey City, NJ, May 2, 36; c 3. ENGINEERING, MATERIALS SCIENCE. *Educ:* Yale Univ, BE, 57, BS, 58; NC State Univ, MS, 60. *Prof Exp:* Dept head, 67-75, prog mgr, 75-77, DEP DIV DIR, NUCLEAR DIV, UNION CARBIDE CORP, 77- *Honors & Awards:* E O Lawrence Award, Dept of Energy, 78. *Mem:* Nat Soc Prof Engrs. *Res:* Materials, applied mechanics, stress analysis, vibration, machine dynamics, and systems engineering. *Mailing Add:* 132 Newport Dr Oak Ridge TN 37830

WATERS, IRVING WADE, b Baldwyn, Miss, June 19, 31; m 54; c 2. PHARMACOLOGY. *Educ:* Delta State Col, BS, 58; Auburn Univ, MS, 60; Univ Fla, PhD(pharmacol), 63. *Prof Exp:* From asst prof to assoc prof, 66-74, PROF PHARMACOL, UNIV MISS, 74- *Mem:* Am Soc Pharmacol & Exp Therapeut; Sigma Xi. *Res:* Drug metabolism; toxicology. *Mailing Add:* Sch Pharm Univ Miss University MS 38677

WATERS, JAMES AUGUSTUS, b Postville, Iowa, June 23, 31; m 57; c 4. ORGANIC MEDICINAL CHEMISTY. *Educ:* Univ Iowa, BS, 53, PhD(pharmaceut chem), 59; Purdue Univ, MS, 57. *Prof Exp:* Res fel org chem, Univ Mich, 59-60; res chemist, 60-89, PHARMACIST, NAT INST ARTHRITIS, METAB & DIGESTIVE DIS, 90- *Mem:* Am Chem Soc; Sigma Xi; fel Am Found Pharmaceut Educ. *Res:* Structure elucidation of natural products; steroid biosynthesis; medicinal chemistry; photochemistry; synthesis nicotinic acetylcholine receptor agonists. *Mailing Add:* Nat Inst Diabetes & Digestive & Kidney Dis Lab Biorg Chem Bldg 8A Rm 1A20 Bethesda MD 20892

WATERS, JAMES FREDERICK, b Oak Park, Ill, Mar 17, 38; m 65. VERTEBRATE ANATOMY, NATURAL HISTORY. *Educ:* Stanford Univ, BA, 59; Univ Wash, PhD(zool), 69. *Prof Exp:* From asst prof to assoc prof, 66-75, PROF ZOOL, HUMBOLDT STATE UNIV, 75- *Concurrent Pos:* Sci translr, Ger to Eng. *Mem:* Soc Study Evolution; Soc Vert Paleont; Am Soc Zoologists; Western Soc Naturalists. *Res:* Functional locomotor anatomy of lizards and snakes. *Mailing Add:* Dept Biol Humboldt State Univ Arcata CA 95221

WATERS, JOHN ALBERT, b Norwich, Eng, Nov 7, 35; c 4. PHYSICAL ORGANIC CHEMISTRY, CATALYSIS. *Educ:* Univ London, BS, 57; Leicester Univ, PhD(org chem), 60. *Prof Exp:* Sr res chemist, Monsanto, 63-75; res specialist, 75-78, RES ASSOC CHEM, MERICHEM CO, 78- *Mem:* Am Chem Soc; fel Am Inst Chemists; Catalyst Soc. *Res:* Organometallics; organic reactions. *Mailing Add:* Merichem Co 1503 Central Ave Houston TX 77012-2743

WATERS, JOSEPH HEMENWAY, b Brockton, Mass, Dec 23, 30. VERTEBRATE ZOOLOGY. *Educ:* Univ Mich, BS, 54, MS, 55; Univ Conn, PhD(zool), 60. *Prof Exp:* Instr biol, Mass State Col Bridgewater, 59-61, Duke Univ, 62-63, Univ RI, 63-64 & Roanoke Col, 64-65; asst prof biol, Villanova Univ, 65-85. *Mem:* Fel AAAS; Am Soc Mammal; Am Soc Ichthyologists & Herpetologists; Sigma Xi. *Res:* Ecology and systematics of vertebrates in eastern North America. *Mailing Add:* 65 Bonney Hill Lane Hanson MA 02341

WATERS, KENNETH LEE, b Monroe, Va, Jan 24, 14; m 39. PHARMACY, MEDICINAL CHEMISTRY. *Educ:* Lynchburg Col, AB, 35; Univ Ga, MS, 37; Univ Md, PhD(pharmaceut chem), 45. *Prof Exp:* Instr chem, Transylvania Col, 36-37 & Univ Ga, 37-39; from jr chemist to asst chemist, US Food & Drug Admin, Md, 39-43; fel, Mellon Inst, 43-47; tech dir, Zemmer Co, 47-48; dean, 48-77, EMER DEAN SCH PHARM, UNIV GA,

77- *Concurrent Pos:* Lectr, Sch Pharm, Univ Pittsburgh, 45-47. *Mem:* AAAS; Am Chem Soc; Am Pharmaceut Asn. *Res:* Drug standardization; preparation of alpha-akloximino acids and their derivatives. *Mailing Add:* 289 Miledge Heights Athens GA 30606

WATERS, LARRY CHARLES, b Glenville, Ga, July 1, 39; m 60; c 1. BIOCHEMISTRY. *Educ:* Valdosta State Col, BS, 61; Univ Ga, MS, 64, PhD(biochem), 65. *Prof Exp:* Am Cancer Soc fel biochem res, 65-67, STAFF BIOCHEMIST, BIOL DIV, OAK RIDGE NAT LAB, 67- *Concurrent Pos:* Lectr, Univ Tenn, Knoxville. *Mem:* Am Soc Biol Chemists. *Res:* Chemistry and biochemistry of nucleic acids; molecular mechanisms of mutagenesis including metabolic activation of mutagens and repair of DNA; viral carcinogenesis. *Mailing Add:* Biol Div Oak Ridge Nat Lab PO Box 2009 Oak Ridge TN 37831-8077

WATERS, MICHAEL DEE, b Charlotte, NC, Apr 17, 42; m 64, 90; c 2. GENETIC TOXICOLOGY. *Educ:* Davidson Col, BS, 64; Univ NC, Chapel Hill, PhD(biochem), 69. *Prof Exp:* Lab chief biochem, Biophys Lab, Edgewood Arsenal, Md, 69-71; unit chief cellular physiol, Cellular Biol Sect, 71-72, sect chief cellular physiol, Pathobiol Res Br, 72-75, chief cellular biochem sect, Biomed Res Br, 75-76, chief, Biochem Br, 76-79, coordr genetic toxicol prog, 78-79, DIR, GENETIC TOXICOL DIV, ENVIRON RES CTR, ENVIRON PROTECTION AGENCY, 79- *Concurrent Pos:* Clin instr, George Washington Univ, 71-72; adj prof, Univ NC, Chapel Hill, 80- *Honors & Awards:* Bronze Medal, Environ Protection Agency, 80 & 88. *Mem:* NY Acad Sci; Am Chem Soc; Tissue Cult Asn; Sigma Xi; Environ Mutagen Soc (pres, 91-92). *Res:* Microbial, mammalian cell, organ culture and whole animal systems for studies of genetic, biochemical and physiological effects of environmental pollutants. *Mailing Add:* Genetic Toxicol Div Environ Protection Agency Research Triangle Park NC 27711

WATERS, NORMAN DALE, b Twin Falls, Idaho, May 1, 22; m 54; c 2. ENTOMOLOGY. *Educ:* Univ Calif, Berkeley, BS, 49, PhD, 55. *Prof Exp:* Mem staff for parasite introd, India, Pakistan, Nepal & Afghanistan, USDA, 49-51; entomologist, Cashewnut Improv Comn, India, 52-53; pest control supvr citrus, Limonera Co, Calif, 55-57; ENTOMOLOGIST, EXP STA, UNIV IDAHO, 57-, ASSOC RES PROF ENTOM, 70- *Mem:* Entom Soc Am; Int Orgn Biol Control; Sigma Xi. *Res:* Biological control; legume forage insects; pollinating insects. *Mailing Add:* Rte 2 Box 2137 R-E Ctr Univ Idaho Parma ID 83660

WATERS, RICHARD C(ABOT), b Framingham, Mass, Apr 8, 50; m 87. AUTOMATED TOOLS. *Educ:* Brown Univ, BS, 72; Harvard Univ, MS, 73; Mass Inst Technol, PhD(computer sci), 78. *Prof Exp:* Res scientist, Mass Inst Technol, 78-82, prin res scientist, 82-91; SR RES SCIENTIST, MITSUBISHI ELEC RES LABS, 91- *Mem:* Sr mem Inst Elec & Electronics Engrs; Asn Comput Mach; Am Asn Artificial Intel. *Res:* Automated tools for the support of the software process. *Mailing Add:* Mitsubishi Elec Res Labs 201 Broadway Cambridge MA 02139

WATERS, ROBERT CHARLES, b Long Beach, Calif, Apr 27, 30; m; c 4. ENGINEERING MANAGEMENT, ENGINEERING ECONOMICS. *Educ:* Univ Calif, Los Angeles, BS, 56, MBA, 63; Univ Southern Calif, DBA(bus econ), 68. *Prof Exp:* Trainee, mfg mgt prog, Gen Elec Co, 56-58, supvr, instrumentation shop, jet engine dept, 58-59; syst analyst admin, TRW Systs Group, 59-61, cost engr, syst lab, 61-63, mkt planning mgr, 63-68; sr res analyst consult, Resource Mgt Corp, 68-69; vpres consult, EMSCO Eng & Mgt Sci Corp, 69-72; assoc prof eng mgt, Univ Mo, Rolla, 72-79; PROF ENG ADMIN, GEORGE WASHINGTON UNIV, 79-, CHMN DEPT, 84- *Concurrent Pos:* Staff specialist, US Water Resources Coun, 76-78; prin investr, Maritime Protection Pract & Proposals, US Dept Transp, 83-85; vis prof, Anderson Grad Sch Mgt, Univ Calif, Los Angeles, 88. *Mem:* Am Soc Eng Educ; Am Soc Eng Mgt (treas, 80-84); Sigma Xi; AAAS; Inst Mgt Sci. *Res:* Economics and management of transportation, particularly maritime; management of technological innovation; impact of technology on society and vice versa; water resources planning. *Mailing Add:* Dept Eng Mgt George Washington Univ Washington DC 20052

WATERS, RODNEY LEWIS, b Long Beach, Calif, July 13, 36; m 58; c 3. LASERS. *Educ:* Univ Calif, Los Angeles, BA, 63; Pepperdine Univ, MBA, 79. *Prof Exp:* Res engr, NAm Rockwell, 63-65; prod mgr, Korad Dept, Union Carbide Corp, 65-63; mkt mgr, Quantrad Corp, 73-74; PRES, FLOROD CORP, 74- *Concurrent Pos:* Lectr, Univ Calif, Los Angeles, 79-87. *Mem:* Semiconductor Equip & Mat Inst; Int Soc Hybrid Microelectronics. *Res:* Developed first Q-switched YAG laser resistor trimmer; laser for semiconductor photomask repair; laser cutter for very large scale integrated failure analysis; sealed excimer; commercial laser chemical liquid deposition. *Mailing Add:* Florod Corp PO Box 213 Torrance CA 90501

WATERS, ROLLAND MAYDEN, b Tucson, Ariz, Apr 7, 26; m 59; c 3. ORGANIC CHEMISTRY. *Educ:* Univ Ark, BS, 48; Yale Univ, PhD(org chem), 54. *Prof Exp:* Sr res chemist gen org chem, Dow Chem Co, 53-63; RES CHEMIST PHEROMONES, USDA, 63- *Concurrent Pos:* Lectr chem, Univ Md-Univ Col, 72- *Mem:* Am Chem Soc; AAAS. *Res:* Isolation, structure elucidation, and synthesis of chemicals affecting the behavior of insects; the goal is alternative methods of controlling insects to reduce our use of pesticides. *Mailing Add:* 20442 Meadow Pond Pl Gaithersburg MD 20879-1130

WATERS, THOMAS FRANK, b Hastings, Mich, May 17, 26; m 53; c 3. AQUATIC ECOLOGY, FISHERIES. *Educ:* Mich State Univ, BS, 52, MS, 53, PhD(fishery biol), 56. *Prof Exp:* Supvr, Pigeon River Trout Res Sta, Mich Dept Conserv, 56-57; from asst prof to assoc prof, 58-68, PROF FISHERY BIOL, UNIV MINN, ST PAUL, 68- *Honors & Awards:* Excellence in Benthic Sci Award, NAm Benthological Soc (pres, 75). *Mem:* Am Fisheries Soc; Ecol Soc Am; Am Inst Fishery Res Biol; NAm Benthological Soc (pres, 75). *Res:* Limnology; ecology of aquatic invertebrates and stream fish populations. *Mailing Add:* Dept Fisheries & Wildlife Univ Minn St Paul MN 55108

WATERS, WILLIAM E, b Springfield, Mass, July 2, 22; m 43, 76; c 4. FOREST ENTOMOLOGY. *Educ:* State Univ NY Col Forestry, Syracuse, BS, 48; Duke Univ, MF, 49; Yale Univ, PhD, 58. *Prof Exp:* Entomologist, Bur Entom & Plant Quarantine, USDA, 49-53, entomologist, Forest Serv, 53-59, chief div forest insect res, 59-63, div forest insect & dis res, 63-65, asst dir insects & dis, 65-66, chief forest insect & dis lab & prin ecologist, Northeastern Forest Exp Sta, Conn, 66-68, chief forest insect res, Div Forest Insect & Dis Res, 68-73, chief scientist, Pac Southwest Forest & Range Exp Sta, 73-75; dean, Col Natural Resources, 75-77, prof, 77-86, EMER PROF FORESTRY & ENTOM, UNIV CALIF, BERKELEY, 86- *Concurrent Pos:* Instr, Quinnipiac Col, 61-64; lectr, Yale Univ, 64-68; mem pop dynamics working party & chmn impact of destructive agents proj group, Int Union Forest Res Orgn; chmn working party on forest insects & dis, Food & Agr Orgn-NAm Forestry Comn; leader, Pine-Beetle IPM Proj, NSF/Environ Protection Agency, 74-84; assoc dir, Calif Agr Res Sta, Univ Calif, Berkeley, 75-77; chmn, statist ecol sect, Int Asn Ecol, 85-; prin investr, US-Korea Coop Res Prog, 87-90. *Honors & Awards:* Distinguished Statist Ecologist Award, Int Asn Ecol, 90. *Mem:* Entom Soc Am; Soc Am Foresters; Sigma Xi; Int Asn Ecol. *Res:* Forest insect ecology; population dynamics; biometrics; insect behavior; forest pest management. *Mailing Add:* Col Natural Resources Univ Calif Berkeley CA 94720

WATERS, WILLIAM EDWARD, physics, for more information see previous edition

WATERS, WILLIAM F, b Dayton, Ohio, Apr 25, 43. PSYCHOLOGY, PSYCHOPHYSIOLOGY. *Educ:* Tulane Univ, AB, 64; Case Western Reserve Univ, MS, 66, PhD(psychol), 69, Am Bd Prof Psychol, dipl, 73. *Prof Exp:* Assoc prof med psychol, Dept Psychiat, Univ Mo, Columbia, 68-79, assoc prof psychol, Dept Psychol, 70-79; dir clin training, 79-81, PROF, DEPT PSYCHOL, LA STATE UNIV, 79- *Concurrent Pos:* Dir psychol serv, MId-Mo Ment Health Ctr, 68-74; consult, Ochsner Clin Baton Rouge Sleep Dis Clin, 88- *Mem:* Am Psychol Asn; Soc Psychophysiol Res; Am Sleep Dis Asn. *Res:* Etidogy and treatment of insomnia; psychophysiology of selective attention, emotion, habituation; psychophysiological disorders. *Mailing Add:* Dept Psychol Audubon Hall La State Univ Baton Rouge LA 70803

WATERS, WILLIAM LINCOLN, organic chemistry, organometallic chemistry, for more information see previous edition

WATERS, WILLIE ESTEL, b Smith Town, Ky, Sept 19, 31; m 52; c 3. HORTICULTURE. *Educ:* Univ Ky, BS, 54, MS, 58; Univ Fla, PhD(veg crops), 60. *Prof Exp:* Asst county agr agent, Ky, 54; asst soils, Univ Ky, 56-58; asst veg crops, 58-60, from asst ornamental horticulturist to assoc ornamental horticulturist, 60-68, horticulturist & head Agr Res Ctr, Apopka, Fla, 68-70, HORTICULTURIST & DIR GULF COAST RES & EDUC CTR, UNIV FLA, 70- *Honors & Awards:* Alex Laurie Award, Am Soc Hort Sci, 68. *Mem:* Fel Am Soc Hort Sci. *Res:* Soil and plant nutrition; weed control and physiology of ornamental crops. *Mailing Add:* Gulf Coast Res & Educ Ctr Univ Fla 5007 60th St E Bradenton FL 34203

WATERSON, JOHN R, b Decatur, Ill, May 11, 44; c 4. HUMAN GENETICS, PEDIATRICS. *Educ:* Univ Mich, BS, 66, MD, 75; Yale Univ, PhD(biophys), 71. *Prof Exp:* Instr, 79, ASST PROF PEDIAT, UNIV MICH, 80-; ASSOC PROF PEDIAT & MICROBIOL & IMMUNOL, NORTHEAST UNIV COL MED. *Mem:* Am Soc Human Genetics; Am Acad Pediat. *Res:* Human gene structure and chromosomal organization. *Mailing Add:* Dept Pediat Children's Hosp Med Ctr 281 Locust Akron OH 44308

WATERWORTH, HOWARD E, b Randolph, Wis, Sept 3, 36; m 66; c 4. PLANT VIROLOGY. *Educ:* Univ Wis, BS, 58, PhD(plant path), 62. *Prof Exp:* Res plant pathologist, USDA, 64-80, res virologist & chief, Germplasm Resources Lab, Res Ctr, 80-82, adminr, 82-89, LEAD SCIENTIST, AGR RES SERV, USDA, BELTSVILLE, 89- *Mem:* Am Inst Biol Sci; Am Phytopath Soc; Am Soc Hort Sci; Coun Agr Sci & Technol. *Res:* Testing of new stone fruits, grasses, tropical crops and woody ornamentals for the presence of viruses; identification of new viruses. *Mailing Add:* Agr Res Serv USDA BARC-E Bldg 580 Beltsville MD 20705

WATKIN, DONALD M, NUTRITION, GERONTOLOGY. *Educ:* Harvard Univ, MD, 46. *Prof Exp:* MGR, OCCUP HEALTH DIV, OFF AVIATION MED, FED AVIATION ADMIN, DEPT TRANSP, 81-; RES PROF, SCH MED & HEALTH SCI, GEORGE WASHINGTON UNIV, 81- *Res:* Internal medicine. *Mailing Add:* Dept Transp FAA Off Aviation Med 800 Independence Ave SW Suite 327 Washington DC 20591

WATKINS, ALLEN HARRISON, b Charlottesville, Va, Apr 25, 38; m 62, 83; c 4. TECHNICAL MANAGEMENT. *Educ:* Va Polytech Inst, BS, 61. *Prof Exp:* Prog mgt space syst, Manned Spacecraft Ctr, NASA, 62-70, prog mgt earth resources, 70-73; DIR, EARTH RESOURCES OBSERV SYST DATA CTR, DEPT INTERIOR, 73- *Honors & Awards:* Pecora Award. *Mem:* Am Soc Photogram; Am Inst Aeronaut & Astronaut; Int Acad Astronaut. *Mailing Add:* 4312 Ash Grove Ave Sioux Falls SD 57103

WATKINS, BRENTON JOHN, b Adelaide, Australia, Aug 5, 46; US citizen; c 1. APPLICATIONS OF RADAR TO ATMOSPHERE SCIENCE. *Educ:* Univ Adelaide, BSc Hons, 69; La Trobe Univ, MSc, 72; Univ Alaska, PhD(geophys), 76. *Prof Exp:* Physicist, Dept Sci, Govt Australia, 69-72; tech officer, La Trobe Univ, 77; staff scientist, Lincoln Lab, Mass Inst Technol, 78-80; sr res asst, 72-76, asst prof physics, 80-85, ASSOC PROF PHYSICS, GEOPHYS INST, UNIV ALASKA, 85-, ASSOC PROF GEOPHYS, 80- *Concurrent Pos:* Prin investr, NSF grants, 81-83, 84-86 & 87-89 & NASA, 84-86; co-prin investr, Air Force Off Sci Res Grant, 85-86. *Mem:* Australian Inst Physics; Am Meterol Soc; Am Geophys Union. *Res:* Development of numerical computer simulations of the earth's ionosphere; use of incoherent scatter radar at high latitudes to study the ionosphere; development of radar techniques to study the lower atmosphere using turbulence-scatter; development of hardware and software for radar systems; real-time computer software and hardware. *Mailing Add:* Dept Physics Univ Alaska Fairbanks AK 99701

WATKINS, CHARLES B, b Petersburg, Va, Nov 20, 42; m 64; c 2. HEAT TRANSFER, FLUID MECHANICS. *Educ:* Howard Univ, BSME, 64; Univ NMex, MS, 66, PhD(mech eng), 70. *Prof Exp:* Staff mem, Sandia Labs, 64-71; from asst prof to assoc prof, 73-79, PROF MECH ENG, HOWARD UNIV, 79-, CHMN DEPT, 73-; DEAN, SCH ENG, CITY COL, CITY UNIV NEW YORK, 86- *Concurrent Pos:* Mem, Eng Chem & Energetics Adv Panel, NSF, 75-78; mem, Selection Panel Nat Res Coun Post Doctoral Fel Prog, 80; consult, US Army, USN & Sandia Labs, 67-; bd on eng educ, Am Soc Mech Engrs, 84-87; mem Mech, Struct & Mat Eng Adv Bd, NSF, 88-89. *Honors & Awards:* Ralph R Teetor Award, Soc Automotive Engrs, 80. *Mem:* Am Soc Mech Engrs; Fel Am Soc Mech Engrs; Soc Automotive Engrs; Sigma Xi. *Res:* Heat transfer; fluid mechanics; gas bearing dynamics; mechanical design; truck wheels; product liability; author or co-author of 57 publications. *Mailing Add:* Sch Eng City Col Convent Ave 140 St New York NY 10031

WATKINS, CHARLES H(ENRY), b Henshaw, Ky, Mar 21, 13; m 40; c 4. CHEMICAL ENGINEERING. *Educ:* Univ Louisville, BS, 35, MS, 36; Purdue Univ, PhD(chem eng), 39. *Prof Exp:* Res engr, Standard Oil Develop Co, NJ, 39-46; div coordr, Universal Oil Prod Co, 46-60, div coordr eng res & develop, 60-77, consult eng res & develop, Universal Oil Prod Process Div, 77-83; RETIRED. *Mem:* Am Chem Soc; Am Inst Chem Engrs. *Res:* Olefin production; aromatics; polymerization of hydrocarbons; alkyllation of isoparaffines with ethylene, propylene and butenes; hydrogenation; inorganic reactions in organic solvents; hydrofining; hydrocracking. *Mailing Add:* 8033 SW 103rd Lane Ocala FL 32676

WATKINS, CHARLES LEE, b Fairfield, Ala, Oct 27, 42. CHEMISTRY. *Educ:* Univ Ala, Tuscaloosa, BS, 64; Univ Fla, MS, 66, PhD(chem), 68. *Prof Exp:* Lab asst gen chem, Univ Ala, Tuscaloosa, 63-64; res assoc, Univ NC, Chapel Hill, 69-70; asst prof, 70-74, ASSOC PROF CHEM, UNIV ALA, BIRMINGHAM, 74- *Mem:* Am Chem Soc; Sigma Xi. *Mailing Add:* Dept Chem Univ Ala Birmingham AL 35294

WATKINS, CLYDE ANDREW, b McKees Rocks, PA, Dec 1, 46; m 84; c 2. RESEARCH ADMINISTRATION. *Educ:* Duquesne Univ, BS, 69; Pa State Univ, PhD(physiol), 77. *Prof Exp:* Res assoc physiol, 77-79, asst prof anesthesia & physiol, M S Hershey Med Ctr, 79-82; asst prof physiol, 82-85, assoc prof physiol & pharmacol, Med Col Ga, 85-87; HEALTH SCIENTIST ADMINR, NIH, 87- *Mem:* Am Physiol Soc; Am Thoracic Soc; Am Heart Asn; AAAS. *Res:* Metabolic response of the lung and vasculature to injury with emphasis on membrane processes and protein turnover; molecular basis of anesthesia. *Mailing Add:* Off Sci Integrity NIH Bldg 31 Rm BlC39 Bethesda MD 20892

WATKINS, DARRELL DWIGHT, JR, b Woodbine, Iowa, Mar 7, 43; m 64; c 2. ORGANOMETALLIC CHEMISTRY. *Educ:* Univ Nebr, Omaha, BA, 68; Univ Nebr, Lincoln, PhD(chem), 73. *Prof Exp:* Asst instr chem, Univ Nebr, Omaha, 68-70; res assoc, Ohio State Univ, 73-76; SR RES CHEMIST, MONSANTO CO, 76- *Mem:* Am Chem Soc. *Res:* Synthesis of new and novel organometallic compounds and the use of those compounds in homogeneous catalysis. *Mailing Add:* 12440 Glengate Dr Maryland Heights MO 63043-2576

WATKINS, DAVID HYDER, thoracic surgery; deceased, see previous edition for last biography

WATKINS, DEAN ALLEN, b Omaha, Nebr, Oct 23, 22; m 44; c 3. ELECTRICAL ENGINEERING. *Educ:* Iowa State Col, BS, 44; Calif Inst Technol, MS, 47; Stanford Univ, PhD(elec eng), 51. *Prof Exp:* Engr, Collins Radio Co, Iowa, 47-48; staff mem, Los Alamos Sci Lab, 48-49; head microwave tube sect, Res & Develop Labs, Hughes Aircraft Co, 51-53; from assoc prof to prof elec eng, Stanford Univ, 53-64, dir electron devices lab, 55-64; pres, 57-67, chief exec officer, 67-79, CHMN BD, WATKINS-JOHNSON CO, 67- *Concurrent Pos:* Lectr, Stanford Univ, 64-69. *Mem:* Nat Acad Eng; Am Phys Soc; fel Inst Elec & Electronics Engrs. *Res:* Microwave devices and systems. *Mailing Add:* Watkins-Johnson Co 3333 Hillview Ave Palo Alto CA 94304

WATKINS, DON WAYNE, b Louisville, Ky, June 9, 40; m 66; c 3. PHYSIOLOGY. *Educ:* Univ Louisville, BChE, 63; Univ Wis-Madison, PhD(physiol), 68. *Prof Exp:* Lectr physiol, Med Sch, Makerere Univ, Uganda, 68-70; asst prof, 70-75, ASSOC PROF PHYSIOL, MED SCH, GEORGE WASHINGTON UNIV, 75-, INTERIM CHAIR PHYSIOL, 89- *Mem:* Am Physiol Soc. *Res:* Membrane transport, physiology, epithelia; kidney salt and water; frog skin; trace element and mineral metabolism. *Mailing Add:* Dept Physiol George Washington Univ Med Sch 2300 Eye St NW Washington DC 20037

WATKINS, DUDLEY T, b Youngstown, Ohio, May 2, 38; m 60; c 3. ANATOMY. *Educ:* Oberlin Col, AB, 60; Western Reserve Univ, MD, 66, PhD(anat), 67. *Prof Exp:* Fel anat, 66-67, asst prof, 67-72, assoc prof, 72-80, PROF ANAT, HEALTH CTR, UNIV CONN, 80- *Mem:* AAAS; Am Diabetes Asn; Am Asn Anatomists; Sigma Xi. *Res:* Mechanism of action of alloxan in the production of diabetes; mechanism of insulin secretion. *Mailing Add:* Dept Anat Univ Conn Health Ctr 263 Farmington Ave Farmington CT 06032

WATKINS, ELTON, JR, b Portland, Ore, Aug 16, 21. SURGERY, CANCER. *Educ:* Reed Col, BA, 41; Univ Ore, MD, 44. *Prof Exp:* Asst path, Med Sch, Univ Ore, 41-43, biochem, 43-44, resident thoracic surg, Hosps & Clins, 45-48, instr physiol, Med Sch, 49-50; Ore Heart Asn res fel, Children's Hosp Med Ctr, Boston, Mass, 51-53; from instr to asst prof surg, Harvard Med Sch, 54-57; sr surgeon, div res, 64-67, SR SURGEON & DIR SIAS SURG LAB, LAHEY CLIN FOUND, 57-, CHMN DIV RES, 64- *Concurrent Pos:* Surgeon, Children's Hosp Med Ctr, Boston, Mass, 56-57; asst surgeon, Peter Bent Brigham Hosp, Boston, Mass, 57; consult, Blood Preservation Lab, US Naval Hosp, Boston, 56-; USPHS grant organ transplantation, Sias Surg Lab, Lahey Clin Found, 59-65, USPHS grant cancer immunother, 72-; lectr, Royal Col Surgeons Eng, 63. *Mem:* AAAS; Am Asn Cancer Res; Transplantation Soc; Am Fedn Clin Res; Int Soc Surgeons. *Res:* Cardiovascular physiology and surgery; pancreatic transplantation; regional cancer chemotherapy; experimental manipulation of host immune response to cancer. *Mailing Add:* Lahey Clin Med Ctr Burlington MA 01805

WATKINS, GEORGE DANIELS, b Evanston, Ill, Apr 28, 24; m 49; c 3. PHYSICS. *Educ:* Randolph-Macon Col, BS, 43; Harvard Univ, AM, 47, PhD(physics), 52. *Hon Degrees:* DSc, Randolph-Macon Col, 76. *Prof Exp:* Physicist, Res & Develop Ctr, Gen Elec Co, 52-75; SHERMAN FAIRCHILD PROF PHYSICS, LEHIGH UNIV, 75- *Concurrent Pos:* Adj prof, Rensselaer Polytech Inst, 62-65; NSF fel, Oxford Univ, 66-67; adj prof, State Univ NY, Albany, 70-71; Alexander von Humboldt Sr US Scientist Award, 83-84. *Honors & Awards:* Oliver E Buckley Prize Solid State Physics, 78. *Mem:* Nat Acad Sci; Fel Am Phys Soc; fel AAAS. *Res:* Nuclear and electron spin resonance studies; solid state physics; radiation effects and point defects in solids. *Mailing Add:* Sherman Fairchild Lab 161 Lehigh Univ Bethlehem PA 18015

WATKINS, IVAN WARREN, b Minneapolis, Kans, Jan 14, 34; m 55; c 4. PHYSICS. *Educ:* Univ Kans, BS, 55, MS, 57; Tex A&M Univ, PhD, 68. *Prof Exp:* Instr physics, Ft Hays Kans State Col, 58-62; from asst prof to prof physics, 63-74, PROF GEOG & EARTH SCI, ST CLOUD STATE COL, 74- *Mem:* Am Asn Physics Teachers. *Res:* Theoretical molecular spectroscopy; student attitudes about sciences and how the attitudes correlate with success in science courses. *Mailing Add:* Dept Earth Sci St Cloud State Univ St Cloud MN 56301

WATKINS, JACKIE LLOYD, b Melvin, Tex, Jan 16, 32; m 55; c 4. GEOLOGY. *Educ:* Southern Methodist Univ, BS, 52, MS, 54; Univ Mich, PhD(geol), 58. *Prof Exp:* Assoc prof, 58-72, PROF GEOL, MIDWEST UNIV, 72-, CHMN DEPT, 77- *Concurrent Pos:* Chmn bd, Watkins Mineral Corp. *Mem:* AAAS; Paleont Soc; Soc Econ Paleont & Mineral; Geol Soc Am; Nat Asn Geol Teachers; Sigma Xi. *Res:* Invertebrate paleontology; tetracorals and tabulate corals. *Mailing Add:* 117 Pembroke Lane Wichita Falls TX 76301

WATKINS, JOEL SMITH, JR, b Poteau, Okla, May 27, 32; m 56, 71; c 2. GEOPHYSICS, SEISMOLOGY. *Educ:* Univ NC, AB, 53; Univ Tex, PhD(geol), 61. *Prof Exp:* Geophysicist, Regional Geophys Br, US Geol Surv, 61-64, Astrogeol Br, 64-66; res assoc geophys, Mass Inst Technol, 66-67; from assoc prof to prof, Univ NC, Chapel Hill, 67-73; prof geophys, Univ Tex Marine Sci Inst, 73-77; sr res assoc, Gulf Res & Develop Co, Houston, 77-79, mgr, Geol Interpreters Dept, Gulf Sci & Tech Co, Pittsburgh, 79-81, mgr geol, Cent Exp Group Int, Gulf Oil E&P Co, Houston, 81-82, expl mgr, Eastern US Frontier, 82-83, vpres, Expl Res, Gulf Res & Develop Co, 83-85; PROF OCEANOG, TEX A&M UNIV, 85-, COOK PROF GEOSCI, 86-, HEAD, DEPT GEOPHYS, 88- *Concurrent Pos:* Co-investr, Apollo Active Seismic Exp, 65-72, Apollo Lunar Seismic Profiling Exp, 71-73; mem adv comt, Int Phase Ocean Drilling, 74-; co-chief, Deep Sea Drilling Proj, 79. *Honors & Awards:* Except Sci Achievement Medal, NASA, 73. *Mem:* Am Geophys Union; Soc Explor Geophysicists; Geol Soc Am; Am Asn Petrol Geol. *Res:* Marine geophysics; explosion seismology. *Mailing Add:* Dept Geophys Tex A&M Univ College Station TX 77843

WATKINS, JULIAN F, II, b Marvell, Ark, Mar 18, 36; m 54; c 2. ENTOMOLOGY. *Educ:* Univ Ark, BSA, 56; Kans State Univ, MS, 62, PhD(entom), 64. *Prof Exp:* Asst county agent, Ark Agr Exten Serv, 56-60; asst prof, 64-68, assoc prof, 68-79, actg chmn, 79-81, PROF BIOL, BAYLOR UNIV, 79-, VCHMN, 81- *Mem:* Entom Soc Am. *Res:* Taxonomy and behavior of army ants. *Mailing Add:* Dept Biol Baylor Univ Waco TX 76798

WATKINS, KAY ORVILLE, b Nunn, Colo, Apr 28, 32; m 61; c 3. INORGANIC CHEMISTRY, PHYSICAL CHEMISTRY. *Educ:* Adams State Col, BA, 55; Vanderbilt Univ, PhD(inorg chem), 61. *Prof Exp:* From asst prof to assoc prof, 61-70, PROF CHEM, ADAMS STATE COL, 70-, CHMN, DIV SCI & MATH, 77- *Concurrent Pos:* NSF acad year exten grant, 63-65 & res grant, 65-68; NSF high sch lectr, 66-67 & fac develop fel, 78; vis scientist, Brookhaven Nat Lab, 68-69 & Argonne Nat Lab, 73; vis prof, Univ Utah, 81 & Univ Hawaii, 87. *Mem:* Am Chem Soc. *Res:* Equilibria and kinetic studies of inorganic systems in solution. *Mailing Add:* Dept Chem Adams State Col Alamosa CO 81101

WATKINS, KENNETH WALTER, b Philadelphia, Pa, Apr 22, 39; m 62, 84; c 2. PHYSICAL CHEMISTRY. *Educ:* Kans State Univ, BS, 61, PhD(chem), 65. *Prof Exp:* Fel chem, Univ Wash, 65-66; ASST PROF CHEM, COLO STATE UNIV, 66- *Concurrent Pos:* Vis asst prof chem, Univ Wis, 79-80. *Mem:* Am Chem Soc; AAAS. *Res:* Chemical education. *Mailing Add:* Dept Chem Colo State Univ Ft Collins CO 80523

WATKINS, LINDA ROTHBLUM, b Norfolk, Va, June 29, 54; m 76. NEUROANATOMY, NEUROPHYSIOLOGY. *Educ:* Va Polytech Inst & State Univ, BS, 76; Med Col Va, PhD(physiol), 80. *Prof Exp:* LECTR NEUROSCI, DEPT PHYSIOL, MED COL VA, 81- *Concurrent Pos:* Fel, Dept Physiol, Med Col Va, 81- *Mem:* Int Asn Study Pain. *Res:* Behavioral, neuroanatomical and neuropharmacological investigations of endogennous opiate and non-opiate pain inhibitory systems which are activated by brain stimulation, morphine or environmental stimuli. *Mailing Add:* Dept Psychol Univ Colo Box 345 Boulder CO 80309

WATKINS, MARK E, b New York, NY, Apr 13, 37; m; c 3. MATHEMATICS. *Educ:* Amherst Col, AB, 59; Yale Univ, MA, 61, PhD(math), 64. *Prof Exp:* Instr math, Univ NC, Chapel Hill, 63-64, asst prof, 64-68; assoc prof, 68-76, PROF MATH, SYRACUSE UNIV, 76- *Concurrent Pos:* Vis assoc prof, Univ Waterloo, 67-68 & 80; vis prof, Vienna Tech Univ, 73-74 & Univ Paris-Orsay, 86; Ger Acad Exchange Serv grants, West Berlin, 80 & 89. *Mem:* Am Math Soc. *Res:* Problems related to vertex-connectivity in graphs; imbedding of graphs; automorphism groups of graphs and systems. *Mailing Add:* Dept Math Syracuse Univ Syracuse NY 13244-1150

WATKINS, MAURICE, b East Chicago, Ind, Aug 8, 46; m 76. METALLURGICAL ENGINEERING. *Educ:* Ill Inst Technol, BS, 68, PhD(metall eng), 73. *Prof Exp:* Engr eng mat, Res & Develop Labs, Continental Can Co, 68-69; res assoc, 73-85, GROUP LEADER, MAT RES, EXXON PROD RES CO, 85- *Concurrent Pos:* Vis lectr, Black Exec Exchange Prog, Urban League, 76-; indust adv & prog coordr for Jets Club, Houston Independent Sch Dist, 78-; Exxon rep, Nat Asn Corrosion Engrs & Am Petrol Inst. *Mem:* Am Soc Metals; Nat Asn Corrosion Engrs. *Res:* Welding, corrosion, stress corrosion cracking, hydrogen embrittlement and fatigue of materials used in the manufacture of equipment for oil and gas exploration and production. *Mailing Add:* 5623 Bent Bough Houston TX 77088

WATKINS, NANCY CHAPMAN, b Bowling Green, Ky, Mar 19, 39; m 62; c 1. PHYSICS, POLYMER SCIENCE. *Educ:* Univ Ky, BS, 61; Rensselaer Polytech Inst, PhD(chem eng), 67. *Prof Exp:* Sr res scientist fiber struct, Am Enka Corp, 67-80; STAFF ENGR, IBM CORP, 80- *Mem:* Am Chem Soc. *Res:* Thermal analysis, scanning electron microscopy, x-ray diffraction and computer applications to fiber structural studies; cross-linking technique applied to thermal studies. *Mailing Add:* One Warwick Lane Frankfort KY 40601

WATKINS, PAUL ALLAN, b Baltimore, Md, Oct 12, 49. BIOCHEMISTRY. *Educ:* Johns Hopkins Univ, BA, 71, MD, 78, PhD(biochem), 79. *Prof Exp:* Res assoc, Nat Heart, Lung & Blood Inst, NIH, 78-82, med staff fel, 82-84; ASST PROF, KENNEDY INST, 84- *Mem:* Am Soc Biol Chemists; Am Fedn Clin Res. *Res:* Regulation of fatty acid metabolism in normal and pathological states; regulation of adenylate cyclase and related systems by adenosine diphosphate-ribosyl transferases. *Mailing Add:* Kennedy Inst 707 N Broadway Baltimore MD 21205

WATKINS, PAUL DONALD, microbiology, biochemistry; deceased, see previous edition for last biography

WATKINS, ROBERT ARNOLD, b Boston, Mass, Aug 3, 26; m 83; c 2. ELECTROOPTICS, MILITARY SYSTEMS. *Educ:* Brown Univ, ScB, 47; Ohio State Univ, MS, 48, PhD(physics), 53. *Prof Exp:* Instr physics, Univ WVa, 48-49; engr, Zenith Radio Corp, 52-54; sr engr, Raytheon Co, Goleta, 54-57, sect mgr, 57-60, dept mgr, 60-63, prin engr, 63-80, dept mgr, 80-89, CONSULT ENGR, RAYTHEON CO, GOLETA, 89- *Honors & Awards:* Centennial Medal, Inst Elec & Electronics Engrs. *Mem:* Inst Elec & Electronics Engrs; Optical Soc Am; Am Inst Aeronaut & Astronaut. *Res:* Optical systems design with associated electronic and mechanical configurations; laser effects on materials; combined radio frequency and optical systems; digital based radio frequency systems. *Mailing Add:* 844 Vereda del Ciervo Goleta CA 93117-5304

WATKINS, SALLIE ANN, b Jacksonville, Fla, June 27, 22. PHYSICS. *Educ:* Notre Dame Col, Ohio, BS, 45; Catholic Univ, MS, 54, PhD(physics), 58. *Prof Exp:* Instr chem & physics, Notre Dame Acad, Ohio, 45-49, Elyria Dist Cath High Sch, Ohio, 49-50; instr physics, Notre Dame Col, Ohio, 50-53, prof, 57-66; teaching asst, Catholic Univ, 55-56; PROF PHYSICS, UNIV SOUTHERN COLO, 66- *Concurrent Pos:* NSF res grant biophys, 60-63; Oak Ridge Asn Univs res partic fel, Savannah River Lab, 66-68; mem exec bd, Am Asn Physics Teachers, 83-86; sr educ fel, Am Inst Physics, Washington, DC, 87-88. *Mem:* Am Phys Soc; Am Asn Physics Teachers. *Res:* Nuclear reactor physics; ultrasonics and biophysics; history and philosophy of science. *Mailing Add:* 108 S Lynx Dr Pueblo CO 81007

WATKINS, SPENCER HUNT, b Mayfield, Ky, Sept 15, 24; m 46; c 3. ORGANIC CHEMISTRY. *Educ:* Univ Ill, BS, 47; Univ Wis, PhD(chem), 50. *Prof Exp:* Res chemist, Hercules, Inc, 50-55, res supvr, 55-57, tech rep, 57-60, asst sales mgr, 60-63, mgr tech serv, 63-65, mgr mkt develop, 65-74, dir develop, pine & paper chem dept, 74-78, dir technol, 78-88; RETIRED. *Mem:* Am Chem Soc; Tech Asn Pulp & Paper Indust; Com Develop Asn; Am Asn Textile Chem & Colorists. *Res:* Paper chemicals, rosin chemistry; urea-formaldehyde resins. *Mailing Add:* 125 Wildwood Lane Naples FL 33942

WATKINS, STANLEY READ, b University Park, NMex, June 22, 29; m 52; c 2. ATOMIC ABSORPTION, SPECTROSCOPY & HPLC. *Educ:* NMex State Univ, BS, 51; Univ Colo, PhD(analytical chem), 58. *Prof Exp:* Chemist, E I du Pont de Nemours & Co, 51-54; PROF CHEM, COE COL, 58- *Mem:* Am Chem Soc. *Res:* Analysis of trace metals in biological systems; trace analysis. *Mailing Add:* Dept Chem Coe Col Cedar Rapids IA 52402

WATKINS, STEVEN F, b Amarillo, Tex, May 14, 40; m 66; c 2. STRUCTURAL CHEMISTRY, CRYSTALLOGRAPHY. *Educ:* Pomona Col, BA, 62; Univ Wis, PhD(phys chem), 67. *Prof Exp:* Res fel, Bristol Univ, 67-68; asst prof, 68-73, ASSOC PROF CHEM, LA STATE UNIV, BATON ROUGE, 73- *Concurrent Pos:* vis prof, Univ Houston, 66-67; vis scientist, Brookhaven Nat Lab, 72; chmn, Orgn Prophet Using Scientist, 85. *Mem:* AAAS; Am Chem Soc; Am Crystallog Asn; Royal Soc Chem; Sigma Xi. *Res:* Molecular structure of organometallic, inorganic and organic biological molecules in the solid state by means of x-ray and neutron crystallography. *Mailing Add:* Dept Chem La State Univ Baton Rouge LA 70803

WATKINS, TERRY ANDERSON, b Brady, Tex, June 29, 38. STATISTICS. *Educ:* WTex State Univ, BS, 61; Ill Inst Technol, MS, 64; Tex Tech Univ, PhD(math statist), 72. *Prof Exp:* Instr math, WTex State Univ, 64-65, asst prof, Angelo State Univ, 66-68, 71-72; instr, Tex Tech Univ, 70-71, asst prof, 72-77; ASSOC PROF MATH STATIST, UNIV NEW ORLEANS, 77- *Mem:* Am Statist Asn. *Res:* Estimation; jackknifing procedures. *Mailing Add:* Dept Math Univ New Orleans New Orleans LA 70148

WATKINS, W DAVID, ANESTHESIOLOGY, PHARMACOLOGY. *Educ:* Univ Mich, PhD(toxicol); Univ Colo, MD, 75. *Prof Exp:* PROF & CHMN DEPT ANESTHESIOL, DUKE UNIV MED CTR, 83- *Mailing Add:* Dept Anesthesiol Duke Univ Med Ctr Box 3081 Durham NC 27710

WATKINS, WILLIAM, b Los Angeles, Calif, July 7, 42; m; c 2. MATHEMATICS, STATISTICS. *Educ:* Univ Calif, Santa Barbara, BA, 64, MA, 68, PhD(math), 69. *Prof Exp:* PROF MATH, CALIF STATE UNIV, NORTHRIDGE, 69- *Concurrent Pos:* Ed, Col Math J. *Mem:* Math Asn Am; Soc Indust & Appl Math. *Res:* Linear and multilinear algebra. *Mailing Add:* Dept Math Calif State Univ Northridge CA 91330

WATLING, LES, b Calgary, Alta, Oct 13, 45. TAXONOMY OF CRUSTACEA, MARINE BENTHIC ECOLOGY. *Educ:* Univ Calgary, BSc, 65; Univ Pac, MS, 68; Univ Del, PhD(marine studies), 74. *Prof Exp:* Res biologist, Univ Del, 74-76; asst prof oceanog, 76-80, actg chmn, dept oceanog & zool, 80-81, ASSOC PROF OCEANOG & ZOOL, UNIV MAINE, 80-, ACTG DIR, DARLING MARINE CTR, 85- *Concurrent Pos:* Assoc ed, J Crustacean Biol, 80-81; mem gov bd, Crustacean Soc, 80-84; sr postdoctoral fel, Smithsonian, 90. *Mem:* Soc Syst Zool; Crustacean Soc; Palaeont Asn; Ecol Soc Am; Am Soc Limnologists & Oceanogr. *Res:* Ecology of marine benthos, particularly of boreal and polar seas; taxonomy of amphipoda and cumacea from Antarctica to tropics; author of 50 articles. *Mailing Add:* Darling Marine Ctr Univ Maine Walpole ME 04573

WATLINGTON, CHARLES OSCAR, b Midlothian, Va, Apr 9, 32; m 55; c 2. MEDICINE, PHYSIOLOGY. *Educ:* Va Polytech Inst, BS, 54; Med Col Va, Va Commonwealth Univ, MD, 58, PhD(physiol), 68. *Prof Exp:* Intern med, Univ Calif, San Francisco, 58-59, asst resident, 60-62; from instr to asst prof, 62-69, assoc prof, 69-76, PROF MED, MED COL VA, VA COMMONWEALTH UNIV, 76- *Concurrent Pos:* Fel endocrinol, Univ Calif, San Francisco. *Mem:* Am Physiol Soc; Am Fedn Clin Res; Endocrine Soc; Am Diabetes Asn; Am Soc Nephrology. *Res:* Regulation of ion transport; sodium and calcium homeostasis; cellular mechanism of action of hormones. *Mailing Add:* Dept Med Endominol Va Commonwealth Univ Med Col PO Box 145 Richmond VA 23298-0145

WATNE, ALVIN LLOYD, b Shabbona, Ill, Jan 13, 27; m 66; c 4. SURGERY. *Educ:* Univ Ill, BS, 50, MD, 52, MS, 56. *Prof Exp:* Intern, Indianapolis Gen Hosp, Ind, 52-53; res asst surg, Univ Ill, 53-54; resident, Res & Educ Hosp, Univ Ill, 54-58; assoc cancer res surgeon, Roswell Park Mem Inst, 58-59, assoc chief cancer res surgeon, 59-62; from assoc prof to prof surg, WVa Univ, 62-73, cancer coordr, 62-73, actg chmn surg, 73-75, chmn dept, 75-86; PROF & CHMN, COL MED, UNIV ILL, PEORIA, 86- *Mem:* Am Asn Cancer Res; Am Col Surgeons; Soc Univ Surgeons; NY Acad Sci; Sigma Xi. *Res:* Cancer metastases, including dissemination of tumor cells via the blood and lymph and their lodgement and growth; etiology and prevention of polyposis and coli and colon cancer. *Mailing Add:* 420 NE Glenn Oak Suite 302 Peoria IL 61603

WATNICK, ARTHUR SAUL, b Brooklyn, NY, Feb 4, 30; m 57; c 2. ENDOCRINOLOGY. *Educ:* Brooklyn Col, BS, 53, MA, 56; NY Univ, PhD(physiol), 63. *Prof Exp:* Biochemist, Sloan-Kettering Inst, 53-54, Beth-El Hosp, Brooklyn, NY, 54-56; from asst chemist to assoc chemist, 56-62, from scientist to sr scientist, 62-70, assoc dir, Dept Allergy & Inflammation, 81-85, ASSOC DIR, BIOL RES, SCHERING CORP, 70-, MGR DEPT PHYSIOL, 77, SECT LEADER, 85- *Concurrent Pos:* Lectr endocrinol, Fairleigh Dickenson Univ, 66- & Rutgers Univ, 71-; lectr physiol anat, Bloomfield Col, 85; adj prof immunol, Fairleigh Dickenson Univ, 88- *Mem:* AAAS; Soc Exp Biol & Med; Am Chem Soc; Endocrine Soc; NY Acad Sci; Reticuloendothelial Soc. *Res:* Reproductive physiology; immunology. *Mailing Add:* 27 Harding Dr South Orange NJ 07079

WATRACH, ADOLF MICHAEL, b Poland, Jan 7, 18; nat US; m 55. VETERINARY PATHOLOGY. *Educ:* Royal (Dick) Vet Col, Scotland, MRCVS, 48; Glasgow Univ, PhD(path), 58. *Prof Exp:* Asst vet path, Vet Sch, Glasgow Univ, 49-51; from instr to prof, 51-85, EMER PROF VET PATH, COL VET MED, UNIV ILL, URBANA, 85- *Mem:* AAAS; Electron Micros Soc Am; Am Col Vet Path; Vet Med Asn; Int Acad Path. *Res:* Ultrastructural pathology; virus-cell relationship; viral oncogenesis; mammary tumor biology; cell pathobiology. *Mailing Add:* 645 Cosmos Way Prescott AZ 86303

WATREL, WARREN GEORGE, b Brooklyn, NY, Jan 5, 35; m 60; c 3. MICROBIOLOGY, BIOCHEMISTRY. *Educ:* Syracuse Univ, BS, 57, MS, 58. *Prof Exp:* Asst to dir of NSF, Syracuse Univ, 57, instr, 57-58; pharmaceut sales & mkt staff, Lederle Labs, Am Cyanamid Co, 60-62; biochem specialist, M&T Chem, Am Can Co, 62-64; sales mgr, Pharmacia Fine Chem, Inc, 64-65, dir mkt & gen mgr, 65-72; vpres, Vineland, Vista Labs, Inc, Ideal & Nickolson Inst, Damon Corp, 72-74; vpres, Pharmachem Corp, 74-75; exec vpres, Newton Industs Inc, 76-79; VPRES, SECOL INC & SELOMAS INC, 79- *Concurrent Pos:* Am Cyanamid Co & NSF grants, 57-58. *Honors & Awards:* Mkt Award, Am Chem Soc; Award, Res & Develop Soc Cosmet Chem. *Mem:* AAAS; Am Chem Soc; Am Mkt Asn; Am Soc Microbiol; NY Acad Sci. *Res:* Application research and development of products biologically active for use in commercial products; research and development of products to be used in gel filtration chromatography; development, manufacture and sale of veterinarian, pharmacy, human, cosmetics industrial chemicals and medical electonic instruments. *Mailing Add:* 506 Collins Ave Hasbrouck Heights NJ 07604

WATROUS, JAMES JOSEPH, b Cleveland, Ohio, July 20, 42; m 70; c 2. PHYSIOLOGY. *Educ:* Univ Dayton, BSEd, 65, MS, 69; Georgetown Univ, PhD(biol), 72. *Prof Exp:* Teacher sec sch, Ohio, 64-67; teaching asst biol, Univ Dayton, 67-69 & Georgetown Univ, 69-72; asst prof, 72-80, ASSOC PROF BIOL, ST JOSEPH'S UNIV, 80-, CHMN DEPT, 77- *Mem:* Am Phys Soc; NY Acad Sci; AAAS; Sigma Xi; Am Soc Zoologists. *Res:* Transport of materials across biological membranes; nerve muscle physiology. *Mailing Add:* Dept Biol St Joseph's Univ Philadelphia PA 19131

WATSCHKE, THOMAS LEE, b Charles City, Iowa, Apr 12, 44; m 65. AGRONOMY, HORTICULTURE. *Educ:* Iowa State Univ, BS, 67; Va Polytech Inst & State Univ, MS, 69, PhD(agron), 71. *Prof Exp:* assoc prof agron, 70-80, PROF TURFGRASS SCI, PA STATE UNIV, UNIVERSITY PARK, 80- *Mem:* Am Soc Agron; Crop Sci Soc Am; Int Turfgrass Soc. *Res:* Turfgrass physiology, weed control and growth regulation. *Mailing Add:* Dept Agron 116 ASI Bldg Pa State Univ University Park PA 16802

WATSON, ALAN KEMBALL, b Vernon, BC, Sept 1, 48; m 71; c 3. WEED SCIENCE, BIOLOGICAL CONTROL. *Educ:* Univ BC, BSc, 70, MSc, 72; Univ Sask, PhD(weed sci), 75. *Prof Exp:* From asst prof to assoc prof, 75-86, PROF, MACDONALD COL, MCGILL UNIV, 86- *Concurrent Pos:* Dir, Biopesticides Res Lab, McGill Univ, 87- *Honors & Awards:* Medal Agron Distinction, Order Agronomists Que, 85. *Mem:* Weed Sci Soc Am; Soc Nematologists; Can Phytopath Soc. *Res:* Biology and control of weeds with emphasis on biological control utilizing plant pathogens. *Mailing Add:* Dept Plant Sci McGill Univ 21111 Lakeshore Rd Ste-Anne-de-Bellevue PQ H9X 1C0 Can

WATSON, ANDREW JOHN, b Mich, Aug 1, 21; m 50, 68. WEED SCIENCE. *Educ:* Mich State Univ, PhD(soil sci), 49. *Prof Exp:* agronomist, Dow Chem Co, 49-86; RETIRED. *Mem:* Weed Sci Soc Am. *Res:* Herbicide development. *Mailing Add:* 5509 Siebert St Midland MI 48640

WATSON, ANDREW SAMUEL, b Highland Park, Mich, May 2, 20; m 67; c 2. PSYCHIATRY. *Educ:* Univ Mich, BS, 42; Temple Univ, MD, 50, MS, 51. *Prof Exp:* From instr to asst prof psychiat, Med Sch, Univ Pa, 45-59, assoc prof, Law Sch, 55-59; from asst prof psychiat & asst prof law to assoc prof psychiat & assoc prof law, 59-66, PROF PSYCHIAT, MED SCH, UNIV MICH, ANN ARBOR, & PROF LAW, LAW SCH, 66- *Concurrent Pos:* Lectr social work, Sch Social Work, Bryn Mawr Col, 55-59; psychiat consult, Mich Dept Corrections, 59-; comnr, Mich Law Enforcement Criminal Justice Comn, 68-72; mem, Adv Comt Divorce, Nat Conf Comnr Uniform State Laws, 69-70; mem, Surgeon's Gen Sci Adv Comn TV & Social Behav, 69-72. *Honors & Awards:* Issac Ray Award, Am Psychiat Asn, 78. *Mem:* Group Advan Psychiat; Am Col Psychiatrists. *Res:* Family treatment; professionalizing process of lawyers; applications of psychiatric concepts to law. *Mailing Add:* 555 E William 21-D Ann Arbor MI 48104

WATSON, ANNETTA PAULE, b Pleasure Ridge Park, Ky, May 2, 48; m 75. TERRESTRIAL ECOLOGY, ENTOMOLOGY. *Educ:* Purdue Univ, West Lafayette, BS, 70; Univ Ky, PhD(entom), 76. *Prof Exp:* Consult res, Oak Ridge Nat Lab, Union Carbide Corp, 75-76; vis scientist, Div Entom, Commonwealth Sci & Indust Res Orgn, Australia, 76; guest scientist, 77, RES STAFF, HEALTH & SAFETY RES DIV, OAK RIDGE NAT LAB, MARTIN MARIETTA ENERGY SYSTS, 77- *Mem:* Nat Asn Environ Prof; AAAS; Sigma Xi. *Res:* Acute and long term toxicity of chemical warfare agents and their breakdown products; assessment of energy technologies on human systems, human ecology, insect ecology, and entomology; litter decomposition; industry development; industry development. *Mailing Add:* Health & Safety Res Div Bldg 4500 S PO Box 2008 Oak Ridge TN 37831-6101

WATSON, BARRY, b Middlesbrough, Eng, Dec 2, 40. BIOPHYSICAL CHEMISTRY. *Educ:* Univ Bradford, BSc, 65, PhD(phys chem), 68. *Prof Exp:* Scientist, Unilever Res Ltd, 66-68; Off Saline Water fel, Mellon Inst Sci, 69-72; res chemist, Owens-Ill Inc, 72-85; RES DIR, ANATRACE INC, 85- *Mem:* Royal Soc Chem; Am Chem Soc; Soc Electroanalytical Chem. *Res:* Thermodynamics of aqueous solutions; electrochemistry; biomedical research; electroanalytical chemistry. *Mailing Add:* Anatrace Inc 1280 Dussel Dr Maumee OH 43537

WATSON, CHARLES S, b Chicago, Ill, Aug 16, 32; c 4. PSYCHOACOUSTICS. *Educ:* Ind Univ, AB, 58, PhD(psychol), 63. *Prof Exp:* Teaching assoc psychol, Ind Univ, 59-61; asst prof, Univ Tex, 62-65; from assoc prof to prof, Wash Univ, 67-76; sr res assoc, Cent Inst for Deaf, St Louis, 66-76; dir res, Boys Town Inst Commun Dis In Children, 76-; prof human commun, Med Sch, Creighton Univ, 76-; AT DEPT SPEECH & HEARING SCI, UNIV IND. *Concurrent Pos:* Ed assoc, Perception & Psychophysics, 68-73; mem, Comt Hearing & Bioacoust, Nat Res Coun-Nat Acad Sci, 70-, chmn, 81-83; assoc ed, J Acoust Soc Am, 73-78; mem, NSF Rev Panel Sensory Physiol & Perception, 78- *Mem:* Fel Acoust Soc Am; fel Am Psychol Asn; AAAS; Sigma Xi; Asn Res Otolaryngol. *Res:* Hearing and deafness; perception of complex sounds; psychophysical methods; learning, memory and selective attention in auditory perception. *Mailing Add:* Dept Speech & Hearing Sci Ind Univ Bloomington IN 47405

WATSON, CLARENCE ELLIS, JR, b Stillwater, Okla, Apr 13, 51; m 72; c 2. PLANT BREEDING, EXPERIMENTAL STATISTICS. *Educ:* NMex State Univ, BS, 72, MS, 74; Ore State Univ, PhD(crop sci), 76. *Prof Exp:* Res asst, NMex State Univ, 73-74; res asst, Ore State Univ, 74-76; from asst prof to prof agron, 76-90, PROF EXP STATIST, MISS STATE UNIV, 89- *Mem:* Am Soc Agron; Crop Sci Soc Am; Biomet Soc. *Res:* Forage grass breeding; genetics; management with particular emphasis on host-plant resistance and stress tolerance. *Mailing Add:* Dept Exp Statist Box NZ Mississippi State MS 39762

WATSON, CLAYTON WILBUR, b Neosho, Mo, Feb 24, 33; m 55; c 3. NUCLEAR ENGINEERING. *Educ:* Wash Univ, BS, 54; Iowa State Univ, MS, 55, PhD(nuclear eng), 60. *Prof Exp:* Engr, Westinghouse Atomic Power, 55-57; res assoc, Rocketdyne, 60-62; asst prof nuclear eng, Univ Fla, 62-63; MEM STAFF, LOS ALAMOS SCI LAB, UNIV CALIF, 63- *Mem:* Am Nuclear Soc. *Res:* Monte Carlo analysis; nuclear reactor physics and applications; radiation environmental analysis; nuclear applications in space; advanced space systems; nuclear weapons; advanced-technology assessment; systems analysis and evaluation; technology development planning. *Mailing Add:* 113 Barranca Rd Los Alamos NM 87544

WATSON, DAVID GOULDING, b Toronto, Ont, May 7, 29; nat US; m 61; c 5. PEDIATRICS. *Educ:* Univ Toronto, MD, 52. *Prof Exp:* From asst prof to assoc prof, 59-75, PROF PEDIAT, MED CTR, UNIV MISS, 75- *Concurrent Pos:* NIH spec fel, Univ Fla, 72-73. *Mem:* Am Acad Pediat; Am Col Cardiol; Royal Col Physicians Can. *Res:* Pediatric cardiology. *Mailing Add:* Dept Pediat Univ Miss Med Ctr Jackson MS 39216

WATSON, DAVID LIVINGSTON, b Burlington, Ont, Oct 22, 26; US citizen; m 51; c 3. ENTOMOLOGY. *Educ:* Univ Guelph, BSA, 51; Cornell Univ, PhD(entom), 56. *Prof Exp:* Res specialist, Chevron Chem Co, 55-59, coordr indust res, 59-61, res supvr, 61-66; assoc dir indust res, 66-70, DIR PROD DEVELOP DIV, VELSICOL CHEM CORP, CHICAGO, 70- *Mem:* AAAS; Entom Soc Am; Am Phytopath Soc; Weed Sci Soc Am. *Res:* Genetic relationship of phosphate resistance in mites; development of crop protection chemicals in agriculture and related fields. *Mailing Add:* 127 Whittington Course St Charles IL 60174-1435

WATSON, DENNIS RONALD, b Overton, Tex, Dec 7, 41; m 70; c 1. CHEMISTRY. *Educ:* Howard Payne Col, BA, 64; Univ Colo, Boulder, MS, 67, PhD(chem), 70. *Prof Exp:* From asst prof to assoc prof, 70-81, PROF CHEM, LA COL, 81-, CHMN DEPT, 70- *Mem:* Am Chem Soc. *Res:* Air and water pollution topics that can be performed by senior level students dealing with local problems and conditions. *Mailing Add:* Dept Chem La Col Pineville LA 71360

WATSON, DENNIS WALLACE, b Morpeth, Ont, Apr 29, 14; nat US; m 41; c 2. MICROBIOLOGY. *Educ:* Univ Toronto, BSA, 34; Dalhousie Univ, MSc, 37; Univ Wis, PhD(bact), 41. *Hon Degrees:* DSc, Univ Wis-Madison, 81. *Prof Exp:* Asst, Biol Bd Can, NS, 35-37, sci asst, 37-38; asst bact, Univ Wis, 38-41, Alumni Res Found fel, 41-42, res assoc, 42; vis investr, Rockefeller Inst, 42; vis investr, Connaught Labs, Toronto, 42-44; asst prof bact, Univ Wis, 46-49; from assoc prof to prof bact, 49-64, Regents prof, 80, prof & head dept, 64-84, REGENTS EMER PROF MICROBIOL, MED SCH, UNIV MINN, MINNEAPOLIS, 84- *Concurrent Pos:* Med consult, Fed Security Agency, Washington, DC, 44; assoc mem comn immunization, Armed Forces Epidemiol Bd, 46-59; vis prof, Med Sch, Univ Wash, 50; mem allergy & immunol study sect, NIH, 54-58, mem bd sci counr, Div Biol Standards, 57-59, mem allergy & immunol training grant comt, Nat Inst Allergy & Infectious Dis, 58-60 & 64-66, chmn, 66, mem nat adv coun, 67-71; USPHS spec res fel, WGer, 60-61; mem ad hoc comt med microbiol, Div Med Sci, Nat Acad Sci; vis investr, Osaka Univ, Japan, 71; vis instr, WHO Immunol Course, High Inst Pub Health, Alexandria, Egypt, 81. *Mem:* AAAS; Am Chem Soc; Am Soc Microbiol (vpres, 67-68, pres, 68-69); Soc Exp Biol & Med (pres, 75-76); Infectious Dis Soc Am; Am Asn Immunologists. *Res:* Host-parasite relationships; chemistry and immunology of microbial toxins; pathogenesis of group A streptococci; mechanisms of nonspecific resistance to infection. *Mailing Add:* Med Sch Dept Microbiol Univ Minn Box 196 Mayo Minneapolis MN 55455

WATSON, DONALD PICKETT, b Port Credit, Ont, May 20, 12; nat US; m 55. HORTICULTURE. *Educ:* Univ Toronto, BSA, 34; Univ London, MSc, 37; Cornell Univ, PhD(plant anat), 48. *Prof Exp:* Instr hort, State Univ NY Agr Inst, Long Island, 37-42; from asst prof to prof hort & head ornamental hort, Mich State Univ, 48-63; prof hort, 63-75, head dept, 63-66, urban horticulturist, 66-75, EMER PROF HORT, UNIV HAWAII, 75- *Mem:* Am Hort Soc. *Res:* Plant science and anatomy; horticultural television and therapy. *Mailing Add:* 5443 Drover Dr San Diego CA 92115

WATSON, DUANE CRAIG, b Enid, Okla, Dec 8, 30; m 51; c 3. ANALYTICAL CHEMISTRY. *Educ:* Eastern NMex Univ, BA, 51, BS, 56, MS, 57. *Prof Exp:* Res chemist, El Paso Natural Gas Prod, Tex, 57-63; prof res chemist, Philip Morris, Inc, 63-74, sr scientist, Philip Morris, USA, 74-90, SEC LEADER, PHILIP MORRIS, USA, 90- *Concurrent Pos:* Lectr, Univ Tex, El Paso, 61-63. *Res:* Smoke chemistry; trace analyses; pollution; process instrumentation; gas chromatography. *Mailing Add:* Philip Morris USA Box 26583 Richmond VA 23261

WATSON, EARL EUGENE, b Sterling, Ill, July 27, 39; m 78; c 5. ACOUSTICS. *Educ:* Fla State Univ, BS, 62, MS, 63; Pa State Univ, PhD(eng acoust), 71. *Prof Exp:* Res engr, Piezeo-Technol, 62; instr eng sci, Fla State Univ, 63-67; res asst, Pa State Univ, 71-74, asst prof acoust, 74-75; mgr acoust, Wolverine Div, UOP Inc, 75-78; PROG MGR, WYLE LAB, 78- *Concurrent Pos:* Res engr, Recon, Inc, 63-65. *Mem:* Audio Eng Soc Inc. *Res:* Acoust holography; noise control. *Mailing Add:* 5645 Mary Lane Dr San Diego CA 90250

WATSON, EDNA SUE, b Batesville, Miss, July 8, 45; m 65; c 3. IMMUNOLOGY. *Educ:* Univ Miss, BA, 67, MS, 70, PhD(biol), 75. *Prof Exp:* RES ASSOC, RES INST PHARMACEUT SCI, SCH PHARM, UNIV MISS, 74- *Mem:* AAAS; Can Soc Immunol; Brit Soc Immunol. *Res:* Poison ivy dermatitis, desensitization and immunoprophylaxis; cellular immunology, manipulation of the immune response by cyclic nucleotides. *Mailing Add:* Chem Dept Coulter Hall Univ Miss University MS 38677-0013

WATSON, EVELYN E, b Corbin, Ky, Dec 15, 28; m 53; c 2. INTERNAL DOSIMETRY, MATHEMATICAL MODELING. *Educ:* Univ Ky, BA, 49. *Prof Exp:* Res assoc, Med Div, Radiopharmaceut Internal Dose Info Ctr, Oak Ridge Assoc Univs, 71-77, scientist, 77-81, prog mgr, Manpower Educ, Res & Training Div, 81-89, PROG DIR, RADIOPHARMACEUT INTERNAL DOSE INFO CTR, MED SCI DIV, OAK RIDGE ASSOC UNIVS, 89- *Concurrent Pos:* Mem, Med Internal Radiation Dose Comt, 80- & Task Group Nat Coun Radiation Protection & Measurements Sci Comt 57 on Placental Transfer, 85-; consult, Food & Drug Admin Radiopharmaceut Adv Comt, 83- *Honors & Awards:* Spec Award, Excellence Technol Transfer, Fed Lab Consortium, 85. *Mem:* Soc Nuclear Med; Health Physics Soc; Sigma Xi. *Res:* Calculation of radiation dose to normal persons and to patients with various diseases; improve estimation of dose; disseminate information to the nuclear medicine community. *Mailing Add:* Radiopharmaceut Internal Dose Info Ctr Oak Ridge Assoc Univs PO Box 117 Oak Ridge TN 37831-0117

WATSON, FLETCHER GUARD, b Baltimore, Md, Apr 27, 12; m 35; c 4. SCIENCE EDUCATION. *Educ:* Pomona Col, AB, 33; Harvard Univ, MA, 35, PhD(astron), 38. *Prof Exp:* Instr & asst astron, Harvard Univ, 33-38, exec secy & res assoc, 38-41; instr, Radcliffe Col, 41; tech aide, Nat Defense Res Corp Radiation Lab, Mass Inst Technol, 42-43; from asst prof to assoc prof sci educ, 46-57, prof educ, 57-66, Henry Lee Shattuck prof, 66-78, EMER PROF EDUC, HARVARD UNIV, 78- *Concurrent Pos:* Ford Found fel, Europe, 64-65; co-dir, Harvard Proj Physics, 64-; prof educ, NY Univ, 78-81. *Honors & Awards:* Distinguished Serv Citation, Nat Sci Teachers Asn, 72. *Mem:* AAAS; Nat Sci Teachers Asn; Am Asn Physics Teachers; Nat Asn Res Sci Teaching; Asn Educ Teachers in Sci (pres, 63-64); AAAS; Sigma Xi. *Res:* Development and evaluation of new high school physics course; studies of development of science teachers and influence of various teacher-types on pupils. *Mailing Add:* 24 Hastings Rd Belmont MA 02178

WATSON, FRANK YANDLE, b Charlotte, NC, May 18, 25; m 49; c 4. PATHOLOGY. *Educ:* Univ Md, MD, 49; Am Bd Path, dipl, 58. *Prof Exp:* Res physician path, Charlotte Mem Hosp, NC, 52-56; from instr to asst prof path, Col Med, State Univ NY Downstate Med Ctr, 56-61; assoc pathologist, 61-72, DIR LABS, MOUNTAINSIDE HOSP, NJ, 72-; CLIN ASST PROF PATH, COL MED, STATE UNIV NY DOWNSTATE MED CTR, 61- *Concurrent Pos:* Surg pathologist, Inst Path, Kings County Med Ctr, NY, 56-61. *Mem:* Fel Col Am Pathologists; fel Am Soc Clin Pathologists; AMA; Asn Am Med Cols. *Res:* General and surgical pathology. *Mailing Add:* 33 Washington St No 7A Newark NJ 07102

WATSON, GARY HUNTER, b Lewisburg, Tenn, June 28, 51; m 81; c 2. REPRODUCTIVE ENDOCRINOLOGY, STEROID HORMONE ACTION. *Educ:* Univ SC, BS, 74; Med Col Ga, PhD(endocrinol), 82. *Prof Exp:* Postdoctoral fel, Tex Tech Health Sci Ctr, 82-85; res scientist endocrinol, Col Osteop Med & Surg, 85-87, asst prof biochem & endocrinol, 87-91, ASSOC PROF BIOCHEM ENDOCRINOL/MOLECULAR BIOL, COL OSTEOP MED, OKLA STATE UNIV, 91- *Concurrent Pos:* Actg dir res, Col Osteop Med & Surg, Okla State Univ, 85-87, dir res, Col Osteop Med, 87-91; adj prof, Col Vet Med, Dept Physiol, Okla State Univ, 88; prin investr, Okla Ctr Advan Sci & Technol, 89. *Mem:* AAAS; NY Acad Sci; Sigma Xi. *Res:* Mechanism of steroid hormone action as it relates to reproduction, cancers of reproductive systems, and adipose accretion and metabolism; steroid control of gene expression. *Mailing Add:* Col Osteop Med Okla State Univ 1111 W 17th St Tulsa OK 74107

WATSON, GEOFFREY STUART, b Bendigo, Australia, Dec 3, 21; m 53; c 4. MATHEMATICAL STATISTICS. *Educ:* Univ Melbourne, BA, 42; NC State Col, PhD, 52. *Hon Degrees:* DSc, Univ Melbourne, 67. *Prof Exp:* Res officer, Commonwealth Sci & Indust Res Orgn, Australia, 43; tutor math, Trinity Col, Univ Melbourne, 44-47; res officer appl econ, Cambridge Univ, 49-51; sr lectr statist, Univ Melbourne, 51-54; sr fel, Australian Nat Univ, 55-58; res assoc math, Princeton Univ, 58-59; assoc prof, Univ Toronto, 59-62; prof statist & chmn dept, Johns Hopkins Univ, 62-68; on leave to inst genetics, Univ Pavia, 68-69; chmn dept, 70-79, PROF STATIST, PRINCETON UNIV, 70- *Concurrent Pos:* Guggenheim fel, 76-77. *Mem:* Fel AAAS; fel Inst Math Statist; fel Am Statist Asn; fel Royal Statist Soc; fel Int Statist Inst. *Res:* Application of mathematics, especially probability theory, stochastic processes and statistics, to science; statistical methods in geophysics. *Mailing Add:* Dept Statist Fine Hall Princeton Univ Princeton NJ 08544

WATSON, GEORGE E, III, b New York, NY, Aug 13, 31; m 66; c 2. ZOOLOGY, ORNITHOLOGY. *Educ:* Yale Univ, BA, 53, MS, 61, PhD(biol), 64. *Prof Exp:* From asst cur to cur ornith, Nat Mus Natural Hist, Smithsonian Inst, 62-67, chmn dept vert zool, 67-72, cur vert zool, 72-85; assoc pathobiol, Sch Pub Health & Hyg, Johns Hopkins Univ, 70-85; RETIRED. *Concurrent Pos:* Mem, Seabird Comt, 66-, Nomenclature Comt, 78-, Int Ornith Cong & Comt Res & Explor, Nat Geog Soc, 75- *Mem:* Fel AAAS; fel Am Ornith Union (secy, 73-77, vpres, 84-85); Brit Ornith Union; corresp mem Ger Ornith Soc; Sigma Xi. *Res:* Marine ornithology, especially Antarctica; systematics of birds of Palearctic and Oriental realms. *Mailing Add:* 4323 Cathedral Ave NW Washington DC 20016

WATSON, HAL, JR, b Jacksonville, Tex, Dec 27, 39; m 63; c 2. ENGINEERING MECHANICS. *Educ:* Columbia Univ, BA, 62; Univ Tex, Austin, MS, 65, PhD(eng mech), 67. *Prof Exp:* Res engr, Eng Mech Res Lab, Univ Tex, Austin, 64-67; from asst prof to assoc prof solid mech, 67-76, ASSOC PROF CIVIL & MECH ENG, INST TECHNOL, SOUTHERN METHODIST UNIV, 77- *Concurrent Pos:* NSF study grant, 68-69; Dept Defense-Off Naval Res consult grant, Dept Statist, Southern Methodist Univ. *Mem:* Am Soc Mech Engrs; Soc Exp Stress Anal; Am Inst Aeronaut & Astronaut; Am Soc Eng Educ; Sigma Xi. *Res:* Acoustics; wave propagation in solids; dynamic properties of materials in high pressure and high temperature environments. *Mailing Add:* Southern Methodist Univ Dallas TX 75205

WATSON, HUGH ALEXANDER, b Ottawa, Ont, Oct 8, 26; US citizen; m 58; c 2. PHYSICS. *Educ:* Univ Toronto, BS, 48; McGill Univ, MS, 49; Mass Inst Technol, PhD(physics), 52. *Prof Exp:* Mem staff div indust coop, Mass Inst Technol, 51-52; mem tech staff, 52-88, DEPT HEAD, PROCESSING TECHNOL & FABRICATION, AT&T BELL LABS, 88- *Mem:* Sr mem, Inst Elec & Electronics Engrs. *Res:* Electron lithography and x-ray lithography. *Mailing Add:* Bell Labs Inc Murray Hill NJ 07974

WATSON, J(AMES) KENNETH, b Ada, Okla, Sept 23, 29; m 58; c 3. APPLICATIONS OF MAGNETISM, POWER ELECTRONICS. *Educ:* Univ Okla, BS, 51; Mass Inst Technol, SM, 55; Rice Univ, PhD, 66. *Prof Exp:* Observer, Seismic Eng Co, 51-52; teaching asst elec eng, Mass Inst Technol, 52-54; electronics engr, Gen Electronic Labs, Inc, 54-56; proj engr, Systron-Donner Co, 56-58; asst prof elec eng & proj engr, OSAGE Comput Lab, Univ Okla, 58-63, consult dept prev med & pub health, Med Ctr, 61-63; chief engr, Rice Comput Proj, Rice Univ, 63-66; assoc prof, 66-79, PROF ELEC ENG, UNIV FLA, 79- *Concurrent Pos:* Vis res physicist, Magnetism & Metall Div, US Naval Ord Lab, Md, 72-73; vis prof elec eng, Calif Inst Technol, 80-81; lectr & consult, E J Bloom Assocs, 85-86. *Mem:* Sr mem Inst Elec & Electronics Engrs. *Res:* Characterization and modeling of magnetic materials for device and circuit applications; ferrite transformers; power electronics. *Mailing Add:* Dept Elec Eng Univ Fla Gainesville FL 32611-2030

WATSON, JACK ELLSWORTH, b Robertsdale, Pa, Apr 17, 38; m 62; c 3. GENETICS, HUMAN GENETICS. *Educ:* Shippensburg State Col, BS, 61; Purdue Univ, MS, 63, PhD(genetics), 67. *Prof Exp:* From asst prof to assoc prof, 65-77, PROF ZOOL & ENTOM, AUBURN UNIV, 77- *Concurrent Pos:* Lectr, St Bernard Col, 70; consult, Pub Sch Sci Prog, Miss. *Mem:* Southern Genetics Group; Genetics Soc Am; Am Genetic Asn; Sigma Xi; AAAS. *Res:* Structure, function and manipulation of human metaphase chromosomes; effects of heterochromatin on chromosome pairing; genetics and cytogenetics of sickle cell families; inheritance of threshold characters in plants and animals. *Mailing Add:* Dept Zool & Entom Auburn Univ Auburn AL 36849

WATSON, JACK SAMUEL, b Oliver Springs, Tenn, Oct 18, 35; m 60; c 2. CHEMICAL ENGINEERING. *Educ:* Univ Tenn, BS, 58, MS, 62, PhD(chem eng), 67. *Prof Exp:* DEVELOP ENGR, CHEM TECHNOL DIV, OAK RIDGE NAT LAB, 58- *Mem:* Am Inst Chem Engrs. *Res:* Mass transfer; ion exchange; solvent extraction; fluid mechanics; heat transfer. *Mailing Add:* 349 Sevenoaks Dr Knoxville TN 37922-3402

WATSON, JACK THROCK, b Casey, Iowa, May 2, 39; m 66; c 2. ANALYTICAL CHEMISTY. *Educ:* Iowa State Univ, BS, 61; Mass Inst Technol, PhD(anal chem), 65. *Prof Exp:* Res chemist, US Air Force Sch Aerospace Med, 65-68; asst prof, Sch Med, Vanderbilt Univ, 69-72, assoc prof pharmacol, 73-79; PROF BIOCHEM & CHEM, MICH STATE UNIV, 80- *Concurrent Pos:* Fel, Univ Strasbourg, 68-69. *Mem:* Am Soc Mass Spectrometry; Am Chem Soc; Sigma Xi. *Res:* Gas chromatography in separation and mass spectrometry in elucidation of structure of biologically significant molecules; prostaglandins; biogenic amines; selective detection of drugs with gas chromatography-mass spectrometry computer systems; biochemical applications of gas chromatography. *Mailing Add:* Biochem Dept Mich State Univ East Lansing MI 48824

WATSON, JAMES DEWEY, b Chicago, Ill, Apr 6, 28; m; c 2. HUMAN GENOME RESEARCH. *Educ:* Univ Chicago, BS, 47; Ind Univ, PhD(zool), 50. *Hon Degrees:* DSc, Univ Chicago, 61, Ind Univ, 63, Long Island Univ, 70, Adelphi Univ, 72, Brandeis Univ, 73, Albert Einstein Col Med, 74, Hofstra Univ, 76, Harvard Univ, 78, Rockefeller Univ, 80, Clarkson Col, 81, State Univ NY, 83, Rutgers Univ, 88, Bard Col, 91; LLD, Notre Dame Univ, 65; MD, Buenos Aires, Arg, 86. *Prof Exp:* Res staff men, Univ Copenhagen, 50-51, Cavendish Lab, Cambridge Univ, 51-52 & 55-56; sr res fel biol, Calif Inst Technol, 53-55; from asst prof to prof biol, Harvard Univ, 56-76; assoc dir, 88-89, DIR, NAT CTR HUMAN GENOME RES, NIH, 89-; DIR COLD SPRING HARBOR LAB, 68- *Honors & Awards:* Nobel Prize in Med, 62; Eli Lilly Award, 60; Albert Lasker Prize, Am Pub Health Asn, 60; John Collins Warren Prize, Mass Gen Hosp, 59; John J Carty Gold Medal, Nat Acad Sci, 71; Presidential Medal of Freedom, 77. *Mem:* Nat Acad Sci; Am Asn Cancer Res; AAAS; Am Soc Biol Chemists; Am Philos Soc. *Res:* Zoology; biology; human genome. *Mailing Add:* Nat Ctr Human Genome Res NIH Bldg 38A Rm 605 Bethesda MD 20892

WATSON, JAMES E, JR, b Red Springs, NC, Jan 10, 38; m 62. HEALTH PHYSICS, RADIATION PROTECTION. *Educ:* NC State Univ, BS, 60, MS, 62; Univ NC, PhD(environ sci & eng), 70. *Prof Exp:* Nuclear engr, US Army Ballistic Res Labs, 62-67; health physicist, Oak Ridge Nat Lab, 67; br chief health physics, Tenn Valley Authority, 70-74; PROF HEALTH PHYSICS & DIR RADIOL HYG PROG, UNIV NC, 74- *Concurrent Pos:* Mem, Task Force Low-Level Radioactive Waste Mgt, US Dept Energy, 80-81; chmn, NC Radiation Protection Comn, 82-83 & Radiol Health Sect, Am Pub Health Asn, 85; nat lectr, Sigma Xi, 91-92. *Mem:* Fel Health Physics Soc (pres, 85); Am Pub Health Asn; Sigma Xi. *Res:* Radiological impact assessments of natural sources of radiation, nuclear power generation and low-level radioactive waste management; environmental and occupational radiation surveillance. *Mailing Add:* Dept Environ Sci & Eng Univ NC Chapel Hill NC 27599-7400

WATSON, JAMES FREDERIC, b Port Huron, Mich, Aug 26, 31; m 52; c 3. METALLURGY, CERAMICS. *Educ:* Univ Mich, BS, 53, MS, 56, PhD(metall eng), 58. *Prof Exp:* Metallurgist, Magnesium Div, Dow Chem Co, Mich, 53 & Ballistic Res Lab, Aberdeen Proving Ground, Md, 54-55; staff scientist, Convair-Astronaut Div, Gen Dynamics Corp, 58-62; asst chmn metall, 62-68, mgr mat & processes lab, 68-70, assoc dir res & develop div, 70-71, DEPT MGR MAT SCI, GULF GEN ATOMIC, 71- *Concurrent Pos:* Lectr, Univ Calif, Los Angeles, 59-60, 61-62; mem mat adv bd, comt eval mat, Nat Acad Sci, 60-61. *Mem:* Am Soc Metals; Brit Inst Metals. *Res:* High temperature materials for nuclear reactors; materials and processes for components of steam power plants; properties of materials for missiles and spacecraft at cryogenic temperatures. *Mailing Add:* 4961 Quincy San Diego CA 92109

WATSON, JAMES RAY, JR, b Anniston, Ala, Dec 6, 35; m 60; c 3. PLANT TAXONOMY. *Educ:* Auburn Univ, BS, 57, MS, 60; Iowa State Univ, PhD(bot), 63. *Prof Exp:* Res asst agron, Auburn Univ, 58-60; from asst prof to assoc prof bot, 63-78, PROF BIOL SCI, MISS STATE UNIV, 78- *Concurrent Pos:* NSF fel, 64-65. *Mem:* Sigma Xi. *Res:* Woody flora of Mississippi. *Mailing Add:* Dept Biol Sci Miss State Univ PO Drawer GY Mississippi State MS 39762-5759

WATSON, JEFFREY, b Butterknowle, Eng, Oct 4, 40. RESOURCE MANAGEMENT, INFORMATION SCIENCE. *Educ:* Univ Durham, BSc, 62, dipl educ, 63, PhD(zool), 66. *Prof Exp:* Scientist, 66-71, dep ed, Off of the Ed, 71-79, ED-IN-CHIEF, CAN J FISH & AQUATIC SCI, FISHERIES RES BD CAN, 79- *Concurrent Pos:* Consult. *Mem:* Am Fisheries Soc; Am Soc Info Sci; Coun Biol Ed (pres, 81-82). *Mailing Add:* Jeffrey Watson Assoc 1331 Brenton Halifax ON B3J 2K5 Can

WATSON, JERRY M, b Independence, Kans, June 26, 42; m 62. ACCELERATOR DEVELOPMENT & LINEAR ACCELERATORS, FREE-ELECTRON LASERS. *Educ:* Univ Chicago, BS, 64, MS, 65, PhD(physics), 71. *Prof Exp:* Asst physicist, Argonne Nat Lab, 70-77, physicist, 77-82, group leader, heavy ion fusion & accelerator develop, 80-82; staff mem, Los Alamos Nat Lab, 82-89, group leader, free electron lasers, 83-89, assoc div leader, accelerator technol, 85-89, group leader, LINAC, 89-90, PHYSICIST, SUPERCONDUCTING SUPER COLLIDER LAB, LOS ALAMOS NAT LAB, 89-, DEP DIV LEADER, ACCELERATOR SYSTS, 90- *Mem:* Am Phys Soc; AAAS. *Res:* Development of accelerators for various applications, high energy physics, free-electron lasers, and heavy ion fusion; design of accelerator systems. *Mailing Add:* 1001 Plantation Dr De Soto TX 75115-5281

WATSON, JOHN ALFRED, b Chicago, Ill, May 21, 40; m 60; c 4. BIOCHEMISTRY. *Educ:* Ill Inst Technol, BS, 64; Univ Ill, Chicago, USPHS fel & PhD(biochem), 67. *Prof Exp:* USPHS fel biochem, Brandeis Univ, 67-69; asst prof, 69-76, assoc dean admis med, 73-80, PROF BIOCHEM, UNIV CALIF, SAN FRANCISCO, 76- *Concurrent Pos:* From asst dean, to assoc dean student affairs, 69-73; estab investr, Am Heart Asn, 80- *Mem:* Am Oil Chem Soc; Am Soc Biol Chemists; Sigma Xi; Tissue Cult Asn; Nat Inst Sci. *Res:* Regulation of sterol and non-sterol isopentenoid synthesis in isolated cultured vertebrate and invertebrate cells. *Mailing Add:* 78 Santa Ana Daly City CA 94015

WATSON, JOHN ERNEST, audiology, for more information see previous edition

WATSON, JOHN H L, b May 27, 16; m; c 3. ELECTRON MICROSCOPY. *Educ:* McMaster Univ, Hamilton, Ont, BA(math & physics), 39; Univ Toronto, MA, 40, PhD(physics), 43. *Prof Exp:* Lectr physics, Univ BC, 43-45; physicist, Shawinigan Chem Co, 45-47; chmn dept physics, Edsel B Ford Inst Med Res, Henry Ford Hosp, Detroit, 47-81; RETIRED. *Concurrent Pos:* Statist officer, Electron Micros Soc Am. *Mem:* Electron Micros Soc Am (pres, 58). *Mailing Add:* 652 Hupp Cross Birmingham MI 48010

WATSON, JOHN THOMAS, b Indianapolis, Ind, Jan 9, 40; m 64, 86; c 6. CARDIOVASCULAR PHYSIOLOGY, BIOMEDICAL ENGINEERING. *Educ:* Univ Cincinnati, BSME, 62; Southern Methodist Univ, MSME, 66; Univ Tex Southwestern Med Sch, PhD(physiol), 72. *Prof Exp:* Student engr, Indianapolis Power & Light Co, 57-59; design consult, Nat Cash Register Co, 59-62; systs engr, Ling-Temco Vought, Inc, 62-66; teaching asst physiol, Univ Tex Health Sci Ctr, Dallas, 66-69, adj instr, 69-71, instr thoracic & cardiovasc surg & physiol, 71-74, asst prof surg & physiol, 74-76, asst prof, Grad Sch Biomed Sci, 73-80. *Concurrent Pos:* Chief devices & technol br, Nat Heart, Lung & Blood Inst, 76-; trustee, Int Soc Artificial Organs. *Honors & Awards:* Spec Recognition Award, Pub Health Serv. *Mem:* Am Heart Asn; Am Soc Artificial Internal Organs; Am Soc Mech Engrs; Int Soc Artificial Organs. *Res:* Circulatory assistance, ischemic heart disease. *Mailing Add:* Nat Heart Lung & Blood Inst Devices & Technol Br Fed Bldg Rm 312A Bethesda MD 20892

WATSON, JOSEPH ALEXANDER, b Pittsburgh, Pa, July 23, 26; m 50; c 2. RADIOBIOLOGY, MICROBIOLOGY. *Educ:* Univ Pittsburgh, BS, 50, MS, 52, PhD(microbiol), 62. *Prof Exp:* Res asst radiation health, Univ Pittsburgh, 51-56; res biochemist, Radioisotope Serv, Vet Admin Hosp, Pittsburgh, Pa, 56-57; res assoc radiation health, 57-59, asst prof, 61-67, assoc prof, 67-74, prof radiobiol, 74-, assoc prof radiol, 72-, dir radiation health training prog, 68-, dir med physics training prog, 74-88, EMER PROF RADIOBIOL, GRAD SCH PUB HEALTH, UNIV PITTSBURGH, 88- *Mem:* Radiation Res Soc; Am Soc Microbiol; Am Pub Health Asn. *Res:* Pulmonary clearance of radioactive dusts; radiation effects on the lungs; metabolic control mechanisms in the cell. *Mailing Add:* 121 Mallard Dr McKees Rock PA 15136

WATSON, KENNETH, b Montreal, Que, July 25, 35; US citizen; m 59; c 2. GEOPHYSICS. *Educ:* Univ Toronto, BA, 57; Calif Inst Technol, MS, 59, PhD(geophys), 64. *Prof Exp:* Geophysicist, 63-76, chief, Br Petrophys & Remote Sensing, 76-79, RES GEOPHYSICIST, US GEOL SURV, 80- *Concurrent Pos:* Lectr, Northern Ariz Univ, 66-67; prin investr, NASA, 77-81; assoc ed, J Geophys Res, 68-70 & Geophys, 83- *Mem:* AAAS; Soc Explor Geophysicists; Am Geophys Union; Sigma Xi. *Res:* Planetary science; behavior of volatiles on the lunar surface; infrared emission and visible light reflection; terrestrial remote sensing investigations; thermal modeling; analysis of satellite thermal infrared data; remote sensing for mineral exploration. *Mailing Add:* 2054 S Moore Ct Denver CO 80227

WATSON, KENNETH DE PENCIER, b Vancouver, BC, July 19, 15; m 41; c 3. GEOLOGY. *Educ:* Univ BC, BASc, 37; Princeton Univ, PhD(geol), 40. *Prof Exp:* Instr econ geol, Princeton Univ, 40-43; assoc mining engr, Dept Mines, BC, 43-46; from assoc prof to prof geol & geog, Univ BC, 46-50; prof geol, Univ Calif, Los Angeles, 50-57; chief geologist, Dome Explor Ltd, Can, 57-58; PROF GEOL, UNIV CALIF, LOS ANGELES, 58- *Concurrent Pos:* Consult, Dome Mines Ltd, 58-; dir, Sigma Mines Ltd, 66- *Mem:* Fel Geol Soc Am; fel Mineral Soc Am; Soc Econ Geol; Mineral Asn Can. *Res:* Petrology and mineral deposits. *Mailing Add:* 2625 S Rimpau Blvd Los Angeles CA 90016

WATSON, KENNETH FREDRICK, b Pasco, Wash, Feb 17, 42; m 64; c 2. BIOCHEMISTRY, MOLECULAR VIROLOGY. *Educ:* Northwest Nazarene Col, AB, 64; Ore State Univ, PhD(biochem), 69. *Prof Exp:* Res assoc molecular virol, Inst Cancer Res, Columbia Univ, 69-71, instr, Col Physicians & Surgeons, 71-72; res assoc virol, Robert Koch Inst, 72-73; from asst prof to assoc prof biochem, Univ Mont, 73-81, prof biochem, 81-83; dir molecular virol, Abbott Labs, 83-85; vpres acad affairs, 85-87, ASST PRES, NORTHWEST NAZARENE COL, 89- *Concurrent Pos:* Fel, Nat Cancer Inst, 69-71; res fel, Int Agency Res Cancer, 72-73; fac res award, Am Cancer Soc, 76-81. *Mem:* AAAS; Sigma Xi. *Res:* Replication of viral nucleic acids; mechanism of RNA tumor virus replication and virus-induced cell transformation; characterization of reverse transcriptase and its use in gene synthesis; role of protein phosphorylation in virus life cycle. *Mailing Add:* Asst Pres Northwest Nazarene Col Nampa ID 83686

WATSON, KENNETH MARSHALL, b Des Moines, Iowa, Sept 7, 21; m 46; c 2. PHYSICS. *Educ:* Iowa State Col, BS, 43; Univ Iowa, PhD(physics), 48. *Hon Degrees:* DSc, Indiana Univ, 76. *Prof Exp:* Lab instr, Iowa State Col, 42-43; radio engr, US Naval Res Lab, Washington, DC, 43-46; instr physics, Univ Iowa, 48 & Princeton Univ, 48; AEC fel, Inst Advan Study & Radiation Lab, Univ Calif, 48-50; asst prof physics, Ind Univ, 51-53; assoc prof, Univ Wis, 53-59; prof physics, Univ Calif, Berkeley, 59-81; PROF OCEANOG, UNIV CALIF, SAN DIEGO, & DIR, MARINE PHYS LAB, SCRIPPS INST OCEANOG, 81- *Concurrent Pos:* Staff mem, Lawrence Berkeley Lab; consult, Phys Dynamics, Inc, Mitre Corp & Sci Applications, Inc. *Mem:* Nat Acad Sci; Am Geophys Union; Am Phys Soc. *Res:* Statistical mechanics; physical oceanography. *Mailing Add:* Scripps Inst Oceanog Univ Calif Mail Code 0213 La Jolla CA 92093-0213

WATSON, MARSHALL TREDWAY, b Blacksburg, Va, Dec 27, 22; m 52; c 2. PHYSICAL CHEMISTRY, TEXTILE CHEMISTRY. *Educ:* Va Polytech Inst, BS, 43; Princeton Univ, MA, 48, PhD(phys chem), 49. *Prof Exp:* Asst, Princeton Univ, 46-48; from res chemist to sr res chemist, Eastman Kodak Co, 49-63, res assoc, 63-68, head fibers res div, 68-76, dir , Eastman Chem Div, 76-86; RETIRED. *Mem:* Am Chem Soc; Am Asn Textile Technol. *Res:* Mechanism of protein denaturation; mechanical and rheological properites of polymers; processing of polymers into fibers; structure and properties of fibers. *Mailing Add:* 1700 Longview St Kingsport TN 37660

WATSON, MARTHA F, b Janesville, Wis, Feb 2, 35. MATHEMATICS. *Educ:* Murray State Col, AB, 56; Univ Ky, MA, 58, PhD(math), 62. *Prof Exp:* Assoc prof, 62-74, PROF MATH, WESTERN KY UNIV, 74- *Mem:* Am Math Soc; Math Asn Am. *Res:* Complex analysis. *Mailing Add:* Dept Math Phys & Comput Sci Georgetown Col 400 E Main St Georgetown KY 40324

WATSON, MAURICE E, SLUDGE ANALYSIS, WATER ANALYSIS. *Educ:* Univ Nebraska, BSc, 63, MSc, 67; Univ Guelph, PhD, 72. *Prof Exp:* Lab dir, North Carolina State Univ, 72-76; LAB HEAD, OHIO STATE UNIV, 76- *Mem:* Am Soc Agron; Soil Sci Soc Am. *Res:* Plant tissue analysis. *Mailing Add:* Ohio State Univ Res-Extension Analytical Lab Wooster OH 44691

WATSON, MAXINE AMANDA, b New Rochelle, NY, May 8, 47. PLANT POPULATION BIOLOGY. *Educ:* Cornell Univ, BS, 68; Yale Univ, MDH, 70, PhD(population biol), 74. *Prof Exp:* Asst prof, Univ Utah, 75-80; asst prof, 80-84, ASSOC PROF POP BIOL, IND UNIV, 84- *Mem:* Soc Study Evolution; Ecol Soc Am; Bot Soc Am; Am Bryological & Lichenological Soc; Aquatic Plant Mgt Soc. *Res:* Regulation of population structure in clonal plants using demographic, morphological and physiological approaches; investigation of patterns of resource allocation in plants designed to identify relevant currencies. *Mailing Add:* Dept Biol Ind Univ Bloomington IN 47405

WATSON, MICHAEL DOUGLAS, b St Thomas, Ont, July 27, 36; m 59; c 3. AERONOMY. *Educ:* Univ Western Ont, BSc, 57, MS, 59, PhD(physics), 61. *Prof Exp:* Asst res officer, Herzberg Inst Astrophysics, 61-67, assoc res officer, 67-77, sr res officer, 77-81; dir asst, Can Ctr Space Sci, Nat Res Coun Can, 81-82, chief, Prog Eval Off, 82-86, mgr, Indust Develop Off, 81-89, PROJ MGR, ADMIN SERV & PROP MGT, NAT RES COUN CAN, 90- *Res:* Shock-tube excitation of powdered solids; plasma jet diagnostics; optical studies of aurora; observational studies of infrasonic waves from meteors. *Mailing Add:* 660 Sandra Ave Ottawa ON K1G 2Z8 Can

WATSON, NATHAN DEE, b Westfield, NC, Oct 14, 35; m 57; c 1. HEAT TRANSFER, DYNAMICS. *Educ:* NC State Univ, BS, 62; Polytech Inst, MS, 68; NC State Univ, PhD, 77. *Prof Exp:* Supvr aerospace engr, Langley Res Ctr, NASA, 62-90; RETIRED. *Res:* Thermal design and analysis of hypersonic aircraft and spacecraft; development of mathematical analysis techniques to predict the rigid and flexible body responses of large aerospace aircraft. *Mailing Add:* 110 Claxton Hampton VA 23696

WATSON, P(ERCY) KEITH, b Staffordshire, Eng, Dec 22, 27; m 53; c 4. ELECTRICAL ENGINEERING, PHYSICS. *Educ:* Univ Birmingham, BSc, 48, PhD(elec eng), 52. *Prof Exp:* Engr, English Elec Co, 53-54; fel, Nat Res Coun Can, 54-56; physicist res lab, Gen Elec Co, 56-62; engr, English Elec Co, 62-63; STAFF MEM, XEROX CORP, 63- *Mem:* Sigma Xi. *Res:* Dielectrics and electrostatics; electrical conduction and breakdown in liquids, solids and gases; electrophotography; electrical insulation. *Mailing Add:* Xerox Corp 800 Phillips Rd 0114-24D Webster NY 14580

WATSON, PHILIP DONALD, b Leeds, Eng, Oct 20, 41; m 65; c 2. MAMMALIAN CARDIOVASCULAR PHYSIOLOGY, CAPILLARY PERMEABILITY. *Educ:* Univ Leeds, BSc, 64; Univ Southern Calif, MS, 71, PhD(biomed eng), 75. *Prof Exp:* Res engr, Western Gear Corp, Calif, 66-69; fel physiol, Univ Calif, Davis, 75-77; asst prof, 77-80, assoc prof, 80-87, PROF PHYSIOL, UNIV SC, 87- *Concurrent Pos:* NIH fel, 72-75, 75-77; Pharmacia travel award, Microcirculatory Soc, 81; mem coun Microcirculatory Soc, 84-86. *Mem:* Microcirculatory Soc; Am Physiol Soc. *Res:* Solute and water movement between blood and tissue. *Mailing Add:* 413 Stanford Bridge Rd Columbia SC 29212-1936

WATSON, RAND LEWIS, b Denver, Colo, Aug 29, 40; m 62; c 2. CHEMICAL PHYSICS. *Educ:* Colo Sch Mines, BS, 62; Univ Calif, Berkeley, PhD(nuclear chem), 66. *Prof Exp:* Res assoc, Lawrence Radiation Lab, Univ Calif, 66-67; from asst prof to assoc prof, 69-77, PROF CHEM, TEX A&M UNIV, 77-, ASST DEAN SCI, 80- *Concurrent Pos:* Vis prof, Univ Calif, Berkeley, 78-79. *Mem:* Am Phys Soc; Am Chem Soc; Sigma Xi. *Res:* Fast electron rearrangement following multiple ionization in heavy-ion-collisions; excited state distributions in fast ions penetrating solids and gases; x-ray spectroscopy of few-electron ions. *Mailing Add:* 1307 Angelina Ct College Station TX 77840

WATSON, RAYMOND COKE, JR, b Anniston, Ala, Aug 31, 26; m 48; c 4. DEFENSE SCIENCES, ELECTRO-OPTICS. *Educ:* Jacksonville State Univ, BS; Univ Ala, MSE; Univ Fla, MS; Calif Coast Univ, PhD(eng sci). *Prof Exp:* Chief engr, Dixie Serv Co, 48-54; head dept physics & eng, Jacksonville State Univ, 54-60; vpres, eng & res, Teledyne Brown Eng, 60-70; dir, continuing educ, eng & math, Univ Ala, Huntsville, 70-76; PRES & PROF ENG & MATH, SOUTHEASTERN INST TECHNOL, HUNTSVILLE, ALA, 76- *Concurrent Pos:* Adj assoc prof, Univ Ala, Huntsville, 61-70; consult, var defense industs, 70- *Mem:* Inst Elec & Electronics Engrs; Am Inst Aeronaut & Astronaut; Inst Mgt Sci; Optical Soc Am; Int Soc Optical Engrs; Inst Indust Engrs; Opers Res Soc Am. *Res:* Defense systems, space systems & electro-optics; sensor technologies; space based defense. *Mailing Add:* Southeastern Inst Technol PO Box 1485 Huntsville AL 35807

WATSON, RICHARD ALLAN, b New Market, Iowa, Feb 23, 31; m 55; c 1. PHILOSOPHY OF HISTORICAL SCIENCES. *Educ:* Univ Iowa, BA, 53, MA, 57, PhD(philos), 61; Univ Minn, MS, 59. *Prof Exp:* Instr philos, Univ Mich, 61-64; from asst prof to assoc prof, 64-74, PROF PHILOS, WASH UNIV, 74-, SR RES ASSOC EARTH & PLANETARY SCI, 77- *Concurrent Pos:* Aerial photog, USAF, 53-55; geologist, Univ Chicago Orient Inst, 59-60, 68 & 70; consult, US Nuclear Regulatory Comn, 76. *Mem:* AAAS; Cave Res Found (pres 65-67); Philos Sci Asn; Am Philos Asn. *Res:* Landslide geomorphology in New Mexico and Iran; paleoclimatology in New Mexico, Kentucky, Iran and Turkey; Karst geomorphology in Kentucky; theoretical articles in geology, anthropology, and archeology; glacial geomorphology in the Yukon; Pleistocene geomorphology. *Mailing Add:* Dept Philos Wash Univ St Louis MO 63130

WATSON, RICHARD E, b New York, NY, Sept 30, 31; m 58; c 4. PROPERTIES OF METALS. *Educ:* Amherst Col, BA, 53; Mass Inst Technol, PhD(physics), 59. *Prof Exp:* Res asst, Mass Inst Technol, 59-60; mem staff, Avco Res & Develop, Wilmington, Mass, 60-61; NSF fel, Atomic Energy Res Estab, Harwell, Eng, 61-62 & Univ Uppsala, Sweden, 62; mem staff, Bell Labs, 63-65; SR RES PHYSICIST, BROOKHAVEN NAT LAB, 65- *Concurrent Pos:* Consult, Nat Bur Standards, 65-83; Nordic Inst Theoret Atomic Physics vis prof, Univ Gothenburg, Sweden, 70, Univ Lund, Sweden & Univ Aarhus, Denmark, 75; mem, Nat Acad Panel on Nat Need for Synchrotron Radiation Facil, 76. *Honors & Awards:* Hume-Rothery Award, Metall Soc of Am Inst Mining, Metall & Petrol Engrs, 85. *Mem:* Fel Am Phys Soc; Mat Res Soc; Metall Soc of Am Inst Mining, Metall & Petrol Engrs. *Res:* Electronic properties of metals and alloys. *Mailing Add:* Physics Dept Brookhaven Nat Lab Upton NY 11973

WATSON, RICHARD ELVIS, b Carterville, Ill, Apr 9, 12; m 36; c 2. PHYSICS. *Educ:* Southern Ill Norm Univ, BEd, 32; Univ Ill, AM, 35, PhD(physics), 38. *Prof Exp:* Instr physics, Eastern Ill Teachers Col, 38-39; sci ed, Coop Test Serv, NY, 39-40; asst prof physics, Southern Ill Teachers Col, 40-42; res technologist, Elec Div, Leeds & Northrup Co, 46-58; PROF PHYSICS, SOUTHERN ILL UNIV, CARBONDALE, 58-, CHMN DEPT, 76- *Mem:* Am Phys Teachers; Inst Elec & Electronics Eng. *Res:* Scattering of neutrons by light nuclei; scattering of fast electrons by Coulomb field; electrometer amplifiers and recorders; automatic electrical controllers; economic loading of power systems; modelling human visual response. *Mailing Add:* 517 N Almond Carbondale IL 62901

WATSON, RICHARD WHITE, JR, b Indiana, Pa, Apr 13, 33; m 58; c 3. CLINICAL LABORATORY MANAGEMENT. *Educ:* Cornell Univ, BS, 59; Rutgers Univ, MS, 61, PhD(bact), 64. *Prof Exp:* Sr lab technician, Rutgers Univ, 61-62; res microbiologist, Anheuser-Busch, Inc, 64-67 & Esso Res & Eng Co, 67-71; LAB DIR, NAT HEALTH LABS, INC, CRANFORD, NJ, 71- *Mem:* Am Soc Microbiol; Am Asn Clin Chem. *Res:* Analytical biochemistry. *Mailing Add:* 16 Walnut St New Providence NJ 07974

WATSON, ROBERT BARDEN, b Champaign, Ill, Apr 14, 14; m 41; c 3. PHYSICS, LASERS. *Educ:* Univ Ill, AB, 34; Univ Calif, Los Angeles, MA, 36; Harvard Univ, PhD(physics), 41. *Prof Exp:* Asst physics, Univ Calif, Los Angeles, 35-36; asst, Harvard Univ, 36-37, instr, 37-38, physics & commun eng, 38-40, res assoc physics, 40-41, spec res assoc, Underwater Sound Lab, 41-45; res physicist, Defense Res Lab, Univ Tex, 45-52, mil physics res lab, 52-57, from asst prof to assoc prof physics, 46-60; chief, Physics, Electronics & Mech Br, Phys & Eng Sci Div, Off Chief Res & Develop, Dept Army, 60-72, chief, Phys & Eng Sci Div, 72-75, staff officer, Directorate Army Res, Off Dept Chief Staff Res, Develop & Acquisition, 75-76; CONSULT, 76- *Mem:* Fel AAAS; fel Acoust Soc Am; Am Asn Physics Teachers; Inst Elec & Electronics Engrs; Optical Soc Am. *Res:* Electronic circuits; architectural and musical acoustics; propagation of high frequency electromagnetic radiation; semiconductors; management of broad research programs in physical and engineering sciences. *Mailing Add:* 1176 Wimbledon Dr McLean VA 22101-2938

WATSON, ROBERT FLETCHER, b Charlottesville, Va, Jan 24, 10; m 46; c 2. MEDICINE. *Educ:* Univ Va, MD, 34. *Prof Exp:* Intern, House of the Good Samaritan, Boston, Mass, 34; intern, Mass Gen Hosp, 35-36; from asst to instr, Med Col, Cornell Univ, 36-39; from asst mem to assoc mem, Rockefeller Inst, 39-46; assoc prof med, 46-50, from assoc prof to prof clin med, 50-75, EMER PROF CLIN MED, MED COL, CORNELL UNIV, 75-; CONSULT, NY HOSP, 75- *Concurrent Pos:* Asst resident, NY Hosp, 36-38, resident physician, 38, assoc attend physician, 46-50, attend physician & chief-of-serv, Vincent Astor Diag Serv, 50-75; asst resident, Hosp, Rockefeller Inst, 39-41, resident physician, 41-46; clin prof med, Med Col, Cornell Univ, 75-78. *Mem:* AAAS; Am Soc Clin Invest; Harvey Soc; Am Heart Asn; Am Rheumatism Asn. *Res:* Cardiovascular physiology; hemolytic streptococcal infection; rheumatic fever and collagen diseases. *Mailing Add:* 93 Biltmore Dr Charlottesville VA 22901

WATSON, ROBERT FRANCIS, b Knoxville, Tenn, Nov 20, 36; m 58; c 3. CHEMISTRY, SCIENCE EDUCATION. *Educ:* Col Wooster, AB, 58; Univ Tenn, Knoxville, PhD(chem), 63. *Prof Exp:* From asst prof to assoc prof chem, Memphis State Univ, 63-68; from asst prog dir to assoc prog dir, Undergrad Educ in Sci, 68-73, prog mgr, Off Exp Proj & Progs, 73-75, prog dir, Undergrad Instrnl Improvement Progs, 75-78, dep dir, 78-81, opers res analyst, US Off Mgt & Budget, 81-83, dep dir, Div Sci Educ Mat Res, 83-84, head, Off Col Sci Instrumentation, 84-87, DIR, DIV UNDERGRAD SCI, ENG & MATH EDUC, NSF, 87- *Concurrent Pos:* Res assoc, Oak Ridge Nat Lab, 65; Nat Acad Sci res assoc, Naval Stores Lab, USDA, Fla, 67-68. *Mem:* Am Chem Soc; AAAS. *Res:* Chemistry of indanes; nonclassical ions; physical properties of sulfoxides; federal programs for support of science education and scientific research; federal science policy, administration and budget analysis; science education policy. *Mailing Add:* Div Undergrad Sci Eng & Math Educ NSF Washington DC 20550

WATSON, ROBERT JOSEPH, geophysics, for more information see previous edition

WATSON, ROBERT LEE, b Scribner, Nebr, Dec 17, 31; m 53; c 4. EPIDEMIOLOGY, PUBLIC HEALTH. *Educ:* Iowa State Univ, DVM, 55; Univ Minn, MPH, 63, PhD, 73. *Prof Exp:* Jr asst vet, Ga State Dept Health, USPHS, 55-57, asst vet, Md State Dept Health, 57, vet epidemiologist, 57-60, sr asst vet, Univ Minn, 60-61, vet epidemiologist, Miss State Bd Health, 61-63; trainee epidemiol, Sch Pub Health, Univ Minn, 63-64; instr med, 64-67, asst prof epidemiol, 67-70, asst prof med & prev med, 70-71, assoc prof prev med, 71-78, chief div epidemiol & biostatist, 77, PROF PREV MED, MED CTR, UNIV MISS, 78- *Concurrent Pos:* Trainee epidemiol, Sch Pub Health, Univ Minn, 64-67; co-investr, NIH grants, 66-73, 77-82, contract, 71-76, 85- *Mem:* Soc Epidemiol Res; Asn Teachers Prev Med. *Res:* Health and health care statistics; socio-cultural-economic factors related to blood pressure levels; toxemia of pregnancy. *Mailing Add:* Dept Prev Med Univ Miss Med Ctr 2500 N State St Jackson MS 39216

WATSON, ROBERT LEE, b Plainview, Ark, Nov 8, 34; m 65; c 3. ENTOMOLOGY. *Educ:* Univ Ark, Fayetteville, BS, 56, MS, 63; Auburn Univ, PhD(entom), 68. *Prof Exp:* Instr zool, Auburn Univ, 66-67; asst prof, 67-71, ASSOC PROF BIOL, UNIV ARK, LITTLE ROCK, 71-, CHMN DEPT, 73- *Mem:* Entom Soc Am; Am Inst Biol Sci. *Res:* Taxonomy and ecology of Tabanidae aquatic ecology. *Mailing Add:* 45 Colony Rd Little Rock AR 72207

WATSON, ROBERT LOWRIE, b Morristown, Tenn, June 15, 46; m 73; c 3. PROCESS INSTRUMENTATION. *Educ:* East Tenn State Univ, BS, 68, MS, 70. *Prof Exp:* Physicist, Tenn Eastman Co, 69-75, sr physicist, 75-79, res assoc 79-81; RES ASSOC, RES LABS, EASTMAN CHEM CO, EASTMAN KODAK CO, 81- *Mem:* Instrument Soc Am; Optical Soc Am. *Res:* Design of instruments for measuring parameters of chemical streams, plastics, and man-made fibers in laboratory and plant floor environments, primarily using optical sensing techniques. *Mailing Add:* Eastman Chem Co Res PO Box 1972 Kingsport TN 37662

WATSON, RONALD ROSS, b Tyler, Tex, Dec 9, 42; m 66; c 4. IMMUNOLOGY, NUTRITION. *Educ:* Brigham Young Univ, BS, 66; Mich State Univ, PhD(biochem), 71. *Prof Exp:* Res fel immunol & microbiol, Sch Pub Health, Harvard Univ, 71-73; asst prof microbiol, Med Ctr, Univ Miss, 73-74; asst prof microbiol & immunol, Sch Med, Ind Univ, Indianapolis, 74-78; assoc prof immunol & nutrit, Purdue Univ, W Lafayette, 78-82; from res assoc prof to res prof, Dept Family & Community Med, Univ Ariz Med Sch, 82-88; DIR, ALCOHOL RES CTR, NAT INST ALCOHOL ABUSE & ALCOHOLISM, 88- *Concurrent Pos:* Collabr & vis scientist, Int Ctr Med Res, Cali, Colombia, 73-80; res adv, NIH Fel Training Prog Sexually Transmitted Dis, Ind Univ, 74-76; vis prof immunol, Brigham Young Univ, 77 & nutrit, Wash State Univ, 79; vis nutritionist & immunologist, Egyption Nutrit Inst, Cairo, Egypt, 79 & 80. *Mem:* Am Asn Immunologists; Am Soc Microbiol; Am Inst Nutrit; Sigma Xi; AAAS. *Res:* Alcohol, cocaine, and immunity; effect of malnutrition on immune systems in mice and humans; vitamins, immunomodulation and cancer prevention; drugs of abuse as cofactors in immunomodulation in AIDS. *Mailing Add:* Dept Family & Community Med Univ Ariz Med Sch Tucson AZ 85724

WATSON, STANLEY ARTHUR, b Los Angeles, Calif, Aug 30, 15; m 42; c 4. AGRICULTURAL AND FOOD CHEMISTRY, CROP TECHNOLOGY. *Educ:* Pomona Col, AB, 39; Univ Ill, AM, 42, PhD(agron), 49. *Prof Exp:* Asst bot, Univ Ill, 38-42, agron, 46-48; from jr chemist to asst chemist Northern Regional Res Lab, Bur Agr & Indust Chem, USDA, 42-46; sect leader, Res Dept, CPC Int, Inc, 48-68, asst dir, Explor Res Dept, 68-75, res scientist, Agron & Milling Dept, 75-78; coordr north-cent regional proj grain qual, Ohio Agr Res & Develop Ctr, Ohio State Univ, 78-86; RETIRED. *Mem:* Am Chem Soc; Am Asn Cereal Chem; Sigma Xi. *Res:* Composition, structure, agronomics and processing of cereal grains; industrial corn wet milled process and products. *Mailing Add:* 811 Buchholz Dr Wooster OH 44691

WATSON, STANLEY W, b Seattle, Wash, Jan 3, 21; m 52. BACTERIOLOGY. *Educ:* Univ Wash, BS, 49, MS, 51; Univ Wis, PhD(bact), 57. *Prof Exp:* Fisheries biologist, US Fish & Wildlife Serv, Wash, 52-54; res assoc, 57-71, sr scientist, 71-86, EMER SCIENTIST, WOODS HOLE OCEANOG INST, 86-; PRES, ASSOCS CAPE COD INC, 72- *Mem:* Am Soc Microbiol; Brit Soc Gen Microbiol; Am Soc Limnol & Oceanog. *Res:* Marine microbiology; nitrification; marine slime molds; fish diseases; myxobacteria. *Mailing Add:* PO Box 224 Woods Hole MA 02543

WATSON, THEO FRANKLIN, b Plainview, Ark, July 2, 31; m 60; c 4. ENTOMOLOGY, ECOLOGY. *Educ:* Univ Ark, BS, 53, MS, 58; Univ Calif, Berkeley, PhD(entom), 62. *Prof Exp:* From asst prof to assoc prof entom, Auburn Univ, 62-66; assoc prof, 66-70, PROF & ENTOMOLOGIST, AGR EXP STA, UNIV ARIZ, 70- *Mem:* Entom Soc Am. *Res:* Agricultural entomology and ecology; ecology of cotton insects and the integrated approach to pest control in cotton. *Mailing Add:* Dept Entom Univ Ariz Tucson AZ 85721

WATSON, VANCE H, b Kennett, Mo, Nov 25, 42; m 64; c 2. AGRONOMY. *Educ:* Southeast Mo State Univ, BS, 64; Univ Mo, Columbia, MS, 66; Miss State Univ, PhD(agron), 69. *Prof Exp:* Soil conservationist, Soil Conserv Serv, USDA, 63-64; res asst agron, Univ Mo, Columbia, 64-66; from asst prof to assoc prof, 66-77, PROF AGRON, MISS STATE UNIV, 77- *Mem:* Am Soc Agron; Crop Sci Soc Am; Am Forage & Grassland Coun. *Res:* Forage crop ecology and management. *Mailing Add:* 14 Pinecrest St Starkville MS 39759

WATSON, VELVIN RICHARD, b Streator, Ill, June 2, 32; m 58; c 3. GAS DYNAMICS, NUMERICAL ANALYSIS. *Educ:* Univ Calif, Berkeley, BS, 59, MS, 61; Stanford Univ, PhD(plasma physics), 69. *Prof Exp:* RES SCIENTIST GAS DYNAMICS & NUMERICAL ANALYSIS, AMES RES CTR, NASA, 61- *Concurrent Pos:* Instr, San Jose State Univ, 74-79. *Mem:* Am Inst Aeronaut & Astronaut; Asn Comput Mach; Inst Elec & Electronics Engrs. *Res:* Improving our understanding of plasma dynamics and gas dynamics by utilizing numerical analysis and computer simulations. *Mailing Add:* Mail Stop 258-2 Ames Res Ctr NASA Moffett Field CA 94035

WATSON, WILLIAM CRAWFORD, b Glasgow, Scotland, Dec 20, 27; m 54; c 4. INTERNAL MEDICINE, GASTROENTEROLOGY. *Educ:* Glasgow Univ, MB, ChB, 50, MD, 60, PhD(med), 64; FRCPS(G), 66; FRCP & FACP, 78. *Prof Exp:* Resident surg, Ballochmyle Hosp, 50-51; resident med, Glasgow Royal Infirmary, 53, sr house officer, 53-54, sr house officer cardiol, 54-55, sr registr med, 55-60, lectr, 61-66; prof, Univ EAfrica, 66-67; consult, Glasgow Royal Infirmary, 67-69; assoc prof, 69-72, PROF MED, UNIV WESTERN ONT, 72- *Concurrent Pos:* EAfrican Med Res Coun fel, Nairobi, Kenyatta Nat Hosp, 66-67; dir gastroenterol, Victoria Hosp, 69-88, chief staff, 84-90. *Mem:* Brit Soc Gastroenterol; Can Asn Gastroenterol (past pres); Can Soc Clin Invest; Am Gastroenterol Asn. *Res:* Biophysical and biochemical aspects of intestinal structure and function. *Mailing Add:* Fac Med Univ Western Ont London ON N6A 4G5 Can

WATSON, WILLIAM DOUGLAS, b Memphis, Tenn, Jan 12, 42; m 69. ASTROPHYSICS. *Educ:* Mass Inst Technol, BS, 64, PhD(physics), 68. *Prof Exp:* Res assoc, Mass Inst Technol, 68-70; res assoc, Cornell Univ, 70-72; from asst prof to assoc prof physics & astron, 72-77, PROF PHYSICS & ASTRON, UNIV ILL, URBANA, 77- *Concurrent Pos:* A P Sloan Res fel, 74. *Mem:* Am Astron Soc; Am Phys Soc; Int Astron Union. *Res:* Theoretical astrophysics; theory of the interstellar medium; atomic and molecular processes. *Mailing Add:* Dept Physics Univ Ill 227 Loomis Lab Urbana IL 61801

WATSON, WILLIAM HAROLD, JR, b Tex, Sept 2, 31; m 56; c 2. STRUCTURAL CHEMISTRY, NATURAL PRODUCTS CHEMISTRY. *Educ:* Rice Univ, BA, 53, PhD(chem), 58. *Prof Exp:* From asst prof to assoc prof, 57-64, PROF CHEM, TEX CHRISTIAN UNIV, 64-, DIR FASTBIOS LAB, 72-, CHMN, 81- *Concurrent Pos:* Vis prof, Inst Technol, Monterrey, 75-; guest prof, Univ Bonn & Heidelberg, 79. *Mem:* Am Soc Pharmacog; Am Chem Soc; Am Phys Soc; NAm Phytochem Soc; Royal Soc Chem; Sigma Xi. *Res:* Structure of biologically active molecules; phytochemical investigations of Central and South America plants; structure and reactivity of molecule exhibiting deformed pi-election system. *Mailing Add:* Dept Chem Tex Christian Univ Ft Worth TX 76129

WATSON, WILLIAM MARTIN, JR, b Annapolis, Md, May 10, 46; m 73; c 2. COATINGS POLYMERS, ORGANIZATIONAL RESOURCES & INTERNAL QUALITY CONSULTING. *Educ:* Ga Inst Technol, BS, 68; Univ Ill, MS, 72, PhD(phys chem), 73. *Prof Exp:* RES CHEMIST, ROHM & HAAS CO, 73- *Mem:* AAAS; Am Chem Soc; Sigma Xi. *Res:* Research management, quality and organizational development coordination (internal); environmental monitoring. *Mailing Add:* Hilltown Pike Hilltown PA 18927

WATSON, WYNNFIELD YOUNG, b Toronto, Ont, Feb 5, 24; m 50; c 3. ENTOMOLOGY, INVERTEBRATE ZOOLOGY. *Educ:* Univ Toronto, BA, 50, PhD(entom), 55. *Prof Exp:* Res officer, Fed Dept Forestry, 50-61; from assoc prof to prof zool, Laurentian Univ, 61-74, chmn dept biol, 67-71, dir grad studies, 69-74; chmn dept biol, 74-80, PROF ZOOL, WILFRID LAURIER UNIV, 80- *Concurrent Pos:* Nat Res Coun Can grants, 65-67. *Mem:* Entom Soc Can; Coleopterists Soc. *Res:* Systematics of Coccinellidae; carabidae of eastern North America. *Mailing Add:* Dept Biol Wilfrid Laurier Univ 75 University Ave W Waterloo ON N2L 3C5 Can

WATT, BOB EVERETT, b Tulsa, Okla, July 20, 17; m 46; c 3. HIGH ENERGY LASER DEVELOPMENT. *Educ:* Rice Univ, BA, 39, MA, 40, PhD, 46. *Prof Exp:* Staff mem radar develop, Radiation Lab, Mass Inst Technol, 41-45; staff mem well logging, Geophys Res Lab, Texaco, 46-47; staff mem nuclear physics & laser res, Los Alamos Nat Lab, 47-51, group leader, 51-62, polarized nuclear reactions, 62-70, laser fusion res, 70-77; PRES & CONSULT, WATTLAB, 77- *Mem:* Fel Am Phys Soc; AAAS; Sigma Xi. *Res:* Nuclear reactions of light elements; energy spectrum of fission neutrons; particle scattering from and nuclear reactions with polarized helium (mass no 3); high speed electronic instrumentation; nuclear reactions involving neutrons and gamma rays. *Mailing Add:* 1447 45th St Los Alamos NM 87544

WATT, DANIEL FRANK, b High River, Alta, Feb 9, 38; m 72. METALLURGY, MECHANICAL PROPERTIES OF POLYMERS. *Educ:* Univ Alta, BSc, 61; McMaster Univ, PhD(metall), 68. *Prof Exp:* Sr scientist, Dept Mines & Tech Surv, Ont, Can, 61-63; Nat Res Coun fel, Cavendish Lab, Eng, 67-69; asst prof eng mat, Univ Windsor, 69-76; gen mgr, Curtis-Hoover Industs, Houston, 77-78; assoc prof, 79-81, PROF ENG MAT, UNIV WINSOR, 81- *Concurrent Pos:* Consult, Dominion Foundries & Steel Ltd, Ont, Can, 71. *Mem:* Am Soc Metals; Am Inst Mining, Metall & Petrol Engrs; Soc Plastics Engrs. *Res:* Metal fatigue; mechanical metallurgy. *Mailing Add:* Dept Mech Eng Mat Group Univ Windsor Windsor ON N9B 3P4 Can

WATT, DAVID MILNE, JR, b Cincinnati, Ohio, July 7, 42; m 67; c 2. SURFACE CHEMISTRY, ENGINEERING MANAGEMENT. *Educ:* Princeton Univ, BSE, 64; Univ Calif, Berkeley, PhD(chem eng), 69. *Prof Exp:* Asst prof chem eng, Cornell Univ, 69-71; prod develop chemist, Procter & Gamble Co, Cincinnati, 71-77; proj leader, Cloriox Tech Ctr, 77-84; sr sect head, res & develop, Exxon Enterprises, Epidemic Div, San Jose, Calif, 84-85; mgr prog develop, Energy Res Ctr, 85, MGR PROG DEVELOP, SCH ENG, MINES DEVELOP OFF & CTR INNOVATION & BUS DEVELOP, UNIV NDAK, 87-, ASSOC PROF ENG MGT, 90- *Concurrent Pos:* Petrol Res Fund initiation grant, Cornell Univ, 69-71, NSF starter grant, 70-71; training adminr, Elec Power Res Inst, Palo Alto, Calif, 85; mgt training consult, 85- *Mem:* Am Chem Soc; Am Soc Training & Develop. *Res:* Chemical engineering; formulation of household cleaning products. *Mailing Add:* Univ NDak SEM Box 8103 Univ Sta Grand Forks ND 58202-8103

WATT, DEAN DAY, b McCammon, Idaho, Sept 21, 17; m 46; c 5. BIOCHEMISTRY. *Educ:* Univ Idaho, BS, 42; Iowa State Col, PhD(bact physiol), 49. *Prof Exp:* Res chemist, Westvaco Chlorine Prod, Inc, 42-44; instr bact, Iowa State Col, 47-49; asst microbiologist, Agr Exp Sta, Purdue Univ, 49-53; assoc prof biochem, Tulane Univ La, 53-60; assoc prof zool, Ariz State Univ, 60-63; mem staff, Midwest Res Inst, Mo, 63-69; PROF BIOCHEM, SCH MED, CREIGHTON UNIV, 69- *Concurrent Pos:* Sr res biochemist & head dept physiol sci, Southeast La Hosp, Mandeville, La, 53-60. *Mem:* AAAS; Int Soc Toxinology; Sigma Xi. *Res:* Metabolism of bacteria; pigments of fungi; biochemistry of mental disease; chemistry of animal venoms. *Mailing Add:* Dept Biochem Creighton Univ Sch Med Omaha NE 68131

WATT, JAMES PETER, b Truro, NS, Sept 22, 49. GEOLOGY, APPLIED PHYSICS. *Educ:* Dalhousie Univ, BSc, 71, MSc, 72; Harvard Univ, PhD(appl physics), 78. *Prof Exp:* Vis res fel, Coop Inst Res Environ Sci, Univ Colo, 78-79; res fel, Seismol Lab, Calif Inst Technol, 79-81; ASST PROF GEOL, RENSSELAER POLYTECH INST, 81- *Mem:* Am Geophys Union; Can Geophys Union; Soc Explor Geophysicists; AAAS; Sigma Xi. *Res:* Mechanical properties of rocks and minerals. *Mailing Add:* Dept Geol Rensselaer Polytech Inst Troy NY 12181

WATT, JOSEPH T(EE), JR, b Honolulu, Hawaii, July 16, 33; m 71; c 2. ELECTRICAL ENGINEERING. *Educ:* Rice Univ, BA, 54, BS, 55; Univ Tex, MS, 63, PhD(elec eng), 65. *Prof Exp:* Eng trainee, Gen Elec Co, 55-56, engr adv eng prog, 57-59, analyst comput sci & info retrieval, 59-60; res asst elec eng, Univ Tex, 61-65; from asst prof to assoc prof, 65-83, PROF ELEC ENG & DIR ENG CO-OP EDUC, LAMAR UNIV, 84- *Mem:* Inst Elec & Electronics Engrs. *Res:* Automatic control theory including system identification and adjustment of adaptive systems; digital systems; operations research; microcomputers. *Mailing Add:* Dept Elec Eng Lamar Univ Box 10029 Beaumont TX 77710

WATT, KENNETH EDMUND FERGUSON, b Toronto, Ont, July 13, 29; m 55; c 2. ECOLOGY. *Educ:* Univ Toronto, BA, 51; Univ Chicago, PhD(zool), 54. *Hon Degrees:* LLD, Simon Fraser Univ, 70. *Prof Exp:* Biometrician, Ont Dept Lands & Forests, 54-57; sr biometrician, Statist Res & Serv, Res Br, Can Dept Agr, 57-61; head statist res & serv, Can Dept Forestry, 61-63; assoc prof zool, 63-64, PROF ZOOL, UNIV CALIF, DAVIS, 65- *Concurrent Pos:* Consult, Sci Secretariat, Can Privy Coun, 66; sr fel, East-West Ctr, Honolulu, 75. *Honors & Awards:* Entom Soc Can Gold Medal, 69. *Mem:* Ecol Soc Am; Soc Gen Systs Res; Soc Comput Simulation; Japanese Soc Pop Ecol. *Res:* Theoretical, experimental and field ecology of fish and insects; biomathematics; applied statistics; computer simulation studies for evaluating resource management strategies; epidemiology; regional and global modelling and simulation. *Mailing Add:* Dept Zool Univ Calif Davis CA 95616

WATT, LYNN A(LEXANDER) K(EELING), b Winnipeg, Man, Oct 25, 24; m 48; c 4. ELECTRICAL ENGINEERING. *Educ:* Univ Man, BSc, 47; Univ Chicago, SM, 51; Univ Minn, PhD(elec eng), 59. *Hon Degrees:* DEng, Carleton Univ, 89. *Prof Exp:* Lectr physics, Univ Man, 48-49, 51-52; asst res off, Atomic Energy Can, Ltd, 52-55; from asst prof to prof elec eng, Univ Wash, 59-66; prof, Univ Waterloo, 66-90, dean grad studies, 72-83, actg dean res, 88-89; COORDR, ONT CTR EXCELLENCE, 87- *Concurrent Pos:* Chmn, Ont Coun Grad Studies, 76-78, exec vchmn, 83- *Mem:* Am Phys Soc; Inst Elec & Electronics Engrs; Can Asn Grad Schs (pres, 77-78); Can Asn Physicists; Sigma Xi. *Res:* Diffusion in III-IV compounds; imperfections in semiconductors; physical properties of semiconductor devices. *Mailing Add:* Dept Elec Eng Univ Waterloo Waterloo ON N2L 3G1 Can

WATT, MAMADOV HAME, b Senegal, Africa, Aug 30, 43; c 2. WATER RESOURCES. *Educ:* Univ Grenoble, France, MS, 70; DSc(fluid mech), 72. *Prof Exp:* Instr math, Lycee Voiron, France, 69-70; res asst fluid mech, Univ Grenoble, France, 70-72; teacher math, Sidewell Friends, 73-74; asst prof phys sc, Wash Tech Inst, 74-76; sr res energy, Planning Res Corp, 76-79; PROF ENVIRON SCI, UNIV DC, 79- *Concurrent Pos:* Dir & prof, DC Water Res Ctr, 79- *Mem:* Soil Conserv Serv; Int Solar Energy Soc; Am Water Works Asn. *Res:* Water resources; energy and environmental projects. *Mailing Add:* 8412 Grubb Rd Chevy Chase MD 20815

WATT, ROBERT DOUGLAS, b Santa Paula, Calif, July 7, 19; m 46; c 2. EXPERIMENTAL HIGH ENERGY PHYSICS, PHYSICS ENGINEERING. *Educ:* Univ Calif, BS, 42. *Prof Exp:* Physicist, Lawrence Radiation Lab, Univ Calif, 46-67; group leader, Stanford Linear Accelerator Ctr, Stanford Univ, 67-86; RETIRED. *Concurrent Pos:* Consult, Argonne Nat Lab & Lawrence Radiation Lab. *Mem:* Am Phys Soc. *Res:* Particle accelerator development and operation; development and use of liquid hydrogen bubble chambers for detection of high energy particles of nuclear physics; research to prove existence of magnetic monopoles. *Mailing Add:* 11117 Palos Verde Dr Cupertino CA 95014

WATT, ROBERT M, b Springfield, Ill, Aug 8, 51; m 78; c 1. IMMUNOCHEMISTRY. *Educ:* Univ Ill, Urbana-Champaign, BS, 73, MS, 76, PhD(microbiol & immunol), 78. *Prof Exp:* Teaching asst immunochem, Univ Ill, Urbana-Champaign, 74-77; NIH fel biochem, Med Ctr, Duke Univ, 78-81; ASST PROF MICROBIOL & IMMUNOL, UPSTATE MED CTR, STATE UNIV NY, 81- *Mem:* Am Chem Soc; NY Acad Sci. *Res:* The structure and organization of apoprotein B of human low density lipoprotein; the role of cell surface immunoglobulins in the activation of B lymphocytes. *Mailing Add:* E I du Pont de Nemours & Co Inc 512 Elmwood Ave Sharon Hill PA 19079

WATT, WARD BELFIELD, b Washington, DC, Oct 21, 40; m 79; c 2. EVOLUTIONARY BIOLOGY. *Educ:* Yale Univ, BA, 62, MS, 64, PhD(biol), 67. *Prof Exp:* From asst prof to assoc prof, 69-85, PROF BIOL, STANFORD UNIV, 85- *Concurrent Pos:* Mem bd trustees, Rocky Mountain Biol Lab, 71-75, 77-90 & vpres, 82-86, pres, 87-88; mem adv panel syst biol, NSF, 73-75. *Mem:* Am Soc Naturalists; Genetics Soc Am; Soc Study Evolution; fel AAAS; Arctic Inst NAm; Int Soc Study Origin Life. *Res:* Study of adaptive mechanisms and microevolutionary processes in natural insect populations from perspectives of biochemistry, physiology, genetics and ecology. *Mailing Add:* Dept Biol Sci Stanford Univ Stanford CA 94305-5020

WATT, WILLIAM JOSEPH, b Carbondale, Ill, Dec 15, 25; m 56; c 3. INORGANIC CHEMISTRY. *Educ:* Univ Ill, BS, 49; Cornell Univ, MS, 51, PhD, 56. *Prof Exp:* Asst prof chem, Davidson Col, 51-53; from asst prof to assoc prof, dean col, 66-84, PROF CHEM, WASHINGTON & LEE UNIV, 65-, DEPT HEAD, 86- *Mem:* AAAS; Am Chem Soc; Chem Soc France. *Res:* Magnesium fluoride gels; inorganic polymers; boron compounds; molten salts. *Mailing Add:* Washington & Lee Univ Lexington VA 24450

WATT, WILLIAM RUSSELL, b Camden, NJ, June 28, 20; m 46; c 3. POLYMER CHEMISTRY, COATINGS. *Educ:* Univ Pa, BA, 49; Univ Del, MS, 52, PhD(chem), 55. *Prof Exp:* Instr gen chem, Philadelphia Col Textiles & Sci, 49-51; res chemist, Am Viscose Corp, 54-60; sr res chemist, Avisun Corp, 60-64; sr res assoc polymer chem, Am Can Co, 64-74, res fel, 77-83; chief chemist, Norland Prod Co, 83-86; consult, Union Carbide & Occidental Chem Co, 86-87; RETIRED. *Mem:* Am Chem Soc. *Res:* Photochemistry; organic coatings; ultraviolet curable coatings; cellulose derivatives; stereospecific polymerization; catalysis. *Mailing Add:* 662 Cascade Dr S Mt Laurel NJ 08054

WATT, WILLIAM STEWART, b Perth, Scotland, Feb 25, 37; US citizen; m 62; c 3. PHYSICAL CHEMISTRY, SPECTROSCOPY. *Educ:* Univ St Andrews, Scotland, BSc, 59; Univ Leeds, PhD(phys chem), 62. *Prof Exp:* Cornell Prof Exp: Fel, Cornell Univ, 62-64; res chemist, Cornell Aeronaut Lab, Buffalo, NY, 64-71; head chem laser sect, Naval Res Lab, 71-73, dep head laser physics br, 73-76, head laser physics br, Optical Sci Div, 76-79; gen mgr, Wash Opers, 79-80, vpres prog develop, 80-90, SR VPRES & DIR PROGS, W J SCHAFER ASSOC, 91- *Concurrent Pos:* Assoc ed, Inst Elec & Electronics Engrs J Quantum Electronics, 78-81. *Honors & Awards:* J B Cohen Res Prize, 62. *Mem:* Am Phys Soc; Combustion Inst; Sigma Xi; Inst Elec & Electronics Engrs. *Res:* Laser physics and development; laser-induced chemistry; energy transfer and reaction rate measurements; optical diagnostics. *Mailing Add:* 6721 Pine Creek Ct McLean VA 22101

WATTENBERG, ALBERT, b New York, NY, Apr 13, 17; m 43; c 3. PHYSICS. *Educ:* City Col New York, BS, 38; Columbia Univ, MA, 39; Univ Chicago, PhD(physics), 47. *Prof Exp:* Spectroscopist, Schenley Prod, Inc, NY, 39-41; asst, Off Sci Res & Develop, Columbia Univ, 41-42; group leader, Metall Lab, Univ Chicago, 42-46; sr physicist & group leader, Argonne Nat Lab, 47-50; vis asst prof, Univ Ill, 50-51; res physicist, Nuclear Sci Lab & lectr, Mass Inst Technol, 51-58; RES PROF PHYSICS, UNIV ILL, URBANA, 66- *Concurrent Pos:* Actg dir nuclear physics div, Argonne Nat Lab, 49-50; NSF fel, Univ Rome, 62-63; vis prof, Stanford Univ, 73 & 80-81. *Honors & Awards:* "Nuclear Pioneer" Award, Soc Nuclear Med, 77; Bronze Medal, Am Nuclear Soc, 77. *Mem:* Am Phys Soc; Sigma Xi; AAAS; Hist Sci Soc. *Res:* Spectroscopy; nuclear chain reactors; photoneutron techniques; photonuclear reactions; elementary particle physics, decay of mesons, history of physics. *Mailing Add:* Dept Physics Univ Ill Urbana IL 61803

WATTENBERG, FRANKLIN ARVEY, b New York, NY, May 16, 43; m 64; c 2. MATHEMATICS. *Educ:* Wayne State Univ, BS, 64; Univ Wis-Madison, MS, 65, PhD(math), 68. *Prof Exp:* Benjamin Peirce asst prof math, Harvard Univ, 68-71; from asst prof to assoc prof, 71-79, PROF MATH, UNIV MASS, 79- *Mem:* Am Math Soc. *Res:* Differential topology; nonstandard analysis; algebraic topology; probability; mathematical economics. *Mailing Add:* Dept Math Univ Mass Amherst MA 01003

WATTENBERG, LEE WOLFF, b New York, NY, Dec 22, 21; m 45; c 6. PATHOLOGY. *Educ:* City Col New York, BS, 41; Univ Minn, Minneapolis, BM, 49, MD, 50; Am Bd Path, dipl, 56. *Prof Exp:* From instr to assoc prof, 56-66, Hill prof, 59-66, PROF PATH, MED SCH, UNIV MINN, MINNEAPOLIS, 66- *Concurrent Pos:* Lederle med fac award, 57-59. *Mem:* Histochem Soc (vpres, 66); Soc Exp Biol & Med; Am Soc Exp Path; Am Asn Pathologists & Bacteriologists; Am Asn Cancer Res. *Res:* Histochemistry; cancer research; experimental pathology. *Mailing Add:* Dept Lab Med & Path/6-133 Jackson Hall Univ Minn Sch Med Minneapolis MN 55455

WATTERS, CHRISTOPHER DEFFNER, b Ironton, Ohio, Dec 7, 39; m 67; c 3. CELL BIOLOGY. *Educ:* Univ Notre Dame, BS, 61; Princeton Univ, MA, 64, PhD(biol), 66. *Prof Exp:* Instr biol, Princeton Univ, 64-66; res assoc cell biol prog, Univ Minn, St Paul, 66-68; from asst prof to assoc prof, 73-78, chmn dept, 76-82, chmn, Div Nat Sci, 82-85, PROF BIOL, MIDDLEBURY COL, 78-, CHMN DEPT, 88- *Concurrent Pos:* Vis scientist, Physiol Lab, Cambridge Univ, 73-74; vis assoc prof, Dartmouth Col, 75; vis prof, Med Sch, Univ Colo, 80-81; Irene Heinz & John LaPorte professorship premed sci, 82-; Vis scientist, Hannah Res Inst, 87-88. *Mem:* Am Soc Cell Biologists; Biochem Soc Gt Brit; Sigma Xi. *Res:* Lactation calcium transport; structural and functional organization of cell membranes; cellular aspects of development. *Mailing Add:* Dept Biol Middlebury Col Middlebury VT 05753

WATTERS, EDWARD C(HARLES), JR, b Monroe, Mich, Feb 16, 23; m 46; c 3. MATHEMATICS, ELECTRONICS. *Educ:* Univ Notre Dame, BSEE, 43, MS, 46; Univ Md, PhD, 54. *Prof Exp:* Asst prof math, US Naval Acad, 46-55; sr engr, Bendix Aviation Corp, 55-57; fel engr, 57, supv engr, 57-58, adv engr, 58-59 & weapon control dept, 59-61, consult engr, Surface Div, 61-70, consult engr, Electronic Systs Div, 70-84, CONSULT ENGR, DEFENSE OPERS DIV, WESTINGHOUSE ELEC CORP, BALTIMORE, 84- *Res:* Radar; weapon systems; signal processing. *Mailing Add:* 133 Spa View Ave Annapolis MD 21401

WATTERS, GARY Z, b Gilson, Ill, Oct 11, 35; m 57; c 3. FLUID MECHANICS, HYDRODYNAMICS. *Educ:* Chico State Col, BS, 57; Stanford Univ, MS, 58, PhD(civil eng), 63. *Prof Exp:* From instr to asst prof civil eng, Chico State Col, 58-61; from asst prof to assoc prof civil eng, Utah State Univ, 63-77, prof, 77-80, asst dean, 72-74, assoc dean, 74-76; DEAN ENG, COMPUT SCI & TECHNOL, CALIF STATE UNIV, CHICO, 80- *Concurrent Pos:* Am Soc Eng Educ/Ford Found resident in eng pract, 71. *Mem:* Am Soc Civil Engrs; Am Soc Eng Educ; Nat Soc Prof Engrs. *Res:* Hydraulic transients; finite element methods in fluid mechanics; economic design of irrigation systems. *Mailing Add:* Col Eng Comput Sci & Technol Calif State Univ Chico CA 95929-0003

WATTERS, GORDON VALENTINE, b Winnipeg, Man, Apr 8, 28; m 57; c 3. NEUROLOGY, PEDIATRICS. *Educ:* Univ Minn, Minneapolis, BA, 51; Univ Man, MD, 57. *Prof Exp:* Asst prof neurol, Winnipeg Children's Hosp, Univ Man, 63-65; asst prof, Children's Med Ctr, Harvard Univ, 65-69; PROF NEUROL & PEDIAT, MONTREAL CHILDREN'S HOSP, MCGILL UNIV, 69- *Mem:* Am Acad Neurol; Can Neurol Soc. *Res:* Degenerative disease of nervous system; cerebrospinal fluid dynamics. *Mailing Add:* Montreal Children's Hosp Montreal PQ H3H 1P3 Can

WATTERS, JAMES I, chemistry; deceased, see previous edition for last biography

WATTERS, KENNETH LYNN, b Iowa City, Iowa, Jan 21, 39; m 64; c 3. INORGANIC CHEMISTRY, MATERIALS CHEMISTRY. *Educ:* Univ Ill, Urbana, BS, 62; Brown Univ, PhD(chem), 70. *Prof Exp:* Res assoc chem, State Univ NY Buffalo, 69-70; from asst prof to assoc prof, 70-88, PROF CHEM, UNIV WIS-MILWAUKEE, 88- *Mem:* Am Chem Soc; Coblentz Soc; AAAS. *Res:* Spectroscopic studies of transition metal complexes; Raman, resonance Raman, and infrared spectroscopies; studies of catalytic properties of transition metal complexes and metal cluster compounds; studies of thin film deposits. *Mailing Add:* Dept Chem Univ Wis Milwaukee WI 53201

WATTERS, ROBERT JAMES, b Glasgow, Scotland, July 22, 46. GEOLOGICAL ENGINEERING, ENGINEERING GEOLOGY. *Educ:* Univ Strathclyde, BS, 69; Univ London, MS, 70, PhD(eng geol), 72. *Prof Exp:* Resident eng geologist, Sir Alexander Gibb & Partners Consult Engrs, London, 72-74; proj geol engr, Dames & Moore Consult Engrs, Los Angeles, 74-77; from asst prof to assoc prof, 78-87, PROF GEOL ENG, MACKAY SCH MINES, UNIV NEV, 87- *Mem:* Int Soc Rock Mech; Inst Civil Engrs; Brit Geotech Soc; Geol Soc London; Asn Eng Geologists. *Res:* Rock mechanics applied to the design of surface and underground excavations; geotechnical documentation and analysis of soil and rock masses; site investigation techniques; ground improvement and instrumentation. *Mailing Add:* Mackay Sch Mines Univ Nev Reno NV 89557

WATTERS, ROBERT LISLE, b Everett, Wash, June 25, 25; m 48; c 2. RADIOCHEMISTRY, HEALTH PHYSICS. *Educ:* Univ Wash, BS, 50, PhD(chem), 63; Harvard Univ, MS, 59; Am Bd Health Physics, dipl, 66; recert, 81. *Prof Exp:* Engr asst, Hanford Atomic Prod Oper, Gen Elec Co, Wash, 50-52, supvr bioassay lab, 52-56, supvr radiation monitoring, 56-58; engr, Boeing Co, 59-60; res specialist radiol hazard eval, Atomic Int Div, NAm Aviation, Inc, Calif, 63-65; assoc prof radiochem, Colo State Univ, 65-72; ENVIRON RADIOACTIVITY SPECIALIST, DIV BIOMED & ENVIRON RES, US DEPT ENERGY, 72-, GROUP LEADER LAND & FRESHWATER RES, 73-, PHYS SCIENTIST, 79- *Concurrent Pos:* Lectr, Mobile Radioisotope Lab, Oak Ridge Inst Nuclear Studies, 66- *Mem:* AAAS; Health Physics Soc. *Res:* Translocation of plutonium and americium in the body; environmental behavior of polonium; environmental behavior of actinides. *Mailing Add:* 826 Flagler Dr Gaithersburg MD 20878

WATTERSON, ARTHUR C, JR, b Ellwood City, Pa, Apr 19, 38. ORGANIC CHEMISTRY, POLYMER CHEMISTRY. *Educ:* Geneva Col, BS, 60; Brown Univ, PhD(org chem), 65. *Prof Exp:* Res assoc chem, Johns Hopkins Univ, 64-65; PROF CHEM, LOWELL TECHNOL INST, 65- *Mem:* Am Chem Soc; Royal Soc Chem. *Res:* Polymer stereochemistry; nuclear magnetic resonance of macromolecules; synthesis of natural products; decomposition of n-nitroso amides; deamination reactions; nitrogen heterocycles. *Mailing Add:* Dept Chem Univ Lowell One University Ave Lowell MA 01854-1996

WATTERSON, D MARTIN, b Pensacola, Fla, Sept 9, 46. MOLECULAR BIOLOGY. *Educ:* Emory Col, BS, 69; Emory Univ, PhD(biochem), 75. *Prof Exp:* Assoc prof, 81-85, PROF PHARMACOL, VANDERBILT UNIV SCH MED, 86- *Concurrent Pos:* Investr, Howard Hughes Med Inst, 81- *Mem:* Am Soc Biol Chemists; Am Soc Cell Biol; Am Chem Soc; Biophys Soc; AAAS; Am Soc Plant Physiol. *Res:* Molecular mechanism of calcium action. *Mailing Add:* Light Hall Rm 823 Vanderbilt Univ Nashville TN 37232-0295

WATTERSON, JON CRAIG, b Kalamazoo, Mich, Nov 25, 44; m 67; c 2. PLANT PATHOLOGY. *Educ:* Carleton Col, BA, 66; Univ Wis, MS, 71. *Prof Exp:* Res asst veg path, Univ Wis, 66-71, res assoc cranberry path, 71-72; HEAD PATH DEPT, PETOSEED CO INC, 72- *Mem:* Am Phytopath Soc. *Res:* Genetics of disease resistance in vegetable crops; breeding for disease resistance in vegetable crops. *Mailing Add:* Petoseed Res Ctr Petoseed Co Inc 37437 State Hwy 16 Woodland CA 95695

WATTERSON, KENNETH FRANKLIN, b London, Eng, July 16, 29; m 68; c 2. LITHOGRAPHY. *Educ:* Univ London, BSc, 52, PhD(chem), 59. *Prof Exp:* Lect demonstr chem, Birkbeck Col, Univ London, 52-58, res asst, Imp Col, 58-60; res assoc, Cornell Univ, 60-62; sr res chemist, Pennsalt Chem Corp, 62-66; engr, Homer Res Labs, Bethlehem Steel Corp, 66-77; SR SCIENTIST PHOTOCHEM, PRINTING PROD DIV, HOECHST CELANESE CORP, 78- *Res:* Metal-gas reactions; surface analysis; ion microprobe spectrometry; auger spectroscopy; corrosion; photochemistry light-sensitive coatings; lithography. *Mailing Add:* RD 1 Box 301A Hellertown PA 18016

WATTERSON, RAY LEIGHTON, embryology; deceased, see previous edition for last biography

WATTERSTON, KENNETH GORDON, b Rockville Centre, NY, Apr 9, 34; m 61; c 2. FOREST SOILS, ENVIRONMENTAL MANAGEMENT. *Educ:* State Univ NY Col Forestry, Syracuse, BS, 59, MS, 62; Univ Wis, PhD(soils), 66. *Prof Exp:* Res asst forest soils, State Univ NY Col Forestry, Syracuse, 59-61; res asst soils, Univ Wis, 61-65; from asst prof to assoc prof forest soils, 65-75, PROF FOREST SOILS, STEPHEN F AUSTIN STATE UNIV, 75-, ASST DEAN, 87- *Concurrent Pos:* Chmn, Forest Soils Div, Soil Sci Soc Am, 80-81, bd dirs, 81-82. *Mem:* Soil Sci Soc Am; Ecol Soc Am; Soc Am Foresters. *Res:* Forest soil-site relationships; forest soil classification; soil pollution and reclamation. *Mailing Add:* Sch Forestry Box 6109 SFA Sta Nacogdoches TX 75962-6109

WATTHEY, JEFFREY WILLIAM HERBERT, b London, Eng, Dec 6, 37; m 61; c 3. ORGANIC CHEMISTRY, MEDICINAL CHEMISTRY. *Educ:* Imp Col, Univ London, BSc, 59; St Catherine's Col, Oxford, DPhil(org chem), 62. *Prof Exp:* Res fel, Dept Chem, Univ Calif, Los Angeles, 62-63; sr staff scientist, 64-83, mgr chem res opers, Pharmaceut Div, 84-89, MGR THERAPEUT AREA DATABASE SUPPORT, CIBA-GEIGY, SUMMIT, NJ, 90- *Mem:* Am Chem Soc. *Res:* Organic synthesis; synthesis of biologically active substances, with emphasis on antihypertensive and central nervous system agents. *Mailing Add:* 19 Franklin Rd Mendham NJ 07945-1807

WATTON, ARTHUR, b Dudley, Eng, Feb 12, 43. SOLID STATE PHYSICS, NUCLEAR MAGNETIC RESONANCE. *Educ:* Univ London, BSc, 65; McMaster Univ, PhD(physics), 71; Univ Chicago, SM, 75. *Prof Exp:* Fel physics, Univ Waterloo, 71-74; asst prof, 75-81, ASSOC PROF PHYSICS, UNIV VICTORIA, 81- *Concurrent Pos:* Nat Res Coun Can Grant, 75- *Mem:* Can Asn Physicists. *Res:* NMR studies of molecular motions in the solid state; solid-solid phase transitions; low temperature rotational quantum effects of molecules. *Mailing Add:* Dept Physics Univ Victoria Box 1700 Victoria BC V8W 2Y2 Can

WATTS, CHARLES D, b Atlanta, Ga, Sept 21, 17; c 4. HEALTH ADMINISTRATION. *Educ:* Morehouse Col, BS, 39; Howard Univ Col Med, MD, 43; Am Bd Surg, dipl, 50. *Hon Degrees:* DSc, Duke Univ, 91. *Prof Exp:* Intern, Freedmen's Hosp, Washington, DC, 43-44; instr surg, Howard Univ Dept Surg, 44-45, dir, Cancer Teaching Proj, Howard Univ & Freedmen's Hosp, 49; dir student health servs, NC Central Univ, 52-59; vpres & med dir, NC Mutual Life Ins Co, 60-70, sr vpres & med dir, 70-89; PVT PRACT PRIMARY CARE, 880. *Concurrent Pos:* Adv Comt on Health, NC Comn Civil Rights, 65; clin instr surg, Duke Med Ctr, 69; chmn bd, Health Systs Agency 76-79; attend surg, Watts Hosp, 68-76, Durham County Gen Hosp, 76-87; Coun Inst Med, Nat Sci Found, 80-83; founder, developer & dir, Lincoln Community Health Ctr, 70-71; pvt pract surg, 50- *Mem:* Inst Med-Nat Acad Sci; fel Am Col Surgeons. *Res:* Medicine and human relations. *Mailing Add:* 510 Simmons St Durham NC 27701

WATTS, CHARLES EDWARD, b Mo, Mar 21, 28; m 57; c 2. MATHEMATICS. *Educ:* Drury Col, MusB, 50; Univ Calif, MA, 56, PhD, 57. *Prof Exp:* Instr math, Univ Chicago, 57-69, NSF fel, 60-61; from asst prof to assoc prof, 61-69, chmn dept, 76-78, PROF MATH, UNIV ROCHESTER, 69- *Mem:* Math Asn Am. *Res:* Algebraic topology; homological algebra. *Mailing Add:* Dept Math Univ Rochester Wilson Blvd Rochester NY 14627

WATTS, DANIEL JAY, b East Cleveland, Ohio, Oct 19, 43; m 73; c 2. CHEMISTRY, BOTANY. *Educ:* Ohio State Univ, BSc, 65; Ind Univ, Bloomington, AM, 68, PhD(org chem), 69. *Prof Exp:* res investr org chem, E R Squibb & Sons, Inc, 69-77; res assoc org chem, Am Can Co, 77-81, sr res assoc org chem, 81-83; dir corp liaison, 83-85, dir opers, 85-88, DEP EXEC DIR, HAZARDOUS SUBSTANCE MGT RES CTR, NJ INST TECHNOL, 88- *Mem:* AAAS; Am Chem Soc; Water Pollution Control Fedn; Bot Soc Am; Royal Chem Soc London; Am Inst Biol Sci. *Res:* Antibiotics; anti-infective agents; organic synthesis; chemotaxonomy; natural products; resource recovery; environmental sciences; pollution prevention. *Mailing Add:* Hazardous Substance Mat Res Ctr NJ Inst Technol Newark NJ 07102

WATTS, DANIEL THOMAS, b Wadesboro, NC, July 31, 16; m 63; c 2. PHARMACOLOGY. *Educ:* Elon Col, AB, 37; Duke Univ, PhD(physiol), 42. *Prof Exp:* Asst zool & physiol, Duke Univ, 39-42; physiologist, Naval Air Exp Sta, Pa, 46-47; from asst prof to assoc prof pharmacol, Med Sch, Univ Va, 47-53; prof & head dept, Med Ctr, WVa Univ, 53-66; prof & dean, 66-82, EMER PROF PHARMACOL, SCH BASIC SCI, MED COL VA, VA COMMONWEALTH UNIV, 82- *Concurrent Pos:* Consult, Walter Reed Army Inst Res, 58-67 & res progs, Med Col Va-Va Coommonwealth Univ, 84- *Mem:* Am Soc Pharmacol & Exp Therapeut; Soc Exp Biol & Med; AAAS. *Res:* Central nervous system depressants; cardiovascular drugs; experimental shock; aerospace physiology. *Mailing Add:* 504 Ridge Top Rd Richmond VA 23229

WATTS, DENNIS RANDOLPH, b Riverside, Calif, Dec 7, 43; m 84; c 3. PHYSICAL OCEANOGRAPHY. *Educ:* Univ Calif, Riverside, BA, 66; Cornell Univ, PhD(physics), 73. *Prof Exp:* Res assoc phys oceanog, Yale Univ, 72-74; asst prof, 74-80, assoc prof phys oceanog, 80-88, PROF PHYS OCEANOG, SCH OCEANOG, UNIV RI, 88- *Mem:* AAAS; Am Geophys Union; Sigma Xi; Oceanog Soc. *Res:* Descriptive and dynamical study of ocean currents, their fluctuations such as eddies, and other processes controlling the oceanic thermocline. *Mailing Add:* Sch Oceanog Univ RI Kingston RI 02881

WATTS, DONALD GEORGE, b Winnipeg, Man, Dec 4, 33; m 58; c 2. APPLIED STATISTICS. *Educ:* Univ BC, BASc, 56, MASc, 58; Univ London, PhD(elec eng), 62. *Prof Exp:* Systs analyst, DeHavilland Aircraft Co, Can, 62-64; vis asst prof math, Univ Wis-Madison, 64-65, assoc prof statist, 65-70; PROF STATIST, QUEEN'S UNIV, ONT, 70- *Concurrent Pos:* Res fel, Rohm & Haas Co, Spring House, Pa, 84-85. *Honors & Awards:* Heaviside Premium Award, Brit Inst Elec Engrs, 61. *Mem:* Fel Am Statist Asn; Int Statist Inst; Royal Statist Soc; Statist Soc Can. *Res:* Time series analysis, control and applications of statistics in many disciplines; nonlinear estimation; teaching methods. *Mailing Add:* Dept Math & Statist Queen's Univ Kingston ON K7L 3N6 Can

WATTS, EXUM DEVER, b Nashville, Tenn, Mar 19, 26; m 48; c 3. ORGANIC CHEMISTRY. *Educ:* George Peabody Col, BS & MA, 48; Vanderbilt Univ, PhD, 54. *Prof Exp:* Instr chem, Florence State Col, 48-49; asst prof, Harding Col, 52-54; from asst prof to assoc prof, 54-60, PROF CHEM, MID TENN STATE UNIV, 60-, PROF PHYSICS, 77- *Mem:* AAAS; Am Chem Soc; Nat Sci Teachers Asn. *Res:* Organic mercurials; ultraviolet spectra; interpretation of organic spectra. *Mailing Add:* Dept Chem Mid Tenn State Univ Murfreesboro TN 37132

WATTS, JEFFREY LYNN, b Homer, La, May 5, 55; m 74; c 2. VETERINARY MICROBIOLOGY. *Educ:* La Tech Univ, Ruston, BS, 78 & MS, 83. *Prof Exp:* Microbiologist, Bienville Gen Hosp, 75-78; res assoc, Hill Farm Res Sta, 78-83, instr, 83-88, asst prof, 88-91. *Mem:* Asn Vet Microbiologists; Nat Reg Microbiologists; Am Soc Microbiol. *Res:* Epidemiology of microorganisms responsible for bovine mastitis, development of identification methods for mastitis pathogens based upon current species descriptions; virulence factors associated with mastitis pathogens. *Mailing Add:* Animal Health Therapeut Res 7923-190-NR Upjohn Co Kalamazoo MI 49001

WATTS, JOHN ALBERT, JR, b Brooklyn, NY, Nov 30, 49; m. CELLULAR PHYSIOLOGY, PHARMACOLOGY. *Educ:* Drew Univ, BA, 71; Univ Md, PhD(zool), 77. *Prof Exp:* Teaching asst marine biol, Drew Univ, 69; teaching asst physiol, Univ Md, 71-77; fel physiol, M S Hershey Med Ctr, 77-79; from asst prof to assoc prof, 79-90, PROF PHYSIOL, UNIV NC, CHARLOTTE, 90- *Concurrent Pos:* Instr seashore life, Children's Sch Sci, 71. *Mem:* Am Physiol Soc; AAAS; Int Soc Heart Res; Sigma Xi. *Res:* Possible roles of calcium and changes in mitochondrial function in the causes of cell death from myocardial ischemic injury; factors limiting reperfusion of heart tissue. *Mailing Add:* Dept Biol Univ NC Charlotte NC 28223

WATTS, MALCOLM S M, b New York, NY, Apr 30, 15; m 47; c 4. INTERNAL MEDICINE. *Educ:* Harvard Univ, AB, 37, MD, 41. *Prof Exp:* From asst clin prof to assoc clin prof med, 53-71, coordr cardiovasc bd, 52-56, actg dir cardiovasc res inst, 56-57, asst dean, 56-66, spec asst to chancellor, 63-71, dir extended progs med educ, 74-85, prof, 71-90, assoc dean, 66-90, EMER CLIN PROF MED, SCH MED, UNIV CALIF, SAN FRANCISCO, 90- *Concurrent Pos:* Ed, Calif Med, 68-73, Western J Med, 74-90 & J Continuing Educ Health Prof; proj dir, Calif Area Health Educ Ctr Syst, 79-90; vpres, Hospice San Francisco, 79-85; dir, Extended Progs Med Educ, Univ Calif, San Francisco, 74-85. *Honors & Awards:* John T McGovern Award, Am Med Writers Asn, 86. *Mem:* Nat Inst Med; AAAS; fel Am Col Physicians; Am Soc Internal Med (pres, 65); Am Acad Political & Social Soc. *Res:* Private practice of internal medicine; examination of the role of the physician in modern society; medical education; health professions education. *Mailing Add:* San Francisco Sch Med Univ Calif San Francisco CA 94143

WATTS, PLATO HILTON, JR, b Florence, SC, May 30, 41; m 75; c 1. CLINICAL CHEMISTRY, PHYSICAL CHEMISTRY. *Educ:* Furman Univ, BS, 65; Univ Md, PhD(chem), 70. *Prof Exp:* Res assoc & fel biophys, Sch Med, Univ Md, 70-71, asst prof chem, 72; res assoc mat sci, Ctr Mat Res, 71-72; teacher chem, Char-Meck Sch, 72-75; assoc dir clin chem, Diag Labs, 75-80; house officer, St Louis Childrens Hosp, 85-87. fel pediat, 87-88; DIR ALLERGY, ROCHE BIOMED, 88- *Mem:* AAAS; Sigma Xi; Am Chem Soc; Am Asn Clin Chemists. *Res:* Application of analytical and radioimmunologic techniques to problems in clinical chemistry. *Mailing Add:* 1312 Dogwood Dr Gibsonville NC 27249

WATTS, SHERRILL GLENN, b Meridian, Miss, July 15, 37; m 61; c 2. PHYSICAL CHEMISTRY. *Educ:* Miss Southern Col, BA, 59, MS, 61; Emory Univ, PhD(phys chem), 65. *Prof Exp:* Res assoc chem, Emory Univ, 65-67; asst prof, Agnes Scott Col, 67-68; assoc prof math & chem, Atlanta Baptist Col, 68-71, chem, Morehouse Col, 73-74; asst to dean sci & libr studies, Ga Inst Technol, 73-76, asst dean, 76-82; actg acad dean, 83-84, ASSOC PROF & CHAIR DEPT MATH & SCI, S GA COL, 82- *Concurrent Pos:* Prin investr, NSF Women in Sci Careers Proj, 80-83. *Mem:* Sigma Xi; AAAS; Am Asn Higher Educ; Am Chem Soc. *Res:* Nuclear magnetic resonance coupling in small molecules. *Mailing Add:* Dept Sci & Math S Ga Col Douglas GA 31533

WATTS, TERENCE LESLIE, b Leicester, Eng, May 5, 35; m 62; c 2. PHYSICS. *Educ:* Univ London, BSc, 57; Yale Univ, PhD(physics), 63. *Prof Exp:* Res asst nuclear physics, Yale Univ, 63; res assoc particle physics, Duke Univ, 63-64; mem res staff, Lab Nuclear Sci, Mass Inst Technol, 64-65, asst prof physics, 65-70; ASSOC PROF PHYSICS, RUTGERS UNIV, NEW BRUNSWICK, 70- *Mem:* Am Phys Soc. *Res:* Experimental particle physics; phenomenology; data processing of bubble chamber photographs; interactive use of computers. *Mailing Add:* Dept Physics Rutgers Univ New Brunswick NJ 08903

WATTSON, ROBERT K(EAN), JR, b Kansas City, Mo, Oct 18, 22; m 43; c 5. MECHANICAL ENGINEERING. *Educ:* Okla Agr & Mech Col, BS, 46; Mass Inst Technol, SM, 48. *Prof Exp:* Under eng aide, US Engrs, Mo, 41-42; instr mech eng, Okla Agr & Mech Col, 46-47; asst aeronaut eng, Mass Inst Technol, 47-48; assoc prof mech eng, NDak Agr Col, 48-53; proj engr res dept, Cessna Aircraft Co, 53-56; assoc prof aeronaut eng, Univ Wichita, 56, chief engr & head eng res dept, 57-60; mgr short take-off & landing aerodyn, Boeing Co, Wichita Div, 60-61, mgr vertical take-off & landing aerodyn, 61-63; chief aerodyn, Lear Jet Industs, Inc, 63-64, chief tech staff, 64, chief res & develop, Aircraft Div, 64-67; design specialist adv design-com, Beech Aircraft Corp, 67-70; assoc chmn dept aeronaut eng, Tri-State Col, 70-78; staff tech specialist, Gates Lear-Jet Corp, 78-84; chmn dept mech eng technol, Oregon Inst Technol, 85-87; ASSOC CHMN AERONAUT ENG & MECH ENG, TRI-STATE UNIV, 87- *Mem:* Soc Automotive Engrs; fel Am Inst Aeronaut & Astronaut; Sigma Xi; Am Soc Eng Ed; Nat Soc Prof Engrs. *Res:* Vertical/short takeoff and landing aerodynamics; preliminary aerodynamic design. *Mailing Add:* 5004 W Robinson Wichita KS 67212

WATWOOD, VERNON BELL, JR, b Opelika, Ala, Sept 24, 35; m 58; c 3. STRUCTURAL ENGINEERING. *Educ:* Auburn Univ, BCE, 57; Cornell Univ, MS, 61; Univ Wash, PhD(eng), 66. *Prof Exp:* Asst prof civil eng, Miss State Univ, 61-62; engr, Boeing Co, Wash, 62-64; sr res engr, Esso Prod Res Co, Tex, 66-67; eng assoc, Pac Northwest Labs, Battelle Mem Inst, 67-70; lab mgr, Res Labs, Franklin Inst, Pa, 70-73; assoc prof, 73-80, PROF CIVIL ENG & CHMN DEPT, MICH TECHNOL UNIV, 80- *Concurrent Pos:* Lectr, Grad Res Ctr, Wash, 69-70; consult, mining indust, 73- *Mem:* Am Soc Civil Engrs. *Res:* Finite element methods in stress analysis and dynamic response of structures; mathematical modeling and design of large surface mining and materials handling equipment such as draglines, shovels, conveyors and bucket-wheels. *Mailing Add:* Dept Civil Eng Mich Technol Univ Houghton MI 49931

WATZKE, ROBERT COIT, b Madison, Wis, Dec 19, 22; m 56; c 2. OPHTHALMOLOGY. *Educ:* Univ Wis, BS, 50, MD, 52; Am Bd Ophthal, dipl, 57. *Prof Exp:* Intern, Med Ctr, Ind Univ, 52-53; resident, Univ Wis, 56; res asst ophthal, Harvard Med Sch, 56-57; from asst prof to assoc prof, Col Med, Univ Iowa, 66-73, prof ophthal, 73-; AT DEPT BIOL, ORE HEALTH SCI UNIV, PORTLAND. *Mem:* Fel Am Acad Ophthal. *Res:* Clinical ophthalmology; retinal detachment and the vitreous humor of the eye. *Mailing Add:* Dept Ophthal Ore Health Sci Univ 3181 SW Sam Jackson Park Rd Portland OR 97201

WATZMAN, NATHAN, b Powhattan Point, Ohio, Feb 15, 26; m 59; c 3. PSYCHOPHARMACOLOGY, PHYSIOLOGY. *Educ:* Univ Pittsburgh, BS, 47 & 55; MS, 57, PhD(muscular dystrophy), 61. *Hon Degrees:* DSc, 77. *Prof Exp:* Asst prof pharmacol, Sch Pharm, Northeast La State Col, 59-63; res assoc prof, Sch Pharm, Univ Pittsburgh, 63-68; health sci adminr, Div Res Grants, NIH, 68-69, Bur Health Manpower Educ, 69-74; assoc dir regional progs, Div Assoc Health Professions, Health Resource Admin, 74-78, chief spec progs, 78-81; exec secy, div res grants, NIH, 81-84, CHIEF CLIN SCI REV SECT, DIV RES GRANTS, NIH, 84- *Mem:* Am Soc Pharmacol & Exp Therapeut. *Res:* Drug interaction with stress and other environmental modifications of behavior. *Mailing Add:* NIH 348 Westwood Bldg Bethesda MD 20892

WAUCHOPE, ROBERT DONALD, b Atlanta, Ga, Aug 31, 42; m 78; c 6. AGRICULTURAL CHEMISTRY, ANALYTICAL CHEMISTRY. *Educ:* Univ NC, Chapel Hill, BS, 65; NC State Univ, MS, 68, PhD(phys chem), 70. *Prof Exp:* Agr chemist, Ore State Univ, 70-72; RES CHEMIST, SOUTHERN WEED SCI LAB, SCI & EDUC ADMIN-AGR RES, USDA, 72- *Concurrent Pos:* Int Union Pure & Appl Chem; Coun Agr Sci & Tech. *Mem:* Am Chem Soc; Sigma Xi; AAAS. *Res:* Aqueous chemistry and soil behavior of pesticides; pesticide residue analysis; environmental behavior of pesticides. *Mailing Add:* Nematodes Weeds & Crops Res USDA-Agr Res Serv PO Box 748 Tifton CA 31793

WAUD, BARBARA E, b Kitchener, Ont, May 18, 31; US citizen; m 56; c 3. ANESTHESIOLOGY. *Educ:* Univ Western Ont, MD, 56, cert, Am Bd Anesthesiol, 66. *Prof Exp:* Instr anesthesia, Sch Med, Boston Univ, 64-66; instr, Harvard Med Sch, 66-71, asst prof, 71-76; prof anesthesiol, 76-77, PROF PHARMACOL & ANESTHESIOL, UNIV MASS MED SCH, 77- *Mem:* Am Soc Anesthesiologists; Asn Univ Anesthetists; AMA. *Res:* Mechanism of action and kinetics of drugs acting at the neuromuscular junction. *Mailing Add:* Dept Anesthesiol Med Sch Univ Mass 55 Lake Ave N Worcester MA 01655

WAUD, DOUGLAS RUSSELL, b London, Ont, Oct 21, 32; m 56; c 3. PHARMACOLOGY. *Educ:* Univ Western Ont, MD, 56; Oxford Univ, DPhil(pharmacol), 64. *Prof Exp:* Intern, St Joseph's Hosp, London, Ont, 56-57; instr pharmacol, Harvard Med Sch, 59-60; demonstr, Oxford Univ, 61-63; from assoc to assoc prof, Harvard Med Sch, 63-74; PROF PHARMACOL, MED SCH, UNIV MASS, 74- *Concurrent Pos:* USPHS career develop award, 66- *Mem:* AAAS. *Res:* Mechanisms of drug action at molecular level; autonomic and cardiovascular pharmacology; neuromuscular junction physiology and pharmacology; anesthetic agents; uptake and distribution of drugs. *Mailing Add:* Dept Pharmacol S7-137 Univ Mass Med Sch 55 Lake Ave N Worcester MA 01655

WAUER, ROLAND H, ecology, ornithology, for more information see previous edition

WAUGH, DOUGLAS OLIVER WILLIAM, b Hove, Eng, Mar 21, 18; m 46. MEDICAL EDUCATION, PATHOLOGY. *Educ:* McGill Univ, MD, CM, 42, MSc, 48, PhD(path), 50; Royal Col Physicians & Surgeons Can, cert path, 54; FRCP, 64. *Prof Exp:* Demonstr & asst surg pathologist, Path Inst, McGill Univ, 46-47, assoc prof path, 51-57, Miranda Fraser assoc prof comp path, 57; assoc prof path, Univ & asst pathologist, Hosp, Univ Alta, 50-51; from assoc prof to prof, Queen's Univ, Can, 58-64; prof & head dept, Dalhousie Univ, 64-70; prof path & dean fac med, Queen's Univ, Ont, 70-75; exec dir, Asn Can Med Cols, 75-83; RETIRED. *Concurrent Pos:* Asst prov pathologist, Alta, 50-51; mem cancer diag clin, Alta Dept Health, 50-51; mem comt consults, Can Tumor Registry, 55-62; med mem adv comt tumor registry, Nat Cancer Inst, 56-62, dir inst, 65, pres, 74-76; dir labs, Hotel Dieu Hosp, Kingston, 58-64; chmn, Can Cytol Coun, 64-65; adj prof path, Univ Ottawa, 76-83; med journalist, editorialist & essayist, 83-; mem, Bd Accreditation, Can Assoc Univ Sch Nursing, 84- *Honors & Awards:* Queen Elizabeth II Jubilee Medal, 77. *Mem:* Am Soc Exp Pathologists; Am Asn Pathologists & Bacteriologists; Can Asn Pathologists; Int Acad Pathologists; sr mem Can Med Asn. *Res:* Renal diseases; hypertension; lesions of experimental hypersensitivity. *Mailing Add:* 183 Marlborough Ave Ottawa ON K1N 8G3 Can

WAUGH, JOHN BLAKE-STEELE, electronics, physics, for more information see previous edition

WAUGH, JOHN DAVID, b Charleston, WVa, Sept 20, 32; m 83; c 3. SOLID MECHANICS. *Educ:* Univ SC, BS, 54; Yale Univ, MS, 62. *Prof Exp:* Stress analyst missiles div, Bendix Corp, 56-58; assoc dean, Col Eng, 68-78, dean, col eng, 77-87, vprovost, 87-89, PROF ENG, UNIV SC, 58- *Concurrent Pos:* Bd dirs, Assoc Media-based Continuing Eng Educ, 76-, Engrs Deans Coun, 83-88, Nat Inst Eng Mgt, 88- *Honors & Awards:* Litman Award. *Mem:* Fel Am Soc Eng Educ; Am Soc Civil Engrs; Nat Soc Prof Engrs. *Mailing Add:* Univ SC Col Eng Columbia SC 29208

WAUGH, JOHN LODOVICK THOMSON, b Avonhead, Scotland, Nov 13, 22; m 49; c 4. PHYSICAL INORGANIC CHEMISTRY. *Educ:* Univ Glasgow, BSc, 43, PhD(inorg chem), 49. *Prof Exp:* Plant supt, Imp Chem Industs, Ltd, 43-46; asst lectr, Univ Glasgow, 46-49; Climax Molybdenum Co res fel, Calif Inst Technol, 49-50, inst res fel, 51-53; asst prof chem, Univ Hawaii, 50-51; res chemist, Pac Coast Borax Co, 53-56; assoc prof chem, 56-72, PROF CHEM, UNIV HAWAII, 72- *Concurrent Pos:* Prog dir, NSF Undergrad Res Partic Prog, 60-63; assoc res officer, Neutron Physics Br, Atomic Energy Can, Ltd, Ont, 62-63. *Mem:* NY Acad Sci; fel The Chem Soc; assoc Royal Inst Chem. *Res:* Isopoly and heteropoly compounds; boron and boron compounds; x-ray crystallography; intermetallic compounds; lattice dynamics. *Mailing Add:* Dept Chem/Bil 219 Univ Hawaii Manoa 2500 Campus Rd Honolulu HI 96822

WAUGH, JOHN STEWART, b Willimantic, Conn, Apr 25, 29; m 83; c 2. PHYSICAL CHEMISTRY. *Educ:* Dartmouth Col, AB, 49; Calif Inst Technol, PhD(chem, physics), 53. *Prof Exp:* From instr to prof, 53-73, AA Noyes prof chem, 73-88, INST PROF, MASS INST TECHNOL, 89- *Concurrent Pos:* Sloan res fel, 58-62; assoc, Retina Found, 61-70; vis scientist, USSR Acad Sci, Moscow, 62 & 75; Guggenheim fel & res assoc physics, Univ Calif, Berkeley, 63-64, mem sci & educ adv comt, Univ Calif, 81-87; consult, Lawrence Radiation Lab, Univ Calif, 64-75; ed, Advan Magnetic Resonance, 65-88; Robert A Welch Found lectr, Univ Tex, 69; mem chem rev panel, Argonne Nat Lab, 70-74, chmn, 73-74; vis prof & Humboldt fel, Max Planck Inst Med Res, 72; Falk-Plaut lectr, Columbia Univ, mem fel adv panel, Alfred P Sloan Found, 77-; mem adv comt, Stanford Magnetic Resonance Lab, 76-, Nat Magnetic Lab, 76-80, Magnetic Resonance Ctr, Univ SC, 78-82 & Lawrence Berkeley Lab, 80-87; Reilly lectr, Univ Notre Dame, 78; Lucy Pickett lectr, Mt Holyoke Col, 78; mem sci & educ adv comt, Univ Calif, 81-; adv prof, East China Normal Univ, Shanghai, 84; chmn, Div Chem Phys, Am Phys Soc, 84; Joliot-Curie prof, Ecolesuperievre Physics & Chem, Paris, 85; Fairchild scholar, Calif Inst Technol, 89. *Honors & Awards:* Irving Langmuir Award, Am Chem Soc, 76; Pittsburgh Award in Spectros, 78; R W Vaughan Mem lectr, Denver, 81; McElvain lectr, Univ Wis, 81; G N Lewis Mem lectr, Univ Calif, Berkeley, 82; Wolf Prize in Chem, 83-84; Pauling Medal, 84; Dreyfus Lectr, Dartmouth Col, 84; Kistiakowsky Lectr, Howard Univ, 85; McDowell Lectr, BC, 88; Baker Lectr, Cornell Univ, 90. *Mem:* Nat Acad Sci; fel Am Acad Arts & Sci; Am Phys Soc; AAAS. *Res:* Magnetic resonance. *Mailing Add:* Mass Inst Technol Rm 6-235 Cambridge MA 02139

WAUGH, MARGARET H, Newark, NJ, Dec 14, 23. CARDIOVASCULAR & GASTRO-INTESTINAL PHARMACOLOGY. *Educ:* NJ Women's Col, BS, 44. *Prof Exp:* Asst res investr, E R Squibb & Sons, 76-86. *Mem:* NY Acad Sci; Am Soc Pharmacol & Exp Therapeut. *Mailing Add:* 15 Emerson Pl Newark NJ 07114

WAUGH, WILLIAM HOWARD, b New York, NY, May 13, 25; m 52; c 3. PHYSIOLOGY, INTERNAL MEDICINE. *Educ:* Tufts Col, MD, 48. *Prof Exp:* Intern internal med, Long Island Col Hosp, 48-49, asst resident, 50-51; asst resident med, Univ Md Hosp, 51-52; cardiovasc trainee, Med Col Ga, 54-55, asst res prof physiol, 55-60, USPHS sr res fel, 59-60, assoc med, 57-60; from assoc prof to prof med, Col Med, Univ Ky, 60-71, head renal div, 60-68, Ky Heart Asn chair cardiovasc res, 63-71; dir clin sci, 71-76, PROF MED & PHYSIOL, SCH MED, ECAROLINA UNIV, 71- *Concurrent Pos:* Estab investr, Ga Heart Asn, Med Col Ga, 58, physician in chg hemodialysis sect, 59-60; chmn, Policy Rev Comt Human Res, ECarolina Univ, 72-90. *Honors & Awards:* Founder's Award, Am Heart Asn, 82. *Mem:* Microcirculatory Soc; Am Soc Nephrology; Am Physiol Soc; Am Heart Asn; fel Am Col Physicians. *Res:* Hemodynamics; circulatory and renal physiology; nephrology. *Mailing Add:* ECarolina Univ Sch Med Greenville NC 27858-4354

WAVE, HERBERT EDWIN, b Portsmouth, NH, Oct 13, 23; m 47; c 1. ENTOMOLOGY, PLANT PATHOLOGY. *Educ:* Univ Maine, BS, 52; Rutgers Univ, MS, 60, PhD(entom), 61. *Prof Exp:* Entomologist, USDA, 52-58; res asst entom, Rutgers Univ, 58-61; entomologist, USDA, 61-62; asst prof entom, Univ Mass, 62-67; assoc prof plant & soil sci, Univ Maine, 67-86; RETIRED. *Concurrent Pos:* Emer ext edur, Fruit Specialist. *Mem:* Sigma Xi. *Res:* Biology and ecology of potato infesting species of aphids; extension pest control for tree and small fruits; orchard herbicides and growth regulators. *Mailing Add:* Box 448 East Winthrop ME 04343

WAVRIK, JOHN J, b New York, NY, Mar 17, 41; m 61. GEOMETRY. *Educ:* Johns Hopkins Univ, AB, 61, MA, 64; Stanford Univ, PhD(math), 66. *Prof Exp:* Joseph Fells Ritt instr math, Columbia Univ, 66-69; ASSOC PROF MATH, UNIV CALIF, SAN DIEGO, 69- *Mem:* Am Math Soc; Math Asn Am. *Res:* Algebraic geometry; computers in pure mathematics. *Mailing Add:* Dept Math C-012 Univ Calif San Diego La Jolla CA 92093

WAWERSIK, WOLFGANG R, b Frankenholz, Ger, Apr 23, 36; US citizen; m 65; c 2. ROCK MECHANICS. *Educ:* Tech Hochsch Aachen, Ger, Dipl-Ing, 61, Univ Minn, MS, 63, PhD(mineral eng/rock mech), 68. *Prof Exp:* Res assoc, Dept Earth & Planetary Sci, Mass Inst Technol, 68-69; from asst prof to assoc prof mech eng, Dept Mech & Indust Eng, Univ Utah, 69-74; mem tech staff, 74-83, distinguished mem tech staff, 83-86, SUPVR, GEOMECH DIV, SANDIA NAT LABS, 86- *Concurrent Pos:* vis scientist, Fed Inst Geosci & Natural Resources, Hannover, Ger, 80. *Mem:* Am Inst Mining, Metall & Petrol Engrs; Am Geophys Union; Sigma Xi. *Res:* Experimental rock mechanics. *Mailing Add:* 7512 Pickard Ave NE Albuquerque NM 87110

WAWNER, FRANKLIN EDWARD, JR, b Petersburg, Va, Dec 12, 33; m 53; c 3. MATERIALS SCIENCE. *Educ:* Randolph-Macon Col, BS, 59; Univ Va, MS, 68, PhD(mat sci), 71. *Prof Exp:* Physicist, Army Eng Res & Develop Lab, 59-61; sr physicist, Texaco Exp Inc, 61-66; RES PROF COMPOSITE MAT, DEPT MAT SCI, UNIV VA, 71- *Concurrent Pos:* Consult, Textron Spec Mat, Lowell, Mass, 69- & various other corps. *Mem:* Metall Soc; Am Ceramic Soc. *Res:* Composite materials; chemical vapor deposition; characterization of mechanical properties; fracture; structure; microstructure. *Mailing Add:* Dept Mat Sci Thornton Hall Univ Va Charlottesville VA 22903

WAWSZKIEWICZ, EDWARD JOHN, b North Smithfield, RI, Feb 10, 33. MICROBIOLOGY, BIOCHEMISTRY. *Educ:* Harvard Univ, AB, 54; Univ Calif, Berkeley, PhD, 61. *Prof Exp:* Asst microbiol, Hopkins Marine Sta, Stanford Univ, 54-55; res asst microbiol, Univ Calif, Berkeley, 55-61; USPHS fel, Max Planck Inst Cell Chem, Ger, 61-63; resident res assoc, Argonne Nat Lab, 64-66; asst mem, Inst Biomed Res, 66-70; ASSOC PROF MICROBIOL, UNIV ILL MED CTR, 70- *Mem:* AAAS; Am Soc Microbiol; fel Am Inst Chemists; Am Inst Biol Sci; Am Chem Soc. *Res:* Metabolism of thiobacilli; erythritol metabolism by propionibacteria; propionate metabolism; erythromycin biosynthesis; mouse salmonellosis, iron metabolism; pacifarins; enology. *Mailing Add:* Dept Microbiol M/C 790 Univ Ill PO Box 6998 Chicago IL 60680

WAX, HARRY, b Boston, Mass, June 17, 18; m 50; c 2. BIOCHEMISTRY. *Educ:* Univ Calif, Los Angeles, MA, 41; Iowa State Col, PhD(chem), 49. *Prof Exp:* Dir fine chems div, Wm T Thompson Co, 40-44; res chemist, Dr Geo Piness Allergy Group, 49-53; res chemist pharmaceuts, Stuart Co, 53-59, mgr prod develop, 59-69, Stuart Pharmaceut Res Dept, Atlas Chem Indust, 69-74; STAFF, ARCHON PURE PROD CORP, 74- *Concurrent Pos:* Instr, Univ Calif, Los Angeles, 51-52. *Mem:* AAAS; Am Chem Soc; Am Pharmaceut Asn; Sigma Xi. *Res:* Enzymatic synthesis of peptide bonds; application of ion exchange resins and filter paper electrophoresis to the separation of allergens; organosynthesis; pharmaceutical manufacturing processes. *Mailing Add:* 1830 Westholme Ave 105 Los Angeles CA 90025-4941

WAX, JOAN, b Detroit, Mich, Sept 14, 21. CLINICAL INVESTIGATION, PHARMACOLOGY. *Educ:* Wayne State Univ, BA, 43; Univ NC, MS, 47. *Prof Exp:* From res asst to asst res pharmacologist to res pharmacologist, Parke, Davis & Co, 47-76, sr scientist & clin pharmacologist, 77-80, CLIN SCIENTIST, PARKE-DAVIS PHARMACEUT RES DIV, WARNER-LAMBERT CO, 80- *Mem:* AAAS; Am Soc Pharmacol & Exp Therapeut; Am Chem Soc; Sigma Xi. *Res:* Analgetic and anti-inflammatory agents; narcotic antagonists; drug-induced gastrointestinal ulcerogenesis; clinical investigation regarding drugs in arthritis, dysmenorrhea. *Mailing Add:* 301 N Roadrunner Pkwy No 607 Las Cruces NM 88001

WAX, NELSON, b Philadelphia, Pa, Apr 2, 17; m 42; c 3. ELECTRICAL ENGINEERING. *Educ:* Univ Pa, BS, 37, MS, 38; Ohio State Univ, PhD(elec eng), 42. *Prof Exp:* Asst, Ohio State Univ, 38-39, instr elec eng, 42; mem tech staff, Bell Tel Labs, Inc, 42-48; from asst prof to prof elec eng & res, Univ Ill, Urbana, 48-87; RETIRED. *Concurrent Pos:* Guggenheim fel, 54-55; consult, Rand Corp, 58-68; assoc mem, Ctr Adv Study, Univ Ill, 64-65; vis prof, Univ Tex, 65-66; vis scientist, Weirman Inst, 78. *Res:* Microwave electronics; communication theory; non-linear oscillations. *Mailing Add:* 802 Centura Champaign IL 61820

WAX, ROBERT LEROY, b Des Moines, Iowa, July 7, 38. SPACE PHYSICS. *Educ:* Calif Inst Technol, BS, 60; Univ Ill, Urbana, MS, 61; Univ Calif, Berkeley, PhD(physics), 65. *Prof Exp:* Mem tech staff, TRW Systs Group, 66-71; physicist, Space Environ Lab, Nat Oceanic & Atmospheric Admin, 71-73; consult physicist, 73-78, SR STAFF SYST ENGR, DEFENSE & SPACE SYSTS GROUP, TRW INC, 78- *Mem:* Am Geophys Union. *Mailing Add:* 201 S Poinsettia Ave Manhattan Beach CA 90266

WAXLER, GLENN LEE, b Olney, Ill, Feb 24, 25; m 46; c 2. VETERINARY PATHOLOGY. *Educ:* Univ Ill, BS, 51, DVM, 53; Mich State Univ, MS, 59, PhD(vet path), 61. *Prof Exp:* Pvt pract, 53-57; from instr to assoc prof vet path, 57-66, prof, 66-90, EMER PROF PATH, MICH STATE UNIV, 90- *Mem:* Am Vet Med Asn; Am Col Vet Path; Asn Gnotobiotics; Conf Res Workers Animal Dis. *Res:* Germ-free research; enteric disease of swine; histopathology of diseases of domestic animals. *Mailing Add:* 1824 Cahill Dr East Lansing MI 48823-4702

WAXMAN, ALAN DAVID, b New York, NY, Mar 9, 38; m 61; c 3. NUCLEAR MEDICINE. *Educ:* Univ Southern Calif, BA, 58, MD, 63. *Prof Exp:* Intern med, Los Angeles County Gen Hosp, 63-64; resident med, Los Angeles Vet Admin Hosp, 64-65; res assoc metab, Metab Serv, Cancer Inst, NIH, 65-67; asst prof, 68-70, ASSOC PROF RADIOL, SCH MED, UNIV SOUTHERN CALIF, 70-; DIR NUCLEAR MED, CEDARS-SINAI MED CTR, 77- *Concurrent Pos:* Staff physician radiol, Los Angeles County-Univ Southern Calif Med Ctr, 78- *Mem:* Soc Nuclear Med; Am Fedn Clin Res; AMA; fel Am Col Physicians; Am Col Nuclear Physicians. *Res:* Applications of nuclear medicine technology in the detection and evaluation of disease processes, primarily in oncology and hepatic disease. *Mailing Add:* Cedars-Sinai Med Ctr 8700 Beverly Blvd Los Angeles CA 90048

WAXMAN, DAVID, b Albany, NY, Feb 7, 18; m 50; c 6. INTERNAL MEDICINE, CARDIOLOGY. *Educ:* Syracuse Univ, BS, 42; Syracuse Col Med, MD, 50. *Prof Exp:* Instr internal med, 61-64, asst prof, 64-69, asst dean, 70-71, assoc dean, 71-72, dean, 72-74, vchancellor students, 74-76, vchancelor, 76-77, dep exec vchancellor, 77, EXEC VCHANCELLOR, MED CTR, UNIV KANS, 77- *Concurrent Pos:* Consult, Vet Admin Hosp, 61- & Muson Army Hosp, 79-; assoc prof internal med, Med Ctr, Univ Kans, 69-77, dir med outpatient serv, 70-74 & prof internal med, 77-; major gen mobilization & augmentee to surgeon gen, USAF, 70-78, nat consult educ to surgeon gen, 80- *Mem:* AMA; Am Soc Internal Med; fel Am Col Physicians; Sigma Xi. *Res:* Non-invasive methods and techniques for the evaluation and assessment of cardiac disfunction; drug inhibition of fatty acid mobilization and catecholamine-induced metabolic changes in humans. *Mailing Add:* Kansas Univ Med Ctr 39th & Rainbow Kansas City KS 66103

WAXMAN, HERBERT SUMNER, b Boston, Mass, Sept 1, 36; m 60; c 3. MEDICINE, HEMATOLOGY. *Educ:* Mass Inst Technol, BS, 58; Harvard Univ, MD, 62; Am Bd Internal Med, dipl, 69, dipl hemat, 74. *Prof Exp:* From intern to resident med, Mass Gen Hosp, Boston, 62-64; res assoc biochem, Nat Cancer Inst, 64-66; resident, Mass Gen Hosp, Boston, 66-67; from asst prof to prof med, Sch Med, Temple Univ, 68-77, dep chmn dept, 75-77; chmn, Dept Med, Baystate Med Ctr, Springfield, Mass, 77-79; prof med, Tufts Univ, 77-79; CHMN, DEPT MED, ALBERT EINSTEIN MED CTR, 79-; DEP CHMN, DEPT MED, SCH MED, TEMPLE UNIV, 79- *Concurrent Pos:* Mead Johnson scholar, Am Col Physicians, 66-67; NIH trainee hemat, Sch Med, Wash Univ, 67-68; advan clin fel, Am Cancer Soc, 68-71. *Mem:* Am Fedn Clin Res; Am Soc Hemat; fel Am Col Physicians; Assoc Prog Dir Int Med. *Res:* Control of protein synthesis in blood cells; clinical studies in sickle cell and related diseases; computer assisted medical diagnosis systems. *Mailing Add:* Albert Einstein Med Ctr York & Tabor Rds Philadelphia PA 19141

WAXMAN, LLOYD H, PROTEASES, PROTEIN CHEMISTRY. *Educ:* Harvard Univ, PhD(biochem), 75. *Prof Exp:* SR RES ASSOC, SCH MED, HARVARD UNIV, 81- *Mailing Add:* Dept Biol & Chem Merck Sharp & Dohme Sumneytown Pike West Point PA 19486

WAXMAN, RONALD, b Newark, NJ, Nov 28, 33; m 55; c 3. DIGITAL SYSTEM SPECIFICATION & ANALYSIS, COMPUTER HARDWARE DESCRIPTION LANGUAGES. *Educ:* NJ Inst Technol, BS, 55; Syracuse Univ, MS, 63. *Prof Exp:* Mem eng staff, Int Bus Mach, 55-87; PRIN SCIENTIST, UNIV VA, 87- *Concurrent Pos:* Chair, Design Automation Standards Subcomt, Inst Elect & Electronics Engrs Computer Soc, 83-88, Tech Comt, 88-90, lectr, 86-88, mem bd gov, 89-93; Mem adv bd, Spec Interest Group Design Automation, Asn Comput Mach, 88- *Mem:* Fel Inst Elec & Electronics Engrs; Inst Elec & Electronics Engrs Computer Soc; Asn Comput Mach. *Res:* Methods of designing digital systems; application of hardware description languages to the design process; system specification and design. *Mailing Add:* Dept Elec Eng Thornton Hall Univ Va Charlottesville VA 22903

WAXMAN, SAMUEL, HEMATOLOGY, ONCOLOGY. *Educ:* State Univ NY Downstate Med Ctr, MD, 63. *Prof Exp:* CLIN PROF MED, MT SINAI SCH MED, 83-, HEAD, CANCER CHEMOTHER FOUND LAB, 72- *Res:* Differentiation therapy. *Mailing Add:* Med Dept Mt Sinai Sch Med One Gustave L Levy Pl New York NY 10029

WAXMAN, SIDNEY, b Providence, RI, Nov 13, 23; m 48; c 3. ORNAMENTAL HORTICULTURE. *Educ:* Univ RI, BS, 51; Cornell Univ, MS, 54, PhD(ornamental hort), 57. *Prof Exp:* PROF ORNAMENTAL HORT, UNIV CONN, 57- *Honors & Awards:* Merit Award, Int Plant Propagator's Soc. *Mem:* Am Soc Hort Sci; hon mem Int Plant Propagators Soc; Am Conifer Soc. *Res:* Photoperiodism; plant propagation; seed and bud dormancy; development of dwarf forms of conifers their introduction and propagation. *Mailing Add:* Dept Plant Sci Univ Conn Storrs CT 06268

WAXMAN, STEPHEN GEORGE, b Newark, NJ, Aug 17, 45; m 68; c 2. NEUROLOGICAL PATHOPHYSIOLOGY, ION CHANNEL BIOLOGY. *Educ:* Harvard Univ, AB, 67; Albert Einstein Col Med, PhD(neurosci), 70, MD, 72. *Hon Degrees:* MA, Yale Univ, 86. *Prof Exp:* Res asst biophys, McLean Hosp, Harvard Med Sch, 65-67, clin fel neurol, 72-75, from asst prof to assoc prof, 75-78; res fel neurosci, Albert Einstein Col Med, 70-72; prof neurol & assoc chmn, Stanford Med Sch & chief, neurol, Vet Admin Hosp, Palo Alto, 81-86; PROF NEUROL & CHMN DEPT, SCH MED, YALE UNIV & NEUROLOGIST-IN-CHIEF, YALE-NEW HAVEN HOSP, 86-; DIR, NEUROSCI RES CTR, VET ADMIN MED CTR, WEST HAVEN, 86- *Concurrent Pos:* Epilepsy Found Chauveau fel, Med Res Coun Cerebral Functions Res Group, Univ Col London, 69; res fel neurosci, Mass Inst Technol, 74-75, from vis asst prof to vis assoc prof biol, 75-78; Nat Inst Neurol & Commun Dis & Stroke res career develop award, 75; mem, adv bd, Regeneration Progs, Vet Admin, 81-91, Comt Decade of Brain, Am Acad Neurol & Bd Sci Counselors, Nat Inst Neurol Dis & Stroke, 90-, Bd Biobehav Sci, Inst Med, 91-; estab investr, Nat Mult Sclerosis Soc, 87. *Honors & Awards:* First Ann Trgyve Tuve Mem Award, NIH, Inst Advan Educ Sci, 75. *Mem:* Inst Med-Nat Acad Sci; Am Acad Neurol; Asn Res Nerv & Ment Dis. *Res:* Molecular and cellular aspects of diseases of the brain and spinal cord; mechanisms underlying functional recovery after injury to nerve cells and glial cells; electron microscopy; immunocytochemistry; neurophysiology; biophysics; computer simulations; molecular biology of ion channels; multiple sclerosis, spinal cord injury and stroke. *Mailing Add:* Dept Neurol LCI 707 Yale Med Sch PO Box 3333 New Haven CT 06510

WAY, E LEONG, b Watsonville, Calif, July 10, 16; m 44; c 2. PHARMACOLOGY, TOXICOLOGY. *Educ:* Univ Calif, BS, 38, MS, 40, PhD(pharmaceut chem), 42. *Prof Exp:* Asst pharmacol, Univ Calif, 42; pharmaceut chemist, Merck & Co, Inc, NJ, 42-43; from instr to asst prof pharmacol, Med Sch, George Washington Univ, 43-48; from asst prof to prof pharmacol & toxicol, med ctr, Univ Calif, San Francisco, 49-87, vchmn dept, 57-67, chmn dept, 73-78, EMER PROF PHARMACOL & TOXICOL, MED CTR, UNIV CALIF, SAN FRANCISCO, 87- *Concurrent Pos:* USPHS spec res fel, Univ Bern, 55-56; China Med Bd res fel, Univ Hong Kong, 62-63; consult, Attorney Gen, Calif, 59-60 & Dept Corrections, Calif, 64-70; mem, Comt on Probs Drug Safety, Nat Acad Sci-Nat Res Coun, 65-71 & Comt on Probs Drug Dependence, 68-74, 77- & chmn bd, 78-82; mem comt on abuse of depressant & stimulant drugs, Dept Health, Educ & Welfare, 66-68, mem pharmacol study sect, 66-70, chmn, 68-70, mem comt on narcotic addiction & drug abuse rev, 70-74, chmn, 71-74; mem sci adv comt drugs, Bur Narcotics & Dangerous Drugs, 68-74; mem alcohol & drug dependence serv adv group & merit rev bd, Vet Admin, 71-76; mem controlled substances adv comt, Food & Drug Admin, 74-78; Sullivan-Sterling distinguished vis prof, Morehouse Sch Med, 81-82; hon prof pharmacol & neurosci, Guangzhou Med Col, 86; vis prof neuropsychopharmacol, Gunma Univ Med Sch, Japan, 89-90; sr staff fel, Nat Inst Drug Abuse, 90; chmn bd, Comt Probs Drug Dependence, 78-82; bd dirs, Li Found, 70-, pres, 85-; nat adv coun, Am Bur Med Advan China, 81-; secy, Int Narcotic Res Conf, 83-87, treas, 87- *Honors & Awards:* Am Pharmaceut Found Award, 62; Ebert Prize Cert, Am Pharmaceut Asn, 62; Kauffman lectr, Ohio State Univ, 66; Forbes lectr, Va Commonwealth Univ, 78; Cultural Citation & Gold Medal, Ministry Educ, Repub China, 78. *Mem:* Fel AAAS; Am Soc Pharmacol & Exp Therapeut (pres, 76-77); Am Pharmaceut Asn; fel Am Col Neuropsychopharmacol; Fed Am Soc Exp Biol (pres, 77-78). *Res:* Drug metabolism; pharmacology analgetics, and endophins; drug tolerance and physical dependence mechanisms. *Mailing Add:* Dept Pharmacol & Exp Therapeut Univ Calif Med Ctr San Francisco CA 94143

WAY, FREDERICK, III, b Sewickley, Pa, Jan 4, 25; m 48; c 2. COMPUTER SCIENCE. *Educ:* Univ Pittsburgh, BS, 50. *Prof Exp:* Physicist, Babcock & Wilcox Co, 51-54, anal eng, Res Lab, 54-56; assoc prof, 56-70, assoc dir, Comput Ctr, 56-80, PROF COMPUT ENG & SCI, CASE WESTERN RESERVE UNIV, 70-, DIR, JENNINGS COMPUT CTR, 80- *Concurrent Pos:* Consult, Thompson Ramo Wooldridge Co, 59 & Bailey Meter Co, 60- *Mem:* Soc Indust & Appl Math; Asn Comput Mach; Math Asn Am; Sigma Xi. *Res:* Investigation and implementation of automatic programming methods for digital computers. *Mailing Add:* 1258 Castleton Rd Cleveland Heights OH 44121

WAY, GEORGE H, b Wheeling, WVa, May 30, 30; m 53; c 1. ENGINEERING. *Educ:* Princeton Univ, BSE, 52. *Prof Exp:* Staff mem, Pa Railroad Co, 54-66; res engr, Planning Dept, The Chessie Syst, 66-73; asst to vpres, 73-83, sr asst vpres, 83-85, VPRES RES & TEST DEPT, ASN AM RAILROADS, 85- *Concurrent Pos:* Fel, Permanent Way Inst, 71. *Mem:* Am Railroad Eng Asn. *Res:* Concrete cross tie design and performance; development of automated railway track geometry and assessment techniques; design and evaluation of epoxy glued insulating rail joints; railway track design, construction and maintenance. *Mailing Add:* Asn Am Railroads Res & Test Dept 50 F St NW Washington DC 20001

WAY, JAMES LEONG, b Watsonville, Calif, Mar 21, 27; m 47; c 3. PHARMACOLOGY, TOXICOLOGY. *Educ:* Univ Calif, Berkeley, BA, 51; George Washington Univ, PhD(pharmacol), 55. *Prof Exp:* USPHS fel pharmacol, Univ Wis, 55-57; USPHS sr fel, 57-58; from instr to asst prof pharmacol, Univ Wis, 59-62; assoc prof, Med Col Wis, 62-67; prof pharmacol, Wash State Univ, 67-82; SHELTON PROF PHARMACOL & TOXICOL, DEPT PHARMACOL & TOXICOL, TEX A&M UNIV, 82- *Concurrent Pos:* USPHS career develop award, 58-62, spec res fel, 73-75; mem fed task group toxicol eval pesticide in mammalian species, Environ Protection Agency, 73-75; mem toxicol study sect, NIH, 74-78, mem toxicol data bank rev comt, NIH, 74-82; vis scientist, Div Molecular Pharmacol, Nat Inst Med Res, London, 73-75; mem sci adv bd, Nat Ctr Toxicol Res, 79-82; vis prof, NSF, 81-82; mem, exec bd, Nat Ctr Toxicol Res, 80-82, Environ Health Sci Comt Study Sect, NIH, 85-89; assoc ed, Ann Rev Pharmacol & Toxicol, 89-94; pres, Western Pharmacol Soc, 76; Distinguished scholar, Nat Acad Sci, 86-87; endowed chair toxicol, Tex A&M Univ, 82. *Mem:* AAAS; Am Chem Soc; Am Soc Pharmacol & Exp Therapeut (secy & treas, 89); Soc Toxicol. *Res:* Cancer; nucleic acid; drug metabolism and anticholinesterase alkylphosphate antagonists; cyanide, nitrite and alkylphosphate hydrogen sulfide, carbon monoxide, PCB poisoning; molecular and marine pharmacology; environmental toxicology; drug carrier cells; molecular biology. *Mailing Add:* Dept Med Pharmacol Med Sci Bldg Tex A&M Univ College Station TX 77843

WAY, JON LEONG, b Madison, Wis, Feb 15, 61. PEDIATRIC DENTISTRY, PUBLIC HEALTH LEAD DENTIST. *Educ:* Wash State Univ, BS, 83; Univ Wash, DDS, 87. *Prof Exp:* Res technician, Sch Pharm, Wash State Univ, 78-83; instr pediat dent, Sch Dent, Univ Wash, 90; lead dentist, Seattle-King County Dept Pub Health, 88-91. *Mem:* Am Dent Asn; Fedn Dentaire Int. *Res:* Cyanide toxicity and its antagonism; selective toxic of squoxin. *Mailing Add:* 11308 Joffre St No 4 Los Angeles CA 90049

WAY, KATHARINE, b Sewickley, Pa, Feb 20, 03. PHYSICS. *Educ:* Columbia Univ, BS, 32; Univ NC, PhD(physics), 38. *Prof Exp:* Huff res fel, Bryn Mawr Col, 38-39; from instr to asst prof physics, Univ Tenn, 39-42; physicist, US Naval Ord Lab, 42, Manhattan Proj, Oak Ridge Nat Lab, 42-48 & Nat Bur Stand, 49-53; dir nuclear data proj, Nat Res Coun, 53-63 & Oak Ridge Nat Lab, 64-68; adj prof physics, Duke Univ, 68-88; RETIRED. *Concurrent Pos:* Ed, Atomic Data & Nuclear Data Tables, 73-82. *Mem:* Fel Am Phys Soc; fel AAAS. *Res:* Nuclear fission; radiation shielding; nuclear constants. *Mailing Add:* Carolina Meadows Apt 2-102 Chapel Hill NC 27514

WAY, KERMIT R, b Detriot, Mich, Sept 21, 39; m 77; c 1. PHYSICAL CHEMISTRY, ATOMIC PHYSICS. *Educ:* Luther Col, BA, 61; Mich State Univ, MSc, 65; Univ Iowa, PhD(phys chem), 76. *Prof Exp:* Res chemist, Chrysler Corp Detroit, Mich, 61-62; instr, Concordia Col Minn, 64-65; lectr, Waterloo Lutheran Univ, Ont, Can, 65-69; asst prof, Drake Univ Iowa, 75-76; res assoc, Mass Inst Technol, 76-78; asst prof, Augustana, SDak, 79-84; SR PROJ ENGR, BENDIX FIELD CORP, COLUMBIA, MD, 84- *Concurrent Pos:* Vis asst prof, Carthage Col, Wis, 78-79. *Mem:* Am Phys Soc; Am Chem Soc; Soc Photo-optical Instrumentation Engrs. *Res:* Optical recording and developmental optical recording systems technical advisors for development from optical recording system. *Mailing Add:* 9713 Summer Park Ct Columbia MD 21046-1812

WAY, LAWRENCE WELLESLEY, b St Louis, Mo, Nov 15, 33; m 71; c 2. SURGERY. *Educ:* Cornell Univ, AB, 55; Univ Buffalo, MD, 59. *Prof Exp:* Clin instr, 67-69, from asst prof to assoc prof, 69-75, PROF SURG, SCH MED, UNIV CALIF, SAN FRANCISCO, 75-, VCHMN DEPT, 72-; CHIEF SURG SERV, VET ADMIN HOSP, SAN FRANCISCO, 72- *Concurrent Pos:* Fel gastrointestinal physiol, Univ Calif, San Francisco, 67-68 & Vet Admin Ctr, Univ Calif, Los Angeles, 68-69; mem, Surg & Bioengineering Study Sect, NIH, 81-85. *Mem:* Am Col Surg; Am Gastroenterol Asn; Am Surg Asn; Soc Surg Alimentary Tract; Soc Univ Surgeons. *Res:* Gastrointestinal secretion; bile formation. *Mailing Add:* Dept Surg Univ Calif 680-A San Francisco CA 94143

WAY, WALTER, b Rochester, NY, June 27, 31; m 55; c 3. ANESTHESIOLOGY, PHARMACOLOGY. *Educ:* Univ Buffalo, BS, 53; State Univ NY, MD, 57; Univ Calif, MS, 62. *Prof Exp:* USPHS trainee pharmacol, 60-61; from instr to asst prof anesthesia, 61-63, from asst prof to assoc prof, 63-74, PROF ANESTHESIA & PHARMACOL, MED CTR, UNIV CALIF, SAN FRANCISCO, 74- *Mem:* Am Soc Pharmacol & Exp Therapeut; Asn Univ Anesthetists. *Res:* Opioid clinical pharmacology. *Mailing Add:* Dept Anesthesia & Pharmacol Univ Calif San Francisco CA 94143

WAYGOOD, EDWARD BRUCE, b Macclesfield, Eng, Dec 15, 45; Can citizen; m 73; c 2. BIOCHEMISTRY, MICROBIOLOGY. *Educ:* Univ Man, BSc, 68, MSc, 69; Univ Toronto, PhD(med biol), 73. *Prof Exp:* Fel biol, Johns Hopkins Univ, 74-77; ASST PROF BIOCHEM, UNIV SASK, 77- *Concurrent Pos:* Med Res Coun Can grant, 74-76 & 76-78. *Res:* Microbial physiology; microbial transport; metabolic regulation; bacterial sugar-phosphotransference system. *Mailing Add:* Dept Biochem Univ Sask Col Med Saskatoon SK S7N 0W0 Can

WAYGOOD, ERNEST ROY, b Bramhall, Eng, Oct 26, 18; m 50; c 1. PLANT PHYSIOLOGY, BIOCHEMISTRY. *Educ:* Ont Agr Col, BSc, 41; Univ Toronto, MSc, 47, PhD, 49. *Prof Exp:* Assoc prof plant physiol, McGill Univ, 49-54; prof, 54-79, EMER PROF BOT, UNIV MAN, 79- *Mem:* Can Soc Plant Physiol (pres, 59-60); fel Chem Inst Can; fel Royal Soc Can; Am Soc Plant Physiol; Sigma Xi. *Res:* Enzyme mechanisms in respiration and photosynthesis; physiology of host parasite relationships. *Mailing Add:* 14836 Beachview Ave White Rock BC V4B 1N7 Can

WAYLAND, BRADFORD B, b Lakewood, Ohio, Dec 14, 39; m 59; c 1. INORGANIC CHEMISTRY. *Educ:* Western Reserve Univ, AB, 61; Univ Ill, PhD(inorg chem), 64. *Prof Exp:* From asst prof to assoc prof, 64-75, PROF CHEM, UNIV PA, 75- *Mem:* Am Chem Soc; Royal Soc Chem. *Res:* Transition metal ion complexes; molecular complexes; thermodynamics; magnetic resonance; metalloporphyrin species; metal ions in biological systems; surface complexes; metal site catalysis. *Mailing Add:* Dept Chem Univ Pa Philadelphia PA 19104

WAYLAND, J(AMES) HAROLD, b Boise, Idaho, July 2, 09; m 33; c 2. MICROCIRCULATION, INTRAVITAL MICROSCOPY. *Educ:* Univ Idaho, BS, 31; Calif Inst Technol, MS, 35, PhD(physics), 37. *Hon Degrees:* DrSci, Univ Idaho, 77. *Prof Exp:* Asst math, Calif Inst Technol, 31-34; instr, Univ Idaho, 34-35; asst prof physics, Univ Redlands, 38-45; physicist, US Naval Ord Test Sta, Calif, 45-48; from assoc prof to prof appl mech, 49-63, prof, 63-79, EMER PROF ENG SCI, CALIF INST TECHNOL, 79- *Concurrent Pos:* Fel, Calif Inst Technol, 38-41, war res fel, 44-45; contract employee, US Naval Ord Lab, 41-42 & 11th Naval Dist, US Dept Navy, 42-44; Guggenheim fel, Univ Strasbourg, 53-54; mem int comn microcirculation & capillary exchange, Int Union Physiol Sci, 78-84; vis prof, Shinshu Univ Midsch, 73, Univ Limburg, Holland, 79 & Univ Tsukuba, Japan, 87; Alexander von Humboldt sr scientist award, 82; distinguished vis prof, Univ Del, 85. *Honors & Awards:* Ehrenmitglied Ges, Mikrozirkultation, 80; Eugene M Landis Award, Microcirculatory Soc, 81; Malpighi Prize, Europ Soc Microcirculation, 88. *Mem:* Microcirculation Soc; Am Soc Eng Educ; Am Soc Enol; Am Phys Soc; Int Soc Biorheology; Am Physiol Soc; AAAS; hon mem Europ Soc Microcirculation. *Res:* Biological engineering science; hemorheology; intrarital microscopy; biophysics and bioengineering of the peripheral circulation. *Mailing Add:* Div Eng & Appl Sci 104-44 Calif Inst Technol Pasadena CA 91125

WAYLAND, JAMES ROBERT, JR, b Plainview, Tex, May 3, 37; m 61; c 2. ASTROPHYSICS & GEOPHYSICS, AGRICULTURAL PHYSICS. *Educ:* Univ of the South, BS, 59; Univ Ariz, PhD(physics), 67. *Prof Exp:* Res assoc astrophys, Univ Md, 67-70; asst prof astrophysics & agr physics, Tex A&M Univ, 70-74; mem, 74-90, SR MEM TECH STAFF, SANDIA LABS, 90- *Honors & Awards:* IR-100 Award. *Mem:* Am Phys Soc. *Res:* Cosmic ray physics; environmental impact of energy generating systems; health physics; petroleum engineering. *Mailing Add:* Sandia Labs Orgn 9114 Bldg 868 Rm 122 Albuquerque NM 87185

WAYLAND, ROSSER LEE, JR, b Charlottesville, Va, Dec 30, 30; m 53; c 3. TEXTILE CHEMISTRY. *Educ:* Univ Va, BS, 49, MS, 50, PhD(chem), 52. *Prof Exp:* Res chemist, Dan River, Inc, 51-54, group leader, Res Div, 54-60, asst dir res, 60-74, mgr, Chem Prod Dept, 74-78, vpres, 78-79, tech dir, Chem

Prod Div, 79-90; TECH DIR, HICKSON DANCHEM CORP, 90- *Honors & Awards:* Olney Medal, Am Asn Textile Chemists & Colorists, 75. *Mem:* Am Asn Textile Chemists & Colorists; Am Chem Soc. *Res:* Organic synthesis; thermosetting textile resins. *Mailing Add:* 319 W Main St Danville VA 24541-2804

WAYLAND, RUSSELL GIBSON, b Treadwell, Alaska, Jan 23, 13; m 43, 65; c 2. MINING & PETROLEUM GEOLOGY, ENGINEERING. *Educ:* Univ Wash, BS, 34; Univ Minn, MS, 35, PhD(econ geol), 39; Harvard Univ, AM, 37. *Prof Exp:* Geologist & engr, Homestake Mining Co, SDak, 30-39; geologist, US Geol Surv, 39-42; minerals specialist, Army-Navy Munitions Bd, 42-45; mining indust control officer, Off Mil Govt for Ger, 45-48; US chmn combined coal control group, Allied High Comn, Ger, 48-52; staff engr, Off Dir, US Geol Surv, 52-58, regional geologist, Condrv Div, Los Angeles, 58-66, asst chief, Conserv Div, DC, 66-67, chief, Conserv Div, Reston, Va, 67-78, res, Off Dir, 78-80; CONSULT ENERGY MINERALS, 80- *Concurrent Pos:* Res asst, Univ Minn, 34-36, instr, 37-39; geologist & engr, Alaska Juneau Gold Mining Co, 37; Wash rep, Am Inst Prof Geologists, Wash, 82-88; comnr, Va Oil & Gas Conserv Comn, 82-90. *Honors & Awards:* Distinguished Serv Award, Dept Interior, 68. *Mem:* Fel Mineral Soc Am; Am Inst Mining, Metall & Petrol Engrs; Geol Soc Am; Soc Econ Geologists; Am Inst Prof Geologists; Asn Eng Geologists; fel AAAS; Sigma Xi. *Res:* Industrial minerals; coal, petroleum and gold; mineral land appraisal. *Mailing Add:* 4660 N 35th Arlington VA 22207-4462

WAYMAN, C(LARENCE) MARVIN, b Wheeling, WVa, Aug 12, 30; m 56; c 2. METALLURGY. *Educ:* Purdue Univ, BS, 52, MS, 55; Lehigh Univ, PhD(metall), 57. *Prof Exp:* From asst prof to assoc prof, 57-64, PROF METALL, UNIV ILL, URBANA, 64- *Concurrent Pos:* Guggenheim fel, 69; overseas fel, Churchill Col, Cambridge Univ, 69. *Honors & Awards:* Zay Jeffries Award, Am Soc Metals, 77; Matthewson Gold Medal, Am Inst Mining & Metall Engrs, 78. *Mem:* Fel Am Soc Metals; fel Am Inst Mining & Metall Engrs; fel Inst Metallurgists Gt Brit; Japan Inst Metals; Mat Res Soc. *Res:* Solid state phase transformations; growth and properties of thin films; transmission electron microscopy; shape memory alloys. *Mailing Add:* 115B Metall & Mining Bldg 1304 W Green St Univ Ill Urbana IL 61801

WAYMAN, MICHAEL LASH, b Kingston, Ont, Feb 1, 43; m 65; c 2. METALLURGY. *Educ:* Univ BC, BASc, 64; McMaster Univ, MSc, 66; Cambridge Univ, PhD(metall), 68. *Prof Exp:* Res assoc metall, Univ Toronto, 68-69; from asst prof to assoc prof, 69-77, PROF METALL, UNIV ALTA, 78- *Mem:* Am Soc Metals; Can Inst Mining & Metall. *Res:* Effects of environment on structure and on mechanical properties of metals and alloys. *Mailing Add:* Dept Metall & Mat Univ Alta Edmonton AB T6G 2M7 Can

WAYMAN, MORRIS, b Can, Mar 19, 15; m 37; c 2. ORGANIC & WOOD CHEMISTRY, BIOTECHNOLOGY. *Educ:* Univ Toronto, BA, 36, MA, 37, PhD(chem), 41. *Prof Exp:* Asst chem, Univ Toronto, 36-40; res chemist, Dye & Chem Co, Can, 41-43 & Can Int Paper Co, 43-48; mem staff, Indust Cellulose Res Ltd, 48-52; tech dir, Columbia Cellulose Co, Ltd, 52-59; res dir, Sandwell & Co, Ltd, 59-63; PROF CHEM ENG & APPL CHEM, UNIV TORONTO, 63-, PROF FORESTRY, 73- *Mem:* Fel AAAS; Am Chem Soc; Tech Asn Pulp & Paper Indust; Can Soc Microbiologists; fel Royal Soc Can; fel Chem Inst Can. *Res:* Chemistry of wood; lignin; pulp and paper; cellulose to alcohol conversion; conversion of municipal solid waste (garbage) to fuels and chemicals; feasibility analysis; author of books. *Mailing Add:* 17 Noel Ave Toronto ON M4G 1B2 Can

WAYMAN, OLIVER, b Logan, Utah, Jan 8, 16; m 43; c 5. ANIMAL PHYSIOLOGY, ANIMAL NUTRITION. *Educ:* Utah State Agr Col, BS, 47; Cornell Univ, PhD, 51. *Prof Exp:* Asst animal husbandman, 51-52, assoc animal scientist, 52-57, chmn dept animal sci, 54-60, animal scientist, 57-80, EMER ANIMAL SCIENTIST, UNIV HAWAII, 80- *Concurrent Pos:* Res assoc psychoenergetic lab, Univ Mo, 60-61; Rockefeller Found grant, Colombian Land & Cattle Inst, 69; USDA Coop State Res Serv grant, Honolulu, 71-74; Dept Planning & Econ Develop grant, Hawaii, 75-77. *Mem:* Am Soc Animal Sci; Soc Study Reproduction. *Res:* Influence of tropical environment upon growth, development and reproduction of cattle; improvement of feeding value of tropical by-products and forage; use of the whole pineapple plant as ruminant feed. *Mailing Add:* 2749 Peter St Honolulu HI 96816

WAYMIRE, JACK CALVIN, b Dayton, Ohio, Jan 10, 41; div; c 1. NEUROBIOLOGY, NEUROCHEMISTRY. *Educ:* Earlham Col, BA, 63; Ohio State Univ, PhD(physiol), 69. *Prof Exp:* Res assoc, Med Ctr, Univ Colo, 69-73; asst prof neurochem, Univ Calif, Irvine, 73-78; chmn dept, 84-88, ASSOC PROF NEUROBIOL, MED SCH, UNIV TEX, HOUSTON, 78- *Concurrent Pos:* Prin investr, Nat Inst Neurol & Commun Dis & Stroke, 74- & Nat Inst Aging, 76-78. *Mem:* Soc Neurosci; Sigma Xi. *Res:* Cellular basis for regulation of monoamine synthesis, secretion and plasticity; aging of the nervous system. *Mailing Add:* Dept Neurobiol & Anat Univ Tex Box 20708 Houston TX 77025

WAYMOUTH, CHARITY, b London, Eng, Apr 29, 15. CELL BIOLOGY, BIOCHEMISTRY. *Educ:* Univ London, BSc, 36; Aberdeen Univ, PhD(biochem), 44. *Hon Degrees:* ScD, Bowdoin Col, 82. *Prof Exp:* Technician, Crumpsall Hosp, Manchester, Eng, 37-39; biochemist, Manchester Gen Hosps, 39-41; mem sci staff & head tissue cult dept, Chester Beatty Res Inst, Royal Cancer Hosp, 47-53; res assoc, 55-57, staff scientist, 57-63, asst dir training, 69-72, sr staff scientist, 63-81, asst dir res, 76-77, assoc dir sci affairs, 77-80, interim dir, 80-81, EMER SR STAFF SCIENTIST, JACKSON LAB, 81- *Concurrent Pos:* Beit mem fel, Aberdeen Univ, 44-47, Nat Inst Med Res, London, 45, Carlsberg Biol Found, Copenhagen, 46 & St Thomas' Hosp Med Sch, Eng, 46-47; Am Cancer Soc-Brit Cancer Campaign exchange fel, Jackson Lab, 52-53, res fel, 53-55; hon lectr, Univ Maine, 64-80; ed in chief, In Vitro, Tissue Cult Asn, 68-75; Rose Morgan vis prof, Univ Kans, 71. *Mem:* AAAS; Tissue Cult Asn (pres, 60-62); NY Acad Sci. *Res:* Design of culture media tailored to the needs of particular cells for growth, differentiation or function, with respect to major and minor nutrients, hormones and growth factors, osmolality, ion balance and gas phase. *Mailing Add:* 16 Atlantic Ave Bar Harbor ME 04609

WAYMOUTH, JOHN FRANCIS, b Ingenio Barahona, Dominican Repub, May 24, 26; m 49; c 4. APPLIED PHYSICS. *Educ:* Univ of the South, BS, 47; Mass Inst Technol, PhD(physics), 50. *Prof Exp:* Lab asst, Univ of the South, 43-44 & 46-47; asst, Mass Inst Technol, 49-50; sr engr, Sylvania Elec Prod Inc Div, Gen Tel & Electronics Corp, 50-58, sect head, 58-65, mgr physics lab, 65-69, dir res Sylvania Lighting Prod Group, 69-88; INDEPENDENT CONSULT, 88- *Concurrent Pos:* Consult, Magnet Lab, Mass Inst Technol, 58-65; mem, US Govt Adv Comt, Tech Electron Prod Radiation Safety Standards Comt, 77-79 & Adv Comt Physics, NSF, 81-83. *Honors & Awards:* W Elenbaas Award, Eindhoven Tech Univ, Dutch Phys Soc & N V Philips Co, 73. *Mem:* Am Phys Soc; Illum Eng Soc. *Res:* Oxide cathodes; electroluminescent phosphors; gaseous electronics. *Mailing Add:* 16 Bennett Rd Marblehead MA 01945

WAYNANT, RONALD WILLIAM, b Gettysburg, Pa, Oct 4, 40. ELECTRO-OPTICS, FIBER OPTICS. *Educ:* Johns Hopkins Univ, BES, 62; Cath Univ, MSEE, 66, PhD(elec eng), 71. *Prof Exp:* Asst engr, laser res & develop, Westinghouse Corp, 62-69; res elec engr, short wave length laser, Naval Res Lab, 69-86; SR OPTICAL ENGR, FOOD & DRUG ADMIN, 86- *Concurrent Pos:* Adj prof, elec eng & fac quantum electronics, Cath Univ; ed-in-chief, Circuits & Devices Mag, 86- *Mem:* Fel Inst Elec & Electronics Engrs; fel Optical Soc Am; Soc Photo-Optical Instrumentation Eng. *Res:* Medical fiber optics; laser surgery including angioplasty and ophthalmology; laser interaction with tissue; non-linear interaction; optical imaging. *Mailing Add:* 13101 Claxton Dr Laurel MD 20708

WAYNE, BURTON HOWARD, b Acton, Mass, Nov 18, 24; m 47; c 3. ELECTRICAL ENGINEERING. *Educ:* Mich State Univ, BS, 51, MS, 54, PhD(elec eng), 60. *Prof Exp:* Instr elec eng, Mich State Univ, 55-64; assoc prof, 64-71, PROF ENG ANALYSIS & DESIGN & CHMN DEPT, UNIV NC, CHARLOTTE, 71- *Mem:* Inst Elec & Electronics Engrs; Am Soc Eng Educ; Nat Soc Prof Engrs. *Res:* Circuit theory. *Mailing Add:* Dept Elec Eng Univ NC Univ Sta Charlotte NC 28223

WAYNE, CLARENCE EUGENE, b Moundsville, WVa, June 5, 56. DYNAMICAL SYSTEMS, STATISTICAL MECHANICS. *Educ:* Univ Va, BS, 78; Harvard Univ, AM, 79, PhD(physics), 82. *Prof Exp:* Postdoctoral fel, Inst Math & Appln, Univ Minn, 82-83; asst prof, 83-86, ASSOC PROF MATH, PA STATE UNIV, 86- *Concurrent Pos:* Res instr, Univ Va, 83-84. *Mem:* Am Phys Soc; Am Math Soc. *Res:* Dynamical systems, particularly hamiltonian systems with many degrees of freedom; classical statistical mechanics. *Mailing Add:* Dept Math Pa State Univ University Park PA 16802

WAYNE, GEORGE JEROME, b New York, NY, Sept 13, 14; m 41; c 2. PSYCHIATRY, PSYCHOANALYSIS. *Educ:* Brooklyn Col, BS, 34; Univ Western Ont, MD, 39; Southern Calif Psychoanal Inst, PhD(psychoanal), 77; Am Bd Psychiat & Neurol, dipl, 50. *Prof Exp:* Med dir, Los Angeles Neurol Inst, 45-56; CLIN PROF PSYCHIAT, SCH MED, UNIV CALIF, LOS ANGELES, 53- *Concurrent Pos:* Consult, Vet Admin Hosp, 50; teaching analyst, Inst Psychoanal Med Southern Calif, 53-; med dir, Edgemont Hosp, 56-81; consult, Camarillo & Metrop State Hosp, 57; US Info Agency, 59-; pres, Nat Asn Pvt Psychiat Hosps, 75. *Mem:* Fel Geront Soc; fel Am Psychiat Asn; Am Psychoanal Asn; Aerospace Med Asn; fel Am Col Physicians; Am Acad Psychiat Law. *Res:* Psychotherapy and somatic therapies in schizophrenia; psychiatric explorations of aged people; cause and treatment of psychoses; research in teaching methods; forensic psychiatry; workers compensation. *Mailing Add:* 25031 Hidden Mesa Ct Monterey CA 93940

WAYNE, LAWRENCE GERSHON, b Los Angeles, Calif, Mar 11, 26; m 48, 62; c 5. MICROBIOLOGY. *Educ:* Univ Calif, Los Angeles, BS, 49, MA, 50, PhD(microbiol), 52. *Prof Exp:* Chief bact res lab, Vet Admin Hosp, San Fernando Calif, 52-71; CHIEF TUBERC RES LAB, VET ADMIN HOSP, LONG BEACH, 71-; ASSOC CLIN PROF MED, UNIV CALIF, IRVINE-CALIF COL MED, 70- *Concurrent Pos:* Mem infectious dis res prog comt, Vet Admin, 61-64, mem pulmonary dis res prog comt, 64; mem lab comt, Vet Admin-Armed Forces Coop Study Chemother Tuberc, 61-66; consult, Calif Dept Pub Health, 63-69; mem mycobacterium taxon subcomt, Int Asn Microbiol Socs, 66-, chmn, 76-; mem adv comt actinomycetes, Bergey's Manual Trust, 67-; mem judicial comn, Int Comt Syst Bact, 73-86, chmn, 78-; mem bact & mycol study sect, Nat Inst Allergy & Infectious Dis, 71-74. *Honors & Awards:* Bergey Award, 88. *Mem:* Fel Am Acad Microbiol; Am Soc Microbiol; Am Thoracic Soc (secy-treas, 72-74). *Res:* Natural history and diagnostic techniques of tuberculosis and fungus diseases; physiology and classification of mycobacteria. *Mailing Add:* Tuberc Res Lab Vet Admin Hosp 5901 E Seventh St Long Beach CA 90822

WAYNE, LOWELL GRANT, b Washington, DC, Nov 27, 18; m 42; c 2. AIR POLLUTION, ENVIRONMENTAL HEALTH. *Educ:* Univ Calif, Berkeley, BS, 37; Calif Inst Technol, PhD(inorg chem), 49; Am Bd Indust Hyg, dipl. *Prof Exp:* Fel petrol refining, Mellon Inst, 49-52; sr phys chemist, Stanford Res Inst, 53-54; indust health engr, Univ Calif, Los Angeles, 54-56; res photochemist, Air Pollution Control Dist, Los Angeles, 56-62; res analyst, Allan Hancock Found, Univ Southern Calif, 62-69, res assoc comput sci, 63-65, res biol sci, 65-69, sector head, Air Pollution Control Inst, 65-72, res assoc, Sch Pub Admin, 69-72; vpres & dir res, 72-85, SR SCIENTIST, PAC ENVIRON SERV, INC, 85-; SR SCIENTIST, VALLEY RES CORP, 87- *Concurrent Pos:* Prof consult, Comt Motor Vehicles Emissions, Nat Acad Sci-Nat Res Coun, 71-73. *Mem:* Fel AAAS; fel Am Inst Chemists; Am Chem Soc; Air Pollution Control Asn; Am Indust Hyg Asn. *Res:* Kinetics and photochemistry of gas phase reactions, especially chemical reactions in polluted urban atmospheres; oxides of nitrogen; air quality modelling; air quality evaluation; atmospheric chemistry. *Mailing Add:* 1763 Pickett Rd McKinleyville CA 95521

WAYNE, WILLIAM JOHN, b Cass Co, Mich, Apr 23, 22; m 46; c 3. GEOMORPHOLOGY, QUATERNARY STRATIGRAPHY. *Educ:* Ind Univ, AB, 43, MA, 50, PhD(geol), 52. *Prof Exp:* Head glacial geologist, State Geol Surv, Ind, 52-68; assoc prof geol, 68-71, PROF GEOL, UNIV NEBR, LINCOLN, 71- *Concurrent Pos:* Vis prof, Univ Wis, 66-67; res, Inst Argentino de Nivologia y Glaciologia, 80; NSF grants, Int Progs, 80, 82-84, Nat Geog Soc res grant, 87-88; Fulbright lectr (Argentina), 87. *Mem:* AAAS; Geol Soc Am; Asn Geol Arg; Ger Quaternary Asn; Am Quaternary Asn; Am Inst Prof Geologists; Asn Eng Geologists. *Res:* Quaternary stratigraphy and paleontology; geomorphology; environmental geology; geomorphology and Pleistocene stratigraphy in Indiana and Nebraska; alpine geomorphology in Nevada and Argentina. *Mailing Add:* Dept Geol 214 Bessey Hall Univ Nebr Lincoln NE 68588-0340

WAYNER, MATTHEW JOHN, b Clifton, NJ, Sept 7, 27; c 3. NEUROSCIENCE, PSYCHOPHARMACOLOGY. *Educ:* Dartmouth Col, AB, 49; Tufts Univ, MS, 50; Univ Ill, PhD(psychol), 53. *Prof Exp:* From asst prof to prof psychol, Brain Res Lab, Syracuse Univ, 53-82; DIR, DIV LIFE SCI, UNIV TEX, 83- *Concurrent Pos:* ed-in-chief, Physiol & Behav, Pharmacol, Biochem & Behav, Brain Res Bulletin & Neurosci Biobehav Rev. *Mem:* Am Psychol Asn; Am Physiol Soc; Int Brain Res Orgn; Soc Neurosci; Am Col Neuropsychopharmacol. *Res:* Hypothalamic mechanisms and ingestive behavior; neural mechanisms of ingestive behavior and drug action. *Mailing Add:* Div Life Sci Univ Tex 700 NW Loop 1604 San Antonio TX 78285

WAYNER, PETER C, JR, b Taunton, Mass, Aug 18, 34; m 63; c 3. CHEMICAL ENGINEERING. *Educ:* Rensselaer Polytech Inst, BSChE, 56; Mass Inst Technol, SM, 60; Northwestern Univ, PhD(chem eng), 63. *Prof Exp:* Res engr, United Aircraft Res Labs, Conn, 63-65; from asst prof to assoc prof, 65-75, PROF CHEM ENG, RENSSELAER POLYTECH INST, 75- *Concurrent Pos:* Consult heat transfer & fluid mech; chmn, Heat Transfer & Energy Conversion Div, Am Inst Chem Engrs, 85. *Honors & Awards:* Elected Fel, Am Inst Chem Engrs. *Mem:* Am Inst Chem Engrs; Am Chem Soc; Am Soc Mech Engrs. *Res:* Use of interfacial phenomena to control transport phenomena; heat and mass transfer; fluid mechanics; boiling. *Mailing Add:* Dept Chem Eng Rensselaer Polytech Inst Troy NY 12180-3590

WAYRYNEN, ROBERT ELLIS, b Lake Norden, SDak, Oct 24, 24; m 46; c 2. GENERAL CHEMISTRY. *Educ:* SDak State Col, BS, 48; Univ Utah, PhD(chem), 53. *Prof Exp:* From chemist to sr chemist, Photo Prod Dept, E I du Pont de Nemours & Co, Inc, 52-62, res supvr, 62-65, tech serv group supvr, 65-66, field sales mgr, 66-69, tech mgr photo prod dept, 69-85; RETIRED. *Mem:* Soc Photog Sci & Eng; Sigma Xi. *Res:* Photosynthesis; photography; photographic chemistry. *Mailing Add:* 1138 Webster Dr Webster Farm Wilmington DE 19803

WAZIRI, RAFIQ, b Afghanistan, Dec 11, 33; US citizen; m 67; c 3. NEUROBIOLOGY, PSYCHIATRY. *Educ:* Am Univ, Beirut, BS, 56, MD, 60. *Prof Exp:* Res fel neurophysiol, Med Sch, Harvard Univ, 64-66; asst prof psychiat, Col Med, Univ Iowa, 66-68; asst prof physiol & assoc prof psychiat, Col Med, Univ Tenn, 70-72; ASSOC PROF PSYCHIAT, COL MED, UNIV IOWA, 72- *Mem:* Soc Neurosci. *Res:* Neurotransmission in the central nervous system of a plysia; lithium effects on the nervous system; alcohol effects on neural tissues in culture. *Mailing Add:* 500 Newton Rd Iowa City IA 52210

WAZZAN, A R FRANK, b Lattakia, Syria, Oct 17, 35; US citizen; m 59; c 3. NUCLEAR ENGINEERING, FLUID MECHANICS. *Educ:* Univ Calif, Berkeley, BS, 59, MS, 61, PhD(eng sci), 63. *Prof Exp:* From asst prof to assoc prof eng, 63-69, PROF ENG & APPL SCI, UNIV CALIF, LOS ANGELES, 74-, DEAN, SCH ENG, 87- *Concurrent Pos:* Guggenheim fel, Copenhagen, 66; consult, McDonnell Douglas Corp, 62-71, Lawrence Radiation Lab, 65-67, Westinghouse Elec Corp, 74-76, NAm Aviation, 75-78, Honeywell, 76-78 & Rand Corp, 75-; reviewer heat & mass transfer, thermodyn & quantum mech, fluid mech & nuclear eng, Appl Mech Rev, 71-; vis scholar with EDF, Paris & Off Comnr Atomic Energy, Saclay, France, 73 & 79. *Mem:* Fel Am Nuclear Soc; Am Inst Aeronaut & Astronaut. *Res:* Modeling of fuel elements for fast breeder reactor; stability and transition of laminar flows; thermodynamics of solids and of dense gases; thermal hydraulics of pressurized water reactors. *Mailing Add:* Deans Off 7400 Boelter Hall Sch Eng & Appl Sci Univ Calif Los Angeles Los Angeles CA 90024-1600

WEAD, WILLIAM BADERTSCHER, b Columbus, Ohio, Mar 11, 40; m 62; c 3. MICROCIRCULATION, VASCULAR SMOOTH MUSCLE CELL CONTRACTION. *Educ:* Wabash Col, BA, 62; Ohio State Univ, MS, 67, PhD(physiol), 69. *Prof Exp:* Res & teaching asst physiol, Ohio State Univ, 63-69; asst prof, 69-81, basic sci coordr nursing & allied health, 75-80, ASSOC PROF PHYSIOL, SCH MED, UNIV LOUISVILLE, 81-, VCHMN, DEPT PHYSIOL & BIOPHYS, 85- *Concurrent Pos:* Prin investr, Am Heart Asn & Am Lung Asn; fel pulmonary pathophysiol, Health Sci Ctr, Univ Tex, Dallas, 82. *Mem:* Am Heart Asn; Am Physiol Soc; Int Union Physiol Sci; Europ Microcirculatory Soc. *Res:* Isolated vascular smooth muscle cell control. *Mailing Add:* Dept Physiol & Biophys Univ Louisville Health Sci Ctr Louisville KY 40292

WEAKLEY, MARTIN LEROY, b Piedmont, WVa, June 5, 25; m 50; c 4. AGRICULTURAL CHEMISTRY. *Educ:* Antioch Col, BS, 51. *Prof Exp:* Lab asst, Nat Cash Register Co, Ohio, 49-51; chemist, Celanese Corp Am, Tex, 51-54 & John Deere Chem Co, Okla, 54-65; chemist, Res & Develop Dept, Nipak Inc, 65-67, mgr chem appln, 67-78; RETIRED. *Concurrent Pos:* Sci instr, Pryor Sch Dist, Okla, 81-91. *Res:* Product and process development and use application in agricultural field; fertilizer chemistry; scientific education. *Mailing Add:* Rte 2 Box 229 Pryor OK 74361

WEAKLIEM, HERBERT ALFRED, JR, b Newark, NJ, Mar 24, 26; m 55; c 4. PHYSICAL CHEMISTRY. *Educ:* Rutgers Univ, BSc, 53; Cornell Univ, PhD(phys chem), 58. *Prof Exp:* mem tech staff, RCA Labs, 58-84; sr scientist, Chronar Corp, 84-90; CONSULT, 90- *Concurrent Pos:* Vis prof physics, Univ Calif, Los Angeles, 72-73. *Mem:* Am Chem Soc; Am Phys Soc; Sigma Xi; Optical Soc. *Res:* Solid state and molecular spectroscopy; plasma chemistry; optics. *Mailing Add:* 132 King George Rd Pennington NJ 08534

WEAKS, THOMAS ELTON, b Cumberland City, Tenn, Sept 12, 34; m 59; c 2. PLANT PHYSIOLOGY. *Educ:* Austin Peay State Univ, BS, 56; George Peabody Col, MA, 60; Univ Tenn, Knoxville, PhD(bot), 71. *Prof Exp:* High sch instr, Fla, 58-65; instr biol, Brevard Jr Col, 66-67; from asst prof to assoc prof bot, 71-82, PROF BIOL SCI, MARSHALL UNIV, 82- *Concurrent Pos:* Sigma Xi res grant, 74; instnl res grant, Marshall Univ, 75; consult, US Army CEngr. *Mem:* Am Soc Plant Physiologists; Sigma Xi; Scand Soc Plant Physiologists. *Res:* Inhibitory action of canavanine in higher plants; phytotoxin effects on fungus diseases of legumes; allelopathic interference as factor affecting periphyton communities. *Mailing Add:* Dept Bot Marshall Univ Huntington WV 25701

WEALE, GARETH PRYCE, b Presteigne, Powys, UK, Mar 22, 63. CCD DETECTORS, IR DETECTOR SYSTEMS. *Educ:* Univ Birmingham, UK, BSc, 84. *Prof Exp:* Res engr, GEC Res, 84-87; RES ENGR, ITRES RES, 88- *Concurrent Pos:* Consult, Heartland Perforating, 89-90. *Res:* CCD's design and application; PC based imaging spectrograph. *Mailing Add:* 6815 Eighth St NE No 110 Calgary AB T2E 7H7 Can

WEAR, JAMES OTTO, b Francis, Okla, Oct 25, 37; m 59; c 2. PHYSICAL CHEMISTRY, CLINICAL ENGINEERING & HOSPITAL SAFETY. *Educ:* Univ Ark, BS, 59, MS, 60, PhD(phys chem), 62. *Prof Exp:* Staff mem, Sandia Corp, 61-65; res chemist, Southern Res Support Ctr, Vet Admin, 65-66, dir opers, 66-68, actg chief, 67-68, chief cent res instrument prog, 66-85, DIR ENG TRAINING CTR, VET ADMIN HOSP, LITTLE ROCK, 72-; PROF BIOMED INSTRUMENTATION TECHNOL & CHMN DEPT, COL HEALTH RELATED PROF, UNIV ARK, LITTLE ROCK, 72- *Concurrent Pos:* Abstractor, Chem Abstr Servs, 62-65; prof, Philander Smith Col, 66-85; asst prof, Grad Inst Technol, Univ Ark, 66-69, asst prof, Med Sch, 68-75, assoc prof, 75-76; chmn, Cert Hosp Safety Exam Comt, 88-90; mem, Cert Bd Examr Biomed Equip Technol, 73-78; mem, Clin Eng Bd Exam, 82-87; pres, Ark Sci Assocs, 69-73; chmn, Ark Sci & Technol Coun, 73-75. *Mem:* AAAS; Am Chem Soc; Asn Advan Med Instrumentation; Am Soc Hosp Engrs; Am Inst Chemists. *Res:* Hospital safety; electrochemistry; radiochemistry; science education; research support and management; hospital instrumentation maintenance; health and social planning and program evaluation. *Mailing Add:* 5104 Randolph Rd North Little Rock AR 72116

WEAR, ROBERT LEE, b Princeville, Ill, Feb 28, 24; m 46; c 2. ORGANIC POLYMER CHEMISTRY. *Educ:* Univ Ill, BS, 46; Univ Nebr, MS, 48, PhD(chem), 50. *Prof Exp:* Sr res chemist, 49-55, group supvr, 55-61, res specialist, 62-67, SR RES SPECIALIST, MINN MINING & MFG CO, 67- *Mem:* Am Chem Soc. *Res:* Condensation polymers. *Mailing Add:* 93 Kraft Rd West St Paul MN 55118-3813

WEARDEN, STANLEY, b Goliad, Tex, Oct 1, 26; m 51; c 5. STATISTICS. *Educ:* St Louis Univ, BS, 50; Univ Houston, MS, 51; Cornell Univ, PhD, 57. *Prof Exp:* Asst biol, Univ Houston, 50-51, instr, 51-53; asst animal husb, Cornell Univ, 53-57; asst prof math, Kans State Univ, 57-59, from assoc prof to prof statist, 59-66; dir div statist, 66-69, chmn, Dept Statist & Comput Sci, 69-73, PROF STATIST, WVA UNIV, 66-, DEAN, GRAD SCH, 72- *Concurrent Pos:* Hon res fel, Birmingham Univ, 63-64; USPHS spec fel, 63-64. *Mem:* Am Statist Asn; Am Soc Animal Sci; Biomet Soc. *Res:* Statistical methods in study of quantitative inheritance; low temperature biology; statistics in agricultural research and genetics. *Mailing Add:* Dept Statist & Comput Sci WVa Univ Knapp Hall Morgantown WV 26506

WEARE, BRYAN C, b York, Maine, Aug 22, 47. CLIMATE DYNAMICS, ATMOSPHERIC SCIENCE. *Educ:* Bates Col, BS, 69; State Univ NY, Buffalo, PhD(biophys sci), 74. *Prof Exp:* Res assoc, dept meteorol, Mass Inst Technol, 74-76; asst prof, 76-82, ASSOC PROF ATMOSPHERIC SCI, UNIV CALIF, DAVIS, 82- *Concurrent Pos:* Vis prof, Meteorol Inst, Univ Munich, 83-84. *Mem:* Am Meteorol Soc; Royal Meteorol Soc. *Res:* Interaction of oceans and atmosphere of tropical regions. *Mailing Add:* Dept Land Air & Water Resources Univ Calif Davis CA 95616

WEARE, JOHN H, b Boston, Mass, Mar 8, 40; m 63; c 1. CHEMISTRY. *Educ:* Harvey Mudd Col, BS, 62; Johns Hopkins Univ, PhD(chem), 67. *Prof Exp:* NSF grant, Air Force Off Sci Res, 68-69; asst prof, 69-80, ASSOC PROF PHYS CHEM, UNIV CALIF, SAN DIEGO, 80- *Res:* Theoretical chemistry, particularly molecular and atomic structure and interactions and the study of irreversible processes. *Mailing Add:* Dept Chem Univ Calif San Diego Box 109 La Jolla CA 92093-0332

WEARLY, WILLIAM L, b Warren, Ind, Dec 5, 15. MINING ENGINEERING. *Educ:* Purdue Univ, BSEE, 37. *Prof Exp:* Engr, Joy Mfg Co, Pittsburgh, Pa, 37-42, vpres sales, 42-55, pres & chief exec officer, 55-62; vpres & dir, Ingersoll Rand, 62-67, chmn & chief exec officer, 67-80; RETIRED. *Concurrent Pos:* Dir, Am-SAfrican Co & Med Care Int. *Mem:* Nat Acad Eng; Am Inst Elec Eng; Am Soc Mining Engrs; sr mem Inst Elec & Electronics Engrs. *Mailing Add:* One Milbank Unit 2F Greenwich CT 06830

WEARN, RICHARD BENJAMIN, b Newberry, SC, Aug 1, 16; m 40; c 5. ORGANIC CHEMISTRY. *Educ:* Clemson Col, BS, 37; Univ Ill, MS, 39, PhD(org chem), 41. *Prof Exp:* Asst, Nat Starch Prod, Inc, NY, 37-38; asst chem, Univ Ill, 39-40; res chemist, Eastman Kodak Co, NY, 41-42; res chemist, Southern Res Inst, 46-47, head org chem div, 47-48; asst res dir, Res & Develop Dept, Colgate-Palmolive Peet Co, 48-52, assoc dir, Colgate-

Palmolive Int, Inc, 52-57, dir res & develop, Household Prod Div, 57-65, dir prod develop, 65-67, tech dir, 67-71, vpres res & develop, 71-77; RETIRED. *Mem:* Am Chem Soc. *Res:* Heterogeneous catalytic reactions; terpene derivatives; diene addition products of diaroyl ethylenes; structure of cannabinol and of cannabidiol; isolation and synthesis of active principle in marihuana; utilization of soaps and synthetic detergents. *Mailing Add:* Still Hopes Box 172 100 Seventh St Ext West Columbia SC 29169-7151

WEART, HARRY W(ALDRON), b Seneca Falls, NY, July 10, 27; m 53; c 3. PHYSICAL METALLURGY. *Educ:* Rensselaer Polytech Inst, BMetE, 51; Univ Wis, MS, 52, PhD(metall), 57. *Prof Exp:* Instr foundry metall, Univ Wis, 53-56; res engr metal physics, Res Lab, Westinghouse Elec Corp, 56-60; asst prof phys metall, Cornell Univ, 60-64; prof metall eng & chmn dept, 64-88, PROF METALL ENG, UNIV MO, ROLLA, 89- *Concurrent Pos:* Am Coun Educ intern acad admin, Univ Calif, Berkeley, 70-71; vis foreign lectr, Inst Appl Physics, Univ Tsukuba, Japan, 89. *Mem:* Am Soc Eng Educ; Am Soc Metals Int; Am Inst Mining, Metall & Petrol Engrs; Mat Res Soc. *Res:* Phase transformations; diffusion, especially surface diffusion; crystal growth; amorphization in layered structures. *Mailing Add:* Dept Metall Eng Univ Mo Rolla MO 65401

WEART, RICHARD CLAUDE, b Brandon, Iowa, July 20, 22; m 46; c 3. STRATIGRAPHY, PALEONTOLOGY. *Educ:* Cornell Col, BA, 43; Syracuse Univ, MS, 48; Univ Ill, PhD(geol), 50. *Prof Exp:* Asst prof geol, Tex Tech Col, 50-52; from paleontologist & paleogeologist to asst chief geologist, Latin Am Div, Sun Oil Co, 52-60, head staff geologist, 60-62, mgr geol res, 63-71, mgr geol, 71-75, mgr strategic explor, 75-78; MGR EXPLOR RES & DEVELOP, SUNMARK EXPLOR CO, 78- *Mem:* Am Asn Petrol Geologists. *Res:* Petroleum geology. *Mailing Add:* 6828 Meadowcreek Dr Dallas TX 75240

WEART, SPENCER RICHARD, b Detroit, Mich, Mar 8, 42; m 71; c 2. PHYSICS HISTORY. *Educ:* Cornell Univ, BA, 63; Univ Colo, PhD(physics & astrophys), 68. *Prof Exp:* Teaching & res asst physics, Joint Inst Lab Astrophys, Univ Colo, 63-68; res asst astron, Hawaii Inst Astrophys, Univ Hawaii, 66-67; fel solar physics, Mt Wilson & Palomar Observ, 68-71; res assoc hist, Univ Calif, Berkeley, 71-74; DIR CTR HIST PHYSICS, AM INST PHYSICS, 74- *Concurrent Pos:* Teaching asst, Calif Inst Technol, 69-70; res apprenticeship, Inst Int Studies, 72-73; vis prof, Princeton, 90. *Mem:* Am Astron Soc; Hist Sci Soc (treas, 83-87); Soc Social Studies Sci. *Res:* Solar chromosphere; origins of sunspots; space telescopes; history of geophysics, solid state and nuclear physics, French science, and modern astrophysics. *Mailing Add:* Am Inst Physics Ctr Hist Physics 335 E 45th St New York NY 10017

WEART, WENDELL D, b Brandon, Iowa, Sept 24, 32; m 54; c 3. GEOPHYSICS. *Educ:* Cornell Col, BA, 53; Univ Wis, PhD(geophys), 61. *Prof Exp:* Geophysicist, Ballistics Res Lab, 56-59; geophysicist, Sandia Corp, 59-69, supvr underground physics div, 69-75, DEPT MGR, NUCLEAR WASTE SYSTS, SANDIA LABS, 75- *Mem:* AAAS; Am Geophys Union. *Res:* Earth physics relating to underground explosion; nuclear waste disposal in geologic media, particular emphasis on salt. *Mailing Add:* Sandia Labs Orgn 6330 Bldg 823 Rm 4096 Albuquerque NM 87185

WEARY, MARLYS E, b Chicago, Ill, Mar 13, 39. PYROGENS, ENDOTOXINS. *Educ:* Valparaiso Univ, BA, 60; Univ Ill, MS, 62; Lake Forest Sch Mgt, MBA, 81. *Prof Exp:* Pharmacologist, Baxter Travenol Labs, Inc, 62-66, supvr biol control, 66-81, mgr Microbiol Tech Serv, 81-82, pyrogen technol, 82-86, microbiol/pyrogen technol, 86-89, sr res scientist, Baxter Healthcare Corp, Inc, 89-90; PRIN, MERIT CONSULT SERV, 90- *Concurrent Pos:* Lectr, Ctr Prof Advan, Sch Pharm, Univ Ill. *Mem:* NY Acad Sci; Parenteral Drug Asn. *Res:* Biologic control and pyrogen testing; limulus lysate research; endotoxin research. *Mailing Add:* Merit Consult Serv 513 S George St Mt Prospect IL 60056

WEARY, PEYTON EDWIN, b Evanston, Ill, Jan 10, 30; m 52; c 3. DERMATOLOGY. *Educ:* Univ Va, MD, 55. *Prof Exp:* Intern, Univ Hosps Cleveland, 55-56; from asst resident to resident, 58-61, from instr to assoc prof, 61-70, PROF DERMAT, SCH MED, UNIV VA, 70-, VCHMN DEPT, 75- *Concurrent Pos:* Chmn, Coun Nat Prog Dermat, 72-75. *Honors & Awards:* Gold Medal, Am Acad Dermat, 90. *Mem:* Am Dermat Asn; Am Acad Dermat; Soc Invest Dermat. *Res:* Exploration of the keratinolytic abilities of various dermatophyte fungal organisms and investigation of ecology of certain lipophilic yeast organisms on the skin surface. *Mailing Add:* Dept Dermat Univ Va Hosp Charlottesville VA 22908

WEAST, CLAIR ALEXANDER, b Modesto, Calif, Oct 13, 13; m 40; c 2. FOOD CHEMISTRY. *Educ:* Univ Calif, BS, 37, MS, 39, PhD(agr chem), 42. *Prof Exp:* Asst, Univ Calif, 37-42; chemist, USDA, 42-44; chief chemist, Pac Can Co, 44-46; res dir, Tillie Lewis Foods, Inc, 46-78; RETIRED. *Res:* Food products; chlorophyllase. *Mailing Add:* J16127 S Cottage Ave Manteca CA 95336

WEATHERBEE, CARL, b Michigan City, Ind, Nov 21, 16; m 50; c 5. ORGANIC CHEMISTRY. *Educ:* Hanover Col, AB, 40; Univ Ill, AM, 46; Univ Utah, PhD(chem), 50. *Prof Exp:* Asst chem, Univ Ill, 46; instr, Reed Col, 50; asst prof, Univ Hawaii, 50-51; prof chem & chmn dept, Millikin Univ, 52-82, emer prof, 82; RETIRED. *Concurrent Pos:* Researcher, Univ Utah, 59-60, NDak State Univ, 69-70. *Mem:* Am Chem Soc; Am Inst Chemists; Sigma Xi. *Res:* Mannich bases; nitrogen mustards. *Mailing Add:* 1360 W Macon St Decatur IL 62522-2704

WEATHERBEE, JAMES A, b Chicago, Ill, Jan 22, 43; m; c 2. MICROTUBULES, GROWTH FACTORS. *Educ:* Ill Inst Technol, PhD(biol), 72. *Prof Exp:* sr staff scientist, 85-89, DIR, RES DEVELOP, R & D SYSTS, 89- *Concurrent Pos:* From adj instr to adj asst prof, Rutgers Med Sch, Univ Med & Dent NJ, 81-85. *Mem:* Am Soc Cell Biol. *Res:* Biochemistry and biological effects of growth factor. *Mailing Add:* R & D Systs 614 McKinley Pl NE Minneapolis MN 55413

WEATHERBY, GERALD DUNCAN, b Neodesha, Kans, Mar 13, 40; m 68; c 2. BIOCHEMISTRY. *Educ:* Univ Kans, BS, 62 & 63, PhD(biochem), 69. *Prof Exp:* Teacher high sch, Kans, 63-65; asst prof chem, Lake Superior State Col, 69-77, assoc prof, 77-81; chmn dept chem, 81, assoc prof, 81-82, PROF CHEM, OKLA CITY UNIV, 83- *Res:* Mechanism of riboflavin catalyzed carbon-carbon bond oxidations; extrapolation of these model system studies to elucidate mechanisms of flavoenzyme catalysis. *Mailing Add:* Dept Chem Okla City Univ 2501 N Blackwelder Oklahoma City OK 73106

WEATHERFORD, THOMAS WALLER, III, b Uriah, Ala, Mar 12, 30; m 57; c 2. PERIODONTOLOGY. *Educ:* Auburn Univ, DVM, 54; Univ Ala, Birmingham, DMD, 61, MSD, 69; Am Bd Periodont, dipl, 76. *Prof Exp:* Intern pedodontics, Sch Dent, 61-62, instr dent, 62-69, from asst prof to assoc prof periodont, 69-77, PROF PERIODONT, SCH DENT, UNIV ALA, 77-, DIR POSTDOCTORAL EDUC, 70-, ASST PROF COMP MED, SCH MED & SR SCIENTIST, INST DENT RES, 77- *Concurrent Pos:* NIH trainee, 61-62; staff dentist, Birmingham Vet Admin Hosp, 62-65, resident periodont, 66-68, res assoc, 68-70; resident periodont, Sch Dent, Univ Ala, 66-68; consult, Dept Animal Serv, Univ Ala, 67-69; investr, Inst Dent Res, 70-; mem, Am Asn Dent Schs; consult, Birmingham Vet Admin Hosp, Tuskegee Vet Admin Hosp, Children's Hosp & Eye Found Hosp (Birmingham). *Mem:* AAAS; Am Dent Asn; Int Asn Dent Res; Am Acad Periodont; Am Asn Lab Animal Sci; Sigma Xi. *Res:* Animal models in periodontal research; dental plaque control; histochemistry of the periodontium. *Mailing Add:* Sch Dent Univ Ala University Station AL 35294

WEATHERFORD, W(ILLIAM) D(EWEY), JR, b Orange, Tex, Nov 20, 23; m 45; c 2. CHEMICAL ENGINEERING, COMBUSTION. *Educ:* Univ Tex, BS, 44; Univ Pittsburgh, MS, 49, PhD(chem eng), 54. *Prof Exp:* Tutor chem eng, Univ Tex, 44-45; asst process engr, Neches Butane Products Co, 45-47; jr fel, Mellon Inst, 47-54, fel, 54-58; sr res engr, Southwest Res Inst, 58-60, sect mgr, 60-68, staff engr, 68-71, sect mgr, 71-83, inst engr, 83-85; CONSULT, CHEM ENGR, 85- *Concurrent Pos:* Consult, Appl Physics Lab, Johns Hopkins Univ, 54-56; mem FAA-Safer Tech Group, Post-crash fire hazard reduction & Nat Res Coun, assessment mat submarine hull insulation. *Mem:* Combustion Inst; Am Inst Chem Engrs. *Res:* Principles and applications of fluid flammability, combustion and fire safety; mechanisms of fretting wear; diffusional processes; fire-resistant fuels; thermophysical properties of alkali metals; properties and physical chemistry of invert microemulsion. *Mailing Add:* 219 Anne Lewis Dr San Antonio TX 78216-6606

WEATHERLEY, ALAN HAROLD, b Sydney, Australia, Mar 28, 28. FISH & POLLUTION BIOLOGY. *Educ:* Univ Sydney, BSc, 49; Univ Tasmania, MSc, 59; Univ Glasgow, PhD(zool), 61. *Prof Exp:* Res fel physiol, Univ Sydney, 49-51; res scientist fishery biol, Commonwealth Sci & Indust Res Orgn, 51-57; asst lectr zool, Univ Glasgow, 58-59, acting lectr, 59-60; lectr zool, Australian Nat Univ, 60-62, sr lectr, 62-71, reader, 71-72; prof fisheries biol, Inst Biol & Geol, Univ Tromsö, 74-75; PROF ZOOL, LIFE SCI DIV, SCARBOROUGH COL, UNIV TORONTO, 75- *Concurrent Pos:* Vis fel, Leverhulme Trust, 70. *Honors & Awards:* Publ Award, Wildlife Soc, 72; Hilary Jolly Award, Australian Soc Limnol, 74. *Mem:* Can Soc Zool; Am Fisheries Soc; Australian Soc Limnol (pres 65-66); fel Int Acad Fishery Scientists. *Res:* Studies on freshwater fish in fields of ecology, distribution, taxonomy, physiology, especially thermal tolerance and somatic growth, conservation and heavy metal pollution. *Mailing Add:* Dept Zool Scarborough Col Univ Toronto Scarborough ON M1C 1A4 Can

WEATHERLY, GEORGES LLOYD, b 1942; div; c 2. OCEANOGRAPHY. *Educ:* Univ Va, BS, 64; Harvard Univ, MA, 66, MEng, 67; Nova Univ, PhD(phys oceanography), 71. *Prof Exp:* Fel, Nova Univ, 71-72; exchange scientist, Inst Oceanog, USSR Acad Sci, 72-73; from asst prof to assoc prof, 73-78, PROF OCEANOG, FLA STATE UNIV, 78- *Mem:* Am Geophys Union; AAAS. *Res:* Near bottom currents and deep circulation in the ocean; oceans bottom boundary layer; suspended sediment transport dynamics in the oceans; turbulent processes in the ocean. *Mailing Add:* Dept Oceanog Fla State Univ Tallahassee FL 32306

WEATHERLY, NORMAN F, b Elkton, Ore, June 22, 32; m 52; c 6. MEDICAL PARASITOLOGY. *Educ:* Ore State Univ, BS, 53, MS, 60; Kans State Univ, PhD(parasitol), 62. *Prof Exp:* NIH trainee parasitol, 62-63, from asst prof to assoc prof, 63-72, PROF PARASITOL, SCH PUB HEALTH, UNIV NC, CHAPEL HILL, 72- *Concurrent Pos:* Consult, Nat Inst Gen Med Sci, 76-80; adj prof, Dept Microbiol, Duke Univ, 81- *Mem:* Am Soc Parasitol; Am Soc Trop Med & Hyg; Am Micros Soc. *Res:* General immunobiology of helminth parasites; cell mediated responses of hosts to helminth parasites. *Mailing Add:* Dept Parasitol 201H Sch Pub Health Univ NC Chapel Hill NC 27599-7400

WEATHERRED, JACKIE G, b Pampa, Tex, Mar 14, 34; m 60; c 3. PHYSIOLOGY. *Educ:* Univ Tex, DDS, 59, PhD(physiol), 65. *Prof Exp:* Dent consult, Tex Inst Rehab & Res, 60-62; instr physiol & res assoc oral path, Med Col Va, 62-63; from asst prof to assoc prof physiol, Dent Sch, Univ Md, Baltimore, 63-69; prof, 69-90, COORDR PHYSIOL & DIR GRAD PROGS, ORAL BIOL, SCH DENT, MED COL GA, 90- *Concurrent Pos:* Mem test construction comt, Coun Nat Bd Dent Examr; basic sci consult, Coun Dent Educ; Am Col Dent fel, 71; mem oral biol & med study sect, NIH, 76-80; chmn coun of fac, Am Asn Dent Sch, 78-79; vpres fac, Am Asn Dent Sch, 79-82. *Mem:* AAAS; Int Asn Dent Res; NY Acad Sci. *Res:* Plasma kinins and oral physiology; circulation in dental pulp; predisposing factors in experimental carcinoma of the hamster cheek pouch; membrane transport in oral epithelium. *Mailing Add:* Sch Dent Med Col Ga Augusta GA 30902-1129

WEATHERS, DWIGHT RONALD, b Milledgeville, Ga, Aug 14, 38; m 75; c 4. ORAL PATHOLOGY. *Educ:* Emory Univ, DDS, 62, MSD, 66. *Prof Exp:* From asst prof to assoc prof, 67-79, PROF ORAL PATH, SCH DENT, 79-, PROF DERM, PROF PATH, EMORY UNIV, 88- *Concurrent Pos:* Dir bd

& dipl, Am Bd Oral Path; prof, Winship Cancer Ctr, Emory Univ, Dean, 85-; Consult, Hospitals Test Construction Com, Am Bd Path. *Honors & Awards:* Fel, Am Col Dentists, 87. *Mem:* Am Acad Oral Path; Am Dent Asn. *Res:* Herpes simplex virus; neoplasia of oral cavity. *Mailing Add:* Emory Univ Sch Dent 1462 Clifton Rd NE Atlanta GA 30322

WEATHERS, LEWIS GLEN, b Sunset, Utah, July 5, 25; m 46; c 4. PLANT PATHOLOGY. *Educ:* Utah State Col, BS, 49, MS, 51; Univ Wis, PhD(phytopath), 53. *Prof Exp:* Asst bot, Utah State Univ, 49-51; asst plant path, Univ Wis, 51-53; vchmn dept plant path, 71-73, chmn dept, 73-77, PROF PLANT PATH & PLANT PATHOLOGIST, UNIV CALIF, RIVERSIDE, 53-, ASSOC DEAN, COL NATURAL & AGR SCI, 77- *Concurrent Pos:* Guggenheim fel, 61-62; Rockefeller fel, 63; NATO fel, 73. *Mem:* Am Phytopath Soc; Int Orgn Citrus Virol; Sigma Xi. *Res:* Citrus and virus diseases; scion and rootstock uncongenialities in citrus; effect of environmental factors on virus diseases; interactions of unrelated viruses in mixed infections. *Mailing Add:* Dept Plant Path Univ Calif Riverside CA 92521

WEATHERS, WESLEY WAYNE, b Homer, Ill, Sept 28, 42. ENVIRONMENTAL PHYSIOLOGY. *Educ:* San Diego State Col, BS, 64; Univ Calif, Los Angeles, MA, 67, PhD(zool), 69. *Prof Exp:* USPHS cardiovasc scholar, Sch Med, Univ Calif, Los Angeles, 69-70; from asst prof to assoc prof physiol, Rutgers Univ, New Brunswick, 70-75; asst prof physiol, 75-76, ASSOC PROF PHYSIOL, UNIV CALIF, DAVIS, 76- *Mem:* Am Physiol Soc; Am Soc Zoologists; Sigma Xi; Cooper Ornith Soc; Am Ornith Union. *Res:* Comparative physiology of temperature regulation and vertebrate ecological energetics. *Mailing Add:* Dept Avian Sci Univ Calif Davis CA 95616

WEATHERSBY, AUGUSTUS BURNS, b Pinola, Miss, May 19, 13; m 45; c 2. MEDICAL ENTOMOLOGY, PARASITOLOGY. *Educ:* La State Univ, AB, 38, MS, 40, PhD(entom), 54. *Prof Exp:* Asst dist entomologist, State Dept Agr, La, 40-42; entomologist-parasitologist, Naval Med Res Inst, 42-62; prof, 62-82, EMER PROF ENTOM, UNIV GA, 82- *Mem:* AAAS; Entom Soc Am; Am Soc Parasitol; Am Soc Trop Med & Hyg; Am Mosquito Control Asn; fel, Royal Soc Trop Med & Hyg. *Res:* Medical entomology; innate immunity of mosquitoes to malaria; malaria survey, control and parasitology; drug action and immunity in malaria; life cycles of malaria; exoerythrocytic stages, tissue culture, and time-lapse cinephotomicrography; cryobiology. *Mailing Add:* 210 Bishop Dr Athens GA 30606

WEATHERSPOON, CHARLES PHILLIP, b Tucson, Ariz, Dec 1, 42; m 64; c 2. SILVICULTURE, TREE SEEDLING PHYSIOLOGY. *Educ:* Univ Ariz, BS, 64; Duke Univ, PhD(plant physiol), 68. *Prof Exp:* Res botanist remote sensing, US Army Eng Topog Labs, 70-73; forester silvicult, Kaibab Nat Forest, 73-76, forester timber inventory, Southwestern Region, 76-78, RES FORESTER SILVICULT & TREE SEEDLING PHYSIOL RES, PAC SOUTHWEST FOREST & RANGE EXP STA, US FOREST SERV, 78- *Concurrent Pos:* Consult, Nat Acad Sci, Comt Effects Herbicides, Vietnam, 72-73. *Mem:* Soc Am Foresters; Am Soc Plant Physiol; Sigma Xi. *Res:* Establishment and growth of forest stands in relation to their environment; effects of cultural treatment, especially prescribed fire, on this environment; assessment of physiological condition of tree seedlings. *Mailing Add:* 8263 Granada Dr Redding CA 96002

WEAVER, ALBERT BRUCE, b Mont, May 27, 17; m 45; c 3. PHYSICS, ACADEMIC ADMINISTRATION. *Educ:* Univ Mont, AB, 40; Univ Idaho, MS, 41; Univ Chicago, PhD(physics), 52. *Prof Exp:* Physicist, US Naval Ord Lab, 42-45; res assoc physics, Univ Chicago, 52-53 & Univ Wash, 53-54; from asst prof to assoc prof physics, Univ Colo, 54-58, chmn dept, 56-58; head physics dept, 58-70, assoc dean, Col Lib Arts, 61-70, provost acad affairs, 70-72, exec vpres, 72-83, PROF PHYSICS, UNIV ARIZ, 58- *Mem:* Fel AAAS; fel Am Phys Soc. *Res:* Cosmic rays. *Mailing Add:* 5726 E Holmes Tucson AZ 85711

WEAVER, ALFRED CHARLES, b Johnson City, Tenn, July 18, 49; c 1. COMPUTER SCIENCE. *Educ:* Univ Tenn, BS, 71; Univ Ill, MS, 73, PhD(comput sci), 76. *Prof Exp:* Vis asst prof, Univ Ill, 76-77; ASSOC PROF COMPUT SCI, UNIV VA, 77- *Concurrent Pos:* Consult, Gen Elec, Lockheed, NASA, Johnson Controls & Johnson & Johnson, 72- *Mem:* Asn Comput Mach; Inst Elec & Electronics Engrs; Sigma Xi. *Res:* Computer networks. *Mailing Add:* Thornton Hall Univ Va Charlottesville VA 22901

WEAVER, ALLEN DALE, b Galesburg, Ill, Nov 15, 11; wid; c 2. PHYSICS. *Educ:* Knox Col, BS, 33; Univ Mich, MS, 47; NY Univ, PhD(sci educ), 54. *Prof Exp:* High sch teacher, Ill, 35-37; jr high sch teacher, 37-40; instr physics, phys sci & math, Md State Teachers Col, Salisbury, 47-55; assoc prof phys sci, Northern Ill Univ, 55-60, prof physics, 60-81; RETIRED. *Mem:* Nat Sci Teachers Asn; Nat Asn Res Sci Teaching. *Res:* Science education. *Mailing Add:* 591 Garden Rd De Kalb IL 60115

WEAVER, ANDREW ALBERT, b Sarasota, Fla, Dec 10, 26; m 51; c 4. ENTOMOLOGY. *Educ:* Col Wooster, BA, 49; Univ Wis, MS, 51, PhD(zool), 55. *Prof Exp:* Asst prof biol, 56-66, PROF BIOL, COL WOOSTER, 66- *Mem:* AAAS; Am Micros Soc; Entom Soc Am; Am Soc Zoologists; Soc Syst Zool. *Res:* Taxonomy of centipedes and copepods. *Mailing Add:* Dept Biol Col Wooster Wooster OH 44691

WEAVER, CHARLES EDWARD, b Lock Haven, Pa, Jan 27, 25; m 46; c 3. CLAY MINERALOGY, SEDIMENTOLOGY. *Educ:* Pa State Univ, BS, 48, MS, 50, PhD(mineral), 52. *Prof Exp:* Res assoc mineral, Pa State Univ, 52; res assoc mineral, Shell Res & Develop Co, 52-55, res scientist, 55-59; res group leader mineral, Continental Oil Co, 59-63; assoc prof, 63-65, dir, Sch Geophys Sci, 70-81, PROF MINERAL, GA INST TECHNOL, 65- *Honors & Awards:* Mineral Soc Am Award, 58; Sigma Xi Res Award, 72; Distinguished Mem Clay Minerals Soc, 85. *Mem:* Clay Minerals Soc (vpres, 66, pres, 67); Mineral Soc Am; Geochem Soc; Geol Soc Am; Soc Econ Paleontologists & Mineralogists. *Res:* Clay mineralogy and petrology; geochemistry of sediments; clay-water chemistry; radioactive dating of sediments; geothermometry of shales; diagenesis-metamorphism; salt mineralogy; envionmental geochemistry, biogeochemistry. *Mailing Add:* Sch Earth & Atmospheric Sci Ga Inst Technol Atlanta GA 30332

WEAVER, CHARLES HADLEY, b Tenn, Jan 27, 20; m 44; c 4. ELECTRICAL ENGINEERING. *Educ:* Univ Tenn, BS, 43, MS, 48; Univ Wis, PhD(elec eng), 56. *Prof Exp:* Engr & supvr, Tenn Eastman Corp, 43-46; from instr to prof elec eng, Univ Tenn, 46-59; Westinghouse prof, Auburn Univ, 59-63, head prof, 63-65; dean eng, 65-68, chancellor, 68-71, vpres continuing educ, 71-81, dean, Space Inst, 75-81, PROF ELEC ENG, UNIV TENN, KNOXVILLE, 82- *Concurrent Pos:* Consult, Sverdrup & Parcel, Inc, 53-54 & Oak Ridge Nat Lab, 55-64. *Mem:* Inst Elec & Electronics Engrs; Am Soc Eng Educ. *Mailing Add:* Dept Elec Eng Univ Tenn Knoxville TN 37916

WEAVER, CHRISTOPHER SCOT, b New York, NY, Feb 6, 51. OPTICAL VIDEODISC, TELECOMMUNICATIONS. *Educ:* Hobart Col, AB, 73; Wesleyan Univ, MA & MS, 76, CAS, 77. *Prof Exp:* Assoc dir news, Nat Broadcasting Co, 77-78; mgr technol res, Am Broadcasting Co, 78-80; vpres sci & technol, Nat Cable TV Asn, 80-81; vpres res & develop, VML Labs, 81-83; PRES, MEDIA TECHNOL ASSOCS, LTD, 82- *Concurrent Pos:* Asst prof aerodyn, Norwich Univ, 78; telecommun ed, Video Mag, 80-81; res fel, Mass Inst Technol, 80-82, vis scholar, 82-; US Nat Comt rep, Int Union Radio Sci, 81; tech adv, subcomt commun, US Cong, 81-83 & Off Technol Assessment, 83-85. *Honors & Awards:* NSF/Navy Award. *Mem:* Inst Elec & Electronics Engrs; Soc Info Display; Soc Motion Picture & TV Engrs. *Res:* Participatory and interactive videodisc systems; computer graphics; idiosyncratic computer systems; optical data storage; man-machine interface. *Mailing Add:* c/o Media Technol 15235 Shady Grove Rd Suite 100 Rockville MD 20850

WEAVER, CONNIE MARIE, b LaGrande, Ore, Oct 29, 50; m 71; c 3. NUTRITION. *Educ:* Ore State Univ, BS, 72, MS, 74; Fla State Univ, PhD(foods & nutrit), 78. *Prof Exp:* PROF FOODS & NUTRIT, PURDUE UNIV, 78- *Concurrent Pos:* Res fel, Kraft Inc. *Mem:* Sigma Xi; Inst Food Technol; Am Chem Soc; Am Inst Nutrit. *Res:* Mineral bioavailability. *Mailing Add:* Dept Foods & Nutrit Stone Hall Purdue Univ West Lafayette IN 47907

WEAVER, DAVID DAWSON, b Twin Falls, Idaho, Feb 12, 39; m 67; c 2. MEDICINE, HUMAN GENETICS. *Educ:* Col Idaho, BS, 61; Univ Ore, MS & MD, 66. *Prof Exp:* Intern med, Milwaukee County Gen Hosp, Wis, 66-67; biogeneticist, Arctic Health Res Ctr, 67-70, pediat residency, Med Sch, Univ Ore, 70-72; USPHS fel human genetics, Med Sch, Univ Wash, 72-74; March of Dimes fel, Genetics & Metab Dis, Dept Pediat, Univ Ore Health Sci Ctr, Portland, 74-76; asst prof, 76-81, assoc prof, dept med genetics, 81-86, PROF, SCH MED, IND UNIV, 86-, DIR CLIN SERV, 76- *Concurrent Pos:* Lectr genetics & cell biol, Univ Alaska, 68-70. *Mem:* Am Soc Human Genetics; Sigma Xi. *Res:* Dysmorphology; birth defects; prenatal diagnosis; microcephaly. *Mailing Add:* Dept Med & Molecular Genetics Ind Univ Sch Med 975 W Walnut St Indianapolis IN 46202-5251

WEAVER, DAVID LEO, b Albany, NY, Apr 18, 37; m 66. THEORETICAL PHYSICS. *Educ:* Rensselaer Polytech Inst, BS, 58; Iowa State Univ, PhD(physics), 63. *Prof Exp:* Res assoc physics, Iowa State Univ, 63-64; from asst prof to assoc prof, 64-77, PROF PHYSICS, TUFTS UNIV, 77- *Concurrent Pos:* NATO fel physics, Europ Orgn Nuclear Res, Switz, 65-66; Nat Nuclear Energy Comt fel, Frascati Nat Lab, Italy, 68-69. *Mem:* Am Phys Soc. *Res:* Theoretical elementary particle physics; mathematical physics; molecular biophysics. *Mailing Add:* Dept Physics Tufts Univ Medford MA 02155

WEAVER, DONALD K(ESSLER), JR, b Great Falls, Mont, July 18, 24; m 48; c 2. ELECTRICAL ENGINEERING. *Educ:* Stanford Univ, BS, 48, MS, 49, EE, 50, PhD(elec eng), 59. *Prof Exp:* Asst, Stanford Univ, 48-50; sr res engr, Stanford Res Inst, 50-56; assoc prof elec eng & dir electronics lab, 56-64, dir eng exp sta, 64-69, PROF ELEC ENG, MONT STATE UNIV, 64- *Mem:* Am Soc Eng Educ; Inst Elec & Electronics Engrs. *Res:* Network theory; communication theory. *Mailing Add:* 2404 Spring Creek Dr Bozeman MT 59715

WEAVER, EDWIN SNELL, b Hartford, Conn, Jan 30, 33; m 55; c 5. PHYSICAL CHEMISTRY. *Educ:* Yale Univ, BS, 54; Cornell Univ, PhD(chem), 59. *Prof Exp:* Asst chem, Cornell Univ, 55-57; from asst prof to assoc prof, 58-71, chmn dept, 72-78, PROF CHEM, MT HOLYOKE COL, 71- *Concurrent Pos:* NSF sci fac fel & vis prof, Univ Calif, San Diego, 64-65; vis prof, Univ Conn, 71-72; vis fel, Yale Univ, 78-79; NSF Sci Fac fel, 78-79. *Mem:* AAAS; Am Chem Soc. *Res:* Physical chemistry of polymers and proteins; neutron activation analysis. *Mailing Add:* 115 Woodridge St South Hadley MA 01075

WEAVER, ELLEN CLEMINSHAW, b Oberlin, Ohio, Feb 18, 25; m 44; c 2. PLANT PHYSIOLOGY, REMOTE SENSING OF OCEAN. *Educ:* Western Reserve Univ, AB, 45; Stanford Univ, MA, 52; Univ Calif, PhD(genetics), 59. *Prof Exp:* Staff mem, Carnegie Inst Dept Plant Biol, 61-62; res assoc, Biophys Lab, Stanford Univ, 62-67; resident res assoc, Ames Res Ctr, NASA, 67-69; dir off sponsored res, 74-77, assoc prof, 69-77, interium exec vpres, 78-79, PROF BIOL SCI, SAN JOSE STATE UNIV, 78- *Concurrent Pos:* Vis prof, Univ Hawaii, 85. *Mem:* AAAS; Am Soc Plant Physiol; Bot Soc Am; Asn Women in Sci. *Res:* Mechanisms of photosynthesis, using wild type and mutant strains of algae; light-induced electron transport as monitored by means of electron paramagnetic resonance spectroscopy; evolution of photosynthesis; remote sensing of chloryphyll in oceans. *Mailing Add:* Dept Biol Sci San Jose State Univ San Jose CA 95192

WEAVER, ERVIN EUGENE, b Centralia, Wash, Mar 12, 23; m 48; c 3. ENVIRONMENTAL CHEMISTRY. *Educ:* Manchester Col, BA, 45; Univ Ill, MA, 47; Case-Western Reserve Univ, PhD(inorg chem), 51. *Prof Exp:* Asst, Univ Ill, 45-47; asst prof chem, Baldwin-Wallace Col, 47-51; from asst prof to assoc prof, Wabash Col, 51-61; res scientist, Chem Dept Ford Motor Sci Lab, 61-66, res scientist, Vehicle Emissions, Prod Res, 66-72, supvr catalysts develop & testing, Engine & Foundry Div, 72-73, emission planning assoc, Automotive Emissions & Fuel Econ Off, Ford Motor Co, 73-83; RETIRED. *Concurrent Pos:* Consult, Argonne Nat Lab, 56-59. *Mem:* Soc Automotive Engrs; Air Pollution Control Asn; Am Chem Soc; Sigma Xi. *Res:* Hexafluorides of heavy metals; xenon fluorides; nonaqueous solvents; air pollution; catalytic control of automotive emissions. *Mailing Add:* Two Unutsi Ct Brevard NC 28712-9276

WEAVER, GEORGE THOMAS, b Anna, Ill, Mar 11, 39; m 60; c 3. FOREST ECOLOGY, SILVICULTURE. *Educ:* Southern Ill Univ, Carbondale, BA, 60, MSEd, 62; Univ Tenn, Knoxville, PhD(bot), 72. *Prof Exp:* Instr bot, 70-71, asst prof forestry, 71-77, ASSOC PROF FORESTRY, SOUTHERN ILL UNIV, CARBONDALE, 77- *Mem:* Ecol Soc Am; Am Inst Biol Sci. *Res:* Dry matter production and nutrient cycling in forest ecosystems; soil-site relationships of forests. *Mailing Add:* Dept Forestry Univ Tenn Knoxville TN 37996

WEAVER, HAROLD FRANCIS, b San Jose, Calif, Sept 25, 17; m 39; c 3. ASTRONOMY. *Educ:* Univ Calif, AB, 40, PhD(astron), 42. *Prof Exp:* Nat Res Coun fel, Yerkes Observ, Chicago & McDonald Observ, 42-43; tech aide, Nat Defense Res Comt, DC, 43-44; physicist, Radiation Lab, 44-45, from asst astronr to assoc astronr, Lick Observ, 45-51, assoc prof, 51-56, dir radio astron lab, 58-72, PROF ASTRON, UNIV CALIF, BERKELEY, 56- *Concurrent Pos:* Mem, US Army Air Force-Nat Geog Soc eclipse exped, Brazil, 47. *Mem:* Int Astron Union; Am Astron Soc; Sigma Xi. *Res:* Spectroscopy of peculiar stars; star clusters; galactic structure; radio astronomy. *Mailing Add:* Dept Astron 601 Campbell Hall Univ Calif Berkeley CA 94720

WEAVER, HARRY EDWARD, JR, b Philadelphia, Pa, Feb 1, 23; m 44; c 2. EXPERIMENTAL PHYSICS. *Educ:* Case Inst Technol, BS, 43, MS, 48; Stanford Univ, PhD(physics), 52. *Prof Exp:* Physicist, Manhattan Proj, Tenn, 44-46; instr physics, Case Inst Technol, 46-48; asst, Stanford Univ, 48-52; asst, Physics Inst, Zurich, 52-54; physicist, Varian Assocs, Calif, 54-69; physicist, Hewlett-Packard Co, 69-88; PHYSICIST, SETS, INC, MILILANI, HAWAII, 88- *Concurrent Pos:* Vis lectr, Univ Zurich, 59-61. *Mem:* Am Phys Soc; Sigma Xi. *Res:* Nuclear magnetic and electron resonance, especially in solids; cryogenic engineering and application of high field superconductive materials in high field solenoids for high resolution nuclear magnetic resonance; trajectory analysis of ions in electric fields produced by quadrupole structures of finite length and with imperfections as applied to the design of mass spectrometers; x-ray photoelectron spectroscopy; super critical fluid, especially carbon dioxide; Raman spectroscopy of simple molecules & solvents effects of ultraviolet spectra of chromophores. *Mailing Add:* 987 Westridge Dr Portola Valley CA 94028

WEAVER, HARRY TALMADGE, b Brewton, Ala, Dec 15, 38; m 59; c 3. SOLID STATE PHYSICS. *Educ:* Auburn Univ, BS, 60, MS, 61, PhD(physics), 69. *Prof Exp:* Mem staff eng, Humble Oil & Refining Co, 62-63; mem staff physics, 68-78, SUPVR, SANDIA LABS, 78- *Mem:* Am Phys Soc; Inst Elec & Electronics Engrs; AAAS. *Res:* Design and development of III-V semiconductor devices and optoelectronic components. *Mailing Add:* Sandia Labs Div 1141 Albuquerque NM 87185

WEAVER, HENRY D, JR, b Harrisonburg, Va, May 5, 28; m 52; c 4. PHYSICAL CHEMISTRY. *Educ:* George Washington Univ, BS, 50; Univ Del, MS & PhD(phys chem), 53. *Prof Exp:* Assoc prof chem, Eastern Mennonite Col, 51-57; assoc prof, Goshen Col, 57-71, actg dean, 70-72, prof chem, 71-79, provost, 72-79; dep dir educ abroad, Univ Calif, Santa Barbara, 79-91, adj lectr chem, 80-91; RETIRED. *Concurrent Pos:* Tech adv, Lima, Peru, 64-65; Fulbright lectr, Tribhuvan Univ, Nepal, 69-70. *Mem:* AAAS; Am Chem Soc; Am Sci Affil (pres, 62). *Res:* Heterogenous kinetics; metal; acid systems; kinetics of complex ion formation. *Mailing Add:* 4985 Old Oak Pl Santa Barbara CA 93111

WEAVER, JAMES B, JR, b Hartwell, Ga, Jan 28, 26; m 49; c 1. GENETICS, AGRONOMY. *Educ:* Univ Ga, BSA, 50, NC State Univ, MS, 52, PhD(genetics, agron), 55. *Prof Exp:* Asst county agt, Univ Ga, 52-53, asst prof agron, 55-58; dir cotton res, DeKalb Agr Res Inc, 58-63; dir cotton res, Cotton Hybrid Res, Inc, 63-65; supt grounds & agronomist, Univ Ga, 65-67, from asst prof to prof agron, 67-87; CONSULT, 87- *Mem:* Am Soc Agron; Crop Sci Soc Am; Entom Soc Am. *Res:* Basic research on cotton improvement; utilization of hybrid vigor in cotton; insect resistance. *Mailing Add:* 155 Hardin Dr Athens GA 30605-1519

WEAVER, JAMES COWLES, b Faribault, Minn, Sept 8, 40; m 66; c 2. BIOPHYSICS, MEDICAL PHYSICS. *Educ:* Carleton Col, BS, 62; Yale Univ, MS, 63, PhD(physics), 69. *Prof Exp:* Fel physics, Res Lab Electronics, 69-72, res assoc, 72-74, staff physicist, 74-78, LECTR PHYSICS, MASS INST TECHNOL, 74-, RES ASSOC BIOPHYS, DEPT NUTRIT & FOOD SCI, 78- *Concurrent Pos:* lectr med physics, Harvard-Mass Inst Technol, Div Health Sci & Technol, 78-86, assoc dir, Biomed Eng Ctr, 85-, prin res scientist, 86- *Mem:* Am Phys Soc; Am Chem Soc; Bioelectrochem Soc; AAAS; Sigma Xi; Int Soc Anal Cytol; Bioelectromagnetics Soc. *Res:* Biophysics and medical physics; electromagnetic field effects on cells and tissue, particularly electroporation; rapid methods of cell analysis; biosensors. *Mailing Add:* Mass Inst Technol Rm 20A-128 Cambridge MA 02139

WEAVER, JEREMIAH WILLIAM, organic chemistry; deceased, see previous edition for last biography

WEAVER, JOHN HERBERT, b Cincinnati, Ohio, Sept 16, 46. SOLID STATE PHYSICS. *Educ:* Univ Mo, BS, 67, MS, 69; Iowa State Univ, PhD(physics), 73. *Prof Exp:* Fel physics, Mat Res Ctr, Univ Mo-Rolla, 73; res assoc physics, 74-75, asst scientist, 75-77, assoc scientist, Synchrotron Radiation Ctr, Phys Sci Lab, Univ Wis-Madison, 77-; DEPT CHEM ENG & MAT SCI, UNIV MINN. *Concurrent Pos:* Assoc, Ames Lab, US Dept Energy, 75-; adj prof mat sci prog, 81. *Mem:* Am Phys Soc; Sigma Xi. *Res:* Optical properties of solids; electronic structure of metals and alloys; hydrogen in metals; photoemission and surface physics. *Mailing Add:* Dept Chem Eng & Sci Univ Minn 151 Amundson Minneapolis MN 55455

WEAVER, JOHN SCOTT, geophysics, for more information see previous edition

WEAVER, JOHN TREVOR, b Birmingham, Eng, Nov 5, 32; m 60; c 3. GEOPHYSICS. *Educ:* Bristol Univ, BSc, 53; Univ Sask, MSc, 55, PhD(physics), 59. *Prof Exp:* From instr to asst prof math, Univ Sask, 58-61; leader appl math group, Defence Res Estab Pac, BC, 61-66; from asst prof to assoc prof physics, 66-72, PROF PHYSICS, UNIV VICTORIA, 72- *Concurrent Pos:* Lectr math, Univ Victoria, 62-64; Nat Sci Eng Res Coun Can res grants, 66-; travel fel, Univ Cambridge, 72-73; actg chmn physics, Univ Victoria, 78-79, chmn, 80-88; vis fel, Univ Edinburgh, 79-80; sci collabr, Cantonal Observ, Neuchâtel, Switz, 84; actg dean sci, Univ Victoria, 88-89, Nat Sci Eng Res Coun Can & Roy Soc Bilateral Exchange, Univ Edinburgh, 89 & NSF Bilateral Exchange Cantoner Observ, Neuchâtel, 90. *Mem:* Am Geophys Union; Can Asn Physicists; fel Royal Astron Soc; Can Geophys Union. *Res:* Electromagnetic induction in the earth; electromagnetic theory; geomagnetic variations. *Mailing Add:* Dept Physics & Astron Univ Victoria Victoria BC V8W 3P6 Can

WEAVER, KENNETH NEWCOMER, b Lancaster, Pa, Jan 16, 27; m 50; c 2. GEOLOGY. *Educ:* Franklin & Marshall Col, BS, 50; Johns Hopkins Univ, MA, 52, PhD(geol), 54. *Prof Exp:* Geologist, Medusa Portland Cement Co, 56-61, mgr geol & quarry dept, 61-63; DIR, MD GEOL SURV, 63- *Concurrent Pos:* Governor's rep, Interstate Oil Compact Comn, 64- & Interstate Mining Compact, 74-, Chmn, 77; mem, Md Mining Coun, 73-, chmn, 77-88; mem, Mid-Atlantic Gov Coastal Resources Coun, 74-79, Subcomt Mgt Major Underground Projs, US Nat Comt Tunneling Technol, Nat Res Coun, 77-79, Comt Surface Mining & Reclamation, 78-80, Comt Disposal Excess Spoil, 80-81, Comt Geol Mapping, 83- & Comt Water Res, 89-; chmn, Md Land Reclamation Comt, 68-76 & 79-, State Mapping Comt, 76-, Comt Abandoned Mined Lands Res Priorities, Nat Res Coun, 87, GEOREF Adv Comt, Am Geol Inst, 91-; mem Md-Del Joint Boundary Comn, 74-, Outer Continental Shelf Policy Comt, 74-79 & 85-, Md Comn Artistic Property, 86- & Delamarva Comt, Nat Water Qual Assessment Prog, US Geol Surv; mem, Groundwater Interagency Taskforce, Sci & Technol Comts, US House Rep, 78-79; liaison mem, Bd Earth Sci, 83-88; mem-at-large, Am Geol Inst Ex Comt, 89-90. *Mem:* Fel AAAS; fel Geol Soc Am; Am Inst Mining, Metall & Petrol Eng; Am Inst Prof Geologists; Asn Am State Geol (vpres, 71, pres, 73). *Res:* Geology of industrial minerals; environmental and structural geology; research administration. *Mailing Add:* Md Geol Surv 2300 St Paul St Baltimore MD 21218

WEAVER, L(ELLAND) A(USTIN) C(HARLES), b Winnipeg, Man, Sept 29, 37; m 58; c 3. ELECTRICAL ENGINEERING. *Educ:* Univ Toronto, BASc, 60; Univ Ill, MS, 62, PhD(elec eng), 66; Univ Pittsburgh, MBA, 76. *Prof Exp:* Instr elec eng, Univ Ill, 62-66; sr engr, Optical Physics Dept, Westinghouse Res Labs, 66-74, mgr gas laser res, 74-77, prog mgr excimer laser technol, 77-79, asst to dir power systs res & develop, 79-82, dir int res & develop, 82-85, managing dir, Advan Prod Tech, Europe, Westinghouse Elec, 85-87, DIR, INDUST & COM RES & DEVELOP, WESTINGHOUSE RES LABS, 87- *Mem:* Sr mem Inst Elec & Electronics Engrs; Am Phys Soc; Sigma Xi. *Res:* Gaseous and quantum electronics; gas lasers; research planning; technology assessment; robotics; environmental tech. *Mailing Add:* 216 Harwick Dr Pittsburgh PA 15235

WEAVER, LAWRENCE CLAYTON, b Bloomfield, Iowa, Jan 23, 24; m 49; c 4. PHARMACOLOGY. *Educ:* Drake Univ, BS, 49; Univ Utah, PhD(pharmacol), 53. *Prof Exp:* Asst pharmacol, Univ Utah, 49-53; pharmacologist, Res Dept, Pitman-Moore Co, 53-58, dir pharmacol labs, 59-60, assoc dir pharmacol res, 60-61, head biomed res, Pitman-Moore Div, Dow Chem Co, 61-64, asst dir res & develop labs, 64-65, asst to gen mgr, 65-66; vpres-prof rel, Pharmaceut Mfrs Asn, 84-89; prof pharmacol & dean, 66-84, EMER DEAN, COL PHARM, UNIV MINN, MINNEAPOLIS, 89- *Honors & Awards:* Am Pharmaceut Asn Found Res Achievement Award, 63; Remington Hon Gold Medal, Am Pharmaceut Asn, 89; Distinguished Serv Award, Bd Dirs, Am Asn Col Pharm, 89. *Mem:* Am Soc Pharmacol; Soc Exp Biol & Med; Am Pharmaceut Asn; Am Pub Health Asn; Acad Pharmaceut Sci. *Res:* Combinations and assay of anticonvulsant drugs; pharmacology of cardiovascular and central nervous system drugs; social and administrative pharmacy; health care delivery systems; discovery, development and distribution of orphan products. *Mailing Add:* Col Pharm 7-107 Health Sci Unit F Univ Minn 308 Harvard St SE Minneapolis MN 55455

WEAVER, LEO JAMES, b Springfield, Mo, Apr 18, 24; m 49. ORGANIC CHEMISTRY. *Educ:* Drury Col, BS, 49; Univ Mich, MS, 50. *Prof Exp:* Res chemist, Org Chems Div, Monsanto Chem Co, 50-54, res chemist, Inorg Chem Div, 54-56, res group leader detergents & surfactants, 56-60, asst dir res, 60-62, dir prod sales, 62-64, dir com develop, 64-65, dir res & develop, 65-69, pres, Monsanto Enviro-Chem Systs, Inc, 69-72; EXEC VPRES, CALGON CORP, 72- *Concurrent Pos:* Mem, Indust Res Inst. *Honors & Awards:* Monsanto du Bois Res Award, 54. *Mem:* AAAS; Am Mgt Asn; Am Chem Soc; NY Acad Sci. *Res:* Alkylation reactions of aromatics and olefins; sulfonation and sulfation of organics; detergents and surfactants. *Mailing Add:* Upper St Clair 2220 Clarmont Dr Pittsburgh PA 15241

WEAVER, LESLIE O, plant pathology; deceased, see previous edition for last biography

WEAVER, LYNN E(DWARD), b St Louis, Mo, Jan 12, 30; m 83; c 5. ELECTRICAL ENGINEERING, NUCLEAR ENGINEERING. *Educ:* Univ Mo, BSEE, 51; Southern Methodist Univ, MSEE, 55; Purdue Univ, PhD(elec eng), 58. *Prof Exp:* Develop engr, McDonnell Aircraft Corp, 52-53; aerophys engr, Convair Corp, 53-55; instr elec eng, Purdue Univ, 55-58; assoc prof, Univ Ariz, 58-59, prof nuclear eng & head dept, 59-69; assoc dean, Univ Okla, 69-70; exec asst to pres & dir off environ studies, Argonne Univs Asn, 70-72; dir, Sch Nuclear Eng & Health Physics, Ga Inst Technol, 72-82; dean eng & distinguished prof, Auburn Univ, 82-87; PRES, FLA INST TECHNOL, MELBOURNE, 87- *Concurrent Pos:* Mem, Govt Comt Atomic Energy, Ariz, 62-64; comnr, Ariz AEC, 64-69; mem adv comt space power & propulsion, NASA, 70-75; exec ed, Ann Nuclear Energy, 78-; chmn coord comt energy, Am Asn Eng Soc, 81-82; chmn pub affairs coun, Am Asn Eng Soc, Washington DC, 84-87. *Mem:* Am Soc Eng Educ; fel Am Nuclear Soc; Inst Elec & Electronics Engrs; Sigma Xi. *Res:* Nuclear reactor dynamics and control. *Mailing Add:* Fla Inst Technol Melbourne FL 32901

WEAVER, LYNNE C, b June 14, 1945; m. PHYSIOLOGY OF THE AUTONOMIC NERVOUS SYSTEM. *Educ:* Mich State Univ, DVM, 68, PhD(pharmacol), 75. *Prof Exp:* Prof physiol, Mich State Univ, 84-86; STAFF, JOHN P ROBARTS RES INST, DEPT STROKE & AGING, 87- *Honors & Awards:* Res Career Develop Award, NIH. *Mem:* Soc Neurosci; Am Physiol Soc. *Res:* Sympathetic reflexes affecting kidney and splanchic circulation; organization of discrete patterns of sympathetic discharge. *Mailing Add:* Dept Stroke & Aging John P Robarts Res Inst PO Box 5015 100 Perth Dr London ON N6A 5K8

WEAVER, MICHAEL JOHN, b London, Eng, Mar 30, 47. ELECTROCHEMISTRY, ANALYTICAL CHEMISTRY. *Educ:* Univ London, BSc, 68, PhD(chem) & DIC, 72. *Prof Exp:* Res fel chem, Calif Inst Technol, 72-75; asst prof chem, Mich State Univ, 75-80, assoc prof, 80-; DEPT CHEM, PURDUE UNIV 82- *Mem:* Am Chem Soc. *Res:* Kinetics of electrode processes; structure of electrode-electrolyte interfaces; measurement of rapid electrochemical reaction rates; theories of electron transfer kinetics; chemistry of metal macrocyles. *Mailing Add:* Dept Chem Purdue Univ West Lafayette IN 47907-9980

WEAVER, MILO WESLEY, b Lufkin, Tex, Feb 16, 13; m 34; c 5. MATHEMATICS. *Educ:* Univ Tex, BA, 35, MA, 50, PhD, 56. *Prof Exp:* Pub sch teacher, Tex, 34-45; from instr to assoc prof math, 45-77, EMER ASSOC PROF MATH, UNIV TEX, AUSTIN, 77- *Concurrent Pos:* NSF res fel, 58-59. *Mem:* Am Math Soc; Math Asn Am. *Res:* Theory of semigroups; mappings on a finite set; associative algebras. *Mailing Add:* 1051 Lonesome Trail Driftwood TX 78619

WEAVER, MORRIS EUGENE, b Morrison, Okla, June 10, 29; m 50; c 4. ANATOMY, ZOOLOGY. *Educ:* York Col, Nebr, BS, 51; Univ Omaha, BS, 53; Ore State Univ, MA, 56, PhD(zool), 59. *Prof Exp:* From instr to prof anat, Dent Sch, Ore Health Sci Univ, 72-89; RETIRED. *Concurrent Pos:* Nat Inst Dent Res fel, Inst Animal Physiol, Eng, 66-67; vis reader, Med Sch, Univ Ibadan, 72-73; NATO fel, Med Sch, Bristol, Eng, 79-80. *Mem:* Am Asn Anatomists; Int Asn Dent Res; Sigma Xi. *Res:* Cell biology and mitosis; dental research using swine as experimental animals; microcirculation and temperature regulation; tooth eruption. *Mailing Add:* 6354 SW Garden Home Rd Portland OR 97201

WEAVER, NEVIN, b Navasota, Tex, Jan 4, 20; m 62; c 2. INSECT PHYSIOLOGY, BEE BEHAVIOR. *Educ:* Southwestern Univ, AB, 41; Tex A&M Univ, MS, 43, PhD(entom), 53. *Prof Exp:* Asst biol, Southwestern Univ, 38-41; asst, Tex A&M Univ, 41-43; beekeeper, 46-48; asst entom, Tex A&M Univ, 48-51, from instr to assoc prof, 51-65; chmn dept biol, 65-68 & 74-76, chmn campus planning, 69-70, prof, 65-87, EMER PROF BIOL, UNIV MASS, BOSTON, 88- *Concurrent Pos:* Secy, Am comt, Bee Res Asn, 58; fel biol, Harvard Univ, 59-60; assoc ed, J Apicult Res, 82-87. *Mem:* AAAS; Int Bee Res Asn; Animal Behav Soc. *Res:* Honeybee and stingless bee dimorphism, development, physiology, biochemistry, especially lipids, pheromones and behavior; the roles of pheromonal and behavioral signals in the interactions of queen and worker honeybees. *Mailing Add:* Dept Biol Univ Mass Boston MA 02125

WEAVER, OLIVER LAURENCE, b Birmingham, Ala, Feb 6, 43; div; c 2. MATHEMATICAL PHYSICS, NUCLEAR PHYSICS. *Educ:* Calif Inst Technol, BS, 65; Duke Univ, PhD(physics), 70. *Prof Exp:* Instr & res assoc, Duke Univ, 69-70; from asst prof to assoc prof, 70-84, PROF PHYSICS, KANS STATE UNIV, 84- *Mem:* Sigma Xi; Am Phys Soc. *Res:* Applications of group theory in nuclear physics; atomic and nuclear scattering theory. *Mailing Add:* Dept Physics Kans State Univ Manhattan KS 66506

WEAVER, PAUL FRANKLIN, b Allentown, Pa, Feb 11, 26; m 64; c 3. ELECTRICAL ENGINEERING. *Educ:* Cornell Univ, PhD(elec eng), 59. *Prof Exp:* Mem tech staff, Bell Tel Labs, 47-52; asst prof elec eng, Cornell Univ, 59-65; assoc prof, 65-72, chmn, 76-79, PROF ELEC ENG, UNIV HAWAII, 72- *Concurrent Pos:* Staff mem, Space Environ Lab, Nat Oceanic & Atmospheric Admin, Boulder, Colo, 71-72; fac fel & consult, GTE Labs, 78-84; vis prof, Naval Res Labs, Washington, DC, 88-89 & 90. *Mem:* Inst Elec & Electronics Engrs. *Res:* Radio wave propagation; ionospheric physics; microwave circuit design. *Mailing Add:* Dept Elec Eng Univ Hawaii Honolulu HI 96822

WEAVER, R(OBERT) E(DGAR) C(OLEMAN), b New Orleans, La, Mar 30, 32; m 65; c 1. CHEMICAL ENGINEERING. *Educ:* Tulane Univ, BS, 53, MS, 55; Princeton Univ, MA, 57, PhD(chem eng), 58. *Prof Exp:* Process design engr, Ethyl Corp, 58-60; from asst prof to assoc prof chem eng, Tulane Univ, 60-66, prof & chmn dept, 66-81; PROF CHEM, METALL & POLYMER ENG & DEAN, COL ENG, UNIV TENN, 81- *Concurrent Pos:* Consult, Ethyl Corp, 60-65, Humble Oil & Refining Co, 65-73 & Dept Nat Resources, State of La, 74- *Mem:* Am Chem Soc; Am Inst Chem Engrs; Soc Petrol Engrs. *Res:* Heterogeneous catalysis; diffusion; automatic control; process dynamics and simulation; biomedical engineering; fluidization; applied econommics; resource management; environmental engineering; coal technology; reservior engineering New Orleans, Louisiana and Knoxville, Tennessee. *Mailing Add:* 321 Saint Charles Ave New Orleans LA 70130

WEAVER, RALPH SHERMAN, b Doaktown, NB, Nov 7, 35. MATHEMATICAL MODELING OF BIOPHYSICAL & BIOMEDICAL CONCERNS, ENVIRONMENTALLY SUSTAINABLE DEVELOPMENT. *Educ:* Univ NB, BSc, 56; McGill Univ, MSc, 59, PhD(nuclear physics), 62. *Prof Exp:* Demonstr, McGill Univ, 56-61, res asst, 61-62; sci officer, Defense Res Bd, 62-73, group head, 73-75; EXEC DIR, ALTA ENVIRON CTR, ALTA DEPT ENVIRON, 75-, ASST DEP MINISTER, 89- *Concurrent Pos:* Dir, Can Asn Physicists, 66-68, & Inst Res Pub Policy, 81-91; mem, Fac Grad Studies, York Univ, 66-69; vpres, Fifth Int Conf Indoor Air Qual, 90- *Mem:* Can Asn Physicists; Sigma Xi; Can Res Mgt Asn. *Res:* Heat, vapor and diffusion modeling; underwater life support systems and decompression models; biomedical and physiological systems; disposal of hazardous wastes; air pollution; environmental impacts. *Mailing Add:* Alta Environ Ctr Bag 4000 Vegreville AB T0B 4L0 Can

WEAVER, RICHARD WAYNE, b Twin Falls, Idaho, June 25, 44; m 64; c 3. SOIL MICROBIOLOGY. *Educ:* Utah State Univ, BS, 66; Iowa State Univ, PhD(soil microbiol), 70. *Prof Exp:* Assoc prof, 70-76, PROF SOIL MICROBIOL, TEX A&M UNIV, 82- *Concurrent Pos:* Fulbright Distinguished Lectr, 86. *Mem:* Fel Soil Sci Soc Am; fel Am Soc Agron; Am Soc Microbiol; fel Crop Sci Soc Am. *Res:* Soil nitrogen; microbial ecology; nitrogen fixation; waste disposal. *Mailing Add:* Dept Soil & Crop Sci Tex A&M Univ College Station TX 77843

WEAVER, ROBERT F, b Topeka, Kans, July 18, 42; m 65; c 2. BIOCHEMISTRY. *Educ:* Col Wooster, BA, 64; Duke Univ, PhD(biochem), 69. *Prof Exp:* NIH fel, Univ Calif, San Francisco, 69-71; from asst prof to assoc prof, 71-81, PROF BIOCHEM, UNIV KANS, 81-, CHMN DEPT, 84- *Concurrent Pos:* Res grants, NIH, 72-89 & Am Cancer Soc, 75-77, 81-82 & 85-88; Am Cancer Soc res scholar, Univ Zurich, Switz, 78-79 & NERC Inst Virol, Oxford, UK, 89-90. *Mem:* Am Soc Biochem & Molecular Biol; Am Soc Virol. *Res:* Structure and function of eucaryotic RNA polymerases; transcription control; molecular biology of baculoviruses. *Mailing Add:* Dept Biochem Univ Kans Lawrence KS 66044

WEAVER, ROBERT HINCHMAN, b Buckhannon, WVa, Dec 2, 31; m 58; c 1. BIOCHEMISTRY. *Educ:* WVa Wesleyan Col, BS, 53; Univ Md, MS, 55, PhD(biochem), 57. *Prof Exp:* Fel, Enzyme Inst, Univ Wis, 57-61; assoc prof, 61-70, PROF CHEM, UNIV WIS-STEVENS POINT, 70- *Mem:* AAAS; Am Chem Soc. *Res:* Amine and carbohydrate metabolism; enzyme chemistry. *Mailing Add:* Dept Chem Univ Wis Stevens Point WI 54481

WEAVER, ROBERT JOHN, b Lincoln, Nebr, Sept 23, 17; m 51; c 2. PLANT PHYSIOLOGY. *Educ:* Univ Nebr, AB, 39, MS, 40; Univ Chicago, PhD(plant physiol), 46. *Prof Exp:* Res assoc bot, Univ Chicago, 46-48; from asst viticulturist to assoc viticulturist, Exp Sta, Univ Calif, Davis, 48-58, viticulturist, 58-85, prof viticulture, 58-85; RETIRED. *Concurrent Pos:* Fulbright sr res scholar, Superior Col Agr, Greece, 56; vis res worker, Res Inst Grapevine Breeding, Ger, 63; Fulbright sr res scholar, Indian Agr Res Inst, New Delhi, 69; res fel, Enol & Viticult Res Inst, SAfrica, 73; res fel grapes, Foreign Agr Orgn, UN, Cyprus, 79. *Mem:* AAAS; Am Soc Plant Physiol; Bot Soc Am; Am Soc Enol; Am Soc Hort Sci. *Res:* Plant hormones and regulators; physiology of the grapevine. *Mailing Add:* Dept Viticult & Enol Univ Calif Agr Exp Sta Davis CA 95616

WEAVER, ROBERT MICHAEL, b Goshen, Ind, June 23, 42; m 75; c 2. CLAY MINERALOGY, SOIL SCIENCE. *Educ:* Berea Col, BA, 64; Mich State Univ, MS, 66; Univ Wis-Madison, PhD(soil sci), 70. *Prof Exp:* Asst prof soil sci, Cornell Univ, 71-78; MEM STAFF, CLAY DIV, J M HUBER CORP, 78- *Mem:* Soil Sci Soc Am; Clay Minerals Soc. *Res:* Genesis of clay minerals; clay mineralogy of tropical soils. *Mailing Add:* Clay Div J M Huber Corp Huber GA 31040

WEAVER, ROBERT PAUL, b Binghamton, NY, July 19, 52; m 73. ASTROPHYSICS, PHYSICS. *Educ:* Colgate Univ, AB, 74; Univ Colo, MS, 76, PhD(astrophys), 77. *Prof Exp:* Res asst physics, Univ Calif Los Alamos Sci Lab, 75; res asst astrophys, Univ Colo, 75-77; res assoc, Nat Acad Sci- Nat Res Coun, 77-78; STAFF MEM PHYSICS, LOS ALAMOS NAT LAB, 78- *Mem:* Am Astron Soc. *Res:* Transfer of radiation, especially in the x-ray regime; interstellar dynamics. *Mailing Add:* 39 Los Arboles Dr Los Alamos NM 87544

WEAVER, SYLVIA SHORT, marine biology, for more information see previous edition

WEAVER, W DOUGLAS, Ft Fairfield, Maine, Mar 14, 45. PREHOSPITAL CARDIAC CARE, CARDIAC TECHNOLOGY. *Educ:* Univ Maine, BA, 67; Tufts Univ Sch Med, MD, 71. *Prof Exp:* asst prof, 79-85, ASSOC PROF CARDIOL, UNIV WASH, 85- *Concurrent Pos:* Prin investr, Myocardial Infraction Triage & Intervention Trial, 87-; mem, Coun Clin Cardiol, Am Heart Asn, 84- *Mem:* Am Heart Asn; fel Am Col Cardiol. *Res:* Epidemiology of sudden death; prehospital emergency cardiac care; thrombolytic therapy shelves; development and testing of automatic external defibrinators. *Mailing Add:* No 205 MITI 1910 Fairview Ave E Seattle WA 98102

WEAVER, WARREN ELDRED, b Sparrows Point, Md, June 5, 21; m 45; c 4. CHEMISTRY. *Educ:* Univ Md, BS, 42, PhD(pharmaceut chem), 47. *Prof Exp:* Asst pharm, Univ Md, 42, asst antigas prep, Off Sci Res & Develop, 42-44; asst insecticides, 44-45; chemist, US Naval Res Lab, 45-50; assoc prof chem & pharmaceut chem, 50-54, from actg chmn to chmn dept, 50-56, dean, Sch Pharm, 56-81, PROF CHEM & PHARMACEUT CHEM, MED COL VA, 54-, EMER DEAN, SCH PHARM, 81- *Concurrent Pos:* Mem pharm rev comt, NIH, 68-72; mem bd dirs, Am Found Pharm Educ, 69-73; mem,

Secretary's Comn to rev Report of President's Task Force on Prescription Drugs, 70; mem, Am Coun Pharmaceut Educ, 74-80. *Mem:* AAAS; Am Chem Soc; Am Pharmaceut Asn; Am Inst Hist Pharm; Am Asn Cols Pharm (vpres, 67-68, pres, 68-69). *Res:* Synthetic peptides; insecticides; correlation of structure with fungicidal activity. *Mailing Add:* 403 Horsepen Rd Richmond VA 23229

WEAVER, WILLIAM BRUCE, b Catskill, NY, Sept 1, 46; c 2. ATOMIC & MOLECULAR PHYSICS, PLASMA PHYSICS. *Educ:* Univ Ariz, BS, 68; Inst Technol, 71; Western Reserve Univ, PhD(astrophys), 72. *Prof Exp:* RES ASTRONR, MONTEREY INST RES ASTRON, 72-, PRES, 86- *Concurrent Pos:* trainee, NSF & Case Western Reserve Univ, 68-72; sr prin scientist, BDM Inc, 73- *Mem:* Am Astron Soc; Royal Astron Soc. *Res:* Spectral classification; statistics and astrophysics of star formation; spectrophotometry of peculiar stars; design & construction of astronomical instrumentation including eclipses, spectrophotometers and solid state detectors. *Mailing Add:* Monterey Inst Res Astron 900 Maj Sherman Lane Monterey CA 93940

WEAVER, WILLIAM JUDSON, b Twin Falls, Idaho, May 7, 36; m 65; c 2. PATHOPHYSIOLOGY. *Educ:* Col Idaho, BS, 58; Univ Ore, MS, 65, PhD(exp path), 71. *Prof Exp:* Res assoc surg, Ore Health Sci Univ, 71-83; ASSOC PROF BIOL, LINFIELD COL, PORTLAND, ORE, 83- *Concurrent Pos:* Sci consult, Res Industs Corp, Utah, 75-83. *Mem:* AAAS; NY Acad Sci. *Res:* Chemistry; pharmacology and toxicology of dimethyl sulfoxide. *Mailing Add:* 24 Del Prado Lake Oswego OR 97035-1312

WEAVER, WILLIAM MICHAEL, b Lima, Ohio, Feb 18, 31; m 66. ORGANIC CHEMISTRY. *Educ:* Johns Carroll Univ, BS, 53; Purdue Univ, MS, , 56, PhD, 58. *Prof Exp:* Asst org chem, Purdue Univ, 53-58; from instr to assoc prof, 58-70, PROF ORG CHEM, JOHN CARROLL UNIV, 70- *Mem:* Am Chem Soc. *Res:* Aliphatic nitrocompound; kinetics; reactions in aprotic solvents. *Mailing Add:* 22279 Douglas Rd Cleveland OH 44122-2038

WEBB, ALAN WENDELL, b Enid, Okla, Sept 20, 39; m 61; c 4. HIGH PRESSURE MATERIALS SCIENCE. *Educ:* Brigham Young Univ, BS, 63, PhD(phys chem), 69. *Prof Exp:* RES CHEMIST, US NAVAL RES LAB, 68- *Concurrent Pos:* Nat Acad Sci-Nat Res Coun resident res associateship, 68-70. *Mem:* Am Chem Soc; Am Soc Home Inspectors; Am Phys Soc. *Res:* High pressure and high temperature (HPHT) synthesis of inorganic compounds; effects of HPHT on potential superconductors; HPHT sintering of flame-grown diamond; use of syncrotron radiation for high pressure structural studies. *Mailing Add:* Phase Transformation Sect US Naval Res Lab Washington DC 20375-5000

WEBB, ALBERT DINSMOOR, b Victorville, Calif, Oct 10, 17; wid; c 2. ENOLOGY. *Educ:* Univ Calif, BS, 39, PhD(chem), 48. *Hon Degrees:* Dr, Univ Bordeaux, 82. *Prof Exp:* From asst prof to prof enol, Univ Calif, 48-60, from asst chemist to chemist, Exp Sta, 48-82, prof enol, Col Agr, Univ Calif Davis, 60-82, chmn dept viticult & enol, 73-81; RETIRED. *Concurrent Pos:* Fulbright res scholar, Australia, 58, Alko Res Labs, Helsinki, Finland, 69; NATO scholar, Bordeaux, 62; vis res prof, Univ Stellenbosch, SAfrica, 70; scholar, res sta, Siebeldingen, WGer, 81; dir, Wine Indust Tech Seminars, 80- *Mem:* Am Chem Soc; Am Soc Enologists (pres, 74-75). *Res:* Identification of aroma and flavor compounds in grapes and wines. *Mailing Add:* Dept Viticult & Enol Univ Calif Davis CA 95616

WEBB, ALFREDA JOHNSON, b Mobile, Ala, Feb 21, 23; m 59; c 3. VETERINARY MEDICINE. *Educ:* Tuskegee Inst, BS, 43, DVM, 49; Mich State Univ, MS, 51. *Prof Exp:* From instr to assoc prof anat, Tuskegee Inst, 50-59; prof biol, 59-78, PROF LAB ANIMAL SCI, NC A&T STATE UNIV, 77- *Mem:* Am Asn Vet Anat; Sigma Xi. *Res:* Histology; cytology; embryology. *Mailing Add:* 137 N Dudley St Greensboro NC 27401

WEBB, ALLEN NYSTROM, b Wichita, Kans, Dec 14, 21; m 43; c 3. PHYSICAL CHEMISTRY. *Educ:* Kans State Univ, BS, 43; Univ Calif, Berkeley, PhD(phys chem), 49. *Prof Exp:* Jr chemist, Boston Consol Gas Co, 43; res chemist, Stand Oil Co, Ind, 44-46; res asst chem, Univ Calif, 46-49; from res chemist to sr res chemist, Texaco Inc, 49-66, res assoc, 66-80; RETIRED. *Concurrent Pos:* Chmn, Gordon Res Conf Catalysis, 70; mem adv bd, Petrol Res Fund, 70-72. *Mem:* Am Chem Soc. *Res:* Catalysis; chemisorption; molecular spectra; fuel cells; exchange reactions; kinetics; hydrodesulfurization. *Mailing Add:* 5901 Cedar Cliff Dr Austin TX 78759

WEBB, ANDREW CLIVE, b Bishop's Stortford, Eng, Feb 17, 47; m 80; c 1. DEVELOPMENTAL BIOLOGY. *Educ:* Southampton Univ, BSc, 69, PhD(biol), 73. *Prof Exp:* Fel biol, Purdue Univ, West Lafayette, 73-75, asst prof, 75-81, ASSOC PROF BIOL SCI, WELLESLEY COL, 81- *Concurrent Pos:* Prin investr, NIH, 80-82. *Mem:* Brit Soc Develop Biol; AAAS; Sigma Xi. *Res:* Ultrastructure and biochemistry of amphibian oogenesis with special reference to the organization and expression of the mitochondrial genome. *Mailing Add:* Dept Biol Sci Wellesley Col Wellesley MA 02181

WEBB, BILL D, b Ralls, Tex, June 13, 28; m 54; c 3. PLANT BIOCHEMISTRY. *Educ:* Tex A&M Univ, BS, 56, MS, 59, PhD(biochem & nutrit), 61. *Prof Exp:* RES CHEMIST, AGR RES SERV, USDA, 61- *Mem:* Am Asn Cereal Chem; Inst Food Technologists; Sigma Xi. *Res:* Physicochemical properties of the rice grain and their relation to rice milling, cooking, processing and nutritive quality. *Mailing Add:* USDA Nat Rice Qual Lab Rte 7 Box 999 Beaumont TX 77713

WEBB, BURLEIGH C, b Greensboro, NC, Jan 9, 23; m 49; c 3. AGRONOMY, PLANT PHYSIOLOGY. *Educ:* Agr & Tech Col, NC, BS, 43; Univ Ill, MS, 47; Mich State Univ, PhD, 52. *Prof Exp:* Instr agron, Tuskegee Inst, 47-49; asst prof, Ala Agr & Mech Col, 51-52; assoc prof, Tuskegee Inst, 52-59; PROF AGRON, NC A&T STATE UNIV, 59-, DEAN SCH AGR, 63- *Mem:* AAAS; Am Soc Agron; Sigma Xi. *Res:* Plant physiology and ecology of forage crop plants; role of growth regulators in developmental and growth phenomena in crop plants. *Mailing Add:* 137 N Dudley St Greensboro NC 27411

WEBB, BYRON H, b New York, NY, May 2, 03; m 32; c 4. DAIRY INDUSTRY. *Educ:* Univ Calif, BS, 25; George Washington Univ, MS, 28; Cornell Univ, PhD(dairy indust), 31. *Prof Exp:* Jr dairy mfg specialist, Bur Dairy Indust, USDA, 26-29, from assoc specialist to sr specialist, 30-48, prin dairy technologist, 48-51; sr scientist, Nat Dairy Res Labs, Inc, 51-53, asst dir res, 53-54, dir, 54-60, chief dairy prods lab, Eastern Utilization Res & Develop Div, Agr Res Serv, USDA, DC, 50-72; RETIRED. *Concurrent Pos:* Consult, dairy & food technol, 72-88. *Honors & Awards:* Borden Award, Am Dairy Sci Asn, 43. *Mem:* AAAS; Am Chem Soc; Am Dairy Sci Asn; Inst Food Technologists. *Res:* Methods of improvement of dairy products; development of new products, by-products from milk. *Mailing Add:* 5426 Burkittsville Rd Burkittsville MD 21718

WEBB, BYRON KENNETH, b Cross Anchor, SC, Feb 2, 34; m 61; c 1. AGRICULTURAL ENGINEERING. *Educ:* Clemson Univ, BS, 55, MS, 62; NC State Univ, PhD(agr eng), 66. *Prof Exp:* From asst prof to assoc prof agr eng, 55-72, head dept, 77-84, assoc dean & dir, 84-87, PROF AGR ENG, CLEMSON UNIV, 72- DEAN & DIR, 87- *Mem:* Am Soc Agr Engrs; Am Soc Eng Educ; Sigma Xi. *Res:* Fruit and vegetable mechanization. *Mailing Add:* 103 Barre Hall Clemson Univ Clemson SC 29631

WEBB, CHARLES ALAN, b Charlottesville, Va, July 23, 47; m 74; c 1. TECHNICAL & BUSINESS, MANAGEMENT. *Educ:* Lehigh Univ, BA, 69; Univ Miami, PhD(phys chem), 75. *Prof Exp:* Chemist, E I Du Pont de nemours & Co, Inc, textile fibers dept, 69-70, res chemist, 75-78, mkt rep, polymer prod dept, 78-80, res supvr, 80-82, tech mfg supt, 82-85, sr planning consult, 85-86, venture mgr, Packaging Prod Div, 86-88, BUS MGR, PACKAGING PROD, E I DU PONT DE NEMOURS & CO, INC, 89- *Res:* Electrochemical corrosion of copper in chloride and amino acid solutions; spunbonded nonwoven polymer technology; stabilization, polyethylene and Surlyn product development; polypropylene stabilization; vinyl acetate and polyvinyl alcohol technology. *Mailing Add:* Polymer Prod Dept D6090-1 E I du Pont de Nemours & Co Inc 1007 N Market St Wilmington DE 19898

WEBB, CYNTHIA ANN GLINERT, b New York, NY, Oct 24, 47; US & Israel citizen; m 70; c 3. DEVELOPMENT BIOLOGY, CELL BIOLOGY. *Educ:* Technion Israel Inst Technol, BSc, 68, MSc, 70; Weizmann Inst Sci, PhD(biochem), 75. *Prof Exp:* Fel, Rockefeller Univ, 75-77; res fel, Weizmann Inst Sci, 77-79, sr scientist, 80-87; sr scientist, Biotechnol Gen, Ltd, 87-89; SR SCIENTIST, TEVA PHARMACEUT INDUSTS, LTD, 90- *Mem:* Israel Immunol Soc; Am Soc Cell Biol; Int Soc Develop Biol. *Res:* Early embryonic development in mammals; cellular and molecular aspects of differentiation of embryonic cells and teratocarcinomas; immunology of embryonic antigens; glycoproteins on gametes and early mammalian embryos. *Mailing Add:* Teva Pharmaceut Industs Ltd PO Box 3190 Petah Tikva 49131 Israel

WEBB, DAVID R, JR, b Taft, Calif, Nov 10, 44; m 66; c 2. IMMUNOLOGY, BIOCHEMISTRY. *Educ:* Calif State Univ, BA, 66, MS, 68; Rutgers Univ, PhD(microbiol), 71. *Prof Exp:* Res immunologist, Univ Calif, San Francisco, 71-73; asst mem, 73-78, ASSOC MEM IMMUNOL, ROCHE INST MOLECULAR BIOL, 78- *Concurrent Pos:* Dernham Jr fel, Am Cancer Soc, 72; adj prof, Grad Dept Biochem, City Univ New York, 76-78 & Dept Zool, Rutgers Univ, 78-; adj assoc prof, Col Physicians & Surgeons, Columbia Univ, 80- *Mem:* Am Asn Immunol; Am Soc Microbiol; Sigma Xi; NY Acad Sci. *Res:* Structure and function of antigen-specific/helper factors; purification and biochemical analysis of immune mediators. *Mailing Add:* Dept Cell Biol Roche Inst Nutley NJ 07110

WEBB, DAVID RITCHIE, b Taft, Calif, Nov 10, 44; m 66; c 2. PROTEIN PURIFICATION & ANALYSIS, ANALYSIS OF CYTOKINE-MEDIATED REGULATION OF CELL FUNCTION. *Educ:* Calif State Col, Fullerton, BA, 66, MA, 68; Rutgers State Univ, Phd(microbiol), 71. *Prof Exp:* Dernham fel immunol, Univ Calif, San Francisco, 71-73; asst mem, Roche Inst Molecular Biol, 73-79, assoc mem, 79-87; DISTINGUISHED SCIENTIST, SYNTEX RES, 87-, DIR, INST IMMUNOL & BIOL SCI, 90- *Concurrent Pos:* Adj prof biochem, City Univ New York, 76-79; adj assoc prof, Dept Zool, Rutgers Univ, 78-87 & Dept Human Genetics, Columbia Univ, 80-87; mem adv comt, Physiol, Cell & Molecular Biol, NSF, 80-82; affil, Cancer Biol Prog, Stanford Univ, 89- *Mem:* Am Asn Immunologists; Sigma Xi; Am Soc Microbiol; NY Acad Sci; AAAS. *Res:* Molecular basis of lymphocyte activation; regulation of immunity at the cellular and molecular level; development of immune modulating drugs. *Mailing Add:* Dept Molecular Immunol S3-6 Syntex Res 3401 Hillview Ave Palo Alto CA 94303

WEBB, DAVID THOMAS, b Darby, Pa, Aug 2, 45; m 75. PHOTOMORPHOGENESIS, PLANT TISSUE CULTURE. *Educ:* West Chester State Col, BA, 67; Univ Mont, PhD(bot), 78. *Prof Exp:* Asst prof biol, Simmons Col, 78-80; asst prof bot, Univ PR, 80-84; asst prof biol, Queens Univ, Ont, 84-87; RES SCIENTIST, BC RES CORP, 88- *Concurrent Pos:* Proj leader, Agrogen Biotech, 88. *Mem:* Bot Soc Am; Can Bot Asn; Can Soc Plant Physiologists. *Res:* Effects of light on cycad root growth and nodulation; cycad-Nostoc cymbiosis in virto; clonal propagation of economically important gymnosperms. *Mailing Add:* Dept Biol Queen Univ Kingston ON K7L 3N6 Can

WEBB, DENIS CONRAD, b Skowheghan, Maine, May 12, 38; m 73; c 1. MICROWAVE ENGINEERING, MAGNETICS. *Educ:* Univ Mich, BSE, 60, MS, 61; Stanford Univ, PhD(appl physics), 71. *Prof Exp:* From assoc to sr engr electromagnetics, Westinghouse Defense & Space Ctr, 61-66; engr magnetics, Phys Electron Lab, 71-72; supvr electron eng acoust, 72-87, SUPVR ELECTRON ENG, MICROWAVE TECH, NAVAL RES LAB, 87- *Mem:* Inst Elec & Electronics Engrs. *Res:* Magnetoelastic propagation in highly magnetostrictive materials; excitation and propagation of magnetostatic surface waves; microwave superconductivity. *Mailing Add:* Naval Res Lab Code 6850 Washington DC 20375

WEBB, DONALD WAYNE, b Brandon, Man, July 12, 39; m 61; c 2. ENTOMOLOGY. *Educ:* Univ Man, BSc, 61, MSc, 63; Univ Ill, PhD, 81. *Prof Exp:* Lectr, St Paul's Col, Man, 62-63; from asst taxonomist to assoc taxonomist, 66-82, TAXONOMIST, ILL NATURAL HIST SURV, 82- *Concurrent Pos:* Assoc prof agr entom, Univ Ill, 81- *Res:* Aquatic insect ecology and the taxonomy of the mature and immature stages of the Chironomidae; systematics of mecoptera and diptera. *Mailing Add:* Ill Natural Hist Surv 607 E Peabody Champaign IL 61820

WEBB, FRED, JR, b Hinton, WVa, Feb 23, 35; m 56; c 2. GEOLOGY. *Educ:* Duke Univ, AB, 57; Va Polytech Inst, MS, 59, PhD(struct geol), 65. *Prof Exp:* Asst geol, Va Polytech Inst, 59-60, instr, Eng Exp Sta, 61-63, instr, Inst, 63-66, resident dir summer field sta, 65; assoc prof geol, Catawba Col, 66-68; from asst prof to assoc prof, 68-73, PROF GEOL, APPALACHIAN STATE UNIV, 73-, CHMN DEPT, 72- *Concurrent Pos:* Vis prof, Va Polytech Inst & State Univ, 71-84; exchange prof geol, Northeast Univ Technol, Shenyang, People's Repub China, 82-83, hon prof, 85- *Mem:* Nat Asn Geol Teachers; Soc Econ Paleontologists & Mineralogists; Geol Soc Am. *Res:* Stratigraphy and tectonics of southern Appalachians; Ordovician paleotopography; stratigraphy and regional geology of northeastern China; Ordovician/ Carboniferous unconformity in northeast China; environments of deposition of Proterozoic rocks in Liaoning Province, China. *Mailing Add:* Dept Geol Appalachian State Univ Boone NC 28608

WEBB, GEORGE DAYTON, b Oak Park, Ill, June 22, 34; m 57; c 3. NEUROPHYSIOLOGY, ION TRANSPORT. *Educ:* Oberlin Col, AB, 56; Yale Univ, MAT, 57; Univ Colo, PhD(physiol), 62. *Prof Exp:* High sch teacher, DC, 57-58; asst prof, 66-69, ASSOC PROF PHYSIOL, UNIV VT, 69- *Concurrent Pos:* Vis fel biochem, Univ Copenhagen, 62-63; vis fel neurol & biochem, Columbia Univ, 63-66. *Mem:* Am Physiol Soc; Am Soc Hypertension; Soc Gen Physiol; Biophys Soc; fel Am Col Nutrit. *Res:* Molecular physiology of synaptic and conducting membranes, especially cholinergic mechanisms and ion transport; sodium and potassium transport in red blood cells; essential hypertension. *Mailing Add:* Dept Physiol & Biophys Given Med Bldg Univ Vt Burlington VT 05405

WEBB, GEORGE N, b Fredericktown, Ohio, Mar 24, 20; m 46; c 2. CLINICAL ENGINEERING, ELECTRICAL SAFETY ENGINEERING. *Educ:* Univ Toledo, BE, 47; NC State Col, MS, 52. *Prof Exp:* Instr elec eng, Univ Toledo, 47-49; instr NC State Col, 49-51, res engr, 51-52; electronics engr, 52-57, asst, 57- 61, instr, 61-64, ASST PROF BIOMED ENG, SCH MED, JOHNS HOPKINS UNIV, 64-, CLIN ENGR, HOSP, 71- *Concurrent Pos:* Secy gen, Int Fedn Med Electronics & Biol Engrs, 67-71, mem conf planning & policies comt, 69-; chmn const comt, Alliance for Engrs in Med & Biol, 69-; mem instrumentation group, Int Soc Comn Heart Dis Resources, 70-; partic, appl physiol & bioeng study sect, NIH, 72-76. *Mem:* AAAS; Inst Elec & Electronics Engrs; Asn Advan Med Instrumentation; Int Fedn Med Electronics & Biol Engrs; Biomed Eng Soc; Sigma Xi. *Res:* Instruments for data acquisition; processing and display for intensive care; electrocardiogram, cardiac pressure, respiratory function; methodology for assuring quality of performance and safe operation of patient related electrical equipment; codes and standards for electrical safety. *Mailing Add:* 5622 Alhambra Ave Baltimore MD 21212

WEBB, GEORGE RANDOLPH, b Norfolk, Va, Feb 25, 38; m 59; c 4. ENGINEERING SCIENCE, STABILITY. *Educ:* Mass Inst Technol, BS, 59; Va Polytech Inst, PhD(eng mech), 64. *Prof Exp:* From asst prof to assoc prof mech eng, Tulane Univ, 64-73; assoc prof physics, 73-76, PROF PHYSICS, CHRISTOPHER NEWPORT COL, 76- *Concurrent Pos:* NSF vis scientist, Univ Va, 71-72, NSF pub serv resident, Smithfield Times, 77-78. *Mem:* Soc Naval Archit & Marine Eng; Soc Natural Philos; Am Soc Mech Engrs; Am Soc Eng Educ; Sigma Xi. *Res:* Theories of stability; bifurcation theory and catastrophe theory applications. *Mailing Add:* Physics Dept Christopher Newport Col Newport News VA 23606

WEBB, GLENN FRANCIS, b Cleveland, Ohio, Sept 30, 42; m 73; c 2. MATHEMATICAL ANALYSIS. *Educ:* Ga Inst Technol, BS, 65; Emory Univ, MS, 66, PhD(math), 68. *Prof Exp:* From asst prof to assoc prof, 68-79, PROF MATH, VANDERBILT UNIV, 79-, CHAIR, MATH DEPT, 87- *Concurrent Pos:* Vis assoc prof math, Univ Ky, 73; Ital Nat Coun Res fel, Univ Rome, 74; vis prof math, Univ Padova, 80 & Univ Graz, 81. *Mem:* Am Math Soc; Math Asn Am. *Res:* Differential equations and functional analysis and mathematical biolgy. *Mailing Add:* Dept Math Vanderbilt Univ Nashville TN 37235

WEBB, GLENN R, b Chicago, Ill, May 22, 18; m 53; c 1. MALACOLOGY. *Educ:* Univ Ill, BS, 48, MS, 49; Univ Okla, PhD, 60. *Prof Exp:* Res asst, Univ Ill, 48-49 & Univ Okla, 53-56; instr biol, Henderson State Teachers Col, 56-57; fishery res biologist, US Fish & Wildlife Surv, Ft Worth, Tex, 58-59; instr biol, Coastal Carolina Jr Col, 59-60 & SC, Florence & Conway Branches, 60-62; asst prof, High Point Col, 62-63; prof, 63-84, EMER PROF BIOL, KUTZTOWN STATE COL, 84- *Mem:* Am Malacol Union. *Res:* Pulmonate land snails life histories with special regard to sexology as a clue to phylogeny; snail autoecology; Polygyridae-Helicoidea; inter-species hybridization. *Mailing Add:* 14 Henry Rd Fleetwood PA 19522

WEBB, HAROLD DONIVAN, b Franklin, Ind, Sept 23, 09; m 37; c 4. ELECTRICAL ENGINEERING, APPLIED PHYSICS. *Educ:* Franklin Col, AB, 31; Ind Univ, AM, 32, PhD(physics), 39. *Prof Exp:* Teacher math & sci, Needham Sch, Ind, 34-35; teacher physics & chem, Baylor Sch, Tenn, 35-36; teacher math, chem & physics, Franklin High Sch, Ind, 36-39; prof math & physics, West Liberty Col, WVa, 39-42; elec eng physicist radar, Evans Signal Lab, NJ, 42-47; from asst prof to prof, 47-77, EMER PROF ELEC ENG DEPT, UNIV ILL, 77- *Mem:* Inst Elec & Electronics Engrs; fel AAAS; Am Geophys Soc; Int Union Radio Sci. *Res:* First radar contact with the moon; work in radio direction finding, chiefly with matched channel receivers; used moon reflection data and satellite data to measure and study the electron content of the ionosphere and the moon surface; radio astronomy. *Mailing Add:* 812 Delaware Ave Urbana IL 61801

WEBB, HELEN MARGUERITE, b Nova Scotia, July 7, 13; nat US. INVERTEBRATE PHYSIOLOGY. *Educ:* Northwestern Univ, BS, 46, MS, 48, PhD(zool), 50. *Prof Exp:* Asst prof biol, Boston Col, 50-52; asst prof physiol, 52-59, assoc prof biol sci, 59-66, prof, 66-79, EMER PROF BIOL SCI, GOUCHER COL, 79- *Concurrent Pos:* Mem corp, Marine Biol Lab, Woods Hole. *Mem:* Soc Gen Physiol. *Res:* Biological rhythms; crustacean endocrinology. *Mailing Add:* Marine Biol Lab Woods Hole MA 02543

WEBB, J(OHN) WARREN, b Austin, Tex, Apr 4, 45; m 67; c 3. FOREST & WETLAND ECOLOGY, WILDLIFE MANAGEMENT & ENDANGERED SPECIES. *Educ:* Univ Tex, Austin, BA, 67; Rhodes Univ, Grahamstown, SAfrica, PhD(insect ecol), 75. *Prof Exp:* Natural sci chmn, Mary Holmes Jr Col, West Point, Miss, 69-71; lectr entomol, Rhodes Univ, SAfrica, 74-75; post doctorate ecol & forest entomol, Univ Ga, Athens, 75-78; RES STAFF, OAK RIDGE NAT LAB, 78- *Concurrent Pos:* Prin investr, Elec Power Res Inst, 80-82; consult forest mitigation, CEngr, Tenn Tombigbee Proj, 83, environ trends, Coun Environ Qual, 91; mem, Task Force on Environ Assessment, Europ Econ Comm, UN, 90-91. *Mem:* Sigma Xi; Nat Asn Environ Professionals. *Res:* Surveys for Gypsy Moth and endangered bat species on the Oak Ridge Reservation; environmental assessments and evaluations for various federal agencies, primarily associated with energy production technologies. *Mailing Add:* Oak Ridge Nat Lab PO Box 2008 Bldg 1505 Oak Ridge TN 37831-6036

WEBB, JAMES L A, b Webb, Miss, Nov 17, 17; m 46; c 3. ORGANIC CHEMISTRY. *Educ:* Washington & Lee Univ, BS, 39; Johns Hopkins Univ, PhD(org chem), 43. *Prof Exp:* Jr instr chem, Johns Hopkins Univ, 39-43, instr, 43-45; from asst prof to prof, Southwestern at Memphis, 45-59; prof chem, Goucher Col, 59-83, chmn dept, 65-83; RETIRED. *Concurrent Pos:* Res chemist, Chapman Chem Co, 51-59. *Mem:* Am Chem Soc. *Res:* Organic heterocyclic compounds; theory and antidotes for heavy metal poisoning; disubstituted pyridines; bipyrryls and pyrrole pigments; asphalt additives. *Mailing Add:* 131 Greenbriar Dr Memphis TN 38117

WEBB, JAMES R, b Lafayette, Ind, Sept 18, 45; m 74; c 3. ELECTRICAL ENGINEERING, COMPUTER SCIENCES. *Educ:* Univ Colo, BS, 69. *Prof Exp:* Technician, Colo Inst Inc, 61-69; group leader, ESSA-NOAA, McMurdo Sta, 69-71; electronic engr, Colo Instr Inc, 71-72; pres, Commodore Eng Corp, 72-73; dir engr prod develop, Commodore Bus Mach, 73-74; staff engr new prod, Data Pathing Inc, 74-76; staff engr space telescope, Ball Brothers Res Corp, 76-77; VPRES ENG, DATA RAY CORP, 77- *Mem:* Inst Elec & Electronics Engrs. *Res:* Patents relative to keyboards and calculators. *Mailing Add:* 1077 Marble Ct Boulder CO 80303-3223

WEBB, JAMES RAYMOND, b Anderson, Ind, Aug 5, 54; m 83. QUASAR VARIABILITY, MULTIFREQUENCY SPECTRA. *Educ:* Ball State Univ, BS, 81; Univ Fla, MA, 84, PhD(astron), 88. *Prof Exp:* Asst prof physics, Stephen F Austin State Univ, 88-89; resident astronr, Computer Sci Corp, 89-90; ASST PROF PHYSICS, FLA INT UNIV, 90- *Mem:* Am Astron Soc; Nat Space Soc. *Res:* Variations of quasars in all frequency ranges, radio through x-ray; optical variations; energetics of quasar outbursts; continuum models of quasars. *Mailing Add:* 4808 SW 138th Ave Miami FL 33175

WEBB, JERRY GLEN, b Rosefield, La, Feb 17, 38; m 61; c 3. THEORETICAL PHYSICS. *Educ:* Northeast La State Col, BS, 61; Univ Ark-Fayetteville, MS, 64; Tex A&M Univ, PhD(physics), 69. *Prof Exp:* Instr physics, Cent State Col, Okla, 62-65; assoc prof, 69-75, head, Dept Phys Sci, 78-80, PROF PHYSICS, UNIV ARK, MONTICELLO, 75- *Concurrent Pos:* Res partic, Savannah River Lab, Aiken, SC, 70. *Mem:* Am Asn Physics Teachers. *Res:* Three particle scattering problem. *Mailing Add:* Dept Physics Univ Ark Box 3598 Monticello AR 71655

WEBB, JOHN DAY, b Wash, DC, Nov 30, 49. POLYMER DEGRADATION PROCESSES, PHOTOCHEMISTRY. *Educ:* Univ Colo, Denver, BA, 73; Univ Colo, Boulder, BS, 77; Univ Denver, MS, 82. *Prof Exp:* Chemist, Protex Industs, Inc, 73-75; teaching asst chem, Univ Colo, 75-77; process engr, Shell Chem Co, 77-78; SR ENGR, SOLAR ENERGY RES INST, 78- *Mem:* Am Chem Soc. *Res:* Fourier transform infrared spectroscopy. *Mailing Add:* Solar Energy Res Inst 1617 Cole Blvd Golden CO 80401

WEBB, JOHN RAYMOND, b Morgantown, WVa, Sept 18, 20. AGRONOMY. *Educ:* WVa Univ, BS, 40, MS, 42; Purdue Univ, PhD(agron), 53. *Prof Exp:* Instr & asst soils, WVa Univ, 41-43; soil surveyor, Soil Conserv Serv, USDA, 46-47; instr & asst soils, WVa Univ, 47-48; asst agron, Purdue Univ, 49-52; asst prof, 52-56, assoc prof soils, Agr Exp Sta, 56-80, PROF AGRON, DEPT AGRON, IOWA STATE UNIV, 80- *Mem:* Am Soc Agron; Soil Sci Soc Am. *Res:* Soil fertility. *Mailing Add:* Dept Agron Iowa State Univ 120 Agronomy Ames IA 50011

WEBB, KENNETH EMERSON, JR, b Hamilton, Ohio, Feb 26, 43; c 2. ANIMAL NUTRITION. *Educ:* Ohio Univ, BS, 65; Univ Ky, MS, 67, PhD(animal sci), 69. *Prof Exp:* From asst prof to assoc prof, 69-82, PROF RUMINANT NUTRIT, VA POLYTECH INST & STATE UNIV, 82- *Mem:* Am Soc Animal Sci; Am Inst Nutrit; Animal Nutrit Res Coun; Agr Res Inst. *Res:* Ruminant nutrition, protein, amino acids and gastrointestinal physiology. *Mailing Add:* 3020 Animal Sci Va Polytech Inst 0306 Blacksburg VA 24061-0306

WEBB, KENNETH L, b Old Fort, Ohio, July 18, 30; m 73. MARINE BIOLOGY, BIOLOGICAL OCEANOGRAPHY. *Educ:* Antioch Col, BA, 53; Ohio State Univ, MSc, 54, PhD(plant physiol), 59. *Prof Exp:* Res assoc, Marine Inst, Univ Ga, 60-65; assoc marine scientist, Va Inst Marine Sci, 65-79, sr marine scientist, 79-83; from asst prof to prof, 65-87, CHANCELLOR PROF MARINE SCI, COL WILLIAM & MARY, 87- *Concurrent Pos:* Vis assoc prof, Ohio State Univ, 65 & dept oceanog, Univ Hawaii, 72-73; mem, Adv Comt Ocean Sci, NSF, 78-81. *Mem:* Fel AAAS; Am Soc Limnol & Oceanog; Am Soc Zoologists; Estuarine Res Fedn; Phycol Soc Am; Sigma Xi.

Res: Intedisciplinary investigations related to energy flow and nutrient cycling in marine environments including estuaries, salt marshes, seagrass systems, and coral reefs; physiology of marine organisms; image analysis. *Mailing Add:* Sch Marine Sci Col William & Mary Gloucester Point VA 23062

WEBB, LELAND FREDERICK, b Hollywood, Calif, July 27, 41; m 63; c 2. MATHEMATICS EDUCATION. *Educ:* Univ Calif, Santa Barbara, BA, 63; Calif State Polytech Univ, San Luis Obispo, MA, 68; Univ Tex, Austin, PhD(math & educ), 71. *Prof Exp:* Lectr math & educ, Calif State Polytech Univ, San Luis Obispo, 67-68; curric writer math, Southwest Educ Develop Lab, 70; teaching asst math, Univ Tex, Austin, 71; res assoc math, Res & Develop Ctr Teacher Educ, Austin, 71; from asst prof to assoc prof, 71-78, PROF MATH & MATH EDUC, CALIF STATE UNIV, BAKERSFIELD, 78-, CHMN DEPT, 82-85, 90- *Concurrent Pos:* Consult, Educ Develop Ctr, Newton, Mass, 71-76, Greenfield Sch Dist, Bakersfield, 73-87 & Maricopa Sch Dist, Calif, 74-75; NSF sec sch math & sci grant, 73-75 & 74-75; sabbatical leave, Agder Reg Col, Kristiansand, Norway, 80; consult, Bishop Union Elem Sch Dist, 80-; vis lectr, NSF Inst, Tokyo, Japan, 75. *Mem:* Sigma Xi; Nat Coun Teachers Math; Am Educ Res Asn; Sch Sci & Math Asn. *Res:* Learning in science and mathematics education; author of kindergarten through eighth grade mathematics textbook series. *Mailing Add:* Dept Math Calif State Univ Bakersfield CA 93311

WEBB, MARY ALICE, b Austin, Tex, Jan 16, 51. PLANT CELL BIOLOGY, IMMUNOCYTOCHEMISTRY. *Educ:* Univ Tex, Austin, BA, 75; Univ Tex, Arlington, MA, 80; Univ Wis-Madison, PhD(plant cell biol), 88. *Prof Exp:* ASST PROF PLANT ANAT & CELL BIOL, PURDUE UNIV, 89- *Mem:* AAAS; Am Soc Plant Physiologists; Bot Soc Am; Electron Micros Soc Am; Sigma Xi. *Res:* Calcification in plants; nitrogen-fixing symbioses in plants; light and electron microscopy; immunological approaches. *Mailing Add:* Bot & Plant Path Dept Purdue Univ West Lafayette IN 47907

WEBB, MAURICE BARNETT, b Neenah, Wis, May 14, 26; m 56; c 2. PHYSICS. *Educ:* Univ Wis, BS, 50, MS, 52, PhD, 56. *Prof Exp:* Proj Lincoln, Mass Inst Technol, 52-53; res lab, Gen Elec Co, 56-61; assoc prof, 61-65, PROF PHYSICS, UNIV WIS-MADISON, 65- *Mem:* Am Phys Soc. *Res:* Solid state physics; x-ray scattering; surface physics; low energy electron diffraction. *Mailing Add:* Dept Physics Univ Wis Madison WI 53706

WEBB, NEIL BROYLES, b Junta, WVa, May 19, 30; m 53; c 5. BIOCHEMISTRY, MICROBIOLOGY. *Educ:* WVa Univ, BS, 53; Univ Ill, MS, 57; Univ Mo, PhD(food sci), 59. *Prof Exp:* Asst prof food sci, Mich State Univ, 59-62; dir food tech, Eckert Packing Co, Ohio, 62-66; assoc prof food sci, NC State Univ, 66-77; PRES, WEBB FOODLAB, INC, 72- *Mem:* Inst Food Technol; Am Meat Sci Asn; Am Chem Soc. *Res:* Food processing systems; product development; nutritional quality foods. *Mailing Add:* 4019 Glen Laurel Dr Raleigh NC 27612

WEBB, NORVAL ELLSWORTH, JR, b Indianapolis, Ind, Oct 17, 27; m 50; c 3. PHARMACY, PHARMACEUTICAL CHEMISTRY. *Educ:* Purdue Univ, BS, 50, MS, 52, PhD(pharm), 56. *Prof Exp:* Asst pharm, Purdue Univ, 50-52; from instr to assoc prof, SDak State Col, 52-62; chemist, Pharmaceut Res & Develop Dept, Merrell-Dow Pharmaceut Div, Dow Chem Co, Inc, 80-90; RETIRED. *Mem:* Am Pharmaceut Asn; Acad Pharmaceut Sci. *Res:* Interaction of pharmaceutical dose forms with packaging materials; dose form development. *Mailing Add:* 605 Branchhill Loveland Rd Cincinnati OH 45241

WEBB, PAUL, b Hemel Hempstead, Eng, Dec 23, 45; m 70; c 2. BIOMECHANICS, PHYSIOLOGICAL ECOLOGY. *Educ:* Univ Bristol, BSc, 67, PhD(zool), 71. *Prof Exp:* Nat Res Coun fel, Pac Biol Sta, 70-72; from asst prof to assoc prof, 72-80, PROF NATURAL RESOURCES, SCH NATURAL RESOURCES, UNIV MICH, 80- *Concurrent Pos:* NSF grants, 75-83; consult, Calif Inst Tech & Detroit Edison. *Mem:* AAAS; Am Fisheries Soc; Am Soc Zool; Can Soc Zool; Soc Exp Biol. *Res:* Functional morphology, biomechanics and behavior of animals, particularly fish; whole animal bioenergetics and energetic correlates of environmental adaptation, including responses to manmade stress. *Mailing Add:* 2081 Natural Sci Biol Univ Mich Main Campus Ann Arbor MI 48109-1048

WEBB, PAUL, b Cleveland, Ohio, Dec 2, 23; m 48; c 2. PHYSIOLOGY. *Educ:* Va, BA, 43, MD, 46; Seattle, Wash, MS, 52. *Prof Exp:* Intern, Mont Gen Hosp, Can, 46-47, asst res med, 47-48; post surgeon, Arctic Training Ctr, US Army, Alaska, 49-50; res assoc, Seattle, Wash, 50-51; asst prof physiol, Okla, 52-54; res physiologist & med officer, Med Lab, USAF, 54-57, chief environ sect, 57-58; PRIN ASSOC, WEBB ASSOCS, INC, 58- *Concurrent Pos:* Pres & chmn, Environ, Inc, 59-66; med C, US Army, 43-50. *Mem:* Fel AAAS; Marine Technol Soc; Human Factors Soc; Physiol Soc; Aerospace Med Asn. *Res:* Environmental physiology; human thermal tolerance; protective clothing; artificial atmospheres; atmospheric conditioning equipment; hygrometry; cutaneous water loss and heat regulation; aviation and space physiology; heat loss and human calorimetry. *Mailing Add:* PO Box 308 Yellow Springs OH 45387

WEBB, PETER NOEL, b Wellington, NZ, Dec 14, 36; m 78; c 2. GEOLOGY, MICROPALEONTOLOGY. *Educ:* Victoria Univ Wellington, BSc, 59, MSc, 60; State Univ Utrecht, PhD(geol), 66. *Prof Exp:* Micropaleontologist, NZ Geol Surv, 60-73; prof micropaleont & chmn dept, Northern Ill Univ, 73-80; PROF MICROPALEONT & CHMN DEPT, OHIO STATE UNIV, 80- *Concurrent Pos:* Vis micropaleontologist, Hebrew Univ Jerusalem, 67 & Fed Geol Surv, Hanover, Ger, 69; secy, Nat Comt Geol, NZ, 70-73; Late Cenozoic Working Group, Sci Comt Antarctic Res, 72-77, convenor, Ice Shelf Drilling Group, 73-78; ad hoc comt, Antarctic Geol, Nat Acad Sci. *Honors & Awards:* Hamilton Prize, Royal Soc NZ, 60; Cotton Prize, Victoria Univ Wellington, 60; Medal, US Polar Serv, 79. *Mem:* Soc Econ Paleontologists & Mineralogists; Geol Soc New Zealand; Geol Soc Am; Paleont Soc; Antarctican Soc; Am Geophys Union. *Res:* Cretaceous, Tertiary and recent Foraminifera from Southern Hemisphere, particularly New Zealand, Antarctica, South America and intervening areas, with special emphasis on biostratigraphy, population compositions and climate relationship. *Mailing Add:* Dept Geol & Mineral 107 Mendenhall Lab Ohio State Univ 125 S Oval Mal Columbus OH 43210

WEBB, PHILIP GILBERT, b Norwich, NY, Oct 17, 43; m 68; c 3. INDUSTRIAL ORGANIC CHEMISTRY. *Educ:* Hamilton Col, AB, 65; Univ Rochester, PhD(org chem), 70. *Prof Exp:* Res chemist org pigments, Pigments Div, Am Cyanamid Co, 69-74; plant tech mgr, Chemetron Corp, 74-76, plant mgr, 76-79, TECHNOL SERV MGR PIGMENTS, BASF CORP, 79- *Mem:* Am Chem Soc; Water Pollution Control Fedn. *Res:* New and improved organic pigments; dispersability and lightfastness of organic compounds in inks, paints and plastics. *Mailing Add:* 110 W 39th St Holland MI 49423

WEBB, R CLINTON, b Evansville, Ind, Dec, 31, 48; m 71; c 2. CARDIOVASCULAR PHYSIOLOGY, HYPERTENSION. *Educ:* Southern Ill Univ, BA, 71, MS, 73; Univ Iowa, PhD(anat), 76. *Prof Exp:* Fel physiol, Univ Mich, 76-78, fel pharmacol, Univ Instelling Antwerpen, 78-79; res scientist, 79-80, asst prof, 80-83, ASSOC PROF PHYSIOL, UNIV MICH, 83- *Concurrent Pos:* Mem, Coun High Blood Pressure Res, Soc Exp Biol & Med. *Honors & Awards:* C & F Demuth Award, Int Soc Hypertension, 82. *Mem:* Am Physiol Soc; Soc Exp Biol Med. *Res:* Physiology of vascular smooth muscle, with particular emphasis placed on: vascular reactivity in hypertension and diabetes, cellular and subcellular mechanisms of contraction and relaxation and adrenerqic neurotransmission in vascular smooth muscle. *Mailing Add:* Dept Physiol Med Sch Univ Mich 7813 Med Sci Bldg II Ann Arbor MI 48109-0622

WEBB, RALPH L, b Parker, Kans, Feb 22, 34; m 61; c 2. MECHANICAL ENGINEERING, HEAT TRANSFER. *Educ:* Kans State Univ, BS, 57; Rensselaer Polytech Inst, MS, 62; Univ Minn, PhD(mech eng), 69. *Prof Exp:* Instr mech eng, Kans State Univ, 57; exp engr, Knolls Atomic Power Lab, 60-62; mgr heat transfer res dept, Trane Co, 63-77; assoc prof, 77-80, PROF MECH ENG, PA STATE UNIV, 80- *Concurrent Pos:* Assoc tech ed, J Heat Transfer, 73-76; tech ed, Heat Transfer Eng, 77- & J Heat Recovery Systs, 82- *Honors & Awards:* Thermotank Gold Medal, Heat Transfer Design Refrigeration Indust, Inst Refrigeration Hall, 87; Heat Transfer Mem Award, Am Soc Mech Engrs, 87. *Mem:* Fel Am Soc Mech Engrs; Am Inst Chem Engrs; Am Soc Heating & Refrig Engrs. *Res:* Applied research on enhanced heat transfer, including convection, two-phase flow, boiling and condensation; design of heat exchangers. *Mailing Add:* Dept Mech Eng Pa State Univ University Park PA 16802

WEBB, RICHARD, b Los Angeles, Calif, Sept 10, 46; m; c 2. MACROSCOPIC QUANTUM TUNNELING. *Educ:* Univ Calif, Berkeley, BA, 68; Univ Calif, San Diego, MS, 70, PhD, 73. *Prof Exp:* Res assoc, Univ Calif, San Diego, 73-75; from asst to assoc res physicist, Argonne Nat Lab, 75-78; RES STAFF MEM & MGR, T J WATSON RES CTR, IBM, 78- *Honors & Awards:* Simon Mem Prize, 89. *Mem:* Fel Am Phys Soc. *Res:* Macroscopic quantum tunneling in Josephson junctions at low temperatures; investigations of the Aharonov-Bohm effect and universal periodic conductance fluctuations in very small semiconducting and normal metal rings; measurement and temperature, magnetic field, and Fermi Energy dependencies of the conduction process of very small Si MOSFET devices in both insulating and metallic regimes. *Mailing Add:* IBM Res Div PO Box 218 Yorktown Heights NY 10598

WEBB, RICHARD C(LARENCE), b Omaha, Nebr, Sept 2, 15; m 41; c 1. ELECTRICAL ENGINEERING. *Educ:* Univ Denver, BS, 37; Purdue Univ, MS, 44, PhD(elec eng), 51. *Prof Exp:* Traffic trainee, Mountain States Tel & Tel Co, 37-39; instr elec eng, Purdue Univ, 39-45; res engr labs, Radio Corp Am, 45-50 & 52; prof elec eng, Iowa State Univ, 50; prof elec eng, Univ Denver & sect head, Denver Res Inst, 52-56; pres, Colo Res Corp, 56-61; pres & tech dir, Colo Instruments, Inc, 61-73; pres, Data Ray Corp, 75-85; PRES, WEBB ENG CO, 73- *Concurrent Pos:* Consult, electronic eng; vis lectr elec eng, Univ Colo. *Mem:* Fel Inst Elec & Electronics Engrs; Sigma Xi; Soc Motion Pictures & TV Engrs. *Res:* Communication and information theory; electro-mechanical and electrooptical instruments; television; digital data systems. *Mailing Add:* PO Box 3078 Estes Park CO 80517-3078

WEBB, RICHARD LANSING, b Mountain Lakes, NJ, July 28, 23; m 49; c 3. POLYMER CHEMISTRY. *Educ:* Mass Inst Technol, BS, 48; Columbia Univ, MA, 49. *Prof Exp:* Jr chemist, Cent Labs, Gen Foods Corp, NJ, 48 & 49-50; res chemist, Cent Res Div, Am Cyanamid Co, 50-60, sr res chemist, 60-69, group leader, Lederle Labs Div, NY, 69-76, sr res chemist, Chem Res Div, Conn, 76-89; RETIRED. *Mem:* Am Chem Soc. *Res:* Reactions of acrylonitrile and derivatives; hydrogen cyanide polymerization; x-ray induced addition reactions of olefins including polymerization; synthesis of polymers for biological applications; photosensitive polymers. *Mailing Add:* PR 01 New London NH 03257-1601

WEBB, ROBERT CARROLL, b Petersburg, Va, Mar 6, 47; c 3. HIGH ENERGY PHYSICS. *Educ:* Univ Pa, BA, 68; Princeton Univ, MA, 70, PhD(physics), 72. *Prof Exp:* Adj asst prof physics, Univ Calif, Los Angeles, 72-74; res assoc, Princeton Univ, 75-76, asst prof physics, 76-80; assoc prof, 80-87, PROF PHYSICS, TEX A&M UNIV, 87- *Mem:* Am Phys Soc; Inst Elec & Electronics Engrs. *Res:* Experiments in high energy particle physics; charge conjugation symmetry in proton-antiproton collisions; production of new particle states in high energy hadron collisions; searches for super heavy GUT magnetic monopoles in cosmic rays; the study of high energy proton-antiproton collisions. *Mailing Add:* Physics Dept Tex A&M Univ College Station TX 77843-4242

WEBB, ROBERT G, b Long Beach, Calif, Feb 18, 27; m 86; c 1. VERTEBRATE ZOOLOGY. *Educ:* Univ Okla, BS, 50, MS, 52; Univ Kans, PhD(zool), 60. *Prof Exp:* Instr biol, WTex State Univ, 57-58; PROF BIOL, UNIV TEX, EL PASO, 62- *Mem:* Am Soc Ichthyol & Herpet; Soc Study Amphibians & Reptiles (pres, 80); Herpetologists League. *Res:* Systematics, zoogeography and evolution of amphibians and reptiles, especially in the southwestern United States and northern Mexico; trionychid turtles worldwide. *Mailing Add:* Univ Tex El Paso TX 79968

WEBB, ROBERT HOWARD, b Burlington, Vt, Oct 17, 34; m 53, 81; c 2. MEDICAL PHYSICS, BIOENGINEERING & BIOMEDICAL ENGINEERING. *Educ:* Harvard Univ, AB, 55; Rutgers Univ, PhD(physics), 59. *Prof Exp:* Res assoc physics, Stanford Univ, 59-63; asst prof, Tufts Univ, 63-69; sr staff scientist, Block Eng Inc, 69-78; sci, 78-86, SR SCI, EYE RES INST RETINA FOUND, 86- *Honors & Awards:* IR-100, 81. *Mem:* Asn Res Vision & Ophthal; Am Asn Physics Teachers; Optic Soc Am. *Res:* Medical diagnostic instrumentation; flow cytometry, scanning microscopy optics; chemical and solid state physics; electron and nuclear spin resonance; low temperatures. *Mailing Add:* Eye Res Inst Retina Found 20 Staniford St Boston MA 02114

WEBB, ROBERT LEE, b Topeka, Kans, Nov 19, 26; m 52; c 2. ORGANIC CHEMISTRY. *Educ:* Washburn Univ, BS, 49. *Prof Exp:* Analytical chemist, I H Milling Co, 49-50; res chemist, Org Chem Dept, Glidden Co, 50-56, mgr res lab, 56-59, prod mgr, 59-61, asst tech dir, 61-62, vpres, Terpene Res Inst, 62-64; asst mgr aromatic chem develop, Union Camp Corp, 64-65, supt mfg tech serv terpenes, 65-68, gen mgr terpene & aromatic chem, 68-75, sr vpres & gen mgr, chem group, 75-87; RETIRED. *Honors & Awards:* Pioneer in Terpene Chem Award, Am Chem Soc. *Mem:* Am Chem Soc. *Res:* Terpene and resin chemistry; synthesis and manufacture of flavor and perfumery chemicals from terpenes; composition and reconstitution of essential oils. *Mailing Add:* 2479 Holly Point Rd E Orange Park FL 32073

WEBB, ROBERT MACHARDY, b Hamilton, Ohio, Dec 2, 15; m 61. PHYSICAL GEOGRAPHY. *Educ:* Memphis State Col, BS, 49; Ohio State Univ, MA, 50; Univ Kans, PhD(geog), 62. *Prof Exp:* Instr geog, Ohio Wesleyan Univ, 52-53; prof geog, Univ Southwestern La, 56-80, coordr, 75-80; RETIRED. *Concurrent Pos:* Consult, Dept Educ, State of La, 67-68. *Mem:* Asn Am Geog; Int Geog Union; Nat Coun Geog Educ. *Res:* Agricultural land use; climatology. *Mailing Add:* 159 Whittington Dr Lafayette LA 70503

WEBB, RODNEY A, b Eng, July 23, 46; Can citizen; m 70. INVERTEBRATE PHYSIOLOGY, NEUROENDOCRINOLOGY. *Educ:* Univ London, BSc, 68; Univ Toronto, PhD(zool), 72. *Prof Exp:* Demonstr zool, Univ Toronto, 68-72; Nat Res Coun Can fel parasitol, Inst Parasitol, MacDonald Col, McGill Univ, 72-74; asst prof zool, Univ NB, 74-75; res assoc zool, 75-77, asst prof, 77-81, ASSOC PROF BIOL, YORK UNIV, 81- *Mem:* Can Soc Zoologists; Am Soc Parasitologists. *Res:* Fine structure, physiology and biochemistry of parasitic helminths; neurobiology and neuroendocrinology of annelids and helminths; endocrine control of spermatogenesis in leeches. *Mailing Add:* Dept Biol York Univ 4700 Keele St Toronto ON M3J 1P6 Can

WEBB, ROGER P(AUL), b Cedar City, Utah, Dec 28, 36; m 57. ELECTRICAL ENGINEERING. *Educ:* Univ Utah, BSEE, 57; Univ Southern Calif, MSEE, 59; Ga Inst Technol, PhD(elec eng), 64. *Prof Exp:* Engr, Douglas Aircraft Corp, 57-59; proj engr, Sperry Phoenix Co, 59-60; from asst prof to assoc prof, 63-77, ASSOC DIR & GA POWER PROF ELEC ENG, GA INST TECHNOL, 78- *Concurrent Pos:* Consult, Lockheed-Ga Co, 66- *Mem:* Inst Elec & Electronics Engrs. *Res:* Automatic control systems. *Mailing Add:* Sch Elec Eng Ga Inst Technol Atlanta GA 30332

WEBB, RYLAND EDWIN, b Dondi, Angola, Africa, Jan 24, 32; US citizen; m 58; c 3. NUTRITIONAL BIOCHEMISTRY. *Educ:* Univ Ill, BS, 54, PhD(nutrit), 61. *Prof Exp:* Res scientist, Lederle Labs, Am Cyanamid Co, 61-63; from asst prof to assoc prof biochem & nutrit, 63-73, head dept, 73-82, PROF HUMAN NUTRIT & FOODS, VA POLYTECH INST & STATE UNIV, 73- *Mem:* Am Inst Nutrit; Am Dietetic Asn. *Res:* Nutrition and toxicant interactions; applied international nutrition programs. *Mailing Add:* Dept Human Nutrit & Foods Va Polytech Inst & State Univ Blacksburg VA 24061

WEBB, SAWNEY DAVID, b Los Angeles, Calif, Oct 31, 36; m 58; c 2. PALEONTOLOGY, ZOOLOGY. *Educ:* Cornell Univ, BA, 58; Univ Calif, Berkeley, MA, 61, PhD(paleont), 64. *Prof Exp:* Instr paleont, Univ Calif, Berkeley, 63-64; from asst cur mus & asst prof zool to assoc cur & assoc prof, 64-70, PROF GEOL & ZOOL & CUR FOSSIL VERT, FLA STATE MUS, UNIV FLA, 70- *Concurrent Pos:* NSF grants; Guggenheim fel, 73-74; vis prof, Yale Univ, 76; ed, Paleobiology, 75-78, Soc Vert Paleont News Bulletin, 79-82 & Quaternary Res, 82-86; pres, Soc Vert Paleont, 78-79; chmn, US Nat Comn Int Quaternary Asn, 82-86; bd mem, Simroe Found. *Mem:* Soc Vert Paleont (pres, 79-); Soc Study Evolution; Am Soc Mammal; Int Quaternary Asn; Sigma Xi. *Res:* Fossil mammals, evolution, paleoecology and functional morphology. *Mailing Add:* Fla State Museum Univ Fla Gainesville FL 32611

WEBB, THEODORE STRATTON, JR, b Oklahoma City, Okla, Mar 4, 30; m 52; c 2. APPLIED PHYSICS, AEROSPACE ENGINEERING. *Educ:* Univ Okla, BS, 51; Calif Inst Technol, PhD(physics), 55. *Prof Exp:* Dir aerospace tech dept, Gen Dynamics, 69-74, vpres res & eng, 74-80, vpres F-16 prog, 80-89, consult, 89-90, VPRES A-12 PROG, GEN DYNAMICS, 91- *Concurrent Pos:* Adj prof physics, Tex Christian Univ, 55-71. *Mem:* Am Phys Soc; AAAS. *Res:* Management of research and development; aircraft design and development; propulsion systems; nuclear shielding and reactor theory; low energy particle physics. *Mailing Add:* Gen Dynamics PO Box 748 Ft Worth TX 76101

WEBB, THOMAS EVAN, b Edmonton, Alta, Mar 4, 32; m 61; c 2. BIOCHEMISTRY. *Educ:* Univ Alta, BSc, 55, MSc, 57; Univ Toronto, PhD(biochem), 61. *Prof Exp:* Nat Res Coun Can fels, Nat Res Coun, Ont, 61-63 & Univ Wis-Madison, 63-65; asst prof biochem, Med Sch, Univ Man, 65-66; asst prof, Cancer Res Unit, McGill Univ, 66-70, actg dir, 69-70; assoc prof, 70-74, PROF PHYSIOL CHEM, COL MED, OHIO STATE UNIV, 74- *Concurrent Pos:* Prin investr, NIH grants. *Mem:* AAAS; Am Asn Cancer Res; Am Soc Biol Sci; Fedn Am Socs Exp Biol. *Res:* The biochemistry of cancer, development of cancer-detection and carcinogens-detection tests and development of anti-carcinogens. *Mailing Add:* Dept Physiol Chem Ohio State Univ Col Med 333 W Tenth Ave Columbus OH 43210

WEBB, THOMAS HOWARD, b Norfolk, Va, July 16, 35; m 57; c 2. ORGANIC CHEMISTRY, LUBRICATION ENGINEERING. *Educ:* Univ Va, BA, 57; Duke Univ, MA, 60, PhD(org chem), 61. *Prof Exp:* Res chemist, Sinclair Res, Inc, 61, proj leader lubricant additives, 63-65, group leader, 65-66, asst sect leader indust oils, 66-69; supvr lubricants, asphalts & waxes, BP Oil Corp, 69-70; sr proj leader indust oils res, 70-72, res assoc lubricants & lubrication, 72-80, RES ASSOC, EXPLOR & PROD, STANDARD OIL CO, OHIO, 81- *Concurrent Pos:* Sr res scientist, Sinclair Oil Corp, 68. *Mem:* AAAS; Am Soc Testing & Mat; Am Soc Lubrication Eng; Am Chem Soc. *Res:* Additives for lubricants; synthesis and structure-property correlations; lubricant formulation and product design. *Mailing Add:* B P Res 4440 Warrensville Center Rd Warrensville Heights OH 44128-2837

WEBB, THOMPSON, III, b Los Angeles, Calif, Jan 13, 44; m 69; c 2. PALYNOLOGY, CLIMATOLOGY. *Educ:* Swarthmore Col, BA, 66; Univ Wis-Madison, PhD(meteorol), 71. *Prof Exp:* Inst Sci & Technol fel, Univ Mich, Ann Arbor, 70-71, asst res paleoecologist, Great Lakes Res Div, 71-72; asst res prof, 72-75, assoc prof, 75-84, PROF GEOL SCI, BROWN UNIV, 84- *Concurrent Pos:* Vis fel, Clare Hall & Bot Sch, Cambridge Univ, Eng, 77-78, Cires, Univ Colo, Boulder, 88-89; ed, Rev Palaeobot & Palynology, 80-, J Veg Sci, 90-, Quaternary Sci Revs, 82-, J Climate, 88-, Ecol & Ecol Monographs, 85-88, J Quaternary Sci, 85-; prin investr, Coop Holocene Mapping Proj, 77-; pres, Subcomn Holecene NAm & Greenland, Int Union Quaternary Res, 82-91; acad adv, Ctr for the First Americans, Orono, Maine, 84- *Mem:* Fel AAAS; Am Meteorol Soc; Am Quaternary Asn; Am Asn Stratig Palynologists; Ecol Soc Am. *Res:* Use of multivariate statistical techniques for calibrating quaternary pollen data in vegetational and climatic terms; production of paleoclimatic and paleovegetation maps; development of global data; sets of paleoclimatic data. *Mailing Add:* Dept Geol Sci Box 1846 Brown Univ Providence RI 02912

WEBB, WATT WETMORE, b Kansas City, Mo, Aug 27, 27; m 50; c 3. CONDENSED MATTER PHYSICS. *Educ:* Mass Inst Technol, ScB, 47, ScD(metall), 55. *Prof Exp:* Res engr, Union Carbide Metals Co, 47-52, res scientist, 55-59, coordr fundamental res, 59-60, asst dir res, 61; assoc prof eng physics, 61-65, dir, 83-88, PROF APPL PHYSICS, CORNELL UNIV, 65- *Concurrent Pos:* Consult; mem, var adv panels Nat Res Coun, NSF, 58-; John Simon Guggenheim Found fel, 74-75; assoc ed, Phys Rev Letters, 75-; exec comt, Div Biol Physics, Am Phys Soc, 75-78, chmn, 89-90; mem, publ comt, Biophys J; chmn, long range planning comt, Cornell Res Found, 85-88; NIH Fogarty Int Scholar 88-; dir, Develop Resource for Biophys Imaging Optoelectronics, NIH. *Honors & Awards:* Biol Physics Prize, Am Phys Soc, 91. *Mem:* Fel Am Phys Soc; fel AAAS; Biophys Soc; Am Soc Cell Biol; Inst Soc Photo Optical Engrs. *Res:* Critical and collective phenomena; fluctuations; superconductivity; membrane and cellular biophysics; biophysical processes; biophysical instrumentation; chemical kinetics; fluid dynamics. *Mailing Add:* Sch Appl & Eng Physics Clark Hall Cornell Univ Ithaca NY 14853-2501

WEBB, WATTS RANKIN, b Columbia, Ky, Sept 8, 22; m 44; c 4. SURGERY. *Educ:* Univ Miss, BA, 42; Johns Hopkins Univ, MD, 45; Am Bd Surg, dipl, 52; Am Bd Thoracic Surg, dipl, 53. *Prof Exp:* Chief surgeon, Miss State Sanatorium, 52-63; prof surg & chmn, Div Thoracic & Cardiovasc Surg, Univ Tex Southwest Med Sch Dallas, 64-70; prof surg & chmn dept, State Univ NY Upstate Med Ctr, 70-77; PROF SURG & CHMN DEPT, SCH MED, TULANE UNIV, 77- *Concurrent Pos:* Prof, Sch Med, Univ Miss, 55-63. *Mem:* Fel Am Col Surgeons; Am Asn Thoracic Surgeons; fel Am Col Chest Physicians; Am Col Cardiol; Am Soc Artificial Internal Organs; Am Surg Asn; Soc Univ Surgeons; Soc Thoracic Surgeons. *Res:* Shock; cardiac-pulmonary transplantation; organ preservation; hypothermia; myocardial physiology. *Mailing Add:* 1430 Tulane Ave New Orleans LA 70112

WEBB, WILLIAM ALBERT, b Paterson, NJ, Mar 10, 44; m 66; c 2. MATHEMATICS. *Educ:* Mich State Univ, BS, 65; Pa State Univ, University Park, PhD(math), 68. *Prof Exp:* Res instr math, Pa State Univ, 68-69; from asst prof to assoc prof math, 69-82, PROF MATH, WASH STATE UNIV, 82- *Mem:* Am Math Soc; Math Asn Am. *Res:* Analytic and combinatorial number theory; number theoretic questions concerning polynomial rings over finite fields; combinatorics and graph theory; cryptology. *Mailing Add:* Dept Math Wash State Univ Pullman WA 99164-3113

WEBB, WILLIAM GATEWOOD, b Charleston, SC, July 17, 25; m 52; c 3. MEDICINAL CHEMISTRY. *Educ:* Univ of South, BS, 50; Univ Rochester, PhD(org chem), 54. *Prof Exp:* Res chemist, Columbia-Southern Chem Corp, WVa, 54-55; res assoc, Res Div, Sterling-Winthrop Res Inst, 55-57, Patent Div, 57-61, PATENT AGENT, STERLING RES GROUP, 61- *Mem:* Am Chem Soc; NY Acad Sci. *Res:* Synthetic organic chemistry. *Mailing Add:* Patent Dept Sterling Res Group Rensselaer NY 12144

WEBB, WILLIAM LOGAN, b Chattanooga, Tenn, Feb 13, 30; c 5. PSYCHIATRY. *Educ:* Princeton Univ, AB, 51; Johns Hopkins Univ, MD, 55. *Hon Degrees:* MA, Univ Pa, 71. *Prof Exp:* From instr to asst prof psychiat, Sch Med, Johns Hopkins Univ, 61-64; from asst prof to prof psychiat, Univ Pa, 64-74; prof psychiat & chmn Dept, Univ Tenn, Memphis, 74-; AT INST LIVING. *Mem:* Am Psychiat Asn; AMA; Acad Psychosom Med; Am Pain Soc; Am Psychosom Soc. *Mailing Add:* Inst Living 400 Washington St Hartford CT 06106

WEBB, WILLIAM PAUL, b Bismarck, NDak, Dec 30, 22; m 47; c 4. ORGANIC CHEMISTRY. *Educ:* Univ Notre Dame, BS, 44 & 47; Univ Minn, PhD(org chem), 51. *Prof Exp:* Sr res assoc, Chevron Res Co, Standard Oil Co Calif, San Francisco, 51-86; RETIRED. *Mem:* Am Chem Soc. *Res:* Petrochemicals; patent liaison. *Mailing Add:* 16 Chestnut Ave San Rafael CA 94901

WEBB, WILLIS KEITH, b McCoy, Va, Apr 21, 28; m 50; c 5. REGULATORY VETERINARY MEDICINE. *Educ:* Tex A&M Univ, DVM, 57. *Prof Exp:* Vet livestock inspector, Animal Dis Eradication Div, USDA, 57; vet, US Food & Drug Admin, 57-62; res assoc, Mead Johnson & Co, Ind, 62-70; VET MED OFFICER, USDA, 70- *Res:* Regulatory veterinary medicine; eradication of disease from domestic animals; animal welfare; interstate movement of livestock. *Mailing Add:* Rte 3 Box 136 Christiansburg VA 24073

WEBB, WILLIS LEE, b Nevada, Tex, July 9, 23; m 42; c 1. METEOROLOGY. *Educ:* Southern Methodist Univ, BS, 52; Univ Okla, MS, 70; Colo State Univ, PhD, 72. *Prof Exp:* Meteorol observer, US Weather Bur, 42-43, meteorologist, 46-52, physicist, 52-55; meteorologist & chief scientist, Atmospheric Sci Lab, US Army Electronics Command, 55-80; MEM STAFF, SCHELLENGER RES LAB, UNIV TEX, EL PASO, 80- *Concurrent Pos:* Mem upper atmosphere rocket res comt, Space Sci Bd, Nat Acad Sci, 59-65; chmn meteorol rocket network comt, Inter-Range Instrumentation Group, 60-; consult, Cath Univ, 60-61; adj prof physics, Univ Tex, El Paso, 63-; mem air blast subcomt, Inter-Oceanic Canal Studies Group, 64-74; Army adv, Nat Comt Clear Air Turbulence, 65-66. *Mem:* Am Meteorol Soc; Am Geophys Union; Am Inst Aeronaut & Astronaut; Am Phys Soc. *Res:* Synoptic exploration of the 25-100 kilometer region with small rocket vehicles to determine the circulation, thermal and composition structure; development of meteorological satellite techniques for mesoscale applications such as severe storm, air pollution, and combat on a global basis. *Mailing Add:* 4929 Blue Ridge Circle El Paso TX 79904

WEBBER, CHARLES LEWIS, JR, b Bay Shore, NY, July 26, 47; m 70; c 2. MEDICAL & RESPIRATORY PHYSIOLOGY, NONLINEAR DYNAMICS. *Educ:* Taylor Univ, AB, 69; Loyola Univ, Chicago, PhD(physiol), 74. *Prof Exp:* Fel physiol, Max Planck Inst Physiol & Clin Res, Bad Nauheim, 73-75; asst prof, 75-81, ASSOC PROF PHYSIOL, LOYOLA UNIV, CHICAGO, 81- *Concurrent Pos:* Sect leader prog proj grant, Nat Heart, Lung & Blood Inst, 75-80; prin investr, NIH res grant; course dir, med physiol, Loyola Univ. *Mem:* Am Physiol Soc; Soc Neurosci. *Res:* Central nervous system regulation of cardiopulmonary mechanisms; structure function correlations in the central nervous system; application of mathematics and laboratory computer systems to biological problem solving; teaching computer literacy to medical students; corelatioin of sympathetic and respiratory neural discharges; application of nonlinear dynamics to physiological systems and rhythms. *Mailing Add:* Dept Physiol Loyola Univ Chicago 2160 S First Ave Maywood IL 60153

WEBBER, DONALD SALYER, b Los Angeles, Calif, Jan 15, 17; m 51; c 2. PHYSICS. *Educ:* Univ Calif, Los Angeles, BA, 38, MA, 41, PhD(physics), 54; Calif Inst Technol, MS, 42. *Prof Exp:* Assoc physics, Univ Calif, Los Angeles, 49-54, instr, 54-55; mem tech staff, Ramo-Wooldridge Div, Thompson Ramo Wooldridge, Inc, 55-60, mem sr staff, 60-61, assoc mgr photo equip dept, 61; mgr solar physics prog, Calif Div, Lockheed Aircraft Corp, 61-63, mgr astron sci lab, 64-65; sr staff engr, TRW Systs Group, 65-80; RETIRED. *Concurrent Pos:* Consult physicist, 80-; prin invest: Study Correlation Guid Ref Mats, 66, Lab Measurements Satellite Optical Signatures, 72, Measurements of Atmospheric Pollutants, 75. *Mem:* Am Phys Soc; Optical Soc Am; Sigma Xi. *Res:* Infrared spectroscopy of crystals; photo-optical instrumentation; solar physics; remote sensing. *Mailing Add:* 3551 Knobhill Dr Sherman Oaks CA 91423

WEBBER, EDGAR ERNEST, b Worcester, Mass, Sept 11, 32; m 58; c 2. BOTANY, PLANT MORPHOLOGY. *Educ:* Univ Mass, BS, 55, PhD(bot), 67; Cornell Univ, MS, 61. *Prof Exp:* Instr bot, Wellesley Col, 61-62; asst prof, Pa State Univ, Behrend Campus, 65-67; from asst prof to assoc prof, 67-74, PROF BIOL, KEUKA COL, 74- *Concurrent Pos:* NSF grant, 65, Am Philos Soc grant, 68, Pa State fac grant, 65-67, Keuka Col fac grants, 87-88; teaching/res, Marine Sci Inst, Nahant, Mass, 68-80. *Mem:* Phycol Soc Am; Int Phycol Soc; Brit Phycol Soc. *Res:* Ecology and systematics of benthic salt marsh algae; culture and life history studies; morphology; 25 published papers in marine biology. *Mailing Add:* Dept Biol Keuka Col Keuka Park NY 14478

WEBBER, GAYLE MILTON, b Sioux City, Iowa, Aug 30, 31; m 60; c 3. ORGANIC CHEMISTRY. *Educ:* Morningside Col, BS, 53; Univ Iowa, MS, 58. *Prof Exp:* Sr res asst, 58-74, RES INVESTR, G D SEARLE & CO, 74- *Mem:* Am Chem Soc. *Res:* Steroids; peptides and amino acids. *Mailing Add:* 2026 Harrison St Evanston IL 60201-2222

WEBBER, GEORGE ROGER, b Toronto, Ont, Nov 2, 26; m 54. GEOLOGY. *Educ:* Queen's Univ, Can, BSc, 49; McMaster Univ, MSc, 52; Mass Inst Technol, PhD(geol), 55. *Prof Exp:* Res assoc, McGill Univ, 55-59, asst prof, 59-66, assoc prof geol, 66-88; RETIRED. *Res:* Geochemistry; x-ray and optical spectrography. *Mailing Add:* 66 13th St Roxboro PQ H8Y 1L7 Can

WEBBER, HERBERT H, b Vancouver, BC, Oct 15, 41; m 63. BIOLOGY. *Educ:* Univ BC, BSc, 63, PhD(zool), 66. *Prof Exp:* NATO fel, Can for Study at Hopkins Marine Sta Stanford, 66-68; asst prof biol, Wake Forest Univ, 68-72; asst prof biol, 72-74, ASSOC PROF MARINE RESOURCES, HUXLEY COL ENVIRON STUDIES, WESTERN WASH UNIV, 74- *Mem:* AAAS. *Res:* Invertebrate physiology; reproductive physiology of Gastropoda Mollusca. *Mailing Add:* Dept Environ Sci Western Wash Univ 516 High St Bellingham WA 98225

WEBBER, J ALAN, b Chicago, Ill, Aug 14, 40; div; c 1. MEDICAL REGULATORY AFFAIRS. *Educ:* Univ Colo, BA, 62; Stanford Univ, PhD(org chem), 67. *Prof Exp:* Sr org chem, 66-70, from res scientist to sr res, 70-88. REGULAR SCIENTIST, LILLY CORP CTR, 88- *Mem:* Am Chem Soc; Am Soc Microbiol. *Res:* Structural modification and synthesis of natural products and complex organic molecules; medicinal chemistry of agents with antibiotic activity. *Mailing Add:* Dept MC 676 Lilly Corp Ctr Indianapolis IN 46285

WEBBER, JOHN CLINTON, b Shreveport, La, Apr 2, 43; m 64; c 2. RADIO ASTRONOMY, COMPUTER CONTROL. *Educ:* Calif Inst Technol, BS, 64, PhD(astron), 70. *Prof Exp:* Res assoc astron, Univ Ill, Urbana-Champaign, 69-71, asst prof, 71-77, sr scientist, 77-80; res staff, 80-83, asst dir astron, Haystack Observ, Mass Inst Technol, 83-88; DIV VPRES, INTERFEROMETRICS INC, 88- *Mem:* Am Astron Soc; Int Astron Union; Int Union Radio Sci,. *Res:* Radio astronomy research; quasars and very-long-baseline interferometry; equipment development; computer control of radio telescopes and data acquisition systems; recorder system for very-long-baseline array; satellite tracking. *Mailing Add:* Interferometrics Inc 8150 Leesburg Pike Vienna VA 22182

WEBBER, LARRY STANFORD, b New Orleans, La, Apr 1, 45; m 67; c 2. BIOMETRICS, BIOSTATISTICS. *Educ:* La State Univ, BS, 67; Yale Univ, MPhil, 70, PhD(epidemiol & pub health), 73. *Prof Exp:* Nat Acad Sci statistician, Atomic Bomb Casualty Comn, 72-74; from asst prof to assoc prof, 74-84, PROF, MED CTR, LA STATE UNIV, 84- *Mem:* Am Statist Asn; Biomet Soc; Am Heart Asn; Am Pub Health Asn; Sigma Xi. *Res:* Design, implementation and analysis of an ongoing longitudinal research study of cardiovascular risk factor variables in an entire community; development of health promotion model for school children. *Mailing Add:* Dept Med Specialized Ctr Res La State Univ 1542 Tulane Ave New Orleans LA 70112-2822

WEBBER, MARION GEORGE, b Golden, Colo, Dec 8, 21; m 45; c 9. PHARMACEUTICAL CHEMISTRY. *Educ:* Univ Colo, BS, 47, MS, 48; Univ Fla, PhD(pharm), 51. *Prof Exp:* Instr pharm & pharmaceut chem, Univ Fla, 50-51; asst prof pharmaceut chem, 51-52, assoc prof pharm, 53-72, PROF PHARM, UNIV HOUSTON, 72- *Mem:* Am Col Apothecaries; Am Pharmaceut Asn. *Res:* Tillandsia usneoides; pharmaceutical formulation and stability studies. *Mailing Add:* 8138 Glenbrook Houston TX 77017

WEBBER, MILO M, b Los Angeles, Calif, Sept 27, 30; m 55; c 2. RADIOLOGY, NUCLEAR MEDICINE. *Educ:* Univ Calif, Los Angeles, BA, 52, MD, 55; Am Bd Radiol, dipl, 61; Am Bd Nuclear Med, dipl, 72. *Prof Exp:* Intern surg serv, 55-56, resident radiol, 56-57 & 59-60, instr, 60-61, lectr, 61-62, from asst prof to assoc prof, 62-74, PROF RADIOL SCI, SCH MED, UNIV CALIF, LOS ANGELES, 74- *Concurrent Pos:* Consult, Queen of Angels Hosp, Los Angeles, 65-77; res radiologist, Lab Nuclear Med & Radiation Biol, Sch Med, Univ Calif, Los Angeles, 65-78. *Mem:* Am Col Radiol; AMA; Soc Nuclear Med; Radiol Soc NAm. *Res:* Development and refinement of organ radioisotope scanning procedures, including development of white blood cell scanning and thrombosis localization techniques; application of telecommunications to the practice of radiology and nuclear medicine. *Mailing Add:* Radiol Sci BL428 CHS 172 115 Univ Calif Los Angeles CA 90024

WEBBER, MUKTA MALA (MAINI), oncology, cell biology, for more information see previous edition

WEBBER, PATRICK JOHN, b Bedfordshire, UK, Feb 24, 38; m 63. ECOLOGY, BIOLOGY. *Educ:* Univ Reading, BSc, 60; Queen's Univ, Ont, MSc, 63, PhD, 71. *Prof Exp:* Asst prof biol, York Univ, 66-69; from asst prof to prof biol, Univ Colo, Boulder, 69-89, mem fac, Inst Arctic & Alpine Res, 69-89; DIR, KELLOG BIOL STA & PROF FORESTRY, MICH STATE UNIV, 90- *Concurrent Pos:* Dir ecol prog, NSF, Washington DC, 87-89. *Mem:* AAAS; Ecol Soc Am; Arctic Inst NAm; Can Bot Asn. *Res:* Primary productivity, phenology and phytosociology of terrestrial communities; long-term ecological research; landscape ecology. *Mailing Add:* Dept Forestry Mich State Univ East Lansing MI 48824

WEBBER, RICHARD HARRY, b Camillus, NY, Jan 2, 24; m 46; c 7. ANATOMY. *Educ:* St Benedict's Col, BS, 48; Univ Notre Dame, MS, 49; St Louis Univ, PhD(anat), 54. *Prof Exp:* Instr biol, St Benedict's Col, 48; instr, Niagara Univ, 49-51; asst anat, Sch Med, St Louis Univ, 51-54; asst prof, Sch Med, Creighton Univ, 54-59; assoc prof, Sch Med, Temple Univ, 59-61; from assoc prof to prof, 61-89, EMER PROF ANAT, SCH MED, STATE UNIV NY BUFFALO, 89- *Concurrent Pos:* Lederle med fac scholar, 56-59; from chmn elect to chmn, Sect Anat Sci, Am Asn Dent Schs, 71-73; mem, Am Benedictine Acad; fel, Human Biol Coun; vis prof, Anat Inst, Univ Heidelberg, 82-83 & 85-87 & 90, guest prof, 84 & 87. *Mem:* Am Asn Anatomists; Sigma Xi; Cajal Club. *Res:* Peripheral autonomic nervous system. *Mailing Add:* 2947 Sunset Dr Grand Island NY 14072

WEBBER, RICHARD JOHN, b Mar 2, 48. CARTILAGE BIOLOGY, JOINT BIOLUBRICATION. *Educ:* State Univ NY, Stony Brook, PhD(path), 81. *Prof Exp:* ASSOC PROF PATH & BIOCHEM, UNIV ARK MED SCI, 82- *Mailing Add:* Dept Path Univ Ark Med Sch Mail Slot 517 4301 W Markham St Little Rock AR 72205

WEBBER, RICHARD LYLE, b Akron, Ohio, Nov 2, 35; div; c 2. PHYSIOLOGICAL OPTICS, DENTISTRY. *Educ:* Albion Col, AB, 58; Univ Mich, Ann Arbor, DDS, 63; Univ Calif, Berkeley, PhD(physiol optics), 71. *Prof Exp:* Intern dent, USPHS Hosp, Seattle, Wash, 63-64, clin investr, Mat & Technol Br, Dent Health Ctr, Div Dent Health, USPHS, 64-68, lectr, Univ Calif, Berkeley, 70-71, staff investr, Diag Systs, Oral Med & Surg Br, 71-75, chief diag methodology sect, 75-77, CHIEF CLIN INVEST BR, NAT INST DENT RES, 75-, CHIEF, DIAG SYST BR, 78- *Concurrent Pos:* Clin investr, Nat Inst Dent Res, NIH, 71-72, chief, Clin Invests & Res Servs Br,

73-75, chief, Clin Invests Br & chief, Methodology Sect, 75-80, chief, Diag Systs Br & chief Diag Methdology Sect, 80-83, Chief, Diag Systs Br, 83-88; chmn, Dept Diag Sci, Sch Dent, Univ Ala, Birmingham, 88-90, Dept Prof Dent & Radiol, Bowman Gray Sch Med, Winston Salem, NC, 90- *Honors & Awards:* Commendation Medal, USPHS. *Mem:* Fel AAAS; Am Acad Dent Radiol; Int Asn Dent Res; Soc Photo-Optical Instrument Eng; Sigma Xi; Am Dent Asn. *Res:* Image processing and the study of factors influencing diagnostic systems. *Mailing Add:* 2970-E Walnut Forest Ct Winston-Salem NC 27103

WEBBER, STANLEY EUGENE, b Boston, Mass, June 8, 19; c 4. ELECTRICAL ENGINEERING. *Educ:* Mass Inst Technol, BS, 41, MS, 42. *Prof Exp:* Res assoc, Gen Elec Co, 42-57, mgr eng, 57-64; mgr eng, Litton Industs, Inc, 64-67, mgr linear beam dept, Electron Tube Div, 67-76, vpres, 72-76, PRES, LITTON INDUSTS, INC, 76- *Mem:* Fel Inst Elec & Electronics Engrs. *Res:* Design and manufacture of ultra high frequency vacuum tubes. *Mailing Add:* Litton Industs 960 Industrial Rd San Carlos CA 94070

WEBBER, STEPHEN EDWARD, b Springfield, Mo, Sept 19, 40; m 62; c 3. PHYSICAL CHEMISTRY, POLYMER SPECTROSCOPY. *Educ:* Wash Univ, AB, 62; Univ Chicago, PhD(chem), 65. *Prof Exp:* NSF fel, Univ Col, Univ London, 65-66; from asst prof to assoc prof, 66-82, PROF CHEM, UNIV TEX, AUSTIN, 83- *Concurrent Pos:* Vis prof, Fed Polytech Lausanne, Switz, 83, Univ Paris, Orsey, 85. *Mem:* Am Chem Soc; Sigma Xi. *Res:* Energy transfer in polymers; polymer photochemistry. *Mailing Add:* Dept Chem Univ Tex Austin TX 78712

WEBBER, THOMAS GRAY, organic chemistry; deceased, see previous edition for last biography

WEBBER, WILLIAM A, b Nfld, Can, Apr 8, 34; m 58; c 3. HISTOLOGY, PHYSIOLOGY. *Educ:* Univ BC, MD, 58. *Prof Exp:* Intern, Vancouver Gen Hosp, 58-59; fel physiol, Med Col, Cornell Univ, 59-61; from asst prof to assoc prof anat, Univ BC, 61-69, assoc dean med, 71-77, dean, 77-90, PROF ANAT, UNIV BC, 69-, ASSOC VPRES ACAD, 90- *Mem:* Am Asn Anatomists; Can Asn Anatomists. *Res:* Renal physiology; kidney structure. *Mailing Add:* Dept Anat Univ BC Vancouver BC V6T 1W5 Can

WEBBER, WILLIAM R, b Bedford, Iowa, June 9, 29; m 61; c 1. PHYSICS, ASTROPHYSICS. *Educ:* Coe Col, BS, 51; Univ Iowa, MS, 54, PhD(physics), 57. *Prof Exp:* Asst prof physics, Univ Md, 58-59; NSF fel, Imp Col, Univ London, 59-61; from asst prof to assoc prof, Univ Minn, Minneapolis, 61-69; PROF PHYSICS & DIR, SPACE SCI CTR, UNIV NH, 69- *Concurrent Pos:* Co-ed, J Geophys Res, Am Geophys Union, 61-; mem Fields & Particles Subcomt, NASA, 63-66; consult, Boeing Aircraft Co & NAm Aviation, Inc, 63-; vis prof, Univ Adelaide, 68-69 & Danish Space Res Ctr, 71; NSF fel, Univ Tasmanes, 76 & von Humboldt, Max Planck Inst Physics & Astrophysics, Garching, 85-86. *Mem:* Am Phys Soc; Am Geophys Union; Am Astron Soc. *Res:* Charge composition and energy spectrum of cosmic rays; motion of charged particles in the earths magnetic field; solar-terrestrial relationships; particle detectors; x-ray and gamma ray astronomy. *Mailing Add:* Dept Physics Univ NH Durham NH 03824

WEBBINK, RONALD FREDERICK, b Hutchinson, Kans, Sept 21, 45; m 70; c 2. ASTROPHYSICS. *Educ:* Mass Inst Technol, BS, 67; Univ Cambridge, PhD(astron), 75. *Prof Exp:* Res assoc astron, Univ Ill, 75-77, res asst prof, 77-78, from asst prof to assoc prof, 78-85, PROF ASTRON, UNIV ILL, 85-, DEPT CHAIR, 89- *Concurrent Pos:* Vis fel, Joint Inst Lab Astrophys, Boulder, Colo, 84-85. *Mem:* Am Astron Soc; Royal Astron Soc; Astron Soc Pac; Int Astron Union. *Res:* Structure and evolution of close binary stars; mass and angular momentum loss from stars; stellar interiors; cataclysmic variable stars; globular star clusters. *Mailing Add:* Astron Dept 103 Astron Bldg 1002 W Green St Urbana IL 61801

WEBBON, BRUCE WARREN, b Bridgeport, Conn, June 7, 45; div; c 2. BIOENGINEERING, MECHANICAL ENGINEERING. *Educ:* Ga Inst Technol, BSME, 68; Univ Fla, MSME, 69; Univ Mo, PhD(mech eng), 78. *Prof Exp:* Turbine aerodyn engr jet engines, Pratt & Whitney Aircraft Co, 67-68; res asst mech eng, Univ Fla, 68-69; sr thermodyn engr life support syst, LTV Aerospace Corp, 69-72; res assoc mech eng, Univ Mo, 72-75; res scientist life sci, Ames Res Ctr, NASA, 75-81; prog dir, Bioinstrumentation Technol, SRI Int, 81-86; BR CHIEF, EXTRAVEHICULAR BR, AMES RES CTR, NASA, 86- *Mem:* Am Inst Aeronaut & Astronaut; AAAS; Aerospace Med Asn. *Res:* Zero gravity fluid mechanics and phase change processes; human thermoregulation; life support systems; physiological instrumentation. *Mailing Add:* 40 Forest Rd Woodside CA 94062

WEBER, ALBERT VINCENT, b Pittsburgh, Pa, Jan 30, 25; m 48; c 3. BOTANY. *Educ:* Duquesne Univ, BEd, 50, MSc, 52; Univ Minn, PhD(bot), 57. *Prof Exp:* Asst bot, Duquesne Univ, 50-52; asst, Univ Minn, 52-56; from instr to assoc prof, 56-63, co-dir summer session, 69-74, PROF BIOL, UNIV WIS-LA CROSSE, 64-, ASSOC DEAN, COL ARTS LETT & SCI, 67-, DIR, SCH HEALTH & HUMAN SERV, 74- *Mem:* Sigma Xi; Bot Soc Am. *Res:* Plant anatomy, morphology and morphogenesis. *Mailing Add:* Col Lett Sci Univ Wis La Crosse WI 54601

WEBER, ALFONS, b Dortmund, Ger, Oct 8, 27; nat US; m 55; c 3. PHYSICS. *Educ:* Ill Inst Technol, BS, 51, MS, 53, PhD(physics), 56. *Prof Exp:* Asst physics, Ill Inst Technol, 51-53, instr, 53-56; fel, Nat Res Coun Can, 56-57; from asst prof to prof physics, Fordham Univ, 57-81, chmn dept, 64-69; physicist, Nat Bur Standards, 77-80, CHIEF, MOLECULAR PHYSICS DIV, NAT INST STANDARDS & TECHNOL, 80- *Mem:* Fel Am Phys Soc; AAAS; Optical Soc Am; Coblentz Soc. *Res:* High resolution Raman and infrared spectroscopy; molecular mechanics; optics. *Mailing Add:* Molecular Physics Div 574 Rm B268 Bldg 221 Nat Inst Standards & Technol Gaithersburg MD 20899

WEBER, ALFRED HERMAN, b Philadelphia, Pa, Jan 15, 06; m 32; c 7. NUCLEAR & SPACE PHYSICS, BIOPHYSICS. *Educ:* St Joseph's Col, Pa, AB, 28, MA, 31; Univ Pa, PhD(physics), 36. *Hon Degrees:* DSc, St Joseph's Col, Pa, 68. *Prof Exp:* Instr physics & math, St Joseph's Col, Pa, 28-34, from asst prof to prof physics & head dept, 34-39; from asst prof to prof, St Louis Univ, 39-74; tech dir, USAAF Radio Sch, 42-43, chmn dept physics, 51-74; lectr Space Physic, Wash Univ St Louis, 60-61; prof physics, Univ Ala, Huntsville, 62-70; EMER PROF PHYSICS & CHMN DEPT, ST LOUIS UNIV, 74- *Concurrent Pos:* Sigma Xi, Res Corp, NSF, Army Res Off, AEC, Am Cancer Soc & Dept Defense grants; consult, Argonne Nat Lab, 47-57, US Army Ballistic Missile Agency, 57-60 & Marshall Space Flight Ctr, NASA, 60-70; vpres, Assoc Midwest Univs, Argonne Nat Lab, 60, pres, 61; lectr space physics, Washington Univ, St Louis, 60-61; consult, 74- *Mem:* Fel Am Phys Soc; Am Asn Univ Prof; Am Asn Physics Teachers; Sigma Xi. *Res:* Electron emission and diffraction; properties of thin metallic films; x-ray diffraction; neutron diffraction and scattering; nuclear spectroscopy; plasma and space physics; nuclear spectroscopy; photoelectricity; atomic and molecular physics; optics. *Mailing Add:* 546 Chevy Chase Dr Sarasota FL 34243

WEBER, ALLEN HOWARD, b Lorenzo, Idaho, May 15, 38; m 59; c 3. ATMOSPHERIC DIFFUSION, MICROMETEOROLOGY. *Educ:* Brigham Young Univ, BS, 60; Univ Ariz, MS, 62; Univ Utah, PhD(meteorol), 66. *Prof Exp:* Asst prof meteorol, Univ Okla, 66-68 & meteorol & environ health, 68-69; consult, Nat Severe Storms Lab, 69; from asst prof to assoc prof meteorol, NC State Univ, 69-77; RES STAFF METEOROLOGIST, SAVANNAH RIVER LAB, 87- *Concurrent Pos:* NSF grant, Univ Okla, 68-69; res meteorologist, Environ Protection Agency, Nat Environ Res Ctr, 70-72; Environ Protection Agency grant, NC State Univ & NC Water Resources Res Inst, 72-77 & turbulence & diffusion res, Savannah River Lab, 77- *Mem:* Am Meteorol Soc; Am Nuclear Soc. *Res:* Atmospheric turbulence; physical meteorology. *Mailing Add:* Environ Technol Div Savannah River Lab Aiken SC 29808

WEBER, ALLEN THOMAS, b Long Prairie, Minn, Sept 7, 43; m 69; c 2. MICROBIOLOGY. *Educ:* Univ Mich, Ann Arbor, BS, 65; Univ Wis-Madison, MS, 67, PhD(bact), 70. *Prof Exp:* Asst prof, 70-73, ASSOC PROF BIOL, UNIV NEBR, OMAHA, 73- *Mem:* AAAS; Am Soc Microbiol; Brit Soc Gen Microbiol; Sigma Xi. *Res:* Morphogenesis and development of microorganisms; cellular slime molds; microbial genetics. *Mailing Add:* Dept Biol Univ Nebr Omaha NE 68182

WEBER, ALVIN FRANCIS, b Hartford, Wis, Mar 13, 18; m 45; c 3. CYTOLOGY. *Educ:* Iowa State Col, DVM, 44; Univ Wis, BA, 46, MS, 48, PhD(vet sci), 49. *Prof Exp:* Instr vet sci, Univ Wis, 44-49; from asst prof to assoc prof, 49-55, head dept, 65-73, PROF VET ANAT, UNIV MINN, ST PAUL, 55- *Concurrent Pos:* USPHS spec fel, Univ Giessen, 59; NIH fel, Univ Bern, 72. *Mem:* Am Asn Anat; Sigma Xi. *Res:* Bovine uterine histology, histological and cytochemical changes of the adrenal gland and structure of secretory components of the udder; electron microscopic studies of the hematopoietic organs of domestic animals. *Mailing Add:* Dept Vet Anat Univ Minn 239 A Vet Sci St Paul MN 55108

WEBER, ANNEMARIE, b Rostock, Ger, Sept 11, 23; US citizen. PHYSIOLOGY, BIOCHEMISTRY. *Educ:* Univ Tubingen, MD, 50. *Prof Exp:* Res asst physiol, Univ Tubingen, 50-51; res fel biophys, Univ Col, London & physiol, Univ Md, 51; Rockefeller fel phys chem, Harvard Med Sch, 52; res fel physiol, Univ Tubingen, 53-54; res assoc neurol, Columbia Univ, 54-59; asst mem physiol, Inst Muscle Dis, New York, 59-63, assoc mem, 63-65; prof biochem, St Louis Univ, 65-72; PROF BIOCHEM, UNIV PA, 72- *Mem:* Am Physiol Soc; Soc Gen Physiol; Biophys Soc; Am Soc Biol Chemists; Sigma Xi. *Res:* Regulation of muscular activity. *Mailing Add:* Sch Med Univ Pa Philadelphia PA 19104

WEBER, ARTHUR GEORGE, chemistry; deceased, see previous edition for last biography

WEBER, ARTHUR L, b June 7, 43; m 74. BIOCHEMISTRY, ORGANIC CHEMISTRY. *Educ:* Univ Miami, PhD(biochem), 73. *Prof Exp:* Postdoctoral, Univ Ala, Birmingham, 73-75, Univ Calif, Berkeley, 75-76 & Salk Inst, 76-79; RES ASSOC, SALK INST BIOG STUDIES, 79- *Mem:* Am Chem Soc; Fedn Am Socs Exp Biol; Int Soc Study Origin of Life. *Res:* study of the chemical origin of life. *Mailing Add:* Salk Inst PO Box 85800 San Diego CA 92138-9216

WEBER, ARTHUR PHINEAS, b Brooklyn, NY, Mar 10, 20; m 42; c 2. CONTINUOUS CHEMICAL REACTOR DESIGN, MIXING. *Educ:* City Col NY, BS, 41; Oak Ridge Inst Reactor Technol, 47. *Prof Exp:* Design engr, Hendrick Mfg Co, 41-42; exec engr, Chemurgy Design Corp, 42-44; dir process develop & design, Kellex Corp, 44-49; tech dir, Int Eng Inc, 49-55; OWNER CONSULT, ARTHUR PHINEAS WEBER ENGRS, 51- *Concurrent Pos:* Instr chem eng, City Col, New York, 43-44; assoc prof nuclear eng, NY Univ, 51-54; prof chem eng, Polytech Univ, Brooklyn, 55-67; consult, Nat Res Coun; regist prof engr, NY, Ohio, Pa. *Mem:* Fel AAAS; Am Inst Chem Engrs; Am Chem Soc; Nat Soc Prof Engrs; Am Soc Safety Engrs; NY Acad Sci. *Res:* Continuous chemical reactor design; mixing; nuclear physics design; drying systems; process equipment optimization; engineering parameters of insurance liability. *Mailing Add:* 1334 Surrey Lane Rockville Centre NY 11570

WEBER, BARBARA C, b Prairie du Chien, Wis, Nov 15, 47; m; c 1. ENTOMOLOGY. *Educ:* Viterbo Col, BA, 69; Univ Minn, St Paul, MS, 71; Southern Ill Univ Carbondale, PhD, 82. *Prof Exp:* Conserv forest pest specialist entom, Minn Dept Natural Resources, 73-74; entomologist Dutch elm dis, Minn Dept Agr, 74-75; proj leader, USDA, 83-86, legis asst, 87-88, staff asst, 88-91, RES ENTOMOLOGIST, FOREST SERV, USDA, 75-, ACTG DIR, 91- *Concurrent Pos:* Ed, Walnut Coun Bull, 81-85; cong fel, Am Polit Sci Asn, 86-87. *Honors & Awards:* Am Black Walnut Res Award, 83.

Mem: Entom Soc Am; Soc Am Foresters; Sigma Xi. *Res:* Insect pests of black walnut; disease transmission by insects; taxonomy of scolytid insects; silvicultural management of black walnut to control insect problems. *Mailing Add:* Pac SW Forest & Range Exp Sta 1960 Addison St Berkeley CA 94704

WEBER, BRUCE HOWARD, b Cleveland, Ohio, June 8, 41. BIOCHEMISTRY. *Educ:* San Diego State Univ, BS, 63; Univ Calif, San Diego, PhD(chem), 68. *Prof Exp:* Am Cancer Soc fel, Molecular Biol Inst, Univ Calif, Los Angeles, 68-70; asst prof, 70-72, assoc prof, 72-76, PROF CHEM, CALIF STATE UNIV, FULLERTON, 76- *Concurrent Pos:* NIH res fel, 75; res scientist, Div Neurosci, City of Hope Nat Med Ctr, 75-77. *Mem:* AAAS; Am Chem Soc; Brit Soc Hist Sci; Am Soc Biol Chemists; Sigma Xi. *Res:* Structure, function, and evolution of proteins; history of biochemistry. *Mailing Add:* Dept Chem Calif State Univ Fullerton CA 92634

WEBER, CARL JOSEPH, b Evanston, Ill, Nov 7, 54; m 77; c 2. RADIATION CROSSLINKED POLYMER SYSTEMS, FLUOROPOLYMERS. *Educ:* Univ Calif, Santa Barbara,BS, 76; Univ Mass, MS, 80, PhD(chem), 81. *Prof Exp:* Develop engr electron mat develop, Thick Film Systs, Subsid Ferro Corp, 76-78; res chemist process res, US Borax Res Corp, Div US Borax Chem Corp, 81-85; res chemist, Polymer Mat Develop, 85-87, develop group leader, 87-90, ENG MGR, THERMOFIT DIV, RAYCHEM CORP, 90-; FEL INORG & POLYMER CHEM, UNIV MASS, AMHERST, 81- *Mem:* Am Chem Soc; AAAS; Sigma Xi. *Mailing Add:* 105 Atherwood Ave Redwood City CA 94061

WEBER, CHARLES L, b Dayton, Ohio, Dec 2, 37. COMMUNICATION SYSTEMS, RADAR SYSTEMS. *Educ:* Univ Dayton, BSEE, 58; Univ Southern Calif, MSEE, 60; Univ Calif, Los Angeles, PhD, 64. *Prof Exp:* Mem tech staff, Hughes Aircraft Co, Calif, 58-62; from asst prof to assoc prof, 64-79, PROF ELEC ENG, UNIV SOUTHERN CALIF, 79- *Concurrent Pos:* Consult. *Mem:* Inst Elec & Electronics Engrs. *Res:* Blind equalization; spread spectrum systems. *Mailing Add:* PO Box 45968 Los Angeles CA 90045

WEBER, CHARLES WALTER, b Harold, SDak, Nov 30, 31; m 61; c 2. BIOCHEMISTRY, NUTRITION. *Educ:* Colo State Univ, BS, 56, MS, 58; Univ Ariz, PhD(biochem, nutrit), 66. *Prof Exp:* Res chemist, Univ Colo, 60-63; from asst prof to assoc prof, 66-73, PROF NUTRIT BIOCHEM, UNIV ARIZ, 73- *Mem:* Inst Food Technologists; Poultry Sci Asn; Am Soc Clin Nutrit; NY Acad Sci; Am Inst Nutrit; Am Asn Cereal Chem. *Res:* Interaction between trace elements and fiber; evaluation of desert plants for nutritional quality. *Mailing Add:* Dept Nutrit & Food Sci Rm 309 Shantz Bldg Univ Ariz Tucson AZ 85721

WEBER, CHARLES WILLIAM, b Streator, Ill, Dec 28, 22; m 48; c 3. ANALYTICAL CHEMISTRY, ENVIRONMENTAL MONITORING. *Educ:* Northwestern Univ, BS, 49; Ind Univ, PhD(anal chem), 53. *Prof Exp:* Analytical chemist, 53-58, Sect Anal Develop, 58-62, dept head chem anal, Oak Ridge Gaseous Diffusion Plant, 62-77, ENVIRON ANALSIS FIVE-PLANT COORDR, MARTIN MARIETTA ENERGY SYSTS, INC, 77- *Mem:* AAAS; Sigma Xi; fel Am Inst Chem; Am Chem Soc; NY Acad Sci; Am Soc Testing & Mat. *Res:* Instrument development; gas analysis; automation; laboratory management; process control. *Mailing Add:* 1021 W Outer Dr Oak Ridge TN 37830

WEBER, CLIFFORD E, b Fresno, Calif, May 20, 18; m 44; c 2. ADVANCED NUCLEAR REACTORS, NUCLEAR ELEMENT. *Educ:* Univ Calif, Berkeley, BS, 41; Johns Hopkins Univ, PhD(chem), 44. *Prof Exp:* Group leader, Manhattan Proj, Johns Hopkins Univ, 42-46; sect head, Knolls Atomic Power Lab, Gen Elec Co, 46-61; dir, Atomics Int, 61-62, proj mgr, 62-68; br chief, AEC, 68-75 & Energy Res & Develop Admin, 75-77; int specialist, 78-89, DEP DIR, ADVAN REACTOR PROGS, DEPT ENERGY, 89- *Mem:* Fel Am Soc Metals. *Res:* Nuclear fuels and materials; nuclear fuel development; fluorins and fluorocarbons; sodium corrosion; fast reactor development; nuclear reactor technology; author of numerous technical publications; awarded one patent. *Mailing Add:* 15317 Carrolton Rd Rockville MD 20853

WEBER, DARRELL J, b Thornton, Idaho, Nov 16, 33; m 62; c 7. BIOCHEMISTRY, PLANT PATHOLOGY. *Educ:* Univ Idaho, BS, 58, MS, 59; Univ Calif, PhD(plant path), 63. *Prof Exp:* Res asst agr chem, Univ Idaho, 57-59; res asst plant path, Univ Calif, 59-63; res assoc biochem, Univ Wis, 63-65; asst prof, Univ Houston, 65-69; assoc prof, 69-73, PROF BOT, BRIGHAM YOUNG UNIV, 73- *Concurrent Pos:* Fels, Univ Wis, 63-65; consult, NASA, 66-67; grants, USDA, 66-68, NASA, 67-68, NIH, 68-71 & NSF, 71-78; fel biochem, Mich State Univ, 75-76, NSF, 76-79, & 84-88, USDA, 80-81 & 88-90. *Honors & Awards:* Maeser Award, 74. *Mem:* NAm Mushroom Soc; Am Phytopath Soc; Am Bot Soc; New Crops Soc; Am Mycol Soc. *Res:* Phytochemistry; mode of action of fungicides; metabolism of fungal spores; biochemistry of host-parasite complexes; salt tolerance of plants; toxic compounds in plants. *Mailing Add:* Dept Bot 285 Widtsoe Brigham Young Univ Provo UT 84602

WEBER, DAVID ALEXANDER, b Lockport, NY, Mar 6, 39; m 61; c 2. MEDICAL PHYSICS, NUCLEAR MEDICINE. *Educ:* St Lawrence Univ, BS, 60; Univ Rochester, PhD(medical physics), 71. *Prof Exp:* Teaching asst physics, Univ Buffalo, 60-61; asst attend physicist, Mem Sloan-Kettering Cancer Ctr, 61-68; AEC grad lab fel radiation biol & biophys, Univ Rochester, 68-70, asst prof radiol, 70-75, asst prof radiation biol & biophys, 70-80, assoc prof radiol, 75-87, assoc prof radiation biol & biophys, Sch Med & Dent, 80-87; HEAD NUCLEAR MED RES GROUP, GROUP LEADER & SR SCIENTIST, BROOKHAVEN NAT LAB, 87-; PROF RADIOL, STATE UNIV NY, STONY BROOK, 88- *Concurrent Pos:* Clin fac, Sch Health Related Prof, Rochester Inst Technol, 76-86; vis scientist, Lund Univ Hosp, Sweden, 78-79; Fogart Sr Int fel, 78-79; tech assistance expert, Int Atomic Energy Agency, 85, 88. *Mem:* Soc Nuclear Med; Am Asn Physicists in Med; Health Physics Soc; fel Am Col Nuclear Physicians; Sigma Xi. *Res:* Development of tracer procedures for the study of bone, brain, heart and lungs; metabolic and functional radionuclide studies to investigate physiological processes and mechanisms of disease; development of single photon emission computed tomography (SPECT) technology, planar imaging techniques and methods of computer assisted image processing. *Mailing Add:* Nuclear Med Res Group Med Dept Brookhaven Nat Lab Upton NY 11973

WEBER, DAVID FREDERICK, b North Terre Haute, Ind, Nov 18, 39; m 63; c 2. CYTOGENETICS. *Educ:* Purdue Univ, BS, 61; Ind Univ, MS, 63, PhD(genetics), 67. *Prof Exp:* From asst prof to assoc prof, 67-78, PROF GENETICS, ILL STATE UNIV, 78- *Concurrent Pos:* Contract, 70-83 & USDA, 86- *Mem:* AAAS; Sigma Xi; Genetics Soc Am. *Res:* Analysis of meiosis in monosomics; cytological behavior of univalents; effects of a monosomic chromosome on recombination; study of the frequency and types of spontaneous chromosome abberations arising in monosomics; determination of effects of monosomy on lipid content in Zea mays; screening for ultrastructural differences in monosomics; determination of free amino acid profiles in monosomics; analysis of restriction fragment length polymorphisms with monosomics; identification and mapping of duplicate RFLP loci in maize. *Mailing Add:* Dept Biol Sci Ill State Univ Normal IL 61761

WEBER, DEANE FAY, b Aberdeen, SDak, May 17, 25; m 50; c 3. SOIL MICROBIOLOGY. *Educ:* Jamestown Col, BS, 50; Kans State Univ, MS, 52, PhD, 59. *Prof Exp:* Biochemist & bacteriologist, Quain & Ramstad Clin, Bismarck, NDak, 52-55; asst vet bact, pathogenic bact & virol, Kans State Univ, 57-58; soil scientist, Soil & Water Conserv Br, Agr Res Serv, USDA, 58-64 & IRI Res Inst, Campinas, Brazil, 64-66, microbiologist soybean invests, Crops Res Div, USDA, 67-72, microbiologist, Cell Cult & Nitrogen Fixation Lab, 72-82, Nitrogen Fixation & Soybean Genetics Lab, 82-89; RETIRED. *Mem:* AAAS. *Res:* Legume microbiology; nitrogen fixation. *Mailing Add:* 14972 Belle Ami Dr Laurel MD 20707

WEBER, DENNIS JOSEPH, b Kalamazoo, Mich, Mar 30, 34; div; c 6. DRUG METABOLISM, PHARMACOKINETICS. *Educ:* Western Mich Univ, BS, 58, MA, 62; Univ Fla, PhD(pharm), 67. *Prof Exp:* Res asst phys & anal chem, Upjohn Co, 58-62; mgr anal chem, Syntex Labs, Calif, 67-70; res scientist phys & anal chem, 70-79, RES SCIENTIST DRUG METAB RES, UPJOHN CO, 79- *Mem:* Am Pharmaceut Asn; Acad Pharmaceut Sci; Royal Soc Chem; Am Chem Soc; Am Asn Pharmaceut Scientists. *Res:* Kinetics of hydrolysis of drugs; correlation of spectra and structure of hydrazones; structure and stability constants of metal complexes of thiouracils; partition chromatography of steroids; high pressure liquid chromatography; pharmacokinetics. *Mailing Add:* Upjohn Co 7256-126-2 Kalamazoo MI 49001

WEBER, EDWARD JOSEPH, b Troy, NY, July 17, 48. AURORAL PHYSICS, SPACE PHYSICS. *Educ:* Union Col, BS, 70; Boston Col, PhD(physics), 75. *Prof Exp:* Res asst physics, Boston Col, 70-74; RES PHYSICIST IONOSPHERIC PHYSICS, GEOPHYS LAB, USAF, 74- *Mem:* Am Geophys Union; Sigma Xi. *Res:* Auroral dynamics; ionospheric structure and dynamics; optical detection of aurora and airglow; magnetospheric dynamics inferred from the aurora. *Mailing Add:* 61 Brooks Ave Newtonville MA 02160

WEBER, EICKE RICHARD, b Munnerstadt, WGer, Oct 28, 49; m 85. SEMICONDUCTOR MATERIALS. *Educ:* Univ Cologne, WGer, BS, 70; MS, 73, PhD(physics), 76, DrS(physics), 83. *Prof Exp:* Sci asst solid state physics, Tech Univ Aachen, WGer, 73-76, Univ Cologne, WGer, 76-82; asst prof mat sci, 83-87, ASSOC PROF MAT SCI, UNIV CALIF, BERKELEY, 87- *Concurrent Pos:* Int fel solid state physics, State Univ NY Albany, 79-80; res assoc solid state physics, Univ Lund, Sweden, 82-83; assoc fac, Lawrence Berkeley Lab, 84-; consult, semiconductor indust co, 83- *Honors & Awards:* Prince distinguished lectr, Ariz State Univ, 83. *Mem:* Am Physical Soc; Ger Physical Soc; Mat Res Soc; Electrochem Soc; Metall Soc. *Res:* Investigations of lattice defects in semiconductors, microscopic identification, electronic properties; influence on the device performance; specific topics include transition metal gettering in silicon, deep level defects in semiconductors, metals GaAs contacts, heteroepitoxial growth of semiconductors; microwave asorption of semiconductors. *Mailing Add:* Dept Mat Sci Univ Calif Berkeley CA 94720

WEBER, ERNST, b Vienna, Austria, Sept 6, 01; nat US; m 36; c 2. ELECTRICAL ENGINEERING. *Educ:* Vienna Tech Univ, dipl, 24, DSc(elec mach), 27; Univ Vienna, PhD(physics), 26. *Hon Degrees:* DSc, Pratt Inst, 58, Long Island Univ, 63; DEng, Newark Col, 59, Univ Mich, 64 & Polytech Inst Brooklyn, 69. *Prof Exp:* Res engr, Oesterreichische Siemens-Schuckert-Werke, Austria, 24-29 & Siemens-Schuckert-Werke, Ger, 29-30; vis prof, 30-31, res prof elec eng, 31-41, off investr, Off Sci Res & Develop contract, 42-45, prof elec eng & head dept & dir, Microwave Res Inst, 45-57, pres, Inst, 57-69, EMER PRES, POLYTECH INST BROOKLYN, 69- *Concurrent Pos:* Secy, Polytech Res & Develop Co, Inc, 44-52, pres, 52-60; chmn div eng, Nat Res Coun, 70-74; mem, Comn Sociotech Systs, Nat Res Coun, 74-78, consult, 78- *Honors & Awards:* Howard Coonley Medal, Am Standards Asn, 66; Founders Award, Inst Elec & Electronics Engrs, 70; L E Grinter Award, Engrs Coun Prof Develop, 78; Nat Medal Sci, 87. *Mem:* Nat Acad Sci; Nat Acad Eng; AAAS; fel Am Phys Soc; fel Inst Elec & Electronics Engrs (pres, 59 & 63). *Res:* Electromagnetic theory; circuit analysis; conformal mapping; microwave measurements. *Mailing Add:* PO Box 1619 28 Vineyard Rd Tryon NC 28782

WEBER, ERWIN WILBUR, b Chicago, Ill, Oct 8, 31. ELECTRICAL ENGINEERING. *Educ:* Ill Inst Technol, BS, 58, MS, 59, PhD(network theory), 64. *Prof Exp:* Res engr, Ill Inst Technol, 60-61, from instr to asst prof, 61-70, res engr, Res Inst, 66-76, ASSOC PROF ELEC ENG, ILL INST TECHNOL, 70-, DIR, DAN F ADAL RICE CAMPUS, 76- *Concurrent Pos:* Assoc engr, Armour Res Found, 62-63. *Mem:* AAAS; Inst Elec & Electronics Engrs; Sigma Xi. *Res:* Network theory; electrical network synthesis; computer aided circuit design. *Mailing Add:* 201 E Loop Rd Wheaton IL 60187

WEBER, EVELYN JOYCE, b Tower Hill, Ill, Nov 9, 28. LIPID BIOCHEMISTRY. *Educ:* Univ Ill, BS, 53; Iowa State Univ, PhD(biochem), 61. *Prof Exp:* Asst biochem, Iowa State Univ, 56-61; res assoc, Univ Ill, 61-65, from asst prof to prof plant biochem, 65-82; res chemist, USDA, 65-87; prof, 82-87, EMER PROF PLANT BIOCHEM, UNIV ILL, URBANA, 87- *Concurrent Pos:* Consult & ed, Univ Ill, Urbana, 87- *Mem:* AAAS; Am Chem Soc; fel Am Inst Chem; Am Oil Chem Soc; Am Soc Plant Physiol. *Res:* Identification and characterization of complex lipids; metabolism of fatty acids and lipids in corn and other plants; metabolism of vitamin E, carotenoids and other natural antioxidants. *Mailing Add:* 808 E Mumford Dr Urbana IL 61801-6325

WEBER, FAUSTIN N, b Toledo, Ohio, Nov 5, 11; m 37; c 4. ORTHODONTICS. *Educ:* Univ Mich, DDS, 34, MS, 36; Am Bd Orthod, dipl. *Hon Degrees:* FACD, FICD. *Prof Exp:* From asst prof to assoc prof, 36-51, chmn dept, 36-78, prof, 78-82, EMER PROF ORTHOD, UNIV TENN CTR HEALTH SCI, MEMPHIS, 81- *Concurrent Pos:* Consult. *Honors & Awards:* Albert H Ketcham Award, 80; Distinguished Serv Scroll, Am Asn Orthod, 82. *Mem:* Am Asn Orthod; Am Cleft Palate Asn; AAAS; fel Am Col Dent; fel Int Col Dent; Int Dent Fedn; Int Asn Dent Res; Am Asn Dent Res; NY Acad Sci; Am Asn Dent Schs; Am Dent Asn; So Asn Orthod. *Res:* Child growth and development. *Mailing Add:* Dept Orthod Univ Tenn Ctr Health Sci Memphis TN 38163

WEBER, FLORENCE ROBINSON, b Milwaukee, Wis, Aug 26, 21; m 59. GEOLOGY. *Educ:* Univ Chicago, BS, 43, MS, 48. *Hon Degrees:* DSc, Univ Alaska, 87. *Prof Exp:* Lab instr & librn, Univ Chicago, 42-43, librn, 47; subsurface geologist, Shell Oil Co, Tex, 43-47 & State Geol Surv, Mo, 47; geologist, Alaska Br, 49-54, DC, 54-57, GEOLOGIST, ALASKA BR, US GEOL SURV, 57-, GEOLOGIST-IN-CHARGE, COL OFF, 59-; DISTINGUISHED LECTR GEOL, UNIV ALASKA, FAIRBANKS, 59- *Concurrent Pos:* Arctic Inst NAm grant, 56. *Mem:* AAAS; Geol Soc Am; Am Asn Petrol Geol; Arctic Inst NAm; Sigma Xi. *Res:* Stratigraphy; structure; geomorphology; glaciology; petrology; paleontology; petroleum geology. *Mailing Add:* PO Box 80745 Fairbanks AK 99708

WEBER, FRANK E, b Chicago, Ill, Feb 1, 35; m 57; c 3. FOOD SCIENCE. *Educ:* Ill Inst Technol, BS, 57; Univ Ill, MS, 63, PhD(food sci), 64. *Prof Exp:* Res scientist-leader, Chem Sect, R T French Co, 64-71, mgr prod develop, 71-73, mgr tech res, 73-76; tech dir, 76-80, DIR RES & DEVELOP, RECKITT & COLMAN NORTH AM INC, 81-; MGR, NEW PROD RES, MILLER BREWING CO, 81- *Mem:* Am Chem Soc; Inst Food Technologists; Am Asn Cereal Chemists; Sigma Xi. *Res:* Product development; quality control; natural and artificial flavorings; isolation and analysis of flavor substances; food analysis methodology product development; industrial waste treatment; brewing technology. *Mailing Add:* 418 E Northpoint Rd Mequon WI 53092

WEBER, FRANK L, b St Louis, Mo, Feb 14, 24; m 47, 77; c 7. CERAMICS ENGINEERING, ANALYTICAL CHEMISTRY. *Educ:* Mo Sch Mines & Metall, BA. *Prof Exp:* Chemist, Johnston Foil Mfg Co, 46-49; analyst, Am Brake Shoe Co, 49-57; chemist, Walsh Refractories Co, 57-68; qual control supvr, Div Combine Eng, C E Refractories Co, 68-72; process engr, Findlay Refractories, Co, 72-88; RETIRED. *Mem:* Am Ceramic Soc. *Res:* Ceramics engineering. *Mailing Add:* RD 1 Washington PA 15301

WEBER, FREDERICK, JR, b Hilgen, Ger, Feb 14, 23; nat US; m 46; c 3. MICROBIOLOGY, PUBLIC HEALTH. *Educ:* Univ RI, BS, 48; Pa State Univ, MS, 50; Mich State Univ, PhD, 56. *Prof Exp:* Instr, Wartburg Col, 50-51 & Wayne State Univ, 51-53; mem res staff, Swift & Co, 56-59; MEM RES STAFF, JOSEPH E SEAGRAM & SONS, INC, 59- *Mem:* Am Soc Microbiol. *Res:* Psychiophiles in dairy products; flavor development by microorganisms in foods; aerobic digestion of wastes; alcoholic fermentations. *Mailing Add:* 9927 Silverwood Lane V Sta Louisville KY 40272

WEBER, GEORGE, b Budapest, Hungary, Mar 29, 22; nat US; m 58; c 3. ONCOLOGY, PHARMACOLOGY. *Educ:* Queen's Univ, Ont, BA, 50, MD, 52. *Hon Degrees:* Dr Med & Surg, Univ Chieti, Italy; Dr Med, Med Univ, Budapest, Hungary & Univ Leipzig, Ger; DSc, Tokushima Univ, Japan. *Prof Exp:* Nat Cancer Inst Can fel, Univ BC, 52-53; Cancer Res Soc sr fel, Montreal Cancer Inst, 53-58, head dept path chem, 56-59; assoc prof biochem & microbiol, 59-60, assoc prof pharmacol, 60-61, PROF PHARMACOL, SCH MED, IND UNIV, INDIANAPOLIS, 61-, DIR LAB EXP ONCOL, 74-, DISTINGUISHED PROF & DIR EXP ONCOL & DISTINGUISHED PROF PHARMACOL & TOXICOL, 90- *Concurrent Pos:* Ed, Adv in Enzyme Regulation, 62-; Oxford Biochem Soc lectr, 69; assoc ed, Cancer Res, 70-; rapporteur, Int Cancer Cong, Tex, 70; mem sci adv comt, Pharmacol B Study Sect, USPHS, 70-74 & 84-86, chmn exp therapeut study sect, Nat Cancer Inst, 76-78; mem, Tiberine Acad, 71-; mem sci adv, Damon Runyon Mem Fund, 71-75; mem adv comt instnl grants, Am Cancer Soc, 72-76; Aaron Brown lectureship, Case Western Reserve Univ, 77; mem, US Nat Organizing Comt & Prog Comt, 13th Int Cancer Cong, 79-82 & Int Adv Comt, 14th Int Cancer Cong, 83-86; outstanding invest award, Nat Cancer Inst, NIH, 86-93. *Honors & Awards:* Alecce Award, Cancer Res, Rome, Italy, 71; G H A Clowes Award, Am Asn Cancer Res, 82; G F Gallanti Prize, Int Soc Clin Chemists, 84; J H Wilkinson Award, Int Soc Clin Enzym, 87. *Mem:* Am Soc Pharmacol & Exp Therapeut; hon mem, All-Union Biochem Soc, USSR Nat Acad Sci; Am Asn Cancer Res; hon mem Hungarian Cancer Soc. *Res:* Oncology; biochemical pharmacology; regulation of enzymes and metabolism; neoplasia; chemotherapy; liver, kidney and colon tumors of different growth rates; hormone action; leukemia treatment. *Mailing Add:* Rm 337 Riley Res Wing Ind Univ Sch Med 702 Barnhill Dr Indianapolis IN 46202-5200

WEBER, GEORGE RUSSELL, b Novinger, Mo, Dec 29, 11; m 47; c 2. BACTERIOLOGY. *Educ:* Univ Mo, BS, 35; Iowa State Col, PhD(sanit & food bact), 40. *Hon Degrees:* DSc, Int Univ Found Malta, 86. *Prof Exp:* Asst chemist, Exp Sta, Univ Mo, 35-36; instr, Iowa State Col, 40-42; bacteriologist, USPHS, 46, sr asst scientist, 47-49, scientist, 49-53, chief sanitizing agents unit, 51-53; proj leader, US Indust Chem Co, Nat Distillers & Chem Corp, 53-73, sr res microbiologist, Res Div, 73-75, res assoc, 75-76; RETIRED. *Concurrent Pos:* Asst, Iowa State Col, 36-40; lectr, Eve Col, Univ Cincinnati, 69-70. *Honors & Awards:* Bronze Medal, Albert Einstein Int Acad Found, 87. *Mem:* AAAS; Am Soc Microbiol; fel Am Pub Health Asn; NY Acad Sci; fel Royal Soc Health; Sigma Xi; fel Int Biog Asn. *Res:* Sanitary and food bacteriology; germicidal efficiency of hypochlorites, chloramines and quaternary ammonium compounds; antibiotics; fermentations; yeast hybridization; ruminant and chinchilla nutrition; biological metal corrosion. *Mailing Add:* 1525 Burney Lane Cincinnati OH 45230

WEBER, GREGORIO, b Buenos Aires, Arg, July 4, 16; m 47; c 3. BIOCHEMISTRY, BIOPHYSICS. *Educ:* Univ Buenos Aires, MD, 43; Cambridge Univ, PhD(biochem), 47. *Prof Exp:* Beit mem fel biochem, Cambridge Univ, 48-52; lectr biochem, Sheffield Univ, 53-56, sr lectr, 56-60, reader biophys, 60-62; prof biochem, 62-86, prof, Ctr Advan Studies, 71-86, EMER PROF BIOCHEM, UNIV ILL, URBANA, 86- *Concurrent Pos:* Vis lectr, Univ Ill, 59 & Stanford Univ, 61 & 79; vis prof, Brandeis Univ, 60, Univ Wash, 64, Harvard Univ, 70 & Univ Calif, Los Angeles, 77; ed, J Biol Chem, Am Soc Biol Chemists, 68; Guggeheim Found fel, 70; coun mem, Am Biophys Soc, 70-73; vis res lectr, Nat Comn Sci & Technol Res, Univ Buenos Aires, 81 & hon prof, 87- *Honors & Awards:* First Nat Lectr, Biophys Soc, 69; NATO Lectr, Europe, 75; Rumford Prize, 80; Repligen Award for Chem of Biol Processes, Am Chem Soc, 86. *Mem:* Nat Acad Sci; Am Soc Biol Chemists; Am Chem Soc; fel Am Acad Arts & Sci; Am Biophys Soc. *Res:* Physical chemistry of proteins; fluorescence methods; excited states. *Mailing Add:* Univ Ill 1209 W California Urbana IL 61801

WEBER, HANS JOSEF, b Aachen, WGer, March 5, 42; m 68. OPERATIONS MANAGEMENT, PROGRAM DEVELOPMENT. *Educ:* Gonzaga Univ, BS, 64; San Diego State Univ, MA, 68. *Prof Exp:* SR VPRES, SCI APPLNS INT CORP, 86- *Mem:* Am Nuclear Soc; Am Soc Non Destructive Testing; Am Defense Preparedness Asn. *Res:* Development of non-destructive assay instrumentation for special nuclear materials; development of explosive detection systems for letter bombs. *Mailing Add:* 7916 Laurel Ridge Rd San Diego CA 92120

WEBER, HANS JURGEN, b Berlin, Ger, May 3, 39; m 66; c 1. THEORETICAL NUCLEAR PHYSICS. *Educ:* Univ Frankfurt, BS, 60, MS, 61, PhD(theoret physics), 65. *Prof Exp:* Res assoc theoret nuclear physics, Duke Univ, 66-67; asst prof, 68-71, assoc prof, 71-76, PROF THEORET NUCLEAR PHYSICS, UNIV VA, 77- *Concurrent Pos:* Sesquicentennial assoc, Max Planck Inst, Mainz & Univ Frankfurt, 72-73, Orsay, 79, Lyon, 78. *Mem:* Am Phys Soc; AAAS; Sigma Xi; Am Asn Physics Teachers. *Res:* Medium energy physics; photonuclear physics; group theory; quark models. *Mailing Add:* Dept Physics Univ Va Charlottesville VA 22901

WEBER, HARRY A, b Indianapolis, Ind, July 13, 21; m 49; c 3. INDUSTRIAL MANAGEMENT, ELECTRICAL ENGINEERING. *Educ:* Purdue Univ, BSEE, 42. *Prof Exp:* Dept mgr, Delco-Remy Div, Gen Motors Corp, 43-44; proj engr, US Naval Ord Plant, 46-50; chief, Prog Sect, US Atomic Energy Comn, 50-54, chief, Opers Br, 54, dir, Develop & Prod Div, 54-56, dir, Non-Nuclear Div, 56-57, dir, Plant Opers, 57-60, dir weapons prod, 60-69, WEAPONS PROG MGR, US DEPT ENERGY, 69- *Concurrent Pos:* Gen partner, Big Bend Co, Colo & NMex. *Mem:* Nat Soc Prof Engrs; Inst Elec & Electronics Engrs; Mensa. *Res:* Technical administration. *Mailing Add:* 7133 Kiowa Ave NE Albuquerque NM 87110

WEBER, HARRY P(ITT), b Pittsburgh, Pa, June 20, 31; m 52; c 3. ELECTRICAL ENGINEERING, PHYSICS. *Educ:* Univ Pittsburgh, BSEE, 58, MSEE, 61, DSc(elec eng), 64. *Prof Exp:* Res asst, Mellon Inst Indust Res, 56-59; asst prof elec eng, Univ Pittsburgh, 59-64; sr syst analyst, Radio Corp Am Missile Test Proj, Patrick AFB, 64-65; head dept elec eng, 66-71, dean eng & sci, 71-79, dean eng & sci, 71-79, dean Grad Sch, 68-82, assoc vpres acad affairs, Fla Inst Technol, 80-82; pres, Hawthorne Col, 82-86; Mgr, Tech Anal, GE Missile Test Proj, 86-88, MGR OPER ANALYTICAL, COMPUTER SCI RAYTHEON, PATRICK AFB, FL, 88- *Concurrent Pos:* Consult, Radio Corp Am Missile Test Proj, Patrick AFB, 66-68. *Mem:* Sigma Xi; Inst Elec & Electronics Engrs. *Res:* Missile test range, instrumentation and tracking, space test range concepts. *Mailing Add:* 310 Greenway Ave Satellite Beach FL 32937

WEBER, HEATHER R(OSS) WILSON, b Passaic, NJ, Mar 7, 43; m 63; c 4. BIOCHEMISTRY. *Educ:* Boston Univ, BA, 65; Univ Southern Calif, PhD(cellular & molecular biol), 74. *Prof Exp:* Res asst hemat, Peter Bent Brigham Hosp, 65-67; res asst biophys, Harvard Med Sch, 67-68; res assoc biochem, Univ Southern Calif, 73-74, NIH cancer training grant, 74-76, NIH fel biochem, 76-79, res scientist, 79-84, RES ASST PROF, MOLECULAR BIOL SECT BIOL DEPT, UNIV SOUTHERN CALIF, 85- *Mem:* AAAS; Am Chem Soc; Am Soc Biochem & Molecular Biol; Sigma Xi; Am Soc Cell Biol. *Res:* Transcriptional regulation of eucaryotic gene expression; mechanism of hormonal effects on cyclic nucleotide metabolism; membrane bound enzymes. *Mailing Add:* Dept Molecular Biol Univ Southern Calif University Park Los Angeles CA 90089-1340

WEBER, J K RICHARD, b London, UK, May 29, 57; m 89. HIGH TEMPERATURE MATERIALS. *Educ:* Sir John Cass Col, London, BS, 83; Imperial Col, London, PhD(metall) & DIC, 86; Inst Metals, CEng, 90. *Prof Exp:* Res consult corrosion, Int Nickel Co, 83-86; res assoc chem, Univ Toledo, Ohio, 86-88; PRIN SCIENTIST RES & DEVELOP, INTERSONICS CORP, 88- *Mem:* Inst Metals; Am Ceramic Soc; Metall Soc; Mat Res Soc; Am Soc Metall Int. *Res:* Containerless processing; high temperature materials; optical properties; non-contact temperature measurement; chemical kinetics and corrosion; ceramics and refractory metals processing research, development and design. *Mailing Add:* Intersonics Inc 3453 Commercial Ave Northbrook IL 60062

WEBER, JAMES ALAN, b Santa Monica, Calif, Mar 16, 44; m 70. BOTANY-PHYTOPATHOLOGY, ECOLOGY. *Educ:* Univ Calif, Berkeley, AB, 66; Univ Mich, AM, 67, PhD(bot), 73. *Prof Exp:* Asst, 73-78, ASST RES SCIENTIST, UNIV MICH, 78- *Concurrent Pos:* Co-ed, Mich Botanist, 84- *Mem:* Sigma Xi; Am Soc Plant Physiologists; Bot Soc Am; Ecol Soc Am; Am Inst Biol Sci; AAAS. *Res:* The interaction of physiological processes with environmental factors; photosynthesis and growth. *Mailing Add:* 2160 NW Beachwood Pl Corvallis OR 97330

WEBER, JAMES EDWARD, b Chicago, Ill, Nov 16, 57. OPTOMETRY, IRIS ULTRASTRUCTURE. *Educ:* Southern Ill Univ, Carbondale, BA, 80, MS, 83, PhD(physiol), 87; New England Col Optom, Boston, OD, 89. *Prof Exp:* Teaching asst anat, Southern Ill Univ, Carbondale, 81-87; consult; MEM STAFF, VISTACON, 87- *Mem:* Soc Study Reproduction; Am Soc Andrology; Am Asn Anatomists; Am Soc Cell Biol; NY Acad Sci; AAAS; Am Optom Asn. *Res:* Iris ultrastructure; open angle glaucoma. *Mailing Add:* Vistacon PO Box 10157 Jacksonville FL 32247

WEBER, JAMES H(AROLD), b Pittsburgh, Pa, Nov 21, 19; m 43; c 3. CHEMICAL ENGINEERING. *Educ:* Univ Pittsburgh, BS, 41, MS, 47, PhD(chem eng), 48. *Prof Exp:* Asst, Univ Pittsburgh, 46-48; from instr to prof chem eng, Univ Nebr, Lincoln, 48-64, chmn dept, 58-71, regents prof chem eng, 64-84, EMER PROF, UNIV NEBR, LINCOLN, 84- *Concurrent Pos:* Consult, Phillips Petrol Co, 52-79, Natural Gas Processors Asn, 59 & 64 & C F Braun & Co, 65. *Mem:* Am Chem Soc; Am Inst Chem Engrs; Soc Hist Technol. *Res:* Non-adiabatic absorption of ammonia in a wetted wall tower; applied thermodynamics; distillation; absorption; kinetics. *Mailing Add:* 1919 S 77th Lincoln NE 68506

WEBER, JAMES HAROLD, b Madison, Wis, July 21, 36; m 61; c 3. ENVIRONMENTAL CHEMISTRY. *Educ:* Marquette Univ, BS, 59; Ohio State Univ, PhD(chem), 63. *Prof Exp:* Asst prof, 63-70, assoc prof, 70-77, PROF CHEM, UNIV NH, 77- *Mem:* Am Chem Soc. *Res:* Chemistry of estuaries; environmental organometallic chemistry; biogeochemistry of tin. *Mailing Add:* Parsons Hall Univ NH Durham NH 03824-3598

WEBER, JAMES R, b May 16, 45. ANALYSIS OF OIL RESERVES. *Educ:* St Louis Univ, BS, 67; Colo Sch Mines, MS, 71. *Prof Exp:* Geol engr, Gulf Oil Corp, 68-72; geol engr, Pennzoil Co, Tex, 72-74; consult, oil industry, 74-86; PRES, CORAL PROD CORP, 86- *Res:* Geological & engineering analysis of the reserves & potential reserves of the Morrow Formation in southwestern Colorado. *Mailing Add:* 1230 Atlantis Dr Lafayette CO 80026

WEBER, JANET CROSBY, b Chicago, Ill, June 1, 23; m 49; c 3. COMPUTER SCIENCE, SOFTWARE SYSTEMS. *Educ:* Iowa State Univ, BS, 45; Univ Ill, Urbana, PhD(chem), 48. *Prof Exp:* Asst prof vet res, Mont State Univ, 48-50; res assoc, 60-63, from instr to asst prof, 63-79, ASSOC PROF OPHTHAL, MED CTR, IND UNIV, INDIANAPOLIS, 79- *Mem:* Mended Hearts Asn Women Sci; Sigma Xi. *Res:* Computer methods in clinical and research studies; development of computer programs for statistical analysis; microcomputers for research; application of statistics and computers to ophthalmic research. *Mailing Add:* Dept Ophthal Rotary Bldg 3rd Fl 702 Rotary Circle Indianapolis IN 46223

WEBER, JEAN ROBERT, b Thun, Switz, Apr 28, 25; Can citizen; m 54; c 3. GEOPHYSICS. *Educ:* Swiss Fed Inst Technol, Prof Eng, 52; Univ Alta, PhD(physics), 60. *Prof Exp:* Res engr, PTT Res Labs, Berne, Switz, 51-53; lectr, Univ Alta, 57-58; geophysicist-in-chg, Oper Hazen, Int Geophys Year, 58-59; RES SCIENTIST, DOM OBSERV, DEPT ENERGY, MINES & RESOURCES, CAN, 60- *Concurrent Pos:* Mem Arctic Inst NAm Baffin Island Exped, 53 & Univ Toronto, Salmon Glacier Exped, 56; leader, Dom Observ NPole Exped, 67. *Mem:* Am Geophys Union; Soc Explor Geophys; fel Arctic Inst NAm; Glaciol Soc. *Res:* Communications electronics; dosimetry of beta radiation and biological effects on allium cepa roots; regional gravity interpretations; continental margins; upper mantle, Arctic Ocean; application of geophysics to glaciology. *Mailing Add:* Geo Surv Can-Geophys Dept One Observatory Crescent Ottawa ON K1A 0Y3 Can

WEBER, JEROME BERNARD, b Claremont, Minn, Sept 19, 33; m 56; c 4. SOIL CHEMISTRY, WEED SCIENCE. *Educ:* Univ Minn, BS, 57, MS, 59, PhD(soil chem), 63. *Prof Exp:* Assoc prof soil pesticide chem, 62-71, PROF SOIL PESTICIDE CHEM & WEED SCI, NC STATE UNIV, 71- *Concurrent Pos:* NSF lectr, Clemson Univ, 79; consult, WHO, Venezuela, 80, Univ SAfrica, 82; mem, Coun Agr Sci & Technol. *Honors & Awards:* Sigma Xi Res Award; Res Award, Weed Sci Soc Am; Wright lectr, Purdue Univ, 80. *Mem:* AAAS; Am Soc Agron; Weed Sci Soc Am; Am Chem Soc; Clay Minerals Soc. *Res:* Chemistry of soil, fate and biological availability of applied organic compounds, especially herbicides; effects of pesticides on environmental quality; weed ecology studies; behavior of gases in soil; behavior of toxic organics in waters and soils. *Mailing Add:* Crop Sci Box 7620 NC State Univ Raleigh NC 27695

WEBER, JOHN DONALD, b Lagon, La, Nov 1, 34; m 62; c 2. PHARMACEUTICAL CHEMISTRY. *Educ:* Xavier Univ La, BS, 57; Univ Notre Dame, MS, 61; Georgetown Univ, PhD(org chem), 72. *Prof Exp:* Instr chem, Southern Univ New Orleans, 62-63; anal chemist, Food & Drug Admin, 63-68, res chem, 68-78; mem staff organic synthesis, USDA, 78-79; RES CHEMIST, FOOD & DRUG ADMIN, 79- *Mem:* Am Chem Soc; Sigma Xi; Asn Off Anal Chemists; NY Acad Sci. *Res:* Pharmaceutical analyses; chromatography; nuclear magnetic resonance fluorimetric techniques; optical purity of drugs; stereochemistry and chemical kinetics; problems of relationships between stereoisomerism and physiological activity. *Mailing Add:* 7204 Seventh St NW Washington DC 20012

WEBER, JOHN R, b Ft Madison, Iowa, Mar 18, 24; m 48; c 2. PLANT PHYSIOLOGY. *Educ:* Univ Iowa, AB, 48, MS, 49. *Prof Exp:* Prin lab technician, Citrus Exp Sta, Univ Calif, 52-56; plant physiologist, FMC Corp, 56-67, mem spec projs egg handling systs, 67-76, int mgr, Citrus Mach Div, 76-82; RETIRED. *Res:* Post-harvest physiology of fruits and vegetables. *Mailing Add:* 2102 Oak Crest Dr Riverside CA 92506

WEBER, JOSEPH, b Paterson, NJ, May 17, 19; m 42, 72; c 4. PHYSICS. *Educ:* US Naval Acad, BS, 40; Cath Univ Am, PhD(physics), 51. *Prof Exp:* PROF PHYSICS, UNIV MD, COLLEGE PARK, 48- *Concurrent Pos:* Vis prof, Univ Calif, Irvine; Nat Res Coun & Guggenheim fels, 55-56; Fel, Inst Advan Study, 55-56, 62-63 & 69-70 & Lorentz Inst Theoret Physics, State Univ Leiden, 56. *Honors & Awards:* First Prize, Gravity Res Found, 59; Sigma Xi Award, 70; Boris Pregel Prize, NY Acad Sci, 73. *Mem:* Fel Am Phys Soc; fel Inst Elec & Electronics Engrs. *Res:* General relativity; microwave spectroscopy; irreversibility; scattering. *Mailing Add:* Dept Physics & Astron Univ Md College Park MD 20742

WEBER, JOSEPH M, b Budapest, Hungary, Oct 10, 39; Can citizen; m 64, 73; c 2. VIROLOGY. *Educ:* Univ BC, BSc, 64, MSc, 66; McMaster Univ, PhD(virol), 69. *Prof Exp:* Nat Cancer Inst res assoc virol, Ohio State Univ, 69-70; asst prof to assoc prof, 70-80, PROF MICROBIOL, UNIV SHERBROOKE, 80-, HEAD DEPT, 88- *Concurrent Pos:* Lectr, Ohio State Univ, 69-70; Med Res Coun Can scholar, 71-76; Nat Cancer Inst Can scholar, 76-80, res assoc, 80-85, Terry Fox cancer res scientist, 85-89. *Mem:* Am Asn Cancer Res; Am Soc Virol. *Res:* Viral oncology; cellular oncogenes; adenovirus genetics; genetic engineering; function of the E1a gene, transformation revertants, drug-resistance and detransformation, cellular factors interacting with E1a; adenovirus endoprotease function in maturation; adenovirus-lymphocyte interactions. *Mailing Add:* Dept Microbiol Univ Ctr Hosp Sherbrooke PQ J1H 5N4 Can

WEBER, JOSEPH T, b Brooklyn, NY, Jan 1, 38; m 77; c 1. NEUROANATOMY, NEURODEVELOPMENT. *Educ:* Univ Calif, Berkeley, AB, 73; Univ Wis, PhD(anat), 78. *Prof Exp:* Fel neuroanat, Med Sch, Univ Wis, 78-80; asst prof, 80-83, assoc prof, 83-86, PROF NEUROANAT, MED SCH, TULANE UNIV, 86- *Concurrent Pos:* Ad hoc reviewer, NSF, 81-; prin investr, NIH, 81- *Mem:* Soc Neurosci; Am Asn Anatomists; AAAS; Sigma Xi. *Res:* Anatomical studies of extrageniculate visual pathways; mechanisms involved in the control of head and eye movements; development of visual centers. *Mailing Add:* Tulane Med Sch 1430 Tulane Ave New Orleans LA 70112

WEBER, JULIUS, b Brooklyn, NY, Apr 8, 14; m 47; c 4. CYTOLOGY, PHOTOGRAPHY. *Hon Degrees:* DSc, Jersey City State Col, 74. *Prof Exp:* Asst, Brooklyn Jewish Hosp, New York, 33-35; chief histol technician, Kingston Ave Hosp Infectious Dis, 35-36 & Israel Zion Hosp, Brooklyn, 36-39; head dept med photog, Columbia-Presby Med Ctr, New York, 39-49; HEAD DEPT PHOTOG RES, BETH ISRAEL HOSP, 49-; EMER RES ASSOC, EINSTEIN SCH MED, 84- *Concurrent Pos:* Mem training div, Inst Inter-Am Affairs, US Dept State, 46; consult, Western Union Tel Co, 46-48, Chem Corps, US Army, 47-50, Ansco Div, Gen Aniline & Film Corp, 47-48, Nat Film Bd Can, 51, US Naval Hosp, St Albans, NY, 51-52, St Francis Hosp, New York, 59, Perkin-Elmer Corp, 60 & Ehrenreich Photo-Optical Industs, Inc, 63; lectr, Sch Med, Univ Calif, Los Angeles, 48; guest lectr, Col Physicians & Surgeons, Columbia Univ, 50, Sch Eng, 60; dir med photog, William Douglas McAdams, Inc, 52-59; head dept med photog, Hosp Joint Dis, 53-73; res assoc, Waldemar Med Res Found, 55-; lectr, Grad Sch Pub Admin, NY Univ, 55-57; head dept med photog, Misericordia Hosp, 56-; med photographer, Knickerbocker Hosp, 58-63; assoc, Dept Mineral, Am Mus Natural Hist, 60; ed, Image Dynamics Sci & Med, 69-71; dir, Wildcliff Natural Sci Ctr, 69-74; res assoc, Dept Mineral, Royal Ont Mus, Toronto, 71-; res assoc, Dept Med, Einstein Sch Med, 76-84. *Mem:* Soc Photog Sci & Eng; assoc Photog Soc Am; fel Biol Photog Asn; fel Royal Micros Soc; fel Royal Photog Soc Gt Brit. *Res:* Photographic instrumentation for endoscopy, micromineralogy; time lapse photomicrography and cinematography; ultraviolet, infrared and interference photomicrography and photomacrography; neurocytology and neuropathology photoimpregnation techniques. *Mailing Add:* 1040 Cove Rd Mamaroneck NY 10543

WEBER, KARL T, INTERNAL MEDICINE, CARDIOLOGY. *Educ:* Temple Univ, MD, 68. *Prof Exp:* DIR, DIV CARDIOL & CARDIOVASC INST, DEPT MED, MICHAEL REESE HOSP, UNIV CHICAGO, 83- *Mailing Add:* Michael Reese Hosp Univ Chicago 5801 Ellis Ave Chicago IL 60616

WEBER, KENNETH C, b Cold Spring, Minn, June 30, 37; m 57; c 4. PULMONARY PHYSIOLOGY. *Educ:* Univ Minn, BSEE, 63, PhD(physiol), 68. *Prof Exp:* USPHS trainee physiol, Univ Minn, 62-68; from asst prof to assoc prof, 68-76, PROF PHYSIOL, WVA UNIV, 76- *Concurrent Pos:* Chief physiol sect, 68-85, chief, Lab Invest Br, Appalachian Lab Occup Respiratory Dis, Nat Inst Occup Safety & Health, 85-; chmn, Gordon Res Conf Non-Ventilatory Lung Function, 75. *Mem:* AAAS; Am Physiol Soc; Am Thoracic Soc; Sigma Xi; Am Heart Asn (vpres, 81-82). *Res:* Cardiopulmonary hemodynamics; respiration physiology; non-ventilatory lung function; lung metabolism. *Mailing Add:* Dept Pharmacol & Toxicol WVa Univ Med Ctr Morgantown WV 26506

WEBER, LAVERN J, b Isabel, SDak, June 7, 33; m 59; c 4. PHARMACOLOGY, ACADEMIC ADMINISTRATION. *Educ:* Pac Lutheran Univ, BA, 58; Univ Wash, MS, 62, PhD(pharmacol), 64. *Prof Exp:* From instr to asst prof pharmacol, Sch Med, Univ Wash, 64-69; assoc prof pharmacol & toxicol, 69-75, asst dean, Grad Sch, 74-77, dir, Marine & Freshwater Biomed Ctr, 78-81, PROF PHARMACOL, TOXICOL & FISHERIES, SCH PHARM, ORE STATE UNIV, 75-, DIR, MARINE SCI CTR, 77- *Concurrent Pos:* Supt, Coastal Ore Marine Exp Sta, 88- *Mem:* Am Soc Pharmacol & Exp Therapeut; Soc Toxicol; Soc Exp Biol & Med; Sigma Xi; Am Asn Lab Animal Sci. *Res:* Marine sciences; biochemistry of autonomic nervous system; comparative pharmacology, physiology and toxicology; comparative pharmacology of the autonomic nervous system; liver toxicology; comparative neuromuscular pharmacology. *Mailing Add:* Ore State Univ Marine Sci Ctr Newport OR 97365

WEBER, LEON, b Detroit, Mich, Feb 4, 31; m 52; c 3. SURFACE CHEMISTRY, PHYSICAL CHEMISTRY. *Educ:* Wayne State Univ, BS, 52, MS, 54; Carnegie-Mellon Univ, MS & PhD(phys chem), 67. *Prof Exp:* Chemist, Wayne County Rd, Comn, 51-54; res chemist, Shell Develop Co, 54-56; scientist nuclear chem, Westinghouse Atomic Power Co, 56-57; sr res chemist, Gulf Res & Develop Co, 57-67; SCIENTIST SURFACE CHEM, SCM CHEMICALS, INC, 67- *Mem:* AAAS; Am Chem Soc; Sigma Xi. *Res:* Colloid and surface chemistry of pigments; gas-solid sorption phenomena; reaction kinetics by thermal analysis. *Mailing Add:* 8246 Streamwood Dr Baltimore MD 21208-2135

WEBER, LESTER GEORGE, b St Louis, Mo, July 23, 24; m 49; c 5. CHEMICAL ENGINEERING. *Educ:* Purdue Univ, BS, 49; Wash Univ, MS, 54. *Prof Exp:* Analyst, Uranium Div, Mallinckrodt Chem Works, Mo, 49-50, engr pilot plant, 50-51, supvr metal pilot plant, 51-54, asst mgr plant design liaison group, Uranium Div, 54-57, tech supt mfg dept, 57-58, supvr prod tech dept, 58-60, asst mgr, 60-62, asst mgr develop dept, 62-63; res engr, Yerkes Res & Develop Lab, NY, 63-65, res engr, Circleville Res & Develop Lab, Ohio, 65-66, staff engr, 66, res supvr, 66-69, tech supt cellophane, Clinton Film Plant, 69-73, customer serv mgr, Plastic Prod & Resins Dept, Packaging Films Div, BIVAC Meat Packaging Systs, 73-76, STAFF ENGR, POLYMER PROD DEPT, MFG DIV, WASHINGTON WORKS, E I DU PONT DE NEMOURS & CO, INC, 76- *Mem:* Am Inst Chem Engrs; Am Soc Plastics Engrs. *Mailing Add:* 2030 MacKenzie Dr Columbus OH 43220

WEBER, LOUIS RUSSELL, experimental physics, for more information see previous edition

WEBER, MARVIN JOHN, b Fresno, Calif, Feb 26, 32; m 57; c 2. LASERS, SPECTROSCOPY. *Educ:* Univ Calif, Berkeley, AB, 54, MA, 56, PhD(physics), 59. *Prof Exp:* Asst physics, Univ Calif, 54-59; prin res scientist & mgr solid state lasers, Res Div, Raytheon Co, 59-73; group leader, 73-87, ASSOC DIV LEADER, LAWRENCE LIVERMORE NAT LAB, 87- *Concurrent Pos:* Vis res assoc, Stanford Univ, 66; consult, Div Mat Res, NSF, 73-76; ed-in-chief, Handbk Series Laser Sci & Technol, CRC, 78-; transfer assignment, US Dept Energy, 84-85; guest worker, Nat Bur Standards, 84-85; assoc ed, J Luminescence, 85-; tech consult, US Dept Energy, 86-87; regional ed, J Non-Crystalline Solids, 88-; spokesman, Nat Labs Synchrotron Radiation Facil, Univ Calif, 88-; consult prof, Stanford Univ, 90- *Honors & Awards:* IR-100 Award, 79; George W Morey Award, Am Ceramics Soc, 83. *Mem:* Fel Am Phys Soc; fel Optical Soc Am; Am Ceramics Soc; Am Asn Crystal Growth; Mat Res Soc; Sigma Xi. *Res:* Magnetic resonance; optical spectroscopy of solids; luminescence materials; lasers; solid state physics; materials science. *Mailing Add:* Lawrence Livermore Nat Lab Univ Calif PO Box 808 Livermore CA 94550

WEBER, MICHAEL JOSEPH, b New York, NY, Aug 23, 42; m 67. VIROLOGY, CELL BIOLOGY. *Educ:* Haverford Col, BSc, 63; Univ Calif, San Diego, PhD(biol), 68. *Prof Exp:* Am Cancer Soc Dernham jr fel, Univ Calif, Berkeley, 68-70; asst prof microbiol, Univ Ill, Urbana, 70-75, assoc prof, 75-; AT DEPT MICROBIOL, SCH MED, UNIV VA, CHARLOTTESVILLE. *Concurrent Pos:* NIH res career develop award, 75. *Mem:* Am Soc Microbiol; Tissue Cult Asn; Soc Gen Physiologists; Am Soc Biol Chemists. *Res:* Control of growth of animal cells and malignant transformation; tumor virus-induced cell surface changes. *Mailing Add:* Dept Microbiol Univ Va Sch Med Charlotte VA 22908

WEBER, MORTON M, b New York, NY, May 26, 22; m 55; c 2. MICROBIOLOGY, BIOCHEMISTRY. *Educ:* City Col New York, BS, 49; Johns Hopkins Univ, ScD(microbiol), 53. *Prof Exp:* Instr zool & parasitol, St Francis Col, 49; instr med microbiol, Johns Hopkins Univ, 51-55; instr bact & immunol, Harvard Med Sch, 56-59; from asst prof to assoc prof, 56-63, chmn dept, 64-87, PROF MICROBIOL, SCH MED, ST LOUIS UNIV, 63-, EMER CHMN DEPT, 87- *Concurrent Pos:* Am Cancer Soc fel, McCollum-Pratt Inst, Johns Hopkins Univ, 53-56; mem microbial chem study sect, NIH, 69-73; vis sci, Microbiol Unit Dept Biochem, Oxford Univ, 70-, Linacre Col, 70- *Mem:* Fel AAAS; Am Soc Biol Chemists & Molecular Biologists; Am Soc Microbiol; Am Acad Microbiol; NY Acad Sci; fel Infectious Dis Soc Am; Soc Gen Microbiol, UK; Sigma Xi. *Res:* Physiology and biochemistry of microorganisms; pathways and mechanisms of electron transport; mode of action of antibiotics and other antimicrobial agents; biochemical regulatory mechanisms. *Mailing Add:* Dept Microbiol Sch Med St Louis Univ 1402 S Grand St Louis MO 63104

WEBER, NEAL ALBERT, b Towner, NDak, Dec 14, 08; m 40; c 3. ENTOMOLOGY, ECOLOGY. *Educ:* Univ NDak, AB, 30, MS, 32; Harvard Univ, AM, 33, PhD(zool), 35. *Hon Degrees:* ScD, Univ NDak, 58. *Prof Exp:* Assoc prof biol, Univ NDak, 36-43 & anat, Sch Med, 43-47; from assoc prof to prof, 47-74, EMER PROF ZOOL, SWARTHMORE COL, 74- *Concurrent Pos:* Mem expeds, WI, 33-36, Trinidad, 34-36, Orinoco Delta, 35, Brit Guiana, 35-36, Barro Colo Island, CZ & Colombia, 38, Anglo-Egyptian Sudan, Uganda & Kenya, 39; biologist, Am Mus Natural Hist Exped, CAfrica, 48, Middle East, 50-52, Trop Am, 54- & Europ Mus, 57; consult, Arctic Res Lab, Alaska, 48-50; mem dept zool, Col Arts & Sci Univ Baghdad, Iraq, 50-52; vis prof, Univ Wis, 55-56; mem panel biol & med sci, Comt Polar Res, Nat Acad Sci, 58-60, panel fels, 64-66; mem & US del spec comt Antarctic res, Int Coun Sci Unions, Australia, 59; sci attache, Am Embassy, US Dept State, Buenos Aires, Arg, 60-62; mem, Latin Am Colloquium, Arg, 65 & Brazil, 68; consult, Venezuelan Univs, 72; adj prof biol sci, Fla State Univ, Tallahassee, 74- *Honors & Awards:* John F Lewis Prize, Am Philos Soc, 73. *Mem:* AAAS; fel Entom Soc Am; Ecol Soc Am; Mycol Soc Am; Am Soc Zool. *Res:* Tropical ecology; fungus-growing ants and their fungi; zoogeography. *Mailing Add:* 1805 Aaron Rd Tallahassee FL 32303

WEBER, NORMAN, b San Luis Obispo, Calif, Nov 25, 34; m 71; c 6. THERMAL SCIENCES. *Educ:* Calif State Polytech Univ, BS, 57; Univ Southern Calif, MS, 67; Mont State Univ, PhD(mech eng), 71. *Prof Exp:* From res engr to sr res engr heat transfer res, Rocketdyne, Div Rockwell Int, 57-68; res asst, Mont State Univ, 68-71; sr develop engr II nuclear safety prog mgr, Westinghouse Hanford Co, 71-74; supvr thermal & hydraul anal, 74-81, ASST HEAD, NUCLEAR SAFEGUARDS & LICENSING DIV, SARGENT & LUNDY ENGRS, 81-, ASSOC, 85- *Mem:* Am Nuclear Soc; Am Soc Mech Engrs; AAAS; NY Acad Sci; Am Solar Energy Soc. *Res:* Natural and forced convection heat transfer. *Mailing Add:* 1041 E Porter Ave Naperville IL 60540

WEBER, PAUL VAN VRANKEN, b Highland Park, Ill, Mar 12, 21; m 48; c 3. PLANT PATHOLOGY. *Educ:* Cornell Univ, BS, 43; Univ Wis, PhD(plant path), 49. *Prof Exp:* Asst plant pathologist, Ohio Agr Exp Sta, 49-52; asst plant pathologist & geneticist, Campbell Soup Co, 52-59; chief, Bur Plant Path, State Dept Agr, NJ, 59-88; RETIRED. *Mem:* Am Phytopath Soc; Am Inst Biol Sci. *Res:* Plant disease and insect surveys; administration. *Mailing Add:* 264 Buckner Ave Haddonfield NJ 08033

WEBER, PETER B, b Berlin, Ger, July 31, 34; c 1. BIOCHEMISTRY. *Educ:* Univ Cologne, DNatSc(biol, chem), 61. *Prof Exp:* NIH grants, Univ Ill, Chicago, 64-65; NSF grants, State Univ NY Buffalo, 65-68; from asst prof to assoc prof, 68-84, PROF BIOCHEM, ALBANY MED COL, 84- *Mem:* Fedn Am Socs Exp Biol; Soc Complex Carbohydrates; Sigma Xi; NY Acad Sci. *Res:* Biochemistry of membrane glycoproteins and bacterial polysaccharides. *Mailing Add:* Dept Biochem Albany Med Col Albany NY 12209

WEBER, RICHARD GERALD, b Newport News, Va, Dec 20, 39. ENTOMOLOGY. *Educ:* Eastern Mennonite Col, BS(biol) & BA(foreign lang), 69; Univ Del, MS, 71; Kans State Univ, PhD(entom), 75. *Prof Exp:* Instr insect morphol, insect taxonomy, gen entomology, Kans State Univ, 75-77; INSTR INSECT STRUCT, INSECT PHOTOGRAPHY, NATURAL HIST INSECTS, DEPT ENTOM & APPL SCI, UNIV DEL, 77- *Mem:* Am Entom Soc; Am Mosquito Control Asn; Entom Soc Am; Biol Photographic Asn. *Res:* Insect morphology in relation to behavior; mosquito oviposition behavior; photographic and electronics applications for entomological research. *Mailing Add:* Dept Entom & Ecol Univ Del Newark DE 19717-1303

WEBER, RICHARD RAND, b Columbia, Pa, July 28, 38; m 65; c 3. RADIO ASTRONOMY, INSTRUMENTATION. *Educ:* Franklin & Marshall Col, AB, 60; Univ Md, MS, 68. *Prof Exp:* Teacher physics & math, Wilson High Sch, West Lawn, Pa, 60-61; RES SCIENTIST RADIO ASTRON, GODDARD SPACE FLIGHT CTR, NASA, 64- *Mem:* Am Astron Soc; Am Geophys Union; Int Union Radio Sci. *Res:* Microwave studies of cosmic background radiation; low frequency; studies of galactic, solar, planetary radio emissions; radio experiments on spacecraft. *Mailing Add:* 10715 Moosberger Ct Columbia MD 21044

WEBER, ROBERT EMIL, b Oshkosh, Wis, Dec 17, 30; m 53; c 3. POLYMER CHEMISTRY. *Educ:* Univ Wis-Oshkosh, BS, 53; Univ Iowa, PhD, 59. *Prof Exp:* Res chemist, Res & Develop Ctr, Wis, 58-66, sr res scientist, 66-68, mgr prod develop lab, Munising Div, Mich, 68-71, group leader, Advan Develop Lab, Res & Eng Ctr, Neenah, 71-75, dir res & develop, Munising Paper Div, 75-82, SR RES FEL, KIMBERLY-CLARK CORP, 82- *Mem:* Am Chem Soc. *Res:* Physical properties of polymers solutions; physical-chemical properties of fiber-elastomer composites. *Mailing Add:* Kimberly-Clark Corp 1400 Holcomb Bridge Rd Roswell GA 30076-9703

WEBER, ROBERT HARRISON, b Wauseon, Ohio, Feb 8, 19; m 41; c 2. GEOLOGY. *Educ:* Ohio State Univ, BSc, 41; Univ Ariz, PhD(geol), 50. *Prof Exp:* Geologist, Shell Oil Co, 41-42; econ geologist, 50-66, SR GEOLOGIST, NMEX BUR MINES & MINERAL RESOURCES, 66- *Concurrent Pos:* Fac assoc, NMex Inst Mining & Technol, 65- *Mem:* Fel AAAS; fel Geol Soc Am; Soc Econ Geol; Soc Am Archaeol; Am Quaternary Asn. *Res:* Mineral deposits; petrography and petrology of volcanic rocks; Quaternary stratigraphy and geomorphology of the Southwest; meteoritics; early man in the New World. *Mailing Add:* 1502 Evergreen Dr Socorro NM 87801

WEBER, THOMAS ANDREW, theoretical chemistry, for more information see previous edition

WEBER, THOMAS BYRNES, b Oklahoma City, Okla, Sept 1, 25; m 58; c 4. BIOCHEMISTRY. *Educ:* Okla State Univ, BS, 48; La State Univ, PhD(biochem), 54. *Prof Exp:* Res scientist, Animal Dis Res Ctr, Agr Res Ctr, USDA, 54-57; head biochem dept, US Navy Dent Res Inst, 57-59; head atmospheric & gas anal, USAF Sch Aerospace Med, 59-62; head adv res med develop, Beckman Instruments, Inc, 62-67; pres, Biosci Planning Inc, 67-69; pres, Med Patents, Inc, 69-74; pres, Weber Dent Prod, 74-77, PRES, GENERICS INT, 74- *Concurrent Pos:* Life sci consult to indust, 58-62; consult, Life Sci & Instrumentation, US Govt, 63-; mem bd dirs, Metab Dynamic Found, 64- *Mem:* AAAS; Am Chem Soc. *Res:* Life support systems; multiphasic screening methods; biochemical and physiological instrumentation; monitoring in closed ecological systems; body fluid analysis; automation of testing tools; handling of ethical pharmaceuticals; heat transfer; high temperature combustion and incineration. *Mailing Add:* 4722 Calle De Vida San Diego CA 92124-2308

WEBER, THOMAS W(ILLIAM), b Orange, NJ, July 15, 30; m 66; c 2. CHEMICAL ENGINEERING. *Educ:* Cornell Univ, BChE, 53, PhD(chem eng), 63; Newark Col Eng, MS, 58. *Prof Exp:* Engr, Esso Res & Eng, Linden, NJ, 55-58; instr chem engr, Cornell Univ, 61-62; from asst prof to assoc prof, State Univ NY, 63-83, assoc chmn dept, 80-82, chmn dept, 82-89, PROF CHEM ENG, STATE UNIV NY, BUFFALO, 83- *Concurrent Pos:* NSF grant, 66-68. *Honors & Awards:* ASEE Award, AT&T Found, 87-88. *Mem:* Fel Am Inst Chem Engrs; Am Soc Eng Educ; Sigma Xi. *Res:* Dynamics and control of chemical engineering equipment and processes; gaseous adsorption in granular solids, especially adiabatic and isothermal adsorption in packed beds; liquid adsorption. *Mailing Add:* Dept Chem Eng Furnas Hall State Univ NY Amherst NY 14260

WEBER, WALDEMAR CARL, b Chicago, Ill, May 4, 37; m 69. MATHEMATICS. *Educ:* US Naval Acad, BSc, 59; Univ Ill, Urbana, MSc, 64, PhD, 68. *Prof Exp:* Asst prof, 68-72, asst chair, Dept Math & Statist, 77-87, ASSOC PROF MATH, BOWLING GREEN STATE UNIV, 72 - *Mem:* Am Math Soc; Math Asn Am. *Res:* Geometry; applied mathematics. *Mailing Add:* Dept Math & Statist Bowling Green State Univ Bowling Green OH 43403

WEBER, WALLACE RUDOLPH, b Murphysboro, Ill, Aug 1, 34; m 60; c 2. SYSTEMATIC BOTANY. *Educ:* Southern Ill Univ, BA, 56, MS, 59; Ohio State Univ, PhD(bot), 68. *Prof Exp:* Instr biol, Otterbein Col, 59-62; teaching assoc bot, Ohio State Univ, 63-67; from asst prof to assoc prof, 67-79, PROF BIOL, SOUTHWEST MO STATE UNIV, 79- *Mem:* Bot Soc Am; Am Soc Plant Taxon; Int Asn Plant Taxon; Sigma Xi; Nat Asn Biol Teachers. *Res:* Floristics of Missouri Ozarks; biosystematic studies. *Mailing Add:* Dept Biol Southwest Mo State Univ Springfield MO 65804

WEBER, WALTER J, JR, b Pittsburgh, Pa, June 16, 34; m; c 4. ENVIRONMENTAL & WATER RESOURCES ENGINEERING. *Educ:* Brown Univ, ScB, 56; Rutgers Univ, ScM, 59; Harvard Univ, AM, 61, PhD(water resources eng), 62; Am Acad Environ Engrs, dipl, 75. *Prof Exp:* Engr, Caterpillar Tractor Co, 56-57; instr civil eng, Rutgers Univ, 57-59; fel water resources eng, Harvard Univ, 62-63; from asst prof to prof civil & water resources eng, Univ Mich, Ann Arbor, 63, 87, dir, Univ Prog Water Resources, 68-91, Earnest Boyce distinguished prof eng, 87, DIR, GREAT LAKE & MID-ATLANTIC HAZARDOUS SUBSTANCE RES CTR, UNIV MICH, ANN ARBOR, 89- *Concurrent Pos:* Engr, Soil Conserv Serv, 59; vis scholar, Univ Calif, Berkeley & Univ Melbourne, 71; Nalco res award, Asn Environ Eng Professors, 79. *Honors & Awards:* Faraday lectr, Univ Mich, 70; James R Rumsey Mem Award, Mich Water Control Asn, 75; Willard F Shephard Award, Mich Water Pollution Control Asn, 80; Rudolph Herning Medal, Am Soc Civil Engrs, 80, Simon W Freese Award, 84, G Brooks Ernest Award, 85; F J Zimmerman Award, Am Chem Soc, 82; Thomas R Camp Award, Boston Soc Civil Engrs, 82; Eng Sci Res Award, Asn Environ Eng Prof, 84; Thomas R Camp Medal, Water Pollution Control Fedn, 88 & Gordon M Fair Medal, 90. *Mem:* Nat Acad Eng; AAAS; Am Chem Soc; Am Inst Chem Engrs; Am Soc Civil Engrs; Water Pollution Control Fedn; Sigma Xi; Am Water Works Asn; Asn Environ Eng Professors; Int Asn Water Pollution Res & Control. *Res:* Water quality and pollution control; water and wastewater treatment; water resources systems design, modeling and water basin management. *Mailing Add:* 181 Water Resources Eng Bldg 1-A Univ Mich Ann Arbor MI 48109

WEBER, WENDELL W, b Maplewood, Mo, Sept 2, 25; m 52; c 2. PHARMACOLOGY, PEDIATRICS. *Educ:* Cent Methodist Col, BA, 45; Northwestern Univ, PhD(phys chem), 50; Univ Chicago, MD, 59. *Prof Exp:* Asst prof chem, Univ Tenn, 49-51; opers res analyst, Off of Chief Chem Officer, Dept Army, Washington, DC, 51-55; from resident to chief resident pediat, Univ Calif, San Francisco, 60-62; NIH spec fel human genetics, Univ Col, Univ London, 62-63; from instr to prof pharmacol, Sch Med, NY Univ, 63-74; PROF PHARMACOL, UNIV MICH, ANN ARBOR, 74- *Concurrent Pos:* NIH spec fel biochem, Sch Med, NY Univ, 63-65; Health Res Coun NY career scientist award, 65-70 & 70-; mem pharmacol-toxicol comt, Nat Inst Gen Med Sci, 69-73. *Mem:* Am Chem Soc; Am Soc Pharmacol & Exp Therapeut; Am Soc Human Genetics; fel NY Acad Sci. *Res:* Physical chemistry; human genetics; biochemical genetics and pharmacogenetics; drug metabolism and toxicity. *Mailing Add:* Dept Pharmacol Univ Mich Ann Arbor MI 48109

WEBER, WILFRIED T, b Rosenheim, Ger, Feb 10, 36; US citizen; m 60; c 2. PATHOLOGY, IMMUNOLOGY. *Educ:* Cornell Univ, BS, 59, DVM, 61; Univ Pa, PhD(path), 66. *Prof Exp:* From asst prof to prof, 66-75, prof path & chmn Dept Pathobiol, Sch Vet Med, Univ Pa, 78-89; CONSULT, 89- *Concurrent Pos:* NIH res grants, 67-83; USDA grants, 83- *Honors & Awards:* Lindback Award, 75. *Mem:* Am Vet Med Asn; Reticuloendothelial Soc; Am Asn Pathologists; Am Asn Immunologists. *Res:* Hematology; immunopathology; cancer research; tissue culture of lymphoid cells and macrophages. *Mailing Add:* Dept Pathobiol Univ Pa Sch Vet Med Philadelphia PA 19174-6049

WEBER, WILLES HENRY, b Reno, Nev, Sept 22, 42; m 65; c 1. OPTICS, SPECTROSCOPY & SPECTROMETRY. *Educ:* Calif Inst Technol, BS, 64; Univ Wis, MS, 65, PhD(physics), 68. *Prof Exp:* Prin res scientist assoc, 68-78, STAFF SCIENTIST, RES STAFF, FORD MOTOR CO, 78- *Concurrent Pos:* Assoc ed, Optics Lett, Optical Soc Am, 77-80; adj assoc prof physics, Univ Mich, Ann Arbor, 77-; topical ed, J Optical Soc Am, 80-86; pres, Ann Arbor Chap, Optical Soc Am, 85-86; ed-in-chief, J Optical Soc AmB, 87-88; vis prof physics, Univ Mich, Ann Arbor, 88-89. *Mem:* Am Phys Soc; fel Optical Soc Am; Sigma Xi. *Res:* Atomic and molecular physics; semiconductor physics, injection phenomena, instabilities, lasers, luminescence; plasma effects in metals and semiconductors; infrared laser spectroscopy; optical effects at surfaces; Raman spectroscopy of solids. *Mailing Add:* Res Staff Physics Dept Ford Motor Co Dearborn MI 48121-2053

WEBER, WILLIAM ALFRED, b New York, NY, Nov 16, 18; m 40; c 3. BOTANY. *Educ:* Iowa State Col, BS, 40; State Col Wash, MS, 42, PhD(bot), 46. *Prof Exp:* From asst prof to assoc prof, 46-82, EMER PROF NATURAL HIST & CUR HERBARIUM, UNIV COLO, BOULDER, 82- *Concurrent Pos:* Cur, Lichen Herbarium, Am Bryol & Lichenological Soc, 54-70. *Mem:* AAAS; Bot Soc Am; Am Bryol & Lichenological Soc; Am Soc Plant Taxon; Int Asn Plant Taxon; Lichenological Soc London; fel Linnean Soc London. *Res:* Lichen and bryophyte flora of Colorado, Galapagos Islands and Australasia; boreal and arctic elements in the Rocky Mountain flora; vascular flora of Rocky Mountains/Altai. *Mailing Add:* Campus Box 350 Univ Colo Mus Boulder CO 80309

WEBER, WILLIAM J, b Watertown, Wis, July 19, 49; div. RADIATION EFFECTS, ADVANCED CERAMICS. *Educ:* Univ Wis-Oshkosh, BS, 71; Univ Wis-Madison, MS, 72, PhD(nuclear energy), 77. *Prof Exp:* SR RES SCIENTIST, PAC NORTHWEST DIV, BATTELLE, 77- *Concurrent Pos:* Vis scientist, Europ Inst Transuranium Elements, Karlsruhe, Ger, 83; lectr, Tri-cities Univ Ctr, 88; prog mgr, special assignment, US Dept Energy, 88-92. *Mem:* Am Ceramic Soc; Mat Res Soc; Sigma Xi. *Res:* Materials science research on radiation effects in materials; defect/property relationships; advanced ceramic processing and properties; electrical transport in ceramics; materials characterization, especially x-ray diffraction and electron microscopy techniques. *Mailing Add:* PO Box 999 MS K2-44 Richland WA 99352

WEBER, WILLIAM MARK, b Great Bend, Kans, Nov 24, 41; m 64; c 2. QUATERNARY GEOLOGY. *Educ:* Colo Col, BS, 63; Mont State Univ, MS, 65; Univ Wash, PhD(geol), 71. *Prof Exp:* Lab technician, Lincoln DeVore Testing Lab, 59-63, consult geologist, 65-67; instr geol, Colo Col, 67-68; asst prof, Minot State Col, 70-74; mem staff, Dept Geol, Univ Mont, 74-77; GEOLOGIST, LEWIS & CLARK NAT FOREST, 77-; PRES, WEBER & ASSOC, INC, 77- *Concurrent Pos:* Instr, Univ Colo, Cragmoor Campus, 66-68; co-dir exp col, Minot State Col, 71-72, NSF grant, 71-72. *Mem:* Geol Soc Am; Asn Prof Geol Scientists. *Mailing Add:* Box 773 Lewiston MT 59457

WEBER, WILLIAM PALMER, b Washington, DC, Nov 7, 40; m 63; c 4. SYNTHETIC INORGANIC & ORGANOMETALLIC CHEMISTRY. *Educ:* Univ Chicago, BS, 63; Harvard Univ, MS, 65, PhD(chem), 68. *Prof Exp:* Res chemist, Dow Chem, 67-68; from asst prof to assoc prof, 68-78, PROF CHEM, UNIV SOUTHERN CALIF, 78- *Concurrent Pos:* Chair, Univ Admis Comt, Univ Southern Calif, 80-84, pres, Fac Senate, 85-86 & chair, Dept Chem, 86-89. *Mem:* Am Chem Soc. *Res:* Organosilicon polymer synthesis. *Mailing Add:* Loker Hydrocarbon Res Inst Univ Southern Calif Los Angeles CA 90089-1661

WEBERG, BERTON CHARLES, b St Paul, Minn, Dec 23, 30; m 58; c 4. ORGANIC CHEMISTRY. *Educ:* Hamline Univ, BS, 54; Univ Colo, PhD(org chem), 58. *Prof Exp:* Sr res chemist, Abrasives Lab, Minn Mining & Mfg Co, 58-64; PROF CHEM, MANKATO STATE UNIV, 64- *Mem:* Am Chem Soc. *Res:* Chemistry of hindered ketones and vinyl ethers; polymer chemistry. *Mailing Add:* Dept Chem Mankato State Univ Mankato MN 56001

WEBERS, GERALD F, b Racine, Wis, Apr 14, 32; m 58; c 2. PALEONTOLOGY. *Educ:* Lawrence Univ, BS, 54; Univ Minn, MS, 61, PhD(geol), 64. *Prof Exp:* Res assoc geol, Univ Minn, 64-66; PROF GEOL, MACALESTER COL, 66- *Mem:* Geol Soc Am; Paleont Soc. *Res:* Evolution, taxonomy and paleoecology of Paleozoic invertebrate fossil faunas, especially trilobites, primitive mollusks and conodonts; antarctic geology. *Mailing Add:* 1757 N Abert St Paul MN 55113

WEBERS, VINCENT JOSEPH, b Racine, Wis, Apr 28, 22; m 49; c 6. ORGANIC CHEMISTRY. *Educ:* Univ Wis, BS, 43; Univ Minn, PhD(org chem), 49. *Prof Exp:* Lab asst, Univ Minn, 43-45; chemist, Wyeth Inst Appl Biochem, 46; lab asst, Univ Minn, 47; chemist, Cent Res Dept, E I du Pont de Nemours & co, Inc, 49-53, res chemist, Photo Prod Dept, 53-65, res assoc, 65-66 & org chem dept, 66-67, res supvr, Photo Prod Dept, 67-73, res assoc, 73-76, res fel, 76-85, CONSULT, E I DU PONT DE NEMOURS & CO, INC, 85- *Honors & Awards:* V F Payne Award, Am Chem Soc, 65. *Mem:* Am Chem Soc; Sigma Xi; Soc Imaging Sci & Technol. *Res:* Polymer chemistry; lithography; photochemistry; photopolymerization; sensitometry; photosensitive systems; metal coating and plating. *Mailing Add:* 2322 Wynnwood Rd Wilmington DE 19810

WEBERT, HENRY S, b New Orleans, La, Jan 21, 29; m 70; c 3. BOTANY, PLANT PHYSIOLOGY. *Educ:* Loyola Univ, Ill, BS, 51, MEd, 53; La State Univ, MS, 62, PhD(bot), 65. *Prof Exp:* Teacher high sch, Ill, 54-59; partic biol, NSF Acad Year Inst, Brown Univ, 59-60; res asst bot, La State Univ, 60-64; from asst prof to assoc prof, 64-70, PROF BIOL, NICHOLLS STATE UNIV, 70- *Mailing Add:* Dept Biol Sci Nicholls State Univ Thibodaux LA 70310

WEBRE, NEIL WHITNEY, b Baton Rouge, La, Feb 9, 38; m 64; c 4. COMPUTER SCIENCE, APPLIED PHYSICS. *Educ:* La State Univ, BS, 60; Harvard Univ, AM, 68. *Prof Exp:* Programmer, Brookhaven Nat Lab, 60-63; systs analyst, Columbia Univ, 63-65 & Max Planck Inst, Munich, Ger, 65-66; appl mathematician, ABT Assocs, Cambridge, Mass, 68-69; chmn dept, 84-86, PROF COMPUTER SCI, CALIF POLYTECH, SAN LUIS OBISPO, 69- *Mem:* Asn Comput Mach; Inst Elec & Electronics Engrs. *Res:* Data structures, file structures, and software engineering; effects of data abstraction and object-oriented approaches on software engineering. *Mailing Add:* Dept Computer Sci Calif Polytech State Univ San Luis Obispo CA 93407

WEBSTER, ALLEN E, b Minneapolis, Minn, Oct 9, 38; m 60; c 2. PROCESS CONTROL SYSTEMS, INSTRUMENTATION. *Educ:* Univ Minn, BSME, 60. *Prof Exp:* Proj engr, DuPont, 60-64, tech engr, 64-67, res design engr, 67-74, sr res design engr, 74-81, res engr, 81-83, sr res engr, 83-86, SR TECH ENGR, DUPONT, 86- *Mem:* Instrument Soc Am. *Res:* Substrates for floppy disks; film processing procedures and equipment; microwave drying; film coating and treating; film handling and slitting; batch processing with PLCs and tristate matrix programming; DCS systems configuration. *Mailing Add:* 5309 Chickadee Circle Orient OH 43146

WEBSTER, BARBARA DONAHUE, b Winthrop, Mass, May 19, 29; m 56; c 1. PLANT MORPHOGENESIS. *Educ:* Univ Mass, BS, 50; Smith Col, MA, 52; Harvard Univ, PhD, 57. *Prof Exp:* Instr plant sci, Vassar Col, 52-54; instr biol, Tufts Univ, 57-58; instr agron, Univ Calif, 58-72, res morphologist & lectr agron & range sci, 72-79, prof & assoc dean biol sci, 79-81, assoc dean grad studies & res, 81-89, ASSOC VCHANCELLOR RES, DEPT AGRON & RANGE SCI, UNIV CALIF, DAVIS, 89- *Concurrent Pos:* Res investr,

Brookhaven Nat Lab, 52-53 & 55; NSF fel, Purdue Univ, 58-60; mgt fel, Univ Calif, Davis, 80; vis prof, Univ Nairobi, Kenya, 83; vis scholar, Univ Tex, Austin, Nat Acad Sci & NSF, 87. *Mem:* Bot Soc Am (pres elect, 82, pres, 83); Am Soc Plant Physiol; fel Am Sci Hort Sci; fel AAAS. *Res:* Plant growth regulators and abscission; morphology and physiology of plant growth; reproductive biology and pollination mechanisms. *Mailing Add:* Dept Agron & Range Sci Univ Calif Davis CA 95616

WEBSTER, BURNICE HOYLE, b Leeville, Tenn, Mar 3, 10; m 39; c 3. THORACIC DISEASES. *Educ:* Vanderbilt Univ, BA, 36, MD, 40. *Hon Degrees:* DSc, Holy Trinity Col, 71; PhD, Fla Res Inst, 72, Dr Humanities, 80. *Prof Exp:* Prof anat & path, Gupton-Jones Col Mortuary Sci, 36-42; intern & resident, St Thomas Hosp, 40-43; ASSOC INTERNAL MED & CHEST DIS, MED SCH, VANDERBILT UNIV, 43-; PROF MED & MED CONSULT, GUPTON SCH MORTUARY SCI, 44- *Concurrent Pos:* Vis physician, Chest Clin, Vanderbilt Univ, 43-48; consult, Vet Admin, 44-; dir, Nashville Chest Clin, 45-48; pres med staff, Protestant Hosp, 47; pres, Holy Trinity Col, 69-; lectr, St Thomas & Baptist Hosps. *Mem:* AAAS; fel Am Thoracic Soc; fel Am Geriat Soc; Am Soc Trop Med & Hyg; fel AMA. *Res:* Parasitic and fungus disease of the lung; changes in fungi following multiple animal passages; pathogenic mutation; thoracology. *Mailing Add:* 2315 Valley Brook Rd Nashville TN 37215

WEBSTER, CLYDE LEROY, JR, b Colorado Springs, Colo, Nov 15, 44; m 65; c 2. INORGANIC CHEMISTRY, GEOCHEMISTRY. *Educ:* Walla Walla Col, BSc, 68; Colo State Univ, PhD(chem), 72. *Prof Exp:* Consult chem, Accu-Labs Res Inc, 73-74; asst mgr chem, Instrument Anal Div, Com Testing & Eng, 74-75; asst prof chem, Loma Linda Univ, 75-78, assoc chmn dept, 77-80; ASSOC PROF CHEM & CHMN DEPT, WALLA WALLA COL, 80- *Concurrent Pos:* Consult, Geosci Res Inst, 75- *Mem:* Am Chem Soc. *Res:* Ore body genesis and processes of fossilization as related to the great deluge theory. *Mailing Add:* 5183 Sierra Vista Riverside CA 92505

WEBSTER, CURTIS CLEVELAND, b Roxbury, Vt, Sept 22, 22; m 47; c 5. MATHEMATICS, NUCLEAR ENGINEERING. *Educ:* Univ Vt, AB, 47; Mo Sch Mines, MS, 50. *Prof Exp:* Lab asst physics, Case Inst Technol, 47-48; instr, Mo Sch Mines, 48-50; scientist, Oak Ridge Nat Lab, 50-56; sr scientist, Radiation & Nucleonics Lab, Mat Eng Dept, Westinghouse Elec Corp, 56-58, sr scientist, Atomic Power Dept, 58-59, proj engr, Testing Reactor, 59-60, hazards eval engr, 60-62; develop engr, Develop Dept Opers Div, 62-70, COMPUT APPLNS SPECIALIST, MATH DIV, OAK RIDGE NAT LAB, 70- *Mem:* AAAS. *Res:* Neutron radiation effects; reactor fuel element development; reactor power measurement by activation of oxygen in coolant; heat transfer problems in water cooled reactor; safety problems in nuclear reactors; reactor physics; application of computers to science and engineering. *Mailing Add:* 122 N Western Ave Oak Ridge TN 37830

WEBSTER, D(ONALD) S(TEELE), b Wellsboro, Pa, Nov 27, 17; m 56; c 4. CHEMICAL ENGINEERING. *Educ:* Pa State Univ, BS, 40. *Prof Exp:* Jr engr, Eng Serv Div, E I du Pont de Nemours & Co, 40-42, engr, US Army contract, Metall Lab, Chicago, 42-43, engr, Clinton Lab, Tenn, 43-44, engr, Hanford Eng Works, Wash, 44-45, engr, Eng Serv Div, Titanium Plant, Del, 45-46, engr, Eng Res Lab, Exp Sta, 46-52, res supvr, Savannah River Lab, SC, 54-56, sr res supvr, 56-62, res mgr chem eng, 62-69; assoc dir, Chem Eng Div, Argonne Nat Lab, 69-76, dep dir, 76-82; RETIRED. *Mem:* Am Inst Chem Engrs; Am Chem Soc; AAAS. *Res:* Fluid dynamics; solvent extraction methods; radiochemical processing; fluidized-bed combustion. *Mailing Add:* 4271 Vaucluse Rd Aiken SC 29801

WEBSTER, DALE ARROY, b St Clair, Mich, Jan 11, 38; m 59; c 1. BIOCHEMISTRY. *Educ:* Univ Mich, BS, 60; Univ Calif, Berkeley, PhD(biochem), 65. *Prof Exp:* Res fel med, Mass Gen Hosp & Harvard Med Sch, 65-68; from asst prof to assoc prof, 68-78, PROF BIOL, ILL INST TECHNOL, 78- *Concurrent Pos:* Vis prof biol, Osaka Univ, Japan, 74-75; guest prof biol, Konstanz Univ, Ger, 79; vis prof, Dept Pub Health, Kyoto Univ, Japan, 83-84. *Mem:* AAAS; Am Soc Microbiol; Am Soc Biochem & Molecular Biol. *Res:* The structure, function and regulation of biosynthesis of heme proteins in bacterial respiration and terminal electron transport. *Mailing Add:* Dept Biol Ill Inst Technol Chicago IL 60616

WEBSTER, DAVID DYER, b Grand Rapids, Minn, May 27, 18; m 46; c 2. MEDICINE, NEUROLOGY. *Educ:* Univ Minn, BS, 42, MD, 51; Am Bd Psychiat & Neurol, dipl, 60. *Prof Exp:* PROF NEUROL, MED SCH, UNIV MINN, MINNEAPOLIS, 57- *Concurrent Pos:* Staff neurologist & dir neurophysiol lab, Minneapolis Vet Hosp, 55-, chief neurol serv, 76- *Mem:* AAAS; AMA; Am Acad Neurol. *Res:* Movement disorders; medical electronics. *Mailing Add:* Dept Neurol Univ Minn Vet Admin Med Ctr 54th St & 48th Ave S Minneapolis MN 55417

WEBSTER, DAVID HENRY, b Berwick, NS, Oct 29, 34; m 60; c 1. HORTICULTURE. *Educ:* Acadia Univ, BSc, 54, MSc, 55; Univ Calif, Davis, PhD(plant physiol), 65. *Prof Exp:* Mem res staff, NS Res Found, 58-62; RES SCIENTIST, CAN DEPT AGR, 65- *Mem:* Am Soc Hort Sci. *Res:* Fruit tree physiology; nutrition; soil physical properties. *Mailing Add:* Can Dept Agr Res Sta Kentville NS B4N 1J5 Can

WEBSTER, DENNIS BURTON, b Covington, Va, Dec 14, 42. INDUSTRIAL ENGINEERING. *Educ:* WVa Univ, BSIE, 65, MSIE, 66; Purdue Univ, Lafayette, PhD(indust eng), 69. *Prof Exp:* Field studies analyst, Am Viscose Div, FMC Corp, 64, indust engr, 66; asst prof, 70-75, assoc prof, 75-80, PROF INDUST ENG, AUBURN UNIV, 81- *Concurrent Pos:* Mem, Col Indust Coun Mat Handling Educ. *Mem:* Am Inst Indust Engrs. *Res:* Materials flow models; simulation models and simulation optimization; scheduling and production control. *Mailing Add:* Dept Indust Eng Univ Ala Box 6316 Tuscaloosa AL 35487

WEBSTER, DOUGLAS B, b Fond du Lac, Wis, Jan 14, 34; m 55; c 2. OTORHINOLARYNGOLOGY. *Educ:* Oberlin Col, AB, 56; Cornell Univ, PhD(zool), 60. *Prof Exp:* Fel psychobiology, Calif Inst Technol, 60-62; from asst prof to prof biol, NY Univ, 62-73, actg chmn dept, 67-68; PROF OTORHINOLARYNGOL & ANAT, LA STATE UNIV MED CTR, NEW ORLEANS, 73-, CLIN PROF AUDIOL & SPEECH PATH, SCH ALLIED HEALTH, 74- *Concurrent Pos:* NIH grants, NY Univ, 63-65, NY Univ & La State Univ Med Ctr, 65-75 & La State Univ Med Ctr, 73-91. *Mem:* AAAS; Am Asn Anatomists; Soc Neurosci; Am Soc Zoologists (secy, 69-75); Sigma Xi. *Res:* Morphology, behavior and physiology of hearing in vertebrates. *Mailing Add:* Kresge Herig Res Lab Med Ctr La State Univ 2020 Gravier St New Orleans LA 70112-2234

WEBSTER, EDWARD WILLIAM, b London, Eng, Apr 12, 22; nat US; m 50, 61; c 6. MEDICAL PHYSICS. *Educ:* Univ London, BSc, 43, PhD(elec eng), 46. *Hon Degrees:* Am, Harvard Univ, 90. *Prof Exp:* Res engr, Eng Elec Co, UK, 45-49; guest researcher, Mass Inst Technol, 49-50, radiation physicist, 50-51; lectr elec eng & nuclear energy, Queen Mary Col, Univ London, 52-53; prof radiol, Div Health Sci & Technol, Harvard-Mass Inst Technol, 78-86, PHYSICIST, MASS GEN HOSP, 53-, PROF RADIOL, HARVARD MED SCH, 75- *Concurrent Pos:* London County Coun Blair traveling fel, Mass Inst Technol, 49-50; from asst to asst prof, Harvard Med Sch, 53-67, assoc clin prof physics in radiol, 67-69, assoc prof radiol, 69-75; USPHS spec fel, 65-66; lectr med radiation physics, Sch Pub Health, Harvard Univ, 71-86, consult, Mass Eye & Ear Infirmary, 57- & Int Atomic Energy Agency, 60-64; mem comt radiol, Nat Acad Sci, 62-68; consult, Children's Hosp Med Ctr, Boston, 62-; mem, Nat Coun Radiation Protection & Measurements, 64-, bd dirs, 81-88; consult, WHO, 65 & 67; consult, USPHS, 61-63; chmn physics credentials comt, Am Bd Radiol, 66-76; mem radiol health study sect, Dept HEW, 69-72, radiol trining grant comt, NIH, 69-73 & adv comt med uses of isotopes, US Nuclear Regulatory Comn, 71-, adv comt comt environ hazards, Dept Vet Affairs, 85-; mem comt biol effects ionizing radiation, Nat Acad Sci, 77-80 & US Nat Comt, Int Union Pure & Appl Biophys, 71-74, Oversight Comt Radioepidemiologic Tables, 83-84, Comt Ionizing Radiation Dosimetry, 85-86; secy-gen, Second Int Conf Med Physics, 69; mem, US deleg, UN Sci Comt Effects Atomic Radiation; consult, Radiation Effects Res Found, Hiroshima, Japan, 88; Sigma Xi Nat lectr, Health Physics, 88-89. *Honors & Awards:* Garland lectr, Calif Radiol Soc, 80; William Coolidge Medal, Am Asn Physicists Med, 83; Gold Medal, Am Col Radiol, 90. *Mem:* Fel Am Asn Physicists in Med (pres, 63-64); Radiol Soc NAm (vpres, 77-78); fel Am Col Radiol; fel Health Physics Soc; Soc Nuclear Med; Radiation Res Soc. *Res:* Radiological physics; application of radiation, radioisotopes and electronic methods to medical diagnosis and therapy; radiation protection. *Mailing Add:* Dept Radiol Mass Gen Hosp 32 Fruit St Boston MA 02114

WEBSTER, ELEANOR RUDD, b Cleveland, Ohio, Oct 11, 20. ORGANIC CHEMISTRY, HISTORY OF SCIENCE. *Educ:* Wellesley Col, AB, 42; Mt Holyoke Col, MA, 44; Radcliffe Col, MA, 50, PhD(chem), 52. *Prof Exp:* Asst chem, Mt Holyoke Col, 42-44; chemist synthetic org res, Eastman Kodak Co, 44-47; from instr to prof, 52-85, dean of freshmen & sophomores, 56-60, chmn dept chem, 64-67, & 79-81, dir inst chem, 64-72, dir continuing educ, 69-70, EMER PROF CHEM, WELLESLEY COL, 85- *Concurrent Pos:* Fulbright grant, Belg, 51-52; NSF sci fac fel, Univ Chicago, Oak Ridge, Univ Calif, Berkeley, 60-61, Univ East Anglia, 81-82; vis prof, New Hall, Cambridge Univ, 68 & Univ East Anglia, 81-82. *Mem:* Am Chem Soc; Hist Sci Soc. *Res:* Physical organic chemistry; dissemination of science, the public's understanding since 1850. *Mailing Add:* 162 Western Ave Sherborn MA 01770

WEBSTER, EMILIA, b Arad, Romania, Oct, 27, 50; US citizen. OPTICAL COMMUNICATIONS, OPTICS. *Educ:* Univ Calif, Los Angeles, BS, 74, MS, 78, PhD(physics), 80. *Prof Exp:* Teaching assoc physics, Univ Calif, Los Angeles, 74-79; mem tech staff electro-optics, 80-81, MEM TECH STAFF ADVAN SPACE COMMUN, AEROSPACE CORP, 81- *Mem:* Am Phys Soc; Optical Soc Am; Inst Elec & Electronics Engrs; Soc Photo-Optical Instrumentation Engrs; Soc Photog Scientists & Engrs. *Res:* Laser communications; lasers; optics; communications theory. *Mailing Add:* PO Box 25307 Los Angeles CA 90025

WEBSTER, FERRIS, b St Boniface, Man, Aug 7, 34; m. PHYSICAL OCEANOGRAPHY. *Educ:* Univ Alta, BSc, 56, MSc, 57; Mass Inst Technol, PhD(geophys), 61. *Prof Exp:* Res asst phys oceanog, Woods Hole Oceanog Inst, 59-62, res assoc, 62-63, asst scientist, 63-65, assoc scientist, 65-70, chmn dept phys oceanog, 71-73, sr scientist, 70-78, assoc dir, 73-78; asst adminr res & develop, Nat Oceanic & Atmospheric Admin, 78-82; sr fel, Nat Acad Sci, 82-83; PROF PHYS OCEANOG, UNIV DEL, 83- *Concurrent Pos:* Asst prof, Mass Inst Technol, 66-68. *Mem:* AAAS; Am Geophys Union; Am Meteorol Soc. *Res:* Ocean currents; the influence of the ocean on climate variability; management of oceanic data, emphasizing needs of World Climate Research Program; time-series analysis. *Mailing Add:* Col Marine Studies Univ Del Lewes DE 19958-1298

WEBSTER, GARY DEAN, b Hutchinson, Kans, Feb 15, 34; m 64; c 2. GEOLOGY, PALEONTOLOGY. *Educ:* Univ Okla, BS, 56; Univ Kans, MS, 59; Univ Calif, Los Angeles, PhD(geol), 66. *Prof Exp:* Geologist, Amerada Petrol Corp, 56-57; geologist, Belco Petrol Corp, 60; geologist, Shell Oil Co, 63; lectr phys geol, Calif Lutheran Col, 63-64; mus scientist, Univ Calif, Los Angeles, 64-65; asst prof geol & paleont, San Diego State Col, 65-68; from asst prof to assoc prof, 68-77, chmn dept, 80-85, PROF GEOL, WASH STATE UNIV, 77- *Concurrent Pos:* Mem, Am Geol Inst, Int Field Inst, Spain, 71. *Mem:* AAAS; Soc Econ Paleont & Mineral; Paleont Soc; fel Geol Soc Am; Brit Palaeont Asn; Am Inst Prof Geol. *Res:* Late Paleozoic paleontology and stratigraphy, especially crinoids and conodonts; stratigraphy sedimentation. *Mailing Add:* Dept Geol Wash State Univ Pullman WA 99164-2812

WEBSTER, GEORGE CALVIN, b South Haven, Mich, July 17, 24; m 60; c 2. BIOCHEMISTRY, MOLECULAR BIOLOGY. *Educ:* Western Mich Univ, BS, 48; Univ Minn, MS, 49, PhD(biol), 52. *Prof Exp:* Res fel, Calif Inst Technol, 52-55; from assoc prof to prof biochem, Ohio State Univ, 55-61; vis prof enzyme chem, Univ Wis, 61-65; chief chemist, Aerospace Serv Div, Cape Kennedy, Fla, 65-70; prof biol sci & head dept, Fla Inst Technol, 70-86, assoc dean, Col Sci & Eng, 85-86; RETIRED. *Concurrent Pos:* USPHS spec res fel, 61-63; Am Heart Asn estab investr, 63-65. *Mem:* AAAS; Am Soc Biol Chemists; Sigma Xi; Am Soc Cell Biol; Brit Biochem Soc; Geront Soc; Am Aging Asn. *Res:* Molecular biology of aging; control of gene expression. *Mailing Add:* 530 Majorca Ct Satellite Beach FL 32937-3266

WEBSTER, GORDON RITCHIE, b Kindersley, Sask, Jan 7, 22; m 50; c 2. SOIL CHEMISTRY. *Educ:* Univ BC, BSA, 49, MSA, 51; Univ Ore, PhD(soils), 58. *Prof Exp:* Res officer soil sci, Can Dept Agr, 49-60; from assoc prof to prof soil sci, 60-91, EMER PROF, UNIV ALTA, 91- *Mem:* Can Soc Soil Sci; Int Soc Soil Sci. *Res:* Reclamation of solonetzic soils; reclamation of salt spills. *Mailing Add:* 11817 76th Ave Edmonton AB T6G 0L1 Can

WEBSTER, GRADY LINDER, b Ada, Okla, Apr 14, 27; m 56; c 1. PLANT TAXONOMY. *Educ:* Univ Tex, BA, 47, MA, 49; Univ Mich, PhD(bot), 54. *Prof Exp:* Lectr trop bot, Harvard Univ, 53, NSF fel biol, 53-55, instr bot, 55-58; from asst prof to assoc prof biol sci, Purdue Univ, 58-66; PROF BOT, UNIV CALIF, DAVIS, 66-; DIR, J M TUCKER HERBARIUM, 87- *Concurrent Pos:* Guggenheim fel bot, State Univ Utrecht, 64-65; prog dir systematic biol, NSF, 81-82. *Mem:* Am Soc Plant Taxon (pres, 82); Soc Study Evolution; Bot Soc Am; Asn Trop Biol; Int Asn Plant Taxon; Sigma Xi; Linnaean Soc London. *Res:* Evolution and systematics of vascular plants, especially Euphorbiaceae; vegetational and floristic plant geography; pollination ecology. *Mailing Add:* Dept Bot Univ Calif Davis CA 95616

WEBSTER, HAROLD FRANK, b Buffalo, NY, June 25, 19; m 51; c 3. PHYSICS. *Educ:* Univ Buffalo, BA, 41, MA, 44; Cornell Univ, PhD(physics), 53. *Prof Exp:* Mem staff, Radiation Lab, Mass Inst Technol, 43-45; asst, Cornell Univ, 45-51; physicist, Res Lab, Gen Elec Co, 51-86; RETIRED. *Concurrent Pos:* US deleg gen assembly, Int Sci Radio Union, 60; mem adv comt, Physics Today, 77-80. *Honors & Awards:* Baker Award, Inst Elec & Electronics Eng, 58. *Mem:* Am Phys Soc; Inst Elec & Electronics Eng. *Res:* Thermionic emission from single crystal surfaces; metal surface wetting; electron beam dynamics; cesium plasma; energy conversion; ultraviolet sensors; solid state science; microelectronics. *Mailing Add:* 77 St Stephens Lane W Schenectady NY 12302

WEBSTER, HARRIS DUANE, b Lansing, Mich, Jan 14, 20; m 43; c 2. VETERINARY PATHOLOGY. *Educ:* Mich State Col, DVM, 43, MS, 51. *Prof Exp:* Vet, US Regional Poultry Res Lab, East Lansing, 44 & 46-47; gen pract, Mich, 47-48; instr animal path, Mich State Col, 48-52; res scientist path & toxicol, Upjohn Co, 52-83; RETIRED. *Mem:* Am Vet Med Asn; Am Col Vet Path; Int Acad Pathologists. *Res:* Toxicologic, experimental pathologic studies and laboratory animal health consultation and service. *Mailing Add:* 715 Jenks Blvd Kalamazoo MI 49007

WEBSTER, HENRY DEFOREST, b New York, NY, Apr 22, 27; m 51; c 5. NEUROLOGY, NEUROPATHOLOGY. *Educ:* Amherst Col, BA, 48; Harvard Med Sch, MD, 52; Am Bd Psychiat & Neurol, dipl & cert neurol, 59. *Prof Exp:* Intern & asst resident med, Harvard Serv, Boston City Hosp, 52-54; asst resident & resident neurol, Mass Gen Hosp, 54-56, res fel neuropath, 56-58; asst instr & assoc neurol, Harvard Med Sch, 58-66, asst prof neuropath, 66; assoc prof neurol, Sch Med, Univ Miami, 66-69, prof, 69; Assoc chief lab neuropath & neuroanat 75-84, Chief LAB EXP NEUROPATH, 84- *Concurrent Pos:* Prin investr, Nat Inst Neurol Dis & Stroke grants, 62-69, partic neuroanat vis scientist prog, 64-65; assoc neurologist & asst neuropathologist, Mass Gen Hosp, 63-66; secy gen, VIII Int Cong Neuropath, 78. *Mem:* Am Asn Neuropath (vpres, 76-77, pres, 78-79); Am Soc Cell Biol; Am Acad Neurol; Am Neurol Asn; Sigma Xi. *Res:* Experimental neuropathology utilizing electron microscopy and immunocytochemistry, especially the formation and breakdown of myelin. *Mailing Add:* NIH Bldg 36 Rm 4A-29 Bethesda MD 20892

WEBSTER, ISABELLA MARGARET, b Chicago, Ill, Jan 22, 11. PUBLIC HEALTH. *Educ:* Northwestern Univ, BA, 32; Univ Minn, PhD(org chem), 36; Woman's Med Col Pa, MD, 47; Univ Liverpool, dipl, 65; Univ Hawaii, MPH, 69. *Prof Exp:* Asst chem, Univ Minn, 32-36, instr org chem, 36-37; intern, Mary Immaculate Hosp, NY, 47-48; intern obstet & gynec, Georgetown Univ Hosp, 48; mem med staff, Holy Family Hosp, Mandar, India, 48-54; doctor chg, Archbishop Attipetty Jubilee Mem Hosp, S India, 54-57; chief med servs, Kokofu Leprosarium, Ghana Leprosy Serv, 58-68; consult pub health, Atat Hosp, Ethiopia, Nangina Hosp, Nangina, Kenya, Virika Hosp, Ft Portal, Uganda & Lower Shire Mobile Health Unit, Chiromo, Malawi, 69-71; dist superior, Med Mission Sisters, Ghana, 71-73, mem, Cent Med Mission Sisters, Rome, 73-79; DIOCESAN MED COORDR, DIOCESE KAKAMEGA, KENYA, 81- *Concurrent Pos:* Consult primary health, Med Mission Sisters; chmn health working group, SEDOS, Rome. *Res:* Organic chemistry; bacteriology; obstetrics and gynecology; leprosy. *Mailing Add:* Diocese Kakamega PO Box 838 Kakamega Kenya

WEBSTER, JACKSON DAN, b Tacoma, Wash, Feb 26, 19; m 44; c 3. PARASITOLOGY, ORNITHOLOGY. *Educ:* Whitworth Col, Wash, BSc, 39; Cornell Univ, MSc, 41; Rice Inst, PhD(parasitol), 47. *Prof Exp:* Field researcher ornith, Alaska, 40 & 46; asst prof biol, Jamestown Col, 47-49; assoc prof, 49-53, prof, 53-84, EMER PROF BIOL, HANOVER COL, 84- *Concurrent Pos:* Field researcher ornith, 50-88. *Mem:* Soc Parasitol; Wilson Ornith Soc; Cooper Ornith Soc; Am Ornith Union. *Res:* Septematics, distribution and populations of birds; systematics of tapeworms. *Mailing Add:* Dept Biol Hanover Col Hanover IN 47243

WEBSTER, JACKSON ROSS, b Brigham City, Utah, May 3, 45; m 68; c 2. STREAM ECOLOGY. *Educ:* Wabash Col, BA, 67; Univ Ga, PhD(ecol, zool), 75. *Prof Exp:* From asst prof to assoc prof, 75-87, PROF BIOL, VA POLYTECH INST & STATE UNIV, 87- *Concurrent Pos:* Vis scientist, Oak Ridge Nat Lab, 81-82. *Mem:* Ecol Soc Am; Am Soc Limnol & Oceanog; NAm Benthological Soc; Int Asn Theoret & Appl Limnol; Am Soc Naturalists. *Res:* Nutrient and energy dynamics in stream ecosystems; ecosystem modeling; effects of watershed disturbance on streams. *Mailing Add:* Dept Biol Va Polytech Inst & State Univ Blacksburg VA 24061

WEBSTER, JAMES ALBERT, b Mineola, NY, June 2, 28; m 51; c 4. ORGANIC CHEMISTRY. *Educ:* Col Wooster, BA, 51; Univ Pittsburgh, MS, 56. *Prof Exp:* Res chemist, Res & Eng Div, Monsanto Chem Co, 56-61, res chemist, Monsanto Res Corp, 61-66, sr res chemist, 66-71, sr group leader, 71-75, SR RES SPECIALIST, MONSANTO RES CORP, 75- *Mem:* Am Chem Soc; AAAS. *Res:* Silicone chemistry; polymer synthesis; fluorine chemistry. *Mailing Add:* 7611 Eagle Creek Dr Dayton OH 45459-3411

WEBSTER, JAMES ALLAN, b Lincoln, Nebr, May 1, 39; m 67; c 2. ENTOMOLOGY. *Educ:* Univ Ky, BS, 61, MS, 64; Kans State Univ, PhD(entom), 68. *Prof Exp:* RES ENTOMOLOGIST, AGR RES SERV, USDA, 68- *Concurrent Pos:* From adj asst prof to adj assoc prof, Dept Entom, Mich State Univ, 68-81; adj assoc prof, Dept Entom, Okla State Univ, 81-89, adj prof, 89- *Mem:* Entom Soc Am; Am Soc Agron. *Res:* Insect resistance in grain and forage crops. *Mailing Add:* 1301 N Western Rd Stillwater OK 74075-2714

WEBSTER, JAMES RANDOLPH, JR, b Chicago, Ill, Aug 25, 31; m 54; c 3. PULMONARY PHYSIOLOGY. *Educ:* Northwestern Univ, Chicago, BS, 53, MS & MD, 56. *Prof Exp:* From asst prof to assoc prof, 67-77, PROF MED, MED SCH, NORTHWESTERN UNIV, CHICAGO, 77- *Concurrent Pos:* USPHS fel pulmonary physiol, Med Sch, Northwestern Univ, Chicago, 62-64; assoc dir inhalation ther, Chicago Wesley Mem Hosp, 65-66, dir pulmonary function lab, 66-; chief med, Northwestern Mem Hosp, 72- *Mem:* Am Thoracic Soc; Am Fedn Clin Res; Am Col Physicians. *Res:* Diseases of the chest. *Mailing Add:* 222 E Superior Chicago IL 60611

WEBSTER, JOHN GOODWIN, b Plainfield, NJ, May 27, 32; m 54; c 4. MEDICAL INSTRUMENTATION, ELECTRODES. *Educ:* Cornell Univ, BEE, 53; Univ Rochester, MSEE, 65, PhD(elec eng), 67. *Prof Exp:* Res engr, NAm Aviation, Inc, 54-55; head instrumentation, Boeing Airplane Co, 55-59; head telemetry, Radiation, Inc, 59-61; staff engr, Mitre Corp, 61-62; staff engr, Int Bus Mach Corp, 62-63; res assoc elec eng, Univ Rochester, 67; from asst prof to assoc prof, 67-73, PROF ELEC & COMPUT ENG, UNIV WIS-MADISON, 73- *Concurrent Pos:* NSF res grant, 68-70 & 84- 91; NIH res career develop award, 71-76; NASA res grant, 72-73; NIH res grant, 76-83 & 88-91; dir, Biomed Eng Ctr, 76-80; assoc ed, Trans Biomed Eng, Inst Elec & Electronics Engrs, 78-85; mem, NIH Surg Bioeng Study Sect, 86-90. *Honors & Awards:* Donald P Eckman Educ Award, Instrument Soc Am, 74; Western Electric Fund Award, Am Soc Eng Educ, 78. *Mem:* Inst Elec & Electronics Engrs; Instrument Soc Am; Biomed Eng Soc; Asn Advan Med Instrumentation; Am Soc Eng Educ. *Res:* Medical devices and instrumentation; electrodes for monitoring and stimulation; biopotential amplifiers and interference; electrosurgical units; sensory substitution systems; tactile sensors for robotics and medicine; electrical impedance tomography. *Mailing Add:* Dept Elec & Comput Eng Univ Wis 1415 Johnson Dr Madison WI 53706

WEBSTER, JOHN H, b Belleville, Ont, Dec 17, 28; m 53; c 3. RADIOTHERAPY. *Educ:* Queen's Univ, Ont, MD, 55. *Prof Exp:* Sr cancer res radiologist, Roswell Park Mem Inst, 59-62, assoc cancer res radiologist, 62-63, assoc chief cancer res radiologist, 63-64, chief cancer res radiologist, 64-74; prof therapeut radiol & chmn dept, McGill Univ, 74-80; PROF, DEPT RADIOL, ONCOL CTR, UNIV PITTSBURGH, 80- *Concurrent Pos:* USPHS grants; therapeut radiologist-in-chief, Montreal Gen Hosp, Royal Victoria Hosp, Montreal Children's Hosp & Jewish Gen Hosp, 74-79; sr consult radiotherapist, Montreal Neurol Hosp, 74-79; consult radiotherapist, Presby Univ Hosp & Mazec Womens Hosp, 79- *Mem:* Am Col Radiol; Am Soc Therapeut Radiol; Can Med Asn; Soc Chmn Acad Radiation Oncol Progs; Can Asn Radiologists. *Res:* Experimental radiotherapy and allied fields; oncologically related research; cancer patient care systems. *Mailing Add:* Joint Radiol & Oncol Ctr Magee Women's Hosp Div Forbes Ave & Halket St Pittsburgh PA 15213

WEBSTER, JOHN MALCOLM, biology, parasitology & nematology, for more information see previous edition

WEBSTER, JOHN ROBERT, b Riverdale, Calif, May 5, 16; m 43; c 3. FOOD SCIENCE. *Educ:* Calif State Univ, Fresno, AB, 39. *Prof Exp:* Dir res & develop, Lindsay Olive Growers, 39-81; RETIRED. *Concurrent Pos:* Fluidize bed combustion engr, 81- *Mem:* Am Chem Soc; Inst Food Technologists. *Res:* Improving nutrition, improving quality, and decreasing process cost of olives, cherries, pimiento and pickled peppers. *Mailing Add:* 386 Alameda St Lindsay CA 93247-1509

WEBSTER, JOHN THOMAS, b Fond du Lac, Wis, Sept 12, 27; m 55; c 3. STATISTICS. *Educ:* Ripon Col, BA, 51; Purdue Univ, MS, 55; NC State Col, PhD(statist), 60. *Prof Exp:* Statistician, Westinghouse Elec Corp, 55-57; asst prof statist, Bucknell Univ, 60-62; assoc prof, 62-76, PROF STATIST, SOUTHERN METHODIST UNIV, 76- *Mem:* Am Statist Asn; Biomet Soc; Am Soc Qual Control. *Res:* Design and analysis of experiments. *Mailing Add:* 1225 Glen Cove Dr Richardson TX 75080

WEBSTER, KARL SMITH, b Orleans, Vt, Aug 18, 24; m 53; c 2. MECHANICAL ENGINEERING, ENGINEERING MECHANICS. *Educ:* Univ Vt, BSME, 49; Pa State Univ, MS, 58. *Prof Exp:* Test engr, Gen Elec Co, NY, 49-50, appln eng, Mass, 50-52; chief engr, W J Nolan Co, 52-53; develop engr, Fels Gear Shaper Co, 53-54; res asst mech eng, Pa State Univ,

54, instr, 54-58; asst prof, Univ NH, 58-62, assoc prof mech design technol, Tech Inst, 62-65; Dept Interior consult, River Syst as Reactor Proj, Univ Water Resources, 67-69, asst to vpres acad affairs, Univ, 70-72, assoc prof mech eng, Univ Maine, Orono, 65-77, assoc prof, 77-80, prof mech eng technol, 80-89, EMER PROF MECH ENG TECHNOL, SCH ENG TECHNOL, COL ENG & TECHNOL, UNIV MAINE, ORONO, 89- *Concurrent Pos:* Sabbatical mfg engr, resources & equip develop, GE, Burlington, Vt, 86-87. *Mem:* Am Soc Eng Educ; Soc Mfg Engrs. *Res:* Internal combustion engine projects; industrial and manufacturing engineering. *Mailing Add:* Rm 221 E Annex Univ Maine Orono ME 04469

WEBSTER, LARRY DALE, b Westfall, Kans, Feb 8, 39; m 63; c 1. MECHANICS, MATERIALS SCIENCE. *Educ:* Colo Sch Mines, MetE, 61; Cornell Univ, PhD(mat sci), 65. *Prof Exp:* Chmn sci dept, Lamar Community Col, 65-67; res scientist weapons effects, Kaman Sci Corp, 67-73; sr engr reactor technol, Bettis Atomic Power Lab, Pa, 73-77; RES SCIENTIST ACOUST DEVICES, KAMAN SCI CORP, 77- *Res:* Finite element analyses of coupled structural and electromagnetic fields; model development for piezoelectric devices and for magnetically driven flyer plates associated with impact test facilities. *Mailing Add:* 9078 Dogwood Ct Monument CO 80132

WEBSTER, LEE ALAN, b Mt Holly, NJ, June 30, 41; m 64; c 2. TRANSPORTATION & TRAFFIC ENGINEERING. *Educ:* Univ Del, BCE, 63; Univ Ill, Urbana, MSc, 65, PhD(traffic eng), 68. *Prof Exp:* Asst prof civil eng, Univ Mass, Amherst, 68-76; transp engr, Curran Assocs, 76-79; PROF & COORDR ENG SCI, GREENFIELD COMMUNITY COL, 79- *Mem:* Am Soc Civil Engrs; Am Soc Eng Educ. *Res:* Transportation and traffic engineering education; transportation planning; computer applications to engineering and education. *Mailing Add:* Dept Natural Sci Greenfield Community Col One College Dr Greenfield MA 01301

WEBSTER, LESLIE T, JR, b New York, NY, Mar 31, 26; m 55; c 4. BIOCHEMISTRY. *Educ:* Amherst Col, BA, 47; Harvard Med Sch, MD, 48; Am Bd Internal Med, dipl, 57. *Hon Degrees:* ScD, Amherst Col, 81. *Prof Exp:* Demonstr med, Sch Med, Case Western Reserve Univ, 55-57, instr, 57-60, from sr instr to asst prof biochem, 58-66, asst prof med, 60-70, from asst prof to assoc prof pharmacol, 66-70; prof pharmacol & chmn dept, Med & Dent Schs, Northwestern Univ, Chicago, 70-76; J H HORD PROF PHARMACOL & CHMN DEPT, SCH MED, CASE WESTERN RESERVE UNIV, 76- *Concurrent Pos:* Nat Vitamin Found Wilder fel, 56-59; sr investr, USPHS, 59-61, res career develop award, 61-69; mem gastroenterol & nutrit training grants comt, Nat Inst Arthritis & Metab Dis, 65-69; consult, NIH, 71-, WHO, 77-83 & Rockefeller Found, 78-86; Macy fac scholar, 80-81; mem, Cellular & Molecular Basis Dis Rev Comt, Nat Inst Gen Med Sci, 84-88. *Mem:* Am Col Physicians; Am Soc Clin Invest; Am Soc Biol Chemists; Am Soc Pharmacol & Exp Therapeut; Am Soc Trop Med & Hyg; Asn Med Sch Pharmacol. *Res:* Enzymology; pharmacoparasitology; immunopharmacology; drug metabolism. *Mailing Add:* Dept Pharmacol Case Western Reserve Univ Sch Med 2109 Adelbert Rd Cleveland OH 44106-4901

WEBSTER, MERRITT SAMUEL, b Cheyney, Pa, June 5, 09; m 36; c 3. MATHEMATICS. *Educ:* Swarthmore Col, AB, 31; Univ Pa, AM, 33, PhD(math), 35. *Prof Exp:* Asst instr math, Univ Pa, 31-33; instr, Univ Nebr, 35-38; from instr to prof, 38-75, EMER PROF MATH, PURDUE UNIV, WEST LAFAYETTE, 75- *Mem:* Am Math Soc; Math Asn Am. *Res:* Orthogonal polynomials; interpolation. *Mailing Add:* 2741 N Salisbury St No 3109 West Lafayette IN 47906

WEBSTER, ORRIN JOHN, b Arkansas City, Kans, June 26, 13; m 36; c 3. AGRONOMY. *Educ:* Univ Nebr, BSc, 34, MSc, 40; Univ Minn, PhD(plant breeding & genetics), 50. *Prof Exp:* Agronomist, Soil Conserv Serv, USDA, 35-36, Dryland Agr Div, 36-43 & Cereal Corps Res Br, Agr Res Serv, 43-63, dir-coordr major cereal proj, Orgn For African Unity, 63-71, proj leader corn, sorghum & millet res, Fed Exp Sta, PR, 71-74; ADJ PROF AGRON, UNIV ARIZ, 74- *Concurrent Pos:* Adv corn & sorghum breeding, Govt Nigeria, Liberia, Sudan & Africa, 51; secy sorghum res comt, USDA, 53-67; mem comt preserv indigenous strains sorghum, Nat Acad Sci, 61- *Mem:* Am Soc Agron. *Res:* Plant breeding of sorghum; genetics and cytogenetics of sorghum. *Mailing Add:* Dept Plant Sci Univ Ariz Tucson AZ 85721

WEBSTER, OWEN WRIGHT, b Devils Lake, NDak, Mar 25, 29; m 53; c 5. POLYMER CHEMISTRY. *Educ:* Univ NDak, BS, 51; Pa State Univ, PhD(chem), 55. *Hon Degrees:* DSc, Univ NDak, 86. *Prof Exp:* Res chemist, 55-74, group leader, 74-79, RES LEADER, E I DU PONT DE NEMOURS & CO, INC, 84- *Mem:* Am Chem Soc; Sigma Xi; AAAS. *Res:* Synthetic organic chemistry; adamantanes; cyanocarbons; hydrogen cyanide; polymers; group transfer polymerization. *Mailing Add:* 2106 Navaro Rd Wilmington DE 19803

WEBSTER, PAUL DANIEL, III, b Mt Airy, NC, Apr 26, 30; m 57; c 2. MEDICINE, GASTROENTEROLOGY. *Educ:* Univ Richmond, BS, 52; Bowman Gray Sch Med, MD, 56. *Prof Exp:* Fel med, Univ Minn, 60-63; fel gastroenterol, Sch Med, Duke Univ, 63-66, asst prof med, Med Ctr, 67-68; assoc prof, 68-71, PROF MED, MED COL GA, 71-, CHMN DEPT, 77- *Concurrent Pos:* USPHS fel, 65-66; chief med serv, Vet Admin Hosp, Augusta, Ga, 73. *Mem:* Am Gastroenterol Asn; Am Physiol Soc; Am Col Gastroenterol; Am Inst Nutrit; Am Soc Clin Invest. *Res:* Hormonal control of pancreatic protein synthesis; pancreatic structure and function; cancer of pancreas. *Mailing Add:* 517 Scottsway Augusta GA 30909

WEBSTER, PETER JOHN, b Stockport, UK, May 30, 42; c 2. METEOROLOGY, CLIMATOLOGY. *Educ:* Mass Inst Technol, PhD(meteorol), 71. *Prof Exp:* Asst prof, Univ Wash, Seattle, 72-77; prin res scientist, Commonwealth Sci & Indust Res Orgn, Aspendale Victoria, Australia, 77-83; PROF METEOROL, DEPT METEOROL, PA STATE UNIV, 83- *Concurrent Pos:* Tata prof res, Indian Inst Sci, Bangalore, India, 82; George Haltiner prof res, Dept Naval Postgrad, Monterey, Calif, 83; chief ed, J Atmospheric Sci, Am Meteorol Soc, 83-86; consult, NASA Goddard Lab Atmospheres, Nat Geog Soc, 83-86. *Honors & Awards:* Wilson Res Award, 89; Jule G Charney Award, 90; Humboldt Award, 90. *Mem:* Fel Am Meteorol Soc. *Res:* Dynamics of low frequencies flow with particular emphasis on the evolution of monsoons and equatorial waves. *Mailing Add:* Dept Meteorol Pa State Univ 518 Walker Bldg University Park PA 16802

WEBSTER, PORTER GRIGSBY, b Wheatley, Ky, Nov 25, 29; m 61. MATHEMATICS. *Educ:* Georgetown Col, BA, 51; Auburn Univ, MS, 56, PhD(math), 61. *Prof Exp:* PROF MATH, UNIV SOUTHERN MISS, 61- *Mem:* Am Math Soc; Math Asn Am. *Mailing Add:* Dept Math Ss Box 5045 Univ Southern Miss Hattiesburg MS 39406

WEBSTER, ROBERT EDWARD, b New Haven, Conn, May 31, 38; m 60; c 3. MOLECULAR BIOLOGY, MOLECULAR GENETICS. *Educ:* Amherst Col, BA, 59; Duke Univ, PhD(microbiol), 65. *Prof Exp:* NSF fel genetics, Rockefeller Univ, 65-66, asst prof, 66-71; assoc prof, 71-76, PROF BIOCHEM, MED CTR, DUKE UNIV, 76- *Mem:* AAAS; Am Soc Cell Biol; Am Soc Biol Chemists; Am Soc Microbiol; Am Soc Virol. *Res:* Phage genetics and morphogenesis; membrane structure and synthesis. *Mailing Add:* Dept Biochem Med Ctr Duke Univ Durham NC 27710

WEBSTER, ROBERT G, b Balclutha, NZ, July 5, 32; Australian citizen; c 3. VIROLOGY, IMMUNOLOGY. *Educ:* Otago Univ, BS, 55, MS, 57; Australian Nat Univ, PhD(microbiol), 62. *Prof Exp:* Virologist, NZ Dept Agr, 58-59; Fulbright scholar, Dept Epidemiol, Sch Pub Health, Univ Mich, Ann Arbor, 62-63; res fel microbiol, John Curtin Med Sch, Australian Nat Univ, 64-66, fel, 66-67; assoc mem, Lab Immunol, St Jude Children's Res Hosp, 68-69; assoc prof, 68-74, CLIN PROF MICROBIOL, UNIV TENN CTR HEALTH SCI MEMPHIS, 74-; MEM LABS VIROL & IMMUNOL, ST JUDE CHILDREN'S RES HOSP, 69- *Concurrent Pos:* Coordr, US-USSR Joint Comt Health Coop, Ecol of Human Influenza & Animal Influenza Rels to Human Infection, 74- *Mem:* Am Soc Microbiol; Am Asn Immunologists. *Res:* Structure and immunology of influenza viruses. *Mailing Add:* Dept Virol St Judes Childrens Hospital 322 N Lauderdale Memphis TN 38101

WEBSTER, ROBERT K, b Solomonville, Ariz, Jan 15, 38; m 59; c 3. PHYTOPATHOLOGY. *Educ:* Utah State Univ, BS, 61; Univ Calif, Davis, PhD(plant path), 66. *Prof Exp:* Res assoc plant path, NC State Univ, 66; from asst prof to assoc prof, 66-75, PROF PLANT PATH, UNIV CALIF, DAVIS, 75- *Mem:* Mycol Soc Am; Am Phytopathological Soc; Am Soc Naturalists. *Res:* Genetics of plant pathogenic fungi; field crop diseases. *Mailing Add:* Dept Plant Path Univ Calif Davis CA 95616

WEBSTER, TERRY R, b Hamilton, Ohio, Feb 10, 38; m 64; c 1. BOTANY, PLANT MORPHOLOGY. *Educ:* Miami Univ, BA, 60; Univ Sask, MA, 62, PhD(bot), 65. *Prof Exp:* Asst prof bot, 65-71, ASSOC PROF BIOL, UNIV CONN, 71- *Mem:* Am Fern Soc (secy, 73-76); 73-78); Soc Am. *Res:* Morphology of the genus Selaginella. *Mailing Add:* Dept Ecol Univ Conn U-43 75 N Eagleville Storrs CT 06268

WEBSTER, THOMAS G, b Topeka, Kans, Jan 23, 24; m 48; c 3. PSYCHIATRY, ACADEMIC ADMINISTRATION. *Educ:* Ft Hays Kans State Col, AB, 46; Wayne State Univ, MD, 49. *Prof Exp:* Commonwealth fel, Col Ment Health, Mass Inst Technol, 54-56; NIMH career teacher grant, Harvard Med Sch, 56-58, instr psychiat, 59-63; training specialist, Psychiat Training Br, NIMH, 63-66, chief, Continuing Educ Br, 66-72; prof psychiat & behav sci & chmn dept, 72-75, prof child develop & health, 75-80, prof, 75-86, EMER PROF PSYCHIAT & CHILD HEALTH & DEVELOP, GEORGE WASHINGTON UNIV, 86- *Concurrent Pos:* Dir, Presch Retard Children's Prog Greater Boston, 58-62; vis prof phychiat, Harvard Med Sch, 80-82. *Mem:* AAAS; Am Psychiat Asn; Am Col Psychiat; Am Acad Child Psychiat; Group Advan Psychiat. *Res:* Psychiatric education; child development; psychopathology; health policy, international; program evaluation in fields of manpower and training. *Mailing Add:* George Washington Univ Med Ctr 2112 F St NW Washington DC 20037

WEBSTER, WILLIAM JOHN, JR, b New York, NY, May 3, 43; m 80; c 1. PLANETARY SCIENCES. *Educ:* Univ Rochester, BS, 65; Case Western Reserve Univ, PhD(astron), 70. *Prof Exp:* Res assoc astron, Nat Radio Astron Observ, Va, 68-70; Nat Acad Sci-Nat Res Coun resident res assoc, Solar Physics Lab, 70-71, staff scientist, Meteorol & Earth Sci Lab, 71-74, STAFF SCIENTIST, GEOL & GEOMAGNETISM BR, LAB TERRESTRIAL PHYSICS, GODDARD SPACE FLIGHT CTR, 74- *Honors & Awards:* Pub Serv Award, NASA, 79, Group Achievement Award, 90. *Mem:* Am Astron Soc; Am Inst Aeronaut & Astronaut. *Res:* Radio interferometry; gaseous nebulae; radio spectroscopy; continuum mapping of radio sources; microwave observation of earth; planetary radio astronomy; asteroids; comets. *Mailing Add:* Geophys Br Code 622 Goddard Space Flight Ctr Greenbelt MD 20771

WEBSTER, WILLIAM MERLE, b Warsaw, NY, June 13, 25; m 47; c 2. PHYSICS. *Educ:* Union Col, NY, BS, 45; Princeton Univ, PhD(elec eng), 54. *Prof Exp:* Res engr, Labs, 46-54, mgr adv develop, RCA Semiconductor & Mat Div, 54-59, dir, Electronic Res Lab, RCA, 59-66, staff vpres mat & devices res, 66-68, vpres labs, 68-85, VPRES & SR TECH ADV, RCA LABS, RCA CORP, 85- *Honors & Awards:* Fredrik Philips Award, Inst Elec & Electronics Engrs, 80. *Mem:* Nat Acad Eng; fel Inst Elec & Electronics Engrs. *Res:* Solid state and gaseous electronics; electron physics. *Mailing Add:* 77 Cleveland Lane Princeton NJ 08540

WEBSTER, WILLIAM PHILLIP, b Mt Airy, NC, Apr 26, 30; m 52; c 4. DENTISTRY. *Educ:* Univ NC, Chapel Hill, BS, 56, DDS, 59, MS, 68. *Prof Exp:* Instr prosthodont, 59-61, asst prof periodont & oral path, 61-65, asst prof periodont, oral path & path, 65-67, trainee, 67-68, assoc prof path, 69-71, assoc prof dent ecol, 69-73, PROF PATH & DENT ECOL, SCHS MED & DENT, UNIV NC, CHAPEL HILL, 72-; CHIEF, DIV ORAL MED, HOSP DENT SERV, NC MEM HOSP, 68- *Concurrent Pos:* Res assoc, Sch Med, Univ NC, Chapel Hill, 59-65; mem med adv bd, Nat Hemophilia Found,

62-72; mem med adv comt, Vet Admin Hosp, Fayetteville, NC & sci adv comt, World Fedn Hemophilia. *Mem:* Am Acad Forensic Sci; Am Dent Asn; Am Soc Exp Path; NY Acad Sci; Int Soc Thrombosis & Haemostasis. *Res:* Transplantation; biology; blood coagulation; pathology; oral medicine; oral oncology; hematology. *Mailing Add:* Rte 1 Box 313 Pittsboro NC 27312

WECHSLER, MARTIN T, b New York, NY, July 27, 21; m 53; c 3. MATHEMATICS. *Educ:* Queen's Col, NY, BS, 42; Univ Mich, MA, 46, PhD(math), 52. *Prof Exp:* Physicist, Nat Bur Standards, 42-45; instr math, Wayne State Univ, 51-52 & Princeton Univ, 52-53; from instr to asst prof, Wash State Univ, 53-56; from inst to prof, 56-75, chmn dept, 68-75, assoc dean, Col Liberal Arts, 75-83, PROF MATH, WAYNE STATE UNIV, 83- *Mem:* Am Math Soc. *Res:* Topology; groups of homeomorphisms. *Mailing Add:* Math 1085 Fac/Admin Bldg Wayne State Univ Detroit MI 48202

WECHSLER, MONROE S(TANLEY), b New York, NY, May 1, 23; m 50; c 3. MATERIALS SCIENCE, SOLID STATE PHYSICS. *Educ:* City Col New York, BS, 44; Columbia Univ, MA, 50, PhD(physics), 53. *Prof Exp:* Elec engr, USN Air Magnetics Lab, 47-48; mem sci staff, Columbia Univ, 51-54; physicist, Oak Ridge Nat Lab, 54-69; prof metall, Univ Tenn, 65-69; chmn, Dept Metall, Iowa State Univ, 70-75, chief, Metall Div, Ames Lab, USAEC, 70-75, sr metallurgist, Ames Lab, US Dept Energy, 70-75, PROF, DEPT MAT SCI & ENG, IOWA STATE UNIV, 70-, PROF NUCLEAR ENG, 90- *Concurrent Pos:* Ed, Nuclear Eng Mat Handbk, 77- *Mem:* Fel Am Phys Soc; Am Inst Mining, Metall & Petrol Engrs; Am Soc Metals; Am Nuclear Soc; Am Soc Eng Educ; Sigma Xi. *Res:* Irradiation effects on metals; materials for nuclear reactors; phase transformations; shape-memory alloy heat engines. *Mailing Add:* 3602 Woodland Ames Iowa City IA 50010

WECHSLER, STEVEN LEWIS, b Bronx, NY, May 30, 48; m 74; c 1. MOLECULAR GENETICS, VIROLOGY. *Educ:* City Col New York, BS, 70; Univ NC, Chapel Hill, PhD(molecular genetics), 75. *Prof Exp:* RES FEL MEASLES VIRUS, DEPT MICROBIOL, HARVARD MED SCH, 75- *Mem:* Am Soc Microbiol. *Res:* Molecular genetics of measles virus and its relationship to persistent infections and chronic diseases such as subacute sclerosing panencephalites and multiple sclerosis. *Mailing Add:* Ophthal Res-Cedars Sinai Med Ctr 8700 Beverly Blvd Halper Bldg Suite 111 Los Angeles CA 90048

WECHTER, MARGARET ANN, b Chicago, Ill, Sept 30, 35. ANALYTICAL CHEMISTRY, RADIOCHEMISTRY. *Educ:* Mundelein Col, BS, 62; Iowa State Univ, PhD(anal chem), 67. *Prof Exp:* Fel chem, Purdue Univ, 67-68, asst prof chem, Calumet Campus, 69-73; ASSOC PROF CHEM, SOUTHEASTERN MASS UNIV, 73-; MEM FAC, DEPT CHEM, SOUTHEASTERN MASS UNIV, 80- *Concurrent Pos:* Mem staff, Ames Lab, Iowa State Univ, 70-80. *Mem:* Am Chem Soc; Sigma Xi. *Res:* Development, analytical applications and electrochemistry of the tungsten bronze electrodes; activation analysis. *Mailing Add:* Dept Chem Southeastern Mass Univ North Dartmouth MA 02747

WECHTER, WILLIAM JULIUS, b Louisville, Ky, Feb 13, 32; m 52, 82; c 3. PHARMACOLOGIC MAINTENANCE OF SKELETAL BONE. *Educ:* Univ Ill, AB, 53, MS, 54; Univ Calif, Los Angeles, PhD, 57. *Prof Exp:* Asst chem, Univ Ill, 54; asst org chem, Univ Calif, Los Angeles, 55-56; res assoc, Dept Chem, Upjohn Co, 57-68, res head, 68-79, res mgr hypersensitivity dis, 79-84; dir res & develop planning, Boehringer Ingelheim Bd, 84; dir, Clin & Pharmaceut Res, Boots Pharmaceut, Inc, 85-88; RES PROF MED, SCH MED, LOMA LINDA UNIV, 88- *Concurrent Pos:* Vis scholar, Depts Chem & Med Microbiol, Stanford Univ, 67-68; vchmn, Gordon Conf Med Chem, 72, chmn, 73; adj prof biochem, Kalamazoo Col, 74-84; vis lectr path, Harvard Med Sch, 77-78; adj prof pharmacol & therapeut, La State Univ Med Ctr, Shreveport, 85-88. *Honors & Awards:* Upjohn Prize Award, 75; W Heinlein Hall Lectureship, 78. *Mem:* AAAS; Am Chem Soc; Transplantation Soc; Am Asn Immunologists; Am Rheumatology Asn; fel Am Col Clin Pharmacol; Royal Soc Med. *Res:* Isolation, structure and synthesis of hormones; chiral drugs; pharmacological maintenance of skeletal bone; clinical pharmacology, inflammation. *Mailing Add:* Loma Linda Univ Med Ctr Loma Linda CA 92350

WECK, FRIEDRICH JOSEF, b Puettlingen, Ger, Nov 10, 18; nat US; m 56; c 2. ORGANIC CHEMISTRY, PHYSICAL CHEMISTRY. *Educ:* Univ Bonn, MS, 50; Univ Saarland, PhD(natural sci), 54. *Prof Exp:* Assistantship, Univ Saarland, 51-52; res assoc, Nat Ctr Sci Res, France, 52-55; head res & develop lab, Saar Water Asn, Ger, 55-57; from group leader new prod & processes sect, to res specialist, Am Potash & Chem Corp, 57-64; dir chem res, Garett Res & Develop Co, 64; pres, F J Weck Co, 64-81, PRES, WECK LABS INC, 81- *Concurrent Pos:* Consult, Union Saarbruecken, 54-57. *Honors & Awards:* Bronze Medal, Chem Soc France, 57. *Mem:* Am Chem Soc; Ger Chem Soc; fel Am Inst Chem. *Res:* Organic-inorganic synthesis; mono-polymers; photosensitive materials; environmental safety; food and water; liquid and solid ion exchangers; saline mineral chemistry; extraction-crystallization of salts; product and process development and testing; instrumental analyses for industry and government. *Mailing Add:* Weck Labs Inc Contract Res 14859 E Clark Ave City of Industry CA 91745-1308

WECKER, LYNN, b New York, NY, Sept 27, 47. NEUROPHARMACOLOGY, ACETYLCHOLINE METABOLISM. *Educ:* State Univ NY, BS, 69; Univ Fla, PhD(pharmacol), 72. *Prof Exp:* Postdoctoral fel, Vanderbilt Univ Sch Med, 73-74, instr pharmacol, 74-75, asst prof, 76-78; asst prof med chem & pharm, Col Pharm, Northeastern Univ, 75-76; from asst prof to prof pharmacol, La State Univ Med Ctr, 78-90; PROF PHARMACOL, COL MED, UNIV SFLA, 90-, CHAIR, 91- *Concurrent Pos:* Mem coun, Am Soc Neurochem. *Mem:* AAAS; Am Inst Nutrit; Am Soc Pharmacol & Exp Therapeut; NY Acad Sci; Sigma Xi; Int Soc Neurochem. *Res:* Regulation of the synthesis and release of acetylcholine by brain and the enhancement of these regulatory processes by pharmacological and nutritional manipulations; neurochemical effects of chronic nicotine administration and plasticity of neuronal nicotinic and GABAergic receptors; neurochemical changes in the aged brain. *Mailing Add:* 12901 Bruce B Downs Blvd Box 9 Tampa FL 33612-4799

WECKER, STANLEY C, b New York, NY, Apr 29, 33; m 64. VERTEBRATE ECOLOGY, APPLIED ECOLOGY. *Educ:* City Col New York, BS, 55; Univ Mich, MS, 57, PhD(zool), 62. *Prof Exp:* Asst prof biol, Hofstra Univ, 62-63; from instr to asst prof, 63-70, assoc prof, 70-82, PROF BIOL, CITY COL NEW YORK, 82-, DIR, ENERGY, ECOL, ENVIRON PROG, 76- *Concurrent Pos:* Res grants, Sigma Xi, 63 & NSF, 64-68; US Dept HEW Off Educ grant & staff ecologist, Environ Educ Proj, North Westchester-Putnam Coop Educ Serv, NY, 71-72; spec consult, NY State Educ Dept, 72; mem, Environ Defense Fund, NASA, 75-77. *Honors & Awards:* Award, Am Soc Mammal, 61. *Mem:* Fel AAAS; Am Inst Biol Sci; Ecol Soc Am; Am Soc Mammal; NY Acad Sci; Sigma Xi. *Res:* Behavioral ecology of deer mice, especially habitat orientation; application of satellite data to coastal zone management. *Mailing Add:* Dept Biol City Col New York New York NY 10031

WECKESSER, LOUIS BENJAMIN, b Baltimore, MD, Feb 22, 28; m 49; c 3. HEAT TRANSFER, MECHANICAL ENGINEERING. *Educ:* Univ MD, BS, 52, MS, 56. *Prof Exp:* GROUP SUPVR ENG APPL PHYSICS LAB, JOHNS HOPKINS UNIV, 52- *Mem:* Am Inst Aeronaut & Astronaut. *Res:* Tactical missile thermal design; thermal insulation development; missile radome design and test. *Mailing Add:* Appl Physics Lab Johns Hopkins Rd Laurel MD 20810

WECKLER, GENE PETER, b San Francisco, Calif, July 3, 32; m 56; c 4. ELECTRICAL ENGINEERING. *Educ:* Utah State Univ, BS, 58; San Jose State Col, MS, 64; Stanford Univ, DEng, 68. *Prof Exp:* Jr engr, Convair Astronaut, 58-59; elec engr, Shockley Transistor Corp. 59-62; develop engr, Opto-Electronic Devices, Inc, 62-63; mem tech staff, Fairchild Semiconductor Div, 63-71; DIR ENG, RETICON CORP, 71- *Mem:* Inst Elec & Electronics Engrs; Int Soc Optical Eng. *Res:* Development of silicon p-n junction photodetectors, especially integrated arrays for image sensing; development and application of solid state devices for signal processing. *Mailing Add:* Eg&G Solid State Prod Group 345 Potrero Ave Sunnyvale CA 94086

WECKOWICZ, THADDEUS EUGENE, b Iskorst, USSR, Oct 10, 18; Can citizen; m 66. PSYCHIATRY. *Educ:* Univ Edinburgh, MB & ChB, 45; Univ Leeds, DPM, 52; Univ Sask, PhD(psychol), 62. *Prof Exp:* Registr psychiat, Univ Leeds, 53-54, sr registr, Bolton Group Hosps, Eng, 54-55; sr resident, Hosp, Univ Sask, 55-56; res psychiatrist, Weyburn Ment Hosp, Sask, 56-59; res assoc psychiat & sessional lectr psychol, 62, from asst prof to assoc prof, 62-74, prof psychiat & psychol. Univ Alta, 74-84, mem staff, Ctr Advan Study Theiret Psychol, 66-84; RETIRED. *Concurrent Pos:* Fel, Hosp, Univ Sask, 55-56. *Mem:* Can Psychiat Asn. *Res:* Psychological and biological aspects of schizophrenia; psychotropic drug research, particularly hallucinogenic drugs; psychopharmacology; multivariate study of depression; studies of learned helplessness retardation in depression. *Mailing Add:* Dept Psychiat Univ Alta Edmonton AB T6G 2E4 Can

WECKSUNG, GEORGE WILLIAM, b Muscatine, Iowa, Oct 31, 31; m 61; c 3. COMPUTER IMAGE PROCESSING, APPLIED MATHEMATICS. *Educ:* Univ Iowa, BA, 58; Calif State Univ, Long Beach, MA, 63. *Prof Exp:* Comput engr inertial navig, Autonetics, 61-64; sci specialist digital signal processing EG&G, Inc, Las Vegas, 64-69; res engr inertial navig, Teledyne Systs Co, 69-70; sci specialist comput image processing EG&G, Inc, Los Alamos, 70-74; STAFF MEM, COMPUT IMAGE PROCESSING, LOS ALAMOS NAT LAB, 74- *Mem:* Soc Photo-optical Instrumentation Engrs. *Res:* Remote sensing; computer image processing; stereo imagery and photogrammetry; digital holography; computed tomography. *Mailing Add:* 161 El Corto St Los Alamos NM 87544

WECKWERTH, VERNON ERVIN, b Herman, Minn, Apr 29, 31; m 55; c 5. BIOSTATISTICS, HOSPITAL ADMINISTRATION. *Educ:* Univ Minn, BS, 54, MS, 56, PhD(biostatist), 63. *Prof Exp:* Teaching asst biostatist, Univ Minn, 54-56, instr, 56-58; head res & statist, Am Hosp Asn, 58-60; from lectr to assoc prof, 60-68, PROF HOSP & HEALTH CARE ADMIN, COORDR CONTINUING HOSP EDUC & PROF FAMILY PRACT, MED SCH, UNIV MINN, MINNEAPOLIS, 68- *Concurrent Pos:* Assoc dir, Hosp Res & Educ Trust, 58-60; mem adv comt to Nat Ctr for Health Servs Res & Develop, 69- *Mem:* Am Statist Asn; Biomet Soc; Am Hosp Asn; Am Pub Health Asn; Sigma Xi. *Res:* Continuing education of health care workers including professionals; health care delivery systems research; teaching research and statistics in health care. *Mailing Add:* 1666 Coffman St Unit 334 St Paul MN 55108-1326

WEDBERG, STANLEY EDWARD, b Bridgeport, Conn, Aug 28, 13; m 41; c 2. MICROBIOLOGY. *Educ:* Univ Conn, BS, 37; Yale Univ, PhD(bact), 40. *Prof Exp:* Instr immunol, Sch Med, Yale Univ, 40-41; from instr to assoc prof bact, 41-59, head dept, 55-66, prof, 59-69, fac coordr educ TV, 67-69, prof biol, 68-69, EMER PROF BIOL, UNIV CONN, 69- *Concurrent Pos:* Consult bacteriologist, Windham Community Mem Hosp, Conn, 46-57; biologist, Southwestern Co!, Calif, 69-80; lectr, San Diego Paramedic Prog, 79- & San Diego Health Dept; mem, Conn Clean Water Task Force, vchmn, Conn Clean Air Task Force, Conn adv comt, Foods, Drugs, Cosmetics & Devices; co-producer & pcrformer, Survival, 68-69. *Mem:* AAAS; Sigma Xi; Am Soc Microbiol; Am Acad Microbiol. *Res:* Microbial thermogenesis; bacterial capsule staining; antihistamines on antibody production of rabbits; germicidal gases; author of three microbiology texts. *Mailing Add:* 3361 Ullman St Point Loma San Diego CA 92106

WEDDELL, DAVID S(TOVER), b Philadelphia, Pa, Mar 8, 17; m 43; c 4. CHEMICAL ENGINEERING. *Educ:* Pa State Col, BS, 38; Mass Inst Technol, ScD(chem eng), 41. *Prof Exp:* Chem engr, Monsanto Co, Ala, 41-44, 45-47, develop dir, Wash, 49-50, develop dept, Mo, 50-52, asst dir, 53-54, Europ tech rep, 54-60, asst dir develop, Overseas Div, 60-62, dir, 63-64, dir proj eval, 65-82, financial analyst, int div, 83-86; RETIRED. *Concurrent Pos:* Mem Nat Defense Res Comt, Ohio, 44-45, Mo, 47-48. *Mem:* Am Inst Chem Engrs. *Res:* Turbulent mixing in flames and its reproduction in liquid models; process design; application of phosphate chemicals; commercial chemical development; investment analysis and planning; international operations. *Mailing Add:* 21 Fair Oaks St Louis MO 63124

WEDDELL, GEORGE G(RAY), b Baltimore, Md, Jan 30, 23; m 56. CHEMICAL ENGINEERING. *Educ:* Pa State Univ, BS, 44. *Prof Exp:* Asst, SAM Labs, Columbia Univ, 44-45; asst, Carbide & Carbon Chem Co, NY, 45-46, chem engr, Tenn, 46-47; fel, Mellon Inst, 47-52; vpres, Geotic Industs, Inc, 52-53, pres, 54-55; asst mgr, O Hommel Co, 55-56; fel engr, Bettis Atomic Power Lab, Westinghouse Elec Corp, 56-66, supvr, 66-71, mgr, 71-89; RETIRED. *Mem:* Am Chem Soc; Am Nuclear Soc. *Res:* Utilization of fine particles; manufacture and application of perlite; radioactive waste disposal; thermal and hydraulic nuclear engineering; nuclear fuel cycle. *Mailing Add:* 349 Dale Rd Bethel Park PA 15122

WEDDELL, JAMES BLOUNT, b Evanston, Ill, Apr 29, 27; m 60. SPACE PHYSICS. *Educ:* Drew Univ, AB, 49; Northwestern Univ, MS, 51, PhD(physics), 53. *Prof Exp:* Asst physics, Northwestern Univ, 49-53; res engr, Westinghouse Elec Corp, 53-57; sr scientist, Martin-Marietta Co, 57-62; supvr, Rockwell Int Corp, 62-80, mgr, 80-85, chief scientist, 85-87, proj mgr, 87-90; RETIRED. *Mem:* Am Phys Soc; Sigma Xi. *Res:* Space shuttle payload and cargo integration; solar physics; magnetosphere; nuclear reactions; radiation shielding; solar-electric propulsion systems. *Mailing Add:* 936 S Peregrine Pl Anaheim CA 92806

WEDDING, BRENT (M), b Walnut, Ill, May 3, 36; m 69; c 2. EXPERIMENTAL PHYSICS. *Educ:* Hamilton Col, AB, 58; Univ Ill, Urbana, MS, 61, PhD(physics), 67. *Prof Exp:* Lectr physics, Southern Ill Univ, Carbondale, 62-63; res supvr, 75-76, SR PHYSICIST, CORNING GLASS WORKS, 67-, DEVELOP ASSOC PHYSICS, 76-, MGR PHYS PROPERTIES RES, 79- *Mem:* Am Phys Soc; Optical Soc Am; Am Ceramic Soc; Sigma Xi. *Res:* Optical glass development. *Mailing Add:* Corning Glass Works Corning NY 14830

WEDDING, RANDOLPH TOWNSEND, b St Petersburg, Fla, Nov 6, 21; m 43; c 2. BIOCHEMISTRY. *Educ:* Univ Fla, BSA, 43, MS, 47; Cornell Univ, PhD(plant physiol), 50. *Prof Exp:* Tech asst, Agr Exp Sta, Univ Fla, 41-43, 46-47; teaching asst bot, Cornell Univ, 47-49, instr plant physiol, 49-50; jr plant physiologist, 50-52, asst plant physiologist, 52-58, assoc plant physiologist & lectr, 58-61, assoc prof biochem & assoc biochemist, 61-63, chmn dept, 66-75, PROF BIOCHEM & BIOCHEMIST, UNIV CALIF, RIVERSIDE, 63- *Concurrent Pos:* Agr Res Coun fel, Oxford Univ, 59-60, Orgn Econ Coop & Develop sr sci fel, 64; sr sci fel, Natural Environ Res Coun, 71-72. *Mem:* Am Soc Plant Physiol; Am Soc Biol Chem; Brit Soc Exp Biol; Brit Biochem Soc; Scand Soc Plant Physiol; Japanese Soc Plant Physiol. *Res:* Metabolic control; mechanisms of enzyme action and regulation. *Mailing Add:* Dept Biochem Univ Calif Riverside CA 92521

WEDDLETON, RICHARD FRANCIS, b Boston, Mass, Oct 10, 39; m 62; c 3. POLYMER CHEMISTRY. *Educ:* Mass Inst Technol, BS, 61; Ind Univ, Bloomington, MS, 63, PhD(org chem), 65. *Prof Exp:* Chemist, Mat & Processes Lab, Gen Elec Co, 65-73; mgr insulation eng, Nat Elec Coil Div, McGraw-Edison Co, 73-77; sr engr, 77-80, ASST TO GEN MGR, WESTINGHOUSE ELEC CORP, 80- *Mem:* Am Chem Soc; Inst Elec & Electronics Eng. *Res:* Project management, development and testing of insulation system for use in electrical rotating machinery. *Mailing Add:* Westinghouse Elec Corp Mc 100 The Quadrangle 4400 Alafaya Trail Orlando FL 32826-2399

WEDEEN, RICHARD P, b New York, NY, Jan 19, 34; m 57; c 1. NEPHROLOGY, ENVIRONMENTAL TOXINS. *Educ:* NY Univ, MD, 59. *Prof Exp:* Prof prev med, dir, Div Occup & Environ Med & prof med, NJ Med Sch, 76-78; clin prof interdisciplinary med, Univ Med & Dent NJ, Grad Sch Health Related Prof, 90; ASSOC CHIEF STAFF RES & DEVELOP, VET ADMIN MED CTR, 78- *Concurrent Pos:* Vis prof med, Univ Antwerp, Belg, 85; consult ed, Am J Indust Med; contrib ed, Arch Environ Health; asst ed, Mt Sinai J Med; actg chief staff, Vet Admin Med Ctr, 88. *Mem:* Emer mem Am Soc Clin Invest; fel Am Col Physicians; Harvey Soc; Sigma Xi; Am Physiol Soc; Am Soc Nephrology. *Res:* The role of lead as a cause of renal disease in the occupational setting and as a cause of hypertension and gout with renal disfunction; animal research demonstrating pathogenic mechanisms of interstitial nephritis including crystal nephropathy; history of occupational medicine; author of one book. *Mailing Add:* Vet Admin Med Ctr East Orange NJ 07019

WEDEGAERTNER, DONALD K, b Kingsburg, Calif, Sept 9, 36; m 58; c 2. ORGANIC CHEMISTRY. *Educ:* Univ Calif, Berkeley, BS, 58; Univ Ill, Urbana, PhD(org chem), 62. *Prof Exp:* Res asst org chem, Iowa State Univ, 62-63; from asst prof to assoc prof, 63-71, chmn dept chem, 74-80, PROF ORG CHEM, UNIV PAC, 71- *Concurrent Pos:* Res asst org chem, Univ Rochester, 69-70, vis prof, 78-79; vis prof, Univ Utah, 83-84. *Mem:* Am Chem Soc. *Res:* Reaction mechanisms. *Mailing Add:* Dept Chem Univ of the Pac Stockton CA 95211

WEDEKIND, GILBERT LEROY, b Zion, Ill, Feb 28, 33; m 53; c 2. MECHANICAL ENGINEERING. *Educ:* Univ Ill, BS, 59, MS, 61, PhD(mech eng), 65. *Prof Exp:* Res assoc mech eng, Univ Ill, 65-66; from asst prof to assoc prof, 66-77, PROF ENG, OAKLAND UNIV, 77- *Concurrent Pos:* NSF eng res grants, 67-69, 72-74, 80-82 & 84-86; consult, Propulsion Systs Lab, US Army Tank-Automotive Command, 67-70, Climate Control Opers, Ford Motor Co, 71-73 & Mfg Develop, Gen Motors Tech Ctr, 73-80. *Mem:* Am Soc Mech Engrs; Am Soc Eng Educ; Sigma Xi. *Res:* Thermal modeling of spacecraft; experimental and theoretical study of transient two-phase evaporating and condensing flow phenomena; modeling components processes and systems which utilize fluid and thermal phenomena for their operation. *Mailing Add:* Sch Eng Oakland Univ Rochester MI 48309

WEDEL, ARNOLD MARION, b Lawrence, Kans, Jan 31, 28; m 54; c 3. MATHEMATICS. *Educ:* Bethel Col, Kans, AB, 47; Univ Kans, MA, 48; Iowa State Col, PhD(math), 51. *Prof Exp:* PROF MATH, BETHEL COL, KANS, 51- *Mem:* Math Asn Am; Am Math Soc. *Res:* Hypergeometric series and volterra transforms. *Mailing Add:* Dept Math Bethel Col North Newton KS 67117

WEDEMEYER, GARY ALVIN, b Fromberg, Mont, Oct 15, 35; m 57; c 2. FISHERIES, BIOCHEMISTRY. *Educ:* Univ Wash, BS, 57, MS, 63, PhD(fisheries), 65. *Prof Exp:* BIOCHEMIST, FISH PHYSIOL, NAT FISHERIES RES CTR, US DEPT INTERIOR, 65- *Concurrent Pos:* Affil prof, Col Fisheries, Univ Wash, 65- *Mem:* Am Inst Fishery Res Biol; Am Fisheries Soc. *Res:* Biochemistry and physiology of fishes; pollution, disease, aquaculture; tolerance of fish and fish populations to environmental stress. *Mailing Add:* Nat Fisheries Res Ctr Dept Interior Bldg 204 Naval Sta Puget Sound Seattle WA 98115

WEDGWOOD, RALPH JOSIAH PATRICK, b Eng, May 25, 24; nat US; m 43; c 3. PEDIATRICS, RHEUMATOLOGY. *Educ:* Harvard Med Sch, MD, 47; Am Bd Pediat, dipl, 55. *Prof Exp:* Res fel pediat, Harvard Med Sch, 49-51; sr instr pediat & biochem, Western Reserve Univ, 53-57, asst prof pediat & prev med, Sch Med, 57-62; assoc prof, 62-63, chmn dept, 62-72, PROF PEDIAT, SCH MED, UNIV WASH, 63- *Concurrent Pos:* Spec consult & mem gen clin res ctr comt, NIH, 62-66, mem nat adv res resources coun, 66-70; mem sci adv comn, Nat Found, 69-; mem sci adv bd, St Jude's Children's Res Hosp, Memphis, 70-76; Markle scholar, 60-65, vis scholar, St John's Col, Cambridge UK, 69-70, Rockefeller Study Ctr, Bellagio, Italy, 85. *Mem:* Am Asn Immunologists; Am Rheumatism Asn; Heberden Soc; Am Pediat Soc; Infectious Dis Soc Am. *Res:* Immunobiology; natural resistance factors; infectious and rheumatic diseases in children; general pediatrics. *Mailing Add:* Dept Pediat RD-20 Univ Wash Sch Med Seattle WA 98195

WEDIN, WALTER F, b Frederic, Wis, Nov 28, 25; m 55; c 2. AGRONOMY. *Educ:* Univ Wis, BS, 50, MS, 51, PhD(agron, soils), 53. *Prof Exp:* Wis Alumni Res Found assistantship agron, Univ Wis, 50-53, proj assoc hort, 53, asst prof, 53-57; res agronomist, Forage & Range Res Br, Crops Res Div, Agr Res Serv, USDA, Minn, 57-61; from assoc prof to prof agron, 61-90, dir, World Food Inst, 73-77, EMER PROF AGRON, IOWA STATE UNIV, 91-; ADJ PROF, DEPT AGRON & PLANT GENETICS, UNIV MINN, 91- *Concurrent Pos:* Asst prof, Inst Agr, Univ Minn, 59-61; consult, US Agency Int Develop-Iowa State Univ, Uruguay, 64; consult, Coop States Res Serv, USDA, 64, 70, 72 & 74, prin agronomist, 68. *Mem:* AAAS; fel Am Soc Agron; Crop Sci Soc Am; Am Soc Animal Sci; Am Forage & Grassland Coun (pres, 85); Coun Agr Sci & Technol; Soil & Water Conserv Soc; Nature Conserv. *Res:* Evaluation of forage crops, especially utilization as pasture, hay or silage; techniques involved in measurement of forage nutritive value; animal intake and digestibility; grassland development, international. *Mailing Add:* Dept Agron & Plant Genetics Univ Minn St Paul MN 55108

WEDLER, FREDERICK CHARLES OLIVER, JR, b Philadelphia, Pa, June 10, 41; m 67; c 3. BIOCHEMISTRY. *Educ:* Univ NC, Chapel Hill, BS, 63; Northwestern Univ, Evanston, PhD(biochem), 68. *Prof Exp:* Molecular Biol Inst fel biochem, Univ Calif, Los Angeles, 68-70; from asst prof to assoc prof chem & biochem, Rensselaer Polytech Inst, 70-78; assoc prof, 79-85, PROF MOLECULAR & CELL BIOL, PA STATE UNIV, UNIVERSITY PARK, 85- *Concurrent Pos:* Petrol Res Fund grant, Rensselaer Polytech Inst, 71-, NSF res grant, 72-81, 89-; NASA res grant, 74-79, NIH, 77-87; res chemist, NIH, 77-78; NIH res grants, 78-82; fac res award, Am Cancer Soc, 80-84; vis scientist, Ctr Neurochem, Strasbourg, France. *Mem:* AAAS; Am Chem Soc; Am Soc Biochem & Molecular Biol. *Res:* Enzymology; kinetics and mechanisms of action of key regulatory enzymes; development of new probe systems for protein structures; thermophilic proteins; neurochemistry. *Mailing Add:* Dept Molecular & Cell Biol Althouse Lab Pa State Univ University Park PA 16802

WEDLICK, HAROLD LEE, b Detroit, Mich, Feb 26, 36; m 59; c 2. HEALTH PHYSICS, ENVIRONMENTAL HEALTH. *Educ:* Wayne State Univ, BS, 61, MS, 63; Am Bd Health Physics, cert, 74. *Prof Exp:* Chemist, City Detroit Health Dept, Mich, 61-63; res assoc environ health, Univ Mich, 64-66; assoc prof radiol sci, Lowell Technol Inst, 66-75; res assoc, Occup & Environ Safety Dept, Pac Northwest Labs, Battelle Mem Inst, 75-80; owner & dir, Wedlick's Educ Develop Systs Safety, 80-86; PRES, WESTINGHOUSE, HANFORD, 86- *Mem:* Health Physics Soc; Conf Radiol Health; Am Nuclear Soc. *Res:* Transfer mechanisms and measurements of radionuclides in the environment. *Mailing Add:* 411 26th Pl Kennewick WA 99336

WEDLOCK, BRUCE D(ANIELS), b Providence, RI, Mar 20, 34; m 58; c 2. SOLID STATE PHYSICS. *Educ:* Mass Inst Technol, SB & SM, 58, ScD(elec eng), 62. *Prof Exp:* From instr to assoc prof elec eng, Mass Inst Technol, 60-71; staff scientist, Block Eng, Inc, Mass, 71-72; DIR, LOWELL INST SCH, 73- *Concurrent Pos:* Ford Found fel, 62-64. *Mem:* Fel Inst Elec & Electronics Engrs; Am Soc Eng Educ. *Res:* Physical electronics of solid state devices; development of two-year technical school curricula in emerging technologies. *Mailing Add:* Mass Inst Technol Rm E32-105 Cambridge MA 02139

WEDMAN, ELWOOD EDWARD, b Harper, Kans, Aug 22, 22; m 46; c 3. VETERINARY MICROBIOLOGY. *Educ:* Kans State Univ, DVM, 46; Univ Minn, MPH, 57, PhD(microbiol), 64. *Prof Exp:* Pvt pract, 45-47, 49-50; vet in charge Mex-US Campaign Eradication Foot & Mouth Dis, Mex, 47-49; vet in charge Wichita Union Stockyards, Animal Dis Eradication Div, Agr Res Serv, USDA, DC, 52-54, staff vet, Lab Serv, 58-61, res assoc, Univ Minn & USDA, 54-58, chief vet diag serv, Animal Dis Labs, USDA, Iowa, 61-64; assoc dir vet med res inst, Col Vet Med, Iowa State Univ, 64-72; dean, Sch Vet Med, Ore State Univ, 72-86. *Concurrent Pos:* Consult, USDA, 63-64, 65-66 & 67 & Pan-Am Health Orgn, 65, 67; mem, Nat Conf Vet Lab Diagnosticians. *Mem:* Am Vet Med Asn; US Animal Health Asn; Am Col Vet Microbiol; Conf Res Workers Animal Dis. *Res:* Microbiology and epidemiology of animal diseases. *Mailing Add:* 8100 NW Ridge Dr Corvallis OR 97330

WEE, WILLIAM GO, b Iloilo City, Philippines, Sept 9, 37; m 67; c 2. INTELLIGENT SYSTEMS. *Educ:* Mapua Inst Technol, BSEE, 62; Purdue Univ, Lafayette, MS, 65, PhD(elec eng), 67. *Prof Exp:* Prin res engr, Honeywell Systs & Res Ctr, 67-71; assoc prof, 71-77, PROF ELEC &

COMPUTER ENG, UNIV CINCINNATI, 77- *Mem:* Sr mem Inst Elec & Electronics Engrs. *Res:* Theory and application of pattern recognition; adaptive and learning systems; artificial neural networks; computer vision; artificial intelligence. *Mailing Add:* Elec & Computer Eng Dept Cincinnati OH 45221

WEED, ELAINE GREENING AMES, b Detroit, Mich, Mar 15, 32; m 53; c 3. GEOLOGY. *Educ:* Mt Holyoke Col, BA, 53. *Prof Exp:* GEOLOGIST, US GEOL SURV, 57- *Mem:* Geol Soc Am; Am Asn Petrol Geologists; AAAS. *Res:* Geology and resources of United States Atlantic Margin; subsurface geology of southeastern Massachusetts; impacts of resource development. *Mailing Add:* 421 S Van Ness Ave Apt 44 Los Angeles CA 90020

WEED, GRANT B(ARG), b Salt Lake City, Utah, Aug 19, 35; m 64; c 4. CERAMICS ENGINEERING, METALLURGY. *Educ:* Univ Utah, BS, 61, PhD(ceramic eng), 67. *Prof Exp:* Source engr, Hercules Powder Co, Utah, 60-61; res chemist, Exp Sta, E I du Pont de Nemours & Co, 66-72; res engr, Cent Res Labs, NL Industs, 72-73, supvr mat equip develop, Tech Dept, Magnesium Div, 73-80; AT MAGNESIUM DIV, AMAX SPECIALITY METALS CORP, 80- *Mem:* Am Ceramic Soc; Nat Inst Ceramic Engrs. *Res:* High temperature ceramics and ceramic-metal composites; refractories for molten salt containment. *Mailing Add:* 1011 E Millbrook Way Bountiful UT 84054

WEED, H(OMER) C(LYDE), b Sun City, Kans, Mar 30, 20; m 58. PHYSICAL CHEMISTRY. *Educ:* Univ Ariz, BS, 42; Ohio State Univ, MS, 48, PhD(chem), 57. *Prof Exp:* Assoc develop chemist, Tenn Eastman Corp, 42-46; asst chem, Ohio State Univ, 46-47, 48, 49, res found, 47-48, 49, 53-57, metall, 50-53; chemist, Inorg Mat Div, 57-76, CHEMIST, EARTH SCI DIV, LAWRENCE LIVERMORE NAT LAB, UNIV CALIF, 76- *Mem:* AAAS; Am Chem Soc; Am Geophys Union. *Res:* High temperature chemistry; high pressure mechanical properties of solids and powders; radionuclide transport in earth materials by water; rheology of silicate melts; IR spectroscopy on silicates. *Mailing Add:* 805 Los Alamos Ave Livermore CA 94550

WEED, HERMAN ROSCOE, b Union City, Pa, Aug 5, 22; m 46; c 3. ELECTRICAL ENGINEERING, BIOMEDICAL ENGINEERING. *Educ:* Pa State Col, BS, 45; Ohio State Univ, MS, 48. *Prof Exp:* Instr elec eng, Pa State Col, 43-46; from instr to assoc prof, 46-59, PROF ELEC ENG, OHIO STATE UNIV, 59-, DIR, BIOMED ENG CTR, 71-, PROF PREVENTIVE MED, 78- *Concurrent Pos:* Mem staff, Westinghouse Elec Corp, 49; mem staff, Res Found, Ohio State Univ, 48-50, mem staff, Exp Sta, 52-53, consult, 51-; mem staff & consult, Robbins & Myers Co, 53-78; US deleg, Am Control Coun, Int Fedn Automatic Control Cong, Russia, 60, London, 66, chmn systs comt, Am Automatic Control Coun; consult, Solid-State Controls, Inc, 61-65 & Am Inst Biol Sci, 77-80; vis prof, Univ Cairo Egypt, Univ Karlsruhe, WGer, 79; hon prof, Zhejiang Med Univ, Hangzhou, People's Repub China, 86; dir, Biomed Eng, Proj Hope Worldwide, 79- *Honors & Awards:* Outstanding Biomed Engr Year, Am Soc Eng Educ, 87. *Mem:* Inst Elec & Electronics Engrs; Am Soc Eng Educ; Asn Advan Med Instrumentation. *Res:* Automatic control; bioelectrical instrumentation; electronics; systems; non-invasive diagnosis; ultrasound; physiological systems; biomedical-clinical engineering education; functional muscle stimulation for handicapped; real time visual information for visually handicapped. *Mailing Add:* Elect Eng Dept Ohio State Univ 2015 Neil Ave Columbus OH 43210

WEED, JOHN CONANT, b Lake Charles, La, July 7, 12; m 39; c 4. OBSTETRICS & GYNECOLOGY. *Educ:* Tulane Univ, BS, 33, MD, 36, MS, 40; Am Bd Obstet & Gynec, dipl, 46. *Prof Exp:* Instr anat, 39-41, from assoc clin prof to clin prof, 53-83, EMER PROF OBSTET & GYNEC, SCH MED, TULANE UNIV, 83- *Concurrent Pos:* Mem staff, Ochsner Clin, 45-; chmn dept obstet & gynec, Ochsner Found Hosp, 63-73; bd trustees, Alton Ochsner Med Found. *Mem:* AMA; Am Fertil Soc (pres, 73-74); Am Col Obstet & Gynec; Am Col Surgeons; Soc Reproductive Surg; Soc Gynec Surg. *Res:* Infertility; investigation of auto-immune factors in cases of moderate to severe endometriosis has revealed C'3 (complement) deposition in uterine endometrium, searching for antibody and serologic proof. *Mailing Add:* 1525 Dufossat St New Orleans LA 70115

WEED, LAWRENCE LEONARD, b Troy, NY, Dec 26, 23; m 52; c 5. COMPUTER SCIENCE, MEDICINE. *Educ:* Hamilton Col, BA, 45, Col Physicians & Surgeons, MD, 47. *Prof Exp:* Asst prof med & pharmacol, Sch Med, Yale Univ, 54-56; dir, Med Educ, Eastern Maine Gen Hosp, Bangor, 56-60; asst prof microbiol, Case Western Reserve Univ, 61-64, assoc prof, 64-69; PROF COMMUNITY MED, COL MED, UNIV VT, 69- *Concurrent Pos:* Prof med & dir, Outpatient Clin, Cleveland Metrop Gen Hosp, 64-69; dir, Promis Lab, 69-81, chief scientist, Promis Info Systs, Inc, 81-82. *Mem:* Am Col Physicians; Am Soc Microbiol. *Res:* Problem oriented medical information system. *Mailing Add:* Irish Settlement Underhill VT 05489

WEED, STERLING BARG, b Salt Lake City, Utah, Mar 25, 26; m 49; c 2. SOIL CHEMISTRY. *Educ:* Brigham Young Univ, AB, 51; NC State Col, MS, 53, PhD(soils), 55. *Prof Exp:* Asst prof soils, Cornell Univ, 55-56; from asst prof to assoc prof, 56-70, PROF SOILS, NC STATE UNIV, 70- *Mem:* Soil Sci Soc Am; Clay Minerals Soc; Int Soc Soil Sci; Int Clay Minerals Soc. *Res:* Clay mineralogy; clay-organic interactions. *Mailing Add:* Dept Soil Sci NC State Univ Box 7619 Raleigh NC 27695-7619

WEEDEN, ROBERT BARTON, b Fall River, Mass, Jan 8, 33; m 59; c 3. ZOOLOGY. *Educ:* Univ Mass, BSc, 53; Univ Maine, MSc, 55; Univ BC, PhD(zool), 59. *Prof Exp:* Asst, Univ Maine, 53-55; asst, Univ BC, 55-58; instr zool, Wash State Univ, 58-59; prof wildlife mgt, 68-74, head dept biol & wildlife 86-88, PROF RESOURCE MGT, UNIV ALASKA; RES BIOLOGIST, ALASKA DEPT FISH & GAME, 59- *Concurrent Pos:* Dir, Policy Develop, Off of Gov, Alaska, 75-76. *Mem:* Arctic Inst NAm; Cooper Ornith Soc; Wildlife Soc; Am Ornith Union. *Res:* Avian ecology, particularly of Tetraonidae and alpine-arctic environments; resource policy and land use. *Mailing Add:* Dept Bio Sci Univ Alaska Fairbanks AK 99701

WEEDMAN, DANIEL WILSON, b Nashville, Tenn, Oct 19, 42; m 68; c 2. OBSERVATIONAL ASTRONOMY, TELESCOPE DESIGN. *Educ:* Vanderbilt Univ, BA, 64; Univ Wis, PhD(astron), 67. *Prof Exp:* Fac assoc astron, Univ Tex, 67-69, asst prof, 69-70; asst prof, Vanderbilt Univ, 70-73, assoc prof, 75-78; assoc prof, Univ Minn, 74-75; PROF ASTRON, PA STATE UNIV, 78- *Concurrent Pos:* Exchange scientist, Nat Acad Sci, USSR, 70 & 78; mem bd dirs, Asn Univ Res Astron, Inc, 78-81; counr, Am Astron Soc, 80-83; vis sr scientist, NASA HQ, 90-92. *Mem:* Am Astron Soc; Int Astron Union. *Res:* Observational research in extragalactic astronomy, primarily galactic nuclei, quasars, and starburst galaxies using optical, ultraviolet, radio and infrared techniques; astrophysics management. *Mailing Add:* Astron Dept 525 Davey Lab Pa State Univ University Park PA 16802

WEEDON, ALAN CHARLES, b Oxford, Eng, Mar 29, 51. ORGANIC PHOTOCHEMISTRY. *Educ:* London Univ, BSc, 73, PhD(chem) & DIC, 76; ARCS, 73. *Prof Exp:* Res assoc, 77-80, from asst prof to assoc prof, 80-91, PROF CHEM, UNIV WESTERN ONT, 91- *Concurrent Pos:* Mem, Photochem Unit, Univ Western Ont, 80-, actg dir, 82, dir, 85- *Honors & Awards:* Merck Frosst Award, Can Inst Chem, 91. *Mem:* Can Inst Chem; Royal Soc Chem; Am Chem Soc. *Res:* Organic photochemistry and synthetic organic chemistry; applications of photochemistry to organic synthesis; mechanistic photochemistry; light induced-electron transfer. *Mailing Add:* Chem Dept Photochem Unit Univ Western Ont London ON N6A 5B7 Can

WEEDON, GENE CLYDE, b Washington, DC, June 11, 36; m 61; c 3. POLYMER CHEMISTRY. *Educ:* Va Polytech Inst & State Univ, BS, 59. *Prof Exp:* Res chemist, Esso Res & Eng Co, 60-66; res chemist, 66-71, RES MGR, FIBERS DIVISIONS CO, ALLIED SIGNAL CORP, 71- *Mem:* Am Chem Soc; Res Soc Am; Soc Plastics Engrs; Sigma Xi. *Res:* Development of modified polymers and processes for applications in films, fibers, and moldings area. *Mailing Add:* 4548 Shanto Ct Richmond VA 23237-2571

WEEGE, RANDALL JAMES, b May 14, 26; US citizen; m 51; c 3. GEOLOGY. *Educ:* Univ Wis-Madison, BS, 51, MS, 55. *Prof Exp:* Geologist, Anaconda Co, 55-56; geologist-engr, Uranium Div, Calumet & Hecla, Inc, 56-60, resident geologist, Calumet Div, 60-61, asst chief geologist, 61-65, dir geol, 65-68; dir geol, 68-74, dir explor, 74, dir mineral develop, Mineral Sci Div, 74-81, DIR MINERAL EXP, UNIVERSAL OIL PROD CO, 82- *Concurrent Pos:* Mineral consult, Parsons-Jurden Corp, 66-, Consol Papers Inc, Coleman Eng Co Nord Inc. *Mem:* Soc Econ Geologists; Am Inst Mining, Metall & Petrol Eng; fel Geol Soc Am; Soc Geol Appl Mineral Deposits. *Res:* Mineral exploration techniques and methods. *Mailing Add:* Rte 2 Spread Eagle WI 54219

WEEKES, TREVOR CECIL, b Dublin, Ireland, May 21, 40; m 64; c 3. ASTROPHYSICS. *Educ:* Nat Univ Dublin, BSc, 62, PhD(physics), 66; Nat Univ Ireland, DSc, 78. *Prof Exp:* Lectr III physics, Univ Col, Dublin, 64-66; fel astrophys, 66-67, resident dir, Mt Hopkins Observ, 69-76, ASTROPHYSICIST, SMITHSONIAN INST, 67- *Concurrent Pos:* Consult, Atomic Energy Res Estab, Harwell, 64-66; vis prof, Dublin Inst Advan Studies, 71 & Univ Ariz, 72 & 74; res assoc, Harvard Col Observ, 67- & Steward Observ, Univ Ariz, 76-; vis prof, Royal Greenwich Observ, 80. *Mem:* AAAS; Am Astron Soc; Royal Astron Soc. *Res:* Gamma ray astronomy; cosmic ray physics; meteor detection; atmospheric Cerenkov and fluorescence radiation; transient astronomy; extragalactic astronomy. *Mailing Add:* Whipple Observ Box 97 Amado AZ 85645-0097

WEEKMAN, GERALD THOMAS, entomology; deceased, see previous edition for last biography

WEEKMAN, VERN W(ILLIAM), JR, b Jamestown, NY, June 28, 31; m 55; c 2. REACTION ENGINEERING, CATALYSIS. *Educ:* Purdue Univ, BS, 53, PhD(chem eng), 63; Univ Mich, MS, 54. *Prof Exp:* Chem engr, 54-55, res chem engr, Res Dept, 57-60, fel, 60-63, sr res chem engr, 63-65, eng assoc, 66-67, mgr, Systs Res, 67-76, mgr, Spec Proc, 76-77, mgr, Catalyst Res, 77-79, mgr, Proc Res & Develop, 79-80, pres, Mobil Solar Energy Corp, 80-85, MGR, CENT RES LAB, MOBIL OIL CORP, 85- *Concurrent Pos:* Ed, Indust & Eng Chem Ann Rev, Am Chem Soc, 71-75; lectr, Am Inst Chem Engrs, 79. *Honors & Awards:* Wilhelm Award, Am Inst Chem Engrs, 82. *Mem:* Nat Acad Eng; Am Inst Chem Engrs; Am Chem Soc; Sigma Xi. *Res:* Chemical reaction kinetics; diffusion and heat transfer in catalysis; process dynamics and control; solar photovoltaics. *Mailing Add:* PO Box 1025 Cent Res Lab Mobil Oil Corp Princeton NJ 08540

WEEKS, CHARLES MERRITT, b Buffalo, NY, Mar 23, 44; m 75; c 2. BIOPHYSICS. *Educ:* Cornell Univ, BS, 66; State Univ NY Buffalo, PhD(biophys), 70. *Prof Exp:* SR RES SCIENTIST, X-RAY CRYSTALLOG, MED FOUND BUFFALO, 70- *Mem:* Am Crystallog Asn; Sigma Xi. *Res:* Direct methods of phase determination in x-ray crystallography; crystal structures of steroids and related biological materials; protein crystallography. *Mailing Add:* Med Found Buffalo 73 High St Buffalo NY 14203

WEEKS, DAVID LEE, b Boone, Iowa, June 24, 30; m 57; c 4. EXPERIMENTAL STATISTICS. *Educ:* Okla State Univ, BS, 52, MS, 57, PhD(statist), 59. *Prof Exp:* Asst math, 54-57, from asst to assoc prof statist, 57-66, PROF STATIST, OKLA STATE UNIV, 66- *Concurrent Pos:* Consult, Phillips Petrol Co, 61-91, Air Force Armament Lab, 69-73 & ASI Systs Int, Inc, 78-91; NSF fac fel, Cornell Univ, 67-68; statist consult, RCA, Albuquerque, 73 & Decisions & Designs, 84; vis prof NMex State Univ, 76-77 & 83-84. *Mem:* Am Statist Asn; Biomet Soc; Sigma Xi. *Res:* Design and analysis of experiments; linear models. *Mailing Add:* Dept Statist MS-301 Okla State Univ Stillwater OK 74078

WEEKS, DENNIS ALAN, b Brainerd, Minn, May 29, 43; m 68; c 1. DATA PARALLEL COMPUTER ALGORITHMS, CRYPTOGRAPHIC ALGORITHMS. *Educ:* Univ Ill, Urbana, BA, 67. *Prof Exp:* Mem tech staff, SCM Corp, 73-74; syst analyst, Xebec Systs Inc, 74-76 & Spectra Med Systs, 76-77; syst engr, Data Gen Corp, 77-83 & Convex Computer Corp, 87-89;

consult, Matrix Assocs, 83-87; COURSE DEVELOPER, MASPAR COMPUTER CORP, 89- *Mem:* Asn Comput Mach; Soc Indust & Appl Math. *Res:* Algorithms for data-parallel computer architectures with emphasis on computer algebra; history of mathematics. *Mailing Add:* MasPar Computer Corp 749 N Mary Ave Sunnyvale CA 94086

WEEKS, DONALD PAUL, b Terre Haute, Ind, Feb 15, 41; m 64; c 4. MOLECULAR BIOLOGY, PHYSIOLOGY. *Educ:* Purdue Univ, BSA, 63; Univ Ill, PhD(agron), 67. *Prof Exp:* Res assoc molecular biol, Inst Cancer Res, 67-68, NIH fel, 68-70, res assoc, 70-74, asst mem, 74-78, assoc mem, 78-82; prin scientist, Zoecon Corp, Palo Alto, Calif, 82-89; DIR, CTR BIOTECHNOL & PROF BIOL SCI & BIOCHEM, UNIV NEBR, LINCOLN, 89- *Concurrent Pos:* Consult outside contracts, 85- *Mem:* AAAS; Am Soc Plant Physiol; Am Soc Cell Biol; Int Soc Plant Molecular Biol. *Res:* Protein synthesis; gene regulation; plant cell transformation; genetic eng plants. *Mailing Add:* Ctr Biotechnol Univ Nebr Lincoln 101 Manter Hall Lincoln NE 68588-0159

WEEKS, DOROTHY WALCOTT, physics; deceased, see previous edition for last biography

WEEKS, GEORGE ELIOT, b Montgomery, Ala, July 10, 39; m 62; c 2. AEROSPACE ENGINEERING. *Educ:* Univ Ala, BS, 60, MS, 61; Va Polytech Inst, PhD(mech), 66. *Prof Exp:* Stress analyst aircraft, Hayes Int, 61; asst prof aerospace eng, Miss State Univ, 61-63; aerospace technologist, NASA Langley Field, 63-67; from asst prof to assoc prof, 67-77, PROF AEROSPACE ENG, UNIV ALA, 77- *Concurrent Pos:* Consult, US Army Missile Command, Hayes Int Corp, Remtech, Inc, Norton Co. *Mem:* Am Inst Aeronaut & Astronaut. *Res:* Composite materials; structural dynamics; stress analysis; numerical mathematics. *Mailing Add:* Univ Ala Box 870280 Tuscaloosa AL 35487-0280

WEEKS, GERALD, b Birmingham, Eng, Feb 5, 41. BIOCHEMISTRY. *Educ:* Birmingham Univ, BSc, 62, PhD(biochem), 66. *Prof Exp:* Res assoc biochem, Duke Univ Med Ctr, 66-69; res fel, Leicester Univ, 69-71; from asst prof to assoc prof, 71-84, PROF MICROBIOL, UNIV BC, 84- *Mem:* Am Soc Microbiol; Can Biochem Soc. *Res:* Molecular basis of cell-cell interaction during the differentiation of the cellular slime mould, Dictyostelium discoideum. *Mailing Add:* Dept Microbiol Univ BC No 300-6174 University Blvd Vancouver BC V6T 1W5 Can

WEEKS, GREGORY PAUL, b Seattle, Wash, July 16, 47. TEXTILE CHEMISTRY, INSTRUMENTATION. *Educ:* Univ Wash, BS, 69; Univ Ill, Urbana, PhD(phys chem), 74. *Prof Exp:* Res chemist, E I du Pont de Nemours & Co, Inc, 74-80. *Res:* Development of instrumentation for on-line automated analysis of polymer solutions used in synthetic textile fiber manufacture; synthetic textile technology generally, with emphasis on infrared engineering technology for instrumentation. *Mailing Add:* 108 Cameron Dr Hockessin DE 19707-9684

WEEKS, HARMON PATRICK, JR, b Orangeburg, SC, Oct 4, 44; m 65, 90; c 2. WILDLIFE ECOLOGY. *Educ:* Univ Ga, BSF, 67, MS, 69; Purdue Univ, PhD(wildlife ecol), 74. *Prof Exp:* Lectr wildlife ecol, Yale Univ, 73-74; asst prof, 74-79, ASSOC PROF WILDLIFE MGT, PURDUE UNIV, WEST LAFAYETTE, 79- *Mem:* Wildlife Soc; Am Soc Mammalogists; Am Ornithologists Union. *Res:* Effects of fragmentation and silvicultural practices on wildlife populations; adaptations of homeothermic vertebrates to sodium deficiencies; avian breeding biology. *Mailing Add:* Dept Forestry & Nat Resources Purdue Univ West Lafayette IN 47907

WEEKS, JAMES ROBERT, b Des Moines, Iowa, Aug 13, 20; m 43; c 3. PHARMACOLOGY. *Educ:* Univ Nebr, BSc, 41, MS, 46; Univ Mich, PhD(pharmacol), 52. *Prof Exp:* From instr pharm to prof pharmacol, 50-57; res assoc, Upjohn Co, 57-84; RETIRED. *Mem:* AAAS; Am Soc Pharmacol & Exp Therapeut; Soc Exp Biol & Med; Sigma Xi. *Res:* Cardiovascular pharmacology; hypertension; prostaglandins; experimental addiction. *Mailing Add:* 1417 Sheridan Kalamazoo MI 49001

WEEKS, JOHN DAVID, b Birmingham, Ala, Oct 11, 43; m 88; c 1. CHEMICAL PHYSICS, MATERIALS PHYSICS. *Educ:* Harvard Col, BA, 65; Univ Chicago, PhD(chem physics), 69. *Prof Exp:* Res fel chem, Univ Calif, San Diego, 69-71; res fel physics, Cambridge Univ, 72; mem tech staff mat physics, Bell Tell Labs 73-90; PROF, INST PHYS SCI & TECHNOL & DEPT CHEM, UNIV MD, 90- *Honors & Awards:* Joel Henry Hildebrand Award, Am Chem Soc, 90. *Mem:* Fel Am Phys Soc; Am Chem Soc; AAAS. *Res:* Statistical physics; theory of crystal growth; pattern formation. *Mailing Add:* Inst Phys Sci & Technol Univ Md College Park MD 20742

WEEKS, JOHN LEONARD, b Bath, Eng, May 10, 26; m 52; c 5. OCCUPATIONAL HEALTH. *Educ:* Univ London, MB, BS, 53, MD, 75; Royal Col Obstet & Gynec, dipl, 54; Royal Col Physicians & Surgeons, DIH, 57; MFOM (Eng), 78; Can Bd Occup Med, cert, 80. *Prof Exp:* Intern, St Thomas's Hosp, London, Eng, 53-54; sr intern, St Helier Hosp, London, 54-55; indust physician, London Transport Exec, 55-58; med officer, Dept Health, Nfld, 58-60; med officer, Int Nickel Co, 60-62; supt, Health & Safety Br, 62-68, dir health & safety div, hi div, 68-86, SR ADV HEALTH & SAFETY, WHITESHELL NUCLEAR RES ESTAB, ATOMIC ENERGY CAN, LTD. *Concurrent Pos:* Hon prof, Univ Man, 66. *Mem:* Can Med Asn. *Res:* Industrial health and toxicology, particularly organic coolants; radiation protection and biology; environmental medicine; epidemiology. *Mailing Add:* Whiteshell Nuclear Res Estab Atomic Energy Can Ltd Pinawa MB R0E 1L0 Can

WEEKS, JOHN R(ANDEL), IV, b Orange, NJ, Oct 30, 27; m 51; c 2. METALLURGY. *Educ:* Colo Sch Mines, EMet, 49; Univ Utah, MS, 50, PhD(metall), 53. *Prof Exp:* Asst metall eng, Univ Utah, 49-52; assoc metallurgist, Brookhaven Nat Lab, 53-59, metallurgist, 59-72; sr metallurgist, US AEC, 72-74; leader, corrosion sci group, 74-83, sr metallurgist & head mat technol div, Brookhaven Nat Lab, 83-87; SR CONSULT, 87- *Concurrent Pos:* Adj assoc prof, State Univ NY, Stony Brook, 62-63; consult, Aerojet-Gen Corp, 65-71 & US Nuclear Regulatory Comn, 75-; adj prof metall & nuclear eng, Polytech Inst NY, 78-84. *Mem:* Fel Am Soc Metals; Am Inst Mining, Metall & Petrol Engrs; Am Nuclear Soc; Nat Asn Corrosion Engrs. *Res:* Corrosion in aqueous solutions; liquid metal corrosion; reactor metallurgy; corrosion and stress corrosion cracking of nuclear reactor materials. *Mailing Add:* Brookhaven Nat Lab Upton NY 11973

WEEKS, L(ORAINE) H(UBERT), b Mt Lake Park, Md, Feb 27, 18; m 41; c 2. ELECTRICAL ENGINEERING. *Educ:* Univ Md, BS, 40. *Prof Exp:* Elec engr, Copper Wire Eng Asn, DC & Mo, 40-43; elec engr, Frank Horton & Co, Mo, 46-48; dist design engr, Monongahela Power Co, WVa, 48-50, planning engr, 50-63, mgr planning eng, 63-68; mgr syst transmission planning, Allegheny Power Serv Corp, 68-69, mgr spec planning studies, 70-73, exec dir, 73-83; RETIRED. *Concurrent Pos:* Elec engr consult, 83-84. *Mem:* Inst Elec & Electronics Engrs; Nat Soc Prof Engrs. *Res:* Electric utility planning of generation and transmission facilities. *Mailing Add:* 3035 McClellan Dr Greensburg PA 15601

WEEKS, LEO, b Norman Park, Ga, June 18, 25; m 54; c 2. GENETICS. *Educ:* Ga Southern Col, BS, 48; George Peabody Col, MA, 51; Univ Nebr, PhD(zool), 54. *Prof Exp:* Teacher & asst prin pub sch, Ga, 48-50; instr biol, George Peabody Col, 50-51; instr & asst, Univ Nebr, 51-54; asst prof biol, Austin Peay State Col, 54-56; prof & head dept, Berry Col, 56-62; prof biol, Ga Southern Col, 62-67; HEAD DEPT BIOL, HIGH POINT COL, 68- *Mem:* Am Genetic Asn; Am Inst Biol Sci; Sigma Xi. *Res:* Drosophila melanica and other species. *Mailing Add:* Dept Biol High Point Col High Point NC 27262

WEEKS, LESLIE VERNON, b Lazear, Colo, July 24, 18; m 44; c 3. SOIL SCIENCE. *Educ:* Univ Calif, Berkeley, BS, 52; Univ Calif, Riverside, PhD(soil sci), 66. *Prof Exp:* Sr lab technician, Univ Calif, Riverside, 51-56, prin lab technician, 56-70, staff res assoc soil sci, 70-78, lectr soil sci, 75-78; RETIRED. *Res:* Water movement in liquid and vapor phases due to water potential and thermal gradients in unsaturated soils; use of computers in water movement studies under transient flow conditions. *Mailing Add:* 5496 Jurupa Ave Riverside CA 92504

WEEKS, MAURICE HAROLD, b Germantown, NY, Nov 9, 21; m 48; c 2. TOXICOLOGY. *Educ:* Union Col, BS, 49, MS, 50. *Prof Exp:* Collab path, Forest Prod Lab, US Forest Serv, Wis, 49; chemist, Hanford Works, Gen Elec Co, 50-55; pharmacologist, Directorate Med Res, Army Chem Ctr, 55-66; PHARMACOLOGIST, US ARMY ENVIRON HYG AGENCY, ABERDEEN PROVING GROUNDS, 66- *Mem:* AAAS; Sci Res Soc Am; Am Indust Hyg Asn; Soc Toxicol. *Res:* Animal metabolism studies of administered chemicals; toxicology and hazard evaluation of inhaled aerosols and vapors. *Mailing Add:* 27 Idlewild Bel Air MD 21014

WEEKS, PAUL MARTIN, b Clinton, NC, June 11, 32; m 57; c 6. SURGERY. *Educ:* Duke Univ, AB, 54; Univ NC, MD, 58. *Prof Exp:* From instr to prof surg, Med Ctr, Univ Ky, 64-70; CHIEF DIV PLASTIC SURG, DEPT SURG, SCH MED, WASH UNIV, 71- *Mem:* Am Col Surgeons; Soc Univ Surgeons; Am Chem Soc; Plastic Surg Res Coun. *Res:* Interrelationships of collagen and mucopolysaccharides in determining tissue compliance; effects of environment in cell synthesis, particularly regarding mucopolysaccharide or collagen synthesis; effects of irradiation on collagen and mucopolysaccharide synthesis. *Mailing Add:* 4949 Barnes Hosp Plaza Suite 17424 St Louis MO 63110

WEEKS, RICHARD W(ILLIAM), b Los Angeles, Calif, Mar 21, 22; m 47; c 1. ELECTRICAL ENGINEERING. *Educ:* Calif Inst Technol, BS, 52; Stanford Univ, MS, 58. *Prof Exp:* Assoc engr, Res Lab, Int Bus Mach Corp, 52-55; lectr elec eng, San Jose State Col, 55-56, assoc prof, 58-60; develop engr, Pick Labs, 56-58; asst prof elec eng, Mont State Univ, 60-63; supv bioeng lab, Presby Med Ctr, 63-65; assoc prof elec eng, Nat Resources Res Inst, 65-71, ASSOC PROF ELEC ENG, UNIV WYO, 71- *Concurrent Pos:* Consult various indust concerns. *Mem:* Am Soc Eng Educ; Inst Elec & Electronics Engrs. *Res:* Applications of bioengineering to wildlife management and domestic livestock husbandry; general electronic instrumentation and telemetry. *Mailing Add:* 2679 Kennedy Laramie WY 82070

WEEKS, RICHARD WILLIAM, b Borger, Tex, April 12, 42; m 68; c 2. AQUEOUS CORROSION, NUCLEAR FUELS. *Educ:* Swarthmore Col, BS, 64; Calif Inst Technol, MS, 65; Univ Ill, PhD(theoret & appl mech), 68; Univ Chicago, MBA, 82. *Prof Exp:* Asst mech engr, 68-71, group leader fatigue & fracture, 71-73, assoc div dir, Mat Sci Div, 73-86, DIV DIR, MAT & COMP TECH DIV, ARGONNE NAT LAB, 86- *Concurrent Pos:* Consult, Elec Power Res Inst, 76-80; sr mech engr, Argonne Nat Lab, 79-; lectr, Short Course, Northwestern Univ, 80-87; assoc ed, J Nuclear Eng & Des, 76- *Mem:* Sigma Xi. *Res:* Aqueous stress corrosion cracking in light water reactor systems; high temperature mechanical properties including creep-fatigue interactions; computer modelling of nuclear fuel element performance; materials science engineering; superconductivity technology. *Mailing Add:* 13743 Janas Pkwy Lockport IL 60441

WEEKS, ROBERT A, b Birmingham, Ala, Aug 23, 24; m 48; c 4. SOLID STATE PHYSICS. *Educ:* Birmingham-Southern Col, BS, 48; Univ Tenn, MS, 51; Brown Univ, PhD(physics), 66. *Prof Exp:* Physicist, Solid State Div, Oak Ridge Nat Lab, 51-84, prin investr, lunar mat, 68-74, prin investr, Libyan desert glass, 68-81; PRIN INVESTR, REDOX EQUILIBRIA IN OPTICAL WAVEGUIDE MAT, VANDERBILT UNIV, 81-, RES PROF, DEPT MME, 84- *Concurrent Pos:* Distinguished vis prof, Am Univ Cairo, 68, 70-71; res fel, Univ Reading, 71; consult, Dept Phys Sci & Mat Eng, Am Univ Cairo, 71-; assoc ed, J Geophys Res, 68-74; Fulbright short term lectr, 80; invite prof, Fed Polytech Sch Lausannne, 81; consult, Kuwait Inst Sci Res, Technol Develop Coop, 76-, Spectran Corp, 82-, Ga Inst Tech, 83-, Gen Elec Corp,

84-85 & UT Corp, 84-; adj prof, Univ Pa, 77- & Vanderbilt Univ, 79-; co-chmn, First Int Conf Effects of Modes of Formation on Struct of Glasses, 84, co-ed proc, 85, prin investr, New Mat by Ion Implantation, NSF grant, 86-89, co-prin investr, Radiation Effects on Low Thermat Expansion Mat, NRL grant & Free Electron Laser Effects on Mat, SDI grant, co-chmn, 2nd Int Conf Effects of Modes of Formation on Struct of Glasses, 87, co-ed proc, 88, ed, J Non-Crystalline Solids, 88. *Mem:* AAAS; Am Phys Soc; Sci Res Soc Am; Am Ceramic Soc. *Res:* Optical and magnetic properties of intrinsic and extrinsic defects of crystalline and glassy solids; effects of photon and particle irradiation upon the properties of diamagnetic non-conducting solids; magnetic properties of extraterrestrial solids; transport in solids; ion implantation effects on solids. *Mailing Add:* Dept Mech & Mat Eng Vanderbilt Univ PO Box 1678 Nashville TN 37235

WEEKS, ROBERT JOE, b Quapaw, Okla, Feb 19, 29; m 48; c 3. MYCOLOGY, MEDICAL MICROBIOLOGY. *Educ:* Southwest Mo State Col, BS, 61. *Prof Exp:* Mycologist, Kans City Field Sta, Commun Dis Ctr, Bur Labs, USPHS, 62-64, microbiologist, 64-67, chief soil ecol unit, Mycoses Sect, Kansas City Labs, Ecol Invests Prog, Ctr Dis Control, 67-74, microbiologist, microbiol sect, Proficiency Testing Br, 74-79; microbiologist, Mycotic Dis Div, Ctr Infectious Dis, Ctr Dis Control, Atlanta, Ga, 79-85; CONSULT ENVIRON MYCOL, 85- *Mem:* Mycol Soc Am. *Res:* Relationship of the pathogenic fungi to their soil environment. *Mailing Add:* Rte 6 Box 135A Rogers AR 72756

WEEKS, STEPHAN JOHN, b Minneapolis, Minn, Apr 13, 50; m 77; c 3. ANALYTICAL SPECTROMETRY. *Educ:* St Olaf Col, BA, 72; Univ Fla, PhD(chem), 77. *Prof Exp:* Res chemist lasers spectros, Nat Bur Standards, 77-79, res chemist, tribology, 79-84; CHEMIST, AMES LAB, 84- *Concurrent Pos:* Res assoc, Nat Res Coun fel, 77-79. *Mem:* Am Chem Soc; Soc Appl Spectros. *Res:* Trace molecularanalysis; method development; laser spectrometry; trace atomic analysis; computer data analysis. *Mailing Add:* Ames Lab 26 Spedding Hall Ames IA 50011

WEEKS, STEPHEN P, b Wilmington, Del, July 3, 49; m 73; c 2. SILICON MATERIALS DEVELOPMENT, ELECTRONIC CERAMICS. *Educ:* Univ Del, BS, 71; Col William & Mary, MS, 73; Univ Pa, PhD(solid state physics), 77. *Prof Exp:* Postdoctoral, AT&T Bell Labs, 77-78, mem tech staff, 79-84, distinguished tech staff, 84-85; res mgr, Standard Oil Ohio, 85-87; res coordr, Brit Petrol, London, 87-90; DIR PLANNING & CONTROL, BP AM RES, 91- *Mem:* Am Phys Soc. *Res:* Integrated circuit process technology development; electronic/optical materials research. *Mailing Add:* BP Am 4440 Warrensville Ctr Rd Cleveland OH 44128

WEEKS, THOMAS F, b Wheeling, WVa, Apr 12, 35; m 55; c 2. PLANT PHYSIOLOGY. *Educ:* West Liberty State Col, AB, 63; Purdue Univ, MS, 67, PhD(develop bot), 70. *Prof Exp:* Mem staff sec educ, Ohio County Bd Educ, WVa, 63-64; guest lectr biol, Purdue Univ, 70-71; assoc prof, 71-80, PROF BIOL, UNIV WIS-LA CROSSE, 80- *Mem:* AAAS; Bot Soc Am; Am Soc Plant Physiol. *Res:* Plant growth regulators; control mechanisms in plants. *Mailing Add:* Dept Biol Univ Wis La Crosse 1725 State St La Crosse WI 54601

WEEKS, THOMAS JOSEPH, JR, b Tarrytown, NY, Aug 31, 41; m 67; c 2. CATALYSIS. *Educ:* Colgate Univ, BA, 63; Univ Colo, PhD(chem), 67. *Prof Exp:* Res chemist, Res & Eng Ctr, Johns-Manville Corp, Manville, NJ, 66-68; sr res chemist, Union Carbide Corp, Tarrytown, NY, 70-72, proj chemist, 72-73, sr staff chemist, 73-75, supvr, 75-78; group leader, Ashland Chem Corp, 78-80, res mgr, 80-83, com develop mgr, 83-86, DEPT HEAD, RES & DEVELOP, ASHLAND CHEM CORP, 87- *Mem:* Am Chem Soc; Catalysis Soc; AAAS. *Res:* Management of research and development on thermosetting polymer systems; waste management and safety-research and development building administration; synthesis, testing and manufacturing of heterogeneous catalysts. *Mailing Add:* Ashland Chem Co PO Box 2219 Columbus OH 43216

WEEKS, WILFORD FRANK, b Champaign, Ill, Jan 8, 29; m 84; c 2. GLACIOLOGY, HYDROLOGY. *Educ:* Univ Ill, BS, 51, MS, 53; Univ Chicago, PhD(geol), 56. *Prof Exp:* Geologist, Mineral Deposits Br, US Geol Surv, 52-55; glaciologist, USAF Cambridge Res Ctr, 55-57; asst prof, Wash Univ, 57-62; glaciologist, Cold Regions Res & Eng Lab, 62-89; CHIEF SCIENTIST, ALASKA SYNTHETIC APERTURE RADAR FACIL, GEOPHYS INST, UNIV ALASKA, FAIRBANKS, 86- *Concurrent Pos:* Adj assoc prof, Dartmouth Col, 62-72, adj prof, 72-85; mem, Polar Res Bd & chmn, Panel on Glaciol, Nat Acad Sci, 71-77; chmn, Div River, Lake & Sea Ice, Int Comn Snow & Ice; Japan Soc Promotion Sci vis prof, Inst Low Temp Sci, Hokkaido Univ, Japan, 73; mem, adv panel, Jet Propulsion Lab; lectr, Advan Study Inst Air, Sea & Ice Interactions, NATO; Off Naval Res Chair, Arctic Marine Sci, 78-79; NASA Earth Orbital Shuttle Synthetic Aperture Radar Team & NASA Earth Syst Sci Comt; Radarsat Sci Group, Can Ctr Remote Sensing; advisor to US Arctic Res Comn. *Honors & Awards:* Seligman Crystal Award, Int Glaciol Soc, 89. *Mem:* Nat Acad Eng; fel Arctic Inst NAm; fel Am Geophys Union; Int Glaciol Soc (vpres, 69-72, pres, 72-75). *Res:* Geophysics of sea, lake and river ice. *Mailing Add:* Geophys Inst Univ Alaska Fairbanks AK 99775-0800

WEEKS, WILLIAM THOMAS, b Portchester, NY, Mar 28, 32; m 54; c 2. ENGINEERING PHYSICS. *Educ:* Williams Col, BA, 54; Univ Mich, MS, 56, PhD(physics), 60. *Prof Exp:* Assoc physicist, Data Systs Div, 60-61, staff physicist, Systs Develop Div, 61-64, adv physicist, Components Div, 64-68, sr engr components div, 68-84, SR TECH STAFF MEM, GEN TECHNOL DIV, IBM CORP, 84- *Mem:* Sigma Xi. *Res:* Scientific computation and numerical analysis as applied to design of electronic digital computers. *Mailing Add:* Innsbruck Blvd Hopewell Junction NY 12533

WEEMS, CHARLES WILLIAM, b Greeneville, Tenn, Mar 1, 41; m 66; c 3. REPRODUCTIVE ENDOCRINOLOGY, GROWTH PHYSIOLOGY. *Educ:* ETenn State Univ, BS, 64, MA, 66; WVa Univ, PhD(endocrinol), 75. *Prof Exp:* Captain chem corps, US Army Biol Labs, Fort Detrick, Md, 66-68; instr, Fairmont State Col, WVa, 68-71, from asst prof to assoc prof biol & physiol, 71-76; from assoc prof to prof agr & endocrinol, Ariz State Univ, 76-83; PROF & CHMN, DEPT ANIMAL SCI, UNIV HAWAII, MANOA, 83- *Concurrent Pos:* prin investr, UpJohn Co, Inc, 76-; adhoc reviewer reproduction & growth, USDA, 85-, biotechnol peer panel reviewer growth, 85-; prog dir, USDA Biotech Grants, 87. *Mem:* NY Acad Sci; Soc Study Reproduction; Endocrine Soc; Am Soc Animal Sci; Sigma Xi; Soc Study Fertil. *Res:* Mechanisms regulating ovarian, corpus luteum, and uterine function; embryonic signals regulating pregnancy and conception; uterine capacity; survival of embryos; mechanisms of action and regulation of hormone receptors. *Mailing Add:* Dept Animal Sci Univ Hawaii 1800 East-West Rd Honolulu HI 96822

WEEMS, HOWARD VINCENT, JR, b Rome, Ga, Apr 11, 22; m 50; c 4. ENTOMOLOGY. *Educ:* Emory Univ, BA, 46; Univ Fla, MS, 48; Ohio State Univ, PhD(entom), 53. *Prof Exp:* Asst entom, Univ Fla, 46-47, assoc prof entom, 66-73; instr biol, Univ Miss, 48-49; res asst, Ohio Biol Surv, Ohio State Univ, 49-53; TAXON ENTOMOLOGIST, DIV PLANT INDUST, FLA DEPT AGR & CONSUMER SERV, 53-; PROF ENTOM, UNIV FLA, 73- *Concurrent Pos:* Head Cur, Fla State Collection Arthropods, 54-; assoc arthropods, Fla State Mus Natural Hist, 59-; ed, Arthropods of Fla & Neighboring Land Areas, Fla State Collection Arthropods, Fla Dept Agr & Consumer Serv, 65-; assoc ed, Fla Entomologist; courtesy assoc prof entom, Fla A&M Univ, 77-; dir, Ctr Syst Entom Inc. *Mem:* Fel AAAS; Asn Trop Biol; Entom Soc Am; assoc Ecol Soc Am; Soc Syst Zool; Am Forestry Asn; Asn Syst Collections; Sigma Xi; Smithsonian Natural Hist Asn; Brazilian Entom Soc; Lepidopterists Soc. *Res:* Taxonomy and ecology of syrphid flies; identifications of Florida arthropods. *Mailing Add:* Fla Dept Agr & Consumer Serv PO Box 1269 Gainesville FL 32602

WEEMS, MALCOLM LEE BRUCE, b Nashville, Kans, Dec 8, 45; m 72; c 1. PHYSICS. *Educ:* Kans State Teachers Col, BSE, 67, MS, 69; Okla State Univ, PhD(physics), 72. *Prof Exp:* Lectr physics, Kans State Teachers Col, 68-69; instr, E Cent State Univ, 72-75, dir sci lab prog for blind, 74-80, asst prof physics, 75-79, assoc prof physics, 79-82, chmn, Dept Physics, 82-899, intern dean, Sch Math & Sci, 90-91, PROF PHYSICS, E CENT STATE UNIV, 82-, CHMN, DEPT PHYS & ENVIRON SCI, 89-, DEAN, 91- *Mem:* Am Inst Physics; Am Asn Physics Teachers. *Res:* Stellar atmospheres; educational innovation. *Mailing Add:* Sch Math & Sci E Cent Univ Ada OK 74820-6899

WEEMS, ROBERT EDWIN, b Richmond, Va, Jan 22, 47; m 83; c 3. GEOLOGY. *Educ:* Randolph-Macon Col, BS, 68; Va Polytech Inst & State Univ, MS, 75; George Washington Univ, PhD(geol), 78. *Prof Exp:* Instr, George Mason Univ, 76-79; RES GEOLOGIST, US GEOL SURV, 78- *Concurrent Pos:* Consult, Dahlgren Naval Res Lab, 77-79. *Mem:* Am Asn Ichthyologists & Herpetologists. *Res:* Vertebrate paleontology; stratigraphy and paleoseismicity of the Maryland, Virginia, West Virginia, North Carolina and South Carolina region from the Precambrian to present; Atlantic coastal plain stratigraphy; Newark supergroup statigraphy. *Mailing Add:* 1065 Burwick Dr Herndon VA 22313

WEEMS, WILLIAM ARTHUR, b Carlsbad, NMex, July 3, 44; m 65; c 2. NEUROPHYSIOLOGY, GASTROINTESTINAL MOTILITY. *Educ:* Baylor Univ, BS, 67, MS, 70; Univ Ill, PhD(physiol), 73. *Prof Exp:* Res fel, Mayo Found, 73-76; asst prof, 76-82, ASSOC PROF PHYSIOL, MED SCH, UNIV TEX, HOUSTON, 82- *Mem:* Sigma Xi; Soc Neurosci; Am Physiol Soc; Asn Comput Mach. *Res:* Information processing in sympathetic prevertebral ganglia and neural control of gastrointestinal motility. *Mailing Add:* Dept Physiol & Cell Biol Univ Tex Med Sch PO Box 20708 Houston TX 77225

WEERTMAN, JOHANNES, b Fairfield, Ala, May 11, 25; m 50; c 2. MATERIAL SCIENCE. *Educ:* Carnegie Inst Technol, BS, 48, DSc(physics), 51. *Prof Exp:* Fulbright fel, Ecole Normal Superieure, Paris, 51-52; solid state physicist, US Naval Res Lab, 52-58. sci liaison officer, US Off Naval Res US Embassy, Eng Lab, 58-59; from assoc prof to prof, 59-68, WALTER P MURPHY PROF MAT SCI, NORTHWESTERN UNIV, EVANSTON, 68-, PROF GEOPHYS, 63- *Concurrent Pos:* Consult, US Army Cold Regions Res & Eng Lab, 60-75, US Naval Res Lab, 60-67, Bain Lab, US Steel Co, 60-62 & Oak Ridge Nat Lab, 63-68; vis prof, Calif Inst Technol, 64; consult, Los Alamos Sci Lab, 67-; vis prof, Scott Polar Res Inst, Cambridge Univ, 71-72, Swiss Fed Reactor Res Inst, 86; Guggenheim fel, 70-71. *Honors & Awards:* Horton Award, Sect Hydrol, Am Geophys Union, 62; Mathewson Gold Medal, Am Inst Mining, Metall & Petrol Eng, 77; Seligman Crystal Award, Int Glaciol Soc, 83; Acta Metallurgica Gold Medal, 80. *Mem:* Nat Acad Eng; AAAS; fel Geol Soc Am; fel Am Phys Soc; fel Am Soc Metals; fel Am Geophys Union; fel Metals Minerals & Mat Soc. *Res:* Creep of crystals; dislocation theory; internal friction; theory of glacier movement; metal physics; glaciology; geophysics; fatigue; fracture. *Mailing Add:* Dept Mat Sci & Eng Northwestern Univ Evanston IL 60208

WEERTMAN, JULIA RANDALL, b Muskegon, Mich, Feb 10, 26; m 50; c 2. SOLID STATE PHYSICS. *Educ:* Carnegie Inst Technol, BS, 46, MS, 47, DSc(physics), 51. *Prof Exp:* Rotary Int fel, Ecole Normale Superieure, Univ Paris, 51-52; physicist, US Naval Res Lab, 52-58; vis asst prof, 72-73, from asst prof to assoc prof, 73-82, PROF MAT SCI, NORTHWESTERN UNIV, EVANSTON, 82-, DEPT CHMN, 87- *Concurrent Pos:* Guest prof, ETH-Zurich, 86. *Honors & Awards:* Creativity Res Award, NSF, 81 & 86; Guggenheim Fel, 86-87. *Mem:* Nat Acad Eng; Am Inst Phys; fel Am Soc Metals Int; Am Crystallog Asn; Am Soc Testing & Mat; Mat Res Soc; Am Phys Soc; Mining Metals & Mat Soc. *Res:* Dislocation theory; high temperature fatigue; small angle neutron scattering; nanocrystalline material. *Mailing Add:* Dept Mat Sci & Eng Northwestern Univ Evanston IL 60208

WEESE, RICHARD HENRY, b Hyer, WVa, Dec 13, 38; m 68. ORGANIC POLYMER CHEMISTRY. *Educ:* WVa State Col, BS, 64; Bucknell Univ, MS, 67. *Prof Exp:* Scientist chem, 67-74, sr scientist chem, 74-78, RES SECT MGR, RES LABS, ROHM & HAAS CO, BRISTOL, 78- *Mem:* Soc Plastic Engrs. *Res:* Synthesis and application of organic polymers, including additives or modifiers. *Mailing Add:* RR 1 Glenwood Dr Washington Crossing PA 18977

WEETALL, HOWARD H, b Chicago, Ill, Nov 17, 36; m 62; c 2. IMMUNOCHEMISTRY, ENZYMOLOGY. *Educ:* Univ Calif, Los Angeles, BA, 59, MA, 61. *Prof Exp:* Scientist, Jet Propulsion Lab, Calif Inst Technol, 61-65; sr res biologist, Space Gen Corp, 65-66; immunochemist, Bionetics Res Corp, 66-67; res biochemist, Corning Glass Works, 67-71, res assoc, 71-73, sr res assoc biochem, 73-78, mgr biosci res, 78-84, MGR BIOMED RES, CORNING GLASS WORKS, 78-, RES FEL, 84- *Concurrent Pos:* Res fel, Ciba-Corning Diag, 85-90; res biol, Nat Inst Standards & Technol, 91- *Mem:* AAAS; Am Chem Soc; Am Soc Microbiol; NY Acad Sci; Am Soc Biol Chemists; Am Asn Immunol. *Res:* Immobilized biologically active molecules, including antigens, antibodies and enzymes and the characteristics of such materials. *Mailing Add:* Nat Inst Standards & Technol Gaithersburg MA 20899

WEETE, JOHN DONALD, b Dallas, Tex, June 14, 42; m; c 1. FUNGAL PHYSIOLOGY, FUNGAL BIOCHEMISTRY. *Educ:* Stephen F Austin State Univ, BS, 65, MS, 68; Univ Houston, PhD(biol), 70. *Prof Exp:* Fel, Baylor Col Med, 69-70; vis scientist, Lunar Sci Inst, 70-71, staff scientist, 71-72, NASA prin investr lunar sample anal, 72; from asst prof to prof, 72-86, ALUMNI PROF BOT & MICROBIOL, AUBURN UNIV, 86- *Concurrent Pos:* Vis prof, Univ Zurich, 80; invited prof, Univ Paul Sabatier, Toulouse, France; assoc ed, Can J Microbiol. *Honors & Awards:* Res Award, Am Phytopath Soc, 68; William Howard Smith Fac Res Award, Auburn Univ, 80; Int Sci Exchange Award, Can, 87. *Mem:* Am Phytopath Soc; Am Soc Microbiologists; Sigma Xi; fel Am Acad Microbiol; Mycol Soc Am; Am Oil Chemists Soc; AAAS. *Res:* Fungus physiology and biochemistry; lipid composition, function and metabolism of fungi; role of sterols in membranes; mode of action of antifungal sterol biosynthesis inhibitors. *Mailing Add:* Dept Bot & Microbiol Auburn Univ Auburn AL 36830

WEETMAN, DAVID G, b Poughkeepsie, NY, Feb 20, 38; m 65; c 2. ORGANIC CHEMISTRY. *Educ:* Pa State Univ, BS, 59; Univ Minn, PhD(org chem), 68. *Prof Exp:* RES CHEMIST, TEXACO RES LAB, TEXACO, INC, 68- *Mem:* Am Chem Soc. *Res:* Synthesis and reactions of dihalocyclopropanes derived from 2- and 3- methyl substituted 4-ethoxy-2H-1-benzothiopyrans. *Mailing Add:* 12 Prentiss Dr Hopewell Junction NY 12533-6015

WEETMAN, GORDON FREDERICK, b York, Eng, Apr 24, 33; Can citizen; c 3. FORESTRY. *Educ:* Univ Toronto, BScF, 55; Yale Univ, MS, 58, PhD(forestry), 62. *Prof Exp:* Res forester, Pulp & Paper Res Inst Can, 55-72; prof silvicult, Fac Forestry, Univ NB, Fredericton, 72-78, PROF SILVICULT, FAC FORESTRY, UNIV BC, VANCOUVER, 78- *Concurrent Pos:* Ed, Forestry Chronicle, Can Inst Forestry, 67-71. *Mem:* Can Inst Forestry (pres, 73-74); Sigma Xi. *Res:* Blockage of the nitrogen cycle by raw humus accumulations in boreal forests; nitrogen fertilization; nutrient losses in logging; silviculture; Canadian forestry problems. *Mailing Add:* Fac Forestry Univ BC Vancouver BC V6T 1W5 Can

WEFEL, JOHN PAUL, b Cleveland, Ohio, Apr 28, 44; m 67; c 2. PHYSICS, ASTROPHYSICS. *Educ:* Valparaiso Univ, BS, 66; Wash Univ, MA, 68, PhD(physics), 71. *Prof Exp:* Nat Acad Sci-Nat Res Coun resident res assoc astrophys, Naval Res Lab, 71-73, res physicist, 73-75; Robert R McCormick fel, Enrico Fermi Inst, Univ Chicago, 75-77, sr res assoc, 77-82; ASST PROF PHYSICS, LA STATE UNIV, 82- *Mem:* AAAS; Am Phys Soc; Sigma Xi. *Res:* Cosmic ray astrophysics; utilized both passive and electronic detectors to measure element and isotopic abundances; studied nuclear fragmentation reactions of importance in cosmic ray propagation calculations; nucleosynthesis studies; instrument development. *Mailing Add:* 689 Castle Kirk Dr Baton Rouge LA 70808

WEFERS, KARL, b Bonn, Ger, Aug 4, 28; m 57; c 2. CRYSTALLOGRAPHY, SURFACE CHEMISTRY. *Educ:* Univ Bonn, Dr rer nat, 58. *Prof Exp:* Group leader crystallog & struct chem, Vereinigte Aluminum Werke, Bonn, Ger, 58-66; sci assoc, Alcoa Res Labs, 67-80, FEL, ALCOA LABS, ALCOA TECH CTR, 80- *Mem:* Sigma Xi; Adhesion Soc; Mat Res Soc. *Res:* Physical chemistry and structural chemistry of extractive metallurgy of aluminum; surface chemistry of aluminum; adhesion; surface finishing. *Mailing Add:* 528 Joyce Ave Apollo PA 15613

WEG, JOHN GERARD, b New York, NY, Feb 16, 34; m 56; c 6. PULMONARY DISEASES. *Educ:* Col Holy Cross, AB, 55; New York Med Col, MD, 59. *Prof Exp:* Actg chief pulmonary, Wilford Hall, USAF Hosp, Lackland AFB, 63-64, chief pulmonary & inhalation ther sect, 64-66, pulmonary & infectious dis, 66-67; asst prof internal med, Baylor Col Med, 67-71, assoc prof, 71; assoc prof internal med, 71-74, physician in charge Pulmonary & Crit Care Med Div, 71-85, PROF INTERNAL MED, MED SCH, UNIV MICH, 74- *Concurrent Pos:* Clin dir pulmonary & internal med, Jefferson Davis Hosp, 67-71; consult, Area Consult Surgeon Gen, Methodist Hosp, Houston, Tex & Med Adv Comt, Assoc Degree Prog Health Sci, South Tex Jr Col, 68-71, Vet Admin Hosp, Ann Arbor, Mich, 72- & Wayne Count Gen Hosp, 73-84, NIH Rev Panel & Study Sect. *Mem:* Am Col Chest Physicians (pres 80-81); Am Col Physicians; Am Thoracic Soc (secy & treas 74-76); Am Bd Internal Med; Am Fedn Clin Res. *Res:* Correlation of clinical, physiologic, biochemical-immunologic, and pathologic mechanisms of pulmonary disease; acute respiratory failure; pulmonary emboli; diffuse interstitial disease; sarcoidosis; occupational and environmental medicine. *Mailing Add:* Box 0026 BIH245 1500 E Medical Center Dr Ann Arbor MI 48109-0026

WEG, RUTH B(ASS), b New York, NY, Oct 12, 20; c 3. GERONTOLOGY. *Educ:* Hunter Col, BA, 40; Univ Southern Calif, MS, 54, PhD(zool), 58. *Prof Exp:* Res assoc, dept biochem, 58-59, biol & physiol, 60-70, biochemist, 62-64, biologist in residence, Air Pollution Control Inst, 66, assoc prof biol, Univ & assoc dir educ & training, 68-74, assoc prof biol & geront, 75-84, PROF BIOL & GERONT, UNIV SOUTHERN CALIF, 84-, RES ASSOC, GERONT RES INST, ETHEL PERCY ANDRUS GERONT CTR, 85- *Concurrent Pos:* Consult, Rossmoor-Cortese Inst Study Retirement & Aging, Univ Southern Calif, 67, mem teaching fac, summer inst study geront, 68-, pre-med adv, Univ, 70-, dean student affairs, Leonard Davis Sch Geront, 74-76; Ida Beam Distinguished prof, Univ Iowa, 86. *Honors & Awards:* Dr George C Griffiths Mem Award, 77. *Mem:* AAAS; Am Inst Biol Sci; Fedn Am Socs Exp Biol; Geront Soc; Am Soc Aging; Asn Geront Higher Educ. *Res:* Normal pathological aging; nutrition and age, health and wellness in the later years; sexuality and sex roles in mid-life and later years; mid-life men and women; education for gerontology; sex differences in morbidity, mortality and longevity. *Mailing Add:* Andrus Geront Ctr Univ Southern Calif Univ Park Los Angeles CA 90089-0191

WEGE, ANN CHRISTENE, b Manitowoc, Wis, Mar 10, 56. RESEARCH ADMINISTRATION. *Educ:* Univ Wis, BS, 78. *Prof Exp:* Food serv mgr, Saga Food Serv Corp, 78-79; lab technician, Bio-Tech Resources, Inc, 79-82, res microbiologist, 82-84, sr res microbiologist, 84-86, group leader, Brewing & Anal Serv, 86-90. *Mem:* Nat Inst Food Technologists; Master Brewers Asn Am; Am Asn Brewing Chemists. *Res:* Developing industrial fermentations for the food & brewing industry primary area, fermented flavors. *Mailing Add:* 125 Cherry Ave Francis Creek WI 54227

WEGE, WILLIAM RICHARD, b Shawano, Wis, Mar 31, 26; m 50; c 3. RADIOLOGY. *Educ:* Marquette Univ, DDS, 52; Univ Ala, MS, 67. *Prof Exp:* Assoc prof radiol, 67-76, PROF ORAL MED & RADIOL, MED COL GA, 76- *Concurrent Pos:* NIH spec fel, Med Col Ga, 67-68; consult, US Army, Ft Jackson, SC, 67- & Vet Admin, 68- *Mem:* Fel Am Acad Dent Radiol (treas, 72); Int Asn Dent Res (pres, 71); fel Am Col Dent. *Res:* Cytogenetics in relation to radiation. *Mailing Add:* Salem Rd Morgantown GA 30560

WEGENER, PETER PAUL, b Berlin, Ger, Aug 29, 17; nat US. FLUID PHYSICS, GAS DYNAMICS. *Educ:* Univ Berlin, Dr rer nat(physics, geophys), 43. *Hon Degrees:* Dr Ing Eh, Univ Karlsruhe, 79. *Prof Exp:* Mem & head, Basic Res Group Wind Tunnels, Ger, 43-45; head Res Group Hypersonic Wind Tunnel Design & Res, US Naval Ord Lab, 46-53; chief, Fluid Mech Sect, Jet Propulsion Lab, Calif Inst Technol, 53-60; prof appl sci, 60-72, chmn dept, 66-71, Harold Hodgkinson prof, 72-87, HAROLD HODGKINSON EMER PROF ENG & APPL SCI, YALE UNIV, 87- *Concurrent Pos:* Consult. *Honors & Awards:* Sr Am Scientist Humboldt Award, 79. *Mem:* Fel Am Phys Soc; fel Inst Advan Studies Berlin. *Res:* Gas dynamics; fluid dynamics; chemical physics related to flow problems such as chemical kinetics and condensation. *Mailing Add:* Mason Lab PO Box 2159 Yale Univ New Haven CT 06520

WEGENER, WARNER SMITH, b Cincinnati, Ohio, June 23, 35; m 58; c 3. MICROBIOLOGY, BIOCHEMISTRY. *Educ:* Univ Cincinnati, BS, 57, PhD(microbiol), 64. *Prof Exp:* Asst mem, Res Labs, Albert Einstein Med Ctr, 64-68; asst prof, 68-72, ASSOC PROF MICROBIOL, SCH MED, IND UNIV, INDIANAPOLIS, 72- *Concurrent Pos:* NIH fel, 65-66. *Mem:* Am Soc Microbiol. *Res:* Intermediary metabolism and cellular regulatory processes; biochemical basis of microbial pathogenicity. *Mailing Add:* Dept Microbiol Ind Univ Sch Med 1120 South Dr Indianapolis IN 46223

WEGGEL, JOHN RICHARD, b Philadelphia, Pa, Nov 29, 41; m 64; c 3. ENGINEERING, HYDRAULICS. *Educ:* Drexel Inst Technol, BS, 64; Univ Ill, MS, 66, PhD(civil eng), 68. *Prof Exp:* Asst prof, Univ Ill, Urbana, 68-71; hydraul engr, Coastal Eng Res Ctr, 71-73, spec asst to dir, 73-77, chief eval br, 77-83; assoc prof, 83-88, PROF & HEAD, DEPT CIVIL & ARCHIT ENG, DREXEL UNIV, 88- *Concurrent Pos:* Mem steering comt, Deep Draft Harbor Study, NAtlantic Div, CEngr, US Army, 72-; mem, PIANC Waves Comn, 73-76; prof lectr, George Washington Univ, 74-83; consult, US Army CEngr & var munic. *Mem:* Am Soc Civil Engrs; Am Geophys Union. *Res:* Coastal engineering; civil engineering; coastal processes and ocean engineering; development of design criteria for coastal works and littoral processes. *Mailing Add:* 627 Rodman Ave Jenkintown PA 19046

WEGGEL, ROBERT JOHN, b Cleveland, Ohio, Mar 16, 43; m 80. ELECTROMAGNETISM, ENGINEERING PHYSICS. *Educ:* Mass Inst Technol, BS, 64; Harvard Univ, BS, 66. *Prof Exp:* MEM SPONSORED RES STAFF MAGNET DESIGN, FRANCIS BITTER NAT MAGNET LAB, MASS INST TECHNOL, 61- *Concurrent Pos:* Consult magnet design, var cos & individuals, 66-76. *Mem:* Sigma Xi. *Res:* Designer of world's most intense continuous field electromagnets employing water-cooled solenoids separately or in combination with superconducting coils (hybrid magnets), or long-duration pulse coils precooled with liquid nitrogen. *Mailing Add:* Mass Inst Technol Rm NW 14-2124 Cambridge MA 02139

WEGMAN, DAVID HOWE, b Baltimore, Md, Mar 13, 40; m 69; c 2. OCCUPATIONAL DISEASE EPIDEMIOLOGY. *Educ:* Swarthmore Col, BA, 63; Harvard Med Sch, MD, 66; Harvard Sch Pub Health, MS, 72. *Prof Exp:* Med intern, Cleveland Metrop Gen Hosp, 66-67; med epidemiologist, Nat Commun Dis Ctr, USPHS, 67-69; dir, indust health, Urban Planning Aid, Inc, 69-71; occup hyg physician, Div Occup Hyg, Commonwealth Mass, 72-78; from asst prof to assoc prof occup health, Harvard Sch Pub Health, 72-83; assoc physician, Brigham & Women's Hosp, 82-83; prof & dir environ & occup health sci, Univ Calif, Los Angeles, 83-87; PROF & CHAIR, DEPT WORK ENVIRON, COL ENG, UNIV LOWELL, 87- *Concurrent Pos:* Mem, Task Group on Surveillance, Nat Inst Occup Safety & Health, 76-77, Study Sect, 80-83, sci comt on epidemiol, Int Comn on Occup Health, 82-, Task Force III, Nat Inst Environ Health Sci, 84; contrib ed, Am J Indust Med, 78-; vis prof, Swed Nat Bd Occup Safety & Health, 84; mem, Bd Sci Counr,

Nat Inst Occup Health & Safety, 88-; secy, Sci Comt, Int Comn Occup Health, 90-93; chair, Occup Health Adv Bd, United Auto Workers, Gen Motors & Epidemiol Rev Bd, DuPont Corp. *Honors & Awards:* Alfred L Frechette Award, Mass Pub Health Asn, 79. *Mem:* Fel Am Col Epidemiol; fel Am Col Prev Med; Am Pub Health Asn; Am Soc Epidemiol; Soc Occup & Environ Health; Soc Epidemiol Res; Am Occup Med Asn; Int Epidemiol Asn. *Res:* Epidemiologic study of occupational cancer and occupational respiratory disease; the development of occupational disease and hazard surveillance strategies; the evaluation of effectiveness of public policy and regulations. *Mailing Add:* Dept Work Environ Univ Lowell Lowell MA 01854

WEGMAN, EDWARD JOSEPH, b Terre Haute, Ind, July 4, 43; m 67; c 2. MATHEMATICAL STATISTICS. *Educ:* St Louis Univ, BS, 65; Univ Iowa, MS, 67, PhD(statist), 68. *Prof Exp:* From asst prof to assoc prof statist, Univ NC, 68-78; dir statist & probability prog, 78-82, head, Math Sci Div, Off Naval Res, 82-86; DUNN PROF & CHAIR, GEORGE MASON UNIV, 86- *Concurrent Pos:* Res assoc, NSF, 68-72 & Off Naval Res, 72-74; consult, Naval Coastal Syst Lab 72-74, NC State Utilities Comn,74, Gov James Hunt NC,75, NC Dept Pub Instr, 76- 77 , US Off Mgt & Budget, 80- 81; prin investr, Air Force Off Sci Res, 74-78, Off Naval Res, 86-91, Army Res Off, 86-91, NSF, 87-91; sr fac fel, NSF, 76- 77. *Honors & Awards:* Meritorious Civilian Serv Medal, USN, 82. *Mem:* Fel Inst Math Statist; Fel Am Statist Asn; Fel Royal Statist Soc; sr mem Inst Elec & Electronics Engrs; Soc Indust Appl Math; Fel AAAS. *Res:* Statistical inference under order restrictions, time series analysis; function estimation; computational statistics; parallel computing. *Mailing Add:* Ctr Comput Statist George Mason Univ Fairfax VA 22030

WEGMAN, MYRON EZRA, b Brooklyn, NY, July 23, 08; m 36; c 4. PUBLIC HEALTH, PEDIATRICS. *Educ:* City Col of City Univ New York, BA, 28; Yale Univ, MD, 32; Johns Hopkins Univ, MPH, 38. *Prof Exp:* From asst to instr pediat, Sch Med, Yale Univ, 32-36; pediat consult, State Dept Health, Md, 36-41; asst prof child hyg, Sch Trop Med, Univ PR, 41-42; dir training & res, Dept Health, New York, 42-46, dir, Sch Health Serv, 43-46; prof pediat & head dept, Sch Med, La State Univ, 46-52; dir div educ & training, Pan Am Sanit Bur, WHO, 52-56, secy gen, Pan Am Health Orgn, 57-60; dean, Sch Pub Health, 60-74, prof pub health & prof pediat, 60-78, EMER DEAN, SCH PUB HEALTH, MED SCH, UNIV MICH, ANN ARBOR, 74-, JOHN G SEARLE EMER PROF PUB HEALTH & EMER PROF PEDIAT & COMMUN DIS, 78- *Concurrent Pos:* From intern to resident, New Haven Hosp, Conn, 32-36; lectr, Maternal & Child Health, Johns Hopkins Univ, 39-46; asst prof, Col Physicians & Surgeons, Columbia Univ, 41-44; asst prof, Med Col, Cornell Univ, 42-46; pediatrician-in-chief, Charity Hosp, New Orleans, 46-52, pres, Vis Staff, 50-51; consult, Children's Hosp, Washington, DC, 53-60; spec lectr, Sch Med, George Washington Univ, 54-60; chief exec bd, Am Pub Health Asn, 65-70, pres, 72; pres, Comprehensive Health Planning Coun, SE Mich, 68-74; WHO vis prof, Univ Malaya, 74; mem, Soc of Scholars, Johns Hopkins Univ, 75; John G Searle prof pub health, Univ Mich, 75-78; chmn comt pediat hospitalization rates, Nat Res Coun, 75-77; adv, Kellogg Nat Fel Prog. *Honors & Awards:* Grulee Award, Am Acad Pediat, 58; Townsend Harris Medal, City Col New York, 61; Bronfman Prize, 67; Walter P Reuther Award, United Automobile Workers, 74; Sedgwick Medal, Am Pub Health Asn, 74. *Mem:* Fel AAAS; Am Pub Health Asn (pres, 72); Am Pediat Soc; Soc Exp Biol & Med; fel AMA; Soc Pediat Res; Asn Sch Pub Health (pres, 63-66); Am Asn World Health (pres, 82-84); Pan Am Health & Educ Found (pres, 84-85); Physicians Soc Responsibility (pres, 87). *Res:* Infant mortality; prepaid health care; international health organization. *Mailing Add:* Sch Pub Health Univ Mich Ann Arbor MI 48109-2029

WEGMAN, STEVEN M, b Sioux Falls, SDak, Apr 29, 53; m 78; c 2. ELECTROMAGNETISM. *Educ:* SDak State Univ, BS, 77. *Prof Exp:* Proj engr, SDak Energy Off, 78-82; dir, Govt Off Energy Policy, 82-90; STAFF ENGR, SDAK PUB UTILITIES COMN, 91- *Concurrent Pos:* Consult, SDak Renewable Energy Asn, 79-, Western Area Power Admin, 80-, SDak Cement Plant, 85- & Govt Off Energy Policy, 90-; lectr, SDak State Univ, 80- *Mem:* Am Soc Heating Refrig & Air Conditioning Engrs. *Res:* Energy use in residential, commercial and institutional buildings. *Mailing Add:* SDak Pub Utilities Comn 125 S Madison Pierre SD 57501

WEGMANN, THOMAS GEORGE, b Milwaukee, Wis, Sept 29, 41; m 65; c 1. GENETICS, IMMUNOLOGY. *Educ:* Univ Wis, BA, 63, PhD(med genetics), 68. *Prof Exp:* From asst prof to assoc prof biol, Harvard Biol Labs, 69-74; assoc prof, 74-76, PROF IMMUNOL & PRIN INVESTR, MED RES COUN IMMUNOBIOL UNIT, UNIV ALTA, 76- *Mem:* Am Asn Immunologists. *Res:* Genetics and the immune response; immunological tolerance and developmental analysis of genetically-marked chimeric systems. *Mailing Add:* Dept Immunol Univ Atla Edmonton AB T6G 2E2 Can

WEGNER, GENE H, b Madison, Wis, Aug 30, 30; m 53; c 4. FERMENTATION TECHNOLOGY, SINGLE CELL PROTEIN. *Educ:* Univ Wis, BS, 53, MS, 57, PhD(bact, biochem), 62. *Prof Exp:* Asst bact, Univ Wis, 55-57; microbiologist, Eli Lilly & Co, 57-59; asst bact, Univ Wis, 59-60, dept fel, 60-61; sr res microbiologist, Phillips Petrol Co, 62-80, biotechnol sect supvr, 80-84; vpres, res & develop, Provesta Corp, 84-87; MGR FERMENTATION PROD RES & DEVELOP, PHILLIPS PETROL CO, 88- *Mem:* Am Soc Microbiol; Am Chem Soc; Soc Indust Microbiol. *Res:* Hydrocarbon microbiology; fatty acid metabolism; microbial lipids; single cell protein; continuous culture; pilot plant development; biopolymers; biotechnology; rDNA fermentations and product isolation; yeast and food fermentations; bioremediation. *Mailing Add:* Phillips Res Ctr Phillips Petrol Co Bartlesville OK 74004

WEGNER, HARVEY E, b Tacoma, Wash, Aug 12, 25; m 49; c 3. NUCLEAR PHYSICS. *Educ:* Univ Puget Sound, BS, 48; Univ Wash, MS, 51, PhD(physics), 53. *Prof Exp:* Asst physicist, Brookhaven Nat Lab, 53-56; physicist, Los Alamos Sci Lab, 56-62; physicist, 62-66, co-dir, Tandem Van de Graff Facil, 70-75, CONSTRUCT MGR, TANDEN VAN DE GRAFF FACIL, BROOKHAVEN NAT LAB, 66-, SR PHYSICIST, 68- *Concurrent Pos:* Consult, Radiation Dynamics, Inc, 63- & Gen Ionex, 74- *Mem:* Fel Am Phys Soc; Am Asn Physics Teachers. *Res:* Accelerator construction; cyclotrons in nuclear physics and machine development; semiconductor detectors; low energy accelerator construction and development; heavy ion reaction and fusion physics. *Mailing Add:* Dept Phys Brookhaven Nat Lab Upton NY 11973

WEGNER, KARL HEINRICH, b Pierre, SDak, Jan 5, 30; m 57; c 3. MEDICINE, PATHOLOGY. *Educ:* Yale Univ, BA, 52; Harvard Med Sch, MD, 59. *Prof Exp:* prof path & chmn dept, Sch Med, Univ SDak, 68-73, dean sch med & vpres health affairs, Univ, 73-79; PATHOLOGIST & MED DIR, LAB CLIN MED & SIOUX VALLEY HOSP, 62- *Mem:* Col Am Path; Am Soc Clin Path; Int Acad Path. *Mailing Add:* Lab Clin Med 1212 S Euclid St Sioux Falls SD 57105

WEGNER, MARCUS IMMANUEL, b South Haven, Mich, Mar 3, 15; m 41; c 4. BIOCHEMISTRY, NUTRITION. *Educ:* St Norbert Col, BS, 36; Univ Wis, PhD(nutrit biochem), 41. *Prof Exp:* Asst chemist, Exp Sta, Agr & Mech Col, Univ Tex, 41-43; asst nutritionist, Exp Sta, NDak Col, 43-44; res chemist, Mead Johnson & Co, Ind, 44-48; asst res dir, Res Div, Oscar Mayer & Co, 48-51; nutritionist, Pet Milk Co, 51-59, sect leader new prod develop, 59-60, group mgr, 60-61; res dir, Ward Foods, Inc, 61-64; asst res dir, Best Foods Div, Corn Prod Co, 64-69; br chief & diet appraisal, Consumer & Food Economics Div, Agr Res Serv, USDA, 69-71, asst dir, Eastern Regional Res Lab, 71-80; CONSULT FOOD SCI & NUTRIT, 80- *Mem:* Am Asn Cereal Chem; Am Chem Soc; Inst Food Technol; Sigma Xi. *Res:* Chemistry and nutrition of protein hydrolysates; bakery products; food and diet appraisal; new and wider uses for American farm commodities; meats and dairy products; convenience frozen foods; new product development. *Mailing Add:* 401 Woodley Rd Santa Barbara CA 93108

WEGNER, PATRICK ANDREW, b South Bend, Ind, Nov 14, 40. ORGANOMETALLIC CHEMISTRY. *Educ:* Northwestern Univ, Evanston, BA, 62; Univ Calif, Riverside, PhD(chem), 66. *Prof Exp:* Res chemist, E I du Pont de Nemours & Co, Inc, 66-68; vis prof chem, Harvey Mudd Col, 68-69; from asst prof to assoc prof, 69-74, PROF CHEM, CALIF STATE UNIV, FULLERTON, 75-, CHMN DEPT, 78- *Concurrent Pos:* Grants, Petrol Res Corp, Calif State Univ, Fullerton, 68-72, 74-78, Res Corp, 70-72, NASA, 72-74; Alexander von Humboldt fel, 77-78. *Mem:* Am Chem Soc. *Res:* Boron hydride chemistry; transition metal organometallic chemistry. *Mailing Add:* Dept Chem Calif State Univ Fullerton CA 92634

WEGNER, PETER, b Aug 20, 32; US citizen; m 56; c 4. COMPUTER SCIENCE. *Educ:* Univ London, BSC, 53, PhD(comput sci), 68; Pa State Univ, MA, 59. *Prof Exp:* Res assoc, Comput Ctr, Mass Inst Technol, 59-60; asst dir statist lab, Harvard Univ, 60-61; lectr comput sci, London Sch Econ, 61-64; asst prof, Pa State Univ, 64-66; assoc prof, Cornell Univ, 66-69; assoc prof comput sci, 69-, PROF, BROWN UNIV, 75- *Concurrent Pos:* Ed-in-chief, Asn Comput Mach Press Bks. *Mem:* Asn Comput Mach; Math Asn Am. *Res:* Programming language theory and implementation; computer science education; semantics of programming languages; software engineering. *Mailing Add:* Dept Comput Sci Brown Univ Brown Sta Providence RI 02912

WEGNER, ROBERT CARL, b New York, NY, Jan 7, 44; m 69; c 2. GEOPHYSICS, GEOLOGY. *Educ:* Queen's Col, NY, BA, 67; Lehigh Univ, MS, 72; Rice Univ, PhD(geophys), 78. *Prof Exp:* Vis lectr geol, Am Geol Inst, 68; geophys interpreter, Exxon Co USA, 69-70, res specialist, 74-80, supvr geophys, Exxon Prod Res Co, 80-86, COORDR EXXON CO, INT, 86- *Concurrent Pos:* Cities Serv fel, 70; adj prof, Rice Univ, 81-82; chmn, Continuing Educ Comt, Soc Explor Geophysicists, 86. *Mem:* Am Geophys Union; Soc Explor Geophysicists; Sigma Xi; AAAS. *Res:* Geophysical research in reflection seismology applied to the exploration of oil and gas reservoirs. *Mailing Add:* Exxon Prod Res PO Box 2186 Houston TX 77252

WEGNER, THOMAS NORMAN, b Cleveland, Ohio, July 8, 32; m 62; c 1. ANIMAL PHYSIOLOGY, BIOCHEMISTRY. *Educ:* Mich State Univ, BS, 54; Colo State Univ, MS, 56; Univ Calif, Davis, PhD(animal physiol), 64. *Prof Exp:* Asst animal pathologist, 64-66, asst prof dairy sci, 67-76, LECTR & RES ASSOC ANIMAL SCI, AGR EXP STA, UNIV ARIZ, 76- *Mem:* Am Dairy Sci Asn. *Res:* Carbohydrate metabolism in ruminant animals; pathogenesis of coccidioidomycosis, biochemistry of the immune responses; physiology of abnormal milk production and heat stress on dairy cows. *Mailing Add:* Dept Dairy Sci Univ Ariz Tucson AZ 85721

WEGST, W(ALTER) F(REDERICK), b Philadelphia, Pa, Nov 20, 09; m 33; c 2. CHEMICAL ENGINEERING. *Educ:* Drexel Inst Technol, BS, 31. *Prof Exp:* Chemist, Philadelphia Quartz Co, 31-41; group leader, BASF Wyandotte Corp, 41-54, res supvr, 54-64, res assoc, 64-73; RETIRED. *Mem:* Am Chem Soc. *Res:* Silicate manufacture; formulation of detergents and sanitizing agents; inhibition alkali action on glass by use of berylliate, zincate and organics. *Mailing Add:* 8151 Wood Grosse Ile MI 48138-1143

WEGST, WALTER F, JR, b Philadelphia, Pa, Dec 26, 34; div; c 2. RADIOLOGICAL HEALTH. *Educ:* Univ Mich, BSE, 56, MSE, 57, PhD(environ health), 63; Am Bd Health Phys, Dipl, 66; Bd Cert Safety Prof, Cert. *Prof Exp:* Reactor health physicist, Phoenix Mem Lab, Univ Mich, 57-58, lab health physicist, 58-59, lab supvr, 59-60; inst health physicist, Calif Inst Technol, 63-68, safety mgr, 68-71, mgr security, 71-73, mgr safety, 71-79; dir res & occup safety, Univ Calif, Los Angeles, 79-88; CONSULT, 88-; MGR ENVIRON SAFETY & HEALTH, RAYTHEON SERV NEV, LAS VEGAS, 90- *Concurrent Pos:* Consult, City of Hope Nat Med Ctr, Atomics Int Div Rockwell Int, 74-, Southern Calif Edison, Genetech & Training Resources Div, Nuclear Support Serv. *Mem:* Health Phys Soc; Am Indust Hyg Asn; Sigma Xi; AAAS; Am Health Physics Soc; Am Conf Govt Indust Hygienists. *Res:* Radiobiological studies on mammalian cells; secondary electron production by charged particle passage through matter; university health physics problems. *Mailing Add:* 7620 Cruz Bay Ct Las Vegas NV 89128

WEGWEISER, ARTHUR E, b New York, NY, Feb 20, 34; wid; c 2. GEOLOGY. *Educ:* Brooklyn Col, BA, 55; Hofstra Univ, MS, 58; Wash Univ, St Louis, PhD(geol), 66. *Prof Exp:* Sci teacher, Island Trees High Sch, 58-61; chmn, Dept Earth Sci, 69-75, PROF GEOL, EDINBORO UNIV PA, 65-, DIR MARINE SCI CONSORTIUM, 68- *Concurrent Pos:* Adj lectr, Hofstra Univ, 58-61; fel, Woods Hole Oceanog Inst & Washington Univ, St Louis, Mo. *Mem:* Soc Econ Paleont & Mineral; Sigma Xi; Nat Asn Geol Teachers. *Res:* Micropaleontology-ecology of recent Foraminifera; environments of deposition. *Mailing Add:* Edinboro Univ Pa 5341 Gibson Hill Rd Edinboro PA 16412

WEHAUSEN, JOHN VROOMAN, b Duluth, Minn, Sept 23, 13; m 38; c 4. WATER WAVE THEORY, SHIP HYDRODYNAMICS. *Educ:* Univ Mich, BS, 34, MS, 35, PhD(math), 38. *Prof Exp:* Instr math, Brown Univ, 37-38, Columbia Univ, 38-40 & Univ Mo, 40-44; consult opers res group, Off Field Serv, Off Sci Res & Develop, USN, 44-46, mathematician, David Taylor Model Basin, 46-49, actg head mech br, Off Naval Res, 49-50; assoc res mathematician, Inst Eng Res, 56-57, res mathematician, Dept Naval Archit, 57-58, assoc prof, 58-59, PROF ENG SCI, UNIV CALIF, BERKELEY, 59- *Concurrent Pos:* Lectr, Univ Md, 45-50; exec ed, Math Rev, Am Math Soc, 50-56; vis prof, Univ Hamburg, 60-61, Flinders Univ SAustralia, 67, Univ Nantes, France, 73 & 84, Univ Grenoble, 79, Technion, 82, Univ Tel Aviv, 82 & Chalmers Univ, Sweden, 82; ed, Ann Rev Fluid Mech, 70-85. *Honors & Awards:* Davidson Medal, Soc Naval Archit Marine Eng, 84. *Mem:* Nat Acad Eng; Soc Naval Archit & Marine Eng; Math Asn Am; Am Math Soc. *Res:* Fluid mechanics, especially theory of water waves and hydrodynamics of ships. *Mailing Add:* Dept Naval Archit & Offshore Eng Univ Calif Berkeley CA 94720

WEHE, ROBERT L(OUIS), b Topeka, Kans, Apr 14, 21; m 42; c 5. MECHANICAL ENGINEERING. *Educ:* Univ Kans, BS, 48; Univ Ill, MS, 51. *Prof Exp:* Instr mech eng, Univ Ill, 48-51; asst prof, 51-57, ASSOC PROF MECH ENG, CORNELL UNIV, 57- *Concurrent Pos:* NSF fac fel, Univ Ill, 59-60; proj engr, Corning Glass Works, 57-58, consult, 55-59; consult, Lycoming Div, Avco, Inc, 56. *Mem:* Am Soc Mech Engrs; Am Soc Eng Educ. *Res:* Hydrodynamic lubrication; design of automatic machinery. *Mailing Add:* 409 Campbell Dr Ithaca NY 14850

WEHINGER, PETER AUGUSTUS, b Goshen, NY, Feb 18, 38; m 67. ASTRONOMY. *Educ:* Union Col, NY, BS, 60; Ind Univ, MA, 62; Case Western Reserve Univ, PhD(astron), 66. *Prof Exp:* Res asst astron, Ind Univ, 60-62; NASA fel, Warner & Swasey Observ, Case Western Reserve Univ, 63-65; from instr to asst prof, Univ Mich, 65-70; assoc prof, Univ Kans, 70-72; vis assoc prof, Tel-Aviv Univ, Israel, 72-75; prin res fel, Royal Greenwich Observ, Eng, 75-78; vis sr scientist, Max Planck Inst Astron, Heidelberg, Ger, 78-80; RES PROF DEPT PHYSICS & ASTRON, ARIZ STATE UNIV, 81- *Concurrent Pos:* Sr res assoc, Ohio State Univ, 78-79; vis astron, Royal Greenwich Observ, Eng, 83; NASA predoctoral fel, 63-65; Smithsonian fel, Wise Observ, 72-75; prin res fel, UK Sci Res Council 75-78; sr fel, Max Planck-Gesellschaft, 78-80; vis res fel, Stromlo & Siding Spring Observ, Australian Nat Univ, 86, 87, 89 & 90; ed, Halley Hotline, Electronic Bull Bd, Int Halley Watch, 85-87, Electronic Newslett on Comets, 89-90; discipline spec spectros Int Halley Watch, NASA, Jet Propulsion Lab, 82-90; dir, Ariz Space Consortium, Ariz State Univ, 90- *Honors & Awards:* Sigma Xi Prize, 60. *Mem:* Am Astron Soc; Royal Astron Soc; Int Astron Union; Sigma Xi; Planetary Soc. *Res:* Astronomical instrumentation; molecular spectroscopy; digital archiving of cometary spectra; cometary spectroscopy; spectroscopy and imaging of active galaxies, quasars and quasar host galaxies; Jupiter-Io sodium cloud. *Mailing Add:* Dept Physics & Astron Ariz State Univ Tempe AZ 85287-1504

WEHLAU, AMELIA W, b Berkeley, Calif, Feb 5, 30; m 50; c 4. ASTRONOMY. *Educ:* Univ Calif, Berkeley, AB, 49, PhD(astron), 53. *Prof Exp:* Lectr, 65-71, asst prof, 71-78, ASSOC PROF ASTRON, UNIV WESTERN ONT, 78- *Concurrent Pos:* Mem, Can Comt, Int Astron Union, 67-70. *Mem:* Am Astron Soc; Astron Soc Pac; Can Astron Soc; Int Astron Union; fel Royal Astron Soc. *Res:* Variable stars in globular clusters and related stellar systems. *Mailing Add:* Dept Astron Univ Western Ont London ON N6A 3K7 Can

WEHLAU, WILLIAM HENRY, b San Francisco, Calif, Apr 7, 26; m 50; c 4. ASTRONOMY. *Educ:* Univ Calif, AB, 49, PhD(astron), 53. *Prof Exp:* Instr astron, Case Inst Technol, 53-55; assoc prof, 55-65, PROF ASTRON, UNIV WESTERN ONT, 65-, HEAD DEPT, 55- *Mem:* Am Astron Soc; Royal Astron Soc Can; Can Astron Soc. *Res:* Stellar spectroscopy and photometry; astrophysics. *Mailing Add:* Dept Astron Univ Western Ont London ON N6A 3K7 Can

WEHLE, LOUIS BRANDEIS, JR, b Washington, DC, Dec 28, 18; m 46; c 4. AERONAUTICAL ENGINEERING. *Educ:* Harvard Univ, BS, 40. *Prof Exp:* Stress analyst, 41-48, head struct methods group, 48-56, head nuclear sect, 56-59, staff engr, 61-72, CHIEF SCIENTIST, GRUMMAN AEROSPACE CORP, BETHPAGE, 72- *Mem:* Am Phys Soc; Am Inst Aeronaut & Astronaut. *Res:* Stress analysis; structural mechanics. *Mailing Add:* 28 Camel Hollow Rd Huntington NY 11743

WEHMAN, ANTHONY THEODORE, organic chemistry, organometallic chemistry; deceased, see previous edition for last biography

WEHMANN, ALAN AHLERS, b New York, NY, Dec 28, 40; m 68; c 2. PARTICLE PHYSICS. *Educ:* Rensselaer Polytech Inst, BS, 62; Harvard Univ, MA, 63, PhD(physics), 68. *Prof Exp:* Res assoc physics, Univ Rochester, 67-69; physicist I, Meson Lab Sect, 69-80, scientiest I, Magnet Test Facil, 80-82, Tevatron I Antiproton Source Sect, 83-85, SCIENTIST I, RES DIV, FERMI ACCELERATOR LAB, 85- *Concurrent Pos:* Vis scientist, Inst Physics, Univ Mex, 83. *Mem:* Am Phys Soc. *Res:* Experimental high energy particle physics. *Mailing Add:* Fermilab MS 220 Batavia IL 60510

WEHNER, ALFRED PETER, b Wiesbaden, Ger, Oct 23, 26; US citizen; m 55; c 4. TOXICOLOGY. *Educ:* Johannes Gutenberg Univ, Can Med, 49, DDS, 51, ScD, 53; Acad Toxicol Sci, dipl, 88. *Prof Exp:* Pvt pract, WGer, 51-53; dr dent, Guggenheim Dent Clin, 53-54; Dr med dent, 7100th Hosp, USAF, Europe, 54-56; res asst microbiol, Field Res Lab, Mobil Oil Co, Tex, 57-62; sr res scientist biomed res, Biomet Instrument Corp, 62-64; dir & pres, Electro-Aerosol Inst, Inc, 64-67; prof biol & chmn dept sci, Univ Plano, 66-67; res assoc inhalation toxicol, Battelle Pac Northwest Labs, 67-77, mgr environ & indust toxicol, 78-80, task leader indust toxicol, 80-89; PRES, BIOMED & ENVIRON CONSULT CORP, 89- *Concurrent Pos:* Fel clin pedodontia, Guggenheim Dent Clin, 53-54; consult, Vet Admin Hosp, McKinney, Tex, 63-65; US rep, Int Soc Biometeorol, 72-80; exec bd, Int Soc Aerosols Med, 70-80; bd dirs, Am Inst Biomed Climatol, 73-, proj dir & prin investr res projs exceeding $15 million; ed, Medicef, Biol Interaction of Inhaled Mineral Fibers & Cigarette Smoke; ed-in-chief, Bull Am Inst Biomed Climat; mem, subcomt D221101, Am Soc Testing & Mat. *Mem:* Int Soc Biometeorol; Int Soc Aerosols Med; Am Inst Biomed Climat (secy, 80-84, pres, 84-); Sigma Xi; AAAS; Int Soc Aerobiol. *Res:* Inhalation toxicology; biological effects of electro-aerosols and air ions; bioclimatology; author of more than 120 science papers and contributions to science books; investigation of biological effects of various air pollutants in animal models. *Mailing Add:* 312 Saint St Richland WA 99352

WEHNER, DONALD C, b Middletown, NY, Apr 1, 29; m 50; c 3. WATER POLLUTION, ENVIRONMENTAL SCIENCES. *Educ:* Univ Bridgeport, BA, 51. *Prof Exp:* Biologist, Lederle Div, 51-54 & 56-58, res biologist, Indust Chem Div, 58-69, asst to dept head, Water Treating Chem Dept, 69-70. microbiologist-field engr, 70-74, dist sales mgr, 74-75, TECH SPECIALIST, PAPER CHEMS DEPT, AM CYANAMID CO, 75- *Concurrent Pos:* Mem subcomt biol anal of waters for sub-surface injection, Am Petrol Inst, 59-64. *Mem:* Soc Indust Microbiol (treas, 66-68, vpres, 69-70, pres, 71-72); Am Inst Biol Sci; Water Pollution Control Fedn; Tech Asn Pulp & Paper Indust. *Res:* Research and development of industrial algicides, bactericides and fungicides. *Mailing Add:* 4059 Westmoreland Dr S Mobile AL 36601

WEHNER, GOTTFRIED KARL, b Ger, Sept 23, 10; nat US; m 39; c 3. PHYSICS. *Educ:* Inst Technol, Munich, Ger, dipl, 36, DrIng, 39. *Prof Exp:* Physicist, Inst Technol, Munich, 39-45, Wright-Patterson AFB, Ohio, 47-55 & Gen Mills, Inc, 55-63; dir appl sci div, Litton Industs, Inc, 63-68; PROF ELEC ENG, UNIV MINN, MINNEAPOLIS, 68- *Honors & Awards:* Welch Medal, Am Vacuum Soc, 71. *Mem:* Fel Am Phys Soc; fel Inst Elec & Electronics Engrs; hon mem Am Vacuum Soc; Ger Phys Soc; Europ Phys Soc. *Res:* Plasma and surface physics; thin films; vacuum physics; sputtering. *Mailing Add:* 6017 Walnut Dr Minneapolis MN 55436

WEHNER, JEANNE M, SEIZURES, RECEPTOR PLASTICITY. *Educ:* Univ Minn, PhD(biochem). *Prof Exp:* ASST PROF PHARM & MOLECULAR GENETICS, INST BEHAV GENETICS, UNIV COLO. *Mailing Add:* Univ Colo PO Box 447 Boulder CO 80303

WEHNER, PHILIP, b Chicago, Ill, July 21, 17; m 43; c 2. INDUSTRIAL ORGANIC CHEMISTRY. *Educ:* Univ Chicago, PhD(org chem), 43. *Prof Exp:* Res chemist, Gen Aniline & Film Corp, Pa, 43-46; assoc chemist, Argonne Nat Lab, 46-52; chemist, Ciba States Ltd, 52-55; chemist, Toms River-Cincinnati Chem Corp, 55-60, mgr res & develop, 60-64, vpres, 64-68, pres, 68-73; vpres prod & tech develop, Ciba-Geigy Corp, 73-78, vpres tech affairs, Dyestuffs & Chem Div, 78-79; RETIRED. *Mem:* Am Chem Soc; Am Asn Textile Chem & Colorists; AAAS. *Mailing Add:* 3404 Old Onslow Rd Greensboro NC 27407

WEHR, ALLAN GORDON, b Brooklyn, NY, July 31, 31; m 51; c 2. CHEMICAL ENGINEERING, MATERIALS ENGINEERING. *Educ:* Mo Sch Mines, BS, 58; Univ Mo, PhD(metall eng), 62. *Prof Exp:* Res engr, P R Mallory Co, 59-60; assoc prof metall eng, 61-62, prof mat eng & head dept, 62-73, PROF CHEM ENG, MISS STATE UNIV, 73- *Mem:* Am Soc Metals; Am Soc Eng Educ; Am Inst Chem Engrs. *Res:* Fracture; magnetohydrodynamics; mechanical behavior of materials; nitrogen oxide generation and control; uses of microwave and radio requency energy. *Mailing Add:* PO Box AW Mississippi State MS 39762

WEHR, CARL TIMOTHY, b San Francisco, Calif, Feb 15, 43; m 66; c 2. BIOCHEMISTRY, GENETICS. *Educ:* Whitman Col, BA, 65; Ore State Univ, PhD(microbial physiol), 69. *Prof Exp:* Fel microbiol, Dept Bact, Univ Calif, Davis, 69-71; res biologist molecular biol, Virus Lab, Univ Calif, Berkeley, 71-75; dir res chem, Agr Sci Labs, 75-78; sr chemist biochem, 78-80, MGR, HIGH PERFORMANCE LIQUID CHROMATOGRAPHY LAB, VARIAN INSTRUMENTS, 80- *Concurrent Pos:* Jane Coffin Childs Mem Fund Med Res fel, 69-71. *Mem:* AAAS; Am Soc Microbiol; Inst Food Technol. *Res:* Development of analytical methods in biomedical research and biotechnology. *Mailing Add:* Box 2093 Walnut Creek CA 95495

WEHR, HERBERT MICHAEL, b San Francisco, Calif, Feb 15, 43; m 67; c 1. FOOD SCIENCE, MICROBIOLOGY. *Educ:* Univ Calif, Berkeley, BS, 66; Ore State Univ, MS, 68, PhD(biochem), 72. *Prof Exp:* Supvr microbiol & food chem, 71-73, asst adminr, 73-77, ADMINR, LAB SERV, ORE DEPT AGR, 78-, ADMINR, EXPORT SERV CTR, 90- *Concurrent Pos:* Mem, Conf on Interstate Milk Shipments, Nat Lab Comn, 77-; chmn, Ore Comt Synthetic Chem Environ, 77-; adminr, Pesticide Anal & Response Ctr, 85-; mem, Nat Dairy Coun, Dairy Res Found Sci Adv Comt, 85-88, Nat Comt, Microbiol Criteria Foods, 88-90; mem bd dirs, Asn Off Anal Chemists, 87-, pres, 90-91. *Honors & Awards:* Distinguished Serv Award, Asn Off Anal Chemists, 88, Gov Mgt Award, 85, 89. *Mem:* Inst Food Technologists; Asn Off Anal Chemists; Int Asn Milk, Food & Environ Sanitarians; Am Chem Soc; Coun Agr Sci & Technol. *Res:* Food microbiology; food biochemistry; methods development in food chemistry and microbiology. *Mailing Add:* Lab Serv Ore Dept Agr Salem OR 97310

WEHR, THOMAS A, b Louisville, Ky, Oct 15, 41; m 63; c 1. PSYCHOBIOLOGY. *Educ:* Yale Univ, BA, 65; Univ Louisville, MD, 69; Am Bd Psychiat & Neurol, 83. *Prof Exp:* Clin assoc, Sect Psychiat, Lab Clin Sci, Intramural Res Prog, NIMH, 73-75, res med officer, 75-76, psychiatrist, 76, chief, Clin Res Unit, 77-84, CHIEF, CLIN PSYCHOBIOL BR, INTRAMURAL RES PROG, NIMH, 82- *Concurrent Pos:* USPHS, 73-75, 76-; instr, Washington Sch Psychiat, DC, 76-78; pvt pract psychiat, Bethesda, Md, 77-; vis prof, Univ Naples Med Sch, Italy, 84; adj prof, Dept Biol Sci, Univ Md, 86-; mem adv bd, Soc Res Biol Rhythms. *Honors & Awards:* Anna-Monika Found Award, 81. *Mem:* Am Psychiat Asn; Am Col Neuropsychopharmacol; Soc Biol Psychiat; Sleep Res Soc. *Res:* Psychobiology; patents. *Mailing Add:* Clin Psychobiol Br NIH NIMH Bldg 10 Rm 45239 9000 Rockville Pike Bethesda MD 20892

WEHRBEIN, WILLIAM MEAD, b Omaha, Nebr, Sept 12, 48. RADIATIVE TRANSFER. *Educ:* Nebr Wesleyan Univ, BS, 70; Univ Colo, PhD(astrogeophysics), 77. *Prof Exp:* Res assoc postdoctoral, Atmospheric Sci, Univ Wash, 77-81; asst prof, 81-85, ASSOC PROF PHYSICS, NEBR WESLEYAN UNIV, 81- *Concurrent Pos:* Chair, Physics Dept, Nebr Wesleyan Univ, 85- *Mem:* Am Asn Physics Teachers; AAAS; Am Meteorol Soc; Union Concerned Scientists. *Res:* Planetary atmospheres, especially IR radiative transfer. *Mailing Add:* Nebr Wesleyan Univ 5000 St Paul Ave Lincoln NE 68504-2796

WEHRENBERG, JOHN P, b Springfield, Ill, Aug 10, 27; m 66. MINERALOGY. *Educ:* Univ Mo, BS, 50; Univ Ill, MS, 52, PhD, 56. *Prof Exp:* From asst prof to assoc prof, 55-66, PROF GEOL, UNIV MONT, 66- *Concurrent Pos:* Vis scientist, FBI Acad, Va. *Mem:* Mineral Soc Am; Geochem Soc; Am Crystallog Asn. *Res:* Forensic mineralogy; crystallography; infrared spectra of minerals. *Mailing Add:* Dept Geol Univ Mont Missoula MT 59812

WEHRING, BERNARD WILLIAM, b Monroe, Mich, Aug 3, 37; m 59; c 4. NUCLEAR ENGINEERING, NUCLEAR PHYSICS. *Educ:* Univ Mich, BSE(physics) & BSE(math), 59; Univ Ill, MS, 61, PhD(nuclear eng), 66. *Prof Exp:* From asst prof to prof nuclear eng, Univ Ill, Urbana, 66-84; PROF NUCLEAR ENG, NC STATE UNIV, 84-, DIR NUCLEAR REACTOR PROG, 84- *Concurrent Pos:* Consult, Construct Eng Res Lab, Champaign, Ill, 70-71, Los Alamos Sci Lab & Argonne Nat Lab, 77-79; sabbatical, Van de Graaff Lab, Oak Ridge Nat Lab, 73-74; mem, Cross Sect Eval Working Group, Fission Prod, Brookhaven Nat Lab, 75- *Mem:* Fel Am Nuclear Soc; Am Soc Eng Educ; Am Phys Soc; Inst Elec & Electronics Engrs; Instrument Soc Am; Sigma Xi. *Res:* Fission physics; interaction of radiation with matter; radiation detection; neutron spectroscopy and dosimetry. *Mailing Add:* Nuclear Eng Teaching Lab 10100 Burnett Rd Austin TX 78758

WEHRLE, LOUIS, JR, food microbiology, for more information see previous edition

WEHRLE, PAUL F, b Ithaca, NY, Dec 18, 21; m 44; c 4. PEDIATRICS, MICROBIOLOGY. *Educ:* Univ Ariz, BS, 47; Tulane Univ, MD, 47. *Prof Exp:* Clin instr pediat, Univ Ill Col Med, 50-51; res assoc epidemiol & microbiol, Grad Sch Pub Health, Univ Pittsburgh, 51-53; res assoc, Poliomyelitis Lab, Johns Hopkins Univ, 53-55; from asst prof to assoc prof pediat, Col Med, State Univ NY Upstate Med Ctr, 55-61, actg chmn dept microbiol, 59-61; prof, 61-71, HASTINGS PROF PEDIAT, SCH MED, UNIV SOUTHERN CALIF, 71- *Concurrent Pos:* Asst med supt, Chicago Munic Contagious Dis Hosp, Ill, 50-51; head physician contagious dis serv, Los Angeles County-Univ Southern Calif Med Ctr, 61-63, chief physician, Children's Div, 63- *Honors & Awards:* Cert Appreciation Smallpoc Irradication, 87; Medal of Merit, Intl Pediat Asn. *Mem:* Am Acad Pediat; Soc Pediat Res; Am Soc Microbiol; Am Pub Health Asn; Am Asn Immunologists. *Res:* Viral infections in man, especially enteroviruses; antibiotic action; epidemiology of infectious diseases. *Mailing Add:* 233 Esplanade San Clemente CA 92672

WEHRLI, PIUS ANTON, b Bazenheid, Switz, July 27, 33; m 62; c 1. CHEMISTRY. *Educ:* Swiss Fed Inst Technol, Dipl Chem Eng, 64, PhD(chem), 67. *Prof Exp:* Sr chemist, 67-69, res fel, 69-70, res group chief, 70-73, res sect chief, 73-78, dir, Kilo Lab & Process Res, Chem Res Dept, 78-85, ASST VPRES & DIR TECH & PROD DEVELOP, HOFFMANN-LA ROCHE INC, 85- *Mem:* Am Chem Soc; Swiss Chem Soc; Soc Ger Chem. *Res:* Development in synthetic organic chemistry. *Mailing Add:* Hoffmann-La Roche Inc Kingsland Rd Nutley NJ 07110-1199

WEHRLY, THOMAS EDWARD, b Richmond, Ind, Nov 5, 47; m 76; c 2. STOCHASTIC MODELLING, NONPARAMETRIC FUNCTION ESTIMATION. *Educ:* Univ Mich, BS, 69; Univ Wis-Madison, MA, 70, PhD(statist), 76. *Prof Exp:* From asst prof statist to assoc prof, 76-88, PROF DEPT STATISTICS, TEX A&M UNIV, 88- *Concurrent Pos:* Vis res fel, Australian Nat Univ, 90. *Mem:* Am Statist Asn; Biometrics Soc; Inst Math Statist; Royal Statist Soc. *Res:* Stochastic modelling for biological systems and statistical inference for these stochastic models; nonparametric function estimation. *Mailing Add:* Dept Statist Tex A&M Univ College Station TX 77843-3143

WEHRMANN, RALPH F(REDERICK), b Williamstown, Mo, June 24, 18. ELECTROCHEMISTRY. *Educ:* Culver-Stockton Col, BA, 39; St Louis Univ, MS, 41, PhD(inorg chem), 43. *Prof Exp:* Group leader, Manhattan Proj, 43-47; asst res dir, Fansteel Co, 48-61; res scientist, LTV Corp, 62-67; res dir, Poco Graphite Div, Unocal Co, 68-72; SPECIALIST EXTEN FAC, UNIV MO, 73- *Mem:* Am Chem Soc; fel Am Inst Chemists; Am Ceramic Soc. *Res:* Coatings for uranium and thorium; isolation and purification of radioactive elements; preparation of and applications for refractory metals, alloys and compounds; refractory oxidation-resistant materials; preparation and properties of synthetic graphite; high temperature chemistry. *Mailing Add:* 121 S Meramee Suite 200 Clayton MO 63105-1725

WEHRMEISTER, HERBERT LOUIS, b Chicago, Ill, Nov 8, 20; m 53; c 3. CHEMISTRY. *Educ:* Ill Inst Technol, BS, 44; Northwestern Univ, MS, 46, PhD(chem), 48. *Prof Exp:* Lab asst, Portland Cement Asn, Ill, 38-40; control chemist, W H Barber Co, 40-44; res chemist, Miner Labs, 44-46; res chemist, Com Solvents Corp, 49-77, IMC Corp, 77-86; RETIRED. *Mem:* Sigma Xi; Am Chem Soc. *Res:* Organic chemistry; derivatives of sulfenic acids; syntheses in the thiazole series; nitoparaffin and hydroxylamine derivatives; zearalenone and derivatives. *Mailing Add:* 2711 Deming St Terre Haute IN 47803

WEHRY, EARL L, JR, b Reading, Pa, Feb 13, 41. ANALYTICAL CHEMISTRY, SPECTROSCOPY. *Educ:* Juniata Col, BS, 62; Purdue Univ, PhD(chem), 65. *Prof Exp:* Instr chem, Ind Univ, 65-66, asst prof, 66-70; from asst prof to assoc prof, 70-77, PROF CHEM, UNIV TENN, KNOXVILLE, 77- *Honors & Awards:* Meggers Award, Soc Appl Spectros, 82. *Mem:* AAAS; Soc Appl Spectros; Am Chem Soc; Optical Soc Am. *Res:* Fluorescence and phosphorescence; photochemistry; laser spectroscopy. *Mailing Add:* Dept Chem Univ Tenn Knoxville TN 37996-1600

WEI, CHIN HSUAN, b Yuanlin, Taiwan, Oct 25, 26; m 54; c 2. X-RAY CRYSTALLOGRAPHY. *Educ:* Cheng Kung Univ, BS, 50; Purdue Univ, MS, 58; Univ Wis-Madison, PhD(phys chem), 62. *Prof Exp:* Asst phys chem, Cheng Kung Univ, 50-55, instr gen chem, 55-56; res assoc struct chem, Univ Wis, Madison, 62-65, sr spectroscopist, 65-66; BIOPHYSICIST, OAK RIDGE NAT LAB, 66- *Concurrent Pos:* Vis prof, Cheng Kung Univ, 70-71. *Mem:* Am Chem Soc; Am Crystallog Asn. *Res:* Structure determination of organometallic complexes and compounds of biological interest; isolation, purification, characterization and crystalliation of proteins. *Mailing Add:* 129 Newhaven Rd Oak Ridge TN 37830

WEI, CHING-YEU, b Tainan, Taiwan, Dec 17, 48; US citizen; m; c 3. MATERIALS SCIENCE ENGINEERING. *Educ:* Nat Taiwan Univ, BS, 70; Cornell Univ, PhD(mat sci), 77. *Prof Exp:* Staff res scientist, 77-89, MGR, GEN ELEC CORP RES & DEVELOP, 89- *Mem:* Sr mem Inst Elec & Electronics Engrs. *Res:* Amorphous-silicon devices; radiation-hardened micron and submicron CMOS process as well as various infrared detector technologies; manager of a medical imager development program. *Mailing Add:* Corp Res & Develop Gen Elec Co Schenectady NY 12301

WEI, CHUNG-CHEN, b Taiwan, Apr 9, 40; m 65; c 1. ORGANIC CHEMISTRY, PHOTOCHEMISTRY. *Educ:* Nat Taiwan Univ, BS, 63; Colo State Univ, PhD(org chem), 69. *Prof Exp:* Fel org chem, Johns Hopkins Univ, 69-71; res assoc bio-org chem, Yale Univ, 71-73; assoc chemist, Midwest Res Inst, 73; SR CHEMIST, HOFFMANN-LA ROCHE INC, 74- *Mem:* Am Chem Soc. *Res:* Synthesis of heterocyclic compounds and natural products; organic photochemistry. *Mailing Add:* Chem Res Dept Hoffmann-La Roche Inc 340 Kingsland St Bldg 76/11 Nutley NJ 07110-1199

WEI, DIANA YUN DEE, b Che-Kiang, China, June 8, 30; m 62; c 1. MATHEMATICS. *Educ:* Taiwan Norm Univ, BS & BEd, 53; Univ Nebr, MS, 60; McGill Univ, PhD(math), 67. *Prof Exp:* Lectr math, Taipei Inst Technol, Taiwan, 53-58; teaching asst, Univ Nebr, 58-60 & Univ Wash, 60-62; lectr, McGill Univ, 62-65; asst prof, Marianopolis Col, 65-68 & Sir George Williams Univ, 68-72; prof, Sch Comn Baldwin-Cartier, 72-75; PROF MATH, NORFOLK STATE UNIV, 75- *Concurrent Pos:* Nat Res Coun Can grant award, 69, 70, 71. *Mem:* Am Math Soc; Can Math Cong. *Res:* Groups, rings and modules; homology; category; linear algebra. *Mailing Add:* Dept Math & Comput Sci Norfolk State Univ 2401 Corprew Ave Norfolk VA 23504

WEI, EDWARD T, b Shanghai, China, Dec 6, 44; m 66; c 2. PHARMACOLOGY, TOXICOLOGY. *Educ:* Univ Calif, Berkeley, AB, 65; Univ Calif, San Francisco, PhD(pharmacol), 69. *Prof Exp:* Nat Inst Arthritis & Metab Dis fel, Stanford Univ, 69-70; from asst prof to assoc prof, 70-80, PROF TOXICOL, SCH PUB HEALTH, UNIV CALIF, BERKELEY, 80- *Concurrent Pos:* Nat Inst Drug Abuse fel, 70-81; Nat Inst Environ Health Sci fel, 74-76; assoc prof toxicol, Dept Pharmacol, Univ Calif, San Francisco, 75-80, prof, 80-; sr Int Fogarty Fel, 84-85. *Honors & Awards:* Merit Award, Nat Inst Drug Abuse, 87. *Mem:* AAAS; Soc Toxicol; Am Soc Pharmacol & Exp Therapeut. *Res:* Anti-inflammatory peptides; toxic chemicals and their mechanisms of action; biological mechanisms of morphine dependence. *Mailing Add:* Dept Biomed & Environ Health Univ Calif 2120 Oxford St Berkeley CA 94720

WEI, ENOCH PING, b Shanghai, China, July 19, 42; US citizen; m 69; c 2. CEREBRAL MICROCIRCULATION, PERIPHERAL CIRCULATION. *Educ:* Univ NC, Chapel Hill, BA, 67, PhD(physiol), 72. *Prof Exp:* Fel, 71-73, res assoc med, 73-85, DIR RES, MED COL VA, 81-, RES ASST PROF MED, 85- *Concurrent Pos:* Mem, Nominating Comt, Microcirculatory Soc, 87-; vis scientist minority insts, 83- *Mem:* Am Physiol Soc; Int Soc Cerebral Blood Flow & Metab; Am Fedn Clin Res; Microcirculatory Soc; AAAS. *Res:* Local regulatory mechanisms of the cerebral circulation in areas related to high blood pressure, brain injury, ischemia, and endothelium-derived relaxing factor (EDRF). *Mailing Add:* Dept Med Med Col Va MCV Sta Box 281 Richmond VA 23298-0001

WEI, GUANG-JONG JASON, b Fu-Chou, China, Mar 14, 46; m 71. PHYSICAL BIOCHEMISTRY. *Educ:* Cheng Kung Univ, Taiwan, BS, 68; Univ Ill, Urbana, PhD(phys chem), 74. *Prof Exp:* Res assoc, 74-75, NIH fel phys biochem, Mich State Univ, 75-77; res specialist, 77-80, SCIENTIST, UNIV MINN, ST PAUL, 80- *Mem:* Am Chem Soc. *Res:* Physical chemistry of macromolecules; dynamic light scattering. *Mailing Add:* ECOLAB 840 Sibley Hwy Roseville MN 55118-1700

WEI, JAMES, b Macao, China, Aug 14, 30; nat US; m 56; c 4. CHEMICAL ENGINEERING. *Educ:* Ga Inst Technol, BChE, 52; Mass Inst Technol, MS, 54, ScD(chem eng), 55. *Prof Exp:* Res assoc, Socony Mobil Oil Co, Inc, 55-65, sr res assoc, Mobil Oil Corp, 65-68, sr scientist, Mobil Res & Develop Corp, 68-69, mgr anal, LRAS Group, Mobil Oil Corp, NY, 69-71; Allan P

Colburn prof chem eng, Univ Del, 71-77; head, 77-88, WARREN K LEWIS PROF DEPT CHEM ENG, MASS INST TECHNOL, 77- *Concurrent Pos:* Consult, Mobil Oil Corp; chmn, catalyst panel, Comt Motor Vehicle Emission, Nat Acad Sci; consult ed, Chem Eng Ser, McGraw-Hill; vis prof, Princeton Univ, 62-63 & Calif Inst Technol, 65; mem, Sci Adv Bd, Environ Protection Agency, 76-79; Fairchild scholar, Calif Inst Technol, 77; chief ed, Adv Chem Eng, 80- *Honors & Awards:* Petrol Chem Award, Am Chem Soc, 66; Prof Progress Award, Am Inst Chem Engrs, 70, William H Walker Award, 80, Warren K Lewis Award, 85. *Mem:* Nat Acad Eng; Am Chem Soc; Am Inst Chem Engrs (pres, 88); Am Acad Arts & Sci. *Res:* Catalysis and kinetics; chemical reactors; applied mathematics; structure of chemical processing industries. *Mailing Add:* Dept Chem Eng Mass Inst Technol Cambridge MA 02139

WEI, L(ING) Y(UN), b Hangyang, China, May 23, 20; Can citizen; m 44; c 4. SOLID STATE PHYSICS, ELECTRICAL ENGINEERING. *Educ:* Northwestern Col Eng, China, BS, 42; Univ Ill, MS, 49, PhD(solid state physics), 58. *Prof Exp:* Engr, Directorate Gen Telecommun, China & Formosa, 42-56; asst prof elec eng, Univ Wash, 58-60; assoc prof, 60-63, PROF ELEC ENG, UNIV WATERLOO, 63- *Mem:* AAAS; Biophys Soc; Inst Elec & Electronics Engrs. *Res:* Semiconductor materials; physics of thin films and physical mechanisms of nerve conduction. *Mailing Add:* Waterloo Univ Univ Ave Waterloo ON N2L 3X8 Can

WEI, LEE-JEN, b China, Apr 2, 48; m; c 1. BIOSTATISTICS. *Educ:* Fu Jen Univ, Taiwan, BS, 70; Univ Wis, Madison, PhD(statist), 75. *Prof Exp:* Asst prof statist, Univ SC, 75-79, assoc prof, 79-80; cancer expert, Nat Cancer Inst, 80-81; PROF STATIST, GEORGE WASHINGTON UNIV, 81- *Concurrent Pos:* Comt mem, Nat Res Coun, 82- *Mem:* Am Statist Asn. *Mailing Add:* Dept Biostatist Harvard Sch Pub Health 677 Huntington Ave Boston MA 02115

WEI, LESTER YEEHOW, b Foochow, China, Sept 4, 44; US citizen; m 72; c 2. ANALYTICAL CHEMISTRY. *Educ:* Cheng Kung Univ, BSE, 67; ETex State Univ, MS, 70, EdD(chem), 74. *Prof Exp:* Res assoc chem, Emory Univ 74-76 & Ill Geol Surv, 76-77; asst prof scientist & res chemist, Ill Natural Hist Surv, 77-91; CONSULT, 91- *Mem:* Am Chem Soc. *Res:* Chemistry and analysis of pesticides; gas chromatography; mass spectroscopy. *Mailing Add:* Ill Natural Hist Surv 607 E Peabody Champaign IL 61820

WEI, LUN-SHIN, b Hou-long, Formosa, Jan 14, 29; m 56; c 4. FOOD SCIENCE. *Educ:* Taiwan Prov Col Agr, BS, 51; Univ Ill, MS, 55, PhD, 58. *Prof Exp:* Asst agron, 55-57, sci analyst, 57-59, res assoc food sci, 59-64, assoc prof, 64-76, PROF FOOD SCI, UNIV ILL, URBANA, 76- *Honors & Awards:* Educ & Res Award, Land of Lincoln Soybean Asn, 73. *Mem:* AAAS; Am Chem Soc; Inst Food Technologists. *Res:* Foods and plant materials analysis; food processing and preservation; product development; food utilization of soybeans. *Mailing Add:* 309 McHenry St Urbana IL 61801

WEI, PAX SAMUEL PIN, b Chungking, China, Nov 11, 38; US citizen; m 64; c 3. PHYSICAL CHEMISTRY. *Educ:* Nat Taiwan Univ, BS, 60; Univ Ill, Urbana, MS, 63; Calif Inst Technol, PhD(chem), 68. *Prof Exp:* Mem tech staff, Bell Tel Labs, NJ, 67-69; res scientist, Boeing Sci Res Labs, 69-71, RES SCIENTIST, BOEING AEROSPACE CO, 72- *Mem:* Am Chem Soc; Am Phys Soc. *Res:* Atomic and molecular spectroscopy; surface science; electron diffraction; laser effects. *Mailing Add:* 6612-129th Pl SE Bellevue WA 98006-4041

WEI, PETER HSING-LIEN, b Shantung, China, Feb 11, 22; m 48; c 4. MEDICINAL CHEMISTRY. *Educ:* St John's Univ, China, BS, 48; Columbia Univ, MA, 53; Univ Pa, PhD, 69. *Prof Exp:* Res chemist, Norwich Pharmacal Col, 52-60; res chemist, Wyeth Labs, Inc, 60-69, sr res chemist, 69-77, group leader, 77-87; RETIRED. *Mem:* Am Chem Soc. *Res:* Pharmaceuticals; synthesis of heterocyclic compounds of biological interest; synthetic approach to studies of chemical compounds as anticancer, antinflammatory and antiallergic agents acting through the immune system of animals and humans. *Mailing Add:* 430 Ridge Lane Springfield PA 19064

WEI, ROBERT, b Apr 16, 39; US citizen; m 76; c 2. BIOCHEMISTRY, IMMUNOLOGY. *Educ:* George Washington Univ, BA, 62, PhD(biochem), 72; Georgetown Univ, MS, 69. *Prof Exp:* Prin res chemist clin chem, Wash Ref Lab Inc, 74-75; sr res scientist clin chem, Electro-Nucleonics, Inc, 75-76, sr res scientist immunol, 76-78; ASSOC PROF BIOCHEM, CLEVELAND STATE UNIV, 78- *Mem:* Am Bd Clin Chem; Am Asn Immunol. *Res:* Oxygen radical mediated tissue damage. *Mailing Add:* Dept Chem Cleveland State Univ 1983 E 24th St Cleveland OH 44115

WEI, ROBERT PEH-YING, b Nanking, China, Sept 16, 31; m 54; c 2. APPLIED MECHANICS, MATERIALS SCIENCE. *Educ:* Princeton Univ, BSE, 53, MSE, 54, PhD(mech eng), 60. *Prof Exp:* Instr mech eng, Princeton Univ, 54-57, res asst aeronaut eng, 58-59; assoc technologist, fracture mech, Appl Res Lab, US Steel Corp, 59-61, technologist, 61-62, sr res engr, 62-64, assoc res consult, 64-66; assoc prof mech, 66-70, PROF MECH, LEHIGH UNIV, 70- *Mem:* Am Soc Testing & Mat; Sigma Xi. *Res:* Fracture mechanics; mechanics and metallurgical aspects of fatigue crack growth and stress corrosion cracking; experimental stress analysis. *Mailing Add:* Dept Mech Lehigh Univ Bethlehem PA 18015

WEI, STEPHEN HON YIN, b Shanghai, China, Sept 17, 37; US citizen; m 63; c 2. PEDODONTICS, HISTOLOGY. *Educ:* Univ Adelaide, BDS, 61, Hons, 62, MDS, 65; Univ Ill, Chicago, MS, 67; Univ Iowa, DDS, 71. *Prof Exp:* Dent surgeon, Royal Adelaide Hosp, Univ Adelaide, 63-64, teaching registr, Queen Elizabeth Hosp, 65; res asst pedodontics, Univ Ill, Chicago, 65-66, instr, 67; from asst prof to assoc prof, 67-74, PROF PEDODONT, COL DENT, UNIV IOWA, 74- *Concurrent Pos:* Dentist, Vet Admin Hosp, 69-, consult, 74; consult, Res Comt, Am Acad Pedodont, 71-, ed, 78-; mem res comt, Off Maternal & Child Health, Bur Community Health Serv, HEW. *Mem:* Am Dent Asn; Am Asn Hosp Dent; Int Asn Dent Res; Am Acad Pedodontics; Am Soc Dent Children. *Res:* Preventive dentistry; remineralization of teeth; systemic and topical fluoride therapy; electron optical studies of dental and other hard tissues; clinical research in pedodontics and dental caries. *Mailing Add:* Dept Children's Dent & Orthod Prince Philip Dent Hosp 34 Hospital Rd Hong Kong Hong Kong

WEI, WEI-ZEN, b Tainan, Taiwan, May 1, 51; US citizen; c 2. TUMOR IMMUNOLOGY, CELLULAR IMMUNOLOGY. *Educ:* Nat Taiwan Univ, BS, 73; State Univ NY, MS, 75; Brown Univ, PhD(biol), 78. *Prof Exp:* Postdoctoral fel, Dept Path, Health Ctr, Univ Conn, 78-79; postdoctoral res assoc, 80-81, instr, 81-82, CLIN INSTR, DEPT PATH, OHIO STATE UNIV, 83- *Concurrent Pos:* Scientist, Dept Immunol, Mich Cancer Found, 83-86, asst mem, 86-91, assoc mem, 91-; adj prof cancer biol, Dept Pharmacol, Wayne State Univ, 90- *Mem:* Am Asn Immunologists; Am Asn Cancer Res. *Res:* Determination of the host cellular and humoral immune reactivities during mouse mammary neoplastic progressions; identification of immune markers for early detection of mammary lesions; modulation of the immune reactivities to prevent tumoripenesis from preneoplastic lesions. *Mailing Add:* Dept Immunol Mich Cancer Found 110 E Warren Ave Detroit MI 48201

WEI, WILLIAM WU-SHYONG, b Shin-Chu, Taiwan, June 2, 40; m 71; c 3. STATISTICS, MATHEMATICS. *Educ:* Nat Taiwan Univ, BA, 66; Univ Ore, BA, 69; Univ Wis-Madison, PhD(statist), 74. *Prof Exp:* From asst prof to assoc prof, 74-85, dir bus statist, 77-82, chmn dept, 82-87, PROF STATIST, TEMPLE UNIV, 86- *Concurrent Pos:* Assoc ed, J Forecasting, 85- *Mem:* Am Statist Asn; Inst Math Statist. *Res:* Time series analysis, forecasting, statistical modelling and applications. *Mailing Add:* Dept Statist Temple Univ Philadelphia PA 19122-2595

WEI, YEN, b Linchuan, Jiangxi Prov, China, Sept 1, 57; m 86; c 1. CHEMISTRY MATERIALS, CHEMICAL EDUCATION. *Educ:* Peking Univ, Beijing, China, BS, 79, MS, 81; City Col New York, MA, 84; City Univ New York, PhD(chem), 86. *Prof Exp:* Postdoctoral assoc polymer sci, Dept Mat Sci & Eng, Mass Inst Technol, 86-87; asst prof chem, 87-91, ASSOC PROF CHEM, DEPT CHEM, DREXEL UNIV, 91- *Concurrent Pos:* Drexel res scholar award, Drexel Univ, 88; DuPont young fac award, DuPont Co, 91. *Mem:* Am Chem Soc; Sigma Xi; AAAS. *Res:* Polymer chemistry; electrically conductive polymers; mechanistic studies of polymerizations; nonlinear optical materials; polymer network and composites; solid state organic chemistry; high temperature superconductors biomedical applications of polymers. *Mailing Add:* Dept Chem Drexel Univ Philadelphia PA 19104

WEIBEL, ARMELLA, b Ewing, Nebr, Feb 7, 20. MATHEMATICS, SCIENCE EDUCATION. *Educ:* Alverno Col, BSE, 46; Univ Wis, MS, 52. *Prof Exp:* Teacher, St Clara Sch, Ill, 38-48; teacher, Frankenstein High Sch, Mo, 48-50; from instr to asst prof, 52-53, ASSOC PROF MATH, ALVERNO COL, 56- *Concurrent Pos:* Dir, Alverno Ctr, NSF Minn Math & Sci Teaching Proj, 63-70. *Mem:* Nat Coun Teachers Math. *Res:* Math education. *Mailing Add:* 3401 S 39th St Milwaukee WI 53215

WEIBEL, CHARLES ALEXANDER, b Terre Haute, Ind, Oct 28, 50; m 86; c 1. ALGEBRAIC K-THEORY. *Educ:* Univ Mich, BS(physics) & BA(math), 72; Univ Chicago, SM, 73, PhD(math), 77. *Prof Exp:* Opers res analyst, Standard Oil, Ind, 70-76; mem, Inst Advan Study, 77-78, 85-86; asst prof math, Univ Pa, 78-80; from asst prof to prof, 80-89, PROF MATH, RUTGERS UNIV, 89- *Concurrent Pos:* Ed, J Pure & Appl Algebra, 83-, managing ed, 89- *Mem:* Am Math Soc. *Res:* Algebraic K-theory: computational methods, structural relationships and applications to other research areas, primarily algebra and topology. *Mailing Add:* Dept Math Rutgers Univ New Brunswick NJ 08903

WEIBEL, DALE ELDON, b De Witt, Nebr, Dec 14, 20; m 42; c 4. AGRONOMY. *Educ:* Univ Nebr, BSc, 42, MSc, 47; Iowa State Univ, PhD(plant breeding, path), 55. *Prof Exp:* Res agronomist, Div Cereal Crops & Dis, USDA, Kans, 47-53, field crops res br, Tex, 53-58; assoc prof, 58-61, PROF AGRON, OKLA STATE UNIV, 61- *Mem:* Crop Sci Soc Am; fel Am Soc Agron. *Res:* Breeding for disease and insect resistance in grain sorghum, forage sorghum and broomcorn. *Mailing Add:* Dept Agron Okla State Univ Stillwater OK 74075

WEIBELL, FRED JOHN, b Murray, Utah, Oct 18, 27; m 49; c 4. BIOMEDICAL ENGINEERING, COMPUTER SCIENCES. *Educ:* Univ Utah, BS, 53; Univ Calif, Los Angeles, MS, 69, PhD(eng), 77. *Prof Exp:* Staff mem eng, Sandia Corp, 53-57, sect supvr, 57-62; chief biomed eng sect, Vet Admin Western Res Support Ctr, 62-67, asst chief biomed eng & comput sci, 67-72, CHIEF, VET ADMIN BIOMED ENG & COMPUT CTR, 72- *Mem:* Biomed Eng Soc (secy-treas, 69-); Instrument Soc Am; Asn Advan Med Instrumentation. *Res:* Electrical safety in the hospital; management information systems covering all medical research in Veterans Administration; medical instrumentation. *Mailing Add:* Biomed Eng & Comput Ctr Dept Vet Affairs 16111 Plummer St Sepulveda CA 91343

WEI-BERK, CAROLINE, b Taipei, Taiwan, China, Oct 10, 56; US citizen; m 80. PHYSICAL PROPERTIES OF POLYMERS, POLYMER COMPOSITES. *Educ:* Nat Taiwan Univ, BS, 78; Yale Univ, MS, 79; Carnegie-Mellon Univ, MS, 84, PhD(chem), 86. *Prof Exp:* Sr scientist, Alcoa, 86-89; SCIENTIST, HIMURET, 89- *Mem:* Am Chem Soc. *Res:* Characterization and processing of liquid crystalline polymers; solution properties of polymers; thermoplastic composites; alloys and blends. *Mailing Add:* Six Dublin Way Ashton PA 19014-1227

WEIBLEN, PAUL WILLARD, b Miller, SDak, Feb 15, 27; m 67; c 1. GEOLOGY, PETROLOGY. *Educ:* Wartburg Col, BA, 50; Univ Minn, MA, 52, MS & PhD(geol), 65. *Prof Exp:* Asst prof, 65-71, assoc prof, 71-80, PROF GEOL, UNIV MINN, MINNEAPOLIS, 80- *Mem:* Electron Probe Anal Soc Am; Sigma Xi. *Res:* Petrology, especially the study of gabbroic rocks and associated mineralization; lunar petrology; Precambrian geology; application of electron probe analysis to problems in mineralogy, geochemistry and petrology. *Mailing Add:* 1519 Brook Ave SE Minneapolis MN 55414

WEIBRECHT, WALTER EUGENE, b New York, NY, June 25, 37. INORGANIC CHEMISTRY. *Educ:* Franklin & Marshall Col, BS, 59; Cornell Univ, PhD(chem), 64. *Prof Exp:* Am Oil fel chem, Harvard Univ, 63-64; asst prof, Mich State Univ, 64-66; asst prof, 66-77, ASSOC PROF CHEM, UNIV MASS, BOSTON, 77- *Mem:* AAAS; Am Chem Soc; The Chem Soc. *Res:* Studies involving borazine as a Lewis acid; B-hydroxyborazines; synthesis; silicon-nitrogen bond cleavage in symmetrically and unsymmetrically alkoxylated silazanes; silicon, germanium and tin transamination equilibria. *Mailing Add:* Dept Chem Univ Mass Boston MA 02125-3386

WEIBUST, ROBERT SMITH, b Newport, RI, May 6, 42. GENETICS, ZOOLOGY. *Educ:* Colby Col, AB, 64; Univ Maine, Orono, MS, 66, PhD(zool), 70. *Prof Exp:* Nat Cancer Inst fel, Jackson Lab, 64-65, Nat Inst Gen Med Sci fel, Jackson Lab & Univ Maine, Orono, 68-70, fel, Jackson Lab, 70; from asst prof to assoc prof, 70-81, PROF BIOL, MOORHEAD STATE UNIV, 81- *Mem:* AAAS; Am Genetic Asn; Sigma Xi; Genetics Soc Am. *Res:* Mammalian genetics. *Mailing Add:* Dept Biol Moorhead State Univ Moorhead MN 56563

WEICHEL, HUGO, b Selz, Ukraine, July 23, 37; US citizen; m 61; c 2. LASERS. *Educ:* Portland State Univ, BS, 60; USAF Inst Technol, MS, 65; Univ Ariz, PhD, 72. *Prof Exp:* USAF, 60-89, res physicist, nuclear rocket propulsion, Edwards AFB, Calif, 61-63, student arc plasmas, Aeronaut Res Lab, Ohio, 64-65, proj officer laser res, Air Force Weapons Lab, NMex, 65-69, assoc prof & dep dept head physics, USAF Inst Technol, 72-78, aeronaut engr, 78-83, dir physics, Air Force Off Sci Res, 83-87, dep dir, Defense Nuclear Agency, 87-89; PHYSICIST, NICHOLS RES CORP, 89- *Mem:* Am Phys Soc; fel Soc Optical Eng. *Res:* Lasers; optics; laser repair; plasmas. *Mailing Add:* 4533 Gilbertson Rd Fairfax VA 22032

WEICHENTHAL, BURTON ARTHUR, b Stanton, Nebr, Nov 7, 37; m 60; c 1. ANIMAL NUTRITION. *Educ:* Univ Nebr, BS, 59; SDak State Univ, MS, 62; Colo State Univ, PhD(nutrit), 67. *Prof Exp:* Beef cattle exten specialist, Univ Ill, Urbana, 67-81, prof animal sci, 80-81; ASSOC DIR, PANHANDLE RES & EXTEN CTR, UNIV NEBR, 81- *Mem:* Am Soc Animal Sci; Coun Agr Sci & Technol. *Res:* Ruminant nutrition and physiology. *Mailing Add:* Panhandle Res & Exten Ctr Univ Nebr-Lincoln 4502 Ave I Scottsbluff NE 69361

WEICHERT, DIETER HORST, b Breslau, Ger, May 2, 32; Can citizen; m 62; c 2. SEISMOLOGY. *Educ:* Univ BC, BASc, 61, PhD(geophys), 65; McMaster Univ, MSc, 63. *Prof Exp:* Res asst geophys, Univ BC, 60-61, elec eng, Nat Res Coun Can, 61 & geophys, Univ Toronto, 63; res scientist, Earth Physics Br, 65-78, actg dir, 88-90, RES SCIENTIST, PAC GEOSCI CTR, GEOL SURV CAN, DEPT ENERGY, MINES & RESOURCES, 78- *Concurrent Pos:* Vis scientist, Geophys Inst, Univ Karlsruhe, 70, UK Atomic Energy Authority, 71, Inst, Frankfurt Univ, 71 & Fed Rep Ger Seismol Ctr, Graefenberg, 76. *Mem:* Seismol Soc Am; Am Geophys Union; Can Geophys Union. *Res:* Geophysics. *Mailing Add:* 8750 Pender Park Dr Sidney BC V8L 3Z5 Can

WEICHLEIN, RUSSELL GEORGE, b Dayton, Ohio, Apr 23, 15; m 48; c 1. MICROBIOLOGY. *Educ:* St Mary's Univ, Tex, BS, 39; Trinity Univ, Tex, MS, 55. *Prof Exp:* Tester, Ref Lab, Tex Co, 47; asst biologist, Found Appl Res, 48-52; assoc res bacteriologist, Sanit Sci Dept, Southwest Res Inst, 53-58, dept chem res, 58-65; from instr to assoc prof, 65-80, EMER ASSOC PROF BIOL SCI, SAN ANTONIO COL, 80- *Res:* Microbial genetics; biology; bacteriology. *Mailing Add:* 306 Waxwood Lane San Antonio TX 78216-6834

WEICHMAN, BARRY MICHAEL, ANTI-INFLAMMATORY DRUGS, PULMONARY PHARMACOLOGY. *Educ:* Univ Rochester, PhD(pharmacol), 79. *Prof Exp:* Sect head, anti-inflammatory & analgesic pharmacol, Ayerst Labs Res, Inc, 84-86, assoc dir, metab dis unit, 86-87, ASSOC DIR, INFLAMMATION/ALLERGY SUBDIV, WYETH-AYERST RES, 87- *Mem:* Am Soc Pharmacol & Exp Therapeut; NY Acad Sci; Inflammation Res Asn; AAAS. *Mailing Add:* Dept Immunopharmacol Wyeth-Ayerst Res Princeton NJ 08543-8000

WEICHMAN, FRANK LUDWIG, b Liegnitz, Ger, Sept 23, 30; Can citizen; m 58; c 2. EXPERIMENTAL SOLID STATE PHYSICS. *Educ:* Brooklyn Col, BS, 53; Northwestern Univ, PhD(physics), 59. *Prof Exp:* From asst prof to assoc prof, 58-70, PROF PHYSICS, UNIV ALTA, 70- *Concurrent Pos:* Mem fac, Univ Strasbourg, 65-66; vis prof, Technion, Haifa, Israel, 73-74; sr indust fel, Bell Northern Res Ltd, Ottawa, 81-82. *Mem:* Can Asn Physicists; Am Phys Soc. *Res:* Optical and electrical properties of gallium arsenide with emphasis on the role of defect structures on photoconductivity, luminescence and electroluminescence; solar energy as applied to photovoltaics and greenhouses. *Mailing Add:* Dept Phys Univ Alta Edmonton AB T6G 2J1 Can

WEICHMAN, PETER BERNARD, b Edmonton, Alta, Mar 6, 59; US citizen; m 89; c 1. PHASE TRANSITION & CRITICAL PHENOMENA, LOW TEMPERATURE PHYSICS. *Educ:* Univ Alta, BSc, 81; Cornell Univ, MS, 84 PhD(physics), 86. *Prof Exp:* Res fel, 86-89, ASST PROF PHYSICS, CALIF INST TECHNOL, 89- *Mem:* Am Phys Soc. *Res:* Phase transitions and critical phenomena, especially in quantum mechanical systems, including super-fluidity of helium in porous media and thin film conductors; turbulence in fluids and ocean surface waves. *Mailing Add:* Caltech 114-36 Pasadena CA 91125

WEICHSEL, MORTON E, JR, b Pueblo, Colo, June 17, 33. PEDIATRICS, NEUROLOGY. *Educ:* Univ Colo, Boulder, BA, 55; Univ Buffalo, MD, 62; Am Bd Pediat, dipl & cert; Am Bd Psychiat & Neurol, dipl & cert child neurol. *Prof Exp:* Intern pediat, Buffalo Children's Hosp, NY, 63; resident, Med Ctr, Stanford Univ, 63-65, fel pediat neurol, 65-67; fel Med Ctr, Univ Colo, 67-68; clin instr pediat, Med Ctr, Stanford Univ, 68-69, fel develop neurochem, 69-71; asst prof human develop & med, Col Human Med, Mich State Univ, 71-74; assoc prof, 74-80, PROF PEDIAT & NEUROL, HARBOR GEN HOSP & SCH MED, UNIV CALIF, LOS ANGELES, 80-, PROF PEDIAT, KING/DREW MED CTR, 89- *Mem:* Soc Neurosci; Am Acad Neurol; Soc Pediat Res; Child Neurol Soc; Am Fedn Clin Res. *Res:* Developmental neurochemistry. *Mailing Add:* Harbor-UCLA Med Ctr Torrance CA 90509

WEICHSEL, PAUL M, b New York, NY, July 22, 31; m 55, 84; c 5. MATHEMATICS. *Educ:* City Col New York, BS, 53; NY Univ, MS, 54; Calif Inst Technol, PhD(math), 60. *Prof Exp:* From instr to asst prof math, Univ Ill, Urbana, 60-65; NATO fel, Math Inst, Oxford, Eng, 61-62; res fel, Inst Advan Studies, Australian Nat Univ, 65-66; from asst prof to assoc prof, 66-75, PROF MATH, UNIV ILL, URBANA, 75- *Concurrent Pos:* Part time consult, Argonne Nat Lab, 63-64; vis prof, Hebrew Univ, Jerusalem, 70-71, Univ Tel Aviv, Israel, 71-72, Weizmann Inst Sci, Rehorot, Israel, 88-89; math coordr, Nat Coord Ctr Curric Develop, Dept Technol & Soc, Col Eng, State Univ NY, Stony Brook, 76-78. *Mem:* Am Math Soc; Math Asn Am. *Res:* Algebra; theory of graphs; theory of finite groups; use of techniques from linear algebra and group theory to study graphs, especially those with many symmetries and those with a high degree of regularity. *Mailing Add:* Dept Math Univ Ill Urbana IL 61801

WEICK, CHARLES FREDERICK, b Buffalo, NY, Jan 19, 31; m 59; c 2. INORGANIC CHEMISTRY. *Educ:* Mt Union Col, BS, 53; Univ Rochester, PhD, 59. *Prof Exp:* Assoc prof, 58-80, PROF CHEM, UNION COL, NY, 80- *Concurrent Pos:* Fel, Univ Kent, Canterbury, Eng, 71-72. *Mem:* Am Chem Soc. *Res:* Coordination chemistry; radiochemistry. *Mailing Add:* Dept Chem & Sci Bldg 5216 Union Col Schenectady NY 12308

WEICK, RICHARD FRED, b 1934; m 57; c 2. NEUROENDOCRINOLOGY, REPRODUCTION. *Educ:* Tex A&M, BS, BA, 56; Stanford Univ, PhD(physiol), 70. *Prof Exp:* Res scientist, NASA, 59-66; PROF PHYSIOL, UNIV WESTERN ONT, 73- *Concurrent Pos:* Vis fel, Corpus Christi Col, Cambridge, UK, 89-90. *Mem:* Soc Study Reprod; Endocrine Soc; Int Soc Neuroendocrinol; Am Physiol Soc; Can Physiol Soc; Soc Neurosci. *Res:* Neuroendocrine regulation of gonadotropin secretion and the reproductive cycle. *Mailing Add:* Dept Physiol Univ Western Ont London ON N6A 5C1 Can

WEIDANZ, WILLIAM P, b Jackson Heights, NY, Jan 30, 35; m 60; c 6. IMMUNOLOGY, MEDICAL MICROBIOLOGY. *Educ:* Rutgers Univ, BS, 56; Univ RI, MS, 58; Tulane Univ, PhD(microbiol), 61. *Prof Exp:* NIH fel, 61-64; asst prof immunol & pathogenic bact, La State Univ, 64-66; from asst prof to assoc prof, 66-77, PROF MICROBIOL, HAHNEMANN MED COL, 77- *Concurrent Pos:* NIH res grant, 64-67 & 75-78; fac res coun fel, La State Univ, 65; Eleanor Roosevelt Int Cancer fel, Fibiger-Laboratoriet, Copenhagen, Denmark, 75-76. *Mem:* Am Asn Immunologists; Am Soc Microbiol. *Res:* immunity to malaria. *Mailing Add:* Prof Microbiol & Immunol Hahnemann Univ Sch Med Broad & Vine Philadelphia PA 19102

WEIDE, DAVID L, b Jan 9, 36. REGIONAL GEOLOGY, GEOMORPHOLOGY. *Educ:* Univ Calif, AB, 58, MA, 68, PhD(geomorphol), 74. *Prof Exp:* Geologist, Calif Well Logging Co, 58-59; geologist, US Army, 59-61; assoc res geologist, Douglas Aircraft Co, 61-64; mus scientist, dept geol, Univ Calif, Los Angeles, 64-73, staff geologist, Archaeol Surv Off, 64-73; asst prof, 73-78, dept chmn, 80-81, ASSOC PROF, DEPT GEOSCI, UNIV NEV, LAS VEGAS, 78- *Res:* Geology applied to archaeology, air photo interpretation and related remote sensing; budget and personnel management; mineralogy and petrology; scientific editing; geologic cartography. *Mailing Add:* Dept Geosci Univ Nev 4505 Maryland Pkwy Las Vegas NV 89154

WEIDEN, MATHIAS HERMAN JOSEPH, b Narrowsburg, NY, Nov 3, 23; m 54; c 4. INSECT TOXICOLOGY. *Educ:* Manhattan Col, BS, 46; Cornell Univ, PhD(entom), 54. *Prof Exp:* Chemist, Lederle Labs, Am Cyanamid Co, 46-47; asst entom, Cornell Univ, 47-48 & 51-52, asst prof insecticide chem, 54-59; entomologist, Union Carbide Agr Prod Co, 59-66, res scientist, Explor Insecticide Res, 66-86; scientist, Discovery Res, Rhone-Poulenc Ag Co, 87-90; RETIRED. *Concurrent Pos:* Exclusive consult, 90- *Mem:* Entom Soc Am; Am Chem Soc. *Res:* Action of insecticides and detoxication mechanisms in insects; insecticide research and development. *Mailing Add:* 525 North St Chapel Hill NC 27514

WEIDENBAUM, SHERMAN S, b New York, July 8, 25; m 48; c 3. CHEMICAL ENGINEERING. *Educ:* Columbia Univ, BS, 47, MS, 48, PhD(chem eng), 53. *Prof Exp:* Asst chem eng, Columbia Univ, 48-49; sr chem engr, Corning Glass Works, 52-56, sr melting engr, 56-58, res chem engr, 58-66, managing dir UN Proj, Israel Ceramic & Silicate Inst, Haifa, 62-65, proj engr, Corning Glass Works, 65-66, mgr tech serv, Latin Am Area, Corning Glass Int, 66-68; prof phys sci, 68-70, proj mgr & dir, 70-77, PROF PHYS SCI, USCG ACAD, 77- *Mem:* Electrochem Soc; Am Soc Qual Control; Am Chem Soc; Am Inst Chem Engrs; Sigma Xi. *Res:* Solids mixing and processing; paste mixing; pelletizing; applications to glass batch preparation; glass melting; theoretical and practical aspects of solids mixing; industrial statistics; international technical assistance. *Mailing Add:* USCG Acad 17 Fifth Ave Waterford CT 06385

WEIDENSAUL, T CRAIG, b Reedsville, Pa, Apr 4, 39; m 65; c 3. PLANT PATHOLOGY, FORESTRY. *Educ:* Gettysburg Col, BA, 62; Duke Univ, MF, 63; Pa State Univ, PhD(plant path), 69. *Prof Exp:* Plant pathologist, USDA Forest Serv, 63-66; res asst, Pa State Univ, 66-69, fel scholar, 69-70; from asst prof to assoc prof, 70-79, PROF PLANT PATH & FORESTRY, OHIO STATE UNIV, 79- & DIR LAB ENVIRON STUDIES, OHIO AGR RES & DEVELOP CTR, 70- *Concurrent Pos:* Consult forester & plant pathologist; fel, Indust Applns Ctr, NASA; asst dir, Sch Natural Resources, Ohio State Univ, 90- *Honors & Awards:* Gov Award, Nat Wildlife Fedn. *Mem:* Soil Sci Soc Am; Am Phytopath Soc; Soc Am Foresters; Am Soc Agron; Sigma Xi. *Res:* Epidemiology of fusarium canker of sugar maple; precipitation quality and effects on terrestrial ecosystems; effects of gaseous and metallic air pollutants on plants and plant disease. *Mailing Add:* 134 Williams Hall Ohio Agr Res & Develop Ctr Wooster OH 44691

WEIDENSCHILLING, STUART JOHN, b Montclair, NJ, Sept 22, 46; m 73; c 2. PLANETARY SCIENCE, ASTRONOMY. *Educ:* Mass Inst Technol, BS, 68, MS, 69, PhD(earth & planetary sci), 76. *Prof Exp:* Fel, Dept Terrestrial Magnetism, Carnegie Inst, 76-78; res scientist, 78-, SR SCIENTIST, PLANETARY SCI INST, SCI APPLN, INC. *Mem:* Am Geophys Union; Am Astron Soc; Int Astron Union. *Res:* Origin and evolution of the solar system. *Mailing Add:* Planetary Sci Inst 2421 E Sixth St Tucson AZ 85719-5234

WEIDHAAS, DONALD E, b Northampton Mass, Feb 12, 28; m 53; c 2. MEDICAL ENTOMOLOGY. *Educ:* Univ Mass, BS, 51; Cornell Univ, NY, PhD(entom), 55. *Prof Exp:* Med entomologist, Sci & Educ Admin-Agr Res, USDA, 56-62, asst chief, 62-67, actg chief, 67, dir insects affecting man & animals res lab, 67-84; RETIRED. *Concurrent Pos:* Consult, 85- *Mem:* Entom Soc Am; Am Mosquito Control Asn; Am Asn Trop Med & Hyg. *Res:* Chemical and alternative methods of control, resistance studies, physiology, toxicology and biology of insects affecting man or of medical importance; integrated control; dynamics and modelling. *Mailing Add:* 1330 NW 25th Terr Gainesville FL 32605

WEIDHAAS, JOHN AUGUST, JR, b Northampton, Mass, Oct 13, 25; m 48, 75; c 5. ENTOMOLOGY. *Educ:* Univ Mass, BS, 49, MS, 52, PhD(entom), 59. *Prof Exp:* Instr entom, Univ Mass, 51 & 53-59; from asst prof to assoc prof, Cornell Univ, 59-67; ASSOC PROF ENTOM & EXTEN SPECIALIST, VA POLYTECH INST & STATE UNIV, 67- *Concurrent Pos:* Collabr, Agr Res Serv, USDA; consult, Elm. *Mem:* Entom Soc Am; Entom Soc Can; Int Soc Arboricult (past pres); hon mem Soc Munic Arborists; Arboricultural Res & Educ Acad (past pres). *Res:* Forest, shade tree, ornamental and beneficial insects; Acarina. *Mailing Add:* Dept Entom Va Polytech Inst Blacksburg VA 24061

WEIDIE, ALFRED EDWARD, b New Orleans, La, July 31, 31; m 60; c 3. GEOLOGY. *Educ:* Vanderbilt Univ, BA, 53; La State Univ, MS, 58, PhD(geol), 61. *Prof Exp:* From asst prof to assoc prof, 61-71, chmn dept earth sci, 67-73, PROF GEOL, UNIV NEW ORLEANS, 71- *Res:* Structural geology and physical stratigraphy. *Mailing Add:* Dept Earth Sci Univ New Orleans New Orleans LA 70148

WEIDIG, CHARLES F, b Houston, Tex, Oct 24, 45; m 67; c 2. MARKETING, PHARMACEUTICALS INTERMEDIATES. *Educ:* Tex Christian Univ, BS, 69, PhD(chem), 74; Sam Houston State Univ, MA, 71. *Prof Exp:* Anal chemist, Indust Labs, Ft Worth, 66-69; res fel, Edsel B Ford Inst Med Res, 74-76; sr res chemist, 76-85, PROD MGR, ETHYL CORP, 85- *Mem:* Am Chem Soc; Sigma Xi. *Res:* Rapid reaction kinetics and mechanisms of enzyme-catalyzed reactions; synthesis, kinetics and mechanism of hydrolysis of nitrogen base adducts of substituted boranes; transition metal chemistry; homogeneous and heterogeneous catalysis; lubricant additives; fuel additives; polymer synthesis; epoxy and polyurethane chemistry; pharmaceutical chemistry; pesticides; antioxidants. *Mailing Add:* Ethyl Corp 451 Florida Blvd Baton Rouge LA 70801

WEIDLER, DONALD JOHN, b Fredericksburg, Iowa, Sept 26, 33; m 58; c 2. INTERNAL MEDICINE, CLINICAL PHARMACOLOGY. *Educ:* Wartburg Col, BA, 59; Univ Iowa, MS, 62, MD, 65, PhD(physiol, biophys), 69. *Prof Exp:* Intern internal med, Boston City Hosp, 65-66; instr physiol & biophys, Univ Iowa, 66-69; from asst prof & mem grad fac to assoc prof physiol & biophys, Univ Nebr Med Ctr, 69-72; resident internal med, Sch Med, Univ Mich, 72-74, instr pharmacol, 73-77, asst prof internal med, 74-79; ASSOC PROF & CHIEF DIV CLIN PHARMACOL, SCH MED, UNIV MIAMI, 79- *Concurrent Pos:* NIH fel, Univ Iowa, 66-69. *Mem:* Am Physiol Soc; Am Col Physicians; Am Soc Clin Pharmacol & Therapeut; Am Soc Internal Med; Am Col Clin Pharmacol; Sigma Xi. *Res:* Clinicial pharmacology; clinical pharmacodynamics; clinical pharmacokinetics. *Mailing Add:* Box 015996 Miami FL 33101

WEIDLICH, JOHN EDWARD, JR, b Akron, Ohio, July 31, 19; m 69; c 1. MATHEMATICS. *Educ:* Stanford Univ, BS, 48, MS, 50; Univ Calif, Berkeley, PhD(math), 61. *Prof Exp:* Instr math, Univ Santa Clara, 49-51; from assoc res scientist to res scientist, Lockheed Missile & Space Co, 59-64; from asst prof to assoc prof, 64-70, actg chmn dept, 66-67, res sabbatical, 71-72, PROF MATH, CALIF STATE UNIV, HAYWARD, 70- *Concurrent Pos:* Consult, Lockheed Missile & Space Co, 64-65. *Mem:* Am Math Soc; Math Asn Am; Soc Indust & Appl Math. *Res:* Pure mathematics in analysis, differential equations and asymptotic; mathematical nature of physical phenomena, including work in theoretical orbit mechanics and trajectory studies in the n-body problem. *Mailing Add:* Dept Math Calif State Univ Hayward CA 94542

WEIDLINGER, PAUL, b Budapest, Hungary, Dec 22, 37. CIVIL ENGINEERING, ENGINEERING MECHANICS. *Educ:* Swiss Polytech Inst, MS, 37. *Prof Exp:* PRES, WEILINGER ASSOCS, 47-, CONSULT ENG, 48- *Concurrent Pos:* Vis lectr civil eng, Harvard Univ, 46-, Mass Inst Technol, 64- *Honors & Awards:* J James Croes Award, 63, Ernest E Howad Award, 85, Brown Medal, 87. *Mem:* Nat Acad Eng; Am Inst Aeronaut & Astronaut; NY Acad Sci; Am Soc Civil Engrs. *Mailing Add:* Weidlinger Assocs 333 Seventh Ave New York NY 10001

WEIDMAN, PATRICK DAN, b Santa Clara, Calif, Nov 18, 41. FLUID MECHANICS. *Educ:* Calif State Polytech Col, BS, 63; Calif Inst Technol, MS, 64; Univ Southern Calif, PhD(aerospace eng), 73. *Prof Exp:* Fel, Univ Southern Calif, 73-75, res assoc & lectr, 75-81; ASSOC PROF, DEPT MECH ENG, UNIV COLO, 81- *Mem:* AAAS; Am Phys Soc; Am Geophys Union; Sigma Xi; Soc Indust & Appl Math. *Res:* Wave theory, stability of fluid flow porus; fluid mechanics; wave propagation theory; stokes flow; porus media flow fluid stability; low-gravity fluid mechanics. *Mailing Add:* 665 Manhattan Dr 205 Boulder CO 80303

WEIDMAN, ROBERT MCMASTER, b Missoula, Mont, Mar 20, 23; m 51; c 4. GEOLOGY. *Educ:* Calif Inst Technol, BS, 44; Univ Ind, MA, 49; Univ Calif, PhD, 59. *Prof Exp:* Geologist, Stand Oil Co, Calif, 44-47 & State Geol Surv, Ind, 48; instr, Fresno State Col, 49-50; from instr to assoc prof, 53-68, dir, Geol Field Sta, 78-86, PROF GEOL, UNIV MONT, 68- *Mem:* Geol Soc Am; Am Asn Petrol Geologists. *Res:* Geologic remote sensing. *Mailing Add:* Dept Geol Univ Mont Missoula MT 59801

WEIDMAN, ROBERT STUART, b Philadelphia, Pa, Feb 17, 54. BAND THEORY, SURFACE PHYSICS. *Educ:* Univ Del, BS, 76; Univ Ill, PhD(physics), 80. *Prof Exp:* Res assoc, Univ Ill, 80; ASST PROF PHYSICS, MICH TECHNOL UNIV, 80- *Mem:* Am Phys Soc; Sigma Xi. *Res:* Band theory and optical properties of semiconductors and insulators; electronic structure of point defects and impurities; theory of heterogeneous catalysis. *Mailing Add:* PO Box 641 South Range MI 49963

WEIDMANN, SILVIO, b Konolfinger Berne, Switz, Apr 7, 21; m 47; c 3. ELECTROMAGNETIC FIELDS IN BIOLOGY. *Educ:* Univ Bern, MD, 47. *Hon Degrees:* Dr hc, Univ Paris, Sud, 76; Med Dr hc, Uppsala, 77; DSc hc, Leicester, 82. *Prof Exp:* Assoc prof, 58-68, chmn, 68-86, EMER PROF PHYSIOL, UNIV BERN, 86- *Concurrent Pos:* Vis prof, Downstate Med Ctr, State Univ, NY, Brooklyn, NY, 54-55 & Univ PR, 76-77; rector, Univ Bern, 74-75. *Mem:* Hon mem Am Physiol Soc. *Res:* Cardiac electrophysiology using intracellular microelectrodes, ionic fluxes in heart, cell-to-cell coupling in heart. *Mailing Add:* Physiol Inst Univ Bern Buehlplatz Five Bern CH-3012 Switzerland

WEIDNER, BRUCE VAN SCOYOC, b Pottstown, Pa, Oct 29, 08; m 34; c 2. CHEMISTRY. *Educ:* Pa State Col, BS, 31, MS, 32, PhD(inorg chem), 35. *Prof Exp:* Asst chem, Pa State Col, 32-35, instr arts & sci, Hazleton Undergrad Ctr, 35-37; instr, Univ Alaska, 37-39, asst prof chem, 39-42; asst prof chem, Middlebury Col, 42-46; assoc prof, Utah State Agr Col, 46-47; from assoc prof to prof chem, Miami Univ, 47-79; RETIRED. *Concurrent Pos:* Dir, NSF Summer Sec-Sci Insts Teachers, 60-73; first sem, Miami, 85-86. *Mem:* Am Chem Soc; Soc Appl Spectros; AAAS; Sigma Xi. *Res:* Quantitative and inorganic chemistry (air pollution). *Mailing Add:* 308 University Ave Oxford OH 45056

WEIDNER, EARL, b Burke, SDak, Nov 13, 35. ZOOLOGY. *Educ:* Univ SDak, BS, 58, MS, 60; Tulane Univ, PhD(biol), 69. *Prof Exp:* Postdoctoral, Rockefeller Univ, 69-72; from asst prof to assoc prof, 72-86, PROF, DEPT ZOOL & PHYSIOL, LA STATE UNIV, 86- *Mem:* Am Soc Cell Biol; Soc Protozool; Invert Path Soc. *Res:* Interactions of intracellular parasites with host cells; interface of host-parasite interaction; microsporidian invasions into host cells. *Mailing Add:* Dept Zool & Physiol La State Univ Baton Rouge LA 70803

WEIDNER, JERRY R, b Bloomington, Ill, July 19, 38; m 62. GEOCHEMISTRY. *Educ:* Miami Univ, BA, 60, MS, 63; Pa State Univ, University Park, PhD(geochem), 68. *Prof Exp:* Res assoc geochem, Stanford Univ, 67-68; from asst prof to assoc prof geol, Univ Md, College Park, 70-80; MEM FAC, BOWIE STATE COL, 80- *Concurrent Pos:* Nat Res Coun fel NASA Goddard Space Flight Ctr, Md, 68-70. *Mem:* AAAS; Am Geophys Union; Mineral Soc Am; Geochem Soc. *Res:* Experimental petrology. *Mailing Add:* Dept Geol Univ Md College Park MD 20742-4211

WEIDNER, MICHAEL GEORGE, JR, b Birmingham, Ala, July 18, 22; m 60; c 1. SURGERY. *Educ:* Vanderbilt Univ, BA, 44, MD, 46; Am Bd Surg, dipl, 56. *Prof Exp:* Instr surg, Sch Med, Vanderbilt Univ, 53-57; from sr instr to asst prof, Sch Med, Western Reserve Univ, 57-62; assoc prof, Med Univ SC, 62-68, asst dean student affairs, 70-72, assoc dean student affairs & proj dir, Area Health Educ Ctr, 72-76, assoc dean, 76-83; chief of staff, Vet Admin Hosp, 76-83; PROF SURG, MED UNIV SC, 68- *Concurrent Pos:* Res fel, Vanderbilt Univ, 54-57; chief surg serv, Vet Admin Hosp, Charleston, SC, 66-72. *Mem:* AAAS; Am Col Surgeons; Soc Univ Surgeons; Soc Exp Biol & Med; NY Acad Sci. *Res:* Hemorrhagic shock; animal and human gastrointestinal physiology, especially ulcer disease. *Mailing Add:* Dept Surg Med Univ SC 171 Ashley Ave Charleston SC 29425

WEIDNER, RICHARD TILGHMAN, b Allentown, Pa, Mar 31, 21; m 47; c 3. PHYSICS, MAGNETIC RESONANCE. *Educ:* Muhlenberg Col, BS, 43; Yale Univ, MS, 43, PhD(physics), 48. *Prof Exp:* Physicist, US Naval Res Lab, 44-46; lab asst & instr physics, Yale Univ, 43-44, asst, 46-47; from instr to prof physics, Rutgers Univ, 47-88, asst dean, Col Arts & Sci, 66-70, assoc dean, Rutgers Col, 70-77 & actg dean, 77-78, EMER PROF PHYSICS, RUTGERS UNIV, NEW BRUNSWICK, 88- *Mem:* Fel Am Phys Soc; Sigma Xi. *Res:* Electron spin resonance; texts in elementary physics; general physics. *Mailing Add:* 56 W Market St No 6 Bethlehem PA 18018

WEIDNER, TERRY MOHR, b Allentown, Pa, May 31, 37; m 62; c 3. PLANT PHYSIOLOGY. *Educ:* West Chester State Col, BSEd, 59; Ohio State Univ, MS, 61, PhD(bot), 64. *Prof Exp:* From asst prof to assoc prof, 64-74, PROF BOT, EASTERN ILL UNIV, 74-, CHMN DEPT, 76- *Mem:* Am Inst Biol Sci; Am Soc Plant Physiol; Sigma Xi. *Res:* Carbohydrate translocation in higher plants; active ion uptake by roots of higher plants; sulfur dioxide effects on mosses. *Mailing Add:* Dept Bot Eastern Ill Univ Charleston IL 61920

WEIDNER, VICTOR RAY, b Clarion, Pa, Jan 4, 32; m 67. OPTICAL PHYSICS. *Educ:* Clarion State Col, BS, 60. *Prof Exp:* PHYSICIST, NAT BUR STANDARDS, 60- *Honors & Awards:* Bronze Metal Award, US Dept Com. *Mem:* Optical Soc Am. *Res:* Infrared spectrophotometry; development of research spectrophotometers and spectrophotometric standards. *Mailing Add:* 24401 Hipsley Mill Rd Gaithersburg MD 20882

WEIDNER, WILLIAM JEFFREY, b Michigan City, Ind, Jan 21, 47; m 71. PHYSIOLOGY, EXPERIMENTAL BIOLOGY. *Educ:* Mich State Univ, BA, 68, MS, 71, PhD(physiol), 73. *Prof Exp:* Res assoc physiol, Mich State Univ, 73-74; fel, Univ Calif, San Francisco, 74-75; ASST PROF PHYSIOL, UNIV CALIF, DAVIS, 75- *Mem:* AAAS; Am Physiol Soc; Sigma Xi; Aerospace Med Asn. *Res:* Cardiopulmonary physiology; acceleration biology; history of science. *Mailing Add:* 1897 Sharon Dr Davis CA 95616

WEIER, RICHARD MATHIAS, b Streator, Ill, July 18, 40; m 69; c 2. MEDICINAL CHEMISTRY. *Educ:* Loras Col, BS, 62; Wayne State Univ, PhD(org chem), 67. *Prof Exp:* Res chemist, Ash-Stevens, Inc, Mich, 66-67; res chemist, Res & Develop Div, Kraftco Corp, 68-69; res investr, 69-77, res scientist I, 77-85, RES SCIENTIST II, GD SEARLE & CO, 85- *Mem:* AAAS; Am Chem Soc. *Res:* Synthesis of steroids and natural products; aldosterone blockade; diuretics; anti-hypertensives; prostaglandin synthesis; gastric antisecretory agents; design and synthesis of antagonists of lipid mediators of hypersensitivity; carbohydrate synthesis. *Mailing Add:* Searle Res & Develop G D Searle & Co 4901 Searle Pkwy Skokie IL 60077

WEIFFENBACH, CONRAD VENABLE, b Oak Park, Ill, Aug 25, 42. NUCLEAR PHYSICS. *Educ:* Univ Mich, BS, 64, PhD(physics), 70. *Prof Exp:* Res fel, Foster Radiation Lab, McGill Univ, 71-74; vol physics, US Peace Corp, King Mongkut Inst Technol, Bangkok, 76-78; asst prof physics, Kean Col NJ, 78-80; MEM FAC, PHYSICS DEPT, UNIV MAINE, 80- *Res:* Experimental nuclear structure physics. *Mailing Add:* Dept Physics Cornell Col Mt Vernon IA 52314

WEIGAND, WILLIAM ADAM, b Chicago, Ill, Oct 26, 38. CHEMICAL ENGINEERING. *Educ:* Ill Inst Technol, BS, 62, MS, 63, PhD(chem eng), 68. *Prof Exp:* Engr res, Esso Res Labs, 63-64; asst prof, 67-72, assoc prof chem eng, Purdue Univ, 72-79; prog dir, NSF, 76-77 & 79-83; prof chem eng, Ill Inst of Tech, 83-88; PROF CHEM ENG, UNIV MD, COLLEGE PARK, 88- *Concurrent Pos:* Prog dir, NSF, 76-77. *Mem:* Am Inst Chem Eng; Am Chem Soc; Sigma Xi. *Res:* Growth kinetics; dynamic modeling; optimal fermentor type and operation; automatic control of bioreactors. *Mailing Add:* Dept Chem Eng Univ Maryland College Park MD 20742

WEIGAND, WILLIS ALAN, cement research, for more information see previous edition

WEIGEL, PAUL H(ENRY), b New York, NY, Aug 11, 46; div; c 1. CELL SURFACE RECEPTORS, GLYCOPROTEINS. *Educ:* Cornell Univ, BA, 68; Johns Hopkins Univ Sch Med, MA, 69, PhD(biochem), 75. *Prof Exp:* Nat Cancer Inst postdoctoral fel, Biol Dept, Johns Hopkins Univ, 75-78; from asst prof to assoc prof, 78-87, PROF BIOCHEM & CELL BIOL, UNIV TEX MED BR, GALVESTON, 87-, VCHMN, 90- *Concurrent Pos:* Consult, TelTech Inc; Pathobiochem Study Sect, NIH 85-87; prin investr NIH grants. *Mem:* Am Chem Soc; Am Soc Cell Biol; Am Soc Biochem & Molecular Biol; AAAS; NY Acad Sci. *Res:* Receptor mediated endocytosis; cell surface receptors for extracellular matrix molecules; cell recognition and responses to external complex carbohydrate molecules; metabolism of asialoglycoproteins and hyaluronic acid; the role of hyaluronic acid in wound healing. *Mailing Add:* Dept Human Biol Chem & Genetics Univ Tex Med Br Galveston TX 77550

WEIGEL, ROBERT DAVID, b Buffalo, NY, Dec 31, 23. VERTEBRATE ZOOLOGY. *Educ:* Univ Buffalo, BA, 49, MA, 55; Univ Fla, PhD(zool), 58. *Prof Exp:* Instr gen biol, Univ Fla, 57; asst prof biol, Howard Col, 58-59; assoc prof, 59-64, PROF BIOL, ILL STATE UNIV, 64- *Mem:* AAAS; Soc Study Evolution; Soc Vert Paleont; Paleont Soc; Am Soc Mammal; Sigma Xi. *Res:* Avian paleontology; ornithology; osteology. *Mailing Add:* Box 487 Bloomington IL 61701

WEIGEL, RUSSELL C(ORNELIUS), JR, b Teaneck, NJ, Dec 10, 40. PLANT TISSUE CULTURE. *Educ:* Univ Del, BA, 62; George Washington Univ, BS, 63; Univ Md, College Park, MS, 68, PhD(plant physiol), 70. *Prof Exp:* Agr res investr, E I du Pont de Nemours & Co, Inc, 64-65; teaching asst bot & plant physiol, Univ Md, 65-69; res biologist, Plant Res Lab, E I du Pont de Nemours & Co, Inc, 69-80, sr res biologist, Biochem Dept, 80-82; adj assoc prof, Dept Bot, Univ Tenn, 82-86; ASSOC PROF, DEPT BIOL SCI, FLA INST TECHNOL, 86- *Mem:* Am Soc Plant Physiol; Sigma Xi; AAAS; Int Asn Plant Tissue Cult; Tissue Cult Asn. *Res:* Various aspects of plant tissue culture; regeneration; mutant selection; micropropagation. *Mailing Add:* Dept Biol Sci Fla Inst Technol 150 W University Blvd Melbourne FL 32901-6988

WEIGENSBERG, BERNARD IRVING, b Montreal, Que, Feb 6, 26; m 50; c 3. EXPERIMENTAL PATHOLOGY. *Educ:* McGill Univ, BSc, 49, MSc, 51, PhD(biochem), 53. *Prof Exp:* Res asst biochem, 49-54, res assoc path chem, 54-62, asst prof, 62-74, ASSOC PROF EXP PATH, MCGILL UNIV, 74- *Concurrent Pos:* Res assoc, Can Heart Found, 58-66; mem coun arteriosclerosis & coun thrombosis, Am Heart Asn. *Mem:* Fel Am Soc Study Arteriosclerosis; Am Soc Exp Path; Can Atherosclerosis Soc. *Res:* Atherosclerosis from cholesterol and lipoproteins; atherosclerosis from white mural non-occlusive thrombisis; myointimal thickenings from injury; lipid and connective tissue metabolism. *Mailing Add:* Lyman Duff Med Sci Bldg McGill Univ 2650 Bedford Montreal PQ H3S 1G1 Can

WEIGER, ROBERT W, medicine; deceased, see previous edition for last biography

WEIGHT, FORREST F, b Waynesboro, Pa, Apr 17, 36; m 61; c 3. NEUROSCIENCE, NEUROPHYSIOLOGY. *Educ:* Princeton Univ, AB, 58; Columbia Univ, Col Physicians & Surgeons, MD, 62. *Prof Exp:* Intern med, Univ NC Hosp, 62-63; resident, Mary Imogene Bassett Hosp, 63-64; vis scientist, Univ Goteborg, Sweden, 68-69; chief sect synaptic pharmacol, NIMH, 69-78; chief, lab preclin studies, 78-86, CHIEF ELECTROPHYSIOL SECT, NAT INST ALCOHOL ABUSE & ALCOHOLISM, 86- *Concurrent Pos:* Vis fel, Swed Med Res Coun, 68-69; secy assembly scientists, NIMH, 77-78; Heritage vis lectr & Int Brain Res Orgn vis lectr. *Honors & Awards:* Outstanding Performance Medal, Pub Health Serv. *Mem:* Am Physiol Soc; Am Soc Pharmacol & Exp Therapeut; Soc Neurosci; Int Brain Res Orgn; Res Soc Alcoholism; Int Soc Biomed Res Alcoholism. *Res:* Cellular mechanisms of communication and information processing in the nervous system; mechanisms of synaptic transmission; mechanisms controlling nerve cell excitability; identification of neurotransmitters; molecular basis of membrane permeability changes; cellular and molecular mechanisms of drug actions. *Mailing Add:* Electrophysiol Sect Nat Inst Alcohol Abuse & Alcoholism 12501 Washington Ave Rockville MD 20852

WEIGL, PETER DOUGLAS, b New York, NY, Nov 9, 39; m 68. ECOLOGY, ANIMAL BEHAVIOR. *Educ:* Williams Col, AB, 62; Duke Univ, PhD(vert ecol), 69. *Prof Exp:* Asst prof, 68-74, assoc prof, 74-80, PROF BIOL, WAKE FOREST UNIV, 80- *Mem:* AAAS; Am Soc Zoologists; Am Soc Mammal; Ecol Soc Am; Soc Study Evolution; Sigma Xi. *Res:* Vertebrate zoology; evolution; energetics of behavior, species interactions, functional morphology. *Mailing Add:* Dept Biol Wake Forest Univ Winston-Salem NC 27109

WEIGLE, JACK LEROY, b Montpelier, Ohio, Sept 5, 25; m 54; c 4. PLANT BREEDING. *Educ:* Purdue Univ, BS, 50, MS, 54; Mich State Univ, PhD, 56. *Prof Exp:* Asst prof hort, Colo State Univ, 56-61; from asst prof to assoc prof, 61-73, PROF HORT, IOWA STATE UNIV, 73- *Mem:* Am Soc Hort Sci; Am Genetic Asn. *Res:* Genetic, cytogenetic and physiological investigations and development of new cultivars of ornamental plants. *Mailing Add:* Dept Hort Iowa State Univ 106 Hort Ames IA 50011

WEIGLE, ROBERT EDWARD, b Shiloh, Pa, Apr 27, 27; m 49; c 1. APPLIED MECHANICS, APPLIED MATHEMATICS. *Educ:* Rensselaer Polytech Inst, BCE, 51, MS, 57, PhD, 59. *Prof Exp:* Eng consult, 52-54; construct supt, Long Serv Co, Inc, 54-55; assoc res scientist, Dept Mech, Rensselaer Polytech Inst, 55-59; res dir, Res Lab, US Army, 59-62, tech dir, Benet Weapons Lab, 62-69, dir lab, Watervliet Arsenal, 69-77, chief scientist, Benet Weapons Lab, 70-77, tech dir, Armament Res & Develop Command, 77-82, dir, Army Res Off, 82-88; DIR, PHYS SCI LAB, NMEX STATE UNIV, 88- *Concurrent Pos:* Consult high pressure technol comt, Nat Res Coun, 60-61, Army mem gun tube technol comt, Mat Adv Bd, 67-69; mem, Adv Panel Weapons & Mat, NATO, 60-67, consult, Adv Group Aeronaut Res & Develop-Struct & Mat Panel, 71; mem, Dept Defense Forum Phys & Eng Sci; mem, Army Res Coun, 68-70, US Army Materiel Command Tech Adv Bd, 77-84, actg dep Res & Tech, 83, chmn, reorganization study panel, 84, dept Defense Comt on Res, 82-88; consult cannon wear & erosion, Mat Adv Bd, 77-78. *Honors & Awards:* Crozier Prize, 85; Presidential Citations, 65, 78 & 82. *Mem:* Nat Soc Prof Eng; Soc Exp Mech; AAAS; Am Defense Preparedness Asn; Am Acad Mech; Am Soc Mech Engrs; Sigma Xi. *Res:* Fatigue and fracture behavior of high strength alloy steels as applied to high performance, large caliber weapons and related equipment; mechanical behavior of hi-performance composite materials. *Mailing Add:* Phys Sci Lab NMex State Univ Box 30002 Las Cruces NM 88003-0002

WEIGLE, WILLIAM O, b Monaca, Pa, Apr, 28, 27; div; c 2. IMMUNOLOGY, CELL BIOLOGY. *Educ:* Univ Pittsburgh, BS, 50, MS, 51, PhD(bact), 56. *Prof Exp:* Res technician, Dept Path, Sch Med, Univ Pittsburgh, 51-52, res assoc, 55-58, asst res prof, 58-59, asst prof immunochem, 59-61; assoc mem, 61-63, mem dept, 63-82, chmn dept immunopathol, 78-82, vchmn dept, 82-85, MEM, DEPT IMMUNOL, SCRIPPS CLIN & RES FOUND, 82-, HEAD, DIV CELLULAR IMMUNOL, 84-, CHMN DEPT, 85- *Concurrent Pos:* USPHS fel, 56-59, sr res fel, 59-61, career res award, 62-; adj assoc prof, Univ Calif, San Diego, 61-67, adj prof biol, 67- *Honors & Awards:* Parke-Davis Award, Am Soc Exp Path, 67. *Mem:* Am Soc Microbiol; Am Asn Immunologists; Am Soc Exp Path; Am Acad Allergy; NY Acad Sci. *Res:* Mechanisms involved in immunity and diseases of hypersensitivity; immunochemistry; immunopathology. *Mailing Add:* Dept Immunol IMM9 Scripps Clin & Res Found 10666 N Torrey Pines Rd La Jolla CA 92037

WEIGMAN, BERNARD J, b Baltimore, Md, Nov 18, 32; m 60; c 4. ENGINEERING PHYSICS, COMPUTER SCIENCE. *Educ:* Loyola Col, Md, BS, 54; Univ Notre Dame, PhD(physics), 59. *Prof Exp:* From instr to assoc prof physics, 58-67, chmn, dept math, 60-64 & chmn dept physics & eng, 64-73, PROF PHYSICS, LOYOLA COL, MD, 67- *Concurrent Pos:* Consult, Martin Co, 59-61 & Westinghouse Corp, 74-, dir, Master Eng Sci Digital Systs. *Mem:* Am Asn Physics Teachers; Asn Comput Mach. *Res:* Optical and electrical properties of thin films; behavior of an aerosol system; thermoelectric, photoelectric and field emission of electrons; properties of molecules; small computer systems; real time data collection. *Mailing Add:* Dept Eng Sci Loyola Col 4501 N Charles St Baltimore MD 21210

WEIHAUPT, JOHN GEORGE, b La Crosse, Wis, Mar 5, 30; m 61. ASTROGEOLOGY, OCEANOGRAPHY. *Educ:* Univ Wis, BS, 52, MS, 53 & 71, PhD(geomorphol), 73. *Prof Exp:* Teaching asst geol, Univ Wis, 52-53, res asst, 53-54; explor geologist, Anaconda Co, 56-57; seismologist, United Geophys Corp, 57-58; geophysicist, Arctic Inst NAm, 58-63; chmn dept phys & biol sci, US Armed Forces Inst, Dept Defense, 63-73; dean acad affairs, Sch Sci, Ind Univ-Purdue Univ, 73-78, asst dean grad sch, 75-78; vchancellor Acad Affairs, 81-86, PROF GEOL, UNIV COLO, DENVER, 86-; ASSOC ACAD VPRES & DEAN GRAD STUDIES & RES, SAN JOSE STATE UNIV, 78- *Concurrent Pos:* Explor geologist, Am Smelting & Ref Co, 53; codiscoverer, USARP Range, Antarctica, 59-60 & Mt Weihaupt named in his honor, 66; lectr, Univ Wis-Madison, 63-73; hon mem, Exped Polaire Francais; mem, Man/Environ Commun Ctr, Community Coun Pub TV & Int Coun Correspondence Educ; sci consult, McGraw-Hill Book Co, 65, geol consult, 68-; sci consult, Holt, Rinehart & Winston, Inc, 65-; ed & consult, John Wiley & Sons, 68; vpres Univ Res Found, San Jose State Univ; liaison rep, AAAS. *Mem:* AAAS; Am Geophys Union; fel Geol Soc Am; Asn Am Geog; Int Soc Study Time; Sigma Xi. *Res:* Antarctic geology and geophysics in regions of Victoria Land and the South Pole in Antarctica; geophysical-biological periodic relationships; marine geology and geophysics; fluvial, glacial and coastal geomorphology; channels and paleoclimate of Mars; meteorite crater and impact phenomena; discoverer of the Wilkes Land Anomaly, Victoria Land, Antartica; author of seven books and 50 research articles. *Mailing Add:* Univ Colo PO Box 173346 1200 Larimer St Box 172 Denver CO 80217-3346

WEIHE, JOSEPH WILLIAM, mathematics; deceased, see previous edition for last biography

WEIHER, JAMES F, b Waverly, Iowa, Mar 30, 33; m. SURFACE CHEMISTRY, CATALYSIS. *Educ:* Carleton Col, BA, 55; Iowa State Univ, PhD(phys chem), 61. *Prof Exp:* Res chemist, Los Alamos Sci Lab, 55; res chemist, Study Group Radiochem, Max Planck Inst, Mainz, Ger, 55-57; res asst, Inst Atomic Res, Iowa State Univ, 57-61; fel, E I du Pont de Nemours & Co, Inc, 61-62, res chemist, Cent Res Dept, 62-86; CONSULT, 86- *Concurrent Pos:* Fulbright fel. *Res:* Nuclear chemistry and radiochemistry; kinetics; molecular structure; solid state; Mossbauer effect; magnetic susceptibility; catalysis; lab automation and control; artificial intelligence. *Mailing Add:* Eight Pinecrest Dr Wilmington DE 19810-1414

WEIHING, JOHN LAWSON, b Rocky Ford, Colo, Feb 26, 21; m 48; c 4. PLANT PATHOLOGY. *Educ:* Colo Agr & Mech Col, BS, 42; Univ Nebr, MS, 49, PhD(bot), 54. *Prof Exp:* Assoc plant path, Agr Exten, Univ Nebr, Lincoln, 49-50, exten plant pathologist, 50-64, dir, Panhandle Sta, 71-84, prof plant path, 60-85; RETIRED. *Concurrent Pos:* Mem Univ Nebr group, Agency Inst Develop, 64- *Mem:* Bot Soc Am; Am Phytopath Soc; AAAS; Am Inst Biol Sci. *Res:* Epidemiology of plant diseases. *Mailing Add:* 1605 Holly Dr Gering NE 69341

WEIHING, ROBERT RALPH, b Ft Collins, Colo, Jan 14, 38. BIOCHEMISTRY, CELL BIOLOGY. *Educ:* Rice Univ, BA, 59; Johns Hopkins Univ, PhD(biochem), 65, MD, 67. *Prof Exp:* Staff assoc, NIH, 68-71; staff scientist, Worcester Found Exp Biol, 71-77, sr scientist cell biol, 77-85; res prof biol, Clark Univ, 85-88; RESIDENT PATH, UNIV MD, 89- *Concurrent Pos:* Am Cancer Soc res scholar, 77-80. *Mem:* Am Soc Cell Biol; Am Soc Biochem & Molecular Biol. *Res:* Molecular basis of movement in non-muscle cells. *Mailing Add:* Dept Path Univ Maryland Syst-Hosp 22 S Greene St Baltimore MD 21201

WEIJER, JAN, b Heerenveen, Neth, Jan 28, 24; m 52. GENETICS. *Educ:* Univ Groningen, BSc, 51, MSc, 52, DSc, 54. *Prof Exp:* Asst genetics, Univ Groningen, 46-52 & exp embryol, 51-52; geneticist, Firestone Bot Res Inst, 52-54, Can, 55-57; prof hort, Univ Alta, 57-59, chmn dept genetics, 67-69, prof genetics, 60-80; RETIRED. *Concurrent Pos:* Lectr, Neth, 59; consult, US AEC, 67; mem coun, Int Orgn Pure & Appl Biophys. *Mem:* AAAS; Radiation Res Soc; Am Genetic Asn; Genetics Soc Am; NY Acad Sci. *Res:* Microbial genetics and cytology, especially fine structure of the gene; biochemical genetics and mutation; biological assessment of radiation damage in humans. *Mailing Add:* PO Box 3015 Sherwood Park AB T8A 2A6 Can

WEIK, MARTIN H, JR, b New York, NY, Oct 5, 22; m 43; c 6. PHYSICS, ELECTRICAL ENGINEERING. *Educ:* City Col New York, BS, 49; Columbia Univ, MS, 51; George Washington Univ, DSc, 79. *Prof Exp:* Asst eng, Columbia Univ, 49-52; designer electronic digital comput, Ballistic Res Labs, Aberdeen Proving Ground, Md, 53-64; chief res & eng systs br & off chief res & develop, US Army Res Off, 64-69, chief data mgt div, 69-73; eng systs analyst, US Mil Commun Electronics Bd, Defense Commun Agency, 73-79; SYSTS ANALYST, DYNAMIC SYSTS, INC, 81- *Concurrent Pos:* Chief US deleg, Int Orgn Standardization Tech Comt 97, subcomt 1 vocabulary, 61-; chmn, Am Nat Standards Inst Tech Comt X3K5 Vocabulary. *Honors & Awards:* Meritorious Civilian Serv Award, US Army. *Mem:* AAAS; Inst Elec & Electronics Engrs; Asn Advan Med Instrumentation. *Res:* Logical design of electronic digital computers; design and development of technical information systems; information processing and computer language and vocabulary development; fiber optics. *Mailing Add:* 2200 Columbia Pike Arlington VA 22204

WEIKEL, JOHN HENRY, JR, b Palmerton, Pa, June 14, 29. TOXICOLOGY. *Educ:* Trinity Col, Conn, BS, 51; Univ Rochester, PhD(pharmacol), 54. *Prof Exp:* Res assoc, AEC Proj, Univ Rochester, 51-54; pharmacologist, Div Pharmacol, Food & Drug Admin, 54-56; sr pharmacologist, 56-58, group leader, 58-59, sect leader, 59-61, asst dept dir, Res Ctr, 61-62, dir chem pharmacol & safety eval, 62-68, DIR PATH & TOXICOL, MEAD JOHNSON & CO, 68- *Mem:* Am Chem Soc; Soc Exp Biol & Med; Am Soc Pharmacol & Exp Therapeut; Soc Toxicol; NY Acad Sci. *Res:* Inorganic metabolism; drug metabolism; toxicology. *Mailing Add:* Res Lab Mead Johnson Co Evansville IN 47721

WEIL, ANDRE, MATHEMATICS. *Prof Exp:* EMER PROF MATH, INST ADVAN STUDY, 76- *Mem:* Nat Acad Sci. *Mailing Add:* Sch Math Inst Advan Study Princeton NJ 08540

WEIL, ANDREW THOMAS, b Philadelphia, Pa, June 8, 42; m 90. MEDICINE, ETHNOPHARMACOLOGY. *Educ:* Harvard Col, AB, 64; Harvard Med Sch, MD, 68. *Prof Exp:* LECTR, UNIV ARIZ, COL MED, 83- *Concurrent Pos:* Inst Current World Affairs fel, 71-75; res assoc ethnopharmacol, Harvard Bot Mus, 71-84; adj prof addiction studies, Health Sci Ctr, Univ Ariz, 78-82; pres, Beneficial Plant Res Inst, 79- *Mem:* Sigma Xi. *Res:* Ethnopharmacology, especially native uses of psychoactive and medicinal plants; drug abuse; herbal medicine; alternative and holistic medicine; altered states of consciousness; mind body interactions. *Mailing Add:* Div Social Perspectives Med Ariz Health Sci Ctr Col Med Tucson AZ 85724

WEIL, BENJAMIN HENRY, b St Joseph, Mo, July 8, 16; m 44; c 3. CHEMICAL ENGINEERING, COMMUNICATIONS. *Educ:* Univ Mo-Columbia, BS, 39; Univ Wis-Madison, MS, 40. *Prof Exp:* Head info sect & chem mkt researcher, Chem Div, Gulf Res & Develop Co, 40-45; mgr, Info Serv Div, Eng Exp Sta, Ga Inst Technol, 45-50; mgr info serv, Ethyl Corp Res Labs, 50-57; head, Tech Info Sect, Exxon Res & Eng Co, 57-72, sr staff adv & monographs ed, 72-82; INFO CONSULT, BEN H WEIL, INC, 82- *Concurrent Pos:* Trustee, Eng Index Inc, 64-79, dir, 65-68 & vpres, 65-67; creator, Copyright Clearance Ctr, Asn Am Publ, 77, vpres, 72-80; mem bd, Int Coun Sci Unions, 77-88, US Nat Rep Abstr Bd, Nat Acad Sci-Nat Res Coun, 78-81. *Honors & Awards:* Patterson-Crane Award, Am Chem Soc, 77, Herman Skolnik Award, 81; Miles Conrad lectr, Nat Fedn Abstracting & Indexing Serv, 78. *Mem:* Am Chem Soc; Nat Fedn Abstr & Indexing Serv (secy, 68-69 & 72-73, treas, 71-72, pres, 75-76); Am Soc Info Sci; fel AAAS; Assoc Info Mgrs. *Res:* Information-center management and design; information systems; copyright-compliance operations; internal-reporting operations and systems; technical writing, editing and publishing. *Mailing Add:* Four Wells Lane Warren NJ 07060-5311

WEIL, CARROL S, b St Joseph, Mo, Dec 16, 17; m 40; c 2. BIOMETRICS-BIOSTATISTICS. *Educ:* Univ Mo, BA, 39, MA, 40. *Prof Exp:* Bacteriologist, Anchor Serum Co, Mo, 41-42; res assoc & toxicologist, Carnegie-Mellon Univ, 42-43; unit head, Manhattan Div, Univ Rochester, 43-45; fel, Carnegie-Mellon Univ, 45-52, chief toxicologist & fel, Carnegie-Mellon Res Inst, 52-80; corp fel, Bushy Run Res Ctr, Union Carbide Corp, 80-82; CONSULT, CARROL S WEIL, INC, 83- *Concurrent Pos:* Ambassador Toxicol, Soc Toxicol, 89. *Honors & Awards:* Herbert E Stokinger Lectr & Award, Am Conf Govt Indust Hygienists, 84; George H Scott Award, Toxicol Forum, 84; Merit Award, Soc Toxicol, 85. *Mem:* Am Chem Soc; Am Indust Hyg Asn; Biomet Soc; Distinguished fel Soc Toxicol (secy, 63-67, pres, 68-69). *Res:* Chemical hygiene and toxicology; biometrics. *Mailing Add:* 4326 McCaslin St Pittsburgh PA 15217

WEIL, CLIFFORD EDWARD, b East Chicago, Ind, Nov 19, 37; m 60; c 3. REAL ANALYSIS. *Educ:* Wabash Col, BA, 59; Purdue Univ, MS, 61, PhD(math), 63. *Prof Exp:* Instr math, Princeton Univ, 63-64 & Univ Chicago, 64-66; from asst prof to assoc prof, 66-78, PROF MATH, MICH STATE UNIV, 78- *Concurrent Pos:* Vis prof math, Philipps-Univ, WGer, 73-74. *Mem:* Am Math Soc; Math Asn Am. *Res:* Functions of a real variable. *Mailing Add:* Dept Math A231 Mich State Univ East Lansing MI 48824-1027

WEIL, EDWARD DAVID, b Philadelphia, Pa, June 13, 28; m 52; c 2. ORGANIC CHEMISTRY. *Educ:* Univ Pa, BS, 50; Univ Ill, MS, 51, PhD(org chem), 53; Pace Univ, MBA, 82. *Prof Exp:* Res & develop chemist, Hooker Electrochem Co, 53-56, supvr agr chem res, Hooker Chem Corp, 56-65; supvr org res, Stauffer Chem Co, Dobbs Ferry, 65-69, sr res assoc, 69-76, sr scientist, 76-86; CONSULT, INTERTECH SERV, 86-; RES PROF, POLYTECH UNIV, BROOKLYN, NY, 88- *Honors & Awards:* IR-100 Award, Indust Res Inst, 74. *Mem:* Am Chem Soc. *Res:* Vinyl polymers; organic sulfur; chlorine derivatives; pesticides; organic phosphorus; rubber chemicals; flame retardants; textile chemicals; polymer additives; research planning; specialty chemicals and polymers. *Mailing Add:* Six Amherst Dr Hastings-on-Hudson NY 10706

WEIL, FRANCIS ALPHONSE, b Selestat, France, Nov 5, 38; Can citizen; m 64; c 2. ENVIRONMENT, ENERGY CONSERVATION. *Educ:* Univ Strasbourg, BSc, 58; Univ Paris, DiplEng, 61; Dalhousie Univ, MSc, 62, PhD(physics), 68. *Prof Exp:* Teaching asst math, Univ Paris at the Sorbonne, France, 60-61; res asst physics, Saclay Nuclear Res Ctr, France, 66-67; PROF PHYSICS, UNIV MONCTON, 68- *Concurrent Pos:* Fel, Killam Found, 68; Nat Res Coun grant, Univ Moncton, 69-70, Univ Res Coun, 70-72; dean, Fac Sci & Eng, Univ Moncton, 80-90; Fisheries & Oceans, Can, grant, 89, 90-91; dir, Sci Peace. *Honors & Awards:* Prix Schlumberger Award, 89. *Mem:* Can Asn Physicists; Fr-Can Asn Advan Sci; Am Asn Physics Teachers. *Res:* Quantum field theory; energy conservation (environment). *Mailing Add:* Dept Physics Univ Moncton Moncton NB E1A 3E9 Can

WEIL, HERSCHEL, b Rochester, NY, July 26, 21; m 43; c 2. ELECTROMAGNETIC SCATTERING. *Educ:* Univ Rochester, BS, 43; Brown Univ, ScM, 45, PhD(appl math), 48. *Prof Exp:* Optical engr, Bausch & Lomb Optical Co, 43-44; asst appl math, Brown Univ, 46-47, res assoc, 47-48; engr, Gen Elec Co, 48-52; res assoc, 52-53, assoc res engr, 53-55, res engr, 55-60, lectr, 58-60, assoc prof elec eng, 60-67, prof, 67-87, EMER PROF ELEC ENG & COMP SCI, UNIV MICH, ANN ARBOR, 87- *Concurrent Pos:* Vis assoc prof, Univ Paris-Orsay, 64-65; sr vis fel, Nuffield Radio Astron Labs, Univ Manchester, 75; sr fel, Nat Ctr Atmospheric Res, 77-78. *Mem:* Inst Elec & Electronics Engrs; Optical Soc Am. *Res:* Electromagnetic theory and applications to engineering and atmospheric optics. *Mailing Add:* Dept Elec Eng & Comput Sci Univ Mich Ann Arbor MI 48109-2122

WEIL, JESSE LEO, b Ann Arbor, Mich, Dec 9, 31; div; c 2. NUCLEAR PHYSICS. *Educ:* Calif Inst Technol, BS, 52; Columbia Univ, PhD(physics), 59. *Prof Exp:* Res asst nuclear physics, Columbia Univ, 54-58; res assoc, Rice Univ, 59-60 & 61-63 & Univ Hamburg, 60-61; from asst to assoc prof, 63-73, PROF NUCLEAR PHYSICS, UNIV KY, 73- *Concurrent Pos:* Res assoc, Rutherford High Energy Lab, 71-72. *Mem:* Am Asn Univ Prof; Am Phys Soc; Sigma Xi. *Res:* Nuclear reactions; scattering of nucleons and nuclei; nuclear energy level studies; radioactivity and fission studies beta decay and nuclear masses; neutron polarization. *Mailing Add:* Dept Physics Univ Ky Lexington KY 40506

WEIL, JOHN A, b Hamburg, Ger, Mar 15, 29; nat US; m 47; c 2. CHEMICAL PHYSICS. *Educ:* Univ Chicago, MS, 50, PhD(chem), 55; Univ Sask, DSc, 85. *Prof Exp:* Res chemist, Inst Study Metals, Univ Chicago, 49-52, Inst Nuclear Studies, 53-54; Corning fel chem, Princeton Univ, 55-56, instr phys chem, 56-59; from assoc scientist to sr scientist, Argonne Nat Lab, Ill, 59-71; PROF

CHEM, UNIV SASK, 71- *Concurrent Pos:* Fulbright scholar physics, Univ Canterbury, NZ, 67; vis prof, Univ Chicago, 80; vis prof physics, Lehigh Univ, 87. *Honors & Awards:* Erskine lectr, Univ Canterbury, 87. *Mem:* Fel Chem Inst Can; Am Phys Soc; Brit Int Physics & Phys Soc; Int Soc Magnetic Res. *Res:* Paramagnetic resonance; quantum chemistry; electronic structure of inorganic complexes; defect structure of silicates and other solids; organic free radicals. *Mailing Add:* Dept Chem Univ Sask Saskatoon SK S7N 0W0 Can

WEIL, JOHN VICTOR, b Detroit, Mich, May 10, 35; m 64; c 1. INTERNAL MEDICINE. *Educ:* Yale Univ, BS, 57, MD, 61. *Prof Exp:* Intern path, Sch Med, Yale Univ, 61-62, intern med, 62-63, asst resident, 63-64 & 67-68; asst prof, 68-72, ASSOC PROF MED, MED CTR, UNIV COLO, DENVER, 72- *Concurrent Pos:* NIH res fel cardiol, Univ Colo, Denver, 66-67. *Res:* Physiological responses to hypoxia; peripheral circulation, erythropoiesis, ventilation, oxygen transport. *Mailing Add:* Dept Med CVP Res Lab Univ Colo Health Sci Ctr 4200 E Ninth Ave Box B-133 Denver CO 80262

WEIL, JON DAVID, b Evansville, Ind, Mar 24, 37; m 85; c 2. HUMAN GENETICS, GENETIC COUNSELING. *Educ:* Swarthmore Col, BA, 58; Univ Calif, Davis, PhD(genetics), 63, Wright Inst, PhD(clin psychol), 84. *Prof Exp:* NSF fel molecular biol, Univ Ore, 63-65; NIH fel, Harvard Univ, 65-67; from asst prof to assoc prof molecular biol, Vanderbilt Univ, 67-77; res geneticist, 77-86, ASST ADJ PROF PEDIAT, UNIV CALIF, SAN FRANCISCO, 86- *Concurrent Pos:* Postdoctoral fel psychol, Dept Pediat, Univ Calif, San Francisco, 84-86; lectr, Calif Sch Prof Psychol, 77, 86-88. *Mem:* Am Soc Human Genetics; AAAS; Am Psychol Asn. *Res:* Activities; pathogenic mechanisms of human aneuploidy; psychological consequences of genetic diseases. *Mailing Add:* 1608 Belvedere Berkeley CA 94702

WEIL, MARVIN LEE, b Gainesville, Fla, Sept 28, 24; m 54; c 3. PEDIATRIC NEUROLOGY, VIROLOGY. *Educ:* Univ Fla, BS, 43; Johns Hopkins Univ, MD, 46; Am Bd Pediat, dipl; Am Bd Psychiat & Neurol, dipl. *Prof Exp:* Intern, Duke Univ Hosp, 46-47; asst resident pediat, 47-48; resident, Children's Hosp, Cincinnati, 50-52, res assoc, Res Found, 52-53; clin asst prof pediat, Sch Med, Univ Miami, 58-65; Nat Inst Neurol Dis & Stroke spec fel, Johns Hopkins Univ & Univ Calif, Los Angeles, 65-68; from asst prof to assoc prof, 68-78, PROF PEDIAT & NEUROL, UNIV CALIF, LOS ANGELES, 78- *Concurrent Pos:* Attend pediatrician, Cincinnati Gen Hosp, 52-53; pvt pract, 53-65; chief div pediat neurol, Harbor Gen Hosp, 68-; sr int fel, Fogarty Int Ctr, Karolinska Inst, NIH, Stockholm, Sweden, 76-77. *Mem:* AAAS; Am Asn Immunologists; fel Am Acad Pediat; fel Am Acad Neurol; Child Neurol Soc; Am Pediat Soc. *Res:* Growth characteristics, genetics, neurotropic behavior of mammalian viruses; immune responses of central nervous system; infections of the central nervous system. *Mailing Add:* Harbor Univ Calif Med Ctr 1000 W Carson St Torrance CA 90509

WEIL, MARVIN LEE, b Gainesville, Fla, Sept 28, 24; m 54; c 3. PEDIATRICS, CHILD NEUROLOGY. *Educ:* Univ Fla, BS, 43; Johns Hopkins Univ, MD, 46. *Prof Exp:* Intern pediat, Duke Univ, 46-47, asst resident, 47-48; asst chief, Neurotrop Virus Sect, Div Virus & Rickettsial Dis, Army Med Dept Res & Grad Sch, 48-50; res pediat, Cincinnati Children's Hosp, 50-52; clin instr pediat, Univ Cincinnati, 52-53; clin asst prof pediat, Univ Miami Sch Med, 54-65; spec fel pediat neurol, Nat Inst Neurol Dis & Blindness, Johns Hopkins Univ, 65-66, Nat Inst Neurol Dis & Blindness, Univ Calif, Los Angeles, 66-68; from asst prof to prof, 68-89, EMER PROF PEDIAT & NEUROL, SCH MED, UNIV CALIF, LOS ANGELES, 89- *Concurrent Pos:* Pvt pract pediat, Miami, Fla, 53-65; mem, Accident Prev Comt, Am Acad Pediat, 64-67, bd dirs, Child Neurol Soc, 80-82; Fogarty Int Ctr, sr int fel, Dept Virol, Karolinska Inst, Stockholm, Sweden, NIH, 76-77; acad visitor, Dept Biochem, Univ Oxford, UK, 89- *Mem:* Child Neurol Soc; Am Pediat Soc; Am Asn Immunologists; Am Acad Pediat; Am Acad Neurol; Int Child Neurol Soc. *Res:* Etiology, immunology and biochemistry of infections and inflammation of the nervous system; effects of intracranial hypertension on brain metabolism; pathogenesis of demyelination. *Mailing Add:* 82 Thames St Oxford 0X1 1SU England

WEIL, MAX HARRY, b Baden, Switz, Feb 9, 27; nat US; m 55; c 2. CARDIOVASCULAR DISEASES. *Educ:* Univ Mich, AB, 48; State Univ NY Downstate Med Ctr, MD, 52; Univ Minn, PhD(med), 57; Am Bd Internal Med, dipl, 62. *Prof Exp:* Intern internal med, Cincinnati Gen Hosp, Ohio, 52-53; resident, Univ Hosps, Univ Minn, 53-55; asst clin prof, Sch Med, Univ Southern Calif, 57-59, from asst prof to assoc prof, 59-71, dir, Ctr Critically Ill, 68-80, prof med, 71-81, prof biomed eng, 72-81; dir, Inst Critical Care Med, Los Angeles, 76-82; DISTINGUISHED PROF MED & PHYSIOL & CHMN, CHICAGO MED SCH, 82-; CHIEF, DIV CARDIOL, 83- *Concurrent Pos:* Vis prof, Univ Pittsburgh; chief cardiol, City of Hope Med Ctr, Duarte, Calif, 57-59, consult, 59-63; attend physician, Los Angeles County Gen Hosp, 58-71, sr atten cardiologist, Children's Div, 58-65; consult physician, Cedars-Sinai Med Ctr, 65-81, hon physician, 81-; chmn, Dept Med, Univ Health Sci, Chicago Med Sch, 82-; past mem comt shock, Nat Acad Sci-Nat Res Coun; fel coun circulation & coun thrombosis, Am Heart Asn; pres, Int Critical Care Med, Inc, 75- *Mem:* Fel Am Col Cardiol; fel Am Col Physicians; Am Physiol Soc; Am Soc Pharmacol & Exp Therapeut; Inst Elec & Electronics Engrs. *Res:* Clinical cardiology and cardiorespiratory physiology; critical care medicine; studies on circulatory shock and cardiopulmonary resuscitation both experimental and clinical; biomedical instrumentation and automation; applications of computer techniques to bedside medicine. *Mailing Add:* 3810 Mission Hills Rd Northbrook IL 60062

WEIL, MICHAEL RAY, b Buffalo, NY, Feb 10, 51; m 84. COMPARATIVE ENDOCRINOLOGY. *Educ:* Univ Mich, BA, 72; St Louis Univ, MS, 76, PhD(biol), 80. *Prof Exp:* ASST PROF BIOL, UNIV WIS-EAU CLAIRE, 79- *Mem:* Soc Study Amphibians & Reptiles; Herpetologists League; Am Soc Ichthyologists & Herpetologists; Sigma Xi; Am Soc Zoologists; Soc Study Evolution. *Res:* Comparative vertebrate endocrinology; seasonal cycles of reptilian and amphibian reproductive hormones; hormone binding proteins in the blood; developmental endocrinology. *Mailing Add:* Dept Biol Univ Wis Eau Claire WI 54701

WEIL, RAOUL BLOCH, b La Paz, Bolivia; US citizen; m 56; c 2. PHYSICS. *Educ:* Univ La Paz, Bachiller, 45; Univ Ill, Urbana, BSEE, 49; Univ Calif, Riverside, MA, 63, PhD(physics), 66. *Prof Exp:* Trainee, Allis Chalmers Mfg Co, 49-51; consult engr, Bolivian Com Corp, 51-53, head tech dept, 53-55; prof elec eng, Univ La Paz, 56-60; res asst cosmic ray physics, Univ Chicago, 60-61; solid state physics, Univ Calif, 61-65; res physicist, Monsanto Co, 65-67, proj leader, 67-69; assoc prof elec eng, Wash Univ, 69-71; assoc prof, 71-90, PROF PHYSICS, ISRAEL INST TECHNOL, 90- *Concurrent Pos:* Res assoc, Cosmic Physics Lab, Chacaltaya, 57-60; secy, Bolivian Nat Comt for the Int Geophys, 57-60; vis assoc prof, Univ Ill, Urbana, 77-78; vis prof, Univ Louis Pasteur, Strasbourg, France, 85 & Solar Energy Res Inst, Golden, Colo, 86; mem Israel rep, Thins Films Div, Int Union Vacuum Sci Tech & Appl Physics Comt, Europ Phys Soc, 86; mem Israel representant to the applied physics comt of the Europ Phys Soc, 86. *Mem:* Am Phys Soc; Inst Elec & Electronics Engrs; Israel Laser & Electrooptics Soc (secy, 73-75, pres, 76-); Israel Phys Soc (secy, 76-77); Sigma Xi; Israel Vacuum Soc. *Res:* Solar cells; optical properties of semiconductors in the infrared; lasers; superionic conductors; amorphous silicon, ferroelectricity-superconductivity. *Mailing Add:* Physics Dept Technion Israel Inst Technol Haifa 32000 Israel

WEIL, ROLF, b Neunkirchen, Ger, Aug 5, 26; nat US. METALLURGY, ELCTROCHEMISTRY. *Educ:* Carnegie Inst Technol, BS, 46, MS, 49; Pa State Univ, PhD(metall), 51. *Hon Degrees:* ME, Stevens Inst Technol, 67. *Prof Exp:* Metallurgist, Duquesne Smelting Corp, 46-48; grad asst, Pa State Univ, 49-51; assoc metallurgist, Argonne Nat Lab, 51-54; from asst prof to prof mat sci & eng, 56-91, EMER PROF MAT SCI & ENG STEVENS INST TECHNOL, 91- *Honors & Awards:* Sci Achievement Award, Am Electroplaters Soc; Res Award, Electrodeposition Div, Electrochem Soc. *Mem:* Electrochem Soc; Am Inst Mining, Metall & Petrol Engrs; Am Electroplaters & Surface Finishers Soc; fel Brit Inst Metal Finishing. *Res:* Structure and properties of electrodeposited metals; electron microscopy; metal strengthening mechanisms; corrosion. *Mailing Add:* Dept Mat & Metall Eng Castle Point Sta Hoboken NJ 07030

WEIL, THOMAS ANDRE, b Ft Riley, Kans, June 27, 44; m 70; c 2. INORGANIC CHEMISTRY. *Educ:* State Univ NY Col Oswego, BA, 66; Univ Cincinnati, PhD(chem), 70. *Prof Exp:* Nat Res Coun fel, US Bur Mines, Pa, 70-71; asst prof chem, Trenton State Col, 71-72; NIH fel, Univ Chicago, 72-74; res chemist, 74-77, res supvr, 77-82, dir explor res, 82-87, MGR, NEW PROD RES & DEVELOP, AMOCO CHEM CORP, 87- *Concurrent Pos:* NSF exchange fel, US-USSR Sci & Technol, 74. *Mem:* Am Chem Soc. *Res:* Catalysis; heterogeneous and homogeneous catalysis; transition metal complexes; metals in organic synthesis; mechanisms of metal catalyzed reactions; energy related research. *Mailing Add:* Amoco Chem Corp Box 400 Naperville IL 60560-0400

WEIL, THOMAS P, b Mt Vernon, NY, Oct 2, 32; m 65. PUBLIC HEALTH, HOSPITAL ADMINISTRATION. *Educ:* Union Col, NY, AB, 54; Yale Univ, MPH, 58; Univ Mich, PhD(med care orgn), 65. *Prof Exp:* S S Goldwater fel hosp admin, Mt Sinai Hosp, New York, 57-58; assoc consult, John G Steinle & Assocs, Mgt Consults, 58-61; asst prof, Sch Pub Health, Univ Calif, Los Angeles, 62-65; assoc dir & consult, Touro Infirmary, New Orleans, 64-66; prof grad studies health serv mgr & dir, Schs Med, Bus & Pub Admin, Univ Mo, Columbia, 66-71; vpres & prin, E D Rosenfeld Assocs, Inc, New York, 71-75; PRES, BEDFORD HEALTH ASSOCS, INC, 75- *Concurrent Pos:* Consult to numerous hosps, planning agencies & med ctrs; vis prof & W K Kellogg Found grant, Univ NSW, 69. *Mem:* Fel Am Pub Health Asn; Am Hosp Asn; Am Col Hosp Adminr; Am Asn Hosp Consult. *Res:* Management, organization and financing of health services, especially hospital and medical care administration. *Mailing Add:* 1400 Town Mountain Rd Asheville NC 28804

WEIL, WILLIAM B, JR, b Minneapolis, Minn, Dec 3, 24; m 49; c 2. PEDIATRICS. *Educ:* Univ Minn, BA, 45, BS, 46, MB, 47, MD, 48. *Prof Exp:* USPHS & univ res fels, Harvard Univ, 51-52; chmn dept, 68-79, PROF PEDIAT & HUMAN DEVELOP, MICH STATE UNIV, 68-; assoc prof, Col Med, Univ Fla, 63-65, E I du Pont prof for handicapped children, 65-68; chmn dept, 68-79, PROF HUMAN DEVELOP, MICH STATE UNIV, 68- *Mem:* AAAS; Soc Pediat Res (secy-treas, 62-69, pres, 69-70); Am Pediat Soc; Am Acad Pediat; NY Acad Sci. *Res:* Renal disease; diabetes; nutrition. *Mailing Add:* Dept Human Develop B140 Life Sci Mich State Univ East Lansing MI 48824-1317

WEILER, EDWARD JOHN, b Chicago, Ill, Jan 15, 49. ASTRONOMY. *Educ:* Northwestern Univ, BA, 71, MS, 73, PhD(astron), 76. *Prof Exp:* Res assoc astron, Avionics Lab, Northwestern Univ, 73-74; RES ASSOC, PRINCETON UNIV-NASA GODDARD SPACE FLIGHT CTR, 76- *Mem:* AAAS. *Res:* Chromospheric activity in late-type binary stars and the evolution of galaxies. *Mailing Add:* Astrophys Div NASA HQ Code E2 Washington DC 20546

WEILER, ERNEST DIETER, b Neuwied, Ger, June 30, 39; US citizen; m 67. ORGANIC CHEMISTRY. *Educ:* Univ Minn, BChem, 62; Univ Nebr, PhD(org chem), 66; Temple Univ, MBA, 74. *Prof Exp:* Fel, Univ Basel, 66-67; chemist, Rohm & Haas Co, 67-73, chemist animal health res, 73-74, proj leader animal health res, 74-75, proj leader process res, 75-77, dept mgr, plastics intermediates, 77-82, dir plastics, Europ Labs, 82-85, dir toxicol, 85-86, DIR, PROD INTEGRITY, ROHM & HAAS CO, 86- *Mem:* Am Chem Soc; Sigma Xi. *Mailing Add:* Rohm & Haas Co Independence Mall West Philadelphia PA 19007

WEILER, JOHN HENRY, JR, b Lincoln, Nebr, July 8, 25; m 58; c 2. PLANT TAXONOMY. *Educ:* Univ Nebr, BSc, 58; Univ Calif, Berkeley, PhD(taxon), 62. *Prof Exp:* PROF PLANT SCI, CALIF STATE UNIV, FRESNO, 62- *Mem:* Am Soc Plant Taxon; Bot Soc Am; Int Asn Plant Taxon. *Res:* Biosystematics; floristics of central California. *Mailing Add:* Dept Plant Sci Calif State Univ 6241 N Maple Ave Fresno CA 93740

WEILER, KURT WALTER, b Phoenix, Ariz, Mar 16, 43; m 79; c 3. RADIO INTERFEROMETRY, OPTICAL INTERFEROMETRY. *Educ:* Univ Ariz, BS, 64; Calif Inst Technol, PhD(physics), 70. *Prof Exp:* Sr sci officer, Neth Found Radio Astron, 70-74; sci collabr, Lab Radio Astron, Bologna, Italy, 75-76; res assoc, Max Planck Inst Radio Astron, 76-79; prog dir, NSF, Washington, DC, 79-85; HEAD, INTERFEROMETRIC RES SECT, NAVAL RES LAB, WASHINGTON, DC, 85- *Concurrent Pos:* Consult, Max Planck Inst Radio Astron, Bonn, Ger, 76; proj dir, Optical Interferometry Proj, US Naval Observ, Naval Res Lab, 89- *Mem:* Am Astron Soc; Int Astron Union; Royal Astron Sci; Int Union Radio Sci. *Res:* Radio emission from supernovae and supernova remnants; stellar research using the techniques of optical interferometry. *Mailing Add:* Naval Res Lab Code 4215 Washington DC 20375-5000

WEILER, LAWRENCE STANLEY, b Middleton, NS, July 11, 42; m 64; c 3. ORGANIC CHEMISTRY. *Educ:* Univ Toronto, BSc, 64; Harvard Univ, PhD(org chem), 68. *Prof Exp:* From asst prof to assoc prof, 68-80, PROF CHEM, UNIV BC, 80- *Mem:* Am Chem Soc; Chem Inst Can; Royal Soc Chem; Swiss Chem Soc. *Res:* Synthesis and study of novel organic compounds; synthesis of natural products and organic metals. *Mailing Add:* Dept Chem 2036 Main Mall Univ BC Vancouver BC V6T 1Y6 Can

WEILER, MARGARET HORTON, b Sewickley, Pa, Apr 30, 41; m 62; c 2. PHYSICS. *Educ:* Radcliffe Col, AB, 62; Univ Maine, MS, 64; Mass Inst Technol, PhD(physics), 77. *Prof Exp:* Instr physics, Univ Maine, 64-65; comput programmer linguistics, Harvard Computation Ctr, Harvard Univ, 65; res staff mem, Francis Bitter Nat Magnet Lab, Mass Inst Technol, 65-74, asst prof physics, 77-83; sr scientist, Res Div, Raytheon Co, 83-88; SR PRIN DEVELOP ENG, LORAL INFRARED & IMAGING SYSTS, 88- *Mem:* AAAS; Am Phys Soc; Int Elec & Electronics Engrs. *Res:* Semiconductor device physics. *Mailing Add:* 356 Lincoln St Apt 14 Waltham MA 02154

WEILER, ROLAND R, b Estonia, Feb 23, 36; Can citizen; m 65; c 3. GEOCHEMISTRY. *Educ:* Univ Toronto, BA, 59, MA, 60; Dalhousie Univ, PhD(oceanog), 65. *Prof Exp:* Res scientist, Bedford Inst Oceanog, Can Dept Energy, Mines & Resources, 64-67 & Can Ctr for Inland Waters, 67-80. *Mem:* AAAS; Am Soc Limnol & Oceanog; Int Asn Gt Lakes Res. *Res:* Chemical limnology and geochemistry of sediments. *Mailing Add:* 37 Mercer St Dundas ON L9H 2N8 Can

WEILER, THOMAS JOSEPH, b St Louis, Mo, May 5, 49. ELEMENTARY PARTICLE PHYSICS. *Educ:* Stanford Univ, BS, 71; Univ Wis, PhD(physics), 76. *Prof Exp:* Sr res asst, Univ Liverpool, 76-78; res assoc theoret physics, Northeastern Univ, 78-81; asst res physicist II, Univ Calif, San Diego, 81-; AT DEPT PHYS & ASTRON, VANDERBILT UNIV. *Concurrent Pos:* Vis theoret physics, Ctr Europ Nuclear Res, Rutherford Lab, 77, Aspen Ctr, 80, 81. *Mem:* Sigma Xi; Am Phys Soc. *Res:* Work towards the description of the ultimate subunits of matter and the forces by which they interact and combine; implications of particle physics for cosmology. *Mailing Add:* Dept Physics & Astron Vanderbilt Univ Box 1807 Sta B Nashville TN 37235

WEILER, WILLIAM ALEXANDER, b Milwaukee, Wis, Nov 8, 41; m 63; c 3. BACTERIOLOGY, MICROBIAL ECOLOGY. *Educ:* Dartmouth Col, AB, 63; Purdue Univ, PhD(microbiol), 69. *Prof Exp:* Instr microbiol, Purdue Univ, 66-69; asst prof, 69-74, assoc prof, 74-80, PROF BOT, EASTERN ILL UNIV, 80- *Mem:* Am Soc Microbiol. *Res:* Herbicide effects on soil microflora; petroleum degradation in soil and water ecosystems. *Mailing Add:* Dept Bot Eastern Ill Univ Charleston IL 61920

WEILL, CAROL EDWIN, b Brooklyn, NY, Dec 12, 18; m 47; c 2. ORGANIC CHEMISTRY. *Educ:* City Col New York, BS, 39; Columbia Univ, AM, 43, PhD(chem), 44. *Prof Exp:* Res chemist, Div War Res, Columbia Univ, 44-45 & Takamine Lab, 45-47; from instr to assoc prof, 47-60, PROF CHEM, RUTGERS UNIV, NEWARK, 60- *Mem:* Am Chem Soc. *Res:* Enzymes; carbohydrates. *Mailing Add:* Dept Chem Rutgers Univ Newark NJ 07102

WEILL, DANIEL FRANCIS, b Paris, France, Nov 29, 31; US citizen; m 57; c 3. PETROLOGY, GEOCHEMISTRY. *Educ:* Cornell Univ, AB, 56; Univ Ill, MS, 58; Univ Calif, Berkeley, PhD(geol), 62. *Prof Exp:* Res assoc geochem, Univ Calif, Berkeley, 62-63; asst prof geol, Univ Calif, San Diego, 63-66; from assoc prof to prof geol, Univ Ore, 66-85, dir, Ctr Volcanology, 69-70, assoc dean arts & sci, 76-78 & 81-83; PROG DIR, NSF, 85- *Concurrent Pos:* Grants, NSF, 65-66, 66-68, & 79, NASA, 68-78, 71-78 & 79-81 & Am Chem Soc, 70-72; Fulbright-Hays sr res fel, UK, 72-73. *Mem:* Mineral Soc Am; Royal Soc Chem. *Res:* Physical chemistry of geological systems; silicate liquid density, viscosity; diffusion; lunar sample analysis; redox equilibria in silicate melts; experimental trace element distribution; thermodynamic properties of mineral and liquid silicate systems; science administration. *Mailing Add:* Div Earth Sci NSF Washington DC 20550

WEILL, GEORGES GUSTAVE, b Strasbourg, France, Apr 9, 26. MATHEMATICAL ANALYSIS, APPLIED MATHEMATICS. *Educ:* Univ Paris, DSc(physics), 55; Univ Calif, Los Angeles, PhD(math), 60. *Prof Exp:* Res scientist, Gen Radio Co, France, 52-56; res fel, Calif Inst Tech, 56-59; res fel math, Harvard Univ, 60-62; lectr & res fel, Yale Univ, 62-64; asst prof, Yeshiva Univ, 64-65; PROF, POLYTECH UNIV NY, 66- *Concurrent Pos:* Consult, Electro-Optical Systs, Inc, 59-60 & Raytheon Co, 61-62. *Mem:* Am Math Soc; sr mem Inst Elec & Electronics Engrs; Math Soc France. *Res:* Complex analysis; Riemann surfaces; diffraction theory; antennas; theoretical and applied electromagnetics. *Mailing Add:* Dept Math Polytech Univ NY 333 Jay St Brooklyn NY 11201

WEILL, HANS, b Berlin, Ger, Aug 31, 33; US citizen; m; c 3. MEDICINE, PHYSIOLOGY. *Educ:* Tulane Univ, BA, 55, MD, 58; Am Bd Internal Med, dipl & Am Bd Pulmonary Dis, dipl, 65. *Prof Exp:* From instr to prof med, 62-85, SCHLIEDER PROF PULMONARY MED, TULANE UNIV MED SCH, 85-, DIR CTR BIOENVIRON RES, 90- *Concurrent Pos:* Attending staff, Vet Admin Hosp, New Orleans, 63-83; consult, USPHS Hosp, 64-76; dir, Specialized Ctr Res & consult task force environ lung dis, Nat Heart, Lung & Blood Inst, 72-, mem, Pulmonary Dis Adv Comt, 81, chmn, 82; mem, Pulmonary Dis Bd, Am Bd Internal Med, 80-86; mem, Nat Heart Lung & Blood Adv Coun, NIH, 86-90; bd gov, Am Bd Internal Med, 85-88. *Mem:* Am Thoracic Soc (pres, 76-77); fel Am Col Chest Physicians; fel Am Col Physicians; Am Fedn Clin Res; fel Royal Soc Med. *Res:* Occupational respiratory diseases; environmental health sciences. *Mailing Add:* Dept Med Tulane Univ Med Sch 1700 Perdido St New Orleans LA 70112

WEIMANN, LUDWIG JAN, b Lipnica, Ger, May 24, 41; m 72; c 2. PHYSICAL CHEMISTRY, PHOTOCHEMISTRY. *Educ:* Univ Poznan, MS, 63, PhD(photochem), 70. *Prof Exp:* Instr photochem & radiochem, Univ Poznan, 63-70; fel theoret chem, Kans Univ, 70-72; fel vision chem, Univ Mo, Kansas City, 72-73, instr phys & gen chem, 73-75, vis asst prof phys chem, 75-76; res & develop dir polymer photochem, K C Coatings Inc, 76-80; TECH DIR RES & DEVELOP, BERTEK, INC, 80- *Mem:* Am Chem Soc; Soc Mfg Eng. *Res:* Radical and cationic photopolymerization; application in ink making area; kinetics of photopolymerization and physicochemical properties of ultraviolet-cured films; kinetics of drug transport through membranes. *Mailing Add:* Bertek Inc 110 Lake St St Albans VT 05478

WEIMAR, VIRGINIA LEE, b Condon, Ore, Oct 23, 22. PHYSIOLOGY. *Educ:* Ore State Col, MS, 47; Univ Pa, PhD(physiol), 51. *Prof Exp:* Physiologist, USN, 51; res assoc biochem, Wills Eye Hosp, Philadelphia, Pa, 52-54; researcher ophthal, Col Physicians & Surgeons, Columbia Univ, 55-58, res assoc, 58-59, asst prof, 59-61; res assoc, 61-63, asst prof, 63-68, ASSOC PROF OPHTHAL, MED SCH, ORE HEALTH SCI UNIV, PORTLAND, 68- *Concurrent Pos:* Nat Cancer Inst fel, Univ Pa, 54-55; NIH travel award; deleg, Int Cong Ophthal, Belg, 58 & Jerusalem Conf Prev of Blindness, 71. *Mem:* Harvey Soc; Am Physiol Soc; Asn Res Vision & Ophthal; Soc Gen Physiol; Am Soc Cell Biol. *Res:* Biochemistry and physiology of trauma; wound healing; inflammation; corneal wound healing; cellular ultramicrochemistry; biomathematics. *Mailing Add:* 305 SE 89th Ave Portland OR 97216

WEIMBERG, RALPH, b San Diego, Calif, Dec 22, 24; m 52; c 2. PHYSIOLOGY, PLANT BACTERIOLOGY. *Educ:* Univ Calif, AB, 49, MA, 51, PhD(bact), 55. *Prof Exp:* Fel, Biol Div, Oak Ridge Nat Labs, 55-56; instr microbiol, Sch Med, Western Reserve Univ, 56-58; biochemist, Northern Utilization Res & Develop Div, 58-65, BIOCHEMIST US SALINITY LAB, AGR RES SERV, US DEPT AGR, 65- *Concurrent Pos:* Vis assoc prof, Univ Calif, Davis, 64; vis res scientist, Bot Dept, Hebrew Univ, Jerusalem, 77 & 81. *Mem:* AAAS; Am Soc Microbiol; Am Soc Plant Physiol; Am Soc Biol Chemists; NY Acad Sci. *Res:* Plant physiology and metabolism; properties and location of enzymes; biochemistry. *Mailing Add:* 5444 Quince St Riverside CA 92506

WEIMER, DAVID, b Marion, Ind, Oct 13, 19; m 44; c 4. GAS DYNAMICS. *Educ:* Ohio State Univ, BS, 41, MS, 46. *Prof Exp:* Res asst physics, Princeton Univ, 41-45; prof, Am Col SIndia, 46-47; res assoc, Princeton Univ, 47-52; res engr, Armour Res Found, Ill, 52-55 & Am Mach & Foundry Co, 55-56; sr staff scientist physics, Lockheed Corp, Calif, 56-58, 60-64; asst prof, Ohio Northern Univ, 64-68; staff scientist, Martin-Marietta Corp, Colo, 68-69; ASSOC PROF PHYSICS, OHIO NORTHERN UNIV, 69- *Concurrent Pos:* Sr staff engr, Martin-Marietta Corp, Colo, 72-73. *Mem:* Am Phys Soc. *Res:* Gas physics and shock wave phenomena; dissociation and ionization of gas at high temperatures; radiation of excited species in planetary atmospheres. *Mailing Add:* Box 234 Ada OH 45810

WEIMER, F(RANK) CARLIN, b Dayton, Ohio, July 27, 17; m 44; c 3. ELECTRICAL ENGINEERING. *Educ:* Univ Ohio, BS, 38; Ohio State Univ, MSc, 39, PhD(elec eng), 43. *Prof Exp:* Asst, 39-41, from instr to prof, 41-83, EMER PROF ELEC ENG, OHIO STATE UNIV, 83- *Concurrent Pos:* Consult, Battelle Mem Inst, 57- *Honors & Awards:* Centennial Medal, Inst Elec & Electronics Engrs, 84. *Mem:* AAAS; Am Soc Eng Educ; Nat Soc Prof Engrs; Inst Elec & Electronics Engrs; Sigma Xi. *Res:* Servomechanisms; magnetic fields in machinery; permeance analysis; automatic control; signal processing. *Mailing Add:* Dept Elec Eng Ohio State Univ 2015 Neil Ave Columbus OH 43210

WEIMER, HENRY EBEN, immunochemistry; deceased, see previous edition for last biography

WEIMER, JOHN THOMAS, b McKeesport, Pa, Mar 14, 30; m 57; c 3. BIOLOGY, TOXICOLOGY. *Educ:* St Vincent Col, BA, 52. *Prof Exp:* Res biologist toxicol, Chem Systs Lab, US Army, 56-88; RETIRED. *Mem:* Soc Toxicol; Sigma Xi. *Res:* Inhalation toxicology, pharmacology, aerosol technology. *Mailing Add:* 1500 Southview Rd Bel Air MD 21014

WEIMER, PAUL KESSLER, b Wabash, Ind, Nov 5, 14; m 42; c 3. PHYSICS. *Educ:* Manchester Col, AB, 36; Univ Kans, AM, 38; Ohio State Univ, PhD(physics), 42. *Hon Degrees:* DSc, Manchester Col, 68. *Prof Exp:* Asst, Univ Kans, 36-37; prof physics, Tabor Col, 37-39; asst, Ohio State Univ, 39-42; res engr, RCA Labs, 42-64, fel tech staff, 64-81. *Concurrent Pos:* Consult, 81- *Honors & Awards:* Award, TV Broadcasters Asn, 46; Zworykin Prize, Inst Elec & Electronics Eng, 59 & Morris Liebmann Mem Prize, 66; Albert Rose Award, Inst Graphic Commun, 87. *Mem:* Nat Acad Eng; fel Inst Elec & Electronics Eng; Ger Photog Soc. *Res:* Nuclear physics; electron optics; photoconductivity; secondary emission; semiconductor devices; television camera tubes and solid state image sensors. *Mailing Add:* 112 Random Rd Princeton NJ 08540

WEIMER, ROBERT FREDRICK, b Wheeling, WVa, Jan 16, 40. CHEMICAL ENGINEERING. *Educ:* Mass Inst Technol, SB, 61; Univ Calif, Berkeley, PhD(chem eng), 65. *Prof Exp:* Staff engr, Air Prod & Chem Inc, 67-73, process mgr, 73-76, develop mgr, 76-78, dir develop, Corp Res &

Develop Dept, 78-81; chief engr, Int Coal Refining Co, 81-82; tech mgr process technol, 82-85, CHIEF ENGR, AIR PROD & CHEM INC, 85- *Mem:* AAAS; Am Inst Chem Engrs; Am Chem Soc. *Res:* Energy conversion; heat transfer; cryogenics; adsorption. *Mailing Add:* Air Prod & Chem Inc 7201 Hamilton Blvd Allentown PA 18195

WEIMER, ROBERT J, b Glendo, Wyo, Sept 4, 26; m 48; c 3. GEOLOGY. *Educ:* Univ Wyo, BA, 48, MA, 49; Stanford Univ, PhD, 53. *Prof Exp:* Dist geologist, Union Oil Co Calif, Mont, 53-54; consult res geologist, 54-57; from asst prof to assoc prof, 57-72, head dept, 64-69, Getty prof geol eng, 78-83, PROF GEOL, COLO SCH MINES, 72-, EMER PROF GEOL ENG, 83- *Concurrent Pos:* Consult, petrol indust, 54-; exchange prof, Univ Colo, 60; Fulbright lectr, Univ Adelaide, 67; vis prof, Univ Calgary, 70; mem res assoc comt, Nat Acad Sci, 70-73; mem, Inst Technol, Bandung, Indonesia, 75. *Honors & Awards:* Sidney Powers Medal, 84; Medal, Colo Sch Medal. *Mem:* AAAS; Geol Soc Am; hon mem Am Asn Petrol Geol; hon mem Soc Econ Paleont & Mineral (secy-treas, 66-68, vpres, 71, pres, 72). *Res:* Stratigraphic research in the application of modern sedimentation studies to the geologic record; regional framework of sedimentation in the Cretaceous, Jurassic and Pennsylvanian rock systems; tectonics and sedimentation; geologic history of the Rocky Mountain region; stratigraphic record of global sea level changes. *Mailing Add:* 25853 Mt Vernon Rd Rte 3 Golden CO 80401

WEIN, ROSS WALLACE, b Exeter, Ont, Oct 29, 40; m 67. PLANT ECOLOGY. *Educ:* Univ Guelph, BSA, 65, MSc, 66; Utah State Univ, PhD(ecol), 69. *Prof Exp:* Can Nat Res Coun fel, Univ Alta, 69-71, vis asst prof plant ecol, 71-72; from asst prof to prof plant ecol, Univ NB, 72-78, dir, Fire Sci Ctr, Frederickton, 78-87; PROF FORESTRY, UNIV ALTA, EDMONTON, 87- *Concurrent Pos:* dir northern studies, Boreal Inst, 87- *Mem:* Ecol Soc Am; Am Soc Range Mgt; Can Bot Asn. *Res:* Plant production and nutrient cycling following wildfire; plant community dynamics; boreal and arctic ecology. *Mailing Add:* Dept Forestry & Sci Univ Alta Edmonton AB T6G 2M7 Can

WEINACHT, RICHARD JAY, b Union City, NJ, Dec 10, 31; m 55; c 6. MATHEMATICS. *Educ:* Univ Notre Dame, BS, 53; Columbia Univ, MS, 55; Univ Md, PhD(math), 62. *Prof Exp:* NSF fel math, Courant Inst Math Sci, NY Univ, 62-63; from asst prof to assoc prof, 63-74, PROF MATH, UNIV DEL, 74- *Concurrent Pos:* Vis assoc prof, Rensselaer Polytech Inst, 69-70; Fulbright fel & vis prof, Darmstadt Tech Univ, 76-77; Ger Acad Exchange Serv vis scientist, Univ Cologne, 80; vis prof, Univ Rome, 84. *Mem:* Am Math Soc; Soc Indust & Appl Math. *Res:* Partial differential equations; singular perturbations. *Mailing Add:* Dept Math Sci Univ Del Newark DE 19716

WEINBACH, EUGENE CLAYTON, b Pine Island, NY, Nov 5, 19; m 38; c 2. BIOCHEMISTRY. *Educ:* Univ Md, BS, 42, PhD(org med chem), 47. *Prof Exp:* Res chemist, US Naval Res Lab, Washington, DC, 47; USPHS & Nat Cancer Inst fel, Sch Med, Johns Hopkins Univ, 47-48, instr physiol chem, 48-50; RES CHEMIST, NIH, 50-, HEAD, SECT PHYSIOL & BIOCHEM, LAB PARASITIC DIS, NAT INST ALLERGY & INFECTIOUS DIS, 69- *Concurrent Pos:* Guest worker, Wenner-Gren Inst, Stockholm, Sweden, 60-62. *Mem:* Am Chem Soc; Am Soc Biol Chemists; Sigma Xi. *Res:* Biochemical mechanisms of drug action; intermediary metabolism of parasites and their hosts; biological oxidations and phosphorylations; biochemistry of mitochondria. *Mailing Add:* Lab Parasitic Dis NIH Bethesda MD 20892

WEINBAUM, CARL MARTIN, b Manchester, Eng, Jan 16, 37; US citizen; m 63, 91; c 4. MATHEMATICS. *Educ:* Queens Col, NY, BS, 58; Harvard Univ, AM, 60; NY Univ, PhD(math), 63. *Prof Exp:* Asst prof math, Univ Calif, Los Angeles, 63-68; assoc prof, Univ Hawaii, 68-70 & Hawaii Loa Col, 70-77; fac mem, Woodmere Acad, 77-78; mem tech staff, Network Anal Corp, 78-79; programmer/analyst, Royal Ins Co, 80-84; software engr, Grumman Aerospace Corp, 84-88; SOFTWARE ENGR, UMECORP, 88- *Mem:* Am Math Soc. *Res:* Infinite groups, particularly word problems and small cancellation and knot groups. *Mailing Add:* 246 Miller Ave Mill Valley CA 94941

WEINBAUM, GEORGE, b Brooklyn, NY, July 27, 32; m 63; c 4. PULMONARY RESEARCH. *Educ:* Univ Pa, AB, 53; Pa State Univ, MS, 55, PhD(biochem), 57. *Prof Exp:* Chief biochem labs, Geisinger Med Ctr, 57-61; asst mem, Res Labs, 61-66, assoc mem, 66-71, bioscientist, Dept Pulmonary Dis, Albert Einstein Med Ctr, 71-82; RES PROF MICROBIOL, HEALTH SCI CTR, TEMPLE UNIV, 73-; res assoc prof, 82-88, RES PROF, DEPT MED, UNIV PA, 88- *Concurrent Pos:* Fulbright res fel, Tokyo, 59-61; NIH res career develop award, 69-74; chief, res div, Dept Med, Grad Hosp, 82- *Mem:* Am Soc Biochem & Molecular Biol; Am Asn Path; Am Soc Microbiol; Am Chem Soc; Sigma Xi; Am Thoracic Soc. *Res:* Membrane structure, synthesis and function; role of glycoproteins in organization of cell surface; etiology of emphysema and pulmonary fibrosis; action of proteinases on cell membranes. *Mailing Add:* Res Labs Grad Hosp 415 S 19th St Philadelphia PA 19146

WEINBAUM, SHELDON, b Brooklyn, NY, July 26, 37; m 62; c 2. FLUID PHYSICS, CHEMICAL ENGINEERING. *Educ:* Rensselaer Polytech Inst, BAE, 59; Harvard Univ, SM, 60, PhD(eng), 63. *Prof Exp:* Mem res staff, Sperry Rand Res Ctr, 63-64; prin scientist, Avco Everett Res Lab, 64; theoret aerodynamicist, Gen Elec Space Sci Lab, 64-67, prin scientist, 67; from asst prof to prof mech eng, 67-80, Herbert G Kayser chair prof eng, 80-86, DISTINGUISHED PROF ENG, CITY COL NEW YORK, 86- *Concurrent Pos:* Consult, Avco Everett Res Lab, 60-61, Wilmer Eye Inst, Sch Med, Johns Hopkins Univ, 67-70, Gen Elec Co, 68-69 & Boeing Sci Res Labs, 68-70; vis prof physiol, Flow Studies Imp Col Sci Tech, London, 73-74; sr fel Sci Res Coun, Gt Brit, 74; Russell Springer vis prof mech eng, Univ Calif, Berkeley, 79; vis prof mech eng, Mass Inst Technol, 80-81; creativity grant award, NSF, 85; Gordon McKay Prize fel, Harvard Univ, 59; NSF fel, 60. *Mem:* Biomed Eng Soc; Am Soc Mech Engrs; fel Am Phys Soc. *Res:* Fluid mechanics; biofluid mechanics; bioheat transfer; biophysics; two phase flow; low Reynolds number flow; high speed gas dynamics; interacting boundary layers and wake; wakes; heat transfer. *Mailing Add:* Dept Mech Eng 140th St & Convent Ave New York NY 10031

WEINBERG, ALVIN MARTIN, b Chicago, Ill, Apr 20, 15; m 40, 74; c 2. NUCLEAR PHYSICS. *Educ:* Univ Chicago, AB, 35, MS, 36, PhD(physics), 39. *Hon Degrees:* LLD, Univ Chattanooga, 63, Alfred Univ & Denison Col, 67; ScD, Univ of the Pac, 66, Worcester Polytech Inst, 71, Univ Rochester & Butler Univ, 73, Wake Forest Univ, 74 & Univ Louisville, 78; EngD, Stevens Inst Technol, 73. *Prof Exp:* Asst math biophys, Univ Chicago, 39-41, physicist, Metall Lab, 41-45; physicist, Clinton Labs, Tenn, 45-48; dir physics div, Oak Ridge Nat Lab, 48-49, res dir, 49-55, dir, 55-74; dir off energy res & develop, Fed Energy Admin, 74; dir, 75-85, DISTINGUISHED FEL, INST ENERGY ANALYSIS, 85- *Concurrent Pos:* Mem sci adv bd, USAF, 56-59; mem, President's Sci Adv Comt, 60-62; Regents' lectr, Univ Calif, 66 & 78; distinguished fel, Inst Energy Anal, 85- *Honors & Awards:* Atoms for Peace Award, 60; Lawrence Mem Award, US AEC, 69; Heinrich Hertz Prize, Univ Karlsruhe, 75; Award, NY Acad Sci; Enrico Fermi Award, 80; Harvey Prize, Technion, 82. *Mem:* Nat Acad Sci; Nat Acad Eng; Am Philos Soc; fel Am Nuclear Soc (pres, 59-60); Royal Neth Acad Sci; Am Acad Arts & Sci. *Res:* Nuclear energy; mathematical theory of nerve function; science policy; energy analysis. *Mailing Add:* 111 Moylan Lane Oak Ridge TN 37830

WEINBERG, BARBARA LEE HUBERMAN, b New York, NY, Nov 28, 34; m 57; c 3. MOLECULAR BIOLOGY. *Educ:* City Col New York, BS, 55; Yale Univ, MS, 58; Duke Univ, PhD(biochem), 64. *Prof Exp:* Asst prof org chem & microbiol, Marymount Col, 79-80; fel endocrine res, Res Inst Hosp Joint Dis, 80-81; fel biochem, NY Med Col, 81-84; consult, 84-86; CONSULT, TECHNOL CTR, SILICON VALLEY, 90- *Mem:* AAAS; Sigma Xi. *Res:* Recombinant plasmids have been constructed to study synthesis and regulation of expression of genes for initiation factor-3 and other synthetases; analysis of molecular mechanisms at level of transcriptional or translational control. *Mailing Add:* 1844 Schooldale Dr San Jose CA 95124-1136

WEINBERG, BERND, b Chicago, Ill, Jan 30, 40; m 65. SPEECH PATHOLOGY. *Educ:* State Univ NY, BS, 61; Ind Univ, MA, 63, PhD, 65. *Prof Exp:* Prof otolaryngol, Sch Med & prof speech path, Sch Dent, Ind Univ, 64-66; res fel speech sci, Nat Inst Dent Res, 66-68; dir speech res lab, Med Ctr, Ind Univ, Indianapolis, 68-74; prof speech path & speech sci, Purdue Univ, West Lafayette, 74-87; DIR INSTNL RELS, RES CORP TECHNOL, 87- *Mem:* Acoust Soc Am; Am Speech & Hearing Asn; Am Cleft Palate Asn. *Res:* Speech acoustics; speech physiology. *Mailing Add:* Res Corp Technol 6840 E Broadway Blvd Tucson AZ 85710-2815

WEINBERG, CRISPIN BERNARD, b Minneapolis, Minn, Jan 13, 51; m 80; c 3. TISSUE ENGINEERING, ARTIFICIAL ORGANS. *Educ:* Univ Chicago, SB & SM, 73; Harvard Univ, PhD(neurobiol), 80. *Prof Exp:* Teaching fel neurobiol, Harvard Univ, 73-76; res fel, Mass Inst Technol, 81-84; dir BVE res, 85-90, DIR GRAFT ARTERY PROG, ORGANOGENESIS INC, 91- *Mem:* Am Soc Cell Biol; Am Soc Artificial Internal Organs; AAAS; Tissue Cult Asn. *Res:* Major research activities directed at fabricating and testing a biological vascular graft based on the principles of guided tissue regeneration; studies have involved cell-matrix interactions, matrix biochemistry, immunology, and experimental surgery; four publications and two patents. *Mailing Add:* Organogenesis Inc 83 Rogers St Cambridge MA 02142

WEINBERG, DANIEL I, b New York, NY, July 11, 28; m 57; c 3. COMPUTER ENGINEERING, ELECTROMAGNETIC FIELDS. *Educ:* Clarkson Col Technol, BEE, 48; Duke Univ, MS, 62, PhD(physiol), 69. *Prof Exp:* Engr, Fairchild Engine & Airplane Corp, 48-51; sr engr, Atomic Prod Div, Gen Elec Co, 51-57; assoc, Astra, Inc, Conn & NC, 57-59, mgr med eng, 59-64; staff engr, Adv Systs Develop Div, IBM, 64-70, adv systs engr, Data Processing Div, 70-75 & Gen Systs Div, 75-78, adv systs anal, 78-80, sr engr, Internal Telecommun, 81-86, sr engr & scientist, Santa Teresa Lab, 87-90, CONSULT, IBM CORP, 90- *Concurrent Pos:* Consult, 57-64; res assoc, Med Ctr, Duke Univ, 60-64; assoc, Col Physicians & Surgeons, Columbia Univ, 65-69; adv lab safety, Bd Educ, Scarsdale, NY, 81-84; Technol Ctr Silicon Valley, 90- *Mem:* Inst Elec & Electronics Engrs. *Res:* Application of engineering to solution of biological research and medical problems; computers in medicine and instrumentation; electrical safety in medicine; cardiac electrophysiology; physiological monitoring; instrumentation and measurement; control of nonionizing electromagnetic fields from industrial facilities; communications systems; biological effects of electromagnetic fields. *Mailing Add:* 1844 Schooldale Dr San Jose CA 95124-1136

WEINBERG, DAVID SAMUEL, b St Louis, Mo, Feb 15, 38; m 61; c 2. ORGANIC ANALYTICAL CHEMISTRY. *Educ:* Univ Ariz, BS, 60, PhD(phys org chem), 65. *Prof Exp:* Res assoc, Stanford Univ, 64-65; res chemist, Phillips Petrol Co, Okla, 65-71; group leader, Owens-Ill Co, Ohio, 71-79; sect mgr, Sverdrup Technol, Inc, Tenn, 79-83; SECT MGR, SOUTHERN RES INST, ALA, 83- *Mem:* Am Chem Soc. *Res:* Structure and mechanism in organic chemistry; trace organic analysis. *Mailing Add:* 600 Belle Terre Circle Birmingham AL 35226

WEINBERG, DONALD LEWIS, b New York, NY, Aug 1, 31; div; c 1. INFRARED DETECTORS, LASERS. *Educ:* Harvard Univ, AB, 52, AM, 53, PhD(appl physics), 59. *Prof Exp:* Res physicist, Res & Develop Div, Corning Glass Works, 59-64; res assoc, Lincoln Lab, Mass Inst Technol, 64-67; res physicist, Electronics Res Ctr, Nat Aeronaut & Space Admin, 67-70; mem tech staff, Eng Res Ctr, Western Elec Co, 70-71; RES PHYSICIST, NAVAL RES LAB, 71- *Mem:* Inst Elec & Electronics Engrs. *Res:* Nuclear magnetic resonance in alloys; small angle x-ray scattering; nonlinear optics; physical electronics of television pick-up tubes, image intensifiers, silicon charge-coupled devices, and infrared charge-injection devices. *Mailing Add:* Code 6552 Naval Res Lab Washington DC 20375-5000

WEINBERG, ELLIOT CARL, b Chicago, Ill, Aug 17, 32; m 62, 79; c 4. MATHEMATICS. *Educ:* Purdue Univ, BA, 56, MS, 58, PhD(math), 61. *Prof Exp:* From instr to asst prof, 60-66, ASSOC PROF MATH, UNIV ILL, URBANA, 66- *Mem:* Am Math Soc; Math Asn Am. *Res:* Ordered algebraic structures. *Mailing Add:* Dept Math Univ Ill 1409 W Green St Urbana IL 61801

WEINBERG, ELLIOT HILLEL, b Duluth, Minn, Dec 28, 24; m 47; c 2. SOLID STATE PHYSICS. *Educ:* Univ Mich, BS & MS, 47; Univ Iowa, PhD(physics), 53. *Prof Exp:* Asst, Univ Mich, 42-43, res assoc, 47-48; res physicist, Mass Inst Technol, 48-49; from instr to assoc prof physics, Univ Minn, 49-58; prof & chmn dept, NDak State Univ, 58-62; chief scientist, San Francisco Br, 62-67, dir div phys sci, 67-71, liaison scientist, London, 65-66, dir res Off Naval Res, 71-79; DIR, NAVY CTR INT SCI & TECHNOL, NAVAL GRAD SCH, MONTEREY, CALIF, 83- *Concurrent Pos:* Mem, Col Physics Comn Rev, 62. *Mem:* AAAS; Am Phys Soc; Am Soc Eng Educ; Am Asn Physics Teachers. *Res:* Solid state physics; meteroology; atmospheric and underwater optics; laser physics and applications; foreign science and technology. *Mailing Add:* Naval Post Grad Sch NCIST Monterey CA 93940

WEINBERG, ERIC S, b New York, NY, May 15, 42; c 1. MOLECULAR GENETICS. *Educ:* Univ Rochester, BA, 63; Rockefeller Univ, PhD(develop biol), 69. *Prof Exp:* Helen Hay Whitney Found res fel, Univ Palermo, Italy, 69-70, Lab Molecular Embryol, Naples, 70-71, Univ Edinburgh, 71-72; asst prof, Dept Biol, Johns Hopkins Univ, 72-79; ASSOC PROF, DEPT BIOL, UNIV PA, 79- *Concurrent Pos:* Mem, molecular biol study sect, NIH, 83-87; chair, Grad Group Molecular Biol, Univ Pa, 87- *Mem:* Am Soc Cell Biol. *Res:* Regulation of genes during early embryonic development; histone genes of the sea urchin. *Mailing Add:* Dept Biol Univ Pa Philadelphia PA 19104

WEINBERG, ERICK JAMES, b Ossining, NY, Aug 29, 47; m 72; c 2. QUANTUM FIELD THEORY. *Educ:* Manhattan Col, BS, 68; Harvard Univ, MA, 69, PhD(physics), 73. *Prof Exp:* Mem, Inst Advan Study, 73-75; from asst prof to assoc prof, 75-87, PROF PHYSICS, COLUMBIA UNIV, 87- *Concurrent Pos:* Alfred P Sloan Found fel, 78-80. *Mem:* Am Phys Soc. *Res:* Theoretical elementary particle physics and quantum field theory, including the study of the very early universe. *Mailing Add:* Physics Dept Columbia Univ New York NY 10027

WEINBERG, EUGENE DAVID, b Chicago, Ill, Mar 4, 22; m 49; c 4. MEDICAL MICROBIOLOGY. *Educ:* Univ Chicago, BS, 42, MS, 48, PhD(microbiol), 50. *Prof Exp:* Asst instr microbiol, Univ Chicago, 47-50; from instr to assoc prof, 50-61, assoc dean res & grad develop, 77-79, PROF MICROBIOL & MED SCI, IND UNIV, BLOOMINGTON, 61- *Mem:* AAAS; Am Soc Microbiol; fel Am Acad Microbiol. *Res:* Roles of trace metals and of metal-binding agents in microbial physiology and in chemotherapy; nutritional immunity; antimicrobial compounds; environmental control of secondary metabolism. *Mailing Add:* Dept Biol Ind Univ Bloomington IN 47401

WEINBERG, FRED, b Poland, Apr 3, 25; nat Can; m 56; c 1. PHYSICS. *Educ:* Univ Toronto, BASc, 47, MA, 48, PhD(metall), 51. *Prof Exp:* Res scientist, Phys Metall Div, Mines Br, Can Dept Energy, Mines & Resources, 51-67, head metal physics sect, 61-67; prof, 67-90, EMER PROF METALL, UNIV BC, 90- *Concurrent Pos:* Vis prof, Cavendish Lab, Cambridge Univ, 62-63 & Hebrew Univ, Jerusalem, 75-76; head, Dept Metall Eng, Univ BC, 80. *Honors & Awards:* Robert Woolson Hunt Award, Am Inst Mech Engrs, 80; Alcan Award, Can Inst Mining & Metall, 80. *Mem:* Am Inst Mining, Metall & Petrol Eng; fel Metall Soc. *Res:* Metal physics; deformation and solidification of metals. *Mailing Add:* Dept Metall Univ BC Vancouver BC V6T 1W5 Can

WEINBERG, I JACK, b New York, NY, May 25, 35; m 58; c 2. COMPUTER SCIENCE, APPLIED MATHEMATICS. *Educ:* Yeshiva Univ, BA, 56; Mass Inst Technol, SM, 58, PhD(math), 61. *Prof Exp:* Mgr math, Avco Corp, 61-70; assoc prof math, Lowell Technol Inst, 70-73, prof, 73-80; PROF, UNIV LOWELL, 80- *Concurrent Pos:* Instr, Northeastern Grad Sch Arts & Sci, 64-69. *Mem:* Am Math Soc; Soc Indust & Appl Math; Asn Comput Mach; Math Asn Am. *Res:* Numerical analysis; differential equations; theory of elasticity. *Mailing Add:* 84 Rawson Rd Brookline MA 02146

WEINBERG, IRVING, b New York, NY, July 3, 18; m 47; c 1. PHYSICS. *Educ:* Stanford Univ, BS, 50; Univ Colo, MS, 52, PhD, 58. *Prof Exp:* Physicist, Midway Labs, Chicago, 52-53; instr physics, Univ Colo, 54-57; sr physicist, Mobil Oil Co, NJ & Tex, 58-59; sr res physicist, Aeronutronic Div, Ford Motor Co, 59-64; res physicist, Jet Propulsion Lab, 64-67; physicist, Div Res, 67-72, PHYSICIST, MAT PROG, NASA, 72- *Concurrent Pos:* Consult, Magnolia Petrol Co, Tex. *Mem:* Am Phys Soc; Sigma Xi. *Res:* Space physics; radio frequency spectroscopy of the solid state; transport phenomena in solids; energy band theory; interaction of radiation with matter. *Mailing Add:* 20695 Beaconsfield Blvd Rocky River OH 44116

WEINBERG, JERRY L, b Detroit, Mich, Dec 2, 31; m 61; c 2. ASTROPHYSICS, ATMOSPHERIC PHYSICS. *Educ:* St Lawrence Univ, BS, 58; Univ Colo, PhD(astrophys, atmospheric physics), 63. *Prof Exp:* Res asst astrophys, High Altitude Observ, Boulder, Colo, 59-63; astrophysicist, Haleakala Observ, Univ Hawaii, 63-68; astrophysicist, Dudley Observ, 68-73, assoc dir, 70-73, res prof astron, State Univ NY, Albany, 73-80, dir, Space Astron Lab, 73-80; RES PROF ASTRON & DIR, SPACE ASTRON LAB, UNIV FLA, 80-, DIR SPACE SCI INST, 87- *Concurrent Pos:* Mem organizing comt, Comn 21, Int Astron Union; mem, Int Comt Space Res, Int Coun Sci Unions; mem working group photom, Int Asn Geomag & Aeronomy. *Honors & Awards:* Exceptional Sci Achievement Medal & Skylab Achievement Award, NASA, 74. *Mem:* AAAS; Am Astron Soc; Int Astron Union; Am Geophys Union; Astron Soc Pac; Sigma Xi; Optical Soc Am. *Res:* Space astronomy; instrumentation, interplanetary medium; night sky research; atmospheric optics. *Mailing Add:* Space Astron Lab 1810 NW Sixth St Gainesville FL 32609

WEINBERG, LOUIS, b New York, NY, July 15, 19; m 51; c 3. ELECTRICAL ENGINEERING, APPLIED MATHEMATICS. *Educ:* Brooklyn Col, AB, 41; Harvard Univ, MS, 47; Mass Inst Technol, ScD(elec eng), 51. *Prof Exp:* Instr elec eng, Mass Inst Technol, 47-51; res physicist & head commun & networks res sect, Hughes Aircraft Co, 51-61; vpres info processing, Conductron Corp, 61-64; vis prof elec eng, Univ Mich, 64-65; prof elec eng, City Col New York, 65-90; RETIRED. *Concurrent Pos:* Lectr, Univ Calif, Los Angeles, 52-54; vis assoc prof, Univ Southern Calif, 55-56; vis prof, Calif Inst Technol, 58-59; vchmn comn VI, Int Union Radio Sci, 60; NSF grant, City Col New York, 66-68, Res Found grant, 70-72; consult, Gen Elec Co & IBM, T J Watson Res Ctr; vis prof math eng, Univ Tokyo, 79. *Mem:* Fel AAAS; Am Soc Eng Educ; fel Inst Elec & Electronics Engrs; fel NY Acad Sci. *Res:* Network analysis; synthesis of lumped networks and lumped-distributed networks; graphs and matroids; games, computers, and switching networks; fault detection and test pattern generation. *Mailing Add:* 11 Woodland St Tenafly NJ 07670-2309

WEINBERG, MARC STEVEN, b Boston, Mass, Aug 9, 48; m 71; c 2. AERONAUTICAL ENGINEERING, INERTIAL INSTRUMENTS. *Educ:* Mass Inst Technol, BS & MS, 71, ME, 73, PhD(mech eng), 74. *Prof Exp:* Res asst ground transport, Mech Eng Dept, Mass Inst Technol, 71-74; mil officer gas turbine engines, Aeronaut Syst Div, US Air Force, 74-75; STAFF ENGR INERTIAL GUID, CHARLES STARK DRAPER LAB, 75- *Concurrent Pos:* Nat Sci Fel, 71-73; Norway Sci fel, 84. *Mem:* Am Soc Mech Engrs; Am Inst Aeronaut & Astronaut. *Res:* Inertial navigation gyroscopes and accelerometers, applied control and estimation; microfabrication. *Mailing Add:* 119 Broad Meadow Rd Needham MA 02192

WEINBERG, MYRON SIMON, b New York, NY, July 18, 30; m 54; c 3. TOXICOLOGY, RESEARCH ADMINISTRATION. *Educ:* NY Univ, BA, 50; Fordham Univ, BS, 54; Univ Md, MS, 56, PhD(med chem, pharmacol), 58. *Prof Exp:* Res assoc med, Sinai Hosp, Baltimore, 56-58; chemist, Ortho Res Found, NJ, 58-59; chief chemist, Norwalk Hosp, Conn, 59-65; assoc dir biol opers, Food & Drug Res Labs, NY, 65-67; dir biol sci lab, Foster D Snell, Inc, 67-68, exec vpres, 69-70, pres, 70-73; vpres, Church & Dwight Co, Inc, 73-77; sr vpres, Booz, Allen & Hamilton, Inc, 77-83; PRES, WEINBERG CONSULT GROUP INC, 83- *Concurrent Pos:* Prof, Med Sch, Georgetown Univ, lectr, City Col New York & Fairfield Univ. *Mem:* Fel Royal Soc; fel Royal Soc Health; Am Soc Clin Pharmacol & Chemother; Am Chem Soc; AAAS; fel Am Inst Chem; fel Am Inst Chem Eng. *Res:* Human and animal testing of chemicals; pharmaceuticals; cosmetics; governmental liaison and regulatory advice; research management; regulatory administration and compliance; biology; new uses for chemicals; new business opportunities; litigation strategy; legal and regulatory strategy. *Mailing Add:* 2828 Pennsylvania Ave NW Suite 301 Washington DC 20007

WEINBERG, NORMAN, electrical engineering, acoustics; deceased, see previous edition for last biography

WEINBERG, PHILIP, b New York, NY, Dec 13, 25; m 47; c 3. COMMUNICATIONS, ELECTRICAL ENGINEERING. *Educ:* Univ Denver, BSEE, 50; Stanford Univ, MSEE, 51. *Hon Degrees:* LHD, Bradley Univ, 71. *Prof Exp:* Instr elec eng, Univ NMex, 52; asst prof, Univ Utah, 53-56; prof & head dept, Bradley Univ, 56-77, dean Col Commun & Fine Arts, 78-86, dean, Col Eng & Technol, 86-89, EMER PROF ENG & TECHNOL, 89- *Concurrent Pos:* Pres, Ill Valley Pub Telecommun Corp, 69-; dir, West Cent Ill Educ Telecommun Corp, 77-85. *Res:* Urban telecommunications; information utilities. *Mailing Add:* Hartmann Ctr Bradley Univ Peoria IL 61625

WEINBERG, ROBERT A, b Pittsburgh, Pa, Nov 11, 42; m 76; c 2. MOLECULAR BIOLOGY. *Educ:* Mass Inst Technol, BS, 64, PhD(biol), 69. *Hon Degrees:* DSc, Northwestern Univ, 84, State Univ NY, Stony Brook, 88; ScD, City Univ New York, 89. *Prof Exp:* Instr biol, Stillman Col, Ala, 65-66; fel, Weizmann Inst, 69-70 & Salk Inst, 70-72; res assoc, 72-73, from asst prof to assoc prof, Dept Biol & Ctr Cancer Res, 73-82, PROF BIOL, MASS INST TECHNOL, CAMBRIDGE, 82-; MEM, WHITEHEAD INST BIOMED RES, CAMBRIDGE, MASS, 82- *Concurrent Pos:* Fel, Helen Hay Whitney Found, 70; res scholar award, Am Cancer Soc, 74-77; Rita Allen Found scholar, 76-80; chmn, Sci Panel Oncogen Res, Inst Med-Nat Acad Sci, 84, Med Consumers Adv Comt, Mass Inst Technol, 88- & Comt Biol Warfare, Fedn Am Scientist, 89-; prof biol, Am Cancer Soc, 85; distinguished basic scientist award, Milken Family Med Found, 90; Lila Gruber cancer res award, Am Acad Dermat, 90; mem adv bd, Nat Coun, Fedn Am Scientists, Inst Molecular Path, Austria, Comt Responsible Conduct Res, US Inst Med & Weizmann Inst, Israel. *Honors & Awards:* Millard Schult lectr, Mass Gen Hosp, 82, Warren Triennial Prize, 83; Robert Koch Found Medal, Bonn, Ger, 83; Hammer Cancer Found Award & US Steel Found Award, Nat Acad Sci, 84; Howard Taylor Ricketts Award, Univ Chicago Med Ctr, 84; Brown-Hazen Award, NY State Dept Health, 84; Antonio Feltrinelli Prize Nat Acad Lincei, Rome, 84; Bristol-Myer Award for Distinguished Achievement in Cancer Res, 84; Katherine Berkann Judd Award, Mem Sloan-Kettering Cancer Ctr, 86; Sloan Prize, Gen Motors Cancer Res Found, 87; Lucy Wortham James Award, Soc Surg Oncol, 89; Res Recognition Award, Samuel Roberts Noble Found, 90. *Mem:* Nat Acad Sci; fel AAAS; fel Am Acad Arts & Sci. *Res:* Gene transfer, allowing the detection and isolation of a series of genes from human tumor cells, each of which is capable of inducing the normal cell to undergo conversion to a tumor cell. *Mailing Add:* Nine Cambridge Ctr Whitehead Inst Rm 367 Cambridge MA 02142

WEINBERG, ROBERT P, b Gary, Ind, Feb 12, 55; m 85; c 1. THERMAL PHYSIOLOGY. *Educ:* Purdue Univ, BS, 77; Ind Univ, PhD(physiol), 82. *Prof Exp:* RES PHYSIOLOGIST, US NAVAL MED RES INST, 83- *Mem:* Sigma Xi; NY Acad Sci; Am Physiol Soc; Inst Elec & Electronics Engrs. *Res:* Thermal protection in cold environments; diving. *Mailing Add:* 13544 Spinning Wheel Dr Germantown MD 20874

WEINBERG, ROGER, b New York, NY, Jan 1, 31. BIOSTATISTICS. *Educ:* Univ Tex, PhD(genetics), 54; Univ Mich, Ann Arbor, PhD(comput sci), 70. *Prof Exp:* USPHS fel microbiol, Calif Inst Technol, 57-58; from instr to asst prof bact, Univ Pittsburgh, 58-65; USPHS spec fel biostatist, Univ Mich, Ann Arbor, 65-68, res assoc logic of comput, 68-70; assoc prof comput sci, Kans State Univ, 70-72; PROF BIOMET, MED CTR, LA STATE UNIV, NEW ORLEANS, 72- *Mem:* Biometric Soc; Asn Comput Mach. *Res:* Computer applications to medicine. *Mailing Add:* Dept Biomet Med Ctr La State Univ New Orleans LA 70112

WEINBERG, SIDNEY B, b Philadelphia, Pa, Sept 13, 23; m 51; c 3. PATHOLOGY. *Educ:* Univ Buffalo, MD, 50; Am Bd Path, dipl, 59. *Prof Exp:* Asst med examr forensic path, New York, 54-59; CHIEF MED EXAMR, SUFFOLK COUNTY, 60-; PROF FORENSIC PATH, STATE UNIV NY, STONY BROOK, 70- *Concurrent Pos:* Attend pathologist, Vet Admin Hosp, Brooklyn, 54-55; instr, Col Med & Post-Grad Med Sch, NY Univ, 55-57, asst prof, 57-70; lectr, Columbia Univ, 57-73; res assoc, Med Sch, Cornell Univ, 58-59; consult, Brookhaven Mem & Huntington Hosps, 61-, Southside & Good Samaritan Hosps, 65- & Cent Suffolk Hosp, 66-; dir labs, Brunswick Hosp. *Mem:* AAAS; fel Am Soc Clin Path; AMA; Col Am Path; fel Am Acad Forensic Sci. *Res:* Forensic pathology, especially sudden death and coronary artery disease; automotive trauma; narcotic addiction and industrial poisoning. *Mailing Add:* Off Med Exam Suffolk County Off Bldg Hauppauge NY 11787

WEINBERG, STEVEN, b New York, NY, May 3, 33; m 54; c 1. THEORETICAL PHYSICS. *Educ:* Cornell Univ, AB, 54; Princeton Univ, PhD(physics), 57. *Hon Degrees:* DSc, Knox Col & Univ Chicago, 78, Univ Rochester & Yale Univ, 79, City Univ New York, 80, Clark Univ, 82 & Dartmouth Col, 84; DLitt, Wash Col, 85; Dr, Weizmann Inst, 85 & Columbia Univ, 90. *Prof Exp:* Instr physics, Columbia Univ, 57-59; res assoc, Lawrence Radiation Lab, Univ Calif, Berkeley, 59-60, from asst prof to prof physics, 60-69; prof, Mass Inst Technol, 69-73; Higgins prof physics, Harvard Univ, 73-83; JOSEY PROF SCI, UNIV TEX, AUSTIN, 82- *Concurrent Pos:* Consult, Inst Defense Anal, 60-73; Sloan Found fel, 61-65; vis prof, Mass Inst Technol, 67-69; consult, US Arms Control & Disarmament Agency, 71-73; counr, Am Phys Soc, 72-75; consult, Stanford Res Inst, 73-; sr scientist, Smithsonian Astrophys Observ, 73-; mem, Bd Overseers of the Superconducting Supercollider, 83-86; Morris Loeb vis prof, Harvard Univ, 83; sr consult, Smithsonian Astrophys Observ, 83-; dir, Jerusalem Sch Theoret Physics, 83-; vis prof, Harvard, 83-; sr consult, Smithsonian Astrophys Observ, 83-; coun scholar, Libr Cong, 82-85; mem, Sci Bk Comt, Sloan Found, 85; mem, Supercollider Site Eval Comt, Nat Acad Sci-Nat Acad Eng, 87, Supercollider Sci Policy Comt, 89- *Honors & Awards:* Nobel Prize in Physics, 79; Nat Medal of Sci, 91; Morris Loeb Lectr, Harvard Univ, 66; J R Oppenheimer Prize, 73; Am Inst Physics-US Steel Found Sci Writing Award, 77; Richtmeyer Lectr, Am Asn Physics Teachers, 74; Scott Lectr, Cavendish Lab, Cambridge Univ, 75; Dannie Heineman Prize, 77; Silliman Lectr, Yale Univ, 77; Elliot Cresson Medal, Franklin Inst, 79, Madison Medal, Princeton Univ, 91; Lauritsen Lectr, Calif Inst Technol, 79; Shalit Lectr, Weizmann Inst, 79; Henry Lectr, Princeton Univ, 79; Schild Lectr, Univ Tex, 79; Bampton Lectr, Columbia Univ, 83; Hilldale Lectr, Univ Wis, 85; Klein Lectr, Univ Stockholm, 89. *Mem:* Nat Acad Sci; Am Acad Arts & Sci; Am Phys Soc; Int Astron Union; Royal Soc London; Am Philos Soc. *Res:* Elementary particles; field theory; cosmology. *Mailing Add:* Dept Physics Univ Tex Austin TX 78712-1081

WEINBERG, WILLIAM HENRY, b Columbia, SC, Dec 5, 44; m 89. CHEMICAL ENGINEERING, CHEMICAL PHYSICS. *Educ:* Univ SC, BS, 66; Univ Calif, Berkeley, PhD(chem eng), 71. *Prof Exp:* NATO fel phys chem, Cambridge Univ, 71-72; from asst prof to assoc prof chem eng, Calif Inst Technol, 72-74, prof chem eng & chem physics, 77-89; PROF CHEM & NUCLEAR ENG, UNIV CALIF, SANTA BARBARA, 89- *Concurrent Pos:* Prin investr, US-USSR Exchange Prog Chem Catalysis, 74-80; Alfred P Sloan Found fel, 76; Camille & Henry Dreyfus Found teacher-scholar, 76; vis prof chem, Harvard Univ, 80; Alexander von Humboldt sr US scientist award, 82; Chevron Distinguished Prof Chem Eng & Chem Physics, Calif Inst Tech, 81-86; vis prof, Univ Pittsburgh, 87-88; ed, Appl Surf Sci, Surface Sci Reports, Langmuir. *Honors & Awards:* Wayne B Nottingham Prize, Am Phys Soc, 72; Victor K Lamer Award, Am Chem Soc, 72; A P Colburn Award, Am Inst Chem Engrs, 81; Alexander von Humboldt Sr US Scientist Award, 82. *Mem:* AAAS; Am Chem Soc; Am Inst Chem Engrs; Am Vacuum Soc; fel Am Phys Soc. *Res:* Physics and chemistry of solids and solid surfaces; gas-surface interactions; chemical adsorption and heterogeneously catalyzed surface reactions. *Mailing Add:* Div Chem & Nuclear Eng Univ Calif Santa Barbara CA 91306

WEINBERGER, ARNOLD, b Czechoslovakia, Oct 23, 24; US citizen; m 56; c 3. MEMORIES, DESIGN AUTOMATION. *Educ:* City Col New York, BEE, 50. *Prof Exp:* Electronic scientist, Nat Bur Standards, 50-60; res staff mem, IBM, Yorktown Heights, 60-66; SR ENGR DEVELOP, IBM, POUGHKEEPSIE, 66- *Mem:* Fel Inst Elec & Electronics Engrs. *Res:* Computer arithmetic and logic including very large scale integration, programmable logic array, design automation, memories and system design. *Mailing Add:* IBM S Rd Box 950 Poughkeepsie NY 12602

WEINBERGER, CHARLES BRIAN, b Macon, Ga, Jan 31, 41; m 76; c 2. CHEMICAL ENGINEERING. *Educ:* Univ Calif, Berkeley, BS, 63; Univ Mich, Ann Arbor, MSE, 64, PhD(chem eng), 70. *Prof Exp:* Engr, Shell Develop Co, 64-66; res engr, Eng Tech Lab, E I du Pont de Nemours & Co, Inc, 70-72; asst prof, 72-78, actg dept head, 81-82, ASSOC PROF CHEM ENG, DREXEL UNIV, 78-, DEPT HEAD, 88- *Mem:* Am Inst Chem Engrs; Soc Rheology; Sigma Xi. *Res:* Extension flow of non-Newtonian fluids; polymer processing; fluid mechanics. *Mailing Add:* Dept Chem Eng Drexel Univ Philadelphia PA 19104

WEINBERGER, DANIEL R, b New York, NY, May 24, 47; m; c 1. BRAIN DISORDER RESEARCH. *Educ:* Johns Hopkins Univ, BA, 69; Univ Pa, MD, 73; Am Bd Psychiat & Neurol, cert psychiat, 78, cert neurol, 84. *Prof Exp:* Grad fel med, Univ Calif, Los Angeles Sch Med, 73-74; clin fel psychiat, Harvard Med Sch, 74-77; res ward dir, Adult Psychiat Br, Intramural Res Prog, NIMH, 77-78, staff psychiatrist, 77-81, head, Clin Neuropsychiat & Neurobehav Unit, St Elizabeths Hosp, 81-82, chief, Sect Clin Neuropsychiat & Neurobehav, Neuropsychiat Br, 83-86, CHIEF, CLIN BRAIN DIS BR, INTRAMURAL RES PROG, NEUROSCI CTR, ST ELIZABETHS, WASHINGTON, DC, 86- *Concurrent Pos:* Gen med prac, Bridgewater Med Ctr, Mass, 74-76; emergency rm physician, Cardinal Cushing Gen Hosp, Mass, 74-77; asst clin prof psychiat, George Washington Univ, 78-81; dir, Movement Dis, Dementia Clin & Nat Inst Neurol & Commun Dis & Stroke, NIH, 83-86, Behav Neurol Serv, St Elizabeths Hosp, 83-88. *Honors & Awards:* Morton Prince Award, Am Psychopath Asn, 84; Judith B Silver Award, Nat Alliance Ment Ill, 85; Joel Elkes Int Award, Am Col Neuropsychopharmacol, 89. *Mem:* AMA; Am Psychiat Asn; AAAS; Am Acad Neurol; Soc Neurosci; Am Neuropsychiat Asn. *Res:* Neurology; psychiatry. *Mailing Add:* NIMH Neurosci Ctr St Elizabeths Clin Brain Dis 2700 Martin Luther King Jr Ave SE Washington DC 20032

WEINBERGER, DOREEN ANNE, b Bethlehem, Pa, Jan 27, 54. OPTICAL PHYSICS, NONLINEAR OPTICS. *Educ:* Mt Holyoke Col, BA, 75; Univ Ariz, MS, 80, PhD(optical sci), 84. *Prof Exp:* Sr tech aide solid state physics, Bell Tel Labs, 74; res student solar physics, Kitt Peak Nat Observ, 75; res asst physics, Mass Inst Technol, 75-78; res asst optical sci, Univ Ariz, 78-84; ASST PROF ELEC ENG & COMPUT SCI, UNIV MICH, 84- *Concurrent Pos:* Trainee physics, Gen Motors Labs, 78; assoc fac instr, Pima Community Col, Tucson, Ariz, 80-81. *Honors & Awards:* Presidential Young Investr Award, NSF, 86. *Mem:* Optical Soc Am; Sigma Xi; Asn Women Sci; AAAS. *Res:* Nonlineamties in optical fibers; second harmonic generation; nonlinear pulse propagation; physics of optical nonlinearities of semiconductors and heterostructures for use in all-optical logic and image processing applications. *Mailing Add:* Dept Elec Eng & Comput Sci Univ Mich Ann Arbor MI 48109-2122

WEINBERGER, EDWARD BERTRAM, b Pittsburgh, Pa, Mar 21, 21; m 44; c 3. COMPUTER SCIENCE. *Educ:* Mass Inst Technol, BS, 41; Univ Pittsburgh, MS, 47, PhD(math), 50. *Prof Exp:* Phys chemist, Gulf Res & Develop Co, 43-51, mathematician, 51-52, head comput anal sect, 53-62, res assoc, 62-66, sr scientist, 66-71; systs analyst computer ctr, Carnegie-Mellon Univ, 72-73, sr systs analyst, Admin Systs Dept, 73-90; RETIRED. *Concurrent Pos:* Lectr, Univ Pittsburgh, 47-73. *Mem:* AAAS; Asn Comput Mach. *Res:* Digital computers; programming languages; numerical analysis; data processing. *Mailing Add:* 6380 Caton St Pittsburgh PA 15217-3031

WEINBERGER, HANS FELIX, b Vienna, Austria, Sept 27, 28; nat US; m 57; c 3. APPLIED MATHEMATICS. *Educ:* Carnegie Inst Technol, BS & MS, 48, ScD, 50. *Prof Exp:* Fel, Inst Fluid Dynamics & Applied Math, Univ Md, 50-51, res assoc, 51-53, from asst res prof to assoc res prof, 53-60; assoc prof, 60-61, head, Sch Math, 67-69, PROF MATH, UNIV MINN, MINNEAPOLIS, 61-, DIR, INST MATH & APPLICATIONS, 81- *Concurrent Pos:* Vis mem, Courant Inst Math Sci, NY Univ, 66-67; vis prof, Univ Ariz, 70-71 & Stanford Univ, 72-73. *Mem:* Am Math Soc; Soc Natural Philos; Soc Indust & Appl Math. *Res:* Approximation of eigenvalues, quadratic functionals and solutions of partial differential equations. *Mailing Add:* Dept Math Vincent Hall 206 Church St SE Minneapolis MN 55455

WEINBERGER, HAROLD, b New York, NY, Mar 24, 10; m 37; c 4. INORGANIC CHEMISTRY. *Educ:* Polytech Inst Brooklyn, BSc, 31, MS, 33, PhD(org chem), 36. *Prof Exp:* Asst chem, Polytech Inst Brooklyn, 31-35; instr, Long Island Univ, 36-42 & City Col New York, 42-44; proj leader, Gen Chem Co, NY, 44-46; chief chemist, Glyco Prods Inc, 46-48 & Heyden Chem Corp, 48-52; dir res, Fine Organics, Inc, 52-59; assoc prof, 59-62, chmn dept, 60-69, dir career ctr, Col Sci & Eng, 71-73, prof, 62-76, fac leader, Teaneck-Hackensack Campus, 75-76, EMER PROF CHEM, FAIRLEIGH DICKINSON UNIV, 76- *Concurrent Pos:* Asst, Long Island Univ, 31-33; instr, Brooklyn Col, 41-42; consult, 76- *Mem:* Fel AAAS; Am Chem Soc; fel Am Inst Chem; NY Acad Sci. *Res:* Resin chemistry; germicides; fatty acid derivatives; aroma materials. *Mailing Add:* 655 Pomander Walk Teaneck NJ 07666

WEINBERGER, LEON WALTER, b New York, NY, Aug 28, 23; m 50; c 3. SANITARY ENGINEERING. *Educ:* Cooper Union, BCE, 43; Mass Inst Technol, MS, 47, ScD, 49. *Prof Exp:* Eng draftsman, NAm Aviation, Inc, 43; asst, Mass Inst Technol, 47-48, res assoc, 48-49; from asst prof to assoc prof civil & sanit eng, Case Inst Technol, 49-62; chief, Basic & Appl Sci Br, Div Water Supply & Pollution Control, USPHS, 63-66; asst comnr res & develop, Fed Water Pollution Control Admin, 66-68; vpres environ control mgt, Zurn Industs Inc, 68-70; PRES, LEON W WEINBERGER & ASSOC, LTD & PRES, ENVIRON QUAL SYSTS INC, 71-; CHIEF ENGR, PEER CONSULT, PC, 78- *Concurrent Pos:* Nat Found Fels Consult engr, 49-62; chmn, Am Sci Team, Am-Ger Coop Exchange Water Pollution; US rep, comt water pollution, Orgn Econ Coop & Develop; mem, Expert Adv Panel Environ Health; WHO; mem comts water & pollution, Nat Acad Sci; US rep, Int Oceanog Comt; mem, Comt Water Resources Res; prof, Nat Grad Univ & univ seminar assoc, Columbia Univ, 68-; mem adv bd, Dept Natural Resources, State Md, 74-77. *Mem:* Fel Am Soc Civil Engrs; Water Pollution Control Fedn; Am Water Works Asn; Am Acad Environ Engrs. *Res:* Water supply and treatment; waste water treatment; water pollution control. *Mailing Add:* 7400 Masters Dr Potomac MD 20854

WEINBERGER, MILES, b McKeesport, Pa, June 28, 38; m 64; c 4. PEDIATRICS, RESPIRATORY DISEASE. *Educ:* Univ Pittsburgh, MD, 65. *Prof Exp:* Assoc prof pediat & pharmacol, 75-80, PROF PEDIAT, UNIV IOWA, 80-, CHMN PEDIAT ALLERGY & PULMONARY DIV & DIR, CYSTIC FIBROSIS CTR, 75- *Mem:* Fel Am Acad Allergy; fel Am Acad Pediat; fel Soc Pediat Res; Am Thoracic Soc; fel Am Col Allergy. *Res:* Pharmacotherapy of asthma; clinical pharmacology of antiasthmatic drugs; pediatric allergy. *Mailing Add:* Dept Pediat Univ Iowa Hosp Iowa City IA 52242

WEINBERGER, MYRON HILMAR, b Cincinnati, Ohio, Sept 21, 37; m 60; c 3. INTERNAL MEDICINE, HYPERTENSION. *Educ:* Ind Univ, AB, 59, MD, 63. *Prof Exp:* Intern internal med, Sch Med, Ind Univ, 63-64, resident, 64-66; res fel endocrinol, Sch Med, Stanford Univ, 66-69; from asst prof to assoc prof, 69-76, PROF MED, SCH MED, IND UNIV, 76- *Concurrent Pos:* Investr, Specialized Ctr Res Hypertension, Ind Univ, 70-75, prin investr-dir, 75-; mem prog comt, Coun High Blood Pressure Res, 78. *Mem:* Am Heart Asn; Am Fedn Clin Res; Endocrine Soc. *Res:* Blood pressure control; salt and water metabolism, hormones; circulation; kidney; adrenal; high blood pressure. *Mailing Add:* Ind Univ Sch Med 541 Clinical Dr Rm 409 Indianapolis IN 46202-5111

WEINBERGER, NORMAN MALCOLM, b Cleveland, Ohio, Aug 10, 35; m 54; c 7. PSYCHOBIOLOGY, NEUROPHYSIOLOGY. *Educ:* Western Reserve Univ, AB, 57, MS, 59, PhD(psychol), 61. *Prof Exp:* Res physiologist, Univ Calif, Los Angeles, 64; from asst prof to assoc prof, 65-74, chmn, 75-78, PROF PSYCHOBIOL, UNIV CALIF, IRVINE, 74- *Concurrent Pos:* Consult ed, Physiol Psychol; assoc ed, Exp Neurol; found mem, Ctr Neurobiol Learning & Memory. *Mem:* AAAS; Soc Neurosci; Int Brain Res Orgn; NY Acad Sci. *Res:* Neurobiological bases of learning and attention. *Mailing Add:* Dept Psychobiol Univ Calif Irvine CA 92717

WEINBERGER, PEARL, plant physiology, environmental biology; deceased, see previous edition for last biography

WEINBERGER, PETER JAY, b New York, NY, Aug 6, 42. MATHEMATICS, COMPUTER SCIENCE. *Educ:* Swarthmore Col, BA, 64; Univ Calif, Berkeley, PhD(math), 69. *Prof Exp:* Mem tech staff, Bellcomm Inc, 69-70; asst prof math, Univ Mich, Ann Arbor, 70-76; mem staff computer sci res, 76-83, dept head, Comput Sci Res, 83-90, DIR, SOFTWARE & SYSTS RES LAB, BELL LABS, 90- *Mem:* Am Math Soc; Asn Comput Mach; Math Asn Am; Sigma Xi. *Res:* Number theory; operating systems and compilers; data bases and distributed computing. *Mailing Add:* Bell Labs Murray Hill NJ 07974

WEINBERGER, STEVEN ELLIOTT, b Philadelphia, Pa, Jan 28, 49; m 70; c 2. PULMONARY DISEASE. *Educ:* Princeton Univ, AB, 69; Harvard Univ, MD, 73. *Prof Exp:* Clin assoc, Pulmonary Br, Nat Heart, Lung & Blood Inst, NIH, 75-78; from instr to asst prof, 78-89, ASSOC PROF MED, HARVARD UNIV SCH MED, 89- *Concurrent Pos:* Clin dir, Pulmonary Unit, Beth Israel Hosp, Boston, 86- *Mem:* Am Col Physicians; Am Thoracic Soc; Am Fedn Clin Res; Am Col Chest Physicians; Am Physiol Soc. *Res:* Control of breathing; neurohumoral aspects of ventilatory control; pathophysiologic mechanisms of dyspnea. *Mailing Add:* Pulmonary Unit Beth Israel Hosp 330 Brookline Ave Boston MA 02215

WEINBRANDT, RICHARD M, b Lexington, Nebr, July 20, 44; m 67, 87; c 3. PETROLEUM ENGINEERING. *Educ:* Univ Calif, Berkeley, BS, 67, MS, 68; Stanford Univ, PhD(petrol eng), 72. *Prof Exp:* Res engr, Chevron Oil Field Res, 68-69 & Union Oil Co, 71-76; mgr eng, Aminoil, 76-81; CONSULT, ENHANCED OIL RECOVERY, 81- *Concurrent Pos:* Teaching asst, Univ Calif, Berkeley, 68 & Stanford Univ, 71; consult, Alcoa, Aminoil, Dow Chem, Gary Energy, PQ Energy Serv Sci Software & Westinghouse. *Mem:* Am Inst Mining, Metall & Petrol Engrs; Soc Petrol Engrs. *Res:* Improved oil recovery methods; reservoir simulation; fluid flow in porous media. *Mailing Add:* Enhanced Oil Recovery Consult Skyline RanchBox 20 Jackson WY 83001

WEINDLING, JOACHIM I(GNACE), b Antwerp, Belg, Feb 18, 27; US citizen; m 54; c 3. OPTIMIZATION, ENGINEERING ECONOMY. *Educ:* City Col New York, BME, 48; Columbia Univ, MS, 49, PhD(opers res), 67. *Prof Exp:* Lectr mech eng, City Col New York, 48-51; chief engr, Korfund Dynamics, Inc, 51-59; vpres eng, Vibration Mountings & Controls, Inc, 59-61; assoc prof mech eng, Drexel Inst, 61-67; prof eng, PMC Cols, 67-69; chmn, Dept Opers Res & Syst Anal, 73-74, PROF SYSTS ENG, POLYTECH INST NEW YORK, 69-, DIR OPERS RES PROG, 70- *Concurrent Pos:* Lectr, City Col New York, 51-61; adj assoc prof, NY Univ, 58-61; indust consult, 61- *Mem:* Am Soc Mech Engrs; Opers Res Soc Am; Am Soc Eng Educ; Am Inst Indust Engrs; Sigma Xi. *Res:* Markov chains; reliability; computer-aided optimum designs; shock and vibration controls; mathematical programming; technology utilization and education in developing countries. *Mailing Add:* 204 Fairfield Dr Wallingford PA 19086

WEINER, BRIAN LEWIS, b London, Eng, June 21, 45; m 80; c 1. MATHEMATICAL STRUCTURE OF REDUCED DENSITY MATRICES COHERENT STATES, GROUP THEORETIC STRUCTURE OF QUANTUM MECHANICS. *Educ:* Univ Leicester, BSc, 66, PhD(math) & PhD(chem), 70. *Prof Exp:* Inst math, Leicester Col Art & Technol, 69-70; postdoctoral fel, Dept Math, Queens Univ, Ont, Can, 70-73; royal soc postdoctoral fel, Quantum Chem Group, Uppsala Univ, Sweden, 73-75, res asst, 75-80, res asst math, Dept Quantum Chem, 79-80; vis assoc prof, Quantum Theory Proj, Univ Fla, 80-81, Dept Computer Sci, 84-85, Dept Math, 85-86; asst prof, 86-90, ASSOC PROF, DEPT PHYSICS, PA STATE UNIV, 90- *Concurrent Pos:* Vis assoc prof, Dept Chem, Univ Fla, Gainsville, 80-86; from asst prof to assoc prof, Dept Physics, Pa State Univ, DuBois, 86- *Mem:* Am Phys Soc; Am Math Soc; Am Chem Soc; Soc Indust & Appl Math. *Res:* Theoretical development of quantum mechanical methods for the description of processes involving atomic, molecular and condensed matter systems, especially the use of lie group methods; quantum chemical calculations of reactions on surfaces molecular spectra and percursor reactions in the formation of soot. *Mailing Add:* 1036 Crabapple Dr State Col PA 16801

WEINER, CHARLES, b Brooklyn, NY, Aug 11, 31; div; c 1. HISTORY OF PHYSICS, HISTORY OF BIOLOGY. *Educ:* Case Inst Technol, BS, 60, MA, 63, PhD(hist of sci), 65. *Prof Exp:* Dir proj hist of recent physics in US, Am Inst Physics, 64-65, dir, Ctr for Hist of Physics, 65-74, PROF HIST SCI & TECHNOL, MASS INST TECHNOL, 74-, DIR, ORAL HIST PROG, 75- *Concurrent Pos:* NSF grants, 65-79 & 81; mem adv comt, Nat Union Catalog of Manuscript Collections, Libr of Cong, 65-71; vis lectr, Polytech Inst Brooklyn, 66; proj dir comt hist of contemp physics, Joint Am Inst Physics-Am Acad Arts & Sci, 66-74; Guggenheim fel, Niels Bohr Inst, Copenhagen, Denmark, 70-71. *Mem:* Fel AAAS; Hist Sci Soc; Soc Hist Technol; Am Hist Asn; Orgn Am Historians. *Res:* History of twentieth century science; social impact of science and technology; development of the physical sciences in the United States; history of genetic engineering and of university-industry relations; ethical issues in science and engineering. *Mailing Add:* Dept Technol & Human Affairs Mass Inst Technol 77 Mass Ave Cambridge MA 02139

WEINER, DANIEL LEE, b Dayton, Ky, Apr 13, 50; m 74; c 2. PHARMACOKINETICS, STATISTICAL SOFTWARE. *Educ:* Univ Ky, BS, 72, MS, 73, PhD(statist), 76. *Prof Exp:* Head, biostatist, Merrell Dow Pharmaceut, Div Dow Chem Co, 82-85; exec vpres, Statist Consult Inc, 85-88; DIR, BIOMETRY SYNTEX RES, 88- *Concurrent Pos:* Adj asst prof, Dept Environ Health, Div Epidemiol & Biostat, Col Med, Univ Cincinnati, 79-81, adj assoc prof, 82-87, adj assoc prof pharmacokinetics, 85- *Mem:* Am Statist Asn; Biomet Soc. *Res:* Bioavailability and pharmacokinetics in relation to biological response; application of statistics to the evaluation of drugs. *Mailing Add:* Syntex Res A4-200 3401 Hillview Ave Palo Alto CA 94303

WEINER, EUGENE ROBERT, b Pittsburgh, Pa, Sept 16, 28; m 52, 81; c 4. PHYSICAL CHEMISTRY. *Educ:* Ohio Univ, BS, 50; Univ Ill, MS, 57; Johns Hopkins Univ, PhD(chem), 63. *Prof Exp:* Sect chief appl physics sect, Interior Ballistics Lab, Aberdeen Proving Grounds, Md, 50-55; instr, McCoy Col, 62-63; sr scientist, Johnston Labs, Inc, 63-65; from asst prof to assoc prof, 65-84, PROF CHEM, UNIV DENVER, 84- *Concurrent Pos:* Consult, US Geol Surv, 70- *Mem:* Am Chem Soc; Am Phys Soc. *Res:* Atomic and molecular collisions; photochemistry at liquid-solid interface; gas phase ion-molecule reactions, free radical reactions, gas kinetics and energetics; laser Raman spectroscopy. *Mailing Add:* Dept Chem Univ Denver Denver CO 80208

WEINER, HENRY, b Cleveland, Ohio, May 18, 37; m 60; c 2. BIOCHEMISTRY, MOLECULAR BIOLOGY. *Educ:* Case Inst Technol, BS, 59; Purdue Univ, PhD(org chem), 63. *Prof Exp:* Res assoc biol, Brookhaven Nat Lab, 63-65; NIH res fel biochem, Karolinska Inst, Sweden, 65-66; from asst prof to assoc prof, 66-76, PROF BIOCHEM, PURDUE UNIV, LAFAYETTE, 76- *Mem:* Res Soc Alcoholism; Am Soc Biol Chem; Am Chem Soc; Am Asn Univ Prof; Int Soc Biomed Res Alcoholism. *Res:* Enzymology; protein chemistry; enzyme precursors; effects of ethanol on metabolism; acetaldehyde metabolism. *Mailing Add:* Dept Biochem Purdue Univ West Lafayette IN 47907-9980

WEINER, HERBERT, b Vienna, Austria, Feb 6, 21; nat US; m 53; c 3. PSYCHIATRY, NEUROLOGY. *Educ:* Harvard Univ, AB, 43; Columbia Univ, MD, 46. *Prof Exp:* Exchange fel neurol, 50; USPHS fel psychiat, 52-53; instr psychiat, Sch Med & Dent, Univ Rochester, 53-54; guest lectr, Wash Sch Psychiat, 54-55; from instr to assoc prof, 55-66, PROF PSYCHIAT, ALBERT EINSTEIN COL MED, 66-, PROF NEUROSCI, 74- *Concurrent Pos:* Asst vis psychiatrist & chief adult psychiat in-patient serv, Bronx Munic Hosp Ctr, 55-56, asst vis physician, 56-61, assoc vis psychiatrist, 56-; consult, Home Care Dept, Montefiore Hosp, 55-59; USPHS fel ment health, 56-58 & res career develop award, 62-65; Stern fel, 59-62; mem bd trustees, Scarborough Country Day Sch, 62-; mem ment health study sect, Div Res Grants, NIH, 62-, chmn, 66-67; attend psychiatrist & dir educ prog, Div Psychiat, Montefiore Hosp & Med Ctr, New York, 65-69, chmn dept psychiat, 69-; fel, Ctr Advan Study Behav Sci, Stanford Univ, Guggenheim Mem Found fel & Commonwealth Fund grant-in-aid, 72-73; ed-in-chief, Psychosom Med, 72-82. *Honors & Awards:* Spec Presidential Commendation from Am Psychiat Asn, 87. *Mem:* AAAS; Am Acad Neurol; Am Psychosom Soc (pres, 71-72); Asn Res Nerv & Dis (pres, 79-80); Psychiat Res Soc; Acad Behav Med Res (pres, 83-84); Int Col Psychosom Med (pres, 87-89). *Res:* Neurophysiology; psychophysiology. *Mailing Add:* Neuropsychiat Inst UCLA Health Sci Ctr Los Angeles CA 90024-1759

WEINER, HOWARD JACOB, b Chicago, Ill, Aug 26, 37; m 71. PROBABILITY. *Educ:* Ill Inst Technol, BSEE, 58; Univ Chicago, MS, 60; Stanford Univ, PhD(statist), 64. *Prof Exp:* Actg asst prof statist, Univ Calif, Berkeley, 64; asst prof, Stanford Univ, 64-65; from asst prof to assoc prof, 65-72, vchmn dept, 76-80, PROF MATH, UNIV CALIF, DAVIS, 72- *Concurrent Pos:* Vis assoc prof statist, Stanford Univ, 76, vis res assoc, 77 & 78. *Mem:* Math Asn Am. *Res:* Age dependent branching processes; sequential random packing. *Mailing Add:* Dept Math Univ Calif Davis CA 95616

WEINER, IRWIN M, b New York, NY, Nov 5, 30; c 2. PHARMACOLOGY. *Educ:* Syracuse Univ, AB, 52; State Univ NY, MD, 56. *Prof Exp:* Fel pharmacol, Sch Med, Johns Hopkins Univ, 56-58, from instr to asst prof pharmacol & exp therapeut, 58-66; assoc prof, 66-68, PROF PHARMACOL, STATE UNIV NY UPSTATE MED CTR, 68-, DEAN, COL MED, 87-, VPRES MED & BIOMED EDUC, 87- *Concurrent Pos:* Fel pharmacol & exp therapeut, Sch Med, Johns Hopkins Univ, 58-60; vis prof molecular biol, Albert Einstein Col Med, 64-65; USPHS res career develop award, 64-66; mem study sect pharmacol & exp therapeut, NIH, 65-69; ed, J Pharmacol & Exp Therapeut, 65-72; consult, Sterling Winthrop Res Inst, 68-82. *Mem:* AAAS; Am Soc Pharmacol & Exp Therapeut; NY Acad Sci. *Res:* Pharmacology of diuretics and uricosuric agents; renal excretion of drugs; intestinal absorption of bile salts; bacterial cell wall biosynthesis. *Mailing Add:* Deans Off State Univ NY Upstate Med Ctr Syracuse NY 13210

WEINER, JACOB, b Brooklyn, NY, Dec 24, 47. PLANT ECOLOGY, BOTANY. *Educ:* Antioch Col, BA, 70; Univ Mich, MS, 74; Univ Ore, PhD(biol), 78. *Prof Exp:* Asst prof, 78-84, ASSOC PROF BIOL, SWARTHMORE COL, 84- *Concurrent Pos:* Trainee syst biol, NIH, 77-78; vis scholar, Harvard Univ, 81; vis scientist, Univ Col North Wales, 81-82, Imp Col, Silwood Park, 89-90, Univ Basel, 90; Smithsonian postdoctoral fel, 84. *Mem:* Brit Ecol Soc; Ecol Soc Am; Bot Soc Am; Sigma Xi. *Res:* Plant population biology. *Mailing Add:* Dept Biol Swarthmore Col 500 College Ave Swarthmore PA 19081-1397

WEINER, JEROME HARRIS, b New York, NY, Apr 5, 23; m 50; c 2. APPLIED MATHEMATICS. *Educ:* Cooper Union, BME, 43; Columbia Univ, AM, 47, PhD(math), 52. *Prof Exp:* Asst tech dir heat & mass flow analyzer lab, Columbia Univ, 51-57, asst prof civil eng & eng mech, 52-56, prof mech eng, 56-68; L HERBERT BALLOU UNIV PROF ENG & PHYSICS, BROWN UNIV, 68- *Mem:* Am Math Soc; Am Phys Soc. *Res:* Thermal stresses; crystal defects. *Mailing Add:* 24 Taber Ave Providence RI 02906

WEINER, JOEL HIRSCH, b Montreal, Que, Dec 23, 46; m 71; c 2. MICROBIOLOGY, MOLECULAR BIOLOGY. *Educ:* McGill Univ, BSc, 68; Cornell Univ, PhD(biochem), 72. *Prof Exp:* Fel biochem, Stanford Univ, 72-76; from asst prof to assoc prof, 76-84, PROF BIOCHEM, UNIV ALTA, 84- *Concurrent Pos:* Med Res Coun scholar, Univ Alta, 76-81. *Mem:* Am Soc Biol Chemists; Am Soc Microbiol; Can Biochem Soc; AAAS. *Res:* Investigation of membrane bioenergetics using molecular biological and biochemical approaches with emphasis on bacterial anaerobic electron transport chain proteins; study of membrane protein structure and function using biochemical and biophysical techniques. *Mailing Add:* Dept Biochem Univ Alta Edmonton AB T6G 2H7 Can

WEINER, JOHN, b Malvern, NY, Apr 14, 43; wid; c 2. CHEMICAL PHYSICS. *Educ:* Pa State Univ, BS, 64; Univ Chicago, PhD(chem), 70. *Prof Exp:* Fel & lectr chem, Yale Univ, 70-73; asst prof chem, Dartmouth Col, 73-78; assoc prof, 78-83, PROF CHEM, UNIV MD, 83- *Concurrent Pos:* Vis fel, Univ Paris, 77-78. *Mem:* Am Chem Soc; Am Phys Soc; AAAS; Sigma Xi; Optical Soc Am. *Res:* Reactive and inelastic collision processes in ion-molecule systems; internal state excitation leading to chemiluminescence phenomena; laser-induced excitation; ionization and reactive collisions in crossed-beam studies; optical/collisional interactions. *Mailing Add:* Dept Chem Univ Md College Park MD 20742

WEINER, LAWRENCE MYRON, b Milwaukee, Wis, May 21, 23; m 44; c 2. MICROBIOLOGY, IMMUNOLOGY. *Educ:* Univ Wis, BA, 47, MS, 48, PhD(med microbiol), 51. *Prof Exp:* From instr to assoc prof, 51-65, chmn dept & assoc dean, 70-72, dep dean, 72-79, dean, 79-81, PROF MICROBIOL & PATH, SCH MED, WAYNE STATE UNIV, 65- *Concurrent Pos:* Pres, Mich State Bd Examr in Basic Sci, 62-68. *Mem:* Am Soc Microbiol; Soc Exp Biol & Med. *Res:* Immunology of infectious diseases; hypersensitivity to chemical agents; clinical microbiology. *Mailing Add:* 540 E Canfield Ave Detroit MI 48201

WEINER, LOUIS I, b Philadelphia, Pa, May 3, 13; m 36; c 1. TEXTILE ENGINEERING. *Educ:* Temple Univ, BS, 36; Lowell Tech Inst, MS, 65. *Prof Exp:* Technologist, Res & Develop Lab, Philadelphia Qm Depot, 43-50, chief, Textile Eng Lab, 50-58; asst chief res, Textile Clothing & Footwear Div, Res & Eng Command, US Dept Army, 58-63, gen phys scientist, 63-67, head fibrous mat res group, Clothing & Org Mat Div, Natick Labs, 67-69, res phys scientist, Clothing & Personal Life Support Equip Lab, 69-72; RETIRED. *Concurrent Pos:* Vis prof, Lowell Technol Inst, 65-72; expert, UN Indust Develop Orgn, 68. *Mem:* Fiber Soc; Sigma Xi. *Res:* Wear resistance; heat transfer; fiber and fabric structure and performance. *Mailing Add:* 36 Morrill St West Newton MA 02165

WEINER, LOUIS MAX, b Chicago, Ill, Nov 11, 26; m 57; c 3. ALGEBRA, MATHEMATICS GENERAL. *Educ:* Univ Chicago, SB, 47, SM, 48, PhD(math), 51. *Prof Exp:* Asst, Univ Chicago, 48; personnel examr, Chicago Civil Serv Comn, 51-52; asst prof math, DePaul Univ, 52-58; res engr, Mech Res Div, Am Mach & Foundry Co, 58-62, supvr, Anal Sect, Gen Am Res Div, 62-64; assoc prof, 64-68, chmn dept, 68-74, PROF MATH, NORTHEASTERN ILL UNIV, 64- *Concurrent Pos:* Instr, Amundsen Br, Chicago City Jr Col, instr, Oakton Community Col, 77- *Mem:* Am Math Soc; Math Asn Am; Sigma Xi; Mensa. *Res:* Algebra; linear algebras; operations research; slide rule type computers. *Mailing Add:* 3144 Greenleaf Wilmette IL 60091-2008

WEINER, MATEI, b Bucharest, Romania, Aug 21, 33; US citizen; m 66; c 2. MEDICAL MYCOLOGY, ANTIBIOTIC SUSCEPTIBILITY STUDIES. *Educ:* Univ Bucharest, Romania, Zoo Nic Engr, 57; Wagner Col, Staten Island, MS, 70. *Prof Exp:* Inspector & technologist, Dept Health, City Bucharest, Romania, 58-66; asst supvr & technologist, Maimonides Med Ctr, Brooklyn, 67-70; SUPVR, MICROBIOL DEPT, METHODIST HOSP BROOKLYN, 70- *Concurrent Pos:* Sr lectr, Sch Med Technol, Methodist Hosp Brooklyn, 71-; clin instr, Health Sci Ctr, State Univ NY, 80-; clin assoc, Tech Col, City Univ New York, 80-82. *Mem:* Am Soc Microbiol. *Res:* Senior or contributing author of 24 publications in clinical microbiology and medical research. *Mailing Add:* Methodist Hosp 506 Sixth St Brooklyn NY 11215

WEINER, MILTON LAWRENCE, polymer chemistry, plastics chemistry; deceased, see previous edition for last biography

WEINER, MURRAY, b New York, NY, Apr 18, 19; m 51; c 1. MEDICINE. *Educ:* City Col New York, BS, 39; NY Univ, MS & MD, 43. *Prof Exp:* From instr to asst prof med, Col Med, NY Univ, 54-72; vpres & dir biol res, Geigy Res Labs, NY, 67-72; vpres res & sci affairs, Merrell-Nat Labs, 72-81; DIR CLIN PHARMACOL, UNIV CINCINNATI, 81- *Concurrent Pos:* Asst vis physician & consult, Bellevue Hosp, NY, 49-72, Univ Hosp, 53-72, Long Island Jewish Hosp, 54-57 & North Shore Hosp, NY, 55-72; assoc vis physician, Willard Parker Hosp, NY, 52-57 & Goldwater Mem Hosp, NY, 56-72; vis staff, Cincinnati Gen Hosp & Christian Holmes Hosp, Ohio, 72-; clin prof med, Univ Cincinnati, 72- *Mem:* Soc Exp Biol & Med; Am Physiol Soc; AMA; Am Heart Asn. *Res:* Clotting mechanism; anticoagulant and anti-inflammatory drugs; drug disposition. *Mailing Add:* Pharmacol Dept Int Med Univ Cincinnati Med Ctr 231 Bethesda Ave Cincinnati OH 45267-0578

WEINER, MYRON, b Baltimore, Md, May 27, 43; m 87; c 2. PHARMACOLOGY. *Educ:* Univ Md, Baltimore City, BS, 66, PhD(pharmacol), 72. *Prof Exp:* Instr pharmacol, Sch Nursing, Univ Md, Baltimore City, 67-68; instr anat & physiol, Sch Nursing, St Agnes Hosp, 68-70; instr, Catonsville Community Col, Md, 70-71; asst prof pharmacol, Sch Pharm, Univ Southern Calif, 71-77; ASSOC PROF PHARMACOL, SCH PHARM, UNIV MD, 77- *Concurrent Pos:* Nat Inst Aging grant, 84-85, basic res support grant, 86-88; Nat Cancer Inst grant, 73-76. *Mem:* AAAS; NY Acad Sci; Am Col Clin Pharmacol; Am Pharmaceut Asn; Am Asn Cols Pharm. *Res:* Alteration of drug biotransformation caused by cyclic nucleotides; mechanisms of altered hepatic and extrahepatic drug metabolism (oxidative and conjugative) caused by cancer, diabetes, and aging. *Mailing Add:* 4528 Stonecrest Dr Ellicott City MD 21043

WEINER, NORMAN, b Rochester, NY, July 13, 28; m 55; c 5. PHARMACOLOGY. *Educ:* Univ Mich, BS, 49; Harvard Med Sch, MD, 53. *Prof Exp:* Intern, Harvard Med Serv, Boston City Hosp, 53-54, instr pharmacol, Harvard Med Sch, 56-58, assoc, 58-61, asst prof, 61-67; chmn dept, 67-77, PROF PHARMACOL, MED CTR, UNIV COLO, DENVER, 67- *Concurrent Pos:* Dir neuropharmacol lab, Mass Ment Health Ctr, 64-67. *Mem:* AAAS; Am Soc Pharmacol & Exp Therapeut; Am Soc Neurochem; Am Col Neuropsychopharmacol; Asn Med Sch Pharmacologists. *Res:* Metabolism of biologically active amines; regulation of synthesis of neurotransmitters; synthesis, storage and release of tissue amines; effect of drugs on energy metabolism of brain; ionization of drugs, drug distribution and relation to biological activity. *Mailing Add:* Dept Pharmacol C-236 Univ Colo Health Sci Ctr 4200 E Ninth Ave Denver CO 80262

WEINER, RICHARD, b Brooklyn, NY, Aug 21, 36; m 60; c 2. PHYSIOLOGY. *Educ:* Long Island Univ, BS, 58; NY Univ, PhD(physiol), 65. *Prof Exp:* USPHS fel microcirc, NY Univ Med Ctr, 65, asst res scientist, 66; from instr to asst prof, 66-72, ASSOC PROF PHYSIOL, NEW YORK MED COL, 73- *Mem:* AAAS; Am Physiol Soc; Microcircuitry Soc. *Res:* Regulation of vascular smooth muscle. *Mailing Add:* Dept Pediat Albert Einstein Hosp 1825 Eastchester Rd Bronx NY 10461

WEINER, RICHARD D, b Brooklyn, NY, Nov 25, 45; m 68; c 2. PSYCHIATRY. *Educ:* Mass Inst Technol, BS, 67; Univ Pa, MSE, 69; Duke Univ, PhD(physiol), 73 & MD, 73. *Prof Exp:* Med res assoc psychiat, Durham Vet Admin Med Ctr, 77-79, res assoc, 77-79, asst prof, 79-83, staff physician, 80-90, ASSOC PROF, DUKE UNIV MED CTR, 84- *Concurrent Pos:* Chmn, Task Force on Electroconvulsive Ther, Am Psychiat Asn, 78. *Mem:* Am Psychiat Asn; Soc Biol Psychiat; Am EEG Soc; Col Int Neuro-Psychopharmacol. *Res:* Beneficial and adverse effects of electroconvulsive and on neurophysiologic correlates of mind. *Mailing Add:* Box 3309 Duke Univ Med Ctr Durham NC 27710

WEINER, RICHARD IRA, b New York, NY, Nov 6, 40. NEUROENDOCRINOLOGY. *Educ:* Pa State Univ, BS, 63, MS, 65; Univ Calif, San Francisco, PhD(endocrinol), 69. *Prof Exp:* Fel, Brain Res Inst, Univ Calif, Los Angeles, 69-71; asst prof physiol, Univ Tenn Med Units, 71-72; asst prof anat, Sch Med, Univ Southern Calif, 72-74; assoc prof, 74-80, PROF OBSTET & GYNEC & PHYSIOL, SCH MED, UNIV CALIF, SAN FRANCISC0, 80- *Concurrent Pos:* Mem and chmn biochem endocrinol study sect, NIH; chmn, grad prog endocrinol, Univ Calif, San Francisco. *Mem:* Endocrine Soc; Int Neuroendocrine Soc; Soc Neurosci; Am Physiol Soc. *Res:* Mechanisms of action of hypothalamic hormones in the neuroendocrine regulation of prolactin. *Mailing Add:* Dept Obstet & Gynec Univ Calif Sch Med San Francisco CA 94143

WEINER, ROBERT ALLEN, b New York, NY, Apr 3, 40; m 61; c 2. REACTOR MATERIALS, FUEL PIN MODELLING. *Educ:* Columbia Univ, AB, 61; Harvard Univ, AM, 62, PhD(physics), 67. *Prof Exp:* Asst res physicist, Univ Calif, San Diego, 67-69; asst prof physics, Carnegie-Mellon Univ, 69-75; sr scientist, Advan Reactors Div, 75-84, sr engr, Nuclear Fuels Div, 84-86, PRIN ENGR, NUCLEAR FUELS DIV, WESTINGHOUSE ELEC CO, 86- *Mem:* Am Nuclear Soc; AAAS; Am Phys Soc. *Res:* Irradiation damage mechanisms in metals and alloys; fast breeder reactor fuel; cladding model development; nuclear fuel rod performance. *Mailing Add:* 1433 Denniston Ave Pittsburgh PA 15217

WEINER, RONALD MARTIN, b Brooklyn, NY, May 7, 42; m 64; c 3. MICROBIOLOGY. *Educ:* Brooklyn Col, BS, 64; LI Univ, MS, 67; Iowa State Univ, PhD(microbiol), 70. *Prof Exp:* Bacteriologist, Greenpoint, Coney Island Hosps, NY, 64-66; teaching asst, Iowa State Univ, 67-68, instr, 69-70; asst prof to assoc prof, 70-87, PROF MICROBIOL & CTR MARINE BIOTECH, UNIV MD, COLLEGE PARK, 87- *Honors & Awards:* Fulbright Award, 90. *Mem:* AAAS; Am Soc Microbiol; Am Inst Biol Sci; fel Am Acad Microbiol. *Res:* Morphogenesis of Hyphomonas; mechanisms of procaryote-invertebrate symbiosis: settlement, aquaculture, polysaccharide films; molecular approaches to the study of microbial ecology. *Mailing Add:* Dept Microbiol Univ Md College Park MD 20742

WEINER, STEPHEN DOUGLAS, b Philadelphia, Pa, Jan 1, 41; m 61; c 2. APPLIED PHYSICS. *Educ:* Mass Inst Technol, BS, 61, PhD(physics), 65. *Prof Exp:* Mem staff, 65-73, assoc group leader, 73-, GROUP LEADER, LINCOLN LAB, MASS INST TECHNOL,. *Res:* Missile defense research; reentry physics; electromagnetic scattering and propagation; operations research. *Mailing Add:* Lincoln Lab Mass Inst Technol 244 Wood St Lexington MA 02173

WEINER, STEVEN ALLAN, b New York, NY, June 6, 42; m 71; c 1. PHYSICAL ORGANIC CHEMISTRY. *Educ:* Columbia Univ, AB, 63; Iowa State Univ, PhD(org chem), 67. *Prof Exp:* Res assoc, Calif Inst Technol, 67-68; staff scientist, Ford Motor Co, 68-74, proj mgr, Res Staff, 74-77 prin staff engr, 77-90, MGR PROCESS DEVELOP, FORD MOTOR CO, 90- *Concurrent Pos:* Lectr, Univ Mich-Dearborn, 70-71. *Honors & Awards:* Leibmann Mem Award, Am Chem Soc, 59. *Mem:* Opers Res Soc Am; Am Chem Soc; AAAS; Sigma Xi. *Res:* Photochemistry; free radical kinetics; combustion products and reactions; sodium-sulfur battery; energy storage and conversion; metal casting and heat treating; energy management; process modeling and control; manufacturing process development. *Mailing Add:* Ford Motor Co Mfg Develop Ctr 24500 Glendale Ave Redford MI 48239

WEINFELD, HERBERT, b New York, NY, Feb 7, 21; m 46; c 1. BIOCHEMISTRY. *Educ:* City Col New York, BS, 42; Univ Mich, MS, 48, PhD(biochem), 52. *Prof Exp:* Asst biochem, Univ Mich, 47-49, asst, Sloan-Kettering Inst, 51-54; instr, Inst Indust Med, NY Univ-Bellevue Med Ctr, 54-55; assoc cancer res scientist, 55-65, prin cancer res scientist, Dept Med, Roswell Park Mem Inst, 65-80; EMER RES PROF BIOCHEM, STATE UNIV NY, BUFFALO, 80- *Concurrent Pos:* Res prof biochem and chmn dept, Roswell Park Grad Div, State Univ NY, Buffalo, 71-80. *Mem:* Am Chem Soc; Am Soc Biochem & Molecular Biol. *Res:* Metabolism of nucleosides, nucleotides and nucleic acids; cellular agents controlling mitotic events. *Mailing Add:* 51 Brookville Dr Tonawanda NY 14150-7165

WEINGART, RICHARD, civil engineering, for more information see previous edition

WEINGARTEN, DONALD HENRY, b Boston, Mass, Feb 16, 45; m 71; c 2. THEORETICAL PHYSICS. *Educ:* Columbia Col, AB, 65; Columbia Univ, PhD(physics), 70. *Prof Exp:* Res assoc theoret particle physics, Fermi Nat Accelerator Lab, 69-71; res fel, Niels Bohr Inst, Univ Copenhagen, 71-73; res assoc, Lab Theoret Physics, Univ Paris XI, 73-74; res assoc, Univ Rochester, 74-76; from asst prof to assoc prof, Ind Univ, 76-83, prof physics, 83-84; RES STAFF MEM, T J WATSON RES CTR, IBM, 83- *Mem:* Fel Am Phys Soc. *Res:* Theoretical particle physics; mathematical physics. *Mailing Add:* IBM TJ Watson Res Ctr PO Box 218 Yorktown Heights NY 10598

WEINGARTEN, NORMAN C, b Buffalo, NY, Jan 11, 47; m 69; c 2. FLIGHT SIMULATION, FLYING QUALITIES. *Educ:* State Univ NY, Buffalo, BS, 68. *Prof Exp:* Intern, Calspan Corp, 68, res engr, 69-83, prin engr, 84-90, SECT HEAD, CALSPAN CORP, 90- *Concurrent Pos:* Mem, Am Inst Aeronaut & Astronaut Flight Simulation Tech Comt, 85-88; prog mgr, Calspan Corp, 88- *Mem:* Sr mem, Am Inst Aeronaut & Astronaut; Sigma Xi. *Res:* Flight simulation directed towards development of flight control systems and aircraft flying qualities, stability and control. *Mailing Add:* Calspan Corp PO Box 400 Buffalo NY 14225

WEINGARTEN, VICTOR I, b New York, NY, Jan 18, 31; m 54; c 2. ENGINEERING MECHANICS, CIVIL ENGINEERING. *Educ:* City Col New York, BME, 52; NY Univ, MSME, 54; Univ Calif, Los Angeles, PhD(eng mech), 64. *Prof Exp:* Stress analyst, Northrop Corp, 56-59; mem tech staff eng mech, TRW Systs, Inc, 59-61; mem tech staff eng mech, Aerospace Corp, 61-64; assoc prof, 64-71, chmn dept, 73-82, PROF CIVIL ENG, UNIV SOUTHERN CALIF, 71- *Concurrent Pos:* Consult, Aerojet-Gen Corp, 64-65, Northrop Norair, 64-71 & Hughes Aircraft Co, 67-83. *Mem:* Am Soc Mech Engrs; Am Soc Civil Engrs. *Res:* Buckling of nuclear containment vessels; elastic stability and free vibrations of plates and shells; fluid-structure interaction problem; application of visco-elastic finite element techniques to biomechanics problems. *Mailing Add:* Dept Civil Eng University Park & Univ Southern Calif Los Angeles CA 90089

WEINGARTNER, DAVID PETER, b Escanaba, Mich, Mar 13, 39; m 64; c 2. PHYTOPATHOLOGY, PLANT NEMATOLOGY. *Educ:* Univ Mich, BS, 62; Mich State Univ, PhD(plant path), 69. *Prof Exp:* Teacher high sch, Inkster, Mich, 62-64; asst prof & asst plant pathologist, 69-75, ASSOC PROF & ASSOC PLANT PATHOLOGIST, AGR RES & EDUC CTR, INST FOOD & AGR, UNIV FLA, 75- *Concurrent Pos:* Res award, Fla Fruit & Veg Asn, 75. *Mem:* Soc Nematol; Am Phytopath Soc; Potato Asn Am. *Res:* Nematode and disease control in vegetables; nematode population dynamics; epidemiology; disease forecasting; pest management; potato corky ringspot; early blight; bacterial wilt. *Mailing Add:* Agr Res Ctr & Educ Ctr Univ Fla Hastings FL 32145

WEINGARTNER, HERBERT, b Karlsruhe, Ger, Apr 4, 35; US citizen; m 67; c 2. PSYCHOBIOLOGY, PSYCHOPHARMACOLOGY. *Educ:* City Univ New York, BS, 56; Johns Hopkins Univ, MA, 62, PhD(exp psychol), 66. *Prof Exp:* Teacher math & sci, Bd Educ, New York City, 56-58; clin psychologist & lectr, Sheppard & Enoch Pratt Hosp, 61-63; from instr to asst prof psychol & assoc dir med psychol, Sch Med, Johns Hopkins Univ, 64-70; from assoc prof to prof psychol, Univ Md, 70-77; res psychologist, NIMH, 77-78, actg chief psychol, Lab Psychol & Psychopath, 78-81, chief, Unit on Cofrutive Studies, 82-86; chair, Dept Psychol, George Washington Univ, 87-90; CHIEF, COGNITION SECT, NAT INST AGING, 91- *Concurrent Pos:* Vis scientist psychol, NIMH, 74-75 & 76-77; Gilman fel, Johns Hopkins Univ; NIH res fel; mem, Huntington's Comn, 77-78. *Mem:* Am Psychol Asn; Sigma Xi; Am Psychol Soc. *Res:* Biology of learning and memory; cognitive psychology; mood disturbance; state dependent learning; impairments on memory. *Mailing Add:* 11213 Debra Dr Potomic MD 20854

WEINGARTNER, KARL ERNST, UTILIZATION OF SOY BEANS. *Educ:* Univ Ill, PhD(food sci). *Prof Exp:* SCIENTIST, SOYBEAN UTILIZATION SPECIALIST, GRAIN LEGUME IMPROV PROG, INT INST TROP AGR, 85- *Mailing Add:* 702 Lyndhurst St Dunedin FL 34698

WEINGOLD, ALLAN BYRNE, b New York, NY, Sept 2, 30; m 52; c 4. OBSTETRICS & GYNECOLOGY. *Educ:* Oberlin Col, BA, 51; New York Med Col, MD, 55; Am Bd Obstet & Gynec, dipl, 64. *Prof Exp:* From asst prof to prof obstet & gynec, New York Med Col, 70-73, asst chmn dept, 71-73; PROF OBSTET & GYNEC & CHMN DEPT, SCH MED, GEORGE WASHINGTON UNIV, 73- *Concurrent Pos:* Am Cancer Soc fel gynec malignancy, 60-61; training dir, USPHS grant, 66-68; chief obstet & gynec, Metrop Hosp, New York, 67-70; consult, NIH Cancer Ctr, Walter Reed Army Med Ctr, Columbia Hosp for Women & Fairfax Hosp; mem sub-spec bd, Maternal Fetal Med, 76- *Mem:* Fel Am Col Obstet & Gynec; Am Gynec Soc; Soc Perinatal Obstetricians; Perinatal Res Soc; fel Am Col Surgeons. *Res:* Studies on monitoring of the fetal environment by endocrine, biochemical and biophysical indices. *Mailing Add:* Dept Obstet & Gynec George Washington Univ Sch Med 2150 Penn Ave NW Washington DC 20037

WEINHOLD, ALBERT RAYMOND, b Evans, Colo, Feb 14, 31; m 52; c 2. PLANT PATHOLOGY. *Educ:* Colo State Univ, BS, 53, MS, 55; Univ Calif Davis, PhD(plantpath), 58. *Prof Exp:* Res asst, Univ Calif, Davis, 55-57, from instr to assoc prof, 59-72, from jr plant pathologist to assoc plant pathologist, 59-72, chmn dept, Berkeley, 76-84, actg dean, Col Nat Resources, 84-86, PROF PLANT PATH & PLANT PATHOLOGIST, EXP STA, UNIV CALIF, BERKELEY, 72- *Concurrent Pos:* Sr ed, Phytopath, 70-73, ed-in-chief, 73-75. *Mem:* AAAS; fel Am Phytopathological Soc (pres 87-88). *Res:* Disease and pathogen physiology; soil-borne pathogens; root diseases; potato diseases. *Mailing Add:* Dept Plant Path Univ Calif Berkeley CA 94720

WEINHOLD, PAUL ALLEN, b Evans, Colo, Sept 23, 35; m 56; c 4. BIOCHEMISTRY. *Educ:* Colo State Univ, BS, 57; Univ Wis, PhD(biochem), 61. *Prof Exp:* Asst prof to assoc prof biochem, Med Sch Univ, Mich, Ann Arbor, 72-84; biochemist, 65-77, SUPV RES BIOCHEMIST, VET ADMIN HOSP, 77-; PROF BIOCHEM, MED SCH, UNIV MICH, ANN ARBOR, 84- *Concurrent Pos:* USPHS res fel biochem, Harvard Med Sch, 63-65; res career scientist, Vet Affairs, Med Res. *Mem:* AAAS; Am Soc Biol Chemists. *Res:* Biochemistry of development, phospholipid metabolism and control mechanisms in metabolism. *Mailing Add:* Dept Biochem Univ Mich Ann Arbor MI 48109

WEINHOUSE, SIDNEY, b Chicago, Ill, May 21, 09; c 3. BIOCHEMISTRY, CANCER. *Educ:* Univ Chicago, BS, 33, PhD(chem), 36; Dr Med & Surg, Unoiv Chieti, Italy, 79. *Hon Degrees:* DSc, Med Col Pa, 72, Temple Univ, 76, Jefferson Med Col, 85. *Prof Exp:* Coman fel org chem, Univ Chicago, 41-44; head biochem res, Houdry Process Corp, Pa, 44-47; biochem res dir, Res Inst, 47-50, prof chem & biol, chmn div biochem, Inst Cancer Res & head dept metab chem, Lankenau Hosp Res Inst, 50-57, dir Fels Res Inst, 63-74, prof, 57-79, emer prof biochem, Sch Med, Temple Univ, 80-87; SR SCIENTIST, LANKENAU MED RES CTR, 87- *Concurrent Pos:* Mem biochem study sect, NIH, 53-58 & nat adv coun, Nat Cancer Inst, Dept HEW, 58-62; assoc dir, Fels Res Inst, 61-63; chmn comt biol chem, Div Chem & Chem Technol, Nat Acad Sci-Nat Res Coun, 62-64; mem sci adv comt, Damon Runyon Mem Fund, 52-60 & Environ Health Sci Adv Comt, 68-72. *Honors & Awards:* Philadelphia Sect Award, Am Chem Soc, 66; G H A Clowes Award, Am Asn Cancer Res, 72; Papanicolaou Award, 76; Lucy Wortham James Award, Soc Surg Oncol, 79; Nat Award, Am Cancer Soc; Achievement Award, Am Cancer Soc, 87. *Mem:* Nat Acad Sci; Am Soc Biol Chemists; Am Soc Biol Chem & Molecular Biol; Soc Exp Biol & Med; Am Chem Soc; Am Cancer Soc; fel NY Acad Sci; hon mem Japanese Asn Cancer Res. *Res:* Carbohydrate and fatty acid metabolism in normal and neoplastic cells; dietary and hormonal regulation of enzymes of carbohydrate and fatty acid metabolism in liver tumors; effects of hormones on gluconeogenesis; comparative studies on control of respiration and glycolysis in liver and liver tumors. *Mailing Add:* Lankenau Med Res Ctr Philadelphia PA 19151

WEINIG, SHELDON, b New York, NY, Jan 15, 28; div; c 3. METALLURGY. *Educ:* NY Univ, BME, 51; Columbia Univ, MS, 53, DEngSc, 55. *Hon Degrees:* LLD, St Thomas Aquinas Col. *Prof Exp:* PRES, CHMN & CHIEF EXEC OFFICER, MAT RES CORP, 57- *Concurrent Pos:* Mem adv comt, US Dept Com, 76-78; dir, Semiconductor Equip & Mat Inst, 76-; mem adv comt, Polytech Univ NY, 77-; mem, US-Japan Coop Sci Comt. *Honors & Awards:* Semmy Award, Semiconductor Equipment & Mat Inst, 80. *Mem:* Nat Acad Sci; AAAS; Am Inst Mining, Metall & Petrol Engrs; Am Soc Metals; Am Phys Soc. *Res:* Electronic materials. *Mailing Add:* Mat Res Corp Rte 303 Orangeburg NY 10962

WEININGER, STEPHEN JOEL, b New York, NY, Mar 28, 37; m 61; c 3. ORGANIC CHEMISTRY. *Educ:* Brooklyn Col, BA, 57; Univ Pa, PhD(org chem), 64. *Prof Exp:* Sr demonstr phys chem, Univ Durham, 64-65; from asst prof to assoc prof, 65-77, PROF CHEM, WORCESTER POLYTECH INST, 77- *Concurrent Pos:* Consult, Natick Res & Develop Command, US Army; vis prof, Colorado State Univ, 76-78; Danforth assoc, 79-85; Mellon fel, Prog Sci Tech & Sci, Mass Inst Technol. *Mem:* Am Chem Soc; Am Asn Univ Professors. *Res:* chemistry of hot intermediates; organic synthesis; polymer chemistry. *Mailing Add:* Dept Chem Worcester Polytech Inst Worcester MA 01609-2280

WEINKAM, ROBERT JOSEPH, b Cincinnati, Ohio, Dec 27, 42; c 4. MEDICINAL CHEMISTRY. *Educ:* Xavier Univ, Ohio, BS, 64; Duquesne Univ, PhD(chem), 68. *Prof Exp:* Fel, Sya Res Inst Calif, 68-69; fel, Sch Med, Univ Calif, San Francisco, 69-70, asst prof, Pharmaceut Chem, 71-80, asst prof neurosurg, 78-80; ASSOC PROF MED CHEM, PURDUE UNIV, WEST LAFAYETTE, 80- *Concurrent Pos:* Res career develop award, NIH, 75-80. *Mem:* Am Chem Soc; Am Soc Mass Spectros; Asn Res Vision & Ophthal. *Res:* Biomedical application of mass spectrometry; drug metabolism; ophthalmic drugs; prodrugs. *Mailing Add:* Allargan Pharmaceut Inc 2525 Dupont Dr Irvine CA 92715-1599

WEINMANN, CLARENCE JACOB, b Oakland, Calif, May 27, 25; m 57; c 2. PARASITOLOGY. *Educ:* Univ Calif, BS, 50, PhD(parasitol), 58. *Prof Exp:* Instr microbiol, Col Med, Univ Fla, 58-60; res fel biol, Rice Univ, 60-62; from asst prof to assoc prof, 62-75, PROF ENTOM & PARASITOL, UNIV CALIF, BERKELEY, 75- *Concurrent Pos:* Fel trop med & parasitol, Univ Cent Am, 59. *Mem:* AAAS; Am Soc Parasitol; Entom Soc Am; Am Soc Trop Med & Hyg; Wildlife Dis Asn. *Res:* Immunity in helminth infections; arthropod-borne helminthiases. *Mailing Add:* Div Entom Univ Calif 2120 Oxford St Berkeley CA 94720

WEINREB, EVA LURIE, b New York, NY; m 50. BIOLOGICAL STRUCTURE, CELL BIOLOGY. *Educ:* NY Univ, BA, 48; Univ Wis, MA, 49, PhD(zool), 55. *Prof Exp:* Asst zool, Univ Wis, 49 & 51-55; res assoc path, Sch Med, Marquette Univ, 55-56; dir basic sci, St Mary's Hosp, Sch Nursing, Milwaukee, 56-57; head animal res, Kolmer Res Ctr, 57-58; asst prof biol, Milwaukee-Downer Col, 59-60; fel cell biol & res assoc anat, Med Col, Cornell Univ, 61-63; asst prof biol, Washington Square Col, NY Univ, 63-69; cell biologist & sr res scientist, Biomed Sect, Geometric Data Corp, 71-72; assoc prof, 73-80, PROF BIOL, COMMUNITY COL PHILADELPHIA, 80- *Concurrent Pos:* Consult, Volu-Sol Med Industs, Inc, 72-73, Trial Advocacy Found Pa. *Mem:* AAAS; Am Asn Anatomists; Am Soc Cell Biol; Am Soc Zool; Am Soc Allied Health Prof. *Res:* Comparative histology; hematology; pathology; electron microscopy; anatomy. *Mailing Add:* Dept Biol Community Col Philadelphia 1700 Spring Garden St Philadelphia PA 19130

WEINREB, MICHAEL PHILIP, b Lakewood, NJ, Feb 2, 39; m 66; c 2. ATMOSPHERIC PHYSICS, REMOTE SENSING. *Educ:* Univ Pa, BA, 60; Brandeis Univ, PhD(physics), 66. *Prof Exp:* Instr physics, Brandeis Univ, 64-65; physicist, Electronics Res Ctr, NASA, Mass, 65-70; PHYSICIST,

NAT ENVIRON SATELLITE, DATA & INFO SERV, NAT OCEANIC & ATMOSPHERIC ADMIN, 70- *Concurrent Pos:* Adj prof math, Am Univ, 84-85. *Mem:* Am Meteorol Soc; Optical Soc Am; Am Geophys Union. *Res:* Radiative transfer theory; atomic and molecular physics; remote sensing of atmospheric temperature and constituent profiles. *Mailing Add:* Nat Environ Satellite, Data & Info Serv Nat Oceanic & Atmospheric Admin Washington DC 20233

WEINREB, ROBERT NEAL, b New York, NY, Nov 23, 49. MEDICAL SCIENCE, SURGERY. *Educ:* Mass Inst Technol, BS, 71; Harvard Med Sch, MD, 75. *Prof Exp:* PROF & VCHMN, DEPT OPHTHAL, UNIV CALIF, SAN DIEGO, 84- *Honors & Awards:* Alcon Award, Alcon Res Inst, 83; Hon Award, Am Acad Ophthal. *Mem:* Asn Res Vision & Ophthal; Am Acad Ophthal; Am Glaucoma Soc; Int Soc Eye Res; Found Eye Res. *Res:* Cell biology of the trabecular meskwork in normal, aging and glaucomatous eyes; imaging of the optic disk and nerve fiber layer; psychophysical investigations in glaucoma; biochemistry of aqueous humor. *Mailing Add:* Ophthal T-014 Univ Calif San Diego La Jolla CA 92093

WEINREB, SANDER, b New York, NY, Dec 9, 36; m 57; c 2. RADIO ASTRONOMY, MICROWAVE ENGINEERING. *Educ:* Mass Inst Technol, BSEE, 58, PhD(elec eng), 63. *Prof Exp:* Staff mem, Lincoln Lab, Mass Inst Technol, 63-65; head electronics div, Nat Radio Astron Observ, 65-77, scientist, 77-84, asst dir electronics, 85-88; MGR, MILLIMETER WAVE DESIGN & TEST, MARTIN MARIETTA LABS, 88- *Concurrent Pos:* Foreign adv, Neth Found Radio Astron, 72-77; adv, Nat Astron & Ionospheric Observ, 72-78. *Mem:* Sigma Xi; fel Inst Elec & Electronics Engrs. *Res:* Low noise receivers; millimeter wave devices. *Mailing Add:* Martin Marietta Labs 1450 S Rolling Rd Baltimore MD 21227

WEINREB, STEVEN MARTIN, b Brooklyn, NY, May 10, 41; m 65; c 2. ORGANIC CHEMISTRY. *Educ:* Cornell Univ, AB, 63; Univ Rochester, PhD(chem), 67. *Prof Exp:* NIH fel, Columbia Univ, 66-67; NIH fel, Mass Inst Technol, 67-68, res assoc, 68-70; from asst prof to assoc prof chem, Fordham Univ, 70-78; from assoc prof to prof, 78-87, MARKER PROF CHEM, PA STATE UNIV, 80- *Concurrent Pos:* Res fel, Alfred P Sloan Found, 75-79; res career develop award, NIH, 75-80; Guggenheim fel, 83-84. *Mem:* Am Chem Soc; Royal Soc Chem. *Res:* Synthesis of natural products; heterocyclic chemistry. *Mailing Add:* Dept Chem Pa State Univ University Park PA 16802

WEINREICH, DANIEL, b Claremont, France, June 6, 42; US citizen; m 36; c 2. NEUROBIOLOGY. *Educ:* Bethany Col, BS, 64; Univ Utah, PhD(pharmacol), 70. *Prof Exp:* Res asst neuropharmacol, Sandoz Pharmaceut, Inc, 64-65; NSF fel, City of Hope Nat Med Ctr, 70-72, NIMH fel, 73-74; asst prof, 74-77, ASSOC PROF PHARMACOL, SCH MED, UNIV MD, BALTIMORE, 74- *Concurrent Pos:* Extramural reviewer, NSF, 73-, Neurobiol Study Sect, 80-81; NSF res grant, 74-83. *Mem:* Neurosci Soc. *Res:* Identification and regulation of neurotransmitter and neuromodulator substances in single nerve cells; cellular autonomic neuropharmacology. *Mailing Add:* Dept Pharmacol & Exp Therapeut Univ Maryland Sch Med 660 W Redwood St Baltimore MD 21201

WEINREICH, GABRIEL, b Vilna, Poland, Feb 12, 28; nat US; m 51, 71; c 5. PHYSICS. *Educ:* Columbia Univ, AB, 48, MA, 49, PhD(physics), 54. *Prof Exp:* Asst physics, Columbia Univ, 49-51; mem tech staff, Bell Tel Labs, 53-60; vis assoc prof, 60, assoc prof, 60-64, col prof, 74-76, PROF PHYSICS, UNIV MICH, ANN ARBOR, 64- *Concurrent Pos:* Assoc ed, J Acoust Soc Am, 87-90. *Mem:* Fel Acoust Soc Am. *Res:* Atomic spectra; solid state theory; electron-phonon interactions; nonlinear optics; thermodynamics; electron-atom scattering; musical acoustics; atomic beam kinetics. *Mailing Add:* Randall Lab Physics Univ Mich Ann Arbor MI 48109-1120

WEINRICH, ALAN JEFFREY, b Passaic, NJ, Aug 24, 53; m 83. INDUSTRIAL HYGIENE, OCCUPATIONAL HEALTH & SAFETY. *Educ:* Rutgers Univ, BS, 75; Univ Iowa, MS, 88. *Prof Exp:* Indust hygienist, Div Occup Health, Tenn State, 75-78; campus ministry assoc, St Mark Lutheran Church, Charlottesville, Va, 78-80; indust hygienist, Col Pharm, Univ Tenn, 80-82; occup health consult, Memphis, Tenn, 81-82; vol teacher var subj, Int Sch, Moshi, Tanzania, 82-84; sr res asst indust hyg, Dept Prev Med, Univ Iowa, 85-89; SR INDUST HYGIENIST, PSI ENERGY, 89- *Mem:* Am Indust Hyg Asn; Am Conf Govt Indust Hygenists; Am Pub Health Asn. *Res:* Evaluating current and developing new environmental sampling methods for measuring air contaminants in agricultural workplaces; evaluating the relationship between levels of contaminants and the health status of workers; developing educational, procedural and engineering methods for minimizing worker exposure to the most important contaminants. *Mailing Add:* PSI Energy 1000 E Main St Plainfield IN 46168

WEINRICH, JAMES D, b Cleveland, Ohio, July 2, 50. HUMAN REPRODUCTIVE STRATEGY, GENDER TRANSPOSITIONS. *Educ:* Princeton Univ, AB, 72; Harvard Univ, PhD(biol), 76. *Prof Exp:* Teaching asst evolution, Harvard Univ, 76-77, jr fel, Soc Fels, 77-80; NIH training grant fel med psychol, Psychohormonal Res Unit, Sch Med, Johns Hopkins Univ, 80-81, instr psychiat, 81-83; res fel, Sch Med, Boston Univ, 82-83, asst res prof psychiat, family studies, 83-87, assoc dir, Family Studies Lab, 85-87; ASST RES PSYCHOBIOLOGIST, SCH MED, UNIV CALIF, SAN DIEGO, 87-; MGR & DATA MGR, HUMAN NUTRIT RES CTR, 89- *Concurrent Pos:* Bd mem, Found for Sci Study Sexuality, 87- *Honors & Awards:* Hugo Beigel Award, Soc for Sci Study of Sex, 87. *Mem:* Soc Sci Study Sex; Sigma Xi. *Res:* The development, evolution and expression of human sex roles and gender transpositions in health and disease, expecially AIDS. *Mailing Add:* HIV Neurolbehav Res Ctr 2760 Fifth Ave No 200 San Diego CA 92103

WEINRICH, MARCEL, b Jendiesow, Poland, July 23, 27; nat US; div; c 1. PHYSICS. *Educ:* Bethany Col, WVa, BS, 46; Univ WVa, MS, 48; Columbia Univ, PhD, 57. *Prof Exp:* Instr physics, Univ WVa, 47-49; physicist, Res & Develop Ctr, Gen Elec Co, 57-69; chmn dept, 69-73, PROF PHYSICS, JERSEY CITY STATE COL, 69- *Concurrent Pos:* Coordr gen studies review, Jersey City State Col, co-chmn, Pres Task Force for the 1980's. *Mem:* Fel AAAS; Am Inst Physics; Am Asn Physics Teachers (pres, 72-74); Am Phys Soc; NY Acad Sci. *Res:* Meson and plasma physics; breakdown of parity conservation in meson decays; controlled fusion reactors. *Mailing Add:* Dept Physics Jersey City State Col 2039 Kennedy Blvd Jersey City NJ 07305

WEINRYB, IRA, b New York, NY, Nov 20, 40; m 67; c 2. BIOCHEMISTRY, PHARMACOLOGY. *Educ:* Columbia Univ, BS, 61; Yale Univ, MEng, 62, MS, 65, PhD(molecular biophys), 67. *Prof Exp:* Nat Acad Sci-Nat Res Coun resident res assoc, Lab Phys Biochem, Naval Med Res Inst, Nat Naval Med Ctr, 67-69; from res investr to sr res investr, Dept Biochem Pharmacol, Squibb Inst Med Res, 69-73, head biochem sect, 73-75, res fel, Dept Pharmacol, 75; dir biochem & drug disposition, USV Pharmaceut Corp, 75-81; assoc div dir biol res, Revion Health Care, 81-84, div dir chem res & develop, 84-89. *Concurrent Pos:* Chmn, Biochem Pharmacol Discussion Group. *Mem:* Am Soc Biol Chem; fel NY Acad Sci; Am Chem Soc; Int Soc Study Xenobiotics. *Res:* Drug discovery and development, drug design, biochemical pharmacology, immunopharmacology, molecular bases of disease; spectroscopy of biological molecules; biochemical pharmacology; mechanism of drug action. *Mailing Add:* 413 Gwynedd Valley Dr PO Box 131 Gwynedd Valley PA 19437-0131

WEINSHANK, DONALD JEROME, b Chicago, Ill, Apr 29, 37; m 59; c 2. HISTORY & PHILOSOPHY OF SCIENCE. *Educ:* Northwestern Univ, BA, 58; Univ Wis-Madison, MS, 61, PhD(biochem), 69. *Prof Exp:* From instr to prof natural sci, 67-81, prof computer sci & natural sci, 81-87, PROF COMPUTER SCI, UNIV COL, MICH STATE UNIV, 87- *Mem:* Asn Comput Mach; Am Chem Soc. *Res:* Computer assisted instruction; computer science education; concordances to works of Charles Darwin. *Mailing Add:* Computer Sci Dept A-732 Wells Hall Mich State Univ East Lansing MI 48824

WEINSHENKER, NED MARTIN, b Brooklyn, NY, Oct 13, 42; m 66; c 2. ORGANIC CHEMISTRY. *Educ:* Polytech Inst Brooklyn, BSc, 64; Mass Inst Technol, PhD(org chem), 69. *Prof Exp:* NIH fel, Harvard Univ, 68-69; asst prof chem, Univ Md, College Park, 69-70; dir phys sci & prin scientist, Alza Corp, 70-72; VPRES RES, DYNAPOL, 72- *Concurrent Pos:* Lectr, Stanford Univ, 71; guest lectr, Dept Human Biol, Stanford Univ, 79, 81. *Mem:* Am Chem Soc; The Chem Soc; NY Acad Sci. *Res:* Synthetic organic chemistry; new synthetic methods and physiologically active compounds; polymeric reagents and membrane structure; creation of new safe non-toxic food additives; new drug delivery systems. *Mailing Add:* Churchill Oaks Consult 541 Churchill Ave Palo Alto CA 94301-8015

WEINSHILBOUM, RICHARD MERLE, b Eldorado, Kans, Mar 31, 40; m 65; c 2. PHARMACOLOGY, INTERNAL MEDICINE. *Educ:* Univ Kans, BA, 62, MD, 67. *Prof Exp:* Intern internal med, Mass Gen Hosp, 67-68, asst resident, 68-69; res assoc pharmacol, Lab Clin Sci, NIMH, 69-71; sr resident internal med, Mass Gen Hosp, 71-72; consult asst prof pharmacol & internal med, 72-75, chief, clin pharmacol unit, 74-90, assoc prof pharmacol, 76-79, internal med, 77-79, prof pharmacol & internal med & William L McKnight-3M prof neurosci, 79-84, dir res, 84-88, CHMN, DEPT PHARMACOL, MAYO FOUND, 89- *Concurrent Pos:* Consult, Mayo Clin, 72-; Pharmaceut Mfg Asn Fedn fac develop award, 73-75; estab investr, Am Heart Asn, 76-81; Burroughs-Wellcome scholar clin pharmacol, 81-86. *Honors & Awards:* Rawls-Palmer Award, Am Soc Clin Pharmacol Therapeut, 79. *Mem:* Am Soc Pharmacol & Exp Therapeut; Am Soc Clin Pharmacol & Therapeut; Am Fedn Clin Res; Am Soc Neurochem; Soc Neurosci. *Res:* Neurochemistry, biogenic amines; pharmacogenetics; biochemical genetics and clinical pharmacology. *Mailing Add:* Dept Pharmacol 200 First St SW Rochester MN 55905

WEINSIER, ROLAND LOUIS, CLINICAL NUTRITION. *Educ:* Harvard Univ, MD & PhD(pub health), 73. *Prof Exp:* PROF NUTRIT & DIR CLIN DIV, MED CTR, UNIV ALA, BIRMINGHAM, 75- *Mailing Add:* Dept Nutrit Sci Univ Ala Birmingham Med Ctr UAB Sta PO Box 188 Birmingham AL 35294

WEINSTEIN, ABBOTT SAMSON, biostatistics; deceased, see previous edition for last biography

WEINSTEIN, ALAN DAVID, b New York, NY, June 17, 43; m 67; c 1. GEOMETRY. *Educ:* Mass Inst Technol, BS, 64; Univ Calif, Berkeley, MA, 66, PhD(math), 67. *Prof Exp:* Vis fel, Inst Advan Sci Studies, France, 67; C L E Moore instr math, Mass Inst Technol, 67-68; NATO fel, Math Inst, Univ Bonn, 68-69; from asst prof to assoc prof, 69-76, PROF MATH, UNIV CALIF, BERKELEY, 76- *Concurrent Pos:* Sloan fel, 71-73; vis prof, Rice Univ, 78-79. *Mem:* Am Math Soc; Math Asn Am. *Res:* Symplectic manifolds; fourier integral operators; Hamiltonian dynamical systems; riemannian geometry. *Mailing Add:* Dept Math Univ Calif 2120 Oxford St Berkeley CA 94720

WEINSTEIN, ALAN IRA, b New York, NY, Apr 7, 40; m 66; c 2. METEOROLOGY, OCEANOGRAPHY. *Educ:* City Col New York, BS, 61; Pa State Univ, MS, 63, PhD(meteorol), 68. *Prof Exp:* Res meteorologist, Meteorol Res Inc, 63-66; res asst meteorol, Pa State Univ, 66-68; res scientist, Meteorol Res Inc, Cohu Electronics Inc, 69-71; res physicist, 71-74, br chief stratiform cloud physics, Air Force Geophys Labs, 74-77; dir res, Naval Environ Prediction Res Facil, 77-84, dir, Ocean Sci Div, 84-88, DIR OCEAN & ATMOSPHERIC PHYSICS DIV, OFF NAVAL RES, 88- *Mem:* Am Meteorol Soc; Royal Meteorol Soc; Sigma Xi; AAAS; Am Geophys Union; Oceanog Soc. *Res:* Marine meteorology; atmospheric physics; physical oceanography. *Mailing Add:* Off Naval Res (Code 1122) Arlington VA 22217

WEINSTEIN, ALAN JAY, b Brooklyn, NY, Dec 22, 57; m 91. COLLIDER DETECTOR DESIGN. *Educ:* Harvard Univ, BA, 78, PhD(physics), 83. *Prof Exp:* Asst res physicist high energy physics, dept physics, Harvard Univ, 83-84; asst res physicist high energy physics, Univ Calif, Santa Cruz, 84-88;

ASST PROF PHYSICS, CALIF INST TECHNOL, 88- *Concurrent Pos:* SSC Fel, Superconducting Supercollider Lab, 91- *Mem:* Am Phys Soc. *Res:* High energy e plus e minus collisions; decays of clepton, charmed and bottom mesons; design of high energy physics collider; tracking detectors. *Mailing Add:* Caltech 256-48 Pasadena CA 91125

WEINSTEIN, ALEXANDER, b Astrakhan, Russia, Nov 22, 93; nat US; m 17; c 1. GENETICS, HISTORY OF SCIENCE. *Educ:* Columbia Univ, BS, 13, AM, 14, PhD(zool), 17. *Prof Exp:* Asst zool, Columbia Univ, 16-17, Sigma Xi res fel, 21-22, lectr, 28; res asst genetics, Carnegie Inst, 17-19; mem ed staff, Am Men Sci, 20-21; Johnston scholar, Johns Hopkins Univ, 22-23, Nat Res Coun fel, 23-24, assoc zool & hist sci, 30-36; Int Ed Bd fel, Cambridge Univ, 24-25; assoc zool, Univ Ill, 26-27; prof genetics, Univ Minn, 28-29; researcher, Am Philos Soc grant, Columbia Univ, 37-42; instr physics, City Col New York, 42-49; res fel biol, 51-54, RESEARCHER, COMMONWEALTH FUND GRANT, HARVARD UNIV, 51- *Mem:* AAAS; Am Soc Nat; Am Soc Zool; Genetics Soc Am; Hist Sci Soc. *Res:* Crossing over and multiple-strand theory; radiation genetics; human genetics; heredity and development; comparative genetics of Drosophila; history of biology and physics; ancient and modern science; environmental and genetic factors in the history of science. *Mailing Add:* 41 Linnaen St Cambridge MA 02138

WEINSTEIN, ALVIN SEYMOUR, b Lynn, Mass, June 12, 28; m 52; c 3. MECHANICAL ENGINEERING & LAW, PREVENTIVE LAWYERING FOR CORPORATIONS IN RISK REDUCTION. *Educ:* Univ Mich, BS, 51; Carnegie Inst Technol, MS, 53, PhD(mech eng), 55; Franklin Pierce Law Ctr, JD, 83. *Prof Exp:* prof mech eng & pub policy, Carnegie Mellon Univ, 55-85; adj prof law, Franklin Pierce Law Ctr & legal-tech consult, 85-90; PARTNER, WEINSTEIN, BOHAN & ASSOC, 90- *Concurrent Pos:* Res award, Am Soc Test & Mat, 65; pres, TEC Consults. *Honors & Awards:* Melville Medal, Am Soc Mech Engrs, 72; Western Elec Teaching Award, Am Soc Eng Educ, 73. *Mem:* Am Soc Mech Engrs; Am Soc Testing & Mat; Asn Iron & Steel Engrs. *Res:* Materials forming; technical aspects of products liability litigation; methodology for developing product safety standards; product safety analysis. *Mailing Add:* Weinstein Bohan & Assoc 371 Fore St Portland ME 04101

WEINSTEIN, ARTHUR, b 1944. RHEUMATOID ARTHRITIS, SYSTEMIC LUPUS ERYTHEMATHOSUS. *Educ:* Univ Toronto, MD, 67. *Prof Exp:* PROF MED & HEAD, DIV RHEUMATIC DIS & IMMUNOL, NY MED COL, 85- *Concurrent Pos:* Chief rheumatology, Westchester Med Ctr; dir, Diag Immunol Lab, NY Med Col, Rheumatology Training Prog; mem, educ coun, Am Col Rheumatology, 90- *Mem:* Fel Can Arthritis Soc; fel Am Col Physicians; fel Am Col Rheumatology; Am Asn Immunologists; Am Asn Path; Am Fedn Clin Res; AAAS. *Res:* Therapeutic studies in rheumatoid arthritis; laboratory studies in SLE and Lyme disease. *Mailing Add:* Div Rheumatic Dis & Immunol New York Med Col Valhalla NY 10595

WEINSTEIN, ARTHUR HOWARD, b Brooklyn, NY, Jan 20, 24; m 46; c 2. ORGANIC POLYMER CHEMISTRY, RUBBER CHEMISTRY. *Educ:* Queens Col, NY, BS, 44; Ohio State Univ, MS, 48, PhD(org chem), 50. *Prof Exp:* Asst path chemist, Bellevue Hosp, New York, 46; anal chemist, Dept Purchase, New York, 50; sr res chemist, chem res & develop div, Goodyear Tire & Rubber Co, 51-87; RETIRED. *Concurrent Pos:* chmn prof activ comt, Akron Div, Am Chem Soc, 52. *Mem:* Emer mem Am Chem Soc. *Res:* Aminocellulose derivatives; synthesis of aromatic sulfur compounds; emulsion polydienes with terminal or internal functionality; polymerization modifiers; castable pre-elastomers; polydiene elastomers self-resistant to oxidation or to combustion; chemically-resistant saturated nitrilated elastomers; urethane-cured polydienes. *Mailing Add:* 2400 Cambridge Dr Hudson OH 44236

WEINSTEIN, BERNARD ALLEN, b Bridgeport, Conn, Nov 15, 46; m 70; c 2. SOLID STATE PHYSICS. *Educ:* Univ Rochester, BS, 68; Brown Univ, PhD(physics), 74. *Prof Exp:* Max Planck Inst Solid State Res fel, Stuttgart, 71-73; res assoc solid state mat, Nat Bur Standards, 73-75; asst prof physics, Purdue Univ, West Lafayette, 75-78; res scientist, Xerox Corp, 78-87; PROF, DEPT PHYSICS & ASTRON, STATE UNIV NY, BUFFALO, 87- *Concurrent Pos:* Alfred P Sloan res fel, 75-77. *Mem:* Am Phys Soc; Fedn Am Scientists. *Res:* Optical properties of semiconductors, with specific work in Raman scattering, visible and infrared spectroscopy and ultra-high pressure research. *Mailing Add:* Dept Physics State Univ NY Fronczak Hall Buffalo NY 14260

WEINSTEIN, BERTHOLD WERNER, b New York, NY, Oct 11, 47; m 70; c 2. GENERAL PHYSICS, MATHEMATICS. *Educ:* Brigham Young Univ, BS, 71; Univ Ill, Urbana, MS, 73, PhD(physics), 75. *Prof Exp:* RES PHYSICIST, GROUP & DIV LEADER, LAWRENCE LIVERMORE NAT LAB, UNIV CALIF, 74- *Mem:* Am Phys Soc; AAAS. *Res:* Inertial fusion target fabrication and measurement; liquid metal ion beam sources; computer simulation and analysis of high energy physics experiments; electrohydrodynamics, particularly ion spraying; high energy physics; nuclear weapons. *Mailing Add:* Lawrence Livermore Nat Lab Div L-377 PO Box 808 Livermore CA 94550

WEINSTEIN, CONSTANCE DE COURCY, b London, Eng, Aug 31, 24; US citizen; m 59; c 3. BIOCHEMISTRY. *Educ:* Univ London, BSc, 48, PhD(biochem), 53. *Prof Exp:* Res biochemist, Hosp Sick Children, London, 53-54; res fel biochem, Jefferson Med Col, Philadelphia, 55-56; sr instr biochem, Med Sch, Western Reserve Univ, Cleveland, 56-61; res scientist cancer, City of Hope, Duarte, Calif, 70-74; health sci adminr, NIH, 75-81, exec secy, div res grants, 81-86, dep chief cardiac dis br, 86-89, CHIEF CARDIAC DIS BR, NAT HEART, LUNG & BLOOD INST, NIH, 89- *Mem:* Am Heart Asn; Int Soc Heart Res. *Mailing Add:* Cardiac Dis Br Nat Heart Lung & Blood Inst Fed Bldg Rm 3C06 7550 Wisconsin Ave Bethesda MD 20298

WEINSTEIN, CURT DAVID, b New York, NY, May 11, 51. COMMUNICATIONS EFFECTIVENESS, MEASUREMENT SCIENCE. *Educ:* Clarkson Col Technol, BA, 73; Univ Mich, MA, 75. *Prof Exp:* Res assoc, Neuro Commun Res Labs Inc, 75-78, vpres, 78-84; dir, Adam Info, 85-87; PRES, PHARMACEUT & COSMETIC EVAL, NEURO COMMUN RES LABS INC, 87- *Concurrent Pos:* Proprieter, Weinstein's Market Res & Analysis, 84- *Res:* Perception, communications-effects, cosmetic-skin, measurement science, biodetection, electrophysiological applications of neuropsychology. *Mailing Add:* 39B Mill Plain Rd Apt 208 Danbury CT 06811

WEINSTEIN, DAVID, bacteriology, cytogenetics, for more information see previous edition

WEINSTEIN, DAVID ALAN, b New York City, NY, Mar 26, 51. NUTRIENT CYCLING, COMPUTER MODELING. *Educ:* Darmouth Col, BA, 73; Univ NH, MS, 76; Univ Tenn, PhD(ecol), 82. *Prof Exp:* Grad res fel, Environ Sci Div, Oak Ridge Nat Lab, 77-81; RES ASSOC, ECOSYST RES CTR, ENVIRON PROTECTION AGENCY CTR EXCELLENCE, CORNELL UNIV, 81- *Mem:* Ecol Soc Am; Sigma Xi. *Res:* Mechanisms of response of ecosystems to perturbations through the analysis of nutrient cycling patterns by computer simulation techniques. *Mailing Add:* 51 Freese Rd Ithaca NY 14850

WEINSTEIN, HAREL, b June 5, 45; US citizen; m 67; c 1. BIOPHYSICAL CHEMISTRY. *Educ:* Israel Inst Technol, BSc, 66, MSc, 68, DSc(quantum chem), 71. *Prof Exp:* Asst chem, Israel Inst Technol, 66-68, sr res asst, 68-71, lectr, 71-73; res assoc, Johns Hopkins Univ, 73-74; from asst prof to assoc prof, 74-79, PROF PHARMACOL, MT SINAI SCH MED, 79-, PROF PHYSIOL & BIOPHYS & CHMN DEPT, 85- *Concurrent Pos:* Consult res assoc biochem, Tel-Aviv Univ, 73-78; vis scientist genetics, Med Ctr, Stanford Univ, 74-75; mem ed & adv bd, Molecular Pharmacol, 77-; Irma T Hirschl Trust res career scientist, 78-82; Alcohol, Drug Abuse & Ment Health Admin res grant, 79-83 & 84-; consult, Merck Sharp & Dohme, 79-81; consult, US Environ Protection Agy, 83-; NSF res grants, 83-86; bd sci adv, NIH-NCI; adv panel, NIH-DRR, 85-; coop agreement grant, US Environ Protection Agency, 84- *Honors & Awards:* Int Soc Quantum Biol Award, 87. *Mem:* Biophys Soc; Am Soc Pharmacol Exp Therapeut; Am Chem Soc; Int Soc Quantum Biol (pres, 85-). *Res:* Development and application of theoretical methods to study molecular structure, reactivity and interactions of biological systems; molecular recognition in hormone and drug action and energy transfer and storage in proteins; receptor theory and molecular mechanisms of neurotransmission and enzyme activity; structure and function of calcium binding protein; protein DNA interactions. *Mailing Add:* Dept Physiol & Biophys Sch Med Fifth Ave at 100th St Box 1218 New York NY 10029

WEINSTEIN, HERBERT, b Brooklyn, NY, Mar 10, 33; m 57; c 3. FLUIDIZATION, CHEMICAL REACTOR ENGINEERING. *Educ:* City Col New York, BEChE, 55; Purdue Univ, MSChE, 57; Case Inst Technol, PhD(eng), 63. *Prof Exp:* Staff mem, Los Alamos Sci Lab, 56-58; res engr, Lewis Lab, NASA, 59-63; asst prof chem eng, Ill Inst Technol, 63-66, assoc prof, 66-72, prof, 72-77; prof, 77-87, HERBERT G KAYSER PROF CHEM ENG, CITY UNIV NEW YORK, 87- *Concurrent Pos:* Vis res assoc, Michael Reese Hosp & Med Ctr, 65-77; vis prof mech eng, Technion Israel Inst Technol, 72-73, biomed eng, Rush Med Col, 73-76; Lady Davis prof mech eng, Technion Israel Inst Technol, 85. *Mem:* Am Inst Chem Engrs; Sigma Xi. *Res:* Chemical reactor engineering: fluidization, tracer methods, heat and mass transfer in catalysts; fluid mechanics; biomedical engineering: tracer methods in the circulation and glucose sensors. *Mailing Add:* Dept Chem Eng City Univ New York New York NY 10031

WEINSTEIN, HOWARD, b New York, NY, Nov 9, 27. NEUROCHEMISTRY, CELL PHYSIOLOGY. *Educ:* Cornell Univ, BA, 49; State Univ Iowa, PhD(zool), 56. *Prof Exp:* Instr zool & physiol, Wis State Univ-Stevens Point, 56-57; USPHS fel neuroendocrinol, Case Western Reserve Univ, 58-61; res scientist, City of Hope Med Ctr, Duarte, Calif, 61-74; STAFF SCIENTIST, NAT INST NEUROL DIS & STROKE, 74- *Concurrent Pos:* USPHS grants, 68- *Mem:* Soc Neurosci; Am Soc Neurochem; Int Soc Neurochem; Am Soc Zoologists. *Res:* Membrane transport processes in central nervous system. *Mailing Add:* 14012 Eagle Ct Rockville MD 20853

WEINSTEIN, HYMAN GABRIEL, b Worcester, Mass, June 15, 20. BIOCHEMISTRY, NUTRITION. *Educ:* Worcester Polytech Inst, BS, 42; Univ Ill, MS, 47; Univ Chicago, PhD, 53. *Prof Exp:* Res assoc, Rheumatic Fever Res Inst, Northwestern Univ Med Sch, 49-53; chief, biochem res, West Side Vet Admin Hosp, 54-64; chief, Res on Aging Lab, 65-88, DIR RES GERIAT/EXTENDED CARE, VET ADMIN MED CTR, NORTH CHICAGO, 88-; ASSOC DIR RES, CTR AGING & PROF BIOCHEM, UNIV HEALTH SCI, CHICAGO MED SCH, NORTH CHICAGO, 90- *Concurrent Pos:* Lectr, Northeastern Ill Univ, 75- *Mem:* AAAS; Am Chem Soc; Soc Complex Carbohydrates; NY Acad Sci; Sigma Xi; Am Soc Biochem & Molecular Biol. *Res:* Complex carbohydrates in development, aging, nutrition and disease; preventive medicine. *Mailing Add:* 270 E Crescent Knoll Rd Libertyville IL 60048

WEINSTEIN, I BERNARD, b Madison, Wis, Sept 9, 30; m 52; c 3. MEDICINE. *Educ:* Univ Wis, BS, 52, MD, 55. *Prof Exp:* Nat Cancer Inst spec res fel bact & immunol, Harvard Med Sch & Mass Inst Technol, 59-61; from asst prof to assoc prof med, Col Physicians & Surgeons, dir, Div Environ Sci, Sch Pub Health, 78-90, PROF MED, COLUMBIA UNIV, 73-, PROF PUB HEALTH, 78-, DIR, DIV ONCOL & PROF GENETICS & DEVELOP, 90-; FRODE JENSEN PROF MED & DIR, COMPREHENSIVE CANCER CTR, 85- *Concurrent Pos:* Career scientist, Health Res Coun, City of New York, 61-72; assoc vis physician, Francis Delafield Hosp, 61-66; from asst attend physician to assoc attend physician, Presby Hosp, 67-81, attend physician, 81-; Europ Molecular Biol Orgn travel fel, 70-71; adv, Lung Cancer Segment, Carcinogenesis Prog, Nat Cancer Inst, 71-74 & Chem & Molecular Biol Segment, 73-76; mem, Interdisciplinary Commun Prog, Smithsonian Inst, 71-74, Pharmacol B Study Sect, NIH, 71-75 & numerous sci & adv comts, Nat Cancer Inst & Am Cancer Soc, 76-88; adv, Roswell Park Mem Inst, Buffalo, NY, Brookhaven Nat Lab, Div Cancer Cause & Prevention, Nat Cancer Inst, Coun on Anal & Projs, Am Cancer Soc, Int Agency for Res on Cancer, WHO, Lyon, France; assoc ed, Cancer Res, 73-76 & 86-, J Environ Path & Toxicol, 77-84 & J Cellular Physiol, 82-89; coun deleg, AAAS, 85-88; Nakasone vis prof, Tokyo, 87; Gen Motors Cancer Res Found vis prof, Int Agency Res Cancer, Lyon, 88. *Honors & Awards:* Meltzer Medal, 64; Louise Weissberger Lectr, Univ Rochester, 81; Mary Ann Swetland Lectr, Case Western Reserve Univ, 83; Daniel Laszlo Mem Lectr, Montefiore Med Ctr, 83; Samuel Kuna Distinguished Lectr, Rutgers Univ, 85; Clowes Award, Am Asn Cancer Res, 87; Ester Langer Lectr, Univ Chicago, 89; Harris Mem Lectr, Mass Inst Technol, 89; Silvio O Conte Award, Environ Health Inst, 90. *Mem:* Inst Med-Nat Acad Sci; Am Soc Microbiol; Int Soc Quantum Biol; Am Asn Cancer Res; AAAS (pres, 90-91); Am Soc Clin Investigation; NY Acad Sci. *Res:* Oncology; cellular and molecular aspects of carcinogenesis; environmental carcinogenesis; control of gene expression. *Mailing Add:* Inst Cancer Res Columbia Univ 701 W 168th St New York NY 10032

WEINSTEIN, IRA, b Oak Park, Ill, Jan 30, 28; m 54; c 3. ENDOCRINOLOGY, BIOCHEMICAL PHARMACOLOGY. *Educ:* Roosevelt Univ, BS, 49; Univ Ill, MS, 52; George Washington Univ, PhD(microbiol), 60. *Prof Exp:* From instr to asst prof pharmacol, Vanderbilt Univ, 60-69; assoc prof, Sch Med, Univ Fla, 69-75; assoc prof pharmacol, Sch Med, Univ Mo-Columbia, 75-80, prof, 80-81; PROF, UNIV TENN CTR HEALTH SCI, MEMPHIS, 81- *Concurrent Pos:* Fel pharmacol, Med Sch, Vanderbilt Univ, 60-63; USPHS fels, 60-64; Olson Mem Fund fel, 65-66; vis lectr, Hebrew Univ Israel, 65-66; Am Heart Asn advan res fel, 65-67; fel coun on arteriosclerosis, Am Heart Asn. *Mem:* Sigma Xi; Am Soc Microbiol; Am Soc Pharmacol & Exp Therapeut. *Res:* Bacterial physiology, endocrines and drug effects upon lipid metabolism. *Mailing Add:* Dept Pharmacol Univ Tenn Ctr Health Sci 800 Madison Ave Memphis TN 38163

WEINSTEIN, IRAM J, b Brooklyn, NY, Oct 17, 36; m 70; c 4. SENSOR SYSTEMS ANALYSIS, SIGNAL PROCESSING. *Educ:* Rensselaer Polytech Inst, BS, 56; Northeastern Univ, MSEE, 61; Stanford Univ, PhD(elec eng), 67. *Prof Exp:* Sr engr, Raytheon Corp, 56-60; sr res engr, Stanford Res Inst, 60-68, dir syst eval, 68-72, dir, Ctr Anal Pub Serv, 72-76; chief scientist & vpres measurement syst, Systs Planning Corp, 78-; SR SCIENTIST, SCI APPLN INT CORP. *Concurrent Pos:* Lectr, Stanford Univ, 67-79. *Mem:* Inst Elec & Electronics Engrs; Opers Res Soc Am; Am Inst Aeronaut & Astronaut; Asn Comput Mach. *Res:* Analysis and evaluation of large scale systems; radar systems engineering; signal processing. *Mailing Add:* Sci Appln Int Corp 4001 N Fairfax Dr Suite 500 Arlington VA 22203

WEINSTEIN, IRWIN M, b Denver, Colo, Mar 5, 26; m 51; c 2. HEMATOLOGY, INTERNAL MEDICINE. *Educ:* Univ Colo Med Ctr Denver, MD, 49. *Prof Exp:* Instr med, Univ Chicago, 53-54, asst prof, 54-55; assoc prof, 55-59, assoc clin prof, 59-70, CLIN PROF MED, UNIV CALIF, LOS ANGELES, 70- *Concurrent Pos:* Chief hemat, Vet Admin Ctr, Wadsworth Hq, Los Angeles, 55-59; chief staff, Cedars-Sinai Med Ctr, Los Angeles, 72-75; gov, Am Col Physicians, Southern Calif Region I, 89. *Honors & Awards:* Israel Cancer Res Fund Founders Award, 87. *Mem:* Inst Med-Nat Acad Sci; Am Soc Hemat; fel Am Col Physicians; Am Fedn Clin Res. *Res:* Mechanisms of anemia; radioactive chromium and iron for studying red cell production and distruction. *Mailing Add:* 8635 W Third St Los Angeles CA 90048

WEINSTEIN, JEREMY SAUL, b Brooklyn, NY, Apr 9, 44; m 67; c 3. OPERATIONS RESEARCH, COMPUTER SCIENCE. *Educ:* City Col New York, BE, 66; Purdue Univ, Lafayette, MSIE, 68, PhD(indust eng), 71. *Prof Exp:* Res engr, 70-72, mgr bus planning, Laundry Group, 72-76, mgr prod scheduling & mis, 76-79, mgr indust eng & mis, Marion Div, 79-82, dir, Mfg Eng, Findlay Div, 82-88, group dir, Mfg Serv, 88-89, VPRES, OPERS & PLANNING, WHIRLPOOL CORP, BENTON HARBOR, MICH, 89- *Concurrent Pos:* Instr, Benton Harbor Exten, Mich State Univ, 70-71. *Mem:* Inst Mgt Sci; Inst Indust Engrs. *Mailing Add:* Whirlpool Corp 2000 M 63 North Benton Harbor MI 49022

WEINSTEIN, LEONARD HARLAN, b Springfield, Mass, Apr 11, 26; m 50; c 2. PLANT PHYSIOLOGY, ENVIRONMENTAL BIOLOGY. *Educ:* Pa State Univ, BS, 49; Univ Mass, MS, 50; Rutgers Univ, PhD(plant physiol), 53. *Prof Exp:* Fel soils, Rutgers Univ, 53-54; from assoc plant physiologist to plant physiologist, 55-63, prog dir plant chem, 63-69; dir ecosysts res ctr, 89-90, ADJ PROF NAT RESOURCES, CORNELL UNIV, 79-; PROG DIR ENVIRON BIOL, BOYCE THOMPSON INST PLANT RES, 69- *Concurrent Pos:* Mem bd dirs, 73-, trustee, 78-, Boyce Thompson Inst Plant Res, mem & dir, Boyce Thompson Southwestern Arboretum; Sci Adv Comn, Environ Sci Div, Oak Ridge Nat Lab, 85-87, Ecosysts Res Ctr Adv Comm, 85-, Environ Protection Agency, Sci Adv Bd Environ Effects, Transp & Rate, 87- , Off Technol & Assessment, Adv Panel Oxidants, 87-, Nat Asn State Univ & Land Grant Cols, Environ Comm, 87- *Mem:* AAAS; Am Soc Plant Physiol; Air Pollution Control Asn; Fedn Am Scientists; Am Inst Biol Sci; Soc Environ Toxicol & Chem. *Res:* Air pollution; plant nutrition; plant senescence; environmental biology; effects of atmospheric pollutants on plant growth, development, productivity and quality; plant toxicology. *Mailing Add:* Boyce Thompson Inst Plant Res at Cornell Tower Rd Ithaca NY 14853

WEINSTEIN, LOUIS, b Bridgeport, Conn, Feb 26, 09; m 34. INTERNAL MEDICINE, INFECTIOUS DISEASES. *Educ:* Yale Univ, BS, 28, MS, 30, PhD(bact), 31; Boston Univ, MD, 43. *Hon Degrees:* ScD, Boston Univ, 73. *Prof Exp:* Instr bact, Med Sch, Yale Univ, 37-39; res assoc immunol, Sch Med, Boston Univ, 39-44, asst med, 43-44, from instr to assoc prof, 44-57; lectr pediat, Med Sch, Tufts Univ, 50-57, prof med, 57-75; chief infectious dis serv, Vet Admin Hosp, West Roxbury, Mass, 75-78; physician, 75-83, SR CONSULT MED, BRIGHAM & WOMEN'S HOSP, BOSTON, 83-; EMER PROF MED, MED SCH, TUFTS UNIV, BOSTON, 85- *Concurrent Pos:* Asst, Harvard Med Sch, 45-46, instr, 46-49, lectr, 49-75, vis prof med, 75-83, lectr med, 83-; chief infectious dis serv, Mass Mem Hosp, 47-57 & New Eng Ctr Hosp & Boston Floating Hosp, 57-75; assoc physician, Med Serv, Mass Gen Hosp, 58-75, mem, bd consult, 75-83, hon physician, 83-; assoc physician in chief, New Eng Med Ctr Hosps, 62-71; mem, bact & mycol study sect, Nat Inst Allergy & Infectious Dis, NIH, 62-66; assoc ed, J Infectious Dis, 68-79; chmn, Infectious Dis Bd, Am Bd, Internal Med, 71-76; mem, Int Adv Comt, Kuvin Ctr Study Infectious & Trop Dis, Jerusalem, Israel, 75-; ed-in-chief, Infectious Dis Pract, 77- *Honors & Awards:* Finland Award & Bristol Award, Am Soc Infectious Dis; Medal lectr, Am Col Chest Physicians. *Mem:* Am Acad Arts & Sci; AAAS; AMA; Am Soc Microbiol; Am Soc Infectious Dis; Am Soc Clin Invest; Am Fedn Clin Res; Asn Am Physicians; Sigma Xi. *Res:* Chemotherapy of infection; host factors in infectious disease. *Mailing Add:* 26 Greylock Rd Newtonville MA 02160

WEINSTEIN, MARK JOEL, b Syracuse, NY, Mar 2, 45; m; c 1. BLOOD COAGULATION PROTEIN. *Educ:* Univ Calif, San Diego, PhD(chem), 71. *Prof Exp:* ASSOC PROF BIOCHEM, SCH MED, BOSTON UNIV & SPEC SCI STAFF, BOSTON CITY HOSP, 84- *Mem:* Am Soc Biol Chem; Am Chem Soc; Am Heart Asn; AAAS; NY Acad Sci. *Res:* von Willebrand Factor/Factor VIII. *Mailing Add:* Boston Univ Sch Med S-402 80 E Concord St Boston MA 02118

WEINSTEIN, MARVIN, b Bronx, NY, June 7, 42; m 67; c 1. THEORETICAL HIGH ENERGY PHYSICS. *Educ:* Columbia Univ, BS, 63, MS, 64, PhD(physics), 67. *Prof Exp:* Physics mem, Inst Advan Study, 67-69; vis asst prof physics, Yeshiva Univ, 69-70 & NY Univ, 70-72; sr res assoc physics, Stanford Linear Accelerator Ctr, Stanford Univ, 72-80. *Mem:* Sigma Xi; Am Chem Soc. *Res:* Current algebra; gauge theories of strong, weak and electromagnetic interactions; non-perturbatic methods in quantum field theory. *Mailing Add:* 24 Maddaket Southwyck Village Scotch Plains NJ 07076

WEINSTEIN, MARVIN STANLEY, b New York, NY, May 24, 27; m 52; c 2. ACOUSTICS. *Educ:* St Louis Univ, BS, 48; Univ Md, MS, 51, PhD, 56. *Prof Exp:* Asst, Univ Md, 48-49; physicist, US Naval Ord Lab, 49-59; vpres, Underwater Systs Inc, 59-62, pres, 62-74, chmn bd, 75-87; RETIRED. *Concurrent Pos:* Dir, Mandex Corp. *Mem:* Acoust Soc Am. *Res:* Underwater acoustics; propagation at long and short range in deep and shallow water, noise, sinusoidal and explosive signals; ultrasonics; ultrasonic modeling; electronics. *Mailing Add:* 14305 Northwyn Dr Silver Spring MD 20904

WEINSTEIN, NORMAN J(ACOB), b Rochester, NY, Dec 31, 29; m 57; c 3. POLLUTION CONTROL SYSTEMS, HAZARDOUS WASTE MANAGEMENT. *Educ:* Syracuse Univ, BChE, 51, MChE, 53; Ore State Univ, PhD(chem eng), 56. *Prof Exp:* Chem engr, Esso Res & Eng Co, 56-61, sr engr, 61-66, eng assoc, 66; eng & develop asst dir, Princeton Chem Res Inc, 66-67, eng & develop dir, 67-69; PRES, RECON SYSTS, INC, 69- *Concurrent Pos:* Adj prof, Newark Col Eng, 63-66; chief tech adv, Indian Pollution Control Inst, 84-87. *Mem:* Am Chem Soc; Am Inst Chem Engrs; Am Soc Testing & Mat; NY Acad Sci. *Res:* Air and water pollution control; petroleum, petrochemical and metallurgical processes; catalysis; fluidized solids; hydrocarbon and coal gasification; economic evaluation; liquid and solid waste disposal and recycling. *Mailing Add:* RR 1 Box 1175 Princeton NJ 08540-9801

WEINSTEIN, PAUL P, b Brooklyn, NY, Dec 9, 19; m 54; c 2. DEVELOPMENTAL BIOLOGY. *Educ:* Brooklyn Col, AB, 41;. *Prof Exp:* Jr parasitologist, USPHS, St Bd Health, Fla, 42-44, sr asst sanitarian, Washington, DC, Ga & PR, 44-46, from scientist to scientist dir & chief lab parasitic dis, NIH, 49-68; chmn dept, 69-75, prof, 69-90, EMER PROF BIOL, UNIV NOTRE DAME, 90- *Concurrent Pos:* Vis scientist, Nat Inst Med Res, Eng, 62-63; mem, Parasitic Dis Panel, US-Japan Coop Med Sci Prog, 65-69, chmn, 69-73; mem, comt int ctr med res & training, NIH, 70-73; mem adv sci bd, Gorgas Mem Inst Trop Prev Med, 72-88; mem, Nat Adv Comt, Primate Res Ctr, Univ Calif, Davis, 73-77; mem, microbiol & infectious dis adv comt, NIH, 78-81; consult, Walter Reed Army Inst Res, US Army, 77-79; vis prof, Juntendo Univ Sch Med, Tokyo, Japan, 89. *Honors & Awards:* Ashford Award, Am Soc Trop Med & Hyg, 57. *Mem:* Fel AAAS; Am Soc Trop Med & Hyg (pres, 85); Am Soc Parasitol (pres, 72); Japanese Soc Parasitol. *Res:* Cultivation and physiology of parasitic helminths; host-parasite relationships; cobalamins and parasite development. *Mailing Add:* Dept Biol Univ Notre Dame Notre Dame IN 46556

WEINSTEIN, ROBERT, b Brooklyn, NY, Aug 10, 49; m 83; c 2. VASCULAR CELL BIOLOGY, HEMOSTASIS & THROMBOSIS. *Educ:* Brandeis Univ, AB, 71; NY Univ, MD, 75. *Prof Exp:* CHIEF, DIV CLIN RES, VASCULAR LAB, SAINT ELIZABETH HOSP, BOSTON, 85-, DIR, TRANSFUSION SERV, 85- *Concurrent Pos:* Investr biomed res, St Elizabeth Hosp, 85-; assoc prof med, Tufts Med Sch. *Mem:* AAAS; Am Soc Cell Biol; Am Soc Hemat; Am Fedn Clin Res; Am Asn Blood Banks; NY Acad Sci; Am Soc Apheresis. *Res:* Erykrocyte-endorelial interaction and the pathogenesis of vascular disease in sickle cell anemia. *Mailing Add:* Dept Biomed Res St Elizabeth's Hosp 738 Cambridge St Boston MA 02135

WEINSTEIN, RONALD S, b Schenectady, NY, Nov 20, 38; m 64; c 2. EXPERIMENTAL PATHOLOGY, ELECTRON MICROSCOPY. *Educ:* Union Col, NY, BS, 60; Tufts Univ, MD, 65; Am Bd Path, dipl. *Prof Exp:* Res asst electron micros, Mass Gen Hosp, 62-63; instr path, Sch Med, Tufts Univ, 67-69, assoc prof, 72-75; PROF PATH & CHMN DEPT, RUSH MED COL, 75- *Concurrent Pos:* From intern to resident path, Mass Gen Hosp, 65-70, head, Mixter Lab Electron Micros, 66-70; teaching fel, Harvard Med Sch, 65-70; investr toxicol, Aerospace Med Res Lab, Wright-Patterson AFB, 70-72; chmn dept, Rush-Presbyterian-St Lukes Med Ctr. *Mem:* Soc Develop Biol; NY Acad Sci; Am Asn Anatomists; Soc Toxicol; Am Asn Path. *Res:* Development and application of high resolution electron microscopy techniques to the study of biological membrane ultrastructure; comparative and functional studies on normal and neoplastic cell membranes; environmental toxicology. *Mailing Add:* 1753 W Congress Pkwy Chicago IL 60612

WEINSTEIN, ROY, b New York, NY, Apr 21, 27; m 54; c 2. ACCELERATOR EXPERIMENTS, UNDERGROUND EXPERIMENTS. *Educ:* Mass Inst Technol, BS, 51, PhD(physics), 54. *Hon Degrees:* Lycoming Col, ScD, 82. *Prof Exp:* From instr to asst prof physics, Brandeis Univ, 54-56; asst prof physics, Mass Inst Technol, 56-59; from assoc prof to prof physics, Northeastern univ, 61-74, prof & chmn, 74-82; prof & dean natural sci, 82-88, PROF DEPT PHYSICS & DIR, INST BEAM PARTICLE DYNAMICS, UNIV HOUSTON, 88- *Concurrent Pos:* Prin investr, Nat Sci Found res at Univs, 61-; ed, McGraw Hill, 59-64; consult, Camb Electronics Accel (Harvard Univ), Stanford Univ Lin Accel & several co, 66-; NSF fel, Bohr Inst, 59-60, Harvard Univ, 69-70 & Guggenheim fel, Stanford Univ, 68-69,. *Honors & Awards:* Nat Triennial Scholar, Phi Kappa Phi, 82-85. *Mem:* Fel Am Phys Soc; NY Acad Sci. *Res:* Particle physics; neutrino astrophysics; nuclear physics; atomic physics; superconducting permanent magnets. *Mailing Add:* Inst Beam Particle Dynamics Rm 632 SR 1 Univ Houston Houston TX 77204-5506

WEINSTEIN, SAM, b Omaha, Nebr, May 24, 16; m 46; c 2. ORTHODONTICS. *Educ:* Creighton Univ, DDS, 41; Northwestern Univ, MSD, 48; Am Bd Orthod, dipl. *Prof Exp:* From asst prof to prof orthod, Col Dent, Univ Nebr, 54-71, chmn dept, 63-71; prof orthod, 71-83, EMER PROF ORTHOD, UNIV CONN, 83- *Concurrent Pos:* Mem dent study sect, NIH, 69-73; consult, Coun Dent Educ, Am Dent Asn, 69-76; examr, Coun on Educ, Can Dent Asn, 74-77. *Honors & Awards:* Louise Ada Jarabak Mem Int Orthod Teachers & Res Award, 83. *Mem:* Fel AAAS; Sigma Xi; Int Asn Dent Res; Int Soc Cranio-Facial Biol. *Res:* Theoretical mechanics application to soft tissue forces and tooth movement; cleft palate embryology; growth. *Mailing Add:* Univ Conn Health Ctr Farmington CT 06032

WEINSTEIN, STANLEY EDWIN, b New York, NY, Apr 26, 42; m 64. MATHEMATICS, COMPUTER SCIENCE. *Educ:* Hunter Col, BA, 62; Mich State Univ, MS, 64, PhD(math), 67. *Prof Exp:* Teaching asst math, Mich State Univ, 62-67; from asst prof to assoc prof math, Univ Utah, 67-75; PROF & CHMN, DEPT MATH & COMPUT SCI, OLD DOMINION UNIV, 75- *Concurrent Pos:* USAF Off Sci Res grant, Univ Utah, 72-73; vis assoc prof, Dept Math, Ariz State Univ, 72-73. *Mem:* Soc Indust & Appl Math; Asn Comput Mach; Am Math Soc. *Res:* Approximation theory; numerical analysis; solution of nonlinear equations. *Mailing Add:* Dept Math & Comput Sci Old Dominion Univ 5215 Hampton Blvd Norfolk VA 23508

WEINSTEIN, STEPHEN B, b New York, NY, Nov 25, 38. MULTIMEDIA COMMUNICATIONS. *Educ:* Mass Inst Technol, SB, 60; Univ Mich, MS, 62; Univ Calif, PhD(elec eng), 66. *Prof Exp:* mem tech staff, Bell Telephone, 68-79; vpres technol strategy, Am Express, 79-84; DIV MEM, MULTIMEDIA COMMUN RES DIV, BELL COMMUN RES, 84- *Concurrent Pos:* Lectr, Data Commun, Polytech Col Brooklyn, 81, Princeton Univ, 86-87. *Honors & Awards:* Centennial Medal, Inst Elec & Electronics Engrs, 84. *Mem:* Fel Inst Elec & Electronics Engrs. *Res:* Network communication of data; modems; information transfer and retrieval. *Mailing Add:* MultiMedia Commun Res Div Bell Commun Res 445 South St Box 1910 Rm 2E 288 Morristown NJ 07960

WEINSTEIN, STEPHEN HENRY, b Bronx, NY, Apr 14, 37; m 66; c 1. DRUG METABOLISM. *Educ:* Queens Col, NY, BS, 58; Adelphi Col, MS, 61; Adelphi Univ, PhD(biochem), 67. *Prof Exp:* Scientist, Warner-Lambert Res Inst, 67-68; sr biochemist, 68-73, group leader, Endo Labs, Inc, 73-77; sr res biochemist, E I du Pont de Nemours & Co, Inc, 77-80; sect head, Squibb Inst Med Res, 80-85, sr group leader, 85- 88, asst dept dir, 88-89, EXEC DIR, METAB & PHARMACOKINETICS DEPT, BRISTOL-MYERS SQUIBB INST PHARMACEUT RES, 89- *Mem:* Am Soc Pharmacol & Exp Therapeut; Int Soc Study Xenobiotics; Am Asn Pharmaceut Scientists. *Res:* Metabolism and function of phosphatides; mechanisms of membrane transport; pharmacokinetics; drug metabolism; biochemical pharmacology. *Mailing Add:* Bristol-Myers Squibb Inst Pharmaceut Res PO Box 4000 Princeton NJ 08543-4000

WEINSTOCK, ALFRED, b Toronto, Ont, May 3, 39; div; c 2. CELL BIOLOGY, PERIODONTOLOGY. *Educ:* Univ Toronto, DDS, 62; McGill Univ, PhD(anat), 69. *Prof Exp:* Res fel dent med, Forsyth Dent Ctr, Boston, 62-63; Nat Res Coun Can res fel periodont, Sch Dent Med, Harvard Univ, 63-66; Nat Res Coun Can res fel anat sci, Sch Med, McGill Univ, 66-67, from lectr to asst prof, 67-70; assoc prof, 70-76, chmn sect periodont, 71-74, CLIN PROF DENT & ANAT, SCH DENT & SCH MED, CTR HEALTH SCI, UNIV CALIF, LOS ANGELES, 77- MEM, DENT RES INST, 73- *Concurrent Pos:* Nat Res Coun Can res scholar, Sch Med, McGill Univ, 67-69, Med Res Coun Can res scholar, 69-70; NIH res grant, Univ Calif, Los Angeles, 71-79; consult, Vet Admin Hosp, Brentwood & Sepulveda, Calif, 72-; mem staff, Univ Calif, Los Angeles Med Ctr Hosp, Cedars-Sinai Med Ctr, 76-; chmn, Div Dent, Cedars-Sinai Med Ctr, 82-86. *Honors & Awards:* Res Award, Can Dent Asn, 71. *Mem:* AAAS; Am Acad Periodont; Am Asn Anatomists; Am Dent Asn; Sigma Xi; Am Soc Cell Biol. *Res:* Structural and functional aspects of secretory cells involved in the elaboration of collagen, enamel and other glycoproteins, mainly in mineralizing tissues; histology; experimental pathology; periodontal disease. *Mailing Add:* 9201 Sunset Blvd Suite 710 Los Angeles CA 90069

WEINSTOCK, BARNET MORDECAI, b Brooklyn, NY, Oct 10, 40; m 66; c 2. SEVERAL COMPLEX VARIABLES, FUNCTIONAL ANALYSIS. *Educ:* Columbia Univ, BA, 62; Mass Inst Technol, PhD(math), 66. *Prof Exp:* From instr to asst prof math, Brown Univ, 66-73; assoc prof math, Univ Ky, 73-77; assoc prof, 77-80, actg chmn, 80-81, chmn dept, 81-85, PROF MATH, UNIV NC, CHARLOTTE, 80- *Mem:* Am Math Soc; Math Asn Am. *Res:* Functional analysis; several complex variables. *Mailing Add:* Dept Math Univ NC Charlotte NC 28223

WEINSTOCK, HAROLD, b Philadelphia, Pa, Dec 25, 34; m 61; c 2. SOLID STATE PHYSICS, LOW TEMPERATURE PHYSICS. *Educ:* Temple Univ, BA, 56; Cornell Univ, PhD(helium three), 62. *Prof Exp:* From res asst to res assoc physics, Cornell Univ, 56-62; asst prof, Mich State Univ, 62-65; assoc prof physics, Ill Inst Technol, 65-73, dir, Educ Technol Ctr, 79-85, prof, 73-86, dir, technol & continuing educ, 85-86; PROG MGR, AIR FORCE OFF SCI RES, 86- *Concurrent Pos:* Vis prof, Cath Univ Louvain, 70 & Cath Univ Nijmegen, 72-73; vis staff mem, Los Alamos Nat Lab, 72-86; consult, Tech Adv Serv for Atty, 74- & Air Force Off Sci Res, 85-86; dir, NATO Advan Study Inst, 76, 85, 88 & 89. *Mem:* Fel Am Phys Soc; Mat Res Soc. *Res:* Thermal, electrical and magnetic properties of solids; radiation damage in solids; superconductivity; computer use in science education; nondestructive evaluation; scanning tunneling microscopy; superconductive magnetometry. *Mailing Add:* Air Force Off Sci Res-NE Bolling 20332-6448

WEINSTOCK, IRWIN MORTON, b New York, NY, July 17, 25; m 56; c 3. BIOCHEMISTRY, SCIENCE ADMINISTRATION. *Educ:* Univ Okla, BS, 47; Univ Ill, MS, 48, PhD(chem), 51. *Prof Exp:* Res assoc biochem, Dept Psychiat, Med Col, Cornell Univ, 51-57; res assoc, New York Med Col, 57-59; from asst mem to assoc mem, Inst Muscle Dis, 64-74; dir spec neurol lab, Nassau County Med Ctr, 74-80; CLIN DIR, RES TEST LABS INC, 80- *Concurrent Pos:* Lectr, Hunter Col, 53-54 & Columbia Univ, 68-75; adj assoc prof med, Health Sci Ctr, State Univ NY, Stony Brook, 76-80. *Mem:* AAAS; Harvey Soc. *Res:* Intermediary metabolism and enzymology of muscle wasting conditions. *Mailing Add:* Res Test Labs Inc 255 Great Neck Rd Great Neck NY 11021

WEINSTOCK, JEROME, b Brooklyn, NY, Sept 12, 33; m 55; c 3. PLASMA PHYSICS, FLUID MECHANICS. *Educ:* Cooper Union, BChE, 55; Cornell Univ, PhD(phys chem), 59. *Prof Exp:* Sr scientist, Nat Bur Standards, 59-65; SR SCIENTIST, NAT OCEANIC & ATMOSPHERIC ADMIN, 65- *Concurrent Pos:* Nat Res Coun-Nat Acad Sci res fel, 59-62; Emil Schweingerg scholarship. *Mem:* Am Phys Soc; Am Geophys Union. *Res:* Statistical mechanics; turbulence theory; transport theory; fluctuation theory; molecular collision theory; basic research in turbulence theory, atmospheric waves, and plasma physics; statistical physics. *Mailing Add:* Plasma Physics Bldg Nat Oceanic & Atmospheric Admin Boulder CO 80302

WEINSTOCK, JOEL VINCENT, GASTROENTEROLOGY, INTERNAL MEDICINE. *Educ:* Wayne State Univ, MD, 73. *Prof Exp:* DIR, DIV GASTROENTEROL, UNIV IOWA, 86- *Mailing Add:* Dept Internal Med Univ Iowa Hosp & Clin Iowa City IA 52242

WEINSTOCK, JOSEPH, b New York, NY, Jan 30, 28; wid; c 10. MEDICINAL CHEMISTRY, PHARMACOLOGY. *Educ:* Rutgers Univ, BS, 49; Univ Rochester, PhD(chem), 52. *Prof Exp:* Res assoc chem, Northwestern Univ, 52-54, instr, 54-56; sr chemist, Smith Kline & French Labs, 56-62, group leader, 62-67, sr investr, 67-76, asst dir chem, 78-87, fel, 87-90, SR FEL, SMITHKLINE BEECHAM PHARMACEUT, 90- *Mem:* AAAS; Am Chem Soc; NY Acad Sci. *Res:* Medicinal and synthetic organic chemistry; organic reaction mechanisms; pteridines; diuretic, anti-inflammatory, antihypertensive agents; drug metabolism and identification of metabolites; dopamine agonists; benzazepines; non-peptide angiotensin antagonists. *Mailing Add:* Med Chem L431 Res & Develop SmithKline Beecham Pharmaceut PO Box 1539 King of Prussia PA 19406-0939

WEINSTOCK, LEONARD M, b Passaic, NJ, Jan 30, 27; m 60; c 2. ORGANIC CHEMISTRY. *Educ:* Rutgers Univ, BS, 50; Ind Univ, PhD(chem), 58. *Prof Exp:* dir process res, Merck & Co, 78-83, sr dir, 83-90; CONSULT, 90- *Mem:* Am Chem Soc. *Res:* Chemistry of heterocyclic compounds and beta-lactam antibiotics; synthetic electrochemistry; new reagents; new synthetic methods. *Mailing Add:* Three Myrtle Bank Lane Hilton Head Island SC 29926

WEINSTOCK, MANUEL, b Philadelphia, Pa, Apr 15, 27; m 54; c 3. MECHANICAL ENGINEERING, AERONAUTICAL ENGINEERING. *Educ:* Drexel Inst, BS, 47; Univ Pa, MS, 50. *Prof Exp:* Struct designer, Widdicombe Eng Co, Pa, 47-50; stress analyst, Piasecki Helicopter Corp, 50-51; aeronaut engr, Naval Air Develop Ctr, 51-54; mech engr, Frankford Arsenal, 54-57 & 58-62, supvry ord engr, 57-58, supvry mech engr, Mech Eng Propellant Actuated Devices Dept, 62-74, supvr ind eng, 74-77, SUPVR MECH ENGR ARTIL & TANK METAL PARTS, US ARMY, 77- *Mem:* Am Soc Mech Engrs; Sigma Xi. *Res:* Interior ballistics; digital and analog simulation of the performance of propellant actuated devices and complete systems used for emergency escape from military aircraft. *Mailing Add:* 1016 City Ave Philadelphia PA 19151

WEINSTOCK, ROBERT, b Philadelphia, Pa, Feb 2, 19; m 50; c 2. MATHEMATICAL PHYSICS. *Educ:* Univ Pa, AB, 40; Stanford Univ, PhD, 43. *Prof Exp:* Instr physics, Stanford Univ, 43-44, math, 46-50, actg asst prof, 50-54; res assoc radar countermeasures, Radio Res Lab, Harvard Univ, 44-45; from asst prof to assoc prof math, Univ Notre Dame, 54-59; vis assoc prof math, 59-60, from assoc prof to prof, 60-83, EMER PROF PHYSICS, OBERLIN COL, 83- *Concurrent Pos:* NSF fel, Oxford Univ, 65-66. *Mem:* Fel AAAS; Am Phys Soc; Am Asn Physics Teachers. *Res:* Mathematical physics; linear algebra; Newton's Principia. *Mailing Add:* Dept Physics Oberlin Col Oberlin OH 44074

WEINSTOCK, W, b Philadelphia, Pa, Aug 18, 25. RADAR & DEFENSE SYSTEM DESIGN. *Educ:* Univ Pa, BSEE, 46, MSEE, 54, PhD(elec eng), 64. *Prof Exp:* Design engr radar design, Philco, 46-49; prin scientist, missile & surface radar div, RCA, 49-87; RETIRED. *Concurrent Pos:* Consult, 87- *Mem:* Fel Inst Elec & Electronics Engrs. *Mailing Add:* Six Beryl Rd Cheltenham PA 19012

WEINSWIG, MELVIN H, b Lynn, Mass, Feb 2, 35; m 60; c 3. PHARMACEUTICAL CHEMISTRY. *Educ:* Mass Col Pharm, BS, 55, MS, 57; Univ Ill, PhD(pharmaceut chem), 61. *Prof Exp:* From asst prof to assoc prof pharmaceut chem, Butler Univ, 61-69; assoc dean, 71-76, PROF PHARM & CHMN EXTEN SERV PHARM, SCH PHARM, UNIV WIS-MADISON, 69- *Concurrent Pos:* Consult, Continuing Co, Educ Health Prof. *Honors & Awards:* Award of Merit Continuing Educ, Am Asn Cols Pharm. *Mem:* Am Chem Soc; Am Pharmaceut Asn; Am Asn Cols Pharm. *Res:* Novel analytical approach to combination pharmaceutical products; drug abuse education and research. *Mailing Add:* Rm 2308 Pharm Bldg Sch Pharm Univ Wis Madison WI 53706

WEINTRAUB, BRUCE DALE, b Buffalo, NY, Sept 19, 40; m 68; c 1. ENDOCRINOLOGY, INTERNAL MEDICINE. *Educ:* Princeton Univ, AB, 62; Harvard Med Sch, MD, 66. *Prof Exp:* Intern med, Peter Bent Brigham Hosp, Boston, Mass, 66-67, resident, 67-68; clin assoc endocrinol, Nat Inst Arthritis & Metab Dis, 68-71; instr med, Harvard Med Sch, 71-72; SR INVESTR ENDOCRINOL, NAT INST ARTHRITIS, METAB & DIGESTIVE DIS, 72- *Concurrent Pos:* Clin & res fel, Nat Inst Arthritis & Metab Dis, 68-71; USPHS res grant & asst med, Mass Gen Hosp, Boston, 71-72. *Mem:* Am Fedn Clin Res; Endocrine Soc. *Res:* Structure and properties of hormones secreted by tumors; applications of affinity chromatography to endocrinology; subunits of glycoprotein hormones. *Mailing Add:* Metab & Digestive Dis Molecular Cellular & Nutrit Endocrinol Br Nat Inst Arthritis Bldg 10 Rm 8N 316 Bethesda MD 20892

WEINTRAUB, HAROLD M, b Newark, NJ, June 2, 45; m; c 2. GENE EXPRESSION, CELLULAR REGULATION. *Educ:* Harvard Univ, BA, 67; Univ Pa, PhD(cell differention), 71, MD, 73. *Prof Exp:* Helen Hay Whitney fel, Lab Molecular Biol, Med Res Coun, 72-73; from asst prof to assoc prof biochem sci, Princeton Univ, 73-77; MEM, FRED HUTCHINSON CANCER RES CTR, 78- *Concurrent Pos:* Rita Allen Found scholar, 76-81; assoc ed, J Cell Biol; mem study sect molecular biol, NIH; investr, Howard Hughes Med Inst; asst ed, Sci & Cell. *Honors & Awards:* Lilley Award, 82; Lounsberry Award, Nat Acad Sci, 91. *Mem:* Nat Acad Sci; Am Acad Arts & Sci. *Res:* Structure, function and replication of eukaryotic chromosomes; cell transformation by avian tumor viruses; gene regulation and development; author of over 180 technical journal articles. *Mailing Add:* Fred Hutchinson Cancer Res Ctr 1124 Columbia St Mail Stop M-723 Genetics Seattle WA 98104

WEINTRAUB, HERBERT D, b New York, NY, Feb 17, 30; m 63; c 3. ANESTHESIOLOGY, MEDICAL EDUCATION. *Educ:* NY Univ, BA, 50; Oxford Univ, MA, 58, BM, BCh, 59. *Prof Exp:* Intern rotating, Strong Mem Hosp, Univ Rochester, 59-61; resident anesthesiol, Columbia Presby Hosp & Med Ctr, New York, 61-63, NIH res fel, 63-64; asst prof, Med Sch, Univ Pa, 64-69; asst prof, Med Sch, Univ Chicago, 69-71; assoc prof, George Washington Univ, 71-74, dir med educ, Dept Anesthesia, Med Ctr, 71-78, dir operating rm, 76-90, interim chmn, 91, PROF ANESTHESIOL, GEORGE WASHINGTON UNIV, 74-, ASSOC CHMN, DEPT ANESTHESIA, MED CTR, 79-,. *Concurrent Pos:* Mem sr staff, Hosp Univ Pa, 64-69; assoc dir, Dept Anesthesiol, Michael Reese Hosp, Chicago, 69-71. *Mem:* Asn Univ Anesthetists; Am Soc Anesthesiol; Int Anesthesia Res Soc; Soc Neurosurg Anesthetists & Neurol Supportive Care; Soc Cardiovascular Anesthesiologists. *Res:* Anesthesia for cardio-thoracic surgery; educational methods as applied to anesthesiology. *Mailing Add:* Dept Anesthesiol George Washington Univ Med Ctr Washington DC 20052

WEINTRAUB, HERSCHEL JONATHAN R, b Cincinnati, Ohio, Aug 19, 48; m 81. MOLECULAR GRAPHICS, DRUG DESIGN. *Educ:* Case Inst Technol, BS, 70; Case Western Reserve Univ, MS, 73, PhD(macromolecular sci), 75. *Prof Exp:* Res assoc, Dept Med Chem, Purdue Univ, 75-76, dir comput based educ, Sch Pharm, 76-77, asst prof med chem, 77-82; sr phys chemist, Lilly Res Labs, Eli Lilly & Co, 82-83; sr chemist, Abbott Lab, Chicago, Ill, 83-86; HEAD, THEORET CHEM DEPT, MARION MERRELL DOW RES INST, 86- *Concurrent Pos:* Nat Library Med fel, Dept Comput Sci, Univ Ill, Urbana, 78; prin investr, NIH grants, dept med chem, Purdue Univ, 79-81, adj prof med chem, 82-; consult, various pharmaceut & chem co, 78-; adj prof med chem, Univ Cincinnati, 86- *Mem:* Am Chem Soc; Inst Elec & Electronics Engrs; Biophys Soc; Int Soc Quantum Biol. *Res:* Drug design, with emphasis on solution conformational properties of drugs; modeling drug-receptor interactions; development of drug design software systems. *Mailing Add:* Marion Merrell Dow Res Inst 2110 E Galbraith Rd PO Box 156300 Cincinnati OH 45215

WEINTRAUB, HOWARD STEVEN, biopharmaceutics, for more information see previous edition

WEINTRAUB, JOEL D, b New York, NY, May 2, 42; m 68; c 2. ANIMAL BEHAVIOR, ENVIRONMENTAL SCIENCES. *Educ:* City Col New York, BS, 63; Univ Calif, Riverside, PhD(zool), 68. *Prof Exp:* Dir environ studies, 72-75, & 85, from asst prof to assoc prof, 68-77, PROF ZOOL, CALIF STATE UNIV, FULLERTON, 77- *Concurrent Pos:* Environ consult. *Mem:* Am Soc Ichthyol & Herpet; Sigma Xi; Ecol Soc Am; Herpetologists League. *Res:* Ecology of amphibians and reptiles; homing and orientation of vertebrates; urban ecology; vertebrate food habits; geographic information systems. *Mailing Add:* Dept Biol Sci Calif State Univ Fullerton CA 92634

WEINTRAUB, LEONARD, b New York, NY, Apr 21, 26; m 50; c 2. ORGANIC CHEMISTRY, PHARMACEUTICAL CHEMISTRY. *Educ:* City Col New York, BS, 48; Polytech Inst Brooklyn, MS, 54, PhD(org chem), 68. *Prof Exp:* Org chemist, Francis Delafield Hosp, NY, 50-54; org chemist, Bristol Myers Co, 54-62, group leader, 62-68, dept head org chem, 68-69, dir chem res, 69-85; CONSULT, 85- *Mem:* AAAS; Am Chem Soc; Royal Soc Chem; NY Acad Sci; Am Pharmaceut Asn. *Res:* Heterocyclic chemistry; molecular complexes in organic chemistry; salicylate chemistry; synthetic methods; pharmaceutical and cosmetic analysis. *Mailing Add:* 88 Greenwood Dr Millburn NJ 07041-1428

WEINTRAUB, LESTER, b New York, NY, Feb 1, 24; m 50; c 2. ORGANIC POLYMER CHEMISTRY. *Educ:* City Col New York, BS, 46; Fordham Univ, MS, 49; NY Univ, PhD(org chem), 54. *Prof Exp:* Sr chemist, Atomic Energy Comn Proj, Columbia Univ, 53-55, group leader, 55-58; group leader, Adv Res Proj Agency Proj, NY Univ, 58-59; sr chemist, Cent Res Labs, Air Reduction Co, Inc, NJ, 59-71; group leader polymer res, 71-73, TECH MGR POLYMERS & CONSULT SPECIALTY POLYVINYL CHLORIDE POLYMERS, PANTASOTE CO, INC, 85-; SR RES CHEM, PERMEABLE CONTACT LENS, INC, 86- *Mem:* Am Chem Soc; Am Inst Chem; Soc Plastics Eng. *Res:* Organic polymer synthesis and structure determination; polyvinyl chloride technology. *Mailing Add:* Ten Pardee Ct Mt Laurel NJ 08054

WEINTRAUB, LEWIS ROBERT, b New York, NY, Aug 15, 34; m 67; c 2. HEMATOLOGY. *Educ:* Dartmouth Col, AB, 55; Harvard Med Sch, MD, 58. *Prof Exp:* Intern med, Hosp, Univ Pa, 58-59; asst resident res in med, Hosp, Univ Mich, 59-61; fel hemat, Mt Sinai Hosp, New York, 61-62; res hematologist, Walter Reed Army Inst Res, 62-65; from asst prof to assoc prof med, Sch Med, Tufts Univ, 65-72; assoc prof, 72-77, PROF MED, SCH MED, BOSTON UNIV, 77- *Concurrent Pos:* Asst chief hemat, Walter Reed Gen Hosp, 62-65; asst hematologist & asst physician, New Eng Med Ctr Hosps, 65-72; assoc vis physician, Boston Univ Hosp, 72-, chief & physician, hemat sect, 77-; NIH res career develop award, 66. *Mem:* AAAS; AMA; Am Soc Hemat; NY Acad Med. *Res:* Clinical hematology and research in field of iron metabolism; chemotherapy trials for hematologic malignancies. *Mailing Add:* 75 E Newton Boston MA 02118

WEINTRAUB, MARVIN, b Radom, Poland, Oct 17, 24; nat Can; m 48; c 4. PLANT VIROLOGY. *Educ:* Univ Toronto, BA, 47, PhD(bot), 50. *Prof Exp:* Demonstr bot, Univ Toronto, 45-50; prin res scientist & head virus chem & physiol sect, Res Br, Can Dept Agr, 50-71, DIR, RES STA, AGR CAN, 71- *Concurrent Pos:* Hon prof, Univ BC, 71- *Honors & Awards:* Queen's Silver Jubilee Medal, 77. *Mem:* AAAS; Am Phytopath Soc; Can Phytopath Soc; fel NY Acad Sci; Int Soc Plant Morphol. *Res:* Metabolism and cytology of virus-infected plants; fine structure and electron microscopy; virus inhibitors; movement in plants. *Mailing Add:* Res Sta Agr Can 6660 NW Marine Dr Vancouver BC V6T 1X2 Can

WEINTRAUB, PHILIP MARVIN, b Cleveland, Ohio, Feb 22, 39; m 83; c 3. ORGANIC CHEMISTRY. *Educ:* Ohio State Univ, BSc, 60, MSc, 63, PhD(chem), 64. *Prof Exp:* Res chemist, Pioneering Res Lab, Textile Fibers Dept, E I du Pont de Nemours & Co, Inc, 64-66; sr chemist, Hess & Clark, 66-70; SR RES CHEMIST, MARION MERRELL DOW RES INST, 70- *Mem:* Am Chem Soc; NY Acad Sci. *Res:* Steroids; heterocycles; strained ring polycyclic systems; prostaglandins. *Mailing Add:* Marion Merrell Dow Res Inst 2110 E Galbraith Rd Cincinnati OH 45215

WEINTRAUB, ROBERT LOUIS, b Washington, DC, May 9, 12; m 38; c 3. PLANT & CELL PHYSIOLOGY. *Educ:* George Washington Univ, BS, 31, MA, 33, PhD(plant physiol), 38. *Prof Exp:* Biochemist, Div Radiation & Organisms, Smithsonian Inst, 37-47; supvry plant physiologist, Army Biol Labs, Ft Detrick, 47-59, phys sci adminr, Army Gen Staff, 59-63; prof bot, 63-77, EMER PROF BOT, GEORGE WASHINGTON UNIV, 77- *Concurrent Pos:* Plant physiologist, Smithsonian Radiation Biol Lab, 66-76; lectr bot, NC State Univ, 79- *Mem:* AAAS; Am Chem Soc; Am Inst Biol Sci; Am Soc Plant Physiol; Sigma Xi. *Res:* Effects of radiant energy on plants; plant growth regulators; microbial physiology. *Mailing Add:* 1657 31st St NW No 305 Washington DC 20007

WEINTRAUB, SOL, mathematics, for more information see previous edition

WEINZIMMER, FRED, b New York, NY, May 29, 25; m 47; c 2. NUCLEAR ENGINEERING, MECHANICAL ENGINEERING. *Educ:* Clarkson Col Technol, BME, 50; Polytech Inst Brooklyn, MME, 55. *Prof Exp:* Draftsman diesel design, US Dept Navy, NY, 43; engr, Am Gas & Elec Co, 50-51; asst proj engr, R R Popham, 51-52; engr, Ebasco Serv, Inc, 52-54; systs engr, Bettis Plant, Westinghouse Elec Corp, 54-55, lead systs engr, 55-56; systs tech supvr, Atomic Power Equip Dept, Gen Elec Co, 56-57, lead systs engr, 57-62, specialist, Vallecitos Exp Superheat Reactor Prog & Design, 62-65, prin engr, ECPL proj, 65-66, mgr proj anal, 67, proj mgr, Shoreham Proj, 67-71, proj mgr, Nine Mile Pt Unit 2, 71-74, proj mgr, Nuclear Energy Div, Somerset 1 & 2 Proj, 74-76, proj mgr, BWRSD-High Flow Hydraul Facil, 76-78, proj mgr, NPD-Clinton 1 & 2 Proj, 78-81; dir, proj eng, GPU Nuclear Corp, 81-82, dir spec projs, 82-84, mgr spec projs, Oyster Creek NGS, GPU nuclear, 84-89; RETIRED. *Res:* Major nuclear power plant operation, design, layout and construction for utility application. *Mailing Add:* Nine Coventry Terr Marlboro NJ 07746

WEINZWEIG, AVRUM ISRAEL, b Toronto, Ont, Apr 22, 26; m 53, 63; c 5. MATHEMATICS. *Educ:* Univ Toron[illegible] BASc, 50; Harvard Univ, AM, 53, PhD(math), 57. *Prof Exp:* Res physici[illegible] Weizmann Inst, 48-49; asst chief geophysicist, Weiss Geophys Corp, 50-5[illegible], instr math, Univ Calif, Berkeley, 57-59, actg asst prof, 59-60; asst prof, Northwestern Univ, 60-65; PROF MATH, STATIST & COMPUT SCI, UNIV ILL, CHICAGO, 65- *Concurrent Pos:* Consult, Solomon Schechter Day Schs, Ill, 64-80 & Second Int Study Math, 76-; dir, Inst Learning & Teaching Math, 84-; Math Adv Comt, Ill Off Educ, 84-86; pres, Int Comn for Study & Improv of Teaching of Math, 88-; consult, Sabin Magnet Sch, 88- *Mem:* Am Math Soc; Math Asn Am; Nat Coun Teachers Math; Math Asn Am; Asn Teachers Math. *Res:* Algebraic topology, particularly fiber spaces; category theory; learning theory; learning and acquisition of mathematical concepts; mathematics education. *Mailing Add:* Dept Math S&CS MC 249 Univ Ill Box 4348 Chicago IL 60680

WEIPERT, EUGENE ALLEN, b Monroe, Mich, Nov 17, 31; m 60; c 8. INDUSTRIAL ORGANIC CHEMISTRY. *Educ:* Univ Detroit, BS, 52, MS, 54; Iowa State Univ, PhD(org chem), 57. *Prof Exp:* Res supvr, Wyandotte Chem Corp, Mich, 58-72; tech mgr, Southern Sizing Co, 72-82; LAB MGR, MAZER CHEM CO, 82-; DIR RES & DEVELOP, PPG, MAZER CHEM CO, 88- *Res:* Alkylene oxides, amines; organometallics; organic reaction mechanisms; surfactants. *Mailing Add:* 211 68th St Kenosha WI 53143

WEIR, ALEXANDER, JR, b Crossett, Ark, Dec 19, 22; m 46; c 3. CHEMICAL ENGINEERING. *Educ:* Univ Ark, BSChE, 43; Polytech Inst Brooklyn, MChE, 46; Univ Mich, PhD, 54. *Prof Exp:* Analyst, W Bauxite Plant, Am Cyanamid & Chem Corp, Ark, 41, chemist, Berger Plant, 42, chem engr, Stanford Res Labs, Am Cyanamid Co, Conn, 43-47; from asst to proj supvr, Aircraft Propulsion Lab, Univ Mich, 48-57, lectr chem & metall eng, 54-56, asst prof, 56-58; from consult to asst mgr, Atlas Prog Off, Space Tech Labs, Ramo-Wooldridge Corp, 56-60; corp sr sci & tech adv res & develop mgt, Northrop Corp, 60-67, dir plans & progs, Corp Labs, 67-70; prin scientist air qual, Southern Calif Edison Co, 70-76, mgr, chem systs res & develop, 76-84, environ res, 84-85, chief scientist, 85-88; CONSULT, 88- *Mem:* AAAS; Am Geophys Union; Am Chem Soc; Combustion Inst; Am Inst Chem Engrs. *Res:* Effects of beta radiation on combustion; spectroscopic investigation of flames, temperatures and compositions behind detonation waves; flow through sonic orifices; location of Mach discs in supersonic jets; development of sulfur dioxide and nitrogen oxide removal systems for electric generating stations; fluegas desulferization; coal gasification (cool water project); solar central thermal power (solar one); solar salt ponds. *Mailing Add:* 8229 Villa Vista Dr Palos Verdas CA 90293

WEIR, BRUCE SPENCER, b Christchurch, NZ, Dec 31, 43; m 71; c 2. POPULATION GENETICS, QUANTITATIVE GENETICS. *Educ:* Univ Canterbury, NZ, BSc Hons, 65; NC State Univ, PhD(statist), 68. *Prof Exp:* Sr lectr statist, Massey Univ, 70-75, reader, 76; assoc prof, 76-81, PROF STATIST & GENETICS, NC STATE UNIV, 81- *Concurrent Pos:* Assoc ed, Genetics, 77-, Theoret Pop Biol, 79-87, Biomet, 84-89 & J Heredity, 90-; Guggenheim Fel, 83. *Mem:* Am Statist Asn; Biomet Soc; Am Soc Human Genetics; Genetics Soc Am; Soc Study Evolution; Am Soc Naturalists. *Res:* Development of statistical methodology for the analysis of genetic data; development of population genetics theory. *Mailing Add:* Dept Statist NC State Univ Box 8203 Raleigh NC 27695-8203

WEIR, EDWARD EARL, II, b St Petersburg, Fla, Mar 9, 45. CHEMISTRY, COMPUTER SOFTWARE. *Educ:* Ga Inst Technol, BS, 67; Fla State Univ, PhD(chem), 72. *Prof Exp:* Res assoc biochem, Fla State Univ, 67-72; res assoc oncol, Univ Ala, 72-73, instr biochem, 73-74; res assoc oncol, Johns Hopkins Hosp, 74-77; anal lab mgr & qual asurance coordr, Hittman Assocs, 77-82; VPRES TECH & ANALYTICAL DATA PROCESSING, 82- *Concurrent Pos:* Fels, NIH, 72-74, Am Cancer Soc, 74-76 & USPHS, 76-77. *Res:* Quality assurance coordinator as well as laboratory manager for organic and inorganic analytical laboratories dealing in coal, synthetic fuels, water and other media; computer programs dealing with data reduction, data base (particularly quality assurance), cost analysis and purchasing; biochemical ecology and pesticide fate and effects. *Mailing Add:* 602 St Johnsbury Rd Baltimore MD 21228-4046

WEIR, EDWARD KENNETH, b Jan 7, 43. PULMONARY VASCULAR RESEARCH, CLINICAL CARDIOLOGY. *Educ:* Univ Oxford, UK, BA, 64, MA, BM, Bch, 67, DM, 76; Royal Col Phys, MRCP, 71; Am Bd Internal Med, dipl, 80,81. *Prof Exp:* Assoc prof med, Univ Minn, 78-85; STAFF PHYSICIAN MED, VET ADMIN MED CTR, MINN, 78-; PROF MED, MED SCH, UNIV MINN, 85- *Mem:* Am Physiol Soc; Am Col Cardiol; Am Col Chest Physicians; Am Soc Exp Biol & Med; Am Thoracic Soc; Am Fedn Clin Res. *Res:* Pulmonary vascular reactivity and the mechanisms involved in hypoxic pulmonary vasoconstriction; sulfg hydryl redox status in cell division and hypertrophy. *Mailing Add:* Dept Med Vet Admin Med Ctr Univ Minn One Veterans Dr Minneapolis MN 55417

WEIR, JAMES HENRY, III, b East Orange, NJ, Sept 25, 32; m 58; c 3. MEDICAL RESEARCH. *Educ:* Princeton Univ, AB, 54; Columbia Univ, MD, 58. *Prof Exp:* From med res assoc to dir med serv, 63-75, DIR MED RES, WARNER-LAMBERT CO, 75-, VPRES REGULATORY & MED AFFAIRS, 81. *Concurrent Pos:* Macy teaching fel obstet-gynec, Col Physicians & Surgeons, Columbia Univ, 62-63. *Mem:* Am Fertil Soc; Am Fedn Clin Res; Sigma Xi. *Res:* Clinical investigation of new drugs and instrumentation devices; administration of government regulatory affairs and medical affairs. *Mailing Add:* Warner-Lambert Co 201 Tabor Rd Morris Plains NJ 07950

WEIR, JAMES ROBERT, JR, b Middletown, Ohio, Dec 29, 32; m 52, 81; c 5. METALLURGICAL ENGINEERING. *Educ:* Univ Cincinnati, BS, 55; Univ Tenn, MS, 61. *Prof Exp:* Metallurgist, 55-57, group leader metall, 60-66, asst sect chief, 67-70, sect chief, 70-73, ASSOC DIR, METALS & CERAMICS DIV, OAK RIDGE NAT LAB, 73- *Concurrent Pos:* Mem bd trustees, Am Soc Metals, 77-80. *Honors & Awards:* E O Lawrence Award, AEC, 73. *Mem:* Fel Am Soc Metals; fel AAAS. *Res:* High-temperature properties of metals; fatigue; radiation damage in metals, fuel element design. *Mailing Add:* Metals & Ceramics Div Oak Ridge Nat Lab PO Box 2008 Oak Ridge TN 37831-6134

WEIR, JOHN ARNOLD, b Saskatoon, Sask, Apr 5, 16; nat US; m 46; c 2. GENETICS. *Educ:* Univ Sask, BSA, 37; Iowa State Col, MS, 42, PhD(genetics), 48. *Prof Exp:* Asst animal breeding, Dom Exp Sta, Alta, 37-40; instr, Univ Sask, 42, assoc prof animal husb, 48-50; from asst prof to prof, 50-84, EMER PROF GENETICS, UNIV KANS, 84- *Concurrent Pos:* Consult, Animal Resources Adv Comt, USPHS, 64-67; USPHS spec fel & hon res assoc hist sci, Harvard Univ, 66-67. *Mem:* Genetics Soc Am; Am Genetic Asn; Sigma Xi. *Res:* Mammalian genetics; sex ratio and behavior of mice; history of genetics; history of aeronautics; history of agriculture. *Mailing Add:* Div Biol Univ Kans Lawrence KS 66045

WEIR, MICHAEL ROSS, B Austin, Tex, Dec 30, 42; m 64; c 3. OTITIS MEDIA, PEDIATRIC NEPHROLOGY. *Educ:* Harvard Univ, BA, 65; Univ Tex, Galveston, MD, 69. *Prof Exp:* Internship residency pediat, Letterman Army Med Ctr, 69-72; chief pediat, Vicenza Army Hosp, Italy, 72-76; staff pediatrician, William Beaumont Army Med Ctr, 76-78, chief outpatient serv, 78-82, asst chief pediat, 82-84, chief Dept Clin Invest, 84-87; CHIEF PEDIAT, MADIGAN ARMY MED CTR, 87- *Mem:* Am Acad Pediat. *Res:* Otitis media; pediatric nephrology; fluid and electrolyte abnormalities; toxicology. *Mailing Add:* Madigan Army Med Ctr Box 406 Tacoma WA 98431-5406

WEIR, ROBERT JAMES, JR, b Washington, DC, Nov 26, 24; m 47; c 3. TOXICOLOGY. *Educ:* Univ Md, College Park, BS, 48, MS, 50, PhD, 55. *Prof Exp:* Asst physiol, Univ Md, 48-51; res assoc toxicol, Hazleton Labs, Va, 51-56, head agr chem dept, 56-58, res applns specialist, 58-62, dir, Hazleton Labs, SA, Lausanne, Switz, 62-65, vpres, Inst Indust & Biol Res, Cologne, Ger, 63-66, vpres, Mkt, Hazleton Labs, Va, 67-69; VPRES, LITTON BIONETICS, INC, 69- *Mem:* AAAS; Am Chem Soc (secy-treas, Agr & Food Div, 67-68); Soc Cosmetic Chemists; Environ Mutagen Soc; Soc Toxicol; Sigma Xi. *Res:* Toxicology; biochemistry; pharmacology; safety evaluation of drug, cosmetic, food chemical and pesticide development. *Mailing Add:* 1606 Simmons Ct McLean VA 22101-5154

WEIR, RONALD DOUGLAS, b St John, NB, Jan 10, 41; m 63; c 2. PHYSICAL CHEMISTRY. *Educ:* Univ NB, BSc, 63; Univ London, DIC & PhD, 66. *Prof Exp:* Nat Res Coun Can fel, 66-68, from asst prof to assoc prof, 68-81, PROF ENG, ROYAL MIL COL CAN, 76-, PROF CHEM ENG, 81-, HEAD DEPT CHEM & CHEM ENG, 90- *Concurrent Pos:* Sr vis inorganic chem, Oxford Univ, 78-79; adj prof chem, Queen's Univ, 86-; vis prof chem, Univ Mich, 85-90. *Mem:* Chem Inst Can; Can Soc Chem Engrs; fel Royal Soc Chem; Am Soc Eng Educ; Int Union Pure & Appl Chem; Calorimetry Conf. *Res:* Thermodynamic properties and orientational disorder in crystals; low temperature calorimetry; dielectrics; equations of state and intermolecular forces. *Mailing Add:* Dept Chem & Chem Eng Royal Mil Col Can Kingston ON K7K 5L0 Can

WEIR, WILLIAM CARL, b Lakeview, Ore, Aug 24, 19; m 46; c 2. RESEARCH ADMINISTRATION. *Educ:* Ore State Col, BS, 40; Univ Wis, MS, 41, PhD(animal husb, biochem), 48. *Prof Exp:* Asst, Univ Wis, 45-46, instr, 47; assoc prof animal sci, Ore State Col, 48; from asst prof to prof, Univ Calif, Davis, 48-73, dean students, 58-65, prof nutrit & chmn dept, 73-81, assoc prog dir, Small Ruminant Prog, 81-88, assoc dean, int progs, 85-90; RETIRED. *Concurrent Pos:* Fulbright res grant, Univ Western Australia, 65-66; mem comt sheep nutrit, Nat Acad Sci-Nat Res Coun; vis scientist & Univ Calif rep, Univ Chile-Univ Calif Coop Prog, Santiago, Chile, 70-72. *Mem:* Am Soc Animal Sci. *Res:* Sheep nutrition; sheep and goat management. *Mailing Add:* 887 Linden Lane Davis CA 95616

WEIR, WILLIAM DAVID, b Oakland, Calif, Mar 15, 41; c 2. PHYSICAL CHEMISTRY. *Educ:* Occidental Col, AB, 62; Princeton Univ, AM, 63, PhD(chem), 65. *Prof Exp:* Instr chem, Harvard Univ, 65-68; from asst prof to assoc prof chem, Reed Col, 68-86. *Concurrent Pos:* Consult, indust & govt. *Mem:* Am Chem Soc; Electrochem Soc; Asn Comput Machinery; Inst Elec & Electronics Engrs. *Res:* Chemical kinetics; kinetics and mechanisms of electrochemical reactions and dynamics of membrane function; relaxation methods; instrumentation and computation in chemical research; theoretical protein dynamics; computer graphics and interactive modelling applications in chemistry. *Mailing Add:* 104 Hilltop Dr Los Gatos CA 95030

WEIR, WILLIAM THOMAS, b Wildwood, NJ, Dec 23, 31; m 54; c 4. SYSTEMS ENGINEERING, OPERATIONS RESEARCH. *Educ:* Drexel Univ, BSEE, 54, MS, 58; Univ Pa, PhD(systms eng, opers res), 72. *Prof Exp:* Component engr elec eng, RCA Corp, 54-56; mgr syst eng, Gen Elec Co, 56-73; PRES & CHMN BD DIRS, EVAL ASSOCS, INC, 73- *Concurrent Pos:* Adj prof physics, Drexel Univ, 58-, adj prof eng mgt, 63-; mem, US Sci Deleg, Peoples Repub China, 79. *Mem:* Sr mem Inst Elec & Electronics Engrs (treas, past vpres); Oper Res Soc Am. *Res:* Reliability. *Mailing Add:* Eval Assocs Inc One Belmont Ave Bala Cynwyd PA 19004

WEIRES, RICHARD WILLIAM, JR, fruit entomology; deceased, see previous edition for last biography

WEIRICH, GUNTER FRIEDRICH, b Eisenach, Ger, Feb 17, 34; div. INSECT ENDOCRINOLOGY. *Educ:* Univ Munich, PhD(zool), 63. *Prof Exp:* Res assoc endocrinol, Philipps Univ, Marburg, Ger, 64-65; res fel insect physiol, Biol Div, Oak Ridge Nat Lab, 65-66; res assoc endocrinol, Philipps Univ, 66-70; sr scientist biol chem, Zoecon Corp, 70-73; res assoc biol chem, Tex A&M Univ, 73-77; RES ENTOMOLOGIST BIOL CHEM, BELTSVILLE AGR RES CTR-EAST, USDA, 77- *Mem:* AAAS; Ger Soc Biol Chem; Europ Soc Comp Endocrinol. *Res:* Metabolism of insect hormones. *Mailing Add:* Insect Neurobiol & Hormone Lab Bldg Beltsville Agr Res Ctr-East USDA Beltsville MD 20705

WEIRICH, WALTER EDWARD, b Saginaw, Mich, Nov 20, 38; m 60; c 2. VETERINARY SURGERY. *Educ:* Mich State Univ, BS, 61, DVM, 63; Univ Wis, MS, 70, PhD(vet sci, cardiol), 71; Am Col Vet Surgeon, dipl, 78. *Prof Exp:* Officer in chg, Vet Corps, US Army, 63-65; pract vet, Madison Vet Clin, 65-68; NIH fel cardiol, Univ Wis-Madison, 68-71; from asst prof to assoc prof cardiol & surg, 71-75, actg head dept cardiol & surg, 75-76, head dept small animal clin, sch vet med, 76-87, PROF CARDIOL & SURG, PURDUE UNIV, WEST LAFAYETTE, 78- *Concurrent Pos:* Mem cardiol comt, Am Animal Hosp Assoc, 73-; reviewer, J Am Vet Med Asn. *Mem:* Am Vet Med Asn; Am Acad Vet Cardiol (pres elect); Am Asn Vet Med Cols; Sigma Xi; Am Asn Vet Clin. *Res:* Hypothermia for cardiac arrest surgery; myocardial infarctions and vascular surgery in the canine. *Mailing Add:* Vet Clin Sch Vet Med Purdue Univ West Lafayette IN 47907

WEIS, DALE STERN, b Cleveland, Ohio, Oct 11, 24; m 70; c 2. MICROBIOLOGY, PROTOZOOLOGY. *Educ:* Western Reserve Univ, BS, 45, MS, 51; Yale Univ, PhD, 55. *Prof Exp:* Instr bact, Albertus Magnus Col, 53-54; res assoc plant physiol, Univ Minn, 55-57; res assoc biochem, Univ Chicago, 57-58; from instr to asst prof biol, Univ Col, 58-64; chmn natural sci 2, Shimer Col, 64-66; from asst prof to assoc prof, 66-77, PROF BIOL, CLEVELAND STATE UNIV, 77- *Concurrent Pos:* Charles F Kettering fel, Univ Minn, 55-57. *Mem:* AAAS; Soc Protozool; Am Soc Microbiol; Am Soc Cell Biol. *Res:* Metabolism of algae; symbiosis; host-symbiote interaction; cell-cell interaction and recognition during infection. *Mailing Add:* 21067 Kenwood Ave Cleveland OH 44116

WEIS, JERRY SAMUEL, b Salina, Kans, Dec 23, 35; m 61; c 3. BIOLOGY. *Educ:* Kans Wesleyan Univ, AB, 58; Univ Kans, MA, 60, PhD(bot), 64. *Prof Exp:* Asst prof biol, Univ Minn, 64-65; NIH fel biol, Yale Univ, 65-66; asst prof, 66-71, asst dir div biol, 69-73, ASSOC PROF BIOL, KANS STATE UNIV, 72-, ASSOC DIR DIV BIOL, 75- *Mem:* AAAS; Inst Soc Ethics & Life Sci; Am Asn Higher Educ. *Res:* Bioethics. *Mailing Add:* Div Biol Kans State Univ Manhattan KS 66506

WEIS, JUDITH SHULMAN, b New York, NY, May 29, 41; m 62; c 2. AQUATIC BIOLOGY, ECOLOGY. *Educ:* Cornell Univ, BA, 62; NY Univ, MS, 64, PhD(biol), 67. *Prof Exp:* Lectr biol, Hunter Col, 64-67; from asst prof to assoc prof, 67-76, assoc dean, 85-86, PROF ZOOL, RUTGERS UNIV, NEWARK, 76- *Concurrent Pos:* Rutgers Res Coun grant, 67-76, NJ sea grant, Nat Oceanic & Atmospheric Admin, 77-; AAAS Cong Sci fel, 83-84; consult, Environ Protection Agency, 84-, NOAA, NJ Dept Environ Protection; mem bd, Am Inst Biol Sci, 87-89, 90-92 & Soc Environ Toxicol & Chem, 90-93; prog dir, NSF, 88-90. *Mem:* Am Soc Zoologists; Am Inst Biol Sci; Estuarine Res Fedn; Soc Environ Toxicol & Chem; AAAS; Am Fisheries Soc; Asn Women Sci; fel AAAS. *Res:* Marine biology; effects of pollutants on aquatic animals; ecology. *Mailing Add:* Dept Biol Sci Rutgers Univ Newark NJ 07102

WEIS, LEONARD WALTER, b New York, NY, June 23, 23; m 55; c 2. GEOLOGY. *Educ:* Harvard Univ, SB, 43; Mass Inst Technol, SM, 47; Univ Wis, PhD(geol), 65. *Prof Exp:* Res observer, Blue Hill Meteorol Observ, 43; asst meteorol, Mass Inst Technol, 44-47; instr geol & geog, Univ RI, 47-49; asst prof geol & actg chmn dept, Coe Col, 53-54; asst prof, Lawrence Univ, 55-65; ASST PROF GEOL, UNIV WIS CTR-FOX VALLEY, 65- *Mem:* Am Meteorol Soc; Geol Soc Am; Am Geophys Union; Geochem Soc; NY Acad Sci; Sigma Xi. *Res:* Igneous and metamorphic petrology; glacial geology, including petrography of sediments; paleoclimatology; meteorological instruments and observations. *Mailing Add:* 120 N Green Bay Rd Appleton WI 54911

WEIS, PAUL LESTER, b Chicago, Ill, June 22, 22; m 45, 69; c 2. GEOLOGY. *Educ:* Univ Wis, BS, 47, PhD(geol), 52. *Prof Exp:* Geologist, US Geol Surv, 50-51 & 53-81; asst prof geol, Univ Va, 51-53; RETIRED. *Concurrent Pos:* Consult geologist, 81- *Mem:* Am Inst Prof Geologists; fel Geol Soc Am; Soc Econ Geol. *Res:* Mineral resources; areal geology of northwestern United States; geochemistry of stable carbon isotopes; metallic mineral deposits in sedimentary rocks; creation "science" and its impact on science education in public schools. *Mailing Add:* S 5106 Sunward Dr Spokane WA 99223

WEIS, PEDDRICK, b South Paris, Maine, June 4, 38; m 62; c 2. ANATOMY, EMBRYOLOGY. *Educ:* NY Univ, DDS, 63. *Prof Exp:* NSF res fel, 63-64; instr anat, Col Dent, NY Univ, 64-67; from asst prof to assoc prof, 67-78, PROF ANAT, NJ MED SCH & GRAD SCH BIOMED SCI, UNIV MED & DENT NJ, 78- *Mem:* AAAS; Am Asn Anat; Soc Environ Toxicol Chem. *Res:* Ultrastructural and biochemical effects of environmental pollutants, especially heavy metals; effects of pollutants on development and behavior of aquatic organisms. *Mailing Add:* Dept Anat NJ Med Sch Univ Med & Dent NJ Newark NJ 07103

WEIS, ROBERT E(DWARD), b Ark, May 11, 18; m 48; c 4. CHEMICAL ENGINEERING. *Educ:* Univ Cincinnati, MS, 41. *Prof Exp:* Asst chem eng, Univ Cincinnati, 40-41; asst chem engr, Phillips Petrol Co, 41-42, assoc chem engr, 42-43, master chem engr, 43-44, sr chem engr, 44-48, pilot plant mgr, 48-51, asst supt, Philtex Exp Sta, 51, supt, 51-53, mgr, Bartlesville Develop Pilot Plant, 53-60, asst to mgr, Process Develop Div, 60-61, systs eng adminr, 61-63, mgr, Process Optimization Br, 63-68; dir, Process Optimization Dept, Appl Automation, Inc, 68-77; res & develop proj coordr, Phillips Petrol Co, Bartlesville, 77-82, facilities, planning, res & develop proj coord, 82-84; RETIRED. *Mem:* AAAS; Am Chem Soc; Am Inst Chem Engrs. *Res:* Process optimization; development and application of digital computer systems with on-line instrumentation and control to test, model, evaluate, and optimize operations and profitability of complex petroleum and chemical processes. *Mailing Add:* Box 553 Bartlesville OK 74005-0553

WEISBACH, JERRY ARNOLD, b New York, NY, Dec 23, 33; m 58; c 3. ORGANIC CHEMISTRY, MEDICINAL CHEMISTRY. *Educ:* Brooklyn Col, BS, 55; Harvard Univ, MA, 56, PhD(chem), 59. *Prof Exp:* Sr med chemist, Smith Kline & French Labs, 60-65, group leader, 65-67, assoc dir chem, 67-71, assoc dir res, US Pharmaceut Prod, 71-75, dep dir res, 75-77, vpres res, 77-79; PRES, PARKE DAVIS PHARMACEUT RES DIV, WARNER LAMBERT CORP, 79-, VPRES, 81- *Mem:* AAAS; Am Chem Soc; Acad Pharmaceut Sci; NY Acad Sci; Am Soc Microbiol. *Res:* Structure, isolation and synthesis of natural products, particularly antibiotics, alkaloids and lipids; organic biochemistry and synthetic medicinal chemistry. *Mailing Add:* Pharmaceut Res Div Warrer Lambert Co 2800 Plymouth Rd Ann Arbor MI 48105

WEISBART, MELVIN, b Toronto, Ont, Dec 28, 38; m 63; c 3. COMPARATIVE ENDOCRINOLOGY, COMPARATIVE PHYSIOLOGY. *Educ:* Univ Toronto, BSc, 61, MA, 63; Univ BC, PhD(physiol), 67. *Prof Exp:* Nat Res Coun Can fel, Fisheries Res Bd Can, 68-69; asst prof biol, Wayne State Univ, 69-76; res assoc, Univ Wash, 76-78; from asst prof to prof biol, St Francis Xavier Univ, 76-89, chmn, dept biol, 83-89; PROF & HEAD, DEPT BIOL, UNIV REGINA, 89- *Concurrent Pos:* Fac res award, Wayne State Univ, 70; pres, Can Soc Zoologists, 88-89. *Honors & Awards:* Fel, AAAS, 88. *Mem:* AAAS; Am Soc Zoologists; Can Soc Zoologists; Am Physiol Soc; Soc Protection Old Fishes. *Res:* Fish physiology and endocrinology; evolution of corticosteroids; role of corticosteroids in salmonid adaptation to marine environment; steroid receptors in salmonid egg maturation. *Mailing Add:* Dept Biol Univ Regina 3737 Wascana Pkwy Regina SK S4S 0A2 Can

WEISBECKER, HENRY B, b Kassel, Ger, July 20, 25; US citizen; m 72; c 2. ELECTRICAL ENGINEERING. *Educ:* Pratt Inst, BEE, 45; NY Univ, MEE, 48; Munich Tech Univ, Dr Ing(elec eng), 57. *Prof Exp:* Sr engr, A B Dumont Labs, 50-51, W L Maxson Corp, 52-54 & Simmons Aerocessories, 54-58; dir res elec eng, Manson Labs, Conn, 58-59 & Loral Inc, 60; sr prin engr, Litton Industs, 60-66; assoc prof elec eng, NJ Inst Technol, 66-72; INDEPENDENT ENG CONSULT, 72- *Concurrent Pos:* Independent consult engr, 72- *Mem:* Sr mem Inst Elec & Electronics Engrs. *Res:* Transistors; servos; infrared techniques; communications; author of textbooks and numerous articles. *Mailing Add:* 712 Tenth St Union City NJ 07087

WEISBERG, HERBERT, b New York, NY, June 30, 31. MEDICINE. *Educ:* City Col New York, BS, 53; Univ Lausanne, MD, 58. *Prof Exp:* From instr to asst prof med, NY Med Col-Flower & Fifth Ave Hosp, 64-71, assoc prof anat & med, 71-76; PRECEPTOR FAMILY MED, UNIV CALIF, DAVIS, 77- *Concurrent Pos:* USPHS trainee gastroenterol, NY Med Col-Flower & Fifth Ave Hosp, 62-64, USPHS spec fel electron micros, 66-68; vis prof, Sch Med, NY Univ, 71. *Mem:* Am Fedn Clin Res; Am Gastroenterol Asn. *Res:* Intracellular pathway of absorption for nutrients in the intestine, especially vitamin B-12. *Mailing Add:* 6525 Seward Park Ave S Seattle WA 98118

WEISBERG, JOSEPH SIMPSON, b Jersey City, NJ, June 7, 37; m 64; c 2. OCEANOGRAPHY, METEOROLOGY. *Educ:* Jersey City State Col, BA, 60; Montclair State Col, MA, 64; Columbia Univ, EdD(earth sci educ), 69. *Prof Exp:* Teacher pub schs, NJ, 60-64; chmn dept geosci, 73-83, PROF GEOSCI, JERSEY CITY STATE COL, 60-, DEAN, SCH ARTS & SCI, 83- *Concurrent Pos:* Sci consult, US Off Educ, 68-69; adv, NJ Dept Environ Protection, 75- *Honors & Awards:* Award Merit, US Environ Protection Agency, 78. *Mem:* AAAS; Geol Soc Am; Am Meteorol Soc; Nat Asn Geol Teachers; Nat Sci Teachers Asn. *Res:* Use of visual aids in science teaching; inquiry and learning; environmental aspects of the geosciences; author of various works on oceanography and meteorology; use of microcomputers in education. *Mailing Add:* Dean Arts & Sci Jersey City State Col 2039 Kennedy Mem Blvd Jersey City NJ 07305

WEISBERG, ROBERT H, b Brooklyn, NY, May 20, 47. PHYSICAL OCEANOGRAPHY, OCEAN CIRCULATION. *Educ:* Cornell Univ, BS, 69; Univ RI, MS, 72, PhD(phys oceanog), 75. *Prof Exp:* Res asst, Grad Sch Oceanog, Univ RI, 69-74, res assoc phys oceanog, 74-76; prof, Dept Marine, Earth & Atmospheric Sci, NC State Univ, 76-88. *Concurrent Pos:* Assoc prof dept marine sci, Univ SFla, 84-85. *Mem:* Am Geophys Union; Am Meteorol Soc. *Res:* Equatorial, coastal, and estuarine circulation dynamics. *Mailing Add:* Dept Marine Sci Univ Fla 140 Seventh Ave S St Petersburg FL 33701

WEISBERG, SANFORD, US citizen; m; c 1. APPLIED STATISTICS. *Educ:* Univ Calif, AB, 69; Harvard Univ, AM, 70, PhD(statist), 73. *Prof Exp:* PROF APPL STATIST, UNIV MINN, ST PAUL, 72- *Honors & Awards:* Youden Prize, 89. *Mem:* Am Statist Asn; Inst Math Statist; Biomet Soc; Royal Statist Soc. *Res:* Data analysis and methods; linear models; statistical computing; graphical methods. *Mailing Add:* Dept Appl Statist Univ Minn St Paul MN 55108

WEISBERG, STEPHEN BARRY, b New York, NY, Apr 19, 54; m 84. POPULATION BIOLOGY, ICHTHYOLOGY. *Educ:* Univ Mich, BGS, 74; Univ Del, PhD(biol), 81. *Prof Exp:* Res asst nutrient dynamics, Univ Del, 77-80; instr ecol & wildlife biol, West Chester State Col, 80-81; RES SCIENTIST FISH ECOL, MARTIN MARIETTA ENVIRON CTR, 81- *Concurrent Pos:* Adj asst prof, Univ Md, 86- *Mem:* Ecol Soc Am; Am Soc Zoologists; Am Fisheries Soc; Atlantic Estuarine Res Soc; Estuarine Res Fedn. *Res:* Population dynamics and feeding ecology of fish; impacts of fish feeding on population dynamics of prey species; effects of hydroelectric operations on aquatic biota. *Mailing Add:* 5474 Delphinium Ct Columbia MD 21045

WEISBERGER, WILLIAM I, b New York, NY, Dec 20, 37; m 61; c 3. THEORETICAL HIGH ENERGY PHYSICS. *Educ:* Amherst Col, BA, 59; Mass Inst Technol, PhD(physics), 64. *Prof Exp:* Res assoc physics, Stanford Linear Accelerator Ctr, 64-66; asst prof, Princeton Univ, 66-70; PROF PHYSICS, STATE UNIV NY, STONY BROOK, 70- *Concurrent Pos:* Sloan Found fel, 67-69; vis scientist, Weizmann Inst Sci, 68-69 & 74-75; Guggenheim fel, 74-75; vis prof, Univ Wash, 78-79. *Mem:* Fel Am Phys Soc; AAAS. *Mailing Add:* Inst Theoret Physics State Univ NY Stony Brook NY 11794-3840

WEISBLAT, DAVID IRWIN, organic chemistry; deceased, see previous edition for last biography

WEISBORD, NORMAN EDWARD, geology, invertebrate paleontology; deceased, see previous edition for last biography

WEISBRODT, NORMAN WILLIAM, b Cleves, Ohio, June 30, 42; m 65; c 3. PHYSIOLOGY, PHARMACOLOGY. *Educ:* Univ Cincinnati, BS, 65; Univ Mich, Ann Arbor, PhD(pharmacol), 70. *Prof Exp:* USPHS res fel, Univ Iowa, 70-71; from instr to asst prof physiol, 71-75, ASSOC PROF PHYSIOL & PHARMACOL, UNIV TEX MED SCH HOUSTON, 75- *Concurrent Pos:* Fel, Univ Iowa, 70-71; assoc prof, Univ Tex Grad Sch Biomed Sci, 71-; res scientist develop, Nat Inst Drug Abuse, 76-81; assoc ed, Am J Physiol, 77-80. *Mem:* Am Asn Clin Res; Am Physiol Soc; Soc Exp Biol & Med; Am Gastroenterol Asn. *Res:* Smooth muscle physiology and pharmacology; gastrointestinal motility. *Mailing Add:* Dept Physiol & Pharmacol Univ Tex Med Sch PO Box 20708 Houston TX 77225

WEISBROT, DAVID R, b Brooklyn, NY, Dec 29, 31; m 60; c 3. POPULATION GENETICS. *Educ:* Brooklyn Col, BS, 53, MA, 58; Columbia Univ, PhD(zool), 63. *Prof Exp:* Substitute instr biol, Brooklyn Col, 56-58; res asst genetics, Univ Conn, 58-59 & Long Island Biol Asn, 59-60; instr biol, City Col New York, 60-62; fel genetics, Univ Calif, Berkeley, 63-64; asst prof biol, Tufts Univ, 64-69; assoc prof, State Univ NY Binghamton, 69-72; assoc prof, 72-78, PROF BIOL, WILLIAM PATERSON COL, NJ, 78- *Concurrent Pos:* Adj assoc prof, Columbia Univ, 76-81. *Mem:* AAAS; Genetics Soc Am; Am Genetic Asn; Sigma Xi. *Res:* Genotypic interactions among competing strains of Drosophila; relationship of genetic and morphological differences among sibling species; cytogenetics. *Mailing Add:* 1103 Sussex Rd Teaneck NJ 07666

WEISBROTH, STEVEN H, b New York, NY, Sept 16, 34; m 58; c 3. LABORATORY ANIMAL MEDICINE, COMPARATIVE PATHOLOGY. *Educ:* Cornell Univ, BS, 58; Wash State Univ, MS, 60, DVM, 64; Am Col Lab Animal Med, dipl. *Prof Exp:* NIH fel lab animal med, NY Univ Med Ctr, 64-66; asst prof path & dir animal facilities, Rockefeller Univ, 66-69; asst prof path & dir, Dept Lab Animal Med, State Univ NY Stony Brook, 69-70, assoc prof path & dir, Div Lab Animal Resources, 70-78; PRES, ANIMAL MED LABS, INC, 70- *Concurrent Pos:* NIH fel lab animal med, Rockefeller Univ, 69-; mem coun, Am Asn Accreditation Lab Animal Care, 75-79. *Honors & Awards:* Res Award, Am Asn Lab Animal Sci, 72. *Mem:* Am Vet Med Asn; Am Asn Lab Animal Sci; Am Soc Exp Path. *Res:* Spontaneous diseases in laboratory animals. *Mailing Add:* 10800 S Glen Rd Rockville MD 20855-2701

WEISBURGER, ELIZABETH KREISER, b Greenlane, Pa, Apr 9, 24; div; c 3. ONCOLOGY, TOXICOLOGY. *Educ:* Lebanon Valley Col, BS, 44; Univ Cincinnati, PhD(org chem), 47. *Hon Degrees:* DSc, Univ Cincinnati, 81; DSc, Lebanon Valley, 89. *Prof Exp:* Res assoc, Univ Cincinnati, 47-49; res fel, Nat Cancer Inst, 49-51, res org chemist, Biochem Lab, 51-61, carcinogen screening sect, 61-72, chief, Carcinogen Metab & Toxicol Br, 73-78, chief, Lab Carcinogen Metab, 78-81, asst dir chem carcinogen, Div Cancer Etiology, 81-89; CONSULT, 89- *Concurrent Pos:* Asst chief ed, J Nat Cancer Inst, 71-87. *Honors & Awards:* Hillebrand Prize & Garvan Medal, Am Chem Soc, 81. *Mem:* AAAS; Am Asn Cancer Res; Am Chem Soc; Royal Soc Chem; Soc Toxicol; Am Soc Biol Chem & Molecular Biol. *Res:* Metabolism of chemical carcinogens, carcinogen testing, chemical carcinogenesis and toxicology. *Mailing Add:* 5309 McKinley St Bethesda MD 20814-1413

WEISBURGER, JOHN HANS, b Stuttgart, Ger, Sept, 15, 21; nat US; c 3. NUTRITION, ENVIRONMENTAL HEALTH. *Educ:* Univ Cincinnati, AB, 47, MS, 48, PhD(org chem & cancer res), 49. *Hon Degrees:* MD, Umea Univ, 80. *Prof Exp:* Fel, Nat Cancer Inst, 49-50, head phys-org chem unit, Lab Biochem, 50-61, head carcinogen screening sect, 61-72, dir bioassay segment, 71-72; vpres res, Am Health Found & dir Naylor Dana Inst Dis Prev, Valhalla, NY, 72-87; RES PROF PATH, NY MED COL, 74-; PRES, WEISBURGER ASSOCS, NORTH WHITE PLAINS, NY, 87- *Concurrent Pos:* assoc ed, J Nat Cancer Inst, 60-62, Cancer Res, 69-80, Arch Toxicol, 77-87, J Am Col Toxicol, 82-, Prev Med, 89-; mem, Working Groups Monagraphs Series, Int Agency Res Cancer; mem, Dept Health Educ Welfare, Fed Drug Admin, US Dept Agr, Pub Health Serv, Interdept Tech Panel on Carcinogens, 62-71; chmn, Approach 1-3, Nat Cancer Prog Strategic Plan, 71-74; chmn, subcomt positive controls, Nat Acad Sci, Nat Res Coun conf on Carcinogen Testing of New Drugs, 73; mem, US Dept Agr Expert Panel on Nitrites & Nitrosamines, 73-77; chmn, Wkshp Colorectal Cancer, Int Union Against Cancer, Geneva, 75; mem, Nat Cancer Inst Clearinghouse Environ Carcinogens, 76-78; mem, orgn comt, Princess Takamatsu Cancer Res Found, Seventh Int Symp, 76; co-chmn, orgn comt, US-Japan Coorp Wkshp GI Tract Cancer, 79; chmn, Symp Large Bowel Cancer, Organ Gastroenterol Seventh Cong, 82; chmn, Int Union Against Cancer Natural Carcinogens Human Cancer Develop, 82; mem, Panel Irritants & Vesicants Comn Life Sci, Nat Res Coun, Nat Acad Sci, 83; co-chmn, Prof Educ Comt Westchester Div, Am Cancer Soc, 83-89 & 85-88; coun, Europ Asn Cancer Res, 85-89; chmn, Sci Rev Panel, NJ State Comn Cancer Res, 88-90; mem, sci adv bd, Am Water Works Asn, 88-; co-chmn, workshop health effects tea, 91; invited lectr numerous nat & int conf. *Honors & Awards:* Merit Award, Soc Toxicol, 81, Ambassador Award, 90; Distinguished Serv Award, Am Soc Prev Oncol, 90. *Mem:* Am Asn Cancer Res; Am Soc Biochem & Molecular Biol; Am Soc Pharmacol & Exp Therapeut; Soc Exp Biol & Med; Soc Toxicol; Am Chem Soc; Am Gastroenterol Asn; fel NY Acad Sci; Sigma Xi; Am Soc Prev Oncol; Biochem Soc. *Res:* Etiology of cancer, mechanisms of carcinogenesis; bioassay and metabolism of carcinogens and drugs; host factors in cancer induction and development; nutrition, endocrinology, immunology, cancers of the endocrine, digestive and excretory organs; preventive medicine; preventive oncology. *Mailing Add:* Am Health Found Naylor Dana Inst Dis Prev Valhalla NY 10595-1599

WEISE, CHARLES MARTIN, b Bridgeville, Pa, July 8, 26; m 51; c 5. POPULATION ECOLOGY, ORNITHOLOGY. *Educ:* Ohio Univ, BS, 50; Univ Ill, MS, 51, PhD(zool), 56. *Prof Exp:* Asst prof biol, Fisk Univ, 53-56; from asst prof to assoc prof, 55-66, chmn dept, 61-63, 78-84, PROF BIOL SCI, UNIV WIS-MILWAUKEE, 66-, CHMN DEPT, 89- *Mem:* AAAS; Cooper Ornith Soc; Wilson Ornith Soc; Am Soc Zoologists; Am Ornith Union; Sigma Xi. *Res:* Annual Physiological and reproductive cycles in birds; field ornithology; behavioral ecology and population ecology. *Mailing Add:* Dept Biol Sci Univ Wis Milwaukee WI 53201

WEISE, JURGEN KARL, b Nov 7, 37; US citizen; m 64; c 3. POLYMER CHEMISTRY, ORGANIC CHEMISTRY. *Educ:* Univ Bonn, BS, 60; Polytech Inst Brooklyn, PhD(polymer chem), 66. *Prof Exp:* Ger Res Asn fel chem, Univ Mainz, 66-67; res chemist, Cent Res Dept, E I du Pont de Nemours & Co, 67-71, res chemist, Elastomer Chem Dept, 71-77; mgr res lab, Bostik Tech Ctr, Europe 77-82; mgr profit ctr, Schill & Seilacher, Hamburg, Europe, 83-84; TECH RES DIR, DENTSPLY, GER, 86- *Mem:* Am Chem Soc. *Res:* Reactions of polymers; polymer structures and their effects on properties; ring-opening polymerization; fluoro-polymers; specialty elastomers; adhesives and sealants development. *Mailing Add:* AM Hohen Berg 30 Usingen 6390 Germany

WEISEL, GEORGE FERDINAND, JR, b Missoula, Mont, Mar 21, 15; m 50; c 2. ZOOLOGY. *Educ:* Univ Mont, BS, 41, MA, 42; Univ Calif, Los Angeles, PhD(zool), 49. *Prof Exp:* Asst zool, Univ Mont, 41-42, Univ Mich, 42-43 & Scripps Inst Oceanog, Univ Calif, San Diego, 47-48; from instr to assoc prof comp anat & gen zool, Univ Mont, 47-69, prof comp anat & ichthyol, 69-90; RETIRED. *Mem:* Am Soc Ichthyol & Herpet; Am Soc Zoologists; Sigma Xi. *Res:* Anatomy, histology, sex organs, life histories, osteology and endocrinology of fish. *Mailing Add:* 615 Pattee Canyon Rd Missoula MT 59801

WEISENBERG, RICHARD CHARLES, b Columbus, Ohio, Apr 2, 41. CELL BIOLOGY. *Educ:* Univ Calif, Santa Barbara, BA, 63; Univ Chicago, PhD(biophys), 68. *Prof Exp:* USDA trainee, Brandeis Univ, 68-70; asst prof, 71-73, ASSOC PROF BIOL, TEMPLE UNIV, 73- *Concurrent Pos:* Investr, Marine Biol Lab, Woods Hole, 71- *Mem:* Am Soc Cell Biol. *Res:* Cell division and motility; biochemistry of microtubules. *Mailing Add:* Dept Biol Temple Univ Philadelphia PA 19122

WEISENBORN, FRANK L, b Portland, Ore, Feb 26, 25; div; c 4. ORGANIC CHEMISTRY, INFORMATION SCIENCE. *Educ:* Reed Col, BA, 45; Univ Wash, PhD(chem), 49. *Prof Exp:* Atomic Energy Comn fel, Harvard Univ, 49-50, USPHS fel, 50-51; sr res chemist, Riker Labs, Inc, 51-53; res assoc, Univ Calif, Los Angeles, 53; sr res chemist, 53-59, from res assoc to sr res assoc, 59-63, dir org chem, 63-81, DIR SCI INFO, SQUIBB INST MED RES, E R SQUIBB & SONS, INC, 81- *Concurrent Pos:* Chmn, Gordon Res Conf Natural Prod, 75-76. *Mem:* Am Chem Soc; Royal Soc Chem. *Res:* Structure and synthesis of antibiotics, steroids and alkaloids; natural and synthetic hypotensive agents; computer assisted molecular modeling; information science. *Mailing Add:* Squibb Inst Med Res PO Box 4000 Princeton NJ 08540

WEISER, ALAN, b Houston, Tex, July 16, 55; m 78; c 2. NUMERICAL ANALYSIS. *Educ:* Rice Univ, BA, 76; Yale Univ, MS, 77, PhD(comput sci), 81. *Prof Exp:* RES MATHEMATICIAN, EXXON PROD RES CO, 81- *Mem:* Soc Indust & Appl Math; Am Math Soc. *Res:* Numerical reservoir simulation; numerical linear algebra; numerical methods for hyperbolic equations. *Mailing Add:* 4050 Grennoch Houston TX 77025

WEISER, CONRAD JOHN, b Middlebury, Vt, June 20, 35. PLANT STRESS PHYSIOLOGY. *Educ:* NDak State Univ, BS, 57; Ore State Univ, PhD(hort), 60. *Prof Exp:* Asst prof, Hort Dept, Univ Minn, 61-63, assoc prof, 63-66, prof, 66-73; PROF HORT & DEPT HEAD, ORE STATE UNIV, 73- *Concurrent Pos:* Chair, Nat Agr Sci Team, 80-, Temperature Stress in Plants, Gordon Conf, 89; mem, adv coun, NSF, Int Div, 81-84, bd agr, Nat Acad Sci, Nat Res Coun, 86-92, Nat Arboretum Adv Coun, 88-99, Coun, AAAS, 92-94; co-chair, Ore State Univ Long-Range Planning Comn, 86-87. *Honors & Awards:* Alex Laurie Award, Am Soc Hort Sci, 66, J H Gourley Award, 73. *Mem:* Fel Am Soc Hort Sci (vpres, 79-80, pres-elect, 79-80, pres, 80-81); Am Soc Plant Physiol; Soc Cryobiol; fel AAAS; Coun Agr Sci Tech. *Res:* Crop stress physiology, plant freezing injury and physiological responses which permit plants to acclimate in response to environmental stimuli. *Mailing Add:* Dept Hort Cordley Hall 2042 Ore State Univ Corvallis OR 97331

WEISER, DAN, b St Louis, Mo, July 10, 33; m 54; c 4. MATHEMATICS. *Educ:* Rice Univ, BA, 54, MA, 56, PhD(math), 58. *Prof Exp:* Asst math, Rice Univ, 57-58; sr res mathematician, Field Res Lab, Mobil Res & Develop Co, 58-77, STATISTICIAN, MOBIL EXPLOR & PROD SERV, INC, 77- *Mem:* Soc Indust & Appl Math. *Res:* Applied mathematics and political science. *Mailing Add:* 3851 Rugged Circle Dallas TX 75224

WEISER, DAVID W, inorganic chemistry; deceased, see previous edition for last biography

WEISER, KURT, b Vienna, Austria, Dec 24, 24; US citizen; m 57; c 3. PHYSICS. *Educ:* Harvard Univ, BA, 49; Cornell Univ, PhD(phys chem), 54. *Prof Exp:* Mem tech staff, RCA, 54-58 & T J Watson Ctr, IBM Corp, 58-73; PROF DEPT ELEC ENG & DIR, SOLID STATE INST, TECHNION, 73- *Mem:* Fel Am Phys Soc. *Res:* Solid state physics with emphasis on semiconductors. *Mailing Add:* Dept Elec Eng Technion Haifa Israel

WEISER, PHILIP CRAIG, b Portland, Ore, Oct 26, 41; m 66; c 1. EXERCISE PHYSIOLOGY, REHABILITATION. *Educ:* Univ Wash, BS, 63; Univ Minn, MS, 67, PhD(physiol), 69. *Prof Exp:* Adj prof biol, Univ Colo, Boulder, 70-76; clin asst prof prev med, Med Ctr, Univ Colo, Denver, 76-80; dir, Dept Physiol, Nat Asthma Ctr, Denver, 77-80; prog dir, Community Fitness Ctr, Inst Health Educ, Wheat Ridge Colo, 80-81; dir, Ore Fitness & Health Ctr, Good Samaritan Hosp, Portland, Ore, 81-83; res physiologist, Pulmonary Div, Va Med Ctr, Philadelphia, 83-86; DIR, CARDIO-PULMONARY FITNESS REHAB CTR, MED COL PA, PHILADELPHIA, 86-, RES ASSOC PROF, DEPT MED, 87- *Concurrent Pos:* Res physiologist, Physiol Div, Med Res & Nutrit Lab, Fitzsimons Army Med Ctr, Med Serv Corps, US Army, 68-71, chief mil performance br, 71-73; res fel, Cardiovasc Pulmonary Res Lab, Med Ctr, Univ Colo, Denver, 73-75; res fel, Dept Clin Physiol, Nat Asthma Ctr, Denver, 75-76; assoc fel, Harvard Univ, 83. *Mem:* Am Heart Asn; Am Physiol Soc; Am Thoracic Soc; Am Col Sports Med. *Res:* Biological adaptation to stress, particularly the psychophysiological factors limiting physical performance in aging individuals and in patients with cardiovascular and pulmonary disease. *Mailing Add:* Pulm Diag & Rehab Dept Med Col Penn 3300 Henry Ave Philadelphia PA 19119

WEISER, ROBERT B(RUCE), b Ashley, Ohio, July 19, 27; m 53, 72; c 5. CHEMICAL ENGINEERING, ORGANIC CHEMISTRY. *Educ:* Ohio State Univ, BS & MS, 51, PhD, 54. *Prof Exp:* Res engr, Exp Sta, Polychem Dept, E I du Pont de Nemours & Co, Inc, Del, 54-58, Res & Develop Lab, Wash Works, 58-90; CONSULT, 90- *Concurrent Pos:* Lectr, Marietta Col, 61-65. *Mem:* Am Inst Chem Engrs. *Res:* Reaction kinetics; diffusional operations; fluid mechanics. *Mailing Add:* One Fox Hill Dr Parkersburg WV 26101

WEISER, RUSSEL SHIVELY, b Grimes, Iowa, Sept 28, 06; m 31; c 1. BACTERIOLOGY, IMMUNOLOGY. *Educ:* NDak Col, BS, 30, MS, 31; Univ Wash, PhD(bact), 34. *Prof Exp:* Assoc bact & path, 34-36, from instr to assoc prof & actg exec officer, 36-45, assoc prof microbiol, 45-49, prof, 49-77, EMER PROF IMMUNOL, SCHS MED & DENT, UNIV WASH, 77- *Concurrent Pos:* Mem leprosy res panel, US-Japan Coop Med Sci Prog. *Mem:* Am Asn Immunologists; Reticuloendothelial Soc; Brit Soc Immunol; Transplantation Soc. *Res:* Immunology of syphilis; immunology of cancer; immunologic tissue injury; macrophages; cell-mediated immunity. *Mailing Add:* 174 Tiburon Ct Aptos CA 95003

WEISFELD, LEWIS BERNARD, b Philadelphia, Pa, July 2, 29; m 66. PLASTICS PRODUCTS LIABILITY. *Educ:* Univ Pa, BA, 52; Univ Del, MS, 53, PhD(phys-org chem), 56. *Prof Exp:* Res engr, Silicone Prod Dept, Gen Elec Co, 56-58; sr res scientist, US Rubber Co, 58-63; dir labs, Advan Div, Cincinnati Milacron Chems, 63-74, vpres & sci dir, 74-75, dir res, Cincinnati Milacron Inc, 75-79; dir res & develop, Plastics Div, Container Corp Am, 79; CONSULT CHEMIST, WEISFELD & ASSOCS, 79- *Concurrent Pos:* Comnr, Nat Cert Comn Chem & Chem Eng, 83-89. *Mem:* Am Chem Soc; Am Inst Chemists; fel Soc Plastics Engrs; NY Acad Sci; AAAS; Am Soc Testing & Mat; Asn Consult Chemists & Chem Engrs (pres, 82-84). *Res:* Government regulatory developments related to plastics industry. *Mailing Add:* One Franklin Town Blvd Suite 1204 Philadelphia PA 19103

WEISFELDT, MYRON LEE, b Milwaukee, Wis, Apr 25, 40; m 63; c 3. CARDIOLOGY, CARDIOPULMONARY RESUSCITATION. *Educ:* Johns Hopkins Univ, BA, 62, MD, 65. *Prof Exp:* House officer med, Columbia Presbyterian Med, 65-67; clin assoc, NIH, 67-69; house officer med, Mass Gen Hosp, 69-70, fel cardiol, 70-72; from asst prof to assoc prof, 72-78, PROF MED, JOHNS HOPKINS UNIV, 78- *Concurrent Pos:* Mem res rev comt A, Nat Heart Lung Blood Inst, 82-86, chmn cardiol adv comt, 87-90, prin investr heart dis, SCOR, 77- *Mem:* Fel AAAS; Am Soc Clin Invest; Am Asn Physicians; Am Heart Asn (pres, 89-90); Am Col Cardiol; Asn Univ Professors. *Res:* Hemodynamics and left ventricular function, particularly using magnetic resonance imaging and spectroscopy; cardiopulmonary resuscitation hemodynamics and age associated changes in the cardiovascular system. *Mailing Add:* Johns Hopkins Hosp Blalock 536 Baltimore MD 21205

WEISGERBER, DAVID WENDELIN, b Delphos, Ohio, May 20, 38; m 65; c 2. INFORMATION SCIENCE. *Educ:* Bowling Green State Univ, BS, 60; Univ Ill, PhD(org chem), 65. *Prof Exp:* Chemist res, E I du Pont de Nemours & Co, Inc, 64-69; assoc indexer, 69-71, group leader compound name data base, 71-72, sr indexer, 72-73, asst to ed, 73-77, mgr chem substance handling, 77-79, dir ed opers, 79-82, ED & DIR, CHEM ABSTRACTS SERV, 82- *Mem:* Am Chem Soc; NY Acad Sci; Am Soc Info Sci. *Res:* Synthetic organic chemistry; chemical nomenclature; scientific information storage and retrieval. *Mailing Add:* 6178 Middlebury Dr E Worthington OH 43085

WEISGERBER, GEORGE AUSTIN, b Philadelphia, Pa, Dec 31, 18; m 47; c 2. PETROLEUM CHEMISTRY. *Educ:* Philadelphia Col Pharm, BSc, 40; Univ Del, MSc, 50, PhD(org chem), 51. *Prof Exp:* Res chemist, Johnson & Johnson, NJ, 40-48; from res chemist to sr chemist, Esso Res & Eng Co, 51-60, from res assoc to sr res assoc, 60-74, SR RES ASSOC, EXXON RES & ENG CO, 74- *Mem:* Am Chem Soc; Tech Asn Pulp & Paper Indust; Am Soc Testing & Mat. *Res:* Petroleum product research; additives, burner fuels, industrial and motor lubricants; wax; asphalt. *Mailing Add:* 208 Oak Lane Cranford NJ 07016

WEISGRABER, KARL HEINRICH, b Norwich, Conn, July 13, 41; m 64; c 2. ORGANIC CHEMISTRY, BIOCHEMISTRY. *Educ:* Univ Conn, BA, 64, PhD(org chem), 69. *Prof Exp:* Staff fel, Nat Inst Arthritis & Metab Dis, 69-70 & Nat Res Coun Res Assoc, USDA, Calif, 70-71; prin scientist, Meloy Labs, Va, 71-72; sr staff fel, Nat Heart & Lung Inst, 72-81. *Mem:* Am Chem Soc; The Chem Soc; NY Acad Sci. *Res:* Experimental atherosclerosis; study of serum lipoproteins; mechanism of transport of serum constituents across aortic endothelium. *Mailing Add:* 1018 Rudgear Rd Walnut Creek CA 94596

WEISHAUPT, CLARA GERTRUDE, botany, for more information see previous edition

WEISHEIT, JON CARLETON, b Mt Vernon, Wash, Oct 10, 44; m 65; c 2. ATOMIC PHYSICS, ASTROPHYSICS. *Educ:* Univ Tex, El Paso, BS, 66; Rice Univ, MS, 69, PhD(physics), 70. *Prof Exp:* Res fel astron, Harvard Univ, 70-72, lectr astron, 71-72; physicist, Lawrence Livermore Nat Lab, Univ Calif, 72-79, group leader atomic physics, 78-79; res physicist, Plasma Physics Lab, Princeton Univ, 79-81, lectr astron, 80-81; staff scientist, Lawrence Livermore Nat Lab, 81-87; PROF SPACE PHYSICS & ASTRON, RICE UNIV, 88- *Mem:* Fel Am Phys Soc; Am Astron Soc; Int Astron Union; Sigma Xi. *Res:* Atomic processes in laboratory and cosmic plasmas; extragalactic astronomy. *Mailing Add:* Space Physics & Astron Dept Rice Univ Houston TX 77251

WEISLEDER, DAVID, b New York, NY, Sept 30, 39; m; c 4. ORGANIC CHEMISTRY, NUCLEAR MAGNETIC RESONANCE. *Educ:* City Col New York, BS, 61; Univ Cincinnati, PhD(chem), 66. *Prof Exp:* RES CHEMIST, NAT CTR AGR UTILIZATION RES, AGR RES, USDA, 66- *Mem:* Am Chem Soc. *Res:* Nuclear magnetic resonance spectroscopy of natural products and their derivatives or analogs. *Mailing Add:* Nat Ctr Agr Utilization Res USDA Agr Res Serv 1815 N University St Peoria IL 61604

WEISLER, LEONARD, b Rochester, NY, Sept 9, 12; m 42; c 4. ORGANIC CHEMISTRY. *Educ:* Univ Rochester, BS, 34, MS, 36, PhD(org chem), 39. *Prof Exp:* Asst, Univ Rochester, 35-38; anal chemist, Distillation Prod, Inc, 42-53; org res & develop chemist, Paper Serv Div, Eastman Kodak Co, 53-78; RETIRED. *Concurrent Pos:* Sr lectr chem, Univ Rochester, 54-84; adj prof chem, Monroe Community Col, Rochester, NY, 79-84. *Mem:* Am Chem Soc. *Res:* Alkylation of nitro compounds; chemistry and biochemistry of vitamin E; synthesis of vitamin A; fats, oils and hydrocarbons; metallurgy; photographic chemistry. *Mailing Add:* Six Eastland Ave Rochester NY 14618

WEISLOW, OWEN STUART, b Cleveland, Ohio, Mar 11, 38. IMMUNOLOGY, MICROBIOLOGY. *Educ:* Delaware Valley Col Sci & Agr, BS, 65; Thomas Jefferson Univ, MS, 68, PhD(microbiol), 70. *Prof Exp:* Fel immunol, Albert Einstein Med Ctr, Philadelphia, 70-71; fel immunol & virol, Thomas Jefferson Univ, 71-73, instr, 73-74; assoc found scientist immunol, Southwest Found Res & Educ, 74-76; scientist III, 76-80, SR SCIENTIST IMMUNOL, FREDERICK CANCER RES CTR, LITTON BIONETICS INC, 80- *Mem:* Am Soc Microbiol; AAAS; Am Asn Immunologists. *Res:* Immunobiology of oncogenic viruses; tumor immunology; regulation of cell proliferation. *Mailing Add:* Head Anti-AIDS Virus Drug Screen Lab NCI Frederick Cancer Res & Develop Ctr Bldg 430 Box B Frederick MD 21701

WEISMAN, GARY RAYMOND, b Cincinnati, Ohio, Apr 17, 49; m 71; c 2. ORGANIC CHEMISTRY, PHYSICAL CHEMISTRY. *Educ:* Univ Ky, BS, 71; Univ Wis, PhD(org chem), 76. *Prof Exp:* Res chemist org chem, Univ Calif, Los Angeles, 76-77; ASST PROF ORG CHEM, UNIV NH, 77- *Concurrent Pos:* Prin investr, Petrol Res Found grant, 78-81 & res corp grant, 79- *Mem:* Am Chem Soc; Sigma Xi. *Res:* Organic host guest chemistry; conformational analysis; applications of nuclear magnetic resonance; structure and chemistry of radical ions; organic electrochemistry and electron spin resonance in mechanistic investigations. *Mailing Add:* Dept Chem Parsons Hall Univ NH Durham NH 03824

WEISMAN, HAROLD, b Brooklyn, NY, Oct 24, 28; c 2. ANESTHESIOLOGY. *Educ:* Univ Calif, Los Angeles, BA, 52, DO, 56, MD, 62. *Prof Exp:* Intern, Pac Hosp, Long Beach, Calif, 56-57; pvt pract gen med, 57-66; resident anesthesia, 66-68, asst prof in residence surg anesthesia, 69-71, ASST PROF ANESTHESIOL, SCH MED, UNIV CALIF, LOS ANGELES, 71- *Concurrent Pos:* Parke-Davis grant, 70; McNeil Labs grant, 71; secy-treas, Community Hosp North Hollywood, Calif, 59-61, from asst chief of staff to chief of staff, 61-63; mem consult staff, Bel Air Mem Hosp, 68, Valley Presby Hosp, Van Nuys, 68-69, Vet Admin Wadsworth Gen Hosp, 71 & Mt Sinai Hosp, Los Angeles, 72-; mem fac, Jules Stein Eye Inst. *Mem:* AMA; AAAS; Am Soc Anesthesiologists; Int Anesthesia Res Soc; fel Am Col Anesthesiologists. *Res:* Effect of various anesthetic agents on the oculocardiac reflex; effect of general anesthetics on intraocular pressure; retinal artery flow during anesthesia and surgery; determination of damage to retina with Doppler microwave energy output in New Zealand albino rabbits exposed chronically. *Mailing Add:* 4159 Adlon Pl Encino CA 91436

WEISMAN, HARVEY, b Winnipeg, Man, Feb 25, 27; m 57; c 2. HISTOLOGY, NEUROPHARMACOLOGY. *Educ:* Univ Man, BSc, 53, MSc, 55, PhD(histol), 69. *Prof Exp:* Lectr zool, 58-65, asst prof, 65-72, ASSOC PROF PHARMACOL, UNIV MAN, 82-, PROF, ASSOC HEAD, DIR GRAD STUDIES FAC MED, 82- *Mem:* Can Asn Anatomists; Pharmacol Soc Can. *Res:* Histological, cytological and biochemical reorganization of the central nervous system in mammals as a result of psychopharmacological drugs; histopathology of primary and secondary cardiomyopathics; pathological effects of maternal smoking on fetal growth and health. *Mailing Add:* Dept Pharmacol & Therapeut Univ Man Fac Med 753 McDermot Ave Winnipeg MB R3E 0W3 Can

WEISMAN, JOEL, b New York, NY, July 15, 28; m 55. CHEMICAL ENGINEERING, NUCLEAR ENGINEERING. *Educ:* City Col New York, BChE, 48; Columbia Univ, MS, 49; Univ Pittsburgh, PhD, 68. *Prof Exp:* Chem engr, Etched Prod Corp, NY, 50-51; assoc chem engr, Brookhaven Nat Lab, 51-54; from engr to sr engr, 54-57, supvry engr, Atomic Power Dept, Westinghouse Elec Corp, 58-59; sr engr, Nuclear Develop Corp Am, NY, 59-60; fel engr, 60-66, mgr thermal & hydraul anal, Atomic Power Dept, Westinghouse Elec Corp, Pa, 67-68; from asst prof to assoc prof, 68-72, dir, Nuclear Energy Prog, 76-86, PROF NUCLEAR ENG, UNIV CINCINNATI, 72- *Concurrent Pos:* Consult, nuclear power indust, 69- *Mem:* Fel Am Nuclear Soc; Am Inst Chem Engrs. *Res:* Nuclear power reactor technology; experimental heat transfer and fluid flow; design of large scale experiments; system optimization. *Mailing Add:* Dept Mech Indust & Nuclear Eng Univ Cincinnati Cincinnati OH 45221

WEISMAN, R(OBERT) BRUCE, b Baltimore, Md, Nov 23, 50; m 86. MOLECULAR PHOTOPHYSICS AND PHOTOCHEMISTRY. *Educ:* Johns Hopkins Univ, BA, 71; Univ Chicago, PhD(chem), 77. *Prof Exp:* Fel, Univ Pa, 77-79; asst prof, 79-84, ASSOC PROF CHEM, RICE UNIV, 84- *Concurrent Pos:* Res Fel, Alfred P Sloan Found, 85. *Mem:* Am Phys Soc; AAAS; Sigma Xi. *Res:* Use of time-resolved laser spectroscopies to study molecular photophysics and photochemistry; laser spectroscopy. *Mailing Add:* Dept Chem Rice Univ PO Box 1892 Houston TX 77251

WEISMAN, ROBERT A, b Kingston, NY, Dec 16, 36; m 63; c 3. BIOCHEMISTRY. *Educ:* Union Univ, NY, BS, 58; Mass Inst Technol, PhD(biochem), 63. *Prof Exp:* Staff fel biochem, Sect Cellular Physiol, Lab Biochem, Nat Heart Inst, 63-66; asst prof biochem, Med Col Pa, 66-68; from asst prof to assoc prof, 68-74, prof biochem, Univ Tex Health Sci Ctr San Antonio, 74-77; chmn dept biomed, Wright State Univ, 87-89, dir, biomed sci PhD prog, 71-86, PROF BIOCHEM, SCH MED & COL SCI & ENG, WRIGHT STATE UNIV, 77-, ASSOC DIR, MAGNETIC RES LAB, 85- *Mem:* Am Soc Biol Chemists. *Res:* Magnetic resonance; position emission tomography. *Mailing Add:* Dept Biochem Wright State Univ Glenn Hwy Dayton OH 45435

WEISMAN, RUSSELL, b Cleveland, Ohio, Jan 20, 22; m 47; c 4. MEDICINE, HEMATOLOGY. *Educ:* Western Reserve Univ, AB, 44, MD, 46. *Prof Exp:* Res fel med, 52-54, instr, 54-55, sr instr clin med & path, 55-57, asst prof, 57-63, assoc prof, 63-80, PROF MED, CASE WESTERN RESERVE UNIV, 80- *Mem:* AAAS; Am Soc Hemat; Am Fedn Clin Res; Sigma Xi; Am Asn Blood Banks; Cent Soc Clin Res. *Res:* Hemolytic anemia; immunohematology; relation of the spleen and blood destruction. *Mailing Add:* 2074 Abington Rd Cleveland OH 44106

WEISMANN, THEODORE JAMES, b Pittsburgh, Pa, Apr 21, 30. PHYSICAL CHEMISTRY. *Educ:* Duquesne Univ, BS, 52, MS, 54, PhD(phys chem), 56. *Prof Exp:* RES DIR, GULF RES & DEVELOP CO, 56- *Mem:* Am Chem Soc; Geochem Soc; Am Phys Soc; Am Soc Mass Spectrometry; AAAS; Sigma Xi. *Res:* Boron chemistry; mass spectrometry; organic and isotopic geochemistry; marine geochemistry; molecular structure of organic radicals; magnetic susceptibilities; geochronometry. *Mailing Add:* 106 Short St Pittsburgh PA 15237

WEISMILLER, RICHARD A, b Elwood, Ind, Feb 23, 42; m 66; c 2. AGRONOMY. *Educ:* Purdue Univ, BS, 64, MS, 66; Mich State Univ, PhD(soil chem, clay mineral), 69. *Prof Exp:* Spec scientist soil stabilization, USAF Weapons Lab, 69-73; res agronomist remote sensing, Purdue Univ, West Lafayette, 73- 83; PROF, SOIL & WATER RESOURCES, UNIV MD, 83- *Mem:* Am Soc Agron; Soil Sci Soc Am; Clay Minerals Soc; Soil & Water Conserv Soc; Sigma Xi. *Res:* Relation of spectral reflectance of soils to their physicochemical properties; application of remote sensing to soils mapping, land use inventories and change detection as related to land use; BMP's for nonpoint source pollution control. *Mailing Add:* Dept Agron Univ Md College Park MD 20742

WEISS, ALAN, b Cleveland, Ohio, Dec 5, 55; m 78; c 2. LARGE DEVIATIONS. *Educ:* Case Western Reserve Univ, BS(math) & BS(physics), 76; NY Univ, MA, 79, PhD(math), 81. *Prof Exp:* MEM TECH STAFF, AT&T BELL LABS, 81- *Concurrent Pos:* Vis lectr, Systs Res Ctr, Univ Md, 86. *Mem:* Oper Res Soc Am. *Res:* Applications of the theory of large deviations to computer and communication systems; efficiency of certain algorithms for parallel computation. *Mailing Add:* AT&T Bell Labs Rm 2C-118 600 Mountain Ave Murray Hill NJ 07974

WEISS, ALVIN H(ARVEY), b Philadelphia, Pa, Apr 28, 28; m; c 2. CHEMICAL ENGINEERING, CATALYSIS. *Educ:* Univ Pa, BS, 49, PhD(phys chem), 65; Newark Col Eng, MS, 55. *Prof Exp:* Chem engr, Fiber Chem Co, 49-51, US Army Chem Ctr, Edgewood Proving Ground, 51-53, Colgate Palmolive Co, 53-55 & Houdry Process Corp, 55-63; res assoc interdisciplinary res, Inst Coop Res, Univ Pa, 63-65, res investr, 65-66; assoc prof, 66-68, PROF CHEM ENG, WORCESTER POLYTECH INST, 68- *Concurrent Pos:* Lectr, Univ Pa, 64-66. *Mem:* AAAS; fel Am Inst Chem Engrs; Am Chem Soc; Catalysis Soc (secy, 68-); Ger Soc Chem Apparatus. *Res:* Petroleum and petrochemical processing; hydrodealkylation; hydrodechlorination; dehydrogenation; kinetics and mechanisms of complex reactions; catalysis. *Mailing Add:* Dept Chem Eng Worcester Polytech Inst Worcester MA 01609

WEISS, ANDREW W, b Streator, Ill, Mar 13, 30; m 61; c 3. ATOMIC PHYSICS, THEORETICAL PHYSICS. *Educ:* Univ Detroit, BS, 52, MS, 54; Univ Chicago, PhD(atomic physics), 61. *Prof Exp:* PHYSICIST, NAT BUR STANDARDS, 61- *Honors & Awards:* Dept Com Silver Medal; Fel, Am Phys Soc. *Mem:* Am Phys Soc. *Res:* Application of electronic computers to the determination of the electronic structure and properties of atoms and simple molecules. *Mailing Add:* Div 531 Nat Bur Standards & Technol Gaithersburg MD 20899

WEISS, ARTHUR JACOBS, b Philadelphia, Pa, Apr 11, 25; m 52; c 3. INTERNAL MEDICINE. *Educ:* Pa State Col, BS, 45; Univ Pa, MD, 50. *Prof Exp:* Instr, 57-62, ASST PROF MED, JEFFERSON MED COL, 62- *Concurrent Pos:* Am Heart Asn advan res fel, 57- *Mem:* Am Asn Cancer Res; Am Soc Hemat; Am Soc Pharmacol & Exp Therapeut; Am Col Physicians; Am Soc Clin Oncol. *Res:* Long-term storage of various tissues; hematology; malignant diseases; oncology. *Mailing Add:* Dept Med Thomas Jefferson Univ Med Col 1025 Walnut St Philadelphia PA 19107

WEISS, BENJAMIN, b Newark, NJ, Nov 16, 22; m 47; c 6. BIOCHEMISTRY, ORGANIC CHEMISTRY. *Educ:* Univ Iowa, BS, 44; Univ Ill, PhD(biochem), 49. *Prof Exp:* Asst inorg chem, Univ Ill, 47-49; res biochemist, Harper Hosp, 49-53; res assoc, Columbia Univ, 53-65, asst prof brain metab, Col Physicians & Surgeons & mem staff, NY State Psychiat Inst, 53-76; RES SCIENTIST, INST NEUROCHEM, WARD'S ISLAND, NY, 76- *Mem:* Am Chem Soc; Am Soc Biol Chemists. *Res:* Chemistry and biochemistry of long chain bases; synthesis of long chain base antimetabolites; lipid metabolism; chemical modification of enzymes; isolation of natural products. *Mailing Add:* Inst Neurochem Ward's Island NY 10035

WEISS, BERNARD, b New York, NY, May 27, 25; m 78; c 2. NEUROTOXICOLOGY. *Educ:* NY Univ, BA, 49; Univ Rochester, PhD(psychol), 53. *Prof Exp:* Res assoc psychol, Univ Rochester, 53-54; exp & physiol psychologist, Sch Aviation Med, US Air Force, 54-56; instr med, Sch Med, Johns Hopkins Univ, 56-65, from instr to asst prof pharmacol, 59-65; assoc prof radiation biol & biophys & brain res, 65-67, prof radiation biol & biophys, psychol & brain res, Sch Med & Dent, 67-79, PROF TOXICOL & DEP DIR, ENVIRON HEALTH SCI CTR, UNIV ROCHESTER, 79- *Concurrent Pos:* Mem behav pharmacol comt, NIMH, 65-67; mem comt biol effects of atmospheric pollutants, Nat Acad Sci-Nat Res Coun, 71-74, Sci Adv Bd, Environ Protection Agency, 83-; partic, US-USSR Environ Health Exchange Prog, 73-79; mem comt, Toxicol Study Sect, NIH, 81-85 & complex mixtures, Nat Acad Sci-Nat Res Coun, 85-88, Comt Neurotoxicol & Risk Assessment, 88-, sci adv bd, Environ Protection Agency, 81- *Honors & Awards:* Stockinger Award, Am Conf Govt Indust Hygienists, 90. *Mem:* AAAS; Am Soc Pharmacol & Exp Therapeut; Am Psychol Asn; Behav Pharmacol Soc (pres, 61-64); Soc Toxicol. *Res:* Chemical influences on behavior. *Mailing Add:* Environ Health Sci Ctr Univ Rochester Sch Med & Dent Rochester NY 14642

WEISS, BERNARD, b New York, NY, July 17, 36; m 66; c 1. DNA REPAIR. *Educ:* Columbia Univ, AB, 56; Col Physicians & Surgeons, Columbia Univ, MD, 60. *Prof Exp:* Fel biochem, Harvard Med Sch, 65-67; from asst prof to assoc prof microbiol, 67-80, PROF MOLECULAR BIOL & GENETICS, SCH MED, JOHNS HOPKINS UNIV, 80- *Mem:* Am Soc Microbiol; Am Soc Biol Chemists. *Res:* Enzymatic repair of DNA in bacteria; microbial genetics. *Mailing Add:* Dept Microbiol Johns Hopkins Univ Sch Med 720 Rutland Ave Baltimore MD 21205

WEISS, C DENNIS, b Oklahoma City, Okla, July 27, 39; m 62. COMPUTER SCIENCE, ENGINEERING. *Educ:* Stanford Univ, BS, 61; Columbia Univ, MS, 62, PhD(elec eng), 66. *Prof Exp:* Asst prof elec eng, Johns Hopkins Univ, 66-72; mem tech staff, Bell Tel Labs, Inc, 72-82; dir, Integrated Systs Develop Lab, AT&T Info Systs, 82-90; DIR, UNIX SYST SOFTWARE, UNIX SYST LABS, 90- *Mem:* Inst Elec & Electronics Engrs. *Res:* Computers; communication systems; software engineering. *Mailing Add:* UNIX Syst Labs 190 River Rd Rm 1-301 Summit NJ 07901

WEISS, CHARLES, JR, b San Francisco, Calif, Dec 20, 37; m 69; c 2. SCIENCE POLICY, ENVIRONMENT. *Educ:* Harvard Univ, AB, 59, PhD(chem physics, biochem), 65. *Prof Exp:* Teaching fel, Harvard Univ, 62-64; NIH fel biophys, Lab Chem Biodyn, Lawrence Radiation Lab, Univ Calif, 67-69; biophysicist, IBM Watson Lab, Columbia Univ, 69-71; sci & technol adv, Int Bank Reconstruct & Develop, 71-86; adj prof, Univ Pa, 86-90; CHMN, INT TECHNOL DEVELOP & FINANCE, INC, 86-; VIS LECTR, WOODROW WILSON SCH, PRINCETON UNIV, 89- *Concurrent Pos:* Mem corp bd, Vols Tech Assistance, Nat Climate Adv Bd, Coun Foreign Rels & World Acad Arts & Sci. *Mem:* AAAS; Am Phys Soc; Am Chem Soc; Soc Int Develop. *Res:* Science and technology in developing countries and Eastern Europe; environmental policy in developing countries and Eastern Europe; role of development assistance organizations in technological research and technology transfer; venture capital in developing countries. *Mailing Add:* 6309 Crathie Lane Bethesda MD 20816

WEISS, CHARLES FREDERICK, b Cohoctah, Mich, Apr 2, 21; m 47; c 3. PEDIATRICS, PHARMACOLOGY. *Educ:* Univ Mich, BA, 42; Vanderbilt Univ, MD, 49; Am Bd Pediat, dipl, 60. *Prof Exp:* Pvt pract, 54-58; med coordr clin invest in pediat & virus res, Parke, Davis & Co, Mich, 58-69; from instr to assoc prof pharm & pharmacol & assoc prof pediat, Col Med, Univ Fla, 69-73; chief of staff, Hope Haven Children's Hosp, 73-77; coordr prof affairs & consult pediat to surgeon gen, USAF, 77-84; vpres med affairs & chmn bd sci adv, himedics, 85-89; MED EXEC DIR, SUNLAND CTR, ORLANDO, 84-; DIR PLANNING & DEVELOP & CONSULT PEDIAT PHARMACOL, DAKLE, INC, 85- *Concurrent Pos:* Staff physician, Children's Hosp, Mich, 54-69 & Receiving Hosp, Detroit; clin asst prof pediat & commun dis, Univ Hosp, Univ Mich, Ann Arbor; med consult to dir div ment retardation, State of Fla, 69-; med dir & vpres, Fla Special Olympics, 76; consult & qual assurance, Dept Health & Rhab Serv, State of FLa, 84- *Honors & Awards:* President's Award, Am Acad Pediat, 84, Outstanding Serv Award, 86. *Mem:* Am Soc Clin Pharmacol & Therapeut; AMA; fel Am Acad Pediat; Am Fedn Clin Res. *Res:* Virus and infectious diseases; pediatric clinical pharmacology; author of many publications. *Mailing Add:* 5011 Ocean Blvd 3rd Floor Siesta Key FL 34242

WEISS, CHARLES MANUEL, b Scranton, Pa, Dec 7, 18; m 42. AQUATIC BIOLOGY. *Educ:* Rutgers Univ, BS, 39; Johns Hopkins Univ, PhD(biol), 50. *Prof Exp:* Asst bacteriologist, Woods Hole Oceanog Inst, 40-42, res assoc & biologist in-chg, Miami Beach Sta, 42-46, res assoc marine biol, 46-47; chemist-biologist, Baltimore Harbor Proj, Dept Sanit Eng, Johns Hopkins Univ, 47-50; basin biologist, Div Water Pollution Control, USPHS, 50-52; biologist, Sanit Chem Br, Med Labs, Army Chem Ctr, Edgewood, Md, 52-56; assoc prof sanit sci, 56-62, dep head, Dept Environ Sci & Eng, 67-77, PROF ENVIRON BIOL, SCH PUB HEALTH, UNIV NC, CHAPEL HILL, 62- *Concurrent Pos:* Consult, Pan Am Health Orgn; Bigelow fel, Woods Hole Oceanog Inst. *Mem:* Fel AAAS; Am Chem Soc; Soc Int Limnol; Am Soc Microbiol; fel Am Pub Health Asn; fel NY Acad Sci. *Res:* Response of aquatic biota to environmental stress; water quality criteria and indices; limnology of lakes and impoundments; stream pollution. *Mailing Add:* Environ Sci Eng Univ NC Sch Pub Health CB 7400 Rosenau Chapel Hill NC 27599-7400

WEISS, DANIEL LEIGH, b Long Branch, NJ, July 27, 23; m 51; c 3. PATHOLOGY, MEDICAL SCIENCE. *Educ:* Columbia Univ, AB, 43, MD, 46. *Prof Exp:* Res asst path & exp med, Beth Israel Hosp, 50-51; res asst path, Mt Sinai Hosp, NY, 51-53; dir inst path & lab med, DC Gen Hosp, 53-63; prof path, Col Med, Univ Ky, 63-77; exec secy, div med sci, Nat Res Coun, Nat Acad Sci, 77-82; dep dir, Affil Educ Progs Serv, Off Acad Affairs, Dept Med & Surg, Vet Admin, 83-87; CONSULT MED SYSTS, RES, CARE & EDUC, 87- *Concurrent Pos:* Clin prof, Georgetown Univ, 53-63, George Washington Univ, 53-63 & 79- & Howard Univ, 59-63; consult, NIH, 56-73; US liasion rep, Comt Genetic Experimentation, WHO, 78-; mem biomech adv comn, US Dept Transp, 79-81 & Health Servs Comn, Am Red Cross, 80; rep, Nat Res Coun, Interagency Comn, Handicapped Res, 80-81 & Comn Nat Standards, Medicolegal Invest Death, Nat Inst Justice, US Dept Justice, 80-; Vet Admin rep, Educ & Pub Info Task Force, Pub Working Group for Health, Nat Disaster Med Serv, Exec Off Pres, 83-87; mem, Comt Interagency Radiation Res & Policy Coord, Off Sci & Technol Policy, Exec Off of Pres, 87-; consult, Booz-Allen Hamilton, Titan Corp & Oakridge Assoc Univs. *Honors & Awards:* Sunderman Clin Scientist Award, Asn Clin Scientists. *Mem:* Am Soc Clin Path; Am Asn Path; Col Am Path; Am Soc Exp Path; Asn Clin Sci; Am Asn Hist Med. *Res:* Experimental virus infection; primary and secondary vasculitis; pathology of infection and immunity; skeletal pathology; paleopathology; science administration; radiation pathobiology; environmental pathobiology; medical education. *Mailing Add:* 7201 Park Terr D Alexandra VA 22307

WEISS, DAVID STEVEN, b Newark, NJ, Mar 3, 44; m 69. PHOTOCONDUCTIVITY, PHOTOCHEMISTRY. *Educ:* Lehigh Univ, BS, 65; Columbia Univ, PhD(chem), 69. *Prof Exp:* NIH fel, Iowa State Univ, 69-72; asst prof chem, Univ Mich, Ann Arbor, 72-78; RES ASSOC, EASTMAN KODAK CO, 78- *Mem:* Inter-Am Photochem Soc; Am Chem Soc; Sigma Xi; Soc Photog Scientists & Engrs. *Mailing Add:* 67 Eastwood Trail Rochester NY 14622

WEISS, DAVID WALTER, b Vienna, Austria, July 6, 27; nat US; m 51; c 3. MICROBIOLOGY, IMMUNOLOGY. *Educ:* Brooklyn Col, BA, 49; Rutgers Univ, PhD(microbiol), 52; Oxford Univ, DPhil, Med, 57. *Hon Degrees:* DSc, Brooklyn Col, City Univ, NY, 83. *Prof Exp:* Asst, Rockefeller Inst 52-55; dir med res coun tuberc unit, Oxford Univ, 56-57; from asst prof to prof bact & immunol, Univ Calif, Berkeley, 57-67, res immunologist, Cancer Res Genetics Lab, 62-67; PROF IMMUNOL & CHMN, LAUTENBERG CTR GEN RES & TUMOR IMMUNOL, HADASSAH MED SCH, HEBREW UNIV, JERUSALEM, 68- *Concurrent Pos:* Merck sr fel, Nat Acad Sci, 55-57; Am Cancer Soc scholar, 62-63; res prof, Miller Inst Basic Res Sci, 66-67; Herbert Abeles Scholar, 74; Herbert Abeles Scholar, 74; vis prof neoplastics dis, Mt Sinai Med Sch, NY, 77-78; mem, Midwinter Conf Immunologists. *Honors & Awards:* Ungerman-Lubin Cancer Res Award, 77; Frank Sinatra Award, Pan Am Cancer Cytol Soc, 73; Prentis Prize & Medal, 83. *Mem:* Am Asn Cancer Res; Transplantation Soc; Am Asn Immunologists; NY Acad Sci; Path Soc Gt Brit & Ireland. *Res:* Pathogenesis and host-parasite relationships; development of nonliving vaccines; relationship of specific and nonspecific immunogenic activities of microorganisms; tumor immunology; mechanisms of immunological unresponsiveness; oncogenic viruses; antibody formation; endotoxins. *Mailing Add:* Dept Immunol Hadassah Med Sch Hebrew Univ PO Box 1172 Jerusalem 91010 Israel

WEISS, DENNIS, b New York, NY, July 2, 40; m 65; c 3. MICROPALEONTOLOGY, ENVIRONMENTAL GEOLOGY. *Educ:* City Col New York, BS, 63; NY Univ, MS, 67, PhD(geol), 71. *Prof Exp:* Lectr geol, 64-71, asst prof, 71-80, ASSOC PROF EARTH & PLANETARY SCI, CITY COL NEW YORK, 80-, CHMN DEPT, 78- *Mem:* AAAS; Geol Soc Am; Soc Econ Paleontologists & Mineralogists; Am Asn Stratig Palynologists; Nat Asn Geol Teachers. *Res:* Quaternary paleo-environments. *Mailing Add:* Dept Earth Sci City Univ NY City Col Covent Ave & 138th New York NY 10031

WEISS, DOUGLAS EUGENE, b Aurora, Ill, July 28, 45. ORGANIC POLYMER CHEMISTRY, RADIATION CHEMISTRY & PROCESSING OF POLYMERS. *Educ:* Univ Kans, BS, 68; Univ Nebr, PhD(org chem), 72. *Prof Exp:* Sr res assoc polymers, Univ E Anglia, 72-73; res assoc org chem, Univ Nebr, 73-74; chemist, Elastomer Chem Dept, E I du Pont de Nemours & Co, Inc, 74-78; res specialist, 3M Co, 78-81, prod develop supvr, 81-85, mat res mgr, 85-88, TECH MGR, CORP RES PROCESS TECHNOL LAB, 3M CO, 88- *Mem:* Am Chem Soc; Sigma Xi. *Res:* Heat-shrinkable technology; radiation processing; olefinic polymers; radiation-grafting; primers; hot melt and pressure sensitive adhesives; optical properties of polymers; elastomers; polyurethanes; C-13 nuclear magnetic reanance. *Mailing Add:* 400 Natchez Ave N Golden Valley MN 55422

WEISS, EARLE BURTON, b Waltham, Mass, Nov 23, 32; m 63; c 2. PULMONARY PHYSIOLOGY. *Educ:* Northeastern Univ, BS, 55; Mass Inst Technol, MS, 57; Albert Einstein Col Med, MD, 61; Am Bd Internal Med, dipl, 69. *Prof Exp:* Nat Heart Inst fel, Tufts Lung Sta, Boston City Hosp, 64-66; from instr to assoc prof, Sch Med, Tufts Univ, 66-77; assoc prof, 71-77, PROF MED, MED SCH, UNIV MASS, 77-; DIR, DEPT RESPIRATORY DIS, ST VINCENT HOSP, 71- *Concurrent Pos:* Assoc, Tufts Lung Sta, Boston City Hosp, 66-71, physician chg, Pulmonary Function & Physiol Sect, Hosp, 66-71 & Cent Arterial Blood Gas Lab, 67-71, assisting physician, Tufts Med Serv, 67-70, assoc dir, 69-70, dir respiratory intensive care unit, 70-71; consult physiol, Norfolk County Sanatorium, Mass, 66-69; mem toxicol info prog, Nat Libr Med, 70-; tuberc consult, Mass Dept Pub Health, 72-; lectr med, Med Sch, Tufts Univ, 78; assoc affil prof life sci, Worcester Polytech Inst, 76- *Mem:* AAAS; Am Fedn Clin Res; fel Am Col Physicians; fel Am Col Chest Physicians; Am Soc Internal Med. *Res:* Mechanisms of airway hyperactivity in toxic oxygen radicals and intracellular calcium. *Mailing Add:* Brigham & Women's Hosp Boston MA 14612

WEISS, EDWIN, b Brooklyn, NY, Aug 23, 27; m 52; c 2. ALGEBRA. *Educ:* Brooklyn Col, BS, 48; Mass Inst Technol, PhD, 53. *Prof Exp:* Instr math, Univ Mich, 53-54; NSF fel, Inst Advan Study, 54-55; Benjamin Peirce instr, Harvard Univ, 55-58; asst prof, Univ Calif, Los Angeles, 58-59; mem staff, Lincoln Lab, Mass Inst Technol, 59-65; PROF MATH, BOSTON UNIV, 65- *Mem:* Am Math Soc. *Mailing Add:* 16 Warwick Rd Brookline MA 02146

WEISS, EMILIO, b Pakrac, Yugoslavia, Oct 4, 18; nat US; m 43; c 2. MEDICAL MICROBIOLOGY. *Educ:* Univ Kans, AB, 41; Univ Chicago, MS, 42, PhD(bact), 48. *Prof Exp:* Asst histol, parasitol & bact, Univ Chicago, 42 & 47-48; res assoc, 48-50; instr bact, Loyola Univ, Ill, 47-48; asst prof, Ind Univ, 50-53; chief virol br, Chem Corps Biol Lab, US Dept Army, Ft Detrick, Md, 53-54; res prof prev med & biometrics, Uniformed Serv Univ Health Sci, 77-84; asst head virol div, Naval Med Res Inst,54-63, dep dir microbiol dept, 63-72, dir, 72-74, chmn microbiol dept, 74-80, chair sci, 80-86; RETIRED. *Concurrent Pos:* Vis scientist, Naval Med Res Inst, 86- *Mem:* Am Soc Microbiol; Soc Exp Biol & Med; Am Asn Immunologists; Am Acad Microbiol; Am Soc Rickettsiol (pres, 80-82). *Res:* Microbiol physiology and pathogenesis; rickettsiae, chlamydiae and other bacterial pathogens. *Mailing Add:* 3612 Raymond St Chevy Chase MD 20815-4152

WEISS, FRED TOBY, b Oakland, Calif, July 24, 16; m 40; c 3. ANALYTICAL CHEMISTRY, ENVIRONMENTAL CHEMISTRY. *Educ:* Univ Calif, Los Angeles, BS, 38; Harvard Univ, MS, 40, PhD(chem), 41. *Prof Exp:* Res chemist, Shell Oil Co, Ill, 41-43, group leader fuels res, 43-46, res chemist, Shell Develop Co, Calif, 46-52, res supvr, 52-72, staff res chemist, Bellaire Res Ctr, Tex, 72-75, sr staff res chemist, Bellaire Res Ctr, Shell Develop Co, Tex, 75-81, sr staff environ scientist, Shell Oil Co, 81-83; RETIRED. *Concurrent Pos:* Mem adv comt, NSF-Res Appl to Nat Needs Study of Petrol Indust in Del Estuary, 74-77; mem outer continental shelf adv bd, US Dept Interior, 79- *Mem:* AAAS; Am Chem Soc. *Res:* Characterization and analysis of organic structures; analytical methods for process control; combined use of chemical and physical methods for analysis of organic compounds; environmental analysis; development of analytical methods for studying fate and effects of petroleum in marine environments; surveys of impacts of offshore petroleum operations in the environment. *Mailing Add:* Three Indian Gulch Piedmont CA 94611

WEISS, GARY BRUCE, b New York, NY, Oct 5, 44; div; c 1. ONCOLOGY, HEMATOLOGY. *Educ:* NY Univ, BA, 65, MD, 71, PhD(biochem), 72. *Prof Exp:* Intern med, Sch Med, Univ Calif, San Francisco, 71-72, resident, 72-73; res assoc hemat, Nat Heart & Lung Inst, Bethesda, Md, 73-76; fel hemat, Univ Wash, 76-77; from asst prof to assoc prof int med, human biol chem & genetics, 77-84, CLIN ASSOC PROF, UNIV TEX MED BR GALVESTON, 84- *Mem:* Am Fedn Clin Res; Asn Comput Mach; Am Soc Hemat; Am Soc Biol Chemists; Am Statist Asn. *Res:* Medical ethics in clinical oncology. *Mailing Add:* Div Hemat Univ Tex Med Br Galveston TX 77550

WEISS, GEORGE B, b Plainfield, NJ, Apr 29, 35; m 60; c 2. PHARMACOLOGY. *Educ:* Princeton Univ, AB, 57; Vanderbilt Univ, PhD(pharmacol), 62. *Prof Exp:* USPHS res fels pharmacol, Vanderbilt Univ, 62 & Univ Pa, 62-64; from asst prof to assoc prof pharmacol, Med Col Va, 64-70; from assoc prof to prof pharmacol, Univ Tex Health Ctr, Dallas, 70-82; sr res fel, 82-88, DIST RES FEL, CIBA-GEIGY CORP, 89- *Mem:* Biophys Soc; Am Soc Pharmacol & Exp Therapeut; Am Physiol Soc. *Res:* Cellular pharmacology; actions of drugs on membrane permeability to ions; excitation-contraction coupling and calcium ion in smooth (especially vascular) and striated muscle; actions of calcium antagonists, especially calcium channel blockers. *Mailing Add:* Res Dept Ciba-Geigy Corp Summit NJ 07901

WEISS, GEORGE HERBERT, b New York, NY, Feb 19, 30; m 61; c 3. APPLIED MATHEMATICS. *Educ:* Columbia Univ, AB, 51; Univ Md, MA, 53, PhD, 58. *Prof Exp:* Physicist, US Naval Ord Lab, 51-54; math asst, Ballistic Res Lab, Aberdeen Proving Ground, US Army, 54-56; asst math, Inst Fluid Dynamics & Appl Math, Univ Md, 56-58, res assoc, 59-60, res asst prof, 60-63; Weizmann fel, Weizmann Inst Sci, 58-59; NIH study grant, Rockefeller Inst, 63-64; study grant, 64-67, CHIEF PHYS SCI LAB, DIV COMPUT RES & TECHNOL, NIH, 67- *Concurrent Pos:* Physicist, US Naval Ord Lab, 56-61; consult, Gen Motors Corp, 60-64; consult, IBM Corp, 64; Fulbright sr fel, Imp Col, Univ London, 68-69; assoc ed, Transp Res, 73- & Cancer Res, 82-; Chemometrics & Intel Lab Systs, 86- *Mem:* Opers Res Soc Am; Soc Indust & Appl Math. *Res:* Traffic theory; statistical mechanics; biometry; biochemical separation techniques, stochastic processes. *Mailing Add:* 1105 N Belgrade Rd Silver Spring MD 20902

WEISS, GERALD S, b Boyertown, Pa, July 26, 34; m 58; c 3. INORGANIC CHEMISTRY. *Educ:* Drexel Inst, BS, 57; Univ Pa, PhD(inorg chem, anal chem), 65. *Prof Exp:* Res chemist, Rohm & Haas Co, 57-59; instr chem, Drexel Inst, 59-65, asst prof, 65-67; assoc prof, 67-69, PROF CHEM, MILLERSVILLE STATE COL, 69-, CHMN DEPT, 75- *Mem:* Am Chem Soc. *Res:* Synthesis of volatile hydrides of group IV elements; infrared spectroscopic analysis of small molecules; heavy metal complex ion synthesis and spectroscopic analysis. *Mailing Add:* Dept Chem Millersville Univ Millersville PA 17551

WEISS, GERSON, b New York, NY, Aug 1, 39; m 59; c 4. REPRODUCTIVE ENDOCRINOLOGY. *Educ:* NY Univ, BA, 60, MD, 64; Am Bd Obstet & Gynec, dipl, 71, cert reproductive endocrinol, 74, 81. *Prof Exp:* From instr to prof obstet & gynec, Sch Med, NY Univ, 69-85, dir, Div Reproduction & Endocrinol, Med Ctr, 80-85; PROF & CHMN DEPT OBSTET & GYNEC, NJ MED SCH, 85- *Concurrent Pos:* Fel reproductive endocrinol, Sch Med, Univ Pittsburgh, 71-73; John Polachek Found Med Res grant, 75; Nat Inst Child Health & Human Develop res grant, 75; United Cerebral Palsy res grant, 77; prin invesr, Mellon Found; res grants, NIH. *Honors & Awards:* Perdue-Frederick Award for Med Res. *Mem:* Soc Gynec Invest; Endocrine Soc; Am Fertil Soc; Soc Study Reproduction; NY Acad Med; NY Acad Sci. *Res:* Control and function of the pregnancy and postpartum corpus luteum; relaxin physiology; control of pituitary gonadotropin secretion. *Mailing Add:* Dept Obstet-Gynec NJ Med Sch 100 Bergen St Newark NJ 07103

WEISS, GUIDO LEOPOLD, b Trieste, Italy, Dec 29, 28; nat US; m 50. MATHEMATICS. *Educ:* Univ Chicago, PhB, 49, MS, 51, PhD, 56. *Prof Exp:* From instr to assoc prof math, DePaul Univ, 55-60; chmn dept, 67-70, PROF MATH, WASH UNIV, 66- *Honors & Awards:* Chauvenet Prize, Math Asn Am, 67. *Mem:* Am Math Soc; Math Asn Am. *Res:* Harmonic analysis; complex and real variables. *Mailing Add:* Dept Math Wash Univ St Louis MO 63130

WEISS, HAROLD GILBERT, b Perth Amboy, NJ, Feb 6, 23; m 44; c 2. INORGANIC CHEMISTRY, PHYSICAL CHEMISTRY. *Educ:* Univ Calif, Los Angeles, BS, 48. *Prof Exp:* Res chemist, US Naval Ord Test Sta, 48-52; supvr catalysis res, Olin Mathieson Chem Corp, 52-59; lab mgr, Nat Eng Sci Co, 59-61; dir chem, Dynamic Sci Corp, 61-66; PRES, WEST COAST TECH SERV, INC, 66- *Mem:* Am Chem Soc; Am Inst Chemists; Inst Environ Sci. *Res:* Boron hydrides; organoboranes; surface chemistry and catalysis; analytical chemistry; mass and infrared spectrometry; high vacuum technology; combustion-fire research; environmental control; analytical instrumentation. *Mailing Add:* 4016 Montego Dr Huntington Beach CA 92649-2494

WEISS, HAROLD SAMUEL, b New York, NY, Sept 10, 22; m 49; c 4. CARDIOPULMONARY, CANCER. *Educ:* Rutgers Univ, BSc, 47, MSc, 49, PhD(physiol), 50. *Prof Exp:* From instr to assoc prof avian physiol, Rutgers Univ, 50-62; from assoc prof to prof, 62-91, EMER PROF PHYSIOL, COL MED, OHIO STATE UNIV, 90- *Concurrent Pos:* Meteorol, Air Traffic Control, USAF, 43-46, Res & Develop, Aviation Physiol, 51-54. *Mem:* Am Physiol Soc; Poultry Sci Asn; Soc Exp Biol & Med; Aerospace Med Asn; Undersea Med Soc; AAAS. *Res:* Cardiopulmonary; blood pressure; atherosclerosis; lung mechanics; environmental physiology; acceleration; temperature control; gaseous environment and respiratory disease; environment and cancer. *Mailing Add:* 4196 Graves Hall Ohio State Univ Col Med 333 W Tenth Ave Columbus OH 43210

WEISS, HARRY JOSEPH, b Pittsburgh, Pa, Feb 15, 23; m 47; c 3. APPLIED MECHANICS, APPLIED MATHEMATICS. *Educ:* Carnegie Mellon Univ, BS, 47, MS, 49, DSc(appl math), 51. *Prof Exp:* Instr math, Carnegie-Mellon Univ, 49-51; asst prof appl math, Brown Univ, 51-53; from asst prof to prof math, 54-64, prof eng mech & head dept, 64-86, PROF ENG MECH, IOWA STATE UNIV, 86- *Concurrent Pos:* Vis scientist, Nat Bur Standards, 61; consult, Gen Dynamics Astronaut, 62-66. *Mem:* Soc Eng Sci (vpres, 77-79); Am Soc Mech Engrs; Sigma Xi; Am Soc Eng Educ. *Res:* Integral transforms; partial differential equations; boundary value problems; elasticity. *Mailing Add:* 2529 Northwood Ames IA 50011

WEISS, HARVEY JEROME, b New York, NY, June 30, 29; m 57; c 2. INTERNAL MEDICINE, HEMATOLOGY. *Educ:* Harvard Univ, AB, 51, MD, 55. *Prof Exp:* Intern, Bellevue Hosp, Columbia Univ, 55-56; resident med, Manhattan Vet Admin Hosp, NY, 56-58; Dazian fel hemat, Mt Sinai Hosp, NY, 58-59; instr, Sch Med, NY Univ, 62-64; asst attend hematologist, Mt Sinai Hosp, 65-69; assoc clin prof, 69-71, assoc prof, 72-74, PROF MED, COL PHYSICIANS & SURGEONS, COLUMBIA UNIV, 75- *Concurrent Pos:* Consult, Walter Reed Army Med Ctr, 65-70; asst prof med, Mt Sinai Sch Med, 66-69; dir, Div Hemat, Roosevelt Hosp, NY, 69-; mem hemat speciality comt, Am Bd Internal Med, 81- *Mem:* Soc Exp Biol & Med; Am Physiol Soc; Am Soc Clin Invest; Am Soc Hemat; Asn Am Physicians. *Res:* Hematology; blood coagulation; disorders of hemostasis; platelet physiology. *Mailing Add:* Dept Med St Luke's Roosevelt Hosp Ctr 428 W 59th St New York NY 10019

WEISS, HARVEY RICHARD, b New York, NY, May 13, 43; m 66; c 2. PHYSIOLOGY, PHARMACOLOGY. *Educ:* City Col New York, BS, 65; Duke Univ, PhD(physiol), 69. *Prof Exp:* Warner-Lambert joint fel pharmacol, Warner-Lambert Res Inst & Col Physicians & Surgeons, Columbia Univ, 69-71; from asst prof to assoc prof, 71-82, PROF PHYSIOL, RUTGERS MED SCH, UNIV MED & DENT NJ, 82- *Mem:* AAAS; Am Physiol Soc. *Res:* Physiologic and pharmacologic control of oxygen transport to tissue; coronary circulation; regional cerebral and myocardial oxygen consumption. *Mailing Add:* Dept Physiol & Biophys Robert Wood Johnson Med Sch Piscataway NJ 08854-5635

WEISS, HERBERT KLEMM, b Lawrence, Mass, June 22, 17; m 45; c 2. SYSTEMS DESIGN & SYSTEMS SCIENCE. *Educ:* Mass Inst Technol, BS, 37, MS, 38. *Prof Exp:* Mech engr, Coast Artillery Bd, Va, 38-42, Anti-aircraft Artillery Bd, NC, 42-44 & Army Ground Forces Bd, Tex, 44-46; chief, Weapons Effect Br & dep chief, Terminal Ballistics Lab, Aberdeen Proving Ground, US Dept Army, Md, 46-50, chief, Weapon Systs Lab, 50-53; chief, Weapon Systs Anal Dept, Northrop Aircraft Corp, 53-58; mgr, Adv Systs Develop, Aeronutronic Div, Ford Motor Co, 58-60, mgr, Mil Systs Planning, 60-62; dir, Systs Anal & Eval, Aerospace Corp, 62-65; sr scientist, Litton Industs, 65-80, scientist, Data Systs Div, 80-82; RETIRED. *Concurrent Pos:* Mem tech adv panel ord, Off Asst Secy Defense Res & Develop, 54-64; consult, Weapons Systs Eval Group, Off Asst Secy Defense, 55-56; lectr, Univ Calif, Los Angeles, 57-58; consult, Sci Adv Bd, USAF, 58-59, mem, 59-63; summer mem study group, Nat Acad Sci-USAF, 58 & Proj Endicott, Opers Eval Group, USN, 60; consult, Opers Eval Group, USN, 60-62; mem, US Army Sci Adv Panel, 65-76; consult, Pres Sci Adv Comt. 66-72. *Mem:* AAAS; Opers Res Soc Am (vpres, 58-59); assoc fel Am Inst Aeronaut & Astronaut; sr mem Inst Elec & Electronics Engrs. *Res:* Fire control; servo-mechanisms; computing devices; weapon systems research and development; operations research; systems analysis; ballistics; management sciences. *Mailing Add:* PO Box 2668 Palos Verdes Peninsula CA 90274

WEISS, HERBERT V, b Brooklyn, NY, Nov 16, 21; m 55; c 2. CHEMISTRY. *Educ:* NY Univ, BA, 42, MS, 49; Univ Cincinnati, PhD(biochem), 52. *Prof Exp:* Toxicologist, Chief Med Exam Lab, NY, 47-49; fel physiol, Sch Med, Univ Rochester, 52-53; instr indust med, Post-grad Med Sch, NY Univ, 53-56; supvry radiochemist, US Naval Radiol Defense Lab, San Francisco, 56-69; res chemist, Naval Oceans Systs Ctr, 69-88; ADJ PROF, SAN DIEGO STATE UNIV, 88- *Mem:* AAAS; Am Chem Soc. *Res:* Nuclear chemistry; analytical chemistry; industrial hygiene and toxicology; environmental chemistry. *Mailing Add:* Dept Chem San Diego State Univ San Diego CA 92182-0328

WEISS, IRA PAUL, b New York, NY, Feb 27, 42; m 67; c 2. NEUROSCIENCES. *Educ:* City Col New York, BS, 65; Syracuse Univ, PhD(physiol, psychol), 69. *Prof Exp:* Instr psychol, Onondaga Community Col, 68; NIH fel neurophysiol, Callier Hearing & Speech Ctr, Dallas, Tex, 69-70, NIH spec fel neurophysiol & consult res design, 70-72; res psychologists neurosci, 72-80, vchmn, Dept Psychol, 75-76, CO-DIR, EVOKED RESPONSE LAB, CHILDRENS HOSP NAT MED CTR, 78- *Concurrent Pos:* Asst prof lectr, Sch Med, George Washington Univ, 72-80, assoc prof lectr, 80-81, assoc prof, 82- *Mem:* Soc Neurosci; Soc Psychophysiol Res; AAAS. *Res:* Relationships of brainstem and cortical sensory evoked potentials to developments, neurological and sensory disorders and to behavior. *Mailing Add:* Evoked Response Lab Childrens Hosp Nat Med Ctr 111 Michigan Ave MW Washington DC 20010

WEISS, IRMA TUCK, b New York, NY, Aug 5, 13; m 38; c 2. BIOCHEMISTRY, ORGANIC CHEMISTRY. *Educ:* NY Univ, BS, 33, MS, 36, PhD, 42. *Prof Exp:* Asst instr chem, Washington Sq Col, 40-42, from instr to assoc prof biochem, Col Dent, 43-76, prof, 76-80, EMER PROF BIOCHEM, COL DENT, NY UNIV, 80- *Mem:* Sigma Xi (secy, Sci Res Soc Am, 59-60, treas, 60-61, vpres, 61-62, pres, 62-63); Am Chem Soc. *Res:* Organic synthesis; application of biochemistry in clinical dentistry; electrophoresis studies of salivary proteins. *Mailing Add:* 401 First Ave Apt 19D New York NY 10010-4007

WEISS, IRVING, b New York, Apr 10, 19; m 44; c 3. MATHEMATICAL STATISTICS. *Educ:* Univ Mich, BS, 41; Columbia Univ, MA, 48; Stanford Univ, PhD(statist), 55. *Prof Exp:* Res asst statist, Stanford Univ, 51-55; instr math, Lehigh Univ, 55-56; tech staff statistician, Bell Tel Labs, 56-59 & Mitre Corp, 59-62; ASSOC PROF MATH, UNIV COLO, BOULDER, 62- *Concurrent Pos:* Consult, State Dept Employ, Colo, 63, Beech Aircraft Corp, 63-64, dept biol, Univ Rochester, 64-65, Nat Ctr Atmospheric Res, 64-72 & 75, Behav Res & Eval Corp, 73-74 & US Environ Protection Agency, 75. *Mem:* Inst Math Statist; Am Statist Asn. *Res:* Probability; applied probability theory; statistical inference; stochastic processes; control charts; tolerance intervals. *Mailing Add:* Dept Math Univ Colo Box 426 Boulder CO 80309

WEISS, JAMES ALLYN, b W Bend, Wis, Apr 16, 43; m 71; c 2. ORGANIC CHEMISTRY. *Educ:* Univ Wis-Madison, BS, 65; Pa State Univ, Univ Park, PhD(chem), 71. *Prof Exp:* ASST PROF CHEM, PA STATE UNIV, WORTHINGTON SCRANTON CAMPUS, 70- *Mem:* AAAS; Am Chem Soc; Sigma Xi; Am Chem Soc. *Res:* Organic synthesis; isolation and structural elucidation of natural products; stereochemistry of organic molecules. *Mailing Add:* West Scranton Campus Pa State Univ Dunmore PA 18512

WEISS, JAMES MOSES AARON, b St Paul, Minn, Oct 22, 21; m 46; c 2. PSYCHIATRY. *Educ:* Univ Minn, AB, 41, BS, 47, MB, 49, MD, 50; Yale Univ, MPH, 51; Am Bd Psychiat & Neurol, dipl, 57. *Prof Exp:* Asst psychol, Col St Thomas, 41-42; intern med, USPHS Hosp, Seattle, 49-50; from instr to asst prof psychiat, Sch Med, Wash Univ, 54-60; assoc prof & founding chmn dept, 60-61, PROF PSYCHIAT & CHMN DEPT, SCH MED, UNIV MO, COLUMBIA, 61-, PROF COMMUNITY MED, 71- *Concurrent Pos:* Res & clin fel psychiat, Sch Med, Yale Univ, 51-53; fac fel, Inter-Univ Coun Inst Social Geront, Univ Conn, 58; dir training, Malcolm Bliss Ment Health Ctr, City of St Louis, 54-60, dir psychiat clin, 54-59, dir div community psychiat serv, 58-59; vis psychiatrist, St Louis City Hosps, Barnes & Affil Hosps & Wash Univ Clins, 54-60; consult to state, fed & nat agencies & orgn, 60-; vis prof, Inst Criminol, Cambridge Univ, 68-69; chancellor's emissary, Univ Mo, Columbia, 79-81; sr res fel, NSF & Am Coun Educ, 84; vis prof, All India Inst Med Sci & Univ Malaya, 84. *Honors & Awards:* Distinguished Serv Commendation, Nat Coun Community Ment Health Centers, 82, 83, & 86; Kohler Distinguished Lectr, St Louis Univ, 88. *Mem:* Fel Royal Col Psychiatrists; fel Am Psychiat Asn; fel Am Pub Health Asn; fel Royal Soc Med; fel Am Col Psychiatrists; fel Am Col Prev Med. *Res:* Social psychiatry and gerontology; psychiatric problems of aging; suicide; homicide; antisocial behavior. *Mailing Add:* Dept Psychiat Univ Mo Med Ctr Columbia MO 65212

WEISS, JAMES OWEN, b Memphis, Tenn, Sept 25, 31; m 55; c 3. ORGANIC CHEMISTRY, POLYMER CHEMISTRY. *Educ:* Duke Univ, BS, 52; Univ Va, MS, 54, PhD(org chem), 57. *Prof Exp:* Res chemist, Shell Oil Co, 57-58 & E I du Pont de Nemours & Co, 58-61; res chemist, Chemstrand Res Ctr, 61-66; res mgr fiber develop, Beaunit Fibers, 66-68; group leader fiber develop, Celanese Fibers Co, 68-72; chem develop mgr, 73-78, qual control dir, 78-81, FILAMENT PROD DIR, HOECHST FIBERS INDUSTS, 81- *Mem:* Am Chem Soc. *Res:* Polyester fiber research and development; nylon, high temperature resistant fiber, spandex fiber, polypropylene fiber and vinyl polymer and fiber research and development. *Mailing Add:* 310 Continental Dr Greenville SC 29615-3418

WEISS, JAY M, b Jersey City, NJ, Mar 20, 41; m 63; c 2. PSYCHOPHYSIOLOGY. *Educ:* Lafayette Col, BA, 62; Yale Univ, PhD(psychol), 67. *Prof Exp:* USPHS fel & guest investr, Rockefeller Univ, 67-68, from asst prof to assoc prof physiol, 69-85; PROF, PSYCHIAT DEPT, DUKE UNIV, 85- *Concurrent Pos:* MacArthur fel. *Res:* Psychological factors influencing physiological effects of stress; psychosomatic disorders; motivation. *Mailing Add:* Dept Psychiat Duke Univ Med Ctr Durham NC 27710

WEISS, JEFFREY MARTIN, b Philadelphia, Pa, July 1, 44; m 67; c 2. ATMOSPHERE DYNAMICS, MARINE SCIENCES. *Educ:* Princeton Univ, AB, 66; Harvard Univ, MA, 67, PhD(physics), 72. *Prof Exp:* Vis scientist high energy physics, Europ Orgn Nuclear Res, Geneva, Switz, 72-74; res assoc, Nevis Labs, Columbia Univ, 74-76; res assoc physics, Stanford Linear Accelerator Ctr, Stanford Univ, 76-81; physicist, Argonne Nat Lab, 81-85; sr res physicist, 85-86, ASST DIR, APPL ELECTROMAGNETICS & OPTICS LAB, SRI INT, MENLO PARK, CALIF, 87- *Concurrent Pos:* NSF fel, Europ Orgn Nuclear Res, 72-73; consult, Dynametrics Co, Philadelphia, Pa, 78-80; secy-treas, Stanford Linear Accelerator Ctr-Lawrence Berkely Lab Users Orgn, 84-85. *Mem:* Am Phys Soc; Inst Elec & Electronics Engrs Antennas & Propagation Soc; Geosci & Remote Sensing Soc; Appl Computational Electromagnetics Soc. *Res:* Electron-positron colliding beam physics; production of charmed baryons and mesons; electromagnetics in geophysical and remote sensing applications; intersecting storage rings; finite-difference time-domain calculations of field propagation in complex geometries. *Mailing Add:* SRI Int 333 Ravenswood Ave Menlo Park CA 94025

WEISS, JERALD AUBREY, b Cleveland, Ohio, June 9, 22; m 49; c 2. MICROWAVE PHYSICS. *Educ:* Ohio State Univ, PhD(physics), 53. *Prof Exp:* Instr math, Univ Wyo, 49-51; mem tech staff magnetic mat, Bell Tel Labs, Inc, 53-61; vpres, Hyletronics Corp, Mass, 61-62; from asst prof to assoc prof, 62-66, PROF PHYSICS, WORCESTER POLYTECH INST, 66- *Concurrent Pos:* Consult, US Army Natick Res & Develop Ctr, 75 & Lincoln Lab, Mass Inst Technol, 62- *Mem:* Am Phys Soc; sr mem Inst Elec & Electronics Engrs; AAAS; Sigma Xi. *Res:* Theory of atomic and molecular structure; magnetic materials and electromagnetic interactions, theory and applications; magnetic resonance spectroscopy and microwave applications. *Mailing Add:* 30 Wayland Hills Rd Wayland MA 01778

WEISS, JEROME, b Brooklyn, NY, Aug 27, 22; m 43; c 2. PHYSICAL CHEMISTRY. *Educ:* Cornell Univ, BA, 48; Ind Univ, PhD(phys chem), 51. *Prof Exp:* Asst chem, Ind Univ, 48-50, instr, Exten, 50; chemist, Brookhaven Nat Lab, 51-73; ASST PROF CHEM, SUFFOLK COUNTY COMMUNITY COL, 73- *Concurrent Pos:* Lectr, Hofstra Col, 55; Dewar res fel, Univ Edinburgh, 57; consult, Am Soc Testing & Mat & Brookhaven Nat Lab, 85; teacher, Smithtown High Sch, 75-77 & Northport High Sch, 78-86. *Mem:* AAAS; Radiation Res Soc; Royal Soc Chem; NY Acad Sci; fel Am Inst Chemists. *Res:* Organic radiation chemistry; chemical dosimetry; health physics; membrane transport. *Mailing Add:* 17 Locust Ave Stony Brook NY 11790

WEISS, JOHN JAY, b Chicago, Ill, May 28, 53; m 82; c 2. ALLERGY. *Educ:* Northwestern Univ, BA, 75, MS, 79, PhD(tumor cell biol), 88. *Prof Exp:* Mgr prod develop, Kallestad Diagnostics, 82-83, mgr res & develop, 86-90; DIR MGR, DIAMED INC, 90- *Res:* Development of immunodiagnostic products for general use; immunology and allergy diagnostic products. *Mailing Add:* 401 Cumberland Ave No 1002 Portland ME 04101

WEISS, JONAS, b New York, NY, Feb 17, 34; m 59; c 3. POLYMER CHEMISTRY. *Educ:* City Col New York, BS, 55; NY Univ, PhD(phys chem), 62. *Prof Exp:* Chemist, Esso Res & Eng Co, 62-64; supvr org & polymer chem, Am Standard Co, 64-70; group leader polymer chem, Nat Patent Develop Corp, 70-74; group leader polymer applns, 74-80, sr staff scientist, 80-84, DIR PROD SAFETY & REGULATORY AFFAIRS, CIBA-GEIGY CORP, 84- *Mem:* Am Chem Soc. *Res:* Polymers and plastics for composites, coatings, sealants, adhesives and binders; permeability and rate of release of polymers; polymers and plastics for building materials; foam insulation; indicators and controls; semi-permeable membranes; ion-exchange polymers. *Mailing Add:* Seven Skyline Dr Ciba-Geigy Hawthorne NY 10532

WEISS, JOSEPH FRANCIS, b Taylor, Pa, Jan 26, 40; m 68; c 2. BIOCHEMISTRY, RADIOBIOLOGY. *Educ:* Univ Scranton, BS, 61; Ohio State Univ, MS, 63, PhD(physiol chem), 66. *Prof Exp:* NIH fel neurochem, Inst Pharmacol, Univ Milan, Italy, 66-68; instr neurosurg in biochem cancer, Med Ctr, NY Univ, 68-72, asst prof exp neurosurg, 72-74; res chemist biochem & radiobiol, 74-76, chief, Physiol Chem Div, 78-88, CHIEF RADIO PROTECTION DIV, ARMED FORCES RADIOBIOL RES INST, DEFENSE NUCLEAR AGENCY, 86-, PROJ MGR, 88- *Concurrent Pos:* Vis prof, Univ Cagliari, Italy, 88, Univ Pisa, Italy, 90. *Mem:* Am Chem Soc; Int Soc Immunopharmacol; Am Asn Cancer Res; Am Soc Clin Oncol; Radiation Res Soc. *Res:* Biochemical markers of cancer and radiation injury; glycoproteins, lipids, trace metals; radiobiology; immunopharmacology; neurochemistry; carcinogenesis; radio protection. *Mailing Add:* Dept Radiation Biochem Armed Forces Radiobiol Res Inst Bethesda MD 20889-5145

WEISS, JOSEPH JACOB, b Detroit, Mich, Mar 22, 34; m 68; c 2. RHEUMATOLOGY. *Educ:* Univ Mich, BA, 55, MD, 61. *Prof Exp:* Asst dir, Rheum Sect, 72, ATTEND PHYSICIAN, WAYNE COUNTY GEN HOSP, 71-; ASST PROF INTERNAL MED, MED SCH, UNIV MICH, 72- *Concurrent Pos:* Rheumatol consult, Vet Admin Hosp, Ann Arbor, Mich, 75- *Mem:* Fel Am Col Physicians; Am Soc Clin Res; Am Fedn Clin Res. *Res:* Investigation of the etiology of frozen shoulder; use of arthrography to delineate the cause and assist the diagnosis of shoulder pain and immobility. *Mailing Add:* 18829 Farmington Rd Livonia MI 48152

WEISS, KARL H, b Hamburg, Ger, June 21, 26; nat US; m 48; c 2. PHYSICAL CHEMISTRY. *Educ:* Columbia Univ, BS, 51; NY Univ, PhD(chem), 57. *Prof Exp:* Res chemist, Color Res Corp, 47-50, tech adminr, 50-54; instr chem, NY Univ, 56-59, asst prof, 59-61; from asst prof to assoc prof, 61-65, chmn dept, 69-79, vprovost, res & grad studies, 79-83, vpres res & vprovost, 83-88, PROF CHEM, NORTHEASTERN UNIV, 65-, VPRES ACAD DEVELOP, 88-, INTERIM VPRES COOP EDUC, 90- *Concurrent Pos:* NSF sr fel, Quantum Chem Group, Univ Uppsala, 68-69; Fulbright-Hayes scholar, 77-78; vis prof, Univ Konstanz, WGer, 77-78; dir, Mass Technol Park Corp, 82-, vchmn bd, 87-; tour lectr, Am Chem Soc, 73-75, adv bd coop educ, 79-85, chmn Task Force Implement Report Chem Educ, 86-89; northeast regional lectr, Sigma Xi, 73-75. *Mem:* Am Chem Soc; NY Acad Sci; Royal Soc Chem; fel Am Inst Chem; Am Soc Photobiol; AAAS. *Res:* Photochemistry of complex molecules; laser photochemistry; quantum chemistry; charge transfer interaction. *Mailing Add:* Off Acad Develop Northeastern Univ Boston MA 02115

WEISS, KENNETH MONRAD, b Cleveland, Ohio, Nov 29, 41; m 81; c 3. GENETIC EPIDEMIOLOGY, BIOLOGICAL ANTHROPOLOGY. *Educ:* Oberlin Col, BA, 63; Univ Mich, MA, 69, PhD(biol anthrop), 72. *Prof Exp:* Res assoc human genetics, Med Sch, Univ Mich, 72-73; prof demog & pop genetics, Univ Tex Grad Sch Biomed Sci, Houston, 73-85; PROF GENETICS & HEAD DEPT ANTHROP, PA STATE UNIV, UNIVERSITY PARK, 85- *Concurrent Pos:* Prof, Univ Tex Sch Pub Health, Houston, 74- *Honors & Awards:* Juan Comas Award, Am Asn Phys Anthropologists, 72; Leigh Lectr, Univ Utah, 85. *Mem:* AAAS; Am Asn Phys Anthropologists; Soc Epidemiol Res; Sigma Xi; Am Asn Human Genetics. *Res:* Demographic evolution of human populations; demographic genetic epidemiology of degenerative diseases; human evolution and biological anthropology. *Mailing Add:* Dept Anthrop Pa State Univ 409 Carpenter University Park PA 16802

WEISS, KLAUDIUSZ ROBERT, b Le Mans, France, June 7, 44. NEUROBIOLOGY. *Educ:* Univ Warsaw, MA, 67; State Univ NY Stony Brook, PhD(psychol), 73. *Prof Exp:* Fel, Col Physicians & Surgeons, Columbia Univ, 73-76, asst prof, Dept Anat & Psychiat, Columbia Univ Col Physicians & Surgeons, 77-; sr res scientist, NY State Psychiat Inst, 76-90; PROF PHYSIOL & BIOPHYS, MT SINAI SCH MED, 90- *Mem:* Soc Neurosci; Am Soc Zool. *Res:* Analysis of the neural basis of behavioral plasticity. *Mailing Add:* 1212 Fifth Ave No 12B New York NY 10029

WEISS, LAWRENCE H(EISLER), b Chicago, Ill, May 14, 38; m 62; c 2. PRODUCT DEVELOPMENT, PROCESS DESIGN & DEVELOPMENT. *Educ:* Ill Inst Technol, BS, 59; Univ Calif, Berkeley, MS, 62; Johns Hopkins Univ, PhD(chem eng), 68. *Prof Exp:* Assoc develop engr, United Tech Ctr Div, United Aircraft Corp, 61-63; res scientist, Res & Develop Div, Union Camp Corp, NJ, 67-72; chem systs res specialist, Consol Edison Co NY, Inc, 72-75; sr technologist, Chem Systs Inc, 75-77, mgr eng projs, 77-82; managing partner, Energy Technol Assoc, 82-86; mgr, Process Indust Com Develop, Airco Indust Gases, 86-87; mgr, Bus Develop, Princeton Combustion Res Labs, 87-88; SR CONSULT, PA CONSULT GROUP, 89- *Concurrent Pos:* Mem, Interagency Flue Gas Desulfurization Task Force, US Environ Protection Agency; mem, Clean Fuels Task Force, Elec Power Res Inst. *Mem:* Am Inst Chem Engrs; Am Chem Soc. *Res:* Chemical and biochemical process research and development; low volume manufacturing; product development; bioabsorbable materials. *Mailing Add:* 301 State Rd Princeton NJ 08540

WEISS, LEON, b Brooklyn, NY, Oct 4, 25; m 49; c 6. ANATOMY. *Educ:* Long Island Col Med, MD, 48. *Prof Exp:* Intern med, Maimonides Hosp, Brooklyn, NY, 48-49, asst resident, 49-50; instr med, Col Med, State Univ NY Downstate Med Ctr, 52-53; lectr, Grad Sch, Univ Md, 54-55; assoc anat, Harvard Med Sch, 55-57, from asst prof to prof anat, 57-76, Sch Med, Johns Hopkins Univ; PROF CELL BIOL & CHMN DEPT ANIMAL BIOL, SCH VET MED, UNIV PA, 76- *Concurrent Pos:* USPHS res fel anat, Harvard Med Sch, 50-52; mem med mission to establish hemat lab, Nat Defense Med Ctr, Formosa, 55. *Mem:* Electron Micros Soc Am; Histochem Soc; Am Asn Anatomists; Tissue Cult Asn. *Res:* Microscopic anatomy; electron microscopy; histochemistry; tissue culture; connective tissues; reticuloendothelial system; histophysiology of the lympho-hematopoietic system. *Mailing Add:* Dept Animal Biol Univ Pa Sch Vet Med 3800 Spruce St 6046 Philadelphia PA 19104

WEISS, LEONARD, b Brooklyn, NY, Mar 14, 34; m 58; c 2. APPLIED MATHEMATICS, ENGINEERING. *Educ:* City Col New York, BEE, 56; Columbia Univ, MSEE, 59; Johns Hopkins Univ, PhD(elec eng), 62. *Hon Degrees:* MA, Brown Univ, 66. *Prof Exp:* Lectr elec eng, City Col New York, 56-59; staff scientist, Res Inst Adv Studies, 62-64; from asst prof to assoc prof appl math & eng, Brown Univ, 64-68; prof elec eng, Inst Fluid Dynamics & Appl Math, Univ Md, College Park, 68-78; staff dir, US Senate Subcomt on Energy, Nuclear Proliferation & Fed Serv, 77-86; STAFF DIR, US SENATE COMT GOVT AFFAIRS, 87- *Concurrent Pos:* Alfred P Sloan res fel, 66-68; res mathematician, Naval Res Lab, Washington, DC, 70-77; legis asst to Sen John Glenn, 76-77; Cong sci fel, Inst Elec & Electronics Engrs, 76-77. *Mem:* AAAS; Am Math Soc; Soc Indust & Appl Math; Math Asn Am; Inst Elec & Electronics Engrs. *Res:* System theory; control theory; signal theory; analysis and structure of dynamical systems described by differential equations; nuclear proliferation. *Mailing Add:* Staff Dir US Senate Comt Govt Affair 340 Dirksen Senate Bldg Washington DC 20510

WEISS, LEONARD, b London, Eng, June 15, 28; m 51; c 3. PATHOLOGY, CELL BIOLOGY. *Educ:* Cambridge Univ, BA, 50, MA, MB, BChir, 53, MD, 58, PhD(biol), 63, ScD, 71. *Prof Exp:* House physician, Westminster Hosp, Univ London, 54-55, resident pathologist, Hosp & res assoc & registr morbid anat, Med Sch, 55-58; mem sci staff, Med Res Coun, Nat Inst Med Res, London, 58-60 & Strangeways Res Lab, Cambridge Univ, 60-64; DIR CANCER RES & CHIEF CANCER RES CLINICIAN, DEPT EXP PATH, ROSWELL PARK MEM INST, 64-; RES PROF DERMAT, MED SCH, STATE UNIV NY, BUFFALO, 74- *Concurrent Pos:* Res prof biophys, State Univ NY Buffalo, 65-74. *Mem:* Fel AAAS; Path Soc Gt Brit & Ireland; fel Royal Col Path; fel Brit Inst Biol; fel Col Am Pathologists. *Res:* Metastasis; biophysics of cell contact phenomena. *Mailing Add:* Dept Exp Path Roswell Park Mem Inst Buffalo NY 14263

WEISS, LIONEL EDWARD, b London, Eng, Dec 11, 27; m 64; c 2. GEOLOGY. *Educ:* Univ Birmingham, BSc, 49, PhD, 53; Univ Edinburgh, ScD, 56. *Prof Exp:* Res assoc, 51-53, assoc, 56-57, from asst prof to assoc prof, 57-64, Miller res prof, 65-67, PROF GEOL, UNIV CALIF, BERKELEY, 64- *Concurrent Pos:* Guggenheim fel, 61 & 69. *Res:* Natural and experimental deformation of rocks and minerals; structural geology. *Mailing Add:* Four Seaview Dr Santa Barbara CA 93108

WEISS, LIONEL IRA, b New York, NY, Sept 5, 23; m 46; c 3. MATHEMATICAL STATISTICS. *Educ:* Columbia Univ, BA, 43, MA, 45, PhD(math statist), 53. *Prof Exp:* From asst prof to assoc prof statist, Univ Va, 49-56; assoc prof math, Univ Ore, 56-57; assoc prof, 57-61, PROF OPERS RES, CORNELL UNIV, 61- *Concurrent Pos:* Mem, Nat Res Coun, 66-69. *Mem:* Inst Math Statist; NY Acad Sci; Sigma Xi. *Res:* Statistical decision theory; asymptotic statistical theory. *Mailing Add:* Upson Hall Cornell Univ Ithaca NY 14850

WEISS, LOUIS CHARLES, electric fields, textile physics; deceased, see previous edition for last biography

WEISS, MALCOLM PICKETT, b Washington, DC, June 28, 21; m 43; c 4. GEOLOGY. *Educ:* Univ Minn, PhD(geol), 53. *Prof Exp:* From instr to assoc prof geol, Ohio State Univ, 52-67; chmn dept, 67-72, prof, 67-88, PROF EMER GEOL, NORTHERN ILL UNIV, 88- *Mem:* Geol Soc Am; Soc Econ Paleontologists & Mineralogists; Am Asn Petrol Geologists; AAAS. *Res:* Stratigraphy; sedimentary petrography; carbonate petrology. *Mailing Add:* Dept Geol Northern Ill Univ De Kalb IL 60115

WEISS, MARK LAWRENCE, b Brooklyn, NY, Nov 1, 45; div; c 2. PRIMATOLOGY, PHYSICAL ANTHROPOLOGY. *Educ:* State Univ NY Binghamton, BA, 66; Univ Calif, Berkeley, MA, 68, PhD(anthrop), 69. *Prof Exp:* Asst prof, 69-74, assoc prof, 74-87, CHMN, DEPT ANTHROPOLOGY, WAYNE STATE UNIV, 83-, PROF, 87- *Concurrent Pos:* Adj prof anat, Sch Med, Wayne State Univ, 69-; fel, William Beaumont Hosp, 74; res assoc genetics, Leicester Univ, UK, 82-83; on leave, NSF, 90-91. *Mem:* AAAS; Am Anthrop Asn; Am Asn Phys Anthropologists; Brit Soc Study Human Biol. *Res:* Biochemical polymorphisms; primate genetics; primate microevolution. *Mailing Add:* Dept Anthropology Wayne State Univ Detroit MI 48202

WEISS, MARTIN, b New York, NY, Jan 21, 19; m 49; c 3. GEOLOGY. *Educ:* City Col New York, BS, 48; Univ Mich, MS, 51, PhD(geol), 54. *Prof Exp:* Asst geol, Mus Paleont, Univ Mich, 51-53; geologist, US Geol Surv, 53-63; oceanogr, Nat Oceanog Data Ctr, 63-72, marine geologist, Nat Geophys & Solar Terrestrial Data Ctr, Nat Oceanic & Atmospheric Admin, 68-75; MEM STAFF, US GEOL SURV, RESTON, 75- *Mem:* Geol Soc Am; Am Geophys Union; Marine Technol Soc. *Res:* Geological oceanography; military geology; Paleozoic ostracoda; review of environmental impact statements. *Mailing Add:* 3710 Prado Pl FAirfax VA 22031

WEISS, MARTIN GEORGE, plant breeding; deceased, see previous edition for last biography

WEISS, MARTIN JOSEPH, b New York, NY, May 4, 23; m 51; c 3. ORGANIC CHEMISTRY. *Educ:* NY Univ, AB, 44; Duke Univ, PhD(chem), 49. *Prof Exp:* Asst, Duke Univ, 44-47; res fel org chem, Hickrill Chem Res Found, 49-50; res chemist, Pharmaceut Res Dept, Calco Chem Div, 50-54, group leader, 54-75, dept head, 75-83, ASSOC DIR PHARMACEUT RES, LEDERLE LABS, AM CYANAMID CO, 83- *Concurrent Pos:* Asst, Comt Med Res, Off Sci Res & Develop, Duke Univ, 44-45. *Mem:* Am Chem Soc. *Res:* Synthetic medicinal chemistry in prostaglandins, antibiotics, steroids, nucleosides and carbohydrates, indoles and other heterocyclics; chemotherapy; anti-arthritic, anti-allergy, anti-atherosclerotic and hypoglycemic agents. *Mailing Add:* Lederle Labs Am Cyanamid Co Pearl River NY 10965

WEISS, MARVIN, b New York, NY, Feb 6, 14; m 40; c 3. PHARMACEUTICAL CHEMISTRY, SCIENCE ADMINISTRATION. *Educ:* Brooklyn Col, BS, 37; Univ Ill, MS, 38. *Prof Exp:* Dir, develop & control labs, Am Pharmaceut Co, 41-57; dir, anal lab, Berkeley Chem Coorp, 57-66; group leader, corp control & anal, Millmaster Onyx Corp, 66, dir, anal lab, Berkeley Chem Dept, 66-78, corp tech dir, 66-78, tech dir, AGross & Co, 78-83; RETIRED. *Concurrent Pos:* Group consult, Mantrose-Hauser Div, US Printing Ink Div, Onyx Div & Carboquimica SA Div, 65-83. *Mem:* Am Chem Soc; Am Oil Chemists Soc. *Res:* Urethanes; nitrogen heterocycles; multiple condensations; acetic acid-ammonium acetate reactions with carbonyl compounds; crossed Cannizzaro syntheses; urea reactions; radiation curable inks and coatings; fatty acids and fatty acid derivates; shellac and shellac derivatives. *Mailing Add:* 227 Elkwood Ave New Providence NJ 07974-1816

WEISS, MAX LESLIE, b Salt Lake City, Utah, Aug 12, 33; c 5. MATHEMATICS. *Educ:* Yale Univ, BA, 55; Cornell Univ, MS, 58; Univ Wash, PhD(math), 62. *Prof Exp:* Instr math, Reed Col, 58-60 & Univ Wash, 62-63; NSF fel, Inst Advan Study, 63-64; from asst prof to assoc prof, 64-72, PROF MATH, UNIV CALIF, SANTA BARBARA, 72- *Concurrent Pos:* Assoc provost, Col Creative Studies, 71-77, 83-84, actg provost, 84-86, provost, 86- *Mem:* Am Math Soc; Asn Mem Inst Advan Study. *Res:* Function algebras and complex variables. *Mailing Add:* Dept Math Univ Calif Santa Barbara CA 93106

WEISS, MAX TIBOR, b Hungary, Dec 29, 22; nat US; m 53; c 4. PHYSICS. *Educ:* Mass Inst Technol, MS, 47, PhD(physics), 51. *Prof Exp:* Engr, Radio Corp Am, 43-44 & US Naval Ord Lab, 45-46; res assoc, Milrowave Spectros Lab, Mass Inst Technol, 46-50; mem tech staff, Bell Tel Labs, Inc, 50-59; assoc dept mgr, Appl Physics Dept, Hughes Aircraft Corp, 59-61; dir electronics lab, Aerospace Corp, 61-63, gen mgr labs div, 63-67, asst mgr eng opers, TRW Systs, 67-68; gen mgr, Electronics & Optics Div, Aerospace Corp, 68-78, vpres & gen mgr lab opers, 78-81, vpres eng group, 81-86; VPRES, TECH, NORTHROP CORP, 86- *Concurrent Pos:* Lectr, City Col New York, 53. *Mem:* AAAS; fel Inst Elec & Electronics Engrs; fel Am Phys Soc. *Res:* Microwaves; magnetics; quantum electronics; communications; microelectronics. *Mailing Add:* Northrop Corp 1840 Century Park E Los Angeles CA 90067

WEISS, MICHAEL DAVID, b Chicago, Ill, Nov 12, 42. MATHEMATICAL ECONOMICS, BEHAVIOR UNDER UNCERTAINTY. *Educ:* Brandeis Univ, BA, 64; Brown Univ, PhD(math), 70; Univ Md, MA, 84. *Prof Exp:* Asst prof math, Wayne State Univ, 69-74; analyst, Ketron Inc, 74-76; math statistician, 76-85, AGR ECONOMIST, ECON RES SERV, USDA, 85- *Mem:* Am Econ Asn; Economet Soc; Am Agr Econ Asn; Am Math Soc; Soc Indust & Appl Math. *Res:* Economic theory of behavior under uncertainty; mathematical economics and econometrics; probability and statistics; ergodic theory; theory of fuzzy sets; mathematical applications. *Mailing Add:* 7797 Heatherton Lane Potomac MD 20854-3264

WEISS, MICHAEL JOHN, b St Paul, Minn, June 13, 55. PEST MANAGEMENT. *Educ:* Purdue Univ, BS, 77; Ohio State Univ, MS, 79; Univ Nebr, PhD(entom), 83. *Prof Exp:* Asst prof res, Mont State Univ, 83-85; ASST PROF ENTOM, NDAK STATE UNIV, 85- *Mem:* Sigma Xi; Entom Soc Am. *Res:* Applied insect ecology; agroecosystem research. *Mailing Add:* Dept Entom NDak State Univ Main Campus Fargo ND 58105

WEISS, MICHAEL KARL, b Hatzfeld, Romania, Nov 11, 28; nat US; m 54; c 3. CHEMISTRY. *Educ:* Western Reserve Univ, MS, 55. *Prof Exp:* Res chemist, Harshaw Chem Co, 52-54; supvr anal chem, Repub Steel Corp Res Ctr, 55-62, chief anal sect, 62-67; head anal div, 67-72, mgr qual control dept, Bunker Hill Co, 72-82; LAB MGR, BP MINERALS AM, 83- *Mem:* Am Chem Soc; Soc Appl Spectros; Am Soc Testing & Mat. *Mailing Add:* 2225 Elkridge Dr Camden SC 29020

WEISS, MICHAEL STEPHEN, b Queens Co, NY, Mar 20, 43; m 65; c 2. SPEECH & HEARING SCIENCES, SPEECH PATHOLOGY. *Educ:* Long Island Univ, BA, 64; Purdue Univ, MS, 68, PhD(speech sci), 70. *Prof Exp:* Coordr res, Cleveland Hearing & Speech Ctr & sr res assoc speech sci, Case Western Reserve Univ, 70-71; asst prof speech, Howard Univ, 71-72; instr laryngol & otol & sci dir, Info Ctr Hearing, Speech & Dis Human Commun, Sch Med, Johns Hopkins Univ, 72-76; coordr continuing educ media, Boys Town Inst Commun Dis Children, 76-78; assoc prof dept speech path & audiol, 78-84, PROF DEPT COMMUN DIS, WEST CHESTER UNIV, 84- *Concurrent Pos:* Lectr, Gallaudet Col, 74-76; assoc prof otolaryngol, Creighton Univ, 76-78. *Mem:* Am Speech & Hearing Asn; Acoust Soc Am; NY Acad Sci. *Res:* Speech science; physiological and acoustical events which underlie speech production and perception. *Mailing Add:* Dept Commun Dis West Chester Univ West Chester PA 19383

WEISS, MITCHELL JOSEPH, b Chicago, Ill, Nov 12, 42. INVERTEBRATE ZOOLOGY. *Educ:* Brown Univ, ScB, 64; Univ Mich, PhD(zool), 70; Rutgers Univ, MLS, 88. *Prof Exp:* Instr zool, Univ Iowa, 73-74; asst prof biol, 74-81, vis asst prof, 81-82, GUEST INVESTR, RUTGERS UNIV, NEW BRUNSWICK, 82- *Concurrent Pos:* NIH res fel, Univ Wash, 70-73; prin investr NSF res grant, 76-78; consult, Oriel Corp, 83. *Mem:* Am Soc Zoologists; AAAS; Am Micros Soc; Int Asn Meiobenthologists. *Res:* Functional and comparative anatomy and development of insect brain centers at microscopic and ultrastructural levels; biology of phylum Gastrotricha; current emphasis on sexuality, life cycles and systematics of freshwater gastrotrichs. *Mailing Add:* 51-B Phelps Ave New Brunswick NJ 08901

WEISS, NOEL SCOTT, b Chicago, Ill, Mar 10, 43. EPIDEMIOLOGY, BIOSTATISTICS. *Educ:* Stanford Univ, AB, 65, MD, 67; Harvard Univ, MPH, 69, DrPH, 71. *Prof Exp:* Epidemiologist, Nat Ctr Health Statist, 71-73; from asst prof to assoc prof, 73-79, PROF, DEPT EPIDEMIOL, SCH PUB HEALTH & COMMUNITY MED, UNIV WASH, SEATTLE, 79-, CHMN DEPT, 84- *Concurrent Pos:* Asst mem, Fred Hutchinson Cancer Res Ctr, Seattle, Wash, 74-76, assoc mem, 76-82, mem, 82-; Nat Cancer Inst res grants, 75-; assoc ed, Cancer Epidemiol & Prev, 87-90; chair, Adv Comt Occup Studies Sect, Environ Epidemiol Br & mem, Bd Sci counselors, Div Cancer Etiology, Nat Cancer Inst, NIH; mem, Adv Comt Study Intermediate Outcome Breast Cancer, Ctr Health Studies, Group Health Coop, 88-90, Cancer Epidemiol Pac Basin, Hawaii, 89 & Workshop Etiology Mult Myeloma, Nat Cancer Inst, 90. *Mem:* Nat Acad Sci. *Res:* Endocrinologic determinants of disease; author of 160 technical publications. *Mailing Add:* Dept Epidemiol SC-36 Univ Wash Seattle WA 98195

WEISS, NORMAN JAY, b Brooklyn, NY, May 28, 42; m 65; c 2. MATHEMATICAL ANALYSIS. *Educ:* Harvard Univ, BA, 63; Princeton Univ, PhD, 66. *Prof Exp:* Instr math, Princeton Univ, 66-68; asst prof, Columbia Univ, 68-71; asst prof, 71-72, ASSOC PROF MATH, QUEENS COL, NY, 73- *Mem:* Am Math Soc. *Res:* Real and Fourier analysis. *Mailing Add:* Dept Math Queens Col 65-30 Kissena Blvd Flushing NY 11367

WEISS, PAUL ALFRED, biology; deceased, see previous edition for last biography

WEISS, PAUL B, physics, computer science, for more information see previous edition

WEISS, PAUL STORCH, b Ithaca, NY, Oct 10, 59; m 82; c 3. SURFACE CHEMISTRY, SURFACE PHYSICS. *Educ:* Mass Inst Technol, SB & SM, 80; Univ Calif, Berkeley, PhD(phys chem), 86. *Prof Exp:* Postdoctoral mem tech staff, AT&T Bell Labs, 86-88; ASST PROF CHEM, PA STATE UNIV, 89- *Concurrent Pos:* Vis scientist, Almaden Res Ctr, IBM Corp, 88-89; NSF presidential young investr, 91. *Mem:* Am Chem Soc; Am Phys Soc; Sigma Xi. *Res:* Analytical and physical chemistry; surface chemistry and physics; low temperature scanning tunneling microscopy; solid state chemistry; novel techniques for imaging, spectroscopy and detection. *Mailing Add:* Dept Chem 152 Davey Lab Pa State Univ University Park PA 16802

WEISS, PETER JOSEPH, b Vienna, Austria, Nov 23, 18; nat US; m 48; c 2. PHARMACEUTICAL CHEMISTRY. *Educ:* George Washington Univ, BS, 51; Georgetown Univ, MS, 53, PhD(biochem), 56. *Prof Exp:* Chemist, Food & Drug Admin, 47-56, chief chem br, Div Antibiotics, 56-70, dep dir, Nat Ctr Antibiotic Anal, 70-71, dir, 72-79; RETIRED. *Mem:* Am Chem Soc; Acad Pharmaceut Sci; Am Pharmaceut Asn. *Res:* Chemical analysis of antibiotics. *Mailing Add:* 1309 Caddington Ave Silver Spring MD 20901-1049

WEISS, PHILIP, b New York, NY, June 12, 16; m 43; c 2. ORGANIC CHEMISTRY. *Educ:* NY Univ, BS, 39, MSc, 41, PhD(org chem), 48. *Prof Exp:* Res chemist, Lederle Labs, Am Cyanamid Co, 41-46; sr res chemist, Wallace & Tiernan Prod, Inc, 46-52; sr proj chemist, Colgate-Palmolive Co, 52-57; sr res scientist, Electrochem Dept, Res Labs, Gen Motors Corp, 57-58, asst head electrochem & polymers dept, 58-59, head, Polymers Dept, Res Labs, 59-81; ADJ PROF, DEPT CHEM, OAKLAND UNIV, 80- *Concurrent Pos:* Instr, Cooper 49-51; ed, J Appl Polymer Sci; mem, Comt Crit & Strategic Mat, Nat Mat Adv Bd, Adv Bd, Col Eng, Univ Detroit & Adv Bd, Polymer Prog, Col Eng, Princeton Univ; US rep, Plastics & High Polymers Sect, Int Union Pure & Appl Chem; proj mgr, Paint Res Inst. *Honors & Awards:* Exner Medal, 69. *Mem:* Am Chem Soc; Soc Plastics Engrs. *Res:* Polymer synthesis, research and development in plastics, rubber, adhesives and surface coatings; graft and block copolymers; mechanisms of finish failure; adhesion and cohesion; mechanical behavior of polymers; polymer flammability, aging, waste disposal and processing; electrically conducting polymers. *Mailing Add:* Dept Chem Oakland Univ Rochester MI 48309-4401

WEISS, RAINER, b Berlin, Ger, Sept 29, 32; US citizen; m 59; c 2. PHYSICS. *Educ:* Mass Inst Technol, BS, 55, PhD(physics), 62. *Prof Exp:* Asst prof physics, Tufts Univ, 60-62; res assoc, Princeton Univ, 62-64; from asst prof to assoc prof, 64-73, PROF PHYSICS, MASS INST TECHNOL, 73- *Concurrent Pos:* Phys sci comt, NASA, 70-74, Mgt Opers Working Group, Shuttle Astron, 73-76, Airborne Astron, 73-86, chmn, Panel Exp Relativity & Gravitation, 74-76, SSSC comt, 79-82, Infrared Detector Panel 78, Space & Earth Sci Adv Comt, 82; chmn subcomt gravitational physics, NSF, 78, coordr panel interferometric observ for gravitational waves, 86; mem, Nat Acad Space Sci Bd, 83-86; mem, Panel Joint Inst Lab Astrophys, Bd Assessment of NBS Progs, Nat Acad Sci, 85- *Honors & Awards:* Achievement Award, NASA, 84, Group Achievement Award, 90. *Mem:* AAAS; Am Phys Soc; Am Astron Soc. *Res:* Experimental atomic physics, atomic clocks, laser physics; experimental gravitation; millimeter and submillimeter astronomy; cosmic background measurements. *Mailing Add:* Dept Physics Mass Inst Technol 77 Massachusetts Ave Cambridge MA 02139

WEISS, RICHARD GERALD, b Akron, Ohio, Nov 13, 42. PHOTOCHEMISTRY, PHYSICAL ORGANIC CHEMISTRY. *Educ:* Brown Univ, ScB, 65; Univ Conn, MS, 67, PhD(chem), 69. *Prof Exp:* NIH res fel chem, Calif Inst Technol, 69-71; vis prof, Inst Chem, Univ Sao Paulo, 71-74; from asst prof to assoc prof, 70-85, PROF CHEM, GEORGETOWN UNIV UNIV, 86- *Concurrent Pos:* Nat Acad Sci overseas fel, 71-74; vis scientist, Max Planck Inst Radiation Chem, 81-82; vis prof, Nat Superior Sch Chem, Strasbourg, 82 & Univ Bordeaux, 82; consult, World Bank, 83-; vis prof, Indian Inst Sci, Bangalore, India, 89-90. *Mem:* Am Chem Soc; InterAm Photochem Soc; Europ Photochem Asn; Am Asn Univ Professors. *Res:* Mechanisms and rates of organic and photochemical reactions; steric effects in electronic energy transfer and in decay of excited states; mechanistic studies in ordered media. *Mailing Add:* 4604 W Virginia Ave Bethesda MD 20814

WEISS, RICHARD JEROME, b New York, NY, Dec 14, 23; m; c 2. PHYSICS. *Educ:* City Col New York, BS, 44; Univ Calif, MA, 47; NY Univ, PhD(physics), 50. *Prof Exp:* Physicist, US Army Mat Res Ctr, 50-80; prof, Univ Surrey, England, 81-85; PROF, KINGS COL, LONDON, 80- *Concurrent Pos:* Vis fel, Cavendish Labs, Cambridge Univ, 56-57; lectr, Mass Inst Technol, 59; Secy Army fel, Imp Col, Univ London, 62-63; chmn, Comn Electron Distributions, Int Union Crystallog, 72-75; ed, Int J Optical Sensors, 85-90; contrib ed, Lasers & Optronics, 90- *Mem:* Am Phys Soc; Am Crystallog Soc. *Res:* Solid state, neutron and x-ray physics; electron structure of solids, materials; lasers; optoelectronics. *Mailing Add:* Four Lawson St Avon MA 02322

WEISS, RICHARD LOUIS, b Evanston, Ill, June 24, 44. BIOCHEMISTRY, GENETICS. *Educ:* Stanford Univ, BS, 66; Univ Wash, PhD(biochem), 71. *Prof Exp:* USPHS fel, Univ Mich, 71-72, Am Cancer Soc fel, 72-73; from asst prof to assoc prof, 74-85, PROF CHEM, UNIV CALIF, LOS ANGELES, 85- *Mem:* Am Chem Soc; Am Soc Microbiol; Genetics Soc Am; Fedn Am Soc Exp Biol. *Res:* Regulation of amino acid metabolism in eucaryotic microorganisms. *Mailing Add:* Dept Chem Univ Calif Los Angeles CA 90024-1569

WEISS, RICHARD RAYMOND, b Takoma Park, Md, Aug 26, 28; m 56; c 2. METEOROLOGY, ELECTRICAL ENGINEERING. *Educ:* Univ Md, BSE, 52; Univ Mich, Ann Arbor, MSE, 55; Univ Wash, PhD(elec eng), 67. *Prof Exp:* Engr, Sperry Gyroscope Co, 52-54; cardiovasc trainee, Univ Wash, 59-60; asst prof elec eng, Seattle Univ, 60-64; lectr, Univ Wash, 64-67, res asst prof, 67-74, sr res assoc atmospheric sci, 74-81; ASSOC PROF ELEC ENG, SEATTLE UNIV, 81- *Mem:* Inst Elec & Electronics Engrs; Am Geophys Union. *Res:* Radar meteorology. *Mailing Add:* 1501 McGilvaa Blvd Seattle WA 98112

WEISS, ROBERT ALAN, b Cleveland, Ohio, Oct 29, 50. POLYMER SCIENCE. *Educ:* Northwestern Univ, BS, 72; Univ Mass, PhD(chem eng), 76. *Prof Exp:* Res engr, Exxon Chem Co, 75-77, staff engr, Exxon Res & Eng Co, 77-81; ASSOC PROF, DEPT CHEM ENG, UNIV CONN, 81- *Concurrent Pos:* Vis prof, Stevens Inst Tech, 80; chmn, Eng Property Struct Div, Am Chem Soc, 83-86, & Plastics Anal Div, 85-86. *Mem:* Soc Plastics Engrs; Am Chem Soc; Soc Rheol; NAm Thermal Anal Soc. *Res:* Polymer structure; property relationships; ionomers; composite materials. *Mailing Add:* Chem Eng Dept U-139 Univ Conn Storrs CT 06268

WEISS, ROBERT JEROME, b West New York, NJ, Dec 9, 17; m 45; c 3. PSYCHIATRY. *Educ:* George Washington Univ, AB, 47; Columbia Univ, MD, 51; Dartmouth Col, MA, 64. *Prof Exp:* Intern, Columbia Div, Bellevue Hosp, 51, asst resident med, 53; resident psychiat, Columbia Psychoanal Clin, 54-59; chief, Mary Hitchcock Mem Hosp, Hanover, NH, 59-70; vis prof, 70-75, assoc dir, Ctr Community Health & Med Care, Harvard Med Sch, 70-75, assoc dean health care progs, 71-75; prof psychiat & social med, Col Physicians & Surgeons, Columbia Univ, 75-86, dir community health systs, 75-86, Delemar prof pub health pract & dean, 80-86, EMER DEAN, SCH PUB HEALTH, COLUMBIA UNIV, 86-, EMER DELEMAR PROF, 86-, EMER PROF SOCIAL MED & PSYCHIAT, 86- *Concurrent Pos:* NIMH career teacher trainee, Columbia Univ, 56-58; resident, NY State Psychiat Inst & Presby Hosp, NY, 57; asst attend, Vanderbilt Clin, 57-58 & Presby Hosp, 58-59; assoc, Col Physicians & Surgeons, Columbia Univ, 57-59; consult, NH Div Ment Health, 59-70, Vet Admin Hosp, White River Junction, Vt, 60-70 & Bur Health Manpower Educ, 72-; prof psychiat & chmn dept, Dartmouth Med Sch, 59-70; chmn adv comn, NH Dept Health & Welfare, 61; coordr panel, NIMH, 65-67, chmn subcomt psychiat, 67-68; psychiatrist, Beth Israel Hosp, 70-75; attend, Presby Hosp, NY, 75-; consult, Nat Ctr Health Serv Res, 75- & NIMH, 77-; vis prof, Community Med, Univ NMex Med Sch, 87-; vis scholar, Univ Calif, 86-87; vis prof psychiat, Harvard Univ, 89- *Honors & Awards:* Bi-Centennial Medal, Columbia Univ Col Physicians & Surgeons, 67. *Mem:* AAAS; fel Am Psychiat Asn; Am Asn Med Cols; Am Acad Psychoanal; NY Acad Sci. *Res:* Epidemiology; health care; preventive psychiatry; community medicine. *Mailing Add:* 34 Allison Rd Princeton NJ 08540-3002

WEISS, ROBERT JOHN, b Pomona, Calif, Apr 9, 37; m 64; c 2. MATHEMATICS. *Educ:* La Verne Col, BA, 58; Univ Calif, Los Angeles, MA & PhD(math), 62. *Prof Exp:* Assoc prof math, Bridgewater Col, 62-68; assoc prof & head dept, 68-74, PROF MATH, MARY BALDWIN COL, 74- *Mem:* Am Math Soc; Math Asn Am; Sigma Xi. *Res:* Algebraic topology. *Mailing Add:* 411 Rainbow Dr Staunton VA 24401

WEISS, ROBERT MARTIN, b New York, NY; m; c 2. UROLOGY, PHARMACOLOGY. *Educ:* Franklin & Marshall Col, BS, 57; State Univ NY Downstate Med Ctr, MD, 60. *Hon Degrees:* MA, Yale Univ, 76. *Prof Exp:* Intern med, Second (Cornell) Med Div, Bellevue Hosp, New York, 60-61; resident gen surg, Beth Israel Hosp, NY, 61-62; resident urol, Columbia-Presby Med Ctr, 63-64; fel, Col Physicians & Surgeons, Columbia Univ, 64-65; resident, Columbia-Presby Med Ctr, 65-67; instr surg-urol, Sch Med, Yale Univ, 67-68, asst prof urol, 68-71, assoc prof surg-urol, 71-76, PROF SURG-UROL, SCH MED, YALE UNIV, 76-, PROF & CHIEF UROL, 88- *Concurrent Pos:* Res assoc pharmacol, Col Physicians & Surgeons, Columbia Univ, 67-75, adj assoc prof, 75-77, adj prof, 77-; attend, Yale-New Haven Hosp, 67-; consult, West Haven Vet Admin Hosp, 67- & Waterbury Hosp, 76-; mem obstruction & neuromuscular dis comt, Nat Inst Arthritis, Metab & Digestive Dis, 74-75; fel, Timothy Dwight Col, Yale Univ, 74-; asst ed, J Urol, 75-82; mem adv panel, US Pharmacopeia & Nat Formulary, 76-81; assoc sect ed, J Urol, 86- *Mem:* Am Physiol Soc; Soc Gen Physiologists; Am Acad Pediat; Am Col Surgeons; Am Urol Asn. *Res:* Mechanical and electrophysiologic properties of ureteral smooth muscle; role of cyclic nucleotides in smooth muscle function. *Mailing Add:* Urol Sect Sch Med Yale Univ 333 Cedar St New Haven CT 06510

WEISS, ROGER HARVEY, b New York, NY, July 13, 26; m 53; c 2. ANALYTICAL CHEMISTRY. *Educ:* Col Holy Cross, BNaval Sci, 46; Cornell Univ, AB, 50; Ga Inst Technol, PhD(chem), 68. *Prof Exp:* Instr chem, Univ Toledo, 54-57; from asst prof to assoc prof, 59-, PROF CHEM, HUMBOLDT STATE UNIV, 74- *Concurrent Pos:* Fulbright lectr, Univ Sind, Pakistan, 73-74. *Mem:* Am Chem Soc. *Res:* Titrimetry with solid titrants; trace analysis; chelation; absorption spectrophotometry; photometric titrations. *Mailing Add:* Dept Chem Humboldt State Univ Arcata CA 95521-4957

WEISS, ROLAND GEORGE, b Milwaukee, Wis, July 11, 49. INDUSTRIAL ENGINEERING. *Educ:* Univ Wis-Milwaukee, BS, 72; Northwestern Univ, MS, 75, PhD(indust eng), 76. *Prof Exp:* Oper res analyst technol policy, Exp Technol Incentives Prog, Ctr Field Methods, Nat Bur Standards, 76-79; systs engr, Bell Labs, 79-83; CHIEF ENGR, CTA INC, ROCKVILLE, MD, 88- *Concurrent Pos:* Prof rels vchmn, Col Pub Progs & Processes, Inst Mgt Sci, 77-81. *Mem:* AAAS; Am Inst Indust Engrs; Am Psychol Asn; Inst Elec & Electronics Engrs; Inst Mgt Sci. *Res:* Development of methodology for analyzing complex, unstructured problems, including systems analysis, decision and architectural design and administrative experimentation; systems engineering of large, advanced systems with hardware, software, organizational market components. *Mailing Add:* 44 Columbia Ave Takoma Park MD 20912

WEISS, RONALD, b Chicago, Ill, Jan 29, 37; m 67; c 3. FOOD SCIENCE. *Educ:* Ariz State Univ, BS, 58, MS, 59; Mich State Univ, PhD(chem), 64, MBA, 67. *Prof Exp:* Res chemist, 64-67, proj coordr, 67-69, mgr prod develop, 69-73, dir growth & develop, 73-75, DIR PLANNING, MILES LABS, INC, 75- *Mem:* Am Chem Soc; Inst Food Technologists; Soft Drink Technologists Asn; Sigma Xi. *Mailing Add:* 1505 Greenbriar Dr Elkhart IN 46515

WEISS, SAMUEL BERNARD, b New York, NY, May 18, 26; m 61; c 2. BIOCHEMISTRY. *Educ:* City Col New York, BS, 48; Univ Southern Calif, PhD(biochem), 54. *Prof Exp:* Res assoc biochem, Mass Gen Hosp, Boston, 56-57; asst prof, Rockefeller Inst, 57-58; asst prof, 58-63, PROF BIOCHEM, UNIV CHICAGO, 63- *Concurrent Pos:* Am Heart Asn fel biochem, Univ Chicago, 54-56; Guggenheim fel, Salk Inst, 70-71; res assoc, Argonne Cancer Res Hosp, 58-, assoc dir, 67-74. *Honors & Awards:* Theobold Smith Award, AAAS, 61; Am Chem Soc Award Enzyme Chem, 66. *Mem:* Am Soc Biol Chemists; Am Acad Arts & Sci. *Res:* Enzymology of reactions in the synthesis of lipids, proteins and nucleic acids; viral transfer RNAs; mechanism of action of polycyclic aromatic hydrocarbons. *Mailing Add:* Franklin McLean Mem Res Inst Univ Chicago Box 433 IB-7 Chicago IL 60637

WEISS, SIDNEY, b New York, NY, Dec 16, 20; m 44; c 3. BIOCHEMISTRY. *Educ:* Queens Col, NY, BS, 42; Fordham Univ, MS, 46, PhD(biochem), 49; Rutgers Univ, MS, 66. *Prof Exp:* Res chemist, Food Res Labs, Inc, NY, 42-46; res assoc biochem, Inst Cancer Res, Philadelphia, Pa, 49-58; sr res chemist, Colgate-Palmolive Co, 58-59, sect head, 59- 63, mgr biol res, 63-74, assoc dir res, 74-79, dir basic res, 79- 85; RETIRED. *Mem:* Am Chem Soc; Am Soc Biol Chem; Am Asn Dent Res; Am Statist Asn. *Res:* Biological oxidations; enzymatic reactions and microbiological transformations. *Mailing Add:* 76 Juniper Dr Levittown PA 19056

WEISS, SOL, b Austria, Apr 19, 13; US citizen; m 40; c 2. MATHEMATICS. *Educ:* Brooklyn Col, BS, 34; Columbia Univ, MA, 36. *Prof Exp:* Teacher, Philadelphia Pub Sch Syst, Pa, chmn dept math, 50-64; from asst prof to assoc prof math, 64-78, EMER PROF MATH SCI, WEST CHESTER UNIV, 78- *Concurrent Pos:* Consult, Wilmington Sch Dist, Del, 64, Philadelphia Pub Sch Syst, 64-65 & 67-, Cecil County Pub Schs, Md, 66, Upward Bound Prog, Franklin & Marshall Col, 67 & Lehigh Univ Social Restoration Prog, Pa State Prisons, 73-74; trustee, West Chester Univ; lectr, Chinese Univ, 86. *Mem:* Math Asn Am. *Res:* Mathematics education for the low achiever; book for parents on how to help their children with math. *Mailing Add:* 102 Crosshill Rd Wynnewood PA 19096

WEISS, STANLEY, b New York, NY, Apr 9, 29; m 60; c 3. METALLURGY. *Educ:* Polytech Inst Brooklyn, BMetE, 51; Mass Inst Technol, SM, 55, ScD(metall), 64. *Prof Exp:* Engr, Gen Elec Co, 51-54; sr engr, Alco Prod, 55-56; mgr eng, Gen Elec Co, Ohio, 56-61; res assoc metall, Mass Inst Technol, 64-68; EMER PROF, MAT ENG, UNIV WIS-MILWAUKEE, 68- *Concurrent Pos:* Consult, Gen Elec Co, 61-, Standard Thomson Corp, 63-68, Walter Kidde-Fenwal Div, 64-68 & Cryogenics Technol, 65-; corp consult, Snap-On Tools Corp, 71- *Mem:* Am Welding Soc; Am Soc Mech Engrs; Am Soc Metals; Am Soc Testing Mat; Sigma Xi. *Res:* Brazing mechanisms; residual stresses in welding; failure analysis; forged powder metals; joining of forged powder metals; corrosion of prosthetic devices. *Mailing Add:* Com Bldg Suite 580 744 N Fourth St Milwaukee WI 53203

WEISS, STANLEY H, b Brooklyn, NY, Jan 28, 54; m 81; c 2. MEDICAL ONCOLOGY, CLINICAL EPIDEMIOLOGY. *Educ:* Yale Univ, BA, 74; Harvard Univ, MD, 78. *Prof Exp:* Res assoc immunol, Robert B Brigham Hosp, 75; dir, path diagnosis regist, Peter B Brigham Hosp, 76-78; intern med, Montefiore Hosp & Med Ctr, 78-79, resident, 79-81, asst attend physician,

81-82; clin assoc oncol, Nat Cancer Inst, 82-87, med staff fel epidemiol, 83-87, ADJ ASST PROF, DEPT PREV MED & BIOMET, UNIFORMED SERV, UNIV HEALTH SCI, NIH, 86-, ASST PROF, DEPT PREV MED & COMMUNITY HEALTH, 87-, DEPT MED, 88- *Concurrent Pos:* Res assoc epidemiol, Ctr Dis Control, 80; special consult, WHO, special prog Res, Develop & Res Training in Human Reproduction, 87, mem, NIH Study Sect, 88; chief aids & retrovirol epidemiol, NJ Sch Med, 87-, fel humanities, 88-; assoc att med, UMDNJ, Univ Hosp, 89-, dir Div Infectious Dis Epidemiol, 89-, AIDS fac, Acad Med, NJ, 89-, exec sci comt, NJ Community Res Initiative, 89-; mem, Physicians Adv Comm, NJ Dept Health, 90-, sci adv comm, clin dir Network, Comm Prog Clinic Res AIDS, 90- *Honors & Awards:* Barge Math Prize, 71; Am Soc Clin Oncol Award, 85. *Mem:* Am Soc Clin Oncol; Am Col Physicians; Am Soc Microbiol; fel Infectious Dis Soc Am; Am Fedn Clin Res; Am Pub Health Asn. *Res:* Epidemiology of the acquired immunodeficiency syndrome and other retroviral associated illnesses; mathematical modeling and decision theory applied to the pathogenesis of neoplasms and chronic diseases; critical evaluation of diagnostic and therapeutic modalities; author of over 60 articles and of 58 scientific abstracts. *Mailing Add:* 42 Ridge Dr Livingston NJ 07039-3716

WEISS, STEPHEN FREDRICK, b Berkeley, Calif, Mar 6, 44; m 68; c 2. COMPUTER SCIENCE. *Educ:* Carnegie-Mellon Univ, BS, 66; Cornell Univ, MS, 69, PhD(comput sci), 70. *Prof Exp:* From asst prof to assoc prof, 70-84, PROF COMPUT SCI, UNIV NC, CHAPEL HILL, 84- *Concurrent Pos:* Consult, Environ Protection Agency, 75-78, Res Triangle Inst & Lipids Res Clinics. *Mem:* Asn Comput Mach; Asn Computational Ling. *Res:* Natural language analysis; information retrieval. *Mailing Add:* Dept Comput Sci Univ NC Chapel Hill NC 27599-3175

WEISS, THEODORE JOEL, b Rochester, NY, Aug 16, 19; m 41; c 2. LIPID CHEMISTRY. *Educ:* Syracuse Univ, AB, 40, PhD(biochem), 53. *Prof Exp:* Asst chemist feed control, State Inspection & Regulatory Serv, Md, 41-43; res chemist powdered milk, Borden Co, 44-49 & Dairymen's League Coop Asn, 49-52; res chemist & div head, Swift & Co, 52-63; tech dir, Capital City Prod Co, 63-64; sr proj leader, Res & Develop Dept, Hunt-Wesson Foods, Inc, 64-68; res chemist, Southern Utilization Res & Develop Div, Agr Res Serv, USDA, 68-70, Dairy Prod Lab, 70-72; tech mgr, Indust Sales Dept, Hunt-Wesson Foods, Inc, 72-85; CONSULT, 85- *Concurrent Pos:* Consult, 85- *Mem:* AAAS; Am Chem Soc; Inst Food Technologists; Am Oil Chem Soc. *Res:* Edible fats and oils; margarine; peanut butter; mayonnaise. *Mailing Add:* 16775 Lake Terrace Way Yorba Linda CA 92686

WEISS, THOMAS E, b New Orleans, La, June 15, 16; m 50; c 2. MEDICINE. *Educ:* Tulane Univ, MD, 40. *Prof Exp:* From instr to assoc prof med, Sch Med, Tulane Univ, 47-64, prof clin med, 64-84, emer prof med, 64-; RETIRED. *Concurrent Pos:* Mem staff, Ochsner Clin; trustee, Alton Ochsner Med Found, chmn, vol support comt, 84- *Mem:* Am Rheumatism Asn (past pres). *Res:* Rheumatic diseases; gout, clinical observations and correlating clinical finds with test and treatments; studies of radioisotope joint scanning in patients with arthritis. *Mailing Add:* Ochsner Clin 1514 Jefferson Hwy New Orleans LA 70121

WEISS, ULRICH, organic chemistry; deceased, see previous edition for last biography

WEISS, VOLKER, b Rottenmann, Austria, Sept 2, 30; nat US; m 57; c 2. METALLURGY. *Educ:* Vienna Tech Univ, dipl, 53; Syracuse Univ, MS, 55, PhD(solid state sci & technol), 57. *Prof Exp:* Asst, Neth Steel Factory, 52; res assoc metall, 54-57, from asst prof to assoc prof, 57-65, PROF MAT SCI, L C SMITH COL ENG, SYRACUSE UNIV, 65-, PROF ENG & PHYSICS, 86- *Concurrent Pos:* Indust consult, US & Ger, 56-; sr sci fel, NATO, 67-68; assoc chem metall, L C Smith Col Eng, Syracuse Univ, 65-72, chmn solid state sci & technol prog, 60-67, assoc dean sponsored progs, 72-78, vpres res & grad affairs, 77-86. *Honors & Awards:* Minor Award, NASA, 68. *Mem:* Fel Am Soc Metals; Am Soc Testing & Mat; Am Inst Mining, Metall & Petrol Engrs; Brit Inst Metals; Ger Metall Soc. *Res:* Metal physics; fracture mechanics; fatigue; residual stresses; solid state reactions; superplasticity; x-ray diffraction; application of Ai to materials science. *Mailing Add:* 449 Link Hall Syracuse Univ Syracuse NY 13244

WEISS, WILLIAM, b New York, NY, June 12, 23; m 56; c 4. BIOSTATISTICS. *Educ:* George Washington Univ, BA, 48. *Prof Exp:* Chief statistician, US Food & Drug Admin, 52-62; asst chief perinatal res br, Nat Inst Neurol & Commun Dis & Stroke, NIH, 62-66, chief, Off Biomet & Field Studies, 66-84; PVT CONSULT STATIST, 84- *Mem:* Biomet Soc; fel Am Statist Asn; Drug Info Asn; fel AAAS. *Res:* Biostatistical applications in neurology; design and analysis of clinical trials in multiple sclerosis. *Mailing Add:* 609 Jerry Lane NW Vienna VA 22180

WEISS, WILLIAM, b Philadelphia, Pa, July 30, 19; wid; c 3. PULMONARY DISEASES, EPIDEMIOLOGY. *Educ:* Univ Pa, BA, 40, MD, 44. *Prof Exp:* Mem staff, Sch Med & Grad Sch Med, Univ Pa & Med Col Pa, 45-66; from assoc prof to prof med, Hahnemann Med Col, 66-84, dir div occup med, 75-84, EMER PROF MED, HAHNEMANN UNIV, 84- *Concurrent Pos:* Chief tuberc, Harbor Gen Hosp, Torrance, Calif, 49-50; clin dir, Pulmonary Dis Serv, Philadelphia Gen Hosp, 50-74; chest consult, Norristown State Hosp, Pa, 51-60; dir, Philadelphia Pulmonary Neoplasm Res Proj, 57-67; Int Agency Res Cancer travel fel, London Mass Radiography Units, 69; ed, Philadelphia Med, 76- *Honors & Awards:* Ann Sci Award, Am Cancer Soc, 79. *Mem:* AMA; Am Thoracic Soc; Am Col Physicians. *Res:* Pulmonary disease, particularly lung cancer. *Mailing Add:* 3912 Netherfield Rd Philadelphia PA 19129

WEISSBACH, ARTHUR, b New York, NY, Aug 27, 27; m 58; c 2. MOLECULAR BIOLOGY. *Educ:* City Col New York, BS, 47; Columbia Univ, PhD(biochem), 53. *Prof Exp:* Nat Found Infantile Paralysis fel, NIH, 53-55; asst prof biochem, Albany Med Col, Union Univ, NY, 55-56; biochemist, NIH, 56-68; head dept cell biol, Roche Inst Molecular Biol, 68-82, assoc dir, 83-89; EXEC ED, ANALYTICAL BIOCHEM, 88- *Concurrent Pos:* Prof lectr, Georgetown Univ, 57-58; NSF fel, 59-60; prof lectr, George Washington Univ, 61-66; adj prof, Dept Human Genetics & Develop, Columbia Univ, 69-80, Univ Med & Dent, NJ, 81- *Mem:* Am Soc Biol Chemists; Int Soc Plant Molecular Biol. *Res:* Biochemistry of plant and animal nucleic acids. *Mailing Add:* PO Box 168 Sanibel FL 33957

WEISSBACH, HERBERT, b New York, NY, Mar 16, 32; c 4. GENE EXPRESSION, MOLECULAR BIOLOGY. *Educ:* George Washington Univ, MS, 55, PhD, 57. *Prof Exp:* Biochemist, Nat Heart Inst, 53-58; lectr, George Washington Univ, 62-68; head sect enzymes & metab & actg chief lab clin biochem, NIH, 68-69; assoc dir, Roche Inst Molecular Biol & dir, dept biochem, 69-83, DIR, ROCHE INST MOLECULAR BIOL, 83- *Concurrent Pos:* NSF travel grant, Int Cong Biol Chem Socs, Moscow, 61; ed, J Pharmacol & Exp Therapeut, 67-72, Int J Neuropharm, 69- 76 & J Biol Chem, 72-77; adj prof, Dept Human Genetics, Columbia Univ, 69- *Honors & Awards:* Am Chem Soc Enzyme Award. *Mem:* Nat Acad Sci; AAAS; Am Soc Biol Chemists; Am Chem Soc; Am Soc Microbiol; Am Soc Pharmacol & Exp Therapeut. *Res:* Protein synthesis; gene expression; mechanism of enzyme & coenzyme action. *Mailing Add:* Roche Inst Molecular Biol Nutley NJ 07110

WEISSBERG, ALFRED, b Boston, Mass, Jan 29, 28; wid; c 3. MATHEMATICS, SCIENCE ADMINISTRATION. *Educ:* Northeastern Univ, BS, 52; Univ NH, MS, 54. *Prof Exp:* Prin mathematician, Battelle Mem Inst, 53-60; math statistician, US Food & Drug Admin, 60-64; supvry opers res analyst, Nat Bur Standards, 64-67; head file orgn & statist, Toxicol Info Prog, Nat Libr Med, 67-70; Sci & Technol Commun Off, 70-80, SCI ADMIN, NAT INST NEUROL COMMUN DIS & STROKE, NIH, 80- *Mem:* AAAS; Am Soc Info Sci; Am Statist Asn. *Res:* Mathematical statistics; information system design and operation; application of computer systems to information retrieval. *Mailing Add:* 1024 Noyes Dr Silver Spring MD 20910

WEISSBERG, ROBERT MURRAY, b Cleveland, Ohio, Mar 3, 40. PHARMACOLOGY, PHYSIOLOGY. *Educ:* Ohio State Univ, PhD(physiol), 72. *Prof Exp:* STAFF RESEARCHER PHARMACOL, SYNTEX RES, 73- *Mem:* Am Soc Pharmacol & Exp Therapeut. *Res:* Respiratory pharmacology. *Mailing Add:* Dept Exp Pharmacol Syntex Labs 3401 Hillview Ave Palo Alto CA 94306

WEISSBLUTH, MITCHEL, b Russia, Jan 7, 15; m 40; c 3. PHYSICS. *Educ:* Brooklyn Col, BA, 36; George Washington Univ, MA, 41; Univ Calif, PhD, 50. *Prof Exp:* Metallurgist, USN Yard, Washington, DC, 37-41; radio engr, Crosley Radio Corp, Ohio, 42-43; sr res engr, Jet Propulsion Lab, Calif Inst Technol, 43-45; asst physics, Univ Calif, 45-49, lectr, 50; res assoc, 50-51, from instr to asst prof radiol physics, 51-66, dir, Biophys Lab, 64-67, assoc prof, 67-76, PROF APPL PHYSICS, STANFORD UNIV, 76- *Concurrent Pos:* Physicist, Western Regional Res Lab, Bur Agr & Indust Chem, USDA, 48; Fulbright grant, Weizmann Inst Sci, Israel, 60-61; liaison scientist, Off Naval Res, London, 67-68, sr liaison scientist, Tokyo, 78-79. *Mem:* Am Phys Soc; Int Soc Quantum Biol (pres, 73-75). *Res:* Rocket motors; pion production; light scattering; x-ray microscopy; medical accelerators; electron spin resonance; thermoluminescence; enzyme reactions in magnetic fields; hypochromism; triplet states; Mossbauer resonance; electronic states in hemoglobin; Jahn-Teller effect; synchrotron radiation; photon-atom interactions. *Mailing Add:* Dept Appl Physics Stanford Univ Stanford CA 94305

WEISSE, ALLEN B, b New York, NY, Dec 6, 29; m 67; c 2. CARDIOLOGY, MEDICINE. *Educ:* NY Univ, BA, 50; State Univ NY, MD, 58; Am Bd Internal Med, dipl, 65, cert cardiovasc dis, 67. *Prof Exp:* Res fel cardiol, Sch Med, Univ Utah, 61-63; instr med, Sch Med, Seton Hall Univ, 63-65; from instr to assoc prof, 65-74, PROF MED, NJ MED SCH, UNIV MED & DENT NJ, 74- *Mem:* Am Fedn Clin Res; Am Physiol Soc; Am Heart Asn; fel Am Col Physicians; fel Am Col Cardiol. *Res:* Cardiovascular physiology and disease; medical history. *Mailing Add:* NJ Med Sch Univ Med & Dent NJ 185 S Orange Ave Newark NJ 07103

WEISSENBERGER, STEIN, b San Francisco, Calif, July 30, 37; m 59; c 2. DECISION ANALYSIS. *Educ:* Mass Inst Technol, BS & MS, 60; Stanford Univ, PhD(aeronaut & astronaut sci), 65. *Prof Exp:* Dynamics engr, Lockheed Missiles & Space Co, 62-65; from asst prof to assoc prof mech eng, Univ Santa Clara, 65-76; group leader & dep div leader, 76-82, DIV LEADER, LAWRENCE LIVERMORE LAB, UNIV CALIF, 82- *Concurrent Pos:* Sr res assoc, Nat Res Coun, NASA-Ames Res Ctr, 73-75. *Mem:* Am Soc Mech Engrs; Inst Elec & Electronics Engrs; Opers Res Soc Am. *Res:* Control and decision theory; control systems. *Mailing Add:* Lawrence Livermore Lab Livermore CA 94550

WEISSENBURGER, DON WILLIAM, b New Brunswick, NJ, May 22, 47. MAGNETODYNAMICS, MAGNETOSTATICS. *Educ:* Colo Col, BS, 69. *Prof Exp:* Mem eng & sci staff, Plasma Physics Lab, Princeton Univ, 75-; STAFF ENGR, GRUMANN SPACE ELECTRONIC DIV. *Res:* Design and analysis of magnetic fields, eddy currents and related phenomena associated with controlled fusion research. *Mailing Add:* PO Box 390 Monmouth NJ 08852

WEISSENBURGER, JASON T, b Wheeling, WVa, Dec 11, 32. NOISE CONTROL, VIBRATIONS. *Educ:* Wash Univ, St Louis, BSME, 55, MS, 59, DSc(appl mech), 66. *Prof Exp:* Tech specialist, McDonnell Aircraft Corp, 59-69; PRES, ENG DYNAMICS INT, 70- *Concurrent Pos:* Adj prof & lectr appl mech, Wash Univ, St Louis, 55-70; adj prof, Univ Mo, Rolla & Cent Mo State Univ. *Mem:* Nat Coun Acoust Consults (pres, 88-90); Acoust Soc Am; Am Soc Mech Engrs; Inst Noise Control Eng. *Res:* Eigenvalue mathematics. *Mailing Add:* 219 Orrick Lane Kirkwood MO 63122

WEISSER, EUGENE P, b Pittsburgh, Pa, Feb 12, 22; m 48; c 5. CHEMICAL ENGINEERING. *Educ:* Univ Pittsburgh, BS, 49. *Prof Exp:* Res asst chem eng, Mellon Inst, 49-51; jr chem engr, Res Dept, Koppers Co, Inc, Monroeville, Pa, 51-55, chem engr, 55-60, sr chem engr, 60-61, group mgr res eng, 61-67, sr res group mgr, 67-77; mgr res eng, Arco Polymers, Inc, Atlantic Richfield Co, Philadelphia, 77-81, MGR, PILOT PLANT SERV, ARCO CHEM CO, PHILADELPHIA, 81- *Mem:* Am Inst Chem Engrs. *Res:* The design, construction and operation of pilot plants, particularly concerning polymers. *Mailing Add:* 6731 W Barivista Dr Verona PA 15147

WEISSGERBER, RUDOLPH E, b Ger, June 2, 21; US citizen; m 50; c 1. MEDICINE. *Educ:* Univ Greifswald, MD, 45; Univ Hamburg, dipl trop dis & med parasitol, 48; Emory Univ, cert bus admin, 65. *Prof Exp:* Pres, Nordmark Chem Co, NY, 52-55; med sci dir, E R Squibb & Sons, Inc, Europe, 55-63, dir prod planning & develop, E R Squibb Int Co, NY, 63-64, dir res & develop, Squibb Int Co, 65-66; vpres & sci dir, Int Div, Bristol-Myers Co, 67-68; vpres, Merrell Int & dir, Sci & Com Develop-Europe, Richardson-Merrell Inc, 69-81, dir, licensing mgt, Merrell Dow Pharmaceut Inc, Europe, 81-90; RETIRED. *Res:* Pharmaceutical companies and product acquisitions; licensing and assessment of pharmaceutical products. *Mailing Add:* PO Box 1406 Graefelfing Munich D 8032 Germany

WEISSKOPF, BERNARD, b Berlin, Ger, Dec 11, 29; US citizen; m 65; c 2. PEDIATRICS, PSYCHIATRY. *Educ:* Syracuse Univ, BA, 51; State Univ Leiden, MD, 55; Am Bd Pediat, dipl, 65. *Prof Exp:* Physician, State Univ Leiden, 58; intern, Meadowbrook Hosp, Hempstead, NY, 58-59, resident pediat, 59-60; asst chief pediat, USAF Hosp, Maxwell AFB, 60-62; fel pediat & child psychiat, Hosp & Sch Med, Johns Hopkins Univ, 62-64; asst prof pediat, Univ Ill Col Med, 64-66; assoc prof, 66-72, PROF PEDIAT, SCH MED, UNIV LOUISVILLE, 72-, ASSOC PSYCHIAT, ASSOC OBSTET & GYNEC & DIR CHILD EVAL CTR, 66- *Concurrent Pos:* Clin coordr, Ill State Pediat Inst, 64-66. *Mem:* Fel Am Acad Pediat; Am Asn Ment Deficiency; Am Soc Human Genetics. *Res:* Behavioral aspects of pediatrics, especially learning disorders and mental retardation genetics. *Mailing Add:* Child Eval Ctr 334 E Broadway Louisville KY 40202

WEISSKOPF, MARTIN CHARLES, b Omaha, Nebr, Apr 21, 42; m 88; c 2. X-RAY ASTRONOMY, ASTROPHYSICS. *Educ:* Oberlin Col, AB, 64; Brandeis Univ, PhD(physics), 69. *Prof Exp:* Res assoc physics, Brandeis Univ, 68-69; res assoc, Columbia Univ, 69-71, lectr, 70-71, asst prof, 71-77; SR X-RAY ASTRON ASTROPHYS & AXAF PROJ SCI, 77-, CHIEF, X-RAY ASTRON BR, 84- *Concurrent Pos:* Co-investr, High Energy Astron Observ-2, NASA, 72- & Columbia Exp NASA's OSO-8, 75-77; guest-investr, NASA's High Energy Astrophys Observ-1, 78- *Mem:* Am Phys Soc; Am Astron Soc. *Res:* X-ray astronomy and high energy astrophysics. *Mailing Add:* Code ES-65 Marshall Space Flight Ctr Huntsville AL 35812

WEISSKOPF, VICTOR FREDERICK, b Vienna, Austria, Sept 19, 08; nat US; m 34; c 2. PHYSICS. *Educ:* Univ Gottingen, PhD(physics), 31. *Hon Degrees:* Twenty five from various US & foreign univs, 61-70. *Prof Exp:* Res assoc, Univ Berlin, 31-32; Rockefeller Found fel, Univs Copenhagen & Cambridge, 32-33; res assoc, Swiss Fed Inst Technol, 33-36 & Univ Copenhagen, 36-37; from instr to asst prof physics, Univ Rochester, 37-43; dep div leader, Los Alamos Sci Lab, 43-46; prof, 46-74, EMER INST PROF PHYSICS, MASS INST TECHNOL, 74- *Concurrent Pos:* Dir gen, Europ Ctr Nuclear Res, 61-65. *Honors & Awards:* Planck Medal, 56; Prix Mondial Cino Del Duca, 72; Nat Medal Sci, 80; Fermi Prize, 89; Pub Serv Medal, Nat Acad Sci. *Mem:* Nat Acad Sci; fel Am Phys Soc (vpres, 59, pres, 60); Fedn Am Sci; corresp mem French, Austrian, Danish, Spanish, Scottish, Ital, Pontifical, Soviet & Bavarian Acad Sci; Am Acad Arts & Sci (pres, 77-80). *Res:* Quantum mechanics; electron theory; theory of nuclear phenomena; particle physics. *Mailing Add:* Mass Inst Technol Bldg 6-303 Cambridge MA 02139

WEISSLER, ARNOLD M, b Brooklyn, NY, May 13, 27; c 4. CARDIOLOGY, INTERNAL MEDICINE. *Educ:* NY Univ, BA, 48; State Univ NY Downstate Med Ctr, MD, 53. *Prof Exp:* Trainee internal med, Maimonides Hosp, 53-54 & Duke Hosp, 54-60, assoc med, Duke Hosp, 59-60; asst prof, Med Br, Univ Tex, Galveston, 60-61; from asst prof to prof, Ohio State Univ, 61-71, dir, Div Cardiol, 63-71; prof med & chmn dept, Wayne State Univ, 71-81, chief, Fib Vstfiol, 81-; PROF & CHMN DEPT MED, ROSE MED CTR. *Concurrent Pos:* Am Heart Asn res fel, Duke Hosp, Durham, NC, 55-57; chief, Dept Med, Harper-Grace Hosps, 71-81; mem cardiovasc bd, Am Bd Internal Med, chmn, 75-77; fel, Coun Clin Cardiol, Am Heart Asn, 78, chmn, 78-80. *Mem:* Fel Am Col Physicians; Am Clin & Climat Asn; fel Am Col Cardiol; Am Soc Clin Invest; Am Soc Pharmacol & Exp Therapeut. *Res:* Cardiovascular physiology; congestive heart failure; noninvasive techniques in cardiology; myocardial metabolism. *Mailing Add:* Mayo Clin 200 First St Rochester MN 55901

WEISSLER, GERHARD LUDWIG, b Eilenburg, Ger, Feb 20, 18; nat US; m 53; c 2. PHYSICS. *Educ:* Univ Calif, MA, 41, PhD(physics), 42; Am Bd Radiol, dipl, 52. *Prof Exp:* Instr radiol, Med Sch, Univ Calif, 42-44; from asst prof to assoc prof, Univ Southern Calif, 44-52, head dept, 51-56, prof, 52-88, emer prof physics, 88; RETIRED. *Mem:* Fel Am Phys Soc; fel Optical Soc Am. *Res:* Gaseous electronics; vacuum ultraviolet radiation physics; photo-absorption and photo-ionization cross sections; photoelectric effect; optical constants and solid state physics; upper atmosphere and astrophysical problems; nuclear accelerator physics; vacuum ultraviolet spectroscopy of hot gaseous plasmas. *Mailing Add:* 16434 Westfall Pl Encino CA 91436

WEISSMAN, ALBERT, b New York, NY, Aug 1, 33; m 57; c 5. PSYCHOPHARMACOLOGY. *Educ:* NY Univ, BA, 54; Columbia Univ, MA, 55, PhD, 58. *Prof Exp:* Instr psychol, Columbia Univ, 57; sr res psychologist, Pfizer Res Labs, 58-61, mgr psychopharmacol, 61-72, ASST DIR PHARMACOL, PFIZER INC, 72- *Mem:* Am Soc Pharmacol & Exp Therapeut. *Res:* Catecholamines and indolylalkylamines; learning and memory; addiction. *Mailing Add:* Asst Dir Pharmacol Pfizer Inc Groton CT 06340

WEISSMAN, CHARLES, b New York, NY, April, 24, 51. CRITICAL CARE MEDICINE, INTERNAL MEDICINE. *Educ:* Yeshiva Univ, BA, 72; Downstate Med Ctr, State Univ NY, MD, 76. *Prof Exp:* Asst prof anesthesiol & med, Col Physicians & Surgeons, Columbia Univ, 83-89; asst attend physician anesthesiol & med, 83-89, ASSOC ATTEND PHYSICIAN ANESTHESIOL, PRESBY HOSP, 89-; ASSOC PROF CLIN ANESTHESIOL & CLIN MED, COL PHYSICIANS & SURGEONS, COLUMBIA UNIV, 89- *Res:* The metabolic and respiratory aspects of the critically ill patients; the respiratory patterns of post operative patients; anesthesiology. *Mailing Add:* 622 W 168th St New York NY 10032

WEISSMAN, DAVID E(VERETT), b New York, NY, Sept 18, 37; m 61; c 4. ELECTRICAL ENGINEERING. *Educ:* NY Univ, BA & BEE, 60, MEE, 61; Stanford Univ, PhD(elec eng), 68. *Prof Exp:* Asst elec eng, NY Univ, 60-61; elec engr, Dorne & Margolin, Inc, 61-63; res engr, Stanford Res Inst, 63-68; from asst prof to assoc prof, 68-82, PROF ENG, HOFSTRA UNIV, 82- *Concurrent Pos:* Ed-in-chief, Inst Elec & Electronics Engrs J Oceanic Eng, 79-82. *Honors & Awards:* Centennial Medal, Inst Elec & Electronics Engrs, 84. *Mem:* AAAS; Inst Elec & Electronics Engrs; Int Sci Radio Union; Am Geophys Union. *Res:* Development of radar remote sensing techniques for ocean surface observation from aircraft and satellites and other random media probing; demonstrated the usefulness of dual frequency radars to the measurement of ocean wave heights and directional spectrum; radar remote sensing. *Mailing Add:* Dept Eng Hofstra Univ Hempstead NY 11550

WEISSMAN, EARL BERNARD, b Detroit, Mich, Feb 21, 42; m 67; c 2. CLINICAL CHEMISTRY. *Educ:* Wayne State Univ, BS, 65, PhD(biochem), 72. *Prof Exp:* Fel clin chem, Buffalo Gen Hosp, 72-74; asst mgr lab opers, Lab Procedures, Upjohn Co, 74-77; TECH DIR, AM REFINING LAB, 78- *Concurrent Pos:* Consult, Ventura County Gen Hosp, 74-78; asst prof, Calif State Univ, 75- *Mem:* Am Asn Clin Chemists. *Res:* Porphyrin metabolism; clinical methods development. *Mailing Add:* 10702 Providence Dr Villa Park CA 92667

WEISSMAN, EUGENE Y(EHUDA), b Bucharest, Romania, Sept 23, 31; US citizen; m 58; c 2. TECHNICAL MANAGEMENT, RESOURCE MANAGEMENT. *Educ:* Israel Inst Technol, BSc, 53, ChemE, 54; Univ Mich, MSc, 59; Case Western Reserve Univ, PhD(chem eng), 63; Univ Chicago, MBA, 72. *Prof Exp:* Chem engr, Israel AEC, 53-57, actg proj dir separation processes & water isotopes, 57-58; asst, Case Western Reserve Univ, 59-63; chem engr, Res & Develop Ctr, Gen Elec Co, 63-65, sr engr direct energy conversion oper, 65, mgr appl res & tech develop, 65-68; mgr electrochem res dept, Johnson Controls, Inc, 68-73; dir inorg-electrol res & develop, BASF Corp, 73-76, tech dir chem spec bus, 76-79, dir res tech support, BASF Wyandotte Corp, 79-86, dir res tech support, 86-89, DIR UNIV TECH LIAISON, BASF CORP, 89- *Concurrent Pos:* Res & develop engr, Hercules Powder Co, 60-61; dir, Heat Transfer & Energy Conversion Div, Am Inst Chem Eng, 74-75; dir, Indust Res Inst, 87-90, Mich Mat & Processing Inst, 90-; mem, adv coun, Univ Akron Col Eng; mem, planning & policy bd, Ctr for Interfacial Eng, Univ Minn; indust assoc prog, Chem Dept, Purdue Univ; indust affil prog, Mass Inst Technol; comt corp assoc, Am Chem Soc; mem, nat res coun, coun chem res; mem, conf comt, Nat Conf Advan of Res. *Honors & Awards:* Gen Elec Co Inventors Award. *Mem:* Am Inst Chem Engrs; Indust Res Inst; Am Chem Soc; Comm Develop Asn; Mat Res Soc; AAAS. *Res:* Electrochemistry and electrochemical engineering; transport phenomena; surface phenomena; research and development management; materials; formulations chemistry; analytical chemistry; information science; university relations; technology transfer; management of technical affairs and resources. *Mailing Add:* 356 N Clifton Rd Birmingham MI 48010

WEISSMAN, HERMAN BENJAMIN, physics, for more information see previous edition

WEISSMAN, IRA, electrical engineering, physics, for more information see previous edition

WEISSMAN, IRVING L, b Great Falls, Mont, Oct 21, 39; m 61; c 4. BIOLOGICAL SCIENCES. *Educ:* Mont State Col, BS, 60; Stanford Univ, MD, 65. *Prof Exp:* NIH fel, Dept Radiol, Stanford Univ, 65-67, res assoc, 67-68, from asst prof to assoc prof, Dept Path, 69-81, PROF PATH, SCH MED, STANFORD UNIV, 81-, PROF DEVELOP BIOL, 89- *Concurrent Pos:* Sr Dernham fel, Calif Div, Am Cancer Soc, 69-73, fac res award, Nat Am Cancer Soc, 74-78; Josiah Macy Found scholar, 74-75; mem, Immunobiol Study Sect, NIH, 76-80, outstanding investr award, 86; mem sci rev bd, Howard Hughes Med Inst, 86-; mem, Steering & Res Panels on AIDS, Nat Acad Sci-Nat Inst Med, 86-; mem, Sci Adv Comt, Irvington House Inst, 87-; Karel & Avice Beekhuis prof cancer biol, 87; Fifth Ann vis prof cancer biol, Univ Tex Health Sci Ctr, 87; distinguished lectr, Western Soc Clin Invest, 90. *Honors & Awards:* James McGinnis Mem Lectr, Duke Univ, 82; George Feigen Mem Lectr, Stanford Univ, 86; Albert Coons Mem Lectr, Harvard Univ, 87; James Stahlman Lectr, Vanderbilt Univ, 87; R E Smith Lectr, Univ Tex Syst Cancer Ctr, 88; Chauncey D Leake Lectr, Univ Calif, 89; Pasarow Award, 89; Harvey Lectr, Rockefeller Univ, 89; Rose Litman Lectr, 90. *Mem:* Nat Acad Sci; fel AAAS; Am Acad Arts & Sci; Am Asn Immunologists; Am Asn Univ Pathologists; Am Asn Pathologists; Am Soc Microbiol; Am Asn Cancer Res; Int Immunol. *Res:* Lymphocyte differentiation; molecular biology of lymphocyte surface receptors; viral oncogenesis; cellular immunology. *Mailing Add:* Dept Path B257 Beckman Ctr Sch Med Stanford Univ Stanford CA 94305

WEISSMAN, MICHAEL BENJAMIN, b St Louis, Mo, Aug 28, 49. FLUCTUATION SPECTROSCOPY. *Educ:* Harvard Univ, AB, 70; Univ Calif, San Diego, MS, 72, PhD(physics), 76. *Prof Exp:* Res fel chem, Harvard Univ, 76-78; ASST PROF PHYSICS, UNIV ILL, URBANA, 78- *Concurrent Pos:* NSF fel, 77-78. *Mem:* Biophys Soc; Am Phys Soc. *Res:* Fluctuation spectroscopy; techniques and applications to biophysics; origins of 1/f noise in electrical systems. *Mailing Add:* Dept Physics Univ Ill 1110 W Green St Urbana IL 61801

WEISSMAN, MICHAEL HERBERT, b New York, NY, Jan 15, 42; m 67; c 5. PEDIATRICS, BIOMEDICAL ENGINEERING. *Educ:* Cooper Union, BME, 63; Northwestern Univ, Evanston, MS, 65, PhD(civil eng), 67; Wash Univ, MD, 76. *Prof Exp:* Res asst, Northwestern Univ, Evanston, 65-67; asst prof bioeng, Carnegie-Mellon Univ, 67-71, assoc prof bioeng & chem eng, 71-73; pediat resident, Bronx Munic Hosp Ctr, 76-79, pediat chief resident, 79-80; PEDIATRICIAN, MT KISCO MED GROUP, 80-; CLIN ASST PROF PEDIAT, ALBERT EINSTEIN COL MED, YESHIVA UNIV, 82- *Concurrent Pos:* NIH grant, Carnegie-Mellon Univ, 69-74. *Honors & Awards:* George F Gill Prize Pediat, Wash Univ, 76. *Mem:* Sigma Xi; AMA; Am Acad Pediat. *Res:* Artificial internal organs; biological transport processes; physiological systems analysis; fluid mechanics; physiological simulation; general pediatrics. *Mailing Add:* Dept Pediat Mt Kisco Med Group Mt Kisco NY 10549

WEISSMAN, MYRNA MILGRAM, b Boston, Mass, Apr 17, 35; m; c 4. EPIDEMIOLOGY, PSYCHIATRY. *Educ:* Brandeis Univ, BA, 56; Univ Pa, MSW, 58; Yale Univ, PhD(chronic dis epidemiol), 74. *Prof Exp:* Psychiat social worker, Inst Juv Res, 58-59 & Dept Psychol Med, Southern Gen Hosp, 59-60; social worker, Clin Ctr, NIH, 60-67; res assoc psychiat, Clin Psychopharmacol Res Unit, Yale Univ, 67-71, asst prof, Conn Ment Health Ctr, 71-74, dir, Depression Res Unit, 81-87, asst prof psychiat, Sch Med, 75-77, from assoc prof to prof psychiat & epidemiol, 75-87; PROF EPIDEMIOL & PSYCHIAT, COL PHYSICIANS & SURGEONS, COLUMBIA UNIV, 87-; CHIEF, DIV CLIN GENETIC EPIDEMIOL, NY STATE PSYCHIAT INST, 87-; RES DIR, CHILD ANXIETY & DEPRESSION CLIN, BABY'S HOSP, COLUMBIA PRESBY MED CTR, 87- *Concurrent Pos:* Chmn utilization rev proj, Yale Univ Suicide Panel, 69-72; consult, Psychopharmacol Res Br, NIMH, 73-76, Clin Res Br, 77 & Nat Inst Drug Abuse, 77; mem, White House Task Force Epidemiol of Ment Health, 77-78, Nat Adv Comt, NIMH, 78 & Res Study Sect, Nat Inst Alcohol Abuse & Alcoholism, 75-80; consult planning epidemiol res, Alcohol, Drug Abuse & Ment Health Admin, 78; consult, Dept HEW, 79, Inst Med-Nat Acad Sci & White House Off Sci & Technol Policy, 80, Soc Sci Res Coun & NIMH, 81, Nat Health Res & Develop Prog, Health & Welfare Can, 81-83, NIMH & WHO, 83; vis sr scholar, Inst Med, Nat Acad Sci & Ment Health Sect, 79-80; vpres, Am Coun Affective Dis, Inc, NY, 83; mem, adv bd, Ctr Alcohol Studies, Rutgers Univ, NJ, 86-90, Nat Depressive & Manic Depressive Asn, 86-; mem, Task Force Res, Am Psychiat Asn, 88-89; mem, Extramural Sci Adv Bd, NIMH, 88-92; mem, comt, Epidemiol & Vet Follow-Up Studies, Inst Med, Nat Acad Sci, 89-; mem, Depression Panel of Coun on Sci Affairs, AMA, 90; mem, Sci Adv Comt, Panic Dis Prev & Pub Educ Prog, NIMH, 90-92; mem, bd dirs, Am Sucide Found, & NY State Res Found Ment Hyg, 90; consult, numerous orgns and insts; Method to Extend Res in Time, NIMH grant, 90- *Honors & Awards:* Found Fund Prize, Res Psychiat, Am Psychiat Asn, 78; Anna Monika Found Award, 85; Rema Lapouse Ment Health Epidemiol Award, Am Pub Health Asn, 85; Maudsley Bequest lectr, Royal Col Psychiat, London & Cambridge Univ, Eng; Outstanding Res in Affective Dis Award, Nat Depressive & Manic Depressive Asn, 89; Res Award, Am Suicide Found, 90; Anne Pollock Lederer Found Res Award study depression in young adults, 90; Sr Investr Award, Nat Alliance Res on Schizophrenia & Depression, 91. *Mem:* AAAS; Soc Epidemiol Res; Soc Life Hist Res Psychopath; Am Pub Health Asn; Int Epidemiol Asn; Am Psychopath Asn; fel Am Col Neuropsychopharmacol; fel Am Psychiat Asn. *Res:* Effectiveness of pharmacotherapy and psychotherapy in the treatment of depression; development of model methodologies for use in cognate studies in the future; epidemiology; genetics of affective, anxiety disorders. *Mailing Add:* Col Physicians & Surgeons Columbia Univ 722 W 168th St Box 14 New York NY 10032

WEISSMAN, NORMAN, b New York, NY, Sept 12, 14; m 37; c 3. CLINICAL CHEMISTRY, TOXICOLOGY. *Educ:* City Col New York, BS, 35; Columbia Univ, PhD(biochem), 41. *Prof Exp:* Res fel dent med, Harvard Sch Dent Med, 41-43, instr, 43-46; lectr, Johns Hopkins Univ, 46-47, asst prof physiol chem & prev med, Sch Med, 47-51; assoc prof med, Col Med, State Univ NY Downstate Med Ctr, 51-56; assoc prof path & biochem, Col Med, Univ Utah, 56-70; asst dir, Chem Dept, 70-74, sr res scientist, Res Dept, 74-80, RES ASSOC, BIO-SCI LABS, 80- *Concurrent Pos:* Chemist, Maimonides Hosp, NY, 51-56 & Univ Hosp, Univ Utah, 57-65; consult, Vet Admin Hosp, Salt Lake City, Utah, 60-70. *Mem:* Am Soc Biol Chemists; Am Asn Clin Chemists; Am Acad Forensic Sci; Harvey Soc; Sigma Xi. *Res:* Amino acid metabolism; bacterial chemistry; histochemistry; copper and connective tissue; toxicology methods. *Mailing Add:* 16901 Mooncrest Dr Encino CA 91436

WEISSMAN, PAUL MORTON, b New York, NY, Oct 17, 36; m 83; c 3. ORGANOMETALLIC CHEMISTRY, BIOINORGANIC CHEMISTRY. *Educ:* City Col New York, BS, 60; Purdue Univ, Lafayette, PhD(chem), 64. *Prof Exp:* Res asst air pollution, NIH, Md, 59-60; res chemist, Gen Aniline & Film, Inc, 64-65; Petrol Res Fund grant, Univ Cincinnati, 65-66; asst prof chem, Brock Univ, 66-68; asst prof, 68-72, ASSOC PROF CHEM, FAIRLEIGH DICKINSON UNIV, 72- *Concurrent Pos:* Nat Res Coun Can grant, Brock Univ, 66-68; NSF fel, Brandeis Univ, 67; vis prof, Hebrew Univ, 75. *Mem:* NY Acad Sci. *Res:* Organometallic synthesis and reaction mechanisms; voltammetry in non-aqueous solvents; transition metal complexes containing a metal-metal bond; food chemistry and nutrition. *Mailing Add:* Dept Chem Fairleigh Dickinson Univ Madison NJ 07940

WEISSMAN, PAUL ROBERT, b Brooklyn, NY, Sept 28, 47. COMETS, PLANETARY SCIENCES. *Educ:* Cornell Univ, AB, 69; Univ Mass, MS, 71; Univ Calif, Los Angeles, MS, 73, PhD(planetary physics), 78. *Prof Exp:* Sr scientist, 74-80, mem tech staff, 80-90, DEP PROJ SCIENTIST, COMET RENDEZVOUS MISSION, JET PROPULSION LAB, CALIF INST TECHNOL, 87- *Concurrent Pos:* Vis mem, Inst Advan Study, 85; Harlow Shapley Vis Lectr, 82- *Mem:* Am Astron Soc; Sigma Xi; AAAS; Int Astron Union; Am Geophys Union. *Res:* Physical and dynamical studies of small bodies in the solar system, comets, meteors and asteroids; theories of origin and implications for formation of the solar system; design of spacecraft and trajectories for solar system exploration; celestial mechanics; planetary sciences. *Mailing Add:* Mail Stop 183 601 Jet Propulsion Lab 4800 Oak Grove Dr Pasadena CA 91109

WEISSMAN, ROBERT HENRY, b Chicago, Ill, July 4, 42; m 87; c 1. SOLID STATE ELECTRONICS. *Educ:* Univ Mich, BS, 64; Stanford Univ, MS, 65, PhD(elec eng), 70. *Prof Exp:* Mem tech staff res & develop light-emitting diode devices, 70-73, proj mgr res & develop mat & processes, 73-80, res & develop sect mgr, 80-84, TECHNOL RES & DEVELOP MGR, HEWLETT PACKARD CO, 84- *Mem:* Electrochem Soc; Int Soc Hybrid Microelectronics; Optical Soc Am. *Res:* Materials and process technology; solid state device physics; optoelectronic devices; optical fiber technology. *Mailing Add:* 3436 Shady Spring Lane Mountain View CA 94040-4543

WEISSMAN, SAMUEL ISAAC, b South Bend, Ind, June 25, 12; m 43; c 2. SPECTROSCOPY, MAGNETIC RESONANCE. *Educ:* Univ Chicago, BS, 33, PhD(phys chem), 38. *Hon Degrees:* Univ Siena, Italy, 86; Wash Univ, 88. *Prof Exp:* Fel, Univ Chicago, 39-41; Nat Res Coun fel, Univ Calif, 41-42; res chemist, Manhattan Proj, Calif, 42-43 & NMex, 43-46; from asst prof to assoc prof, 46-55, prof, 55-80, EMER PROF CHEM, WASH UNIV, 80- *Mem:* Nat Acad Sci; Am Chem Soc. *Res:* Chemical spectroscopy; fluorescence; electrical conductivity; paramagnetic resonance. *Mailing Add:* Dept Chem Wash Univ St Louis MO 63130

WEISSMAN, SHERMAN MORTON, b Chicago, Ill, Nov 22, 30; m 59; c 4. PHYSIOLOGY, BIOCHEMISTRY. *Educ:* Northwestern Univ, BS, 50; Univ Chicago, MS, 51; Harvard Univ, MD, 55. *Prof Exp:* Intern, Boston City Hosp, 55-56; clin assoc metab serv, Gen Med Br, Nat Cancer Inst, NIH, 56-58; asst resident med, Ill Educ & Res Hosp, Chicago, 58-59; spec res fel, Dept Biochem, Univ Glasgow, Scotland, Nat Cancer Inst, 59-60; sr investr metab serv, Nat Cancer Inst, NIH, 60-67; from assoc prof to prof med, molecular biophys & biochem, 67-72, prof, 72-87, STERLING PROF HUMAN GENETICS MED, MOLECULAR BIOPHYS & BIOCHEM, YALE UNIV SCH MED, 87- *Concurrent Pos:* Assoc dir, Comprehensive Cancer Ctr, Yale Univ, 75-81; mem virol & cell biol study sect, Am Cancer Soc, 72-74, co-chmn, 74-75, chmn, 76; mem, Ad Hoc Comt Synthetic Nucleic Acids, Nat Res Coun Assembly Life Sci, 74, Bd Sci Counselors, Nat Cancer Inst, 75-79, chmn, 77-79, mem, Cancer Spec Progs Adv Comt, 79-83; assoc ed, Cancer Res, 80-81. *Mem:* Nat Acad Sci; Am Soc Biol Chemists; Am Soc Clin Invest; Asn Am Physicians; Brit Biochem Soc; fel AAAS; Am Physiol Soc; Am Soc Hemat; Am Soc Human Genetics. *Res:* Nucleic acid metabolism; molecular genetics; author of over 170 technical journal articles. *Mailing Add:* Dept Human Genetics Yale Univ Sch Med 333 Cedar St New Haven CT 06510

WEISSMAN, SUZANNE HEISLER, b Dalles ,Ore, June 20, 49; m 76; c 1. ANALYTICAL CHEMISTRY. *Educ:* Ore State Univ, BS, 71; Univ Ill, MS, 73, PhD(chem), 75. *Prof Exp:* Vis lectr chem, Univ Ill, 75, vis asst prof, 76; chemist, Lovelace Biomed & Environ Res Inst, Inc. 76-80; mem tech staff, 80-86, SUPVR, SANDIA NAT LABS, 86- *Mem:* Am Chem Soc; Soc Appl Spectros; Anal Lab Mgrs Asn. *Res:* Determination of trace elements, ultratrace analyses; inductively coupled plasma-atomic emission spectroscopy; atomic absorption spectroscopy; plasma diagnostics. *Mailing Add:* Sandia Labs Org No 1821 PO Box 5800 Albuquerque NM 87185-5800

WEISSMAN, WILLIAM, b New York, NY, Oct 12, 18; m 48; c 2. SURFACE CHEMISTRY. *Educ:* Univ Ala, Tuscaloosa, BA, 39; St John's Univ, NY, dipl chem, 44 & Newark Col Eng, dipl chem, 48. *Hon Degrees:* PhD, Sci Inst Caracas, 52. *Prof Exp:* Head insulation dept, Inslx Co, 39-46; lacquer chemist, Maas & Waldstein Co, Inc, 46-53; chief chemist, Nat Foil Co, Inc, 53-58; res chemist, Metro Adhesives Inc, 58-61; teacher chem, Passaic County Tech & Voc High Sch, 61-69; res chemist, US Rubber Reclaiming Co, Inc, 69-73; adhesive chemist, Cataphote Div, Ferro Inc, 73-74; sr scientist, Vicksburg Chem Co, 74-75; CONSULT, COMPLEX CHEMS, INC, 75- *Concurrent Pos:* Consult, Standard Chem Co, 53-69; Sussex County Bd Educ fel, Sussex County Tech Sch, Newton, NJ, 64-65; Passaic County Bd Educ fel, Montclair State Col, 67-68; environ consult. *Mem:* Am Chem Soc; AAAS; Nat Geog Soc. *Res:* Adhesion and cohesion, organic and inorganic gases, liquids and solids; atomic micro and macro parameters. *Mailing Add:* 108 Katherine Dr Vicksburg MS 39180

WEISSMANN, BERNARD, b New York, NY, Dec 2, 17. BIOCHEMISTRY. *Educ:* City Col New York, BS, 38; Univ Mich, MS, 39, PhD, 51. *Prof Exp:* Res assoc, Col Physicians & Surgeons, Columbia Univ, 50-53 & Mt Sinai Hosp, NY, 53-57; from asst prof to assoc prof, 58-67, PROF BIOL CHEM, UNIV ILL COL MED, 67- *Mem:* Am Soc Biol Chem; Am Chem Soc; Brit Biochem Soc; Soc Complex Carbohydrates. *Res:* Enzymology and structure of acid mucopolysaccharide catabolism; synthetic carbohydrate chemistry. *Mailing Add:* 1124 & Dobson St Evanston IL 60202-3819

WEISSMANN, GERALD, b Vienna, Austria, Aug 7, 30; US citizen; m 53; c 2. CELL BIOLOGY, INTERNAL MEDICINE. *Educ:* Columbia Univ, BA, 50; NY Univ, MD, 54; Am Bd Internal Med, dipl, 63. *Prof Exp:* Res fel biochem, Arthritis & Rheumatism Found, 58-59; from instr to assoc prof, 59-70, PROF MED, SCH MED, NY UNIV, 70-, DIR DIV RHEUMATOLOGY, 74- *Concurrent Pos:* USPHS spec res fel, Strangeways Res Lab, Cambridge Univ, 60-61; sr investr, Arthritis & Rheumatism Found, 61-65; career investr, Health Res Coun, NY, 66-70; consult, US Food & Drug Admin & Nat Heart & Lung Inst, NIH; investr & instr physiol, Marine Biol Lab, Woods Hole, Mass, 70-, trustee, 85 & 89; Guggenheim fel, Ctr Immunol & Physiol, Paris, 73-74; ed-in-chief, Inflammation, Advan Inflammation Res. *Honors & Awards:* Alessandro Robecchi Prize for Rheumatology, 72; Marine Biol Lab Prize, 74 & 79, Centennial Award, 88; Lila Gruber Award, 79. *Mem:* Am Soc Cell Biol; Am Soc Exp Path; Am Soc Clin Invest; Am Rheumatism Asn (pres, 81-82); Soc Exp Biol & Med; Harvey Soc (pres, 80-81); Asn Am Physicians. *Res:* Study of lysosomes as they relate to cell injury; physiology and pharmacology of lysosomes and of artificial lipid structures; neutrophil activation; prostaglandins; leukotrienes; liposomes. *Mailing Add:* Dept Med NY Univ Med Ctr 550 First Ave New York NY 10016

WEISSMANN, GERD FRIEDRICH HORST, b Leipzig, Ger, Nov 26, 23; US citizen; m 54; c 3. MATERIALS & SOIL SCIENCE. *Educ:* Tech Univ, Berlin, Dipl Ing, 50, Dr Ing, 70; Pa State Univ, MS, 53. *Prof Exp:* supvr mech mat, Bell Tel Labs, 53-81; PRES, DAMPING MEASUREMENT SYSTS, 81- *Mem:* Am Soc Civil Engrs; Am Soc Testing & Mat; Soc Exp Stress Anal. *Res:* Materials research; classification of metals; soil dynamics; foundation engineering; experimental mechanics. *Mailing Add:* 36 Circle Rd Florham Park NJ 07932

WEISSMANN, SIGMUND, b Vienna, Austria, July 1, 17; nat US; m 45; c 2. CRYSTALLOGRAPHY. *Educ:* Polytech Inst Brooklyn, PhD(chem), 52. *Prof Exp:* Res specialist x-ray diffraction, Col Eng, 49-87, PROF MAT SCI & DIR MAT RES LAB, RUTGERS UNIV, NEW BRUNSWICK, 63- *Concurrent Pos:* Consult, Lawrence Livermore Lab & Univ Calif, Savannah River Lab, US Steel Corp, Monroeville, Pa; ed, Metals & Alloys, Joint Comt Powder Diffraction Standards, Int Ctr Diffraction Data; Lady Davis vis prof, Hebrew Univ, Jerusalem, 80; nat lectr, Sigma Xi, 62. *Honors & Awards:* Howe Medal, Am Soc Metals, 62; NSF Res Creativity Award, 83. *Mem:* Am Crystallog Asn; Am Inst Mining, Metall & Petrol Engrs; Am Soc Metals; Am Soc Testing & Mat; NY Acad Sci. *Res:* X-ray crystallography; crystal imperfections of metals and alloys; irradiation of solids; recrystallization; recovery deformation; creep and fatigue of metals and alloys; displacive phase transformation, fracture, stress-corrosion. *Mailing Add:* Dept Mech & Mat Sci Col Eng Rutgers Univ Piscataway NJ 08855

WEIST, WILLIAM GODFREY, JR, b New York, NY, July 6, 31; m 56; c 4. GEOLOGY, HYDROLOGY. *Educ:* Amherst Col, BA, 53; Univ Colo, MS, 56. *Prof Exp:* Geologist, Ground Water Br, US Geol Surv, Colo, 56-63, Ariz, 63-67, hydrologist, Water Resources Div, US Geol Surv, NY, 67-70, supvry hydrologist, Ind, 70-73, sub dist chief, Ind Dist, 72-73, chief, Hydrol Studies Sect, Ind Dist, 73-75, staff hydrologist, 75-79; hydrologist, Environ Affairs Off, Denver, 79-82; hydrologist, 82-84, SR PROJ MGR, OFF SURFACE MINING, DENVER, 84- *Mem:* Geol Soc Am; Am Inst Prof Geologists; AAAS; Fedn Am Scientists. *Res:* Ground-water hydrology. *Mailing Add:* Off Surface Mining 1020 15th St Denver CO 80202

WEISTROP, DONNA ETTA, b New York, NY, June 10, 44; m. ASTRONOMY. *Educ:* Wellesley Col, BA, 65; Calif Inst Technol, PhD(astron), 71. *Prof Exp:* Vis lectr astron, Tel Aviv Univ, 71-73; univ fel astron, Ohio State Univ, 73-74; asst astronr, Kitt Peak Nat Observ, 74-77; res fel, Univ Ariz, 77-78; astrophysicist, Goddard Space Flight Ctr, 78-85; sr scientist, Appl Res Corp, 85-90; ASSOC PROF ASTRON, UNIV NEV, LAS VEGAS, 90- *Concurrent Pos:* Vis scientist, Lunar & Planetary Lab, Univ Ariz, 78-80. *Mem:* Am Astron Soc; Int Astron Union; AAAS; Astron Soc Pac. *Res:* Characteristics of the nebulosity associated with BL lac objects and quasi-stellar objects; development of and applications for electronic detectors; luminosity functions and density distributions of intrinsically faint stars; QSO's and BL lac objects; identification of faint radio sources; characteristics of emission line galaxies. *Mailing Add:* 1742 Saddleback Ct Henderson NV 89014

WEISZ, JUDITH, b Budapest, Hungary, Aug 6, 26; Brit citizen. REPRODUCTIVE PHYSIOLOGY, NEUROENDOCRINOLOGY. *Educ:* Newnham Col, Eng, BA, 48, MB, BCh, 51. *Prof Exp:* Clin training, London Hosp, Eng, 48-51; intern, Hadassah Med Sch, Hebrew Univ, Jerusalem, 52-54; second asst internal med, Tel Hashomer Govt & Mil Hosp, Israel, 54-56; registr internal med, London Hosp, Eng, 56-57; first asst internal med, Tel Hashomer Govt & Mil Hosp, Israel, 57-59; res fel endocrinol, Mt Sinai Hosp, New York, 60-62; fel, Training Prog Steroid Biochem, Worcester Found Exp Biol, Mass, 62-63, staff scientist, Training Prog Reproductive Physiol, 63-70, assoc dir, Training Prog Physiol of Reproduction, 68-72, sr scientist, Worcester Found Exp Biol, 70-72; PROF, DIV REPRODUCTIVE BIOL, DEPT OBSTET & GYNEC, MILTON S HERSHEY MED CTR, PA STATE UNIV, 76-, HEAD DIV, 75-, SR MEM, GRAD SCH FAC, UNIV, 73- *Concurrent Pos:* Chmn, pub bd enquiry, Depo-provera, Food & Drug Admin, 82-85. *Mem:* Brit Med Asn; Endocrine Soc; Int Soc Neuroendocrinol; Soc Study Reproduction. *Res:* Neuroendocrine regulation of reproductive function; steroid biochemistry. *Mailing Add:* Dept Obstet & Gynec Milton S Hershey Med Ctr Hershey PA 17033

WEISZ, PAUL B, b Vienna, Austria, Nov 3, 21; nat US; m 45; c 3. EMBRYOLOGY. *Educ:* McGill Univ, BSc, 43, MSc, 44, PhD(zool), 46. *Prof Exp:* Demonstr zool, McGill Univ, 43-46, lectr, 46-47, lectr biol, Sir George Williams Col, 43-44, lectr, 44-47; from instr to assoc prof, 47-57, PROF BIOL, BROWN UNIV, 57- *Concurrent Pos:* Mem, Ed Policies Comn & Comn Undergrad Educ Biol Sci. *Mem:* Soc Develop Biol; Am Soc Zool; Nat Asn Biol Teachers; Nat Sci Teachers Asn; Am Chem Soc. *Res:* Embryology and development of amphibians and crustaceans; morphogenesis; cytochemistry and nuclear functions in Protozoa; comparative embryology; science writing. *Mailing Add:* 135 Freeman Pkwy Providence RI 02906

WEISZ, PAUL BURG, b Pilsen, Czech, July 2, 19; m 43; c 2. CHEMISTRY, CHEMICAL ENGINEERING. *Educ:* Auburn Univ, BSc, 40; Swiss Fed Inst Technol, ScD, 66. *Hon Degrees:* ScD, Swiss Fed Inst Technol, 80. *Prof Exp:* Asst, Univ Berlin, 38-39 & Bartol Res Found, Pa, 40-46; res assoc, Mobil Res & Develop Corp, 46-61, sr scientist, 61-67, mgr process res, 67-69, mgr, Cent Res Div, 69-82, sr scientist & sci adv, 82-84; DISTINGUISHED PROF CHEM & BIOENG, UNIV PA, 84- *Concurrent Pos:* Instr, Swarthmore Col, 42-43; vis prof, Princeton Univ, 74-76; agent, US Patent Off; chmn policy bd, Ctr Catalysis Sci & Technol, Univ Del, 77-81; mem, Energy Res Adv Bd, 85-; consult, res & develop strategy, 84- *Honors & Awards:* Murphee Award, Am Chem Soc, 72, Chem Contemp Technol Probs Award, 86 & Carothers Award, 87; Leo Friend Award, Am Chem Soc, 77; Wilhelm Award, Am Inst Chem Eng, 78; Lavoisier Medal, Soc Chem France, 83; Perkin Medal, Soc Chem Indust, 85; Pioneer Award, Am Inst Chemists, 74; DGKM Kolleg Award, 88. *Mem:* Fel Am Phys Soc; Am Chem Soc; Nat Acad Eng; fel Am Inst Chemists; Am Inst Chem Engrs. *Res:* Cosmic rays; Geiger counters; electric discharge phenomena; catalysis; diffusion phenomena; petroleum; petroleum processes; energy technology; basic and interdisciplinary phenomena in the sciences; angiogenesis; cell proliferation *Mailing Add:* Chem Eng Dept Univ Pa Philadelphia PA 19104

WEISZ, ROBERT STEPHEN, b New York, NY, May 24, 18; m 56; c 3. INORGANIC CHEMISTRY. *Educ:* Cornell Univ, AB, 39, PhD(chem), 42. *Prof Exp:* Asst instr chem, Cornell Univ, 39-41; res fel, Westinghouse Elec & Mfg Co, 42, res engr, 42-46; res engr, Thomas A Edison, Inc, 46-49 & RCA Labs, 49-56; res dir, Telemeter Memories, Inc, 56-60, Ampex Comput Prod Co, 60-61 & Electronics Memories, Inc, 61-69; indust consult, 69-77; RETIRED. *Mem:* Am Chem Soc. *Res:* Reactions in solids; ceramics; electrical properties of inorganic compounds; magnetic materials. *Mailing Add:* 2109 Mt Calvary Rd Santa Barbara CA 93105

WEISZ-CARRINGTON, PAUL, PATHOLOGY, IMMUNOLOGY. *Educ:* Nat Univ, Mex, MD, 71. *Prof Exp:* DIR LABS, PATH & IMMUNOL, SHREVEPORT VET MED CTR, 80- *Res:* Anatomic pathology. *Mailing Add:* Dept Path Vet Admin West Side Med Ctr 820 Damen Ave Chicago IL 60612

WEITH, HERBERT LEE, NUCLEIC ACID CHEMISTRY, CHEMICAL SYNTHESIS. *Educ:* Purdue Univ, PhD(biochem), 69. *Prof Exp:* ASSOC PROF BIOCHEM, PURDUE UNIV, 76- *Res:* Physical chemistry. *Mailing Add:* Dept Biochem Purdue Univ Lafayette IN 47907

WEITKAMP, LOWELL R, b Lincoln, Nebr, June 13, 36; m 73; c 1. BIOCHEMICAL GENETICS. *Educ:* Reed Col, BA, 58; Univ Rochester, MD, 63; Univ Mich, MS, 66. *Prof Exp:* Internship, Univ Wis Hosp, Madison, 63-64; fel, Univ Mich, 64-66, res assoc, 66-69; from asst prof to assoc prof, 69-78, PROF GEN, SCH MED, UNIV ROCHESTER, 78- *Concurrent Pos:* NIH res career develop award, 71-76; vis prof, Univ Inst Med Genetics, Univ Copenhagen, 73; vis fel, Dept Human Biol, John Curtin Sch Med Res, Australian Nat Univ, 75-76; NIMH res scientist develop award, 76-81. *Honors & Awards:* Clarke Inst Award. *Mem:* Am Genetic Asn; Am Soc Human Genetics; Int Soc Animal Genetics. *Res:* Biochemical genetic markers in susceptibility to nonmendelian familial diseases and decreased reproductive performance. *Mailing Add:* Div Genetics Box 641 Med Ctr Univ Rochester 601 Elmwood Ave Rochester NY 14642

WEITKAMP, WILLIAM GEORGE, b Fremont, Nebr, June 22, 34; m 56; c 3. NUCLEAR PHYSICS. *Educ:* St Olaf Col, BA, 56; Univ Wis, MS, 61, PhD(physics), 65. *Prof Exp:* Res asst physics, Univ Wis, 59-64; res asst prof, Univ Wash, Seattle, 64-67; asst prof, Univ Pittsburgh, 67-68; sr res assoc, 68-73, res assoc prof, 73-78, RES PROF PHYSICS, UNIV WASH, 78-, TECH DIR, NUCLEAR PHYSICS LAB, 68- *Concurrent Pos:* Acad guest, Fed Polytech, Zurich, Switz, 74-75. *Mem:* Am Phys Soc. *Res:* Polarization phenomena in reactions involving light nuclei; time reversal invarience in nuclear reactions; isobaric analog states in heavy nuclei; nuclear instrumentation. *Mailing Add:* Nuclear Physics Lab Univ Wash Seattle WA 98195

WEITLAUF, HARRY, b Seattle, Wash, July 8, 37; m 64. REPRODUCTIVE PHYSIOLOGY. *Educ:* Univ Wash, BS, 59, MD, 63. *Prof Exp:* Fel reproductive physiol, Med Ctr, Univ Kans, 66-68, from asst prof to assoc prof anat, 68-74; assoc prof, Univ Ore Med Ctr, 74-80, prof, 80-; HEAD DEPT ANAT, TEX TECH UNIV HEALTH SCI CTR. *Mem:* Am Asn Anat; Soc Study Reproduction. *Res:* Blastocyst metabolism during the preimplantation phase of development. *Mailing Add:* Dept Cell Biol & Anat Tex Tech Univ Health Sci Ctr Lubbock TX 79430

WEITSMAN, ALLEN WILLIAM, b Greenbelt, Md, July 11, 40; div; c 4. MATHEMATICS. *Educ:* Syracuse Univ, BS, 62, MS, 64, PhD(math), 68. *Prof Exp:* From asst prof to assoc prof, 68-74, PROF MATH, PURDUE UNIV, LAFAYETTE, 74- *Concurrent Pos:* NSF grants, 69-81; Sloan Found grant, 72. *Res:* Classical function theory. *Mailing Add:* Dept Math Purdue Univ West Lafayette IN 47907

WEITZ, ERIC, b New York, NY, Sept 18, 47; m; c 1. CHEMICAL PHYSICS, PHYSICAL CHEMISTRY. *Educ:* Mass Inst Technol, BS, 68; Columbia Univ, PhD(chem), 72. *Prof Exp:* Res assoc chem, Univ Calif, Berkeley, 72-74; from asst prof to assoc prof, 74-84, PROF CHEM, NORTHWESTERN UNIV, EVANSTON, 84- *Concurrent Pos:* Alfred P Sloan fel; vis fel, Joint Inst Lab Astrophys, 83-84. *Mem:* Fel Am Phys Soc; Am Chem Soc. *Res:* Vibrational energy transfer in small molecules and the relation of internal energy to chemical reactivity; applications to laser-induced chemistry; studies of matrix isolated molecules; laser induced surface processes; transient spectroscopy. *Mailing Add:* Dept Chem Northwestern Univ 2145 Sheridan Rd Evanston IL 60208

WEITZ, JOHN HILLS, b Cleveland, Ohio, Sept 20, 16; m 45; c 3. ECONOMIC GEOLOGY. *Educ:* Wesleyan Univ, BA, 38; Lehigh Univ, MS, 40; Pa State Univ, PhD(geol), 54. *Prof Exp:* Asst, Johns Hopkins Univ, 40-42; from jr mineral economist to mineral economist, US Bur Mines, Washington, DC, 42-46; coop geologist, State Geol Surv, Pa, 46-52; asst prof geol, Lehigh Univ, 47-52; geologist & secy, Independent Explosives Co, 52-61, pres, 61-88; RETIRED. *Mem:* AAAS; Geol Soc Am; Mineral Soc Am; Am Inst Mining, Metall & Petrol Engrs; Soc Explosive Engrs. *Res:* Economic geology; clay minerals. *Mailing Add:* 12 Chippenham Ct Rocky River OH 44116

WEITZ, JOSEPH LEONARD, b Cleveland, Ohio, June 2, 22; m 49; c 3. GEOLOGY. *Educ:* Wesleyan Univ, BA, 44; Yale Univ, MS, 46, PhD(geol), 54. *Prof Exp:* Geologist, Fuels Br, US Geol Surv, 48-55 & Independent Explosives Co, Pa, 55-58; asst prof geol, Wesleyan Univ, 58-60 & Colo State Univ, 60-61; assoc prof, Hanover Col, 61-62; from assoc prof to prof, 62-83, EMER PROF GEOL, COLO STATE UNIV, 83- *Concurrent Pos:* Geologist, US Geol Surv, 66-88; dir, Earth Sci Curric Proj, Colo, 67-69; mem, Coun Educ Geol Sci, 67-72; ed, J Geol Educ, 71-74. *Mem:* Geol Soc Am; Am Asn Petrol Geologists; Sigma Xi; Nat Asn Geol Teachers. *Res:* Geology of mineral fuels; Mesozoic stratigraphy; structural geology; geologic compilation. *Mailing Add:* Box 9 Timnath CO 80547-0009

WEITZMAN, ELLIOT D, b Newark, NJ, Feb 4, 29; m; c 2. NEUROLOGY, NEUROPHYSIOLOGY. *Educ:* Univ Iowa, BA, 50; Univ Chicago, MD, 55. *Prof Exp:* Instr neuroanat, Univ Chicago, 51-52; instr med physiol, Univ Ill, 52; asst neurol, Columbia Univ, 58-59; assoc, 61-63, from asst prof to assoc prof, 63-69, PROF NEUROL, ALBERT EINSTEIN COL MED, 69-, CHMN DEPT, 71-, PROF NEUROSCI, 74-; CHIEF DIV NEUROL, MONTEFIORE HOSP & MED CTR, 69- *Concurrent Pos:* Consult, Jacobi Hosp, 61-69, Bronx Vet Admin Hosp, 63-67 & Health Res Coun, NY, 64-69; ed, Sleep Rev. *Mem:* Am Neurol Asn; Soc Neurosci; Am Acad Neurol; Asn Res Nerv & Ment Dis; Asn Psychophysiol. *Res:* Study of sleep. *Mailing Add:* 2410 Barker Ave Bronx NY 10467

WEITZMAN, STANLEY HOWARD, b Mill Valley, Calif, Mar 16, 27; m 48; c 2. ICHTHYOLOGY. *Educ:* Univ Calif, AB, 51, AM, 53; Stanford Univ, PhD(biol), 60. *Prof Exp:* Sr lab technician, Univ Calif, 50-56; instr anat, Sch Med, Stanford Univ, 57-62; assoc cur, 63-67, CUR, DIV FISHES, DEPT VERT ZOOL, SMITHSONIAN INST, 67- *Mem:* Am Soc Ichthyologists & Herpetologists; Am Fisheries Soc; Soc Syst Zool; Soc Study Evolution. *Res:* Evolution, systematics and morphology of fishes, especially South American freshwater fishes. *Mailing Add:* Div Fishes Stop 159 Smithsonian Inst Washington DC 20560

WEITZNER, HAROLD, b Boston, Mass, May 19, 33; m 62; c 2. APPLIED MATHEMATICS, PLASMA PHYSICS. *Educ:* Univ Calif, AB, 54; Harvard Univ, AM, 55, PhD(physics), 58. *Prof Exp:* NSF fel, 58-59; assoc res scientist, 59-60, res scientist, 60-62, from asst prof to assoc prof, 62-69, PROF MATH, COURANT INST MATH SCI, NY UNIV, 69-, DIR, MAGNETOFLUID DYNAMICS DIV, 87-, CHAIR, DEPT MATH, 88- *Concurrent Pos:* Consult, Los Alamos Nat Lab, Univ Calif, 62-, Lawrence Livermore Lab, Livermore, 77-81 & Oak Ridge Nat Lab, 78-; Magnetic Fusion Energy Adv Comt, 87-90. *Mem:* Fel Am Phys Soc. *Res:* Wave propagation problems in magnetohydrodynamics; kinetic theory and plasma oscillation problems; magnetohydrodynamic equilibrium; stability theory. *Mailing Add:* Courant Inst Math Sci NY Univ 251 Mercer St New York NY 10012

WEITZNER, STANLEY, b New York, NY, Mar 11, 31; m 61; c 3. PATHOLOGY. *Educ:* NY Univ, BA, 51; Univ Geneva, MD, 56. *Prof Exp:* Intern, Queens Gen Hosp, Jamaica, NY, 57-58; resident path, Vet Admin Hosp, New York, 58-62; assoc pathologist, Southside Hosp, Bay Shore, 63-64; dir lab, Monticello Hosp, 65-66; from instr to assoc prof path, Sch Med, Univ NMex, 67-74; assoc prof, Univ Tex Health Sci Ctr, San Antonio, 74-75; from assoc prof to prof path, Univ Miss Med Ctr, Jackson, 75-78, dir surg path, 75-78; prof path, Northeastern Ohio Univs Col Med, 78-81; dir labs & chief path, Timken Mercy Hosp, 78-81; dir autopsy path, Herman Hosp, Houston, Tex, 81-85, assoc dir surg path & cytol, 85-87, dir cytol, 87-89, DIR SURG & PATH, HERMAN HOSP, HOUSTON, TEX, 87-; PROF PATH, MED SCH, UNIV TEX, HOUSTON, 81- *Concurrent Pos:* Instr, Sch Med, NY Univ, 60-61; dep med examr, Suffolk County, NY, 64; assoc, Col Physicians & Surgeons, Columbia Univ, 65-67; asst chief lab, Vet Admin Hosp, Albuquerque, NMex, 66-74; chief anat path, Vet Admin Hosp, San Antonio, 74-75; att pathologist, Hermann Hosp, Houston Tex, 81- *Mem:* Int Acad Path; Am Soc Clin Path; Am Acad Oral Path; NY Acad Sci; Col Am Pathologists. *Res:* Laboratory medicine with prime interest in surgical and anatomical pathology. *Mailing Add:* Dept Path & Lab Med 6431 Fannin St Univ Tex Med Sch Houston TX 77030

WEJKSNORA, PETER JAMES, b New York, NY, Jan 12, 50; c 1. GENE EXPRESSION, SOMATIC CELL HYBRIDS. *Educ:* Brooklyn Col, BS, 72; Brandeis Univ, PhD(biol), 77. *Prof Exp:* Fel, Albert Einstein Col Med, 77-81; asst prof genetics, 81-87, ASSOC PROF BIOL SCI, DEPT BIOL SCI, UNIV WIS, MILWAUKEE, 87- *Mem:* Am Soc Cell Biol; Am Soc Microbiol; Am Soc Biol Chem. *Res:* Molecular biology; control of eukaryotic gene expression particularly ribosomal RNA and protein genes; gene control in somatic cell hybrids. *Mailing Add:* Dept Biol Sci Univ Wis PO Box 413 Milwaukee WI 53201

WEKELL, MARLEEN MARIE, b Spokane, Wash, Sept 26, 42; m 64; c 1. FOOD MICROBIOLOGY, BIOCHEMISTRY. *Educ:* Seattle Univ, BS, 64; Univ Wash, MS, 72, PhD(environ microbiol, 75. *Prof Exp:* Teaching asst Dept Chem, Seattle Univ, 61-65; res technician biochem, Sch Med, Univ Wash, 65-70, teaching asst marine biochem, Sch Fisheries, 70-72, res asst environ microbiol, Dept Food Sci, 72-75; res chemist environ biochem, Nat Marine Fisheries Serv, Seattle, Wash, 75-76; DIR CHEM & MICROBIOL, US FOOD & DRUG ADMIN, SEATTLE, WASH, 82- *Concurrent Pos:* Chmn Local Sect, Inst Food Technologist, 82-83; adj res prof Dept Chem, Seattle Univ, 86-; adj res assoc prof, Dept Food Sci, Univ Wash, 86-; gen referee seafood, Asn Official Anal Chemists, Int Asn Milk, Food & Environ Sanitarians, 87. *Mem:* AAAS; Am Soc Microbiol; Am Chem Soc. *Res:* Environmental and food microbiology; food chemistry. *Mailing Add:* 21431 NE Sixth St Redmond WA 98053-3906

WEKSLER, BABETTE BARBASH, b New York, NY, Jan 18, 37; m 58; c 2. HEMATOLOGY, INTERNAL MEDICINE. *Educ:* Swarthmore Col, BA, 58; Col Physicians & Surgeons Columbia Univ, MD, 63. *Prof Exp:* Intern med, Bronx Munic Hosp, 63-64; resident, Georgetown Univ Hosp, 65-67; USPHS fel microbiol, Wright Fleming Inst St Mary's Hosp Med Sch, London, 67-68; Am Cancer Soc Clin fel hemat, NY Hosp Cornell Med Ctr, 68-69; asst prof, 70-75, assoc prof, 75-80, PROF MED, CORNELL UNIV MED COL, 81- *Concurrent Pos:* Assoc attend physician, NY Hosp Cornell Med Ctr, 75-80, attend physician, 81- *Mem:* Fel NY Acad Sci (gov, 78-80); Am Fedn Clin Res; Am Soc Hemat; Am Physiol Soc; Am Soc Clin Invest. *Res:* Platelet physiology and biochemistry; inflammation and hemostasis; prostaglandins. *Mailing Add:* Dept Med Cornell Univ Med Col 1300 York Ave Rm C-608 New York NY 10021

WEKSLER, MARC EDWARD, b New York, NY, Apr 16, 37; m 58; c 2. MEDICINE, IMMUNOLOGY. *Educ:* Swarthmore Col, BA, 58; Columbia Univ, MD, 62. *Prof Exp:* From asst prof to assoc prof, 70-78, WRIGHT PROF MED, MED COL, CORNELL UNIV, 78-, DIR DIV GERIAT & GERONT, 78- *Concurrent Pos:* USPHS fel, Transplantation Unit, St Mary's Hosp, London, Eng, 67-68 & spec fel, NY Hosp, 68-70; asst attend physician, NY Hosp, 70-75, assoc attend physician, 75-78, attend physician, 78 & dir geront & geriat; adj physician, Mem Hosp, NY; law lectr, Cornell Univ, 80; consult, World Health Org, Pontifical Acad Sci, Nat Res Coun. *Mem:* Am Soc Clin Invest; fel Am Col Physicians; Am Asn Immunologists; Geront Soc; Am Asn Physicians. *Res:* Gerontology and cellular immunology. *Mailing Add:* Dept Med Cornell Univ Med Col 1300 York Ave New York NY 10021

WEKSTEIN, DAVID ROBERT, b Boston, Mass, Feb 26, 37; m 58; c 4. PHYSIOLOGY, GERONTOLOGY. *Educ:* Boston Univ, AB, 57, MA, 58; Univ Rochester, PhD(physiol), 62. *Prof Exp:* From instr to assoc prof, 62-76, PROF PHYSIOL, COL MED, UNIV KY, 78-, ASSOC DIR, SANDERS BROWN CTR, AGING, 79- *Mem:* AAAS; Soc Exp Biol & Med; Geront Soc; Am Soc Zoologists; Am Physiol Soc. *Res:* Developmental physiology; physiology of temperature regulation; biology of aging; circadian rhythms. *Mailing Add:* 101 Sanders Brown Bldg Univ Ky Col Med Lexington KY 40536

WELBER, IRWIN, b New York, NY, Mar 3, 24; m 46; c 3. TELECOMMUNICATIONS, NATIONAL SECURITY RESEARCH & DEVELOPMENT. *Educ:* Union Col, BS, 48; Rennselaer Polytech Inst, MS, 50. *Prof Exp:* Staff mem transmission systs, Bell Tel Labs, 50-65, dir microwave transmission, 65-71, from assoc exec dir to exec dir transmission syst, 71-81, vpres transmission systs, AT&T Bell Labs, 81-85; sr exec vpres, Nat Lab, Sandia Corp, 85-86, pres, 86-89; RETIRED. *Mem:* Nat Acad Eng; AAAS; NY Acad Sci; fel Inst Elec & Electronics Engrs; Sigma Xi. *Res:* Telecommunications systems including first active communication satellite-Telstar; research and activity in submarine cable systems and microwave radio. *Mailing Add:* 1604 La Tuna Pl SE Albuquerque NM 87123

WELBY, CHARLES WILLIAM, b Bakersfield, Calif, Oct 9, 26; m 48; c 2. HYDROGEOLOGY, STRATIGRAPHY. *Educ:* Univ Calif, BS, 48, MS, 49; Mass Inst Technol, PhD(geol), 52. *Prof Exp:* Geologist, Calif Co, 52-54; asst prof geol, Middlebury Col, 54-58, Trinity Col, Conn, 58-61 & Rensselaer Polytech Inst, 61-62; assoc prof geol & chmn dept, Southern Miss Univ, 62-65; assoc prof, 65-77, PROF GEOL, NC STATE UNIV, 77- *Concurrent Pos:* Consult groundwater, environ geol, remote sensing, shoreline erosion. *Mem:* Am Asn Petrol Geologists; fel Geol Soc Am; Soc Econ Paleont & Mineral; Asn Eng Geologists; Nat Water Well Asn; Sigma Xi. *Res:* Occurrence and management of ground water; structure and stratigraphy; occurrence and exploration for petroleum; ground water pollution; environmental geology; remote sensing. *Mailing Add:* Dept Marine Earth Atmospheric Sci NC State Univ PO Box 8208 Raleigh NC 27695-8208

WELCH, AARON WADDINGTON, b Georgetown, Md, July 25, 16; m 41; c 3. BOTANY. *Educ:* Univ Md, BS, 37; Iowa State Col, PhD(plant path), 42. *Prof Exp:* Mgr, Southeastern Exp Sta, Iowa, 40-42; assoc pathologist, Div Forage Crops, USDA, 45-47; plant pathologist, Exp Sta, E I du Pont de Nemours & Co, Inc, 47-60, mgr res & develop farm, Clayton, 60-73, dir res & develop farm, Clayton, 73-88. *Mem:* Am Phytopath Soc. *Res:* Plant pathology; helminthology; herbicides. *Mailing Add:* 5922 Carmel Lane Raleigh NC 27609

WELCH, ANNEMARIE S, b Knoxville, Tenn, Aug 20, 37; m 59. MEDICINE. *Educ:* Duke Univ, BS, 59, MS, 60; Johns Hopkins Sch Med, MD, 74. *Prof Exp:* Res asst, dept pharm, Duke Univ, 60-62 & dept pharm, Med Col Va, 62-64; res assoc, dept biol, Col William & Mary, 64-66, Univ Tenn Mem Res Hosp, 66-69 & Md Psychiat Res Ctr, 69-70; NIH fel, dept pharmacol, Johns Hopkins Univ, 70-71; intern, Yale Univ Hosp, 74-75, jr resident, Yale New Haven Hosp, 75-76; staff physician, Gaylord Hosp, 76-77; sr resident, Univ Conn Health Ctr, 77-78; emergency rm physician, 78-79, asst attend med, 79-80, jr attend med, 80-83, ASSOC ATTEND MED, NEW BRIT GEN HOSP, CONN, 83-; SR PARTNER, WELCH ASSOCS, 76- *Concurrent Pos:* Secy, Conn State Med Soc, 86-; deleg, Am Med Asn, 88- *Mem:* Am Soc Pharmacol & Exp Therapeut; Am Col Physicians; AMA; Am Soc Internal Med; AAAS. *Res:* Neurotransmitter medicine; neurochemistry; neuropharmacology; biochemistry and pharmacology of behavior; biochemistry and therapy of sleep disorders; neuropsychoimmunology. *Mailing Add:* Dept Med New Brit Gen Hosp 100 Grand St New Britain CT 06050

WELCH, ARNOLD D(EMERRITT), b Nottingham, NH, Nov 7, 08; m 33, 66; c 3. PHARMACOLOGY, CHEMOTHERAPY. *Educ:* Univ Fla, BS, 30, MS, 31; Univ Toronto, PhD(pharmacol), 34; Wash Univ, MD, 39. *Hon Degrees:* DSc, Univ Fla, 73. *Prof Exp:* Asst physiol, Exp Sta, Univ Fla, 29-31; asst pharmacol, Sch Med, Univ Toronto, 31-35; asst pharmacol, Sch Med, Wash Univ, 35-36, instr, 36-40; dir pharmacol res, Sharp & Dohme, Inc, 40-44, asst dir res, 42-43, dir, 43-44; prof pharmacol & dir dept, Sch Med, Western Reserve Univ, 44-53; prof & chmn dept, Sch Med & Grad Sch, Yale Univ, 53-67, Eugene Higgins prof, 57-67; dir, Squibb Inst Med Res, 67-72, pres, 72-74, vpres res & develop, E R Squibb & Sons, Inc, 67-74; scientist & cancer export, Nat Cancer Inst, 83-87; emer scientist, NIH, 87; chmn biochem & clin pharmacol, 75-81, scientist, 81-83, EMER CHMN, ST JUDE CHILDREN'S RES HOSP, 81- *Concurrent Pos:* Fulbright scholar, Oxford Univ, 52-53; Commonwealth scholar, Inst Therapeut Biochem, Univ Frankfurt, Inst Org Chem & Biochem, Acad Sci Prague, 64-65; mem, Sci Adv Comt, Leonard Wood Mem Found, 48-53; mem comts, Nat Res Coun, 48-56, chmn comt growth, 52-54; consult to Surgeon Gen, USPHS, 52-64; mem, Div Comt Biol & Med Res, NSF, 53-55; mem, Pharmcol & Exp Therapeut Study Sect, NIH, 59-63, chmn, 60-63, chmn chemother study sect, 63-65; mem, Sci Adv Bd, St Jude Hosp, 69-72; mem, Sci Adv Bd Biol Sci, Princeton Univ, 69-75. *Honors & Awards:* Sollmann Award, Am Soc Pharmacol & Exp Therapeut, 66; Mem Medal, Sloan-Kettering Inst, 88; Heyrorsky Gold

Medal, Czech Acad Sci, 90. *Mem:* Asn Am Physicians; Am Soc Hemat; Brit Biochem Soc; Am Soc Clin Pharmacol & Therapeut; Am Asn Cancer Res; Am Soc Pharmacol & Therapeut; Am Chem Soc; Am Soc Biol Chem & Molecular Biol. *Res:* Cellular localization of pressor amines; sulfonamides; filariasis; structure and action of choline-like compounds; biosynthesis of labile methyl group; biosynthesis and antagonism of utilization of nucleic acid precursors; metabolic approaches to cancer and virus chemotherapy; inhibition of enzyme induction; mechanisms of resistance to cytotoxic nucleosides; biochemical approaches to cancer chemotherapy. *Mailing Add:* 8101 Connecticut Ave No C-410 Chevy Chase MD 20815

WELCH, ASHLEY JAMES, b Ft Worth, Tex, May 3, 33; m 52; c 3. ELECTRICAL & BIOMEDICAL ENGINEERING. *Educ:* Tex Tech Col, BS, 55; Southern Methodist Univ, MS, 59; Rice Univ, PhD(elec eng), 64. *Prof Exp:* Aerophys engr, Gen Dynamics, Ft Worth, 57-60; instr elec eng, Rice Univ, 60-64; prof elec eng & biomed engr, 75-86, MARRON E FORSMAN PROF ELEC & COMPUTER ENG & BIOMED ENG, UNIV TEX, AUSTIN, 86- *Mem:* Fel Inst Elec & Electronics Engrs; Am Soc Laser Surg & Med. *Res:* Laser-tissue interaction; application of lasers in medicine. *Mailing Add:* Dept Elec Eng Univ Tex Austin TX 78712

WELCH, CLARK MOORE, b Mountain Grove, Mo, Mar 3, 25; m 66; c 3. CELLULOSE CHEMISTRY, COTTON TEXTILE CHEMISTRY. *Educ:* Univ Tenn, BS, 46, MS, 47, PhD(chem), 52. *Prof Exp:* Res chemist, Monsanto Chem Co, 47-48; fel, Univ Tenn, 52; asst prof, La State Univ, 52-56; res chemist, 56-66, res leader, 66-82, RES CHEMIST, SOUTHERN REGIONAL RES CTR, 82- *Honors & Awards:* Fed Inventors Award, US Dept Com, 85. *Mem:* Am Chem Soc; Am Asn Textile Chemists & Colorists; AAAS; Sigma Xi. *Res:* Polycarboxylic acids as formaldehyde-free durable press reagents for cotton fabrics; development of glyoxal glycol coreactants as formaldehyde-free crosslinking agents for cotton; metal complexes of hydrogen peroxide as durable antibacterial agents for cotton textiles. *Mailing Add:* 5704 Ruth St Metairie LA 70003

WELCH, CLAUDE ALTON, b Flint, Mich, Oct 24, 21; m 49; c 2. ZOOLOGY. *Educ:* Mich State Univ, BS, 48, PhD(zool), S7. *Prof Exp:* From instr to prof natural sci, Mich State Univ, 53-69; prof, 69-73, chmn dept, 69-77, O T WALTER PROF BIOL, MACALESTER COL, 73- *Concurrent Pos:* Supvr blue version writing team, Biol Sci Curric Study, 61; consult, Adaptation Comt, Japan, 63-65 & Turkey, 65. *Mem:* AAAS; Nat Asn Biol Teachers (pres, 72); Nat Sci Teachers Asn. *Res:* Cell physiology. *Mailing Add:* 148 El Viento Green Valley AZ 85614

WELCH, CLETUS NORMAN, b Convoy, Ohio, Feb 2, 37; m 60; c 2. PHYSICAL INORGANIC CHEMISTRY. *Educ:* Bowling Green State Univ, BS, 61; Ohio State Univ, MS, 64, PhD(chem), 66. *Prof Exp:* Res assoc, 66-81, sr res assoc, 81-86, SCIENTIST, PPG INDUSTS, INC, 86- *Mem:* Am Chem Soc; Soc Vac Coaters. *Res:* Halogen and halogen-oxygen chemistry; synthesis of electrocatalysts; electrode processes; electrochemical cell design and materials of construction; electroless, electrolytic and vapor deposition of metallic coatings; photochronics, electrochromics, thin films. *Mailing Add:* 3317 Hermar Dr Murrysville PA 15668-1602

WELCH, DAVID O(TIS), b Richmond, Va, Mar 9, 38; m 78; c 2. MATERIALS SCIENCE, SOLID STATE PHYSICS. *Educ:* Univ Tenn, BS, 60; Mass Inst Technol, SM, 62; Univ Pa, PhD(metall), 64. *Prof Exp:* Res assoc metal physics, Res Inst Advan Studies, Martin Co, 62; NATO res fel, Solid State Physics Div, Atomic Energy Res Estab, Harwell, Eng, 64-65; asst prof metal phys, Solid State & Mat Prog, Princeton Univ, 66-72, res fel, 72; from assoc physicist to physicist, Physics Dept, 72-77, Mat Sci Div, 77-90, asst head, 82-90, HEAD, MAT SCI DIV DEPT APPL SCI, BROOKHAVEN NAT LAB, 90-, SR PHYSICIST, 90- *Concurrent Pos:* Consult, Metals & Ceramics Div, Oak Ridge Nat Lab, 66-68; consult & vis scientist, Mat Sci Div, Argonne Nat Lab, 68; vis prof, Univ Sao Paulo, 70, Univ Campinas, 88; vis scientist, KFA Jülich, 74; adj prof mat sci, State Univ NY, Stony Brook, 80-; mem solid state sci panel, Nat Res Coun, 80-84. *Mem:* AAAS; Am Phys Soc; Sigma Xi; Mats Res Soc; Metals Minerals & Mat Soc. *Res:* Theoretical materials science. *Mailing Add:* Bldg 480 Brookhaven Nat Lab Upton NY 11973

WELCH, DEAN EARL, b Aledo, Ill, Aug 5, 37; m 58; c 6. ORGANIC CHEMISTRY. *Educ:* Monmouth Col, BA, 59; Mass Inst Technol, PhD(org chem), 63. *Prof Exp:* Chemist, Escambia Chem Corp, 63-64; assoc scientist, 64-66, scientist, 66-68, qual assurance dir, 68-69, chem res dir, 69-71, res dir, 71-72, vpres res, 72-84, VPRES OPERS, SALSBURY LABS, 84- *Mem:* Am Chem Soc. *Res:* Organic synthesis; research and development administration. *Mailing Add:* 128 Park Lane Dr Charles City IA 50616

WELCH, FRANK JOSEPH, b Fresno, Calif, Aug 5, 29; m 54; c 5. ORGANIC CHEMISTRY. *Educ:* Fresno State Univ, BS, 51; Stanford Univ, PhD(org chem), 54. *Prof Exp:* Res chemist polymers, Union Carbide Chem Co, 54-62, group leader latex polymerization, Res & Develop Dept, Union Carbide Corp, 62-71; dir res, Avery Label Div, Avery Int, 71-88, SR RES ASSOC, AVERY DENNISON CORP, 88- *Mem:* Tech Asn Pulp & Paper Indust; Am Chem Soc; Tech Asn Graphic Arts. *Res:* Ionic and free radical polymerization kinetics; organo-metallic and organophosphorus chemistry; latex polymerization; coatings and printing inks; adhesives. *Mailing Add:* 9339 Amelga Dr Whittier CA 90603

WELCH, GARTH LARRY, b Brigham City, Utah, Feb 14, 37; m 60; c 6. ACADEMIC ADMINISTRATION, INORGANIC CHEMISTRY. *Educ:* Univ Utah, BS, 59, PhD(inorg chem), 63. *Prof Exp:* Fel, Univ Calif, Los Angeles, 62-64; from asst prof to assoc prof, Weber State Univ, 64-72, dean, Sch Natural Sci, 74-83, exec dir bus affairs, 83-89, assoc vpres phys fac, 90-92, PROF CHEM, WEBER STATE UNIV, 72- *Concurrent Pos:* Vis prof, Brigham Young Univ, 80. *Mem:* Am Chem Soc; Sigma Xi. *Res:* Kinetics of complex ions; reversion rates of polymers. *Mailing Add:* 3910 N 800 W Ogden UT 84404

WELCH, GARY ALAN, b Santa Monica, Calif, June 19, 42. ASTRONOMY. *Educ:* Harvey Mudd Col, BSc, 64; Univ Wash, MSc, 67, PhD(astron), 69. *Prof Exp:* Postdoctoral res assoc, Van Vleck Observ, Wesleyan Univ, 69-72, Mich State Univ, 72-73; asst prof astron, Wheaton Col, 73-74; asst prof, 74-78, ASSOC PROF ASTRON, SAINT MARY'S UNIV, HALIFAX, 78- *Concurrent Pos:* Hon fel astron, Univ Wis, Madison, 80-81. *Mem:* Int Astron Union; Am Astron Soc; Can Astron Soc; Royal Astron Soc Can; Astron Soc Pac. *Res:* Structure stellar content and evolution of ES0 galaxies; photometric studies of peculiar early-type galaxies; digital imaging techniques and processing of digital data in astronomy. *Mailing Add:* Dept Astron Saint Mary's Univ Halifax NS B3H 3C3 Can

WELCH, GARY WILLIAM, b Buffalo, NY, Jan 4, 43; m 72; c 5. ANESTHESIOLOGY, CRITICAL CARE MEDICINE. *Educ:* Univ Va, BA,64, MD, 70, PhD(anat), 70, JD, 87. *Prof Exp:* Chief anesthesiol sect, US Army Inst Surg Res, 74-76; asst prof, 77-79, dir, Surg Intensive Care Unit, 78-82, vchmn, Dept Anesthesiol, 80-81, actg chmn, 81-82, ASSOC PROF ANESTHESIOL & SURG, UNIV MASS MED CTR, 79-, CHMN DEPT ANETHESIOL & ASSOC DIR, SURG INTENSIVE CARE UNIT, 82- *Mem:* Soc Critical Care Med; Am Shock Soc; Am Burn Asn; Am Soc Anesthesiologists. *Res:* High frequency; high pressure; jet ventilation; metabolic changes following major surgery; use of epidural narcotics for pain relief; fluid therapy for treatment of shock; genetic analysis malignant hyperthermia. *Mailing Add:* 93 Cherry St Shrewsbury MA 09545

WELCH, GEORGE BURNS, b Soso, Miss, Nov 23, 20; m 54; c 1. AGRICULTURAL ENGINEERING. *Educ:* Miss State Univ, BS, 47; Tex Agr & Mech Col, MS, 50; Okla State Univ, PhD(eng), 65. *Prof Exp:* Instr agr, Jones County Jr Col, 47-48; instr agr eng, Miss State Univ, 48-49, assoc agr engr, Miss Agr Exp Sta, 51-62, assoc prof agr eng, 52-62; res asst, Okla State Univ, 62-65; prof agr eng & agr engr, 65-80, PROF AGR & BIOL ENG, MISS STATE UNIV, 80- *Concurrent Pos:* Miss State Univ-US Agency Int Develop seed processing engr, Brazil, 70-72. *Mem:* Am Soc Agr Engrs. *Res:* Design of agricultural machinery and agricultural buildings; design of seed processing and storage facilities for tropical countries. *Mailing Add:* Dept Biol & Agr Eng Miss State Univ Box 5465 Mississippi State MS 39762

WELCH, GEORGE RICKEY, b Rockwood, Tenn, May 29, 47; m 67; c 2. THEORETICAL BIOCHEMISTRY, BIOPHYSICAL CHEMISTRY. *Educ:* Univ Tenn, BS, 70, PhD(biochem), 75. *Prof Exp:* Res fel theoret biophys, Univ Libre Bruxelles, 75-77; res scientist biochem, Univ Tex Med Sch, 77-78; ASST PROF BIOL SCI, UNIV NEW ORLEANS, 78- *Concurrent Pos:* Res fel, Solvay Int Inst Physics & Chem, 75-77. *Mem:* AAAS; Am Phys Soc; Biophys Soc. *Res:* Enzymology and metabolic regulation; theoretical biophysical chemistry; modeling of biochemical processes; bioenergetics. *Mailing Add:* Dept Biol Sci Univ New Orleans New Orleans LA 70148

WELCH, GORDON E, b Sabinal, Tex, Aug 20, 33; m 57; c 3. MICROBIOLOGY. *Educ:* Southwest Tex State Univ, BS, 60, MA, 62; Tex A&M Univ, PhD(microbiol), 66. *Prof Exp:* Chmn dept biol & chem, Southwest Tex Jr Col, 60-62; assoc prof biol, San Antonio Col, 66-67; assoc prof, 67-73, PROF BIOL, ANGELO STATE UNIV, 73-, DEAN COL SCI, 77- *Concurrent Pos:* Consult, Bexar Co Hosp Dist, San Antonio, Tex, 66. *Mem:* AAAS; Am Soc Microbiol; Am Inst Biol Sci. *Res:* Virology, particularly myxoviruses, their nature, properties and pathogenicity. *Mailing Add:* Dean Sci Angelo State Univ 2601 West Ave N San Angelo TX 76909

WELCH, GRAEME P, b Los Angeles, Calif, Nov 25, 17; m 45; c 4. BIOPHYSICS. *Educ:* Univ Calif, Los Angeles, AB, 40; Univ Calif, Berkeley, MA, 50, PhD(biophys), 57. *Prof Exp:* Assoc physicist, Div War Res, Univ Calif, 41-45, physicist, Donner Lab, 48-59; engr biophys, Saclay Nuclear Res Ctr, France, 59-61; biophysicist, Donner Lab, Berkeley Radiation Lab, Univ Calif, Berkeley, 61-80; RETIRED. *Mem:* Radiation Res Soc. *Res:* Biological effects of radiation; physics and dosimetry of heavy-particle radiations. *Mailing Add:* 1020 Oxford St Berkeley CA 94707

WELCH, H(OMER) WILLIAM, b Beardstown, Ill, Oct 21, 20; m 42; c 2. ELECTRONIC & ELECTRICAL ENGINEERING. *Educ:* DePauw Univ, BA, 42; Univ Mich, MS, 49, PhD, 52. *Prof Exp:* Asst physics & electronics, Univ Wis, 42-43; res assoc, Radio Res Lab, Harvard Univ, 43-45; instr physics, Purdue Univ, 45-46; res physicist, Eng Res Inst, Univ Mich, 46-50, proj leader, Electronic Defense Group, 51-53, from assoc prof to prof elec eng, 53-57; dir res & develop, Motorola Mil Electronics Div, 57-62, gen mgr control syst div, 62-66, asst to chief tech off, 66-67; asst dean, Col Eng Sci, 68-80, PROF ENG, ARIZ STATE UNIV, 67-, DIR, PROG SOCIETY, VALUES & TECHNOL, 79- *Concurrent Pos:* Mem adv groups, US Dept Defense. *Mem:* Fel Inst Elec & Electronics Engrs; Am Soc Eng Educ; Soc Hist Technol; fel AAAS. *Res:* Microwave tubes and solid state devices; communications systems; integrated microelectronics; remote monitoring and control systems; social impacts of technology. *Mailing Add:* Col Eng & Appl Sci Ariz State Univ Tempe AZ 85287

WELCH, HUGH GORDON, b Memphis, Tenn, Oct 30, 37; m 59; c 2. EXERCISE PHYSIOLOGY, RESPIRATORY PHYSIOLOGY. *Educ:* Lambuth Col, BA, 59; Univ Fla, PhD(physiol), 66. *Prof Exp:* Asst prof, Univ Mich, Ann Arbor, 66-68; assoc prof, 68-76, PROF PHYSIOL, UNIV TENN, KNOXVILLE, 76- *Concurrent Pos:* Guest scientist, Univ Copenhagen, Denmark, 74-75. *Mem:* Am Physiol Soc; Am Col Sports Med. *Res:* Physiology of exercise-in particular in the study of those factors that limit human work capacity in health and in certain disease states. *Mailing Add:* Dept Zool Univ Tenn Rte 3 Box 409 Gatlinburg TN 37738

WELCH, J PHILIP, b Macclesfield, Eng, June 18, 33; m 58; c 4. HUMAN GENETICS. *Educ:* Univ Edinburgh, MB & ChB, 58; Johns Hopkins Univ, PhD, 69. *Prof Exp:* Fel med, Johns Hopkins Hosp, 63-67; asst prof, 67-72, ASSOC PROF PEDIAT, DALHOUSIE UNIV, 72- *Concurrent Pos:* Consult, Children's Hosp, Victoria Gen Hosp, Grace Maternity Hosp & Halifax Infirmary, 67- *Mem:* Am Soc Human Genetics; Am Fedn Clin Res;

Am Asn Advan Aging Res; Can Genetics Soc; Soc Study Social Biol. *Res:* Biochemical, behavioral and cytogenetic aspect of mental abilities and aberrations. *Mailing Add:* Dept Biol Sci Dalhousie Univ Halifax NS B3H 4H6 Can

WELCH, JAMES ALEXANDER, b Versailles, Ky, May 25, 24; m 51; c 2. ANIMAL NUTRITION, ANIMAL PHYSIOLOGY. *Educ:* Univ Ky, BS, 47; Univ Ill, PhD(animal sci), 52. *Prof Exp:* From asst prof to assoc prof animal husb, 52-60, actg chmn dept animal sci, 62, prof & animal scientist, 60-87, EMER PROF ANIMAL SCI, UNIV WVA, 88- *Concurrent Pos:* Chief party, WVa Univ-AID contract team, Uganda & lectr, Vet Training Inst, Entebbe, 63-66. *Mem:* Fel AAAS; Am Soc Animal Sci. *Res:* Estrus control in cattle; non-protein nitrogen utilization by ruminants. *Mailing Add:* 1370 Headlee Ave Morgantown WV 26505

WELCH, JAMES EDWARD, b San Rafael, Calif, July 19, 11; m 37; c 3. GENETICS. *Educ:* Univ Calif, BS, 34, MS, 35; Cornell Univ, PhD(genetics), 42. *Prof Exp:* Trainee agron, Soil Conserv Serv, USDA, 35, jr agr aide, 35-36, agr aide, 36; jr olericulturist, Exp Sta, Univ Hawaii, 36-38, asst olericulturist, 38-39; asst, Maize Genetics Coop, Cornell Univ, 40-42; asst horticulturist, Exp Sta, La State Univ, 42-43; assoc horticulturist, Regional Veg Breeding Lab, USDA, SC, 43-47; asst olericulturist, 47-52, assoc olericulturist, 52-79, EMER ASSOC OLERICULTURIST, EXP STA, UNIV CALIF, DAVIS, 79- *Concurrent Pos:* Lectr veg crops, Col Agr, Univ Calif, Davis, 49-59 & 76-79; collabr, Plant Sci Res Div, USDA, 58-74. *Honors & Awards:* Western Res Man of the Year Award, Pac Seedmon's Asn, 65. *Mem:* Fel AAAS; Am Inst Biol Sci; Am Soc Hort Sci; Am Genetic Asn. *Res:* Plant morphological characters associated with earworm resistance in maize; genetic linkage in autotetraploid maize; genetics of male sterility in the carrot; breeding improved varieties of vegetable crops, especially sweet corn, lima beans, carrots, celery, tomatoes and lettuce. *Mailing Add:* 739 A St Davis CA 95616-3604

WELCH, JAMES GRAHAM, b Ithaca, NY, Aug 16, 32; m 56; c 4. ANIMAL NUTRITION. *Educ:* Cornell Univ, BS, 55; Univ Wis, MS, 57, PhD(biochem, animal husb), 59. *Prof Exp:* Asst, Univ Wis, 55-59; asst prof animal husb, Rutgers Univ, 59-63, assoc prof nutrit, 63-67; assoc prof, 67-70, PROF ANIMAL SCI, UNIV VT, 70- *Mem:* Am Soc Animal Sci; Am Dairy Sci Asn; Am Inst Nutrit. *Res:* Ruminant nutrition. *Mailing Add:* Dept Animal Sci Univ Vt Burlington VT 05405

WELCH, JAMES LEE, b Medford, Ore, June 23, 46. HEALTH SCIENCES. *Educ:* SOre State Col, BS, 68; Loma Linda Univ, BS, 69, MPH, 72, DHSC, 74. *Prof Exp:* Chief technologist med technol, Park Ave Hosp, 69-70; supvr lab, Hollywood Presby Hosp, 70-72, Temple Hosp, 72-; CHMN & ASSOC PROF MED TECHNOL HEALTH SCI, CALIF STATE UNIV, DOMINGUEZ HILLS, 75- *Concurrent Pos:* Mem, Am Bd Bioanal, 75-; consult exten allied health, Univ Calif, Riverside, 74- *Mem:* AAAS; Am Soc Clin Pathologists; Am Pub Health Asn. *Mailing Add:* Dept Med Technol Calif State Univ 1000 E Victoria Dominquez Hills Carson CA 90747

WELCH, JANE MARIE, b Springfield, Mass, May 16, 50; m 77. NUCLEAR WASTE FORMS. *Educ:* Univ Mass, BS, 72; Pa State Univ, MS, 74, PhD(geochem & mineral), 78. *Prof Exp:* SCIENTIST & ENVIRON MGR, EG&G IDAHO, INC, 80- *Mem:* Am Ceramic Soc; Mat Res Soc. *Res:* Phase equilibria of silicate and titanate systems; development and characterization of nuclear waste forms. *Mailing Add:* 750 Brandon Dr Idaho Falls ID 83402

WELCH, JASPER ARTHUR, JR, b Baton Rouge, La, Jan 5, 1931; m 53, 85; c 3. OPERATIONS RESEARCH. *Educ:* La State Univ, BS, 52; Univ Calif, Berkeley, MA, 54, PhD(physics), 58. *Prof Exp:* USAF, 52-, group leader, Lawrence Livermore Lab, 53-57, Weapons Lab, 57-62; sr staff, RAND Corp, 62-64; USAF, asst to comdr, systs command, 64-65, tech dir, WCoast Study Facil, 65-68, chief anal, Studies & Analyses, 69-71, asst dir defense, Res & Eng, Off Secy Defense, 71-73, sr asst to Secy Defense for Atomic Energy, 73-74, asst strategic initiatives Hq, 74-75, asst chief of staff, Studies & Analyses, 75-79, coordn, Defense Pol, Nat Security Coun, 79-81, spec asst chief staff, Hq, 81, asst dep chief of staff, 81-83, DEFENSE CONSULT, RES & DEVELOP, 83- *Concurrent Pos:* Mem, NASA Adv Comt Fluid Mech, 64-66; consult, President's Sci Adv Comt, 62-65, NATO Adv Group Aerospace, Res & Develop, 71-80, Defense Sci Bd, 68-, Air Force Sci Adv Bd, 57- & Army Sci Bd, 80-; mem adv coun, NASA, 84-88. *Mem:* Nat Acad Eng; Am Geophys Union; Coun Foreign Rel; Am Phys Soc. *Res:* Aerospace system requirements, design and program management. *Mailing Add:* 1901 Ft Myer Dr Ste 1120 Arlington VA 22209

WELCH, JEROME E, b Portland, Maine, Jan 11, 45; c 2. SOIL GEOCHEMISTRY. *Educ:* Calif Polytech State Univ, BS, 72; Univ Calif, Riverside, PhD(soil chem), 80. *Prof Exp:* Res assoc, Dept Soil & Environ Sci, Univ Calif, Riverside, 72-78; ASST SCIENTIST, KANS GEOL SURV, UNIV KANS, 78- *Mem:* Soil Sci Soc Am; Soc Environ Geochem & Health; Int Soc Soil Sci. *Res:* Determining the trace element geochemistry of the major soils of Kansas; determining the trace element geochemistry of Kansas soils that have been disturbed or affected as a result of mineral and coal mining; geochemical and hydrological investigations of solid and hazardous waste disposal sites. *Mailing Add:* 2606 Bonanza St Lawrence KS 66046-2598

WELCH, JOHN F, JR, b Salem, Mass, Nov 11, 35. SCIENCE ADMINISTRATION. *Educ:* Univ Mass, BS, 57; Univ Ill, MS, 58, PhD(chem eng), 60. *Prof Exp:* Var mgt positions, Gen Elec Co, 60-68, gen mgr, Plastics Bus Dept, 68-72, vpres, 72-79, vchmn, 79-81, CHMN & CHIEF EXEC OFFICER, GEN ELEC CO, 81- *Concurrent Pos:* Chmn, Bus Coun; mem, Bus Roundtable. *Mem:* Nat Acad Eng. *Mailing Add:* Gen Elec Co 3135 Eastern Turnpike Fairfield CT 06431

WELCH, LIN, b Tahoka, Tex, Dec 9, 27; m 50; c 3. SPEECH PATHOLOGY. *Educ:* WTex State Univ, BS, 48; Baylor Univ, MA, 49; Univ Mo, PhD(speech path), 60. *Prof Exp:* Instr speech, Blue Mountain Col, 52-53; instr, Univ Mo, 53-56; PROF SPEECH PATH & AUDIOL, CENT MO STATE UNIV, 56-, CHMN, 68- *Concurrent Pos:* Chmn, Continuing Educ Comt, Am Speech & Hearing Asn, 72-73; mem, Educ & Standards Bd, Am Speech & Hearing, 80-82; mem, Adv Comn Commun Dis, Mo Bd Med & Healing Arts. *Mem:* Fel Am Speech & Hearing Asn. *Res:* Cleft palate; stuttering. *Mailing Add:* Dept Speech Path & Audiol Cent Mo State Univ Warrensburg MO 64093

WELCH, LLOYD RICHARD, b Detroit, Mich, Sept 28, 27; m 53; c 3. MATHEMATICS, COMBINATORICS. *Educ:* Univ Ill, BS, 51; Calif Inst Technol, PhD(math), 58. *Prof Exp:* Sr res engr, Jet Propulsion Lab, Calif Inst Technol, 57-59; staff mathematician, Inst Defense Anal, 59-65; assoc prof elec eng, 65-68, PROF ELEC ENG, UNIV SOUTHERN CALIF, 68- *Mem:* Nat Acad Eng; Math Asn Am; Soc Indust & Appl Math; Am Math Soc; fel Inst Elec & Electronics Engrs. *Res:* Communication theory; combinatorics; probability theory. *Mailing Add:* Dept Elec Eng Univ Southern Calif Los Angeles CA 90089-0272

WELCH, MELVIN BRUCE, b Hood River, Ore, Feb 24, 45; m 67; c 2. INORGANIC CHEMISTRY, POLYMER CHEMISTRY. *Educ:* Linfield Col, BA, 67; Univ Utah, PhD(inorg chem), 74. *Prof Exp:* Res chemist, 74-80, SR RES CHEMIST, PHILLIPS PETROL CO, 80- *Mem:* Am Chem Soc. *Res:* Olefin polymerization and organometallic chemistry. *Mailing Add:* 4716 Barlow Dr Bartlesville OK 74006-6915

WELCH, MICHAEL JOHN, b Stoke-on-Trent, Eng, June 28, 39; div; c 2. RADIOCHEMISTRY, NUCLEAR MEDICINE. *Educ:* Cambridge Univ, BA, 61, MA, 64; Univ London, PhD(radiochem), 65. *Prof Exp:* Res assoc chem, Brookhaven Nat Lab, 65-67; from asst prof to assoc prof, 67-74, PROF RADIATION CHEM, SCH MED, WASH UNIV, 74- *Honors & Awards:* Paul C Aebersold Award, Soc Nuclear Med, 80; St Louis Award, Am Chem Soc, 88. *Mem:* Royal Soc Chem; Am Chem Soc; Soc Nuclear Med (pres, 84-85); Radiation Res Soc. *Res:* Hot atom chemistry; isotopes in medicine. *Mailing Add:* One Spoede Lane St Louis MO 63141-7763

WELCH, PETER D, b Detroit, Mich, May 19, 28; m 74; c 2. MATHEMATICAL STATISTICS. *Educ:* Univ Wis, MS, 51; NMex State Univ, MS, 56; Columbia Univ, PhD(math statist), 63. *Prof Exp:* Assoc mathematician, Phys Sci Lab, NMex State Univ, 51-56; RES STAFF MEM PROBABILITY & STATIST, IBM RES CTR, 56- *Concurrent Pos:* Adj prof, Columbia Univ, 64-65, 74-75, 84-90. *Mem:* Opers Res Soc. *Res:* Queueing theory; time series analysis; signal processing. *Mailing Add:* 85 Croton Ave Mt Kisco NY 10549

WELCH, RAYMOND LEE, b Emporia, Kans, Nov 2, 43; m 64; c 2. ORGANIC CHEMISTRY, ORGANIC COATINGS. *Educ:* Kans State Teachers Col, BA, 65; Iowa State Univ, PhD(org chem), 69. *Prof Exp:* Group leader, Hercules Res Ctr, Wilmington, 69-86; DIR TECH SERV, MAGNOX INC, 86- *Mem:* Am Chem Soc. *Res:* Product development; organic coatings; magnetic recording. *Mailing Add:* Tatnall Estates 374 Springhouse Lane Hockessin DE 19707-9689

WELCH, RICHARD MARTIN, b Brooklyn, NY, Nov 4, 33; m 54; c 3. BIOCHEMICAL PHARMACOLOGY. *Educ:* St John's Univ, NY, BS, 57; Jefferson Med Col, MS, 60, PhD(pharmacol), 62. *Prof Exp:* Asst prof pharmacol, Jefferson Med Col, 62-63; GROUP LEADER MED BIOCHEM, BURROUGHS WELLCOME RES LABS, 65-, SR PHARMACOLOGIST, 77-, PRIN SCIENTIST & ASST DIR PHARMACOKINETIC & DRUG METAB. *Concurrent Pos:* Fel, Albert Einstein Col Med, 63-65. *Mem:* Am Soc Pharmacol & Exp Therapeut; Soc Toxicol. *Res:* Pharmacodynamics; absorption, distribution and metabolism of drugs. *Mailing Add:* Med Biochem Burroughs Wellcome Co 3030 Cornwalis Rd Research Triangle Park NC 27709

WELCH, ROBIN IVOR, b Douglas, Ariz, Mar 13, 30; m 68; c 4. EARTH SCIENCE, ECOLOGY. *Educ:* Univ Calif, Berkeley, BS, 55, MS, 56, PhD(wildland resource sci), 71. *Prof Exp:* Sr eng photo interpretation res, Mark Systs, Inc, 63-66; sr eng syst anal res, Stanford Res Inst, 66-70; res eng remote sensing res & teaching, Earth Satellite Corp, 70-75; assoc res scientist & vis assoc prof, Tex A&M Univ, 75-77; prog mgr & dir training remote sensing res Airview spec corp contract, Ames Res Ctr, NASA, Humboldt State Univ, 77-81; PRES & FOUNDER AERIAL PHOTO RES, AIRVIEW SPECIALISTS CORP, 54-; OPERS RES SPECIALIST & SYSTS ENG, LOCKHEED MISSLES & SPACE CO, 81- *Concurrent Pos:* Consult, State Alaska Dept Econ Develop, 69-72, NASA Johnson Space Ctr, 75-77, Cent Intel Agency, 73-75, Greek Govt, 72-73 & Argentine Govt, 70-71. *Mem:* Am Soc Photogram; Am Forestry Asn. *Res:* Remote sensing of earth agriculture and water resources; data analysis by interactive man and machine systems of remote sensing data; teaching methods and curriculum development in remote sensing applications. *Mailing Add:* 3419 Churin Dr Mountain View CA 94040

WELCH, RONALD MAURICE, b Chicago, Ill, Dec 30, 43; m 67; c 2. ATMOSPHERIC PHYSICS, METEOROLOGY. *Educ:* Calif State Univ, Long Beach, BS, 65, MA, 67; Univ Utah, PhD(physics), 71, PhD(meteorol), 76. *Prof Exp:* Engr-scientist, Missile & Space Systs Div, Douglas Aircraft Co, Calif, 66; res assoc geophys, Univ Utah, 71-72, res assoc physics, 72, teaching assoc, 72-73, assoc instr, Meteorol Dept, 73-74, res assoc, 74-76; res assoc, Dept Atmospheric Sci, Colo State Univ, 76-78; mem staff, Meteorol Inst, Univ Mainz, Ger, 78-81; assoc prof meteorol, Old Dominion Univ, 81-82; sr scientist, 82-90, PROF, DEPT METEOROL, SDAK SCH MINES TECHNOL, 90- *Concurrent Pos:* Vis scientist, Inst Meterol, Univ Mainz, 78-81, Naval Oceanog & Atmospheric Res Lab, Univ Calif, Monterey, 89-90; adj prof, Colo State Univ, 87-; Am Soc Environ Educ fac fel, NASA, Langley, 83, 84 & 88; mem sci adv panel, Earth Observing Syst Data & Info Syst. *Mem:* AAAS; Am Phys Soc; Am Geophys Union; Sigma Xi; Am Meteorol Soc; Inst

Elec & Electronics Engrs. *Res:* Radiative transfer in planetary atmospheres; climatology; electronic properties of materials, remote sensing; artificial intelligence. *Mailing Add:* Dept Meteorol SDak Sch Mines Technol Rapid City SD 57701

WELCH, ROSS MAYNARD, b Lancaster, Calif, May 8, 43; m 65; c 2. PLANT NUTRITION, PLANT PHYSIOLOGY. *Educ:* Calif State Polytech Univ, San Luis Obispo, BS, 66; Univ Calif, Davis, MS, 69, PhD(soil sci), 71. *Prof Exp:* Res assoc plant mineral nutrit, Dept Agron & US Plant, Soil & Nutrit Lab, 71-72, from asst prof to assoc prof, 74-87, PLANT PHYSIOLOGIST, AGR RES SERV, US PLANT, SOIL & NUTRIT LAB, USDA, 72-, PROF AGRON, 87-; assoc prof, 81-87, PROF AGRON, CORNELL UNIV, 87- *Concurrent Pos:* Asn prof agron, Cornell Univ, 74-; vis prof, Sch Environ & Life Sci, Murdoch Univ, WAustralia, 80-81. *Mem:* NY Acad Sci; Sigma Xi; Am Soc Plant Physiol; fel Am Soc Agron; Soc Environ Geochem & Health; fel Soil Sci Soc Am; Crop Sci Soc Am. *Res:* Plant mineral nutrition; ion transport by whole plants, plant tissues and cells; trace element physiology and biochemistry; physiological form and bioavailability of mineral elements in plants to animals and humans. *Mailing Add:* US Plant Soil & Nutrit Lab Cornell Univ Tower Rd Ithaca NY 14853

WELCH, ROY ALLEN, b Waukesha, Wis, Nov 14, 39; m 67. GEOGRAPHY, PHOTOGRAMMETRY. *Educ:* Carroll Col, BS, 61; Univ Okla, MA, 65; Univ Glasgow, PhD(geog, photogram), 68. *Prof Exp:* Photo-analyst geog, US Govt, 62-64; mgr earth sci dept, Itek Corp, 68-69; Nat Acad Sci-Nat Res Counc res assoc, US Geol Surv, 69-71; assoc prof geog, 71-77, PROF GEOG, UNIV GA, 77- *Concurrent Pos:* Consult, AID, 72; mem working group image qual & optical transfer function-modulation transfer function, Comn I, Int Soc Photogram, 72-76; mem remote sensing comt, Asn Am Geogrs, 73-75; remote sensing specialist, US Geol Surv, 73- *Honors & Awards:* III Talbert Abrams, Am Soc Photogram, 71, Presidential Citation, 75. *Mem:* Am Soc Photogram; Brit Photogram Soc; Asn Am Geogrs; Brit Cartog Soc; Sigma Xi; Inst Elec & Electronics Engrs. *Res:* Analyses of the quality and applications of aircraft and satellite imagery for geographic tasks, with particular reference to land use mapping and cartography. *Mailing Add:* 440 Cherokee Ridge Athens GA 30601

WELCH, STEPHEN MELWOOD, b San Diego, Calif, July 1, 49; m 73; c 2. AGROMETEROLOGY, SYSTEMS SCIENCE. *Educ:* Mich State Univ, BS, 71, PhD(zool), 77. *Prof Exp:* From asst prof to prof entom, Kans State Univ, 78-84, tech develop coordr, 85-86, exten computer systs coordr, 87-89, PROF AGRON, KANS STATE UNIV, 90- *Concurrent Pos:* Mem bd dir, Asn Agr Comput Co. *Mem:* Am Soc Agron; Asn Comput Mach; Am Soc Agr Engrs; Asn Agr Comput Co. *Res:* The application of computers and systems science to agriculture. *Mailing Add:* Dept Agron Kans State Univ Manhattan KS 66506-5501

WELCH, STEVEN CHARLES, b Inglewood, Calif, Feb 18, 40; div; c 2. CHEMISTRY. *Educ:* Univ Calif, Los Angeles, BS, 64; Univ Southern Calif, PhD(chem), 68. *Prof Exp:* NIH fel, Calif Inst Technol, 68-70; from asst prof to prof chem & pharm, Univ Houston, 70-90; PROF & CHAIR, CALIF STATE UNIV, SAN MARCOS, 90- *Concurrent Pos:* Nat Inst Gen Med Sci grant, Univ Houston, 71-77, Welch Found grant, 72-90; Nat Cancer Inst contract, 77-80, Nat Cancer Inst grant, 87-90. *Mem:* Am Chem Soc. *Res:* Synthetic organic chemistry; synthesis of terpene natural products and development of new synthetic methods and reagents. *Mailing Add:* Dept Chem Calif State Univ San Marcos CA 92096

WELCH, WALTER RAYNES, b Rumford, Maine, Oct 25, 20; m 44; c 3. MARINE BIOLOGY. *Educ:* Univ Maine, BS, 47, MS, 50. *Prof Exp:* Fisheries res biologist, Clam Prog, US Nat Marine Fisheries Serv, 49-58, prog leader, 58-64, proj leader lobster prog, 64-67, prog leader, 67-71, asst lab dir, 71-73, consult div water resources mgt & marine adv serv, 71-73; marine resource scientist & proj leader, Maine Dept Marine Resources, 73-78, div chief, 78-81, asst res dir, 81-85; RETIRED. *Concurrent Pos:* Pvt consult molluscan & environ probs, 78-86. *Honors & Awards:* Unit Citation, Nat Oceanic & Atmospheric Admin, 73. *Res:* Molluscan and crustacean ecology; environmental measurement and interpretation. *Mailing Add:* Six Factory Core Rd Boothbay Harbor ME 04538

WELCH, WAYNE WILLARD, b Clinton, Iowa, May 20, 34; m 51; c 3. EDUCATIONAL PSYCHOLOGY, PROGRAM EVALUATION. *Educ:* Univ Wis, La Crosse, BS, 56; Univ Pa, MS, 60; Purdue Univ, MS, 63; Univ Wis, PhD(sci educ), 66. *Prof Exp:* Res assoc, Harvard Univ, 65-69; from asst prof to assoc prof, 69-74, asst dean, 70-74, PROF EDUC PSYCHOL, UNIV MINN, 74- *Concurrent Pos:* Prin investr eval teacher educ proj, NSF, 71-77, nat assessment educ progress, 80-84, prog officer, 72-73; vis prof, Univ Wash, 76-77; Fulbright lect, Weizemann Inst Sci, Israel, 79; mem comt sci indicators, Nat Res Coun, 83-87; vis scholar, Western Australia Inst Technol, 84; Fulbright Res Scholar, Univ Waikato, NZ, 85; head Off Studies & Prog Assessment, NSF, 88-89, expert consult, 90-91. *Mem:* Nat Asn Res Sci Teaching (pres, 73-74); fel AAAS; Am Educ Res Asn; Nat Sci Teachers Asn; Am Eval Asn. *Res:* improvement of science education; identification and measurement of factors that influence science learning; evaluation of innovative programs; national assessment of science achievement; science education indicators. *Mailing Add:* 210 Burton Hall Univ Minn Minneapolis MN 55455

WELCH, WILLARD MCKOWAN, JR, b Frankfort, Ky, Mar 1, 44; m 66; c 2. ORGANIC CHEMISTRY. *Educ:* Mass Inst Technol, BS, 65; Rice Univ, PhD(org chem), 69. *Prof Exp:* RES CHEMIST, MED RES LABS, PFIZER, INC, 70- *Res:* Synthetic organic chemistry; central nervous system drugs; heterocyclic chemistry. *Mailing Add:* Med Res Labs Pfizer Inc Groton CT 06340

WELCH, WILLIAM HENRY, JR, b Los Angeles, Calif, Dec 13, 40; m 65; c 4. BIOCHEMISTRY. *Educ:* Univ Calif, Berkeley, BA, 63; Univ Kans, PhD(biochem), 69. *Prof Exp:* NIH fel, Brandeis Univ, 69-70; asst prof, 70-76, ASSOC PROF BIOCHEM, UNIV NEV, RENO, 76- *Mem:* Sigma Xi; AAAS; Am Chem Soc. *Res:* Enzymology; molecular basis of adaptation to temperature stress; mechanisms by which metal ions alter enzyme action and protein conformation. *Mailing Add:* Dept Biochem Univ Nev Reno NV 89557

WELCH, WILLIAM JOHN, b Chester, Pa, Jan 17, 34; m 55; c 3. RADIO ASTRONOMY. *Educ:* Stanford Univ, BS, 55; Univ Calif, Berkeley, PhD(eng sci), 60. *Hon Degrees:* Dr, Univ Bordeaux I, 79. *Prof Exp:* From asst prof to prof elec eng, 60-72, PROF ELEC ENG & ASTRON & DIR RADIO ASTRON LAB, UNIV CALIF, BERKELEY, 72- *Concurrent Pos:* Mem, NSF Atron Adv Panel, 73-76; trustee at large, Assoc Univ Inc, 74-81; mem, Arecibu Adv Bd, 77-80. *Mem:* Am Astron Soc; Inst Elec & Electronics Engrs; Int Sci Radio Union; AAAS; Int Astron Union. *Res:* Radio astronomical studies of the planets, interstellar medium and extragalactic radio sources; instrumentation for radio astronomy at millimeter wave lengths. *Mailing Add:* Radio Astron Lab Univ Calif Berkeley CA 94720

WELCH, WINONA HAZEL, botany; deceased, see previous edition for last biography

WELCH, ZARA D, b North Manchester, Ind, June 28, 15. ORGANIC CHEMISTRY. *Educ:* Manchester Col, AB, 37; Purdue Univ, MS, 39, PhD(org chem), 42. *Prof Exp:* Asst, Purdue Univ, 37-41; res chemist, Va Smelting Co, 42; res fel chem, 42-49, from instr to assoc prof, 49-83, admin asst to head, 49-68, asst to head dept, 68-83, EMER PROF CHEM, PURDUE UNIV, 83- *Concurrent Pos:* Off Sci Res & Develop fel, 42-43. *Mem:* Am Chem Soc; fel AAAS. *Res:* Dehydrohalogenation; halogenation; fluorine chemistry; recovery of uranium. *Mailing Add:* Dept Chem Purdue Univ West Lafayette IN 47907

WELCHER, RICHARD PARKE, b Hartford, Conn, July 21, 19; m 50; c 5. CHEMISTRY. *Educ:* Trinity Col, Conn, AB, 41; Mass Inst Technol, BS, 43, PhD(org chem), 47. *Prof Exp:* Control supvr anal dept, Tenn Eastman Corp, Tenn, 44-45; sr res chemist, Am Cyanamid Co, Stamford, 47-81; RETIRED. *Mem:* Am Chem Soc; Sigma Xi. *Res:* Heterocyclic compounds; organophosphorus, nitrogen and sulfur chemistry; nitriles, amides, polyamines, cyanogen, s-triazines, dicyandiamide; biocides, ion and electron exchange resins; surfactants; mining chemicals; chemistry of cationic polymers. *Mailing Add:* 16 Watch Tower Lane Old Greenwich CT 06870-1109

WELD, CHARLES BEECHER, b Vancouver, BC, Feb 3, 99; m 30; c 3. HUMAN PHYSIOLOGY, NUTRITION. *Educ:* Univ BC, BA, 22, MA, 24; Univ Toronto, MD, 29. *Hon Degrees:* LLD, Dalhousie Univ, 70. *Prof Exp:* Res asst, Connaught Labs, Univ Toronto, 24, res assoc & lectr physiol, 30-36; prof, 36-69, EMER PROF PHYSIOL, DALHOUSIE UNIV, 69- *Concurrent Pos:* Physiologist, Hosp Sick Children, Toronto, 31-36; mem, bd dirs, NS Mus, 50- *Honors & Awards:* Starr Gold Medal, 33. *Mem:* Am Physiol Soc; Can Physiol Soc (pres, 46); fel Royal Soc Can. *Res:* Diphtheria toxoid; freezing of fish; parathyroid; acute intestinal obstruction; pneumothorax; aqueous humor; lipemia; intestinal secretion; interstitial fluid pressures. *Mailing Add:* 6550 Waegwoltic Ave Halifax NS B3H 2B4 Can

WELDEN, ARTHUR LUNA, b Birmingham, Ala, Jan 27, 27; m 50; c 2. MYCOLOGY. *Educ:* Birmingham-Southern Col, AB, 50; Univ Tenn, MS, 51; Univ Iowa, PhD(bot), 54. *Prof Exp:* Asst prof biol, Milliken Univ, 54-55; from instr to prof bot, 55-79, chmn, dept biol, 85-90, IDA RICHARDSON PROF BOT, TULANE UNIV, 80- *Concurrent Pos:* Am Philos Soc grant, 57, PI NSF grants, 69-73. *Honors & Awards:* Socio Honorario, Soc Mycol Mex. *Mem:* Mycol Soc Am; Sigma Xi; Int Asn Plant Taxon; Asn Trop Biol; AAAS. *Res:* Myxomycetes; tropical fungi; Thelephoraceae. *Mailing Add:* Dept Ecol/ Evolution & Org Biol Tulane Univ New Orleans LA 70118

WELDES, HELMUT H, b Munich, Fed Repub Ger, July 4, 28; US citizen. TECHNICAL MANAGEMENT, CHEMISTRY. *Educ:* Univ Karlsruhe, Fed Repub Ger, BS, 51, MS, 54; Univ Aachen, Fed Repub Ger, PhD(chem), 56. *Prof Exp:* Res assoc, Max Planck Inst Coal Res, 54-57; res & develop group leader, P Q Corp, 57-64, res mgr, 64-66, dir res & develop, 66-72, vpres, P Q Int, Inc, 72-78; res dir in Egypt, Nat Acad Sci, 78-81; owner, Tech Transfer Int, 81-85; PRES, PERMETHYL CORP, 85- *Concurrent Pos:* Partner, R D Systs, 81- *Mem:* Am Chem Soc; Am Inst Chem Engrs; Am Inst Chemists; Ger Chem Soc. *Res:* Small ring and multi-ring organic compounds; organometallics; soluble silicates; finally divided silica; aliphatic and olefinic hydrocarbon oligomers; detergents and cleaners; silicones and related compounds. *Mailing Add:* 514 Charleston Greene Malvern PA 19355

WELDON, EDWARD J, JR, b New York, NY, Apr 8, 38; m 65; c 3. INFORMATION SCIENCE. *Educ:* Manhattan Col, BSEE, 58; Univ Fla, MSEE, 60, PhD(elec eng), 63. *Prof Exp:* Engr, Bell Tel Labs, 63-66; assoc prof, 66-71, PROF ELEC ENG, UNIV HAWAII, 71- *Concurrent Pos:* Pres, Adtech, Inc. *Mem:* AAAS; Inst Elec & Electronics Engrs. *Res:* Information theory; coding error control. *Mailing Add:* Dept Elec Eng Univ Hawaii Manoa 2500 Campus Rd Honolulu HI 96822

WELDON, HENRY ARTHUR, b Atlanta, Ga, July 11, 47; m 69. ELEMENTARY PARTICLE PHYSICS, ASTROPHYSICS. *Educ:* Mass Inst Technol, SB, 69, PhD(physics), 74. *Prof Exp:* Res assoc, Stanford Linear Accelerator Ctr, 74-76; res assoc physics, Univ Pa, 76-79, asst prof, 79-86; ASST PROF PHYSICS, WVA UNIV, 86- *Res:* High temperature quantum field theory; quark-gluon plasma; ultrarelativistic heavy ion collisions; applications to the early universe; unified gauge theories; CP violation. *Mailing Add:* Dept Physics WVa Univ Morgantown WV 26506

WELDON, VIRGINIA V, b Toronto, Ont, Sept 8, 35; US citizen; div; c 2. PEDIATRIC ENDOCRINOLOGY, MEDICAL ADMINISTRATION. *Educ:* Smith Col, AB, 57; State Univ NY Buffalo, MD, 62. *Hon Degrees:* LHD, Rush Univ, 85. *Prof Exp:* Intern, resident & fel, Sch Med, Johns Hopkins Univ & Hosp, 62-67, instr pediat, Univ, 67-68; from instr to prof pediat, Sch Med, Wash Univ, 68-89, asst vchancellor med affairs, 75-81, assoc vchancellor, 81-83, vpres Med Ctr, 81-89, dep vchancellor med affairs, 83; VPRES PUB POLICY, MONSANTO CO, 89- *Concurrent Pos:* Consult, Adv Comt Endocrinol & Metab, Food & Drug Admin, 73-76; mem, State of Mo Health Manpower Planning Task Force, 76-; co-dir div metab & endocrinol, St Louis Children's Hosp, 73- 77; gen clin res ctr adv comt, NIH, 76-80, nat adv res resources coun, 80-84; comnr, St Louis Zool Park & vpres comn, 83-; chmn, Asn Am Med Cols, 85-86; bd dirs, Southwest Bell Corp, 85-91, G D Searle & Co & Nutrasweet Co; adv dir, Monsanto Bd Dirs, mem adv comts, Monsanto Agr Co & Monsanto Chem Co. *Mem:* Inst Med-Nat Acad Sci; fel AAAS; Sigma Xi; Endocrine Soc; Soc Pediat Res; Am Pediat Soc; Asn Am Med Cols. *Res:* Aldosterone secretion in children; disorders of growth and growth hormone secretion in children. *Mailing Add:* Monsanto Co 800 N Lindberg Blvd St Louis MO 63167

WELDON, WILLIAM FORREST, b San Marcos, Tex, Jan 12, 45; US citizen; m 68; c 2. ELECTROMAGNETIC LAUNCH, PULSED POWER. *Educ:* Trinity Univ, Tex, BS, 67; Univ Tex, Austin, MS, 70. *Prof Exp:* Engr, Cameron Iron Works, 67-68; proj engr, Glastron Boat Co, 70-71; chief proj engr, Nalle Plastics Co, 71-73; res engr, 73-77, tech dir, 77-85, DIR CTR ELECTROMECH, UNIV TEX, AUSTIN, 85-, PROF ELEC & MECH ENG, 86- *Concurrent Pos:* Mem, Nat Dept Defense Adv Panel Electromagnetics Tech, 78-83 & Nat Res Coun Comt Mobile Elec Power, 87- *Honors & Awards:* Peter Mark Medal, 86. *Mem:* Inst Elec & Electronics Engrs; Am Soc Mech Engrs; Nat Soc Prof Engrs. *Res:* Development of rotating electrical machines for pulsed power applications; industrial applications of pulsed power; development of electromagnetic launch technology. *Mailing Add:* Ctr Electromech Univ Tex Austin 10100 Burnet Rd Austin TX 78758-4497

WELFORD, NORMAN TRAVISS, b London, Eng, Feb 5, 21; nat US; m 44; c 3. MEDICAL DEVICE REGULATION, LEGAL TESTIMONY. *Educ:* Cambridge Univ, BA, 41, MA & MB, BCh, 45. *Prof Exp:* Intern & resident, Addenbrooks Hosp, Cambridge, Eng, 45-46; res assoc biophys, Univ Western Ont, 55-56; assoc prof psychophysiol, Antioch Col & res assoc psychophysiol-neurophysiol, Fels Res Inst, 56-66; dir biomed eng & fac assoc psychiat, Univ Tex Med Br, Galveston, 66-78; med officer, Ctr for Devices & Radiol Health, Food & Drug Admin, 78-87; RETIRED. *Concurrent Pos:* USPHS res fel, Univ Western Ont, 55-56. *Mem:* Inst Elec & Electronics Engrs; Sigma Xi; Asn Advan Med Instrumentation. *Res:* Automation of data gathering and handling and stimulus presentation; psychophysiology of human sensory-motor performance and fetal heart rate behavior. *Mailing Add:* 19 Lakeshore Dr NE Alberqueque NM 87112

WELGE, HENRY JOHN, b St Louis, Mo, Aug 15, 07; m 40; c 2. FLUID DYNAMICS. *Educ:* Univ Ill, BSc, 29; Calif Inst Technol, MSc, 32, PhD(phys chem), 36. *Prof Exp:* Anal chemist, Richfield Oil Co, Calif, 29-31; res chemist, Tex Co, 37; from instr to asst prof chem, Agr & Mech Col, Tex, 37-44; sr res chemist, Exxon Prod Res Co, Houston, Tex, 44-72. *Concurrent Pos:* Consult petrol prod, 72- *Honors & Awards:* Anthony Incas Gold Medal, Soc Petrol Engrs. *Mem:* Am Inst Mining, Metall & Petrol Eng; Soc Petrol Engrs. *Res:* Photochemistry; reaction kinetics and equilibria and diffusion; capillarity in petroleum production; single phase and multiphase fluid flow in porous media. *Mailing Add:* 286 St Josephs Long Beach CA 90803

WELHAN, JOHN ANDREW, b Winnipeg, Man, Aug 24, 50; m 78; c 2. ISOTOPE GEOCHEMISTRY, GEOTHERMAL CHEMISTRY. *Educ:* Univ Man, BSc, 72; Univ Waterloo, MSc, 74; Univ Calif, San Diego, PhD(geol sci), 81. *Prof Exp:* Staff res assoc III, 74-81, FEL RES GEOLOGIST VI, SCRIPPS INST OCEANOG, 81- *Concurrent Pos:* Consult, Geothermal Div, Union Oil Co & Harding-Lawson Assocs, 80. *Mem:* Am Geophys Union. *Res:* Origins and geochemical significance of hydrothermal methane in mid-ocean ridges and continental geothermal systems; carbon hydrogen and nitrogen isotopic anomalies in hydro-thermal gases and correlations with gas and helium isotope compositions. *Mailing Add:* Dept Earth Sci Mem Univ St Johns NF A1B 3X5 Can

WELIKY, IRVING, b Mt Vernon, NY, Aug 29, 24; m 51; c 4. CLINICAL PHARMACOLOGY, PHARMACOKINETICS. *Educ:* Ill Wesleyan Univ, BS, 48; Columbia Univ, PhD(biochem), 58. *Prof Exp:* Res asst, Sloan-Kettering Inst Cancer Res, 49-52; res fel, Mass Gen Hosp, 57-60, asst biochemist, 60-62; asst prof biochem, Sch Med, Univ Pittsburgh, 62-68; sr res investr, Squibb Inst Med Res, 68-69, res group leader drug metab, 69-73, asst clin pharmacol dir, 73-76, assoc clin pharmacol dir, 76-77; assoc dir, Wallace Labs, 77-79, dir clin pharmacol, 79-89; CONSULT CLIN PHARMACOL, 90- *Concurrent Pos:* Res fel biol chem, Harvard Med Sch, 57-60, res assoc, 60-62; asst prof obstet & gynec & mem grad fac, Univ Pittsburgh, 62-68. *Mem:* Am Soc Clin Pharmacol & Therapeut; Am Col Clin Pharmacol. *Res:* Intermediary metabolism and biosynthesis of nucleic acids, porphyrins and steroid hormones; clinical pharmacology; pharmacokinetics; drug metabolism. *Mailing Add:* 851 Black Canyon Dr Estes Park CO 80527

WELIKY, NORMAN, b New York, NY, Nov 1, 19; m 55; c 2. SEPARATION TECHNIQUES, BIOMOLECULAR PROCESSES. *Educ:* City Col New York, BChE, 39; Polytech Inst Brooklyn, PhD(chem), 57. *Prof Exp:* Chemist, Mineral Pigments Corp, 46-47 & Reichhold Chem, Inc, 50-52; Nuclear Magnetic Resonance res fel chem, Harvard Univ, 56-57; res fel hemoglobins, Dept Chem & Chem Eng, Calif Inst Technol, 57-59, group supvr molecular struct & synthesis, Jet Propulsion Lab, 59-65; mem tech staff, Biosci Dept, TRW Systs, 66-67, asst mgr biosci & electrochem dept, 68-71, sr scientist, 71-75; res scientist div allergy & immunol, dept pediat, Los Angeles County Harbor-Univ Calif Los Angeles Med Ctr, 76-81; asst res scientist, Dept Cytol & Cytogenetics, City of Hope Nat Med Ctr, 83-85; CONSULT, 86- *Concurrent Pos:* Consult, separation techniques, immunochemistry. *Mem:* AAAS; Am Chem Soc; Sigma Xi. *Res:* Physical chemistry; enzyme chemistry; specific insoluble adsorbents; biological specificity; immunology-allergy; environmental quality criteria; cancer; autoimmune disease, phagocytosis; autoimmune disease in mouse model; phagocytics in macrophages, endocytic and exocytic steps. *Mailing Add:* 1072 Ridge Crest St Monterey Park CA 91754

WELKER, EVERETT LINUS, b Greenview, Ill, Mar 30, 11; m 33; c 2. MATHMATICAL STATISTICS. *Educ:* Univ Ill, AB, 30, AM, 31, PhD(math statist), 38. *Prof Exp:* Asst math, Univ Ill, 35-38, instr, 38-42, assoc, 42-44, from asst prof to assoc prof, 44-47, assoc math, Bur Med Econ Res, AMA, 47-52; statistician, Dept Defense, 52-57; mgr adv studies dept, ARINC Res Corp, 57-63; mgr syst effectiveness anal prog, Tempo, Gen Elec Ctr Advan Studies, Calif, 63-71; staff scientist, TRW Systs, 71-76; CONSULT, 76- *Mem:* Inst Math Statist. *Res:* Correlation theory; vital and engineering statistics; biometrics; reliability. *Mailing Add:* 1171 Northwood Lake Northport AL 35476-1901

WELKER, GEORGE W, b Cumberland City, Tenn, July 4, 23; m 50; c 3. PARASITOLOGY, BACTERIOLOGY. *Educ:* Mid Tenn State Univ, BS, 44; George Peabody Col, MA, 50; Ohio State Univ, PhD, 62. *Prof Exp:* Teacher high sch, Ohio, 46-49; asst biol, George Peabody Col, 49-50; from asst prof to assoc prof, Ball State Univ, 50-65, prof biol, 65-86, chmn, Biol Dept 80-86; RETIRED. *Concurrent Pos:* Danforth teacher study grant, 57-58. *Mem:* AAAS; Am Sci Affil; Sigma Xi. *Res:* Helminth parasites and microbial physiology. *Mailing Add:* 1732 N Colson Dr Muncie IN 47304

WELKER, J(OHN) REED, b Rexburg, Idaho, Dec 1, 36; m 58; c 3. CHEMICAL ENGINEERING. *Educ:* Univ Idaho, BS, 59, MS, 61; Univ Okla, PhD(chem eng), 65. *Prof Exp:* Instr, Univ Idaho, 60-61; group leader, Oil Recovery Corp, 62-63; res engr, Res Inst, Univ Okla, 65-70; vpres, Univ Engrs, Inc, 70-77; pres, Appl Technol Corp, 77-83; PROF ENG, UNIV ARK, 83- *Mem:* Am Inst Chem Engrs; Am Chem Soc; Combustion Inst; Am Gas Asn. *Res:* Fire research; fire safety; atmospheric dispersion; liquefied natural gas plant safety; fire extinguishment and control. *Mailing Add:* PO Box 116 Fayetteville AR 72702-0116

WELKER, NEIL ERNEST, b Batavia, NY, Apr 21, 32; m 76; c 6. BIOCHEMISTRY. *Educ:* Univ Buffalo, BA, 58; Western Reserve Univ, PhD(microbiol), 63. *Prof Exp:* Fel, Univ Ill, Urbana, 63-64; from asst prof to assoc prof biol sci, 64-74, PROF BIOCHEM, MOLECULAR & CELL BIOL, NORTHWESTERN UNIV, 74- *Mem:* AAAS; Am Soc Microbiol; Brit Soc Gen Microbiol; Am Soc Biol Chemists. *Mailing Add:* Dept Biochem Molecular Biol & Cell Biol Northwestern Univ 633 Clark St Evanston IL 60201

WELKER, WALLACE I, b Batavia, NY, Dec 17, 26; c 3. NEUROPHYSIOLOGY, NEUROANATOMY. *Educ:* Univ Chicago, PhD(psychol), 54. *Prof Exp:* Asst, Yerkes Labs Primate Biol, 52-54; PROF NEUROPHYSIOL, MED SCH, UNIV WIS-MADISON, 67- *Concurrent Pos:* NIH fel neurophysiol, Univ Wis-Madison, 54-56, Sister Kenny Found scholar, 57-62, NIH career develop fel, 62-67. *Mem:* Am Asn Anatomists; Soc Neurosci. *Res:* Comparative neurobiology; neurophysiology and neuroanatomy; analysis of somato sensory circuits of mammalian brain. *Mailing Add:* Dept Neurophysiol 281 Med Sci Bldg Med Sch Univ Wis 1300 University Madison WI 53706

WELKER, WILLIAM V, JR, b Milwaukee, Wis, Nov 12, 28; m 51; c 4. WEED SCIENCE, HORTICULTURE. *Educ:* Univ Wis, BS, 52, PhD(hort, plant physiol), 62. *Prof Exp:* RES HORTICULTURIST, SCI & EDUC ADMIN-AGR RES, USDA, 59-; RES PROF SOILS & CROPS, RUTGERS UNIV, NEW BRUNSWICK, 59- *Mem:* Weed Sci Soc Am; Am Soc Hort Sci. *Res:* Chemical control of weeds in horticultural crops; influence of long term use of herbicides upon crops; fate of herbicides in soil. *Mailing Add:* USDA-Agr Res Serv Appalachian Fruit Res Sta Rte 2 Box 45 Kearneysville WV 25430

WELKIE, GEORGE WILLIAM, b Hazleton, Pa, Apr 11, 32; m 57; c 2. VIROLOGY, PLANT PHYSIOLOGY. *Educ:* Pa State Univ, BS, 52, MS, 54; Univ Wis, PhD(plant path), 57. *Prof Exp:* Asst prof, 57-62, ASSOC PROF BOT & PLANT PATH, UTAH STATE UNIV, 62- *Concurrent Pos:* NSF fel virol, Rothamsted Exp Sta, Eng. *Mem:* Am Phytopath Soc; Sigma Xi. *Res:* Virus infection and synthesis; effect of virus infection on host metabolism; mineral nutrition of plants with relation to metabolism. *Mailing Add:* Dept Bot Utah State Univ Logan UT 84321-5305

WELKOWITZ, WALTER, b Brooklyn, NY, Aug 3, 26; m 51; c 3. ELECTRICAL ENGINEERING. *Educ:* Cooper Union, BS, 48; Univ Ill, MS, 49, PhD, 54. *Prof Exp:* Res assoc elec eng, Bioacoust Lab, Univ Ill, 48-54; lectr acoust, Grad Sch Eng, Columbia Univ, 54-55; gen mgr, Gulton Industs, Inc, 55-64; prof elec eng, 64-86, chmn dept, 72-86, PROF BIOMED ENG, CHEM DEPT, RUTGERS UNIV, NEW BRUNSWICK, 86- *Concurrent Pos:* Adj prof, Rutgers Med Sch, Col Med & Dent, NJ, 72- *Honors & Awards:* Centennial Medal, Inst Elec & Electronics Engrs, 84. *Mem:* Fel Inst Elec & Electronics Engrs; Am Soc Artificial Internal Organs; Soc Math Biol; NY Acad Sci. *Res:* Biomedical engineering; instrumentation. *Mailing Add:* Dept Biomed Eng Rutgers Univ PO Box 909 Piscataway NJ 08854

WELLAND, GRANT VINCENT, b Toronto, Ont, June 25, 40; m 66. MATHEMATICS. *Educ:* Purdue Univ, BS, 63, MS, 65, PhD(math), 66. *Prof Exp:* Teaching asst math, Purdue Univ, 63-65; from asst prof to assoc prof, 66-77, PROF MATH & COORDR PROBS & STATIST, UNIV MO, ST LOUIS, 77- *Concurrent Pos:* Dir, NSF undergrad res prog, 67-68; NSF res grant, 68-70; vis prof, Univ Madrid, 70. *Mem:* AAAS; Am Math Soc. *Res:* Harmonic analysis and differentiation. *Mailing Add:* Dept Math Univ Mo 8001 Nat Bridge Rd St Louis MO 63121

WELLAND, ROBERT ROY, b Toronto, Ont, Jan 31, 33; m 57; c 3. MATHEMATICS. *Educ:* Univ Okla, BS, 56, MA, 57; Purdue Univ, PhD, 60. *Prof Exp:* Instr math, Univ Okla, 56-57; NSF asst, Purdue Univ, 57-58, from instr to asst prof, 58-60; asst prof, Ohio State Univ, 60-63; asst prof, 63-74, ASSOC PROF MATH, NORTHWESTERN UNIV, EVANSTON, 74- *Concurrent Pos:* Vis instr, Univ Chicago, 61. *Mem:* Am Math Soc. *Res:* Functional analysis; special spaces; nonlinear analysis. *Mailing Add:* Dept Math Northwestern Univ 633 Clark St Evanston IL 60208

WELLDON, PAUL BURKE, b Nashua, NH, May 10, 16; m 41; c 2. POLYMER CHEMISTRY, RESEARCH ADMINISTRATION. *Educ:* Dartmouth Col, AB, 37, AM, 39; Univ Ill, PhD(org chem), 42. *Prof Exp:* Instr chem, Dartmouth Col, 37-39; res chemist, Hercules Inc, 42-51, mgr, Cellulose Prod Res Div, 52-60, tech asst to dir res, 60-62, mgr, Appln Res Div, 62-65, mgr, Personnel Div, Res Ctr, 67-79; RETIRED. *Mem:* Am Chem Soc. *Res:* Chlorination of organic compounds; chemistry of cellulose derivatives; polyolefins; polymer applications; organic chemistry. *Mailing Add:* Harbor Lights CR344 Islesboro ME 04848

WELLE, STEPHEN LEO, ENERGY METABOLISM, NOREPINEPHRINE METABOLISM. *Educ:* Northern Ill Univ, PhD(bio-psychol), 78. *Prof Exp:* ASST PROF MED, SCH MED & DENT, UNIV ROCHESTER, 78- *Mailing Add:* Dept Med & Physiol Monroe Community Hosp Endocrine Unit Rochester NY 14620

WELLENREITER, RODGER HENRY, b Bloomington, Ill, Oct 23, 42; m 63; c 2. POULTRY NUTRITION. *Educ:* Ill State Univ, BS, 64; Mich State Univ, MS, 67, PhD(animal husb), 70. *Prof Exp:* Sr scientist, 70-80, RES SCIENTIST POULTRY NUTRIT, LILLY RES LABS, ELI LILLY & CO, 80- *Concurrent Pos:* Sigma Xi res award, 70. *Mem:* Poultry Sci Asn; World Poultry Sci Asn; Sigma Xi. *Res:* Means of improving the efficiency of conversion of animal feedstuffs into products for human comsumption. *Mailing Add:* Eli Lilly Co Greenfield Lab Greenfield IN 46140

WELLER, CHARLES STAGG, JR, b Nashville, Tenn, Dec 28, 40; m 66; c 2. SPACE SCIENCE, RADAR. *Educ:* Mass Inst Technol, BS, 62; Univ Pittsburgh, PhD(physics), 67. *Prof Exp:* Res physicist, Naval Res Lab, 67-86, sect head, 74-86; SR RES SCIENTIST, DAVID TAYLOR RES CTR, 86- *Mem:* Am Phys Soc; Am Geophys Union; Am Astron Soc; Int Astron Union. *Res:* Space science; upper atmospheric studies; interplanetary medium; ultraviolet optics; ocean sciences; electromagnetic signatures. *Mailing Add:* Code 1402 David Taylor Res Ctr Bethesda MD 20084-5000

WELLER, DAVID LLOYD, b Munfordville, Ky, Sept 28, 38; m 68; c 1. BIOCHEMISTRY, MOLECULAR BIOLOGY. *Educ:* Rochester Inst Technol, BS, 62; Iowa State Univ, PhD(bio-chem), 66. *Prof Exp:* Res fel molecular biol, Children's Cancer Res Found, Boston, Mass, 66-67; asst prof agr biochem, 67-71, chmn biol sci prog, 71-75, chmn cell biol, 72-75, asst dean, Col Agr & assoc dir, Agr Exp Sta, 75-77, assoc prof, 71-77, PROF BIOCHEM & MICROBIOL, UNIV VT, 77- *Mem:* Am Chem Soc; Biophys Soc; NY Acad Sci; Am Inst Chem; Soc Protozoologists. *Res:* Ribosomes and RNAases of Entamoeba; isoelectric focusing of proteins. *Mailing Add:* 369 S Union Burlington VT 05401

WELLER, EDWARD F(RANK), JR, b Baltimore, MD, Nov 30, 1919; m 43; c 3. ELECTRONICS ENGINEERING. *Educ:* Univ Cincinnati, EE, 43. *Prof Exp:* Res engr instrumentation, Gen Motors Res Labs, 46-52, asst head, Physics Dept, 52-62, head, Electronics & Instrumentation Dept, 62-74, head, Electronics Dept, 74-82; RETIRED. *Mem:* Fel Inst Elec & Electronics Engrs; Sigma Xi. *Res:* Ferroelectricity. *Mailing Add:* 14230 Boliver Dr Sun City AZ 85351

WELLER, GLENN PETER, b New Orleans, La, Dec 7, 43; m 69; c 2. MATHEMATICS. *Educ:* Tulane Univ, BA, 64; Univ Chicago, SM, 65, PhD(math), 68. *Prof Exp:* Asst prof math, Roosevelt Univ, 68-69 & Univ Ill, Chicago Circle, 69-75; ASST PROF, KENNEDY-KING COL, 75- *Mem:* Math Asn Am. *Res:* Geometric topology. *Mailing Add:* Dept Data Processing Harold Washington Col 30 E Lake St Chicago IL 60601

WELLER, GUNTER ERNST, b Haifa, June 14, 34; m 63; c 3. POLAR METEOROLOGY. *Educ:* Univ Melbourne, BSc, 62, MSc, 65, PhD(meteorol), 67. *Prof Exp:* Meteorologist, Commonwealth Bur Meteorol, 60-61; glaciologist, Australian Nat Antarctic Res Exped, 62-66; NSF res fel, Univ Melbourne, 67; from asst prof to assoc prof geophys, 68-73, assoc dir planning, 78-83, PROF GEOPHYS, GEOPHYS INST, UNIV ALASKA, 73-, DEP DIR, 84- *Concurrent Pos:* Prog mgr polar meteorol, Off Polar Progs, NSF, 72-74; US rep, working group meteorol, Sci Comt Antartic Res, 74-; proj mgr, Arctic Proj Off, Outer Continental Shelf, Nat Oceanic & Atmospheric Admin, 75-81; mem, Polar Res Bd, Nat Acad Sci, 75-79, chmn, 85-; pres, Int Comn Polar Meterol, Int Union Geol & Geophys, 80-83; mem, Climate Res Comt, Nat Acad Sci, 81-84; dir, NASA-Univ Alaska, SAR Facil, 86-, NOAA Panel on Climate & Global Change, 87-, Ctr Global Change & Arctic Syst Res, 90- *Honors & Awards:* Polar Medal, Commonwealth of Australia, 69; Antarctic Serv Medal, US Govt, 74. *Mem:* Am Meteorol Soc; Am Geophys Union; Arctic Inst NAm; Int Glaciol Soc; Int Arctic Comt. *Res:* Polar meteorology and climatology; micrometeorology; glacio-meteorology; problems of resource recovery and remote sensing in polar regions. *Mailing Add:* Geophys Inst Univ Alaska Fairbanks AK 99775-0800

WELLER, HENRY RICHARD, b East Rutherford, NJ, Mar 15, 41; m 64; c 3. EXPERIMENTAL NUCLEAR PHYSICS. *Educ:* Fairleigh Dickinson Univ, BS, 62; Duke Univ, PhD(nuclear physics), 68. *Prof Exp:* Fel physics, Univ Fla, 67-68, from asst prof to prof, 68-80; PROF PHYSICS, DUKE UNIV, 80- *Mem:* Am Phys Soc. *Res:* Experimental nuclear structure studies, especially capture reactions involving polarized and unpolarized projectiles; studies of D-state effects in A = 3 and 4 nuclei using tensor polarized deuteron capture reactions. *Mailing Add:* Dept Physics Duke Univ Durham NC 27706

WELLER, JOHN MARTIN, internal medicine, nephrology; deceased, see previous edition for last biography

WELLER, LAWRENCE ALLENBY, b New York, NY, July 3, 27; m 57; c 2. ELECTROMAGNETIC PULSE, AIRCRAFT ELECTRONICS. *Educ:* City Col New York, BS, 49; Univ Toronto, MA, 52; Ohio State Univ, PhD(physics), 73. *Prof Exp:* Physicist, Warner & Swasey Res Corp, 53-56, Foster Wheller Corp, 56-68; sr res physicist, Monsanto Co, 58-71; res asst & fel, Dept Physics, Ohio State Univ, 71-74; engr, Cincinnati Electronics Corp, 74-75; physicist, Kornylak Corp, 75-76; sr scientist, Pedco Environ, Inc, 76-77; eng technician, Nat Inst Occup Safety & Health, 77-80; SPECIALIST ENGR, BOEING CO, 80- *Concurrent Pos:* Res assoc, Dept Physics, Ohio State Univ, 74-80. *Mem:* Am Phys Soc; Am Nuclear Soc; Inst Elec & Electronics Engrs. *Res:* Infrared spectra of molecules; lasers; materials handling machinery; isotope separation; nuclear engineering; protective equipment for workers; electromagnetic pulse effects on electrical and electronic systems; radar. *Mailing Add:* 8612 Chalet Dr Wichita KS 67207

WELLER, LOWELL ERNEST, b Continental, Ohio, Apr 17, 23; m 44; c 2. ORGANIC CHEMISTRY, BIOCHEMISTRY. *Educ:* Bowling Green State Univ, BS, 48; Mich State Univ, MS, 51, PhD(chem), 56. *Prof Exp:* Asst chem, Bowling Green State Univ, 46-48; from instr to asst prof biochem, Mich State Univ, 48-57; from assoc prof to prof, 57-89, head dept, 57-89, EMER PROF CHEM, UNIV EVANSVILLE, 89- *Concurrent Pos:* NSF, AEC & Energy Res & Develop Admin grants. *Mem:* AAAS; Am Chem Soc; Royal Soc Chem; Sigma Xi. *Res:* Organic synthesis, reaction mechanisms and structure determination especially by spectroscopic methods. *Mailing Add:* Dept Chem Univ Evansville Evansville IN 47722

WELLER, MILTON WEBSTER, b St Louis, Mo, May 23, 29; m 47; c 1. ECOLOGY, FISH & WILDLIFE SCIENCES. *Educ:* Univ Mo, AB, 51, MA, 54, PhD(zool), 56. *Prof Exp:* Instr zool, Univ Mo, 56-57; from asst prof to prof, Iowa State Univ, 57-74, chmn fisheries & wildlife sect, 67-74; prof & head Dept Entom, Fisheries & Wildlife, Univ Minn, St Paul, 74-82; KLEBERG CHAIR PROF, WILDLIFE ECOL, TEX A&M UNIV, 82- *Concurrent Pos:* NSF res grants, 64-65, 70-74 & 76-77, Dept Interior res grants, 76-81, 86-92. *Mem:* Fel AAAS; Ecol Soc Am; Cooper Ornith Soc; Wildlife Soc; fel Am Ornith Union; Soc Wetland Sci. *Res:* Wetland and waterbird ecology, including habitat restoration and evaluation. *Mailing Add:* Dept Wildlife & Fisheries Sci Tex A&M Univ College Station TX 77843

WELLER, PAUL FRANKLIN, b Kankakee, Ill, Aug 30, 35; m 58; c 2. PHYSICAL CHEMISTRY. *Educ:* Univ Ill, BS, 57; Cornell Univ, PhD(chem), 62. *Prof Exp:* Res scientist, T J Watson Res Ctr, Int Bus Mach Corp, 61-65; from asst prof to prof chem, State Univ NY Col Fredonia, 65-75, chmn dept, 74-75; prof chem & dean arts & sci, Western Ill Univ, 75-79; acad vpres, Calif Polytech Univ, Pomona, 79-85; PRES, FRAMINGHAM STATE COL, 85- *Mem:* Am Asn Higher Educ; Am Chem Soc; Sigma Xi. *Res:* Wide band gap semiconductors; defect oxides; solid state materials characterization; crystal growth. *Mailing Add:* Framingham State Col 100 State St Framingham MA 01701

WELLER, RICHARD IRWIN, b Newark, NJ, Mar 3, 21; m 53; c 2. PHYSICS, ACADEMIC ADMINISTRATION. *Educ:* City Col New York, BEE, 44; Union Col, BS, 48; Fordham Univ, MS, 50, PhD(physics), 53. *Prof Exp:* Elec engr, NAm Phillips Co, 44, Guy F Atkinson Co, 44-45, NAm Phillips Co, 45-46, Crow, Lewis & Wick, 47, Edward E Ashley, 48, Allied Processes Co, 50, V L Falotico & Assocs, 50-51 & Singmaster & Breyer, 51; instr physics, Brooklyn Col, 52-53; asst prof, State Univ, NY Maritime Col, 53-54; med physicist, Brookhaven Nat Lab, 54-57; prof physics, Franklin & Marshall Col, 57-70, chmn dept, 57-61; prof physics & dean, Sch Sci & Math, Edinboro State Col, 70-81; CONSULT, 81- *Concurrent Pos:* Instr, Manhattan Col, 49-50, Fordham Univ, 50-52, Newark Col Eng, 52-53, Broward Community Col, 81-83, Brevard Community Col, 83-84 & Fla Inst Technol, 84-85; consult, Brookhaven Nat Lab, Fairchild Camera & Instrument Corp & Nuclear Sci & Eng Corp; mem Nat Res Coun subcomt, Nat Acad Sci. *Mem:* AAAS; Am Asn Univ Prof; Inst Elec & Electronics Engrs; Sigma Xi; Am Soc Eng Educ; Health Physics Soc; Am Asn Physics Teachers; Am Physics Soc. *Res:* Atmospheric electricity; radioactivity; radioisotopes dosimetry and instrumentation; biophysics. *Mailing Add:* 750 N Atlantic Ave-PH-1 Cocoa Beach FL 32931-3154

WELLER, ROBERT ANDREW, b Boston, Mass, July 27, 50; m 72; c 2. OCEANOGRAPHY. *Educ:* Harvard Univ, BA, 72; Univ Calif, San Diego, PhD(oceanog), 78. *Prof Exp:* Res asst, Harvard Univ, 70-72; Res asst, Scripps Inst Oceanog, 72-78, res oceanographer, Woods Hole Oceanog Inst, 78-79; scholar, 79-80, asst scientist, 80-84, assoc scientist, 84-88, ASSOC SCIENTIST WITH TENURE, WOODS HOLE OCEANOG INST, 88- *Honors & Awards:* James B Macelwane Award, Am Geophys Union, 86. *Mem:* AAAS; fel Am Geophys Union; Am Meteorol Soc. *Res:* Experimental work in observing and understanding the response of the upper ocean to atmospheric forcing. *Mailing Add:* Clark Lab Woods Hole Oceanog Inst Woods Hole MA 02543

WELLER, S(OL) W(ILLIAM), b Detroit, Mich, July 27, 18; m 43; c 4. PHYSICAL CHEMISTRY, CHEMICAL ENGINEERING. *Educ:* Wayne State Univ, BS, 38; Univ Chicago, PhD(chem), 41. *Prof Exp:* Fel, NY Univ, 41; res assoc, Nat Defense Res Comt, Chicago, 41-43; proj leader, Gen Foods Corp, 43-44; res scientist, Manhattan Proj, Columbia Univ, 44-45; phys chemist, US Bur Mines, Pa, 45-47, asst chief, Coal Hydrogenation Sect, 47-50; chief, Fundamental Res Sect, Houdry Process Corp, 50-58; mgr, Propulsion Dept Aeronutronic Div, Ford Motor Co, 58-60, mem sr staff, 60, mgr chem & mat, 60-63, dir, Chem Lab, 63-65; actg chmn dept chem eng, State Univ NY, 68-69, 72-73 & 76-77, prof chem eng, 65-89, C C Furnas Mem prof, 83-89, EMER PROF CHEM ENG, STATE UNIV NY, BUFFALO, AMHERST, 89- *Concurrent Pos:* Civilian, AEC, 44; vis prof, Univ Calif, Berkeley, 67; UN tech expert, Israel, 71-72; sr Fulbright lectr, Univ Madrid, 75, Istanbul Tech Univ, 81. *Honors & Awards:* H H Storch Award, 81; E V Murphree Award, 82; J F Schoellkopf Medal, 84. *Mem:* AAAS; Am Chem Soc; Am Inst Chem Engrs. *Res:* Catalysis; kinetics and mechanisms of catalytic reactions; synthetic liquid fuels. *Mailing Add:* Dept Chem Eng State Univ NY at Buffalo Amherst NY 14260

WELLER, THOMAS HUCKLE, b Ann Arbor, Mich, June 15, 15; m 45; c 4. TROPICAL MEDICINE, INFECTIOUS DISEASES. *Educ:* Univ Mich, AB, 36, MS, 37; Harvard Med Sch, MD, 40. *Hon Degrees:* LLD, Univ Mich, 56; ScD, Gustavus Adolphus Col, 75; LHD, Lowell Univ, 77; ScD, Univ Mass, 85. *Prof Exp:* Res fel comp path & trop med, Harvard Med Sch, 40, teaching fel bact, 40-41, Milton fel pediat, 47-48; USPHS fel, 48; instr comp path & trop med, 48-49, from asst prof to assoc prof trop pub health, 49-54, head dept, 54-81, Strong prof, 54-85, dir, Ctr Prev Infectious Dis, 66-81, EMER PROF TROP PUB HEALTH, SCH PUB HEALTH, HARVARD UNIV, 85- *Concurrent Pos:* Intern, Children's Hosp, 41-42, asst res, 46, asst dir res, Div Infectious Dis, Children's Med Ctr, 49-56, assoc physician, 49-55; area consult, US Vet Admin, 49-64; consult & mem trop med & parasitol study sect, USPHS, 53-55; dir, Comn Parasitic Dis, Armed Forces Epidemiol Br, 53-59, mem, 59-72. *Honors & Awards:* Nobel Prize Physiol & Med, 54; Mead Johnson Award, Am Acad Pediat, 53; Kimble Methodol Award, 54; Ledlie Prize, 63; Weinstein Award, 73; Bristol Award, Infectious Dis Soc Am, 80. *Mem:* Nat Acad Sci; Am Soc Trop Med & Hyg (pres, 64); Asn Am Physicians; hon mem Royal Soc Trop Med & Hyg, 87. *Res:* In vitro cultivation of viruses, especially poliomyelitis, mumps and Coxsackie viruses; etiology of epidemic pleurodynia, varicella and Herpes zoster; cytomegalic inclusion and Rubella; helminth infections, especially schistosomiasis and enterobiasis. *Mailing Add:* 56 Winding River Rd Needham MA 02192

WELLERSON, RALPH, JR, b New York, NY, Dec 12, 24; m 51; c 2. MICROBIOLOGY. *Educ:* Hobart Col, BS, 49; Rutgers Univ, MS, 51; Purdue Univ, PhD(bact), 54. *Prof Exp:* Sr scientist, 54-62, RES FEL MICROBIOL, DIV MICROBIOL, ORTHO RES FOUND, ORTHO PHARMACEUT CORP, RARITAN, NJ, 62- *Mem:* Am Soc Microbiol. *Res:* Metabolism and nutrition of microorganisms; microbial fermentations; chemotherapy of Trichomonas vaginalis; immunological control of fertility. *Mailing Add:* 14 Oxbow Lane Basking Ridge NJ 07920

WELLES, HARRY LESLIE, b Rockville, Conn, Jan 2, 45; m 66; c 1. PHARMACEUTICS. *Educ:* Cent Conn State Col, BS, 67; Univ Conn, MS, 69, PhD(pharmaceut), 74. *Prof Exp:* Res assoc pharmaceut, Sch Pharm, Univ Wis-Madison, 72-74; asst prof pharmaceut, Col Pharm, Dalhousie Univ, 74-80; res scientist, 80-84, GROUP LEADER, NORWICH-EATON PHARMACEUT, NORWICH, NY, 84- *Mem:* Acad Pharmaceut Sci; Am Asn Pharmaceut Sci. *Res:* Parenteral formulations; dissolution; drug-excipient interactions; controlled release dosage formulations. *Mailing Add:* Norwich-Eaton Pharmaceut Box 191 Norwich NY 13815

WELLES, SAMUEL PAUL, b Gloucester, Mass, Nov 9, 07; wid; c 3. VERTEBRATE PALEONTOLOGY. *Educ:* Univ Calif, AB, 30, PhD(vert paleont), 40. *Prof Exp:* Field & lab asst, 31-39, asst cur reptiles & amphibians, 39-42, sr mus cur, 42-46, prin mus paleontologist & lectr, 46-76, actg dir, 47-48, RES ASSOC, MUS PALEONT, UNIV CALIF, BERKELEY, 76- *Concurrent Pos:* Mem, US Nat Mus exped, 30 & Univ Calif expeds, 31-42, 45, 47, 49, 56, 58, 64-68, 72, 73, 76, 81; Fulbright scholar, Univ Canterbury, 69-70. *Mem:* Fel Geol Soc Am; Paleont Soc; Soc Vert Paleont. *Res:* Triassic labyrinthodonts; Cretaceous Plesiosaurs; dinosaurs. *Mailing Add:* Mus Paleont Univ Calif Berkeley CA 94720

WELLING, DANIEL J, b Kansas City, Mo, May 1, 37; m; c 3. THEORETICAL PHYSICS, BIOPHYSICS. *Educ:* Rockhurst Col, BS, 58; St Louis Univ, PhD(physics), 63. *Prof Exp:* Asst prof physics, Southern Ill Univ, 64; AEC fel theoret physics, Argonne Nat Lab, 64-66; asst prof & assoc scientist, Univ Wis-Milwaukee, 66-72; vis assoc prof path, 72-73, assoc prof path & physiol, 73-80, RES PROF PATH & PHYSIOL, MED CTR, UNIV KANS, 80- *Concurrent Pos:* Assoc scientist, McDonnell Aircraft Corp, 63; res scientist, VA Med Ctr, Kans City, Mo. *Mem:* Am Phys Soc; Am Soc Nephrology. *Res:* Medical physics; mathematical modeling of transport in the nephron. *Mailing Add:* Dept Physiol Univ Kans Med Ctr Kansas City KS 66103

WELLING, LARRY WAYNE, RENAL PHYSIOLOGY, PATHOLOGY. *Educ:* Univ Kans, PhD(physiol), 72. *Prof Exp:* STAFF PATHOLOGIST, VET ADMIN MED CTR, KANSAS CITY, MO, 72- *Mailing Add:* 220 Westover Rd Kansas City MO 64113

WELLINGTON, GEORGE HARVEY, b Springport, Mich, Sept 19, 15; m 39; c 3. ANIMAL SCIENCE. *Educ:* Mich State Univ, BS, 37, PhD, 54; Kans State Univ, MS, 40. *Prof Exp:* Asst, Kans State Univ, 38-40; agr exten agent, Mich, 45-46; from asst prof to prof animal sci, 47-77, EMER PROF ANIMAL, CORNELL UNIV, 78- *Concurrent Pos:* Ford Found consult livestock in Syria, 62; vis prof, Univ Aleppo, Syria, 65-66; consult, Fed Univ Minas Gerias, Brazil, 78, Govt Botswana, 79, Govt Malawi, 82, Va State Univ, 84, & Govt Bangladesh, 85- *Honors & Awards:* Signal Serv Award, Am Meat Sci Asn, 66. *Mem:* Am Soc Animal Sci; Am Meat Sci Asn. *Res:* Meat animal growth development and composition; meat quality factors and processing. *Mailing Add:* Morrison Hall Cornell Univ Ithaca NY 14850

WELLINGTON, JOHN SESSIONS, b Glendale, Calif, Sept 28, 21; m 44; c 2. PATHOLOGY. *Educ:* Univ Calif, Berkeley, AB, 42; Univ Calif, San Francisco, MD, 45. *Prof Exp:* Instr path, Sch Med, Univ Calif, 53-55, lectr, 55-56; vis asst prof, Univ Indonesia, 56-58; assoc prof path, Sch Med, Univ Calif, San Francisco, 58-69, prof, 69-77, assoc dean, 65-77; PROF PATH & ASSOC DEAN, JOHN A BURNS SCH MED, UNIV HAWAII, HONOLULU, 77- *Mem:* Fel Col Am Path; Am Soc Exp Path; Int Acad Path. *Res:* Leukemia; injury and repair in hematopoietic tissue. *Mailing Add:* Dept Path Univ Calif San Francisco 11980 Henno Rd Glen Ellen CA 95442

WELLINGTON, WILLIAM GEORGE, b Vancouver, BC, Aug 16, 20; m 59; c 2. INSECT ECOLOGY, BIOMETEOROLOGY. *Educ:* Univ BC, BA, 41; Univ Toronto, MA, 45, PhD(zool, entom), 47. *Prof Exp:* Meteorol officer, Meteorol Serv, Can Dept Transport, 42-45; forest biologist, Forest Biol Div, Can Dept Agr, 46-60; head bioclimat sect, Can Dept Forestry, 60-64, prin scientist insect ecol, 64-68; prof ecol, Univ Toronto, 68-70; dir, Inst Animal Resource Ecol, Univ BC, 73-79, prof plant sci & resource ecol, 70-86, chmn, Resource Ecol & Planning Coun, 76-78, EMER PROF, UNIV BC, 86- *Concurrent Pos:* Vis prof, NC State Univ, 72, 75, 81, San Diego State Univ, 75, Laval Univ, 81, Univ Calgary, 83 & Simon Fraser Univ, 87; mem, Exec Biol Coun Can, 77-78; Killam sr res fel, 80-81; grad prog appraiser, Ont Coun Grad Studies, 82; assoc ed, Can J Zool, 83-86. *Honors & Awards:* Gold Medal for Outstanding Achievement in Can Entom, Entom Soc Can, 68; Outstanding Achievement Award Bioclimat, Am Meteorol Soc, 69; C J Woodworth Award, Entom Soc Am, 79. *Mem:* Am Meteorol Soc; Can Soc Zoologists; Entom Soc Am; fel Entom Soc Can (pres, 77-78); Japan Soc Pop Ecol; fel Explorer Club; fel Royal Soc Can. *Res:* Micro- and synoptic climatology; interdisciplinary associations of meteorology and the life sciences; population biology; physiological, behavioral and social variations affecting population dynamics; insect behavior; human ecology. *Mailing Add:* Resource Ecol Univ BC Vancouver BC V6T 1W5 Can

WELLIVER, ALBERTUS, ENGINEERING ADMINISTRATION. *Prof Exp:* SR VPRES ENG & TECHNOL, BOEING CO, 90- *Mem:* Nat Acad Eng. *Mailing Add:* Boeing Co PO Box 3707 MS 13-35 Seattle WA 98124

WELLIVER, PAUL WESLEY, b Danville, Pa, Mar 28, 31; m 56; c 2. SCIENCE EDUCATION. *Educ:* Western Md Col, BA, 52; Pa State Univ, MEd, 58, PhD(sci educ), 65. *Prof Exp:* Teacher gen sci, Roosevelt Jr High, Pa, 55-58; lectr nuclear energy, Oak Ridge Inst Nuclear Studies, 59-60; TV studio teacher phys sci, NC In-Sch TV, 61-66; consult sci educ, NC Dept Educ, 66-67; dir educ instr TV, Miss Authority Educ TV, 67-69; PROF EDUC SCI EDUC, PA STATE UNIV, 69-, DIR, REGIONAL COMPUT RESOURCE CTR, 84- *Concurrent Pos:* NSF fel, Pa State Univ, 58-59; consult nat proj improv TV instr, Nat Asn Educ Broadcasters, 64-65 & invest sci proj, Pa Dept Educ, 70-78; dir, Comprehensive Sch Improv ITV Proj, 66-67, Miss ITV Curric Lab, 67-68, instr develop inst proj, Pa State Univ, 73-77 & sci ITV proj, Pa Dept Educ, 72-78; pres, Asn Educ Commun & Technol, 83-84. *Mem:* AAAS; Nat Asn Res Sci Teaching; Nat Sci Teachers Asn; Nat Soc Study Educ; Asn Educ Commun & Technol. *Res:* The application of communications media and technology to the development, dissemination, implementation and maintenance of science instruction over a large geographic region. *Mailing Add:* Dept Curric Instr Pa State Univ 166 Chambers Bldg University Park PA 16802

WELLMAN, ANGELA MYRA, b Sudbury, Suffolk, Eng, June 13, 35; m 59. MICROBIAL PHYSIOLOGY. *Educ:* Bristol Univ, BSc, 56, PhD(mycol), 61. *Prof Exp:* Demonstr bot, Bristol Univ, 56-59; Nat Res Coun Can fel, Univ Western Ont, 59-61, instr, 61-62, lectr, 62-64, asst prof, 64-68, asst dean sci, 74-75, assoc dean sci, 75-82, assoc prof bot, 68-82; RETIRED. *Mailing Add:* Spring St Po Box 112 Port Stanley ON N0L 2A0 Can

WELLMAN, DENNIS LEE, b Freeport, Ill, Mar 18, 42; m 68; c 2. SYSTEMS ENGINEERING, RESEARCH PLATFORMS. *Educ:* Univ Colo, BS, 71; Colo State Univ, BS, 80. *Prof Exp:* Electronics engr aerospace, Ball Bros Res Corp, 72; ELECTRONICS ENGR ATMOSPHERIC RES, NAT OCEANIC & ATMOSPHERIC ADMIN, 72- *Mem:* Am Geophys Union. *Res:* Inadvertent alterations of precipitation distribution and amount; physical and chemical characteristics of natural and man-made aerosols; chemical reactions of pollutant gases; atmospheric transmissivity; air quality, acid precipitation and deposition, pollutant transport. *Mailing Add:* NOAA/ERL/ARL/ARS 325 Broadway R/E/ARX1 Boulder CO 80303

WELLMAN, HENRY NELSON, b Kansas City, Kans, Nov 10, 33; m 57; c 6. INTERNAL MEDICINE. *Educ:* Rockhurst Col, BS, 56; Sch Med, St Louis Univ, MD, 60. *Prof Exp:* Intern med, St Louis Univ Hosps, 60-61, postgrad radiol biol & med, 61; resident med, Univ Cincinnati, 63-65, fel, 65-66, asst prof med & radiol, Sch Med, 66-69, assoc prof, 69-71; PROF MED & RADIOL, SCH MED, IND UNIV, 71- *Concurrent Pos:* Asst attend physician, Cincinnati Gen Hosp, 67-70; chief nuclear med sect, Pop Studies Prog, Dept Health, Educ & Welfare, 69-71; sr scientist award, Alexander Von Humboldt Found, 79-80; US tech advisor, Int Electrotech Comt, 77-81; dir nuclear med, Sch Med, Ind Univ, 72- & staff radiol med, Univ Med Ctr, 71-; consult, Bur Drugs, Div Oncol & Radiopharm, 75- & WHO, 77- *Mem:* Am Med Asn; fel Am Col Physicians; fel Am Col Radiol; fel Am Col Nuclear Physicians (secy, 85-87); Soc Nuclear Med. *Res:* Development of newer diagnostic uses of radiolabelled substances, especially in pulmonary, skeletal cerebral and neoplastic processes for clinical nuclear medicine. *Mailing Add:* Nuclear Med Uh P-16 Ind Univ Sch Med 1100 W Michigan St Indianapolis IN 46223

WELLMAN, RUSSEL ELMER, b New Berlin, NY, July 16, 22; m 48; c 2. XEROGRAPHY, THERMOGRAPHY. *Educ:* Howard Col, BA, 50; Univ Rochester, PhD(phys chem), 55. *Prof Exp:* Res chemist, Callery Chem Co, 55, group leader appl chem, 56-60; sr chemist, Inorg & Phys Chem Sect, Southern Res Inst, 60-66; unit mgr polymer technol, Xerox Corp, Rochester, NY, 66-67, scientist, 67-73, sr chemist, 73-76; consult, 76-77; SR CHEMIST, PITNEY BOWES, DANBURY, CONN, 78- *Mem:* Am Chem Soc; NY Acad Sci; Am Inst Chem; Soc Photog Sci & Eng; Am Soc Testing & Mat. *Res:* Viscometry; physical chemistry of polymers; adhesion of polymers; microencapsulation; materials for non-impact printing processes. *Mailing Add:* 206 Grasslands Rd Southbury CT 06488

WELLMAN, WILLIAM EDWARD, b Ft Wayne, Ind, Oct 22, 32; m 66; c 2. ORGANIC CHEMISTRY. *Educ:* Col Wooster, BA, 54; Ohio State Univ, PhD(org chem), 60. *Prof Exp:* Chemist, Esso Res & Eng Co, 60-63, sr chemist, 63-64, proj leader, 64-69, res assoc, 68-73, sect head, 69-73, mgr oxygen solvents & lower alcohols, Solvents Technol Div, 73-80, MGR TECHNOL DEVELOP & APPLN, EXXON CHEM CO, 80- *Res:* Petrochemicals, solvents and fiber intermediates; process development. *Mailing Add:* 1900 E Linden Ave Linden NJ 07036-1111

WELLMANN, KLAUS FRIEDRICH, anatomic pathology, clinical pathology; deceased, see previous edition for last biography

WELLNER, DANIEL, b Antwerp, Belg, June 9, 34; US citizen; m 62; c 1. BIOCHEMISTRY. *Educ:* Harvard Col, AB, 56; Tufts Univ, PhD(biochem), 61. *Prof Exp:* Instr biochem, Sch Med, Tufts Univ, 62-63, sr instr, 63-65, asst prof, 65-67; asst prof, 67-69, ASSOC PROF BIOCHEM, MED COL, CORNELL UNIV, 69- *Concurrent Pos:* NATO fel biochem, Weizmann Inst Sci, 61-62; Lederle med fac award, 64-67; ad hoc mem, Pathobiol Chem Study Sect, NIH, 79, mem, Med Biochem Study Sect, 81-85. *Mem:* Sigma Xi; Am Chem Soc; NY Acad Sci; Harvey Soc; Am Soc Biochem & Molecular Biol. *Res:* Ribonuclease; amino acid oxidases; flavoproteins; mechanism of enzyme action; biochemistry of cancer; inborn errors of metabolism. *Mailing Add:* Cornell Univ Med Col 1300 York Ave New York NY 10021

WELLNER, MARCEL, b Antwerp, Belg, Feb 8, 30; US citizen; m 61; c 2. PHYSICS. *Educ:* Mass Inst Technol, BS, 52; Princeton Univ, PhD(physics), 58. *Prof Exp:* Instr physics, Princeton Univ, 56-58 & Brandeis Univ, 58-59; mem, Inst Advan Study, 59-60; res assoc physics, Univ Ind, 60-63; NSF fel, Atomic Energy Res Estab, Eng, 63-64; from asst prof to assoc prof, 64-71, PROF PHYSICS, SYRACUSE UNIV, 71- *Concurrent Pos:* Physicist, Cavendish Lab, Cambridge Univ, Eng, 68-69; vis scientist, Inst Advan Studies, Dublin, Ireland, 80. *Mem:* Am Phys Soc. *Res:* Quantum field theory and mathematical methods. *Mailing Add:* Dept Physics Syracuse Univ Syracuse NY 13244-1130

WELLNER, VAIRA PAMILJANS, b Aluksne, Latvia, Jan 28, 36; US citizen; m 62; c 1. BIOCHEMISTRY. *Educ:* Boston Univ, AB, 58; Tufts Univ, PhD(biochem), 64. *Prof Exp:* Res assoc biochem, Sch Med, Tufts Univ, 66-67; res assoc biochem, Col Med, 72-87, RES ASSOC, DEPT SURG, BURN CTR, CORNELL UNIV MED COL, 87- *Concurrent Pos:* Res fel biochem, Sch Med, Tufts Univ, 63-66 & Col Med, Cornell Univ, 67-72. *Mem:* Am Chem Soc; Sigma Xi; Am Soc Biochem Molecular Biol; AAAS. *Res:* Mechanisms of enzyme action. *Mailing Add:* Cornell Univ Med Col 1300 York Ave New York NY 10021

WELLONS, JESSE DAVIS, III, b Roanoke, Va, Apr 4, 38; m 58; c 4. WOOD TECHNOLOGY, POLYMER SCIENCE. *Educ:* Duke Univ, BS, 60, MF, 63, PhD(wood technol, polymer sci), 66. *Prof Exp:* Res chemist, Res Triangle Inst, NC, 62-65, res assoc wood chem, Iowa State Univ, 65-66, , from asst prof to assoc prof, 66-70; assoc prof forest prod chem, Ore State Univ, 70-77, prof, 77-81; MGR RES & DEVELOP, CHEM DIV, GA PAC CORP, 81- *Mem:* Forest Prod Res Soc; Soc Wood Sci & Technol; Am Chem Soc. *Res:* Wood adhesives; tannin adhesives; plywood processing; use of plastics to modify properties of wood; sorption and diffusion of monomers in wood. *Mailing Add:* Ga Pac Corp 133 Peachtree St NE Floor 22 Atlanta GA 30303-1847

WELLS, ADONIRAM JUDSON, b Chicago, Ill, Apr 1, 17; m 37; c 6. PHYSICAL CHEMISTRY. *Educ:* Harvard Univ, SB, 38, AM, 40, PhD(phys chem), 41. *Prof Exp:* Res chemist, Ammonia Dept, E I du Pont de Nemours & Co, Del, 41-46, res supvr, 46-48, mgt asst, 48-50, mgr film develop, 50-53, dir, Yerkes Res Lab, NY, 53-55, asst res dir film dept, Del, 55-59, res dir electrochem dept, 59-69, dir indust prod div, 69-77, dir specialty prods div, Fabrics & Finishes Dept, 77-80; CONSULT, AM LUNG ASN, 81- *Mem:* Am Chem Soc; Am Thoracic Soc; Soc Epidemiol Res. *Res:* Infrared and Raman spectroscopy; thermodynamics; process and market development; polymer chemistry; high temperatures; epidemiology of passive smoking. *Mailing Add:* 41 Windermere Way Kennett Square PA 19348

WELLS, BARBARA DURYEA, b Birmingham, Ala, Feb 4, 39. SPECTROSCOPY, NUCLEIC ACID CHEMISTRY. *Educ:* Univ Miami, BS, 60; Univ Calif, San Francisco, MS, 69, PhD(biochem & biophysics), 73. *Prof Exp:* Instr chem, Polk Jr Col, 65; technician, Univ Calif, San Francisco, 66-68; fel, Fla State Univ, 73-75, Columbia Univ, 75-78; res scientist, Johns Hopkins Univ, 78-79; ASST PROF BIOCHEM, UNIV WIS-MILWAUKEE, 79- *Mem:* Am Soc Biol Chemists; Biophys Soc; Am Chem Soc; Sigma Xi; AAAS. *Res:* Spectroscopic techniques; fluorescence electron paramagnetic resonance, nuclear magnetic resonance for studying the structure of transfer RNA in order to correlate structural changes with biological functions. *Mailing Add:* 1815 E Olive Milwaukee WI 53211

WELLS, BENJAMIN B, JR, b Rochester, Minn, May 31, 41; m 67. MATHEMATICAL ANALYSIS. *Educ:* Univ Mich, BS, 61, MS, 62; Univ Calif, Berkeley, PhD(math), 67. *Prof Exp:* From instr to asst prof math, Univ Ore, 67-70; Fulbright lectr, Univ Santiago, Chile, 70-71; assoc prof, 71-76, PROF MATH, UNIV HAWAII, 76- *Mem:* Am Math Soc. *Res:* Measure theory; harmonic analysis. *Mailing Add:* 9480 Par Dr Nokesville VA 22123

WELLS, BOBBY R, b Wickliffe, Ky, July 30, 34; m 60; c 1. SOIL CHEMISTRY, SOIL FERTILITY. *Educ:* Murray State Univ, BS, 59; Univ Ark, MS, 61; Univ Mo, PhD(soil chem), 64. *Prof Exp:* Asst soils, Univ Ark, 59-60; asst county agent, Univ Mo, 60-61, asst soils, 61-64; asst prof agr, Murray State Univ, 64-66; asst prof, Rice Br Exp Sta, 66-70, assoc prof, 70-75, PROF AGRON, RICE EXP STA, UNIV ARK, FAYETTEVILLE, 75- *Mem:* Am Soc Agron; Soil Sci Soc Am; Sigma Xi. *Res:* Investigations into soil-plant relationships for rice. *Mailing Add:* Agron 115 Pl Sci Univ Ark Fayetteville AR 72701

WELLS, CHARLES EDMON, b Dothan, Ala, May 19, 29; m 62; c 3. PSYCHIATRY, NEUROLOGY. *Educ:* Emory Univ, AB, 48, MD, 53. *Prof Exp:* From intern to asst resident med, New York Hosp, 53-54; clin assoc neurol, NIH, 54-56; resident, New York Hosp, 57-58; assoc prof neurol, 61-75, psychiat, 68-72, PROF NEUROL, SCH MED, VANDERBILT UNIV, 75-, PROF PSYCHIAT & VCHMN DEPT, 72- *Concurrent Pos:* NIH fel neurol, New York Hosp-Cornell Med Ctr, 58-59; mem coun aging, Am Psychiat Asn, 82- *Mem:* Am Psychiat Asn; Am Col Psychiat; Am Neurol Asn; Asn Res Nerv & Ment Dis. *Res:* Dementia; neuropsychiatry; use of literature in teaching of psychiatry. *Mailing Add:* 310 25th Ave N Suite 309 Nashville TN 37203

WELLS, CHARLES FREDERICK, b Atlanta, Ga, May 4, 37; m 62; c 2. CATEGORY THEORY, PROGRAMMING LANGUAGE MODEL THEORY. *Educ:* Oberlin Col, AB, 62; Duke Univ, PhD(math), 65. *Prof Exp:* From asst prof to assoc prof, 65-80, PROF MATH, CASE WESTERN RESERVE UNIV, 80- *Concurrent Pos:* Prin investr, NSF grants, 65-67 & 87-89; guest, Math Res Inst, Swiss Fed Inst Technol, Zurich, 75-76 & 83; guest, Oxford Univ Computer Lab, 90- *Mem:* Math Asn Am; Am Math Soc; Asn Comput Mach. *Res:* Category theory, theoretical computer science. *Mailing Add:* Dept Math Case Western Reserve Univ Cleveland OH 44106-1712

WELLS, CHARLES HENRY, b Chicago, Ill, Dec 6, 31; m 56; c 2. PHYSIOLOGY. *Educ:* Randolph-Macon Col, BA, 54; Mich State Univ, MS, 60, PhD(physiol), 63. *Prof Exp:* Instr physiol, Univ Tex Med Br, 62-64; asst prof, Med Col SC, 64-67; asst prof, Univ Tex Med Br, Galveston, 67-73, assoc prof physiol, 73-80; WITH CRITIKON CORP, 80- *Concurrent Pos:* Consult, USAF, 69-; chief, Physiol Div, Shriners Burns Inst. *Mem:* Fedn Am Socs Exp Biol; Undersea Med Soc; Am Physiol Soc; Am Burn Asn; Int Soc Burn Injury. *Res:* Blood rheology, microcirculation in stress; thermal injury; decompression sickness; computer aided instructional systems development. *Mailing Add:* Critikon Inc PO Box 31800 Tampa FL 33622

WELLS, CHARLES VAN, b Summerton, SC, July 1, 37; m 66; c 2. BOTANY. *Educ:* Presby Col, SC, BS, 59; Appalachian State Univ, MA, 63; Univ Ariz, PhD(bot), 69. *Prof Exp:* Teacher & coach, high sch, SC, 59-62; teaching asst biol, Appalachian State Univ, 62-63; asst prof biol, Newberry Col, 63-64; teaching asst biol, Univ Ariz, 64-69; assoc prof, 64-80, PROF BIOL, LENOIR RHYNE COL, 80- *Mem:* Sigma Xi. *Res:* Effects of radiation on vegetative morphology of sirogonium. *Mailing Add:* Dept Biol Lenoir Rhyne Col Box 473 Hickory NC 28601

WELLS, DANIEL R, b New York, NY, May 2, 21; m 43; c 2. PHYSICS. *Educ:* Cornell Univ, BME, 42; NY Univ, MS, 55; Stevens Inst Technol, PhD(physics), 63. *Prof Exp:* Res assoc plasma physics, Princeton Univ, 55-64; assoc prof physics, 64-67, PROF PHYSICS, UNIV MIAMI, 67- *Concurrent Pos:* Res assoc, Stevens Inst Technol, 61-64; assoc prof, Seton Hall, 62-64; res grants, AEC & Air Force Off Sci Res, 64-74. *Mem:* Am Phys Soc. *Res:* Controlled thermonuclear research. *Mailing Add:* 6950 SW 62nd St Miami FL 33143

WELLS, DARRELL GIBSON, b Pierre, SDak, Feb 21, 17; m 46; c 3. PLANT BREEDING, PLANT PATHOLOGY. *Educ:* SDak State Col, BS, 41; State Col Wash, MS, 43; Univ Wis, PhD, 49. *Prof Exp:* Asst agronomist, State Col Wash, 43-45; asst, Univ Wis, 45-49; from assoc prof to prof agron, Miss State Univ, 49-62; prof wheat breeding, SDak State Univ, 62-83; RETIRED. *Concurrent Pos:* Mem tech asst mission, Western Region, Nigeria, Int Develop Serv, 58-60. *Mem:* Am Soc Agron. *Res:* Small grain genetics and breeding; cowpea and lima bean breeding. *Mailing Add:* Rte 4 Box 233 Brookings SD 57006

WELLS, DARTHON VERNON, b Saline Co, Ark, Oct 11, 29; m 50; c 2. ORGANIC CHEMISTRY. *Educ:* Univ Northern Ala, BS, 54; Univ Miss, MS, 57; Univ SC, PhD(chem), 60. *Prof Exp:* Res assoc chem, Brown Univ, 59-61; from asst prof to assoc prof, 61-72, PROF CHEM, LA STATE UNIV, ALEXANDRIA, 72- *Mem:* Am Chem Soc; Sigma Xi. *Res:* Chemistry of organic peroxide decomposition; nucleophilic displacement of phosphorus; carbanion rearrangement reactions. *Mailing Add:* Dept Chem La State Univ Alexandria LA 71302

WELLS, DAVID ERNEST, b Montreal, Que, June 29, 39; m 64; c 3. GEODESY, OCEAN MAPPING. *Educ:* Mt Allison Univ, BSc, 61; Univ BC, BASc, 63, MASc, 66; Univ NB, PhD(geod), 74. *Prof Exp:* Res scientist, Dept Fisheries & Oceans, Bedford Inst Oceanog, 74-80; PROF HYDROGRAPHY GEOD, UNIV NB, 80- *Concurrent Pos:* Nat deleg, Int Asn Geod, 86-88; pres, Can GPS Assocs, 86-; mem, Nat Marine Coun Can, 88- & Int Adv Bd, Int Hydrographic Orgn & Int Cong Surv, 90- *Mem:* Am Geophys Union; Am Cong Surv & Mapping; Hydrographic Soc; Int Asn Geod. *Res:* Ocean mapping data cleaning, interpretation and management software tool development; kinematic applications of the global positioning system. *Mailing Add:* 538 Squires St Fredericton NB E3B 3V4 Can

WELLS, EDDIE N, b Mar 13, 40; US citizen. PLANETOLOGY, REMOTE SENSING. *Educ:* Murray State Univ, BS, 62; Univ Pittsburgh, PhD(earth & planets), 77. *Prof Exp:* Res assoc, Lab Planetary Studies, Cornell Univ, 77-79; res assoc, Univ Pittsburgh, 79-81, res asst prof geol & planetary sci, 81-82; OPERS ASTROM, COMPUTER SCI CORP, SPACE TELESCOPE SCI INST, BALTIMORE, MD, 82- *Mem:* Am Geophys Union; Am Astron Soc. *Res:* Remote sensing of the earth; astronomical observations of planets, satellites and asteroids; laboratory studies of optical properties of rocks, minerals, glasses, and ices. *Mailing Add:* 3700 San Martin Dr Space Telescope Sci Inst Baltimore MD 21218

WELLS, EDWARD JOSEPH, b Sydney, Australia, Oct 10, 36; m 62; c 2. PHYSICAL CHEMISTRY. *Educ:* Univ Sydney, BSc, 58, MSc, 60; Oxford Univ, DPhil(magnetic resonance), 62. *Prof Exp:* Gowrie travelling scholar, 59-61; fel magnetic resonance, Univ BC, 61-63; instr phys chem, 63-64; res assoc magnetic resonance, Univ Ill, Urbana, 64-65; asst prof chem, 65-67, actg chmn dept, 75-76, chmn dept, 76-79, ASSOC PROF CHEM, SIMON FRASER UNIV, 67-, ASSOC MEM PHYSICS DEPT, 69- *Concurrent Pos:* With Australian Nat Serv, 55-58. *Mem:* Fel Chem Inst Can; fel Royal Soc Chem; Am Inst Physics; fel Royal Soc Arts; Can Asn Physicists; Sigma Xi. *Res:* Chemical and biological applications of nuclear magnetic resonance. *Mailing Add:* Dept Chem Simon Fraser Univ Burnaby BC V5A 1S6 Can

WELLS, ELIZABETH FORTSON, b Shreveport, La, March 4, 43. PLANT TAXONOMY, FLAVONOID CHEMOTAXONOMY. *Educ:* Agnes Scott Col, Decataur, Ga, BA, 65; Univ NC, Chapel Hill, MA, 70, PhD (bot), 77. *Prof Exp:* Instr biol, Univ Richmond, 70-72; fel, Univ BC, 77-79; ASST PROF

BOT, GEORGE WASHINGTON UNIV, 79- *Mem:* Am Soc Plant Taxonomists; Bot Soc Am; Int Asn Plant Taxon; Asn Southeastern Biologists; Am Inst Biol Sci; Sigma Xi. *Res:* Revising the genus Heuchera (Saxifragaceae) using breeding studies, flavonoid analysis and morphometrics; flavonoids in Leucanthemum (Asterceae). *Mailing Add:* 8023 Washington Ave Alexandria VA 22308

WELLS, FRANK EDWARD, b Granby, Mo, Mar 29, 25; m 47; c 2. RAPID ANALYTICAL METHODOLOGY. *Educ:* Southwest Mo State Univ, BS, 51; Purdue Univ, MS, 58, PhD(bact), 61. *Prof Exp:* Med technologist, Vets Admin Hosp, Springfield, 51-53; lab dir, Henningsen Foods Inc, 53-56; instr food prods technol, Purdue Univ, 56-61; tech dir, Monarck Egg Corp, 61; sr microbiologist, 61-65, PRIN MICROBIOLOGIST, MIDWEST RES INST, 65- *Mem:* Soc Invert Path; Wildlife Dis Asn. *Res:* Control agents for food spoilage; microbial insect control agents; disease monitoring of laboratory animals; biodegradation of xenobiotics in soil; detection of aerosolized biological agents; industrial fermentations; enzyme immobilization. *Mailing Add:* 3200 NE 64th Terr Kansas City MO 64119

WELLS, FREDERICK JOSEPH, b Dayton, Ohio, Nov 13, 44. SEISMOLOGY. *Educ:* Univ Dayton, BS, 66; Brown Univ, ScM, 68, PhD(seismol), 72. *Prof Exp:* Res geophysicist, 72-75, advan res geophysicist, Denver Res Ctr, Marathon Oil Co, 75-77, sr geophysicist, 77-80, MGR GEOPHYS PROJ, MARATHON INT OIL CO, 80- *Mem:* Soc Explor Geophysicists; Am Asn Petrol Geologists. *Mailing Add:* Marathon Oil Co PO Box 3128 Houston TX 77253

WELLS, GARLAND RAY, b El Dorado, Ark, Oct 10, 36; m 56; c 1. FORESTRY, ECONOMICS. *Educ:* La Polytech Inst, BS, 58; NC State Univ, MF, 61; Duke Univ, DF, 68. *Prof Exp:* From instr to asst prof forestry, La Polytech Inst, 62-65; asst prof, 65-75, ASSOC PROF FORESTRY, UNIV TENN, KNOXVILLE, 75- *Mem:* Soc Am Foresters; Am Econ Asn. *Res:* Land ownership research especially the practice of forestry by private, non-industrial forest owners. *Mailing Add:* Dept Forestry Univ Tenn Knoxville TN 37916

WELLS, GARY NEIL, b Springfield, Ill, Nov 19, 41; m 65; c 2. PLANT PHYSIOLOGY, BIOCHEMISTRY. *Educ:* Western Ill Univ, BA, 65, MS, 67; Univ Ill, PhD(plant biochem), 71. *Prof Exp:* Chem eng tech chem, Ill State Dept, 61-63; res assoc molecular biol, Univ Okla, 71-73; from asst prof to assoc prof, 73-83, assoc head, 85-86, PROF BIOL, FLA INST TECHNOL, 83-, HEAD DEPT BIOL, 86- *Concurrent Pos:* NIH fel, Univ Okla, 72-73. *Mem:* AAAS; Am Soc Plant Physiol; Am Soc Photo Biol. *Res:* Molecular biology of development; regulatory mechanisms operating at the level of transcription and translation and their role in plant growth and development. *Mailing Add:* Dept Biol Sci Fla Inst Technol Melbourne FL 32901

WELLS, HARRINGTON, b Columbia, Mo, June 20, 52; m 74, 88; c 3. POPULATION ECOLOGY, POPULATION GENETICS. *Educ:* Occidental Col, BA, 74; Univ Calif, Santa Barbara, PhD(biol), 79. *Prof Exp:* Asst prof, 80-85, ASSOC PROF BIOL, UNIV TULSA, 86- *Concurrent Pos:* Co-dir & co-prin investr, Tulsa Teacher Enhancement, NSF grant, 85-92; vis assoc prof, Dept Entom, G B Pant Univ, Pantnagar, India, 90-91; dir & prin investr, Okla Statewide Teacher Sci Lit, 91-96. *Mem:* AAAS. *Res:* Population biology based reproduction kinetics; factors altering population growth at low densities and resulting kinetic models for conservation biology; foraging ecology of nectivores; potential evolutionary and agricultural implications of alternate foraging behaviors. *Mailing Add:* Dept Biol Sci Univ Tulsa Tulsa OK 74104

WELLS, HENRY BRADLEY, b Ridgeland, SC; m 47; c 4. BIOSTATISTICS. *Educ:* Emory Univ, BA, 50; Univ NC, MSPH, 53, PhD(biostatist), 59. *Prof Exp:* Chief statistician, Ga State Dept Pub Health, 50-55; statistician, NC State Bd Health, 56-58, consult, 58-64; from instr to assoc prof biostatist, 58-69, prof biostatist, 69-80, EMER PROF, SCH PUB HEALTH, UNIV NC, CHAPEL HILL, 81-; PROF BIOSTATIST, BOWMAN GRAY SCH MED, WAKE FORREST UNIV, 81- *Mem:* Fel Am Statist Asn; Biomet Soc; fel Am Pub Health Asn; Int Statist Inst; Int Asn Surv Statisticians; AAAS. *Res:* Design of clinical trials, survivorship analysis; survey methods in demographic research; evaluation of health programs. *Mailing Add:* Dept Pub Health Sci Bowman Gray Sch Med 300 S Hawthorne Rd Winston-Salem NC 27103

WELLS, HERBERT, b New Haven, Conn, July 27, 30; m 59; c 3. PHARMACOLOGY, ORTHODONTICS. *Educ:* Yale Univ, BA, 52; Harvard Univ, DMD, 56. *Prof Exp:* Res assoc orthod, Sch Dent Med, Harvard Univ, 59-60, assoc pharmacol, 60-63, asst prof dent & dir, Dent Clin, 63-68; PROF PHARMACOL, HENRY M GOLDMAN SCH GRAD DENT, BOSTON UNIV, 68-, ASST DEAN, PREDOCTORAL PROG, 80- *Concurrent Pos:* Res fel orthod & pharmacol, Sch Dent Med, Harvard Univ, 56-59; mem, Panel Drugs Dent, Nat Res Coun, 66- & Dent Study Sect, NIH, 74- *Honors & Awards:* Lord-Chaim Res Award, 60; Oral Sci Prize, Int Asn Dent Res, 64. *Mem:* Am Soc Pharmacol & Exp Therapeut; Int Asn Dent Res; Sigma Xi. *Res:* Growth and secretion of exocrine and endocrine glands; salivary glands; parathyroid glands; experimental teratology; cleft palate formation. *Mailing Add:* Oral Pharmacol Lab Boston Univ Sch Grad Dent 20 E Concord St Boston MA 02112

WELLS, HERBERT ARTHUR, b Jersey City, NJ, Aug 4, 21; m 46; c 3. MECHANICAL ENGINEERING. *Educ:* Cooper Union, BME, 47; Newark Col Eng, MSME, 52. *Prof Exp:* Mem tech staff, 47-56, SUPVR, BELL TEL LABS, INC, 56- *Mem:* Am Soc Mech Engrs; Nat Soc Prof Engrs. *Res:* Ship design and cable laying methods; hardening of structures to resist nuclear blast effects; antenna structural design for arctic and other applications. *Mailing Add:* 772 Norman Pl Westfield NJ 07090-3466

WELLS, HOMER DOUGLAS, b Blaine, Ky, Nov 11, 23; m 42; c 1. PLANT PATHOLOGY. *Educ:* Univ Ky, BS, 48, MS, 49; NC State Col, PhD(plant path), 54. *Prof Exp:* Tech asst agron agr exp sta, Univ Ky, 48-49, asst, 49-50; asst plant path, NC State Col, 50-52; asst agronomist, Ga Coastal Plain Exp Sta, USDA, 52-53, pathologist forage crops, 53-88; RETIRED. *Honors & Awards:* Cert Merit, USDA, 86. *Mem:* Am Phytopath Soc; Sigma Xi; AAAS. *Res:* Turf and forage crop disease problems. *Mailing Add:* Rte 5 Box 262 Tifton GA 31794

WELLS, IBERT CLIFTON, b Fayette, Mo, Apr 12, 21; m 48; c 6. BIOCHEMISTRY. *Educ:* Cent Methodist Col, AB, 42; St Louis Univ, PhD(biochem), 48. *Prof Exp:* From instr to assoc prof biochem, Col Med, State Univ NY Upstate Med Ctr, 50-61; chmn dept, 61-76, PROF BIOCHEM, SCH MED, CREIGHTON UNIV, 61-, PROF MED, 76- *Concurrent Pos:* Nat Res Coun fel, Calif Inst Technol, 48-50. *Honors & Awards:* Com Solvents Corp Award, 52. *Mem:* Am Soc Biol Chemists; Soc Exp Biol & Med. *Res:* Cholesterol metabolism; choline and one-carbon metabolism; biochemistry of disease. *Mailing Add:* Dept Biomed Sci Div Biochem Dept Med Creighton Univ Sch Med 2500 California St Omaha NE 68178

WELLS, J GORDON, b Salt Lake City, Utah, Sept 18, 18; m 41; c 1. PHYSIOLOGY. *Educ:* Pepperdine Col, BS, 46; Univ Southern Calif, PhD, 51. *Prof Exp:* Mem fac phys educ, Pepperdine Col, 46-48; asst aviation physiol, Univ Southern Calif, 48-49; aviation physiologist, USAF Sch Aviation Med, 49-56; supvr human eng, Norair Div, Northrop Corp, 56-61; asst dir life sci, Appl Sci, Space & Info Systs Div, NAm Aviation, Inc, 61-62, mgr Apollo Crew Systs, Apollo Eng, 62-66, asst to vpres & gen mgr life sci opers, 66-67, dir space progs, Life Sci Opers, 67, mgr life sci & systs, Sci & Technol, Res Eng & Test, Space Div, 67-68, mgr life sci & systs, 68-71, supvr life sci, Systs Eng & Technol, Res & Eng & Test, Space Div, NAm Rockwell Corp, 71-75; assoc dean admin, Grad Sch Educ, 75-83, asst prof, 75-77, ASSOC PROF, GRAD SCH EDUC & PSYCHOL, PEPPERDINE UNIV, 83- *Mem:* AAAS; Human Factors Soc; Am Astronaut Soc; assoc fel Am Inst Aeronaut & Astronaut; fel Aerospace Med Asn (vpres, 59-60). *Res:* Physiological aspects of aerospace medicine; human factors and bioastronautics. *Mailing Add:* Attn Personnel Pepperdine Univ 3415 Sepulveda Blvd Los Angeles CA 90034

WELLS, JACK NULK, b McLouth, Kans, May 17, 37; m 60; c 2. MEDICINAL CHEMISTRY, PHARMACOLOGY. *Educ:* Park Col, BA, 59; Univ Mich, MS, 62, PhD(med chem), 63. *Prof Exp:* From asst prof to assoc prof med chem, Purdue Univ, 63-67; vis scholar, 72-73, asst prof physiol, 73-75, asst prof pharmacol, 75-77, FROM ASSOC PROF TO PROF PHARMACOL, SCH MED, VANDERBILT UNIV, 77- *Concurrent Pos:* Fel, Ohio State Univ, 63. *Mem:* AAAS; Am Soc Pharmacol & Exp Therapeut; Am Chem Soc; Sigma Xi. *Res:* Phosphodiesterase; smooth muscle physiology and airchemistry. *Mailing Add:* 3604 Saratoga Nashville TN 37205

WELLS, JACQUELINE GAYE, b Pittsburgh, Pa, May 17, 31; m 51; c 2. ALGEBRA. *Educ:* Univ Pittsburgh, BS, 52, MS, 64, PhD(math), 72. *Prof Exp:* ASST PROF MATH, PA STATE UNIV, MCKEESPORT, 64- *Mailing Add:* 1005 Fawcett Ave McKeesport PA 15132

WELLS, JAMES EDWARD, physical inorganic chemistry, for more information see previous edition

WELLS, JAMES HOWARD, b Howe, Tex, June 20, 32; m 53; c 2. MATHEMATICS. *Educ:* Tex Tech Col, BS, 52, MS, 54; Univ Tex, PhD(math), 58. *Prof Exp:* Instr math, Univ Tex, 57-58; from instr to asst prof, Univ NC, 58-59; vis asst prof, Univ Calif, Berkeley, 60-61; assoc prof, 62-69, PROF MATH, UNIV KY, 69- *Mem:* Am Math Soc; Math Asn Am. *Res:* Analysis. *Mailing Add:* Dept Math Univ Ky Lexington KY 40506

WELLS, JAMES RAY, b Delaware, Ohio, May 28, 32; m 58; c 2. PLANT TAXONOMY, PLANT ECOLOGY. *Educ:* Univ Tenn, BS, 54, MS, 56; Ohio State Univ, PhD(bot), 63. *Prof Exp:* Asst prof biol, E Carolina Univ, 63-64 & Old Dominion Univ, 64-66; BOTANIST, CRANBROOK INST SCI, 66- *Concurrent Pos:* Vis prof, Stephen F Austin State Col, 64; NSF grant, 67-69; adj prof biol, Oakland Univ, 69- & Wayne State Univ, 73-; chmn, Mich Natural Areas Coun, 70-72; pres, Mich Bot Club, 81-83. *Mem:* Bot Soc Am; Am Soc Nat; Sigma Xi. *Res:* Plant ecology; biosystematics of the genus Polymnia; botany of Michigan Islands. *Mailing Add:* Cranbrook Inst Sci Box 801 Bloomfield Hills MI 48303-0801

WELLS, JAMES ROBERT, b Moundsville, WVa, Apr 5, 40; m 81; c 2. CHEMISTRY, STRATEGIC PLANNING. *Educ:* Wheeling Col, BS, 62; Univ Pittsburgh, PhD(chem), 67. *Prof Exp:* Chemist plastics dept, Exp Sta, 67-70, sr res chemist, Sabine River Lab, 70-72, res supvr, Polymer Intermediates Dept, 72-74, lab dir, 74-76, prod supt nylon intermediates-methanol-power, Sabine River Works, Plastics Prod & Resins Dept, 76-77, res mgr, Feedstocks Res & Develop Div, Cent Res & Develop Dept, 77-79, prin consult, Corp Plans Dept, Wilmington, 79-81, PRIN CONSULT, CENT RES & DEVELOP DEPT, E I DU PONT DE NEMOURS & CO, INC, 81- *Mem:* Am Chem Soc. *Mailing Add:* 24 Drake Rd Chesapeake City MD 21915-1709

WELLS, JANE FRANCES, b Davenport, Iowa, Feb 24, 44; m 81; c 1. ALGEBRA. *Educ:* Marycrest Col, BA, 66; Univ Iowa, MS, 67, PhD(math), 70, Univ Il, MS, 90. *Prof Exp:* Asst prof math, Purdue Univ, Ft Wayne, 70-74; UNIV PROF, GOV STATE UNIV, 74- *Concurrent Pos:* Vis appointment, US Environ Protection Agency, 80-81. *Mem:* Am Math Soc; Math Asn Am; Am Statist Asn; Asn Women Math; Asn Comput Mach; Inst Elec & Electronics Engrs Comput Soc. *Res:* Information retrieval, data base systems. *Mailing Add:* Div Sci Gov State Univ Park Forest IL 60466

WELLS, JOHN ARTHUR, plant chemistry; deceased, see previous edition for last biography

WELLS, JOHN CALHOUN, JR, b Tampa, Fla, May 12, 41; m 63; c 2. EXPERIMENTAL NUCLEAR PHYSICS. *Educ:* Fla State Univ, BS, 61; Johns Hopkins Univ, PhD(physics), 68. *Prof Exp:* Nat Acad Sci-Nat Res Coun assoc & res physicist, US Naval Ord Lab, Md, 68-70; from asst prof to assoc prof, 70-80, PROF PHYSICS, TENN TECHNOL UNIV, 80- *Concurrent Pos:* Consult physics div, Oak Ridge Nat Lab, 71- *Mem:* Am Phys Soc; Sigma Xi; Am Asn Univ Prof; Am Asn Physics Teachers. *Res:* Nuclear charged-particle reactions; heavy-ion reactions; gamma-ray spectroscopy; neutron reactions. *Mailing Add:* Dept Physics Tenn Technol Univ Cookeville TN 38505

WELLS, JOHN MORGAN, JR, b Hopewell, Va, Apr 12, 40. DIVING PHYSIOLOGY, HYPERBARIC. *Educ:* Randolph-Macon Col, BS, 62; Univ Calif, San Diego, PhD(marine biol), 69. *Prof Exp:* Res physiologist, Wrightsville Marine Bio-Med Lab, NC, 69-72; asst prof physiol, Sch Med, Univ NC, 70-72; sci coordr marine biol, Manned Undersea Sci & Technol Off, 72-79, dir, diving prog, 79-91, DIR, EXP DIVING UNIT, NAT OCEANIC & ATMOSPHERIC ADMIN, 91- *Concurrent Pos:* Guest Scientist, Naval Med Res Inst, 84-85. *Mem:* Undersea & Hyperbaric Med Soc; Am Acad of Underwater Sci; Nat Asn Diver Med Tech. *Res:* Blood function at high hydrostatic and inert gas pressures; physiological symbiosis between algae and invertebrates; community metabolism of marine benthic communities; protection of divers in polluted waters. *Mailing Add:* Nat Oceanic & Atmospheric Admin Exp Diving Unit Bldg 1519 Ft Ustus VA 23604-5544

WELLS, JOHN WEST, b Philadelphia, Pa, July 15, 07; m 32; c 1. PALEONTOLOGY, MARINE ZOOLOGY. *Educ:* Univ Pittsburgh, BS, 28; Cornell Univ, MA, 30, PhD(paleont), 33. *Prof Exp:* Instr geol, Univ Tex, 29-31; Nat Res Coun fel, Brit Mus, Paris & Berlin, 33-34; asst sci, NY State Norm Sch, 37-38; from instr to prof geol, Ohio State Univ, 38-48; prof geol, 48-74, EMER PROF GEOL, CORNELL UNIV, 74- *Concurrent Pos:* Geologist, US Geol Surv, 46-70, Off Strategic Serv, 44-45, Bikini Sci Resurv, 47 & Pac Sci Bd, Arno Atoll Exped, 50; Fulbright lectr, Univ Queensland, 54; pres, Paleont Res Inst, 61-64. *Honors & Awards:* Paleont Medal, Paleont Soc, 74. *Mem:* Nat Acad Sci; fel Geol Soc Am; Paleont Soc (pres, 61-62). *Res:* Invertebrate paleontology and paleoecology; vertebrate paleontology; stratigraphy; invertebrate zoology. *Mailing Add:* Dept Geol Sci Cornell Univ Ithaca NY 14853

WELLS, JOSEPH, b Boston, Mass, Oct 6, 34; m 56; c 5. NEUROANATOMY, NEUROBIOLOGY. *Educ:* Univ RI, BS, 56; Duke Univ, PhD(anat), 59. *Prof Exp:* Instr anat, Duke Univ, 59-61 & Yale Univ, 61-63; asst prof, Sch Med, Univ Md, Baltimore, 63-68; ASSOC PROF ANAT, COL MED, UNIV VT, 68- *Concurrent Pos:* NIH fel, 59-61; NIH res grant, 64-67, 83-; Lederle Med Fac Award, Univ Md, 68, Sandoz Found fel neurosci, 78-80. *Mem:* Am Asn Anatomists; Soc Neurosci. *Res:* Neuroanatomy using silver stains; systems neurobiology using electron microscopy; neuronal plasticity of somatosensory system; neural transplantation. *Mailing Add:* Dept Anat Given Bldg Univ Vt Col Med Burlington VT 05405

WELLS, JOSEPH S, b Meade, Kans, Mar 19, 30; m 56; c 4. PHYSICS. *Educ:* Kans State Univ, BS, 56, MS, 58; Univ Colo, PhD(physics), 64. *Prof Exp:* Proj leader microwave noise, Electronics Calibration Ctr, 59-62, physicist, Radio & Microwave Mat Sect, 62-67, res physicist, Quantum Electronics Div, 67-73, RES PHYSICIST, TIME & FREQUENCY DIV, NAT BUR STANDARDS, 74- *Concurrent Pos:* Res assoc, Physics Dept, Univ Colo, 64-66, lectr, 66-73, adj prof, 73-79. *Honors & Awards:* Gold Medal, Dept Com, 74. *Mem:* Am Phys Soc; Sci Res Soc Am. *Res:* Paramagnetic and antiferromagnetic resonance; microwave measurements; infrared frequency synthesis; laser stabilization and frequency measurements; tunable lasers and spectroscopy; infrared frequency standards and molecular spectroscopy. *Mailing Add:* 2385 Kohler Dr Boulder CO 80303

WELLS, KENNETH, b Portsmouth, Ohio, July 24, 27; m 54; c 2. MYCOLOGY. *Educ:* Univ Ky, BS, 50; Univ Iowa, MS & PhD(bot), 57. *Prof Exp:* Asst bot, Univ Iowa, 54-57; instr & jr botanist, 57-58, from asst prof & asst botanist to assoc prof & assoc botanist, 59-72, chmn dept bot & agr bot, 78-82, PROF BOT & BOTANIST, UNIV CALIF, DAVIS, 72- *Concurrent Pos:* Fulbright res fel, Inst Bot, Sao Paulo, Brazil, 63-64; Nat Acad Sci exchange scientist, Romania, 71-72; Fulbright grant, Univ Tübingen, Ger, 83-84. *Mem:* Mycol Soc Am; Brit Mycol Soc; Mycol Soc Japan; AAAS. *Res:* Taxonomy of the saprobic Heterobasidiomycetes; compatibility and mating tests as a guide to speciation; ultrastructure of the fungi, especially of the basidium and spindle apparatus. *Mailing Add:* Dept Bot Univ Calif Davis CA 95616

WELLS, KENNETH LINCOLN, b Lone Mountain, Tenn, May 28, 35; m 60. SOIL SCIENCE, AGRONOMY. *Educ:* Univ Tenn, BS, 57, MS, 59; Iowa State Univ, PhD(soil sci), 63. *Prof Exp:* Res assoc soils, Iowa State Univ, 59-63; agriculturist, Tenn Valley Authority, 63-65, agronomist, 65-69; from asst prof to assoc prof agron, prof, 69-77, EXTEN PROF AGRON, UNIV KY, 77- *Mem:* Am Soc Agron; Soil Sci Soc Am. *Res:* Soil fertility, crop production, genesis and classification. *Mailing Add:* 2684 Walnut Hill Rd Lexington KY 40515

WELLS, KENTWOOD DAVID, b Alexandria, Va, Mar 23, 48; m 88. HERPETOLOGY, ANIMAL BEHAVIOR. *Educ:* Duke Univ, AB, 70; Cornell Univ, PhD(vert zool), 76. *Prof Exp:* From asst prof to assoc prof biol, 73-89, PROF BIOL, UNIV CONN, 89- *Concurrent Pos:* Fel, Smithsonian Trop Res Inst, 75-76. *Mem:* Am Soc Ichthyol & Herpet; Soc Study Amphibians & Reptiles; Ecol Soc Am; Animal Behav Soc. *Res:* Social behavior, mating systems and behavioral ecology of vertebrates, especially amphibians; history of science, especially evolutionary biology and early behavioral studies. *Mailing Add:* Dept Ecol Evolution Biol Univ Conn Storrs CT 06268

WELLS, LARRY GENE, b Covington, Ky, July 28, 47; m 71; c 2. AGRICULTURAL ENGINEERING. *Educ:* Univ Ky, BS, 69, MS, 71; NC State Univ, PhD(biol, agr eng & math), 75. *Prof Exp:* From asst prof to assoc prof, 74-86, PROF AGR ENG, UNIV KY, 86- *Concurrent Pos:* Prin investr, NSF Res Initiation Grant, 76-78, Philip Morris Res Grant, 82-88, Coop State Res Serv, 87-88; investr, Sci & Educ Admin Grant, 78-80. *Mem:* Am Soc Agr Eng; Int Soc Terrain Vehicle Systs. *Res:* Design and development of agricultural machinery; simulation of machinery systems; dynamic soil - machinery interactions; limiting vehicular soil compaction in surface mine reclamation. *Mailing Add:* Dept Agr Eng Univ Ky Lexington KY 40506-0075

WELLS, MARION ALVA, electrical engineering, for more information see previous edition

WELLS, MARION ROBERT, b Jackson, Miss, Feb 9, 37; m 59; c 4. MOLECULAR BIOLOGY. *Educ:* Memphis State Univ, BS, 60, MA, 63; Miss State Univ, PhD(zool), 71. *Prof Exp:* Instr biol, Troy State Col, 63-64; assoc prof, 64-77, PROF BIOL, MID TENN STATE UNIV, 77- *Mem:* Sigma Xi; Am Physiol Soc. *Res:* Binding of insecticides to cell particulate. *Mailing Add:* Dept Biol Mid Tenn State Univ Murfreesboro TN 37132

WELLS, MARK BRIMHALL, b Pocatello, Idaho, June 10, 29; m 74; c 6. COMPUTER SCIENCE, MATHEMATICS. *Educ:* Denver Univ, BA, 50; Univ Calif, Berkeley, PhD(math), 61. *Prof Exp:* Staff mem prog, Los Alamos Sci Lab, Univ Calif, 51-68; group leader computer sci, 68-73; vis prof math sci, Rice Univ, 70; vis prof math & computer sci, Univ Denver, 77-78; staff mem computer sci, Los Alamos Nat Lab, Univ Calif, 73-80; DEPT HEAD, COMPUTER SCI DEPT, NMEX STATE UNIV, 80- *Mem:* Asn Comput Mach; Am Math Soc; Math Asn Am. *Res:* Programming languages; combinatorial algorithm design. *Mailing Add:* Computer Sci Dept NMex State Univ Las Cruces NM 88003

WELLS, MICHAEL ARTHUR, b Los Angeles, Calif, Nov 8, 38; m 58; c 3. BIOCHEMISTRY. *Educ:* Univ Southern Calif, BA, 61; Univ Ky, PhD(biochem), 65. *Prof Exp:* Am Cancer Soc fel biochem, Univ Wash, 65-67; from asst prof to prof biochem, 67-86, HEAD BIOCHEM, UNIV ARIZ, 86- *Concurrent Pos:* Macy fac scholar, Josiah Macy Found, 75-76. *Mem:* AAAS; Am Chem Soc; Am Soc Biol Chemists. *Res:* Lipid and lipoprotein metabolism in insects. *Mailing Add:* Dept Biochem Biosci W Univ Ariz Tucson AZ 85721-0002

WELLS, MICHAEL BYRON, b Kansas City, Mo, July 3, 22; m 48; c 4. OPTICAL PHYSICS, RADIATION PHYSICS. *Educ:* Univ Mo, Kansas City, BA, 48, MA, 50. *Prof Exp:* Instr, Univ Mo, Kansas City, 50-56; proj nuclear engr, Gen Dynamics, Ft Worth, Tex, 56-63; vpres, Radiation Res Assocs, Inc, 63-89; RETIRED. *Concurrent Pos:* Instr, Tex Christian Univ, 58-63. *Mem:* Am Nuclear Soc; Optical Soc Am. *Res:* Nuclear radiation transport calculations; radiation shielding; atmospheric optics; ultraviolet, visible and infrared radiation transport in planetary atmospheres; nuclear engineering. *Mailing Add:* Well Consults Inc 3812 Glenmont Dr Ft Worth TX 76113

WELLS, MILTON ERNEST, b Calera, Okla, Nov 28, 32; m 55; c 2. ANIMAL BREEDING. *Educ:* Okla State Univ, BS, 55, MS, 59, PhD(animal breeding), 62. *Prof Exp:* Asst prof animal sci, Imp Ethiopian Col Agr, 61-65; asst prof dairy sci, Okla State Univ, 65-74, assoc prof, 74-77, prof animal sci, 77-80; SELF EMPLOYED. *Mem:* Am Soc Animal Sci; Am Dairy Sci Asn. *Res:* Reproductive physiology; vibriosis; international animal agriculture; acrosome of sperm cells. *Mailing Add:* Rte 1 Box 645 Stillwater OK 74074

WELLS, OTHO SYLVESTER, b Burgaw, NC, Sept 15, 38; m 68; c 2. HORTICULTURE. *Educ:* NC State Univ, BS, 61; Mich State Univ, MS, 63; Rutgers Univ, PhD(hort), 66. *Prof Exp:* Asst prof, 66-71, ASSOC PROF PLANT SCI, UNIV NH, 71-, EXTEN HORTICULTURIST, VEG, 74- *Mem:* Am Soc Hort Sci. *Res:* Crop production under environmentally controlled conditions. *Mailing Add:* Dept Plant Sci Univ NH Durham NH 03824

WELLS, OUIDA CAROLYN, b Atlanta, Ga, July 23, 33. ZOOLOGY. *Educ:* Agnes Scott Col, BA, 55; Emory Univ, MS, 56, PhD(genetics), 58. *Prof Exp:* Asst, Emory Univ, 55-57; res assoc, Oak Ridge Nat Lab, 58-60; asst prof biol, 60-65, assoc prof natural sci, 65-68, from asst dean to dean, 69-81, PROF BIOL, LONGWOOD COL, 68- *Mem:* Genetics Soc Am; Soc Protozool. *Res:* Genetics, cytology, physiology and biochemistry of microorganisms; radiation biology of ciliate Tetrahymena pyriformis, especially survival and death of irradiated cells using high levels of ionizing radiation. *Mailing Add:* Dept Nat Sci Longwood Col Farmville VA 23901

WELLS, PATRICK HARRINGTON, b Palo Alto, Calif, June 19, 26; m 51; c 3. BEE LANGUAGE CONTROVERSY. *Educ:* Univ Calif, AB, 48; Stanford Univ, PhD(biol), 51. *Prof Exp:* Asst prof zool, Univ Mo, 51-57; from asst prof to assoc prof biol, 57-71, PROF BIOL, OCCIDENTAL COL, 71- *Concurrent Pos:* Res assoc, Univ Calif, 67-72; ed, Southern Calif Acad Sci Bull, 72-75. *Mem:* Fel AAAS; Bee Res Asn; Am Soc Zool; Lepidop Soc; Sigma Xi; Hist Sci Soc. *Res:* Foraging behavior and recruitment in honey bees; pollination biology; natural history; population biology, ecology and physiology of butterflies and other arthropods. *Mailing Add:* Dept Biol Occidental Col Los Angeles CA 90041

WELLS, PATRICK ROLAND, b Liberty, Tex, Apr 1, 31. PHARMACOLOGY. *Educ:* Tex Southern Univ, BS, 57; Univ Nebr, MS, 59, PhD(pharmaceut sci), 61. *Prof Exp:* Asst prof pharmacol, Fordham Univ, 61-63; from asst prof to assoc prof & actg chmn dept, Univ Nebr, 63-70; PROF PHARMACOL & DEAN SCH PHARM, TEX SOUTHERN UNIV, 70- *Mem:* Nat Pharmaceut Asn; Sigma Xi; Am Pharmaceut Asn. *Res:* Cardiovascular screening of plant tissue culture. *Mailing Add:* 5330 Fairgreen Lane Houston TX 77048

WELLS, PHILIP VINCENT, b Brooklyn, NY, Apr 24, 28; m 59; c 4. BOTANY. *Educ:* Brooklyn Col, BA, 51; Univ Wis, MS, 56; Duke Univ, PhD(bot), 59. *Prof Exp:* Asst, Univ Wis, 54-55 & Duke Univ, 55-58; instr bot, Univ Calif, Santa Barbara, 58-59 & Calif Polytech Col, 59-60; res assoc biol, NMex Highlands Univ, 60-62; from asst prof to assoc prof bot, 62-71, PROF BOT & SYST ECOL, UNIV KANS, 71- *Concurrent Pos:* Actg dir, Bot Garden & vis assoc prof, Univ Calif, Berkeley, 66-67. *Mem:* AAAS; Soc Study Evolution; Bot Soc Am; Ecol Soc Am. *Res:* Pleistocene paleobotany; systematics, ecology and evolution in Arctostaphylos; vegetation of North America; physiological ecology. *Mailing Add:* Dept Biol Univ Kans 7026 Haw Lawrence KS 66045

WELLS, PHILLIP RICHARD, b Northampton, Mass, May 23, 36; m 62; c 3. FOOD SCIENCE, DAIRY SCIENCE. *Educ:* Univ Mass, BS, 57; Pa State Univ, MS, 59; Rutgers Univ, PhD(dairy sci), 62. *Prof Exp:* Proj leader, Food Prod Develop, Colgate Palmolive Co, 62-66; sect leader, New Prod Develop, Corn Prod Co, 66-72, asst to dir res, 72-75, asst dir tech serv, 75-78, asst to dir nutrit, 78-79, NUTRIT RES ASSOC, BEST FOODS, CPC INT, 79- *Mem:* Nutrit Today Soc; Inst Food Technol. *Res:* Dried and concentrated dairy products; snack products; stabilizers and emulsifiers; protein based foods; dried eggs; research management; food laws; advertising and labeling. *Mailing Add:* Best Foods Res Ctr CPC Int 1120 Commerce Ave Union NJ 07083

WELLS, RALPH GORDON, b Newark, Ohio, Sept 24, 15; m 42. MATERIALS SCIENCE. *Educ:* Muskingum Col, BA, 39; Ohio State Univ, MSc, 47; Univ Mich, PhD(mineral), 51. *Prof Exp:* Res metallurgist, US Steel Corp, 47-51, supvry technologist, 51-55; assoc res engr, Univ Mich, 56-59; sr res scientist, Res Ctr, Crucible Inc, 59-80; RETIRED. *Concurrent Pos:* Res assoc earth Sci, Carnegie Mus Natural Hist, 76- *Mem:* Am Phys Soc; sr mem Inst Elec & Electronics Engrs; Am Inst Mining Metall & Petrol Eng; Sigma Xi. *Res:* Application of tools and techniques of mineralogy and crystallography to metallurgy and ceramics. *Mailing Add:* 5253 Sherwood Dr Pittsburgh PA 15236-1838

WELLS, RAYMOND O'NEIL, JR, b Dallas, Tex, June 12, 40; m 63; c 2. MATHEMATICS. *Educ:* Rice Univ, BS, 62; NY Univ, MS, 64, PhD(math), 65. *Prof Exp:* From asst prof to assoc prof, 65-74, chmn dept, 76-79, PROF MATH, RICE UNIV, 74- *Concurrent Pos:* Vis asst prof, Brandeis Univ, 67-68; Fulbright grant, 68; mem, Inst Advan Study, 70-71 & 79-80; mem, Regional Conf Bd Math Sci, 74-77; Guggenheim fel, John Simon Guggenheim Mem Found, 74-75; US sr scientist award, Alexander von Humboldt Found, 74-75; vis prof math, Univ Gottingen, 74-75; mem, Coun Am Math Soc, 79-, US Nat Comt Math, 84-87 & US Comn Math Instr; ed, Trans & Mem Am Math Soc, 79-82, managing ed, 83-, Math Surv & Monogr, 83-85, managing ed, 85- & Contemp Math, managing ed, 83-; Ulam vis prof, Univ Colo, 83-84; Nat Acad Sci exchange vis, Bulgaria, 84; mem adv bd, Math Sci, 85-; mem, US Comm Math Inst, 85-; dir, Am Math Proj, Berkeley, 87. *Mem:* Am Math Soc; Math Asn Am; fel AAAS; Am Phys Soc; Asn Mem Inst Advan Study. *Res:* Analytic continuation and approximation theory in several complex variables; theory of complex manifolds and spaces; automorphic functions; algebraic geometry; mathematical physics. *Mailing Add:* Dept Math Rice Univ Houston TX 77251

WELLS, ROBERT DALE, b Uniontown, Pa, Oct 2, 38; m 60; c 2. BIOCHEMISTRY, MOLECULAR BIOLOGY. *Educ:* Ohio Wesleyan Univ, BS, 60; Univ Pittsburgh, PhD(biochem), 64. *Prof Exp:* NIH fel, Univ Pittsburgh, 64; NIH fel, enzyme Inst, Univ Wis-Madison, 64-66; from asst prof to assoc prof biochem, 66-73, prof biochem, 73-81, PROF & CHMN, DEPT BIOCHEM, SCHS MED & DENT, UNIV ALA, BIRMINGHAM, 82- *Concurrent Pos:* Mem regional coun, Am Inst Biol Sci, 66-68; Guggenheim Award, 76-77; ed, J Biol Chem, 77- *Mem:* AAAS; Am Chem Soc; Am Soc Biol Chem; Sigma Xi; Am Soc Microbiol. *Res:* Polynucleotides synthesis; DNA physical chemistry, synthesis and replication; polypeptide synthesis; tumor virology. *Mailing Add:* Dept Biochem Univ Ala University Sta Birmingham AL 35294

WELLS, RONALD ALLEN, b Norton, Va, Sept 12, 42; div; c 2. ARCHAEOASTRONOMY, EGYPTOLOGY. *Educ:* Univ Calif, Berkeley, AB, 64; Univ London, dipl, 66, PhD(astron), 67. *Prof Exp:* Assoc res astronr I, Space Sci Lab, Univ Calif, Berkeley, 67-69, II, 69-71 & III, 71-72 & res assoc, 73-79, res assoc, Dept Near Eastern Studies, 81-91, MICROCOMPUTER SPECIALIST, DEPT PLANT BIOL, PLANT GENE EXPRESSION CTR, UNIV CALIF, BERKELEY, 89- *Concurrent Pos:* Proj dir, NASA grant, 71-72; prin investr, Nat Endowment Humanities grant, 83-84; sr fel, Binat Fulbright Comn, Cairo, 83-84; lectr physics & astron, The Crowden Sch, Berkeley, 84-87; publ asst, Freeman-Cooper & Co, San Francisco, 85-86; sr fel, Binat Fulbright Comn, Hamburg, WGer, 87-88; prin invest, Michela Schiff Giorgini Found grant, 87. *Mem:* Fel Royal Astron Soc; Am Astron Soc; Am Res Ctr Egypt. *Res:* Origins and applications of astronomy in ancient Egypt from hieroglyphic texts, inscriptions, precise temple orientations and stellar-lunar rise and set positions; chronology and calendars of ancient Egypt. *Mailing Add:* Dept Plant Biol Univ Calif Berkeley CA 94720

WELLS, RUSSELL FREDERICK, b Brooklyn, NY, Oct 24, 37; div; c 2. EXERCISE PHYSIOLOGY, SPORTS MEDICINE. *Educ:* Lafayette Col, BA, 59; Springfield Col, MS, 62; Univ NC, Chapel Hill, MA, 66; Purdue Univ, PhD(biol educ), 70. *Prof Exp:* Teacher & coach pvt sch, NC, 62-65; NSF acad year fel zool, Univ NC, Chapel Hill, 65-66; asst prof biol, Montclair State Col, 66-68; asst prof biol sci, Purdue Univ, 70-71; asst prof, 71-74, assoc dean col, 81-83, ASSOC PROF BIOL, ST LAWRENCE UNIV, 74-,. *Concurrent Pos:* Adj prof, San Diego State Univ, 78-79; vis assoc res physiologist, Univ Calif, San Diego, 78-79; vis physiologist, Australian Inst Sport, Canberra, 80-87; vis fel, Fac Sci, Australian Nat Univ, 86-87. *Mem:* Am Inst Biol Sci; Nat Asn Biol Teachers; Am Col Sports Med. *Res:* Development of time-lapse cline studies of prolonged biological phenomens; physiology of exercise as it pertains to adult fitness and intercollegiate athletes. *Mailing Add:* Dept Biol St Lawrence Univ Canton NY 13617

WELLS, SAMUEL ALONZO, JR, b Cuthbert, Ga, Mar 16, 36; m 64; c 2. GENERAL SURGERY. *Educ:* Emory Univ, Atlanta, Ga, MD, 61. *Prof Exp:* Intern & resident internal med, Johns Hopkins Hosp, Baltimore, Md, 61-63; guest investr, Dept Tumor Biol, Karolinska Inst, Stockholm, Sweden, 63-64; clin assoc, Surg Br, Pub Health Serv, Nat Cancer Inst, NIH, Bethesda, Md, 64-66; asst resident & resident surg, Duke Univ, Durham, NC, 66-70; sr investr, Surg Br, Nat Cancer Inst, NIH, 70-72; from assoc prof to prof surg, Duke Univ Med Ctr, 72-81; BIXBY PROF SURG & CHMN DEPT, SCH MED, WASH UNIV, ST LOUIS, MO, 81- *Concurrent Pos:* Mem treatment comt, Breast Cancer Task Force, Nat Cancer Inst, NIH, 74-77, chmn, 77-78; mem bd sci counrs, Div Cancer Treatment, Nat Cancer Inst, NIH, 82-86, chmn, 84-86; ed-in-chief, World J Surg, 83-86. *Mem:* Soc Clin Surg (treas, 80-84); Am Surg Asn; Am Soc Clin Invest; Am Asn Cancer Res; Int Surg Soc; Endocrine Soc. *Res:* Cancer; endocrinology. *Mailing Add:* Dept Surg Wash Univ Sch Med 4960 Audubon Ave St Louis MO 63110

WELLS, STEPHEN GENE, b Linton, Ind, Mar 4, 49; m 74; c 2. GEOMORPHOLOGY OF ARID-SEMIARID LANDS, QUATERNARY GEOLOGY & LANDSCAPE EVOLUTION. *Educ:* Ind Univ, BS, 71; Univ Cincinnati, MS, 73, PhD(geol), 76. *Prof Exp:* From asst prof to prof geol/geomorphol, Dept Geol, Univ NMex, 76-91, asst chair, 86-89, chair, 89-91; PROF GEOMORPHOL, DEPT EARTH SCI, UNIV CALIF, RIVERSIDE, 91- *Concurrent Pos:* Fac res assoc, USAF Weapons Lab, Kirtland AFB, 77; fac appointment geologist, Western Minerals Br, US Geol Surv, 82-85; panel mem, Geol Soc Am, Quaternary Geol & Geomorphology Div, 83-85; vis lectr, Dept Geog, Univ Liverpool, 84; prin investr, NSF grants, 86-; pres, Neotec, Inc, 88-; pres lectr, Univ NMex, 88-90; vis affil, Earth & Environ Sci Div, Los Alamos Nat Labs, NMex, 89; Gladys Cole res award, Geol Soc Am, 91. *Mem:* Fel Geol Soc Am; Am Geophys Union; Int Asn Sedimentologists; Cave Res Found. *Res:* Geomorphology, quaternary geology, and environmental geology with an emphasis on surficial processes of dry (arid-semiarid) lands and Quaternary stratigraphy and landscape evolution within the southwestern United States; geomorphology and stratigraphy of alluvial fans and eolian landforms in arid and humid environments; arroyo/valley-floor processes and watershed management with emphasis on the variations in geomorphic processes related to changes in land management practices, vegetation conditions, and climate; reconstructions and semi-quantitative models of geomorphic and hydrologic responses to late Quaternary climatic changes in the Mojave Desert of southern California; quantitative assessments of differential uplift rates related to complex plate tectonic geometries of Central America and Spain; tectonic geomorphology of the eastern Mojave Desert Block and northern Rio Grande Rift in New Mexico; volcanic geomorphology/hazards in southwestern United States and quantifying the periodicity and character of small basaltic volcanic eruptions related to the proposed radioactive waste repository in southern Nevada; research problems are applied to practical problems of hazardous waste siting, reclamation, and land management problems. *Mailing Add:* Dept Earth Sci Univ Calif Riverside NM 92521

WELLS, WARREN F, b Des Moines, Iowa, May 16, 26; m 50; c 3. BIOCHEMISTRY. *Educ:* Univ Northern Iowa, BA, 50, MA, 55; Univ Ill, PhD(biochem), 59. *Prof Exp:* Instr sci, Clear Lake High Sch, Iowa, 50-52; instr chem, Undergrad Div, Univ Ill, 53-54, res asst biochem, Col Med, 54-59; from instr to asst prof, 59-70, ASSOC PROF BIOCHEM, MED & DENT SCHS, NORTHWESTERN UNIV, CHICAGO, 70-, DIR STUDENT SERV, 77- *Concurrent Pos:* Consult, Col Am Path, 62-64 & Vet Admin Hosp, 64-74; biochemist, Vet Admin Hosp, 63-65; lectr, Dept Ment Health, State of Ill, 67-69. *Mem:* NY Acad Sci; Am Asn Dent Schs. *Res:* Nucleic acid and protein metabolism of normal and pathologic nervous tissue; biochemical changes accompanying periodontal disease. *Mailing Add:* Dept Cell Molecular & Struct Biol Northwestern Univ McGaw Med Ctr Chicago IL 60611

WELLS, WILLARD H, b Austin, Tex, Feb 23, 31; m 56; c 2. OPTICAL PHYSICS. *Educ:* Univ Tex, BS, 52; Calif Inst Technol, PhD(physics), 59. *Prof Exp:* Scientist, Jet Propulsion Lab, Calif Inst Technol, 59-62, res group supvr, 62-67; CHIEF SCIENTIST, TETRA TECH, INC, HONEYWELL, 67- *Mem:* Am Phys Soc. *Res:* Quantum electronics and mechanics; applied mathematics; spacecraft mechanics and space communications; underwater optical systems; optical communications, especially fiber optics; radiactive transfer; coherent fiberoptics and interferometry. *Mailing Add:* Data Systs 630 N Rosemead San Diego CA 91720

WELLS, WILLIAM LOCHRIDGE, b Mayfield, Ky, Oct 12, 39. FLUE GAS DESULFURIZATION. *Educ:* Univ Ky, BSChE, 62; Univ Ill, Urbana, MS, 64, PhD(chem), 67; Univ SC, MSChE, 74; Univ Tenn, MBA, 82. *Prof Exp:* Asst, Univ Ill, 62-67; asst prof, Murray State Univ, 67-69; NSF grant & res assoc, Wayne State Univ, 69-71; asst prof, Southwest Baptist Col, 71-72; res asst chem eng, Univ SC, 72-73; prof, Midlands Tech Col, 73-75; chem engr, Tenn Valley Authority, 75-77, projs mgr air res, 77-80, prog mgr, gaseous emission control, 80-83; dir, Ctr Res Sulfur Coal, 83-91; CONSULT ENGR, ENERGY & ENVIRON AFFAIRS, 91- *Concurrent Pos:* Consult, var pvt & pub orgs, 83- *Mem:* Am Chem Soc; Royal Soc Chem; Am Inst Chem Engrs; Am Inst Indust Engrs. *Res:* Comparative economic evaluations; coal cleaning; coal conversion; flue gas desulfurization. *Mailing Add:* PO Box 407 Mayfield KY 42066

WELLS, WILLIAM RAYMOND, b Winder, Ga, Nov 28, 36; m 56; c 3. AEROSPACE ENGINEERING, APPLIED MATHEMATICS. *Educ:* Ga Inst Technol, BS, 59; Va Polytech Inst, MS, 61, PhD(aerospace eng), 68; Harvard Univ, MA, 64. *Prof Exp:* Aerospace technologist, Langley Res Ctr, NASA, 59-68; from asst prof to prof aerospace eng, Univ Cincinnati, 68-77; prof eng & chmn dept, Wright State Univ, 77-84; DEAN COL ENG, UNIV NEVADA, LAS VEGAS, 84- *Concurrent Pos:* NASA grant, Langley Res Ctr, 70-; consult, Flight Dynamics Lab, USAF, 72- *Mem:* Assoc fel Am Inst Aeronaut & Astronaut; Am Soc Eng Educ; Am Acad Mech; Am Asn Univ Prof; Sigma Xi. *Res:* Systems identification, flight mechanics and control theory. *Mailing Add:* 3561 Gallup Ct Las Vegas NV 89121-5901

WELLS, WILLIAM T, b Wytheville, Va, Aug 9, 33; m 55; c 3. APPLIED STATISTICS, APPLIED MATHEMATICS. *Educ:* Col William & Mary, BS, 54; NC State Univ, MS, 56, PhD(exp statist), 59. *Prof Exp:* Res asst appl math, NC State Col, 58-59; statistician, Res Triangle Inst, 59-62; engr guided missiles range div, Pan Am World Airways, Inc, 62-65; sr consult appl statist, 65-67, mgr appl sci dept, 67-68, vpres & dir appl sci div, Wolf Res & Develop Corp, 68-77; VPRES ENG, EG&G-WASH ANALYTICAL SERV CTR, INC, 77- *Mem:* Inst Math Statist; Am Statist Asn. *Res:* Statistical methods in trajectory determination; geodesy; instrumentation evaluation; biomedical data analysis. *Mailing Add:* 5211 Auth Rd Suite 204 Suitland MD 20746

WELLS, WILLIAM WOOD, b Traverse City, Mich, June 8, 27; m 50; c 4. BIOCHEMISTRY. *Educ:* Univ Mich, BS, 49, MS, 51; Univ Wis, PhD, 55. *Prof Exp:* Res assoc, Upjohn Co, 51-52; asst biochem, Univ Wis, 52-55; from instr to assoc prof, Univ Pittsburgh, 55-66; PROF BIOCHEM, MICH STATE UNIV, 66- *Concurrent Pos:* Mem, Metab Study Sect, NIH, 66-70. *Mem:* Am Chem Soc; Am Soc Biol Chem; Am Soc Neurochem; Int Soc Neurochem; Am Inst Nutrit. *Res:* Sterol structure and metabolism; relationship of sterol metabolism to experimental atherosclerosis; galactose metabolism and mental retardation; brain energy metabolism; metabolic regulations of lysosomes; microtubule associated enzymes; myoinositol and polyphosphoinositide metabolism; thioltransferase function. *Mailing Add:* Dept Biochem 413 Biochem Bldg Mich State Univ East Lansing MI 48824

WELLSO, STANLEY GORDON, b Oshkosh, Wis, Feb 13, 35; m 57; c 2. CEREAL INSECTS & BUPRESTIDAE. *Educ:* Univ Wis, BS, 57; Tex A&M Univ, MS, 62, PhD(entom), 66. *Prof Exp:* Asst prof entom, Colo State Univ, 65-67; entomologist, Agr Res Serv, Sci & Educ Admin-Fed Res, USDA, Mich State Univ, 67-73, res leader entom & small grains, 73-85, adj prof entom, 77-86; RES ENTOMOLOGIST & ADJ ASSOC PROF, PURDUE UNIV, 86- *Mem:* AAAS; Entom Soc Am; Sigma Xi; Coleopterists Soc. *Res:* Diapause induction and termination in insects; influence of photoperiod on insect's growth and development; rearing insects on artificial diets; host plant resistance of cereal insects; taxonomy and ecology of buprestids; small grain insects; insect behavior and feeding during oviposition; buprestid taxonomy; sugar depletion and cold hardiness of wheat affected by Hessian fly larval feeding; Hessian fly diapause; systematics of buprestids in the Chrysobthris femorata complex from North America and Actenodes especially from Central America. *Mailing Add:* Entom Dept Purdue Univ West Lafayette IN 47907-1158

WELMERS, EVERETT THOMAS, b Orange City, Iowa, Oct 27, 12; m 38; c 2. MATHEMATICS. *Educ:* Hope Col, AB, 32; Univ Mich, PhD(math), 37. *Hon Degrees:* ScD, Hope Col, 66. *Prof Exp:* From instr to asst prof math, Mich State Col, 37-44; flight res engr, Bell Aircraft Corp, 44, flutter engr, 44-46, group leader dynamic anal, 46-49, chief dynamics, 49-57, dir, L D Bell Res Ctr, 57-60, asst to pres, Corp, 58-60; group dir satellite systs, Aerospace Corp, 60-63, asst for tech oper, Manned Systs Div, 64-67, asst to gen mgr, El Segundo Tech Opers, 67-68, asst to pres, 68-77, aerospace historian, 78-80; dean, Col Eng, Northrop Univ, 80-82; RETIRED. *Concurrent Pos:* Prof, Millard Fillmore Col, 45-59; mem, Air Training Command Adv Bd, 57-68; on leave to Inst Defense Anal & Adv Res Proj Agency, 59-60; comnr, Community Redevelop Agency, City Los Angeles, 68-83. *Mem:* Am Math Soc; Math Asn Am; Am Inst Aeronaut & Astronaut; Inst Elec & Electronics Engrs; Opers Res Soc Am. *Res:* Integration theory; jet propulsion; flutter; applied mathematics; aircraft and helicopter dynamics; operations analysis; computers; system analysis; satellites. *Mailing Add:* 1626 Old Oak Rd Los Angeles CA 90049

WELNA, CECILIA, b New Britain, Conn. MATHEMATICS. *Educ:* St Joseph Col, Conn, BS, 49; Univ Conn, MA, 52, PhD(ed), 60. *Prof Exp:* Instr, Mt St Joseph Acad, 49-50; asst instr math, Univ Conn, 50-55; instr, Univ Mass, 55-56; from asst prof to prof, 56-68, prof math & chmn dept, 68-82, DEAN, COL EDUC, NURSING & HEALTH PROFESSIONS, UNIV HARTFORD, 82- *Mem:* Math Asn Am; Sigma Xi; Nat Coun Teachers Math. *Mailing Add:* 176 Smith St New Britain CT 06053-3621

WELPLY, JOSEPH KEVIN, b New York, NY, Jan 2, 53; m 84; c 2. BIOCHEMISTRY. *Educ:* Univ Calif, Santa Barbara, BA, 75, Los Angeles, PhD(biochem), 81. *Prof Exp:* Fel biochem, Johns Hopkins Sch Med, 81-83, M D Anderson Hosp & Tumor Inst, 83-85, Oxford Univ, 85-86; sr res chemist, 86-87, res specialist, 87-89, SR RES SPECIALIST, MONSANTO CO, 89- *Concurrent Pos:* Assoc ed, Glycobiol, 91. *Mem:* AAAS. *Res:* Role of cell surface carbohydrate-recognizing proteins in cell adhesion, particularly as adhesion relates to inflammation; development of agents which prevent cell-adhesion and result in anti-inflammatory therapeutics. *Mailing Add:* 800 N Lindbergh Blvd T30 St Louis MO 63167

WELSCH, CLIFFORD WILLIAM, JR, b St Louis, Mo, Sept 10, 35; m 58; c 3. PHYSIOLOGICAL CHEMISTRY, ONCOLOGY. *Educ:* Univ Mo, BS, 57, MS, 62, PhD(physiol chem), 65. *Prof Exp:* Instr physiol chem, Univ Mo, 63-65; asst prof natural sci, 65-66, from asst prof to assoc prof anat, 68-76, PROF ANAT, COL HUMAN MED, MICH STATE UNIV, 76- *Concurrent Pos:* Nat Cancer Inst res fel oncol, Mich State Univ, 66-68; NIH career development award, 71-76; assoc ed, J Cancer Res, 80- *Mem:* AAAS; Am Asn Cancer Res; Soc Exp Biol & Med; Am Physiol Soc. *Res:* Investigations in mammary gland carcinogenesis. *Mailing Add:* Dept Pharmacol & Toxicol 357 Giltner Hall Mich State Univ East Lansing MI 48824

WELSCH, FEDERICO, b Sevilla, Spain, Dec 26, 33; US citizen; m 59; c 4. MOLECULAR BIOLOGY, MEDICINE. *Educ:* Univ Barcelona, BA, 50; Univ Valencia. MD, 55, DMedSci, 57; Ctr Res & Advan Study, Mex, MS, 65; Dartmouth Col, PhD(molecular biol), 68. *Prof Exp:* Intern internal med, Univ Hamburg, 54-55; from instr to asst prof physiol chem, Univ Valencia, 55-57; prof physiol, Univ Guadalajara, 58-60; prof biochem, Univ Chihuahua, 60-63; instr, Ctr Advan Res & Study, Mex, 63-65; asst prof, Dartmouth Med Sch, 68-70; assoc dir, 70-74, EXEC DIR, WORCESTER FOUND EXP BIOL, 74-; ASSOC RES PROF BIOCHEM, MED SCH, UNIV MASS, 70- *Concurrent Pos:* USPHS & Am Heart Asn grants; hon collabr, Span Res Coun, 58; sr Fulbright scholar, Peru, 75; pub policy scholar, Off Dir, NIH, 77; mem IV prog, health syst mgt, Bus Sch, Harvard Univ, 75; govt affairs liaison, Asn Independent Res Insts; budget analyst, Deleg Basic Biomed Res, 78- *Mem:* AAAS; Am Chem Soc; Asn Am Med Cols; Asn Independent Res Insts (pres, 83-85). *Res:* Molecular biology and medicine. *Mailing Add:* Calderwood Neck Rd Vinalhaven ME 04863

WELSCH, FRANK, b Berlin, Ger, Apr 14, 41; div; c 1. PRENATAL TOXICOLOGY, TERATOGENESIS. *Educ:* Free Univ, Berlin, DVM, 65; Am Bd Toxicol, dipl, 81. *Prof Exp:* From instr pharmacol to prof pharmacol & toxicol, Mich State Univ, East Lansing, 71-82; STAFF SCIENTIST, DEPT EXP PATH & TOXICOL, CHEM INDUST INST TOXICOL, 82- *Concurrent Pos:* Prin investr, Extramural Prog, NIH, Nat Inst Child Health & Human Develop, 72-80, Nat Found March Dimes, 75-78 & Nat Inst Environ Health Sci, 80-82; res fel award, Alexander Humboldt Found, 77-78; mem, Reproduction & Develop Toxicol Prog Rev Subcomt, Bd Sci Counr, Nat Toxicol Prog, 87-, Sci Rev Panel Health Res, Environ Protection Agency, 86-, Rev Comt Health Effects, Elec Power Res Inst, 87- *Mem:* Am Soc Pharmacol & Exp Therapeut; Soc Toxicol; Teratol Soc; Ger Soc Pharmacol & Toxicol. *Res:* Effects of chemicals on structural, biochemical and functional development of mammalian species; mechanisms of teratogenesis; placenta function. *Mailing Add:* Chem Indust Inst Toxicol PO Box 12137 Research Triangle Park NC 27709

WELSCH, GERHARD EGON, b Saarland, Ger, Sept 19, 44; m 73; c 4. MATERIALS SCIENCE & ENGINEERING. *Educ:* Aachen Tech Univ, Dipl Ing, 68; Case Western Reserve Univ, MS, 70, PhD(mat sci), 74. *Prof Exp:* Res scientist aircraft mat, Deutsche Forschungs, Versuchsanstalt for Luft, Raumfahrt, Cologne, Ger, 74-76; sr res engr refractory metals, Gen Elec Co, 76-79; ASSOC PROF, CASE WESTERN RES UNIV, 79- *Mem:* Ger Metall Soc; Am Soc Metals; Metall Soc-Am Inst Mining Engrs; Mat Res Soc. *Res:* Refractory metals and titanium alloys; heat treatment and processing of metals; electron microscopy; ion implantation into metal surfaces; metallurgy of high temperature materials. *Mailing Add:* 2514 Edge Hill Rd Cleveland Heights OH 44106

WELSCH, ROY ELMER, b Kansas City, Mo, July 31, 43. STATISTICS. *Educ:* Princeton Univ, AB, 65; Stanford Univ, MS, 66, PhD(math), 69. *Prof Exp:* Asst prof opers res, 69-73, assoc prof, 73-79, PROF MGT SCI & STATIST, SLOAN SCH MGT, MASS INST TECHNOL, 79- *Concurrent Pos:* Assoc ed, J Am Statist Asn. *Mem:* Fel Am Statist Asn; Inst Math Statist. *Res:* Data analysis; robust statistics; multiple comparisons; graphics. *Mailing Add:* Dept Mgt & Statist Mass Inst Technol 77 Massachusetts Ave Cambridge MA 02139

WELSH, BARBARA LATHROP, b New London, Conn; m; c 5. MARINE ECOLOGY. *Educ:* Mt Holyoke Col, BA, 57; Univ Md, MS, 70; Univ RI, PhD(oceanog), 73. *Prof Exp:* Instr zool, Univ Md, 68-69; sr ecologist marine systs, Vast, Inc, 72-73; asst prof biol sci, 73, ASSOC PROF MARINE SCI & BIOL SCI, UNIV CONN, 73- *Concurrent Pos:* Appointment, comt hazardous mat, Nat Res Coun, Nat Acad Sci, 73-76; consult, Normandean Assocs, 75- & Environ Qual Bd, Commonwealth PR, 76-77. *Mem:* Estuarine Res Fedn (pres, 81-83); Am Geophys Union; Sigma Xi. *Res:* Ecology of marine coastal systems, particularly detrital based systems with tidal interaction. *Mailing Add:* 52 Shore Dr Waterford CT 06385

WELSH, DAVID ALBERT, b Pittsburgh, Pa, Oct 25, 42; m 64; c 3. ORGANIC CHEMISTRY, POLYMER CHEMISTRY. *Educ:* Carnegie Mellon Univ, BS, 64, MS, 68, PhD(org chem), 69. *Prof Exp:* Res chemist, Kippers Co Res Labs, 68-69, Edgewood Arsenal, 69-71 & Koppers Co Res Labs, 71-74; SR RES SCIENTIST, RES DEPT, ARCO/POLYMERS INC, 74- *Mem:* Am Chem Soc; Sigma Xi. *Res:* Free radical polymerization and copolymerization; polymer properties; organic synthesis. *Mailing Add:* 1042 Harvard Rd Monroeville PA 15146

WELSH, DAVID EDWARD, b Chicago, Ill, July 20, 42; m 85; c 1. INTERCONNECTION ENGINEERING, FIBER OPTICS INTERCONNECTION TECHNOLOGY. *Educ:* Calif Coast Univ, BS, 76, MS, 88. *Prof Exp:* Proj engr IC sockets, Scanbe Div Zero Mfg, 69-73; prog mgr keyswitches, Hi-Tek, 75-76; mgr eng serv elec connectors, Malco, A Microdot Co, 76-82, mgr res & develop fiber optics, 82-87; sr prod engr elec connectors, 73-75, MGR NEW PROD DEVELOP ELEC CONNECTORS, ITT CANNON, 87- *Concurrent Pos:* Chmn, US Chess Fedn Computer Chess Comt, 82-90. *Mem:* AAAS; Inst Elec & Electronics Engrs; Int Inst Connector & Interconnection Technol. *Res:* Interconnection technology; fiber optics; computer simulation of optical interconnections, wargaming; author of various publications. *Mailing Add:* 17512 Chatham Dr Tustin CA 92680

WELSH, GEORGE W, III, b New York, NY, Aug 7, 20; m 50; c 2. INTERNAL MEDICINE, ENDOCRINOLOGY. *Educ:* Yale Univ, BA, 42; Univ Rochester, MD, 50; Am Bd Internal Med, dipl, 57. *Prof Exp:* Asst med, Med Col, Cornell Univ, 51-52, Dartmouth Med Sch, 52-54 & Sch Med, Univ Wash, 54-55; from instr to asst prof med, 56-64, dir continuing med educ, 65-68, dir off continuing educ health sci, 68-74, ASSOC PROF MED, COL MED, UNIV VT, 64- *Concurrent Pos:* Fel med, Mary Hitchcock Hosp, Hanover, NH, 52-53; Nat Inst Arthritis & Metab Dis trainee, Sch Med, Univ Wash, 54-55, USPHS res fel, 55; USPHS grants, 62-66; chmn interdept coun aging, State of Vt, 63-69; mem adv comt, Northeast Regional Med Libr Serv, 70-74, chmn, 72; adv coun, Off Health Care Educ, Northeast Ctr Continuing Educ, 71- *Mem:* Fel Am Col Physicians; Am Diabetes Asn; Endocrine Soc; Am Fedn Clin Res; Am Soc Internal Med. *Mailing Add:* Univ Assocs Med Univ Health Ctr One S Prospect St Burlington VT 05401

WELSH, JAMES FRANCIS, b Pittsburgh, Pa, June 21, 30; m 55; c 5. ZOOLOGY. *Educ:* State Univ NY, BA, 53; Univ Calif, Los Angeles, PhD, 59. *Prof Exp:* Instr biol, St Mary's Col, 57-58; instr zool & human anat, Univ Calif, Los Angeles, 58-59; asst prof anat & physiol, 59-67, assoc prof physiol,

67-70, PROF ZOOL, HUMBOLDT STATE UNIV, 70- *Mem:* AAAS; Am Soc Parasitol; Wildlife Dis Asn; Genetics Soc Am. *Res:* Immunological research on Hymenolepis nana; enzyme systems of trematodes. *Mailing Add:* Dept Biol Humboldt State Univ Arcata CA 95521

WELSH, JAMES P, b Buffalo, NY. THERMAL SYSTEMS MANAGEMENT. *Educ:* Carnegie Tech, Pittsburg, BS, 38. *Prof Exp:* PRES THERMAL TECH LAB, BUFFALO, NY, 71- *Mem:* Fel Inst Elec & Electronics Engrs. *Res:* Light weight magnetic and thermal design of lightweight power systems. *Mailing Add:* 504 Brantwood Dr Snyder NY 14226

WELSH, JAMES RALPH, b Langdon, NDak, Sept 4, 33; m 52; c 4. PLANT BREEDING, PLANT GENETICS. *Educ:* NDak State Univ, BS, 56; Mont State Univ, PhD(plant genetics), 63. *Prof Exp:* Exten agt agron, NDak Exten Serv, 56-60; from asst prof to assoc prof, Mont State Univ, 63-68; assoc prof, Colo State Univ, 68-72, prof agron, 72-80; DEAN AGR & DIR, AGR EXP STA, MONT STATE UNIV, 80- *Mem:* Fel Am Soc Agron; Crop Sci Soc Am; Sigma Xi. *Res:* Wheat breeding and genetics; drought tolerance in winter wheat; high yielding cultivars. *Mailing Add:* Dean Agr Mont State Univ Bozeman MT 59717

WELSH, JOHN ELLIOTT, SR, b Berea, Ky, Nov 4, 27; m 65; c 3. GEOLOGY. *Educ:* Berea Col, AB, 50; Univ Wyo, MA, 51; Univ Utah, PhD(geol), 59. *Prof Exp:* Jr geologist, Magnolia Petrol Corp, 51; geologist, Shell Oil Co, 53-56; asst prof geol, Western State Col Colo, 56-61 & Colo State Univ, 61-62; res geologist, Kennecott Explor Inc, Salt Lake City, 70-79; CONSULT GEOLOGIST, 79- *Concurrent Pos:* Consult geologist, 56-70. *Mem:* Am Asn Petrol Geol; Soc Econ Geol. *Res:* Structural and stratigraphic analyses of mining districts; stratigraphy and structure of the overthrust belt and the basin and range province in the western United States; nonmetallic geology. *Mailing Add:* 4780 Bonair St Holladay UT 84117

WELSH, LAWRENCE B, b Santa Barbara, Calif, Oct 21, 39; m 61; c 2. SOLID STATE PHYSICS. *Educ:* Pomona Col, BA, 61; Univ Calif, Berkeley, PhD(physics), 66. *Prof Exp:* Fel physics, Univ Pa, 66-68; asst prof, Northwestern Univ, Evanston, 68-74; group leader mat sci, 74-81, RES SCI, UOP, INC, 81- *Concurrent Pos:* Vis asst prof physics, Northwestern Univ, 74-75. *Mem:* Am Phys Soc; AAAS. *Res:* Ceramics; thin film depositions; fuel cell technology; nuclear magnetic resonance in metals; zeolites; catalysts. *Mailing Add:* 2203 Lincolnwood Dr Evanston IL 60201

WELSH, MICHAEL JAMES, MEMBRANE PHYSIOLOGY, EPITHELIAL TRANSPORT. *Educ:* Univ Iowa, MD, 74. *Prof Exp:* ASSOC PROF MED, DEPT INTERNAL MED, UNIV IOWA, 81- *Res:* Airway smooth muscle. *Mailing Add:* Dept Med Univ Iowa Hosp Iowa City IA 52242

WELSH, RICHARD STANLEY, b Philadelphia, Pa, Nov 15, 21; m 56; c 2. BIOCHEMISTRY. *Educ:* Harvard Col, SB, 43; Univ Pa, MS, 46; Stanford Univ, PhD(biochem, phys chem), 52. *Prof Exp:* Asst instr inorg anal, Univ Pa, 44-45; technician virus res, Rockefeller Inst, 46-47; res asst, Stanford Univ, 57-59; Am Heart Asn advan res fel, Univ Redlands, 59-61, Am Heart Asn estab investr, 61-64; Am Heart Asn estab investr, Univ Calif, Riverside, 64-65, Lab Molecular Biol, NIH, 65-66 & Brookhaven Nat Lab, 66-67; assoc scientist, Div Microbiol, Brookhaven Nat Lab, 67-68; res assoc, Inst Med, Atomic Res Inst, WGer, 68-86; RETIRED. *Concurrent Pos:* Am Heart Asn res fel virus res, Univ Redlands, 57-59. *Mem:* Biophys Soc; NY Acad Sci. *Res:* Molecular characterization of DNA subunit fractions isolated nondegradatively from calf thymus, liver and other sources; polymerization reactions of the subunits with specific phosphopeptides, enzymes, adenosinetriphosphate and their biological significance; characterization of the phosphopeptides split out of DNA during its cleavage into subunits by reaction with chelating reagents and mechanism of the reaction; amino acid and sewuence analysis of the phosphopeptides. *Mailing Add:* Inst Med Res Ctr Julich 517 Germany

WELSH, ROBERT EDWARD, b Pittsburgh, Pa, Oct 1, 32; m 56, 88; c 3. PHYSICS. *Educ:* Georgetown Univ, BS, 54; Pa State Univ, PhD(physics), 60. *Prof Exp:* Res physicist, Carnegie Inst Technol, 60-63, asst dir nuclear res ctr, 62-63; assoc prof physics, Col William & Mary, 63-68, asst dir, Space Radiation Effects Lab, 67-72, chair, Dept Physics, 88-91, PROF PHYSICS, COL WILLIAM & MARY, 68- *Concurrent Pos:* Consult, Langley Res Ctr, NASA, 64-66 & Los Alamos Sci Lab 77-80; guest scientist, Argonne Nat Lab & Brookhaven Nat Lab, 71-; res physicist, Rutherford Lab, Eng, 72-73; sci assoc, Europ Coun Nuclear Res, Geneva, Switz, 83, 84. *Mem:* Fel Am Phys Soc; Am Asn Univ Prof. *Res:* Experimental particle physics; muon physics; exotic atoms studies with muons, pions, kaons, and sigma hyperons; antiproton-proton atomic systems; rare decay searches with neutral K mesons. *Mailing Add:* Dept Physics Col William & Mary Williamsburg VA 23185

WELSH, RONALD, b Houston, Tex, Oct 13, 26; m 50; c 3. PATHOLOGY. *Educ:* Univ Tex, BA, 47, MD, 50; Am Bd Path, dipl, 55. *Prof Exp:* Asst prof path, Univ Tex Med Br, Galveston, 55-57; from asst prof to assoc prof, 57-63, PROF PATH, SCH MED, LA STATE UNIV, NEW ORLEANS, 63- *Concurrent Pos:* Pathologist, Univ Tex Med Br Hosps, Galveston, 55-57; sr vis pathologist, Charity Hosp, New Orleans, 57-, sect dir surg path; path consult, Vet Admin Hosp, New Orleans, Children's Hosp New Orleans. *Honors & Awards:* Am Cancer Soc Award, 81. *Mem:* AMA; Am Soc Clin Path; Col Am Path; Int Acad Path. *Res:* Thyroid disease; electron microscopy, particularly of basic activities of inflammatory cells; oncology; surgical pathology. *Mailing Add:* Dept Path La State Univ Med Ctr 1901 Perdido St New Orleans LA 70112

WELSH, SUSAN, FOOD & NUTRIENT CONSUMPTION PATTERNS. *Educ:* Univ Md, PhD(nutrit), 74. *Prof Exp:* DIR, DIV EDUC, HUMAN NUTRIT INFO SERV, USDA, 85- *Mailing Add:* 12105 Darnestown Rd Gaithersburg MD 20878

WELSH, THOMAS LAURENCE, b Chicago, Ill, Nov 8, 32; m 60; c 3. FOOD SCIENCE, PHARMACEUTICAL CHEMISTRY. *Educ:* Univ Ill, BS, 59, PhD(pharmaceut chem), 63. *Prof Exp:* Res pharmacist, Miles Labs, Inc, 62-67; tech dir, Xttrium Labs, 67-69; vpres res & develop, 69-87, VPRES OPERS, MILES LABS, INC, 87- *Mem:* Soap & Detergent Asn; Sigma Xi. *Res:* Protein food products; effervescent pharmaceutical products; micronutrient delivery systems; human nutrition; household insecticides. *Mailing Add:* Miles Inc 7123 W 65th St Chicago IL 60638

WELSH, WILLIAM JAMES, b Philadelphia, Pa, Dec 24, 47. THEORETICAL & PHYSICAL CHEMISTRY. *Educ:* St Joseph Col, BS, 69; Univ Pa, PhD(theoret phys chem), 75. *Prof Exp:* Instr chem, Univ Pa, 69-74; chemist, Procter & Gamble Co, 75-78; ASST PROF DEPT CHEM & PHYSICS, EDGECLIFF COL, 78-; asst prof, Univ Cincinnati, 80-85. *Concurrent Pos:* Fel theoret polymer chem, Univ Cincinnati, 78-; extramural assoc, NIH, 86. *Mem:* Am Chem Soc; Am Phys Soc. *Res:* Configuration, dependent properties and conformational energies of long chain molecules; statistical mechanics; intermolecular forces; theory of liquids; physical adsorption; structure-property relationships of anticancer drugs and other bioactive materials; molecular computer graphics. *Mailing Add:* Chemistry Dept Univ Mo 8001 Natural Bridge Rd St Louis MO 63121-4401

WELSHIMER, HERBERT JEFFERSON, b West Mansfield, Ohio, Feb 23, 20; m 46; c 3. BACTERIOLOGY. *Educ:* Ohio State Univ, BSc, 43, PhD(bact), 47. *Prof Exp:* Asst med bact, Ohio State Univ, 44-46; instr bact, Ind Univ, 47-49; from asst prof to prof bact, 49-85, EMER PROF MICROBIOL & IMMUNOL, MED COL VA, VA COMMONWEALTH UNIV, 85- *Concurrent Pos:* USPHS fel, Ohio State Univ, 47; attend bacteriologist, Johnston-Willis Hosp, 54-80; chmn subcomt listeria & related organisms, Int Comt Syst Bact, 74-82. *Mem:* AAAS; Am Soc Microbiol; fel Am Acad Microbiol; NY Acad Sci. *Res:* Clinical bacteriology; immunology; lysozyme; bacteriophage; bacterial cytology; listeriosis. *Mailing Add:* 7400 Biscayne Rd Richmond VA 23294

WELSHONS, WILLIAM JOHN, b Pitcairn, Pa, July 18, 22; m 49; c 4. GENETICS. *Educ:* Univ Calif, BA, 49, MA, 52, PhD(zool), 54. *Prof Exp:* Res assoc, Biol Div, Oak Ridge Nat Lab, 54-55, mem staff, 55-65; head dept, 65-75, PROF GENETICS, IOWA STATE UNIV, 65- *Concurrent Pos:* NSF sr fel, Santiago, Chile, 63-64; mem adv comt, Chem Rev Bd, State Iowa, 70-71. *Mem:* Genetics Soc Am; Am Inst Biol Sci; Sigma Xi. *Res:* Gene structure recombination and cytogenetic analysis in Drosophila; mammalian sex determination and cytogenetics. *Mailing Add:* Dept Genetics Iowa State Univ Ames IA 50011

WELSTEAD, WILLIAM JOHN, JR, b Newport News, Va, July 17, 35; m 57; c 3. ORGANIC CHEMISTRY. *Educ:* Univ Richmond, BS, 57; Univ Va, PhD(chem), 62. *Prof Exp:* Org chemist, Army Chem Ctr, Md, 62; NSF grant, Iowa State Univ, 63; res chemist, 64-72, assoc dir chem res, 72-73, DIR CHEM RES, A H ROBINS CO, 73- *Mem:* AAAS; Am Chem Soc; Sigma Xi. *Res:* Synthetic organic and medicinal chemistry; heterocyclics. *Mailing Add:* 8306 Brookfield Rd Richmond VA 23227

WELSTED, JOHN EDWARD, b Norwich, Eng, Dec 6, 35; div; c 2. PHYSICAL GEOGRAPHY. *Educ:* Bristol Univ, BSc, 58, cert educ, 61, PhD(geog), 71; McGill Univ, MSc, 60. *Prof Exp:* Asst master geog, Maidenhead Grammar Sch, Eng, 61-62; teacher, Oromocto High Sch, NB, 62-64; demonstr geog, Bristol Univ, 64-65; from asst prof to assoc prof, 65-78, PROF GEOG, BRANDON UNIV, 78- *Honors & Awards:* Autometric Award, Am Soc Photogram, 80. *Mem:* Can Asn Geog; Can Water Resources Asn; Geol Asn Can; Am Soc Photogram & Remote Sensing. *Res:* Rate of meander migration on rivers in southwest Manitoba; flooding by the rivers of southwest Manitoba; legal implications of shifting miler channels in Southwest Manitoba. *Mailing Add:* Dept Geog Brandon Univ Brandon MB R7A 6A9 Can

WELT, ISAAC DAVIDSON, b Montreal, Que, May 13, 22; nat US; m 45; c 3. INFORMATION SCIENCE, DOCUMENTATION. *Educ:* McGill Univ, BSc, 44, MSc, 45; Yale Univ, PhD(physiol chem), 49. *Prof Exp:* Instr chem, Sir George Williams Col, 46; lab asst anat, Yale Univ, 46-47, asst, Nutrit Lab, 47-48; instr chem, New Haven YMCA Jr Col, Conn, 48-49; asst, Div Physiol & Nutrit, Pub Health Res Inst New York, Inc, 49-51; asst prof biochem, Col Med, Baylor Univ, 51-53; res assoc pharmacol, Chem-Biol Coord Ctr, Nat Res Coun, Washington, DC, 53-55, dir cardiovasc lit proj, Div Med Sci, Nat Acad Sci-Nat Res Coun, 55-61; assoc dir & chief, Wash Br, Inst Advan Med Commun, 61-64; prog dir sci & tech info systs, 64-67, prof info sci, Ctr Technol & Admin, 64-87, PROF, COMPUTER SCI INFO SYST, AM UNIV. *Concurrent Pos:* Asst dir radioisotope unit, Vet Admin Hosp, Tex, 51-53; prof lectr chem, Am Univ, 56-61. *Honors & Awards:* Info Sci & Technol Coun Award. *Mem:* AAAS; Am Soc Info Sci. *Res:* Endocrine influences and isotopes in intermediary; metabolism; nutrition; medical and biological literature research; chemical-biological correlations; research administration; education in information science and documentation; information storage and retrieval systems; computers and society. *Mailing Add:* Dept Computer Sci & Info Systs Am Univ Washington DC 20016

WELT, MARTIN A, b Brooklyn, NY, Oct 7, 32; m 62; c 3. NUCLEONICS. *Educ:* Clarkson Col, BChE, 54; Iowa State Univ, MS, 55; Mass Inst Technol, SM, 57; NC State Univ, PhD(physics), 64. *Prof Exp:* Reactor physicist, US AEC, Washington, DC, 57-58; supvr energy conversion sect, Chance Vought Corp, 59-61; pres, Int Sci Corp, NC, 61-67; pres, Radiation Technol, Inc, 68-86; FOUNDER & CHMN, ALPHA OMEGA TECHNOL, INC, 86- *Concurrent Pos:* Lectr, George Washington Univ, 58; adj prof, Southern Methodist Univ, 60-62; asst prof, NC State Univ, 64-67; mem, Ames Lab, US AEC, 54-55; aeronaut res scientist, Lewis Lab, Nat Adv Comt Aeronaut, 56; aeronaut res scientist, Union Carbide Co, Tenn, 56-66; dir, Adv Res Assocs, 62-63; mem, Working Comt, Proj Starfire, Southern Interstates Nuclear Bd, 64-65; dir, Nuclear Reactor Proj, NC State Univ, 64-66. *Mem:* AAAS; Am Nuclear Soc; Am Chem Soc; Am Phys Soc; Inst Food Technologists. *Res:*

Radiation preservation of food; radioisotope and radiation physics; plasma oscillations; radiation processing; design and analysis of nuclear facilities; hazards evaluation; thermoelectric energy conversion; nuclear research administration. *Mailing Add:* 1279 Rte 46 Parsippany NJ 07054

WELTER, ALPHONSE NICHOLAS, b Dudelange, Luxembourg, Apr 8, 25; US citizen; m 54; c 5. ANATOMY, HYDROBIOLOGY. *Educ:* Loras Col, AB, 52; Univ Ill, MS, 57, PhD(physiol), 59. *Prof Exp:* Instr physiol, Sch Med, Marquette Univ, 59-62; res physiologist, Lederle Div, Am Cyanamid Co, 62-67; RES SPECIALIST, 3M CO, 67- *Concurrent Pos:* USPHS fel, 59-61; Nat Heart Inst res fel, 61-62; grants, Wis Heart Asn, 60-61 & Am Heart Asn, 60-62. *Mem:* AAAS; assoc mem Am Physiol Soc; NY Acad Sci; Soc Environ Toxicol & Chem. *Res:* Cardiovascular physiology and pharmacology, specifically pulmonary circulation; respiratory and renal physiology. *Mailing Add:* 7181 County Rd Woodbury MN 55125

WELTER, C JOSEPH, b Tiffin, Ohio, June 11, 32; m 55; c 3. MICROBIOLOGY, IMMUNOLOGY. *Educ:* King's Col, BS, 54; Univ Notre Dame, MS, 56; Mich State Univ, PhD, 59. *Prof Exp:* Asst zool, Univ Notre Dame, 54-56; asst parasitol, Mich State Univ, 56-59; dir res, Diamond Labs, Inc, 59-69, vpres res, 69-74; PRES, AMBICO, INC, 74- *Mem:* AAAS; Am Soc Parasitol; US Animal Health Asn; Am Soc Trop Med & Hyg; Am Soc Microbiol. *Res:* Protozoology; parasitology; virology. *Mailing Add:* 3906 SW 28 Pl Des Moines IA 50321

WELTER, DAVE ALLEN, b Lorain, Ohio, Aug 7, 36; m 64; c 2. CYTOGENETICS, CHROMOSOME STRUCTURE. *Educ:* Univ Ga, BS, 61; Med Col Ga, MS, 62, PhD(anat), 70. *Prof Exp:* Dir cytogenetics lab, Gracewood Hosp, 62-69; instr, 70-72, asst prof, 72-79, ASSOC PROF ANAT, MED COL GA, 79- *Concurrent Pos:* Cytogenetic consult, Ft Gordon Hosp, 65 & Gracewood Hosp, Ga, 69- *Res:* Birth defects; neuroanatomy; gross anatomy; embryology. *Mailing Add:* 505 Henderson Dr Augusta GA 80909

WELTMAN, CLARENCE A, b New York, NY, Mar 17, 19; m 43; c 2. PHYSICAL CHEMISTRY, ORGANIC CHEMISTRY. *Educ:* NY Univ, BA, 40. *Prof Exp:* Assoc chemist, Explosives Res Lab, Nat Defense Res Comt, 41-45; res chemist, ALOX Corp, 45-49, chief chemist, 49-54, exec vpres & tech dir, 54-60, pres, 60-88; CONSULT, 88- *Concurrent Pos:* Vpres, RPM Inc, 80-, corp dir res & develop, 81-88. *Mem:* AAAS; Am Chem Soc; Am Soc Testing & Mat; Am Soc Lubrication Eng; Nat Asn Corrosion Eng. *Res:* Development of organic surface active agents and their application to problems of lubrication and corrosion prevention. *Mailing Add:* 5068 Woodland Dr Lewiston NY 14092

WELTMAN, JOEL KENNETH, b New York, NY, May 22, 33; m 56; c 2. IMMUNOLOGY, BIOCHEMISTRY. *Educ:* State Univ NY, MD, 58; Univ Colo, PhD(microbiol), 63. *Hon Degrees:* MA, Brown Univ, 72. *Prof Exp:* Intern, Ind Univ, 58-59; instr microbiol, Univ Colo, 63; asst prof, 66-70, ASSOC PROF MED, BROWN UNIV, 70 - *Mem:* Am Assoc Cancer Res; Am Soc Biol Chemists; Am Acad Allergy Immunol. *Res:* Cancer immunology. *Mailing Add:* Div Biomed Sci Brown Univ Providence RI 02912

WELTNER, WILLIAM, JR, b Baltimore, Md, Dec 8, 22; m 47; c 3. PHYSICAL CHEMISTRY. *Educ:* Johns Hopkins Univ, BE, 43; Univ Calif, PhD(chem), 50. *Prof Exp:* Res asst, Hercules Powder Co, Del, 43-44 & Manhattan Proj, Columbia Univ, 44-46; fel, Univ Minn, 50; instr chem, Johns Hopkins Univ, 50-54; fel, Harvard Univ, 54-56; res chemist, Union Carbide Res Inst, Tarrytown, NY, 56-66; PROF CHEM, UNIV FLA, 66- *Concurrent Pos:* Mem opers res group, US Army Chem Ctr, 51-52; consult, Nat Bur Stand, 54. *Mem:* Am Chem Soc; Am Phys Soc. *Res:* Quantum and high temperature chemistry; molecular spectroscopy and structure; electron-spin resonance. *Mailing Add:* Dept Chem Univ Fla Gainesville FL 32611-2046

WELTON, ANN FRANCES, b Evanston, Ill, Oct 6, 47; M 86. MOLECULAR PHARMACOLOGY. *Educ:* Lake Forest Col, BA, 69; Mich State Univ, PhD(biochem), 74. *Prof Exp:* Asst biochem dept, Mich State Univ, 69-74; fel lab nutrit & endocrinol, Nat Inst Health, 74-77; sr scientist pharmacol, Hoffmann-La Roche, NC, Nutley, NJ, 77-79, res group chief, 80-82, sect head, Dept Pharmacol II, 82-83, asst dir, 83-85, dir Allergy & Inflammation Res, Dept Pharmacol & Chemother, 85-88, sr dir Allergy & Inflammation Res, Dept Pharmacol & Chemother, 88-90, sr dir pharmacol, 89-90; adj asst prof, 83-88, ADJ ASSOC PROF, DEPT BIOCHEM, UNIV MED & DENT NJ, NEWARK, NJ, 88-; ASST VPRES & SR DIR PHARMACOL, HOFFMANN-LA ROCHE, INC, NUTLEY, NJ, 90- *Concurrent Pos:* NIH fel, Nat Res Serv Award, 74-77. *Mem:* AAAS; Biophys Soc; NY Acad Sci; Am Soc Pharm & Exp Therapeuts; Am Women Sci; Sigma Xi. *Res:* Membrane biochemistry; development of antiallergy agents; immunology; therapeutics for pulmonary and allergic diseases; araichidlonic acid metabolism. *Mailing Add:* 37 Maple Dr North Caldwell NJ 07006

WELTON, THEODORE ALLEN, b Saratoga Springs, NY, July 4, 18; m 43; c 4. THEORETICAL PHYSICS. *Educ:* Mass Inst Technol, BS, 39; Univ Ill, PhD(physics), 43. *Prof Exp:* Instr physics, Univ Ill, 43-44; jr scientist theoret physics, Los Alamos Sci Lab, 44-45; res assoc, Mass Inst Technol, 46-48; asst prof, Univ Pa, 48-50; prin physicist, 50-59, SR PHYSICIST, OAK RIDGE NAT LAB, 59-; FORD FOUND PROF PHYSICS, UNIV TENN, 63- *Mem:* Fel AAAS; fel Am Phys Soc; Electron Micros Soc Am. *Res:* Quantum theory of fields; theoretical nuclear physics; quantum theory of irreversible processes; theory of nuclear reactors and shielding; theory of particle accelerators; theory of lasers; theory of electron microscopy. *Mailing Add:* 121 Clark Lane Oak Ridge TN 37830

WELTON, WILLIAM ARCH, b Fairmont, WVa, June 21, 28; m 56; c 2. DERMATOLOGY. *Educ:* Harvard Univ, AB, 50; Univ Md, MD, 54. *Prof Exp:* CHMN DIV DERMAT, SCH MED, WVA UNIV, 61- *Concurrent Pos:* Osborne fel dermal path, 59-60. *Mem:* Am Acad Dermat. *Res:* Skin pathology. *Mailing Add:* Dept Med WVa Univ Sch Med Morgantown WV 26505

WELTY, JAMES RICHARD, b Garden City, Kans, Oct 23, 33; m 53; c 5. MECHANICAL ENGINEERING, CHEMICAL ENGINEERING. *Educ:* Ore State Univ, BS, 54, MS, 59, PhD(chem eng), 62. *Prof Exp:* Test engr, Pratt & Whitney Aircraft, 54; instr mech eng, 58-61, from asst prof to assoc prof, 62-67, PROF MECH ENG, ORE STATE UNIV, 67-, HEAD DEPT, 70- *Concurrent Pos:* Res engr, US Bur Mines, Ore, 62-64; mem tech staff, Bell Tel Labs, Pa, 64; vis prof, Thayer Sch Eng, Dartmouth Col, 67; res grants, US Environ Protection Agency, 68-71, US AEC, US Dept Energy & NSF, 69- *Mem:* Am Soc Mech Engrs. *Res:* Heat transfer; natural convection in liquid metals; non-Newtonian fluids in natural and forced flows; numerical modeling of thermal plumes; fluidized bed heat transfer. *Mailing Add:* Dept of Mech Eng Ore State Univ Corvallis OR 97331

WELTY, JOSEPH D, b Marion, Ind, Nov 22, 31; m 76; c 4. PHARMACOLOGY, PHYSIOLOGY. *Educ:* Purdue Univ, BS, 58; Univ SDak, MA, 62, PhD(pharmacol), 63. *Prof Exp:* Asst scientist, Dr Salisbury Labs, 58-61; instr pharmacol, Sch Med, Univ SDak, Vermillion, 63-64, from asst prof to assoc prof physiol, 64-72, prof, 72-; PROF, DEPT PHYSIOL, FAC MED, KUWAIT UNIV. *Concurrent Pos:* Consult staff, Sacred Heart Hosp, Yankton, SDak, 67-; mem, Int Study Group Res Cardiac Metab. *Mem:* Soc Exp Biol & Med; Am Physiol Soc. *Res:* Cardiovascular physiology, contractile proteins in congestive heart failure and antiarrhythmias. *Mailing Add:* Fac Med Kuwait Univ PO Box 24923 Safat 13110 Kuwait

WELTY, RONALD EARLE, b Winona, Minn, Dec 7, 34; m 62; c 2. AGRICULTURE, BOTANY. *Educ:* Winona State Univ, BS, 56; Univ Minn, MS, 61, PhD(plant path), 65. *Prof Exp:* Teacher high sch, Minn, 56-57 & 58-59; res asst plant path, Univ Minn, 59-61 & 62-65; instr bot & plant path, La State Univ, 62; res assoc, 65-66, from asst prof to prof plant path, NC State Univ, 66-82; SUPVRY PLANT PATHOLOGIST, AGR RES SERV, USDA, 66-; PROF PLANT PATH, OREGON STATE UNIV, 82- *Mem:* Sigma Xi; Mycol Soc Am; Am Phytopath Soc; Am Forage & Grassland Coun. *Res:* Diseases of forage crops grown for seed and seedborne plant pathogens; general phytopathology. *Mailing Add:* Nat Forage Seed Prod Res Ctr Ore State Univ 3450 SW Campus Way Corvallis OR 97331-7102

WEMMER, DAVID EARL, b Sacramento, Calif, Aug 27, 51; m 75; c 2. MAGNETIC RESONANCE. *Educ:* Univ Calif, Davis, BS, 73, Berkeley, PhD(chem), 79. *Prof Exp:* Fel physics, Univ Dortmund, Fed Repub Ger, 78-79; res asst biophysics, Stanford Magnetic Resonance Lab, Stanford Univ, 79-82; res asst prof, Univ Wash, 82-85; asst prof, 85-89, ASSOC PROF CHEM, UNIV CALIF, BERKELEY, 90- *Mem:* Am Chem Soc. *Res:* Applications of magnetic resonance in biophysics; problems of molecular structure and dynamics. *Mailing Add:* Dept Chem Univ Calif Berkeley CA 94720

WEMPE, LAWRENCE KYRAN, b Hutchinson, Kans, Oct 3, 41; m 65; c 1. POLYMER SYNTHESIS, DIELECTRIC MATERIALS. *Educ:* Rockhurst Col, BA, 63; Univ Kans, PhD(org chem), 68. *Prof Exp:* Lab asst water treatment chem, Deady Chem Co, Kans, 62-63; sr chemist, Rohm & Haas Co, 68-77; sr prin res chemist, Air Prod & Chem, Inc, 77-83; FROM SUPVR TO GROUP SUPVR, PENNWALT CORP, 83- *Concurrent Pos:* Instr, Montgomery County Community Col, Blue Bell, Pa, 73-76. *Res:* Polymer chemistry, polymer synthesis and polymer structure/property relationships; organic synthesis; dielectric materials; polymers for drug delivery; free radical polymerization, emulsion polymerization and polymerization mechanisms; textile and paper chemistry; coatings and composites; polymer crosslinking mechanisms; electrical properties of polymers. *Mailing Add:* Pennwalt Corp 900 First Ave PO Box 1536 King of Prussia PA 19406-0018

WEMPLE, STUART H(ARRY), b Rockford, Ill, July 27, 30; m 67; c 3. SOLID STATE PHYSICS, ELECTRICAL ENGINEERING. *Educ:* Northwestern Univ, BS, 53; Calif Inst Technol, MS, 54; Mass Inst Technol, PhD(elec eng), 63. *Prof Exp:* Mem tech staff, 54-85, DEPT HEAD, BELLABS, 85- *Mem:* Am Phys Soc; Inst Elec & Electronics Engrs; Sigma Xi. *Res:* Transport and optical properties of wide band gap semiconductors, including the effects of phase transition; gallium arsenide field-effect transistors and integrated circuits. *Mailing Add:* 112 Deborah Dr Wyomissing Hills PA 19610

WEMPNER, GERALD ARTHUR, b Waupun, Wis, Nov 11, 28; m 52; c 2. MECHANICAL ENGINEERING. *Educ:* Univ Wis, BS, 52, MS, 53; Univ Ill, PhD(eng mech), 57. *Prof Exp:* From instr to asst prof mech, Univ Ill, 53-59; assoc prof mech & civil eng, Univ Ariz, 59-62; vis prof struct mech, Univ Calif, Berkeley, 62-63; NSF fel eng mech, Stanford Univ, 63-64; prof, Univ Ala, Huntsville, 64-73; PROF ENG MECH, GA INST TECHNOL, 73- *Concurrent Pos:* Humboldt sr fel, Ruhr Univ, Ger, 73-74; Gillam fel, Univ Calgary, Alta, 83. *Mem:* Fel Am Soc Mech Engrs; fel Am Soc Civil Engrs; fel Am Acad Mech. *Res:* Contributions to theory and approximation of solids and shells with particular theories and methods for finite deformations and related nonlinear problems. *Mailing Add:* Dept Civil Eng Ga Inst Technol Atlanta GA 30332

WEMYSS, COURTNEY TITUS, JR, b Arlington, NJ, Dec 30, 22; m 51; c 3. ZOOLOGY. *Educ:* Swarthmore Col, AB, 47; Rutgers Univ, PhD, 51. *Prof Exp:* Asst zool, Rutgers Univ, 47-51; res fel bact & immunol, Harvard Med Sch, 51-52; asst prof biol, Loyola Univ, 53-54; instr physiol, NY Med Col, 54-60; assoc prof, 60-70, PROF BIOL, HOFSTRA UNIV, 70- *Concurrent Pos:* Guest investr, Rockefeller Inst, 60- *Mem:* NY Acad Sci. *Res:* Invertebrate immunity; comparative serology; tissue specificity. *Mailing Add:* 62 Oldfield Rd Greenlawn NY 11790

WEN, CHIN-YUNG, chemical engineering; deceased, see previous edition for last biography

WEN, RICHARD YUTZE, b Shanghai, China, Mar 17, 30; m 62; c 2. ORGANIC POLYMER CHEMISTRY. *Educ:* Wesleyan Univ, BA, 51; Univ Mich, MS, 53; Ind Univ, PhD(org chem), 62. *Prof Exp:* Chemist, Nalco Chem Co, 53-56; res chemist, Dow Chem Co, Mich, 62-69; sr chemist, 69-74, res specialist, Cent Res Labs, 74-78, RES SPECIALIST, MAGNETIC TAPE DIV, 3M CO, 78- *Mem:* AAAS; Am Chem Soc. *Res:* Coating technology. *Mailing Add:* 1900 Fredeen Ct New Brighton MN 55112-2412

WEN, SHIH-LIANG, b China; US citizen; m 30; c 2. KORTEWEG-DEVRIES EQUATIONS. *Educ:* Nat Taiwan Univ, BS, 56; Univ Utah, MS, 61; Purdue Univ, MS, 65, PhD(math), 68. *Prof Exp:* Assoc res engr, Boeing Co, Seattle, Wash, 61-63; teaching asst, Purdue Univ, 63-68; from asst prof to assoc prof, 68-81, PROF, OHIO UNIV, 81-, CHMN, 85- *Concurrent Pos:* Res analyst, Appl Math Res Lab, USAF, Wright-Patterson, 72; vis res scientist, Courant Inst Math Sci, NY Univ, 78-79; hon prof, Jiangxi Univ, Jiangxi, China, 85- & Lanzhou Univ, Lanzhou, China, 88- *Mem:* Am Math Soc; Math Asn Am; Soc Indust & Appl Math. *Res:* Asymptotic evaluation of multiple fourier integrals; water wave problems; eigenvalue problems in fluid mechanics; boundary layer equations; classical toroidal plasma; two-dimensional Korteweg-deVries equations. *Mailing Add:* Math Dept Ohio Univ 321 Morton Hall Athens OH 45701-2979

WEN, SUNG-FENG, b Hsinchu, Taiwan, Mar 3, 33; US citizen; m 66; c 2. MEDICINE. *Educ:* Nat Taiwan Univ, MB, 58. *Prof Exp:* Intern med, Univ Louisville, 62-63; resident, Chicago Med Sch, 63-64; res fel nephrol, Univ Wis-Madison, 64-66, instr, 66-67; res fel renal physiol, McGill Univ, 67-70; asst prof med, 70-74, assoc prof, 74-79, Rennebohm Prof, 75-80, PROF MED, UNIV WIS-MADISON, 79- *Concurrent Pos:* Mem coun kidney cardiovasc dis, Am Heart Asn. *Mem:* Am Fedn Clin Res; Am Soc Nephrol; Int Soc Nephrol; Nat Kidney Found. *Res:* Renal physiology and pathophysiology, especially related to renal transport of sodium, potassium, phosphate and glucose under normal and abnormal conditions using micropuncture techniques; hemodynamic alterations and diabetic nephropathy. *Mailing Add:* Dept Med Univ Wis Sch Clin 523 Clin Sci Ctr Madison WI 53792

WEN, WEN-YANG, b Hsin-tsu, Taiwan, Mar 7, 31; m 59; c 2. PHYSICAL CHEMISTRY. *Educ:* Nat Taiwan Univ, BS, 53; Univ Pittsburgh, PhD(chem), 57. *Prof Exp:* Res assoc, Univ Pittsburgh, 57-58; res fel, Northwestern Univ, 58-60; asst prof chem, DePaul Univ, 60-62; from asst prof to assoc prof, 62-73, PROF CHEM, CLARK UNIV, 73- *Concurrent Pos:* Humboldt scholar, Univ Karlsruhe, 70-71 & 73, Univ Goettingen, 76 & Morgantown Energy Technol Ctr, Dept Energy, 78-79. *Mem:* Am Chem Soc; AAAS; Sigma Xi. *Res:* Structure of water; thermodynamic properties of large ions in solutions; tetraalkylammonium salts and hydrophobic bonds; nuclear magnetic resonance; alkali metal catalysis on coal gasification; thermal and catalytic cracking of coal tar; polymer-gas interaction by NMR. *Mailing Add:* Dept Chem Clark Univ Worcester MA 01610

WENCLAWIAK, BERND WILHELM, b Du-Hamborn, WGer, Aug 11, 51; m 76; c 4. SUPERCRITICAL FLUIDS. *Educ:* Westphalian Wilhelms Univ, Münster, dipl chem, 75, PhD(chem), 78. *Prof Exp:* Sci asst anal chem, Westphalian Wilhelms Univ, Münster, 78-84; res assoc, Univ Colo, Boulder, 82 & 84-85; ASSOC PROF ANALYTICAL CHEM, UNIV TOLEDO, 85- *Concurrent Pos:* Consult, Ger Prod Testing Found, 79-80. *Mem:* Soc Ger Chemists; Am Chem Soc; Am Asn Mass Spectros. *Res:* Analytical and environmental chemistry; chromatography; mass spectrometry; metal-containing compounds, such as metal chelates; reactions of antitumor drugs of the cisplatinum type with DNA. *Mailing Add:* Dept Chem Univ Toledo 2801 W Bancroft St Toledo OH 43606

WEND, DAVID VAN VRANKEN, b Poughkeepsie, NY, Oct 18, 23; m 53; c 3. MATHEMATICS. *Educ:* Univ Mich, BS, 45, MA, 46, PhD(math), 55. *Prof Exp:* Instr math, Reed Col, 49-51 & Iowa State Univ, 52-55; from asst prof to assoc prof, Univ Utah, 55-66; from assoc prof to prof math, Mont State Univ, 66-91; RETIRED. *Mem:* Math Asn Am. *Res:* Functions of a complex variable; differential equations. *Mailing Add:* Dept Math Mont State Univ Bozeman MT 59717-0240

WENDE, CHARLES DAVID, b Wilmington, Del, Dec 4, 41; m 65; c 1. SPACE PHYSICS. *Educ:* Mass Inst Technol, BS, 63; Univ Iowa, MS, 66, PhD(physics), 68. *Prof Exp:* Res assoc space physics, Univ Iowa, 68-69; astrophysicist, 69-81, SCI DATA SYST MGR, HST PROF, GODDARD SPACE FLIGHT CTR, NASA, 81- *Mem:* AAAS; Am Geophys Union; Int Union Radio Sci; Sigma Xi. *Res:* Oversee implementation of portions of Hubble Space Telescope ground system; application of interactive computing to modeling experiment hardware and to data reduction and analysis. *Mailing Add:* 8700 Nightingale Dr Seabrook MD 20706

WENDEL, CARLTON TYRUS, b Fredericksburg, Tex, Oct 6, 39; m 63; c 2. ANALYTICAL CHEMISTRY. *Educ:* Tex Lutheran Col, BS, 62; Tex Tech Col, MS, 65, PhD(chem), 67. *Prof Exp:* From instr to assoc prof, 67-89, chmn dept, 80-89, PROF CHEM & CHAIR, TEX WOMAN'S UNIV, 89- *Mem:* Am Chem Soc; Sigma Xi. *Res:* Measurement of conceptual skills development. *Mailing Add:* Dept Chem Tex Woman's Univ Denton TX 76204

WENDEL, JAMES G, b Portland, Ore, Apr, 18, 22; m 44; c 6. COMPUTER SCIENCE, GENERAL. *Educ:* Reed Col, BA, 43; Calif Inst Technol, PhD(math), 48. *Prof Exp:* Asst Nat Defense Res Comt, Calif Inst Technol, 42-45, instr, 45-48; instr math, Yale Univ, 48-51; assoc mathematician, Rand Corp, 51-52; from asst prof to assoc prof math, La State Univ, 52-55; from asst prof to prof math, 55-87, assoc chmn dept, 68-70, 73-77 & 79-84, EMER PROF MATH, UNIV MICH, ANN ARBOR, 87- *Concurrent Pos:* Vis prof, Aarhus Univ, 62-64, Univ London, 70-71, Univ WAustralia, 78, Calif Inst Technol, 79, Weizmann Inst, 85 & Univ Hawaii, 86. *Mem:* Am Math Soc; Math Asn Am; Asn Comput Mach. *Res:* Probability theory. *Mailing Add:* 437 Ferne Ave Palo Alto CA 94306-4621

WENDEL, OTTO THEODORE, JR, b Philadelphia, Pa, Mar 21, 48; m 69; c 1. NEUROPHARMACOLOGY, CARDIOVASCULAR PHARMACOLOGY. *Educ:* St Andrews Col, BA, 69; Wake Forest Univ, MS, 73, PhD(pharmacol), 74. *Prof Exp:* Instr neuropharmacol, Dept Neurol, Bowman Gray Sch Med, 74-76, instr pharmacol, 76-78, asst prof, 78-79; asst prof pharmacol, Kirksville Col Osteop Med, 79-86; asst dean med educ, 86-90, DEAN, ALLIED HEALTH PROFESSIONS, COL OSTEOP MED PAC, 90- *Concurrent Pos:* Assoc community med, Bowman Gray Sch Med, 74-79. *Mem:* Sigma Xi; Soc Neurosci; Am Soc Pharmacol & Exp Therapeut. *Res:* The relationship between endogneous opioid activity and the genesis and or maintenance of hypertension. *Mailing Add:* Col Osteop Med Pac College Plaza Pomona CA 91766-1889

WENDEL, SAMUEL REECE, b Charleston, Ill, Sept 1, 44; m 67; c 3. BIOINORGANIC CHEMISTRY, ORGANOMETALLIC CHEMISTRY. *Educ:* Univ Ill, Urbana, BS, 66; Univ Mont, PhD(org chem), 73. *Prof Exp:* Chemist, 66-69, SR PROJ CHEMIST ORGANOSILICON CHEM, DOW CORNING CORP, 73- *Mem:* AAAS; Am Chem Soc. *Res:* Design and synthesis of bioactive organosilicon compounds; silicone biomaterials. *Mailing Add:* Eli Lilly & Co IC742 Bldg 110 Lilly Corp Ctr Indianapolis IN 46285

WENDELKEN, JOHN FRANKLIN, b Lexington, Ky, Nov 12, 45; m 68; c 2. SURFACE & SOLID STATE PHYSICS. *Educ:* Univ Ill, BS, 68, MS, 70, PhD(physics), 75. *Prof Exp:* PHYSICIST SURFACE PHYSICS, OAK RIDGE NAT LAB, 74- *Mem:* Am Phys Soc; Am Vacuum Soc; Am Chem Soc. *Res:* Geometric, electronic and vibrational properties of clean and absorbate covered single crystal surfaces. *Mailing Add:* Solid State Div MS 6024 Bldg 3025 Oak Ridge Nat Lab PO Box 2008 Oak Ridge TN 37831

WENDER, IRVING, b New York, NY, June 19, 15; m 42; c 3. FUEL SCIENCE, ORGANOMETALLIC CHEMISTRY & CATALYSIS. *Educ:* City Col New York, BS, 36; Columbia Univ, MA, 45; Univ Pittsburgh, PhD(chem), 50. *Prof Exp:* Chemist & res assoc, Manhattan Proj, Univ Chicago, 44-46; org chemist, Pittsburgh Coal Res Ctr, US Bur Mines, 46-53, chief chem sect, 53-71, res dir, Pittsburgh Energy Res Ctr, Energy Res & Develop Admin, 71-75, dir, 75-77, dir, Pittsburgh Energy Technol Ctr, Dept Energy, 77-78, spec adv to asst dir fossil energy, 78-79, dir, Off Advan Res & Technol, Off Fossil Energy, 79-81; RES PROF, DEPT CHEM/PETROL ENG, UNIV PITTSBURGH, 81- *Concurrent Pos:* Lectr, Univ Pittsburgh, 63-69, adj prof, 69-78, 82- *Honors & Awards:* Bituminous Coal Res Award, 56, 60; H H Storch Award, 64; Pittsburgh Award, Am Chem Soc, 68; K K Kelley Award, 69; Career Serv Award, Nat Civil Serv League, 76; Award Petrol Chem, Am Chem Soc, 82; H H Lowry Award, Dept Energy, 88. *Mem:* Am Chem Soc; Am Inst Chem Eng; AAAS. *Res:* Chemistry of carbon monoxide, metal carbonyls, coal conversion; catalysis; reactions at high pressures; synthetic fuels from coal; carbon monoxide chemistry. *Mailing Add:* Dept Chem/Petrol Eng 1249 Benedum Hall Univ Pittsburgh Pittsburgh PA 15261

WENDER, PAUL ANTHONY, PHOTOCHEMISTRY, BIOLOGICAL ACTIVITY. *Educ:* Wilkes Col, BS, 69; Yale Univ, PhD, 73; Columbia Univ, Postdoc, 74. *Prof Exp:* From asst prof to assoc prof, Harvard Univ, 74-81; PROF, STANFORD UNIV, 81- *Concurrent Pos:* Consult, Eli Lilly & Co, 80-, Chevron, 84-; Cope Scholar Award, Am Chem Soc; Dreyfus Teacher-Scholar Award. *Honors & Awards:* Guenther Award, Am Chem Soc; ICI Am Chem Award, Stuart Pharmaceut. *Mem:* Am Chem Soc; AAAS. *Res:* Organic synthesis; organic photochemistry; computer modelling; tumor promotion; organic synthesis. *Mailing Add:* Dept Chem Stanford Univ Stanford CA 94305

WENDER, PAUL H, b New York, NY, May 12, 34; m 70; c 3. PSYCHIATRY, CHILD PSYCHIATRY. *Educ:* Harvard Univ, AB, 55; Columbia Univ, MD, 59. *Prof Exp:* Intern, Barnes Hosp, St Louis, 59-60; resident adult psychiat, Mass Ment Health Ctr, 60-62; resident, St Elizabeth's Hosp, 62-63; resident child psychiat, Johns Hopkins Univ, 64-67, asst prof pediat & psychiat, 67-73; PROF PSYCHIAT, COL MED, UNIV UTAH, 73- *Concurrent Pos:* NIMH fel, NIH, Bethesda, Md, 64-66, res psychiatrist, 67-73. *Honors & Awards:* Hofheimer Award, Am Psychiat Asn, 74. *Mem:* Am Psychiat Asn; Am Acad Child Psychiat; Psychiat Res Soc. *Res:* Genetics and schizophrenia; minimal brain dysfunction in children. *Mailing Add:* Dept Psychiat Univ Utah Col Med Salt Lake City UT 84132

WENDER, SIMON HAROLD, b Dalton, Ga, Sept 4, 13; m 42; c 3. BIOCHEMISTRY. *Educ:* Emory Univ, AB, 34, MS, 35; Univ Minn, PhD(agr biochem), 38. *Prof Exp:* Res assoc, Med Sch, Emory Univ, 38-39; assoc chemist, Exp Sta, Agr & Mech Col, Tex, 39-41; from instr to asst prof chem, Univ Ky, 41-46; from assoc prof to prof, 46-53, res prof, 53-83, GEORGE L CROSS EMER PROF BIOCHEM, UNIV OKLA, 83- *Concurrent Pos:* Former mem bd dirs, Oak Ridge Assoc Univs; chmn coun, Oak Ridge Inst Nuclear Studies, Okla, rep to coun, 52-64 & 71-80; vis res assoc, Argonne Nat Lab, 54-64; vis prof, Univ Wis, 66 & Univ Calif at Davis, 73. *Honors & Awards:* Okla Chemist Award, 83. *Mem:* Fel AAAS; Am Chem Soc; Am Soc Biol Chem; Soc Exp Biol & Med; Am Soc Plant Physiol; Phytochem Soc NAm. *Res:* Chromatography; plant phenolics and plant and animal oxidoreductases. *Mailing Add:* Dept Chem 620 Parrington Oval Univ Okla Norman OK 73019

WENDLAND, RAY THEODORE, b Minneapolis, Minn, July 11, 11; m 46; c 1. ORGANIC CHEMISTRY. *Educ:* Carleton Col, BA, 33; Iowa State Univ, PhD(chem), 37. *Prof Exp:* Res chemist, Universal Oil Prod Co, 38-39; instr org chem & biochem, Coe Col, 39-42; asst prof, Middlebury Col, 42-43; res chemist synthetic rubber, War Prod Bd, Univ Minn, 43-44; asst prof org & biol chem, Lehigh Univ, 44-47; prof org chem, NDak State Univ, 47-55; res fel petrol chem, Mellon Inst, 55-58; prof chem & chmn div sci & math, Winona State Col, 58-63; prof chem, 63-76, EMER PROF CHEM, CARROLL COL, WIS, 76- *Concurrent Pos:* Consult. *Mem:* AAAS; Am Chem Soc; Sigma Xi. *Res:* Organic synthesis; polymer and petroleum chemistry. *Mailing Add:* 2315 Morningside Dr Waukesha WI 53186

WENDLAND, WAYNE MARCEL, b Beaver Dam, Wis, Aug 9, 34; m 56; c 4. METEOROLOGY, PALEOCLIMATOLOGY. *Educ:* Lawrence Col, BA, 56; Univ Wis-Madison, MS, 65, PhD(meteorol), 72. *Prof Exp:* Weather forecaster, USAF, 56-64; proj supvr meteorol, Univ Wis-Madison, 66-70, instr climat, 70, asst prof geog & meteorol, 70-76; assoc prof geog, Univ Ill, Urbana, 76-80; head, climat sect, 80-84, PRIN SCIENTIST & STATE CLIMATOLOGIST, ILL STATE WATER SURV, 84- *Concurrent Pos:* Adj prof geog, Univ Ill, Urbana. *Mem:* Am Meteorol Soc; Am Quaternary Asn; Am Asn State Climatologists. *Res:* Past climatic circulation patterns; climatic episodes of the Holocene; climatic reconstructions from tree rings; climatic variability. *Mailing Add:* Ill State Water Surv 2204 Griffith Dr Champaign IL 61820

WENDLAND, WOLFGANG LEOPOLD, b Poznan, Poland, Sept 20, 36; Ger citizen; m 64; c 2. MATHEMATICS, MECHANICAL ENGINEERING. *Educ:* Tech Univ, Berlin, BS(mech eng), 58, BS(math), 58, dipl ing, 61, Dr Ing, 65, Habilitation, 69. *Prof Exp:* Sci collabr, Tech Univ Berlin, 61-63, sci asst, 63-64; sci asst dept numerical math, Hahn Meitner Inst, Berlin, 64-69, prof, 69-70; prof, Tech Univ Darmstadt, 70-86; PROF, UNIV STUTTGART. *Concurrent Pos:* Vis unidel chair prof, Univ Del, 73-74; chmn math dept, Tech Univ Darmstadt, 74-75; Fulbright Stipendium, Univ Del, 77-; vis prof, Ore State Univ, 77-, Univ Md, 81, Univ Del, 81, Univ Concepcion, Chile, 81, Australian Nat Univ, Canberra, 85, Univ Del, 89 & 90. *Mem:* Am Math Soc; Ger Asn Appl Math Mech; Soc Appl Math Mech; Int Soc Interaction Mech Math; Int Soc Comp Mech Eng; Int Asn Boundary Element Methods. *Res:* Applied mathematics; partial differential equations; mathematical physics. *Mailing Add:* Iu Himmel 62a Stuttgart 80 7000 Germany

WENDLANDT, WESLEY W, b Galesville, Wis, Nov 20, 27. INORGANIC CHEMISTRY, ANALYTICAL CHEMISTRY. *Educ:* Wis State Col, River Falls, BS, 50; Univ Iowa, MS, 52, PhD(chem), 54. *Prof Exp:* From asst prof to prof chem, Tex Tech Col, 54-66; chmn dept, 66-72, PROF CHEM, UNIV HOUSTON, 66- *Concurrent Pos:* Vis prof, NMex Highlands Univ, 61; ed-in-chief, Thermochimica Acta. *Honors & Awards:* Mettler Award, 70. *Mem:* AAAS; Am Chem Soc; The Chem Soc; NAm Thermal Anal Soc; Int Confedn Thermal Anal. *Res:* Coordination compounds; metal chelates; thermogravimetry; differential thermal analysis; solid state chemistry; reflectance spectroscopy. *Mailing Add:* 7818 Sands Point Dr Houston TX 77036

WENDLER, GERD DIERK, b Hamburg, WGer, June 16, 39; m 69; c 2. METEOROLOGY. *Educ:* Innsbruck Univ, PhD(meteorol), 64. *Prof Exp:* Data process asst meteorol, Inst Meteorol, Innsbruck Univ, 60-64, res asst, 65-66; asst geophysicist, 66-67, asst prof meteorol, 67-70, ASSOC PROF METEOROL, GEOPHYS INST, UNIV ALASKA, FAIRBANKS, 70- *Concurrent Pos:* NSF grant, McCall Glacier, Geophys Inst, Univ Alaska, Fairbanks, 69-, Sea grant Arctic Ocean, 71- & NASA satellite grant cent Alaska, 72- *Mem:* Am Meteorol Soc; Am Geophys Union; Glaciol Soc; Arctic Inst NAm; Ger Soc Polar Res. *Res:* Meteorology in the arctic, especially of Alaska. *Mailing Add:* Dept Physics Univ Alaska 116 Bunnell Fairbanks AK 99701

WENDRICKS, ROLAND N, b Casco, Wis, July 26, 30; m 52; c 4. PHYSICAL CHEMISTRY. *Educ:* St Norbert Col, BS, 52; Northwestern Univ, MS, 61. *Prof Exp:* Group supvr blow molding process, 52-67, supvr blow molding plastics res & develop, 67-70, MGR PLASTICS EQUIP ENG, AM CAN CO, 70- *Concurrent Pos:* Chmn, SPE-Blow Molding Div, 82. *Mem:* Soc Plastics Eng. *Res:* Processing of thermoplastic polymers. *Mailing Add:* Am Nat Can Co 2301 Industrial Dr Neenah WI 54957

WENDROFF, BURTON, b New York, NY, Mar 10, 30. MATHEMATICS. *Educ:* NY Univ, BA, 51, PhD(math), 58; Mass Inst Technol, SM, 52. *Prof Exp:* Staff mem, Los Alamos Sci Lab, 52-66; from assoc prof to prof math, Univ Denver, 66-74; GROUP LEADER & STAFF MEM, LOS ALAMOS SCI LAB, 73- *Mem:* Am Math Soc; Soc Indust & Appl Math. *Res:* Applied mathematics; numerical analysis. *Mailing Add:* Los Alamos Nat Lab Group T-7 MS-B 284 Los Alamos NM 87544

WENDT, ARNOLD, b Red Bud, Ill, Jan 14, 22; m 43; c 1. MATHEMATICS. *Educ:* Univ Wis, PhD(math), 52. *Prof Exp:* PROF MATH, WESTERN ILL UNIV, 52- *Mem:* AAAS; Am Math Soc; Math Asn Am. *Res:* Analysis and applied mathematics. *Mailing Add:* Western Ill State Col Macomb IL 61455

WENDT, CHARLES WILLIAM, b Plainview, Tex, July 12, 31; m 55; c 4. SOIL PHYSICS. *Educ:* Tex A&M Univ, BS, 51, PhD(soil physics), 66; Tex Tech Col, MS, 57. *Prof Exp:* Res asst agron, Tex Tech Col, 53-55, from instr to asst prof, 57-63; res asst soil physics, 63-65, res assoc, 65-66, from asst prof to assoc prof, 66-74, PROF SOIL PHYSICS, TEX A&M UNIV, 74- *Concurrent Pos:* Consult, cotton prod, Ministry Agr, Repub Sudan, Dryland Agr, Cong of US, Technol & Assement Off, Int Irrigation Asn, proj eval, Environ Protection Agency, Nat Water Resources Coun, Nat Res Coun, Southeast Consortium, Int Develop. *Honors & Awards:* Outstanding Res Scientist, High Plains Res Found, 82. *Mem:* AAAS; Am Soc Agron; Soil Sci Soc Am; Am Soc Plant Physiol; Brit Plant Growth Regulator Group; Plant Growth Regulation Soc Am. *Res:* Efficient utilization of rainfall and limited irrigation water through plant modification (breeding, growth regulators, antitranspirants) and soil modification (irrigation, furrow, diking, and evaporation suppressants), in US; semi-arid agriculture in Africa. *Mailing Add:* Tex Agr Exp Sta Tex A&M Univ Agr Res & Exten Ctr Rte 3 Lubbock TX 79401

WENDT, JOST O L, b Berlin, Ger, July 2, 41; m 61; c 2. CHEMICAL ENGINEERING, COMBUSTION. *Educ:* Glasgow Univ, BSc, 63; Johns Hopkins Univ, MSE, 66, PhD(chem eng), 68. *Prof Exp:* Instr thermodyn, Johns Hopkins Univ, 66-67; engr, Emeryville Res Ctr, Shell Develop Co, Calif, 68-72; from asst prof to assoc prof, 72-79, PROF CHEM ENG, UNIV ARIZ, 79- *Concurrent Pos:* Consult combustion appln, Nat Acad Sci/Nat Acad Engrs; sr vis scientist, Environ Proctection Agency 84-86; prin investr, res grants on combustion. *Mem:* Am Inst Chem Engrs; Combustion Inst; Int Flame Res Found. *Res:* Combustion generated air pollution; pollution aspects of burner design; coal combustion; kinetics; combustion science; hazardous waste incineration. *Mailing Add:* Dept Chem Eng Univ Ariz Tucson AZ 85721

WENDT, RICHARD P, b St Louis, Mo, Oct 6, 32; m 70; c 2. PHYSICAL CHEMISTRY, TRANSPORT PROCESSES. *Educ:* Washington Univ, AB, 54; Univ Wis, PhD(phys chem), 61. *Prof Exp:* Asst prof chem, La State Univ, 62-66; assoc prof, 66-73, PROF CHEM, LOYOLA UNIV, NEW ORLEANS, 73- *Concurrent Pos:* Consult, Ironite Products Co; NIH res fel, 71. *Mem:* AAAS. *Res:* Diffusion in liquids; nonequilibrium thermodynamics; mass transfer across synthetic membranes; iron oxide/H_2S reactions. *Mailing Add:* Dept Chem Loyola Univ 6363 St Charles Ave New Orleans LA 70118

WENDT, ROBERT CHARLES, b Aurora, Ill, July 5, 29; m 53; c 4. SURFACE & POLYMER CHEMISTRY. *Educ:* NCent Col, Ill, BA, 51; Univ Ill, PhD(phys chem), 55. *Prof Exp:* Res chemist, Yerkes Lab, Film Dept, 55-64, staff scientist, 64-69, staff scientist, Exp Sta Lab, Film Dept, 69-75, sr res chemist, 76-81, RES ASSOC, EXP STA LAB, POLYMER PRODS DEPT, E I DU PONT DE NEMOURS & CO, INC, 81- *Mem:* Am Chem Soc; Am Vacuum Soc; Elec Micros Soc Am. *Res:* Diffusion phenomena; polymer physical properties; adhesion; polymer surface chemistry and physics; electron spectroscopy. *Mailing Add:* 3316 Coachman Rd Wilmington DE 19803-1943

WENDT, ROBERT L(OUIS), b Chicago, Ill, July 10, 20; m 46. ENGINEERING. *Educ:* Harvard Univ, AB, 40. *Prof Exp:* Eng, Sperry Gyroscope Co, 40-52, mgr eng, 52-57, mgr sales & subcontracts, 57-62, mgr B-58 bomb navig syst prog, 62-63, dir prog control, 63-64, mgr, Polaris/Poseidon, 64-69, group mgr ship & mil systs, 69-71, vpres & gen mgr, Systs Mgt, 71-75, vpres & gen mgr, Gyroscope, 75-80, PRES, SPERRY DIV, SPERRY CORP, 80- *Mem:* Sr mem Inst Elec & Electronics Engrs; Inst Navig; Am Soc Naval Engrs. *Res:* Management of major programs, especially development and production of complex electronic systems. *Mailing Add:* One Hillcrest Lane Woodbury NY 11797

WENDT, ROBERT LEO, HYPERTENSION, CARDIAC DISEASES. *Educ:* Univ Cincinnati, PhD(pharmacol), 68. *Prof Exp:* ASSOC DIR, DIV EXP THERAPEUT, WYETH LABS, INC, PHILADELPHIA, 82- *Mailing Add:* Wyeth Labs Inc PO Box 8299 Philadelphia PA 19101

WENDT, THEODORE MIL, b Ft Collins, Colo, Sept 14, 40; m 63; c 2. MICROBIOLOGY, CHEMISTRY. *Educ:* Colo State Univ, BS, 64, MS, 66, PhD(microbiol), 68. *Prof Exp:* Res microbiologist, US Army Natick Labs, 68-78; mgr microbiol, Arbook, Inc, 78-80; mgr microbiol, 81-86, DIR RES, SURGIKOS, INC, 87- *Mem:* Am Soc Microbiol; Sigma Xi; AAAS; Soc Indust Microbiol; Am Soc Testing & Mat; Asn Off Anal Chemists. *Res:* Microbiological deterioration of materials, especially polymers; water pollution abatement through biological activity; relationship of chemical structure to biological susceptibility; aquatic microbial ecology; disinfectants; biocides in health care; clinical environments; clinical research and toxicology. *Mailing Add:* 5400 Overridge Dr Arlington TX 76017

WENESER, JOSEPH, b New York, NY, Nov 23, 22; m 56. THEORETICAL PHYSICS. *Educ:* City Col, BS, 42; Columbia Univ, MA, 48, PhD(physics), 52. *Prof Exp:* Asst physics, Manhattan Proj, Columbia Univ, 42-46; assoc physicist, Brookhaven Nat Lab, 52-55; asst prof, Univ Ill, 55-57; chmn dept physics, 70-75, SR PHYSICIST, BROOKHAVEN NAT LAB, 57- *Mem:* Fel Am Phys Soc; fel AAAS. *Res:* Theoretical nuclear physics. *Mailing Add:* Dept Physics Bldg 510A Brookhaven Nat Lab Upton NY 11973

WENG, LIH-JYH, b Fukien, China, Dec 3, 37; wid; c 2. ELECTRICAL ENGINEERING, COMPUTER SCIENCE. *Educ:* Cheng Kung Univ, Taiwan, BS, 59; Northeastern Univ, MSEE, 63, PhD(elec eng), 66. *Prof Exp:* Res engr, Adcom, Inc, 65-66; sr res assoc coding theory, Northeastern Univ, 66-67, from asst prof to assoc prof elec eng, 67-73; sr eng specialist, CNR, Inc, 73-78; consult engr, 78-81, SR CONSULT, DIGITAL EQUIP CORP, 81- *Concurrent Pos:* NSF initiation grant, 68-69; consult, Honeywell Info Systs Inc, Mass, 70-71; lectr, Northeastern Univ, 73-82. *Mem:* Inst Elec & Electronics Engrs. *Res:* Algebraic coding theory; digital communications; mass storage techniques in computers; error-control; communication theory; cryptology. *Mailing Add:* Digital Equip Corp SHR 1-3/E29 333 South St Shrewsbury MA 01545-4112

WENG, TU-LUNG, engineering mechanics, mechanical engineering, for more information see previous edition

WENG, TUNG HSIANG, b Fukien, China, Jan 16, 33; m 61; c 3. ELECTRICAL ENGINEERING. *Educ:* Nat Taiwan Univ, BS, 56; Univ Iowa, MS, 59; Univ Mo, Columbia, PhD(elec eng), 67. *Prof Exp:* Design engr, Oak Mfg Co, 60-61; proj engr, Sula Elec Co, 61-63 & Simpson Elec Co, 63-64; asst prof elec eng, Univ Mo, Columbia, 67-69; from asst prof to assoc prof, 69-83, PROF ELEC ENG, OAKLAND UNIV, 83- *Mem:* Inst Elec & Electronics Engrs; Sigma Xi. *Res:* Solid state properties and devices. *Mailing Add:* Sch Eng Oakland Univ Rochester MI 48309-4401

WENG, (FRANK) TZONG-RUEY, b Jakarta, Indonesia, Jan 5, 34; US citizen; m 72. MEDICINE, PEDIATRICS. *Educ:* Nat Taiwan Univ, MD, 60. *Prof Exp:* Asst prof pediat, Sch Med, NY Univ, 70-75; assoc prof pediat, Sch Med, Wayne State Univ, 75-80; MEM STAFF, CHILDREN'S HOSP PITTSBURGH, 80-; STAFF MEM, CHILDREN'S HOSP DETROIT. *Mem:* Fel Am Acad Pediat; fel Am Col Chest Physicians; Am Thoracic Soc. *Res:* Pulmonary diseases in children. *Mailing Add:* 125 De Soto St Pittsburgh PA 15213

WENG, WU TSUNG, b Taiwan, China, Aug 11, 44; m 67; c 2. ELECTROMAGNETISM, ENGINEERING PHYSICS. *Educ:* Nat Taiwan Univ, BS, 66; Nat Tsing Hua Univ, MS, 68; State Univ NY, Stony Brook, PhD(physics), 74. *Prof Exp:* Accelerator physicist, Stanford Linear Accelerator Ctr, 83-87; accelerator physicist, Accelerator Dept, 77-83, mgr, Booster Proj, AGS Dept, 87-90, ACCELERATOR DIV HEAD, AGS DEPT, BROOKHAVEN NAT LAB, 90- *Concurrent Pos:* Chmn, Accelerator Sci & Technol Comt, Nuclear & Plasma Sci Soc, Inst Elec & Electronics Engrs. *Mem:* Am Phys Soc; Inst Elec & Electronics Engrs. *Res:* Accelerator physics; non-linear dynamits; beam control and feedback. *Mailing Add:* Bldg 911B Brookhaven Nat Lab Upton NY 11973

WENGER, BYRON SYLVESTER, b Russell, Kans, Oct 13, 19; m 47; c 4. DEVELOPMENTAL BIOLOGY, TERATOLOGY. *Educ:* Univ Wyo, BS, 40, MS, 41; Washington Univ, PhD(zool), 49. *Prof Exp:* NIH fel pharmacol, Wash Univ, 49-51; from asst prof to assoc prof anat, Univ Kans, 51-62, from assoc prof comp biochem & physiol to prof biochem, 62-69; prof anat, Univ Sask, 69-88; PROF ANAT, ROSS UNIV, 89- *Concurrent Pos:* Vis assoc prof, Wash Univ, 64-66; res assoc, Univ Calif, Davis, 79-80. *Mem:* Am Asn Anatomists; Am Soc Zoologists; Soc Develop Biol; Int Soc Develop Biologists; Teratology Soc. *Res:* Experimental and biochemical studies of differentiation in normal chick and mouse embryos and during genetic and drug induced teratogenesis; microsurgery on chick embryos; ultramicrochemical analysis of samples dissected from lyophilized microtome sections. *Mailing Add:* Ross Univ Sch Med Box 266 Roseau Commonwealth Dominica West Indies

WENGER, CHRISTIAN BRUCE, b Philadelphia, Pa, July 24, 42. THERMOREGULATION, CIRCULATION. *Educ:* Col Wooster, AB, 64; Yale Univ, MD, 70, PhD(environ physiol), 73. *Prof Exp:* Instr physics, Col Wooster, 67-68; vis asst fel, John B Pierce Found Lab, 74-75, asst fel, 75-82, assoc fel physiol, 82-84, vis assoc fel, 84-86; RES PHARMACOLOGIST, THERMAL PHYSIOL & MED DIV, US ARMY RES INST, ENVIRON MED, 84- *Concurrent Pos:* Guest referee ed, Am Physiol Soc, 73-; post-doc fel, Dept Epidemiol & Pub Health, Yale Univ, 74-76, asst prof, 76-, sr res assoc & lectr, 82-83, res scientist & lectr, 83-84; res assoc physiol, Sch Pub Health, Harvard Univ, 85-89. *Mem:* Am Physiol Soc; Sigma Xi. *Res:* Thermoregulatory physiology; circulatory effects of heat and cold stress and exercise; heat acclimatization; heat and cold disorders. *Mailing Add:* Thermal Physiol & Med Div US Army Res Inst Environ Med Kans St Natick MA 01760-5007

WENGER, DAVID ARTHUR, biochemical genetics, pediatrics, for more information see previous edition

WENGER, FRANZ, b Bern, Switz, Nov 28, 25; nat US; m 55. PHYSICAL CHEMISTRY. *Educ:* Univ Bern, Lic phil nat, 53, PhD(chem), 54. *Prof Exp:* Chemist, Lonza, Inc, Switz, 55; fel photochem, Nat Res Coun Can, 55-57; fel polymer sci, Mellon Inst, 58-63; sr staff assoc, Cent Res Lab, Celanese Corp, NJ, 63-66; mgr spec prod res, Polaroid Corp, Mass, 66-69; GROUP V PRES-IN-CHG RES & DEVELOP & ENG, ENGELHARD INDUSTS DIV, ENGELHARD MINERALS & CHEM CORP, NEWARK, 69- *Mem:* Am Chem Soc. *Res:* Reaction kinetics; structure-properties relationship of polymeric materials; photographic technology; process research and development. *Mailing Add:* 363 Cherry Hill Rd Mountainside NJ 07092-2032

WENGER, GALEN ROSENBERGER, b Sellersville, Pa, May 16, 46; m 72; c 2. OPERANT BEHAVIOR. *Educ:* Goshen Col, Ind, BA, 68; WVa Univ, PhD(pharmacol), 71. *Prof Exp:* Fel, Univ Col Med Ctr, 72-73; fel, Harvard Med Sch, 73-75, instr pharmacol, 75-78; from asst prof to assoc prof, 78-86, PROF PHARMACOL, UNIV ARK MED SCI, LITTLE ROCK, 86- *Concurrent Pos:* Ad hoc consult, Nat Inst Environ Health Sci; consult, Nat Ctr Toxicol Res, 83-89; prin investr, res grant, Nat Inst Drug Abuse, 83-; toxicol study sect mem, NIH, 90- *Mem:* Am Soc Pharmacol & Exp Therapeut; Soc Toxicol; Behav Pharmacol Soc; Behav Toxicol Soc. *Res:* Behavioral effects of drugs of abuse affecting the central nervous system; behavioral effects of toxic agents; toxicology; experimental psychology. *Mailing Add:* Dept Pharmacol Slot 611 Univ Ark Med Sci 4301 W Markham Little Rock AR 72205

WENGER, JOHN C, b Manhattan, NY, Jan 13, 41; m 63; c 2. MEASURE THEORY. *Educ:* Univ Mich, BS, 63; Univ Chicago, SM, 66; Ill Inst Technol, PhD(math), 79. *Prof Exp:* ASSOC PROF MATH, LOOP COL, 68- *Concurrent Pos:* Teaching asst, Ill Inst Technol, 77-79, instr, 79- *Mem:* Math Asn Am; Am Math Soc; Nat Coun Teachers Math. *Res:* Generalizations of the Riesz-Markov representation theorem in both topological and more abstract settings. *Mailing Add:* 198 Bloom Highland Park IL 60035

WENGER, LOWELL EDWARD, b Middlebury, Ind, Nov 17, 48; m 76; c 2. PHYSICS. *Educ:* Purdue Univ, BS, 71, MS, 73, PhD(physics), 75. *Prof Exp:* Res assoc physics, Purdue Univ, 75-76; asst prof, 76-81, assoc prof, 81-85, PROF PHYSICS, WAYNE STATE UNIV, 85- *Concurrent Pos:* Grants, Cottrell Res Corp, 77-79 & Wayne State Fac Res Award, 77-78, 84-85; fel, Alfred P Sloan Res Found, 78-82, Fulbright Res, 82-83; career develop chmn, Wayne State, 84-85. *Mem:* Am Phys Soc; Sigma Xi; Mat Res Soc. *Res:* Study of magnetic properties of magnetic alloys; heat capacity and magnetic susceptibilities of solids at very low temperatures; characterization of high-temperature superconducting oxides. *Mailing Add:* Dept Physics Wayne State Univ Detroit MI 48202

WENGER, NANETTE KASS, b New York, NY, Sept 3, 30. MEDICINE, CARDIOLOGY. *Educ:* Hunter Col, BA, 51; Harvard Med Sch, 54. *Prof Exp:* Intern, Mt Sinai Hosp, New York, 54-55, resident med, 55-56, chief resident cardiol, 56-57; sr asst resident med, Grady Mem Hosp, Atlanta, 58; instr med, 59-62, assoc, 62-64, from asst prof to assoc prof, 64-71, PROF MED, SCH MED, EMORY UNIV, 71-, DIR, CARDIAC CLINS, GRADY MEM HOSP, ATLANTA, 60-; DIR CARDIAC CLINS GRADY MEM HOSP, ATLANTA, 60- *Concurrent Pos:* Fel cardiol, Sch Med, Grady Mem Hosp, Emory Univ, 58-59; fel, Coun Clin Cardiol, Am Heart Asn, 70; dir, Proj Cardiac Eval & Med & Voc Rehab, 66-; mem, Rehab Comt, Inter-Soc Comn Heart Disease Resources, 69-75; mem, Nat Thrombosis Adv Comt, 71-74 & Heart Panel, Heart, Lung & Blood Vessel Dis Act, 72; consult, Int Div Social & Rehab Serv, Dept Health, Educ & Welfare, 70- & J Chest, 70-; chmn, Prog Comt, Am Heart Asn, 75-76; mem, Clin Trials Comn, Nat Heart, Lung & Blood Inst, 78-81; assoc ed, J Behav Med, 77; ed, J Cardiol, 82-; consult ed, J Cardiovasc & Pulmonary Med, 83- *Honors & Awards:* Myrtle Wreath Award, 67. *Mem:* AMA; fel Am Col Cardiol; Am Heart Asn (past vpres); Am Fedn Clin Res; Am Thoracic Asn (vpres, 75-). *Res:* Urokinase pulmonary embolism trial; ischemic heart disease in young adults; clinical evaluation of myocardial infarction patients; evaluation of patient education programs; author of over 500 scientific books. *Mailing Add:* 864 Somerset Dr Atlanta GA 30327

WENGER, RONALD HAROLD, b Dayton, Ohio, Nov 30, 37; m 63. MATHEMATICS. *Educ:* Miami Univ, Ohio, AB, 59; Mich State Univ, MS, 61, PhD(math), 65. *Prof Exp:* ASSOC PROF MATH, UNIV DEL, 65-, ASST TO PROVOST FOR ACAD PLANNING, 69- *Concurrent Pos:* Assoc dean, Col Arts & Sci, Univ Del, 68-69, 72-; Am Coun Educ fel acad admin, 70-71. *Mem:* Am Math Soc; Math Asn Am. *Res:* Semigroup rings. *Mailing Add:* 20 Winslow Rd Newark DE 19711-5210

WENGER, THOMAS LEE, b New York, NY, Feb 20, 45; c 2. MEDICINE. *Educ:* Princeton Univ, AB, 66; Boston Univ, MD, 71. *Prof Exp:* House officer, Harlem Hosp Med Ctr, 71-75; res assoc, Columbia Presby Med Ctr, 74-75; cardiol fel, Duke Univ Med Ctr, 75-78; SR CLIN RES SCIENTIST, BURROUGHS WELLCOME CO, 78-, DIR, PROF SER DIV, MED. *Concurrent Pos:* Adj assoc prof med, Duke Univ Med Ctr, 78- *Mem:* Am Heart Assoc; fel Am Col Cardiol; AAAS. *Res:* Cardiovascular electrophysiology and pharmacology. *Mailing Add:* Burroughs Wellcome Co 3030 Cornwallis Rd Research Triangle Park NC 27709

WENGERD, SHERMAN ALEXANDER, b Berlin, Ohio, Feb 17, 15; m 40; c 4. CIVIL ENGINEERING, GEOGRAPHICAL EXPLORATION. *Educ:* Wooster Col, AB, 36; Harvard Univ, MA, 38, PhD(geol), 47. *Prof Exp:* Mining geologist, Ramshorn Mining, Challis, Idaho, 38; geologist, Shell Oil Co, 40-42; hydrographic engr, photogram engr, geophysicist & photogeologist, Lt, USN, 42-45; res geologist, Shell Oil Co, 45-47; prof geol, Univ NMex, 47-76, emer prof, 76-; RETIRED. *Concurrent Pos:* Vpres, Oil Recovery Corp, 56-61; ed, Am Asn Petrol Geol Bull, 57-59, pres Asn, 71-72; adv bd, Energy Equities Inc, 71-77; dir, Thompson Int Corp, 72; dir, Found Advan Paleont & Sedimentology, 77-80. *Honors & Awards:* Pres Award, Am Asn Petrol Geol, 48. *Mem:* Hon mem Four Corners Geol Soc; hon mem Am Asn Petrol Geol; fel Geol Soc Am; Am Geog Soc; Sigma Xi; Am Asn Univ Prof; Nat Soc Engrs. *Res:* Exploration research on petroleum geology; published numerous articles in various journals and chapters in books. *Mailing Add:* 1040 Stanford Dr NE Albuquerque NM 87106

WENIG, HAROLD G(EORGE), b New York, NY, Sept 24, 24; m 49; c 2. MECHANICAL ENGINEERING, ELECTROMAGNETISM. *Educ:* City Col New York, BME, 45; Yale Univ, ME, 48; NY Univ, ScD(mech eng), 52. *Prof Exp:* Stress analyst, Otis Elevator Co, 45; mech designer, H K Ferguson Co, 47; mech engr, Sanderson & Porter Co, 48-49; dir eng projs, Bulova Res & Develop Labs, Inc, 51-58; consult engr, 58-59; PRES, WRIGHT INDUSTS, INC, 59- *Concurrent Pos:* Dir, Krystinel Corp. *Honors & Awards:* I B Laskowitz Gold Medal, NY Acad Sci, 68. *Mem:* AAAS; fel NY Acad Sci; Tech Asn Pulp & Paper Indust; Inst Elec & Electronics Engrs; Soc Photog Scientists & Engrs; Am Phys Soc; Am Soc Mech Engrs; Nat Soc Prof Engrs. *Res:* Aerothermodynamics; dynamics of rigid bodies in extreme force fields; hydrology; research and development management; magnetism and magnetic materials. *Mailing Add:* 375-08E S End Ave New York NY 10280

WENIS, EDWARD, b Linden, NJ, May 21, 19; m 42; c 2. CHEMICAL ENGINEERING. *Educ:* Newark Col Eng, BS, 40; Stevens Inst Technol, MS, 42. *Prof Exp:* Salesman, Sun Oil Co, 38-39; chemist & county supvr Dutch elm disease, USDA, Pa, 39-42; org chemist, Hoffman-LaRoche Inc, 42-66, mgr, Thin Layer Chromatogram Labs & Res Serv, 67-79; RETIRED. *Mem:* Am Chem Soc; fel Am Inst Chemists. *Res:* Synthesis of new drugs; development of riboflavin, lyxoflavin, folic acid and khellin; antituberculosis, hypoglycemic agents and psychoenergizers. *Mailing Add:* 104 Hillcrest Ave Leonia NJ 07605

WENK, EDWARD, JR, b Baltimore, Md, Jan 24, 20; m 41; c 3. SCIENCE POLICY, CIVIL ENGINEERING. *Educ:* Johns Hopkins Univ, BE, 40, DEng, 50; Harvard Univ, MSc, 47. *Hon Degrees:* DSc, Univ RI, 68; LHD, Johns Hopkins, 90. *Prof Exp:* Ship struct designer, Boston Navy Shipyard, Mass, 41-42; supvr turret test sect, David Taylor Model Basin, US Naval Dept, 42-45, supt struct dynamics sect, 45-48, head submarine struct br, 48-50, chief struct div, 50-56; chmn dept eng mech, Southwest Res Inst, 56-59; sr specialist sci & technol, Legis Reference Serv, Libr Cong, 59-61; tech asst to President's Sci Adv, White House, 61-62; tech asst to dir, Off Sci & Technol & exec secy, Fed Coun Sci & Technol, 62-64, chief, Sci Policy Res Div & spec adv to librn, Libr Cong, 64-66; exec secy, Nat Coun Marine Resources & Eng Develop, Exec Off of President, 66-70; dir, prog social mgt technol, 73-78, prof, 70-83, EMER PROF ENG & PUB AFFAIRS, UNIV WASH, 83- *Concurrent Pos:* Mem pressure vessel res comt & chmn, Design Div, Welding Res Coun, 53-59; lectr, Univ Md, 52 & 54-56, Seattle Univ, 84; consult comt undersea warfare, Nat Acad Sci, 57-58, mem panel naval vehicles, 60-71; chmn comt pub eng policy, Nat Acad Eng, 69-75; vis scholar, Woodrow Wilson Int Ctr Scholars, 70-71; Ford Found fel, 70-72; mem bd dirs, URS Systs Corp, 71-88; mem, Nat Adv Comt Oceans & Atmosphere, 71-72; vchmn, US Cong Technol Assessment Adv Coun, 74-79; consult to White House, US Cong, Nat Oceanic & Atmospheric Admin & UN Secretariat, UK, Sweden, Australia & Philippines, States of Alaska & Wash. *Mem:* Nat Acad Eng; Nat Acad Pub Admin; Am Soc Civil Engrs; Soc Exp Stress Anal (pres, 57-58); fel Am Soc Mech Engrs; Sigma Xi; Int Asn Impact Assessment (pres,

82-83); fel AAAS. *Res:* Applied mechanics; strength of ships and deep-diving submarines; experimental stress analysis; thin shell structures; ocean engineering; marine affairs; technology assessment; science policy research; public administration; futures; decision theory; technology-intensive public policy; public involvement; technology as social process; impact analysis; decision behavior; futures; principles for social management of technology; technological literacy, techno- ethics; education reform. *Mailing Add:* Univ Wash 255 Wilcox Hall FX-10 Seattle WA 98195

WENK, EUGENE J, b New York, NY, Oct 21, 27; m 54; c 3. HORMONE RECEPTORS, OCULAR TISSUE. *Educ:* Columbia Univ, AB, 50, AM, 51; New York Med Col, PhD(anat), 72. *Prof Exp:* From instr to asst prof, 72-79, ASSOC PROF ANAT, NY MED COL, 79- *Mem:* NY Acad Sci; Am Asn Anatomists; Am Asn Clin Anatomists. *Res:* Ultrastructure and function of ocular tissue. *Mailing Add:* Dept Cell Biol & Anat New York Med Col Valhalla NY 10595

WENK, HANS-RUDOLF, b Zurich, Switz, Oct 25, 41; m 70. TEXTURE ANALYSIS, ELECTRON MICROSCOPY. *Educ:* Univ Basel, BA, 63; Univ Zurich, PhD(crystallog), 65. *Prof Exp:* From asst prof to assoc prof geol, 67-73, PROF GEOL & GEOPHYS, UNIV CALIF, BERKELEY, 73- *Concurrent Pos:* Co-worker, Swiss Geol Comn, 66-; vis prof, Univ Frankfurt, 74, Univ Metz, 79, Univ Kiel, 81, Univ Nanjing, 86, Univ Hamburg, 88; collabr, LANL-CMS, 84- *Honors & Awards:* Homboldt Sr US Scientist Award, 87. *Mem:* Mineral Soc Am; Am Geophys Union; Am Crystallog Asn. *Res:* Structural geology of metamorphic belts; texture analysis of deformed rocks, ceramics and metals; electron microscopy and x-ray diffraction of rockforming minerals. *Mailing Add:* Dept Geol & Geophys Univ Calif 497 Earth Sci Bldg Berkeley CA 94720

WENK, MARTIN LESTER, CELL BIOLOGY. *Educ:* Columbia Univ, PhD(cell biol), 71. *Prof Exp:* PRIN INVESTR & HEAD DEPT, MICROBIOL ASSOCS, INC, BETHESDA, 85- *Res:* In vivo experimental carcinogenesis. *Mailing Add:* 9740 Wightman Rd Gaithersburg MD 20879

WENKERT, ERNEST, b Vienna, Austria, Oct 16, 25; nat US; m 48; c 4. ORGANIC CHEMISTRY. *Educ:* Univ Wash, BS, 45, MS, 47; Harvard Univ, PhD(chem), 51. *Hon Degrees:* Dr, Univ Paris-Sud, 78. *Prof Exp:* Instr chem, Lower Columbia Jr Col, 47-48; from asst prof to prof org chem, Iowa State Univ, 51-61; prof, Ind Univ, Bloomington, 61-69, Herman T Briscoe prof, 69-73; E D Butcher prof chem, Rice Univ, 73-80, chmn dept, 76-80; PROF CHEM, UNIV CALIF, SAN DIEGO, 80- *Concurrent Pos:* Lectr & vis prof, var US & foreign orgn & acad insts, 51-; actg head dept org chem, Weizmann Inst, 64-65; Guggenheim fel, 65-66; mem NIH med chem B study sect, 71-72, chmn, 72-75; chmn, Gordon Conf Steroids & Other Natural Prod, 64 & 65; mem, NIH med chem B fel rev comn, 67-70; mem, Comt Direction, Inst Chem Substances Naturelles, France, 74-77; mem, Oak Ridge Nat Lab Chem Div Adv Comn, 75-; chief tech adv, UNESCO, 78. *Honors & Awards:* Ernest Guenther Award, Am Chem Soc. *Mem:* Am Chem Soc; Royal Soc Chem; corresp mem Acad Brazileira Ciencias; Swiss Chem Soc. *Mailing Add:* Chem Dept D-006 Univ Calif San Diego La Jolla CA 92093

WENNBERG, JOHN E, b June 2, 34; m. EPIDEMIOLOGY, PUBLIC POLICY. *Educ:* Stanford Univ, BA, 56; McGill Med Sch, MD, 61; Johns Hopkins Univ, MPH, 66. *Prof Exp:* Intern med, DC Gen Hosp, 61-62; assoc med resident, Johns Hopkins Hosp, Baltimore, Md, 62-63, fel, Renal Dis & Pharmacol, 63-65; resident chronic dis, Baltimore City Hosp, Md, 66-67; dir, Northern New Eng Regional Med Prog, Burlington, Vt, 67-71; interim dir, Coop Health Info Ctr, Vt, 72-73; sr assoc, Harvard Ctr Community Health & Med Care, Harvard Med Sch, Boston, Mass, 73-75, asst prof, Dept Prev & Social Med, 73-78; assoc prof, 79, PROF EPIDEMIOL, DEPT COMMUNITY & FAMILY MED, DARTMOUTH MED SCH, HANOVER, NH, 80-, DIR, CTR EVAL CLIN SCI, 89- *Concurrent Pos:* Asst prof, Dept Med, Univ Vt, Burlington, 67-72; mem, Comt Health Data Systs, Inst Med, 73-74, Comt Study Resources Clin Invest, 88-, Health Sci Policy Bd, 89-; mem, Coun Res & Develop, Am Hosp Asn, 73-75; mem, Ctr Anal Health Pract, Harvard Sch Pub Health, 75-78; mem bd dirs, Am Med Rev Res Ctr, 85-, Collab Ctr Small Area Anal, Univ Copenhagen, 86-88; physicians adv group, NY Health & Hosps Corp, 88- *Mailing Add:* Dept Community & Family Med Dartmouth Med Sch Hanover NH 03756

WENNER, ADRIAN MANLEY, b Roseau, Minn, May 24, 28; m 57; c 2. ZOOLOGY. *Educ:* Gustavus Adolphus Col, BS, 51; Chico State Col, MA, 55; Univ Mich, MS, 58, PhD(zool), 61. *Prof Exp:* Prin elem sch, Ore, 54-55; teacher high sch, Calif, 55-56; fel zool, Univ Mich, 56-60; from asst prof biol sci to assoc prof biol, 60-73, PROF NATURAL HIST, UNIV CALIF, SANTA BARBARA, 73- *Concurrent Pos:* Consult, Teledyne, Inc, 62-64 & Autonetics Div, NAm Aviation, Inc, 64-65. *Mem:* AAAS; Am Soc Zoologists; Sigma Xi; Am Soc Naturalists; Crustacean Soc. *Res:* Problems of growth and egg production in marine crustaceans as they occur in nature; natural history of marine crustacea; animal communication. *Mailing Add:* Biol Sci Dept Univ Calif Santa Barbara CA 93106

WENNER, BRUCE RICHARD, b Lancaster, Pa, Apr 25, 38; m 65; c 3. TOPOLOGY. *Educ:* Col Wooster, BA, 60; Duke Univ, PhD(math), 64. *Prof Exp:* Asst prof math, Univ Vt, 64-68; from asst prof to assoc prof, 68-76, PROF MATH, UNIV MO-KANSAS CITY, 76- *Concurrent Pos:* NASA res grant, 65-66. *Mem:* Am Math Soc; Math Asn Am. *Res:* Dimension theory with regard to topological dimension functions, especially metrizable spaces. *Mailing Add:* Dept of Math Univ of Mo Kansas City MO 64110

WENNER, CHARLES EARL, b Lattimer, Pa, May 2, 24; m 48; c 4. BIOCHEMISTRY. *Educ:* Temple Univ, BA, PhD(chem), 53. *Prof Exp:* Res fel, Lankenau Hosp Res Inst & Inst Cancer Res, 50-54, res assoc, 54-55; sr scientist, 56-61, assoc scientist, 61-65, ASSOC CHIEF SCIENTIST, DEPT EXP BIOL, ROSWELL PARK MEM INST; RES PROF BIOCHEM, GRAD SCH, STATE UNIV NY BUFFALO, 70-, CHMN, DEPT BIOCHEM, ROSWELL PARK DIV, 80- *Concurrent Pos:* Runyon fel, Inst Cancer Res, 52-54; from asst prof to assoc prof biochem, Grad Sch, State Univ NY Buffalo, 58-70; Johnson Res Found vis res prof, Univ Pa, 65-66. *Mem:* Am Chem Soc; Am Asn Cancer Res; Biophys Soc; Fedn Am Soc Exp Biol. *Res:* Energy control mechanisms; transforming growth factors and signal transduction mechanisms; mechanism of action of cocarcinogens; membrane transport and cell proliferation. *Mailing Add:* Dept Exp Biol Roswell Park Mem Inst 666 Elm St Buffalo NY 14263

WENNER, DAVID BRUCE, b Flint, Mich, May 28, 41; m 68, 83; c 2. GEOCHEMISTRY, GEOLOGY. *Educ:* Univ Cincinnati, BS, 63; Calif Inst Technol, MS, 66, PhD(geochem), 71. *Prof Exp:* Vis asst prof, 71, asst prof, 71-80, ASSOC PROF GEOL, UNIV GA, 80- *Mem:* AAAS; Geol Soc Am; Am Geophys Union. *Res:* Stable isotope geochemistry with applications to petrology, hydrology and archaeology. *Mailing Add:* Dept Geol Univ Ga Athens GA 30602

WENNER, HERBERT ALLAN, b Drums, Pa, Nov 14, 12; m 42; c 4. MEDICINE. *Educ:* Bucknell Univ, BSc, 33; Univ Rochester, MD, 39; Am Bd Microbiol, dipl, 62; Am Bd Pediat, 49. *Prof Exp:* Instr prev med, Sch Med, Yale Univ, 44-46; from asst prof to assoc prof pediat & bact, Univ Kans, 46-51, res prof pediat, 51-69; distinguished prof pediat, 69-81, clin prof, Sch Med, 70-88, EMER PROF, UNIV MO, KANSAS CITY, 88- *Concurrent Pos:* Babbott fel, Sch Med, Yale Univ, 41-42; Nat Res Coun fel, Yale Univ & Johns Hopkins Univ, 43-44; NIH res career award, 62-69; assoc physician, Dept Internal Med, New Haven Hosp, Conn, 44-46; consult, Mo State Bd Health & Nat Commun Dis Ctr; mem, Echovirus Subcomt, Picornavirus Study Group, NIH; clin prof pediat, Univ Kans Med Ctr, Kansas City, 73- *Honors & Awards:* Presidential & Distinguished Serv Awards, Nat Found Infantile Paralysis; Cert Serv Award, Panel Picornaviruses, NIH. *Mem:* Fel AAAS; fel Am Acad Pediat; fel Am Pub Health Asn; Soc Pediat Res; Am Pediat Soc. *Res:* Etiology, pathogenesis and epidemiology of infectious diseases. *Mailing Add:* Children's Mercy Hosp 24th at Gillham Rd Kansas City MO 64108

WENNERBERG, A(LLAN) L(ORENS), b Chicago, Ill, Jan 20, 32; m 53; c 2. ELECTRONICS ENGINEERING. *Educ:* Ind Inst Technol, BSEE, 56; Univ Notre Dame, MSEE, 61. *Prof Exp:* Res engr, Sylvania Microwave Res Lab, 56-57; res engr, 57-65, mgr electronics res, 65-67, dir electronics res, 67-69, DIR RES, WHIRLPOOL CORP, 69- *Mem:* Inst Elec & Electronics Engrs; Sci Res Soc Am; Sigma Xi. *Res:* Precision control systems involving digital techniques implemented with microcircuitry. *Mailing Add:* 2804 Sunnydale Ave St Joseph MI 49085-2426

WENNERSTROM, ARTHUR J(OHN), b New York, NY, Jan 11, 35. FLUID MECHANICS. *Educ:* Duke Univ, BS, 56; Mass Inst Technol, MS, 58; Swiss Fed Inst Technol, DScTech(compressor aerodyn), 65. *Prof Exp:* Sr engr, Aircraft Armaments, Inc, 58-59; res engr, Sulzer Bros, Ltd, Switz, 60-62; proj engr, Northern Res & Eng Corp, 65-67; group leader, 67-75, CHIEF, COMPRESSOR RES GROUP, AIR FORCE AERO PROPULSION LAB, 75- *Concurrent Pos:* US coordr, Propulsion & Energetics Panel, Adv Group Aerospace Res & Develop, NATO, 77-85; chmn, Gas Turbine Div, Am Soc Mech Engrs, 80-81; consult UNESCO, Nat Aeronaut Lab, India, 81; ed, Am Soc Mech Engrs J Eng Gas Turbines & Power, 83-88, Am Soc Mech Engrs J Turbomachinery, 86-88. *Honors & Awards:* Airbreathing Propulsion Award, Am Inst Aeronaut & Astronaut, 79; Cliff Garrett Turbomachinery Award, Soc Automotive Engrs, 86. *Mem:* Fel Am Soc Mech Engrs; fel Am Inst Aeronaut & Astronaut. *Res:* Gas turbines; aerodynamics of transonic and supersonic axial compressors; experimental aerodynamics; aerodynamic instrumentation and measurement techniques. *Mailing Add:* 4300 Fair Oak Rd No 5 Dayton OH 45405

WENNERSTROM, DAVID E, b Glendale, Calif, June 19, 45; m 73; c 2. MICROBIAL PHYSIOLOGY, MICROBIAL PATHOGENESIS. *Educ:* Univ Ark, BS, 68, MS, 70; Univ Tenn, PhD(microbiol), 73. *Prof Exp:* Fel lipid biochem, Hormel Inst, Univ Minn, 73-76; ASST PROF MICROBIAL PHYSIOL, DEPT MICROBIOL & IMMUNOL, UNIV ARK MED SCI, 76- *Concurrent Pos:* Prin invest, Inst Allergy & Infectious Dis, NIH, Univ Ark Med Sci, 80-83. *Mem:* Am Soc Microbiol; AAAS; Sigma Xi. *Res:* Pathogenesis of group B streptococci in lung tissue is studied using mice which have been shown to be an appropriate model of early-onset group B streptococcal disease of the human newborn. *Mailing Add:* Dept Microbiol & Immunol Univ Ark Med Sci 4301 W Markham Little Rock AK 72205

WENRICH, KAREN JANE, b Lebanon, Pa, Apr 9, 47; m 69, 83. GEOLOGY. *Educ:* Pa State Univ, BS, 69, MS, 71, PhD(geol), 75. *Prof Exp:* Geologist, Molybdenum Corp Am, 69; instr geol, Bucknell Univ, 73-74; GEOLOGIST, US GEOL SURV, 74- *Concurrent Pos:* US Geol Surv adv, Energy Res & Develop Admin Nat Uranium Resource Eval Prog, 75-78. *Mem:* Geol Soc Am; Sigma Xi; Am Geophys Union; Asn Explor Geochemists. *Res:* Uranium exploration, specifically developing exploration guides for uranium silicic volcanic rocks and the use of uranium in water, stream sediments, soils and modern decaying plant materials as a tool for exploration; trace element geochemistry in volcanic rocks; mineralized breccia pipes in northern Arizona. *Mailing Add:* Mail Stop 939 Fed Ctr US Geol Surv Denver CO 80225

WENSCH, GLEN W(ILLIAM), b Chicago, Ill, Nov 15, 17; m 42; c 2. PHYSICAL METALLURGY. *Educ:* Univ Ill, BS, 46, MS, 47, PhD(metall eng), 49. *Prof Exp:* Staff metall engr, Los Alamos Sci Lab, 49-51; sr res metallurgist, Fansteel Metall Co, 51-52; chief reactor mat br, Savannah River Proj, US AEC, 52-54; sr metall engr, Vitro Corp, 54-55; chief, Liquid Metals Projs Br, US AEC, 55-70, dir reactor develop, 70-76, exec asst to asst adminr, Energy Res & Develop Admin, 76-78; consult engr, 78-85; RETIRED. *Concurrent Pos:* Chief, Fast Reactor Team, Europe, 55, 57, 59, 60 & 63; pres, Int Plutonium Conf, France, 60; mem & leader, Power Reactor Deleg, USSR, 64 & 70; US mem, Int Working Group Fast Reactors, Int Atomic Energy Agency; pres, Glen W Wensch Corp; consult, Argonne Nat Lab, Dept Energy, 89- *Mem:* AAAS; fel Am Soc Metals; Am Nuclear Soc; fel NY Acad Sci; fel Am Inst Chemists. *Res:* Plutonium; sodium components and technology; fast breeder reactors. *Mailing Add:* RR 1 Box 54 Champaign IL 61821

WENSLEY, CHARLES GELEN, b Niagara Falls, NY, Feb 19, 49; m 77; c 2. POLYMER PHYSICAL CHEMISTRY, FORMULATIONS DEVELOPMENT. *Educ:* Univ Calif, San Diego, BA, 71; Univ Wis, Madison PhD(chem), 77. *Prof Exp:* MGR SPEC MEMBRANE DEVELOP, ENVIROGENICS SYSTS CO, 78- *Res:* Develop synthetic semi-permeable membranes that have the capactity to separate mixtures of gases, or salts from water, or components of an azeotrope using pressure as the driving force. *Mailing Add:* 18092 Stratford Circle Villa Park CA 92667

WENT, FRITZ, botany; deceased, see previous edition for last biography

WENT, HANS ADRIAAN, b Bogor, Indonesia, Dec 3, 29; nat US; m 51; c 2. PHYSIOLOGY. *Educ:* Univ Calif, AB, 51, MA, 53, PhD(zool), 58. *Prof Exp:* From instr to asst prof, 59-69, ASSOC PROF ZOOL, WASH STATE UNIV, 69- *Mem:* Am Soc Zool; Sigma Xi. *Res:* Cell division, especially molecular origin of mitotic apparatus; cell physiology. *Mailing Add:* Dept Zool Sci Hall Wash State Univ Pullman WA 99164-4220

WENTE, HENRY CHRISTIAN, b New York, NY, Aug 18, 36. MATHEMATICS. *Educ:* Harvard Univ, BA, 58, MA, 59, PhD(math), 66. *Prof Exp:* From instr to asst prof, Tufts Univ, 63-70, lectr, 70-71; from asst prof to assoc prof math, 71-77, PROF MATH, UNIV TOLEDO, 77- *Concurrent Pos:* Univ grant & res assoc, Math Inst, Univ Bonn, 72-73. *Mem:* AAAS; Am Math Soc; Sigma Xi. *Res:* Existence theorems in the calculus of variations, especially those arising from two-dimensional parametric surfaces immersed in Euclidean space; surfaces minimizing area subject to a volume constraint. *Mailing Add:* Dept Math Univ Toledo 2801 W Bancroft St Toledo OH 43606

WENTINK, TUNIS, JR, b Paterson, NJ, Feb 3, 20; m 46, 68. PHYSICAL CHEMISTRY. *Educ:* Rutgers Univ, BS, 41; Cornell Univ, PhD(chem), 54. *Prof Exp:* Res chemist, Photoprods Div, E I du Pont de Nemours & Co, 41-43; from res assoc to mem staff, Div Indust Co-op, Mass Inst Technol, 47-48; from res assoc to specialist microwave spectros, Brookhaven Nat Lab, 48-50; asst, Cornell Univ, 50-54; physicist, Gen Elec Co, 53-55; prin res scientist & supvr, Chem Lab, Avco-Everett Res Lab, Avco Corp, 55-59, from prin scientist to sr consult scientist, Adv Res & Develop Div, 59-67; prin scientist, GCA Corp, 67-68; head exp physics dept, Panametrics, Inc, 68-70; assoc dir, dir Inst Arctic Environ Eng, 72-73, 72-, PROF PHYSICS, GEOPHYS INST, UNIV ALASKA, 70- *Concurrent Pos:* NSF vis prof, Geophys Inst, Univ Alaska, 68; consult. *Mem:* Hon mem Sigma Xi. *Res:* Spectroscopy; radiation chemistry of high temperature gases and solids; ultraviolet to microwave regions and instrumentation; molecular absolute intensities and radiative lifetimes; wind energy; energy systems; re-entry physics and signatures; ablation; arctic environmental technology. *Mailing Add:* 1718 Bridgewater Dr Fairbanks AK 99709

WENTLAND, MARK PHILIP, b New Britain, Conn, Jan 22, 45; m 70; c 2. MEDICINAL CHEMISTRY, ORGANIC CHEMISTRY. *Educ:* Cent Conn State Univ, BS, 66; Rice Univ, PhD(chem), 70. *Prof Exp:* Res investr, 70-81, sr res investr, 81-84, prin res investr, 84-87, res leader , 87-89, ASST RES DIR, STERLING RES GROUP, 89- *Concurrent Pos:* Lectr, Rensselaer Polytech Inst, 71, adj assoc prof, 72-81, adj prof, 81- *Mem:* Am Chem Soc; Sigma Xi; Am Asn Cancer Res. *Res:* The design and synthesis of potentially useful medicinal agents. *Mailing Add:* Sterling Res Group Rensselaer NY 12144

WENTLAND, STEPHEN HENRY, b New Britain, Conn, May 1, 40; m 65; c 2. BIOCHEMISTRY, ORGANIC CHEMISTRY. *Educ:* Rensselaer Polytech Inst, BS, 62; Yale Univ, MS, 64, PhD(chem), 68. *Prof Exp:* NIH fel chem, Ind Univ, Bloomington, 68-70; sr org chemist, Smith Kline & French, Inc, 70-72; res assoc biochemist, Univ Colo Med Ctr, Denver, 72-74, instr med, 74-77; PROF CHEM, HOUSTON BAPTIST UNIV, 77-, CHMN DEPT, 83- *Mem:* Am Chem Soc. *Res:* Synthesis and assay of biochemical and organic compounds; organic and biochemistry textbook and chemical technical report writing. *Mailing Add:* 11426 Carvel Lane Houston TX 77072-2934

WENTORF, ROBERT H, JR, b Wis, May 28, 26; m 49; c 3. PHYSICAL CHEMISTRY. *Educ:* Univ Wis, BSChE, 48, PhD(chem), 52. *Hon Degrees:* DSc, Univ Wis, 81. *Prof Exp:* Asst chem, Univ Wis, 46; RES ASSOC CHEM, CORP RES & DEVELOP CTR, GEN ELEC CO, 52- *Concurrent Pos:* Brittingham vis prof, Univ Wis, 67-68; dept chem eng distinguished res prof, RPI, Troy, NY, 88- *Honors & Awards:* Ipatieff Prize, 65; New Mat Prize, Am Phys Soc, 77. *Mem:* Nat Acad Eng; Am Chem Soc; Sigma Xi. *Res:* High pressure chemistry and physics, diamond synthesis, cubic BN, energy systems, solar energy utilization. *Mailing Add:* RR 3 Box 154A Greenwich NY 12834-9803

WENTWORTH, BERNARD C, b Freedom, Maine, Feb 16, 35; m 60; c 4. AVIAN PHYSIOLOGY. *Educ:* Univ Maine, Orono, BS, 57; Univ Mass, MS, 60, PhD(avian physiol), 63. *Prof Exp:* Physiologist, US Dept Interior, 63-69; assoc prof, 69-73, PROF POULTRY SCI, UNIV WIS-MADISON, 73- *Mem:* AAAS; Poultry Sci Asn; Soc Study Reproduction; Endocrine Soc. *Res:* Basic physiology of birds and comparative endocrinology of animals as related to applied benefits to biomedicine and agriculture; primodial germ cell differentiation. *Mailing Add:* Dept Poultry Sci Animal Sci Bldg Univ Wis Madison WI 53706

WENTWORTH, BERTTINA BROWN, microbiology, for more information see previous edition

WENTWORTH, CARL M, JR, b New York, NY, Feb 8, 36; m 68; c 2. TECTONICS, ENVIRONMENTAL GEOLOGY. *Educ:* Dartmouth Col, AB, 58; Stanford Univ, MS, 60, PhD(geol), 67. *Prof Exp:* GEOLOGIST, US GEOL SURV, 63- *Mem:* Geol Soc Am; AAAS. *Res:* Neotectonics; major geologic hazards of United States; active faults and movement histories; landslides and slope stability; engineering character of geologic materials; geology of California Coast Ranges. *Mailing Add:* US Geol Surv Mail Stop 975 345 Middlefield Rd Menlo Park CA 94025

WENTWORTH, GARY, b Orange, Mass, Aug 3, 39; m 61; c 2. ORGANIC CHEMISTRY, POLYMER CHEMISTRY. *Educ:* Rensselaer Polytech Inst, BS, 61; Ga Inst Technol, PhD(org chem), 66. *Prof Exp:* Res chemist, Union Carbide Corp, 66-68; sr res chemist, Monsanto Develop Ctr, 68-73, res specialist, 73-76; res group leader, Monsanto Polymers & Resins Co, 76-82; DIR RES, SHERWIN-WILLIAMS RES CTR, 82- *Mem:* Am Chem Soc; The Chem Soc. *Res:* Fiber chemistry; polymerization; polymer structure/property relationships; coatings science. *Mailing Add:* Sherwin Williams Co 10909 S Cottage Grove Ave Chicago IL 60628

WENTWORTH, JOHN W, b Greenville, Maine, Nov 3, 25. TELEVISION ENGINEERING. *Educ:* Univ Maine, BS, 49. *Prof Exp:* Mgr, Broadcast Tech Training, RCA Corp, 73-82; PROPRIETOR, WENTWORTH TRAINING SERV, CHERRY HILL, NJ, 83- *Mem:* Fel Inst Elec & Electronics Engrs; fel Soc Motion Picture & TV Engrs. *Res:* Development and design of television broadcast equipment. *Mailing Add:* Wentworth Training Service 137 Colwick Rd Cherry Hill NJ 08002

WENTWORTH, RUPERT A D, b Hattiesburg, Miss, Nov 5, 34; m 56, 72; c 3. INORGANIC CHEMISTRY. *Educ:* Fordham Univ, BS, 55; Mich State Univ, PhD(inorg chem), 63. *Prof Exp:* Fel with Prof T S Piper, Univ Ill, 63-65; from asst prof to assoc prof, 65-74, PROF CHEM, COL ARTS & SCI, GRAD SCH, IND UNIV, BLOOMINGTON, 74- *Mem:* Am Chem Soc. *Res:* The chemistry of simple molybdenum complexes with substrates for molybdoenzymes. *Mailing Add:* Dept Chem Ind Univ Bloomington IN 47405-4001

WENTWORTH, STANLEY EARL, b Natick, Mass, July 13, 40; m 67; c 2. ORGANIC POLYMER CHEMISTRY. *Educ:* Northeastern Univ, BS, 63, PhD(org chem), 67. *Prof Exp:* Res chemist, Army Natick Labs, 67-68, RES CHEMIST, EMERGING MAT DIV, ARMY MAT TECH LAB, 70- *Concurrent Pos:* Army liaison rep on several nat mat adv bd study comts. *Mem:* Am Chem Soc; Adhesion Soc; SAMPE. *Res:* Organic synthesis in the areas of organofluorine compounds, diazoalkanes and monomers for high temperature resins; thermal analysis of organic materials; synthesis of polyphenylquinoxalines and polyurethanes; studies of adhesives and adhesive bonding; synthesis of semiconductive polymers; composite materials technology; studies in adhesion science. *Mailing Add:* Emerging Mat Div Army Mat Tech Lab Watertown MA 02172-0001

WENTWORTH, THOMAS RALPH, b Boston, Mass, Sept 4, 48. PLANT ECOLOGY. *Educ:* Dartmouth Col, AB, 70; Cornell Univ, PhD(plant ecol), 76. *Prof Exp:* ASST PROF BOT, NC STATE UNIV, 76- *Mem:* Ecol Soc Am; Brit Ecol Soc; Sigma Xi. *Res:* Plant community ecology. *Mailing Add:* Dept Bot Box 7612 NC State Univ Raleigh NC 27695-7612

WENTWORTH, WAYNE, b Rochester, Minn, May 29, 30; m 54; c 4. ANALYTICAL CHEMISTRY, PHYSICAL CHEMISTRY. *Educ:* St Olaf Col, BA, 52; Fla State Univ, PhD(chem), 57. *Prof Exp:* Res mathematician, Radio Corp Am, 56-59; from asst prof to assoc prof, 59-68, PROF CHEM, UNIV HOUSTON, 68- *Mem:* Am Chem Soc. *Res:* Electron attachment to molecules; molecular complexes; molecular spectroscopy. *Mailing Add:* Dept Chem Univ Houston Cullen Blvd Houston TX 77004

WENTZ, WILLIAM BUDD, b Philadelphia, Pa, Aug 9, 24; m 45; c 3. OBSTETRICS & GYNECOLOGY, ONCOLOGY. *Educ:* Univ Pa, BA, 51, MA, 53; Western Reserve Univ, MD, 58; Am Bd Obstet & Gynec, dipl, 66. *Prof Exp:* Prin investr physiol, US Naval Aviation Med Acceleration Lab, 53-54; intern med, Lankenau Hosp, Philadelphia, Pa, 48-49, prin investr gynec & oncol, 59-63; asst prof obstet & gynec, Hahnemann Med Col, 63-66; assoc prof obstet & gynec, 66-71, PROF REPRODUCTIVE BIOL, SCH MED, CASE WESTERN RESERVE UNIV, 71- *Concurrent Pos:* Am Cancer Soc grants, Lankenau Hosp, Philadelphia, 59-63; Nat Cancer Inst grants, 64- *Mem:* Am Col Obstet & Gynec; Soc Gynec Oncol; Am Soc Cytol. *Res:* Experimental gynecological pathology; carcinogenesis; cancer research treatment of malignant and premalignant disease. *Mailing Add:* 2105 Adelbert Rd Cleveland OH 44106

WENTZ, WILLIAM HENRY, (JR), b Wichita, Kans, Dec 18, 33; m 55; c 2. AERONAUTICAL ENGINEERING, MECHANICAL ENGINEERING. *Educ:* Univ Wichita, BS, 55, MS, 61; Univ Kans, PhD(eng mech), 69. *Prof Exp:* Instr mech eng, Univ Wichita, 57-58; res engr, Boeing Co, 58-63; from asst prof to prof aeronaut eng, 63-83, DISTINGUISHED PROF, WICHITA STATE UNIV, 83-, EXEC DIR, INST AVIATION RES, 88- *Concurrent Pos:* NSF sci fac fel, 67-68; prin investr NASA res grants, Wichita State Univ, 70- *Honors & Awards:* Teetor Award, Soc Automotive Engrs, 73; Gen Aviation Award, Am Inst Aeronaut & Astronaut, 81. *Mem:* Assoc fel Am Inst Aeronaut & Astronaut; Am Soc Eng Educ; Soc Automotive Engrs. *Res:* Aerodynamics of wings and bodies; low speed delta wing vortex flows; trailing vortices and wake turbulence; computer analysis of airfoil sections; airfoil, flap and control surface design; separated flows; wind tunnel testing techniques and instrumentation; wind turbine blade and control system design. *Mailing Add:* Inst Aviation Res Wichita State Univ Box 93 Wichita KS 67208

WENTZEL, DONAT GOTTHARD, b Zürich, Switz, June 25, 34; US citizen; m 59; c 1. ASTROPHYSICS. *Educ:* Univ Chicago, BA, 54, BS, 55, MS, 56, PhD(physics), 60. *Prof Exp:* From instr to assoc prof astron, Univ Mich, 60-66; assoc prof, 67-74, PROF ASTRON, UNIV MD, COLLEGE PARK, 74- *Concurrent Pos:* Alfred P Sloan res fel, 62-66; vis lectr, Princeton Univ, 64; vis prof, Tata Inst Fundamental Res, Bombay, 73; acad guest, Fed Inst Technol, Zurich, 78; President Comn on Teaching Astron, Int Astron Union, 79-82; prog dir, NSF, 86-88. *Mem:* Fel AAAS; Am Astron Soc; Int Astron Union; Sigma Xi. *Res:* Effects of magnetic fields on fluid dynamics and charged particles on the sun and in interplanetary and interstellar space; astronomy education. *Mailing Add:* Astron Dept Univ Md College Park MD 20742

WENZEL, ALAN RICHARD, b Port Chester, NY, Feb 8, 38. APPLIED MATHEMATICS. *Educ:* NY Univ, BAE, 60, MS, 62, PhD(math), 70. *Prof Exp:* Mem res staff, Wyle Labs, 65-68; res assoc, Univ Miami, 70-73; Nat Res Coun assoc, Ames Res Ctr, NASA, 73-75, vis scientist, Inst Comput Appln Sci & Eng, Langley Res Ctr, 75-80; Naval Ocean Res & Develop Activ, 80-87; SR STAFF SCIENTIST, PLANNING SYSTS INC, SLIDELL, LA, 87- *Concurrent Pos:* Adj assoc prof, Dept Physics, Univ New Orleans, 91- *Mem:* Soc Indust & Appl Math; Am Math Soc. *Res:* Theoretical research in fluid dynamics and wave propagation. *Mailing Add:* 210 Anthony Dr Slidell LA 70458

WENZEL, ALEXANDER B, b Mexico City, Mex, Aug 12, 36; US citizen; m 60; c 3. MECHANICAL ENGINEERING, PHYSICS. *Educ:* NMex Mil Inst, BS, 56; NMex State Univ, MS, 59; St Mary's Univ, MBA, 72. *Prof Exp:* Instr physics & math, NMex State Univ, 56-57, Univ Md, 57-60 & Univ Del, 60-62; head terminal ballistics, Lab Ballistics & Explosives, Gen Motors Defense Labs, 62-66; mgr mat dynamics, Lab Ballistics & Shock Physics, Cleveland Army Tank-Auto Comn, 66-68; sr engr ballistics & explosives, 68-71, group leader, 71-73, mgr, 73-77, DIR BALLISTICS & EXPLOSIVES, SOUTHWEST RES INST, 77- *Concurrent Pos:* Sci adv, Phys Sci Lab, NMex, 56-57. *Mem:* Am Defense Preparedness Asn; Instrument Soc Am; Am Inst Aeronaut & Astronaut. *Res:* Explosive and ballistic technology; impulsive loading of structures; vulnerability and survivability analyses; hazards and safety analysis; response of materials to high strains and pressures; weapons effects; penetration mechanics; shaped charge technology. *Mailing Add:* 3211 Hitching Post San Antonio TX 78217

WENZEL, BERNICE MARTHA, b Bridgeport, Conn, June 22, 21; m 52. NEUROBEHAVIOR. *Educ:* Beaver Col, AB, 42; Columbia Univ, AM, 43, PhD(psychol), 48. *Prof Exp:* Instr psychol, Newcomb Col, Tulane Univ, 45-46; from instr to asst prof, Barnard Col, Columbia Univ, 46-55; from asst prof to assoc prof physiol, Univ Calif, Los Angeles, 59-69, vchmn dept, 71-73, prof physiol, 69-89, prof psychiat, 71-89, asst dean, 74-89, EMER PROF, SCH MED, UNIV CALIF, LOS ANGELES, 89- *Concurrent Pos:* Fel, Ment Health Training Prog, Sch Med, Univ Calif, Los Angeles, 57-59, asst dean, 74-89. *Mem:* Fel AAAS; Am Physiol Soc; Int Brain Res Orgn; Soc Neurosci; Am Psychol Asn; Am Chemoreception Sci. *Res:* Behavioral physiology; olfaction; ethology and evolution of chemoreception, especially olfaction in birds. *Mailing Add:* 3334 Scadlock Lane Sherman Oaks CA 91403

WENZEL, BRUCE ERICKSON, b New Orleans, La, Jan 14, 38; m 64; c 3. PHYSICAL & ANALYTICAL CHEMISTRY, ENVIRONMENTAL SCIENCES. *Educ:* Tulane Univ, BS, 59; Mich State Univ, PhD(chem), 69. *Prof Exp:* Instrument analyst chem, Univ Mich, 61-63; proj chemist, Standard Oil Co, 68-70, anal chemist, 70-76; group leader anal, Ashland Petrol Co, 76-80; Lago Oil & Transp, 80-82; sr chemist, Henkel Corp, 83-86; chem lab mgr, Minn Dept Health, 86-87; environ group leader, Fed Cartridge Co, 88-89; SR ANALYTICAL CHEMIST, CHEM WASTE MGT, 89- *Concurrent Pos:* Indust adv comt, dept chem, Marshall Univ, 77-80; adv comt, Mat Anal Dept, Inst Mining & Minerals Res, Ky Ctr Energy Res, 78-80. *Mem:* Sigma Xi; Am Chem Soc. *Res:* Sulfur specific detectors in gas chromatography; gas chromatographic analysis of petroleum, shale oil and coal liquifaction products; general analysis of petroleum, shale oil and coal liquifaction fractions; environmental analysis; hazardous waste remediation; quality assurance; quality control laboratory; analysis of hazardous waste. *Mailing Add:* 1450 190th St Homewood IL 60430-4006

WENZEL, DUANE GREVE, pharmacology; deceased, see previous edition for last biography

WENZEL, FREDERICK J, b Marshfield, Wis, Aug 5, 30; m 52; c 6. BIOCHEMISTRY, BIOLOGY. *Educ:* Univ Wis-Stevens Point, BS, 56; Univ Chicago, MBA, 79. *Prof Exp:* Res asst, St Joseph's Hosp, Marshfield, 50-53; dir labs, Marshfield Clin, 53-65, secy found, 58-64, exec dir, Marshfield Med Found, 65-76, EXEC DIR, MARSHFIELD CLIN, 76- *Mem:* Am Pub Health Asn; NY Acad Sci; Am Fedn Clin Res. *Res:* Hypersensitivity phenomenon in the lung, such as farmer's lung and maple bark disease; health services; natural history of pulmonary thromboembolism, especially diagnosis and treatment and studies of the fibrinolytic process in this disease. *Mailing Add:* 610 S Sycamore Ave Marshfield WI 54449

WENZEL, HARRY G, JR, b Pittsburgh, Pa, Sept 4, 37; m 63; c 1. CIVIL ENGINEERING. *Educ:* Carnegie Inst Technol, BS, 59, MS, 61, PhD(civil eng), 64. *Prof Exp:* Instr civil eng, Carnegie Inst Technol, 62-64; from asst prof to assoc prof, 64-84, PROF CIVIL ENG & ASST DEAN, UNIV ILL, URBANA, 84- *Honors & Awards:* Walter L Huber Res Prize, Am Soc Civil Engrs, 77. *Mem:* Am Soc Civil Engrs; Am Geophys Union; Am Soc Eng Educ. *Res:* Hydrology; hydraulic engineering; urban water resources. *Mailing Add:* 207 Eng Hall Univ Ill 1308 W Green St Urbana IL 61801

WENZEL, JAMES GOTTLIEB, b Springfield, Minn, Oct 16, 26; m 50; c 4. OCEAN ENGINEERING, OCEAN MINING & THERMOENERGY. *Educ:* Univ Minn, BAeroEng, 48, MS, 50. *Hon Degrees:* PhD, Calif Lutheran Col, 85. *Prof Exp:* Aerodynamic engr, Convair, San Diego, 48-55, proj mgr anti-submarine warfare & ocean systs, 56-57, asst to vpres eng, 58-59; asst to sr vpres eng, Gen Dynamics, San Diego, 59-61; mgr govt planning, USN, 61-62; mgr cruise missiles systs, Lockheed Missles & Space Craft Co, Inc, 62-63, mgr ocean systs, 63-70, asst gen mgr res & develop div & dir ocean systs, 70-72, vpres ocean systs, 72-84; PRES & CHMN, MARINE DEVELOP ASSOCS, INC, 84- *Concurrent Pos:* Instr, Univ Minn, 49-50; lectr eng, Univ Calif, Los Angeles, 50-57, lectr ocean syst develop planning, 66, lectr deep submergence systs develop, 68; vpres, Lockheed Petrol Systs Ltd, Vancouver, BC, 70-75; pres & chmn, Maine Develop Asn Inc, 84, pres Maine Develop Asn, Sci & Tech Inc, 85; chmn, Ore Resource Expos, 86- *Mem:* Nat Acad Eng; fel & assoc mem Am Inst Aeronaut & Astronaut; assoc mem Royal Aeronaut Soc; Soc Naval Architects & Marine Engrs; Marine Tech Soc. *Res:* Pioneer in development of Deep Quest research submarine system, 8,000 feet, deep submergence rescue system, 5,000 feet, deep submergence search vehicle, 20,000 feet, offshore petroleum system and deep ocean mining system. *Mailing Add:* PO Box 3409 Saratoga CA 95070

WENZEL, JOHN THOMPSON, b Philadelphia, Pa, June 23, 46; m 71; c 2. PLASTICS RECYCLING. *Educ:* Stanford Univ, BS, 67; Univ Chicago, MS, 70, PhD(chem), 75. *Prof Exp:* Chemist, Nat Bur Standards, 76-81; engr, St Gobain Res, 81-83; dir res, Co St Gobain, 83-87; PROF CERAMICS ENG, RUTGERS UNIV, 87-, DIR, CTR PLASTICS RECYCLING RES, 89- *Concurrent Pos:* Vis scientist, Danish AEC, 71-74; SRC res fel physics, Univ Kent & AERE Harwell, UK, 74-76; sci coun, Centre Nat Res Sci, 85-87; course dir glass technol, Ctr Prof Advan, 89- *Honors & Awards:* IR-100 Award, Indust Res Inst, 79. *Mem:* Am Chem Soc; Am Phys Soc; Fel Am Ceramic Soc; Inst Elec & Electronics Engrs. *Res:* Technical, economic and social aspects of plastics recycling, including collection schemes, sorting systems, resin recovery, plastics reprocessing and end-use market studies. *Mailing Add:* Ctr Plastics Recycling Res Rutgers Univ PO Box 1179 Piscataway NJ 08855-1179

WENZEL, LEONARD A(NDREW), b Palo Alto, Calif, Jan 21, 23; m 44; c 4. CHEMICAL ENGINEERING. *Educ:* Pa State Col, BS, 43; Univ Mich, MS, 48, PhD(chem eng), 49. *Prof Exp:* Develop engr, Colgate-Palmolive-Peet Co, 49-51; from asst prof to assoc prof, 51-61, chmn dept, 62-84, PROF CHEM ENG, LEHIGH UNIV, 61-; CHIEF SCI, ARENCIBIA TECH, 87- *Concurrent Pos:* Mem exec comt, Cryogenic Eng Conf, 64-67; lectr, Univ Ala, Huntsville, 64- & Esso Res & Eng Co, 65-66; expert chem eng & proj coord, Proj COL-5, UNESCO, Bucaramanga, Colombia, 69-70; consult, Air Prod & Chem, Pearsall Chem Co & United Aircraft Corp; vis sr engr, Exxon Chem Co, 84-85. *Honors & Awards:* Hillman Award, Lehigh Univ, 84, Stabler Award, 87. *Mem:* Am Chem Soc; Am Inst Chem Engrs; Am Soc Eng Educ. *Res:* Low temperature processing; heat transfer; thermodynamics; fluidization. *Mailing Add:* Dept Chem Eng Lehigh Univ Bethlehem PA 18015

WENZEL, RICHARD LOUIS, b Marietta, Ohio, Sept 4, 21; wid; c 3. PUBLIC HEALTH ADMINISTRATION, PREVENTIVE MEDICINE. *Educ:* Marietta Col, AB, 43; Ohio State Univ, MD, 46; Univ Mich, MPH, 47; Am Bd Prev Med, cert pub health, 63. *Prof Exp:* Intern, Jersey City Med Ctr, 46-47; resident obstet & gynec, St Ann's Maternity Hosp, Columbus, Ohio, 47; chief commun dis & dep health off, Columbus Dept Health, 53-58; health officer, Marietta & Washington County, 58-60; assoc prof pub health admin, Sch Pub Health, Univ Mich, Ann Arbor, 60-70; HEALTH COMNR, TOLEDO & LUCAS COUNTY HEALTH DEPTS, 70- *Concurrent Pos:* Asst prof, Col Med, Ohio State Univ, 54-57; consult, Div Health Mobilization, USPHS, 61-72, Div Indian Health, 64-67 & Bur Med Serv, 66-67; mem, Emergency Health Preparedness Adv Comt, 67-72; consult, Nat Comn Community Health Serv, 64-66; assoc clin prof, Dept Med, Med Col Ohio, 72-; adj prof pub health admin, Sch Pub Health, Univ Mich, Ann Arbor, 70-73; mem, courtesy med staff, Toledo Hosp, 83- *Mem:* Fel Am Pub Health Asn; fel Am Col Prev Med; US Conf City Health Offs. *Res:* Survey, assessment, and evaluation of community health services. *Mailing Add:* Toledo Health Dept 635 N Erie St Toledo OH 43624

WENZEL, ROBERT GALE, b Terra Bella, Calif, June 23, 32; m 73; c 2. PHYSICS, OPTICS. *Educ:* Univ Calif, Berkeley, BA, 59; Univ NMex, MS, 66, PhD(physics), 79. *Prof Exp:* STAFF MEM PHYSICS, LOS ALAMOS NAT LAB, 59- *Concurrent Pos:* Int Atomic Energy Agency vis scientist, Inst Atomic Energy, Sao Paulo, Brazil, 66-67; sr vis fel, UK Sci Res Coun, Heriot-Watt Univ, 81. *Mem:* Optical Soc Am. *Res:* Tunable lasers, nonlinear optics and picosecond optical devices. *Mailing Add:* Rte 1 Box 11 Cundiyo Rd Santa Fe NM 87501

WENZEL, RUPERT LEON, b Owen, Wis, Oct 16, 15; m 40; c 3. ZOOLOGY. *Educ:* Cent YMCA Col, AB, 38, Univ Chicago, PhD(zool), 62. *Prof Exp:* Asst zool, Cent YMCA Col, 36-38; res asst, Univ Chicago, 37-40; asst cur, Field Mus Natural Hist, 40-50, chmn, Dept Zool, 70-77, cur, 51-80, EMER CUR INSECTS, FIELD MUS NATURAL HIST, 81- *Concurrent Pos:* Lectr, Roosevelt Col, 46-47 & Univ Chicago, 62-80; res assoc biol, Northwestern Univ, 59-80, vis prof, 63. *Honors & Awards:* Order of Vasco Nunez de Balboa, Panama, 67. *Mem:* AAAS; Entom Soc Am; Soc Study Evolution; Soc Syst Zool; Coleopterists Soc. *Res:* Taxonomy of streblid and nycteribiid batflies and histerid beetles; zoogeography; evolution of ectoparasites of terrestrial vertebrates. *Mailing Add:* Dept Zool Field Mus Natural Hist Chicago IL 60605

WENZEL, WILLIAM ALFRED, b Cincinnati, Ohio, Apr 30, 24; m 55; c 2. ELEMENTARY PARTICLE PHYSICS. *Educ:* Williams Col, AB, 44; Calif Inst Technol, MS, 48, PhD(physics), 52. *Prof Exp:* Res fel physics, Calif Inst Technol, 52-53; assoc dir physics, 70-73, RES PHYSICIST, LAWRENCE BERKELEY LAB, UNIV CALIF, 53- *Mem:* Fel Am Phys Soc; AAAS; Sigma Xi. *Res:* Experimentation in elementary particle physics; study of weak and electromagnetic interactions and of hadron phenomena at high transverse momenta; development of electronic instrumentation and accelerator facilities. *Mailing Add:* Lawrence Berkeley Lab Univ Calif Berkeley CA 94720

WENZEL, ZITA MARTA, biological rhythms, aging, for more information see previous edition

WENZINGER, GEORGE ROBERT, b Newport News, Va, July 24, 33. CHEMISTRY. *Educ:* Washington Univ, AB, 55; Univ Rochester, PhD(chem), 60. *Prof Exp:* NSF fel, Yale Univ, 62-63; asst prof, 63-76, ASSOC PROF ORG CHEM, UNIV SFLA, 76- *Mem:* AAAS; Am Chem Soc. *Res:* Conjugate elimination reactions; synthesis. *Mailing Add:* Dept Chem Univ SFla Tampa FL 33620

WENZL, JAMES E, b Greenleaf, Kans, Mar 24, 35; c 4. PEDIATRIC NEPHROLOGY. *Educ:* Creighton Univ, MD, 59; Univ Minn, Minneapolis, MS, 63. *Prof Exp:* Asst prof, 67-71, assoc prof, 71-75, PROF PEDIAT, UNIV OKLA HEALTH SCI CTR, OKLAHOMA CITY, 75- *Concurrent Pos:* Chief pediat, Nephrology Serv, Okla Children's Mem Hosp, Olkahoma City, 73-; interim chmn pediatrics, Health Sci Ctr, Univ Okla, Oklahoma City, 76-77, vchmn, 77-; J Pediat Prof, Eastern Va Med Sch, Norfolk & Med Col Va, Richmond, 76; vis prof, Hosp Nat DeNinos, Sch Med, Univ Costa Rica, San Jose, 77; assoc ed, Contemporary Dialysis, 81- *Mem:* Am Soc Nephrology; Am Soc Pediat Nephrology; Int Soc Nephrology; AMA. *Mailing Add:* 4816 NW 62nd Terr Oklahoma City OK 73122

WEPFER, WILLIAM J, b Milwaukee, Wis, Dec 17, 52; m 79; c 1. MECHANICAL ENGINEERING. *Educ:* Marquette Univ, BSE, 74; Stanford Univ, MS, 76; Univ Wis-Madison, PhD(mech eng), 79. *Prof Exp:* Asst prof, 80-86, ASSOC PROF MECH ENG, GEORGE W WOODRUFF SCH MECH ENG, GA TECH, 86-, ASSOC DIR GRAD STUDIES, 89- *Concurrent Pos:* Dow Young fac award, Am Soc Eng Educ, 86, AT&T Found Award, 88. *Honors & Awards:* Teetor Award, Soc Automotive Engrs, 85; Delos-Fluke Award, Am Soc Eng Educ, 88; E K Campbell Award, Am Soc Heating Refrig & Air Conditioning Engrs, 89. *Mem:* Am Soc Mech Engrs; Am Soc Heating Refrig & Air Conditioning Engrs; Instrument Soc Am; Am Soc Eng Educ; Sigma Xi; Soc Automotive Engrs. *Res:* Thermal systems analysis; thermal performance of solid-vapor heat pump systems; heat transfer analysis of engine cooling systems; vapor-compression heat pumps hvac systems; heat and mass transfer studies of textile drying processes; analysis of high temperature solid-oxide fuel cell systems. *Mailing Add:* 2995 Randolph Rd NE Atlanta GA 30345

WEPPELMAN, ROGER MICHAEL, b Pittsburgh, Pa, Nov 4, 44; m 71. BIOCHEMISTRY, ENDOCRINOLOGY. *Educ:* Univ Pittsburgh, BS, 65, PhD(microbiol), 70. *Prof Exp:* Instr microbiol, Univ Pittsburgh, 70-71; Am Cancer Soc fel biochem, Univ Calif, Berkeley, 71-73; sr res biochemist, Dept Basic Animal Sci Res, Merck & Co, 73-77, res fel, 77-80, sr res fel, 80-84; mgr, Microbiol & Agr Res Div, Merck, Sharp & Dohme Res Labs, 84-87, assoc dir, Regulatory Affairs Animal Health, 87-89; MGR, REGULATORY AFFAIRS, ANIMAL SCI DIV, MONSANTO, 89- *Concurrent Pos:* Assoc mem, Grad Fac, Rutgers Univ, 82-89. *Mem:* Sigma Xi; Am Chem Soc. *Res:* Endocrinology of Avian growth and reproduction; genetics of drug resistance of parasitic protozoa; registration of animal health products. *Mailing Add:* Ten Saddle Creek Ct Chesterfield MO 63005

WERBACH, MELVYN ROY, b New York, NY, Nov 11, 40; m 67; c 2. BIOFEEDBACK, CLINICAL NUTRITION. *Educ:* Columbia Col, BA, 62; Tufts Univ Sch Med, MD, 66. *Prof Exp:* Consult, div clin neurol, City of Hope Nat Med Ctr, 72-75; dir psychol serv, Pain Control Unit, Univ Calif, Los Angeles Hosp & Clins, 75-80; ASST CLIN PROF, DEPT PSYCHIAT, SCH MED, UNIV CALIF, LOS ANGELES, 78- & DEPT ANESTHESIOL, 80- *Concurrent Pos:* Chmn, dept ment health, Ross-Loos Med Co, Cigna Health Plans, 71-75, dir, Biofeedback Med Clin, 72- *Honors & Awards:* Clarke lect, Am Col Advan Med, 89. *Mem:* Am Psychiat Asn; Asn Appl Psychophysiol & Biofeedback; Am Col Nutrit. *Res:* Nutritional medicine; third line medicine (writer). *Mailing Add:* 18663 Ventura Blvd No 221 Tarzana CA 91356

WERBEL, LESLIE MORTON, b New York, NY, Mar 31, 31; m 58; c 3. ORGANIC CHEMISTRY, MEDICINAL CHEMISTRY. *Educ:* Queens Col, NY, BS, 51; Columbia Univ, AM, 52; Univ Ill, PhD(chem), 57. *Prof Exp:* Res chemist, 57-67, sr res chemist, 67-75, sr res scientist, 75-76, sr res assoc, 76-77, dir chem prod contract res, Parke-Davis & Co, 77-80, dir chem contract res, 80-82, DIR CANCER CHEM SYNTHESIS, PARKE-DAVIS PHARMACEUT RES DIV, WARNER-LAMBERT CO, 82- *Concurrent Pos:* Lectr col pharm, Univ Mich, 67, adj prof, 71. *Mem:* Am Chem Soc; Am Asn Cancer Res. *Res:* Medicinal chemistry; chemotherapy of parasitic infections; cancer chemotherapy; relation of intermediary metabolism of host and invading organism to drug action; novel heterocyclic ring systems. *Mailing Add:* Parke Davis & Co 2800 Plymouth Rd Ann Arbor MI 48106

WERBER, ERNA ALTURE, virology; deceased, see previous edition for last biography

WERBER, FRANK XAVIER, b Vienna, Austria, Apr 8, 24; nat US; m 50; c 2. ORGANIC CHEMISTRY, POLYMER CHEMISTRY. *Educ:* Queens Col, NY, BS, 44; Univ Ill, MS, 47, PhD(chem), 49. *Prof Exp:* Asst, Queens Col, NY, 46; res chemist, Org Res Dept, Res Ctr, B F Goodrich Co, 49-55, sr res chemist, 55-56; dir polymer res, Res Div, W R Grace & Co, 57-63, vpres org & polymer res, 63-67; nat prog staff, Dept Agr, Agr Res Serv, Beltsville, Md, 67-84; VPRES RES & DEVELOP, J P STEVENS & CO, INC, 67- *Concurrent Pos:* Chmn bd, Textile Res Inst, 73-75; mem bd, Indust Res Inst, 76-82, 83-86. *Mem:* Fel AAAS; Am Chem Soc; NY Acad Sci; Am Asn Textile Technol; Am Asn Textile Chemists & Colorists; Textile Res Inst. *Res:* Industrial organic chemistry; modification of polymers; adhesives; condensation polymers; textile chemistry, technology, machinery and engineering. *Mailing Add:* Dept Agr-Agr Res Serv Nat Prog Staff Bldg 005 BARC Beltsville MD 20705

WERBIN, HAROLD, b New York, NY, Oct 2, 22; m 77. BIOLOGICAL CHEMISTRY, CANCER. *Educ:* Brooklyn Col, BS, 44; Polytech Inst Brooklyn, MS, 47, PhD(chem), 50. *Prof Exp:* Asst, Brooklyn Jewish Hosp, 44-46; res assoc, Polytech Inst Brooklyn, 47-48; dir labs, Hillside Hosp, 50-53; res assoc biochem, Argonne Cancer Res Hosp, Chicago, 53-56; res biochemist dept physiol & soils & plant nutrit, Univ Calif, Berkeley, 57-66; from assoc prof to prof biol, Univ Tex, Dallas, 66-89; ADJ PROF, DEPT CELL BIOL & NEUROSCI, SOUTHWESTERN MED CTR, UNIV TEX, 89- *Concurrent Pos:* AEC contract, 67-, Energy Res & Develop Admin contract, 75-76; NSF grant, 67-71 & 78-80; Robert A Welch Found grant, 71-82 & Meadows Found grant, 84-86; NIH sr res fel, Dept Cell Biol, Univ Tex Health Sci Ctr, 82-83; NIH grant, subcontractor, 85-88. *Mem:* Am Soc Biol Chem. *Res:* Role of the cytoplasm in chemical carcinogenesis. *Mailing Add:* Dept Cell Biol & Neurosci Southwestern Med Ctr 5323 Harry Hines Blvd Dallas TX 75235

WERBLIN, FRANK SIMON, b New York, NY, Jan 24, 37; c 1. BIOENGINEERING. *Educ:* Mass Inst Technol, BS, 58, MS, 62; Johns Hopkins Univ, PhD(bioeng), 68. *Prof Exp:* From asst prof to assoc prof, 70-75, PROF ELEC ENG, UNIV CALIF, BERKELEY, 75- *Concurrent Pos:* Guggenheim fel, 74; Miller prof, Miller Found, 76. *Res:* Neurophysiological and biochemical basis for function of the vertebrate retina; mechanisms of light transduction, contrast detection, gain control studied in terms of molecular events in photoreceptors, ionic events at synapses, membrane events in neurons. *Mailing Add:* Dept Cell & Molecular Biol Univ Calif 2120 Oxford St Berkeley CA 94720

WERDEGAR, DAVID, b New York, NY, Sept 16, 30; m 61; c 2. COMMUNITY HEALTH. *Educ:* Cornell Univ, AB, 51, MA, 52; NY Med Col, MD, 56; Univ Calif, Berkeley, MPH, 70. *Prof Exp:* Consult, Calif State Dept Pub Health, 64-65; from asst prof to assoc prof, family & Community Med Sch Med, Univ Calif, Fresno 65-75, PROF MED, FAMILY COMMUNITY MED & ASSOC DEAN, DEPT FAMILY & COMMUNITY MED SCH MED UNIV CALIF, SAN FRANCISCO, 75; DIR, SAN FRANCISCO DEPT HEALTH, 85- *Concurrent Pos:* NIMH spec fel, Univ Calif, San Francisco, 63-64, Dept Health, Educ & Welfare Div Chronic Dis fel, 64-67; co-dir, Regional Med Prog Cardiovasc Dis Prev Northwest Calif, 67-70; planning officer, Regional Med Prog, Calif; med dir, Univ Calif Home Care Serv, 65- *Mem:* Fel AAAS; Am Pub Health Asn; Asn Teachers Prev Med; Am Fedn Clin Res; Am Col Physicians. *Res:* Family medicine; health policy; social aspects of health care; organization of health care services; evaluation of quality of health care. *Mailing Add:* Univ Calif Med Ctr 405A San Francisco CA 94143

WERDEL, JUDITH ANN, b Lackawanna, NY, June 22, 37. INFORMATION SCIENCE, SCIENCE POLICY. *Educ:* State Univ NY Buffalo, BA, 58; Mt Holyoke Col, MA, 59. *Prof Exp:* Sci info specialist, Shell Develop Co, Calif, 59-63; prof asst info sci, Off Doc, 63-70, prof assoc, Bd Int Orgn & Progs & Bd Sci & Technol for Int Develop, 70-74, staff officer, Comt Int Sci & Tech Info Progs, 74-81, prof assoc, Bd Sci & Technol, Int Develop, Comn Int Rels, Nat Acad Sci, 81-83; proj adminr, Develop Info Res & Ref Serv, USAID, 83-86; INDEPENDENT CONSULT, 86- *Concurrent Pos:* Secy bd dirs, Doc Abstr, Inc, 66-68. *Mem:* Am Soc Info Sci. *Res:* National and international policies and programs for the development of scientific and technical information systems; technical assistance in scientific and technical information systems; promotion and development of the field of information science and technology. *Mailing Add:* 1817 Corcoran St NW Washington DC 20009

WERDER, ALVAR ARVID, b Sweden, Mar 12, 17; nat US; m 44; c 2. MICROBIOLOGY. *Educ:* Univ Minn, BA, 45, MS, 47, PhD, 49. *Prof Exp:* From instr to asst prof bact, Univ Minn, 49-52; from assoc prof to prof microbiol, Sch Med, Univ Kans, 52-84, chmn dept, 61-82; RETIRED. *Mem:* Am Soc Microbiol; Am Asn Pathologists. *Res:* Oncogenic viruses; studies on germfree animals. *Mailing Add:* 3311 W 74th Terr Prairie Village KS 66280

WERGEDAL, JON E, b Eau Claire, Wis, Feb 19, 36. BIOCHEMISTRY. *Educ:* St Olaf Col, BA, 58; Univ Wis, MS, 60, PhD(biochem), 63. *Prof Exp:* biochemist, Vet Admin Hosp, Seattle, 62-; AT DEPT BIOCHEM, LOMA LINDA UNIV, CALIF. *Concurrent Pos:* Res instr med, Sch Med, Univ Wash, 63-71, res asst prof, 71-76. *Mem:* AAAS. *Res:* Bone metabolism. *Mailing Add:* 1355 Parker Ct Redlands CA 92373

WERGIN, WILLIAM PETER, b Manitowoc, Wis, Apr 20, 42; m 62; c 2. CYTOLOGY. *Educ:* Univ Wis-Madison, BS, 64, PhD(bot), 70. *Prof Exp:* Res cytologist plant path, USDA, 70-72, res cytologist weed sci, 72-74, res cytologist nematol & proj leader animal reproduction, Agr Res Serv, 74-79, res leader, Plant Stress Lab, 79-88, RES LEADER, ELECTRON MICROS LAB, USDA, 88- *Honors & Awards:* Diamond Award, Bot Soc Am, 75; prin investr, Agr Competitive Grants Prog, 78 & co-prin investr, US-Israel Agr Res & Develop Fund, 79. *Mem:* Am Soc Cell Biol; Bot Soc Am; Electron Micros Soc Am; Am Soc Plant Physiologists; Soc Nematologists; AAAS; Sigma Xi; fel Royal Micros Soc. *Res:* Current transmission and scanning electron microscopic examinations include host-parasite interactions between higher plants and nematodes and effects of environmental stress on crop plants; environmental stress effects caused by air pollutants, mineral deficiencies or toxicities and temperature, water and light extremes in plants. *Mailing Add:* Beltsville Agr Res Ctr-E Agr Res Serv USDA Beltsville MD 20705

WERKEMA, GEORGE JAN, b Vancouver, Wash, Nov 24, 36; m 69; c 2. PHYSICAL CHEMISTRY, HEALTH PHYSICS. *Educ:* Calvin Col, BS, 58; Univ Colo, PhD(chem), 65. *Prof Exp:* From res chemist to sr res chemist, Dow Chem Co, 63-72, res mgr, Rocky Flats Div, 72-75; health physicist, US Energy Res & Develop Admin, 75-81; chief, Health Protection Br, 81-83, CHIEF, NUCLEAR MGT BR, US DEPT ENERGY, ALBUQUERQUE, 83- *Mem:* Am Chem Soc; Sigma Xi. *Res:* X-ray crystallography; computer programming; instrument development; health physics; environmental control and monitoring; industrial hygiene. *Mailing Add:* 3504 Embudito Dr NE Albuquerque NM 87111

WERKHEISER, ARTHUR H, JR, b Easton, Pa, May 2, 35; m 62; c 2. LASERS. *Educ:* Lafayette Col, BS, 57; Univ Tenn, MS, 59, PhD(physics), 65. *Prof Exp:* Instr physics, Univ Tenn, 64; physicist, Res & Develop Directorate, 65-75, br chief, Modeling & Anal Div, Army High Energy Lasers Res & Eng Directorate, 75-82, actg div chief, Concepts & Prog Mgt, Directed Energy Directorate, US Army Missile Command, Redstone Arsenal, 82-88; assoc prof, Physics Dept, 70-89, MGR LASER APPLICATIONS RES LAB SR RES SCIENTIST, CTR APPL OPTICS, UNIV ALA, HUNTSVILLE, 89- *Mem:* Am Phys Soc. *Res:* Isomer shifts of solid solutions; conduction electron polarization; Mössbauer effect physics; mathematical modeling of high energy laser systems; laser propagation; high power laser processing of materials. *Mailing Add:* 1602 Drake Ave Huntsville AL 35802

WERKING, ROBERT JUNIOR, b Richmond, Ind, Jan 21, 30; m 51; c 2. SCIENCE EDUCATION. *Educ:* Hillsdale Col, BS, 61; Syracuse Univ, MS, 65; Ind Univ, Bloomington, EdD(sci educ), 71. *Prof Exp:* Asst prof physics, 65-69, asst prof sci educ, 71-73, CHMN DIV NATURAL SCI & MATH, MARION COL, 73- *Mem:* Am Asn Physics Teachers; Nat Sci Teachers Asn; fel Am Sci Affil. *Res:* Science instruction; instructional objectives; teaching effectiveness; instructional methods. *Mailing Add:* Div Natural Sci & Math Ind Wesleyan Univ Marion IN 46953-9980

WERKMAN, JOYCE, b Ill, 40. CHEMISTRY, INSTRUMENTATION. *Prof Exp:* PRES & OWNER, INSTRUMENTS RES & IND, 65- *Res:* Development of instruments to detect toxic gases, and for research and plant operations; design and manufacture of safety products for laboratory sciences and instruments for automatizing tedious laboratory tasks; design and manufacture of other devices that facilitate laboratory work in fields of chemistry and biochemistry. *Mailing Add:* 108 Franklin Ave Cheltenham PA 19012

WERKMAN, SIDNEY LEE, b Washington, DC, May 3, 27; c 1. MEDICINE. *Educ:* Williams Col, AB, 48; Cornell Univ, MD, 52. *Prof Exp:* Assoc prof psychiat, Med Sch, George Washington Univ, 61-69; assoc prof, 69-72, PROF PSYCHIAT, UNIV COLO MED CTR, DENVER, 72- *Concurrent Pos:* Commonwealth Fund fel, Florence, Italy, 63-64; res consult, USPHS, 60-68; assoc dir, Joint Comn Ment Health Children, 67-68; consult, NIMH, 74- *Mem:* AAAS; Am Psychiat Asn; Am Acad Child Psychiat; Am Orthopsychiat Asn; Group Advan Psychiat. *Res:* Psychological factors in nutrition and development; brain dysfunction in children; attitude studies; psychological adjustment of Americans overseas. *Mailing Add:* Dept Psychiat Univ Colo Med Ctr 4200 E Ninth Ave Denver CO 80262

WERMAN, ROBERT, b Brooklyn, NY, May 2, 29; m 54; c 4. NEUROPHYSIOLOGY. *Educ:* NY Univ, AB, 48, MD, 52. *Prof Exp:* Intern, Montefiore Hosp, New York, 52-53; resident neurol, Mt Sinai Hosp, 53-54 & 56-58; asst prof, Col Physicians & Surgeons, Columbia Univ, 60-61; prof psychiat, Sch Med, Ind Univ, Indianapolis, 61-69, prof anat & physiol, Ind Univ Bloomington, 64-69; PROF NEUROPHYSIOL, HEBREW UNIV, JERUSALEM, 69- *Concurrent Pos:* Nat Inst Neurol Dis & Blindness trainee neurophysiol, Columbia Univ, 58-60; vis scientist, Cambridge Univ, 60-61. *Mem:* AAAS; Soc Gen Physiologists; Am Physiol Soc. *Res:* Electrophysiology; neuromuscular junction; membrane properties and ionic movements; synaptic physiology; spinal cord; death of neurons; Ca2. *Mailing Add:* Dept Neurobiol Hebrew Univ Jerusalem Jerusalem 91904 Israel

WERMUND, EDMUND GERALD, JR, b Arlington, NJ, Apr 15, 26; m 76; c 2. GEOLOGY. *Educ:* Franklin & Marshall Col, BS, 48; La State Univ, PhD(geol), 61. *Prof Exp:* Instr geol, La State Univ, 52-57; sr res technologist, Field Res Lab, Mobil Oil Corp, 57-68, res assoc, Mobil Res & Develop Corp, 68-70; tech mgr, Remote Sensing, Inc, 70-71; res scientist, Bur Econ Geol, 71-73, assoc dir, 73-88, RES SCIENTIST, BUR ECON GEOL, UNIV TEX, AUSTIN, 88- *Mem:* AAAS; Am Asn Petrol Geologists; fel Geol Soc Am; Am Soc Photogram; Sigma Xi; Asn Geoscientists Int Develop. *Res:* Petroleum geology, environmental geology; regional geology. *Mailing Add:* Box X Bur Econ Geol Univ Tex Austin TX 78713-7508

WERMUS, GERALD R, b St Paul, Minn, May 8, 38; m 65; c 2. BIOCHEMISTRY, CLINICAL CHEMISTRY. *Prof Exp:* Formulation chemist, Econ Lab, Inc, 60-61; res chemist, 61-72, res supvr, 72-74, res mgr, 74-78, tech serv mgr, Instrument Prod Div, 78-81, mgr, new methods, 81-83, proj mgr, 84-86, PROD MGR, MED PROD DEPT, CLIN SYSTS DIV, E I DU PONT DE NEMOURS & CO, INC, 87- *Concurrent Pos:* Mem, Nat Comt Clin Lab Standards. *Mem:* Am Asn Clin Chem; Am Chem Soc. *Res:* Development of clinical laboratory methodology; bacterial metabolism; lipid peroxidation; laboratory administration; product management; project management. *Mailing Add:* 114 Venus Dr Newark DE 19711

WERMUTH, JEROME FRANCIS, b Madison, Wis, Oct 19, 36; m 64; c 5. DEVELOPMENTAL BIOLOGY. *Educ:* Univ Wis-Madison, BS, 57, MS, 60; Ind Univ, PhD(zool), 68. *Prof Exp:* Instr biol, Rockhurst Col, 61-64; asst prof, St Joseph's Col, Ind, 65-66; res assoc, Univ Notre Dame, 68-69; asst prof, 69-78, asst dean, 84-86, ASSOC PROF BIOL, PURDUE UNIV, CALUMET CAMPUS, 78-, EXEC ASST TO CHANCELLOR, 87- *Concurrent Pos:* Vis asst prof & Nat Cancer Inst spec res fel life sci, Ind State Univ, Terre Haute, 72-73; vis asst prof zool, Ind Univ, Bloomington, 75-76. *Res:* Effects of x-irradiation on the developmental physiology of colonial marine cnidarians. *Mailing Add:* Exec Asst to Chancellor Purdue Univ Calumet Hammond IN 46323

WERNAU, WILLIAM CHARLES, b Flushing, NY, Jan 22, 47; m 68; c 4. STREPTOMYCES FERMENTATION, ANTIBIOTICS. *Educ:* Cooper Union, BS, 68; Univ Calif, Berkeley, PhD(chem eng), 72. *Prof Exp:* Res scientist, 72-76, sr res scientist, 76-77, proj leader, 77-81, mgr, 81-82, asst dir, 82-87, DIR FERMENTATION RES, PFIZER, INC, 87- *Mem:* Am Soc Microbiol; Am Chem Soc. *Res:* Continuous fermentation, organic acid biosynthesis, biopolymer synthesis (patents), enzyme production, pilot plant scale-up, fermentor design, fermentation and recovery process development; mass transfer in aerobic systems; antibiotics, anticoccidials, antiparasitics and streptomyces genetics. *Mailing Add:* 27 Quinley Way Waterford CT 06385

WERNER, ARNOLD, b Brooklyn, NY, June 8, 38; m 66; c 2. PSYCHOSOMATIC MEDICINE, MEDICAL EDUCATION. *Educ:* Brooklyn Col, BS, 59; Univ Rochester, MD, 63. *Prof Exp:* Instr psychiat, Sch Med, Univ Rochester, 66-67, Temple Univ, 67-69; from asst prof to assoc prof, 69-78, PROF PSYCHIAT, MICH STATE UNIV, 78- *Mem:* Fel Am Psychiat Asn; Am Psychosom Soc. *Res:* Adaptation to cancer; medical education; health concerns of normal people. *Mailing Add:* A228 E Fee Hall Dept Psychiat Mich State Univ East Lansing MI 48824-1316

WERNER, DANIEL PAUL, b Philadelphia, Pa, Dec 12, 38; m 63; c 2. MECHANICAL ENGINEERING. *Educ:* Drexel Univ, BSME, 61; Univ Denver, MSME, 63; Univ Mich, Ann Arbor, PhD(mech eng), 68. *Prof Exp:* Asst prof mech eng, Duke Univ, 68-71; chief compressor engr, Freezing Equip Sales, Inc, 71-77; mgr, airconditioning eng, Bohn Heat Transfer Div, Gulf & Western Ind, Inc, 77-84; VPRES ENG, HYDROTHERM DIV, AUTOMATION IND, INC, 84- *Mem:* Am Soc Mech Engrs; Am Soc Heating, Refrig & Air-Conditioning Eng. *Res:* Pressure drop and heat transfer in two phase liquid-gas flows; thermal stresses in thick walled heat exchanger tubing; mechanical refrigeration systems development; mechanical air conditioning system development; pulse combustion system. *Mailing Add:* 190 Brookside Ave Allendale NJ 07401

WERNER, EARL EDWARD, b Aliquippa, Pa, July 11, 44; m 71. ECOLOGY, ZOOLOGY. *Educ:* Columbia Univ, AB, 66; Mich State Univ, PhD(ecol), 72. *Prof Exp:* Asst prof, Univ Iowa, 72-73; assoc prof zool, Mich State Univ, Kellogg Biol Sta, 78-84; PROF BIOL, UNIV MICH, 86- *Honors & Awards:* Mercer Award, Ecol Soc Am, 78. *Mem:* AAAS; Ecol Soc Am; Brit Ecol Soc; Int Soc Theoret & Appl Limnol; Am Soc Naturalists. *Res:* Community ecology; population biology; competition; foraging strategies. *Mailing Add:* Biol Dept Univ Mich Ann Arbor MI 48109

WERNER, ERVIN ROBERT, JR, b Philadelphia, Pa, May 2, 32; m 53; c 5. ORGANIC CHEMISTRY. *Educ:* Haverford Col, BS, 54; Univ Md, MS, 58; Univ Pa, PhD(org chem), 60. *Prof Exp:* Chemist, Rohm & Haas Co, 57-60; res chemist, 60-69, staff chemist, 69-72, RES ASSOC, E I DU PONT DE NEMOURS & CO, INC, 72- *Mem:* Am Chem Soc; Fedn Soc Paint Technol. *Res:* Organic synthesis; flame retardant polyesters; organophosphorus chemistry; water solution polymers; aqueous wire enamels; emulsion systems and house paints; interior emulsion paints; automotive clear finishes; automotive waterborne basecoats. *Mailing Add:* E I du Pont de Nemours & Co Inc 3500 Grays Ferry Ave Philadelphia PA 19146

WERNER, F(RED) E(UGENE), b Mansfield, Ohio, Sept 22, 27; m 51; c 3. PHYSICAL METALLURGY, MAGNETIC MATERIALS. *Educ:* Mass Inst Technol, SB, 50, ScD(metall), 55. *Prof Exp:* Instr metall, Mass Inst Technol, 50-53, asst, 53-55; res engr, Res Labs, Westinghouse Elec Corp, 55-60, sect mgr alloy studies, 60-62, mgr, Metall Dept, 62-68, mat consult, 69, mgr, Magnetics Dept, 70-82, mgr patents & libr, 82-88; CONSULT, 88- *Concurrent Pos:* Gen chem, Magnetism Conf, 83; chmn, Magnetism adv comt, 84. *Mem:* Soc Metals; Inst Mining, Metall & Petrol Engrs; Inst Elec & Electronics Engrs; fel Am Soc Metals. *Res:* Metallic alloys; application of metals and alloys to practical uses; metals processing, research and development in physical and mechanical metallurgy; metals joining; soft and permanent magnetic materials. *Mailing Add:* 619 Cascade Rd Pittsburgh PA 15221

WERNER, FLOYD GERALD, b Ottawa, Ill, June 1, 21; m 52; c 3. ENTOMOLOGY. *Educ:* Harvard Univ, SB, 43, PhD(biol), 50. *Prof Exp:* From asst prof to assoc prof zool, Univ Vt, 50-54; from asst prof to prof, 54-89, EMER PROF ENTOM, UNIV ARIZ, 89- *Mem:* AAAS; Entom Soc Am; Coleopterists Soc. *Res:* Taxonomy of Coleoptera, Meloidae and Anthicidae. *Mailing Add:* Dept Entom Univ Ariz Tucson AZ 85721

WERNER, FRANK D(AVID), b Junction City, Kans, Mar 14, 22; m 46; c 3. AERONAUTICAL ENGINEERING. *Educ:* Kans State Col, BS, 43; Univ Minn, MS, 48, PhD(aeronaut eng), 55. *Prof Exp:* Jr physicist, Appl Physics Lab, Johns Hopkins Univ, 43-47; res assoc, Rosemount Aeronaut Lab, Univ Minn, 47-56; pres & dir, Rosemount, Inc, 56-68; pres, Park Energy Co, 76-82; PRES & DIR, ORIGIN INC, 68- & TECH LINE CORP, 90- *Concurrent Pos:* Consult, 48-57. *Mem:* Am Phys Soc; Instrument Soc Am. *Res:* Instrumentation, particularly temperature and pressure measurement; platinum resistance thermometry and measurement of air data parameters for airplane flights; solar heat collectors. *Mailing Add:* Box SR9 Jackson WY 83001

WERNER, GERHARD, b Vienna, Austria, Sept 28, 21; m 58; c 2. PHARMACOLOGY, PSYCHIATRY. *Educ:* Univ Vienna, MD, 45. *Prof Exp:* From instr to asst prof pharmacol, Univ Vienna, 43-52; prof, Univ Calcutta, 52-54 & Med Sch, Univ Sao Paulo, 55-57; assoc prof, Med Col, Cornell Univ, 57-61 & Med Sch, Johns Hopkins Univ, 61-65; chmn dept, Univ Pittsburgh, 65-75, dean, Sch Med, 74-78, prof psychiat, 78-89, FS Cheever Distinguished Prof, 80-81, PROF PHARMACOL, SCH MED, UNIV PITTSBURGH, 65-, PROF PSYCHOL, 70-, EMER PROF PSYCHIAT, 89- *Concurrent Pos:* Mem adv bd, Indian Coun Med Res, 52-54; mem pharmacol study sect, NIH, 65-69, mem chem-biol info handling prog, 68-71; NSF consult, 69-71; pres, Medcoup, Inc, 90- *Honors & Awards:* Humboldt Prize, 84. *Mem:* Asn Computer Mach; Int Brain Res Orgn; Asn Res Nerv & Ment Dis; Sigma Xi; Am Asn Artificial Intel; Am Psychiat Asn. *Res:* Neuropharmacology; psychoanalysis; neurophysiology; computer applications; expert systems for medical diagnosis; artificial intelligence. *Mailing Add:* Psychiat Dept Vet Admin Med Ctr Highland Dr Pittsburgh PA 15206

WERNER, HARRY EMIL, b Brooklyn, NY, June 7, 21; m 43; c 3. GEOLOGY. *Educ:* Syracuse Univ, AB, 47, PhD(geol), 56; Wash Univ, AM, 49. *Prof Exp:* Field geologist, State Geol Surv, Va, 48-51; asst prof geol, St Lawrence Univ, 51-55; res geologist, Pan Am Petrol Corp, 55-60, res sect supvr, 60-63; assoc prof geol, Univ Pittsburgh, 63-80; Seminole Energy, 80-89; RETIRED. *Mem:* Geol Soc Am; Am Asn Petrol Geologists. *Res:* Stratigraphy; petrology; structural geology; carbonate sedimentation; underwater drilling techniques. *Mailing Add:* 1510 SW 28th Ct Ft Lauderdale FL 33315

WERNER, JOAN KATHLEEN, neuroanatomy, gross anatomy; deceased, see previous edition for last biography

WERNER, JOHN ELLIS, b Erie, Pa, Oct 25, 32; m 57; c 4. METALLURGY, CHEMICAL ENGINEERING. *Educ:* Pa State Univ, BS, 54, MS, 60. *Prof Exp:* Metall engr, Lackawanna Plant, Bethlehem Steel Corp, 54-58, res engr, Res Dept, 60-65, supvr steelmaking processes, Steelmaking Sect, 65-66, asst sect mgr, 66-69, sect mgr phys metall, Homer Res Labs, 69-74, asst div mgr prod res, 74-76, asst div mgr primary processes, 76-78, div mgr raw mat & chem processes, Res Dept, 78-82, dir res, 82-84, dir technol transfer & ventures, diversified group, Bethlehem Steel Corp, 84-88; PRES, BEN FRANKLIN TECHNOL CTR CENT/NORTHERN PA, INC, 89- *Mem:* Am Inst Mining, Metall & Petrol Engrs; Am Soc Metals. *Res:* Steelmaking processes; microstructural control to enhance properties; research management; technical forecasting. *Mailing Add:* Ben Franklin Tech Ctr Cent & NPa Rider Bldg 5th Floor 120 S Burrowes St Union Park PA 16801

WERNER, JOHN KIRWIN, b Conrad, Mont, Sept 14, 41; m 67; c 2. HERPETOLOGY, PARASITOLOGY. *Educ:* Carroll Col, Mont, BA, 63; Univ Notre Dame, PhD(biol), 68. *Prof Exp:* Chief dept med zool, 406th Med Lab, US Med Command, Japan, 68-71; asst prof biol, 71-75, head dept, 75-78, ASSOC PROF BIOL, NORTHERN MICH UNIV, 75- *Concurrent Pos:* Assoc investr, US Army Res & Develop Protocol, 68-71; US Forest Serv grants amphibian ecol, 71-75; scientist, Int Ctr Med Res, Cali, Colombia, 78-81. *Mem:* Herpetologists League; Am Soc Parasitologists; Am Soc Ichthyologists & Herpetologists. *Res:* Ecology and reproduction of amphibians and reptiles; blood parasites of lower vertebrates. *Mailing Add:* Dept Biol Northern Mich Univ Marquette MI 49855

WERNER, LINCOLN HARVEY, b New York, NY, Feb 19, 18; m 44; c 2. ORGANIC CHEMISTRY. *Educ:* Swiss Fed Inst Technol, dipl, 41, DrTechSci, 44. *Prof Exp:* Asst, Swiss Fed Inst Technol, 44-46; res chemist, Ciba-Geigy Ltd, Basel, Switz, 46-47, res chemist, Ciba Pharmaceut Prods, Inc, NJ, 47-68, mgr chem res admin, Ciba Pharmaceut Co, 68-84; RETIRED. *Mem:* Am Chem Soc. *Res:* Pharmaceuticals; cardiovascular drugs; diuretics. *Mailing Add:* 94 Larned Rd Summit NJ 07901

WERNER, MARIO, b Zurich, Switz, Aug 21, 31; US citizen; m 68; c 1. LABORATORY MEDICINE. *Educ:* Univ Zurich, MD, 56, DrMed, 60. *Prof Exp:* Asst physician, Kantonsspital St Gallen, Switz, 57-58 & Univ Basel, 58-61; NIH res fel metab, Swiss Acad Med Sci, 61-62; NIH res fel physiol, Rockefeller Univ, 62-64; head physician, Med Sch Essen, WGer, 64-66; asst prof lab med, Univ Calif, San Francisco, 67-70; assoc prof med path, Wash Univ, 70-72; PROF LAB MED, GEORGE WASHINGTON UNIV, 72- *Concurrent Pos:* Consult, Univ Tex M D Anderson Hosp & Tumor Inst, 70-80 & Nat Heart & Lung Inst, 73-83; chmn health care delivery comt, Bi-State Regional Med Prog, 71-72; trustee, Found Interdisciplinary Biocharacterizations Pop, 71-75; mem, Path A Study Sect, NIH, 75-80; mem clin chem device classification panel, Food & Drug Admin, 77-79; dir, Am Bd Clin Chem, 79-85; ed-in-chief, Clinica Chimica Acta, 85-; chmn, Nat Coun Health Lab Serv, 87-89. *Honors & Awards:* CCE Commissioners Medal, Am Soc Clin Path. *Mem:* Am Soc Clin Path; Col Am Path; Acad Clin Lab Physicians & Sci; Am Asn Clin Chem; Am Fedn Clin Res; Nat Acad Clin Biochem; Int Soc Clin Enzym (pres, 86-90); hon mem Ital Soc Lab Med; corresp mem Ger Soc Lab Med. *Res:* Laboratory data processing and diagnostic discrimination by computer; blood protein and lipoprotein metabolism. *Mailing Add:* Div Lab Med George Washington Univ Med Ctr Washington DC 20037

WERNER, MICHAEL WOLOCK, b Chicago, Ill, Oct 9, 42; m 67; c 2. ASTROPHYSICS, INFRARED ASTRONOMY. *Educ:* Haverford Col, BA, 63; Cornell Univ, PhD(astron), 68. *Prof Exp:* Res physicist, US Naval Res Lab, 63-64; vis fel, Inst Theoret Astron, Cambridge, Eng, 68-69; lectr physics, Univ Calif, Berkeley, 69-72; asst prof physics, Calif Inst Technol, 72-79, staff assoc, Hale Observ, 72-79; RES SCIENTIST, AMES RES CTR, NASA, 79- *Concurrent Pos:* Proj scientist, Space Infrared Telescope Facil, 84- *Honors & Awards:* Alfred P Sloan Fel, 72. *Mem:* AAAS; Am Astron Soc. *Res:* Observational infrared astronomy; astrophysics of the interstellar medium; development of telescopes and systems for infrared space astronomy. *Mailing Add:* NASA Ames Res Ctr MS 245-6 Moffett Field CA 94035

WERNER, PATRICIA ANN SNYDER, b Flint, Mich, July 7, 41; m 71, 91. ECOLOGY, BOTANY. *Educ:* Mich State Univ, MS, 68, PhD(plant ecol), 72. *Prof Exp:* Res assoc ecol, Univ Iowa, 72-73; from asst prof to prof bot & zool, Mich State Univ, 81-86; dir, Trop Ecosysts Res Ctr, Commonwealth Sci & Indust Res Orgn, Australia, 85-90; DIR, DIV BIOTIC SYSTS & RESOURCES, NSF, WASHINGTON, DC, 90- *Concurrent Pos:* Prin investr grants, NSF, 73-86, USDA, 82-86; vis prof, Univ Otago, Dunedin, NZ, 81, Harvard Univ, 82 & Australian Nat Univ, 82. *Mem:* Ecol Soc Am; Bot Soc Am; Brit Ecol Soc; Int Soc Plant Pop Biol (secy-treas, 77-85, pres, 85-); fel AAAS. *Res:* Plant population ecology; life histories and competition; weeds; goldenrods; community structure; niche theory and empirical quantification; species relationships in disturbed communities; succession; seeds. *Mailing Add:* Rm 215 BBS BSR NSF Washington DC 20550

WERNER, RAYMOND EDMUND, b Cincinnati, Ohio, Apr 18, 19; m 41; c 2. ORGANIC CHEMISTRY. *Educ:* Univ Cincinnati, ChE, 41, MS, 43, ScD(org chem), 45; Xavier Univ, Ohio, MBA, 69. *Prof Exp:* Res chemist, Interchem Corp, Ohio, 45-46; develop chemist, Sterling Drug, Inc, 46-55, dir develop, 55-75, vpres res & develop, Tech Ctr, Hilton Davis Div, 75-80; RETIRED. *Mem:* Am Chem Soc; Sigma Xi. *Res:* Administration of research and development in areas of fine organics, dyes, pharmaceuticals, pigments and graphic arts materials. *Mailing Add:* 9400 Southgate Dr Cincinnati OH 45241

WERNER, RICHARD ALLEN, b Reading, Pa, Feb 20, 36. INSECT ECOLOGY. *Educ:* Pa State Univ, BS, 58 & 60; Univ Md, MS, 66; NC State Univ, PhD(entom), 71. *Prof Exp:* Forester, US Forest Serv, USDA, 57-59, res entomologist, Forestry Sci Lab, NC, 60-74, res entomologist, Inst Northern Forestry, 74-84, prog leader, 85-91; CONSULT, 91- *Concurrent Pos:* NIH/NSF res grant, Univ Md, 64-65; affil prof forestry, Univ Alaska, 76-, prin res assoc, Inst Arctic Biol, 82- *Mem:* Soc Am Foresters; Entom Soc Can; Entom Soc Am; Soc Chem Ecol. *Res:* Aggregation behavior of bark beetles to semiochemicals; host susceptibility to bark beetles in white spruce ecosystems; plant chemical defensive systems in relation to hardwood defoliators; spruce bark beetles. *Mailing Add:* USDA Forest Serv Inst Northern Forestry Fairbanks AK 99775-5500

WERNER, ROBERT GEORGE, b Plymouth, Ind, Mar 6, 36; m 58; c 2. ZOOLOGY. *Educ:* Purdue Univ, BS, 58; Univ Calif, Los Angeles, MA, 63; Ind Univ, PhD(zool), 66. *Prof Exp:* Asst prof zool, State Univ NY Col Forestry, Syracuse, 66-69; asst prof fisheries, Cornell Univ, 69-70; assoc prof, 70-76, PROF ZOOL, COL ENVIRON SCI & FORESTRY, STATE UNIV NY, 76-; CO-DIR, GREAT LAKES RES CONSORTIUM, 86- *Concurrent Pos:* Vis scientist, Scottish Marine Biol Asn, 78; pres, Early Life Hist Sect, Am Fisheries Soc. *Honors & Awards:* Fulbright Fel, 88. *Mem:* Am Fisheries Soc; Am Soc Limnol & Oceanog; Ecol Soc Am. *Res:* Ecology of larval freshwater fish. *Mailing Add:* Dept Environ & Forest Biol Col Environ Sci & Forestry State Univ NY Syracuse NY 13210-2788

WERNER, RUDOLF, b Königsberg, Ger, Dec 17, 34; m 61; c 2. CHEMISTRY. *Educ:* Univ Freiburg, Dipl(chem), 60, Dr rer nat(chem), 63. *Prof Exp:* Res asst, Inst Macromolecular Chem, Univ Freiburg, 61-65; res assoc, Carnegie Inst Genetics Res Unit, 65-67; sr staff investr, Cold Spring Harbor Lab Quant Biol, 68-70; assoc prof, 70-77, PROF BIOCHEM, SCH MED, UNIV MIAMI, 77- *Concurrent Pos:* Estab investr, Am Heart Asn, 69-74. *Mem:* AAAS; Am Soc Biol Chemists. *Res:* Molecular biology; DNA replication. *Mailing Add:* Dept Biochem Sch Med Univ Miami 1600 N Miami & Tenth Ave Miami FL 33101

WERNER, SAMUEL ALFRED, b Elgin, Ill, Jan 5, 37; m 61; c 1. SOLID STATE PHYSICS, NUCLEAR ENGINEERING. *Educ:* Dartmouth Col, AB, 59, MS, 61; Univ Mich, Ann Arbor, PhD(nuclear eng), 65. *Prof Exp:* Instr solid state physics & mat, Thayer Sch Eng, Dartmouth Col, 60-61; staff scientist, Physics Dept, Sci Lab, Ford Motor Co, 64-70, sr scientist, 70-75; chmn dept physics & astron, 81-83, PROF PHYSICS, UNIV MO, COLUMBIA, 75-, MILLSAP DISTINGUISHED PROF PHYSICS, 85- *Concurrent Pos:* Vis scientist, Argonne, Oak Ridge & Brookhaven Nat Labs, 64-, Aktiebolaget Atomenergi, Sweden, 70 & Inst Laue-Langevin, France, 77-; consult, Argonne Nat Lab, 68-; adj assoc prof, Univ Mich, 69-74; fel Swed Res Coun, 70; mem neutron scattering adv comt, Oak Ridge Nat Lab, 81-; mem panel access current status facil & res neutron scattering, Nat Acad Sci, 83-84; vis scientist, Nat Bur Standards, 83-84. *Mem:* AAAS; fel Am Phys Soc; Sigma Xi; NY Acad Sci. *Res:* Neutron scattering; magnetism; phase transitions; fundamental neutron physics; neutron interferometry; charge density and spin density waves. *Mailing Add:* Dept Physics Univ Mo Columbia MO 65211

WERNER, SANFORD BENSON, b Newark, NJ, Jan 5, 39; m 68; c 2. PREVENTIVE MEDICINE. *Educ:* Rutgers Univ, BA, 60; Wash Univ, MD, 64; Univ Calif, MPH, 70. *Prof Exp:* Intern internal med, Vanderbilt Univ Hosp, Nashville, Tenn, 64-65; med epidemiologist, Epidemic Intel Serv, Ctr Dis Control, Ga, 65-67; resident internal med, Univ Wash Hosps, Seattle, 67-69; internist, Alaska Clin, Anchorage, 69; MED EPIDEMIOLOGIST INFECTIOUS DIS CONTROL, CALIF STATE DEPT HEALTH, 70- *Concurrent Pos:* Resident prev med, Sch Pub Health, Univ Calif, Berkeley, 70-71, lectr infectious dis epidemiol, 70-; WHO fel, communicable dis Asia; consult, Nat Acad Sci. *Mem:* Fel Am Col Prev Med; fel Am Col Epidemiol. *Res:* Epidemiology of botulism, typhoid, salmonellosis and other enteric diseases. *Mailing Add:* Gen Infectious Dis Unit Calif Dept Health Serv 2151 Berkeley Way Berkeley CA 94704

WERNER, SIDNEY CHARLES, b New York, NY, June 29, 09; m 47; c 4. THYROID, MEDICINE. *Educ:* Columbia Univ, AB, 29, MD, 32, ScD(med), 37. *Prof Exp:* Asst med, 34-78, prof, 74-78, EMER PROF CLIN MED, COL PHYSICIANS & SURGEONS, COLUMBIA UNIV, 78- *Concurrent Pos:* From intern to attend physician, Presby Hosp, New York, 32-74, consult med, 74-84, emer consult med, 84-; spec consult, Endocrine Study Sect, NIH, 52-55, endocrine panel, Nat Cancer Inst, 55-58, Cancer Chemother Serv Ctr, 58-62; Jacobaeus lectr, Finland, 68; consult, var hosps, NY & Conn; vis prof, Univ Ariz, 78-84, vis emer prof, 84- *Honors & Awards:* Wilson Award, 60; Stevens Triennial Award, 66; Distinguished Serv Award, Am Thyroid Asn, 69; Distinguished Thyroid Scientist, Int Thyroid Cong, 75; Sidney C Werner ann lectr, thyroid res, dept med, Col Physicians & Surgeons, Columbia Univ, 77. *Mem:* Fel AAAS; Am Thyroid Asn (pres, 72-73); Am Soc Clin Invest; Endocrine Soc; Asn Am Physicians. *Res:* Thyroid physiology and disease. *Mailing Add:* 4528 E Calle Del Conde Tucson AZ 85718

WERNER, THOMAS CLYDE, b York, Pa, June 19, 42; m 65; c 2. ANALYTICAL CHEMISTRY. *Educ:* Juniata Col, BS, 64; Mass Inst Technol, PhD(anal chem), 69. *Prof Exp:* Fel chem, Harvard Med Sch & Mass Gen Hosp, Boston, 69-70; fel med sch, Tufts Univ, 70-71; from asst prof to assoc prof, 71-85, PROF CHEM & DEPT CHMN, UNION COL, NY, 85- *Mem:* Am Chem Soc. *Res:* Application of absorption and luminescence spectroscopy to the study of molecular structure. *Mailing Add:* Dept Chem Union Col Schenectady NY 12308

WERNER, WILLIAM ERNEST, JR, b Mt Marion, NY, June 30, 25; m 47; c 2. ECOLOGY. *Educ:* State Univ NY, BA, 50, MA, 51; Cornell Univ, PhD(mammal), 54. *Prof Exp:* Instr biol, State Univ NY Col Teachers, Albany, 51-52; PROF BIOL, BLACKBURN COL, 54- *Concurrent Pos:* Nat Heart Inst fel, Marine Inst, Miami, 63-64. *Mem:* AAAS; Am Soc Mammalogists; Am Soc Ichthyologists & Herpetologists; Ecol Soc Am; Sigma Xi. *Res:* Ecology of mammals, reptiles, amphibians and marine invertebrates. *Mailing Add:* 333 E Cherry Carlinville IL 62626

WERNICK, JACK H(ARRY), b St Paul, Minn, May 19, 23; m 47; c 2. PHYSICAL METALLURGY, SOLID STATE CHEMISTRY. *Educ:* Univ Minn, BS, 47, MS, 48; Pa State Univ, PhD(metall chem), 54. *Prof Exp:* Metallurgist, Manhattan Proj, Los Alamos Sci Lab, Calif, 44-46; instr metall, Pa State Univ, 49-54; mem tech staff, 54-64, head, Physical Metall Res Dept, 64-73, head, Solid State Chem Res Dept, 73-81, head, device mat res dept, Bell Labs, 81-83, MGR MAT SCI RES, BELL COMMUN RES, 84- *Mem:* Nat Acad Eng; fel Am Soc Metals; fel Am Inst Mining, Metall & Petrol Engrs; fel Am Phys Soc; fel NY Acad Sci; Electrochem Soc; AAAS. *Res:* Constitution and physical chemistry of materials, particularly thermodynamic, magnetic and semiconducting properties. *Mailing Add:* Bell Commun Res 331 Newman Springs Rd Red Bank NJ 07701-7020

WERNICK, ROBERT J, b New York, NY, June 19, 28; m 54; c 3. MATHEMATICS. *Educ:* Univ Mich, Ann Arbor, BSE, 49; Stevens Inst Technol, MS, 52; Rensselaer Polytech Inst, PhD(math), 69. *Prof Exp:* Res engr, Stevens Inst Technol, 51-53; tech asst, Panel Hydrodyn Submerged Bodies, Comt Undersea Warfare, Nat Res Coun, Washington, DC, 54; mathematician, Gen Eng Labs, Gen Elec Co, NY, 59-62 & Mech Technol, Inc, 62-64; asst prof math, State Univ NY, Albany, 64-68; ASSOC PROF MATH, STATE UNIV NY, OSWEGO, 69- *Concurrent Pos:* Consult, Advan Technol Labs, Gen Elec Co, NY, 63-66 & Mech Technol, Inc, 65; referee, J Appl Mech, 69- *Mem:* Am Math Soc; Soc Indust & Appl Math; Math Asn Am. *Res:* Numerical analysis; game theory applied to determining the spectra of linear operators. *Mailing Add:* 4500 Ulloa St San Francisco CA 94116

WERNICK, WILLIAM, b New York, NY, Dec 18, 10; m 33; c 2. MATHEMATICS. *Educ:* NY Univ, BS, 33, MS, 34, PhD(math), 41. *Prof Exp:* Teacher, New York Bd Educ, 37-46, chmn dept math, 46-65; assoc prof math, City Col New York, 65-77; RETIRED. *Concurrent Pos:* Lectr, Hunter Col, 46-50; assoc, Columbia Univ, 50-65. *Mem:* Am Math Soc; Math Asn Am; Asn Symbolic Logic. *Res:* Symbolic logic; general topology; calculus of operators; complete sets of logical functions; functional dependence in the calculus of propositions; distributive properties of set operators. *Mailing Add:* 615 Palmer Rd Yonkers NY 10701

WERNIMONT, GRANT (THEODORE), analytical chemistry, applied statistics, for more information see previous edition

WERNSMAN, EARL ALLEN, b Vernon, Ill, Nov 4, 35; m 59; c 1. CROP BREEDING. *Educ:* Univ Ill, Urbana, BS, 58, MS, 60; Purdue Univ, PhD(genetics), 63. *Prof Exp:* Res assoc genetics, Iowa State Univ, 63-64; from asst prof to assoc prof, 64-72, PROF GENETICS, NC STATE UNIV, 72- *Mem:* fel Am Soc Agron; Am Genetic Asn; Genetics Soc Can; AAAS; Sigma Xi; Crop Sci Soc Am. *Res:* Genetics of Nicotiana and alkaloid production; physiology of cytoplasmic male-sterility in Nicotiana; interspecific hybridization in Nicotiana, tissue and anther culture. *Mailing Add:* Box 7620 NC State Univ Raleigh NC 27695-7620

WERNTZ, CARL W, b Washington, DC, Aug 7, 31; m 58; c 2. THEORETICAL NUCLEAR PHYSICS, ASTROPHYSICS. *Educ:* George Washington Univ, BS, 53; Univ Minn, MS, 55, PhD(physics), 60. *Prof Exp:* Res assoc physics, Univ Wis, 60-62; assoc prof, 62-70, chmn dept, 75-78, PROF PHYSICS, CATH UNIV AM, 70- *Mem:* Am Phys Soc; Am Asn Physics Teachers; Sigma Xi. *Res:* Theory of interaction of pions with nuclei; study of solar nuclear reactions; three body problem. *Mailing Add:* Dept Physics Cath Univ Am Washington DC 20064

WERNTZ, HENRY OSCAR, b Atlantic City, NJ, Jan 19, 30; m 57; c 3. ZOOLOGY. *Educ:* Rutgers Univ, BS, 52; Yale Univ, PhD(zool), 58. *Prof Exp:* Instr biol, Harvard Univ, 57-60; asst prof biol, 60-64, actg chmn dept, 66-67, chmn dept, 75-81, actg resident dir, Marine Sci & Maritime Studies Ctr, 82-84, ASSOC PROF BIOL, NORTHEASTERN UNIV, 64- *Mem:* AAAS; Am Soc Zoologists. *Res:* Osmotic and ionic regulation in invertebrates; environmental physiology. *Mailing Add:* Dept of Biol Northeastern Univ Boston MA 02115

WERNTZ, JAMES HERBERT, JR, b Wilmington, Del, Sept 3, 28; m 55; c 4. EDUCATIONAL ADMINISTRATION, PHYSICS. *Educ:* Oberlin Col, BA, 50; Univ Wis, MS, 52, PhD(physics), 57. *Prof Exp:* From asst prof to prof physics, Univ Minn, Minneapolis, 56-81, dir, Ctr Educ Develop, 67-81 & Univ Col, 78-81; VCHANCELLOR ACAD AFFAIRS & PROF PHYSICS, UNIV NC, CHARLOTTE, 81- *Honors & Awards:* Distinguished Serv Citation, Am Asn Physics Teachers, 68. *Mem:* Fel AAAS; Am Asn Higher Educ; Am Phys Soc; Am Asn Physics Teachers; Sigma Xi. *Res:* Physics of very low temperatures, especially liquid helium phenomena; science education. *Mailing Add:* VChancellor Acad Affairs Univ NC Charlotte NC 28223

WERNY, FRANK, b Ger, June 6, 36; US citizen; c 5. NATURAL PRODUCTS CHEMISTRY, TEXTILE TECHNOLOGY. *Educ:* Univ Puget Sound, BS, 58; Univ Hawaii, PhD(org chem), 63. *Prof Exp:* Ford Found fel, Free Univ Berlin, 62-63; res chemist, Carothers Res Lab, 63-67, sr res chemist, Textile Res Lab, 67-72, Christina Labs, 72-76, Chestnut Run Textile Res Lab, 76-85, RES ASSOC, E I DU PONT DE NEMOURS & CO, INC, 85- *Mem:* Am Chem Soc; Am Soc Testing & Mat. *Res:* Polyamide polymers and fibers; flammability of polymers and textiles; irradiation grafting and melt blending of polymers; polymer rheology; carpet yarns and carpets; nonwoven fabric research and development; fiber light scattering, carpet luster. *Mailing Add:* 334 Old Bailey Lane West Chester PA 19382-8437

WERSHAW, ROBERT LAWRENCE, b Norwalk, Conn, Sept 17, 35; m 63; c 2. HYDROLOGY, GEOCHEMISTRY. *Educ:* Tex Western Col, BS, 57; Calif Inst Technol, MS, 59; Univ Tex, PhD(geol), 63. *Prof Exp:* HYDROLOGIST, US GEOL SURV, 63- *Mem:* Geochem Soc; Am Chem Soc. *Res:* Geochemistry of naturally occurring polyelectrolytes and their interaction with water pollutants. *Mailing Add:* 1566 Winona Ct Denver CO 80204-1143

WERSTIUK, NICK HENRY, b Dec 31, 39; Can citizen. ORGANIC CHEMISTRY, PHYSICAL CHEMISTRY. *Educ:* Univ Alta, BSc, 62; Johns Hopkins Univ, MA, 63, PhD(chem), 66. *Prof Exp:* From asst prof to assoc prof, 67-78, PROF CHEM, MCMASTER UNIV, 78- *Mem:* Chem Inst Can; Am Chem Soc. *Res:* Base-catalyzed H-D exchange and rearrangement of polycyclic betones and other carbon acids; synthesis of hydrogen-isotope labelled organic compounds; photoelectron spectroscopy of transients. *Mailing Add:* Dept Chem McMaster Univ 1280 Main St W Hamilton ON L8S 4L8 Can

WERT, CHARLES ALLEN, b Battle Creek, Iowa, Dec 31, 19; m 42; c 2. FUEL TECHNOLOGY & PETROLEUM ENGINEERING. *Educ:* Morningside Col, BA, 41; Univ Iowa, MS, 43, PhD(physics), 48. *Prof Exp:* Mem staff radiation lab, Mass Inst Technol, 43-45; asst, Univ Iowa, 46-48; instr, inst study metals, Univ Chicago, 48-50; from assoc prof to prof mining & metall eng, 50-56, head dept, 66-86, RACHEFF PROF MAT ENG, UNIV ILL, 86- *Honors & Awards:* Alexander von Humboldt Found sr scientist award, WGer, 81, 87. *Mem:* Fel Am Phys Soc; Am Inst Mining, Metall & Petrol Engrs; fel Am Soc Metals; fel AAAS; Sigma Xi; fel Metall Soc; fel Am Soc Mats. *Res:* Internal friction of metals; diffusion in solids; electron microscopy of solids; alloying behavior of metals with carbon, nitrogen, oxygen and carbon; chemistry and physical structure of coal. *Mailing Add:* 201 Metall Mining Bldg Univ Ill 1304 W Green St Urbana IL 61801

WERT, JAMES J, b Barron County, Wis, Jan 9, 33; m 58; c 2. ENGINEERING, MATERIALS SCIENCE. *Educ:* Univ Wis, BS, 57, MS, 58, PhD(metall eng), 61. *Prof Exp:* Asst engr, Bettis Atomic Power Lab, 58-59; res scientist, A O Smith Corp, 61; from asst prof to prof metall eng, 61-72, dir mat sci & eng div, 67-69, chmn dept mat sci & eng, 69-72, chmn mat, mech & struct div, 72-76, chmn dept mech eng & mat sci, 76-82, GEORGE A SLOAN PROF METALL, VANDERBILT UNIV, 72- *Concurrent Pos:* Westinghouse fel, Carnegie Inst Technol; consult, Nat Acad Sci; vis prof, Cambridge Univ, 74-75; Fulbright lectr, Middle East, 82. *Mem:* Am Soc Eng Educ; Brit Inst Metals; Am Inst Mining, Metall & Petrol Engrs; Am Welding Soc; Royal Micros Soc. *Res:* X-ray diffraction studies of the effects of deformation on superlattice structures; kinetics of antiphase domain growth; residual stresses in metals, friction and wear; fracture, fatigue and failure analysis. *Mailing Add:* 2510 Ridgewood Dr Nashville TN 37215

WERT, JONATHAN MAXWELL, JR, b Port Royal, Pa, Nov 8, 39; m 83; c 5. ENVIRONMENTAL SCIENCES. *Educ:* Austin Peay State Univ, BS, 66, MS, 68; Univ Ala, PhD(educ admin), 74. *Prof Exp:* Chief naturalist, Pa Dept Environ Resources, 68-69; chief naturalist, Bays Mountain Park, 69-71; supvr environ educ sect, Tenn Valley Authority, 71-75; consult energy conserv, environ ctr, Univ Tenn, Knoxville, 75-77; sr prof assoc, Energy Exten Serv, Pa State Univ, 77-81; chief div energy prog develop, Pa Dept Community Affairs, 81-83; PRES, MGT DIAGNOSTICS, MGT CONSULTS, MGT ANALYSES, SAFETY ASSESSMENTS, ORGN DEVELOP & HUM RESOURCES MGT, 83- *Concurrent Pos:* Mem adv bd, Environ Educ Report, Inc, 74-78; mem nat adv coun environ educ, Dept HEW, 75-78; mem, Energy Exten Serv Adv Coun, US Dept of Energy, 87-88. *Honors & Awards:* Am Motors Conserv Award, 76. *Mem:* Conserv Educ Asn; Am Soc Ecol Educ; Nat Audubon Soc; Nat Educ Asn. *Res:* Development of methodologies for carrying out comprehensive environmental planning, including technological assessment, cost-benefit analysis and environmental impact assessment; energy conservation; systems planning; development of numerous models and survey instruments for assing organization and management effectiveness, organization culture and climate including safety culture. *Mailing Add:* PO Box 194 Port Royal PA 17082

WERTH, GLENN CONRAD, b Denver, Colo, July 21, 26; m 50; c 3. GEOPHYSICS. *Educ:* Univ Colo, BS, 49; Univ Calif, Los Angeles, MS, 50, PhD(physics, acoust), 53. *Prof Exp:* Physicist, Arthur D Little, Inc, 53-54 & Calif Res Corp, 54-59; physicist, 59-66, assoc dir, Lawrence Livermore Lab, Univ Calif, 66-81; dir new prog develop, Woodward Clyde Consults, 81-90; RETIRED. *Mem:* Seismol Soc; Am Nuclear Soc; Am Geophys Union; Corporate Planning Asn. *Res:* Seismology; geophysical exploration for oil; use of nuclear explosives in industry and commerce; development of alternate energy supplies; analysis of energy issues and role for research and development; corporate planning. *Mailing Add:* 11473 Ghirardelli Ct Gold River CA 95670

WERTH, JEAN MARIE, b Rochester, NY, Jan 21, 43. MICROBIAL PHYSIOLOGY. *Educ:* Nazareth Col Rochester, BS, 64; Syracuse Univ, MS, 69, PhD(molecular biol), 73. *Prof Exp:* Asst prof, 72-77, assoc prof biol, 77-82, PROF BIOL, WILLIAM PATERSON COL, 82- *Concurrent Pos:* Vis scientist, Roche Inst Molecular Biol, 80-81. *Mem:* AAAS; Am Soc Microbiol; Sigma Xi; NY Acad Sci. *Res:* Gene expression in mormoniella vitripennis during various developmental stages. *Mailing Add:* 24 Reality Dr Kinnelon NJ 07405

WERTH, JOHN, electrochemical & electrical engineering, for more information see previous edition

WERTH, JOHN ST CLAIR, JR, mathematics, for more information see previous edition

WERTH, RICHARD GEORGE, b Markesan, Wis, Feb 5, 20; m 43; c 1. ORGANIC CHEMISTRY. *Educ:* Wartburg Col, BA, 42; Univ Wis, MS, 48, PhD, 50. *Prof Exp:* Jr chemist, Electrochem Dept, E I du Pont de Nemours & Co, 42-44 & 46; Alumni Res Found asst, Univ Wis, 48-50; head dept chem, 61-69, chmn dept, 74-77, from assoc prob to prof chem, 50-90, EMER PROF CHEM, CONCORDIA COL, MOORHEAD, MINN, 90- *Concurrent Pos:* Res partic, Oak Ridge Nat Lab, 62-63; vis fel, Cornell Univ, 70-71; vis scientist, Univ Hyg Lab, Iowa City, Iowa, 83. *Mem:* AAAS; Am Chem Soc; Am Inst Chemists; Soc Appl Spectros; Am Radio Relay League. *Res:* Methods of organic synthesis of polycyclic compounds; formaldehyde products; reaction mechanism for dehydration of bicycloheptanediols; mass spectra of vinylogous imides. *Mailing Add:* Chem Dept Concordia Col Moorhead MN 56562-0001

WERTH, ROBERT JOSEPH, b Hays, Kans, Apr 4, 40; c 1. ZOOLOGY. *Educ:* St Benedict's Col, Kans, BS, 61; Univ Mo, Kansas City, MS, 65; Univ Colo, Boulder, PhD(zool), 69. *Prof Exp:* Asst prof, 69-76, ASSOC PROF BIOL, PURDUE UNIV, CALUMET CAMPUS, 76- *Mem:* AAAS; Am Soc Ichthyologists & Herpetologists. *Res:* Reptilian ecology and physiology; correlation of environmental requirement with physiological adaptation. *Mailing Add:* Dept Biol Purdue Univ Calumet Campus Hammond IN 46323

WERTHAMER, N RICHARD, b Milwaukee, Wis, Feb 9, 35; div; c 1. PHYSICS. *Educ:* Harvard Col, BA, 56; Univ Calif, PhD(physics), 61. *Prof Exp:* Res assoc, Univ Calif, San Diego, 61-62; mem tech staff, Bell Labs, 62-75; chmn, NY State Energy Res & Develop Authority, 76-78; DIR, BECTON DICKINSON & CO, 83- *Mem:* Fel Am Phys Soc; fel AAAS; Am Asn Clin Chem. *Mailing Add:* Becton Dickinson & Co One Becton Dr Franklin Lakes NJ 07417

WERTHEIM, ARTHUR ROBERT, b Newark, NJ, July 5, 15; m 47. MEDICINE. *Educ:* Dartmouth Col, AB, 35; Jefferson Med Col, MD, 39; Am Bd Internal Med, dipl, 50, recertification, 74. *Prof Exp:* Intern, Philadelphia Gen Hosp, 39-41, resident, 41-43; resident, Goldwater Mem Hosp, 46-47; asst, Long Island Col Med, 48; instr, Col Med, State Univ NY Downstate Med Ctr, 49-51; from asst prof to assoc prof, 51-69, prof, 69-81, EMER PROF MED, COL PHYSICIANS & SURGEONS, COLUMBIA UNIV, 81-; AT MED SERV, PRESBY HOSP. *Concurrent Pos:* Res fel, Goldwater Mem Hosp, 47; asst med, Col Physicians & Surgeons, Columbia Univ, 47; from clin assoc to dir med serv, Maimonides Hosp, Brooklyn, 48-51; res assoc, Goldwater Mem Hosp, 51-54, vis physician, 54-68; actg chief, Med Serv, Francis Delafield Hosp, New York, 71-74, chief, 74-75; attending physician, Presby Hosp, NY, 81, consult med serv, 81- *Mem:* Harvey Soc; AMA; fel Am Col Physicians. *Res:* Clinical and laboratory research in chronic diseases; metabolic aspects of hypertension and atherosclerosis. *Mailing Add:* Med Serv Presby Hosp 630 W 168th St New York NY 10032

WERTHEIM, GUNTHER KLAUS, b Berlin, Ger, Feb 26, 27; nat US; m 56; c 3. SURFACE SCIENCE & PHOTOEMISSION. *Educ:* Stevens Inst Technol, ME, 51; Harvard Univ, AM, 52, PhD(physics), 55. *Prof Exp:* Assoc phys oceanog, Woods Hole Oceanog Inst, 54; mem tech staff, Bell Tel Labs, Inc, 55-62, head, Crystal Physics Res Dept, 62-84, DISTINGUISHED MEM TECH STAFF, BELL LABS, 84- *Concurrent Pos:* Adj prof, Stevens Inst Technol, 66-68; mem vis comts physics, Harvard Univ, 69-75 & Bartol Res Found, 72-87. *Honors & Awards:* Humboldt Sr Scientist Award, 86. *Mem:* Am Phys Soc; Sigma Xi. *Res:* X-ray photo-electron spectroscopy; Mössbauer effect; angular correlations; magnetism; semiconductors; synchrotron radiation. *Mailing Add:* 175 Woodland Ave Convent Station NJ 07961

WERTHEIM, ROBERT HALLEY, b Carlsbad, NMex, Nov 9, 22; m 46; c 2. NUCLEAR PHYSICS. *Educ:* US Naval Acad, BS, 45; Mass Inst Technol, SM, 54. *Prof Exp:* Staff officer, Armed Forces Spec Weapons Proj, USN, 48-49, spec asst nuclear applns, Spec Projs Off, 56, head re-entry body sect, 56-61, weapons develop dept, Naval Ord Test Sta, 61-62, mem staff, Dir of Defense Res & Eng, 62-65, head missile br, Spec Projs Off, 65-67, dep tech dir, 67-68, tech dir, 68-77, dir, Spec Projs Off, 77-80; sr vpres sci & eng, Lockheed Corp, 81-88; CONSULT, 88- *Concurrent Pos:* Consult, Off Secy Defense Sci Bd, Inst Def Anal & Ctr Naval Anal; mem, Comt Int Security & Arms Control, Nat Acad Sci; mem, Nat Security Adv Group, Los Alamos Nat Lab, Charles Stark Draper Lab Corp, Sci Adv Group, Joint Strategic Target Planning Staff, Defense Advan Res Proj Agency. *Honors & Awards:* William S Parsons Award, 71; Gold Medal, Am Soc Naval Engrs, 73. *Mem:* Nat Acad Eng; Sigma Xi; Am Soc Naval Engrs; fel Am Inst Aeronaut & Astronaut. *Res:* Weapons systems research, development and testing, especially ballistic missiles and nuclear applications. *Mailing Add:* 177705 Devereux Rd San Diego CA 92128

WERTHEIMER, ALAN LEE, b Cleveland, Ohio, Dec 22, 46; m 69; c 2. OPTICAL ENGINEERING, MICROPROCESSORS. *Educ:* Univ Rochester, BS, 68, PhD(optics), 74. *Prof Exp:* Optical syst & lens design, Itek Corp, 68-69; SR SCIENTIST OPTICS RES, LEEDS & NORTHRUP CO, 74- *Mem:* Optical Soc Am; Soc Photo-Optical Instrumentation Eng. *Res:* Light scattering to analyze particle size distributions in air and liquid media; engineering and project management of optically based, microprocessor controlled measuring and recording instruments; optical processing of photographic images; fiber optics communications. *Mailing Add:* Nine Kalleston Dr Pittsford NY 14534

WERTHEIMER, ALBERT I, b Buffalo, NY, Sept 14, 42; m 65; c 2. PUBLIC HEALTH ADMINISTRATION. *Educ:* Univ Buffalo, BS, 65; State Univ NY, Buffalo, MBA, 67; Purdue Univ, PhD(med sociol, pharm), 69. *Prof Exp:* Asst prof pharm, Sch Pharm, State Univ NY, Buffalo, 69-71, asst prof mkt, Sch Mgt, 70-71; researcher med care, Social Security Admin, Dept HEW, 72 & USPHS, 72-73; PROF PHARM ADMIN & DIR GRAD PROG, UNIV MINN, MINNEAPOLIS, 73- *Concurrent Pos:* WHO fel, 75. *Mem:* Am Pub Health Asn; Am Pharmaceut Asn; Am Sociol Asn; AMA. *Res:* Drug use process; drug ecology studies. *Mailing Add:* Pharm 5-130 H S Unit F Univ Minn 308 Harvard St SE Minneapolis MN 55455-0343

WERTHEIMER, MICHAEL ROBERT, b Capetown, SAfrica, Jan 22, 40; Can citizen; m 73; c 2. SOLID STATE PHYSICS, MATERIALS SCIENCE ENGINEERING. *Educ:* Univ Toronto, BASc, 62, MA, 63; Univ Grenoble, Dr es Sc, 67. *Prof Exp:* Process engr thermodyn, Can Liquid Air Ltd, 63-64, res engr cryog, 67-72; PROF ENG PHYSICS, ECOLE POLYTECH, MONTREAL, 73- *Concurrent Pos:* Pres, Polyplasma Inc, 82- *Mem:* Fel Inst Elec & Electronics Engrs; Elec Insulation Soc; Am Phys Soc; Can Asn Physicists. *Res:* Electrical insulation; dielectrics; polymers; plasma chemistry; materials science; surface science; engineering physics. *Mailing Add:* Dept Eng Physics, Ecole Polytechnique Box 6079 Sta A Montreal PQ H3C 3A7 Can

WERTMAN, LOUIS, b New York, NY, Oct 2, 25; m 47; c 3. BIOMEDICAL ENGINEERING. *Educ:* City Col New York, BME, 50. *Prof Exp:* Assoc prof, 51-65, PROF MECH TECHNOL, NEW YORK CITY COMMUNITY COL, 65-, ADMINR ELECTROMECH, 66- *Concurrent Pos:* Consult engr, 60- *Mem:* AAAS; Am Soc Eng Educ; Soc Mfg Engrs; Nat Soc Prof Engrs. *Res:* Automated biomedical engineering utilizing electromechanical components and systems; educational research in new program development. *Mailing Add:* 18616 Aberdeen Rd Jamaica NY 10467

WERTMAN, WILLIAM THOMAS, b Franklin, Pa, Dec 25, 20; m 42; c 3. PETROLEUM ENGINEERING, CHEMICAL ENGINEERING. *Educ:* Grove City Col, BS, 42. *Prof Exp:* Chem engr, Joseph E Seagram & Sons, Inc, Md, 42-45 & Bradford Labs, Inc, Pa, 46-48; chem engr, Petrol & Natural Gas Div, US Bur Mines, Pa, 48-54, proj leader, Morgantown Petrol Res Lab, WVa, 54-63, proj cooodr, 63-65, actg chief lab, 65-66, chief lab, 66-70, dep res dir, Morgantown Energy Res Ctr, 70-75, sr staff specialist, Morgantown Energy Technol Ctr, US Dept Energy, 75-81; sr petrol engr, Sci Applns Inc, Morgantown, Va, 81-82; RETIRED. *Concurrent Pos:* Consult eng, 82- *Mem:* Am Chem Soc; Soc Petrol Engrs; Am Petrol Inst; Am Gas Asn. *Res:* Petroleum and natural gas production; reservoir engineering and rock analysis; subsurface formation temperatures; fluid flow through porous media; water analysis and treatment; coal chemistry and gasification research. *Mailing Add:* 433 Washington St Morgantown WV 26505

WERTZ, DAVID LEE, b Hammond, Ind, Feb 3, 40; m 61; c 2. INORGANIC CHEMISTRY, PHYSICAL CHEMISTRY. *Educ:* Ark State Univ, BS, 62; Univ Ark, PhD(chem), 66. *Prof Exp:* From asst prof to assoc prof, 66-74, PROF CHEM, UNIV SOUTHERN MISS, 74- *Mem:* Am Chem Soc. *Res:* X-ray diffraction and spectral studies of solute-solvent interactions in concentrated solutions; x-ray diffraction studies of liquid structure. *Mailing Add:* Univ Southern Miss Box 5043 Southern Sta Hattiesburg MS 39406

WERTZ, DENNIS WILLIAM, b Reading, Pa, Mar 21, 42; m 62; c 2. PHYSICAL CHEMISTRY. *Educ:* Univ Md, BS, 64; Univ SC, PhD(chem), 68. *Prof Exp:* Res assoc spectros, Mass Inst Technol, 68-69; asst prof, 69-74, ASSOC PROF CHEM, NC STATE UNIV, 74- *Mem:* AAAS. *Res:* Resonance Raman and electronic absorption spectroscopy of metal dimines and their reduction properties; localization of the redox electron. *Mailing Add:* Dept Chem NC State Univ Raleigh NC 27607

WERTZ, GAIL T WILLIAMS, b Washington, DC, Oct 31, 43; m 66, 87. VIROLOGY, MOLECULAR BIOLOGY. *Educ:* Col William & Mary, BS, 66; Univ Pittsburgh, PhD(microbiol), 70. *Prof Exp:* NIH fel, Med Sch, Univ Mich, Ann Arbor, 70-71, sr res assoc virol, 71-73; from asst prof to assoc prof, Bact & Immunol, Univ NC Med Sch, 73-83, prof microbiol, 83-87; PROF MICROBIOL, UNIV ALA MED SCH, 87- *Concurrent Pos:* NSF & NIH res grants; vis prof, Univ Wis, 83-84; consult, WHO Vacine Prog, 83-91; ASM Div lectr, Am Soc Microbiol, 87, found lectr, 91. *Mem:* AAAS; Am Soc Microbiol; Am Soc Virol; Sigma Xi. *Res:* Virus-host interactions; viral replication; mechanism of viral nucleic acid synthesis and replication; effect of virus on host macromolecular synthesis and host response to infection-interference phenomenon. *Mailing Add:* Univ Ala BHS Box 70 UAB Sta Birmingham AL 35294

WERTZ, HARVEY J, b Muskogee, Okla, May 1, 36. ELECTRICAL ENGINEERING, APPLIED MATHEMATICS. *Educ:* Univ Kans, BSEE, 58, MSEE, 59; Univ Wis, PhD(elec eng), 62. *Prof Exp:* Instr elec eng, Univ Kans, 58-59; from asst prof to assoc prof, Univ Wis-Madison, 62-69; consult, 64-65, 68-69, mem tech staff, 66-68, 69-73, head, Orbital Systs Prog Dept, 73-79, PRIN DIR COMPUT SYSTS SUBDIV, AEROSPACE CORP, 79- *Concurrent Pos:* Asst prof, Math Res Ctr, US Army. *Mem:* Soc Indust & Appl Math; Asn Comput Mach; Inst Elec & Electronics Engrs. *Res:* Application of analytical and computer-oriented mathematics to the analysis and synthesis of physical systems; coordinated development of algorithms and custom hardware to achieve highly efficient processing, specification and development of computer software. *Mailing Add:* 1004 Centinela Ave Santa Monica CA 90403

WERTZ, JAMES RICHARD, b Kingman, Ariz, Feb 20, 44; m 67; c 1. THEORETICAL ASTROPHYSICS, ASTRONAUTICS. *Educ:* Mass Inst Technol, SB, 66; Univ Tex, Austin, PhD(physics), 70; George Washington Univ, MS, 78. *Prof Exp:* Asst prof physics & astron, Moorhead State Col, 70-73, NSF inst grant, 71-73; sr analyst, Comput Sci Corp, 73-78; dir spacecraft eng, Western Union Space Comt, TRW, 78-80, sr systs engr, TRW, 80-84; PRES, MICROCOSM, INC, TORRANCE, CALIF, 84- *Concurrent Pos:* Mem, Task Group Educ Astron, Am Astron Soc, 74- *Mem:* AAAS; Brit Interplanetary Soc; Am Phys Soc; Am Astron Soc; Am Inst Aeronaut & Astronaut; Sigma Xi. *Res:* Spacecraft attitude determination, interstellar travel and navigation; hierarchical cosmology. *Mailing Add:* 2601 Airport Dr-230 Torrance CA 90505

WERTZ, JOHN EDWARD, b Denver, Colo, Dec 4, 16; m 43; c 3. CHEMISTRY. *Educ:* Univ Denver, BS, 37, MS, 38; Univ Chicago, PhD(chem), 48. *Prof Exp:* Asst prof chem, Augustana Col, 41-44; instr physics, Gustavus Adolphus Col, 44-45; instr mech eng, 45-47, from asst prof to prof phys chem, 48-74, PROF CHEM, UNIV MINN, MINNEAPOLIS, 74- *Concurrent Pos:* Fulbright res scholar & Guggenheim fel, Clarendon Lab, Oxford Univ, 57-58. *Mem:* Fel Am Phys Soc; Brit Inst Physics; Am Chem Soc; The Chem Soc. *Res:* Fluorescence of crystals; surface energy of solids; adsorption of vapors on solids; magnetic susceptibility of adsorbed layers; nuclear and paramagnetic resonance; structure and electronic properties of defects in solids; infrared detectors. *Mailing Add:* 700 Arbogast St St Paul MN 55126-4123

WERTZ, PHILIP WESLEY, b Lykens, Pa, Oct 25, 49; m 73; c 2. DERMATOLOGY. *Educ:* Rutgers Univ, AB, 71; Univ Wis-Madison, PhD(biochem), 76. *Prof Exp:* Fel, McArdle Lab Cancer Res, 76-79, res assoc, 79-81; asst scientist, Univ Iowa Col Med, 84-90; ASSOC PROF ORAL

PATH, DOWS INST DENT RES, 90- *Mem:* Soc Investigative Dermat; Am Oil Chemists Soc; Am Soc Biochem & Molecular Biol. *Res:* Investigation of the structure, composition and metabolism of epithelial lipids and attempting to define relationships between molecular structure and function of lipids in normal and diseased skin and oral mucosa. *Mailing Add:* Dows Inst Dent Res Univ Iowa Col Dent Iowa City IA 52242

WERTZ, RONALD DUANE, b Loveland, Colo, Jan 29, 29; m 51; c 5. INDUSTRIAL & MANUFACTURING ENGINEERING. *Educ:* Univ Colo, BS, 52. *Prof Exp:* Mem tech staff, Hughes Aircraft Corp, 52-55; sect engr, Lockheed Missiles & Space Div, 55-61; DIR, PRODUCTIVITY LAB, BALL CORP, 61- *Mem:* Inst Elec & Electronics Engrs; Soc Mfg Engrs; Am Soc Nondestructive Testing. *Res:* Specialized instruments and equipment to improve manufacturing productivity; electro-optics; ultrasonics; computer science; electronics; mechanisms; robotics; machine design. *Mailing Add:* 2005 Vassar Dr Boulder CO 80303

WESCHLER, CHARLES JOHN, b Youngstown, Ohio, Jan 29, 48; m 71. INDOOR AIR CHEMISTRY. *Educ:* Boston Col, BS, 69; Univ Chicago, MS, 72, PhD(chem), 74. *Prof Exp:* Fel phys inorg chem, Northwestern Univ, 74-75; mem tech staff chem, Bell Labs, 75-84, mem tech staff chem, Bell Communs Res, 84-87, DISTINGUISHED MEM PROF STAFF ENVIRON & CONTAMINANTS RES GROUP, BELLCORE, 87- *Concurrent Pos:* Lab grad partic, Argonne Nat Lab, 72-73; mem Nat Acad Sci comt, Advan Assessing Human Exposure Airborne Pollutants, 87-90. *Mem:* AAAS; Am Chem Soc; Am Geophys Union; Am Soc Testing & Mat; Air & Waste Mgt Asn. *Res:* Indoor aerosols; elemental and chemical composition of aerosols as a function of particle size; chemistry occurring within atmospheric aerosols; comparisons between indoor and outdoor airborne pollutants; organic constituents of airborne particles. *Mailing Add:* Bell Commun Res 331 Newman Springs Rd Red Bank NJ 07701-7040

WESCOTT, EUGENE MICHAEL, b Hampton, Iowa, Feb 15, 32; div; c 3. SPACE PHYSICS, EXPLORATION GEOPHYSICS. *Educ:* Univ Calif, Los Angeles, AB, 55; Univ Alaska, MS, 60, PhD(geophys), 64. *Prof Exp:* Geophysicist, Geophys Res & Develop Br, US AEC, 55-58; sr res asst, 58-64, from asst prof to assoc prof, 64-74, PROF RES GEOPHYS, GEOPHYS INST, UNIV ALASKA, 74- *Concurrent Pos:* Nat Acad Sci resident res assoc, NASA-Goddard Space Flight Ctr, 66-69; vis prof inst geophys & planetary physics, Univ Calif, Los Angeles, 76-77. *Mem:* Am Geophys Union; Int Asn Geomagnetism & Aeronomy; Soc Explor Geophys; Geothermal Resources Council. *Res:* Electric and magnetic fields of the upper atmosphere and of magnetospheric plasmas; auroral mechanisms; geothermal resources exploration research. *Mailing Add:* Dept Geol Univ Alaska Fairbanks AK 99701

WESCOTT, LYLE DUMOND, JR, b Hackensack, NJ, Jan 27, 37; m 85; c 2. ORGANIC CHEMISTRY, POLYMER CHEMISTRY. *Educ:* Ga Inst Technol, BS, 59; Pa State Univ, PhD(chem), 63. *Prof Exp:* Res fel, Pa State Univ, 63-64; res chemist, Baytown Labs, Esso Res & Eng Co, Tex, 64-68; from asst prof to assoc prof, 68-77, head dept, 74-83, PROF CHEM, CHRISTIAN BROS COL, 77- *Mem:* Am Chem Soc; Royal Soc Chem. *Res:* Low temperature reactions of vapor species of refractory materials; stabilization of polymeric materials; flame and smoke suppressants for polymeric materials adhesives. *Mailing Add:* Dept Chem Christian Bros Univ 650 E Pkwy S Memphis TN 38104

WESCOTT, RICHARD BRESLICH, b Chicago, Ill, July 8, 32; m 54; c 3. VETERINARY PARASITOLOGY. *Educ:* Univ Wis, BS, 54, MS, 64, PhD(vet med), 65; Univ Minn, DVM, 58; Am Col Lab Animal Med, dipl, 67. *Prof Exp:* USPHS fel parasitol, Univ Wis, 62-65; assoc prof vet microbiol, Sch Vet Med, Univ Mo, Columbia, 65-71; PROF VET PATH, WASH STATE UNIV, 71- *Concurrent Pos:* Actg chmn dept vet path, Wash State Univ, 75-77. *Mem:* Am Vet Med Asn; Sigma Xi. *Res:* Laboratory animal medicine; nematode-virus interactions in gnotobiotic host animals. *Mailing Add:* Dept Vet Path Wash State Univ Pullman WA 99163

WESCOTT, WILLIAM B, b Pendleton, Ore, Nov 10, 22; m 69; c 2. ORAL PATHOLOGY. *Educ:* Univ Ore, DMD, 51, MS, 62. *Prof Exp:* Dir oral tumor registry, Dent Sch, Univ Ore, 53-54; pvt practr gen dent, Ore, 53-59; asst gen path, Med Sch, Univ Ore, 59-60; from asst prof to assoc prof path, 62-69, prof path & assoc dean admin affairs, 69-72; co-dir res & educ training prog, Oral Dis Res Lab, Vet Admin Hosp, Houston, 72-75; chief dent serv, Vet Admin Hosp, Durham, NC, 75-78; co-dir, Southeastern Regional Med Educ Ctr, Birmingham, 78-79; prof oral diag radiol path, Dent Sch, Loma Linda Univ, 79-85; prof oral diagnosis, oral med & oral path, Sch Dent, Univ Calif, Los Angeles, 79-85; dir, Dent Educ Ctr, Los Angeles, 79-85; CHIEF, DENT SERV, VET ADMIN MED CTR, SAN FRANCISCO, 85-; PROF, RESTORATIVE DENT, SCH DENT, UNIV CALIF, SAN FRANCISCO, 85- *Concurrent Pos:* Am Cancer Soc fel, 59-61; USPHS fel, 61-63; clin assoc oral & dent med, Med Sch, Univ Ore, sr clin instr, 67-69; prof path, Univ Tex Dent Br, Houston, 72-75 & Duke Univ Med Sch, 76-78; prof dent, Univ NC, 75-78; consult, US Naval Regional Dent Ctr, 82-; prof, Golden Gate Univ, 86- *Mem:* Am Dent Asn; Int Asn Dent Res; fel Am Acad Oral Path; Int Asn Microbiol; fel Am Col Dent; Asn Mil Surgeons US. *Res:* Correlation of clinical and histopathologic findings; fluorescent antibody technic; foreign body reaction; bacteriologic and fungal changes under bizarre atmospheres and increased pressures; immunology; salivary glands (xerostomia). *Mailing Add:* Vet Admin Med Ctr 4150 Clemente St 160 San Francisco CA 94121

WESELOH, RONALD MACK, b Los Angeles, Calif, June 30, 44. ENTOMOLOGY. *Educ:* Brigham Young Univ, BS, 66; Univ Calif, Riverside, PhD(entom), 70. *Prof Exp:* Asst agr scientist, 70-75, assoc agr scientist, 75-81, AGR SCIENTIST, CONN AGR EXP STA, 81- *Mem:* Entom Soc Am; Entom Soc Can; Sigma Xi; Int Orgn Biol Control; Ecol Soc Am. *Res:* Control of insect pests by means of parasites and predators; insect behavior and ecology. *Mailing Add:* Dept Entom Conn Agr Exp Sta New Haven CT 06504

WESELY, MARVIN LARRY, b Cedar Bluffs, Nebr, May 5, 44; m 68; c 2. MICROMETEOROLOGY. *Educ:* Univ Nebr, Lincoln, BS, 65; Univ Wis-Madison, MS, 68, PhD(soil sci), 70. *Prof Exp:* Physicist, US Army Ballistic Res Labs, 70-73; asst meteorologist, 73-76, sect head, Atmospheric Physics Sect, Radiol & Environ Res Div, 81-83, prog mgr, Atmospheric Physics Prog, Environ Res Div, 83-87, METEOROLOGIST, ARGONNE NAT LAB, 76-, SECT HEAD, ATMOSPHERIC RES SECT, 87- *Mem:* Am Meteorol Soc; Am Soc Agron; Am Geophys Union; Royal Meteorol Soc; AAAS. *Res:* Studies of turbulent transfer of heat, momentum and pollutants in the lower atmosphere; remote sensing of turbulence with ground-based systems; atmospheric boundary layer physics. *Mailing Add:* Argonne Nat Lab Bldg 203 ER Argonne IL 60439

WESENBERG, CLARENCE L, b Bradley, SDak, May 15, 20; m 42; c 6. ELECTRONICS ENGINEERING. *Educ:* SDak State Univ, BS, 43. *Prof Exp:* Radial-physics surveyor, Argonne Nat Lab, 44-46, electronics designer, 46-53; electronics engr, Stromberg-Carlson Co, 53-58, reliability engr, Gen Dynamics/Electronics, 58-62; prof reliability engr, Honeywell, Inc, 62-67; SR RELIABILITY ENGR, 3M CO, 67- *Mem:* Inst Elec & Electronics Engrs. *Res:* Metrology engineering involving calibration and measurements via mechanical and electromechanical instruments; maintaining accurate parameter measurement control for all magnetic tape parameters affecting performance and usage. *Mailing Add:* 8617 W River Rd N Minneapolis MN 55444

WESENBERG, DARRELL, b Madison, Wis, Dec 6, 39; m 68; c 3. AGRONOMY, GENETICS. *Educ:* Univ Wis-Madison, BS, 62, MS, 65, PhD(agron), 68. *Prof Exp:* RES AGRONOMIST, RES & EXTEN CTR, AGR RES SERV, USDA, 68- *Mem:* Crops Sci Soc Am; Am Soc Agron; AAAS; Am Genetic Asn; Sigma Xi. *Res:* Plant breeding and plant genetics in cereal crops. *Mailing Add:* USDA-ARS Nat Small Grains Germplasm Res Facil PO Box 307 Aberdeen ID 83210

WESER, DON BENTON, b Wellsburg, WVa, Feb 7, 42; m 65; c 3. BIOCHEMISTRY. *Educ:* Bethany Col, BS, 64; WVa Univ, MS, 66; Ga Inst Technol, PhD(org chem), 71. *Prof Exp:* Asst prof chem, Clemson Univ, Sumter, 66-67 & 71-73; assoc prof, Univ SC, Sumter, 73-77; PROF CHEM, FROSTBURG STATE UNIV, 77- *Concurrent Pos:* Consult, Fibred Co, Lavale & med prof, 87; prof, Goucher Col, 88-90. *Mem:* Sigma Xi; Am Chem Soc. *Res:* Synthesis of certain alkaloids and precursors to such compounds; reaction mechanisms for nucleophilic substitution reactions. *Mailing Add:* Rte 2 Box 236 Frostburg MD 21532

WESER, ELLIOT, b New York, NY, Jan 12, 32; m 55; c 1. MEDICINE, GASTROENTEROLOGY. *Educ:* Columbia Univ, AB, 53, MD, 57; Am Bd Internal Med, dipl, 64; Am Bd Nutrit, cert, 71. *Prof Exp:* Med intern & asst resident, Sch Med & King County Hosp, Univ Wash, 57-59; sr resident med, Bronx Munic Hosp Ctr, Albert Einstein Col of Med, 59-60; clin assoc gastroenterol, Nat Inst Arthritis & Metab Dis, 61-63; from instr to asst prof med, Med Col, Cornell Univ, 63-67; prof physiol & med, Univ Tex Med Sch, San Antonio, 67-76, dep chmn dept, 69-76, head sect gastroenterol, 67-76; CHIEF MED SERV, AUDIE L MURPHY VET ADMIN HOSP, SAN ANTONIO, 76- *Concurrent Pos:* Res fel med, New York Hosp-Cornell Med Ctr, 60-61; New York Res Coun career scientist award, 63-67; asst attend physician, New York Hosp, 64-67; asst vis physician, Bellevue Hosp, 64-67; attend physician, Bexar County Hosps, 67- *Mem:* AAAS; fel Am Col Physicians; Am Fedn Clin Res; Am Gastroenterol Asn; AMA. *Mailing Add:* Dept Med Univ Tex Health Sci Ctr 7703 Floyd Curl Dr San Antonio TX 78284-7870

WESLER, OSCAR, b Brooklyn, NY, July 12, 21. MATHEMATICS, STATISTICS. *Educ:* City Col New York, BS, 42; NY Univ, MS, 43; Stanford Univ, PhD(math statist), 55. *Prof Exp:* Asst math, NY Univ, 42-43; asst & instr, Princeton Univ, 43-46; res assoc math statist, Stanford Univ, 52-55, actg asst prof statist, 55-56; from asst prof to assoc prof math, Univ Mich, 56-64; PROF STATIST & MATH, NC STATE UNIV, 64- *Concurrent Pos:* Consult inst sci & technnol, Univ Mich, 56-64; vis assoc prof, Stanford Univ, 60, vis prof, 62-63, 73, 74 & 78; NSF nat lectr, 63-; prof on-site studies prog, Int Bus Mach Corp, 66-; vis scholar, Univ Calif, Berkeley, 72-73. *Mem:* Am Math Soc; Inst Math Statist. *Res:* Probability and statistics; stochastic processes; statistical decision theory; functional analysis. *Mailing Add:* Dept Statist NC State Univ Raleigh NC 27695-8203

WESLEY, DEAN E, b Flint, Mich, Feb 26, 37; m 59; c 4. SOIL FERTILITY. *Educ:* Mich State Univ, BS, 61; SDak State Univ, PhD(soil fertil), 65. *Prof Exp:* Exten soil specialist & asst SCI & Educ Admin-Agr Res, Univ Nebr, 65-66; from asst prof to assoc prof agron, 66-74, assoc prof, 74-80, PROF AGR, WESTERN ILL UNIV, 80- *Mem:* Am Soc Agron; Soil Sci Soc Am; Soil Conserv Soc Am. *Res:* Plant nutrition; nutrients such as phosphorus, zinc, iron and sulfur. *Mailing Add:* Eight Indian Trail Rd Macomb IL 61455

WESLEY, ROBERT COOK, b Jamestown, Ky, Aug 19, 26; m 53; c 2. DENTISTRY. *Educ:* Berea Col, AB, 50; Univ Louisville, DMD, 54. *Prof Exp:* Pvt practr, 54-67; asst prof prosthodont, Univ Ky, 67-72, actg dir, Gen Pract Residency Dent, 80, dir, 78-90, assoc prof family pract & dir, Div Oral Health, Dept Family Pract, Col Med, 75-78, ASSOC PROF PROSTHODONT, COL DENT, UNIV KY, 72- *Concurrent Pos:* Consult, Vet Admin Hosp, Lexington, Ky, 72; ed, Southeastern Acad Newslett, 73-81; assoc ed, J Prosthetic Dent, 73-81. *Mem:* Int Asn Dent Res; Am Dent Asn; Fed Prosthetic Organs (secy, 84-, pres, 89); Am Prosthodont Soc. *Res:* Complete dentures especially related to geriatric patients. *Mailing Add:* Dept Oral Health Pract Univ Ky Med Ctr Lexington KY 40536

WESLEY, ROY LEWIS, b Liberty, Ky, May 1, 29; m 54; c 5. FOOD SCIENCE. *Educ:* Berea Col, BS, 55; Purdue Univ, MS, 58, PhD, 66. *Prof Exp:* From asst prof to assoc prof poultry sci, Va Polytech Inst & State Univ, 58-74, prof food sci, 74-89; RETIRED. *Mem:* Poultry Sci Asn; Inst Food Technol. *Res:* Physiology. *Mailing Add:* 704 Southgate Dr Blacksburg VA 24060

WESLEY, WALTER GLEN, b Ft Worth, Tex, Oct 12, 38; m 68; c 5. THEORETICAL PHYSICS. *Educ:* Tex Christian Univ, BA, 61; Univ NC, PhD(physics), 70. *Prof Exp:* From asst prof to assoc prof physics, 66-75, PROF PHYSICS, MOORHEAD STATE UNIV, 75- *Mem:* Hist Sci Soc; Royal Astron Soc Can; Am Asn Physics Teachers; Am Phys Soc. *Res:* Gravitational theory and quantum gravitation; interaction of science and society; history of astronomy and physics: 17-19th century. *Mailing Add:* Dept Physics Moorhead State Univ Moorhead MN 56560

WESNER, JOHN WILLIAM, b Newark, NJ, July 14, 36; m 65; c 2. PHYSICAL DESIGN, CONSUMER PRODUCT DESIGN. *Educ:* Carnegie Inst Technol, BS, 58; Calif Inst Technol, MS, 59; Carnegie-Mellon Univ, PhD(mech eng), 68. *Prof Exp:* Sr engr, Atomic Power Div, Westinghouse Elec Corp, 64-68; from mem tech staff to supur, Bell Tel Lab, 68-82; supvr, IS Res & Develop, 83-86, SUPVR, GBS, AT&T, 87- *Concurrent Pos:* Chairperson, Design Educ Comt, Am Soc Mech Engrs, 83-87, invited participant, Decennial Educ Conf, 86, chairperson, Design for Manufacturability Comt, 87-89, chair, Design Eng Div, 91-; assoc ed, Mfg Rev, 88-, adv ed, Res Eng Design, 89-; chairperson, ad hoc adv comt, Prog in Design Theory & Methodology, NSF, 87-90. *Mem:* Fel Am Soc Mech Engrs; Sigma Xi. *Res:* Physical design (mechanical design and development) of small business telecommunications systems. *Mailing Add:* 63 Glendale Dr Freehold NJ 07728-1357

WESOLOWSKI, WAYNE EDWARD, b Cicero, Ill, July 25, 45; m 68; c 2. PHYSICAL & ENVIRONMENTAL CHEMISTRY. *Educ:* St Procopius Col, BS, 67; Univ Ariz, PhD(chem), 71. *Prof Exp:* Res scientist, Freeman Lab, Inc, 71-73; vpres, Chicago Sci, Inc, 73-74; ASSOC PROF, ILL BENEDICTINE COL, 74- *Concurrent Pos:* Fel, Res Corp, 74-75. *Mem:* Am Chem Soc; Air Pollution Control Asn. *Res:* Paramagnetic behavior of transition metals in anisotropic ligand fields; isokinetic particulate sampling of stationary environmental pollution sources. *Mailing Add:* Dept Chem Ill Benedictine Col Lisle IL 60532-0900

WESSEL, GUNTER KURT, b Berlin, Ger, Mar 29, 20; US citizen; m 53; c 3. ASTRONOMY. *Educ:* Tech Hochsch, Berlin, BS, 40; Univ Gottingen, dipl, 47, PhD(physics), 48. *Prof Exp:* Physicist, Lorenz Radio AG, Berlin, 43-45; fel, Nat Res Coun Can, 51-53; consult physics, Gen Elec Co, Syracuse NY, 53-61; prof physics, Syracuse Univ, 61-86; RETIRED. *Res:* Atomic physics; spectroscopy; masers; lasers; magnetic resonance. *Mailing Add:* Four Jamar Dr Fayetteville NY 13066

WESSEL, HANS U, b Duisburg, Ger, Apr 18, 27; US citizen; m 55; c 5. CARDIOLOGY, PHYSIOLOGY. *Educ:* Univ Freiburg, MD, 53. *Prof Exp:* Instr med, 60-63, assoc, 63-65, asst prof, 65-67, asst prof pediat, 67-71, assoc prof eng sci, 73-76, PROF PEDIAT, MED SCH, NORTHWESTERN UNIV, CHICAGO, 76-, PROF ENG SCI, 76- *Concurrent Pos:* NIH cardiovasc res trainee, Med Sch, Northwestern Univ, Chicago, 61-63; Am Heart Asn estab investr, 65-70, fel coun circulation, 65- *Mem:* AAAS; assoc fel Am Col Cardiol; assoc mem Inst Elec & Electronics Engrs; sr mem Instrument Soc Am; Biomed Eng Soc. *Res:* Bioengineering; effect of pulmonary vascular disease on pulmonary gas exchange; instrumentation; indicator dilution techniques; thermal velocity probes; patient monitoring system; exercise physiology in children. *Mailing Add:* Children's Mem Hosp 2300 Children's Plaza Chicago IL 60614

WESSEL, JOHN EMMIT, b Los Angeles, Calif, Mar 8, 42. CHEMICAL PHYSICS, SOLID STATE PHYSICS. *Educ:* Univ Calif, Los Angeles, BS, 65; Univ Chicago, PhD(chem), 70. *Prof Exp:* Fel chem, Univ Pa, 69-72, instr, 72-74; mem tech staff, 74-78, res scientist, 78-80, SR SCIENTIST, CHEM & PHYSICS LAB, AEROSPACE CORP, 80- *Concurrent Pos:* Prin investr on Dept of Energy & Aerospace Corp Sponsored Prog. *Honors & Awards:* Aerospace Corp Pres Award. *Mem:* Am Phys Soc; Sigma Xi; AAAS; Am Chem Soc. *Res:* Laser spectroscopy applied to atomic and molecular detection, laser communications, and infrared detection. *Mailing Add:* 919 Duncan Ave Manhattan Beach CA 90266

WESSEL, PAUL ROGER, ocean acoustics, solid state physics, for more information see previous edition

WESSEL, WILLIAM ROY, b Louisville, Ky, Nov 20, 37. GLOBAL WEATHER FORECASTING, PLANETARY ATMOSPHERE MODELING. *Educ:* Univ Notre Dame, BS, 59; Univ Mich, MS, 61, PhD(physics), 65. *Prof Exp:* Fel, Univ Chicago, 64-66; res scientist, Argonne Nat Lab, 66-72; res assoc, Fla State Univ, 72-74; vis scientist, Nat Ctr Atmospheric Res, 74-79; sr consult, Control Data Corp, 79-87; SCI CONSULT, W ROY WESSEL & ASSOCS, 87- *Mem:* Soc Indust & Appl Math; Am Math Soc. *Res:* Construction and adaptation of fluid dynamic models in the form of computer programs; development of numerical algorithms and writing of scientific software. *Mailing Add:* 3545 Arthur Ct No 3 Boulder CO 80304

WESSELLS, NORMAN KEITH, b Jersey City, NJ, May 11, 32; c 4. DEVELOPMENTAL BIOLOGY. *Educ:* Yale Univ, BS, 54, PhD(zool), 60. *Prof Exp:* Am Cancer Soc fel, 60-62; from asst prof to assoc prof biol sci, 62-70, chmn dept, 72-78, actg dir, Hopkins Marine Sta, 72-76, assoc dean humanities & sci, 77-81, dean humanities & sci, 81-88, PROF BIOL SCI, STANFORD UNIV, 71- *Concurrent Pos:* Am Cancer Soc scholar cancer res, Dept Biochem, Univ Wash, 68-69; Guggenheim Found fel, 75-76; chmn, Yale Coun Comt Biol Sci. *Honors & Awards:* Herman Beerman Award, Soc Investigative Dermat, 71. *Mem:* Am Soc Zoologists; Soc Develop Biol (pres, 78). *Res:* Embryonic induction; cytodifferentiation; chemistry, ultrastructure of skin, pancreas development; development of nerve cells, axons. *Mailing Add:* 28015 Stonehenge Lane Eugene OR 97402

WESSELS, BRUCE WARREN, b New York, NY, Oct 18, 46; m 68; c 2. MATERIALS SCIENCE, ELECTRONIC MATERIALS. *Educ:* Univ Pa, BS, 68; Mass Inst Technol, PhD(mat sci), 73. *Prof Exp:* Mem tech staff, Gen Elec Res Ctr, 72-77; from asst prof to assoc prof, 77-83, PROF MAT SCI, NORTHWESTERN UNIV, 84- *Concurrent Pos:* Prin investr, NSF grant, DOE grant; consult, NSF, ARO; chmn, Elec Mat Comn, Metall Soc; mem ed bd, J Electronic Mats. *Mem:* Electrochem Soc; Am Inst Mining, Metall & Petrol Eng; Mat Res Soc. *Res:* Semiconductor physics; defects in semiconductors; thin films. *Mailing Add:* Dept Mat Sci Northwestern Univ Evanston IL 60201

WESSENAUER, GABRIEL OTTO, electrical engineering; deceased, see previous edition for last biography

WESSINGER, WILLIAM DAVID, b Honolulu, Hawaii, Nov 8, 51. BEHAVIORAL PHARMACOLOGY, PSYCHOPHARMACOLOGY. *Educ:* Rutgers Univ, BS, 73; Med Col Va, PhD(pharmacol & toxicol), 83. *Prof Exp:* Res asst environ sci, Rutgers Univ, 71-73; teaching asst oceanog, Univ Mass, Amherst, 73-74; res assoc pharmacol, 83-85, instr pharmacol & toxicol, 85-86, ASST PROF PHARMACOL & TOXICOL, UNIV ARK MED SCI, 86- *Mem:* Am Soc Pharmacol & Exp Therapeut; Soc Stimulus Properties Drugs; Int Study Group Investigating Drugs as Reinforcers; Soc Neurosci; Behav Pharmacol Soc; Behav Toxicol Soc. *Res:* Behavioral pharmacology of narcotics, phencyclidine and other psychoactive drugs; discriminative stimulus properties of drugs, tolerance, dependence and drug interactions; drugs of abuse. *Mailing Add:* Dept Pharmacol & Toxicol Univ Ark Med Sci Slot 611 Little Rock AR 72205

WESSLER, MAX ALDEN, b Jacksonville, Ill, Jan 16, 31; m 53; c 3. THERMODYNAMICS, COMBUSTION. *Educ:* Bradley Univ, BSME, 52; Univ Southern Calif, MSME, 54; Purdue Univ, PhD, 66. *Prof Exp:* Mech engr, Guided Missiles Labs, Hughes Aircraft Co, Calif, 52-54; first lieutenant, USAF, 54-56; from asst prof to assoc prof, 56-71, PROF MECH ENG, BRADLEY UNIV, 71-, CHMN DEPT, 74- *Concurrent Pos:* Mem tech staff, Electron Dynamics Dept, Hughes Res Labs, 60-62; mem bd dirs, Soc Automotive Engrs, 76-78; mem bd dirs, Accreditation Bd Eng & Technol, 77-83 & Eng Accreditation Comn, 84-88; chmn, Mech Eng Div, Am Soc Eng Educ, 86-87. *Mem:* Soc Automotive Engrs; Am Soc Eng Educ; Am Soc Mech Engrs. *Res:* Thermodynamics and combustion; fluid mechanics. *Mailing Add:* 923 N Maplewood Ave Peoria IL 61606

WESSLER, STANFORD, b New York, NY, Apr 20, 17; m 42; c 3. MEDICINE. *Educ:* Harvard Univ, BA, 38; NY Univ, MD, 42; Am Bd Internal Med, dipl. *Prof Exp:* Asst med, Harvard Med Sch, 49-51, instr, 51-54, assoc, 54-57, asst prof & tutor, 57-64; prof med, Sch Med, Wash Univ, 64-74, John E & Adeline Simon prof, 66-74; PROF MED & ASSOC DEAN POST-GRAD, MED SCH, NY UNIV, 74- *Concurrent Pos:* Res fel med, Harvard Med Sch, 46-49; Nat Heart Inst trainee, 49-51; Am Heart Asn estab investr, 54-59; James F Mitchell Award heart & vascular res, 72-; asst, Beth Israel Hosp, Boston, 46-49, assoc, 49-56, physician, Vasc Clin, 54-64, assoc vis physician, 57-58, vis physician, 59-64, head, Anticoagulation Clin, 60-64, dir, Clin Res Ctr Thrombosis & Atherosclerosis, Harvard Med Sch, 61-64, assoc dir, 64-; mem, Comt Thrombosis & Hemorrhage, Nat Res Coun, 60-64, Inst Med, 86-88; physician-in-chief, Jewish Hosp, St Louis & assoc physician, Barnes Hosp, 64-75; mem, Med Adv Bd, Coun Circulation, Am Heart Asn, 64- & Coun Stroke, 67-76, vchmn, Coun Thrombosis, 71-74, chmn, 74-76, chmn, Publ Comt, 72-77, vpres coun, 74-76; vchmn heart training comt, Nat Heart Inst, 65-67 & mem, Thrombosis Adv Comt, 67-71; chmn, Subgroup Thromboembolism, Inter-Soc Comn Heart Dis Resources, 69-72; dir, Nat Heart & Lung Inst Thrombosis Ctr, Wash Univ Sch Med, 71-74; mem, Comn Stroke, Nat Inst Neurol Dis & Stroke, 72-74; attend physician, NY Univ Med Ctr, Univ Hosp, New York, 74- & Bellevue Hosp Ctr, 74- *Honors & Awards:* James F Mitchell Award, Heart & Vascular Res, 72; Award of Merit, Am Heart Asn, 78. *Mem:* Am Fedn Clin; Am Soc Hemat; Int Soc Internal Med; Am Soc Clin Invest; fel Am Col Physicians; Am Physiol Soc; Asn Am Physicians. *Res:* Blood coagulation; peripheral vascular disease. *Mailing Add:* Dept Med NY Univ Sch Med 60 Rye Rd Rye NY 10580

WESSLING, RITCHIE A, b Iowa, Sept 15, 32; m 61; c 5. PHYSICAL CHEMISTRY, POLYMER SCIENCE. *Educ:* Mich State Univ, BS, 57, MS, 59; Univ Pa, PhD(phys chem), 62. *Prof Exp:* Chemist, Wyandotte Chem Co, 59; chemist, Polymer Res Lab, Dow Chem Co, 62, assoc scientist, polymer sci group, 68-78, res scientist, Cent Res, Phys Res Lab, 78-80, AG Prod Dept, 80-82, res scientist, 82-87, SR RES SCIENTIST, CENT RES, DOW CHEM CO, 87- *Mem:* Am Chem Soc; AAAS; Am Inst Chemists. *Res:* Relationship between physical properties of polymeric materials and their chemical structure; membranes. *Mailing Add:* Cent Res Dow Chem Co 2800 Mitchell Dr Walnut Creek CA 94598

WESSMAN, GARNER ELMER, bacteriology; deceased, see previous edition for last biography

WESSON, JAMES ROBERT, b Jackson Gap, Ala, Nov 1, 21; m 43; c 4. MATHEMATICS. *Educ:* Birmingham-Southern Col, BS, 49; Vanderbilt Univ, MA, 49, PhD(math), 53. *Prof Exp:* Instr math, Univ Tenn, 49-50; from asst prof to assoc prof, Birmingham-Southern Col, 51-57; from asst prof to assoc prof, Vanderbilt Univ, 57-66, assoc dean, prof math, 61-90; RETIRED. *Concurrent Pos:* Dir undergrad studies, Vanderbilt Univ, 70-77. *Mem:* Math Asn Am. *Res:* Projective planes; abstract algebra; numerical solutions of differential equations. *Mailing Add:* Dept Math Vanderbilt Univ Box 1595 Sta B Nashville TN 37235

WESSON, LAURENCE GODDARD, JR, b Midland, Mich, Oct 18, 17; m 48; c 4. PHYSIOLOGY, INTERNAL MEDICINE. *Educ:* Haverford Col, AB, 38; Harvard Univ, MD, 42. *Prof Exp:* From instr to asst prof physiol, Col Med, NY Univ, 46-50, from instr to assoc prof med, Postgrad Med Sch, 50-62; prof med, Jefferson Med Col, 62-88; RETIRED. *Mem:* Am Physiol Soc; Am Fedn Clin. *Res:* Renal and cellular physiology; renal diseases. *Mailing Add:* Tinker Hill Rd RD 2 Phoenixville PA 19460

WESSON, PAUL STEPHEN, b Nottingham, Eng, Sept 11, 49; Can citizen; m 80; c 2. COSMOLOGY, SOLAR SYSTEM. *Educ:* Univ London, BSc, 71; Univ Cambridge, MSc, 72, PhD(astron), 79. *Prof Exp:* Asst prof physics, Univ Alta, Can, 80-84; assoc prof, 84-88, PROF PHYSICS, UNIV WATERLOO, CAN, 88- *Concurrent Pos:* Vis prof space sci, Univ Calif, Berkeley & Gravity Group, Stanford Univ, 90-91. *Mem:* Int Astron Union; Royal Astron Soc; Am Astron Asn; Can Inst Theoret Astrophys. *Res:* Cosmology; solutions of general relativity; clusters of galaxies; fundamental constants of physics; the solar system and its origin; formation and rotation of earth. *Mailing Add:* Dept Physics Univ Waterloo Waterloo ON N2L 3G1 Can

WESSON, ROBERT LAUGHLIN, b San Francisco, Calif, Feb 26, 44; m 66; c 2. SEISMOLOGY, TECTONOPHYSICS. *Educ:* Mass Inst Technol, SB, 66; Stanford Univ, MS, 68, PhD(geophys), 70. *Prof Exp:* Res assoc, 70-72, chief, Off Earthquake Studies, 78-80, asst dir res, 80-81, asst dir res progs, 81-82, GEOPHYSICIST, US GEOL SURV, 72- *Concurrent Pos:* US chmn, US-USSR Working Group Earthquake Prediction, 78-82. *Mem:* Am Geophys Union; Seismol Soc Am; Geol Soc Am; Earthquake Engr Res Inst; AAAS. *Res:* Earthquake prediction; seismology; tectonophysics. *Mailing Add:* 905 Nat Ctr US Geol Surv Reston VA 22092

WEST, A(RNOLD) SUMNER, b Philadelphia, Pa, Jan 12, 22; m 46; c 2. CHEMICAL ENGINEERING, CHEMICAL PRODUCT SAFETY & REGULATIONS. *Educ:* Univ Pa, BS, 43; Pa State Univ, MS, 46. *Prof Exp:* Asst, Pa State Univ, 43-46; process engr, Rohm & Haas Co, 46-52, process group leader, 52-62, semi-works supt & head res comput lab, 62-72, mgr petrol chem res, 72-77, sr tech specialist govt & regulatory affairs, 78-87; OWNER & PRIN, A S WEST ASSOCS, 87- *Concurrent Pos:* Pres, United Eng Trustees, Inc, 86-87. *Honors & Awards:* Founders Award, Am Inst Chem Engrs, 79, T J Hamilton Mem Award, 82 & F J Van Antwerpen Award, 83. *Mem:* Am Chem Soc; Nat Soc Prof Engrs; Am Inst Chem Engrs (pres, 77); Soc Automotive Engrs; Water Pollution Control Fedn. *Mailing Add:* A S West Assocs 3896 Sidney Rd Huntingdon Valley PA 19006-2347

WEST, ANITA, b New York, NY, Oct 21, 30; m 55; c 2. TECHNOLOGY TRANSFER, OPERATIONS RESEARCH. *Educ:* Univ Denver, BA, 60, MS, 62, PhD(math educ), 69. *Prof Exp:* Eng, Martin Marietta, 60-63; systs anal, IBM, 63-64; res math, Denver Res Instit, 64-88, div head, 78-88, chief oper officer, 86-88; prin scientist, Appl Res Assocs, 88-90; SR DIR, MCREL, 88- *Concurrent Pos:* Instr, civil eng, Univ Denver, 64-68; instr math, Arapahoe Community Col, 70-74; Gov Sci & Adv Coun, Colo, 81-85. *Mem:* Sigma Xi; NY Acad Sci; AAAS. *Res:* Training technology, technology transfer and evaluation with cost benefit analysis; criminal justice research; math evaluation. *Mailing Add:* 3235 S St Paul St Denver CO 80210

WEST, ARTHUR JAMES, II, b Boston, Mass, Dec 14, 27; div; c 3. PARASITOLOGY, MARINE BIOLOGY. *Educ:* Suffolk Univ, BS, 51, MA, 56; Univ NH, MS, 62, PhD(zool), 64. *Prof Exp:* From instr to assoc prof biol, Suffolk Univ, 52-65, prof & co-chmn dept, 65-68; dean div natural sci, New Eng Col, 68-70; chmn dept biol, 70-73, prog dir, Pre-Col Educ in Sci, NSF, 72-73, PROF BIOL, SUFFOLK UNIV, 70-, CHMN DEPT, 78-, DIR, ROBERT S FRIEDMAN COBSCOOK BAY LAB, 75- *Concurrent Pos:* Instr sci & chmn dept, Emerson Col, 56-59; asst prof biol & chmn dept, Mass Col Optom, 57-60; mem, Gov Comn Ocean Mgt, Mass; NSF dir var prog, NH Col & Univ Coun, 68-78 & 78- *Mem:* Am Soc Parasitol; Am Inst Biol Scientists. *Res:* Embryology and histochemistry of larval forms of Acanthocephala; morphology of priapulids; life cycle studies of marine parasites. *Mailing Add:* Suffolk Univ 41 Temple St Boston MA 02114

WEST, BOB, b Ellenville, NY, Mar 7, 31; m 57, 82; c 3. TOXICOLOGY. *Educ:* Union Univ, BS, 52; Purdue Univ, MS, 54, PhD, 56. *Prof Exp:* Vpres, Rosner-Hixson Labs, Chicago, 60-68; dir, sci & regulatory affairs, Vick Chem Co, Mt Vernon, NY, 68-75; PRES, BOB WEST ASSOCS, INC, STAMFORD, CONN, 75- & PRIN, FOOD, DRUG, CHEM SERV, FAIRFAX, VA. *Mem:* Am Soc Pharmacol & Exp Therapeut; Soc Toxicol; Acad Pharmaceut Sci; Drugs Info Asn (past pres); Am Col Clin Pharmacol. *Res:* Pharmaceuticals; chemicals, diagnostics and devices. *Mailing Add:* Food Drug Chem Serv 3771 Center Way Fairfax VA 22033

WEST, BRUCE DAVID, b Madison, Wis, July 10, 35; m 57; c 2. BIOCHEMISTRY. *Educ:* Univ Wis, BS, 57, MS, 61, PhD(biochem), 62. *Prof Exp:* Fel biochem, Univ Wis, 62-63; asst prof chem, Univ NMex, 63-69; asst prof, 69-74, ASSOC PROF CHEM, EASTERN MICH UNIV, 74- *Mem:* Am Chem Soc; The Chem Soc. *Res:* Coumarin anticoagulants; structure-activity relationship, synthesis, biodegradation. *Mailing Add:* Dept Chem Eastern Mich Univ Ypsilanti MI 48197

WEST, CHARLES ALLEN, b Greencastle, Ind, Nov 4, 27; m 52; c 2. PLANT BIOCHEMISTRY. *Educ:* DePauw Univ, AB, 49; Univ Ill, PhD(chem), 52. *Hon Degrees:* DSc, DePauw Univ, 81. *Prof Exp:* Instr chem, Univ Calif, Los Angeles, 52 & 55, from asst prof to assoc prof, 56-67, vchmn dept, 70-75, chmn dept, 81-84, actg dean phys sci, 88, PROF CHEM, UNIV CALIF, LOS ANGELES, 67- *Concurrent Pos:* Guggenheim fel, 61-62. *Mem:* Am Chem Soc; Am Soc Biol Chemists; Am Soc Plant Physiol. *Res:* Chemistry and biosynthesis of natural products of physiological importance, including gibberellins and other plant growth regulators; metabolic regulation; molecular basis of plant disease resistance. *Mailing Add:* Dept Chem & Biochem Univ Calif Los Angeles CA 90024-1569

WEST, CHARLES DAVID, b Riverside, Calif, July 25, 37; m 63; c 3. ANALYTICAL CHEMISTRY. *Educ:* Pomona Col, BA, 59; Mass Inst Technol, PhD(analytical chem), 64. *Prof Exp:* From res chemist to sr res chemist, Beckman Instruments, Inc, 64-67; from asst prof to assoc prof, 67-80, chmn dept, 80-83, PROF CHEM, OCCIDENTAL COL, 80- *Mem:* AAAS; Am Chem Soc; Soc Appl Spectros. *Res:* Emission spectroscopy; flame photometry; atomic absorption instrument design; atomic and molecular fluorescence instrumentation. *Mailing Add:* Dept Chem Occidental Col 1600 Campus Rd Los Angeles CA 90041-3397

WEST, CHARLES DONALD, b Ogden, Utah, Oct 25, 20; m 46; c 4. BIOCHEMISTRY. *Educ:* Univ Utah, BA, 41, MD, 44, PhD, 50. *Prof Exp:* Instr med, Med Col, Cornell Univ, 50-54, res assoc, Sloan-Kettering Div, 52-53, asst prof, 53-57; from asst res prof to assoc res prof biochem, 57-71, assoc prof med, 65-69, PROF MED, MED CTR, UNIV UTAH, 69-, PROF BIOCHEM, 71-, CO-DIR, CLIN RES CTR, 66- *Concurrent Pos:* From asst to assoc, Sloan-Kettering Inst Cancer Res. 50-57; asst dir prof serv res & assoc chief staff, Vet Admin Hosp, Salt Lake City, 57-65. *Mem:* AAAS; Endocrine Soc; Asn Cancer Res; Fedn Clin. *Res:* Harvey Soc. Res: Endocrinology. *Mailing Add:* 50 N Medical Dr Salt Lake City UT 84132

WEST, CHARLES HUTCHISON KEESOR, b Wheeling, WVa, Aug 9, 48; m 72; c 2. PHYSIOLOGY, NEUROPHYSIOLOGY. *Educ:* Ohio Univ, BS, 70, MS, 72; Mich State Univ, PhD(physiol), 77. *Prof Exp:* Res asst zool, Ohio Univ, 71, teaching asst, 71-72; res asst physiol, Mich State Univ, 72-77; fel neurophysiol, Univ Wis, 77-80; RES SCIENTIST, GA MENT HEALTH INST, 80-; ASST PROF, DEPT PSYCHIAT, EMORY UNIV, 80- *Mem:* AAAS; Am Soc Zool; Soc Neurosci. *Res:* Basic motivation and attention controlling brain mechanisms and their relationship to mental illness; electrophysiology and intracranial self-stimulation paradigm with various physiological, pharmacological and behavioral manipulations. *Mailing Add:* Dept Educ Psychiat Emory Univ 1364 Clifton Rd NE Atlanta GA 30322

WEST, CHARLES P, organic chemistry; deceased, see previous edition for last biography

WEST, CHARLES PATRICK, b Minneapolis, Minn, June 13, 52. FORAGE PHYSIOLOGY & QUALITY. *Educ:* Univ Minn, BS, 74, MS, 78; Iowa State Univ, PhD(agron), 81. *Prof Exp:* Res fel, Ministry Agr & Fisheries, NZ, 82-84; asst prof, 84-90, ASSC PROF, UNIV ARK, 90- *Mem:* Am Soc Agron; Am Forage & Grassland Coun; Crop Sci Soc Am; Brit Grassland Soc. *Res:* Physiological and pest interactions in endophyte infected grasses. *Mailing Add:* Univ Ark 276 Altheimer Dr Fayetteville AR 72703-5586

WEST, CHRISTOPHER DRANE, b Norfolk, Va, April 29, 43; m 74; c 3. NEUROANATOMY. *Educ:* Brown Univ, AB, 64; Boston Univ, AM, 66, PhD(exp & physiol psychol), 73. *Prof Exp:* NIMH trainee psychobiol, Worcester Found, 72-75; researcher psychol, Vet Admin Hosp, Mass, 75-82; ASST PROF ANAT & NEUROSCI, SCH MED, BOSTON UNIV, 78- *Concurrent Pos:* NIH prin investr, Div Dept Neurol, Harvard Med Sch, 76-77; lectr neurol, Harvard Neurol Unit, Beth Israel Hosp, 75- & Harvard Med Sch, 75-; psychologist & behav consult, Dementia Study Unit, Rodger's Vet Admin Hosp, 85- *Mem:* Am Asn Anatomists; AAAS; Am Psychol Asn; Asn Adv Behav Theory; Geront Soc Am. *Res:* Comparative anatomy of the auditory system and comparative auditory behavior; brain changes in Down's Syndrome with age; operant behavior management in Alzheimeas disease; relation of dietary restriction to aging, receptor function and neuron morphology. *Mailing Add:* Ten Remington St Cambridge MA 02138

WEST, CLARK DARWIN, b Jamestown, NY, July 4, 18; m 44; c 3. IMMUNOLOGY, NEPHROLOGY. *Educ:* Col Wooster, AB, 40; Univ Mich, MD, 43. *Prof Exp:* Intern surg, Univ Mich Hosp, 43-44, resident pediat, 44-46; from asst prof to assoc prof, 51-62, PROF PEDIAT, COL MED, UNIV CINCINNATI, 62- *Concurrent Pos:* Children's Hosp Res Found scholar, 48-49, fel, 53-; Nat Res Coun sr fel pediat, Children's Hosp Res Found, 49-50 & Cardiopulmonary Lab, Bellevue Hosp, New York, 50-51; res assoc, Children's Hosp Res Found, 51-53, assoc dir, 63-, supv biochemist, Hosp, 51-65, attend pediatrician, 51-; attend pediatrician, Cincinnati Gen Hosp, 53-; mem, Gen Clin Res Ctr Comt, Div Res Facilities & resources, NIH, 65-69; mem, Urol & Renal Dis Training Comt, Nat Inst Arthritis, Metab & Digestive Dis, 72-73. *Mem:* Am Physiol Soc; Soc Pediat Res (secy-treas, 58-62, pres, 63-64); Am Soc Nephrology; Am Asn Immunologists; Am Pediat Soc. *Res:* Abnormalities of the immune system related to the pathogenesis of glomerulonephritis. *Mailing Add:* Children's Hosp Res Found Elland Ave & Bethesda Cincinnati OH 45229

WEST, COLIN DOUGLAS, b Rochdale, Eng, June 21, 41; m 65; c 1. CLASSICAL PHYSICS. *Educ:* Univ Liverpool, BSc, 61, MSc, 64, PhD(physics), 65. *Prof Exp:* Res fel, Res Lab, Harwell, 65-68; sci officer, UK Atomic Energy Authority, 68-73; vis scientist, Oak Ridge Nat Lab, 73-74; proj mgr physics, UK Atomic Energy Res Lab, 74-77; res assoc planning, 77-80, mgr & systs technol group, 81-85, DIR ADVAN NEUTRON SCI PROJ, OAK RIDGE NAT LAB, 86- *Concurrent Pos:* Nuclear physics res fel, Univ Liverpool, 63-65. *Honors & Awards:* Jointly Awarded Sci, Educ & Mgt Premium, Inst Elec Engrs, London, 75. *Res:* Application of long range planning techniques to laboratory and research organizations; small scale heat engines for irrigation and other applications; irradiation engineering. *Mailing Add:* Rte 3 Box 262A Oliver Springs TN 37840

WEST, DAVID ARMSTRONG, b Beirut, Lebanon, Apr 9, 33; US citizen; m 58; c 3. GENETICS. *Educ:* Cornell Univ, BA, 55, PhD(vert zool), 59. *Prof Exp:* Asst prof zool, Cornell Univ, 59-60; NATO fel, 60-61; USPHS fel, 61-62; asst prof, 62-68, ASSOC PROF ZOOL, VA POLYTECH INST & STATE UNIV, 68- *Concurrent Pos:* Sci Res Coun sr vis fel, 66; ed, Va J Sci, 74-76. *Mem:* Lepidopterist Soc; Lepidoptera Res Found; Soc Study Evolution; Am Soc Naturalists. *Res:* Ecological genetics; genetics of natural populations of butterflies; polymorphisms. *Mailing Add:* Dept Biol Va Polytech Inst & State Univ Blacksburg VA 24061

WEST, DENNIS R, b Kennett, Mo, Nov 22, 46; m 69; c 1. CROP BREEDING, QUANTITATIVE GENETICS. *Educ:* Miss State Univ, BS, 69, MS, 75; Univ Nebr, PhD(agron), 78. *Prof Exp:* Res assoc, NC State Univ, 78-79; asst prof, 79-84, ASSOC PROF PLANT & SOIL SCI, UNIV TENN, 84- *Mem:* Am Soc Agron; Crop Sci Soc Am; AAAS; Coun Agr Sci & Technol. *Res:* Basic and applied research in plant breeding and genetics; quantitative genetics; maize breeding; host plant resistance; development of maize germplasm. *Mailing Add:* Dept Plant Sci Univ Tenn Knoxville TN 37996

WEST, DONALD K, b Providence, RI, May 14, 29; m 57; c 3. ASTROPHYSICS. *Educ:* Univ RI, BS, 57; Rutgers Univ, MS, 60; Univ Wis, PhD(astron), 64. *Prof Exp:* ASTROPHYSICIST, GODDARD SPACE FLIGHT CTR, NASA, 64- & OBS ADMINR, 78- *Mem:* Am Astron Soc. *Res:* Physics of emission line stars; astronomical observations from space telescopes. *Mailing Add:* Code 672 Goddard Space Flight Ctr Greenbelt MD 20715

WEST, DONALD MARKHAM, b Pasadena, Calif, Apr 22, 25; m 48. ANALYTICAL CHEMISTRY. *Educ:* Stanford Univ, BS, 49, PhD(chem), 58. *Prof Exp:* Actg instr chem, Stanford Univ, 54-55; from asst prof to assoc prof, 56-65, PROF CHEM, SAN JOSE STATE UNIV, 65- *Mem:* Am Chem Soc. *Res:* Analysis of organic compounds. *Mailing Add:* Dept Chem San Jose State Univ 125 S Seventh St San Jose CA 95192-0101

WEST, DOUGLAS BRENT, b Queens, NY, Nov 14, 53; m 86. GRAPH THEORY, OPTIMIZATION. *Educ:* Princeton Univ, AB, 74; Mass Inst Technol, PhD(math), 78. *Prof Exp:* Vis lectr & res asst, dept comput sci, Stanford Univ, 78-79; asst prof, dept math, Princeton Univ, 79-82; asst prof, dept math & asst res prof, Coord Sci Lab, 82-85, ASSOC PROF, DEPT MATH & ASSOC RES PROF, COORD SCI LAB, UNIV ILL, 85- *Concurrent Pos:* Co-prin investr, Off Naval Res res grant, 85-91; prin investr, NSF res grant, 85-87 & NSA res grant, 90-; assoc ed, Am Math Monthly, 86-; mem res staff, IDA/CCR, Univ Calif, Berkeley, 87-88, vis assoc prof, comp sci, 89-90. *Mem:* Math Asn Am; Soc Indust & Appl Math; Asn Comput Mach. *Res:* Graph theory; partially ordered sets; extremal, structural and algorithmic questions involving discrete mathematical and computational structures. *Mailing Add:* Dept Math Univ Ill 1409 W Green St Urbana IL 61801

WEST, DOUGLAS XAVIER, b Tacoma, Wash, June 11, 37; m 64; c 2. INORGANIC CHEMISTRY. *Educ:* Whitman Col, AB, 59; Wash State Univ, PhD(chem), 64. *Prof Exp:* Instr chem, Upsala Col, 64-65; from asst prof to prof, Cent Mich Univ, 65-75, dir univ honors progs, 70-72; chmn dept, 75-86, PROF INORG CHEM, ILL STATE UNIV, 86- *Concurrent Pos:* Vis prof, Univ Leicester, 72, 74 & 81 , Wash State Univ, 80, Univ Poona, 90; Indo Am fel, 90. *Mem:* Am Chem Soc; Sigma Xi; Royal Soc Chem. *Res:* Transition metal complexes of n-oxides, thiosemicarbazones and thioureas; electron spin resonance and other forms of spectroscopy. *Mailing Add:* Dept Chem Ill State Univ Normal IL 61761

WEST, ERIC NEIL, b Montreal, Que, Mar 28, 41; m 65; c 2. STATISTICS, COMPUTER SCIENCE. *Educ:* Royal Mil Col Can, BSc, 63; Iowa State Univ, MS, 67, PhD(statist), 70. *Prof Exp:* Res assoc statist & comput, Iowa State Univ, 67-70; asst prof comput sci, Univ Alta, 70-72; assoc prof statist & comput sci, 72-73, assoc prof quant methods, Sir George Williams Campus, Concordia Univ, 73-; AT DEPT BUS ADMIN, UNIV WINDSOR, CAN. *Concurrent Pos:* Consult, Pro Data Serv, 70-; pvt consult, 70- *Mem:* Am Statist Asn; Inst Math Statist; Sigma Xi. *Res:* Statistical inference; computation systems analysis; business decision making; forecasting. *Mailing Add:* Dean Bus Admin Univ Windsor Windsor ON N9B 3P4 Can

WEST, FELICIA EMMINGER, b Chicora, Pa, Sept 14, 26; m 48; c 2. SCIENCE EDUCATION. *Educ:* J B Stetson Univ, BS, 48; Univ Fla, MEd, 65, EdD(sci educ), geol), 71. *Prof Exp:* Teacher high sch, Fla, 60-64; instr physics & earth sci, Miami Dade Jr Col, 65-66; instr earth sci & phys sci, St Johns River Jr Col, 66-67; teacher & student gen sci, Lab Sch, Univ Fla, 67-69, teacher & asst prof gen sci & earth sci, 69-72; staff assoc, AAAS, 72-75; CHMN DIV NATURAL SCI, MATH & PHYS EDUC, FLA JR COL, S CAMPUS, 75- *Concurrent Pos:* Coord ed, Fla Asn Sci Teachers J, 75-; mem bd dirs, Fedn Unified Sci Educ. *Res:* Development of unified science curriculum materials for use at community college level; development of field guides to specific sites in Florida for use by secondary and community college instructors. *Mailing Add:* Dept Math & Sci Fla Community Cols 11901 Beach Blvd Jacksonville FL 32216

WEST, FRED RALPH, JR, b Pittsburgh, Pa, June 7, 25. PHARMACOLOGY. *Educ:* Hampton Inst, BS, 47; Tuskegee Inst, MS, 48; Univ Chicago, PhD(pharmacol), 56; Howard Univ, MD, 63. *Prof Exp:* Instr chem, St Augustine's Col, 48-50; asst, George W Carver Found, 50-53; instr pharmacol, 56-65, ASST PROF PHARMACOL, SCH MED, HOWARD UNIV, 65- *Mem:* AAAS. *Res:* Cultivation of animal and plant cells in vitro. *Mailing Add:* 2251 Pimmit Dr No 405 Falls Church VA 22043-2813

WEST, GEORGE CURTISS, b Newton, Mass, May 13, 31; c 4. PHYSIOLOGICAL ECOLOGY, ORNITHOLOGY. *Educ:* Middlebury Col, BA, 53; Univ Ill, Urbana, MS, 56, PhD(physiol ecol), 58. *Prof Exp:* Fel div biosci, Nat Res Coun Can, 59-60; asst prof zool, Univ RI, 60-63; from asst prof to prof zoophysiol, Inst Arctic Biol, Univ Alaska, 63-83, actg dean, Col Biol Sci & Renewable Resources, 74-75, actg dir, Div Life Sci & Inst Arctic Biol, 74-77, dir Biomed Ctr, 72-80, prof zoophysiol, 80-83; RETIRED. *Concurrent Pos:* Alexander von Humboldt Found fel, Aschoff Div, Max Planck Inst Physiol of Behav, 71-72; mem US-USSR bilateral exchange working group protection northern ecosysts, 75- *Mem:* Am Ornith Union; Am Physiol Soc; Ecol Soc Am; Wildlife Soc; Wilson Ornith Soc. *Res:* Bioenergetics and temperature regulation of birds; migration, fat deposition, food habits of birds; fatty acid analysis of plant and animal lipids. *Mailing Add:* PO Box 841 Homer AK 99603

WEST, GORDON FOX, b Toronto, Ont, Apr 21, 33. GEOPHYSICS. *Educ:* Univ Toronto, BASc, 55, MA, 57, PhD(geophys), 60. *Prof Exp:* Geophysicist, Dom Gulf Co, 55-56; lectr geophys, 58-66, assoc prof, 66-72, PROF PHYSICS, UNIV TORONTO, 72- *Mem:* Soc Explor Geophysicists; Am Geophys Union; Geol Asn Can; Can Asn Physicists. *Res:* Electromagnetic geophysical methods; geophysical studies of precambrian shields. *Mailing Add:* Dept Physics Univ Toronto 60 St George St Toronto ON M5S 1A7 Can

WEST, HARRY IRWIN, JR, b Foley, Ala, Dec 3, 25; m 56; c 4. SPACE PHYSICS, NUCLEAR PHYSICS. *Educ:* Auburn Univ, BS, 46, MS, 47; Stanford Univ, PhD(physics), 55. *Prof Exp:* PHYSICIST, LAWRENCE LIVERMORE NAT LAB, UNIV CALIF, 55- *Mem:* Fel Am Phys Soc; Am Geophys Union. *Res:* Nuclear spectroscopy and measurements of charged particles in the earth's radiation belts. *Mailing Add:* L-233 Lawrence Livermore Nat Lab Univ Calif Livermore CA 94550

WEST, HERSCHEL J, b Portsmouth, Va, Nov 13, 37; m 63; c 2. HIGH VOLTAGE PHENOMENA. *Educ:* Brigham Young Univ, BS, 63; Kinsington Univ, MS, 80; Donsbach Univ, PhD(nutrit), 82. *Prof Exp:* Eng officer, USN, 63-66; ELEC ENGR, DIV LABS, BONNEVILLE POWER ADMIN, 66- *Concurrent Pos:* Consult unexplained elec flashovers, 75-, res consult new human growth factor, 82- *Res:* High voltage phenomena; unexplained outages. *Mailing Add:* 3801 Edgewood Dr Vancouver WA 98661

WEST, JAMES EDWARD, b Grinnell, Iowa, May 1, 44; m 65; c 2. MATHEMATICS. *Educ:* La State Univ, BS, 64, PhD(math), 67. *Prof Exp:* Asst prof math, Univ Ky, 68-69; from asst prof to assoc prof, 69-76, PROF MATH, CORNELL UNIV, 76- *Concurrent Pos:* Mem, Inst Advan Study, 67-68, 77-78 & 84-85, Math Sci Res Inst, 85; vis prof, La State Univ, 72, Univ KY, 80; lectr, Stefan Bunach Inst, Warsaw, 74; exchange scientist, Nat Acad Sci, USSR & Poland, 78; NSF res grant, 70-92; invited address, Int Cong Math, 78. *Mem:* Am Math Soc; NY Acad Sci; AAAS. *Res:* Topology of infinite-dimensional spaces and manifolds; geometric and point-set topology. *Mailing Add:* Dept Math White Hall Cornell Univ Ithaca NY 14853

WEST, JERRY LEE, b North Wilkesboro, NC, Nov 1, 40; m 65; c 2. FISH BIOLOGY, ZOOLOGY. *Educ:* Appalachian State Univ, BS, 62; NC State Univ, MS, 65, PhD(zool), 68. *Prof Exp:* From asst prof to assoc prof, 67-90, actg head dept biol, 74-77 & 88-89, PROF BIOL, WESTERN CAROLINA UNIV, 90- *Mem:* Am Fisheries Soc; Am Inst Fishery Res Biologists. *Res:* Ecology of fishes in Southern Appalachian streams. *Mailing Add:* Dept Biol Western Carolina Univ Cullowhee NC 28723

WEST, JOHN B(ERNARD), b Elliott, Iowa, Feb 23, 25; m 47; c 4. CHEMICAL ENGINEERING, NUCLEAR ENGINEERING. *Educ:* Iowa State Univ, BS, 48, PhD, 54. *Prof Exp:* Jr engr, Gen Elec Co, NY, 48-50; from asst prof to prof, Sch Chem Eng, Okla State Univ, 54-76; Mgr, Black Fox Sta Proj, Pub Serv Co Okla, 81-83, mgr, Black Fox Sta Eng, 76-81, dir fuels, 83-89; RETIRED. *Concurrent Pos:* Consult, 89- *Mem:* Am Inst Chem Engrs. *Res:* Fossil fuel price and availability forecasting; project management. *Mailing Add:* 7901 S Yukon Tulsa OK 74132-2644

WEST, JOHN B, b Adelaide, SAustralia, Dec 27, 28; US citizen. PHYSIOLOGY, MEDICINE. *Educ:* Univ Adelaide, BS & MB, 52, MD, 58, DSc, 80; Univ London, PhD(appl physiol), 60. *Hon Degrees:* Dr, Univ Barcelona, 88. *Prof Exp:* Res assoc respiratory physiol, Royal Postgrad Med Sch, London, 54-60; physiologist, Himalayan Sci & Mountaineering Exped, 60-61; asst prof physiol, Univ Buffalo, 61-62; lectr med, Royal Postgrad Sch, London, 63-68; PROF MED, UNIV CALIF, SAN DIEGO, 69- *Concurrent Pos:* Mem, Cardiovasc Study Sect, NIH, 71-75, chmn, 73-75; chmn fac, Univ Calif, San Diego Sch Md, 72-73; mem, physiol comt, Nat Bd Med Examrs, 73-76, life sci adv comt, NASA, 85-86, comt space biol & med, Nat Acad Sci, 86- & Int Union Physiol Sci, 84-87. *Honors & Awards:* Ernst Jung Prize Med, 77; George C Griffith lectr, Am Heart Asn, 78, Dickinson W Richards lectr, 80; I S Ravdin lectr, Am Col Sur, 82; Telford Mem lectr, Manchester Univ, UK, 83; Harry G Armstrong Lectr, Aerospace Med Asn, 84; J Burns Amberson lectr, Am Thoracic Soc, 84; Suzanne Kronheim lectr, Undersea Med Soc, 84; Orr Reynolds Prize, Am Physiol Soc, 87. *Mem:* Am Physiol Soc; Am Soc Clin Invest; Am Thoracic Soc; Brit Physiol Soc; Royal Col Physicians London; Am Soc Gravitation & Space Biol; Asn Am Physicians; Asn Chmn Dept Physiol; Fleischner Soc (pres, 85); Royal Geog Soc. *Res:* Respiratory function in health and disease. *Mailing Add:* Dept Med Univ Calif San Diego La Jolla CA 92093

WEST, JOHN M(AURICE), b Long Branch, NJ, June 10, 27; m 51; c 2. FERMENTATION, BIOCHEMICAL ENGINEERING. *Educ:* Columbia Univ, BA, 50, BS, 51, MS, 54. *Prof Exp:* Res chem engr, E R Squibb Div, Olin Mathieson Chem Corp, 53-60; prod mgr enzymes, Nopco Chem Co, 60-63; group leader, Hoffmann-La Roche, Inc, 63-77, dir fermentation process develop, 77-80, dir biotechnol, 80-85; mgr biotechnol, John Brown Engrs & Constructors, Stamford, Conn, 85-86; SR PROJ MGR, AM CYANAMID AGR R&D, PRINCETON, NJ, 86- *Mem:* Am Soc Microbiol; Am Inst Chem Engrs; Am Chem Soc. *Res:* Fermentation process development and engineering; continuous sterilization and automatic foam control; automatic chemical analysis and control; pH control; deeptank and semi-solid fermentations; isolation recovery, purification of fermentation and natural products; biotechnology process evaluations; process design, project management; plant trouble shooting and manufacturing cost reduction. *Mailing Add:* 87 Cooper Ave Montclair NJ 07043

WEST, JOHN M, b Stillwell, Okla, Jan 18, 20. NUCLEAR ENGINEERING. *Educ:* Northeastern Okla State Univ, BS, 35; Univ Iowa, MS, 41. *Prof Exp:* Assoc dir, Reactor Eng Div, Argonne Nat Lab, 49-57; exec vpres, Gen Nuclear Eng Corp, 57-65; sr vpres, Combustion Eng Inc, 65-85; RETIRED. *Honors & Awards:* Charter A Coffin, Gen Elec Corp, 49; Walter H Zinn, Am Nuclear Soc, 82. *Mem:* Nat Acad Eng; Am Nuc Soc. *Mailing Add:* 1608 SE 40th Terr Cape Coral FL 33904

WEST, JOHN WYATT, b Decaturville, Tenn, Oct 18, 23; m 47; c 3. POULTRY NUTRITION. *Educ:* Univ Tenn, BSA, 47, MS, 48; Purdue Univ, PhD(poultry nutrit), 51. *Prof Exp:* Asst dir feeds res, Security Mills, Inc, Tenn, 51; from assoc prof to prof poultry husb, Miss State Univ, 52-56; prof poultry sci & head dept, Okla State Univ, 56-68; ASSOC DEAN, SCH AGR, CALIF POLYTECH STATE UNIV, SAN LUIS OBISPO, 68- *Mem:* Poultry Sci Asn. *Res:* Arsenic compounds and vitamin-amino acid interrelationships in poultry nutrition; nutritional value of cottonseed meal in broiler and turkey rations; antibiotic-protein interrelationships in broiler rations. *Mailing Add:* Dept Animal Sci Calif Polytech State Univ San Luis Obispo CA 93407

WEST, KEITH P, b Simla, Colo, Aug 20, 20; m 46; c 3. ZOOLOGY, RADIATION BIOLOGY. *Educ:* Chico State Col, AB, 42; Stanford Univ, MA, 48. *Prof Exp:* Instr biol & chem, Vallejo Col, 47-48; from instr to prof biol sci, Drexel Univ, 48-86; RETIRED. *Concurrent Pos:* Asst dean col eng & sci, Drexel Univ, 67-68, actg head dept biol sci, 71-72, & 83; consult. *Mem:* AAAS; Health Physics Soc. *Res:* Biological effects of radiation; sanitary quality control of food products. *Mailing Add:* 2803 Dogwood Lane Broomare PA 19008

WEST, KENNETH CALVIN, b Broken Bow, Nebr, Apr 1, 35; m 60; c 3. ANALYTICAL CHEMISTRY. *Educ:* Wheaton Col, Ill, BS, 56; Ind Univ, PhD(anal chem), 67. *Prof Exp:* Asst prof, 67-75, ASSOC PROF CHEM, ST LAWRENCE UNIV, 75- *Mem:* Am Chem Soc. *Res:* Instrumentation. *Mailing Add:* Dept Chem St Lawrence Univ Canton NY 13617

WEST, KEVIN J, TRACE METAL ION DETERMINATION, STUDIES METAL-LIGAND COMPLEXES. *Educ:* Ripon Col, BA, 80; Univ Iowa, PhD(anal chem), 86. *Prof Exp:* Instr, 84-86, ASST PROF ANAL CHEM, IND UNIV-PURDUE UNIV, FT WAYNE, 87- *Mem:* Am Chem Soc; Sigma Xi. *Res:* Studies involving substituted hydrazone ligands as complexing agents for the ultraviolet, visible or fluorimetric determination of trace metal ions; use of computers as teaching aids in general and analytical chemistry. *Mailing Add:* Dept Chem Univ Wis Whitewater Whitewater WI 53190

WEST, KEVIN JAMES, COMPLEX FORMATION. *Educ:* Ripon Col, BA, 80; Univ Iowa, PhD(chem), 86. *Prof Exp:* Instr anal chem, Ind Univ-Purdue Univ, Ft Wayne, 84-86, asst prof, 86-89; ASST PROF ANAL CHEM, UNIV WIS-WHITEWATER, 89- *Mem:* Am Chem Soc; Sigma Xi. *Res:* Ultraviolet-visible spectrophotometry of metal-ligand complexes for the determination of metal ion concentration; adaptation and development of instrumental measurements to undergraduate education. *Mailing Add:* Chem Dept Univ Wis 800 W Main St Whitewater WI 53190

WEST, LOUIS JOLYON, b New York, NY, Oct 6, 24; m 44; c 3. PSYCHIATRY, NEUROLOGY. *Educ:* Univ Minn, BS, 46, MB, 48, MD, 49; Am Bd Psychiat & Neurol, dipl, 54. *Prof Exp:* Intern med, Univ Hosps, Univ Minn, 48-49; asst psychiat, Med Col, Cornell Univ, 49-52; prof psychiat & head dept psychiat, neurol & behav sci, Sch Med, Univ Okla, 54-69; PROF PSYCHIAT & CHMN DEPT, SCH MED, UNIV CALIF, LOS ANGELES, 69-, MED DIR, NEUROPSYCHIAT INST, 69-, PSYCHIATRIST IN CHIEF, UNIV CALIF HOSPS & CLINS, 69- *Concurrent Pos:* Resident, Payne Whitney Clin, New York Hosp, 49-52; res coordr, Okla Alcoholism Asn & chief behav sci, Okla Med Res Found, 56-69; consult, Oklahoma City Vet Admin Hosp, 56-69, USAF Hosp, Tinker AFB, Okla, 56-66,USAF Aero-Space Med Ctr, 61-66 & Peace Corps, 62-63; nat consult, Surgeon Gen, USAF, 57-62, mem adv coun, Behav Sci Div, Air Force Off Sci Res, 56-58; mem prof adv coun, Nat Asn Ment Health, 59-64; consult, US Info Agency, 60-61; mem exec coun adv comt behav res, Nat Acad Sci-Nat Res Coun, 61-63; mem nat adv comt psychiat, neurol & psychol, Spec Med Adv Group, US Vet Admin; mem, Nat Adv Ment Health Coun, NIMH, 65-69, White House Conf Civil Rights, 66 & Nat Adv Comt Alcoholism, Dept HEW; consult ed, Med Aspects Human Sexuality, 67-; mem bd dir, Kittay Found, 72-; mem residency rev comt psychiat & neurol, AMA, 73-; mem adv panel res & develop, US Army, 74- *Mem:* Fel Am Col Neuropsychopharmacol; fel Am Col Psychiat; fel Am Psychiat Asn; Pavlovian Soc NAm (pres, 74-75); Soc Biol Psychiat. *Res:* Experimental psychopathology, especially relating to disturbances of perception and altered states of consciousness; psychophysiological correlates in clinical practice; alcohol and drug abuse; life-threatening behavior; interaction of biological, psychological and sociocultural factors in personality development and function. *Mailing Add:* Neuropsychiat Inst Univ Calif Los Angeles CA 90024

WEST, MARTIN LUTHER, b Waco, Tex, Dec 25, 36; m 68; c 2. RADIATION CHEMISTRY, RADIATION PHYSICS. *Educ:* Univ Tex, BS, 60, MA, 62, PhD(physics), 67. *Prof Exp:* SR RES SCIENTIST PHYSICS, PAC NORTHWEST LAB BATTELLE NORTHWEST, 67- *Mem:* AAAS; Radiation Res Soc. *Res:* Reaction mechanisms and fast kinetic measurements for charged particle impact; pulsed radioluminescence and UV spectroscopy instrumentation. *Mailing Add:* 1628 Davison Ave Richland WA 99352

WEST, MIKE HAROLD, b Lewiston, Idaho, Feb 29, 48; m 73; c 2. ANALYTICAL CHEMISTRY. *Educ:* Western Wash State Col, BA, 70; Wash State Univ, PhD(chem), 76. *Prof Exp:* Res assoc, Univ New Orleans, 75-76; asst prof chem, Tougaloo Col, 76-77; res assoc, Tex A&M Univ, 77-78; sr chemist, Coors Spectro Chem Lab, Coors Porcelain Co, 78-81; sr chemist, energy systs group, Rockwell Hanford Opers, 81-86; STAFF MEM, LOS ALAMOS NAT LAB, 86- *Mem:* Am Chem Soc; Soc Appl Spectros. *Res:* Use of molten salts for the processing of octinide elements research into mechanisms for the purification and recovery of actinides with molten salt systems. *Mailing Add:* 116 Aragon St Los Alamos NM 87544

WEST, NEIL ELLIOTT, b Portland, Ore, Dec 17, 37; m 63; c 1. ECOLOGY. *Educ:* Ore State Univ, BS, 60, PhD(plant ecol), 64. *Prof Exp:* From asst prof to assoc prof plant ecol, Dept Range Sci & Ecol Ctr, 64-75, PROF RANGE SCI, UTAH STATE UNIV, 75- *Concurrent Pos:* Forest ecologist, Ore Forest Res Lab, Ore State Univ, 63; NSF fel & vis prof, Inst Ecol, Univ Ga, 70-71; Mellon vis lectr, Yale Sch Forestry & Environ Studies, 78-79; vis scholar, Adelarde Univ, 85; vis prof, Hebrew Univ, 86; consult, Nat Park Serv, Occidental Petrol, Argonne Nat Lab & Environ Protection Agency. *Honors & Awards:* W R Chapline Res Award, Soc Range Mgt, 91. *Mem:* Int Asn Ecol; Ecol Soc Am; Brit Ecol Soc; Int Asn Veg Sci; Int Asn Landscape Ecol. *Res:* Plant ecology theory and its application to wildland resource management, particularly desertification, synecology and soil-vegetation relationships; systems ecology; community structure succession, productivity, nutrient cycling in arid, semi-arid, riparian and woodland ecosystems. *Mailing Add:* Dept Range Sci Utah State Univ UMC 52 Logan UT 84322-5230

WEST, NORMAN REED, b Oak Park, Ill, Aug 13, 43; m 68; c 3. NEUROBIOLOGY, PSYCHOPHARMACOLOGY. *Educ:* Judson Col, BA, 67; Thomas Jefferson Univ, PhD(pharmacol), 74. *Prof Exp:* ASST PROF PATH & ANAT, STATE UNIV NY, HEALTH SCI CTR, SYRACUSE, NY, 77. *Concurrent Pos:* PMAF fel, Dept Anat & Neurobiol, Sch Med, Wash Univ, Mo, 73-75, NIH fel, 75-77; pres res award, Psychopharmacol Prog, Am Soc Pharmacol & Exp Therapeut. *Mem:* Soc Neurosci; Am Asn Anat; Soc Exp Neuropath. *Res:* Neurobiology of cryogenic lesions in the central nervous system; tissue culture modelling of neuropathologic lesions in human tissue; response of neurons to endogenous and exogenous agents in vivo and in tissue culture. *Mailing Add:* Dept Path 766 Irving Ave Syracuse NY 13210

WEST, PHILIP WILLIAM, b Crookston, Minn, Apr 12, 13; m 35, 64; c 3. ENVIRONMENTAL SCIENCES. *Educ:* Univ NDak, BS & MS, 35, 36; Univ Iowa, PhD(chem), 39. *Hon Degrees:* DSc, Univ NDak, 58. *Prof Exp:* Asst chemist, State Geol Surv, NDak, 35-36; asst sanit chem, Univ Iowa, 36-37; asst chemist, State Dept Health, Iowa, 37-40; res chemist & microchemist, Econ Lab, Inc, Minn, 40; from instr to prof, 40-53, Boyd prof, 53-80, EMER BOYD PROF CHEM, LA STATE UNIV, BATON ROUGE, 80-; CHMN & DIR, WEST-PAINE LABS, BATON ROUGE, 80- *Concurrent Pos:* Smith lectr, Okla State Univ, 55; consult, Ethyl Corp, A D Little Co & USPHS; consult, Kem-Tech Labs, Inc & chmn bd, 66-73; ed, Analytica Chimica Acta, 58-77; co-ed, Sci Total Environ, 73-77; mem working party 1, Sci Comt Probs Environ, 71-74; adj prof, Environ Protection Agency; mem chem panel, NSF; pres comt new reactions, Int Union Pure & Appl Chem & pres anal chem div, 66-70, mem sect toxicol & indust hyg, 71-73; mem study sect, USPHS, 60-65; vis prof, Rand Afrikaans Univ, SAfrica, 80, Univ Colo, 63; mem, tech adv comt La Air Pollution Control Comm, 79-85; mem, Gov Task Force Environ Health, 83-85. *Honors & Awards:* Southwest Award, Am Chem Soc, 54; Coates Award, 67; Fisher Award, 74 & Creative Advan Environ Sci & Technol Award, Am Chem Soc, 81. *Mem:* Am Chem Soc; hon mem Brit Soc Anal Chem; hon mem Austrian Asn Microchem & Anal Chem; hon men Japanese Soc Anal Chem. *Res:* Water treatment and analysis; polarized light microscopy; spot tests; organic reagents; complex ions; analysis of petroleum; polarography; chromatography; high frequency titrations; inorganic extractions; catalyzed and induced reactions; air pollution; industrial hygiene; personal monitors. *Mailing Add:* West-Paine Labs Inc 7979 GSRI Ave Baton Rouge LA 70820

WEST, RICHARD LOWELL, b Quincy, Fla, Mar 20, 34; m 56; c 4. ORGANIC CHEMISTRY, ORNITHOLOGY. *Educ:* Univ of the South, BS, 55; Univ Rochester, PhD(chem), 61; Univ Del, MBA, 73. *Prof Exp:* Sr chemist, Atlas Chem Ind, Inc, 59-66; res chemist, ICI Am Inc, 66-70, res supvr chem, 70-74, mgr chem & polymer res, 75-79, mgr herbicides & PGR chem, 79-81, asst dir corp res, 81-83, mgr anal chem & info sci, 84-86; RES ASSOC, DEL MUS NATURAL HIST, 87- *Concurrent Pos:* dir, Del Breeding Bird Atlas Proj, 83-87; ornithologist, 87-; ed, American Birds, 89- *Mem:* Am Ornithol Union. *Res:* Research management; organic synthesis; industrial chemical development; avian population studies. *Mailing Add:* 2808 Rabbit Hill Rd Tallahassee FL 32312-3137

WEST, ROBERT A, b Valparaiso, Ind, June 14, 51; m 79; c 1. PLANETARY ATMOSPHERES, RADIATIVE TRANSFER. *Educ:* Calif Inst Technol, BS, 73; Univ Ariz, PhD(planetary sci), 77. *Prof Exp:* Res assoc, Univ Colo, Boulder, 78-84, lectr astron, 79-84; mem tech staff, 84-89, OUTER PLANETS GROUP LEADER, JET PROPULSION LAB, CALIF INST TECHNOL, 89- *Concurrent Pos:* Co-invest, Voyager Photopolarimeter Exp, 80-90, Galileo Ultraviolet Spectrometer Exp, 87-, Cassini Imaging Team, 90-, Cassini Ultraviolet Spectrometer Exp, 90-, Cassini Huygens DISR Exp, 90- *Honors & Awards:* Group Achievement Awards, Voyager, Pioneer & Solar Mesosphere Explorer, NASA. *Mem:* Am Astron Soc; Am Geophys Union; Int Astron Union. *Res:* Radiative transfer in planetary atmospheres including observation and interpretation of multiply-scattered light observed by spacecraft and ground-based instruments. *Mailing Add:* MS 169-237 JPL 4800 Oak Grove Dr Pasadena CA 91109

WEST, ROBERT ELMER, b Blackfoot, Idaho, Apr 2, 38; div; c 3. EXPLORATION GEOPHYSICS. *Educ:* Univ Idaho, BS, 61; Univ Ariz, MS, 70, PhD(geosci), 72. *Prof Exp:* Physicist, Phillips Petrol Co, 61-62, 65; res assoc geophys, Univ Ariz, 68-69, 71-72; geophysicist, Humble Oil & Refining Co, 72-74; geophysicist, Mining Geophys Surv, 74-85; CONSULT GEOPHYSICIST, 85- *Res:* Application of gravity, magnetics and electrical methods to engineering studies and to exploration for ore deposits and ground water. *Mailing Add:* 2821 N Fontana Tucson AZ 85705

WEST, ROBERT MACLELLAN, b Appleton, Wis, Sept 1, 42; m 65; c 1. VERTEBRATE PALEONTOLOGY. *Educ:* Lawrence Col, BA, 63; Univ Chicago, SM, 64, PhD(evolutionary biol), 68. *Prof Exp:* Res assoc geol & geophys sci, Princeton Univ, 68-69; asst prof biol, Adelphi Univ, 69-74; cur geol, Milwaukee Pub Mus, 74-83; dir, Carnegie Mus Natural Hist, 83-87; DIR, CRANBROOK INST SCI, 87-, VPRES, CRANBROOK EDUC COMMUNITY, 87- *Concurrent Pos:* Adj assoc prof, Dept Geol Sci, Univ Wis-Milwaukee, 74-83; adj prof, dept earth & planetary sci, Univ Pittsburgh, 83-; hon cur, Milwaukee Pub Mus, 84; adj prof, Dept Biol Sci & Anthrop & Sociol, Oakland Univ, 87, 88-; Indo-US Subcomt Educ & Cult, 90- *Honors & Awards:* Guyot Award, Nat Geog Soc, 81. *Mem:* Geol Soc Am; Asn Sci Mus Dir; Am Soc Mammal; Am Asn Mus; Soc Vert Paleont. *Res:* Asiatic mammalian evolution; paleontologic aspects of plate tectonics; evolution of early Tertiary mammals and mammalian communities; biostratigraphy of Tertiary deposits of intermontane basins in North America. *Mailing Add:* Cranbrook Inst Sci Box 801 500 Lone Pine Rd Bloomfield Hills MI 48303-0801

WEST, RONALD E(MMETT), b Rosebush, Mich, Sept 7, 33; m 55; c 3. CHEMICAL ENGINEERING. *Educ:* Univ Mich, BSE, 54, MSE, 55, PhD(chem eng), 58. *Prof Exp:* From asst prof to assoc prof, 57-74, PROF CHEM ENG, UNIV COLO, BOULDER, 74- *Mem:* Am Inst Chem Engrs; Water Pollution Control Fedn. *Res:* Water pollution control; renewable energy sources. *Mailing Add:* Cohlear Corp 61 Inverness Dr E Englewood CO 80112

WEST, RONALD ROBERT, b Centralia, Ill, Nov 14, 35; m 58, 78, 82; c 2. PALEOBIOLOGY, PALEOECOLOGY. *Educ:* Univ Mo, Rolla, BS, 58; Univ Kans, MS, 62; Univ Okla, PhD(paleoecol geol), 70- *Prof Exp:* Stratigrapher, Shell Oil Co, Okla, 56, micropaleontologist, La, 58-59; invert paleontologist, Kans Geol Surv, 60, geologist, 61; paleobiologist & paleoecologist, Humble Oil & Refining Co, Tex & Okla, 61-67; instr geol, Univ Okla, 67-68; from asst prof to assoc prof paleobiol, 69-79, ancillany prof, Biol Div, 74-79, PROF PALEOBIOL, DEPT GEOL, KANS STATE UNIV, 79- *Concurrent Pos:* Am Chem Soc-Petrol Res Fund grant paleobiol; mem adv coun, Friends of Woodrow Wilson Nat Fel Found; consult res lab, Amoco Prod Co, Okla, 74-; NSF grant, Nat Mus Natural Hist, 77-78; vis prof, Oxford, Univ Kans, 80; geologist, Tell Qargen, Syria, 84; fel Woodrow Wilson, 59-61; ACS-PRF grant, chaetetids reefs, 85-87; collabr, Dept Paleobiol, Sovi Hisonia Inst, 78-; corresp mem subcommission conb strat found, Int Geophys Union, 81-; tech ed, Paleontol Soc, 85- *Honors & Awards:* Geol Soc Am Award, 72. *Mem:* Int Paleont Asn; Paleont Soc; Soc Econ Mineralogists & Paleontologists; Paleont Asn; Geol Soc Am; Sigma Xi. *Res:* Paleoecology and paleobiology of upper paleozoic invertebrates; structure and dynamics of benthic fossil communities; carbonate sedimentation (reefs); recent marine invertebrate ecology; organism substrate relationships and functional morphology of marine invertebrates especially cocalline sponges (chaetetids) and chaetetids taxonomy and biology. *Mailing Add:* Geol Dept Kans State Univ Thompson Hall Manhattan KS 66506

WEST, ROSE GAYLE, b Pascagoula, Miss, Oct 31, 43; m 62; c 1. PHYSICAL CHEMISTRY. *Educ:* Univ Southern Miss, BA, 65, PhD(phys chem), 69. *Prof Exp:* Asst prof chem, 68-73, actg chmn dept, 72-74, ASSOC PROF CHEM, WILLIAM CAREY COL, 73-, CHMN DEPT, 74- *Mem:* Am Chem Soc. *Res:* Thermo chemistry, heats of combustion and resonance energies of aromatic hydrocarbons and 5- and 6-membered aromatic nitrogen heterocyclic compounds; special projects for undergraduate physical chemistry laboratories. *Mailing Add:* RFD 3 Box 795 Sumrall MS 39482-9236

WEST, SEYMOUR S, biophysics, anatomy; deceased, see previous edition for last biography

WEST, SHERLIE HILL, b Pall Mall, Tenn, Feb 18, 27; m 49; c 1. PLANT PHYSIOLOGY, AGRONOMY. *Educ:* Tenn Polytech Univ, BS, 49; Univ Ky, MS, 54; Univ Ill, PhD(agron, bot), 58. *Prof Exp:* Asst agronomist, Univ Ky, 54-55; res agronomist, USDA, 58-60; from asst agronomist to assoc agronomist, 58-70, asst dean res, Inst Food & Agr Sci, 72-79, AGRONOMIST, UNIV FLA, 70-, PROF SEED TECHNOL, 79- *Concurrent Pos:* Consult, AID Progs, 59-64; plant physiologist, USDA, 60-72; res leader, USDA. *Mem:* Am Chem Soc; Am Soc Plant Physiol; fel Am Soc Agron; fel Crop Sci Am; fel AAAS. *Res:* Nucleic acid metabolism and growth due to environmental factors; mechanism of hormone action; drought and cold tolerance; genetic criteria of selection of superior plants; cool temperature effects on carbohydrate metabolism; mechanisms of cool temperature dormancy in tropical grasses; seed quality and deterioration. *Mailing Add:* Dept Agron Bldg 661 Univ Fla Gainesville FL 32611

WEST, TERRY RONALD, b St Louis, Mo, Aug 15, 36; m 57; c 2. ENGINEERING GEOLOGY, CIVIL ENGINEERING. *Educ:* Wash Univ, AB & BS, 59, MA, 62; Purdue Univ, Lafayette, MSCE, 64, PhD(eng geol), 66. *Prof Exp:* Teaching asst geol, Wash Univ, 59-61; staff engr, H M Reitz Consult Engr, Mo, 61; from instr to asst prof eng geol, 61-71, ASSOC PROF ENG GEOL, PURDUE UNIV, WEST LAFAYETTE, 71- *Concurrent Pos:* Consult, ATEC Assoc Inc, 69-80; team leader eng soils group, lab appl remote sensing, 69-77; consult, 80- *Mem:* Geol Soc Am; Am Soc Civil Engrs; Asn Eng Geol; Am Soc Testing & Mat. *Res:* Evaluation of geological materials for engineering uses; remote sensing of earth materials; subsurface geology and ground water. *Mailing Add:* Dept Earth & Atmospheric Sci Purdue Univ West Lafayette IN 47907

WEST, THEODORE CLINTON, b Central, SC, May 17, 19; m 42; c 3. PHARMACOLOGY. *Educ:* Univ Wash, BS, 48, MS, 49, PhD(pharmacol), 52. *Prof Exp:* From instr to prof pharmacol, Univ Wash, 49-68, asst chmn dept, 63-68, asst dean planning, 66-68; prof med educ & pharmacol & dir off med educ, Univ Calif, Davis, 68-77, prof med sch med, 78-88. *Mem:* Am Soc Pharmacol & Exp Therapeut; Sigma Xi. *Res:* Pharmacology of cardiac and smooth muscle; medical education. *Mailing Add:* 8807 72nd St SE Snohomish WA 98290

WEST, THOMAS PATRICK, b Peabody, Mass, Dec 9, 53. GENE EXPRESSION, CORN UTILIZATION. *Educ:* Purdue Univ, BS, 74; Tex A&M Univ, MS, 76, PhD(biochem), 80. *Prof Exp:* Res assoc microbiol, Mich State Univ, 80-81 & Univ Ariz Health Sci Ctr, 81-82; NIH postdoctoral chem, Boston Col, 82-83; asst prof biol, Univ Southern Miss, 83-87; ASSOC PROF CHEM, SDAK STATE UNIV, 88- *Concurrent Pos:* Prin investr, NIH grant, 88-90. *Mem:* Am Chem Soc; AAAS; Am Soc Microbiol; Sigma Xi; Genetics Soc Am; Soc Exp Biol & Med. *Res:* Study of pyrimidine biosynthesis and utilization by psuedomonads, fungal synthesis of biopolymers, fungal xylanoytic catabolism of hemicellulose and microbial biodeterioration of plastics. *Mailing Add:* Sta Biochem SDak State Univ Box 2170 Brookings SD 57007

WEST, WALTER SCOTT, b Fayette, Wis, Mar 12, 12; m 40; c 3. ECONOMIC GEOLOGY, GEOCHEMISTRY. *Educ:* Cornell Col, AB, 34; Univ Wis-Platteville, BE, 35; Univ Tenn, MS, 37. *Prof Exp:* Prin & basketball coach , high sch, Mo, 37-38; instr geol, Univ NC, 39-42; eng aide & cartographer, Alaskan Geol Br, 42-46, geologist, 46-54, secy geol names comt, Geol Div, 54-67, CHIEF, WIS LEAD-ZINC PROJ, EASTERN MINERAL RESOURCES BR, GEOL DIV, US GEOL SURV, 66- *Concurrent Pos:* Dir, Citizens Nat Bank, Darlington, Wis, 68- *Mem:* Am Inst Mining, Metall & Petrol Eng; Arctic Inst NAm; Soc Econ Geologists. *Res:* Radioactive mineral and bse metaldeposits, pumice and riprap in Alaska; trace element and lead isotope studies; mineralogy and genesis of lead and zinc deposits in Wisconsin and Tennessee; geologic and geochemical mapping and topical studies in Upper Mississippi Valley zinc-lead district, Wisconsin, Illinois, Iowa and Minnesota; stratigraphy. *Mailing Add:* US Geol Surv Warner Hall Univ Wis Platteville WI 53818

WEST, WARWICK REED, JR, b Evington, Va, Feb 9, 22; m 46; c 3. INVERTEBRATE ZOOLOGY. *Educ:* Lynchburg Col, BS, 43; Univ Va, PhD(biol), 52. *Prof Exp:* Instr biol, Lynchburg Col, 46-49; from asst prof to prof, 52-88, chmn dept, 65-85, EMER PROF BIOL, UNIV RICHMOND, 88- *Mem:* Sigma Xi. *Res:* Milipede anatomy. *Mailing Add:* Dept Biol Univ Richmond Richmond VA 23173

WEST, WILLIAM LIONEL, b Charlotte, NC, Nov 30, 23; m 72. ZOOLOGY, PHARMACOLOGY. *Educ:* J C Smith Univ, BS, 47; Univ Iowa, PhD, 55. *Prof Exp:* Asst zool, Univ Iowa, 49-55, res assoc radiation, Col Med, 55-56; from instr to assoc prof pharmacol, 56-73, PROF PHARMACOL & RADIOL & CHMN DEPT PHARMACOL, COL MED, HOWARD UNIV, 73- *Mem:* Am Soc Zool; Am Soc Pharmacol & Exp Therapeut; Am Physiol Soc; Am Nuclear Soc; NY Acad Sci; Sigma Xi. *Res:* Biochemical and endocrine pharmacology; cellular physiology. *Mailing Add:* Dept Pharmacol Howard Univ Washington DC 20059

WEST, WILLIAM T, b Holyoke, Mass, June 26, 25; m 50; c 2. HISTOLOGY, ANATOMY. *Educ:* Am Int Col, BA, 49; Univ Rochester, PhD(anat), 56. *Prof Exp:* Fel anat, Univ Rochester, 56-57, instr, 57-58; assoc staff scientist, Jackson Mem Lab, 58-62; from asst prof to assoc prof, State Univ NY Downstate Med Ctr, 62-90, emer assoc prof anat, 90-; RETIRED. *Mem:* Sigma Xi. *Res:* Histopathology; human anatomy; pathology; endocrinology. *Mailing Add:* Two Malcolm Ct Oceanside NY 11572

WESTALL, FREDERICK CHARLES, b Pasadena, Calif, Nov 6, 43; m 68; c 4. BIOCHEMISTRY. *Educ:* Univ Calif, Los Angeles, BS, 64; San Diego State Col, MS, 66; Univ Calif, San Diego, PhD(chem), 70. *Prof Exp:* Multiple Sclerosis Soc res fel biochem, Salk Inst, 70-72, res assoc, 72-73, from asst res prof to assoc res prof, 73-84; PRES, INST DIS RES, 84-; PROF, CALIF POLYTECH, POMONA, 89- *Concurrent Pos:* Adj prof, Harvey Mudd Col, 87-89. *Mem:* Soc Neurosci; Am Chem Soc. *Res:* Biochemistry of neurological diseases; solid phase peptide synthesis; aging; immunological effects of adjuvants; biochemistry of neurotransmitters. *Mailing Add:* Dept Chem Calif State Polytech Univ Pomona CA 91768-4032

WESTAWAY, KENNETH C, b Hamilton, Ont, Aug 14, 38; m 62; c 3. PHYSICAL ORGANIC CHEMISTRY, ENVIRONMENTAL HEALTH. *Educ:* McMaster Univ, BSc, 62, PhD(phys org chem), 68. *Prof Exp:* Asst prof chem, Laurentian Univ, 68-75, chmn dept, 74-76, assoc prof, 75-82, PROF CHEM, LAURENTIAN UNIV, 82- *Concurrent Pos:* Vis prof, Univ Wis-Madison, 76-77, Univ Sask, 85-86. *Honors & Awards:* Fel Chem Inst Can. *Mem:* Chem Inst Can; Am Chem Soc. *Res:* Mechanisms of organic reactions; nucleophilic substitution reactions and elimination reactions; kinetic isotope effects; solvent effects and substituent effects on transition state structure; acid mists and gaseous pollutants in dieselized underground mines; identification and analysis of organic compounds. *Mailing Add:* Dept Chem Laurentian Univ Sudbury ON P3E 2C6 Can

WESTBERG, KARL ROGERS, b Norwalk, Conn, Dec 17, 39; m 71; c 3. CHEMISTRY, PHYSICS. *Educ:* Bowdoin Col, BA, 61; Brown Univ, PhD(chem), 69. *Prof Exp:* Engr, Perkin-Elmer Corp, 61; MEM TECH STAFF, AEROSPACE CORP, 68- *Concurrent Pos:* Vis res fel, Univ Birmingham, Eng, 85. *Mem:* Am Phys Soc. *Res:* Chemical kinetics; spacecraft contamination; aerospace sciences; air-pollution chemistry. *Mailing Add:* Aerospace Corp PO Box 92957 Los Angeles CA 90009

WESTBROOK, DAVID REX, b London, Eng, May 12, 37; m 60, 74; c 4. APPLIED MATHEMATICS. *Educ:* Univ London, BSc, 58, PhD, 61. *Prof Exp:* Lectr math, Univ Singapore, 61-64; sr lectr, Univ Melbourne, 64; lectr, Univ Nottingham, 64-65; vis mem, NY Univ, 65-66; asst prof, 66-71, ASSOC PROF MATH, UNIV CALGARY, 71- *Concurrent Pos:* Nat Res Coun Can res grant, 67-68. *Mem:* Soc Indust & Appl Math; Inst Math & Appl UK; Can Appl Math Soc. *Res:* Applications of applied maths to industrial problems; numerical methods for PDE. *Mailing Add:* Dept Math Statist & Math Univ Calgary Calgary AB T2N 1N4 Can

WESTBROOK, EDWIN MONROE, b San Juan, PR, June 30, 48; m 74; c 4. PROTEIN CRYSTALLOGRAPHY, STRUCTURAL BIOLOGY. *Educ:* Univ Calif, Berkeley, AB, 69; Stanford Univ, MS, 71; Univ Chicago, MD & PhD(biophys), 81. *Prof Exp:* Teaching fel biochem, Univ Calif, Los Angeles, 81-83; asst scientist biochem, 83-87, SCIENTIST BIOPHYS, ARGONNE NAT LAB, 87- *Concurrent Pos:* Asst prof biochem & molecular biol, Univ Chicago, 83-88; assoc prof biochem, molecular biol & cell biol, Northwestern Univ, 88- *Mem:* Am Crystallog Asn. *Res:* Application of protein crystallographic methods to the study of biological macromolecules; steroid-protein interaction; enzyme kinetics; antimicrobial proteins; microbial toxins. *Mailing Add:* Div Biol & Med Res Argonne Nat Lab 9700 S Cass Ave Argonne IL 60439

WESTBROOK, J(ACK) H(ALL), b Troy, NY, Aug 19, 24; m 47; c 5. PHYSICAL METALLURGY, CERAMICS. *Educ:* Rensselaer Polytech Inst, BMetE, 44, MMetE, 47; Mass Inst Technol, ScD(metall), 49. *Prof Exp:* Asst metall, Rensselaer Polytech Inst, 44, 46-47 & Mass Inst Technol, 47-49; res assoc, Res & Develop Ctr, Gen Elec Co, 49-71, mgr eng mat & processes info oper, 71-74, mgr mat info serv, Corp Res & Develop, 74-81, mgr spec projs, 81-85; PRES & PRIN CONSULT, SCI TECH KNOWLEDGE SYSTS, 85- *Concurrent Pos:* Adj assoc prof, Rensselaer Polytech Inst, 57-59; mem subcomts, Nat Mat Adv Bd, Nat Acad Sci, 59-63, 68, 75-77 & 79-82; chmn, Gordon Conf Solid State Studies in Ceramics, 60; Nat Acad Sci US-USSR exchange fel, 71; consult, USAF, US Army, Nat Sci Fedn, NASA & Advan Res Proj Agency, Dept Defense; chmn, Mat Info Comn, Fedn Mat, 74-; mem mat adv panel, Off Technol Assessment, US Cong, 75-77; chmn adv panel on data for indust needs, CODATA, 76- *Honors & Awards:* Turner Award, Electrochem Soc, 57; Templin Award, Am Soc Testing & Mat, 59; Geisler Award, Am Soc Metals, 59, Campbell Mem lectr, 76; Am Inst Mining, Metall & Petrol Engrs Award, 63; Am Ceramic Soc Award, 67; Hofmann Prize, Lead Develop Asn, 71; Jeffries Mem lectr, Am Soc Metals, 79. *Mem:* Fel Am Soc Metals; Am Inst Mining, Metall & Petrol Engrs; fel Am Ceramic Soc; Electrochem Soc; fel AAAS; Sigma Xi; fel Am Inst Chemists. *Res:* Intermetallic compounds; mechanical properties of refractory materials; hardness measurement techniques; grain boundaries; materials information; research and development planning; technological forecasting and assessment; history of metallurgy. *Mailing Add:* Brookline Rd RFD 2 Ballston Spa NY 12020

WESTBY, CARL A, b Los Angeles, Calif, Feb 8, 36; m 58; c 6. MICROBIAL PHYSIOLOGY. *Educ:* Univ Calif, Riverside, AB, 58; Univ Calif, Davis, PhD(bact), 64. *Prof Exp:* Fel bact physiol, Sch Med, Univ Pa, 64-67; asst prof bact, Utah State Univ, 67-73; assoc prof, 73-80, PROF MICROBIOL, SDAK STATE UNIV, 80- *Concurrent Pos:* Med prod consult, Med Prod Div, 3M Co, 74- *Mem:* Am Soc Microbiol; Sigma Xi. *Res:* Bacterial physiology, genetics, and gene engineering. *Mailing Add:* Dept Microbiol SDak State Univ Brookings SD 57006

WESTCOTT, KEITH R, b Salt Lake City, Utah, Sept 30, 52. PROTEIN CHEMISTRY. *Educ:* Univ Calif, Berkeley, AB, 74; Univ Ill, Urbana-Champaign, MS, 78, PhD(biochem), 80. *Prof Exp:* Res assoc, Dept Biochem, Duke Univ Med Ctr, 80-83; researcher, Dept Biol Chem, Univ Calif, Los Angeles, 83-85; scientist, Alpha Therapeut Corp, 85-86; SCIENTIST, AMGEN, INC, 86- *Mem:* Am Soc Biochem & Molecular Biol; Am Chem Inst; AAAS; fel Am Inst Chemists; Sigma Xi; NY Acad Sci. *Res:* Development of protein purification methods for human therapeutics; growth factor and cytokine isolation; protein folding. *Mailing Add:* Amgen Inc Amgen Ctr Thousand Oaks CA 91320

WESTCOTT, PETER WALTER, ecology, zoology; deceased, see previous edition for last biography

WESTDAL, PAUL HAROLD, b Wynyard, Sask, Nov 5, 21; m 47; c 3. ENTOMOLOGY, VIROLOGY. *Educ:* Univ Man, BSc, 47, MSc, 50, PhD(entom), 69. *Prof Exp:* Res entomologist, 46-74, sr res entomologist econ entom, Agr Can, 75-82; CONSULT, WESTDAL AGRI CONSULTS, 82- *Concurrent Pos:* Adj prof, Univ Man, 71-82; ed, The Man Entomologist, 77-81. *Mem:* Entom Soc Can; Sigma Xi. *Res:* Biology and control of insect pests of sunflowers and rapeseed. *Mailing Add:* 40 Garnet Bay Winnipeg MB R3T 0L6 Can

WEST-EBERHARD, MARY J, b Pontiac, Mich, Aug 20, 41. ENTOMOLOGY. *Educ:* Univ Mich, BA, 63, MS, 64, PhD(zool), 67. *Prof Exp:* Teaching fel, Dept Zool, Univ Mich, 63-65; postdoctoral fel biol, Harvard Univ, 67-69; assoc, 73-75, ENTOMOLOGIST, DEPT ENTOM, SMITHSONIAN TROP RES INST, 75- *Concurrent Pos:* Milton fel, Harvard Univ, 68-69; staff mem, Dept Biol, Univ Valle, Cali, Colombia, 72-78; distinguished vis scientist, Mus Zool, Univ Mich, 82; mem, Int Comt, Int Union Study Social Insects; mem bd dirs, Orgn Trop Studies, 85-87; mem adv comt, Monteverde Conserv League Comt Human Rights, Nat Acad Sci, 90- *Mem:* Nat Acad Sci; Am Soc Naturalists; Sigma Xi; Int Union Study Social Insects; Soc Study Evolution (vpres II, 87-88, pres-elect, 91); Int Soc Hist, Philos & Social Studies Biol. *Res:* Author or co-author of over 50 publications. *Mailing Add:* Dept Entom Smithsonian Trop Res Inst APO Miami FL 34002-0011

WESTENBARGER, GENE ARLAN, b Lancaster, Ohio, July 25, 35; m 58; c 4. PHYSICAL CHEMISTRY. *Educ:* Ohio Univ, BS, 57; Univ Calif, Berkeley, PhD(phys chem), 63. *Prof Exp:* Chemist, Battelle Mem Inst, 57; res & develop coordr, Qm Food & Container Inst, 58-59; asst prof, 63-67, ASSOC PROF PHYS CHEM, OHIO UNIV, 67- *Mem:* Am Phys Soc; Am Chem Soc. *Res:* Thermodynamic and magnetic studies at low temperature. *Mailing Add:* Dept Chem Ohio Univ Athens OH 45701-2979

WESTENBERG, ARTHUR AYER, b Menomonie, Wis, Mar 1, 22; m 45; c 1. CHEMICAL PHYSICS. *Educ:* Carleton Col, AB, 43; Harvard Univ, AM, 48, PhD(chem), 50. *Prof Exp:* Chemist, Mayo Clin, 50; asst prof chem, Lafayette Col, 50-52; sr staff mem, Appl Physics Lab, Johns Hopkins Univ, 52-58, prin staff mem, 58-77, supvr chem physics res, 63-77; RETIRED. *Concurrent Pos:* Consult, Proj Squid, Off Naval Res, 60-65; mem adv comt, Army Res Off, 66-71; mem comt assess environ effects of supersonic transport, US Dept Com, 71. *Honors & Awards:* Hillebrand Award, Am Chem Soc, 66; Silver Medal, Combustion Inst, 66. *Res:* Chemical kinetics; high temperature gas properties; combustion; air pollution chemistry; electron spin resonance spectroscopy. *Mailing Add:* Box 295 Manchester VT 05254

WESTENFELDER, CHRISTOF, b Stuttgart, WGer, July 1, 42. NEPHROLOGY, PHYSIOLOGY. *Educ:* Univ Munich, BS, 68; Univ Kiel-Lubek, WGer, MD, 69. *Prof Exp:* Intern, Rittbegkrankenhaus, W Berlin, 69-70 & Cook County & Suburban Hosps, Chicago, 70-71; resident internal med, Abraham Lincoln Sch Med, Univ, 71-73, fel nephrology, 73-75, asst prof, 75-81; dir hemodialysis, Univ Ill Hosp, 75-80 & Vet Admin Med Ctr, W Side Hosp, 81-82; fac biomed eng, Ctr Artificial Hearts Devices, 85, PROF MED, SCH MED, UNIV UTAH, 83-; CHIEF NEPHROLOGY, VET ADMIN MED CTR, SALT LAKE CITY, 83-, DIR, TRANSPLANT & ACUTE CARE UNIT, 84- *Concurrent Pos:* Vis prof, Univ Chicago, 79-82, Univ Freiburg, Fed Repub Ger, 80, Univ Vienna, Austria, 81, Univ Calif, San Francisco, 85, Cleveland Clin, 84-85, Univ Winniped, Can, 85, Univ BC, Vancouver, 85. *Mem:* Am Physiol Soc; Am Soc Renal Biochem & Metab; fel Am Col Physicians; Am Fedn Clin Res; Nat Kidney Found; AMA; Am Soc Nephrology; Int Soc Nephrology; Am Soc Artificial Internal Organs; Am Heart Asn; Ger Med Asn; AAAS. *Res:* Atrial natriuretic factor release in animals with artificial hearts; acute renal failure, bioenergetics; contratility of glomeruli; renal converting enzyme, enzymology. *Mailing Add:* Div Nephrology Univ Utah 50 Medical Dr Salt Lake City UT 84132

WESTENSKOW, DWAYNE R, b LaGrande, Ore, Apr 16, 47; m; c 5. ANESTHESIOLOGY. *Educ:* Brigham Young Univ, BS, 72; Univ Utah, ME, 75, PhD(bioeng), 76. *Prof Exp:* Bioengr, Pulmonary Functions Lab, LDS Hosp, Salt Lake City, Utah, 73; proj engr, Health Sci Res Div, Sandoz, Inc, 75-76; res instr, Dept Anesthesiol, Univ Utah, 75-78, from asst prof to assoc prof, 79-88, res instr, Dept Surg, 78-85, res asst prof, 79-88, res assoc prof, Dept Bioeng, 82-89, PROF, DEPT ANESTHESIOL, UNIV UTAH, 88-, RES PROF, DEPT BIOENG, 89- *Concurrent Pos:* Consult, var corp, foreign & US, 76-; grants, foreign & US, corp & educ, 77-91; mem bd dirs, Instrumentation Res Labs, 77-80, subcomt, Equip Monitoring & Eng Technol, Am Soc Anesthesiologists, 85, 90 & 91; mem, Int Adv Comt, Comput in Anesthesia, 82-; sect ed, J Annals Biomed Eng, 82-85, field ed, Anesthesia & Analgesia, 82-, Inst Elec & Electronics Engrs, 81-, consult ed, J Crit Care, 85- *Honors & Awards:* Career Achievement Award, Asn Advan Med Instrumentation, 79 & Award for Excellence in Commun, 85. *Mem:* Asn Univ Anesthetists; sr mem Biomed Eng Soc; Asn Advan Med Instrumentation; sr mem Inst Elec & Electronics Engrs; Int Anesthesia Res Soc; Am Soc Anesthesiologists; Crit Care Soc. *Res:* Development continuous oxygen consumption measuring system and computer control of IV fluid infusion rate; computerization of anesthesia monitoring; control of mechanical ventilation; over 100 technical papers published; 13 patents. *Mailing Add:* Dept Anesthesiol Univ Utah 50 N Medical Dr Salt Lake City UT 84132

WESTER, DERIN C, b Oakland, Calif, Nov 21, 49; c 3. PEDIATRIC AUDIOLOGY. *Educ:* Brigham Young Univ, BS, 72, MCD, 79; Univ Utah, PhD(commun dis), 90. *Prof Exp:* ASST PROF AUDIOL, UNIV ALA, 90- *Mem:* Am Speech Lang & Hearing Asn; Am Acad Audiol; Am Auditory Soc. *Res:* Audiological research on central auditory processing; ultra high frequency audiometry; hereditary hearing impairments. *Mailing Add:* Dept Commun Dis Univ Ala PO Box 870242 Tuscaloosa AL 35487-0242

WESTER, RONALD CLARENCE, b Gardner, Mass, Aug 26, 40; m 67; c 2. DRUG METABOLISM, PHARMACOKINETICS. *Educ:* Clark Univ, AB, 62; Univ Ill, Urbana, MS, 69, PhD(physiol), 70. *Prof Exp:* Res asst metab, Univ Rochester Sch Med, 64-66, res asst, Worcester Fedn Exp Biol, 62-64; res investr, G D Searle & Co, 72-76, res scientist drug metab, 76-81; PROF PHARM, SCH PHARM & RES DERMATOLOGIST, SCH MED, UNIV CALIF, SAN FRANCISCO, 81- *Concurrent Pos:* Fel, Univ Ill, 66-70, Cornell Univ, 70-72; vis scientist, Ore Regional Primate Ctr, 72. *Mem:* Am Soc Clin Pharmacol & Therapeut. *Res:* Bioavailability and drug metabolism in experimental animals and man; percutaneous absorption in experimental animals and man; clinical dermatology and pharmaceutical sciences. *Mailing Add:* Dept Pharmacol Univ Calif San Francisco Med Sci 513 Parnassus Ave San Francisco CA 94143-0446

WESTERBERG, ARTHUR WILLIAM, b St Paul, Minn, Oct 9, 38; m 63; c 2. CHEMICAL ENGINEERING. *Educ:* Univ Minn, Minneapolis, BSc, 60; Princeton Univ, MSc, 61; Univ London, PhD(chem eng) & Imp Col, dipl, 64. *Prof Exp:* Sr analyst software eng, Control Data Corp, 65-67; from asst prof to prof chem eng, Univ Fla, 67-76; dir, Design Res Ctr, 78-80, dept head, 80-83, PROF CHEM ENG, CARNEGIE-MELLON UNIV, 76-, SWEARINGEN PROF CHEM ENG, 82- *Concurrent Pos:* Chem Eng Div lectureship, Am Soc Eng Educ, 81; inst lectr, Am Inst Chem Engrs, 89. *Honors & Awards:* Comput Syst Technol Div Award, Am Inst Chem Engrs, 83, William H Walker Award, 87, McAfee Award, Pittsburgh Sect, 90. *Mem:* Nat Acad Eng; Am Inst Chem Engrs. *Res:* Developing computer software systems to aid in process analysis and optimization; process synthesis; expert systems in process design. *Mailing Add:* Dept Chem Eng Carnegie Mellon Univ Pittsburgh PA 15213

WESTERDAHL, CAROLYN ANN LOVEJOY, b Oklahoma City, Okla, Apr 16, 35; m 61. SURFACE SCIENCE, ENERGETIC MATERIALS. *Educ:* Univ Chicago, BA, 55, BS, 57; Univ Calif, Berkeley, PhD(chem), 61. *Prof Exp:* CHEMIST, PICATINNY ARSENAL, 67- *Mem:* Sigma Xi; Am Chem Soc. *Res:* Surface studies. *Mailing Add:* Seven Comanche Trail Denville NJ 07834

WESTERDAHL, RAYMOND P, b Chicago, Ill, Mar 22, 29; m 61. PHYSICAL CHEMISTRY, POLLUTION CONTROL. *Educ:* Univ Ill, BS, 51; Univ Chicago, MS, 59, PhD(phys chem), 62. *Prof Exp:* Res chemist, Esso Res & Eng Co, Standard Oil Co, NJ, 62-67; res chemist, Feltman Res Lab, Picatinny Arsenal, 67-77, CHEM ENGR, US ARMY ARMAMENT RES & DEVELOP CTR, 77- *Mem:* AAAS; fel Am Inst Chemists; Am Chem Soc; Soc Appl Spectros. *Res:* Mechanisms of pyrotechnic reactions; Raman spectroscopy; chemiluminescent reactions; reactions in fused salts; analysis and control of air and water pollutants. *Mailing Add:* Seven Comanche Trail Denville NJ 07834

WESTERFELD, WILFRED WIEDEY, b St Charles, Mo, Dec 13, 13; m 38; c 5. MOLYBDENUM XANTHINE OXIDAZE, CARBOLIGASE REACTIONS. *Educ:* Mo Univ, BS, 34; St Louis Univ, PhD(biochem), 38. *Hon Degrees:* DSc, State Univ NY, 89. *Prof Exp:* Assoc prof biochem, Harvard Med Sch, 40-45; prof biochem, Upstate Med Ctr, State Univ NY, 45-79; RETIRED. *Honors & Awards:* Am Chem Soc, 68. *Mailing Add:* 7607 Hunt Lane Fayetteville NY 13066

WESTERFIELD, CLIFFORD, veterinary anatomy; deceased, see previous edition for last biography

WESTERHOF, NICOLAAS, b DeBilt, Utrecht, May 4, 37; m 61; c 2. ANIMAL PHYSIOLOGY. *Educ:* Univ Utrecht, Neth, BS, 57, MS, 62; Univ Pa, PhD(biomed eng), 68. *Prof Exp:* Res assoc physics, Univ Utrecht, 60-64; res assoc physiol, Georgetown Univ, Washington, DC, 64-66; res assoc biomed eng, Univ Pa, Philadelphia, 66-69; researcher physiol, 69-72, lectr, 72-80, PROF PHYSIOL, FREE UNIV AMSTERDAM, 80- *Mem:* Biomed Eng Soc; Inst Elec & Electronics Engrs. *Res:* Cardiovascular physiology, especially hemodynamics; cardiac pump function and coronary pressure-flow relations. *Mailing Add:* Lab Physiol Free Univ Amsterdam Amsterdam 1081 BT Netherlands

WESTERHOUT, GART, b The Hague, Neth, June 15, 27; m 56; c 5. ASTRONOMY. *Educ:* State Univ Leiden, Drs, 54, PhD(astron), 58. *Prof Exp:* Res asst astron, Univ Observ, State Univ Leiden, 52-54, sci officer, 54-59, chief sci officer, 59-62; dir astron prog, 62-73, chmn div math, phys & eng sci, 72-73, prof astron, Univ Md, College Park, 62-77; SCI DIR, US NAVAL OBSERV, 77- *Concurrent Pos:* NATO fel, 59; mem user's comt, Nat Radio Astron Observ, 65-78; mem, NSF Dicke Panel Radio Astron Facil, 67-69; vchmn, Div Phys Sci, Nat Res Coun, 69-73, mem comt radio frequencies, 72-82; trustee-at-large, Assoc Univs Inc, 71-74; mem US nat comt, Int Astron Union, 71-74; mem US nat comt, Int Sci Radio Union, 72-78, chmn comn radio-astron, 75-78; Humboldt Found Award, Ger, 73-75; mem, Inter-Union Comn for Allocation of Frequencies, 75-82; mem vis comn, Max Planck Inst Radion Astron, Bonn, WGermany, 76-81; mem vis comn, Haystack Observ, Mass Inst Technol, 76-81 & chmn, 79-81; mem, Arecibo Adv Bd, Cornell Univ, 77-80 & chmn, 79; mem, Sci Coun, Stellar Data Ctr, Strasbourg, Fr, 78-84; counr, Am Astron Soc, 78. *Mem:* Am Astron Soc (vpres, 85-); Royal Astron Soc; Int Astron Union; Int Sci & Radio Union; Sigma Xi. *Res:* Radio astronomy; 21-centimeter line research; fundamental astronomy; galactic structure; optical and radio astrometry; very long baseline interferometry, optical interferometry; calibration curve data arrays; Earth rotation; timekeeping. *Mailing Add:* Off Sci Dir US Naval Observ Washington DC 20392

WESTERMAN, ARTHUR B(AER), b Pittsburgh, Pa, June 29, 19; m 42; c 3. METALLURGY. *Educ:* Carnegie Inst Technol, BS, 39. *Prof Exp:* Asst metallurgist, Crucible Steel Co Am, Pa, 39-42; res metallurgist, Metals Res Lab, Carnegie Inst Technol, 42-43; res engr, Battelle Develop Corp, 43-50, asst div chief, 50-58, proj coordr, 58-62, asst res mgr, 62-75, coordr large proj, 75-80, mem staff, Battelle Mem Inst, 80-90; RETIRED. *Mem:* Am Soc Metals; AAAS. *Res:* Scientific and technical information. *Mailing Add:* 71 N Merkle Rd Columbus OH 43209

WESTERMAN, DAVID SCOTT, b Ann Arbor, Mich, July 12, 46; m 68, 84; c 2. TECTONIC GEOLOGY. *Educ:* Allegheny Col, BS, 69; Lehigh Univ, MS, 71, PhD(geol), 72. *Prof Exp:* Asst prof earth sci, Northeastern Univ, 72-78; vis asst prof, Univ Southern Maine, 78-79; asst prof geol, Univ Maine, Orono, 80; asst prof, Colloy Col, 80-82; asst prof, 82-83, ASSOC PROF GEOL, NORWICH UNIV, 83- *Concurrent Pos:* Field geologist bedrock mapping, Maine Geol Surv, 75-83 & Vt Geol Surv, 83-; co-ed, Geol Sci Maine, 78-; vis asst prof geol, Unity Col, 82. *Mem:* Geol Soc Am; Sigma Xi; Planetary Soc. *Res:* Reconstruction of ancient global-scale tectonic events which are recorded in Central Vermont and Eastern Maine; resolution of ancient stress fields in Maine. *Mailing Add:* Dept Geol Norwich Univ 65 S Main St Northfield VT 05663

WESTERMAN, EDWIN J(AMES), b USA, Jan 18, 35; m 59, 81; c 8. METALLURGICAL ENGINEERING, METALLURGY. *Educ:* Mont Sch Mineral Sci & Technol, BS, 56; Rensselaer Polytech Inst, PhD(metall), 59. *Prof Exp:* Metallurgist, Res Lab, Gen Elec Co, 59-61; metallurgist, Dept Metall Res, Kaiser Aluminum, Spokane Wash, 61-69, head alloy metall sect, 70-73, mgr, Alloy & Properties Res Dept, 73-79, mgr, Reduction Res Dept, 80-83, TECH MGR, CTR TECHNOL, KAISER ALUMINUM & CHEM CORP, PLEASANTON, CALIF, 69-, MGR CAN STOCK RES PROG, 84- *Mem:* Am Soc Metals; Am Inst Mining, Metall & Petrol Engrs. *Res:* Powder metallurgy; aluminum alloy development; x-ray diffraction; electron and ion microprobes; nondestructive testing of metals; can stock metallurgy and technology. *Mailing Add:* 164 Woodview Terr Dr San Ramon CA 94583

WESTERMAN, HOWARD ROBERT, b Jersey City, NJ, Feb 17, 26; m 50; c 2. SYSTEMS ENGINEERING. *Educ:* St Peter's Col, BS, 48; Univ Chicago, SM, 49; Columbia Univ, PhD(chem physics), 52. *Prof Exp:* Analyst oper res, Off Chief Naval Oper, 52-54; mem tech staff, 54-66, DEPT HEAD SYST ENG, AT&T BELL LABS, 66- *Concurrent Pos:* Allied Chem Co fel, Columbia Univ, 52-53. *Mem:* Sigma Xi. *Res:* Quantum mechanics; operations research; orbital mechanics. *Mailing Add:* 37 Cheshire Sq Little Silver NJ 07739

WESTERMAN, IRA JOHN, b Louisville, Ky, Oct 10, 45; m 79. ORGANIC CHEMISTRY, POLYMER CHEMISTRY. *Educ:* Univ Ky, BS, 67; Duke Univ, MS, 69, PhD(org chem), 73. *Prof Exp:* Res fel, Univ Ariz, 73-74; sr res chemist polymer chem, B F Goodrich Chem Div, 74-80; SR RES CHEMIST POLYMER CHEM, PHILLIPS PETROL CO, 80- *Mem:* Am Chem Soc; Soc Plastics Engrs. *Res:* Polyelectrolytes; water-soluble polymers; heterocyclic chemistry; polar cycloadditions; oil recovery chemicals. *Mailing Add:* 299 Pheasant Run Wadsworth OH 44281-2347

WESTERMAN, RICHARD EARL, b Great Falls, Mont, Jan 18, 35; m 58; c 4. METALLURGY. *Educ:* Mont Col Mineral Sci & Technol, BS, 56; Rensselaer Polytech Inst, PhD(metall), 60. *Prof Exp:* Engr, Hanford Labs, Gen Elec Co, 56-62, sr scientist, 62-64; res assoc corrosion res, 64-66, mgr, 66-69, res assoc, Metall Sect, 69-80, TECH LEADER ENVIRON MECH PROPERTIES, PAC NORTHWEST LAB, BATTELLE MEM INST, 80- *Concurrent Pos:* Lectr, Richland Joint Ctr Grad Study. *Mem:* Am Soc Metals. *Res:* Kinetics of high temperature oxidation and evaporation of superalloys; diffusion and solubility of hydrogen in iron, zirconium and titanium alloys; gas-metal reactions; aqueous corrosion; thermodynamics; powder metallurgy; dental implant development; nuclear waste disposal. *Mailing Add:* 1804 Marshall Richland WA 99352

WESTERMAN, WILLIAM JOSEPH, II, b St Louis, Mo, June 27, 37; m 63; c 3. MECHANICAL ENGINEERING, PHYSICS. *Educ:* Vanderbilt Univ, BSME, 59, MS, 61; Washington Univ, DSc, 64. *Prof Exp:* Lectr elec eng, Washington Univ, 60-63; res scientist mech eng, Martin Co, 64-67; mgr special eng prog, McDonnell Douglas, 67-74; gen mgr res & develop, Chamberlain Mfg Co, 74-80; PRES, COGSDILL TOOL PROD CORP, 80- *Concurrent Pos:* Adj assoc prof, Univ Fla, 64-73; fel, NSF, 62-63. *Mem:* Am Soc Mech Eng; Soc Mfg Engr; Soc Automotive Engrs; Cutting Tool Mfrs Am (dir, 84-85). *Res:* Fluidics; automatic controls; fluid systems contamination; metal cutting; burnishing; reaming. *Mailing Add:* 207 East Springs Rd Columbia SC 29223

WESTERMANN, D T, b July 4, 41; US citizen; c 3. SOIL SCIENCE, SOIL CHEMISTRY. *Educ:* Colo State Univ, BS, 63; Ore State Univ, MS, 65, PhD, 68. *Prof Exp:* SOIL SCIENTIST, SNAKE RIVER CONSERV RES CTR, SCI & EDUC ADMIN-AGR RES, USDA, 68- *Concurrent Pos:* Affil prof, Univ Idaho, 71-, Utah State Univ, 87- *Mem:* Sigma Xi. *Res:* Plant nutrition. *Mailing Add:* Rte 1 Box 186 USDA Sea Ar Kimberly ID 83341

WESTERMANN, FRED ERNST, b Cincinnati, Ohio, Mar 14, 21; m 49; c 3. METALLURGICAL & MATERIALS ENGINEERING. *Educ:* Univ Cincinnati, ChE, 43, MS, 47, PhD(metall eng), 57. *Prof Exp:* Jr engr develop elastomers, Inland Mfg Div Gen Motors Corp, 43-44; FAC METALL ENG, COL ENG, UNIV CINCINNATI, 48- *Concurrent Pos:* Consult, Gen Elec Aircraft Nuclear Propulsion, 57-61, Metcut Res Assoc, Inc, 65-74 & Delhi Foundry Sand Co, 75-78; co investr grant, NSF, 73- 76, mem mat adv team educ modules mat sci & eng, 76-80, Fusite Div, Emerson Elec, 77-89. *Mem:* Fel Am Soc Metals; Am Soc Eng Educ; Am Foundrymen's Soc; Sigma Xi. *Res:* Powder metallurgy; powder characterization, compaction and sintering; foundry engineering; molding materials; cast iron structure property relationships; welding metallurgy. *Mailing Add:* 9575 Millbrook Dr Cincinnati OH 45231

WESTERMANN, GERD ERNST GEROLD, b Berlin, Ger, May 11, 27; m 56; c 3. GEOLOGY. *Educ:* Brunswick Tech Univ, BSc, 50; Univ Tubingen, MSc & PhD(geol, paleont), 53. *Prof Exp:* Paleontologist, Geol Surv Ger, 53-57; from lectr to assoc prof, 57-69, PROF GEOL, MCMASTER UNIV, 69- *Concurrent Pos:* Consult, 58-62; mem, Leader Int Geol Correl Prog. *Mem:* Paleont Res Inst; Am Paleont Soc; Soc Econ Paleont & Mineral; UK Palaeont Asn; Int Paleont Asn (secy-gen, 74-82); Can Geol Asn. *Res:* Mesozoic Mollusca, especially Jurassic Ammonoidea and Triassic Pectinacea; functional morphology of cephalopods; taxonomy; intercontinental biochronology and biogeography. *Mailing Add:* Dept Geol McMaster Univ 1280 Main St W Hamilton ON L8S 4L8 Can

WESTERN, ARTHUR BOYD, b Detroit, Mich, Feb 29, 44; m 66; c 3. DESIGN OF MOBILE ROBOTS. *Educ:* Rollins Col, BS, 65; Mont State Univ, MS, 72, PhD(physics), 76. *Prof Exp:* Asst prof physics, Mont Col Mineral Sci & Technol, 76-81, head, dept physics & geophys eng, 78-86, assoc prof physics, 81-86; PROF PHYSICS & APPL OPTICS, ROSE-HULMAN INST TECHNOL, 86- *Concurrent Pos:* Staff engr-scientist, MSE Inc, 85- *Mem:* Am Asn Physics Teachers; Int Soc Optical Eng. *Res:* Holographic interferometry, phase transitions, ultrasonic measurements, dielectric measurements, refractory ceramics, robotics; design of high-interaction magnetohydrodynamics generators. *Mailing Add:* Dept Physics & Appl Optics Rose-Hulman Inst Technol 5500 Wabash Ave Terre Haute IN 47803

WESTERN, DONALD WARD, b Poland, NY, May 7, 15; m 43; c 5. MATHEMATICS. *Educ:* Denison Univ, BA, 37; Mich State Col, MA, 39; Brown Univ, PhD(math), 46. *Prof Exp:* Instr math, Mich State Col, 37-39; from assoc prof to prof math, 48-74, chmn, Dept Math & Astron, 52-72, Charles A Dana prof, 74-80, EMER PROF MATH, FRANKLIN & MARSHALL COL, 80- *Concurrent Pos:* NSF fac fel, 60. *Mem:* Am Math Soc; Math Asn Am. *Res:* Inequalities in the complex plane. *Mailing Add:* Franklin-Marshall Col Lancaster PA 17604

WESTERVELT, CLINTON ALBERT, JR, b Portland, Ore, June 15, 36; m 65. INVERTEBRATE ZOOLOGY, PARASITOLOGY. *Educ:* Lewis & Clark Col, BA, 58; Univ Ariz, MS, 61, PhD(zool), 66. *Prof Exp:* From asst prof to assoc prof, 65-82, PROF BIOL, CHAPMAN COL, 82- *Mem:* Am Soc Parasitologists. *Res:* Biology and systematics of rhabdocoel turbellarians. *Mailing Add:* Chapman Col 333 N Glassell St Chapman Col Orange CA 92666

WESTERVELT, FRANKLIN HERBERT, b Benton Harbor, Mich, Mar 26, 30; m 48; c 2. ENGINEERING, COMPUTER SCIENCE. *Educ:* Univ Mich, Ann Arbor, BSE(mech eng) & BSE(math), 52, MSE, 53, PhD(mech eng), 61. *Prof Exp:* Instr eng graphics, Univ Mich, Ann Arbor, 53-56, res assoc comput ctr, 61-66, assoc dir ctr, 66-71, from asst prof to prof mech eng, 61-71; prof eng & comput sci & dir comput serv ctr, 71-83, PROF ENG & COMPUT SCI, WAYNE STATE UNIV, 83-; EXEC VPRES RES, ST SYSTS DEVELOP INC, 83- *Concurrent Pos:* Proj dir conversational use of comput, Advan Res Projs Agency-US Dept Defense, 65-70; trustee, Argonne Univ Asn, 71-75, 81-82. *Mem:* Asn Comput Mach; Nat Soc Prof Engrs. *Res:* Computing systems; very large databases; interactive computing; systems architecture; micro-programming; information management systems; parallel distributed processors. *Mailing Add:* Dept Elec & Comp Eng Wayne State Univ 3148 Eng Detroit MI 48202-9960

WESTERVELT, FREDERIC BALLARD, JR, b Washington, DC, June 11, 31; m; c 2. NEPHROLOGY. *Educ:* Univ Va, MD, 55. *Prof Exp:* HEAD, RENAL DIV, DEPT MED, UNIV VA, 64-, DIR, DEPT RENAL SERV, 65-, PROF MED, 78- *Mem:* Am Col Physicians; Am Soc Artificial Internal Organs. *Res:* Hermodialysis; uremia. *Mailing Add:* Sch Med Univ Va Box 133 Charlottesville VA 22908

WESTERVELT, PETER JOCELYN, b Albany, NY, Dec 16, 19. THEORETICAL PHYSICS. *Educ:* Mass Inst Technol, BS, 47, MS, 49, PhD(physics), 51. *Prof Exp:* Mem staff, Radiation Lab, Mass Inst Technol, 40-41 & Underwater Sound Lab, 41-45, asst physics, 46-47, res assoc, 48-50; from asst prof to assoc prof, 51-70, PROF PHYSICS, BROWN UNIV, 70- *Concurrent Pos:* Consult to asst attache for res, USN, Am Embassy, London, 51-52, Bolt, Beranek & Newman, Inc & Appl Res Labs, Univ Tex, Austin, 71-, mem, Subcomt Aircraft Noise, NASA, 54-59; mem comt hearing & bio-acoust, Armed Forces-Nat Res Coun, 57-, mem exec coun, 60-61, chmn, 67-68; mem sonic boom comt, Nat Acad Sci, 68-71. *Honors & Awards:* Recipient Rayleish Medal, Inst Acoust, UK, 85- *Mem:* Fel Acoust Soc Am; fel Am Phys Soc; Am Astron Soc; Am Phys Soc. *Res:* Physical effects of high amplitude sound waves; air acoustics; underwater sound; general relativity. *Mailing Add:* Dept Physics Brown Univ Providence RI 02912

WESTFAHL, PAMELA KAY, b Miami, Okla. REPRODUCTIVE PHYSIOLOGY. *Educ:* Wash State Univ, BSc, 72; Univ Okla, PhD(physiol), 80. *Prof Exp:* Fel, Ore Regional Primate Res Ctr, 80-83; asst prof, Dept Zool, Miami Univ, 83-88; ASST PROF, BASIC SCIENTIST DEPT, CALIF COL PODIAT MED, 88- *Mem:* Soc Study Reproduction; Endocrine Soc; Am Physiol Soc. *Res:* Factors involved in regulation of the corpus luteum during the menstrual or estrous cycle and pregnancy; feedback relationships of gonadal hormones on hypothalamic-pituitary function. *Mailing Add:* Basic Scientist Dept Calif Col Podiat Med 1210 Scott St San Francisco CA 94115

WESTFALL, DAVID PATRICK, b Harrisburg, WVa, June 9, 42; m 65; c 2. PHARMACOLOGY. *Educ:* Brown Univ, BA, 64; WVa Univ, MS, 66, PhD(pharmacol), 68. *Prof Exp:* Demonstr pharmacol, Oxford Univ, 68-70; from asst prof to prof pharmacol, Med Sch, WVa Univ, 70-82; PROF & CHMN PHARMACOL, SCH MED, UNIV NEV, 82- *Concurrent Pos:* J H Burn fel pharmacol, Oxford Univ, 68-70. *Mem:* AAAS; Am Soc Pharmacol & Exp Therapeut; Soc Neurosci. *Res:* Pharmacology and physiology of smooth and cardiac muscle; factors governing the sensitivity of muscle to drugs. *Mailing Add:* Dept Pharmacol Sch Med Univ Nev Reno NV 89557

WESTFALL, DWAYNE GENE, b Aberdeen, Idaho, Nov 21, 38; m 61; c 3. SOIL CHEMISTRY, AGRONOMY. *Educ:* Univ Idaho, BS, 61; Wash State Univ, PhD(soils), 68. *Prof Exp:* Res asst soils, Wash State Univ, 66-67; from asst prof to assoc prof soil chem, Tex A&M Univ, 67-74; sr plant nutritionist, Great Western Sugar Co Agr Res Ctr, Longmont, 74-78; PROF AGRON, COLO STATE UNIV, 78- *Concurrent Pos:* Mem staff soil fertil, Colo State Univ Water Mgt Res Prog, Lahore, Pakistan, 78-80, Int Develop. *Mem:* Sigma Xi; fel Am Soc Agron; Soil Sci Soc Am; Int Soil & Sci Soc. *Res:* Soil chemistry of submerged soils; fertility of rice, sugarbeets, small grains and other agramomic crops; micronutrients and nitrogen efficiency and utilization; research and training programs in developing countries; crop production systems. *Mailing Add:* Dept Agron Plant Sci Bldg Colo State Univ Ft Collins CO 80523

WESTFALL, HELEN NAOMI, b Grafton, WVa, June 23, 33; m 52; c 3. BACTERIAL PHYSIOLOGY, PATHOGENICITY. *Educ:* Old Dominion Univ, BS, 71, MS, 74; WVa Univ, PhD(med microbiol), 80. *Prof Exp:* Res asst, Old Dominion Univ, 71-72, asst, Gen Biol & Life Sci Labs, 73-74; sci instr, Portsmouth Cath High Sch, 74-75; instr biol, Alderson-Broaddus Col, 75-77; Benedum fel, WVa Univ, 77-80; nat res coun assoc, Naval Med Res Inst, 80-82, res microbiologist, 82-85; asst prof, 85-89, ASSOC PROF, SDAK STATE UNIV, 89- *Mem:* Am Soc Microbiol; Sigma Xi; Am Women Sci. *Res:* Characterization of chemotaxis by Campylobacter species; isolation, purification and characterization of nitrate reductase from species of campylobacter; expression of Leptospira antigens by Escherichia coli, potential for use as diagnostic reagents or vaccines. *Mailing Add:* 1734 Orchard Dr Brookings SD 57006-3521

WESTFALL, JANE ANNE, b Berkeley, Calif, June 21, 28. NEUROCYTOLOGY, CELL ULTRASTRUCTURE. *Educ:* Col Pac, AB, 50; Mills Col, MA, 52; Univ Calif, Berkeley, PhD(zool), 65. *Prof Exp:* Res asst zool, Univ NC, 52-53 & Univ Calif, Berkeley, 55-56; lab technician cancer res, Univ Calif, Berkeley, 57-58, lab technician zool, 58-65, asst res zoologist, 65-67; asst prof anat, 67-70, assoc prof physiol sci & dir ultrastruct res lab, 70-76, PROF ANAT & PHYSIOL, KANS STATE UNIV, 76- *Concurrent Pos:* Vis prof molecular, cellular & develop biol, Univ Colo, Boulder, 74-75. *Mem:* Electron Micros Soc Am; Am Soc Cell Biol; Am Soc Zoologists; Soc Neurosci; Am Asn Anat; Sigma Xi. *Res:* Electron microscopy, scanning, conventional and high voltage, of sensory receptor cells, synapses and neuromuscular junctions in simple nervous systems; ultrastructure of lung tissues with stress and respiratory disease. *Mailing Add:* Dept Anat & Physiol Kans State Univ Manhattan KS 66506

WESTFALL, MINTER JACKSON, JR, b Orlando, Fla, Jan 28, 16; m 45; c 3. BIOLOGY. *Educ:* Rollins Col, BS, 41; Cornell Univ, PhD(nature study), 47. *Prof Exp:* Wildlife technician, Ala Coop Wildlife Res Unit, 37; asst dir mus, Rollins Col, 37-40; from asst to sr asst biol, Cornell Univ, 42-47; from asst prof to assoc prof, 47-69, PROF ZOOL & ENTOM, UNIV FLA, 69- *Concurrent Pos:* US dep game warden, 36-70; Howell fel, Highlands Biol Sta, NC, 53. *Mem:* Entom Soc Am. *Res:* Wildlife management of mourning dove; bird migration and movement; taxonomy, ecology, zoogeography, life histories of the Odonata. *Mailing Add:* Dept Zool Flint Hall Univ Fla Gainesville FL 32603

WESTFALL, RICHARD MERRILL, b Denver, Colo, Dec 17, 56. EXTRATERRESTRIAL RESOURCE DEVELOPMENT, SUPERCONDUCTIVITY POWER SUPPLY DESIGN. *Prof Exp:* Res asst, Nat Oceanic & Atmospheric Admin, Boulder, Colo, 78-79; chemist photovoltaic res, Solar Energy Res Inst, Golden, Colo, 79-80; res dir, Galactic Prod Inc, Denver, 80-82; assoc engr process chem & optoelectronic device fabrication, Tex Med Instruments, 86-87; FOUNDER & RES DIR, CEL SYSTS CORP, DENVER, 82-, GALACTIC MINING INDUSTS INC, DENVER, 88- *Concurrent Pos:* Founder, Galactic Educ Develop Inst, 89- *Mem:* Air Force Asn. *Res:* Solid-state physics device fabrication; optoelectronic device experience; spacecraft propulsion system design; ultra-high power to weight ratio superconducting power supply design; orbital metallurgical satellite design; author of one publication; granted two patents. *Mailing Add:* Galactic Mining Industs Inc 4838 Stuart St Denver CO 80212-2922

WESTFALL, THOMAS CREED, b Latrobe, Pa, Oct 31, 37; m 61; c 1. PHARMACOLOGY. *Educ:* WVa Univ, AB, 59, MS, 61, PhD(pharmacol), 62. *Prof Exp:* From instr to asst prof pharmacol, WVa Univ, 62-65; from asst prof to assoc prof pharmacol, Sch Med, Univ Va, prof, 69-80; MEM FAC, DEPT PHARMACOL, SCH MED, ST LOUIS UNIV, 80- *Concurrent Pos:* Nat Heart Inst fel physiol, Karolinska Inst, Sweden, 63-64; dir grad studies, Dept Pharmacol, Sch Med, Univ Va, 68-; IUPHAR int fel, 74; vis fac scholar, Group Biochem Neuropharmacol, Lab Molecular Biol, Col France, Paris, 74-75. *Mem:* AAAS; Am Soc Pharmacol & Exp Therapeut; Soc Exp Biol & Med; Soc Neurosci; Am Heart Asn. *Res:* Autonomic, cardiovascular, biochemical and neuropharmacology; influence of drugs on the syntheses, uptake, storage, metabolism and receptor interaction of biogenic amines, particularly catecholamines; cardiovascular and autonomic actions of nicotine; neurotransmitter physiology and pharmacology. *Mailing Add:* Dept Pharmacol Sch Med St Louis Univ 1402 S Grand Blvd St Louis MO 63104

WESTFIELD, JAMES D, b Nov 21, 37; US citizen; m 61; c 4. ENVIRONMENTAL ENGINEERING, TECHNOLOGY ASSESSMENT. *Educ:* Univ Nev, Reno, BS, 61; Univ Mich, Ann Arbor, MPH, 64, PhD(environ sci), 66. *Prof Exp:* Chemist, Nev State Health Dept, 61-62; consult engr sanit eng, Industs, 62-66; asst prof civil eng, Ga Inst Technol, 66-68; sr sanit engr & off mgr, Eng Sci Inc, Calif, 68-72; dir environ eng group, Tetra Tech Inc, 72-73; vpres, Develop Sci Inc, Sagamore, 73-83; VPRES, INT RESOURCES GROUP, WASHINGTON, DC, 83- *Concurrent Pos:* Dir solid waste training prog, Ga Inst Technol; consult, Agua Tech Inc & prin, Eco Sci Labs, 67-68; Ford Found residency in eng pract grant, 68-69. *Mem:* Am Soc Civil Engrs; Am Chem Soc; Sigma Xi. *Res:* Energy systems design; chemical and physical water and wastewater treatment; resource management; technology assessment. *Mailing Add:* Agency Int Develop 8100 USAID Islamabad Washington DC 20090-6950

WESTGARD, JAMES BLAKE, b Billings, Mont, Feb 12, 35; m 66; c 3. ELEMENTARY PARTICLE PHYSICS. *Educ:* Reed Col, BA, 57; Syracuse Univ, PhD(physics), 63. *Prof Exp:* Res fel physics, Carnegie Inst Technol, 63-66; asst prof, 66-73, assoc prof, 66-80, PROF PHYSICS, IND STATE UNIV, TERRE HAUTE, 80- *Concurrent Pos:* Vis prof, Dartmouth Col, 83. *Mem:* Am Phys Soc; AAAS. *Res:* Particle physics; biomathematics; computer applications; electromagnetism. *Mailing Add:* Dept Physics Ind State Univ 217 N Sixth St Terre Haute IN 47809

WESTHAUS, PAUL ANTHONY, b St Louis, Mo, Dec 10, 38. ATOMIC PHYSICS, MOLECULAR PHYSICS. *Educ:* St Louis Univ, BS, 61; Washington Univ, PhD(physics), 66. *Prof Exp:* Proj assoc, Theoret Chem Inst, Univ Wis, 66-67; ASSOC PROF PHYSICS, OKLA STATE UNIV, 68- *Concurrent Pos:* Res staff scientist, Yale Univ, 69; NIH career develop award, 72-77. *Mem:* Am Phys Soc; Sigma Xi. *Res:* Atomic structure and radiative transitions; electronic structure of molecules and intermolecular forces; quantum biology; many-electron problem. *Mailing Add:* Dept Physics Okla State Univ Stillwater OK 74074

WESTHEAD, EDWARD WILLIAM, JR, b Philadelphia, Pa, June 19, 30; c 2. BIOCHEMISTRY. *Educ:* Haverford Col, BA, 51, MA, 52; Polytech Inst Brooklyn, PhD(polymer & phys chem), 55. *Prof Exp:* Am Scand Found res fel, Biochem Inst, Univ Uppsala, 55-56, NSF res fel, 56-57; res assoc physiol chem, Univ Minn, 58-60; asst prof biochem, Dartmouth Med Sch, 61-66; assoc prof & actg head dept, 66-71, PROF BIOCHEM, UNIV MASS, AMHERST, 71- *Concurrent Pos:* NIH career develop award, 61-66; NIH spec fel, Oxford Univ, 72-73. *Mem:* Am Chem Soc; Am Soc Biol Chemists. *Res:* Enzyme mechanisms and control; biochemistry of catecholamine secretion. *Mailing Add:* Dept Biochem Univ Mass Amherst MA 01003

WESTHEIMER, FRANK HENRY, b Baltimore, Md, Jan 15, 12; m 37; c 2. ORGANIC CHEMISTRY, ENZYMOLOGY. *Educ:* Dartmouth Col, AB, 32; Harvard Univ, MA, 33, PhD(chem), 35. *Hon Degrees:* DSc, Dartmouth Col, 61, Univ Chicago, 73, Univ Cincinnati, 76, Tufts Univ, 78, Univ NC, 83 & Bard Col, 83, Weizmann Inst, 87 & Univ Ill, Chicago, 88. *Prof Exp:* Nat Res Coun fel chem, Columbia Univ, 35-36; res assoc org chem, Univ Chicago, 36-37, from instr to asst prof chem, 37-44; res supvr, Explosives Res Lab, Nat Defense Res Comt, Pa, 44-45; from assoc prof to prof chem, Univ Chicago, 46-54; prof, Harvard Univ, 54-60, chmn dept, 59-62, Loeb prof, 60-82, sr prof, 82-83, EMER LOEB PROF CHEM, HARVARD UNIV, 83- *Concurrent Pos:* Vis prof chem, Harvard Univ, 53-54; chmn comt surv chem, Nat Acad Sci, 64-65, mem coun, 72-75 & 76-79; mem, President's Sci Adv Comt, 67-70; vis prof, Univ Calif, Los Angeles, 55, Berkeley, 58, Boston Univ, 84, Ohio State Univ, 85 , Univ Calif, San Diego, 86 & 88; mem coun, Am Philos Soc, 81-84. *Honors & Awards:* Willard Gibbs Medal, 70; James Flack Norris Award, Am Chem Soc, 70; Theodore William Richards Medal, 76; Richard Kokes Award, Nat Acad Sci, 80; Chas Frederick Chandler Award, 80; Lewis C Rosenstiel Award, 81; Robert A Welch Award, 82; Arthur C Cope Award, 82; Nichols Medal, 82; Ingold Medal, 83; Nat Medal of Sci, 86; Paracelsus Medal, 88; Priestley Medal, 88. *Mem:* Nat Acad Sci; Am Philos Soc; Am Chem Soc; fel AAAS (secy, 85-90); Am Acad Arts & Sci; foreign mem Royal Soc London. *Res:* Electrostatic effects in organic chemistry; mechanism of nitration; chemical and biochemical oxidation; molecular mechanics; enzymic and chemical decarboxylation; phosphate esters; photoaffinity labeling. *Mailing Add:* Converse Lab Harvard Univ Cambridge MA 02138

WESTHEIMER, GERALD, b Berlin, Ger, May 13, 24. PHYSIOLOGICAL OPTICS. *Educ:* Univ Sydney, BSc, 47; Ohio State Univ, PhD(physics), 53. *Hon Degrees:* DSc, Univ NSW, Australia, 88; ScD, State Univ NY, 90. *Prof Exp:* Optometrist, Australia, 45-51; assoc optom, Ohio State Univ, 51-52; from asst prof to assoc prof physiol optics, 54-60; prof, Univ Houston, 53-54; assoc prof physiol optics & optom, Univ Calif, Berkeley, 60-63, prof physiol optics, 63-68, chmn physiol optics group, 64-67, prof physiol, 68-89, PROF HEAD, DIV NEUROBIOL, UNIV CALIF, BERKELEY, 89- *Concurrent Pos:* Vis researcher, Physiol Lab, Cambridge Univ, 58-59; mem, Nat Acad Sci-Nat Res Coun Comt Vision, 57-72, mem exec coun, 69-72; mem visual sci study sect, NIH, 66-70, mem vision res training comt, Nat Eye Inst, 70-74; ed, Vision Res, 72-79 & 85-89, chmn, 86-91; assoc ed, Invest Ophthal, 73-74, Exp Brain Res, 73-89, Optics Lett, 77-79, J Optical Soc Am, 80-82, Human Neurobiol, 81-87, Spatial Visions, 85 & Ophthal & Physiol Optics, 85, J Physiol, 87-; mem commun sci cluster, President's Biomed Res Panel, 75; chmn, Visual Sci B Study Sect, NIH, 77-79; mem, Bd Sci Counrs, Nat Eye Inst, NIH, 80-84, chmn, 81-83; corresp ed, J Proc Royal Soc London, B, 90- *Honors & Awards:* Tillyer Medal, Optical Soc Am, 78; Proctor Medal, Asn Res Vision & Ophthal, 79; von Sallman Prize, Columbia Univ, 86; Prentice

Medal, Am Acad Optom, 86; Bicentennial Medal, Australia Optom Asn, 88; Sackler lectr, Tel Aviv Univ, 89; Hebb lectr, McGill Univ, 91-; Ferrier lectr, Royal Soc London, 92. *Mem:* Fel AAAS; Sigma Xi; fel Optical Soc Am; Soc Neurosci; Brit Physiol Soc; Int Brain Res Orgn; fel Royal Soc London. *Res:* Biophysics and physiology of visual system; neurophysiology. *Mailing Add:* Dept Molecular & Cell Biol Univ Calif Div Neurobiol Berkeley CA 94720

WESTHOFF, DENNIS CHARLES, b Jersey City, NJ, Nov 20, 42; m 63, 78; c 2. FOOD MICROBIOLOGY. *Educ:* Univ Ga, BSA, 66; NC State Univ, MS, 68, PhD(food sci), 71. *Prof Exp:* From asst prof to assoc prof, 70-79, PROF DAIRY SCI, UNIV MD, COLLEGE PARK, 79-, CHMN DEPT ANIMAL SCI, 85- *Concurrent Pos:* Sci adv, Food & Drug Admin, 76- *Mem:* Am Soc Microbiol; Am Dairy Sci Asn. *Res:* Bioprocessing of foods; food safety. *Mailing Add:* Dept Animal Sci Univ Md College Park MD 20742-2311

WESTINE, PETER SVEN, b Boston, Mass, Apr 8, 40; m 62; c 2. PENTRATION MECHANICS, ARMOR. *Educ:* Swarthmore Col, Pa, BS, 62; Cornell Univ, MCE, 64. *Prof Exp:* Res engr, Southwest Res Inst, 64-71, sr engr, 71-79, staff engr, 79-87; TECH CONSULT, ALCOA TECH LAB, 87- *Mem:* Am Soc Civil Engrs; Am Defense Preparedness Asn. *Res:* The effects of explosion (ground shock, air blast, fragments, and thermal fireballs) on shelters, buildings, people, aircraft and vehicles; aircraft and vehicle armors; KE and CE threats with passive and reactive systems including metals, composites, ceramics, explosives, etc. *Mailing Add:* 603 Jefferson Dr Apollo PA 15613

WESTING, ARTHUR H, b New York, NY, July 18, 28; m 56; c 2. ECOLOGY, CONSERVATION. *Educ:* Columbia Univ, AB, 50, Yale Univ, MF, 54, PhD, 59. *Hon Degrees:* DSc, Windham Col, 73. *Prof Exp:* Res forester, US Forest Serv, 54-55; asst prof forestry, Purdue Univ, 59-64; assoc prof tree physiol, Univ Mass, 64-65; assoc prof biol, Middlebury Col, 65-66; assoc prof bot, Windham Col, 66-71, chmn dept biol, 66-74, prof bot, 71-76; sr res fel, Stockholm Int Peace Res Inst, 76-78 & 83-87; prof ecol & dean, Sch Natural Sci, Hampshire Col, 78-83; sr res fel, Int Peace Res Inst, Oslo, 88-90; CONSULT, 90- *Concurrent Pos:* Fel forest biol, NC State Col, 60; Bullard fel, Harvard Univ, 63-64, res fel, 70; fel bot, Univ Mass, 66; trustee, Vt Wild Land Found, 66-75, Vt Acad Arts & Sci, 67-71 & 91- & Rachel Carson Coun, 79-87, World Coun Biosphere, 88-; fel nuclear sci, St Augustine's Col, 68; vis scholar, Stockholm Int Peace Inst, 75 & 81; adj prof ecol, Hampshire Col, 83- *Honors & Awards:* NY Acad Sci Award, 83; Bulgarian Protection of Nature Medal, 84; Global 500 Award, UN, 90. *Mem:* Fel AAAS; Fauna & Flora Protection Soc; fel Scientists' Inst Pub Info; Int Peace Res Asn; Sigma Xi; Int Primate Protection League. *Res:* Forest ecology; environmental effects of war. *Mailing Add:* Westing Associates RFD 1 Box 126 Putney VT 05346

WESTKAEMPER, JOHN C(ONRAD), b San Antonio, Tex, Dec 5, 23; m 47; c 2. AERONAUTICAL ENGINEERING, MECHANICAL ENGINEERING. *Educ:* Univ Tex, BS, 47, MS, 59, PhD, 67. *Prof Exp:* Aerodynamicist, Consol Vultee Aircraft Corp, 47-50; from sr engr to asst proj engr, Sverdrup & Parcel, Inc, 50-52; sr res engr, Convair Div, Gen Dynamics Corp, 52-55; res engr, Defense Res Lab, Univ Tex, 55-60; res engr, Res Br, Engine Test Facil, ARO, Inc, 60; spec instr, Appl Res Lab, 57-60, asst prof aerospace eng, 60-67, ASSOC PROF AEROSPACE ENG, UNIV TEX, AUSTIN, 67-, RES ENGR, APPL RES LAB, 60- *Mem:* Am Inst Aeronaut & Astronaut; Sigma Xi; Am Soc Eng Educ. *Res:* Fluid mechanics and heat transfer, including testing and test facility design. *Mailing Add:* 3402 Foothills Terr Austin TX 78731

WESTLAKE, DONALD G(ILBERT), b Aurora, Ill, July 9, 28; m 50; c 1. METALLURGY. *Educ:* Northern Ill Univ, BS, 50; Iowa State Univ, PhD(metall), 59. *Prof Exp:* Teacher high sch, Ill, 50-51; jr chemist, Argonne Nat Lab, 51-52; asst, Iowa State Univ, 54-59; asst metallurgist, Mat Sci Div, Argonne Nat Lab, 59-64, assoc metallurgist, 64-72, metallurgist, 72-83, sr metallurgist, 83-84; RETIRED. *Mem:* Sigma Xi. *Res:* Metal-hydrogen alloys; structure and properties of metal hydrides; reaction rates and diffusion; hydrogen embrittlement of metals; interstitial occupancy in hydrides of intermetallic compounds; deformation twinning. *Mailing Add:* 611 Plamondon Ct Wheaton IL 60187

WESTLAKE, DONALD WILLIAM SPECK, b Woodstock, Ont, Feb 27, 31; m 54; c 2. MICROBIOLOGY. *Educ:* Univ BC, BS, 53, MS, 55; Univ Wis, PhD, 58. *Prof Exp:* From asst res officer to assoc res officer, Nat Res Coun Can, 58-66; assoc prof microbiol, 66-69, PROF MICROBIOL & CHMN DEPT, UNIV ALTA, 69- *Honors & Awards:* Can Soc of Microbiol Award, 86. *Mem:* Can Soc Microbiol; Chem Inst Can; Can Soc Biochem; Brit Soc Gen Microbiol. *Res:* Microbial biochemistry; environmental microbiology; metabolic pathways; fermentations. *Mailing Add:* Dept Microbiol Fac Sci Univ Alta Edmonton AB T6G 2M7 Can

WESTLAKE, ROBERT ELMER, SR, b Jersey City, NJ, Oct 2, 18; m 44; c 3. INTERNAL MEDICINE, CARDIOLOGY. *Educ:* Princeton Univ, AB, 40; Columbia Univ, MD, 43. *Prof Exp:* From instr to asst prof med, Col Med, State Univ NY Upstate Med Ctr, 49-52, from clin asst prof to clin prof, 52-67; dir prof serv, Community-Gen Hosp, 67-81; EMER CLIN PROF, STATE UNIV NY UPSTATE MED CTR, 81- *Concurrent Pos:* Mem, Adv Comt Heart, Cancer & Stroke, Surgeon Gen, 66-67; mem, Adv Coun, US Dept Defense, 69-74; electrocardiographer, Syracuse Med Ctr. *Mem:* AAAS; Am Col Physicians; Am Fedn Clin; Am Soc Internal Med (pres, 65-66); fel Soc Advan Med Systs. *Res:* Electrocardiography. *Mailing Add:* 5217 SE Sea Island Way Stuart FL 34997

WESTLAKE, WILFRED JAMES, mathematics, statistics, for more information see previous edition

WESTLAND, ALAN DUANE, b Toledo, Ohio, Dec 29, 29; m 56. INORGANIC CHEMISTRY. *Educ:* Univ Toronto, BA, 53, MA, 54, PhD(chem), 56. *Prof Exp:* Fel, Univ Muenster, 56-58; from asst prof to assoc prof, 58-69, PROF INORG CHEM, UNIV OTTAWA, 69- *Mem:* Am Chem Soc; Chem Inst Can; Royal Soc Chem. *Res:* Analytical and physical inorganic chemistry of the transition elements. *Mailing Add:* Dept Chem Univ Ottawa Ottawa ON K1N 6N5 Can

WESTLAND, ROGER D(EAN), b Winnebago Co, Iowa, June 26, 28; m 56; c 2. PHARMACEUTICAL CHEMISTRY, ORGANIC CHEMISTRY. *Educ:* St Olaf Col, AB, 50; Univ Kans, MS, 52; Univ Mich, PhD(chem), 59. *Prof Exp:* Asst res chemist, Parke, Davis & Co, 52-55; asst, Univ Mich, 57-58; from assoc res chemist to sr res chemist, 59-71, res scientist, 71-73, mgr, 73-85, DIR CHEM-BIOL INFO, PARKE-DAVIS PHARMACEUT RES, 85- *Mem:* Am Chem Soc; Sigma Xi. *Res:* Medicinal chemistry and information science. *Mailing Add:* Parke-Davis Pharmaceut Res 2800 Plymouth Rd Ann Arbor MI 48106

WESTLER, WILLIAM MILO, b Grove City, Pa, Dec 28, 50; m 80. PHYSICAL BIOCHEMISTRY. *Educ:* Grove City Col, BS, 72; John Carroll Univ, MS, 74; Purdue Univ, PhD(chem), 80. *Prof Exp:* Oper dir, Biochem Magnetic Resonance Lab, Purdue Univ, 79-88; OPER DIR, NAT MAGNETIC RESONANCE FACIL, UNIV WIS, MADISON, 88- *Mem:* Am Chem Soc; Sigma Xi. *Res:* One dimensional and two dimensional nuclear magnetic resonance spectroscopic techniques applied to the investigation of the structural and the dynamic properties of biological macromolecules. *Mailing Add:* Dept Biochem/Univ Wis 420 Henry Mall Madison WI 53706

WESTLEY, JOHN LEONARD, b Wilsonville, Nebr, Aug 29, 27; m 56; c 3. ENZYMOLOGY, SULFUR BIOCHEMISTRY. *Educ:* Univ Chicago, PhB, 48, PhD(biochem), 54. *Prof Exp:* NSF fel biochem, Calif Inst Technol, 54-55, res fel, 55-56; from instr to assoc prof, 56-64, PROF BIOCHEM, UNIV CHICAGO, 64- *Concurrent Pos:* USPHS res career develop award, 62-72. *Mem:* Am Soc Biol Chemists. *Res:* Mechanisms of enzyme action; sulfur metabolism; kinetic analysis; enzyme regulation; cyanide detoxication. *Mailing Add:* Dept Biochem CLSC 761B Univ Chicago 920 E 58th St Chicago IL 60637

WESTLEY, JOHN WILLIAM, b Cambridge, Eng, Feb 5, 36; div; c 2. ORGANIC CHEMISTRY, BIOCHEMISTRY. *Educ:* Univ Nottingham, BSc, 58, PhD(org chem), 61. *Prof Exp:* Res assoc org chem, Stanford Univ, 61-62 & Med Ctr, Univ Calif, San Francisco, 62-63; res assoc org chem & biochem, Sch Med, Stanford Univ, 63-68; sr chemist, 68-71, res fel, Dept Microbiol, 71-80, asst dir microbiol, Chem Res Dept, Hoffman-La Roche Inc, 80-85; ASSOC DIR, BIOMOLECULAR DISCOVERY, SMITHKLINE & FRENCH LABS, 85- *Concurrent Pos:* Squibb Inst Med Res & NIH fels, 61-62; NIH fel, 62-63; NASA grant, 63-68. *Mem:* Am Chem Soc; Royal Soc Chem; NY Acad Sci; Sigma Xi. *Res:* Structure determination, chemistry, activity and biosynthesis of natural products; optical resolution of asymmetric compounds; the polyether antibiotics; ionophores. *Mailing Add:* SmithKline Beecham Pharmaceut PO Box 1539/MCL-950 King Prussia PA 19406-0939

WESTMACOTT, KENNETH HARRY, b Wantage, Eng, Nov 22, 29; m 64; c 2. PHYSICAL METALLURGY, MATERIALS SCIENCE. *Educ:* Univ Birmingham, PhD(phys metall, mat sci), 69. *Prof Exp:* Exp officer metall, Atomic Energy Res Estab, Eng, 51-62; res assoc mat sci, Stanford Univ, 63; res physicist metal physics, Michelson Lab, US Naval Weapons Ctr, 64-76; PRIN INVESTR, LAWRENCE BERKELEY LAB, 76- *Mem:* Assoc Brit Inst Metall; Am Inst Mining, Metall & Petrol Engrs. *Res:* Structure of metals; crystal lattice defects; mechanical properties; relation of microstructure to properties; quench and irradiation damage; dislocation theory; transmission electron microscopy. *Mailing Add:* Mat & Molecular Res Div Lawrence Berkeley Lab Berkeley CA 94720

WESTMAN, JACK CONRAD, b Cadillac, Mich, Oct 28, 27; m 53; c 3. CHILD PSYCHIATRY. *Educ:* Univ Mich, BS, 49, MD, 52, MS, 59. *Prof Exp:* Intern, Duke Univ Hosp, 52-53; resident psychiat, Univ Mich, 56-57; from instr to assoc prof, Med Sch, Univ Mich, 58-65; coordr diag & treat unit, Ctr Ment Retardation & Human Develop, 66-74, dir child psychiat div, Univ Hosp, 65-78, PROF PSYCHIAT, MED SCH, UNIV WIS-MADISON, 65- *Concurrent Pos:* Fel child psychiat, Univ Mich, 57-59; lectr, Sch Social Work, Univ Mich, 59-64; mem, Sr Staff, Children's Psychiat Hosp, Ann Arbor, Mich, 59-65, dir outpatient serv, 61-62, dir outpatient & day care serv, 62-65. *Mem:* Fel Am Orthopsychiat Asn; Asn Am Med Cols; fel Am Acad Child Psychiat; Am Asn Ment Deficiency; Soc Prof Child Psychiatrists; Sigma Xi. *Res:* Efficacy of medication of hyperkinesis; psychiatric aspects of learning disabilities and mental retardation; the role of child psychiatry in divorce; predicting later adjustment from nursery school behavior; individual differences in children; child advocacy. *Mailing Add:* 1234 Dartmouth Rd Madison WI 53705

WESTMAN, WALTER EMIL, b New York, NY, Nov 5, 45. PLANT ECOLOGY, ENVIRONMENTAL POLICY. *Educ:* Swarthmore Col, BA, 66; Macquarie Univ, MSc, 69; Cornell Univ, PhD(ecol), 71. *Prof Exp:* Congressional fel ecol, US Senate Subcomt Air & Water Pollution, 71; lectr bot, Univ Queensland, 72-74; vis prof environ planning, Univ Calif, Los Angeles, 75-76, prof geog, 76-84; res scientist, NASA Ames Res Ctr, 84-87; STAFF SCIENTIST, LAWRENCE BERKELEY LAB, UNIV CALIF, 87- *Concurrent Pos:* Consult, Off Technol Assessment, 76-79, Global 2000 Study; prin investr, NSF, 77-81; Nat Res Coun associateship, NASA Ames Res Ctr, 85-87. *Mem:* Ecol Soc Am; Am Asn Geographers. *Res:* Plant ecology; remote sensing; environmental impact assessment; general environmental sciences. *Mailing Add:* Appl Sci Div Lawrence Berkeley Lab One Cyclotron Rd Bldg 90 Rm 4000 Berkeley CA 94720

WESTMANN, RUSSELL A, b Fresno, Calif, May 20, 36; m 58; c 2. CIVIL ENGINEERING, APPLIED MECHANICS. *Educ:* Univ Calif, Berkeley, BS, 59, MS, 60, PhD(civil eng), 62. *Prof Exp:* Actg asst prof civil eng, Univ Calif, Berkeley, 62; NSF fel math, 62-63; res fel aeronaut, Calif Inst Technol, 63-65, asst prof civil eng, 65-66; from asst prof to assoc prof mech & struct, 66-73, actg chmn dept, Sch Eng & Appl Sci, 69-70, assoc dean, 76-81, PROF MECH & STRUCT, SCH ENG & APPL SCI, UNIV CALIF, LOS ANGELES, 73- *Concurrent Pos:* Mem transp res bd, Nat Res Coun, 65-75; sr vis fel, Sci Res Coun, Eng, 74. *Mem:* Am Soc Civil Engrs. *Res:* Solid mechanics; engineering mathematics; fracture mechanics. *Mailing Add:* Dept Mech Eng Univ Calif Los Angeles 405 Hilgard Ave Los Angeles CA 90024

WESTMEYER, PAUL, b Dillsboro, Ind, Dec 9, 25; m 47; c 6. SCIENCE EDUCATION. *Educ:* Ball State Univ, BS, 49, MS, 53; Univ Ill, EdD, 60. *Prof Exp:* Pub sch teacher, Ind, 49-54; instr chem & sci, Univ Ill High Sch, 54-61; from asst prof to assoc prof sci educ, Univ Ill, 60-63; assoc prof, Univ Tex, 63-66; prof, Fla State Univ, 66-73; PROF EDUC, UNIV TEX, SAN ANTONIO, 73- *Concurrent Pos:* Instr, Exten, Purdue Univ, 51-54; consult, Chicago Sch Bd, Ill, 54; proj assoc, Earlham Col, 59-62; partic, Int Seminar Chem, Dublin, Ireland, 60; consult, CBA Insts, 61-63; NSF in-serv grant, 64-65, coop col-sch sci grants, 64-66, summer inst grants & in-serv grants, 67-73 & 81-82; vis scientist, Uniformed Servs Univ Health Scis, 87-90. *Mem:* AAAS; Am Chem Soc; Nat Asn Res Sci Teaching; Nat Sci Teachers Asn; Asn Educ Teachers Sci (pres, 71-72). *Res:* Course development and evaluation; computer-assisted-instruction in chemistry; development of laboratory materials, tests and instructional procedures in chemistry; better utilization of staff in science teaching; computer programs for teaching statistics; higher education; technology assisted instruction in health sciences. *Mailing Add:* 6900 Loop 1604 W San Antonio TX 78249

WESTMORE, JOHN BRIAN, b Welling, Eng, Apr 23, 37; m 61; c 2. PHYSICAL CHEMISTRY, INORGANIC CHEMISTRY. *Educ:* Univ London, BSc, 58, PhD(phys chem), 61. *Prof Exp:* Fel chem, Nat Res Coun Can, 61-63; asst prof, 63-69, assoc prof, 69-76, PROF CHEM, UNIV MAN, 76- *Mem:* Chem Inst Can; The Chem Soc; Am Soc Mass Spectrometry. *Res:* Ionization and dissociation of molecules; chromatography and mass spectrometry of metal chelates, nucleosides, nucleotides and steroids; analytical chemistry; thermodynamics of metal chelation. *Mailing Add:* Dept Chem Univ Man Winnipeg MB R3T 2N2 Can

WESTMORELAND, BARBARA FENN, b New York, NY, July 22, 40. ELECTROENCEPHALOGRAPHY, NEUROSCIENCE. *Educ:* Mary Washington Col, BS, 61; Univ Va, MD, 65. *Prof Exp:* Intern, Vanderbilt Univ, 65-66; resident neurol, Univ Va, 66-70; fel, EEG, Mayo Clin, 70-71; asst prof neurol, 73-78, assoc prof neurol, 78-85, PROF NEUROL, MAYO MED SCH, 85- *Concurrent Pos:* Consult, EEG, Mayo Clin, 71-; epilepsy adv comt, NIH. *Honors & Awards:* Colgate Darden Award, Mary Washington Col, 61. *Mem:* Am Epilepsy Soc (treas, 78-80, pres, 87-88); Am EEG Soc (secy, 85-87); Central EEG Assoc (secy & treas, 76-78, pres, 79-80); Sigma Xi (pres, Mayo Chapter, 87-88); NIH; Am Acad Neurol. *Res:* Clinical electroencephalography as it relates to various types of EEG patterns and the EEG findings in various disease entities as in epilepsy, metabolic derangements, and coma. *Mailing Add:* Mayo Clin Electroncephalog Sect 200 First St SW Rochester MN 55905

WESTMORELAND, DAVID GRAY, b Mooresville, NC, Aug 5, 46; m 80; c 2. ANALYTICAL & PHYSICAL CHEMISTRY. *Educ:* Univ NC, Chapel Hill, BS, 68; Stanford Univ, PhD(chem), 73. *Prof Exp:* Sr scientist chem, 73-83, RES FEL, ROHM & HAAS CO, 83- *Mem:* Am Chem Soc; Am Soc Mass Spectrometry; AAAS. *Res:* Structure determination of unknown compounds; trace analysis for organic compounds. *Mailing Add:* Rohm & Haas Co 727 Norristown Rd Spring House PA 19477

WESTMORELAND, PHILLIP R, b Asheboro, NC, Mar 20, 51; m 79; c 1. COMBUSTION KINETICS, GAS KINETICS FOR MATERIALS SYNTHESIS. *Educ:* NC State Univ, BS, 73; La State Univ, MS, 74; Mass Inst Technol, PhD(chem eng), 86. *Prof Exp:* Res engr, Chem Technol Div, Oak Ridge Nat Lab, 74-79; ASST PROF, CHEM ENG DEPT, UNIV MASS, AMHERST, 86- *Concurrent Pos:* Consult, Oak Ridge Nat Lab, 80-, Mass Inst Technol, 86-89; NSF presidential young investr, 90. *Mem:* Am Inst Chem Engrs; Am Chem Soc; Am Soc Eng Educ; Combustion Inst; Electrochem Soc; AAAS. *Res:* Quantum theories of elementary-reaction Kinetics; experiments and modeling of chemical reactions in microelectronics fabrication (particularly plasma-enhanced CVD), flame combustion, pyrolysis and steam cracking of hydrocarbons. *Mailing Add:* Dept Chem Eng 159 Goessmann Lab Univ Mass Amherst MA 01003

WESTMORELAND, WINFRED WILLIAM, b Santa Maria, Calif, Feb 7, 19; m 42; c 3. DENTISTRY. *Educ:* Col Physicians & Surgeons San Francisco, DDS, 42; Univ Calif, MPH, 57; Am Bd Dent Pub Health, dipl. *Prof Exp:* Extern oral surg, San Francisco Hosp, 46-47; clin instr oper dent, Col Physicians & Surgeons, Univ of Pac, 49-53, asst clin prof prosthetic dent & lectr dent mat, 53-58, lectr pub health dent, 58-69, asst clin prof dent pub health, 66-69; sr dent consult, Calif Dept Health, Oakland, 67-81; dent health prog consult, 81-84; RETIRED. *Concurrent Pos:* Pub health dent officer, Calif State Dept Health, 55-67; consult, Resident Training Prog, US Army, Ft Ord, Calif, 67-72; pvt practr. *Mem:* Am Dent Asn; Am Pub Health Asn; Int Asn Dent Res. *Res:* Epidemiology of dental caries; periodontal disease; cleft lip and palate; radiation exposure; dental fluorosis; dental health administration and education. *Mailing Add:* 644 S N St Livermore CA 94550

WESTNEAT, DAVID FRENCH, b Oradell, NJ, June 18, 29; m 58; c 3. ANALYTICAL CHEMISTRY. *Educ:* Allegheny Col, BS, 50; Univ Pittsburgh, PhD(chem), 56. *Prof Exp:* Res chemist, E I du Pont de Nemours & Co, 56-60; asst prof chem, Akron Univ, 60-65; chmn dept, 68-72, assoc prof, 65-77, PROF CHEM, WITTENBERG UNIV, 77- *Concurrent Pos:* Vis prof, Sci & Soc Prog, Cornell Univ, 72-73; vis scientist, Univ Hyg Lab, Univ Iowa, 80-81. *Mem:* AAAS; Sigma Xi. *Res:* Instrumental analysis, especially absorption spectroscopy and gas chromatography. *Mailing Add:* Dept Chem Wittenburg Univ PO Box 720 Springfield OH 45501

WESTNEAT, DAVID FRENCH, JR, b Wilmington, Del, June 16, 59; m 85; c 1. BEHAVIORAL ECOLOGY, ORNITHOLOGY. *Educ:* Carleton Col, BA, 81; Univ NC, PhD(zool), 86. *Prof Exp:* Vis lectr behav, Univ NC, Chapel Hill, 87; NSF fel res, Cornell Univ, Ithaca, NY, 87-89, res assoc, 89-90; ASST PROF ZOOL, UNIV KY, LEXINGTON, 90- *Concurrent Pos:* Prin investr, Cornell Univ/Univ Ky, 89- *Honors & Awards:* Young Investr Award, Am Soc Naturalists, 89. *Mem:* Sigma Xi; Am Ornithologists Union; Animal Behav Soc; Int Soc Behav Ecologists; Am Soc Naturalists; Soc Study Evolution. *Res:* Ecological causes and evolutionary consequences of animal social behavior; use of genetic techniques to analyze the outcome of alternative mating behavior in birds and to study the costs and benefits of parental behavior. *Mailing Add:* Univ Ky 101 Morgan Bldg Lexington KY 40506-0225

WESTOFF, CHARLES F, POPULATION RESEARCH. *Prof Exp:* FAC MEM, POPULATION RES OFF, PRINCETON UNIV. *Mem:* Inst Med-Nat Acad Sci. *Mailing Add:* Princeton Univ Pop Res Off 21 Prospect Ave Princeton NJ 08544

WESTON, ARTHUR WALTER, b Smiths Falls, Ont, Feb 13, 14; nat US; m 40; c 3. ORGANIC CHEMISTRY. *Educ:* Queen's Univ, Ont, BA, 34, MA, 35; Northwestern Univ, PhD(org chem), 38. *Prof Exp:* Asst chem, Northwestern Univ, 35-37, res assoc, 38-40; from res chemist to asst head org res, Abbott Labs, 40-54, asst to dir develop, 54-55, asst dir, 55-57, dir res, 57-59, dir res & develop, 59-61, dir co, 59-69, vpres res & develop, 61-68, vpres sci affairs, 68-78, vpres corp licensing, 78-79, consult, 79-85; PRES, ARTHUR W WESTON & ASSOCS, 79- *Concurrent Pos:* Mem War Manpower Comn, Off Sci Res & Develop, 42-45; mem ad hoc comt chem agts, Dept of Defense, 61-65; mem, Indust Res Inst & Dirs Indust Res; mem indust panel sci & technol, NSF, 74-79. *Mem:* Am Chem Soc; Pharmaceut Mfrs Asn; Chem Inst Can; Sigma Xi. *Res:* Organic medicinals; antibiotics; plant processes. *Mailing Add:* 349 E Hilldale Pl Lake Forest IL 60045

WESTON, CHARLES RICHARD, b South Gate, Calif, Apr 24, 33; m 53; c 4. DEVELOPMENTAL BIOLOGY, MICROBIAL ECOLOGY. *Educ:* Univ Calif, Santa Barbara, BA, 57; Princeton Univ, PhD(biol), 65. *Prof Exp:* Res scientist, E R Squibb Inst Med Res, 60-61; res assoc & asst prof, Univ Rochester, 62-66; Nat Acad Sci-Nat Res Coun resident res assoc, Jet Propulsion Lab, Calif Inst Technol, 66-68; lectr, 68-69, ASSOC PROF BIOL, CALIF STATE UNIV, NORTHRIDGE, 70- *Mem:* AAAS; Bot Soc Am; Am Soc Microbiol. *Res:* Morphogenesis of fungal mycelium; development of instrumentation for remote life detection; interactions and distribution of soil microorganisms. *Mailing Add:* Dept Biol Calif State Univ 18111 Nordhoff St Northridge CA 91330

WESTON, HENRY GRIGGS, JR, b Hemet, Calif, Apr 7, 22; m 47; c 3. ECOLOGY, VERTEBRATE ZOOLOGY. *Educ:* San Diego State Col, BA, 43; Univ Calif, Berkeley, MA, 47; Iowa State Col, PhD(zool), 50. *Prof Exp:* Asst prof biol, Grinnell Col, 50-55; prof biol, San Jose State Univ, 55-87; RETIRED. *Mem:* Wildlife Soc; Cooper Ornith Soc; Am Ornith Union; Am Soc Mammal. *Res:* Birds of California; bird-banding; field ecology. *Mailing Add:* 8650 Brockway Vista Kings Beach CA 95717

WESTON, JAMES A, b Washington, DC, June 20, 36; m 58; c 2. DEVELOPMENTAL BIOLOGY, CELL BIOLOGY. *Educ:* Cornell Univ, AB, 58; Yale Univ, PhD(biol), 62. *Prof Exp:* USPHS fel zool, Univ Col, London, 62-64; from asst prof to assoc prof biol, Case Western Reserve Univ, 64-70; assoc prof, 70-74, PROF BIOL, UNIV ORE, 74- *Concurrent Pos:* Mem med adv bds, Nat Neurofibromatosis Found, Inc & Familial Dysautonomia Found, Inc. *Mem:* Fel AAAS; Soc Develop Biol (secy, 73-76); Int Soc Develop Biol; Am Soc Cell Biol. *Res:* Cellular control of morphogenetic movements in vertebrate development; regulation of cellular phenotypic expression of neural crest cells in vivo and in vitro; properties of cell surfaces in normal and transformed states. *Mailing Add:* Inst Neurosci Univ Ore Eugene OR 97403

WESTON, JAMES T, medicine, pathology, for more information see previous edition

WESTON, JOHN COLBY, b Dover-Foxcroft, Maine, Dec 31, 26; m 51; c 2. HISTOLOGY, EMBRYOLOGY. *Educ:* Bowdoin Col, AB, 51; Syracuse Univ, MS, 54, PhD(zool), 56. *Prof Exp:* From instr to assoc prof anat, Col Med, Ohio State Univ, 55-67; assoc prof, 67-69, PROF BIOL, MUHLENBERG COL, 69-, CHIEF HEALTH PROF ADV, 76- *Concurrent Pos:* Mem, Working Group on Essential Knowledge (GPEP), Asn Am Med Cols, adv group, Assessing Change Med Educ; col marshall, Muhlenberg Col, 86- *Res:* Biochemical, histochemical and electron microscopic analyses of development; cytology. *Mailing Add:* Dept Biol Muhlenberg Col 24th & Chew St Allentown PA 18104

WESTON, KENNETH CLAYTON, b Buffalo, NY, Jan 8, 32; m 63; c 3. COMPUTER AIDED ENGINEERING. *Educ:* Cornell Univ, BME, 55; Rice Univ, MS, 65, PhD(mech eng), 69. *Prof Exp:* Res scientist aerodyn & thermodyn, NASA Lewis Res Ctr, Ohio, 55-58, res engr, Space Task Group, Va, 58-60, sect head aerothermodyn, Manned Spacecraft Ctr, Tex, 60-68; assoc prof mech & aerospace eng, 68-80, PROF MECH ENG, UNIV TULSA, 80- *Concurrent Pos:* Vis prof, US Mil Acad, 85-86. *Honors & Awards:* Teetor Award, Soc Automotive Engrs, 78; Robert W Cox Award, Am Soc Mech Engrs. *Mem:* Am Soc Eng Educ; Am Inst Aeronaut & Astronaut; Am Soc Mech Engrs; Soc Automotive Engrs. *Res:* Combined radiative and conductive energy transfer; thermodynamics; mechanics of continua; fluid mechanics; energy conversion; computer aided design and graphics; gas turbines; engineering software tools. *Mailing Add:* Dept Mech Eng Univ Tulsa 600 S College Ave Tulsa OK 74104

WESTON, KENNETH W, b Milwaukee, Wis, Feb 1, 29; m 66; c 2. ALGEBRA. *Educ:* Univ Wis, BS, 53, MS, 55, PhD(math), 63. *Prof Exp:* Instr math, Univ Wis-Milwaukee, 61-63; asst prof, Univ Notre Dame, 63-69; assoc prof, Marquette Univ, 69-71; assoc prof, 71-89, PROF, UNIV WIS-

PARKSIDE, 89- *Mem:* Am Math Soc; Math Asn Am. *Res:* Groups satisfying Engel condition and connections between ring and group theory; model theory. *Mailing Add:* Dept Math Univ Wis-Parkside Box 2000 Kenosha WI 53141

WESTON, RALPH E, JR, b San Francisco, Calif, Nov 9, 23; m 51; c 3. CHEMICAL KINETICS, PHOTOCHEMISTRY. *Educ:* Univ Calif, BS, 46; Stanford Univ, PhD(chem), 50. *Prof Exp:* Asst, Harvard Univ, 49-51; from assoc chemist to chemist, 51-65, dep dept chmn, 82-89, SR CHEMIST, BROOKHAVEN NAT LAB, 65- *Concurrent Pos:* Vis scientist, Saclay Nuclear Res Ctr, France, 60-61; vis lectr, Univ Calif, Berkeley, 68-69; lectr, Columbia Univ, 71-72; mem, comt chem kinetics, Nat Res Coun, 75-77, Army Basic Res Comt, 75-81, adv bd, Off Chem & Chem Technol, 80-81. *Mem:* Am Chem Soc; Sigma Xi. *Res:* Kinetics and dynamics of gas phase reactions; photochemistry; using laser and synchrotron radiation. *Mailing Add:* Chem Dept Brookhaven Nat Lab Upton NY 11973

WESTON, RAYMOND E, b Chicago, Ill, July 15, 17. CARDIOVASCULAR DISEASE, NUTRITION. *Educ:* Univ Chicago, MD & PhD(physiol), 41. *Prof Exp:* CLIN PROF MED, UNIV CALIF, LOS ANGELES, 78- *Mem:* Am Bd Int Med; Am Fedn Clin Res; Am Physiol Soc; Am Soc Clin Invest; Soc Exp Biol & Med. *Res:* Salt and water metabolism. *Mailing Add:* 9201 Sunset Blvd 512 Los Angeles CA 90069

WESTON, ROY FRANCIS, b Reedsburg, Wis, June 25, 11; m 34; c 2. ENVIRONMENTAL ENGINEERING. *Educ:* Univ Wis, BCE, 33; NY Univ, MCE, 39; Environ Eng Intersoc, dipl. *Hon Degrees:* DEng, Drexel Univ, 81. *Prof Exp:* Jr hwy engr, Wis Hwy Dept, 34-36; dist engr, Wis Dept Health, 36-37; sanit eng res fel, NY Univ, 37-39; sanit engr, Atlantic Refining Co, Philadelphia, 39-55; PRES & CHMN BD, ROY F WESTON, INC, ENVIRON CONSULTS, 55- *Concurrent Pos:* Mem vis comt, Dept Civil & Urban Eng, Univ Pa & Ctr Marine & Environ Studies, Lehigh Univ; pres, Am Acad Environ Engrs, 73-74; mem, US Environ Control Seminar, Neth & Eastern Europe, 72; mem, Comt Environ, US Chamber Com, 75; mem, Indust & Prof Adv Comt, Pa State Univ, 75. *Honors & Awards:* Indust Wastes Medal, Water Pollution Control Fedn, 50, Arthur Sidney Bedell Award, 59; Gordon Maskew Fair Award, Am Acad Environ Engrs, 77. *Mem:* Nat Acad Eng; fel Am Soc Civil Engrs; Am Pub Health Asn; Am Chem Soc; Am Inst Chem Engrs. *Res:* Environmental control. *Mailing Add:* Roy F Weston Inc One Weston Way West Chester PA 19380-1499

WESTON, VAUGHAN HATHERLEY, b Parry Sound, Ont, May 1, 31; m 54; c 4. APPLIED MATHEMATICS, THEORETICAL PHYSICS. *Educ:* Univ Toronto, BA, 53, MA, 54, PhD, 56. *Prof Exp:* Lectr math, Univ Toronto, 57-58; res assoc, Radiation Lab, Univ Mich, 58-59, from assoc res mathematician to res mathematician, 59-69; PROF MATH, PURDUE UNIV, WEST LAFAYETTE, 69- *Mem:* Am Math Soc; Soc Indust & Appl Math; Am Phys Soc. *Res:* Electromagnetic theory; diffraction; plasmas; inverse scattering. *Mailing Add:* Div Math Sci Purdue Univ West Lafayette IN 47907

WESTON, WILLIAM LEE, b Grand Rapids, Minn, Aug 13, 38; m 64; c 2. PEDIATRIC DERMATOLOGY, CUTANEOUS DERMATOLOGY. *Educ:* Whitman Col, AB, 60; Univ SDak, BMS, 63; Univ Colo, MD, 65. *Prof Exp:* Resident pediat, Univ Colo, 65-67; resident, Univ Calif, San Francisco, 67-68; med officer, Ft George, Minn, 68-70; resident dermat, 70-72, fel immunol, 72-73, from asst prof to assoc prof, 73-79, CHMN DEPT DERMAT, UNIV COLO, 76-, PROF DERMAT & PEDIAT, 79- *Mem:* Soc Pediat Dermat (secy/treas, 76-80, pres, 83-84); Am Acad Dermat; Soc Invest Dermat. *Res:* Skin diseases in children, especially birthmarks. *Mailing Add:* Univ Colo Sch Med 4200 E Ninth St Denver CO 80220

WESTOVER, JAMES DONALD, b Clarkdale, Ariz, Sept 22, 34; m 59; c 4. ORGANIC CHEMISTRY. *Educ:* Ariz State Univ, BS, 50, MS, 62; Brigham Young Univ, PhD(chem), 66. *Prof Exp:* Res chemist, Dacron Res Lab, E I du Pont de Nemours & Co, Inc, NC, 65-70; from asst prof to assoc prof, 70-82, PROF CHEM, CALIF POLYTECH STATE UNIV, 82- *Res:* Synthetic organic chemistry in area of nitrogen heterocyclic compounds; polyester fibers. *Mailing Add:* 141 Del Norte Way San Luis Obispo CA 93405-1507

WESTOVER, LEMOYNE BYRON, b Curwensville, Pa, Aug 12, 28; m 52; c 3. ANALYTICAL CHEMISTRY. *Educ:* Dickinson Col, BS, 50; Pa State Univ, MS, 52. *Prof Exp:* Chemist anal chem, Hercules, 52-54, US Army, 54-56; chemist anal chem, Dow Chem Co, 56-69, group leader, mass spectros, 59-70, res mgr anal chem, 70-89; RETIRED. *Mem:* Am Chem Soc; Sigma Xi. *Res:* Mass spectrometry; liquid chromatography; gas chromatography; infrared spectroscopy. *Mailing Add:* 318 Mt Everest Ct Clayton CA 94517

WESTOVER, THOMAS A(RCHIE), b Westover, Pa, June 1, 09; m 40; c 3. ELECTRICAL ENGINEERING. *Educ:* Carnegie Inst Technol, BSc, 36; Stevens Inst Technol, MSc, 45. *Prof Exp:* Tester & inspector, Allis Chalmers Mfg Co, Pa, 36-40; tester, Pub Serv Elec & Gas Co, NJ, 40-42; sr engr, Fairchild Camera & Instrument Corp, 42-46, proj engr, 46-47; staff engr, Servo Corp Am, 47-54, engr in charge control systs, Eng Dept, 54-61; asst chief electronics engr, Missile Systs Div, Repub Aviation Corp, 61-63; asst chief engr, Electronic Prod Div, Fairchild Hiller Corp, 63-65; eng mgr, RR Prods Div, Servo Corp Am, 65-70, sr consult engr, Advan Eng Dept, 70-72; RETIRED. *Concurrent Pos:* Instr, State Univ NY Agr & Tech Inst, Long Island, 50-60. *Mem:* Sr mem Inst Elec & Electronics Engrs. *Res:* Servomechanisms; electricity and magnetism; instruments and equipment; electronic and infrared systems; meteorological measurement systems. *Mailing Add:* 22 Parkway Dr Sag Harbor NY 11963

WESTPFAHL, DAVID JOHN, b Scranton, Pa, Aug 19, 53. ASTROPHYSICS. *Educ:* Dartmouth Col, AB, 75; Mont State Univ, MS, 78 & PhD(physics), 85; Yale Univ, MS, 79. *Prof Exp:* Instr physics, Mont State Univ, 85-86; res assoc astrophys, Dominion Astrophys Observ, 86-88; asst scientist, Nat Radio Astron Observ, 88-89; ASST PROF ASTROPHYS, NMEX INST MINING & TECHNOL, 89- *Concurrent Pos:* Fel, Kahlmeyer Found, 89, dir, 90-; dir & prin investr, Joint Observ Cometary Res, 90-; collaborating scientist, Nat Radio Astron Observ, 90- *Mem:* Am Astron Soc. *Res:* Internal kinematics and dynamics of disk galaxies; evolution of gas in spiral galaxies. *Mailing Add:* Dept Physics NMex Inst Mining & Technol Socorro NM 87801

WESTPHAL, HEINER J, b Seesen, Ger, Feb 13, 35; m; c 2. MAMMALIAN GENE RESEARCH. *Educ:* Med Sch, Bonn, MD, 62, Venia legendi, Ulm, 75. *Prof Exp:* Mem staff, Inst Biochem, Freiburg, Ger, 62-65, Inst Hyg & Microbiol, 65-66, Salk Inst Biol Sci, San Diego, Calif, 67-70 & Cold Spring Harbor Lab, NY, 72-; CHIEF, LAB MAMMALIAN GENES & DEVELOP, NAT INST CHILD HEALTH & HUMAN DVELOP, NIH, 72- *Concurrent Pos:* Assoc ed, Molecular Biol & Med; adv bd mem, Experientia; instr, FAES Grad Sch, NIH. *Honors & Awards:* Gerhard-Domagk Prize. *Mem:* Soc Develop Biol. *Res:* Enzyme biochemistry and regulation in yeast; induction of cell DNA synthesis during lytic infection of monkey cells with SV40; integration of SV40 DNA in transformed cells; mammalian gene regulation. *Mailing Add:* Nat Inst Child Health & Human Develop NIH Bldg 6 Rm 338 Bethesda MD 20892

WESTPHAL, JAMES ADOLPH, b Dubuque, Iowa, June 13, 30; m 67; c 1. PLANETARY SCIENCES. *Educ:* Univ Tulsa, BS, 53. *Prof Exp:* Sr res fel, 66-71, assoc prof planetary sci, Calif Inst Technol, 71-77; PROF PLANETARY SCI, CALIF INST TECHNOL, 78- *Concurrent Pos:* Prin investr, wide field/planetary camera, space telescope. *Mem:* Am Astron Soc. *Res:* Infrared astronomy; infrared atmospheric properties; planetary astronomy; astronomical instrumentation; space astronomy. *Mailing Add:* M5170-25 Planetary Sci Calif Inst Technol Pasadena CA 91125

WESTPHAL, MILTON C, JR, b Philadelphia, Pa, June 2, 26; m 78; c 6. PEDIATRICS, GASTROENTEROLOGY. *Educ:* Yale Univ, BS, 47; Univ Pa, MD, 51; Am Bd Pediat, dipl, 57. *Prof Exp:* Intern, Univ Pa Hosp, 51-52; resident, Children's Hosp Philadelphia, 54-56; from asst instr to instr pediat, Sch Med, Univ Pa, 55-61; from asst prof to assoc prof, State Univ NY-Buffalo, 61-67; chmn dept, 67-76, PROF PEDIAT, COL MED, MED UNIV SC, 67-, CHIEF, SECT GASTROENTEROL, 76- *Concurrent Pos:* Chief resident, Children's Hosp Philadelphia, 56-57, from asst physician to assoc physician, 57-61, mem, Res Dept, 57-61; asst pediatrician to outpatients, Pa Hosp, 57-58, hosp, 58-59, assoc pediatrician, 59-61; asst attend, Children's Hosp Buffalo, 61-64, assoc attend, 64-; dir, Buffalo Poison Control Ctr, 61-67; proj dir & chmn, Comt Prin Investrs, Collab Study Cerebral Palsy, Ment Retardation & Other Neurol & Sensory Dis Infancy & Childhood, 63-64; asst dean admis, Univ SC, 86- *Mem:* Am Pediat Soc. *Res:* Gastroenterology; ion transport across intestinal muscosain in cystic fibrosis. *Mailing Add:* 171 Ashley Ave Charleston SC 29425

WESTPHAL, WARREN HENRY, b Easton, Pa, Feb 19, 25; m 46; c 3. ENERGY MINERAL DEVELOPMENT. *Educ:* Columbia Univ, AB, 47. *Prof Exp:* Jr mining engr, NJ Zinc Co, NJ, 47-48, geologist, 48-49, res geologist, Pa, 49-50, geophysicist, Colo, 50-55; sr geologist, Tidewater Assoc Oil Co, NMex, 55; chief geophysicist, Utah Construct & Mining Co, 56-59; sr geophysicist, Stanford Res Inst, 59-66, chmn earth sci dept, 66-69; vpres mining, Intercontinental Energy Corp, 69-79; PRES, WESTPHAL ASSOCS INC, 79- *Honors & Awards:* Founder Award, Energy Minerals Div, Am Asn Petrol Geologists. *Mem:* AAAS; Am Inst Prof Geologists; Am Asn Petrol Geologists; Am Inst Mining, Metall & Petrol Engrs; Soc Economic Geologists. *Res:* Earthquake seismology; uranium geology; in-situ mining; underground coal gasification; geochemistry of gold deposition. *Mailing Add:* 4398 S Akron Englewood CO 80111

WESTRUM, EDGAR FRANCIS, JR, b Albert Lea, Minn, Mar 16, 19; m 43; c 4. PHYSICAL CHEMISTRY, THERMODYNAMICS. *Educ:* Univ Minn, BChem, 40; Univ Calif, PhD(phys chem), 44. *Prof Exp:* Res chemist, Metall Lab, Univ Chicago, 44-46 & Radiation Lab, Univ Calif, 46; from asst prof to assoc prof, 46-57, prof, 58-89, EMER PROF PHYS CHEM, UNIV MICH, ANN ARBOR, 89- *Concurrent Pos:* Chmn comt data for sci & technol, Nat Acad Sci, 73; chmn comn & phys chem div, Int Union Pure & Appl Chem, 73-77; ed, J Chem Thermodyn, 68-79 & Bull Thermodyn & Thermochem, 55-75; secy gen, Comt on Data for Sci & Technol, 74-81, ed-in- chief, 82- *Honors & Awards:* Hugh Huffman Award, F D Rossini Lectr, Bausch & Lomb Hon Sci Award, Distinguished Fac Achievement Award. *Mem:* Fel AAAS; fel Am Inst Chemists; Am Chem Soc; Royal Soc Chem; fel Am Phys Soc. *Res:* Thermodynamics, actinide, lanthanide, and transition compounds; thermochemistry; cryogenic calorimetry; molecular dynamics plastic crystals; thermophysics of phase ordering, Schottky, transitions. *Mailing Add:* Dept Chem Univ Mich Ann Arbor MI 48109-1055

WESTRUM, LESNICK EDWARD, b Tacoma, Wash, Oct 19, 34. NEUROSURGERY, BIOLOGICAL STRUCTURE. *Educ:* Wash State Univ, BS, 58; Univ Wash, MD, 63; Univ London, PhD(anat), 66. *Prof Exp:* NIH fel, Univ London, 63-66, hon res asst anat, Univ Col, 64-66; res asst prof, 66-67, asst prof, 67-77, ASSOC PROF SURG & BIOL STRUCT, SCH MED, UNIV WASH, 77- *Mem:* AAAS; Am Asn Anatomists; Soc Neurosci. *Res:* Studies of synapses in normal and experimental conditions; emphasis on trigeminal and limbic systems. *Mailing Add:* Dept Biol Struct Sch Med Univ Wash Seattle WA 98195

WESTWATER, EDGEWORTH RUPERT, b Denver, Colo, Oct 29, 37; div; c 3. ATMOSPHERIC PHYSICS. *Educ:* Western State Col Colo, BA, 59; Colo Univ, MS, 62, PhD(physics), 70. *Prof Exp:* Chemist cement chem, Ideal Cement Co, 59-60; physicist atmospheric physics, Nat Bur Standards, 60-65 & Environ Sci Serv Admin, 65-70; physicist atmospheric physics, 70-84, SUPV PHYSICIST, NAT OCEANIC & ATMOSPHERIC ADMIN, 84- *Concurrent Pos:* Mem ad hoc working group inversion methods, Radiation Comn, Int Asn Meteorol & Atmospheric Physics, 66-; comt on radio frequencies, Nat Acad Sci; mem, comn F, URSI. *Mem:* Am Meteorol Soc; Optical Soc Am; Soc Indust & Appl Math; Math Asn Am; Sigma Xi. *Res:* Remote sensing of the atmosphere; radiative transfer in the atmosphere; mathematics of ill-posed problems. *Mailing Add:* RJE/WPS Wave Propagation Lab 325 Broadway Boulder CO 80303-3328

WESTWATER, J(AMES) W(ILLIAM), b Danville, Ill, Nov 24, 19; m 42; c 4. CHEMICAL ENGINEERING, HEAT TRANSFER. *Educ:* Univ Ill, BS, 41; Univ Del, MChE, 43, PhD(chem eng), 48. *Prof Exp:* From asst prof to assoc prof, 48-59, head dept, 62-80, PROF CHEM ENG, UNIV ILL, 59- *Concurrent Pos:* Lectr, Am Inst Chem Engrs, 64; Donald L Katz lectureship, Univ Mich, 78. *Honors & Awards:* William H Walker Award, Am Inst Chem Engrs, 66; Max Jakob Award, Am Soc Mech Engrs, 72; Vincent Bendix Award, Am Soc Eng Educ, 74; Reilly lect, Notre Dame Univ, 58. *Mem:* Nat Acad Eng; Am Chem Soc; Am Soc Eng Educ; fel Am Inst Chem Engrs; Am Soc Mech Engrs. *Res:* Heat transfer; phase changes; boiling; condensation. *Mailing Add:* Dept Chem Eng Univ Ill Urbana IL 61801

WESTWICK, CARMEN R, medical-surgical nursing, gerontology, for more information see previous edition

WESTWICK, ROY, b Vancouver, BC, May 23, 33; m 59; c 3. MATHEMATICS. *Educ:* Univ BC, BA, 56, MA, 57, PhD(math), 60. *Prof Exp:* Nat Res coun Can overseas fel, math, Univ Col, London, 60-62; from asst prof to assoc prof, 62-70, PROF MATH, UNIV BC, 70- *Concurrent Pos:* Can Coun sr fel, 67-68. *Mem:* Can Math Soc. *Res:* Linear and multilinear algebra. *Mailing Add:* Dept Math Univ BC 2075 Westbrook Hall Vancouver BC V6T 1W5 Can

WESTWOOD, ALBERT RONALD CLIFTON, b Birmingham, Eng, June 9, 32; nat US; m 56; c 2. RESEARCH MANAGEMENT, MATERIALS SCIENCE. *Educ:* Univ Birmingham, BSc, 53, PhD(phys metall), 56, DSc(mat sci), 68; Coun Eng Inst, UK, CEng, 78. *Prof Exp:* Tech officer, Res Dept, Imp Chem Industs, Eng, 56-58; scientist, Res Inst Advan Studies, Martin Marietta Labs, Baltimore, Md, 58-61, sr scientist, 61-64, assoc dir & head, Mat Sci Dept, 64-69, dep dir, 69-74, dir, 74-84, corp dir res & develop, Martin Marietta Corp, Bethesda, Md, 84-87, vpres res & develop, 87-90, vpres sci, 90, VPRES RES & TECHNOL, MARTIN MARIETTA CORP, BETHESDA, MD, 90- *Concurrent Pos:* Lectr, Chance Tech Col, Eng, 54 & Handsworth Tech Col, 54-56; fel, Johns Hopkins Univ, 65-; guest lectr, Acad Sci, USSR, 69, 76, 78 & 81; mem adv comt, Inorg Mat Div, Nat Bur Standards, 72-75; mem adv bd, J Mat Sci, UK, 74-, Colloids & Surfaces, 79-83 & Mech & Physics of Surfaces, 79-; mem rev comt, Mat Sci Div, Argonne Nat Labs, 76-82; mem, Nat Mat Adv Bd, Nat Res Coun, 80-85 & Comn Eng & Tech Studies, 85-; mem mat res adv comn, NSF, 80-84; bd dirs, Metall Soc, Inc, 80-83, financial officer, 85-; Md Humanities Coun, 84-; adv bd, Oak Ridge Nat Lab, 84-; bd dir, Martin Marietta Energy Systs, Inc, 84-; adv coun, Sch Arts & Sci, Johns Hopkins Univ, 84-; adv comt, Sch Eng, Univ Md, 84-; bd dirs, US Advan Ceramics Asn, 85-88; chmn, Sect P Indust Sci, AAAS, 89-90; distinguished lectr, Am Soc Mech Engrs, 89-91; trustee, Am Inst Mining, Metall & Petrol Engrs, 90- *Honors & Awards:* Beilby Gold Medal & Prize, Royal Inst Chem, Soc Chem Indust & Inst Metals UK, 70; Tewksbury Lectr, Univ Melbourne, Australia, 74; Burgess Lectr, Am Soc Metals, 84; Campbell Mem Lectr, Am Soc Metals Int, 87; Henry Krumb Lectr, Metall Soc & Soc Mining Engrs, 88. *Mem:* Nat Acad Eng; fel Am Soc Metals; fel Brit Inst Metallurgists; fel Brit Inst Physics; Metall Soc Inc; fel AAAS; foreign assoc Royal Swed Acad Eng Sci; Indust Res Inst (pres, 89-90); fel Minerals, Metals & Mat Soc (pres, 90-91). *Res:* Research and development management theory; mechanical behavior, chemomechanical effects, surface and environmental effects; metals, ceramics, semiconductors. *Mailing Add:* 20430 Owen Brown Rd Columbia MD 21044

WESTWOOD, MELVIN (NEIL), b Hiawatha, Utah, Mar 25, 23; m 46; c 4. POMOLOGY, PLANT PHYSIOLOGY. *Educ:* Utah State Univ, BS, 53; Wash State Univ, PhD(pomol, plant physiol), 56. *Prof Exp:* Asst field botanist, Utah State Univ, 51-52, supt, Hort Res Field Sta, 52-53; asst, Wash State Univ, 53-55; from asst res horticulturist to res horticulturist, USDA, 55-60; assoc prof, 60-67, PROF HORT, ORE STATE UNIV, 67- *Honors & Awards:* Gourley Award, Am Soc Hort Sci, 58 & 77 & Stark Award, 69 & 77; Paul Howe Shepard Award, Am Pomol Soc, 68; Wilder Medal Award, Am Pomol Soc, 81. *Mem:* AAAS; fel Am Soc Hort Sci (pres, 74-75); Am Soc Plant Physiologists; Scand Soc Plant Physiol. *Res:* Deciduous fruit tree and rootstock physiology; growth dynamics; plant hormones; chemical thinning; high density orchard systems. *Mailing Add:* 2130 NW Elmwood Pl Corvallis OR 97330

WESTWOOD, WILLIAM DICKSON, b Kirkcaldy, Scotland, Jan 4, 37; m 61; c 1. PHYSICS. *Educ:* Univ Aberdeen, BSc, 59, PhD(physics), 62, DSc, 86. *Hon Degrees:* PhD(physics), Univ Adelaide, 66. *Prof Exp:* Scientist, Northern Elec Co Ltd, 62-65; lectr physics, Flinders Univ, Australia, 66-69; scientist, Northern Elec Co Ltd, 69-70; mgr thin film physics, 71-79, MGR MAT & DEVICE RES, BELL-NORTHERN RES LTD, 79- *Mem:* Fel Brit Inst Physics; Can Asn Physicists; Am Vacuum Soc (secy, 85-); Sigma Xi. *Res:* Thin film physics; sputtering; spectroscopy of gas discharges; integrated optics; surface analysis; optical recording devices; optoelectronic devices; gallium-arsenide devices communication systems. *Mailing Add:* Dept 5C10 Bell-Northern Res Ltd Box 3511 Sta C Ottawa ON K1X 4H7 Can

WESWIG, PAUL HENRY, b St Paul, Minn, July 13, 13; m 40; c 2. ANIMAL NUTRITION. *Educ:* St Olaf Col, BA, 35; Univ Minn, MS, 39, PhD(biochem), 41. *Prof Exp:* Instr chem, St Olaf Col, 36-37; asst biochem, Univ Minn, 38-41; asst prof biochem, Univ & asst chemist, Exp Sta, 41-42, assoc prof biochem, 46-47, PROF BIOCHEM, ORE STATE UNIV, 57- *Concurrent Pos:* Nutrit Surv, Ethiopia, 58, Malaya, 62, Paraguay, 65, 72, Sri Lanka, 79. *Mem:* Am Chem Soc; Am Inst Nutrit; Am Dairy Sci Asn; Am Soc Animal Sci. *Res:* Trace mineral metabolism in ruminants and laboratory animals including requirement, interrelationship, toxicity, tissue enzyme activity and element concentration including selenium, copper and heavy metals. *Mailing Add:* 2112 NW Polk St Corvallis OR 97330

WETEGROVE, ROBERT LLOYD, b San Diego, Calif, Jan 25, 48; m 71. MICROBIOLOGY, INDUSTRIAL WATER TREATMENT. *Educ:* Univ Tex, Austin, BA, 70, MA, 72, PhD(microbiol), 78; Ill Inst Technol, MBA, 88. *Prof Exp:* Res scientist microbiol, Nalco Chem Co, 78-81, res group leader waste treat chemicals, 81-82, cooling water chem, 82-84, mkt develop mgr, 84-87, res develop mgr biotechnol, 87-89, CORP RES ASSOC, NALCO CHEM CO, 89- *Mem:* Am Soc Microbiol; Soc Indust Microbiol; Water Pollution Control Fedn; Sigma Xi. *Res:* Biotechnology relating to industrial water treatment. *Mailing Add:* One Nalco Ctr Nalco Chem Co Naperville IL 06563-1198

WETHERALD, RICHARD TRYON, b Plainfield, NJ, Mar 28, 36; m 63; c 3. GENERAL CIRCULATION MODELING, MATH & COMPUTER ANALYSIS. *Educ:* Univ Mich, BS, 62, MS, 63. *Prof Exp:* Assoc engr computer develop, Westinghouse Elec, 63-64; RES ASSOC CLIMATE RES, GEOPHYS FLUID DYNAMICS LAB, NAT OCEANIC & ATMOSPHERIC ADMIN, 64- *Concurrent Pos:* Investr, Dept Energy, 87-, Prog Climate Model Diag & Intercomparison, 90-; contribr, Intergovt Panel on Climate Change, 89- *Res:* Climate research; climate sensitivity; greenhouse warming; changes in solar insulation; cloud feedback; oxygen-induced changes of temperature, hydrology and clouds. *Mailing Add:* 78 Taylor Terr Hopewell NJ 08525

WETHERELL, DONALD FRANCIS, b Manchester, Conn, Nov 25, 27; m 49; c 4. PLANT PHYSIOLOGY. *Educ:* Univ Conn, BA, 51; Univ Md, MS, 53, PhD(plant physiol), 56. *Prof Exp:* Res assoc & lectr plant physiol, Univ Md, 56-58; from asst prof to assoc prof plant physiol, 58-67, actg chmn dept bot, 61-64, chmn regulatory biol sect, 67-68, PROF PLANT PHYSIOL, UNIV CONN, 67- *Concurrent Pos:* Guggenheim fel, 65-66. *Mem:* Am Soc Plant Physiologists; Int Asn Plant Tissue Cult. *Res:* Regulatory mechanisms of growth and development in plants. *Mailing Add:* Dept Molecular & Cell Biol U-125 Univ Conn Main Campus 75 N Eaglevil Storrs CT 06268

WETHERELL, HERBERT RANSON, JR, b Chicago, Ill, Jan 25, 27; m 62; c 1. PHARMACOLOGY, TOXICOLOGY. *Educ:* Yale Univ, BS, 49, PhD(pharmacol), 54. *Prof Exp:* Asst prof physiol & pharmacol, Med Col, Univ Nebr, 53-61; toxicologist, Wayne County Med Examr Off, 61-72; toxicologist, Crime Lab, Mich Dept Pub Health, 72-77; TOXICOLOGIST, CRIME LAB, MICH STATE POLICE, 77- *Mem:* Am Chem Soc; Am Acad Forensic Sci; Royal Soc Chem London; Sigma Xi. *Res:* Relationship between chemical structure and pharmacologic activity; toxicology and drug metabolism; chlorophyll chemistry; infrared spectrophotometry. *Mailing Add:* Toxicol Lab Mich State Police 714 S Harrison Rd East Lansing MI 48823

WETHERHOLD, ROBERT CAMPBELL, b Wilmington, Del, Nov 19, 51. COMPOSITE MATERIALS. *Educ:* Univ Del, BME & BA, 74, MMAE, 76, PhD(mech eng), 83. *Prof Exp:* Engr, E I du Pont Co, 76-78, spec engr, 80-81; res assoc, Ctr Composite Mat, Univ Del, 78-80; asst prof, 83-89, ASSOC PROF MECH & MAT, STATE UNIV NY, BUFFALO, 89- *Concurrent Pos:* Am Soc Eng Educ/NASA-Lewis fac fel, 85 & 86; Air Force Off Sci Res fac fel, Wright-Patterson AFB, 87 & 88; vis scientist, Air Force Mat Lab, Wright-Patterson AFB, 89 & Rockwell Int Sci Ctr, 90. *Mem:* Am Soc Mech Eng; Am Inst Aeronaut & Astronaut; Am Ceramic Soc; Am Soc Composite. *Res:* All aspects of composite materials, from their fabrication through their end use; durability issues such as fracture of composites and performance of high-temperature composites. *Mailing Add:* Dept Mech & Aerospace Eng State Univ NY Buffalo NY 14260

WETHERILL, GEORGE WEST, b Philadelphia, Pa, Aug 12, 25; m 50. GEOPHYSICS. *Educ:* Univ Chicago, PhB, 48, SB, 49, MS, 51, PhD(physics), 53. *Prof Exp:* Mem staff, Carnegie Inst, Wash Dept Terrestrial Magnetism, 53-60; prof geophys & geol, Univ Calif, Los Angeles, 60-75, chmn dept planetary & space sci, 68-72; DIR DEPT TERRESTRIAL MAGNETISM, CARNEGIE INST WASH, 75- *Concurrent Pos:* Vis prof, Calif Inst Technol, 59. *Honors & Awards:* Leonard Medal, Meteoritical Soc, 81; Gilbert Award, Geol Soc Am, 84; Kuiper Prize, Am Astron Soc, 86. *Mem:* Nat Acad Sci; fel Am Acad Arts & Sci; fel Am Geophys Union; Meteoritical Soc (vpres, 72-74 & 80-82, pres, 82-84); Geochem Soc (vpres, 73-74, pres, 74-75); Int Astron Union; Int Asn Geochem & Cosmochem (pres, 77-80). *Res:* Planetology; geochronology; meteorites; origin and evolution of solar system; lunar history; Precambrian geology; kinetics of human lead metabolism. *Mailing Add:* Dept Terrestrial Magnetism Carnegie Inst 5241 Broad Br Rd Washington DC 20015

WETHERINGTON, RONALD K, b St Petersburg, Fla, Nov 27, 35; m 72; c 5. BIOLOGICAL ANTHROPOLOGY. *Educ:* Tex Tech Univ, BA, 58; Univ Mich, MA, 60, PhD(anthrop), 64. *Prof Exp:* PROF ANTHROP, SOUTHERN METHODIST UNIV, 64- *Mem:* Sigma Xi; Soc Med Anthrop; AAAS; Am Asn Phys Anthrop. *Res:* Skeletal growth and development in man; human ecology and evolutionary theory; adaptive aspects of human demography. *Mailing Add:* Dept Anthrop Southern Methodist Univ Dallas TX 75275

WETHERN, JAMES DOUGLAS, b Minneapolis, Minn, July 12, 26; m 48; c 4. CHEMICAL ENGINEERING. *Educ:* Univ Wis, BS, 47; Inst Paper Chem, MS, 49, PhD(paper chem eng), 52. *Prof Exp:* Sr engr, Crown Zellerbach Corp, 51-53, chief, Pulping Sect, 53-55 & Paper Sect, 55-56, coordr appl res, 56-59; tech dir, Pulp & Paperboard Div, Riegel Paper Co, 59, mgr mfg serv, 59-65, vpres & res mgr, La Opers, 65-68, vpres mfg, Paper Div, 68-72; vpres mfg, 72-80, pres, 80-82, MGR PLANNING & DEVELOP, BRUNSWICK PULP & PAPER CO, GA PAC, 82- *Mem:* Fel Tech Asn Pulp & Paper Indust; Am Mgt Asn; Can Pulp & Paper Asn. *Res:* Pulp and paper. *Mailing Add:* Ga Pac PO Box 1438 Brunswick GA 31520

WETHINGTON, JOHN A(BNER), JR, b Tallahassee, Fla, Apr 18, 21; m 43; c 1. NUCLEAR ENGINEERING, NUCLEAR CHEMISTRY. *Educ:* Emory Univ, AB, 42, MS, 43; Northwestern Univ, Evanston, PhD(chem & physics), 50. *Prof Exp:* Vis res asst chem, Princeton Univ, 43-44; scientist, Fercleve Corp, Tenn, 44-45; assoc scientist, Oak Ridge Nat Lab, 45-46, sr scientist, 49-53; from asst prof chem eng to assoc prof nuclear eng, 53-60, prof, 60-85, EMER PROF, NUCLEAR ENG, UNIV FLA, 85- *Concurrent*

Pos: US deleg, Geneva Atomic Energy Conf, Switz, 58; consult, PR Nuclear Ctr, 61-63; vis scientist, Oak Ridge Nat Lab, 58, 79-80 & Lawrence Livermore Lab, 71-72; tech adv, Fla Power & Light, 89-91. *Honors & Awards:* Award, Am Soc Eng Educ, 57. *Mem:* AAAS; Am Chem Soc; Am Nuclear Soc. *Res:* Surface physics; exchange reactions; use of nuclear explosions; processing of nuclear reactor fuel; isotope separation; effect of ionizing radiation on fluorocarbons; radioactive waste disposal; technologically enhanced natural radiation. *Mailing Add:* Dept Nuclear Eng Univ Fla Gainesville FL 32611-2055

WETLAUFER, DONALD BURTON, b New Berlin, NY, Apr 4, 25; m 50; c 2. BIOCHEMISTRY. *Educ:* Univ Wis, BS, 46, PhD(biochem), 54. *Prof Exp:* Jr chemist, Argonne Nat Lab, 44 & 46-47; res chemist, Bjorksten Res Labs, Inc, 48-50; asst anal chem, Univ Wis, 44-46, asst biochem, 50-52, res assoc enzymol, Inst Enzyme Res, 54; res assoc, Children's Cancer Res Found & Dept Biol Chem, Harvard Med Sch, 58-61; asst prof biochem, Sch Med, Ind Univ, Indianapolis, 61-62; from assoc prof to prof biochem, Med Sch, Univ Minn, Minneapolis, 62-75; chmn dept, 75-85, DU PONT PROF CHEM, UNIV DEL, 75- *Concurrent Pos:* Nat Found Infantile Paralysis fel protein chem, Carlsberg Lab, Denmark, 55-56; Am Heart Asn fel, Biol Labs, Harvard Univ, 56-58; tutor, Harvard Univ, 58-61; prin investr, USPHS, 61-; consult, Nat Inst Gen Med Sci, 64-e; investr, Max Planck Inst Ernahrungsphysiologie, 74-78; consult, 80-84, NSF, prin investr, 80-; Res Career Develop Award, USPHS, 61-66. *Mem:* Am Chem Soc; Am Soc Biochem & Molecular Biol; Biophys Soc; fel Am Inst Chemists. *Res:* Chemical and physical basis of structure, stability and reactivity of proteins; acquisition of three-dimensional structure of macromolecules; high performance protein chromatography. *Mailing Add:* Dept Chem & Biochem Univ Del Newark DE 19711

WETMORE, CLIFFORD MAJOR, b Akron, Ohio, June 18, 34; m 59; c 2. LICHENOLOGY. *Educ:* Mich State Univ, BS, 56, MS, 59, PhD(bot), 65. *Prof Exp:* From instr to assoc prof biol, Wartburg Col, 64-70; from asst prof to assoc prof, 70-84, PROF BOT, UNIV MINN, ST PAUL, 84- *Concurrent Pos:* NSF res grants, 66-68 & 71-73, fel, 70; nat park serv contracts, 82-; cur cryptograms, Univ Minn, 70- *Mem:* Am Bryol & Lichenological Soc; Int Asn Plant Taxonomists; Brit Lichen Soc; Sigma Xi. *Res:* Lichens of the Black Hills, South Dakota, and Minnesota; desert lichens; lichen genera; distributions of lichens; herbarium computer techniques; lichens and air pollution. *Mailing Add:* Dept Plant Biol Univ Minn St Paul MN 55108

WETMORE, DAVID EUGENE, b Stella, Nebr, Dec 18, 35; m 59; c 2. COMPUTER LITERACY. *Educ:* Park Col, BA, 58; Univ Kans, MA, 62; Tex A&M Univ, PhD(org chem), 65. *Prof Exp:* Res chemist, Sun Oil Co, 65-67; from asst prof to assoc prof chem, St Andrew Presby Col, 67-81, chmn chem prog, 68-77, chmn, sci div, 77-81, prof chem & computer sci, 82-84; PROF CHEM & COMPUTER SCI, BREVARD COL, 85- *Concurrent Pos:* Consult, Coun Independent Cols. *Mem:* Asn Comput Mach. *Res:* Computer education. *Mailing Add:* Dept Computer Sci Brevard Col Brevard NC 28718

WETMORE, RALPH HARTLEY, morphology, developmental biology; deceased, see previous edition for last biography

WETMORE, STANLEY IRWIN, JR, b Queens, NY, June 1, 39; m 61; c 2. ORGANIC CHEMISTRY. *Educ:* Rensselaer Polytech Inst, BS, 60, MS, 62; State Univ NY Buffalo, PhD(org chem), 73. *Prof Exp:* Chemist, Mobil Oil Corp, 62-63; from instr to assoc prof, Va Mil Inst, 64-69, head, Dept Chem, 83-89, PROF CHEM, VA MIL INST, 79-, DIR RES LABS, 85- *Concurrent Pos:* NSF sci fac fel, 70-71. *Mem:* Am Chem Soc; Sigma Xi. *Res:* Synthesis and properties of unique carbenes; photochemistry of small ring heterocyclic compounds; development of ethanol as farm fuel. *Mailing Add:* Dept Chem Va Mil Inst Lexington VA 24450

WETMUR, JAMES GERARD, b New Castle, Pa, July 1, 41; m 65; c 3. BIOPHYSICAL CHEMISTRY. *Educ:* Yale Univ, BS, 63; Calif Inst Technol, PhD(chem), 67. *Prof Exp:* Asst prof chem & biochem, Univ Ill, Urbana, 69-74; assoc prof, 74-82, PROF MICROBIOL, MT SINAI SCH MED, CITY UNIV NEW YORK, 82- *Concurrent Pos:* Vpres, NY Acad Sci, 86-89. *Mem:* Am Chem Soc; Am Soc Biol & Molecular Biol; Am Soc Microbiol; Sigma Xi; NY Acad Sci. *Res:* Kinetics of renaturation of DNA; DNA-protein interactions; molecular biology - gene expression. *Mailing Add:* 994 Post Rd Scarsdale NY 10583

WETS, ROGER J B, b Uccle, Belg, Feb 20, 37; m 61; c 2. MATHEMATICS. *Educ:* Free Univ Brussels, BA, 59; Univ Calif, Berkeley, PhD(appl math), 64. *Prof Exp:* Staff mem, Boeing Sci Res Lab, Wash, 64-70; prof math, Univ Chicago, 70-71; PROF MATH, UNIV KY, 71- *Concurrent Pos:* Vis lectr, Univ Wash, 66, Univ Calif, Berkeley, 67 & Inst Info & Automation Res, Paris, 69; vis res, Math Ctr, Montreal, 70; Guggenheim fel, 81-82. *Mem:* Am Math Soc; Soc Indust & Appl Math. *Res:* Mathematical programming; stochastic optimization. *Mailing Add:* 2121 Pinehurst Ct El Cerrito CA 94530

WETSTONE, HOWARD J, b Hartford, Conn, Apr 27, 26; m 47; c 4. MEDICINE, HOSPITAL ADMINISTRATION. *Educ:* Wesleyan Univ, BA, 47; Tufts Univ, MD, 51. *Prof Exp:* Intern med, New Eng Ctr Hosp, 51-52, jr asst resident, 52-53; asst resident, 53-54, resident, 54-55, dir liver enzyme lab, 58-60, dir biochem res lab, 60-63, dir med res, 63-66, asst dir dept med, 66-70, dir outpatient dept, 70-72, sr physician internal med, 70-88, dir ambulatory serv, 72-83, VPRES MED AFFAIRS, CONN HEALTH SYST, HARTFORD HOSP, 83- *Concurrent Pos:* Assoc prof med & community med & health care, Med Sch, Univ Conn, 71- *Mem:* Sigma Xi; Am Col Physicians. *Res:* Pharmacogenetics; hypertension; enzymology; community health systems; primary care. *Mailing Add:* 77 Kenwood Circle Bloomfield CT 06002

WETTACH, WILLIAM, b Pittsburgh, Pa, Mar 13, 10; m 40; c 3. CHEMICAL ENGINEERING. *Educ:* Princeton Univ, BS, 32, ChE, 33; Univ Pittsburgh, PhD(chem eng), 41. *Prof Exp:* Vpres, W W Lawrence & Co, 33-35 & Wettach Paint & Chem Co, 35-38; chemist, Peerless Paint & Chem Co, 38-40; managing dir, Indust Paint Co, 40-53, pres, 53-69; vpres & tech dir, Sterling Div, Reichhold Chem Inc, 69-80; RETIRED. *Concurrent Pos:* Vpres, Sterling Varnish Co, 41-69, Mercury Varnish & Sterling Varnish of Can, 67-75. *Mem:* AAAS; Am Chem Soc; Am Inst Chem Engrs; fel Am Inst Chem; Sigma Xi. *Res:* Development of water soluble alkyd resins; formulation of industrial coatings for metallic substances. *Mailing Add:* 659 Grove St Sewickley PA 15143-1234

WETTACK, F SHELDON, b Coffeyville, Kans, Dec 5, 38; m 56; c 4. PHYSICAL CHEMISTRY. *Educ:* San Jose State Col, AB, 60, MA, 62; Univ Tex, Austin, PhD(chem), 68. *Prof Exp:* High sch teacher, Calif, 61-64; from asst prof to assoc prof, 67-72, PROF CHEM, HOPE COL, 72-, DEAN NATURAL SCI, 74- *Concurrent Pos:* Camille & Henry Dreyfus Found Teacher-Scholar Award, 70-75. *Mem:* Am Chem Soc. *Res:* Photochemistry; energy transfer; fluorescence spectroscopy. *Mailing Add:* Dean Fac Univ Richmond Richmond VA 23173

WETTE, REIMUT, b Mannheim, Ger, May 12, 27; US citizen; m 51; c 5. BIOSTATISTICS, BIOMATHEMATICS. *Educ:* Univ Heidelberg, MS, 52, DSc(natural sci), 55. *Prof Exp:* Sci asst biomet, Zool Inst, Univ Heidelberg, 52-61; asst biometrician, Univ Tex, 61-64, mem grad fac, 65-66, assoc prof biomath, Univ Tex M D Anderson Hosp & Tumor Inst, 64-66; prof, 66-90, EMER PROF BIOSTATIST, SCH MED, WASH UNIV, 90- *Mem:* Am Statist Asn; Biomet Soc; Inst Math Statist. *Res:* Mathematical approaches to basic science and medical aspects of neoplasia; application and problem-oriented development of mathematical-statistical methodology in biomedical research; methods of mathematical-genetical epidemiology. *Mailing Add:* Div Biostatist Wash Univ Sch Med St Louis MO 63110

WETTEMANN, ROBERT PAUL, b New Haven, Conn, Nov 12, 44; m 68; c 3. REPRODUCTIVE PHYSIOLOGY. *Educ:* Univ Conn, BS, 66; Mich State Univ, MS, 68, PhD(dairy), 72. *Prof Exp:* From asst prof to prof animal sci, 72-80, REGENTS PROF, OKLA STATE UNIV, 84- *Concurrent Pos:* Vis prof, Univ Fla, 80-81. *Honors & Awards:* Richard Hoyt Award, Am Dairy Sci Asn, 71. *Mem:* Am Soc Animal Sci; Am Dairy Sci Asn; Soc Study Fertil; Soc Study Reproduction. *Res:* Influence of the environment on endocrine and reproductive function in animals. *Mailing Add:* Dept Animal Sci Okla State Univ Stillwater OK 74078

WETTERHAHN, KAREN E, b Plattsburgh, NY, Oct 16, 48. INORGANIC BIOCHEMISTRY, PHYSICAL BIOCHEMISTRY. *Educ:* St Lawrence Univ, BS, 70; Columbia Univ, PhD(chem), 75. *Prof Exp:* Chemist formulations, Mearl Corp, 70-71; res fel chem, Columbia Univ, 71-75; fel biochem, Inst Cancer Res, 75-76; asst prof, Dept Chem, 76-82, asst prof biochem prog, 78-82, ASSOC PROF, DARTMOUTH COL, 82- *Concurrent Pos:* Alfred P Sloan fel, 81-85. *Mem:* AAAS; Am Chem Soc; Am Asn Cancer Res. *Res:* Mechanisms of chemical carcinogenesis; metabolism and nucleic acid interactions of carcinogens; inorganic (metal) carcinogenesis; structure-function relationships of modified nucleic acids. *Mailing Add:* Dept Chem Dartmouth Col Hanover NH 03755

WETTSTEIN, FELIX O, b Uerikon, Switz, Jan 1, 32; m 60; c 3. MOLECULAR BIOLOGY. *Educ:* Swiss Fed Inst Technol, BS, 56, PhD(agr chem), 60. *Prof Exp:* Fel, Univ Pittsburgh, 60-64; asst res biochemist, Univ Calif, Berkeley, 64-67; assoc prof, 67-74, PROF MOLECULAR BIOL, MED MICROBIOL & IMMUNOL, UNIV CALIF, LOS ANGELES, 74-, VCHMN DEPT, 77- *Mem:* AAAS; Am Soc Microbiol. *Res:* Regulation of RNA and protein biosynthesis in differentiating and transformed animal cells; viral carcinogenesis. *Mailing Add:* Dept Microbiol & Immunol Univ Calif Los Angeles Sch Med 405 Hilgard Ave Los Angeles CA 90024

WETTSTEIN, JOSEPH G, b San Francisco, Calif, Sept 5, 54; m 83; c 2. NEUROPHARMACOLOGY, PSYCHOPHARMACOLOGY. *Educ:* Univ Calif San Diego, BA, 76; Univ Ky, PhD(pharmacol), 84. *Prof Exp:* Res fel, Harvard Med Sch, 84-86, instr, 87-88; DIR, INST RES JOUVEINAL, 88- *Concurrent Pos:* Travel award, Am Soc Pharmacol & Exp Therapeut, 79 & 82; guest lectr, Univ Ky, 81-83; nat res serv award, NIH, 84-87. *Mem:* Am Soc Pharmacol & Exp Therapeut; France Soc Pharmacologists; Behav Pharmacol Soc; Europ Behav Pharmacol Soc; Soc Stimulus Properties Drugs; Soc Neurosci. *Res:* Mechanisms of actions of drugs in the brain; evaluate how drugs may be useful as therapeutic agents for brain disorders. *Mailing Add:* Dept CNS Pharmacol Inst Res Jouveinal 3-9 rue de la Loge BP 100 Fresnes 94265 France

WETTSTEIN, PETER J, IMMUNOGENETICS, TRANSPLANTATION. *Educ:* Univ NC, Chapel, Hill, PhD(genetics), 77. *Prof Exp:* Assoc prof, Wistar Inst, Philadelphia, 83-88; sr scientist, McLaughlin Res Inst, 88-91; CONSULT, DEPT SURG, MAYO CLIN, 91- *Mailing Add:* Dept Surg Mayo Clin Rochester MN 55905

WETZEL, ALBERT JOHN, b New Orleans, La, Dec 29, 17; m 46; c 4. MISSILE & SPACE TECHNOLOGY, AIRCRAFT FLIGHT TESTING. *Educ:* Tulane Univ, BEng, 39; Johns Hopkins Univ, MS, 50. *Prof Exp:* Exp test pilot, USAF, 43-45, tech asst to secy defense, US Dept Defense, 50-55, comdr, 40th Bombardment Wing-Jet, USAF Strategic Air Command, 55-56, prog dir, Space & Missile Div, 56-62, dir, Strategic Prog, 62-65; vpres univ develop & res, 65-80, ASST TO PRES, TULANE UNIV, 80- *Concurrent Pos:* Mem, Rocket & Space Panel, President's Sci Adv Comt, 65-71, bd dirs, Gulf S Res Inst & Inst Defense Anal; deleg, Nat Conf Advan Res. *Mem:* Fel Am Inst Aeronaut & Astronaut; AAAS; Sigma Xi. *Res:* Space and missile technology; aircraft flight testing; missile guidance systems. *Mailing Add:* Seven Richmond Pl New Orleans LA 70115

WETZEL, ALLAN BROOKE, b Dayton, Ohio, May 29, 33; m 59; c 3. NEUROPSYCHOLOGY. *Educ:* Univ Ky, BS, 54; Ohio State Univ, MA, 63, PhD(psychol), 65. *Prof Exp:* Trainee biosci, Stanford Univ, 65-67; trainee neurophysiol, Univ Wis-Madison, 67-69, Nat Inst Neurol Dis & Stroke spec fel, 68-69; instr surg, 70-74, ASST PROF SURG & OTO-

MAXILLOFACIAL SURG, MED SCH, NORTHWESTERN UNIV, CHICAGO, 74- *Concurrent Pos:* Res investr neuropsychol, Neurosurg Res Lab, Northwestern Mem Hosp, 70-74. *Mem:* AAAS; Am Psychol Asn; Soc Neurosci; Sigma Xi. *Res:* Brain function; neurophysiology; neuroendocrinology. *Mailing Add:* 5222 N Sawyer Ave Chicago IL 60625

WETZEL, JOHN EDWIN, b Hammond, Ind, Mar 6, 32; m 62. MATHEMATICS. *Educ:* Purdue Univ, Lafayette, BS, 54; Stanford Univ, PhD(math), 64. *Prof Exp:* From instr to asst prof, 61-68, assoc prof, 68-87, PROF MATH, UNIV ILL, URBANA-CHAMPAIGN, 87- *Mem:* Am Math Soc; Math Asn Am. *Res:* Combinational geometry. *Mailing Add:* Dept Math Univ Ill Urbana-Champaign 1409 W Green St Urbana IL 61801

WETZEL, KARL JOSEPH, b Waynesboro, Va, May 29, 37; m 68; c 2. EXPERIMENTAL NUCLEAR PHYSICS. *Educ:* Georgetown Univ, BS, 59; Yale Univ, MS, 60, PhD(physics), 65. *Prof Exp:* NSF fel, Inst Tech Nuclear Physics, Darmstadt, Ger, 65-66, Ger Govt guest res fel, 66-67; fel, Argonne Nat Lab, 67-69; from asst prof to assoc prof physics, 72-81, PROF PHYSICS, UNIV PORTLAND, 81- *Concurrent Pos:* Vis assoc prof neurol, Univ Ore, Health Sci Ctr, 76-77, chmn physical & life sci, 80-86; NSF fac fel, 76-77; dean, Grad Sch, 87- *Honors & Awards:* Gov Award, Ore Mus Sci & Indust, 72; Culligan Award, Univ Portland, 85. *Mem:* Am Phys Soc; Soc Tech Commun; Am Asn Univ Profs. *Res:* Neutron capture gamma rays; electron and photon scattering; photonuclear processes; electroneurological measurements; pionic atoms. *Mailing Add:* Dept Physics Univ Portland Portland OR 97203

WETZEL, NICHOLAS, b Jacksonville, Fla, July 17, 20; m 45; c 6. NEUROSURGERY. *Educ:* Princeton Univ, AB, 42; Northwestern Univ, MD, 46, MS, 50, PhD, 58. *Prof Exp:* Clin asst, 52-54, instr, 54-55, assoc, 55-57, asst prof, 57-63, ASSOC PROF SURG, MED SCH, NORTHWESTERN UNIV, CHICAGO, 63- *Mem:* AAAS; Am Asn Neurol Surg; AMA; Am Col Surgeons. *Res:* Human stereotaxic surgery for movement disorders; intractable pain; human olfactory system. *Mailing Add:* 333 E Huron St Chicago IL 60611

WETZEL, ROBERT GEORGE, b Ann Arbor, Mich, Aug 16, 36; m 59; c 4. LIMNOLOGY. *Educ:* Univ Mich, BSc, 58, MSc, 59; Univ Calif, Davis, PhD(limnol), 62. *Hon Degrees:* PhD, Univ Uppsala, Sweden, 83. *Prof Exp:* Tech asst, Univ Mich, 58-59; res technician, US Fish & Wildlife Serv, Mich, 59; res & tech asst, Univ Calif, 59-62; res assoc, Ind Univ, 62-65; from asst prof to prof bot, Mich State Univ, 65-86, adj prof zool, 79-86; prof biol & res scientist, Univ Mich, 86-90; BISHOP PROF BIOL, UNIV ALA, 90- *Concurrent Pos:* Res fel, Aquatic Res Unit, Ind Dept Natural Resources, 63-64; NSF res fel, 63-65 & 67-69, travel award, Int Asn Limnol Cong, 65 & 68; partic, AEC Contract, 65-; co-ed, Commun, 68- & Archiv fuer Hydrobiologie, 81-; Off Water Resources Res res fel, 69-71; int consult, Int Biol Prog, Int Coun Sci Unions, 69-; numerous NSF grants, 64-76; mem bd dirs, Am Soc Limnol & Oceanog, 68-71 & Ecol Soc Am, 70-72; US rep, Aquatic Ecol Comn, Int Asn Ecol, 68-; external examr, var schs & univs, 70-88; adv, Aqua Fennica, Water Asn Finland & Finnish Limnol Soc, 90-, Aquatic Plant Mgt Res, Univ Fla, 75, Ecol Sci Improv, Ark, 79, Elec Power Res Inst, 80, Environ Protection Agency, 82 & other insts. *Mem:* Fel AAAS; Am Inst Biol Sci; Am Soc Limnol & Oceanog (vpres, 79-80, pres, 80-81); Ecol Soc Am; Int Asn Theoret & Appl Limnol (gen secy & treas, 68-); Sigma Xi; Aquatic Plant Mgt Soc; Freshwater Biol Assn UK; Int Asn Ecol; Int Asn Great Lakes Res. *Res:* Biological productivity of California, Indiana and Michigan lakes; physiological ecology of algae and aquatic macrophytes; author of numerous publications. *Mailing Add:* Dept Biol Univ Ala Tuscaloosa AL 35487-0344

WETZEL, ROLAND H(ERMAN), b Wis, Apr 29, 23; m 45; c 2. CHEMICAL ENGINEERING. *Educ:* Univ Wis, BS, 45, PhD(chem eng), 51. *Prof Exp:* Asst chem eng, Univ Wis, 46-50; res engr process develop, E I du Pont de Nemours & Co, Inc, 51-53, res supvr, 54-66, res mgr, 67-69, asst lab dir, Pigments Dept, 69-73, tech supt, 73-75, eng assoc, 76-77; RETIRED. *Mem:* Sigma Xi. *Res:* Unit operations; process development. *Mailing Add:* RD 4 Box 253 Landenberg PA 19350

WETZEL, RONALD BURNELL, b Hanover, Pa, May 26, 46; m 76; c 2. PROTEIN CHEMISTRY, PROTEIN ENGINEERING. *Educ:* Drexel Univ, BS, 69; Univ Calif, Berkely, PhD(org chem), 73. *Prof Exp:* Fel, Max Planck Inst Exp Med, 73-75 & Dept Molecular Biophys & Biochem, Yale Univ, 75-78; sr scientist, Dept Protein Biochem, Genentech Inc, 78-82, sr scientist, Biomolecular Chem Dept, 82-89; RES FEL, MACROMOLECULAR SCI DEPT, SMITHKLINE BEECHAM, 89- *Concurrent Pos:* Adj assoc prof, Dept Chem, Univ Calif, Santa Cruz, 85-89; ed, Protein Eng, 86- *Mem:* Am Chem Soc; AAAS; Peptide Soc. *Res:* Protein structure/function relationships studied by chemical and molecular biological approaches; medical applications of protein engineering; roles of disulfide bonds in globular proteins. *Mailing Add:* Macromolecular Sci Dept SmithKline Beecham Inc PO Box 1539 King of Prussia PA 19406

WETZSTEIN, H(ANNS) J(UERGEN), b Wuppertal-E, Ger, June, 1920; nat US; m 54; c 2. ELECTRICAL ENGINEERING. *Educ:* Univ Cape Town, BSc, 47; Harvard Univ, MS, 49, DSc(elec eng), 52. *Prof Exp:* Dir res, Sci Specialties Corp, 53-55; sr eng scientist, Missile Electronics & Controls Dept, Radio Corp Am, 55-61; mem staff, Inst Naval Studies, Inst Defence Anal, 61-65; sr mem staff, Arthur D Little, Inc, Mass, 65-67; sr eng scientist, Aerospace Syst Div, RCA Corp, 67-71; prin elec engr, Optical Syst Div, ITEK Corp, Lexington, 71-77; sr staff engr, W J Schafer Assocs Inc, 77-83; CONSULT, ITEK OPTICAL SYSTS, 86- *Mem:* Inst Elec & Electronics Engrs, London. *Res:* Servomechanism; system analysis; transistor measurements; analog-digital conversion and computers; passive detection physics and techniques; submarine and antisubmarine warfare; laser systems; active optics; military detection systems. *Mailing Add:* 33 Bayfield Rd Wayland MA 01778-4205

WEWERKA, EUGENE MICHAEL, b St Paul, Minn, Nov 7, 38; m 59; c 4. PHYSICAL ORGANIC CHEMISTRY. *Educ:* Univ Minn, BA, 62, PhD(org chem), 65. *Prof Exp:* STAFF MEM, LOS ALAMOS SCI LAB, 65- *Concurrent Pos:* Adj prof, Univ NMex, 72- *Mem:* Fel Am Inst Chemists; Am Chem Soc. *Res:* Chemical kinetics; mechanism studies; polymer characterization; analytical methods; physical chemistry of polymers; chemistry of coal and oil shale conversion processes; fossil fuels environmental studies; environmental chemistry. *Mailing Add:* MS A102 Los Alamos Nat Lab Los Alamos NM 87545-0001

WEXELL, DALE RICHARD, b Corning, NY, Apr 10, 43; m 78; c 2. INORGANIC CHEMISTRY, GLASS & CERAMICS. *Educ:* Fordham Univ, BS, 64; Georgetown Univ, MS, 69, PhD(inorg chem), 71. *Prof Exp:* Instr chem, Georgetown Univ, 64-70; fel, Corning Glass Works, 71-72, res scientist glass chem, 72-80; res assoc, 81-91, SR RES ASSOC, CORNING INC, 91- *Concurrent Pos:* Mem, Nat Mat Res Coun, 81-84; lectr, Elmira Col; mem State Univ NY Genesco Col Coun. *Mem:* AAAS; Am Chem Soc; Royal Soc Chem; Am Ceramic Soc; Sigma Xi; NY Acad Sci. *Res:* Aqueous silicates; heteropoly electrolytes; high-temperature materials; surface chemistry of glass; inorganic coatings technology; opal glasses; microwave processing; photochromism; electrical, magnetic and optical behavior in glass and ceramics; fiber reinforced composites. *Mailing Add:* Corning Glass Works SP-FR-5-1 Corning NY 14831

WEXLER, ARTHUR SAMUEL, b New York, NY, Oct 27, 18; m 47; c 3. ANALYTICAL CHEMISTRY. *Educ:* Kans State Col, BS, 40; Polytech Inst Brooklyn, PhD(chem), 51. *Prof Exp:* Chemist, C F Kirk & Co, 47-49 & Nopco Chem Co, 49-51; sr chemist, Pepsi-Cola Co, 51-60; mgr, Anal Lab, Dewey & Almy Chem Div, W R Grace & Co, Cambridge, 60-82; RETIRED. *Mem:* Am Chem Soc. *Res:* Analytical methods for monomers and polymers; physical organic chemistry; biochemistry; infrared spectroscopy. *Mailing Add:* 9312 Greyrock Rd Silver Spring MD 20910

WEXLER, BERNARD CARL, b Boston, Mass, May 1, 23; m 46; c 3. EXPERIMENTAL MEDICINE. *Educ:* Univ Ore, BS, 47; Univ Calif, MA, 48; Stanford Univ, PhD(anat, biochem), 52. *Prof Exp:* Asst anat, Sch Med, Stanford Univ, 49-52; mem res staff, Baxter Labs, 52-55; res assoc, May Inst Med Res, 55-61, asst dir, 61-64, dir, 64-81; from asst prof to assoc prof exp path, 55-71, exp med, 71-75, PROF EXP MED & PATH, COL MED, UNIV CINCINNATI, 75- *Concurrent Pos:* Am Heart Asn advan res fel, 60-62; Nat Heart Inst res career develop award, 62-72; lectr, Dominican Col, 50-52; res assoc, Stanford Res Inst, 51-52; mem, Coun Arteriosclerosis & Coun Basic Sci, Am Heart Asn, 62-, Coun Stroke & Coun High Blood Pressure Res; mem, Coun Arteriosclerosis & Ischemic Heart Dis, Int Soc Cardiol; mem med staff, Jewish Hosp, Cincinnati. *Mem:* Am Soc Physiologists; AAAS; Am Diabetes Asn; Asn Am Med Cols; Endocrine Soc. *Res:* Pituitary-adrenal physiology; experimental pathology. *Mailing Add:* 7640 De Mar Rd Cincinnati OH 45243

WEXLER, BERNARD LESTER, b Newton, Mass, April 20, 45; m 68; c 2. QUANTUM ELECTRONICS. *Educ:* Yale Col, BS, 66, MS, 68, PhD(quantum electronics), 73. *Prof Exp:* RES PHYSICIST, LASER PHYSICS BR, OPTICAL SCI DIV, NAVAL RES LAB, 73- *Res:* Infrared laser, particularly 16 micron laser; gas lasers and related areas in atomic and molecular physics; excimer laser development; phase conjugation; nonlinear optics. *Mailing Add:* 5314 North Second St Arlington VA 22203

WEXLER, JONATHAN DAVID, b Phoenix, Ariz, Dec 12, 37. COMPUTER SCIENCE. *Educ:* Ariz State Univ, BS, 59; Univ Wis-Madison, MS, 65, PhD(comput sci), 70. *Prof Exp:* Asst mathematician, Ill Inst Technol Res Inst, 61-64; asst prof, Computer Sci Dept, State Univ NY Buffalo, 70-75; software eng specialist, Western Develop Labs, Ford Aerospace & Commun Corp, Palo Alto, 76-79; software develop engr, BTI Computer Systs, Sunnyvale, Calif, 79-82; pres, Starflower Technol Inc, 82-83; sr software engr, MedaSonics Inc, 84-85; mem tech staff, DocuGraphix Inc, Cupertino, Calif, 85-86; STAFF ENGR, LOCKHEED MISSLES & SPACE CO, SUNNYVALE, CALIF, 87- *Mem:* Asn Comput Mach. *Res:* Data structure representations of knowledge; software engineering methodologies; artificial intelligence; knowledge-based remote diagnostic systems for computers; microcomputer-based merged text-graphics systems; expert systems for computer software debugging. *Mailing Add:* 986 Starflower Ct Sunnyvale CA 94086

WEY, ALBERT CHIN-TANG, b Kaohsiung, Taiwan, July 28, 55; US citizen. ACOUSTIC MICROSCOPY, ELECTRO-OPTICS. *Educ:* Nat Chiao-Tung Univ, Taiwan, BS, 77; State Univ NY, Stony Brook, MS, 80, PhD(elec eng), 86. *Prof Exp:* Lectr commun, Chinese Army Commun & Elec Sch, 77-79; res engr lasers, Quantronix Corp, 80-84; res assoc magneto-optics, US Naval Res Lab, 84-86; sr engr electro-optics, Amphenol Fiber Optic Prod, Allied-Signal Inc, 86-88; RES DIR ACOUST MICROS, SONOSCAN, INC, 88- *Concurrent Pos:* Dir, Chinese Inst Engrs, USA, 80-83. *Honors & Awards:* IR-100 Award, Res & Develop Mag, 86-87. *Mem:* Inst Elec & Electronics Engrs Ultrasonics Soc; Soc Mfg Engrs; Am Soc Nondestructive Testing; Am Ceramic Soc; Am Soc Metals Int; Soc Photo-Optical Instrumentation Engrs. *Res:* Acoustic microscopy technologies and their applications in nondestructive testing and evaluation of microelectronic components and structural materials, such as ceramics, metals, polymers, and other advanced composite materials; author of 38 publications. *Mailing Add:* Sonoscan Inc 530 E Green St Bensenville IL 60106

WEY, JONG-SHINN, b Taiwan, Oct 26, 44; m 66; c 2. CHEMICAL ENGINEERING. *Educ:* Nat Taiwan Univ, BS, 67; Clarkson Col Technol, MS, 70, PhD(chem eng), 73. *Prof Exp:* Res asst chem eng, Clarkson Col Technol, 68-73; sr res chemist, 73-78, res assoc, 78-84, LAB HEAD, RES LABS, EASTMAN KODAK, 84- *Concurrent Pos:* Nat crystallization comt, Am Inst Chem Eng, 77. *Mem:* Am Inst Chem Eng; fel Soc Photog Sci & Eng. *Res:* Crystallization; precipitation; nucleation; growth; size-distribution control; solid liquid separation; photographic emulsion. *Mailing Add:* Res Labs Eastman Kodak Co Rochester NY 14650

WEYAND, JOHN DAVID, b Faulkton, SDak, Aug 13, 39; m 63; c 5. CERAMIC ENGINEERING, GEOLOGICAL ENGINEERING. *Educ:* SDak Sch Mines & Technol, BS, 61; Univ Mo, Rolla, MS, 64, PhD(ceramic eng), 71. *Prof Exp:* Jr geologist field geol, Shell Oil Co, 61; ceramic engr, Minn Mining & Mfg Co, 62-66 & Battelle Mem Inst, 66-68; supvr refractories, Interpace Corp, 71-73; STAFF ENGR CERAMICS, ALUMINUM CO AM, 73- *Mem:* Am Ceramic Soc; Nat Inst Ceramic Eng, ASTM. *Res:* Boride, nitride and oxide ceramic development for placement in chloride and fluoride electroylsis cells as corrosion resistant refractories or electrodes; development refractories, substrate, infrared and light transmitting ceramics and glass elements; mechanical property determination of advanced ceramics. *Mailing Add:* 2187 Brandon Idaho Falls ID 83402

WEYBREW, JOSEPH ARTHUR, b Wamego, Kans, July 13, 15; m 42; c 2. PLANT CHEMISTRY. *Educ:* Kans State Col, BS, 38, MS, 39; Univ Wis, PhD(plant physiol), 42. *Prof Exp:* Res dir, W J Small Co, 42; asst nutritionist, Exp Sta, Kans State Col, 42-43; chief chemist, Indust Hyg Div, State Bd Health, Kans, 43; assoc res prof animal nutrit, 46-49, assoc res prof agron, 49-51, res prof, 51-56, res prof chem, 56-60, actg head chem res, 57-60, WILLIAM NEALS REYNOLDS DISTINGUISHED PROF AGR, NC STATE UNIV, 57- *Concurrent Pos:* Res prof field crops, 60-76. *Mem:* AAAS; Am Chem Soc. *Res:* Tobacco biochemistry, especially fluecuring and quality evaluation; tobacco biogenetics. *Mailing Add:* 112 Pineland Circle Raleigh NC 27607

WEYENBERG, DONALD RICHARD, b Gelvil, Nebr, July 11, 30; m 55; c 2. ORGANOSILICON CHEMISTRY, SILICONE MATERIALS. *Educ:* Univ Nebr, BS, 51; Pa State Univ, PhD(chem), 58. *Prof Exp:* Chemist, Chem Labs, 51-65, res mgr chem, 65-68, dir, Corp Develop, 68-70, silicone res, 70-71, mgr, Resins & Chem Bus, 71-74, New Venture Bus, 74-77, dir res, 77-79, VPRES RES & DEVELOP, DOW CORNING CORP, 79- *Honors & Awards:* Award, Am Chem Soc, 83. *Mem:* Am Chem Soc; NY Acad Sci; Am Ceramic Soc; Soc Chem Indust Brit; Sigma Xi; AAAS. *Res:* Research and development management in organosilicon chemistry, siloxane polymers and the application of silicone materials. *Mailing Add:* 4601 Arbor Dr Midland MI 48640

WEYH, JOHN ARTHUR, b Havre, Mont, Sept 9, 42; m 62; c 4. ANALYTICAL CHEMISTRY, INORGANIC CHEMISTRY. *Educ:* Col Great Falls, BA, 64; Wash State Univ, MS, 66, PhD(chem), 68. *Prof Exp:* Instr chem, Wash State Univ, 66-67; from asst prof to assoc prof, 68-78, PROF CHEM, WESTERN WASH UNIV, 78- *Mem:* Am Chem Soc. *Res:* Synthesis, characterization and kinetic studies on coordination compounds. *Mailing Add:* Dept of Chem Western Wash Univ Bellingham WA 98225

WEYHENMEYER, JAMES ALAN, b Hazelton, Pa, Feb 12, 51; m 73; c 2. NEUROCHEMISTRY, NEUROANATOMY. *Educ:* Knox Col, BA, 73; Ind Univ, PhD(cell biol), 77. *Prof Exp:* Assoc instr anat, Sch Med, Ind univ, 73-77; fel, Col Med, Univ Iowa, 77-79; asst prof, 79-85, ASSOC PROF NEUROSCI, COL MED, UNIV ILL, 85- *Mem:* Soc Neurosci; AAAS; Am Chem Soc; Am Soc Cell Biol. *Res:* Cell and molecular biology of neuropeptides and their receptors in the mammalian central nervous system; physicochemical characterization of angiotensin II and its receptor in the central nervous system and their relationship to the pathogenesis of essential hypertension. *Mailing Add:* Dept Cell Struct Biol, Univ Ill Urbana IL 61801

WEYHMANN, WALTER VICTOR, b Roanoke, Va, Nov 27, 35; m 57; c 1. SOLID STATE PHYSICS, LOW TEMPERATURE PHYSICS. *Educ:* Duke Univ, BS, 57; Harvard Univ, AM, 58, PhD(physics), 63. *Prof Exp:* Res fel physics, Harvard Univ, 63-64; from asst prof to assoc prof 64-75, PROF PHYSICS & HEAD SCH PHYSICS & ASTRON, UNIV MINN, MINNEAPOLIS, 75- *Mem:* Am Phys Soc. *Res:* Nuclear magnetic resonance measurement of sublattice magnetizations; weak magnetic phenomena and nuclear ordering at very low temperatures; production of very low temperatures. *Mailing Add:* Dept Physics Univ Minn Minneapolis MN 55455

WEYL, PETER K, b Ger, May 6, 24; nat US; m 47; c 3. ENVIRONMENTAL MANAGEMENT. *Educ:* Univ Chicago, ScM, 51, PhD(physics), 53. *Prof Exp:* Lectr physics, Roosevelt Col, 51-53; asst prof, Brazilian Ctr Phys Res, 53-54; physicist, Explor & Prod Res Labs, Shell Develop Co, 54-59, sr physicist, 59-63; prof oceanog, Ore State Univ, 63-66; PROF OCEANOG, STATE UNIV NY STONY BROOK, 66- *Concurrent Pos:* Lectr, Univ Houston, 55-62; vis prof, Hebrew Univ, Jerusalem, 72-73 & Univ Concepcion, Chile, 79. *Mem:* AAAS; Am Geophys Union. *Res:* Chemical and physical oceanography; ocean-climate interaction; environmental stability; coastal zone management. *Mailing Add:* 90 Christian Ave Stony Brook NY 11790

WEYLAND, JACK ARNOLD, b Butte, Mont, June 12, 40; m 65; c 5. PHYSICS, SCIENCE COMMUNICATIONS. *Educ:* Mont State Univ, BS, 62, PhD(solid state physics), 69. *Prof Exp:* Asst prof, 68-77, PROF PHYSICS, SDAK SCH MINES & TECHNOL, 78- *Concurrent Pos:* Grants, US Dept Transp, 69-72 & Res Corp, 70-77. *Mem:* AAAS; Am Phys Soc; Am Asn Physics Teachers. *Res:* High pressure diffusion and magnetic susceptibility studies; adhesion of ice to concrete surfaces; science errors in the movies. *Mailing Add:* Dept Physics SDak Sch Mines & Technol Rapid City SD 57701

WEYLER, MICHAEL E, b Boston, Mass, 40; m 62; c 3. FAILURE ANALYSIS, FORENSIC ENGINEERING. *Educ:* Tufts Univ, BSME, 62; Univ Mich Ann Arbor, MSE, 63, PhD(mech eng), 69. *Prof Exp:* Asst prof, Univ Mich, Dearborn, 63-69; construct officer, US Navy, 70-71, ocean engr, 71-78, pub works officer, Antarctica, 78-80; instr mech eng, US Naval Acad, Annapolis, 80-82; sr res specialist, Exxon Prod Res, Houston, 82-86; dir facil progs, US Naval Reserve, 86-90; CONSULT ENGR, CH&A CORP, HOUSTON, 90- *Concurrent Pos:* Capt, Civil Eng Corps, US Naval Reserve. *Mem:* Nat Soc Prof Engrs. *Res:* Application of field experience in engineering and construction to the analysis of failure in mechanical systems; specializes in analysis of failures due to extreme environmental conditions, such as arctic and offshore conditions. *Mailing Add:* 4303 Birchcroft Dr Houston TX 77088

WEYMANN, RAY J, b Los Angeles, Calif, Dec 2, 34; m 56; c 3. ASTRONOMY, SPECTROSCOPY. *Educ:* Calif Inst Technol, BS, 56; Princeton Univ, PhD(astron), 59. *Prof Exp:* Res fel astron, Calif Inst Technol, 59-61; from asst prof to prof astron, Univ Ariz, 61-86, head, Dept Astron & Dir, 70-75, astrom, Steward Observ, 70-86; dir, 86-88, STAFF, OBSERV CARNEGIE INST WASH, 88- *Mem:* Nat Acad Sci; Am Astron Soc. *Res:* Theoretical astrophysics; stellar spectroscopy. *Mailing Add:* Observ Carnegie Inst Wash 813 Santa Barbara St Pasadena CA 91101

WEYMOUTH, JOHN WALTER, b Palo Alto, Calif, Jan 14, 22; m 66; c 3. ARCHAEOMETRY, GEOPHYSICS. *Educ:* Univ Calif, AB, 43, MA, 50, PhD(physics), 51. *Prof Exp:* Instr physics, Vassar Col, 52-54; from asst prof to assoc prof, Clarkson Col Technol, 54-58; from asst prof to prof, 58-89, EMER PROF PHYSICS, UNIV NEBR-LINCOLN, 89- *Mem:* Am Phys Soc; Soc Am Archaeol; Soc Historical Archaeol; Soc Archaeol Sci (vpres, 81-82, pres 82-83). *Res:* Physical methods in archaeology; geophysical surveying of archaelogical sites. *Mailing Add:* Dept Physics Univ Nebr Lincoln NE 68588-0111

WEYMOUTH, PATRICIA PERKINS, b Birmingham, Mich, Dec 31, 18; div; c 3. NATURAL SCIENCE, HISTORY OF SCIENCE. *Educ:* Russell Sage Col, AB, 40; Univ Cincinnati, PhD(biochem), 44. *Prof Exp:* Asst biochem, Armored Med Res Lab, Ft Knox, 44-46; res assoc med physics, Univ Calif, 46-49; res assoc radio, Stanford-Lane Hosp, San Francisco, 49-52, Vassar Col, 52-54 & Clarkson Col Technol, 54-58; res assoc biochem & nutrit, Univ Nebr, 58-67; from asst to assoc prof, 69-75 prof natural sci, 75-86, EMER PROF, MICH STATE UNIV, 86. *Concurrent Pos:* Dir, Kedzie Lects, for high sch sci students, 88. *Mem:* Fel AAAS; NY Acad Sci; Sigma Xi. *Res:* Bacteriological and cancer biochemistry; nucleic acid and enzyme studies; interpenetrance of science and other disciplines. *Mailing Add:* Dept Natural Sci Mich State Univ 100 N Kedzie Lane East Lansing MI 48823

WEYMOUTH, RICHARD J, b Brewer, Maine, July 19, 28; c 1. ANATOMY, ENDOCRINOLOGY. *Educ:* Univ Maine, BS, 50; Univ Mich, MS & PhD, 55; Marquette Univ, MD, 63. *Prof Exp:* Instr anat, Miami Univ, 55-59; asst prof, Sch Med, Marquette Univ, 59-61; intern, Univ Mich Hosps, 63-64; from assoc prof to prof anat, Med Col Va, 64-75; chmn dept, 75-84, assoc dean student affairs & admis, 76-82, PROF ANAT, SCH MED, UNIV SC, 75-, ASSOC DEAN ACAD AFFAIRS, 82- *Concurrent Pos:* Lectr med, Med Col Va, 64-75; Fulbright scholar, Ankara, Turkey, 85-86. *Mem:* Am Asn Anatomists; Sigma Xi; Transplantation Soc. *Res:* Electron microscopy, endocrine glands and kidney. *Mailing Add:* Dept Anat Sch Med Univ SC Columbia SC 29208

WEYNA, PHILIP LEO, b Chicago, Ill, May 11, 32; m 54; c 4. ORGANIC CHEMISTRY, POLYMER CHEMISTRY. *Educ:* Loyola Univ, Ill, BS, 54; Univ Wis, PhD(org chem), 58. *Prof Exp:* Res chemist, Morton Chem Co, 58-64, supvr polymer res, 64-66, dir, 66-73, mgr chem specialties bus group, 73-76, dir com develop, 76-82; CONSULT, 82- *Mem:* Am Chem Soc; Tech Asn Pulp & Paper Indust. *Res:* Polymeric coatings and adhesives. *Mailing Add:* 6604 Rhode Island Trail Crystal Lake IL 60012-3118

WEYNAND, EDMUND E, b San Antonio, Tex, Nov 21, 20; m 50; c 4. MECHANICAL ENGINEERING, AERODYNAMICS. *Educ:* Univ Tex, BS, 49; Mass Inst Technol, SM, 50, MechE, 51, ScD(mech eng), 53. *Prof Exp:* Asst eng & draftsman, San Antonio Air Depot, 42; asst, Proj Squid, Mass Inst Technol, 52-53; sr & proj propulsion engr, Convair Div, Gen Dynamics Corp, Tex, 53-56; assoc prof mech eng, 56-62, PROF MECH ENG, SOUTHERN METHODIST UNIV, 62- *Concurrent Pos:* Consult, Convair Div, Gen Dynamics Corp, 57-59 & Ling-Tempco-Vought, Vought Aeronautics. *Mem:* Am Inst Aeronaut & Astronaut; Am Soc Eng Educ; Sigma Xi. *Res:* Internal aerodynamics; nozzles; jet mixing; base drag; inlets; wind tunnel testing; ejectors; optical instrumentation; thermodynamics; heat transfer; fluid mechanics. *Mailing Add:* 4195 Lively Lane Dallas TX 75220

WEYTER, FREDERICK WILLIAM, b Philadelphia, Pa, Oct 7, 34; m 65; c 2. BIOCHEMICAL GENETICS. *Educ:* Univ Pa, AB, 56; Amherst Col, MA, 58; Univ Ill, PhD(biochem), 62. *Prof Exp:* From instr to asst prof, 62-73, ASSOC PROF BIOL, COLGATE UNIV, 73- *Concurrent Pos:* Fulbright lectr, Afghanistan, 65-66. *Mem:* Am Chem Soc. *Res:* Drug metabolism in microorganisms; regulation of arginine biosynthesis in E coli. *Mailing Add:* Dept Biol Colgate Univ Hamilton NY 13346

WHALEN, CAROL KUPERS, PSYCHOLOGY. *Educ:* Stanford Univ, BS, 63; Univ Calif, Los Angeles, MA, 65 & PhD(clin develop psychol), 67. *Prof Exp:* Res asst, Stanford Univ, 61-63; intern, Psychol Clin, Univ Calif, Los Angeles, 64-65, Long Beach Vet Admin Hosp & Fernald Sch, Univ Calif, Los Angeles, 65-66, Fairview State Hosp & S Coast Child Guid Clin, 66-67; res specialist & co-prin investr, Therap Pyramids Proj, Fairview State Hosp, 67-70; asst res psychologist, 70-71, asst prof social ecol, 70-73, assoc prof psychiat & human behav, 74-79, assoc prof soc ecol, 73-79, actg dir, Prog Social Ecol, 85-86, PROF PSYCHIAT & HUMAN BEHAV & SOCIAL ECOL, UNIV CALIF IRVINE, 79- *Concurrent Pos:* Res asst, Stanford Univ, 61-63, teaching asst, 62; ed consult, J Appl Behav Anal, J Appl Develop Psychol, J Child Psychol & Psychiat, J Consult & Clin Psychol & J Personality & Social Psychol; USPHS predoc fel, Univ Calif, Los Angeles, 64-67; res consult, Educ Assessment Ctr Handicapped Children, 70-71; Nat Regist Health Serv Providees Psychol; Panel Deinstitutionalized Children & Youth, NAS, 78-81; Res Review Panel, NIMH, 80-84; comt child develop res & pub policy, NAS, 82-87; data & safety monitoring comt, Dietary Intervention Study Children, Nat Heart, Lung & Blood Inst, 87- *Mem:* Fel Am Psychol Asn. *Res:* Author of over 25 publications. *Mailing Add:* Social Ecol Univ Calif Irvine Irvine CA 92717

WHALEN, JAMES JOSEPH, b Meriden, Conn, Feb 16, 35; m 59; c 3. ELECTROMAGNETIC COMPATIBILITY, MICROWAVES. *Educ:* Cornell Univ, BEE, 58; Johns Hopkins Univ, MSE, 62, PhD(elec eng), 69. *Prof Exp:* Res staff asst, Carlyle Barton Lab, Johns Hopkins Univ, 62-69, assoc res scientist, 69-70; res scientist, Nat Oceanic & Atmospheric Admin, 70; from asst prof to assoc prof, 70-81, PROF ELEC ENG, STATE UNIV NY, BUFFALO, 81- *Concurrent Pos:* Consult, McDonnell Douglas Corp, St Louis, Mo, 76-78, Southeast Ctr Elec Eng Educ, 77-79 & 82-, Universal Energy Systs, 79-82 & Digital Equip Corp, 84-85. *Mem:* Inst Elec & Electronics Engrs; Am Soc Eng Educ. *Res:* Semiconductor devices; microwave overstressing & damage analysis of gallium arsenide metal-semiconductor field-effect transistors and integrated circuits; electronic circuit analysis program applications; electromagnetic compatibility, RFI effects in integrated circuits; microwaves; millimeter waves; prediction, measurement and suppression of electromagnetic interference in microelectronic circuits; determination of the microwave electrical overstress properties of gallium arsenide metalized semiconductor field-effect microwave transistors; measurements. *Mailing Add:* Dept Elec & Comput Eng State Univ NY Buffalo 215 B Bonner Hall Buffalo NY 14260

WHALEN, JAMES WILLIAM, b Enid, Okla, Mar 16, 23; m 46; c 2. SURFACE CHEMISTRY. *Educ:* Univ Okla, BS, 46, MS, 47, PhD(chem), 51. *Prof Exp:* Res assoc, Mobil Oil Corp, 50-68; chmn dept chem, Univ Tex, El Paso, 68-72, dean col sci, 72-75, prof, 68-69, EMER PROF CHEM, UNIV TEX, EL PASO, 89- *Concurrent Pos:* Vis prof, Univ Heidelberg, 84, Univ Regensburg, 84; consult, Ballard Res Inc, Vancouver, BC, 76-86; vis sr scientist, Ballard Res, 86; lectr, Tex Univ Consortium/Inst Technol Malaysia, Shah Alam, Malaysi, 87-88. *Mem:* Am Chem Soc; Int Asn Colloid & Interface Scientists. *Res:* Surface phenomena; adsorption; calorimetry. *Mailing Add:* Dept Chem Univ Tex El Paso TX 79968

WHALEN, JOSEPH WILSON, b Battle Creek, Mich, May 27, 23; m 54; c 2. BIOCHEMISTRY, MICROBIOLOGY. *Educ:* Mich State Univ, BS, 49, MS, 51, PhD(microbiol), 55. *Prof Exp:* Bacteriologist, Arthur S Kimball Sanatorium, Battle Creek, 48-50, lab dir, 54-55; bacteriologist, Calhoun County Health Dept, Mich, 50-52; bacteriologist, Biol Labs, Pitman-Moore Co, 55-56, head bact dept, 56-64, mgr bact & immunochem depts, 64-71; asst to dir, Biol Labs, Dow Chem Co, 72-73, res specialist, 73-76, sr res specialist, 76-90; RES SCIENTIST, WHALEN ENTERPRISES, 90- *Concurrent Pos:* Tech adv, LIFE Labs, Quito, Ecuador, 64-73; consult, Mich Dept Health, 88- *Mem:* Sigma Xi; Am Soc Microbiol; AAAS. *Res:* Bacteriological and immunochemical investigations in tuberculosis; bacterins and vaccines; antimicrobial agents; antibiotics; chemotherapy; microbial genetics; environmental microbiology. *Mailing Add:* 6014 Sturgeon Creek Pkwy Midland MI 48640

WHALEN, THOMAS J(OHN), b Rochester, NY, Sept 1, 31; m 53; c 8. PHYSICAL METALLURGY, CERAMICS. *Educ:* Alfred Univ, BS, 53; Pa State Univ, MS, 55, PhD(metall), 57, Oakland Univ, cert, 90. *Prof Exp:* Asst, Pa State Univ, 53-55; PRIN RES SCIENTIST, SCI RES STAFF, FORD MOTOR CO, 57- *Mem:* Fel Am Ceramic Soc; Am Soc Metals; Soc Automotive Engrs; Am Statist Asn. *Res:* Powder metallurgy; surface energies of solids, liquids and glasses; mechanical strength of ceramics; processing and mechanical properties of ceramics; statistical design of experiments. *Mailing Add:* Ford Motor Co 20000 Rotunda Dr Box 2053 Dearborn MI 48121

WHALEN, WILLIAM JAMES, b Ft Dodge, Iowa, July 9, 15; m 46; c 3. PHYSIOLOGY. *Educ:* Stanford Univ, BA, 48, MA, 49, PhD(physiol), 51. *Prof Exp:* Asst physiol, Stanford Univ, 48-49, res assoc, 49-51, instr, 50-51; from instr to asst prof, Univ Calif, Los Angeles, 51-60; assoc prof, Col Med, Univ Iowa, 60-67; dir res, St Vincent Charity Hosp, 67-; prof, Case-Western Res Univ, 67-80; CONSULT, 80- *Concurrent Pos:* Fulbright scholar, 58-59; adj prof, Case Western Reserve Univ, 67-80, emer prof, 80- *Mem:* AAAS; Microcirc Soc; Am Physiol Soc; Cardiac Muscle Soc (pres, 66-68); NY Acad Sci. *Res:* Cardiovascular research; cardiac function in isolated preparations; respiratory control mechanisms; autonomic pharmacology; tissue oxygen tension and cell metabolism; chemoreceptors. *Mailing Add:* 2805 Bellamah Dr Santa Fe NM 87505

WHALEY, HOWARD ARNOLD, b Iroquois, Ill, Sept 1, 34; m 54; c 4. ORGANIC CHEMISTRY. *Educ:* Univ Ill, Urbana, BS, 56; Univ Wis-Madison, PhD(org chem), 61. *Prof Exp:* Res chemist, Lederle Labs, Am Cyanamid Co, 61-66; RES CHEMIST, UPJOHN CO, 66- *Mem:* Am Chem Soc. *Res:* Finding, isolating and studying new antibiotics and modifying known antibiotics. *Mailing Add:* Chem & Biol Screening Upjohn Co Kalamazoo MI 49001-0199

WHALEY, JULIAN WENDELL, b Parkersburg, WVa, Aug 12, 37; m 61, 70; c 4. PLANT PATHOLOGY, PLANT SCIENCE. *Educ:* W Liberty State Col, BS, 59; WVa Univ, MS, 61; Univ Ariz, PhD(plant path), 64. *Prof Exp:* Sr plant pathologist, Eli Lilly & Co, 64-70; PROF PLANT SCI, CALIF STATE UNIV, FRESNO, 70- *Mem:* Am Phytopath Soc; Coun Agr Sci & Technol. *Res:* Plant protection; pesticides; soil fungi; grape diseases. *Mailing Add:* Dept Plant Sci Calif State Univ 6241 N Maple Ave Fresno CA 93740

WHALEY, KATHARINE BIRGITTA, b Barnehurst, Eng, 1956. CHEMICAL DYNAMICS, CONDENSED PHASE CHEMISTRY. *Educ:* Oxford Univ, BA, 78, MA, 80; Univ Chicago, MSc, 81, PhD(chem), 84. *Prof Exp:* Goldmeir fel, Hebrew Univ Jerusalem, 84-85; postdoctoral fel, Tel Aviv Univ, 85-86; ASST PROF CHEM, UNIV CALIF, BERKELEY, 86- *Concurrent Pos:* A P Sloan fel, 91-93. *Mem:* Am Chem Soc; Am Phys Soc; Royal Soc Chem. *Res:* Chemical dynamics of interacting quantum systems; description of clusters and development of techniques for study of strongly correlated finite systems. *Mailing Add:* Dept Chem Univ Calif Berkeley CA 94720

WHALEY, PETER WALTER, b Baltimore, Md, June 27, 37; m 60; c 3. GEOLOGY. *Educ:* Ohio Wesleyan Univ, BA, 59; Univ Ky, MS, 64; La State Univ, PhD(geol), 69. *Prof Exp:* From asst prof to assoc prof, 68-77, PROF GEOL, MURRAY STATE UNIV, 77- *Mem:* Soc Econ Paleontologists & Mineralogists; Geol Soc Am; Sigma Xi. *Res:* Modern depositional environments; carboniferous system of the Eastern United States. *Mailing Add:* 802 Guthrie Dr Murray KY 42071

WHALEY, RANDALL MCVAY, physics; deceased, see previous edition for last biography

WHALEY, ROSS SAMUEL, b Detroit, Mich, Nov 7, 37; m 58; c 3. FOREST ECONOMICS, NATURAL RESOURCE & ENVIRONMENTAL POLICY. *Educ:* Univ Mich, BS, 59; Colo State Univ, MS, 61; Univ Mich, PhD(natural resource econ), 69. *Prof Exp:* Instr forestry, Colo State Univ, 60-61 & assoc dean, Col Forestry & Natural Resources, 70-73; res forester, Southern Forest Exp Sta, US Forest Serv, 61-63; asst prof natural resource econ, Utah State Univ, 65-67, prof forest sci & head dept, 67-69; head dept landscape archit & regional planning, Univ Mass, 73-76, dean, Col Food & Natural Resources, 76-78; dir forest econ, Forest Serv, USDA, 78-84; PRES, STATE UNIV NY, COL ENVIRON SCI & FORESTRY, 84- *Concurrent Pos:* Consult, Intermountain Forest & Range Exp Sta, USDA Forest Serv, 66-67, Rocky Mountain Forest & Range Exp Sta, 67, Pub Land Law Rev Comn, 70, Wallace, McHarg, Roberts & Todd, Joe Meheen Eng & Geddes, Brecher, Qualls, Cunningham, Architects. *Mem:* Soc Am Foresters; Sigma Xi. *Res:* Application of economic theory to problems of natural resources policy and regional planning. *Mailing Add:* Pres/Chancellor State Univ NY Col Environ Sci & Forestry Syracuse NY 13210

WHALEY, THOMAS PATRICK, b Atchison, Kans, Jan 13, 23; m 46, 69; c 2. APPLIED SOLAR THERMAL ENERGY. *Educ:* St Benedicts Col, BS, 42; Univ Kans, PhD(chem), 50. *Prof Exp:* Proj leader, Res Lab, Ethyl Corp, 50-55, res supvr, Develop Lab, 55-58, sr res assoc, 58-62; mgr inorg & phys chem, Int Minerals & Chem Corp, 62-69, dir, Anal & Tech Serv, 69-74; tech dir, Sipi Metals Corp, 74-76; assoc dir solar energy, 76-81, SR ADV, INST GAS TECHNOL, 81- *Concurrent Pos:* Consult, Dearborn Chem Co, 74; instr, Oakton Community Col, 74; pres, Consanal Corp, 74-; ed, chem Bull, 76-81. *Mem:* Am Chem Soc. *Res:* Alkali metals; organometallics; inorganic compounds of alkali and alkaline earth metals; refractory metals; metal plating; inorganic phosphates; non-ferrous metals; precious metals; solar energy; fossil fuels; solar production of chemicals; energy planning and systems analysis. *Mailing Add:* Inst Gas Technol IIT Ctr 3424 S State St Chicago IL 60616

WHALEY, THOMAS WILLIAMS, b Albuquerque, NMex, June 13, 42. ORGANIC CHEMISTRY. *Educ:* Univ NMex, BS, 67, MS, 69, PhD(chem), 71. *Prof Exp:* MEM STAFF CHEM, LOS ALAMOS SCI LAB, UNIV CALIF, 71- *Concurrent Pos:* Adj asst prof chem, Univ NMex, Los Alamos Grad Ctr, 73-; ed, J Labelled Compounds & Radiopharmaceut, 74- *Mem:* Am Chem Soc; Sigma Xi; AAAS. *Res:* Organic synthesis with stable isotopes. *Mailing Add:* Los Alamos Sci Lab Los Alamos Nat Lab PO Box 1663 MS M880 Los Alamos NM 87545

WHALEY, WILSON MONROE, b Baltimore, Md, July 21, 20; m 56; c 3. TEXTILE CHEMISTRY. *Educ:* Univ Md, BS, 42, MS, 44, PhD(chem), 47. *Prof Exp:* Org chemist, Naval Res Lab, 44-47; fel chem, Univ Ill, 47-49; asst prof, Univ Tenn, 49-53; asst dir res labs, Pabst Brewing Co, 53-55; sect head chem, Res Ctr, Gen Foods Corp, 55-59; asst tech dir, Midwest Div, Arthur D Little, Inc, 59-62; mgr indust develop, IIT Res Inst, 62-65, mgr org chem, 63-65; dir res & planning, Burlington Industs, Inc, 65-71; pres, Whaley Assocs, NY, 71-75; prof textile chem & head dept, NC State Univ, 75-85; RETIRED. *Concurrent Pos:* Consult, Oak Ridge Nat Lab, 51-55; adj assoc prof, Cornell Univ, 74-75. *Mem:* Am Chem Soc; Am Inst Chemists. *Res:* Synthesis and chemistry of dyes; heterocyclic and organophosphorus compounds; polymer syntheses; textile fibers, finishes and processes; plastics, resins and composite structures; textile chemicals, polymers and processes; mutagenicity & carcinogenicity of dyes and intermediates. *Mailing Add:* PO Box 1009 St Michaels MD 21663-1009

WHALIN, EDWIN ANSIL, JR, b Barlow, Ky, Mar 6, 24; m 48; c 4. PHYSICS. *Educ:* Univ Ill, BS, 45, MS, 47, PhD(physics), 54. *Prof Exp:* From asst prof to prof physics, Univ NDak, 54-66; assoc prof, 66-70, PROF PHYSICS, EASTERN ILL UNIV, 70- *Mem:* Am Phys Soc; Sigma Xi. *Res:* Nuclear physics. *Mailing Add:* Nine Heather Dr Charleston IL 61920

WHALING, WARD, b Dallas, Tex, Sept 29, 23. ATOMIC PHYSICS. *Educ:* Rice Inst, BA, 44, MA, 47, PhD(physics), 49. *Prof Exp:* Fel, 49-52, from asst prof to assoc prof, 52-62, PROF PHYSICS, CALIF INST TECHNOL, 62- *Mem:* Am Phys Soc. *Res:* Penetration of charged particles through matter; atomic spectroscopy. *Mailing Add:* Kellogg Radiation Lab Calif Inst Technol Pasadena CA 91125

WHALLEY, EDWARD, b Darwen, Eng, June 20, 25; m 56; c 3. PHYSICAL CHEMISTRY. *Educ:* Univ London, BSc & ARCS, 45, PhD(phys chem), 49, DSc, 63; Imp Col, dipl, 49. *Prof Exp:* Lectr chem, Royal Tech Col, Salford, Eng, 48-50; fel, Nat Res Coun Can, 50-52, asst res officer, Pure Chem Div, 52-53, from asst res officer to sr res officer, Appl Chem Div, 53-61, PRIN RES OFFICER, CHEM DIV, NAT RES COUN CAN, 62-, HEAD HIGH PRESSURE SECT, 62- *Concurrent Pos:* Vis prof, Univ Western Ont, 67 & Kyoto Univ, Japan, 74-75; chmn, Phys Chem Div, Chem Inst Can, 71-72; ed, Physics & Chem Ice, 73; assoc mem comn thermodyn & thermochem, Int Union Pure & Appl Chem, 73; chmn, Can Nat Comt for Int Asn Properties of Steam, 74-83, secy, 83-90; hon treas, Int Asn Advan High Pressure Sci & Technol, 77-85; mem, Arctic Circle, Ottawa, 70-, exec comt, 90- *Honors & Awards:* Silver Jubilee Medal, 77; Centennial Medal, Royal Soc Can, 83. *Mem:* Am Phys Soc; fel Chem Inst Can; fel Royal Soc Can (assoc hon treas, 69-71, assoc hon secy, 71-74, hon secy, 74-77); Int Glacial Soc. *Res:* High-pressure physical chemistry; far-infrared and Raman spectroscopy. *Mailing Add:* Div Chem Steacie Inst Molecular Sci Ottawa ON K1A 0R6 Can

WHAN, GLENN A(LAN), b North Lima, Ohio, Aug 8, 30; m 55; c 3. CHEMICAL & NUCLEAR ENGINEERING. *Educ:* Ind Inst Technol, BSChE, 51; Mont State Univ, MSChE, 53; Carnegie Inst Technol, PhD(chem eng), 57. *Prof Exp:* Asst prof chem eng, Univ NMex, 57-59; mem staff, Nuclear Propulsion Div, Los Alamos Sci Lab, Univ Calif, 59-60; from asst prof to assoc prof, 60-66, prof nuclear eng & chmn dept, 66-71, chmn dept chem & nuclear eng, 71-75, PROF CHEM & NUCLEAR ENG, UNIV NMEX, 71-, ASSOC DEAN, COL ENG, 76- *Concurrent Pos:* Consult, ACF

Indust, Inc, 58-63, Sandia Lab, 63-68, US AEC Int Div, Lisbon, Portugal, 68, Western Interstate Nuclear Bd, Denver, Colo, 71, Los Alamos Sci Lab, 75 & State of NMex, 78; vis staff mem, Eng Div, Los Alamos Sci Lab, 74-75. *Mem:* Fel Am Nuclear Soc; Am Inst Chem Engrs; Am Soc Eng Educ; Sigma Xi; Nat Soc Prof Engrs. *Res:* Energy resources and systems analysis; nuclear energy systems and safety; nuclear fuel cycles; radioactive waste management; benefit-risk-cost analysis; technology assessment. *Mailing Add:* Dept Chem & Nuclear Eng Univ NMex Albuquerque NM 87131

WHAN, RUTH ELAINE, physical chemistry; deceased, see previous edition for last biography

WHANG, ROBERT, b Honolulu, Hawaii, Mar 7, 28; m 56; c 4. INTERNAL MEDICINE, NEPHROLOGY. *Educ:* St Louis Univ, BS, 52, MD, 56; Am Bd Internal Med, dipl, 65, 74, cert nephrol, 72; Am Bd Nutrit, cert, 81. *Prof Exp:* Intern med, Johns Hopkins Univ Hosp, 56-57; asst resident, Baltimore City Hosps, 57-59, resident, 59-60; from instr to assoc prof med, Sch Med, Univ NMex, 63-71; prof, Sch Med, Univ Conn, 71-73, assoc dean, Vet Admin Hosp Affairs, 72-73; prof med, Sch Med, Ind Univ, Indianapolis, 73-78; PROF MED & VHEAD DEPT MED, COL MED, UNIV OKLA, OKLAHOMA CITY, 78-; CHIEF, MED SERV, VET ADMIN HOSP, 78- *Concurrent Pos:* Life Ins Med res fel, Univ NC, 60-62, USPHS trainee renal dis, 62-63; chief metab, Vet Admin Hosp, Albuquerque, 66-71; chief staff, Vet Admin Hosp, Newington, Conn, 71-73 & Indianapolis, Ind, 73-78. *Mem:* Fel Am Col Physicians; Am Fedn Clin Res; Int Soc Nephrology; Am Soc Nephrology; Am Col Nutrit. *Res:* Magnesium deficiency, interrelationship of magnesium and potassium, electrolyte changes in uremia. *Mailing Add:* 921 NE 13th St Oklahoma City OK 73104

WHANG, SUKOO JACK, b Seoul, Korea, Feb 3, 34; US citizen; m 63; c 3. MEDICAL MICROBIOLOGY, IMMUNOLOGY. *Educ:* Univ Calif, Los Angeles, MS, 60, PhD(med microbiol, immunol), 63, MD, 72; Am Bd Med Microbiol, dipl, 75; Am Bd Path, dipl, 77. *Prof Exp:* Asst prof microbiol, Calif State Polytech Univ, 63-64; chief microbiol & serol dept, Providence Hosp, Southfield, Mich, 64-65; chief microbiol & immunol dept, Ref Lab, Div Abbott Labs, Calif, 65-69; CHIEF MICROBIOL & IMMUNOL DIV, CLIN LAB, CHMN INFECTION CONTROL COMT & PROG DIR SCH MED TECHNOL, WHITE MEM MED CTR, LOS ANGELES, 77 - *Concurrent Pos:* Chief microbiol & Immunol Div, Clin Lab, White Mem Med Ctr, 69-70; adj prof, Pac Union Col, Angwin,77- *Mem:* NY Acad Sci; Am Soc Microbiol; Sigma Xi; fel Am Col Physicians; fel Am Soc Clin Pathologists; Am Med Asn. *Res:* Clinical microbiology, pathology and serology; syphilis serology; diagnostic tests for the detection of inborn errors of metabolism; fluorescent antibody testing. *Mailing Add:* 1325 Via Del Ray South Pasadena CA 91030

WHANG, SUNG H, b Suh-Byuck, SKorea, Feb 17, 36; US citizen; m 30; c 4. INTERMETALLIC MATERIALS, SUPERCONDUCTING OXIDES PROCESSING. *Educ:* Seoul Nat Univ, BS, 62; Columbia Univ, MS, 75, DEngsci, 78. *Prof Exp:* Lectr metall eng, Seoul Nat Univ, 69-72; staff scientist, Northeastern Univ, 79-81, sr scientist mat sci, 81-85; assoc prof, 85-91, PROF METALL & MAT SCI, POLYTECH UNIV, 91- *Concurrent Pos:* Mem, Bd Rev Met Trans A, 89-; ed, Conf Proc on Superconducting Mat, TMS, 89- & vchmn, 90- *Mem:* Am Soc Metals; Mat Res Soc; Sigma Xi; Metall Soc Am Inst Mech Engrs; NY Acad Sci. *Res:* Processing of high Jc superconducting oxide materials; deformation and properties of titanium aluminides for high temperature applications; composite materials processing for structural damping applications; rapid solidification processing. *Mailing Add:* Dept Mat Sci Polytech Univ Six Metrotech Ctr Brooklyn NY 11201

WHANG, YUN CHOW, b Foochow, China, Dec 13, 33; m 59; c 3. SPACE SCIENCE, FLUID MECHANICS. *Educ:* Taiwan Col Eng, BS, 54; Univ Minn, Minneapolis, PhD(fluid mech), 61. *Prof Exp:* Asst prof aerospace eng, Univ Fla, 61-62; from asst prof to assoc prof space sci, 62-67, chmn dept, 71-84, PROF MECH ENG, CATH UNIV AM, 67- *Concurrent Pos:* NASA grant, Cath Univ Am, 69-, NSF grant, 71-, USAF Off Sci Res grant, 86- *Mem:* Assoc fel Am Inst Aeronaut & Astronaut; Am Geophys Union; Am Soc Mech Engrs; Am Soc Eng Educ. *Res:* Solar wind and its interaction with the earth, the moon and other planets; interplanetary shocks; coronal slow shocks. *Mailing Add:* Dept Mech Eng Cath Univ Am Washington DC 20064

WHANGBO, MYUNG HWAN, b Korea, Oct 21, 45; US citizen; c 2. SOLID STATE CHEMISTRY. *Educ:* Seoul Univ, BSc, 68, MSc, 70; Queens Univ, PhD(chem), 74. *Prof Exp:* Fel, Queens Univ, 75-76; assoc chem, Cornell Univ, 76-77; from asst prof to assoc prof chem, 78-87, PROF CHEM, NC STATE UNIV, 87- *Concurrent Pos:* Camille & Henry Dreyfus teacher-scholar, 80-85; vis prof, Bell Labs, Murray Hill, 81, Univ de Nantes, 84, Argonne Nat Lab, 85, 86 & Univ Paris-Sud, 87, 89. *Honors & Awards:* Sigma Xi Res Award, 81. *Mem:* Am Chem Soc; Sigma Xi. *Res:* Molecular orbital interpretation of structures and reactivities of organic and inorganic systems; structure property relationships of crystalline materials; theoretical chemistry. *Mailing Add:* Dept Chem NC State Univ Raleigh NC 27695-8204

WHANGER, PHILIP DANIEL, b Lewisburg, WVa, Aug 30, 36; m 64; c 2. NUTRITIONAL BIOCHEMISTRY. *Educ:* Berry Col, BS, 59; WVa Univ, MS, 61; NC State Univ, PhD(nutrit), 65. *Prof Exp:* Res assoc biochem, Mich State Univ, 65-66; from asst prof to assoc prof, 66-78, PROF NUTRIT & BIOCHEM, ORE STATE UNIV, 78- *Concurrent Pos:* NIH res fel, Mich State Univ, 66-67, res grants selenium & myopathies, Ore State Univ, 68-; NIH spec fel, 72; assoc staff, Harvard Med Sch, 72-73; vis scientist, Gen Acad Exchange Serv, Univ Tubingen, 86; NSF Int fel, 80-81; vis scientist, Commonwealth Sci & Indust Res Orgn, Wembley, Western Australia & Acad Prev Med, Beijing, China, 88. *Mem:* Am Inst Nutrit; Am Soc Animal Sci; Int Bioinorg Scientists; Soc Environ Geochem & Health. *Res:* Altered metabolic pathways under selenium deficiency; relationships of vitamin E and selenium in myopathies; biochemical properties of selenium and cadmium metallo-proteins; metabolic pathways for incorporation of selenium into proteins; selenium and glutathione peroxidise in human blood fractions; selenium deficiencies in primates; selenium intake on human blood and urine fractions. *Mailing Add:* Dept Agr Chem Ore State Univ Corvalis OR 97330

WHARRY, STEPHEN MARK, b Dalheart, Tex, Oct 23, 55. COMPUTER PROGRAMMING, CRUDE OIL ANALYSES. *Educ:* Iowa State Univ, BS, 77; Northwestern Univ, MS, 79, PhD(phys org chem), 81. *Prof Exp:* Res chemist, 81-87, SR RES CHEMIST, PHILLIPS PETROL, 87- *Mem:* Am Chem Soc; Sigma Xi; Soc Appl Spectros; AAAS. *Res:* Applications of nuclear magnetic resonance spectroscopy to petroleum fractions and engineering plastics; method development and automation. *Mailing Add:* 4842 Clearview Circle Bartlesville OK 74006-5501

WHARTON, CHARLES BENJAMIN, b Gold Hill, Ore, Mar 29, 26; m 53; c 3. PLASMA PHYSICS, MICROWAVE TECHNOLOGY. *Educ:* Univ Calif, Berkeley, BSEE, 50, MS, 52. *Prof Exp:* Proj engr, Lawrence Radiation Lab, Univ Calif, 50-62; staff mem exp physics, Gen Atomic Div, Gen Dynamics Corp, Calif, 62-67; dir lab plasma studies, 72-73, PROF PLASMA PHYSICS, CORNELL UNIV, 67- *Concurrent Pos:* Tech advisor, UN Conf on Peaceful Uses of Atomic Energy, Geneva, Switz, 58; sci engr, Max Planck Inst Physics & Astrophys, Ger, 59-60; consult, Aerojet-Gen Nucleonics Div, Gen Tire & Rubber Co, 60-62, US Naval Res Lab, Washington, DC, 70-, Lawrence Livermore Lab, 75-77, Power Conversion Technol, Inc, 79-81, Occidental Res Corp, 79-82, Sandia Nat Lab, Albuquerque, NMex, 82-86 & Los Alamos Nat Lab, NMex, 83-84; controlled fusion res mem eval panel on quantum electronics & plasma physics, Nat Res Coun, 70-73; vis scientist, Max Planck Inst Plasma Physics, Munich, Ger, 73-74; dir courses, Int Sch Plasma Physics, Varenna, Italy, 78, 82 & 86; vis prof, Univ Calif, Irvine, 79-80 & Occidental Res Corp, 80-81; US participating scientist, Joint Prog Plasma Physics, Cornell Univ & Physical Res Lab, Ahmedabad, India, 81-84. *Honors & Awards:* Alexander von Humboldt sr scientist award, 73. *Mem:* Fel Am Phys Soc; fel Inst Elec & Electronics Engrs; Nuclear & Plasma Sci Soc (vpres, 75). *Res:* Plasma diagnostics; waves in plasmas; plasma instabilities; microwave technology; electronic circuitry; nonlinear waves; relativistic electron beams; plasma heating; controlled fusion research; intense ion beams. *Mailing Add:* 303 N Sunset Dr Ithaca NY 14850

WHARTON, DAVID CARRIE, b Avoca, Pa, Nov 3, 30; m 61; c 3. BIOCHEMISTRY. *Educ:* Pa State Univ, BS, 52, MS, 54, PhD(plant biochem), 56. *Prof Exp:* Asst plant biochem, Pa State Univ, 52-56; fel enzyme chem, Enzyme Inst, Univ Wis, 59-61; res scientist, E I du Pont de Nemours & Co, 61-62; asst prof biochem, Univ Wis, 62-64 & Sch Med, Univ Va, 64-66; from asst prof to assoc prof, Cornell Univ, 66-73; prof biochem, Univ Tex Health Sci Ctr, San Antonio, 73-81; PROF & CHMN BIOL, NORTHEASTERN UNIV, BOSTON, 81- *Mem:* Am Soc Biol Chemists; Am Chem Soc; Brit Soc Gen Microbiol; Am Soc Microbiol. *Res:* Electron transport; metalloenzymes. *Mailing Add:* Dept Biol Northeastern Univ Boston MA 02115

WHARTON, H(ARRY) WHITNEY, b Watertown, NY, May 4, 31; m 55; c 2. ANALYTICAL CHEMISTRY. *Educ:* Iowa State Univ, BS, 53, MS, 58, PhD(anal chem), 60. *Prof Exp:* Res chemist, Rath Packing Co, 56; asst anal chem, Iowa State Univ, 56-60; res chemist, 60-61, group leader anal chem, 61-64, group leader, Soap Prod Div, 65, sect head anal chem, Food Prod Div, 65-70, sect head new prod res, 70-74, SECT HEAD FOODS ANALYSIS FOOD PROD DIV, PROCTER & GAMBLE CO, 74- *Mem:* Asn Off Anal Chemists; Am Chem Soc; Am Oil Chem Soc. *Res:* Micro methods of analysis involving spectrophotometry, microdiffusion, spectrophotometric and nonaqueous titrations, polarography and inorganic oxidation-reduction reactions; managing the development of analytical methods for food products. *Mailing Add:* Procter & Gamble Co Food Prod Div 6071 Center Hill Rd Cincinnati OH 45224

WHARTON, JAMES HENRY, b Mangum, Okla, July 23, 37; m 56; c 2. PHYSICAL CHEMISTRY. *Educ:* Northeast La Univ, BS, 59; La State Univ, PhD(phys chem), 62. *Prof Exp:* Asst prof, 62-63 & 65-69, assoc dean, Col Chem & Physics, 69-71, assoc prof, 69-80, PROF CHEM, LA STATE UNIV, BATON ROUGE, 80-, DEAN, GEN COL, 71-, CHANCELLOR. *Concurrent Pos:* Consult, Univ Tex, San Antonio, 70-71. *Mem:* Am Chem Soc. *Res:* Molecular spectroscopy; electron spin resonance. *Mailing Add:* 934 Kenilworth Pkwy Baton Rouge LA 70808

WHARTON, LENNARD, b Boston, Mass, Dec 10, 33; m 57; c 3. PHYSICAL CHEMISTRY. *Educ:* Mass Inst Technol, BS, 55; Univ Cambridge, MA, 57; Harvard Univ, PhD(chem), 63. *Prof Exp:* From asst prof to assoc prof, Univ Chicago, 63-78, prof chem, 78; vpres eng & technol, Worthington Group, McGraw Edison Co, 78-; AT DEPT CHEM, UNIV CHICAGO, ILL. *Concurrent Pos:* Alfred P Sloan res fel, 64-66; res assoc prof, Univ Chicago, 78-83; consult, Northrop Electronics Div, 77-78; vpres technol, Studebaker Worthington Corp, 78- *Mem:* Am Phys Soc; sr mem Inst Elec & Electronics Engrs; Am Inst Chem Engrs; AAAS; Am Soc Mech Engrs. *Res:* Molecular beams and structure; spectroscopy; chemical kinetics; experimental physical chemistry; scattering phenomena; surface sciences; solar energy conversion; electrical power transmission and distribution. *Mailing Add:* Packer Eng PO Box 353 Naperville IL 60566

WHARTON, MARION A, b Cayuga, Ont, Nov 17, 10; nat US. NUTRITION. *Educ:* Univ Toronto, BA, 33; Univ Western Ont, MS, 34; Mich State Univ, PhD(nutrit), 47; Am Bd Nutrit, dipl. *Prof Exp:* Instr, Univ Toronto, 36-37; dietician, Prov Dept Health, Ont, 37-39; asst, Mich State Univ, 40-46; res nutritionist, Univ WVa, 46-48; asst prof foods & nutrit, Ohio State Univ, 48-52; from asst prof to assoc prof, NDak Agr Col, 52-55; prof, Univ Southern Ill, 55-61 & Univ RI, 61-63; assoc prof, Iowa State Univ, 63-65; from assoc prof to prof, 65-78, EMER PROF FOODS & NUTRIT, CALIF STATE UNIV, LONG BEACH, 78- *Mem:* Am Dietetic Asn; Am Home Econ Asn; Sigma Xi. *Mailing Add:* 13650 Del Monte Dr Seal Beach CA 90740

WHARTON, PETER STANLEY, b Oxford, Eng, May 9, 31; m 55; c 4. ORGANIC CHEMISTRY. *Educ:* Cambridge Univ, BA, 52, MA, 57; Yale Univ, MS, 57, PhD, 59. *Prof Exp:* Fel, Columbia Univ, 58-60; from instr to prof org chem, Univ Wis-Madison, 60-68; PROF CHEM, WESLEYAN UNIV, 68- *Honors & Awards:* Frederick Gardner Cottrell Award, 61. *Mem:* Am Chem Soc; The Chem Soc. *Res:* Synthetic and mechanistic alicyclic chemistry. *Mailing Add:* Dept Chem Wesleyan Univ Middletown CT 06457-3262

WHARTON, WALTER WASHINGTON, b Boone, Ky, Mar 25, 26; m 47; c 5. PHYSICAL CHEMISTRY. *Educ:* Georgetown Col, AB, 50; Univ Ky, MS, 52, PhD, 55. *Prof Exp:* Res chemist, Redstone Arsenal, 54-59, supvr res chem & chief adv technol br, 59-79, sr exec serv, Propulsion Dir, 79-90, DIR, WEAPONS SCI DIRECTORATE, US ARMY MISSILE COMMAND, REDSTONE ARSENAL, 90- *Concurrent Pos:* Instr, Exten Ctr, Univ Ala, 56- *Res:* Propulsion technology; propellant chemistry; combustion kinetics; laser technology; optic and photonics research. *Mailing Add:* 2811 Barcody Rd SE Huntsville AL 35801

WHARTON, WILLIAM RAYMOND, b Knoxville, Tenn, Mar 30, 43; m 67; c 3. EXPERIMENTAL NUCLEAR PHYSICS. *Educ:* Stanford Univ, BS, 65; Univ Wash, PhD(nuclear physics), 72. *Prof Exp:* Res assoc exp nuclear physics, Argonne Nat Lab, 72-74 & Rutgers Univ, 74-75; asst prof, 75-80, ASSOC PROF PHYSICS, CARNEGIE-MELLON UNIV, 80- *Mem:* Am Phys Soc. *Res:* Experimental medium energy nuclear physics involving the study of nuclei or nucleons interacting with pions, kaons and antiprotons. *Mailing Add:* Dept Physics Wheaton Col Wheaton IL 60187

WHATLEY, ALFRED T, b Denver, Colo, Apr 20, 22; div; c 4. PHYSICAL CHEMISTRY. *Educ:* Princeton Univ, AB, 48, AM, 50, PhD(chem), 52. *Prof Exp:* Chemist, Hanford Works, Gen Elec Co, 52-55, eng consult, Aircraft Nuclear Propulsion, 55-57, physicist, Vallecitos Atomic Lab, 57-61; staff engr, Martin Co, 61-62; sr sci specialist, EG&G, 62-70; exec dir, Western Interstate Nuclear Bd, Lakewood, Colo, 70-76; RETIRED. *Concurrent Pos:* Mem, Colo Air Pollution Control Comn, chmn, 78- *Mem:* Am Nuclear Soc; Sigma Xi. *Res:* Nuclear science. *Mailing Add:* PO Box 540 Breckenridge CO 80424-0540

WHATLEY, BOOKER TILLMAN, b Alexandria, Ala, Nov 5, 15; m 43. HORTICULTURE, PLANT PHYSIOLOGY. *Educ:* Ala Agr & Mech Col, BS, 41; Rutgers Univ, PhD, 57. *Prof Exp:* Agr exten agent, Butler County, Ala, 46-47; prin high sch, Ala, 47-50; tech oper officer, Chofu Hydroponic Farm, Japan, 50-54; assoc prof & head dept hort, Southern Univ, 57-60; adv hort, US Opers Mission, Ministry Agr, Ghana, 60-62; prof hort, Southern Univ, 62-68; prof plant & soil sci, Tuskegee Inst, 68-81; PRES, WHATLEY FARMS, INC, 81- *Mem:* AAAS; Am Soc Hort Sci; Am Soc Plant Physiol; Sigma Xi. *Res:* The effect of budding methods, wrapping materials and hormones on Myristica fragrans and its vegetative propagation. *Mailing Add:* Seven Square Haardt Dr Montgomery AL 36105

WHATLEY, JAMES ARNOLD, b Calvert, Tex, Feb 26, 16; m 39; c 2. ANIMAL BREEDING. *Educ:* Agr & Mech Col, Tex, BS, 36; Iowa State Col, MS, 37, PhD(animal breeding), 39. *Prof Exp:* From asst prof to prof animal husb, Okla State Univ, 39-64, from assoc dir to dir, Agr Exp Sta, 64-74, dir agr res, 66-68, dean agr, 68-74, assoc dir, Agr Exp Sta, 74-81; RETIRED. *Concurrent Pos:* With bur animal indust, USDA, 44. *Mem:* Sigma Xi; Am Soc Animal Sci. *Res:* Swine breeding. *Mailing Add:* Agr Exp Sta Okla State Univ Stillwater OK 74075

WHATLEY, MARVIN E, chemical engineering, for more information see previous edition

WHATLEY, THOMAS ALVAH, b Midland, Ark, Aug 23, 32; m 54; c 4. PHYSICAL CHEMISTRY, INSTRUMENTATION. *Educ:* Fresno State Col, BS, 53; Univ Ore, PhD(phys chem), 61. *Prof Exp:* Sr scientist, Lockheed Aircraft Corp, 58-61; inorg res group head, United Tech Div, United Aircraft Corp, 61-65; sr res chemist, F&M Div, Hewlett-Packard Co, Pa, 65-68, eng specialist, Appl Res Labs, 68-74, mgr applns, 74-78; CONSULT MEM ENG STAFF, XEROX CORP, 78- *Mem:* Am Chem Soc; Am Soc Mass Spectrometry; Microbeam Anal Soc; Am Vacuum Soc. *Res:* Ion probe mass spectrometry; analytical instrumentation development; material science; solid state device structure analysis; anti-counterfeiting technology. *Mailing Add:* Xerox Corp 250 N Halstead St Pasadena CA 91107

WHATMORE, GEORGE BERNARD, b Seattle, Wash, Aug 31, 17; m 42; c 3. NEUROPHYSIOLOGY OF CONDITIONING & LEARNING, PHYSIOPATHOLOGY & TREATMENT OF FUNCTIONAL DISORDERS. *Educ:* Univ Wash, BS, 40, MS, 41; Univ Chicago, PhD(neurophysiol), 46, MD, 48. *Prof Exp:* Intern & resident med, King County Hosp, Seattle, 48-50; resident physician clin neurophysiol, Lab Clin Physiol, Chicago, 50-51; PVT PRACT PHYSICIAN INTERNAL MED & FUNCTIONAL DIS, 51-; PRIN INVESTR NEUROPHYSIOL FUNCTIONAL DIS, PAC NORTHWEST RES FOUND, SEATTLE, 66- *Concurrent Pos:* Univ fel physiol, Univ Chicago, 42-43, Rawson fel, 43-44, Sheldon fel, 45-46, teaching asst, 42-45. *Mem:* AAAS; Behav Ther & Res Soc; Acad Psychosomatic Med; Asn Appl Psychophysiol & Biofeedback; Sigma Xi; Int Stress & Tension Control Asn. *Res:* Investigation of the neurophysiologic mechanisms and pathways involved in conditioning and learning investigation of the role of action-potential output from the premotor and motor cortex in the etiology, physiopathology and treatment of functional disorders; investigation of the role of action- potential output from the premotor and motor cortex in the etiology, physiopathology and treatment of functional disorders. *Mailing Add:* 10524 SE 27th St Bellevue WA 98004-7231

WHAYNE, TOM FRENCH, b Columbus, Ky, Dec 26, 05; m 34; c 2. MEDICINE. *Educ:* Univ Ky, AB, 27; Wash Univ, MD, 31; Harvard Univ, MPH, 49, DrPH, 50; Am Bd Prev Med, dipl, 49. *Prof Exp:* Intern, Mo Baptist Hosp, St Louis, 31-32; house physician, Mo Pac Hosp, 32-33; surgeon, Civilian Conserv Corps, 33-34; physician, Fitzsimons Gen Hosp, Denver, 34; physician, CZ, Med Corps, 38-41, chief, Med Int Div, Off Surgeon Gen, 41-43, asst mil attach med, US Embassy, London, 43-44, chief prev med sect, 12th Army Group, US Army, 44-45; chief prev med sect, Off Chief Surgeon, Europe, 45-46, dep chief prev med div, Off Surgeon Gen, Washington, DC, 46-47, chief, 47-48, chief dept training doctrines, Army Med Serv Grad Sch, Walter Reed Army Med Ctr, 50-51, chief prev med div, Off Surgeon Gen, DC, 51-55; prof prev med & pub health, Sch Med, Univ Pa, 55-63, assoc dean, 58-63; asst vpres med ctr, 63-67, actg dean col med, 66-67, prof community med & assoc dean col med, 63-72, EMER PROF COMMUNITY MED, UNIV KY, 74- *Concurrent Pos:* Consult, Surgeon Gen, US Army, 55-75; adv, US Deleg, World Health Assembly, Geneva, 48 & 53; mem, Bd Dirs, Gorgas Mem Inst Trop & Prev Med, 50-85. *Mem:* Fel Am Soc Trop Med & Hyg; Am Epidemiol Soc; fel Am Pub Health Asn; fel NY Acad Med; fel Am Col Prev Med; Sigma Xi. *Res:* Epidemiology. *Mailing Add:* 623 Tateswood Dr Lexington KY 40502

WHEALTON, JOHN HOBSON, b Brooklyn, NY, Apr 27, 43; m 72. PLASMA PHYSICS. *Educ:* Univ Lowell, BS, 66; Univ Del, MS, 68, PhD(physics), 71. *Prof Exp:* Res assoc Div Eng, Brown Univ, 71-72, res assoc, Dept Chem, 72-73; res assoc, Joint Inst Lab Astrophys, Univ Colo-Nat Bur Stand, 73-75, MEM STAFF, THERMONUCLEAR DIV, OAK RIDGE NAT LABS, 75- *Mem:* Fel Am Phys Soc; Inst Elec & Electronics Engrs; Am Nuclear Soc. *Res:* Analysis of drift tube swarm experiments; kinetic theory of diffusion and mobility in collision-dominated weakly ionized gases in presence of strong fields; space charge ion extraction optics; R F accelerator modeling; 3-dimensional analysis of Maxwell equations. *Mailing Add:* 185 Outer Dr Oak Ridge TN 37830

WHEASLER, ROBERT, b Indianapolis, Ind, Dec 26, 24; m 46; c 2. AERONAUTICAL ENGINEERING, AERODYNAMICS. *Educ:* Purdue Univ, BS, 53, MS, 54, Univ Okla, PhD(eng sci), 64. *Prof Exp:* Res engr, Boeing Airplane Co, 54-55; res engr, Aircraft Gas Turbine Div, Gen Elec Co, 55; instr aeronaut eng, Purdue Univ, 55-58; asst prof, 58-65, PROF AERONAUT & MECH ENG, UNIV WYO, 65- *Concurrent Pos:* Instr, Univ Okla, 60; vis prof, US Naval Acad, 82-83; brig gen, USAF (retired). *Mem:* Sigma Xi; Air Force Hist Found; Air Force Asn. *Res:* Thermodynamics; heat transfer; aircraft and missile propulsion; gas dynamics. *Mailing Add:* Dept Mech Eng Univ Wyo Laramie WY 82070

WHEAT, JOHN DAVID, b Ranger, Tex, July 12, 21; m 50; c 3. ANIMAL GENETICS. *Educ:* Agr & Mech Col, Tex, BS, 42, MS, 51; Iowa State Col, PhD(animal breeding, genetics), 54. *Prof Exp:* From asst prof to prof animal sci, Kans State Univ, 54-88; RETIRED. *Concurrent Pos:* Beef cattle breeding adv, Ministry of Animal & Forest Resources, US Agency Int Develop-Kans State Univ Contract Team, Northern Nigeria, 66-68; mem fac, Ahmadu Bello Univ, Nigeria, 66-68, livestock breeding consult, Taiwan, 72 & 75, Costa Rica, 80 & 81 & Nigeria, 84. *Mem:* Am Soc Animal Sci; Am Genetic Asn. *Res:* Population genetics; muscling selection research in swine. *Mailing Add:* 3501 Dickens Ave Manhattan KS 66502

WHEAT, JOSEPH ALLEN, b Charlottesville, Va, Mar 31, 13; m 42; c 1. CHEMISTRY. *Educ:* Univ Va, BS, 34; Cornell Univ, MS, 36, PhD(inorg chem), 39. *Prof Exp:* Instr chem, Trinity Col, Conn, 39-40; microchemist, Biochem Res Found, Del, 40-41; chemist, Celanese Corp Am, NJ, 41-49; spectroscopist, Air Reduction Co, 49-53; chemist, Savannah River Lab, Atomic Energy Div, E I du Pont de Nemours & Co, Inc, 53-78; PROF COMPUTER SCI, VOORHEES COL, 67-, TRUSTEE, 74- *Mem:* Am Chem Soc; Soc Appl Spectros; Am Inst Chemists. *Res:* Instrumental analysis; infrared, emission and atomic absorption spectroscopy; application of computers and programmable calculators to reduction of spectroscopic data; programming of computer to interact simultaneously with from one to four data gathering instruments, such as 4K channel pulse height analyzer, infrared spectrophotometer, atomic absorption spectrometer. *Mailing Add:* 1478 Canterbury Ct SE Aiken SC 29801-5118

WHEAT, ROBERT WAYNE, b Springfield, Mo, Nov 10, 26; m 48; c 3. MICROBIOLOGY, BIOCHEMISTRY. *Educ:* Wash Univ, PhD(microbiol), 55. *Prof Exp:* USPHS fel biochem, NIH, Md, 55-56; instr biochem, 56-58, assoc, 58-60, assoc prof microbiol, 66-74, ASST PROF BIOCHEM, SCH MED, DUKE UNIV, 60-, PROF MICROBIOL, 74- *Concurrent Pos:* Sabbatical vis prof Freilurg, Breisgan, West Ger, Max Planck Inst Immunobiology, 65; NIH consult, 69-72, 85-89; vis scientist, Rocky Mountain Lab, Nat Inst Allergy & Infectious Dis, NIH, 78-79. *Mem:* Am Chem Soc; Am Soc Biol Chem; Am Soc Microbiol. *Res:* Biochemistry of microorganisms, amino sugars and polysaccharides; cell surface antigens. *Mailing Add:* Dept Microbiol & Biochem 18 Rp4 Duke Univ Med Ctr Durham NC 27710

WHEATLAND, DAVID ALAN, b Boston, Mass, Aug 27, 40; m 65; c 1. INORGANIC CHEMISTRY. *Educ:* Brown Univ, ScB, 63; Univ Md, PhD(inorg chem), 67. *Prof Exp:* Asst prof chem, Bowdoin Col, 67-73; RES CHEMIST, S D WARREN RES LAB, 73- *Concurrent Pos:* Petrol Res Fund grant, 68-70. *Res:* Reprographic research and development. *Mailing Add:* 20 Storn Cmb Frsde Portland ME 04110

WHEATLEY, VICTOR RICHARD, b London, Eng, Nov 4, 18; m 43; c 2. BIOCHEMISTRY. *Educ:* Univ London, BSc, 47, PhD(org chem), 50, DSc(chem, biochem), 68. *Prof Exp:* Biochemist, St Bartholomew's Hosp Med Col, London, 48-57; res assoc dermat, Univ Chicago, 57-59; sr res assoc, Stanford Univ, 59-62; ASSOC PROF DERMAT, MED CTR, NY UNIV, 62- *Concurrent Pos:* NIH fels, Med Ctr, NY Univ, 62-68. *Honors & Awards:* Bronze Medal, Am Acad Dermat, 58; Spec Award, Soc Cosmetic Chem, 62. *Mem:* Am Soc Biol Chemists; NY Zool Soc; NY Acad Sci; Am Inst Chemists. *Res:* Biochemistry of skin, especially the lipid metabolism of skin. *Mailing Add:* Dept Dermatol NY Univ Sch Med New York NY 10016

WHEATLEY, W(ILLIAM) A(RTHUR), b Deming, NMex, Nov 9, 23; m 45; c 4. ELECTRONICS. *Educ:* Univ Mich, BS, 47, MS, 48. *Prof Exp:* Res assoc electronics, Willow Run Res Ctr, Univ Mich, 48-50, res engr res admin, 50-55; vpres, Strand Eng Co, 55-60; vpres, Electronic Assistance Corp, 60-64, mgt consult, 64-68; mkt mgr, United Telecontrol Electronics, 68-70; gen mgr, Wave Energy Systs, Inc, Newtown, Pa, 70-72; COMMUN CONSULT, HARVEY J KRASNER ASSOCS INC, 72- *Mem:* Inst Elec & Electronics Engrs. *Res:* Analog computers; complex military system design; electronic equipment prototype design; field test; research administration; engineering management; management consulting. *Mailing Add:* Harvey J Krasner Assoc Inc 87 Water Mill Lane Great Neck NY 11021

WHEATON, BURDETTE CARL, b Mankato, Minn, July 3, 38; m 68; c 3. ALGEBRA. *Educ:* Mankato State Col, BS, 59; Univ Iowa, MS, 61, PhD(math), 65. *Prof Exp:* Instr math, Univ Iowa, 59-63; asst prof, Western Ill Univ, 63-65; asst prof, 65-72, PROF MATH, MANKATO STATE UNIV, 72- *Mem:* Am Math Soc; Math Asn Am; Sigma Xi; Nat Coun Teachers Math. *Res:* Abstract algebra, particularly group theory and group representations. *Mailing Add:* Dept Math Mankato State Univ PO Box 8400 Mankato MN 56002-8400

WHEATON, ELMER PAUL, b Elyria, Ohio, Aug 15, 09; m 33; c 2. ENGINEERING, OCEANOGRAPHY. *Educ:* Pomona Col, BA, 33. *Prof Exp:* Sound technician, Columbia Motion Picture Studios, 33; riveter & assembler, Douglas Aircraft Co, 34-36, res engr, 36-40, asst chief res sect, 40-43 & eng labs, 43, spec asst to vpres eng, 43-44, on loan to radiation lab, Mass Inst Technol, 44, on loan to Rand Corp, 45, mgr appl physics lab, 45, chief dynamics & sound control, Missile Projs, 45-55, chief missiles engr, 55-58, dir missiles & space systs, 58, vpres eng, 58-60, vpres eng technol, 60-61, vpres eng & corp vpres, 61, dir, Astropower, Inc, 61; asst to pres, Lockheed Missiles & Space Co, Lockheed Aircraft Corp, 62, vpres & gen mgr, Space Prog Div, 62-63, vpres & gen mgr, Res & Develop Div, 63-74, vpres, Corp, 62-74, pres, Lockheed Petrol Serv Ltd, 72-74; DIR, MARINE DEVELOP ASSOCS, INC, 85- *Concurrent Pos:* Lectr, Guggenheim Aeronaut Lab, Calif Inst Technol, 41-54; mem spec indust comt missiles for res & develop, Off Secy USAF, 54; consult, Adv Panel Aeronaut Res & Eng, Off Dir Defense, 57-59; mem comt ocean eng, Nat Acad Eng, 67-; consult & mem panel ocean eng, Nat Coun Marine Resources & Eng Develop, 68-69; mem, Calif State Marine Res Comt, 72-78, marine bd, Mat Acad Sci-Nat Acad Eng, adv panel, Int Decade Ocean Explor, NSF, 70-71 & Sea Grant Coord Coun, Univ Calif, 70-; consult, 74- *Honors & Awards:* Robert M Thompson Award for outstanding civilian leadership, Navy League US, 71; Aerospaces Contrib to Soc Award, Am Inst Aeronaut & Astronaut, 78. *Mem:* Nat Acad Eng; fel Am Inst Aeronaut & Astronaut; fel Am Astronaut Soc; fel Marine Technol Soc; AAAS. *Res:* Acoustics; electronics; aeronautical and systems engineering; flutter and vibration; missile and space systems engineering; aerospace management; ocean systems engineering and management; research and development management. *Mailing Add:* 501 Portola Rd PO Box 8087 Portola Valley CA 94028-7603

WHEATON, GREGORY ALAN, b Muskegon, Mich, July 18, 47; m 73. ORGANIC CHEMISTRY. *Educ:* State Univ Iowa, BS, 69, MS, 73, PhD(org chem), 76. *Prof Exp:* Res chemist org chem, Atlantic Richfield Co, 76-81; RES SCIENTIST, PENNWALT CORP, 81- *Mem:* Am Chem Soc; Sigma Xi. *Res:* Homogeneous and heterogeneous catalysis in the conversion of petrochemical feedstocks to chemical intermediates, particularly partial oxidation of 2, 3 and 4 carbon olefins. *Mailing Add:* 130 Harvest Rd Swedesboro NJ 08085

WHEATON, JONATHAN EDWARD, b Fullerton, Calif, Jan 22, 47; m 67; c 2. NEUROENDOCRINOLOGY. *Educ:* Univ Calif, Davis, BS, 69; Ore State Univ, MS, 70, PhD(animal physiol), 73. *Prof Exp:* Fel neuroendocrinol, Southwestern Med Sch, 73-75; asst prof, 75-80, ASSOC PROF PHYSIOL, UNIV MINN, ST PAUL, 80- *Mem:* Am Soc Animal Sci; Sigma Xi; Soc Study Reprod. *Res:* Control and effects of neurohormones; reproductive endocrinology. *Mailing Add:* Dept Animal Sci 122 Peters Hall Univ Minn 1404 Gortner Ave St Paul MN 55108

WHEATON, ROBERT MILLER, b Danbury, Ohio, Oct 11, 19; m 43, 67; c 5. INDUSTRIAL CHEMISTRY. *Educ:* Oberlin Col, AB, 41. *Prof Exp:* Chemist, Celotex Corp, 41; chemist & process specialist, Trojan Powder Co, 42-43, head process specialists, 44-45; chemist, Dow Chem Co, 46-49, group leader res, 50-55, div leader, 56-71, assoc scientist, Western Div Res, 72-82; RETIRED. *Concurrent Pos:* Chmn, Gordon Res Conf Ion Exchange, 63. *Mem:* Am Chem Soc; Sigma Xi. *Res:* Synthesis, applications and properties of ion exchange resins. *Mailing Add:* 156 Warwick Dr Walnut Creek CA 94598

WHEATON, THOMAS ADAIR, b Orlando, Fla, Apr 5, 36; m 61; c 2. PLANT PHYSIOLOGY, HORTICULTURE. *Educ:* Univ Fla, BS, 58, MS, 60; Univ Calif, Davis, PhD(plant physiol), 63. *Prof Exp:* Asst horticulturist, 63-70, assoc prof hort & assoc horticulturist, 70-79, PROF & HORTICULTURIST, INST FOOD & AGR SCI, AGR RES & EDUC CTR, UNIV FLA, LAKE ALFRED, 79- *Mem:* Am Soc Plant Physiol; Am Soc Hort Sci; Sigma Xi. *Res:* Chilling injury in plants; nitrogen metabolism and growth regulation in citrus. *Mailing Add:* Univ Fla Agr Res & Educ Ctr 700 Experimental Station Rd Lake Alfred FL 33850

WHEBY, MUNSEY S, b Roanoke, Va, Nov 19, 30; m 55; c 3. INTERNAL MEDICINE. *Educ:* Roanoke Col, BS, 51; Univ Va, MD, 55. *Prof Exp:* Asst chief hemat, Walter Reed Army Inst Res & Walter Reed Gen Hosp, 59-61, chief gastroenterol, Walter Reed Army Inst Res, 61-62, chief med div, US Army Trop Res Med Lab, 62-65; assoc prof med, Rutgers Med Sch, 65-66; assoc prof, 66-72, PROF MED, SCH MED, UNIV VA, 72- *Mem:* Am Fedn Clin Res; AMA; Am Soc Hemat. *Res:* Gastrointestinal absorption of iron; folic acid and B-12 metabolism. *Mailing Add:* Dept Internal Med Univ Va Sch Med Box 502 Charlottesville VA 22908

WHEEDEN, RICHARD LEE, b Baltimore, Md, Nov 29, 40; m 62; c 2. MATHEMATICS. *Educ:* Johns Hopkins Univ, AB, 61; Univ Chicago, MS, 62, PhD(math), 65. *Prof Exp:* Instr math, Univ Chicago, 65-66; mem, Inst Adv Study, 66-67; from asst prof to assoc prof, 67-74, PROF MATH, RUTGERS UNIV, NEW BRUNSWICK, 74- *Concurrent Pos:* NSF fel, 66-67; NSF grants, 67- *Mem:* Am Math Soc. *Res:* Harmonic analysis. *Mailing Add:* Dept Math Rutgers Univ New Brunswick NJ 08903

WHEELER, ALBERT HAROLD, b St Louis, Mo, Dec 11, 15; m 38; c 3. BACTERIOLOGY. *Educ:* Lincoln Univ, AB, 36; Iowa State Col, MS, 37; Univ Mich, MSPH, 38, DrPH, 44. *Prof Exp:* Clin technician, Col Med, Howard Univ, 38-40; asst, 41-44, res assoc, Univ Hosp, 44-52, asst prof bact, 52-58, ASSOC PROF BACT, MED SCH, UNIV MICH, ANN ARBOR, 59-, ASSOC PROF MICROBIOL, 74- *Concurrent Pos:* Consult, Serol Lab, Univ Hosp, Univ Mich. *Mem:* Am Asn Immunol. *Res:* Active and passive immunity in experimental syphilis; serodiagnosis of syphilis; serology of biologic false positive reactions in syphilis; treponemicidal activity of various animal sera and complements. *Mailing Add:* 234 Eighth St Ann Arbor MI 48103

WHEELER, ALFRED GEORGE, JR, b Nebraska City, Nebr, Apr 11, 44. ENTOMOLOGY. *Educ:* Grinnell Col, BA, 66; Cornell Univ, PhD(insect ecol), 71. *Prof Exp:* ENTOMOLOGIST, BUR PLANT INDUST, PA DEPT AGR, 71- *Concurrent Pos:* Consult, Dames & Moore, 73-74; adj asst prof, Pa State Univ, 73-81, adj assoc prof, 81-88, prof, 88-; ed, Regulatory Hort, 75-, Melsheimer Entomol Series, 78-83; vis fel, Cornell Univ, 86, 87 & 88. *Honors & Awards:* Distinguished Achievement Award Regulatory Entomol, Entomol Soc Am, 86. *Mem:* Entom Soc Am. *Res:* Life history studies of Hemiptera-Heteroptera, especially Miridae; biology of insects affecting ornamental plants; study of insect-plant associations. *Mailing Add:* Dept Agr Bur Plant Indust 2301 N Cameron St Harrisburg PA 17110-9408

WHEELER, ALFRED PORTIUS, b Brooklyn, NY, Sept 16, 47; m 69; c 4. PHYSIOLOGY, CELL BIOLOGY. *Educ:* Butler Univ, BS, 69; Duke Univ, PhD(zool), 75. *Prof Exp:* Instr, Duke Univ, 74-76; asst prof, 76-81, ASSOC PROF ZOOL, CLEMSON UNIV, 82- *Concurrent Pos:* Vis prof, Duke Univ, 77- *Mem:* Sigma Xi; Am Soc Zoologists. *Res:* Physiology of biomineralization, especially in molluscs; physiology and subcellular localization of carbonic anhydrase; matrix function in biomineralization; the biology and industrial applications of mineralization inhibitors. *Mailing Add:* Dept Biol Sci Clemson Univ 201 Sikes Hall Clemson SC 29634

WHEELER, ALLAN GORDON, b Gary, Ind, July 12, 23; m 49; c 4. PHARMACOLOGY, PHYSIOLOGY. *Educ:* Valparaiso Univ, BA, 48; Univ Wis, MA, 50. *Prof Exp:* Asst pharmacol & anesthesiol, Univ Wis, 50-54; assoc pharmacologist, Res Ctr, Mead Johnson & Co, Ind, 54-58, sr pharmacologist, 54-59, group leader toxicol, 59-68; supvr indust toxicol, ICI US, Inc, 68-89, toxicol coordr, Bio-Med Res Lab, 88-89; RETIRED. *Mem:* Am Indust Hyg Asn; Drug Info Asn; Environ Mutagen Soc; Sigma Xi; Am Asn Lab Animal Sci. *Res:* Anesthesiology; toxicology. *Mailing Add:* Seven Brandywine Blvd Edgewood Hills Wilmington DE 19809

WHEELER, BERNICE MARION, b Winsted, Conn, June 30, 15. ZOOLOGY. *Educ:* Conn Col, AB, 37; Smith Col, MA, 39; Yale Univ, PhD(zool), 48. *Prof Exp:* Asst zool, Smith Col, 37-39; instr, Westbrook Jr Col, 39-42; asst, Yale Univ, 42-47; from instr to emer prof zool, Conn Col, 47-80; RETIRED. *Concurrent Pos:* Ford Found fel, 54-55. *Mem:* Sigma Xi. *Res:* Genetics; ecology; evolution. *Mailing Add:* 35 W Main St Niantic CT 06357

WHEELER, CLAYTON EUGENE, JR, b Viroqua, Wis, June 30, 17; m 52; c 3. DERMATOLOGY. *Educ:* Univ Wis, BA, 38, MD, 41; Am Bd Dermat, dipl, 51. *Prof Exp:* Resident & instr internal med, Med Sch, Univ Mich, 42-44, resident & instr dermat, 49-51; from asst prof to prof, Sch Med, Univ Va, 51-62; chmn dept, 72-87, chief div, 62-72, PROF DERMAT, SCH MED, UNIV NC, CHAPEL HILL, 62- *Concurrent Pos:* Res fel endocrinol & metab, Univ Mich, 47-48; chmn, Residency Rev Comt Dermat; rep, Am Bd Med Specialties; pres, Am Bd Dermat, 77-78. *Honors & Awards:* Rothman Award, 79. *Mem:* Soc Invest Dermat (pres, 73-74); AMA; Am Acad Dermat (pres, 84-85); Asn Prof Dermat (pres, 75-76); Am Dermat Asn (pres, 82-83). *Res:* Viral diseases of skin, especially Herpes simplex. *Mailing Add:* Dept Dermat NC Mem Hosp Chapel Hill NC 27514

WHEELER, DARRELL DEANE, b West Liberty, Ky, Feb 24, 39; m 63; c 2. MEMBRANE PHYSIOLOGY, TRANSPORT. *Educ:* Transylvania Col, AB, 62; Univ Ky, PhD(physiol), 67. *Prof Exp:* Asst prof, 68-75, assoc prof, 75-81, PROF PHYSIOL, MED UNIV SC, 81- *Concurrent Pos:* NIH fel physiol & biophys, Univ Ky, 67-68; NIH res grants (PI), 69-72 & 75-86 & 90-; Instr, Transylvania Univ, 68. *Mem:* Am Physiol Soc; AAAS. *Res:* Cell physiology; membrane transport in the nervous system. *Mailing Add:* Dept Physiol Med Univ SC Charleston SC 29425

WHEELER, DESMOND MICHAEL SHERLOCK, b Northwich, Cheshire, Eng, Apr 18, 29; m 53. ORGANIC CHEMISTRY. *Educ:* Nat Univ Ireland, BSc, 50, PhD(chem), 55, DSc, 77; Univ Dublin, MA, 54. *Prof Exp:* Dep lectr phys chem, Trinity Col, Dublin, 52-53, asst lectr org chem, 53-55; res fel chem, Harvard Univ, 55-58; asst prof chem, Univ Nebr, 58-59 & Univ SC, 59-61; from asst prof to assoc prof, 61-66, PROF CHEM, UNIV NEBR, LINCOLN, 66- *Concurrent Pos:* Vis lectr, Univ Ill, 64; res fel, Univ Sussex, 67-68; NATO sr fel, Sch Pharm, Univ London, 70; vis scholar, Columbia Univ, 78. *Mem:* Am Chem Soc; Royal Soc Chem; Inst Chem Ireland. *Res:* Synthesis of naturally occurring guinones; synthesis of diterpenoid acids; structures of plant extractives. *Mailing Add:* Dept Chem Univ Nebr Lincoln NE 68588-0304

WHEELER, DONALD ALSOP, b Philadelphia, Pa, Aug 16, 31; m 53; c 4. HUMAN GENETICS, EVOLUTION. *Educ:* Mich State Univ, BS, 53, MS, 56; Cornell Univ, PhD(plant breeding), 61. *Prof Exp:* Instr biol, Delta Col, 61-65, head dept, 63-65; assoc prof, 65-73, asst head dept, 71-73, PROF BIOL, EDINBORO UNIV PA, 73- *Concurrent Pos:* Post-doctoral study, Johns Hopkins, 76. *Mem:* Sigma Xi; AAAS. *Mailing Add:* 5471 Sherrod Hill Rd Edinboro PA 16412-1864

WHEELER, DONALD BINGHAM, JR, b Cleveland, Ohio, May 24, 17. PHYSICS. *Educ:* Lehigh Univ, BS, 38; Calif Inst Technol, PhD(physics), 47. *Prof Exp:* Instr physics, Occidental Col, 41-42; asst prof, 47-57, ASSOC PROF PHYSICS, LEHIGH UNIV, 57- *Mem:* Am Phys Soc. *Res:* Electric dipole moment determinations; microwave propagation; dispersion and absorption of electromagnetic waves in fatty acids. *Mailing Add:* 1806 Main St Bethlehem PA 18018

WHEELER, EDWARD NORWOOD, b Yancey, Tex, Oct 11, 27; m 50; c 5. ORGANIC CHEMISTRY. *Educ:* Tex Col Arts & Indust, BS, 47, BSCE, 49; Univ Tex, MA, 51, PhD(org chem), 53. *Prof Exp:* Res chemist, Celanese Chem Co, 53-55, group leader, 55-62, sect head, 62-67, dir chem res, 67-69, dir res, Tech Ctr, 69-72, dir develop, 72-74, dir planning, 74-75, dir res, develop & planning, 75-76, vpres res, develop & planning, 76-79, vpres res & develop, 79-83; RETIRED. *Concurrent Pos:* Consult Petrochem Process Litigation. *Mem:* Am Chem Soc. *Res:* Acrylic acid; vinyl monomers; propiolactone reactions; palladium-olefin reactions; liquid phase oxidation of carbonyl compounds and olefins; process development and synthesis of petrochemicals. *Mailing Add:* 9238 Moss Haven Dallas TX 75231-1412

WHEELER, EDWARD STUBBS, b Philadelphia, Pa, June 3, 27; m 52; c 4. RESEARCH ADMINISTRATION. *Educ:* Haverford Col, AB, 48; Cornell Univ, PhD(chem), 52. *Prof Exp:* Asst chem, Cornell Univ, 48-51; assoc chemist org chem res, Atlantic Ref Co, 52-53, supv chemist, 53-59; mgr adhesives div, Amchem Prods, Inc, 59-62; mgr thermosetting polymer develop, Insulating Mat Dept, Gen Elec Co, 63-66, mgr-engr, Insulator Dept, 66-71, consult, Corp Exec Staff, 71-75; vpres technol, 75-85, vpres & gen mgr, Organic Insulators Div, Lapp Insulator Co, 85-86; INDEPENDENT MGT CONSULT, 86- *Concurrent Pos:* Mem bd dirs, Am Nat Metric Coun, 74-75. *Mem:* Fel AAAS; Am Chem Soc; Inst Elec & Electronics Engrs; Sigma Xi. *Res:* Insulators; polymers; electrical insulation; synthetic organic chemistry; engineering standards; metric conversion. *Mailing Add:* 33 Longate Rd Clinton CT 06413-1343

WHEELER, FRANK CARLISLE, b Millinocket, Maine, Jan 26, 17; m 46; c 2. PHARMACEUTICAL CHEMISTRY. *Educ:* Mass Col Pharm, BS, 40, MS, 42; Purdue Univ, PhD(pharmaceut chem), 49. *Prof Exp:* Anal chemist, Burroughs Wellcome & Co, 42-43; pharmaceut chemist, Eli Lilly & Co, 49-58, chief ampoule pilot plant, 58-65, head, 65-66, dir, Parenteral Opers Div, 66-75, dir qual control & tech serv, 75-79; RETIRED. *Mem:* Am Chem Soc; Am Pharmaceut Asn; Pharmaceut Mfrs Asn; Parenteral Drug Asn. *Res:* Pharmaceutical development; manufacture and control of chiefly parenteral products. *Mailing Add:* 23 Southern Pine Trail Ormond Beach FL 32174-5988

WHEELER, GEORGE CARLOS, b Bonham, Tex, Apr 10, 97; m 21, 41; c 2. ENTOMOLOGY. *Educ:* Rice Inst, AB, 18; Harvard Univ, MS, 20, ScD(entom), 21. *Hon Degrees:* LLD, Univ NDak, 70. *Prof Exp:* Instr zool, Syracuse Univ, 21-24, asst prof, 24-26; prof, 26-65, head dept, 26-63, univ prof, 65-67, Emer Univ Prof Biol, Univ NDak, 67-; adj res assoc entom, Desert Res Inst, Univ Nev, 67-85; res assoc, Natural Hist Mus, Los Angeles, 76-82, FLA DEPT AGR, 86- *Mem:* AAAS; Entom Soc Am; Am Entom Soc; Am Asn Univ Prof; Sigma Xi. *Res:* Ant larvae; biogeography of ants of Western United States. *Mailing Add:* 3358 NE 58th Ave Silver Springs FL 32688

WHEELER, GEORGE LAWRENCE, b Rockville Center, NY, June 16, 44; m 68; c 2. PHYSICAL CHEMISTRY, BIOCHEMISTRY. *Educ:* Cath Univ Am, AB, 67; Univ Md, PhD(phys chem), 73. *Prof Exp:* Res assoc phys chem, Yale Univ, 73-75, NIH fel, Med Sch, 75-77; from asst prof to prof, 77-85, J F BUCKMAN PROF CHEM, UNIV NEW HAVEN, 85- *Concurrent Pos:* NSF/Inst Sci Equip Prog grant, Univ New Haven, 77-80; staff mem, Life Sci Div, Los Alamos Nat Lab, 81-83. *Mem:* Am Chem Soc; Am Crystallog Asn; NY Acad Sci. *Res:* Biochemistry of light activated enzymes in the retina; intermolecular interactions in molecular crystals; oscillating chemical reactions. *Mailing Add:* Dept Chem Univ New Haven 300 Orange Ave West Haven CT 06516-1916

WHEELER, GILBERT VERNON, b Sour Lake, Tex, July 5, 22; m 42; c 2. CHEMISTRY, SPECTROSCOPY. *Educ:* Millikin Univ, BS, 44; Univ Ill, MS, 47. *Prof Exp:* Instr physics, Millikin Univ, 43-44; physicist, Res & Develop Dept, Phillips Petrol Co, 48-51, physicist, Atomic Energy Div, 52-53, reactor engr, 51-52, supvr spectrochem lab, 53-63, supvr spectros sect, 63-65; supvr spectros sect, Allied Chem Corp, 65-78; staff scientist, Exxon Nuclear Idaho, 78-81; RETIRED. *Concurrent Pos:* Pres, Soc Appl Spectros, 73; int del, Soc Appl Spectros, 75, 77 & 79. *Mem:* Soc Appl Spectros (pres, 73). *Res:* Graphite furnaces; inductive coupled plasmas and sputter sources for spectroscopy; isotopic and isotope dilution mass spectrometry. *Mailing Add:* Rte 1 Box 235 A Salmon ID 83467

WHEELER, GLYNN PEARCE, b Milan, Tenn, Oct 13, 19; m 43; c 2. BIOCHEMISTRY. *Educ:* Vanderbilt Univ, AB, 41; Univ Akron, MS, 47; Vanderbilt Univ, PhD(org chem), 50. *Prof Exp:* Anal chemist, Tenn Coal, Iron & RR Co, Ala, 41; shift supvr, Ala Ord Works, 42; res chemist, B F Goodrich Co, Ohio, 42-46; chemist, Southern Res Inst, 46-48, biochemist, 50-56, head intermediary metab sect, 56-66, head cancer biocehm div, 66-85; RETIRED. *Mem:* AAAS; Am Chem Soc; Am Asn Cancer Res; Am Soc Biol Chem. *Res:* Cancer biochemistry; chemotherapy of cancer; nucleic acids. *Mailing Add:* PO Box 19986 Birmingham AL 35219-0986

WHEELER, HAROLD A(LDEN), b St Paul, Minn, May 10, 03; m 26; c 3. RADIO ENGINEERING. *Educ:* George Washington Univ, BS, 25. *Hon Degrees:* DSc, George Washington Univ, 72; DEng, Stevens Inst Technol, 78. *Prof Exp:* Engr, Hazeltine Corp, NJ, 24-29, engr, Hazeltine Serv Corp, NY, 29-39, vpres & chief consult engr, Hazeltine Electronics Corp, 39-46; pres, Wheeler Labs, Inc, 47-68; vpres & dir, Hazeltine Corp, 59-65, chmn bd, 65-77, chief scientist, 68-87, emer chmn bd, 77-87; RETIRED. *Concurrent Pos:* Consult, Off Secy Defense, 50-64 & Defense Sci Bd, 63-64; dir, Inst Radio Engrs, 34, 40-45. *Honors & Awards:* Modern Pioneer Award, Nat Asn Mfrs, 40; Liebmann Prize, Inst Radio Engrs, 40,; Medal of Honor, Inst Elec & Electronics Engrs, 64; Armstrong Medal, Radio Club Am, 64. *Mem:* Fel Inst Elec & Electronics Engrs; Brit Inst Elec Engrs; Sigma Xi; Am Inst Eng. *Res:* Radio receivers and transmitters; television; radar; antennas; microwaves; communication network theory; conformal mapping of fields; tracking radar for guided missile systems; submarine and subsurface radio communication; radio guidance of aircraft. *Mailing Add:* 416 Atascadero Dr Santa Barbara CA 93110-2004

WHEELER, HARRY ERNEST, b WCharleston, Vt, Jan 25, 19; m 44. PHYTOPATHOLOGY. *Educ:* Univ Vt, BS, 41; La State Univ, MS, 47, PhD(bot), 49. *Prof Exp:* Lab asst bot & bact, Univ Vt, 39-41; asst bot & plant path, La State Univ, 46-49, from asst prof to prof bot & plant path, 49-67; PROF PLANT PATH, UNIV KY, 67- *Concurrent Pos:* Vis investr & res partic, Biol Div, Oak Ridge Nat Lab, 49-50; Guggenheim fel, Biol Labs, Harvard Univ, 58. *Mem:* AAAS; Bot Soc Am; Am Phytopath Soc; Mycol Soc Am. *Res:* Genetics and cytology of fungi; host relations of plant pathogens; electron microscopy. *Mailing Add:* 3293 Bellefonte Dr Lexington KY 40502

WHEELER, HENRY ORSON, b Los Angeles, Calif, Apr 7, 24; m 47; c 2. MEDICINE. *Educ:* Harvard Med Sch, MD, 51. *Prof Exp:* Assoc prof med, Col Physicians & Surgeons, Columbia Univ, 62-68; PROF MED, SCH MED, UNIV CALIF, SAN DIEGO, 68- *Mem:* Fedn Am Socs Exp Biol; AAAS; Am Soc Clin Invest; Am Physiol Soc. *Res:* Hepatic physiology; bile formation; gallbladder; ion transport. *Mailing Add:* Dept Med Univ Hosp Univ Calif San Diego 225 W Dickinson St San Diego CA 92103

WHEELER, JAMES DONLAN, b St Louis, Mo, July 19, 23. BIOCHEMISTRY, EXERCISE PHYSIOLOGY. *Educ:* St Louis Univ, AB, 47, PhL, 48, MS, 52, STL, 56; Univ Mo, Kansas City, PhD(pharmaceut chem), 65. *Prof Exp:* Instr chem, St Louis Univ High Sch, 50-51; from instr to assoc prof, 56-74, head dept, 67-74, PROF CHEM, ROCKHURST COL, 74- *Mem:* AAAS; Am Chem Soc; Nat Sci Teachers Asn. *Res:* Biochemistry and physiology of the effects of training and exercise; learning theory as applied to freshman chemistry students. *Mailing Add:* Rockhurst Col 1100 Rockhurst Rd Kansas City MO 64110-2508

WHEELER, JAMES ENGLISH, b Durham, NC, May 5, 38; m 66; c 3. PATHOLOGY. *Educ:* Harvard Univ, AB, 58; Johns Hopkins Univ, MD, 62. *Prof Exp:* Intern med, Johns Hopkins Univ, 62-63, resident path, 63-66; resident path, State Univ NY Upstate Med Ctr, 69-70; assoc, 70-72, from asst prof to assoc prof, 72-86, PROF PATH & OBSTET & GYNEC, SCH MED, UNIV PA, 86- *Concurrent Pos:* USPHS cancer control sr clin trainee, Mem Hosp Cancer & Allied Dis, New York, 66-67. *Mem:* Fel US-Can Acad Path; AAAS; Int Soc Gynec Pathologists. *Res:* Surgical and gynecological pathology. *Mailing Add:* Dept Path Hosp Univ Pa Philadelphia PA 19104

WHEELER, JAMES WILLIAM, JR, b Clarksburg, WVa, Oct 2, 34; m 57; c 1. ORGANIC CHEMISTRY. *Educ:* Antioch Col, BS, 57; Stanford Univ, MS, 59, PhD(chem), 62. *Prof Exp:* NSF fel chem, Cornell Univ, 63-64, NIH trainee, 64; from asst prof to assoc prof, 64-71, NIH spec fel, 71-72, PROF CHEM, HOWARD UNIV, 71- *Mem:* AAAS; Am Chem Soc; Am Soc Mass Spectros. *Res:* Chemistry of arthropod and mammalian pheromones, small ring compounds and monoterpenes. *Mailing Add:* Dept Chem Howard Univ Washington DC 20059

WHEELER, JEANETTE NORRIS, b Newton, Iowa, May 21, 18; m 41; c 1. ENTOMOLOGY. *Educ:* Univ NDak, BA, 39, MS, 56, PhD, 62. *Prof Exp:* From instr biol to asst prof, Univ NDak, 46-65, res assoc, 65-67; res assoc, Desert Res Inst, Univ Nev Syst, Reno, 67-85. *Concurrent Pos:* Res assoc entom, Natural Hist Mus, Los Angeles County, 76-82; Res Assoc, Fla Dept Agr, 86- *Mem:* Entom Soc Am; Sigma Xi. *Res:* Taxonomy and morphology of the ant larvae; desert ants; Western & Florida ants. *Mailing Add:* 3358 NE 58th Ave Silver Springs FL 32688

WHEELER, JOE DARR, b Dallas, Tex, Dec 29, 30; m 55; c 2. CHEMICAL ENGINEERING, THERMODYNAMICS. *Educ:* Rice Univ, BA, 52, BS, 53; Purdue Univ, PhD(chem eng), 64. *Prof Exp:* Foreman chem eng, Am Cyanamid Co, 53; salesman houses, S R Franck & Co, 54; chem engr, Humble Oil & Refining Co, 57-59; RES ADV, EXXON PROD RES CO, 63- *Res:* Thermodynamics of non-ideal mixtures; applied statistics; fluid mechanics; permafrost drilling. *Mailing Add:* 803 Atwell Dr Bellaire TX 77401

WHEELER, JOHN ARCHIBALD, b Jacksonville, Fla, July 9, 11; m 35; c 3. THEORETICAL PHYSICS. *Educ:* Johns Hopkins Univ, PhD(physics), 33. *Hon Degrees:* ScD, Western Reserve Univ, 58, Univ NC, 59, Univ Penn, 68, Middlebury Col, 69, Yeshiva Univ, 73, Yale Univ, 74, Univ Md, 77, Gusbarus Adolphus Univ, 81, Cath Univ Am, 82, Univ Newcastle upon Tyme, 83, Princeton Univ, 86, Univ Conn, 89; PhD, Univ Uppsala, 75; LLD, Johns Hopkins Univ, 77; LittD, Drexel Univ, 87. *Prof Exp:* Nat Res Coun fel, NY Univ & Copenhagen Univ, 33-35; from asst prof to assoc prof physics, Univ NC, 35-38; asst prof, Princeton Univ, 38-42, physicist atomic energy proj, 39-42; physicist, Metall Lab, Univ Chicago, 42-43, E I du Pont de Nemours & Co, Del, 43-44, Hanford Eng Works, Wash, 44-45 & Los Alamos Sci Lab, 50-53; from assoc prof to prof, 45-66, Joseph Henry prof, 66-76, EMER PROF PHYSICS, PRINCETON UNIV, 76-; dir Ctr Theoret Physics, 76-86, Ashbel Smith prof, 79-86, Roland Blumberg prof, 81-86, EMER PROF PHYSICS, UNIV TEX, AUSTIN, 86- *Concurrent Pos:* US rep cosmic ray comn, Int Union Pure & Appl Physics, Poland, 47, vpres, Union, 51-54; Guggenheim fel, Univ Paris & Copenhagen Univ, 49-50; dir proj Matterhorn, Princeton Univ, 51-53; Lorentz prof, Univ Leiden, 56; mem adv comt, Oak Ridge Nat Lab, 57-67; sci adv, US Senate Del, Conf NATO Parliamentarians, France, 57; adv, Joint Cong Comt Atomic Energy; chmn proj 137, forerunner present proj Jason, Dept Defense Advan, Res Proj Agency, 58; consult, AEC; trustee, Battelle Mem Inst, 59-89; Fulbright prof, Kyoto Univ, 62; chmn joint comt on hist theoret physics in 20th century, Am Phys Soc-Am Philos Soc, 62-; vis fel, Clare Col, Cambridge Univ, 64; mem, US Gen Adv Comt Arms Control & Disarmament, 69-76; Battelle Mem prof, Univ Wash, 75. *Honors & Awards:* Morrison Prize, NY Acad Sci, 46; Einstein Prize, Strauss Found, 65; Enrico Fermi Award, AEC, 68; Franklin Medal, Franklin Inst, 69; Nat Medal Sci, 71; J Robert Oppenheimer Mem Prize, 84. *Mem:* Nat Acad Sci; fel Am Phys Soc (pres, 66); Am Philos Soc (vpres, 71-73); Am Math Soc; Am Acad Arts & Sci; Int Astron Union; Danish Royal Acad Sci; NY Acad Sci; Int Acad Philos Sci, (vpres, 87-90). *Res:* Atomic and nuclear physics; scattering theory; fission; nuclear chain reactors; direct electromagnetic interaction between particles; mathematics of semiclassical analysis of physical processes; mu-meson; relativity; space-time and geometrodynamics. *Mailing Add:* Dept Physics Princeton Univ Princeton NJ 08544-0708

WHEELER, JOHN C, b Urbana, Ill, Mar 26, 41; div; c 1. THEORETICAL CHEMISTRY, CHEMICAL PHYSICS. *Educ:* Oberlin Col, BA, 63; Cornell Univ, PhD(theoret chem), 68. *Prof Exp:* NSF fel chem, Harvard Univ, 67-69; from asst prof to assoc prof 69-81, PROF CHEM, UNIV CALIF, SAN DIEGO, 81- *Concurrent Pos:* Alfred P Sloan Found fel, 72-76; John Simon Guggenheim fel, 83-84. *Mem:* Sigma Xi; fel Am Phys Soc; Am Chem Soc. *Res:* Statistical mechanics and thermodynamics of single and multi component systems, phase transitions and critical phenomena; rigorous bounds in statistical mechanics and thermodynamics; equilibrium polymerization, polymer solutions, micellar solutions and microemulsions; reconstruction of densities from modified moments; surface properties of solids from modified moments. *Mailing Add:* Dept Chem 0340 Univ Calif La Jolla CA 92093-0340

WHEELER, JOHN CRAIG, b Glendale, Calif, Apr 5, 43; m 66; c 2. THEORETICAL ASTROPHYSICS. *Educ:* Mass Inst Technol, BS, 65; Univ Colo, PhD(physics), 69. *Prof Exp:* Res fel, Calif Inst Technol, 69-71; asst prof astron, Harvard Univ, 71-74; from assoc prof to prof astron, 74-85, dept chmn, 86-90, SAMUEL T & FERN YANAGISAWA REGENTS PROF ASTRON, UNIV TEX, AUSTIN, 85- *Honors & Awards:* Fulbright Travel Award, Italy, 91. *Mem:* Am Astron Soc; Sigma Xi; Int Astron Union. *Res:* High energy and relativistic astrophysics; supernova hydrodynamics; black hole physics; active nuclei of galaxies; compact objects in binary systems. *Mailing Add:* Dept Astron Univ Tex Austin TX 78712

WHEELER, JOHN OLIVER, b Mussoorie, India, Dec 19, 24; m 52; c 2. GEOLOGY. *Educ:* Univ BC, BASc, 47; Columbia Univ, PhD(geol), 56. *Prof Exp:* Asst geol, Columbia Univ, 49-51; tech officer, 52-55, geologist, 56-67, head cordilleran & Pac margins sect, 67-70, chief regional & econ geol div, 70-73, dep dir, 73-79, res scientist, 79-90, EMER RES SCIENTIST, GEOL SURV CAN, 90- *Concurrent Pos:* Vis prof, Univ Toronto, 72; Can coordr, Decade NAm Geol, Geol Soc Am, 81-; chmn, Steering Comt, Proj Lithoprobe, 82-84. *Honors & Awards:* Logan Medal, Geol Asn Can, 83. *Mem:* Fel Royal Soc Can; fel Geol Soc Am; Geol Asn Can (pres, 70-71); Can Inst Mining & Metall; Can Geosci Coun (pres, 81). *Res:* Geological mapping in Central and Southern Yukon and Southeastern British Columbia; glacial geology in Southern Yukon; tectonics and structure of southern part of Western Canadian cordillera; recent glacier fluctuations of Selkirk Mountains; tectonics of Canadian cordillera and Canada. *Mailing Add:* Geol Surv Can 100 W Pender St Vancouver BC V6B 1R8 Can

WHEELER, KEITH WILSON, b Iowa City, Iowa, Jan 9, 18; m 40; c 2. INFORMATION SCIENCE. *Educ:* Knox Col, AB, 38; Purdue Univ, MS, 40, PhD(org chem), 44. *Prof Exp:* Asst, Purdue Univ, 39-43; res chemist, William S Merrell Co, Ohio, 43-56, head records off, 57-64, head sci info dept, 64-71; sr prin investr, Tech Info Serv Dept, Mead Johnson Res Ctr, 71-80, sr res assoc, planning dept, 80-83; RETIRED. *Mem:* Fel AAAS; fel Am Inst Chemists; Sigma Xi; Am Chem Soc; Drug Info Asn. *Res:* Documentation. *Mailing Add:* 100 Wilson Point Dr Hot Springs AK 71913

WHEELER, KENNETH THEODORE, JR, b Dover, NH, Sept 11, 40; m 80. BIOPHYSICS, RADIATION BIOLOGY. *Educ:* Harvard Univ, BA, 62; Wesleyan Univ, MAT, 63; Univ Kans, PhD(radiation biophys), 70; Brown Univ, MA, 83. *Prof Exp:* Asst radiation biologist, Colo State Univ, 70-72, asst prof radiation biol, 72; from asst prof to assoc prof neurol surg & radiol, Med Sch, Univ Calif, San Francisco, 72-76; assoc prof radiation oncol, Univ Rochester, 76-81; prof radiation med, RI Hosp, Brown Univ, 81-83; prof & sr scientist, Radiation Biophys, Univ Kans, 83-86; PROF RADIOL & DIR, EXP RADIATION ONCOL, BOWMAN GRAY SCH MED, WAKE FOREST UNIV, 86- *Mem:* Radiation Res Soc; Biophys Soc; Am Asn Cancer Res; Sigma Xi; Am Aging Asn. *Res:* In vivo DNA damage and repair in normal nondividing tissue and tumor tissue; development of combined modality therapy for brain tumors; mechanisms of chemosensitization; molecular mechanisms of tumor cell heterogeneity. *Mailing Add:* Dept Radiol Bowman Gray Sch-Wake Forest 300 S Hawthorne Rd Winston-Salem NC 27103

WHEELER, LAWRENCE, environmental psychology, psychophysics; deceased, see previous edition for last biography

WHEELER, LEWIS TURNER, b Houston, Tex, Sept 28, 40. MECHANICS. *Educ:* Univ Houston, BS, 63, MS, 64; Calif Inst Technol, PhD, 69. *Prof Exp:* From asst prof to assoc prof, 68-76, PROF MECH ENG & MATH, UNIV HOUSTON, 76- *Concurrent Pos:* NSF grants, 69-71, 72-74 & 74-76; assoc ed, J Applied Mech. *Mem:* Am Soc Mech Engrs; Soc Indust & Appl Math; Soc Natural Philos; Am Acad Mech. *Res:* Mathematical theory of elasticity; wave propagation in solids. *Mailing Add:* Cullen Col Eng 4800 Calhoun Rd Houston TX 77204-4792

WHEELER, MARSHALL RALPH, b Carlinville, Ill, Apr 7, 17; m 44, 66; c 3. ZOOLOGY, GENETICS. *Educ:* Baylor Univ, BA, 39; Univ Tex, PhD(genetics), 47. *Prof Exp:* From instr to prof, 47-77, EMER PROF ZOOL, UNIV TEX, AUSTIN, 77- *Concurrent Pos:* Gosney fel, Calif Inst Technol, 49-50. *Mem:* Genetics Soc Am; Am Soc Nat; Soc Study Evolution; Soc Syst Zool; Entom Soc Am. *Res:* Speciation and taxonomy in Drosophila; biology of acalyptrate Diptera; insect cytogenetics. *Mailing Add:* Univ Texas Patterson Bldg Rm 140 Austin TX 78712

WHEELER, MARY FANETT, b Cuero, Tex, Dec 21, 38; m 63; c 1. NUMERICAL ANALYSIS. *Educ:* Univ Tex, BA, 60, MA, 63, PhD(math), 71. *Prof Exp:* Programmer math, Univ Tex Comput Ctr, 61-65; from programmer to instr, 65-73, from asst prof to assoc prof, 73-81, PROF MATH SCI, RICE UNIV, 81- *Concurrent Pos:* Noah Harding prof, Rice Univ. *Mem:* Am Math Soc; Soc Indust & Appl Math. *Res:* Numerical solution of partial and ordinary differential equations; parallel computation; flow in porous media. *Mailing Add:* Dept Math Sci Rice Univ Box 1892 Houston TX 77251

WHEELER, MICHAEL HUGH, b Rolla, Mo, Nov 13, 40; m 69; c 2. BIOCHEMISTRY, BIOLOGY. *Educ:* Tex A&M Univ, BS, 65, MS, 70. *Prof Exp:* Microbiologist, Wadley Inst Molecular Biol, 71-72; RES CHEMIST, NAT COTTON PATH RES LAB, AGR RES SERV, USDA, 72- *Mem:* Electron Microscope Soc Am; Am Soc Microbiol; Am Phytopath Soc. *Res:* Biochemical and utrastructural aspects of fungal physiology and morphogenesis; melanogenesis; cell wall composition; nuclear behavior and host-plant, fungal-parasite interactions. *Mailing Add:* 1003 Timm Dr College Station TX 77840

WHEELER, NED BRENT, b Ogden, Utah, May 4, 36; m 56; c 4. ELECTRO-OPTICS, INTERFEROMETER DESIGN & MEASUREMENTS. *Educ:* Utah State Univ, BS, 64, MS, 65. *Prof Exp:* Instr elec eng, Weber State Univ, 65-67; engr, Stewart Radiance Lab, 67-68; ENGR, PHILLIPS LAB, HANSCOM AFB, 68- *Concurrent Pos:* Lectr elec eng, Univ Lowell, 67- *Res:* Development of hardware, techniques and procedures for training astronauts for operating payloads in orbit; infrared radiation measuring sensors and equipment for upper atmosphere measurements; radiometer and photometer design and measurements; author of 20 papers and publications. *Mailing Add:* 19 Porter Rd Chelmsford MA 01824

WHEELER, NICHOLAS ALLAN, b The Dalles, Ore, May 24, 33; m 62; c 2. MATHEMATICAL PHYSICS. *Educ:* Reed Col, BA, 55; Brandeis Univ, PhD(physics), 60. *Prof Exp:* NSF fel, State Univ Utrecht & Europ Orgn Nuclear Res, 60-62; res assoc physics, Brandeis Univ, 62-63; asst prof, 63-65, assoc prof, 65-77, PROF PHYSICS, REED COL, 77- *Res:* Structure and interconnections among physical theories, especially classical and quantum dynamics, classical field theories, statistical mechanics and thermodynamics. *Mailing Add:* Dept Physics Reed Col 3203 SE Woodstock Blvd Portland OR 97202

WHEELER, ORVILLE EUGENE, b Memphis, Tenn, Dec 31, 32; m 56; c 1. STRUCTURAL MECHANICS, COMPUTER APPLICATIONS. *Educ:* Vanderbilt Univ, BE, 54; Univ Mo, MSCE, 56; Tex A&M Univ, PhD(civil eng), 66. *Prof Exp:* Stress analyst, Chance Vought Aircraft, 59-60 & Hayes Aircraft Brown Eng, 60-62; sect chief, Marshall Space Flight Ctr, NASA, 62-66; design specialist, Gen Dynamics, 66-72; chief struct engr, Bucyrus Erie Co, 72-78; dean, 78-87, HERF PROF STRUCT MECH, MEMPHIS STATE UNIV, 87- *Concurrent Pos:* Vis prof, Univ Tex, 67-70, Southern Methodist Univ, 67-72; consult, Wheeler Engrs Inc, 79- *Mem:* Am Soc Civil Engrs; Am Soc Testing Mat; Am Inst Steel Construct; Nat Soc Prof Engrs; Am Soc Eng Educ; NY Acad Sci; Asn Computer Mach. *Res:* Structural mechanics particularly computer applications, fracture mechanics, design methodology and fatigue. *Mailing Add:* PO Box 241396 Memphis TN 38124

WHEELER, RALPH JOHN, b Devine, Tex, Sept 14, 29; m 57; c 1. ANALYTICAL CHEMISTRY. *Educ:* Trinity Univ, San Antonio, Tex, BS, 63. *Prof Exp:* Res chemist, Southwest Res Inst, 63-68; anal chemist, Gulf South Res Inst, 68-73, mgr anal chem, 73-74, assoc dir, Life Sci Div, 74-83, dir mkt, Intox Labs, 83-85; toxicologist, 86-89, PRES, TPS, INC, 89- *Mem:* Soc Toxicol. *Res:* Carcinogenesis bioassay of pesticides and other environmental chemicals; development of new chromatographic instrumentation and methodology for the analysis of airborne polynuclear arenes; pharmaceutical and chemical safety evaluations through animal research. *Mailing Add:* 4315 Joyce Lane Mt Vernon IN 47620-9624

WHEELER, RICHARD HUNTING, b Brooklyn, NY, Jan 30, 31; m 54; c 3. FOREST HYDROLOGY, WATERSHED MANAGEMENT. *Educ:* Univ Maine, Orono, BS, 53; Colo State Univ, MF, 69. *Prof Exp:* Forester, Savannah River Proj, US Forest Serv-AEC, SC, 57-59; forester, Ouachita Nat Forest, US Forest Serv, 59-62, forester & staff consult water resources, Roosevelt Nat Forest, 62-64 & Arapaho Nat Forest, 64-66, hydrologist & staff consult, Northern Region, Div Soil, Air & Water Mgt, 66-74; forest hydrologist & consult, Food & Agr Orgn UN, Mae Sa Watershed Proj, Chiang Mai, Thailand, 74-77; hydrologist & staff consult, Northern Region, USDA Forest Serv, Missoula, Mont, 77-79; hydrologist & staff consult, Mt Hood Nat Forest, 79-86; RETIRED. *Concurrent Pos:* Consult to UN Environ Prog, Asia & Pac Region, Bangkok, Thailand, 76; fac affil, Sch Forestry, Univ Mont, 78-79. *Mem:* Soc Am Foresters. *Res:* Wildlife water quality and water resource management; general forest management. *Mailing Add:* 5013 SE 22nd St Gresham OR 97080

WHEELER, ROBERT FRANCIS, b Austin, Tex, Nov 3, 43; m 76. FUNCTIONAL ANALYSIS, GENERAL TOPOLOGY. *Educ:* Rice Univ, BS, 65; Univ Mo, Columbia, MA, 68, PhD(math), 70. *Prof Exp:* Vis asst prof math, La State Univ, 71-72; from asst prof to assoc prof, 72-83, PROF MATH, NORTHERN ILL UNIV, 83- *Concurrent Pos:* Math res grants, NSF, 77-78 & 80-81. *Mem:* Am Math Soc; Sigma Xi. *Res:* Measures on topological spaces; the strict topology on spaces of continuous functions; Banach space theory. *Mailing Add:* 18 Golfview Pl De Kalb IL 60115

WHEELER, ROBERT LEE, b Minneapolis, Minn, Jan 17, 44; m 67; c 3. INTEGRAL EQUATIONS, VISCOELASTICITY. *Educ:* Univ Minn, BS, 66; Univ Wis-Madison, MA, 69, PhD(math), 71. *Prof Exp:* From asst prof to assoc prof math, Univ Mo, Columbia, 71-80; assoc prof, 80-83, PROF MATH, VA POLYTECH INST & STATE UNIV, 83- *Concurrent Pos:* Vis asst prof math, Iowa State Univ, 74-75. *Mem:* Am Math Soc; Soc Indust & Appl Math; Math Asn Am. *Res:* Volterra integral equations; integro-partial differential equations; dynamics of viscoelastic structures. *Mailing Add:* Dept Math Va Polytech Inst & State Univ Blacksburg VA 24061-0123

WHEELER, RURIC E, b Clarkson, Ky, Nov 30, 23; div; c 2. MATHEMATICAL STATISTICS, NUMBER THEORY. *Educ:* Western Ky Univ, AB, 47; Univ Ky, MS, 48, PhD(math, statist), 52. *Prof Exp:* Instr math & statist, Univ Ky, 48-52; asst prof statist, Fla State Univ, 52-53; from assoc prof to prof math, 53-65, head dept math & eng, 55-65, chmn div natural sci, 65-67, asst to acad dean, 67-68, dean, Howard Col Arts & Sci, 68-70, vpres acad affairs, 70-87, UNIV PROF MATH, SAMFORD UNIV, 87-

Concurrent Pos: Consult, Dynamics Dept, Hayes Int Corp, 56-67; trustee, Mid-South Technol Inst, 58-67; dir, NSF vis sci prog, 63-67 & coop prog, 65-67; trustee, Gorgas Found, 68-; mem, Am Conf Acad Deans. *Honors & Awards:* Humboldt Award, Fed Repub Ger, 86. *Mem:* Am Math Soc; Math Asn Am; Asn Math Teachers; Am Educ Asn; Am Asn Higher Educ; Am Asn Univ Adminr (vpres, 74-76, pres, 76-77). *Res:* Statistical distributions; stochastic processes; mathematics education; author of ten college textbooks. *Mailing Add:* Dept Math Samford Univ Birmingham AL 35209

WHEELER, RUSSELL LEONARD, b Freeport, NY, June 12, 43. STRUCTURAL GEOLOGY, TECTONICS. *Educ:* Yale Univ, BS, 66; Princeton Univ, PhD(geol), 73. *Prof Exp:* Asst prof geol, WVa Univ, 71-77, assoc prof, 77-80; WITH US GEOLOGICAL SURVEY, 80- *Concurrent Pos:* Grants, WVa Univ Fac Senate, 74 & Petrol Res Fund, Am Chem Soc, 74-77; mem fac, US Geol Surv, 75-77; contract, US Dept Energy, 78. *Mem:* AAAS; Geol Soc Am; Am Geophys Union. *Res:* Structural analysis of fold and thrust belts; applications of robust statistical methods to structural geology and tectonics and economic applications thereof. *Mailing Add:* US Geological Survey Fed Ctr Mail Stop 966 Box 25046 Denver CO 80225

WHEELER, SAMUEL CRANE, JR, b Montclair, NJ, June 3, 13; m 42; c 6. ASTRONOMY. *Educ:* Miami Univ, AB, 42; Univ Ill, MS, 43; Ohio State Univ, PhD(physics), 60. *Prof Exp:* Asst, Miami Univ, 39-42, instr physics, 43-44; asst, Univ Ill, 42-43; asst physicist, Div War Res, Univ Calif, San Diego, 44-46; physicist, US Navy Electronics Lab, 46-48; from instr to prof, 48-78, chmn, dept physics & astron, 60-70, EMER PROF PHYSICS, DENISON UNIV, 78- *Concurrent Pos:* NSF fac fel, 59-60 & prog dir, Div Instnl Progs, 63-64, consult, Div Undergrad Educ Sci, 66-; mem exam bd, NCent Asn Cols & Sec Schs, Comn Higher Educ, 70- *Mem:* Am Phys Soc; Am Asn Physics Teachers; Am Astron Soc. *Res:* Underwater sound calibration techniques; sonic properties of materials; theory of infrared spectra of polyatomic molecules; theoretical molecular physics. *Mailing Add:* 2342 Silver St SW Granville OH 43023

WHEELER, THOMAS NEIL, b Ocala, Fla, Feb 6, 43; m 67. PESTICIDES, SYNTHETIC ORGANIC CHEMISTRY. *Educ:* Univ Fla, BS, 64; Cornell Univ, PhD(org chem), 69. *Prof Exp:* From asst prof to assoc prof chem, Fla Technol Univ, 69-75; res chemist, 75-80, RES SCIENTIST/GROUP LEADER, UNION CARBIDE CORP, 80- *Concurrent Pos:* Petrol res fund type B grant, Fla Technol Univ, 70-72. *Mem:* Am Chem Soc. *Res:* Exploratory synthesis of pesticides. *Mailing Add:* 8605 Woodlawn Dr Raleigh NC 27612-2673

WHEELER, WALTER HALL, vertebrate paleontology; deceased, see previous edition for last biography

WHEELER, WILLIAM HOLLIS, b Akron, Ohio, Feb 10, 46. MATHEMATICAL LOGIC. *Educ:* Vanderbilt Univ, BA, 68; Yale Univ, PhD(math), 72. *Prof Exp:* Asst prof, 72-80, ASSOC PROF MATH, IND UNIV, BLOOMINGTON, 80- *Concurrent Pos:* Vis lectr math, Bedford Col, Univ London, 73-74. *Mem:* Am Math Soc; Asn Symbolic Logic. *Res:* Model theory; metamathematics of algebra; applications of logic to algebra. *Mailing Add:* Dept Math Ind Univ Bloomington IN 47405

WHEELER, WILLIAM JOE, b Flora, Ind, Oct 14, 40; m 62; c 2. MEDICINAL CHEMISTRY. *Educ:* Purdue Univ, BS, 62; Butler Univ, MS, 66; Purdue Univ, PhD(med chem), 70. *Prof Exp:* Teacher math & sci, Northwestern Sch Corp, Henry County, Ind, 62-63; anal chemist, Allison Div, Gen Motors Corp, Ind, 63-65; org chemist, Eli Lilly & Co, 65-67; asst, Purdue Univ, Lafayette, 67-68, fel, 68-70; sr pharmaceut chemist, 70-80, RES SCIENTIST, ELI LILLY & CO, 80- *Mem:* Am Chem Soc; Am Soc Microbiol; Sigma Xi; Int Isotope Soc. *Res:* Synthesis of both stable and radioactive isotopically labeled compounds for drug metabolism studies; new methodology for isotopic labeling. *Mailing Add:* 12030 Emerald BLF Indianapolis IN 46236-8970

WHEELER, WILLIS BOLY, b Oakland, Calif, June 13, 38; m 64; c 2. BIOCHEMISTRY. *Educ:* George Washington Univ, BS, 61, MS, 63; Pa State Univ, PhD(biochem), 66. *Prof Exp:* Res asst cancer, George Washington Univ, 61-63; instr chem pesticides, Pa State Univ, 64-66; asst prof 66-72, assoc prof, 72-78, PROF PESTICIDES, UNIV FLA, 78- *Mem:* AAAS; Am Chem Soc; Sigma Xi. *Res:* Disappearance of chemical from plant and plant environment; behavior of pesticides in soil ecosystems; movement through soils; effects on soil microbial populations; metabolism; residue detection methodology. *Mailing Add:* Univ Fla 359 Food Sci Bldg IFAS 0163 Gainesville FL 32611-0163

WHEELESS, LEON LUM, JR, b Jackson, Miss, Nov 6, 35; m 57; c 3. ANALYTICAL CYTOLOGY. *Educ:* Mass Inst Technol, SB, 58; Univ Rochester, MS, 62, PhD(elec eng), 65. *Prof Exp:* Sect head, Electronics Dept, Bausch & Lomb, Inc, 58-61, tech specialist, Biophys Dept, 61-65, res scientist, Cent Res Lab, 65-68, sr res scientist, Biomed Res Dept, Anal Systs Div, 68-69, dir biomed res, 69-71; assoc prof, 71-81, PROF PATH, LAB MED & ELEC ENG, MED CTR, UNIV ROCHESTER, 81-, DIR, ANALYTICAL CYTOL UNIT, DEPT PATH, 75-, PROF UROL, 87- *Honors & Awards:* Inst Elec & Electronics Engrs Centennial Medal, 84. *Mem:* Inst Elec & Electronics Engrs; Am Soc Cytol; Soc Anal Cytol. *Res:* Biomedical instrumentation; systems for automatic recognition of abnormal cells; automated cytopathology instrumentation; pattern recognition; image and flow cytometry. *Mailing Add:* Dept Path Univ Rochester Med Ctr Rochester NY 14642

WHEELIS, MARK LEWIS, b Chelsea, Mass, Jan 8, 44; m 65; c 2. MICROBIOLOGY, GENETICS. *Educ:* Univ Calif, Berkeley, AB, 65, MA, 67, PhD(bacteriol), 69. *Prof Exp:* Lab technician bacteriol, Univ Calif, Berkeley, 65-66; res assoc, Univ Ill, Urbana-Champaign, 69-70; asst prof, 70-76, ASSOC PROF BACTERIOL, COL LETT & SCI, UNIV CALIF, DAVIS, 76- *Concurrent Pos:* USPHS res grant, Univ Calif, Davis, 72-73. *Mem:* Am Soc Microbiol; Genetics Soc Am; Brit Soc Gen Microbiol; Am Soc Biol Chemists. *Res:* bacterial metabolism; dissimilation of aromatic acids. *Mailing Add:* Dept Microbiol Univ Calif Davis CA 95616

WHEELOCK, EARLE FREDERICK, b New York, NY, Feb 19, 27; m 55; c 3. ONCOLOGY. *Educ:* Mass Inst Technol, BS, 50; Columbia Univ, MD, 55; Rockefeller Inst, PhD(biol), 61. *Prof Exp:* Intern med, Clins, Univ Chicago, 55-56; resident, Strong Mem Hosp, Rochester, NY, 56-57; fel biol & virol, Rockefeller Inst, 57-61; from asst prof to assoc prof prev med, Sch Med, Western Reserve Univ, 61-71; prof microbiol, Jefferson Med Col, 71-81; PROF PATH, HAHNEMANN MED COL, 81- *Concurrent Pos:* USPHS res career develop award, 66-71. *Mem:* Am Soc Clin Invest; Am Asn Immunol; Soc Exp Biol & Med; Am Soc Microbiol; Am Asn Cancer Res. *Res:* Animal virology; mechanism of host resistance to viral infections; role of leucocytes and interferon in human viral infections; effect of nontumor viruses on virus-induced leukemia in mice; suppression of leukemia viral infections; tumor dormant states in animals and man. *Mailing Add:* Dept Path Hahnemann Univ Broad & Vine Sts Philadelphia PA 19102

WHEELOCK, KENNETH STEVEN, b Kansas City, Mo, Sept 18, 43; m 72; c 2. PETROLEUM CHEMISTRY, INORGANIC CHEMISTRY. *Educ:* Univ Mo, Kansas City, BS, 65; Tulane Univ, PhD(chem), 70. *Prof Exp:* Chemist, 69-72, res chemist, 72-77, staff chemist, 77-83, sr staff chemist, Exxon Res & Develop Labs, 83-87; sr patent chemist, 87-89, SR RES CHEMIST, PHILLIPS PETROL, BARTLEVILLE, OKLA, 89- *Concurrent Pos:* Consult, Dept Chem, Tulane Univ, 70- *Honors & Awards:* Award, Am Chem Soc, 65. *Mem:* Am Chem Soc; AAAS; NY Acad Sci; Sigma Xi. *Res:* Theoretical aspects of transition metal chemistry and catalysis; low valent complexes of transition metals with unsaturated ligands; quantum chemistry of catalysis; theory of finely divided metals; perovskite catalysts and fluid catalytic cracking. *Mailing Add:* 2917 Monticello Dr Bartlesville OK 74006

WHEELOCK, THOMAS DAVID, b Chihuahua, Mex, May 15, 25; m 52; c 2. CHEMICAL ENGINEERING. *Educ:* Iowa State Univ, BS, 49, PhD(chem eng), 58. *Prof Exp:* Sales engr, Chem Equip Co Calif, 49-51; chem engr, Westvaco Chlor-Alkali Div, Food Mach & Chem Corp, 51-54; from instr to assoc prof, 57-64, PROF CHEM ENG, IOWA STATE UNIV, 64- *Concurrent Pos:* Masua hon lectr, 80-81. *Mem:* Am Chem Soc; Am Soc Eng Educ; Soc Mining Engrs, Am Inst Chem Engrs; Sigma Xi; fel Am Soc Chem Engrs. *Res:* Process thermodynamics and kinetics; fluidized bed reactors; coal and other mineral utilization processes. *Mailing Add:* Dept Chem Eng Iowa State Univ Ames IA 50011

WHEELON, ALBERT DEWELL, b Moline, Ill, Jan 18, 29; m 53; c 2. THEORETICAL PHYSICS. *Educ:* Stanford Univ, BSc, 49; Mass Inst Technol, PhD(physics), 52. *Prof Exp:* Asst, Res Lab Electronics, Mass Inst Technol, 51-52; sr mem tech staff, Ramo-Wooldridge Corp, 53-62; with US Govt, 62-66; vpres & group exec, 66-80, chmn & chief exec officer, 86-88, SR VPRES & PRES SPACE & COMMUN GROUP, HUGHES AIRCRAFT CO, 80- *Concurrent Pos:* Vis prof, Mass Inst Technol, 89-; trustee, Calif Inst Technol, Santa Fe Inst & Aerospace Corp. *Mem:* Nat Acad Eng; Am Phys Soc; fel Inst Elec & Electronics Engrs. *Res:* Meson theory; general relativity; turbulence theory; analysis of ballistic missile and space systems; electromagnetic propagation and radio signal statistics. *Mailing Add:* 320 S Canyon View Dr Los Angeles CA 90049

WHEELWRIGHT, EARL J, b Rexburg, Idaho, Mar 26, 28; m 47; c 6. NUCLEAR CHEMISTRY. *Educ:* Brigham Young Univ, BS, 50; Iowa State Univ, PhD, 55. *Prof Exp:* Chemist, Hanford Atomic Prod Oper, Gen Elec Co, 55-60, sr scientist, 60-64; res assoc, 65-77, SR STAFF SCIENTIST, PAC NORTHWEST LABS, BATTELLE MEM INST, 77-, MGR, CHEM TECHNOL PROGS, 85- *Mem:* Am Nuclear Soc. *Res:* Separation chemistry, ion exchange and chelation chemistry as applied to lanthanides, actinides and fission products. *Mailing Add:* 1416 Sunset Richland WA 99352

WHEILDON, W(ILLIAM) M(AXWELL), JR, b Ashland, Mass, June 16, 08; m 51; c 3. MECHANICAL ENGINEERING. *Educ:* Mass Inst Technol, BS, 30, MS, 31. *Prof Exp:* Res engr, Norton Co, 37-66, chief protective prod, Div Res & Develop, 66-67, res assoc indust ceramics, 67-73; RETIRED. *Concurrent Pos:* Mech engr consult, 73- *Mem:* Assoc fel Am Inst Aeronaut & Astronaut; Am Chem Soc; Am Welding Soc; Am Soc Testing & Mat. *Res:* Abrasion and corrosion resistant ceramic materials beyond the range of metals; high temperature ceramic protective and insulating coatings; flame-sprayed ceramic coatings; ceramic cutting tools; ceramic armor and hot pressed industrial ceramics. *Mailing Add:* 84 Gates St Framingham MA 01701

WHELAN, ELIZABETH M, b New York, NY, Dec 4, 43; m 71; c 1. EPIDEMIOLOGY, PUBLIC HEALTH. *Educ:* Conn Col, AB, 65; Yale Univ, MPH, 67; Harvard Univ, MS, 68, ScD, 71. *Prof Exp:* EXEC DIR AM COUN SCI & HEALTH, 72- *Concurrent Pos:* Res assoc, Sch Pub Health, Harvard Univ, 74- *Mem:* Am Pub Health Asn; Nutrit Today Asn. *Res:* Author or coauthor of over 100 publications. *Mailing Add:* Am Coun Sci & Health 1995 Broadway New York NY 10023

WHELAN, JAMES ARTHUR, b Steele Co, Minn, Sept 25, 28; m 50; c 3. ECONOMIC GEOLOGY. *Educ:* Univ Minn, BMinE, 49, MS, 56, PhD, 59. *Prof Exp:* Instr mining eng, Univ Minn, 57-59; from asst prof to prof mineral, Univ Utah, 59-68, prof mining & geol eng, 68-69; dep off in chg construct, US Navy, Marianas, 69-71; PROF GEOL & GEOPHYS SCI, UNIV UTAH, 71- *Mem:* Geol Soc Am. *Res:* Geochemistry; mineralogy. *Mailing Add:* 2312 Sunnyside Ave Salt Lake City UT 84108

WHELAN, JEAN KING, b Reno, Nev, Nov 12, 39; m 70. ORGANIC CHEMISTRY. *Educ:* Univ Calif, Davis, BS, 61; Mass Inst Technol, PhD(org chem), 65. *Prof Exp:* NIH fel chem, Brandeis Univ, 65-67; from asst prof to assoc prof, Fairleigh Dickinson Univ, 67-75; res assoc, 75-78, res specialist, 78-83, SR RES SPECIALIST, WOODS HOLE OCEANOL INST, 83- *Concurrent Pos:* Shipboard org chemist, deep sea drilling proj. *Mem:* Am Asn Petrol Geologists; Geol Soc Am; AAAS; Am Chem Soc; Sigma Xi. *Res:* Marine chemistry, organic geochemistry; petroleum genesis and migration; chemical and microbiological transformations of organic compounds in marine sediments; pyrolysis-GC-MS-in organic geochemistry; development of new organic geochemical methods. *Mailing Add:* Dept Chem Fye Bldg Woods Hole Oceanog Inst Woods Hole MA 02543

WHELAN, JOHN MICHAEL, b Lyndhurst, NJ, Sept 12, 21; m 43; c 3. POLYMER CHEMISTRY. *Educ:* Stevens Inst Technol, ME, 41, MS, 43; Polytech Inst Brooklyn, PhD(org chem), 59. *Prof Exp:* Res chemist, Union Carbide Corp, 41-53, group leader, 53-55, sect head, 55-63, asst dir res & develop, 63-72, res assoc, 72-83; RETIRED. *Mem:* Am Chem Soc. *Res:* Synthetic polymers; organometallic polymerization catalysts. *Mailing Add:* 38 Colony Ct Murray Hill NJ 07974

WHELAN, THOMAS, III, b Houston, Tex, Dec 21, 44; m 68; c 1. MARINE GEOCHEMISTRY. *Educ:* Austin Col, BA, 66; Univ Tex, Austin, MA, 68; Tex A&M Univ, PhD(chem oceanog), 71. *Prof Exp:* Asst prof marine sci, Coastal Studies Inst, La State Univ, Baton Rouge, 71-75, assoc prof marine sci, 75-79; PRES, CARBON SYSTS, INC, 79-; PRES, WHELAN & ASSOC INC. *Mem:* Geochem Soc; AAAS; Am Asn Plant Physiologists; Sigma Xi. *Res:* Organic geochemistry of marine environments; geochemistry of natural gases; effects of oil in coastal environment. *Mailing Add:* 12979 Kingsbridge Lane Houston TX 77077-2268

WHELAN, WILLIAM JOSEPH, b Salford, UK, Nov 14, 24; m 51. BIOCHEMISTRY. *Educ:* Univ Birmingham, BSc, 45, PhD, 48, DSc(org chem), 55. *Prof Exp:* Sr lectr, Univ Col NWales, 48-55; sr mem, Lister Inst Prev Med, London, Eng, 56-64; prof biochem, Royal Free Hosp Sch Med, Univ London, 64-67; PROF BIOCHEM & CHMN DEPT, SCH MED, UNIV MIAMI, 67- *Concurrent Pos:* Secy-gen, Fedn Europ Biochem Socs, 65-67 & Pan-Am Asn Biochem Socs, 69-72; mem physiol chem study sect, NIH, 71-75, chmn, 73-75; gen-secy, Int Union Biochem, 73-83; ed-in-chief, Trends Biochem Sci, 76-78; chmn, Comt Genetic Exp, 77-81; mem exec bd, Int Coun Sci Unions, 78-80; ed-in-chief, J Bio Essays, 84-89, J Fedn Am Soc Exp Biol, 87- *Honors & Awards:* Carl Lucas Alsberg lectr, 67; Ciba Medal & lectr, 69; Diplome d'Honneur, Fedn Europ Biochem Socs, 74; Saare Medal, Asn Cereal Res, 79. *Mem:* Brit Biochem Soc; Am Soc Biol Chem; Am Chem Soc; Sigma Xi; Am Soc Cell Biol; fel AAAS. *Res:* Glycogen and starch, structure and metabolism. *Mailing Add:* PO Box 016129 Miami FL 33101

WHELAN, WILLIAM PAUL, JR, b Brooklyn, NY, Sept 22, 23; wid; c 3. ORGANIC POLYMER CHEMISTRY. *Educ:* Holy Cross Col, AB, 43, MS, 47; Columbia Univ, PhD(chem), 52. *Prof Exp:* Res scientist, Uniroyal Inc, World Hq, Middlebury, 52-66, sr res scientist, Corp Res & Develop, 66-85; RETIRED. *Mem:* Am Chem Soc; Sigma Xi. *Res:* Flame and smoke inhibition in polymers; rocket motor insulators; rubber and plastic product formulation; cellular products; polyurethane synthesis; vinyl polymerization; blowing agents; ablatives; organic synthesis; reaction mechanisms; solvolysis theory; crystallization kinetics; photopolymers; low profile additives. *Mailing Add:* 27 Orchard Lane Woodbury CT 06798

WHELLY, SANDRA MARIE, b Fall River, Mass, Aug 8, 45. BIOCHEMISTRY. *Educ:* Salve Regina Col, BA, 68; Univ Nebr, PhD(biochem), 73. *Prof Exp:* Res asst chem, Salve Regina Col, 67-68; res asst microbiol, Peter Bent Brigham Hosp, 68-69; NIH trainee biochem, Eppley Inst, Col Med, Univ Nebr, 69-72, grad res asst, 72- 73; fel, Temple Univ, Sch Med, 74-76, res assoc 76-77; res asst prof, 77-81, ASST PROF BIOCHEM, TEXAS TECH UNIV HEALTH SCI CTR, 81- *Concurrent Pos:* Res assoc grant, Am Cancer Soc, 74-75; NIH res grant, Child Health & Develop, 81-84 & 85-89. *Res:* Transcriptional and translational control mechanisms of cellular proliferation; biochemistry of hormone action. *Mailing Add:* Tex Tech Health Sci Ctr Lubbock TX 79430

WHELTON, ANDREW, b Cork, Ireland, Oct 6, 40; m 74; c 2. MEDICINE. *Educ:* Nat Univ Ireland, MB & MD, 63, FCP, 75, FACP, 83. *Prof Exp:* Mem staff, Renal Metal Unit, Walter Reed Army Inst Res, 67-68; ASSOC PROF MED, SCH MED, JOHNS HOPKINS UNIV, 76- *Concurrent Pos:* Consult renal dis, USPHS Hosp, Baltimore, 71- & to Surgeon Gen, USAF, Washington, DC, 71- & Union Mem Hosp, 73-; consult med, Dept Obstet & Gynec, Johns Hopkins Univ, 75- *Mem:* Int Soc Nephrology; AMA; Am Soc Nephrology; Am Fedn Clin Res; fel Am Col Clin Pharmacol (pres elect); fel Am Col Physicians. *Res:* Drug metabolism in renal failure; drug nephrotoxicity; acute renal failure. *Mailing Add:* Med Nephrol Div Johns Hopkins Univ Sch Med 720 Rutland Ave Baltimore MD 21205

WHELTON, BARTLETT DAVID, b San Francisco, Calif, Dec 2, 41. MEDICINAL CHEMISTRY, INTERMEDIARY METABOLISM. *Educ:* Univ San Francisco, BS, 63; Univ Wash, PhD(med chem), 69. *Prof Exp:* Fel med chem, Univ Alta, 69-71; asst prof, Univ of the Pac, 71-74; from asst prof to assoc prof, 74-85, PROF MED CHEM, EASTERN WASH UNIV, 85- *Concurrent Pos:* Fac res fel, Argonne Nat Lab, 82-83. *Mem:* Am Chem Soc; Sigma Xi. *Res:* Synthesis of antineoplastic agents; heavy metal toxicology; owl distribution and behavior. *Mailing Add:* Dept Chem & Biochem Eastern Wash Univ Cheney WA 99004

WHEREAT, ARTHUR FINCH, b New York, NY, June 30, 27; m 53; c 3. CARDIOVASCULAR DISEASES. *Educ:* Williams Col, BA, 47; Univ Pa, MD, 51. *Prof Exp:* Intern & med resident internal med, Hosp Univ Pa, 51-54, fel cardiol, 54-56; assoc in biochem, 56-60, asst prof med, 60-68, ASSOC PROF MED, SCH MED, UNIV PA, 68-; CHIEF OF STAFF, VET ADMIN HOSP, PHILADELPHIA, 76- *Concurrent Pos:* Mem coun arteriosclerosis, Am Heart Asn, 64 & coun clin cardiol, 69. *Mem:* Fel Am Col Physicians; fel Am Col Cardiol; Am Soc Biol Chemists; Am Physiol Soc; Am Heart Asn. *Res:* Biochemical changes in arterial wall associated with atherogenesis; biochemical changes in heart muscle during hypoxia and ischemia. *Mailing Add:* Univ Pa Hosp 3400 Spruce St Philadelphia PA 19104

WHERRETT, JOHN ROSS, b Regina, Saskatchewan, Nov 28, 30; m 58; c 2. NEUROLOGY, NEUROCHEMISTRY. *Educ:* Queen's Univ, MDCM, 55; Royal Col Physicians & Surgeons, FRCP, 63; Univ London, PhD(biochem), 66. *Prof Exp:* Clin teacher, Univ Toronto, 63-65, assoc, 65-68, from asst prof to prof & dir Neurol prog med, 68-89, ACTG DIR, INST MED SCI, UNIV TORONTO, 91- *Concurrent Pos:* Travelling fel, R S McLaughlin Found, 58-60; scholar acad med, Markle Found, 63-68; res fel, Am Col Physicians, 63-66; staff physician, Toronto Gen Hosp, 63-; consult neurol, Clarke Inst Psychiat, Univ Toronto, 69-75; mem & chmn grant comt neurosciences, Med Res Coun Can, 67-72; mem, Inst Med Sci, Univ Toronto, 69-; mem & chmn comt fel review, Ont Ministry Health, 71-77; mem, Bd & Med Res Comt, Ont Heart Found, 77-82; mem bd & chmn adv bd, Parkinson Found Can, 86- *Mem:* Am Acad Neurol; Am Neurol Asn; Am Soc Neurochem; Can Neurol Soc (pres, 78-79); Can Soc Clin Invest. *Res:* Investigation of the structure and metabolism of glycosphingolipids and phospholipids and of the disturbances occurring in inherited degenerative and in inflammatory diseases of the nervous system. *Mailing Add:* Toronto Hosp 200 Elizabeth St Toronto ON M5G 2C4 Can

WHETSEL, KERMIT BAZIL, b Tenn, Dec 9, 23; m 49; c 4. MOLECULAR SPECTROSCOPY, COLOR MEASUREMENT. *Educ:* ETenn State Univ, BS, 43; Univ Tenn, MS, 47, PhD(chem), 50. *Prof Exp:* From chemist to sr chemist, Tenn Eastman Co Div, Eastman Kodak Co, 50-66, develop assoc, 66-73, sr develop assoc 73-81, develop fel, 81-86; RETIRED. *Concurrent Pos:* Nat Acad Sci-Nat Res Coun resident res assoc, US Naval Res Lab, 60-61. *Mem:* Am Chem Soc; Soc Appl Spectros; Coblentz Soc; fel Am Soc Testing & Mat; Sigma Xi; Am Inst Chemists. *Res:* Instrumental analysis of organic compounds; correlation of absorption spectra with structure; solvent effects on infrared spectra; spectroscopic study of hydrogen bonded complexes. *Mailing Add:* 1501 Dobyns Dr Kingsport TN 37664

WHETSTONE, STANLEY L, JR, b Newark, NJ, Aug 30, 25; m 52; c 4. EXPERIMENTAL NUCLEAR PHYSICS. *Educ:* Williams Col, BA, 49; Univ Calif, Berkeley, PhD(physics), 55. *Prof Exp:* Staff mem & physicist, Los Alamos Sci Lab, 55-70 & 75-76; physics sect head, Int Atomic Energy Agency, Vienna, Austria, 70-75; PHYSICIST, US DEPT ENERGY, 76- *Concurrent Pos:* Vis lectr, Univ Wash, 67-68. *Mem:* Am Phys Soc. *Mailing Add:* ER-23/GTN US Dept Energy Washington DC 20545

WHETTEN, JOHN T, b Willimantic, Conn, Mar 16, 35; m 60; c 3. GEOLOGY, OCEANOGRAPHY. *Educ:* Princeton Univ, AB, 57; Univ Calif, Berkeley, MA, 59; Princeton Univ, PhD(geol), 62. *Prof Exp:* Fulbright fel, Australia & NZ, 62-63; res instr oceanog, Univ Wash, 63-64, res asst prof, 64-65, from asst prof to prof geol, 65-80, assoc dean grad sch, 68-69, chmn dept, 69-74; geologist, US Geol Surv, Seattle, 75-80; asst div leader, 80-81, dep div leader, 81-84, div leader, Earth & Space Sci Div, 84-86, ASSOC DIR, ENERGY & RES APPLN, LOS ALAMOS NAT LAB, 86- *Mem:* AAAS; Geol Soc Am; Soc Econ Paleontologists & Mineralogists. *Res:* Sedimentology; sedimentary petrology; marine geology. *Mailing Add:* MS A107 Los Alamos Nat Lab PO Box 1663 Los Alamos NM 87545

WHETTEN, NATHAN REY, b Provo, Utah, Aug 11, 28; m 53; c 3. PHYSICS. *Educ:* Yale Univ, BS, 49, MS, 50, PhD(physics), 53. *Prof Exp:* PHYSICIST, RES LAB, GEN ELEC CO, 53- *Concurrent Pos:* Vis lectr, Union Col, 64-65, adj prof, 67- *Mem:* Am Phys Soc; hon life mem Am Vacuum Soc (pres-elect, 75, pres, 76, treas, 84-). *Res:* Cosmic rays; secondary electron emission; surface physics; high vacuum; mass spectrometry; electron physics; medical physics. *Mailing Add:* Res & Develop Ctr Gen Elec Co PO Box 8 Schenectady NY 12301

WHETTEN, ROBERT LLOYD, b Mesa, Ariz, Oct 15, 59. ATOMIC & MOLECULAR CLUSTERS. *Educ:* Weber State Col, BA, 80; Cornell Univ, MS, 82, PhD(phys chem), 84. *Prof Exp:* Res fel, NSF, 81-84 & Exxon Res & Eng Co, 84-85; ASST PROF CHEM & BIOCHEM, UNIV CALIF, LOS ANGELES, 85- *Mem:* Am Chem Soc; Am Phys Soc. *Res:* Investigation of metal-atom and other clusters by laser and radiofrequency spectroscopy and chemical kinetics; spectroscopy of molecular excited states; quantum and semiclassical theory of molecular states; molecular interactions with electromagnetic fields. *Mailing Add:* Dept Chem & Biochem Univ Calif 405 Hilgard Ave Los Angeles CA 90024

WHICKER, DONALD, b Noblesville, Ind, Nov 23, 44. STRUCTURAL ANALYSIS. *Educ:* Purdue Univ, BS, 67, MS, 68, PhD(eng), 73. *Prof Exp:* Assoc sr res engr, 73-79, sr res engr, 79-80, STAFF RES ENGR, GEN MOTORS RES LABS, 80- *Mem:* Am Soc Mech Engrs; Soc Automotive Engrs. *Res:* Lubrication; tire traction; tire rolling resistance; development and application of analytical techniques for engine structural analysis and design. *Mailing Add:* Eng Mech Dept Gen Motors Res Labs Warren MI 48098

WHICKER, FLOYD WARD, b Cedar City, Utah, July 24, 37; m 57; c 3. RADIATION BIOLOGY, ECOLOGY. *Educ:* Colo State Univ, BS, 62, PhD(radiation biol), 65. *Prof Exp:* From asst prof to assoc prof, 65-80, PROF RADIATION BIOL, COLO STATE UNIV, 80- *Concurrent Pos:* Consult, Western Rad Consult, Inc & Rockwell Int. *Mem:* Fel AAAS; Ecol Soc Am; Wildlife Soc; Health Physics Soc; hon scientist, Sigma Xi. *Res:* Radiation ecology; radionuclide behavior in natural ecosystems and radiation effects on plant and animal populations. *Mailing Add:* Dept Radiol & Radiation Biol Colo State Univ Ft Collins CO 80523

WHICKER, LAWRENCE R, b Bristol, Va, Oct 3, 34; m 58; c 2. MICROWAVE PHYSICS, ELECTROMAGNETISM. *Educ:* Univ Tenn, BS, 57, MS, 58; Purdue Univ, PhD(elec eng), 64. *Prof Exp:* Teaching asst, Univ Tenn, 58; sr engr, Microwave Electronics Div, Sperry Rand Corp, 58-61; fel engr, Surface Div, Westinghouse Elec Corp, 64-65, assoc dir appl physics, 65-66, mgr microwave physics group, Aerospace Div, 66-68, adv engr microwave & antenna group, 68-69; head microwave techniques br, electronics div, Val Res Lab, 69-; MGR, GAAS PROGS, WESTINGHOUSE ELEC CO, BALTIMORE, MD. *Concurrent Pos:* Lectr, Univ Md, 64-; vpres res & develop, I-Tel, Inc, 67-68. *Mem:* Fel Inst Elec & Electronics Engrs. *Res:* Millimeter-coupled mode techniques; microwave filters; electromagnetic propagation studies; microwave solid state techniques, including microwave latching phasers, acoustics and integrated circuits. *Mailing Add:* 1218 Balfour Dr Arnold MD 21012

WHIDBY, JERRY FRANK, b Baltimore, Md, Oct 29, 43; m 67; c 3. ANALYTICAL CHEMISTRY, PHYSICAL CHEMISTRY. *Educ:* NGa Col, BS, 65; Univ Ga, PhD(anal chem), 70. *Prof Exp:* Res chemist, Gen Elec Co, Mo, 71-72; res assoc anal chem, 72-75, res chemist, 75-80, sr scientist, 80-81, MGR RES & DEVELOP, PHILIP MORRIS INC, 81- *Mem:* Am Chem Soc. *Res:* Proton exchange kinetics; environmental research-sensors; kinetics of filter action. *Mailing Add:* Philip Morris USA Res Ctr PO Box 26583 Richmond VA 23261

WHIDDEN, STANLEY JOHN, b Bayshore, NY, Oct 10, 47. DIVING PHYSIOLOGY & MEDICINE, FORENSIC MEDICINE. *Educ:* Southeastern La Univ, Hammond, BS, 71, MS, 73; Auburn Univ, PhD(physiol), 79; Univ Autonoma de Ciudad Juarez, Mex, MD, 84. *Prof Exp:* Res fel shock physiol, LSUMC, 80-82; chief researcher diving med, Baromed Res Inst, JESMC, 84-86; fac fel aerospace med, HNC, SAM, USAF, Brooks AFB, Tex, 86-87 & SBRI, Johnson Space Ctr, NASA, Tex, 87-88; major, Army Reserve Civil Affairs, Oper Just Cause, Panama, 89-90 & Desert Shield/Storm, Saudi Arabia, 90-91; ASST DIR SCI FORENSIC MED, NAT INST JUSTICE, US DEPT JUSTICE, 89- *Concurrent Pos:* Lectr, Physiol Dept, Univ Wis, Madison, 78-79; vis asst prof biol sci, Univ New Orleans, La, 79-80; NIH shock & trauma fel grant, 80, USPHS cardiovasc training grant, 80 & shock & trauma training grant, 81; asst prof physiol, LSUMC, 85- & asst prof, Nat Defense Univ, Washington, DC, 88-; dir, Technol Assessment Prog, Nat Inst Justice, 88- *Mem:* AAAS; Am Chem Soc; Am Physiol Soc; Shock Soc; Soc Neurosci; Undersea & Hyperbaric Med Soc. *Res:* Underlining cardiovascular and metabolic responses during shock resuscitation with hyperbaric oxygen medical treatment. *Mailing Add:* Nat Inst Justice 633 Indiana Ave NW Washington DC 20531

WHIFFEN, JAMES DOUGLASS, b New York, NY, Jan 16, 31; m 60; c 1. SURGERY, BIOENGINEERING. *Educ:* Univ Wis, BS, 52, MD, 55; Am Bd Surg, dipl, 63. *Prof Exp:* Intern, Ohio State Univ, 55-56; resident, 56-57 & 59-62, from instr to assoc prof, 62-71, actg chmn dept, 72-74, PROF SURG, MED SCH, UNIV WIS-MADISON, 71-, ASST DEAN MED SCH, 75- *Concurrent Pos:* Nat Heart Inst res fel, 62-64; res career develop award, 65-75; Markle Scholar, 66- *Mem:* Am Col Surg; Am Soc Artificial Internal Organs; Am Soc Test & Mat. *Res:* Cardiovascular prostheses; cardiopulmonary support devices; biomaterials. *Mailing Add:* Univ Hosps 600 Highland Ave Madison WI 53705

WHIGAN, DAISY B, b Nueva Vizcaya, Philippines, Oct 5, 47; US citizen. ANALYTICAL CHEMISTRY, PHARMACEUTICAL CHEMISTRY. *Educ:* Adamson Univ, Philippines, BS, 66; Univ Hawaii, MS, 69. *Prof Exp:* Sr res investr, Squibb Inst Med Res, 69-90; RES PROJ LEADER, BRISTOL-MYERS SQUIBB PHARM RES INST, 90- *Mem:* Am Chem Soc; NY Acad Sci; AAAS. *Res:* Analytical method development for pharmaceuticals in bulk formulations and body fluids; supervision of atomic spectroscopy laboratory research involving inductively coupled plasma atomic emission spectroscopy, graphite furnace and flame atomic absorption; robotics/automation. *Mailing Add:* Bristol-Myers Squibb Res Inst PO Box 191 New Brunswick NJ 08903

WHIGHAM, DAVID KEITH, b Blanchard, Iowa, Aug 15, 38; m 64; c 3. AGRONOMY. *Educ:* Iowa State Univ, BS, 66, MS, 69, PhD(crop prod), 71. *Prof Exp:* Instr agron, Iowa State Univ, 68-71; agronomist, USDA, 71-73; asst prof agron, Univ Ill, Urbana, 73-77; from asst prof to assoc prof, 77-82, PROF AGRON, IOWA STATE UNIV, 82- *Concurrent Pos:* Dir, Int Affairs, Iowa State Univ; cert prof, Agron Crops & Soils, Am Reg; assoc, Int Agr & Rnr Develop. *Mem:* Fel Am Soc Agron; fel Crop Sci Soc Am; Nat Asn Col & Teachers Agr; Coun Agr Sci Tech; Soc Int Develop; Nat Asn Foreign Student Affairs. *Res:* Crop production of economic crops; cropping systems research. *Mailing Add:* Dept Agron Iowa State Univ Ames IA 50011-1050

WHIKEHART, DAVID RALPH, b Pittsburgh, Pa, Aug 21, 39; m 69; c 2. PLASMA MEMBRANES, TRANSPORT ENZYMES. *Educ:* WVa Univ, PhD(biochem), 69. *Prof Exp:* Res assoc, Harvard Med Sch, 71-72; asst biochemist, McLean Hosp, Belmont, Mass, 71-72; spec fel, Nat Eye Inst, 72-74, sr staff fel ophthalmic biochem, 74-78; from asst prof to assoc prof, 78-90, PROF, SCH OPTOM, UNIV ALA, BIRMINGHAM, 90- *Concurrent Pos:* Res fel neurochem, Harvard Med Sch, 69-71, fel neurochem, McLean Hosp, Belmont, Mass, 69-71; asst prof, dept biochem, Sch Med, Univ Ala, Birmingham, 79- 82, assoc prof, 82-90. *Mem:* Asn Res Vision & Ophthal. *Res:* Metabolism and transport in the cornea; biochemistry of alkali burned corneas; biochemistry of corneal tissue cultures; ocular diabetes. *Mailing Add:* Vision Sci Res Ctr Sch Optom Univ Ala Birmingham Birmingham AL 35294

WHILLANS, IAN MORLEY, b Toronto, Ont, Feb 25, 44; m 74. GLACIOLOGY, GLACIAL GEOLOGY. *Educ:* Univ Bristol, BSc, 66; Ohio State Univ, PhD(geol), 75. *Prof Exp:* Vis scientist glaciol, Geophys Isotope Lab, Univ Copenhagen, 75-76; res assoc, Inst Polar Studies, 76-77; ASST PROF GEOL, OHIO STATE UNIV, 77- *Concurrent Pos:* Mem, comt snow & ice, Am Geophys Union, 76-77 & comt glaciol, Polar Res Bd, Nat Acad Sci, 78- *Mem:* Int Glaciol Soc; Am Geophys Union. *Res:* Dynamics of polar ice sheets; mechanics of quaternary ice sheets. *Mailing Add:* Dept Geol 107 Meadenhall Lab Ohio State Univ Main Campus Columbus OH 43210

WHINNERY, JAMES ELLIOTT, b Amarillo, Tex, May 1, 46; m 85; c 1. FIGHTER AVIATION MEDICINE, ACCELERATION PHYSIOLOGY. *Educ:* WTex State Univ, BS, 68; Tex Christian Univ, PhD(phys chem) & MAT, 72; Univ Tex, Galveston, MD, 75. *Prof Exp:* Flight surgeon, USAF Sch Aerospace Med, 75-80, chief, Biodynamics, 83-85; dir, Cardiovasc & Renal Med, Merck Sharp & Dohme Res Labs, 80-83; CHIEF AEROSPACE MED SCIENTIST, NAVAL AIR DEVELOP CTR, 86- *Concurrent Pos:* Flight surgeon, Tex Air Nat Guard, 75-80, clin comdr, 83-85; sr flight surgeon, Pa Air Nat Guard, 80-83 & Ala Air Nat Guard, 85-86; spec res fel, Air Power Res Inst, Air War Col, 85-86; adv air surgeon, Nat Guard Bur, 86-90; asst command surgeon, USAF, Europe & Air Force Surgeon, Opers Desert Shield-Desert Storm, 90-91. *Honors & Awards:* Arnold D Tuttle Award, Aerospace Med Asn, 79 & 86. *Mem:* Fel Am Col Cardiol; fel Aerospace Med Asn; fel Am Inst Chemists; Sigma Xi; Am Chem Soc. *Res:* Expert fighter aviation medicine; acceleration cardiovascular; neurophysiology; acceleration induced loss of consciousness. *Mailing Add:* Off Chief Aeromed Sci Naval Air Develop Ctr Warminster PA 18974

WHINNERY, JOHN R(OY), b Read, Colo, July 26, 16; m 44; c 3. OPTICS. *Educ:* Univ Calif, BS, 37, PhD(elec eng), 48. *Prof Exp:* From test engr to res assoc, Gen Elec Co, NY, 37-46; from lectr to prof elec eng, Univ Calif, Berkeley, 46-87, head dept, 56-59, dean col eng, 59-63, EMER PROF ELEC ENG, UNIV CALIF, BERKELEY, 87- *Concurrent Pos:* Head microwave tube res, Hughes Aircraft Co, 51-52, Guggenheim fel, 59; vis mem tech staff, Bell Tel Labs, 63-64; mem sci & technol comt manned space flight, NASA, 64-70; mem standing comt controlled thermonuclear res, Atomic Energy Comn, 70-; mem, President's Comt Nat Medal Sci, 70-72 & 79-81. *Honors & Awards:* Educ Medal, Inst Elec & Electronics Engrs, 67, Medal Honor, 85; Lamme Award, Am Soc Eng Educ, 75; Founders Award, Nat Acad Eng, 86. *Mem:* Nat Acad Sci; Nat Acad Eng; Am Phys Soc; fel Inst Elec & Electronics Engrs; fel Optical Soc Am; fel Am Acad Arts & Sci; fel AAAS. *Res:* Microwave and quantum electronics, including microwave electron devices, wave guiding systems, optical guiding systems and lasers for communication purposes. *Mailing Add:* Dept Elec Eng & Comput Sci Univ Calif Berkeley CA 94720

WHIPKEY, KENNETH LEE, b Cortland, Ohio, June 5, 32; m 62. MATHEMATICS, STATISTICS. *Educ:* Kent State Univ, AB, 53, MA, 58; Univ Evansville, MS, 84; Case Western Reserve Univ, PhD(educ statist), 69. *Prof Exp:* Instr high sch, 54-57; asst math, Kent State Univ, 57-58; instr high sch, 58-59; asst prof, Youngstown Univ, 59-67; from assoc prof to prof math, Westminster Col, Pa, 68-83; Stevens prof computer sci, Birmingham Southern Col, 83-84; ASSOC PROF MATH, UNIV NC, WILMINGTON, 84- *Concurrent Pos:* Instr, In-Serv Insts Teachers, NSF, 64-67; lectr & vis scientist, Ohio Acad Sci, 64-66; lectr, Holt, Rinehart & Winston, 65. *Mem:* Math Asn Am. *Res:* Factor analysis; calculus; computer science. *Mailing Add:* 805 Seapath Tower Wrightsville Beach NC 28480

WHIPP, BRIAN JAMES, b Tredegar, Wales, Mar 3, 37. RESPIRATORY PHYSIOLOGY. *Educ:* Stanford Univ, PhD(physiol), 67; Loughborough Univ, Eng, DSc, 82. *Prof Exp:* PROF PHYSIOL & MED, HARBOR-UNIV CALIF LOS ANGELES MED CTR, 70- *Mem:* Am Physiol Soc; Am Thoracic Soc; Am Col Sports Med; Physiol Soc. *Mailing Add:* Dept Physiol Univ Calif Los Angeles CA 90024-1751

WHIPP, SHANNON CARL, b Jacksonville, Fla, May 3, 31; m 56; c 5. VETERINARY PHYSIOLOGY, VETERINARY MICROBIOLOGY. *Educ:* Univ Minn, BS, 57, DVM, 59, PhD(physiol), 65. *Prof Exp:* Field vet, Minn Livestock Bd, 59-60; instr physiol, Univ Minn, 60-65; res vet, 65-77, res leader, 77-80, CHIEF PHYSIOPATH DIV, NAT ANIMAL DIS CTR, AGR RES SERV, USDA, 80- *Concurrent Pos:* NIH fel, 62-65. *Mem:* Am Soc Vet Physiol & Pharmacol; Am Vet Med Asn; Comp Gastroenterol Soc; Conf Res Workers Animal Dis; NY Acad Sci; Sigma Xi. *Res:* Mechanisms of diarrhea; secretory diarrhea; enteric colibacillosis. *Mailing Add:* Nat Animal Dis Ctr PO Box 70 Ames IA 50010

WHIPPEY, PATRICK WILLIAM, b Reading, UK, Feb 18, 40; m; c 3. IMAGE PROCESSING. *Educ:* Univ Reading, BSc, 62, PhD(physics), 66. *Prof Exp:* Asst lectr physics, Univ Reading, 65-66; lectr, 66-67, asst prof, 67-72, ASSOC PROF PHYSICS, UNIV WESTERN ONT, 72- *Mem:* Am Asn Physics Teachers. *Mailing Add:* Dept Physics Univ Western Ont London ON N6A 3K7 Can

WHIPPLE, CHRISTOPHER GEORGE, b Columbus, Ohio, Feb 17, 49; m 70; c 2. RISK ANALYSIS, ENERGY ECONOMICS. *Educ:* Purdue Univ, BS, 70; Calif Inst Technol, MS, 71, PhD(eng sci), 74. *Prof Exp:* Mem tech staff, Elec Power Res Inst, 74-90; VPRES & DIR, WESTERN OPERS, CLEMENT INT, 90- *Concurrent Pos:* Lectr mech eng, Stanford Univ, 78-79; course dir, Chautauqua-Type Short Course, Col Teachers, NSF-AAAS, 78-81, Adv Study Inst Technol Risk Assessment, NATO, 81; mem Adv Comt, NSF Proj Risk Assessment, 78, 81-82, 85; mem, Comt Health & Ecol Effects Synfuel Industs, Nat Acad Sci, 82-83 & Bd Radioactive Waste Mgt, 85-; mem, comt on Nuclear Safety Res, 85-86, Risk Perception & Commun, 88-90, Waste Isolation Pilot Plant, 89-, Radioactive Waste Mgt with USSR, 90. *Honors & Awards:* Outstanding Serv Award, Soc Risk Analysis, 90. *Mem:* Soc Risk Anal (pres, 82-83); AAAS; Sigma Xi; Nat Asn Environ Prof. *Res:* Analysis and management of technological risks; risk communication; radioactive waste management. *Mailing Add:* Clement Int 160 Spear St Suite 1380 San Francisco CA 94105-1535

WHIPPLE, EARL BENNETT, b Thomson, Ga, Apr 9, 30; m 51; c 4. PHYSICAL CHEMISTRY. *Educ:* Emory Univ, BS, 51, PhD(phys chem), 59. *Prof Exp:* Chemist, Va-Carolina Chem Corp, Va, 52-53; res scientist chem, Union Carbide Res Inst, 59-66, group leader chem, 66-70, mgr res, Cent Sci Lab, Tarrytown Tech Ctr, Union Carbide Corp, 70-74; RES ADV, CENT RES, PFIZER, INC, 74- *Concurrent Pos:* Adj prof, Rockefeller Univ, 69-74. *Mem:* NY Acad Sci; Am Chem Soc; AAAS; Int Soc Magnetic Resonance. *Res:* Chemical applications of nuclear and electron spin resonance; structure and electronic properties of molecules. *Mailing Add:* Seven Forest Hills Dr Madison CT 06443

WHIPPLE, FRED LAWRENCE, b Red Oak, Iowa, Nov 5, 06; m 46; c 3. ASTRONOMY, SPACE PHYSICS. *Educ:* Univ Calif, Los Angeles, AB, 27; Univ Calif, Berkeley, PhD(astron), 31; Harvard Univ, MA, 45. *Hon Degrees:* DSc, Am Int Col, 48, Temple Univ, 61 & Univ Ariz, 79; DLitt, Northeastern Univ, 61. *Prof Exp:* Mem staff, Observ, 31-77, instr, Univ, 32-38, lectr, 38-45, from assoc prof to prof, 45-70, chmn dept, 49-56, Phillips prof, 70-77, EMER PHILLIPS PROF ASTRON, HARVARD UNIV, 77- *Concurrent Pos:* Res assoc radio res lab, Off Sci Res & Develop, 42-45; leader, Harvard proj upper atmospheric & meteor res, Bur Ord, US Navy, 46-51, Air Res & Develop Command, US Air Force, 48-62, Off Naval Res, 51-57, Off Ord Res, US

Army, 53-57, dir, Harvard Radio Meteor Proj, Nat Bur Standards, 57-61, NSF, 60-63 & NASA, 63-; Lowell lectr, Lowell Technol Inst, 47; dir, Astrophys Observ, Smithsonian Inst, 55-73, sr scientist, 73-77; ed, Smithsonian Contrib Astrophys, 56-; regional ed, Planetary & Space Sci, 58-, mem, US Rocket & Satellite Res Panel, 46-; mem subcomt, Nat Adv Comt Aeronaut, 46-52; mem panel upper atmosphere, US Res & Develop Bd, 47-52; adv panel astron, NSF, 52-55, chmn, 54-55, mem div comt math & phys sci, 64-; mem sci adv bd, USAF, 53-62, assoc adv, 63-67, mem geophys & space tech panels; mem comts meteorol & atmospheric sci, Nat Acad Sci-Nat Res Coun, 58-, space sci bd, 58- & subcomt potential contamination & interference from space exp, 63-; dir optical satellite tracking proj & proj dir orbiting astron observ, NASA, 58-72, mem space sci working group orbiting astron observ, 59-69, consult aeronomy subcomt, 61-63, comt planetary atmospheres, 62-63, dir meteorite photog & recovery prog, 62-73, mem working group geod satellite prog, 63- & mem comet & asteroid sci adv comt, 71-72, chmn, 73-74; mem joint bio-astronaut comt, Armed Forces-Nat Res Coun, 59-61; spec consult comt sci & astronaut, US House Rep, 60-73; mem working groups geod satellites & tracking, telemetry & dynamics, Comt Space Res, 60-, chmn sci coun geod uses artificial satellites, 65-; chmn, Gordon Res Conf Chem & Physics Space, 63; trustee-at-large, Univ Corp Atmospheric Res, Colo, 64-68 & mem, Comt Nat Ctr Atmospheric Res Staff-Univ rels, 65-68, Mem, vpres & pres comns, Int Astron Union, 32-, voting rep, 52 & 55; deleg, Inter-Am Astrophys Cong, Mex, 42; mem comn 3, U S nat comt, Int Sci Radio Union, 49-61; Int Astron Fedn, 55-; mem working group satellite tracking & comput & chief investr proj optical tracking artificial earth satellites, Int Geophys Yr, 55-58, mem tech panel earth satellite prog & tech panel rocketry, 55-59. *Honors & Awards:* Donohue Medal; Smith Medal, Nat Acad Sci, 49; Am Astronaut Soc Award, 61; Space Pioneers Medallion, 68; Leonard Medal, Meteoritical Soc, 70; Kepler Medal, AAAS, 71; Henry Medal, Smithsonian Inst, 73; Gold Medal, Royal Arts Soc, Eng, 83. *Mem:* Nat Acad Sci; fel Am Astron Soc (vpres, 48-50); fel Am Astronaut Soc (vpres, 62-64); fel Am Geophys Union; fel Am Inst Aeronaut & Astronaut; Am Acad Arts & Sci; Am Philos Soc; Sigma Xi; AAAS; Am Meteorol Soc. *Res:* Photometry; comet discoveries and theory; colors of external galaxies; novae; meteor orbits; earth's upper atmosphere; stellar and pla. *Mailing Add:* Smithsonian Astrophys Observ 60 Garden St Cambridge MA 02138

WHIPPLE, GERALD HOWARD, b Calif, Feb 6, 23; m 47; c 5. MEDICINE. *Educ:* Harvard Univ, SB, 43; Univ Calif, MD, 46; Am Bd Internal Med, dipl, 60. *Prof Exp:* Instr, Sch Med, Boston Univ, 56-57, assoc, 57-60, from asst prof to assoc prof, 60-68; from assoc prof to prof med, Col Med, Univ Calif, Irvine, 71-78, chief cardiol div, 74-78; PROF CARDIOL MED & DIR, DIV CARDIOL, UNIV NEV, RENO, 78- *Concurrent Pos:* Res fel med, Harvard Univ, 53-56; vol asst med, Congenital Heart Clin, Children's Med Ctr, Boston, 54-56; physician in chg, EKG Lab, Univ Hosp, Boston Univ, 56-58, physician, Cardiac Care Univ, 65-68, assoc vis physician, 56-68, assoc mem, Evans Mem Res Found, 56-68, sect head clin cardiol res; consult cardiol, Congenital Heart Clin, Boston City Hosp, 58-59; res consult, Providence Vet Admin Hosp, RI, 60-68; res assoc, Mass Inst Technol, 63-66; fel coun clin cardiol, Am Heart Asn; heart coordr area VIII, Regional Med Prog Cancer, Heart & Stroke, 68-; physician & dir intensive cardiac care unit, Orange County Med Ctr, 69-70; chief med serv, Vet Admin Hosp, Long Beach, 70-74 & Reno, 81- *Mem:* Fel Am Col Physicians; fel Am Col Cardiol. *Res:* Academic cardiology; cardiology; cardiac arrhythmias; epidemiologic evaluation of acute ischemic heart disease and sudden death. *Mailing Add:* 1000 Locust St Reno NV 89520

WHIPPLE, ROYSON NEWTON, b Buffalo, NY, May 28, 12; m 35; c 2. FOOD TECHNOLOGY. *Educ:* Univ Mich, BS, 35; Cornell Univ, MS, 39. *Prof Exp:* Instr high sch, NY, 35-45; prof & head div food technol, State Univ NY Agr & Tech Col, Morrisville, 45-57, pres, 57-78; RETIRED. *Mem:* Inst Food Technologists. *Res:* Background needs for food technologists engaged in fruit and vegetable processing. *Mailing Add:* 330 Bahama Dr Indialantic FL 32903-3006

WHIRLOW, DONALD KENT, b Pittsburgh, Pa, May 2, 38; m 62; c 3. FLUID DYNAMICS, AERODYNAMICS. *Educ:* Carnegie Inst Technol, BS, 60, MS, 61, PhD(mech eng), 64. *Prof Exp:* Sr engr, 66-71, mgr fluid dynamics, 71-76, ADV ENGR, FLUID DYNAMICS RES, WESTINGHOUSE SCI & TECHNOL CTR, 76- *Mem:* Am Soc Mech Engrs; Sigma Xi. *Res:* Fluid dynamics and heat transfer; turbomachinery; acoustic flowmetering; transformers; transonic flow; unsteady aerodynamics; two-phase steam flow; finite element analysis. *Mailing Add:* 4040 W Benden Dr Murrysville PA 15668

WHISLER, FRANK DUANE, b Burton, WVa, Nov 20, 34; m 56; c 4. SOIL PHYSICS. *Educ:* Univ WVa, BS, 57, MS, 58; Univ Ill, PhD(soil physics), 64. *Prof Exp:* Soil scientist, Agr Res Serv, USDA, Ill, 58-69, soil scientist, Water Conserv Lab, 69-73; PROF AGRON & AGRONOMIST, MISS STATE UNIV, 73- *Honors & Awards:* Award, Soil Sci Soc Am, 63. *Mem:* Am Soc Agron; Soil Sci Soc Am; Sigma Xi. *Res:* Water movement into and through soils or other porous material from both a theoretical and experimental point of view. *Mailing Add:* Dept Agron Miss State Univ Box 5248 Mississippi State MS 39762

WHISLER, HOWARD CLINTON, b Oakland, Calif, Feb 4, 31; m 53; c 2. BOTANY, MICROBIOLOGY. *Educ:* Univ Calif, Berkeley, BSc, 54, PhD(bot), 61. *Prof Exp:* NATO fel, Univ Montpellier, 60-61; asst prof bot, McGill Univ, 61-63; from asst prof to assoc prof, 63-73, PROF BOT, UNIV WASH, 73- *Concurrent Pos:* NSF fel, Univ Geneva, 68-69. *Mem:* Mycol Soc Am; Soc Invert Path. *Res:* Development of the aquatic phycomycetes; insect microbiology. *Mailing Add:* Dept Bot Univ Wash Seattle WA 98195

WHISLER, WALTER WILLIAM, b Davenport, Iowa, Feb 9, 34; m 59; c 2. BIOCHEMISTRY, NEUROSURGERY. *Educ:* Augustana Col, AB, 55; Univ Ill, MD, 59, PhD(biochem), 69; Am Bd Neurol Surg, dipl, 67. *Prof Exp:* USPHS fel, 64-65; PROF BIOCHEM & NEUROSURG & CHMN DEPT NEUROSURG, RUSH MED COL, 70- *Concurrent Pos:* Attend neurosurgeon & chmn dept neurosurg, Presby-St Luke's Hosp, 70-; clin assoc prof, Univ Ill Col Med, 70- *Mem:* AMA; Am Asn Neurol Surg; Cong Neurol Surg; Int Soc Res Stereoencephalotomy. *Res:* Mechanisms of catecholomine oxidation; metabolism of the psychotomimetic amines; biochemistry of brain tumors. *Mailing Add:* Dept Neurosurg RSH-Presby St Lukes Med 1753 W Congress Pkwy Chicago IL 60612

WHISNANT, JACK PAGE, b Little Rock, Ark, Oct 26, 24; m 44; c 3. MEDICINE, NEUROLOGY. *Educ:* Univ Ark, BS, 48, MD, 51. *Prof Exp:* From instr to assoc prof, 56-69, PROF NEUROL, MAYO MED SCH, UNIV MINN, 69-, CHMN DEPT, 71- *Concurrent Pos:* Consult, Mayo Clin, 55-, head sect neurol, 63-71. *Mem:* Am Acad Neurol; Am Neurol Asn; Sigma Xi. *Res:* Clinical neurology, especially vascular diseases of the nervous system. *Mailing Add:* Mayo Clin Rochester MN 55901

WHISONANT, ROBERT CLYDE, b Columbia, SC, Apr 20, 41; m 63; c 2. STRATIGRAPHY, SEDIMENTOLOGY. *Educ:* Clemson Univ, BS, 63; Fla State Univ, MS, 65, PhD(geol), 67. *Prof Exp:* Petrol geologist, Humble Oil & Ref Co, 67-71; from asst prof to assoc prof geol, 71-81, PROF & CHMN GEOL, RADFORD UNIV, 81- *Concurrent Pos:* Consult, Humble Oil & Refining Co, 72 & Nat Geog Soc, 81; grantee, Geol Soc Am, 72, Petrol Res Fund, Am Chem Soc, 83-85 & Jeffres Mem Trust, 83-85, 86 & 89-91; mem, Va Oil & Gas Conserv Bd, 82-90; mem, Nat Asn Geol Teachers. *Mem:* Geol Soc Am; Am Asn Petrol Geol; Soc Econ Paleontologists & Mineralogists; Sigma Xi. *Res:* Paleozoic rocks of the southern Appalachians, specifically, sedimentary petrography and paleocurrent features; paleoenvironmental determinations; stratigraphic analysis. *Mailing Add:* Dept Geol Radford Univ Radford VA 24142

WHISSELL-BUECHY, DOROTHY Y E, b St Louis, Mo, Apr 12, 26; m 56; c 3. HUMAN GENETICS. *Educ:* Wellesley Col, AB, 48; Univ Tex Southwestern Med Sch, Dallas, MD, 56; Univ Calif, Berkeley, PhD(genetics), 68. *Prof Exp:* INSTR GENETICS, UNIV CALIF, BERKELEY, 64-, ASST RES GENETICIST, INST HUMAN DEVELOP, 69- *Concurrent Pos:* Instr, Univ Exten, 71-78. *Mem:* Behav Genetics Asn; Sigma Xi. *Res:* Assessment of health at various stages of development, especially middle age; genetics of olfaction and gustation; genetics of personality. *Mailing Add:* 1203 Tolman Hall Univ Calif Inst Human Develop Berkeley CA 94720

WHISTLER, DAVID PAUL, b Summit, NJ, June 15, 40; m 64; c 2. VERTEBRATE PALEONTOLOGY. *Educ:* Univ Calif, Riverside, BA, 63, MA, 65; Univ Calif, Berkeley, PhD(paleont), 69. *Prof Exp:* Mus scientist paleont, Univ Calif Mus Paleont, 67-69; asst prof geol, Tex Tech Univ, 69-70; CUR VERT PALEONT, NATURAL HIST MUS, LOS ANGELES COUNTY, 69- *Mem:* Soc Vert Paleont (secy & treas, 85-89); Paleont Soc; Geol Soc Am; Am Soc Mammalogists; Soc Econ Paleontologists & Mineralogists; Nat Asn Geol Teachers. *Res:* Evolution, taxonomy and biostratigraphy of smaller vertebrates, amphibians, reptiles rodents and insectivores in later Tertiary. *Mailing Add:* 900 Expos Blvd Los Angeles CA 90007

WHISTLER, ROY LESTER, b Morgantown, WVa, Mar 21, 12; m 35; c 1. CARBOHYDRATES. *Educ:* Heidelberg Col, BS, 34; Ohio State Univ, MS, 35; Iowa State Univ, PhD, 38. *Hon Degrees:* DSc, Heidelberg Univ, 53; DLitt, St Thomas Inst Advan Study, 82; DAgr, Purdue Univ, 85. *Prof Exp:* Instr chem, Iowa State Col, 35-38; fel, Nat Bur Stand, 38-40; head, Starch Struct Sect, USDA, 40-45 & Northern Regional Res Lab, Bur Agr Chem & Eng, 45-46; prof, 46-75, asst dept head, 48-60, chmn, Inst Agr Utilization Res, 61-78, Hillenbrand distinguished prof, 75-82, ADJ PROF, WHISTLER CTR CARBOHYDRATE RES, PURDUE UNIV, WEST LAFAYETTE, 82- *Concurrent Pos:* Vis lectr, Univ Witwatersrand, 61, & 85, Cape Town, 65, NZ & Australia, 67 & 74, Czech Acad Sci & Hungarian Acad Sci, 68, Taiwan, 70 & France & Poland, 75; Far East, Vladivostok Acad Sci lectr to USSR, 76; Nat Res Coun lectr, Brazil, 77; Welsh Found lectr, Tex, 77; guest lectr, Repub SAfrica, 77 & 85; guest, Polish Acad Sci, 78 & 85, People's Repub China, 84, 85. *Honors & Awards:* Hudson Award, Am Chem Soc, 60, Payen Award, 67; Annual Res Award, Japanese Tech Soc Starch, 67; Alsberg Schoch Award, Am Asn Cereal Chem, 70, Osborn Medal, 74; Ger Saare Medal, 74; Spencer Award, 75; Hendricks Award, USDA, 91. *Mem:* AAAS (pres, 73); Am Chem Soc; Am Soc Biochemists; Am Asn Cereal Chem (pres, 72-73); Am Inst Chemists (pres, 80-81); Int Carbohydrate Orgn (pres, 60); hon mem Argentine Chem Soc; Sigma Xi. *Res:* Chemistry and biochemistry of carbohydrates, both fundamental and practical; simple sugars; polysaccharides; carbohydrates in industry; basic and industrial medicine and foods. *Mailing Add:* Dept Biochem Purdue Univ Lafayette IN 47907

WHITACRE, CAROLINE C, b Cincinnati, Ohio, Nov 4, 49. NEUROIMMUNOLOGY, AUTOIMMUNE DISEASE. *Educ:* Ohio State Univ, PhD(med microbiol), 75. *Prof Exp:* ASST PROF MED, OHIO STATE UNIV, 81- *Mem:* AAAS; Am Soc Microbiol; Am Asn Immunologists. *Mailing Add:* Dept Med Microbiol & Immunol 5072 Graves Hall Ohio State Univ 333 W Tenth Ave Columbus OH 43210

WHITACRE, DAVID MARTIN, b Mariemont, Ohio, Dec 4, 43; m 66; c 2. ENVIRONMENTAL SCIENCES. *Educ:* Wilmington Col, Ohio, AB, 65; Ohio State Univ, MSc, 66; Univ Ariz, PhD(entom, zool), 69; Am Bd Toxicol, dipl, 82. *Prof Exp:* Asst prof biol, Univ PR, Mayaguez, 69-71; State of Ariz res grant entom, Univ Ariz, 71-72; proj mgr entom toxicol, Velsicol Chem Corp, 72-73, dir environ res, 73-76, dir environ sci & toxicol, 76-85; vpres res, 86, head agr res Switz, 86-87, VPRES, DEVELOP CROP, SANDOZ PROTECTION CORP, 87- *Concurrent Pos:* Post doctoral fel, Univ Ky, 72. *Mem:* Entom Soc Am; Am Chem Soc. *Res:* Metabolism of pesticides in plants and animals. *Mailing Add:* Sandoz Crop Protection Crop 1300 E Touhy Ave Des Plaines IL 60018

WHITACRE, GALE R(OBERT), b Salem, Ohio, June 23, 33; m 57; c 3. MECHANICAL ENGINEERING. *Educ:* Univ Cincinnati, BS, 56; Purdue Univ, MS, 58. *Prof Exp:* Res mech engr, 57-65, sr mech engr, 65-71, fel, 71-77, PRIN RESEARCHER, BATTELLE MEM INST, 77- *Mem:* Am Soc Mech Engrs. *Res:* Heat transfer; boundary layer flow; ablation; re-entry thermal analysis. *Mailing Add:* 3042 Midgard St Columbus OH 43202

WHITAKER, CLAY WESTERFIELD, b Greenville, Ky, Apr 17, 24; c 4. OTOLARYNGOLOGY. *Educ:* Berea Col, BA, 48; Western Reserve Univ, MD, 52; Am Bd Otolaryngol, dipl, 58. *Prof Exp:* Asst otolaryngologist, Highland View Hosp, Cleveland, Ohio, 52; asst to dir Ear, Nose & Throat Dept, St Luke's Hosp, Cleveland, 56; asst otolaryngologist, Univ Hosp & demonstr & clin instr otolaryngol, Med Sch, Western Reserve Univ, 56-61; from asst prof to assoc prof, 64-70, Prof surg, Sch Med, Univ Southern Calif, 70-; prof & dir ear, nose & throat res training, Los Angeles County-Univ Southern Calif Med Ctr, 64-83, chief physician & dir Dept Otolaryngol, 65-83; RETIRED. *Concurrent Pos:* Consult, Highland View Hosp, Cleveland, 56-61, Dept Pub Social Serv-Med Sci Div Hearing Aid Prog, Los Angeles County, 65-, Porterville State Hosp, Calif, 66-, tech adv comt on hearing aids, Dept Health Care Serv, State of Calif, 67-, Children's Hosp, Los Angeles, 67- & Hollywood Presby Hosp, Los Angeles, 67- *Mem:* AMA; Am Acad Ophthal & Otolaryngol. *Res:* Otorhinolaryngology; experimental surgery for device as practical prosthetic larynx; etiologic factors influencing Bell's Palsy. *Mailing Add:* 25 Briarcliff Dr Asheville NC 28804

WHITAKER, ELLIS HOBART, b Salem, Mass, Dec 2, 08; m 35, 66; c 2. PLANT PHYSIOLOGY. *Educ:* Worcester Polytech Inst, BS, 30; Cornell Univ, MS, 36, PhD(plant physiol), 49. *Prof Exp:* Engr, Gilbert & Barker Mfg, Mass, 30-31; teacher, Monson Acad, 33-35 & Westover Sch, Conn, 36-39; asst gen bot, Cornell Univ, 40-41; instr phys sci & gen biol, State Univ NY Col Oneonta, 41-48, asst prof chem, 48-54, assoc prof biol, 54-64; from assoc prof to prof biol, 64-76, EMER PROF BIOL, SOUTHEASTERN MASS UNIV, 76- *Concurrent Pos:* Consult, W W Norton Co, 79-80. *Mem:* Fel AAAS. *Res:* General biology; enzymes in insectivorous plants. *Mailing Add:* 32 Prospect St South Dartmouth MA 02748

WHITAKER, EWEN A, b London, Eng, June 22, 22; m 46; c 3. PLANETARY SCIENCES, HISTORY OF SELENOGRAPHY. *Educ:* Brit Inst Mech Eng, cert, 42. *Prof Exp:* Lab asst chem anal & phys testing, Siemens Bros & Co Ltd, 40-41, lab asst spectrochem anal, 41-49; sci asst, Royal Greenwich Observ, 49-53, from asst exp officer to exp officer, 53-58; res assoc, Yerkes Observ, Univ Chicago, 58-60; res fel, Lunar & Planetary Lab, Univ Ariz, 60-87, assoc res scientist, 82-87; RETIRED. *Concurrent Pos:* Co-experimenter, Ranger moonshots, 62-66; mem TV exp team, Surveyor Spacecraft, 64-68; mem site selection team, Orbiter 5, 67-68; mem, Apollo Orbital Sci Photo Team, 69-73; lectr, Lunar & Planetary Lab, Univ Ariz, 83-87. *Honors & Awards:* Walter Goodacre Prize & Gold Medal, Brit Astron Asn, 82. *Mem:* Int Astron Union; Am Astron Soc; fel Royal Astron Soc. *Res:* Study of moon, particularly of surface features and properties by earthbased telescopic observations and research and spacecraft; history of selenography; standardization of lunar nomenclature. *Mailing Add:* Lunar & Planetary Lab Univ Ariz Tucson AZ 85721

WHITAKER, JOHN O, JR, b Oneonta, NY, Apr 22, 35; m 57; c 3. VERTEBRATE ECOLOGY, MAMMALOGY. *Educ:* Cornell Univ, BS, 57, PhD(vert zool), 62. *Prof Exp:* From asst prof to assoc prof, 62-70, PROF LIFE SCI, IND STATE UNIV, TERRE HAUTE, 70- *Mem:* Fel AAAS; Am Soc Mammal; Ecol Soc Am. *Res:* Food habits, parasites, habitats and interrelations of species. *Mailing Add:* Dept Life Sci Ind State Univ Terre Haute IN 47809

WHITAKER, JOHN ROBERT, b Lubbock, Tex, Sept 13, 29; m 52; c 4. FOOD & SCIENCE TECHNOLOGY, AGRICULTURAL & FOOD CHEMISTRY. *Educ:* Berea Col, AB, 51; Ohio State Univ, PhD(agr biochem), 54. *Prof Exp:* Lab asst chem, Berea Col, 50-51; asst agr biochem, Ohio State Univ, 51-52, asst instr, 53-54, fel, 54; from instr to assoc prof, 56-67, assoc dean, Col Agr & Environ Sci, 75-83, 86-90, prof & chair, dept biochem & biophys, 84-85, PROF & BIOCHEMIST FOOD SCI & TECHNOL, UNIV CALIF, DAVIS, 67-, ASSOC DEAN AFFAIRS, COL AGR & ENVIRON SCI, 86- *Concurrent Pos:* Enzyme Group Med Res Lab, US Army, Army Chem Ctr, Md, 54-56; NIH spec fel, Northwestern Univ, 63-64; vis prof, Nat Univ, Mexico City, 68, Vet Col, Norway, 72, Univ Bristol, Eng, 72-73, Norwegian Food Res Inst, 79, Nat Polytech Inst, Mexico, 80, Univ Campinas, Brazil, 80, Metropolitan Univ Mexico, 80, People's Repub China, 85, 87; Fulbright fel, Univ Sao Paulo, Brazil, 85. *Honors & Awards:* William V Cruess Award, Inst Food Technologists, 73; spec fel, NIH, 72-73; fel, Inst Food Technologists; Fulbright-Hays Award, 85; Agr & Food Chem Res Award, Am Chem Soc, 85; Agr & Food Chem Distinguished Serv Award, Am Chem Soc, 89; Agr & Food Chem Fel Award, Am Chem Soc, 90. *Mem:* Am Chem Soc; Am Soc Biol Chemists; Am Soc Plant Physiologists. *Res:* Relationship between structure and function in enzymes; chemical and enzymatic modification of food proteins; naturally occurring enzyme inhibitors; over 190 publications on enzymes and enzyme inhibitors important to food science, including much work on beans, and on chemical and enzymatic modification of protein. *Mailing Add:* Dept Food Sci & Technol Univ Calif Davis CA 95616-8598

WHITAKER, JOHN SCOTT, b Oroville, Calif, Nov 10, 48; m 83; c 2. HIGH ENERGY PHYSICS. *Educ:* Univ Calif, Berkeley, BA, 70, PhD(physics), 76. *Prof Exp:* Asst, Lawrence Berkeley Lab, 76; sci assoc, Europ Coun Nuclear Res, 77-78; from asst prof to assoc prof physics, Mass Inst Technol, 78-85; ASSOC PROF PHYSICS, BOSTON UNIV, 85- *Mem:* Am Phys Soc; Am Asn Physics Teachers; Am Asn Univ Prof. *Res:* High energy particle physics. *Mailing Add:* Dept Physics Boston Univ 590 Commonwealth Ave Boston MA 02215

WHITAKER, MACK LEE, b Forest City, NC, Dec 2, 31; m 56; c 2. MATHEMATICS. *Educ:* Appalachian State Teachers Col, BS, 53, MA, 56; Fla State Univ, EdD, 61. *Prof Exp:* Teacher, Piedmont High Sch, 56-58; from assoc prof to prof math, Radford Col, 60-68, chmn dept, 62-66; assoc prof math educ, Auburn Univ, 68-69; PROF MATH, RADFORD UNIV, 69- *Mem:* Math Asn Am. *Res:* Mathematics education; abstract algebra; foundations of mathematics. *Mailing Add:* Dept Math Radford Univ Box 5803 Radford VA 24142

WHITAKER, ROBERT DALLAS, b Tampa, Fla, Mar 5, 33; m 60; c 3. INORGANIC CHEMISTRY. *Educ:* Univ Washington & Lee, BS, 55; Univ Fla, PhD(inorg chem), 59. *Prof Exp:* Asst prof chem, Wash & Lee Univ, 59-62; from asst prof to assoc prof, 62-74, PROF CHEM, UNIV SFLA, TAMPA, 74- *Mem:* Am Chem Soc. *Res:* Molecular addition compounds. *Mailing Add:* Dept Chem Univ SFla Tampa FL 33620

WHITAKER, SIDNEY HOPKINS, b Spring Valley, Ill, Apr 7, 40. GEOLOGY, HYDROGEOLOGY. *Educ:* Oberlin Col, BA, 62; Univ Ill, PhD(geol), 65. *Prof Exp:* Fel, Sask Res Coun, 65-67, asst res officer, 67-71, assoc res officer geol, 71-75, sr res scientist geol, 75-77; pres, Silverspoon Res & Consult Ltd, 77-81; SCI ADV, ATOMIC ENERGY CAN LTD, 81- *Concurrent Pos:* Mem geosci working group, Can Adv Comt Remote Sensing, Can Ctr Remote Sensing, 75-81; mem, Coord Group Geol Disposal, Nuclear Energy Agency, Orgn Econ Cooperation & Develop, 82-85 & tech subgroup, Stripa Proj, 87-92. *Mem:* Glaciol Soc; Geol Asn Can; Geol Soc Am; Sigma Xi. *Res:* Review Canadian geoscience programs for deep disposal of radioactive waste; techniques of groundwater exploration; hydrogeology; field geology; mapping; subsurface exploration and stratigraphy; lignite exploration. *Mailing Add:* Box 46 River Hills Manitoba MB R0E 1T0 Can

WHITAKER, THOMAS BURTON, b Asheville, NC, May 16, 39; m 60; c 3. AGRICULTURAL ENGINEERING. *Educ:* NC State Univ, BS, 62, MS, 64; Ohio State Univ, PhD(agr eng), 67. *Prof Exp:* From asst prof to assoc prof, 67-76, PROF BIOL & AGR ENG, NC STATE UNIV, 76-, AGR ENGR, AGR RES SERV, USDA, 67- *Honors & Awards:* Bailey Award, Am Peanut Res & Educ Asn, 76; Golden Peanut Res Award, Nat Peanut Coun, 80. *Mem:* Am Soc Agr Engrs; Am Peanut Res & Educ Soc; Sigma Xi; Asn Off Anal Chemists. *Res:* Quality control of agricultural commodities with main emphasis concerning the detection, control, and elimination of mycotoxins in food products. *Mailing Add:* Agr Res USDA NC State Univ Campus Box 7625 Raleigh NC 27695-7625

WHITAKER, THOMAS WALLACE, b Monrovia, Calif, Aug 13, 05; m 31; c 2. GENETICS. *Educ:* Univ Calif, BS, 27; Univ Va, MS, 29, PhD(genetics, cytol), 31. *Prof Exp:* Asst bot, Bussey Inst & Arnold Arboretum fel, Harvard Univ, 31-34; assoc prof bot, Agnes Scott Col, 34-36; from assoc geneticist to geneticist, Bur Plant Indust, 36-52, sr geneticist, Hort Crops Res Br, 52-56, prin geneticist & invest leader, Crops Res Div, 56-61, res geneticist & invest leader, 61-73, COLLABR CROPS, SCI & EDUC ADMIN-FED RES, WESTERN REGION, USDA, 73- *Concurrent Pos:* Guggenheim Mem Found fel, Wash Univ, 46-47 & Univ Calif, Davis, 59; consult, Peabody Archaeol Proj, Tehuacan, Mex, 62; lectr, Tulane Univ, 67, Univ Ariz, 71, Univ Ragistan, 72, Purdue Univ, 72 & Univ Ill, 73, Univ Nacional de la Plata, 80-81; ed, Hortsci, 69, Jour, 69-75; mem comt genetic vulnerability of major crops, Nat Res Coun, 71; res assoc, Univ Calif, San Diego, 73-; mem, Knight-Whitaker plant explor exped, 80 & leader, Whitaker-Providenti exped, 83. *Mem:* Fel Am Soc Hort Sci (pres, 74); hon life mem Can Soc Hort Sci; fel Torrey Bot Club; Soc Econ Bot (pres, 69); Am Plant Life Soc (exec secy, 55-); Am Phytopathological Soc; fel AAAS; Bot Soc Am; Soc Study Evolution; Am Soc Archeol. *Res:* Plant breeding; vegetable crops; genetics and cytology of Lactuca and the Cucurbitaceae; origin and domestication of cucurbits as related to cultural history; development of lettuce cultivars. *Mailing Add:* 2534 Ellentown Rd La Jolla CA 92037

WHITAKER, WILLIAM ARMSTRONG, b Little Rock, Ark, Jan 10, 36; m 57; c 2. COMPUTER SCIENCE, NATURAL SCIENCE. *Educ:* Tulane Univ, BS, 55, MS, 56; Univ Chicago, PhD(physics), 63. *Prof Exp:* Asst physics, Tulane Univ, 54-56; USAF, 56-, prof off res & develop, 57-68, chief high altitude group, Weapons Lab, 68-70, chief scientist, 70-72, mil asst res, Off Dir Defense Res & Eng, 73-75, spec asst to dir defense advan res proj agency, 75-80, TECH DIR FOR DIGITAL APPLNS, AIR FORCE ARMAMENT DEVELOP LAB, ELGIN AFB, 80- *Mem:* Am Phys Soc; Am Astron Soc; Am Geophys Union; Am Meteorol Soc; Inst Elec & Electronics Engrs. *Res:* Direction of advanced research, defense software management, common high order computer programming languages. *Mailing Add:* PO Box 3036 McLean VA 22103

WHITAKER-AZMITIA, PATRICIA MACK, b Winnipeg, Man, Can, Jan 24, 53; m 83; c 2. NEUROPHARMACOLOGY, DEVELOPMENTAL PHARMACOLOGY. *Educ:* Univ Man, BSc, 75; Univ Toronto, MSc, 76, PhD(pharmacol), 79. *Prof Exp:* Vis scientist biol psychiat, Clin Res Centre, Harrow, UK, 79-80; asst prof psychiat, Univ Toronto, 81-83; asst prof, 83-90, ASSOC PROF PSYCHIAT, STATE UNIV NY, STONY BROOK, 90- *Concurrent Pos:* Adj assoc prof, Dept Biol, NY Univ, 83-; assoc prof, Dept Psychol, State Univ NY, Stony Brook, 90-; chmn, Biomed Sect, NY Acad Sci, 91-92. *Mem:* Soc Neurosci; Int Soc Develop Neurosci; NY Acad Sci; Int Brain Res Orgn. *Res:* Developmental pharmacology; function and pharmacology of serotonin receptors; pharmacology of astroglial cells; role of neurotransmitters in brain development. *Mailing Add:* Dept Psychiat State Univ NY Stony Brook NY 11794-8101

WHITBY, KENNETH T(HOMAS), mechanical engineering; deceased, see previous edition for last biography

WHITBY, OWEN, b Luton, Eng, Feb 24, 42; Can & UK citizen; m 70. STATISTICS, ACTUARIAL METHODS. *Educ:* McMaster Univ, BSc Hons, 64; Stanford Univ, MS, 66, PhD(statist), 72. *Prof Exp:* From asst prof to assoc prof statist, Teachers Col, Columbia Univ, 71-78; Consult, 78-79; from actuarial assoc to secy, Swiss Re Holding, 79-80; secy, North Am Reins Corp, 80-84, asst vpres, 84-87, vpres, 87-89; vpres, 84-88, sr vpres, 88-90, EXEC VPRES, ATRIUM CORP, 90- *Mem:* AAAS; Am Statist Asn; Asn Comput Mach; Biomet Soc; Inst Math Statist; Soc Actuaries; Int Actuarial Asn. *Res:* Biostatistics; mathematical statistics; actuarial methods. *Mailing Add:* 106 Morningside Dr New York NY 10027-6026

WHITCOMB, CARL ERWIN, b Independence, Kans, Oct 26, 39; m 63; c 2. HORTICULTURE, PLANT ECOLOGY. *Educ:* Kans State Univ, BSA, 64; Iowa State Univ, MS, 66, PhD(hort, plant ecol), 69. *Prof Exp:* Asst prof ornamental hort, Univ Fla, 67-72; prof hort, Okla State Univ, 72-86; PRES, LACEBARK INC. *Mem:* Am Soc Hort Sci; Ecol Soc Am; Am Soc Agron; Soc Am Foresters. *Res:* Plant interactions in man-made or man-managed landscapes; production, establishment and maintenance of landscape plants. *Mailing Add:* Lacebark Inc Rte 5 Box 174 Stillwater OK 74074

WHITCOMB, DONALD LEROY, b Ilion, NY, Feb 1, 25; m 48; c 4. ANALYTICAL CHEMISTRY. *Educ:* Univ Rochester, BS, 52, PhD(phys chem), 56. *Prof Exp:* Res chemist, Res Labs, Eastman Kodak Co, NY, 46-53; mem tech staff, Bell Tel Labs, Pa, 56-63; mem tech staff, Microwave Tube Div, Hughes Aircraft Co, 63-64 & Space Systs Div, 64-66; chemist, Motorola Semiconductor Prod Inc, 66-70; chief chemist, Ariz State Dept Health, 70-81; chem supvr, WRF Opers, Ariz Pub Serv Co, 81-83; SR SCIENTIST, NUCLEAR PROCESS CHEM, ARIZ NUCLEAR POWER PROJ, 83- *Mem:* Am Chem Soc; Am Phys Soc. *Mailing Add:* 7420 N 179th Ave Waddell AZ 85355-9712

WHITCOMB, RICHARD T, b Evanston, Ill, Feb 21, 21. ENGINEERING ADMINISTRATION. *Educ:* Worcester Polytech Inst, BS, 43. *Hon Degrees:* DSc, Old Dominion Univ, 85. *Prof Exp:* Assoc, Transonic Aerodyn Br, Langley Res Ctr, NASA, 43-58, head br, 58-80, distinguished res assoc, Langley Res Ctr, 80-90; RETIRED. *Concurrent Pos:* Consult, 80- *Honors & Awards:* Sci Achievement Medal, NASA, 59; Sylvanus Albert Reed Award, Am Inst Aeronaut & Astronaut, 69; Nat Med Sci, Pres US, 73; Wright Bros Mem Trophy, Nat Aeronaut Asn, 74; Meritorious Serv Aviation Award, Nat Bus Aircraft Asn, 78. *Mem:* Nat Acad Eng; fel Am Inst Aeronaut & Astronaut. *Res:* Author of various publications. *Mailing Add:* 46 Lakeshore Dr Apt 1B Hampton VA 23666

WHITCOMB, STANLEY ERNEST, b Denver, Colo, Jan 23, 51; m 77. GRAVITATIONAL WAVE ASTRONOMY, PRECISION OPTICAL MEASUREMENTS. *Educ:* Calif Inst Technol, BS, 73; Univ Chicago, PhD(physics), 80. *Prof Exp:* Nat Needs fel, Univ Chicago, 80; asst prof physics, Calif Inst Technol, 80-85; res engr, Electronics Div, Northrop Corp, 85-89; sr systs specialist, Loral Electro-optical Systs, 89-91; DEP DIR, LIAO PROJ, CALIF INST TECHNOL, 91- *Mem:* Optical Soc Am. *Res:* Development of ultra-high precision optical interferometers for gravitational wave detection, gravitational wave astrophysics, precision optical measurements. *Mailing Add:* Calif Inst Technol MS 102-33 Pasadena CA 91125

WHITCOMB, WALTER HENRY, b Enid, Okla, Jan 26, 28; m 46; c 3. INTERNAL MEDICINE, NUCLEAR MEDICINE. *Educ:* Univ Okla, BA, 50, MD, 53; Am Bd Nuclear Med, dipl, 72. *Prof Exp:* Clin asst, Sch Med & chief res med, Med Ctr, Univ Okla, 56-57, chief exp med group, Radiobiol Lab, Univ Tex-USAF, 58-60; investr & instr radiobiol, Bionucleonics Dept, USAF Sch Aerospace Med, 60-62; from asst prof to assoc prof, 62-79, PROF MED, SCH MED, UNIV OKLA, 79-; DIR, VET ADMIN MED CTR, 81- *Concurrent Pos:* Res fel hemat, Univ Okla, 57-58; clin investr, Southwest Cancer Chemother Study Group, Vet Admin Hosp, 62-66, chief radioisotopes serv & hemat sect, 62-70, assoc chief staff res & educ, 63-67; from asst prof to assoc prof radiol, Sch Med, Univ Okla, 64-79, asst dean vet affairs, 70-79; chief staff, Vet Admin Hosp, 70-79; mgt support staff, Dept Med & Surg, Vet Admin Cent Off, 79-81. *Mem:* Am Col Physicians; Cent Soc Clin Res; Am Soc Hemat. *Res:* Biological effects of radiation; control of erythropoiesis; physiology of erythropoietin and erythropoietin inhibitor factors; operations research. *Mailing Add:* Vet Admin Hosp 921 NE 13th St Oklahoma City OK 73104

WHITCOMB, WILLARD HALL, b Manchester, NH, July 2, 15; m 43. TROPICAL ENTOMOLOGY. *Educ:* Bates Col, BS, 38; Agr & Mech Col, Tex, MS, 42; Cornell Univ, PhD(entom), 47. *Prof Exp:* Entomologist, Ministry Agr, Venezuela, 47-52 & Shell Co, Venezuela, 52-56; prof entom, Univ Ark, 56-67; prof entom, 67-84, EMER PROF ENTOM, BIG BEND HORT LAB, UNIV FLA, 84-; PRES, FITO TECHNICA FLORIDANA CORP, 84- *Mem:* Entom Soc Am; Int Orgn Biol Control; Int Palm Soc. *Res:* Biological control of arthropods; ecology and population dynamics; pest management; tropical entomology; cotton pests; Formicidae and Araneida in natural biological control; tropical entomology pheromone use; habitat management for biological control. *Mailing Add:* 4013 NW 39th Way Gainesville FL 32606

WHITE, ADDISON HUGHSON, b Clovis, Calif, Oct 13, 09; m 34; c 1. SCIENCE ADMINISTRATION. *Educ:* Occidental Col, AB, 30. *Prof Exp:* Mem staff, Bell Telephone Labs, 30-53, dir chem physics res, 53-58, exec dir res phys sci, 58-67; RETIRED. *Concurrent Pos:* Dir semiconductor res, Bell Telephone Labs, 54-58. *Mem:* Am Phys Soc. *Res:* Dielectric properties of solids; electron diffraction of thin films; thermionic emission. *Mailing Add:* 47 Woodland Ave Summit NJ 07901

WHITE, ALAN GEORGE CASTLE, bacteriology; deceased, see previous edition for last biography

WHITE, ALAN WAYNE, b Kingsport, Tenn, Feb 18, 54; m 75; c 3. BIOLOGICALLY ACTIVE MOLECULES. *Educ:* Univ Tenn, BS, 76; Harvard Univ, PhD(chem), 81. *Prof Exp:* Teaching fel org chem, Harvard Univ, 76-78, res asst, 78-81; from res chemist to sr res chemist, 81-88, PRIN RES CHEMIST, EASTMAN CHEM DIV, EASTMAN KODAK CO, 89- *Mem:* Am Chem Soc; Sigma Xi; Drug Info Asn. *Res:* Design and synthesis of novel organic compounds and polymers. *Mailing Add:* Eastman Chem Div PO Box 1972 Kingsport TN 37662

WHITE, ALAN WHITCOMB, b Norwood, Mass, March 15, 45; m 71; c 3. PHYTOPLANKTON ECOLOGY & PHYSIOLOGY. *Educ:* Col William & Mary, BS, 66; Harvard Univ, MA, 69, PhD(biol), 72. *Prof Exp:* Fel, Dept Microbiol & Chem, Hadassah Med Sch, Jerusalem, 72-73; res scientist, Can Dept Fisheries & Oceans, 73-86; marine sci adv, Sea Grant Prog, Woods Hole Oceanog Inst, 86-90; MARINE RESOURCES MGT SPECIALIST, NAT MARINE FISHERIES SERV, WOODS HOLE, MASS, 90- *Concurrent Pos:* Asst ed, J Fisheries Res Bd, Can, 78; vis scientist, Nansei Regional Fisheries Res Lab, Hiroshima, Japan, 82-83; leader, Nat Coord Ctr Harmful Algae Blooms; chmn, Third Int Conf Toxic Dinoflagellates, 85. *Mem:* Phycol Soc Am; Am Soc Limnol & Oceanog. *Res:* Ecology, physiology and toxicology of toxic dinoflagellate blooms and red tides; the fate of dinoflagellate toxins in the marine food web; consequences of toxins for fisheries resources. *Mailing Add:* Nat Marine Fisheries Serv NE Fisheries Ctr Woods Hole MA 02543

WHITE, ALBERT CORNELIUS, b Clearwater, Fla, July 17, 27; m 49; c 3. ENTOMOLOGY. *Educ:* Clemson Col, BS, 51; Univ Wis, MS, 53. *Prof Exp:* Res entomologist, Ortho Div, Chevron Chem Co, 53-69, int res specialist, 69-76; INDEPENDENT CONSULT, 78- *Concurrent Pos:* Entomologist, WFla Arthroprod Res Lab, 77-78; pres, Fla Entom Soc. *Mem:* Entom Soc Am; Mex Entom Soc. *Res:* Contract research and development on all pesticides, citrus production and pest control consultation; investigations of pesticide damage and environmental studies. *Mailing Add:* 817 W Fairbanks Ave Orlando FL 32804

WHITE, ALBERT GEORGE, JR, b Centralia, Ill, July 16, 40; m 67; c 4. MATHEMATICS. *Educ:* Southern Ill Univ, Edwardsville, BA, 62; Univ Mo, Columbia, MA, 64; St Louis Univ, PhD(math), 68. *Prof Exp:* Asst prof math, Ill State Univ, 67-69; assoc prof, 69-77, PROF MATH, ST BONAVENTURE UNIV, 77-, CHMN DEPT, 70- *Mem:* Am Math Soc; Math Asn Am. *Mailing Add:* Dept Math St Bonaventure Univ St Bonaventure NY 14778

WHITE, ALBERT M, b Derby, Conn, June 12, 26; m 55; c 6. CLINICAL PHARMACY, INSTITUTIONAL. *Educ:* Univ Conn, BS, 48, MS, 52. *Prof Exp:* Asst chem & pharm, Univ Conn, 50-52; from instr to assoc prof, 56-72, PROF PHARM, ALBANY COL PHARM, 72-, ASSOC DEAN, 74-,. *Concurrent Pos:* Clin assoc prof admin med, State Univ NY Upstate Med Ctr; assoc clin prof, Albany Vet Admin Hosp; consult, Whitney M Young Health Ctr & Villa Mary Immaculate Nursing Home & NY Dept Health, Parsons Family Care Ctr; state dir, Am Bd Dipl in Pharm. *Mem:* Am Pharmaceut Asn; Am Soc Hosp Pharmacists; Am Soc Consult Pharmacists; Am Asn Cols Pharm; Pres Pharm Leadership Soc. *Res:* Delivery of clinical pharmaceutical services to institutionalized and health center patients; practice of clinical pharmacy by community pharmacists; evaluation of practice experience programs. *Mailing Add:* Albany Col Pharm Albany NY 12208

WHITE, ALICE ELIZABETH, b Glen Ridge, NJ, Apr 5, 54; m 90. SOLID STATE PHYSICS, TECHNICAL MANAGEMENT. *Educ:* Middlebury Col, BA, 76; Harvard Univ, MA, 78, PhD(physics), 82. *Prof Exp:* Postdoctoral mem tech staff, 82-84, mem tech staff, 84-88, DEPT HEAD, AT&T BELL LABS, 88- *Concurrent Pos:* Consult, Lincoln Labs, Mass Inst Technol, 79-82. *Honors & Awards:* Maria Goeppert-Mayer Award, Am Phys Soc, 91. *Mem:* Am Phys Soc; Mat Res Soc. *Res:* High dose ion implantation for compound formation; defects in superconductors. *Mailing Add:* AT&T Bell Labs Rm 1D-360 600 Mountain Ave Murray Hill NJ 07974-2070

WHITE, ALVIN MURRAY, b New York, NY, June 21, 25; m 46; c 2. MATHEMATICS, SCIENCE EDUCATION. *Educ:* Columbia Univ, AB, 49; Univ Calif, Los Angeles, MA, 51; Stanford Univ, PhD, 61. *Prof Exp:* Asst prof math, Univ Santa Clara, 54-61; mem math res ctr, US Army, Wis, 61-62; assoc prof, 62-80, PROF MATH, HARVEY MUDD COL, 80- *Concurrent Pos:* Fac fel, Danforth Found, 75-76; vis scientist, Div Study & Res in Educ, Mass Inst Technol, 76; mem blue ribbon comt on writing standards, Calif Comn Post Sec Educ, 80; initiator & proj dir, New Interdisciplinary Holistic Approaches to Teaching/Learning, Fund Improv Post Sec Educ, 77-81; ed, Humanistic Math Network Newslet. *Mem:* Am Math Soc; Math Asn Am; Fedn Am Scientists; AAAS; Am Asn Univ Profs; Sigma Xi. *Res:* Function theoretical aspects of partial differential equations; quasiconformal mapping; nature of scientific creativity; nurture of scientific creativity; interdisciplinary teaching; science education. *Mailing Add:* Dept Math Harvey Mudd Col Claremont CA 91711-5990

WHITE, ANDREW MICHAEL, b Elyria, Ohio, Mar 17, 42; m 65; c 2. ICHTHYOLOGY, LIMNOLOGY. *Educ:* Ohio State Univ, BS, 66, PhD(zool), 73. *Prof Exp:* Res assoc parasitol, US Dept Interior Dis Invest Lab, 63-64; teaching assoc zool, Ohio State Univ, 65-70; trustee ecol, Cleveland Environ Res, 73-77; PROF BIOL, JOHN CARROLL UNIV, 70- *Concurrent Pos:* Res dir grant, USEPA Study Lake Erie Fisheries, 72-75; assoc cur fishes, Cleveland Mus Natural Hist, 75-; mem endangered species comt, Ohio Biol Surv, 76-; consult, Nat Comn Water Quality, 76-77; vpres ecologist, Environ Resource Asn, 77-; res dir grant, Fish Degradation NE Ohio Streams, Ohio Dept Natural Resources, 72-82, sea lamprey grant, US Fish & Wildlife Serv, 83-, endangered species grant, 85-; pres, Ohio Acad Sic, 87. *Mem:* Am Asn Parasitol; Am Fisheries Soc; Am Asn Limnol Oceanog; Int Asn Great Lakes Res. *Res:* Freshwater fisheries research, especially concerned with the Great Lakes ecosystem; major emphasis of ecology of non-game species in relation to spawning, growth, zoogeography, subspeciation and niche utilization. *Mailing Add:* Dept Biol John Carroll Univ University Heights Cleveland OH 44118

WHITE, ANDREW WILSON, JR, b Thomaston, Ga, Aug 1, 27; m 50; c 2. SOIL CONSERVATION, SOIL FERTILITY. *Educ:* Univ Ga, BS, 49, MS, 58, PhD, 69. *Prof Exp:* Soil scientist, Soil & Water Conserv Res Div, USDA, 51-61, res soil scientist, Southern Piedmont Conserv Res Ctr, 61-75, soil scientist, Southeastern Fruit & Nut Tree Lab, Sci & Educ Admin-Agr Res, 75-86; CONSULT, 86- *Mailing Add:* PO Box 386 Watkinsville GA 30677

WHITE, ARLYNN QUINTON, JR, b Norfolk, Va, Jan 17, 46; m 83; c 2. INVERTEBRATE PHYSIOLOGY, MARINE ECOLOGY. *Educ:* NC Wesleyan Col, BS, 68; Univ Va, MS, 72; Univ SC, PhD(biol), 76. *Prof Exp:* Instr biol, Univ SC, 76; vis asst prof, 76-78, asst prof biol & assoc dir environ ctr, 78-82, assoc prof biol, 83-88, PROF BIOL, JACKSONVILLE UNIV, 88- *Concurrent Pos:* Ed, newslett, Estuarine Res Soc, 80-82. *Mem:* AAAS; Am Soc Zoologists; Sigma Xi; Am Inst Biol Sci; Estuarine Res Soc; Nat Marine Educators Asn (treas, 86-90). *Res:* Behavioral physiology; locomotion in decapods; rhythmicity in marine invertebrates; effects of pollutants on behavior and physiology. *Mailing Add:* Dept Biol & Marine Sci Jacksonville Univ Jacksonville FL 32211

WHITE, ARLYNN QUINTON, JR, b Norfolk, Va, Jan 17, 46; m 83; c 2. BENTHIC MARINE ORGANISMS, WATER QUALITY. *Educ:* NC Wesleyan Col, BS, 68; Univ Va, MS, 72; Univ SC, PhD(biol), 76. *Prof Exp:* From asst prof to assoc prof, 76-88, PROF & CHAIR BIOL & MARINE SCI, JACKSONVILLE UNIV, 88- *Concurrent Pos:* Consult, A Quinton White & Assoc, 76-; assoc dir, Environ Ctr, 78- & Charter Marine Sci Ctr, 80- *Mem:* Nat Marine Educr Asn (treas, 86-90); Estuarine Res Fedn; Sigma Xi; AAAS; Am Inst Biol Sci. *Res:* Marine science-pollution and ecology of St Johns River, estuarines and artificial reefs; behavior and physiology of benthic marine organism; water quality-pollution impacts on marine organisms. *Mailing Add:* Dept Biol & Marine Sci Jacksonville Univ Jacksonville FL 32211

WHITE, ARNOLD ALLEN, b New York, NY, Oct 13, 23; m 53; c 5. BIOCHEMISTRY. *Educ:* Univ Iowa, AB, 47, MS, 49; Georgetown Univ, PhD(biochem), 54. *Prof Exp:* From instr to asst prof biochem, Georgetown Univ, 52-56; from asst prof to assoc prof, 56-77, PROF BIOCHEM, UNIV MO, COLUMBIA, 77-, INVESTR, DALTON RES CTR, 66- *Mem:* AAAS; Am Soc Biol Chemists; Sigma Xi. *Res:* signal translation mechanisms; cyclic nucleotide research. *Mailing Add:* Dalton Res Ctr Univ Mo Columbia MO 65211

WHITE, ARTHUR C, b Williamsburg, Ky, Aug 1, 25; m 49; c 3. INTERNAL MEDICINE, INFECTIOUS DISEASES. *Educ:* Univ Ky, BS; Harvard Univ, MD, 52. *Prof Exp:* Instr med, Vanderbilt Univ, 53-58; from instr to asst prof, Univ Louisville, 58-63; assoc prof, Med Col Ga, 63-67; PROF MED, SCH MED, IND UNIV, INDIANAPOLIS, 67- *Concurrent Pos:* Consult, Vet Admin Hosps, Louisville, Ky, 59-63 & Augusta, Ga, 63-67; drug efficacy study, Nat Acad Sci, 66. *Mem:* Am Soc Microbiol; Am Fedn Clin Res; Am Col Physicians; Infectious Dis Soc Am. *Res:* Staphylococcal epidemiology and immunology; immunoglobulins and their activity; immunology of gram negative infections; histamine release. *Mailing Add:* Infectious Dis Div Ind Univ Sch Med Indianapolis IN 46223

WHITE, ARTHUR THOMAS, II, b Orange, NJ, Oct 7, 39; m 61; c 2. MATHEMATICS. *Educ:* Oberlin Col, AB, 61; Mich State Univ, MS, 66, PhD(math), 69. *Prof Exp:* Actuarial trainee, Home Life Ins Co, 61-62; from asst prof to assoc prof, 69-79, PROF MATH, WESTERN MICH UNIV, 79- *Concurrent Pos:* Commun electronics officer, USAF, 62-65; asst & fel, Mich State Univ, 65-69; NSF grant, 73-74; vis prof, Royal Holloway Col, Univ London, 77-78; managing ed, J Graph Theory, 78-80; vis lectr, Clemson Univ, 79; vis fel, Wolfson Col, Oxford Univ & sr vis, Math Inst, 84-85; fac teaching fel, Western Mich Univ, 85-86; Clarke Benedict Williams Lectureship Math, Kalamazoo Col, 88. *Mem:* Am Math Soc; London Math Soc; Math Asn Am; Sigma Xi; Nat Coun Teachers Math. *Res:* Topological graph theory; imbedding of graphs; genus of graphs and of groups; block designs; symmetrical maps; change ringing; random topological graph theory; enumerative topological graph theory. *Mailing Add:* 2502 Law Ave Kalamazoo MI 49008

WHITE, AUGUSTUS AARON, III, b Memphis, Tenn, June 4, 36; m 74; c 3. ORTHOPEDIC SURGERY, BIOMEDICAL ENGINEERING. *Educ:* Brown Univ, BA, 57; Stanford Univ, MD, 61. *Hon Degrees:* Dr, Karolinska Inst, 69. *Prof Exp:* Intern, Univ Hosp, Ann Arbor, Mich, 61-62; from instr to assoc prof orthop surg, Sch Med, Yale Univ, 65-78, dir biomech res, Sect Orthop Surg, 73-78; PROF ORTHOP SURG, HARVARD MED SCH, 78-; ORTHOP SURGEON-IN-CHIEF, BETH ISRAEL HOSP, 78- *Concurrent Pos:* Chief resident orthop surg, Vet Admin Hosp, W Haven, Conn, 66; attend orthop surgeon, Yale-New Haven Hosp, 69-78; consult, Vet Admin Hosp, W Haven & Hill Health Ctr, New Haven, 69-78; mem bioeng res comt, Int Coun Sports & Phys Educ, 74; Am Brit Can traveling fel award, Am Orthop Asn, 75; mem adv coun, Nat Arthritis, Metab & Digestive Dis, 79-82; bd fel, Brown Univ, 81-82. *Mem:* Orthop Res Soc; Cervical Spine Res Soc (pres, 88); Int Soc Study Lumbar Spine; Am Acad Orthop Surgeons; Nat Med Asn. *Res:* Mechanical studies on the entire human spine designed to provide knowledge and technology applicable to clinical problems; development of an engineering system which will accelerate fracture healing. *Mailing Add:* Beth Israel Hosp 330 Brookline Ave Boston MA 02215

WHITE, BENJAMIN STEVEN, b Boston, MA, Sept 29, 45; m 66; c 2. STOCHASTIC PROCESSES, WAVE PROPAGATION. *Educ:* Mass Inst Technol, SB, 67; Univ Ariz, MA, 68; New York Univ, PhD(math), 74. *Prof Exp:* Vis mem, Courant Inst, New York Univ, 74-75; instr, appl math, Calif Inst Technol, 75-78; mem tech staff, Jet Propulsion Lab, 78-81; sr staff mathematician, 81-84, head, applied math group, 86-89, RES ASSOC, EXXON RES & ENG CO, 85- *Concurrent Pos:* Instr, Math Dept, New York Univ, 71-72; vpres, Perceptive Systs Inc, Pasadena, Calif, 81. *Mem:* Soc Indust & Appl Math; Am Math Soc; AAAS. *Res:* Applications of stochastic processes; stochastic differential equations; wave propagation in random media. *Mailing Add:* Exxon Res & Eng Co Rte 22 E Annandale NJ 08801

WHITE, BERNARD HENRY, b Chicago, Ill, Oct 15, 47; m 68; c 3. PHYSICAL CHEMISTRY, CHEMICAL ENGINEERING. *Educ:* Univ Cincinnati, BS, 69; Univ Wash, MS, 71; Univ Houston, PhD(phys chem), 76. *Prof Exp:* Res scientist, 76-78, sr scientist, 78-81, sect head, Exxon Res & Eng Co, 81-86, PRECIOUS METALS CATALYSTS, EXXON CO, USA, 86- *Concurrent Pos:* Robert A Welch res fel, dept chem, Univ Houston, 73-76. *Mem:* Am Chem Soc; Am Phys Soc. *Res:* Synthetic fuels, both liquids and gases, derived from coal and shale; chemistry of coal liquefaction and gasification; process designs for coal and shale; oil refining processes; reforming and hydrocracking. *Mailing Add:* Exxon Co USA PO Box 2812 Houston TX 77252

WHITE, BERNARD J, b Portland, Ore, Jan 8, 37; m 63; c 5. BIOCHEMISTRY. *Educ:* Univ Portland, BS, 58; Univ Ore, MA, 61, PhD(biochem), 63. *Prof Exp:* Asst prof chem, Loras Col, 63-68; asst prof, 68-74, ASSOC PROF BIOCHEM, IOWA STATE UNIV, 74- *Concurrent Pos:* Vis prof, Univ Md, 76-77. *Mem:* AAAS; Am Chem Soc. *Res:* Protein structure; biochemical evolution; enzymology. *Mailing Add:* 135 N Russell Ames IA 50010-5963

WHITE, BLANCHE BABETTE, b Cumberland, Md, July 25, 05. CHEMISTRY. *Educ:* Goucher Col, AB, 25; Univ Chicago, MS, 27. *Prof Exp:* Asst chem, Goucher Col, 25-26; chemist, Celanese Corp Am, 27-45, sect head cellulose derivatives, 46-56; head tech info, Charlotte Develop Labs, NC, 56-59; asst librn, Res Div, W R Grace & Co, 59-63, lit scientist, 63-70; chem lit consult, 70-80; RETIRED. *Mem:* Am Chem Soc. *Res:* Scientific information retrieval; cellulose chemistry. *Mailing Add:* 1316 Fenwick Lane Apt 813 Silver Spring MD 20910-3504

WHITE, BRIAN, b Brigg, Eng, Feb 19, 36; m 62; c 2. CARBONATE SEDIMENTOLOGY. *Educ:* Univ Wales, BSc, 63, PhD(geol), 66. *Prof Exp:* Fel, Dalhousie Univ, 66-68; from instr to asst prof, 68-73, assoc prof geol & chmn dept, 73-82, PROF GEOL, SMITH COL, 82- *Mem:* Soc Econ Paleontologists & Mineralogists; Geol Soc Am; Sigma Xi; Int Asn Sedimentologists. *Res:* Stratigraphy, sedimentary petrology and micropaleobotany of precambrian sedimentary rocks; onaternary carbonates; fossil coral reefs. *Mailing Add:* Dept Geol Smith Col Northampton MA 01063

WHITE, BRUCE LANGTON, b Wellington, NZ, Mar 2, 31; m 54; c 2. PHYSICS. *Educ:* New Zealand, BSc, 52; Univ London, DIC & PhD, 56. *Prof Exp:* Res fel, 56-59, res assoc, 59-60, from asst prof to assoc prof, 60-70, PROF PHYSICS, UNIV BC, 70- *Mem:* Am Phys Soc. *Res:* Experimental low energy nuclear physics; experimental cosmology and gravitation; Möbauer effect. *Mailing Add:* Dept Physics Univ BC Univ Campus 6224 Agr Rd Vancouver BC V6T 2A6 Can

WHITE, CHARLES A, JR, b San Diego, Calif, Aug 1, 22; m 60; c 3. OBSTETRICS & GYNECOLOGY. *Educ:* Colo Agr & Mech Col, DVM, 45; Univ Utah, MD, 55; Am Bd Obstet & Gynec, dipl, 64. *Prof Exp:* Pvt pract vet med, 45-51; intern, Salt Lake County Gen Hosp, 55-56; resident obstet & gynec, Dee Mem Hosp, Ogden, Utah, 56-57; resident, Univ Hosp, Univ Iowa, 59-61, assoc, Col Med, 61-62, from asst prof to prof, 62-74; prof obstet & gynec & chmn dept, WVa Univ, 74-80; PROF OBSTET & GYNEC & HEAD DEPT, SCH MED, LA STATE UNIV, 80- *Concurrent Pos:* Examr, Am Bd Obstet & Gynec, 71- *Mem:* AMA; Am Gynec & Obstet Soc; Am Col Obstet & Gynec; Soc Gynec Surgeons; Obstet & Gynec Travel Club; Am Col Surgeons. *Mailing Add:* Dept Obstet & Gynec La State Univ Med Sch 1542 Tulane Ave New Orleans LA 70112-2822

WHITE, CHARLES F(LOYD), b Columbus, NMex, Aug 4, 13; m 51; c 3. ELECTRONICS, SYSTEMS ENGINEERING. *Educ:* Univ Calif, BS, 35, MS, 38. *Prof Exp:* Electronics lab asst, Farnsworth TV Lab, Calif, 38; instr elec eng, Univ Calif, 38-39 & Yale Univ, 39-40; from physicist to elec engr, US Dept Navy, DC & Calif, 40-43, sect head spec electronic instrumentation develop, US Naval Res Lab, 46-53, consult & coordr radar systs res, 53-59, consult electronic eng, res & develop, 59-66, systs analyst naval anal staff, 66-68, electronics scientist, Satellite Commun Br, 69-75; CONSULT, 75- *Mem:* Sigma Xi; Inst Elec & Electronics Engrs. *Res:* Signal and data processing; electrical networks; technical writing. *Mailing Add:* 4216 Dorris Rd Irving TX 75038-3909

WHITE, CHARLES HENRY, b Birmingham, Ala, Mar 15, 43; m 65; c 1. DAIRY MICROBIOLOGY. *Educ:* Miss State Univ, BS, 65, MS, 69; Univ Mo, PhD(dairy microbiol), 71. *Prof Exp:* Sr food scientist, Archer Daniels Midland Co, 71-72; asst prof dairy microbiol, Univ Ga, 72-76; dir, qual assurance, Dean Foods Co, 76-80; prof dairy sci, La State Univ, 80-85; PROF & E W CUSTER CHAIR DAIRY SCI, MISS STATE UNIV, 85- *Concurrent Pos:* Dairy consult; partic, Comt to Revise Stand Methods for Exam Dairy Prod, 74- *Mem:* Am Dairy Sci Asn; Inst Food Technol; Int Asn Milk & Food Sanitarians; Nat Environ Health Asn; Cult Dairy Prod Inst. *Res:* Psychrotrophic bacteria and relationship with shelf-life of dairy products, including measurement of proteolytic activity of raw milk as well as determination of heat-stable protease from the psychrotrophs; diacetyl reductases. *Mailing Add:* Dairy Sci Dept Drawer Dd Miss State Univ Mississippi State MS 39762

WHITE, CHARLES RAYMOND, b Wabash, Ind, Dec 23, 33; c 1. OPERATIONS RESEARCH, INDUSTRIAL ENGINEERING. *Educ:* Purdue Univ, BSME, 55, MSIE, 57, PhD(opers res), 63. *Prof Exp:* Opers res analyst, Armour & Co, 63-66; ASSOC PROF INDUST ENG, AUBURN UNIV, 66- *Mem:* Inst Mgt Sci; Am Inst Indust Engrs; Sigma Xi. *Res:* Maintenance engineering and energy conservation. *Mailing Add:* Dept Indust Eng Auburn Univ Auburn AL 36830

WHITE, CHARLEY MONROE, b Rose Hill, Ill, Dec 9, 32; m 60; c 3. ECOLOGY, WILDLIFE BIOLOGY. *Educ:* Eastern Ill Univ, BS, 60; Purdue Univ, MS, 62, PhD(ecol), 68. *Prof Exp:* Asst prof, 66-71, assoc prof, 71-78, PROF BIOL, UNIV WIS-STEVENS POINT, 78- *Mem:* Wildlife Soc; Am Soc Mammal. *Res:* Productivity of White-tailed deer; population dynamics. *Mailing Add:* 1509 Treder Ave Stevens Point WI 54481

WHITE, CHRISTOPHER CLARKE, b Haverhill, Mass, June 24, 37. MATHEMATICS. *Educ:* Bowdoin Col, AB, 59; Miami Univ, MA, 63; Univ Ore, PhD(math), 67. *Prof Exp:* Asst prof math, Univ NH, 67-70; ASSOC PROF MATH, CASTLETON STATE COL, 70-, CHMN DEPT, 74- *Mem:* Sigma Xi. *Res:* Banach algebras; harmonic analysis. *Mailing Add:* Box 1554 Castleton VT 05735

WHITE, CLARK WOODY, b Rome, Ga, May 4, 40; m 70; c 2. PHYSICS. *Educ:* Mass Inst Technol, BS, 62; Duke Univ, PhD(physics), 67. *Prof Exp:* Mem tech staff physics, Bell Labs, 67-75; MEM RES STAFF PHYSICS, OAK RIDGE NAT LAB, 75- *Honors & Awards:* IR-100 Award, 83; Woody Award, Mat Res Soc, 84. *Mem:* Mat Res Soc (pres, 84); fel Am Phys Soc. *Res:* Solid state physics; ion-solid collisions; surface physics; ion implantation; laser annealing. *Mailing Add:* Solid State Div Oak Ridge Nat Lab Bldg 3137 MS-6057 Oak Ridge TN 37830

WHITE, CLAYTON M, b Afton, Wyo, Apr 19, 36; m 59; c 5. VERTEBRATE ZOOLOGY, ECOLOGY. *Educ:* Univ Utah, AB, 61, PhD(zool), 68. *Prof Exp:* Instr zool & cur birds, Univ Kans, 65-66; instr zool & res fel, Cornell Univ, 68-70; from asst prof to assoc prof, 70-78, PROF ZOOL, BRIGHAM YOUNG UNIV, 78- *Concurrent Pos:* Consult, Columbus Labs, Battelle Mem Inst, 72-, Bechtel Group, 80-; NAm coordr, Int Coun Bird Preservation, 81; mem adv comt, div polar progs, NSF, 82- *Honors & Awards:* Francis B Roberts Award, 68. *Mem:* AAAS; Am Ornith Union; Soc Syst Zool; Cooper Ornith Union; Wilson Ornith Soc. *Res:* Avian evolution and systematics; ecology of raptorial birds; impact of environmental pollution in avian populations. *Mailing Add:* 1146 S 300 Orem UT 84058

WHITE, COLIN, b Australia, Aug 25, 13; m 43; c 2. BIOMETRY. *Educ:* Univ Sydney, BSc, 35, MSc, 36, MB & BS, 40. *Prof Exp:* Intern, Sydney Hosp, Australia, 41; lectr physiol, Univ Sydney, 42; med officer, Australian Inst Anat, 43-46; lectr physiol, Univ Birmingham, 46-48; asst prof, Univ Pa, 48-50; lectr physiol, Univ Birmingham, 50-53; from asst prof to assoc prof, 53-62, PROF PUB HEALTH, SCH MED, YALE UNIV, 62- *Mem:* Am Statist Asn; Biomet Soc; Royal Statist Soc; fel Int Statist Inst; Sigma Xi. *Res:* Epidemiology of chronic diseases; vital statistics. *Mailing Add:* 107 Thornton St Hamden CT 06517

WHITE, DAVID, b Russia, Jan 14, 25; nat US; m 45; c 3. PHYSICAL CHEMISTRY. *Educ:* McGill Univ, BSc, 44; Univ Toronto, PhD(chem), 47. *Prof Exp:* Asst, Univ Toronto, 44-47; fel, Ohio State Univ, 48-50, asst dir cryogenic lab, 50-53; asst prof chem, Syracuse Univ, 53-54; from asst prof to prof, Ohio State Univ, 54-66; dir, Lab for Res Struct of Matter, 81-87, PROF CHEM & CHMN DEPT, UNIV PA, 66- *Concurrent Pos:* Vis prof, Technion & Weizmann Insts, Israel, 63-64; Fulbright fel & vis prof, Univ Kyoto & Univ Tokyo, Japan, 65; Nat Ctr Sci Res fel & vis prof, Inst Appl Quantum Mech, France, 74-75. *Mem:* Am Chem Soc; Sigma Xi; Am Phys Soc. *Res:* Low temperature thermodynamics and solid state nuclear magnetic resonance; molecular structure from Nuclear Magnetic Resonance studies of matrix isolated spectra; optical coherence and relaxation studies of solids at low temperatures. *Mailing Add:* Dept Chem Univ Pa Philadelphia PA 19104

WHITE, DAVID, b Boston, Mass, Apr 26, 36; m 59; c 3. MICROBIOLOGY. *Educ:* Brandeis Univ, AB, 58, PhD(biol), 65. *Prof Exp:* Res scientist, Exobiol Div, Ames Res Ctr, NASA, 63-65; res assoc microbial physiol, Med Sch, Univ Minn, 65-67; asst prof, 67-74, ASSOC PROF MICROBIOL, IND UNIV, BLOOMINGTON, 74-, ASSOC PROF GEOL, 80- *Concurrent Pos:* Res grants, Am Cancer Soc, 68-70, NSF, 68-70, 70-72. *Mem:* Am Soc Microbiol. *Res:* Microbial physiology; microbial development; myxobacteria. *Mailing Add:* Dept Biol Ind Univ Bloomington IN 47401

WHITE, DAVID ARCHER, b Philadelphia, Pa, Jan 22, 27; m 52; c 3. GEOLOGY. *Educ:* Dartmouth Col, BA, 50; Univ Minn, MS, 51, PhD(geol), 54. *Prof Exp:* Sr res adv. Exxpm Prod Res Co, 54-86; PVT CONSULT, 86- *Mem:* Geol Soc Am; Am Asn Petrol Geol. *Res:* Geology of the Mesabi range, Minnesota; geochemistry; stratigraphy; hydrocarbon assessment. *Mailing Add:* 3202 W Anderson Suite 208-148 Austin TX 78757

WHITE, DAVID CALVIN, b Sunnyside, Wash, Feb 18, 22; m 49, 66; c 2. ELECTRICAL POWER TECHNOLOGY. *Educ:* Stanford Univ, BS, 46, MS, 47, PhD, 49. *Prof Exp:* Elec engr, Kaiser Co, Inc, 42-45; assoc prof elec eng, Univ Fla, 49-52; from asst prof to prof, 52-62, dir, Energy Lab, 72-89, FORD PROF, MASS INST TECHNOL, 62-, DEP DIR, ENERGY LAB, 90- *Concurrent Pos:* Lectr, Univ London, 61; vis prof & consult, Purdue Univ, 64-68; sr adv & vis prof, Birla Inst Technol & Sci, Pilani, India, 68-70; coun mem, Univ Benin, Nigeria, 70-72; trustee, Lowell Technol Inst, 72-74; mem adv coun, Elec Power Inst, 80-87, chmn, 84-86; mem res coord panel, Gas Res Inst, 80-87, chmn, 84-86; mem corp, Woods Hole Oceanog Inst, 77-83. *Honors & Awards:* Westinghouse Award, Am Soc Eng Educ, 61. *Mem:* Nat Acad Eng; fel Am Acad Arts & Sci; Am Soc Eng Educ; fel Inst Elec & Electronics Engrs. *Res:* Energy supply and demand analysis; energy conversion devices and systems; research and development planning. *Mailing Add:* Mass Inst Technol Energy Lab One Amherst St Bldg E40 AC3-455 Cambridge MA 02139-4307

WHITE, DAVID CLEAVELAND, b Moline, Ill, May 18, 29; m 56; c 3. TOXICOLOGY, ANALYTICAL CHEMISTRY. *Educ:* Dartmouth Col, AB, 51; Tufts Univ, MD, 55; Rockefeller Univ, PhD(biochem), 62. *Prof Exp:* Intern, Univ Hosp, Univ Pa, 55-56, instr physiol, 56-58, res assoc med, 58; from asst prof to prof biochem, Univ Ky, 62-72; prof biol & assoc dir, Prog Med Sci, Fla State Univ, 72-86; PROF MICROBIOL & ECOL, UNIV TENN, 86-, DIR INST APPL MICROBIOL, 86-; DISTINGUISHED SCIENTIST, UNIV TENN KNOXVILLE/OAK RIDGE NAT LAB, 86- *Concurrent Pos:* Prof community health & family med, Med Sch, Univ Fla, Gainesville, 75-86; adj prof, Interdept Toxicol Prog, Med Ctr, Univ Ark, Little Rock & Nat Ctr Toxicol Res, Jefferson, 81-; vis prof, Am Soc Microbiol. *Honors & Awards:* P R Edwards Award, Am Soc Microbiol, 81; Sci & Tech Achievement Award, US Environ Protection Agency, 87. *Mem:* Am Soc Biol Chem; Am Soc Limnol & Oceanog; Am Diabetic Asn; Gulf Estuarine Res Soc; Soc Toxicol. *Res:* Microbial ecology of microbially influenced corrosion; toxicant biodegradation; groundwater bioremediation; antarctle benthic ecology; fermentation technology; application of analytical chemistry to microbiol system. *Mailing Add:* Inst Appl Microbiol 10515 Research Dr Suite 300 Knoxville TN 37932-2567

WHITE, DAVID EVANS, b Syracuse, NY, Dec 13, 32; m 52; c 4. FOREST ECONOMICS. *Educ:* State Univ NY Col Forestry, Syracuse Univ, BS, 59, MS, 60, PhD(econ), 65. *Prof Exp:* Forester, Crown-Zellerbach Corp, 60-61; instr forest econ, State Univ NY Col Forestry, Syracuse Univ, 61-64; from asst prof to assoc prof, 64-71, dir div forestry, 66-76, PROF FOREST ECON & POLICY, WVA UNIV, 71- *Mem:* Fel, Soc Am Foresters. *Res:* Forest resources policy and administration; multi-disciplinary studies in environmental decision-making; natural resources economics; land use planning. *Mailing Add:* Dept Forestry WVa Univ Box 6125 Morgantown WV 26506-6125

WHITE, DAVID GOVER, b Woodbury, NJ, Sept 21, 27; m 59. INORGANIC CHEMISTRY. *Educ:* Cornell Univ, BChE, 50; Harvard Univ, PhD(chem), 54. *Prof Exp:* From asst prof to assoc prof, 53-62, PROF CHEM, GEORGE WASHINGTON UNIV, 62- *Concurrent Pos:* NSF fel, 60. *Mem:* Am Chem Soc; Royal Soc Chem. *Res:* Organometallic chemistry; boron-nitrogen compounds; metal complexes. *Mailing Add:* Dept Chem George Washington Univ Washington DC 20052

WHITE, DAVID HALBERT, physical organic chemistry, chemical evolution; deceased, see previous edition for last biography

WHITE, DAVID HYWEL, b Cardiff, Wales, June 4, 31; m 54; c 2. EXPERIMENTAL PARTICLE PHYSICS. *Educ:* Univ Wales, BSc, 53; Univ Birmingham, PhD(physics), 56. *Prof Exp:* Res fel physics, Univ Birmingham, 56-58, asst lectr, 58-59; res assoc, Univ Pa, 59-61, asst prof, 61-64; assoc prof, 64-69, prof physics, Cornell Univ, 69-78; sr physicist, Brookhaven Nat Lab, 78-86; STAFF, LOS ALAMOS NAT LAB, 86- *Mem:* Am Phys Soc. *Res:* Experimental particle physics; weak interactions; strong interactions; electromagnetic interactions. *Mailing Add:* MP-4 H846 Los Alamos Nat Lab Los Alamos NM 36830

WHITE, DAVID RAYMOND, b Oak Park, Ill, Sept 20, 40; m 64; c 2. ORGANIC CHEMISTRY. *Educ:* St John's Univ, BA, 62; Univ Wis, PhD(org chem), 66. *Prof Exp:* Grant, NIH, 66-68; MEM STAFF, UPJOHN CO, 68- *Mem:* Am Chem Soc; The Chem Soc. *Res:* Biogenetic type synthesis; steroid reactions; new synthetic methods; prostaglandin synthesis; antibiotics chemistry. *Mailing Add:* 1815 Greenlawn Ave Kalamazoo MI 49007-4324

WHITE, DAVID SANFORD, b Ashburnham, Mass, Sept 16, 45; m 66; c 1. LIMNOLOGY, AQUATIC ENTOMOLOGY. *Educ:* DePauw Univ, AB & MS, 70; Univ Louisville, PhD (biol), 74. *Prof Exp:* Res biol ecol, Univ Okla, Biol Sta, 74-77; res scientist limnol, 77-80, ASSOC RES LIMNOLOGIST, GREAT LAKES RES DIV & ASST PROF NATURAL RESOURCES, UNIV MICH, 80- *Concurrent Pos:* Vis scientist, Ill Natural Hist Surv, 77. *Mem:* Entom Soc Am; Am Entom Soc; Ecol Soc Am; Sigma Xi. *Res:* Benthic ecology, the distribution and abundance of aquatic invertebrates in relation to sediment and water quality; ecology and systematics of riffle beetles. *Mailing Add:* Dept Biol Sci Murray State Univ Murray KY 42071

WHITE, DEAN KINCAID, b Tulsa, Okla, Aug 27, 44; m 67; c 1. DENTAL PATHOLOGY. *Educ:* Univ Okla, BS, 66; Univ Mo-Kansas City, DDS, 70; Ind Univ, MSD, 72; Am Bd Oral Path, dipl, 75. *Prof Exp:* Asst prof path, Sch Dent, Temple Univ, 72-77; PROF ORAL PATH & CHMN DEPT, COL DENT, UNIV KY, 77- *Concurrent Pos:* Consult, Vet Admin. *Mem:* Am Dent Asn; Am Acad Oral Path. *Res:* Clinical research in oral neoplasia. *Mailing Add:* Dept Oral Path Univ Ky Col Dent Lexington KY 40506

WHITE, DENIS NALDRETT, b Bristol, Eng, June 10, 16; m 38; c 4. MEDICINE. *Educ:* Cambridge Univ, BA, 37, MA, MB & BCh, 40, MD, 56; FRCP, FACP. *Prof Exp:* First asst med, London Hosp, 43-46; sr registr, Univ London, 46-48; from asst prof prof, 48-81, EMER PROF, QUEENS'S UNIV, ONT, 81- *Concurrent Pos:* Chief ed, Ultrasound in Med & Biol, 73- *Honors & Awards:* Pioneer Award, Am Inst Ultrasonics Med. *Mem:* Fel Am Col Physicians; Am Electroencephalog Soc; Am Acad Neurol; Can Soc Electroencephalog; Am Inst Ultrasonics Med (past pres); hon mem World Fedn Ultrasound Med; hon mem Japanese Soc Ultrasonics Med; hon mem Mex Asn Ultrasonics Med; hon mem Yugoslav Asn Soc Ultrasonics Med Biol; Am Neurol Asn; Japanese Soc Ultrasonics Med. *Res:* Neurology; medical ultrasonics; ultrasonic doppler techniques. *Mailing Add:* 230 Alwington Pl Kingston ON K7L 4P8 Can

WHITE, DONALD BENJAMIN, b Framingham, Mass, Feb 15, 30; m 53; c 6. ORNAMENTAL HORTICULTURE, GENETICS. *Educ:* Univ Mass, BS, 56; Iowa State Univ, PhD(hort genetics, breeding), 61. *Prof Exp:* Res assoc hort, Iowa State Univ, 56-59, res asst, 59-61; from asst prof to assoc prof, 61-69, PROF HORT, UNIV MINN, ST PAUL, 69-, PROF LANDSCAPE ARCHIT, 74- *Mem:* Am Soc Hort Sci; Am Soc Agron; Soil Sci Soc Am. *Res:* Physiology of cold acclimation and dwarfing of woody plants; breeding and genetics of grasses; physiology of chemical growth regulation of monocots. *Mailing Add:* Dept Hort Sci 305 Alderman Hall Univ Minn St Paul 1970 Folwell Ave St Paul MN 55108

WHITE, DONALD EDWARD, b Dinuba, Calif, May 7, 14; m 41; c 3. GEOLOGY. *Educ:* Stanford Univ, AB, 36; Princeton Univ, PhD(econ geol, petrol), 39. *Prof Exp:* Assoc geologist, Geol Surv Nfld, Can, 37-38; geologist, US Geol Surv, 39-63, res geologist, 63-81; RETIRED. *Concurrent Pos:* Asst chief mineral deposits br, US Geol Surv, DC, 58-60. *Honors & Awards:* Penrose Medal, Geol Soc Am, 84. *Mem:* Nat Acad Sci; fel Geol Soc Am; Fel Soc Econ Geologists (pres, 82); fel Mineral Soc Am; Geochem Soc. *Res:* Origin and geochemistry of thermal and mineral springs and their relations to volcanism and ore deposits; geothermal energy; origin and nature of ore-forming fluids; origin and characteristics of geysers; isotope geology of waters and associated rocks alteration; abnormal geothermal gradients conductive and convective. *Mailing Add:* 222 Blackburn Ave Menlo Park CA 94025

WHITE, DONALD GLENN, b Charleston, WVa, Mar 16, 46; m 68; c 2. PLANT PATHOLOGY. *Educ:* Marshall Univ, BA, 68, MS, 70; Ohio State Univ, PhD(plant path), 73. *Prof Exp:* Lectr plant path, Ohio State Univ, 73-74; ASST PROF PLANT PATH, UNIV ILL, URBANA, 74- *Mem:* Am Phytopath Soc; Crop Sci Soc Am; Am Soc Agron. *Res:* Fungal diseases of field crops; stalk rot, ear rot, storage molds of corn; mycotoxins. *Mailing Add:* Dept Agron Univ Ill N 425 Turner Urbana IL 61801

WHITE, DONALD HARVEY, b Berkeley, Calif, Apr 30, 31; m 53; c 5. NUCLEAR SPECTROSCOPY. *Educ:* Univ Calif, Berkeley, BA, 53; Cornell Univ, PhD(physics), 60. *Prof Exp:* Asst Cornell Univ, 53-57, DuPont Scholar, 58-59; res physicist, Lawrence Livermore Lab, Univ Calif, 60-71; PROF PHYSICS, WESTERN ORE STATE COL, 71- *Concurrent Pos:* Lectr, Univ Calif, Berkeley, 70; consult, Lawrence Livermore Lab, 71-; vis res physicist, Inst Laue-Langevin, Grenoble, France, 77-78, 84-85; fel, Minna-Heineman, 77-78; pres, Ore Acad Sci, 79-80. *Mem:* Am Phys Soc; Am Asn Physics Teachers. *Res:* Experimental nuclear physics; thermal-neutron capture processes; nuclear spectroscopy. *Mailing Add:* Div Natural Sci & Math Western Ore State Col Monmouth OR 97361

WHITE, DONALD HENRY, b W Monroe, La, May 9, 40. ENVIRONMENTAL CONTAMINANTS ON WILDLIFE & THEIR HABITATS. *Educ:* Northeast La Univ, BS, 70, MS, 71; Univ Ark, PhD(zool), 75. *Prof Exp:* Zoologist, US Fish & Wildlife Serv, Patuxent Wildlife Res Ctr, Laurel, Md, 74-76, res zoologist, Gulf Coast Field Sta, Victoria, Tex, 76-83, RES ZOOLOGIST, US FISH & WILDLIFE SERV, PATUXENT WILDLIFE RES CTR, SOUTHEAST RES STA, ATHENS, GA, 83-; ADJ ASSOC PROF, SCH FOREST RESOURCES, UNIV GA, ATHENS, 83- *Mem:* Am Ornithologists Union; Asn Field Ornithologists; Nat Wildlife Fedn. *Res:* Study the effects of environmental contaminants on wildlife and their habitats, primarily birds in aquatic habitats. *Mailing Add:* US Fish & Wildlife Serv Sch Forest Resources Univ Ga Athens GA 30602

WHITE, DONALD PERRY, b New York, NY, May 19, 16; m 46; c 3. FORESTRY, SOIL SCIENCE. *Educ:* State Univ NY, BS, 37; Univ Wis, MS, 40, PhD(forest soils), 51. *Prof Exp:* Asst soils, Univ Wis, 38-40; forester, US Indian Serv, 40-42 & 46-47; asst soils, Univ Wis, 47-48; from instr to asst prof silvicult, State Univ NY Col Forestry, Syracuse Univ, 48-56; assoc prof forestry, 57-64, mem adv bd forest sci, 59-65, prof, 65-80, EMER PROF, MICH STATE UNIV, 83- *Concurrent Pos:* Collabr, US Forest Serv, 59-64; comnr, NAm Foreign Soils Conf, 68-78. *Mem:* Soc Am Foresters; Soil Sci Soc Am; fel Am Soc Agron. *Res:* Forest soils and fertilization; forest hydrology; watershed management; herbicides; reforestation techniques; Christmas trees. *Mailing Add:* 2870 College Rd Holt MI 48842

WHITE, DONALD ROBERTSON, b Schenectady, NY, Sept 27, 24; m 47; c 4. OPTICAL PHYSICS. *Educ:* Union Col, BS, 48; Princeton Univ, MA, 50, PhD(physics), 51. *Prof Exp:* Res asst physics, Princeton Univ, 51-52; physicist, Gen Elec Co, 52-68, mgr optical physics br, 68-79, tech adminr, Corp Res & Develop Ctr, 79-85; RETIRED. *Concurrent Pos:* Coffin fel, 48; Adj prof, Rensselaer Polytech Inst, 60-65. *Mem:* Fel Am Phys Soc; Combustion Inst. *Res:* Shock tubes and shock wave phenomena; gaseous detonation; optically pumped lasers; light scattering sensors. *Mailing Add:* 16 Garnsey Rd Rexford NY 12148

WHITE, DWAIN MONTGOMERY, b Minneapolis, Minn, Feb 16, 31; m 56; c 4. ORGANIC CHEMISTRY, POLYMER CHEMISTRY. *Educ:* Univ Wis, BS, 53; Mass Inst Tech, PhD (chem), 56. *Prof Exp:* RES CHEMIST, RES & DEVELOP CTR, GEN ELEC CO, 56- *Concurrent Pos:* Coolidge fel, 88. *Mem:* Am Chem Soc; Sigma Xi. *Res:* Organic synthesis and structure determination; oxidative coupling reactions; heterocyclics; synthesis and reactions of polyphenylene oxides. *Mailing Add:* Gen Elec Res & Develop Ctr PO Box 8 Schenectady NY 12301

WHITE, EDMUND W(ILLIAM), b Philadelphia, Pa, July 8, 20; m 48; c 4. CHEMICAL ENGINEERING, FUELS CHEMISTRY. *Educ:* Columbia Univ, AB, 40, BS, 41, ChE, 42; Lehigh Univ, PhD(chem eng), 52; Am Univ, cert opers res, 71. *Prof Exp:* Chem engr, Westvaco Chlorine Prods Corp, 42-44; design engr, C L Mantell, 46-47; tech staff engr, Diamond Alkali Co, 47-49; asst chem eng & res asst heat transfer, Inst Res, Lehigh Univ, 49-50 & teach asst, 50-51; pilot plant engr & proj leader, Cities Serv Res & Develop Co, NJ, 51-56, staff engr, NY, 56-60, staff engr & budget officer, Cities Serv Athabasca, Inc, 60-65; oper res analyst, Naval Supply Systs Command, 65-66, FUEL RES ENGR, DAVID TAYLOR RES CTR, ANNAPOLIS LAB, 66- *Concurrent Pos:* Instr indust stoichoimetry, Drexel Univ, 56; chmn, Petrol Prods & Lubricants, 88-91; mem, Steering Comt, Int Asn Stability & Handling Liquid Fuels; chmn, Am Soc Testing & Mat. *Honors & Awards:* Award of Merit, Am Soc Testing & Mat. *Mem:* Am Chem Soc; Am Inst Chem Engrs; Sigma Xi; Am Soc Testing & Mat; Int Asn Stability & Handling Liquid Fuels. *Res:* Fuels storage and stability; ship fuels and fuel systems; fuel purification; liquid-solid and liquid-liquid separations; filters and filter/separators; synfuels properties and composition; fuels at low temperatures. *Mailing Add:* 908 Crest Park Dr Silver Spring MD 20903-1307

WHITE, EDWARD, b Florence, SC, Nov 23, 33; m 55; c 5. IMMUNOLOGY, ENDODONTICS. *Educ:* Emory Univ, DDS, 58; Med Univ SC, MS, 66; Univ Calif, Los Angeles, PhD(microbiol, immunol), 69. *Prof Exp:* Asst prof microbiol & immunol, Sch Dent, Univ Southern Calif, 69-72; chmn dept, 72-77, PROF ENDODONTICS, COL DENT MED, MED UNIV SC, 72- *Mem:* Transplantation Soc; Am Soc Microbiol. *Res:* Immunology of transplantation; etiology of dental pulpal disease. *Mailing Add:* 505 W Cheves St Florence SC 29501-4449

WHITE, EDWARD AUSTIN, b Brooklyn, NY, Nov 28, 15; m 48; c 2. BIOCHEMISTRY, NUTRITION. *Educ:* Fordham Univ, BS, 37, MS, 40, PhD(biochem), 46. *Prof Exp:* Instr analytical chem, Fordham Univ, 37-40; analytical res chemist, Calco Chem Co, NJ, 40-42 & Winthrop Chem Co, NY & DC, 42-43; prof biochem, Col Mt St Vincent, 43-46; actg head dept chem, Inst Appl Arts & Sci, NY, 46-47; chief chem gen lab, Japan, 47-50, adv med sci, 50-57, RUSS/GER TRANSLR, US DEPT ARMY, WASHINGTON, DC, 57- *Mem:* Fel AAAS; Am Chem Soc; fel Am Inst Chem. *Res:* Nutrition in animals; analytical methods; pharmaceuticals; scientific translation. *Mailing Add:* 5307 Sangamore Rd Bethesda MD 20816

WHITE, EDWARD JOHN, b Haverhill, Iowa, Jan 26, 32; m 59; c 5. ELECTRICAL ENGINEERING. *Educ:* Iowa State Univ, BS, 58; Univ Va, MEE, 62, DSc(elec eng), 66. *Prof Exp:* Electronic engr, US Govt, 58-59; instr, 59-65, lectr, 65-66, asst prof, 66-69, asst to dean, Sch Eng & Appl Sci, 74-77, actg asst dean, 76-77, asst dean, 77-79, asst chmn, Dept Elec Eng, Univ Va, 80-86; ASSOC DEAN & RES PROF ELEC ENG, VANDERBILT UNIV, 87- *Mem:* Inst Elec & Electronics Engrs. *Res:* Computer graphics; computer aided circuit design. *Mailing Add:* 510 Clementis Dr Nashville TN 37205

WHITE, EDWARD LEWIS, b Boston, Mass, Jan 8, 47; m 70. NEUROANATOMY. *Educ:* Clark Univ, AB, 68; Georgetown Univ, PhD(anat), 72. *Prof Exp:* Premier asst anat, Inst Anat Normale, Switz, 73-75; asst prof, 75-81, ASSOC PROF ANAT, SCH MED, BOSTON UNIV, 81- *Mem:* Am Asn Anatomists. *Res:* Ultrastructure and synaptic organization in mammalian central nervous systems. *Mailing Add:* Three Richfield Park Boston MA 02125

WHITE, EDWARD RODERICK, physical chemistry, chromatography, for more information see previous edition

WHITE, EDWIN HENRY, b Gouverneur, NY, Dec 22, 37; m 61; c 3. FORESTRY, SOIL SCIENCE. *Educ:* State Univ NY Col Forestry, Syracuse Univ, BS, 62, MS, 64; Auburn Univ, PhD(soils), 69. *Prof Exp:* Technician forest soils, State Univ NY Col Forestry, Syracuse Univ, 61-62; instr forestry, Auburn Univ, 64-65, res asst, 65-68; fel forestry & soils, Univ Fla, 68-69; res soil scientist, US Forest Serv, Miss, 69-70; asst prof forestry, Univ Ky, 70-74; assoc prof forest resources, Univ Minn, 74-78, prof, 78-80; PROF FOREST RESOURCES, COL ENVIRON SCI & FORESTRY, SCH FORESTRY, STATE UNIV NY, SYRACUSE, 80- *Mem:* Soil Sci Soc Am; Soil Conserv Soc Am; fel Soc Am Foresters. *Res:* Forest soils and silviculture; soil-site-species relationships; tree planting research; acidic deposition. *Mailing Add:* Fac Forestry Col Environ Sci & Forestry Syracuse NY 13210

WHITE, ELIZABETH LLOYD, b Norfolk, Va, Sept 28, 16. EXPERIMENTAL EMBRYOLOGY, MOLECULAR BIOLOGY. *Educ:* Goucher Col, AB, 37; Bryn Mawr Col, MA, 38, PhD(embryol), 47. *Prof Exp:* Researcher, Wistar Inst, Univ Pa, 40-42; chemist, Res Dept, Gen Refractories Co, 42-46; instr zool, Wash Univ, 47-49; from asst prof to assoc prof zool, 49-61, prof biol, 61-78, EMER PROF BIOL, WHEATON COL, MASS, 78- *Concurrent Pos:* NSF fel hist of sci, Johns Hopkins Univ & Cambridge Univ, 58-59; NSF fel & vis prof biol, Johns Hopkins Univ, 66-67, sr res assoc, 73-74. *Mem:* Fel AAAS; Soc Develop Biol; Am Asn Anat. *Res:* Experimental embryology of teleosts; history of science; specialized nuclear RNA in chick limb-bud differentiation. *Mailing Add:* Fairhaven Health Ctr 7200 Third Ave No 0613 Sykesville MD 21784

WHITE, ELIZABETH LOCZI, b McKees Rocks, Pa, Mar 9, 36; m 59; c 2. MATERIALS SCIENCE ENGINEERING. *Educ:* Univ Pittsburgh, BS, 58; Pa State Univ, MS, 69, PhD(civil eng), 75. *Prof Exp:* Civil engr II highway design, Pa Dept Transp, 57, civil engr IV bridge design, 58-59; res assoc, 75-83, SR RES ASSOC, STORMWATER MODELING & ENG PROBS CARBONATE ROCKS, PA STATE UNIV, 83- *Concurrent Pos:* Anna Rose Hawkes fel Award, Am Asn Univ Women, 74-75. *Mem:* Am Soc Civil Eng; Am Asn Univ Women; Grad Women Sci; Nat Soc Prof Engrs. *Res:* Surface water hydrology; sediment transport, soil properties and flood frequency statistics; chemistry and properties of high temperature cements; geothermal wells and nuclear waste disposal; engineering problems in carbonate rock terrains; computer application to civil engineering; kinetic modeling of cement and cementitious materials; storm water management. *Mailing Add:* Dept Civil Eng 212 Sackett Bldg University Park PA 16802

WHITE, EMIL HENRY, b Akron, Ohio, Aug 17, 26. ORGANIC CHEMISTRY. *Educ:* Univ Akron, BS, 47; Purdue Univ, MS, 48, PhD(chem), 50. *Prof Exp:* Fel, Univ Chicago, 50-51 & Harvard Univ, 51-52; instr org chem, Yale Univ, 52-56; from asst prof to prof org chem, 57-80, D MEAD JOHNSON PROF CHEM, JOHNS HOPKINS UNIV, 80- *Concurrent Pos:* Guggenheim fel, 58-59; NIH sr fel, 65-66 & 72-73. *Mem:* AAAS; Am Chem Soc. *Res:* Mechanism of reaction in organic chemistry; active site mapping of enzymes; chemiluminescence and bioluminescence; deamination reactions. *Mailing Add:* Dept Chem Johns Hopkins Univ Baltimore MD 21218-2680

WHITE, EUGENE L, b Savannah, Ga, Oct 14, 51. COMPUTER PROGRAMMER. *Educ:* Savannah State Col, BSEE, 73. *Prof Exp:* Test engr, automatic test equip, 73-80, test engr, Engr Planning Group, 80-82, SR MFG ENGR, MFG SYST & TECHNOL, WESTINGHOUSE, BALTIMORE, MD, 82- *Mem:* Inst Elec & Electronics Engrs. *Res:* Try to develop new ways with work stations for hardware and software, information systems, data base and also robonic applications, all dealing with computers. *Mailing Add:* 9200 Berger Rd Ms6145c Columbia MD 21045

WHITE, EUGENE WILBERT, b Indiana, Pa, Jan 23, 33; m 52; c 3. INSTRUMENTATION, MINERALOGY. *Educ:* Pa State Univ, BS, 55, MS, 58, PhD(solid state tech), 65. *Prof Exp:* Res asst mineral, Pa State Univ, 55-59; head, X-ray Diffraction Applns Lab, Picker X-ray Co, Ohio, 59-61; design engr, Tem-Pres Res, Inc, Pa, 61; res asst electron microprobe, 62-65, from asst prof to assoc prof solid state sci, 65-74, PROF SOLID STATE SCI, MAT RES LAB, PA STATE UNIV, 74- *Mem:* Am Crystallog Asn; Sigma Xi. *Res:* Electron microprobe research; x-ray spectroscopy. *Mailing Add:* RD 1 Box 182 Rossiter PA 15772

WHITE, FRANKLIN ESTABROOK, b Denver, Colo, Mar 26, 22; m 44; c 2. INSTRUMENTATION. *Educ:* Univ Denver, BS, 48, MA, 51; Univ Mich, MS, 55. *Prof Exp:* Res assoc atmospheric infrared studies, Univ Denver, 49-51; infrared instrumentation, Univ Mich, 51-55; teacher, 59-62, res physicist, 55-83, SR RES PHYSICIST, UNIV DENVER, 83- *Concurrent Pos:* Consult, Air Force Opers Anal Off, 56-71. *Mem:* Am Phys Soc; Sigma Xi. *Res:* Atmospheric infrared absorption; balloon and rocket instrumentation; radio propagation and telemetry; operations analysis; electronics. *Mailing Add:* 4801 S Ogden St Englewood CO 80110

WHITE, FRANKLIN HENRY, b Alton, Ill, Feb 11, 19; m 47; c 2. VETERINARY MICROBIOLOGY. *Educ:* Shurtleff Col, BS, 42; Univ Ill, MS, 48, PhD(bact), 55; Am Bd Med Microbiol, cert pub health & med lab microbiol, 74. *Prof Exp:* Bacteriologist, State Dept Pub Health, Ill, 46-49; instr bact, Col Vet Med, Univ Ill, 49-55; from asst bacteriologist to assoc bacteriologist, Univ Fla, 55-67, assoc prof bact, 61-67, prof bact & bacteriologist, 67-85, EMER PROF BACT, UNIV FLA, 85- *Mem:* Am Soc Microbiol; Conf Res Workers Animal Dis; Wildlife Dis Asn; US Animal Health Asn. *Res:* Pathogenic microbiology and immunology; leptospirosis; vibriosis; wildlife diseases; epizootiology. *Mailing Add:* 3525 NW 12th Ave Gainesville FL 32605

WHITE, FRED D(ONALD), b Charleroi, Pa, Oct 11, 18; m 43; c 3. METEOROLOGY. *Educ:* Miami Univ, Ohio, AB, 41; Univ Wis, PhD(meteorol), 63. *Prof Exp:* Meteorologist, USAF, 41-46, US Weather Bur, 46-58; prog dir meteorol, NSF, 58-66, head atmospheric sci, 66-76; CONSULT, 76- *Concurrent Pos:* Staff officer, Nat Res Coun, Nat Acad Sci, 78-90; ed, Newslett, Am Meteorol Soc, 80-87. *Mem:* Fel Am Meteorol Soc; fel Am Geophys Union; fel AAAS. *Res:* Research and administration in all fields of atmospheric sciences. *Mailing Add:* 3631 N Harrison St Arlington VA 22207

WHITE, FRED G, b Spanish Fork, Utah, Jan 19, 28; m 54; c 7. PLANT BIOCHEMISTRY. *Educ:* Brigham Young Univ, BS, 52, MS, 56; Univ Calif, PhD(biochem), 61. *Prof Exp:* From asst prof to assoc prof, 61-72, PROF CHEM, BRIGHAM YOUNG UNIV, 72- *Concurrent Pos:* Estab investr, Am Heart Asn, 65-70. *Mem:* AAAS; Am Chem Soc; Am Inst Biol Sci; NY Acad Sci; Sigma Xi. *Res:* Mechanisms of enzymes and co-enzymes; biological nitrogen fixation; plant growth regulators; plant biochemistry; plant genetic engineering. *Mailing Add:* 10285 SW 139th Pl Miami FL 33186

WHITE, FRED NEWTON, b Yelgar, La, June 17, 27; m 51. CARDIOVASCULAR PHYSIOLOGY, COMPARATIVE PHYSIOLOGY. *Educ:* Univ Houston, BS, 50, MS, 51; Univ Ill, PhD(zool), 55. *Prof Exp:* Asst prof biol, Univ Houston, 55-57; asst prof exp med, Southwestern Med Sch, Univ Tex, 58-59 & 62-63; assoc prof physiol, Am Univ Beirut, 59-62; prof physiol, Sch Med, Univ Calif, Los Angeles, 63-76; PROF PHYSIOL & DIR PHYSIOL RES LAB, SCRIPPS INST OCEANOG, UNIV CALIF, SAN DIEGO, 76- *Concurrent Pos:* Am Physiol Soc cardiovasc training fels, 57 & 58. *Mem:* Am Physiol Soc; Fedn Am Socs Exp Biol; Soc Exp Biol & Med. *Res:* Control of renin secretion; peripheral circulation; comparative aspects of vertebrate circulation; environmental physiology. *Mailing Add:* PO Box 633 Fredericksburg TX 78624

WHITE, FREDERICK ANDREW, b Detroit, Mich, Mar 11, 18; m 42; c 3. MASS SPECTROMETRY. *Educ:* Wayne State Univ, BS, 40; Univ Mich, MS, 41; Univ Wis, PhD, 59. *Prof Exp:* Res asst physics, Manhattan Proj, Univ Rochester, 43-46; res assoc, Gen Elec Co, 47-62; prof nuclear eng & eng sci, Rensselaer Polytech Inst, 62-81; CONSULT, NASA, 75- *Concurrent Pos:* Sci adv, Rochester Gas & Elec, 80-; adj prof physics, State Univ NY, Albany, 81-88. *Mem:* Acoust Soc Am; Am Chem Soc; Am Phys Soc; Am Soc Mass Spectrometry; Inst Elec & Electronics Engrs; Optical Soc Am. *Res:* Application of isotopic abundance measurements in the physical sciences; acoustics; mass spectrometry; industrial research. *Mailing Add:* 2456 Hilltop Rd Schenectady NY 12309

WHITE, FREDERICK HOWARD, JR, b Washington, DC, Jan 19, 26. PROTEIN CHEMISTRY. *Educ:* Univ Va, BS, 49; Univ Md, MS, 52; Univ Wis, PhD(biochem), 57. *Prof Exp:* Asst chem, Univ Md, 51-52; chemist, Nat Heart Inst, 52-53; asst biochem, Univ Wis, 53-56; chemist, Lab Biochem, Nat Heart, Lung & Blood Inst, 56-75 & Lab Cell Biol, 75-83; RETIRED. *Concurrent Pos:* Vis fel, Australian Nat Univ, Canberra, 84-87; courtesy res scientist, Fla State Univ, 87- *Mem:* Am Soc Biol Chem; Radiation Res Soc. *Res:* Chemistry of sulfur in proteins; protein conformation; radiolysis of proteins; lysozyme, milk proteins. *Mailing Add:* Chem Dept Fla State Univ Tallahassee FL 32306

WHITE, FREDRIC PAUL, b New York, NY, July 24, 42; m 64; c 1. BIOCHEMISTRY, NEUROSCIENCES. *Educ:* Purdue Univ, BChE, 64; Ind Univ, PhD(biol chem), 71. *Prof Exp:* Chem engr, E I du Pont de Nemours & Co, Inc, 64-66; biochem trainee, NIH, 67-71; res scientist neurochem, Med Sch, Univ Colo, 72-73; vis asst prof neurophys, Ind State Univ, 73-74; asst prof, 74-77, ASSOC PROF BIOCHEM, MEM UNIV NFLD, 78- *Concurrent Pos:* Vis scientist, Med Res Coun, 81-82. *Mem:* Am Soc Neurochem; Soc Neurosci. *Res:* Cellular physiology of cerebral endothelial cells and pericytes; synthesis of the stress protein traumin by cells in response to trauma. *Mailing Add:* Col Vet Med Wash State Univ Pullman WA 99164

WHITE, GEORGE CHARLES, JR, b Coatesville, Pa, Sept 14, 18; m 43; c 2. PHYSICS. *Educ:* Villanova Univ, BS, 39. *Prof Exp:* Physicist, Physics & Math Br, Pitman-Dunn Res Labs, 46-52, chief, 53-57, asst dir, Physics Res Lab, 57-58, asst spec missions off, 59-60, tech asst, Inst Res, 60-64, tech asst, Off of Dir, Pitman-Dunn Res Labs, 65-71, dep dir, Pitman-Dunn Res Labs, Frankford Arsenal, US Dept Army, 72-77; CONSULT, PHYS SCI & ADMIN RES, 77- *Mem:* Am Phys Soc; Am Nuclear Soc; Sigma Xi. *Res:* Nuclear, radiation and solid state physics; materials research. *Mailing Add:* 705 Avondale Rd Philadelphia PA 19118

WHITE, GEORGE MATTHEWS, b Salt Lake City, Utah, Dec 7, 41; m; c 4. COMMUNICATION SCIENCE. *Educ:* Mich State Univ, BS, 64; Univ Ore, PhD(chem phys), 68. *Prof Exp:* NIH res fel, Dept Comput Sci, Stanford Univ, 68-70; res scientist comput sci, Xerox Palo Alto Res Ctr, 70-77; res scientist, ITT Defence Commun Div, 77-80; mgr, Speech Recognition Unit, Auricle Inc, 80-81; pres & chmn bd, Texticon, Inc, 81-85; chmn bd, Koala Technologies, 85-88; CONSULT, APPLE COMPUT ENGR, 88- *Mem:* Inst Elec & Electronics Engrs; Asn Comput Mach; Acoust Soc Am; Pattern Recognition Soc. *Res:* Pattern recognition, machine perception, artificial intelligence, automatic speech recognition, signal processing and perceptual psychology. *Mailing Add:* Apple Comput Inc 70 Valley Green M/S 65f Cupertino CA 95014

WHITE, GEORGE MICHAEL, b Toronto, Ont, June 14, 39; m 64; c 3. COMPUTER SCIENCE. *Educ:* Univ Toronto, BASc, 61; Univ Alta, MSc, 65; Univ Calgary, PhD, 68. *Prof Exp:* Systs engr, CAE Electronics, 61-62; Irish Govt fel, Univ Col, Dublin, 68-70; ASST PROF COMPUT SCI, UNIV OTTAWA, 70- *Mem:* Asn Comput Mach; Can Info Processing Soc. *Mailing Add:* Dept Comput Sci Univ Ottawa Ottawa ON K1N 6N5 Can

WHITE, GEORGE NICHOLS, JR, b Concord, Mass, July 1, 19; m 48; c 3. APPLIED MATHEMATICS. *Educ:* Harvard Univ, BS, 41; Brown Univ, MS, 48, PhD(appl math), 50. *Prof Exp:* Trainee, Phys Test Lab, T Mason Co, 41-42; technician radar, US Civil Serv, 42-45; technician electronics, Oceanog Inst, Woods Hole, 45-46; res assoc appl math, Brown Univ, 48-50; mem staff appl math, Los Alamos Nat Lab, 50-85; RETIRED. *Concurrent Pos:* Prof, Univ NMex, 57-60. *Mem:* Am Math Soc; Soc Indust Appl Math. *Res:* Mathematics theory of plasticity; hydrodynamics; elasticity. *Mailing Add:* 119 Tunyo Los Alamos NM 87544

WHITE, GEORGE ROWLAND, b Niagara Falls, NY, Feb 22, 29; m 57. HIGH TECHNOLOGY, IMAGING SYSTEMS. *Educ:* Wesleyan Univ, BA, 50; Iowa State Univ, PhD(physics), 55; Univ Calif, Los Angeles, MS, 67. *Prof Exp:* From engr to dept mgr, Sperry Gyroscope Co, 55-64; from div mgr to chief scientist, 64-67, eng dept mgr, 67-68, div vpres, 68-72, staff vpres, 72-73, vpres prod planning, 74-76, vpres advan develop, 77-79, VPRES RES & DEVELOP & ENG, XEROX CORP, 79- *Concurrent Pos:* Carroll-Ford Found vis prof bus admin, Grad Sch Bus Admin, Harvard Univ, 76-77; chmn, Human Relations Comn, Monroe Co, 77-79; chmn planning bd, Polytech Inst NY, Inst Imaging Sci, 79- *Mem:* Optical Soc Am; Am Phys Soc; Inst Elec & Electronics Engrs; Indust Res Inst. *Res:* Management of technological innovation; development of xerographic and electronic imaging systems; lasers; microwave tubes; particle physics. *Mailing Add:* 911 William Pitt Union Univ Pittsburgh Pittsburgh PA 15260

WHITE, GERALD M(ILTON), b Detroit, Mich, Dec 6, 29; m 56; c 4. COMPUTER SCIENCE. *Educ:* Univ Mich, BS, 51; Harvard Univ, MS, 53, PhD(appl physics), 58. *Prof Exp:* Res assoc, Res Lab, 57-64, mgr comput oper, Res & Develop Ctr, 64-69, res assoc info systs, 69-77, INFO SYSTS ENGR, RES & DEVELOP CTR, GEN ELEC CO, 77- *Mem:* Inst Elec & Electronics Engrs; Sigma Xi. *Res:* Information processing; computer systems and languages. *Mailing Add:* 1274 Hawthorne Rd Schenectady NY 12309

WHITE, GIFFORD, b San Saba, Tex, Feb 17, 12; m 35; c 2. PHYSICS. *Educ:* Univ Tex, BA & MA, 39. *Prof Exp:* Geophysicist, Humble Oil & Refining Co, 34-38; res engr, Sperry Gyroscope Co, 41-47; vpres, Statham Instruments, Inc, 47-52; pres, White Instruments, Inc, 53-79, chmn bd, 79-84; CHMN BD, WHITE PROPERTIES, INC, 84- *Honors & Awards:* Cert of Appreciation, US War Dept, 45. *Mem:* Am Phys Soc; Soc Explor Geophys; fel Inst Elec & Electronics Engrs. *Res:* Instrumentation for physical measurements; circuit theory. *Mailing Add:* 1034 Capital Pkwy No 223 Austin TX 78746

WHITE, GILBERT FOWLER, b Chicago, Ill, Nov 26, 11; wid; c 3. WATER RESOURCES, NATURAL HAZARDS. *Educ:* Univ Chicago, SB, 32, SM, 34, PhD(geog), 42. *Prof Exp:* Geographer, Miss Valley Comt, Pub Work Admin, 34-35, Nat Resources Planning Bd, 35-40, Bur Budget, 40-42; volunteer, Am Friends Serv Comt, 42-46; pres, Haverford Col, 46-55; prof geog, Univ Chicago, 56-69, Univ Colo, 70-78; dir, Natural Hazards Info Ctr, 76-84; EMER DISTINGUISHED PROF & EMER DIR, INST BEHAV SCI, 78- *Concurrent Pos:* VChmn, Pres Water Resources Policy Comn, 50-51; vis prof, Univ Oxford, 62-63; chmn, Bur Budget Task Force Fed Flood Policy, 65-66; chmn bd, Resources Future, 74-79; chmn, Comn Natural Resources, Nat Res Coun, 77-80, pres, Sci Comt on Problems Environ, 76-82; exec ed, Environment, 83- *Honors & Awards:* Daly Medal, Am Geog Soc, 71; Eben Award, Am Water Resources Asn, 72; Environ Award, Nat Acad Sci, 80; Sasakawa Prize, UN, 85; Tyler Prize, 87; Laureat d'Honneur, Int Geog Union, 88; Caulfield Prize, Am Water Resources Asn, 89. *Mem:* Nat Acad Sci; Asn Am Geographers (pres, 61-62); Am Geophys Union; hon mem Am Planning Asn; Soviet Acad Sci; Royal Geog Soc. *Res:* Natural resources management; environmental policy; natural hazards. *Mailing Add:* Inst Behav Sci Campus Box 482 Univ Colo Boulder CO 80309

WHITE, GLENN E, b Sasebo-Shu, Nagusaki-Ken, Japan, Aug 21, 57. REGULATION OF MUSCLE GENE EXPRESSION. *Educ:* Harvard Univ, PhD(anat & cell biol), 83. *Prof Exp:* Res fel med, Beth Israel Hosp, Boston, 83-85; res fel cardiol, Children's Hosp Med Ctr, 85-89; ASST PROF MED RES, DUKE UNIV, 89- *Mem:* Am Soc Cell Biol; NY Acad Sci. *Mailing Add:* 1947 Southwood Dr Apt 1 Durham NC 27707

WHITE, GORDON ALLAN, b Vancouver, BC, Nov 8, 32; m 59; c 3. PLANT PHYSIOLOGY, BIOCHEMISTRY. *Educ:* Univ BC, BA, 54, MA, 55; Iowa State Univ, PhD, 59. *Prof Exp:* Fel plant biochem, Prairie Regional Lab, Nat Res Coun Can, 59-60; asst prof chem, Ore State Univ, 60-62; plant biochemist, Res Ctr, Agr Can, 62-77, res scientist, 77-91; RETIRED. *Res:* Oxidative enzymes and their role in metabolism; biochemistry of fungi, including obligate parasites; carbohydrate catabolism in microorganisms. *Mailing Add:* Agr Can Res Ctr 1400 Western Rd London ON N6G 2V4 Can

WHITE, GORDON JUSTICE, immunochemistry, biochemistry, for more information see previous edition

WHITE, HAROLD BANCROFT, III, b Hartford, Conn, Feb 26, 43; m 66; c 3. VITAMIN METABOLISM, PROTEIN STRUCTURE & FUNCTION. *Educ:* Pa State Univ, BS, 65; Brandeis Univ, PhD(biochem), 70. *Prof Exp:* Res fel chem, Harvard Univ, 70-71; from asst prof to assoc prof, 71-83, PROF CHEM, UNIV DEL, 83- *Concurrent Pos:* Res career develop award, NIH, 77-81; ed bd, J Molecular Evolution, 77-86; vis res scientist genetics, Univ Calif, Davis, 77-78; vis scientist, AFRC-Poultry Res Ctr, Edinburgh, Scotland, 84-85. *Mem:* AAAS; Am Soc Biochem & Molecular Biol; Am Entom Soc; Soc Study Evolution; Am Inst Nutrit. *Res:* Vitamin and coenzyme metabolism; biotin; riboflavin; vitamin binding proteins; egg yolk deposition; molecular evolution; glycoproteins; odonata; entomology; science education. *Mailing Add:* Dept Chem & Biochem Univ Del Newark DE 19716

WHITE, HAROLD BIRTS, JR, b Little Rock, Ark, Mar 13, 29; m 58; c 2. BIOCHEMISTRY. *Educ:* Columbia Univ, AB, 51, MA, 53, PhD(physiol), 57. *Prof Exp:* Fel, Purdue Univ, 57-59, asst prof biochem, 59-61; from asst prof to assoc prof, 61-68, PROF BIOCHEM, SCH MED, UNIV MISS, 68- *Concurrent Pos:* Vis scientist, Univ Milan, 68-69. *Mem:* Am Chem Soc; Am Soc Biol Chem; Sigma Xi. *Res:* Brain lipid modification; poxvirus influence on lipid metabolism. *Mailing Add:* Dept Biochem Univ Miss Sch Med Jackson MS 39216

WHITE, HAROLD D(OUGLAS), b Sugar Valley, Ga, Aug 29, 10; m 35; c 2. AGRICULTURAL ENGINEERING. *Educ:* Univ Ga, BS, 34; Iowa State Col, MS, 38. *Prof Exp:* Agr engr, Abraham Baldwin Agr Col, 34-37; asst, Exp Sta, Univ Iowa, 37; instr agr eng, Iowa State Col, 38-41; asst agr engr, Univ Ga, 41-43; spec supvr, State Dept Educ, Ga, 43-45; from assoc prof to prof, 45-74, EMER PROF AGR ENG, UNIV GA, 74- *Concurrent Pos:* Pres, Prof Engrs & Assocs P C, 76- *Mem:* Sigma Xi; fel Am Soc Agr Engrs; Nat Soc Prof Engrs; Poultry Sci Asn. *Res:* Food and feed processing facilities and equipment; materials handling; dairy and poultry engineering. *Mailing Add:* 561 University Dr Athens GA 30605

WHITE, HAROLD J, b New York, NY, Jan 4, 20. PATHOLOGY. *Educ:* Harvard Univ, BS, 41; Univ Geneva, MD, 52. *Prof Exp:* Instr path, Sch Med, Yale Univ, 57-58; asst prof, 58-61, PROF PATH, SCH MED, UNIV ARK, LITTLE ROCK, 66- *Concurrent Pos:* Mem staff, Vet Admin Hosp, 60-; sr res scientist & consult, Dept Biomed Sci, Gen Motors Res Lab, Warren, Mich, 80- *Mem:* NY Acad Med; Int Acad Path. *Res:* Role of mucopolysaccharides and collagen in pathologic conditions; role of oxidants on the lung. *Mailing Add:* 24 Bass Rocks Rd Gloucester MA 01930

WHITE, HAROLD KEITH, b Straughn, Ind, July 11, 23; m 47; c 3. ORGANIC CHEMISTRY, BIOCHEMISTRY. *Educ:* Butler Univ, BS, 47; Purdue Univ, MS, 50; Ind Univ, PhD(org chem), 54. *Prof Exp:* Asst chemist, State Chem Off Ind, 47-49; res chemist, Mead Johnson & Co, 53-55; assoc prof, 55-57, PROF CHEM, HANOVER COL, 57- *Mem:* Am Chem Soc. *Res:* Stereochemistry; synthesis of polycyclics; medicinal chemistry. *Mailing Add:* 257 Garritt St Hanover IN 47243-9680

WHITE, HAROLD MCCOY, b Camden, SC, Feb 1, 32; m 55; c 3. ORGANIC CHEMISTRY. *Educ:* Clemson Univ, BS, 54, PhD(org chem), 62. *Prof Exp:* Chemist, SC Agr Res Sta, 54-55; fel ozone chem, Univ Tex, 62-64; from asst prof to assoc prof chem, 64-71, PROF CHEM, SOUTHWESTERN STATE COL, 71- *Concurrent Pos:* Welch Found fel, 63-64. *Mem:* Am Chem Soc; Sigma Xi. *Res:* Reactions of ozone with organic compounds; mechanism of the ozonation of hydrocarbons and reactions of ozone in basic media. *Mailing Add:* Dept Chem Southwestern Okla State Col Weatherford OK 73096

WHITE, HARRIS HERMAN, b Ft Worth, Tex, June 24, 49; m 73; c 5. BIOLOGICAL OCEANOGRAPHY. *Educ:* Univ Wash, BS(oceanog) & BS(zool), 71; Univ RI, PhD (oceanog), 76. *Prof Exp:* Asst prof oceanog, Old Dominion Univ, 76-79; ECOLOGIST, NAT OCEANIC & ATMOSPHERIC ADMIN, 80- *Res:* Marine pollution. *Mailing Add:* 6102 Biltmore Ave Baltimore MD 21215

WHITE, HARRY JOSEPH, b Philadelphia, Pa, Feb 19, 31; m 56; c 4. ORGANIC CHEMISTRY. *Educ:* LaSalle Col, BA, 54; Univ Notre Dame, PhD(org chem), 58. *Prof Exp:* Res chemist, Rohm & Haas Co, 58-67, coordr PhD recruiting, 67-68, asst mgr manpower & employment, 68-72, mgr recruiting & placement, 72-85, res personnel dir, 85-88, DIR UNIV REL, ROHM & HAAS CO, 88- *Mem:* Am Chem Soc. *Res:* Technical recruiting, placement and manpower planning; petroleum additives; polymer chemistry. *Mailing Add:* Rohm & Haas Co 727 Norristown Rd Spring House PA 19477

WHITE, HELEN LYNG, b Oceanside, NY, Oct 25, 30; m 55; c 2. BIOCHEMISTRY, ENZYMOLOGY. *Educ:* Russell Sage Col, BA, 52; Univ Del, MS, 63; Univ NC, Chapel Hill, PhD(biochem), 67. *Prof Exp:* Chemist, E I du Pont de Nemours & Co, Inc, 52-56; res assoc med chem, Univ NC, 67-70; sr res pharmacologist, 70-85, PRIN SCIENTIST, WELLCOME RES LABS, 85- *Mem:* Am Soc Pharmacol & Exp Therapeut; Am Soc Biochem & Molecular Biol; Am Chem Soc; Soc Neurosci. *Res:* Enzyme mechanisms and inhibitors; neurochemistry. *Mailing Add:* Div Pharmacol Wellcome Res Labs Research Triangle Park NC 27709

WHITE, HENRY W, b Blytheville, Ark, Dec 20, 41; m 62; c 2. PHYSICS. *Educ:* Pepperdine Col, BA, 63; Univ Calif, Riverside, MS, 65, PhD(physics), 69. *Prof Exp:* ASSOC PROF PHYSICS, UNIV MO, COLUMBIA, 69- *Mem:* Am Phys Soc; Am Asn Physics Teachers. *Res:* Low temperature thermal properties; inelastic electron tunneling spectroscopy. *Mailing Add:* Dept Physics Univ Mo Columbia MO 65211

WHITE, HERMAN BRENNER, JR, b Tuskegee, Ala, Sept 28, 48. PARTICLE PHYSICS. *Educ:* Earlham Col, BA, 70; Mich State Univ, MS, 74; Fla State Univ, PhD, 91. *Prof Exp:* Resident res assoc nuclear physics, Argonne Nat Lab, 71; Alfred P Sloan Found fel accelerator & particle physics, Europ Lab Particle Physics, Geneva Switz, 72; staff physicist particle physics, Univs Res Asn, Fermi Nat Accelerator Lab, 74-76; teaching fel physics, Yale Univ, 76-78; STAFF PHYSICIST PARTICLE PHYSICS, FERMI NAT ACCELERATOR LAB, 78- *Concurrent Pos:* Jr res assoc, Brookhaven Nat Lab. *Mem:* Am Inst Physics. *Res:* Study of high energy hadrons production and neutrino production; hadronic constituent scattering. *Mailing Add:* Res Div MS-220 Fermilab PO Box 500 Batavia IL 60510

WHITE, HORACE FREDERICK, b Fresno, Calif, Apr 25, 25; m 52; c 3. PHYSICAL CHEMISTRY. *Educ:* Fresno State Col, AB, 47; Ore State Col, MS, 50; Brown Univ, PhD(phys chem), 53. *Prof Exp:* Fel, Univ Minn, 52-54; spectroscopist, Res Dept, M W Kellogg Co Div, Pullman, Inc, 54-56, instrumental methods supvr, 56-57; spectroscopist, Res Dept, Union Carbide Chem Co, 57-65; from asst prof to assoc prof, 65-75, actg chmn dept, 66-68, PROF CHEM, PORTLAND STATE UNIV, 75- *Mem:* Am Chem Soc; Sigma Xi. *Res:* Molecular structure using infrared spectroscopy and nuclear magnetic resonance spectrometry techniques; mass spectrometry; x-ray crystallography for structural determinations. *Mailing Add:* Portland State Univ Dept Chem PO Box 751 Portland OR 97207

WHITE, HOWARD DWAINE, b Des Moines, Iowa, Oct 24, 46. BIOCHEMISTRY. *Educ:* Univ Colo, BA, 69; Brandeis Univ, PhD(biochem), 73. *Prof Exp:* Fel biophys, Med Res Coun, London, 73-76, mem sci staff, 76-78; ASST PROF BIOCHEM, UNIV ARIZ, 78- *Mem:* Biophys Soc; Brit Biophys Soc. *Res:* The mechanism by which ATP hydrolysis is coupled to the production of mechanical work in muscle and other contractile systems. *Mailing Add:* 731 Graydor Ave Norfolk VA 23507

WHITE, HOWARD JULIAN, JR, b Batavia, NY, Nov 20, 20; m 49; c 2. SCIENCE ADMINISTRATION. *Educ:* Princeton Univ, AB, 42, PhD(chem), 47; Univ Wis, MS, 44. *Prof Exp:* Asst chem, Univ Wis, 42-44; from res chemist to assoc dir res, Textile Res Inst, NJ, 47-57, dir, 57-60; sr phys chemist, Stanford Res Inst, 60-61; spec asst res to Asst Secy Navy, Res & Develop, 61-64; asst chief phys chem div, 64-66, PROG MGR, OFF STANDARD REFERENCE DATA, NAT BUR STANDARDS, 66-, GUEST SCIENTIST, 88- *Mem:* AAAS; Am Chem Soc; Fiber Soc; Calorimetry Conf; Am Soc Mech Engrs; Am Inst Chem Engrs. *Res:* Surface chemistry; solutions, swelling, adsorption and dyeing of fibers; reference data on thermodynamics and transport properties and colloid and surface properties. *Mailing Add:* 8028 Park Overlook Dr Bethesda MD 20817

WHITE, IRVIN LINWOOD, b Hertford, NC, Mar 15, 32; m 78; c 2. ENERGY RESEARCH & DEVELOPMENT. *Educ:* Pa State Univ, BA, 54; Univ Ariz, PhD(govt), 67. *Prof Exp:* Officer-aviator, US Navy, 54-63; vis asst prof, Univ Ariz, 67-68; asst prof polit sci, Purdue Univ, 68-70; asst dir sci & pub policy prog, Univ Okla, 70-78 & from assoc prof to prof polit sci, 70-80; asst dir energy & mineral resources, Bur Land Mgt, US Dept Interior, 80-81; PRES, NY STATE ENERGY RES & DEVELOP AUTHORITY, 81- *Concurrent Pos:* Co-prin investr, Technol Assessment OCS Oil & Gas Opers, NSF, 71-73 & Inter-Agency study NSea Oil & Gas Develop, 73-74; prin investr coal & oil shale develop, NSF, 74-75 & Technol Assessment Western Energy Resource Develop, US Environ Protection Agency, 75-78; mem var comt, Nat Res Coun, 77-78, 78-79, 82 & 85-86; dir off assessment & spec projs, Off Res & Develop, US Environ Protection Agency, 78-80, actg dir off explor res, 80. *Mem:* Int Asn Impact Assessment (pres, 85-86). *Res:* Development management; radioactive waste management; science; technology; public policy. *Mailing Add:* NY State Energy Res & Develop Authority Two Rockefeller Plaza Albany NY 12223

WHITE, J COURTLAND, b Philadelphia, Pa, May 25, 48; m. PHARMACOLOGY. *Educ:* Univ Va, PhD(biochem), 73. *Prof Exp:* ASSOC PROF BIOCHEM, BOWMAN GRAY SCH MED, WAKE FOREST UNIV, 80- *Mem:* Am Asn Cancer Res; Am Soc Biol Chem; AAAS. *Res:* Biochemical pharmacology of anticancer agents, especially for leukemia. *Mailing Add:* Dept Biochem Bowman Gray Sch Med Wake Forest Univ Winston-Salem NC 27103

WHITE, JACK LEE, b Los Angeles, Calif, Oct 29, 25; m 50; c 1. MATERIALS SCIENCE, CARBON & GRAPHITE. *Educ:* Calif Inst Technol, BS, 49; Carnegie Inst Technol, BS, 50; Imp Col Univ London, dipl, 55; Univ Calif, PhD(metall), 55. *Prof Exp:* Res engr, Univ Calif, 55; Nat Acad Sci res assoc, US Naval Res Lab, 55-57; mem staff, Gen Atomic, 58- 67; res off, Petten Ctr, Europ AEC, 67-69; vis scientist, Gulf Gen Atomic, 69-70; assoc prof mat sci, Univ Calif, Davis, 71-72; staff scientist, Aerospace Corp, 73-88; RES CHEM ENGR, UNIV CALIF, SAN DIEGO, 88- *Concurrent Pos:* Consult, Gulf Gen Atomic, Europe AEC, 70-73. *Honors & Awards:* Graffin Lectr, 84; Skakel Award, Am Carbon Soc, 87. *Mem:* AAAS; Am Chem Soc; Am Ceramic Soc; Am Inst Mining, Metall & Petrol Eng; Brit Inst Metals; Am Carbon Soc. *Res:* Carbonaceous and graphitic materials; high-temperature materials; high-temperature physical chemistry; metallurgical thermodynamics. *Mailing Add:* 690 Rimini Rd Del Mar CA 92014

WHITE, JAMES CARL, b Ft Wayne, Ind, Mar 1, 22; m 46; c 4. ANALYTICAL CHEMISTRY. *Educ:* Ind Univ, BS, 43; Ohio State Univ, MS, 48, PhD(chem), 50. *Prof Exp:* Chemist, Joslyn Mfg Co, 46; asst, Ohio State Univ, 46-50; asst div dir, Oak Ridge Nat Lab, 50-67, from assoc dir to dir, Anal Chem Div, 67-76; gen plant serv mgr, Martin Marietta Energy Systs, Inc, 76-86; RETIRED. *Mem:* AAAS; Am Chem Soc; Int Union Pure & Appl Chem; Fedn Anal Chem & Spectros. *Res:* Research administration; molten salts; separations; reference materials. *Mailing Add:* 5425 Shenandoah Trail Knoxville TN 37919

WHITE, JAMES CARRICK, b Scobey, Mont, Oct 29, 16; m 41; c 3. FOOD MICROBIOLOGY. *Educ:* Cornell Univ, PhD(bact), 44. *Prof Exp:* Dir res, Borden Cheese Co, 44-46; assoc prof dairy indust, 46-51, prof food sci, 77-80, asst dean hotel admin, 79-82, PROF DAIRY INDUST, CORNELL UNIV, 51-, PROF HOTEL ADMIN, 72- *Concurrent Pos:* Sci adv, Ctr Environ Info. *Mem:* Inst Food Technol; Int Asn Milk, Food & Environ Sanitarians. *Res:* Food poisoning; waste technology; food sanitation; acid rain; climate change. *Mailing Add:* Statler Hall Cornell Univ No W209 Ithaca NY 14851

WHITE, JAMES CLARENCE, b Hodge, La, July 7, 36; m 57; c 2. PLANT PATHOLOGY. *Educ:* La Polytech Inst, BS, 59; La State Univ, MS, 61, PhD(plant path), 63. *Prof Exp:* Asst prof bot, Southeastern La Col, 63-65; asst prof, 65-70, PROF BOT, LA TECH UNIV, 70-, PROF BACT, 77- *Mem:* Am Phytopath Soc. *Res:* Pathological histology and studies of Tabasco pepper plants infected with tobacco etch virus. *Mailing Add:* Dept Bot La Tech Univ Ruston LA 71272

WHITE, JAMES DAVID, b Bristol, Eng, June 14, 35; m 60; c 2. ORGANIC CHEMISTRY. *Educ:* Cambridge Univ, BA, 59; Univ BC, MSc, 61; Mass Inst Technol, PhD(org chem), 65. *Prof Exp:* From instr to asst prof chem, Harvard Univ, 65-71; assoc prof, 71-76, PROF CHEM, ORE STATE UNIV, 76- *Concurrent Pos:* Consult med chem, NIH; assoc ed, J Am Chem Soc; Guggenheim fel, 88. *Mem:* Am Chem Soc; Royal Soc Chem; Swiss Chem Soc. *Res:* Organic synthesis and photochemistry; chemistry of natural products; heterocyclic compounds. *Mailing Add:* Dept Chem Ore State Univ Corvallis OR 97331-4003

WHITE, JAMES EDWARD, b Cherokee, Tex, May 10, 18; m 41; c 4. ACOUSTICS. *Educ:* Univ Tex, BS, 40, MA, 46; Mass Inst Technol, PhD(physics), 49. *Prof Exp:* Physicist, Underwater Sound Lab, Mass Inst Technol, 41-45, Defense Res Lab, Univ Tex, 45-46, Mobil Oil Co, 49-55, Marathon Oil Co, 55-69 & Globe Universal Sci, Inc, 69-72; mem fac, Colo Sch Mines, 72-73; L A Nelson prof geol sci, Univ Tex, El Paso, 73-76; Charles Henry Green prof explor geophys, 76-88, EMER PROF, COLO SCH MINES, 87- *Concurrent Pos:* Nat Acad Sci exchange scientist, USSR & Yugoslavia, 73-74; mem space appl bd, Nat Acad Eng, 73-78; Esso vis prof, Univ Sydney, Australia, 75; vis prof, Mass Inst Technol, 82, Univ Tex Austin, 85, Macquarie Univ, Sydney, NSW, 88. *Honors & Awards:* Maurice Ewing Medal, 86; Halliburton Award, 87. *Mem:* Nat Acad Eng; fel Acoust Soc Am; Europ Asn Explor Geophys; hon mem Soc Explor Geophys (past pres); Am Geophys Union. *Res:* Seismic prospecting; waves in solids; engineering geophysics; earthquake dynamics. *Mailing Add:* Dept Geophys Colo Sch Mines Golden CO 80401

WHITE, JAMES EDWIN, b Pittsburgh, Pa, June 4, 35; m 60; c 2. BIOLOGY, ECOLOGY. *Educ:* Dartmouth Col, AB, 57; Rutgers Univ, PhD(zool), 61. *Prof Exp:* Asst prof biol, Parsons Col, 61-62; from instr to assoc prof, 62-74, PROF BIOL, KEUKA COL, 74-, CHMN DEPT, 69- *Concurrent Pos:* Consult, NY State Scholar Exam, 64-65; actg acad dean, Keuka Col, 77-78. *Mem:* AAAS; Am Soc Mammal; Ecol Soc Am. *Res:* Mammal population ecology; animal behavior; small mammal parasites. *Mailing Add:* Div Natural Sci & Math Keuka Col Keuka Park NY 14478

WHITE, JAMES GEORGE, b Aug 28, 29; m; c 5. HEMATOLOGY, PEDIATRICS. *Educ:* Univ Minn, MD, 55. *Prof Exp:* REGENTS PROF HEMAT-PEDIAT, DEPTS LAB MED, PATH & PEDIAT, SCH MED, UNIV MINN, 84-, ASSOC DEAN RES. *Res:* Platelet and other blood cells, hematosis, thrombosis. *Mailing Add:* Depts Lab Med Path & Pediat Sch Med Univ Minn Mayo Bldg Box 490 Minneapolis MN 55455

WHITE, JAMES L(INDSAY), b Brooklyn, NY, Jan 3, 38; m 66. POLYMER PROCESSING. *Educ:* Polytech Inst Brooklyn, BChE, 59; Univ Del, MChE, 62, PhD(chem eng), 65. *Prof Exp:* Res engr, Res Ctr, US Rubber Co, NJ, 63-66, Mich, 66-67; from assoc prof to prof eng, Univ Tenn, Knoxville, 67-70, alumni distinguished serv prof, 74-, prof-in-charge polymer eng, 76-83; DIR POLYMER ENG CTR, UNIV AKRON, OHIO, 83-, HEAD DEPT POLYMER ENG, 83- *Concurrent Pos:* Ed, J Polymer Eng, 81-87. *Honors & Awards:* Bingham Medal, Soc Rheology, 81. *Mem:* Polymer Processing Soc; Soc Rheology; Soc Plastics Engrs; Soc Rheology Japan; Soc Fiber Sci & Technol Japan; Am Chem Soc. *Res:* Rheology of polymer systems; extrasion especially twin screw extrasion, rotational molding; characterization of structure and orientation in solid polymers in polymer processing; filled polymer systems and rubber technology; history of the polymer industry. *Mailing Add:* Polymer Eng Ctr Univ Akron Akron OH 44325-0001

WHITE, JAMES PATRICK, b Indianapolis, Ind, Sept 13, 39; m 62; c 4. MICROBIAL PHYSIOLOGY. *Educ:* Marian Col, Ind, BS, 62; Univ Ark, MS, 65, PhD(microbiol), 67. *Prof Exp:* From asst prof to assoc prof microbiol, 70-77, PROF BIOL, ST BONAVENTURE UNIV, 77- *Mem:* AAAS; Am Soc Microbiol; Mycol Soc Am; Sigma Xi. *Res:* Physiology and nutrition of pigment formation in Helminthosporium species; effects of trace elements in nitrogen metabolism of microorganisms. *Mailing Add:* Dept Biol St Bonaventure Univ St Bonaventure NY 14778

WHITE, JAMES RUSHTON, b Ft Benning, Ga, July 28, 23; m 55; c 2. BIOCHEMISTRY. *Educ:* Stanford Univ, BS, 48, PhD(chem), 53. *Prof Exp:* Res chemist, Pioneering Res Lab, E I du Pont de Nemours & Co, Inc, 53-59; res assoc biochem, Univ Pa, 59-62; from asst prof to prof, 62-85, EMER PROF BIOCHEM, SCH MED, UNIV NC, CHAPEL HILL, 85- *Concurrent Pos:* NSF fel, 60-62. *Mem:* AAAS. *Res:* Macromolecular metabolism in bacteria; antibacterial action of antibiotics and other inhibitors; complexes of nucleic acids with low molecular weight ligands. *Mailing Add:* 210 Ridgecrest Dr Chapel Hill NC 27514-2101

WHITE, JAMES RUSSELL, b Elgin, Ill, July 13, 19; m 44; c 1. PHYSICAL CHEMISTRY. *Educ:* Ind Univ, BS, 42; Yale Univ, PhD(phys chem), 44. *Prof Exp:* Phys chemist, Tenn Eastman Corp, 44-47; res assoc physics, Socony-Vacuum Oil Co, 47-59, supvr nuclear res group, Mobil Res & Develop Corp, 59-69, MGR RESOURCES & PROD RES SECT, MOBIL RES & DEVELOP CORP, 69- *Mem:* AAAS; Am Chem Soc; Am Nuclear Soc. *Res:* Mass spectrometry; radio chemical tracers; phycial chemistry of solutions; ionization potentials; lubricants and lubrication; thermal diffusion; radiation chemistry. *Mailing Add:* 20 Milyko Dr Washington Crossing PA 18977

WHITE, JAMES VICTOR, b Hammond, Ind, May 20, 41; m 66. ACOUSTICS, DYNAMICS. *Educ:* Northwestern Univ, Evanston, BS, 64; Harvard Univ, SM, 65, PhD(acoust), 70. *Prof Exp:* Engr, Jensen Mfg Co, 63-64; asst prof mech eng, Stevens Inst Technol, 70-74; staff scientist, Sound Reproduction Dept, CBS Tech Ctr, 74-80; MEM TECH STAFF, ANALYTICAL SCI CORP, 80- *Concurrent Pos:* Fel, Harvard Univ, 70. *Mem:* Inst Elec & Electronics Engrs; Am Soc Mech Engrs; Acoust Soc Am; Audio Eng Soc. *Res:* Stylus-groove interaction in phonographs; noise control; modeling and identification of dynamic systems. *Mailing Add:* TASC 55 Walkers Brook Dr Reading MA 01867

WHITE, JAMES WILSON, b Salisbury, NC, May 29, 14; m 42; c 3. PHYSICS. *Educ:* Davidson Col, BS, 34; Univ NC, MS, 36, PhD(physics), 38. *Hon Degrees:* DSc, King Col, 65. *Prof Exp:* Instr physics, Emory Jr Col, 38-39; prof, King Col, 39-42; instr, Univ Tenn, 42-44; res physicist, Fulton Sylphon Co, 44-45; from asst prof to prof physics, Univ Tenn, Knoxville, 45-84; RETIRED. *Mem:* Am Phys Soc; Am Asn Physics Teachers. *Res:* Instrumentation. *Mailing Add:* Dept Physics-Astron Univ Tenn Knoxville TN 37996

WHITE, JANE VICKNAIR, b Houma, La, Feb 10, 47; m 68; c 2. NUTRITION. *Educ:* St Mary's Dominican Col, BS, 68; Univ Tenn, Knoxville, PhD(nutrit), 75. *Prof Exp:* Asst prof, 75-80, ASSOC PROF NUTRIT, DEPT FAMILY PRACT MED, UNIV TENN, KNOXVILLE, 75- *Concurrent Pos:* Clin nutrit consult, Vet Admin Cent Off, Washington, DC, 79- *Mem:* Am Dietetic Asn; Am Home Econ Asn; Nat Educ Asn. *Res:* Effect of sulfur nutrition on the enzymes of carbohydrate and fat metabolism; nutrition education in family practice residency programs. *Mailing Add:* 1924 Alcoa Hwy-U67 Knoxville TN 37920

WHITE, JERRY EUGENE, b Mt Vernon, Ill, Oct 6, 46; m 68; c 2. ORGANIC CHEMISTRY. *Educ:* Southern Ill Univ, Carbondale, BA, 68; Vanderbilt Univ, PhD(chem), 72. *Prof Exp:* Res chemist polymer synthesis, Army Mat & Mech Res Ctr, 73-76; sr res chemist, New Prod Lab, Granville Res & Develop Ctr, Dow Chem USA, 76-81, res leader, Cent Res-Polymer Mat Lab, 81-88, RES ASSOC, CENT RES-ORG CHEM & POLYMERS LAB, DOW CHEM CO, MIDLAND, MICH, 88- *Mem:* Am Chem Soc. *Res:* Synthetic, structural and mechanistic studies in the chemistry of novel organic polymers for special service applications; polymer modification reactions. *Mailing Add:* Cent Res Org Chem & Polymer Labs Bldg 1707 Dow Chem Co Midland MI 48674

WHITE, JESSE EDMUND, b Indianapolis, Ind, June 9, 27; m 50; c 3. HISTORY OF CHEMISTRY, SCIENCE EDUCATION. *Educ:* Va Mil Inst, BS, 49; Ind Univ, PhD(chem), 58. *Prof Exp:* Asst prof chem, Lafayette Col, 55-59; from asst prof to assoc prof, 59-71, PROF CHEM, SOUTHERN ILL UNIV, EDWARDSVILLE, 71- *Mem:* Am Chem Soc; Nat Sci Teachers Asn; Asn Univ Professors. *Res:* History of chemistry. *Mailing Add:* Dept Chem Southern Ill Univ Edwardsville IL 62026-1652

WHITE, JESSE STEVEN, b Cleveland, Miss, May 9, 17. PARASITOLOGY. *Educ:* Delta State Col, BS, 40; Miss State Col, MS, 49; Univ Ala, PhD, 59. *Prof Exp:* Asst prof, 46-59, head div sci, 59-70, prof, 59-80, EMER PROF BIOL, DELTA STATE UNIV, 80- *Concurrent Pos:* NSF fel, 58-59. *Mem:* Sigma Xi. *Res:* Medical entomology. *Mailing Add:* 118 W Sunflower St Cleveland MS 38732

WHITE, JOE LLOYD, b Pierce, Okla, Nov 8, 21; m 45; c 5. SOIL MINERALOGY, SOIL CHEMISTRY. *Educ:* Okla State Univ, BS, 44, MS, 45; Univ Wis, PhD(soil chem), 47. *Prof Exp:* From asst prof to assoc prof, 47-57, PROF AGRON, PURDUE UNIV, WEST LAFAYETTE, 57- *Concurrent Pos:* Rockefeller fel natural sci, Nat Res Coun, 53-54; NSF sr fel, Louvain, 65-66; Soil Sci Soc Am Rep, Earth Sci Div, Nat Res Coun, 70-73; Fulbright res scholar, Athens, 72-73; Guggenheim fel, Versailles, 72-73; consult, William H Rorer Co, Pa, 78-; Alexander von Humboldt Found fel, Munich Tech Univ, 80-81. *Honors & Awards:* Charles Medal, Charles Univ, Prague, 61; Soil Sci Award, Am Soc Agron, 69; Sr US Scientist Award, Alexander von Humboldt Found, 80. *Mem:* Fel AAAS; Am Chem Soc; fel Am Soc Agron; fel Mineral Soc Am; Clay Minerals Soc (treas, 69-72); fel Royal Soc Chem. *Res:* Weathering of micaceous minerals; pesticide-soil colloid interactions; application of infrared spectroscopy to study of aluminosilicates; structure and properties of aluminum hydroxide gels and aluminum chlorohydrates; clay-drug interactions. *Mailing Add:* 2505 Roselawn Ave West Lafayette IN 47904

WHITE, JOE WADE, b Dill City, Okla, Aug 22, 40; m 62; c 2. PHYSICAL ORGANIC CHEMISTRY. *Educ:* Okla State Univ, BS, 63; Univ Ariz, PhD(chem), 67. *Prof Exp:* Sr chemist, 67-71, res specialist chem, Indust Tape Lab, 71-72, res supvr, Indust Specialties Lab, 72-73, res mgr, 73-78, lab mgr, Struct Prod Dept, 78-81, tech dir, Decorative Prod Div, 81-87, TECH DIR, INDUST SPECIALITIES DIV, 3M CO, 87- *Mem:* Am Chem Soc; Soc Plastics Eng. *Res:* Kinetics and rheology of gelling polymerizations. *Mailing Add:* Indust Specialties Lab 3M Co Bldg 230-3N-02 St Paul MN 55144-1000

WHITE, JOHN ANDERSON, b Bahia, Brazil, Oct 18, 19; m 76; c 1. VERTEBRATE ZOOLOGY, PALEONTOLOGY. *Educ:* William Jewell Col, AB, 42; Univ Kans, PhD(zool), 53. *Prof Exp:* Instr biol, William Jewell Col, 46-47 & Univ Ill, 53-55; prof, Calif State Col Long Beach, 55-66; cur vert paleont, Idaho Mus Natural Hist & prof biol, Idaho State Univ, 66-85; RES ASSOC, DEPT GEO SCI, UNIV ARIZ, 85- *Mem:* Fel AAAS; Am Soc Mammal; Soc Syst Zool; Soc Vert Paleont; Paleont Soc. *Res:* Systematics, evolution and ecology of late Tertiary and Quaternary rodents and logomorphs. *Mailing Add:* 4831 N Via Entrada Tuscon AZ 85718

WHITE, JOHN ARNOLD, b Chicago, Ill, Jan 30, 33; m 64; c 3. THERMODYNAMIC FLUCTUATIONS, PHASE TRANSITIONS. *Educ:* Oberlin Col, BA, 54; Yale Univ, MS, 55, PhD(physics), 59. *Prof Exp:* Instr physics, Yale Univ, 58-59; instr, Harvard Univ, 59-62; res assoc, Yale Univ,

62-63; physicist, Nat Bur Standards, 63-64; res assoc physics, Univ Md, 65-66; assoc prof, 66-68, PROF PHYSICS, AM UNIV, 68- *Concurrent Pos:* Consult, Nat Bur Standards, 66-72 & 81; NSF grants, Am Univ, 66, 67, 69 & 71; vis scientist, Mass Inst Technol, 72, Nat Bur Standards, 81 & 86; res contracts, Off Naval Res, 73 & 74; fel, Am Soc Eng Educ, Naval Res Lab, 85; Dept Energy, Off Basic Energy Sci res grants, 86, 88, 90. *Honors & Awards:* Boyden Premium, Franklin Inst, 80. *Mem:* AAAS; Sigma Xi; fel Am Phys Soc. *Res:* Atomic beams; magnetism of rare earth ions in solids; lasers; spontaneous emission in external fields; speed of light; unified time-length standardization; relativity; laser light scattering; critical point phenomena; theory of thermodynamic fluctuations and phase transitions in fluids. *Mailing Add:* Dept Physics Am Univ Washington DC 20016-8058

WHITE, JOHN AUSTIN, JR, b Portland, Ark, Dec 5, 39; m 63; c 2. MATERIAL HANDLING, FACILITIES PLANNING. *Educ:* Univ Ark, Fayetteville, BS, 61; Va Polytech Inst, MS, 66; Ohio State Univ, PhD(indust eng), 69. *Hon Degrees:* DEng, Cath Univ Leuven, 85. *Prof Exp:* Indust engr, Tenn Eastman Co, 61-63; instr indust eng, Va Polytech Inst, 63-66; teaching assoc, Ohio State Univ, 66-69; from asst prof to assoc prof indust eng & opers res, Va Polytech Inst & State Univ, 70-74; from assoc prof to prof, 75-84, REGENTS' PROF INDUST & SYSTS ENG, GA INST TECHNOL, 84-, EUGENE C GWALTNEY PROF MFG, 88-; ACTG DEP DIR, NSF, 90- *Concurrent Pos:* Mem bd dir, Mat Handling Educ Found, 77-81, 88-90; dir, Mat Handling Res Ctr, Ga Inst Technol, 82-87; exec consult, Coopers & Lybrand, 84-; mem Mfg Studies Bd, Nat Res Coun, 86-89; sr ed, Inst Indust Engrs Transactions, 84-88; chmn bd dirs, Am Asn Eng Studies, 86; mem, Coun Logistics Mgt, Warehouse Educ & Res Coun; asst dir eng, NSF, 88-90. *Honors & Awards:* Outstanding Indust Eng Award, Inst Indust Engrs, 80, Albert G Holzman Distinguished Educ Award, 88, David F Baker Distinguished Res Award, 90; Reed-Apple Award, Mat Handling Educ Found, 85; Kenneth Andrew Roe Award, Am Asn Eng Socs, 89. *Mem:* Nat Acad Eng; Opers Res Soc Am; Am Soc Eng Educ; fel Am Inst Indust Engrs (pres, 83-84); Int Mat Mgt Soc; Nat Soc Prof Engrs; Soc Mfg Engrs. *Res:* Development of design algorithums for facilities layout, material handling and warehousing systems; author & co-author of numerous publications, books and handbooks. *Mailing Add:* NSF 1800 G St NW Washington DC 20550

WHITE, JOHN DAVID, b Newark, NJ, Feb 14, 28; m 51; c 1. PATHOGENESIS, INFECTIOUS DISEASES. *Educ:* Univ Buffalo, BA, 48, MA, 50; Vanderbilt Univ, PhD(bact), 53. *Prof Exp:* Asst bact, Univ Buffalo, 48-50; asst bot, Vanderbilt Univ, 50-53; res bacteriologist, US Army Hosp, Camp Kilmer, 54-55; bacteriologist, Armed Forces Inst Path, 55-56; bacteriologist, Path Div, US Dept Army, Ft Detrick, 56-59, chief clin path br, 59-68, actg chief path div, 68-71; MICROBIOLOGIST, PATH DIV, US ARMY MED RES INST INFECTIOUS DIS, 71- *Mem:* Sigma Xi; Am Asn Pathologists; NY Acad Sci; Electron Micros Soc Am. *Res:* Immunology; fluorescent antibody methods; electron microscopy. *Mailing Add:* US Army Med Res Inst Infect Dis Path Div Ft Detrick Frederick MD 21701

WHITE, JOHN FRANCIS, b New Orleans, La, Feb 9, 21; m 50; c 5. GEOLOGY, GROUND WATER. *Educ:* Univ Calif, BS, 47, PhD, 55. *Prof Exp:* Geologist, Mining Co Guatemala, 47-48 & Consol Coppermines Corp, 48-51; consult, Hydrothermal Res Proj, 55-59, from asst prof to assoc prof, 55-71, prof geol, Antioch Univ, 71-85; CONSULT, 80- *Res:* Economic geology; geomorphology; petrology; ground water in competent rocks. *Mailing Add:* Dept Geol Antioch Col Yellow Springs OH 45387

WHITE, JOHN FRANCIS, b Indianapolis, Ind, July 21, 44; c 3. CELL PHYSIOLOGY, BIOPHYSICS. *Educ:* Marian Col, BS, 66; Ind Univ, PhD(physiol), 70. *Prof Exp:* asst prof, 73-80, ASSOC PROF PHYSIOL, EMORY UNIV, 80- *Concurrent Pos:* NSF training fel, Univ Rochester, 71-72; NIH fel, Univ BC, 72-73; res career develop award, NIH. *Mem:* Biophys Soc; Am Physiol Soc. *Res:* Mechanisms and regulation of intestinal transport of ions and solutes by absorptive cells; intracellular ionic activities and compartmentalization of ions. *Mailing Add:* Dept Physiol Emory Univ 1364 Clifton Rd NE Atlanta GA 30322

WHITE, JOHN FRANCIS, b Madison, Wis, Dec 2, 29; div; c 1. FOOD & DAIRY SCIENCE. *Educ:* Univ Wis, BS, 51; Harvard Bus Sch, AMP, 74. *Prof Exp:* Food scientist, Kraft Inc, 54-66, div res coordr, 66-68, tech asst, 68-74, vpres & dir res & develop, 74-85, gen mgr spec projs, Retail Venture Div, 85-87; RETIRED. *Mem:* AAAS; Inst Food Technol; Am Chem Soc. *Mailing Add:* 1439 Pebble Creek Dr Glenview IL 60025

WHITE, JOHN FRANCIS, b Boston, Mass, Oct 31, 45; m 70; c 1. CHEMISTRY, CATALYSIS. *Educ:* Amherst Col, BA, 67; Mass Inst Technol, PhD(inorg chem), 72. *Prof Exp:* Group leader res & develop, Emery Industs, 72-76; sr res chemist res & develop, Halcon Res & Develop, 76-78; mgr new ventures res, Oxirane Int, 78-81; MGR PROCESS RES, ARCO CHEM RES & DEVELOP, 81- *Mem:* Am Chem Soc; Sigma Xi. *Res:* Petrochemical process development; homogeous and heterogeneous catalysis; aroma chemical process development. *Mailing Add:* 33 Knob Hill Rd Summit NJ 07901-3024

WHITE, JOHN GREVILLE, b Saltcoats, Scotland, Mar 27, 22; nat US; m 53; c 3. CHEMISTRY. *Educ:* Glasgow Univ, BSc, 44, PhD(chem), 47. *Prof Exp:* Asst chem, Glasgow Univ, 45-47; from instr to asst prof, Princeton Univ, 47-55; mem tech staff, Radio Corp Am, 56-66; prof phys, 66-86, EMER PROF CHEM, FORDHAM UNIV, 87- *Mem:* Am Crystallog Asn. *Res:* X-ray crystal structure analysis; complex organic structures; accurate small organic structures; inorganic structures. *Mailing Add:* 50 Scott Lane Princeton NJ 08544

WHITE, JOHN JOSEPH, III, b Arlington, Mass, Apr 24, 39; m 68; c 2. MECHANICAL ENGINEERING, BALLISTIC SCIENCE. *Educ:* Col William & Mary, BS, 60; Univ NC, PhD(physics), 65. *Prof Exp:* Res assoc physics, Univ NC, 65; asst prof physics, Univ Ga, 67-73; sr engr, BDM Corp, 73-74; res scientist, 74-78, prin res scientist, 78-81, GROUP LEADER, COLUMBUS DIV, BATTELLE MEM INST, 81- *Concurrent Pos:* Dir, Ga State Sci Fair, 73; mem, Landing Vehicle Assault Design Review Panel, 76; mem physics res eval group, Air Force Off Sci Res, 78-; mem pub rels comt, Nat Soc Prof Engrs, 78. *Honors & Awards:* Order of the Engr, Sigma Xi. *Mem:* Am Phys Soc; Am Soc Mech Engrs; Am Defense Preparedness Asn; Epigraphic Soc. *Res:* Applied mechanics; impact phenomena; explosion containment; assessment of advanced defense technology; optical properties of silver halides, high resolution specific heat measurements in antiferromagnets; superconductivity in quenched alloys; methods of data analysis; military and space vehicle design. *Mailing Add:* Battelle Columbus Div 505 King Ave Columbus OH 43201-2693

WHITE, JOHN MARVIN, b Martin, Tenn, June 9, 37; m 56; c 2. GENETICS, ANIMAL BREEDING. *Educ:* Univ Tenn, BS, 59; Pa State Univ, MS, 64; NC State Univ, PhD(animal breeding), 67. *Prof Exp:* From asst prof to prof dairy sci, 67-78, HEAD, DEPT DAIRY SCI, VA POLYTECH INST & STATE UNIV, 78- *Mem:* Biomet Soc; Am Dairy Sci Asn; Am Soc Animal Sci; Sigma Xi. *Res:* Quantitative genetics; measurement of response to single and multiple trait selection and correlated responses in mice and dairy cattle. *Mailing Add:* Dept Dairy Sci Va Polytech Inst Blacksburg VA 24061-0315

WHITE, JOHN MICHAEL, b Danville, Ill, Nov 26, 38; m 60; c 3. CHEMICAL PHYSICS. *Educ:* Harding Col, BS, 60; Univ Ill, MS, 62, PhD(chem), 66. *Prof Exp:* From asst prof to prof, 66-85, NORMAN HACKERMAN PROF CHEM, UNIV TEX, AUSTIN, 85- *Concurrent Pos:* Vis staff mem, Los Alamos Sci Lab. *Mem:* Am Chem Soc; Am Phys Soc. *Res:* surface chemistry. *Mailing Add:* Dept Chem Univ Tex Austin TX 78712-1104

WHITE, JOHN S(PENCER), industrial engineering, quality control, for more information see previous edition

WHITE, JOHN THOMAS, b El Paso, Tex, Aug 23, 31; m 58; c 3. MATHEMATICS. *Educ:* Univ Tex, BA, 52, MA, 53, PhD(math), 62. *Prof Exp:* Spec instr math, Univ Tex, 59-62; asst prof, Univ Kans, 62-65; assoc prof, 65-87, chmn dept 79-87, PROF MATH, TEX TECH UNIV, 87- *Concurrent Pos:* Fel, Tex Ctr Res, 66-; assoc dir, Comt Undergrad prog math, Univ Calif, Berkeley, 70-71. *Mem:* Math Asn Am; Am Math Soc; Soc Indust & Appl Math; Sigma Xi. *Res:* Distribution theory and transform analysis; integral transform theory. *Mailing Add:* Math Dept Tex Tech Univ MS 1042 Lubbock TX 79401

WHITE, JOHN W, b Ardmore, Okla, Aug 9, 33; m 58; c 3. GREENHOUSE STRUCTURES, ENERGY CONSERVATION. *Educ:* Okla State Univ, BS, 55; Colo State Univ, MS, 57; Pa State Univ, PhD(hort), 64. *Prof Exp:* From asst prof to assoc prof 64-75, PROF FLORICULT & ASSOC DIR UNIV OFF INDUST RES & INNOVATION, PA STATE UNIV, UNIVERSITY PARK, 75- *Concurrent Pos:* Rev ed, J Am Soc Hort Sci, 71-; consult, Gulf Res Corp, Fafard Peat Co, Lombardo Assocs, Gen Mills Corp, Gov's Waste Heat Energy Coun, Wellsley Col, St Mary's Col, Gen Elec, Dow Corning. *Honors & Awards:* Garland Award, Am Carnation Soc; Alex Laurie Award Educ & Res; Int Award, Dow Corning. *Mem:* Fel Am Soc Hort Sci; Soil Sci Soc Am; Int Solar Energy Soc; Soc Am Floricult. *Res:* Physical and chemical properties of soil; experimental designs and glazings for greenhouse structures; effects of the environment on plant growth; energy conservation for greenhouses; passive solar heating systems; water quality management and waste water treatment. *Mailing Add:* Dept Hort Pa State Univ University Park PA 16802

WHITE, JON M, GEOPHYSICS. *Educ:* Colo Sch Mines, BS, 68; SDak Sch Mines & Technol, MS & DPhil, 75. *Prof Exp:* Regulatory affairs coordr, Exxon Coal USA, Inc, 78-80; asst prof mining eng, Univ Wyo, 80-82 & Univ Petrol & Minerals, Dharan, Saudi Arabia, 82-85; DIR, ROCK MECH & EXPLOSIVE RES CTR, UNIV MO, ROLLA, 85- *Mem:* Soc Mining Engrs; Am Inst Prof Geol. *Res:* Improved high-technology blasting detonator; gun propulsion systems; novel explosives. *Mailing Add:* Dept Petrol & Mining Eng Univ Mo Rolla MO 65401

WHITE, JUNE BROUSSARD, b Elizabeth, La, Aug 27, 24; div; c 2. INORGANIC CHEMISTRY, ANALYTICAL CHEMISTRY. *Educ:* La Polytech Univ, BS, 44; Univ Southwestern La, BS, 59, MS, 61; La State Univ, Baton Rouge, PhD(inorg chem), 70. *Prof Exp:* Anal chemist, Cities Serv Refining Corp, La, 44-45; chemist, Esso Lab, Standard Oil Co, NJ, La, 45-48; teacher, Parish Sch Bd, La, 53-59; asst prof chem, Univ Southwestern La, 61-66; NSF res partic, La State Univ, Baton Rouge, 64 & 66, instr, 66-68; chmn div natural sci, 69-71, PROF CHEM & PHYSICS & CHMN DEPT, UNION UNIV, TENN, 68- *Mem:* Am Chem Soc. *Res:* Transition metal complexes which have d-2 electronic system; electron spin resonance; electronic transitions; magnetic properties. *Mailing Add:* 104 Brookdale Carriere MS 39426

WHITE, KERR LACHLAN, b Winnipeg, Man, Jan 23, 17; nat US; m 43; c 2. EPIDEMIOLOGY, INTERNAL MEDICINE. *Educ:* McGill Univ, BA, 40, MD & CM, 49; Am Bd Internal Med, dipl, 57. *Hon Degrees:* Dr Med, Univ Leuven, 78; DSc, McMaster Univ, 83. *Prof Exp:* Personnel asst, RCA Victor Co, Can, 41-42; intern & resident med, Mary Hitchcock Mem Hosp, Hanover, NH, 49-52; from asst prof to assoc prof med & prev med, Sch Med, Univ NC, 53-62; prof epidemiol & community med & chmn dept, Col Med, Univ Vt, 62-64; chmn dept, 64-72, prof health care orgn, Sch Hyg & Pub Health, Johns Hopkins Univ, 64-76; dir, Inst Health Care Studies, United Hosp Fund, 76-78; dep dir health sci, Rockefeller Found, NY, 78-84; RETIRED. *Concurrent Pos:* Hosmer fel med & psychiat, Royal Victoria Hosp, McGill Univ, 52-53; Commonwealth Fund advan fel, Med Res Coun Gt Brit & Sch Hyg & Trop Med, Univ London, 59-60; consult, Nat Ctr Health Statist & Health Resources Admin, Dept HEW, 66-78; chmn, US Nat Comn Vital & Health Statist, 75-80; mem expert panel orgn med care, WHO, 67-83; mem bd dirs, Found Child Develop, NY, 69-80; trustee, Case Western

Reserve Univ, 74-78; mem health adv panel, Off Technol Assessment, US Cong, 75-82; consult, China, Australia, Switz, NZ & Sask, Can govt; mem adv comt, Population Health, Can Inst Advan Res. *Honors & Awards:* Distinguished Career Award, Asn Health Serv Res, 87; Robert J Glaser Award, Soc Gen Internal Med, 90. *Mem:* Inst Med-Nat Acad Sci; AAAS; AMA; fel Am Pub Health Asn; fel Am Col Physicians; hon mem Int Epidemiol Asn, 84; hon mem NAm Primary Care Res Group, 85. *Res:* Medical education, health services research and epidemiology. *Mailing Add:* 2401 Old Ivy Rd No 1410 Charlottesville VA 22901

WHITE, KEVIN JOSEPH, b Queens, NY, Aug 28, 36; m 66. MOLECULAR PHYSICS. *Educ:* Georgetown Univ, BS, 58; Duke Univ, PhD(physics), 65. *Prof Exp:* Physicist, Naval Ord Lab, 58; res assoc physics, Duke Univ, 65; PHYSICIST, BALLISTIC RES LABS, 65- *Mem:* Am Phys Soc. *Res:* Millimeter wave microwave spectroscopy; Stark effect in rotational spectra; electron spin resonance; radiation damage in oxidizers; radical formation by atom addition and abstraction reactions; supersonic molecular beams for studying high pressure chemical reactions; combustion of propellants. *Mailing Add:* 406 Fowler Ct Joppa MD 21085

WHITE, LARRY DALE, b Sayre, Okla, Nov 24, 40; m 64; c 2. RANGE MANAGEMENT. *Educ:* Northern Ariz Univ, BS, 63; Univ Ariz, MS, 65, PhD(range ecol), 68. *Prof Exp:* Asst forester, US Forest Serv, 63; range adv, Near East Found, Kenya Govt, 67-69; asst prof range ecosyst mgt, Univ Fla, 70-75, assoc prof, 75-78; RANGE EXTEN SPECIALIST, TEX A&M UNIV, 78- *Honors & Awards:* Outstanding Achievement Award, Soc Range Mgt, 91. *Mem:* Australian Rangeland Soc; Soc Range Mgt. *Res:* Educational programs for county range needs; total range planning; grazing systems; use of prescribed fire; producer training and demonstration; brush control; range seeding; livestock-wildlife habitat relationships; effects of forestry practices on understory vegetation; manipulation of range ecosystems; range watershed. *Mailing Add:* Tex A&M Res & Exten Ctr PO Drawer 1849 Uvalde TX 78802-1849

WHITE, LAWRENCE KEITH, b Lafayette, Ind, Sept 16, 48; m 77; c 3. INTEGRATED CIRCUIT PROCESSING, PHYSICAL CHEMISTRY. *Educ:* Earlham Col, AB, 70; Univ Ill, Urbana, PhD(phys chem), 75. *Prof Exp:* Res assoc bio-inorg chem, Univ NH, 75-76; res scientist pulp & paper chem, Union Camp Corp, 77-78; SEN MEM TECH STAFF, ADVAN SILICON TECHNOL, DAVID SARNOFF RES CTR, SUBSID OF SRI INT, 78- *Mem:* Am Chem Soc; Sigma Xi; Electrochem Soc. *Res:* Solid state science and technology; integrated circuit processing; magnetic resonance; metalloproteins; transition metal chemistry; resist technology; lithography. *Mailing Add:* David Sarnoff Res Ctr RCA Labs Princeton NJ 08543-5400

WHITE, LAWRENCE S, b Chelsea, Mass, Mar 9, 23; m 46; c 2. PHYSICS. *Educ:* Mass Inst Technol, BS, 47. *Prof Exp:* Physicist, Res Lab, Titanium Div, Nat Lead Co, South Amboy, 48-62, sr technologist, 62-70; assoc physicist, Hoffman-La Roche Inc, Nutley, 70-73, sr scientist, 74-80, group leader & mgr, 80-85; RETIRED. *Mem:* Am Chem Soc; AAAS; Acad Pharmaceut Sci; Sigma Xi. *Res:* Physical properties and colorimetry of titania pigments; electron microscopy and diffraction; light scattering; surface properties of pharmaceutical solids; scanning electron microscopy. *Mailing Add:* PO Box 142 South Orleans MA 02662-0142

WHITE, LEE JAMES, b Saginaw, Mich, Apr 9, 39; m 65; c 2. COMPUTER SCIENCE, ELECTRICAL ENGINEERING. *Educ:* Univ Cincinnati, BSEE, 62; Univ Mich, MSEE, 63, PhD(elec eng), 67. *Prof Exp:* Coop engr, Dow Chem Co, 57-61; asst prof eng, Wright State Univ, 67-68; from asst prof to assoc prof, 68-77, PROF COMPUT & INFO SCI, OHIO STATE UNIV, 77-, CHMN DEPT, 80-; AT DEPT COMPUT SCI, UNIV ALTA, CAN. *Concurrent Pos:* Consult, Rockwell Int, 73-74 & Monsanto Res Corp, 77- *Mem:* Inst Elec & Electronics Engrs; Asn Comput Mach. *Res:* Analysis of algorithms and software analysis and testing; pattern recognition, automatic document classification, combinational computing and graph theory. *Mailing Add:* Dept Comput Sci Univ Alta 18507 92nd Ave Edmonton AB T5T 1P2 Can

WHITE, LENDELL AARON, b Sabetha, Kans, Nov 10, 26; m 48; c 2. MICROBIOLOGY. *Educ:* Univ Kans, BA, 51, MA, 55. *Prof Exp:* Bacteriologist, State Pub Health Lab, Kans, 51-53; asst, Virol Lab, Univ Kans, 53-54; bacteriologist, Spec Res Unit, Lab Br, 55-57, virus diag unit, 57-59, encephalitis sect, Tech Br, 59-60, venereal dis res lab, 60-63, viral reagents unit, 63-85, SUPVRY RES MICROBIOLOGIST, CTR DIS CONTROL, USPHS, 71- *Mem:* AAAS; Am Soc Microbiol; Sigma Xi. *Res:* Virology; isolation, identification, typing and determination of antigenic relationships of viruses with established strains; serologic and antigenic relationships among the arthropod-borne encephalitides; infectivity and fluorescent antibody studies with Neisseria gonorrhoeae; production of viral reagents; immunization procedures for reference antisera; inactivation procedures for use with viral antigens; stability of viral cultures and reagents; susceptibility of cell cultures to viral infection, large volume suspension culture. *Mailing Add:* Ctr Dis Control 1600 Clifton Rd Bldg 1 Rm 6430 Atlanta GA 30333

WHITE, LEROY ALBERT, b New York, NY, June 24, 29; m 58; c 2. PHYSICAL CHEMISTRY. *Educ:* Mass Inst Technol, BS, 50; Columbia Univ, MS, 51. *Prof Exp:* Res chemist, Monsanto Chem Co, 51-55; PROJ MGR, SPRINGBORN LAB, CONSULTS, HAZARDVILLE, 55- *Mem:* Am Chem Soc. *Res:* Organic synthesis; vinyl polymerization; nylon and epoxy reactions; general polymer development; coatings; membrane technology; photodegradable plastics; adhesives. *Mailing Add:* Four Ocala St New Haven CT 06516

WHITE, LOWELL ELMOND, JR, b Tacoma, Wash, Jan 16, 28; m 47; c 3. NEUROSURGERY, MEDICAL EDUCATION. *Educ:* Univ Wash, BS, 51, MD, 53. *Prof Exp:* Asst neurosurg, Sch Med, Univ Wash, 54-57, from instr to assoc prof, 57-70, assoc dean, 65-68; prof neurol surg & chief div, Univ Fla, 70-72; chmn dept, 72-80, PROF NEUROSCI, UNIV S ALA, 72- *Concurrent Pos:* Guggenheim Found fel, Univ Oslo, 57-58; consult, Div Res Resources & chmn nat adv comt animal resources, NIH, 65-69; consult, USPHS, 65-, grants admin adv comt, Dept HEW, 66-70. *Mem:* Am Asn Neuropath; Am Asn Anat; Am Asn Neurol Surg; AMA; Asn Am Med Cols; Sigma Xi. *Res:* Neuroanatomy and neurological surgery. *Mailing Add:* Dept Neurosci Col Med Univ S Ala Mobile AL 36688

WHITE, MALCOLM LUNT, b Schenectady, NY, Aug 16, 27; m 51; c 3. PHYSICAL CHEMISTRY. *Educ:* Colgate Univ, BA, 49; Northwestern Univ, PhD(chem), 53. *Prof Exp:* Investr geochem, NJ Zinc Co, 53-59; res chemist, Am Cyanamid Co, 59-61; mem tech staff, Bell Tel Labs, 61-82; res scientist, Lehigh Univ, 83-88; RETIRED. *Mem:* Am Chem Soc; Sigma Xi; Nat Asn Corrosion Engrs. *Res:* Nucleation; geochemistry; physical chemistry of colloid systems; surface chemistry; materials for electron device technology; coating and encapsulation of solid state devices; integrated circuit processing development; organic coatings for corrosion control. *Mailing Add:* 1830 Wilson Ave Bethlehem PA 18018

WHITE, MARK GILMORE, b Galveston, Tex, Jan 15, 49; m 85; c 1. CHEMICAL ENGINEERING. *Educ:* Univ Tex, Austin, BSChE, 71; Purdue Univ, MSChE, 73; Rice Univ, PhD(chem eng), 78. *Prof Exp:* Res engr, Amoco Oil Co, 73-74; asst prof, 77-82, ASSOC PROF CHEM, GA INST TECHNOL, 82-, ASSOC DIR, 88- *Mem:* Am Inst Chem Eng; Sigma Xi; Am Chem Soc. *Res:* Heterogeneous catalysis, kinetics and reactor design; research with the elucidation of reaction mechanisms over catalysts by a study of the kinetics; characterization of the catalysts and the use of isotopic compounds. *Mailing Add:* Sch Chem Eng Ga Inst Technol Atlanta GA 30332

WHITE, MARVIN HART, b Bronx, NY, Sept 6, 37; m 65; c 1. SOLID STATE ELECTRONICS & PHYSICS. *Educ:* Univ Mich, Ann Arbor, BSE(physics) & BSE(math), 60, MS, 61; Ohio State Univ, PhD(elec eng), 69. *Prof Exp:* Adv engr, Advan Technol Labs, Westinghouse Elec Corp, 61-81; SHERMAN FAIRCHILD PROF SOLID STATE STUDIES, ELEC & COMPUT ENG DEPT, LEHIGH UNIV, 81- *Concurrent Pos:* Fulbright res vis prof, Catholique Universite, Louvain-la-Nueve, Belgium, 70-79; ed, Electron Device Nat Newslett, 73-76. *Honors & Awards:* Electron Device Nat lectr, Inst Elec & Electronics Engrs, 82. *Mem:* Fel Inst Elec & Electronics Engrs; Sigma Xi. *Res:* Solid state electron devices and systems; semiconductor surfaces; integrated circuits; solid state electron device modeling and characterization. *Mailing Add:* Sherman Fairchild Lab Bldg 161 Lehigh Univ Bethlehem PA 18015

WHITE, MARY ANNE, b London, Ont, Dec 28, 53; m 77; c 2. SOLID STATE CHEMISTRY, THERMAL PROPERTIES OF SOLIDS. *Educ:* Univ Western Ont, BSc, 75; McMaster Univ, PhD(chem), 80. *Prof Exp:* Postdoctoral fel chem, Natural Sci & Res Coun, Oxford Univ, 79-81; NSERC res fel, Univ Waterloo, 81-83; asst prof, 83-87, ASSOC PROF CHEM, DALHOUSIE UNIV, 87- *Concurrent Pos:* Jr res fel, St Hilda's Col, Oxford Univ, 79-81; mem bd dirs, Calorimetry Conf, 84-86, Discovery Centre, 87- *Mem:* Chem Inst Can; Can Asn Physicists; Am Inst Physics. *Res:* Measurement of thermal properties of solids; investigation of polymorphism, especially in disordered solids and in inclusion compounds. *Mailing Add:* Chem Dept Dalhousie Univ Halifax NS B3H 4J3 Can

WHITE, MAURICE LEOPOLD, b New York, NY, Sept 30, 28; m 51; c 2. MICROBIOLOGY. *Educ:* Univ Calif, Los Angeles, BA, 51, PhD, 57; Am Bd Med Microbiol, dipl, 66. *Prof Exp:* Chief dept bact, Cedars of Lebanon Hosp, Los Angeles, Calif, 57-63; instr, med microbiol & immunol, Sch Med, Univ Calif, Los Angeles, 63-73; CHIEF, MICROBIOL UNIT, LAB SERV, VET AFFAIRS MED CTR, WEST LOS ANGELES, 63-; LECTR MED MICROBIOL & IMMUNOL, SCH MED, UNIV CALIF, LOS ANGELES, 73- *Mem:* AAAS; Am Soc Microbiol. *Res:* Staphylococcal phosphatase; nutrition and bacteriophage studies of Bordetella pertussis; taxonomy of Brucellaceae and Enterobacteriaceae; clinical microbiology. *Mailing Add:* Lab Serv W 113 Vet Admin Med Ctr Los Angeles CA 90073

WHITE, MERIT P(ENNIMAN), b Whately, Mass, Oct 25, 08; m 65; c 4. ENGINEERING MECHANICS. *Educ:* Dartmouth Col, AB, 30, CE, 31; Calif Inst Technol, MS, 32, PhD(civil eng), 35. *Prof Exp:* Engr, Soil Conserv Serv, USDA, 35-37; res assoc, Grad Sch Eng, Harvard Univ, 37-38 & Calif Inst Technol, 38-39; asst prof, Ill Inst Technol, 39-42; sr tech aide, Nat Defense Res Comt, Princeton & London, 42-45; bomb damage analyst, US War Dept, 45, sci consult & supvr prep physics rev, Field Info Agency Tech, 46-47; sci consult, USN Dept, 47; prof, 48-61, head dept, 50-77, COMMONWEALTH PROF CIVIL ENG, UNIV MASS, AMHERST, 61- *Concurrent Pos:* NSF fel, Polish Acad Sci, 64-65. *Mem:* Am Soc Civil Engrs; Am Soc Mech Engrs; Inst Mech Engrs; Int Union Testing & Res Lab Mat & Struct. *Res:* Structural dynamics. *Mailing Add:* Dept Civil Eng Univ Mass Amherst MA 01003

WHITE, MICHAEL GEORGE, b Oakpark, Ill, May 29, 53; m 75; c 1. IONIZATION DYNAMICS, ELECTRON SPECTROSCOPY. *Educ:* Univ Pittsburgh, BS, 74; Univ Calif, Berkeley, PhD(chem), 79. *Prof Exp:* Res fel, dept chem, Univ BC, Vancouver, 79-80; from asst chemist to assoc chemist, 80-84, CHEMIST, BROOKHAVEN NAT LAB, 84- *Mem:* Am Phys Soc. *Res:* Photoelectron studies of the photoionization dynamics of free molecules; non-radiative decay of super-excited neutral states lying in the ionization continuum; spectroscopy and ionization dynamics of optically prepared excited states via multiphoton ionization. *Mailing Add:* Chem Dept Brookhaven Nat Lab Upton NY 11973

WHITE, MORENO J, b Evergreen, Ala, Dec 19, 48; m 78; c 2. POLYMER MATRIX COMPOSITES. *Educ:* Univ Ala, BS, 72; Calif State Univ, MS, 75. *Prof Exp:* Jr engr, McDonnel Douglas Astronaut Corp, 72-73; staff engr, Rockwell Int, 73-75; eng assoc, P D A Eng, 75-76; staff eng, Sci Applications, Inc, 76-80; sr engr, Gen Res Corp, 80-82; sr engr, 82-88, TECH DIR, AEROSPACE SCI, SPARTA, INC, 88- *Honors & Awards:* Technol Transfer Award, Am Defense Preparedness Asn & Strategic Defense Initiation Orgn,

89. *Mem:* Am Soc Mech Engrs; Am Soc Metals; Am Defense Preparedness Asn. *Res:* Advanced materials research and development; structural analysis; advanced composites design; bioengineering; hypervelocity projects design and analysis. *Mailing Add:* Sparta Inc 9455 Towne Centre Dr San Diego CA 92121

WHITE, MYRON EDWARD, b Boston, Mass, May 1, 20; m 48; c 4. MATHEMATICS, COMPUTER SCIENCE. *Educ:* Wesleyan Univ, AB, 41; Columbia Univ, AM, 50, PhD(math), 62. *Hon Degrees:* MEng, Stevens Inst Technol, 83. *Prof Exp:* From instr to assoc prof math, 53-73, dir sci training progs, 63-73, PROF MATH, STEVENS INST TECHNOL, 73-, DIR MOVE AHEAD PROG, 74- *Concurrent Pos:* Teacher, NSF Math Insts; adv bd, math, sci & comp sci, NJ Dept Higher Educ; consult, US Agency Int Develop, India. *Mem:* Am Math Soc; Math Asn Am. *Res:* Computer programming languages. *Mailing Add:* Dept Math Stevens Inst Technol Hoboken NJ 07030

WHITE, N(IKOLAS) F(REDERICK), b Seattle, Wash, Oct 25, 39; m 61; c 3. PETROLEUM ENGINEERING, CIVIL ENGINEERING. *Educ:* Colo Sch Mines, GeolE, 61; Univ Wyoming, MS, 64; Colo State Univ, PhD(civil eng), 68. *Prof Exp:* Assoc petrol engr, 64-65, res civil engr, 68-74, sr res civil eng, 74-79, supvr, 79-80, sr coordr, 80-85, MGR FORMATION EVAL, TEXACO, INC, BELLAIRE, 85- *Mem:* Soc Petrol Engrs. *Res:* Porous media flow; well test design; pressure transient techniques in performance and analysis; petroleum reserve evaluation; enhanced recovery techniques; resevior simulation. *Mailing Add:* 322 Wilchester Blvd Houston TX 77024

WHITE, NATHANIEL MILLER, b Providence, RI, Feb 28, 41; m 67; c 3. ASTRONOMY. *Educ:* Earlham Col, AB, 64; Ohio State Univ MSc, 67, PhD(astron), 71; Northern Ariz Univ, BScE, 84. *Prof Exp:* Res assoc astron, 69-71, astronr, 72-78, SR ASTRONR, LOWELL OBSERV, 78- *Concurrent Pos:* Rac astron, Yavapai Community Col, 77-78; prin investr, NSF, 72-90; elected coun mem & vice-mayor, city Flagstaff, 88-92. *Mem:* Am Astron Soc; Int Astron Union; Inst Elec & Electronics Engrs; Sigma Xi; Soc Photo-optical Instrumentation Engrs. *Res:* Basic data on the atmospheres of cool stars; lunar occultation observations; absolute flux measurements of stars; instrumentation; telescope design and control. *Mailing Add:* Lowell Observ Mars Hill Rd 1400 W Flagstaff AZ 86001

WHITE, NILES C, b Saragossa, Ala, Feb 14, 22; m 52; c 2. CHEMICAL ENGINEERING. *Educ:* Univ Ala, BS, 50. *Prof Exp:* Chem engr, US Naval Ord Sta, 50-51 & Redstone Arsenal, 51-56; supv chemist, Army Rocket & Guided Missile Agency, US Army Missile Command, 56-62, supv res chemist, 62-64, chief, Solid Propellant Chem Br, 64-80; TECH ADV, PROP DIV, ATLANTIC RES CORP, 80- *Honors & Awards:* Res & Develop Achievement Award, Dept Army, 61. *Mem:* AAAS; Am Chem Soc; Am Inst Chem Engrs; Am Inst Aeronaut & Astronaut. *Res:* Rocket propulsion; propellants; combustion; polymer crystallinity; physical properties of elastomers. *Mailing Add:* 823 Watts Dr Huntsville AL 35801-2057

WHITE, NOEL DAVID GEORGE, b Simcoe, Ont, Dec 16, 51; m 73; c 2. STORED PRODUCTS, ECOSYSTEM ANALYSIS. *Educ:* Univ Guelph, BScAgr, 74, MSc, 76; Univ Man, PhD(entomol), 79. *Prof Exp:* Res assoc, Dept Agr Eng, Univ Man, 79-81; RES SCIENTIST, AGR CAN RES STA, 81-, PROJ LEADER, 89- *Concurrent Pos:* Adj prof, Dept Agr Eng, Univ Man, 87- *Mem:* Entom Soc Can; Entom Soc Am; Agr Inst Can; Sigma Xi; Can Fedn Biol Soc. *Res:* Improved basis to manage stored cereals, oil seeds and their products with minimal quality loss from insects, mites and molds, using a multi-disciplinary ecosystem approach. *Mailing Add:* Agr Can Res Sta 195 Dafoe Rd Winnipeg MB R3T 2M9 Can

WHITE, NORMAN EDWARD, b Springfield, Ohio, Jan 20, 17; m 54; c 1. PHYSICAL CHEMISTRY. *Educ:* Wittenberg Univ, AB, 38; Univ Pa, MS, 41, PhD(phys chem), 54. *Prof Exp:* From instr to prof chem, Drexel Univ, 47-65; prof chem, Bloomsburg State Col, 65-85, chmn dept, 65-71; RETIRED. *Mem:* AAAS; Am Chem Soc; Torch Club. *Res:* Hydrogen bond association; molecular weights in solution. *Mailing Add:* Six Kent Rd Bloomsburg PA 17815-8553

WHITE, PAUL A, b Hollywood, Calif, Aug 21, 15; m 39; c 4. MATHEMATICS. *Educ:* Univ Calif, Los Angeles, AB, 37, MA, 39; Univ Va, PhD(math), 42. *Prof Exp:* Asst math, Univ Calif, Los Angeles, 37-39; instr, Univ Va, 39-42; asst prof, La State Univ, 42-46; from asst prof to assoc prof, 46-53, PROF MATH, UNIV SOUTHERN CALIF, 53- *Concurrent Pos:* Asst prof, Tulane Univ, 44; mathematician, Ballistics Res Lab, Aberdeen Proving Ground, 45; vis prof, Univ Innsbruck, 60-61; writer & lectr, African Ed Proj, 66-68; NSF lectr, India, 67; writer, UNESCO Arab Math Proj, 69-70; writer & adv bd mem, Sec Sch Math Curric Improv Study, 70-72; mem adv bd, Sch Math Study Group, 70-72. *Mem:* Am Math Soc. *Res:* Topology; R-regular convergence spaces. *Mailing Add:* 1019 W 52nd Los Angeles CA 90037

WHITE, PAUL C, b Boston, Mass, Oct 2, 41; m 64; c 2. PHYSICS. *Educ:* Harpur Col, BA, 63; State Univ NY Binghamton, MA, 66; Univ Tex, Austin, PhD(physics), 70. *Prof Exp:* from asst to assoc prof physics, St Edward's Univ, 69-75, chmn, Div Physics & Biol Sci, 72-74,; mem staff, Ctr Nat Security Studies, Los Alamos Nat Labs, 75-80, group leader, thermonuclear appln group, 80-83, assoc div leader, appl theoret phys div, 83-85, dep dir, 85-88, actg dir, 88-89, DIV LEADER, APPL THEORET PHYS DIV, CTR NAT SECURITY STUDIES, LOS ALAMOS NAT LABS, 89- *Mem:* AAAS; Am Asn Physics Teachers; Int Inst Strategic Studies. *Res:* General relativistic astrophysics; inhomogeneous cosmologies; relativistic transport theory; radiation biophysics; radiation transport; national security policy and arms control technology. *Mailing Add:* Los Alamos Nat Lab PO Box 1663 MS-B218 Los Alamos NM 87545

WHITE, PETER, b Philadelphia, Pa, June 12, 30; m 53; c 4. HEMATOLOGY. *Educ:* Yale Univ, BA, 51; Univ Pa, MD, 55. *Prof Exp:* From assoc to asst prof med, Sch Med, Univ Pa, 63-69, assoc dir clin res ctr, 67-69; assoc prof med, Med Col Ohio, 69-72, prof med & chief div hemat, 72-77, dep chmn dept med, 69-75; dir med & prof med, Presby Univ Pa Med Ctr, 77-85; ASSOC DEAN & PROF MED, MED COL OHIO, 85- *Concurrent Pos:* USPHS res fel hemat, Sch Med, Univ Pa, 63-65; mem res in nursing in patient care rev comt, Bur Health Prof Educ & Manpower Training, NIH, 70-75; mem coun thrombosis, Am Heart Asn. *Mem:* Am Col Physicians; AAAS; Am Soc Hemat; Am Fedn Clin Res; Int Soc Haemostasis and Thrombosis; Am Geriat Soc. *Res:* Hemoglobin metabolism; platelet metabolism; Alzheimer's disease; cancer in the elderly. *Mailing Add:* Dept Med Med Col Ohio Toledo OH 43699

WHITE, PHILIP CLEAVER, b Chicago, Ill, May 10, 13; m 39; c 3. RESEARCH ADMINISTRATION, CHEMISTRY & FOSSIL FUEL TECHNOLOGY. *Educ:* Univ Chicago, BS, 35, PhD(org chem), 38. *Prof Exp:* Res chemist, Standard Oil Co, Ind, 38-45, group leader, 45, chief chemist, 46-50, div dir, 50-51, mgr res, 56-58, gen mgr res & develop, 58-60, gen mgr res, Amoco Res Ctr, 69-75; mgr res & develop, Pan Am Refining Corp, 51-56, gen mgr res & develop, Am Oil Co, 61-65, vpres res & develop, 66-69; gen mgr res, Amoco Res Ctr, Standard Oil Co, Ind, 69-75; asst adminr, ERDA, US Govt, 75-77; sr tech adv, Dept Energy, 77-78; pres & owner, Energy Consults Inc, 79-88; RETIRED. *Concurrent Pos:* Pres, Indust Res Inst, 71-72, pres coord res coun, 73-75; consult, 78-79. *Mem:* Am Chem Soc; fel Am Inst Chem Engrs. *Res:* Petroleum products, processes and analysis; research administration. *Mailing Add:* 1812 Kalorama Sq Washington DC 20008-4022

WHITE, R MILFORD, physical chemistry; deceased, see previous edition for last biography

WHITE, RALPH E, b Clovis, NMex, Nov 6, 42; m 84; c 4. ELECTROCHEMICAL SYSTEMS, MATHEMATICAL MODELLING. *Educ:* Univ SC, BS, 71; Univ Calif, Berkeley, MS, 73, PhD(chem eng), 77. *Prof Exp:* From asst prof to assoc prof, 77-85, PROF CHEM ENG, TEX A&M UNIV, 85-, ASSOC DEPT HEAD & DIR, CTR ELECTROCHEM ENG, 90- *Concurrent Pos:* Prin investr, Tex A&M Univ, 77-; consult, Dow Chem, 79- *Mem:* Electrochem Soc (treas, 90-); Am Inst Chem Engrs. *Res:* Electrochemical systems; mathematical modelling; fuel cells; plating; corrosion. *Mailing Add:* 2912 Arroyo Ct S College Station TX 77840

WHITE, RALPH LAWRENCE, JR, b Troy, NC, June 19, 41; m 68. SYNTHETIC ORGANIC CHEMISTRY. *Educ:* Univ NC, Chapel Hill, BS, 63; Ind Univ, Bloomington, PhD(org chem), 67. *Prof Exp:* Fel, 67-68; instr med chem, Sch Pharm, Univ NC, Chapel Hill, 68-69; sr res chemist, 69-80, RES ASSOC, NORWICH-EATON PHARMACEUT, 80- *Mem:* Am Chem Soc. *Res:* Chemistry and synthesis of thiophene compounds; synthesis of potential biologically active compounds. *Mailing Add:* Two Hillview Dr Norwich NY 13815-1007

WHITE, RAY HENRY, b Lakewood, Ohio, Apr 28, 36; m 62; c 3. ARTIFICIAL NEURAL NETWORKS. *Educ:* Calif Inst Technol, BS, 57; Univ Calif, Berkeley, PhD(physics), 64. *Prof Exp:* Res asst physics, Univ Calif, 58-63; lectr, Univ Singapore, 63-67; asst prof, Calif State Polytech Col, 67-68; asst prof physics, 68-70, chmn dept, 70-72 & 84-85, chmn dept sci & math, 72-75 & 76-78, assoc prof, 70-81, PROF PHYSICS & COMPUT SCI, UNIV SAN DIEGO, 81- *Concurrent Pos:* Vis fel, Univ Singapore, 75-76; vis scholar, Univ Calif, San Diego, 82-83, 89-90. *Mem:* Int Neural Network Soc; Comput Soc; Inst Elec & Electronics Engrs. *Res:* Learning algorithms in neural networks; superconductivity; physics of music. *Mailing Add:* Dept Physics Univ San Diego San Diego CA 92110

WHITE, RAYMOND E, b Freeport, Ill, May 6, 33; m 56; c 3. ASTRONOMY. *Educ:* Univ Ill, Urbana, BS, 55, PhD(astron), 67. *Prof Exp:* From instr to asst prof astron, 64-72, asst dir, 72-74, res assoc & lectr, 74-81, ASSOC PROF & ASSOC ASTRONR, STEWARD OBSERV, UNIV ARIZ, 81- *Concurrent Pos:* Sr vis scholar, Inst Astron, Cambridge Univ, Eng, 80; fac fel, Univ Ariz, 88- *Honors & Awards:* Harlow Shapley Lectr, Am Astron Soc, 78. *Mem:* AAAS; Am Astron Soc; Royal Astron Soc; Int Astron Union; Sigma Xi; Am Asn Univ Professors. *Res:* Observational astronomy; structure of the Milky Way Galaxy, particularly with respect to the identification and distribution of Population II stellar component; archaeo-astronomy especially of Inca culture in Peru. *Mailing Add:* Steward Observ Univ Ariz Tucson AZ 85721

WHITE, RAYMOND GENE, b Elana, WVa, Oct 10, 30; m 52; c 2. LABORATORY ANIMAL MEDICINE. *Educ:* Okla State Univ, BS, 58, DVM, 60; Univ Nebr, MS, 71. *Prof Exp:* Pvt vet pract, 60-64; res vet, Chemagro Corp, 64-69; res vet, Univ Nebr, 69-76, dir N Platte Sta, 76-80; assoc dean, Col Vet Med, Miss State Univ, 80-82; coordr Reg Off Vet Med, 82-85, DIR INST ANIMAL CARE PROG, UNIV NEBR, 85- *Concurrent Pos:* Consult, Vet Med Mgt. *Mem:* Am Vet Med Asn; Am Asn Exten Vet; Am Asn Bovine Practitioners; AAAS; Am Soc Animal Sci; fel Am Acad Vet Pharmacol & Therapeut. *Res:* Initiating and supervising field research activities involving animal health products, pesticides and anthelmintics; author or co-author of over 50 scientific and non-scientific publications. *Mailing Add:* 107 Vet Basic Sci Lincoln NE 68583-0944

WHITE, RAYMOND PETRIE, JR, b New York, NY, Feb 13, 37; m 61; c 2. ANATOMY, ORAL SURGERY. *Educ:* Med Col Va, DDS, 62, PhD(anat), 67; Am Bd Oral Surg, dipl, 74. *Prof Exp:* Intern oral surg, Med Col Va, 64-65, from asst resident to resident, 65-67; from asst prof to assoc prof, Col Dent, Univ Ky, 67-71, chmn dept, 69-71; asst dean admin affairs & prof oral surg, Va Commonwealth Univ, 71-74; dean sch dent, 74-81, PROF ORAL & MAXILLOFACIAL SURG, 74-, ASSOC DEAN SCH MED, 81- *Concurrent Pos:* Mem adv comt, Am Bd Oral Surg, 74-77, Fayetteville Vet Admin Hosp, NC, 74-; assoc chief staff, NC Mem Hosp, 81- *Mem:* Inst Med-Nat Acad Sci; Am Acad Oral Path; Int Asn Dent Res; Am Asn Oral & Maxillofacial Surg; Sigma Xi. *Res:* Correction facial deformity with surgery-orthodontic therapy; dental health policy & health care delivery. *Mailing Add:* Dept Oral Maxillofacial Surg Univ NC Sch Dent Chapel Hill NC 27599-7450

WHITE, RICHARD ALAN, b Philadelphia, Pa, Oct 25, 35; m 65; c 3. DEVELOPMENTAL ANATOMY, MORPHOLOGY. *Educ:* Temple Univ, BS & MEd, 57; Univ Mich, MA, 59, PhD(bot), 62. *Prof Exp:* NSF fel, Univ Manchester, 62-63; from asst prof to assoc prof, 63-73, prof plant anat, Duke Univ, 73-, chmn dept, 76-85, DEAN ART & SCI & TRINITY COL, 85- *Concurrent Pos:* Treas, orgn trop studies, 85- *Mem:* AAAS; Bot Soc Am; Torrey Bot Club; Am Fern Soc; Int Soc Plant Morphol; Am Phys Soc. *Res:* Comparative morphology of tracheary cells of ferns; developmental studies of fern stellar patterns; comparative and developmental studies of lower vascular plants. *Mailing Add:* Dept Bot Duke Univ Durham NC 27706

WHITE, RICHARD ALLAN, b Boston, Mass, June 9, 46. ASTROPHYSICS. *Educ:* Univ Calif, Berkeley, AB, 68; Univ Chicago, MS, 71, PhD(astron), 78. *Prof Exp:* Res asst astron, Univ Chicago, 69-74; res assoc radio astron, Nat Radio Astron Observ, 77-80; NAT ACAD SCI & NAT RES COUN ASSOC, GODDARD SPACE FLIGHT CTR, NASA, 80- *Mem:* Am Astron Soc. *Res:* Galaxies; clusters of galaxies. *Mailing Add:* 8006 Lakecrest Dr Greenbelt MD 20770

WHITE, RICHARD EARL, b Akron, Ohio, Aug 23, 33; m 77; c 2. SYSTEMATIC ENTOMOLOGY. *Educ:* Univ Akron, BS, 57; Ohio State Univ, MSc, 59, PhD(entom), 63. *Prof Exp:* Asst prof zool, Union Col, Ky, 64-65; RES ENTOMOLOGIST, USDA, 65- *Mem:* AAAS; Entom Soc Am; Coleopterists Soc. *Res:* Taxonomy of Coleoptera, especially the families Anobiidae and Chrysomelidae; insect illustrations. *Mailing Add:* Syst Entom Lab c/o US Nat Mus Natural Hist Washington DC 20560

WHITE, RICHARD EDWARD, b Chicago, Ill, Jan 18, 44; m 70, 88. INTERSTELLAR MATTER, OBSERVATIONAL ASTRONOMY. *Educ:* St Joseph's Col, BS, 65; Columbia Univ, PhD(astron), 71. *Prof Exp:* Carnegie fel, Hale Observ, 70-72; resident res assoc, Goddard Inst Space Studies, NASA, 72-74; asst prof, 74-75 & 76-82, lectr, 75-76, ASSOC PROF ASTRON, SMITH COL, 82- *Concurrent Pos:* Mem, Users Comt, Kitt Peak Nat Observ, 77-79, NASA Proposal Rev Comt, Int Ultraviolent Explorer, 81 & Rev Comt, Dept Physics & Astron, Bowdoin Col, 86; proj dir, NSF Grant Instrnl Sci Equip, 78-80; vis astronr, Smithsonian Inst, Moent Hopkins Observ, 80, F L Whipple Observ, 82-83, Smithsonian Astrophys Observ, 84 & Wyo Infrared Observ, 87-88; assoc chair, Undergrad Prog, Five Col Astron Dept, 84-87 & 89-90; prin investr, NASA res grant, 86-87; supply assoc prof, Univ Wyo, 87; asst to pres, Smith Col, 88-89. *Mem:* Am Astron Soc; Astron Soc Pac; AAAS; Int Astron Union; Sigma Xi. *Res:* Observational investigation of the interstellar medium, primarily at visual and infrared wavelengths, with particular attention to the region near the Pleiades star cluster. *Mailing Add:* Five Col Astron Dept Clark Sci Ctr Smith Col Northampton MA 01060

WHITE, RICHARD JOHN, biochemistry, microbiology, for more information see previous edition

WHITE, RICHARD KENNETH, b Mercer, Pa, Mar 20, 30; m 72; c 2. ORGANIC RESIDUES, WATER QUALITY. *Educ:* Pa State Col, BS, 52; NY Theological Sem, STB, 59; Pa State Univ, MS, 66; Ohio State Univ, PhD(agr eng), 69. *Prof Exp:* From asst prof to prof agr eng, Ohio State Univ, 70-85; NEWMAN PROF NATURAL RESOURCES ENG, CLEMSON UNIV, 85- *Honors & Awards:* Gunlogson Countryside Eng Award, Am Soc Agr Engrs, 80. *Mem:* Am Soc Agr Engrs; Soil Conserv Soc Am. *Res:* Application and utilization of organic residues (livestock manures, municipal and industrial sludges) on cropland; planning and design of municipal solid waste systems; control of odors from livestock facilities; water resource management including surface and groundwater. *Mailing Add:* Dept Agr Eng Clemson Univ 106 McAdams Hall Clemson SC 29634-0357

WHITE, RICHARD MANNING, b Denver, Colo, Apr 25, 30; div; c 2. APPLIED PHYSICS, ELECTRICAL ENGINEERING. *Educ:* Harvard Col, BA, 51; Harvard Univ, AM, 52, PhD(appl physics), 56. *Prof Exp:* Mem tech staff microwave electronics, Gen Elec Microwave Lab, 56-62; PROF ELEC ENG & COMPUT SCI, UNIV CALIF, BERKELEY, 62- *Concurrent Pos:* Fel, John Simon Guggenheim Found, 68-69. *Honors & Awards:* Cleo Brunetti Award, Inst Elec & Electronics Engrs, 86. *Mem:* Fel Inst Elec & Electronics Engrs; Am Inst Physics; AAAS. *Res:* Sensors, ultrasonics, chiefly surface and plate acoustic waves; solar energy. *Mailing Add:* Dept Elec Eng & Comput Sci Univ Calif Berkeley CA 94720

WHITE, RICHARD NORMAN, b Chetek, Wis, Dec 21, 33; m 57; c 2. STRUCTURAL ENGINEERING. *Educ:* Univ Wis, BS, 56, MS, 57, PhD(plate stability), 61. *Prof Exp:* Engr, John A Strand, 58-59; instr struct, Univ Wis, 58-61; from asst prof to assoc prof, Cornell Univ, 61-72, dir, sch civil & environ eng, 78-84, assoc dean engr, 87-90, PROF STRUT ENG, CORNELL UNIV, 72-, JAMES A FRIEND FAMILY DISTINGUISHED PROF ENG, 88- *Concurrent Pos:* Consult, Oak Ridge Nat Labs, 66-68; staff assoc, Gen Atomic Div, Gen Dynamics Corp, Calif, 67-68; vis prof, Univ Calif, Berkeley, 74-75; consult, Sandia Nat Lab, 81- *Honors & Awards:* Collingwood Prize, Am Soc Civil Engrs, 67. *Mem:* Fel Am Soc Civil Engrs; fel Am Concrete Inst; Am Soc Eng Educ; Nat Soc Prof Engrs. *Res:* Structural engineering and model analysis; nuclear power plant structures; behavior of reinforced concrete structures; earthquake engineering. *Mailing Add:* 54 Sunnyslope Rd Ithaca NY 14850

WHITE, RICHARD PAUL, b Gary, Ind, May 27, 25; m 47; c 2. PHARMACOLOGY. *Educ:* Ind State Univ, BS, 47; Univ Kans, MS, 49, PhD(physiol), 54. *Prof Exp:* Instr physiol, Univ Kans, 49-54; psychophysiologist, Galesburg State Res Hosp, Ill, 54-56; from instr to assoc prof, 56-70, PROF PHARMACOL, MED UNITS, UNIV TENN, MEMPHIS, 70- *Concurrent Pos:* NIH career develop award, 59-69. *Mem:* Soc Biol Psychiat; Int Col Neuropsychopharmacol; Soc Neurosci; Int Soc Biochem Pharmacol; Sigma Xi. *Res:* Neuropharmacology. *Mailing Add:* Dept Pharmacol Univ Tenn Med Ctr 800 Madison Ave Memphis TN 38163

WHITE, RICHARD WALLACE, underwater acoustics, for more information see previous edition

WHITE, ROBERT ALLAN, b Chicago, Ill, Dec 16, 34; m 57; c 2. AERODYNAMICS, HEAT TRANSFER. *Educ:* Univ Ill, Urbana, BS, 57, MS, 59, PhD(mech eng), 63. *Prof Exp:* Instr mech eng, Univ Ill, 59-63; aeronaut engr, Aeronaut Res Inst Sweden, 60-61, sr res scientist, 63-65; from asst prof to assoc prof, 65-72, PROF MECH ENG, UNIV ILL, URBANA, 72- & DIR AUTOMOTIVE SYSTS LAB, 79- *Concurrent Pos:* NATO sr fel & Thord-Gray fel, Aeronaut Res Inst Sweden, 68, consult, 69. *Honors & Awards:* Ralph R Teetor, Soc Automotive Engrs; Assoc fel, Am Inst Aeronaut & Astronaut, 86. *Mem:* Sigma Xi. *Res:* Aerodynamics of propulsion systems and vehicle integration; separated flows at subsonic and supersonic mach numbers; dynamics and aerodynamics of automotive systems. *Mailing Add:* 2009 S Cottage Grove Urbana IL 61801

WHITE, ROBERT ALLEN, b Las Cruces, NMex, Nov 19, 44; m 71; c 2. BIOMATHEMATICS, CHEMICAL PHYSICS. *Educ:* NMex State Univ, BS, 66; Univ Chicago, PhD (chem physics), 70. *Prof Exp:* Fel chem, Rice Univ, 70-75; proj investr, 75-76, asst prof, 76-81, ASSOC PROF BIOMATH, M D ANDERSON HOSP & TUMOR INST, 81- *Mem:* Am Phys Soc. *Res:* Mathematical biology; estimation of cell kinetics parameters, demography and the statistical mechanics of DNA denaturation. *Mailing Add:* M D Anderson Cancer Ctr Dept Biomath Univ Tex 6723 Bertner Ave Houston TX 77030

WHITE, ROBERT B, b Ennis, Tex, Jan 5, 21; m 42; c 3. PSYCHIATRY. *Educ:* Tex A&M Univ, BS, 41; Univ Tex, MD, 44; Western New Eng Inst Psychoanal, cert, 59. *Prof Exp:* From asst psychiatrist to sr psychiatrist, Austen Riggs Ctr, Inc, 51-62; assoc prof psychiat, 62-67, PROF PSYCHIAT, UNIV TEX MED BR GALVESTON, 67- *Concurrent Pos:* Teaching analyst, New Orleans Psychoanal Inst, 62-66, training analyst psychoanal, 66-; consult, Alcoholism Res Proj, Col Med, Baylor Univ, 63-65 & Hedgecroft Hosp, Houston, 63-65; mem gen planning comt comprehensive statewide ment health prog planning, Tex State Dept Health, Austin, 63-65; mem, Residency Rev Comt Psychiat & Neurol, 71-77; training analyst, Houston-Galveston Psychoanalytic Sch, 74- *Honors & Awards:* David Rapaport Prize, Western New Eng Inst Psychoanal, 59. *Mem:* AAAS; Am Psychoanalytic Asn; fel Am Psychiat Asn; fel Am Col Psychiatrists; fel Am Col Psychoanal. *Res:* Psychoanalysis. *Mailing Add:* Dept Psychiat Univ Tex Med Br Galveston TX 77550

WHITE, ROBERT E(DWARD), b Jersey City, NJ, Aug 31, 17; m 43; c 5. CHEMICAL ENGINEERING. *Educ:* Polytech Inst Brooklyn, BChE, 38, MChE, 40, DChE, 42. *Prof Exp:* Asst chem eng, Polytech Inst Brooklyn, 38-42; jr engr, Vulcan Copper & Supply Co, Ohio, 39-40; res engr, York Corp, Pa, 42-47; asst prof chem eng, Bucknell Univ, 47-49; head dept, 49-83, from assoc prof to prof, 49-86, EMER PROF CHEM ENG, VILLANOVA UNIV, 86- *Concurrent Pos:* Indust consult, 51- *Mem:* fel Am Inst Chem Engrs. *Res:* Drying, pressing and formation in papermaking; heat and mass transfer; waste treatment. *Mailing Add:* Dept Chem Eng Villanova Univ Villanova PA 19085

WHITE, ROBERT J, b Duluth, Minn, Jan 21, 26; m 50; c 10. SURGERY, NEUROPHYSIOLOGY. *Educ:* Univ Minn, BS, 51, PhD(neurosurg physiol), 62; Harvard Univ, MD, 53. *Hon Degrees:* DSc, John Carroll Univ, 79 & Cleveland State Univ, 80. *Prof Exp:* Intern surg, Peter Bent Brigham Hosp, 53-54; resident, Boston Children's Hosp & Peter Bent Brigham Hosp, 54-55; asst to staff, Mayo Clin, 58-59, res assoc neurophysiol, 59-61; from asst prof to assoc prof, 61-66, co-chmn dept, 72-83, PROF NEUROSURG, SCH MED, CASE WESTERN RESERVE UNIV, 66-; DIR NEUROSURG & BRAIN RES LAB, CLEVELAND METROP GEN HOSP, 61- *Concurrent Pos:* Fel neurosurg, Mayo Clin, 55-58; assoc neurosurgeon, Univ Hosps & sr attend neurosurgeon, Vet Admin Hosp, 61-; vis prof, var univs; consult, Burdenko Neurosurg Inst, Moscow, Polenov Inst, Leningrad, Neurosurg Inst, Kiev & Univs, Rome, Naples, Milan & Palermo; fifteen lectrships, USSR, 61-81 & People's Repub China, 77-81; mem bd dirs, Allen Mem Med Library & Int Ctr Artificial Organs & Transplantation. *Honors & Awards:* L W Freeman Award, Nat Paraplegia Found, 77; Svien lectr, Mayo Clin, 77. *Mem:* Int Soc Transplantation; Am Physiol Soc; Soc Univ Surgeons (past pres); Soc Univ Neurosurg (pres, 77); Am Col Surg. *Res:* Special neurosurgical techniques for vascular disease and tumors of the brain utilizing low temperature and extracorporeal perfusion systems; treatment for spinal cord and head injury involving chemotherapy and low temperature; isolation of the subhuman primate brain and perfect brain isolation; neurochemical and circulatory studies; mind/brain relationship, bioethics and dynamics in health care delivery in the United States and communist countries. *Mailing Add:* Dept Neurosurg Cleveland Metrop Gen Hosp Cleveland OH 44109

WHITE, ROBERT KELLER, b Greeneville, Tenn, Mar 3, 30; m 52; c 2. PSYCHOPHYSIOLOGY, RADIOBIOLOGY. *Educ:* Milligan Col, BA, 52; Univ Tex, PhD(psychol), 62. *Prof Exp:* Res scientist, Radiobiol Lab, Balcones Res Ctr, Univ Tex, 56-61; asst prof psychol, Tex Tech Col, 61-65; proj dir, Armed Forces Radiobiol Res Lab, Defense Atomic Support Agency, 65-67; mem tech staff space res, Bellcomm, Inc, 67-70; chmn dept, 70-71, PROF PSYCHOL, WILLIAM PATERSON COL NJ, 70- *Concurrent Pos:* Lectr, USDA Grad Sch, 65- & Col Gen Studies, George Washington Univ, 66-67. *Mem:* Am Psychol Asn; Simulation Coun; Am Soc Cybernet. *Res:* Effects of whole body irradiation upon the physiology and behavior of various species; in uterine irradiation of rats; gamma neutron pulse irradiation upon the psychophysiology of monkeys. *Mailing Add:* Dept Psychol William Paterson Col NJ 300 Pompton Rd Wayne NJ 07470

WHITE, ROBERT LEE, b Plainfield, NJ, Feb 14, 27; m 52; c 4. ELECTRICAL ENGINEERING. *Educ:* Columbia Univ, BA, 49, MA, 51, PhD(physics), 54. *Prof Exp:* res physicist, Res Labs, Hughes Aircraft Co, 54-61; head magnetics dept, Labs, Gen Tel & Electronics Corp, 61-63; dir, Inst Electronics Med, 73-81, chmn dept elec eng, 81-87, PROF ELEC ENG & MAT SCI, STANFORD UNIV, 63- *Concurrent Pos:* Indust consult, 63-; Guggenheim fel, Oxford Univ, 69-70; vis prof, Tokyo Univ, 75; Guggenheim fel, Swiss Fed Inst Technol, Zurich, 78, Christensen fel, St Catherine's Col, Oxford, 86; spec ltd partner, Mayfield Fund, Mayfield II, 69-80; consult ltd partner, Alpha Partners, 84-; dir Analog Design Tools, 84-88; dir Biostim, Inc, 79-85; consult, IBM, Varian, Novacor, Ampex & Lockheed; William E Ayer prof elec eng, 85-; dir, Exploratorium, 87-89, Ctr Res Info Storage Mat, 90- *Mem:* Fel Am Phys Soc; fel Inst Elec & Electronics Engrs. *Res:* Microwave spectroscopy; solid state physics, especially magnetics; neurophysiology; neural prostheses, especially auditory. *Mailing Add:* Dept Mat Sci & Eng Stanford Univ Stanford CA 94305

WHITE, ROBERT LESTER, b Halifax, NS, Apr 25, 53; m 77; c 2. BIOSYNTHESIS OF NATURAL PRODUCTS, ENZYMES OF SECONDARY METABOLISM. *Educ:* Dalhousie Univ, BSc Hons, 74; McMaster Univ, PhD(chem), 80. *Prof Exp:* Nat Sci & Eng Res Coun postdoctoral biochem, Univ Oxford, UK, 79-81; Nat Sci & Eng Res Coun res fel, Syntex Inc, Ont, 81-83 & Dalhousie Univ, 83-84; asst prof chem, Acadia Univ, 84-90; ASST PROF CHEM, DALHOUSIE UNIV, 90- *Mem:* Am Chem Soc; Chem Inst Can; Royal Soc Chem. *Res:* Bioorganic chemistry; biosynthesis of amino acids and antibiotics; enzymes of secondary metabolism; analysis of amino acids and aminosugars by HPLC; production of amino acids by fermentation; chemical synthesis of amino acids. *Mailing Add:* Dept Chem Dalhousie Univ Halifax NS B3H 4J3 Can

WHITE, ROBERT M, b Boston, Mass, Feb 13, 23; m 48; c 2. METEOROLOGY. *Educ:* Harvard Univ, BA, 44; Mass Inst Technol, MS, 49, PhD, 50. *Hon Degrees:* DSc, Long Island Univ, 76, Rensselaer Polytech Inst, 77 & Univ Wis, 78. *Prof Exp:* Asst meteorol, Mass Inst Technol, 48-50; chief large scale processes sect, Air Force Cambridge Res Ctr, 52-58 & meteorol develop lab, 58-59; res assoc, Mass Inst Technol, 59; assoc dir res, Travelers Ins Co, 59-60, pres, Travelers Res Ctr, Inc, 60-63; chief weather bur, 63-65 & Environ Sci Serv Admin, 65-70, adminr, Nat Oceanic & Atmospheric Admin, US Dept Com, 71-77; chmn climate res bd, Nat Acad Sci, 77-79, adminr, Nat Res Coun & exec officer, Nat Acad Sci, 79-80; pres, Univ Corp Atmospheric Res, 80-83; PRES, NAT ACAD ENG, 83- *Concurrent Pos:* US permanent rep, World Meteorol Orgn, 63-78, mem exec comt, 63-78; fed coordr, Meteorol Serv & Supporting Res, 64-70; co-chmn, Dept Com Meteorol Satellite Prog Rev Bd, NASA, 64-73; mem, President's Comn Marine Sci Eng & Resources, 67-68; mem comt water resources res, Fed Coun Sci & Technol, 67-75; chmn, Interagency Comt Marine Sci & Eng, 70-76; chmn fed comt, Meteorol Serv & Supporting Res, 70-77; chmn, Nat Marine Fisheries Adv Comt, 70-77; mem US deleg, UN Conf Human Environ, Stockholm, 71; chief US deleg, Intergovt Oceanog Comn, UNESCO, Paris, 72-73; mem US deleg, Gov Coun, UN Environ Prog, Nairobi, 73-74; US comnr, Int Whaling Comn, London, 73-77; US chmn, US-France Coop Prog Oceanog, 73-77; chief US deleg, Conf Global Environ Monitoring Systs, UN Environ Prog, 74; US chmn, US-USSR Joint Comn Explor World Oceans, 74-77; chmn comt climate change, White House Domestic Coun, 75, chmn comt weather modification, 76; mem US deleg, UN Conf Desertification, Nairobi, 77; mem coun, Nat Acad Eng, 77-; chmn comt atmosphere & oceans, Fed Coord Coun Sci, Eng & Technol, 77; mem, Comt Atmospheric Sci, Nat Res Coun, 78-, mem, Space Appln Bd, 78- & vchmn; mem US deleg, ICSU Gen Assembly, Athens, 78. *Honors & Awards:* Cleveland Abbe Award, Am Meteorol Soc, 69, Fiftieth Anniversary Medal, 70; Matthew Fontaine Maury Award, Smithsonian Inst, 76; Int Conserv Award, Nat Wildlife Fedn, 77; Neptune Award, Am Oceanic Orgn, 77; Spec Award, Marine Technol Soc, 77; Charles E Lindbergh Award; Int Meteorol Orgn Prize, World Meteorol Orgn. *Mem:* Nat Acad Eng (pres, 83-); Am Geophys Union; Am Meteorol Soc; AAAS; Marine Technol Soc (vpres, 75-77); Royal Meteorol Soc. *Res:* Atmospheric and ocean sciences; environmental science. *Mailing Add:* Nat Acad Eng 2101 Constitution Ave NW Washington DC 20418

WHITE, ROBERT MARSHALL, b Reading, Pa, Oct 2, 38. SOLID STATE PHYSICS. *Educ:* Mass Inst Technol, BS, 60; Stanford Univ, PhD(physics), 64. *Prof Exp:* Res physicist, Lincoln Lab, Mass Inst Technol, 60; res assoc, Microwave Lab, Stanford Univ, 63-64; NSF fel, Univ Calif, Berkeley, 65-66; asst prof physics, Stanford Univ, 66-70; NSF sr fel, Cambridge Univ, 70-71; mgr, Solid State Res Area, Xerox PARC, 71-78 & Storage Technol, Xerox Corp Strategy Off, 78-83, prin scientist, 83-84; vpres res & technol, Data Storage Prod Group, Control Data Res Inc, chief tech officer & vpres res & eng, Control Data Corp & pres, 86-90; UNDER SECY TECHNOL, US DEPT COM, WASHINGTON, DC, 90- *Concurrent Pos:* Vis scientist, Osaka Univ, 63, Ecole Polytech, Paris, 76 & Univ Pernambuco, Brazil, 78; lectr, Dept Appl Physics, Stanford Univ, 71-81; prog comt, MMM Conf, 73-75 & chmn, 81, prog chmn, Int Conf Magnetism, 85; adv comt, Conf Magnetism & Magnetic Mat, 76-78 & 80-85; Humboldt fel, Max Planck Inst, Stuttgart, 81; consult prof appl physics, Stanford Univ & coprin investr, Magnetic Thin Film Res Prog, 82-; adv bd, Ctr Magnetic Rec Res, Univ Calif, San Diego, 84-86; bd dirs & tech adv bd, Microelectronics & Comput Technol Corp, 86-; mem, comt mat sci & eng, Nat Res Coun, 86, Nat Steering Comt, Advan Steady State Neutron Source, 86-, Bd Assessment, Radiation Res Panel, NBS, 87-, Minn Comt Sci & Technol Res & Develop & Nat Mat Adv Bd, 88-; adj prof, Dept Physics, Univ Minn, 87-; mem, Nat Adv Comt on Semiconductors, Mfg Forum, Nat Acad Eng. *Mem:* Fel Inst Elec & Electronics Engrs; fel Am Phys Soc. *Res:* Theory of nonlinear phenomena in ferrites; spinwave theory; magneto-optical phenomena; magnetic properties of amorphous materials; disk storage technology. *Mailing Add:* Under Secy Technol US Dept Com Rm 4824 14th St & Constitution Ave NW Washington DC 20230

WHITE, ROBERT STEPHEN, b Elsworth, Kans, Dec 28, 20; m 42; c 4. ASTROPHYSICS, SPACE PHYSICS. *Educ:* Southwestern Col, Kans, AB, 42; Univ Ill, MS, 43; Univ Calif, PhD(physics), 51. *Hon Degrees:* DSc, Southwestern Col, Kans, 71. *Prof Exp:* Asst physics, Univ Ill, 42-44; asst, Univ Calif, 46-48; physicist, Lawrence Radiation Lab, Univ Calif, 48-61; physicist & head particles & fields dept, Space Physics Lab, Aerospace Corp, 61-67; chmn dept physics, 70-73, PROF PHYSICS, UNIV CALIF, RIVERSIDE, 67-, ASSOC DIR, INST GEOPHYS & PLANETARY PHYSICS, 67- *Concurrent Pos:* Lectr, Univ Calif, 53-54 & 57-59; NSF sr fel, 61-62. *Mem:* AAAS; fel Am Phys Soc; Am Geophys Union; Am Astron Soc. *Res:* Space physics and astrophysics; radiation belts of Earth; albedo neutrons from Earth; solar neutrons and gamma-rays; gamma-rays from astrophysical sources. *Mailing Add:* Dept Physics Univ Calif Riverside CA 92521

WHITE, ROBERTA JEAN, b Loup City, Nebr, Dec 8, 26. VIROLOGY. *Educ:* Univ Nebr, AB, 48, MS, 54; Univ Calif, PhD(bact), 61. *Prof Exp:* Technician, Mayo Clin, 48-51; technician, Col Med, Univ Nebr, 51-52, instr microbiol, 54-56; asst bact, Univ Calif, 56-60; asst prof, 61-68, ASSOC PROF VIROL, COL MED, UNIV NEBR, OMAHA, 68- *Mem:* AAAS; Am Soc Microbiol; Sigma Xi. *Res:* Pathogenesis of virus diseases and multiplication. *Mailing Add:* Dept Med Microbiol Univ Nebr Med Ctr Omaha NE 68105

WHITE, RONALD, b Talladega, Ala, Feb 10, 43; m 67; c 2. COMPUTER SCIENCE, ELECTRONICS. *Educ:* Auburn Univ, BEE, 65, MSEE, 66, PhD(elec eng), 71. *Prof Exp:* Res asst digital res, Auburn Univ, 65- 66; sr engr eng design, Sperry Rand Space Support Div, Sperry Rand Corp, 67-69; res assoc digital res, Auburn Univ, 69-71; engr mgr data process, USAF, Elgin AFB, 71-78; PROF COMPUT SCI, JACKSONVILLE STATE UNIV, 87- *Res:* Application of mini computers and microprocessors. *Mailing Add:* Dept Comput Sci Jacksonville State Univ Jacksonville AL 36265

WHITE, RONALD E, DeSoto, Mo, Feb 21, 48; m; c 2. ENZYMOLOGY, DRUG METABOLISM. *Educ:* Univ Wis, PhD(org chem), 74. *Prof Exp:* From asst prof to assoc prof biochem, Univ Conn, 79-87; ASSOC DIR DRUG METAB, BRISTOL-MYERS SQUIBB, 87- *Concurrent Pos:* Mem, Pharmacol Study Sect, NIH, 90- *Mem:* Am Chem Soc; Am Soc Biol Chemists; AAAS. *Res:* Catalytic chemistry of cytochrome P-450; bioorganic chemistry; drug metabolism; biological oxidation mechanisms; predictive models in biology. *Mailing Add:* Dept Metab & Pharmacokinetics Bristol-Myers Squibb PO Box 4000 Princeton NJ 08543-4000

WHITE, RONALD JEROME, b Wibaux, Mont, Oct 31, 36; m 61; c 2. ZOOLOGY, PHYSIOLOGY. *Educ:* Calif State Polytech Col, BS, 59; Ore State Univ, MS, 61, PhD(physiol), 68. *Prof Exp:* From asst prof to assoc prof, 69-76, PROF BIOL, EASTERN WASH UNIV, 76- *Concurrent Pos:* Chmn, Dept Biol, Eastern Wash Univ, 80-82. *Mem:* AAAS; Am Soc Zool; Sigma Xi. *Res:* Physiology of reproduction in the pigtail macaque. *Mailing Add:* Dept Biol Eastern Wash Univ Cheney WA 99004

WHITE, RONALD JOSEPH, b Opelousas, La, Dec 4, 40; m 63; c 3. THEORETICAL CHEMISTRY, APPLIED MATHEMATICS. *Educ:* Univ Southwestern La, BS, 63; Univ Wis, PhD(phys chem), 68. *Prof Exp:* NSF fel, Oxford Univ, 67-68; assoc, Bell Tel Labs, 68-70; asst prof math, Univ Southwestern La, 70-73; res assoc physiol & biophys, Med Ctr, Univ Miss, 73-75; assoc prof math, Univ Southwestern La, 75-77, prof, 77-81, dir univ honors prog, 75-80; MGR SYSTS ANALYSIS, MGT & TECH SERV CO DIV, GEN ELEC CO, HOUSTON, 80- *Mem:* Am Phys Soc; Sigma Xi; Soc Indust Appl Math; Asn Gifted/Talented Students. *Res:* Perturbation theory; quantum mechanics of small atoms and molecules; mathematical models of physical systems; physiological models. *Mailing Add:* 138 Hickory St Pittsburg TX 75686

WHITE, RONALD PAUL, SR, b Coral Gables, Fla, Mar 1, 35; m 60; c 4. SOIL CHEMISTRY, SOIL FERTILITY. *Educ:* Mass State Col, Bridgewater, BS, 61; Univ Mass, MS, 63; Mich State Univ, PhD(soil sci), 68. *Prof Exp:* Instr soil sci, Mich State Univ, 65-66; RES SCIENTIST, RES STA, CAN DEPT AGR, 68- *Mem:* Am Soc Agron; Soil Sci Soc Am; Agr Inst Can; Can Soc Agron. *Res:* Soil fertility requirements and crop management practices of corn and potatoes; soil manganese levels and plant manganese toxicity; soil phosphorus; plant root cation exchange capacity; agronomy; corn hybrid evaluation; seed potato production. *Mailing Add:* PO Box 1210 Charlottetown PE C1A 7M8 Can

WHITE, ROSCOE BERYL, b Freeport, Ill, Dec 20, 37; m 66; c 1. PLASMA PHYSICS. *Educ:* Univ Minn, BS, 59; Princeton Univ, PhD(physics), 63. *Prof Exp:* Res asst, Princeton Univ, 62, instr, 62-63; res assoc, Univ Minn, 63; US Acad Sci exchange scientist, Lebedev Inst, Moscow, 63-64; vis scientist, Int Ctr Theoret Physics, Italy, 64-66; asst prof physics, Univ Calif, Los Angeles, 66-72; mem, Inst Advan Study, 72-74; RES PHYSICIST, PRINCETON UNIV, 74- *Mem:* Am Phys Soc. *Res:* Theoretical plasma physics. *Mailing Add:* Plasma Physics Lab James Forrestal Campus Princeton Univ PO Box 451 Princeton NJ 08530

WHITE, ROSEANN SPICOLA, b Tampa, Fla, Aug 4, 43; m 65; c 2. BIOCHEMISTRY, IMMUNO-CHEMISTRY. *Educ:* Univ Fla, BS, 65; Univ Tex Southwestern Med Sch, Dallas, PhD(biochem), 70. *Prof Exp:* From asst prof to assoc prof, 72-86, PROF MICROBIOL, UNIV CENT FLA, 86- *Concurrent Pos:* Clin chemist, Orange Mem Hosp, 69-72; consult, Plant Fingerprinting for Patent Atty, 78-, allergy & immunol, Found Cent Fla, 80-; chmn med sci sect, Fla Acad Sci, 85-; lectr path, Residency Prog & Med Technol Prog. *Mem:* AAAS; Am Soc Microbiol; Sigma Xi. *Res:* Control vitamin B-6 biosynthesis; vitamin B-6 regulation of apoenzyme levels; isolation and immuno-chemical analysis of allergens; chemical characterization of H capsulatum antigens; lymphocytes subsets in acquired immune deficiency syndrome and acquired related complex; plant fingerprinting. *Mailing Add:* Dept Biol Sci Univ Cent Fla Orlando FL 32816-0368

WHITE, SAMUEL GRANDFORD, JR, b Anniston, Ala, Feb 6, 45; m 68; c 2. EDUCATION ADMINISTRATION, ELECTRONICS ENGINEERING. *Educ:* Tuskegee Univ, BS, 67, MS, 70; Univ Ill, Urbana, PhD(elec eng), 78. *Prof Exp:* Engr, Westinghouse Elec Corp, 67-68; staff engr, IBM, 70-72; assoc prof & chair elec eng, Digital Electronics & Solid State Electronics, NC A&T State Univ, 82-85; assoc prof elec eng & chair eng, Solid State Electronics, Purdue Univ, Ft Wayne, 85-87; ENG DIR EDUC & TRAINING, MAGNAVOX GOVT & INDUST ELECTRONICS, 87- *Concurrent Pos:* Consult, AT&T, 83-85; prog reviewer, NSF, Presidential Young Investigators, 84; reviewer, NSF, Res Initiatives for Minority Insts, 84; prog evaluator, Eng Accreditation Comn, Inst Elec & Electronics Engrs, 90-95. *Mem:* Sr mem Inst Elec & Electronics Engrs; Am Soc Eng Educ. *Mailing Add:* 5707 Hartford Dr Ft Wayne IN 46835

WHITE, SANDRA L, b Columbia, SC, Aug 30, 41; m 84; c 3. CELL BIOLOGY. *Educ:* Hampton Univ, BA, 63; Univ Mich, MS, 71, PhD(microbiol), 74. *Prof Exp:* Asst prof microbiol, 74-76, asst prof, 79-85, ASSOC PROF MICROBIOL & ONCOL, MEM CANCER CTR, HOWARD UNIV COL MED, 85. *Concurrent Pos:* Health Sci Consult, Curber Assoc, Washington DC, 75; scientific proj adv, Verve Res, Inc, Rockville, MD, 75-76; staff fel, Lab Immunodiagnosis, Nat Cancer Inst, NIH, 76-79; mem NIH Path B Study Sect, 80-84; bd Scientific Counrs, Div Cancer Biol & Diagnosis, Nat Cancer Inst, NIH, 85-89; Nat Bd Med Examiners, 89-93. *Mem:* Am Soc Cell Biol; Reticuloendothelial Soc; Am Soc Microbiologists; Sigma Xi; Am Asn Cancer Res. *Res:* Cellular and tumor immunobiology (656) role of immune effector cells (primarily macrophage and natural killer cells); biological response modifiers in cancer metastasis. *Mailing Add:* Cancer Ctr Dept Microbiol 2041 Georgia Ave NW Washington DC 20060

WHITE, SIDNEY EDWARD, b Manchester, NH, Mar 14, 16; m 46; c 1. GEOLOGY. *Educ:* Tufts Col, BS, 39; Harvard Univ, MA, 42; Syracuse Univ, PhD(geol), 51. *Prof Exp:* Lab asst, Tufts Col, 37-40, lab instr, 41-42, instr geol, 47-48; lab instr, Harvard Univ, 42; asst instr geol, Syracuse Univ, 48-51; from asst prof to prof, 51-85, EMER PROF GEOL, OHIO STATE UNIV, 85- *Concurrent Pos:* Recorder, US Geol Surv, 41-42, geologist, US Geol Surv, 46-48; ed, Geol Soc Am, Geomorph Div, 62-71; geologist, NSF Projs, Univ Colo, 64-66, assoc prof, 65-66, 72, 73; Univ Colo Men & Women Scholastic Honoraries award, 65-66. *Mem:* Geol Soc Am; Am Quaternary Asn; Mex Geol Soc. *Res:* Glacial and Pleistocene geology; volcanology; alpine and periglacial mass movement studies. *Mailing Add:* Dept Geol & Mineral Ohio State Univ Columbus OH 43210

WHITE, SIMON DAVID MANTON, b Kent, Eng, Sept 30, 51; m 84. GRAVITATIONAL DYNAMICS, PHYSICAL COSMOLOGY. *Educ:* Univ Cambridge, BA, 72; Univ Toronto, MSc, 74 & PhD(Castron), 77. *Prof Exp:* Lindemann fel, Astron Dept, Univ Calif Berkeley, 77-78; res fel astron, Churchill Col, 79-80; sr fel astron, Space Sci Lab, Univ Calif Berkeley, 80-84; assoc prof, 84-87, PROF ASTRON, STEWARD OBSERV, UNIV ARIZ, 87- *Concurrent Pos:* Vis astronr, Nat Radio Astron Observ, 78; sci attache, Astrophys Inst Paris, 80; presidential young investr, NSF, 84. *Honors & Awards:* Helen B Warner Prize, Am Astron Soc, 86. *Mem:* Am Astron Soc; Royal Astron Soc. *Res:* Structure, dynamics and evolution of galaxies and systems of galaxies, their formation and their relation to the large scale structure of the universe. *Mailing Add:* Dept Astron Univ Ariz Tucson AZ 85721

WHITE, STANLEY A, b Providence, RI, Sept 25, 31; US citizen; m 56; c 4. DIGITAL SIGNAL PROCESSING, NEURAL NETWORKS & LEARNING SYSTEMS. *Educ:* Purdue Univ, BS, 57, MS, 59, PhD(elec eng), 65. *Prof Exp:* Electronics technician, Allison Div, Gen Motors Corp, 56, circuit designer, 57; engr, Radiochem Corp, 58; mem tech staff, Autonetics Navig Systs Div, NAm Rockwell Corp, 59-61, staff to mgr advan technol, 65-67, group scientist digital systs group, Info Sci Br, 67-70, supvr advan inertial instrument res, Navig & Comput Div, 70, mem tech staff prod eng group, NAm Rockwell Microelectronic Co, 70-72, group leader digital systs group, Info Sci Br, Advan Technol Dept, Res & Technol Div, 72-77, mgr digital systs & signal processing, Electronics Res Ctr, 78-82, sr scientist, Autonetics Sensors & Aircraft Systs Div, Rockwell Int, 82-90; PRES, SIGNAL PROCESSING & CONTROLS ENG CORP, 90- *Concurrent Pos:* Mem fac, Purdue Univ, Univ Calif, Los Angeles, Irvine & Davis; lectre, Univ Southern Calif; chmn & fac mem, Nat Electronics Conf; adj prof elec eng, Univ Calif, Irvine; vis lectr, Davis & Univ Southern Calif. *Honors & Awards:* Centennial Medal, Inst Elec & Electronics Engrs, 84; Leonardo da Vinci Medal, Soc History Technol, 86. *Mem:* Fel AAAS; Sigma Xi; fel Inst Elec & Electronics Engrs; fel NY Acad Sci; fel Inst Advan Eng. *Res:* Digital signal processing architectures, algorithms, & devices; neural and learning systems; microelectronics; digital filtering; systems engineering; author or coauthor of over 100 publications; recipient of over 30 US patents. *Mailing Add:* 433 E Avenida Cordoba San Clemente CA 92672

WHITE, STANTON M, sedimentary petrology, marine geology, for more information see previous edition

WHITE, STEPHEN EDWARD, b Frankfort, Ky, Apr 15, 47; m 69; c 2. POPULATION GEOGRAPHY. *Educ:* Univ Ky, BS, 69, MA, 72, PhD(geog), 74. *Prof Exp:* Planner, Dir Planning, Ky Dept Transp, 69; lieutenant, Mil Intel, US Army, 69-71; planner & sect head, Statewide Transp Systs Planning, 74-75; from asst prof to assoc prof, 75-85, head dept, 79-87, PROF GEOG, KANS STATE UNIV, 85- *Concurrent Pos:* Prin investr, NSF, 78-80, US Dept Interior, 79-81, Gen Serv Found, 83-85 & Ford Found, 88-89 & 90-91; fel Appalachian Studies, Mellon Found, 81; secy-treas, Pop Specialty Group, Asn Am Geographers, 82-84, vpres, 84-85, pres, 85-86; hons res lectr, Mid-Am State Univs Asn, 88-89. *Honors & Awards:* Award Excellence Scholar, Nat Coun Geog Educ, 69, Jour Geog Award, 87. *Mem:* Asn Am Geographers; Nat Coun Geog Educ; Pop Asn Am; Pop Ref Bur. *Res:* Interregional migration research methods; return migration to Appalachia; groundwater depletion in the American High Plains; environmental perception. *Mailing Add:* Dept Geog Dickens Hall Kans State Univ Manhattan KS 66506

WHITE, STEPHEN HALLEY, b Wewoka, Okla, May 14, 40; m 61, 84; c 6. BIOPHYSICS, PHYSIOLOGY. *Educ:* Univ Colo, BA, 63; Univ Wash, MS, 65, PhD(physiol, biophys), 69. *Prof Exp:* Asst prof, 72-75, vchmn dept, 74-75, assoc prof physiol, 75-78, chmn dept physiol & biophys, 77-89, PROF PHYSIOL & BIOPHYS, UNIV CALIF, IRVINE, 78- *Concurrent Pos:* USPHS grant biochem, Univ Va, 71-72; guest biophysicist, Brookhaven Nat Lab, 77-; spec comn cell & membrane biophys, 78-82; mem coun, Biophys Soc, 81-84, prog chmn, 85, secy, 86-94; assoc chmn, Dept Physiol Coun, 81-84, pres, 86-87; res grants, NSF & NIH. *Mem:* Biophys Soc; Biophys Soc; Soc Gen Physiologists (treas, 85-88); Am Physiol Soc. *Res:* Structure of biological membranes and the physical chemistry of lipid bilayer membranes; utilizing x-ray and neutron diffraction and NMR. *Mailing Add:* Dept Physiol & Biophys Univ Calif Irvine CA 92717

WHITE, STUART COSSITT, b Pasadena, Calif, July 20, 42; m 68; c 2. DENTAL RADIOLOGY. *Educ:* Univ Calif, Berkeley, AB, 64; Univ Calif, Los Angeles, DDS, 68; Univ Rochester, PhD(radiation biol), 73. *Prof Exp:* Asst prof, 73-75, assoc prof, 75-80, PROF DENT, 80-, ASSOC DEAN, UNIV CALIF, LOS ANGELES, 79- *Concurrent Pos:* USPHS trainee, Univ Rochester, 68-73; prin investr NIH grant, 77-80; co-investr clin cancer educ prog, Univ Calif, Los Angeles, 76-79; prin investr grant, Am Fund Dent Health, 77-78; consult, Xerox Corp, 77-; fel, Am Acad Dent Radiol, 78-; mem bd dirs, Am Bd Oral & Maxillofacial Radiol, 80- *Honors & Awards:* Edward H Hattan Award, Int Asn Dent Res, 67. *Mem:* Radiation Res Soc; Am Acad Dent Radiol. *Res:* Radiation dosimetry; radiographic imaging technique; utility of radiographic examinations. *Mailing Add:* Sch Dent Univ Calif Los Angeles CA 90024

WHITE, THOMAS DAVID, b Sarnia, Ont, Apr 8, 43. NEUROCHEMISTRY, NEUROPHARMACOLOGY. *Educ:* Univ Western Ont, BSc, 65, MSc, 67; Bristol Univ, PhD(pharmacol), 70. *Prof Exp:* From asst prof to assoc prof, 71-82, PROF PHARMACOL, FAC MED, DALHOUSIE UNIV, 82- *Concurrent Pos:* Med Res Coun Can fel, Univ Alta, 70-71. *Mem:* Can Pharmacol Soc; Int Soc Neurochem; Soc Neurosci. *Res:* Physiology and pharmacology of brain synapses; functions of extracellular nucleotides and nucleosides. *Mailing Add:* Dept Pharmacol Dalhousie Univ Fac Med Halifax NS B3H 4H7 Can

WHITE, THOMAS GAILAND, b Artesia, NMex, Oct 12, 32; m 58; c 2. PLANT BREEDING. *Educ:* NMex State Univ, BS, 54; Tex A&M Univ, MS, 58, PhD(plant breeding), 62. *Prof Exp:* From instr to asst prof cotton genetics, Tex A&M Univ, 60-65; res geneticist, USDA, 65-67; dir biol res, Occidental Petrol Corp, 67-71; crops develop mgr, Gilroy Foods, Inc, 71-74, planning coordr, 74-76, mgr agr res & develop, 76-79, dir tech serv, 79; dir, Agr Res Develop, Mccormick, 79-84; RETIRED. *Concurrent Pos:* Leader monosomic res proj, Nat Cotton Coun Grant, Found Cotton Res & Educ, 63-66; mem, Agr Res Inst. *Mem:* Am Soc Hort Sci; Am Mgt Asn; Sigma Xi. *Res:* Basic genetic and cytogenetic research in cotton; applied research in controlled atmosphere use in transport of perishable commodities; genetics and improvement of onions; management, research and development. *Mailing Add:* 7790 Santa Theresa Dr Gilroy CA 95020

WHITE, THOMAS JAMES, b Stamford, Conn, Oct 5, 45; m 87. BIOCHEMISTRY, MICROBIOLOGY. *Educ:* Johns Hopkins Univ, BA, 67; Univ Calif, Berkeley, PhD(biochem), 76. *Prof Exp:* Fel biochem, G W Hooper Found, Univ Calif, San Francisco, 76-77 & Univ Wis-Madison, 77-78; scientist, Recombinant Molecular Res, Cetus Corp, 71-80, dir, 81-84; vpres res, Cetus Corp, 84-87, vpres & assoc dir res & develop, 87-88; sabbatical, Univ Calif, Berkeley, 88; SR DIR, ROCHE DIAGNOSTICS RES, 89- *Concurrent Pos:* NIH, fel, 77-78. *Mem:* Am Soc Microbiol; Soc Study Evolution; Am Inst Biol Sci; Genetics Soc Am. *Res:* Molecular evolution. *Mailing Add:* Roche Diagnostics Res 1145 Atlantic Ave No 100 Alameda CA 94501

WHITE, THOMAS TAYLOR, surgery, physiology; deceased, see previous edition for last biography

WHITE, THOMAS WAYNE, b Caldwell Co, Ky, Dec 9, 34; m 60; c 4. RUMINANT NUTRITION. *Educ:* Univ Ky, BS, 56, MS, 57; Univ Mo-Columbia, PhD(animal nutrit), 60. *Prof Exp:* Asst dist salesman, Nat Oats Co, 61-62; assoc prof, 62-72, PROF ANIMAL SCI, LA STATE UNIV RICE EXP STA, 72- *Concurrent Pos:* On leave, Purdue Univ, 71. *Mem:* Am Soc Animal Sci; Am Dairy Sci Asn; Am Registry Prof Animal Sci. *Res:* Level and source of roughage in ruminant rations as measured by digestibility and feedlot performance; whole shelled corn with various protein supplement for finishing cattle; influence of type and variety of sorghum grain on ration digestibility; forage evaluation and influence of age, breed of cattle and shade on forage utilization; rice and by-pass protein for ruminants. *Mailing Add:* La State Univ Rice Exp Sta PO Box 1429 Crowley LA 70526

WHITE, TIMOTHY LEE, b San Diego, Calif, May 5, 51; m 79; c 1. QUANTITATIVE FOREST GENETICS. *Educ:* Univ Calif, Berkeley, BS, 73; NC State Univ, Raleigh, MS, 75; Ore State Univ, PhD(forest genetics), 80. *Prof Exp:* RES FORESTER GENETICS & BIOMET, INT PAPER CO, 79- *Mem:* Sigma Xi. *Res:* Population dynamics and genetics of forest ecosystems. *Mailing Add:* Dept Forest Sci Univ Fla Gainesville FL 32611

WHITE, TIMOTHY P, b Buenos Aires, Arg, July 9, 49. MUSCLE PHYSIOLOGY. *Educ:* Univ Calif, Berkeley, PhD(phys educ), 77. *Prof Exp:* ASSOC PROF KINESIOL, UNIV MICH, 83- CHMN DEPT, 84- *Concurrent Pos:* Assoc res scientist, Inst Geol, Univ Mich, 86-; bd trust, Amer Col Sports Med, 88- *Mem:* Am Heart Asn; Am Physiol Soc; fel Am Col Sports Med. *Mailing Add:* 4711 Whitman Circle Ann Arbor MI 48103

WHITE, W(ILLIAM) ARTHUR, b Sumner, Ill, Dec 9, 16; m 41. MINERALOGY, PETROLOGY. *Educ:* Univ Ill, BS, 40, MS, 47, PhD(geol), 55. *Prof Exp:* Spec asst chem, Ill State Geol Surv, 43-44, res asst, 44-47, asst geologist, 47-48, assoc geologist, 48-54, head clay resources & clay mineral tech sect, 58-73, geologist, 54-79, EMER GEOLOGIST, ILL STATE GEOL SURV, 79- *Concurrent Pos:* Prof appl clay mineral technol, Fed Univ Rio Grande de sol, Porto Alegre, Brazil, 70; consult, 79-89. *Mem:* Fel AAAS; fel Mineral Soc Am; Am Chem Soc; fel Am Geol Soc; Geochem Soc; Clay Mineral Soc. *Res:* Physical properties of clays as related to soil mechanics; ceramic properties of clays and sediments; clay mineralogy of sediments and the environments in which they were accumulated; the role of clay minerals in environmental geology. *Mailing Add:* 603 E Colorado Ave Urbana IL 61801

WHITE, WARREN D, b Springfield, Mo, July 7, 15; m 40; c 2. ELECTRONICS ENGINEERING, MATHEMATICS. *Educ:* Drury Col, BS, 36; Univ Mo, Rolla, BSEE, 38. *Prof Exp:* Consult radio engr, DC, 39-41; engr in charge radio frequency div, Columbia Broadcasting Syst, 41-46; eng consult, Airborne Instruments Lab, Cutler-Hammer Inc, 46-65; staff mem, Inst Defense Anal, 65-67; eng consult, Cutler Hammer Inc, 67-74, tech asst to pres, Airborne Instrument Lab, 74-80; RETIRED. *Concurrent Pos:* Spec res assoc, Radio Res Lab, Harvard Univ, 42-45; consult electronics, 80-86. *Honors & Awards:* Barry Carleton Award, Inst Elec & Electronics Engrs Aerospace & Electronic Systs Soc, 83. *Mem:* Fel Inst Elec & Electronics Engrs. *Res:* Radar systems; information theory; antenna and propagation adaptive processes. *Mailing Add:* Laclede Oaks Manor Apt 471 705 S Laclede Station Rd Webster Groves MO 63119

WHITE, WILLARD WORSTER, III, b Perth Amboy, NJ, July 6, 44; m 72. MAGNETOHYDRODYNAMICS, PLASMA PHYSICS. *Educ:* Univ Del, BS, 66; Rensselaer Polytech Inst, PhD(physics), 70. *Prof Exp:* Res physicist, 72-77, sr scientist & asst atmospheric phenomenology div leader, 74-84, SR SCIENTIST & MHD GROUP LEADER, MISSION RES CORP, 84- *Concurrent Pos:* Res assoc, Rensselaer Polytech Inst, 70-72. *Mem:* Am Phys Soc. *Res:* Electromagnetic phenomena in the ionosphere; surface physics of materials; radiation damage in solids; laser propagation phenomena. *Mailing Add:* Mission Res Corp One Tara Blvd Suite 302 Nashua NH 03062

WHITE, WILLIAM, b Millbrook, Ont, Apr 1, 28; m 60; c 3. MEDICAL PHYSICS, RESEARCH ADMINISTRATION. *Educ:* Queen's Univ, Ont, BSc, 50; McGill Univ, PhD(physics), 61. *Prof Exp:* Res physicist, Bldg Prod Ltd, Montreal, Can, 51-56 & Am Radiator & Standard Sanit Res Lab, NJ, 61-63; dir res, Searle Med Instrumentation Group, Searle Anal, Inc Div, G D Searle & Co, 63-79; PRIN, RES & DEVELOP, WHITE CONSULTS, 79-; DIR RES, SIEMENS GAMMASONICS, INC, 81- *Concurrent Pos:* Consult res & develop mgt, 80- *Mem:* Soc Nuclear Med. *Res:* Nuclear spectroscopy and instrumentation in the physical and biological sciences; penetration of charged particles; blocking and channeling of atomic particles in crystals; nuclear medical gamma-ray cameras; methods for the organization of research and development. *Mailing Add:* Siemens Gammasonics Inc 2501 N Barrington Rd Hoffman Estates IL 60195-7372

WHITE, WILLIAM ALEXANDER, b Paterson, NJ, June 15, 06; c 2. GEOLOGY. *Educ:* Duke Univ, AB, 30; Univ NC, MA, 31, PhD(geol), 38; Mont Sch Mines, MS, 34. *Prof Exp:* Petrogr, Lago Petrol Corp, Venezuela, 38-40; from assoc prof to prof geol, Univ NC, Chapel Hill, 44-78; RETIRED; assoc prof geol, 44-50, PROF GEOL, UNIV NC, CHAPEL HILL, 50- *Mem:* Fel Am Geol Soc. *Res:* Geomorphology; glacial geology. *Mailing Add:* Dept Geol Univ NC Chapel Hill NC 27514

WHITE, WILLIAM BLAINE, b Huntingdon, Pa, Jan 5, 34; m 59; c 2. GEOCHEMISTRY, MATERIALS SCIENCE. *Educ:* Juniata Col, BS, 54; Pa State Univ, PhD(geochem), 62. *Prof Exp:* Res assoc chem physics, Mellon Inst, 54-58; res assoc geochem, 62-63, from asst prof to assoc prof, 63-72, PROF GEOCHEM, PA STATE UNIV, 72-, CHMN, SOLID STATE SCI, 90- *Mem:* AAAS; Am Geophys Union; Am Ceramic Soc; Am Mineral Soc; Nat Speleol Soc (exec vpres, 65-67). *Res:* High temperature chemistry; infrared and optical spectroscopy of solids; mineralogy; ground water hydrogeology; solid state chemistry; glass science; infrared, optical and luminescence spectroscopy; geomorphology. *Mailing Add:* 210 Mat Res Lab Pa State Univ University Park PA 16802

WHITE, WILLIAM CHARLES, b Jacksonville, Fla, May 12, 22; m 52. PHYSICS, ASTRONOMY. *Educ:* Ohio Wesleyan Univ, BA, 48; Ohio State Univ, MS, 50. *Prof Exp:* Physicist, Naval Weapons Ctr, 50-73, opers res analyst, 73-78, consult, 78-82; RETIRED. *Concurrent Pos:* Consult, Astrophys Observ, Smithsonian Inst, 59-60, Dearborn Observ, Northwestern Univ, 60-62, Ketron, 79-80 & Mammoth Mt SR area, 81-82; sci observer, Stargazer Balloon Flight, 61; partic, Aerial Photog Eclipse of Quiet Sun, 62, NASA Mobile Launch Exped, 65, Sandia Eclipse Exped, 65 & Oceanog & Geophys Exped, SAm, 67. *Mem:* Am Astron Soc; NY Acad Sci; Sigma Xi. *Res:* Astrophysical research with infrared detectors and balloon-borne observatories; atmospheric physics research with high altitude balloons; ozone as a function of latitude; electronic warfare. *Mailing Add:* PO Box 707 Clinton WA 98236-0707

WHITE, WILLIAM MICHAEL, b Allentown, Pa, Aug 10, 48; m 70; c 4. GEOCHEMISTRY. *Educ:* Univ Calif, Berkeley, BA, 71; Univ RI, PhD(oceanog), 77. *Prof Exp:* Fel geochem, Dept Terrestrial Magnetism, Carnegie Inst, Washington, 77-79; asst res prof, Col Sch Mines, Golden, Col, 79-80; staff mem, Max-Planck Inst Chem, 80-85; assoc prof oceanog, Ore State Univ, Corvallis, Ore, 85-86; ASSOC PROF GEOL SCI, CORNELL UNIV, ITHACA, NY, 86- *Mem:* Geol Soc Am; Am Geophys Union; Geochem Soc. *Res:* Isotope and trace element geochemistry of igneous rocks; composition and chemical evolution of earth's mantle; geochemistry of marine sediments. *Mailing Add:* Dept Geol Sci Snee Hall Cornell Univ Ithaca NY 14853-1504

WHITE, WILLIAM NORTH, b Walton, NY, Sept 16, 25; m 51; c 2. PHYSICAL ORGANIC CHEMISTRY. *Educ:* Cornell Univ, AB, 50; Harvard Univ, MA, 51; PhD(org chem), 53. *Prof Exp:* Nat Res Coun fel, Crellin Labs, Calif Inst Technol, 53-54; from asst prof to assoc prof chem, Ohio State Univ, 54-63; prof chem, Univ Vt, 63-76, chmn dept, 63-71 & 75-76; prof & chmn dept chem, Univ Tex, Arlington, 76-77; PROF CHEM, UNIV VT, 77- *Concurrent Pos:* NSF sr fel biol, Brookhaven Nat Labs, 63-64; NSF sr fel chem, Harvard Univ, 65; vis scholar biochem, Brandeis Univ, 74-75. *Mem:* AAAS; Am Chem Soc; Royal Soc Chem. *Res:* Reaction mechanisms; rearrangements; structure reactivity correlations; electrophilic and nucleophilic substitution mechanisms; carbanion and carbonyl group chemistry. *Mailing Add:* Dept Chem Univ Vt Burlington VT 05405-0125

WHITE, WILLIAM WALLACE, b Cleveland, Ohio, Sept 7, 39; m 64; c 2. OPERATIONS RESEARCH, COMPUTER SCIENCE. *Educ:* Princeton Univ, AB, 61; Univ Calif, Berkeley, MS, 63, PhD(eng sci), 66. *Prof Exp:* Staff mem mgt sci, Philadelphia Sci Ctr, 66-74, advan appl adv, Advan Syst Develop Div, 74-75, sr systs anal, syst prod div, 75-77, RES STAFF MEM, T J WATSON RES CTR, IBM CORP, 77- *Concurrent Pos:* Adj assoc prof statist, 72-74, adj assoc prof math, Columbia Univ, 74-79. *Mem:* Opers Res Soc Am; Asn Comput Mach; Math Prog Soc; Inst Elec & Electronics Engrs Computer Soc; Computer Measurement Group. *Res:* Computer measurement evaluation; mathematical programming and computer performance; analysis, applied to computer design. *Mailing Add:* Six Greenmeadow Rd Pleasantville NY 10570-1716

WHITE, WILLIS S, JR, b Dec 17, 26; m; c 3. ELECTRICAL ENGINEERING. *Educ:* Va Polytech Inst & State Univ, BS; Mass Inst Technol, MS. *Prof Exp:* Elec engr elec design & syst planning & oper, Am Elec Power Serv Corp, NY, 48-52, asst to pres, 52-54, off mgr, 54-58, asst to oper vpres, 58-61, asst mgr, Lynchburg Dist, Appalachian Power Co, Va, 61-62, mgr, 62-66, asst gen mgr, 66-67, asst vpres, 67-69, vpres, 69, exec vpres & oper head, 69-73, sr exec vpres, NY, 73-75, vchmn, 75-76, chief exec officer, 76-90, CHMN BD, AM ELEC POWER, 76- *Concurrent Pos:* Chmn bd trustees, Greater Columbus Conv Ctr; trustee, Battelle Mem Inst; dir, Bank NY, Riverside Methodist Hosp & Methodist Theol Sch Ohio. *Mem:* Nat Acad Eng; sr mem Inst Elec & Electronics Engrs. *Mailing Add:* Am Elec Power One River Plaza Columbus OH 43216-6631

WHITED, DEAN ALLEN, b Nebraska City, Nebr, Mar 28, 40; m 64; c 2. GENETICS, AGRONOMY. *Educ:* Univ Nebr, Lincoln, BS, 62, MS, 64; NDak State Univ, PhD(agron), 67. *Prof Exp:* Agency Int Develop grant & res assoc wheat qual, Univ Nebr, Lincoln, 67-68; from asst prof to assoc prof, 68-80, PROF AGRON, NDAK STATE UNIV, 80-, CHMN, GENETICS INST, 71- *Mem:* Am Soc Agron; Am Genetic Asn. *Res:* Genetics; genetic counseling at Muscular Dystrophy Clinic; soybean genetics and soybean production. *Mailing Add:* Hwy Contract 0 Bismarck ND 58501

WHITEFIELD, PHILIP DOUGLAS, b Taunton Somerset, England, Apr 10, 53; US citizen; m 83; c 2. SPECTROSCOPY, DISCHARGE-FLOW KINETICS. *Educ:* Queen Mary Col, Univ London, BSc, 75, PhD(phys chem), 79. *Prof Exp:* Res asst, Airforce Weapons Lab, 79-81; scientist, McDonnell Douglas Res Lab, 81-90; RES ASSOC PROF, UNIV MO, ROLLA, 90- *Concurrent Pos:* Asst prof, Univ Mo, 84-85. *Mem:* Am Chem Soc; Combustion Inst; Am Inst Aeronaut & Astronaut. *Res:* Hypersonic propulsion. *Mailing Add:* Norwood Hall Univ Mo Rolla MO 65401

WHITEFIELD, RODNEY JOE, b Lewiston, Idaho, Jan 21, 36. OPTICS, SURFACE PHYSICS. *Educ:* Wash State Univ, BS, 57; San Diego State Univ, MS, 59; Ore State Univ, PhD(physics), 70. *Prof Exp:* Mem staff exp physics, General Atomics, 59-60; STAFF ENGR, IBM CORP, 60- *Mem:* Optical Soc Am; Inst Elec & Electronics Engrs. *Res:* Surface measurement and characterization; photo electric effect; ultra high vacuum. *Mailing Add:* 6794 Heathfield Dr San Jose CA 95120

WHITEFORD, ROBERT DANIEL, veterinary medicine; deceased, see previous edition for last biography

WHITEHAIR, CHARLES KENNETH, b Abilene, Kans, Mar 3, 16; m 58; c 2. NUTRITION, PATHOLOGY. *Educ:* Kans State Univ, DVM, 40; Univ Wis, MS, 43, PhD(nutrit), 47. *Prof Exp:* Instr animal dis, Univ Wis, 40-47; from asst prof to prof animal nutrit, Okla State Univ, 47-53, head vet res & prof physiol & nutrit, 54-56; PROF PATH & NUTRIT, MICH STATE UNIV, 56- *Concurrent Pos:* Assoc prof, Univ Ill, 52-53. *Mem:* Am Asn Path; Soc Exp Biol & Med; Am Inst Nutrit; Conf Res Workers Animal Dis (vpres, 64, pres, 65); Am Vet Med Asn. *Res:* Nutritional pathology; swine nutrition and diseases; metabolic diseases of livestock; relation of nutrition to diseases; malnutrition; nutritional therapy. *Mailing Add:* Dept Path Mich State Univ East Lansing MI 48823

WHITEHAIR, LEO A, b Abilene, Kans, June 13, 29; m 58; c 3. VETERINARY MEDICINE, FOOD SCIENCE. *Educ:* Kans State Univ, BS & DVM, 53; Univ Wis, MS, 54, PhD(food sci), 62. *Prof Exp:* Vet off nutrit br, Aeromed Lab, Wright-Patterson AFB, Ohio, 54-58; lab vet nutrit br, Food Inst Armed Forces, Ill, 61-62; vet food technologist, Biol Br, Div Biol & Med, US AEC, 62-67; health scientist adminr, 68-75, dir Primate Res Ctr Prog, Animal Resources Br, Div Res Resources, NIH, 75-85; assoc, Res Admin Am Red Cross Biomed Res & Develop Labs, 85- 87; DIR, LAB ANIMAL SCI PROG, ANIMAL RESOURCES BR, DIV RES RESOURCES, NIH, 87- *Concurrent Pos:* Consult adv, Food & Agr Orgn-UN-WHO-Int Atomic Energy Agency joint expert comt meeting, Rome, 64; tech adv int prog irradiation fruit & fruit juices, Inst Biol & Agr, Seibersdorf Reactor Ctr, Austria, 65; vet off dir, USPHS; bd councillors, Am Col Vet Prev Med, 79-80. *Honors & Awards:* Helwig-Jennings Award, 81; Commendation Medal, USPHS, 82. *Mem:* Am Vet Med Asn; Am Col Vet Prev Med (secy-treas, 75-78, pres, 84-85); Inst Food Technologists; Am Asn Vet Nutritionists; Sigma Xi. *Res:* Animal nutrition; wholesomeness and public health safety aspects of irradiated foods; laboratory animal resources; animal models for biomedical research. *Mailing Add:* 707 Wilson Ave Rockville MD 20850

WHITEHEAD, ANDREW BRUCE, b Quebec, Que, Oct 18, 32; m 62; c 3. PLANETOLOGY, PHYSICS. *Educ:* Univ NB, BSc, 53; McGill Univ, MSc, 55, PhD(physics), 57. *Prof Exp:* Res fel nuclear physics, Atomic Energy Res Estab, Harwell, Eng, 57-60; res fel nuclear physics, 61-62, res specialist, 63-65, group supvr physics, 65-67, sect mgr physics, 67-69, mgr lunar & planetary sci sect, 69-71, asst proj scientist, Mariner 9, 71-73, actg asst mgr, Space Sci Div, 73-75, staff scientist, Space Sci Div, Jet Propulsion Lab, 75-78; mgr progs, 78-79, mgr sensors & controls dept, Honeywell Corp Physical Sci Ctr, 80-81, prin staff scientist, Honeywell Sensors & Signal Processing Lab, 82-88. *Mem:* Am Phys Soc. *Res:* Nuclear structure; particle detection; secondary electron emission; atomic stopping; photovoltaics; computer science; sensors; controls. *Mailing Add:* 3045 Lakeshore Ave Maple Plain MN 55359

WHITEHEAD, ARMAND T, b Reno, Nev, May 19, 36; m 54; c 5. ENTOMOLOGY. *Educ:* Brigham Young Univ, BS, 65; Univ Calif, Berkeley, PhD(entom), 69. *Prof Exp:* Asst prof, 69-78, ASSOC PROF ZOOL, BRIGHAM YOUNG UNIV, 78- *Concurrent Pos:* Vis asst prof, Univ Ill, 75-76; vis prof, Univ Alta, 88-89. *Mem:* Entom Soc Am. *Res:* Neurophysiology and morphology of insect sensory receptors. *Mailing Add:* Dept Zool Brigham Young Univ Provo UT 84602

WHITEHEAD, DANIEL L(EE), b Walland, Tenn, Dec 25, 15; m 37; c 3. ELECTRICAL ENGINEERING. *Educ:* Univ Tenn, BS, 39; Cornell Univ, MS, 40. *Prof Exp:* Elec engr, Aluminum Co Am, 34-39; engr, Cent Sta, Westinghouse Elec Corp, 41-47, supvr analog comput, 47-51, mgr, High Voltage Labs, 51-68, mgr eng labs, 68-81; CONSULT HIGH VOLTAGE MEASURING TECHNOL & SAFETY TECHS, 81- *Concurrent Pos:* Lectr, Univ Pittsburgh, 44- & Carnegie-Mellon Univ, 57-58; lectr high voltage test tech, Inst Elec & Electronics Engrs, 81- *Honors & Awards:* Instrumentation & Measurements Comt Award, Inst Elec & Electronics Engrs, 87. *Mem:* Fel Inst Elec & Electronics Engrs; Am Nat Standards Inst. *Res:* High voltage and high power phenomena and measuring techniques. *Mailing Add:* RD 3 Roundtop Rd Export PA 15632

WHITEHEAD, DONALD REED, b Quincy, Mass, Sept 14, 32; m 55; c 2. BOTANY PALEOECOLOGY. *Educ:* Harvard Univ, AB, 54, AM, 55, PhD(biol), 58. *Prof Exp:* From instr to assoc prof biol, Williams Col, 59-67; assoc prof biol, Ind Univ, Bloomington, 67-74, prof zool, 74-77; res entomologist, Syst Entom Lab, US Nat Mus, USDA, 77-; AT DEPT BIOL, IND UNIV. *Concurrent Pos:* Fulbright fel, Geol Surv Denmark, 58-59, res prof, 63; consult, Jersey Prod Res Co, 60; res prof bot mus, Univ Bergen, 63; NSF grants, 63. *Mem:* AAAS; Ecol Soc Am; Am Ornith Union. *Res:* Pleistocene environmental changes in unglaciated regions; pollen morphology. *Mailing Add:* Dept Bot Ind Univ Bloomington IN 47401

WHITEHEAD, EUGENE IRVING, b Canton, SDak, Mar 4, 18; m 55; c 2. PLANT BIOCHEMISTRY. *Educ:* SDak State Col, BS, 39, MS, 41. *Prof Exp:* Lab asst, 40-42, sta analyst, 42-43, asst agr chemist, 43-46, assoc chemist, 46-60, assoc prof sta biochem, Grad Fac, 60-67, PROF STA BIOCHEM, GRAD FAC, SDAK STATE UNIV, 67- *Mem:* Am Soc Plant Physiol; Am Chem Soc. *Res:* Nitrogen metabolism of cereal crops; winter hardiness of cereal plants. *Mailing Add:* HC 55 Box 853 Sturgis SD 57785-9290

WHITEHEAD, FLOY EUGENIA, b Athens, Ga, Feb 10, 13. NUTRITION. *Educ:* Univ Ga, BS, 36, MS, 42; Harvard Univ, DSc, 51; Am Bd Nutrit, dipl, 52. *Prof Exp:* Teacher, High Schs, Ga, 36-40; asst prof home econ, WGa Col, 40-42; assoc dir health educ, State Dept Pub Health, Ga, 42-43; assoc prof home econ, La State Univ, 44-48 & Miss State Col, 48-49; fel, Sch Pub Health, Harvard Univ, 49-52; dir nutrit, Wheat Flour Inst, Ill, 52-53; dir nutrit educ, Nat Dairy Coun, 53-55; chmn, dept home econ, 55-71, prof, 55-78, EMER PROF HOME ECON, UNIV IOWA, 78- *Concurrent Pos:* Vis lectr, Harvard Univ, 52-54; co-dir nutrit educ res pub schs, Mo, 52-55; pres elect, Nat Coun Adminr Home Econ, 66-67, pres, 67-68; res grant off nutrit, AID, US Dept State. *Honors & Awards:* Roberts Award, Am Dietetic Asn, 56. *Mem:* AAAS; Am Pub Health Asn; Am Home Econ Asn; Am Dietetic Asn. *Res:* Dietary surveys; nutrition education; analysis of nutrition education research, 1900-1970. *Mailing Add:* 306 Ferson Ave Iowa City IA 52240

WHITEHEAD, FRED, b Walland, Tenn, Aug 24, 05; m 35; c 2. CHEMISTRY. *Educ:* Ky Wesleyan Col, AB, 31; Univ Tenn, MS, 33; Univ Mich, PhD(chem), 45. *Prof Exp:* Prof chem, Ky Wesleyan Col, 32-50, dean & registr, 44-50; prof chem, Huntingdon Col, 50-75; RETIRED. *Mem:* Am Chem Soc. *Res:* Rate of dissociation of pentaarylethanes; physics; mathematics. *Mailing Add:* 3496 Cloverdale Rd Montgomery AL 36111

WHITEHEAD, GEORGE WILLIAM, b Bloomington, Ill, Aug 2, 18; m 47. TOPOLOGY. *Educ:* Univ Chicago, SB, 37, SM, 38, PhD(math), 41. *Prof Exp:* Instr math, Univ Tenn, 39, Purdue Univ, 41-45 & Princeton Univ, 45-47; from asst prof to assoc prof, Brown Univ, 47-49; from asst prof to prof math, Mass Inst Technol, 49-85; RETIRED. *Concurrent Pos:* Guggenheim fel & Fulbright res scholar, 55-56; vis prof, Princeton Univ, 58-59; NSF sr fel, 65-66; vis res fel, Birkbeck Col, Univ London, 73. *Mem:* Nat Acad Sci; fel Am Acad Arts & Sci; Am Math Soc; Math Asn Am; London Math Soc. *Res:* Algebraic topology, especially homotopy theory. *Mailing Add:* 25 Bellevue Rd Arlington MA 02174

WHITEHEAD, HOWARD ALLAN, b Toronto, Ont, Apr 17, 27; m 52; c 4. BACTERIOLOGY, INNOVATION. *Educ:* Mt Allison Univ, BSc, 49; McGill Univ, MSc & PhD(bact), 53. *Prof Exp:* Res mgr, 53-73, SR RES FEL, KIMBERLY-CLARK CORP, 73- *Honors & Awards:* K C Mahler Award. *Mem:* Sigma Xi. *Res:* Lethal and mutagenic effects of ultraviolet irradiation of bacterial microorganisms; menstruation, physiology and feminine hygiene; international new product and business development. *Mailing Add:* Res Div Ctr Kimberly-Clark Corp Neenah WI 54956

WHITEHEAD, JAMES RENNIE, b Clitheroe, Eng, Aug 4, 17; m 44; c 2. PHYSICS. *Educ:* Univ Manchester, BSc, 39; Cambridge Univ, PhD(physics), 49. *Prof Exp:* Scientist, Telecommun Res Estab, Eng, 39-51; assoc prof physics, McGill Univ, 51-55; dir res, RCA Victor Co, Ltd, 55-65; dep dir sci secretariat, Govt Can, 65-67, prin sci adv, 67-71, asst secy int affairs, Ministry of State for Sci & Technol, 71-73, spec adv, 73-75; sr adv, Int Develop Res Ctr, Govt Can, 75-76; sr vpres, Philip A Lapp Ltd, 76-82; SR ASSOC & DIR, LAPP-HANCOCK ASSOCS, 82- *Concurrent Pos:* Sr sci officer, Brit Air Comn, Washington, DC, 44-45 & Cambridge Univ, 46-49; consult, Defence Res Bd Can, 52-54; consult sci policy in Venezuela and Guyana, UNESCO, 70-75; Can deleg to sci & technol policy comt, Orgn Econ Coop & Develop, vchmn comt sci policy, 73-75; mem sci comt, NATO, 69-75. *Mem:* Fel Royal Soc Can; sr mem Inst Elec & Electronics Engrs; fel Brit Inst Elec Eng; fel Brit Inst Physics; fel Can Aeronaut & Space Inst. *Res:* Physical electronics; circuits; systems propagation; electron microscopy; friction; science policy. *Mailing Add:* 1368 Chattaway Ave Ottawa ON K1H 7S3 Can

WHITEHEAD, JOHN ANDREWS, b Amesbury, Mass, Apr 21, 41; m 64; c 3. OCEANOGRAPHY, GEOPHYSICS. *Educ:* Tufts Univ, BS, 63; Yale Univ, MS, 65, PhD(appl sci), 68. *Prof Exp:* Fel, Inst Geophys & Planetary Physics, Univ Calif, Los Angeles, 68-69, asst res geophysicist, 69-71; asst scientist, 71-73, assoc scientist, 73-88, SR SCIENTIST, WOODS HOLE OCEANOG INST, 88- *Concurrent Pos:* Mem comt geodesy, NAS-Nat Res Coun, 76-; sr fel, Nat Ctr Atmospheric Res, 77-78 & fel John Simon Guggenheim Mem Found, 82; mem space & terrestrial applications comt, NASA, 78-81. *Mem:* Fel Am Phys Soc; Am Geophys Union; NY Acad Sci; AAAS; Sigma Xi; Am Meteorol Soc. *Res:* Geophysical fluid dynamics of oceans, atmospheres and planetary interiors. *Mailing Add:* Dept Phys Oceanog Woods Hole Oceanog Inst Woods Hole MA 02543

WHITEHEAD, KENNETH E, b Niagara Falls, Ont, 1928; US citizen; m 57; c 3. CHEMICAL ENGINEERING. *Educ:* Univ Toronto, BASc, 52; Ohio State Univ, MSc, 53. *Prof Exp:* Supvr process eng, Atlantic Richfield Co, 62-69; supvr, Union Oil Co, Calif, 72-75, mgr process eng res dept, 75-90; RETIRED. *Concurrent Pos:* Chmn subcomt liquid wastes, Am Petrol Inst, 76-77. *Mem:* Am Inst Chem Eng. *Res:* Process development and design in petrochemicals and petroleum refining. *Mailing Add:* 566 N Lincoln Ave Fullerton CA 92631

WHITEHEAD, MARIAN NEDRA, b Calif, Sept 5, 22. NUCLEAR PHYSICS, PARTICLE PHYSICS. *Educ:* Reed Col, AB, 44; Columbia Univ, MS, 45; Univ Calif, PhD(physics), 52. *Prof Exp:* Physicist, US Naval Ord Test Sta, 45-46 & Radiation Lab, Univ Calif, 49-60; Fulbright sr res fel, Inst Physics, Bologna, Italy, 61-62; physicist, Stanford Linear Accelerator Ctr, Stanford Univ, 62-64; assoc prof, 64-67, chmn dept, 69-75, PROF PHYSICS, CALIF STATE UNIV, HAYWARD, 67- *Mem:* Fel Am Phys Soc; Am Asn Physics Teachers (treas, 74-78). *Res:* Meson and cosmic physics. *Mailing Add:* 12012 Goshen Dr Apt 311 Los Angeles CA 90049

WHITEHEAD, MARVIN DELBERT, b Paoli, Okla, Dec 18, 17; m 40; c 1. PHYTOPATHOLOGY, MYCOLOGY. *Educ:* Okla State Univ, BS, 39, MS, 46; Univ Wis, PhD(plant path, mycol), 49. *Prof Exp:* Asst agr aide, Soil Conserv Serv, USDA, Okla, 36-38; asst agron, Okla State Univ, 39-40; sr seed analyst, Fed State Seed Lab, Ala, 40-42; asst plant path, Univ Wis, 46-48; asst prof, Tex A&M Univ, 49-55; assoc prof, Univ Mo, 55-60; prof bot, Edinboro State Col, 60-63; prof plant path, Ga Southern Col, 63-68; prof bot & plant path, Ga State Univ, 68-75; RETIRED. *Concurrent Pos:* Consulting plant pathologist, US Army, Ft McPherson, Ga, 74-, Ft Riley, Kans, 75- & Ft Campbell, Ky, 78-; owner & dir, Marvern Plant Health Inc, Atlanta, 78-; ed, Ga J Sci, 74- *Mem:* AAAS; Am Phytopath Soc; Mycol Soc Am; Bot Soc Am; Am Inst Biol Sci. *Res:* Field crop disease pathology; soil borne and seed borne diseases; phytopathological histology and techniques; fungus and smut taxonomy; yield loss from plant disease; disease resistance; antibiotics and fungicides in control of Dutch elm disease, oak wilt, verticillium wilt of maple, decline of oak, and hackberry. *Mailing Add:* 817 Clifton Rd NE Atlanta GA 30307

WHITEHEAD, MICHAEL ANTHONY, b London, Eng, June 30, 35; m 77. THEORETICAL & QUANTUM CHEMISTRY, NUCLEAR QUADRUPOLE RESONANCE. *Educ:* Univ London, BSc, 56, PhD(phys chem), 60, DSc(theoret & phys chem), 74. *Prof Exp:* Asst lectr chem, Queen Mary Col, Univ London, 58-60; Fulbright scholar, Univ Cincinnati, 60-62, fel, 60-61, asst prof, 61-62; from asst prof to assoc prof, 62-75, PROF CHEM, MCGILL UNIV, 75- *Concurrent Pos:* Nat Res Coun Can travel fel & vis prof, Cambridge Univ, 71-72; vis prof theoret chem, Oxford Univ, 72-74, dept chem & physics, Univ Geneva, 83-84; vis prof fel, Univ Wales, Aberystwyth, 80; vis prof, Oxford Univ, 90-91. *Mem:* Am Phys Soc; Am Chem Soc; Can Inst Chem; Royal Soc Chem; Sigma Xi; Royal Soc Arts. *Res:* Nuclear quadrupole resonance; electronegativity theory; molecular orbital calculations; beyond Hartree-Fock calculations; surface absorption; chemical absorption; theoretical chemistry; spin density functional theory. *Mailing Add:* Dept Chem McGill Univ 801 Sherbrooke St W Montreal PQ H3A 2K6 Can

WHITEHEAD, WALTER DEXTER, JR, b San Diego, Calif, Nov 30, 22; m 49; c 2. NUCLEAR PHYSICS, ATOMIC PHYSICS. *Educ:* Univ Va, BS, 44, MS, 46, PhD(physics), 49. *Prof Exp:* Asst, Univ Va, 43-45; physicist, Bartol Res Found, 49-53; from asst prof to assoc prof physics, NC State Col, 53-56; from asst prof to assoc prof, 56-61, chmn dept, 68-69, dean fac arts & sci, 71-72, PROF PHYSICS, UNIV VA, 61-, DIR CTR ADVAN STUDIES, 65-, DEAN GRAD SCH ARTS & SCI, 69- *Concurrent Pos:* Vis scientist, Inst Nuclear Physics, 59-60; mem bd admin, Va Inst Marine Sci, 71-78; mem Nat Res Coun eval panel, Ctr Radiation Res, Inst Basic Standards, 71-74. *Mem:* AAAS; fel Am Phys Soc; Am Asn Physics Teachers; Sigma Xi. *Res:* Nuclear spectroscopy; neutron scattering; photonuclear reactions; x-ray interactions. *Mailing Add:* 444 Cabell Hall Univ Va Charlottesville VA 22903

WHITEHEAD, WILLIAM EARL, b Martin, Tenn, May 24, 45; m 68; c 1. PSYCHOPHYSIOLOGY. *Educ:* Ariz State Univ, BA, 67; Univ Chicago, PhD(psychol), 73. *Prof Exp:* Asst prof psychiat & phys med, Col Med, Univ Cincinnati, 73-77, assoc prof, 77-; AT NIH, WASHINGTON, DC. *Concurrent Pos:* Adj asst prof psychol, Univ Cincinnati, 75- *Mem:* AAAS; Soc Psychophysiol Res; Am Psychosom Soc; Am Biofeedback Soc. *Res:* Biofeedback treatment of psychophysiologic disorders; etiology of psychosomatic disorders; psychotropic drug evaluation. *Mailing Add:* 1516 Clayton Rd Joppe MD 21085

WHITEHORN, DAVID, neurophysiology, for more information see previous edition

WHITEHORN, WILLIAM VICTOR, b Detroit, Mich, Oct 3, 15; m 38; c 2. PHYSIOLOGY, MEDICINE. *Educ:* Univ Mich, AB, 36, MD, 39. *Prof Exp:* Asst physiol, Univ Mich, 40-42; res assoc, Ohio State Univ, 42-44, from instr to asst prof physiol & med, 44-47; asst prof physiol, Col Med, Univ Ill, Chicago, 47-50, prof, 54-70; dir div health sci & spec asst to pres med affairs,

Univ Del, 70-74; asst comnr prof & consumer prog, Food & Drug Admin, 74-80; EMER PROF, UNIFORMED SERV UNIV HEALTH SCI, 88- *Concurrent Pos:* Vis prof physiol, Uniformed Serv Univ Health Sci, 80-87. *Mem:* AAAS; Am Physiol Soc; Soc Exp Biol & Med; Am Heart Asn; Cent Soc Clin Res. *Res:* Cardiac mechanics and function; applied physiology of respiration and circulation. *Mailing Add:* 13612 Sherwood Forest Dr Silver Spring MD 20904

WHITEHOUSE, BRUCE ALAN, b Henderson, Ky, Sept 6, 39; m 61; c 2. POLYMER CHEMISTRY, TEXTILE CHEMISTRY. *Educ:* Col Charleston, BS, 63; Ga Inst Technol, PhD(phys chem), 67. *Prof Exp:* Res chemist, Plastics Dept, Polyolefins Div, Orange Tex, 67-72, res chemist, Textile Fibers Dept, Nylon Tech Div, Chattanooga, Tenn, 72-73, sr res chemist, 73, res supvr, Dacron Res Lab, 73-76, res supvr, Textile Res Lab, 76-78, mgr prod strategy, Indust Fibers Div, 78-79, res mgr, Textile Res Lab, Textile Fibers Dept, E I du Pont de Nemours & Co, Inc, 79-83; tech mgr, Tyvek & Spanborded Div, 83-88; lab dir, Pioneering Res Lab, 89-90; TECH DIR ADVAN MAT, DUPONT FIBERS, 91- *Mem:* Am Chem Soc. *Res:* Solid state structure and properties; structure and properties of polymers; chromatography; polymer synthesis composites. *Mailing Add:* Dupont Co Chestnut PO Box 80702 Wiliminton DE 19880-0702

WHITEHOUSE, DAVID R(EMPFER), b Evanston, Ill, Nov 13, 29; m 56; c 3. ELECTRICAL ENGINEERING. *Educ:* Northwestern Univ, BS, 52; Mass Inst Technol, SM, 54, DSc, 58. *Prof Exp:* From asst prof to assoc prof elec eng, Mass Inst Technol, 58-65; prin res scientist, Res Div, 65-67, mgr, Laser Advan Develop Ctr, Raytheon Co, 67-85; PRES, WHITEHOUSE ASSOC, 85- *Concurrent Pos:* Consult, 59-65 & 85-; chmn of the bd, Dymed Corp, 87-; bd dir, Surgilase Inc, 87- *Mem:* Am Phys Soc; Inst Elec & Electronics Engrs; Laser Inst Am (pres, 81); Sigma Xi. *Res:* Lasers; laser and plasma physics. *Mailing Add:* 99 South Ave Weston MA 02193

WHITEHOUSE, FRANK, JR, b Ann Arbor, Mich, Nov 20, 24; m 51; c 4. SCIENCE EDUCATION. *Educ:* Univ Mich, BA & MD, 53. *Prof Exp:* Intern, Blodgett Mem Hosp, Grand Rapids, Mich, 53-54; from instr to asst prof, 54-67, ASSOC PROF MICROBIOL, UNIV MICH, ANN ARBOR, 67- *Concurrent Pos:* Lectr, Ohio State Univ, 59; pre-prof counsr, Univ Mich, 60-; exec dir, Nat Asn Adv Health Professions. *Honors & Awards:* Sr Fulbright lectr microbiol, 79-80. *Mem:* Am Soc Microbiol. *Res:* Enzymatic degradation of antibodies; science education. *Mailing Add:* Dept Microbiol & Immunol Med Sch Bldg II Univ Mich Ann Arbor MI 48109-0620

WHITEHOUSE, GARY E, b Trenton, NJ, Aug 13, 38; m 63; c 2. INDUSTRIAL ENGINEERING, OPERATIONS RESEARCH. *Educ:* Lehigh Univ, BS, 60, MS, 62; Ariz State Univ, PhD(indust eng), 66. *Prof Exp:* Instr indust eng, Lehigh Univ, 62-63 & Ariz State Univ, 63-65; from asst prof to prof, Lehigh Univ, 65-78; prof indust eng & mgt systs & chmn dept, 78-85, DEAN ENG, UNIV CENT FLA, 87- *Concurrent Pos:* Consult, Air Prod & Chem Inc, Martin Marietta, USN. *Honors & Awards:* Maynard Award, Inst Indust Engrs. *Mem:* Am Inst Indust Engrs; Inst Mgt Sci; Opers Res Soc Am; Am Soc Eng Educ; Sigma Xi. *Res:* Theory and applications of networks; mathematical programming; decision theory; production and inventory control. *Mailing Add:* 1298 Hillstream Rd Geneva FL 32732

WHITEHOUSE, GERALD D(EAN), b Sapulpa, Okla, May 17, 36; m 58; c 3. MECHANICAL ENGINEERING. *Educ:* Univ Mo-Rolla, BS, 58; Okla State Univ, MS, 64, PhD(mech eng), 67. *Prof Exp:* Vibration engr, Douglas Aircraft Co, 58-59; from asst prof to assoc prof, 66-77, PROF MECH ENG & CHMN DEPT, LA STATE UNIV, BATON ROUGE, 77- *Mem:* Acoust Soc Am; Am Soc Mech Engrs. *Res:* Research activity in the mechanical design area, particularly in stress analysis and vibrations. *Mailing Add:* Dept Mech Eng & Aerospace La State Univ Mech Eng Bldg Baton Rouge LA 70803

WHITEHOUSE, RONALD LESLIE S, b Birmingham, Eng, Aug 9, 37; Can citizen; div; c 2. ELECTRON MICROSCOPY, MICROBIOLOGY. *Educ:* Univ Nottingham, BSc, 59; Univ Alta, MSc, 61, PhD(plant physiol, biochem), 65. *Prof Exp:* Nat Res Coun Can overseas fel, Bot Lab, Univ Bergen, 66; prof assoc food microbiol, 67-68, asst prof med bact & electron micros, 68-74, ASSOC PROF BACT, UNIV ALTA, 74- *Concurrent Pos:* Sabbatical leaves, Dept Molecular Biol, Pasteur Inst, Paris, 75, 83. *Mem:* Micros Soc Can; NY Acad Sci. *Res:* Electron microscopic methods for biological materials; electron microscopic investigation of chromosomal activity during sporulation in Bacillus Subtilis; development of preparative techniques; bacterial ultrastructure; simonsiella sp; oral microbiology. *Mailing Add:* Dept Med Microbiol Univ Alta Fac Med Edmonton AB T6G 2G3 Can

WHITEHOUSE, WALTER MACINTIRE, SR, radiology; deceased, see previous edition for last biography

WHITEHURST, BROOKS M, b Reading, Pa, Apr 9, 30; m 51; c 3. CHEMICAL ENGINEERING. *Educ:* Va Polytech Inst, BS, 51. *Prof Exp:* Sr tech asst, Am Enka Corp, 51-56; sr engr, Ind Chem Div, Mobil Chem Co, 56-63; proj engr, Texaco Exp, Inc, 63-66; process engr, Tex Gulf Sulphur Co, 67-70; supt tech serv, 70-75, mgr eng serv, Tex Gulf, Inc, Aurora, 75-81, PRES, BROOKS WHITEHURST ASSOC, INC, 81- *Concurrent Pos:* Chmn, Indust Adv Coun, Elizabeth City State Univ. *Mem:* Am Inst Chem Engrs; fel Am Inst Chem; Int Solar Energy Soc; Nat Soc Prof Engrs; Royal Soc Chem. *Res:* Rayon yarns; phosphate, fertilizer, fluorine and environmental processes in air and water; alternate energy systems (solar, wood, alcohol), peat-oil slurry fuels, industrial products from sweet potatoes, carbohydrates and chemical aluminum polishing; the development of chelates from carbohydrates, and the development of chelated metals for agriculture, and industrial waste treatment. *Mailing Add:* 1983 Hoods Creek Dr New Bern NC 28564

WHITEHURST, CHARLES A(UGUSTUS), b Cottondale, Fla, June 27, 29; m 56; c 3. MECHANICAL ENGINEERING, ENVIRONMENTAL ENGINEERING. *Educ:* La State Univ, BS, 56; Southern Methodist Univ, MS, 59; Tex A&M Univ, PhD(mech eng), 62. *Prof Exp:* Aerodyn engr, Gen Dynamics /Ft Worth, 56, propulsion engr, 56-59; assoc prof, La Polytech Inst, 62-63; assoc prof mech & aerospace eng, 63-66, from assoc prof to prof, Div Eng Res, 66-77, PROF COASTAL ENG & ASSOC DEAN RES & GRAD ACTIV, LA STATE UNIV, BATON ROUGE, 77- *Concurrent Pos:* Dir & prin investr, Heat, Mass & Momentum Transfer Studies at Low Temperatures Proj, NSF, 62-63; lectr, Manned Spacecraft Ctr, Tex, 65-66; prin investr, Flow Losses in Flexible Hose Proj, NASA, 65-66, prin investr, Jet Shock Interactions Proj, 66-67, prog mgr & prin investr sustaining univ grant, 66-, prin investr related multidiscipline res, NASA Ctr, 69-, prin investr remote sensing studies La Delta, 72-; consult, La Joint Legis Comt Environ Qual, 71-72, prog mgr La environ mgt syst, 72; prog mgr & prin investr res, Off Water Resources Res, 72- *Mem:* Am Soc Mech Engrs; Am Soc Eng Educ; Sigma Xi. *Res:* Thermodynamics; fluid mechanics; heat transfer; environmental engineering, water resource management and environmental impact assessments. *Mailing Add:* 1090 Longwood Dr Baton Rouge LA 70806

WHITEHURST, DARRELL DUAYNE, b Vernon, Ill, July 8, 38; m 67; c 2. ORGANIC CHEMISTRY. *Educ:* Bradley Univ, AB, 60; Univ Iowa, MS, 63, PhD(org chem), 64. *Prof Exp:* Res chemist, 64-65, sr res chemist, 65-68, group leader catalysis, 68-73, res assoc, 74-75, PRIN INVESTR THREE EPRI CONTRACTS, MOBIL RES & DEVELOP CORP, 75-, GROUP MGR COAL & HEAVY LIQUIDS RES, 80- *Concurrent Pos:* Mem comt task force motor fuel & photochem smog, Am Petrol Inst; res assoc, BPRI, 73, prin investr fundamental coal chem study, 75; res scientist, 83. *Honors & Awards:* Richard A Glen Award & Henry H Storch Award, Am Chem Soc. *Mem:* Carbon Soc; Am Chem Soc. *Res:* Organic syntheses; acetylene oxidations and coordination compounds of platinum; catalysis by ion exchange resins; catalysis by transition metals and compounds thereof; catalysis by zeolites; homogeneous-heterogeneous catalysts interconversion; metal plating; petrochemicals. *Mailing Add:* RD 1 Titusville NJ 08560-9801

WHITEHURST, ELDRIDGE AUGUSTUS, b Norfolk, Va, May 26, 23; m 69; c 5. CIVIL ENGINEERING. *Educ:* Va Mil Inst, BSCE, 47; Purdue Univ, MSCE, 51. *Prof Exp:* Assoc res engr, Portland Cement Asn, 47-49; res asst, Joint Hwy Res Proj, Purdue Univ, 50-51, res engr, 51-52; res engr, Tenn Hwy Res Prog, Univ Tenn, Knoxville, 52-62, dir, 52-72, assoc dir, Eng Exp Sta, 54-72, res prof, 62-72; assoc dir, Transp Res Ctr, 72-80, PROF CIVIL ENG, OHIO STATE UNIV, 72-, DIR, TRANSPLEX, 80- *Concurrent Pos:* Consult var firms; mem, Hwy Res Bd, Nat Acad Sci-Nat Res Coun. *Mem:* Fel Am Soc Testing & Mat; Am Rd Builders. *Res:* Highway materials; nondestructive testing of concrete, particularly by pulse velocity techniques; slipperiness of pavements, stabilization of pavement base courses; durability of concrete; performance of aggregates; bituminous materials and mixes. *Mailing Add:* 2620 Helen St Augusta GA 30904

WHITEHURST, GARNETT BROOKS, b Whitepine, Tenn, Sept 26, 52. BIOCHEMISTRY, ANALYTICAL CHEMISTRY. *Educ:* NC State Univ, BS, 75; Iowa State Univ, PhD(biochem), 80. *Prof Exp:* Fac teaching, Barton Col, Atlantic Christian, 82-89; RES DIR RES & DEVELOP, BROOKS WHITEHURST ASSOCS INC, 89- *Concurrent Pos:* Adj fac, Webster Univ, Pope AFB, 87-; vis asst prof, ECarolina Univ, 91- *Mem:* Am Chem Soc; AAAS; Asn Comput Mach; Inst Elec & Electronics Engrs Computer Soc; Math Asn Am; Asn Off Anal Chemists. *Res:* Development of products for chelation of metalion; development of processes for production of chelating agents. *Mailing Add:* 1983 Hoods Creek Dr New Bern NC 28564

WHITEHURST, HARRY BERNARD, b Dallas, Tex, Sept 13, 22; m 48; c 2. PHYSICAL CHEMISTRY. *Educ:* Rice Inst, BA, 44, MA, 48, PhD(chem), 50. *Prof Exp:* Fel, Univ Minn, 50-51; res chemist, Owens-Corning Fiberglas Corp, 51-59; assoc prof, 59-71, PROF CHEM, ARIZ STATE UNIV, 71- *Mem:* Fel AAAS; Am Chem Soc; fel Am Inst Chemists; Sigma Xi. *Res:* Radiochemistry; adsorption; surface chemistry of glass; electrical properties of oxides; investigation of ionizing radiation effects in solids and photoexcitation of electrons in solids. *Mailing Add:* Dept Chem Ariz State Univ Tempe AZ 85287-1604

WHITEHURST, VIRGIL EDWARDS, b Dunkirk, Ind; 32; m 55; c 1. PHARMACOLOGY, TOXICOLOGY. *Educ:* Anderson Col, BA, 53; Butler Univ, MS, 62; Ind Univ, PhD(biochem), 68. *Prof Exp:* Asst nutrit fluoride chem & preventive dent, Ind Univ, 67-68; assoc prof microbiol & biochem, Howard Univ, 68-73; PHARMACOLOGIST & TOXICOLOGIST, FOOD & DRUG ADMIN, WASHINGTON, DC, 73- *Concurrent Pos:* Vis prof, Howard Univ, 73-78; consult, Commun Progress, Inc, New Haven, Ct, 75-77, Murtis H Taylor Multi-Serv Ctr, Cleveland, Ohio, 78-80. *Mem:* Sigma Xi; Soc Black Scientists; Soc Toxicol. *Res:* Investigate the cardiotoxic effects of beta adrenergic agonists methylxanthenes when used separately and concurrently; cardiotoxic effects of the concurrent use of steroids and methylxanthines; propanolol as an antidote for throphylline; the role of phospholipids, arachidonic acid in isoproterenol induced sudden death in rats. *Mailing Add:* 5600 Fishers Lane Rockville MD 20857

WHITEKER, MCELWYN D, b Harrison Co, Ky, Aug 4, 29; m 50; c 3. ANIMAL SCIENCE. *Educ:* Univ Ky, BS, 51, MS, 57, PhD(nutrit, biochem), 61. *Prof Exp:* From asst prof to assoc prof animal sci, Iowa State Univ, 61-67; LIVESTOCK EXTEN SPECIALIST, UNIV KY, 67-, PROF ANIMAL SCI, 69- *Mem:* Am Soc Animal Sci. *Res:* Nutrition; animal breeding. *Mailing Add:* 705 Cromwell Way Lexington KY 40503

WHITEKER, ROY ARCHIE, b Long Beach, Calif, Aug 22, 27; m 60; c 1. GENERAL CHEMISTRY, ANALYTICAL CHEMISTRY. *Educ:* Univ Calif, Los Angeles, BS, 50, MS, 52; Calif Inst Technol, PhD(chem), 56. *Prof Exp:* Instr chem, Mass Inst Technol, 55-57; from asst prof to prof, Harvey Mudd Col, 57-74, actg chmn dept, 69-71; dep exec secy, Coun Int Exchange

Scholars, 71-72, exec secy, 72-75, dir, 75-76; dean, 76-89, PROF, UNIV PAC, 76- *Concurrent Pos:* NSF sci fac fel, Royal Inst Technol, Sweden, 63-64; vis assoc prof, Univ Calif, Riverside, 67; assoc dir fel off, Nat Res Coun, 67-68; dir summer session, Claremont Grad Sch, 69-70. *Mem:* Am Chem Soc. *Res:* Electroanalytical chemistry; complex ions. *Mailing Add:* 3734 Portsmouth Circle N Stockton CA 95219-3843

WHITELAW, R(OBERT) L(ESLIE), b S China, Apr 24, 17; nat US; m 42; c 3. SYSTEMS DESIGN. *Educ:* Univ Toronto, MSc, 40. *Prof Exp:* Liaison officer, Brit Air Comn, DC, 40-43; sr tech officer, Winnipeg Test Sta, Nat Res Coun Can, 43-45; sr designer, A V Roe Can Ltd, 45-46; chief gas turbine res & develop, De Laval Co, NJ, 46-48; staff engr, Res Ctr, Babcock & Wilcox Co, 48-55, proj engr, NS Savannah, 55-60; spec asst to dir res, Allison Div, Gen Motors Corp, 60-66; prof, 66-88, EMER PROF MECH & NUCLEAR ENG, VA POLYTECH INST & STATE UNIV, 88- *Concurrent Pos:* Assoc prof, US Naval Postgrad Sch, 63-64; vis expert, Int Atomic Energy Agency, Cent Atomic Bariloche, Arg, 81; pub policy expert, Heritage Found, 83-91. *Mem:* Am Soc Mech Engrs; Am Nuclear Soc; Am Soc Heating, Refrig & Air-Conditioning Engrs. *Res:* Thermodynamics; power cycles; cable-suspended transportation systems; geothermal power; underwater vehicles; advanced reactor design; flywheel energy storage; gas turbines and jet propulsion; steam power generation; network analysis. *Mailing Add:* 111 Alleghany St Blacksburg VA 24060

WHITELAW, WILLIAM ALBERT, b Halifax, NS, Sept 3, 41; m 70; c 2. RESPIRATION. *Educ:* Univ Toronto, BSc, 64; McGill Uinv, MDCM, 68, PhD(physiol), 78. *Prof Exp:* Intern, Montreal Gen Hosp, 68-69; resident med, Royal Victoria Hosp, Montreal, 69-71; res fel cardiol, Que Med Res Ctr, Makerere Univ, 71-72; res fel respiration, Can Med Res Ctr, Dept Physiol, McGill Univ, 72-75; clin fel chest med, McLaughlin Found, Edinburgh, 75-76; from asst prof to assoc prof, 76-85, PROF MED, UNIV CALGARY, 85- *Concurrent Pos:* Med med staff, Foothills Hosp, 76-; assoc ed-in-chief, French Rev Respiratory/Illness, 79-; chief, div pulmonary med, Univ Calgary, 81-; mem comt respiration, Can Med Res Ctr, 85-88; mem, res grant rev comt, Am Thoracic Soc, 85-87. *Mem:* Royal Col Physicians & Surgeons Can; Am Physiol Soc; Can Soc Clin Invest; AAAS; Can Thoracic Soc; Am Thoracic Soc. *Res:* Control of muscles of respiration in man and experimental mammals; mechanics of the diaphragm, dyspnea, breath holding and upper airway receptors. *Mailing Add:* Dept Med Heritage Med Res Bldg Univ Calgary Fac Med 3330 Hospital Dr NW Calgary AB T2N 4N1 Can

WHITELEY, HELEN RIABOFF, b Harbin, China, June 8, 22; nat US; m 44. MICROBIAL PHYSIOLOGY. *Educ:* Univ Calif, Berkeley, BA, 42; Univ Tex, MA, 46; Univ Wash, PhD, 51. *Prof Exp:* AEC fel, 51-53, res assoc microbiol, 53-57, from res asst prof to res assoc prof, 57-64, PROF MICROBIOL, UNIV WASH, 64- *Concurrent Pos:* Mem panel, NIH, 65-70 & 75-78. *Mem:* Am Soc Microbiol (vpres, 74-75, pres, 75-76); Am Acad Microbiol; Am Soc Biol Chem. *Res:* Control of viral transcription in B Subtilis phages; properties of RNA polymerase; echinoderm development. *Mailing Add:* Dept Microbiol Univ Wash Seattle WA 98195

WHITELEY, PHYLLIS ELLEN, b Manhasset, NY, Dec 4, 57; m 84; c 2. CELLULAR IMMUNOLOGY. *Educ:* Wash Univ, St Louis, BA, 79, PhD(pharmacol), 84. *Prof Exp:* Postdoctoral fel immunol, Med Sch, Wash Univ, 84-87, res asst prof path, Jewish Hosp, 87-89; SR RES IMMUNOLOGIST AUTOIMMUNITY, MERCK, SHARP & DOHME, 89- *Mem:* Am Asn Immunologists; Am Asn Women in Sci. *Res:* Development, initiation, and regulation of T lymphocytes in autoimmune diseases; develop therapeutics to inhibit T cell activation in diseases such as diabetes and rheumatoid arthritis. *Mailing Add:* Merck Sharp & Dohme Res Lab RY 80W-107 PO Box 2000 Rahway NJ 07065

WHITELEY, ROGER L, b Trenton, NJ, Jan 30, 30. MECHANICAL METALLURGY. *Educ:* Rensselaer Polytech Inst, BS, 52, MS, 53. *Prof Exp:* Engr, supvr & asst sect mgr, 53-64, sect mgr mech processing, 64-68, asst mgr forming & finishing res, 68-69, MGR CONTROL SYSTS RES, BETHLEHEM STEEL CORP, 69- *Honors & Awards:* Grossman Award, Am Soc Metals, 60. *Mem:* Am Soc Metals; Am Inst Mining & Metall Engrs; Am Iron & Steel Engrs; Am Iron & Steel Inst; Am Inst Physics. *Res:* Sheet steel metallurgy, forming and fabrication of metals, fracture and fatigue; instrumentation and automation of steel processes, rolling and rolling mill analysis, systems analysis. *Mailing Add:* Rd 2 Box 2188 Emmaus PA 18049-9802

WHITEMAN, ALBERT LEON, b Philadelphia, Pa, Feb 15, 15; m 45. MATHEMATICS. *Educ:* Univ Pa, AB, 36, AM, 37, PhD(math), 40. *Prof Exp:* Asst instr math, Univ Pa, 38-40; Benjamin Peirce instr, Harvard Univ, 40-42; instr, Purdue Univ, 46; res mathematician, US Dept Navy, 46-48; from asst prof to prof, 48-80, EMER PROF MATH, UNIV SOUTHERN CALIF, 80- *Concurrent Pos:* Mem, Inst Advan Study, 52-54, 59-60 & 67-68; chmn res conf theory numbers, NSF, 55 & 63; ed, Pac J Math, 57-62; res mathematician, Inst Defense Anal, 60-61; found fel, Inst Combinatorics & its Applications, 90. *Mem:* Am Math Soc; Math Asn Am. *Res:* Theory of numbers; combinatorial analysis. *Mailing Add:* Dept Math Univ Southern Calif Los Angeles CA 90089-1113

WHITEMAN, CHARLES E, b Eldred, Ill, Sept 28, 18; m 43; c 3. VETERINARY PATHOLOGY. *Educ:* Kans State Col, DVM, 43; Iowa State Univ, PhD(vet path), 60; Am Col Vet Pathologists, dipl. *Prof Exp:* Assoc prof vet path, Mich State Univ, 60-61; assoc prof, 61-71, PROF VET PATH, COLO STATE UNIV, 71- *Mem:* Am Vet Med Asn; Am Asn Avian Pathologists. *Res:* Placental pathology; respiratory diseases; poultry diseases. *Mailing Add:* 2401 East Ridge Ct Ft Collins CO 80524

WHITEMAN, ELDON EUGENE, b Tarentum, Pa, May 5, 13; m 39; c 3. ZOOLOGY. *Educ:* Greenville Col, BS, 36; Mich State Univ, MS, 41, PhD(zool), 65. *Prof Exp:* Asst dir, Kellogg Bird Sanctuary, 39-41; prof biol, 46-80, chmn, Natural Sci Div, 63-71, dir environ studies, 72-80, EMER PROF BIOL, SPRING ARBOR COL, 80- *Mem:* Nat Audubon Soc; Nat Wildlife Soc. *Res:* Development of a summer travel course for the college student in the area of environmental studies. *Mailing Add:* 170 Harmony Rd PO Box 136 Spring Arbor MI 49283

WHITEMAN, JOE V, b Walkerville, Ill, July 13, 19; m 45; c 1. ANIMAL BREEDING. *Educ:* NMex State Univ, BS, 43; Okla State Univ, MS, 51, PhD, 52. *Prof Exp:* From asst prof to assoc prof, 52-63, PROF ANIMAL SCI, OKLA STATE UNIV, 63- *Mem:* AAAS; Am Soc Animal Sci; Biomet Soc. *Res:* Genetic and environmental factors governing the growth and development of meat animals. *Mailing Add:* 724 S Ridge Rd Stillwater OK 74074

WHITEMAN, JOHN DAVID, b Darby, Pa, May 24, 43; m 69; c 2. COATINGS RESEARCH & DEVELOPMENT. *Educ:* LaSalle Col, BA, 65; Univ Pa, PhD(phys chem), 71. *Prof Exp:* Sr res chemist anal chem, Rohn & Haas Co, 72-75, sr res chemist pharmaceut, 75-76, sr res chemist pioneering coatings, 77-79, res sect mgr indust coatings, 80-84, res mgr Europe polymer, 84-87, res sect mgr pressure sensitive adhesives, 87-89, RES DEPT MGR LEATHER CHEM, ROHM & HAAS CO, 89- *Concurrent Pos:* Fel, Dept Chem, Univ Pa, 71-72. *Mem:* Am Phys Soc; Am Chem Soc. *Res:* Energy band structure of molecular crystals; polymer physics; structure activity relationships in agricultural and pharmaceutical chemicals; organic coatings; analytical chemistry; high solids coatings; pressure sensitive adhesives; leather chemicals. *Mailing Add:* 24 Sassafras Dr Churchville PA 18966

WHITENBERG, DAVID CALVIN, b Duffau, Tex, Feb 6, 31; div; c 1. PLANT PHYSIOLOGY, BIOCHEMISTRY. *Educ:* Tex A&M Univ, BS, 57, MS, 59, PhD(plant physiol, biochem), 62. *Prof Exp:* Res plant physiologist, USDA, 61-65; asst prof, 65-67, assoc prof, 67-80, PROF BIOL, SOUTHWEST TEX STATE UNIV, 80- *Mem:* Am Soc Plant Physiologists; Sigma Xi; Sci Res Soc. *Res:* Seed physiology and biochemistry. *Mailing Add:* Dept Biol Southwest Tex State Univ San Marcos TX 78666-4605

WHITESELL, JAMES JUDD, b Philadelphia, Pa, Oct 14, 39; m 65; c 1. ENTOMOLOGY. *Educ:* Dickinson Col, BS, 62; Univ Fla, MEd, 67, MS, 69, PhD(entom), 74. *Prof Exp:* Teacher sci, James S Rickards Jr High Sch, 63-67; res assoc lovebug res, Dept Entom, Univ Fla, 73-74; teacher biol & zool, Snead State Jr Col, Ala, 74-76, chmn sci & math div, 75-76; assoc prof entom & zool, dept biol, 76-80, assoc prof, 81- 84, PROF SCI EDUC & BIOL, SEC EDUC DEPT, VALDOSTA STATE COL, 84- *Mem:* Sigma Xi; Entom Soc Am. *Res:* Insect behavioral ecology. *Mailing Add:* Educ Ctr Valdosta State Col Valdosta GA 31698

WHITESELL, JAMES KELLER, b Philadelphia, Pa, Nov 2, 44; m 66; c 2. SYNTHETIC ORGANIC CHEMISTRY. *Educ:* Pa State Univ, BS, 66; Harvard Univ, PhD(chem), 71. *Prof Exp:* Fel chem, Woodward Res Inst, 70-73; from asst prof to assoc prof, 73-87, PROF CHEM, UNIV TEX, AUSTIN, 87- *Mem:* Am Chem Soc; Royal Soc Chem. *Res:* Total synthesis of naturally occurring and theoretically interesting molecules; asymmetric induction. *Mailing Add:* Dept Chem Univ Tex Austin TX 78712

WHITESELL, WILLIAM JAMES, b Newnan, Ga, Dec 23, 27; m 60; c 5. THEORETICAL PHYSICS. *Educ:* Univ SC, BS, 48; Purdue Univ, MS, 51, PhD(physics), 59. *Prof Exp:* From instr to asst prof physics, Brooklyn Col, 58-63; asst prof, 63-69, assoc prof, 69-81, PROF PHYSICS, ANTIOCH COL, 81- *Concurrent Pos:* Sr lectr, Victoria Univ, Wellington, 70-72. *Mem:* Am Asn Physics Teachers. *Mailing Add:* Dept Gen Sci Antioch Col Yellow Springs OH 45387

WHITESIDE, BOBBY GENE, b Keota, Okla, June 16, 40; m 64; c 2. FISHERIES. *Educ:* Okla State Univ, BS, 62, MS, 64, PhD(fisheries), 67. *Prof Exp:* Assoc prof, 67-77, PROF BIOL, SOUTHWEST TEX STATE UNIV, 77- *Concurrent Pos:* Mem several univ & sch comts, 67-86; Tex Water Develop Bd grant, 72-73, Soil Conserv Serv grant, 74, US Fish & Wildlife Serv grants, 76, 86-87, 87-88, 88-89 & 90, Am Fishing Tackle Mfrs Asn grant, 77, Pro Bass Mag grant, 77, TERA Corp grant, 78, Tex Parks & Wildlife grants, 87-88, NSF grant, 88-89; US Fish & Wildlife Serv grants, 76; Am Fishing Tackle Mfrs Asn grant, 77; Pro Bass Mag grant, 77; TERA Corp grant, 78. *Mem:* Am Fisheries Soc. *Res:* Fisheries management; population dynamics; ecology; numerous papers presented at professional meetings. *Mailing Add:* Dept Biol Southwest Tex State Univ San Marcos TX 78666-4616

WHITESIDE, CHARLES HUGH, b Grapevine, Tex, June 25, 32; m 56; c 2. ANALYTICAL CHEMISTRY, ENVIRONMENTAL SCIENCES. *Educ:* Tex A&M Univ, BS, 53, MS, 58, PhD(biochem), 60. *Prof Exp:* Robert A Welch Found res fel, Dept Biochem & Nutrit, Tex A&M Univ, 60-61; sr scientist, Mead Johnson & Co, 61-64; teacher chem, Kilgore Col, 64-67, chmn dept, 67-71; PRES, ANA-LAB CORP, 67- *Mem:* Am Chem Soc; Am Oil Chemists' Soc; Am Soc Testing Mat. *Res:* Water quality and waste water technology; animal nutrition; solar energy. *Mailing Add:* 2600 Dudley Rd Kilgore TX 75662

WHITESIDE, EUGENE PERRY, b Champaign, Ill, Oct 18, 12; m 36; c 3. SOIL SCIENCE. *Educ:* Univ Ill, BS, 34; Univ Mo, PhD(soils), 44. *Prof Exp:* Asst soil physics & soil surv, Exp Sta, Univ Ill, 34-38; asst soils, Univ Mo, 38-39; assoc soil surv, Exp Sta, Univ Ill, 39-43; assoc soils, Univ Tenn, 43-44; asst chief soil surv & asst prof soil physics, Exp Sta, Univ Ill, 44-49; from assoc prof to prof, 49-78, EMER PROF SOIL SCI, MICH STATE UNIV, 78- *Concurrent Pos:* Assoc soil surv, Emergency Rubber Proj, US Forest Serv, Calif, 43; consult, Natural Resources Sect, Agr Div, Gen Hqs, Supreme Comdr Allied Powers, Japan, 46-47, Rockefeller Found, Mex, 61, Agency Int Develop, Taiwan, 62-63, Arg, 72 & Food & Agr Orgn, UN, 80-81. *Mem:* Fel Soil Sci Soc Am; Int Soil Sci Soc; Soil Conserv Soc Am; fel Am Soc Agron; Sigma Xi. *Res:* Soil mineralogy, chemistry, geography, classification and genesis. *Mailing Add:* Mich State Univ A 560 Plant & Soil Sci Bldg East Lansing MI 48824

WHITESIDE, JACK OLIVER, b Barnstaple, Eng, June 5, 28; m 51; c 2. PLANT PATHOLOGY. *Educ:* Univ London, BSc, 48, PhD(plant physiol), 53. *Prof Exp:* Plant physiologist, Ministry Agr, Zimbabwe, Africa, 48-53, from plant pathologist to chief plant pathologist, 53-67; assoc plant pathologist, Citrus Exp Sta, 68-73, prof plant path, 73-90, EMER PLANT PATH, UNIV FLA, 90- *Mem:* Int Soc Citricult; Am Phytopath Soc. *Res:* Identification and control of plant diseases present in Zimbabwe; behavior and control of fungus diseases of citrus in Florida. *Mailing Add:* Citrus Res & Educ Ctr Univ Fla Lake Alfred FL 33850

WHITESIDE, JAMES BROOKS, b Tyler, Tex, Jan 5, 42; m 66; c 3. COMPOSITE MATERIALS, STRUCTURAL MECHANICS. *Educ:* Tulane Univ, BS, 64; Sheffield Univ, PhD(mech eng), 68. *Prof Exp:* Sr engr struct mech, 68-74, group leader composite structures, 74-79, RES LAB HEAD APPL MECH, GRUMMAN AEROSPACE CORP, 79- *Concurrent Pos:* Mem, Fed Aviation Admin Panel Independent Experts Struct, 81-; prof engr, 73- *Mem:* Soc Advan Mat Processing Eng; assoc fel Am Inst Aeronaut & Astronaut. *Res:* Design allowables and stress analysis of composites; moisture diffusion in polymers; fatigue and fracture; experimental stress analysis; mechanical and environmental behavior of composite materials. *Mailing Add:* A08-035 Grumman Aerospace Corp Bethpage NY 11714-3580

WHITESIDE, MELBOURNE C, b Washington, DC, Dec 16, 37; m 61; c 2. AQUATIC ECOLOGY. *Educ:* Willamette Univ, BA, 62; Ariz State Univ, MS, 64; Ind Univ, Bloomington, PhD(zool), 68. *Prof Exp:* Res fel limnol, Univ Minn, 68-69; asst prof ecol & limnol, Calif State Univ, Fullerton, 69-72; from asst prof to assoc prof, 72-80, PROF ZOOL, UNIV TENN, KNOXVILLE, 80- *Mem:* Am Soc Limnol & Oceanog; Ecol Soc Am. *Res:* Paleolimnology; community ecology and population dynamics of aquatic organisms; sampling problems in aquatic environments. *Mailing Add:* Dept Biol Univ Minn Duluth MN 55812

WHITESIDE, THERESA L, b Katowice, Poland, Mar 10, 39; US citizen; m 61; c 1. IMMUNOLOGY, IMMUNOPATHOLOGY. *Educ:* Columbia Univ, BS, 62, MA, 64, PhD(microbiol), 67; Am Bd Med Lab Immunol, dipl, 79. *Prof Exp:* NIH fel, Sch Med, NY Univ, 67-69, lectr microbiol & assoc res scientist, 69-70; res assoc ophthal, Col Physicians & Surgeons, Columbia Univ, 70-73; asst prof, 73-79, ASSOC PROF PATH, MED SCH, UNIV PITTSBURGH, 79-, ASSOC DIR CLIN IMMUNOPATH, 73- *Concurrent Pos:* NIH spec fel ophthal, Col Physicians & Surgeons, Columbia Univ, 72-73; Fogarty Int sr fel, Ludwig Inst, Lausanne, Switz, 84-85; dir clin immunopath, Pittsburg Cancer Inst, 85-88, dir immunol monitoring lab, 85- *Mem:* AAAS; Am Asn Immunologists; Am Asn Pathologists; Am Soc Microbiol. *Res:* Immunology of surface-associated antigens; tumor immunology; tumor infiltrating lymphocytes in human tumors; lymphocyte membrane receptors; clinical immunology. *Mailing Add:* Med Sch Dept Path Univ Pittsburgh Pittsburgh Cancer Inst 203 DeSoto St Rm 7201 Pittsburgh PA 15213

WHITESIDE, WESLEY C, b Milan, Ill, Aug 22, 27. BOTANY. *Educ:* Augustana Col, BA, 51; Univ Ill, MS, 56; Fla State Univ, PhD, 59. *Prof Exp:* Instr bot & gen biol, Montgomery Jr Col, 59-60; from asst prof to assoc prof, 60-70, PROF BOT, EASTERN ILL UNIV, 70- *Concurrent Pos:* Res grant, Highlands Biol Sta, 59. *Mem:* Mycol Soc Am. *Res:* Morphology; cytology and taxonomy of the ascomycete fungi, especially the Pyrenomycetes; taxonomy of lichens. *Mailing Add:* RR 3 No 56 Charleston IL 61920

WHITESIDES, GEORGE MCCLELLAND, b Louisville, Ky, Aug 3, 39; m 69; c 2. ORGANIC CHEMISTRY. *Educ:* Harvard Univ, AB, 60; Calif Inst Technol, PhD(chem), 64. *Prof Exp:* From asst prof to assoc prof, Mass Inst Technol, 63-74, prof chem, 74-82; PROF CHEM, HARVARD UNIV, 82- *Mem:* Nat Acad Sci; Am Chem Soc; Am Acad Arts & Sci; fel AAAS. *Res:* Mechanisms and structure; use of enzymes in organic synthesis; applied biochemistry; structure-property relations in organic materials science; heterogeneous catalysis. *Mailing Add:* Chem Dept Harvard Univ Cambridge MA 02138

WHITESIDES, JOHN LINDSEY, JR, b San Antonio, Tex, Feb 27, 43; div; c 2. FLUID MECHANICS, AERONAUTICS. *Educ:* Univ Tex, Austin, BS, 65, PhD(aerospace eng), 68. *Prof Exp:* Asst res prof mech eng, 68-74, assoc prof, 74-80, PROF ENG & APPL SCI, GEORGE WASHINGTON UNIV, 80-; ASSOC DIR, JOINT INST ADVAN FLIGHT SCI, LANGLEY RES CTR, NASA, 84- *Concurrent Pos:* Coordr, George Washington Univ-NASA Prog, Langley Res Ctr, NASA, 68-75; asst dir, George Washington Univ, 75-84; nat fac adv, Am Inst Aeronaut & Astronaut. *Mem:* Assoc fel Am Inst Aeronaut & Astronaut; Soc Eng Sci; Am Soc Eng Educ; AAAS. *Res:* Analytical methods in fluid mechanics and aeronautics; graduate engineering education. *Mailing Add:* Langley Res Ctr J1AFS MS 269 NASA Hampton VA 23665

WHITESITT, JOHN ELDON, b Stevensville, Mont, Jan 15, 22; m 44; c 3. MATHEMATICS. *Educ:* Mont State Univ, AB, 43; Univ Ill, AM, 49, PhD(math), 54. *Prof Exp:* From instr to assoc prof, 46-61, head dept, 61-66, PROF MATH, MONT STATE UNIV, 61- *Mem:* Am Math Soc; Math Asn Am. *Res:* Ring theory; linear algebra; Boolean algebra. *Mailing Add:* Dept Math Mont State Univ Bozeman MT 59717

WHITEWAY, STIRLING GIDDINGS, b Stellarton, NS, May 17, 27; m 52; c 2. PHYSICAL CHEMISTRY, INORGANIC CHEMISTRY. *Educ:* Dalhousie Univ, BSc, 47, dipl, 48, MSc, 49; McGill Univ, PhD(phys chem), 53. *Prof Exp:* Fel photochem, Pure Chem Div, Atlantic Res Lab, Nat Res Coun Can, 52-53, from asst res officer to sr res officer, 53-83, prin res officer, 83-89; RETIRED. *Concurrent Pos:* Spec lectr, Dalhousie Univ, 54-55; adj prof, Tech Univ NS, 84- *Mem:* Chem Inst Can; Can Inst Mining & Metall. *Res:* Chemistry of high temperature reactions; chemistry of coal; ceramics. *Mailing Add:* 20 Day Ave Dartmouth NS B2W 2V6 Can

WHITFIELD, CAROL F(AYE), b Altoona, Pa, May 14, 39; m 87. PHYSIOLOGY, MOLECULAR BIOLOGY. *Educ:* Juniata Col, BS, 61; Syracuse Univ, MS, 64; George Washington Univ, PhD(physiol), 68. *Prof Exp:* Teaching asst zool, Syracuse Univ, 61-63; from res assoc to asst prof, 68-83, ASSOC PROF PHYSIOL, MILTON S HERSHEY MED CTR, COL MED, PA STATE UNIV, HERSHEY, 83-, DIR, MULTIDISCIPLINE LABS, 88- *Concurrent Pos:* Mem bd trustees, Juniata Col, 83-86; prin investr, Nat Heart, Lung & Blood Inst, NIH, Am Heart Asn, 77- *Mem:* Biophys Soc; Am Physiol Soc; Am Soc Cell Biol; Am Soc Hemat. *Res:* Sickle cell anemia, red cell membrane structure, regulation of carrier-mediated sugar transport in erythrocytes and muscles; regulation of membrane permeability; membrane structure; cardiac muscle cytoskeleton. *Mailing Add:* Dept Physiol Milton S Hershey Med Ctr Pa State Univ PO Box 850 Hershey PA 17033

WHITFIELD, CAROLYN DICKSON, b Indianapolis, Ind, Aug 21, 41; m 65. BIOCHEMISTRY. *Educ:* Wellesley Col, AB, 63; Univ Chicago, MS, 65; George Washington Univ, PhD(biochem), 69. *Prof Exp:* Wellcome Found fel, Univ Edinburgh, 69-70; Am Cancer Soc fel, 70-72, asst res biol chemist, 72-74, scholar human genetics, 74-76, asst prof, dept biol chem, Med Sch, Univ Mich, Ann Arbor, 76-; AT COL MED, HOWARD UNIV, WASHINGTON, DC. *Concurrent Pos:* Estab investr, Am Heart Asn, 77-82. *Mem:* Am Soc Biol Chemists. *Res:* Mechanism of action of flavoproteins; isolation and biochemical characterization of Chinese hamster cell mutants in tissue culture; methionine biosynthesis. *Mailing Add:* Dept Biol Chem Col Med Howard Univ Washington DC 20059

WHITFIELD, GEORGE BUCKMASTER, JR, b Newark, NJ, Dec 4, 23; m 44; c 3. BIOCHEMISTRY. *Educ:* Cornell Col, BA, 46; Univ Ill, MS, 51, PhD(chem), 53. *Prof Exp:* Jr res scientist, Upjohn Co, 47-51, sr res scientist, 53-59, sect head microbiol, 59-66, mgr, 66-68, infectious dis res mgr, 68-78, coor prod group adminr, Infectious Dis & Cardiovasc Dis, 78-85; RETIRED. *Mem:* Am Chem Soc; Am Soc Microbiol; NY Acad Sci; Sigma Xi. *Res:* Isolation and characterization of new antibiotics and antitumor agents; paper chromatography; microbiological assay; tissue culture; in vitro, in vivo and clinical evaluation of new antibiotics, antifungal and antiparasitic agents. *Mailing Add:* 109 Candlewyck Dr Kalamazoo MI 49001-5414

WHITFIELD, HARVEY JAMES, JR, b Chicago, Ill, Apr 10, 40; m 65. MOLECULAR BIOLOGY. *Educ:* Univ Ill, Urbana, BS, 61; Univ Ill Col Med, MD, 64. *Prof Exp:* Intern, Res & Educ Hosp, Chicago, 64-65; staff asst molecular biol, NIH, 65-69; USPHS spec fel, Med Res Coun Microbial Genetics Unit, Univ Edinburgh, 69-70; asst prof, 70-75, ASSOC PROF BIOCHEM, MED SCH, UNIV MICH, 75-; MED STAFF FEL, BIOL PSYCH BR, NIMH. *Concurrent Pos:* USPHS grant, Univ Mich, Ann Arbor, 71- *Mem:* AAAS; Am Soc Biol Chemists; Am Soc Microbiol; Genetics Soc Am. *Res:* Replication and segregation of episomal DNA; microbial genetics. *Mailing Add:* NIMH Bldg 36 Rm 2D15 9000 Rockville Pike Bethesda MD 20892

WHITFIELD, JACK D, b Paoli, Okla, May 16, 28; m 49; c 3. AERONAUTICAL ENGINEERING, GAS DYNAMICS. *Educ:* Univ Okla, BS, 51; Univ Tenn, MS, 60; Royal Inst Technol, Sweden, DSc, 72. *Prof Exp:* Test engr, Gen Dynamics-Convair, 51-54; engr, Von Karman Gas Dynamics Facil, SVerdrup Corp, 54-60, asst mgr hypervelocity br, 60-64, mgr, 64-68, dir, Von Karman Facil, 68-75, dir, Engine Test Facil, 75-76, corp vpres & corp prin advan technol, 76-80, exec vpres, 85-88, PRES, SVERDRUP CORP, 80-85 & 89- *Concurrent Pos:* Consult, Off Nat Studies & Aeronaut Res, sponsored by adv group for Aeronaut Res & Develop, Paris, 60; consult arc-driven shock tubes, Vitro/Smith Corp, NY, 63; prof aerospace eng, Univ Tenn, 76- *Honors & Awards:* Gen H H Arnold Award, Am Inst Aeronaut & Astronaut, 68; Nat Ground Testing Award, Am Inst Aeronaut & Astronaut. *Mem:* Fel Am Inst Aeronaut & Astronaut; Nat Soc Prof Engrs. *Res:* Experimental and theoretical supersonic and hypersonic aerodynamics, especially studies of viscous flow phenomena; development and advancement of supersonic and hypersonic gas dynamic test facilities; technical and administrative research management. *Mailing Add:* Sverdrup Corp PO Box 884 Tullahoma TN 37388

WHITFIELD, JAMES F, b Sarnia, Ont, July 1, 31; m 51; c 4. CELL PHYSIOLOGY, CANCER. *Educ:* McGill Univ, BSc, 51; Univ Western Ont, MSc, 52, PhD(bact, immunol), 55. *Prof Exp:* Res officer bact & viral genetics, Atomic Energy Can Ltd, 55-58, res officer cellular radiobiol, 58-62; sect chief, Europ Joint Res Ctr, Europ AEC, Italy, 62-65; head cell physiol sect, Radiation Biol Div, Nat Res Coun Can, 65-72, head cell physiol group, 72-86, head cellular oncol group, Biol Sci Div, 86-90, HEAD CELL SYSTS SECT, INST BIOL SCI, NAT RES COUN CAN, 90- *Mem:* Tissue Cult Asn; Am Soc Cell Biol; NY Acad Sci; AAAS. *Res:* Control of cell proliferation; in vivo and in vitro effects of calcium, hormones, cyclic nucleotides and oncogenes. *Mailing Add:* Biol Sci Div Nat Res Coun Can Ottawa ON K1A 0R6 Can

WHITFIELD, JOHN HOWARD MERVYN, b Thessalon, Ont, Sept 11, 39; m 60; c 4. MATHEMATICS. *Educ:* Abilene Christian Col, BA, 61; Tex Christian Univ, MA, 62; Case Inst Technol, PhD(math), 66. *Prof Exp:* From asst prof to assoc prof, 65-82, chmn dept, 72-75, dean arts & sci, 86-90, PROF MATH, LAKEHEAD UNIV, 82-, VPRES ACAD, 91- *Concurrent Pos:* Vis scholar, Univ Wash, 71-72; vis prof, Univ Waterloo, 78-79. *Mem:* Am Math Soc; Math Asn Am; Can Math Soc; Sigma Xi. *Res:* Functional analysis; differentiable functions and norms; geometry of Banach spaces. *Mailing Add:* Dept Math Sci Lakehead Univ Thunder Bay ON P7B 5E1 Can

WHITFIELD, RICHARD GEORGE, b Philadelphia, Pa, Nov 11, 51; m 73; c 1. PHYSICAL & ANALYTICAL CHEMISTRY. *Educ:* Glassboro State Col, BA, 73; Mich State Univ, PhD(chem), 77. *Prof Exp:* SR RES CHEMIST ANALYSIS, OLIN CHEM GROUP, 78- *Mem:* Am Chem Soc; Sigma Xi. *Res:* Analytical applications of infrared, raman, ultraviolet and visible spectroscopy; instrumental and method development; investigations of the solid state via vibrational spectroscopy; office systems; analytical chemistry as well as materials support. *Mailing Add:* PO Box 2805 Kalamazoo MI 49003

WHITFIELD, ROBERT EDWARD, chemistry; deceased, see previous edition for last biography

WHITFILL, DONALD LEE, b Madill, Okla, Mar 13, 39; m 60; c 2. PHYSICAL INORGANIC CHEMISTRY. *Educ:* Southeastern State Col, BS, 61; Univ Okla, PhD(chem), 66. *Prof Exp:* Instr chem, Univ Okla, 66-67; res scientist, Plant Foods Res Div, Continental Oil Co, 67-70, res scientist, 70-78, sr res scientist, 78-79; res group leader, 79-88, SECT MGR, PROD RES DIV, CONOCO INC, 88- *Honors & Awards:* Distinguished Lectr, Soc Petrol Engrs. *Mem:* Soc Petrol Engrs. *Res:* Transition metal chemistry; electrochemistry; drilling fluid and cement technology; drilling and completions; reservoir and enhanced recovery. *Mailing Add:* 1700 Cedar Lane Ponca City OK 74604

WHITFORD, ALBERT EDWARD, b Milton, Wis, Oct 22, 05; m 37; c 3. ASTROPHYSICS. *Educ:* Milton Col, BA, 26; Univ Wis, MA, 28, PhD(physics), 32. *Prof Exp:* Asst, Washburn Observ, Univ Wis, 32-33; Nat Res Coun fel, Mt Wilson Observ & Calif Inst Technol, 33-35; res assoc astron, Washburn Observ, Univ Wis, 35-38, asst prof astrophys, 38-46, assoc prof astron, 46-48, prof & dir observ, 48-58; astronr & dir, 58-68, astronr & prof, 68-73, EMER PROF ASTRON, LICK OBSERV, UNIV CALIF, SANTA CRUZ, 73- *Concurrent Pos:* Mem staff, Radiation Lab, Mass Inst Technol, 41-46. *Mem:* Nat Acad Sci; Am Astron Soc (vpres, 65-67, pres, 67-70); Am Acad Arts & Sci. *Res:* Photoelectric instrumentation; interstellar absorption; spectrophotometry of stars and galaxies; stellar population of galaxies. *Mailing Add:* 220 Morrissey Blvd Santa Cruz CA 95062

WHITFORD, GARY M, b Gouveneur, NY, Mar 9, 37; m 85; c 3. TOXICOLOGY, PHYSIOLOGY. *Educ:* Univ Rochester, BS, 65, MS, 69, PhD(toxicol), 72; Med Col Ga, DMD, 75. *Prof Exp:* Instr oral biol, NJ Dent Sch, 71-72; from asst prof to prof, 72-85, REGENTS PROF ORAL BIOL, MED COL GA, 85- *Honors & Awards:* H Trendley Dean Award, 86. *Mem:* Sigma Xi; Soc Exp Biol Med; Int Asn Dent Res; Europ Orgn Caries Res; Am Physiol Soc. *Res:* Metabolism, biological effects and toxicology of fluoride. *Mailing Add:* Dept Oral Biol Med Col Ga Augusta GA 30912-1129

WHITFORD, HOWARD WAYNE, b Benavides, Tex, Apr 6, 40; m 65; c 2. VETERINARY MICROBIOLOGY. *Educ:* Tex A&M Univ, BS, 63, DVM, 64, PhD(vet microbiol), 76; Am Col Vet Microbiologists, dipl, 73. *Prof Exp:* Vet lab officer res, US Army Med Unit, Frederick, Md, 65-68; vet officer, Rocky Mountain Lab, Nat Inst Allergy & Infectious Dis, USPHS, 68-70; NIH fel vet microbiol, Sch Vet Med, Tex A&M Univ, 70-71, from grad asst to instr, 71-74; bacteriologist, Diag Serv, 74-87, HEAD, DIAG MICROBIOL, TEX VET MED DIAG LAB, 87- *Concurrent Pos:* Mem, sheep & goat comt, US Animal Health Asn, 74-; consult, Stauffer Chem Co, Houston, Tex facil, 80-82; mem, mycoplasmosis comt, Am Asn Vet Lab Diagnosticians, 81-, chmn, 84-90; chmn, small ruminant pract comt, Tex Vet Med Asn, 84-86 & 88-90, peer asst comt, 87- *Mem:* Am Col Vet Microbiologists; Am Asn Vet Lab Diagnosticians; Am Vet Med Asn; US Animal Health Asn; Am Asn Small Ruminant Practrs. *Res:* Bacteriologic diagnostic techniques; infectious diseases of sheep and goats; pathogenic bacteriology and mycology; epidemiology of anthrax. *Mailing Add:* Drawer 3040 College Station TX 77841

WHITFORD, LARRY ALSTON, b Ernul, NC, Apr 11, 02; m 28; c 4. BOTANY. *Educ:* NC State Col, BS, 25, MS, 29; Ohio State Univ, PhD(bot), 41. *Prof Exp:* High sch teacher, 25-26; from instr to prof, 26-68, EMER PROF BOT, NC STATE UNIV, 68- *Concurrent Pos:* Vis prof, Univ Va, 52 & Univ Fla, 53; consult, Duke Power Co Environ Labs, 73-81 & Aquatic Control, 75-81; nat distinguished lectr, Phycol Soc Am, 67. *Mem:* AAAS; Phycol Soc Am (vpres, 56, pres, 57). *Res:* Floristics and ecology of fresh-water algae. *Mailing Add:* Dept Bot NC State Univ Raleigh NC 27695-7612

WHITFORD, PHILIP BURTON, b Argyle, Minn, Jan 9, 20; m 46; c 2. PLANT ECOLOGY. *Educ:* Northern Ill State Teachers Col, BEd, 41; Univ Ill, MS, 42; Univ Wis, PhD(bot), 48. *Prof Exp:* Asst bot, Univ Wis, 46-48; ed asst conserv, State Bd Natural Resources, Md, 48-49; from asst prof to assoc prof, 49-61, chmn dept, 66-69, prof, 61-82, EMER PROF BOT, UNIV WIS-MILWAUKEE, 82- *Mem:* Fel AAAS; Ecol Soc Am; Am Inst Biol Sci; Sigma Xi. *Res:* Population and distribution of plants of the prairie-forest border; successions of native plants; resource management; applied ecology and conservation. *Mailing Add:* PO Box 57 Fair Water WI 53931

WHITFORD, WALTER GEORGE, b Providence, RI, June 12, 36; m 59, 69; c 3. ECOLOGY. *Educ:* Univ RI, BA, 61, PhD(zool), 64. *Prof Exp:* From asst prof to assoc prof, 64-72, PROF BIOL, NMEX STATE UNIV, 72- *Concurrent Pos:* Ecol consult, Pub Serv Co NMex, 71- & Union Oil Co Calif, 74-; ed, Ecol Soc Am, 75-79, Biol Fertil Soils; prin investr, Journada Long Term Ecol Res Prog, 81- *Mem:* AAAS; Ecol Soc Am; Herpetologists' League. *Res:* Desert ecology; ecology of social insects; soil biology; plant ecology. *Mailing Add:* Dept Biol NMex State Univ Las Cruces NM 88070

WHITFORD-STARK, JAMES LESLIE, b London, Eng, Sept 28, 48; m 83; c 2. REMOTE SENSING-PLANETS, VOLCANOLOGY. *Educ:* Keele Univ, BA, 71; Univ Lancaster, MS, 75; Brown Univ, PhD(geol), 80. *Prof Exp:* Field geologist res, Radiogeol & Rare Minerals Unit, Inst Geol Sci, London, 69; res asst, Dept Environ Sci, Univ Lancaster, 71-76; teaching asst, Dept Geol Sci, Brown Univ, 76-77, res asst, 77-80; vis asst prof teaching, Geol Dept, Univ Mo, Columbia, 80-82; from asst prof to assoc prof, 82-87, CHMN TEACHING, GEOL DEPT, SUL ROSS STATE UNIV, 87- *Concurrent Pos:* Consult, Open Univ, Eng, 75, Volcres, 80-; prin investr, Planetary Geol Prog, NASA, 81, Galilean Satellite Mapping Prog, 82-86; mem, Bd Sci, Chihuahua Desert Res Inst, 87-; dir, NSF China-US Coop Res Proj, 88-90. *Mem:* Fel Geol Soc Am; Am Geophys Union; Am Soc Photogrammetry & Remote Sensing; Int Asn Volcanology & Chem Earth's Interior; fel Geol Soc London. *Res:* Mapping and remote sensing of volcanic materials on all of the terrestrial planets; specializing in the volcanic products on the Moon, Io, and mainland Asia. *Mailing Add:* Geol Dept Sul Ross State Univ Alpine TX 79832

WHITHAM, GERALD BERESFORD, b Halifax, Eng, Dec 13, 27; m 51; c 3. FLUID MECHANICS. *Educ:* Univ Manchester, BSc, 48, MSc, 49, PhD, 53. *Prof Exp:* Res assoc, Inst Math Sci, NY Univ, 51-53; lectr math, Univ Manchester, 53-56; assoc prof, Inst Math Sci, NY Univ, 56-59; prof, Mass Inst Technol, 59-62; PROF APPL MATH, CALIF INST TECHNOL, 62. *Honors & Awards:* Wiener Prizer Appl Math, 80. *Mem:* Fel Am Acad Arts & Sci; fel Royal Soc. *Res:* Fluid dynamics; wave propagation. *Mailing Add:* Dept Appl Math Calif Inst Technol Firestone 217-50 Pasadena CA 91125

WHITING, ALLEN R, US citizen. NONDESTRUCTIVE TESTING. *Educ:* Univ Tex, BS, 61. *Prof Exp:* X-ray lab technician, Univ Tex, 61; spray dept supvr, Aztec Tile Co, 61-62; asst res engr, Dept Mat Eng, Southwest Res Inst, 62-64, res engr, 64-67, sr res engr, 67-69, mgr appl eng, 69-70, mgr appl eng, Dept Spec Eng Serv, 70-72, asst dir, 72-74, dir, Dept Res & Eng, 74-75, dir, Dept Energy Serv, 75-78, exec dir, Qual Assurance Systs & Eng Div, 78-87, DIR, SYSTS ENG & INTEGRATION, CTR NUCLEAR WASTE REGULATORY ANALYSIS, SOUTHWEST RES INST, 87- *Mem:* Am Soc Nondestructive Testing; Am Soc Testing & Mat; Am Soc Mech Engrs. *Res:* Nondestructive testing including X- and gamma-radiography, ultrasonics, magnetic particle and penetrants as they are applied to solve industry problems and also in the research and development area. *Mailing Add:* Dept Systs Eng & Integration PO Drawer 28510 San Antonio TX 78284

WHITING, ANNE MARGARET, b Morrisville, Vt, May 17, 41. VERTEBRATE ANATOMY, EMBRYOLOGY. *Educ:* Eastern Nazarene Col, AB, 63; Univ Ill, Urbana, MS, 65; Pa State Univ, PhD(zool), 69. *Prof Exp:* From asst prof to assoc prof, 68-73, PROF BIOL, HOUGHTON COL, 73- *Mem:* Sigma Xi; Am Soc Zoologists; Am Inst Biol Sci; Am Sci Affil. *Res:* Morphology, histology and histochemistry of squamate cloacal glands. *Mailing Add:* Dept Biol Houghton Col Houghton NY 14744

WHITING, FRANK M, b Tucson, Ariz, Dec 5, 32; m 58; c 2. NUTRITION, BIOCHEMISTRY. *Educ:* Univ Ariz, BS, 56, MS, 68, PhD(agr biochem, nutrit), 71. *Prof Exp:* Field man qual control, United Dairymen Ariz, 56-60, res technician pesticide residues, 60-65; res asst, 65-71, asst prof & asst animal scientist, 71-76, assoc prof & assoc animal scientist, 76-81, PROF ANIMAL SCI & ANIMAL SCIENTIST, UNIV ARIZ, 81- *Mem:* Am Dairy Sci Asn; Sigma Xi. *Res:* Pesticide chemistry; ruminant nutrition; lipid metabolism; pesticide residues in feeds and animal products. *Mailing Add:* 1873 Paseo Reforma Tucson AZ 85705

WHITING, JOHN DALE, JR, b New Castle, Pa, Mar 23, 47. FORENSIC TOXICOLOGY, DRUG ABUSE DETECTION. *Educ:* Westminister Col, BS, 69; Duke Univ, PhD(biochem), 74. *Prof Exp:* Res assoc biol, Princeton Univ, 74-77; res assoc biochem, George Washington Univ, 77-79; res chemist toxicol, Armed Forces Inst Path, 79-85; PRES, TOXICHEM LABS, INC, 85- *Concurrent Pos:* Mem, Subcomt Chromatographic Methods, Nat Comn Clin Lab Studies, 82- *Mem:* AAAS; Am Chem Soc; Am Acad Forensic Sci. *Res:* Isolation and identification of drugs and drug metabolites from tissues using high performance liquid chromatography, gas chromatography, gel chromatography and mass spectrometry and fourier transform infrared; development of analytical procedures to detect tetrahydrocannabinol and its metabolites and other substances of abuse in biological fluids and tissues. *Mailing Add:* 5956 Fulton Dr Mt Airy MD 21771

WHITING, R(OBERT) L(OUIS), b San Antonio, Tex, Feb 25, 18; m 44; c 3. PETROLEUM ENGINEERING. *Educ:* Univ Tex, BS, 39, MS, 42. *Prof Exp:* Instr petrol eng, Univ Tex, 39-43; assoc prof, Mo Sch Mines, 45-46; assoc prof, 46-50, head dept, 46-76, PROF PETROL ENG, TEX A&M UNIV, 50- *Concurrent Pos:* Int consult, US Fed Govt, FTC, var petrol co & foreign govts, 46-; dir, Tex Petrol Res Comt, 51-76. *Honors & Awards:* Mineral Indust Educ Award, Am Inst Mech Engrs, 73. *Mem:* Am Inst Mining, Metall & Petrol Engrs; Am Asn Univ Professors; Am Petrol Inst; Am Asn Eng Educ; Sigma Xi. *Res:* Drilling, production, transportation and marketing in petroleum and natural gas. *Mailing Add:* Dept Petrol Eng Tex A&M Univ College Station TX 77843

WHITLA, WILLIAM ALEXANDER, b Galt, Ont, Oct 16, 38; m 64; c 3. COMPUTER ASSISTED INSTRUCTION. *Educ:* McMaster Univ, BSc, 60, PhD(inorg chem), 65. *Prof Exp:* Nat Res Coun overseas fel x-ray crystallog, Oxford Univ, 65-66; teaching fel, Univ BC, 66-67; asst prof, 67-75, ASSOC PROF CHEM, MT ALLISON UNIV, 75- *Mem:* Chem Inst Can. *Res:* Development of computer programs for chemistry instruction. *Mailing Add:* Dept Chem Mt Allison Univ Sackville NB E0A 3C0 Can

WHITLATCH, ROBERT BRUCE, b Boise, Idaho, July 18, 48. MARINE & POPULATION ECOLOGY. *Educ:* Univ Utah, BS, 70; Univ of the Pac, MS, 72; Univ Chicago, PhD(evolutionary biol), 76. *Prof Exp:* Scholar biol, Woods Hole Oceanog Inst, 76-77; asst prof biol, 77-83, assoc prof marine sci, 83-89, PROF, UNIV CONN, 90- *Concurrent Pos:* Vis scientist, Neth Inst Sea Res, 84-85; sr fel, Nat Res Coun, 85-86. *Mem:* AAAS; Am Inst Biol Sci; Ecol Soc Am; Estuarine Res Fedn. *Res:* Population community ecology of marine benthic systems; role of disturbance agents on community structure; resource partitioning in deposit feeding organisms. *Mailing Add:* Dept Marine Sci Univ Conn Groton CT 06340

WHITLEY, JAMES HEYWARD, electric contact theory; deceased, see previous edition for last biography

WHITLEY, JAMES R, b Jamesport, Mo, Apr 21, 21; m 42. BIOCHEMISTRY, NUTRITION. *Educ:* Univ Mo, AB, 42, MS, 47, PhD(agr chem), 52. *Prof Exp:* Pvt herbicide bus, 52-62; SUPT FISHERIES RES, MO DEPT CONSERV, 62- *Mem:* Am Fisheries Soc; Weed Sci Soc Am; Water Pollution Control Fedn. *Res:* Ecology of fish and other aquatic organisms. *Mailing Add:* 303 S Glenwood Columbia MO 65201

WHITLEY, JOSEPH EFIRD, b Albemarle, NC, Mar 22, 31; m 58; c 2. RADIOLOGY. *Educ:* Wake Forest Univ, BS, 51; Bowman Gray Sch Med, MD, 55; Am Bd Radiol, dipl, 60, cert nuclear med, 72. *Prof Exp:* Intern, Pa Hosp, Philadelphia, 55-56; resident radiol, NC Baptist Hosp, Winston-Salem, 56-69; from asst prof to prof radiol, Bowman Gray Sch Med, 62-78; PROF DIAG RADIOL & CHMN DEPT, UNIV MD SCH MED, 78- *Concurrent Pos:* James Picker Found scholar radiol res, Bowman Gray Sch Med, 59-61, advan fel acad radiol, Karolinska Inst, Sweden & Mass Inst Technol, 61-62; consult, Epilepsy Prog, Nat Inst Neurol Dis & Blindness, 67-79. *Mem:* Am Col Radiol; Asn Univ Radiol; Radiol Soc NAm; Soc Nuclear Med; Am Roentgen Ray Soc. *Res:* Description and evaluation of cardiovascular phenomena by angiographic and radioisotopic techniques, particularly renovascular hypertension and pulmonary embolism; evaluation of modes of medical education. *Mailing Add:* Dept Diag Radiol Univ Md Hosp 22 S Greene St Baltimore MD 21201

WHITLEY, LARRY STEPHEN, b Mattoon, Ill, Jan 30, 37; m 58; c 3. ENVIRONMENTAL BIOLOGY. *Educ:* Eastern Ill Univ, BS, 58; Purdue Univ, MS, 60, PhD(ecol), 63. *Prof Exp:* Instr environ biol, Purdue Univ, 63; from asst prof to assoc prof, 63-71, PROF ZOOL, EASTERN ILL UNIV, 71- *Concurrent Pos:* NIH res grants, 65-67; Fed Water Qual Admin grant, Dept Interior, 69-71. *Mem:* AAAS; Ecol Soc Am; Soc Syst Zool. *Res:* Physiology and systematics of tubificid worms; biology of polluted aquatic ecosystems and the tolerance mechanisms of aquatic organisms. *Mailing Add:* Dept Zool Eastern Ill Univ Charleston IL 61920

WHITLEY, NANCY O'NEIL, b Winston-Salem, NC, Feb 21, 32; m 58; c 2. RADIOLOGY. *Educ:* Bowman Gray Sch Med, MD, 57. *Prof Exp:* Intern, Jefferson Davis Hosp, Houston, Tex, 57-58; cardiovasc trainee, Bowman Gray Sch Med, 59-61; physician, Med Dept, Western Elec Co, 63-66; resident radiol, 66-69, from instr to assoc prof, Bowman Gray Sch Med, 69-78; PROF RADIOL, UNIV MD SCH MED, 78- *Concurrent Pos:* Fel cardiol, Bowman Gray Sch Med, 58-59. *Honors & Awards:* Fel, Am Col Radiol. *Mem:* Am Roentgenol Ray Soc; AMA; Asn Univ Radiologists; Radiol Soc NAm; Am Col Radiol. *Res:* Techniques and procedures of angiography. *Mailing Add:* Dept Radiol 22 Greene St Baltimore MD 21201

WHITLEY, WILLIAM THURMON, b Deland, Fla, Oct 24, 41; m 68; c 1. MATHEMATICS. *Educ:* Stetson Univ, BS, 63; Univ NC, Chapel Hill, MA, 66; Va Polytech Inst & State Univ, PhD(math), 69. *Prof Exp:* Instr math, Va Polytech Inst & State Univ, 69-70; assoc prof math, Marshall Univ, 70-79; assoc prof, 79-82, PROF, UNIV NEW HAVEN, 79 - *Mem:* Am Math Soc; Math Asn Am; Am Statist Asn. *Res:* Deleted products of topological spaces; rings of continuous real-valued functions. *Mailing Add:* Dept Math Univ New Haven West Haven CT 06516

WHITLOCK, CHARLES HENRY, b Richmond, Va, Mar 24, 39; m 62; c 2. CIVIL & AERONAUTICAL ENGINEERING. *Educ:* Univ Va, BAE, 61, MAE, 65; Col William & Mary, MBA, 70; Old Dominion Univ, PhD(civil eng), 77. *Prof Exp:* Aerospace res engr, flight mechanics, 61-70, head syst dynamics sect, 70-72, asst head marine anal sect, 72-74, head data anal sect, 74-76, head wave modeling group, 74-77, HEAD SPECTRAL SIGNATURE & OPTICAL MODELING GROUP, MARINE ENVIRON, NASA LANGLEY RES CTR, 76- *Concurrent Pos:* Instr math, Hampton Inst, 66-67. *Mem:* Am Soc Civil Eng; Am Soc Photogram. *Res:* Remote sensing including optical modeling of spectral signals. *Mailing Add:* 207 Brook Blvd Quinton VA 23141

WHITLOCK, DAVID GRAHAM, b Portland, Ore, Aug 26, 24; m 48; c 3. NEUROANATOMY, NEUROPHYSIOLOGY. *Educ:* Ore State Col, BS, 46; Univ Ore, MD, 49, PhD, 51. *Prof Exp:* Instr anat, Med Sch, Univ Ore, 50-51; asst prof, Univ Calif, Los Angeles, 51-54; from asst prof to prof, State Univ NY Upstate Med Ctr, 55-67, chmn dept, 66-67; PROF ANAT & CHMN DEPT, UNIV COLO MED CTR, DENVER, 67- *Concurrent Pos:* Fulbright res scholar, Inst Physiol, Pisa, Italy, 51-52; consult, US Sci Exhibit, 61 & Neurol Study Sect, USPHS, 60-64; chmn, Neurol B Study Sect, Nat Inst Neurol Dis & Blindness, 66-67. *Mem:* Am Asn Anatomists; Int Brain Res Orgn. *Res:* Anatomy and physiology of peripheral and central nervous system pathways. *Mailing Add:* Dept Anat Univ Colo Med Sch 4200 E Ninth Ave Denver CO 80262

WHITLOCK, GAYLORD PURCELL, b Mt Vernon, Ill, July 7, 17; m 41; c 2. AGRICULTURE, BIOCHEMISTRY. *Educ:* Southern Ill Univ, BEd, 39; Pa State Col, MS, 41, PhD(agr, biochem), 42. *Prof Exp:* Res asst, Iowa State Col, 43-46, asst prof, 46-47; specialist nutrit serv, Merck & Co, Inc, 47-56; dir health ed, Nat Dairy Coun, 56-61; prog leader family & consumer sci, Agr Exten Serv, 61-74, agriculturist coop exten & vpres agr sci, 73-74, exten nutritionist, 74-80, EMER EXTEN NUTRITIONIST, UNIV CALIF, BERKELEY, 80- *Concurrent Pos:* Nutrit educ. *Mem:* Am Chem Soc; Soc Nutrtit Educ; NY Acad Sci; Sigma Xi. *Res:* Vitamins; human and animal nutrition; foods. *Mailing Add:* 1641 Rockville Rd Suisun CA 94585

WHITLOCK, JOHN HENDRICK, b Medicine Hat, Alta, Sept 10, 13; US citizen; m 35; c 2. VETERINARY PARASITOLOGY, PARASITOLOGY. *Educ:* Iowa State Univ, DVM, 34; Kans State Univ, MS, 35. *Prof Exp:* Asst zool, Kans State Univ, 34-35, from instr to asst prof path, 35-44; from asst prof to assoc prof, 44-51, prof, 51-79, EMER PROF PARASITOL, NY STATE COL VET MED & DIV BIOL SCI, CORNELL UNIV, 79- *Concurrent Pos:* Mem bd trustees, Cornell Univ, 71-76. *Mem:* Fel AAAS; Am Soc Parasitologists; Am Vet Med Asn; Biomet Soc. *Res:* Population biology of parasitisms and disease; experimental epidemiology. *Mailing Add:* NY State Vet Col Cornell Univ Ithaca NY 14853

WHITLOCK, L RONALD, b Canton, Pa, July 6, 44; m 68; c 2. ANALYTICAL CHEMISTRY, POLYMER CHEMISTRY. *Educ:* Pa State Univ, BS, 66; Univ Mass, PhD(anal chem), 71. *Prof Exp:* Fel polymer sci, Univ Mass, 70-72; res assoc, 72-85, LAB HEAD, EASTMAN KODAK CO RES LABS, 86- *Mem:* Am Chem Soc; Sigma Xi. *Res:* Development of new methods for chemical analysis of polymers and chemicals using modern analytical instruments. *Mailing Add:* Four Cavan Way Pittsford NY 14534

WHITLOCK, LAPSLEY CRAIG, b Lebanon, Ky, Aug 31, 42; m 62; c 3. EXPERIMENTAL NUCLEAR PHYSICS, SPECTROSCOPY. *Educ:* Georgetown Col, BS, 64; Vanderbilt Univ, PhD, 69. *Prof Exp:* From asst prof to assoc prof, 69-81, PROF PHYSICS, MISS COL, 81-, HEAD DEPT, 70- *Honors & Awards:* Pegram Award, Southeast Sect Am Phys Soc. *Mem:* Am Phys Soc; Am Asn Physics Teachers. *Res:* Gamma ray spectroscopy in decay of radioactive nuclides. *Mailing Add:* Dept Physics Miss Col Clinton MS 39058-4004

WHITLOCK, RICHARD T, b Columbus, Ohio, July 8, 31; m 60; c 1. THEORETICAL PHYSICS. *Educ:* Capital Univ, BS, 58; Western Reserve Univ, MS, 61, PhD(physics), 63. *Prof Exp:* Asst physics, Western Reserve Univ, 58-60, instr, 62-63; from asst prof to assoc prof, Thiel Col, 63-67; ASSOC PROF PHYSICS, UNIV NC, GREENSBORO, 67-, DIR RESIDENTIAL COL, 77- *Mem:* Am Asn Physics Teachers. *Res:* Many-body boson problem with applications to the theory of liquid helium; two-fluid hydrodynamics with applications to the theory of liquid helium; light and sound interactions. *Mailing Add:* Dept Physics Univ NC Greensboro NC 27412

WHITLOCK, ROBERT HENRY, b Canton, Pa, July 28, 41; m 63; c 3. VETERINARY MEDICINE, PATHOLOGY. *Educ:* Cornell Univ, DVM, 65, PhD(nutrit path), 70. *Hon Degrees:* MA, Univ Penn, 77. *Prof Exp:* Intern vet med, NY State Vet Col, Cornell Univ, 65-67, NIH spec fel, 69-70, asst prof, 70-76; assoc prof vet med, Col Vet Med, Univ Ga, 76-78; chief large animal med, 78-88, PROF MED, UNIV PA VET COL, 83- *Concurrent Pos:* Ed, Large Animal Portion on Compendium of Continuing Ed. *Mem:* Comp Gastroenterol Soc; Am Soc Vet Clin Path; Am Vet Med Asn; Am Col Vet Internal Med; Am Asn Bovine Practrs; Nat Acad Practr. *Res:* Pathogenesis of infectious metabolic diseases in domestic animals. *Mailing Add:* New Boltan Ctr Univ Pa Vet Col Kennett Square PA 19348

WHITLOW, GRAHAM ANTHONY, b Cardiff, Wales, May 12, 38; m 62; c 2. METALLURGY. *Educ:* Univ Manchester, BScTech, 59; Univ Wales, PhD(metall), 62. *Prof Exp:* Sci officer metall, Atomic Weapons Res Estab, UK Atomic Energy Authority, 62-67; fel engr, Advan Reactors Div, 67-79, FEL ENGR, METALL SCI DEPT RES & DEVELOP CTR, WESTINGHOUSE ELEC CORP, 79- *Mem:* Am Inst Mining, Metall & Petrol Engrs; Metall Soc; Am Inst Mech Engrs; Nat Asn Corrosion Engrs. *Res:* Energy materials development; effects of corrosive environments on high temperature materials; turbine materials development. *Mailing Add:* Westinghouse Res & Develop Ctr 1310 Beulah Rd Pittsburgh PA 15235

WHITLOW, LON WEIDNER, b Scottsville, Ky, Aug 26, 50; m 75; c 2. DAIRY CATTLE NUTRITION AND MANAGEMENT. *Educ:* Univ Ky, BS, 72; Univ Fla, MS, 74; Univ Wis, PhD(dairy sci), 79. *Prof Exp:* Res asst, Univ Fla, 72-74, asst exten dairyman, 74-75; res asst, Univ Wis, 74-79; asst prof, 79-84, ASSOC PROF DAIRY SCI, NC STATE UNIV, RALEIGH, 84- *Mem:* Am Dairy Sci Asn; Am Soc Animal Sci; Sigma Xi. *Res:* The optimum feeding and nutrition of dairy cattle; effects of mycotoxins on dairy cattle health and productivity; dairy cattle nutrition, toxicology, and management. *Mailing Add:* Box 7621 NC State Univ Raleigh NC 27695-7621

WHITMAN, ALAN B, b Joliet, Ill, Apr 7, 41. CONTINUUM MECHANICS, MECHANICS. *Educ:* Univ Ill, Urbana, BS, 63; Univ Minn, Minneapolis, MS, 66, PhD(eng mech), 68. *Prof Exp:* Res asst mech, Univ Minn, 63-64; engr, Honeywell, Inc, 64-65; res asst mech, Univ Minn, 65-66; res asst biomech, 66-68, asst prof mech, 68-71, assoc prof, 71-79, PROF MECH, WAYNE STATE UNIV, 79- *Concurrent Pos:* NSF initiation grant, Wayne State Univ, 69-70, res grants, 71-73 & 74-77; res fel, Univ Man, 78-79. *Mem:* Soc Eng Sci; Am Acad Mech. *Res:* Theories of rods and shells, stability theory of continuous systems. *Mailing Add:* Dept Mech Eng Wayne State Univ Detroit MI 48202

WHITMAN, ALAN M, b Philadelphia, Pa, Jan 26, 37; m 78; c 3. COHERENCE THEORY. *Educ:* Univ Pa, BSME, 58, MSME, 59, PhD(mech eng), 66. *Prof Exp:* Sr res engr, Power Transmission Div, Gen Elec Co, 65-67; asst prof mech eng, Univ Pa, 67-72, assoc prof, 72-80; assoc prof, Tel Aviv Univ, 80-83, prof interdisciplinary studies, Sch Eng, 83-88; PROF & CHMN, DEPT MECH ENG, VILLANOVA UNIV, 88- *Mem:* Fel Optical Soc Am; Am Soc Mech Engrs. *Res:* Wave propagation in random and layered media; rail vehicle dynamic stability; structural acoustics; fluid-solid interaction. *Mailing Add:* Dept Mech Eng Villanova Univ Villanova PA 19085

WHITMAN, ANDREW PETER, b Detroit, Mich, Feb 28, 26. NON-RIEMANNIAN GEOMETRY. *Educ:* Tulane Univ, BS, 45; Cath Univ, MS, 58, PhD(math), 61; Woodstock Col, Md, STL, 64. *Prof Exp:* Instr civil eng, Tulane Univ, 46-51; asst prof math, Loyola Univ, La, 65-66, actg chmn dept, 66-67; from asst prof to assoc prof math, Univ Houston, 67-74; ASSOC PROF MATH, CATH UNIV RIO DE JANEIRO, BRAZIL, 74- *Concurrent Pos:* NSF res grant, 66-68. *Mem:* Am Math Soc; Math Asn Am; Soc Brasileria Math; Sigma Xi. *Res:* Differential topology and geometry; harmonic maps in non-riemannian manifolds. *Mailing Add:* Rua Marques Sao Vicente-293 PUC/Rj Rio de Janeiro 22451 Brazil

WHITMAN, CHARLES INKLEY, physical chemistry, for more information see previous edition

WHITMAN, DONALD RAY, b Ft Wayne, Ind, Nov 7, 31; div; c 2. THEORETICAL CHEMISTRY, PHYSICAL CHEMISTRY. *Educ:* Case Western Reserve Univ, BS, 53; Yale Univ, PhD(phys chem), 57. *Prof Exp:* From asst prof to assoc prof phys chem, 57-70, assoc vpres, 72-74, VPRES, CASE WESTERN RESERVE UNIV, 74-, PROF PHYS CHEM, 70- *Mem:* Am Phys Soc; AAAS. *Res:* Molecular quantum mechanics. *Mailing Add:* Dept Chem Case Western Reserve Univ Cleveland OH 44106

WHITMAN, GERALD MARTIN, b New York, NY, Jan 18, 41; div; c 2. ANTENNAS, MILLIMETER WAVES. *Educ:* Queens Col, BS, 63; Columbia Univ, BSEE, 63; Polytech Inst NY, MS, 67, PhD(electroph), 69. *Prof Exp:* Fel electroph, Dept Elec Eng, Polytech Inst NY, 69-70; from asst prof to prof elec eng, 70-85, asst vpres acad affairs, Grad Studies & Res, 85-87, DIR, CTR MICROWAVE & LIGHTWAVE ENG, NJ INST TECHNOL, 85- *Concurrent Pos:* Consult, Antenna Team, US Army, Ft Monmouth, 73-76, Millimeter Wave Team, 80-, Microwave Res Inst, Polytech Inst, NY, 79, County Newark, 84-85, County Franklin Lakes, 87 & Bellcore, NJ, 88; delegate, NATO Advan Study Inst, Univ East Anglia, Eng, 79; vis prof, Dept Elec Eng, Polytech Inst NY, 80-81. *Honors & Awards:* Region I Award, Inst Elec & Electronics Engrs, 87. *Mem:* Inst Elec & Electronics Engrs; Sigma Xi. *Res:* Electromagnetics: scattering from periodic surfaces, radiation by integrated dielectric waveguides and antenna devices, transmission and compression of signals in plasma media, scattering in random media using transport theory, microstrip antennas, numerical methods and propagation in buildings. *Mailing Add:* Dept Elec Eng NJ Inst Technol 323 High St Newark NJ 07102

WHITMAN, LAWRENCE C, engineering, for more information see previous edition

WHITMAN, PHILIP MARTIN, b Pittsburgh, Pa, Dec 23, 16. MATHEMATICS. *Educ:* Haverford Col, BS, 37; Harvard Univ, AM, 38, PhD(math), 41. *Prof Exp:* Instr math, Harvard Univ, 38-41 & Univ Pa, 41-44; scientist, Los Alamos Sci Lab, Univ Calif, 44-46; asst prof math, Tufts Col, 46-48; mathematician, Appl Physics Lab, Johns Hopkins Univ, 48-61; prof math, 61-86, chmn dept, 61-67, EMER PROF MATH, RI COL, 87- *Concurrent Pos:* Consult, Weapons Systs Eval Group, Off Secy Defense, 51-55; Parsons fel, Johns Hopkins Univ, 58-59; consult, Opers Eval Group, Off Chief Naval Opers, 60-61. *Mem:* AAAS; Am Math Soc; Opers Res Soc Am; Math Asn Am; Nat Coun Teachers Math. *Res:* Lattice theory; operations research; authored college algebra and trigonometry texts. *Mailing Add:* 1010 Waltham St D427 Lexington MA 02173-8044

WHITMAN, ROBERT V(AN DUYNE), b Pittsburgh, Pa, Feb 2, 28; m 54; c 2. EARTHQUAKE ENGINEERING, GEOTECHNICAL ENGINEERING. *Educ:* Swarthmore Col, BS, 48; Mass Inst Technol, SM, 49, ScD(civil eng), 51. *Hon Degrees:* Dr, Swarthmore Col, 90. *Prof Exp:* Res engr, 51-53, from asst prof to assoc prof civil eng, 53-63, PROF CIVIL ENG, MASS INST TECHNOL, 63- *Concurrent Pos:* Consult, govt & indust, 54-; engr, Stanford Res Inst, 63-64; mem, Soil Dynamics Panel, Comt Earthquake Eng Res, Nat Acad Eng, 66-68; chmn, Risk Assessment Comt, Appl Technol Coun, 74-77, Panel Local Effects, Int Workshop Strong Motion Instrument Arrays, 78 & Seismic Loads Subcomt, Am Nat Standards Inst, 78-85; vis scholar, Cambridge Univ, Eng, 76-77; mem, Earthquake Eng Res Inst, dir, 78-81 & 84-88; mem, Comt Nat Disasters, Nat Res Coun, 80-83, Comt Seismol, 84-87 & Comt Earthquake Eng, 85-; vis prof & sr fel, Norweg Geotech Inst, Univ Calif, Davis, 84- *Honors & Awards:* Huber Res Prize, Am Soc Civil Engrs, 64, Terzaghi Lectr, 81, Terzaghi Award, 87. *Mem:* Nat Acad Eng; Am Soc Civil Engrs; Seismol Soc Am; Int Soc Soil Mech & Found Engrs; Earthquake Eng Res Inst (vpres, 79-81, pres 85-87). *Res:* Soil mechanics, especially dynamic problems; author of various publications. *Mailing Add:* Dept Civil Eng Rm 1-342 Mass Inst Technol Cambridge MA 02139

WHITMAN, ROLLIN LAWRENCE, b Pittsfield, Mass, Aug 25, 47; m 70; c 4. ELECTRICAL ENGINEERING, COMPUTER SCIENCE. *Educ:* Univ Wis, BSEE, 70; Univ Colo, MSEE, 74. *Prof Exp:* Engr sci programmer, Martin Marietta Co, 70-76; SECT LEADER, SIGNAL PROCESS & RADIOGRAPHIC IMAGE ANALYSIS, LOS ALAMOS NAT LAB, 76- *Honors & Awards:* Award Excellence, Nuclear Weapons Prog, 86, Res & Develop 100 Award, 88. *Mem:* Inst Elec & Electronics Engrs. *Res:* Computer image processing and signal analysis in laser fusion target inspection, modeling reactor safety coolant problems; two-dimensional digital filtering on both large scale computers and interactive minicomputer systems; 1D and 2D fitting by modeling and incorporation of system modulation transfer function; computed tomography; radiographic deblurring; 2D point spread function estimation of a high energy x-ray source; quantitative radiography edges and density. *Mailing Add:* Los Alamos Nat Lab Box 1663 MS P 940 Los Alamos NM 87545

WHITMAN, ROY MILTON, b New York, NY, June 16, 25; m 68; c 5. PSYCHIATRY. *Educ:* Ind Univ, BS, 44, MD, 46. *Prof Exp:* Intern, Kings County Hosp, 46-47; resident psychiat, Duke Hosp, 47-48; from instr to asst prof, Univ Chicago, 52-54; from asst prof to assoc prof neurol & psychiat, Med Sch, Northwestern Univ, 54-57, assoc prof, 57; assoc prof, 57-67, actg chmn dept, 80-83, chmn dept, 83-89, PROF PSYCHIAT, COL MED, UNIV CINCINNATI, 67- *Concurrent Pos:* USPHS fel, Clins, Univ Chicago, 50-52; chief neurol & psychiat, Vet Admin Res Hosp, Chicago, Ill, 54-57; consult, Vet Admin Hosp, Cincinnati, Ohio, 57-, Ill State Psychiat Inst, 63-68, Cent Clin, Cincinnati, Ohio & Vet Admin Res Hosp, Chicago; clinician, Cincinnati Univ Hosp. *Mem:* Clin Psychoanal Soc; Am Psychiat Asn; Am Psychoanal Asn; Int Psychoanal Asn; Sigma Xi. *Res:* Psychophysiology of dreaming; sex research; psychoanalytic methods; techniques in psychoanalysis; psychosomatic medicine; leadership and organizational behavior. *Mailing Add:* 7137 Fair Oaks Dr Cincinnati OH 45237

WHITMAN, VICTOR, PEDIATRIC CARDIOLOGY. *Educ:* Univ Tex, MD, 64. *Prof Exp:* CARDIOLOGIST, MIAMI CHILDREN'S HOSP, FLA, 86- *Mailing Add:* Dept Pediat Cardiol Miami Children's Hosp 6125 SW 31st St Miami FL 33155

WHITMARSH, CLIFFORD JOHN, b San Diego, Calif, Apr 11, 46; m 80; c 5. BIOPHYSICS. *Educ:* Harvard Univ, MA, 70, PhD(physics), 75. *Prof Exp:* Fel biophys, Ctr Nuclear Studies Saclay, France, 74-75; fel biophys, Purdue Univ, 75-79; asst prof, Queens's Univ, Can, 80; ASSOC PROF & PLANT PHYSIOLOGIST, UNIV ILL, 81- *Concurrent Pos:* NSF fel. *Mem:* Biophys Soc; Am Soc Plant Physiol; Am Soc Plant Physiol. *Res:* Investigate the function and regulation of the photosynthetic apparatus of plants. *Mailing Add:* Dept Plant Biol Univ Ill Urbana Campus 505 S Goodwin St Urbana IL 61801

WHITMER, JEFFREY THOMAS, b Mar 22, 48; m. CARDIOMYOPATHY, CARDIOPLEGIA. *Educ:* Pa State Univ, PhD(physiol), 74, MD, 77. *Prof Exp:* ASSOC PROF PEDIAT, CHILDREN'S HOSP MED CTR, 82- *Mem:* Am Heart Asn; Biophys Soc; Am Physiol Soc; Int Soc Heart Res; Cardiac Muscle Soc. *Res:* Use of the isolated experimental animal heart perfusion technique to study hermodynamics and relate this to intracellular metabolic regulation in health and disease, as well as during the ischemia of cardioplegic arrest in immature and mature hearts. *Mailing Add:* Dept Med Bristol-Myers USPG 2400 W Lloyd Expwy Evansville IN 47721

WHITMER, JOHN CHARLES, b Kingfisher, Okla, Jan 28, 39; m 67; c 2. PHYSICAL CHEMISTRY. *Educ:* Univ Rochester, BS, 60; Univ Mich, MS, 62, PhD(chem), 65. *Prof Exp:* Asst prof chem, Western Wash Univ, 65-66; lectr, Univ EAfrica, 67-69; from asst prof to assoc prof, 69-76, PROF CHEM, WESTERN WASH UNIV, 76- *Concurrent Pos:* Vis prof, Inst Phys Chem, Univ Trondheim, Norway, 76-77 & Univ York, Eng, 90. *Mem:* AAAS; Am Chem Soc. *Res:* Molecular dynamics and structure. *Mailing Add:* Dept Chem Western Wash Univ Bellingham WA 98225

WHITMER, ROBERT MOREHOUSE, b Battle Creek, Mich, June 14, 08; m 36, 51; c 2. PHYSICS, ELECTRICAL ENGINEERING. *Educ:* Univ Mich, BA, 28, MA, 34, PhD(physics), 38. *Prof Exp:* Mem staff, Bell Tel Labs, Inc, 28-32; mem fac, Amherst Col, 32-33; engr, Philco Radio & TV Corp, Pa, 35-36; physicist, Hercules Powder Co, 36; instr physics, Purdue Univ, 37-41; mem staff, Radiation Lab, Mass Inst Technol, 41-46; prof physics, Rensselaer Polytech Inst, 46-56; sr staff physicist, TRW Systs, 56-73; CONSULT, 73- *Concurrent Pos:* Consult, USAF, 52-53; mem, Security Resources Panel, Off Defense Mobilization, 57; mem, Reentry Body Identification Group, Off Secy Defense, 58. *Mem:* AAAS; Am Phys Soc; sr mem Inst Elec & Electronics Engrs. *Res:* Electromagnetics; wave propagation; electromagnetic shielding; military systems and policy studies. *Mailing Add:* 724 Tenth St Manhattan Beach CA 90266

WHITMIRE, CARRIE ELLA, b Electra, Tex, Oct 17, 26. BACTERIOLOGY, TOXICOLOGY. *Educ:* Univ Tex, BA, 46; Univ Kans, MA, 53, PhD(bact), 55. *Prof Exp:* Bacteriologist, Parkland Hosp, Dallas, Tex, 46-47, Vet Admin Hosp, 47-48 & US Army Chem Ctr, Ft Detrick, Md, 48-52; asst virologist, Univ Kans, 52-55; assoc scientist, Ortho Res Found, NJ, 55-61; proj supvr, Merck Sharp & Dohme Biol Div, Pa, 61-64; tech asst, Winthrop Labs, Biol, NY, 64-67; proj dir & viral oncologist, Microbiol Assocs, Inc, 67-75, dir dept exp oncol, 75-78; toxicologist, Nat Cancer Inst, 79-81; toxicologist, Qual Assurance, GLP Compliance, Toxicol Res Testing Prog Br, Nat Toxicol Prog, Nat Inst Environ Health Sci, 81-87; RETIRED. *Mem:* Am Soc Microbiol; Soc Toxicol. *Res:* Medical human, animal and oncology virology and bacteriology; cancer and toxicology research. *Mailing Add:* 101 Forest Hills Ct Cary NC 27511

WHITMIRE, KENTON HERBERT, b Roanoke, Va, July, 20, 55; m 81; c 2. CLUSTER CHEMISTRY. *Educ:* Roanoke Col, BS, 77; Northwestern Univ, MS, 78, PhD, 82. *Prof Exp:* Fel chem, Cambridge Univ, 81-82; asst prof, 82-88, ASSOC PROF CHEM, RICE UNIV, 88- *Concurrent Pos:* NATO fel & Alexander von Humboldt res fel. *Mem:* Sigma Xi; Am Chem Soc. *Mailing Add:* Dept Chem Rice Univ Houston TX 77251

WHITMORE, BRADLEY CHARLES, b Minneapolis, Minn, Jan 14, 53; m 77; c 2. GALACTIC STRUCTURE, GALACTIC DYNAMICS. *Educ:* Univ Mich, BS, 75, MS, 77, PhD(astron), 80. *Prof Exp:* Instr, Univ Mich, 80; Carnegie fel astron, Dept Terrestrial Magnetism, Carnegie Inst, Washington, DC, 80-82; asst prof astron, Ariz State Univ, 82-83; ASST ASTRONR, SPACE TELESCOPE SCI INST, 83- *Mem:* Am Astron Soc; Int Astron Union. *Res:* Structure, dynamics, and evolution of galaxies; comparison of the dynamics of elliptical galaxies with the spheroidal component of spiral galaxies; observations of polar ring galaxies. *Mailing Add:* Space Telescope Sci Inst Johns Hopkins Univ 3700 San Martin Dr Baltimore MD 21218

WHITMORE, DONALD HERBERT, JR, b Buffalo, NY, May 6, 44; m 70. COMPARATIVE PHYSIOLOGY. *Educ:* Ind Univ, BA, 66; Northwestern Univ, PhD(biol sci), 71. *Prof Exp:* NIH fel insect physiol, Northwestern Univ, 71-73; asst prof, 73-79, ASSOC PROF BIOL, UNIV TEX, ARLINGTON, 79- *Mem:* Sigma Xi; Am Fisheries Soc; Am Soc Zoologists; Entom Soc Am. *Res:* The role of environmental influences on the physiology and biochemistry of animals, particularly how animals adapt to environmental stress. *Mailing Add:* 2723 Westridge Arlington TX 76012

WHITMORE, EDWARD HUGH, b Ottawa, Ill, Feb 26, 26; m 49; c 2. GEOMETRY. *Educ:* Ill State Univ, BS, 49, MS, 51; Ohio State Univ, PhD(math educ, math), 56. *Prof Exp:* Instr high sch, Ill, 48-51; asst prof math, Northern Ill Univ, 55-56; from asst prof to assoc prof, San Francisco State Col, 56-6S; chmn dept, Cent Mich Univ, 65-74 & 76-82, prof math, 65-86; RETIRED. *Mem:* Math Asn Am; Nat Coun Teachers Math. *Res:* Mathematics education; sequences in elementary geometry and their history. *Mailing Add:* Dept Math Cent Mich Univ Mt Pleasant MI 48858

WHITMORE, FRANK CLIFFORD, JR, b Cambridge, Mass, Nov 17, 15; m 39; c 4. SYSTEMATICS OF FOSSIL CETACEA, TERTIARY PALEOGEOGRAPHY. *Educ:* Amherst Col, AB, 38; Pa State Univ, MS, 39; Harvard, AM, 41, PhD, 42. *Prof Exp:* Instr, RI State Col, 42-44; GEOLOGIST, US GEOL SURV, 44- *Concurrent Pos:* Sci consult, US Army, 45-46; vchmn, Comt Res & Explor, Nat Geol Soc, 70- *Honors & Awards:* Meritorious Serv Award, US Dept Interior. *Mem:* Geol Soc Am; AAAS; Soc Vert Paleont; Paleont Soc. *Res:* Endocranial anatomy of Oligocene Artiodactyla; Pleistocene mammal faunas of Kentucky and tertiary mammals of the Atlantic and Gulf Coastal Plain and Panama; whale evolution. *Mailing Add:* Nat Mus Nat Hist NHB 121 Washington DC 20560

WHITMORE, FRANK WILLIAM, b Ponca City, Okla, May 15, 32; m 55; c 2. PLANT PHYSIOLOGY, BIOCHEMISTRY. *Educ:* Okla State Univ, BS, 54; Univ Mich, MF, 56, PhD(forestry), 64. *Prof Exp:* Res forester, US Forest Serv, 57-61, plant physiologist, 64-65; res assoc forestry, Univ Mich, 65-67; from asst prof to assoc prof, 67-76, PROF FORESTRY, OHIO AGR RES & DEVELOP CTR, OHIO STATE UNIV, 76- *Concurrent Pos:* Agr Biotechnol Res Adv Comt, USDA, 88- *Mem:* AAAS; Soc Am Foresters; Am Soc Plant Physiologists; Int Soc Plant Molecular Biol. *Res:* Protoplast technology; lignin biochemistry; tissue culture; conifer embryology. *Mailing Add:* Div Forestry Ohio Agr Res & Develop Ctr Ohio State Univ Wooster OH 44691

WHITMORE, GORDON FRANCIS, b Saskatoon, Sask, June 29, 31; m 54; c 2. RADIOBIOLOGY, CANCER. *Educ:* Univ Sask, BA, 53, MA, 54; Yale Univ, PhD(biophys), 57. *Prof Exp:* From asst prof to assoc prof, 60-65, PROF BIOPHYS, UNIV TORONTO, 65-, HEAD DEPT, 71-, ASSOC DEAN FAC MED, 74-; PHYSICIST, ONT CANCER INST, 56-, ASSOC DIR PHYS DIV, 57-, CHMN DEPT MED BIOPHYS, 80- *Concurrent Pos:* Vis prof, Pa State Univ, 63; mem, Nat Cancer Inst Grants Panel, 63- & Nat Res Coun Assoc Comt Radiobiol, 64-; mem radiation study sect, NIH, 65- *Honors & Awards:* Ernest Berry-Anderson Prize, Royal Soc Edinburgh, 66. *Mem:* Biophys Soc; Radiation Res Soc; Can Soc Cell Biol; Can Asn Physicists; Royal Soc Can. *Res:* Radiation physics; radiobiology of mammalian cells in vitro; action of chemotherapeutic agents; mammalian cell genetics. *Mailing Add:* Dept Med Biophys Univ Toronto Toronto ON M5S 1A8 Can

WHITMORE, HOWARD LLOYD, b Dallas, Wis, Dec 3, 35; m 62; c 2. VETERINARY MEDICINE. *Educ:* Okla State Univ, BS, 58, DVM, 60; Univ Wis, PhD(vet sci), 73; Am Col Theriogenologists, dipl. *Prof Exp:* Vet pvt pract, 60-69; assoc prof, Col Vet Med, Univ Minn, St Paul, 74-80; MEM FAC, DEPT VET CLIN MED, COL VET MED, UNIV ILL, 80- *Honors & Awards:* Burr Beach Award, Dept Vet Sci, Univ Wis, 73. *Mem:* Am Vet Med Asn; Sigma Xi. *Res:* Fertility; abortion and pregnancy diagnosis in dairy cattle. *Mailing Add:* 2807 Woodhaven Dr Champaign IL 61821-7528

WHITMORE, MARY (ELIZABETH) ROWE, b Oakland, Calif, Oct 26, 36; m 61; c 3. ANATOMY, ZOOLOGY. *Educ:* Univ Calif, Berkeley, BA, 59; Smith Col, MA, 61; Univ Minn, Minneapolis, PhD(anat), 69. *Prof Exp:* Teaching asst zool, Smith Col, 59-61; teaching asst anat, Univ Minn, Minneapolis, 61-66; vis asst prof, 70-71, ASST PROF ZOOL, UNIV OKLA, 72- *Mem:* AAAS; Am Soc Zool; Electron Micros Soc Am; Sigma Xi. *Res:* Comparative morphology and histology of endocrine glands in the lower vertebrates; biology of cyclostomes. *Mailing Add:* Dept Zool Univ Okla Norman OK 73019

WHITMORE, RALPH M, b Medina, Tex, Oct 22, 17; m 42; c 2. MATHEMATICS, PROBABILITY. *Educ:* Trinity Univ, BA, 38; Univ Tex, Austin, MA, 41, PhD(math), 64. *Prof Exp:* Instr math, Peacock Acad, 38-39; statistician electronics, City San Antonio, 39-40; instr math statist, Trinity Univ, 40-41; chief instr electronics math, Air Force Tech Sch, 42-44; prof math, physics & comput sci & chmn dept math, Southwestern Univ, 44-89; ADJ PROF MATH, TEMPLE JR COL, 89- *Concurrent Pos:* Statistician & consult, Sandia Corp, 53-61; NSF fel, Univ Okla, 64. *Mem:* Math Asn Am; Soc Indust & Appl Math. *Mailing Add:* 3411 Primrose Georgetown TX 78628-2815

WHITMORE, ROY ALVIN, JR, b Baltimore, Md, Aug 14, 28; m 53; c 3. FORESTRY. *Educ:* Univ Mich, BSF, 52, MF, 54. *Prof Exp:* Forest economist, Cent States Forest Exp Sta, USDA, 52-58; PROF FORESTRY, UNIV VT, 58- *Mem:* Soc Am Foresters; Forest Prod Res Soc. *Res:* Forest products utilization and marketing. *Mailing Add:* Dept Forestry Univ Vt Burlington VT 05405

WHITMORE, ROY WALTER, b San Antonio, Tex, Jan 21, 47; m 69; c 2. STATISTICS. *Educ:* Tex Tech Univ, BS, 69, MS, 71; Tex A&M Univ, PhD(statist), 78. *Prof Exp:* Instr math, Stephen F Austin State Univ, 69-71; lectr statist, Tex A&M Univ, 77-78; asst prof math, Univ NC, Greensboro, 78-80; STATISTICIAN, RES TRIANGLE INST, 80- *Mem:* Sigma Xi; Am Statist Asn; Biomet Soc. *Res:* Survey research; environmental surveys; health surveys; education surveys. *Mailing Add:* PO Box 12194 Res Triangle Inst Research Triangle Park NC 27709

WHITMORE, STEPHEN CARR, b Holyoke, Mass, Oct 17, 31; m 61; c 3. PHYSICS. *Educ:* Amherst Col, BA, 54; Univ Minn, PhD(liquid helium), 66. *Prof Exp:* Res assoc physics, Univ Mich, 66-69; asst prof, 69-85, ASSOC PROF PHYSICS, UNIV OKLA, 85- *Mem:* Am Phys Soc. *Res:* Low temperature physics; liquid helium; superconducters; semiconductors. *Mailing Add:* Dept Physics & Astron Univ Okla Nielsen Hall 440 West Brooks Norman OK 73019

WHITMORE, WILLIAM FRANCIS, b Boston, Mass, Jan 6, 17; m 46; c 4. MATHEMATICS, OCEANOGRAPHY. *Educ:* Mass Inst Technol, SB, 38; Univ Calif, PhD(math), 41. *Prof Exp:* With US Naval Ord Lab, Washington, DC, 41-42; instr physics, Mass Inst Technol, 42-46; opers analyst, Opers Eval Group, USN, 46-57, chief scientist, Spec Projs Off, Bur Ord, 57-59; consult scientist, Chief Scientist's Staff, Lockheed Missiles & Space Co, 59-62, dep chief scientist, 62-64, asst to pres, 64-69, sr consult scientist, 69-74, chief scientist ocean systs, Lockheed Advan Marine Systs Co, 74-84; RETIRED. *Concurrent Pos:* Spec consult, Air Forces Eval Bd, 45; sci analyst, Commanding Gen, 1st Marine Air Wing, Korea, 53; consult, Defense Sci Bd, 66-68 & Marine Bd, Nat Acad Eng, 72-; chmn, Navy Lab Adv Bd, Ord, 67-75; vis comt dept math, Mass Inst Technol, 71-78; tech ed consult, Lockheed Missiles & Space Co, 84- *Mem:* Am Math Soc; Opers Res Soc Am; Math Asn Am; assoc fel Am Inst Aeronaut & Astronaut. *Res:* Classical boundary value problems in physics; weapon systems analysis; oceanography. *Mailing Add:* 14120 Miranda Ave Los Altos Hills CA 94022-2045

WHITNEY, ARTHUR EDWIN, JR, b St Louis, Mo, Apr 21, 38. HEAT TRANSFER, THERMODYNAMICS. *Educ:* Wash Univ, BSME, 59, MS, 60, DSc(mech eng), 64. *Prof Exp:* Asst prof mech eng, Wash Univ, 64-65; SR GROUP ENGR, MCDONNELL DOUGLAS CORP, 65- *Mem:* Am Soc Mech Engrs; Am Inst Aeronaut & Astronaut. *Res:* Heat transfer methods and computer applications applied to aircraft, missiles and spacecraft. *Mailing Add:* 15890 Richborough Rd Chesterfield MO 63017

WHITNEY, ARTHUR SHELDON, b Oberlin, Ohio, Oct 31, 33; m 64; c 2. AGRONOMY. *Educ:* Ohio State Univ, BS, 55; Cornell Univ, MS, 58; Univ Hawaii, PhD(soil sci), 66. *Prof Exp:* Res instr agron, Univ Philippines, 59-60; from asst agronomist to assoc agronomist, 65-76, AGRONOMIST, HAWAII AGR EXP STA, 76- *Concurrent Pos:* Prin investr, Nitrogen Fixation by Legumes Prog, AID, 75-79. *Mem:* Crop Sci Soc Am; Am Soc Agron; Trop Grassland Soc Australia. *Res:* Pasture management; legume agronomy; plant nutrition. *Mailing Add:* Box 11286 Lahaina HI 96761

WHITNEY, CHARLES ALLEN, b Milwaukee, Wis, Jan 31, 29; m 51; c 5. ASTROPHYSICS. *Educ:* Mass Inst Technol, BS, 51; Harvard Univ, AM, 53, PhD(astron), 55. *Prof Exp:* Assoc prof, 63-68, PROF ASTRON, HARVARD UNIV, 68-; PHYSICIST, SMITHSONIAN ASTROPHYS OBSERV, 56- *Concurrent Pos:* Guggenheim Found fel, 71. *Mem:* Int Astron Union; Am Astron Soc; Am Acad Arts & Sci. *Res:* History of astronomy; theory of variable stars and associated problems of gas dynamics. *Mailing Add:* 29 Ingalls Rd Cheshire MA 01225

WHITNEY, CHARLES CANDEE, JR, b Newfane, Vt, Oct 12, 39; m 86; c 2. DRUG METABOLISM, PHARMACOKINETICS. *Educ:* Northeastern Univ, AB, 62; Middlebury Col, MS, 64; Univ Calif, Davis, PhD(org chem), 68. *Prof Exp:* Res chemist, E I du Pont de Nemours & Co, Inc, 68; asst chief clin chem, 3rd US Army Med Lab, 68-70; sr res biochemist, 70-80, res assoc, 80-82, res supvr, 82-84, res mgr drug metab, 85-88, RES MGR, ANALYSIS RES & DEVELOP, EXP STA, E I DU PONT DE NEMOURS & CO, WILMINGTON, DEL, 88- *Concurrent Pos:* Mem, Drug Metab Discussion Group. *Mem:* Am Chem Soc; Sigma Xi; Am Soc Clin Pharm & Therapeut. *Res:* Analytical procedures for drug substance and dosage forms; pharmacokinetics, bioavailability and metabolic fate of drugs in the body; analytical methods for the determination of drugs and their metabolites. *Mailing Add:* 335 Steeplechase Circle Wilmington DE 19808

WHITNEY, CYNTHIA KOLB, b Cumberland, Md, July 11, 41; m 63; c 2. MATHEMATICAL PHYSICS, OPERATIONS RESEARCH. *Educ:* Mass Inst Technol, SB, 63, SM, 65, PhD(physics), 68. *Prof Exp:* Staff physicist, 67-80, assoc div leader & proj mgr, Charles Stark Draper Lab Inc, 80-87; VIS PROF, TUFTS UNIV, 87-, STAFF, W J SHAFER ASSOC INC, 87- *Concurrent Pos:* Consult, Advan Appln Flight Exp Prog, NASA, 75. *Honors & Awards:* David Rist Prize, Mil Opers Res Soc. *Mem:* AAAS; Oper Res Soc Am; Sigma Xi. *Res:* Statistical description of complex physical systems; decision analysis for flexible manufacturing systems and nuclear power generation systems; mathematical formalisms in fundamental physics. *Mailing Add:* Tufts Univ EOTC Medford MA 02155

WHITNEY, DANIEL EUGENE, b Chicago, Ill, June 8, 38; m 63; c 2. MECHANICAL ENGINEERING, AUTOMATION & ROBOTICS. *Educ:* Mass Inst Technol, SB, 60 & 61, MS, 65, PhD(mech eng), 68. *Prof Exp:* From asst prof to assoc prof mech eng, Mass Inst Technol, 68-74; SECT CHIEF, CHARLES STARK DRAPER LAB, 74- *Concurrent Pos:* Consult, Charles Stark Draper Lab, 68-74; mem, Automation Res Coun, 71-78; NSF grant, Mass Inst Technol, 72-73 & Charles Stark Draper Lab, 78-80, 79-81, 81-84, 85-87 & 88-91. *Mem:* Sr mem Inst Elec & Electronics Engrs; fel Am Soc Mech Engrs; Robotics Inst Am. *Res:* Application of computers to control systems and to engineering design and manufacturing; robotics. *Mailing Add:* 141 Rhinecliff St Arlington MA 02174

WHITNEY, DONALD RANSOM, b Cleveland Heights, Ohio, Nov 27, 15; m 39; c 4. STATISTICS, MATHEMATICAL STATISTICS. *Educ:* Oberlin Col, BA, 36; Princeton Univ, MA, 39; Ohio State Univ, PhD(math), 48. *Prof Exp:* Instr math, Mary Washington Col, 39-42; prof math, 48-70, chmn dept, 70-80, prof statist, 70-82, EMER PROF STATIST, OHIO STATE UNIV, 82- *Concurrent Pos:* Consult discrimination cases, Ohio Bell Tel Co, Pub Utilities Comn, Cincinnati Bell Tel Co, NAm Aviation & Goodyear Atomic Corp. *Mem:* Fel AAAS; fel Am Statist Asn; Inst Math Statist; Biomet Soc; Am Math Soc. *Res:* Non-parametric statistics; general statistical methodology. *Mailing Add:* 388 Westview Ave Columbus OH 43214-1428

WHITNEY, ELLSWORTH DOW, b Buffalo, NY, Sept 17, 28; m 54; c 2. PHYSICAL CHEMISTRY. *Educ:* Univ Buffalo, BA, 50; NY Univ, PhD(phys chem), 54. *Prof Exp:* Chemist, Charles C Kawin Co, 46-50; res chemist, Olin Mathieson Chem Corp, 54-57, chem res proj specialist, 57-59; sr res chemist, Carborundum Co, 59-62, sr res assoc, 62-70; assoc prof, 70-75, founding dir, Ctr Res Mining & Mineral Resources, 72-82, affil prof, Dept Nuclear Eng Sci, 76-90, PROF MAT SCI & ENG, UNIV FLA, 75- *Concurrent Pos:* Asst prof, Erie County Technol Inst, 63-68; lectr, Eve Sch, State Univ NY, Buffalo, 66-69; partner, Mat Consult Inc, Gainesville, Fla, 72- *Mem:* Am Ceramic Soc; Am Chem Soc; Am Inst Mining, Metall & Petrol Engrs; Soc Mfg Engrs; Am Asn Univ Prof. *Res:* Crystal growth; kinetics of surface exchange; heterogeneous catalysis; boron and metal hydrides; borohydrides; fluorine oxidizers; high energy propellants; ultrahigh pressure solid state phenomena; phase transformations in solids; ceramic cutting tools and abrasives; solid state reaction kinetics, hard materials; mining and mineral research. *Mailing Add:* 165 Rhines Hall Univ Fla Gainesville FL 32611

WHITNEY, ELVIN DALE, b West Bountiful, Utah, Mar 23, 28; m 58; c 3. PLANT PATHOLOGY, PLANT BREEDING. *Educ:* Utah State Univ, BS, 50; Cornell Univ, PhD(plant path), 65. *Prof Exp:* PLANT PATHOLOGIST, AGR RES SERV, USDA, 65- *Concurrent Pos:* Assoc ed, Phytopath. *Mem:* Am Phytopath Soc; Int Soc Plant Pathologists; Am Soc Sugar Beet Technologists; Sigma Xi. *Res:* Fungal, bacterial and nematode diseases of sugar beet; breeding for disease resistance. *Mailing Add:* 2144 Kennedy Pl St George UT 84770-8061

WHITNEY, EUGENE C, b Columbus, Ohio, Aug 26, 13. ELECTRICAL ENGINEERING, MOTORS & HYDROGENATORS. *Educ:* Univ Mich, BS, 35. *Prof Exp:* Mgr sychron condenser & hydrogenerator, Westinghouse Corp, 49-72, consult elec eng, 72-75; INDEPENDENT CONSULT, 75- *Honors & Awards:* Tesla Award, Inst Elec & Electronics Engrs, 84. *Mem:* Nat Acad Eng; Power Eng Soc; Inst Elec & Electronics Engrs. *Res:* Responsible for the design of the world's largest water generators and synchronizing motors. *Mailing Add:* 249 Cascade Rd Forest Hills Pittsburgh PA 15221

WHITNEY, GEORGE STEPHEN, b Wheatland, Wyo, Feb 5, 34; m 59; c 3. ORGANIC CHEMISTRY, BIOCHEMISTRY. *Educ:* Univ Colo, BA, 55; Northwestern Univ, PhD(org chem), 62. *Prof Exp:* Asst prof org chem, Wabash Col, 61-62; from asst prof to assoc prof, 62-68, PROF ORG CHEM, WASHINGTON & LEE UNIV, 73- *Concurrent Pos:* Fulbright fel, 55-56; Swiss-Am Found fel, Univ Basel, 64-65; Sloan-Washington & Lee fel, Univ Bristol, 70-71. *Mem:* Am Chem Soc; Am Asn Univ Prof. *Res:* Synthetic organic chemistry and mechanisms; free-radical additions of organic sulfur compounds; cycloalkenes; bicyclic compounds; stereochemistry; organic synthesis; carbene insertions in cage compounds. *Mailing Add:* Dept Chem Washington & Lee Univ Lexington VA 24450

WHITNEY, GINA MARIE, b Berea, Ohio, Sept 3, 58. CORROSION, ELECTROCHEMICAL PROCESS. *Educ:* Case Western Res Univ, BS, 80; Univ Calif, PhD, 87. *Prof Exp:* ADV ENGR, IBM CORP, 87- *Mem:* Am Inst Chem Engrs; Am Electroplaters & Surface Finishers Soc; Electrochem Soc. *Res:* Electrodeposition and corrosion of magnetic and electronic materials; transport phenomena in electrolytic processes. *Mailing Add:* IBM Corp 5600 Cottle Rd San Jose CA 95193

WHITNEY, HARVEY STUART, b Langdon, Alta, Oct 14, 35; m 62; c 2. PHYTOPATHOLOGY, MYCOLOGY. *Educ:* Univ Sask, BSA, 56, MSc, 58; Univ Calif, Berkeley, PhD(plant path), 63. *Prof Exp:* Res officer seedling dis, Forest Biol Div, Can Dept Agr, 58-61; RES SCIENTIST, CAN FOREST SERV, CAN DEPT ENVIRON, 61- *Concurrent Pos:* Can Forest Serv fel, Univ Calif, Berkeley, 70-71. *Mem:* AAAS; Soc Invert Path; Mycol Soc Am; Can Phytopath Soc; Can Inst Foresters; Sigma Xi. *Res:* Taxonomy and heterokaryosis in Rhizoctonia; insect-fungus-tree relationships in conifers attacked by bark beetles; tree resistance and predisposition; symbiology; insect pathology; biological control. *Mailing Add:* 1925 Casa Marcia Cres Victoria BC V8N 2X4 Can

WHITNEY, HASSLER, analysis, topology; deceased, see previous edition for last biography

WHITNEY, J(OHN) BARRY, III, b Ft Benning, Ga, Oct 4, 44; m 67; c 3. MOLECULAR GENETICS, BIOCHEMICAL GENETICS. *Educ:* Emory Univ, BS, 67; Univ NC, Chapel Hill, PhD(genetics), 72. *Prof Exp:* NIH postdoctoral fel, Univ Wis, 72-76; res assoc, Jackson Lab, 76-78; expert, NIH, 78-80; asst prof, 80-86, ASSOC PROF MOLECULAR GENETICS, MED COL GA, 87- *Mem:* Am Soc Human Genetics; Genetics Soc Am; Sigma Xi; AAAS. *Res:* Detection and characterization of genetic mutations and variations that affect protein structure or expression in laboratory animals; silent amino acid substitutions; gene mapping; comparative genetics; inborn errors of metabolism; genetic therapy. *Mailing Add:* Dept Biochem & Molecular Biol Med Col Ga Augusta GA 30912-2100

WHITNEY, JAMES ARTHUR, b Middlebury, Vt, Apr 19, 46; m 75; c 2. EDUCATION & ADMINISTRATION, VOLCANOLOGY. *Educ:* Mass Inst Technol, SB & MS, 69; Stanford Univ, PhD(geol), 72. *Prof Exp:* Instr geol, Chabot Jr Col, 71-72; from asst prof to assoc prof, 72-83, PROF GEOL & HEAD DEPT, UNIV GA, 83- *Concurrent Pos:* Woodrow Wilson fel, 68; NSF fel, 68-72; geologist, US Geol Surv, Reston, Va, 79. *Mem:* Fel Mineral Soc Am; fel Geol Soc Am; Am Geophys Union; AAAS; Sigma Xi; Soc Econ Geol. *Res:* Origin, fractionation, volatile components and tectonic significance of granitic magmatism with application to the understanding of tectonic processes, large volume ash-flow tuffs and the generation of ore deposits. *Mailing Add:* 145 Gibbons Way Athens GA 30605

WHITNEY, JAMES MARTIN, b Owosso, Mich, Sept 6, 36; m 63; c 3. MATERIALS SCIENCE ENGINEERING. *Educ:* Ill Col, BA, 59; Ga Inst Technol, BSTE, 59, MSTE, 61; Ohio State Univ, MS, 64, PhD(eng mech), 68. *Prof Exp:* Mat engr, 61-66, MAT RES ENGR, NONMETALLIC MAT DIV, USAF MAT LAB, 66- *Concurrent Pos:* Air Force liaison rep to ad hoc comt micromech fibrous composites, Mat Adv Bd, 63-64. *Honors & Awards:* Award of Merit, Am Soc Testing & Mat, 83. *Mem:* Fiber Soc; Am Soc Mech Engrs; assoc fel Am Inst Aeronaut & Astronaut; Am Acad Mech; fel Am Soc Testing & Mat. *Res:* Determination of the mechanical behavior of fibrous composites as a function of constituent properties and geometry, using principles of mechanics and applied mathematics; authored or coauthored over 60 publications including two texts. *Mailing Add:* 4371 Roundtree Dr Dayton OH 45432

WHITNEY, JODIE DOYLE, b Bosque Co, Tex, Oct 14, 37; m 60; c 2. AGRICULTURAL ENGINEERING. *Educ:* Tex A&M Univ, BS, 59; Pa State Univ, MS, 62; Okla State Univ, PhD(agr eng), 66. *Prof Exp:* From asst agr engr to assoc agr engr, 65-79, AGR ENGR, CITRUS RES & EDUC CTR, UNIV FLA, 79- *Mem:* Am Soc Agr Engrs. *Res:* Mechanization of low volume spraying and citrus harvesting; management of close spaced citrus plantings. *Mailing Add:* Citrus Res & Educ Ctr 700 Experiment Station Rd Lake Alfred FL 33850

WHITNEY, JOEL GAYTON, b Cambridge, Mass, Oct 13, 37; m 71. ORGANIC CHEMISTRY. *Educ:* Harvard Univ, AB, 59; Mass Inst Technol, PhD(org chem), 63. *Prof Exp:* Sr res chemist, 63-80, res supvr biochem dept, 80-84, RES ADMINR PHARMACEUT, E I DU PONT DE NEMOURS & CO, INC, 84- *Mem:* Am Chem Soc. *Res:* Amino acid syntheses; medicinal chemistry, especially heterocyclic chemistry; synthesis of central nervous system agents. *Mailing Add:* 111 Marlbrook Way Kennett Square PA 19348-1719

WHITNEY, JOHN BARRY, JR, b Augusta, Ga, June 25, 16; m 41; c 3. PLANT PHYSIOLOGY. *Educ:* Univ Ga, BS, 35; NC State Col, MS, 38; Ohio State Univ, PhD(plant physiol), 41. *Prof Exp:* Tutor, Univ Ga, 35-36; asst, Ohio State Univ, 38-41; plant physiologist, Cent Fibre Corp, 41-43 & 46; from asst prof to prof bot, Clemson Univ, 55-80, head dept, 77-80; RETIRED. *Concurrent Pos:* Oak Ridge Inst Nuclear Studies fel, Univ Tenn AEC Agr Res Prog, 52-53; area consult, Biol Sci Curric Study SC, 69-72. *Mem:* Am Soc Plant Physiol. *Res:* Water relations of plants; structure and microchemistry of plant cell walls; plant microchemistry; nutrition of microorganisms; radioisotope tracer applications. *Mailing Add:* 215 Wyatt Ave Clemson SC 29631

WHITNEY, JOHN GLEN, b Ponca City, Okla, June 4, 39; m 58; c 4. MICROBIOLOGY. *Educ:* Okla State Univ, BS, 61, PhD(microbiol), 67. *Prof Exp:* Sr microbiologist, 67-71, head fermentation prod res & microbiol res depts, 71-73, dir microbiol & fermentation prod res, 73-77, exec dir, 77-78, VPRES, LILLY RES LABS, ELI LILLY & CO, 78- *Mem:* Soc Indust Microbiol; Am Soc Microbiol. *Res:* Discovery, isolation and evaluation of fermentation products; control and regulation of secondary metabolism. *Mailing Add:* Eli Lilly Corp Ctr Indianapolis IN 46285

WHITNEY, KENNETH DEAN, b San Diego, Calif, Oct 14, 52; m 77; c 2. BOTANY-PHYTOPATHOLOGY. *Educ:* Calif State Univ, Chico, BA, 78, MA, 80; Univ NC, Chapel Hill, PhD(bot), 83. *Prof Exp:* Teaching fel & lect biol, 83-84, asst prof biol, 84-85, NSF RES FEL BIOL, UNIV TEX, ARLINGTON, 85- *Concurrent Pos:* Grad fel, Mycol Soc Am, 81; res fel, NSF, 84. *Mem:* AAAS; Bot Soc Am; Mycol Soc Am; Sigma Xi; Soc Protozoologists. *Res:* Morphology, development and ultrastructure of calcium oxalate crystals in fungi; ultrastructure and development of Mycetozoans; biosystematics of Myxomycetes. *Mailing Add:* 118 Lucas Park Dr San Rafael CA 94903

WHITNEY, LESTER F(RANK), b New Bedford, Mass, Mar 21, 28; m 50; c 7. FOOD ENGINEERING, MACHINE DESIGN. *Educ:* Univ Maine, BS, 49; Mich State Univ, MSAE, 51, PhD(agr eng), 64. *Prof Exp:* Design & develop engr, Ariens Co, Wis, 51-53 & Maine Potato Growers, Inc, 53-54; develop engr, Wirthmore Feed Div, Corn Prod, Inc, 54-56, asst chief engr, Mass, 56-59; asst prof agr eng, Univ Mass, 59-62; NSF fel, Mich State Univ, 62-63; PROF FOOD ENG, UNIV MASS, AMHERST, 63- *Concurrent Pos:* Consult engr food mach processing. *Mem:* Am Soc Agr Engrs; Int Food Technologists; Sigma Xi; fel NSF. *Res:* Agricultural processes; stress analysis; systems analysis; water resources; forage dehydration; fish processing; shellfish depuration; fish waste byproducts. *Mailing Add:* 48 Jeffery Lane Amherst MA 01002-2532

WHITNEY, MARION ISABELLE, b Austin, Tex, Apr 23, 11. GEOLOGY. *Educ:* Univ Tex, BA, 30, MA, 31, PhD(geol, paleont), 37. *Prof Exp:* Teacher pub sch, 33-36; asst prof geol, Kans State Teachers Col, 37-42; teacher geol & biol, Kilgore Col, 42-46; asst prof geol, Tex Christian Univ, 46-51 & Sul Ross State Col, 51-52; prof geol & biol, Ark Polytech Col, 52-54; assoc prof geol, Tulane Univ, 54-55; assoc prof, La Tech Inst, 55-60; teacher biol, Texarkana Col, 60-61; from assoc prof to prof biol, Cent Mich Univ, 61-81; RETIRED. *Mem:* Am Asn Petrol Geol; Soc Econ Paleont & Mineral; Geol Soc Am; Sigma Xi. *Res:* Description of the fauna of the Glen Rose formation of Texas; development of new data concerning the method of aerodynamic erosion of rock, dunes and snow. *Mailing Add:* Box 277 Shepherd MI 48853

WHITNEY, NORMAN JOHN, b Langdon, Alta, July 24, 25; m 51; c 4. MYCOLOGY, PLANT PATHOLOGY. *Educ:* Univ Alta, BSc, 47; Univ Western Ont, MSc, 49; Univ Toronto, PhD(mycol, plant path), 53; McGill Univ, BD, 64. *Prof Exp:* Lectr bot, Univ Toronto, 50-52; plant pathologist, Res Sta, Can Dept Agr, 52-61; lectr bot, McGill Univ, 61-64; lectr biol, 65-73, student counr, 66-73, assoc prof, 73-80, PROF BIOL & COUNR STUDENT SERV, UNIV NB, 80- *Mem:* AAAS; Can Phytopath Soc. *Res:* Soil-borne diseases of plants; marine mycology; science and religion; spore germination in the phyllosphere. *Mailing Add:* Dept Biol Sci Univ NB Box 4400 Fredericton NB E3B 5A3 Can

WHITNEY, PHILIP ROY, b Providence, RI, Nov 10, 35; m 59; c 4. GEOCHEMISTRY, PETROLOGY. *Educ:* Mass Inst Technol, BS, 56, PhD(geol), 62. *Prof Exp:* Asst prof geochem, State Univ NY Col Ceramics, Alfred Univ, 62-67; asst prof geol, Rensselaer Polytech Inst, 67-70; ASSOC SCIENTIST GEOCHEM, NY STATE MUS & SCI SERV, 70- *Concurrent Pos:* Adj assoc prof, Rensselaer Polytech Inst, 75- *Mem:* Geol Soc Am; Geochem Soc. *Res:* Geochemistry and petrology of anorthosite; geology of the Adirondack area; geochemistry of freshwater manganese oxides. *Mailing Add:* Luther Rd East Greenbush NY 12061

WHITNEY, RICHARD RALPH, b Salt Lake City, Utah, June 29, 27; m 50; c 4. FISHERY BIOLOGY. *Educ:* Univ Utah, BA, 49, MS, 51; Iowa State Col, PhD(fisheries mgt), 55. *Prof Exp:* Res biologist, Salton Sea Invest, Univ Calif, 54-57; proj leader, Susquehanna Fishery Study, State Dept Res & Educ, Md, 58-60; chief tuna behav invests, Tuna Resources Lab, US Bur Com Fisheries, 61-67; unit leader, Wash Coop Fishery Res Unit, 67-83; PROF FISHERIES, UNIV WASH, 67- *Concurrent Pos:* Tech adv & chmn fisheries adv bd, George H Boldt, Sr Judge US Dist Court, Tacoma, 74-79; consult, Conn Yankee Atomic Power Co, 65-74; mem sci & statist comt, Pac Fishery Mgt Coun, 76-80; coordr & chmn, mid-Columbia studies comt, 80-; comnr, salmon & stealhead adv comn, 81-84; chmn, fish propagation panel, Northwest Power Planning Coun, 83-84. *Mem:* Fel Am Inst Fishery Res Biol; Am Fisheries Soc; Sigma Xi. *Res:* Aquatic ecology; fisheries. *Mailing Add:* 16500 River Rd Leavenworth WA 98826

WHITNEY, RICHARD WILBUR, b Osawatomie, Kans, Nov 1, 38; m 59; c 4. AGRICULTURAL ENGINEERING. *Educ:* Kans State Univ, BS, 61; Okla State Univ, MS, 67, PhD(agr eng), 72. *Prof Exp:* Instr, Kans State Univ, 61-62; asst prof, La State Univ, Baton Rouge, 72-77; instr, 62-69, asst prof, 77-80, ASSOC PROF AGR ENG, OKLA STATE UNIV, 80- *Concurrent*

Pos: Consult, Charles Machine Works, Perry, Okla, 75, Kincaid Equip Co, Haven, Kans, 77-79, US Pollution Control, Oklahoma City, 80, Kahrs, Nelson, Fanning, Hite & Kellogg, 80 & Eagle Aircraft Co, Boise, Idaho, 81. *Mem:* Am Soc Agr Engrs; Sigma Xi. *Res:* Mechanization of food and fiber production of cotton and horticultural crops; low volume pesticide application equipment and delivery techniques for tick control; development of improved production equipment for forage grasses; determination of potential human inhalation exposure to airborne pentachlorophenol within treated structures. *Mailing Add:* Dept Agr Eng Okla State Univ 522 N Washington Stillwater OK 74075-5021

WHITNEY, ROBERT ARTHUR, JR, b Oklahoma City, Okla, July 27, 35; m 86; c 4. LABORATORY ANIMAL MEDICINE, COMPARATIVE MEDICINE. *Educ:* Okla State Univ, BS, 58, DVM, 59; Ohio State Univ, MS, 65. *Prof Exp:* US Army fel & resident lab animal med, Ohio State Univ, 63-65, chief animal resources br, US Army Edgewood Arsenal, 65-70, dir lab animal training prog, US Army Vet Corps, 68-70, commanding officer, 4th Med Detachment, Viet Nam, 70-71; proj officer, Animal Resources Br, Nat Ctr Res Resources, 71-72, chief, Vet Resources Br, 72-85, DIR, NAT CTR RES RESOURCES, NIH, 85-, DIR, OFF ANIMAL CARE & USE, 87- *Concurrent Pos:* Consult lab animal med, US Army Surgeon Gen Off, 67-70; exec dir, US Govt Interagency Primate Steering Comt, 80-81; chief vet officer, USPHS, 84-; chmn, US Govt Interagency Res Animal Comt, 84- *Mem:* Am Col Lab Animal Med; Am Vet Med Asn; Am Asn Lab Animal Sci; Am Asn Lab Animal Practitioners; Sigma Xi. *Res:* Diseases of laboratory animals; primatology. *Mailing Add:* Nat Ctr Res Resources NIH Bldg 12A Rm 4007 Bethesda MD 20892

WHITNEY, ROBERT BYRON, b Minneapolis, Minn, July 28, 05; wid; c 4. ORGANIC CHEMISTRY. *Educ:* Univ Minn, BA, 24, PhD(org chem), 27. *Prof Exp:* Instr chem, Harvard Univ, 28-30; from instr to prof & actg dean fac, 66, EMER PROF CHEM, AMHERST COL, 71- *Concurrent Pos:* Pratt fel, Univs Heidelberg & Halle, 36-37; consult, Radiation Lab, Mass Inst Technol, 45-46; staff mem, Radiation Lab, Univ, Calif, 54; Petrol Res Fund fel. Oxford Univ, 60-61; coordr, Amherst, Smith & Mt Holyoke Cols & Univ Mass, 63-66. *Mem:* AAAS; Am Chem Soc. *Res:* Kinetics of disproportionation of free radicals; reactions of alkali metals with organic substances; thermodynamics of electrocapillarity; acid-base catalysis. *Mailing Add:* 92 Sunset Ave Amherst MA 01002-2018

WHITNEY, ROBERT C, b Seattle, Wash, July 20, 19; m 42; c 2. SCIENCE EDUCATION, PHYSICS. *Educ:* Univ Wash, BS, 47; Cornell Univ, MS, 58, PhD(sci educ, physics), 63. *Prof Exp:* Teacher, Wash High Sch, 47-55, 56-57 & 58-59; assoc dir shell merit fels, Shell Found, Cornell Univ, 59-61, assoc dir acad year inst, NSF, 61-63; assoc prof, 63-66, PROF PHYS SCI, CALIF STATE UNIV, HAYWARD, 66- *Concurrent Pos:* Consult, Murray, Fremont & Palo Alto Sch Dist, Calif, 65-66 & Livermore Sch Dist, 67; NSF fel, Univ Wash, 71-72. *Mem:* AAAS; Am Asn Physics Teachers; Nat Sci Teachers Asn. *Res:* Improvement of high school physics facilities; improvement in the teaching of high school physics and elementary science. *Mailing Add:* Dept Earth Sci Calif State Univ Hayward CA 94542

WHITNEY, ROBERT MCLAUGHLIN, b St Paul, Minn, Sept 28, 11; m 34, 67; c 2. FOOD CHEMISTRY. *Educ:* Augustana Col, SDak, AB, 36; Univ Ill, PhD(phys chem), 44. *Prof Exp:* Instr math, Augustana Col, SDak, 36-37; high sch teacher, NDak, 37-38 & Ill, 38-40; asst chemist, Ill State Water Surv, 40-42; instr chem, Univ Ill, 42-44; res chemist, Dean Milk Co, 44-46; assoc prof dairy mfg res, 46-50, assoc prof dairy technol, 50-59, prof dairy technol, 59-73, PROF FOOD CHEM, UNIV ILL, URBANA, 73- *Concurrent Pos:* Vis prof, Univ PR, Mayaguez, 73-74 & Univ Baghdad, Iraq, 79. *Honors & Awards:* Borden Co Found Res Award, Am Dairy Sci Asn, 61. *Mem:* Am Chem Soc; Am Dairy Sci Asn; fel Am Inst Chemists; Inst Food Technologists. *Res:* Chemical analysis of dairy products; investigation of flavors in dairy products; ultrasonic bactericidal effects; physical-chemical state of milk proteins; investigation of the proteins in the milk fat-globule membrane. *Mailing Add:* Univ Ill 202 Dairy Mfg Urbana IL 61803

WHITNEY, ROY DAVIDSON, b Langdon, Alta, Dec 30, 27; m 53; c 4. FOREST PATHOLOGY. *Educ:* Univ BC, BSF, 51; Yale Univ, MF, 54; Queen's Univ, Ont, PhD(forest path), 60. *Prof Exp:* RES SCIENTIST, CAN FORESTRY SERV, 51- *Mem:* Am Phytopath Soc; Can Phytopath Soc. *Res:* Investigations of root rots of conifers, including identification of causal fungi, symptomatology, infection courts, damage appraisal and spore germination; determination of pathogenic potentials by inoculations. *Mailing Add:* 83 Thorneloe Cr Sault Ste Marie ON P6A 4J4 Can

WHITNEY, ROY P(OWELL), b Milo, Maine, May 30, 13; m 41; c 2. CHEMICAL ENGINEERING. *Educ:* Mass Inst Technol, SB, 35, SM, 37, ScD(chem eng), 45; Lawrence Univ, MS, 79. *Prof Exp:* Asst, Mass Inst Technol, 35-36, asst dir sch chem eng practice, Bangor Sta, 36-38, dir, 38-42, asst prof chem eng, Univ, 39-45; dir, dept indust coop, Univ Maine, 45-47, prof chem eng & actg head dept, 46-47; prof chem eng, 47-79, res assoc & group leader, 47-57, dean, 56-76, vpres, 58-77, asst to pres, 78-79, EMER PROF CHEM ENG, INST PAPER CHEM, 79- *Concurrent Pos:* Tech adv, Chem Warfare Serv Develop Lab, US Dept Army, 42-45, consult, Chem Corps, 50-52; chmn, comt paper base mat, Nat Acad Sci-Nat Res Coun, 60-66. *Honors & Awards:* Colburn Award, Am Inst Chem Engrs, 48; Div Award, Tech Asn Pulp & Paper Indust, 69, Gold Medal, 80; Pro Bono Labore Award, Finnish Paper Engrs Asn, 78. *Mem:* Am Chem Soc; Am Soc Eng Educ; fel Tech Asn Pulp & Paper Indust; fel Am Inst Chem Engrs; fel Am Inst Chemists. *Res:* Heat and mass transfer, particularly gas absorption and drying; pulp and paper technology. *Mailing Add:* 1709 S Douglas St Appleton WI 54914

WHITNEY, THOMAS ALLEN, b Toledo, Ohio, June 22, 40; m 62; c 3. ORGANIC CHEMISTRY. *Educ:* Northwestern Univ, Evanston, BA, 62; Univ Calif, Los Angeles, PhD(chem), 67. *Prof Exp:* Sr staff chemist, 67-80, RES ASSOC, CORP RES LAB, EXXON RES & ENG CO, 80- *Mem:* Am Chem Soc. *Res:* Homogeneous catalysis; asymmetric synthesis; organic reactions. *Mailing Add:* Duracell Inc 37 A St Needham MA 02194-2806

WHITNEY, WENDELL KEITH, b Miltonvale, Kans, Nov 27, 27; m 45; c 2. ENTOMOLOGY, AGRICULTURE. *Educ:* Kans State Univ, BS, 56, MS, 58, PhD(entom, zool), 62. *Prof Exp:* Biol aide, Stored Prod Insect Br, USDA, Kans, 51-56, entomologist, 56-58; instr entom, Kans State Univ, 58-62; entomologist, Bioprod Dept, Dow Chem Co, 62-68; Ford Found entomologist, Int Inst Trop Agr, Nigeria, 68-73; chief entomologist, Plant Prod Res & Develop, Am Cyanamid Co, 74-82, prin scientist, 82-87; RETIRED. *Concurrent Pos:* Res grants, 58-62; consult, Industs & USDA, 59-62. *Mem:* Entom Soc Am; Nigerian Soc Plant Protection. *Res:* Plant pest control; effects of chemicals on insects. *Mailing Add:* 3005 S Burma Rd Salina KS 67401

WHITNEY, WILLIAM MERRILL, b Coeur d'Alene, Idaho, Dec 5, 29; m 50, 78; c 2. PHYSICS, VLSI TECHNOLOGY. *Educ:* Calif Inst Technol, BS, 51; Mass Inst Technol, PhD(physics), 56. *Prof Exp:* From instr to asst prof physics, Mass Inst Technol, 56-63; mem tech staff, Jet Propulsion Lab, 63-67, mgr guid & control res sect, 67-70, mgr info systs res sect, 70-84, tech leader, Robot Res Prog, 71-78, mgr microelectronics technol sect, 84-86, sr mem tech staff, 86-87, DIV TECHNOLOGIST, OBSERVATIONAL SYSTS DIV, JET PROPULSION LAB, 87- *Honors & Awards:* Except Serv Medal, NASA, 82. *Mem:* AAAS; Am Phys Soc. *Res:* Low temperature and semiconductor physics; computer science. *Mailing Add:* Observational Systs Div Jet Propulsion Lab Pasadena CA 91109

WHITSEL, BARRY L, b Mt Union, Pa, Aug 26, 37; m 60; c 2. NEUROPHYSIOLOGY, NEUROPHARMACOLOGY. *Educ:* Gettysburg Col, AB, 59; Univ Pa, MS, 63; Univ Ill, PhD(pharmacol), 66. *Prof Exp:* Res asst psychopharmacol, Wyeth Inst, 59-61; res assoc pharmacol, Sch Med, Univ Pittsburgh, 65-66, from instr to asst prof, 66-74; ASSOC PROF DENT RES & PHYSIOL, SCH MED, UNIV NC, CHAPEL HILL, 74- *Concurrent Pos:* Res scientist develop award, NIMH, 68-73. *Res:* Pharmacology. *Mailing Add:* Dept Physiol Sch Med Univ NC Chapel Hill NC 27514

WHITSELL, JOHN CRAWFORD, II, b St Joseph, Mo, Dec 21, 29; m 65. SURGERY. *Educ:* Grinnell Col, AB, 50; Wash Univ, MD, 54; Am Bd Surg, dipl, 62; Am Bd Thoracic Surg, dipl, 64. *Prof Exp:* Instr surg, Med Col, Cornell Univ, 63-66, asst attend surgeon, New York Hosp, 63-68, from asst prof to assoc prof, 66-70, surg dir, Renal Transplant Unit, 68-75, PROF SURG, MED COL, CORNELL UNIV, 70-, SURG CONSULT RENAL TRANSPLANT UNIT, NEW YORK HOSP-CORNELL MED CTR, 75- *Concurrent Pos:* Assoc attend surgeon, New York Hosp, 68-70, attend surgeon, 70- *Mem:* AMA; Am Col Surg; Transplantation Soc; NY Acad Sci; Harvey Soc. *Res:* Renal transplantation. *Mailing Add:* 449 E 68th St New York NY 10021

WHITSETT, CAROLYN F, b Portsmouth, Va, Nov 21, 45. PATHOLOGY, MEDICINE. *Educ:* Howard Univ, BS, 66, MD, 70. *Prof Exp:* Med internship, 70-71; jr asst resident, Downstate Med Ctr, 71-72; sr asst resident, New York Hisp, 72-73; fel hemat, Montefiore Hosp & Med Ctr, 73-74; fel immunohemat, New York Blood Ctr, 74-75; fel med hemat div, Mem Hosp, 74-75; asst prof med, Cornell Univ, Med Col, 75-77; ASSOC PROF PATH & MED, EMORY UNIV, SCH MED, 77- *Concurrent Pos:* Res assoc tissue typing lab, Mem Sloan Kettering Cancer Ctr, 75-77; asst attending physician hemat serv, Mem Hosp, 75-77; asst med dir, Mem Hosp Blood Bank, 76-77; med dir, Emory Univ Hosp Blood Bank & Sch Blood Banking, 77- *Mem:* Am Asn Blood Banks; Am Asn Clin Histocompatibility Testing. *Mailing Add:* Dept CP Emory Univ Hosp Atlanta GA 30322

WHITSETT, JOHNSON MALLORY, II, b San Antonio, Tex, Jan 26, 41; m 64; c 1. ANIMAL BEHAVIOR, REPRODUCTIVE PHYSIOLOGY. *Educ:* Univ Tex, Austin, BA, 63, PhD(psychol), 70; New Sch Social Res, MA, 66. *Prof Exp:* Fel zool, Univ Tex, Austin, 69-71; from asst prof to assoc prof, 71-81, PROF ZOOL, NC STATE UNIV, 81- *Concurrent Pos:* Res assoc, NC Dept Ment Health, 71-73; mem physiol fac, NC State Univ, 74-; NSF res grant, 78. *Mem:* AAAS; Animal Behav Soc; Am Ornithol Union; Soc Study Reproduction. *Res:* Hormonal and stimulus control of sexual and aggressive behavior in birds and mammals; environmental influence on reproduction; behavioral aspects of sexual development; photoperiodism. *Mailing Add:* 60004 Dartmouth St Amarillo TX 79109

WHITSETT, THOMAS L, b Tulsa, Okla, July 14, 36; m 59; c 2. INTERNAL MEDICINE, CLINICAL PHARMACOLOGY. *Educ:* Pasadena Col, BA, 58; Univ Okla, MD, 62. *Prof Exp:* Clin asst, Med Ctr, Univ Okla, 67-68; vis asst prof med, Sch Med, Emory Univ, 69-70; from asst prof med to assoc prof, 70-78, asst prof pharmacol, 70-77, PROF MED, MED CTR, UNIV OKLA, 78-, ASSOC PROF PHARMACOL, 77- *Concurrent Pos:* Found fac develop award, Pharmaceut Mfr Asn, 71; trainee clin pharmacol, Med Ctr, Univ Okla, 67-68 & Sch Med, Emory Univ, 68-70. *Mem:* Am Heart Asn; Am Fedn Clin Res; Am Soc Pharmacol & Exp Therapeut; Sigma Xi. *Res:* Early phases of new drug investigation, especially cardiovascular and respiratory agents. *Mailing Add:* Dept Med & Cardiol Univ Okla Med Ctr 920 Stanton L Young Blvd Rm 5 Sp300 Oklahoma City OK 73104

WHITSON, PAUL DAVID, b Gravette, Ark, Mar 20, 40; m 78. BOTANY, PLANT ECOLOGY. *Educ:* Baylor Univ, BS, 62, MS, 65; Univ Okla, PhD(bot), 71. *Prof Exp:* Grad asst biol, Baylor Univ, 63-65; grad asst bot, Univ Okla, 65-71; asst prof biol, Baylor Univ, 70-71; from asst prof to assoc prof, 72-81, PROF BIOL, UNIV NORTHERN IOWA, 81- *Concurrent Pos:* Res consult, Nat Park Serv Univ Okla Res Inst, 69-70; Int Biol Prog fel biol, NMex State Univ, 71-72; prin investr desert biome, Int Biol Prog, 72-75; staff assoc environ biol, NSF, 76-77; exec secy, Fed Comt Ecol Reserves, 76-77; investr endangered species, US Fish & Wildlife Serv, 78-81 & Iowa Dept Natural Resources, 82- *Mem:* Ecol Soc Am; Brit Ecol Soc; Sigma Xi; Cactus & Succulent Soc Am; Am Orchid Soc. *Res:* Structure, dynamics and human influences upon woodland and desert vegetation; phenology and productivity of desert annuals; species biology, Southern Appalachian and Midwestern endangered plants; decide forest and prairie reconstruction; terrestial orchid ecology. *Mailing Add:* Dept Biol Univ Northern Iowa Cedar Falls IA 50614-0421

WHITSON, ROBERT EDD, b Spearman, Tex, Apr 30, 42; m 63; c 2. AGRICULTURAL ECONOMICS, RANGE MANAGEMENT. *Educ:* Tex Tech Univ, BS, 65, MS, 67, PhD(agr econ), 74. *Prof Exp:* Area economist, Tex A&M Univ, 69-71, asst prof, 74-77, assoc prof range econ, Tex Agr Exp Sta, 77-81; SR VPRES, FIRST NAT BANK, 81- *Mem:* Am Soc Agr Econ; Soc Range Mgt. *Res:* Examination of risk management alternatives for ranchers and an evaluation of changing feed price relationships on efficient ranch organizations. *Mailing Add:* First Nat Bank PO Box 1600 San Antonio TX 78296

WHITT, DARNELL MOSES, b Greensboro, NC, Apr 30, 13; m 36; c 1. SOIL PHYSICS, FIELD CROPS. *Educ:* NC State Univ, BS, 34; Univ Mo, AM, 35, PhD(crops), 52. *Hon Degrees:* FEI Grad. *Prof Exp:* Soil surveyor, Soil Conserv Serv, USDA, 35-36, res agronomist, 36-42 & 46-52, res soil conservationist, Agr Res Serv, 52-55, regional liaison officer, Agr Res Serv & Soil Conserv Serv, 55-56, nat liaison officer, 56-59, dir conserv planning, Soil Conserv Serv, 59-63, dir plant sci div, 63-72, dep adminr, 72-75; coordr river basin studies, Int Joint Comn, 75-78; CONSULT, NATURAL RESOURCE CONSERV, 78- *Concurrent Pos:* Mem, Nat Comt Res Needs in Soil & Water Conserv, 58-59; consult, Repub of Nauru & SPac Comn, New Caledonia. *Mem:* Am Soc Agron; Soil Sci Soc Am; Soc Range Mgt; Soil Conserv Soc Am; Int Soc Soil Sci. *Res:* Water pollution. *Mailing Add:* PO Box 82 Green Valley AZ 85622

WHITT, DIXIE DAILEY, b Longmont, Colo, Mar 9, 39; m 63. MICROBIAL ECOLOGY, MICROBIAL PHYSIOLOGY. *Educ:* Colo State Univ, BS, 61, PhD(zool), 65. *Prof Exp:* USPHS fel biochem genetics, Yale Univ, 65-69; res assoc microbiol, 69-87, LECTR MICROBIOL, UNIV ILL, URBANA, 87- *Mem:* Am Genetics Asn; Asn Gnotobiotics; AAAS; Soc Microbiol Ecol & Dis; Am Soc Microbiol; Am Acad Microbiol. *Res:* Host-parasite interactions; biochemical genetics of microorganisms; host-parasite relationships as an expression of the host's environmental conditions; host-intestinal microflora interactions. *Mailing Add:* Dept Microbiol 131 Burrill Hall Univ Ill 407 S Goodwin Ave Urbana IL 61801

WHITT, GREGORY SIDNEY, b Detroit, Mich, June 13, 38; m 63. EVOLUTIONARY GENETICS, MOLECULAR EVOLUTION. *Educ:* Colo State Univ, BS, 62, MS, 65; Yale Univ, PhD(biol), 70. *Prof Exp:* From asst prof to assoc prof zool, 69-72, prof genetics & develop, 77-87, PROF ECOL, ETHOLOGY & EVOLUTION, UNIV ILL, URBANA, 87- *Concurrent Pos:* Assoc ed, J Exp Zool, 74-78 & 84-85, & Develop Genetics, 78-83; co-ed, Isozymes: Current Topics Biol & Med Res, 77-87; ed, Isozyme Bull, 78-81; mem adv bd, Biochem Genetics, 75-; affil, Ctr Aquatic Ecol, Ill Natural Hist Surv, Urbana, 81- *Mem:* Am Genetic Asn; Soc Study Evolution; Genetics Soc Am; Am Soc Ichthyologists & Herpetologists; Am Fisheries Soc; Am Soc Biochem & Molecular Biol; fel AAAS; Am Soc Naturalists; Am Inst Biol Sci; Am Soc Zoologists. *Res:* Biochemical and molecular evolutionary genetics and systematics of fishes; biochemical, developmental, and evolutionary genetics of fishes; evolutionary and systematic analyses using isozyme gene structure and regulation; isozymes as probes of gene structure, function and evolution. *Mailing Add:* Dept Ecol Ethology & Evolution Univ Ill 505 S Goodwin Ave Urbana IL 61801-3799

WHITT, WARD, b Buffalo, NY, Jan 29, 42; m 83; c 2. QUEUEING THEORY, STOCHASTIC PROCESSES. *Educ:* Dartmouth Col, AB, 64; Cornell Univ, PhD(opers res), 69. *Prof Exp:* Vis asst prof opers res, Stanford Univ, 68-69; asst prof admin sci, Sch Orgn & Mgt, Yale Univ, 69-73, assoc prof admin sci & statist, 73-77; MEM TECH STAFF, AT&T BELL LABS, 77- *Concurrent Pos:* NSF res initiation grant admin sci, Yale Univ, 71-73, jr fac fel, 72-73, res grant, 73-75. *Mem:* Oper Res Soc Am; Inst Math Statist. *Res:* Probability theory and its applications; mathematical models in the social sciences; queueing theory; stochastic processes. *Mailing Add:* 86 Hill Top Rd Basking Ridge NJ 07920

WHITTAKER, FREDERICK HORACE, b Columbus, Ohio, Mar 9, 28; m 52; c 2. PARASITOLOGY. *Educ:* Otterbein Col, BA, 51; Univ Ga, MSc, 56; Univ Ill, PhD(zool, parasitol), 63. *Prof Exp:* Instr biol & chem, Spartanburg J- Col, 57-58; res biologist, Abbott Labs, Ill, 63-64; from asst prof to assoc prof, 64-72, PROF ZOOL, UNIV LOUISVILLE, 72- *Concurrent Pos:* Consult, Abbott Labs, 64-65. *Mem:* Sigma Xi; Am Soc Parasitologists. *Res:* Effects of fermentation liquors on invertebrates; taxonomy and life cycles of trematodes and cestodes; scanning electron microscopy of cestodes and trematodes of sharks, skates and rays; systematics and ecology of helminths of cavefishes. *Mailing Add:* Dept Biol Univ Louisville Louisville KY 40292

WHITTAKER, J RICHARD, b Cornwall, Ont, Aug 19, 34; div. EMBRYOLOGY, CELL BIOLOGY. *Educ:* Queen's Univ, Ont, BA, 58, MSc, 59; Yale Univ, PhD(develop biol), 62. *Prof Exp:* Asst prof zool, Univ Calif, Los Angeles, 62-67; assoc prof, Wistar Inst Anat & Biol, 67-81; prof biol & dir marine prog, Boston Univ, 81-85; SR SCIENTIST, MARINE BIOL LAB, 85- *Concurrent Pos:* Investr, 69-, trustee, 78-86, actg dir, Marine Biol Lab, Woods Hole, 86-87; assoc prof anat, Sch Med, Univ Pa, 71-81; mem, Alpha Helix Philippine Exped, 79; vis prof, Kewalo Marine Lab, Univ Hawaii, 81; adj prof biol, Boston Univ, 86-; vis assoc, Calif Inst Technol, 88, 91. *Honors & Awards:* MBL Award, Marine Biol Lab, Woods Hole, 71. *Mem:* Am Soc Zool; Soc Develop Biol; Int Soc Develop Biol; Sigma Xi. *Res:* Localization and segregation of morphogenetic determinants in ascidian embryos; gene regulation in early embryonic development of marine invertebrates; melanocyte differentiation; developmental genetics of Tunicate and vertebrate evolution. *Mailing Add:* Lab Develop Genetics Marine Biol Lab Woods Hole MA 02543

WHITTAKER, JAMES VICTOR, b Los Angeles, Calif, Aug 1, 31. MATHEMATICS. *Educ:* Univ Calif, Los Angeles, BA, 53, MA, 54, PhD, 58. *Prof Exp:* Assoc math, Univ Calif, Los Angeles, 57-58; from instr to assoc prof, 58-69, PROF MATH, UNIV BC, 69- *Mem:* Am Math Soc; Math Asn Am; Can Math Cong; Sigma Xi. *Res:* Geometric topology; probability. *Mailing Add:* Dept Math Univ BC Vancouver BC V6T 1W5 Can

WHITTAKER, PAUL, b Spokane, Wash, Jan 10, 43. TOXICOLOGY, BIOCHEMISTRY. *Educ:* Utah State Univ, PhD(nutrit), 83. *Prof Exp:* CHIEF, NUTRIENT INTERACTION SECT, FOOD & DRUG ADMIN, 86- *Mem:* Am Soc Clin Nutrit; Am Inst Nutrit; Sigma Xi. *Res:* Nutrient interactions of trace elements, with a primary interest in iron metabolism, toxicology and bioavailability. *Mailing Add:* Food & Drug Admin 200 C St SW HFF-268 Washington DC 20204-0001

WHITTAM, JAMES HENRY, b New York, NY, Apr 23, 49; m. PHYSICAL CHEMISTRY, CHEMICAL ENGINEERING. *Educ:* City Col New York, BS, 72, PhD(phys chem), 75; Boston Univ, MBA, 78. *Prof Exp:* Instr chem, City Col New York, 72-75; proj chemist, Gillette Co, 75-78; mgr res chem & chem eng, 78-80, DIR HEALTH SCI, SHAKLEE CORP RES LAB, 80- *Concurrent Pos:* Consult, Gen Foods Corp, 73-74; adv ed, Cosmetic Tech, 78- *Mem:* Am Chem Soc; Soc Cosmetic Chem; Inst Food Technol; Am Oil Chem Soc. *Res:* Surface and colloid science pertaining to the fields of hair and skin cosmetology and food science, nutrition and engineering. *Mailing Add:* Shaklee Corp 444 Market St San Francisco CA 94111-5378

WHITTEMBURY, GUILLERMO, b Trujillo, Peru, Nov 17, 29; m 61; c 3. BIOPHYSICS. *Educ:* San Marcos Univ, Lima, BM, 55; Univ Cayetano Heredia, Peru, MD, 65. *Prof Exp:* Instr anat, San Marcos Univ, Lima, 49-50, asst prof med, 55-57, asst prof biophys, 60-62; sr scientist, 62-67, head dept gen physiol, 67-70, MEM STAFF, VENEZUELAN INST SCI RES, 61- *Concurrent Pos:* Res fel, Biophys Lab, Harvard Med Sch, 57-60; Rockefeller Found fel, 57-59; Helen Hay Whitney Found fel, 59-60; vis prof, Yale Univ, 70; mem, Int Union Pure & Appl Biophys, 63; dir, Latin Am Ctr Biol, 73-84; fel, Churchill Col, Cambridge, 76-; vis scientist, Max Planck Inst Biophys, Frankfurt, 86. *Honors & Awards:* Daniel Carrion Prize, Peru, 65. *Mem:* Am Soc Nephrology; Biophys Soc; Peruvian Nephrology Soc; Int Soc Nephol; Soc Gen Physiologists. *Res:* Transport processes across membranes; kidney physiology. *Mailing Add:* Venezuelan Inst Sci Res PO Box 21827 Caracas 1020A Venezuela

WHITTEMORE, ALICE S, b New York, NY, July 5, 36; m 58; c 2. BIOMATHEMATICS, BIOSTATISTICS. *Educ:* Marymount Manhattan Col, BS, 58; Hunter Col, MA, 64; City Univ New York, PhD(math), 67. *Prof Exp:* From asst prof to assoc prof math, Hunter Col, 67-74; adj assoc prof environ med, Med Ctr, NY Univ, 74-76; FAC MEM, DEPT STATIST, STANFORD UNIV, 76- *Concurrent Pos:* City Univ New York res grants, 69 & 70; Sloan Found res grant, Soc Indust & Appl Math Inst Math & Soc, 74-76; Rockefeller Found res grant, 76-77. *Mem:* AAAS; Soc Indust & Appl Math; Sigma Xi; Am Math Soc; Math Asn Am. *Res:* Environmental carcinogenesis. *Mailing Add:* Dept Health Res & Policy Stanford Univ Sch Med Stanford CA 94305-1024

WHITTEMORE, CHARLES ALAN, b Grand Junction, Colo, Dec 14, 35; m 63; c 2. ORGANIC CHEMISTRY. *Educ:* Stanford Univ, BSc, 57; Univ Colo, PhD(org chem), 63. *Prof Exp:* Sr chemist, Cent Res Labs, Minn Mining & Mfg Co, 63-69; from asst prof to assoc prof chem, Colo Women's Col, 69-77; sr develop chemist, 77-82, group leader, 82-87, SCIENTIST, GA-PAC, 87- *Mem:* Am Chem Soc. *Res:* Organic reaction mechanisms; Friedel-Crafts reactions; phenolic resins; organic synthesis. *Mailing Add:* PO Box 1068 Albany OR 97321

WHITTEMORE, DONALD OSGOOD, b Pittsburgh, Pa, May 4, 44; m 71; c 3. GEOCHEMISTRY, HYDROGEOLOGY. *Educ:* Univ NH, BS, 66; Pa State Univ, University Park, PhD(geochem), 73. *Prof Exp:* Asst prof geol, Kans State Univ, 72-78; ASSOC SCIENTIST, KANS GEOL SURV, 78- *Concurrent Pos:* Courtesy assoc prof geol, Univ Kans. *Mem:* Nat Water Well Asn; Soil Sci Soc Am; Geochem Soc; Am Geophys Union. *Res:* Geochemical identification of saltwater pollution sources in water resources; factors controlling spatial and temporal variations in groundwater quality; geochemistry of oilfield brines; groundwater quality protection using geographic information system. *Mailing Add:* Kans Geol Surv Univ Kans Lawrence KS 66047

WHITTEMORE, IRVILLE MERRILL, petroleum chemistry, for more information see previous edition

WHITTEMORE, O(SGOOD) J(AMES), JR, b Clear Lake, Iowa, Jan 24, 19; m 41; c 3. CERAMIC ENGINEERING. *Educ:* Iowa State Univ, BS, 40, CerE, 50; Univ Wash, Seattle, MS, 41. *Prof Exp:* Fel refractories, Mellon Inst, 41-44; group leader Manhattan Proj, Mass Inst Technol, 44-46; sr engr, Norton Co, 46-56, chief ceramic engr, 56-59, res assoc explor res, 59-64; from assoc prof to prof, 64-87, EMER PROF CERAMIC ENG, UNIV WASH, 87- *Concurrent Pos:* Mem ad hoc comt, Mat Adv Bd, 56 & 58; NASA ceramic mat res grant, 64-78; vis prof, Univ BC, Can, 74-75 & 90; prof, Univ Fed de Sao Carlos, Brazil, 76; NSF sintering grant, 79-85; dir, Wash Mining & Mineral Resources Res Inst, 82-87. *Honors & Awards:* Admiral Earle Award, Worcester Eng Soc, 49; Trinks Indust Heating Award, 55; Azevedo Prize, Brazil, 79 & 82. *Mem:* Nat Inst Ceramic Engrs; fel Am Ceramic Soc (vpres, 75-76); Brit Ceramic Soc; Assoc Brasileira de Ceramica; Am Inst Mining & Metall Engrs; fel Inst Ceramics; Int Inst Sci Sintering. *Res:* Refractories; processing; minerals. *Mailing Add:* 10015 Lakeshore Blvd NE Seattle WA 98125

WHITTEMORE, RUTH, b Cambridge, Mass, June 11, 17. PEDIATRICS, CARDIOLOGY. *Educ:* Mt Holyoke Col, BA, 38; Johns Hopkins Univ, MD, 42; Am Bd Pediat, dipl, 53, cert pediat cardiol, 61. *Hon Degrees:* DSc, Mt Holyoke Col, 83. *Prof Exp:* Intern & resident pediat, New Haven Hosp, 42-44; resident, Johns Hopkins Hosp, 44-45, asst physician, Harriet Lane Cardiac Clin, 45-47; physician, Div Crippled Children, 47-59; dir, New Haven Rheumatic Fever & Cardiac Prog, State Dept Health, Conn, 47-60, sr pediatrician, New Haven Pediat Cardiac Res Prog, 59-66, pediat cardiologist & dir, 66-83; CLIN PROF PEDIAT, SCH MED, YALE UNIV, 66- *Concurrent Pos:* From asst clin prof to assoc clin prof, Sch Med, Yale Univ, 47-66; vchmn, Am Heart Dis Youth, Am Heart Asn, 56-60, chmn, Comt

Congenital Heart Dis, 56-60; chmn task force heart dis & youth, Conn Heart Asn, 75-77. *Mem:* Fel Am Acad Pediat; fel Am Col Cardiol; NY Acad Sci; Sigma Xi; Am Pediat Soc. *Res:* Rheumatic fever; etiology and prevention of congenital heart defects; diagnostic services and care of the pediatric cardiac patient; pregnancy in the congenital cardiac, growth and development of offspring; hyperlipemia; thirty year follow-up of blood pressure in childhood; incidence of cognitive heart disease in the offspring of affectal parents. *Mailing Add:* Dept Pediat Yale Univ Sch Med New Haven CT 06510

WHITTEMORE, WILLIAM LESLIE, b Skowhegan, Maine, Sept 25, 24; m 50. PHYSICS. *Educ:* Colby Col, AB, 45; Harvard Univ, MA, 47, PhD(physics), 49. *Prof Exp:* Assoc scientist, Brookhaven Nat Lab, 48-50, physicist, 50-56; physicist, Gen Atomic Div, Gen Dynamics Corp, 57-67, STAFF PHYSICIST, TRIGA REACTORS FACIL, GEN ATOMIC CO, 67-, SR SCI ADV, 78- *Concurrent Pos:* Vis lectr, Harvard Univ, 50-51; sci consult, Korean Atomic Energy Res Inst, 60-, Indonesian Atomic Agency, 65 & NSF, 70 & 75. *Mem:* Am Phys Soc; Am Nuclear Soc; Archaeol Inst Am; Sigma Xi. *Res:* Utilization of research reactors; neutron research; neutron radiography, isotopes for nuclear medicine. *Mailing Add:* Gen Atomic Co Box 81608 San Diego CA 92138

WHITTEN, BARBARA L, b Minneapolis, Minn, Sept 26, 46; m 82; c 2. ATOMIC & MOLECULAR PHYSICS, MATHEMATICAL PHYSICS. *Educ:* Carleton Col, BA, 68; Univ Rochester, MA, 71, PhD(physics), 76. *Prof Exp:* Instr, Western Col, Miami Univ, 74-76, asst prof interdisciplinary studies, 76-80; res assoc, Physics Dept, Rice Univ, 80-81; physicist, Lawrence Livermore Nat Lab, 81-87; ASSOC PROF PHYSICS, COLO COL, 87- *Mem:* Sigma Xi; Am Phys Soc; Am Asn Physics Teachers. *Res:* Physics of soft x-ray lasers; theoretical and computational studies of atomic and molecular processes in plasmas. *Mailing Add:* Physics Dept Colo Col Colorado Springs CO 80903

WHITTEN, BERTWELL KNEELAND, b Boston, Mass, Apr 1, 41; m 62; c 3. ENVIRONMENTAL PHYSIOLOGY, COMPARATIVE PHYSIOLOGY. *Educ:* Middlebury Col, AB, 62; Purdue Univ, MS, 64, PhD(environ physiol), 66. *Prof Exp:* Res physiologist, US Army Med Res & Nutrit Lab, Fitzsimons Gen Hosp, 66-68, res physiologist, Res Inst Environ Med, Army Natick Labs, 68-72; assoc prof, 72-74, dept head, 81-86, PROF BIOL SCI, MICH TECHNOL UNIV, 74-, DEAN RES & GRAD SCH, 86- *Concurrent Pos:* Dir, Biosource Res Inst, Mich Technol Univ. *Mem:* AAAS; Am Soc Zoologists; Am Physiol Soc. *Res:* Cardiovascular adaptations to hypoxia; exercise physiology. *Mailing Add:* VP Student Serv Mich Technol Univ Houghton MI 49931

WHITTEN, CHARLES A, JR, b Harrisburg, Pa, Jan 20, 40; m 65; c 1. NUCLEAR PHYSICS, INTERMEDIATE ENERGY PHYSICS. *Educ:* Yale Univ, BS, 61; Princeton Univ, MA, 63, PhD(physics), 66. *Prof Exp:* Res physicist, A W Wright Nuclear Struct Lab, Yale Univ, 65-68; asst prof nuclear physics, 68-74, assoc prof, 74-80, PROF PHYSICS, UNIV CALIF, LOS ANGELES, 80- *Concurrent Pos:* Vis scientist, Ctr Nuclear Studies, Saclay, France, 80-81 & 86-87. *Mem:* Am Phys Soc; Sigma Xi. *Res:* Direct reaction spectroscopy; isobaric analogue studies; nuclear structure studies with intermediate energy probes; nucleon-nucleon scattering at intermediate energies. *Mailing Add:* Dept Physics Univ Calif Los Angeles CA 90024

WHITTEN, DAVID G, b Washington, DC, Jan 25, 38; m 60; c 2. BIOPHYSICAL & PHYSICAL ORGANIC CHEMISTRY. *Educ:* Johns Hopkins Univ, BA, 59, MA, 61, PhD(org chem), 63. *Prof Exp:* Sr scientist, Jet Propulsion Lab, Calif Inst Technol, 63-65, NIH fel chem, Inst, 65-66; from asst prof to prof chem, Univ NC, Chapel Hill, 66-80, M A Smith prof, 80-83; C E KENNETH MEES PROF CHEM, UNIV ROCHESTER, 83-, CHMN DEPT CHEM, 88-, DIR, CTR PHOTOINDUCED CHARGE TRANSFER, 89- *Concurrent Pos:* Consult, Sci Data Systs, Inc, 66, Tenn Eastman Co, 66-79, Polaroid Corp, 81-83, Eastman Kodak, 84- & L D Caulh, 83-; Alfred P Sloan Found fel, 70-; Alexander von Humboldt fel, Max Planck Inst Biophys Chem, 72-73; Alexander von Humboldt sr scientist award, 74-75; fel, Japan Soc Prom Sci; Humboldt award, 83. *Mem:* Am Chem Soc; Royal Soc Chem; Am Soc Photobiol; Interam Photochem Soc (pres, 83-86). *Res:* Photobiology; photochemistry in microheterogeneous media; solid state and interfacial chemistry; electron transfer photochemistry; porphyrins and organometallic compounds. *Mailing Add:* Univ Rochester Rochester NY 14627

WHITTEN, ELMER HAMMOND, b Stoughton, Mass, Feb 18, 27; m 50; c 2. MEDICAL PHYSIOLOGY. *Educ:* Northeastern Univ, BS, 52; Mass State Col, Bridgewater, MEd, 67; Colo State Univ, PhD(physiol), 70. *Prof Exp:* Med serv rep drug sales, Pitman-Moore Co, Dow Chem Co, 54-56; admin asst sales, Metals & Controls, Inc, 56-58; head customer serv, Tex Instruments Inc, 58-66; instr human physiol, Colo State Univ, 70; from asst prof med physiol to assoc prof physiol & pharmacol, Univ Health Sci, 70-72, prof physiol, 79, assoc dean acad affairs, 72-88, chmn dept physiol, 71-88, PRES, UNIV HEALTH SCI, 88-; PRES, UNIVERSITY TOWERS INC, 88- *Concurrent Pos:* Pres, Lakeside Hosp Asn, 88-; bd gov, AACOM, 88- *Mem:* NY Acad Sci; Sigma Xi. *Res:* Neonatal enteritis; transport phenomena across the intestinal wall during stages in the progress of enteritis as it affects electrolytes and water. *Mailing Add:* Univ Health Sci Kansas City MO 64124

WHITTEN, ERIC HAROLD TIMOTHY, b Ilford, Eng, July 26, 27; m 53, 76; c 5. GEOLOGY. *Educ:* Univ London, BSc, 48, PhD(geol), 52, DSc(geol), 68. *Prof Exp:* Managerial chief clerk, Rex Thomas, Ltd, 43-45; lectr geol, Queen Mary Col, Univ London, 48-58; from assoc prof to prof geol, Northwestern Univ, Evanston, 58-81, dept chair, 77-81; VPRES ACAD AFFAIRS & PROF GEOL, MICH TECH UNIV, HOUGHTON, 81- *Concurrent Pos:* Vis assoc prof, Univ Calif, Berkeley, 57 & 60, Univ Calif, Santa Barbara, 59 & Univ Colo, 61 & 63. *Mem:* Fel AAAS; fel Geol Soc Am; fel Geol Soc London; Brit Geol Asn; Int Asn Math Geol (pres, 80-84). *Res:* Structural geology and petrology of granitic and deformed rocks; application of statistical analysis to quantitative geology problems. *Mailing Add:* Admin Bldg Mich Technol Univ Houghton MI 49931

WHITTEN, JERRY LYNN, b Bartow, Fla, Aug 13, 37; m 80; c 2. THEORETICAL CHEMISTRY. *Educ:* Ga Inst Technol, BS, 60, PhD(chem), 64. *Prof Exp:* Res assoc chem, Princeton Univ, 63-65, instr, 65; asst prof, Mich State Univ, 65-67; from asst prof to assoc prof, 67-73, chmn dept, 85-89, PROF CHEM, STATE UNIV NY, STONY BROOK, 73-; PROF CHEM & DEAN PHYS & MATH SCI, NC STATE UNIV, 89- *Concurrent Pos:* Res grants, Petrol Res Fund, 66-67, 74-76 & 77-81 & NSF, 67-72; Dept Energy res grants, 77-88; Alfred P Sloan fel, 69-71; vis prof, Univ Bonn & Wuppertal, 79 & Swiss Fed Inst Technol, Zurich, 84; Alfred P Sloan fel, 69-71. *Honors & Awards:* Alexander von Humboldt, Sr Scientist Award, 79. *Mem:* Am Phys Soc; Am Chem Soc; Sigma Xi; NY Acad Sci. *Res:* Theoretical studies of molecular structure and bonding; ab initio many-electron theory; theory of excited electronic states, metallic surfaces and chemisorption. *Mailing Add:* Phys & Math Sci NC State Univ Box 8201 Raleigh NC 27695

WHITTEN, KENNETH WAYNE, b Collinsville, Ala, Feb 4, 32. INORGANIC CHEMISTRY. *Educ:* Berry Col, AB, 53; Univ Miss, MS, 58; Univ Ill, PhD(inorg chem), 65. *Prof Exp:* Instr chem, Univ Miss, 55-56; asst prof, Berry Col, 56-58; instr, Univ Southwestern La, 58-59; asst prof, Miss State Col Women, 59-60 & Univ Ala, 63-66; asst prof chem & coord gen chem, 67-70, ASSOC PROF CHEM, UNIV GA, 70- *Mem:* Am Chem Soc. *Res:* Synthesis in fused salt media; chemical education; theories of testing. *Mailing Add:* Dept Chem Univ Ga Athens GA 30602

WHITTEN, MAURICE MASON, b Providence, RI, Oct 1, 23; m 83. HISTORY OF SCIENCE & TECHNOLOGY. *Educ:* Colby Col, AB, 45; Columbia Univ, MA, 49; Ohio State Univ, PhD, 71. *Prof Exp:* Sci teacher, Wilton Acad, 45-48 & Lewiston Maine High Sch, 48-55; instr phys sci, Gorham State Teachers Col, 55-59; TV sci teacher, State Dept Educ, Maine, 59-60; from asst prof to prof phys sci & chem, Univ Maine, Portland-Gorham, 61-77; prof, 78-83, EMER PROF CHEM, UNIV SOUTHERN MAINE, 83- *Concurrent Pos:* Lectr, Cent Maine Gen Hosp, Lewiston, 52-53. *Honors & Awards:* Elizabeth Thompson Award, Am Acad Arts & Sci, 54. *Mem:* AAAS; Am Chem Soc; Nat Sci Teachers Asn; Hist Sci Soc. *Res:* Science education; history of the gunpowder mills of Maine. *Mailing Add:* 11 Lincoln St Gorham ME 04038-1703

WHITTEN, ROBERT CRAIG, JR, b Bristol, Va, Dec 6, 26; m 53; c 2. AERONOMY. *Educ:* US Merchant Marine Acad, BS, 47; Univ Buffalo, BA, 55; Duke Univ, MA, 58, PhD(theoret physics), 59; San Jose State Univ, MS, 71. *Prof Exp:* Asst, Duke Univ, 55-57, instr, 57-58, asst, 58-59; from physicist to sr physicist, Stanford Res Inst, 59-67; res scientist, NASA-Ames Res Ctr, 67-89; RETIRED. *Concurrent Pos:* Lectr, Stanford Univ, 61-62 & 64-66, Univ Santa Clara, 64 & 69 & San Jose State Univ, 72 , 79 & 84; consult sci, eng & tech documentation. *Honors & Awards:* NASA Group Achievement Award, 80 & 81. *Mem:* Am Geophys Union; assoc fel Am Inst Aeronaut & Astronaut. *Res:* Structure, chemistry and dynamics of planetary atmospheres and ionospheres; chemistry and meteorology of the stratosphere; the quantum mechanical three body problem. *Mailing Add:* 1117 Yorkshire Dr Cupertino CA 95014

WHITTENBERGER, JAMES LAVERRE, b Dahina, Ill, Feb 12, 14; m 43; c 3. AIR POLLUTION HEALTH EFFECTS, OCCUPATIONAL HEALTH. *Prof Exp:* Internship, Cincinnati Gen Hosp, 38-39; res fel surg, Univ Chicago, 39-40; res fel med, Harvard Univ, 40-42, from assoc prof to prof physiol, 46-82; res fel physiol, NY Univ, 43; prof & dir environ med, Univ Calif, Irvine, Los Angeles, 82-88; RETIRED. *Concurrent Pos:* Dept head physiol, Sch Pub Health, Harvard Univ, 48-80, James Stevens Simmons prof pub health, 58-82, dir, Kresge Ctr Environ Health, 58-82, assoc dean, 66-78; sci adv bd, US Environ Protection Agency, 73-81; dir, Occup Health Ctr, Univ Calif, Irvine & Los Angeles, 82-88; chair, Dept Commun & Environ Med, Col Med, Univ Calif Irvine, 82-88; chair, Bd Sci Counselors, US Dept Hosp Health Serv, Nat Inst Occup Safety & Health, 84- *Res:* Air pollution health effects. *Mailing Add:* 1312 Dover Dr Newport Beach CA 92660

WHITTIER, ANGUS CHARLES, b Ottawa, Ont, Oct 17, 21; m 48; c 4. PHYSICS. *Educ:* Queen's Univ, Ont, BSc, 48; McGill Univ, MSc & PhD(physics), 52. *Prof Exp:* Asst res officer, Atomic Energy Can, Ltd, 52-55; supv physicist, Atomic Power Dept, Can Gen Elec, 55-67, mgr reactor anal, 67-70; supt shielding & comput br power projs, 70-75, mgr shielding & reactor physics, Eng Co, 77-84, mgr physics, Atomic Energy Can Ltd, Candu Opers, Sheridan Park, 85-86; RETIRED. *Mem:* Am Nuclear Soc; Can Nuclear Soc. *Res:* Nuclear physics, particularly reactor physics. *Mailing Add:* 2493 Vineland Rd Mississauga ON L5K 2A3 Can

WHITTIER, DEAN PAGE, b Worcester, Mass, July 2, 35; m 58; c 2. PLANT MORPHOLOGY. *Educ:* Univ Mass, BS, 57; Harvard Univ, AM, 59, PhD(biol), 61. *Prof Exp:* Asst prof bot, Va Polytech Inst, 61-64; NIH fel biol, Harvard Univ, 64-65; from asst prof to assoc prof, 65-77, chmn dept gen biol, 75-78, PROF BIOL, VANDERBILT UNIV, 77- *Mem:* Bot Soc Am; Am Fern Soc (treas, 74 & 75, vpres, 80 & 81, pres, 82 & 83); Int Soc Plant Morphologists; Int Asn Pteridologists. *Res:* Morphogenesis; apomixis in lower vascular plants. *Mailing Add:* Dept Gen Biol Vanderbilt Univ Nashville TN 37235

WHITTIER, HENRY O, b Schenectady, NY, Sept 1, 37; m 59; c 1. BOTANY, BRYOLOGY. *Educ:* Miami Univ, BS, 59, MA, 61; Columbia Univ, PhD(biol), 68. *Prof Exp:* Res asst bot, Miami Univ Schooner Col Rebel Exped to SPac, 60; instr, Univ Hawaii, 62-64; res asst bryol, NY Bot Garden, 64-68; from asst prof to prof biol sci, Fla Technol Univ, 68-79; PROF BIOL, UNIV CENT FLA, 79-, DIR ARBORETUM, 86- *Mem:* Am Bryol & Lichenological Soc; Am Inst Biol Sci; Sigma Xi. *Res:* Plant systematics, tropical botany, taxonomy, ethnobotany, ecology and biogeography, especially Pacific islands Bryophyta; conservation biology. *Mailing Add:* Dept Biol Univ Cent Fla Orlando FL 32816-0368

WHITTIER, JAMES S(PENCER), b Farmington, Minn, June 19, 35; m 61; c 1. SYSTEMS ENGINEERING. *Educ:* Univ Minn, BS, 57, MS, 58, PhD(mech, mat), 61. *Prof Exp:* Mem tech staff, Aerospace Corp, 61-65, sect mgr, 65-67, dept head, 67-87, spec assignment to Off of Secy Defense, 87-89, PRIN ENGR, SPACE LAUNCH OPERS, AEROSPACE CORP, 90- *Concurrent Pos:* Sr res fel, Appl Phys Dept, Cornell Univ, 74-75; chmn, Plasmadynamics & Lasers Tech Comt, Am Inst Aeronaut & Astronaut, 83-85. *Mem:* Optical Soc Am; Am Inst Aeronaut & Astronaut; Soc Exp Mech. *Res:* Space launch systems; chemical lasers; laser effects; remote sensing of motions; stress wave propagation. *Mailing Add:* Space Launch Opers M5-703 Aerospace Corp PO Box 92957 Los Angeles CA 90009

WHITTIER, JOHN RENSSELAER, neurology, psychiatry; deceased, see previous edition for last biography

WHITTIG, LYNN D, b Meridian, Idaho, Jan 16, 22; m 45; c 3. SOIL CHEMISTRY, SOIL MINERALOGY. *Educ:* Univ Wis, BS, 49, MS, 50, PhD(soil sci), 54. *Prof Exp:* Soil scientist, Soil Conserv Serv, USDA, 54-56; from asst prof to assoc prof, 57-70, res assoc, 63-64, vchmn Dept Land, Air & Water Resources, 75-79, PROF SOIL SCI, UNIV CALIF, DAVIS, 70-, ASSOC DIR, INST ECOL, 79- *Mem:* Fel Am Soc Agron; Soil Sci Soc Am; Clay Minerals Soc. *Res:* Clay mineralogy and mineral weathering processes; chemistry, morphology and genesis of salt-affected soils; x-ray diffraction and fluorescence methods of analysis. *Mailing Add:* 800 Colby Dr Davis CA 95616

WHITTINGHAM, M(ICHAEL) STANLEY, b Nottingham, Eng, Dec 22, 41; nat US; m 69; c 2. SOLID STATE CHEMISTRY. *Educ:* Oxford Univ, BA, 64, MA, 67, DPhil(chem), 68. *Prof Exp:* Res assoc mat sci, Stanford Univ, 68-72; mem sci staff, Exxon Res & Eng Co, 72-75, head chem physics group, 75-78, dir, Solid State & Catalytic Sci Lab, Corp Res Labs, 78-80, mgr chem eng, Technol Div, 80-84; dir phys sci, Schlumberger, 84-88; PROF CHEM & DIR, INST MAT RES, STATE UNIV NY, BINGHAMTON, 88- *Concurrent Pos:* Demonstr, Dept Inorg Chem, Oxford Univ, 65-67; prin ed, J Solid State Ionics & assoc ed, Chem Mat. *Mem:* Am Chem Soc; Electrochem Soc; Am Phys Soc; Mat Res Soc; Am Inst Chem Engrs. *Res:* Chemical properties of highly non-stoichiometric materials; fast ion transport in solids; electrochemical control of the properties of materials; solid state electrochemistry; high energy-density batteries; synthetic fuels technology. *Mailing Add:* Dept Chem State Univ NY Binghamton NY 13902-6000

WHITTINGHILL, MAURICE, b St Joseph, Mo, May 15, 09; m 32, 55; c 2. GENETICS. *Educ:* Dartmouth Col, AB, 31; Univ Mich, PhD(zool), 37. *Prof Exp:* Instr, Dartmouth Col, 31-33; asst, Univ Mich, 35; Nat Res Coun fel biol sci, Calif Inst Technol, 36-37; fel biol, Bennington Col, 37-42; assoc prof zool, Univ NC, 42-52, prof, 52-74, vis prof, 73, EMER PROF ZOOL, UNIV NC, CHAPEL HILL, 74- *Concurrent Pos:* Prof, Univ Mich, 46; sr biologist, Oak Ridge Nat Lab, 49; Wachtmeister vis prof biol, Va Mil Inst, 76; vis prof biol, Univ NC, Wilmington, 77; T E Powell Jr prof biol, Elon Col, 79-81. *Mem:* Genetics Soc Am; Am Soc Zool; Am Soc Nat; Am Soc Human Genetics; Am Genetic Asn (vpres, 72, pres, 73). *Res:* Genetics of Drosophila; irradiation and temperature effects; mutation and crossing over; spondylitis. *Mailing Add:* 1905 S Lake Shore Dr Chapel Hill NC 27514

WHITTINGTON, BERNARD W, July 19, 20; m; c 3. ENGINEERING. *Educ:* WVa Univ, BSEE, 51. *Prof Exp:* CONSULT, WHITTINGTON ENG, 82- *Concurrent Pos:* Chmn code making panel, Nat Elect Code; eng consult, Union Carbide Corp. *Honors & Awards:* Centennial Award, Inst Elec & Electronics Engrs, Outstanding Achievement Award, Kaufmann Award, Standard Medallion Award. *Mem:* Fel Inst Elec & Electronics Engrs; Nat Fire & Protection Asn. *Mailing Add:* Whittington Engr Inc 100 Whittingshire Lane Charleston WV 25312

WHITTINGTON, STUART GORDON, b Chesterfield, Eng, Apr 16, 42; Can & UK citizen; m 64; c 2. THEORETICAL CHEMISTRY. *Educ:* Cambridge Univ, BA, 63, PhD(chem), 72. *Prof Exp:* Scientist chem, Unilever Res Lab, UK, 63-66; res fel, Univ Calif, San Diego, 66-67; res fel, Univ Toronto, 67-68; scientist, Unilever Res Lab, UK, 68-70; asst prof, 70-75, assoc prof, 75-80, chmn, 85-88, PROF CHEM, UNIV TORONTO, 80- *Concurrent Pos:* Vis prof dept math, Univ Newcastle, Australia, 77; sr res fel, Sci Res Coun, Univ Bristol, 77-78; vis fel, Trinity Col, Oxford, 83-84, vis res fel, Dept Math, Univ Melbourne, Australia, 90. *Res:* Statistical mechanics; Monte Carlo methods; excluded volume effect in polymers; polymer adsorption and colloid stability; percolation theory; phase transitions and critical phenomena. *Mailing Add:* Dept Chem Univ Toronto Toronto ON M5S 1A1 Can

WHITTINGTON, WESLEY HERBERT, b Basil, Ohio, Dec 1, 33; m 53; c 2. ESTER PLASTICIZERS FOR POLYMERS. *Educ:* Capital Univ, BS, 61. *Prof Exp:* Asst to chief chem, Precision Rubber Prod Corp, 61-68; develop scientist, B F Goodrich Chem, 68-78; TECH SR MGR, C P HALL CO, 78- *Mem:* Rubber Div Am Chem Soc. *Res:* Test, study and communicate the use and application of ester plasticizers in rubber and plastics compounds. *Mailing Add:* C P Hall Co 5851 W 73rd St Chicago IL 60638

WHITTLE, CHARLES EDWARD, JR, b Brownsville, Ky, Mar 8, 31; m 52; c 10. PHYSICS, APPLIED MATHEMATICS. *Educ:* Centre Col, AB, 49; Washington Univ, PhD(nuclear physics), 53. *Prof Exp:* Fulbright & Res Corp grants, State Univ Leiden, 53-54; res scientist, Union Carbide Corp, 54-56; asst & assoc prof physics, Western Ky Univ, 56-60, prof & chmn dept, 60-62; coordr res, Centre Col Ky, 62-64, from assoc dean to dean, 64-72, prof physics, 62-74, Matton chair appl math, 72-74; ASST DIR, INST ENERGY ANALYSIS, OAK RIDGE ASSOC UNIVS, 74- *Mem:* Am Asn Physics Teachers; Sigma Xi; Phys Soc. *Res:* Nuclear and optical spectroscopy; applied mathematics and geophysics; energy policy analysis and modeling; energy data analysis and validation; geothermal energy assessment; geophysics. *Mailing Add:* 109 Trevose Lane Oak Ridge TN 37830

WHITTLE, FRANK, b Coventry, Eng, June 1, 07; m 30, 76; c 2. AERONAUTICAL ENGINEERING. *Hon Degrees:* Numerous from foreign insts & univs. *Prof Exp:* Tech adv, Jet Aircraft, Brit Overseas Airway Corp, 48-52, Shell Gp, 53-57; consult, Rolls Royce on Turbo Drill Proj, Bristol Siddeley Engines, 61-70; MEM FAC, US NAVAL ACAD, ANNAPOLIS, MD, 77- *Concurrent Pos:* Adj res prof, US Naval Acad. *Honors & Awards:* James Alfred Ewing Medal, ICE, 44; Gold Medal, RAeS, 44; James Clayton Prize, Inst Mech Engrs, 46, James Watt Int Gold Medal, 77; Daniel Guggenheim Medal, US, 46; Kevin Gold Medal, 47 & Melchett Medal, 49; Knighted by King George VI, 48; Gold Medal, Fedn Aeronaut Int, 51; Churchill Gold Medal, Soc Engrs & Albert Gold Medal, Soc Arts, 52; Franklin Medal, US, 56; John Scott Award, 57; Goddard Award, US, 65; Coventry Award of Merit, 66; Int Commun-Christopher Columbus Prize, City of Genoa, 66; Order of Merit, Queen Elizabeth II, 86; Charles Stark Draper Prize, Nat Acad Eng, 91. *Mem:* Foreign assoc Nat Acad Eng; fel UK Fel Eng; foreign mem Am Acad Arts & Sci; hon fel Soc Exp Test Pilots. *Res:* Author of two publications. *Mailing Add:* 10001 Windstream Dr No 706 Columbia MD 21044

WHITTLE, GEORGE PATTERSON, b Eufaula, Ala, July 1, 25; m 63; c 1. WATER TREATMENT, HAZARDOUS WASTE MANAGEMENT. *Educ:* Ga Inst Technol, BChE, 46, BIE, 47; Univ Fla, MS, 64, PhD(chem), 66. *Prof Exp:* Chem engr, Hercules Powder Co, 47-49; self-employed, Whittle Lumber Co, 50-53; chemist, Swift & Co, 53-55; chief chemist, Allied Chem Co, 55-57; res engr, Tenn Corp, 57-63; PROF CIVIL ENG, UNIV ALA, 67- *Honors & Awards:* Bedell Award, Water Pollution Control Fedn, 77. *Mem:* Am Water Works Asn; Am Soc Civil Eng; Sigma Xi. *Res:* Water quality modeling, hazardous waste management; water treatment and chemistry, pollution control; analytical chemistry of water and wastewater; reaction kinetics of halogen residuals in water. *Mailing Add:* PO Box 1468 Tuscaloosa AL 35487

WHITTLE, JOHN ANTONY, b Settle, Yorks, Eng, Mar 13, 42. ORGANIC CHEMISTRY, BIOCHEMISTRY. *Educ:* Univ Glasgow, BSc, 64; Imp Col, dipl, & Univ London, PhD(org chem), 67. *Prof Exp:* Fel, Rutgers Univ, NJ, 67-69; from asst prof to assoc prof, 69-83, PROF CHEM, LAMAR UNIV, 83- *Mem:* Am Chem Soc. *Res:* Biosynthesis of sesquiterpenoids and other natural products; synthesis of sesquiterpenoid ring systems. *Mailing Add:* Dept Chem Lamar Univ PO Box 10022 Lamar Univ Sta Beaumont TX 77710-0022

WHITTLE, PHILIP RODGER, b Russell Springs, Ky, July 11, 43; m 67; c 2. ORGANIC CHEMISTRY, FORENSIC CHEMISTRY. *Educ:* Univ Ky, BS, 65; Iowa State Univ, PhD(org chem), 69. *Prof Exp:* NIH fel, Univ Colo, Boulder, 69-70; assoc prof, 70-80, PROF CHEM, MO SOUTHERN STATE COL, 80-, DIR, REGIONAL CRIMINALISTICS LAB, 72- *Concurrent Pos:* Consult, Anal Labs. *Mem:* Am Chem Soc (secy-treas, 66-); Am Soc Crime Lab Dirs; Am Acad Forensic Scientists; Sigma Xi. *Res:* Electrocyclic cyclopropane ring openings; toxicology; modern drug analysis; trace evidence; forensic applications. *Mailing Add:* Dept Chem Mo Southern State Col Joplin MO 64801

WHITTLESEY, EMMET FINLAY, b Winchester, Mass, Oct 9, 23; m 66; c 3. MATHEMATICS. *Educ:* Princeton Univ, AB, 48, MA, 55, PhD(math), 56. *Prof Exp:* Instr math, Pa State Univ, 50-51 & Bates Col, 51-54; from instr to assoc prof, 54-65, PROF MATH, TRINITY COL, CONN, 65- *Concurrent Pos:* NSF fel, 62-63. *Mem:* Am Math Soc; Math Asn Am; Sigma Xi. *Res:* Combinatorial topology; functional analysis; integration. *Mailing Add:* 89 Walbridge Rd West Hartford CT 06119

WHITTLESEY, JOHN R B, b Los Angeles, Calif, July 21, 27; m 66; c 2. MARINE SEISMIC EXPLORATION. *Educ:* Calif Inst Technol, BS, 48, MS, 50. *Prof Exp:* Instr physics & astron, Univ Nev, 50; Ford Found behav sci grant & res asst, Univ NC, 52-54; res mathematician res clin neuropsychiat inst, Med Ctr, Univ Calif, Los Angeles, 57-62, data processing analyst brain res inst, 62-64; sr mem res staff seismic explor data processing, Ampex Corp, Ray Geophys Co & Petty-Ray Geophys Co, Houston, Tex, 64-80; sr res scientist, Marine Seismic Data Acquisition & Signal Processing, Geosource Inc, 80-86; RETIRED. *Concurrent Pos:* Statist consult numerous behav scientists, Calif, 58-71; NIMH spec res fel brain res inst, Univ Calif, Los Angeles, 63-64. *Honors & Awards:* Award, Soc Explor Geophys, 65. *Mem:* Fel AAAS. *Res:* Mathematics and digital computers applied to psychiatry, brain research, exploration geophysics; time-series analysis; seismic signal processing and air gun signature analysis; laser fusion; digital signal processing; differential geometry and writing chapters for computer-use texts. *Mailing Add:* 5439 Del Monte Dr Houston TX 77056

WHITTON, LESLIE, b New Bedford, Mass, Sept 1, 23; m 47; c 4. PLANT CYTOLOGY, PLANT GENETICS. *Educ:* Utah State Univ, BS, 49; Univ Calif, MS, 53; Cornell Univ, PhD, 64. *Prof Exp:* Asst prof hort, Univ Maine, Orono, 56-62; asst prof, 62-64, asst prof bot, 64-68, ASSOC PROF BOT, BRIGHAM YOUNG UNIV, 68- *Mem:* Bot Soc Am; Sigma Xi. *Res:* Cytology, genetics and breeding of small fruit species and native shrub species of the Rocky Mountain region. *Mailing Add:* 1812 N 1450 E Provo UT 84604

WHITTOW, GEORGE CAUSEY, b Milford Haven, UK, Feb 28, 30; m 55; c 1. PHYSIOLOGY. *Educ:* Univ London, BSc, 52; Univ Malaya, PhD(physiol), 57. *Prof Exp:* Asst lectr physiol, Univ Malaya, 52-54, lectr, 54-59; sr sci officer, Hannah Res Inst, Ayr, Scotland, 59-65; assoc prof physiol, Rutgers Univ, New Brunswick, 65-68; PROF PHYSIOL, SCH MED, UNIV HAWAII, 68-, CHMN DEPT, 85- *Mem:* Am Physiol Soc; fel Linnean Soc; Brit Inst Biol. *Res:* Physiology of thermoregulation; thermal ecology. *Mailing Add:* Sch Med Univ Hawaii Honolulu HI 96822

WHITTUM-HUDSON, JUDITH ANNE, INFECTIOUS DISEASES, MUCOSAL IMMUNITY. *Educ:* Univ Conn PhD(immunol), 80. *Prof Exp:* ASST PROF PHARMACOL, SCH MED, JOHNS HOPKINS UNIV, 82- *Mailing Add:* 457 Wilmer Woods Sch Med Johns Hopkins Univ 600 N Wolfe St Baltimore MD 21205

WHITTY, ELMO BENJAMIN, b Lee, Fla, Mar 6, 37. AGRONOMY. *Educ:* Univ Fla, BSA, 59, MSA, 61; NC State Univ, PhD(soil sci), 65. *Prof Exp:* From asst prof to assoc prof, 66-77, PROF AGRON, UNIV FLA, 77- *Mem:* Am Soc Agron; Plant Growth Regulator Soc Am; Am Peanut Res & Educ Soc. *Res:* Tobacco production; cultural practices; plant growth regulators; plant nutrition. *Mailing Add:* Dept Agron 301 Newell Hall Univ Fla Gainesville FL 32601

WHITWELL, JOHN C(OLMAN), math statistics, for more information see previous edition

WHITWORTH, CLYDE W, b Paulding Co, Ga, Oct 9, 26; m 51; c 3. PHARMACY. *Educ:* Univ Ga, BS, 50, MS, 56; Univ Fla, PhD(pharm), 63. *Prof Exp:* From instr to asst prof pharm, Univ Ga, 54-60; asst prof, Northeast La State Col, 63-66; assoc prof, 66-80, PROF PHARM, UNIV GA, 80- *Concurrent Pos:* Mead Johnson res award, 65-66; William A Webster Co grant prod stability, 72- *Mem:* Am Pharmaceut Asn. *Res:* Factors influencing drug absorption and drug release from external preparations. *Mailing Add:* Dept Pharm Univ Ga Athens GA 30602

WHITWORTH, WALTER RICHARD, b La Crosse, Wis, Feb 22, 34; m 57; c 4. AQUATIC BIOLOGY. *Educ:* Wis State Col, Stevens Point, BS, 58; Okla State Univ, MS, 61, PhD(zool), 63. *Prof Exp:* Fish biologist, Southeastern Fish Control Lab, 63-64; from asst prof to assoc prof, 64-76, PROF FISHERIES, UNIV CONN, 76- *Concurrent Pos:* Ore Fish Comn, 73; Bulgarian Acad Sci, 81-82. *Mem:* Am Fisheries Soc; Am Soc Ichthyol & Herpet. *Res:* Effects of the environment on fish; primary productivity; fish taxonomy and toxicology. *Mailing Add:* U-87 Univ Conn 1376 Storrs Rd Rm 308 Storrs CT 06269-4087

WHORTON, CHESTER D, b Grundy Co, Mo, Nov, 3, 03. RESEARCH & PRODUCTION. *Educ:* Univ Mo, BS, 25, MA, 26. *Prof Exp:* Producer & consult, Chester D Whorton, 46-80; RETIRED. *Mem:* Sigma Xi; Geol Soc Am; Am Asn Prof Geologist. *Mailing Add:* 1454 W Camino Lucientes Green Valley AZ 85614

WHORTON, ELBERT BENJAMIN, b Stamford, Tex, Nov 10, 38; m 62; c 2. RESEARCH DESIGN & ANALYSIS. *Educ:* Baylor Univ, BS(math) & BS(physics), 62; Tulane Univ, MS, 64; Okla Univ, PhD(biostatist & med comput), 68. *Prof Exp:* Surv statistician dir, La State Health Dept, 64-65; from asst prof to assoc prof & dir biomet, Univ Vt, 68-72, asst prof math, 69-72; assoc prof & dir biostatist, Univ Tex Med Br Galveston, 72-82; RES COLLABR, BROOKHAVEN NAT LAB, 82- *Concurrent Pos:* Statist consult, Bur Manpower Intel, NIH, 68-73, Nat Cancer Inst, 73-, Dow Chem Co, 73-80 & Ethyl Corp, 81-82; dir, Educ & Res Comput Ctr, Univ Tex Med Br Galveston, 75-81, assoc dean, Grad Sch Biomed Sci, 76-81; vis prof, Med Sch, Univ Vt, 81. *Mem:* Am Statist Asn; Environ Mutagen Soc; Biomet Soc; Sigma Xi. *Res:* Development and evaluation of experimental and non-experimental designs in environmental toxicology mutagenesis research; development of improved methods for statistical evaluation of research results, particularly in genetic toxicology. *Mailing Add:* Univ Tex Med Br Galveston TX 77550

WHORTON, M DONALD, b Las Vegas, NMex, Jan 25, 43; m 72; c 3. OCCUPATIONAL HEALTH & EPIDEMIOLOGY. *Educ:* Univ NMex, MD, 68; Johns Hopkins Univ, MPH, 73; Am Bd Internal Med, cert, 75; Am Bd Prev Med Occup Med, cert, 75. *Prof Exp:* Instr anat, Sch Med, Univ NMex, 70-71; staff physician, Family Planning Clin, Washington, DC, 71-72; instr, Sch Health Serv, Johns Hopkins Univ, 73-75, instr med, Sch Med, 74-75; assoc dir, Div Emergency Med, Baltimore City Hosps, 74-75; clin asst prof med, Div Ambulatory & Community Med, Sch Med, Univ Calif, San Francisco, 75-77, lectr, Sch Pub Health, Berkeley, 75-79, dir, Labor Occup Health Prog, Inst Indust Relations, Ctr Labor Res & Educ, 75-77, med dir, 77-79, assoc clin prof occup med, Sch Pub Health, 79-87; prin & sr occup physician/epidemiologist, 78-88, VPRES & CHIEF MED SCIENTIST, ENSR HEALTH SCI, 88- *Concurrent Pos:* Resident, Dept Med, Baltimore City Hosp, 71-74; consult, Occup Health Prog, Inst Indust Relations, Univ Calif, Berkeley; mem, Comt Occup Health & Safety, Am Pub Health Asn, 73-74; mem, Study Sect Occup Safety & Health, Ctr Dis Control/Nat Inst Occup Safety & Health, 80-83; mem, Occup Med Pract Comt, Am Occup Med Asn, 82-; corresp, Comt on Human Rights, Nat Acad Sci, 83-; occup med staff, Dept Med, Franklin Hosp, Ralph K Davies Med Ctr, 84-; mem, Comt Nat Statist, Nat Res Coun, 85-, Panel on Occup Safety & Health Statist, 85-87; mem, Sci Oversight Panel on Ctr Dis Control Select Cancer Study of Vietnam Vets, Inst Med, 88-90. *Mem:* Inst Med-Nat Acad Sci; Am Pub Health Asn; Am Occup Med Asn; Am Col Occup Med; fel Am Col Epidemiol; Soc Occup & Environ Health; AAAS; Int Asn Occup Health. *Res:* Environmental and occupational health issues; occupational reproductive issues. *Mailing Add:* ENSR Health Sci 1320 Harbor Bay Pkwy Suite 100 Alameda CA 94501

WHORTON, RAYBURN HARLEN, b London, Ark, Apr 28, 31; m 56; c 3. PAPER PROCESS CHEMISTRY. *Educ:* Ark Polytech Col, BS, 53; Univ Ark, MS, 56. *Prof Exp:* Instr chem, Ark Polytech Col, 55-56; res chemist, Crossett Co, 57-62; proj supvr, Ga-Pac Corp, 62-63; sr proj chemist, Int Paper Co, 64-69, sect leader paper develop, 69-77, group mgr paper structure & chem, Erling Riis Res, 77-87; RETIRED. *Mem:* Tech Asn Pulp & Paper Indust. *Res:* Papermaking; surface and internal sizing; printability coatings; paper-plastic combinations; converting processes. *Mailing Add:* 6167 Challen Circle Mobile AL 36608

WHYBROW, PETER CHARLES, b Hertfordshire, Eng, June 13, 39; nat US; div; c 2. PSYCHIATRY. *Educ:* Univ London, MB & BS, 62; Royal Col Physicians, dipl, 62; Conjoint Bd Physicians & Surgeons Eng, dipl psychol med, 68. *Hon Degrees:* MA, Dartmouth Col, 74, MA, Univ Pa, 84. *Prof Exp:* House physician, Med Res Coun-Univ Col Hosp, London, 62; house surgeon, St Helier Hosp, Surrey, Eng, 63; sr house physician, Univ Col Hosp, London, 63-64; house physician, Prince of Wales Hosp, London, 64; resident psychiat, Univ NC, 65-67, instr, 67-68; sci officer, Med Res Coun, Eng, 68-69; from asst prof to assoc prof, 69-71, chmn dept, 71-78, prof psychiat, 71-79, exec dean, Dartmouth Med Sch, 80-83; PROF PSYCHIAT, CHMN DEPT PSYCHIAT, UNIV PA, 84- *Concurrent Pos:* NIMH res fel, Univ NC, 67-68; lectr, Univ Col Hosp Med Sch, London, 68-69; dir res training, Dartmouth Hitchcock Affil Hosps, 69-71, dir psychiat, 70-; consult, Vet Admin Hosp, 70-; Josiah Macy Jr fac scholar, 78-79; vis scientist div psychobiol, NIMH; chmn psychiat test comt, Nat Bd Med Examrs, 78-; Joseph Macy Jr Fac Scholar, 78-79. *Honors & Awards:* Sr Investr Award, Nat Alliance Res on Schizophrenia & the Depressions, 89. *Mem:* Brit Med Asn; fel Royal Col Psychiat; Am Asn Chmn Depts Psychiat (pres, 77-78); Brit Soc Psychosom Res; fel Am Psychiat Asn; fel Am Col Psychiatrists. *Res:* Psychobiology of affective disorders, particularly pharmacologic and endocrinologic aspects of bipolon disorders. *Mailing Add:* Dept Psychiat Univ Pa 305 Blockley Hall Philadelphia PA 19104

WHYTE, DONALD EDWARD, b Regina, Sask, Jan 22, 18; nat US; c 4. ORGANIC CHEMISTRY. *Educ:* Univ Sask, BA, 39, MA, 41; Columbia Univ, PhD(chem), 43. *Prof Exp:* Chemist naval stores, Hercules Powder Co, 43-46; head org res sect, 46-52, res serv mgr, 52-57, tech serv mgr, 57-59, dir serv prod develop, 59-61, appl res dir, 61-64, res mgr, 65-71, prod res mgr int opers, dir res & develop int opers, 74-77, vpres Int Res & Develop, S C Johnson & Son, Inc, 77-84; PRES, CONSUMER CHEM INC, 85- *Concurrent Pos:* Fel, Columbia Univ. *Mem:* Am Chem Soc; Am Oil Chem Soc. Am Soc Qual Control. *Res:* Analytical and market research; new product evaluation; new product development of polishes, coating and porelon; polymers; insecticides; insect repellants; synergists and microbiology; development of new consumer and industrial products for overseas introduction; developing overseas research and development laboratories. *Mailing Add:* 2720 Mich Blvd Racine WI 53402-4252

WHYTE, MICHAEL PETER, b New York, NY, Dec 19, 46; m 74. INTERNAL MEDICINE. *Educ:* Washington Square Col, NY Univ, BA, 68; State Univ NY, MD, 72. *Prof Exp:* From intern to resident internal med, Dept Med, Bellevue Hosp, New York, 72-74; clin assoc metab neurol, Nat Inst Neurol & Commun Dis & Stroke, NIH, 74-76; fel endocrinol, 76-79, asst prof med, 80-91, asst prof pediat, 82-91, PROF MED & PEDIAT, SCH MED, WASH UNIV, 91- *Res:* Calcium metabolism and metabolic bone disease; alkaline phosphatase; bone dysplasias. *Mailing Add:* Jewish Hosp St Louis 216 S Kingshighway St Louis MO 63110

WHYTE, THADDEUS E, JR, b Washington, DC, Dec 8, 37; m 59; c 2. GENERAL CATALYSIS. *Educ:* Georgetown Univ, BS, 60; Howard Univ, MS, 62, PhD(phys chem), 65. *Prof Exp:* Phys chemist, Nat Bur Standards, 62-63; res chemist, Howard Univ, 63-64; nuclear chemist, USAF, 64-67; sr res chemist & group leader, Mobil Res & Develop Corp, 67-76; dir indust chem res & develop, Air Prod & Chem, 76-79; managing dir, Catalytic Assocs, 79-85; tech mgr, 85-87, TECH DIR RES & DEVELOP, PQ CORP, 87- *Concurrent Pos:* Lectr, grad sch, Sacramento State Col, 65-67 & Gloucester County Col, 69-70. *Mem:* Am Chem Soc; Am Inst Chemists; Am Inst Chem Eng. *Res:* Industrial chemistry; zeolite synthesis and catalysis; petrochemicals; petroleum research. *Mailing Add:* 280 Cedar Grove Rd Conshohocken PA 19423-2240

WIANT, HARRY VERNON, JR, b Burnsville, WVa, Nov 4, 32; m 54; c 2. FORESTRY. *Educ:* Univ WVa, BSF, 54; Univ Ga, MF, 59; Yale Univ, PhD(forestry), 63. *Prof Exp:* Jr forester, US Forest Serv, 57; asst prof forestry, Humboldt State Col, 61-65; prof & asst to dean, Stephen F Austin State Col, 65-72; PROF FORESTRY, WVA UNIV, 72- *Mem:* Soc Am Foresters. *Res:* Concentration of carbon dioxide near forest floor; ecology and silviculture of redwood; chemical and mechanical control of undesirable hardwoods; prediction of site quality; volume determinations and forest inventory; silviculture of southern forest trees; dendrological techniques. *Mailing Add:* Div Forestry WVa Univ Morgantown WV 26506

WIATROWSKI, CLAUDE ALLAN, b Chicago, Ill, Dec 27, 46. INSTRUMENTATION AUTOMATION. *Educ:* Ill Inst Technol, BS, 68; Univ Ariz, MS, 70, PhD(elec eng), 73. *Prof Exp:* Consult geophys, 71-73; design engr microcomput, Burr Brown Res Corp, 73-75; asst prof elec eng, Univ Colo, 75-81; res & develop, Scott Sci, 82, chief scientist, 83-84; PRES, MT AUTOMATION CORP, 76-; VPRES, PARKCON INC, 86- *Concurrent Pos:* Res asst, Dept Physics, Ill Inst Technol, 67-; res asst, Dept Elec Eng, Univ Ariz, 68-73. *Mem:* Inst Elec & Electronics Engrs; Instrument Soc Am. *Res:* Personal computer applications in instrumentation, automation communication and control. *Mailing Add:* PO Box 6020 Woodland Park CO 80866-6020

WIBERG, DONALD M, b Battle Creek, Mich, Sept, 20, 36; div; c 5. CONTROL & BIOMEDICAL ENGINEERING. *Educ:* Calif Inst Technol, BS, 59, MS, 60, PhD(eng), 65. *Prof Exp:* Sr design engr, Gen Dynamics/Convair, 64-65; asst prof, 65-72, assoc prof, 72-79, PROF ENG, UNIV CALIF, LOS ANGELES, 79-, PROF ANESTHESIA, 80- *Concurrent Pos:* Consult, Douglas Aircraft Co, 66-69 & R & D Assocs, 72-74; Fulbright scholar, 77; consult, Aerospace Corp, 81-90, Rockwell Sci, 89- *Mem:* Inst Elec & Electronics Engrs; Sigma Xi. *Res:* System modelling and parameter identification; biomedical systems, especially respiratory and cardiovascular modelling; optimum control of distributed parameter systems; nuclear reactor kinetics and control; aerospace control. *Mailing Add:* Elec Eng Dept Univ Calif Los Angeles CA 90024-1594

WIBERG, GEORGE STUART, toxicology; deceased, see previous edition for last biography

WIBERG, JOHN SAMUEL, b Plaistow, NH, Dec 4, 30; m 52; c 3. BACTERIOPHAGE, TRANSLATIONAL REGULATION. *Educ:* Trinity Col, Conn, BS, 52; Univ Rochester, PhD(pharmacol), 56. *Prof Exp:* Clin lab officer, Wright-Patterson AFB, Ohio, 57-58; res assoc biochem, Mass Inst Technol, 59-63; asst prof, 63-70, ASSOC PROF BIOPHYS, SCH MED &

DENT, UNIV ROCHESTER, 70- *Concurrent Pos:* NIH res fel, Mass Inst Technol, 59-60. *Mem:* AAAS; Am Soc Biochem & Molecular Biol; Am Soc Microbiol. *Res:* Metal ion interactions with nucleic acids and enzymes; biochemical genetics of bacterial viruses; nucleic acid function and metabolism; regulation of protein synthesis. *Mailing Add:* Dept Biophys Univ Rochester Sch Med & Dent Rochester NY 14642

WIBERG, KENNETH BERLE, b Brooklyn, NY, Sept 22, 27; m 51; c 3. ORGANIC CHEMISTRY. *Educ:* Mass Inst Technol, BS, 48; Columbia Univ, PhD(chem), 50. *Prof Exp:* Instr chem, Univ Wash, Seattle, 50-52, from asst prof to assoc prof, 52-57; vis prof, Harvard Univ, 57-58; prof, Univ Wash, Seattle, 58-60; prof, Yale Univ, 60-68, chmn dept, 68-71, Whitehead prof, 68-90, EUGENE HIGGINS PROF CHEM, YALE UNIV, 90- *Concurrent Pos:* Sloan fel, 58-62; Boomer Mem lectr, Univ Alta, 59; Guggenheim fel, 61-62. *Honors & Awards:* Award, Am Chem Soc, 62, J F Norris Award Phys Org Chem, 73; Cope Scholar Award, Am Chem Soc, 88. *Mem:* Nat Acad Sci; AAAS; Am Chem Soc; Royal Soc Chem; Am Acad Arts & Sci. *Res:* Stereochemistry and kinetics of organic reactions, particularly oxidation reactions and molecular rearrangements; synthesis and reactions of highly strained compounds. *Mailing Add:* Dept Chem Yale Univ PO Box 6666 New Haven CT 06511

WIBERLEY, STEPHEN EDWARD, b Troy, NY, May 31, 19; m 42; c 2. ANALYTICAL CHEMISTRY. *Educ:* Williams Col, AB, 41; Rensselaer Polytech Inst, MS, 48, PhD(chem), 50. *Prof Exp:* Sr chemist, Congoleum Nairn, Inc, 41-44; anal chemist, Gen Elec Corp, 46-48; instr chem, 46-48, res assoc, US AEC contract, 48-50, from asst prof to assoc prof anal chem, 50-57, assoc dean, Grad Sch, 64-65, dean, Grad Sch, 65-79, vprovost grad prog & res, 69-79, chair, Dept Chem, 83-88, PROF ANALYTICAL CHEM, RENSSELAER POLYTECH INST, 57- *Concurrent Pos:* Vis physicist, Brookhaven Nat Labs, 50; consult, Imp Color Chem & Paper Corp, Socony-Mobil Oil Co, Inc, Huyck Felt Co, Schenectady Chem, Inc & Nat Gypsum Co. *Mem:* AAAS; Am Chem Soc; Sigma Xi; Am Asn Univ Profs. *Res:* Instrumental analysis; analysis of radioactive elements; oil shale hazardous wastes. *Mailing Add:* 1676 Tibbitts Ave Troy NY 12180

WICANDER, EDWIN REED, b San Francisco, Calif, July 15, 46; m 75. PALEONTOLOGY, GEOLOGY. *Educ:* San Diego State Univ, BS, 69; Univ Calif, Los Angeles, PhD(geol), 73. *Prof Exp:* ASST PROF GEOL, CENT MICH UNIV, 76- *Mem:* Soc Econ Paleontologists & Mineralogists; Paleont Soc; Palaeont Asn; Sigma Xi; Am Asn Stratig Palynologists. *Res:* Micropaleontology. *Mailing Add:* Dept Geol Cent Mich Univ Mt Pleasant MI 48859

WICHER, KONRAD J, b Siemianowice, Poland, Feb 20, 24; m; c 2. IMMUNOLOGY. *Educ:* Med Sch Rokitnica, MD, 57, PhD(immunol), 62; Am Bd Med Microbiol, dipl, 74. *Prof Exp:* Prof, 74-86, EMER PROF MICROBIOL, MED SCH, STATE UNIV NY, BUFFALO, 86- *Concurrent Pos:* Dir clin microbiol & immunol, Erie County Lab, Buffalo, 67-79; dir clin microbiol, Wadsworth Res Ctr, NY State Dept Health, 79-83. *Mem:* Am Acad Allergy; Am Soc Microbiol; Am Asn Immunologists; Am Soc Clin Pathologists. *Res:* Immunopathology of syphilis. *Mailing Add:* Wadsworth Res Ctr NY State Dept Health Albany NY 12201

WICHER, VICTORIA, b Cordoba, Arg, Mar 19, 33. CLINICAL IMMUNOLOGY. *Educ:* Nat Univ Bordova, Arg, PhD(biochem), 60. *Prof Exp:* From instr to asst prof, 60-63, dir clin biochem, Dept Pediat, Sch Med Cordoba, Arg, 63-67; res assoc, Sch Med, 67-72, asst prof, Dept Microbiol Sch Med, prof staff, Ctr Immunol, State Univ NY, Buffalo, 76-79; RES SCIENTIST, NY STATE DEPT HEALTH, 79- *Mem:* Am Asn Immunologist; Am Soc Microbiol; NY Acad Sci. *Res:* Immunopathology of sexually transmitted diseases. *Mailing Add:* Div Lab & Res Rm 100 W NY State Dept Health 120 New Scotland Ave Albany NY 12201

WICHERN, DEAN WILLIAM, b Medford, Wis, Apr 29, 42; m 68; c 2. STATISTICS. *Educ:* Univ Wis-Madison, BS, 64, MS, 65, PhD(statist), 69. *Prof Exp:* From asst to assoc prof bus, Univ Wis-Madison, 69-76, chmn, Dept Quant Anal, 75-78, prof bus, 76-84; prof bus, 84-85, JOHN E PEARSON PROF BUS, TEX A&M UNIV, 85-, ASSOC DEAN, COL BUS ADMIN, 88- *Concurrent Pos:* Vis mem, US Army Math Res Ctr, Madison, Wis, 78-79. *Mem:* Am Statist Asn; Inst Mgt Sci; Royal Statist Soc. *Res:* Time series analysis; experimental design; applications of statistical methods in business. *Mailing Add:* Tex A&M Univ Col Bus Admin College Station TX 77843-4217

WICHMANN, EYVIND HUGO, b Stockholm, Sweden, May 30, 28; nat US; m 51; c 2. THEORETICAL PHYSICS. *Educ:* Inst Tech, Finland, AB, 50; Columbia Univ, AM, 53, PhD, 56. *Prof Exp:* Mem staff physics, Inst Advan Study, 55-57; from asst prof to assoc prof, 57-67, PROF PHYSICS, UNIV CALIF, BERKELEY, 67- *Mem:* Am Phys Soc. *Res:* Quantum field theory and quantum electrodynamics. *Mailing Add:* Dept Physics Univ Calif Berkeley CA 94704

WICHNER, ROBERT PAUL, b Pecs, Hungary, Apr 29, 33; US citizen; m 54; c 2. ENGINEERING SCIENCE. *Educ:* City Col New York, BS, 54; Univ Cincinnati, MS, 55; Univ Tenn, PhD(eng sci), 64. *Prof Exp:* DEVELOP RES ENGR, OAK RIDGE NAT LAB, 55- *Res:* Experimental and theoretical turbulence research; two-phase flow fluid dynamics and heat transfer; thermal-hydraulic analysis of nuclear reactors; analysis and modeling of thermal discharges. *Mailing Add:* 104 Burgess Lane Oak Ridge TN 37830

WICHOLAS, MARK L, b Lawrence, Mass, June 11, 40; div; c 2. INORGANIC CHEMISTRY. *Educ:* Boston Univ, AB, 61; Mich State Univ, MS, 64; Univ Ill, PhD(chem), 67. *Prof Exp:* From asst prof to assoc prof, 67-79, PROF CHEM, WESTERN WASH UNIV, 79-; DEPT CHMN, 82- *Mem:* Am Chem Soc. *Res:* Physical inorganic chemistry; coordination chemistry of transition metals; bioinorganic chemistry. *Mailing Add:* Dept Chem Western Wash Univ Bellingham WA 98225

WICHTERMAN, RALPH, b Philadelphia, Pa, Sept 8, 07; m 33; c 2. ZOOLOGY. *Educ:* Temple Univ, BS, 30; Univ Pa, MA, 32, PhD(zool), 36. *Prof Exp:* From asst instr to prof, 29-74, EMER PROF BIOL, TEMPLE UNIV, 75- *Concurrent Pos:* Guest investr, Dry Tortugas Lab, 39; vis prof, Univ Pa, 51; AEC protozoologist, Acad Natural Sci, Philadelphia, 52; lectr, Nenski Inst Exp Biol, Warsaw, Poland, 62; Am Philos Soc-NSF award, Zool Sta, Naples, Italy, 62 & 69; life mem, Corp Marine Biol Lab, Woods Hole. *Mem:* Fel AAAS; Am Soc Zool; Am Soc Parasitol; Micros Soc Am; Soc Protozool; Sigma Xi. *Res:* Protozoology; parasitology; cytology; histology; parasitic protozoa; sexual processes in ciliates; biology of Paramecium; gamma and x-radiation of protozoa. *Mailing Add:* 31 Buzzards Bay Ave Woods Hole MA 02543

WICK, EMILY LIPPINCOTT, b Youngstown, Ohio, Dec 9, 21. ORGANIC CHEMISTRY, ACADEMIC ADMINISTRATION. *Educ:* Mt Holyoke Col, AB, 43, MA, 45; Mass Inst Technol, PhD(org chem), 51. *Hon Degrees:* ScD, Mt Holyoke Col, 72. *Prof Exp:* Instr chem, Mt Holyoke Col, 45-46; res assoc org chem, Mass Inst Technol, 51-53; org chemist flavor lab, Arthur D Little, Inc, 53-57; res assoc food sci, Mass Inst Technol, 57-59, asst prof food chem, 59-63, from assoc prof to prof, 63-73, assoc dean student affairs, 65-72; prof chem & dean fac, Mt Holyoke Col, 73-80, asst to the pres, Long Range Planning, 81-86; RETIRED. *Mem:* AAAS; Am Chem Soc; Inst Food Technol. *Res:* Chemistry of food and natural products. *Mailing Add:* 37 Atlantic Ave Rockport MA 01966-1651

WICK, GIAN CARLO, b Torino, Italy, Oct 15, 09; nat US; m 43; c 2. PHYSICS. *Educ:* Univ Torino, PhD(physics), 30. *Prof Exp:* Asst prof theoret physics, Univ Palermo, 37-38; assoc prof, Univ Padova, 38-40; prof, Univ Rome, 40-45; prof, Univ Notre Dame, 46-48; prof, Univ Calif, 48-50; prof, Carnegie Inst Technol, 51-57; sr physicist, Brookhaven Nat Lab, 57-65; prof, 65-77, EMER PROF PHYSICS, COLUMBIA UNIV, 77- *Mem:* Nat Acad Sci; fel Am Phys Soc; Am Acad Arts & Sci. *Res:* Nuclear physics; elementary physics. *Mailing Add:* Via Maria Vittoria 35 Torino 10123 Italy

WICK, JAMES ROY, b Henry Co, Iowa, Dec 17, 12; m 42; c 2. ENTOMOLOGY. *Educ:* Iowa Wesleyan Col, BS, 48; Kans State Col, MS, 50; Iowa State Col, PhD(entom), 54. *Prof Exp:* Instr biol, Iowa State Col, 52-54, asst prof, 54-59; assoc prof, 59-64, PROF ZOOL & CHMN DEPT BIOL SCI, NORTHERN ARIZ UNIV, 64- *Mem:* AAAS; Am Inst Biol Sci; Am Entom Soc; Sigma Xi. *Res:* Insect histology and developmental anatomy. *Mailing Add:* 1701 N Kutch Dr Flagstaff AZ 86001

WICK, O(SWALD) J, b Fargo, NDak, July 15, 14; m 41; c 3. METALLURGICAL & MINING ENGINEERING. *Educ:* Mont Sch Mines, BS, 36, MS, 37. *Prof Exp:* Assoc metall, Col Mines, Univ Wash, Seattle, 37-42; gen supt mercury mine, Pac Mining Co, Wash, 42; metallurgist, Metall Lab, Puget Sound Naval Shipyard, 42-50; metallurgist pile technol, Gen Elec Co, 50-54, sr engr, 54-56, head prod metall, 56, mgr plutonium metall, Hanford Labs, 56-62, mgr metall develop, 62-65; mgr, 65-66, dep mgr metall dept, 66-68, assoc mgr chem & metall div, 68-71, staff engr, 71-85, CONSULT, MAT & CHEM APPLNS DEPT, PAC NORTHWEST LAB, BATTELLE MEM INST, 85- *Concurrent Pos:* Tech adv, US Deleg Second Int Conf Peaceful Uses Atomic Energy, Geneva, 58. *Mem:* Am Inst Mining, Metall & Petrol Engrs; Am Soc Metals. *Res:* Plutonium metallurgy; metal fabrication; nuclear fuel; mineral dressing. *Mailing Add:* Mat & Chem Applns Dept PO Box 999 Battelle Mem Inst Richland WA 99352

WICK, ROBERT S(ENTERS), b Port Washington, NY, Dec 4, 25; m 47; c 5. MECHANICAL ENGINEERING. *Educ:* Rensselaer Polytech Inst, BME, 46; Stevens Inst Technol, MS, 48; Univ Ill, PhD(mech eng), 52. *Prof Exp:* Res engr, Standard Oil Develop Co, 46-47; instr mech eng, Syracuse Univ, 48-50; sr res engr, Jet Propulsion Lab, Calif Inst Technol, 52-55, res group supvr, 55; sr engr, Bettis Atomic Power Lab, Westinghouse Elec Corp, 55-57, supvr, 57-59, mgr adv core nuclear design, 59-62, mgr power physics dept & power reactor eng, 62-66; prof nuclear & aerospace eng, Tex A&M Univ, 66-88; RETIRED. *Mem:* AAAS; Am Nuclear Soc; Am Soc Mech Engrs; Am Inst Aeronaut & Astronaut. *Res:* Nuclear reactor technology; rocket propulsion and combustion; applied and fluid mechanics; thermodynamics. *Mailing Add:* 310 Fairway Dr Bryan TX 77801

WICK, WILLIAM QUENTIN, b Eau Claire, Wis, Oct 24, 27; m 54; c 2. COASTAL ECOSYSTEM MANAGEMENT. *Educ:* Ore State Univ, BS, 50, MS, 52. *Prof Exp:* Wildlife biologist, Nev Fish & Game Comn, 52-53 & Wash Dept Game, 53-60; area exten agent, Exten Serv, 60-68, dir, Marine Exten, Sea Grant Ext Prog, 68-73, PROF & DIR, SEA GRANT COL, ORE STATE UNIV, 73- *Concurrent Pos:* Prof exten, Ore State Univ, 69-, prof wildlife ecol, 72-; vis prof, Australia, Great Barrier Reef, 82; app mem, Marine Div, Nat Asn State Univs & Land Grant Col, 83-; comnr, S Slough Nat Estuarine Sanctuary, 84-; vchair, Nat Sea Grant Dir's Coun, 87; chair & bd gov, Nat Coastal Resources Res & Develop Inst, 85-88. *Honors & Awards:* Superior Serv Award, USDA, 71; Einarsen Award, Wildlife Soc, 72; Compass Award, Marine Technol Soc, 76. *Mem:* Sea Grant Asn (pres, 77-78); Wildlife Soc; Sigma Xi. *Res:* Ocean resources development and conservation, including fisheries, aquaculture, economics, engineering, law, food technology, seabed minerals and estuarine; coastal processes; human use conflicts in the marine environment. *Mailing Add:* Ore State Univ AdS 320 Corvallis OR 97331

WICKE, BRIAN GARFIELD, b Berea, Ohio, July 24, 44; m 67; c 2. PHYSICAL CHEMISTRY. *Educ:* DePauw Univ, BA, 66; Harvard Univ, MA, 71, PhD(phys chem), 71. *Prof Exp:* Res assoc chem, Quantum Inst, Univ Calif, Santa Barbara, 74-75, vis scholar, 75-76; scientist, TRW Systs Group, Redondo Beach, Calif, 76-78; assoc sr res scientist, 78-80, staff res scientist, 80-90, SR STAFF RES SCIENTIST, GEN MOTORS, 90- *Honors & Awards:* John M Cambell Award, 88. *Mem:* InterAm Photochem Soc; Sigma Xi; Combustion Inst. *Res:* Chemical kinetics; gas phase kinetics; soot oxidation; emission control; CFC & air conditioning. *Mailing Add:* Phys Chem Dept Gen Motors Res Lab Warren MI 48090

WICKE, HOWARD HENRY, b Chicago, Ill, Aug 29, 24; m 45; c 4. TOPOLOGY. *Educ:* Univ Iowa, PhD(math), 52. *Prof Exp:* Instr math, Lehigh Univ, 52-54; mem staff, Sandia Corp, 54-61, supvr, 61-70; PROF MATH, OHIO UNIV, 70- *Mem:* Am Math Soc; Math Asn Am. *Res:* General topology; point-set topology; set theory; applied mathematics. *Mailing Add:* Dept Math Ohio Univ Athens OH 45701-2979

WICKELGREN, WARREN OTIS, b Munster, Ind, Oct 15, 41; m 65; c 2. NEUROPHYSIOLOGY. *Educ:* Univ Mich, Ann Arbor, AB, 63; Yale Univ, PhD(psychol), 67. *Prof Exp:* NIH trainee, Yale Univ, 67-69, asst prof physiol, 69-70; asst prof, 70-76, ASSOC PROF PHYSIOL, UNIV COLO MED CTR, DENVER, 76- *Concurrent Pos:* NIH res career develop award, Univ Colo Med Ctr, Denver, 71. *Mem:* Soc Neurosci; Am Physiol Soc. *Res:* Organization of simple vertebrate nervous systems; neurophysiology of learning. *Mailing Add:* Dept Physiol Med Sch Univ Colo 4200 E Ninth Ave Denver CO 80262

WICKER, EVERETT E, b Lockport, NY, Apr 6, 19; m 43; c 2. NUCLEAR SCIENCE. *Educ:* Univ Pittsburgh, BS, 41; Carnegie-Mellon Univ, MS, 57. *Prof Exp:* Physicist, Kennametal, Inc, 41-42 & 46-47; from asst technologist to technologist, 47-54, from supvry technologist to res technologist, 54-64, ASSOC RES CONSULT, RES LAB, US STEEL CORP, 64- *Mem:* Am Phys Soc; Am Nuclear Soc. *Res:* Neutron and charged particle activation analysis; nuclear reactor materials; nuclear and reactor physics; industrial and research uses of radioisotopes and radiation. *Mailing Add:* 815 William Penn Ct Pittsburgh PA 15221

WICKER, ROBERT KIRK, b Altoona, Pa, Mar 4, 38; m 61; c 3. PHYSICAL INORGANIC CHEMISTRY, SCIENCE EDUCATION. *Educ:* Juniata Col, BS, 60; Univ Del, MS, 63, PhD(phys chem), 66. *Prof Exp:* Asst prof chem, Davis & Elkins Col, 65-67; from asst prof to assoc prof, 67-78, PROF CHEM, WASHINGTON & JEFFERSON COL, 78- *Concurrent Pos:* Consult, US Dept Energy & Pittsburgh Energy Technol Ctr, Pa Dept Educ, NSF & pvt indust. *Mem:* AAAS; Am Chem Soc; Sigma Xi. *Res:* Thermodynamic properties of nonaqueous electrolyte solutions; preparation and structure determinations of copper complexes; surface area measurements of catalysts; receptor models of acid precipitation; conversion of waste methane gas. *Mailing Add:* Dept Chem Washington & Jefferson Col Washington PA 15301

WICKER, THOMAS HAMILTON, JR, b Orlando, Fla, Nov 19, 23; m 49; c 3. ORGANIC CHEMISTRY, POLYMER CHEMISTRY. *Educ:* Univ Fla, BS, 44, MS, 48, PhD(org chem), 51. *Prof Exp:* From assoc res chemist to sr res chemist, Tenn Eastman Co, 51-76, res assoc, 77-86; RETIRED. *Mem:* Am Chem Soc; Sigma Xi. *Res:* 2-cyanoacrylate adhesives; condensation polymers; organic chemistry. *Mailing Add:* 4619 Mitchell Rd Kingsport TN 37664

WICKERHAUSER, MILAN, b Zemun, Yugoslavia, Aug 28, 22; m 56; c 2. BIOCHEMISTRY. *Educ:* Chem engr, Univ Zagreb, 46, PhD, 61. *Prof Exp:* Develop chemist, Inst Immunol, Yugoslavia, 46-53, head dept serum & toxoid purification, 53-57, head dept human plasma fractionation, 57-62; immunochemist, Immunol, Inc, Ill, 63; res assoc fractionation plasma protein & blood coagulation studies, Hyland Labs, Calif, 64-66; dir, Am Red Cross Nat Fractionation Ctr, Blood Res Lab, 70-79, sr res scientist, 66-85, head plasma fractionation sect, 68-85, CONSULT PLASMA PROD DEVELOP, AM RED CROSS, 85- *Concurrent Pos:* WHO fel, Wellcome Physiol Res Lab, Eng, Lister Inst Prev Med, London & State Serum Inst, Copenhagen, 50. *Mem:* Am Asn Blood Banks; Int Soc Blood Transfusion; Int Soc Thrombosis & Haemostasis; AAAS. *Res:* Isolation and characterization of plasma proteins with special emphasis on the large scale methodology; blood coagulation; immunoglobulins; development of large-scale plasma fractionation methods. *Mailing Add:* 5021 Acacia Ave Bethesda MD 20814

WICKERSHAM, CHARLES EDWARD, JR, b Terre Haute, Ind, Oct 26, 51; m 74; c 3. SPUTTER DEPOSITION OF THIN FILMS, IC METALLIZATION. *Educ:* Rose-Hulman Inst Technol, BS, 73; Univ Ill, MS, 76, PhD(metall eng), 78. *Prof Exp:* Sr assoc eng, Int Bus Mach, 77; res scientist, Battelle Mem Inst, 77-80, sr res scientist, 80-83; res mgr, Specialty Metals Div, Varian Assocs, 83-86, eng mgr, 86-90; VPRES ENG, TOSOH SMD, INC, 90- *Concurrent Pos:* Consult, Battelle Mem Inst, 83-84. *Mem:* Am Vacuum Soc; Am Soc Metals Int. *Res:* Sputtering target metallurgical structure effects on thin film properties; computer simulation of sputtering processes; thin film properties and sputtering target manufacturing. *Mailing Add:* 571 Arden Rd Columbus OH 43214

WICKERSHAM, EDWARD WALKER, b Kelton, Pa, Apr 26, 32; m 59; c 3. REPRODUCTIVE PHYSIOLOGY, HUMAN SEXUALITY. *Educ:* Pa State Univ, BS, 57, MS, 59; Univ Wis, PhD(dairy physiol), 62. *Prof Exp:* NIH trainee endocrinol, Univ Wis, 62-63; asst prof biol, WVa Univ, 63-64; asst prof zool, 64-68, ASSOC PROF BIOL, PA STATE UNIV, 68- *Concurrent Pos:* Cert sex educator, Am Asn Sex Educators, Counselors & Therapists. *Mem:* Brit Soc Study Fertil; Soc Study Reproduction; Am Asn Sex Educators, Counselors & Therapists. *Res:* Mammalian reproductive physiology and endocrinology; physiology of fertility regulation; biological and health aspects of human sexuality. *Mailing Add:* Dept Biol 417 Mueller Lab Pa State Univ University Park PA 16802

WICKERSHEIM, KENNETH ALAN, b Fullerton, Calif, Mar 4, 28; m 52, 67; c 2. SOLID STATE PHYSICS, SPECTROSCOPY. *Educ:* Univ Calif, Los Angeles, AB, 50, MA, 56, PhD(physics), 59. *Prof Exp:* Mem staff, Los Alamos Sci Lab, 53-55; asst physics, Univ Calif, Los Angeles, 55-58; mem staff res labs, Hughes Aircraft Co, 58-61; res physicist, Palo Alto Labs, Gen Tel & Electronics Lab, Inc, 61-63; assoc prof mat sci, Stanford Univ, 63-64; staff scientist solid state physics, Lockheed Palo Alto Res Labs, 64-65, sr mem, 65-66, head adv electronics, 66-70; pres, Spectrotherm Corp, 70-75; vpres res & develop, UTI Corp, 75-77; vpres & gen mgr, Quantex Corp, 77-78; pres,78-85, CHMN BD, LUXTRON CORP, 85- *Concurrent Pos:* Res assoc sch earth sci, Stanford Univ, 64-; consult, Appl Physics Corp, 63-64, Quantic Industs, 77-, AGA Corp, 78-79. *Mem:* AAAS; fel Am Phys Soc; NAm Hyperthermia Group; Bio Electromagnetics Soc. *Res:* Optical properties of materials; spectra of rare earth ions in solids; infrared spectra of crystals; optical and infrared instrumentation; medical and industrial thermography; fiberoptic sensors and associated instrumentation. *Mailing Add:* 1160 B Pine St Menlo Park CA 94025-3407

WICKES, HARRY E, b Portland, Ore, June 24, 25; m 49; c 3. MATHEMATICS. *Educ:* Brigham Young Univ, BS, 50, MEd, 54; Harvard Univ, MEd, 62; Colo State Col, EdD, 67. *Prof Exp:* Teacher high sch, Mont, 50-51, Idaho, 51-54, prin elem & high sch, 54-57; instr math, 57-63, form asst prof to assoc prof, 64-75, PROF MATH, BRIGHAM YOUNG UNIV, 75- *Mem:* Math Asn Am; Nat Coun Teachers Math. *Res:* Mathematics education. *Mailing Add:* 1733 W 80 S Provo UT 84601

WICKES, WILLIAM CASTLES, b Lynwood, Calif, Nov 25, 46; m 71; c 2. PHYSICS, ASTRONOMY. *Educ:* Univ Calif, Los Angeles, BS, 67; Princeton Univ, MA, 69, PhD(physics), 72. *Prof Exp:* From res assoc to instr physics, Princeton Univ, asst prof, 75-78; asst prof physics, Univ Md, 78-81; MEM TECH STAFF & PROJ MGR, CORVALLIS DIV, HEWLETT-PACKARD CO, 81- *Mem:* Am Astron Soc; Sigma Xi; Am Math Soc; Math Asn Am. *Res:* Double star interferometry; experimental and theoretical cosmology and quantum mechanics; computer science. *Mailing Add:* Hewlett-Packard 1000 NE Circle Blvd Corvallis OR 97330

WICKHAM, DONALD G, b Beaverton, Ore, Feb 24, 22; m 54; c 2. INORGANIC CHEMISTRY. *Educ:* Univ Denver, BS, 47, MS, 50; Mass Inst Technol, PhD(inorg chem), 54. *Prof Exp:* Chemist, Lincoln Lab, Mass Inst Technol, 54-57, sect leader, 57-60; mem staff, Res Labs, Hughes Aircraft Co, 60-61; mgr mat res & develop, Components Div, 61-65, MGR FERRITE MEMORY-CORE DEVELOP, AMPEX COMPUT PROD CO, 65- *Mem:* Am Crystallog Asn; Inst Elec & Electronics Engrs. *Res:* Inorganic solid state chemistry; magnetic materials; inorganic syntheses. *Mailing Add:* Ampex Memory Prods Div 200 N Nash St El Segundo CA 90245

WICKHAM, JAMES EDGAR, JR, b Glen Allen, Va, Apr 7, 33; m 53; c 2. ANALYTICAL CHEMISTRY, INORGANIC CHEMISTRY. *Educ:* Randolph-Macon Col, BS, 57. *Prof Exp:* Asst chemist, Philip Morris, Inc, 57-59, group leader, 59-62, supvr, 62-69, head cigarette testing, 69-71, sr scientist, 71-74, MGR CIGARETTE TESTING SERV, PHILIP MORRIS USA, 74- *Mem:* AAAS; Am Chem Soc. *Res:* Wet and instrumental methods development; gas-liquid chromatography tobacco and smoke chemistry; quality control operations; smoking technology; air flow; research and development management. *Mailing Add:* Philip Morris USA Opers Ctr PO Box 26603 Richmond VA 23261

WICKHAM, M GARY, b Ft Morgan, Colo, Dec 23, 42; m 65. HISTOLOGY, ELECTRON MICROSCOPY. *Educ:* Colo State Univ, BS, 64, MS, 67; Wash State Univ, PhD(zool), 72. *Prof Exp:* Fisheries aide, Colo Game & Fish Dept, 62-65; res asst, Colo Coop Fish Unit, 65-67; teaching asst morphogenesis, dept zool, Washington State Univ, 67-71, res asst, 68-70; res physiologist, Vet Admin Med Ctr, Gainesville, Fla, 71-74 & dir, Core EM Facil, San Diego, Calif, 74-79; assoc prof ocular anat, Northeastern State Univ, Tahlequah, Okla, 79-82, assoc prof biol, 79-85, dir biosci res facil, 81-88, asst res & vpres acad affairs, 85-88, prof biol, 85-88, PROF OPTOM, NORTHEASTERN STATE UNIV, TAHLEQUAH, OKLA, 88- *Concurrent Pos:* NSF fel, Friday Harbor Marine Lab, Univ Wash, 70; Nat Defense Educ Act fel, Wash State Univ, 71; asst prof, dept ophthal, Univ Fla, Gainesville, 72-74; asst res biologist & dir, ophthal res, dept surg, Univ Calif, San Diego, 74-79. *Mem:* AAAS; Am Asn Zool Parks & Aquariums; Am Inst Biol Sci; Am Soc Zoologists; Asn Res Vision & Ophthal. *Res:* Comparative morphology and functional morphology of the structures of the anterior segment of the mammalian eye; cornea, iris, sclera, trabecular meshwork, ciliary body and lens as they relate to intraocular pressure; computer-based morphometry. *Mailing Add:* Col Optom Northeastern State Univ Tahlequah OK 74464

WICKHAM, WILLIAM TERRY, JR, b Cleveland, Ohio, May 28, 29; m 52; c 3. TECHNICAL MANAGEMENT. *Educ:* Heidelberg Col, AB, 51; Case Inst Technol, MS, 54, PhD(org chem), 56. *Prof Exp:* Instr, Case Inst Technol, 55-56; res chemist, Owens-Ill Glass Co, 56-58; group leader, Dow Chem Co, 58-62; tech mgr, Celanese Plastics Co, 62-67; dir res, Southern Div, Dayco Corp, 67-72, vpres res & develop, 72-76; tech dir, Crosby Chem Inc, Picaynne, 76-77; PROF, HEIDELBERG COL, TIFFIN, OHIO, 77- *Concurrent Pos:* Consult tech mgt, Gen Motors Mercurs Inc, Dayco Corp & Hercules Inc, 77- *Mem:* Am Chem Soc; Am Phys Soc; Am Inst Chemists; Sigma Xi. *Res:* Research administration; science adminstration; technical management; structure; manufacture; research managment. *Mailing Add:* 54 Glen View Terr Tiffin OH 44883

WICKLER, STEVEN JOHN, b Volga, SDak, May 17, 52. ENERGETICS, THERMOGENESIS. *Educ:* Univ Calif, Riverside, BA, 74; Univ Mich, MS, 77, PhD(zool), 79. *Prof Exp:* VIS ASST PROF, DEPT ANIMAL PHYSIOL, UNIV CALIF, DAVIS, 79- *Mem:* Am Physiol Soc; Am Soc Zoologists. *Res:* Examine whole animal, tissue, and biochemical adaptations with particular emphasis on energetics and thermogenesis. *Mailing Add:* Dept Animal Sci Calif Polytech Univ Pomona CA 91768

WICKLIFF, JAMES LEROY, b Knoxville, Iowa, Nov 14, 31; m 56; c 3. PLANT BIOCHEMISTRY, PHOTOBIOLOGY. *Educ:* Iowa State Univ, BS, 55, PhD(plant physiol), 62. *Prof Exp:* Assoc bot & plant path, Iowa State Univ, 62-65; asst prof bot & bact, 65-69, ASSOC PROF BIOL SCI, UNIV ARK, FAYETTEVILLE, 69- *Mem:* AAAS; Am Soc Plant Physiol; Am Inst Biol Sci; Am Soc Photobiol. *Res:* Chlorophyll biochemistry; photosynthesis; photophysiology of higher plants. *Mailing Add:* Dept Biol Sci Univ Ark Fayetteville AR 72701

WICKLOW, DONALD THOMAS, b San Francisco, Calif, June 22, 40; div; c 2. MYCOLOGY, ECOLOGY. *Educ:* San Francisco State Col, BA, 62, MA, 64; Univ Wis, PhD(bot), 71. *Prof Exp:* Instr biol, Univ Wis Ctr-Waukesha, 69-70; asst prof, Univ Pittsburgh, 70-76; res scientist mycol, 77-85, LEAD SCIENTIST, NORTHERN REGIONAL RES CTR, USDA, 85- *Concurrent Pos:* Consult, Natural Resource Ecol Lab, Ft Collins, Colo, 74-76; chmn steering comt microbiol & ecol, Argonne Nat Labs, 78-80; mem adv panel ecol & ecosysts studies, NSF, 78-81; organizing comt third Int Symp on Microbiol & Ecol, 83. *Honors & Awards:* Alexopoulous Prize, Mycol Soc Am, 80; fel, Am Acad Microbiol, 88. *Mem:* Brit Mycol Soc; Mycol Soc Am; Ecol Soc Am; Mycol Soc Am. *Res:* Ecology of fungal communities, their organization and role in both native and man-managed ecosystems; ecology of mycotoxin producing fungi; chemistry of fungus and insect interaction. *Mailing Add:* Northern Regional Res Ctr 1815 N University St Peoria IL 61604

WICKLUND, ARTHUR BARRY, b Dec 8, 42; US citizen; m 74. EXPERIMENTAL HIGH ENERGY PHYSICS. *Educ:* Harvard Univ, BA, 64; Univ Calif, Berkeley, PhD(physics), 70. *Prof Exp:* Res asst high energy physics, Univ Calif, Berkeley, 65-70; fel, 70-73, asst physicist, 73-76, PHYSICIST, ARGONNE NAT LAB, 76- *Mem:* Am Phys Soc. *Res:* Strong interaction phenomenology; production mechanisms in few body reactions. *Mailing Add:* Argonne Nat Lab HEP Bldg 362 9700 S Cass Ave Argonne IL 60439

WICKMAN, HERBERT HOLLIS, b Omaha, Nebr, Sept 30, 36; m 85; c 2. COLLOID CHEMISTRY. *Educ:* Univ Omaha, AB, 59; Univ Calif, Berkeley, PhD(chem), 64. *Prof Exp:* Mem tech staff chem physics res lab, Bell Tel Labs, NJ, 64-70; assoc prof, Ore State Univ, 70-79, prof chem, 80-87; PROG DIR, DIV MAT RES, NSF, 87- *Mem:* Am Chem Soc; Am Phys Soc; Biophys Soc; Sigma Xi. *Res:* Magnetism, liquid crystals, membrane biophysics. *Mailing Add:* Div Mat Res NSF 1800 G St NW Washington DC 20550

WICKREMA SINHA, ASOKA J, b Colombo, Ceylon, Sept 8, 37; US citizen. BIOCHEMISTRY, ORGANIC CHEMISTRY. *Educ:* Univ London, BSc, 61; Univ Birmingham, MSc, 64, PhD(org chem), 66. *Prof Exp:* Fel biochem, Univ Birmingham, 66-67; staff scientist, Worcester Found Exp Biol, 67-69; res scientist chem & biochem, 69-75, SR RES SCIENTIST, RES DIV, UPJOHN CO, 75- *Mem:* Am Chem Soc; The Chem Soc; Brit Biochem Soc. *Res:* Chemical synthesis; carbonium ion chemistry; biosynthesis and metabolism of steroids; in vivo and in vitro metabolism; analytical methods and assay development; metabolism absorption, distribution and excretion of drugs; isolation and structure elucidation; radiotracer techniques. *Mailing Add:* Res Div Upjohn Co Kalamazoo MI 49001-0199

WICKS, CHARLES E(DWARD), b Prineville, Ore, July 9, 25; m 48; c 3. CHEMICAL ENGINEERING. *Educ:* Ore State Col, BS, 50; Carnegie Inst Technol, MS, 52, PhD, 54. *Prof Exp:* From asst prof to assoc prof, 54-60, PROF CHEM ENG, ORE STATE UNIV, 60-, HEAD DEPT, 70- *Concurrent Pos:* Chem engr, US Bur Mines, 56-58, proj leader, 58-68; consult, US Bur Mines, 56-68, Pac Power & Light Co, 58-60, Year-in-Indust, E I du Pont de Nemours & Co, 64-65 & Ore Metall Corp, 66-; NSF fel, Univ Wis, 60-61; expert witness prod reliability, various law firms; distiguished scholar exchange to People's Repub China, Nat Acad Sci, 84. *Mem:* Am Inst Chem Engrs; Am Soc Eng Educ. *Res:* Waste water treatment; simultaneous heat and mass transfer phenomena. *Mailing Add:* Dept Chem Eng Ore State Univ Corvallis OR 97331-2702

WICKS, FREDERICK JOHN, b Winnipeg, Man, Nov 22, 37; m 67. MINERALOGY. *Educ:* Univ Man, BSc, 60, MSc, 65; Oxford Univ, DPhil(mineral), 69. *Prof Exp:* Geologist, Giant Yellowknife Mines Ltd, 60 & 61; consult mineralogist, Govt & Indust, 62; mineralogist, Man Hwys Br, 63-65 & Geol Surv Can, 67; asst cur, 70-75, assoc cur, 75-80, CUR MINERAL, ROYAL ONT MUS, 80-, CUR-IN-CHARGE, DEPT MINERAL, 87- *Concurrent Pos:* Adj prof, dept earth sci, Univ Man, Winnipeg, 77; assoc prof, dept geol, Univ Toronto, 80-; mineral consult, Royal Comn Matters Health & Safety Arising from Use of Asbestos, Ont, 82-83. *Honors & Awards:* Hawley Award, Mineral Asn Can, 77 & 78. *Mem:* Mineral Asn Can (secy, 73-75); fel Geol Asn Can; fel Mineral Soc Am; Clay Minerals Soc. *Res:* Structure, chemistry and paragenesis of the serpentine minerals; asbestos deposits; geochemistry and paragenesis of Colombian emerald deposits. *Mailing Add:* 74 Milbrook Crescent Toronto ON M4R 1H4 Can

WICKS, GEORGE GARY, b Copaigue, NY, June 26, 45; m 69; c 1. MATERIALS SCIENCE. *Educ:* Fla State Univ, BS, 67, MS, 69; Harvard Univ, SM, 71; Mass Inst Technol, PhD(eng mat), 75. *Prof Exp:* Engr, 69-70, RES METALLURGIST, SAVANNAH RIVER LAB, E I DU PONT DE NEMOURS & CO, INC, 75- *Res:* Structure and science of glass and amorphous materials; immobilization of radioactive waste in glass matrices. *Mailing Add:* 1586 Citation Dr N Aiken SC 29801

WICKS, WESLEY DOANE, b Providence, RI, Feb 13, 36; m 59; c 3. BIOCHEMISTRY, PHARMACOLOGY. *Educ:* Bates Col, BS, 57; Harvard Univ, MA, 59, PhD(med sci), 64. *Prof Exp:* Staff mem biochem, Biol Div, Oak Ridge Nat Lab, Tenn, 65-69; staff mem, Div Res, Nat Jewish Hosp, Denver, 69-72; assoc prof, 72-78, prof pharmacol, med ctr, Univ Colo, Denver, 78-; AT DEPT BIOCHEM, UNIV TENN. *Concurrent Pos:* Am Cancer Soc fel, Biol Div, Oak Ridge Nat Lab, Tenn, 63-65; hon fac mem, Dept Biosci, Fed Univ Pernambuco, Recife, Brazil, 68- *Mem:* AAAS; Am Soc Biol Chem; Am Soc Pharmacol & Exp Therapeut; Am Chem Soc; Endocrine Soc. *Res:* Regulation of specific protein synthesis by hormones and cyclic adenosine phosphate; use of cultured cells for studies of biochemical regulatory mechanisms. *Mailing Add:* Dept Biochem Waller Life Sci Bldg Rm 407 Univ Tenn Knoxville TN 37996-0840

WICKS, ZENO W, JR, b Port Jervis, NY, July 24, 20; m 41; c 6. POLYMER CHEMISTRY. *Educ:* Oberlin Col, AB, 41; Univ Ill, PhD(org chem), 44. *Prof Exp:* Res chemist, Interchem Corp, 44-47, dist tech dir finishes div, 48-49, res dir textile colors div, 49-51, assoc dir cent res labs, 51-54, dir, 54-59, mgr com develop, 59-61, vpres planning, 62-63, vpres & dir corp, 64-69; mem staff, Inmont Corp, NY, 69-72; prof, 72-80, DISTINGUISHED PROF, DEPT POLYMERS & COATINGS, NDAK STATE UNIV, 81-, CHMN DEPT, 72- *Mem:* Am Chem Soc; Soc Coatings Technol; Oil & Colour Chemists Asn. *Res:* Organic surface coatings and related polymer research. *Mailing Add:* 1345 Branson Apt 4C Las Cruces NM 88001-5380

WICKSON, EDWARD JAMES, b New York, NY, Jan 25, 20; m 52; c 5. PLASTICS FORMULATING. *Educ:* Univ Calif, Berkeley, BS, 42. *Prof Exp:* Anal chemist, Gen Chem Co, Calif, 42-43; chemist, Celanese Corp Am, NJ, 46-50; admin asst to lab dir, Vitro Corp Am, 51-54; sr chemist, Chicopee Mfg Co, 54-55; sr chemist, Enjay Labs, Esso Res & Eng Co, 55-56, group leader, 56-60, res assoc, 60-61, head chem sect, 61-69, vinyl indust assoc, Enjay Chem Co, 69-71, sr res assoc, Enjay Chem Lab, 71-75, chief scientist plasticizers, Esso Chem Europe, 75-78, chief prod applns scientist, Exxon Chem Co, 78-86; PRES, WICKSON PROD RES, LTD, 86- *Mem:* Am Chem Soc; fel Soc Plastics Eng. *Res:* Plasticizers; chemical specialties; oxo alcohols; trialkylacetic acids; propylene polymers. *Mailing Add:* 7973 Walden Rd Baton Rouge LA 70808

WICKSTEN, MARY KATHERINE, b San Francisco, Calif, Mar 17, 48. BIOLOGY, INVERTEBRATE ZOOLOGY. *Educ:* Humboldt State Col, BA, 70, MA, 72; Univ Southern Calif, PhD(biol), 77. *Prof Exp:* Teaching asst biol, Humboldt State Col, 72 & Univ Southern Calif, 73-75 & 76-77; scientist marine biol, Bur Land Mgt Southern Calif Benthic Studies & Anal, 75-76; res assoc marine biol, Los Angeles Harbor Proj, 77-80; asst prof, 80-88, ASSOC PROF BIOL, TEX A&M UNIV, 88- *Concurrent Pos:* Consult systematist, King Harbor Proj, 74-, Gulf Alaska Offshore Surv, 76-77, Bur Land Mgt Southern Calif Benthic Studies & Anal Seasonal Study, 76- & Bur Land Mgt Southern Calif Mussel Bed Surv, 77- *Mem:* Am Soc Zoologists. *Res:* Behavior, systematics and zoogeography of decapod crustaceans of the eastern Pacific Ocean. *Mailing Add:* Dept Biol Tex A&M Univ College Station TX 77843

WICKSTROM, CONRAD EUGENE, b Modesto, Calif, Sept 3, 43; m 77; c 3. MICROBIAL ECOLOGY, PHYCOLOGY. *Educ:* Calif State Univ, Chico, BA, 65, MA, 68; Univ Ore, PhD(biol), 74. *Prof Exp:* Instr, Emory Univ, 73-74, asst prof biol, 74-81; ASST PROF BIOL SCI, KENT STATE UNIV, 81- *Concurrent Pos:* Prin investr NSF grant, 78-80 & 82-83. *Mem:* AAAS; Am Soc Microbiol; Am Soc Limnol & Oceanog; Ecol Soc Am; Phycol Soc Am. *Res:* Biotic components of nitrogen cycle, especially asymbiotic nitrogen fixation; biotic and abiotic control of microbial community structure and function; natural thermal systems and thermal enrichments; aquatic biology. *Mailing Add:* Dept Biol Sci Kent State Univ Kent OH 44242

WICKSTROM, ERIC, b Chicago, Ill, Dec 21, 46; m 67; c 2. BIOPHYSICAL CHEMISTRY. *Educ:* Calif Inst Technol, BS, 68; Univ Calif, Berkeley, PhD(chem), 72. *Prof Exp:* Res asst chem, Univ Calif, Berkeley, 68-72; res assoc molecular, cellular & develop biol, Univ Colo, Boulder, 73-74; asst prof chem, Univ Denver, 74-81; asst prof chem, 82-87, ASSOC PROF CHEM & BIOCHEM MOLECULAR BIOL & SURG, UNIV SFLA, 87- *Concurrent Pos:* Europ molecular biol orgn fel, Univ Leiden, 84; guest researcher, Nat Cancer Inst, 86; guest researcher, Max Planck Inst Molecular Genetics, Berlin, 86; res grants, NIH, NSF, Am Cancer Soc, Leukemia Soc Am, US Army Res & Develop Command, Am Found AIDS Res; site visit team mem, Nat Cancer Inst, 89-90; consult, Genta Ins, 89-, Life Scis, 89-; organizer, Int Union Biochem, Nat Cancer Inst Conf Nucleic Acid Therapeut, 91. *Mem:* Am Chem Soc; Am Soc Biochem & Molecular Biol; Biophys Soc; Am Asn Cancer Res. *Res:* Antisense DNA therapeutics for oncogenes and viral genes; protein synthesis initiation, RNA-protein interactions. *Mailing Add:* Dept Chem Univ SFla Tampa FL 33620

WICKSTROM, JACK, medicine; deceased, see previous edition for last biography

WIDDEN, PAUL RODNEY, b London, Eng, Sept 23, 43; m 67; c 3. SOIL MICROBIOLOGY. *Educ:* Univ Liverpool, BSc Hons, 65; Univ Calgary, PhD(mycol), 71. *Prof Exp:* Asst prof, 73-76, ASSOC PROF MICROBIAL ECOL, CONCORDIA UNIV, SIR GEORGE WILLIAMS CAMPUS, 76-, CHAIR, 89- *Concurrent Pos:* Nat Res Coun Can operating grant, 75-91; Univ res grant, Imp Oil Ltd, 78-81, Que Agr & Fisheries grant, 89-91, Que Govt Infrastructure grant, 89-92; vis scientist, Inst Terestrial Ecol, Merlewood, UK, 81-82. *Mem:* Can Soc Microbiologists; Ecol Soc Am; Mycol Soc Am. *Res:* The distribution of fungi in tundra and temperate forest soils; effects of environment on the distribution of soil fungi; crude oil degradation by arctic soil fungi; fungal decomposition of cellulose in forest soils; vesicular-arbuscular Mycorrhizae in Maple Forest soils. *Mailing Add:* Dept Biol Sir George Williams Campus Concordia Univ 1455 De Maisoneuve Blvd W Montreal PQ H3G 1M8 Can

WIDDER, JAMES STONE, b Cleveland, Ohio, Feb 28, 35; m 60; c 2. IMMUNOLOGY, BIOCHEMISTRY. *Educ:* Ohio State Univ, BS, 57, MS, 59, PhD(immunol), 62. *Prof Exp:* Res microbiologist, Miami Valley Lab, 62-64, prod researcher, Winton Hill Tech Ctr, 64-74, sect head basic skin res & proj, 74-76, ASSOC DIR, MIAMI VALLEY LAB, PROCTER & GAMBLE CO, 76-; PROF, DEPT MOLECULAR BIOL, UNIV CINCINNATI, 86- *Concurrent Pos:* Lectr, Univ Cincinnati, 62- *Mem:* AAAS; Am Soc Microbiol; Int Asn Dent Res (secy, 66-67); Sigma Xi; Am Asn Clin Immunol & Allergy. *Res:* Product development of biologically oriented health care products and associated governmental issues. *Mailing Add:* Miami Valley Labs Procter & Gamble PO Box 39175 Cincinnati OH 45247

WIDEBURG, NORMAN EARL, b Chicago, Ill, Mar 8, 33; m 58; c 4. BIOCHEMISTRY. *Educ:* Ill Inst Technol, BS, 54; Univ Wis, MS, 56. *Prof Exp:* BIOCHEMIST, ABBOTT LABS, 58- *Mem:* AAAS; Am Chem Soc; Am Soc Microbiol. *Res:* Microbial transformations; fermentation; antimicrobial agents; enzymology. *Mailing Add:* Abbott Labs Abbott Park Ap-9A Abbott Park IL 60064

WIDEMAN, CHARLES JAMES, b Walkermine, Calif, Feb 7, 36; m 63; c 2. GEOPHYSICS. *Educ:* Colo Sch Mines, BSc, 58, MSc, 67, PhD(geophys), 75. *Prof Exp:* Sr geophysicist, Westinghouse Elec Corp, 67-68; asst prof, 68-73, ASSOC PROF GEOPHYS & CHMN DEPT, MONT COL MINERAL SCI & TECHNOL, 73- *Mem:* Seismol Soc Am. *Res:* Local seismicity; seismic risk analysis and earth strain studies; gravity investigations over and near the Boulder Batholith of Southwestern Montana. *Mailing Add:* Mont Col Mineral Sci Butte MT 59701

WIDEMAN, CYRILLA HELEN, b Toledo, Ohio. PHYSIOLOGY, BIOCHEMISTRY. *Educ:* Notre Dame Col, Ohio, BS, 49; Univ Notre Dame, MS, 60; Ill Inst Technol, PhD(biol), 70. *Prof Exp:* High sch teacher natural sci, Notre Dame Acad, Elyria Cath High Sch, 49-56; instr biol & chem, Notre Dame Col, 56-61, asst prof biol, 61-67; grad asst, Ill Inst Technol, 67-70; fel biochem & physiol, Cleveland Clin, 70-72; assoc prof biol, 72-77, PROF BIOL, JOHN CARROLL UNIV, 77- *Concurrent Pos:* NIH fel, 70-72; NSF & John Carroll Univ grant, 74. *Mem:* AAAS; Am Inst Biol Sci; NY Acad Sci; Soc Neurosci. *Res:* Neuroendocrinological and biochemical relationships underlying brain behavior patterns with special emphasis on the limbic system, especially hippocampal formation. *Mailing Add:* Dept Biol John Carroll Univ 20700 N Park Blvd Cleveland OH 44118

WIDEMAN, LAWSON GIBSON, b Morrelton, Mo, July 17, 43; m 65; c 3. CATALYSIS, POLYMER CHEMISTRY. *Educ:* Univ Mo, Rolla, BS, 66, MS, 67; Univ Akron, PhD(chem), 71. *Prof Exp:* Staff res chemist, 67-71, assoc scientist, 84-87, SR RES CHEMIST, RES DIV, GOODYEAR TIRE & RUBBER CO, 71-, GROUP LEADER, 83-, RES & DEVELOP ASSOC, 87- *Concurrent Pos:* Mem, Goodyear Sci Adv Coun, 80-, chmn, Chem Control Adv Coun, 85-; part-time mem fac, Univ Akron, 81- *Mem:* Am Chem Soc; Sigma Xi; Catalysis Soc. *Res:* Homogeneous and heterogeneous catalysis; polymer modifications; new monomers; homo- and heterogeneous hydrogenation reactions. *Mailing Add:* Goodyear Tire & Rubber Co 142 Goodyear Blvd Akron OH 44316

WIDEMAN, ROBERT FREDERICK, JR, b Dallas, Tex, June 16, 49; m 71; c 1. AVIAN PHYSIOLOGY, RENAL PHYSIOLOGY. *Educ:* Univ Del, BA, 71; Univ Conn, MS, 74, PhD(physiol), 78. *Prof Exp:* Fel physiol, Col Med, Univ Ariz, 78-81; ASST PROF AVIAN PHYSIOL, POULTRY SCI DEPT, PA STATE UNIV, 81- *Mem:* AAAS; Am Soc Zoologists; Am Physiol Soc; Poultry Sci Asn; Sigma Xi. *Res:* Avian renal and endocrine physiology; endocrinological regulation of renal calcium and phosphate transport; renal and endocrine microanatomy. *Mailing Add:* Pa State Univ 206 William L Henning Bldg University Park PA 16802

WIDENER, EDWARD LADD, SR, b Madison, SDak, Dec 23, 26; m 52; c 4. MATERIALS SCIENCE, INDUSTRIAL & MANUFACTURING ENGINEERING. *Educ:* Purdue Univ, West Lafayette, BS, 49, BS, 51; Univ Kans, MS, 62. *Prof Exp:* Fuel engr, US Steel Corp, 51-52; design engr, Union Carbide Corp, 52-60; instr eng mech, Univ Kans, 60-62; develop engr, E I du Pont de Nemours Co, 62-68; process engr, Kimberly-Clark Corp, 68-75; proj engr, Continental Group, Inc, 75-78; ASSOC PROF MECH ENG TECHNOL, PURDUE UNIV, 78- *Concurrent Pos:* Lab instr, United Asn Welders & Pipefitters, 79-89; plastics consult, Steel Parts Co & Insilco Corp, 80-81; prog evaluator, Accreditation Bd Eng & Technol, 84-90; panelist, NSF, 90-91. *Mem:* Am Soc Eng Educ; Am Soc Mech Engrs; Instrument Soc Am; Soc Mat Int; Tech Asn Pulp & Paper Indust. *Res:* Materials labs; waste disposal and packaging; energy conservation; materials recycling. *Mailing Add:* Mech Eng Technol Dept Knoy 119 Purdue Univ West Lafayette IN 47907

WIDERA, GEORG ERNST OTTO, b Dortmund, WGer, Feb 16, 38; US citizen; m 74; c 2. ENGINEERING MECHANICS. *Educ:* Univ Wis-Madison, BS, 60, MS, 62, PhD(eng mech), 65. *Prof Exp:* From asst prof to prof eng mech, 65-82, PROF MECH ENG, UNIV ILL, CHICAGO, 82-, HEAD DEPT, 83- *Concurrent Pos:* Alexander von Humboldt fel, Univ Stuttgart, 68-69, vis prof, 68; vis scientist, Argonne Nat Lab, 68; vis prof, Univ Wis-Milwaukee, 73-74 & Marquette Univ, 78-79; assoc ed, J Pressure Vessel Technol, 77-81, tech ed, 83-, assoc ed, Appl Mech Revs, 87-; consult various indust orgn; chmn, Pressure Vessel Res Comt, Am Soc Mech Engrs, 82-87, mem bd, Pressure Technol Codes & Standards; chmn, Subcomt Design Procedures Cylindrical Shells, Pressure Vessel Res Coun, Welding Res Coun, 84-87, Comt, Reinforced Openings & External Leads, 87-90, prog chmn, Pressure Vessel Piping Div, Am Soc Mech Engrs, 85-89, vchmn, 89-90, chmn, 90-91. *Mem:* Am Acad Mech; fel Am Soc Mech Engrs; Ger Soc Appl Math & Mech; Am Soc Civil Engrs; Soc Mfg Engrs; Am Soc Eng Educ; Soc Plastics Engrs. *Res:* Plates and shells; composite materials; mechanics of deformation processing; pressure vessels and piping. *Mailing Add:* Dept Mech Eng Box 4348 Univ Ill Chicago IL 60680

WIDERQUIST, V(ERNON) R(OBERTS), b Ft Myers, Fla, Sept 21, 22; m 49; c 3. ELECTRICAL ENGINEERING. *Educ:* Ga Inst Technol, BS, 43, MS, 48. *Prof Exp:* Res prof, Eng Exp Sta, Ga Inst Technol, 46-56; proj mgr, Defense Systs Group, TRW INC, 56-86; RETIRED. *Mem:* Inst Elec & Electronics Engrs; Sigma Xi. *Res:* Program and general management. *Mailing Add:* PO Box 58333 Houston TX 77258-8333

WIDESS, MOSES B, b Sverdlovsk, Russia, Sept 21, 11; nat US; m 35; c 2. GEOPHYSICS. *Educ:* Calif Inst Technol, BS, 33, MS, 34, PhD(elec eng), 36. *Prof Exp:* Party chief, Western Geophys Co, Calif, 36-42; consult geophysicist, Amoco Prod Co, 42-73. *Concurrent Pos:* Consult, 73- *Honors & Awards:* Kauffman Gold Medal Award, Soc Explor Geophys, 77. *Mem:* Soc Explor Geophys; Am Geophys Union; Europ Asn Explor Geophys. *Res:* Geophysical interpretation and methods. *Mailing Add:* 11617 Monica Lane Houston TX 77024

WIDGER, WILLIAM RUSSELL, PHOTOSYNTHESIS, MEMBRANE PROTEINS. *Educ:* State Univ NY, Albany, PhD(chem), 79. *Prof Exp:* Res assoc, Lilly Hall Life Sci, Purdue Univ, 84-86; ASST PROF BIOENERGETICS, DEPT BIOCHEM & BIOPHYS, UNIV HOUSTON, UNIVERSITY PARK, 86- *Res:* Bioenergetics. *Mailing Add:* Dept Biochem & Biophys Sci Univ Houston SR No 1 4800 Calhoun Ave Houston TX 77204

WIDGOFF, MILDRED, b Buffalo, NY, Aug 24, 24; c 2. ELEMENTARY PARTICLE PHYSICS. *Educ:* Univ Buffalo, BA, 44; Cornell Univ, PhD(physics), 52. *Prof Exp:* Asst physics, Manhattan Proj, 44-45; assoc physics, Brookhaven Nat Lab, 52-54; res fel, Harvard Univ, 55-58; from res asst prof to res assoc prof, 58-74, exec officer dep, 68-81, PROF PHYSICS, BROWN UNIV, 74- *Mem:* Fel Am Phys Soc. *Res:* Cosmic rays; medium and high energy particle physics; muons and neutrinos from astrophysical sources. *Mailing Add:* Dept Physics Brown Univ Box 1843 Providence RI 02912

WIDHOLM, JACK MILTON, b Watseka, Ill, Mar 11, 39; m 64; c 3. PLANT PHYSIOLOGY, GENETICS. *Educ:* Univ Ill, BS, 61; Calif Inst Technol, PhD(biochem), 66. *Prof Exp:* Res chemist, Int Minerals & Chem Corp, 65-68; asst prof physiol dept agron, 68-73, assoc prof, 73-77, PROF PLANT PHYSIOL DEPT AGRON, UNIV ILL, URBANA, 77- *Mem:* Am Soc Plant Physiol; Tissue Cult Asn; Scand Soc Plant Physiol; Am Soc Agron; AAAS; Int Asn Plant Tissue Cult; Int Asn Plant Molecular Biol. *Res:* Plant biochemistry and genetics, especially genetic manipulation, control of amino acid biosynthesis and photorespiration. *Mailing Add:* Dept Agron Univ Ill 1102 S Goodwin Urbana IL 61801

WIDIN, KATHARINE DOUGLAS, b Cleveland, Ohio, Oct 1, 52; m 74. BIOLOGY. *Educ:* Kenyon Col, AB, 74; Univ Minn, MS, 77, PhD(phytopath), 80. *Prof Exp:* Lectr biol & microbiol, 80-81, ASST PROF BIOL, MICROBIOL & BOT, CURRY COL, MILTON, MASS, 81- *Concurrent Pos:* Lectr entom & plant dis, Mass Bay Community Col, 80-82; consult, Plant Insect & Dis Clins, Regional Garden Ctrs, 81- *Mem:* Am Phytopath Soc; Mycol Soc Am; Sigma Xi. *Mailing Add:* 13457 Sixth St N Stillwater MN 55082

WIDMAIER, ERIC PAUL, b New York, NY, May 14, 57; m 79; c 1. PHYSIOLOGICAL STRESS, METABOLISM. *Educ:* Northwestern Univ, BA & MS, 79, Univ Calif, San Francisco, PhD(endocrinol), 84. *Prof Exp:* Postdoctoral fel, endocrinol, Worchester Found Exp Biol, 84-86, The Salk Inst, 86-88; ASST PROF BIOL, BOSTON UNIV, 88- *Mem:* AAAS; Am Diabetes Soc; Am Physiol Soc; Endocrine Soc. *Res:* Endocrine and nervous compensatory responses to stress; interaction between stress and metabolism; neuroendocrinology; control of hypothalamic/pituitary/adrenal axis in normal and metabolically-challenged animals; effects of glucose and fats on hormone secretion. *Mailing Add:* Dept Biol Boston Univ Five Cummington St Boston MA 02215

WIDMAIER, ROBERT GEORGE, b Riverside, NJ, June 18, 48; m 77. BIOCHEMISTRY, PHARMACEUTICAL CHEMISTRY. *Educ:* ECarolina Univ, BS, 70; Purdue Univ, PhD(biochem), 76. *Prof Exp:* Res asst biochem, Purdue Univ, 71-74, teaching asst, 74-76; res biochemist basic food sci, Res & Develop, Kraft Inc, 76-79; vpres, tech dir & res & develop, Kurth Malting Co, 79-83; RES BIOCHEMIST, GREAT WESTERN MALTING CO, 83- *Mem:* Am Chem Soc; Inst Food Technologists. *Res:* Flavor development in cultured food products; enzyme analysis; clinical diagnosis; drug metabolism; immunology; agricultural biotechnology research in small grains and cereal crops, applications of enzymes and biopolymers in product development for foods and food ingredients. *Mailing Add:* Penwest 777 108th Ave NE Suite 2390 Bellevue WA 98004-5193

WIDMANN, FRANCES KING, b Boston, Mass, July 23, 35; div; c 2. PATHOLOGY. *Educ:* Swarthmore Col, BA, 56; Western Reserve Univ, MD, 60; Am Bd Path, dipl & cert anat & clin path, 65, cert immunohemat, 73. *Prof Exp:* Intern, Cleveland Metrop Gen Hosp, Ohio, 60-61; resident anat & clin path, Sch Med, Univ NC, Chapel Hill, 61-64; resident clin path, Norfolk Gen Hosp, Va, 64-65, staff pathologist, 65-66; from instr to asst prof path, Sch Med, Univ NC, Chapel Hill, 66-70; asst prof, 71-73, assoc dir, 71-73, DIR, SCH MED TECHNOL, DUKE UNIV, 73-, ASSOC PROF PATH, SCH MED, 73-; ASST CHIEF LAB SERV, VET ADMIN HOSP, 72- *Honors & Awards:* John Elliott Award, Am Asn Blood Banks, 84. *Mem:* Am Asn Blood Banks. *Res:* Blood banking and medical education, especially in clinical pathology and medical technology training. *Mailing Add:* Dept Path-1504 Duke Univ Durham NC 27710

WIDMAYER, DOROTHEA JANE, b Washington, DC, Oct 10, 30. ZOOLOGY, MICROBIAL GENETICS. *Educ:* Wellesley Col, BA, 53, MA, 55; Ind Univ, PhD(zool), 62. *Prof Exp:* Instr biol, Simmons Col, 55-57; instr zool, 61-63, from asst prof to assoc prof biol, 63-74, prof, 74-75, KENAN PROF BIOL, WELLESLEY COL, 75-, CHMN DEPT BIOL SCI, 72- *Concurrent Pos:* NSF sci fac fel inst animal genetics, Univ Edinburgh, 67-68; grant, Ascent of Man, Res Corp, 74. *Mem:* AAAS; Sigma Xi; Am Soc Zool; Soc Protozool. *Res:* Gene action and cytoplasmic inheritance in Paramecium aurelia. *Mailing Add:* Dept Biol Sci Wellesley Col Wellesley MA 02181

WIDMER, ELMER ANDREAS, b Dodge, NDak, Apr 27, 25; m; c 2. HELMINTHOLOGY, MEDICAL PARASITOLOGY. *Educ:* Union Col, Nebr, BA, 51; Univ Colo, MA, 56; Colo State Univ, PhD(zool), 65; Univ NC, MPH, 74. *Prof Exp:* Teacher, High Sch, 52-53; from instr to assoc prof biol, La Sierra Col, 53-67; assoc prof environ & trop health, Loma Linda Univ, 67-71, chmn dept, 67-78, interim assoc dean, 70-80, assoc dean, acad affairs, Sch Health, 80-82, chmn dept, 84-88, PROF ENVIRON & TROP HEALTH, LOMA LINDA UNIV, 71- *Concurrent Pos:* Fel, Sch Med, La State Univ, 66; WHO fel, Africa, 71. *Mem:* Am Inst Biol Sci; Nat Environ Health Asn; Am Soc Parasitol; Am Soc Trop Med & Hyg; Wildlife Dis Asn; Sigma Xi; Soc Vector Ecol. *Res:* Reptilian parasitology; tropical helminthology; host-parasite interactions. *Mailing Add:* Dept Environ Sch Pub Health Loma Linda Univ Loma Linda CA 92350

WIDMER, KEMBLE, b New Rochelle, NY, Feb 26, 13; m 39; c 2. GEOLOGY. *Educ:* Lehigh Univ, AB, 37; Princeton Univ, MA, 47, PhD(geol), 50. *Prof Exp:* From instr to asst prof geol, Rutgers Univ, 48-50; assoc prof & chmn dept, Champlain Col, NY, 50-53; prin geologist, Div Sci, Bur Geol & Topog, NJ Dept Environ Protection, 53-58, nuclear indust coordr, 63-68, state geologist, 58-80, nuclear indust coordr, 63-74; RETIRED. *Concurrent Pos:* Tech consult, US Mil Acad, 63-81; seminar assoc, Columbia Univ; mem adv comt on water data for pub use, US Geol Surv, 76-80. *Mem:* AAAS; Geol Soc Am; Am Inst Mining, Metall & Petrol Eng. *Res:* Areal, economic, Pleistocene and engineering geology. *Mailing Add:* 228 King George Rd Pennington NJ 08534

WIDMER, RICHARD ERNEST, b West New York, NJ, June 19, 22; m 49; c 3. HORTICULTURE. *Educ:* Rutgers Univ, BS, 43, MS, 49; Univ Minn, PhD(hort), 55. *Prof Exp:* Instr hort, 49-55, from asst prof to assoc prof, 55-64, prof hort, Univ Minn, St Paul, 64-88; RETIRED. *Concurrent Pos:* Fulbright study grant, Agr Inst, Ireland, 68-69; AID consult, Hassan II Inst Agron & Vet Med, Rabat, Morocco, 73; sr res fel, NZ Nat Res Adv Coun, Levin Hort Res Ctr, 80-81; consult, Integrated Agr Prod & Mkt Proj, Philippines, 83. *Honors & Awards:* Alex Laurie Award, Soc Am Florists, 81. *Mem:* Fel Am Soc Hort Sci; Int Hort Soc. *Res:* Physiological studies of commercial greenhouse crops; breeding of garden chrysanthemums; ornamental horticulture. *Mailing Add:* 1275 Raymond Ave St Paul MN 55108-1817

WIDMER, ROBERT H, b Hawthorne, NJ, May 17, 16; m 45; c 2. AERONAUTICAL ENGINEERING, SYSTEMS ENGINEERING. *Educ:* Rensselaer Polytech Inst, BS, 38; Calif Inst Technol, MS, 39. *Hon Degrees:* ScD, Tex Christian Univ, 67. *Prof Exp:* Chief aerodyn, Ft Worth Div, Gen Dynamics Corp, 49-51, asst chief engr tech design, 51-59, chief engr, 59-61, vpres res & eng, 61-71, vpres res & eng, Convair Div, Ft Worth & San Diego, 71-74, corp vpres sci & eng, St Louis, Gen Dynamics Corp, 74-81; CONSULT, 81- *Concurrent Pos:* Mem comts aerodyn & propulsion, Nat Adv Comt Aeronaut, 48-58; consult res & eng, Off Asst Secy Defense, 58-64 & USAF Sci Adv Bd, 54-58; mem bd dirs, Univ Tex Eng Found, 56-68, Tex Christian Univ Res Found, 73-77 & Southern Methodist Univ Found Sci & Eng, 73-; chmn, Peer Comn Aero/Astro, Nat Acad Eng, 81-84. *Honors & Awards:* Field of Sci Award, Air Force Asn, 49; Spirit of St Louis Medal, Am Soc Mech Engrs, 63; Reed Aeronaut Medal, Am Inst Aeronaut & Astronaut, 83. *Mem:* Nat Acad Eng; fel Am Inst Aeronaut & Astronaut; Nat Soc Prof Engrs; Air Force Asn. *Mailing Add:* 4765 Overton Woods Dr Ft Worth TX 76109

WIDMER, WILBUR JAMES, b West New York, NJ, Oct 20, 18; m 50; c 3. ENVIRONMENTAL ENGINEERING, LIMNOLOGY. *Educ:* Cooper Union, BCE, 46; Mass Inst Technol, SM, 48. *Prof Exp:* Technician, Gibbs & Cox, Inc, NY, 43-47; from instr to prof, 48-88, EMER PROF, CIVIL ENG, UNIV CONN, 88- *Concurrent Pos:* Consult, Conn State Water Resources Comn, 50-52, C W Riva Co, RI, 50-63 & J M Minges Assocs, Conn, 59-63; res asst, Mass Inst Technol, 56-57; NSF sci fac fel biol oceanog, Narragansett Marine Lab, RI, 64-66; WHO assignment as prof sanit eng, WPakistan Univ Eng & Technol, Lahore & sanit eng adv, Govt Pakistan, 68-70; sanit eng consult in Brazil, Pan-Am Health Orgn, 72; sanitary eng consult, WHO, Egypt, Saudia Arabia & Sudan, 79. *Honors & Awards:* Bedell Award Water Pollution Control, Water Pollution Control Fedn, 72. *Mem:* Fel Royal Soc Health; fel Am Soc Civil Engrs; Am Soc Limnol & Oceanog; Am Water Works Asn; Am Acad Environ Engrs. *Res:* Eutrophication and pollution ecology of fresh and marine waters; waste water treatment systems. *Mailing Add:* Dept Civil Eng Box U-37 Univ Conn Storrs CT 06268

WIDMOYER, FRED BIXLER, b Grandfield, Okla, Nov 25, 20; m 61; c 3. HORTICULTURE. *Educ:* Tex Tech Col, BA, 42, MS, 50; Mich State Univ, PhD(ornamental hort), 54. *Prof Exp:* Instr bot, Tex Tech Col, 46-50; asst prof & exten specialist hort, Mich State Univ, 54-60; assoc prof landscape design & nursery mgt, Univ Conn, 60-63; prin horticulturist, Sci & Educ Admin-Coop Res, USDA, 77-78, prof hort & head dept, 63-77 & 78-82; prof hort/exten hort, 82-86, EMER PROF HORT, NMEX STATE UNIV, 86- *Concurrent Pos:* Mem adv comt, Nat Arboretum; mem rev team hort, Coop State Res Serv, USDA; tech adv, Int Water & Boundry Comn; reviewer, gen operating support grants, IMS; adv comt, NMex Parks & Recreation; Paul Harris fel, 89. *Honors & Awards:* Esther Longyear Murphy Medal, 61; Am Hort Soc Sci Citation, 86; Hon Am Farmer, 85; Am Hort Soc Sci Citation, 86. *Mem:* Am Asn Bot Gardens & Arboretums (pres, 82-83); fel Am Soc Hort Sci; Garden Writers Asn Am. *Res:* Ornamental horticulture; plant anatomy; growth regulators and plant propagation; developmental morphology. *Mailing Add:* Dept Hort NMex State Univ Box 3530 Las Cruces NM 88003

WIDNALL, SHEILA EVANS, b Tacoma, Wash, July 13, 38; m 60; c 2. AERONAUTICAL ENGINEERING, FLUID MECHANICS. *Educ:* Mass Inst Technol, BS, 60, MS, 61, ScD(aeronaut eng), 64. *Hon Degrees:* ScD, New Eng Col, 75, Lawrence Univ, 87, Cedar Crest Col, 88, Smith Col, 90. *Prof Exp:* Res asst aerodyn, Mass Inst Technol, 61-64, Ford fel, 64-66, asst prof, 64-70, assoc prof, 70-74, chairperson, women fac, 76-77, head, Dynamics Div, Dept Aeronaut & Astronaut, 78-79, PROF AERONAUT, MASS INST TECHNOL, 74- *Concurrent Pos:* Consult several industs; dir univ res, US Dept Transp, Washington, DC, 74-75; mem space & aeronaut bd, Nat Acad Eng, 75-78; mem adv comt, NSF, USAF & US Dept Transp; mem bd gov, USAF Acad, 78-; assoc ed, J Appl Mech, 83- *Honors & Awards:* Lawrence Sperry Award, Am Inst Aeronaut & Astronaut, 72; Outstanding Achievement Award, Soc Woman Engrs, 75. *Mem:* Nat Acad Eng; fel Am Inst Aeronaut & Astronaut; fel Am Phys Soc; Soc Women Engrs; fel AAAS; Am Soc Mech Engrs. *Res:* Unsteady aerodynamics; aeroelasticity; aerodynamic noise; turbulence; applied mathematics; vortex flows; numerical analysis; aerospace; transportation; aerodynamics and fluid mechanics; acoustics; noise and vibration. *Mailing Add:* Dept Aeronaut & Astronaut Rm 33-218 Mass Inst Technol 77 Massachusetts Ave Cambridge MA 02139

WIDNELL, CHRISTOPHER COURTENAY, b London, Eng, May 19, 40; m 65; c 2. CELL BIOLOGY, BIOCHEMISTRY. *Educ:* Cambridge Univ, BA, 62; Univ London, PhD(biochem), 65. *Prof Exp:* Res assoc cell biol, Rockefeller Univ, 66-68; staff mem biochem, Nat Inst Med Res, London, Eng, 68-69; assoc prof, 69-77, PROF ANAT & CELL BIOL, SCH MED, UNIV PITTSBURGH, 77- *Concurrent Pos:* Jane Coffin Childs fel, Univ Chicago, 65-66; ed, Arch Biochem & Biophys, 72-; sr Fogarty int fel, Int Inst Cellular & Molecular Path, Brussels, 78-79. *Mem:* Am Soc Biol Chem; Am Soc Cell Biol; Brit Biochem Soc. *Res:* Membrane structure and function; cytochemical localization of membrane proteins; synthesis and assembly of membrane components; cellular aging; endocytosis and membrane recycling. *Mailing Add:* Dept Neurobiol Anat & Cell Sci Univ Pittsburgh Sch Med 3550 Terrace St Pittsburgh PA 15261

WIDNER, JIMMY NEWTON, b Clovis, NMex, Feb 10, 42; m 64; c 2. CROP BREEDING. *Educ:* NMex State Univ, BS, 64; NDak State Univ, PhD(agron), 68. *Prof Exp:* Plant breeder, Great Western Sugar Co, Colo, 68-72; res mgr, Northern Ohio Sugar Co, 72-75; SR PLANT BREEDER, GREAT WESTERN SUGAR CO, 75- *Mem:* Am Soc Agron; Crop Sci Soc Am; Am Soc Sugar Beet Technol. *Res:* Development of improved varieties and hybrids of sugar beets. *Mailing Add:* 11939 Sugarmill Rd Longmont CO 80501

WIDNER, WILLIAM RICHARD, b Baxter Co, Ark, Apr 24, 20; m 43; c 1. PHYSIOLOGY, BACTERIOLOGY. *Educ:* Eastern NMex Univ, AB, 42; Univ NMex, MS, 48, PhD, 52. *Prof Exp:* Lab asst biol, Eastern NMex Univ, 42; asst, Univ NMex, 46-48; biomed researcher, Los Alamos Sci Lab, 48-50; asst, Univ NMex, 50-52; indust hygienist, Sandia Corp, 52-55; teacher, Albuquerque Indian Sch, NMex, 55-56; prof biol & head dept, Howard Payne Col, 56-59; asst prof biol & bact, 59-64, PROF BIOL, BAYLOR UNIV, 64- *Mem:* AAAS; Am Soc Microbiol; Sigma Xi (treas, 73-74). *Res:* Cell mitoses and growth of normal and malignant tissues; effects of ionizing radiations on living cells; radiation produced cataracts; bacterial metabolism. *Mailing Add:* Dept Biol Baylor Univ Waco TX 76798

WIDOM, BENJAMIN, b Newark, NJ, Oct 13, 27; m 53; c 3. STATISTICAL MECHANICS, THERMODYNAMICS. *Educ:* Columbia Univ, AB, 49; Cornell Univ, PhD(chem), 53. *Hon Degrees:* DSc, Univ Chicago, 91. *Prof Exp:* Res assoc chem, Univ NC, 52-54; from instr to prof, 54-83, GOLDWIN SMITH PROF CHEM, CORNELL UNIV, 83- *Concurrent Pos:* Guggenheim & Fulbright fels, 61-62; NSF sr fel, 65; Guggenheim fel, 69; van der Waals prof, Univ Amsterdam, 72; IBM prof, Oxford Univ, 78; Lorentz prof, Leiden Univ, 85. *Honors & Awards:* Boris Pregel Award, NY Acad Sci, 76; Langmuir Award, Am Chem Soc, 82. *Mem:* Nat Acad Sci; Am Phys Soc; Am Chem Soc; NY Acad Sci; Am Acad Arts & Sci. *Res:* Phase transitions; statistical mechanics. *Mailing Add:* Dept Chem Cornell Univ Ithaca NY 14853

WIDOM, HAROLD, b Newark, NJ, Sept 23, 32; m 55; c 3. MATHEMATICS. *Educ:* Univ Chicago, SM, 52, PhD(math), 55. *Prof Exp:* From instr to prof math, Cornell Univ, 55-68; PROF MATH, UNIV CALIF, SANTA CRUZ, 68- *Concurrent Pos:* Res fels, NSF, 59-60 & Sloan Found, 61-63; Guggenheim res fel, 67-68 & 72-73. *Mem:* Am Math Soc. *Res:* operator theory. *Mailing Add:* Dept Math Univ Calif Santa Cruz Santa Cruz CA 95064

WIDRA, ABE, b Philadelphia, Pa, Jan 17, 24; m 52; c 4. MICROBIOLOGY. *Educ:* Brooklyn Col, BA, 48; Univ Fla, MS, 52; Univ Pa, PhD(med microbiol), 54; Am Bd Med Microbiol, dipl, 69. *Prof Exp:* Tech asst bact & serol, Philadelphia Gen Hosp, 49-50; res assoc cytol & cytogenetics, Univ Pa, 54-55; instr bact & immunol, Univ NC, 55-59, asst prof, 59-64; ASSOC PROF MICROBIOL, MED CTR, UNIV ILL, CHICAGO, 64- *Concurrent Pos:* Consult, Presby-St Luke's Hosp, Chicago, 66-80. *Mem:* Am Soc Microbiol. *Res:* Medical mycology; wound healing; synthetic skin. *Mailing Add:* 9273 Fairway Dr No 201 Des Plaines IL 60016-1717

WIDSTROM, NEIL WAYNE, b Hecla, SDak, Nov 11, 33; m 60; c 2. QUANTITATIVE GENETICS, PLANT BREEDING. *Educ:* SDak State Univ, BS, 59, PhD(plant breeding), 62. *Prof Exp:* Fel genetics, NC State Univ, 63-64; RES GENETICIST PLANTS, AGR RES SERV, USDA, 64- *Mem:* AAAS; fel Am Soc Agron; fel Crop Sci Soc Am; Genetics Soc Am; Am Genetic Asn. *Res:* Plant genetics; genetics of resistance to insects by plants; resistance in corn to aflatoxin contamination. *Mailing Add:* Insect Biol & Pop Mgt Res Lab PO Box 748 Tifton GA 31793

WIE, CHU RYANG, b Jangheung, Korea, Apr 25, 57; m 85; c 1. III-V SEMICONDUCTOR MATERIALS & DEVICES, X-RAY ROCKING CURVE. *Educ:* Chonnam Nat Univ, Korea, BS, 80; Seoul Nat Univ, Korea, MS, 82; Calif Inst Technol, PhD(appl physics), 85. *Prof Exp:* Res fel physics, Calif Inst Technol, 85; asst prof, 85-89, ASSOC PROF ELEC ENG, STATE UNIV NY, BUFFALO, 89- *Concurrent Pos:* Vis assoc, Calif Inst Technol, 86; NSF eng initiation award, 87 & presidential young investr award, 88. *Mem:* Am Phys Soc; Mat Res Soc; Sigma Xi; NY Acad Sci; Soc Photo-Optical Instrumentation Engrs; Inst Elec & Electronics Engrs. *Res:* X-ray rocking curve analysis of semiconductor heterostructures; materials and devices of lattice-mismatched III-V heterostructures; semi-insulating III-V materials; strained resonant tunneling structures; electrical and optical characterization of semiconductors. *Mailing Add:* Dept Elec & Computer Eng State Univ NY 201 Bonner Hall Buffalo NY 14260

WIEBE, DONALD, b Indicott, Nebr, June 30, 23; m 45; c 2. ENGINEERING. *Educ:* WVa Univ, BS, 49, MS, 59. *Prof Exp:* Res engr, Joy Mfg Co, 49-51; asst prof mining eng, WVa Univ, 51-53; mgr exp sta, Joy Mfg Co, 53-63; mgr eng mech, Astronuclear Lab, Westinghouse Elec Corp, 63-64; mgr, 64-76, vpres res & eng, 76-86, CONSULT, A STUCKI CO, 86- *Honors & Awards:* Arnold Stucki Award, Am Soc Mech Engrs, 89. *Mem:* Soc Exp Mech; Air Pollution Control Asn; Am Inst Mining, Metall & Petrol Engrs; fel Am Soc Mech Engrs; Instrument Soc Am. *Res:* Product engineering and development; mining, industrial-dust collection; applied research; experimental mechanics; railcar suspension developments; holder of 30 US letters patents and 49 foreign patents relating to mining, industrial and rail transport equipment. *Mailing Add:* 106 Woodland Rd Sewickley PA 15143

WIEBE, H ALLAN, precipitation scavenging, analytical methods, for more information see previous edition

WIEBE, JOHN, b Sask, June 3, 26; m 47; c 6. HORTICULTURE. *Educ:* Ont Agr Col, BSA, 51; Cornell Univ, MS, 53, PhD, 55. *Prof Exp:* Res scientist, Hort Res Inst, Ont Dept Agr, 55-76; DIR, PLANT INDUST DIV, ALTA DEPT AGR, EDMONTON, 76- *Res:* Viticulture and physiology. *Mailing Add:* Dept Practical Theol Can Mennonite Bible Col 600 Shaftesbury Blvd Winnipeg MB R3P 0M4 Can

WIEBE, JOHN PETER, b Neu-Scönsee, Ukraine, Aug 28, 38; Can citizen; m 64; c 2. ENDOCRINOLOGY, CONTRACEPTION. *Educ:* Univ BC, BSc, 63, PhD(physiol), 67. *Prof Exp:* Nat Res Coun Can res fel zool, Univ Leeds, 68-69; asst prof endocrinol, Tex A&M Univ, 70-72; asst prof physiol, 72-77, from asst prof to assoc prof, 74-85, PROF ZOOL & PHYSIOL, UNIV WESTERN ONT, 85- *Mem:* AAAS; Can Soc Cell Biol; Can Biochem Soc; Soc Study Reproduction; Am Soc Andrology; Soc Adv Contraception. *Res:* Steriods; contraception; reproductive endocrinology. *Mailing Add:* Dept Zool Univ Western Ont London ON N6A 5B9 Can

WIEBE, LEONARD IRVING, b Swift Current, Sask, Oct 14, 41; m; c 3. BIONUCLEONICS. *Educ:* Univ Sask, BSP, 63, MSc, 66; Univ Sydney, PhD(drug metab), 69. *Prof Exp:* Asst prof pharmaceut chem bionucleonics, Univ Alta, 70-73, assoc prof, 73-78, chmn res reactor comt, 74-89, PROF RADIOPHARM CHEM, UNIV ALTA, 78-, CHMN BIONUCLEONICS DIV, 74-75 & 87- *Concurrent Pos:* Sessional lectr, Univ Sask, 65-66; fel, Univ Alta, 69-70; sessional lectr, Univ Sydney, 73; AUH fel, 76; von Humboldt fel, Ger Cancer Res Ctr, 76-77; vis prof, PMAC, Searle & MRC, 86; res assoc, Cross Cancer Inst, 78-; dir, biomed & health, Australian Nuclear Sci Technol Orgn, 90. *Honors & Awards:* McNeil Res Prize, 87. *Mem:* Asn Fac Pharm Can; Am Pharm Asn; Can Asn Radiopharm Sci; Soc Nuclear Med; Am Asn Pharm Sci; Can Radiopharm Sci Asn. *Res:* Production of short-lived radionuclides for incorporation into radiopharmaceuticals for use in diagnostic nuclear medicine; emphasis is on radiohalogenated pyrmidine nucleosides and nitroimidazole radiosensitizers for oncology and virology. *Mailing Add:* Div Bionucleonics & Radiopharm Univ Alta Edmonton AB T6G 2N8 Can

WIEBE, MICHAEL EUGENE, b Newton, Kans, Oct 1, 42; m 65; c 1. CELL BIOLOGY, VIROLOGY. *Educ:* Sterling Col, BS, 65; Univ Kans, PhD(microbiol), 71. *Prof Exp:* Fel microbiol, Duke Univ Med Ctr, 71-73; asst prof, Med Col, Cornell Univ, 73-81, adj assoc prof, 81-84; assoc dir, Blood Derivatives Prog, New York Blood Ctr, 80-83, assoc investr, Lindsley F Kimball Res Inst, 80-84, dir, Leukocyte Prod, 83-84; sr scientist, Cell Banking Res & Develop, 84-88, assoc dir, Med & Anal Chem, 88-90, DIR QUAL CONTROL MICROBIOL, GENENTECH INC, 90- *Concurrent Pos:* Mem, Subcomt Interrelationships among Catalogued Arboviruses, Am Comt Arthropod-born Viruses, 77-80; prin investr contracts, US Army Med Res & Develop Command, 77-81; mem, Crit Rev Microbiol Adv Bd, 82-86; mem bd trustees, Sterling Col, Kans, 90- *Mem:* Am Soc Microbiol; Am Soc Virol; AAAS; Soc Exp Biol & Med; Parenteral Drug Asn. *Res:* Molecular virology and cell biology; production of human therapeutics by genetically engineered mammalian cells; quality control of genetically engineered biopharmaceuticals. *Mailing Add:* Qual Control Dept Genentech Inc 460 Pt San Bruno Blvd South San Francisco CA 94080

WIEBE, PETER HOWARD, b Salinas, Calif, Oct 2, 40; m 68; c 2. BIOLOGICAL OCEANOGRAPHY, MARINE BIOLOGY. *Educ:* Ariz State Col, Flagstaff, BS, 62; Scripps Inst Oceanog, Univ Calif, San Diego, PhD(biol oceanog), 68. *Prof Exp:* Fel biol oceanog, Hopkins Marine Sta, Stanford Univ, 68-69; asst scientist, 69-74, assoc scientist, 74-84, SR SCIENTIST, BIOL OCEANOG, WOODS HOLE OCEANOG INST, 84- *Concurrent Pos:* Numerous NSF grants & Off of Naval Res contracts. *Mem:* Am Soc Limnol & Oceanog; AAAS. *Res:* Quantitative ecology of zooplankton with emphasis on the biological and physical-chemical factors which act to regulate the distribution and abundance of oceanic populations and communities. *Mailing Add:* 11 Boulder Circle North Falmouth MA 02556

WIEBE, RICHARD PENNER, b Pittsburgh, Pa, Jan 5, 28; m 56; c 2. MATHEMATICAL LOGIC. *Educ:* Univ Ill, Urbana, BS, 49, MS, 51; Univ Calif, Berkeley, PhD(philos), 64. *Prof Exp:* Instr philos, Johns Hopkins Univ, 60-62; asst prof math, St Mary's Col, Calif, 63-88; RETIRED. *Res:* Foundations of mathematics; semantics; philosophy of science and mathematics. *Mailing Add:* 852 Contra Costa Ave Berkeley CA 94707-1920

WIEBE, ROBERT A, b San Meteo, Calif, Nov 2, 39; m 65. PETROLOGY. *Educ:* Stanford Univ, BS, 61; Univ Wash, Seattle, MS, 63; Stanford Univ, PhD, 66. *Prof Exp:* Asst prof geol, 66-73, ASSOC PROF GEOL, FRANKLIN & MARSHALL COL, 73-, CHMN DEPT, 76- *Concurrent Pos:* NATO fel, Univ Edinburgh, 72-73. *Mem:* Geol Soc Am; Mineral Soc Am. *Res:* Igneous and metamorphic petrology; mineralogy; plutonic igneious rocks of the Northern Appalachians; the Nain Anorthosite-Adamellite complex. *Mailing Add:* Dept Geol Franklin & Marshall Col Box 3003 Lancaster PA 17604

WIEBE, WILLIAM JOHN, b San Mateo, Calif, Mar, 14, 35; m 60; c 3. MICROBIAL ECOLOGY, ANAEROBIC METABOLISM. *Educ:* Stanford Univ, BA, 57; Univ Wash, PhD(fisheries), 65. *Prof Exp:* Fel microbiol & electron micros, Georgetown Univ, 65-67; asst prof, 67-72, assoc prof, 72-77, PROF MICROBIOL, UNIV GA, 77- *Concurrent Pos:* Vis prof, Div Fisheries & Oceanog, Commonwealth Sci & Indust Res Orgn, 75, officer-in-chg, 79-82; mem, Panel Adv Comt Ecol & Ecosyst, NSF, 78, Sci Comt Coral Reefs, Pac Sci Asn, 79. *Mem:* Am Soc Microbiol; Am Soc Limnol & Oceanog; Australian Soc Microbiol; Western Naturalist Asn; Int Soc Coral Reefs. *Res:* Marine microbial ecology in nearshore coastal and estuarine environments; salt marsh sediment fermentation; sulfate reduction and methanogenesis, in coastal zones and coral reefs; macrophyte decomposition and nitrogen fixation. *Mailing Add:* Dept Microbiol Univ Ga Athens GA 30601

WIEBOLD, WILLIAM JOHN, b Vinton, Iowa, Oct 27, 49; m 73; c 1. AGRONOMY, CROP PHYSIOLOGY. *Educ:* Iowa State Univ, BS, 71, MS, 74; Univ Ga, PhD(agron), 78. *Prof Exp:* Res asst soybean breeding, Iowa State Univ, 69-71, teaching asst, 71, res asst soybean physiol, 72-74; res asst, Univ Ga, 74-78, instr crop sci, 77; ASST PROF SOYBEAN PHYSIOL, UNIV MD, 78- *Mem:* Am Soc Agron; Crop Sci Soc Am. *Res:* Physiological and morphological limitations to soybean yield. *Mailing Add:* Dept Agron 202 Kottman Hall Ohio State Univ Main Campus Columbus OH 43210

WIEBUSCH, CHARLES FRED, b Perry, Tex, Mar 11, 03; m 26. ACOUSTICS. *Educ:* Univ Tex, BA, 24, MA, 25. *Prof Exp:* Instr physics, Univ Tex, 25-27; mem tech staff, Bell Tel Labs, Inc, 27-45, sta apparatus engr, 45-51, dir Underwater Systs Lab, 51-62, exec dir Outside Plant & Underwater Systs Div, 62-67. *Concurrent Pos:* Consult acoust, 67-; mem, comt Undersea Warfare, Nat Acad Sci, 53-74, chmn, Comt Underwater Commun, 68-70. *Honors & Awards:* Emile Berliner Award, Audio Eng Soc, 69; VAdmiral Charles B Martell Tech Award, Nat Security Indust Asn, 79. *Mem:* Fel Acoust Soc Am; Inst Elec & Electronics Engrs. *Res:* Sources, transmission and measurement of underwater sounds; recording and reproduction of speech and music; architectural acoustics; management of research and engineering. *Mailing Add:* 700 John Ringling Blvd Apt N206 Sarasota FL 34236-1500

WIEBUSCH, F B, b Brenham, Tex, Aug 26, 23. PERIODONTOLOGY. *Educ:* Univ Tex, BBA, 43, DDS, 47; Am Bd Periodont, dipl. *Prof Exp:* Pub health dent consult, State Health Dept, Tex, 47-51; prof oral diag & therapeut & chmn dept, 54-71, PROF PERIODONT & ASST DEAN CONTINUING EDUC, MED COL VA, 71- *Concurrent Pos:* Consult, Vet Admin Hosps, Richmond & Salem, Va. *Mem:* Am Dent Asn; Am Acad Periodont; fel Int Col Dent; fel Am Col Dent; Am Acad Dent Med. *Res:* Periodontics. *Mailing Add:* MCV Box 566 Richmond VA 23298

WIEBUSH, JOSEPH ROY, b Lancaster, Pa, Oct 18, 20; m 43; c 1. ANALYTICAL CHEMISTRY, MARINE CHEMISTRY. *Educ:* Franklin & Marshall Col, BS, 41; Univ Md, MS, 51, PhD(chem), 55. *Prof Exp:* Supvr explosives dept, Hercules Powder Co, 41-43, chemist, 46-48; asst chem, Univ Md, 48-51; res chemist, Mead Corp, Ohio, 55-56; dir res, Nat Inst Drycleaning, 56-60; from assoc prof to prof chem, US Naval Acad, 60-81, chmn dept, 66-77; CONSULT, 77-; DIR, GASSER ASSOC, INC. *Concurrent Pos:* Adj prof, Univ Cent Fla, 81-89. *Mem:* Am Chem Soc; Sigma Xi. *Res:* Fluorescence analysis; toxicity of organic solvents; analytical techniques for trace elements; marine corrosion and fouling; oceanographic applications; environmental pollution; general chemistry texts. *Mailing Add:* 1830 Ramie Rd Clermont FL 34711

WIECH, NORBERT LEONARD, b Chicago, Ill, Mar 13, 39; m 61, 85; c 3. BIOCHEMICAL PHARMACOLOGY. *Educ:* Univ Notre Dame, BS, 60, MS, 63; Tulane Univ, PhD(biochem), 66. *Prof Exp:* Res assoc nutrit, Sch Pub Health, Harvard Univ, 66-67; sect head biochem pharm, Merrell Int, Strasbourg, France, 72-74, sect head biochem pharmacol, Merrell Res Ctr, 67-85, ASSOC SCIENTIST PROJ ADMIN & RES DIRECTION, MERRELL-DOW, CINCINNATI, 85- *Concurrent Pos:* Adj assoc prof exp med, Univ Cincinnati, 78-, asst prof chem, 81- *Mem:* AAAS; Am Oil Chemists Soc; Am Chem Soc; NY Acad Sci; Am Diabetes Asn. *Res:* Membrane receptors; neuropharmacology; carbohydrate-lipid metabolism in health and disease. *Mailing Add:* Innovaphram Ten Overshot Ct Phonix MD 21131-1851

WIECHELMAN, KAREN JANICE, b Central City, Nebr, Apr 30, 47; m 70; c 1. BIOPHYSICAL CHEMISTRY. *Educ:* Univ Nebr, BS, 69, PhD(biochem), 73. *Prof Exp:* Res assoc biophys, Univ Pittsburgh, 73-76; from asst prof to assoc prof chem, 76-90, CHMN DEPT, UNIV SOUTHWESTERN LA, 88-, PROF CHEM, 90- *Concurrent Pos:* Coun, Undergrad Res. *Res:* Use of fluorescence techniques to investigate various biological systems; cryoprotection of proteins. *Mailing Add:* Dept Chem Univ Southwestern La Lafayette LA 70504

WIECZOREK, GERALD FRANCIS, b Schenectady, NY, Oct 25, 49; m 75; c 1. GEOTECHNICAL ENGINEERING, GEOLOGIC HAZARD & RISK ASSESSMENT. *Educ:* Univ Calif, Berkeley, BS, 71, MS, 72, ME, 74, PhD(geol eng), 78. *Prof Exp:* Res civil engr, Br Geol Risk Assessment, Menlo Park, Calif, 78-87, dep chief eng, Off Earthquakes, Volcanoes & Eng, Reston, Va, 87-90, RES CIVIL ENGR, BR GEOL RISK ASSESSMENT, US GEOL SURV, RESTON, VA, 90- *Mem:* Int Asn Eng Geologists; Asn Eng Geologists; Am Soc Civil Engrs. *Res:* Models ground water fluctuation and reactivation of landslides for improved means of assessing probability of landsliding. *Mailing Add:* US Geol Surv Nat Ctr Mail Stop 922 Reston VA 22092

WIED, GEORGE LUDWIG, b Carlsbad, Czech, Feb 7, 21; wid. OBSTETRICS & GYNECOLOGY. *Educ:* Charles Univ, Prague, MD, 44. *Prof Exp:* Intern, County Hosp, Carlsbad, Czech, 45; resident obstet & gynec, Univ Munich, 46-48; asst, Univ Berlin, 48-52, co-chmn dept, 53; from asst prof to assoc prof, Univ Chicago, 54-65, actg chmn dept, 74-75, prof obstet & gynec, 65-, prof path, 67-, dir schs cytotechnol & cytocybernet, 59-, EMER PROF, SCH MED, UNIV CHICAGO. *Concurrent Pos:* Ed-in-chief, Acta Cytologica, 57-; ed, Monogr Clin Cytol, 64-; ed-in-chief, J Reproductive Med, 67-; ed-in-chief, Anal & Quantitative Cytol, 79- *Honors & Awards:* Surgeon Gen Cert Merit, 52; Goldblatt Cytol Award, 61; George N Papanicolaou Cytol Award, 70. *Mem:* Am Soc Cytol (pres, 65-66); Am Soc Cell Biol; Int Acad Cytol (pres elect, 74-77, pres, 77-80); Ger Soc Obstet & Gynec; Ger Soc Cytol. *Res:* Cytopathology; exfoliative cytology; biological image processing. *Mailing Add:* Dept Obstet & Gynec Box 449 Sch Med Univ Chicago 5801 Ellis Ave Chicago IL 60637

WIEDEMAN, VARLEY EARL, b Oklahoma City, Okla, Mar 14, 33; m 63; c 2. BOTANY, ECOLOGY. *Educ:* Univ Okla, BS, 57, MS, 60; Univ Tex, PhD(bot), 64. *Prof Exp:* Chemist-biologist, USPHS, 59-61; from asst prof to assoc prof, 64-74, PROF PLANT ECOL, UNIV LOUISVILLE, 64- *Mem:* Am Water Resources Asn; AAAS; Sigma Xi. *Mailing Add:* Dept Biol Univ Louisville Louisville KY 40208

WIEDEMANN, ALFRED MAX, b Chicago, Ill, Nov 24, 31; c 2. PLANT ECOLOGY. *Educ:* Utah State Univ, BS, 60, MS, 62; Ore State Univ, PhD(bot), 66. *Prof Exp:* Asst prof bot, Ore State Univ, 65-67, asst prof range mgt, 70; sci fac gen sci, N Geelong High Sch, Victoria, Australia, 69-70; MEM FAC BIOL, EVERGREEN STATE UNIV, 70- *Concurrent Pos:* Lectr bot, Univ Malaya, 67-68; fel, Commonwealth Sci & Indust Res Orgn, Australia-NSF, 68-69. *Mem:* Ecol Soc Am; Brit Ecol Soc. *Res:* Vegatation of interior and coastal sand dunes; identification and description of natural areas. *Mailing Add:* Dept Biol Evergreen State Col 13095 Yerba Buena Rd Olympia WA 98505

WIEDEMEIER, HERIBERT, b Steinheim, WGer, Aug 4, 28; nat US. INORGANIC CHEMISTRY. *Educ:* Univ Muenster, BS, 54, MSc, 57, DSc, 60. *Prof Exp:* Asst inorg & phys chem, Univ Muenster, 56-58, instr, 58-60; res assoc chem, Univ Kans, 60-62; res assoc, Univ Muenster, 62-63; res assoc, Univ Kans, 63-64; asst prof, 64-67, assoc prof phys chem, 67-72, PROF CHEM, RENSSELAER POLYTECH INST, 72- *Honors & Awards:* Medal Except Sci Achievement, NASA, 74. *Mem:* AAAS; Am Chem Soc; Ger Chem Soc. *Res:* Growth of single crystals of metal chalcogenides; crystal growth mechanism and morphology; thermodynamic and kinetic studies of condensation and vaporization processes of inorganic materials at elevated temperatures; crystal growth in zero-gravity. *Mailing Add:* Dept Phys Chem Rensselaer Polytech Inst Troy NY 12180

WIEDENBECK, MARCELLUS LEE, b Lancaster, NY, Oct 11, 19; m 46; c 6. NUCLEAR PHYSICS. *Educ:* Canisius Col, BS, 41; Univ Notre Dame, MS, 42, PhD(physics), 45. *Prof Exp:* Instr physics, Univ Notre Dame, 44-46; from asst prof to prof physics, Univ Mich, Ann Arbor, 46-87; RETIRED. *Mem:* Am Phys Soc. *Res:* Nuclear spectroscopy; beta ray and alpha ray spectra; coincidence methods; spectroscopy of some heavy nuclei. *Mailing Add:* 3786 Elizabeth Rd Ann Arbor MI 48103

WIEDENHEFT, CHARLES JOHN, b Sandusky, Ohio, Oct 23, 41. CHEMISTRY. *Educ:* Capital Univ, BS, 63; Case Western Reserve Univ, MS, 65, PhD(chem), 67. *Prof Exp:* RES SPECIALIST, MONSANTO RES CORP, 67- *Mem:* AAAS; Am Chem Soc. *Res:* Coordination compounds of the actinide ions; thermal analysis. *Mailing Add:* 85 Springwood Dr Springboro OH 45066-1038

WIEDENMANN, LYNN G, b Moline, Ill, Apr 21, 28; m 56; c 4. POLYMER CHEMISTRY, ORGANIC CHEMISTRY. *Educ:* Ill Wesleyan Univ, BS, 50; Univ Iowa, MS, 52, PhD(org chem), 55. *Prof Exp:* Asst chemist, Rocky Mountain Arsenal, 55-57; res chemist, Tex-US Chem Corp, 57-60; res chemist, Rock Island Arsenal, 60-69; PROF CHEM, BLACK HAWK COL, 69- *Mem:* Am Chem Soc; Sigma Xi. *Res:* Polymer synthesis; high temperature resistant elastomers; boron and stereoregular butadiene polymers; butadiene derivatives; antioxidants; organic phosphorus compounds; dibenzopyrylium compounds. *Mailing Add:* 2387 Fifth St East Moline IL 61244

WIEDER, GRACE MARILYN, b New York, NY, May 10, 28. INFRARED & RAMAN SPECTROSCOPY. *Educ:* Univ Vt, BA, 49; Mt Holyoke Col, AM, 51; Polytech Inst Brooklyn, PhD(phys chem), 61. *Prof Exp:* Res assoc chem, Univ Southern Calif, 60-62; instr, 62-65, asst prof, 66-77, ASSOC PROF CHEM, BROOKLYN COL, 78- *Concurrent Pos:* Vis scientist, Univ Wash, 70-71. *Mem:* Am Chem Soc; Am Phys Soc; Sigma Xi. *Res:* Stability constants and spectra of donor-acceptor complexes; Raman and infrared spectra of crystals. *Mailing Add:* Dept Chem Brooklyn Col Brooklyn NY 11210

WIEDER, HAROLD, b Cleveland, Ohio, July 18, 27; m 63; c 3. OPTICAL PHYSICS. *Educ:* Univ Rochester, BS, 50, MA, 57, MS, 58; Case Inst Technol, PhD(physics), 64. *Prof Exp:* Engr, Sarnoff Res Ctr, RCA Labs, 50-54; physicist, Parma Res Ctr, Union Carbide Corp, 57-61; physicist, Watson Res Ctr, 63-68, San Jose Lab, 68-82, & IBM Corp, 63-82. *Concurrent Pos:* Consult, 82- *Res:* Optical, magneto-optic, photoconductive, and structural properties of ordered and disordered films; transient thermal and thermomagnetic techniques; mode coupling and intra-cavity laser effects; level crossing and anticrossing spectroscopy. *Mailing Add:* 20175 Knollwood Dr Saratoga CA 95070

WIEDER, IRWIN, b Cleveland, Ohio, Sept 26, 25; m 53; c 4. MOLECULAR BIOPHYSICS, SPECTROSCOPY. *Educ:* Case Inst Technol, BS, 50; Stanford Univ, PhD(physics), 56. *Prof Exp:* Asst, Stanford Univ, 51-56; res physicist, Westinghouse Elec Corp, 56-60 & Varian Assocs, 60-61; dir res, Interphase Corp, 61-66; prin scientist, Carver Corp, 66-69; NIH spec fel, Dept Biol Sci, Stanford Univ, 70-71; vis prof, dept electronics, Weizmann Inst Sci, Israel, 71-73; pres & tech dir, Anal Radiation Corp, 74-81; SCI CONSULT, BAXTER HEALTH CARE, 82- *Mem:* Am Phys Soc; Inst Elec & Electronics Engrs. *Res:* Magnetic resonance; microwave-optical effects; optical pumping in gases, liquids and solids; masers and lasers; energy transfer in biological systems; immunofluorescent, laser induced fluorescent and fluorescent antibody spectroscopy; time gated fluorescent spectroscopy. *Mailing Add:* 459 Panchita Way Los Altos CA 94022

WIEDER, SOL, b Bronx, NY, Jan 6, 40; m 63; c 3. PHYSICS. *Educ:* City Col New York, BS, 61; NY Univ, MS, 62, PhD(physics), 66. *Prof Exp:* Lectr physics, City Col New York, 62-64; instr, NY Univ, 64-65; instr, Bronx Community Col, 65-66; mem tech staff, Bell Tel Labs, Inc, 66-67; asst prof, NY Univ, 67; assoc prof, 67-75, PROF PHYSICS, FAIRLEIGH DICKINSON UNIV, 75- *Mem:* Am Phys Soc; Am Geophys Union; Am Asn Physics Teachers. *Res:* Many-particle physics; geophysics; solar energy. *Mailing Add:* Dept Math & Comput Sci Fairleigh Dickinson Univ Teaneck Campus Teaneck NJ 07666

WIEDERHOLD, EDWARD W(ILLIAM), b Clermont Co, Ohio, Nov 4, 21. CHEMICAL & NUCLEAR ENGINEERING. *Educ:* Ohio State Univ, BChE, 49. *Prof Exp:* Chem engr, AEC, 50-52; chem engr, Mound Lab, Monsanto Co, 52-58, res chemist, 58-59, sr res chemist, 59-68; first officer, Div Nuclear Safety & Environ Protection, Int Atomic Energy Agency, Vienna, Austria, 68-72; pvt pract, 72-78; CONTRACT ENGR & SYSTS DEVELOP ENGR, GEN DEVICES INC, 78- *Mem:* Am Chem Soc. *Res:* Industrial waste disposal; reactor coolants; cryogenics. *Mailing Add:* 1348 Baldwin Rd Milford OH 45150

WIEDERHOLD, MICHAEL L, b Milwaukee, Wis, Aug 2, 39. SENSORY PHYSIOLOGY. *Educ:* Mass Inst Technol, PhD(commun & biophys), 67. *Prof Exp:* DIR, RES OTORHINOLARYNGOL & ASSOC PROF SURG & PHYSIOL, UNIV TEX HEALTH SCI CTR, SAN ANTONIO, 82- *Mailing Add:* Dept ORL Univ Tex Health Sci Ctr 7703 Floyd Curl Dr San Antonio TX 78284-7777

WIEDERHOLD, PIETER RIJK, b Malang, Indonesia, Jan 24, 28; US citizen; m 56; c 3. PHYSICS, ELECTRICAL ENGINEERING. *Educ:* Delft Univ Technol, Ir, 53. *Prof Exp:* Sr proj engr, Sylvania Elec Prod, Inc, 53-61; mgr energy conversion, Ion Physics Corp, Inc, 61-64; mgr cryogenics, Magnion, Inc, 64-66; div mgr space physics, Comstock & Wescott, Inc, 66-74; PRES, GEN EASTERN CORP, 74- *Mem:* Inst Elec & Electronics Engrs; Am Meteorol Soc; Instrument Soc Am. *Res:* Superconductivity; magnetics; energy conversion; humidity instruments. *Mailing Add:* Gen Eastern Corp 20 Commerce Way Woburn MA 01801

WIEDERHOLT, WIGBERT C, b Warmbrunn, Ger, Apr 22, 31; US citizen; m 60; c 3. NEUROLOGY, NEUROEPIDEMIOLOGY. *Educ:* Univ Freiburg, MD, 55. *Prof Exp:* Asst to staff neurol, Mayo Clin, 65; from asst prof to assoc prof med, Ohio State Univ, 66-72, chief clin neurophysiol, 69-72; neurologist-in-chief, Dept Neurosci, 73-83, chmn dept, 78-83, PROF NEUROSCI, UNIV CALIF, SAN DIEGO, 72- *Honors & Awards:* S Weir Mitchell Award, Am Acad Neurol, 56. *Mem:* AAAS; fel Am Acad Neurol; Am Neurol Asn; Am EEG Soc; Am Asn Electromyog & Electrodiag (secy-treas, 71-76, pres, 77-78). *Res:* Neuroepidemiology. *Mailing Add:* Dept Neurosci M-024 Univ Calif San Diego La Jolla CA 92093

WIEDERHORN, SHELDON M, b May 4, 33. MECHANICAL PROPERTIES OF CERAMICS. *Educ:* Columbia Univ, BS, 56; Univ Ill, MS, 58, PhD(chem eng), 60. *Prof Exp:* Res engr, E I du Pont de Nemours & Co, 60-63; res chemist, Inorg Mat Div, Nat Bur Standards, Washington, DC, 63-68, sect chief, Phys Properties Sect, Gaithersburg, Md, 68-76, chief, Fracture & Deformation Div, 76-77, dep div chief, 77-81, group leader, Mech Properties Group, Ceramics Div, 81-88; INST SCIENTIST, MAT SCI & ENG LAB, NAT INST STANDARDS & TECHNOL, 88- *Concurrent Pos:* Pres young investr award panelist, 85. *Honors & Awards:* Silver Medal, Dept Com, 70, Gold Medal, 82; Ross Coffin Purdy Award, Am Ceramic Soc, 71, Morey Award, 77; Sosman Lectr, Am Ceramic Soc, 85; Dow Distinguished Lectr Mat Sci & Eng, 88. *Mem:* Nat Acad Eng; Am Ceramic Soc. *Res:* Elucidation of the mechanisms of fracture in glass and other brittle ceramics; structural materials. *Mailing Add:* Nat Inst Standards & Technol Bldg 223 Rm B309 Gaithersburg MD 20899

WIEDERICK, HARVEY DALE, b Wetaskiwin, Can, May 10, 37; m 63; c 3. PHYSICS. *Educ:* Royal Mil Col Can, BSc, 59; Johns Hopkins Univ, MAT, 64; Queen's Univ, Ont, PhD(physics), 68. *Prof Exp:* Fel physics, Univ Toronto, 68-69; lectr, Royal Mil Col Can, 69-73, asst prof, 74-77; exchange teacher, Royal Mil Col Sci, Eng, 73-74; Nuffield Found res fel, Univ Kent, Canterbury, Eng, 77-78; PROF PHYSICS, ROYAL MIL COL CAN, 78- *Concurrent Pos:* Defence Res Bd Can grant, 70- *Res:* Electrical properties of ferroic materials. *Mailing Add:* Dept Physics Royal Mil Col Kingston ON K7K 5L0 Can

WIEDERSICH, H(ARTMUT), b Glatz, Ger, Apr 22, 26; US citizen; m 60. MATERIALS SCIENCE, SOLID STATE PHYSICS. *Educ:* Univ Gottingen, BS, 50, Dr rer nat(physics & metall), 54. *Prof Exp:* Res engr, Res Labs, Westinghouse Elec Corp, 54-60; res specialist, Atomics Int Div, NAm Aviation, Inc, 60-62, mem tech staff, Sci Ctr, NAm Rockwell Int, 62-70, group leader, 70-71; group leader, 71-82, assoc div leader, 82-89, SR METALLURGIST, MAT SCI DIV, ARGONNE NAT LAB, 71- *Concurrent Pos:* Adj prof, Dept Mat Sci, Univ Southern Calif, 66-70; assoc ed, Am Inst Physics Appl Physics Lett, 89-90, ed, 90- *Mem:* Fel Am Soc Metals; Mat Res Soc; Am Phys Soc. *Res:* Defects and transport processes in solids, radiation effects on processes in, and microstructure and properties of materials; dislocation theory; defects in solids; magnetic structures by the Mossbauer effect; radiation damage; oxidation processes. *Mailing Add:* Mat Sci Div Bldg 212 Argonne Nat Lab Argonne IL 60439

WIEDMAN, HAROLD W, b Palermo, Calif, Jan 11, 30. PHYTOPATHOLOGY. *Educ:* Chico State Col, AB, 52; Ore State Col, PhD(bot), 56. *Prof Exp:* Asst, Ore State Col, 53-56; asst plant pathologist, NMex State Univ, 56-58 & State Dept Agr, Calif, 58-59; asst prof bot, Humboldt State Col, 59-61; from asst prof to assoc prof, 61-69, PROF BIOL SCI, CALIF STATE UNIV, SACRAMENTO, 69- *Mem:* Am Phytopath Soc; Bot Soc Am. *Res:* Soil-borne diseases; biological control; diseases of vegetables and cotton. *Mailing Add:* Dept Biol Sci Calif State Univ 6000 J St Sacramento CA 95819

WIEDMEIER, VERNON THOMAS, b Harvey, NDak, Jan 10, 35; m 57; c 4. PHYSIOLOGY. *Educ:* NDak State Teachers Col, Valley City, BS, 59; NDak State Univ, MS, 61; Marquette Univ, PhD(physiol), 68. *Prof Exp:* Instr biol, NPark Col, 60-61; asst prof, St Ambrose Col, 61-64; asst prof, 71-75, ASSOC PROF PHYSIOL, MED COL GA, 75- *Concurrent Pos:* NIH trainee, Univ Va, 69-70 & fel, 70-71. *Mem:* Am Physiol Soc. *Res:* Myocardial metabolism and the regulation of coronary blood flow. *Mailing Add:* Dept Physiol Med Col Ga Augusta GA 30912

WIEDOW, CARL PAUL, b Pasadena, Calif, Dec 3, 07; m 47. HIGH VOLTAGE PHENOMENA, MAGNETICS. *Educ:* Occidental Col, AB, 33; Calif Inst Technol, MS, 45 & 46; Ore State Univ, PhD(elec eng), 56. *Prof Exp:* Asst prof electronics, US Naval Postgrad Sch, 56-59; design specialist, Gen Dynamics Astronaut, 59-61 & Ryan Aerospace Div, 61-62; prof & chmn physics, Calif Western Univ, 62-66; staff engr, Marine Advisors, 66-67; CHIEF RES, HUMPHREY INC, 67- *Concurrent Pos:* Consult engr, Elgin Nat Watch Co, W Coast Micronics Div, 59-60, Gen Dynamics Astronaut, 63-64, Havens Industs, 62-64; consult, Solar Div Int Harvester, 64-66 & Anka Industs, 79-; acad asst, NSF, 66-68. *Mem:* Optical Soc; Sigma Xi. *Res:* Ionization and tracking over solid dielectrics; prevention of breakdown in high voltage x-ray units; transmission of very low frquency radio through the earth near the surface for geophysical data telemetry. *Mailing Add:* 3023 Alcott St San Diego CA 92106

WIEGAND, CRAIG LOREN, b Santa Rosa, Tex, Jan 11, 33; m 62; c 2. IRRIGATION, SALINITY. *Educ:* Tex A&M Univ, BS, 55, MS, 56; Utah State Univ, PhD(soil physics), 60. *Prof Exp:* Res soil scientist, 60-78, dir Rio Grande Soil & Water Res Ctr, 69-78, tech adv, Sci & Educ Admin-Agr Res, 73-80, SR SCIENTIST, AGR RES SERV, USDA, 80- *Concurrent Pos:* Mid-career fel, Woodrow Wilson Sch Pub & Int Affairs, Princeton Univ, 74-75; foreign specialist in Japan, 88. *Honors & Awards:* Super Performance Award, 70. *Mem:* AAAS; Fel Soil Sci Soc Am; Am Soc Agron; Int Soil Sci Soc. *Res:* Plant physiology; micrometeorology; plant-water relations; crop modeling; earth observation satellite data applications to agriculture; remote sensing. *Mailing Add:* Remote Sensing Res Unit Agr Res Serv USDA 2413 E Bus Hwy 83 Weslaco TX 78596

WIEGAND, DONALD ARTHUR, b Rochester, NY, July 21, 27; m 59; c 2. SOLID STATE PHYSICS. *Educ:* Cornell Univ, BEE, 52, MEE, 53, PhD(eng physics), 56. *Prof Exp:* Asst & assoc physics, Cornell Univ, 55-56; res physicist, Carnegie Mellon Univ, 56-59, from asst prof to assoc prof physics, 59-68; res physicist, Feldman Res Lab, Picatinny Arsenal, 68-79; supvr res physicist, 79-90, RES PHYSICIST, ENERGETIC MAT DIV, ARMAMENT RES DEVELOP & ENG CTR, 90- *Concurrent Pos:* Fulbright grant, Darmstadt Tech Univ, WGer, 60-61. *Mem:* Am Phys Soc; Mat Res Soc. *Res:* Imperfections in solids; luminescence; photo-conductive processes; optical absorption; x-ray diffraction; x-ray photoelectron spectroscopy; metastable solids; mechanical properties of solids. *Mailing Add:* Energetic Mats Div Armament Res Develop & Eng Ctr Dover NJ 07801

WIEGAND, GAYL, b Estherville, Iowa, July 18, 39. ORGANIC CHEMISTRY. *Educ:* Univ Iowa, BS, 61; Univ Mass, PhD(org chem), 65. *Prof Exp:* From asst prof to assoc prof, 65-77, PROF CHEM, IDAHO STATE UNIV, 77-, DEPT CHAIR, 90- *Mem:* Am Chem Soc; Sigma Xi. *Res:* Mechanisms of organic like reactions occurring at elements other than carbon; organosulfur chemistry; kinetics of slow chemical reactions; geothermal energy prospecting; science for the non-specialist. *Mailing Add:* Dept Chem Idaho State Univ Pocatello ID 83209-0009

WIEGAND, OSCAR FERNANDO, b Mex, Nov 3, 21; US citizen; m 49; c 5. PLANT PHYSIOLOGY. *Educ:* Univ Tex, BA, 50, MA, 52, PhD(cell physiol, chem), 56. *Hon Degrees:* Dr, Univ Guadalajara, 65. *Prof Exp:* Asst prof biol, ETex State Col, 56-57; asst prof pharmacol, Univ Tex Southwest Med Sch, 57-60, vis lectr, Univ Tex, Austin, 60-62, from asst prof to prof zool, 77-87; RETIRED. *Concurrent Pos:* Smith-Mundt fel & vis prof, Univ Guadalajara, 61-62, distinguished prof, 62, gen coord model univ develop prog, 63-67; mem study group for reform & improv educ, Latin Am Univ, 65-; head consult, Univ Reform Model, Cath Univ Rio de Janeiro, 66-; consult univ reform prog, Agency Int Develop-Govt Brazil, 66- *Mem:* Am Soc Pharmacol & Exp Therapeut; Soc Gen Physiol. *Res:* Photomorphogenesis, growth physiology and tropisms; water and electrolyte equilibria in tissues; histamine reactions; tracer technique; respiromtery; growth methods for plant tissues. *Mailing Add:* PO Box 27433 Austin TX 78755

WIEGAND, RONALD GAY, pharmacology, biochemistry; deceased, see previous edition for last biography

WIEGAND, SYLVIA MARGARET, b Cape Town, SAfrica, Mar 8, 45; US citizen; m 66; c 2. ALGEBRA. *Educ:* Bryn Mawr Col, AB, 66; Univ Wash, MA, 67; Univ Wis, PhD(algebra), 72. *Prof Exp:* Teaching asst math, Univ Wis, 67-72; from instr to prof, 72-87, PROF MATH, UNIV NEBR, LINCOLN, 87- *Concurrent Pos:* Vis assoc prof, Univ Conn, 78-79 & Univ Wis, 85-86. *Mem:* Am Math Soc; Math Asn Am; Asn Women Math. *Res:* Commutative algebra work on the prime spectrum of a Noetherian ring; direct sum decompositions; cancellation problem for one-dimensional rings. *Mailing Add:* Dept Math Univ Nebr Lincoln NE 68588-0323

WIEGANDT, HERBERT F(REDERICK), b Newaygo, Mich, Jan 4, 17; m 44; c 2. CHEMICAL ENGINEERING. *Educ:* Purdue Univ, BSChE, 38, MSE, 39, PhD(chem eng), 41. *Prof Exp:* Asst process develop, Eng Exp Sta, Purdue Univ, 38-41; chem engr, Standard Oil Co, Ind, 41-44 & Armour Res Found, Ill Inst Technol, 44-47; from asst prof to assoc prof chem eng, 47-60, prof, 60-87, EMER PROF CHEM ENG, CORNELL UNIV, 87- *Concurrent Pos:* Fulbright Award to France, 60-61; Monsanto Chem Co, 52, & French Petrol Inst, 61 & 64; tech adv, Compagnie Francaise Raffinage, Paris, 72-82. *Mem:* AAAS; Am Chem Soc; Am Inst Chem Engrs. *Res:* Desalination; petroleum processes; extractions; distillation; crystallization. *Mailing Add:* Dept Chem Eng Cornell Univ Ithaca NY 14853

WIEGEL, ROBERT L, b San Francisco, Calif, Oct 17, 22; m 48; c 3. OCEAN ENGINEERING. *Educ:* Univ Calif, BS, 43, MS, 49. *Prof Exp:* Jr res engr, 46-52, lectr civil eng, 56-60, assoc prof, 60-63, asst dean, 63-72, assoc res engr, 52-60, actg dean, 72-73, prof, 63-87, EMER PROF CIVIL ENG, UNIV CALIF, BERKELEY, 87- *Concurrent Pos:* Dir, Calif State Tech Serv Prog, 65-68; pres, Int Comt Oceanic Resources, 72-75; mem comt earthquake eng res, Nat Acad Eng, mem, Marine Bd, 75-81; consult; mem, Comt on Eng, Nat Res Coun, 84-87. *Honors & Awards:* Res Prize, Am Soc Civil Engrs, 62; Moffatt-Nichol Harbor Am Coastal Eng Award, 78; Outstanding Civilian Serv Medal, US Army CEngrs, 85; Int Coastal Eng Award, 85. *Mem:* Nat Acad Eng; hon mem & fel Am Soc Civil Engrs; fel AAAS; PIANC; Sigma Xi; hon mem Int Eng Comt on Oceanic Resources. *Res:* Ocean and coastal engineering; technology transfer. *Mailing Add:* 412 O'Brien Hall Univ Calif Berkeley CA 94720

WIEGERT, PHILIP E, b Antigo, Wis, Apr 7, 27; m 59; c 6. BIOMEDICAL ENGINEERING, ORGANIC CHEMISTRY. *Educ:* Univ Wis, BS, 50; Univ Ill, MS, 51, PhD, 54. *Prof Exp:* Chemist, Mallinckrodt Chem Works, 54-61, group leader, 61-66, asst dir pharmaceut chem, 66-74, plant mgr, 74-77, dir res & develop, Mallinckrodt Crit Care, 77-86, dir qual control & regulatory affairs, 86-89, CONSULT, MALLINCKRODT INC, 89- *Mem:* Am Chem Soc; The Chem Soc; Am Soc Testing & Mat; Asn Advan Med Instrumentation; Soc Plastic Engrs. *Res:* X-ray contrast media; opium alkaloids; pharmaceutical chemicals; medical devices; standards development. *Mailing Add:* Two Horicon Ave Glens Falls NY 12801-2655

WIEGERT, RICHARD G, b Toledo, Ohio, Sept 9, 32; div; c 2. ECOLOGY. *Educ:* Adrian Col, BS, 54; Mich State Univ, MS, 58; Univ Mich, PhD(zool), 62. *Prof Exp:* Instr zool, Univ Mich, 61-62; from asst prof to assoc prof, 62-71, PROF ZOOL, UNIV GA, 71- Prof Exp: NSF grants, Yellowstone Nat Park, 68-78, Sapelo Island Salt Marsh, 75-78; Environ Prediction Res Fac grant, microcosms, 78- *Mem:* AAAS; Am Soc Mammal; Ecol Soc Am; Brit Ecol Soc; Am Soc Naturalists. *Res:* Plant and animal ecology, particularly problems of population and community energy utilization; population density regulation; interspecies competition; systems ecology and modeling the dynamics of thermal spring and estuarine communities. *Mailing Add:* Dept Zool Univ Ga Athens GA 30601

WIEGMAN, DAVID L, HYPERTENSION, EXERCISE. *Educ:* Ind Univ, PhD(physiol), 73. *Prof Exp:* ASST DEAN & ASSOC PROF PHYSIOL, HEALTH SCI CTR, UNIV LOUISVILLE, 81- *Mailing Add:* Dept Physiol & Biophys Health Sci Ctr Univ Louisville Louisville KY 40292

WIEGMANN, NORMAN ARTHUR, algebra, for more information see previous edition

WIEGNER, ALLEN W, b Bethlehem, Pa, July 22, 47; m 78; c 1. SPINAL CORD INJURY. *Educ:* Mass Inst Technol, SB & SM, 70, PhD(elec eng), 78. *Prof Exp:* Comn officer, USPHS, 70-72; assoc med, 78-80, assoc neurol, 80-87, ASST PROF NEUROL, HARVARD MED SCH, 87-; BIOMED ENGR, WEST ROXBURY VET ADMIN MED CTR, 87- *Concurrent Pos:* Lectr, Dept Elec Eng & Computer Sci, Mass Inst Technol, 79-; asst biomed engr, Mass Gen Hosp, 80- *Mem:* Inst Elec & Electronics Engrs; Biomed Eng Soc; Soc Neurosci. *Res:* Role of muscle and reflex in movement and movement disorders, tremor; nonlinear behavior of resting and active muscle; functional electrical stimulation, especially in persons with spinal cord injury; rehabilitation aids. *Mailing Add:* Spinal Cord Injury Serv Vet Admin Med Ctr 1400 VFW Pkwy Boston MA 02132

WIEHE, IRWIN ANDREW, chemical engineering, for more information see previous edition

WIEHE, WILLIAM ALBERT, b Sewickley, Pa, Apr 15, 47; m 68; c 2. WIRE PROCESS ENGINEERING, QUALITY SYSTEMS MANAGEMENT. *Educ:* Mich State Univ, BS, 69. *Prof Exp:* Mgr, Sun Shipbuilding Div, Sun Oil, 69-75; mgr, welding, Rexarc Inc, 75-81; DIR RES, ARCOS ALLOYS DIV, ARMANDA CORP, 81- *Mem:* Am Welding Soc; Am Soc Metals; Am Soc Qual Control. *Res:* Coatings research for shielded metal arc welding at hardsurfacing alloys, stainless steel nickel alloys, cobalt alloys and copper alloys. *Mailing Add:* One Arcos Dr Mt Carmel PA 17851

WIELAND, BRUCE WENDELL, b Carroll, Iowa, Apr 15, 37. MECHANICAL ENGINEERING. *Educ:* Iowa State Univ, BS, 60; Ohio State Univ, PhD(nuclear eng), 73. *Prof Exp:* Engr, Oak Ridge Nat Lab, 60-66; res engr, Battelle Mem Inst, 67-68; NIH spec res fel, Ohio State Univ, 69-73; scientist med radioisotopes, Oak Ridge Assoc Univs, 74-80; scientist, Brookhaven Nat Lab, 81-84; SR RES ENGR, COMPUT TECHNOL & IMAGING, INC, 85- *Concurrent Pos:* Res assoc, Sch Med, Washington Univ, 71-73; consult, Oak Ridge Assoc Univs, 72-73; assoc prof, Univ Calif, Los Angeles, 81. *Mem:* Soc Nuclear Med; AAAS. *Res:* Applications of accelerator-produced radioisotopes in nuclear medicine. *Mailing Add:* Comput Technol & Imaging Inc 950 Gilman St Berkeley CA 94710

WIELAND, DENTON R, b Yorktown, Tex, Oct 28, 27; m 54; c 2. PETROLEUM ENGINEERING. *Educ:* Agr & Mech Col, Tex, BS, 53, MS, 56, PhD, 58. *Prof Exp:* Asst prof petrol eng, Univ Tulsa, 57-61, assoc prof & actg head dept, 61-64; dir tech develop, Dowell Div, Dow Chem Co, 64-68, supvr customer serv, 68-69, mgr sales develop dept, 69-72, mgr eng, 72-76; consult, 76-78; proj leader well completions, Osco, 78-79; sr prod mgr, 79-80, GEN SUPVR, CHEVRON, 80- *Mem:* Am Inst Mining, Metall & Petrol Egnrs; Sigma Xi. *Res:* Physical chemistry of petroleum engineering. *Mailing Add:* PO Box 267 Calvert TX 77837

WIEMER, DAVID F, b Burlington, Wis, Mar 17, 50; m 72; c 3. CHEMICAL ECOLOGY. *Educ:* Marquette Univ, BS, 72; Univ Ill, PhD(org chem), 76. *Prof Exp:* NIH fel, Cornell Univ, 76-78; from asst to assoc prof, 78-89, PROF CHEM, UNIV IOWA, 89- *Concurrent Pos:* AP Sloan Found Fel, 85-89; vis scientist, Scripps Inst Oceanogr, 86. *Mem:* Am Chem Soc; Sigma Xi; AAAS. *Res:* Isolation, characterization, and synthesis of biologically active natural products; development of synthetic methodology based on organo phosphorus chemistry; chemical ecology. *Mailing Add:* Dept Chem Univ Iowa Iowa City IA 52242

WIEMEYER, STANLEY NORTON, b Santa Rosa, Calif, Nov 7, 40; m 68; c 1. WILDLIFE TOXICOLOGY. *Educ:* Humboldt State Col, BS, 63, MS, 67. *Prof Exp:* RES WILDLIFE BIOLOGIST, PATUXENT WILDLIFE RES CTR, US FISH & WILDLIFE SERV, 66- *Mem:* Wildlife Soc; Am Ornithologist's Union. *Res:* Effects of environmental contaminants, including pesticides, metals and PCBs, on birds with major emphasis on birds of prey; captive breeding of birds of prey. *Mailing Add:* Patuxent Wildlife Res Ctr US Fish & Wildlife Serv Laurel MD 20708

WIEN, RICHARD W, JR, b Bay Co, Fla, May 17, 45; m 68; c 2. PHYSICAL CHEMISTRY, PHOTOGRAPHIC SCIENCE. *Educ:* Purdue Univ, BS, 67; Stanford Univ, PhD(phys chem), 71. *Prof Exp:* NSF fel phys chem, Stanford Univ, 67-71; sr res chemist photog sci, Eastman Kodak, 71-79, proj leader, Photog Technol Div, 79-85, tech assoc, 86-89, PROD QUAL DIR, PROF PHOTOG DIV, EASTMAN KODAK, 89- *Concurrent Pos:* Nat tour speaker, Am Chem Soc, 78- *Mem:* Am Chem Soc. *Res:* Improvement of photographic speed of color reversal films; development of reversal color papers; development of new high speed color photographic systems; development of professional negative-positive photographic systems; development of measures of product quality as used by customers. *Mailing Add:* 44 Hilltop Dr Pittsford NY 14534

WIENER, EARL LOUIS, b Shreveport, La, May 30, 33; m 55, 80; c 2. INDUSTRIAL ENGINEERING, PSYCHOLOGY. *Educ:* Duke Univ, BA, 55; Ohio State Univ, MA, 59, PhD(psychol, indust eng), 61. *Prof Exp:* Asst, Aviation Psychol Lab, Ohio State Univ, 58-59, opers res group, 59-60, res assoc systs res group, 60-61; asst prof psychol & indust eng, 62-66, PROF MGT SCI & ADJ PROF PSYCHOL, UNIV MIAMI, 66- *Mem:* Fel Human Factors Soc (pres, 88-89); fel Am Psychol Asn; Soc Eng Psychol; Am Inst Indust Engrs. *Res:* Human factors; aviation safety; human vigilance and monitoring; effect of human performance on systems performance; traffic safety. *Mailing Add:* Dept Mgt Sci Univ Miami Univ Sta Box 248237 Coral Gables FL 33124

WIENER, HOWARD LAWRENCE, b Portland, Ore, Mar 16, 37; m 62; c 1. STATISTICS, OPERATIONS RESEARCH. *Educ:* Univ Ore, BS, 59; Northwestern Univ, Evanston, MS, 61; Cath Univ Am, PhD(math), 71. *Prof Exp:* Sci analyst opers eval group, Ctr Naval Anal, Va, 61-63, opers res analyst-mathematician, Naval Ord Lab, MD, 63-71, opers res analyst, 71-75, head opers anal & planning sect, 75-81, SUPVRY OPERS RES ANALYST, COMBAT MGT BR, NAVAL RES LAB, WASHINGTON, DC, 81- *Mem:* Am Statist Asn; Opers Res Soc Am. *Res:* Applications of statistics and probability to operational problems; data analysis; time series analysis. *Mailing Add:* 3824 Porter St Washington DC 20016

WIENER, JOSEPH, b Toronto, Ont, Sept 21, 27; m 54; c 2. PATHOLOGY, CELL BIOLOGY. *Educ:* Univ Toronto, MD, 53. *Prof Exp:* Assoc path, Col Physicians & Surgeons, Columbia Univ, 60-63, asst prof, 63-68; prof path & attend pathologist, NY Med Col, 68-78; prof & chmn dept path, Sch Med, Wayne State Univ, 78-90; chmn, dept path, Detroit Gen Hosp, 78-90; RETIRED. *Mem:* Am Soc Cell Biol; Am Soc Exp Path; Am Asn Path & Bact. *Res:* Experimental pathology. *Mailing Add:* Dept Path 9374 Scott Wayne State Univ 5950 Cass Ave Detroit MI 48202

WIENER, L(UDWIG) D(AVID), b Nashville, Tenn, Aug 10, 26; m 51; c 2. CHEMICAL ENGINEERING. *Educ:* Vanderbilt Univ, BE, 45, MS, 46; Univ Cincinnati, PhD(chem eng), 49. *Prof Exp:* ENG ASSOC, MOBIL RES & DEVELOP CORP, 49-; Retired. *Mem:* Soc Petrol Engrs. *Res:* Hydrotropic solutions; drilling fluids; thermodynamic properties of hydrocarbon systems. *Mailing Add:* 10730 Bushire Dr Dallas TX 75229

WIENER, ROBERT NEWMAN, b New York, NY, Aug 27, 30; m 54; c 3. CHEMISTRY. *Educ:* Harvard Univ, AB, 51; Univ Pa, MS, 53, PhD, 56. *Prof Exp:* Asst instr chem, Univ Pa, 51-54; instr, Rutgers Univ, 55-58; asst prof, 58-62, ASSOC PROF CHEM, NORTHEASTERN UNIV, 62- *Mem:* Am Phys Soc. *Res:* Physical chemistry; molecular spectroscopy. *Mailing Add:* Dept Chem Northeastern Univ Boston MA 02115

WIENER, SIDNEY, b New York, NY, Nov 17, 22; m 44; c 2. MATERIALS SCIENCE. *Educ:* Univ Calif, Los Angeles, BS, 47; Univ Calif, Berkeley, PhD(biochem), 52. *Prof Exp:* Exploitation engr, Prod Lab, Shell Oil Co, 52-56; mem, tech staff, Airborne Systs Labs, space & commun group, Hughes Aircraft Co, 57-60, group head org mat, Res & Develop Div, 60-62, sect head, 62-65, asst dept mgr mat tech, 65-70, mgr space & commun group support activ, 70-73, mgr proj control, 74-78, sr staff engr, Components & Mat Labs, 78, sr scientist prod assurance eng, 78-80, mgr mat, processes & radiation eng, 80-82; RETIRED. *Mem:* AAAS; Am Chem Soc; Sigma Xi; Planetary Soc. *Res:* Physical chemistry and elucidation of structure of nucleic acid using enzymatic reactions and acid-base relations; technical administration in materials. *Mailing Add:* 5609 Edgemere Dr Torrance CA 90503

WIENER, STANLEY L, b New York, NY, Nov 5, 30; m 53; c 3. INTERNAL MEDICINE, EXPERIMENTAL PATHOLOGY. *Educ:* Univ Rochester, AB, 52; Sch Med, Univ Rochester, MD, 56. *Prof Exp:* From asst prof to prof med, State Univ NY, Stony Brook, 61-72; assoc dir res & educ, Long Island Jewish-Hillside Med Ctr, 72-73, assoc dir med, 73-78; chmn, Dept Med, ETenn State Univ, 78-81, Col Med, 81-89; CHIEF SECT GEN MED, SCH MED, UNIV ILL, CHICAGO, 89- *Concurrent Pos:* Chmn res comt, Am Heart Asn, NY State Affil, 7375. *Mem:* Am Soc Exp Path; Am Soc Exp Biol & Med; Am Fedn Clin Res; Am Rheumatism Asn. *Res:* In vivo studies of fibroblast activation and growth control; studies of neutrophil chemotaxis and enzyme release into inflammatory liquid. *Mailing Add:* Dept Med Univ Ill Sch Med 840 S Wood St Chicago IL 60612

WIENKE, BRUCE RAY, b Chicago, Ill, Sept 21, 40; m 81. THEORETICAL PHYSICS, COMPUTATIONAL PHYSICS. *Educ:* Univ Wis, BS, 63; Marquette Univ, MS, 65; Northwestern Univ, PhD(physics), 71. *Prof Exp:* Teaching asst physics, Northwestern Univ & Marquette Univ, 63-67; res asst theoret physics, Northwestern Univ, 70-71; staff mem, 71-72, staff mem comput physics, 72-78, staff mem comput math, Los Alamos Sci Lab, 79, physicist, Mission Res Corp, 80, SECT LEADER COMPUT PHYSICS, LOS ALAMOS NAT LAB, 81- *Concurrent Pos:* Staff mem, Argonne Nat Lab, 66-68; consult, Square D Co, Milwaukee, 66-72 & Prof Asn Diving Instrs, 77-; instr, Col Santa Fe, 76- *Honors & Awards:* Bausch & Lomb Sci Award, 58. *Mem:* Am Phys Soc; Am Nuclear Soc; Soc Indust & Appl Math; Am Acad Mech; Int Oceanog Soc. *Res:* Theoretical particle and nuclear physics; transport theory and applications for neutral and charged particles; computational physics and numerical methodology; mathematical physics and computing science. *Mailing Add:* Los Alamos Nat Lab MS F664 Los Alamos NM 87545

WIENKER, CURTIS WAKEFIELD, b Seattle, Wash, Feb 3, 45; div; c 1. PHYSICAL ANTHROPOLOGY. *Educ:* Univ Wash, BA, 67; Univ Ariz, MA, 70, PhD(anthrop), 75. *Prof Exp:* From lectr to asst prof, 72-78, ASSOC PROF ANTHROP, UNIV SFLA, 78-, ASSOC DEAN, 88. *Concurrent Pos:* Prin investr, 75-78. *Mem:* Am Asn Phys Anthrop; Sigma Xi; Human Biol Coun; Am Acad Forensic Sci. *Res:* Living human biological variation; human evolution; Black population biology; cultural influences on human biology; biomedical anthropology; forensic anthropology. *Mailing Add:* Soc 107 Univ SFla Tampa FL 33620-8100

WIENS, DELBERT, b Munich, NDak, July 9, 32; m 55; c 3. SYSTEMATIC BIOLOGY, EVOLUTIONARY BIOLOGY. *Educ:* Pomona Col, BA, 55; Univ Utah, MS, 57; Claremont Grad Sch, PhD(bot), 61. *Prof Exp:* Instr biol, Univ Colo, 60-62, asst prof, 62-64; asst prof bot, 64-66, assoc prof, 67-74, PROF BIOL, UNIV UTAH, 74- *Concurrent Pos:* Fulbright lectr & hon prof, Univ Guayaquil, 64-65; mem, Flora of Ceylon Proj, 68; vis lectr, Flinders Univ SAustralia, 72. *Mem:* AAAS; Am Soc Plant Taxon; Bot Soc Am; Int Asn Plant Taxon; Soc Study Evolution. *Res:* Systematics, biogeography, chromosome systems and pollination ecology of flowering plants, particularly the mistletoe family. *Mailing Add:* Dept Biol Univ Utah 1201 S Biol Bldg Salt Lake City UT 84112

WIENS, JOHN ANTHONY, b Moscow, Idaho, Sept 29, 39; m 61, 84; c 4. DESERT ECOLOGY, LANDSCAPE ECOLOGY. *Educ:* Univ Okla, BS, 61; Univ Wis, MS, 63, PhD(zool), 66. *Prof Exp:* From asst prof to prof zool, Ore State Univ, 66-78; distinguished prof, Univ NMex, 78-86; PROF ECOL, COLO STATE UNIV, 86- *Concurrent Pos:* NSF res grant, 67-69 & 74-; Am Philos Soc res grant, 72-75; vis prof, Colo State Univ, 73-77; Nat Oceanic Atmospheric Admin res contract, 75-81; US Forest Serv res contract, 76-81; ed, The Auk, 76-84; Fulbright Sr Scholar, Australia, 84-85; Dept Energy grant, 88-; US MAB grant, 90-; vis prof, Univ Oslo, 89. *Mem:* Am Soc Naturalists; fel Am Ornith Union (treas, 74-78); Ecol Soc Am; Animal Behav Soc; Brit Ecol Soc; Copper Ornith Soc; Int Asn Ecol; Wilson Ornith Soc. *Res:* Vertebrate community structure and behavioral ecology; population modeling and analysis; methods of habitat description and analysis; landscape ecology; scaling in ecological systems; desert ecology; seabird ecology. *Mailing Add:* Dept Biol Colo State Univ Ft Collins CO 80523

WIER, CHARLES EUGENE, b Jasonville, Ind, May 15, 21; m 49; c 3. ECONOMIC GEOLOGY. *Educ:* Ind Univ, AB, 43, AM, 50, PhD(econ geol), 55. *Prof Exp:* Geologist & head coal sect, Ind Geol Surv, 49-75; assoc prof geol, Ind Univ, Bloomington, 65-75; mgr coal explor, Amax Int Coal Co, 75-76, vpres, 76-85; VPRES, HOOSIER MINING CO, 85- *Honors & Awards:* Gordon H Wood Mem Award, 90. *Mem:* Geol Soc Am; Soc Econ Geol; Am Asn Petrol Geol; Am Inst Mining, Metall & Petrol Engrs; Sigma Xi. *Res:* Pennsylvanian stratigraphy; coal resources and coal petrology; environmental geology. *Mailing Add:* PO Box 217 Bloomington IN 47402

WIER, DAVID DEWEY, b Sidon, Miss, Sept 3, 23; m 44; c 2. ELECTRICAL ENGINEERING. *Educ:* Miss State Univ, BS, 44; La Polytech Inst, MS, 64. *Prof Exp:* Design engr, Reliance Elec & Eng Co, 44-46; distribution engr, Miss Power Co, 46-52; self employed, 52-57; from instr to asst prof elec eng, 57-64, assoc prof elec eng, Miss State Univ, 64-; RETIRED. *Mem:* Nat Soc Prof Engrs; Inst Elec & Electronics Engrs. *Res:* Magnetomotive forming of metals for use in construction of space vehicles. *Mailing Add:* Dept Elec Eng PO Drawer EE Mississippi State MS 39762

WIER, JACK KNIGHT, b Cairo, Nebr, Aug 31, 23; m 47. PHARMACOGNOSY. *Educ:* Univ Wis, PhB, 45; Univ Nebr, BS, 56; Univ Wash, Seattle, MS, 59, PhD(pharmacog), 61. *Prof Exp:* From asst prof to assoc prof pharmacog, Univ NC, Chapel Hill, 61-87; RETIRED. *Concurrent Pos:* Consult, F W Dodge Co Div, McGraw-Hill, Inc, 65-69. *Mem:* Am Soc Pharmacog (secy, 70-79, pres, 79-81); Am Pharmaceut Asn; Acad Pharmaceut Sci; Sigma Xi. *Res:* Metabolic products of macrofungi; biosynthesis of indole alkaloids in higher plants. *Mailing Add:* 415 Long Leaf Dr Chapel Hill NC 27515

WIER, JOSEPH M(ARION), b Amsterdam, Mo, Mar 2, 24; m 48; c 2. ELECTRICAL ENGINEERING. *Educ:* Iowa State Col, BS, 49, MS, 50; Univ Ill, PhD(elec eng), 56. *Prof Exp:* Instr elec eng, Iowa State Col, 50; asst digital comput lab, Univ Ill, 50-51, res assoc, 51-56; mem tech staff, AT&T, 56-59, head switching systs study dept, 59-72, data mgt systs dept, 72-77, qual theory & systs dept, 77-80, head customer equip qual dept, Bell Tel Labs, 80-82, customer equip qual dept, AT&T, 83-84, consult, qual, 84-87; RETIRED. *Mem:* AAAS; Inst Elec & Electronics Engrs. *Res:* Electronic digital computers; data communications; data management systems; systems theory. *Mailing Add:* 41 E Larchmont Dr Colts Neck NJ 07722

WIER, WITHROW GIL, b San Diego, Calif, Oct 10, 50. ION CONCENTRATIONS. *Educ:* Utah State Univ, BS, 72; Univ Utah, PhD(physiol), 78. *Prof Exp:* Res fel, Dept Pharmacol, Mayo Found, 78-79; from instr to asst prof pharmacol, Mayo Med Sch, 79-82; asst prof, 82-87, ASSOC PROF PHYSIOL, UNIV MD SCH MED, 87- *Concurrent Pos:* Established investr, Am Heart Asn, 85-90; mem, Coun Basic Sci, AMA. *Honors & Awards:* Louis N Katz Prize, AMA, 79. *Mem:* AMA; Biophys Soc;

Soc Gen Physiologists; Am Physiol Soc; foreign mem Physiol Soc UK. *Res:* Measurement of intracellular ion concentrations in living cells; excitation-contraction coupling in mammalian heart. *Mailing Add:* Dept Physiol Univ MD School Med 660 W Redwood St Baltimore MD 21201

WIERENGA, PETER J, b Uithuizen, Neth, June 27, 34; m 63; c 3. SOIL PHYSICS, AGRONOMY. *Educ:* State Agr Univ, Wageningen, BS, 61, MS, 63; Univ Calif, Davis, PhD(soil sci), 68. *Prof Exp:* Res water scientist, Univ Calif, Davis, 65-68; from asst prof to prof agron, NMex State Univ, 68-88; PROF & DEPT HEAD, SOIL & WATER SCI, UNIV ARIZ, 88- *Concurrent Pos:* Consult, Battelle Northwest, Los Alamos Sci Lab, Sandia Labs, EGG, Off Technol Assessment US Cong & Environ Protection Res Inst; assoc ed, Soil Sci Soc Am Water Resource Res; vis scientist, Nat Ctr Sci Res, Mech Inst, Grenoble, France, 75-76; vis prof, ETH, Zurich & Switz; res award, Col Agr, NMex State Univ, 77. *Honors & Awards:* Westhafer Award, NMex State Univ, 82. *Mem:* Fel Am Soc Agron; fel Soil Sci Soc Am; Am Geophys Union; Neth Royal Soc Agr Sci; fel AAAS; Int Soc Soil Sci. *Res:* Measurement and simulation of transfer processes in soils, such as movement of water, heat and salts; quality of irrigation return flow; irrigation management; trickle irrigation; characterization of vadose zone processes. *Mailing Add:* Soil & Water Sci Dept 429 Shantz Bldg Univ Ariz Tucson AZ 85721

WIERENGA, WENDELL, b Hudsonville, Mich, Feb 5, 48; m 68; c 2. ORGANIC CHEMISTRY. *Educ:* Hope Col, BA, 70; Stanford Univ, PhD(org chem), 73. *Prof Exp:* Res scientist org chem, Exp Chem Res, Upjohn Co, 74-78, head cancer res, 81-82, dir, cancer & viral res, 82-90; SR VPRES RES, PHARMACEUT RES DIV, WARNER-LAMBERT CO, 90- *Concurrent Pos:* Fel, Am Cancer Soc, Dept Chem, Stanford Univ, 73-74. *Mem:* Am Chem Soc; Am Asn Cancer Res. *Res:* Design and synthesis of biologically and medicinally important organic compounds. *Mailing Add:* Pharmaceut Res Div Warner-Lambert Co Ann Arbor MI 48106-1047

WIERENGO, CYRIL JOHN, JR, b Picayune, Miss, Mar 7, 40; m 62; c 2. ORGANIC CHEMISTRY. *Educ:* Univ Southern Miss, BA, 62, MS, 64; Miss State Univ, PhD(chem), 74. *Prof Exp:* Res chemist, Dow Chem Co, 64-67; PROF CHEM, MISS UNIV FOR WOMEN, 67- *Mem:* Am Chem Soc; Sigma Xi. *Res:* The synthesis and chemistry of bis-heterocyclic compounds. *Mailing Add:* Miss Univ Women Box W100 Columbus MS 39701

WIERMAN, JOHN C, b Prosser, Wash, June 30, 49; m 71; c 1. PERCOLATION, RANDOM GRAPHS. *Educ:* Univ Wash, BS, 71, PhD(math), 76. *Prof Exp:* Asst prof math, Univ Minn, 76-81; from asst prof to assoc prof, 81-87, PROF MATH SCI, JOHNS HOPKINS UNIV, 87-, CHAIR, 88- *Concurrent Pos:* Sr res fel, Inst Math & Its Appln, Univ Minn, 87-88. *Mem:* Fel Inst Math Statist; Am Statist Asn; Am Math Soc; Math Asn Am. *Res:* Exact determination and bounding methods for critical probabilities and critical exponents in percolation models; random graphs; probabilistic methods in combinatorics. *Mailing Add:* Math Sci Dept Johns Hopkins Univ 34th & Charles St Baltimore MD 21218

WIERSMA, DANIEL, b Volga, SDak, Nov 4, 16; m 43; c 2. SOIL SCIENCE. *Educ:* SDak State Univ, BS, 42; Univ Wyo, MS, 52; Univ Calif, PhD, 56. *Prof Exp:* County agr exten agent, Butte County, SDak, 46-52; asst, Univ Calif, 52-55; from asst prof to assoc prof agron, 56-64, PROF AGRON, PURDUE UNIV, 64-, DIR WATER RESOURCES RES CTR, 65- *Concurrent Pos:* Consult Rockefeller agr prog, Colombia, SAm, 63 & 65. *Mem:* Am Soc Agron; Soil Sci Soc Am; Soil Conserv Soc Am. *Res:* Water resources development; plant, soil and water relations; response of plants to water conditions; use of water by plants as affected by soil, water and climate. *Mailing Add:* Rte 2 Box 8A Volga SD 57071

WIERSMA, JAMES H, b Beaver Dam, Wis, Jan 4, 40; m 61; c 1. ANALYTICAL CHEMISTRY, GENERAL ENVIRONMENTAL SCIENCES. *Educ:* Wis State Univ, Oshkosh, BS, 61; Univ Mo, Kansas City, MS, 65, PhD(chem), 68. *Prof Exp:* Clin chemist, Mercy Hosp, Oshkosh, 61-62; USPHS traineeship water chem, 67-68; asst prof chem, 68-72, ASSOC PROF CHEM, UNIV WIS-GREEN BAY, 72- *Mem:* AAAS; Sigma Xi; Am Chem Soc. *Res:* Environmental sciences especially related chemistry and development of analytical methods; groundwater quality. *Mailing Add:* Nat & Appl Sci Univ Wis Green Bay WI 54311-7001

WIERWILLE, WALTER W(ERNER), b Cincinnati, Ohio, July 3, 36; m 61; c 2. HUMAN FACTORS & INDUSTRIAL ENGINEERING. *Educ:* Univ Ill, Urbana, BSEE, 58; Cornell Univ, PhD(elec eng), 61. *Prof Exp:* Res asst comput ctr, Cornell Univ, 60-61, assoc electronics engr, Avionics Dept, Physics Div, Cornell Aeronaut Lab, 61-63, res electronics engr, 63-65, prin electronics engr, 65-67, head dynamic systs sect, 67-69; supvry scientist, Sanders Assocs, 69-70, mgr, Electronic Counter-Measures Systs Group, 70-71; assoc prof elec & indust eng & opers res, 71-73, PROF ELEC & INDUST ENG & OPERS RES, VA POLYTECH INST & STATE UNIV, 73- *Concurrent Pos:* Consult, NY Transit Authority, Gen Motors Corp, 74- & USN, 77- *Mem:* Sr mem Inst Elec & Electronics Engrs; fel Human Factors Soc; Soc Indust & Appl Math; sr mem Am Inst Indust Engrs. *Res:* Command and control; workspace layout; human performance modeling; man-machine system simulation; human operator workload; vehicle handling; operator/system interface design. *Mailing Add:* Dept Indust Eng & Oper Res Va Polytech Inst & State Univ Whittemore Hall Blacksburg VA 24061

WIESBOECK, ROBERT A, b Frankfurt, Ger, Jan 19, 30; m 50; c 2. ORGANOMETALLIC CHEMISTRY. *Educ:* Munich Tech Univ, BS, 55, PhD(chem), 57. *Prof Exp:* NSF fel phys org chem, Ga Inst Technol, 58-59; res chemist, Redstone Res Div, Rohm and Haas Co, 59-63; staff scientist, 63-68, res scientist, 68-70, mgr chem, 70-74, MGR ATLANTA RES CTR, US STEEL CORP, 74- *Mem:* AAAS; Am Chem Soc. *Res:* Organic and inorganic fluorine chemistry of nitrogen, phosphorous and sulphur. *Mailing Add:* 5912 Oakleaf Dr Stone Mountain GA 30087-5799

WIESE, ALLEN F, b Eyota, Minn, Dec 16, 25; m 48; c 3. WEED SCIENCE. *Educ:* Univ Minn, BS, 49, MS, 51, PhD(agron), 53. *Prof Exp:* Prof, 53-90, EMER PROF WEED SCI, AGR EXP STA, TEX A&M UNIV, 90- *Honors & Awards:* Res Award, Weed Sci Soc Am, 80, Tex A&M Univ, 80. *Mem:* AAAS; fel Am Soc Agron; fel Weed Sci Soc Am; Weed Sci Soc Am; fel Crop Sci Soc Am. *Res:* Weed control methods in crop production including no-tillage. *Mailing Add:* Tex Agr Exp Sta Bushland TX 79012

WIESE, ALVIN CARL, b Milwaukee, Wis, Aug 13, 13; m 44; c 2. NUTRITION. *Educ:* Univ Wis, BS, 35, MS, 37, PhD(biochem), 40. *Prof Exp:* Asst biochem, Univ Wis, 35-40; instr chem, Okla Agr & Mech Col, 40-42; spec res assoc, Univ Ill, 42-45, spec asst animal nutrit, 45-46; prof agr biochem & head dept, 46-72, prof biochem, 72-78, EMER PROF BIOCHEM, UNIV IDAHO, 78- *Mem:* AAAS; Am Chem Soc; Soc Exp Biol & Med; Poultry Sci Asn; Am Inst Nutrit. *Res:* Nutritional biochemistry; enzymology; effect of fluorides on enzymes; air pollution; trace minerals. *Mailing Add:* 721 S Lynn Moscow ID 83843

WIESE, HELEN JEAN COLEMAN, b San Antonio, Tex, Dec 10, 41; m 75. MEDICAL ANTHROPOLOGY. *Educ:* Univ Wis-Milwaukee, BA, 63; Stanford Univ, MA, 64; Univ NC, Chapel Hill, PhD(anthrop), 72. *Prof Exp:* Asst prof, 72-80, ASSOC PROF BEHAV SCI, COL MED, UNIV KY, 80- *Mem:* Soc Appl Anthrop; Soc Med Anthrop; Am Anthrop Asn; Asn Behav Sci Med Educ; Inst Soc Ethics Life Sci. *Res:* Cross-cultural variation in acceptable body image; effects of pharmaceutical counseling on patient compliance with chemotherapy for congestive heart failure; attitudes toward various contraceptive devices; rates of gonorrhea in two Kentucky counties. *Mailing Add:* Dept Behav Sci Univ Ky Col Med 800 Rose St Lexington KY 40536

WIESE, JOHN HERBERT, b Los Angeles, Calif, Jan 15, 17; m 90; c 2. GEOLOGY. *Educ:* Univ Calif, Los Angeles, AB, 40, MA, 41, PhD(struct geol), 47. *Prof Exp:* Geologist, US Geol Surv, 41-48; geologist, Richfield Oil Co, 48-59, supvr explor res, 59-66, sr geologist, Atlantic Richfield Co, 67-73; consult geologist, 73-87; RETIRED. *Concurrent Pos:* Vis indust prof, Southern Methodist Univ, 70-73. *Res:* Geology of Nevada; petroleum exploration; sedimentology; landslides; continental shelf resources; geology of central California coast. *Mailing Add:* 1331 Dixie Downs Rd No 35 St George UT 84770

WIESE, MAURICE VICTOR, b Columbus, Nebr, Sept 22, 40; m 63; c 3. PLANT PATHOLOGY, CROP LOSS ASSESSMENT. *Educ:* Univ Nebr, BS, 63, MS, 65; Univ Calif, PhD(plant path), 69. *Prof Exp:* Asst prof plant path, 69-74, assoc prof & wheat pathologist, Mich State Univ, 74-78; RES PROF & CROP LOSS COORDR, UNIV IDAHO, 78- *Mem:* Am Phytopath Soc; Am Soc Agron; Crop Sci Soc Am; Sigma Xi. *Res:* Pathogenesis, etiology and control of wheat diseases; assessment of losses in crops; comprehensive yield models; crop management. *Mailing Add:* Dept Plant & Soil Sci Univ Idaho Moscow ID 83843

WIESE, RICHARD ANTON, b Howells, Nebr, Apr 3, 28; m 54; c 8. SOIL FERTILITY. *Educ:* Univ Nebr, BS, 54, MS, 56; NC State Univ, PhD, 61. *Prof Exp:* From asst prof to assoc prof soil fertil, Univ Wis, 61-67; assoc prof agron, 67-74, PROF AGRON, UNIV NEBR, LINCOLN, 74- *Mem:* Am Soc Agron; Soil Sci Soc Am. *Res:* Plant nutrition as effected by soil release of nutrients. *Mailing Add:* Dept Agron Univ Nebr Lincoln NE 68583

WIESE, ROBERT GEORGE, JR, b Boston, Mass, Sept 14, 33; m 58; c 4. GEOLOGY. *Educ:* Yale Univ, BS, 55; Harvard Univ, AM, 57, PhD(geol), 61. *Prof Exp:* Explor geologist, New Park Mining Co, 60-63; explor geologist, US Smelting Refining & Mining Co, 63-64; mem staff geol dept, 64-70, PROF GEOL & CHMN DEPT, MT UNION COL, 70- *Concurrent Pos:* Consult geologist. *Mem:* Am Inst Prof Geologists; Geol Soc Am; Am Inst Mining, Metall & Petrol Eng; Mineral Asn Can; Soc Econ Geol; Nat Asn Geol Teachers; assoc Sigma Xi. *Res:* Petrology and geochemistry of White Pine copper deposit, Michigan; petrology of wallrock alteration; genesis of mineral deposits; coal geology, exploration, development; trace elements in coal; x-ray analysis of raw materials for ceramics. *Mailing Add:* 135 Overlook Dr Alliance OH 44601

WIESE, WARREN M(ELVIN), b Rochester, Minn, Apr 14, 29; m 48; c 4. ENGINEERING MANAGEMENT, REFRIGERATION & AIR CONDITIONING. *Educ:* Univ Minn, BS, 50, MS, 52. *Prof Exp:* Teaching asst, Univ Minn, 50-52; res engr, Gen Motors Res Labs, 52-55, sr res engr, 56-65, sr liaison engr, 66-67, mgr air conditioning & automotive prod eng, Frigidaire Div, 67-72, asst chief engr, 72-75, chief engr, Delco Air Conditioning Div, 75-81, CHIEF ENGR, HARRISON RADIATOR DIV, GEN MOTORS CORP, 81- *Concurrent Pos:* Pres, Delco Air Recreation Asn, 75-83; Comn Chloroflurocarbon in Stratosphere, Nat Acad Sci, 79. *Honors & Awards:* Springer Award, Soc Automotive Engrs, 59. *Mem:* Soc Automotive Engrs. *Res:* Engine combustion; fuel antiknock characteristics; deposit-induced ignition; engine rumble; vehicle exhaust emission; technical liaison; residential and automotive air conditioning systems; air conditioning compressors. *Mailing Add:* Eng Dept Harrison Radiator Div Gen Motors Corp 200 Upper Mountain Rd Lockport NY 14094

WIESE, WOLFGANG LOTHAR, b Tilsit, Ger, Apr 21, 31; nat US; m 57; c 2. ATOMIC PHYSICS, PLASMA PHYSICS. *Educ:* Univ Kiel, BS, 54, PhD(physics), 57. *Prof Exp:* Res assoc physics, Univ Md, 58-59; res physicist, 60-62, chief plasma spectros div, Nat Bur Standards, 62-77, CHIEF ATOMIC & PLASMA RADIATION DIV, NAT INST STANDARDS & TECHNOL, 78- *Concurrent Pos:* Lectr, Univ Calif, Los Angeles, 63-64; Guggenheim fel, Max Planck Inst, Munich, Ger, 66-67; Humboldt award, Ger, 86. *Mem:* Fel Optical Soc Am; Int Astron Union; fel Am Phys Soc. *Res:* Experimental plasma spectroscopy; determination of atomic transition probabilities; measurements of plasma line broadening; evaluation and compilation of spectroscopic data. *Mailing Add:* 8229 Stone Trail Dr Bethesda MD 20817

WIESEL, TORSTEN NILS, b Upsala, Sweden, June 3, 24; div. NEUROBIOLOGY, VISUAL PROCESSING. *Educ:* Karolinska Inst, Sweden, MD, 54. *Hon Degrees:* Various from US & foreign univs, 67-90. *Prof Exp:* Instr physiol, Royal Caroline Medico-Surg Inst & asst, Dept Child Psychiat, Hosp, 54-55; fel, 55-58, asst prof ophthal-physiol, Johns Hopkins Univ Med Sch, 58-59; assoc neurophysiol & neuropharmacol, Harvard Med Sch, 59-60, asst prof, 60-64, asst prof neurophysiol, Dept Psychiat, 64-67, prof physiol, 67-68, prof neurobiol, 68-74, chmn dept, 73-84, Robert Winthrop prof, 74-84; HEAD, LAB NEUROBIOL, ROCKEFELLER UNIV, 83-, VINCENT & BROOKE ASTRO PROF, 83- *Concurrent Pos:* Fel ophthal, Med Sch, Johns Hopkins Univ, 55-58. *Honors & Awards:* Nobel Prize in Med, 81; Jules Stein Award, Trustees Prev of Blindness, 71; Dr Jules C Stein Award, Res to Prev Blindness, 71; Rosenstiel Award, 72; Friedenwald Award, Asn Res Vision & Ophthal, 75; Karl Spencer Lashley Prize, Am Philos Soc, 77; Louisa Gross Horwitz Prize, Columbia Univ, 78; Dickson Prize, Univ Pittsburgh, 79; Ledlie Prize, Harvard Univ, 80. *Mem:* Nat Acad Sci; Am Physiol Soc; Am Acad Arts & Sci; Swed Physiol Soc; AAAS; Am Neuro Soc. *Res:* Neurophysiology, especially the visual system. *Mailing Add:* Rockefeller Univ 1230 York Ave New York NY 10021

WIESENDANGER, HANS ULRICH DAVID, b Zurich, Switz, Jan 13, 28; nat US; m 54; c 4. PHYSICAL CHEMISTRY, TECHNICAL MANAGEMENT. *Educ:* Swiss Fed Inst Technol, dipl, 51, DrScTech, 54. *Prof Exp:* Tech adv inst phys ther, Zurich Univ, 53-55; fel, Univ Calif, Los Angeles, 55-56; sr res chemist, Kaiser Aluminum & Chem Corp, 57-59; phys chemist, Stanford Res Inst, 59-66; mkt mgr sci instrument dept, Electronics Assocs, Inc, 66-70; dir mkt, Uthe Technol Int, 70-72; dir int mkt; Barnes-Hind Pharmaceut, Inc, 72-74; consult, 74-75; dir mkt, Plessy Environ Systs, Inc, 75-77; dir int mkt, Chemetrics Corp, 77-81; pres, Orbiotech, Inc, 81-83; SR ASSOC, STANFORD UNIV, 84- *Concurrent Pos:* Consult, 54-55, 71-77 & 81- *Mem:* Am Chem Soc. *Res:* Surface chemistry; ultra high vacuum; radiochemistry; isotopes; tracer methods; catalysis; instrumentation; mass spectrometry; semiconductor processing equipment; process control; environmental monitoring; clinical laboratory instrumentation; clinical chemistry; technoeconomics; international marketing; long range planning; technology assessment and transfer; new ventures; acquisitions and business opportunities analysis; technology licensing; science communications. *Mailing Add:* 1151 Buckingham Dr Los Altos CA 94024

WIESENFELD, JAY MARTIN, b New Brunswick, NJ, Sept 24, 50; m 79; c 2. ULTRA-FAST OPTOELECTRONICS, OPTICAL COMMUNICATIONS TECHNOLOGY. *Educ:* Harvard Univ, AB & AM, 72; Univ Calif, Berkeley, PhD(chem), 78. *Prof Exp:* Postdoctoral fel, Bell Labs, 78-80, MEM TECH STAFF, AT&T BELL LABS, 80- *Mem:* Inst Elec & Electronics Engrs; Am Phys Soc; Optical Soc Am; Sigma Xi. *Res:* Generation of ultrashort laser pulses; ultra-high-speed optoelectronic devices; optical communications; transient behavior of optically excited semiconductor and molecular systems. *Mailing Add:* AT&T Bell Labs Crawford Hill Lab Holmdel NJ 07733

WIESENFELD, JOEL, b New York, NY, Apr 9, 18; m 45; c 2. CIVIL ENGINEERING, MECHANICS. *Educ:* City Col New York, BCE, 40; Mass Inst Technol, SM, 41; Polytech Inst Brooklyn, PhD(appl mech), 53. *Prof Exp:* Stress analyst, Curtiss-Wright Corp, NJ, 41-45; sr stress analyst, Repub Aviation Corp, NY, 45-46; from instr to prof, Rutgers Univ, 46-88, chmn, Dept Civil & Environ Eng, 70-80, asst dean freshman, 80-88, EMER PROF CIVIL ENG, RUTGERS UNIV, 88- *Mem:* Am Soc Civil Engrs; Am Soc Eng Educ; Nat Soc Prof Engrs; Am Water Works Asn. *Res:* Structural design and analysis; computer techniques applied to structural problems; construction engineering; water distribution system operation. *Mailing Add:* Dept Civil & Environ Eng Rutgers State Univ Piscataway NJ 08903

WIESENFELD, JOHN RICHARD, b New York, NY, July 26, 44. CHEMICAL KINETICS, PHOTOCHEMISTRY. *Educ:* City Col New York, BS, 65; Case Inst Technol, PhD(chem), 69; Cambridge Univ, MA, 70. *Prof Exp:* USAF fel phys chem, Cambridge Univ, 69-70, NSF fel, 70-71, Stokes res fel, Pembroke Col, 70-72; from asst prof to assoc prof, Cornell Univ, 72-84, chmn, 85-88, dep vpres res, 88-90, PROF CHEM, CORNELL UNIV, 84-, VPRES PLANNING, 90- *Concurrent Pos:* US Hon Ramsay fel, Ramsay Mem Trust, UK, 71; Henry & Camille Dreyfus Teacher-Scholar, 77-82; Alfred P Sloan Found Res Fel, 77-79; coun chem res, gov bd, 87-91. *Mem:* AAAS; Am Chem Soc. *Res:* Gas phase kinetics of atoms and molecules in defined quantum states; energy storage and transfer in chemical lasers; environmental chemistry. *Mailing Add:* Dept Chem Cornell Univ Ithaca NY 14853

WIESER, HELMUT, b Austria, July 4, 35; Can citizen; m 67; c 2. SPECTROSCOPY, PHEROMONE ACTION CHEMISTRY. *Educ:* Univ BC, BSc, 62; Univ Alta, PhD(chem), 66. *Prof Exp:* Session instr, 66-68, from asst prof to assoc prof, 68-80, PROF CHEM, UNIV CALGARY, 80-, ASSOC DEPT HEAD, 88- *Concurrent Pos:* Dozent fel, Alexander von Humboldt Found, Ger, 74, 75. *Mem:* Fel Chem Inst Can; Sigma Xi. *Res:* Molecular spectroscopy and structure; infrared and Raman spectroscopy; vibrational circular dichroism spectroscopy, ab initio molecular force fields; mode of pheromone action pheromones in forest pest management. *Mailing Add:* Dept Chem Univ Calgary Calgary AB T2N 1N4 Can

WIESMEYER, HERBERT, b Chicago, Ill, Jan 12, 32; m 54; c 2. MICROBIOLOGY. *Educ:* Univ Ill, BS, 54; Wash Univ, St Louis, PhD, 59. *Prof Exp:* NSF fel, Johns Hopkins Univ, 59-61; fel, McCollum-Pratt Inst, 61; NATO fel, 61-62; asst prof, 62-67, ASSOC PROF MOLECULAR BIOL, VANDERBILT UNIV, 67- *Mem:* Am Soc Microbiol; Genetics Soc Am; Sigma Xi. *Res:* Bacterial physiology. *Mailing Add:* Dept Molecular Biol Vanderbilt Univ Nashville TN 37235

WIESNER, J(EROME) B, b Detroit, Mich, May 30, 15; m 40; c 4. ELECTRICAL ENGINEERING, ARMS CONTROL. *Educ:* Univ Mich, BS, 37, MS, 38, PhD(eng), 50. *Hon Degrees:* EngD, Polytech Inst Brooklyn, 61; DSc, Univ Mich & Lowell Tech Inst, 62, Univ Mass, 64, Brandeis Univ & Lehigh Univ, 65 & Northwestern Univ, 66; DEng, Rensselaer Polytech Inst, 72. *Prof Exp:* Assoc dir broadcasting, Univ Mich, 37-40; chief engr, acoust & rec lab, Libr of Cong, 40-42; mem staff, radiation lab, Mass Inst Technol, 42-45; mem staff & group leader, Los Alamos Sci Lab, Univ Calif, 45-46; from asst prof to prof elec eng, Mass Inst Technol, 46-61, asst dir res lab electronics, 47-50, assoc dir, 50-52, dir, 52-61, head, elec eng dept, 59-60; spec asst sci & technol to President, 61-64, dir, Off Sci & Technol, 62-64; dean sci, 64-66, provost, 66-71, pres, 71-80, INST PROF & EMER PRES, MASS INST TECHNOL, 80- *Concurrent Pos:* Mem, President's Sci Adv Comt, chmn, 61-64; mem, technol adv comn, Off Technol Assessment, US Cong, 74-81, chmn, 76-81. *Honors & Awards:* Migel Medal, Am Found Blind, 71; Centennial Medal, Inst Elec & Electronics Engrs, 84; Arthur M Bueche Award, Nat Acad Eng, 85. *Mem:* Nat Acad Sci; Nat Acad Eng; AAAS; Am Philos Soc; Am Geophys Union; fel Inst Elec & Electronics Engrs; Acoust Soc Am; Am Geophys Union; Am Asn Univ Professors; Sigma Xi. *Res:* Electronics; radar; acoustics; theory of communications; author of 2 books and numerous articles. *Mailing Add:* 20 Ames St E15-207 Cambridge MA 02139

WIESNER, LEO, b Vienna, Austria, May 3, 13; nat US; m 58; c 1. PHYSICS, ELECTRONICS ENGINEERING. *Educ:* Univ Vienna, PhD, 37. *Prof Exp:* Res physicist, Harlem Hosp, NY, 39-43 & Int Electronics Indust, Inc, 43-45; electronics engr, Tuck Electronic Corp, 45-47 & Devenco Inc, 47-51; sr gyro engr, Reeves Instrument Corp, 51-62; mgr appl sci, Technol Group, Timex Corp, Waterbury, 62-78; ENG CONSULT, 78- *Mem:* Am Phys Soc; Inst Elec & Electronics Engrs. *Res:* Electro-optical displays; electronic watches; gyros, accelerometers and related inertial devices. *Mailing Add:* 115-01 Grosvenor Rd Kew Gardens NY 11418

WIESNER, LOREN ELWOOD, b Estelline, SDak, Nov 13, 38; m 59; c 3. SEED PHYSIOLOGY. *Educ:* SDak State Univ, BS, 60, MS, 63; Ore State Univ, PhD(agron), 71. *Prof Exp:* Asst agron, SDak State Univ, 63, asst county agt, 63-64; asst prof seed technol, Mont State Univ, 64-68; res asst agron, Ore State Univ, 68-70; assoc prof, 70-80, PROF SEED PHYSIOL, MONT STATE UNIV, 80- *Mem:* Am Soc Agron; Crop Sci Soc Am; Asn Off Seed Analysts; Asn Off Seed Cert Agencies; Sigma Xi. *Res:* Seed research related to production, technology, physiology and ecology. *Mailing Add:* Dept Plant & Soil Sci Mont State Univ Bozeman MT 59717

WIESNER, RAKOMA, b New York City, NY, May 21, 20; m 58; c 1. BIOCHEMISTRY. *Educ:* Brooklyn Col, BA, 40, MA, 50; Columbia Univ, PhD(biochem), 62. *Prof Exp:* Technician clin lab, Brooklyn Jewish Hosp, 42-47; res asst zool, Columbia Univ, 47-50, res worker biochem, 50-62; res assoc enzymol, Inst Muscle Dis, 62-67; res assoc biochem, Mt Sinai Sch Med, 69-74; res assoc biochem, Columbia Univ, 74-76; RES SCIENTIST, DEPT ENVIRON MED, NY UNIV MED CTR, 76- *Mem:* Am Chem Soc. *Res:* Nucleic acid biochemistry; control mechanisms of enzymes; mechanism of tumor promotion. *Mailing Add:* 115-01 Grosvenor Rd Kew Gardens NY 11418

WIESNET, DONALD RICHARD, b Buffalo, NY, Feb 7, 27; m 52; c 4. HYDROLOGY, GEOLOGY. *Educ:* State Univ NY Buffalo, BA, 50, MA, 51. *Prof Exp:* Asst, Univ Buffalo, 50-51; geologist, US Geol Surv, 52-54, chief manuscript rev sect, 54-55, geophys br, 56-57, asst to geol map ed, 57-59, asst chief br tech illustrations, 59-61, geohydrol map ed, 61-64, proj geologist, 65-67; oceanogr, Naval Oceanog Off, 67-68, res hydrologist, Nat Environ Satellite Data & Info Serv, 68-71, sr res hydrologist, 71-80, chief, Land Sci Br, 80-82; exec dir, Satellite Hydrol Inc, 82-90; CONSULT, 90- *Concurrent Pos:* Mem comt hydrol, US Water Resources Coun, 68-71; rapporteur remote sensing of hydrol elements, Comn Hydrol, World Meteorol Orgn, 72-76; mem work group remote sensing in hydrol, US Nat Com, Int Hydrol Decade, 72-76; mem remote sensing comt, Int Field Year on Great Lakes, 72-75; int comt remote sensing & data transmission for hydrol, Int Asn Hydrol Res, 81-; prin investr, Landsat-1 & 2, NASA, 72-76 & Heat Capacity Map Mission Satellite, 78-80; bd dir, Am Water Resources Asn, 80-82, Antarctican Soc, 83- & Potomac Red, Am Soc Photogram & Remote Sensing, 90-; vpres, Eastern Snow Conf, 82, pres, 83. *Mem:* Fel Geol Soc Am; Am Soc Photogram & Remote Sensing; Int Glaciol Soc; Antarctican Soc; Am Geophys Union; Am Water Resources Asn; Soc Am Mil Engrs. *Res:* Satellite hydrology; remote sensing of hydrologic parameters such as snow, ice, soil moisture, floods, coastal hydrology and ground water; hydrologic maps; disaster studies such as floods, earthquakes, hurricanes and droughts. *Mailing Add:* 601 McKinley St NE Vienna VA 22180

WIEST, STEVEN CRAIG, b Harrisburg, Pa, Aug 4, 51. ORNAMENTAL HORTICULTURE, PLANT PHYSIOLOGY. *Educ:* Cornell Univ, BS, 73, MS, 75, PhD(agron), 79. *Prof Exp:* Res asst ornamental hort, Cornell Univ, 73-77, res asst agron, 77-78; asst prof hort, Rutgers Univ, New Brunswick, 78-80; Asst prof, 80-83, ASSOC PROF, DEPT HORT, KANS STATE UNIV, 83- *Honors & Awards:* Kenneth Post Award, Am Soc Hort Sci, 78, Alex Laurie Award, 81. *Mem:* Am Soc Plant Physiologists; Am Soc Hort Sci. *Res:* Biophysical and biochemical responses of plant cells to environmental stresses, especially those stresses induced directly by such meteorological conditions as low and high temperatures and drought. *Mailing Add:* Dept Hort Waters Hall Kans State Univ Manhattan KS 66506

WIEST, WALTER GIBSON, b Price, Utah, Feb 16, 22; m 48; c 7. BIOCHEMISTRY. *Educ:* Brigham Young Univ, AB, 48; Univ Wis, MS, 51, PhD(biochem), 52. *Prof Exp:* Asst biochem, Univ Wis, 48-52; from instr to assoc prof, Univ Utah, 52-64; assoc prof, 64-68, PROF BIOCHEM OBSTET & GYNEC, SCH MED, WASH UNIV, 68- *Concurrent Pos:* USPHS spec fel, Univ Cologne, 59-60. *Mem:* Am Soc Biol Chem; Endocrine Soc; Soc Gynec Invest. *Res:* Biosynthesis, metabolism and mode of action of steroid hormones, especially progesterone; application of radioisotopic techniques to steroid biochemistry. *Mailing Add:* Dept Obstet & Gynec Wash Univ St Louis MO 63110

WIETING, TERENCE JAMES, b Chicago, Ill, Sept 4, 35; m 70; c 3. PHYSICS. *Educ:* Mass Inst Technol, BS, 57; Harvard Univ, BD, 62; Cambridge Univ, PhD(physics), 69. *Prof Exp:* Res staff mem physics, Naval Supersonic Lab, Mass Inst Technol, 58-60; res scientist, Mithras, Inc, Mass, 62-63; Nat Acad Sci-Nat Res Coun res assoc, 69-71; head, Optical Interactions Sect, Naval Res Lab, 71-80. *Concurrent Pos:* Invited prof, Fed Polytech, Lausanne, Switz, 77-78. *Mem:* Am Phys Soc; Sigma Xi; AAAS. *Res:* Optical properties of metals; physics of low-dimensional materials; Raman scattering; lattice dynamics. *Mailing Add:* US Naval Res Lab Code 4650 Washington DC 20375

WIEWIOROWSKI, TADEUSZ KAROL, b Sopot, Poland, Nov 3, 35; US citizen; m 62; c 2. FERTILIZER TECHNOLOGY, SULFUR TECHNOLOGY. *Educ:* Loyola Univ, La, BS, 59; Tulane Univ, PhD(chem), 65. *Prof Exp:* Asst mgr res & develop, Freeport Minerals Co, 59-81; vpres & dir res & develop, Freeport Res & Eng Co, 81-90. *Mem:* Am Chem Soc. *Res:* Inorganic and physical chemistry; process development; hydrometallurgy; management of chemical research and development; solvent extraction. *Mailing Add:* 2620 Danbury Dr New Orleans LA 70131

WIFF, DONALD RAY, b Youngstown, Ohio, Feb 19, 36; m; c 3. POLYMER PHYSICS, MARKETING RESEARCH. *Educ:* Capital Univ, BS, 58; Kent State Univ, MA, 60; Tex A&M Univ, PhD(physics), 67. *Prof Exp:* Instr physics, Tex A&M Univ, 60-67; res assoc prof physics, Univ Dayton Res Inst, 67-70, res physicist, 70-73, res assoc, 73-76, prin investr, 76-85; SECT HEAD, GENCORP INC, 85. *Mem:* Am Phys Soc; Am Mgt Asn; Soc Advan Mat & Process Eng; Brit Inst Physics; Europ Phys Soc; Rheology Soc; Crystallog Asn; Soc Plastics Eng; Am Chem Soc. *Res:* Theoretical polymer physics; electronic energy band calculation of cubic boron nitride via APW method; molecular weight distribution from ultracentrifugation via ill pased problem regularization methods; molecular composites; in-situ composites; dynamic mechanical recalculation spectra using linear viscoelasticity theory. *Mailing Add:* Gencorp Inc 2990 Gilchrist Rd Akron OH 44305

WIGEN, PHILIP E, b LaCrosse, Wash, May 11, 33; m 54; c 3. SOLID STATE PHYSICS, MAGNETISM. *Educ:* Pac Lutheran Col, BA, 55; Mich State Univ, PhD(physics), 60. *Prof Exp:* Assoc res scientist, Lockheed Res Labs, Calif, 60-63, res scientist, 63-65; assoc prof physics, 65-71, PROF PHYSICS, OHIO STATE UNIV, 71- *Concurrent Pos:* Consult, Res Labs, Battelle Mem Inst, 67-70, Drackett Co, 71-74, A F Avionics Lab, WPAB, 74-78 & Airtron/Litton, 81-; vis prof, Univ Osaka, Japan, 73, Univ Osnabruck, W Ger,86; Univ Zurich, Switz, 87; vis scientist, Phillips Res Lab, Hamburg, WGer, 79. *Mem:* Fel Am Phys Soc; Am Asn Physics Teachers; Inst Elec & Electronics Eng; AAAS; Am Asn Univ Professors; Sigma Xi. *Res:* Magnetism in metals and insulators; paramagnetic resonance; chaos in magnetism. *Mailing Add:* Dept Physics Ohio State Univ 174 W 18th Ave Columbus OH 43210

WIGFIELD, DONALD COMPSTON, b Godalming, Eng, June 13, 43; m 66; c 2. CHEMICAL TOXICOLOGY. *Educ:* Univ Birmingham, BSc, 64; Univ Toronto, PhD(org chem), 67. *Hon Degrees:* DSc, Univ Birmingham, 87. *Prof Exp:* Fel, Univ BC, 67-68, teaching fel, 68-69; asst prof chem, 69-72, assoc prof, 72-78, dir, Ottawa-Carleton Inst Res & Grad Studies Chem, 81-84, PROF CHEM, CARLETON UNIV, 78- *Concurrent Pos:* Vis assoc prof, Univ Victoria, 75-76. *Mem:* Am Chem Soc; fel Royal Soc Chem; fel Chem Inst Can. *Res:* Chemical reaction mechanisms. *Mailing Add:* Dept Chem Carleton Univ Ottawa ON K1S 5B6 Can

WIGGANS, DONALD SHERMAN, b Lincoln, Nebr, July 14, 25; m 51; c 4. BIOCHEMISTRY. *Educ:* Univ Nebr, BSc, 49; Univ Ill, PhD(chem), 52. *Prof Exp:* Instr biochem, Yale Univ, 52-54; from asst prof to assoc prof, 54-61, PROF BIOCHEM, UNIV TEX HEALTH SCI CTR, DALLAS, 61- *Concurrent Pos:* Vis prof, Southern Methodist Univ, 64-66 & Univ Tex, Arlington, 67-74. *Mem:* Am Chem Soc; Am Soc Biol Chemists; Soc Exp Biol & Med; Am Inst Nutrit. *Res:* Intermediary metabolism of amino acids and peptides; mechanism of protein synthesis. *Mailing Add:* Dept Biochem Southwest Med Sch Dallas TX 75235

WIGGANS, SAMUEL CLAUDE, b Lincoln, Nebr, Sept 2, 22; m 57; c 2. PLANT PHYSIOLOGY, AGRONOMY. *Educ:* Univ Nebr, BS, 47; Univ Wis, MS, 48, PhD(plant physiol), 51. *Prof Exp:* Asst agron, Univ Wis, 47-49, asst bot, 49-51; asst prof agron & bot, Iowa State Univ, 51-58; assoc prof hort, Okla State Univ, 58-62; chmn, Dept Hort Sci, Univ Vt, 62-65, prof hort, 62-80, chmn, Dept Plant & Soil Sci, 65-80; agronomist, plant sciences, 80-81, PRIN HORTICULTURIST, COOP STATE RES SERV, USDA, WASHINGTON, DC, 81- *Honors & Awards:* Cert Appreciation, Nat Coun Ther & Rehab through Hort. *Mem:* Am Soc Hort Sci; Am Soc Agron; Am Soc Plant Physiol; Coun Agr Sci & Technol; Am Hort Ther Asn; Nat Hort Indust Coun. *Res:* Growth regulation; fertilizer placement and management techniques; photoperiod and temperature. *Mailing Add:* Plant & Animal Sci Coop State Res Serv USDA 330-C Aerospace Ctr Bldg 901 D St SW Washington DC 20250-2200

WIGGER, H JOACHIM, b Hagen, Ger, May 29, 28; m 57; c 2. MEDICINE, PATHOLOGY. *Educ:* Johanneum Col, Lueneburg, Ger, BA, 49; Univ Hamburg, DMSc, 54. *Prof Exp:* Assoc dir labs, Children's Hosp, Washington, DC, 62-64; assoc, 64-67, asst prof, 67-69, ASSOC PROF PATH, COLUMBIA UNIV, 69- *Concurrent Pos:* Consult, USPHS Hosp, Staten Island, 66-; asst attend pathologist, Presby Hosp, New York, 67-69, assoc attend pathologist, 69- *Mem:* NY Acad Sci. *Res:* Pediatric and developmental pathology. *Mailing Add:* Dept Pediat Path Columbia Univ Col Phys Surg 630 W 168th St New York NY 10032

WIGGERS, HAROLD CARL, PHYSIOLOGY. *Educ:* Western Reserve Univ, PhD(physiol), 35. *Prof Exp:* Prof physiol, Albany Med Col, 47-53, pres & dean, 53-77; RETIRED. *Mailing Add:* 711 Iris Lane Vero Beach FL 32960

WIGGERT, BARBARA NORENE, b Cleveland, Ohio, Jan 7, 38; m 58; c 4. BIOCHEMISTRY, VISUAL SCIENCE. *Educ:* Univ Wis-Madison, BA, 59; Harvard Univ, PhD(biochem), 63. *Prof Exp:* Fel biochem, Dept Physiol Chem, 63-65; NIH fel, 75-76, staff fel, 76-78, res chemist biochem, 78-85, SECT CHIEF, NAT EYE INST, NIH, 85- *Mem:* Am Chem Soc; Asn Res Vision & Ophthal; Sigma Xi; Am Soc Biol Chemists. *Res:* Uptake, binding and translocation of retinoids, such as vitamin A and its analogs, into ocular tissues. *Mailing Add:* NIH Lab Retinal Cell & Molecular Biol Bethesda MD 20892

WIGGILL, JOHN BENTLEY, operations management, oilfield exploration & aquisition, for more information see previous edition

WIGGIN, EDWIN ALBERT, b Exeter, NH, Aug 11, 21; m 47; c 2. CHEMISTRY. *Educ:* Univ NH, BS, 43. *Prof Exp:* Asst, SAM Labs, Columbia Univ, 43-45; res chemist, Carbide & Carbon Chem Co, 45-48; chief tech develop br, Isotopes Div, AEC, 48-54; EXEC VPRES, ATOMIC INDUST FORUM, 54- *Concurrent Pos:* Mem Adv Comt Indust Info, AEC, 53-54, mem Adv Comt Isotope & Radiation Develop, 58-60 & 70-72. *Mem:* AAAS; Am Chem Soc; Am Nuclear Soc; Inst Nuclear Mat Mgt. *Res:* Application of atomic energy and radioactive by-products. *Mailing Add:* 17 Meadowcroft Ct Gaithersburg MD 20879

WIGGINS, ALVIN DENNIE, b Harrisburg, Ill, May 5, 22; m 50; c 6. MATHEMATICAL STATISTICS. *Educ:* Univ Calif, Berkeley, AB, 51, MA, 53, PhD(statist), 58. *Prof Exp:* Res asst statist & assoc biostatist, Sch Pub Health, Univ Calif, Berkeley, 54-57; instr math, Ctr Grad Studies, Univ Wash, 58-63; assoc res biostatistician & lectr biostatist, Sch Pub Health, Univ Calif, Berkeley, 63-69; from asst prof to assoc prof biostatist, 69-82, PROF STATIST, UNIV CALIF, DAVIS, 82- *Concurrent Pos:* Sr statistician, Hanford Labs Oper, Gen Elec Co, 57-63; dir, Stat Lab, Univ Calif, Davis, 87-89; pres, San Francisco Bay Area Chap, Am Stat Asn, 87-88. *Mem:* Inst Math Statist; Biomet Soc; Am Statist Asn; fel Royal Statist Soc. *Res:* Mathematical models of biological phenomena; statistical theory of estimation; application of stochastic processes to problems of health, medicine and biology; design and analysis of experiments; stochastic differential equations in biology. *Mailing Add:* Div Statist Univ Calif Davis CA 95616

WIGGINS, CARL M, b Jackson, Miss, Aug 5, 41; m 65; c 1. TRANSIENT ELECTROMAGNETIC INTERFERENCE, LASER & OPTICAL SYSTEMS ANALYSIS. *Educ:* Lamar Univ, BS, 64; Sam Houston State Col, MS, 66. *Prof Exp:* Staff mem, BDM Int, Inc, 73-75, assoc mgr, 75-80, sr staff, 80-85, prin staff, 85-89, SR PRIN, BDM INT, INC, 89- *Mem:* Sr mem Inst Elec & Electronics Engrs Power Eng Soc; sr mem Nat Asn Radio & Telecommun Engrs, 90. *Res:* Transient electromagnetic interference in electric power systems including measurment, analysis, modeling and simulation; fault location; laser, optical, and electro-optical systems analysis. *Mailing Add:* 11600 Academy Rd NE No 4615 Albuquerque NM 87111

WIGGINS, EARL LOWELL, b Ringwood, Okla, July 11, 21; m 45; c 4. PHYSIOLOGY. *Educ:* Okla State Univ, BS, 47, MS, 48; Univ Wis, PhD(physiol of reproduction), 51. *Prof Exp:* Asst, Univ Wis, 48-50; animal geneticist, Sheep Exp Sta, USDA, Idaho, 50-56; from assoc prof to prof animal sci, Auburn Univ, 56-81; RETIRED. *Mem:* AAAS; Am Soc Animal Sci; Am Genetic Asn; Soc Study Reproduction. *Res:* Puberty and related phenomena in sheep and swine; artificial insemination in swine; causes of reproductive failure in ewes; factors affecting the breeding season in ewes; reproduction, breeding and genetics in farm animals. *Mailing Add:* 300 Shell Toomer Pkwy Auburn AL 36830

WIGGINS, EDWIN GEORGE, b Palo Alto, Calif, June 12, 43; m 65; c 2. MARINE ENGINEERING, MECHANICAL ENGINEERING. *Educ:* Purdue Univ, BS, 65, MS, 68, PhD(mech eng), 76. *Prof Exp:* Asst engr, USS Massey, USN, 65-66, ship supt, Charleston Naval Shipyard, 71-74, chief engr, USS Sampson, 74-76, maintenance plan officer, USN, 76-78; dept head marine eng, Moody Col, 78-81; DEPT HEAD ENG, US MERCHANT MARINE ACAD, 82- *Mem:* Am Soc Mech Engrs; Soc Naval Architects & Marine Engrs; Am Soc Eng Educ. *Res:* Measurement of characteristics of incompressible turbulent boundary layers. *Mailing Add:* Webb Inst Naval Architect Crescent Beach Rd Glen Cove NY 11542

WIGGINS, ERNEST JAMES, b Trenton, Ont, Nov 25, 17; m 45; c 2. PHYSICAL CHEMISTRY, CHEMICAL ENGINEERING. *Educ:* Queen's Univ, Ont, BSc, 38; McGill Univ, PhD(phys chem), 46. *Prof Exp:* Supt eng develop sect, Atomic Energy Proj, Nat Res Coun, 46-48; head munitions & eng sect, Suffield Exp Sta, Defence Res Bd, 48-52; sr chemist, Stanford Res Inst, 52-58; head chem div, Sask Res Coun, 58-61; asst dir, Chem Div, Ont Res Found, 61-62; dir res, Res Coun Alta, 62-77, CONSULT MEM, ALTA OIL SANDS TECHNOL & RES AUTHORITY, 77- *Mem:* Am Chem Soc; Am Inst Aeronaut & Astronaut; Arctic Inst NAm; Chem Inst Can; Brit Soc Chem Indust. *Res:* Chemical process development; environmental studies; energy resource development. *Mailing Add:* 8208 117th St Edmonton AB T6G 1R2 Can

WIGGINS, GLENN BLAKELY, b Toronto, Ont, Jan 29, 27; m 49; c 3. ENTOMOLOGY, FRESHWATER BIOLOGY. *Educ:* Univ Toronto, BA, 49, MA, 50, PhD, 58. *Prof Exp:* Asst biologist, Nfld Fisheries Res Sta, Fisheries Res Bd Can, 50-51; asst cur, Dept Entom, 52-61, from asst cur in chg to cur in chg, 61-76, CUR, DEPT ENTOM, ROYAL ONT MUS, 64-; PROF ZOOL, UNIV TORONTO, 68- *Concurrent Pos:* Vis prof, Univ Minn, 70, 72 & 74 & Univ Mont, 81; mem sci comt, Biol Surv Can, 76-; vpres, Biol Coun Can, 82-84. *Mem:* Fel Entom Soc Can (vpres, 79-80, pres, 81-82); Soc Syst Zool; NAm Benthol Soc; Entom Soc Am. *Res:* Systematic entomology, especially trichoptera; aquatic entomology; biology of temporary pools; domiciliary invertebrates; evolution. *Mailing Add:* Dept Entom Royal Ont Mus 100 Queen's Park Toronto ON M5S 2C6 Can

WIGGINS, IRA LOREN, taxonomic botany & plant distribution; deceased, see previous edition for last biography

WIGGINS, JAMES WENDELL, b Fayette, Ala, May 9, 42; m 64; c 2. SEISMIC IMAGING, IMAGE PROCESSING. *Educ:* Univ Ala, BS, 63; Johns Hopkins Univ, PhD(physics), 68. *Prof Exp:* Res assoc physics, Johns Hopkins Univ, 68-69, biophys, 69-73, NIH spec fel, 73-74, asst prof biophys, 74-80, assoc prof biophys, 80-81; sr res geophysicist, Gulf Res & Develop Co, 82-84; sr res geophysicist, 84-88, MGR GEOPHYS RES & DEVELOP, WESTERN GEOPHYS DIV, WESTERN ATLAS INT, INC, 88- *Mem:* Soc Explor Geophysicists; Am Phys Soc; AAAS; Europ Asn Explor Geophysicists. *Res:* Methods of data collection and processing in seismic imaging; computer modeling of geological structure; image processing techniques applicable to geological, biological and other areas. *Mailing Add:* 12914 Elmington Ct Cypress TX 77429

WIGGINS, JAMES WILLIAM, b Paris, Ark, Mar 5, 40. INORGANIC CHEMISTRY. *Educ:* Univ Ark, Fayetteville, BS, 62; Univ Fla, PhD(chem), 66. *Prof Exp:* Res grant, Univ Calif, Riverside, 66-68; interim asst prof chem, Univ Fla, 68-69; asst prof, 69-73, PROF CHEM & INORG CHEM, UNIV ARK, LITTLE ROCK, 73-, ASSOC DEAN, COL SCI, 80-, ASSOC DEAN, COL HEALTH RELATED PROFESSIONS, 83- *Mem:* Am Chem Soc. *Res:* Boron-nitrogen-carbon chemistry, synthesis of compounds; mechanisms of reactions leading to unusual structures; water quality in Arkansas and the effect of changing the stream beds on the water quality. *Mailing Add:* Dept Chem Univ Ark Little Rock AR 72204

WIGGINS, JAY ROSS, b Baltimore, Md, Apr 12, 47; m 87; c 3. CARDIAC PHARMACOLOGY. *Educ:* Bucknell Univ, BS, 69; Columbia Univ, PhD(pharmacol), 75. *Prof Exp:* Res assoc, Rockefeller Univ, 73-75; from asst prof to assoc prof pharmacol, Univ SFla, 76-81; asst dir pharmacol, Berlex Labs, Inc, 82, assoc dir cardiac & biochem pharmacol, 82-84, pharmacol, 84-88, ASST DIR DIAG IMAGING, BERLEX LABS, INC, 88- *Mem:* AAAS; Am Soc Pharmacol & Exp Therapeut; Am Heart Asn; Int Soc Heart Res; NY Acad Sci; Biophys Soc. *Res:* Contrast echocardiography; electrophysiology and pharmacology of cardiac arrhythmias; mechanisms of excitation-contraction coupling in cardiac muscle. *Mailing Add:* Berlex Labs Inc 300 Fairfield Rd Wayne NJ 07470-7358

WIGGINS, JOHN, b Bellevue, Nebr, Oct 25, 49; m 77. HEAVY ION PHYSICS. *Educ:* Univ Ga, BS, 75; Ind Univ, Bloomington, MS, 80, PhD(physics), 81. *Prof Exp:* Res asst, Cyclotron Facil, Ind Univ, 75-81, res assoc, 81; res staff, lab nuclear sci, Mass Inst Technol, 81-; AT BOLT, BERANEK & NEWMAN INC, CAMBRIDGE. *Mem:* Sigma Xi; NY Acad Sci; Am Phys Soc; AAAS; Am Asn Physics Teachers. *Res:* Experimental intermediate energy nuclear physics; reaction mechanism of protons with nuclei, excitation and decay of giant resonances in nuclei and decay of nuclei at high excitation; nuclear structure and properties from proton and heavy ion induced reactions. *Mailing Add:* A8 Hallowell House 668A Washington St Wellesley MA 02181

WIGGINS, JOHN H(ENRY), JR, b Tulsa, Okla, May 12, 31; m 78; c 2. STRUCTURAL DYNAMICS, GEOPHYSICS. *Educ:* Stanford Univ, BS, 53; St Louis Univ, MS, 55; Univ Ill, PhD(civil eng), 61. *Prof Exp:* Physicist spec weapons ctr, US Air Force, 55-58; sr res engr, Jersey Prod Res Co, 61-64; tech prog dir sonic boom & earthquake effects, res div, John A Blume & Assocs, 64-66; tech dir environ res, Datacraft Inc, 66; pres, J H Wiggins Co, 66-85; PRES, CRISIS MGT CORP, 87- *Honors & Awards:* Moisseiff Award, Am Soc Civil Engrs, 65. *Mem:* Am Inst Aeronaut & Astronaut; Soc Explor Geophysicists; Am Soc Mech Engrs; Am Inst Mining, Metall & Petrol Engrs; Am Geophys Union; Earthquake Eng Res Inst. *Res:* Nuclear weapons effects; earthquake engineering and seismology; oil well drilling and exploration geophysics; sonic boom effects; risk assessment; crisis management. *Mailing Add:* Crisis Mgt Corp 1650 S Pacific Coast Hwy Suite 311 Redondo Beach CA 90277

WIGGINS, JOHN SHEARON, b Chicago, Ill, Feb 8, 15. SPACE PHYSICS, SYSTEMS ENGINEERING. *Educ:* Earlham Col, AB, 36; Calif Inst Technol, MS, 38; Univ Southern Calif, PhD(physics), 56. *Prof Exp:* Lectr physics, Univ Southern Calif, 41-43; from instr to asst prof, Univ Redlands, 44-46; asst prof, Univ Okla, 46-50; lectr physics, Univ Southern Calif, 50-56, asst prof, 57-58; mem tech staff, Semiconductor Div, Hughes Aircraft Co, 58-63; mem tech staff, Space Sci Dept, 65-80, MEM TECH STAFF, SPACE & TECHNOL GROUP, TRW SYSTS, 80- *Concurrent Pos:* UNESCO vis prof, Concepcion Univ, Chile, 64. *Mem:* Am Phys Soc; Sigma Xi. *Res:* Photoelectricity; electron microscopy; optical, beta-ray and gamma-ray spectroscopy; linear accelerator; semiconductor devices; space physics; space science instrumentation; payload design, test and integration; radiation damage and measurement; spacecraft charging; spacecraft operations engineering. *Mailing Add:* 7181 Chelan Way Hollywood CA 90068-2624

WIGGINS, PETER F, b New York, NY, July 18, 35; m 65; c 2. NUCLEAR ENGINEERING. *Educ:* State Univ NY Maritime Col, BMarE, 58; NY Univ, MME, 61; Univ Md, PhD(nuclear eng), 70. *Prof Exp:* Asst inst eng, State Univ NY Maritime Col, 58-61; asst prof eng sci, State Univ NY Agr & Technol Col, Farmingdale, 61-62; asst prof nuclear eng, 62-71, assoc prof naval systs eng, 71-76, chmn dept, 76-81, PROF NAVAL SYSTS ENG, US NAVAL ACAD, 76- *Concurrent Pos:* NSF sci fac fel, 69-70. *Mem:* Am Soc Eng Educ; Soc Naval Archit & Marine Engrs; Am Nuclear Soc; Am Soc Nuclear Engrs; Sigma Xi. *Res:* Neutron activation analysis; capture gamma ray studies using isotopic source californium-252 for mineral exploration. *Mailing Add:* 1016 Harbor Dr Annapolis MD 21403

WIGGINS, RALPHE, b Broadwater, Nebr, Apr 4, 40. GEOPHYSICS. *Educ:* Colo Sch Mines, GpE, 61; Mass Inst Technol, PhD(geophys), 65. *Prof Exp:* Proj dir geophys, Geoscience, Inc, 65-66; res assoc, Mass Inst Technol, 66-70; asst prof, Univ Toronto, 70-73; assoc prof, Univ BC, 73-75; sr res geophysicist, Western Geophys Co, 75-77; prin geophysicist, Del Mar Tech Assocs, 77-78; SR RES ASSOC, MOBIL RES & DEVELOP CORP, 78- *Mem:* Am Geophys Union; Soc Explor Geophys; Seismol Soc Am; Geol Soc Am; fel Royal Astron Soc. *Res:* Seismology; computer applications for interpretation and inversion of geophysical observations. *Mailing Add:* Schlumberger Doll Res Old Quarry Rd Ridgefield CT 06877

WIGGINS, RICHARD CALVIN, b Portsmouth, Va, June 26, 45; m 75; c 2. NEUROCHEMISTRY, DEVELOPMENTAL NEUROBIOLOGY. *Educ:* Duke Univ, BS, 67, PhD(anat), 73. *Prof Exp:* Res assoc neurol, Sch Med, Univ Miami, 72-73; res assoc neurochem, Med Sch, Univ NC, 73-75; from asst prof to assoc prof, 76-88, PROF NEUROBIOL & ANAT, MED SCH, UNIV TEX, 88- *Concurrent Pos:* NIH res career develop award, 79. *Mem:* Am Soc Neurochem; Int Soc Neurochem; Am Soc Neurosci; Am Asn Anatomists; Int Soc Develop Neurosci. *Res:* Biological chemistry of myelin; effects of environmental perturbation and drug abuse on brain development. *Mailing Add:* Dept Neurobiol & Anat Univ Tex Med Sch PO Box 20708 Houston TX 77025

WIGGINS, THOMAS ARTHUR, b Indiana, Pa, Feb 24, 21; m 53; c 2. OPTICS. *Educ:* Pa State Univ, BS, 42, PhD(physics), 53; George Washington Univ, MS, 49. *Prof Exp:* Instr physics, George Washington Univ, 48-50; from asst prof to prof physics, 53-86, EMER PROF PHYSICS, PA STATE UNIV, 87- *Mem:* Fel Am Phys Soc; fel Optical Soc Am. *Res:* Light beam propagation; physical optics; spontaneous and stimulated light scattering. *Mailing Add:* 104 Davey Lab Dept Physics Pa State Univ University Park PA 16802

WIGGINS, VIRGIL DALE, b Tulsa, Okla, June 25, 31; m 52; c 3. PALYNOLOGY. *Educ:* Univ Okla, BS, 57, MS, 62. *Prof Exp:* Sr palynological technician, Sun Oil Co Prod Res, 58-59; explor palynologist, 59-69, sr explor palynologist, Alaskan Div, 69-81, STAFF EXPLOR PALYNOLOGIST, WESTERN REGION, CHEVRON USA, 81- *Concurrent Pos:* Alaskan mem, Int Palynological Comn Working Group P3, 74-; mem bd dirs, Am Asn Stratig Palynologists, 83-85. *Mem:* Am Asn Stratig Palynologists. *Res:* Application of palynology to arctic petroleum exploration. *Mailing Add:* 3048 Naranja Dr Walnut Creek CA 94598

WIGH, RUSSELL, b Weehawken, NJ, Nov 17, 14; m 39; c 3. MEDICINE, RADIOLOGY. *Educ:* Rutgers Univ, BS, 35; Harvard Med Sch, MD, 39. *Prof Exp:* Asst demonstr radiol, Jefferson Med Col, 46-49, instr, 49-50, assoc, 50, asst prof, 50-52; asst prof, Col Physicians & Surgeons, Columbia Univ, 52-54, assoc prof, 54-56; prof & chmn dept, Med Col Ga, 56-63; dir Dept Radiol, Bartholomew County Hosp, 63-71; assoc prof radiol, Sch Med, Ind Univ, Indianapolis, 72-77; from prof to emer prof radiol, Med Col Ga, 77-85; RETIRED. *Concurrent Pos:* Consult, Vet Admin Hosps, New York, 55-56 & Augusta, Ga, 56-63 & Battey State Hosp, Rome, 56-62; clin prof, Sch Med, Univ Louisville, 67-72. *Honors & Awards:* Cert of Merit, Am Roentgen Ray Soc, 51. *Mem:* Radiol Soc NAm; AMA; fel Am Col Radiol. *Res:* Photofluorographic detection of silent gastric neoplasms; clinical radiological investigations of various body systems. *Mailing Add:* 3601 Burning Tree Ct Augusta GA 30907

WIGHT, HEWITT GLENN, b Murray, Utah, Feb 8, 21; m 43, 81; c 4. SYNTHETIC ORGANIC CHEMISTRY. *Educ:* Univ Utah, BS, 43; Univ Calif, PhD(chem), 55. *Prof Exp:* PROF CHEM, CALIF POLYTECH STATE UNIV, SAN LUIS OBISPO, 52- *Mem:* Am Chem Soc. *Res:* Chemical education; organic syntheses; synthesis of potential pharmaceuticals. *Mailing Add:* Dept Chem Calif Polytech State Univ San Luis Obispo CA 93407

WIGHT, JERALD ROSS, b Brigham City, Utah, Oct 5, 31; m 54; c 7. RANGE SCIENCE. *Educ:* Utah State Univ, BS, 53, MS, 59; Univ Wyo, PhD(range sci), 66. *Prof Exp:* Lab technician olericult, Univ Calif, 58-63; RANGE SCIENTIST, AGR RES SERV, USDA, 65- *Mem:* Soil Conserv Soc Am; Am Soc Agron; Soc Range Mgt. *Res:* Plant, soil, climate and animal relationships in range ecosystems, with emphasis on development and evaluation of rangeland simulation models. *Mailing Add:* 11864 Reutzel Boise ID 83709

WIGHTMAN, ARTHUR STRONG, b Rochester, NY, Mar 30, 22; m 45, 77; c 2. MATHEMATICAL PHYSICS. *Educ:* Yale Univ, BA, 42; Princeton Univ, PhD(physics), 49. *Hon Degrees:* DSc, Swiss Fed Inst Technol, 69; Gmttingen, 86. *Prof Exp:* Instr physics, Yale Univ, 43-44; from instr to assoc prof, 49-60, prof math physics, 60-71, THOMAS D JONES PROF MATH PHYSICS, PRINCETON UNIV, 71- *Concurrent Pos:* Nat Res Coun fel, Inst Theoret Physics, Copenhagen, 51-52; NSF fel, Copenhagen & Naples, 56-57; vis prof, Paris, 57; vis prof, Inst Advan Study Sci, Bures-sur-Yvette, 63-64 & 68-69; vis prof, Ecole Polytechnique, 77-78, Univ Adelaide, Australia, 82. *Mem:* Nat Acad Sci; AAAS; Am Math Soc; Am Phys Soc; Fedn Am Sci (treas, 54-56); Royal Soc Arts & Sci. *Res:* Elementary particle physics; quantum field theory; mathematical physics; functional analysis. *Mailing Add:* Dept Physics Princeton Univ Box 708 Princeton NJ 08544

WIGHTMAN, FRANK, b Padiham, Eng, Jan 22, 28; Can citizen; m 56; c 5. PLANT PHYSIOLOGY. *Educ:* Univ Leeds, BSc, 48, PhD(plant physiol), 54. *Hon Degrees:* DSc, Univ Leeds, 81. *Prof Exp:* Sr sci officer plant physiol, Agr Res Coun Unit, Wye Col, Univ London, 52-58; Nat Res Coun Can fel, Nat Res Coun Lab, Univ Sask, Can, 58-59; assoc prof, 60-66, chmn dept, 68-71, PROF BIOL, CARLETON UNIV, 66- *Concurrent Pos:* Nuffield Found vis fel, Univ Col, Univ London, 65-66; vis prof, Univ Lausanne, Switz, 66, Univ Calif, Santa Cruz, 74-75 & Commonwealth Sci & Indust Res Orgn Lab, Adelaide, Australia, 82-83. *Mem:* Can Soc Plant Physiol (pres, 79-80); Am Soc Plant Physiol; Brit Soc Exp Biol; Scand Soc Plant Physiol; fel Royal Soc Can. *Res:* Biosynthesis and physiological activity of indole and phenyl plant growth hormones; characterization of enzymes catalyzing aromatic amino acid metabolism; characterization of hormonal substances regulating flower formation and lateral root initiation. *Mailing Add:* Dept Biol Carleton Univ Ottawa ON K1S 5B6 Can

WIGHTMAN, JAMES PINCKNEY, b Ashland, Va, May 14, 35; m 56; c 4. COLLOID & SURFACE CHEMISTRY. *Educ:* Randolph-Macon Col, BS, 55; Lehigh Univ, MS, 58, PhD(chem), 60. *Prof Exp:* Res assoc fuel sci, Pa State Univ, 60-62; PROF CHEM, VA POLYTECH INST & STATE UNIV, 62- *Concurrent Pos:* Vis res prof, Univ Bristol, 75-76. *Mem:* Am Chem Soc;

Sigma Xi; Adhesion Soc; Int Asn Colloid & Interface Scientists. *Res:* Thermodynamics of adhesion; electron spectroscopic chemical analysis of solids surfaces; surface chemistry focuses on the interaction of liquids with solid surfaces and on the characterization of those solids, applications include adhesion, wetting of coal, reprography, surface migration and metal sorption on polymers. *Mailing Add:* Dept Chem 0212 Va Polytech Inst Blacksburg VA 24061-0212

WIGHTMAN, ROBERT HARLAN, b Ottawa, Ont, Jan 24, 37; m 61; c 3. ORGANIC CHEMISTRY. *Educ:* Univ NB, BSc, 58, PhD(org chem), 62. *Prof Exp:* Nat Res Coun Can overseas fel org chem, Imp Col, Univ London, 62-63; res assoc, Stanford Univ, 63-65; asst prof, 65-69, ASSOC PROF ORG CHEM, CARLETON UNIV, 69- *Mem:* Am Chem Soc; Chem Inst Can. *Res:* New synthetic organic methods; syntheses of organic compounds of biological and theoretical interest. *Mailing Add:* Carleton Univ Ottawa ON K1S 5B6 Can

WIGHTON, JOHN L(ATTA), b Vancouver, BC, June 15, 15. MECHANICAL ENGINEERING. *Educ:* Univ BC, BA, 35, BASc, 44; Univ Mich, MSE, 52, PhD(mech eng), 55. *Prof Exp:* Design engr, Hudson Bay Mining & Smelting Co, Man, 44-47; dist engr, B F Sturtevant Co, Ont, 47-49; design engr, Consol Mining & Smelting Co, BC, 49-51; proj engr, Standard Oil Co, Ind, 55-58; dir labs mech eng, Univ BC, 58-67; dir labs, dept mech eng, Ahmadu Bello Univ, Nigeria, 67-69; dir eng labs, fac eng, Univ Regina, 69-82; RETIRED. *Concurrent Pos:* Lab consult, mech eng dept, Chulalongkorn Univ, Bangkok, Nat Polytech Sch, Ecuador, 74-75, Tehran Polytech, Iran, 76-77, Univ Americas, Puebla, Mex, 81 & Chulalongkorn Univ, Bangkok, 83. *Mem:* Am Soc Eng Educ; Sigma Xi. *Res:* Heating and ventilating; fluid dynamics; particle technology; laboratory management and development. *Mailing Add:* 4453 W Eighth Ave Vancouver BC V6R 2A3 Can

WIGINGTON, RONALD L, b Topeka, Kans, May 11, 32; m 51; c 4. INFORMATION SYSTEMS. *Educ:* Univ Kans, BS, 53, PhD(elec eng), 64; Univ Md, MS, 62; Harvard Bus Sch, AMP, 77. *Prof Exp:* Mem tech staff, Bell Tel Labs, Inc, 53-54; supvry electronic engr, Nat Security Agency, US Dept Defense, 54-65, sr engr, 65-68; dir res & develop, Chem Abstr Serv, 68-84, dep exec dir Wash oper, Am Chem Soc, 84-86, DIR CHEM ABSTRACTS SERV, 86- *Concurrent Pos:* Adj assoc prof comput info sci, Ohio State Univ, 70-78; mem Nat Acad Sci Comput Sci & Eng Bd, 69-72, comt on data for sci & technol, Int Coun Sci Unions Task Group on Comput Use, 68-77 & comn for coord Nat Bibliog Control, 75-; Dept Defense consult, 68-; trustee, Online Comput Libr Ctr, 78-, chmn bd, 84-87; pres, Nat Fedn Abstracting & Indexing Serv, 82. *Mem:* Inst Elec & Electronics Engrs; Am Chem Soc; Sigma Xi. *Res:* High speed instrumentation; transmission line theory; electronic devices,techniques for switching,storage of information; computer system organization; design evaluation; displays and man-machine communications; simulation; computerized information systems; research and development management; general management. *Mailing Add:* Chem Abstracts Serv 2540 Olentangy Rd PO Box 3012 Columbus OH 43210

WIGINTON, DAN ALLEN, b Coleman, Tex, Mar 12, 49; m 74; c 3. GENE REGULATION & EXPRESSION, GENE STRUCTURE. *Educ:* Abilene Christian Univ, BS, 72; Univ Tex, Austin, PhD(biochem), 78. *Prof Exp:* Fel, Univ Ky, 78-80; chemist, Vet Admin Hosp, San Antonio, Tex, 80-84; res asst prof, 84-86, ASST PROF, UNIV CINCINNATI & CHILDREN'S HOSP RES FOUND, 86- *Concurrent Pos:* Fel, Univ Tex Health Sci Ctr, 80-84; fac mem, develop biol prog, grad prog, Univ Cincinnati, 89- *Mem:* Am Soc Biochem & Molecular Biol. *Res:* Gene regulation related to human adenosine deaminase- enhancer, promoter, structure and interaction; relationship of adenosine deaminase and severe combined immunodeficiency disease and its treatment-mutations, expression, gene theory. *Mailing Add:* Children's Hosp Res Found Elland & Bethesda Aves CHRF 2032 Cincinnati OH 45229

WIGLE, ERNEST DOUGLAS, b Windsor, Ont, Oct 30, 28; m 58; c 5. MEDICINE, CARDIOLOGY. *Educ:* Univ Toronto, MD, 53; FRCP(C), 58. *Prof Exp:* McLaughlin Found fel cardiol, Univ Toronto, 59-60; sr res assoc, Ont Heart Found, 63-66; from asst prof to assoc prof, 66-72, PROF MED, UNIV TORONTO, 72- *Concurrent Pos:* Dir cardiovasc unit, Toronto Gen Hosp, 64-72, dir div cardiol, 72-; fel coun clin cardiol, Am Heart Asn, 65- *Mem:* Fel Am Col Physicians; Asn Am Physicians; Am Soc Clin Invests; Am Fedn Clin Res; Can Soc Clin Invest. *Res:* Muscular subaortic stenosis, hemodynamics, pharmacology and electrocardiography; hemodynamics of acute valvular insufficiency; cardiomyopathy; ventricular aneurysm; heart catheterization; automated assessment of left ventricular function; ritral valve prolapse. *Mailing Add:* Gen Div Eaton N12-217 Toronto Hosp 200 Elizabeth St Toronto ON M5G 2C4 Can

WIGLER, MICHAEL H, b New York, NY, Sept 3, 47. MEDICAL GENETICS. *Educ:* Princeton Univ, BA, 70; Rutgers Univ, MMS, 72; Columbia Univ, PhD(microbiol), 78. *Prof Exp:* Staff assoc, Inst Cancer Res, Columbia Univ, 78; HEAD, MAMMALIAN CELL GENETICS SECT, COLD SPRING HARBOR LAB, NY, 78- *Concurrent Pos:* Am Bus for Cancer res award, 82; outstanding investr award, NIH, 85; adj prof, Dept Genetics, Col Physicians & Surgeons, Columbia Univ, 88- *Honors & Awards:* Pfizer Biomed Award, 85; Lifetime Res Prof Award, Am Chem Soc, 86; Drew Award in Biomed Res, Ciba-Geigy, 86; G H A Clowes Mem Award for Cancer Res, 91. *Mem:* Nat Acad Sci. *Res:* Mammalian cell genetics; cancer research; author of numerous publications. *Mailing Add:* Cold Spring Harbor Lab PO Box 100 Cold Spring Harbor NY 11724

WIGLER, PAUL WILLIAM, b New York, NY, Aug 26, 28; m 85; c 4. CANCER CHEMOTHERAPY, PHARMACEUTICAL CHEMISTRY. *Educ:* Queens Col, NY, BS, 50; Brooklyn Col, MA, 52; Univ Calif, Berkeley, PhD(biochem), 58. *Prof Exp:* Jr res biochemist, Virus Lab, Univ Calif, Berkeley, 58; NIH fel chem Univ Wis, 58-60; from asst prof to assoc prof, Sch Med, Univ Okla, 60-66; assoc prof res, 66-68, res prof, 68-78, PROF MED BIOL, MEM RES CTR, UNIV TENN, KNOXVILLE, 78- *Concurrent Pos:* Biochemist, Okla Med Res Found, 60-63, assoc mem, 63-66; Pub Health Serv res career develop award, 66. *Mem:* Am Soc Biol Chemists; Am Chem Soc. *Res:* Kinetics of cell membrane transport; reversal mechanisms in multidrug resistance of cancer cells. *Mailing Add:* Dept Med Biol Mem Res Ctr Univ Tenn Med Ctr Knoxville TN 37920

WIGLEY, NEIL MARCHAND, b Mt Vernon, Wash, Feb 16, 36; c 4. MATHEMATICAL ANALYSIS. *Educ:* Univ Calif, Berkeley, BA, 59, PhD(math), 63. *Prof Exp:* Staff mem math, Los Alamos Sci Lab, 63-65; asst prof, Univ Ariz, 65-67; assoc prof, Univ NC, 67-68; fel, Alexander von Humboldt Found, Univ Bonn, 68-70; assoc prof, 70-74, PROF MATH, UNIV WINDSOR, 74- *Mem:* Am Math Soc. *Res:* Partial differential equations. *Mailing Add:* Dept Math Univ Windsor Windsor ON N9B 3P4 Can

WIGLEY, ROLAND L, b Blawenburg, NJ, Oct 4, 23; m 56; c 2. MARINE ECOLOGY. *Educ:* Univ Maine, BS, 49; Cornell Univ, PhD(ichthyol), 53. *Prof Exp:* Fishery biologist, Dept Conserv & Econ Develop, NJ, 52-53; supervisory fishery res biologist, Nat Marine Fisheries Serv, 53-80; CONSULT, 80- *Concurrent Pos:* Consult, Aquatic Sci Info Retrieval Ctr, Taft Lab, RI, 61-64 & John Wiley & Sons, NY, 66-; US deleg, Protein Resources Panel, US-Japan Coop Prog Natural Resources, 75-76; shellfisheries consult, Food & Agr Orgn, UN; mem adv panel 5, Int Comn, Northwest Atlantic Fisheries Comn; US rep shellfish & benthos comt, Int Coun Explor Sea, 73-75, US mem introd nonindigenous marine organisms panel, 74-; adv, New Eng Fishery Mgt Coun, 78. assoc ed, Nat Shellfisheries Asn Proc; assoc ed, Fishery Bull. *Mem:* Am Soc Ichthyologists & Herpetologists; Soc Syst Zool; Am Fisheries Soc; Am Soc Limnol & Oceanog. *Res:* Ecological aspects of offshore marine benthonic animal communities; taxonomy and geographic distribution of marine fishes and invertebrate organisms. *Mailing Add:* 35 Wilson Rd Woods Hole MA 02543

WIGNALL, GEORGE DENIS, b Bradford, Eng, July 16, 41; m 70; c 2. POLYMER PHYSICS, SMALL ANGLE SCATTERING. *Educ:* Univ Sheffield, BS, 62, PhD(physics), 66. *Prof Exp:* Fel, Harwell Atomic Energy Ctr, 66-68, Calif Inst Technol, 68-69; lab mgr, Imperial Chem Industs, Ltd, 69-79; SR RES SCIENTIST, POLYMER PLASTICS, OAK RIDGE NAT LAB, 79- *Concurrent Pos:* Lectr, Univ Tenn, Knoxville, 81- *Mem:* Am Chem Soc; fel Am Phys Soc; Am Crystallog Asn. *Res:* Structure of synthetic polymers and blends including molecular configuration and domain structure; small angle x-ray and neutron scattering. *Mailing Add:* Solid State Div Bldg 7962 Oak Ridge Nat Lab PO Box X Oak Ridge TN 37831-6393

WIGNER, EUGENE PAUL, b Budapest, Hungary, Nov 17, 02; nat US; m 36, 41; c 2. INVARIANTS, PHILOSOPHICAL IMPLICATIONS OF PHYSICS. *Educ:* Tech Univ Berlin, Dr, 25. *Hon Degrees:* Twenty-one from US & foreign cols & univs, 49-73. *Prof Exp:* Asst, Polytech, Berlin, 26-27, privatdocent, 28-30, N B extraordinary prof theoret physics, 30-33; asst, Univ Göttingen, 27-28; lectr math physics, Princeton Univ, 30, prof, 30-36; prof physics, Univ Wis, 37-38; Thomas D Jones prof math physics, Palmer Phys Lab, 38-71, THOMAS D JONES EMER PROF MATH PHYSICS, PRINCETON UNIV, 71- *Concurrent Pos:* Sci guest, Kaiser Wilhelm Inst Berlin, 31 & Metall Lab, Chicago, 42-45; dir res & develop, Clinton Labs, Tenn, 46-47; Lorentz lectr, Inst Lorentz, Leiden, 57; dir harbor proj civil defense, Nat Acad Sci, 63; dir course 29, Int Sch Physics Enrico Fermi, 63; dir, Civil Defense Res Proj, Oak Ridge, Tenn, 64-65; Kramers prof, State Univ Utrecht, 75. Consult, Off Sci Res & Develop, 41-42, Oak Ridge Nat Lab & Exxon Nuclear Co; mem vis comt, Nat Bur Stand, 47-51; gen adv comt, AEC, 52-57, 59-64; vis prof, La State Univ, 71- *Honors & Awards:* Nobel Prize in Physics, 63; Franklin Medal, Franklin Inst, 50; Fermi Award, 58; Atoms for Peace Award, 60; Max Planck Medal, Ger Phys Soc, 61; George Washington Award, Am Hungarian Studies Found, 64; Semmelweiss Medal, Am Hungarian Med Asn, 65; Nat Medal Sci, 69; Albert Einstein Award, 72. *Mem:* Nat Acad Sci; Am Math Soc; fel Am Phys Soc (vpres, 55, pres, 66); Am Acad Arts & Sci; Am Philos Soc; Am Nuclear Soc; Ger Phys Soc. *Res:* Application of group theory of quantum mechanics; rate of chemical reactions; theory of metallic cohesion; nuclear structure and reactions; philosophical implications of quantum mechanics. *Mailing Add:* Eight Oper Rd Princeton NJ 08540

WIGODSKY, HERMAN S, b Sioux City, Iowa, June 12, 15. BIOETHICS, EPIDEMIOLOGY. *Educ:* Northwestern Univ, PhD(physiol), 40, MD, 41. *Prof Exp:* CLIN PROF PATH, HEALTH SCI CTR, UNIV TEX, 70- *Mem:* Am Med Asn; Am Physiol Soc; Soc Exp Biol Med. *Mailing Add:* 300 Primera Dr San Antonio TX 78212

WIGTON, ROBERT SPENCER, b Omaha, Nebr, Nov 1, 11; m 37; c 2. MEDICINE. *Educ:* Univ Nebr, BSc, 32, MA & MD, 35. *Prof Exp:* Instr, Sch Med, Univ Pa, 38-42; prof, 46-77, PROF EMER NEUROL & PSYCHIAT, COL MED, UNIV NEBR, OMAHA, 77- *Concurrent Pos:* Fel neurol, Hosp Univ Pa, 37-40; resident, Pa Hosp, 40-42; consult, Union Pac RR, 53-77. *Mem:* AAAS; AMA; Am Psychiat Asn. *Res:* Clinical neuropsychiatry; neurophysiology in relation to behavior. *Mailing Add:* 6626 Cuming St Omaha NE 68132

WIIG, ELISABETH HEMMERSAM, b Esbjerg, Denmark, May 22, 35; US citizen; m 58; c 2. SPEECH PATHOLOGY. *Educ:* State Sem Enmdrupborg, Denmark, BS, 56; Case-Western Reserve Univ, MA, 60, PhD(speech path), 67. *Prof Exp:* Instr phonetics, Univ Bergen, Norway, 60-64; asst prof speech path, Univ Mich, 68-70; PROF COMMUN DIS, BOSTON UNIV, 70- *Concurrent Pos:* NIH fel, Univ Mich, 67-68. *Mem:* Am Speech & Hearing Asn; Coun Except Children; Acad Aphasia. *Res:* Language disorders and learning disabilities; acquired aphasia in adults; congenital language disorders in children. *Mailing Add:* 7101 Lake Powell Dr Arlington TX 76016

WIITA, PAUL JOSEPH, b Bronx, NY, Feb 18, 53; m 78; c 2. ASTROPHYSICS, THEORETICAL PHYSICS. *Educ:* Cooper Union, NY, BS, 72; Princeton Univ, MA, 74, PhD(physics), 76. *Prof Exp:* Res asst physics, Princeton Univ, 75-76; res assoc, Enrico Fermi Inst, Univ Chicago, 76-79; asst prof astron, Univ Pa, 79-86; ASSOC PROF PHYSICS, GA

STATE UNIV, 86- *Concurrent Pos:* Instr, Adler Planetarium, Chicago, 77; Compton lectr, Enrico Fermi Inst, Univ Chicago, 77; NSF-NATO fel, Inst Astron, Univ Cambridge, 77-78; vis fel, Copernicus Astron Ctr, Warsaw, 78 & Tata Inst Fundamental Res, India, 81, 82 & 85; vis prof, Indian Inst Sci, 86-87, Tata Inst Fund Res, India, 89 & 90. *Mem:* Am Phys Soc; Am Astron Soc; Royal Astron Soc; Int Astron Union. *Res:* Problems in theoretical astrophysics, including radio galaxies, quasars, relativistic beams, black holes, rotating stars, star formation and planetary system formation. *Mailing Add:* Dept Physics & Astron Ga State Univ Atlanta GA 30303

WIITALA, STEPHEN ALLEN, b Vancouver, Wash, Oct 3, 46; m 68; c 1. ALGEBRA, COMPUTER SCIENCE. *Educ:* Western Wash Univ, BAEd, 68, MA, 71; Dartmouth Col, PhD(math), 75. *Prof Exp:* asst prof math, Nebr Wesleyan Univ, 75-80; asst prof, 80-82, ASSOC PROF MATH, NORWICH UNIV, 82- *Mem:* Am Math Soc; Math Asn Am. *Res:* Quadratic forms on vector spaces of characteristic two; representation theory of finite groups; mathematics and computer science education. *Mailing Add:* Dept Math Norwich Univ Mil Col Northfield VT 05663

WIITANEN, WAYNE ALFRED, b Detroit, Mich, May 6, 35; m 73; c 3. BIONICS, COMPUTER SCIENCES. *Educ:* Harvard Univ, AB, 68, MA, 69, PhD(biol), 72. *Prof Exp:* Consult comput sci, 67-68; vpres, Mgt Eng Inc, 69-71; asst prof biol, Univ Ore, 71-77, asst prof comput sci, 73-77, assoc prof biol, 77-80, interim dir univ comput, 77-80; STAFF RES SCIENTIST, COMPUT SCI DEPT, GEN MOTORS RES LABS, 80- *Mem:* AAAS. *Res:* Applications of computers and mathematics to biological problems with special emphasis on the mammalian nervous system; dynamical biological systems simulation; neural networks. *Mailing Add:* Comput Sci Dept Gen Motors Res Labs Warren MI 48090-9055

WIJANGCO, ANTONIO ROBLES, b Manila, Philippines, Apr 6, 44; m 67; c 2. HIGH ENERGY PHYSICS. *Educ:* Ateneo de Manila Univ, AB, 65; Columbia Univ, MS, 71, PhD(physics), 76. *Prof Exp:* Res assoc, Nevis Labs, Columbia Univ, 75-78; RES ASSOC PHYSICS, BROOKHAVEN NAT LAB, 78- *Concurrent Pos:* Vis scientist, Polytech, France & Europ Coun Nuclear Res, Geneva, Switz, 76- *Mem:* Inst Elec & Electronics Engrs; Philippine-Am Asn Scientists & Engrs (pres). *Res:* Photon beams of energy, 100-400 Giga-electron volts as a probe of nuclear and sub-nuclear matter; designing advanced instrumentation for molecular biology. *Mailing Add:* Biotechnica Diags Inc 61 Moulton St Cambridge MA 02138

WIJNEN, JOSEPH M H, b Wittem, Neth, Sept 22, 20; US citizen; m 67; c 2. PHYSICAL CHEMISTRY, PHOTOCHEMISTRY. *Educ:* Cath Univ Louvain, Lic Sci, 46, Dr Sci(chem), 48. *Prof Exp:* Lectr chem, Cath Univ Louvain, 48-49; Nat Res Coun Can fel photochem, 49-51; res assoc, NY Univ, 51-53; Nat Res Coun Can res officer, 53-55; res assoc chem, Celanese Corp Am, 55-58; sr fel photochem, Mellon Inst, 58-63; chmn dept, 81-83 & 85-88, PROF CHEM, HUNTER COL, 63-, ACTG ASSOC PROVOST, 88- *Concurrent Pos:* Consult, US Bur Mines, Pittsburgh, Pa, 61-63; NSF res grants, 67-69 & 70-72; vis prof, Univ Bonn, 69-70, Univ Amsterdam, 76-77 & Algerian Petrol Inst, 83-84. *Mem:* Am Chem Soc; Nat Combustion Inst; fel Am Inst Chemists. *Res:* Primary processes in photochemical reactions; free radical reactions; kinetics of free radical induced polymerization reactions. *Mailing Add:* Dept Chem Hunter Col City Univ New York 695 Park Ave New York NY 10021

WIJSMAN, ROBERT ARTHUR, b Hague, Netherlands, Aug 20, 20; US citizen; m 53; c 3. MATHEMATICAL STATISTICS. *Educ:* Delft Inst Technol, Netherlands, Ir, 45; Univ Calif, PhD(physics), 52. *Prof Exp:* Lectr med physics, Univ Calif, 52-53, instr math, 53-54, res statistician, 54-55, lectr statist & pub health, 55-56, actg asst prof statist, 56-57; from asst prof to assoc prof, 57-65, prof statist, 65-90, PROF EMER, UNIV ILL, URBANA, 90- *Concurrent Pos:* Vis prof, Columbia Univ, 67-68. *Mem:* Inst Math Statist; Am Math Soc; Am Statist Asn. *Res:* Sequential and multivariate analysis. *Mailing Add:* Dept Statist Univ Ill Urbana IL 61801

WIKEL, STEPHEN KENNETH, b Apr 11, 45; c 2. PARASITOLOGY, INFECTIOUS DISEASES. *Educ:* Univ Sask, Can, PhD(immunol), 77. *Prof Exp:* PROF MICROBIOL & IMMUNOL, SCH MED, UNIV NDAK, 83-, CHESTER FRITZ DISTINGUISHED PROF 87- *Honors & Awards:* Fac Res Award, Sigma Xi. *Mem:* Sigma Xi. *Res:* Immunoparasitology. *Mailing Add:* Dept Microbiol & Immunol Sch Med Univ NDak Grand Forks ND 58202

WIKJORD, ALFRED GEORGE, b Flin Flon, Man, July 15, 43; m 68; c 2. ENVIRONMENTAL IMPACT ANALYSES, WASTE MANAGEMENT. *Educ:* Univ Man, BSc, 64, MSc, 65; McGill Univ, PhD(chem), 69. *Prof Exp:* NATO sci fel, Strasbourg Macromolecular Res Ctr, France, 69-70; res officer, Anal Sci Br, 70-80, head, Nuclear Waste Immobilization Sect, 80-86, MGR ENVIRON & SAFETY ASSESSMENT BR, WHITESHELL LABS, ATOMIC ENERGY CAN LTD, 86- *Mem:* Chem Inst Can; Can Nuclear Soc; Soc Risk Anal. *Res:* Chemistry of nuclear reactors; heavy water production; management of nuclear wastes; environmental impact assessment of waste disposal. *Mailing Add:* Environ Assessment Br AECL Res Whiteshell Lab Pinawa MB R0E 1L0 Can

WIKMAN-COFFELT, JOAN, b Chicago, Ill, June 9, 29; m 70. BIOCHEMICAL CARDIOLOGY. *Educ:* Alverno Col, Milwaukee, BS, 59; St Mary's Col, Winona, Minn, MS, 65; St Louis Univ, PhD(biochem), 69. *Prof Exp:* Instr, St Mary's Col, Winona, Minn, 62-63; teaching asst, St Louis Univ, 64-68; postdoctoral res, Baylor Col Med, 68-70; asst prof biochem, Univ Okla Med Sch, 70-72; asst prof biol chem, Univ Calif Med Sch, Davis, 72-77, asst res biochemist, 77-80, ASSOC RES BIOCHEMIST, CARDIOVASC RES INST & DEPT MED, UNIV CALIF, SAN FRANCISCO, 80-, PROF CARDIOL, 89- *Concurrent Pos:* NIH res fel, Baylor Univ, 68-70; NIH grant, Centro Mex & Univ Mex, 70-72; co-dir & coordr, res proj, NIH, 72-77. *Mem:* AAAS; Am Physiol Soc; Am Col Cardiol; Am Chem Soc; Sigma Xi; Am Heart Asn. *Res:* Preservation of the heart for transplantation; long-term preservation of organs for transplantation; awarded two patents. *Mailing Add:* 1610 Redwood Lane Davis CA 95616

WIKSWO, JOHN PETER, JR, b Lynchburg, Va, Oct 6, 49; m 70; c 2. BIOPHYSICS, MEDICAL PHYSICS. *Educ:* Univ Va, BA, 70; Stanford Univ, MS, 73, PhD(physics), 75. *Prof Exp:* Res fel cardiol, Med Sch, Stanford Univ, 75-77; from asst prof to assoc prof physics, 77-88 PROF PHYSICS VANDERBILT UNIV 88. *Concurrent Pos:* Res fel, Bay Area Heart Res Comt, 75-77; Alfred P Sloan res fel, 80-82. *Honors & Awards:* IR-100 Award, 84. *Mem:* Am Phys Soc; Am Heart Asn; Biophys Soc; Inst Elec & Electronics Engrs; Soc Neuroscience; Sigma Xi. *Res:* Application of electric and magnetic measurements and electromagnetic theory to study the propagation of electrical activity in muscle and nerves; development of instrumentation and analysis techniques for neuromagnetism, magnetocardiography and electrocardiography. *Mailing Add:* Dept Physics & Astron Vanderbilt Univ Box 1807 Sta B Nashville TN 37235

WIKTOROWICZ, JOHN EDWARD, b Nairobi, Kenya, Dec 23, 49; US citizen; m 72; c 3. INSTRUMENTATION. *Educ:* Ill Inst Technol, BS, 74; Univ Tex, Galveston, PhD(human genetics & biol chem), 78. *Prof Exp:* Res fel, Calif Inst Technol, 78-81; res assoc, Scripps Clin & Res Inst, 81-82; asst prof biochem, Va Polytech Inst & State Univ, 82-89; scientist, 89-90, STAFF SCIENTIST & GROUP LEADER BIOSEPARATIONS, APPL BIOSYSTS, INC, 90- *Concurrent Pos:* Consult, Automated Dynamics Corp, 82-85 & Unigen, 89. *Mem:* Am Soc Biochem & Molecular Biol; Protein Soc. *Res:* Protein structure-function relationships; separation and analytical chemistry; biochemical analyses; capillary electrophoresis. *Mailing Add:* Appl Biosysts Inc 3745 N First St San Jose CA 95134

WIKUM, DOUGLAS ARNOLD, b Stoughton, Wis, Oct 3, 33; m 58; c 4. ECOLOGY, BIOLOGY. *Educ:* Univ Wis-Stevens Point, BS, 61, Univ SDak, MA, 65; Univ NDak, PhD(biol), 72. *Prof Exp:* Prof biol, Univ Wis-Stout, 66-74; ecologist, Stone & Webster Eng Corp, Boston, 74-76; assoc prof, 76-80, PROF BIOL, UNIV WIS-STOUT, 80- *Concurrent Pos:* Consult ecologist, NUS Corp, Pittsburgh, Pa, 73; res, US Forest Serv, 7-81; co-investr, ELF Commun Syst Ecol Monitoring Prog Wetland Studies, 83-89; investr, Water Qual Assessment Fresh-water Cult Atlantic Salmon, 88-89. *Mem:* Am Inst Biol Sci; Ecol Soc Am; Sigma Xi. *Res:* Chemical and physical properties of soils; plant community structure; wetlands; black spruce growth; waste water disposal in bog ecosystems. *Mailing Add:* Dept Biol Univ Wis-Stout Menomonie WI 54751

WILANSKY, ALBERT, b St John's, Nfld, Sept 13, 21; nat US; m 69; c 5. MATHEMATICS. *Educ:* Dalhousie Univ, MA, 44; Brown Univ, PhD(math), 47. *Prof Exp:* Demonstr physics, Dalhousie Univ, 42-44; instr math, Brown Univ, 46-48; from asst prof to prof, 48-78, UNIV DISTINGUISHED PROF MATH, LEHIGH UNIV, 78- *Concurrent Pos:* Consult, Frankford Arsenal, 57-58; Fulbright vis prof, Reading Univ, 72-73. *Honors & Awards:* Ford Prize, Math Asn Am, 69. *Mem:* Math Asn Am. *Res:* Pure mathematics; analysis; summability; linear topological space; Banach algebra; functional analysis. *Mailing Add:* Dept Math Lehigh Univ Bethlehem PA 18015

WILBANKS, JOHN RANDALL, b Foreman, Ark, June 10, 38; m 62; c 2. PETROLOGY, STRUCTURAL GEOLOGY. *Educ:* NMex Inst Mining & Technol, BS, 60; Tex Tech Univ, MS, 66, PhD(geol), 69. *Prof Exp:* Geologist, NMex State Hwy Dept, 61-63; teaching asst geol, Tex Tech Univ, 63-65, res assoc geol, Marie Byrd Land, 67-69, co-investr & NSF Antarctic grant, 69-70, vis asst prof geol, 70-71; chmn dept, 71-74, ASSOC PROF & CHMN DEPT GEOSCI, UNIV NEV, LAS VEGAS, 71- *Mem:* Geol Soc Am; Nat Asn Geol Teachers. *Res:* Petrology and structure of metamorphic complexes; Marie Byrd Land, Antarctica; morphology of metamorphic zircons; tectonics of the southern great basin. *Mailing Add:* Dept Geosci Univ Nev 4505 Maryland Pkwy Las Vegas NV 89154

WILBARGER, EDWARD STANLEY, JR, b Billings, Mont, Feb 21, 31; m 59; c 2. PHYSICS, FLUIDS. *Educ:* Va Mil Inst, BS, 52; US Naval Postgrad Sch, MS, 56, Univ Calif, Santa Barbara, PhD, 80. *Prof Exp:* Chief indust hyg sect, Off Surgeon Gen, US Army, 56-58; engr physicist, Proj Res Aviation Med, US Naval Med Res Inst, 58-59; engr physicist, Bioastronaut Res Unit, Ord Missile Command, US Army, Redstone Arsenal, 59, chief inspections, Health & Safety Br, Off Chief Engrs, 59-60; head bioinstrumentation group, Gen Motors Corp, 60-62, sr res physicist, Aerospace Opers Dept, AC Electronics Defense Res Labs, 62-71, sr res physicist, 71-75, head Anal & Des, Delco Electronics div, 75-83, mgr, Aerophysics Range Facil, Aerophys Dept, Delco Systs Opers, 83-88; RES & ENG CONSULT, 88- *Concurrent Pos:* Lectr fluid mech & heat transfer, Univ Calif, Santa Barbara, 79-81 & 88; mem, Nat Res Coun, Nat Acad Sci, 86- *Mem:* Sigma Xi; Am Acad Mech; NY Acad Sci. *Res:* Control engineering; measurement methods; physiological response to stress; design and fabrication of control systems; mobility systems analysis for off-road vehicles; analysis and design of auto safety systems; computational fluid dynamics; hypervelocity interior ballistics and fluid dynamics; hypervelocity impact. *Mailing Add:* 3830 Center Ave Santa Barbara CA 93110

WILBER, CHARLES GRADY, b Waukesha, Wis, June 18, 16; m 52; c 6. PHYSIOLOGY. *Educ:* Marquette Univ, BSc, 38; Johns Hopkins Univ, MA, 41, PhD(gen physiol), 42. *Prof Exp:* Lab asst zool, Marquette Univ, 38-39; asst, Johns Hopkins Univ, 40-42; from instr to assoc prof physiol & div biol labs, St Louis Univ, 42-52; asst prof, Fordham Univ, 46-49; chief animal ecol br, Chem Corps Med Labs, US Army Chem Ctr, Md, 52-56, comp physiol br, Chem Res & Develop Labs, 56-60; prof physiol & dean, Grad Sch, Kent State Univ, 61-64; dir marine labs & prof, Univ Del, 64-67; prof zool, 67-86, DIR FORENSIC SCI LAB, COLO STATE UNIV, 74- *Concurrent Pos:* Leader, Fordham Arctic Exped, 48; assoc, Univ Pa, 53-60; prof lectr, Loyola Col, 57-60; mem corp, Marine Biol Lab, Woods Hole; mem panel environ physiol, US Dept Army; mem life sci comt, Nat Acad Sci-Air Res & Develop Command; consult, USPHS; dep coroner, Larimer County, Colo; dir, Ecol Consults, Inc, 72-74; toxicologist, Thorne Ecol Inst, 72-74; Wellcome vis prof basic med sci, Ohio Med Sch, 84-85. *Mem:* Fel Am Acad Forensic Sci; Am Physiol Soc; fel NY Acad Sci; Sigma Xi. *Res:* Biochemistry of body fluids; chemistry of metabolism; comparative aspects of environmental physiology;

climatic adaption; forensic biology; wound ballistics; environmental quality; environmental pathology; comparative toxicology; oceanography; application of science to the needs of the law; wound ballistics and environmental and forensic toxicology; biological impact of noxious ambient factors on Homo Sapiens. *Mailing Add:* 900 Edwards Ft Collins CO 80524

WILBER, JOE CASLEY, JR, b Jonesboro, Ark, Feb 28, 29; m 51. CHEMISTRY, SCIENCE EDUCATION. *Educ:* Memphis State Col, BS, 50, MA, 53; Univ Ga, EdD, 61. *Prof Exp:* Teacher pub schs, Tenn, 50-51 & 53-56; teacher & chmn Dept Sci, pub sch, Ga, 56-59; from asst prof to assoc prof chem, Ga Southern Col, 60-70; teacher chem & chmn Dept Sci, Wingate Col, 70-73; assoc prof, 73-80, PROF CHEM, PAUL D CAMP COMMUNITY COL, 80- *Mem:* AAAS. *Res:* General chemistry; qualitative analysis, a non-sulfide scheme. *Mailing Add:* Dept Math & Sci Paul D Camp Community Col PO Box 737 Franklin VA 23851

WILBER, LAURA ANN, b Memphis, Tenn, May 26, 34. AUDIOLOGY, SPEECH PATHOLOGY. *Educ:* Univ Southern Miss, BS, 55; Gallaudet Col, MS, 58; Northwestern Univ, PhD(audiol), 64. *Prof Exp:* Teacher hard of hearing & deaf, McKinley Elem Sch, Bakersfield, Calif, 55-57; speech therapist & coordr spec educ, US Army Dependent Sch Syst, Heidelberg, Ger, 57-61; res asst audiol, Northwestern Univ, 61-64; asst res audiologist, Univ Calif, Los Angeles, 64-70; asst prof, Albert Einstein Col Med, 70-75, dir hearing & speech serv, 75-76, assoc prof rehab med, 71-77; PROF AUDIOL, NORTHWESTERN UNIV, EVANSTON, ILL, 78- *Concurrent Pos:* Spec instr, Calif State Col, Los Angeles & Univ Southern Calif, 65-70; dir audiol clin, Hosp, Univ Calif, Los Angeles, 68-69; chmn, Clin Sch-Coun NY, 72-73; mem, Dir Hosp Speech & Hearing, Prog Asn & Soc Ear, Nose & Throat Advan in Children; US rep, Int Stand Orgn; mem, Am Nat Stand Inst; adj assoc prof, City Univ New York, 74-76; actg chmn commun dis, Northwestern Univ, 81. *Mem:* Fel Am Speech & Hearing Asn; fel Acoust Soc Am; Am Auditory Soc; Am Acad Audiol. *Mailing Add:* Dept Commun Sci Northwestern Univ 2299 Sheridan Rd Evanston IL 60208

WILBERGER, JAMES ELDRIDGE, b Richmond, Va, May 5, 52. HEAD INJURY. *Educ:* Univ Richmond, BA, 74; Med Col Va, MD, 78. *Prof Exp:* ASSOC PROF SURG-NEUROSURG, MED COL PA, 88-; CLIN ASSOC PROF NEUROSURG, WVA UNIV, 87- *Concurrent Pos:* Dir neurosurg, Allegheny Gen Hosp, 84-; consult neurosurg, New Medico Rehab Facil, 86-, Harmcoville Rehab Inst, 86- *Mem:* Am Asn Neurol Surgeons; Am Col Sports Med; Nat Head Injury Found. *Res:* Neurotransmitted in head injury, studying the derangement in neurotransmitted following severe head injury and correlating with clinical outcome. *Mailing Add:* Dept Neurosurg Allegheny Gen Hosp 320 E N Ave Pittsburgh PA 15212

WILBORN, WALTER HARRISON, b Arbyrd, Mo, May 20, 35; m; c 3. HUMAN ANATOMY. *Educ:* Harding Col, BA, 57; St Louis Univ, MS, 62; Univ Tenn, PhD(anat), 67. *Prof Exp:* From asst prof to assoc prof anat, Med Ctr, Univ Ala, Birmingham, 67-73; assoc prof, 73-75, PROF ANAT, COL MED, UNIV S ALA, & DIR EM CTR, 76- *Concurrent Pos:* NIH & others res grants, 67- *Mem:* Am Asn Anat; Am Soc Cell Biol; Histochem Soc; Am Asn Path. *Res:* Ultrastructure and cytochemistry; cutaneous pathology; secretory mechanisms; reproductive biology. *Mailing Add:* Electron Micros Ctr Univ S Ala Col Med 1206 MSB Mobile AL 36688

WILBRAHAM, ANTONY CHARLES, b Chester, Eng, July 26, 36; m 65; c 1. HAZARDOUS WASTE MANAGEMENT. *Educ:* Carlett Park Col, Eng, cert chem & physics, 59; Royal Soc Chem, grad chem, 62, res dipl chem, 65, FRSC, 72. *Prof Exp:* Analytical technician chem, Shell Refining Co, Eng, 53-60; res assoc prof, Eckard Col, 67-68; vis asst prof, Eckerd Col, 67-68; from asst prof to assoc prof, 68-79, PROF CHEM, SOUTHERN ILL UNIV, EDWARDSVILLE, 79- *Concurrent Pos:* Res fel chem, Univ Manchester, 73-74; vis scientist, Environ Safety Group, Harwell, Eng, 84. *Mem:* Fel Royal Soc Chem; Am Chem Soc. *Res:* chemical hazardous waste disposal methods. *Mailing Add:* Dept Chem Southern Ill Univ Edwardsville IL 62026

WILBUR, DANIEL SCOTT, b Bath, NY, Mar 18, 50; m 74; c 1. RADIOPHARMACEUTICAL DEVELOPMENT, NUCLEAR MEDICINE. *Educ:* Portland State Univ, BS, 73; Univ Calif, Irvine, PhD(chem), 78. *Prof Exp:* Mem res staff, Los Alamos Nat Lab, 78-84; head, Radiochem Sect, Neorx Corp, 84-86, dir, Radiopharmaceut Chem, 86-90; ASSOC PROF, UNIV WASH, 90- *Concurrent Pos:* Adj prof, Chem Dept, Univ NMex, 81-82. *Mem:* Am Chem Soc; Soc Nuclear Med; Am Asn Cancer Res. *Res:* Development of new radiopharmaceuticals, including design and synthesis of potential radiopharmaceuticals with concurrent studies involving new radiolabeling techniques; development of radiolabeled monoclonal antibodies for tumor diagnosis and treatment. *Mailing Add:* Dept Radiation Oncol RC 08 Univ Wash Med Ctr 1959 NE Pacific St Seattle WA 98195

WILBUR, DAVID WESLEY, b Hinsdale, Ill, Dec 15, 37; m 68; c 4. BIOPHYSICS, MEDICINE. *Educ:* Pac Union Col, BA, 61; Univ Calif, Berkeley, PhD(biophys), 65; Loma Linda Univ, MD, 71; Am Bd Internal Med, cert, 75. *Prof Exp:* Biophysicist, Lawrence Radiation Lab, Univ Calif, 65-67; asst prof physiol & biophys, Sch Med, Loma Linda Univ, 67-69, med intern, 71-72, med resident, 72-74, res assoc biomath, 69-74; Am Cancer Soc med oncol clin fel, Roswell Park Mem Inst, 74-76; MEM, MED ONCOL, DEPT INTERNAL MED, LOMA LINDA UNIV, 76-; CHIEF ONCOL & HEMAT SECT, PETTIS MEM VET HOSP, LOMA LINDA, 80- *Concurrent Pos:* Donner fel, 61-62. *Mem:* Am Col Physicians; Am Soc Clin Oncol; Sigma Xi; AMA. *Res:* Cancer chemotherapy; management of infection in neutropenic patients; role of immunology in cancer. *Mailing Add:* Dept Internal Med Loma Linda Univ Loma Linda CA 92354

WILBUR, DONALD LEE, b Chicago, Ill, Apr 28, 42; c 2. NEUROENDOCRINOLOGY. *Educ:* Ind State Univ, BS, 68, MA, 70; Med Univ SC, PhD(anat), 74. *Prof Exp:* Asst prof anat, Sch Med, Tex Tech Univ, 74-76; ASSOC PROF ANAT, MED UNIV SC, 76- *Concurrent Pos:* Consult, Environ Protection Agency, 78- *Mem:* Am Asn Anatomists; Histochem Soc; Sigma Xi. *Res:* Electron microscopic, immunocytochemical studies of the pituitary gland and circumventricular organs, correlated with radioimmunoassayable levels of circulating hormones; mechanisms of hormone synthesis and release. *Mailing Add:* Dept Anat Med Univ SC 171 Ashley Ave Charleston SC 29425

WILBUR, DWIGHT LOCKE, b Harrow-on-the-Hill, Eng, Sept 18, 03; m 28; c 3. GASTROENTEROLOGY. *Educ:* Stanford Univ, AB, 23; Univ Pa, MD, 26; Univ Minn, MS, 33. *Hon Degrees:* DSc, Dartmouth Col, 73. *Prof Exp:* Resident physician, Univ Pa Hosp, 26-28; 1st asst, Sect Path Anat, Mayo Clin, Rochester, Minn, 29-30, 1st asst, Div Med, 31-33, assoc, Sect Path Anat & from instr to asst prof med, 33-37; from asst clin prof to clin prof, 37-69, EMER CLIN PROF, SCH MED, STANFORD UNIV, 69- *Concurrent Pos:* Consult physician, Mayo Clin, 31-37; asst vis physician, Stanford Serv, San Francisco Hosp, 37-60; assoc ed, Gastroenterology, 43-51; ed, Calif Med, 46-67; chief med serv, French Hosp, 46-73, emer chief, 73-; expert consult, Letterman Gen Hosp, Dept Army, 46-76, emer consult, 76-; assoc clin prof med, Col Physicians & Surgeons, Sch Dent, 48-51; consult, US Naval Hosp, Oakland & SPac Hosp, 50-70; mem bd, Mayo Asn, 51-64; mem civilian health & med adv coun, Dept Defense, 53-59; trustee, Mayo Found, 64-71, emer pub trustee, 71-79; mem, Nat Adv Comn Health Manpower, 66-67; ed-in-chief, Post Grad Med, 69-73. *Honors & Awards:* Julius Friedenwald Medal, Am Gastroenterol Asn, 66; Alfred Stengel Mem Award, Am Col Physicians, 70. *Mem:* Sr mem Inst Med-Nat Acad Sci; master Am Col Physicians; Am Gastroenterol Asn (secy, 47-52, 2nd vpres, 52-53, 1st vpres, 53, pres, 54-55); AMA; fel & hon mem Int Col Dentists. *Mailing Add:* 140 Sea Cliff Ave San Francisco CA 94121

WILBUR, HENRY MILES, b Bridgeport, Conn, Jan 25, 44; m 67, 81; c 3. ZOOLOGY, ECOLOGY. *Educ:* Duke Univ, BS, 66; Univ Mich, Ann Arbor, PhD(zool), 71. *Prof Exp:* Univ Mich Soc Fels jr fel, Div Reptiles & Amphibians, Mus Zool, Univ Mich, Ann Arbor, 71-73; from asst prof to assoc prof, Duke Univ, 73-82, prof zool & chmn, 82-91; COMMONWEALTH PROF BIOL, PROF ENVIRON SCI & DIR, MOUNTAIN LAKE BIOL STA, UNIV VA, 91- *Concurrent Pos:* NSF grad fel, 67-69; Edwin S George scholar, Edwin S George Reserve, Mich, 68-69. *Honors & Awards:* Stoye Award, Am Soc Ichthyol & Herpet, 70. *Mem:* AAAS; Ecol Soc Am; Soc Study Evolution; Brit Ecol Soc; Am Soc Ichthyol & Herpet; Am Soc Naturalists; Soc Study Reptiles & Amphibians. *Res:* Evolutionary ecology; evolution of species interactions and life histories. *Mailing Add:* Dept Biol Univ Va Charlottesville VA 22901

WILBUR, JAMES MYERS, JR, b Philadelphia, Pa, Oct 31, 29; m 60; c 3. ORGANIC CHEMISTRY. *Educ:* Muhlenberg Col, BS, 51; Univ Pa, PhD, 59. *Prof Exp:* Res chemist, J T Baker Chem Co, NJ, 51-53; NIH fel cancer chemother, Univ Minn, 58-60; res chemist, E I du Pont de Nemours & Co, 60-62; fel, Univ Ariz, 62-63; assoc prof chem, 63-66, PROF CHEM, SOUTHWEST MO STATE UNIV, 66- *Mem:* Am Chem Soc. *Res:* Medicinal chemistry; cancer chemotherapy; organic mechanisms; polymers. *Mailing Add:* Dept Chem Southwest Mo State Univ Springfield MO 65802

WILBUR, KARL MILTON, b Binghamton, NY, May 7, 12; m 46; c 2. PHYSIOLOGY. *Educ:* Ohio State Univ, BA, 35, MA, 36; Univ Pa, PhD(zool), 40. *Prof Exp:* Asst zool, Ohio State Univ, 35-36; instr, Univ Pa, 39-40; Rockefeller fel, NY Univ, 40-41; instr zool, Ohio State Univ, 41-42, asst prof physiol, Med Sch, Dalhousie Univ, 42-44; assoc prof zool, 46-50, PROF ZOOL, DUKE UNIV, 50- *Concurrent Pos:* Physiologist, AEC, 52-53. *Mem:* Am Physiol Soc; Am Soc Naturalists; Soc Gen Physiol; Am Soc Zoologists. *Res:* Cellular physiology; calcification in marine organisms; cell division. *Mailing Add:* Dept Zool 239 Biol Sci Duke Univ Durham NC 27706

WILBUR, L(ESLIE) C(LIFFORD), b Johnston, RI, May 12, 24; m 50; c 4. MECHANICAL ENGINEERING. *Educ:* Univ RI, BS, 48; Stevens Inst Technol, MS, 49. *Prof Exp:* From instr to asst prof mech eng, Duke Univ, 49-57; assoc prof, Worcester Polytech Inst, 57-61, prof, 61-86, dir, Nuclear Reactor Facil, 59-86, EMER PROF MECH ENG, WORCESTER POLYTECH INST, 86-, EMER CHMN, NUCLEAR REACTOR FACIL, 86- *Mem:* Am Soc Mech Engrs; Am Soc Eng Educ; Am Nuclear Soc. *Res:* Nuclear reactor technology; thermodynamics. *Mailing Add:* PO Box 105 Sebasco Estates ME 04565

WILBUR, LYMAN D, b Los Angeles, Calif, April 27, 00; m 25. RIVER DIVERSION, CONSTRUCTION. *Educ:* Stanford Univ, BA, 21. *Hon Degrees:* LLD, Col Idaho, 62; DSc, Univ Idaho, 67. *Prof Exp:* Draftsman & asst eng field eng, City San Francisco, 21-24; designer, Merced Irrig Dist, 24-26; design eng, East Bay Munic Utility Dist, 26-29; asst to chief consult eng, Mid Asia Water Econ Serv, 29-31; dist engr eng, Morrison-Knudsen Co, Inc, 32-39, div engr, 40-42, dist mgr, 42-47, chief eng, 47-52, vpres eng, 53-60, vpres foreign oper, 60-65, vpres, 65-70; CONSULT, 71- *Concurrent Pos:* Exec vpres, pres & chmn, Int Eng Co, Inc, Div Morrison-Knudsen Co, Inc, 56-70, constuct mgr, 39-41, resident partner, 65-66. *Mem:* Nat Acad Eng; hon mem Am Soc Civil Engrs; Nat Soc Prof Engrs; Soc Am Mil Engrs. *Mailing Add:* 4502 Hillcest Dr Boise ID 83705-2857

WILBUR, PAUL JAMES, b Ogden, Utah, Nov 8, 37; m 60; c 2. MECHANICAL ENGINEERING. *Educ:* Univ Utah, BS, 60; Princeton Univ, PhD(aeronaut & mech sci), 68. *Prof Exp:* From asst prof to assoc prof, 68-75, PROF MECH ENG, COLO STATE UNIV, 75- *Concurrent Pos:* Nuclear Power engr, US AEC. *Mem:* Am Soc Mech Engrs; Am Inst Aeronaut & Astronaut. *Res:* Electric propulsion in space applications; ion implantation. *Mailing Add:* Dept Mech Eng Colo State Univ Ft Collins CO 80523

WILBUR, RICHARD SLOAN, b Boston, Mass, Apr 8, 24; m 51; c 3. MEDICINE. *Educ:* Stanford Univ, BA, 43, MD, 46; John Marshall Law Sch, JD, 90. *Prof Exp:* Intern, San Francisco County Hosp, 46-47; resident, Stanford Hosp, 49-51 & Univ Pa Hosp, 51-52; mem staff, Palo Alto Med Clin, 52-69; dep exec vpres, AMA, Chicago, 69-71; asst secy defense for health & environ, 71-73; dep exec vpres, AMA, 73-74; sr vpres, Baxter Labs, Inc,

Deerfield, Ill, 74-76; EXEC VPRES, COUN MED SPECIALTY SOCS, 76- *Concurrent Pos:* Assoc prof med, Med Sch, Stanford Univ, 52-69 & Med Sch, Georgetown Univ, 71-; chmn bd dirs, Medic Alert Found & Nat Adv Cancer Coun & secy, accreditation coun, Cont Med Educ, 81-; mem bd visitors, Postgrad Med Sch, Drew Univ; pres, Nat Resident Matching Plan, 91-92. *Mem:* Inst Med-Nat Acad Sci; hon fel Int Col Dent; fel Am Col Physicians; Am Gastroenterol Asn; fel Am Col Physician Execs (pres, 88-89). *Mailing Add:* Coun Med Specialty Socs PO Box 70 Lake Forest IL 60045

WILBUR, ROBERT DANIEL, b Glendale, Calif, May 7, 31; m 52; c 3. AGRICULTURAL RESEARCH MANAGEMENT, PLANT PROTECTION. *Educ:* Calif State Polytech Col, BS, 54; Iowa State Univ, PhD(animal nutrit, bact), 59. *Prof Exp:* Asst nutrit & bact, Iowa State Univ, 54-59, fel, 59; res nutritionist, 59-67, group leader nutrit & physiol, 67-76, mgr pesticides res, 76-80, mgr animal res, 80-81, DIR INT PLANT INDUST, AM CYANAMID CO, 81- *Concurrent Pos:* Consult, tech mgt. *Mem:* AAAS; Am Soc Animal Sci. *Res:* Nutrition and physiology of domesticated animals; crop physiology. *Mailing Add:* Maddock Rd Wash Crossing Titusville NJ 08560

WILBUR, ROBERT LYNCH, b Annapolis, Md, July 4, 25; m 55; c 6. PLANT TAXONOMY. *Educ:* Duke Univ, BS, 46, AM, 48; Univ Mich, PhD, 52. *Prof Exp:* Asst bot, Duke Univ, 46-47, Univ Hawaii, 47-48 & Univ Mich, 48-52; asst prof, Univ Ga, 52-53; asst prof & cur herbarium, NC State Col, 53-57; from asst prof to assoc prof bot, 57-70, chmn dept, 71-78, PROF BOT, DUKE UNIV, 70- CUR, HERBARIUM, 57- *Mem:* Am Soc Plant Taxon; Int Asn Plant Taxon; Torrey Bot Club. *Res:* Systematics and phytogeography of vascular plants; flora of the southeastern United States and Central America; campanulaceae; gentianaceae; ericaceae. *Mailing Add:* 265 Biol Sci Bldg Duke Univ Dept Bot Durham NC 27706

WILBURN, NORMAN PATRICK, b Whittier, Calif, Mar 28, 31; m 56; c 4. CHEMICAL & ELECTRICAL ENGINEERING. *Educ:* Calif Inst Technol, BS, 53, MS, 54, PhD, 58. *Prof Exp:* Engr, Hanford Labs, Gen Elec Co, 58-65; res assoc, Pac Northwest Labs, Battelle Mem Inst, 65-70; res assoc, Hanford Eng Develop Labs, Westinghouse Co, 70-75, mgr, 76-80, adv engr, 81-87, Battelle-Northwest, 87-88; RETIRED. *Concurrent Pos:* Consult. *Res:* Development of mathematical models of chemical and thermohydraulic processes; development of on-line digital computer systems; software engineering; large scale scientific software development; consultant in software quality assurance. *Mailing Add:* PO Box 248 Waxhaw NC 28173-0248

WILCE, ROBERT THAYER, b Carbondale, Pa, Dec 9, 24; m 56. BOTANY. *Educ:* Univ Scranton, BS, 50; Univ Vt, MS, 52; Univ Mich, PhD(bot), 57. *Prof Exp:* Instr bot, Univ Mich, 57-58, fel, Horace Rackham Grad Sch, 58-59; from instr to prof, 59-76, PROF BOT, UNIV MASS, AMHERST, 76- *Mem:* Phycol Soc Am. *Res:* Systematic morphology, distribution and ecology of the attached algae of arctic and subarctic areas, especially the Canadian eastern arctic and northwest Greenland. *Mailing Add:* Dept Bot Univ Mass Amherst Campus Amherst MA 01003

WILCHINSKY, ZIGMOND WALTER, b New York, NY, Aug 26, 15; m 40; c 1. POLYMER PHYSICS. *Educ:* Rutgers Univ, BS, 37, MS, 39; Mass Inst Technol, PhD(physics), 42. *Prof Exp:* Asst physics, Rutgers Univ, 37-39; mem staff radiation lab, Mass Inst Technol, 42-45; sect head, US Naval Res Lab, Washington, DC, 43-46; sr res assoc, Exxon Chem Co, 46-79; consult, 79-82; RETIRED. *Mem:* Am Phys Soc; Am Chem Soc; Am Crystallog Asn. *Res:* Structure of plastics; rubber technology; x-ray diffraction; physical chemistry of catalysts; adsorption; development of microwave generators; vacuum tube development. *Mailing Add:* PO Box 188 Linden NJ 07036

WILCOCK, DONALD F(REDERICK), b Brooklyn, NY, Sept 24, 13; m 38; c 1. ENGINEERING. *Educ:* Harvard Univ, BS, 34; Univ Cincinnati, DEngSci, 40. *Prof Exp:* Res chemist, Sherwin-Williams Co, Ill, 39-42 & Res Lab, Gen Elec Co, NY, 42-45; eng group leader, Thomson Lab, 45-53; mgr mat & chem process eng serv dept, Gen Eng Lab, Gen Elec Co, 53-60, consult engr, Ord Dept, 60-65; dir bearings, lubricant & seal technol, Mech Technol Inc, 65-68, dir technol develop, 69-78; PRES, TRIBOLOCK INC, 78- *Concurrent Pos:* Ed, J Lubrication Technol. *Honors & Awards:* Nat Hersey Award, Am Soc Mech Engrs, 73, Centennial Medal, 81; Wilcock Award, Am Soc Mech Engrs, 89. *Mem:* Fel Am Soc Mech Engrs; fel Am Soc Lubrication Engrs. *Res:* Air bearings; magnetic bearings; bearing design and testing; lubricant testing and development; cryogenics. *Mailing Add:* Tribolock Inc 1949 Hexam Rd Schenectady NY 12309

WILCOX, BENJAMIN A, b Anaconda, Mont, June 18, 34; m 55; c 3. MATERIALS SCIENCE, METALLURGY. *Educ:* Wash State Univ, BS, 56; Stanford Univ, MS, 58, PhD(mat sci), 62. *Prof Exp:* Phys metallurgist, Stanford Res Inst, 56-58; NSF fel, Cambridge Univ, 62-63; fel metals sci, Battelle Mem Inst, 63-71, div chief, High Temperature Mat & Processes Div, 71-73, mgr metals sci sect, Battelle Columbus Labs, 73-74; ceramics prog dir, 74-79, HEAD, METALL POLYMERS & CERAMICS SECT, NSF, 79- *Concurrent Pos:* Vis scientist, Imp Col Sci & Technol & Nat Phys Lab, Eng, 73. *Mem:* Am Soc Metals; Am Ceramic Soc; Am Inst Mining, Metall & Petrol Engrs. *Res:* Deformation and fracture of crystalline solids; high temperature creep, substructural strengthening; dispersion strengthened metals; laser-materials interactions; high temperature corrosion. *Mailing Add:* Mat Sci Div Defense Sci Off Defense Advan Res Projs Agency 3702 N Fairfax Dr Arlington VA 22203-1714

WILCOX, BENSON REID, b Charlotte, NC, May 26, 32; m 59; c 4. CARDIOVASCULAR SURGERY, THORACIC SURGERY. *Educ:* Univ NC, Chapel Hill, BA, 53, MD, 57. *Prof Exp:* From instr to assoc prof, 63-71, PROF SURG, UNIV NC, CHAPEL HILL, 71-, CHIEF DIV CARDIOTHORACIC SURG, 69- *Concurrent Pos:* NIH fel, Bethesda, MD, 60-62 & grant, 68-74; Markle scholar, 67-72. *Mem:* Am Asn Thoracic Surg; Am Col Surg; Am Surg Asn; Soc Thoracic Surg; Soc Univ Surg; Thoracic Surg Dir Asn. *Res:* Application of biomathematical and engineering principles to the study of the circulation; pulmonary circulation in heart disease; surgical anatomy of congenital heart disease. *Mailing Add:* Dept Surg Univ NC CB7065 Chapel Hill NC 27599-7065

WILCOX, BRUCE ALEXANDER, b Hackensack, NJ, May 21, 48; m 89; c 2. CONSERVATION BIOLOGY, SUSTAINABLE DEVELOPMENT. *Educ:* Univ Calif, San Diego, AB, 73, PhD(biol), 79; Yale Univ, MS, 75. *Prof Exp:* From res assoc to sr res assoc, dept biol sci, Stanford Univ, 80-88, exec dir, Ctr Conserv Biol, 83-88, consult assoc prof, human biol, 86-88; PRES, INST SUSTAINABLE DEVELOP/ENVIRONVENTURES INC, 88- *Concurrent Pos:* Mem, comt ecol, Int Union Conserv Nature; consult, UNESCO Man & the Biosphere Prog, US Nat Park Serv, US Forest Serv, US Fish & Wildlife Serv & UN Develop Plan; convenor, First Int Conf Conserv Biol, La Jolla, 78 & Workshop on Mgt Endangered Populations, Stanford Univ, 86; partic, India/US Biosphere Reserve Prog, 82. *Mem:* AAAS; Ecol Soc Am; Soc Conserv Biol. *Res:* Island biogeography; population ecology and genetics; application of population biology to biological conservation; conservation policy; international development; environmental technology development biological diversity; global environmental issues. *Mailing Add:* Inst Sustainable Develop 3000 Sand Hill Rd 1-102 Menlo Park CA 94025

WILCOX, CALVIN HAYDEN, b Cicero, NY, Jan 29, 24; m 47; c 3. MATHEMATICS. *Educ:* Harvard Univ, AB, 51, AM, 52, PhD(math), 55. *Prof Exp:* Mathematician, Air Force Cambridge Res Ctr, 53-55; from instr to assoc prof math, Calif Inst Technol, 55-61; prof math & mem US Army Math Res Ctr, Univ Wis, 61-66; prof math, Univ Ariz, 66-69 & Univ Denver, 69-71; PROF MATH, UNIV UTAH, 71- *Concurrent Pos:* Vis prof, Inst Theoret Physics, Univ Geneva, 70-71, Univ Liege, 73, Univ Stuttgart, 74, 76-77, Kyoto Univ, 75. ed, Rocky Mountain J Math, 75-78; Alexander von Humboldt Found US sr scientist award, 76-77; Fed Polytech, Lausanne, 79, Univ Bonn, 80. *Mem:* AAAS; Am Math Soc; Soc Indust & Appl Math. *Res:* Applied mathematics and mathematical physics, especially theories of wave propagation and scattering in classical and quantum physics; boundary value problems for partial differential equations. *Mailing Add:* Dept Math Univ Utah Salt Lake City UT 84112

WILCOX, CHARLES FREDERICK, JR, b Providence, RI, July 20, 30; m 57; c 3. PHYSICAL CHEMISTY, ORGANIC CHEMISTRY. *Educ:* Mass Inst Technol, BS, 52; Univ Calif, Los Angeles, PhD(org chem), 57. *Prof Exp:* NSF fel, Harvard Univ, 57; from instr to assoc prof, 57-74, PROF CHEM, CORNELL UNIV, 74- *Concurrent Pos:* Guggenheim fel, 66-67; vis prof, Calif Inst Technol, 67; asst ed, J Am Chem Soc, 65-66. *Mem:* Am Chem Soc; The Chem Soc. *Res:* Physical aspects of organic chemistry. *Mailing Add:* Dept Chem Cornell Univ Ithaca NY 14853-0001

WILCOX, CHARLES HAMILTON, b Rochester, NY, May 21, 29. THEORETICAL PHYSICS, ENGINEERING MANAGEMENT. *Educ:* Duke Univ, BS, 50; Univ Ill, MS, 52; Univ Southern Calif, 70. *Prof Exp:* Res physicist & lectr physics, Eng Exp Sta, Ga Inst Technol, 52-53; sr mem tech staff & assoc mgr theoret studies dept, Res Labs, 53-67, mgr tech planning, Aerospace Group, 67-70, dir corp independent res & develop, 70-74, DIR ENG & PROG DEVELOP, AEROSPACE GROUPS, HUGHES AIRCRAFT CO, 74- *Concurrent Pos:* Lectr, Univ Southern Calif, 55-59 & Univ Calif, Los Angeles, 61-63; consult, Stanford Res Inst, 73-76. *Mem:* AAAS; Am Phys Soc; Inst Elec & Electronics Engrs; Sigma Xi; Inst Mgt Sci. *Res:* Scattering and diffraction of electromagnetic waves; radiowave propagation and geophysics; technology planning and development. *Mailing Add:* 10520 Draper Ave Los Angeles CA 90064

WILCOX, CHARLES JULIAN, b Harrisburg, Pa, Mar 28, 30; m 55; c 2. DAIRY SCIENCE. *Educ:* Univ Vt, BS, 50; Rutgers Univ, MS, 55, PhD(animal genetics), 59. *Prof Exp:* Res asst dairy sci, Rutgers Univ, 50 & 53-55; owner & mgr dairy farm, 55-56; res asst dairy sci, Rutgers Univ, 56-59; asst prof dairy sci & assoc geneticist, 59-71, PROF DAIRY SCI & GENETICIST, UNIV FLA, 71- *Mem:* Am Dairy Sci Asn; Am Soc Animal Sci; Latin Am Asn Animal Prod; Am Registry Prof Animal Scientists. *Res:* Quantitative genetics of productive traits of farm animals, including milk yield and composition, reproductive performance, birth weights, gestation lengths, heat tolerance, type conformation, disease resistance, maternal and fetal effects, life span and livability. *Mailing Add:* Dairy Sci Dept Univ Fla Gainesville FL 32611

WILCOX, CHRISTOPHER STUART, b UK, Sept 15, 42; m 64; c 2. NEPHROLOGY, CLINICAL PHARMACOLOGY. *Educ:* Oxford Univ, BA, 66, BMBCh & MA, 68; London Univ, PhD(med & physiol), 74. *Prof Exp:* House physician gen med, Middlesex Hosp, 69; house surgeon surg & urol, Cent Middlesex Hosp, 69-70; clin asst med hypertension & nephrology, Middlesex Hosp, 70-75; ASST PROF MED CLIN PHARMACOL, BRIGHAM & WOMEN'S HOSP, 80-; DIV NEPHROLOGY & HYPERTENSION, VET ADMIN MED CTR, GAINESVILLE. *Concurrent Pos:* House physician neurol, Middlesex Hosp, 70, lectr neurol studies, 70-71; house physician chest dis, Brompton Hosp, 70; asst prof med, Harvard Med Sch, 80- *Mem:* Physiol Soc; Brit Pharmacol Soc; Renal Asn; Am Fedn Clin Res; Am Soc Nephrology. *Res:* Regulation of body fluids and electrolytes and renal vascular resistance; hypertension; autonomic insufficiency; chronic kidney disease. *Mailing Add:* Dept Med & Pharmacol Univ Fla Col Med Div Nephrology & Hypertension Box J-224 JHMHC Gainesville FL 32602

WILCOX, CLIFFORD LAVAR, b Archer, Idaho, Apr 15, 25; m 45; c 5. DAIRY HUSBANDRY. *Educ:* Utah State Univ, BS, 51; Univ Minn, MS, 57, PhD(dairy husb), 59. *Prof Exp:* Asst dairying, 56-59, instr dairy husb, 60, exten dairy specialist, 60-65, asst dir agr exp sta, 68-72, SUPT, AGR EXP STA, UNIV MINN, 65- *Mem:* Am Dairy Sci Asn. *Res:* Dairy cattle breeding. *Mailing Add:* RR 1 No 299 Laporte MN 56461

WILCOX, DONALD BROOKS, b Walden, NY, Feb 23, 11; m 35; c 3. INDUSTRIAL ENGINEERING. *Educ:* Pa State Univ, BS, 33; Ga Inst Technol, MS, 39; Emory Univ, LLB, 52. *Prof Exp:* Instr mech eng, Ga Inst Technol, 36-40, instr indust eng, 46-52; assoc prof & acting head dept, Univ Ala, 41-42; prof, Univ Fla, 52-75; PVT CONSULT & EXPERT WITNESS, 75- *Concurrent Pos:* Consult, Fla Indust Comn. *Mem:* Am Soc Eng Educ; Am

Inst Indust Engrs; Am Soc Safety Engrs; Soc Advan Mgt; Am Soc Prof Engrs. *Res:* Accident prevention engineering; engineering contracts and specifications; engineering economy and law. *Mailing Add:* 2431 NW 41st St No 5102 Gainesville FL 32606

WILCOX, ETHELWYN BERNICE, b Wyoming, Iowa, Mar 19, 06. NUTRITION. *Educ:* Iowa State Univ, BS, 31, MS, 37, PhD(nutrit), 42. *Prof Exp:* Teacher high sch, 32-36; asst & supvr animal nutrit lab, Iowa State Univ, 37-42; asst home economist, Exp Sta, Wash State Univ, 42-43; from asst to prof nutrit, 43-71, EMER PROF NUTRIT, UTAH STATE UNIV, 71- *Concurrent Pos:* Fac honres lectr, Univ Utah, 59, head dept food & nutrit, 65-71. *Mem:* Am Home Econ Asn; Am Dietetic Asn; Inst Food Technologists; Am Inst Nutrit. *Res:* Nutritional status of population groups; lipid metabolism; chemical components of venison flavor. *Mailing Add:* 788 Hillcrest Ave Logan UT 84321

WILCOX, FLOYD LEWIS, science education, for more information see previous edition

WILCOX, FRANK H, b Norwich, Conn, June 15, 27; m 60; c 2. GENETICS. *Educ:* Univ Conn, BS, 51; Cornell Univ, MS, 53, PhD(animal genetics), 55. *Prof Exp:* Assoc prof poultry physiol, Univ Md, 55-67; PROF LIFE SCI, IND STATE UNIV, TERRE HAUTE, 67- *Concurrent Pos:* Poultry Sci travel award to World Poultry Cong, Australia, 62; USPHS res fel, 75-76; vis investr, Jackson Lab, 75-76. *Mem:* World Poultry Sci Asn. *Res:* Biochemical genetics, especially electrophoretic variants in vertebrates. *Mailing Add:* Dept Life Sci Ind State Univ Main Campus Terre Haute IN 47809

WILCOX, GARY LYNN, b Ventura, Calif, Jan 7, 47; m 70; c 2. BIOCHEMICAL GENETICS. *Educ:* Univ Calif, Santa Barbara, BA, 69, MA, 72, PhD(molecular biol), 72. *Prof Exp:* Res assoc biol, Univ Calif, Santa Barbara, 72-74; asst prof, Univ Calif, Los Angeles, 74-77, assoc prof bact, 77-80, prof microbiol, 80-84; pres & chief exec officer, Int Genetic Eng, Inc, 82-89; VCHMN, XOMA CORP, 89- *Concurrent Pos:* Am Can Soc fac res award, 77. *Mem:* Am Soc Microbiol; Genetics Soc Am; Am Soc Biol Chemists; Sigma Xi. *Res:* Protein nucleic acid interactions; genetic engineering; expression of heterologous proteins in microorganisms. *Mailing Add:* Xoma Corp 1545 17th St Santa Monica CA 90404

WILCOX, GEORGE LATIMER, NEUROPHARMACOLOGY, NEUROCHEMISTRY. *Educ:* Univ Colo, PhD(aerospace eng sci), 75. *Prof Exp:* ASSOC PROF PHARMACOL, UNIV MINN, 77- *Mailing Add:* Dept Pharmacol 3-249 Millard Hall Univ Minn 435 Delaware St SE Minneapolis MN 55455

WILCOX, GERALD EUGENE, b Wautoma, Wis, July 17, 25; m 49; c 3. SOIL FERTILITY. *Educ:* Univ Wis, BS, 49, MS, 51, PhD, 53. *Prof Exp:* Asst agronomist, Northern La Hill Farm Exp Sta, La State Univ, 53-57; assoc prof, 57-71, PROF HORT & AGRON, PURDUE UNIV, LAFAYETTE, 71- *Concurrent Pos:* Plant nutritionist, US Agency Int Develop, Brazil, 75-77. *Mem:* Soil Sci Soc Am; Am Soc Hort Sci; Int Soil Sci Soc. *Res:* Mineral nutrition and fertilization of vegetable crops, especially tomatoes and potatoes; soil fertility; culture and mechanization of tomato production; nutrient film technique for tomato, lettuce and cucumber production. *Mailing Add:* Dept Hort Purdue Univ Main Campus West Lafayette IN 47907

WILCOX, HAROLD KENDALL, b Wichita, Kans, Aug 9, 42; m 66; c 2. METHODS DEVELOPMENT, HAZARDOUS WASTE TESTING. *Educ:* Sterling Col, BA, 64; Univ Southern Calif, PhD(chem), 71. *Prof Exp:* Prof chem, Calif Baptist Col, 70-73; chemist, San Bernadino County Air Pollution Control Dist, 73-75; prog mgr, Northrop Serv, Inc, 75-79; SECT HEAD, MIDWEST RES INST, 79- *Concurrent Pos:* Lectr, Riverside City Col, 74-77. *Mem:* Air Pollution Control Asn. *Res:* Test methods for stationary source emissions; development of analysis methods; testing of specialized air pollution sources. *Mailing Add:* Midwest Res Inst 425 Volker Blvd Kansas City MO 64110

WILCOX, HARRY HAMMOND, b Canton, Ohio, May 31, 18; m 41; c 3. ANATOMY. *Educ:* Univ Mich, BS, 39, MS, 40, PhD(zool), 48. *Prof Exp:* Assoc prof biol, Morningside Col, 47-48; assoc anat, Sch Med, Univ Pa, 48-52; from asst prof to prof anat, 52-67, Goodman prof, 67-83, EMER PROF ANAT, UNIV TENN, MEMPHIS, 83- *Mem:* Am Soc Zool; Am Asn Anat; Sigma Xi. *Res:* Effects of aging on the nervous system; internal ear; central nervous system pathways. *Mailing Add:* 1031 Marcia Rd Memphis TN 38117

WILCOX, HENRY G, b Hornell, NY, Jan 26, 33; m 66; c 2. BIOCHEMISTRY, PHARMACOLOGY. *Educ:* Univ Fla, PhD(biochem), 64. *Prof Exp:* Asst prof, 68-74, assoc prof pharmacol, Vanderbilt Univ, 74-77; assoc prof pharmacol, Univ Mo, 77-; PROF PHARMACOL, UNIV TENN. *Concurrent Pos:* NIH fel pharmacol, Vanderbilt Univ, 64-67. *Res:* Plasma lipoprotein metabolism, structure and function; methodology for lipoprotein isolation and analysis; hormonal control of lipid metabolism and transport. *Mailing Add:* Dept Pharmacol Univ Tenn 874 Union Ave Memphis TN 38163

WILCOX, HOWARD ALBERT, b Minneapolis, Minn, Nov 9, 20; m 43; c 3. PHYSICS, ENVIRONMENTAL MANAGEMENT. *Educ:* Univ Minn, BA, 43; Univ Chicago, MA & PhD(physics), 48. *Prof Exp:* Instr physics, Harvard Univ & Radcliffe Col, 43-44; jr scientist, Los Alamos Sci Lab, NMex, 44-46; asst, Inst Nuclear Studies, Univ Chicago, 46-48; from instr to asst prof & mem staff radiation lab, Univ Calif, 48-50; res physicist & head guided missile develop div, US Naval Ord Test Sta, 50-55, head weapons develop dept, 55-58, asst tech dir res & head res dept, 58-59; dep dir defense res & eng, Off Secy Defense, Washington, DC, 59-60; dir res & eng defense res labs, Gen Motors Corp, 60-66, tech dir adv power systs, Res Labs, 66-67; physicist, US Naval Weapons Ctr, China Lake, 71-74; mgr ocean food & energy farm proj, US Naval weapons ctr & undersea ctr, 72-77; staff scientist, Environ Sci Dept, Naval Ocean Syst Ctr, 77-84; TECH & MGT CONSULT, 67- *Concurrent Pos:* Vpres, Minicars, Inc, Goleta, 68-74. *Honors & Awards:* Outstanding Tech Achievement Award, Inst Elec & Electronics Engrs, 77. *Mem:* AAAS; fel Am Phys Soc. *Res:* Production of mesons in nuclear collisions; guided missile system engineering; oceanography; hypervelocity flight physics; lunar and terrestrial vehicles; advanced power systems; technical management; earth's energy balance. *Mailing Add:* 882 Golden Park Ave San Diego CA 92106

WILCOX, HOWARD JOSEPH, b Plattsburgh, NY, Oct 20, 39. MATHEMATICS. *Educ:* Hamilton Col, AB, 61; Univ Rochester, PhD(math), 66. *Prof Exp:* Asst prof math, Univ Conn, 65-67; asst prof, Amherst Col, 67-70; from asst prof to assoc prof, 72-78, chmn dept, 76-80, PROF MATH, WELLESLEY COL, 78- *Mem:* Am Math Soc; Math Asn Am. *Res:* Topological groups; general topology. *Mailing Add:* Dept Math Wellesley Col Wellesley MA 02181

WILCOX, HUGH EDWARD, b Manchester, Calif, Sept 2, 16; m 38; c 5. PLANT PHYSIOLOGY. *Educ:* Univ Calif, BS, 38, PhD, 50; Syracuse Univ, MS, 40. *Prof Exp:* Asst, State Univ NY Col Forestry, Syracuse, 38-40; technician, Dept Forestry, Univ Calif, 41-42; physicist, Radiation Lab, 42-45; physicist & ord engr, US Naval Ord Test Sta, 45-46; assoc prof forest prod & wood technologist, Ore State Col, 46-50; res assoc & proj leader, Res Found, State Univ NY, 50-54; from assoc prof to prof, 54-85, EMER PROF FORESTRY, STATE UNIV NY COL ENVIRON SCI & FORESTRY, 86- *Mem:* AAAS; Soc Am Foresters; Bot Soc Am; Am Soc Plant Physiol. *Res:* Growth periodicity; dormancy; physiology of cambial activity; wound healing and regeneration; growth and differentiation of roots; mycorrhiza. *Mailing Add:* Dept Environ & Forest Biol State Univ NY Col Environ Sci & Forestry Syracuse NY 13210

WILCOX, JAMES RAYMOND, b Minneapolis, Minn, Jan 20, 31; m 55; c 2. GENETICS, FORESTRY. *Educ:* Univ Minn, BA, 53, MS, 59; Iowa State Univ, PhD(plant genetics), 61. *Prof Exp:* Res geneticist, Inst Forest Genetics, Forest Serv, USDA, 61-66, res geneticist, 66-76, SUPV RES GENETICIST, AGR RES SERV, USDA, PURDUE UNIV, WEST LAFAYETTE, 76- *Honors & Awards:* Prosoja Award, World Soybean Res Conf IV, 89. *Mem:* fel, AAAS; fel Am Soc Agron; Am Genetic Asn; fel, Crop Sci Soc Am; Am Soybean Asn. *Res:* Soybean breeding and genetics; genetic control of fatty acid biosynthesis and protein accumulation in soybeans; genetic control of pathogen resistance in soybean. *Mailing Add:* Room 2-310 Lilly Hall Purdue Univ Dept Agron West Lafayette IN 47907

WILCOX, JAROSLAVA ZITKOVA, physics, for more information see previous edition

WILCOX, JOSEPH CLIFFORD, b McLean, Ill, June 18, 30; m 52; c 2. FOOD SCIENCE, FOOD MICROBIOLOGY. *Educ:* Univ Ill, BS, 52, MS, 54. *Prof Exp:* Bacteriologist, 56-58, sect head sausage develop, Res & Develop Dept, Food Res Div, Armour & Co, 58-76; mgr food prod develop, Grocery Prods Div, Miles Lab, Inc, 76-82; OWNER & PRES, J C WILCOX ASSOC, 82- *Mem:* AAAS; Inst Food Technologists. *Res:* Bacteriological, chemical and radiological warfare decontamination; application of bacteriological principles in development and study of food products and associated problems; fresh, semidry and dry sausage items; utilization of nonmeat proteins; cholesterol-free food analog products; product and process development and sanitation audits; microbiological and sanitation audits; sensory evaluation; code dating and shelf-life determination; quality control; industrial real estate; technical service; nutritional labeling. *Mailing Add:* 6864 Halligan Ave E Worthington OH 43085-2618

WILCOX, KENT WESTBROOK, b NC, 1945. VIROLOGY. *Educ:* Duke Univ, BS, 67; Johns Hopkins Univ, MA, 69, PhD(microbiol), 74. *Prof Exp:* Asst prof, 79-86, ASSOC PROF MICROBIOL, MED COL WIS, 86- *Mem:* Am Soc Microbiol; AAAS. *Res:* Regulation of viral gene expression in cells infected by herpes simplex virus. *Mailing Add:* Dept Microbiol Med Col Wis 8701 Watertown Plank Rd Milwaukee WI 53226

WILCOX, LEE ROY, b Chicago, Ill, June 8, 12; m 40; c 2. ALGEBRA, SCIENCE EDUCATION. *Educ:* Univ Chicago, SB, 32, SM, 33, PhD(math), 35. *Prof Exp:* Mem sch math, Inst Advan Study, 35-36, asst, 36-38; instr math, Univ Wis, 38-40; from asst prof to prof, 40-77, dir Ctr Educ Develop, 69-77, EMER PROF MATH, ILL INST TECHNOL, 77- *Mem:* Am Math Soc; Math Asn Am; Sigma Xi. *Res:* Theories of semi-modular and topological lattices; foundations of mathematics; abstract algebra; mathematics education. *Mailing Add:* 1404 Forest Ave Wilmette IL 60091-1634

WILCOX, LOUIS VAN INWEGEN, JR, b Orange, NJ, Aug 24, 31; m 56; c 3. ECOLOGY. *Educ:* Colgate Univ, AB, 53; Cornell Univ, MS, 58, PhD(plant path), 61. *Prof Exp:* Asst plant physiol, Cornell Univ, 55-57; asst plant path, 57-61; asst prof biol, Lycoming Col, 61-65; assoc prof, Earlham Col, 65-71; dir, Fahkahatchee Environ Studies Ctr, 71-73; dir environ qual prog, Hampshire Col, 73-76; prof & chmn, Ctr Environ Sci, Unity Col. 76-78, dean, 78-80, pres, 80-86; RETIRED. *Concurrent Pos:* Consult, Coastal Enterprises, Inc, 80-81 & Maine Audubon Farm Study Proj, 79; mem, bd dirs, Ctr Human Ecol Studies, 79-81. *Mem:* Am Asn Geol; Sigma Xi. *Res:* Mangrove ecology. *Mailing Add:* Rte 1 Box 320 Thorndike ME 04986

WILCOX, LYLE C(HESTER), b Lansing, Mich, Aug 8, 32; m 52; c 3. SYSTEMS ENGINEERING, COMPUTER SCIENCE. *Educ:* Tri-State Col, BSEE, 54; Mich State Univ, MSEE, 58, PhD(elec eng), 63. *Prof Exp:* Fac mem elec eng, Tri-State Col, 52-54 & Mich State Univ, 55-63; dir opers, Vet Admin, Ark, 64-65; assoc prof elec eng, & Draper prof elec & mech eng, Clemson Univ, 65-66, head dept elec eng, 66-72, assoc dean prof studies, 70-73, prof elec & comput eng, 66-80, dean eng, Col Eng, 73-80; pres, Univ Southern Colo, 80-84; VPRES RES & DEVELOP. *Concurrent Pos:* NSF fac fel, 59-61; Ford Found fel, 62-63. *Mem:* AAAS; Inst Elec & Electronics Engrs; Am Soc Mech Engrs. *Res:* Application of systems theory to problems in control and biomedical and operations research; design and use of specialized instrumentation for data acquisition systems used in the study of multi-terminal components; general digital analog simulation from the hybrid point of view. *Mailing Add:* Telex Corp 6422 E 41st St Tulsa OK 74135

WILCOX, MARION WALTER, b Broken Arrow, Okla, Aug 17, 22; m 48; c 2. ENGINEERING MECHANICS. *Educ:* Univ Notre Dame, BSCE, 48, ScD(eng sci), 61; Ill Inst Technol, MSCE, 56. *Prof Exp:* Assoc res engr, Armour Res Found, Ill Inst Technol, 51-54; designer, Kaiser Aluminum Corp, 54-55; tech supvr dynamics, Bendix Prod Div, Bendix Aviation Corp, 55-58; sr engr, Gen Dynamics/Ft Worth, Tex, 58-61; assoc prof mech eng, Univ Ariz, 61-62; assoc prof, 62-66, PROF SOLID MECH & MECH ENG, SOUTHERN METHODIST UNIV, 66- *Concurrent Pos:* Consult, Ling-Temco-Vought Corp, 62-, Off Res Anal, Holloman AFB, NMex, 64-65, Socony Mobil Oil Co, Inc, 65-66 & Koelling Universal Joints, Inc, 65-72. *Mem:* Am Soc Eng Educ; Am Acad Mech; Am Inst Aeronaut & Astronaut; Soc Eng Sci; Sigma Xi. *Res:* Thermoelasticity and elastodynamics. *Mailing Add:* C-ME Dept Southern Methodist Univ Dallas TX 75275

WILCOX, MERRILL, b Milwaukee, Wis, Oct 10, 29; m 62; c 2. PLANT PHYSIOLOGY. *Educ:* Univ Md, BS, 52, MS, 54; NC State Univ, PhD(agron, plant physiol), 61. *Prof Exp:* Biol aide marine biol, Chesapeake Biol Lab, Univ Md, 56-57; from asst prof to assoc prof agron, 60-72, PROF AGRON, UNIV FLA, 72- *Concurrent Pos:* Grants, Am Cancer Soc, 63-66, NSF, 63-65 & Geigy Chem Corp, 70-78. *Mem:* AAAS; Am Soc Plant Physiol; Weed Sci Soc Am; Scand Soc Plant Physiol; Am Chem Soc; Int Palm Soc. *Res:* Structure-activity relationships and metabolism of herbicides and plant growth regulators; abscission by Glyoxime; tobacco growth regulation by flumetralin; hybridization in cocosoid palms. *Mailing Add:* Dept Agron Univ Fla Herbicide Met Lab IFAS Gainesville FL 32601

WILCOX, PAUL DENTON, b Salt Lake City, Utah, Mar 4, 35; m 61; c 1. CERAMIC ENGINEERING, METALLURGY. *Educ:* Univ Utah, BS, 58, PhD(ceramic eng), 62. *Prof Exp:* Supvr active ceramic mat div, 62-80, SUPVR INITIATING & PYROTECHNICS, SANDIA LABS, 80- *Concurrent Pos:* Adj prof, Univ NMex, 69-70. *Honors & Awards:* Dept Energy Award, Improving Qual, 88. *Mem:* Am Ceramic Soc; Int Pyrotech Soc. *Res:* Piezoelectrics; ferroelectrics; glass; glass ceramics; thermoelectrics; acoustic surface waves; ceramic varistors; materials science technology; explosives and propellants; pyrotechnics. *Mailing Add:* 1501 Cedar Ridge Dr NE Albuquerque NM 87112

WILCOX, RAY EVERETT, b Janesville, Wis, Mar 31, 12; m 42; c 4. GEOLOGY. *Educ:* Univ Wis, PhB, 33, PhM, 37, PhD(geol), 41. *Prof Exp:* Geologist, State Geol Surv, Wis, 35-39 & Jones & Laughlin Steel Corp, 41-42; geologist, US Geol Surv, 46-84; RETIRED. *Mem:* AAAS; fel Mineral Soc Am. *Res:* Igneous petrology; volcanology; volcanic ash chronology; petrographic methods; optical crystallography. *Mailing Add:* 3590 Estes St Wheat Ridge CO 80033

WILCOX, ROBERTA ARLENE, b Hopkinton, RI, Nov 12, 32. MEDICAL STATISTICS. *Educ:* Univ RI, BS, 54; Johns Hopkins Univ, ScM, 58. *Prof Exp:* Biostatistician, State Dept Health, NY, 57-59; res statistician, Lederle Labs, Am Cyanamid Co, 59-66; sr biostatistician, Med Div, Ciba-Geigy Corp, 66-72; sr res scientist, Pfizer Cent Res, Pfizer, Inc, 72-77; PRIN BIOSTATISTICIAN, ALCON LABS, INC, 79- *Mem:* Am Statist Asn; Biomet Soc; NY Acad Sci. *Res:* Experimental design and analysis applicable to medical and drug research. *Mailing Add:* 6908 Wilton Dr Ft Worth TX 76133-6131

WILCOX, RONALD BRUCE, b Seattle, Wash, Sept 23, 34; m 58; c 2. BIOCHEMISTRY, ENDOCRINOLOGY. *Educ:* Pac Union Col, BS, 57; Univ Utah, PhD(biochem), 62. *Prof Exp:* Res fel med, Mass Gen Hosp & Harvard Med Sch, 62-65; from asst prof to assoc prof, 65-73, PROF BIOCHEM, SCH MED, LOMA LINDA UNIV, 73- *Mem:* AAAS; Sigma Xi; Endocrine Soc; Am Asn Cancer Res. *Res:* Biochemistry and metabolism of hormones; hormone related carcinogenesis. *Mailing Add:* Dept Biochem Loma Linda Univ Sch Med Loma Linda CA 92354

WILCOX, RONALD ERWIN, b Ft Wayne, Ind, Jan 6, 29; m 59; c 3. GEOLOGY. *Educ:* Iowa State Univ, BS, 50, MS, 52; Columbia Univ, PhD(petrol), 58. *Prof Exp:* Asst geol, Iowa State Univ, 50-52 & Columbia Univ, 52-54; res geologist, Humble Oil & Refining Co, 56-64; sr res geologist, Esso Prod Res Co, 64-72; lectr, Univ Houston, 72-75, adj prof, 75-76; assoc prof, Inst Environ Studies, La State Univ, 77-78; CONSULT, 88- *Concurrent Pos:* Consult geologist, 72- *Honors & Awards:* President's Award, Am Asn Petrol Geologists, 75. *Mem:* Fel AAAS; fel Geol Soc Am; Am Asn Petrol Geologists; Am Geophys Union; Sigma Xi. *Res:* Structural geology; petrology; structure of continental margins; salt tectonics; orogenic belts; metamorphism. *Mailing Add:* PO Box 25096 Baton Rouge LA 70894-5096

WILCOX, ROY CARL, b Alexandria, Va, Feb 4, 33; m 65; c 2. PHYSICAL METALLURGY. *Educ:* Va Polytech Inst, BS, 55, MS, 59; Univ Mo-Rolla, PhD(metall eng), 62. *Prof Exp:* Metallurgist, Naval Ord Lab, 55-56; instr metall, Va Polytech Inst, 57-59, 62, assoc prof, 62-68, assoc dir, Continuing Educ Ctr, 68-69; ASSOC PROF MAT, AUBURN UNIV, 69- *Mem:* Am Soc Metals; Am Inst Mining, Metall & Petrol Engrs; Sigma Xi. *Res:* Deformation of textures of cobalt; study of titanium-aluminum alloys; fracture studies of adhesive bonded joints. *Mailing Add:* Dept Mech Eng Auburn Univ Auburn AL 36830

WILCOX, THOMAS JEFFERSON, b San Francisco, Calif, Oct 2, 42; m 72; c 2. PHYSICS. *Educ:* Univ Calif, Berkeley, BA, 64; Univ Calif, Los Angeles, MS, 66, PhD(physics), 72. *Prof Exp:* Res physicist plasma/particle physics, Univ Calif, Los Angeles, 72-73; mem tech staff plasma physics, TRW Systs Group, 73-75; mem tech staff physics, R&D Assocs, 75-82; sr scientist, Sci Applications, Inc, 82-84, Res & Develop Labs, 84-86, Pac Sierra Res Corp, 86-89, SR SCIENTIST, TRW SPACE & TECHNOL GROUP, 89- *Mem:* Am Phys Soc. *Res:* Plasma physics; radiation transport; optics; electromagnetism; mathematical physics. *Mailing Add:* 235 N Kenter Ave Los Angeles CA 90049

WILCOX, W(ILLIAM) R(OSS), b Manhattan, Kans, Jan 14, 35; m 68; c 4. CHEMICAL ENGINEERING, MATERIALS SCIENCE. *Educ:* Univ Southern Calif, BEng, 56; Univ Calif, Berkeley, PhD(chem eng), 60. *Prof Exp:* Instr chem eng, Univ Calif, Berkeley, 60; mem tech staff, Pac Semiconductors Inc, 60-62; mem tech staff, Aerospace Corp, 62-65, head crystal technol sect, 65-68; assoc prof mat sci & chem eng, Univ Southern Calif, 68-74, prof, 74-75; prof chem eng & chmn dept, 75-86, DIR, CTR ADVAN MAT PROCESSING, 85-, DEAN ENG, 87-, DIR, CTR CRYSTAL GROWTH IN SPACE, CLARKSON UNIV, 86- *Concurrent Pos:* Lectr, Eve Col, Univ Calif, Los Angeles, 62-65; res asst, Lawrence Radiation Lab, Univ Calif, Berkeley, 57-59; consult, NASA, var industs & univs; vis prof, Dept Physics, Univ Estadual de Campinas, Brazil, 75. *Mem:* Fel Am Inst Chem Engrs; Am Soc Eng Educ; Am Asn Crystal Growth (vpres, 84-87); fel AAAS; Am Ceramic Soc. *Res:* Crystal growth, materials processing in space. *Mailing Add:* Clarkson Univ Potsdam NY 13676

WILCOX, W WAYNE, b Berkeley, Calif, Oct 28, 38; m 60; c 2. FOREST PRODUCTS PATHOLOGY, WOOD BIODETERIORATION. *Educ:* Univ Calif, Berkeley, BS, 60; Univ Wis-Madison, MS, 62, PhD(plant path), 65. *Prof Exp:* Plant pathologist, US Forest Prod Lab, Wis, 60-64; from asst to assoc forest prod pathologist, 64-77, lectr, 64-75, FOREST PROD PATHOLOGIST & PROF FORESTRY, UNIV CALIF, BERKELEY, 77- *Concurrent Pos:* Fulbright-Hays sr fel, Ger, 73-74. *Honors & Awards:* Forest Prod Res Soc Award, 65. *Mem:* Forest Prod Res Soc; fel Int Acad Wood Sci; Soc Wood Sci & Technol; Am Inst Biol Sci; Int Asn Wood Anatomists. *Res:* Wood deterioration; microscopy of wood decay; ability to detect, diagnose and evaluate early stages of decay in structures. *Mailing Add:* Forest Prod Lab Univ Calif 1301 S 46th St Richmond CA 94804

WILCOX, WESLEY C, b St Anthony, Idaho, July 19, 25; m 48; c 4. MICROBIOLOGY. *Educ:* Univ Utah, BA, 51, MS, 55; Univ Wash, PhD(microbiol), 58. *Prof Exp:* Donner fel med res, Western Reserve Univ, 58-59, USPHS fel prev med, 59-60; assoc microbiol, Univ Pa, 60-62, asst prof, 62-63; asst prof, Univ Vt, 63-65; PROF MICROBIOL, UNIV PA, 65-, CHMN DEPT, 76-, HEAD LAB MICROBIOL, 80- *Concurrent Pos:* Res career develop award, Univ Pa, 60-63. *Mem:* Am Asn Immunol; Am Soc Microbiol. *Res:* Virology; immunology; biochemistry. *Mailing Add:* 2302 Bridgewater Ct Chester Springs PA 19104

WILCOX, WESLEY CRAIN, b Bloomington, Ill, Apr 8, 26; m 48; c 5. AGRONOMY, BOTANY. *Educ:* Univ Ill, BS, 50, MS, 51. *Prof Exp:* Field supvr, Found Dept, 51-55, corn breeder, Res Dept, 55-66, mgr spec proj res, 66-71, DIR, QUAL ASSURANCE DEPT, CIBA-GEIGY SEED DIV, 71- *Mem:* Soc Com Seed Technologists; Am Soc Agron; Crop Sci Soc Am; Sigma Xi. *Res:* High quality seed of hybrid corn, sorghum, soybeans and farm seeds. *Mailing Add:* CIBA-GEIGY Seed Div Qual Assurance Dept 1301 W Washington St Bloomington IL 61701

WILCOX, WILLIAM JENKINS, JR, b Harrisburg, Pa, Jan 26, 23; m 46; c 3. ISOTOPE SEPARATION, STRATEGIC PLANNING. *Educ:* Washington & Lee Univ, BA, 43; Univ Tenn, MS, 58. *Prof Exp:* Chemist, Tenn Eastman Corp, 43-48; chemist, Union Carbide Corp, 48-49, tech asst to lab dir, 49-55, head dept physics, 55-67, prog mgr, 67-69, prod plants tech dir, Nuclear Div, 69-81; sr staff consult, Martin Marietta Energy Systs, Inc, 81-86; RETIRED. *Concurrent Pos:* Mgt consult, 86- *Mem:* AAAS; Am Chem Soc; Sigma Xi; fel Am Inst Chemists; NY Acad Sci. *Res:* Isotope separation processes, research and development; structure of porous materials; materials development. *Mailing Add:* 412 New York Ave Oak Ridge TN 37830

WILCOXSON, ROY DELL, b Columbia, Utah, Jan 12, 26; m 49; c 4. PLANT PATHOLOGY. *Educ:* Utah State Univ, BS, 53; Univ Minn, MS, 55, PhD(plant path), 57. *Prof Exp:* Asst prof, 57-66, PROF PLANT PATH, UNIV MINN, ST PAUL, 66- *Concurrent Pos:* Spec staff mem, Rockefeller Found; vis prof, Indian Agr Res Inst, New Delhi; dir, Morocco Proj, Univ Minn, 83-87. *Mem:* Am Phytopath Soc; Indian Phytopath Soc; AAAS. *Res:* Diseases of forage crops and cereal crops; cereal rust diseases. *Mailing Add:* Dept Plant Path Univ Minn St Paul MN 55101

WILCZEK, FRANK ANTHONY, b Queens, NY, May 15, 51; m 73; c 1. THEORETICAL PHYSICS. *Educ:* Univ Chicago, BS, 70; Princeton Univ, MA, 72, PhD(physics), 74. *Prof Exp:* Asst prof physics, Princeton Univ, 74-77; mem, Inst Advan Studies, 77-78; assoc prof physics, Princeton Univ, 78-80, prof, 80-; astrophysicist, Inst Theoret Physics, Univ Calif, Santa Barbara, 80-88; PROF SCH NATURAL SCI, INST ADVAN STUDY, 88- *Concurrent Pos:* Adv comt, Brookhaven Nat Lab, 78-81. *Honors & Awards:* Sakurai Prize, Am Phys Soc, 86. *Res:* High energy physics; quantum field theory. *Mailing Add:* Sch Natural Sci Inst Advan Study Princeton NJ 08540

WILCZYNSKI, JANUSZ S, PACKAGING TECHNOLOGY. *Educ:* Mining Acad, Cracow, Inz dipl, 54; Jagellonian Univ, MSc, 57; Univ London, PhD(physics), 61. *Prof Exp:* Mgr tech optics, 62-83, sr mgr, Lithography & Packaging Eng Group, 83-86, DIR PACKAGING TECHNOL, IBM T J WATSON RES LAB, 86-, DIR, ADVAN PACKAGING TECHNOL LAB, 88- *Honors & Awards:* Richardson Medal, Optical Soc Am, 88. *Mem:* Nat Acad Eng; fel Optical Soc Am. *Res:* Sub-half micron optical stepper; thin film fabrication; lithographic processes; normal incidence x-ray telescope; author of various publications; granted several patents. *Mailing Add:* IBM Thomas J Watson Res Ctr PO Box 218 Yorktown Heights NY 10598

WILCZYNSKI, WALTER, b Trenton, NJ, Sept 18, 52. NEUROETHOLOGY, SENSORY PROCESSING. *Educ:* Lehigh Univ, BS & BA, 74; Univ Mich, PhD(neurosci), 78. *Prof Exp:* Postdoctoral fel neurobiol, Cornell Univ, 79-83; asst prof, 83-89, ASSOC PROF PSYCHOL, UNIV TEX, 89- *Concurrent Pos:* Prin investr, Univ Tex, NSF & NIMH grants, 84-; vis scientist, Smithsonian Trop Res Inst, 87 & 90. *Mem:* Soc Neurosci; Am Soc Zoologists; Int Soc Neuroethol; AAAS. *Res:* Investigate the neural mechanism of acoustic communication and reproductive social behavior in anuran amphibians; research combines anatomical and physiological techniques to determine how acoustic information is represented and used by the peripheral auditory system and central nervous system. *Mailing Add:* Dept Psychol Univ Tex Austin TX 78712

WILD, BRADFORD WILLISTON, b Fall River, Mass, Dec 5, 27; m 77. OPTOMETRY, OPTICS. *Educ:* Brown Univ, AB, 49; Columbia Univ, BS, 51, MS, 52; Ohio State Univ, PhD(physiol optics), 59. *Hon Degrees:* DOS, Southern Calif Col Optom. *Prof Exp:* From instr to assc prof optom & physiol optics, Ohio State Univ, 59-69; dean, Col Optom, Pac Univ, 69-74; PROF & DEAN, SCH OPTOM, MED CTR, UNIV ALA, BIRMINGHAM, 74- *Concurrent Pos:* Res optometrist, Gen Vision Sect, US Naval Med Res Lab, Conn. *Mem:* Am Optom Asn; Am Acad Optom (pres, 78-80). *Res:* Physiological optics, especially retinal interaction, border phenomena and problems of visibility. *Mailing Add:* Sch Optom Univ Ala Med Ctr Birmingham AL 35294

WILD, GAYNOR (CLARKE), b Winner, SDak, Nov 10, 34; m 58, 73, 87; c 2. NEUROCHEMISTRY, BIOCHEMISTRY OF FERTILIZATION. *Educ:* SDak Sch Mines & Technol, BS, 55; Tulane Univ, PhD(biochem), 62. *Prof Exp:* Fel biochem, Clayton Found Biochem Inst, Univ Tex, 62-63; res assoc, Rockefeller Univ, 63-65, asst prof, 65-67; asst prof, 67-85, ASSOC PROF BIOCHEM, SCH MED, UNIV NMEX, 85-, ASSOC PROF NEUROL, 87- *Concurrent Pos:* Consult, Los Alamos Nat Lab, 83-87. *Mem:* Am Soc Biol Chemists; Soc Neurosci. *Res:* Neurochemistry; enzymology; lipid biochemistry; sperm acrosome autoantigens and enzymes. *Mailing Add:* Dept Biochem Univ NMex Sch Med Albuquerque NM 87131

WILD, GENE MURIEL, b Fremont, Nebr, Oct 15, 26; m 48; c 4. BIOCHEMISTRY. *Educ:* Iowa State Univ, BS, 48, MS, 50, PhD(biochem), 53. *Prof Exp:* Sr biochemist, Eli Lilly & Co, 53-73, res scientist, 73-89; RETIRED. *Res:* Purification process research in antibiotics; chemical analysis and chromatography of antibiotics and related materials. *Mailing Add:* 7455 Jewel Lane Indianapolis IN 46285

WILD, JAMES ROBERT, b Sedalia, Mo, Nov 24, 45; m 73; c 1. MOLECULAR BIOLOGY. *Educ:* Univ Calif, Davis, BA, 67; Univ Calif, Riverside, PhD(biol), 71. *Prof Exp:* Res biochemist, Univ Calif, Riverside, 72; microbiologist, Naval Med Res Inst, Nat Naval Med Ctr, 72-75; asst prof genetics, Tex A&M Univ, 75-80, assoc prof biochem & genetics, 80-84, prof chmn, Biochem, 87-90; CONSULT, 90- *Mem:* Am Soc Microbiol; Genetics Soc Am; Am Soc Biol Chemists; Sigma Xi. *Res:* Pyrimidine biosynthesis; nucleotide biosynthesis; regulation of gene expression and gene structure; function relationships. *Mailing Add:* Dept Biochem & Biophys Tex A&M Univ College Station TX 77843-2128

WILD, JOHN FREDERICK, b Erie, Pa, June 20, 42; m 66; c 2. NUCLEAR CHEMISTRY. *Educ:* Pa State Univ, BS, 64; Mass Inst Technol, PhD(nuclear chem), 68. *Prof Exp:* Res chemist, Knolls Atomic Power Lab, 68-69; RES CHEMIST, LAWRENCE LIVERMORE LAB, 69- *Mem:* Am Chem Soc. *Res:* Nuclear chemistry with emphasis on decay and chemical properties of isotopes of elements above Z 96; spontaneous fission; heavy-ion reaction mechanisms. *Mailing Add:* Lawrence Livermore Lab Livermore CA 94551

WILD, JOHN FREDERICK, b Wallingford, Conn, Nov 30, 26; m 65; c 3. PHYSICS. *Educ:* Yale Univ, BS, 50, MS, 51, PhD(physics), 58. *Prof Exp:* From instr to asst prof physics, Trinity Col, Conn, 57-62; asst prof, 62-67, ASSOC PROF PHYSICS, WORCESTER POLYTECH INST, 67- *Mem:* Am Phys Soc; Am Asn Physics Teachers; Sigma Xi. *Res:* Wave functions for valence electron of neutral caesium for Fermi-Thomas central field; quantum mechanics; color vision; Foucault knife-edge test; tuned percussion instruments; solar heating. *Mailing Add:* 16 Cavour Circle West Boylston MA 01583

WILD, JOHN JULIAN, b Syndenham, Eng, Aug 11, 14; US citizen; m 68; c 3. CLINICAL MEDICINE, ULTRASOUND. *Educ:* Cambridge Univ, BA, 36, MA, 40, MB, MD, 42, PhD(investigative med), 71. *Prof Exp:* Res assoc, Dept Surg, Univ Minn, 46-51, res assoc med diag ultrasound, Dept Elec Eng, 51-53; dir res, St Barnabas Hosp, Minneapolis, 53-60; dir medico-technol res unit, Minn Found, St Paul, 60-63; DIR MED DIAG ULTRASOUND, MEDICO-TECHNOL RES INST, MINNEAPOLIS, 65- *Concurrent Pos:* Res fel, Marion Ordway Found, St Paul, 46-47 & USPHS, 47-49; prin investr, Nat Adv Cancer Coun, 50-60, Nat Heart Inst, 57-60 & Gen Med Sci Div, 62-63. *Honors & Awards:* Pioneer Award, Am Inst Ultrasound Med, 78. *Mem:* Fel Am Inst Ultrasound Med; AMA; hon mem Brit Inst Radiol. *Res:* Physical detection of disease and deteriorative processes; ultrasonic pulse-echo tissue characterization, cancer detection and diagnosis of the breast and colon; pulse-echo measurement of biological tissues. *Mailing Add:* Medico-Technol Res Inst 4262 Alabama Ave S St Louis Park MN 55416-3105

WILD, ROBERT LEE, b Sedalia, Mo, Oct 9, 21; m 43; c 3. SOLID STATE PHYSICS. *Educ:* Cent Mo State Univ, BS, 43; Univ Mo, Columbia, MA, 48, PhD(physics), 50. *Prof Exp:* Asst instr physics, Univ Mo, 49; asst prof, Univ NDak, 50-53; from asst prof to prof, 53-88, chmn dept, 63-68, EMER PROF PHYSICS, UNIV CALIF, RIVERSIDE, 88- *Concurrent Pos:* NSF fel, Univ Ill, 59-60; vis prof, Tech Univ Denmark, 67-68 & Univ Munster, Ger, 75; Fulbright lectr, Univ Philippines, 81-82; pres, S Calif Sect, Am Asn Physics Teachers, 85-87. *Honors & Awards:* Outstanding Physics Achievement Award, Phil Phy Soc, 86. *Mem:* Am Phys Soc; Am Asn Physics Teachers; Sigma Xi. *Res:* Small angle x-ray scattering by liquids and solids; optical and transport properties of solids. *Mailing Add:* Dept Physics Univ Calif Riverside CA 92521

WILD, WAYNE GRANT, b Waterville, Kans, Aug 9, 17; m 39; c 4. PHYSICS, MATHEMATICS. *Educ:* SDak State Univ, BS, 40; Univ Wis, MS, 48; Univ Ill, MA, 67. *Prof Exp:* Prof physics & head dept, Buena Vista Col, 48-67, chmn natural sci div, 53-67; assoc prof physics, 67-69, assoc prof math, Univ Wis-Stevens Point, 69-82; RETIRED. *Mem:* Sigma Xi; Am Math Asn. *Res:* Thermionic emission. *Mailing Add:* 22 Little Dr Bella Vista AR 72714

WILDASIN, HARRY LEWIS, b York Co, Pa, Oct 10, 23; m 45, 70; c 2. BIOCHEMISTRY. *Educ:* Pa State Univ, PhD(dairy), 50. *Prof Exp:* Asst prof dairying, Univ Conn, 49-52; dir qual control, Whiting Milk Co, Boston, 52-57; dir qual control & govt rels, H P Hood, Inc, Boston, 57-; AT WILDASIN ASSOCS. *Mem:* AAAS; Am Dairy Sci Asn; NY Acad Sci; Nat Environ Health Asn; Am Pub Health Asn; Sigma Xi. *Res:* Frozen milk; lactose; milk proteins; surface active agents; antibiotics; salmonella; radioactive elements in milk. *Mailing Add:* Wildasin Assocs 23 Oxbow Rd Lexington MA 02173

WILDBERGER, WILLIAM CAMPBELL, medical administration, psychiatry; deceased, see previous edition for last biography

WILDE, ANTHONY FLORY, b New York, NY, May 16, 30; m 72. PHYSICAL CHEMISTRY. *Educ:* Yale Univ, BS, 52; Ind Univ, PhD(phys chem), 59. *Prof Exp:* Res chemist, Monsanto Res Corp, 59-63; res chemist, US Army Natick Labs, 63-68, RES CHEMIST, US ARMY MAT TECHNOL LAB, 68- *Mem:* Am Chem Soc; AAAS; NY Acad Sci. *Res:* Polymer rheology, especially dynamic mechanical and optical properties of organic polymers and elastomers; stress wave propagation, fracture and energy dissipation in materials; dielectric and piezoelectric properties of organic polymers; sorption and diffusion of liquids in polymers. *Mailing Add:* US Army Mat Technol Lab Polymer Res Br Watertown MA 02172

WILDE, BRYAN EDMUND, b Salford, Lancashire, UK, Nov 8, 34; US citizen; m 56; c 4. FORENSIC ENGINEER, MATERIALS SELECTION. *Educ:* Royal Inst Chem, AGRIC, 59, 61, PhD(phys chem), 64; Rennselaer Polytech, NY, PhD(math), 66. *Prof Exp:* Mgr mat sci, Phys Chem Div, Exide Batteries, 59-64; supvr mat sci, Corrosion Res Lab, RPI, 64-66; lead scientist mat sci, Gen Elect Nucleonics Lab, 66-68; head mat sci, Corrosion Tech Div, US Steel, 68-84; PROF & DIR MAT SCI, FONTANA CORROSION CTR, OHIO STATE UNIV, 84- *Concurrent Pos:* Nat chmn, Nat Asn Corrosion Engrs, Res in Progress, 78, Gordon Res Conf-Corrosion, 86; pres, NAm Corrosion Construct, Inc, 83-; distinguished vis scholar, UN Develop Prog, China, 86. *Mem:* Am Soc Metals; Mat Res Soc; Electro Chem Soc; fel Inst Corrosion Sci & Tech. *Res:* Corrosion, electro chemistry, environmental by induced degradation of materials, composite corrosion, hydrogen obsorption into materials, fracture. *Mailing Add:* Dept Met Eng Fontana Corrosion Ctr 116 W 19th Ave Columbus OH 43210

WILDE, CARROLL ORVILLE, b Elmhurst, Ill, June 5, 32; m 52, 71; c 3. MATHEMATICS. *Educ:* Ill State Univ, BS, 58; Univ Ill, Urbana, PhD(math), 64. *Prof Exp:* Instr math, SDak Sch Mines & Technol, 58-59 & Col Wooster, 59-61; asst prof, Univ Minn, Minneapolis, 64-68; assoc prof, Naval Postgrad Sch, 68-75, chmn dept, 76-83, fac chmn, 90-91, PROF MATH, NAVAL POSTGRAD SCH, 75- *Concurrent Pos:* Vis prof, US Mil Acad, West Point, 79-80, 84-85; prog dir, NSF, 89-90. *Mem:* Nat Coun Teachers Math; Math Asn Am; Sigma Xi; Soc Indust & Appl Math. *Res:* Scientific computation. *Mailing Add:* Dept Math Code MA/Wm Naval Postgrad Sch Monterey CA 93943

WILDE, CHARLES EDWARD, JR, b Boston, Mass, Nov 5, 18; m 44; c 3. BIOLOGY. *Educ:* Dartmouth Col, AB, 40; Princeton Univ, MA, 47, PhD(biol), 49. *Hon Degrees:* MA, Univ Pa, 72. *Prof Exp:* Instr zool, Dartmouth Col, 40-41 & Princeton Univ, 46-49; from asst prof to prof zool, Sch Dent Med, Univ Pa, 49-75, prof embryol, dept biol & dept path, 57-75; prof zool, 75-86, CHMN DEPT, UNIV RI, 75-, EMER PROF, 86- *Concurrent Pos:* Guggenheim Mem Found fel, 57-58; guest investr, Strangeways Res Lab, Cambridge Univ, 57-58; consult, Dept Animal Genetics, Univ Edinburgh, 57; trustee, Mt Desert Island Biol Lab, dir, 67-70, pres, 77-78. *Mem:* AAAS; Am Soc Zoologists; Soc Develop Biol; Soc Cell Biol; Int Inst Embryol. *Res:* Differentiation of organs in vitro; tissue culture; metabolite control of cell differentiation; embryology of the head; cytochimeras of muscle; temporal relations of energy flow; RNA and protein synthesis in morphogenesis and differentiation; molecular and genomic control of symmetry in embryogenesis; the ontogeny of euryhalinity. *Mailing Add:* 40 Woodbine Rd Wakefield RI 02879

WILDE, CHARLES EDWARD, III, b 1946. PROTEIN STRUCTURE, INFECTIOUS DISEASES. *Educ:* Univ Calif, Berkerley, PhD(molecular biol), 75. *Prof Exp:* Asst prof, 78-86, ASSOC PROF MICROBIOL & IMMUNOL, SCH MED, UNIV IND, 86- *Mem:* Am Soc Microbiol; Am Asn Immunologists; AAAS. *Mailing Add:* Dept Microbiol & Immunol Sch Med Ind Univ 635 Barnhill Dr Indianapolis IN 46202-5120

WILDE, D(OUGLASS) J(AMES), b Chicago, Ill, Aug 1, 29; m 56; c 1. OPTIMIZATION, COMPUTATIONAL GEOMETRY. *Educ:* Carnegie Inst Technol, BS, 48; Univ Wash, MS, 56; Univ Calif, PhD, 60. *Prof Exp:* Chem engr, Pittsburgh Coke & Chem Co, 48-50 & Union Oil Co, 54-56; asst, Univ Calif, 57-58, instr chem eng, 58-59, lectr, 59-60; Fulbright lectr, Nat Advan Sch Chem Industs, 60-61; asst prof, Univ Tex, 61-62; assoc prof, 63-67, prof chem eng, 67-72, assoc dean 78-80, PROF MECH ENG DESIGN, STANFORD UNIV, 72- *Concurrent Pos:* Vis assoc prof, Yale Univ, 63; vis prof, PUC, Rio de Janeiro,77, Univ Sydney, Australia, 84 & Cent Sch Arts & Mfrs, Paris, 84. *Honors & Awards:* Lanchester Prize, Opers Res Soc Am, 68; Maynard Prize, Am Inst Indust Engrs; Design Automation Award, Am Soc Mech Engrs, 88. *Mem:* Am Soc Mech Engrs. *Res:* Optimization theory; optimal design. *Mailing Add:* Dept Mech Eng Stanford Univ Stanford CA 94305

WILDE, GARNER LEE, b Spring Creek, Tex, Sept 29, 26; m 51; c 2. GEOLOGY. *Educ:* Tex Christian Univ, BA, 50, MA, 52. *Hon Degrees:* DSc, Tex Christian Univ, 76. *Prof Exp:* Jr geologist, Humble Oil & Refining Co, Exxon Co, USA, 52-53, from assoc paleontologist to paleontologist, 53-63, sr res geologist, 63-67, prof geologist, 67-71, prof geologist, Hq Staff, 71-76, sr explor geologist, 76-81; mgr, Explor Div, Permian Basin, Harper Oil Co, 81-85; CONSULT GEOLOGIST, 86- *Concurrent Pos:* Lectr, Case Western Reserve Univ, 63; vis lectr, Tex Tech Univ, 67, 78 & Univ Mo, 69; mem bd dirs, Cushman Found Foraminiferal Res, 70-75, pres, 74-75; vis lectr, Kans

State Univ, 71 & Rensselaer Polytech Inst, 71; adj prof geol, Tex Christian Univ, 76-; hon mem, Permian Basin Sect, Soc Econ Paleontologists & Mineralogists. *Mem:* Fel Geol Soc Am; Soc Econ Paleontologists & Mineralogists; Am Asn Petrol Geologists. *Res:* Stratigraphic and paleontological studies on late Paleozoic Fusulinid Foraminifera, Calcareous algae and Mesozoic Calcareous Microfossils; carbonate facies; over 40 publications in fusulinid biostratigraphy, carbonate sedimentation. *Mailing Add:* Five Auburn Ct Midland TX 79705

WILDE, GERALD ELDON, b Ballinger, Tex, Dec 7, 39. ENTOMOLOGY. *Educ:* Tex Tech Col, BS, 62; Cornell Univ, PhD(entom), 66. *Prof Exp:* Res asst entom, Cornell Univ, 62-66; ASSOC PROF ENTOM & RES ENTOMOLOGIST AGR RES STA, KANS STATE UNIV, 66- *Mem:* Entom Soc Am. *Res:* Economic entomology; field crops insects. *Mailing Add:* Dept Entom Waters Hall Kans State Univ Manhattan KS 66506

WILDE, KENNETH ALFRED, b Cedar City, Utah, Mar 4, 29; m 61; c 2. PHYSICAL CHEMISTRY. *Educ:* Univ Utah, BS, 50, PhD(chem), 53. *Prof Exp:* Res chemist, Redstone Res Labs, Rohm and Haas Co, 53-70, res chemist, Res Div, 70-75; sr scientist, Radian Corp, 75-78; consult, 78-88; CONSULT, RADCAN CORP, 88- *Mem:* Am Chem Soc. *Res:* Chemical kinetics in electric discharges; combustion wave theory; high temperature thermodynamics and kinetics; aerothermodynamics; mass transfer process simulation; solution thermodynamics and process simulation. *Mailing Add:* 3604 Laurel Ledge Lane Austin TX 78731

WILDE, PAT, b Chicago, Ill, Sept 25, 35. OCEANOGRAPHY. *Educ:* Yale Univ, BS, 57; Harvard Univ, AM, 61, PhD(geol), 65. *Prof Exp:* Geologist, Shell Oil Co, 57-59; res geologist, Scripps Inst Oceanog, 60-62; lectr ocean eng, Univ Calif, Berkeley, 64-68, res engr, 64-66, res oceanogr, 66-75, asst prof, 68-75, res scientist & oceanogr, Lawrence Berkeley Lab, 75-82, res marine scientist, Dept Paleont, 82-89, lectr ocean eng, 75-88; CONSULT, 88- *Concurrent Pos:* Consult, Coastal Res Panel Earthquake Eng, Nat Acad Eng, 65-67; adv tech adv bd, Dept Eng, City & County of San Francisco, 70-74; vis prof, Inst Geol, Tech Univ Berlin, Ger, 89-90; Humboldt Sr Award, 89-90. *Mem:* AAAS; Geol Soc Am; Am Geophys Union; Marine Technol Soc; Geochem Soc. *Res:* Marine electrochemistry; sediment transport in marine environments; chemostratigraphy of black shales. *Mailing Add:* 1735 Highland Pl No 28 Berkeley CA 94709

WILDE, RICHARD EDWARD, JR, b Los Angeles, Calif, Jan 7, 31; m 60; c 3. VIBRATIONAL RELAXATION, HETEROGENEOUS PHOTOCATALYSIS. *Educ:* Univ Calif, Los Angeles, BS, 56; Univ Wash, PhD(chem), 61. *Prof Exp:* Res assoc, Johns Hopkins Univ, 61-63; from asst prof to assoc prof, 63-79, PROF CHEM, TEX TECH UNIV, 79- *Concurrent Pos:* Vis prof, Univ Durham, Eng, 86. *Mem:* Am Chem Soc; Am Phys Soc; Chem Soc. *Res:* Infrared and Raman spectroscopy of nonmetal hydrides; vibrational relaxation in liquids and solids; heterogeneous photocatalysis. *Mailing Add:* Dept Chem Tex Tech Col Lubbock TX 79409

WILDE, WALTER SAMUEL, b Toronto, Ont, Feb 12, 09; US citizen; m 36; c 3. MEDICAL PHYSIOLOGY. *Educ:* Miami Univ, AB, 31; Univ Minn, MA, 33, PhD(zool), 37. *Prof Exp:* Asst zool & physiol, Univ Minn, 31-33; instr zool, Miami Univ, 33-34; asst zool & physiol, Univ Minn, 34-37; teaching fel physiol, Univ Rochester, 37-38; instr zool, Univ Wyo, 38-39; from instr to asst prof physiol, Sch Med, La State Univ, 39-45; res assoc, Carnegie Inst, 45-47; sr physiologist, NIH, 47; from assoc prof to prof physiol, Tulane Univ, 47-56; prof, 56-75, EMER PROF PHYSIOL, UNIV MICH, ANN ARBOR, 75- *Concurrent Pos:* Guest lectr, Mt Desert Island Biol Lab, 56. *Mem:* Fel AAAS; fel Soc Exp Biol & Med; fel Am Physiol Soc. *Res:* Interstitial and capillary albumin; ion transport and kidney; renal stop flow method; blood-brain barrier and aqueous humor; potassium outflux from heart during single systole (effluogram). *Mailing Add:* 622 Bentley Dr Naples FL 33963-8093

WILDEMAN, THOMAS RAYMOND, b Madison, Wis, May 11, 40; m 65; c 2. ANALYTICAL CHEMISTRY, GEOCHEMISTRY. *Educ:* Col St Thomas, BS, 62; Univ Wis, PhD(phys chem), 67. *Prof Exp:* Lectr chem, Univ Wis, 66-67; asst prof, 67-73, assoc prof, 73-79, PROF CHEM, COLO SCH MINES, 79- *Concurrent Pos:* Consult, US Geol Surv, 70-; chmn, Geochem Div, 81, 82, prog chmn, 84; consult environ chem, 84- *Honors & Awards:* Nat Eng Excellence Award, Consult Engrs Coun Am, 90. *Mem:* AAAS; Am Chem Soc; Geochem Soc; Am Soc Environ Educ; AAAS. *Res:* Properties of trace elements in solids and liquids; isotopic, radiochemical and atomic absorption analysis; geochemistry of trace elements in rocks and waters. *Mailing Add:* Dept Chem Colo Sch Mines Golden CO 80401

WILDENTHAL, BRYAN HOBSON, b San Marcos, Tex, Nov 4, 37; c 5. PHYSICS. *Educ:* Sul Ross State Col, BA, 58; Univ Kans, PhD(physics), 64. *Prof Exp:* Res assoc physics, Rice Univ, 64-66; US Atomic Energy Comn fel, Oak Ridge Nat Lab, 66-68; asst prof, Tex A&M Univ, 68-69; from assoc prof to prof physics, Mich State Univ, 69-83; prof & dept head, physics & atmospheric sci, Drexel Univ, 83-87; DEAN, COL ARTS & SCI, UNIV NMEX, 87- *Concurrent Pos:* Sr US fel, Humboldt Found, Univ Munich, 73; vis scientist, Brookhaven Nat Lab, 74, Max Planck Inst Nuclear Physics, Heidelberg, 76, Soc Heavy Ion Res, Darmstadt, 77; vis prof, Univ Paris, 77; fel, John Simon Guggenheim Mem Found, 77; exec secy, Nuclear Sci Adv Comt, NSF, 78; vis prof, Univ Oxford, 79 & Univ Manchester, 80; vis scientist, Los Alamos Nat Lab, 79. *Mem:* Fel Am Phys Soc; Sigma Xi. *Res:* Study of the low lying quantum states of atomic nuclei via direct reaction experiments and shell model theory. *Mailing Add:* Dept Physics & Astron Univ NMex Albuquerque NM 87131

WILDENTHAL, KERN, b San Marcos, Tex, July 1, 41; m 64; c 2. PHYSIOLOGY, INTERNAL MEDICINE. *Educ:* Sul Ross Col, BA, 60; Univ Tex Southwestern Med Sch Dallas, MD, 64; Cambridge Univ, PhD(cell physiol), 70. *Prof Exp:* Intern, Bellevue Hosp-NY Univ, 64-65; resident & cardiol fel, Parkland Hosp-Univ Tex Southwestern Med Sch Dallas, 65-66; vis scientist, Strangeways Res Lab, Cambridge Univ, 68-70; from asst prof to prof physiol & internal med, 70-75, dean, Grad Sch Biomed Sci, 76-80, dean med sch, 80-86, PROF INTERNAL MED & PHYSIOL & PRES, UNIV TEX SOUTHWESTERN MED CTR, DALLAS, 86- *Concurrent Pos:* Guest scientist, Nat Heart Inst, Bethesda, 67-68; Guggenheim Found fel, Univ Cambridge, Eng, 75-76. *Mem:* Am Soc Clin Invest; Int Soc Heart Res; Royal Soc Med, Gt Brit; Am Physiol Soc; Am Col Cardiol; Asn Am Physicians. *Res:* Cardiac physiology and metabolism. *Mailing Add:* Off Pres Univ Tex Southwestern Med Ctr 5323 Harry Hines Blvd Dallas TX 75235-9002

WILDER, CLEO DUKE, b Macon, Ga, Sept 24, 25; m 50; c 2. VERTEBRATE ZOOLOGY. *Educ:* Univ NC, AB, 48; Univ Tenn, MS, 51; Univ Fla, PhD, 62. *Prof Exp:* Instr biol, Presby Col, SC, 51-53; asst, Univ Fla, 55-57 & 58-59; asst prof, Memphis State Univ, 59-62; asst zool, Va Polytech Inst, 62-69; ASSOC PROF BIOL, MURRAY STATE UNIV, 69- *Mem:* Am Soc Ichthyologists & Herpetologists; Soc Study Amphibians & Reptiles; Sigma Xi; Herpetologists' League. *Res:* Taxonomy, ecology, distribution, behavior and evolution of amphibians and reptiles; ecology of stream drainage systems and cypress swamps in western Kentucky. *Mailing Add:* Box 2484 Univ Sta Murray KY 42071

WILDER, DAVID RANDOLPH, b Lorimor, Iowa, June 11, 29; m 51; c 4. CERAMIC ENGINEERING, METALLURGY. *Educ:* Iowa State Univ, BS, 51, MS, 52, PhD(ceramic eng), 58. *Prof Exp:* From instr to assoc prof, 55-61, chmn dept, 61-64, head dept, 64-75, PROF CERAMIC ENG, IOWA STATE UNIV, 61-, CHMN DEPT MAT SCI & ENG, 75- *Concurrent Pos:* Jr ceramic engr, Ames Lab, Iowa State Univ, 52-55, ceramic engr, 55-57, res assoc, 57-58, engr, 58-61, sr engr, 61-66, div chief, Ceramic & Mech Eng Div, 66-73, sr engr, Ames Lab, Dept Energy, 73-81. *Mem:* Am Soc Eng Educ; fel Am Ceramic Soc; Nat Inst Ceramic Engrs. *Res:* High temperature properties and processing of ceramic materials. *Mailing Add:* 1214 Ridgewood Ames IA 50012

WILDER, HARRY D(OUGLAS), b Westfield, Wis, Aug 22, 32; m 59; c 5. PULP & PAPER SCIENCE, CHEMICAL ENGINEERING. *Educ:* Univ Wis, BS, 55; Inst Paper Chem, MS, 57, PhD, 60. *Prof Exp:* Res aide chem eng, Inst Paper Chem, 59-65; asst dir res & develop, Albemarle Paper Co, Va, 66, dir, 67-68; dir pulp & paper res, Ethyl Corp, 68-76; sr sci assoc, 77-79, CHIEF RES, ASSOC SCOTT PAPER CO, 79- *Mem:* Am Inst Chem Engrs; Tech Asn Pulp & Paper Indust. *Res:* Pulping methods and rates; pulp bleaching; pulping and bleaching chemical generation; pulping research; fiber properties research. *Mailing Add:* Scott Paper Co Scott Plaza Philadelphia PA 19113

WILDER, JAMES ANDREW, JR, b Washingtin, DC, Dec 19, 50; m 73. MATERIALS SCIENCE, GLASS SCIENCE. *Educ:* Cath Univ Am, BSE, 73, MSE, 75, Phd(mat sci), 78. *Prof Exp:* MEM STAFF GLASS CERAMIC RES, SANDIA LABS, 78- *Mem:* Am Ceramic Soc; Inst Elec & Electronics Engrs. *Res:* Phase separation and crystallization in glass; glass-to-metal sealing. *Mailing Add:* Sandia Labs Div 2565 PO Box 5800 Albuquerque NM 87185

WILDER, JOSEPH R, b Baltimore, Md, Oct 5, 20; c 5. SURGERY. *Educ:* Dartmouth Col, BS, 42; Columbia Univ, MD, 45. *Prof Exp:* Chief & dir, Surg Serv, Wright Patterson Hosp, 52-54; fel cardiovasc res, Karolinska Inst, Sweden, 54-55; asst prof surg, NY Med Col, 55-58; dir gen surg, Hosp Joint Dis & Med Ctr, 59-80; PROF SURG, MT SINAI SCH MED, 67- *Concurrent Pos:* Med adv, NY State Legis, 65-75; examr, Am Bd Surg, 70-75; consult, US Off Econ Opportunity, 75-80. *Mem:* Am Bd Surg; Fel Am Col Surgeons; AMA. *Res:* Selective surgical intervention in management of penetrating wounds of the abdomen. *Mailing Add:* One Gustave L Levy Pl New York NY 10029

WILDER, MARTIN STUART, b Brooklyn, NY, May 20, 37; m 64; c 1. MICROBIOLOGY. *Educ:* Brooklyn Col, BS, 60; Univ Kans, MA, 63, PhD(microbiol), 66. *Prof Exp:* Res aide microbiol, Sloan-Kettering Inst Cancer Res, 60-61; Nat Acad Sci-Nat Res Coun res assoc, Microbiol Div, US Dept Army, 66-68; asst prof microbiol, 68-74, ASSOC PROF MICROBIOL, UNIV MASS, AMHERST, 74- *Concurrent Pos:* Nat Acad Sci grant, 74. *Mem:* Am Soc Microbiol; Reticuloendothelial Soc; NY Acad Sci. *Res:* Pathogenesis and pathology of infectious diseases; interactions of platelets, bacteria and leukocytes. *Mailing Add:* Dept Microbiol Univ Mass Amherst MA 01003

WILDER, PELHAM, JR, b Americus, Ga, July 20, 20; m 45; c 3. ORGANIC CHEMISTRY. *Educ:* Emory Univ, AB, 42, MA, 43; Harvard Univ, MA, 47, PhD(org chem), 50. *Prof Exp:* From instr to prof chem, 49-68, prof chem & pharmacol, 68-87, UNIV DISTINGUISHED SERV PROF CHEM, DUKE UNIV, 87- *Concurrent Pos:* Consult, NSF, 60-68, E I du Pont de Nemours & Co, Inc, 66-69 & Res Triangle Inst, 68-; Gov Sci Adv Comt, 62-64; mem advan placement chem comt, Col Entrance Exam Bd, 68-74, chmn, 69-74, mem advan placement standing comt, 69-72; assoc, Comt Prof Training, Am Chem Soc. *Mem:* Am Chem Soc; Sigma Xi. *Res:* Stereochemical studies; kinetic, thermodynamic control and mechanism of organic reactions; quantitative structure-activity relationship studies in pharmacology. *Mailing Add:* 2514 Wrightwood Ave Durham NC 27705

WILDER, RONALD LYNN, b Long Beach, Calif, Feb 10, 47; m 69; c 2. ANIMAL MODELS OF ARTHRITIS, RHEUMATOID ARTHRITIS. *Educ:* Univ Calif, Los Angeles, BS, 69, MD & PhD(molecular biol), 74. *Prof Exp:* Intern med, Univ Calif, San Diego, 74-75, resident, 75-76; res assoc immunol, Lab Immunol, Nat Inst Allergy & Infectious Dis, 76-79, clin assoc, 79-81, SR INVESTR RHEUMATOLOGY, ARTHRITIS BR, NAT INST ARTHRITIS, MUSCULOSKELETAL & SKIN DIS, NIH, 81- *Mem:* Am Asn Immunologists; Am Col Physicians; Am Col Rheumatology; Sigma Xi; Clin Immunol Soc. *Res:* Etiology and pathogenesis of chronic erosive forms of arthritis; animal models of arthritis. *Mailing Add:* Arthritis & Rheumatism Br Nat Inst Arthritis Musculoskeletal & Skin Dis NIH Bldg 10 Rm 9N228 Bethesda MD 20892

WILDER, VIOLET MYRTLE, b Granville, Iowa, Apr 8, 08. BIOCHEMISTRY. *Educ:* Univ Nebr, BA, 28, MA, 34, PhD(biochem), 38. *Prof Exp:* Res chemist, Dr G A Young, Omaha, 38-40; instr biochem, Univ Ark, 40; dir, Lab Maternal & Child Health, Univ Nebr, 40-43 & 46-47; instr physiol chem, Woman's Med Col Pa, 43-46; from asst prof to assoc prof, 47-73, ASSOC PROF BIOCHEM, COL MED, UNIV NEBR MED CTR, OMAHA, 73- *Mem:* AAAS; Am Chem Soc; fel Am Asn Clin Chem. *Res:* Enzyme-hormone relationships; clinical chemistry. *Mailing Add:* 1229 SW Dyer Point Rd Palm City FL 33490

WILDES, PETER DRURY, photochemistry, solar energy, for more information see previous edition

WILDEY, ROBERT LEROY, b Los Angeles, Calif, Aug 22, 34; m 59; c 3. ASTRONOMY, ASTROPHYSICS. *Educ:* Calif Inst Technol, BS, 57, MS, 58, PhD(astron), 62. *Prof Exp:* Res engr, Jet Propulsion Lab, Calif Inst Technol, 59-60, res fel & lectr astron & geol, Mt Wilson & Palomar Observ & div geol sci, Calif Inst Technol, 62-65; from assoc prof to prof astophys & astron, 72-80, PROF MATH PHYSICS & ASTRON, NORTHERN ARIZ UNIV, 82-; ASTRONR & ASTROPHYSICIST, BR ASTROGEOL, US GEOL SERV, 65- *Concurrent Pos:* Consult, United Electrodynamics Corp, 62-63, Aeronutronics Div, Ford Motor Co, 63-64, World Book Encycl Sci Serv, 63-64 & US Geol Surv, 72-; vis prof, Univ Calif, Berkeley, 66; mem planetary astron panel, Space Sci Bd, Nat Acad Sci, 67- *Honors & Awards:* Cert of Appreciation, NASA Apollo Prog, 69- *Mem:* Am Astron Soc; Am Geophys Union; fel Geol Soc Am; Int Astron Union; fel Royal Astron Soc; fel Explorers Club. *Res:* Observational approach to stellar and galactic evolution; co-pioneer (with B C Murray and J A Westphal) of cryogenic far-infrared, infrared studies substantiating Jupiter as a star, discovered hot satellite shadow phenomenon, first detection of far-infrared radiation from a star; Apollo and Mariner-Mars experimenter; gravitation and cosmology; photoclinometry; automated digital photogrammetry and photoclinometry; radiative transfer theory; synthetic aperture radar signal processing; invention of radarclinometry. *Mailing Add:* Observ Astron & Astrophys Northern Ariz Univ Box 6010 Flagstaff AZ 86011

WILDFEUER, MARVIN EMANUEL, b Bronx, NY, Apr 16, 36; m 67; c 2. FERMENTATION PRODUCTS, ANTIBIOTIC PURIFICATION DEVELOPMENT. *Educ:* Queen's Col, NY, BS, 57; Iowa State Univ, MS, 59; Univ Del, PhD(chem), 63. *Prof Exp:* NIH fel molecular biol, Univ Calif, San Diego, 63-65; res scientist, Sansum Clin & Res Found, 65-67; sr chemist, 68-80, RES SCIENTIST, ELI LILLY & CO, 80- *Mem:* Am Chem Soc. *Res:* Responsible for production scale antibiotic purification technology, especially macrolide, beta-lactam and polyether antibiotics; development of new isolation procedures from fermentation broth; antibiotic derivatization. *Mailing Add:* Dept TL924 Eli Lilly Co Pharmaceut Lafayette IN 47902

WILDI, BERNARD SYLVESTER, b Columbus, Ohio, May 23, 20. ORGANIC CHEMISTRY. *Educ:* Ohio State Univ, BSc, 43, PhD(org chem), 48. *Prof Exp:* Res chemist, Nat Defense Res Comt, Ohio State Univ, 43-44 & Manhattan Proj, Los Alamos Sci Lab, NMex, 44-47; Nat Res Coun fel, Harvard Univ, 48-49; mem fac, Fla State Univ, 49-50; mem staff life sci, Monsanto Co, 50-53, group leader, 53-65, mgr, 65-69, distinguished sci fel, 69-82; CONSULT, 82- *Mem:* AAAS; Am Chem Soc. *Res:* Structure of natural products; chemical spectroscopy; organic synthesis. *Mailing Add:* 1234 Folger Kirkwood MO 63122

WILDIN, MAURICE W(ILBERT), b Hutchinson, Kans, June 24, 35; m 58; c 2. MECHANICAL ENGINEERING, THERMAL STORAGE. *Educ:* Univ Kans, BSME, 58; Purdue Univ, MSME, 59, PhD(mech eng), 63. *Prof Exp:* From asst prof to assoc prof, 61-72, chmn dept, 68-73, PROF MECH ENG, UNIV NMEX, 72- *Concurrent Pos:* Staff mem, Jet Propulsion Lab, Pasadena, 67-68 & Sandia Nat Lab, Albuquerque, 84-85; assoc ed, J Solar Energy Eng, 89-92; consult, T Y Lin, Int Southland Indust Honeywell, Inc. *Mem:* Am Soc Heating Refrig & Air Conditioning Engrs; Am Soc Mech Engrs; Int Solar Energy Soc; Am Solar Energy Soc. *Res:* Thermal storage in stratified water tanks; thermocline formation; building energy use. *Mailing Add:* 720 Montclaire Dr NE Albuquerque NM 87110

WILDING, LAWRENCE PAUL, b Winner, SDak, Oct 1, 34; m 56; c 4. SOIL SCIENCE, AGRONOMY. *Educ:* SDak State Univ, BSc, 56, MSc, 59; Univ Ill, PhD(soils), 62. *Prof Exp:* Asst agron, SDak State Univ, 56-59; Campbell Soup Co fel plant sci, Univ Ill, 59-62; from asst prof to prof agron, Ohio State Univ, 62-76; PROF SOIL SCI, TEX A&M UNIV, 76- *Concurrent Pos:* Fel, Univ Guelph, 72; consult, USAID, El Salvador, Sudan, Niger & Cameroon, 78-85, Rockefeller Found China, 87 & 89. *Honors & Awards:* Res Award, Soil Sci Soc Am, 87. *Mem:* Fel Am Soc Agron; fel Soil Sci Soc Am; Int Soil Sci Soc; Soil Conserv Soc Am. *Res:* Soil classification and genesis among different climatic, chronologic and topographic sequences; origin, depth distributions, properties and radiocarbon age of soil opal phytoliths; statistical variability in soil physical and chemical parameters; clay mineralogy; sediment mineralogy and soil erosion; microfabric and micropedology of soil habitats; international agriculture; land evaluation; environmental quality; wetlands quantification. *Mailing Add:* Dept Soil & Crop Sci Tex A&M Univ College Station TX 77843-2474

WILDMAN, GARY CECIL, b Middlefield, Ohio, Nov 25, 42; m 65; c 2. POLYMER CHEMISTRY. *Educ:* Thiel Col, AB, 64; Duke Univ, PhD(phys chem), 70. *Prof Exp:* Res chemist, Hercules Inc, 68-71; assoc prof polymer sci, 71-78, chmn dept, 71-75, PROF POLYMER SCI, UNIV SOUTHERN MISS, 78-, DEAN COL SCI & TECHNOL, 76- *Mem:* Am Chem Soc; Am Crystallog Asn; Sigma Xi; Fedn Socs Paint Technol; Soc Plastics Engrs. *Res:* Structure-property relationships of synthetic polymers; x-ray diffraction studies of single crystals and polymeric materials; surface coatings. *Mailing Add:* 8857 Aldershot Dr Germantown TN 38138

WILDMAN, GEORGE THOMAS, b Grasmere, NH, Nov 14, 35. COST CONTROL, RESOURCE MANAGEMENT. *Educ:* Univ NH, BS, 57; NY Univ, MS, 62; Mass Inst Technol, ScD(chem eng), 73. *Prof Exp:* Chem engr, Merck & Co, Inc, 57-62, sr chem engr, 62-65, engr assoc, 65-68, res fel, 68-72, sect mgr, chem eng res & develop dept, 72-77, tech serv mgr, 77-80, mfg mgr, 80-84, tech oper mgr, 84-87, TECH OPER DIR, MFG DIV, MERCK & CO, INC, 87- *Concurrent Pos:* Educ counr, Mass Inst Technol, 75-; tech steering comt, Ctr Chem Process Safety, 88- *Mem:* Am Inst Chem Engrs; Am Chem Soc. *Res:* Organic chemical process research and development; heterogeneous catalysis; synthesis of heterocyclic compounds; antibiotic synthesis; natural products isolation and purification; chemical reactivity evaluation; process safety management; regulatory affairs; new manufacturing technology. *Mailing Add:* 2068 Old Raritan Rd Westfield NJ 07090

WILDMAN, PETER JAMES LACEY, b London, Eng, Sept 7, 36; m 63; c 1. SPACE PHYSICS, SENSORS. *Educ:* Univ Durham, Eng, BSc, 58, PhD(physics), 68. *Prof Exp:* Res physicist space physics, Air Force Geophys Lab, 68-79; MEM STAFF, BRIT AEROSPACE, ENG, 79- *Mem:* Inst Physics (UK). *Res:* Instrumentation for electrical and optical measurements from space vehicles; investigation of processes in ionosphere and magnetosphere. *Mailing Add:* 32 Bannetts Tree Crescent Alveston Avon BS12 2LY England

WILDMANN, MANFRED, b Karlsruhe, Ger, Apr 16, 30; US citizen; m 54; c 3. MECHANICAL ENGINEERING, COMPUTER SYSTEMS. *Educ:* City Col NY, BS, 54; Univ Calif, Los Angeles, MS, 57. *Prof Exp:* Res specialist inertial guidance, Autonetics Div, NAm Aviation, Inc, Calif, 54-62; mgr mech sect, Sunnyvale, 62-69, mgr Terabit Memory Systs Dept, 69-73, MGR RES DEPT, AMPEX CORP, REDWOOD CITY, 73- *Mem:* Am Soc Mech Engrs. *Res:* Lubrication, handling and control of flexible media; large computer memory systems; gas lubrication; foil bearing; inertial components. *Mailing Add:* 1860 Camino De Los Robles St Menlo Park CA 94025

WILDNAUER, RICHARD HARRY, b New Kensington, Pa, Feb 14, 40; m 66; c 1. DERMATOLOGY, PHARMACEUTICALS. *Educ:* St Vincent Col, BS, 62; WVa Univ, PhD(biochem), 66; Rider Col, MBA, 74. *Prof Exp:* Fel, Univ Kans, 66-67; sr scientist, Johnson & Johnson Res Labs, 67-73, sr group leader, 73-77; new prod dir, McNeil Pharmaceut, 77-79; dir new prod develop, Janssen Pharmaceut, 79-82, vpres res & develop, 82-88; VPRES TECHNOL & BUS DEVELOP, JOHNSON & JOHNSON, 88- *Mem:* NY Acad Sci; Soc Invest Dermat; Med Mycol Soc Am; Am Acad Dermat; Sigma Xi. *Res:* Skin physiology and biochemistry; membrane transport properties; wound healing; physical polymer characterizations; medical mycology; pharmaceutical new product development; clinical trials. *Mailing Add:* Technol & Bus Develop Johnson & Johnson New Brunswick NJ 08903

WILDS, ALFRED LAWRENCE, b Kansas City, Mo, Mar 1, 15; m 37. ORGANIC CHEMISTRY. *Educ:* Univ Mich, BS, 36, MS, 37, PhD(org chem), 39. *Prof Exp:* Asst chem, Univ Mich, 37-39, DuPont fel, 39-40; from instr to prof 40-85, EMER PROF CHEM, UNIV WIS-MADISON, 85- *Concurrent Pos:* Co-off investr, Nat Defense Res Comt, Univ Wis-Madison, 42-45; Guggenheim fel, 57. *Mem:* AAAS; Am Chem Soc; Royal Soc Chem. *Res:* Organic synthesis; integrated syntheses, stereochemistry of catalysis hydrogenations, metal reductions; synthesis of natural products, steroids, hormone analogs; reactions of diazoketones; nuclear magnetic resonance studies. *Mailing Add:* 302 Robin Pkwy Madison WI 53705

WILDS, PRESTON LEA, b Aiken, SC, Dec 18, 26; m 50, 63; c 4. OBSTETRICS & GYNECOLOGY. *Educ:* Yale Univ, BA, 49; Univ Pa, MD, 53; Am Bd Obstet & Gynec, dipl, 62. *Prof Exp:* Asst obstet & gynec, Sch Med, La State Univ, 54-57; pvt pract, SC, 57-59; clin instr, Med Col Ga, 59, from instr to prof, 59-78; PROF OBSTET & GYNEC, EASTERN VA MED SCH, 78- *Mem:* Fel Am Col Obstet & Gynec; AMA. *Res:* Programmed instruction; fetal physiology. *Mailing Add:* Dept Obstet & Gynec Eastern Va Med Sch 600 Gresham Dr Norfolk VA 23507

WILDT, DAVID EDWIN, b Jacksonville, Ill, Mar 12, 50; m 70. REPRODUCTIVE PHYSIOLOGY. *Educ:* Ill State Univ, BS, 72; Mich State Univ, MS, 73, PhD(animal husb, physiol), 75. *Prof Exp:* Res asst reproductive physiol, Endocrine Res Univ, Mich State Univ, 72-75; fel reproductive physiol, Inst Comp Med, Baylor Col Med, 75-79; fel comp reproductive endocrinol, NIH, 79-; AT NAT ZOOL PARK, SMITHSONIAN INST. *Mem:* Sigma Xi; Am Soc Animal Sci; Soc Study Reproduction; Am Soc Vet Physiol & Pharmacol; Int Primatological Soc. *Res:* Relationships of reproductive behavior; gonadotropin and steroid hormone concentrations and time of ovulation in canine, feline and nonhuman primate species; laparoscopy; development and use of frozen semen techniques for the preservation of captive wild mammal species. *Mailing Add:* Div Res Serv NIH Vet Resources Br Bethesda MD 20205

WILDUNG, RAYMOND EARL, b Van Nuys, Calif, Feb 24, 41; m 61; c 2. GEOCHEMISTRY, ENVIRONMENTAL SCIENCES. *Educ:* Calif State Polytech Col, San Luis Obispo, BS, 62; Univ Wis-Madison, MS, 64, PhD(soil sci), 66. *Prof Exp:* NIH fel, Univ Wis-Madison, 66-67; sr res scientist, 67-71, prog leader, mgr environ chem sec, 75-85, assoc mgr environ, 85-86, MGR ENVIRON SCI DEPT, BATTELLE PAC NORTHWEST LABS, 86- *Concurrent Pos:* Grants, USDA, 68-70, Environ Protection Agency, 68-71, US Dept Energy, 68-; Nat Inst Environ Health Sci, 71-82; affil prof, Wash State Univ & Calif State Univ; comt accessory elements, chmn, oil shale panel, comt, soil as mineral resource, Nat Acad Sci; mem exec comt coord solid waste mgt, US Dept Energy; Nuclear Regulatory Comn, 84-87. *Honors & Awards:* E O Lawerence Award. *Mem:* AAAS; Am Chem Soc; Am Soc Agron; Int Soc Soil Sci; Soil Sci Soc Am; Soc Environ Geochem & Health. *Res:* Soil-sediment science; over 250 publications on developing a fundamental understanding of geochemical and metabolic processes controlling pollutant behavior as a basis for assessing environmental impacts of energy development, focus on intergrated effects; fate and behavior of pesticides, residuals, trace metals and metal complex behaviour in soils plants radionuclides, subsurface environment and waters. *Mailing Add:* 1629 George Washington Richland WA 99352

WILE, HOWARD P, b New York, NY, Jan 4, 11; m 35; c 2. RESEARCH ADMINISTRATION. *Educ:* Dartmouth Col, AB, 32; Harvard Univ Grad Bus Sch, cert, 44. *Prof Exp:* Mem staff admin, Mass Inst Technol, 44-46; adminr res, Polytech Inst Brooklyn, 46-65; exec dir comt govt rels, Nat Asn Col & Univ Bus Officers, 65-76; RETIRED. *Concurrent Pos:* Consult, Am Coun Educ; chmn grant admin adv comt, Dept Health, Educ & Welfare, 67-73; mem res & develop study group, Comn Govt Procurement, 71-72; consult, Com Fed Paperwork, 76-77 & NSF, 77-79; instr, Surv Course Grants Admin, HEW, 77-79 & Dept Energy, 78-79; prof adv, Nat Coun Univ Res Adminr, 78-85. *Res:* Governmental relations. *Mailing Add:* 2515 Q St NW Washington DC 20007

WILEMSKI, GERALD, b Dunkirk, NY, Oct 15, 46; div; c 1. PHYSICAL CHEMISTRY, STATISTICAL MECHANICS. *Educ:* Canisius Col, BS, 68; Yale Univ, PhD(chem), 72. *Prof Exp:* Res assoc, Dept Eng & Appl Sci, Yale Univ, 72-74; res assoc & vis asst prof chem, Dartmouth Col, 74-77; prin scientist, 77-85, PRIN RES SCIENTIST, PHYS SCI INC, 85- *Concurrent Pos:* Vis prof, Inst CNR TAE, Messina, Italy, 86, 87, 88, 89 & 90. *Mem:* Am Phys Soc; Electrochem Soc; AAAS; Sigma Xi; Am Chem Soc; NY Acad Sci. *Res:* Statistical mechanics; thermodynamics; electrochemical systems; polymer solutions; nucleation phenomena, aerosols, colloidal suspensions. *Mailing Add:* Phys Sci Inc 20 New Eng Bus Ctr Andover MA 01810-0805

WILEN, SAMUEL HENRY, b Brussels, Belg, Mar 6, 31; nat US; m 60; c 2. ORGANIC CHEMISTRY. *Educ:* City Col New York, BS, 51; Univ Kans, PhD(chem), 56. *Prof Exp:* Asst instr chem, Univ Kans, 51-52, asst, 53-55; res assoc, Univ Notre Dame, 55-57; from instr to assoc prof, 57-71, chmn dept, 84-87, PROF CHEM, CITY COL NEW YORK, 71- *Concurrent Pos:* Sci assoc, State Univ Groningen, 68-69; guest researcher, Free Univ Brussels, 75-76; vis prof, Univ NC, Chapel Hill, 84 & 90. *Mem:* AAAS; Am Chem Soc; Royal Soc Chem. *Res:* Chemistry of heterocyclic compounds; resolving agents and optical resolutions; stereochemistry. *Mailing Add:* Dept Chem City Col New York New York NY 10031

WILENSKY, JACOB T, b New Orleans, La, Aug 16, 42; m 76; c 3. GLAUCOMA, OPHTHALMOLOGY. *Educ:* Tulane Univ, BA, 64, MD, 68. *Prof Exp:* Intern med, Mt Sinai Hosp NY, 68-69; res ophthal, Tulane Univ, Affil Hosp, 69-72,; fel, glaucoma, Wash Univ, Sch Med, 72-73; from asst to assoc prof ophthal, 74-86, PROF OPHTHAL, UNIV ILL, COL MED, 86- *Concurrent Pos:* Dir, Glaucoma Serv, Univ Ill, Eye & Ear Infirmary, 77-; vis prof, Hadassah Univ Hosp, Jerusalem, 79; prin investr, Natural Hist Angle-Closure Glaucoma Suspects Study, 80-86. *Mem:* Am Acad Ophthal; Am Glaucoma Soc; Asn Res Vision & Ophthal. *Res:* Diagnosis and therapy of glaucoma with an emphasis on laser techniques and the testing of new drugs. *Mailing Add:* 1855 W Taylor Chicago IL 60612

WILENSKY, ROBERT, b Brooklyn, NY, Mar 26, 51. COMPUTER SCIENCE. *Educ:* Yale Univ, BA, 72, PhD(comput sci), 78. *Prof Exp:* ASSOC PROF COMPUT SCI, UNIV CALIF, BERKELEY, 78- *Mem:* Asn Comput Mach. *Res:* Artificial intelligence and the simulation behavior; natural language process; computer situation of human thought processes. *Mailing Add:* Dept Comput Sci Univ Calif Berkeley CA 94720

WILENSKY, SAMUEL, b Savannah, Ga, July 9, 37; m 70; c 2. DATA CONVERSION DESIGN, RADIATION EFFECTS ON ELECTRONICS. *Educ:* Mass Inst Technol, BS, 59, PhD(nuclear eng), 64. *Prof Exp:* Asst prof nuclear eng, Mass Inst Technol, 64-65; res assoc, nuclear eng, Mass Gen Hosp, 65-68; DIR ENG, HYBRID SYSTS, 68- *Concurrent Pos:* Asst prof, Harvard Med Sch, 65-68. *Res:* Design and development of data conversion products. *Mailing Add:* 22 Linell Circle Billerica MA 01821

WILES, DAVID M, b Springhill, NS, Dec 28, 32; m 57; c 2. POLYMER CHEMISTRY. *Educ:* McMaster Univ, BSc, 54, MSc, 55; McGill Univ, PhD(phys chem), 57. *Prof Exp:* Nat Res Coun Can & Can Ramsay Mem fels, Univ Leeds, 57-59; asst res officer, High Polymer Sect, 59-61, assoc res officer, 61-67, sr res officer, 67-74, head textile chem sect, 66-86, DIR DIV CHEM, NAT RES COUN CAN, 75- *Honors & Awards:* Dunlop Lectr Award, Chem Inst Can, 81; Textile Sci Award, Textile Tech Fedn Can, 80. *Mem:* Am Chem Soc; Fiber Soc; Chem Inst Can; Can Inst Textile Sci; Royal Soc Chem; fel Royal Soc Can. *Res:* Polymerization kinetics and mechanisms; synthesis of stereoregular polymers; polymer structure; photodegradation of fiber forming macromolecules; polymer stabilization; fiber morphology; modification of fibers; polymer surface studies; microbiological deterioration; composites; high temperature thermoplastics. *Mailing Add:* Div Chem Nat Res Coun Can Ottawa ON K1A 0R9 Can

WILES, DONALD ROY, b Truro, NS, Aug 30, 25; m 52; c 3. INORGANIC CHEMISTRY. *Educ:* Mt Allison Univ, BSc, 46, BEd, 47; McMaster Univ, MSc, 50; Mass Inst Technol, PhD(chem), 53. *Prof Exp:* Chemist, Eldorado Mining & Refining Ltd, Can, 47-48; res assoc radiochem, Chem Inst, Oslo, Norway, 53-55; res assoc metall chem, Univ BC, 55-59; from asst prof to assoc prof nuclear inorg chem, Carleton Univ, 59-69, prof chem, 69-90; CONSULT, RADIOACTIVE WASTE DISPOSAL, 90- *Concurrent Pos:* Vis scientist, Inst Hot Atom Chem, Nuclear Res Ctr, Karlsruhe, Ger, 69-70; chmn, chem dept, Carleton Univ, 79-87. *Mem:* Am Chem Soc; fel Chem Inst Can; Royal Soc Chem; Norweg Chem Soc. *Res:* Dissolution kinetics of metals and oxides; hot atom chemistry in organic solids; nuclear fission; radiochemistry; environmental radiochemistry of radium, thorium; Mössbauer spectroscopy; analytical radiochemistry of radium and thorium; radioanalytical chemistry of radium and thorium, especially in environmental soil, mine waters and water. *Mailing Add:* Dept Chem Carleton Univ Ottawa ON K1S 5B6 Can

WILES, MICHAEL, b Sheffield, Eng, May 8, 40; m 63; c 2. PARASITOLOGY, FRESHWATER ECOLOGY. *Educ:* Univ Leeds, BSc, 62, PhD(zool), 65. *Prof Exp:* Res scientist, Fisheries Res Bd, Can, 65-67; asst prof freshwater ecol, 67-72, assoc prof, 72-80, PROF FRESHWATER ECOL, ST MARY'S UNIV, NS, 80- *Mem:* Can Soc Zool; Brit Soc Parasitol. *Res:* parasites of freshwater and marine fishes of Eastern Canada; diseases and parasites of reef fishes and turkles in Bermuda; marine ecology. *Mailing Add:* Dept Biol St Mary's Univ Robie St Halifax NS B3H 3C3 Can

WILES, ROBERT ALLAN, b Quincy, Mass, Apr 6, 29; m 51; c 5. INDUSTRIAL ORGANIC CHEMISTRY. *Educ:* Univ NH, BS, 51, MS, 55; Mass Inst Technol, PhD(chem), 58. *Prof Exp:* Res chemist, Sun Oil Co, 57-59; res chemist, Solvay Process Div, Allied Chem Co, Allied Corp, 59-64, res supvr, 64-70, res supvr indust chem div, 66-68, mgr process res, 71-80, mgr specialty chem div, 60-80, sr res assoc, 81-82; RETIRED. *Mem:* Am Chem Soc. *Res:* fluorochemicals and process research and development. *Mailing Add:* 151 Deer Lake Circle Ormond Beach FL 32174-4275

WILETS, LAWRENCE, b Oconomowoc, Wis, Jan 4, 27; m 76; c 3. THEORETICAL PHYSICS, NUCLEAR PHYSICS. *Educ:* Univ Wis, BS, 48; Princeton Univ, MA, 50, PhD(physics), 52. *Prof Exp:* Res assoc theoret physics, Proj Matterhorn, Princeton Univ, 51-53; res assoc, Lawrence Livermore Lab, Univ Calif, 53; NSF fel, Inst Theoret Physics, Copenhagen, 53-55; mem staff, Los Alamos Sci Lab, 55-58; mem, Inst Advan Study, Princeton Univ, 57-58; assoc prof, 58-62, PROF THEORET PHYSICS, UNIV WASH, 62- *Concurrent Pos:* Consult, Los Alamos Sci Lab, 58-, Lawrence Livermore Lab, Univ Calif, 58- & Oak Ridge Nat Lab, 65-; NSF sr fel, Weizmann Inst Sci, 61-62; vis prof, Princeton Univ, 69 & Calif Inst Technol, 71; J S Guggenheim fel, Univ Lund & Weizmann Inst Sci, 76-77; Nordita prof, Univ Lund, 76; Alexander von Humboldt sr scientist award; sabbatical vis, Lawrence Berkeley Lab & Stanford Linear Acc, 87-88; Sir Thomas Lyle Res Fel, Univ Melbourne, 89. *Mem:* Fel Am Phys Soc; fel AAAS; Am Asn Univ Prof; Fedn Am Scientists. *Res:* Atomic and nuclear structure and reactions; nuclear substructure, especially the role of quarks and quantum chromodynamics in nuclear physics; meson physics; manybody theory; heavy ions and fission; soliton bag model. *Mailing Add:* Dept Physics FM-15 Univ Wash Seattle WA 98195

WILEY, ALBERT LEE, JR, b Forest City, NC, June 9, 36; m 60; c 4. RADIATION MEDICINE. *Educ:* NC State Univ, BN, 58; Univ Rochester, MD, 63; Univ Wis-Madison, PhD(radiobiol & nuclear eng), 72; Am Bd Radiol, cert, 68; Am Bd Nuclear Med, cert, 75; Am Bd Sci Nuclear Med, cert, 80. *Prof Exp:* Nuclear engr, Nuclear Prod Div, Lockheed Aircraft, Ga, 58; intern med & surg, Med Ctr, Univ Va, 63-64; Nat Cancer Inst fel radiation ther, Med Ctr, Stanford Univ, 64-65, radiation ther & nuclear med, Univ Wis Hosps, 65-68; med dir, US Naval Radiol Defense Lab, San Francisco, Calif, 68-70; asst prof radiation therapy, Univ Tex, M D Anderson Hosp, Houston, 72-73; assoc prof radiol & human oncol, Med Sch, Univ Wis-Madison, 76-79; assoc dir & clin dir radiation oncol & prof human oncol, radiol & med physics, Med Sch, Univ Wis-Madison, 79-88; PROF & CHMN, DEPT RADIATION ONCOL & DIR, CANCER CTR, EASTERN CAROLINA UNIV SCH MED, GREENVILLE, 88- *Concurrent Pos:* Vis prof, Cent Hosp & Radiation Clins, Univ Helsinki, 79 & Univ Linköping, Sweden, 86; consult, Nat Cancer Inst, 81 & Adv Comt Reactor Safeguards, Nuclear Regulatory Comn, 81-; tech adv, Dept Health & Human Serv, State Wis, 81-; mem, bd dirs, Am Bd Sci Nuclear Med, 87-, Wis Gov Biotechnol Bd, Wis Radioactive Waste Bd & US Dept Vet Affairs Oncol Bd, 90- *Mem:* Soc Nuclear Med; Health Physics Soc; Inst Elec & Electronics Engrs; Am Soc Law & Med; fel Am Col Prev Med; Am Soc Radiation Oncologists. *Res:* New radiation techniques in the treatment of cancer of pancreas and biliary tract; the use of nuclear medicine, nuclear magnetic resonance and computerized tomography for improving the quality control and optimization of cancer treatment; use of nuclear reactors in medical research and health physics of radiation accidents; biological dosimetry. *Mailing Add:* Dept Radiation Oncol & Jenkins Cancer Ctr ECarolina Univ Med Sch Greenville NC 27858-4354

WILEY, BILL BEAUFORD, b St Joseph, Mo, Nov 12, 23; m 45; c 2. MICROBIOLOGY, IMMUNOLOGY. *Educ:* Univ Kans, BA, 49, MA, 50; Univ Rochester, PhD(microbiol), 56. *Prof Exp:* Asst dir Rochester Health Bur Labs, Med Ctr, Univ Rochester, 50-56; asst prof microbiol, Univ Sask, 56-62; from asst prof, to assoc prof, 62-78, PROF MICROBIOL, MED CTR, UNIV UTAH, 78- *Concurrent Pos:* Med Res Coun Can fel, Univ Sask, 57-62; Nat Inst Allergy & Infectious Dis fel, Univ Utah, 63- *Mem:* AAAS; Can Soc Microbiol; Am Soc Microbiol; NY Acad Sci; Sigma Xi. *Res:* Encapsulation and virulence of Staphylococcus aureus staphylococcal scalded skin syndrome; sphingomyelinases of staphylococcal toxins. *Mailing Add:* 4041 Lisa Dr Salt Lake City UT 84124

WILEY, DON CRAIG, b Akron, Ohio, Oct 21, 44. BIOPHYSICS. *Educ:* Tufts Univ, SB, 66; Harvard Univ, PhD(biophys), 71. *Prof Exp:* Asst prof biochem & molecular biol, 71-75, assoc prof biochem, 75-80, PROF BIOCHEM & BIOPHYS, HARVARD UNIV, 80-, CHMN BIOPHYS, 81- *Concurrent Pos:* Jane Sloan Coffin Fund grant, Harvard Univ, 72-73; fel, Europ Molecular Biol, 76. *Mem:* Nat Acad Sci; AAAS. *Res:* X-ray diffraction; structure of macromolecules and assemblies of macromolecules; viral membrane glycoproteins. *Mailing Add:* Dept Biochem Harvard Univ Seven Divinity Ave Cambridge MA 02138

WILEY, DOUGLAS WALKER, b Shanghai, China, Apr 8, 29; m 50; c 4. ORGANIC CHEMISTRY, SYNTHETIC ORGANIC CHEMISTRY. *Educ:* Univ Richmond, BS, 49; Columbia Univ, MA, 52; Yale Univ, PhD(chem), 55. *Prof Exp:* RES CHEMIST, CENT RES DEPT, E I DU PONT DE NEMOURS & CO, INC, 55- *Concurrent Pos:* Fel, Hickerell Res Found, 53-54 & Harvard Univ, 54-55. *Mem:* Am Chem Soc; AAAS. *Res:* Synthesis, structure and mechanism in area of fluorocarbons, cyanocarbons, exploratory process chemistry, photochemical processes, organoferromagnets, and solar energy. *Mailing Add:* Cent Res Dept Exp Sta E I du Pont de Nemours & Co Inc Wilmington DE 19898

WILEY, E(DWARD) O(RLANDO), III, b Corpus Christi, Tex, Aug 15, 44; m 77; c 2. ICHTHYOLOGY, SYSTEMATIC BIOLOGY. *Educ:* Southwest Tex State Univ, BS, 66; Sam Houston State Univ, MS, 72; City Univ New York, PhD(biol), 76. *Prof Exp:* From asst cur to cur fishes, 76-88, CUR, MUS NATURAL HIST, UNIV KANS, 88-, PROF BIOL SCI, 88- *Concurrent Pos:* Assoc prof biol sci, Mus Natural Hist, Univ Kans, 78-88; res assoc, US Mus Natural Hist, 80- *Honors & Awards:* Stoye Award Ichthyol, Am Soc Ichthyologists & Herpetologists, 76. *Mem:* Am Soc Ichthyologists &

Herpetologists; Soc Syst Zool. *Res:* Phylogenetic relationships of fishes; theory and practice of phylogenetic systematics; evolutionary theory and its relationship to systematics and biogeography. *Mailing Add:* Mus Natural Hist Univ Kans Lawrence KS 66045

WILEY, JACK CLEVELAND, b Evansville, Ind, Mar 17, 40; m 65; c 2. COMPUTER-AIDED ENGINEERING. *Educ:* Purdue Univ, Lafayette, BS, 62, PhD(eng sci), 68; Univ Ill, Urbana, MS, 64. *Prof Exp:* Asst prof theoret & appl mech, Univ Ill, Urbana, 67-72; sr res engr, Tech Ctr, Deere & Co, 72-79, MGR ENG ANALYSIS, TECH CTR, DEERE & CO, 80- *Concurrent Pos:* Presidential exchange exec, US Dept Com, 79-80. *Mem:* Am Soc Mech Engrs; Soc Mfg Engrs; Am Acad Mech. *Res:* Mechanical system analysis; mechanism and vehicle mechanics; mechanical computer aided engineering; computer graphics and geometric modeling; computer aided design and manufacturing. *Mailing Add:* Deere & Co Tech Ctr 3300 River Dr Moline IL 61265

WILEY, JAMES C, JR, b Higginsville, Mo, Mar 4, 38; m 62; c 2. RADIO-CHEMICAL SYNTHESIS. *Educ:* Univ Mo, Kansas City, BS, 68. *Prof Exp:* Technician, 62-65, jr chemist, 65-68, asst chemist, 68-71, assoc chemist, 71-76, sr chemist, 76-81, prin chemist, 81-85, GROUP LEADER, 85-, TECH OPERS MGR, CHEMSYN SCI LAB, MIDWEST RES INST, 85- *Concurrent Pos:* Prin investr, Nat Cancer Inst, 76- *Mem:* Am Chem Soc; AAAS; Sigma Xi. *Res:* Synthesis of labeled and unlabeled carcinogenic polycyclic aromatic hydrocarbon metabolites. *Mailing Add:* Chemsyn Sci Labs 13605 W 96th Terr Lenexa KS 66215-1253

WILEY, JOHN DUNCAN, b Nashville, Tenn, Mar 23, 42. SOLID STATE PHYSICS. *Educ:* Ind Univ, BS, 64; Univ Wis, MS, 65, PhD(physics), 68. *Prof Exp:* Mem tech staff, Optical & Magnetic Mat Dept, Bell Tel Labs, 68-74; res assoc, Max Planck Inst Solid State Res, 74-75; PROF ELEC & COMPUT ENG, UNIV WIS-MADISON, 76-, CHMN, MAT SCI PROG, 82- *Concurrent Pos:* Fel, NSF, 64-66. *Honors & Awards:* Alexander Von Humboldt Award, 74 & 75. *Mem:* Am Vacuum Soc; Am Phys Soc; Inst Elec & Electronics Engrs. *Res:* Transport properties of semiconductors; optical properties of semiconductors; growth and characterization of semiconductor crystals. *Mailing Add:* 415 Eng Res Bldg 1500 Johnson Dr Madison WI 53706

WILEY, JOHN ROBERT, b San Angelo, Tex, Oct 10, 46; m 68; c 2. NUCLEAR CHEMISTRY. *Educ:* Univ Houston, BS, 69; Purdue Univ, PhD(nuclear chem), 74. *Prof Exp:* Res Chem separations, Savannah River Plant, E I du Pont de Nemours & Co, Inc, 74-89; ASSOC PROF CHEM, PERMIAN BASIN, UNIV TEX, 89- *Concurrent Pos:* Int Atomic Energy Agency, Vienna, 85-87. *Mem:* Am Chem Soc; AAAS; Sigma Xi. *Res:* Management of nuclear plant wastes; toxic chemical waste; gas phase negative ions. *Mailing Add:* Univ Tex Permian Basin Odessa TX 79763-0001

WILEY, JOHN W, b Portland, Ore, Aug 26, 49. MEDICINE. *Educ:* Claremont-McKenna Col, BA, 71; Ore Health Sci Univ, MD, 80. *Prof Exp:* ASST PROF INTERNAL MED & GASTROENTEROL, UNIV MICH, 87- *Mem:* Am Gastroenterol Soc; Am Col Physicians; Am Fedn Clin Res; Am Motility Soc; Cent Soc Clin Res. *Res:* Neurohormonal regulation of neural transmission; motility problems of the gastrointestinal tract. *Mailing Add:* Univ Mich Med Ctr 3912 Taubman Ann Arbor MI 48109

WILEY, LORRAINE, b Sacramento, Calif. PLANT PHYSIOLOGY. *Educ:* Sacramento State Col, AB, 64; Univ Calif, Davis, MS, 66, PhD(plant physiol), 71. *Prof Exp:* Asst prof bot, Howard Univ, 71-72; asst prof, 72-76, assoc prof, 76-81, PROF BIOL, CALIF STATE UNIV, FRESNO, 81- *Concurrent Pos:* Res botanist, Univ Calif, Davis, 71. *Mem:* Am Soc Plant Physiol. *Res:* Plant protein metabolism; seed physiology; stress physiology. *Mailing Add:* Dept Biol Calif State Univ Fresno CA 93740

WILEY, LYNN M, b Tucson, Ariz, Feb 24, 47; div. DEVELOPMENTAL BIOLOGY. *Educ:* Univ Calif, Irvine, BS, 68, MS, 71; Univ Calif, San Francisco, PhD(anat), 75. *Prof Exp:* Cell biologist mammalian develop, San Francisco Med Ctr, Univ Calif, 75-78; mem fac, Univ Va, 78-80; MEM FAC, UNIV CALIF, DAVIS, 80- *Concurrent Pos:* Res Career Develop Award, NIH, 80-84. *Mem:* Am Soc Gravitational & Space Biol; Soc Develop Biol; Am Soc Cell Biol. *Res:* Cell surface in early mammalian development; origin of primary germ layers; regulation of cell determination; cell polarity; effect of ionizing radiation on gametes. *Mailing Add:* Dept Obstet & Gynec Calif Primate Res Ctr Univ Calif Davis CA 95617-8542

WILEY, MARILYN E, DIABETES, GERIATRICS. *Educ:* Colo State Univ, PhD(nutrit), 82. *Prof Exp:* ASST PROF NUTRIT, HEALTH SCI CTR, UNIV TEX, 82- *Mailing Add:* Prog Nutrit & Dietetics PO Box 20708 Health Sci Ctr Univ Tex Houston TX 77225

WILEY, MARTIN LEE, b Pittsburg, Kans, May 12, 35; m 63; c 3. ICHTHYOLOGY. *Educ:* Pittsburg State Univ, BS, 59, MS, 60; George Washington Univ, PhD(zool), 69. *Prof Exp:* Sci asst fisheries biol, Inter-Am Trop Tuna Comn, 61-62; mus technician ichthyol, Div Fishes, US Nat Mus, 66-67; fisheries res biologist, Bur Com Fisheries Ichthyol Lab, 67-68; managing ed, Chesapeake Sci, 68-77, ASST PROF ICHTHYOL, CHESAPEAKE BIOL LAB, 68- *Concurrent Pos:* Co-investr res contract, Naval Ord Lab, 73-74 & 75-76; ed proc, Estuarine Res Fedn, 75-78; managing ed, Estuaries, Estuarine Res Fedn, 78-80. *Mem:* Am Soc Ichthyologists & Herpetologists; Atlantic Estuarine Res Soc; Estuarine Res Fedn. *Res:* Structure and physiology of fish swimbladders; how fish adapt to changes in hydrostatic pressure and effects of underwater explosions; biology of fishes. *Mailing Add:* Chesapeake Biol Lab Box 38 Solomons MD 20688

WILEY, MICHAEL DAVID, b Long Beach, Calif, Nov 28, 39; m 86; c 2. ORGANIC CHEMISTRY. *Educ:* Univ Southern Calif, BS, 61; Univ Wash, PhD(org chem), 69. *Prof Exp:* From asst prof to assoc prof, 68-84, PROF & CHEM, CALIF LUTHERAN UNIV, 84- *Concurrent Pos:* SRC res assoc, Univ Liverpool, Eng, 81. *Mem:* AAAS; Am Chem Soc; Royal Soc Chem. *Res:* Reaction mechanisms; carbcations. *Mailing Add:* Dept Chem Calif Lutheran Univ Thousand Oaks CA 91360

WILEY, PAUL FEARS, organic chemistry; deceased, see previous edition for last biography

WILEY, RICHARD G, b Windridge, Pa, Aug 25, 37; m 60; c 6. ELECTRONIC INTELLIGENCE, RADARS. *Educ:* Carnegie Mellon Univ, BSEE, 59, MSEE, 60, Syracuse Univ, PhD(elec engr), 75. *Prof Exp:* Staff consult, Syracuse Res Corp, Syracuse, NY; FOUNDER & VPRES, RES ASN SYRACUSE, 86- *Mem:* Fel Inst Elec & Electronics Engrs. *Res:* Radar and electronic intelligence. *Mailing Add:* Res Asn Syracuse 510 Stewart Dr North Syracuse NY 13212

WILEY, RICHARD HAVEN, b Mattoon, Ill, May 10, 13; m 40; c 2. CHEMISTRY. *Educ:* Univ Ill, AB, 34, MS, 35; Univ Wis, PhD(chem), 37; Temple Univ, LLB, 43. *Prof Exp:* Res chemist, E I du Pont de Nemours & Co, Inc, 37-45; assoc prof chem, Univ NC, 45-49; prof & chmn dept, Univ Louisville, 49-65; prof chem, 65-78, exec officer doctoral prog, 65-68, EMER PROF CHEM, HUNTER COL, 78- *Concurrent Pos:* NSF sr fel, Imp Col, Univ London, 57-58; vis prof, City Col New York, 63-64; consult; res proj dir with NSF, AEC, NIH, Off Naval Res, Off Ord Res, Bur Naval Ord & NASA; vis scholar, Stanford Univ, 78-79; vol employee, San Jose State Univ, 80-85. *Honors & Awards:* Award, Am Chem Soc, 65. *Mem:* AAAS; Am Chem Soc. *Res:* Polymer chemistry; organic synthesis; mass spectrometry; over 475 publications in scientific journals; patents in field; editor treatises in field. *Mailing Add:* Eight Roosevelt Circle Palo Alto CA 94306

WILEY, RICHARD HAVEN, JR, b Wilmington, Del, June 14, 43; m 71; c 2. SOCIOBIOLOGY, ETHOLOGY. *Educ:* Harvard Univ, BA, 65; Rockefeller Univ, PhD(animal behav), 70. *Prof Exp:* Fel animal behav, Rockefeller Univ, 70-71; asst prof, 71-76, assoc prof, 76-81, PROF BIOL, UNIV NC, CHAPEL HILL, 81- *Concurrent Pos:* Res grants, NIMH & NSF, 69-83; mem bd dir, NC Bot Garden & Orgn Trop Studies, 78-; mem Pop Biol Adv Panel, NSF, 80-83; mem, Scholarly Studies Rev Panel, Smithsonian Inst, 87-88. *Mem:* Animal Behav Soc; Int Soc Behav Ecol; Soc Study Evolution; Am Soc Naturalists; Am Ornithologists Union. *Res:* Evolution and ecology of vertebrate social organization; behavioral mechanisms of aggression and affiliation; acoustic communication by birds and primates. *Mailing Add:* Dept Biol Univ NC Chapel Hill NC 27599-3280

WILEY, ROBERT A, b Ann Arbor, Mich, Sept 5, 34; m 55; c 3. MEDICINAL CHEMISTRY. *Educ:* Univ Mich, BS, 55; Univ Calif, San Francisco, PhD(pharmaceut chem), 62. *Prof Exp:* From asst prof to prof med chem, Univ Kans, 62-84; DEAN, COL PHARM, UNIV IOWA, 84- *Mem:* Am Chem Soc; Am Pharmaceut Asn; Soc Toxicol. *Res:* Relationship between biological activity and chemical properties among drugs; chemical aspects of drug metabolism. *Mailing Add:* Col Pharm Univ Iowa Iowa City IA 52242-1185

WILEY, ROBERT CRAIG, b Washington, DC, Nov 14, 24; m 51; c 3. FOOD PROCESSING, FOOD ENGINEERING. *Educ:* Univ Md, BS, 49, MS, 50; Ore State Univ, PhD(food tech), 53. *Prof Exp:* Asst, Ore State Univ, 51-52; food specialist, US Dept Navy, 53; from asst prof to assoc prof, 53-69, chmn, food sci interdepartmental prog, 84-89, PROF HORT, UNIV MD, COLLEGE PARK, 69- *Concurrent Pos:* Fulbright-Hays sr lectr, Univ Belgrade, Yugoslavia, 79; chair, Fruit & Veg Technol Div, Inst Food Technol, 88; vis prof, Mid East Tech Univ, Turkey, 89; consult, USAID, Morocco, 90. *Honors & Awards:* Woodbury Res Award co-recipient, 61, 62; H W Wiley Medal, US Food & Drug Admin, 81. *Mem:* Am Soc Hort Sci; fel Inst Food Technol. *Res:* Measurement of polysaccharides, fatty acids and enzymes in fruits and vegetables; aroma analyses of fruits; thermal processing of foods; ultrafiltration; reverse osmosis of fruit and vegetable juices; minimally processed refrigerated foods. *Mailing Add:* Dept Hort Food Sci Prog Univ Md Col Agr College Park MD 20742-5611

WILEY, RONALD GORDON, b Akron, Ohio, Mar 21, 47; m 70; c 3. NEUROSCIENCE. *Educ:* Northwestern Univ, BS, 72, MD, 75, PhD(pharmacol), 75; Am Bd Psychiat & Neurol, dipl, 81; Am Bd Internal Med, dipl, 81. *Prof Exp:* Intern & resident internal med, Peter Bent Brigham Hosp, 75-77; resident neurol, NY Hosp, 77-80; fel neurol, Cornell Univ Sch Med, 80-82; instr pharmacol & asst prof neurol, 82-84, CHIEF NEURO-ONCOL SERV, MED CTR & VIS STAFF NEUROL, VANDERBILT UNIV SCH MED, UNIV HOSP, 84-; STAFF PHYSICIAN & CHIEF LAB EXP NEUROL, VET ADMIN MED CTR, NASHVILLE, 82-, ASSOC PROF NEUROL, 87-, ASSOC PROF PHARMACOL, 88- *Concurrent Pos:* Assoc attend neurol, LaGuardia Hosp, NY, 80-; Ed neurol, Med Info Systs, 84-; asst prof pharmacol, Vet Admin Med Ctr, Nashville, 84-88; assoc ed, J Neurocytol. *Honors & Awards:* Sigma Xi Res Award, 72; G D Searle Res Award, 74; Roche Award in Neurosci, 72. *Mem:* Soc Neurosci; NY Acad Sci; AAAS; Am Acad Neurol; Cent Soc Neurol Res; Int Soc Neuroimmunol. *Res:* Suicide-transport; development and testing of axonally transported cytotoxins as tools for making experimental neural lesions; reaction of the central nervous system to selective loss of motor neurons. *Mailing Add:* Dept Neurol & Pharmacol Nashville TN 37232

WILEY, RONALD LEE, b Dayton, Ohio, Oct 4, 37; m 81; c 1. RESPIRATORY PHYSIOLOGY, PULMONARY PHYSIOLOGY. *Educ:* Miami Univ, Ohio, BS, 59; Univ Ky, PhD(physiol, biophys), 66. *Prof Exp:* Teacher, Talawanda High Sch, Ohio, 59-62; instr physiol & NIH fel, Marquette Univ, 66-67; asst prof, 67-71, assoc prof zool & physiol, 71-76, PROF ZOOL, MIAMI UNIV, 76- *Concurrent Pos:* Mem med staff, McCullough-Hyde Mem Hosp, Oxford, Ohio, 73-; from asst prof to prof, Sch Med, Wright State Univ, Dayton, 74-80. *Mem:* Am Physiol Soc; Am Thoracic Soc. *Res:* Control of respiration; perception of respiratory sensations. *Mailing Add:* Dept Zool Miami Univ Oxford OH 45056

WILEY, SAMUEL J, b Philadelphia, Pa, May 27, 39; m; c 5. COMPUTER APPLICATIONS IN INSTRUCTION. *Educ:* St Josephs Univ, BS, 61; Villanova Univ, MA, 63; Temple Univ, PhD, 70. *Prof Exp:* Chmn Dept Math, 82-90, ASSOC PROF, LASALLE UNIV, PHILADELPHIA, 70- *Mem:* Math Asn Am; Inst Elec & Electronics Engrs Comput Soc; Asn Comput Mach. *Res:* Computer applications in instruction. *Mailing Add:* Dept Math LaSalle Univ Olney Ave & 20th St Philadelphia PA 19141

WILEY, WILLIAM CHARLES, b Monmouth, Ill, Aug 7, 24; m 44; c 2. ENGINEERING, PHYSICS. *Educ:* Univ Ill, BS, 49. *Prof Exp:* Dir, Appl Physics Lab, Bendix Corp, 49-68, assoc dir, Planning Res Labs, 68-69, asst gen mgr, Sci Instruments & Equip Div, 69-71; vpres & chief tech officer, Leeds & Northrup Co, 71-85; RETIRED. *Concurrent Pos:* Dir, Univ City Sci Ctr. *Mem:* Fel Instrument Soc Am; AAAS; fel Inst Elec & Electronic Engrs; Ind Res Inst. *Res:* Advanced sensing techniques; electronics; instrumentation; mass spectrometry; electron multipliers. *Mailing Add:* 357 Lake Almanor W Dr Chester CA 96020-9703

WILEY, WILLIAM RODNEY, b Oxford, Miss, Sept 5, 32; m 52; c 1. MICROBIOLOGY, BIOCHEMISTRY. *Educ:* Tougaloo Col, BS, 54; Univ Ill, MS, 60; Wash State Univ, PhD(bact), 65. *Hon Degrees:* LLD, Gonzaga Univ, 88, DrS, Whitman Col, Walla Walla, Wash. *Prof Exp:* Coord Life Sci Prog, 65-74, mgr, biol dept, 74-79, dir res, 79-84, DIR, PAC NORTHWEST DIV, BATTELLE MEM INST, 84- *Concurrent Pos:* Adj prof, 68-75. *Mem:* AAAS; Fedn Am Scientists; Am Soc Microbiol; Am Soc Biol Chem. *Res:* Microbial metabolism, particularly the factors which control intracellular pool formation and membrane transport of amino acids and sugars in microorganisms and mammalian cells; develoment environmental engineering; biotechnology; geoscience; analytical chemistry. *Mailing Add:* Pac Northwest Lab Battelle Mem Inst PO Box 999 Richland WA 99352

WILFONG, ROBERT EDWARD, b Wayne Co, Ill, Jan 3, 20; m 38; c 5. CHEMISTRY, FIBER SCIENCE. *Educ:* Univ Wis, BS, 41, MS, 42, PhD(phys chem), 44. *Prof Exp:* Res chemist, E I du Pont de Nemours & Co, Inc, Va, 44-48, res supvr, 48-51, res mgr, 51-53, tech supt, 53-59, lab dir, 59-64, tech mgr, 64-71, tech dir, 71-83; RETIRED. *Mem:* Am Chem Soc. *Res:* Photosynthesis; submarine detection; kinetics of rocket propellant decomposition; modified rocket propellants; infrared spectroscopy; new textile fibers, orlon, nylon and dacron; aromatic polyamides. *Mailing Add:* 117 Deer Cove Rd Hampstead NC 28443

WILFRET, GARY JOE, b Sacramento, Calif, Oct 13, 43; m 68; c 2. PLANT BREEDING, GENETICS. *Educ:* Univ Hawaii, BS, 65, PhD(hort), 68. *Prof Exp:* Asst prof biol, Ga South Col, 68-69; grom asst geneticist to assoc geneticist, 69-79, GENETICIST, AGR RES & EDUC CTR, UNIV FLA, BRADENTON, 79- *Mem:* Am Soc Hort Sci; Am Asn Trop Biol; Bot Soc Am; Tissue Cult Asn; Am Hort Soc. *Res:* Breeding of ornamental plants for disease resistance and adaptation to subtropical conditions; hybridizing Dendrobium species to determine sexual compatability, to examine meiotic behavior and to clarify genome relationships. *Mailing Add:* Dept Ornamental Hort Univ Fla Gainesville FL 32611

WILGRAM, GEORGE FRIEDERICH, b Vienna, Austria, Apr 12, 24; nat US; m 56; c 3. PHYSIOLOGY, DERMATOLOGY. *Educ:* Univ Vienna, MD, 51; Univ Toronto, MA, 53, PhD, 57. *Prof Exp:* Lectr physiol, Univ Toronto, 57-58; asst prof exp path, Univ Chicago, 58-59; res assoc dermat, Harvard Med Sch, 59-60, asst prof, 61-67; PROF DERMAT, TUFTS UNIV, 67- *Mem:* Am Soc Exp Path; Soc Exp Biol & Med; Am Heart Asn. *Res:* Genetics of keratinization and pigmentation. *Mailing Add:* Ste 630 JBC Bldg Emerson Hosp Tufts Univ Med Sch Concord MA 01742

WILGUS, DONOVAN RAY, b La Plata, Mo, Aug 11, 21; m 52; c 3. ORGANIC CHEMISTRY. *Educ:* Northeast Mo State Teachers Col, AB, 42; Univ Colo, PhD, 51. *Prof Exp:* From assoc res chemist to supv res chemist, 51-66, SR RES ASSOC, CHEVRON RES CO, 66- *Mem:* Am Chem Soc. *Res:* Diels-Alder reaction; lubricating oil additives; synthetic oils. *Mailing Add:* 2912 Cindy Ct Richmond CA 94803-3230

WILHEIT, THOMAS TURNER, b Dallas, Tex, Apr 10, 41; m 66; c 1. MICROWAVE PHYSICS, ATMOSPHERIC PHYSICS. *Educ:* Univ South, BA, 63; Wash Univ, St Louis, MA, 67; Mass Inst Technol, PhD(physics), 70. *Prof Exp:* Physicist Microwave, Goddard Space Flight Ctr, 71-89; PROF METEROL, TEX A&M UNIV, 89- *Concurrent Pos:* Resident res assoc, Nat Acad Sci, 70-71. *Mem:* Inst Elec & Electronics Engrs; Am Meterol Soc; AAAS. *Res:* Passive microwave remote sensing of the earth's surface and atmosphere. *Mailing Add:* Dept Meterol Tex A&M Univ College Station TX 77843

WILHELM, ALAN ROY, b Buffalo, NY, Oct 30, 36. MICROBIOLOGY, VIROLOGY. *Educ:* Stanford Univ, AB, 58; Univ Wis, Madison, MS, 65, PhD(bact), 67. *Prof Exp:* Microbiologist, US Army, 67-69; from asst prof to assoc prof, 69-77, PROF BIOL SCI, CALIF STATE UNIV, CHICO, 77- *Mem:* Sigma Xi; AAAS; Am Soc Microbiol. *Res:* Arboviruses; herpes viruses; viral infection of poikilothermic cells. *Mailing Add:* Dept Biol Sci Chico State Col Chico CA 95929

WILHELM, DALE LEROY, b Greenview, Ill, June 20, 26; m 51; c 4. INORGANIC CHEMISTRY. *Educ:* Univ Ill, BS, 51; Univ Tenn, MS, 52, PhD(chem), 54. *Prof Exp:* From asst prof to assoc prof chem, Univ WVa, 54-63; assoc prof, Cornell Col, 63-66; asst dean, Liberal Arts Col, 73-78, chmn chem dept, 74-78, PROF CHEM, OHIO NORTHERN UNIV, 66-, VPRES ACAD AFFAIRS, 78- *Mem:* AAAS; Am Chem Soc; Royal Soc Chem; Sigma Xi. *Res:* Heteropolyanions; electrophoresis in stabilized media. *Mailing Add:* Dept Chem Ohio Northern Univ Ada OH 45810

WILHELM, EUGENE J, JR, b St Louis, Mo, July 25, 33. BIOGEOGRAPHY, HUMAN ECOLOGY. *Educ:* St Louis Univ, BS, 59; La State Univ, Baton Rouge, MA, 61; Tex A&M Univ, PhD(geog), 71. *Prof Exp:* Instr geog, DePaul Univ, 62-63; asst prof, St Louis Univ, 63-65; asst prof, McGill Univ, 65-68; vis lectr, Univ Va, 68-69 & 71-72; assoc prof, 72-76, PROF GEOG, SLIPPERY ROCK STATE COL, 76- *Concurrent Pos:* Vis prof ecol, Univ Sierra Leone, sr Fulbright-Hays Awardee, 76-77. *Mem:* Asn Am Geog. *Res:* Cultural and natural history of Appalachia; ecological problems in national park areas; ethnobiology; human ecology; folk geography. *Mailing Add:* Dept Geog RD 2 Box 2120 Slippery Rock PA 16057

WILHELM, HARLEY A, b Iowa, Aug 5, 1900. METALLURGY. *Educ:* Drake Univ, AB, 23; Iowa State Univ, PhD(phys chem), 31. *Prof Exp:* EMER PROF, DEPT CHEM & METALL, IOWA STATE UNIV, 70- *Concurrent Pos:* Mem, Atomic Energy Prog, Iowa State Univ, 42 & Ames Lab, Dept Energy. *Honors & Awards:* Gold Medal, Am Soc Mech Engrs, 90. *Mem:* Am Chem Soc; Am Soc Metals Int. *Mailing Add:* 513 Hayward Ave Ames IA 50010

WILHELM, JAMES MAURICE, b Redfield, SDak, May 20, 40; m 69. MICROBIOLOGY, MOLECULAR BIOLOGY. *Educ:* SDak Sch Mines & Technol, BS, 62; Case Western Reserve Univ, PhD(biochem), 68. *Prof Exp:* Am Cancer Soc fel biophys, Univ Chicago, 68-70; asst prof microbiol, Sch Med, Univ Pa, 70-73; asst prof, 73-79, ASSOC PROF MICROBIOL, SCH MED, UNIV ROCHESTER, 79- *Mem:* Am Soc Cell Biol; Am Soc Microbiol. *Res:* Protein synthesis; control of viral replication; cell biology. *Mailing Add:* 1107 Westage at the Harbor Rochester NY 14617

WILHELM, RUDOLF ERNST, b Hanover, Ger, Dec 26, 26; US citizen; m 52; c 3. ALLERGY, IMMUNOLOGY. *Educ:* Univ Ill, Chicago, MD, 51; Am Bd Internal Med, dipl, 61. *Prof Exp:* Asst resident internal med, Detroit Receiving Hosp, Mich, 52-53; resident allergy. Roosevelt Hosp Inst, NY, 57; resident internal med, Henry Ford Hosp, Detroit, 58-59; instr med, Sch Med, La State Univ, New Orleans, 59-60; asst prof med, Sch Med, Wayne State Univ, 60-64, from asst prof to assoc prof dermat, 64-77; CHIEF, ALLERGY SECT, OAKWOOD HOSP, 77-; ALLERGY SPECIALIST, ALLERGY ASSOCS, DEARBORN, 77- *Concurrent Pos:* Consult, USPHS Hosp, Detroit, 62-69; chief, Allergy Sect, Vet Admin Hosp, Allen Park, 60-77; secy, Dept Med, Oakwood Hosp, Dearborn, 65-67. *Mem:* Fel Am Col Physicians; fel Am Acad Allergy. *Res:* Delayed-type allergic skin reactions such as atopic eczema and contact dermatitis; methods and mechanics of allergy hyposensitization injections; anti-allergic drug treatment; methods of medical education in allergy and internal medicine. *Mailing Add:* 751 S Military Rd Dearborn MI 48124

WILHELM, STEPHEN, b Imperial Co, Calif, Apr 19, 19; m 44; c 2. PLANT PATHOLOGY. *Educ:* Univ Calif, AB, 42, PhD(plant path), 48. *Prof Exp:* From instr & jr plant pathologist to prof plant path & plant pathologist, 48-84, EMER PROF PLANT PATH, UNIV CALIF, BERKELEY, 84- *Concurrent Pos:* Guggenheim fel, 58-59. *Mem:* Sigma Xi; fel Am Phytopath Soc; Am Soc Hort Sci. *Res:* Verticilium wilt; diseases of small fruit; root infecting fungi; soil fumigation. *Mailing Add:* 12770 Skyline Blvd Oakland CA 94619

WILHELM, WALTER EUGENE, b St Louis, Mo, May 16, 31; m 61; c 3. PARASITOLOGY, PROTOZOOLOGY. *Educ:* Harris Teachers Col, BA, 55; Univ Ill, MS, 59; Univ Southern Ill, PhD(zool), 65. *Prof Exp:* Lectr embryol, Univ Southern Ill, 62-63; from asst prof to assoc prof, 64-85, PROF BIOL, MEMPHIS STATE UNIV, 85- *Concurrent Pos:* NSF fel parasitol, Univ Ill, 67. *Mem:* Soc Protozool; Am Soc Parasitol. *Res:* Parasitic helminths and protozoa, particularly those which may be used as agents of biological control; ecology of free-living protozoa; biology of limax amoebae. *Mailing Add:* Dept Biol Memphis State Univ Memphis TN 38152

WILHELM, WILBERT EDWARD, b Pittsburgh, Pa, Oct 24, 42; m 63; c 2. APPLIED STOCHASTIC PROCESSES. *Educ:* WVa Univ, BS, 64; Va Polytech Inst, MS, 70, PhD (indust eng & opers res), 72. *Prof Exp:* Mfg prog, Gen Elec Co, 64-67, spec mfg admin, 67-69; NSF fel & instr, Va Polytech Inst, 69-72; asst prof, 72-78, ASSOC PROF, OHIO STATE UNIV, 78- *Concurrent Pos:* NSF fel, 69-72; consult, 73-; prin investr, NSF, 83-, Grumman Aerospace Corp, 85-, NSF, 88, IBM, 88; dir, OR Div IIE, 88-89. *Mem:* Am Inst Indust Engrs; Opers Res Soc Am; Soc Mfg Engrs. *Res:* Developed methods to approximate transient performance of queuing networks, and models to manage material flow in assembly systems including robotic cells; computer aided manufacturing. *Mailing Add:* Dept Indust Tex A&M Univ College Station TX 77843

WILHELM, WILLIAM JEAN, b St Louis, Mo, Oct 5, 35; m 57; c 5. STRUCTURAL & CIVIL ENGINEERING. *Educ:* Ala Polytech Inst, BME, 58; Auburn Univ, MS, 63; NC State Univ, PhD(struct eng), 68. *Prof Exp:* Struct engr, Palmer & Baker Engrs, Inc, 58-59 & 59-60; instr eng graphics, Auburn Univ, 60-61 & 62-64; teaching asst civil eng, NC State Univ, 66-67; from asst prof to prof civil eng, WVa Univ, 67-79, assoc chmn dept, 70-74, chmn dept, 74-79; DEAN, COL ENG, WICHITA STATE UNIV, 79- *Concurrent Pos:* NSF grants, 69-73; Am Iron & Steel Inst res grant, 68-73; Expanded Shale Clay & Slate Inst res grant, 71 & 72; chmn, Civil Eng Div, Am Soc Eng Educ, 77-78 & Rels Indust Div, 84-85; chmn comt curricula & accreditation, Am Soc Civil Engrs, 82-83; chmn educ activ comt, Am Concrete Inst, 83-88, mem bd dir, 86-89 & chmn conv comt, 89- *Honors & Awards:* Joe W Kelly Award, Am Concrete Inst, 86. *Mem:* Fel Am Concrete Inst; fel Am Soc Civil Engrs; Am Soc Eng Educ; Prestressed Concrete Inst; Nat Soc Prof Engrs; sr mem Soc Women Engrs. *Res:* Reinforced and prestressed concrete with particular emphasis on torsional and bond behavior for both normal weight and light-weight aggregate concrete; engineering education. *Mailing Add:* Dean Col Eng Wichita State Univ Wichita KS 67208

WILHELMI, ALFRED ELLIS, metabolism, growth hormones, for more information see previous edition

WILHELMY, JERRY BARNARD, b Sewickley, Pa, July 31, 42; m 64. NUCLEAR CHEMISTRY, NUCLEAR PHYSICS. *Educ:* Univ Ariz, BSChE, 64; Univ Calif, Berkeley, PhD(nuclear chem), 69. *Prof Exp:* Fel nuclear chem, Lawrence Berkeley Lab, 69-72; MEM STAFF NUCLEAR CHEM, LOS ALAMOS SCI LAB, 72- *Concurrent Pos:* Vis res scientist, 78-79, assoc group leader res, 79-82, fel, Max Planck Inst Kernphysik, 82- *Mem:* Am Chem Soc; Am Phys Soc. *Res:* Properties of nuclear fission; fission barriers; fission product spectroscopy, fission produced neutrons; neutrino physics; heavy ion reactions. *Mailing Add:* Los Alamos Nat Lab MS J514 Los Alamos NM 87545

WILHITE, DOUGLAS LEE, b Owensboro, Ky, July 29, 44; m 67; c 3. PHYSICAL CHEMISTRY, QUANTUM CHEMISTRY. *Educ:* Univ Mo-Columbia, BS, 66; State Univ NY Stonybrook, PhD(phys chem), 71. *Prof Exp:* Fel quantum chem, Aerospace Res Labs, Wright-Patterson AFB, Ohio, 71-73; RES ASSOC PHOTOG SCI, IMAGING SYSTS DEPT, RES & DEVELOP DIV, E I DU PONT DE NEMOURS & CO, 73- *Concurrent Pos:* Vis res chemist, Technol Inc, 71-72; Nat Res Coun assoc, 72-73. *Mem:* Am Phys Soc. *Res:* Development of new precipitation processes for AgX microcrystals used in photographic systems; design of more efficient photographic systems; study of fundamental phenomena associated with the photographic process. *Mailing Add:* 1105 Piper Rd Wilmington DE 19803

WILHITE, ELMER LEE, b Owensboro, Ky, July 29, 44; m 66; c 3. PLUTONIUM CHEMISTRY, OFF-GAS. *Educ:* Univ Mo, BS, 66; Wash Univ, MA, 69. *Prof Exp:* Chemist, Savannah River Lab, E I du Pont de Nemours & Co, Inc, 69-79, res chemist, 79-80, tech supvr, Environ Monitoring, 80-81, res supvr, Anal Develop Div, 81-84, RES STAFF CHEMIST, SAVANNAH RIVER LAB, WESTINGHOUSE SAVANNAH RIVER CO, 84- *Mem:* Am Chem Soc; Health Physics Soc; Am Nuclear Soc. *Res:* Environmental assessment of transuranic element migration from buried solid waste; melting and off-gas processing of high-level defense nuclear waste; solidification of low-level and hazardous waste; performance assessment of low-level waste disposal. *Mailing Add:* 773-43A Savannah River Lab Westinghouse Savannah River Co Aiken SC 29808

WILHM, JERRY L, b Kansas City, Kans, Apr 27, 30; m 55; c 2. LIMNOLOGY, ECOLOGY. *Educ:* Kans State Teachers Col, BS, 52, MS, 56; Okla State Univ, PhD(zool), 65. *Prof Exp:* Teacher high sch, Kans, 56-62; US AEC fel, Oak Ridge, Tenn, 65-66; PROF ZOOL, DEPT ZOOL, OKLA STATE UNIV, 66-, HEAD DEPT, 81- *Concurrent Pos:* Biol consult, Am Inst Biol Sci Film Series, 61-62; Fulbright fel, NZ, 79. *Mem:* Am Soc Limnol & Oceanog; Ecol Soc Am. *Res:* Biological effects of oil refinery effluents. *Mailing Add:* Dept Zool Okla State Univ Stillwater OK 74078

WILHOFT, DANIEL C, b Newark, NJ, Nov 16, 30; m 51; c 2. ZOOLOGY. *Educ:* Rutgers Univ, AB, 56; Univ Calif, Berkeley, MA, 58, PhD(zool), 63. *Prof Exp:* From instr to assoc prof, 62-69, chmn dept zool & physiol, 69-75, PROF ZOOL, RUTGERS UNIV, NEWARK, 69- *Mem:* AAAS; Am Soc Zool; fel Zool Soc London. *Res:* Ecology of fresh-water turtles; reptilian endocrinology. *Mailing Add:* 240 Walton Ave South Orange NJ 07079

WILHOIT, EUGENE DENNIS, b Frankfort, Ky, Jan 28, 31; m 58; c 1. PHYSICAL CHEMISTRY. *Educ:* Univ Ky, BS, 53, PhD(phys chem), 56. *Prof Exp:* From res chemist to sr res chemist, Polychem Dept, 56-67, admin asst technol dept, 67, asst div supt, 67-69, div supt res, 69-75, gen tech supt, 75-81, TECH FEL, E I DU PONT DE NEMOURS & CO, INC, 81- *Mem:* AAAS; Am Chem Soc. *Res:* Electrochemistry and electrolytic conductance; reactions and synthesis of polymer intermediates; nonaqueous solutions; oxidation mechanisms. *Mailing Add:* 213 Tracy Lane Victoria TX 77904-1525

WILHOIT, JAMES CAMMACK, JR, b Tulsa, Okla, Dec 22, 25; c 4. MECHANICAL ENGINEERING. *Educ:* Rice Inst Technol, BS, 48; Tex A&M Univ, MS, 51; Stanford Univ, PhD(eng mech), 54. *Prof Exp:* Instr mech eng, Tex A&M Univ, 49-51; sr aerophys eng, Convair, Tex, 53-54; from asst prof to assoc prof mech eng, 54-70, PROF MECH ENG, RICE UNIV, 70- *Mem:* Am Soc Mech Engrs. *Mailing Add:* Rte 2 McCowans Ferry Rd Versailles KY 40383

WILHOIT, RANDOLPH CARROLL, b San Antonio, Tex, Oct 16, 25; m 48; c 3. EVALUATION THERMODYNAMIC PROPERTIES. *Educ:* Trinity Univ, Tex, AB, 47; Univ Kans, MA, 49; Northwestern Univ, PhD, 52. *Prof Exp:* Fel phys chem, Univ Ind, 52-53; asst prof, Tex Tech Col, 53-57; assoc prof, NMex Highlands Univ, 57-60, prof, 60-64; ASSOC PROF CHEM & ASSOC DIR THERMODYN RES CTR, TEX A&M UNIV, 64- *Concurrent Pos:* Assoc ed, J Chem & Eng Data, Am Chem Soc, 71- *Mem:* AAAS; Am Chem Soc. *Res:* Thermochemistry; molecular structure; energetics of biochemical reactions; estimation of thermodynamic properties. *Mailing Add:* Thermodyn Res Ctr Tex A&M Univ College Station TX 77843

WILHOLD, GILBERT A, b East St Louis, Ill, Dec 9, 34; m 82; c 6. PHYSICS, MECHANICS. *Educ:* St Louis Univ, BS, 57. *Prof Exp:* Engr, McDonnell Aircraft Corp, Mo, 57-60 & Chrysler Corp, Ala, 60-62; engr, 62-63, tech asst anal & theoret acoust, 63, tech asst to lab dir, 63-66, dep br chief unsteady fluid mech, 66-69 & Unsteady Gas Dynamics Br, 69-77, AEROPHYS DIV ASST, MARSHALL SPACE FLIGHT CTR, NASA, 77- *Concurrent Pos:* Consult to pvt firms, fed, state & local govt & univs. *Res:* Acoustic noise generated by rocket exhausts and air flow over space vehicle surfaces during flight; random process theory and application; data reduction and analysis; aeroelasticity; fluid mechanics; structural dynamics and vibrations; sonic boom and its environmental effects; unsteady fluid flow in high performance pumps and aeroelastically induced loads; computational fluid dynamics and numerical analysis. *Mailing Add:* Aerophys Div ED 31 Marshall Space Flight Ctr Huntsville AL 35812

WILIMOVSKY, NORMAN JOSEPH, b Chicago, Ill, Sept 9, 25; m 47; c 4. ICHTHYOLOGY, FISHERIES. *Educ:* Univ Mich, BS, 48, MA, 49; Stanford Univ, PhD, 56. *Prof Exp:* Head fish & game off, Mil Govt, Bavaria, Ger, 46; assoc ichthyologist, Fisheries Surv Brazil, 50-51; prin investr, Arctic Invests, Stanford Univ, 51-54, res assoc, 55-56; chief marine fisheries invests, US Fish & Wildlife Serv, Alaska, 56-60; assoc prof, Fisheries & Zool, 60-64, dir, Inst Fisheries, 63-66, PROF FISHERIES, UNIV BC, 64- *Concurrent Pos:* Mem comt proj Chariot, AEC, 60-66; staff specialist, US Coun Marine Resources & Eng Develop, 67-68; mem, Environ Protection Bd, 70-76. *Mem:* Fel AAAS; Am Soc Ichthyol & Herpet; Am Fisheries Soc; Am Soc Limnol & Oceanog; fel Arctic Inst NAm. *Res:* Systematics of fishes; fishery population dynamics; ecology of ice; development and management of fisheries; history of biological exploration; resource policy formulation. *Mailing Add:* Inst Resource Ecol Univ BC Vancouver BC V6T 1W5 Can

WILK, LEONARD STEPHEN, b Adams, Mass, Sept 29, 27; div; c 3. ELECTRICAL ENGINEERING, INSTRUMENTATION. *Educ:* Mass Inst Technol, SB & SM, 55. *Prof Exp:* Group leader missile guid, Instrumentation Lab, Mass Inst Technol, 55-60, asst dir space guid, 60-67, assoc dir instrumentation, Measurement Systs Lab, 67-74; STAFF ENGR, CHARLES STARK DRAPER LAB, INC, 74- *Concurrent Pos:* Lectr, Dept Aeronaut & Astronaut, Mass Inst Technol, 85- *Mem:* Am Inst Aeronaut & Astronaut. *Res:* Inertial navigation and guidance; system analysis and testing; gravity gradiometry; magnetic suspension systems; control systems; experimental tests of gravitation theories; configuration management; detectors for high energy physics; fiber optic alignment; arms control. *Mailing Add:* 555 Tech Sq MS 6E Cambridge MA 02139

WILK, SHERWIN, b New York, NY, Aug 25, 38; m 63; c 2. BIOCHEMISTRY, PHARMACOLOGY. *Educ:* Syracuse Univ, BS, 60; Purdue Univ, MS, 62; Fordham Univ, PhD(biochem), 67. *Prof Exp:* Res asst biochem, Mt Sinai Hosp, 62-67; assoc prof, 69-80, PROF PHARMACOL, MT SINAI SCH MED, 80- *Concurrent Pos:* NIH fel, Sch Med, Cornell Univ, 67-69; NIH res career develop award, Mt Sinai Sch Med, 69- *Mem:* AAAS; Am Soc Pharmacol & Exp Therapeut; Am Soc Neurochem; Am Chem Soc; Sigma Xi. *Res:* Metabolism of gamma glutamyl compounds; peptidases and proteinases in the central nervous system; metabolism of catecholamines in central nervous system. *Mailing Add:* 345 W 88th St New York NY 10024

WILK, STANISLAS FRANCOIS JEAN, mathematical physics, nuclear physics, for more information see previous edition

WILK, WILLIAM DAVID, b Pittsburgh, Pa, Mar 6, 42; m 70; c 3. INORGANIC CHEMISTRY. *Educ:* Thiel Col, BA, 64; Northwestern Univ, PhD(chem), 68. *Prof Exp:* From asst prof to assoc prof, 68-78, PROF CHEM, CALIF STATE COL, DOMINGUEZ HILLS, 78- *Mem:* Royal Soc Chem; Nat Sci Teachers Asn. *Res:* Ligand substitution effects on cobalt III complexes. *Mailing Add:* Dept Chem Calif State Univ Dominguez Hills 1000 E Victoria Carson CA 90747

WILKE, CHARLES R, b Dayton, Ohio, Feb 4, 17; m 45. CHEMICAL ENGINEERING. *Educ:* Univ Dayton, BS, 40; State Col Wash, MS, 42; Univ Wis, PhD(chem eng), 44. *Prof Exp:* Assoc engr, Union Oil Co, Calif, 44-45; instr chem eng, State Col Wash, 45-46; from instr to assoc prof, 46-53, chmn dept, 53-63, PROF CHEM ENG, UNIV CALIF, BERKELEY, 53-, RES ASSOC, LAWRENCE BERKELEY LAB, 50- *Concurrent Pos:* Indust consult, 52-; commencement speaker, Univ Dayton, 61; mem adv bd, Petrol Res Fund, 64-67; mem Calif Bd Registr Prof Engrs, 64-72, pres, 67-68. *Honors & Awards:* Colburn Award, Am Inst Chem Engrs, 51, Walker Award, 65, Founders Award, 86. *Mem:* Nat Acad Eng; AAAS; Am Inst Chem Engrs; Am Chem Soc; Am Soc Eng Educ. *Res:* Mass transfer operations; separation and purification of materials; biochemical engineering; kinetics and scale-up of microbial processes. *Mailing Add:* Dept Chem Eng 110 Gilman Hall Univ Calif Berkeley CA 94720

WILKE, FREDERICK WALTER, b Pana, Ill, Sept 13, 33; m 56; c 4. MATHEMATICS. *Educ:* Drury Col, AB, 54; Wash Univ, MA, 59; Univ Mo, Columbia, MA, 60, PhD(math), 66. *Prof Exp:* Instr math, Southwest Mo State Col, 63-65; asst prof, 65-73, ASSOC PROF MATH, UNIV MO, ST LOUIS, 73- *Mem:* Am Math Soc; Math Asn Am. *Res:* Finite projective planes, especially translation planes. *Mailing Add:* Dept Math & Comput Sci Univ Mo St Louis St Louis MO 63121

WILKE, ROBERT NIELSEN, b San Diego, Calif, July 7, 41; m 73; c 2. CHEMICAL INFORMATION, INFORMATION ANALYSIS. *Educ:* San Diego State Univ, BS, 64; Case Western Reserve Univ, PhD(org chem), 71. *Prof Exp:* Res assoc electrochem, Youngstown State Univ, 71-72, cancer res, Univ Chicago, 72-74; sr chemist, Velsicol Chem Corp, 74-80; RES INFO SCIENTIST, AMOCO CORP, 80- *Mem:* Am Chem Soc; Royal Soc Chem. *Res:* Retrieval and analysis of information related to petroleum industry. *Mailing Add:* Amoco Corp PO Box 3011 Naperville IL 60566

WILKEN, DAVID RICHARD, b Amarillo, Tex, Feb 20, 34; m 55; c 3. BIOCHEMISTRY. *Educ:* Blackburn Col, BA, 55; Univ Ill, MS, 58; Mich State Univ, PhD(biochem), 60. *Prof Exp:* From asst prof to assoc prof biochem, Okla State Univ, 62-66; asst prof, 66-72, ASSOC PROF PHYSIOL CHEM, UNIV WIS-MADISON, 72-; RES CHEMIST, LAB EXP PATH, VET ADMIN HOSP, 66- *Concurrent Pos:* Nat Found fel, Inst Enzyme Res, Univ Wis, 60-62. *Mem:* Am Soc Biol Chem; Am Chem Soc. *Res:* Diabetes; glycoprotein biosynthesis; pantothenic acid metabolism; enzymology. *Mailing Add:* William S Middleton Mem Vet Admin Hosp 2500 Overlook Terr Madison WI 53705

WILKEN, DIETER H, b Los Angeles, Calif, Apr 12, 44. SYSTEMATIC BOTANY. *Educ:* Calif State Univ, Los Angeles, BA, 67; Univ Calif, Santa Barbara, PhD(biol), 71. *Prof Exp:* Res asst, Los Angeles State & County Arboretum, 66-67; asst prof biol, Occidental Col, 71-73; asst prof biol, 73-76, ASSOC PROF BOT, COLO STATE UNIV, 76- *Concurrent Pos:* Cur, Colo State Univ Herbarium, 73- *Mem:* Bot Soc Am; Am Soc Plant Taxon; Int Soc Plant Taxon. *Res:* Systematics of higher plants within field of cytology, biochemistry, anatomy and breeding behavior; ecology and evolutionary dynamics of populations. *Mailing Add:* Dept Biol Colo State Univ Ft Collins CO 80523

WILKEN, DONALD RAYL, b New Orleans, La, Apr 25, 38; m 58; c 1. PURE MATHEMATICS. *Educ:* Tulane Univ, BS, 58, PhD(math), 65; Univ Calif, Los Angeles, MA, 62. *Prof Exp:* NSF fel, Brandeis Univ, 65-66; instr math, Mass Inst Technol, 66-68; from asst prof to assoc prof, 68-75, PROF MATH, STATE UNIV NY, ALBANY, 75- *Res:* Functional and complex analysis. *Mailing Add:* Dept Math State Univ NY Albany NY 12203

WILKEN, LEON OTTO, JR, b Waterbury, Conn, Oct 21, 24; m 46. PHARMACEUTICS, BIOPHARMACEUTICS. *Educ:* Loyola Univ, La, BS, 51, PhD(pharm), 82; Univ Tex, MS, 53, PhD(pharm), 63. *Prof Exp:* Spec instr pharm, Univ Tex, 53-63; assoc prof, 63-72, PROF PHARM, AUBURN UNIV, 72-, HEAD PHARMACEUT DIV, 73- *Concurrent Pos:* Pharm consult, Vet Admin Hosp, Tuskegee, Ala, 74- *Mem:* Am Asn Cols Pharm; Am Pharmaceut Asn; Am Chem Soc; Acad Pharm Sci. *Res:* Sustained release delivery systems, assay of pharmaceuticals from drug delivery systems and biological fluids by high-pressure liquid chromatography; methods of enhancing bioavailability of difficult soluble drugs; development of oral vaccines, stability studies. *Mailing Add:* Dept Pharm Auburn Univ Auburn AL 36849

WILKENFELD, JASON MICHAEL, b Brooklyn, NY, May 28, 39; m 62; c 2. SOLID STATE PHYSICS. *Educ:* Columbia Col, AB, 60; NY Univ, MS, 65, PhD(physics), 72. *Prof Exp:* Programmer analyst, Syst Develop Corp, 63; res asst physics, Radiation & Solid State Lab, NY Univ, 65-70; fel physics, New Eng Inst, 70-72; staff physicist, 73-77, group leader, 77-78, DEPT MGR, IRT CORP, 78- *Concurrent Pos:* Lectr physics, Hunter Col, City Univ New York, 66-69 & NY Univ, 69. *Mem:* Am Phys Soc. *Res:* Nuclear and space radiation effects in materials and systems; electrical properties of dielectrics; positron annihilation as a morphological probe. *Mailing Add:* S Cubed PO Box 1620 La Jolla CA 92038

WILKENING, DEAN ARTHUR, b Raleigh, NC, Sept 21, 50. NUCLEAR STRATEGIES, ARMS CONTROL. *Educ:* Univ Chicago, BA, 72; Harvard Univ, PhD(physics), 81. *Prof Exp:* Ford Found fel, Ctr Sci & Int Affairs, Harvard Univ, 81-83; DIR, FORCE EMPLOY PROG, NAT DEFENSE RES INST, 83- *Mem:* Am Phys Soc. *Res:* Strategic theater forces; arms control proposals. *Mailing Add:* 930 12th St Apt B Santa Monica CA 90403

WILKENING, GEORGE MARTIN, b New York, NY, Dec 31, 23; m 50; c 5. ENVIRONMENTAL HEALTH, ENVIRONMENTAL SCIENCES GENERAL. *Educ:* Queen's Col, NY, BS, 49; Columbia Univ, MS, 50. *Prof Exp:* Indust hygienist, State Health Dept, Va, 50-51; indust hygienist, Esso Res & Eng Co, 52-56, sr indust hygienist, Esso Standard Oil Co, 56-61; asst chief indust hygienist, Humble Oil & Ref Co, 61-63; dir environ health, Mgt & Safety Ctr, AT&T Bell Labs, 80-90; EXEC DIR, ENVIRON & OCCUP HEALTH SCI INST, 90- *Concurrent Pos:* Lectr, Columbia Univ, 59-; chmn, Laser Hazards Stand Comt, 68-; mem, Environ Radiation Adv Comt, Environ Protection Agency, 67-; tech electronic prod radiation stand comt mem, Dept HEW, 68-; mem, Nat Coun Radiation Protection & Measurements, 73-; chmn, Tech Comt Lasers, Int Electrotech Comn, 73-; chmn, Sci Comt Microwaves, Nat Coun Radiation Protection & Measurements, 75; mem study group on non-ionizing radiations, Int Radiation Protection Asn, 75, Nat Acad Sci-Nat Res Coun comt on biosphere effects extremely low frequency radiation, 76- & panel on effects radiation from Pave Paws Radar, 78; adj prof environ med, NY Univ, 83-; consult armed forces epidemiol bd, Dept Defense, Washington, DC, 77; mem, Electromagnetic Radiation Mgt Adv Coun, Nat Telecommun & Info, 79- *Mem:* AAAS; Sigma Xi; Acoust Soc Am; Am Indust Hyg Asn; fel NY Acad Sci; Bioelectromagnetics Soc. *Res:* Dosimetry of exposure to electromagnetic radiations, including acoustical noise; industrial toxicology; biological effects of chemical, physical and biological agents in the environment; development of standards for permissible levels of exposure to environmental agents. *Mailing Add:* 12 Skyline Dr Warren NJ 07060

WILKENING, LAUREL LYNN, b Richland, Wash, Nov 23, 44. METEORITICS, PLANETARY SCIENCES. *Educ:* Reed Col, BA, 66; Univ Calif, San Diego, PhD(chem), 70. *Prof Exp:* Am Asn Univ Women fel, Tata Inst Fundamental Res, India & Max Planck Inst Chem, 71; res assoc chem, Enrico Fermi Inst, Univ Chicago, 72-73; from asst prof to prof, 73-88, head, dept planetary sci & dir, Lunar & Planetary Lab, 81-83, vprovost, 83-85, vpres res & dean grad col, Univ Ariz, 85-88; PROF GEOL SCI, UNIV WASH, 88-, PROVOST & VPRES ACAD AFFAIRS, 88- *Concurrent Pos:* Vchmn, Nat Comn Space, 85-86. *Honors & Awards:* Nininger Meteorite Award, Ariz State Univ, 70. *Mem:* AAAS; fel Meteoritical Soc; Am Geophys Union; Am Astron Soc; Asn Women Sci. *Res:* Chemistry and mineralogy of meteorites, asteroids and comets; cosmochemistry; meteorites and their relationship to asteroids, comets and the formation of the solar system; nature of asteroidal surfaces through theoretical studies and studies of meteorites which probably once resided on the surface of asteroids. *Mailing Add:* Admin 301 Univ Wash Seattle WA 98195

WILKENING, MARVIN C, b Malone, Tex, July 1, 20; m 43; c 4. ANIMAL NUTRITION, BIOCHEMISTRY. *Educ:* Univ Tex, AB, 41; Agr & Mech Col, Tex, MS, 47. *Hon Degrees:* ScD, Athens Col, 56. *Prof Exp:* Res chemist, Dow Chem Co, Tex, 41-44; asst to dir res, Security Mills, Tenn, 47-50; dir res, Ala Flour Mills, 50-66; dir tech serv, Nebr, 67-75, NONRUMINANT NUTRITIONIST & CONSULT, CONAGRA, INC, 75- *Mem:* Fel AAAS; Am Chem Soc; Am Soc Animal Sci; Poultry Sci Asn. *Res:* Application of basic animal nutrition and management research. *Mailing Add:* 4312 Autum Leaves Trail SE Decatur AL 35603

WILKENING, MARVIN H, b Oak Ridge, Mo, Mar 13, 18; m 42; c 2. RADON & DAUGHTER PRODUCTS. *Educ:* Southeast Mo State Univ, BS, 39; Ill Inst Technol, MS, 43, PhD(nuclear physics), 49. *Prof Exp:* Teacher high sch, Mo, 39-41; physicist, Manhattan Proj, 42-45; instr physics, Ill Inst Technol, 46-48; from assoc prof to prof physics & geophys & head dept & dean grad studies, 48-83, EMER PROF PHYSICS, NMEX INST MINING & TECHNOL, 84- *Concurrent Pos:* Mem, Fermi group at first nuclear reactor, Chicago, 42; mem subcomn ions, Aerosols & Radioactiv, Int Comt Atmospheric Elec, 76-80; mem, Nat Coun Radiation Protection, SC61, 79-88; mem, NMex Radiation Tech Adv Coun, 81-87; Chicago Pile Exp, Soc Nuclear Med, 77. *Mem:* Fel AAAS; fel Am Phys Soc; Am Geophys Union; Am Meteorol Soc; Health Physics Soc; NY Acad Sci. *Res:* Radon and its daughter products; atmospheric electricity-ions and aerosols related to natural atmospheric radioactivity. *Mailing Add:* 1218 South Dr Socorro NM 87801

WILKENS, GEORGE A(LBERT), b North Bergen, NJ, June 29, 09; m 46; c 2. CHEMICAL ENGINEERING. *Educ:* Columbia Univ, AB, 29, BS, 30, CE, 31, PhD(chem eng), 33. *Prof Exp:* From res chemist to chief chemist, Arlington Plant, Polychem Dept, E I du Pont de Nemours & Co, Inc, Wilmington, 33-57, plastics consult, Chestnut Run Lab, 57-74; CONSULT ACRYLIC PLASTICS, 74- *Mem:* Am Chem Soc; Am Inst Chem Engrs; Am Soc Testing & Mat. *Res:* Agitation; organic synthesis; cellulose plastics; acrylic and dental resins; dyes; pigments; polymerization. *Mailing Add:* 213 Sypherd Dr Newark DE 19711

WILKENS, JERREL L, b Lorraine, Kans, Aug 5, 37; m 68; c 2. INVERTEBRATE NEUROPHYSIOLOGY. *Educ:* Univ Ottawa, Kans, BA, 59; Tulane Univ, MSc, 61; Univ Calif, Los Angeles, PhD(zool), 67. *Prof Exp:* NIMH fel, Brain Res Inst, Univ Calif, Los Angeles, 67-68; from asst prof to assoc prof, 69-81, PROF BIOL, UNIV CALGARY, 81- *Mem:* Soc Neurosci; Can Soc Zool. *Res:* Neuronal activity associated with hormone release; neurophysiology of crustacean motor systems; brachiopod neuromuscular physiology; invertebrate muscle physiology. *Mailing Add:* Dept Biol Univ Calgary Calgary AB T2N 1N4 Can

WILKENS, JOHN ALBERT, b New York, NY, Oct 28, 47; m 77. CHEMICAL ENGINEERING. *Educ:* Cornell Univ, BS, 69, MChE, 71; Mass Inst Technol, PhD(chem eng), 77. *Prof Exp:* Chem engr emission testing, US Environ Protection Agency, 70-72; RES ENGR, E I DU PONT DE NEMOURS & CO, INC, 77- *Mem:* Am Inst Chem Engrs; Am Chem Soc; Sigma Xi; Catalysis Soc. *Res:* Catalysis; reactor engineering; surface chemistry; filtration. *Mailing Add:* 138 Round Hill Rd Kennett Square PA 19348-2608

WILKENS, LON ALLAN, b Ellsworth, Kans, Sept 7, 42; m 65; c 3. NEUROBIOLOGY. *Educ:* Univ Kans, Lawrence, BA, 65; Fla State Univ, PhD(physiol), 70. *Prof Exp:* Fel neurobiol, Univ Tex, Austin, 70-73; asst prof biol, Bryn Mawr Col, 73-75; from asst prof to assoc prof, 75-87, PROF BIOL, UNIV MO, ST LOUIS, 88-, CHMN DEPT, 89- *Concurrent Pos:* Res fel, dept neurobiol, Res Sch Biol Sci, Australian Nat Univ, 80-82. *Mem:* Am Soc Zoologists; Soc Neurosci; AAAS. *Res:* Neurobiology and behavior of invertebrates, including sensory physiology and central processing of tactile and extraretinal information in decapod crustaceans; visual system physiology and escape behaviors in bivalve molluscs. *Mailing Add:* Dept Biol Univ Mo 8001 Natural Bridge Rd St Louis MO 63121

WILKENS, LUCILE SHANES, b Kansas City, Mo, May 19, 50; m 77. CHEMICAL ENGINEERING. *Educ:* Wash Univ, BS, 72; Mass Inst Technol, PhD(chem eng), 77. *Prof Exp:* RES ENGR, E I DU PONT DE NEMOURS & CO, INC, 77- *Mem:* Am Inst Chem Engrs; Am Chem Soc; Sigma Xi. *Res:* Colloid and surface chemistry; rheology; cryogenics. *Mailing Add:* 138 Round Hill Rd Kennett Square PA 19348-2608

WILKERSON, CLARENCE WENDELL, JR, b Laredo, Tex, Aug 12, 44; m 65; c 2. TOPOLOGY. *Educ:* Rice Univ, BA, 66, PhD(math), 70. *Prof Exp:* Asst prof math, Univ Hawaii, Monoa, 70-72; res assoc, Swiss Fed Inst Technol, 72-73; res assoc, Carleton Univ, 73-74; instr, Univ Pa, 74-75, asst prof math, 75-77; PROF MATH, WAYNE STATE UNIV, 77- *Concurrent Pos:* Alfred P Sloan fel, 78-82. *Mem:* Am Math Soc. *Res:* Algebraic topology and homotopy theory of Lie groups, H-spaces and associated spaces. *Mailing Add:* Dept Math Purdue Univ West Lafayette IN 47907

WILKERSON, JAMES EDWARD, b Kingsville, Tex, Oct 2, 45. CARDIOLOGY, SPORTS MEDICINE. *Educ:* Rice Univ, BS, 67; Univ Ore, MA, 68, MS, 69, PhD(physiol), 70; Univ Miami, MD, 87. *Prof Exp:* Asst res physiologist, Inst Environ Stress, Univ Calif, Santa Barbara, 70-75; dir exercise physiol, Univ Ind, 75-85; OWNER, AMERIFIT CORP, 75-; VPRES RES, HYDROTONICS THER CORP, 86-; FEL, DIV CARDIOVASC DIS, UNIV MIAMI, 90- *Concurrent Pos:* NIH fel, 70-71; sr fel, Heart Asn, 71-72; vis prof, Sch Med, Kuwait Univ, 79, Schs Med & Phys Educ & Sport, Univ Helwan, Egypt, 78; fac mem, Safety Eng Sch, USN, 76-83; chmn, sci subcomt, US Olympic Track & Field Develop Comt, 79-80; vis assoc prof, Dept Environ Sci, Johns Hopkins Univ, 83-84; res physiologist, Sch Aerospace Med, USAF, 84-85. *Mem:* Am Physiol Soc; NY Acad Sci; Am Col Sports Med; Am Col Physicians; Sigma Xi. *Res:* Exercise and environmental effects on cardiovascular and pulmonary functions of humans as they age, including endocrine, plasma volume, muscle biochemical and thermoregulatory effects. *Mailing Add:* 10675 NE 11th Ave Miami Shores FL 33138

WILKERSON, JOHN CHRISTOPHER, b Washington, DC, Mar 15, 26; m 58; c 2. PHYSICAL OCEANOGRAPHY. *Educ:* Univ Md, BS, 51. *Prof Exp:* Phys oceanogr, US Naval Oceanog Off, 60-62, from proj scientist to sr proj scientist, 62-67, proj mgr, 67-77, sr proj scientist, 77-85; SR PROJ SCIENTIST, NAT OCEANIC & ATMOSPHERIC ADMIN, 85- *Concurrent Pos:* Proj mgr data mgt, US Naval Oceanog Off, 74-77. *Res:* Remote sensing; aircraft platforms; data management; oceanographic satellites validation. *Mailing Add:* 4834 Butterworth Pl NW Washington DC 20016

WILKERSON, ROBERT C, b Orange, Tex, June 25, 18; m 42; c 1. PHYSICS, MATHEMATICS. *Educ:* Univ Okla, BS, 41. *Prof Exp:* Seismic comput, Geophys Party, Stanolind Oil & Gas Co, Okla, 41-43; spectroscopist, Sinclair Rubber Inc, Tex, 43-47; res & develop div, 47-49, group leader phys instruments, 49-55, anal res, 55, head anal sect, 55-65, ADMIN MGR, TECH CTR, CELANESE CHEM CO, NY, 65- *Mem:* Coblentz Soc. *Res:* Application of physical instruments in analytical support of organic chemistry research and development; development of application of computer systems for management information data and for technical information storage and retrieval systems; thirty-four publications. *Mailing Add:* 221 Rosebud Corpus Christi TX 78404

WILKERSON, ROBERT DOUGLAS, b Wilson, NC, Aug 5, 44; m 65; c 2. PHARMACOLOGY. *Educ:* Univ NC, Chapel Hill, BS, 67; Med Univ SC, MS, 69, PhD(pharmacol), 72. *Prof Exp:* From asst prof to assoc prof, Col Med, Univ South Ala, 73-79; PROF, MED COL OHIO, 79- *Mem:* AAAS; Am Soc Pharmacol & Exp Therapeut; Am Heart Asn. *Res:* Cardiovascular pharmacology. *Mailing Add:* Dept Pharmacol CS 10008 Med Col Ohio Toledo OH 43699

WILKERSON, THOMAS DELANEY, b Detroit, Mich, Feb 18, 32; div; c 3. PLASMA PHYSICS, SPACE PHYSICS. *Educ:* Univ Mich, BS, 53, MS, 54, PhD(physics), 62. *Prof Exp:* Consult, Proj Matterhorn, Princeton Univ, 59-60, mem proj res staff plasma, 60-61; from asst prof to assoc prof, 61-68, actg dir dept, 68-69, PROF PLASMA & SPACE PHYSICS, INST FLUID DYNAMICS & APPL MATH, UNIV MD, COLLEGE PARK, 68- *Concurrent Pos:* Consult, Radiation Div, US Naval Res Lab, 62-64; physics consult, Atlantic Res, 68-69; lectr, von Karman Inst, Brussels, 69; vpres, Versar, Inc, 69-70, consult, 70-; vis prof, Stanford Univ, 70; consult, Stanford Res Inst, 71; vis, Desert Res Inst, 71. *Mem:* Am Phys Soc; Am Astron Soc. *Res:* Laser-guided electrical discharges in gases, high-resolution absorption spectra of atmospheric gases; laser sounding of the atmosphere, infrared schlieren and interferometry; collision-induced spectroscopy in dense gases. *Mailing Add:* Inst Phys Sci & Technol Univ Md College Park MD 20742

WILKES, CHARLES EUGENE, b Worcester, Mass, Oct 9, 39; m 61; c 4. CHEMISTRY. *Educ:* Worcester Polytech Inst, BS, 61; Princeton Univ, PhD(phys chem), 64. *Prof Exp:* From res chemist to sr chemist, BF Goodrich Co, 64-69, sect leader, 69-73, sr res assoc, 73-74, sect mgr, Corp Res New Technol, 74-78, dir, Technol Assessment & Planning, 78-81, DIR CORP RES, RES CTR, B F GOODRICH CO, 81- *Mem:* Am Chem Soc. *Res:* Chemical physics; polymer characterization; x-ray diffraction; molecular spectroscopy; lab automation. *Mailing Add:* PO Box 365 Bath OH 44210-0365

WILKES, GARTH L, b Syracuse, NY, May 22, 42; div; c 2. POLYMER SCIENCE. *Educ:* State Univ NY Col Environ Sci & Forestry, BS, 64; Univ Mass, Amherst, MS, 67, PhD(phys chem), 69. *Prof Exp:* From asst to assoc prof chem eng, Princeton Univ, 69-78; prof chem eng, 78-81, FWB PROF, VA POLYTECH INST & STATE UNIV, 81- *Concurrent Pos:* Res assoc, Textile Res Inst, 74- *Mem:* Am Inst Chem Eng; Am Chem Soc. *Res:* Property of synthetic and biopolymers. *Mailing Add:* Dept Chem Eng Va Polytech Inst Blacksburg VA 24061

WILKES, GLENN RICHARD, b Houtzdale, Pa, Mar 25, 37; m 61; c 2. INORGANIC CHEMISTRY. *Educ:* Pa State Univ, BS, 60; Univ Wis, PhD(inorg chem), 65. *Prof Exp:* Scholar, Univ Calif, Los Angeles, 65-66; RES SCIENTIST, EASTMAN KODAK CO, 66- *Mem:* Am Chem Soc; Am Crystallog Asn; Royal Soc Chem. *Res:* Crystal and molecular structure determination of organometallic compounds by single crystal x-ray diffraction studies; synthesis of metal carbonyls and their derivatives. *Mailing Add:* 1102 Shoemaker Rd Webster NY 14580-8719

WILKES, HILBERT GARRISON, JR, b Los Angeles, Calif, Oct 2, 37; m 78. ECONOMIC BOTANY. *Educ:* Pomona Col, BA, 59; Harvard Univ, PhD(biol), 66. *Prof Exp:* Asst prof biol, Tulane Univ, 66-70; from asst prof to assoc prof, 70-83, PROF BIOL, UNIV MASS, BOSTON, 83- *Concurrent Pos:* Mem, World Maize Germplasm Comn, Rockefeller Found & Food & Agr Orgn; Indo-Am fel, India, 78-79; Maize res fel, Int Maize & Wheat Improvement Ctr, Mex, 85-86; chmn, NRC Ctr Vulnerability Comn, 87-89. *Mem:* Soc Study Econ Bot (secy, 73-75 & 75-77); Bot Soc Am; Soc Econ Bot (pres, 85); Indian Bot Soc; AAAS. *Res:* Evolution under domestication in cultivated plants, especially maize and its wild relatives, teosinte and tripsacum; conservation biology. *Mailing Add:* Dept Biol Col II Univ Mass Boston MA 02125

WILKES, JAMES C, b Mar 13, 21; US citizen; m 45; c 6. BIOLOGY. *Educ:* Troy State Col, BS, 48; Univ Tenn, MS, 50; Univ Ala, PhD(bot), 54. *Prof Exp:* Prof biol, Tenn Wesleyan Col, 50-51; prof & head dept, Jacksonville State Univ, 52-56; ed adv med sci sch aviation med, Air Univ, 56-57; prof & head dept biol, Huntingdon Col, 57-60 & Miss State Col Women, 60-66; head dept, 68-77, PROF BIOL, TROY STATE UNIV, 66- *Res:* Bryophytes; radiation biology. *Mailing Add:* 300 College St Troy AL 36081

WILKES, JAMES OSCROFT, b Southampton, Eng, Jan 24, 32; m 56. CHEMICAL ENGINEERING, DIGITAL COMPUTING. *Educ:* Cambridge Univ, BA, 54, MA, 60; Univ Mich, MS, 56, PhD(chem eng), 63. *Prof Exp:* Demonstr chem eng, Cambridge Univ, 57-60; from instr to assoc prof chem eng, Univ Mich, Ann Arbor, 60-70, chmn dept, 71-77, Arthur F Thurnau prof, 89-92, PROF CHEM ENG, UNIV MICH, ANN ARBOR, 70-, ASST DEAN ENG, 90- *Concurrent Pos:* Assoc, Trinity Col Music, London, Eng, 51. *Mem:* Soc Petrol Engrs; Am Inst Chem Engrs; Sigma Xi. *Res:* Applied numerical methods; fluid mechanics and heat transfer; metal casting; polymer processing; gas storage; two-phase flow. *Mailing Add:* 805 Colliston Rd Ann Arbor MI 48105

WILKES, JOHN BARKER, b Berkeley, Calif, Jan 17, 16; m 38; c 2. PETROLEUM CHEMISTRY. *Educ:* Univ Calif, BS, 37; Stanford Univ, PhD(chem), 48. *Prof Exp:* Chemist, Poultry Prod of Cent Calif, 37-40; res chemist, Chevron Res Co, 48-62, sr res chemist, 62-65, sr res assoc, 65-83; RETIRED. *Concurrent Pos:* US Army Ord dept, 40-45. *Mem:* Am Chem Soc; AAAS. *Res:* Kinetics; thermodynamics; catalysis; chemicals from petroleum; solubility theory. *Mailing Add:* 1570 Kingwood Dr Hillsborough CA 94010

WILKES, JOHN STUART, b Panama, CZ, Mar 6, 47; m 71. ELECTROCHEMISTRY, ORGANIC CHEMISTRY. *Educ:* State Univ NY Buffalo, BA, 69; Northwestern Univ, Evanston, MS, 71, PhD(chem), 73. *Prof Exp:* Tech dir, Frank J Seiler Res Lab, USAF Acad, 73-75; asst prof chem, Univ Colo, Denver, 76-78; RES CHEMIST, FRANK J SEILER RES LAB, USAF ACAD, 78- *Mem:* AAAS; Am Chem Soc; Electrochem Soc. *Res:* Electrochemistry in molten salts; electroorganic synthesis, thermal decompostion of explosives and propellants. *Mailing Add:* Frank J Seiler Res Lab-NC USAF Acad Colorado Springs CO 80840

WILKES, JOSEPH WRAY, b Ft Worth, Tex, May 10, 22; m 47; c 3. COMPUTER SCIENCE, INDUSTRIAL ENGINEERING. *Educ:* Southern Methodist Univ, BS, 44; Wash Univ, St Louis, MS, 51, DSc(indust eng), 54. *Prof Exp:* Instr mech eng, Southern Methodist Univ, 48-49; lectr indust eng, Wash Univ, St Louis, 49-50, instr, 50-54; from asst prof to prof, 54-84, PROF COMPUT SCI ENG, UNIV ARK, FAYETTEVILLE, 84- *Concurrent Pos:* Asst to vpres & provost, Univ Ark, Fayetteville, 58-59, assoc dir, Comput Ctr, 60-62, actg dir, 62-65, dir, 65-72, chmn dept comput sci, 71-83, head instr & res comput serv, 72-81; comput consult consult, Hwy Dept Ark, 61-65; ed consult, State Data Processing Comt, 62-63. *Mem:* Am Inst Indust Engrs; Asn Comput Mach; Am Soc Eng Educ; Sigma Xi. *Res:* Method and statistical analysis; production planning; operations research. *Mailing Add:* 1923 E Joyce No 247 Fayetteville AR 72703

WILKES, MAURICE V, b Dudley, Eng, June 26, 13. COMPUTER SCIENCE. *Educ:* Cambridge Univ, MA, 34, PhD, 37. *Prof Exp:* Sr consult engr, Digital Equip Corp, 80-86; head, Comput Lab, 46-80, EMER PROF COMPUT TECHNOL, CAMBRIDGE UNIV, 86- *Concurrent Pos:* Mem res strategy, Olivetti Res Bd, 86-90, consult, 90- *Honors & Awards:* Turing Award, Asn Comput Mach; Harry Goode Mem Award, Am Fedn Info Processing Soc, Eckert-Mauchly Award; McDowell Award, Inst Elec & Electronics Engrs; Faraday Medal, IEE, London. *Mem:* Foreign assoc Nat Acad Sci; foreign assoc Nat Acad Eng; fel Royal Soc; foreign hon mem AAAS. *Mailing Add:* Comput Lab Cambridge Univ Cambridge CB2-3QG England

WILKES, RICHARD JEFFREY, b Chicago, Ill, Oct 18, 45; m 70; c 2. EXPERIMENTAL HIGH ENERGY PHYSICS. *Educ:* Univ Mich, BSE, 67; Univ Wis, MS, 69, PhD(physics), 74. *Prof Exp:* Res assoc & instr physics, Univ Wash, 74-79, res sci & tech dir, 79-84, sr res assoc, 84-88, res assoc prof, 88-91, RES PROF, UNIV WASH, 91- *Mem:* Am Phys Soc. *Res:* High energy cosmic ray spectra and interactions; elementary particle physics using nuclear emulsion analysis; image analysis; acoustics. *Mailing Add:* Dept Physics FM-15 Univ Wash Seattle WA 98195

WILKES, STANLEY NORTHRUP, b Corvallis, Ore, Jan 3, 27; m 58; c 3. PARASITOLOGY, MARINE ZOOLOGY. *Educ:* Ore State Univ, BS, 50, MS, 57, PhD(zool), 66. *Prof Exp:* Aquatic biologist, Res Div, Ore Fish Comn, 50-51 & 56-60; asst prof biol, E Carolina Col, 65-66; asst prof, 66-71, ASSOC PROF ZOOL, NORTHERN ARIZ UNIV, 71- *Mem:* Am Soc Parasitol; Am Micros Soc; Sigma Xi. *Res:* Taxonomy, life history studies and distribution of parasitic copepods of fishes, elasmobranchs and invertebrates; taxonomy of monogenetic Trematoda; taxonomy and ecology of marine invertebrates and fishes. *Mailing Add:* Dept Biol Sci Box 5640 Northern Ariz Univ Flagstaff AZ 86011

WILKES, STELLA H, ENZYMOLOGY. *Educ:* La State Univ, MS, 50. *Prof Exp:* SR LECTR, INST OCCUP MED, TEX A&M UNIV, 85- *Mailing Add:* Biochem & Biophys Tex A&M Univ College Station TX 77843

WILKES, WILLIAM ROY, b Harvey, Ill, Feb 15, 39; m 59; c 3. ISOTOPE SEPARATION, FUSION TECHNOLOGY. *Educ:* DePauw Univ, AB, 59; Univ Ill, MS, 61, PhD(physics), 66. *Prof Exp:* Asst prof physics, Wake Forest Univ, 65-67; sr res physicist, Monsanto Res Corp, 67-74, res specialist, 74-77, group leader, 77-79, MGR ISOTOPE SEPARATION, MOUND LAB, MONSANTO RES CORP, 79- *Concurrent Pos:* Adj assoc prof physics, Wright State Univ, 89-90. *Mem:* AAAS; Am Phys Soc; Am Nuclear Soc; Soc Nuclear Med. *Res:* Cryogenic isotope separation; Tritium technology; liquid helium; superconductivity. *Mailing Add:* Mound Lab Monsanto Res Corp Miamisburg OH 45343-3000

WILKIE, BRUCE NICHOLSON, b Perth, Scotland, Jan 19, 41; Can citizen; m 66; c 1. IMMUNOLOGY, VETERINARY MEDICINE. *Educ:* Univ Guelph, DVM, 65; Cornell Univ, PhD(immunol), 71. *Prof Exp:* Res asst path, NY State Vet Col, Cornell Univ, 71; asst prof immunol, Univ Bern, 71-73; from asst prof to assoc prof, 73-80, PROF IMMUNOL, UNIV GUELPH, 80-, CHMN DEPT MICROBIOL & IMMUNOL, 87-, DIR, ANIMAL RES CTR, GUELPH-WATERLOO RES, 87- *Mem:* Can Soc Immunol; Am Asn Immunologists; Int Soc Animal Genetics. *Res:* Veterinary immunology and immunopathology. *Mailing Add:* Dept Vet Microbiol & Immunol Univ Guelph Guelph ON N1G 2W1 Can

WILKIE, CHARLES ARTHUR, b Detroit, Mich, Nov 21, 41; m 64; c 3. POLYMER DEGRADATION AND STABILIZATION. *Educ:* Univ Detroit, BS, 63; Wayne State Univ, PhD(inorg chem), 67. *Prof Exp:* from asst prof to assoc prof, 67-86, PROF CHEM, MARQUETTE UNIV, 86- *Mem:* Am Chem Soc; AAAS; Sigma Xi. *Res:* flame retardants; polymer chemistry; nuclear magnetic resonance spectroscopy; thermal decomposition and flame retardation. *Mailing Add:* Dept Chem Marquette Univ Milwaukee WI 53233

WILKIE, DONALD W, b Vancouver, BC, June 20, 31; m 56, 80; c 3. MARINE BIOLOGY. *Educ:* Univ BC, BA, 60, MSc, 66. *Prof Exp:* Asst res biologist, BC Fish & Game Dept, 60-61; asst cur, Vancouver Pub Aquarium, 61-63; cur, Philadelphia Aquarium, Inc, 63-65; AQUARIUM DIR, SCRIPPS INST OCEANOG, 65- *Concurrent Pos:* Aquarium consult. *Mem:* Int Asn Aquatic Animal Med; Nat Marine Educ Asn; Am Asn Zool Parks & Aquariums; Am Cetacean Soc; Am Asn Ichthyol & Herpet; Am Asn Mus. *Res:* Pigmentation and coloration of fishes; aquariology and methods of public education in aquaria and museums. *Mailing Add:* 4548 Cather Ave San Diego CA 92122

WILKIN, JONATHAN KEITH, b Columbus, Ohio, Oct 08, 45; m 67; c 2. CLINICAL PHARMACOLOGY, DERMA PHARMACOLOGY. *Educ:* Ohio State Univ, BA, 67, MS, 71, MD, 74. *Prof Exp:* Tech asst, Franz Theodore Stone Inst Hydrobiol, Ohio, 65-68; teaching asst zool, Ohio State Univ, 67-71; asst prof dermat, Univ Tex Med Sch, 78-82 & MD Anderson Hosp & Tumor Inst, 80-82; from assoc prof to prof dermat, pharmacol & toxicol, Med Col Va, 82-88; DIR DIV DERMAT, OHIO STATE UNIV, 88-

Concurrent Pos: Chief dermat, Vet Admin Med Ctr, 82-88. *Res:* Cutaneous vascular pharmacology and physiology; dermatopharmacology; transdermal drug delivery systems; effects of topical and systemic drugs on the skin. *Mailing Add:* Div Dermat Ohio State Univ Rm 4731 UHC 456 W Tenth Ave Columbus OH 43210

WILKIN, LOUIS ALDEN, b Bath, Maine, Mar 5, 39; m 60; c 3. ORGANIC CHEMISTRY. *Educ:* The Citadel, BS, 60; Clemson Univ, MS, 62, PhD(org chem), 65. *Prof Exp:* From chemist to sr chemist, 65-75, develop assoc, 75-76, GROUP LEADER, TENN EASTMAN CO, 76- *Mem:* Am Chem Soc. *Res:* Novel rearrangements of acetylenic alcohols and esters upon treatment with basic alumina. *Mailing Add:* 1122 Watauga St Kingsport TN 37660

WILKIN, PETER J, b Yeoville, Eng, Nov 28, 43; m; c 2. PHYSIOLOGY. *Educ:* Univ Ill, PhD(physiol & biophys), 71. *Prof Exp:* HEAD, DEPT BIOL, PURDUE UNIV, CALUMET, 79- *Mem:* Am Physiol Soc; Entom Soc Am; Nature Conservancy (treas, 86-); Sierra Club. *Res:* Human and insect physiology. *Mailing Add:* Dept Biol Purdue Univ Calumet Hammond IN 46323-2094

WILKINS, BERT, JR, b Hattiesburg, Miss, Oct 8, 34; m 57; c 2. CHEMICAL ENGINEERING. *Educ:* Ga Inst Technol, BChE, 58, MS, 61, PhD(chem eng), 65. *Prof Exp:* Sr engr, Lockheed-Ga Co div, Lockheed Aircraft Corp, 63-64; sr engr, Humble Oil & Ref Co, 65-67; asst prof chem eng, Univ Mo, Columbia, 67-68; ASSOC PROF CHEM ENG, LA STATE UNIV, BATON ROUGE, 68-, COORDR ENERGY PROGS, 75- *Concurrent Pos:* NASA-Am Soc Eng Educ fac fel, NASA Manned Spacecraft Ctr, Tex, 68, 69; consult, 70-; mem environ adv panel, Gulf Universal Res Corp, 70-; NASA grant, La State Univ, Baton Rouge, 70-, NSF grant, 71- *Mem:* AAAS; Am Inst Chem Engrs. *Res:* Applied mathematics; transport phenomena; process design; bioengineering; ecological systems analysis. *Mailing Add:* 10341 Parkview Dr Baton Rouge LA 70815

WILKINS, BRIAN JOHN SAMUEL, b London, Eng, Feb 28, 37; Can citizen; m 61; c 3. PHYSICAL METALLURGY. *Educ:* Univ London, BSc, 58, PhD(metall), 61. *Prof Exp:* Sci off fuel element develop, Dounreay Exp Reactor Estab, UK Atomic Energy Auth, 61-64; assoc res off mat develop, 64-80, sr res off mat & mech, 80, SECT HEAD MAT & ENG ANALYSIS, ATOMIC ENERGY CAN LTD RES, WHITESHELL LABS, 80- *Res:* Materials development; mathematical modelling (thermomechanical behavior, diffusion, deformation and fracture of solids) applied to reactor safety and nuclear waste management programs. *Mailing Add:* Atomic Energy Can Ltd Res Whiteshell Labs Pinawa MB R0E 1L0 Can

WILKINS, BRUCE TABOR, b Greenport, NY, June 21, 31; m 56; c 3. RESOURCE EXTENSION ADMINISTRATION. *Educ:* Cornell Univ, BS, 52, PhD(resource planning), 67; Mont State Univ, MS, 56. *Prof Exp:* Res biologist, Mont Fish & Game Dept, 56-59; county agr agent, Broome County Exten Serv, NY, 59-63; exten specialist, 63-67, asst prof, 67-73, assoc prof, 73-80, PROF NATURAL RESOURCES, CORNELL UNIV, 80- *Concurrent Pos:* Prog leader, NY Sea Grant Adv Serv, 72-86, assoc dir, 75-86; vis prof, Univ BC, 74; vis Sea Grant prof, Ore State Univ, 75 & Univ Hawaii, 81. *Mem:* Wildlife Soc; Am Fisheries Soc; Am Asn Univ Profs. *Res:* Relationships of human demands and natural resources, especially fish and wildlife resources; resource policy, particularly marine and recreational policy issues. *Mailing Add:* Dept Natural Resources Fernow Hall Cornell Univ Ithaca NY 14853

WILKINS, CHARLES H(ENRY) T(ULLY), physical metallurgy; deceased, see previous edition for last biography

WILKINS, CHARLES LEE, b Los Angeles, Calif, Aug 14, 38; m 66. ANALYTICAL CHEMISTRY. *Educ:* Chapman Col, BS, 61; Univ Ore, PhD(chem), 66. *Prof Exp:* From asst prof to assoc prof, Univ Nebr, Lincoln, 67-76, prof chem, 76-81; PROF CHEM, UNIV CALIF, RIVERSIDE, 81- *Concurrent Pos:* Vis assoc prof, Univ NC, Chapel Hill, 74-75. *Mem:* Am Chem Soc; Am Soc Test & Mat; Am Soc Mass Spectros; Soc Appl Spectros. *Res:* Fourier transform infrared spectrometry; Fourier transform mass spectrometry; nuclear magnetic resonance spectroscopy and computer applications to chemical problems and laboratory automation. *Mailing Add:* Dept Chem Univ Calif Riverside CA 92521

WILKINS, CLETUS WALTER, JR, b Asheville, NC, June 23, 45; m 69; c 2. ORGANIC CHEMISTRY, POLYMER CHEMISTRY. *Educ:* Morgan State Univ, BS, 69; Pa State Univ, MS, 74, PhD(chem), 76. *Prof Exp:* MEM TECH STAFF MAT RES, BELL LABS, 76- *Mem:* Am Chem Soc; AAAS. *Res:* Photochemistry in thin polymer films; solid state photochemistry; organometallic chemistry. *Mailing Add:* Bell Labs 555 Union Blvd Rm 1826 Allentown PA 18031

WILKINS, CURTIS C, b La Crosse, Wis, Oct 28, 35; m 54; c 3. PHYSICAL CHEMISTRY. *Educ:* Wis State Univ, BS, 57; Mich State Univ, PhD(chem), 64. *Prof Exp:* Asst prof chem, WVa Wesleyan Col, 62-65; assoc prof, 65-70, PROF CHEM, WESTERN KY UNIV, 70- *Concurrent Pos:* NSF res participation fel, Univ Tenn, 65. *Mem:* Am Chem Soc. *Res:* Dilute solution properties of stereo-regular polymers; flame photometry studies of trace elements. *Mailing Add:* Dept Chem Western Ky Univ Bowling Green KY 42104

WILKINS, EBTISAM A M SEOUDI, b Monofia, Egypt, Mar 10, 45; US citizen; m 74; c 2. BIOENGINEERING, CHEMICAL ENGINEERING. *Educ:* Cairo Univ, BSc, 65, MSc, 68; Univ Va, MSc, 73, PhD(chem eng), 76. *Prof Exp:* Res eng metall, Nat Res Ctr, Cairo, Egypt, 65-68; res specialist biomed eng, Div Biomed Eng, Univ Va, 69-73, res asst chem eng, Dept Chem Eng, 73-76; fel bioenergetics, Dept Kinesiology, Simon Fraser Univ, 76-77; fel, dept chem eng, Univ BC, 77-78; MEM FAC, UNIV NMEX, 78-, PROF CHEM ENG. *Concurrent Pos:* Fulbright Fel, 88. *Mem:* Am Inst Chem Engrs. *Res:* Waste treatment (bioreactors); solar energy; biomedical sensors (glucose). *Mailing Add:* Univ NMex Albuquerque NM 87131

WILKINS, HAROLD, b Cobden, Ill, Nov 3, 33. HORTICULTURE, PLANT PHYSIOLOGY. *Educ:* Univ Ill, BS, 56, MS, 57, PhD(hort, plant physiol), 65. *Prof Exp:* Res horticulturist, Gulf Coast Exp Sta, Univ Fla, 65-66; from asst prof to assoc prof hort, Univ Minn, St Paul, 66-74, prof hort sci & landscape archit, 74-89; RETIRED. *Concurrent Pos:* Consult, Gas Chromatography Sch, Fisk Univ, 65. *Honors & Awards:* Lauri Award, 67 & 78; Port Award, 80. *Mem:* Am Soc Hort Sci. *Res:* Post harvest physiology; respiration and ethylene emanation in aging floral tissue; physiology of lilies in response to photoperiod and cold treatments; physiology of flowering. *Mailing Add:* Nurserymen's Exchange 2651 Cabrillo Hwy N Half Moon Bay CA 94019

WILKINS, J ERNEST, JR, b Chicago, Ill, Nov 27, 23; m 84; c 2. NUCLEAR ENGINEERING. *Educ:* Univ Chicago, SB, 40, SM, 41, PhD(math), 42; NY Univ, BME, 57, MME, 60. *Prof Exp:* Instr math, Tuskegee Inst, 43-44; from assoc physicist to physicist, Manhattan Proj, Metall Lab, Univ Chicago, 44-46; mathematician, Am Optical Co, 46-50; sr mathematician, Nuclear Develop Corp Am, 50-55, mgr, Physics & Math Dept, 55-57, asst mgr res & develop, 58-59 & mgr, 59-60; asst chmn, Theoret Physics Dept, Gen Atomic Div, Gen Dynamics Corp, 60-65, asst dir lab, 65-70; distinguished prof appl math physics, Howard Univ, 70-77; assoc gen mgr, EG&G Idaho, Inc, 77-80, dep gen mgr, 80-84; RETIRED. *Mem:* Nat Acad Eng; Am Math Soc; Optical Soc Am; Am Nuclear Soc; Soc Indust & Appl Math; AAAS; Math Asn Am. *Res:* Differential and integral equations; Bessel functions; nuclear reactors; calculus of variation. *Mailing Add:* 587 Virginia Ave NE No 612 Atlanta GA 30306

WILKINS, JUDD RICE, b Chicago, Ill, Dec 12, 20; m 50; c 2. MICROBIOLOGY, BIOENGINEERING. *Educ:* Univ Ill, BS, 46, MS, 47, PhD(bact), 50. *Prof Exp:* Asst prof, Med Sch, Univ SDak, 50-51; res investr, Upjohn Co, 51-57; sr res scientist, Booz-Allen Appl Res, 57-64; dept head bact, Eye Res Found, Bethesda, Md, 64-66; res microbiologist, NASA Langley Res Ctr, 66-82; RETIRED. *Honors & Awards:* Spec Achievement & Outstanding Performance Awards, NASA, 76. *Mem:* Am Soc Microbiol. *Res:* Pollution monitoring; microbial detection methods; instrumentation development; life support systems; man in closed environments; electrochemistry; operations research; mathematical models. *Mailing Add:* 281 LittleTown Quarter Williamsburg VA 23185

WILKINS, MICHAEL GRAY, computer modelling, computer applications, for more information see previous edition

WILKINS, PETER OSBORNE, microbiology, for more information see previous edition

WILKINS, RALPH G, b Southampton, Eng, Jan 7, 27; m 84; c 2. INORGANIC CHEMISTRY. *Educ:* Univ Southampton, BSc, 47, PhD(chem), 50; Univ London, DSc(chem), 61. *Prof Exp:* Res chemist, Imp Chem Indust, Eng, 49-52; res assoc inorg chem, Univ Southern Calif, 52-53; from lectr to sr lectr, Sheffield Univ, 53-62; guest prof, Max Planck Inst Phys Chem, 62-63; prof, State Univ NY Buffalo, 63-73; head dept, 73-76, PROF CHEM, NMEX STATE UNIV, 76- *Mem:* Am Chem Soc; Royal Soc Chem. *Res:* Mechanisms of transition metal complexes and metalloenzyme reactions. *Mailing Add:* Dept Chem Univ Warwick Coventry CV4 7AL England

WILKINS, RAYMOND LESLIE, b Boston, Mass, Jan 13, 25; m 50; c 2. RESEARCH MANAGEMENT. *Educ:* Univ Chicago, AB, 51, MS, 54, PhD(chem), 57. *Prof Exp:* Sr scientist, 56-68, head instrument technol lab, 68-73, mgr, Chem Process Res Dept, 74-78, mgr, Process Control Anal Dept, 79-81, mgr spec proj, Rohm & Hass Co, 82-2-; RETIRED. *Concurrent Pos:* Mem, Pa Gov Sci Adv Comt, 69-75, chmn health care delivery panel, 69-71. *Mem:* Electron Micros Soc Am; NY Acad Sci; fel Royal Micros Soc; Am Chem Soc. *Res:* Correlation of the microstructure of heterogeneous organic plastics, polymers and emulsions with their gross properties; mechanisms of polymer formation; sustainable strategies for applying advanced control systems to chemical processes. *Mailing Add:* 63 Golf Club Dr Langhorne PA 19047-2162

WILKINS, ROGER LAWRENCE, b Newport News, Va, Dec 14, 28; m 55; c 1. CHEMICAL PHYSICS. *Educ:* Hampton Inst, BS, 51; Howard Univ, MS, 52; Univ Southern Calif, PhD(chem physics), 67. *Prof Exp:* Aeronaut scientist, NASA, Ohio, 52-55; sr tech specialist, Rocketdyne Div, NAm Aviation, Inc, 55-60; sr staff scientist, Aerophys Dept, Aerodyn & Propulsion Res Lab, 60-80, SR STAFF SCIENTIST, CHEM KINETICS DEPT, AEROPHYSICS LAB, AEROSPACE CORP, 80- *Mem:* Combustion Inst. *Res:* Chemical lasers; application of quantum, statistical and classical mechanics to treatment of energy transfer processes in chemical reactions; application of computers to calculate properties of molecules from first principles. *Mailing Add:* Aerospace Corp P O Box 92957 M5-747 Los Angeles CA 90009-2957

WILKINS, RONALD WAYNE, b Roscoe, Tex, Aug 29, 43; m 66; c 1. PHYSICS, COMPUTER SCIENCE. *Educ:* Harvard Univ, AB, 65; Univ Ill, Urbana, MS, 67, PhD(physics), 73. *Prof Exp:* Fel physics, Univ Kans, 73-75; mem staff, Ind Univ, 75-77; MEM STAFF PLASMA PHYSICS, LOS ALAMOS SCI LAB, 77- *Mem:* Am Phys Soc. *Res:* Solid state physics, especially equation of state computer control and data acquisition. *Mailing Add:* Los Alamos Nat Lab PO Box 1663 MS E534 Los Alamos NM 87545

WILKINS, TRACY DALE, b Sparkman, Ark, July 25, 43. MOLECULAR BIOLOGY, BIOTECHNOLOGY. *Educ:* Univ Ark, BS, 65; Univ Tex, Austin, PhD(microbiol), 69. *Prof Exp:* Fel pharmacol, Univ Ky Med Ctr, 69-71; from asst prof to prof, 72-85, DEPT HEAD MICROBIOL, VA POLYTECH INST & STATE UNIV, 85- *Mem:* Am Soc Microbiol; Soc Intestinal Microbiol & Dis (pres-elect, 88-89, pres, 89-91). *Res:* Anaerobic microbiology of the intestine; colitis; toxins; degradation of xenobiotics and production of carcinogens by intestinal bacteria. *Mailing Add:* Dept Anaerobic Microbiol Va Polytech Inst & State Univ Blacksburg VA 24061

WILKINSON, BRIAN JAMES, b Huddersfield, Eng, Aug 10, 46; m 87; c 2. MICROBIAL BIOCHEMISTRY, MEDICAL MICROBIOLOGY. *Educ:* Univ Col Wales, BSc, 67; Univ Sheffield, PhD(microbiol), 71. *Prof Exp:* Res assoc biochem & microbiol, Univ Ky, 70-73; res fel biochem, Cambridge Univ, 73-76; asst prof med & microbiol, Univ Minn, 76-78; from asst prof to assoc prof, 79-85, PROF MICROBIOL & CHEM, ILL STATE UNIV, 85- *Concurrent Pos:* Broodbank res fel, Cambridge Univ, 73-76; mem spec UK NIH study section, 85; vis prof, Hull Univ, 86. *Mem:* Am Soc Microbiol; Soc Gen Microbiol; fel Am Acad Microbiol. *Res:* Bacterial cell surface, nature and role in pathogenicity; staphylococcal methicillin resistance; staphylococcal osmoregulation. *Mailing Add:* Dept Biol Sci Ill State Univ Normal IL 61761

WILKINSON, BRUCE H, b Lancaster, Pa, June 2, 42. SEDIMENTOLOGY. *Educ:* Univ Wyo, BS, 65, MS, 67; Univ Tex, PhD(geol), 74. *Prof Exp:* Geologist asst oil shale, US Geol Surv, 65; geologist petrol, Gulf Oil Co, 67-69; asst prof, 73-79, ASSOC PROF GEOL, UNIV MICH, ANN ARBOR, 79- *Mem:* Geol Soc Am; Sigma Xi. *Res:* Source and distribution of Holocene sediments of the Texas Gulf Coast and of Michigan; source and distribution of contemporary lacustrine carbonates; evolution; oceanic chemistry. *Mailing Add:* Dept Geol Sci CC Little Bldg Univ Mich Ann Arbor MI 48109

WILKINSON, BRUCE W(ENDELL), b Shelby, Ohio, Aug 9, 28; m 53; c 2. CHEMICAL & NUCLEAR ENGINEERING. *Educ:* Ohio State Univ, BChE, 51, PhD(chem eng), 58. *Prof Exp:* Chem engr, Dow Chem Co, 54-59, staff asst, 59-63, proj coordr, 63-65; from asst prof to prof, 65-90, assoc dir, Div Eng Res, 81-89, EMER PROF, CHEM ENG, MICH STATE UNIV, 90- *Mem:* Am Inst Chem Engrs; Am Nuclear Soc; Am Soc Eng Educ; Sigma Xi; Nat Soc Prof Engrs. *Res:* Radioisotope applications; nuclear fuel processing; nuclear power; energy; environmental effects of energy. *Mailing Add:* Dept Chem Eng Mich State Univ East Lansing MI 48824

WILKINSON, CHARLES BROCK, b Richmond, Va, Jan 16, 22; m 45; c 1. MEDICINE, PSYCHIATRY. *Educ:* Va Union Univ, BS, 41; Howard Univ, MD, 44; Univ Colo, MS, 50. *Prof Exp:* From intern to resident internal med, Freedmen's Hosp, Washington, DC, 45-47; resident psychiat, Univ Colo, 47-50; from instr to asst prof neuropsychiat, Col Med, Howard Univ, 50-55; staff physician, Rollman Receiving Ctr, Cincinnati, Ohio, 58-59; dir adult outpatient serv, Greater Kansas City Ment Health Found, 59-60, dir training, 60-69, assoc dir found, 63-68; clin assoc prof, Sch Med, Univ Mo, Kansas City, 59-65, chmn dept, 67-69, asst dean, 71-80, assoc dean, 81-87, prof psychiat, 65-91, EMER PROF PSYCHIAT, UNIV MO, KANSAS CITY, 91- *Concurrent Pos:* Lectr, Sch Social Work, Howard Univ, 53-55; asst prof, Med Ctr, Univ Kans, 59-65; consult, Family & Children's Serv, Kans, 62-65; mem, Nat Adv Coun, NIMH, 66-70 & US Nat Comt Vital & Health Statist, 72-; exec dir, Greater Kansas City Ment Health Found, 68-89; consult to ed staff, Psychiat Annals, 70-; consult, Region IV, Fed Aviation Admin, 70-; mem bd trustees, Am Psychiat Asn, 71-74, treas, 76-80; mem bd govs, Group Advan Psychiat, 73-75; consult, Psychiat Educ Br, NIMH, 74-; mem, Panel Behav Sci, Nat Res Coun, Nat Acad Sci, 75-; coord, Task Panel, Orgn & Struct Ment Health Serv, President's Comn Ment Health, 77-78; consult, Western Mo Ment Health Ctr, 90-; E Y Williams distinguished scholar award, Nat Med Asn. *Mem:* Fel Am Psychiat Asn; AMA; Nat Med Asn; fel Am Col Psychiat; Group for Advan Psychiat (pres-elect, 91). *Res:* Family dynamics and family therapy; psychiatry and the community; evaluation of psychiatric efforts in community mental health; racism and its effects; studies of black single parent female headed families; human ecology; post traumatic stress disorder. *Mailing Add:* Dept Psychiat Kansas City MO 64108

WILKINSON, CHRISTOPHER FOSTER, b Yorkshire, Eng, Feb 9, 38; m 76; c 1. ENTOMOLOGY, ORGANIC CHEMISTRY. *Educ:* Univ Reading, BSc, 61; Univ Calif, Riverside, PhD(entom), 65. *Prof Exp:* UK Civil Serv Comn sr res fel insecticide chem, Pest Infestation Lab, Agr Res Coun, Eng, 65-66; from asst prof to prof insect toxicol, Cornell Univ, 66-89, dir, Inst Comparative & Environ Toxicol, 80-86; MANAGING TOXICOLOGIST, VERSAR, INC, 89- *Mem:* Soc Toxicol; Am Chem Soc; Soc Environ Toxicol & Chem; Inst Soc Regulatory Toxicol & Pharmacol. *Res:* Structure-activity relationships and mode of action of insecticide synergists; biochemistry; comparative biochemistry of microsomal drug metabolism; toxicology. *Mailing Add:* Risk Focus Versar Inc 6850 Versar Ctr Springfield VA 22151

WILKINSON, DANIEL R, b Glasgow, Ky, May 30, 38; m 61; c 2. PLANT PATHOLOGY, PLANT BREEDING. *Educ:* Western Ky Univ, BS, 61; Clemson Univ, MS, 63; Univ Ill, PhD(plant path), 67. *Prof Exp:* PLANT PATHOLOGIST, PIONEER HI-BREED INT, INC, 67- *Mem:* Am Phytopath Soc. *Res:* Breeding for disease and insect resistance; genetics. *Mailing Add:* Dept Corn Breeding Pioneer Hi-Bred Int Inc Johnston IA 50131-1733

WILKINSON, DAVID IAN, b Cookstown, Northern Ireland, Dec 17, 32; m 57; c 2. BIOCHEMISTRY. *Educ:* Queens Univ, Belfast, BS, 54, PhD(chem), 57. *Prof Exp:* USPHS res fel chem, Wayne State Univ, 57-58 & Univ Calif, Los Angeles, 58-59; res chemist, Brit Drug Houses, Eng, 59-61; NIH res grant & res fel chem, res assoc biochem, 63-73, ADJ PROF DERMAT, SCH MED, STANFORD UNIV, 73- *Concurrent Pos:* Fulbright fel, 57-59; NIH res grant dermat, Sch Med, Stanford Univ, 71-73. *Mem:* The Chem Soc; Am Chem Soc; Soc Invest Dermat. *Res:* Skin lipids; prostaglandins; metabolism of fatty acids in skin; polyunsaturated fatty acids. *Mailing Add:* 1029 Vernier Pl Stanford CA 94305-1006

WILKINSON, DAVID TODD, b Hillsdale, Mich, May 13, 35; div; c 2. PHYSICS, COSMOLOGY. *Educ:* Univ Mich, BSE, 57, MSE, 59, PhD(physics), 62. *Prof Exp:* Lectr physics, Univ Mich, 62-63; from instr to prof, 63-87, chmn dept, 87-90, PROF PHYSICS, PRINCETON UNIV, 90- *Concurrent Pos:* Alfred P Sloan Found fel, 66-68; John Simon Guggenheim fel, 77-78. *Mem:* Nat Acad Sci; Am Acad Arts & Sci; fel Am Phys Soc; Am Astron Soc. *Res:* Atomic physics, properties of electrons and positrons; gravitation and relativity; primeval galaxies; cosmic microwave radiation. *Mailing Add:* Jadwin Hall Princeton Univ Princeton NJ 08544-0708

WILKINSON, EUGENE P DENNIS, b Long Beach, Calif, Aug 10, 18; m; c 4. ENGINEERING. *Educ:* San Diego State Col, BA, 38. *Prof Exp:* From ensign to vadm, USN, 40-74; exec vpres, Data Design Labs, 76-80; pres & chief exec officer, Inst Nuclear Power Opers, 80-84; CONSULT, 84- *Concurrent Pos:* Chmn, Mgt Anal Co; mem bd, Advan Resource Develop Corp, 84-, Nuclear Oversight Comt, 84-; adv, Nuclear Comt Bd, Philadelphia Elec Co, 84-, Nuclear Oversight Comt, 84- *Honors & Awards:* Golden Fleece Award, 55; George Westinghouse Gold Medal, Am Soc Mech Eng, 83; Oliver Townsend Medal, 84; Uranium Inst Gold Medal, 89. *Mem:* Nat Acad Eng; Am Nuclear Soc; Am Soc Naval Engrs. *Res:* Design, construction, operation and management of nuclear electric generating plants. *Mailing Add:* 1449 Crest Rd Del Mar CA 92014

WILKINSON, GEOFFREY, b Todmorden, Eng, July 14, 21. CHEMISTRY. *Prof Exp:* Prof inorg chem, Mass Inst Tech, 50-51; prof, 56-88, EMER PROF, INORGANIC CHEM, IMP COL, 88- *Honors & Awards:* Nobel Prize in Chem, 73; Churchill Gold Medal, 76; Gairdner Found Award, 76. *Mem:* Nat Acad Sci; Am Acad Arts & Sci; Royal Danish Acad Sci. *Mailing Add:* Chem Dept Imp Col Sci & Technol S Kensington London SW7 2AY England

WILKINSON, GRANT ROBERT, b Derby, Eng, Aug 27, 41; US citizen; m; c 4. PHARMACOLOGY. *Educ:* Univ Manchester, BSc, 63; Univ London, PhD(pharmaceut chem), 66. *Prof Exp:* Asst prof pharm, Col Pharm, Univ Ky, 68-71; assoc prof, 71-77, PROF PHARMACOL, SCH MED, VANDERBILT UNIV, 77- *Concurrent Pos:* USPHS fel, Univ Calif, San Francisco, 66-68. *Mem:* Fel AAAS; Am Soc Pharmacol & Exp Therapeut; NY Acad Sci; Am Pharmaceut Asn; fel Acad Pharmaceut Sci; Am Soc Clin Pharmacol Ther. *Res:* Clinical pharmacology; application of analytical methodology, drug metabolism and pharmacokinetics. *Mailing Add:* Dept Pharmacol Vanderbilt Univ Sch Med Nashville TN 37232-6600

WILKINSON, HAROLD L, b Santa Barbara, Calif, Apr 12, 41; m 66; c 6. MEMBRANE TRANSPORT. *Educ:* Univ Ill, PhD(physiol), 76. *Prof Exp:* Asst prof, 78-89, ASSOC PROF BIOL, MILLIKIN UNIV, 89- *Mem:* Sigma Xi; Am Physiol Soc. *Mailing Add:* Dept Biol Millikin Univ Decatur IL 62522

WILKINSON, HAZEL WILEY, BACTERIAL IMMUNOLOGY. *Educ:* Univ Ga, PhD(microbiol), 72. *Prof Exp:* CHIEF, IMMUNOL LAB, CTRS DIS CONTROL, ATLANTA, GA, 78- *Mailing Add:* Ctr Dis Control-5-313-G05 Atlanta GA 30333

WILKINSON, JACK DALE, b Ottumwa, Iowa, Jan 27, 31; m 53; c 5. MATHEMATICS, EDUCATION. *Educ:* Univ Northern Iowa, BA, 52, MA, 58; Iowa State Univ, PhD(math, educ), 70. *Prof Exp:* PROF MATH, UNIV NORTHERN IOWA, 62- *Concurrent Pos:* Consult, 64-; mem, bd dir, Nat Coun Teachers Math & Exec Comt, bd, Nat Coun Accreditation Teacher Educ; rep, Nat Col Athletic Asn. *Mem:* Am Educ Res Asn. *Res:* Activity learning; attitudes toward mathematical learning; learning styles in mathematics education; problem solving and applications of mathematics in the elementary and junior high schools; attracting under represented groups in to mathematics; developing and implementing a better set of linkages between preservice teacher education; the student teaching experience; the initial year(s) of teaching after graduation. *Mailing Add:* Dept Math Univ Northern Iowa Cedar Falls IA 50614

WILKINSON, JOHN EDWIN, b Tacoma, Wash, Nov 11, 42; m 66; c 2. ANALYTICAL & ENVIRONMENTAL CHEMISTRY, GOVERNMENT & REGULATORY AFFAIRS. *Educ:* Univ Puget Sound, BS, 64, MS, 76. *Prof Exp:* Mgr environ relations & sr anal chemist, tech dir, dept sci & technol, Reichhold Chem Inc, 66-85; consult, environ & regulatory affairs, 85-87; DIR, GOVT AFFAIRS, VULCAN CHEM, 87- *Concurrent Pos:* Chlorodioxin specialist, Am Wood Preservers Inst, 73-, chmn, Environ Progs Task Group, 78-; assoc, Chem Mfrs Asn, Halogenated Solvents Indust Alliance, Chlorine Inst, 85- *Mem:* Am Chem Soc; Am Wood Preservers Inst; AAAS. *Res:* Chlorodioxins present in chlorophenols; kemetic and analytical chemical study in the reduction of chlorodioxins in pentachlorophenol; phenol formaldehyde; urea formaldehyde; polyester resins. *Mailing Add:* 1225 19th St NW Suite 300 Washington DC 20036-2410

WILKINSON, JOHN PETER DARRELL, b Englewood, NJ, Nov 24, 38. INFORMATION SYSTEMS. *Educ:* Cambridge Univ, BA Hons, 60, MA, 64; Yale Univ, MEng, 61, DEng, 64. *Prof Exp:* Specialist dynamics, NAm Aviation, 64-67; consult struct, Space Div, 67-68; mech engr solid mech, Corp Res & Develop, 68-72, mgr solid mech, 72-79, mgr liaison oper, 79-80, mgr res & develop appln oper, 80-83, MGR INFO SYSTEM OPER, GEN ELEC CO, 83- *Concurrent Pos:* Adj prof, Polytech Inst NY, 71-72. *Mem:* Am Soc Mech Engrs. *Res:* Structural vibrations; dynamics; mechanical behavior of materials; fracture mechanics; information systems. *Mailing Add:* 12 Englehart Dr Schenectady NY 12302

WILKINSON, JOHN WESLEY, b Bexley, Ont, Nov 1, 28; m 53; c 2. MATHEMATICAL STATISTICS. *Educ:* Queen's Univ, Ont, BA, 50, MA, 52; Univ NC, PhD(math statist), 56. *Prof Exp:* Statistician, Can Industs, Ltd, 52-53; asst prof math, Queen's Univ, Ont, 56-58; res mathematician, Res Labs, Westinghouse Elec Corp, 58-64, fel mathematician, 64-65; prof statist, 65-70; chmn opers res & statist, 70-76, PROF MGT, RENSSELAER POLYTECH INST, 70- *Concurrent Pos:* Consult, Can Industs, Ltd, 56-58, Watervliet Arsenal, Bendix Corp, Kamyr, Inc, 72-, Shaker Res Corp, 72-, NY State Depts Transp, Health, Budget, 72- & NY State Legis Comn Expenditure Rev; assoc ed, Technometrics, 70-77, ed, 78-80. *Mem:* Inst Math Statist; Inst Mgt Sci; Am Inst Decision Sci; Am Soc Qual Control; fel Am Statist Asn. *Res:* Statistical design of experiments; statistical inference; mathematical modeling; statistical applications to problems of environment and energy. *Mailing Add:* Sch Mgt Rensselaer Polytech Inst Troy NY 12181

WILKINSON, MICHAEL KENNERLY, b Palatka, Fla, Feb 9, 21; m 44; c 3. SOLID STATE PHYSICS. *Educ:* The Citadel, BS, 42; Mass Inst Technol, PhD(physics), 50. *Prof Exp:* Res assoc, Res Lab Electronics, Mass Inst Technol, 48-50; res physicist, Oak Ridge Nat Lab, 50-64, assoc dir, 64-72, dir Solid State Div, 72-86, SR ADV, OAK RIDGE NAT LAB, 86- *Concurrent Pos:* Neely vis prof, Ga Inst Technol, 61-62 & adj prof, 62- *Mem:* Fel AAAS; fel Am Phys Soc; Am Crystallog Asn; Sigma Xi. *Res:* Neutron diffraction and spectrometry; magnetic properties of solids; dynamical properties of crystal lattices; x-ray diffraction; physical electronics. *Mailing Add:* 124 E Morningside Dr Oak Ridge TN 37830

WILKINSON, PAUL KENNETH, b Oneonta, NY, Oct 19, 45; m 67; c 3. BIOPHARMACEUTICS, PHARMACODYNAMICS. *Educ:* Univ Conn, BS Pharm, 69; Univ Mich, Ann Arbor, MS & PhD(pharm), 75. *Prof Exp:* Dep chief pharmacist, USPHS, 69-71; asst, Sch Pharm, Univ Mich, Ann Arbor, 71-75; asst prof, Sch Pharm, Auburn Univ, 75-76; asst prof pharmaceut, Sch Pharm, Univ Conn, 76-82; from res fel to sr res fel, MSDRL, Merck & Co, Rahway, NJ, 82-88; mgr pharmaceut develop, R P Scherer Corp, Ann Arbor, Mich, 88-89; MGR PHARMACEUT DEVELOP, MEDIVENTURES, ANN ARBOR, MICH, 89- *Mem:* Am Pharmaceut Asn; Acad Pharmaceut Sci; Am Asn Cols Pharm; Sigma Xi. *Res:* Study of ethyl alcohol concentrations in various segments of the vascular system and evaluation of the kinetics of the oral absorption of ethanol in fasting subjects. *Mailing Add:* Mediventures Inc 655 Phoenix Dr Ann Arbor MI 48108

WILKINSON, PAUL R, b Calcutta, India, Apr 16, 19; m 47; c 5. ECOLOGY. *Educ:* Cambridge Univ, BA, 41, MA, 45, PhD, 68. *Prof Exp:* Entomologist, Colonial Insecticide Res Unit, Uganda, 46-49; res officer, Commonwealth Sci & Indust Res Orgn, Australia, 50-62; res scientist, Can Dept Agr, 62-84; RETIRED. *Mem:* Acarolog Soc Am; Entomol Soc Am; Entom Soc Can. *Res:* Acarology, ecology, physiology and control of ticks and biting flies. *Mailing Add:* 305 Ortona St Lethbridge AB T1J 4K9 Can

WILKINSON, RALPH RUSSELL, b Portland, Ore, Feb 20, 30; m 56. ECONOMICS AND SMALL BUSINESS. *Educ:* Reed Col, BA, 53; Univ Ore, Eugene, PhD(phys chem), 62; Univ Mo, Kansas City, MBA, 74. *Prof Exp:* Sr res chemist, Sprague Elec Co, Tektronix Inc, MacDermid Inc & Chemagro Agr Div, Mobay Chem Corp, 61-72; res chemist, US Vet Hosp, 73-75; assoc chemist technol assessment, 75-80, sr scientist, Midwest Res Inst, 80-84,; asst prof, Rockhurst Col, 85-86; ASST PROF, CLEVELAND CHIROPRACTIC COL, 87- *Mem:* Am Chem Soc; Sigma Xi; NY Acad Sci. *Res:* Technology and risk assessment; organotins; aryl phosphates; chemical economics; pesticides; PCBs; biodegradation; hazardous wastes. *Mailing Add:* 7911 Charlotte St Kansas City MO 64131

WILKINSON, RAYMOND GEORGE, b Duluth, Minn, June 2, 22; m 48; c 3. ORGANIC CHEMISTRY. *Educ:* Harvard Univ, BS, 43; Univ Mich, MS, 48, PhD(org chem), 52. *Prof Exp:* From res chemist to sr res chemist, Lederle Div, Am Cyanamid Co, 51-62, group leader process improv, 62-66; asst to managing ed, Sub Index Div, Chem Abstracts Serv, Ohio State Univ, 66-68; sr res chemist, Lederle Div, Am Cyanamid Co, 68-84, consult, 84-86; RETIRED. *Mem:* Am Chem Soc; Am Humanist Asn. *Res:* Synthesis of steroids, tetracyclines and their degradation products; antituberculosis agents; antimalarials; antitumor; immunomodulating agents. *Mailing Add:* Seven Surrey Lane Montvale NJ 07645

WILKINSON, ROBERT CLEVELAND, JR, b Grand Rapids, Mich, Oct 2, 23; m 48; c 3. ENTOMOLOGY, ECOLOGY. *Educ:* Mich State Univ, BS, 49, MS, 50; Univ Wis, PhD(entom), 61. *Prof Exp:* Entomologist, State Dept Agr, Mich, 51-53; supvr, Off State Entomologist, Wis, 53-57; res asst entom, Univ Wis, 57-60; from asst entomologist to assoc entomologist, 60-70, PROF ENTOM, UNIV FLA, 71- *Concurrent Pos:* Res grants, St Regis Paper Co, Prosper Energy Corp, US Navy, Buckeye Cellulose Corp, Ford Found, SE Coastal Plains Comn, Southern Forest Dis & Insect Res Coun, Ctr Trop Agr, US Forest Serv. *Mem:* AAAS; Entom Soc Am; Soc Am Foresters; Entom Soc Can. *Res:* Forest entomology; bionomics of pine sawflies and bark beetles. *Mailing Add:* 2224 NW 15th St Gainesville FL 32601

WILKINSON, ROBERT EUGENE, b Oilton, Okla, Oct 24, 26; m 51; c 2. PLANT PHYSIOLOGY, WEED SCIENCE. *Educ:* Univ Ill, BS, 50; Univ Okla, MS, 52; Univ Calif, Davis, PhD(plant physiol), 56. *Prof Exp:* Plant physiologist, Crops Res Div, Agr Res Serv, USDA, Ark, 57-62 & NMex, 62-65, assoc agronomist, 65-73, AGRONOMIST, EXP STA, UNIV GA, 73- *Concurrent Pos:* Sr Fulbright-Hayes Lectr appl ecol, Univ Turku, Finland, 74-75; consult, ESAWQ, Piricicicaba, Sao Paulo, Brazil, 78. *Mem:* AAAS; Weed Sci Soc Am; Am Soc Plant Physiol. *Res:* Absorption; translocation; herbicide response; environmental response of plants and effect of herbicides on metabolism. *Mailing Add:* Dept Agron Ga Exp Sta Griffin GA 30223-1797

WILKINSON, ROBERT HAYDN, b Keighley, Eng, Feb 10, 26; m 56; c 4. INSTRUMENTATION. *Educ:* Univ London, BSc, 48; Syracuse Univ, MEE, 61; Mass Inst Technol, ScD(instrumentation), 65. *Prof Exp:* Apprentice engr, Keighley Lifts, Ltd, Eng, 41-47, jr engr, 47-48; sci officer, Radar Res Estab, 49-51; engr, Eng Elec Co, 51-53; engr, Short Bros & Harland, Northern Ireland, 53-54; sr engr, air arm div, Westinghouse Elec Corp, 54-59; prin engr, Link div, Gen Precision Inc, 58-61; PRIN ENGR, C S DRAPER LAB, MASS INST TECHNOL, 61- *Concurrent Pos:* Lectr, Northeastern Univ, 65-67. *Mem:* Inst Elec & Electronics Engrs; Brit Inst Elec Eng. *Res:* Thermal errors in instruments; methods of measurement of instrument parameters; analog function generator techniques; design of electromechanical sensors; design of servos and controls; measurement and modelling of instrument noise processes. *Mailing Add:* C S Draper Lab Inc 555 Technology Sq MS63 Cambridge MA 02139

WILKINSON, RONALD CRAIG, b Augusta, Ga, Oct 1, 43; m 67. FOREST GENETICS, PHYSIOLOGY. *Educ:* Univ Wash, BS, 65; Yale Univ, MF, 66; Mich State Univ, PhD(forest genetics), 70. *Prof Exp:* RES PLANT GENETICIST, NORTHEASTERN FOREST EXP STA, US FOREST SERV, 70- *Mem:* Phytochem Soc NAm. *Res:* Comparative physiology of species, ecological races and hybrids; natural variation and adaptation; genetic and physiological resistance to insects and diseases; biochemical systematics. *Mailing Add:* US Forest Serv Box 968 Burlington VT 05402

WILKINSON, STANLEY R, b West Amboy, NY, Mar 28, 31; m 57; c 3. AGRONOMY, SOIL SCIENCE. *Educ:* Cornell Univ, BS, 54; Purdue Univ, MS, 56, PhD(soil fertil, plant nutrit), 61. *Prof Exp:* Instr soil fertil & plant nutrit, Purdue Univ, 57-60; res soil scientist, Pasture Res Lab, 60-65, RES SOIL SCIENTIST, SOUTHERN PIEDMONT CONSERV RES CTR, AGR RES SERV, USDA, 65- *Mem:* Am Soc Agron; Soil Sci Soc Am; Int Soc Soil Sci; Sigma Xi. *Res:* Mineral nutrient requirements of forage grasses and legumes; biuret toxicity to corn; root growth response to fertilizer; competitive phenomena between forage species; grazing systems research; land application of wastes and environmental quality; grass tetany fescue toxicosis research. *Mailing Add:* Southern Piedmont Conserv Res Ctr Box 555 Watkinsville GA 30677

WILKINSON, THOMAS LLOYD, JR, b Richmond, Va, Dec 14, 39; m 72; c 2. ADHESIVE BONDING TECHNOLOGY, STRUCTURAL TESTING. *Educ:* Va Commonwealth Univ, BS, 63; Univ Richmond, MC, 76. *Prof Exp:* Test engr, Reynolds Metals Co, 63-66, develop engr, 66-67, sr develop engr, 67-73, develop proj dir, 73-80, sr develop proj dir, 80-85, supvr, Composites & Surface Technol, 85-87, SUPVR, ENG TEST, REYNOLDS METALS CO, 87- *Concurrent Pos:* Chmn, D14 Comt Adhesives, Am Soc Testing & Mat, 89- *Mem:* Am Soc Testing & Mat; Am Soc Metals Int; fel Am Inst Chemists. *Res:* Adhesives technology; organic coatings; structural testing; developments in plastics and composite structures; author of numerous publications and three booklets on adhesive bonding. *Mailing Add:* Reynolds Metal Co PO Box 27003 Richmond VA 23261

WILKINSON, THOMAS PRESTON, b Gisburn, Eng, Mar 14, 41; m 66; c 2. EDUCATION ADMINISTRATION, RESOURCE MANAGEMENT. *Educ:* Univ Durham, BSc, 63; Univ Newcastle, PhD(geomorphol), 72. *Prof Exp:* From lectr to asst prof, 67-73, ASSOC PROF GEOG, CARLETON UNIV, 73-, DIR CONTINUING EDUC, 85- *Concurrent Pos:* Vis lectr, Univ Liverpool, 72-73. *Res:* Fluvial geomorphology; geography curricula; environmental management. *Mailing Add:* Dept Geog Carleton Univ Ottawa ON K1S 5B6 Can

WILKINSON, THOMAS ROSS, b Baltimore, Md, Aug 20, 37; m 65; c 3. MICROBIOLOGY. *Educ:* Univ Notre Dame, BS, 59; Univ Md, College Park, MS, 62; Wash State Univ, PhD(microbiol), 70. *Prof Exp:* Technician aerobiol, Naval Biol Lab, Univ Calif, Berkeley, 65-66; from asst prof to prof path & immunol, SDak State Univ, 70-81, head, Dept Microbiol, 75-81; ASSOC DEAN, COL AGR, ASSOC DIR, AGR EXP STA & PROF MICROBIOL, NDAK STATE UNIV, 81- *Mem:* Sigma Xi; Am Soc Microbiol; NY Acad Sci. *Res:* Rapid isolation technique for Listeria monocytogenes; survival of pathogens on metal surfaces; miniature cell systems for virus isolation and epidemiology of enclosed environments; epidemiology of Listeria and pathogenesis of Listeria L-forms; alcohol fuel production by a small farm scale plant. *Mailing Add:* 118 Ft Rutledge Clemson SC 29631

WILKINSON, W(ILLIAM) C(LAYTON), b Tefte, Ind, Dec 17, 14; m 46; c 4. ELECTROMAGNETIC ENGINEERING. *Educ:* Purdue Univ, BSEE, 41. *Prof Exp:* Res engr, RCA Victor Co, RCA Corp, 41-42 & RCA Labs, 42-61, eng supvr, Missile & Surface Radar Div, 61-77, eng supvr astro electronics, 77-80; PRES, SPACE ANTENNA TECH INC, 80- *Mem:* Sr mem Inst Elec & Electronics Engrs. *Res:* Computer aided design of antennas and microwave systems for space communications satellites and systems. *Mailing Add:* 55 Littlebrook Rd N Princeton NJ 08540

WILKINSON, WILLIAM H(ADLEY), b Galt, Ont, Dec 1, 27; US citizen; c 5. MECHANICAL ENGINEERING. *Educ:* Rensselaer Polytech Inst, BME, 49, MME, 52. *Prof Exp:* Instr mech eng, Rensselaer Polytech Inst, 49-55, asst prof, 55-56; proj engr, 56-58, asst consult, 58-60, assoc staff engr, 60-64, fel, 64-70, prin mech engr, 70-80, SR RES SCIENTIST, COLUMBUS LABS, BATTELLE MEM INST, 80- *Concurrent Pos:* Engr, Am Locomotive Co, 53-54. *Mem:* Soc Automotive Engrs; Am Soc Mech Engrs; Am Soc Heating, Refrig & Air-Conditioning Engrs. *Res:* Thermodynamic cycles and conversion systems; flow phenomena; comfort conditioning; advanced system synthesis and modeling; mechanisms; gearing; advanced vehicle propulsion systems. *Mailing Add:* Battelle Mem Inst Columbus Labs 505 King Ave Columbus OH 43201

WILKINSON, WILLIAM KENNETH, b Newcastle, Ind, Jan 17, 18; m 42; c 5. ORGANIC CHEMISTRY, POLYMER CHEMISTRY. *Educ:* DePauw Univ, AB, 40; Northwestern Univ, MS, 47, PhD(org chem), 48. *Prof Exp:* Training supvr, Trojan Powder Co, 41-43; res chemist, Firestone Tire & Rubber Co, 43-45; from res chemist to res assoc, 48-65, RES FEL, BENGER LAB, TEXTILE FIBERS DEPT, E I DU PONT DE NEMOURS & CO, 65- *Mem:* Am Chem Soc. *Res:* Rubber and textile fibers; explosives; acetylene chemistry. *Mailing Add:* 1010 Glenwood Blvd Waynesboro VA 22980-3411

WILKINSON, WILLIAM LYLE, b Sikeston, Mo, Jan 18, 21; m 45, 69, 89; c 6. OPERATIONS RESEARCH. *Educ:* US Naval Postgrad Sch, BS & MS, 55. *Prof Exp:* Sr staff scientist, George Washington Univ, 65-83; RETIRED. *Concurrent Pos:* Lectr, Univ Calif, Los Angeles, 65-66. *Mem:* Opers Res Soc Am. *Res:* Transportation networks in logistics research; computer-based management information systems; quantitative evaluation of weapon systems and tactics; man-computer systems; military sciences. *Mailing Add:* 1309 Alps Dr McLean VA 22102-1501

WILKNISS, PETER EBERHARD, b Berlin, Ger, Sept 28, 34; US citizen; m 63; c 2. OCEANOGRAPHY, RADIOCHEMISTRY. *Educ:* Munich Tech Univ, MS, 59, PhD(radiochem), 61. *Prof Exp:* Res asst, Tech Univ Munich, Ger, 59-61; res chemist, radiol protection officer, US Naval Ordiance Station, 61-64, head nuclear chem br, 64-66, res oceanographer, us Naval Res Lab, 66-70, head geochem sect, 70-75; prog mgr, Nat Ctr Atmosphere Res Prog, NSF, 75-76, prog mgr Ocean Sediment Coring Prog, 76-80, team mgr, Ocean Drilling proj team, 80, dir, Div Ocean Drilling Progs, 80-81, sr sci assoc, 81-82, dep asst dir, Sci, Technol & Int Affairs Directorate, 82-84, DIR, DIV POLAR PROGS, NSF, 84- *Concurrent Pos:* Nat Res Coun, Polar Res bd liasion mem, 84-; head chem oceanog br, US Naval Res Lab, 71-73, chmn radiol comt, 74-75; mem, comt atmospheric chem & radioactivity, Am Meteorol Soc, 75-78; mem, Interagency Comt Atmosphereic Sci, 75-76; mem, NASA Space Sta Adv Comt, 88- *Mem:* AAAS; Sigma Xi; Am Geophys Union. *Res:* Radio and nuclear chemistry, solid propellants, oceanography, marine geochemistry, satellite images applied to meteorology, air/sea interactions; anthropogenic impact on the global environment, natural radioactivity in the troposphere; radiochemistry applied to oceanography. *Mailing Add:* Polar Prog NSF NSF 1800 G St NW Washington DC 20550

WILKOFF, LEE JOSEPH, b Youngstown, Ohio, Oct 17, 24; m 53; c 1. MICROBIOLOGY, BIOCHEMISTRY. *Educ:* Roosevelt Univ, BS, 48; Univ Chicago, PhD(microbiol), 63. *Prof Exp:* Chemist, H Kramer & Co, 48-49; res asst biochem, Ben May Lab Cancer Res, Univ Chicago, 49-52 & Dept Med, 54-60; biochemist, Vet Admin Hosp, Hines, Ill, 52-54; dir microbiol lab, Woodard Res Corp, 63-64; sr microbiologist, 64-70, HEAD CELL BIOL DIV, SOUTHERN RES INST, 70- *Mem:* AAAS; Am Asn Cancer Res; Soc Exp Biol & Med; Am Soc Microbiol; Tissue Cult Asn; Sigma Xi. *Res:* Cell biology and chemotherapy of tumor cells; effect of anticancer drugs on the kinetic behavior of tumor cells; cellular sites of action of anticancer agents. *Mailing Add:* Southern Res Inst Birmingham AL 35205

WILKOV, ROBERT SPENCER, b New York, NY, Feb 2, 43; m 64; c 2. COMPUTER SCIENCE. *Educ:* Columbia Univ, BS, 64, MS, 66, PhD(elec eng), 68. *Prof Exp:* Res asst elec eng, Columbia Univ, 67-68; mem staff comput sci, 68-72, MGR NETWORK ANALYSIS, IBM RES CTR, 72-, ACCT DEVELOP DIR, 90- *Concurrent Pos:* Consult, Otis Elevator Co, 63-68; lectr, Univ Conn, 69-70; adj asst prof, City Col New York, 70- *Mem:* AAAS; Inst Elec & Electronics Engrs. *Res:* Applications of graph theory and queuing theory in the analysis and synthesis of voice and data communication systems. *Mailing Add:* 7259 Scarsdale Pl San Jose CA 95120

WILKOWSKE, HOWARD HUGO, b Zachow, Wis, Sept 10, 17; m 48; c 3. DAIRY BACTERIOLOGY. *Educ:* Tex Tech Col, BS, 40, MS, 42; Iowa State Col, PhD, 49. *Prof Exp:* Assoc prof dairy mfg & assoc dairy technologist, Agr Exp Sta, Inst Food & Agr Sci, Univ Fla, 50-57, asst dir, Agr Exp Sta, 57-68, asst dean res, 68-79, dir internal energy mgt, 80-81; RETIRED. *Concurrent Pos:* Mem adv coun, Ctr Trop Agr, US AID dairy specialist, Costa Rica, 58, Ghana, 69 & Venezuela, 71. *Mem:* AAAS; Am Soc Microbiol; Am Dairy Sci Asn. *Res:* Antibiotics in dairy products; continuous and automatic manufacture of fermented dairy products; bacteriophage of dairy microorganisms; agricultural research administration. *Mailing Add:* 1040 SW 11th St Gainesville FL 32601

WILKS, JOHN WILLIAM, b Kenosha, Wis, July 5, 44; m 84. ENDOCRINOLOGY, CELL BIOLOGY. *Educ:* Univ Wis-Madison, BS, 66; Cornell Univ, PhD(physiol), 71. *Prof Exp:* Sr res scientist reproductive endocrinol, Fertil Res, UpJohn Co, 70-85, SR SCIENTIST, CANCER & INFECTIOUS DIS RES, UPJOHN CO, 85- *Concurrent Pos:* Vis assoc prof, Dept Cell Biol, Baylor Col Med, 81. *Mem:* Tissue Culture Asn; Endocrine Soc; AAAS; Am Asn Cancer Res; Am Soc Cell Biol. *Res:* Discovery of anti-cancer pharmaceutical therapies; tumor angiogenesis. *Mailing Add:* Cancer & Infectious Dis Res Upjohn Co Kalamazoo MI 49001

WILKS, LOUIS PHILLIP, b Dayton, Ohio, June 28, 13; m 38; c 3. CHEMISTRY. *Educ:* Univ Dayton, BS, 35. *Prof Exp:* Instr chem, Univ Dayton, 35; res chemist, Thomas & Hochwalt Lab, 35-38; RETIRED. *Concurrent Pos:* Consult, 78- *Mem:* AAAS; Am Chem Soc. *Res:* Products from petrochemical by-products; agricultural chemicals; research management and corporate planning. *Mailing Add:* 1906 Mirmar Munster IN 46321-2719

WILL, CLIFFORD MARTIN, b Hamilton, Ont, Can, Nov 13, 46; m 70; c 2. THEORETICAL ASTROPHYSICS, GENERAL RELATIVITY. *Educ:* McMaster Univ, BSc, 68; Calif Inst Technol, PhD(physics), 71. *Prof Exp:* Instr physics, Calif Inst Technol, 71-72; fel, Enrico Fermi Inst, Univ Chicago, 72-74; asst prof physics, Stanford Univ, 74-81; assoc prof, 81-85, PROF PHYSICS, WASH UNIV, ST LOUIS, 85-, CHMN PHYSICS, 91- *Concurrent Pos:* Res fel, Calif Inst Technol, 71-72; Sloan Found res fel, 75-79; fel, Mellon Found, 78-79; chmn, comt accuracy time transfer in satellite systs, Nat Acad Sci, 84-86; mem exec comt, Am Phys Soc, 88-90; div assoc ed, Phys Rev Letters, 89- *Mem:* Fel Am Phys Soc; Am Astron Soc; Sigma Xi; Int Astron Union; Int Soc Gen Relativity & Gravitation; Am Asn Physics Teachers. *Res:* General relativity theory and its applications to astrophysics; gravitational radiation; black holes; cosmology; experimental tests of general relativity. *Mailing Add:* Dept Physics Wash Univ One Brookings Dr Campus Box 1105 St Louis MO 63130

WILL, FRIT GUSTAV, b Breslau, Ger, Jan 12, 31; m 58; c 3. ELECTROCHEMISTRY. *Educ:* Munich Tech Univ, BS, 53, MS, 57, PhD(phys chem), 59. *Prof Exp:* Res scientist, Eng Res & Develop Labs, US Dept Army, Va, 59-60; electrochemist, 60-69, mgr electrochem mat & reactions unit, 69-72, MEM RES STAFF, RES & DEVELOP CTR, GEN ELEC CO, 73- *Concurrent Pos:* Vis prof, Univ Bonn, WGer, 73; div ed, J Electrochem Soc, 74-84; chmn phys elec chem div, Electrochem Soc, 79-80. *Honors & Awards:* Battery Res Award, Electrochem Soc, 64; Indust Res Award IR-100, 75. *Mem:* Electrochem Soc (vpres, 84-86, pres, 87-88). *Res:* Electrocatalysis; electrode kinetics; fuel cells; batteries; solid electrolyte; zinc-halogen; electrochemical sensors; conducting polymers; conduction in oxides and nitrides. *Mailing Add:* Gen Elec Res & Develop Ctr PO Box 8 Schenectady NY 12301

WILL, FRITZ, III, b Richmond, Va, Oct 24, 26; m 54; c 2. ANALYTICAL CHEMISTRY. *Educ:* Univ Va, BS, 49, MS, 51, PhD(chem), 53. *Prof Exp:* Asst chem, Univ Va, 47-51; res chemist, Res Labs, Aluminium Co Am, 53-65; res chemist, Phillip Morris, Inc, 65-69, mgr, Anal Chem Div, Philip Morris Res Ctr, 69-79, mgr beverage res & develop, 80-81, coordr anal chem tobacco process, 82-85; RETIRED. *Honors & Awards:* Medal, Am Inst Chemists, 49. *Mem:* Am Chem Soc; Soc Appl Spectros; Sigma Xi. *Res:* Analytical methods; spectrophotometry; ultraviolet and visual absorption spectroscopy; nuclear magnetic resonance spectroscopy. *Mailing Add:* 2301 Astoria Dr Richmond VA 23235

WILL, JAMES ARTHUR, b Wauwatosa, Wis, Nov 2, 30; m 53; c 3. PHYSIOLOGY. *Educ:* Univ Wis, BS, 52, MS, 53, PhD(vet sci), 67; Kans State Univ, DVM, 60. *Prof Exp:* Vet, Columbus Vet Hosp, Wis, 60-67; from asst prof to assoc prof, 67-74, chmn dept, 74-78, PROF VET SCI, UNIV WIS-MADISON, 74-, PROF ANESTHESIOL, MED SCH, 80-, DIR GRAD SCH RES ANIMAL RESOURCES CTR, 82- *Concurrent Pos:* NIH spec fel, New Med Sch, Univ Liverpool, 72-73; mem coun basic sci & circulation, Am Heart Asn; mem comt primary pulmonary hypertension, WHO; consult to domestic & foreign co. *Mem:* AAAS; Am Vet Med Asn; fel Royal Soc Med, London; Asn Am Vet Med Col (pres); Am Physiol Soc; Am Thoracic Soc; Soc Exp Biol & Med; Am Asn Lab Animal Sci. *Res:* Cardiopulmonary physiopathology, particularly relationship between function and disease under natural and altered environmental conditions or with impairment of function by a disease process. *Mailing Add:* Dept Vet Sci Univ Wis 1655 Linden Dr Madison WI 53706

WILL, JOHN JUNIOR, b Cincinnati, Ohio, Aug 13, 24; m 46; c 3. HEMATOLOGY. *Educ:* Univ Cincinnati, MD, 47. *Prof Exp:* Intern med, Presby Hosp, New York, 47-48; resident internal med, Cincinnati Gen Hosp, 48-50; asst chief med serv, Travis AFB, 50-52; from instr to prof med, 52-76, PROF INTERNAL MED, COL MED, UNIV CINCINNATI, 76- *Concurrent Pos:* Fel, Nutrit & Hemat Lab, Cincinnati Gen Hosp, 52-55, co-dir, Hemat Lab, 57-, clinician, Outpatient Dept & chief clinician, Hemat Clin. *Mem:* Am Fedn Clin Res. *Res:* Nutrition; relationship of vitamin B-12 folic acid and ascorbic acid to the absorption and utilization of versene in iron deficiency anemia; nuclease inhibitors in human white blood cells in leukemia and other disease states; effect of antimetabolites on leukemia in rats and humans; cancer chemotherapy. *Mailing Add:* Cincinnati Health Dept 3101 Burnet Ave Cincinnati OH 45227

WILL, PAUL ARTHUR, b Weslaco, Tex, Feb 9, 46; m 70; c 2. FOOD SCIENCE & TECHNOLOGY. *Educ:* Tex A&M Univ, BS, 70; Okla State Univ, MS, 74, PhD(food sci), 78. *Prof Exp:* Asst prof, 78-85, ASSOC PROF & DIR, SUL ROSS STATE UNIV, 85-; INSTR MEAT SCI, OKLA STATE UNIV, 78- *Mem:* Am Meat Sci Asn; Inst Food Technologists; Am Animal Sci Asn; Sigma Xi. *Res:* Efficiency of red meat production; palatibility of the produced product; method of fabrication and processing; physical and chemical properties of meat. *Mailing Add:* PO Box C 158 Alpine TX 79830

WILL, PETER MILNE, b Peterhead, Scotland, Nov 2, 35; m 59; c 3. COMPUTER SCIENCE, ELECTRICAL ENGINEERING. *Educ:* Aberdeen Univ, BScEng, 58, PhD(elec eng), 60. *Prof Exp:* Mem res staff automatic control, Res Labs Assoc Elec Industs, Ltd, Eng, 61-62; proj leader indust electronics & control, AMF Brit Res Labs, 62-64; sr res physicist, Morehead Patterson Res Ctr, Am Mach & Foundry Co, Conn, 64-65; mem res staff comput sci, T J Watson Res Ctr, IBM Corp, 65-80; dir prod systs eng, Schlumberger Well Serv, 80-85; DIR DESIGN STRATEGY, H P CORP, 85- *Concurrent Pos:* Lectr, Univ Conn, Stamford Exten, 66-71; mem, Eng Res Ctr Eval Group, 89; mem, Info Sci & Tech Comn, DARPA, 87-; mem, Computer Sci & Tech Bd, Nat Acad Sci, 82-86. *Honors & Awards:* Joseph Engelberger Award Robotics, 89. *Mem:* Inst Elec & Electronics Engrs. *Res:* Application of computers to non-traditional fields; image processing; bandwidth compression; robotry; robot vision; multispectral imagery. *Mailing Add:* Schlumberger Well Serv 5000 Gulf Freeway Houston TX 77001

WILL, THEODORE A, b Orange, NJ, Aug 9, 37; m 78; c 2. SOLID STATE PHYSICS, SCIENCE EDUCATION. *Educ:* Johns Hopkins Univ, AB, 59; Univ Chicago, SM, 61; Case Western Reserve Univ, PhD(physics), 68. *Prof Exp:* Asst prof physics, Heidelberg Col, 62-64 & Grinnell Col, 68-72; mem fac, Mat Res Ctr, Nat Univ Mex, 74-83, head dept mat sci, 77-81, vis prof, Univ Alta, 81-83; PROF & CHAIR, DEPT PHYSICS, CALIF STATE UNIV, 83- *Concurrent Pos:* Vis prof, Univ Alta, 81-83; vis prof, Mat Res Ctr, Nat Univ Mex, 72-74. *Mem:* Am Phys Soc; Mex Phys Soc; Am Asn Physics Teachers; Sigma Xi. *Res:* Electron phonon interaction in PbTl and PbBi from superconductor tunneling measurements; electron tunneling. *Mailing Add:* Dept Physics Calif State Univ Dominque Hills Carson CA 90747

WILLARD, DANIEL, b Baltimore, Md, Aug 22, 26; m 58; c 2. PHYSICS. *Educ:* Yale Univ, BS, 49, MS, 50; Mass Inst Technol, PhD(physics), 54. *Prof Exp:* Res assoc physics, Brookhaven Nat Lab, 54-55; instr, Swarthmore Col, 55-58; assoc prof, Va Polytech Inst, 58-61; opers res analyst, Opers Res Off, Johns Hopkins Univ, 61 & Res Anal Corp, Va, 61-64; OPERS RES ANALYST, OFF UNDER SECY US ARMY, 64- *Concurrent Pos:* Consult, Langley Res Ctr, NASA, 59-61. *Mem:* Am Phys Soc; Mil Opers Res Soc. *Res:* Cosmic rays; heavy unstable particles; radio astronomy; mathematical models of combat; operations research; systems analysis. *Mailing Add:* Off Under Secy Army SAUS-OR Washington DC 20310-0102

WILLARD, DANIEL EDWARD, b Cincinnati, Ohio, Oct 24, 34; m 78; c 3. ENVIRONMENTAL ECOLOGY. *Educ:* Stanford Univ, AB, 60, Univ Calif, Davis, PhD(zool), 66. *Prof Exp:* Lectr biol, Univ Tex, Austin, 66-67, asst prof bot & zool, 67-70; asst prof, Univ Wis-Madison, 70-72, assoc scientist, 72-77; ASSOC PROF, SCH PUB & ENVIRON AFFAIRS, IND UNIV, BLOOMINGTON, 77- *Concurrent Pos:* Lectr, San Diego Zool Soc, 60; vis asst prof, Inst Marine Sci, Univ Ore, 72-73; consult, Ore Pub Utility Comn, 75-76 & Wis Attorney Gen Off, 73- *Mem:* Ecol Soc Am; Human Ecol Soc; AAAS; Am Behav Soc; Am Inst Biol Sci. *Res:* Impacts of energy development on biological systems; dynamics of wetlands; human impact on wildlife; management of endangered species; ecological regulation. *Mailing Add:* Dept Pub Admin Ind Univ Bloomington IN 47405

WILLARD, HARVEY BRADFORD, nuclear physics, for more information see previous edition

WILLARD, JAMES MATTHEW, b St Johnsbury, Vt, Nov 18, 39; m 61; c 3. HYDROLOGY & WATER RESOURCES. *Educ:* St Michael's Col, Vt, AB, 61; Cornell Univ, PhD(biochem), 67. *Prof Exp:* Res assoc biochem, Case Western Reserve Univ, 66-69; asst prof biochem, Col Med, Univ Vt, 69-75; asst prof, 75-77, asst dean arts sci, 81-84, ASSOC PROF BIOL, CLEVELAND STATE UNIV, 77- *Honors & Awards:* Fulbright Award, 88-89. *Mem:* Am Soc Biol Chemists. *Res:* Preparation and effect of certain analogues of phosphoribosyl pyrophosphate on de novo purine synthesis; zeolites as dietary supplement; dietary use of fructose; predictive method of catalysis (intermedion theory); restoration of biotic condition with zeolite of acid water. *Mailing Add:* Dept Biol Cleveland State Univ Cleveland OH 44115

WILLARD, JOHN ELA, b Oak Park, Ill, Oct 31, 08; m 37; c 4. PHYSICAL CHEMISTRY. *Educ:* Harvard Univ, SB, 30; Univ Wis, PhD(phys chem), 35. *Prof Exp:* Instr chem, Avon Sch, 30-32 & Haverford Col, 35-37; from instr to prof, 37-47, dean, Grad Sch, 58-63, Vilas res prof, 63-79, chmn, dept chem, 70-72, EMER PROF CHEM, UNIV WIS-MADISON, 79- *Concurrent Pos:* Assoc sect chief, plutonium chem sect, Metall Lab, Univ Chicago, 42-44, dir pile chem div, 45-46; area supvr, Hanford Eng Works, E I du Pont de Nemours & Co, Inc, 44-45; consult, Oak Ridge Nat Lab, 46-49; mem, Phys Chem Panel, Off Naval Res, 48-50, surv comt, AEC, 49 & isotope distrib adv comt, AEC, 53-57; secy, Nuclear Chem Sect, Int Cong Pure & Appl Chem, 51; mem partic inst exec bd, Argonne Nat Lab, 50-53, chmn, 52-53, chem div vis comt, 58-64; mem adv bd, Gordon Res Confs, 55-60, chmn conf radiation chem, 68; assoc ed, Chem Reviews, 55-58 & Radiation Res, 65-68; mem bd vis chem div, Brookhaven Nat Lab, 56-59; mem, Panel on Basic Res & Grad Educ, Pres' Sci Adv Comt, 59-60 & Panel Basic Res & Nat Goals, Nat Acad Sci, 64-65. *Honors & Awards:* Am Chem Soc Award, 59. *Mem:* Am Chem Soc; Am Phys Soc; Radiation Res Soc; AAAS. *Res:* Radiation chemistry; photochemistry; chemical effects of nuclear transformations; nature and reactions of trapped intermediates. *Mailing Add:* Dept Chem Univ Wis Madison WI 53706

WILLARD, MARK BENJAMIN, b Chicago, Ill, Feb 18, 43. NEUROBIOLOGY. *Educ:* Oberlin Col, BA, 65; Univ Wis, PhD(biochem), 71. *Prof Exp:* Teaching fel anat & biochem, 71-74, from asst prof to assoc prof, 74-85, PROF ANAT NEUROBIOL & BIOCHEM, SCH MED, WASH UNIV, 85- *Res:* Role of regulation of neuronal gene expression in controlling the development and regeneration of the cells of the nervous system. *Mailing Add:* Dept Anat Sch Med Univ Wash 660 S Euclid Ave St Louis MO 63110

WILLARD, PAUL EDWIN, b Ogdensburg, NY, Sept 11, 19; m 43; c 2. ORGANIC CHEMISTRY. *Educ:* Univ Chicago, BS, 41. *Prof Exp:* Res chemist, Gen Labs, US Rubber Co, NJ, 41-44 & Celanese Corp Am, 44-47; res chemist, Ohio-Apex, Inc, 47-50, from asst res dir to res dir, Ohio-Apex Div, FMC Corp, 50-57, tech mgr, 57- 58, mgr applns, Tech Serv Lab, Org Chem Div, 58-62, asst dir govt liaison, NJ, 62-65, mgr ceramic fibers res & develop, 65-69, sr res chemist, 69-74, res assoc, 74-81; RETIRED. *Mem:* Am Chem Soc; Nat Asn Corrosion Engrs. *Res:* Hydrogen peroxide; thermosetting plastics, especially rheology; plastics testing and applications; composite materials; ceramic fibers; thermal analysis. *Mailing Add:* Ten Skillman Rd Skillman NJ 08558

WILLARD, PAUL W, b Marshalltown, Iowa, Mar 21, 33; m 52; c 4. HAZARD COMMUNICATION, PHARMACOLOGY. *Educ:* Iowa State Univ, BS, 55; Univ Iowa, PhD(physiol), 59; Ind Univ, MBA, 68; Am Bd Toxicol, dipl, 81. *Prof Exp:* Nat Heart Inst fel physiol, Lankenau Hosp, Philadelphia, Pa, 59-61; sr scientist, Div Pharmacol Res, Eli Lilly & Co, 61-69; clin res coordr, Med Prods Div, 3M Co, 69-75; dir regulatory affairs, Medtronic Inc, 75-77; supvr toxicol serv, 78-80, MGR PROD REGULATORY TOXICOL, 3M CO, 80- *Concurrent Pos:* Adj prof environ health, Univ Minn, 81- *Honors & Awards:* Allderdices Award. *Mem:* Am Soc Pharmacol & Exp Therapeut; Soc Toxicol; Am Col Toxicol; Am Indust Hyg Asn; Am Conf Chem Labeling. *Res:* Cardiovascular pharmacology, physiology and toxicology; regulatory affairs; industrial toxicology; product toxicology. *Mailing Add:* Toxicol Serv 220-2E 3M Ctr St Paul MN 55144

WILLARD, ROBERT JACKSON, b Brockton, Mass, Mar 21, 29; m 54; c 2. GEOLOGY, RESEARCH ADMINISTRATION. *Educ:* Boston Univ, AB, 51, AM, 53, PhD(geol), 58. *Prof Exp:* Instr geol, Wellesley Col, 56-57; from instr to asst prof, Univ Ark, 57-63; geologist, 63-67, head, Fabric Anal Lab, 68-74, geologist & contract tech proj officer, Advan Mining Div, 74-76, staff engr, Minerals Environ Tech, Headquarters, 77-78, geologist & contract tech proj off, Environ Assessment & Ground Control Div, 78-79, STAFF SCIENTIST, OFF RES DIR, TWIN CITIES RES CTR, US BUR MINES, 80- *Concurrent Pos:* Nat Park Serv study grant, 60-61; mem, Collegium of Distinguished Alumni, Boston Univ, Col Liberal Arts. *Mem:* Fel Geol Soc Am. *Res:* Petrofabrics and electron fractography applications in rock mechanics; improvements in longwall mining techniques and equipment; seabed mining technology; relation of rock fabric to various deformation and fragmentation tests; analysis of Vermont & New York Slate industry; evaluation of coal deposits in the Narragansett Basin of Massachusetts and Rhode Island. *Mailing Add:* US Bur Mines 5629 Minnehaha Ave S Minneapolis MN 55417

WILLARD, THOMAS MAXWELL, b Beaumont, Tex, Aug 30, 37; m 59; c 2. INORGANIC CHEMISTRY, ANALYTICAL CHEMISTRY. *Educ:* Lamar Univ, BS, 59; Tulane Univ, PhD(inorg chem), 64. *Prof Exp:* Fac mem, 64-65, PROF CHEM, FLA SOUTHERN COL, 81-, CHMN DEPT, 65- *Concurrent Pos:* Rotary Int group study exchange fel, Japan, 73; Danforth Found assoc, 80- *Mem:* Am Chem Soc; Sigma Xi. *Res:* Interactions between very weak acids and very weak bases in non-polar media. *Mailing Add:* Dept Chem Fla Southern Col 111 Lake Hollingsworth Dr Lakeland FL 33801-5698

WILLARD, WILLIAM KENNETH, b Hagerstown, Md, Nov 5, 29; m 61; c 2. ECOLOGY. *Educ:* Univ Ga, BSF, 57, MS, 60; Univ Tenn, PhD(zool), 65. *Prof Exp:* From asst prof to assoc prof zool, Clemson Univ, 65-75; PROF ZOOL & CHMN DEPT BIOL, TENN TECHNOL UNIV, 75- *Mem:* AAAS; Ecol Soc Am; Am Inst Biol Sci; Sigma Xi. *Res:* Effects of radiations on populations; fate of radioactive materials in the environment; population dynamics; ecosystem analysis; radiation ecology; bioenergetics of food chain relationships; ecological strategies in mammalian population dynamics. *Mailing Add:* Dept Biol Tenn Technol Univ Cookeville TN 38501

WILLARD-GALLO, KAREN ELIZABETH, b Oak Ridge, Tenn, July 8, 53; m 81. RETROVIROLOGY, PROTEIN BIOCHEMISTRY. *Educ:* Randolph-Macon Women's Col, AB, 75; Va Polytech Inst & State Univ, MS, 78 PhD(molecular biol), 81. *Prof Exp:* Teaching asst immunol, Va Polytech Inst & State Univ, 76-78; student assoc, Argonne Nat Lab, 78-81, post-doctoral fel, 81-82; res assoc, Ludwig Inst Cancer Res, 82-85; SCIENTIST, INT INST CELLULAR & MOLECULAR BIOL, 86- *Concurrent Pos:* Vis scientist, dept clin biochem, Rikshopitalet, Univ Oslo, Norway, 80-81, Basel Inst Immunol, Basel, Switz, 86; temp adv immunol, WHO, 87; consult, Baxter-Travenol, 86-88 & Market Intel Res Co, 91- *Mem:* Am Soc Cell Biol; Electrophoresis Soc. *Res:* Analysis of gene expression in normal, leukemic and HIV-infected human lymphocytes by high-resolution two-dimensional gel electrophoresis; investigation into the role of human retroviruses in regulation of the T cell receptor; CD3 complex and its role in AIDS; US patent on method for early detection of infectious mononucleosis. *Mailing Add:* Dept Biochem ICP-UCL75-39 Ave Hippocrate 75 B1200 Brussels Belgium

WILLARDSON, LYMAN S(ESSIONS), b Ephraim, Utah, May 10, 27; m 48; c 6. AGRICULTURAL ENGINEERING. *Educ:* Utah State Univ, BSCE, 50, MSCE, 55; Ohio State Univ, PhD(agr eng), 67. *Prof Exp:* Irrig engr, United Fruit Co, 52-54; irrig engr, Agr Exp Sta, Univ PR, 54-57; agr engr, Agr Res Serv, USDA, Ohio, 57-67, res leader drainage, Imp Valley Conserv Res Ctr, 67-74; PROF AGR & IRRIG ENG, UTAH STATE UNIV, LOGAN, 74- *Concurrent Pos:* Proj leader, On-Farm Water Mgt, Dominican Repub; mem comt, F17.65 Land Drainage, Am Soc Testing & Mat; mem bd dirs, US Comt Irrigation & Drainage & Int Comn Irrigation & Drainage; dir, Rocky Mountain Sect, Am Soc Agr Engrs, chmn comt, SW-231 Drainage Res; chmn, I&D Div Res Comt, Am Soc Civil Engrs. *Mem:* AAAS; Am Soc Civil Engrs; Am Soc Agr Engrs; Nat Soc Prof Engrs; Int Comn Irrig & Drainage; Am Soc Testing Mat. *Res:* Engineering research on problems associated with irrigation and drainage of agricultural lands; drainage and salinity of irrigated lands. *Mailing Add:* Dept Agr & Irrig Eng Utah State Univ Logan UT 84322-4105

WILLARDSON, ROBERT KENT, b Gunnison, Utah, July 11, 23; m 47; c 3. SOLID STATE PHYSICS. *Educ:* Brigham Young Univ, BS, 49; Iowa State Col, MS, 51. *Prof Exp:* Instr physics, Brigham Young Univ, 47-48 & Iowa State Col, 48-49; res physicist, Ames Lab, AEC, 49-51; prin physicist, Battelle Mem Inst, 51-56, asst chief, Phys Chem Div, 56-60; chief scientist, Res Ctr, Bell & Howell Co, 60-64, dir solid state res, 64-67, dir mat res, 67-69, gen mgr electronic mat div, 69-73, pres, Electronic Mat Corp, 73; asst to gen mgr, Electronic Mat Div, Cominco Am Inc, 73-77, mgr electronic mat, Div Sales, 77-81, mgr planning, 81-; pres, Willardson Consult, 83 & 87; pres, Cryscon Techs, 84-86; TECH DIR, ENIMONT AM, 88- *Mem:* Am Phys Soc; Electrochem Soc; Am Chem Soc; Inst Elec & Electronics Eng; Int Soc Hybrid Microelectronics. *Res:* Preparation, electrical and optical properties of high purity metals, alloys and semiconductors; analysis, control and effects of impurities and lattice defects in these materials; electronic transport phenomena in semiconductors. *Mailing Add:* E 12722 23rd Ave Spokane WA 99216

WILLBANKS, EMILY WEST, b Ft Lauderdale, Fla, Nov 25, 30; m 59. COMPUTER SCIENCE, MATHEMATICS. *Educ:* Duke Univ, BS, 52; Univ NMex, MS, 57. *Prof Exp:* Eng aide math, Pratt-Whitney Aircraft Co, 52-54; STAFF MEM MATH & COMPUT, LOS ALAMOS NAT LAB, 54- *Mailing Add:* Los Alamos Nat Lab PO Box 1663 Los Alamos NM 87545

WILLCOTT, MARK ROBERT, III, b Muskogee, Okla, July 23, 33; m 55; c 4. ORGANIC CHEMISTRY. *Educ:* Rice Univ, BA, 55; Yale Univ, MS, 59, PhD(chem), 63. *Prof Exp:* Asst prof chem, Emory Univ, 62-64; from asst prof to assoc prof, 65-73, PROF CHEM, UNIV HOUSTON, 73- *Concurrent Pos:* NIH fel, Univ Wis, 64-65; Guggenheim fel, 72-73; consult, Upjohn Co, 65- & Aldrich Chem Co, 72-; adj prof med, Baylor Col Med, 78-; adj prof chem, Rice Univ, 81- *Mem:* Am Chem Soc; Royal Soc Chem. *Res:* Thermal rearrangements of organic compounds; nuclear magnetic resonance spectroscopy; magnetic resonance imaging; nuclear magnetic resonances in medicine. *Mailing Add:* Dept Res & Develop Radiol Sci Vanderbilt Univ Nashville TN 37232-2625

WILLCOX, ALFRED BURTON, b Sioux Rapids, Iowa, Sept 18, 25; m 48; c 4. MATHEMATICS. *Educ:* Yale Univ, MA, 49, PhD(math), 53. *Prof Exp:* From instr to prof math, Amherst Col, 53-68, exec dir comt undergrad prog math, 63-64; EXEC DIR, MATH ASN AM, 68- *Concurrent Pos:* Vis asst prof, Univ Chicago, 57-58; vis lectr, Uppsala Univ, Sweden, 67-68. *Mem:* Fel AAAS; Am Math Soc; Math Asn Am (vpres, 64-66); Nat Coun Teachers Math; Soc Indust & Appl Math. *Res:* Banach algebras. *Mailing Add:* 9911 Julliard Dr Bethesda MD 20817

WILLDEN, CHARLES RONALD, b Neola, Utah, Sept 30, 29; m 50; c 3. GEOLOGY. *Educ:* Univ Utah, BS, 51, MS, 52; Stanford Univ, PhD(geol), 60. *Prof Exp:* Geologist, US Geol Surv, Colo, 52-68; sr geologist, Vanguard Explor Co, 68-72; GEOLOGIST, SILVER RESOURCES CORP, 72- *Mem:* Geol Soc Am; Soc Econ Geol; Am Asn Petrol Geol. *Res:* Structural and economic geology. *Mailing Add:* 8750 Kings Hill Dr Salt Lake City UT 84121

WILLE, JOHN JACOB, JR, b New York, NY, June 24, 37; m 61; c 4. CELL BIOLOGY. *Educ:* Cornell Univ, BA, 60; Univ Ind, PhD(genetics), 65. *Prof Exp:* Resident res assoc cell biol, Biol Div, Argonne Nat Lab, 65-66, fel, 66-68; asst prof biol sci, Univ Cincinnati, 68-72; NIH spec fel, Univ Chicago, 72-73, res assoc, Dept Biophys, 73-75; asst prof zool & physiol, La State Univ, 75-80; sr res assoc & asst prof, Dept Cell Biol, Mayo Clin, 80-85; SR SCIENTIST, DEPT BIOCHEM, SOUTHERN RES INST, 85- *Mem:* Am Asn Cancer Res; NY Acad Sci; Am Soc Cell Biol. *Res:* Developmental genetics of unicellular organisms; molecular biology of biological rhythms; cancer cell biology; pathogenic mechanisms of uroepithal cells; cancer chemo prevention rehioid cell function. *Mailing Add:* Conda Tec 200 Headquarters Park Dr Skillman NJ 08558

WILLE, LUC THEO, b Ghent, Belg, Dec 8, 56; m 83; c 2. STATISTICAL PHYSICS. *Educ:* State Univ Ghent, BSc, 80, PhD(physics), 83, MS, 85. *Prof Exp:* Postdoctoral asst physics, State Univ Ghent, 83-85; sr res assoc, Daresbury Lab, Sci & Eng Res Coun, UK, 85-87; postdoctoral asst mat sci, Univ Calif, Berkeley, 87-88; ASST PROF PHYSICS, FLA ATLANTIC UNIV, 88- *Concurrent Pos:* Consult, Lawrence Livermore Nat Lab, 89-; vis prof, Univ Nancy, France, 91. *Honors & Awards:* Laureate, Royal Acad Sci, Belg, 88. *Mem:* Am Phys Soc; NY Acad Sci; Mat Res Soc; Soc Indust & Appl Math; Sigma Xi. *Res:* Phase transformations and structural stability of solids, especially superconductors; electronic structure of solids and clusters; self-organization in complex systems. *Mailing Add:* Dept Physics Fla Atlantic Univ Boca Raton FL 33431

WILLEBOORDSE, FRISO, b Bandung, Indonesia, July 31, 35; m 58; c 4. ANALYTICAL CHEMISTRY. *Educ:* Univ Indonesia, BS, 54; Univ Amsterdam, Drs, 57, PhD(anal chem), 69. *Prof Exp:* Lectr inorg chem, Univ Natal, 60-61, sr lectr anal chem, 61-62; res chemist, WVa, Union Carbide Corp, Bound Brook, NJ, 62-69, res scientist, Bound Brook, 69-72, group leader anal res chem & plastics, 72-80; DIR, ANALYTICAL RES & COMPUT SERV, W R GRACE & CO, COLUMBIA, MD, 81- *Concurrent Pos:* Vis prof, Rutgers Univ, 73-74 & 79-80. *Mem:* Am Chem Soc; Royal Neth Chem Soc. *Res:* Polymer characterization, polarography; differential chemical kinetics. *Mailing Add:* 5468 Wingborne Ct Columbia MD 21045

WILLEFORD, BENNETT RUFUS, JR, b Greenville, SC, Oct 28, 21. INORGANIC CHEMISTRY. *Educ:* Emory Univ, BA, 43; Univ Wis, MS, 49, PhD(phys chem), 50. *Prof Exp:* Jr chemist, Shell Develop Co, 43-46; asst, Univ Wis, 46-50; from asst prof to prof, 50-84, EMER PROF CHEM, BUCKNELL UNIV, 84- *Concurrent Pos:* Res fel, Univ Minn, 56-57; consult, US Fish & Wildlife Serv, 60-64; NSF sci fac fel, Univ Munich, 62-63; res assoc, Univ NC, Chapel Hill, 69-70; guest prof, Inorganic Chem Lab, Univ Oxford, Eng, 77-78, Anorganisch-Chem Inst, Tech Univ Munchen, 78, 83, 84 & Univ Cambridge, Eng, 85; Fulbright lectr, Univ Poona, India, 85-86; vis prof, Earlham Col, Richmond, Ind, 86-87, Baylor Univ, Waco, Tex, 88; Fulbright Lectr, Cottington Univ Col, Liberia, WAfrica, 89-90. *Mem:* Am Chem Soc. *Res:* Structure of metal coordination compounds; organometallic chemistry. *Mailing Add:* Dept Chem Bucknell Univ Lewisburg PA 17837

WILLEKE, KLAUS, b Essen, Ger, Mar 26, 41; m 68; c 1. AEROSOL SCIENCE & TECHNOLOGY, INDUSTRIAL HYGIENE. *Educ:* Univ NH, BS, 63; Stanford Univ, MS, 64, PhD(aeronaut & astronaut), 69; Von Karman Inst, dipl, 66. *Prof Exp:* Fel, Max Planck Inst Plasma Physics, Ger, 69-70; asst prof mech eng, Univ Minn, 70-76; assoc prof environ health & adj prof chem eng, 76-81, PROF ENVIRON HEALTH, UNIV CINCINNATI, 81-, ADJ PROF INDUST ENG, 87- *Concurrent Pos:* Vis asst prof, Dept Chem Eng, Kyoto Univ, Japan, 73; mem, Occup Safety & Health Study Sect, Nat Inst Occup Safety & Health, 78-80; assoc dir dept, environ health, Univ Cincinnati, 86- *Mem:* Am Indust Hyiene Asn; AAAS; Air Pollution Control Asn; Am Asn Aerosol Res. *Res:* Particle classification and measurement; aerosol generation and sampling; asbestos fibers; particle deposition in lung; therapeutic aerosols; respirator protection of workers; dust control. *Mailing Add:* Dept Environ Health Univ Cincinnati Cincinnati OH 45267

WILLEMOT, CLAUDE, b Ghent, Belg, Dec 26, 33; Can citizen; m 66; c 4. PLANT LIPID METABOLISM, STRESS PHYSIOLOGY. *Educ:* McGill Univ, MSc, 63, PhD(plant physiol), 64. *Prof Exp:* Nat Res Coun Can fel, Nat Inst Agr Res, Versailles, France, 64-65 & Univ Calif, Davis, 65-67; RES SCIENTIST PLANT PHYSIOL, RES STA, CAN DEPT AGR, 67- *Concurrent Pos:* Lectr, Fac Agr, Laval Univ, 68-; assoc ed, Can J Biochem, 75-79; mem staff, Dept Sci & Indust Res, NZ, 79-80. *Mem:* Can Soc Plant Physiol (secy-treas, 71-72, secy, 72-73); Am Soc Plant Physiol; Soc Cryobiol. *Res:* Mechanism of plant frost hardiness; plant lipid metabolism; winter hardiness of crop plants, mainly alfalfa and wheat. *Mailing Add:* Univ Laval, FSAA Dept STA Pavillon Comtois, Cite Univ Quebec PQ G1K 7P4 Can

WILLEMS, JAN C, b Bruges, Belg, Sept 18, 39; m 65; c 2. MATHEMATICAL SYSTEMS THEORY, SYSTEMS ENGINEERING. *Educ:* Univ Ghent, Electromech Engr, 63; Univ RI, MSc, 65; Mass Inst Technol, PhD(elec eng), 68. *Prof Exp:* Asst prof elec eng, Mass Inst Technol, 68-73; PROF, UNIV GRONINGEN, NETH, 73- *Concurrent Pos:* Sr vis fel, dept appl math & theoret physics, UK Sci Res Coun, 70-71. *Mem:* Fel Inst Elec & Electronics Engrs; Soc Indust & Appl Math; Dutch Math Soc. *Res:* Control theory; stability theory; optimal control; mathematical system theory. *Mailing Add:* Univ Groningen Math Inst Postbox 800 Groningen 9700 AV Netherlands

WILLEMS, NICHOLAS, b Aardenburg, Netherlands, Jan 6, 24; US citizen; m 48; c 5. CIVIL & STRUCTURAL ENGINEERING. *Educ:* Delft Univ Technol, MSc, 46; Univ Pretoria, MComm, 53; Univ Kans, PhD(eng mech), 63. *Prof Exp:* Asst engr, Hague Munic, 46-48; engr, Transvaal Prov Rd Dept, 48-51; construct engr, SAfrican Coal & Oil Corp, 51-52; resident engr, van Niekerk, Kleyn & Edwards, 52-54, jr partner & consult, 54-60; asst prof civil eng, 60-61, from instr to prof, 61-75, chmn dept, 72-75, PROF CIVIL ENG, UNIV KANS, 75- *Concurrent Pos:* Consult, indust & state & fed govt, 60-67; Ford Found grant, 63. *Mem:* Am Soc Civil Engrs; Sigma Xi. *Res:* Structural analysis; plates; shells; dynamic loading; vibrations and matrix analysis. *Mailing Add:* Dept Civil Eng Univ Kans Lawrence KS 66045

WILLEMSEN, HERMAN WILLIAM, b Huissen, Holland, Dec 23, 45; Can citizen; m 80; c 2. PHYSICS. *Educ:* Univ Waterloo, BSc, 70; Univ Toronto, MSc, 72, PhD(physics), 75. *Prof Exp:* Res assoc physics, Univ Toronto, 75-76; res assoc, Argonne Nat Lab, 76-79; SCIENTIST PHYSICS, BELL NORTHERN RES, 79- *Mem:* Can Asn Physicists; Am Phys Soc. *Res:* Superconductivity; phase transitions; defects in insulators; mass memory systems; opto electronic devices. *Mailing Add:* Bell Northern Res 3500 Carling Ave Ottawa ON K1Y 4H7 Can

WILLEMSEN, ROGER WAYNE, b Oskaloosa, Iowa, Jan 14, 44; m 66; c 3. PHYSIOLOGICAL ECOLOGY, BOTANY. *Educ:* Cent Col, Iowa, BA, 66; Kans State Col, Pittsburg, Kans, MS, 68; Univ Okla, PhD(bot), 71. *Prof Exp:* Grants, 72-73, asst prof bot, Rutgers Univ, New Brunswick, 71-78; mem staff, Lilly Res Labs, 78-85; MEM STAFF, RHONE POULENC INC, 85- *Mem:* Entom Soc Am; Am Phytopath Soc; Weed Sci Soc Am. *Res:* The physiology and ecology of weed seed germination; allelopathy; old-field succession; agricultural pest control. *Mailing Add:* Rhone Pouleuc Ag Co Fresno CA 93720

WILLENBERG, HARVEY JACK, b New York, NY, Sept 8, 45; m 71; c 3. MICROGRAVITY MATERIALS SCIENCE. *Educ:* Harvey Mudd Col, BS, 67; Univ Wash, MS, 71, MSE, 72, PhD(nuclear eng), 76. *Prof Exp:* Sr engr, Westinghouse-Hanford, 72-76; staff scientist, Battelle-Northwest Labs, 76-79; scientist, Math Sci Northwest, 79-82; CHIEF SCIENTIST, BOEING DEFENSE & SPACE GROUP, 82- *Mem:* Am Inst Aeronaut & Astronaut. *Res:* Physics and technology development and applications of magnetically-confined thermonuclear plasmas to power generation; two-phase compressible fluid dynamics with heat and mass transfer; materials science and applications under microgravity conditions. *Mailing Add:* 10021 Bluff Dr Huntsville AL 35803

WILLENBROCK, FREDERICK KARL, b New York, NY, July 19, 20; m 44. ENGINEERING. *Educ:* Brown Univ, ScB, 42; Harvard Univ, MA, 47, PhD(appl physics), 50. *Prof Exp:* Res fel & lectr, Harvard Univ, 50-55, from asst dir labs, Div Eng & Appl Physics to dir labs & assoc dean div, 54-67; prof & provost fac eng & appl sci, State Univ NY, Buffalo, 67-70; dir, Inst Appl Technol, Nat Bur Standards, 70-76; dean, Sch Eng & Appl Sci, Southern Methodist Univ, 76-81, Cecil & Ida Green prof eng, 76-86; exec dir, Am Soc Eng Educ, 86-89; ASST DIR, SCI, TECHNOL & INT AFFAIRS, NSF, 89- *Concurrent Pos:* Dir, Inst Appl Technol, Nat Bur Standards, 70-76; consult, Nat Bur Standards, NSF, Nat Res Coun, Off Technol Assessment, Dept Energy, Nat Aeronaut & Space Admin, Sperry Rand Res Ctr, Sperry Gyroscope Great Neck, Bell Aerospace Div, Dresser Indust, SAIC, Western NY Nuclear Res Ctr, Sage Labs Inc, SMU Found Sci & Eng & SW Res Inst; mem coun, Nat Acad Eng, 80-86. *Honors & Awards:* CIBA Medal & Prize, Biochem Soc, 73; Louis & Bert Freedman Found Award, NY Acad Sci, 74; Wilhelm Feldberg Found Prize, Feldberg Found Anglo/Ger Sci Exchange, 76; Sir Hans Krebs Lectr & Medal, Fedn Europe Biochem Soc, 78; Humphry Davy Mem Lectr, Royal Inst Chem, 80; James Rennie Bequest Lectr, Univ Edinburgh, 80; Copley Medal, Coun Royal Soc, 81; Croonian Lectr, Royal Soc, 87. *Mem:* Nat Acad Eng; AAAS; Inst Elec & Electronics Engrs; Sigma Xi; Am Soc Testing & Mats. *Res:* Engineering education, technology and public policy. *Mailing Add:* 1740 New Hampshire Ave NW Unit E Washington DC 20009

WILLENS, RONALD HOWARD, physical metallurgy, solid state physics, for more information see previous edition

WILLER, RODNEY LEE, b Albany, Ore, Nov 27, 48; m 74. STEREOCHEMISTRY, ENERGETIC MATERIALS. *Educ:* East Carolina Univ, BS, 70, MS, 74; Univ NC, PhD(org chem), 76. *Prof Exp:* Res assoc, Mich State Univ, 76-78; asst prof, Tex Tech Univ, 78-80; RES CHEMIST, ORG CHEM, NAVAL WEAPONS CTR, 80- *Mem:* Am Chem Soc; Sigma Xi; AAAS; NY Acad Sci. *Res:* Synthesis and characterization of high density energetic materials with potential application as propellant and explosive ingrediants. *Mailing Add:* B Scotch Pine Dr Newark DE 19711

WILLERMET, PIERRE ANDRE, b Mineola, NY, Aug 21, 41; m 63; c 2. LUBRICATION, TRIBOLOGY. *Educ:* Widener Univ, BS, 67; Univ Pa, PhD(phys chem), 72. *Prof Exp:* RES SCIENTIST, ENG & RES STAFF, FORD MOTOR CO, 72- *Concurrent Pos:* Assoc ed, J Tribology & J Synthetic Lubrication. *Mem:* Sigma Xi; Am Chem Soc; Soc Tribologists & Lubrication Engrs. *Res:* Tribology; chemistry of friction and wear in lubricated contacts; thermo oxidative degradation of hydrocarbons and lubricant additives; friction losses in automotive engines and drive trains. *Mailing Add:* 11400 Melrose Ave Livonia MI 48150

WILLERSON, JAMES THORNTON, CARDIOVASCULAR DISEASES. *Educ:* Baylor Col Med, MD, 65. *Prof Exp:* PROF MED, HEALTH SCI CTR, UNIV TEX, 76-, DIR, DIV CARDIOL, 77- *Mailing Add:* Dept Internal Med Univ Tex Med Sch 6431 Fannin Rm 1-150 Houston TX 77030

WILLETT, COLIN SIDNEY, b Danbury, Eng, Jan 11, 35; US citizen; c 2. PHYSICAL SCIENCE, LASERS. *Educ:* City Univ London, BS, 63; Univ London, PhD(physics), 67. *Prof Exp:* Physicist lasers, Harry Diamond Labs, Dept Army, Adelphi, 67-88; MEM STAFF, DEFENSE RES TECHNOL, INC, 88- *Res:* Atomic collision processes; gas discharge physics; laser systems research and analysis. *Mailing Add:* 12351 Frederick Rd W Friendship Ellicott City MD 21794

WILLETT, DOUGLAS W, b Adams County, NDak, May 25, 37; m 59; c 6. MATHEMATICS. *Educ:* SDak Sch Mines & Technol, BS, 59; Calif Inst Technol, PhD(math), 63. *Prof Exp:* From asst prof to assoc prof math, Univ Alta, 62-66; assoc prof, 66-72, PROF MATH, UNIV UTAH, 72- *Concurrent Pos:* Vis prof, Univ Alta, 71-72. *Mem:* Soc Indust & Appl Math; Can Math Cong. *Res:* Ordinary differential equations and mathematical analysis. *Mailing Add:* Dept Math Univ Utah Salt Lake City UT 84112

WILLETT, HILDA POPE, b Decatur, Ga, July 15, 23; m 56; c 2. MICROBIOLOGY. *Educ:* Woman's Col Ga, AB, 44; Duke Univ, MA, 46, PhD(microbiol), 49. *Prof Exp:* Instr microbiol, 48-50, assoc, 50-52, from asst prof to assoc prof, 52-64, PROF BACT, SCH MED, DUKE UNIV, 64-, DIR, GRAD STUDIES, 80- *Mem:* Am Soc Microbiol; Am Acad Microbiol. *Res:* Physiology of Mycobacterium tuberculosis. *Mailing Add:* Dept Microbiol & Immunol Duke Univ Sch Med Box 3322 Med Ctr Durham NC 27710

WILLETT, JAMES DELOS, b Stockton, Calif, Jan 16, 37; c 2. CHEMISTRY. *Educ:* Univ Calif, Berkeley, BA, 59; Mass Inst Technol, PhD(org chem), 65. *Prof Exp:* Jr chemist res, Merck, Sharp & Dohme Res Labs, 59-61; fel chem, Stanford Univ, 65-68; asst prof chem, 68-73; assoc prof chem, Univ Idaho, 73-78, prof chem & biochem, 78-80; grants assoc, 80-81, staff asst dep dir, 81-82, spec asst to dir, 82-84, health sci adminr, biomed res model develop, animal resources prog, 84-85, CHIEF, BIOL MODELS & MAT RESOURCES SECT, ANIMAL RESOURCES PROG, 85- & CHIEF, OFF PROG PLANNING & EVAL, DIV RES RESOURCES, NIH, 87- *Concurrent Pos:* NIH fel, 65-68; NIH career develop award, Nat Inst Aging, 75-80. *Mem:* Am Chem Soc; Soc Nematologists; Am Aging Asn; AAAS; Tissue Cult Asn. *Res:* Nematodes as models for the study of the effects of senescence on hormonal control systems. *Mailing Add:* Dept Biol George Mason Univ 4400 University Dr Fairfax VA 22030

WILLETT, JOSEPH ERWIN, b Albany, Mo, June 9, 29; m 55; c 3. PLASMA PHYSICS. *Educ:* Univ Mo, BA, 51, MA, 53, PhD(physics), 56. *Prof Exp:* From asst to instr physics, Univ Mo, 53-55, Stewart fel, 55; physicist & aeronaut res engr, US Naval Ord Lab, Md, 56-58; res scientist, McDonnell Aircraft Corp, 58-61, instr, McDonnell Eve Sch, 60-61; staff scientist, Gen Dynamics Ft Worth, Tex, 61-65; assoc prof, 65-81, PROF PHYSICS, UNIV MO, COLUMBIA, 81- *Concurrent Pos:* Instr & adj prof, Tex Christian Univ, 62-64. *Mem:* Am Phys Soc. *Res:* Theoretical studies of the interaction of electromagnetic waves with plasmas; electrostatic and hydromagnetic waves; stimulated raman and brillouin scattering; instabilities in magnetized plasmas; plasma heating; controlled fusion; free-electron lasers. *Mailing Add:* Dept Physics & Astron Univ Mo Columbia MO 65211

WILLETT, LYNN BRUNSON, b Colorado Springs, Colo, Aug 2, 44; m 66; c 2. ANIMAL PHYSIOLOGY, DAIRY SCIENCE. *Educ:* Colo State Univ, BS, 66; Purdue Univ, Lafayette, MS, 68, PhD(animal physiol), 71. *Prof Exp:* From asst prof to assoc prof, 71-84, PROF DAIRY SCI, OHIO AGR RES & DEVELOP CTR, OHIO STATE UNIV, 84-, ASSOC CHAIR, DEPT DAIRY SCI, 86- *Concurrent Pos:* Ed, Pharmacol & Toxicol Sect, J Animal Sci. *Mem:* Am Dairy Sci Asn; Am Soc Animal Sci; Soc Toxicol. *Res:* Elimination; metabolism and toxicity of halogenated hydrocarbons, particularly polybrominated biphenyls and polychlorinated biphenyls by cattle; steroid hormone relationships in cattle. *Mailing Add:* Ohio Agr Res & Develop Ctr Ohio State Univ 1680 Madison Ave Wooster OH 44691-4096

WILLETT, NORMAN P, b Paterson, NJ, Apr 13, 28; m 56; c 3. MICROBIOLOGY. *Educ:* Rutgers Univ, BS, 49; Syracuse Univ, MS, 52; Mich State Univ, PhD(microbiol), 55. *Prof Exp:* Asst dent med, Harvard Univ, 55-57; sr res microbiologist, Squibb Inst Med Res, 57-60; res microbiologist, Bzura, Inc, 60-62; sr res assoc, Lever Brothers, 62-63; chief microbiologist, Food & Drug Res Inc, 63; res assoc, Sch Vet Med, Univ Pa, 63-66; assoc prof, Sch Pharm, 66-73, head dept, 67-87, PROF MICROBIOL, SCH DENT, TEMPLE UNIV, 73-, DENT EDUC COORDR, 87- *Mem:* AAAS; Am Soc Microbiol; Am Chem Soc; Soc Indust Microbiol; NY Acad Sci; Am Asn Dent Schs; Am Inst Biol Sci; Soc Infectious Control in Dent; Int Asn Dent Res; Sigma Xi. *Res:* Biochemical basis of pathogenicity; physiology of streptococci; antibiotic biosynthesis; biochemistry and microbiology of saliva; organic acid and antibiotic fermentation; chemotherapy; streptococcal toxins; infection control. *Mailing Add:* Dept Microbiol & Immunol Temple Univ Sch Med Philadelphia PA 19140

WILLETT, RICHARD MICHAEL, b Louisville, Ky, May 2, 45; m. MATHEMATICS. *Educ:* USAF Acad, BS, 67; NC State Univ, MA, 69, PhD(math), 71. *Prof Exp:* Teaching asst math, NC State Univ, 67-71, instr, 71-72; asst prof, 72-77, ASSOC PROF MATH, UNIV NC, GREENSBORO, 77- *Mem:* Sigma Xi. *Res:* Finite field theory; error-correcting codes. *Mailing Add:* 808 Brookfield Rd Raleigh NC 27609

WILLETT, ROGER, b Northfield, Minn, July 13, 36; m 57; c 6. PHYSICAL CHEMISTRY, CHEMICAL PHYSICS. *Educ:* St Olaf Col, BA, 58; Iowa State Univ, PhD(chem, physics), 62. *Prof Exp:* From instr to assoc prof, 62-72, chmn dept, 74-80, Chmn Chem Physics Prog, 84-86, PROF CHEM, WASH STATE UNIV, 72- *Concurrent Pos:* Vis prof, Univ Zürich, 80; Fulbright scholar, Leiden, Neth, 81; dir, Advan Study Inst, NATO, 83, ed, proc, 85. *Mem:* Am Chem Soc; Am Crystallog Asn; Sigma Xi. *Res:* X-ray diffraction and crystallography; magnetic susceptibility and interactions; chemical bonding; molecular and electronic structure; electronic spectra and electron spin resonance studies of transition metal ions. *Mailing Add:* Dept of Chem Wash State Univ Pullman WA 99164-4360

WILLETTE, GORDON LOUIS, b Dighton, Mass, Dec 19, 33. PETROLEUM CHEMISTRY. *Educ:* Brown Univ, ScB, 55; Univ Minn, PhD(org chem), 59. *Prof Exp:* Chemist, Res Labs, Rohm & Haas Co, 59-76; dir res & develop, Hatco Chem Corp, 76-80; NEW VENTURE MGR SYNTHETIC LUBRICANTS, UNIROYAL CHEM CO, 80- *Mem:* Am Chem Soc; Am Soc Lubrication Engrs; Soc Advan Educ; Sigma Xi. *Res:* Polymer synthesis; ATF and motor oil formulation, synthetic lubricants and plasticizers. *Mailing Add:* 12A Heritage Crest Southbury CT 06488

WILLETTE, ROBERT EDMOND, b Grand Rapids, Mich, Aug 15, 33; div; c 3. MEDICINAL CHEMISTRY, ORGANIC CHEMISTRY. *Educ:* Ferris State Col, BS, 55; Univ Minn, PhD(pharmaceut chem), 60. *Prof Exp:* Spec instr med chem, Ferris State Col, 59-61; NIH res fel, Australian Nat Univ, 61-63; res officer, Div Org Chem, Commonwealth Sci & Indust Res Orgn, 63-64; res assoc med chem, Univ Mich, 65-66; from asst prof to assoc prof, Sch Pharm, Univ Conn, 66-72; chemist, Div Res, 72-81, PRES, DUO RES, NAT INST DRUG ABUSE, 81- *Mem:* AAAS; Am Pharmaceut Asn; Am Chem Soc; Royal Soc Chem. *Res:* Isolation and structure determination of natural products; synthesis of heterocyclic compounds of medicinal interest, particularly analgetics and narcotic antagonists; drug abuse and testing. *Mailing Add:* 160 Green St Annapolis MD 21401-2502

WILLEY, CLIFF RUFUS, b Hornell, NY, Nov 20, 35; m 57; c 5. SOIL PHYSICS. *Educ:* Cornell Univ, BS, 57, MS, 59; Univ Wis-Madison, PhD(soil physics), 62. *Prof Exp:* Soil scientist, Agr Res Serv, USDA, 62-73; chief solid waste serv, 73-74, CHIEF TECH SERV, DEPT NATURAL RESOURCES, MD ENVIRON SERV, 74- *Concurrent Pos:* Assoc prof agr eng, NC State Univ, 67-73. *Mem:* Sigma Xi. *Mailing Add:* 1028 Old Bay Ridge Rd Annapolis MD 21403

WILLEY, JOAN DEWITT, b Summit, NJ, May 10, 48. MARINE CHEMISTRY, ANALYTICAL CHEMISTRY. *Educ:* Duke Univ, BSc, 69; Dalhousie Univ, PhD(chem oceanog), 75. *Prof Exp:* Fel geochem, Mem Univ Nfld, 74-75; vis scientist marine geol, Bedford Inst, 76-77; from asst prof to assoc prof, 77-86, PROF CHEM & MARINE SCI, UNIV NC, WILMINGTON, 86- *Concurrent Pos:* Consult, Am Inst Chemists. *Mem:* Am Chem Soc; AAAS. *Res:* Sediment and interstitial water chemistry in estuaries, including seasonal variations; marine chemistry of silica; reactions between sediments and seawater; rainwater composition in coastal areas. *Mailing Add:* Dept Chem Univ NC 601 S College Rd Wilmington NC 28403-3297

WILLEY, ROBERT BRUCE, b Long Branch, NJ, Sept 15, 30; m 56. ANIMAL BEHAVIOR, INSECT BIOGEOGRAPHY. *Educ:* NJ State Teachers Col, BA, 52; Harvard Univ, PhD(biol), 59. *Prof Exp:* Teacher high sch, NJ, 52-54; from asst prof to assoc prof biol, Ripon Col, 59-65; ASSOC PROF BIOL, UNIV ILL, CHICAGO, 65- *Concurrent Pos:* Res grants, Sigma Xi, 62 & 67, NSF, 64-66, 72-74 & 75-76; mem bd trustees, Rocky Mountain Biol Lab, 63-88, vpres, 72-76, pres, 77-84. *Honors & Awards:* President's Stewardship Award, Nature Conservancy, 88. *Mem:* Am Soc Zool; Soc Study Evolution; Entom Soc Am; Orthopterists'Soc; Animal Behav Soc; Soc Am Naturalists. *Res:* Invertebrate behavior; interspecific behavior and evolution of sympatric insect populations; animal communication systems; insect biogeography and speciation. *Mailing Add:* Dept Biol Sci M/C 066 Univ Ill PO Box 4348 Chicago IL 60680

WILLEY, RUTH LIPPITT, b Wickford, RI, May 11, 28; m 56. PHYCOLOGY, LIMNOLOGY. *Educ:* Wellesley Col, BA, 50; Radcliffe Col, PhD(biol), 56. *Prof Exp:* Docent mus educ, Peabody Mus Natural Hist, Yale Univ, 50-52; instr zool, Wellesley Col, 56-57; res fel ophthal, Mass Eye & Ear Infirmary, 57-58; comm histologist, Triarch Prod, 59-60; asst prof, 65-71, ASSOC PROF BIOL SCI, UNIV ILL, CHICAGO, 71- *Concurrent Pos:* Mem, Rocky Mountain Biol Lab, Colo, 58-, secy, 66-68, mem bd trustees, 75-78, environ officer, 79-85. *Honors & Awards:* President's Stewardship Award, Nature Conservancy, 88. *Mem:* Am Soc Limnol Oceanog; Am Soc Cell Biol; Am Micros Soc; Entom Soc Am; Phycol Soc Am. *Res:* ultrastructure, ecology and phylogeny of euglenoid algae; ecology of epibiotic interactions among freshwater organisms; distribution of salamanders and incidence of paedomorphy in high altitude ponds. *Mailing Add:* Dept Biol Sci MC 066 Univ Ill Box 4348 Chicago IL 60680

WILLHAM, RICHARD LEWIS, b Hutchinson, Kans, May 4, 32; m 54; c 2. ANIMAL BREEDING. *Educ:* Okla State Univ, BS, 54; Iowa State Univ, MS, 55, PhD(animal breeding), 60. *Prof Exp:* Asst prof animal sci, Iowa State Univ, 59-63; assoc prof animal sci, Okla State Univ, 63-66; from assoc prof to prof animal sci, 66-79, C F CURTISS DISTINGUISHED PROF AGR, IOWA STATE UNIV, 79- *Concurrent Pos:* AEC grant, Iowa State Univ, 59-63. *Honors & Awards:* Am Soc Animal Sci Breeding & Genetics Award, 78; Am Soc Animal Sci Indust Serv Award, 86; Nat Cattleman's Asn Res Award, 86. *Mem:* Fel Am Soc Animal Sci. *Res:* Beef cattle breeding; evaluation of the results of selection and crossbreeding; Beef Improvement Federation work with national sire evaluation program development in beef industry; guest curator of art exposition titled - Art About Livestock. *Mailing Add:* Dept Animal Sci Iowa State Univ Ames IA 50011

WILLHITE, CALVIN CAMPBELL, b Salt Lake City, Utah, Apr 27, 52; m 82. TOXICOLOGY, TERATOLOGY. *Educ:* Utah State Univ, BS, 74, MS, 77; Dartmouth Col, PhD(pharmacol), 80. *Prof Exp:* Albert J Ryan Found fel, Med Sch, Darmouth Col, 77-80; teaching fel toxicol, dept pharmacol, Col Med, Health Sci Ctr, Univ Ariz, Tucson, 80; res pharmacol, Toxicol Res Unit, Western Regional Res Ctr, USDA, Berkeley, 80-85; STAFF TOXICOLOGIST, DEPT HEALTH SERV STATE OF CALIF, BERKELEY, 85- *Concurrent Pos:* Adj asst prof, Toxicol Prog, Utah State Univ, Logan, 84-87, adj assoc prof, 88-; prin investr, Nat Inst Child Health & Human Develop grant, 85-88; US Environ Protection Agency Health Effects Task Group, Nat Sanitation Found, 86-88; mem pub affairs comt, Teratology Soc, 87-90; co-prin investr, March of Dimes Res grant, 87-91; mem Human Health Toxics Rev Comt, Ariz Dept Environ Quality, 89-90, TLV Comt, Am Conf Govt Indust Hygienists, 89-; invited lectr, Instituto di Chimica Biologica, Universita di Palermo Policlinico, Palermo, Mondelo, Italy, 91; consult, SRI Int. *Honors & Awards:* Frank R Blood Award, Soc Toxicol, 86. *Mem:* Soc Toxicol; Teratology Soc; Am Soc Pharmacol & Exp Therapeut; Am Conf Govt Indust Hygienists. *Res:* Embryotoxicity, pharmacokinetics, & placental transfer of cancer chemopreventive retinoids, antiviral nucleosides, industrial chemicals and environmental pollutants; pathogenesis of congenital malformations; identification and risk assessment of tetratogenic agents. *Mailing Add:* Toxic Substances Control Prog Calif Dept Health Serv 700 Heinz St Bldg F Suite 200 Berkeley CA 94710

WILLHITE, GLEN PAUL, b Waterloo, Iowa, July 18, 37; m 59; c 5. CHEMICAL & PETROLEUM ENGINEERING. *Educ:* Iowa State Univ, BS, 59; Northwestern Univ, PhD(chem eng), 62. *Prof Exp:* From res scientist to sr res scientist, Continental Oil Co, 62-70; from assoc prof, 69-88, ROSS H FORNEY DISTINGUISHED PROF CHEM & PETROL ENG, UNIV

KANS, 88-, CHMN DEPT, 88- *Concurrent Pos:* Co-dir, Tertiary Oil Recovery Proj, 74-; consult, Off Technol Assessment, 76-78. *Honors & Awards:* Distinguished Achievement Award for Petrol Engr Fac, Soc Petrol Engrs, 81; Lester C Uren Award, 86; Distinguished Mem Soc Petrol Engrs, 86. *Mem:* Am Inst Chem Engrs; Soc Petrol Engrs; Am Chem Soc. *Res:* Transport processes in porous media; environmental heat transfer; numerical solutions of partial differential equations; enhanced oil recovery processes. *Mailing Add:* Dept Chem & Petrol Eng Univ Kans Lawrence KS 66045

WILLHOIT, DONALD GILLMOR, b Kansas City, Mo, Feb 5, 34; m 56; c 4. RADIATION HEALTH. *Educ:* William Jewell Col, AB, 56; Univ Wash, 58; Univ Pittsburgh, ScD(radiation health), 64. *Prof Exp:* Assoc scientist & health physicist, Westinghouse Testing Reactor, 58-60; radiation safety off, Univ Pittsburgh, 60-61, teaching fel, 61-64; asst prof radiol health, Sch Pub Health, 64-68, ASSOC PROF ENVIRON SCI, UNIV NC, CHAPEL HILL, 68-, DIR HEALTH & SAFETY, 74- *Mem:* Health Physics Soc; Radiation Res Soc; Am Indust Hyg Asn. *Res:* Public policy of low-level radioactive and hazardous waste disposal. *Mailing Add:* Dept Environ Sci Univ NC Chapel Hill NC 27514

WILLIAM, JAMES C, JR, RENAL PHYSIOLOGY EPITHELIAL TRANSPORT. *Educ:* Cornell Univ, PhD(physiol), 83. *Prof Exp:* Teaching fel, 83-86, instr nephrology, Univ Ala, Birmingham, 86-; DEPT ANAT & CELL BIOL, MED UNIV SC. *Mailing Add:* Dept Anat & Cell Biol Med Univ SC Charleston SC 29425

WILLIAMS, AARON, JR, b Newark, NJ, Jan 29, 42. PHYSICAL GEOGRAPHY. *Educ:* Fla State Univ, BS, 65; Univ Mo, MA, 67; Univ Okla, PhD(geog), 71. *Prof Exp:* Meteorologist, US Weather Bur, 66; instr geog, 67-69, asst prof, 71-77, ASSOC PROF GEOG, UNIV SALA, 77- *Mem:* Am Asn Geog; Am Meteorol Soc. *Res:* Climatology of coastal and tropical environments; radar climatology. *Mailing Add:* Dept Geol & Geog Life Sci Bldg Rm 341 Univ SAla Mobile AL 36688

WILLIAMS, ALAN EVAN, b Mechanicsburg, Pa, Oct 4, 52. ISOTOPE GEOCHEMISTRY, GEOTHERMICS. *Educ:* Juniata Col, ScB, 74; Brown Univ, MSc, 76, PhD(geol), 80. *Prof Exp:* Res geochemist, 79-80, ASST RES GEOCHEMIST, INST GEOPHYSICS & PLANETARY PHYSICS, UNIV CALIF, RIVERSIDE, 80- *Concurrent Pos:* Vis lectr, Univ Redlands, 80; adj lectr, Univ Calif, Riverside, 81- *Mem:* Am Geophys Union; Geol Soc Am; Geothermal Resources Coun. *Res:* Stable isotopic constraints on systems of water-rock interaction: geothermal systems, ore bodies, shallow intrusives, and ophiolites. *Mailing Add:* 3390 Santacruz Dr Riverside CA 92507

WILLIAMS, ALAN K, b Harrisburg, Pa, Dec 19, 28; m 88; c 3. PROCESS CHEMISTRY-ACTINIDE ELEMENTS, NUCLEAR FUEL REPROCESSING. *Educ:* Univ Northern Colo, BA, 52. *Prof Exp:* Sr res mgr, Dow Chem Co, 52-74; vpres, Allied Gen Nuclear Serv, 74-83; PROJ MGR, BECHTEL NAT INC, 83- *Concurrent Pos:* Consult, US Dept Energy, TMI Recovery, 79-83, Bret Proj, 82-84, Idaho Nat Engr Lab, Off Gas Systs, 80-82, Los Alamos Nat Lab, 82-83; bd dir, Am Chem Soc, 82-85. *Mem:* fel Am Nuclear Soc; AAAS; AM Chem Soc. *Res:* Principal activities are design and operation of nuclear facilities for the reprocessing of spent nuclear fuels from commercial power reactors and processing of actinide elements. *Mailing Add:* PO Box 12 PO Box 3965 45 Fremont St Hathaway Pines CA 95233-0012

WILLIAMS, ALBERT J, JR, b Media, Pa, Feb 28, 03; m 37; c 2. ELECTRICAL ENGINEERING. *Educ:* Swarthmore Col, AB, 24. *Prof Exp:* Engr transformers, Bell Tel Labs, 24-27; res engr measurement & control, Leeds & Northrup Co, 27-34, chief elec div res dept, 34-51, assoc dir res, 51-55, sci adv, 55-68; RETIRED. *Concurrent Pos:* Consult, Leeds & Northrup Co, 68-76; consult, 68-71. *Honors & Awards:* John Price Wetherill Medal, Franklin Inst, 52; Morris E Leeds Award, Inst Elec & Electronics Engrs, 68; Rufus Oldenburger Medal, Am Soc Mech Engrs, 72. *Mem:* Fel Inst Elec & Electronics Engrs; fel Franklin Inst; fel Am Soc Mech Engrs; Am Phys Soc; Sigma Xi. *Res:* Measurements; recorders; control; measurements with extended range and precision; science of golf; aids to functioning of man's body, especially breathing, hearing, seeing, coma and perceiving; aiding man's perception of numbers. *Mailing Add:* 2464 Trewigton Rd Colmar PA 18915

WILLIAMS, ALBERT JAMES, III, b Philadelphia, Pa, Oct 17, 40; m 63; c 1. OCEANOGRAPHY, OCEAN ENGINEERING. *Educ:* Swarthmore Col, BA, 62; Johns Hopkins Univ, PhD(physics), 69. *Prof Exp:* Investr, 69-71, asst scientist, 71-75, ASSOC SCIENTIST, OCEAN ENG, WOODS HOLE OCEANOG INST, 75- *Mem:* AAAS; Am Geophys Union. *Res:* Ocean microstructure, mixing and thermohaline convection; stress and velocity structure in the benthic boundary layer; oceanographic, optical, acoustic, and electronic instrumentation. *Mailing Add:* 12 Nobska Circle Woods Hole MA 02543

WILLIAMS, ALBERT SIMPSON, b York Co, SC, Jan 23, 24; m 46; c 1. PLANT PATHOLOGY, HORTICULTURE. *Educ:* Emory Univ, AB, 48; Univ Tenn, MS, 49; NC State Univ, PhD(plant path), 54. *Prof Exp:* Asst prof biol, Athens Col, 50; plant pathologist, State Plant Bd, Miss, 50-51; asst, NC State Univ, 51-54; assoc prof plant path, Va Polytech Inst, 54-68; exten prof plant path, 68-75, PROF & CHMN HORT DEPT, UNIV KY, 75- *Mem:* Am Phytopath Soc; Am Soc Nematol; Soc Europ Nematol; Am Soc Hort Sci. *Mailing Add:* Dept Hort Univ Ky Lexington KY 40506

WILLIAMS, ANN HOUSTON, b Red Bank, NJ, Dec 18, 43; m 76. MARINE COMMUNITY ECOLOGY, BENTHIC ECOLOGY. *Educ:* Univ SC, BS, 65; Duke Univ, MAT, 72; Univ NC, PhD(zool), 77. *Prof Exp:* Res assoc, Marine Lab, Duke Univ, 77-78; asst prof biol, Southwestern Univ , Memphis, 78-80; from asst prof to assoc prof marine biol, Auburn Univ, 80-90; PROG ADMINR, SIGMA XI, 90- *Concurrent Pos:* Prin investr, NSF grant, 75-77, NSF EPSCOR grant, 86-88; vis prof, Dauphin Island Sea Lab, 81-85; res fel, AAAS-Environ Protection Agency, 83; Ala Exp Sta grants, 85-90 & 89-94. *Mem:* Sigma Xi; Am Soc Zoologists; Ecol Soc Am; Gulf Estuarine Res Soc; AAAS; Estuarine Res Fedn. *Res:* Competitive interactions in shallow coral reef communities; predator-prey interactions in salt marsh communities; predator-prey interactions of seagrass communities. *Mailing Add:* Sigma Xi Hq PO Box 13975 Research Triangle Park NC 27709

WILLIAMS, ANNA MARIA, b Tampa, Fla, June 29, 27. MICROBIOLOGY. *Educ:* Univ Ala, BS, 48; Univ Wis, MS, 51, PhD(bact), 54. *Prof Exp:* Antibiotics lab supvr, Merck & Co, Inc, 48-50; res asst bact, Univ Wis, 50-54; Fulbright res grant, biophys res group, Univ Utrecht, 54-55; proj assoc, McArdle Mem Inst Cancer Res, Univ Wis-Madison, 55-57, res assoc med, Sch Med, 57-64, from asst prof to assoc prof biol sci, 64-89, PROF BIOL SCI, UNIV WIS-PARKSIDE, 89- *Mem:* AAAS; Am Chem Soc; Am Soc Microbiol; Am Inst Biol Sci; Sigma Xi. *Res:* Nucleic acid metabolism; leukemia; cancer chemotherapy; bacterial enzymes; science education. *Mailing Add:* Univ Wis-Parkside Kenosha WI 53141-2000

WILLIAMS, ARDIS MAE, b Boston, Mass, June 8, 19; m 49; c 1. PHYSICAL CHEMISTRY. *Educ:* Mt Holyoke Col, AB, 41; Vassar Col, AM, 46. *Prof Exp:* Chemist, Harvard Med Sch, 41-42; teacher, Low Heywood Sch, 42-44; lectr chem, Vassar Col, 44-46; lectr, Barnard Col, 46-48; res chemist, Merck & Co, 48-49; res chemist, Med Col Va, 50-54; teacher, Tatnall Sch, 63-67; ASSOC PROF CHEM, WEST CHESTER STATE COL, 67- *Mem:* AAAS; Am Chem Soc. *Res:* Surface area and adsorption by activated carbon; relationship of structure and adsorption of organic molecules. *Mailing Add:* RR 1 No 152 Landenberg PA 19350

WILLIAMS, ARTHUR LEE, b Sawyerville, Ala, July 16, 47; m 69; c 1. MOLECULAR BIOLOGY, MOLECULAR GENETICS. *Educ:* Ala State Univ, BS, 69; Atlanta Univ, MS, 71; Purdue Univ, PhD(molecular biol), 75. *Prof Exp:* Lab instr biol, Atlanta Univ, 69-70; teaching asst cell biol, Purdue Univ, 74-75; asst prof biol, Ala State Univ, 75-77, chairperson gen biol, 76-77; asst prof biol, Murray State Univ, 77-80; ASST PROF BIOL, UNIV KY, 80- *Concurrent Pos:* Prin investr, NIH res grant, 77-78. *Mem:* Am Soc Biog Res; Am Soc Microbiol; Nat Inst Sci; NY Acad Sci. *Res:* Biochemical and genetic investigation of the substrate induction of the ilvC gene expression of Escherichia coli. *Mailing Add:* RR 2 No 744 Foster KY 41043

WILLIAMS, ARTHUR OLNEY, JR, b Providence, RI, Apr 7, 13; m 38; c 1. PHYSICS. *Educ:* Mass Inst Technol, BS, 34; Brown Univ, ScM, 36, PhD(physics), 37. *Prof Exp:* From instr to asst prof physics, Univ Maine, 37-42; from asst prof to prof, Brown Univ, 42-75, chmn dept, 56-60, 62-63, Hazard prof, 75-78, emer prof, 78-; RETIRED. *Concurrent Pos:* Mem, RI AEC, 55-68; vis res prof, Underwater Acoust, Naval Postgrad Sch, Monterey, Calif, 81-82. *Mem:* Fel Am Phys Soc; fel Acoust Soc Am; Sigma Xi. *Res:* Physical acoustics. *Mailing Add:* Brown Univ Providence RI 02902

WILLIAMS, ARTHUR ROBERT, b Feb 20, 41; US citizen; m 62; c 3. SOLID STATE PHYSICS. *Educ:* Dartmouth Col, AB, 62; Harvard Univ, PhD(solid state physics), 69. *Prof Exp:* Appl mathematician, Info Res Inc, Mass, 67-68; fel, 68-69, STAFF PHYSICIST, WATSON RES CTR, IBM CORP, 69- *Mem:* Am Phys Soc. *Res:* Effective one-electron theory of Fermi surface and optical data for solids. *Mailing Add:* IBM T J Watson Res Ctr Box 218 Yorktown Heights NY 10598

WILLIAMS, AUSTIN BEATTY, b Plattsburg, Mo, Oct 17, 19; m 46; c 1. SYSTEMATIC ZOOLOGY. *Educ:* McPherson Col, AB, 43; Univ Kans, PhD(zool), 51. *Prof Exp:* Asst genetics, Univ Wis, 43-44; teacher pub schs, Kans, 44-46; shrimp investr, Inst Fisheries Res, Univ NC, 51-52, asst prof, 52-55; asst prof, Univ Ill, 55-56; assoc prof, Inst Fisheries Res, Univ NC, 56-63, prof, Inst Marine Sci, 63-71; SYST ZOOLOGIST, NAT SYSTS LAB, NAT MARINE FISHERIES SERV, 71- *Concurrent Pos:* Adj prof, Univ NC, Chapel Hill, 71-80. *Mem:* Fel AAAS; Am Soc Zool; Ecol Soc Am; Soc Syst Zool (secy, 85-88); Estuarine Res Fedn (secy, 72-73, vpres, 80-81, pres, 83-85). *Res:* Taxonomy; ecology; life histories of decapod crustacea; estuarine ecology. *Mailing Add:* Nat Systs Lab Nat Mus Nat Marine Fisheries Serv Washington DC 20560

WILLIAMS, BENJAMIN HAYDEN, b Davenport, Iowa, Dec 18, 21; div; c 5. ANATOMY, ORTHODONTICS. *Educ:* Ohio State Univ, DDS, 46, MSc, 49; Univ Ill, MS, 51; Am Bd Orthod, dipl. *Prof Exp:* Fel, Univ Ill, 50-51; from instr to assoc prof, 51-64, PROF ORTHOD & CHMN DEPT, COL DENT, OHIO STATE UNIV, 64- *Concurrent Pos:* Staff mem, Children's Hosp, Columbus, Ohio, 61-; mem bd med, State Crippled Children's Servs, Ohio, 66-; mem orthod sect, Am Asn Dent Schs; staff mem, Vet Hosp, Dayton, Ohio, 76- *Mem:* Int Asn Dent Res; Am Asn Orthod; Angle Soc. *Res:* Cranio-facial growth and development; growth predictions; dental development and eruption; normal and abnormal growth; temporo-mandibular joint disorders. *Mailing Add:* Dept Orthod Ohio State Univ Col Dent Columbus OH 43210

WILLIAMS, BENNIE B, b Scranton, Tex, Jan 16, 22; m 48; c 4. MATHEMATICS. *Educ:* Howard Payne Col, BA, 48; Univ Tex, MA, 53, PhD(math), 66. *Prof Exp:* From instr to asst prof math, Howard Payne Col, 48-61; asst prof, 66-71, ASSOC PROF MATH, UNIV TEX, ARLINGTON, 71- *Mem:* Am Math Soc; Math Asn Am. *Res:* Foundations of mathematics; ordinary differential equations. *Mailing Add:* 3405 Halifax Dr Arlington TX 76013

WILLIAMS, BETTY L, COMPOUNDS ON THE CARDIOVASCULAR SYSTEM. *Educ:* Emory Univ, PhD, 67. *Prof Exp:* PROF PHARMACOL, MED BR, UNIV TEX, 82- *Mailing Add:* Dept Pharmacol Univ Tex Med Br Galveston TX 77550

WILLIAMS, BOBBY JOE, b Idabel, Okla, Nov 3, 30; m 57; c 4. PHYSICAL ANTHROPOLOGY, POPULATION GENETICS. *Educ:* Univ Okla, BA, 53, MA, 57; Univ Mich, PhD(anthrop, human genetics), 65. *Prof Exp:* Asst prof anthrop, Univ Wis, Milwaukee, 63-65; from asst prof to assoc prof, 65-77, PROF ANTHROP, UNIV CALIF, LOS ANGELES, 77- *Concurrent Pos:*

Vpres, Am Asn Phys Anthrop, 87-88. *Mem:* AAAS; Am Anthrop Asn; Am Asn Phys Anthrop. *Res:* Human population genetics; human evolution; population processes in simple societies. *Mailing Add:* Dept Anthrop Univ Calif Los Angeles CA 90024

WILLIAMS, BROWN F, b Evanston, Ill, Dec 22, 40; c 2. SOLID STATE PHYSICS. *Educ:* Univ Calif, Riverside, BA, 62, MA, 64, PhD(physics), 66. *Prof Exp:* Mem tech staff, 66-68, leader electron emission, 68-70, mgr electro-optics lab, 70-73, head quantum electronics res, 73-77, dir energy systs res lab, 77-79, staff vpres, display & energy systs, RCA labs, 79-87; vpres, Solid State Res, 87. *Mem:* Fel Inst Elec & Electronics Engrs; Sigma Xi; AAAS; Am Phys Soc. *Res:* Electro-optic devices; optical information recording; solar energy; television picture tubes; electron optics. *Mailing Add:* 27 Honeybrook Dr Princeton NJ 08540

WILLIAMS, BRYAN, b Longview, Tex, July 28, 25. INTERNAL MEDICINE. *Educ:* Southwestern Univ, MD, 47. *Prof Exp:* Asst resident, Mass Mem Hosp, 50-51, chief resident, 53-54; clin asst, Harvard Univ, 55-56; clin asst prof, 63-66, assoc dean student affairs, 66-90, ASSOC DEAN ALUMNI AFFAIRS, SOUTHWESTERN UNIV, 90- *Mem:* Inst Med Nat Acad Sci. *Mailing Add:* Dept Internal Med Univ Tex Health Sci Ctr 5323 Harry Hines Blvd Dallas TX 75235

WILLIAMS, BYRON BENNETT, JR, pharmacology, for more information see previous edition

WILLIAMS, BYRON LEE, JR, b Guantanamo Bay, Cuba, Aug 29, 20; m 42; c 3. ORGANIC CHEMISTRY. *Educ:* ETex State Teachers Col, BS, 40, MS, 42; Univ Okla, PhD(org chem), 53. *Prof Exp:* Teacher pub sch, Tex, 40-42; chem engr, Chem Warfare Serv, US Dept Army, 42-44, chem engr, Chem Corps, 46-47; assoc prof chem, ETex State Teachers Col, 47-41; asst, Res Found, Univ Okla, 51-53; assoc prof chem, ETex State Teachers Col, 53-54; res chemist, 54-59, Monsanto Co, from asst dir res to assoc dir res, Plastics Div, 59-64, dir process tech & eng dept, Hydrocarbons Div, 64-65, dir res, Hydrocarbons & Polymers Div, 65-67, dir corp res dept, 67-76; gen mgr, Technol Div, Monsanto Textiles Co, 76-81; dir, feedstock transition, Monsanto Co, 81-82; TEACHER, MARYVILLE COL, 82- *Mem:* Am Chem Soc. *Res:* Isolation, chemical characterization and identification of flavonoid type chemical compounds from selected natural products using ion exchange, adsorption and paper chromatography; synthesis of pigments and demethylation studies; hydrolytic enzyme studies of natural glycosides. *Mailing Add:* 609 Mosley Rd Creve Coeur MO 63141-7634

WILLIAMS, CALVIT HERNDON, JR, b Houston, Tex, Dec 28, 36; div; c 4. INDUSTRIAL HYGIENE. *Educ:* Univ St Thomas, Tex, BA, 58; Brown Univ, PhD(chem), 64; Am Bd Indust Hyg, dipl, 79. *Prof Exp:* Fel chem, Rice Univ, 64-66; tech staff mem, Sandia Labs, 66-70; asst prof chem, State Univ Campinas, Brazil, 71-76; lab dir, Aer-Aqua Labs, 76-77; SR STAFF SCIENTIST, RADIAN CORP, 77- *Concurrent Pos:* Financial planning, IDS-Am Express, 84-85. *Mem:* Sigma Xi; Am Chem Soc; Am Indust Hyg Asn; Am Inst Chem; Am Soc Mass Spectrometry. *Res:* High energy crossed-molecular-beam reaction kinetics; thermodynamics of high temperature processes; mass spectrometry and gas chromatography applied to environmental and biomedical analysis; chemical aspects and comprehensive practice of industrial hygiene; occupational and environmental health. *Mailing Add:* Radian Corp PO Box 201088 8501 Mo-Pac Blvd Austin TX 78720

WILLIAMS, CAROL ANN, b Stratford, NJ, Oct 3, 40. CELESTIAL MECHANICS, NON-LINEAR DYNAMICS. *Educ:* Conn Col, BA, 62; Yale Univ, PhD(astron), 67. *Prof Exp:* Assoc res engr & consult, Jet Propulsion Lab, 64, 65, 84; part-time instr physics, Conn Col, 66-67; res staff astronr, Yale Univ, 67-68; from asst prof to assoc prof astron, 68-79, ASSOC PROF MATH, UNIV SFLA, TAMPA, 79- *Concurrent Pos:* Adj assoc prof astron, Univ Fla, 73-; ed, Celestial Mech, 80-87; lectr, Am Astron Soc, 82-86; consult, Space Telescope Sci Inst, 83; mathematician, Nat Bur Standards, 85- *Mem:* Am Astron Soc; Sigma Xi; Int Astron Union. *Res:* Celestial mechanics: lunar and planetary theory, three body problem, resonance problem; applied mathematics: perturbation theory with special functions; astrometry: systematic errors in plate reduction and in star catalogs. *Mailing Add:* Dept Math & Physics 114 Univ SFla 4808 E Fowler Ave Tampa FL 33620

WILLIAMS, CAROLE A, b New Haven, Conn, July 31, 47. CARDIOVASCULAR PHYSIOLOGY. *Educ:* Albertus Magnus Col, AB, 69; St Louis Univ, PhD(physiol), 77. *Prof Exp:* Asst prof, 79-85, ASSOC PROF PHYSIOL, EAST TENN STATE UNIV, 85- *Mem:* Am Physiol Soc; AAAS. *Mailing Add:* Dept Physiol East Tenn State Univ Col Med PO Box 19780A Johnson City TN 37614

WILLIAMS, CARROLL BURNS, JR, b St Louis, Mo, Sept 24, 29; m 58; c 3. FOREST ENTOMOLOGY. *Educ:* Univ Mich, BS, 55, MS, 57, PhD(forestry), 63. *Prof Exp:* Entomologist, Pac Northwest Forest & Range Exp Sta, US Forest Serv, Ore, 57-58, res forester, 58-60, forestry sci lab, 61-65, res entomologist, Pac Southwest Forest & Range Exp Sta, Calif, 65-68, leader insect impact proj, Forest Insect & Dis Lab, Northeastern Exp Sta, Conn, 68-72, res entomologist, Pac Southwest Forest & Range Exp Sta, Calif, 72-84, proj leader, Pioneering Res Unit, Integrated Mgt Systs Forest Insect & Dis, 75-84, proj leader, Pest Impact Assessment Technol, 85-88; ADJ PROF, DEPT FOREST & RESOURCE MGT, COL NATURAL RESOURCES, UNIV CALIF, BERKELEY, 88- *Concurrent Pos:* Lectr, Sch Forestry, Yale Univ, 69-71; consult, NSF, 71-74. *Mem:* Soc Am Foresters; AAAS; Entom Soc Am. *Res:* Evaluate and predict impact of forest insect and disease on forest resources; modelling pest management systems; computer simulation experiments of insect and disease control techniques and strategies; forest management decision models. *Mailing Add:* 145 Mulford Hall Univ Calif Berkeley CA 94720

WILLIAMS, CARROLL MILTON, biology, insect physiology & metamorphosis; deceased, see previous edition for last biography

WILLIAMS, CHARLES HADDON, JR, b Washington, DC, June 29, 32; m 62; c 2. BIOCHEMISTRY. *Educ:* Univ Md, BS, 56; Duke Univ, PhD(biochem), 61. *Prof Exp:* Am Cancer Soc fel, Dept Biochem, Sheffield Sci Sch, 61-63; from instr to assoc prof, Univ Mich, Ann Arbor, 63-79; res chemist, 63-74, coordr res, 77-79, SUPVRY RES CHEMIST, GEN MED RES, DEPT VET AFFAIRS MED CTR, ANN ARBOR, 74-; PROF BIOL CHEM, UNIV MICH, ANN ARBOR, 79- *Concurrent Pos:* Res career scientist, Vet Admin. *Mem:* AAAS; Am Chem Soc; Am Soc Biol Chemists; Sigma Xi. *Res:* Mechanism of action and structure of flavoproteins; roles of various amino acid residues in catalysis by flavoproteins and in their structures. *Mailing Add:* Dept Biochem Univ Mich Ann Arbor MI 48109

WILLIAMS, CHARLES HERBERT, b Aurora, Mo, Jan 21, 35; m 56; c 3. MALIGNANT HYPERTHERMIA. *Educ:* Univ Mo-Columbia, BS, 57, MS, 67, PhD(agr chem), 68. *Prof Exp:* Grad res asst agr chem, Univ Mo, Columbia, 63-68, assoc prof biochem, dept biochem & med, 73-76, res assoc, Sinclair Exp Med Res Farm, 76-78, fel toxicol, 79, res assoc human nutrit, 81, Cancer Res Ctr, 82; fel enzymol, Inst Enzyme Res, Univ Wis-Madison, 68-70, asst prof bichem, 70-72, dept anesthesiol, 70-73; dir qual control, Medico Industs, Elmwood, Kans, 79-80; dir anesthesiol res, Dept Anesthesiol & Biochem, 82-86; ASSOC PROF BIOCHEM, DEPT BIOCHEM, SCH MED, TEX TECH UNIV, 82- *Concurrent Pos:* Adj assoc prof biochem, dept chem, Univ Tex, 83-; dir, surg res, Dept Surg & Biochem, 86- *Honors & Awards:* Fulbright Lectr, Univ San Marcos, Peru, 71. *Mem:* NY Acad Sci; Am Physiol Soc; Am Soc Biol Chemists; Int Anesthesia Res Soc. *Res:* Physiology and pharmacology of temperature regulation; metabolism and endocrinology of catecholamines; biophysics of membrane calcium regulation, for example, calcium channels, gene mapping of the MH gene. *Mailing Add:* Dept Surg & Biochem Health Sci Ctr Tex Tech Univ 4800 Alberta Ave El Paso TX 79905

WILLIAMS, CHARLES MELVILLE, b Regina, Sask, Mar 18, 25; m 53; c 3. PHYSIOLOGY, GENETICS. *Educ:* Univ BC, BSA, 49, MSA, 52, Ore State Col, PhD(genetics), 55. *Prof Exp:* Assoc prof animal physiol, 55-67, head dept, 75-83, PROF ANIMAL SCI, UNIV SASK, 77- *Honors & Awards:* Order of Can. *Mem:* Am Soc Animal Sci; Can Soc Animal Prod; fel Agr Inst Can. *Res:* Effect of low environmental temperatures on farm animals. *Mailing Add:* One Moxon Crescent Saskatoon SK S7N 3B8 Can

WILLIAMS, CHARLES WESLEY, b Palestine, Ark, Feb 18, 31; m 59; c 2. ELECTRONICS ENGINEERING. *Educ:* Univ Tenn, BSEE, 59, MSEE, 63. *Prof Exp:* Design engr, Mead Res Labs, Ohio, 59-60; develop engr, Oak Ridge Nat Lab, 60-63; sr develop engr, 63-67, engr & mgr electronics res & develop, 67-80, DIR RES, ORTEC, LTD, 80- *Mem:* Inst Elec & Electronics Engrs; Nat Soc Prof Engrs. *Res:* Nuclear research instrumentation, especially sub-nanosecond time measurements and linear circuit design. *Mailing Add:* Ortec Ltd Mail Stop 8211 Bldg 9201-1 Oak Ridge TN 37830

WILLIAMS, CHRISTINE, b Miami, Fla, Mar 20, 43; c 3. PREVENTIVE CARDIOLOGY. *Educ:* Univ Pittsburgh, BS, 63, MD, 67; Harvard Univ, MPH, 69. *Prof Exp:* Resident, Johns Hopkins Sch Hyg & Pub Health, 70-71, vis scientist, Karolinska Inst, Stockholm, 71-72; resident pediat, Hosp Med Col Pa, 72-73, pediatrician, Children & Youth Comprehensive Care Clin, 74-75; dir child health, Am Health Found, 75-80; dep comnr health, Westchester County Dept Health, 81-85; ASSOC PROF PEDIAT & CHIEF SECT PEDIAT EPIDEMIOL, NY MED COL, 85- *Honors & Awards:* Preventive Cardiol Acad Award, NIH, 87-92. *Mem:* Am Acad Pediat; Am Public Health Asn. *Res:* Preventive cardiology; Lyme disease. *Mailing Add:* New York Med Col Munger Pavilion Valhalla NY 10595

WILLIAMS, CHRISTOPHER NOEL, b York, Eng, Dec 25, 35; Can citizen; m 60; c 4. GASTROENTEROLOGY. *Educ:* Royal Col Physicians & Surgeons London, MRCS & LRCP, 60; Royal Col Physicians & Surgeons Can, FRCP(C), 68; FACP, 76. *Prof Exp:* Fel med, Univ Pa, 69-71; lectr, 69-72, asst prof, 72-76, assoc prof, 76-82, PROF MED, DALHOUSIE UNIV, 82-, ASSOC PHYSICIAN MED, VICTORIA GEN HOSP, 76- *Concurrent Pos:* MacLaughlin fel, Univ Pa, 69-70, Med Res Coun fel, 70-71; asst physician med, Victoria Gen Hosp, 69-76; dir, Gastrointestinal Res Lab, Dalhousie Univ, 71-; consult gastroenterol, Camp Hill Hosp, Halifax, 73-; grants in aid, Med Res Coun Can, 71- & Nat Health & Welfare Can, 73-; consult, Halifax Infirmary, 78-; head prog dir gastroenterol, Dalhousie Univ, 81. *Mem:* Can Med Asn; Can Asn Gastroenterol (pres, 86-); Can Soc Clin Res; Am Gastroenterol Asn; Am Asn Study Liver Dis; Can Asn Study Lives. *Res:* Detailed kinetic studies of bile acid metabolism in health and disease, particularly liver and inflammatory bowel disease; application of 3-alpha, 7-alpha, 12-alpha hydroxysteroid dehydrogenases to bile analysis; dietary and biochemical factors and prevalence of gallstones in Caucasians and Canadian Indians. *Mailing Add:* Dept Med Gastroent Dalhousie Univ Fac Med Sir Chas Tupper Bldg Halifax NS B3H 4H7 Can

WILLIAMS, CHRISTOPHER P S, b Medford, Ore, Oct 12, 31; m 57; c 3. PEDIATRICS, MEDICINE. *Educ:* Univ Ore, BA, 53, MD, 58. *Prof Exp:* Asst prof pediat, Sch Med, Univ Wash, 62-68; ASSOC PROF PEDIAT, CRIPPLED CHILDREN'S DIV, MED SCH, UNIV ORE, 68- *Res:* Medical education. *Mailing Add:* Crippled Childrens Div Ore Health Sci Univ PO Box 574 Portland OR 97207

WILLIAMS, CLAYTON DREWS, b St Louis, Mo, Oct 22, 35; m 59; c 4. THEORETICAL PHYSICS, SOLID STATE PHYSICS. *Educ:* Rice Inst, BA, 57; Wash Univ, PhD(physics), 61. *Prof Exp:* Asst prof, 61-64, ASSOC PROF PHYSICS, VA POLYTECH INST & STATE UNIV, 64- *Mem:* Am Phys Soc. *Res:* Many-body problem; solid-state theory. *Mailing Add:* Dept Physics Va Polytech Inst & State Univ Blacksburg VA 24061

WILLIAMS, CLYDE MICHAEL, b Marlow, Okla, Oct 8, 28; m 53; c 4. PHYSIOLOGY. *Educ:* Rice Inst, BA, 48; Baylor Univ, MD, 52; Oxford Univ, DPhil(physiol), 54. *Prof Exp:* Tech dir radioisotope lab, Vet Admin Hosp, Pittsburgh, Pa, 58-60; res radiol, 60-63, from asst prof to assoc prof, 63-65, PROF RADIOL & CHMN DEPT, COL MED, UNIV FLA, 65- *Mem:* Am

Physiol Soc; Soc Nuclear Med; Am Roentgen Ray Soc; Radiol Soc NAm. *Res:* Metabolism of aromatic amines; gas chromatography of aromatic acids; use of computer for medical diagnosis. *Mailing Add:* 2721 NW 37th Terr Gainesville FL 32605

WILLIAMS, COLIN JAMES, b London, Eng, May 12, 38. PHYSICAL CHEMISTRY, INORGANIC CHEMISTRY. *Educ:* Univ London, BSc, 60, PhD(phys chem, inorg chem), 60 & Imp Col, dipl, 63. *Prof Exp:* Res chemist, Cent Res Div, Mobil Oil Corp, NJ, 63-66; RES CHEMIST, CAHN INSTRUMENT CO, 67- *Mem:* Am Chem Soc; assoc mem Royal Soc Chem. *Res:* Physical and inorganic chemical properties of zeolites and their spectroscopic and catalytic properties; thermoanalysis; diffuse reflectance spectroscopy; physical adsorption. *Mailing Add:* Perkin-Elmer Corp Ten Faraday Irvine CA 92718-2714

WILLIAMS, CONRAD MALCOLM, b Warsaw, NC, Mar 1, 36; m 67; c 1. SOLID STATE PHYSICS. *Educ:* Morgan State Col, BS, 58; Howard Univ, MS, 65, PhD(physics), 72. *Prof Exp:* Res solid state physicist, Div Math Sci, US Naval Res Lab, 60-80; div sci personnel improv, 80-81, prog dir solid state phys, div mat res, NSF, 81-82; HEAD APPL MAGNETICS, MAT SCI DIV, US NAVAL RES LAB, 82- *Concurrent Pos:* Adj prof physics, Howard Univ, 74-80, adj prof elec eng, 80- *Honors & Awards:* Sigma Xi Pure Sci Award, NRL, 79. *Mem:* Sigma Xi; Am Phys Soc; Inst Elec & Electronics Engrs. *Res:* Solid state physics; physics of metals; ferromagnetism (thin films and bulk materials); low temperature; low temperature properties; irradiation effects metals; metal alloys; semiconductors; ion implantation in the ferromagnetic films. *Mailing Add:* Code 6636 US Naval Res Lab Washington DC 20375

WILLIAMS, CURTIS ALVIN, JR, b Moorestown, NJ, June 26, 27; m 60; c 3. BIOLOGY, EDUCATIONAL ADMINISTRATION. *Educ:* Pa State Univ, BS, 50; Rutgers Univ, PhD(zool), 54. *Prof Exp:* Waksman fel & USPHS fel, Pasteur Inst, Paris, 52-54, USPHS fel, Carlsberg Lab, Copenhagen, 54-55; res assoc microbiol, Rockefeller Inst, 55-57; with Nat Inst Allergy & Infectious Dis, 57-60; from asst prof to assoc prof biochem genetics, Rockefeller Univ, 60-69; dean natural sci, 69-80, chmn dept biol, 80-90, PROF BIOL, STATE UNIV NY, PURCHASE, 69- *Concurrent Pos:* Adj prof, Rockefeller Univ, 70-78 & Sch Med, NY Univ, 76-; hon res fel neuroimmunol, Univ Col, London, 78; vis res fel cell biol, Albert Einstein Col Med, 79-80; vis prof neurobiol, Univ Southern Calif, Los Angeles, 85. *Honors & Awards:* Founders Award, Electrophoresis Soc, 82. *Mem:* Fel AAAS; Soc Neurosci; Sigma Xi; Am Soc Microbiol; Am Asn Immunol; Int Soc Neuroimmunol. *Res:* Neurobiology; immunology; cellular biology; neural and behavioral effects of inflammation in CNS. *Mailing Add:* Dept Biol State Univ NY Purchase NY 10577

WILLIAMS, CURTIS CHANDLER, III, b New York, NY, Sept 30, 26; m 55; c 2. CHEMICAL ENGINEERING. *Educ:* Yale Univ, BEng, 48; Mass Inst Technol, SM, 50, ScD(chem eng), 53. *Prof Exp:* Engr, Emeryville Res Ctr, Shell Develop Co, 53-59, supvr process eng, 59-64, sr technologist, Mfg Res Dept, Shell Oil Co, 64-65, head petrol processing dept, Emeryville Res Ctr, Shell Develop Co, 65-67, chief technologist, Wood River Refinery, Shell Oil Co, Ill, 67-71, mgr facilities, 71-74, mgr process eng-refining, 74-82, MGR PROCESS ENG-SEPARATIONS, SHELL OIL CO, 82- *Concurrent Pos:* Chmn tech data comt, API Refining Dept, 65-; mem bd dirs, Fractionation Res Inc, 75- & Heat Transfer Res, Inc, 75- *Honors & Awards:* Cert of Appreciation, Am Petrol Inst, 77. *Mem:* AAAS; Am Chem Soc; fel Am Inst Chem Engrs. *Res:* Design of petroleum refining and petrochemical processes. *Mailing Add:* 13606 Pinerock Lane Houston TX 77079-5914

WILLIAMS, DALE GORDON, b Chicago, Ill, Aug 9, 29; m 52. PHYSICAL CHEMISTRY. *Educ:* Beloit Col, BS, 51; Univ Minn, MS, 54; Univ Iowa, PhD, 57. *Prof Exp:* Engr, Linde Co Div, Union Carbide Corp, 56-58; res aide, Inst Paper Chem, 59-62, res assoc & assoc prof chem, 62-75; res scientist, 75-81, SR RES SCIENTIST, UNION CAMP CORP, 81- *Mem:* Am Chem Soc; Tech Asn Pulp & Paper Indust; Sigma Xi. *Res:* The application of physical chemistry to paper science and technology. *Mailing Add:* 35 Springwood Dr Lawrenceville NJ 08648

WILLIAMS, DANIEL CHARLES, b Compton, Calif, Dec 15, 44; m 64; c 2. CELL BIOLOGY, BONE BIOLOGY. *Educ:* Calif State Univ, Long Beach, BS, 67; Iowa State Univ, PhD(cell biol), 72. *Prof Exp:* Asst microbiologist, Purex Corp, 66-67; res engr microbiol, NAm Aviation, Inc, 67; instr cell & develop biol, Kans State Univ, 72-74; asst prof develop biol, Univ Notre Dame, 74-76; sr scientist, 76-81, GROUP LEADER BONE BIOL, LILLY RES LABS, 86-, RES SCIENTIST, 82- *Concurrent Pos:* NIH fel, 67-71 & 72-74; consult, BioInfo Assocs, 72-74. *Mem:* Am Soc Bone & Mineral Res; Am Soc Cell Biol; Electron Micros Soc Am. *Res:* Structure-function relationships and control mechanisms associated with cellular and developmental processes especially skeletal tissue; bone biology. *Mailing Add:* Bone Biol Res Group MC620 Eli Lilly & Co Indianapolis IN 46285

WILLIAMS, DANIEL FRANK, b Redmond, Ore, Nov 20, 42; c 2. MAMMALOGY, NONGAME WILDLIFE MANAGEMENT. *Educ:* Cent Wash State Col, BA, 66; Univ NMex, MS, 68, PhD(zool), 71. *Prof E .p:* From asst prof to assoc prof biol sci, 71-80, PROF ZOOL SCI, CALIF STATE UNIV, STANISLAUS, 80- *Concurrent Pos:* Res assoc, Carnegie Mus Natural Hist; assoc ed, Mammalian Species, 78-82. *Mem:* Am Soc Mammal; Ecol Soc Am; Wildlife Soc; Soc Conserv Biol. *Res:* Systematics and evolution of mammals; ecology of mammals; evolution of chromosome morphology in mammals; conservation of mammals; plant/animal interactions - effects on plant productivity. *Mailing Add:* Dept Biol Sci Calif State Univ Stanislaus Turlock CA 95380

WILLIAMS, DARRYL MARLOWE, b Denver, Colo, Apr 3, 38; m 66; c 3. HEMATOLOGY. *Educ:* Baylor Univ, MS & MD, 64. *Prof Exp:* House officer med, Affil Hosps, Baylor Univ, 64-66; Univ Utah, 66-68, fel hemat, 68-73, asst prof med, 73-77; from assoc prof to prof med, La State Univ, 77-90; PROF MED, TEX TECH UNIV, LUBBOCK, 90- *Concurrent Pos:* Chief hemat & oncol, La State Univ, Shreveport, 77-85, asst dean res, 81-85, assoc dean acad affairs, 85-86, dean, 86-90; dean, HSC Sch Med, Tex Tech Univ, 90- *Mem:* Am Col Physicians; Am Soc Hemat; Am Inst Nutrit; Am Soc Clin Nutrit; Am Fedn Clin Res; Sigma Xi. *Res:* Biological effects of copper; manifestations of copper deficiency; interactions of copper, iron, and hemoglobin synthesis. *Mailing Add:* Texas Tech Univ Sch Med 3601 Fourth St Lubbock TX 79430

WILLIAMS, DAVID ALLEN, b Wakefield, Mass, Dec 22, 38; m 62; c 4. MEDICINAL CHEMISTRY. *Educ:* Mass Col Pharm, BS, 60, MS, 62; Univ Minn, Minneapolis, PhD(med chem), 68. *Prof Exp:* Sr scientist, Med Chem Div, Mallinckrodt Chem Works, 67-69; asst prof biochem, Mass Col Pharm, 69-77, assoc prof, 77-84, chmn, Dept Chem & Physics, 81-84, PROF MED CHEM, MASS COL PHARM, 84-, ASST DEAN GRAD STUDIES, 88- *Concurrent Pos:* Vis prof, Univ Strathclyde, Glasgow, 86. *Mem:* Am Chem Soc; NY Acad Sci; Am Asn Pharmaceut Scientists. *Res:* Stereochemistry of drug action; application of structure activity relationships to derivatives of biogenic amines; analgesics; high performance liquid chromatography analytical methods development. *Mailing Add:* Mass Col Pharm 179 Longwood Ave Boston MA 02115

WILLIAMS, DAVID BERNARD, b Leeds, Eng, July 25, 49; m 76; c 3. ENGINEERING, MATERIALS SCIENCE. *Educ:* Cambridge Univ, BA, 70, MA, 73, PhD(mat sci), 74. *Prof Exp:* Sci Res Coun fel mat sci, Cambridge Univ, 74-76; from asst prof to assoc prof, 76-83, PROF METALL & MAT ENG, LEHIGH UNIV, 83- *Honors & Awards:* Burton Medal, Electron Micros Soc Am, 84. *Mem:* Electron Micros Soc Am; Am Soc Metals; Am Inst Metals Engrs; Meteoritical Soc; fel Royal Micros Soc; fel Inst Metals UK; Microbeam Anal Soc. *Res:* Application of transmission and scanning transmission electron microscopy to the study of phase transformations in metals and ceramics. *Mailing Add:* Dept Mat Sci & Eng Lehigh Univ Whitaker Lab No 5 Bethlehem PA 18015-3195

WILLIAMS, DAVID CARY, b Santa Monica, Calif, June 22, 35; m 62. NUCLEAR CHEMISTRY, NUCLEAR PHYSICS. *Educ:* Harvard univ, AB, 57; Mass Inst Technol, PhD(nuclear chem), 62. *Prof Exp:* Fel nuclear chem, Princeton Univ, 62-64; mem staff, Los Alamos Sci Lab, 64-66; MEM TECH STAFF, SANDIA LABS, 66- *Mem:* Am Nuclear Soc; AAAS; Am Phys Soc; Am Chem Soc; Sigma Xi. *Res:* Nuclear reactions, nuclear decay schemes and nuclear reaction spectroscopy; fast reactor safety research; statistical models of nuclear reactions; atmospheric tracer studies. *Mailing Add:* Sandia Labs 6424 Albuquerque NM 87185

WILLIAMS, DAVID DOUGLAS F, b Bushey, Eng, May 13, 30; US citizen; m 52; c 6. HORTICULTURE. *Educ:* Univ Reading, BSc & dipl hort, 52; Univ Wis, Madison, PhD(bot & hort), 62. *Prof Exp:* Technician hort, Cent Exp Farm, Ottawa, Can, 52-53; res off fruit breeding, Exp Sta, Morden, Man, 53-56; proj asst hort, Univ Wis, Madison, 56-61; asst horticulturalist & supt hort, Maui Br Sta, Univ Hawaii, 62-67; plant breeder, Pineapple Res Inst Hawaii, 68-71, dir, 71-73; RES DIR, MAUI LAND & PINEAPPLE CO, 73- *Concurrent Pos:* Lectr, Maunaolu Col Maui, 67. *Res:* Effect of cultural, handling and cannery practices on the yield and quality of pineapple products. *Mailing Add:* 517 Olinda Rd Makawao HI 96768

WILLIAMS, DAVID FRANCIS, b New Orleans, La, Sept 4, 38; m 64; c 2. MEDICAL ENTOMOLOGY. *Educ:* Univ Southwest La, BS, 64, MS, 67; Univ Fla, PhD(entom), 69. *Prof Exp:* Asst prof biol, Greensboro Col, 69-71; res entomologist, WFla Arthropod Res Lab, State Fla, 71-74; location & res leader, Fed Exp Sta, Sci & Educ Admin-Agr Res, USDA, St Croix, 74-77; RES ENTOMOLOGIST, MED & VET ENTOM RES LAB, USDA, GAINESVILLE, FLA, 77- *Concurrent Pos:* Adj asst prof entomol, Univ Fla, 79-; expert witness, 87- *Honors & Awards:* Cert Merit, USDA, 77, Super Serv, 82, Invention Award, 85. *Mem:* Entom Soc Am; Sigma Xi. *Res:* Development and evaluation of chemicals, bait formulations, equipment for methods of control of the imported fire ant and the pharaoh's ant; ecology, population dynamics and developmental biology of the imported fire ant; biology, population dynamics and control of biting flies affecting man and animals. *Mailing Add:* USDA-Agr Res Serv PO Box 14565 Gainesville FL 32604

WILLIAMS, DAVID G(ERALD), b Hackensack, NJ, Feb 6, 35; m 76; c 2. SYSTEMS ANALYSIS, UNDERWATER ACOUSTICS. *Educ:* Univ Mich, BSE, 58, MA 62, PhD(physics), 66. *Prof Exp:* Sr scientist underwater acoust, Gen Dynamics/Elec Boat, 65-69; systs analyst, Mystic Oceanog Co, 69-71; PHYSICIST SYSTS ANAL, NAVAL UNDERWATER SYSTS CTR, 71- *Concurrent Pos:* Instr physics, SE Br, Univ Conn, 67-69. *Mem:* Acoust Soc Am; Inst Elec & Electronics Engrs. *Res:* Military systems analysis; war gaming and tactical development; systems performance modelling. *Mailing Add:* 129 Elm Noank CT 06340

WILLIAMS, DAVID JAMES, b Syracuse, NY, Feb 20, 43; m 65; c 3. PHYSICAL CHEMISTRY. *Educ:* Le Moyne Col, NY, BS, 64; Univ Rochester, PhD(phys chem), 68. *Prof Exp:* Scientist, Xerox Corp, 68-75, mgr phys chem, Corp Res Labs, 75-83; HEAD LAB, EASTMAN KODAK CO, 83- *Mem:* Am Chem Soc; Am Phys Soc. *Res:* Mechanistics of photogeneration and transport of electronic charge in organic and polymeric materials; pulsed nuclear magnetic resonance; electron spin resonance; electrical measurements; optical spectroscopy. *Mailing Add:* Corp Res Lab Eastman Kodak Co Rochester NY 14650-2110

WILLIAMS, DAVID JAMES, b Glendora, NJ, Mar 19, 47; m 70; c 3. HORTICULTURE, PLANT PHYSIOLOGY. *Educ:* Del Valley Col, BS, 69; Rutgers Univ, MS, 71, PhD(hort), 74. *Prof Exp:* Res asst hort, Rutgers Univ, 69-74, teaching asst, 70-74; from asst prof to assoc prof, 74-86, PROF HORT, UNIV ILL, 86- *Concurrent Pos:* Res grants, J M Rhoades Co, 77-78, NCent Region Pesticide Impact Assessment Prog, 78-79, 79-80 & 80-81, Abbott Labs, ICI & Stauffer Chem Co, Monsanto, Valent, 84-, Ill Dept Agr, 85-, Ill Dept Energy & Natural Resources, 89-90. *Mem:* Am Soc Hort Sci; Weed Sci Soc Am; Int Plant Propagators Soc; Int Soc Arboricult. *Mailing Add:* Hort Dept Univ Ill Urbana IL 61801

WILLIAMS, DAVID JOHN, b Salem, Mass, Mar 2, 37; div; c 2. GENERAL POLYMER SCIENCE & TECHNOLOGY, HIGH TECHNOLOGY POLYMERS. *Educ:* Lehigh Univ, BS, 59; Case Western Reserve Univ, MS, 62, PhD(polymer sci), 64. *Prof Exp:* Prof chem eng, City Col, City Univ New York, 64-74; tech asst Res & Develop Div, prod mgr, Plastics Div, Am Hoechst Corp, 74-84; res prof & proj mgr, Polymer Sci & Eng Div, Univ Mass, 84-87; CONSULT, POLYMER SCI & TECHNOL, 84-; PRES, PST GROUP, 89- *Concurrent Pos:* Part time teaching polymer sci & technol, Tufts Univ, 77-79, Clark Univ, 83, Northeastern Univ, 84, Worcester Polytech Inst, 88, 90. *Mem:* Soc Plastic Indust; Am Chem Soc; Soc Plastics Engrs; Soc Advan Mat & Process Eng; Sigma Xi; AAAS. *Res:* High performance polymers; polymer blends; rubber toughened polymers; expandable polymers; free radical polymerization technology; structure-property relationships; general electrical & aerospace applications. *Mailing Add:* 1050 Bay Rd Amherst MA 01002

WILLIAMS, DAVID JOHN, III, b Cordele, Ga, Sept 22, 27; m 48; c 3. THERIOGENOLOGY. *Educ:* Univ Ga, DVM, 53, BSA, 61; Auburn Univ, MS, 63; Royal Vet Col, Sweden, FRVC, 65; Am Col Theriogenologists, dipl, 71. *Prof Exp:* Pvt pract vet med, 53-60; instr vet med & surg, Sch Vet Med, Univ Ga, 61; from instr to assoc prof, Auburn Univ, 61-66; from assoc prof to prof, 66-89, EMER PROF VET MED & SURG, COL VET MED, UNIV GA, 90- *Concurrent Pos:* Am Vet Med Asn fel, 64-65; Auburn Univ grant, 65-66; Animal Dis res grants, 67-70, 71-72. *Mem:* Am Vet Med Asn; Soc Theriogenologists; Am Col Theriogenologists. *Res:* Bovine and equine reproduction; fat necrosis of bovine as influenced by ecological system. *Mailing Add:* 1361 Robin Hood Rd Watkinsville GA 30677

WILLIAMS, DAVID LEE, b Oakland, Calif, Oct 11, 39; div; c 1. GEOPHYSICS, OCEANOGRAPHY. *Educ:* Univ Tex, BA, 62; Mass Inst Technol & Woods Hole Oceanog Inst, PhD(marine geophys), 74. *Prof Exp:* Fel, Woods Hole Oceanog Inst, 74; GEOPHYSICIST, US GEOL SURV, 74- *Mem:* Sigma Xi; Am Geophys Union. *Res:* Terrestial heat flow; thermal evolution of the Earth; young volcanic systems; geothermal energy. *Mailing Add:* 615 Newport St Denver CO 80220

WILLIAMS, DAVID LLEWELYN, b Hawarden, Wales, Apr 25, 37; m 62; c 2. METAL PHYSICS. *Educ:* Univ Col NWales, BSc, 57; Cambridge Univ, PhD(superconductivity), 60. *Prof Exp:* Nat Res Coun Can fel, 60-62, from instr to assoc prof, 62-71, assoc dean grad studies, 75-81, head dept, 82-87, PROF PHYSICS, UNIV BC, 71- *Concurrent Pos:* Nat Res Coun sr fel, Copenhagen Univ & Bristol Univ, 69-70; Killam fel, SIN, Switz, 78-79. *Mem:* Can Asn Physicists. *Res:* Nuclear magnetic resonance; positron annihilation in metal single crystals; muon spin rotation. *Mailing Add:* Dept Physics Univ BC Vancouver BC V6T 2A6 Can

WILLIAMS, DAVID LLOYD, b Springfield, Mass, Aug 15, 35; m 64. DRUG DELIVERY, ELECTROCHEMISTRY. *Educ:* Trinity Col, BS, 57; Northwestern Univ, PhD(anal chem), 62. *Prof Exp:* Sr res chemist, Monsanto Co, 61-69; sr res chemist, Am Hosp Supply Corp, 69-70; res assoc dent, Tufts Univ, 70; prog mgr biomed, Abcor Inc, 70-80; res dir, Biotek Inc, 80-84; staff consult, A D Little, 84-85; sr scientist, Giner Inc, 85-88; SR SCIENTIST, COPLEY PHARMACEUT, 88- *Concurrent Pos:* Consult, Joslin Diabetes Found, 74-80, Moleculon, 85; vis lectr, Salem State Col, 90- *Mem:* Electrochem Soc; Am Chem Soc; Int Asn Dent Res; Sigma Xi. *Res:* Dental drug delivery (enzymes, antibiotics, and fluoride) caries measurement instrumentation; wound dressings; microencapsulation; electrochemical instrumentation (lithium, carbon monoxide, enzyme electrodes). *Mailing Add:* 258 Haverhill St Reading MA 01867

WILLIAMS, DAVID NOEL, b Lewisburg, Tenn, Oct 10, 34; m 56; c 2. MATHEMATICAL PHYSICS. *Educ:* Maryville Col, BA, 56; Univ Calif, Berkeley, PhD(theoret physics), 64. *Prof Exp:* Engr, Lockheed Missile Systs Div, Calif, 56-58; fel, Swiss Fed Inst Technol, 63-65, Nuclear Res Ctr, Saclay, France, 65-66 & Inst Adv Study, Princeton Univ, 66-67; asst prof theoret physics, 67-74, ASSOC PROF PHYSICS, UNIV MICH, ANN ARBOR, 74- *Mem:* Am Phys Soc. *Res:* Holomorphic, Lorentz covariant functions; analytic S matrix theory; analytic parametrization of higher spin scattering amplitudes; applications of functional analysis in the scattering theory and quantum field theory of elementary particles. *Mailing Add:* Dept Physics Univ Mich Ann Arbor MI 48104

WILLIAMS, DAVID TREVOR, b Slough, Eng, Oct 4, 40. ENVIRONMENTAL CHEMISTRY. *Educ:* Univ Bristol, BSc, 61; Queen's Univ, MSc, 63, PhD(chem), 66. *Prof Exp:* Res scientist food chem, Foods Directorate, 69-75, RES SCIENTIST ENVIRON CHEM, ENVIRON HEALTH DIRECTORATE, HEALTH & WELFARE, CAN, 75- *Mem:* Royal Soc Chem; Chem Inst Can. *Res:* Identification of organic contaminants in air and drinking water; effects of water treatment procedures on the organic contaminants of drinking water. *Mailing Add:* Environ Health Directorate Tunneys Pasture Ottawa ON K1A 0L2 Can

WILLIAMS, DEAN E, b Iowa City, Iowa, Feb 18, 24; m 44; c 3. SPEECH PATHOLOGY, AUDIOLOGY. *Educ:* Univ Iowa, BA, 47, MA, 49, PhD(speech path), 52. *Prof Exp:* From asst prof to assoc prof speech path, Fla State Univ, 49-53; res assoc, Univ Iowa, 51-52; asst prof, Ind Univ, 53-58; PROF SPEECH PATH, UNIV IOWA, 58- *Concurrent Pos:* Mem adv panel speech path & audiol, US Voc Rehab Admin, 60-62; consult rev panel speech & hearing, Neurol & Sensory Dis Serv Proj, 64-67 & perinatal res, Nat Inst Neurol Dis & Blindness; mem bd dirs, Am Bd Exam Speech Path & Audiol; pres, academic affairs, Am Speech Found. *Mem:* Fel Am Speech & Hearing Asn; Int Soc Gen Semantics. *Res:* Stuttering, especially onset and development of the problem and improved remedial procedures. *Mailing Add:* Johnson Speech & Hearing Ctr Woolf Ave Iowa City IA 52242

WILLIAMS, DENIS R, theoretical chemistry, for more information see previous edition

WILLIAMS, DONALD BENJAMIN, b New York, NY, Aug 8, 33; m 56; c 2. BIOLOGY. *Educ:* Maryville Col, BA, 55; Emory Univ, MS, 57, PhD(biol), 59. *Prof Exp:* From asst prof to assoc prof biol, Maryville Col, 58-61; from asst prof to assoc prof, 61-81, PROF VASSAR COL, 81- *Mem:* Soc Protozool; Am Micros Soc; Sigma Xi. *Res:* Ecology, physiology, ultrastructure and genetics of ciliated protozoa; factors influencing ciliate cyst induction and development. *Mailing Add:* 21 Cedar Ave Poughkeepsie NY 12603

WILLIAMS, DONALD ELMER, b Kansas City, Mo, Mar 7, 30. X-RAY CRYSTALLOGRAPHY. *Educ:* William Jewell Col, AB, 50; Iowa State Univ, PhD(chem), 64. *Prof Exp:* Res asst chem, Iowa State Univ, 57-62, from asst chemist to assoc chemist, 62-67; assoc prof, 67-71, PROF CHEM, UNIV LOUISVILLE, 71- *Concurrent Pos:* Vis prof, Univ Auckland, NZ, 73, State Univ Utrecht, Neth, 80-81; prog officer, Chem Div, NSF, 88-89. *Mem:* Am Chem Soc; Am Crystallog Asn; Am Phys Soc. *Res:* Intermolecular forces in crystals; crystal mechanics; molecular clusters. *Mailing Add:* Dept Chem Univ Louisville Louisville KY 40292

WILLIAMS, DONALD HOWARD, b Ellwood City, Pa, Mar 9, 38; m 60; c 2. INORGANIC CHEMISTRY, SCIENCE ADMINISTRATION. *Educ:* Muskingum Col, BS, 60; Ohio State Univ, PhD(inorg chem), 64. *Prof Exp:* Asst prof chem, Univ Ky, 64-69; dir summer sch, 72-78, col trustee, 76-80, chmn chem dept, 79-82, PROF CHEM, HOPE COL, 69-, DIR INST ENVIRON QUAL, 70- *Concurrent Pos:* Grants, Res Corp, NY, 65-; water resources inst, US Dept Interior, 66-68; Petrol Res Found, Joyce Found & W K Kellogg Found; consult, local industries, Ottawa County, US Dept Energy. *Mem:* Am Chem Soc; Inst Environ Sci; Nat Sci Teachers Asn; AAAS; Am Nuclear Soc. *Res:* Stereochemistry of transition metal complexes; photochemistry; energy sources. *Mailing Add:* Dept Chem Hope Col Holland MI 49423

WILLIAMS, DONALD J, b Fitchburg, Mass, Dec 25, 33; m 53; c 3. SPACE PHYSICS, NUCLEAR PHYSICS. *Educ:* Yale Univ, BS, 55, MS, 58, PhD(nuclear physics), 62. *Prof Exp:* Sr staff physicist, Appl Physics Lab, Johns Hopkins Univ, 61-65; sect head auroral & trapped radiation, Goddard Space Flight Ctr, NASA, 65-68, head particle physics sect, 68-69, head particle physics br, 69-70; dir, Space Environ Lab, Environ Res Labs, Nat Oceanic & Atmospheric Admin, 70-82; prin staff physicist, 82-89, DIR, MILTON S EISENHOWER RES CTR, APPL PHYSICS LAB, JOHNS HOPKINS UNIV, 90- *Concurrent Pos:* Lectr, Univ Colo, Boulder; prin investr, several satellite experiments including Int Sun Earth Explorer Prog, Galileo Prog & ISTP Prog, NASA. *Honors & Awards:* Leigh Page Mem Prize, Yale Univ, 58; Nat Oceanic & Atmospheric Admin Res & Achievement Award, 74. *Mem:* Am Geophys Union; Am Phys Soc. *Res:* Nuclear scattering and excitations; earth's trapped particle population and magnetic field configuration; solar flares and cosmic rays; interplanetary physics; interaction of interplanetary medium with Earth's environment; space plasma instabilities; published numerous papers in professional journals; editor of two books; co-author of book on Quantitative Aspects of Magnetospheric Physics. *Mailing Add:* Appl Physics Lab Johns Hopkins Univ Johns Hopkins Rd Laurel MD 20707

WILLIAMS, DONALD ROBERT, b Morristown, NJ, June 12, 48; m 70; c 3. ORGANIC CHEMISTRY, POLYMER CHEMISTRY. *Educ:* Brown Univ, ScBChem, 70; Mass Inst Technol, ScM, 72; Colo State Univ, PhD(chem), 78. *Prof Exp:* Res chemist dental polymers, Kendall Co, 72-74; res chemist plastics, 78-87, PLASTICS PLANT DEVELOP CHEMIST, ROHM & HAAS CO, 87- *Res:* Asymmetric synthesis; polymer research (continuous flow, emulsion polymerizations, extrusion). *Mailing Add:* Rohm & Haas Delaware Valley Inc PO Box 219 Rte 413 & Old Rte 13 Bristol PA 19007

WILLIAMS, DONALD SPENCER, b Pasadena, Calif, May 28, 39. NETWORK DESIGN. *Educ:* Harvey Mudd Col, BS; Carnegie-Mellon Univ, MS, PhD(computer sci). *Prof Exp:* Supv, Learning Res & Develop Ctr, Univ Pittsburgh, 65-67; consult, 67-69; staff engr, Computer Systs Div, RCA Corp, 69-72; prin investr, Jet Propulsion Lab, Calif Inst Technol, 72-80; CHIEF ENGR, SPACE & DEFENSE DIV, TRW INC, 80- *Mem:* AAAS; Asn Comput Mach; Audio Eng Soc; Nat Fire Protection Asn; Inst Elec & Electronics Engrs. *Res:* Design and management of computer centers, national communications facilities, voice and data, and conference facilities; network design; man-machine communications; robotics and theater operations. *Mailing Add:* PO Box 40700 Pasadena CA 91114-7700

WILLIAMS, DOUGLAS FRANCIS, b Long Branch, NJ, Dec 14, 48; m 75. GEOLOGICAL OCEANOGRAPHY. *Educ:* Brown Univ, BA, 71; Univ RI, PhD(oceanog), 76. *Prof Exp:* Res assoc geochem, Dept Geol Sci, Brown Univ, 76-77; asst prof 77-85, PROF GEOL & MARINE SCI & CHMN DEPT GEOL, UNIV SC, 85- *Mem:* Am Asn Petrol Geologists; Am Geophys Union; Geol Soc Am; Sigma Xi. *Res:* Marine micropaleontology; stable isotope geochemistry of carbonates; paleoclimatology. *Mailing Add:* Dept Geol Univ SC Columbia SC 29208

WILLIAMS, DUANE ALWIN, b Marshfield, Wis, Apr 6, 35; div; c 2. CHEMICAL ENGINEERING. *Educ:* Univ Wis-Madison, BS, 56, MS, 57, PhD(chem eng), 61. *Prof Exp:* Tech serv engr lubricant additives, Enjay Labs, 57-58; fel, Univ Wis-Madison, 61-62; res chem engr pulp & paper, Kimberly-Clark Corp, 62-64; sr res engr & dir explor res, Rocket Res Corp, 64-71; sr res scientist pulp & paper, Kimberly-Clark Corp, 71-76, mgr, 76-84, dir res & develop, 84-91, DIR MAJOR PROJ, KIMBERLY-CLARK CORP, 91- *Mem:* Am Inst Chem Engrs. *Res:* Solid waste management; reaction control systems and gas generation systems for aerospace and commercial applications; pulp mill unit operations; thermal radiation; solar energy utilization; consumer products research and development. *Mailing Add:* 1976 Marathon Ave Neenah WI 54956

WILLIAMS, DUDLEY, b Covington, Ga, Apr 12, 12; m 37; c 2. MOLECULAR SPECTROSCOPY, PLANETARY ATMOSPHERES. *Educ:* Univ NC, AB, 33, MA, 34, PhD(physics), 36. *Prof Exp:* Instr physics, Univ Fla, 36-38, asst prof phys sci, 38-41; staff mem, Radiation Lab, Mass Inst Technol, 41-43; asst prof physics, Univ Okla, 43-44; staff mem, Los Alamos Sci Lab, Calif, 44-46; from assoc prof to prof physics, Ohio State Univ, 46-63, actg chmn dept, 52-53 & 58-59; prof & head dept, NC State Univ, 63-64; regent's distinguished prof, 64-82, EMER PROF PHYSICS, KANS STATE UNIV, 82- *Concurrent Pos:* Guggenheim fel, Univ Amsterdam & Oxford Univ, 56; NSF sr fel, Univ Liege, 61-62. *Mem:* Fel Am Phys Soc; fel Optical Soc Am (vpres, 77, pres-elect, 78, pres, 79); Am Asn Physics Teachers. *Res:* Infrared spectroscopy; microwave transmission; mass spectroscopy; nuclear and atmospheric physics; planetary atmospheres; determination of nuclear magnet moments. *Mailing Add:* 120 Longview Dr Manhattan KS 66502

WILLIAMS, E(DGAR) P, b Pierpont, Ohio, Aug 17, 18; div; c 4. MISSILE DESIGN, HYPERSONIC AERODYNAMICS. *Educ:* Oberlin Col, AB, 40; Calif Inst Technol, MS & AeroEng, 42. *Prof Exp:* Asst wind tunnel, Calif Inst Technol, 40-42; aerodyn engr, Douglas Aircraft Co, 42-48; head missiles aerodyn, Rand Corp, 49-55, head aerodyn, 56-63; chief aeromech br, McDonnell Douglas Astronaut Co, 63, chief engr aero-thermodyn dept, 64-66, staff asst to dir advan missile & reentry systs, 67-72, sr staff engr, 72-74; engr, CDI Corp, 77-78; design specialist, Pomona Div, Gen Dynamics, 78-84; ENGR SPECIALIST, B-2 DIV, NORTHROP CORP, 84- *Concurrent Pos:* Lectr, Univ Calif, Los Angeles, 56-58. *Mem:* Assoc fel Am Inst Aeronaut & Astronaut; Sigma Xi. *Res:* Aerodynamics, particularly hypersonic aerodynamics, glide and reentry vehicles. *Mailing Add:* 6721 Cory Dr Huntington Beach CA 92647

WILLIAMS, EBENEZER DAVID, JR, b Nanticoke, Pa, June 30, 27; m 54; c 1. TEXTILE CHEMISTRY. *Educ:* Swarthmore Col, AB, 47; Univ Pa, MA, 49, PhD(org chem), 52. *Prof Exp:* Asst instr chem, Univ Pa, 47-52; from res chemist to sr res chemist, E I Du Pont de Nemours & Co, Inc, 52-63, tech supvr, 63-66, res assoc, 66-74, sr res fel, 86, DU PONT FEL, TEXTILE RES LAB, FIBERS, E I DU PONT DE NEMOURS & CO, INC, 88- *Mem:* Am Chem Soc; Sigma Xi; AAATC; Fiber Soc. *Res:* Polymer chemistry; synthetic textiles; mechanism of dyeing; dyeing technology of synthetic fibers. *Mailing Add:* 820 Summerset Dr Hockessin DE 19707

WILLIAMS, EDDIE ROBERT, b Chicago, Ill, Jan 6, 45; c 2. MATHEMATICS. *Educ:* Ottawa Univ, BA, 66; Columbia Univ, PhD(math), 71. *Prof Exp:* Instr math, Intensive Summer Studies Prog, Columbia Univ, 70; ASST PROF MATH, NORTHERN ILL UNIV, 70- *Concurrent Pos:* Asst to vpres acad affairs, San Diego State Univ. *Mem:* NY Acad Sci; Am Math Soc. *Res:* Pure mathematics; several complex variable theory; mathematics education; mathematics for the disadvantaged student. *Mailing Add:* Div Finance & Planning Northern Ill Univ De Kalb IL 60115

WILLIAMS, EDMOND BRADY, b Charlotte, NC, Aug 12, 43; m 67; c 2. BIOCHEMISTRY, ORGANIC CHEMISTRY. *Educ:* Duke Univ, BS, 65; Univ NC, Chapel Hill, PhD(org chem), 70. *Prof Exp:* NIH fel microbiol, Med Ctr, Univ Calif, San Francisco, 70-72; fel chem, Univ Ariz, 72-73; instr, Baylor Univ, 73-74; asst prof chem, Univ Wis-Oshkosh, 74-80; ASSOC PROF CHEM, COL ST CATHERINE, 80-, CHAIR, 90- *Mem:* Am Chem Soc; Sigma Xi; AAAS. *Res:* Synthesis of peptides; protein isolation and modification; enzyme kinetics. *Mailing Add:* Chem Dept Mail No 4245 Col St Catherine St Paul MN 55105

WILLIAMS, EDWARD ASTON, b UK, Oct 20, 47. PLASMA PHYSICS, INERTIAL CONFINEMENT FUSION. *Educ:* Cambridge Univ, BA, 68; Princeton Univ, PhD(physics), 73. *Prof Exp:* Res assoc, Univ Colo, 73-75; mem, Inst Advan Study, Princeton, NJ, 75-77; scientist, Univ Rochester, 77-83; PHYSICIST, LAWRENCE LIVERMORE LAB, 83- *Mem:* Fel Am Phys Soc; Am Asn Physics Teachers. *Res:* Plasma physics interactions. *Mailing Add:* Lawrence Livermore Labs L-472 Box 5508 Livermore CA 94550

WILLIAMS, EDWARD JAMES, b Denver, Colo, June 25, 26; m 60; c 2. BIOPHARMACEUTICS, PHARMACOKINETICS. *Educ:* Regis Col, BS, 48; Univ Notre Dame, MS, 50; Purdue Univ, PhD(phys chem), 59. *Prof Exp:* Chief test design & eval serial chem munitions, US Army Chem Corps, 52-54; fel biochem, Med Sch, Marquette Univ, 59-60 & biophys chem, Purdue Univ, 60-63; assoc prof chem, St Norbert Col, 63-71; RES ASSOC PHARM & DIR DRUG DEVELOP LAB, PURDUE UNIV, 79- *Mem:* Sigma Xi. *Res:* Design of new drug dosage forms. *Mailing Add:* 1933 Vinton St Lafayette IN 47904

WILLIAMS, EDWIN BRUCE, b Ladoga, Ind, Nov 3, 18; m 40; c 1. PHYTOPATHOLOGY. *Educ:* Wabash Col, AB, 50; Purdue Univ, MS, 52, PhD(plant path), 54. *Prof Exp:* From asst prof to assoc prof, 54-69, PROF PLANT PATH, PURDUE UNIV, LAFAYETTE, 69-, PLANT PATHOLOGIST, 54- *Mem:* Am Phytopath Soc; Am Pomol Soc. *Res:* Genetics of Venturia inaequalis; breeding apples for disease resistance. *Mailing Add:* 801 S 12th Lafayette IN 47907

WILLIAMS, ELEANOR RUTH, b Ropesville, Tex, Apr 23, 24. NUTRITION, FOODS. *Educ:* Tex Woman's Univ, BS, 45; Iowa State Univ, MS, 47; Cornell Univ, PhD(nutrit), 63. *Prof Exp:* Res assoc nutrit, Tex Agr Exp Sta, 48-49; instr food & nutrit, Southern Methodist Univ, 49-51; from instr to asst prof, Cornell Univ, 51-59; assoc prof, Univ Nebr, 63-65; assoc prof nutrit, Teachers Col, Columbia Univ, 65-72; assoc prof, State Univ NY Col Buffalo, 72-74; ASSOC PROF HUMAN NUTRIT & FOOD SYSTS, UNIV MD, 74- *Mem:* AAAS; Soc Nutrit Educ. *Res:* B vitamins and reproduction in the rat; non-specific nitrogen intake; adequacy of cereal protein for man; nutrition education. *Mailing Add:* Dept Nutrit & Food Univ Md College Park MD 20742-7521

WILLIAMS, ELIOT CHURCHILL, b Chicago, Ill, Nov 9, 13; m 45; c 4. INVERTEBRATE ZOOLOGY, ANIMAL ECOLOGY. *Educ:* Cent YMCA Col, BA, 35; Northwestern Univ, PhD(zool), 40. *Prof Exp:* Instr zool, Cent YMCA Col, 35-36; asst dir, Chicago Acad Sci, 40-47; asst prof biol, Roosevelt Univ, 47-48; from assoc prof to prof, 48-83, chmn, Div Sci, 76-79, EMER PROF ZOOL, WABASH COL, 83- *Concurrent Pos:* Lectr, Roosevelt Univ, 46-47; Ford Found fel, Johns Hopkins Marine Sta, Stanford Univ, 54-55; consult, Fed Chem Co, 54-76. *Mem:* AAAS; Am Soc Zool; Ecol Soc Am; Am Inst Biol Sci; Sigma Xi. *Res:* Animal populations; cave animals; pigmentation in cave planarians; taxonomy and ecology of Symphyla; radioisotope cycling; energy relationships in ecosystems; meiofauna of coral reefs. *Mailing Add:* Dept Biol Wabash Col Crawfordsville IN 47933

WILLIAMS, ELLEN D, b Oshkosh, Wis, Dec 5, 53; m 80; c 2. SURFACE STRUCTURE, SURFACE DIFFUSION. *Educ:* Mich State Univ, BS, 76; Calif Inst Technol, PhD(chem), 82. *Prof Exp:* Res assoc Physics Dept, 81-83, asst prof, 83-87, ASSOC PROF PHYS DEPT, UNIV MD COL PARK, 87-, ASSOC PROF INST PHYS SCI & TECHNOL, 90- *Concurrent Pos:* Exec comt, Surface Sci Div, Am Vaccuum Soc, 89-90; comt mem, Status Women Physics, Am Phys Soc, 90-92. *Honors & Awards:* Presidential Young Investr, NSF, 84; Young Investr Award, Off Naval Res, 86; Maria Goeppert Mayer Award, Am Phys Soc, 90. *Mem:* Am Vaccuum Soc; Am Phys Soc; Sigma Xi; Am Chem Soc. *Res:* Experimental studies of solid surfaces; determination of mechanism of surface diffusion, defect structure, two-dimensional phase transition & equilibrium crystal shape. *Mailing Add:* Dept Physics Univ Md College Park MD 20742-4111

WILLIAMS, ELMER LEE, b Ironton, Ohio, Apr 14, 29; m 57; c 3. PHYSICAL CHEMISTRY. *Educ:* Ohio Univ, BS, 51, MS, 55; Ind Univ, PhD(phys chem), 59. *Prof Exp:* Asst chem, Ind Univ, 55-58; develop engr, Sylvania Elec Prods, Inc Div, Gen Tel & Electronics Corp, 58-60; phys chemist, Owens-Ill, Inc, 60-79; proj mgr, Midland-Ross Corp, 79-84; DIR GAS CHROMATOGRAPHY/MASS SPECTROMETRY FACIL, BOWLING GREEN STATE UNIV, 86- *Concurrent Pos:* Consult, 84- *Mem:* Am Chem Soc; Am Ceramic Soc; Electrochem Soc. *Res:* Diffusion of ions and atoms in glass and molten silicates; oxygen of mass 18 work and tracer work in the solid state; semiconductors; gas lasers; gas discharge displays; propose, plan and manage research and development engineering projects in energy and glass furnace areas; gas chromatography/mass spectrometry analysis. *Mailing Add:* 3615 Maple Way Dr Toledo OH 43614

WILLIAMS, EMMETT LEWIS, b Lynchburg, Va, June 6, 33; m 57; c 3. SOLID STATE PHYSICS, SCANNING ELECTRON MICROSCOPY. *Educ:* Va Polytech Inst, BS, 56, MS, 62; Clemson Univ, PhD(mat eng), 66. *Prof Exp:* Assoc aircraft engr, Lockheed-Ga Co, 56-57; mat engr, Atomic Energy Div, Babcock & Wilcox Co, 57-59; asst prof metall eng, Va Polytech Inst, 59-64; res asst ceramic eng, Clemson Univ, 64-65; res scientist, Union Carbide Nuclear Corp, 65-66; prof physics, Bob Jones Univ, 66-79, chmn dept, 73-79; mat engr, Continental Tel Labs, 79-81; scientist, Lockheed-Ga Co, 81-90; CONSULT, 90- *Concurrent Pos:* Consult, Inland Motors, 61-64, Leaders Am Sci, 66, Continental Tel Labs, 70-72, Polysci Corp & Electrotech Corp; ed, Creation Res Soc Quart, 83-88. *Mem:* Creation Res Soc (vpres, 72-83); Sigma Xi. *Res:* Solid state, surface physics; thermodynamics; formation of limestone stalactites in laboratory; thermodynamics of living organisms; failure analysis of metallic structures. *Mailing Add:* 5093 Williamsport Dr Norcross GA 30092-2124

WILLIAMS, ERNEST EDWARD, b Easton, Pa, Jan 7, 14. HERPETOLOGY, ECOLOGY. *Educ:* Lafayette Col, BS, 33; Columbia Univ, PhD(zool), 49. *Prof Exp:* Asst zool, Columbia Univ, 40-42 & 46-48; from instr to assoc prof, Harvard Univ, 49-70, cur reptiles & amphibians, Mus Comp Zool, 57-80, prof biol, 70-84, Alexander Agassiz prof zool, 72-84; RETIRED. *Concurrent Pos:* Guggenheim fel, 53-54 & 81-82. *Mem:* AAAS; Am Soc Ichthyol & Herpet; Soc Syst Zool; Soc Study Evolution; fel Am Acad Arts & Sci. *Res:* Taxonomy, paleontology and morphology of reptiles; West Indian paleontology and zoogeography; evolution. *Mailing Add:* 28 Wendell St Cambridge MA 02138

WILLIAMS, EUGENE G, b New Haven, Conn, June 9, 25; m 52; c 2. GEOLOGY, MINERALOGY. *Educ:* Lehigh Univ, BA, 50; Univ Ill, MS, 52; Pa State Univ, PhD, 57. *Prof Exp:* Instr geol, Kent State Univ, 52-53; instr, 54-55, res assoc, 56-57, from asst prof to assoc prof, 57-71, PROF GEOL, PA STATE UNIV, 71- *Mem:* Geol Soc Am; Am Asn Petrol Geol. *Res:* Stratigraphy and petrography of upper Paleozoic rocks of eastern United States. *Mailing Add:* 628 Outer Dr State College PA 16801

WILLIAMS, EUGENE H(UGHES), b New Bern, NC, Jan 13, 12; m 40; c 2. CHEMICAL ENGINEERING. *Educ:* NC State Col, BS, 34; Inst Paper Chem, Lawrence, MS, 37. *Prof Exp:* Chem engr, Munising Paper Co, 37-49; supt printing, Milprint, Inc, 49-52; chem engr, Pulp & Paper, Detroit Sulphite Pulp & Paper Co, 52-54; tech dir, Detroit Div, Scott Paper Co, 54-65, tech specialist, 65-72; tech specialist, S D Warren Co, 72-77; RETIRED. *Mem:* Am Chem Soc; Am Inst Chem; Tech Asn Pulp & Paper Indust; Paper Indust Mgt Asn. *Res:* Pulp and paper chemistry; quality control and development. *Mailing Add:* 1920 Raymond Dearborn MI 48124-4340

WILLIAMS, EVAN THOMAS, b New York, NY, May 17, 36; m 59; c 2. ANALYTICAL CHEMISTRY. *Educ:* Williams Col, BA, 58; Mass Inst Technol, PhD(chem), 63. *Prof Exp:* Civil engr, Res Estab Res, Roskilde, Denmark, 63-65; from asst prof to assoc prof, 65-75, chmn dept, 81-84, PROF CHEM, BROOKLYN COL, 76-, DEAN UNDERGRAD STUDIES, 89- *Concurrent Pos:* Consult, Geosci Instruments Corp, 66-72; mem exec bd, New York City Coun on the Environ, 73-76; chmn, Brooklyn Col Fac Coun, 77-85; bd trustees, Packer Col Inst, 78-87; mem, Citizen's Adv Comt, Brooklyn Navy Yard Resource Recovery Plant, 81- *Mem:* AAAS; Am Chem Soc; Am Phys Soc. *Res:* Trace element analysis; environmental applications; proton-induced x-ray emission. *Mailing Add:* Dept Chem Brooklyn Col Brooklyn NY 11210

WILLIAMS, F(ORD) CAMPBELL, b Nanaimo, BC, Dec 28, 21; m 55; c 2. CHEMICAL ENGINEERING. *Educ:* Univ BC, BASc, 43, MASc, 46; Univ Iowa, PhD(chem eng), 48. *Prof Exp:* Instr chem eng, Univ Iowa, 47-48; asst prof, Univ Calif, 48-52; prof chem eng & tech consult, Nat Petrol Coun, Brazil, 52-55; head prof & head res, Petroleo Brasileiro SA, 55-65; consult chem engr, Consultores Industriais Associados, 65-70; VPRES, NATRON CONSULTORIA E PROJETOR SA, 70-, VPRES TECHNOL, 80- *Mem:* Am Inst Chem Engrs. *Res:* Phase equilibria; extraction; petroleum process and product development; engineering design; process research. *Mailing Add:* Natron-Consultoria e Projetos S A Teofilo Otoni 63-11 Rio de Janeiro Brazil

WILLIAMS, FLOYD JAMES, b Electra, Tex, Jan 9, 20; m 55; c 2. GEOLOGY. *Educ:* Univ Calif, BS, 43; Colo Sch Mines, MS, 51; Columbia Univ, PhD(geol), 58. *Prof Exp:* Mining engr, Bradley Mining Co, Idaho, 43; mining engr, Idaho-Md Mines Corp, Calif, 46-47; explosives engr, Hercules Powder Co, 47-48; chief reconnaissance sect, Salt Lake Explor Br, Div Raw Mat, USAEC, 52-54; geologist, Standard Oil Co, Calif, 56-58; assoc prof geol, Univ Redlands, 58-66; supvr spectrog, Kaiser Steel Corp, Calif, 66-72; head dept geol, 72-76, assoc prof, 76-80, CHMN, DIV SCI, SAN BERNARDINO VALLEY COL, 80-, PROF GEOL, 80- *Concurrent Pos:* NSF fel & res assoc geol & geophys, Univ Calif, Berkeley, 63-64; consult geol, City of San Bernardino, Calif, 74- *Mem:* Geol Soc Am; AAAS; Am Geophys Union; Sigma Xi. *Res:* Criteria for active faults, earthquake prediction. *Mailing Add:* Dept of Geol San Bernardino Valley Col San Bernardino CA 92410

WILLIAMS, FORMAN A(RTHUR), b New Brunswick, NJ, Jan 12, 34; m 55, 78; c 6. COMBUSTION, FLUID DYNAMICS. *Educ:* Princeton Univ, BSE, 55; Calif Inst Technol, PhD(eng sci), 58. *Prof Exp:* Asst prof mech eng, Harvard Univ, 58-64; mem tech staff, Inst Defense Anal, 63-64; from assoc prof to prof aerospace eng, Univ Calif, San Diego, 64-81; Robert H Goddard prof, Dept Mech & Aerospace Eng, Princeton, Univ, 81-88; PROF ENG PHYSICS & COMBUSTION, DEPT APPL MECH & ENG SCI, UNIV CALIF, SAN DIEGO, 89-, DIR CTR ENERGY & COMBUSTION RES, 91- *Concurrent Pos:* NSF fel, Imp Col, Univ London, 62; Guggenheim fel, Univs Sydney & Madrid, 70-71; vis prof, Sydney Univ, 70, Univ Colo, 77, lectr, Assoc Physics, Univ Provence, Marseille, 77. *Honors & Awards:* Silver Combustion Medal, Combustion Inst, 78; Alexander von Humboldt US Sr Scientist Award, 82; Bernard Lewis Gold Medal, Combustion Inst, 90. *Mem:* Nat Acad Eng; Am Phys Soc; fel Am Inst Aeronaut & Astronaut; Soc Indust & Appl Math; Combustion Inst; foreign corresp mem Nat Acad Eng Mex; Sigma Xi. *Res:* Aerothermochemistry; combustion theory; heat and mass transfer; fire research; mathematical methods. *Mailing Add:* Dept Appl Mech & Eng Sci Univ Calif San Diego La Jolla CA 92093-0310

WILLIAMS, FRANCIS, b Whitley Bay, Eng, May 14, 27; m 57. FISHERIES, BIOLOGICAL OCEANOGRAPHY. *Educ:* Univ Durham, BSc, 51, MSc, 55; Univ Newcastle, Eng, DSc(marine fisheries), 64. *Prof Exp:* Prin sci officer, Pelagic Fish & Fisheries, EAfrican Marine Fisheries Res Orgn, Zanzibar, 51-62; dir Guinean Trawling Surv, Orgn African Unity, Lagos, Nigeria, 62-66; mgr pelagic fisheries surv WAfrica, dept fisheries, Food & Agr Orgn UN, Rome, Italy, 66-68; assoc res biologist, Scripps Inst Oceanog, Univ Calif, San Diego, 68-73; chmn, Div Fisheries & Appl Estuarine Ecol, Univ Miami, 73-74, chmn, Div Biol & Living Resources, 74-78, prof, 73-89, EMER PROF MARINE SCI, ROSENSTIEL SCH MARINE & ATMOSPHERIC SCI, UNIV MIAMI, 89- *Concurrent Pos:* Exec secy, Gulf & Caribbean Fisheries Inst, 84-; sci fel, Zool Soc London. *Mem:* Marine Biol Asn UK; Challenger Soc; Am Fisheries Soc; fel Am Inst Fishery Res Biol; Am Soc Ichthyol Herpetologists; Am Soc Limnol & Oceanog; Am Inst Biol Sci. *Res:* Exploratory fishing surveys; fish taxonomy; correlation of fish and fisheries with environment; marine affairs; international fisheries development. *Mailing Add:* 16 Otter Burn Way Prudhoe Northumberland NE42 6RD England

WILLIAMS, FRANK LYNN, b Peoria, Ill, Oct 6, 45. CHEMICAL ENGINEERING, CATALYSIS. *Educ:* Northwestern Univ, BS, 68; Stanford Univ, MS, 70, PhD(chem eng), 73. *Prof Exp:* Res engr catalysis, Gen Motors Res Labs, 72-77; asst prof, 77-80, ASSOC PROF CHEM ENG, UNIV NMEX, 80- *Concurrent Pos:* Consult, Sandia Nat Labs, 77- *Honors & Awards:* Kokes Award, NAm Catalysis Soc, 74. *Mem:* Am Inst Chem Engrs; Am Chem Soc; Am Vacuum Soc; NAm Catalysis Soc. *Res:* Methane recovery from coalbeds; heterogenous surface chemistry during catalytic reactions; in-situ energy production; methane recovery from coalbeds. *Mailing Add:* Dept Chem & Nuclear Eng Univ NMex Albuquerque NM 87131-0001

WILLIAMS, FRED DEVOE, b New York, NY, Dec 16, 36; div; c 3. MICROBIOLOGY, EDUCATION. *Educ:* Rutgers Univ, BA, 60, MS, 62, PhD(bact), 64. *Prof Exp:* Instr bact, Rutgers Univ, 63-64; from asst prof to assoc prof, 64-75, PROF BACT, IOWA STATE UNIV, 75-, CHMN, DEPT MICROBIOL, 81- *Concurrent Pos:* Consult, SC Comt on High Educ, 88; chmn adv bd, Soil Technol Corp, 87-88. *Mem:* AAAS; Am Soc Microbiol; Sigma Xi. *Res:* Ecology of microorganisms and the fuction of extracellular polysaccharides; physiology and biochemistry of microbial behavior. *Mailing Add:* Dept Microbiol Iowa State Univ 205 Sci Ames IA 50011

WILLIAMS, FRED EUGENE, b Wichita Falls, Tex, Oct 23, 41; m 64; c 2. FOOD INTAKE CONTROL, DENTAL EDUCATION. *Educ:* Arlington State Col, BS, 66; Baylor Univ, PhD(physiol), 72. *Prof Exp:* Asst prof, 72-77, ASSOC PROF PHYSIOL, BAYLOR COL DENT, 77- *Mem:* Sigma Xi; AAAS; Am Physiol Soc; Am Asn Dental Schs. *Res:* Effects of liver denervation, hypothalamic lesions, hormones and other chemical agents on feeding behavior. *Mailing Add:* Baylor Col Dent 3302 Gaston Ave Dallas TX 75246

WILLIAMS, FREDERICK MCGEE, b Washington, DC, Jan 10, 34; m; c 4. THEORETICAL ECOLOGY. *Educ:* Stanford Univ, AB, 55; Yale Univ, PhD(biol), 65. *Prof Exp:* Asst prof biol, Lehigh Univ, 63-64; from asst prof to assoc prof zool, Univ Minn, Minneapolis, 64-70; ASSOC PROF BIOL, PA STATE UNIV, UNIVERSITY PARK, 70- *Concurrent Pos:* NASA-Am Inst Biol Sci fel, 65-66; chmn grad prog ecol, Pa State Univ, 74-80; enological consult, Nittany Valley Winery, 84-89; chmn, Quant Ecol Option, Ecol Prog, Pa State Univ, 85; pres, Nittany EcoSysts Group Consult, 90-; impact consult, Pa Power & Light. *Mem:* Am Soc Naturalists; AAAS; Ecol Soc Am; Am Soc Zool; Am Inst Biol Sci; Int Soc Ecol Model; Sigma Xi. *Res:* Theoretical population dynamics; environmental impact consulting; statistical estimation of environment impact; mathematical biology; mechanisms of competition and predation; modelling environmental effects on fish populations. *Mailing Add:* Dept Biol Pa State Univ University Park PA 16802

WILLIAMS, FREDERICK WALLACE, b Cumberland, Md, Sept 24, 39; m 64; c 3. ANALYTICAL CHEMISTRY. *Educ:* Univ Ala, BS, 61, MSc, 63, PhD(chem), 65. *Prof Exp:* Nat Acad Sci-Nat Res Coun fel, 65-66, res chemist, 66-73, SUPV RES CHEMIST, US NAVAL RES LAB, DC, 73- *Concurrent Pos:* Tech dir, USS Shadwell. *Honors & Awards:* E O Hulbert Award, US Naval Res Lab Sci & Eng, 81, E E O Award, 83. *Mem:* AAAS; Am Chem Soc; Combustion Inst. *Res:* Fundamental mechanisms of combustion, fire safety, sealing and modeling. *Mailing Add:* 13408 Colwyn Rd Ft Washington MD 20022

WILLIAMS, FREDRICK DAVID, b Winnipeg, Man, Sept 1, 37; m 64; c 3. POLYMER CHEMISTRY, COMPUTER SCIENCES. *Educ:* Univ Man, BSc, 59, MSc, 61, PhD(chem), 62. *Prof Exp:* Ital Govt res scholarship, Inst Indust Chem, Milan Polytech Inst, 62-63; res chemist, Allis-Chalmers Mfg Corp, 63-65; asst prof, 65-69, ASSOC PROF CHEM, MICH TECHNOL UNIV, 69-, DIR, CTR TEACHING EXCELLENCE, 90- *Mem:* Am Chem Soc. *Res:* Polymer chemistry. *Mailing Add:* Dept Chem Mich Technol Univ Houghton MI 49931

WILLIAMS, G(EORGE) BRYMER, b Denver, Colo, Oct 17, 13; m 40; c 3. CHEMICAL ENGINEERING. *Educ:* Univ Mich, BS, 36, PhD(chem eng), 49. *Prof Exp:* Chem engr, M W Kellogg Co, NY, 40-47; from asst prof to assoc prof chem eng, 48-56, PROF CHEM & METALL ENG, UNIV MICH, ANN ARBOR, 56- *Mem:* Am Chem Soc; Soc Mining Engrs; Am Inst Chem Engrs; Chem Inst Can; Sigma Xi. *Res:* Process design; natural resource utilization; petroleum processing. *Mailing Add:* Dept Chem Eng Univ Mich Ann Arbor MI 48109-2136

WILLIAMS, GARETH, b Rhos, Wales, Apr 28, 37; US citizen; m 65; c 2. APPLIED MATHEMATICS. *Educ:* Univ Wales, BSc, 59, PhD(math), 62. *Prof Exp:* Asst prof math, Univ Fla, 62-65; from asst prof to assoc prof, Univ Denver, 65-73; assoc prof, 73-76, PROF MATH, STETSON UNIV, 76- *Mem:* Am Math Soc; Tensor Soc Gt Brit; Am Math Asn; Soc Indust Appl Math. *Res:* Relativity; differential geometry; mathematical models; computer science; linear algebra. *Mailing Add:* Dept Math Stetson Univ 421 N Woodland Deland FL 32720

WILLIAMS, GARETH PIERCE, b Llandudno, Wales, Oct 16, 39; m 64; c 2. PLANETARY ATMOSPHERES, METEOROLOGICAL MODELING. *Educ:* Univ Wales, BSc, 61, PhD(appl math), 64, DSc, 76. *Prof Exp:* Res meteorologist, 64-74, SR RES SCIENTIST, GEOPHYS FLUID DYNAMICS LAB, NAT OCEANIC & ATMOSPHERIC ADMIN, 74- *Concurrent Pos:* Vis fel, Dept Appl Math & Theoret Physics, Cambridge Univ, UK, 79-80; vis prof, Geophys Fluid Dynamics Prog, Princeton Univ, 80-83; Comn Planetary Atmospheres, Int Asn Meteorcol Atmospheric Physics, 83-90; vis fel, Japan Soc Promotion Sci, 89. *Mem:* Fel Am Meteorol Soc. *Res:* Dynamics of planetary atmospheres, particular Earth and Jupiter; computer modeling of planetary atmospheres; theories for great red spot of Jupiter. *Mailing Add:* Geophys Fluid Dynamics Lab Nat Oceanic & Atmospheric Admin PO Box 308 Princeton NJ 08542

WILLIAMS, GARY LYNN, b Carlsbad, NMex, Feb 25, 50. REPRODUCTIVE ENDOCRINOLOGY, REPRODUCTIVE PHYSIOLOGY. *Educ:* NMex State Univ, BS, 72, MS, 74; Univ Ariz, PhD(animal physiol), 78. *Prof Exp:* Asst prof physiol & endocrinol, Dept Animal Sci, NDak State Univ, 78-; PROF PHYSIOL OF REPRODUCTION & GROWTH, AGR EXP STA, TEX A&M UNIV, 79- *Mem:* Am Soc Animal Sci; Soc Study Reproduction; Endocrine Soc; Am Physiol soc. *Res:* Reproductive endocrinology of the postpartum bovine; neuroendocrine-ovarian relationships. *Mailing Add:* Tex A&M Univ Agr Res Sta HCR-2 Box 43C Beeville TX 78102

WILLIAMS, GARY MURRAY, b Regina, Sask, May 7, 40; US citizen; m 66; c 3. PATHOLOGY, TOXICOLOGY. *Educ:* Washington & Jefferson Col, BA, 63; Univ Pittsburgh, MD, 67. *Prof Exp:* Instr path, Med Sch, Harvard Univ, 67-69; staff assoc carcinogenesis, Nat Cancer Inst, 69-71; asst prof path, Fels Res Inst & Med Sch, Temple Univ, 71-75; res assoc prof, 75-80, RES PROF PATH, NEW YORK MED COL, 80-; DIR MED SCI & CHIEF DIV PATH & TOXICOL, AM HEALTH FOUND, 75- *Concurrent Pos:* Int Agency Res on Cancer res training fel, Wenner-Gren Inst, Stockholm, Sweden, 71-72; from intern to resident path, Mass Gen Hosp, 67-69. *Honors & Awards:* Sheard-Sanford Award, Am Soc Clin Path, 67; Arnold J Lehman Award, Am Soc Toxicol, 82. *Mem:* Am Asn Cancer Res; Am Asn Pathologists; Int Acad Path; Soc Exp Biol & Med; Soc Toxicol. *Res:* The genetic and carcinogenic effects of chemicals with emphasis on cell culture techniques. *Mailing Add:* Am Health Found One Dana Rd Valhalla NY 10595

WILLIAMS, GENE R, b Yuba City, Calif, Nov 10, 32; m 54, 69; c 2. PLANT PHYSIOLOGY, PLANT BIOCHEMISTRY. *Educ:* Univ Calif, BS, 57, MS, 59, PhD(plant physiol), 63. *Prof Exp:* Lectr bot, Univ Calif, 61-62; Am Cancer Soc fel biochem, Biol Div, Oak Ridge Nat Lab, 63-65; asst prof, 65-68, ASSOC PROF BOT, IND UNIV, BLOOMINGTON, 68- *Mem:* AAAS; Am Soc Plant Physiol. *Res:* Plant metabolism, protein and nucleic acid synthesis, amino acid activation and chloroplast development; effects of light on plant development. *Mailing Add:* Dept Biol Ind Univ Bloomington Bloomington IN 47401

WILLIAMS, GEORGE, JR, b Benton, La, June 15, 31; m 57. BOTANY. *Educ:* Southern Univ, BS, 57; Univ NH, MS, 59, PhD(bot), 63. *Prof Exp:* Asst bot, Univ NH, 57-63; PROF BIOL, SOUTHERN UNIV, BATON ROUGE, 63- *Mem:* Am Soc Plant Physiol; Sigma Xi. *Res:* Plant growth as modified by light quality and chemical factors, particularly growth hormones. *Mailing Add:* Southern Univ PO Box 10144 Baton Rouge LA 70813

WILLIAMS, GEORGE ABIAH, b Brooklyn, NY, Apr 1, 31; div; c 3. SOLID STATE PHYSICS, NUCLEAR MAGNETIC RESONANCE. *Educ:* Colgate Univ, BA, 52; Univ Ill, PhD, 56. *Prof Exp:* Res assoc physics, Stanford Univ, 56-59; mem tech staff, Bell Tel Labs, NJ, 59-63; vis asst prof physics, Cornell Univ, 63-64; assoc prof, 64-70, assoc chmn dept, 74-83, PROF PHYSICS, UNIV UTAH, 70- *Concurrent Pos:* NSF res grant, 72-74; USAF Off Sci Res grant, 65-70; vis prof physics, Univ Minn, 79 & Univ Calif, Berkeley, 80. *Mem:* Am Phys Soc. *Res:* Wave propagation in solid state plasmas; plasma effects in solids; superconductivity. *Mailing Add:* Dept Physics Univ Utah 2016 J Fletcher Bldg Salt Lake City UT 84112

WILLIAMS, GEORGE ARTHUR, b Wilcox, Ariz, Jan 14, 18; m 43; c 3. GEOLOGY, ENGINEERING. *Educ:* Tex Col Mines & Metal, BS, 43; Univ Ariz, PhD(geol), 51. *Prof Exp:* Engr, Asarco Mining Co, Inc, Mex, 46; geologist, Peru Mining Co, 46-48, supt, Kearney Mine, 48; geologist, US Geol Surv, 51-57; from assoc prof to prof, 57-88, head geol dept, 65-82, EMER PROF GEOL ENG, UNIV IDAHO, 88- *Concurrent Pos:* Gov appointed, Control Bd, Idaho Bur Mines & Geol. *Mem:* Fel AAAS; fel Geol Soc Am; Soc Econ Paleont & Mineral; Am Asn Petrol Geol. *Res:* Economic geology; sedimentation; stratigraphy; gemology. *Mailing Add:* Dept Geol Univ Idaho Col Mines Moscow ID 83844

WILLIAMS, GEORGE CHRISTOPHER, b Charlotte, NC, May 12, 26; m 50; c 4. ZOOLOGY. *Educ:* Univ Calif, Berkeley, AB, 49, MA, Los Angeles, 52, PhD, 55. *Prof Exp:* From instr to asst prof natural sci, Mich State Univ, 55-60; assoc prof, 60-66, PROF BIOL SCI, STATE UNIV NY, STONY BROOK, 66- *Mem:* AAAS; Soc Study Evolution; Am Soc Ichthyol & Herpet; Am Soc Naturalists; Arctic Inst NAm. *Res:* Evolution; marine ecology; ichthyology; animal behavior; population genetics. *Mailing Add:* Seven Yorktown Rd Setauket NY 11785

WILLIAMS, GEORGE HARRY, b Schenectady, NY, Nov 7, 42; m 67; c 1. COMPUTER SCIENCE. *Educ:* Union Col, BSEE & BA, 65; Yale Univ, MS, 66, PhD(eng, appl sci), 70. *Prof Exp:* PROF ELEC ENG & COMPUTER SCI, UNION COL, 70- *Concurrent Pos:* NSF res initiation grant, Union Col, 71-72; prin investr grant, NSF. 72-73. *Mem:* Inst Elec & Electronics Engrs; Asn Computer Mach; Sigma Xi. *Res:* Automata theory, logic design, computer-aided design and artificial intelligence. *Mailing Add:* Dept Elec Eng & Computer Sci Union Col Schenectady NY 12308

WILLIAMS, GEORGE KENNETH, b Detroit, Mich, July 8, 32; m 54; c 4. MATHEMATICS. *Educ:* Univ Ky, BAE, 55, MA, 58; Univ Va, PhD(math), 64. *Prof Exp:* Teacher high sch, Mich, 55-56; asst prof math, Madison Col, 58-60 & Univ Notre Dame, 64-68; assoc prof, 68-72, PROF MATH, SOUTHWESTERN AT MEMPHIS, 72- *Mem:* Am Math Soc; Math Asn Am. *Res:* Complex analysis; topology. *Mailing Add:* Dept Math Rhodes Col Memphis TN 38112

WILLIAMS, GEORGE NATHANIEL, b Kingsland, Ga, May 17, 47; m 70; c 1. INORGANIC CHEMISTRY, ORGANIC CHEMISTRY. *Educ:* Savannah State Col, BS, 69; Tuskegee Inst, MS, 72; Howard Univ, PhD(inorg chem), 78. *Prof Exp:* Lab technician chem, Union Camp Corp, 70; instr, 72-75, ASST PROF CHEM, SAVANNAH STATE COL, 78- *Mem:* Am Chem Soc; Am Inst Chemists. *Res:* Metal incorporation and the anation reactions of porphyrins. *Mailing Add:* Savannah State Col Box 20016 Savannah GA 31404-9716

WILLIAMS, GEORGE RAINEY, b Atlanta, Ga, Oct 25, 26; m 50; c 4. MEDICINE. *Educ:* Northwestern Univ, BS, 47, BMed, 50, MD, 51. *Prof Exp:* Instr surg, Johns Hopkins Hosp, 57-58; from asst prof to assoc prof surg, 58-63, interim dean, Col Med, 81-82, 85-86, & 88-89, PROF SURG, UNIV OKLA, 63-, CHMN DEPT SURG, 74- *Concurrent Pos:* Markle scholar, 60. *Mem:* Soc Univ Surg; Am Surg Asn; Soc Vascular Surg; Am Asn Thoracic Surg; Am Col Surgeons. *Mailing Add:* Dept Surg Univ Okla Health Sci Ctr PO Box 26307 Oklahoma City OK 73126

WILLIAMS, GEORGE RONALD, b Liverpool, Eng, Jan 4, 28; Can citizen; m 52; c 3. BIOCHEMISTRY. *Educ:* Univ Liverpool, PhD(biochem), 51, DSc, 69. *Prof Exp:* Worshipful Co Goldsmith's traveling fel, Banting & Best Dept Med Res, Univ Toronto, 52-53; res assoc, Johnson Found Med Biophys, Univ Pa, 53-55; Med Res Coun res assoc path, Oxford Univ, 55-56; asst prof, Banting & Best Dept Med Res, Univ Toronto, 56-61, assoc prof biochem, 61-66, chmn dept, 70-77, chmn, Div Life Sci, 78-83, prin, Scarborough Campus, 84-89, PROF BIOCHEM, UNIV TORONTO, 66- *Concurrent Pos:* Vis prof geochem, Lamont-Doherty Geol Observ, Columbia Univ, 77-78. *Mem:* Can Biochem Soc (pres, 71-72); fel Royal Soc Can. *Res:* Geobiochemistry; environmental homeostasis; control systems in biochemical and geochemical reaction networks. *Mailing Add:* Develop Life Sci Univ Toronto Scarborough Col West Hill ON M1C 1A4 Can

WILLIAMS, GEORGE W, b Nashville, Tenn, Oct 31, 46; m; c 3. BIOSTATISTICS. *Educ:* Bucknell Univ, BS, 68; George Washington Univ, MA, 70; Univ NC, PhD(biostatist), 72. *Prof Exp:* Assoc prof, 78-81, PROF BIOSTATIST, UNIV MICH, ANN ARBOR, 81-; CHMN, DEPT BIOSTATIST, CLEVELAND CLIN, OHIO, 80- *Mem:* Soc Clin Trials; Am Statist Asn; Am Pub Health Asn; Biomet Soc; Soc Epidemiol Res; Sigma Xi. *Res:* Clinical trials; statistical methods in epidemiology. *Mailing Add:* 5660 Parkwood Circle Chagrin Falls OH 44022

WILLIAMS, GERALD ALBERT, b Plankinton, SDak, Apr 1, 21; m 50; c 3. MEDICAL RESEARCH, ENDOCRINOLOGY. *Educ:* SDak State Col, BS, 45; George Washington Univ, MD, 49. *Prof Exp:* Instr med, Sch Med, Univ Va, 57-59; from asst prof to prof, 59-86, EMER PROF MED, UNIV ILL COL MED, 86- *Concurrent Pos:* Chief nuclear med serv & endocrinol sect, Vet Admin West Side Hosp, Chicago, 59-86; attend physician, Univ Ill Hosp, 59-86, chief endocrinol, 67-86; fac mem, AOA, student selection, 78-86. *Mem:* Am Fedn Clin Res; fel Am Col Physicians; Endocrine Soc; Soc Exp Biol & Med; Cent Soc Clin Res. *Res:* Parathyroid physiology; calcium metabolism; thyroid disorders. *Mailing Add:* Rte One PO Box 136A Keezletown VA 22832

WILLIAMS, GLEN NORDYKE, b Port Arthur, Tex, Nov 15, 38; m 60; c 5. COMPUTER SCIENCE, CIVIL ENGINEERING. *Educ:* Tex A&M Univ, BS, 60, MEng, 61, PhD(civil eng), 65. *Prof Exp:* Systs engr, IBM Corp, 65; from asst prof to assoc prof, 69-87, PROF COMPUTER SCI, TEX A&M UNIV, 87- *Mem:* Am Soc Civil Engrs; Inst Elec & Electronics Engrs; Asn Computer Mach. *Res:* Fluid networks; slope stability; numerical analysis; computer applications; information systems. *Mailing Add:* 3509 Midwest Bryan TX 77801

WILLIAMS, GLENN C(ARBER), b Princeton, Iowa, Oct 9, 14; m 39; c 2. CHEMICAL ENGINEERING. *Educ:* Univ Ill, BS, 37, MS, 38; Mass Inst Technol, ScD(chem eng), 42. *Prof Exp:* Res chem engr, Univ Ill, 37; from asst instr to instr, 38-42, from asst prof to assoc prof, 42-54, PROF CHEM ENG, MASS INST TECHNOL, 54- *Concurrent Pos:* Mem, Nat Adv Comt Aeronaut, 44 & 49; mem adv comt, Army Ord Res & Develop; res adv comt chem energy systs, NASA, 60-; mem comt, Motor Vehicle Emissions, Nat Acad Sci, 71-74. *Honors & Awards:* Egerton Medal, Combustion Inst. *Mem:* Am Chem Soc; Am Inst Chem Engrs; Am Inst Aeronaut & Astronaut; Combustion Inst (pres); Am Acad Arts & Sci. *Res:* Mass and heat transfer; high-output combustion. *Mailing Add:* Dept Chem Eng Mass Inst Technol Cambridge MA 02139

WILLIAMS, GRAHEME JOHN BRAMALD, b Auckland, NZ, Jan 8, 42; m 69; c 1. STRUCTURAL CHEMISTRY, CRYSTALLOGRAPHY. *Educ:* Univ Auckland, BSc, 66, MSc Hons, 67; Univ Alta, PhD(biochem), 72. *Prof Exp:* Fel chem, Univ Montreal, 72-73; res assoc chem, 73-78, assoc chemist, 78-80, CHEMIST, BROOKHAVEN NAT LAB, 81-; prod mgr, 81-86, vpres, 86-89, PRES, ENRAF-NONIUS SERV CO, 89- *Mem:* Am Crystallog Asn; Am Chem Soc; Am Mineral Soc. *Res:* Application of crystallography to structural problems in chemistry and biochemistry, enzymes, drugs and metabolites; crystallographic technique, improved computational methods and instrumentation. *Mailing Add:* ENRAF-NONIUS Co 390 Central Ave Bohemia NY 11716

WILLIAMS, GWYNNE, b Wrexham, Clwyd, UK, Mar 17, 47; m 69; c 2. SQUEEZE CASTING TO PRODUCE ALUMINUM ALLOY & METAL MATRIX COMPOSITE PRODUCTS, SYSTEMS APPROACH TO MATERIAL & DESIGN TRADE- OFFS IN PRODUCT ENGINEERING. *Educ:* Univ Manchester, UK, BS, 68. *Prof Exp:* Mat scientist high pressure diecasting, GKN Group Technol Ctr, Wolverhampton, UK, 68-70, proj leader, 71-72; sect head metal forming, GKN Technol Ltd, Wolverhampton, UK, 72-81, mgr aluminum process develop, squeeze casting & metal matrix comp, 84-86; mgr squeeze form, squeeze cast process & prod, GKN Sarkey Ltd, Tetford, UK, 82-83; dir prod & mfg develop, suspensions & new automotive prod, GKN Technol, Auburn Hills, Mich, 86-91; PRES CONSULT MAT, PROCESS & PROD, MPP TECH SERV INC, ROCHESTER HILLS, MICH, 91- *Concurrent Pos:* Vis lectr process technol, Univ Birmingham, UK, 76-81; referee tech papers, Inst Metals, UK, 76-86. *Mem:* Inst Metals UK; Soc Automotive Engrs; Soc Advan Mat & Process Eng. *Res:* Squeeze casting for aluminum alloys; engineering of materials, manufacturing processes and product technology to improve existing, and develop new, products with improved cast and time efficiency. *Mailing Add:* 3128 Walton Blvd Suite 237 Rochester Hills MI 48309

WILLIAMS, HAROLD, b St John's, Nfld, Mar 14, 34; m 58; c 3. GEOLOGY. *Educ:* Mem Univ Nfld, BSc, 56, MSc, 58; Univ Toronto, PhD(geol), 61. *Prof Exp:* Res scientist, Geol Surv Can, Ont, 61-68; from assoc prof to prof Geol, 68-84, UNIV RES PROF, MEM UNIV NFLD, 84- *Concurrent Pos:* Killam scholar & fel, 76-79. *Honors & Awards:* Gov Gens Medal, 56; Past Pres Medal, Geol Asn, Can, 76; R J W Douglas Medal, Can Soc Petrol Geologists, 81; Willet G Miller Medal, Royal Soc, Can, 87; Logan Medal, Geol Asn of Can, 88. *Mem:* Royal Soc Can; Geol Soc Am; fel Geol Asn Can; Am Asn Petrol Geologists; Can Soc Petrol Geologists. *Res:* Regional geology; ophiolite suites; continental margins; Appalachian geology; tectonics of Atlantic border lands and North American continent. *Mailing Add:* Dept Earth Sci Mem Univ Nfld St John's NF A1B 3X5 Can

WILLIAMS, HAROLD HENDERSON, b Blanchard, Pa, Aug 29, 07; m 35; c 3. NUTRITION. *Educ:* Pa State Univ, BS, 29; Cornell Univ, PhD(nutrit), 33. *Prof Exp:* Asst, Cornell Univ, 29-33; Sterling fel, Yale Univ, 33-35; res assoc, Children's Fund Mich, 35-39, from asst dir to assoc dir res lab, 39-45; prof, 45-73, head dept, 55-64, EMER PROF BIOCHEM, CORNELL UNIV, 73- *Concurrent Pos:* Mem nutrit res adv comt, USDA, 51-61; comt amino acids, Food & Nutrit Bd, Nat Acad Sci-Nat Res Coun, 52-72; nutrit study sect, NIH, 58-62, study sect comt nutrit & med res, Brazil, 62; spec organizing comt, conf fish & nutrit, Food & Agr Orgn, UN, Italy, 61, expert panel milk qual, 63-73; grad educ grants panel, US Bur Com Fisheries, 65-67 & exec comt, Off Biochem Nomenclature, 65-73; vis comt biol & phys sci, Western Reserve Univ, 65-67; overseas corresp, Nutrit Abstr & Rev, 57-71; Am Soc Biol Chem rep, div biol & agr, Nat Res Coun, 65-67. *Honors & Awards:* Borden Award, Am Inst Nutrit, 53. *Mem:* Am Soc Biochem & Molecular Biol; Am Chem Soc; Am Inst Nutrit. *Res:* Amino acid and protein metabolism; selenium metabolism in microorganisms. *Mailing Add:* 1060 Highland Rd Ithaca NY 14850

WILLIAMS, HARRY EDWIN, b Los Angeles, Calif, Mar 11, 30; m 55; c 4. MECHANICAL ENGINEERING. *Educ:* Univ Santa Clara, BME, 51; Calif Inst Technol, MS, 52, PhD(mech eng), 56. *Prof Exp:* Fulbright scholar math, Univ Manchester, 56-57; res engr, Jet Propulsion Lab, Calif Inst Technol, 57-60; from asst prof to assoc prof eng, Harvey Mudd Col, 60-66; liaison scientist, Off Naval Res, London, 66-67; assoc prof eng, 67-71, PROF ENG, HARVEY MUDD COL, 71- *Concurrent Pos:* Consult, Jet Propulsion Lab, Calif Inst Technol, 60-66 & 68-71 & Naval Weapons Ctr, China Lake, 72- *Mem:* Am Soc Mech Engrs. *Res:* Analysis of linear and elastic shells and thin rings. *Mailing Add:* Dept Eng Harvey Mudd Col Claremont CA 91711

WILLIAMS, HARRY LEVERNE, b Watford, Can, Nov 16, 16; m 46; c 2. POLYMER SCIENCE, POLYMER ENGINEERING. *Educ:* Univ Western Ont, BA, 39, MSc, 40; McGill Univ, PhD(phys chem), 43. *Prof Exp:* Res chemist, London Asn War Res, 40-41 & Res & Develop, Polysar Ltd, 46-55, supvr to asst mgr, 55-59 & 61-64, projs mgr, 59-61, prin scientist, 64-67; prof, 67-82, EMER PROF APPL CHEM, DEPT CHEM ENG, UNIV TORONTO, 82- *Concurrent Pos:* Fel, Univ Western Ont, 43-46. *Honors & Awards:* Dunlop Award, Macromolecular Sci Div, Chem Inst Can, 77. *Mem:* Fel Royal Soc Chem; fel AAAS; fel Soc Plastics Engrs; fel Chem Inst Can; fel Plastics & Rubber Inst; fel NY Acad Sci. *Res:* Structure, properties and uses of synthetic high polymers, specifically rheological, viscoelastic, optical, acoustical, electrical, and thermal; foams, blends and composites. *Mailing Add:* Dept Chem Eng & Appl Chem Univ Toronto Toronto ON M5S 1A4 Can

WILLIAMS, HARRY THOMAS, b Hampton, Va, July 22, 41; m 75; c 5. THEORETICAL PHYSICS. *Educ:* Univ Va, BS, 63, PhD(physics), 67. *Prof Exp:* Res assoc nuclear physics, Nat Bur Stand, 67-69; guest prof, Univ Erlangen-N renberg, 70; staff scientist, Kaman Sci Div, Kaman Sci Corp, 71-73; from asst prof to assoc prof, 74-83, PROF PHYSICS, WASHINGTON & LEE UNIV, 83- *Concurrent Pos:* Assoc dean col, Wash & Lee Univ, 86-89. *Mem:* Sigma Xi. *Res:* Effect of baryon resonance admixtures in nuclear wave function upon nuclear properties and reactions; response of circular loop antennae to electromagnetic fields; relativistic spin 3/2 propagator; CLEBSCH-Gordon algebra. *Mailing Add:* Dept Physics Washington & Lee Univ Lexington VA 24450

WILLIAMS, HENRY WARRINGTON, b Dallas, Tex, July 10, 34; m 58; c 4. ZOOLOGY, ANIMAL BEHAVIOR. *Educ:* Southern Methodist Univ, BS, 55; Utah State Univ, MS, 61, PhD(behavior), 66. *Prof Exp:* Assoc prof, 64-70, chmn div natural sci & math, 73-79, PROF & CHAIR BIOL, WESTMINSTER COL, MO, 70- *Concurrent Pos:* Vis prof ecol, Col Natural Resources, Utah State Univ, Logan, 78- *Mem:* AAAS; Animal Behav Soc; Am Ornith Union; Cooper Ornith Soc; Am Asn Univ Prof (pres, 67-68 & 74-75). *Res:* Investigations in the field of animal behavior with particular concern for sound communication in avian species. *Mailing Add:* Dept Biol Westminster Col Fulton MO 65251

WILLIAMS, HIBBARD E, b Utica, NY, Sept 28, 32; c 2. MEDICAL GENETICS. *Educ:* Cornell Univ, AB, 54, MD, 58. *Prof Exp:* Intern & asst resident med, Mass Gen Hosp, 58-60; clin assoc arthritis & metab dis & sr asst surgeon, NIH, 60-62; resident med, Mass Gen Hosp, 62-63, chief resident & teaching asst, Sch Med, Harvard Univ, 63-64, instr, 64-65; from asst prof to assoc prof, 65-72, prof med, Sch Med, Univ Calif, San Francisco, 72-78; prof med & chmn dept, Cornell Med Col, 78-80; PROF MED & DEAN, SCH MED, UNIV CALIF, DAVIS, 80- *Concurrent Pos:* Markle scholar, 68-73; chief med serv, San Francisco Gen Hosp; physician-in-chief, NY Hosp, 78- *Mem:* Am Soc Clin Invest (secy-treas); Asn Am Physicians; Am Fedn Clin Res. *Res:* Inborn errors of metabolism. *Mailing Add:* Sch Med Univ Calif Davis CA 95616

WILLIAMS, HUGH COWIE, b London, Ont, July 23, 43; m 67; c 1. MATHEMATICS, COMPUTER SCIENCE. *Educ:* Univ Waterloo, BSc, 66, Math, 67, PhD(math), 69. *Prof Exp:* Nat Res Coun Can fel, York Univ, 69-70; ASSOC PROF COMPUTER SCI, UNIV MAN, 70- *Res:* Application of the computer to problems arising in the theory of numbers. *Mailing Add:* Dept Computer Sci Univ Man Winnipeg MB R3T 2N2 Can

WILLIAMS, HUGH HARRISON, b Boston, Mass, Dec 4, 44; m 70; c 2. EXPERIMENTAL HIGH ENERGY PHYSICS. *Educ:* Haverford Col, BS, 66; Stanford Univ, PhD(physics), 72. *Prof Exp:* Res assoc physics, Brookhaven Nat Lab, 71-73, assoc physicist, 73-74; from asst prof to assoc prof, 74-82, PROF PHYSICS, UNIV PA, 82- *Concurrent Pos:* Mem high energy physics adv panel, 80-84; mem Fermilab physics adv comt, 82-86, chmn, 84-86; chmn, SSC Detector Res & Develop Comt, 87-91; mem, SSC Policy Adv Comt, 90-; Alfred P Sloan Fel. *Mem:* Am Phys Soc. *Res:* Experimental study of elementary particles, their nature and interactions, with particular emphasis on the study of weak interactions; high energy proton antiproton interactions. *Mailing Add:* Dept Physics Univ Pa Philadelphia PA 19104

WILLIAMS, HULEN BROWN, b Lauratown, Ark, Oct 8, 20; m 42, 71; c 2. PHYSICAL CHEMISTRY. *Educ:* Hendrix Col, AB, 41; La State Univ, MS, 43, PhD(chem), 48. *Prof Exp:* From instr to assoc prof, 43-57, admin asst to dean, 52-56, dean Col Chem & Physics, 68-82, PROF CHEM, LA STATE UNIV, BATON ROUGE, 57- HEAD DEPT, 56- *Concurrent Pos:* Consult, chem indust, govt, legal prof & educ. *Honors & Awards:* Coates Award, Am Chem Soc, 63. *Mem:* Am Chem Soc; Am Soc Plastics Engrs. *Res:* Light scattering of latices; proteins; protein metal complexes; organic reaction mechanisms; geochemistry. *Mailing Add:* 470 Castle Kirk Dr Baton Rouge LA 70808-6011

WILLIAMS, JACK A, b Wichita, Kans, June 29, 26; m 49; c 4. ORGANIC GEOCHEMISTRY. *Educ:* Univ Kans, AB, 50, PhD(org chem), 54. *Prof Exp:* Chemist, Standard Oil Co, Ind, 53-57; CHEMIST, RES CTR, AMOCO PROD CO, 57- *Res:* Organic geochemistry of petroleum and associated sedimentary substances. *Mailing Add:* 7317 E 59th St Tulsa OK 74145

WILLIAMS, JACK L R, b Namoa, Alta, Oct 25, 23; nat US; m 50; c 5. ORGANIC CHEMISTRY. *Educ:* Univ Alta, BSc, 46; Univ Ill, PhD(org chem), 48. *Prof Exp:* Spec asst, Off Rubber Reserve, Univ Ill, 46-48; Du Pont fel, Univ Wis, 48-49; res chemist, 49-55, from res assoc to sr res assoc, 55-68, SR LAB HEAD, EASTMAN KODAK CO, 68- *Mem:* Am Chem Soc. *Res:* Organic synthesis; rubber chemistry; high pressure reactions; oxo synthesis; catalytic hydrogenation; high polymer chemistry; organic photochemistry; photochemistry of boron. *Mailing Add:* RR Namoa AB T0A 2N0 Can

WILLIAMS, JACK MARVIN, b Delta, Colo, Sept 26, 38; m 58; c 3. INORGANIC CHEMISTRY, STRUCTURAL CHEMISTRY. *Educ:* Lewis & Clark Col, BS, 60; Wash State Univ, MS, 64, PhD(phys-inorg chem), 66. *Prof Exp:* Resident res assoc neutron & x-ray diffraction anal, 66-68, from asst chemist to chemist, 68-77, SR CHEMIST & GROUP LEADER, CHEM & MAT SCI DIV, ARGONNE NAT LAB, 77- *Concurrent Pos:* Guest prof, Univ Copenhagen, Denmark, 80 & 85, Univ Mo, 80 & 81; chmn, Gordon Res Conf Inorg Chem, 80. *Mem:* Am Crystallog Asn; Am Chem Soc (treas, Inorg Div, 82-85); Am Phys Soc; AAAS. *Res:* Inorganic chemistry and neutron and x-ray diffraction as applied to the elucidation of the nature of chemical bonding; synthesis and characterization of synthetic metals and superconductors; chemical bonding. *Mailing Add:* 4801 Montgomery St Downers Grove IL 60515

WILLIAMS, JACK RUDOLPH, b Goldsboro, NC, Sept 24, 29; m 61; c 1. SONAR & UNDERWATER ACOUSTICS, RADAR RESEARCH. *Educ:* NC State Univ, BSEE, 56; Univ Calif, Los Angeles, MSEE, 67. *Prof Exp:* Res engr, Naval Res Lab, 54-56; dept mgr, TRW, Inc & Space Technol Labs, 56-63; dir, anti-submarine warfare progs, Interstate Electronics Corp, 63-79; VPRES, RES & TECHNOL, DIAG RETRIEVAL SYSTS, INC, 79-, PRES, DIAG RETRIEVAL SYSTS CALIF, INC, 82- *Concurrent Pos:* mem Frigate Study Comt, NATO Armaments Group, 81-84. *Mem:* sr mem Inst Elec & Electronics Engrs; Acoust Soc Am; Soc Motion Picture & TV Engrs; Navy League; Naval War Col Found; Nat Security Indust Asn. *Res:* Research and equipment development in sonar and radar systems; statistical methods of signal processing; systems based on fourier and modern processing techniques. *Mailing Add:* 351 S Margarita Ave Anaheim CA 92807

WILLIAMS, JAMES C(LIFFORD), III, b Ocala, Fla, Oct 11, 28; m 51; c 2. AEROSPACE ENGINEERING. *Educ:* Va Polytech Inst, BS, 51, MS, 55; Univ Southern Calif, PhD(eng), 62. *Prof Exp:* Aeronaut res intern, Nat Adv Comt Aeronaut, 51; teaching fel fluid mech, Va Polytech Inst, 53-54; aeronaut engr, NAm Aviation Co, Inc, 54-57; res scientist, Univ Southern Calif, 57-62; prof aerospace eng, NC State Univ, 62-80, assoc head, Dept Mech & Aerospace Eng, 72-80; PROF AEROSPACE ENG & HEAD DEPT, AUBURN UNIV, 80- *Concurrent Pos:* Consult,Systs Corp Am, Calif, 58-, Tech Prod Div, Waste King Corp, 60-61, Marquardt Corp, 61-64, Guid & Control Div, Litton Systs, Inc, 62-64, Corning Glass Co, NC, 64-65, Missile & Space Systs Div, Douglas Aircraft Co, Calif, 65-66 & Northrop Space Lab, Ala, 66- *Mem:* Assoc fel Am Inst Aeronaut & Astronaut; Am Soc Mech Engrs; Am Soc Eng Educ. *Res:* Boundary layer theory including internal viscous flows; gas dynamics; magnetohydrodynamics; aerodynamics. *Mailing Add:* Dept Aerospace Eng Auburn Univ Auburn AL 38649

WILLIAMS, JAMES CARL, b Covington, La, 35; m 63; c 4. ANIMAL PARASITOLOGY. *Educ:* Southeastern La Col, BS, 57; La State Univ, Baton Rouge, MS, 62; La State Univ Med Ctr, New Orleans, PhD(med parasitol), 69. *Prof Exp:* From instr to assoc prof, 57-78, PROF VET PARASITOL, LA STATE UNIV, BATON ROUGE, 78- *Mem:* Am Asn Vet Parasitologists; World Asn Advan Vet Pathol; Am Soc Parasitol. *Res:* Epidemiology of parasitism in ruminants; chemotherapy and management control of parasitism in ruminants. *Mailing Add:* Dept Vet Sci La State Univ Baton Rouge LA 70803-6002

WILLIAMS, JAMES CASE, b Salina, Kans, Dec 7, 38; m 60; c 2. METALLURGY, MATERIALS SCIENCE. *Educ:* Univ Wash, BS, 62, MS, 64, PhD(metall eng), 68. *Prof Exp:* Res engr metall, Boeing Co, 62-68; mem tech staff, Rockwell Sci Ctr, 68-70, group leader, 70-73; prog mgr technol, Aerospace Group Staff, Rockwell Int, 73-75; pres, Mellon Inst, 81-83; from assoc prof to prof metall, Carnegie-Mellon Univ, 75-81, dean eng, 83-88; GEN MGR, ENG MAT TECHNOL LABS, GEN ELEC AIRCRAFT ENGINES, 88- *Concurrent Pos:* Consult, USAF Mat Lab & Los Alamos Nat Lab, 75-, Westinghouse Co, 77-, Gen Elec Co, 81- & Hoeganaes Corp, 84-; adv, USAF Off Sci Res, 76-; chmn US deleg, Int Ti Conf, Moscow, 76; co-chmn, US deleg, Int Ti Conf, Kyoto, 80; comt mem, First Fusion Mat Panel, Int Energy Agency; chmn, Nat Mat Adv Bd, Pa State Univ; mem adv bd, Sch Eng, Univ Va; dir, Wheeling Pittsburgh Steel Co, 87-91; trustee, Ore Grad Inst Sci & Technol. *Honors & Awards:* Adams Award, Am Welding Soc, 79; Albert Sauveur lectr, 83. *Mem:* Nat Acad Eng; Am Inst Mining, Metall & Petrol Engrs; Am Soc Metals Int. *Res:* Physical metallurgy of Ti alloys; phase transformations; fracture and fatigue; electron microscopy; strengthening mechanisms; microstructure especially property relationships, powder metallurgy, welding. *Mailing Add:* Gen Elec Aircraft Engines MDH85 Cincinnati OH 45215

WILLIAMS, JAMES D, b Pratt, Kans, June 8, 32; m 56; c 4. ELECTRICAL ENGINEERING, SOLID STATE PHYSICS. *Educ:* Mass Inst Technol, BS & MS, 60; Purdue Univ, PhD(elec eng), 63. *Prof Exp:* Engr, Gen Radio Co, Mass, 57-59; asst elec eng, Mass Inst Technol, 59-60; instr, Purdue Univ, 60-63; staff mem, 63-66, proj leader adv develop, 66-67, proj leader hybrid microcircuits, 67-68, proj leader semiconductor devices, 68-69, div supvr semiconductor circuits, 69-75, DIV SUPVR, INTRUSION DETECTION SYSTS, SANDIA CORP, 75- *Concurrent Pos:* Instr, Franklin Inst, Boston, 59-60; assoc prof, Univ NMex, 67-68. *Mem:* Inst Elec & Electronics Engrs; Am Phys Soc; Am Soc Testing & Mat; Am Vacuum Soc; Sigma Xi. *Res:* Hybrid microcircuits; development of thin film processes and devices for application of components to hybrid microcircuits; development of semiconductor devices and integrated circuits for use in complex weapon systems; intrusion detection systems. *Mailing Add:* 9101 Aspen Ave NE Albuquerque NM 87112

WILLIAMS, JAMES EARL, JR, b Freeport, Pa, June 1, 38; m 58; c 3. PHYSICAL CHEMISTRY. *Educ:* Univ Pittsburgh, BS, 65, MS, 72. *Prof Exp:* Scientist chem, Aluminum Co Am, 65-73, sr scientist, 73-75, group leader chem, 73-79, sect head, 79-90; OPERS DIR, DSSC, CARNEGIE MELLON UNIV, 91- *Mem:* Am Chem Soc; Electrochem Soc; Aluminum Asn; Steel Struct Painting Coun; Sigma Xi; Fedn Socs Coating Technol. *Res:* Leading the development of understanding materials and processes for advanced data storage systems; magnetic recording substrates, media and wearlayers; recording head/media tribology; thin films, magnetic recording. *Mailing Add:* Carnegie Mellon Univ DSSC Pittsburgh PA 15213-3890

WILLIAMS, JAMES G, b Atascedero, Calif, Nov 4, 44; m 66, 77, 85. COMPUTER SCIENCES. *Educ:* Carleton Col, BA 66; Univ Calif, Berkeley, PhD(math), 71. *Prof Exp:* From asst prof to assoc prof, Bowling Green State Univ, 72-76; MEM TECH STAFF, MITRE CORP, 79- *Mem:* Am Math Soc; Asn Comput Sci; Asn Symbolic Logic. *Res:* Program verification including automated theorem proving, formal verification theory, development of formal logical systems and application-independent user-interface translation software; software development methodology; computer security. *Mailing Add:* Mailstop B330 Burlington Rd Bedford MA 01730

WILLIAMS, JAMES GERARD, b New Kensington, Pa, April 12, 41. GEODYNAMICS, SOLAR SYSTEM DYNAMICS. *Educ:* Calif Inst Technol, BS, 63; Univ Calif, Los Angeles, PhD(planetary & space sci), 69. *Prof Exp:* Mem tech staff, NAm Rockwell, 62-68; RES SCIENTIST, JET PROPULSION LAB, CALIF, 69- *Mem:* Int Astron Union; Am Geophys Union; Am Astron Soc. *Res:* Lunar laser range data; orbit of moon; rotations of earth and moon; dynamical evolution of asteroid orbits; asteroid families; main belt morphology; planet crossing asteroids. *Mailing Add:* Jet Propulsion Lab 238-332 4800 Oak Grove Dr Pasadena CA 91109-8099

WILLIAMS, JAMES HENRY, JR, b Los Angeles, Calif, July 14, 18; m 39; c 2. AGRONOMY. *Educ:* Ore State Col, BS, 49; Iowa State Col, MS, 50, PhD(agron), 52. *Prof Exp:* Res assoc, Iowa State Col, 50-52; from asst prof to prof, 52-85, EMER PROF AGRON, UNIV NEBR, LINCOLN, E CAMPUS, 85- *Mem:* Am Soc Agron; Crop Sci Soc Am; Soc Econ Bot; AAAS; Coun Agr Sci & Technol. *Res:* Soybean breeding. *Mailing Add:* 5300 Earl Dr Lincoln NE 68505

WILLIAMS, JAMES HENRY, JR, b Newport News, Va, Apr 4, 41; c 2. MECHANICAL ENGINEERING. *Educ:* Mass Inst Technol, SB, 67, SM, 68; Univ Cambridge, PhD(mech eng), 70. *Prof Exp:* Apprentice machinist, Newport News Shipbuilding & Dry Dock Co, 60-61, apprentice designer, 61-65, mech designer, 65, sr design engr, 68-70; assoc prof, 70-80, PROF MECH ENG, MASS INST TECHNOL, 80- *Concurrent Pos:* NSF res initiation grant, Mass Inst Technol, 72-74, du Pont-Young fac grant, 72-73, Edgerton Professorship, 73-75. *Honors & Awards:* Charles F Bailey Awards, Bronze, 61, Silver, 62, Gold, 63; Teetor Award, Soc Automotive Engrs, 74; Den Hartog Award. *Res:* Applied mechanics and materials, shell theory; earthquake isolation research, nondestructive evaluation and composite materials. *Mailing Add:* Dept Mech Eng Rm 3-360 77 Massachusetts Ave Cambridge MA 02139

WILLIAMS, JAMES HUTCHISON, b Westerville, Ohio, Feb 20, 22; m 43; c 4. OBSTETRICS & GYNECOLOGY. *Educ:* Otterbein Col, AB, 44; Ohio State Univ, MD, 46, MMSc, 52; Am Bd Obstet & Gynec, dipl. *Prof Exp:* From instr to assoc prof, 55-70, assoc dir, Inst Perinatal Studies, 60-64 & Ctr Perinatal Studies, 65-70, PROF OBSTET & GYNEC, OHIO STATE UNIV, 70-, ASSOC DEAN COL MED, 61- *Mem:* Fel Am Col Surg; fel Am Col Obstet & Gynec. *Res:* Perinatal morbidity and mortality; selection of medical students; medical student evaluation in education. *Mailing Add:* Med Surg 1 Rm 125 Univ Calif Irvine CCM Irvine CA 92717

WILLIAMS, JAMES LOVON, JR, b Salem, Ind, May 16, 29; m 52; c 3. WEED SCIENCE, PLANT PHYSIOLOGY. *Educ:* Purdue Univ, Lafayette, BS, 57, MS, 59, PhD(plant path), 61. *Prof Exp:* From instr to prof weed sci, Purdue Univ, Lafayette, 60-86; RETIRED. *Concurrent Pos:* Hon mem, NCent Weed Control Conf. *Mem:* Weed Sci Soc Am. *Res:* Weed control systems to minimize pollution potential from weeds and their control; effects of soil properties and climatic factors on control systems; fate of herbicides in soil and water. *Mailing Add:* 2504 Sequoya Dr Lafayette IN 47905-2766

WILLIAMS, JAMES MARVIN, b Denver, Colo, Apr 27, 34; m 61; c 2. NUCLEAR & CHEMICAL ENGINEERING. *Educ:* Univ NMex, BS, 57, MS, 64. *Prof Exp:* Staff mem, Los Alamos Sci Lab, 60-69; chief, Systs Studies Br, Off Safeguards & Mat Mgt, US AEC, 69-72; group leader, Laser Div, Los Alamos Sci Lab, 72-74; asst dir develop & technol, Off Magnetic Fusion Energy, Dept Energy, 74-78; div leader, systs, Anal & Assessment Div, 78-79, asst dir planning & anal, 79-86, DEP DIR, INDUST APPLNS OFF, LOS ALAMOS NAT LAB, 86- *Mem:* Am Nuclear Soc. *Res:* Development of fusion plasma heaters, magnets and materials; interdisciplinary policy research and analysis; technology assessment; energy systems modeling and economic analysis; strategic planning; technology transfer; economic development. *Mailing Add:* 5630 Cletsoway Dr Albuquerque NM 87105

WILLIAMS, JAMES STANLEY, statistics; deceased, see previous edition for last biography

WILLIAMS, JAMES THOMAS, b Martinsville, Va, Nov 10, 33; m 62; c 2. MEDICINE. *Educ:* Howard Univ, BS, 54, MD, 58; Am Bd Internal Med, dipl, 67, 74 & 80, cert endocrinol & metab, 72. *Prof Exp:* Intern, Philadelphia Gen Hosp, 58-59; resident internal med, DC Gen Hosp, 59-60 & Freedmen's Hosp, 60-62 & 64-65; fel endocrinol, 65-67, asst prof, 67-74, assoc prof, 74-85, PROF MED, HOWARD UNIV, 85- *Mem:* Fel Am Col Physicians; Endocrine Soc; Am Diabetes Asn. *Res:* Clinical endocrinology and metabolic diseases. *Mailing Add:* Dept Med Howard Univ Hosp 2041 Georgia Ave NW Washington DC 20060

WILLIAMS, JEAN PAUL, b New York, NY, Dec 29, 18; m 45. ANALYTICAL CHEMISTRY. *Educ:* Kent State Univ, BS, 40; Univ NC, PhD(chem), 50. *Prof Exp:* Chemist, Nat Bur Standards, 42-45; instr chem, Univ NC, 45-50; res chemist, Corning Glass Works, 50-57, mgr, Tech Serv Res Dept, 57-64, mgr, Instrumental Anal Res Dept, 64-73, mgr anal serv res, 73-81; RETIRED. *Mem:* AAAS; Am Chem Soc; Soc Appl Spectros; Microbeam Anal Soc; Sigma Xi. *Res:* Inorganic chemical, instrumental and x-ray analysis; classical wet methods; polarography; flame spectrophotometry; glass property measurements; microscopy; mass spectrometry; electron microprobe analysis; atomic absorption and emission. *Mailing Add:* 729 Piedra Dr Canon City CO 81212

WILLIAMS, JEFFREY F, b Bristol, Eng, Aug 28, 42; m 64. PARASITOLOGY, IMMUNOLOGY. *Educ:* Univ Bristol, BVSc, 64; Univ Pa, PhD(parasitol), 68. *Prof Exp:* Parasitologist, Pan-Am Health Orgn, Buenos Aires, Arg, 68-71; from asst prof to assoc prof microbiol, 71-77, dean res vet med, 77-79, PROF MICROBIOL & PUB HEALTH ASST, MICH STATE UNIV, 77- *Concurrent Pos:* Dir, Sudan Proj on Collaborative Res on Trop Dis, 79- *Mem:* Brit Soc Immunol; Am Soc Parasitol; Royal Soc Trop Med; Am Soc Trop Med. *Res:* Mechanisms of resistance to helminth infections in domestic animals and man. *Mailing Add:* Dept Microbiol 178 Giltner Hall Mich State Univ East Lansing MI 48824

WILLIAMS, JEFFREY TAYLOR, b Honolulu, Hawaii, Apr 10, 53; m 75; c 1. SYSTEMATIC ICHTHYOLOGY, BIOGEOGRAPHY. *Educ:* Fla State Univ, BS, 75; Univ S Ala, MS, 79; Univ Fla, PhD(zool), 86. *Prof Exp:* COLLECTION MGR, NAT MUS NATURAL HIST, SMITHSONIAN INST, 83- *Concurrent Pos:* Adj prof, Collin County Community Col, 90- *Mem:* AAAS; Am Soc Ichthyologists & Herpetologists; Sigma Xi; Soc Syst Biol. *Res:* Systematics, biogeography and biodiversity in marine ecosystems with emphasis on tropical marine fishes. *Mailing Add:* Div Fishes NHB Smithsonian Inst Washington DC 20560

WILLIAMS, JEFFREY WALTER, b Monroe, Wis, Oct 4, 51; m 73; c 1. ENZYMOLOGY, PROTEIN CHEMISTRY. *Educ:* Univ Wis-Madison, BS, 73, PhD(pharmacol biochem), 77. *Prof Exp:* Postdoctoral fel biochem, Australian Nat Univ, 77-79; res fel chem, Univ Sussex, 79-80; asst prof med chem, Ohio State Univ, 80-85; sr scientist, 85-87, DIR MFG, PROMEGA CORP, 87- *Mem:* Am Soc Biochem & Molecular Biol; AAAS; Soc Ind Microbiol. *Res:* Mechanisms of enzymatic microbiol detoxification of toxic metals; enzymology of DNA modification. *Mailing Add:* 1477 Ravenoaks Trail Oregon WI 53575

WILLIAMS, JEROME, b Toronto, Ont, July 15, 26; m 53; c 2. PHYSICAL OCEANOGRAPHY. *Educ:* Univ Md, BS, 50; Johns Hopkins Univ, MA, 52. *Prof Exp:* Res staff asst, Chesapeake Bay Inst, Johns Hopkins Univ, 52-56; physicist, Vitro Labs, 56-57; asst prof physics, US Naval Acad, 57-64, assoc prof oceanog, 64-72, res prof environ protection, 72-74, liaison scientist, Off Naval Res, London, 86-88, prof, 74-90, EMER PROF OCEANOG, US NAVAL ACAD, 90-; EXEC DIR, ESTUARINE RES FEDN, 90- *Concurrent Pos:* Res assoc, Chesapeake Bay Inst, Johns Hopkins Univ, 57-71; tech ed, Naval Inst, 66-75. *Mem:* Am Asn Physics Teachers; Marine Technol Soc; Am Geophys Union; Am Soc Limnol & Oceanog; Estuarine Res Fedn (vpres, 71-73, secy, 73-75). *Res:* Underwater transparency; oceanographic instrumentation; environmental protection; publications in area of general oceanography, marine optics, oceanographic instruments and fluid physics. *Mailing Add:* Estuarine Res Fedn PO Box 544 Crownsville MD 21032-0544

WILLIAMS, JESSE BASCOM, b Lone Oak, Tex, Oct 24, 17; m 44; c 3. ANIMAL HUSBANDRY. *Educ:* Okla State Univ, BS, 47; Pa State Univ, MS, 48, PhD(dairy husb), 50. *Prof Exp:* Res asst dairy husb, Pa State Univ, 48-50; asst prof, NDak State Univ, asst dairy husbandman, Agr Exp Sta, 50-55; from asst prof to prof animal sci, Univ Minn, St Paul, 55-84; RETIRED. *Concurrent Pos:* Consult, India, Nepal, Syria, UK, Australia, Belg, Zimbabwe, Somalia, Repub Guinea & Sahel. *Mem:* Fel AAAS; Am Soc Animal Sci; Am Dairy Sci Asn. *Res:* Infant ruminant nutrition; synthetic diets; mechanical feeding devices; rural development in third world; immunoglobulin absorption patterns. *Mailing Add:* Dept Animal Sci Univ Minn St Paul MN 55108

WILLIAMS, JIMMY CALVIN, b Palestine, Tex, Oct 26, 43; div; c 5. MICROBIOLOGY & BIOCHEMISTRY, INFECTIOUS DISEASES. *Educ:* Tex A&M Univ, BS, 69, MS, 71, PhD(biochem), 73. *Prof Exp:* Res assoc cancer, Lab Exp Oncol, Riley Cancer Wing, Ind Sch Med, 73-74; microbiologist, Naval Med Res Inst, Nat Naval Med Ctr, 74-78; biochemist, NIH, Nat Inst Allergy & Infectious Dis, Rocky Mountain Lab, 78-83; chief, Rickettsial Dis Lab, 83-89, CHIEF, INTRACELLULAR PATHOGENS BR, BACT DIV, US ARMY MED RES INST INFECTIOUS DIS, 89- *Concurrent Pos:* Vis instr, Lab Exp Oncol, Ind Sch Med, 74, vis asst prof, 75- *Mem:* Sigma Xi; Nat Am Soc Microbiol; NY Acad Sci; AAAS. *Res:* Molecular biology of obligate intercellular bacteria entailing detailed analysis of structure-function immunology and the biochemical strategy of obligate intracellular growth of bacteria in the cytoplasm, phagosome and phagolysosome. *Mailing Add:* 7014 Summerfield Dr Frederick MD 21702

WILLIAMS, JOEL LAWSON, b Sarecta, NC, Nov 10, 41; m 62; c 2. POLYMER CHEMISTRY. *Educ:* NC State Univ, BS, 65, MS, 67, PhD(polymer sci), 70. *Prof Exp:* Sr chemist polymer res, Camille Dreyfus Lab, Research Triangle Inst, 62-74; HEAD MAT SCI DEPT, BECTON DICKINSON RES CTR, RESEARCH TRIANGLE PARK, NC, 74- *Concurrent Pos:* Adj prof chem eng, NC State Univ, 72-, biomed eng, Duke Univ, 83- *Mem:* Am Chem Soc. *Res:* Permeability and diffusion in membranes, polymer synthesis and characterization with special emphasis on the utilization of radiation chemistry as a tool for graft modification of polymeric substrates, ionic polymerization, high-energy irradiation applications, irradiation grafting and blood compatibility of polymers. *Mailing Add:* Becton Dickinson Res Ctr Box 12016 Research Triangle Park NC 27709

WILLIAMS, JOEL MANN, JR, b Suffolk, Va, Apr 6, 40; m 62; c 2. POLYMER SCIENCE, FUEL SCIENCE. *Educ:* Col William & Mary, BS, 62; Northwestern Univ, Evanston, PhD(org chem), 66. *Prof Exp:* NSF fel, Univ Minn, Minneapolis, 66-67, asst prof chem, 67-68; res chemist, Benger Lab, E I du Pont de Nemours & Co, Inc, Va, 68-72; MEM STAFF, LOS ALAMOS NAT LAB, 72- *Concurrent Pos:* Consult, GV Med Inc, Minn, 86- & Shell Dev, Tex, 91- *Honors & Awards:* Fed Lab Consortium Technol Transfer Award, 88. *Mem:* Sigma Xi. *Res:* Environmental chemistry associated with the disposal of energy related wastes; trace element release from coals, oil shales, uranium and their associated coal and coal wastes, especially trace elements of environmental concern; wastes; geothermal energy and uranium mill tailings; microcellular organic foams, deposition of organic films. *Mailing Add:* MST-7 E549 Los Alamos Nat Lab Univ Calif Los Alamos NM 87545

WILLIAMS, JOEL QUITMAN, b Lake Charles, La, Mar 6, 22; m 47; c 2. PHYSICS. *Educ:* Centenary Col, BS, 43; Ga Inst Technol, MS, 48; Duke Univ, PhD(physics), 52. *Prof Exp:* From asst prof to assoc prof, 46-49, 51-70, prof, 70-83, EMER PROF PHYSICS, GA INST TECHNOL, 83- *Mem:* Am Phys Soc; Sigma Xi. *Res:* Microwave spectroscopy. *Mailing Add:* 2792 Dover Rd Atlanta GA 30327

WILLIAMS, JOHN A(RTHUR), b San Bernardino, Calif, Aug 24, 29; m 59. HYDROMECHANICS, CIVIL ENGINEERING. *Educ:* Univ Calif, Berkeley, BS, 52, ME, 54, PhD(civil eng), 65. *Prof Exp:* From asst prof to assoc prof civil eng, 63-72, PROF CIVIL ENG, UNIV HAWAII, 72- *Concurrent Pos:* Asst engr, Hawaii Inst Geophys, 64-66, assoc researcher, 69; proj mgr, Water Resources Res Ctr, 68-76. *Mem:* Am Soc Civil Engrs; Am Geophys Union. *Res:* Hydromechanics; water waves; groundwater movement; geohydrology. *Mailing Add:* Dept Civil Eng Univ Hawaii Manoa Honolulu HI 96822

WILLIAMS, JOHN ALBERT, b Springfield, Ill, Mar 28, 37; m 59; c 2. ASTRONOMY. *Educ:* Univ Mich, AB, 49; Univ Calif, Berkeley, PhD(astron), 63. *Prof Exp:* NSF fel, Univ Calif, Berkeley & Princeton Univ, 63-64; from instr to asst prof astron, Univ Mich, Ann Arbor, 64-70; ASSOC PROF PHYSICS, ALBION COL, 70- *Mem:* AAAS; Am Astron Soc. *Res:* Photometry of astronomical objects; quantitative spectral classification; interstellar matter. *Mailing Add:* Dept Physics Albion Col Albion MI 49224

WILLIAMS, JOHN ANDREW, b Des Moines, Iowa, Aug 3, 41; m 65; c 2. PHYSIOLOGY. *Educ:* Cent Wash State Col, BA, 63; Univ Wash, MD & PhD(physiol, biophys), 68. *Prof Exp:* Staff assoc, Clin Endocrinol Br, Nat Inst Arthritis & Metab Dis, 69-71; from asst prof to prof physiol, Univ Calif, San Francisco, 79-87; PROF PHYSIOL & MED & CHMN DEPT PHYSIOL, UNIV MICH, ANN ARBOR, 87- *Concurrent Pos:* NIH fel, Dept Pharmacol, Univ Utah, 68-69; Helen Hay Whitney Found fel, Univ Cambridge, 71-72; USPHS grants, Univ Calif, San Francisco, 73-88, Univ Mich, 88-; assoc dir, Cell Biol Res Lab, Mt Zion Hosp & Med Ctr, San Francisco, 79-; ed, Am J Physiol, 85-91. *Honors & Awards:* Hoffman La Roche Prize, 85. *Mem:* Endocrine Soc; Am Soc Cell Biol; Am Physiol Soc; Am Soc Clin Investr; Am Gastroenterol Asn. *Res:* Cellular physiology; endocrinology. *Mailing Add:* Dept Physiol Univ Mich Ann Arbor MI 48109

WILLIAMS, JOHN C, b Hazard, Ky, June 18, 25; m 57; c 1. ZOOLOGY. *Educ:* Mich State Univ, BS, 53; Univ Ky, MS, 57; Univ Louisville, PhD(biol), 63. *Prof Exp:* Instr biol, Mary Washington Col, Univ Va, 56-57; asst prof, Transylvania Col, 57-59; from asst prof to prof, Murray State Univ, 62-69; from assoc prof to prof, 69-85, EMER PROF BIOL, EASTERN KY UNIV, 85- *Concurrent Pos:* Mussel Fishery Invests grant, Tenn, Ohio & Green Rivers, 66-69; Com Fisheries Invests of Ky River grant, 72-74; fishery biologist, Ky Dept Fish & Wildlife Res, 86-87. *Mem:* Am Fisheries Soc; Am Soc Ichthyol & Herpet. *Res:* Fisheries biology; mammalogy; herpetology; natural history of freshwater mussels. *Mailing Add:* 1812 Gayle Dr Lexington KY 40505

WILLIAMS, JOHN COLLINS, JR, b Jackson, Tenn, Jan 19, 45; c 2. ORGANIC CHEMISTRY, PHYSICAL CHEMISTRY. *Educ:* Millsaps Col, BS, 67; Tulane Univ, PhD(chem), 72. *Prof Exp:* Instr chem, Sch Arts & Sci, Tulane Univ, 68-71; asst prof, 72-77, ASSOC PROF CHEM, RI COL, 77- *Mem:* Am Chem Soc; Sigma Xi. *Res:* Synthetic, physical and theoretical chemistry and cytotoxicity of organophosphorus heterocycles; P-31 nuclear magnetic resonance; polarography; photochemistry of aromatic phosphines; group theory and isomerization reactions; educational approaches to introductory spectroscopy. *Mailing Add:* 29 Ann St Providence RI 02903

WILLIAMS, JOHN DELANE, b Ordway, Colo, Oct 26, 38; m 80; c 3. STATISTICS, EDUCATIONAL PSYCHOLOGY. *Educ:* Univ Northern Colo, BA, 59, MA, 60, PhD(appl statist), 66. *Prof Exp:* Instr math, Western Wyo Community Col, 62-65; from asst prof to assoc prof, 66-71, PROF STATIST, UNIV NDAK, 71- *Concurrent Pos:* Statist consult, Proj Reclamation, 76-81, sr consult, Computer Ctr, Univ NDak, 72-82. *Honors & Awards:* Sigma Xi Award. *Mem:* Am Statist Asn; Am Educ Res Asn; Am Psychol Asn. *Res:* Appl statist in multiple linear regression; statistical application in educational psychology; life span development and gerontology. *Mailing Add:* PO Box 8158 Univ NDak PO Box 8158 Grand Forks ND 58202

WILLIAMS, JOHN ERNEST, b Dublin, Ireland, Mar 27, 35; m 57; c 2. METALLURGY, MECHANICAL ENGINEERING. *Educ:* Col Advan Technol, Birmingham, Eng, ACT, 61; Univ Birmingham, PhD(metall), 64. *Prof Exp:* Res asst metall, Int Nickel Corp, Eng, 52-58; res fel metal cutting, Univ Birmingham, 61-64, sr res fel, 64-67; lectr, Brunel Univ, 67-70; ASSOC PROF MECH ENG, UNIV CONN, 70- *Concurrent Pos:* Consult, Cornell Univ, 69 & Gen Elec Corp, 71-72. *Mem:* Am Soc Mech Engrs; Brit Inst Metallurgists. *Res:* Materials and mechanical engineering with particular reference to manufacturing and design engineering. *Mailing Add:* Dept Mech Eng Univ Conn Main Campus U- 139 191 Auditorium Storrs CT 06268

WILLIAMS, JOHN F, JR, b Louisville, Ky, Oct 25, 31; m 66. INTERNAL MEDICINE, CARDIOLOGY. *Educ:* Ind Univ, Indianapolis, MD, 56. *Prof Exp:* Intern med, Univ Minn Hosps, 56-57; resident internal med, Med Ctr, Ind Univ, Indianapolis, 57-59, from asst prof to assoc prof internal med, Sch Med, 65-70; PROF MED & DIR DIV CARDIOL, UNIV TEX MED BR GALVESTON, 70- *Concurrent Pos:* Am Heart Asn res fel cardiol, Med Ctr, Ind Univ, Indianapolis, 59-61; USPHS res fel, 61-63; fel, Cardiol Br, Nat Heart Inst, 63-65; chief cardiovasc res lab, Vet Admin Hosp, Indianapolis, 65-70. *Mem:* Am Soc Clin Invest; Am Physiol Soc; Am Fedn Clin Res; fel Am Col Cardiol; fel Am Col Physicians. *Res:* Cardiovascular physiology and pharmacology. *Mailing Add:* Cardiol Div Tex Med Br Galveston TX 77550

WILLIAMS, JOHN FREDERICK, b York, SC, May 14, 23; m 45; c 2. CHEMISTRY. *Educ:* Univ SC, BS, 44; Clemson Univ, MS, 51; Univ Va, PhD(chem), 54. *Prof Exp:* Instr chem, Clemson Univ, 49-51; sr chemist anal chem, Res Dept, Liggett & Myers Tobacco Co, 54-60, res supvr anal chem, Res Dept, Liggett & Myers Inc, 60-80; STATE CHEMIST & DIR LAB, GA DEPT AGR, 80- *Mem:* Am Chem Soc; Coblentz Soc; Soc Appl Spectros; Asn Food & Drug Officials; Asn Anal Chemists. *Res:* Development and application of chromatographic, spectrophotometric, automatic and classical methods of analysis in the study of natural products. *Mailing Add:* Lab Div Ga Dept Agr Atlanta GA 30334

WILLIAMS, JOHN PAUL, b Laramie, Wyo, Aug 11, 46; m 67; c 2. INORGANIC CHEMISTRY, CHEMICAL EDUCATION. *Educ:* Univ Wyo, BS, 69, MS, 70; Ohio State Univ, PhD(inorg chem), 75. *Prof Exp:* Fel inorg chem, Univ Wis, 75-76; mem staff chem, Univ Cincinnati, 76-77; mem staff inorg chem, Ind Univ-Purdue Univ, Indianapolis, 77-79; asst prof chem, Univ Cincinnati, 79-85; asst prof, 85-90, ASSOC PROF CHEM, MIAMI UNIV, HAMILTON, 90- *Mem:* Am Chem Soc. *Res:* Hands-on pre-college science activities. *Mailing Add:* 8774 Constance Lane Cincinnati OH 45231

WILLIAMS, JOHN PETER, plant biochemistry, cytology, for more information see previous edition

WILLIAMS, JOHN RODERICK, b Birmingham, Eng, July 5, 40; m 74; c 2. ORGANIC CHEMISTRY, ORGANIC SYNTHESIS. *Educ:* Univ Western Australia, BSc, 62, PhD(org chem), 66. *Prof Exp:* Vis fel org photochem, NIH, 66-67; NIH fel & res assoc, Columbia Univ, 67-68; from asst prof to assoc prof, 68-81, PROF ORG CHEM, TEMPLE UNIV, 81- *Concurrent Pos:* Vis prof, State Univ Ghent, Belg, 81. *Mem:* Am Chem Soc; fel The Chem Soc. *Res:* Synthesis of natural products; synthesis of enzyme inhibitors; synthesis of affinity labels; new synthetic methods; photochemistry; marine and steroid chemistry. *Mailing Add:* 273 Winding Way Merion Station PA 19066-1225

WILLIAMS, JOHN RUSSELL, b Hartford, Conn, Oct 1, 48; m 71; c 4. ELECTRICAL ENGINEERING, MECHANICAL ENGINEERING. *Educ:* Boston Univ, BA, 70; Brandeis Univ, PhD(org chem), 77. *Prof Exp:* SECT CHIEF, CHARLES STARK DRAPER LAB, 77- *Mem:* Am Chem Soc. *Res:* Organic synthesis; fluorine chemistry; ball bearing lubricants; slip ring lubricants; gas bearing lubrication; the friction, wear and electrical properties of lubricants for sliding electrical contacts. *Mailing Add:* 70 Outlook Dr Lexington MA 02173

WILLIAMS, JOHN T, physical chemistry, for more information see previous edition

WILLIAMS, JOHN WARREN, biochemistry, macromolecules; deceased, see previous edition for last biography

WILLIAMS, JOHN WATKINS, III, b Alexandria, La, Mar 11, 42; m 70. CYTOGENETICS. *Educ:* Univ Southwestern La, BS, 65; La State Univ, MS, 68, PhD(genetics zool), 71. *Prof Exp:* Res assoc genetics, La State Univ, 68-70; from asst prof to assoc prof, 71-79, PROF GENETICS & EMBRYOL, TUSKEGEE UNIV, 79- *Mem:* AAAS; Int Soc Differentiation; Sigma Xi. *Res:* Amphibian cytogenetics and chromosomal banding patterns in the Rana pipiens complex with emphasis on chromosomal markers in species from different geographic regions; banding of amphibian chromosomes. *Mailing Add:* Dept Biol Tuskegee Inst Tuskegee Inst AL 36088

WILLIAMS, JOHN WESLEY, b Mobile, Ala, Dec 31, 44; m 67; c 2. ORGANIC CHEMISTRY. *Educ:* Univ Ala, BS, 67; Univ Ill, MS, 70, PhD(org chem), 73. *Prof Exp:* Rockefeller Found fel, Dept Entom, Univ Ill, 73-75; sr chemist org chem, Abbott Labs, 75-79; SR RES CHEMIST, STAUFFER CHEM CO, 79- *Mem:* Am Chem Soc; AAAS. *Res:* Synthetic organic chemistry especially industrial and agricultural fungicides and bactericides. *Mailing Add:* 1021 Crestview Dr San Carlos CA 94070

WILLIAMS, JOHN WHARTON, b Wichita, Kans, May 3, 45; m 68; c 1. GEOLOGY. *Educ:* Col William & Mary, BS, 67; Stanford Univ, MS, 68, PhD(geol), 70. *Prof Exp:* Geologist, Calif Div Mines & Geol, 71-76; from asst prof to assoc prof, 76-84, PROF & CHMN DEPT GEOL, SAN JOSE STATE UNIV, 84- *Mem:* AAAS; Geol Soc Am; Asn Eng Geologists. *Res:* Detection, analysis and delineation of geologic hazards as to provide for the proper location and construction of engineering works. *Mailing Add:* Dept Geol San Jose State Univ San Jose CA 95192-0102

WILLIAMS, JOSEPH BURTON, b San Angelo, Tex, June 11, 46; m 76. ECOLOGY, BIOLOGY. *Educ:* David Lipscomb Col, BA, 69; Univ Ill, MA, 72, PhD(ecol), 76. *Prof Exp:* Vis instr biol, Univ Ill, 75-76; ASST PROF ECOL, PEPPERDINE UNIV, 77- *Mem:* Ecol Soc Am; Am Ornithologists Union; Cooper Ornith Soc; Sigma Xi. *Res:* Community organization in birds; ecological energetics; pollination ecology. *Mailing Add:* Dept Natural Sci Pepperdine Univ Malibu CA 90265

WILLIAMS, JOSEPH FRANCIS, b Indianapolis, Ind, April 7, 38; m 67; c 2. IMMUNOPHARMACOLOGY, PHARMACOKINETICS. *Educ:* Ind Univ, AB, 62; Univ Utah, PhD(pharmacol), 71. *Prof Exp:* Fel pharmacol, Univ Minn, 70-72; asst prof, 72-78, ASSOC PROF PHARMACOL, UNIV SFLA, 78- *Mem:* Sigma Xi; Int Soc Study Xenobiotics; Am Soc Pharmacol & Exp Ther; Interferon Soc. *Res:* Mechanisms involved in the regulation of the hepatic mixed function oxidase system; the interrelationship between the hepatic parenchyma and kupffer cell; effects of immunoactive substances on hepatic cellular metabolic activity. *Mailing Add:* Dept Pharmacol & Therapeut Col Med Univ SFla Tampa FL 33612

WILLIAMS, JOSEPH LEE, b New Bern, NC, Nov 2, 36; m 62; c 4. FOOD CHEMISTRY, LIPID CHEMISTRY. *Educ:* Morehouse Col, BS, 60; Tuskegee Inst Technol, MS, 62; Univ Ill, Urbana, PhD(food sci), 70. *Prof Exp:* George Washington Carver fel, Carver Found, Tuskegee Inst Technol, 60-62; chemist, Monsanto Co, 63-66; USPHS fel, Burnsides Res Lab, Univ Ill, 68-70; dir multidisciplinary labs, Sch Vet Med, Tuskegee Inst Technol, 70-72; SR SCIENTIST, RES & DEVELOP DIV, KRAFTCO CORP, 72- *Concurrent Pos:* Consult & mgr audiovisual & multimedia learning resource ctr for sci & med stud & individualized study progs, Tuskegee Inst Technol, 70-72; Ninth Annual George Washington Carver lectr, 71. *Mem:* Fel AAAS; Am Chem Soc; Am Oil Chem Soc; fel Am Inst Chemists; Inst Food Technologists. *Res:* Food science and lipid chemistry as it relates to the feeding of the public; biochemical utilization by man and the nutritional impact upon man; flavor constituents in edible oils, shelf life of products; correlation of physical and sensory method for evaluation of flavor components in edible oils; application of chemometrics to edible oil quality. *Mailing Add:* 1279 Arbor Ave Highland IL 60035

WILLIAMS, JOSEPHINE LOUISE, b Bowling Green, Ky, May 23, 26. PHYSICAL CHEMISTRY, SURFACE CHEMISTRY. *Educ:* Western Ky Univ, BS, 47; Northwestern Univ, MS, 50; Univ Cincinnati, BS, 58. *Prof Exp:* Res assoc heterocyclics, Dept Chem, Western Ky Univ, 47-48; sr chemist, Cimcool Div, Cincinnati Milacron Inc, 50-55, sr res chemist, Cent Res Div, 55-57, sr res supvr abrasives, Metal Working Fluids, 57-71, sr res supvr, Com Develop Dept, 71-75, sr res assoc, 75-77, mgr, Com Develop Dept, 77-79, dir appl sci res & develop, 79-85; RETIRED. *Mem:* Am Chem Soc. *Res:* Chemistry of friction, lubrication and wear; mechanisms of wear of abrasives and bonded abrasives; metal working fluids and processes; corrosion, surfactants, emulsions, electrode processes; plastics processing and equipment research and development. *Mailing Add:* 73737 Ashworth Dr Cincinnati OH 45208

WILLIAMS, JOY ELIZABETH P, b Blackshear, Ga, June 12, 29; m 59; c 2. BACTERIOLOGY, BIOCHEMISTRY. *Educ:* Univ Ga, BS, 50, MS, 58, PhD(bact), 61. *Prof Exp:* Teacher high sch, Ga, 52-57; USPHS fel biol div, Oak Ridge Nat Lab, 61-63; res assoc bact, Univ Ga, 63-66, asst prof, 66-68, mgr biol automated info, Computer Ctr, 68-70, ASST PROF MICROS & ASSOC DIR HONS PROG, UNIV GA, 70-, COORDR, ADVAN PLACEMENT PROG, 85- *Concurrent Pos:* Mem, Nat Collegiate Hons Coun. *Mem:* AAAS; Am Soc Microbiol; Sigma Xi. *Res:* Bacterial metabolism of alginic and mannuronic acids; formic hydrogenlyase systems in bacteria; bacterial degradation of organic synthetic sulfur compounds; sulfur metabolism in bacteria. *Mailing Add:* Academic Bldg Univ Ga Athens GA 30601

WILLIAMS, KENNETH BOCK, b Petersburg, Tex, Jan 18, 30; m 52; c 2. PLANT TAXONOMY, ZOOLOGY. *Educ:* Abilene Christian Col, BS, 50; Univ Tex, MA, 59; Univ Ariz, PhD(bot), 67. *Prof Exp:* Asst prof, 67-73, ASSOC PROF BIOL, ABILENE CHRISTIAN COL, 73- *Res:* Biosystematic studies in the Gramineae. *Mailing Add:* Dept Biol Abilene Christian Univ Box 7593 Abilene TX 79601

WILLIAMS, KENNETH L, b Saybrook, Ill, Sept 4, 34; m 54; c 3. ZOOLOGY. *Educ:* Univ Ill, Urbana, BS, 60, MS, 61; La State Univ, Baton Rouge, PhD(zool), 70. *Prof Exp:* Instr comp anat & biol, Millikin Univ, 62-64; assoc prof, 66-79, PROF ZOOL & BIOL, NORTHWESTERN UNIV, 79- *Concurrent Pos:* Sigma Xi grant, La State Univ, 66; NSF fel, Northwestern State Univ, 68, Sigma Xi grant, 71; US Forest Serv grants, 79 & 80. *Mem:* Am Soc Ichthyol & Herpet; Soc Study Amphibians & Reptiles; Soc Syst Zool; Herpet League. *Res:* Systematics and anatomy. *Mailing Add:* Dept Biol Sci Northwestern State Univ Natchitoches LA 71457

WILLIAMS, KENNETH STUART, b Croydon, Eng, Aug 20, 40; m 62; c 3. MATHEMATICS. *Educ:* Univ Birmingham, BSc, 62, DSc, 79; Univ Toronto, MA, 63, PhD(math), 65. *Prof Exp:* Lectr math, Univ Manchester, 65-66; from asst prof to assoc prof, 66-75, chmn dept, 80-84, PROF MATH, CARLETON UNIV, 75- *Mem:* Math Asn Am; Can Math Soc; Am Math Soc. *Res:* Theory of numbers. *Mailing Add:* Dept Math & Statist Carleton Univ Ottawa ON K1S 5B6 Can

WILLIAMS, LANSING EARL, b Spencer, WVa, Aug 8, 21; m 46; c 2. PLANT PATHOLOGY. *Educ:* Morris Harvey Col, BSc, 50; Ohio State Univ, MSc, 52, PhD(bot, plant path), 54. *Prof Exp:* Lab asst, Morris Harvey Col, 49-50; lab asst, Ohio State Univ, 50-52, from instr to assoc prof, 54-65, PROF BOT & PLANT PATH, OHIO STATE UNIV & OHIO AGR RES & DEVELOP CTR, 65-, ASSOC CHMN DEPT, 68- *Mem:* AAAS; Am Phytopath Soc; Sigma Xi; Nat Res Soc. *Res:* Corn viruses and stalk rot; mycotoxins; relation of soil fungal flora to soil-borne plant pathogens. *Mailing Add:* Ohio Res & Develop Ctr Wooster OH 44691

WILLIAMS, LARRY G, b Moscow, Idaho, Jan 8, 35; m 56; c 2. AGRICULTURAL ENGINEERING. *Educ:* Univ Idaho, BS, 56, MS, 59. *Prof Exp:* Asst prof, 56-71, assoc prof, 71-80, PROF AGR ENG, UNIV IDAHO, 80- *Mem:* Am Soc Agr Eng. *Res:* Agricultural mechanization and automation; materials handling and agricultural processing. *Mailing Add:* Dept Agr Eng Univ Idaho Moscow ID 83843

WILLIAMS, LARRY GALE, b Lincoln, Nebr, Sept 28, 39; m 62; c 2. MOLECULAR BIOLOGY. *Educ:* Univ Nebr, Lincoln, BS, 61, MS, 63; Calif Inst Technol, PhD(biochem), 68. *Prof Exp:* NIH fel bot, Univ Mich, Ann Arbor, 67-71; asst prof, 71-86, ASSOC PROF BIOL, KANS STATE UNIV, 87- *Mem:* Sigma Xi. *Res:* Biology education. *Mailing Add:* Div Biol Kans State Univ Manhattan KS 66506

WILLIAMS, LARRY MCCLEASE, CHEMICAL VAPOR DEPOSITION, ELECTRONIC MATERIALS PROCESSING. *Educ:* NC State Univ, BS, 78; Univ Calif, Berkeley, PhD(chem eng), 82. *Prof Exp:* Res asst, Lawrence Berkeley Lab, 78-82; mem tech staff, AT&T Bell Labs, 82-86. *Mem:* Am Chem Soc; Am Vacuum Soc; Electrochem Soc. *Res:* Research and development of novel chemical processes for depositing thin films of inorganic materials that are used in electronic device fabrication. *Mailing Add:* AT&T Bell Labs Murray Hill NJ 07974

WILLIAMS, LAWRENCE ERNEST, b Youngstown, Ohio, Nov 29, 37; m 66; c 2. NUCLEAR MEDICINE, BIOPHYSICS. *Educ:* Carnegie-Mellon Univ, BS, 59; Univ Minn, Minneapolis, MS, 62, PhD(physics), 65; Am Bd Radiol, cert radiol physics, 74. *Prof Exp:* Sr sci officer, Rutherford High Energy Lab, Eng, 65-68; asst prof physics, Western Ill Univ, 68-70; from asst prof to assoc prof radiol, univ Minn, Minneapolis, 73-80; IMAGING PHYSICIST, CITY OF HOPE, DUARTE, CALIF, 80- *Concurrent Pos:* NIH spec fel nuclear med, Nuclear Med Clin, Univ Minn, Minneapolis, 71-73; NIH grant, 74, 88-91; consult, Jet Propulsion Lab; adj assoc prof, med physics, UCLA; bd dir, Epidaurus Corp, 85-87; prof, Eurotech Res Univ, Palo Alto, Ca, 85-; ed, Nuclear Med Physics, CRC Press, 87. *Honors & Awards:* Exhibit Gold Medalist, Soc Nuclear Med, 83; Fel, Am Col Angiology, 88. *Mem:* Soc Nuclear Med; Sigma Xi; Am Asn Physicists in Med; Am Col Radiol; NY Acad Sci. *Res:* Immunological imaging; phospholipid vesicles; biodistributions of radiolabeled compounds; image enhancement. *Mailing Add:* Diag Radiol City of Hope Duarte CA 91010-0269

WILLIAMS, LEAH ANN, b Clarksburg, WVa, July 20, 32. DEVELOPMENTAL BIOLOGY. *Educ:* WVa Univ, AB, 54, MS, 58, PhD(biol), 70. *Prof Exp:* Instr anat & physiol, Exten, Pa State Univ, 58-59; instr gen zool, anat & physiol, W Liberty State Col, summer 59; from instr to asst prof, 59-73, ASSOC PROF BIOL, WVA UNIV, 73-, CHAIRPERSON, DEPT BIOL, 86- *Concurrent Pos:* NSF sci fac develop grant, 77-78; PI-NEI NIH, 82-85. *Mem:* AAAS; Am Soc Zool; Soc Develop Biol; Sigma Xi. *Res:* Regeneration; control mechanisms in the regenerative processes in the eyes of newts; evolutionary studies of the lens proteins. *Mailing Add:* Dept Biol WVa Univ PO Box 6057 Morgantown WV 26506-6057

WILLIAMS, LEAMON DALE, b Flippin, Ark, Sept 28, 35; m 61; c 3. FOOD SCIENCE, BIOCHEMISTRY. *Educ:* Univ Ark, BS, 58, MS, 61; Mich State Univ, PhD(food sci), 63. *Prof Exp:* Res chemist, Foods Div, Anderson, Clayton & Co, Tex, 63-67; sect head, CPC Int Inc, 67-69; dir food res, Cent Soya Co, Inc, 69-78, vpres res, 78-85, vpres, Chernurgy Div, 85-87, sr vpres refined soya prod, 87-90, oil seed prod group, 90-91; RETIRED. *Honors & Awards:* MacGee Award, Am Oil Chem Soc, 63. *Mem:* Poultry Sci Asn; Inst Food Technologists; Am Oil Chem Soc; Am Chem Soc. *Res:* Organic chemistry of lipids; esterifiability of hydroxyls; interesterification and development of fat based derivatives; protein chemistry; food research; animal nutrition. *Mailing Add:* Cent Soya Co Inc PO Box 1400 Ft Wayne IN 46801-1400

WILLIAMS, LELAND HENDRY, b Columbia, SC, Feb 24, 30; m 52; c 2. MATHEMATICS, COMPUTER SCIENCE. *Educ:* Univ SC, BS, 50; Univ GA, MS, 51; Duke Univ, PhD(math), 61. *Prof Exp:* Mathematician, Redstone Arsenal, 51-53; res assoc math & vis asst prof, Duke Univ, 60-62; math consult comput, Fla State Univ, 62-64, asst dir comput ctr, 64-66, asst prof math, Univ, 62-66; dir comput ctr & assoc prof math, Auburn Univ, 66-70; pres & dir, Triangle Univs Comput Ctr, 70-88; COMPUTER RESOURCES ARCHITECT, NAVAL RES LAB, 88- *Concurrent Pos:* Assoc dir & lectr, NSF comput inst, Fla State Univ, 66; adj assoc prof, Duke Univ, Univ NC, Chapel Hill & NC State Univ, 70-; dep dir, Edinburgh Regional Comput Ctr & vis prof, Univ Edinburgh, Scotland, 76-77. *Mem:* Am Sci Affil; Asn Comput Mach; Sigma Xi. *Res:* Numerical analysis; nonnumeric mathematical computation; computation center management. *Mailing Add:* Computer Resources Architect Code 1003-9 Naval Res Lab 4555 Overlook Ave Washington DC 20375-5000

WILLIAMS, LEO, JR, electrical engineering, electronics, for more information see previous edition

WILLIAMS, LESLEY LATTIN, b New Bedford, Mass, Aug 10, 39. PHYSICAL CHEMISTRY. *Educ:* Hollins Col, AB, 61, Univ Wis-Madison, PhD(chem), 68. *Prof Exp:* From asst prof to assoc prof, 68-78, PROF CHEM, CHICAGO STATE UNIV, 78- *Concurrent Pos:* Lectr, Univ Md, Munich Campus, 71-72. *Mem:* Sigma Xi; Am Phys Soc. *Res:* Nuclear magnetic resonance relaxation mechanisms in inorganic fluorides, including solvent effects; computer assisted instruction in chemistry; hexafluorides. *Mailing Add:* Dept Phys Sci Chicago State Univ 95th & King Dr Chicago IL 60628

WILLIAMS, LEWIS DAVID, b Hopkinsville, Ky, Apr 2, 44; div. CHEMISTRY. *Educ:* Univ Chicago, BS, 66; Harvard Univ, PhD(org chem), 71. *Prof Exp:* Atholl McBean fel chem, Stanford Res Inst, 70-71; Presidential Intern, Western Regional Res Lab, Agr Res Serv, USDA, Albany, 72-73; asst lab dir, 73-80, LAB DIR, DIAG DATA INC, MOUNTAIN VIEW, 80- *Mem:* Am Chem Soc. *Res:* Physical organic chemistry; structure-reactivity relationships; biochemistry. *Mailing Add:* 37709 Arlene Ct Fremont CA 94536-3714

WILLIAMS, LORING RIDER, b Buckhannon, WVa, Jan 6, 07; m 41; c 2. INORGANIC CHEMISTRY. *Educ:* WVa Wesleyan Col, BS, 27; WVa Univ, MS, 32; Univ Ill, PhD(inorg chem), 39. *Prof Exp:* Teacher high sch, WVa, 27-31 & 34-38; instr chem, Alderson-Broaddus Col, 32-34; teacher chem, univ high sch, Univ Ill, 38-39; from instr to prof chem, 39-72, chmn dept, 57-61, EMER PROF CHEM, UNIV NEV, RENO, 72- *Mem:* Am Chem Soc. *Res:* Distribution of selenium in plants and soils. *Mailing Add:* 4975 Malapi Rd Sparks NV 89431-1130

WILLIAMS, LOUIS GRESSETT, b Owensboro, Ky, Oct 28, 13; m 42; c 2. FRESH WATER ECOLOGY, ALGOLOGY. *Educ:* Marshall Univ, AB, 37; Duke Univ, MA, 40, PhD(biol), 48. *Prof Exp:* Asst, Marshall Col, 37-38; teacher high sch, Fla, 39-40 & NC, 40-41; asst, Duke Univ, 46-47; instr bot, Univ NC, 48; assoc prof biol, Furman Univ, 48-58; in charge, USPHS Plankton Prog, Nat Water Qual Network, Ohio, 58-65; in charge plankton prog, Nat Water Qual Lab, Minn, 65-67; prof biol, 67-79, EMER PROF BIOL, UNIV ALA, TUSCALOOSA, 79- *Concurrent Pos:* Carnegie grant, 49; Ford Found fel, Univ Calif, 51-52. *Honors & Awards:* Jefferson Award, 51. *Mem:* Fel AAAS; Bot Soc Am; Am Soc Limnol & Oceanog; Ecol Soc Am; Phycol Soc Am. *Res:* Water quality assessment by species diversity and toxicity bioassay on the Great Lakes and major rivers of the United States; pollution assessment from radionuclides in the environment. *Mailing Add:* 1246 Northwood Lake Northport AL 35476

WILLIAMS, LUTHER STEWARD, b Sawyerville, Ala, Aug 19, 40; m 63. MOLECULAR BIOLOGY. *Educ:* Miles Col, BA, 61; Atlanta Univ, MS, 63; Purdue Univ, PhD(molecular biol), 68. *Prof Exp:* Lab instr biol, Spelman Col, 61-62; lab instr, Atlanta Univ, 62-63, instr, 63-64; teaching asst, Purdue Univ, 64-66; Am Cancer Soc fel, State Univ NY Stony Brook, 68-69; asst prof, Atlanta Univ, 69-70; asst prof biol sci, Purdue Univ, West Lafayette, 70-73, assoc prof, 73-80, asst provost, 76-, prof, 80-; AT DEPT BIOL, WASH UNIV, ST LOUIS, MO. *Concurrent Pos:* NSF teaching asst, 62-63; NIH career develop award, Purdue Univ, 71-75; assoc prof biol, Mass Inst Technol, 73-74; mem, Microbiol Training Comt, Nat Inst Gen Med Sci, 71-74; chmn, MARC Prog, Nat Inst Gen Med Sci, 75-76. *Mem:* AAAS; Am Soc Microbiol; NY Acad Sci; Am Chem Soc; Am Soc Biol Chemists. *Res:* Physiological role of aminoacyl-transfer RNA synthetases and transfer RNA's in bacterial metabolism. *Mailing Add:* NSF 1800 G St Washington DC 20550

WILLIAMS, LYMAN O, b State College, Pa, Apr 1, 34; m 63; c 2. STRUCTURAL GEOLOGY. *Educ:* Univ Ga, BS, 56, Univ Iowa, MS, 59, PhD(geol), 62. *Prof Exp:* Explor geologist, Calif Co, 61-63; asst prof geol, Monmouth Col, 63-64; assoc prof, Eastern Tenn State Univ, 64-69; assoc prof, 69-73, PROF GEOL, MONMOUTH COL, 73-, DEPT CHMN, 77-; PROF GEOL, PHILLIPS UNIV, ENID, OKLA. *Mem:* Geol Soc Am. *Res:* Petrology and structure of crystalline rock terranes; remote sensing of environment. *Mailing Add:* Dept Geol Phillips Univ 100 S University Enid OK 73701

WILLIAMS, LYNN DOLORES, b Seattle, Wash, Mar 20, 44. MATHEMATICAL ANALYSIS. *Educ:* Lewis & Clark Col, BS, 66; Univ Ore, MS, 70, PhD(math), 72. *Prof Exp:* Vis asst prof math, Univ Ore, 72-73; asst prof, La State Univ, Baton Rouge, 73-77; asst prof, Converse Col, Spartanburg, SC, 77-78; RES ANALYST, BOEING AEROSPACE CO, 78- *Mem:* Am Math Soc; Sigma Xi; Math Asn Am; Asn Women in Math. *Res:* Functional analysis, especially Banach algebras of operators, Fredholm operator theory and approximate identities in Banach algebras. *Mailing Add:* 3455 Belvidere SW Seattle WA 98126

WILLIAMS, LYNN ROY, b Detroit, Mich, Apr 23, 45; m 84; c 2. MATHEMATICS. *Educ:* King Col, BA, 67; Univ Ky, MA, 68, PhD(math), 71. *Prof Exp:* Asst prof math, La State Univ, Baton Rouge, 71-75; asst prof, 75-77, ASSOC PROF MATH, IND UNIV, SOUTH BEND, 77- *Mem:* Am Math Soc. *Res:* Functional analysis; Hp theory; harmonic analysis; statistics. *Mailing Add:* Dept Math Ind Univ 1700 Mishawaka Ave South Bend IN 46634

WILLIAMS, M COBURN, b Osage City, Kans, Jan 28, 29; m 53; c 2. PLANT PHYSIOLOGY. *Educ:* Kansas State Univ, BS & MS, 51; Univ Ill, PhD(agron), 56. *Prof Exp:* plant physiologist, Agr Res Serv, USDA, 56-89; RETIRED. *Mem:* Weed Sci Soc Am; Soc Range Mgt; Coun Agr Sci & Technol. *Res:* Biochemical and physiological research on poisonous range weeds, especially methods of chemical control and identification toxic compounds; Astragalus, Delphinium, Lupinus, Lotus. *Mailing Add:* 1427 E 800 N Logan UT 84321

WILLIAMS, MARION PORTER, b Salem, Ind, Jan 24, 46; m 68. FOOD SCIENCE. *Educ:* Purdue Univ, BS, 68, PhD(food sci), 73. *Prof Exp:* From sr food scientist to sr res scientist, 73-75, mgr int res & prod develop, 75-78, asst dir prod develop, Res Labs, Carnation Co, 78; dir new prod develop, Res Ctr, Anderson Clayton Foods, 78-81; dir prod develop, 81-85, VPRES PROD DEVELOP, RES & DEVELOP, KRAFT, INC, 85- *Mem:* Inst Food Technologists; Indust Res Inst. *Res:* Development of new products and maintenance/improvement of current product lines. *Mailing Add:* 1310 Edgewood Lane Northbrook IL 60062

WILLIAMS, MARSHALL HENRY, JR, b New Haven, Conn, July 15, 25; m 48; c 4. PHYSIOLOGY, INTERNAL MEDICINE. *Educ:* Yale Univ, BS, 45, MD, 47. *Prof Exp:* Intern, Presby Hosp, New York, 47-48, asst resident med, 48-49; asst resident, New Haven Hosp, Conn, 49-50, asst, 50; chief respiratory sect, Dept Cardiorespiratory Dis, Army Med Serv Grad Sch, Walter Reed Army Hosp, 52-55; dir cardiorespiratory lab, Grasslands Hosp, Valhalla, NY, 55-59; vis asst prof physiol, 55-59, assoc prof med & physiol, 59-66, PROF MED, ALBERT EINSTEIN COL MED, 66- *Concurrent Pos:* NIH trainee, New Haven Hosp, Conn, 50; dir chest serv, Bronx Munic Hosp Ctr, New York, 59- *Mem:* AAAS; Am Physiol Soc; Am Thoracic Soc; Am Soc Clin Invest; Am Heart Asn. *Res:* Respiratory and clinical cardiopulmonary physiology. *Mailing Add:* Albert Einstein Col Med New York NY 10461

WILLIAMS, MARSHALL VANCE, b Memphis, Tenn, Mar 22, 48; m 70; c 2. TUMOR BIOLOGY, VIROLOGY. *Educ:* Memphis State Univ, BS, 70, MS, 73; Univ Ga, PhD(microbiol), 76. *Prof Exp:* Cancer res scientist I, Roswell Park Mem Inst, 76-78; asst prof microbiol, Kirksville Col Osteop Med, Mo, 78-82. *Concurrent Pos:* Adj asst prof, Northeast Mo State Univ, 79-82; consult, Bio-Diesel Fuels Iowa, Inc, 81-82. *Mem:* Am Soc Microbiol; AAAS; Am Soc Pharmacol & Exp Therapeut; Int Orgn Mycoplasmology; Am Soc Biochem & Molecular Biol. *Res:* Deoxyuridine metabolism in neoplastic cells and cells infected with herpes simplex virus; development of antiviral agents. *Mailing Add:* Dept Med Microbiol & Immunol 5072 Graves Hall Ohio State Univ 333 W Tenth Ave Columbus OH 43210

WILLIAMS, MARTHA E, b Chicago, Ill. INFORMATION SCIENCES. *Educ:* Barat Col, AB, 55; Loyola Univ, MA, 57. *Prof Exp:* Assoc chemist, Chem Dept, ITT Res Inst, 61, asst supvr, 61-68, mgr, Info Sci, 62-72, Comput Search Ctr, 68-72; DIR, INFO RETRIEVAL RES LAB, & PROF INFO SCI, COORD SCI LAB, UNIV ILL, 72-, AFFIL, COMPUTER SCI DEPT, 79- *Concurrent Pos:* Mem, comt chem info, 70-73, chmn & mem, large data base subcomt, 71-73, ad hoc panel, Info Storage & Retrieval, 77, Numerical Data Adv Bd, 79-82, Nat Res Network Rev Comt, 87-88, Nat Res Coun & Nat Acad Sci; ed, Ann Rev Info Sci & Technol, 75-, Comput Readable Databases: A Directory & Data Sourcebook, 76-87 & Online Rev, 77-; vpres, Eng Info, Inc, 78-80, chmn bd, 80-88; mem bd regents, Nat Libr Med, 78-82, chmn bd, 81-82; hon fel, Inst Info Scientists, London, Eng, 85. *Mem:* Fel AAAS; Am Chem Soc; Am Soc Info Sci; Asn Comput Mach; Asn Sci Info Dissemination Ctr, (vpres, 71-73, pres, 75-77, past pres, 77-79). *Res:* Online retrieval systems; computer readable databases; systems analysis and design; chemical information systems. *Mailing Add:* RR 1 Box 194 Monticello IL 61856-0194

WILLIAMS, MARY ANN, b Albany, NY, May 18, 25. NUTRITION, BIOCHEMISTRY. *Educ:* Iowa State Col, BS, 46; Cornell Univ, MS, 50; Univ Calif, PhD, 54. *Prof Exp:* Asst pathologist, Univ Ky, 49-51; asst nutrit, Univ Calif, 51-54; res assoc, McCollum-Pratt Inst, Johns Hopkins Univ, 54-55; from instr to asst prof, 55-63, assoc prof, 63-75, PROF NUTRIT, UNIV CALIF, BERKELEY, 75- *Concurrent Pos:* Guggenheim fel, 63-64. *Mem:* AAAS; Am Chem Soc; Am Inst Nutrit; Soc Exp Biol & Med; Biochem Soc; Am Soc Biochem & Molecular Biol. *Res:* Essential fatty acid metabolism and functions. *Mailing Add:* Dept Nutrit Sci 119 Morgan Hall Univ Calif Berkeley CA 94720

WILLIAMS, MARY BEARDEN, b Lexington, Ky, Aug 29, 36. EVOLUTIONARY BIOLOGY, PHILOSOPHY OF SCIENCE. *Educ:* Reed Col, BA, 58; Univ Pa, MA, 61; Univ London, PhD(math biol) & DIC, 67. *Prof Exp:* Res assoc biomath, Univ Tex M D Anderson Hosp & Tumor Inst, 63-64; asst prof, NC State Univ, 67-73; vis asst prof hist & philos sci, Ind Univ, 73-74; asst prof philos, Ohio State Univ, 74-76; mem hon fac, Freshman Honors Prog, 76-78, dir, Ctr Sci & Cult, 84-89, ASSOC PROF LIFE & HEALTH SCI, UNIV DEL, 78- *Mem:* Soc Study Evolution; Philos Sci Asn; Soc Syst Zool; Sigma Xi; Am Philos Asn. *Res:* Axiomatization of evolutionary theory; logical status of evolutionary predictions; evolution of population self-regulation; philosophy of biology; bioethics; medical ethics. *Mailing Add:* 28 W Delaware Ave Newark DE 19716

WILLIAMS, MARY CAROL, PULMONARY CYTOLOGY, CELL BIOLOGY. *Educ:* Univ Calif, San Francisco, PhD(anat), 71. *Prof Exp:* PROF ANAT, UNIV CALIF, SAN FRANCISCO, 76- *Mailing Add:* Cardiovasc Res Inst Univ Calif San Francisco CA 94143

WILLIAMS, MARY CAROL, b Norfolk, Va, Nov 29, 42; m 62; c 2. PLASMA SPECTROCHEMISTRY, ENVIRONMENTAL REGULATION & COMPLIANCE. *Educ:* Univ NMex, BUS, 82, MS, 85. *Prof Exp:* Med technologist, St Mary's Hosp, Minneapolis, 66-68 & E I DuPont, Waynesboro, Va, 68-72; mus guide, 77-78, chem technician, 78-85, STAFF MEM, LOS ALAMOS NAT LAB, 85- *Mem:* Soc Appl Spectros. *Res:* Characterization of environmental, biological, and radioactive materials; enhancing techniques for sample preparation to increase efficiency and to reduce contamination, volatility, and interferences; modification of sample introduction systems and instrumental parameters to improve detection limits and reduce interferences. *Mailing Add:* 51 Zuni Los Alamos NM 87544

WILLIAMS, MARY CARR, b Port Arthur, Tex, Dec 25, 26; m 51; c 1. STEROID CHEMISTRY. *Educ:* Tex Woman's Univ, BA & BS, 49; St Mary's Univ, MS, 75. *Prof Exp:* Res chemist lipid metab, Dept Biochem & Biophys, Tex A&M Univ, 49-63; res scientist, Southwest Found Res & Educ, 63-78; assoc found scientist, steroid metab, Dept Clin Sci & Reprod Biol, 78-81, assoc found scientist lipid metab, Dept Cardiopulmonary Dis, 82-83, SR RES ASSOC, DEPT PHYSIOL & MED, SOUTHWEST FOUND BIOMED RES, 83- *Mem:* Am Chem Soc. *Res:* Metabolism of natural and synthetic steroid hormones; chromatography of steroids and lipoproteins; metabolism of lipoproteins. *Mailing Add:* 116 Elm Spring Lane San Antonio TX 78232

WILLIAMS, MARYON JOHNSTON, JR, b Griffin, Ga, Jan 14, 46; m; c 2. CLINICAL ENGINEERING. *Educ:* Ga Inst Technol, BEE, 68; Rutgers Univ, MS, 70, PhD(biomed eng), 72. *Prof Exp:* From instr to asst prof biomed eng & med, Med Col Ga, 72-77; tech dir, Biomed Eng Shared Technol, affil, NJ Hosp Asn, 77-82 & Hosp Eng Logistics & Planning Inc, 82-90; EXEC DIR, AM TEKDYNE, INC, 91- *Concurrent Pos:* Lectr physiol, Med Col Ga, 75-77; sr engr, Condition Analyzing Corp, 85-; tech dir, Am Tekdyne, 90-91. *Mem:* Sr mem Biomed Eng Soc; sr mem Inst Elec & Electronics Engrs. *Res:* Developed uniquely powered piezo-electric heart assist device; studied control characteristics of artificial heart assist devices; consult on pulmonary function and exercise testing equipment; develop computerized database systems for hospitals. *Mailing Add:* 21 Quaker Rd Princeton Junction NJ 08550-1615

WILLIAMS, MAX L(EA), JR, b Aspinwall, Pa, Feb 22, 22; m 67; c 3. MECHANICS OF FRACTURE. *Educ:* Carnegie Inst Technol, BS, 42; Calif Inst Technol, MS, 47, AeE, 48, PhD, 50. *Prof Exp:* Lectr aeronaut, Calif Inst Technol, 48-50, res fel, 50-51, from asst prof to prof, 51-65; prof eng & dean eng, Univ Utah, 65-73, distinguished prof eng, 73; prof eng & dean sch eng, 73-85, EMER DEAN & DISTINGUISHED SERV PROF ENG, UNIV PITTSBURGH, 85- *Concurrent Pos:* Ed in chief, Int J Fracture, 65-; mem exec comt, Int Cong Fracture, 65-; sci dir, NATO Advan Study Inst, Italy, 67; mem biomat adv comt, Nat Inst Dent Res, 67-70; mem chem rocket adv comt, NASA, 68-73; pres, Utah Eng & Develop Corp, 69-79; mem eng adv comn, NSF, 69-72; NSF sr fel, Imp Col, Univ London, 71-72; nat lectr, Sigma Xi, 72; mem nat mat adv bd, Nat Res Coun & chmn, Comt on Mat Struct & Design, 75-77; assoc mem, Defense Sci Bd, 75-76, 82-85; consult & lectr in field, co-founder & dir Terra Tek, Inc, Salt Lake City, 73-77; adv, Regional Indust Develop Corp, Pittsburgh, 73-83; chmn, Pa Adv Comn I-79 Bridge Failure, 77; dir, MPC Corp, Pittsburgh, 74-86; chmn struct design task group, Coop Automotive Res Prog, Off Sci & Tech Policy, 78-79; dir, US Nat Comn World Energy Conf, 79-83; founding chmn bd dirs, BCR Nat Lab, Univ Pittsburgh, 83-85; Gen Lew Allen vis prof aeronaut, USAF Inst Technol, 85-87; mem, USAF Sci Adv Bd, 85-89, adv, 90-; Sci Adv, USAF Acquisition Logistics Ctr, 87-88; consult, Off Under Secy Defense, 70-, & Dept State, 73-81; mem, Nat Eng Adv Bd, Mercer Univ, 88- *Honors & Awards:* Adhesion Res Award, Am Soc Testing & Mat, 75; Solid Rocket Tech Achievement Award, Am Inst Aeronaut Astronaut, 88. *Mem:* AAAS; fel Soc Exp Mech; Am Chem Soc, Div Rubber Chem; assoc fel Am Inst Aeronaut & Astronaut; Am Soc Eng Educ; Soc Rheology; hon fel Int Cong Fracture, 89. *Res:* Continuum mechanics with application to fracture of solids and interaction with chemical structure of materials. *Mailing Add:* Benedum Eng Hall Univ Pittsburgh Pittsburgh PA 15261

WILLIAMS, MAX W, b Cardston, Alta, Aug 24, 30; US citizen; m 54, 82; c 5. PLANT PHYSIOLOGY, HORTICULTURE. *Educ:* Utah State Univ, BSc, 54, MSc, 57; Wash State Univ, PhD(hort), 61. *Prof Exp:* Res asst, Utah State Univ, 54-55; actg supt, Utah Tree Fruit Exp Sta, 55-58; res asst, Wash State Univ, 58-61; RES LEADER, PLANT PHYSIOLOGIST, AGR RES SERV, USDA, 61- *Concurrent Pos:* Adv, fruit prod, Chile, Arg, Australia, NZ, France, Eng, Holland, Italy, Romania, Poland, Yugoslavia, Israel, Mex. *Mem:* Am Soc Hort Sci; Int Soc Hort Sci. *Res:* Chemical thinning of apples; growth retardants; cytokinins; auxins. *Mailing Add:* USDA Agr Res Serv 1104 N Western Ave Wenatchee WA 98801

WILLIAMS, MERLIN CHARLES, b Howard, SDak, July 20, 31; m 59; c 4. METEOROLOGY, ENGINEERING. *Educ:* SDak State Univ, BS, 53; Univ Chicago, cert, 54; Univ Wyo, MS, 62; Stanford Univ, MSA, 81. *Prof Exp:* Instr civil eng, SDak State Univ, 57-58; instr & res asst weather eng, 58-59; engr, US Bur Reclamation, 59-61; asst civil eng, Univ Wyo, 61-62, proj dir weather modification res, 62-66; dir weather modification res, Fresno State Col Found, 66-71; dir, SDak State Weather Control Comn, 71-74; DIR, OFF WEATHER MODIFICATION PROGS, ENVIRON RES LABS, NAT OCEANIC & ATMOSPHERIC ADMIN, 74- *Concurrent Pos:* Consult adv bd weather modification, NSF, 75- *Mem:* Am Soc Civil Eng; Am Meteorol Soc; Am Geophys Union; Weather Modification Asn (pres, 69). *Res:* Water resources research to investigate increasing water supplies, including weather modification research, fluid mechanics and hydrology; basic hydrometeorological studies; mountain meteorology; snow physics; hurricane modification (abatement), boundary layer dynamics. *Mailing Add:* 6387 Niwot Rd Longmont CO 80503

WILLIAMS, MICHAEL, b London, Eng, Jan 3, 47. NEUROCHEMISTRY, PHARMACOLOGY. *Educ:* Univ London, BSc, 71, PhD(neurochem), 74. *Prof Exp:* Res assoc neurochem, Univ NC, Chapel Hill, 74-76; sr res neurochemist, Merck Inst, 76-80, res fel pharmacol, 80-; staff, Ciba-Geigy Pharmaceut; AREA HEAD NEUROSCI, ABBOTT LABS, 89- *Mem:* Am Soc Pharmacol & Exp Therapeut; Soc Neurosci; AAAS; Am Soc Neurochem. *Res:* Synaptic transmission; receptor function; cyclic nucleotides and protein phosphorylation. *Mailing Add:* Dept Neurosci Abbott Labs Abbott Park IL 60064-3500

WILLIAMS, MICHAEL C(HARLES), b Milwaukee, Wis, June 11, 37. RHEOLOGY, VISCOELASTICITY. *Educ:* Univ Wis-Madison, BS, 59, MS, 60, PhD(rheology), 64. *Prof Exp:* Fel polymer solutions, Inst Theoret Sci, Univ Ore, 64-65; from asst prof to assoc prof chem eng, 65-72, PROF CHEM ENG, UNIV CALIF, BERKELEY, 72- *Concurrent Pos:* mem, career guid comt, 68-, San Francisco Bay Area Eng Coun, 68-, chmn, 70-73; vis prof rheol, Univ Nat Del Sur, Bahia Blanca, Arg, 70; consult, pvt indust, govt & litigation; prin investr, numerous res grants & contracts, NSF, NIH, NASA, Off Naval Res & Petrol Res Fund; fac scientist, Ctr Advan Mat, Lawrence Berkeley Lab, 83- *Mem:* Am Inst Chem Engrs; Am Chem Soc; Soc Rheol; Am Soc Eng Educ; Soc Plastics Engrs; NY Acad Sci. *Res:* Viscoelastic fluid phenomena; biomedical engineering; rheology of polymer melts and solutions; processing fluid mechanics of viscoelastic liquids and inelastic slurries; mechanical and dynamic properties of solid and liquid block copolymers; hemolytic and other damage to flowing blood; biomaterials. *Mailing Add:* Dept Chem Eng Univ Alta Edmonton AB T6G 2G6 Can

WILLIAMS, MICHAEL D, b Covina, Calif, May 19, 39; m 71. NUCLEAR ENGINEERING. *Educ:* Univ Calif, Los Angeles, BS, 61, MS, 63, PhD(nuclear eng), 66. *Prof Exp:* Assoc astrodynamics, Systs Develop Corp, 61; mem tech staff, Space Tech Labs, Inc, 63; res coord, John Muir Inst Environ Studies, 70-75; res assoc magnetohydrodynamics, 66-68, mem staff, 68-70, MEM STAFF, LOS ALAMOS SCI LAB, 75- *Mem:* AAAS. *Res:* Air contaminant dispersion and effects. *Mailing Add:* Rte 5 Box 229A Santa Fe NM 87501

WILLIAMS, MICHAEL EUGENE, b Ina, Ill, Aug 4, 40; m 68. VERTEBRATE PALEONTOLOGY. *Educ:* Mo Sch Mines & Metall, Rolla, BS, 63; Univ Kans, MS, 72; Univ Kans, PhD, 79. *Prof Exp:* CUR VERT PALEONT, CLEVELAND MUS NATURAL HIST, 76- *Mem:* Soc Vert Paleont; Paleont Soc; Int Paleont Union. *Res:* Paleozoic fishes with special emphasis on chondrichthyans; sedimentation and environments of deposition of various black shale units; cyclic events in the geological record. *Mailing Add:* Cleveland Mus Natural Hist Wade Oval University Circle Cleveland OH 44106

WILLIAMS, MICHAEL LEDELL, b Paragould, Ark, Sept 11, 43; m 63; c 2. ENTOMOLOGY, SYSTEMATICS. *Educ:* Ark State Univ, BS, 67; Va Polytech Inst & State Univ, MS, 69, PhD(entom), 72. *Prof Exp:* Asst entomologist, Md Dept Agr, 71-73; ASSOC PROF, DEPT ZOOL-ENTOM, AUBURN UNIV, 73-, ASSOC PROF, DEPT ENTOM, 86- *Concurrent Pos:* Chmn, Entom Sect, Auburn Univ, 83-86. *Mem:* Entom Soc Am; Sigma Xi. *Res:* Insular speciation of scale insects of the Galapagos Islands; systematics and morphology of New World Coccidae (Homoptera: Coccoidea); natural host plant resistance to scale insects; scale insects of Alabama; insects of ornamental plants. *Mailing Add:* Dept Entom Auburn Univ 328 Funchess Hall Auburn AL 36849-5413

WILLIAMS, MICHAEL MAURICE RUDOLPH, b Croydon, Eng, Dec 1, 35; Brit citizen; m 58; c 2. VIBRATIONS IN NUCLEAR SYSTEMS. *Educ:* Univ London, BSc, 58, PhD(nuclear eng), 62, DSc(nuclear eng), 68. *Prof Exp:* Res engr, nuclear eng, Cent Elec Generating Bd, 61-62; res assoc, Brookhaven Nat Lab, 62-63; lectr nuclear eng, Univ Birmingham, Eng, 63-65; reader nuclear eng, Univ London, Eng, 66-70, prof, 70-86, chmn, 80-86, EMER PROF, UNIV LONDON, ENG, 86-; PROF NUCLEAR ENG, UNIV MICH, 86-; PRIN SCIENTIST, ELECTROWALT ENG SERV (UK) LTD, 89- *Concurrent Pos:* Ed, Annals Nuclear Energy; mem, Adv Bd Safety of Nuclear Installations; nuclear engr, Univ Mich, 86-89. *Mem:* Fel Am Nuclear Soc; Inst Nuclear Engrs (vpres, 72-75); fel Inst Physics; fel Royal Soc Arts. *Res:* Neutron transport theory; random processes in nuclear systems; aerosol physics with special references to nuclear reactor safety; radioactive waste; Coagulation and deposition aerosols on closed compartments. *Mailing Add:* 2A Lytchgate Close South Croydon Surrey CR2 0DX England

WILLIAMS, MYRA NICOL, b Dallas, Tex, June 8, 41; m 68; c 2. MOLECULAR BIOPHYSICS, MOLECULAR MODELING. *Educ:* Southern Methodist Univ, BS, 64; Yale Univ, MS, 65, PhD(molecular biophys), 68. *Prof Exp:* Res fel biophys, 69-76, asst to pres, 76-78, dir sci planning, 78-80, sr dir sci & strategic planning, 80-85, EX DIR INFO RESOURCES & STRATEGIC PLANNING, MERCK SHARP & DOHME RES LABS, 85- *Mem:* Am Soc Biol Chemists; Drug Info Asn. *Res:* Structure and function of proteins; molecular modeling and strategic planning; computer and information systems for research and development. *Mailing Add:* Merck Sharp & Dohme Res Labs Rahway NJ 07065

WILLIAMS, NEAL THOMAS, b East Orange, NJ, Mar 16, 21; m 48; c 2. ENGINEERING PHYSICS. *Educ:* Cornell Univ, AB, 48. *Prof Exp:* Supvr magnetron eng, Westinghouse Elec Co, 42-44, develop engr, 48-51; res assoc, Radiation Lab, Columbia Univ, 44-48, assoc res physicist, 52; mem tech staff, Bell Tel Labs, Inc, 51-52; chief engr, L L Constantin & Co, 52-53; res engr, T A Edison, Inc, 53-60; chief engr, Seal-A-Metic, Inc, 60-65; div mgr, Platronics, Inc, 65-66; PRES, PLATRONICS-SEALS, INC, CLIFTON, 66- *Mem:* Am Phys Soc; sr mem Inst Elec & Electronics Engrs; NY Acad Sci. *Res:* Microwave magnetrons and electronics; traveling wave tubes and backward oscillators; radar duplexers; gas discharges; metal-ceramic seals; low voltage x-rays. *Mailing Add:* PO Box 1002 Five Skyline Dr Hopatcong NJ 07843

WILLIAMS, NOREEN, b Brunswick, Maine, June 21, 55; m 81; c 2. BIOENERGETICS, MOLECULAR BIOLOGY. *Educ:* Univ Maine Orono, BS, 77; NY Univ, PhD(biol & biochem), 81. *Prof Exp:* Fel, Sch Med, 81-84, RES ASSOC, MCCOLLUM PRATT INST, JOHNS HOPKINS UNIV, 84- *Mem:* Biophys Soc; NY Acad Sci; AAAS; US Bioenergetics Group. *Res:* Molecular cloning and mechanism studies of energy-linked systems; mitochondrial and bacterial AT Pases; enzyme II mannose of the phosphotransferase system in Escherichia coli. *Mailing Add:* Dept Biochem Uniformed Serv Univ Health Sci 4301 Jones Bridge Rd Bethesda MD 20814

WILLIAMS, NORMAN DALE, b Nebr, Nov 4, 24; m 47; c 2. PLANT GENETICS. *Educ:* Univ Nebr, BS, 51, MS, 54, PhD(agron), 56. *Prof Exp:* Assoc genetics, Argonne Nat Lab, 54-56, res assoc, 56; GENETICIST, SCI & EDUC-AGR RES SERV, USDA, 56- *Concurrent Pos:* Adj Prof, NDak State Univ, 61- *Mem:* Fel AAAS; Am Soc Agron; Am Genetic Asn; fel Crop Sci Soc Am; Genetics Soc Am. *Res:* Genetic studies of host-parasite relationships, especially wheat and wheat stem rust; mutation induction. *Mailing Add:* USDA Northern Crop Sci Lab Box 5677 University Sta Fargo ND 58105-5677

WILLIAMS, NORMAN EUGENE, b Grove City, Pa, July 29, 28; m 53, 72; c 3. CELL BIOLOGY. *Educ:* Youngstown Univ, AB, 52; Brown Univ, ScM, 54; Univ Calif, Los Angeles, PhD(zool), 58. *Prof Exp:* Instr, 57-59, from asst prof to assoc prof, 59-67, PROF BIOL, UNIV IOWA, 67- *Concurrent Pos:* NIH ser fel, Carlsberg Found, Denmark, 63-64 & Dept Biol Struct, Univ Wash, 66-67. *Mem:* Soc Protozool; Am Soc Cell Biol. *Res:* Cellular development; synthesis and assembly of cell surface. *Mailing Add:* Dept Biol Univ Iowa Iowa City IA 52240

WILLIAMS, NORMAN S W, Can citizen. ADVANCED LASER-BASED DIAGNOSTICS, ENVIRONMENTAL TECHNOLOGY. *Educ:* Univ Waterloo, BASc Hons, 74, MASc, 76; Univ BC, PhD(chem eng & fluid mech), 83. *Prof Exp:* Res scientist & engr, Dept Nat Defense Res Estab Valcartien, 83-87; PROF ENVIRON HEALTH & CHEM ENG, APPL SCI & ENG TECHNOL, SENECA COL APPL ARTS & TECHNOL, 88-; PROF SPECTROS & SPECTROMETRY, SOLID STATE PHYSICS, FAC APPL SCI & ENG, UNIV TORONTO, 90- *Concurrent Pos:* Dir, Eng & Sci Advan Technol Consult Serv, 87-; prog consult, Continuing Educ Div, Technol Dept, Seneca Col Appl Arts & Technol, 90- *Mem:* Asn Prof Engrs Ont. *Res:* Advanced laser-based diagnostic techniques; aerospace biomedical and engineering; fluid dynamics; combustion processes. *Mailing Add:* 37 Barrymore Rd Scarborough ON M1J 1W1 Can

WILLIAMS, NORRIS HAGAN, b Birmingham, Ala, Mar 31, 43; m 70; c 2. PLANT TAXONOMY, CHEMICAL ECOLOGY. *Educ:* Univ Ala, BS, 64, MS, 67; Univ Miami, PhD(biol), 71. *Prof Exp:* From asst prof to assoc prof biol, Fla State Univ, 73-81; assoc cur, 81-84, CUR VASCULAR PLANTS, UNIV FLA, 84-, CHMN, NATURAL SCI, 85- *Mem:* Am Chem Soc; Am Soc Plant Taxonomists; Soc Study Evolution; Asn Trop Biol; Int Asn Plant Taxon; Int Soc Chem Ecol; Am Orchid Soc. *Res:* Systematics and evolution of Orchidaceae; chemical attraction of insects to flowers. *Mailing Add:* Fla Mus Natural Hist Univ Fla Gainesville FL 32611-2035

WILLIAMS, OREN FRANCIS, inorganic chemistry; deceased, see previous edition for last biography

WILLIAMS, OWEN WINGATE, geodesy, geophysics, for more information see previous edition

WILLIAMS, PATRICIA BELL, b Detroit, Mich. PHARMACOLOGY. *Educ:* Col Pharm, Univ Mich, BS, 68; Med Col Va, Va Commonwealth Univ, PhD(pharmacol), 72. *Prof Exp:* Lab asst pharmacol, Health Sci, Med Col Va, Va Commonwealth Univ, 68-70, teaching asst, 70-71; asst prof, 72-78, assoc prof, 78-86, PROF PHARMACOL, EASTERN VA MED SCH, 86- *Concurrent Pos:* Consult, United Drug Abuse Coun & Health Adv Coord Comt, Model Cities Comprehensive Health Sci Proj, 72; lectr, Sch Continuing Educ, Univ Va, 72; asst prof nursing & dent hyg, 72-74, assoc prof chem sci, Old Dom Univ, 79-; Tidewater Heart Asn res grant, 75; Am Heart Asn/Va Affil res grant, 76-77 & 87-; Nat Inst Heart Lung & Blood Inst grant, 76-82; res grants, Am Heart Asn, 83-87, Lions Eye Bank & Res Ctr, 84-, Alzheimer's Res Found Tidewater, 85-89; Basic Sci Coun, Am Heart Asn; Alzheimer's & Related Dis Res Fund/Commonwealth Va, 86; Alzheimer's & Related Dis Assoc/Nat Hq, 86-90. *Mem:* Fel Am Col Clin Pharmacol; Asn Women in Sci; Am Fedn Clin Res; Am Heart Asn; Am Soc Pharmacol & Exp Therapeut; Soc Clin Trials. *Res:* Cardiovascular pharmacology and physiology of vascular smooth muscle with particular interest in the etiology and treatment of hypertension, peripheral vascular disease and the cell biology of vascular endothelium and calcium-mediated events. *Mailing Add:* Dept Pharmacol Eastern Va Med Sch PO Box 1980 Norfolk VA 23501

WILLIAMS, PATRICK KELLY, b San Angelo, Tex, July 31, 43; m 68. ECOLOGY. *Educ:* Univ Tex, Austin, BA, 66; Univ Minn, Minneapolis, MS, 69; Ind Univ, Bloomington, PhD(zool), 73. *Prof Exp:* Asst prof, 73-80, ASSOC PROF BIOL, UNIV DAYTON, 80- *Mem:* Ecol Soc Am; Am Soc Mammalogists; Am Soc Ichthyologists & Herpetologists; Sigma Xi. *Res:* Experimental population ecology on rodents with emphasis on natural regulation and management. *Mailing Add:* Dept Biol Univ Dayton 300 College Park Dayton OH 45469

WILLIAMS, PAUL HUGH, b Vancouver, BC, May 6, 38; m 63. PLANT PATHOLOGY, PLANT GENETICS. *Educ:* Univ BC, BSA, 59; Univ Wis, PhD(plant path), 62. *Prof Exp:* From asst prof to assoc prof, 62-71, PROF PLANT PATH, UNIV WIS-MADISON, 71- *Concurrent Pos:* J S Guggenheim fel, 77-78. *Honors & Awards:* Jakob Eriksson Medal, Swed Acad Sci, 81. *Mem:* Am Phytopath Soc; Am Genetics Asn; Am Soc Hort Sci. *Res:* Genetics and cytology of host-parasite relations and resistance breeding for disease resistance in vegetables; crucifer genetics. *Mailing Add:* Five Straubel Ct No 508 Madison WI 53704

WILLIAMS, PETER J, b Croydon, Eng, Sept 27, 32; m 57; c 3. SOIL PHYSICS, GEOTECHNICAL SCIENCE. *Educ:* Cambridge Univ, BA, 54, MA, 58; Univ Stockholm, Fil Lic & Fil Dr, 69. *Prof Exp:* Res off soil mech sect, Div Bldg Res, Nat Res Coun Can, 57-69; assoc prof geog, 69-71, PROF GEOG, CARLETON UNIV, 71-, DIR GEOTECH SCI LABS, 79- *Concurrent Pos:* Royal Norweg Coun Sci & Indust Res fel, Norweg Geotech Inst, 63-65; lectr several univs in UK, Sweden, Norway, US & Can; consult & geotech adv, Northern Pipelines, Can Govt; vis scholar, Scott Polar Res Inst, Univ Cambridge, UK, 75-76, 82-83 & 89-90. *Mem:* Can Asn Geog; Norweg Geotech Soc; Can Geotech Soc. *Res:* Physics of freezing soils and application to engineering, especially Northern pipelines; geomorphology, especially frost action; author of textbooks and general interest book. *Mailing Add:* Geotech Sci Labs Carleton Univ Ottawa ON K1S 5B6 Can

WILLIAMS, PETER M, b New York, NY, July 17, 27; m 58; c 3. CHEMICAL OCEANOGRAPHY. *Educ:* Washington & Lee Univ, BS, 49; Univ Calif, Los Angeles, MS, 58, PhD(oceanog), 60. *Prof Exp:* Asst res chemist, Smith, Kline & French Labs, 49-51; lab technician, Citrus Exp Sta, Univ Calif, Riverside, 53-54; asst prof marine chem, Inst Oceanog, Univ BC, 60-63; from asst to assoc res chemist, 63-76, RES CHEMIST, INST MARINE RESOURCES, UNIV CALIF, SAN DIEGO, 76- *Mem:* Am Geophys Union; Am Soc Limnol & Oceanog. *Res:* Organic and carbon isotope chemistry of sea water with respect to dissolved and particulate organic matter derived from marine organisms. *Mailing Add:* Scripps Inst Oceanogr Univ Calif San Diego La Jolla CA 92093-0218

WILLIAMS, PHILIP CARSLAKE, b Mountain Ash, Wales, May 26, 33; Can citizen; m 70; c 4. ANALYTICAL CHEMISTRY, CEREAL PULSE UTILIZATION & EVALUATION. *Educ:* Univ Wales, BS, 54, PhD(agr biochem), 58. *Prof Exp:* Res officer, Agr Res Inst, Wagga Wagga, NSW, 58-64; fel cereal chem, Nat Res Coun Can, 64-65; RES SCIENTIST, GRAIN RES LAB, CAN GRAIN COMN, 65- *Concurrent Pos:* Consult, Int Ctr Agr Res Dry Areas, Syria, 75-, Int Develop Res Ctr, 74 - *Mem:* Am Asn Cereal Chemists. *Res:* Near-infrared reflectance spectroscopic analysis of cereal grains, oilseals, pulses, and derived products; planning and design of large-scale analytical operations; applied statistical analysis; imunograph; 117 scientific ppublications. *Mailing Add:* Grain Res Lab 1404-303 Main St Winnipeg MB R3C 3G8 Can

WILLIAMS, PHLETUS P, b Junior, WVa, Aug 3, 33; m 60; c 3. MICROBIOLOGY, BIOCHEMISTRY. *Educ:* Davis & Elkins Col, BS, 55; Univ Md, MS, 59; NDak State Univ, PhD(animal nutrit), 68. *Prof Exp:* Microbiologist beef cattle res br, Animal Husb Res Div, Md, 59-60, dairy cattle res br, 60-61, beef cattle res br, 61-64, prof bact, 72-73, MICROBIOLOGIST, METAB & RADIATION RES LAB, NDAK STATE UNIV, SCI & EDUC ADMIN-AGR RES, USDA, 64- ADJ PROF BACT, 73- *Mem:* AAAS; Am Soc Animal Sci; Am Soc Microbiol; Brit Soc Gen Microbiol. *Res:* Development of rumen protozoal controlled bovines; chemical, physiological, cultural and metabolical study of rumen bacteria and protozoa; microbial metabolic fate studies with lipoidal and pesticidal compounds. *Mailing Add:* Nat Animal Dis Ctr Ames IA 50010

WILLIAMS, RALPH C, JR, b Washington, DC, Feb 17, 28; m 51; c 4. INTERNAL MEDICINE, IMMUNOLOGY. *Educ:* Cornell Univ, AB, 50, MD, 54. *Prof Exp:* Guest investr immunol, Rockefeller Inst, 61-63; from asst prof to prof med, Med Sch, Univ Minn, Minneapolis, 63-69; PROF MED & CHMN DEPT, SCH MED, UNIV NMEX, 69- *Concurrent Pos:* Consult, Bur Hearings & Appeals, Soc Security Admin, 65- *Mem:* Am Rheumatism Asn; Am Fedn Clin Res; Am Soc Clin Invest; Am Asn Immunol; Soc Exp Biol & Med; Am Chem Soc. *Res:* Rheumatic diseases; immunopathology; immunoglobulin abnormalities and their relation to disease. *Mailing Add:* Dept Med Univ Fla CG91 JHMHC Gainesville FL 32610

WILLIAMS, RALPH EDWARD, b Ontario, Ore, July 20, 43; m 63. PHYTOPATHOLOGY. *Educ:* Univ Idaho, BS, 65, MS, 69; Wash State Univ, PhD(plant path), 72. *Prof Exp:* Res asst plant path, Dept Plant Sci, Univ Idaho, 65-66, Wash State Univ, 66-68; res plant pathologist, Forest Serv, USDA, 67-70; weed control specialist, Latah County, Idaho, 70; PLANT PATHOLOGIST, FOREST PEST MGT, FOREST SERV, USDA, 70- *Mem:* Optical Soc Am. *Res:* Developed growth impact methodology; conducted survey and analyzed survey data for root disease centers in forests of northern Idaho and western Montana; developed models for root disease center occurrence; designed and established root disease management evaluations. *Mailing Add:* 507 Goodwin Dr Richardson TX 75081

WILLIAMS, RAY CLAYTON, b Louisville, Ky, July 17, 44. PERIODONTOLOGY, MICROBIOLOGY. *Educ:* Samford Univ, AB, 66; Univ Ala, Birmingham, DMD, 70; Harvard Univ, cert periodont, 73. *Prof Exp:* Res fel periodont, Sch Dent Med, Harvard Univ, 70-73; res fel microbiol, Forsyth Dent Ctr, Boston, Mass, 70-74; from instr to asst prof, 74-85, actg chmn dept, 82-83, ASSOC PROF PERIODONT, SCH DENT MED, HARVARD UNIV, 85-, CHMN DEPT, 83- *Concurrent Pos:* Consult, WRoxbury Vet Admin Hosp, Mass, 77-80, attend physician, 80-; consult, Children's Hosp Med Ctr, 81- *Mem:* Am Acad Periodont; Int Asn Dent Res; Am Soc Microbiol. *Res:* Pharmacologic interception of periodontal diseases; nuclear medicine. *Mailing Add:* Dept Periodont Harvard Sch Dent Med 188 Longwood Ave Boston MA 02115

WILLIAMS, RAYMOND CRAWFORD, b Kansas City, Mo, Sept 22, 24; m 59; c 1. VETERINARY ANATOMY. *Educ:* Kans State Col, DVM, 46; Cornell Univ, MS, 55, PhD, 61. *Prof Exp:* Instr, 46-54, from asst prof to assoc prof, 54-64, PROF ANAT & HISTOL, SCH VET MED, TUSKEGEE UNIV, 64- *Concurrent Pos:* Vis prof, Cornell Univ, 75; external examr, Univ Ibadan, Nigeria, 76-77; Fulbright fel, Life Adv Comt, 82-85. *Mem:* AAAS; Am Vet Med Asn; Am Asn Vet Anat; World Asn Vet Anat; Southern Soc Anatomists. *Res:* Descriptive vertebrate anatomy; fetal size and age relationships; dentition development; anatomical museum methods. *Mailing Add:* Dept Anat Sch Vet Med Tuskegee Univ Tuskegee AL 36088

WILLIAMS, REDFORD BROWN, JR, b Raleigh, NC, Dec 14, 40; m 63; c 2. BEHAVIORAL MEDICINE, PSYCHOSOMATIC MEDICINE. *Educ:* Harvard Univ, AB, 63; Yale Univ Sch Med, MD, 67. *Prof Exp:* Intern & resident internal med, Yale-New Haven Hosp & Med Ctr, 67-70; clin assoc clin psychophysiol, NIH, 70-72; from asst prof psychiat & med to assoc prof psychiat, 72-78, PROF PSYCHIAT, DUKE UNIV MED CTR, 78-, DIR, BEHAV MED RES CTR, 85- *Concurrent Pos:* Assoc prof med, Duke Univ Med Ctr, 81-; prin investr, NIMH res scientist develop award, 74-84 & 84-89, Nat Heart Lung & Blood Inst, 76-91; mem Behav Med Study Sect, NIH, 79-82; consult, President's Biomed & Behav Res Panel, 75-76; lectr, Japanese Psychosomatic Soc, 83; vis scientist, USSR Cardiol Res Ctr, 87. *Mem:* Fel Soc Behav Med (pres, 83-84); Am Psychosomatic Soc; Acad Behav Med Res; Am Col Neuropsychopharmacol; Soc Psychophysiol Res; Am Heart Asn. *Res:* Behavioral medicine, identification of biobehavioral factors that play a role in the etiology , pathogenesis and course of coronary heart disease, with particular emphasis on type A behavior, hostility and anger. *Mailing Add:* Box 3926 Duke Univ Med Ctr Durham NC 27710

WILLIAMS, REED CHESTER, b Chicago, Ill, June 10, 41. ANALYTICAL CHEMISTRY. *Educ:* Lawrence Univ, BA, 63; Univ Wash, PhD(chem), 68. *Prof Exp:* RES CHEMIST, E I DU PONT DE NEMOURS & CO, INC, 68- *Mem:* Am Chem Soc; Sigma Xi. *Res:* Analytical chemistry; application of high speed liquid column chromatography to the separation and quantitation of complex mixtures. *Mailing Add:* 51 Fox Den Rd Newark DE 19898

WILLIAMS, RICHARD, b Chicago, Ill, Aug 5, 27; m 61; c 3. PHYSICAL CHEMISTRY. *Educ:* Miami Univ, AB, 50; Harvard Univ, PhD(phys chem), 54. *Prof Exp:* Instr chem, Harvard Univ, 55-58; MEM TECH STAFF, RCA LABS, 58- *Concurrent Pos:* Fulbright lectr, Sao Carlos Sch Eng, 69. *Honors & Awards:* Callinan Prize, Electrochem Soc. *Mem:* Fel Am Phys Soc; Brazilian Acad Sci. *Res:* Electrical properties of insulators; liquid crystals; luminescence of organic molecules; physical chemistry of surfaces. *Mailing Add:* David Sarnoff Res Ctr Princeton NJ 08543-5300

WILLIAMS, RICHARD ALVIN, b Canton, Ohio, July 21, 36; m 72; c 4. ELECTRICAL ENGINEERING, COMPUTER SCIENCE. *Educ:* Ohio State Univ, BSEE, 59, MSc, 61, PhD(elec eng), 65. *Prof Exp:* Staff mem reliability eng, Sandia Corp, NMex, 59-60; assoc supvr Electrosci Lab, Ohio

State Univ, 60-68, asst prof elec eng, 66-68; eng specialist, goodyear Aerospace Corp, 79-86; ASSOC PROF ELEC ENG, UNIV AKRON, 66-; ENG SPECIALIST, LORAL CORP, 87- *Res:* Communications; communication satellite systems; computers, simulation programming and application; environmental engineering; electric power systems; transportation; aerospace electronics; geographic information systems; geophysical data bases. *Mailing Add:* Dept Elec Eng Univ Akron Akron OH 44325.

WILLIAMS, RICHARD ANDERSON, b Akron, Ohio, July 21, 31; m 51; c 5. ADVANCED STRUCTURAL MATERIALS & COMPOSITES. *Educ:* Wabash Col, AB, 53; Univ Rochester, PhD(chem), 57. *Prof Exp:* Res chemist, Patent Div, 56-60, sr res chemist, Nylon Tech Div, 60-68, supvr res, Qiana Tech Div, 68-71, supvr res & develop, Orlon-Lycra Tech Div, 71-72, patent supvr, Patent Liaison Div, Textile Fibers Dept, 72-77, develop assoc, Carpet Fibers Tech Div, 77-84, patent assoc, 84-86, SR PATENT ASSOC, TEXTILE FIBERS DEPT, E I DU PONT DE NEMOURS & CO, 87- *Res:* Olefin-forming elimination reactions; polymer chemistry; synthetic fibers and applications. *Mailing Add:* 819 Morris Rd Hockessin DE 19707

WILLIAMS, RICHARD JOHN, b Hazleton, Pa, May 24, 44; m 66. SCIENCE ADMINISTRATION, PLANETARY & EARTH SCIENCES. *Educ:* Lehigh Univ, BA, 66; Johns Hopkins Univ, MA, 68, PhD(geochem), 70; Univ Houston, MA (pub admin), 83. *Prof Exp:* Space scientist lunar studies, 70-73, sr space scientist lunar and planetary studies, 73-78, supvry space scientist, 79-87, SR OPER MGT ENGR, NASA HQ, 88- *Honors & Awards:* Sigma Xi. *Mem:* Am Inst Aeronaut & Astronaut; Am Geophys Union; AAAS; Nat Mgt Asn. *Res:* Theoretical and experimental petrology; space industrialization; science and technology policy. *Mailing Add:* 11907 Winterthur No 105 Reston VA 22091-1954

WILLIAMS, RICHARD KELSO, b Chattanooga, Tenn, Oct 20, 38; m 66. MATHEMATICS. *Educ:* Vanderbilt Univ, BA, 60, MA, 62, PhD(math), 65. *Prof Exp:* From asst prof to assoc prof, 65-77, chmn dept, 78-80, PROF MATH, SOUTHERN METHODIST UNIV, 77- *Mem:* Math Asn Am; Am Math Soc. *Res:* Complex function theory; topology. *Mailing Add:* Dept Math Southern Methodist Univ Dallas TX 75275

WILLIAMS, RICHARD STANLEY, b Kodiak, Alaska, Oct 27, 51; m. SURFACE SCIENCE, THIN-FILM MATERIALS. *Educ:* Rice Univ, BA, 74; Univ Calif, Berkeley, MS, 76, PhD(chem), 78. *Prof Exp:* Mem tech staff, AT&T Bell Labs, 78-80; from asst prof to assoc prof, 80-86, PROF PHYS CHEM, UNIV CALIF, LOS ANGELES, 86- *Concurrent Pos:* Consult ed, Chem Physics Lett, 86-, Chem Mats, 88-90; mem Defense Study Group, Inst Defense Anal, Washington, DC, 86-89; guest researcher, Frontier Mat Res Prog, Rikagaku Kenkyusho (Inst Physics & Chem Res Japan) Waho-shi, Saitania, Japan, 87-; res award, Camille & Henry Dreyfus Found, 83, Alfred P Sloan Found, 84. *Mem:* Am Chem Soc; Am Phys Soc; Am Vacuum Soc; Mat Res Soc; Int Union Pure & Appl Chem. *Res:* Chemistry of solid surfaces and interfaces with the goal of creating novel materials systems with unusual and/or useful electrical, optical, mechanical or chemical properties. *Mailing Add:* Dept Chem Univ Calif Los Angeles CA 90024-1569

WILLIAMS, RICHARD SUGDEN, JR, b New York, NY, Dec 6, 38; m 60; c 2. SATELLITE GLACIOLOGY, PLANETARY VOLCANIC GEOMORPHOLOGY. *Educ:* Univ Mich, Ann Arbor, BS, 61, MS, 62; Pa State Univ, PhD(geol), 65. *Prof Exp:* Proj scientist geol, Air Force Cambridge Res Labs, 65-68, res geologist, 68-69, br chief, 69-71; GEOLOGIST, US GEOL SURV, 71- *Concurrent Pos:* Assoc ed, J Photogram Eng & Remote Sensing, 76-77; 2nd dep, 1st dep & dir, Remote Sensing Appln Div, Am Soc Photogram & Remote Sensing, 76-79; expert consult, World Glacier Monitoring Serv, UNESCO, 82-; sci corresp, Dagens Nyheter, Stockholm, Sweden, 83-; mem, Gov Coun, Int Glaciological Soc, 83-87; chief sci ed, Annals of Glaciol, 86-87; mem, Joint Satellite Mapping & Remote Sensing Comt, Am Soc Photogram Eng & Remote Sensing, Am Congr Surv & Mapping, 87-; chair, Solid Earth Processes Task Group, 87-91; prog leader, Sea Level Change Integrating Theme, 90-91; mem, US Global Change Res Prog, Working Group Global Changes, Comt Earth & Environ Sci, Fed Coord Coun Sci Educ & Technol, Off Sci & Tech Policy & Exec Off Pres; mem, Earth Sci Educ, K-12, Framework Steering Comt, Am Geol Inst, 88- & Comt Res & Explor, Nat Geol Surv, 90-; mem-at-large, Sect E, Geol & Geog, AAAS, 89- *Honors & Awards:* Alan Gordon Mem Award, Am Soc Photogram, 78. *Mem:* Fel Geol Soc Am; fel Iceland Sci Soc; Am Geophys Union. *Res:* Satellite and aerial remote sensing of dynamic geomorphic processes; volcanoes and glaciers with particular emphasis on Iceland; landsat data to monitor global changes on the Earth's surface; author of over 150 publications. *Mailing Add:* US Geol Surv Br Atlantic Marine Geol Quissett Campus Woods Hole MA 02543

WILLIAMS, RICHARD TAYLOR, b Tarboro, NC, May 27, 46. SOLID STATE PHYSICS. *Educ:* Wake Forest Univ, BS, 68; Princeton Univ, MA, 71, PhD(physics), 74. *Prof Exp:* PHYSICIST, NAVAL RES LAB, 69- *Mem:* Am Phys Soc; AAAS. *Res:* Effects of ionizing radiation in insulating solids, particularly time-resolved studies of exciton self-trapping and defect formation in halide crystals; vacuum-ultraviolet spectroscopy of solids. *Mailing Add:* Dept Physics Wake Forest Univ Box 7507 Reynolds Sta Winston-Salem NC 27109

WILLIAMS, RICKEY JAY, b Muskogee, Okla, May 13, 42. PHYSICAL INORGANIC CHEMISTRY. *Educ:* Tex Christian Univ, BA, 64, MD, PhD(phys chem), 68. *Prof Exp:* Fel Los Alamos Sci Lab, 68-70 & Baylor Univ, 70-71; asst prof, 71-74, assoc prof, 74-80, PROF & HEAD DEPT CHEM, PHYSICS GEOL & GEOPHYSICS, MIDWESTERN STATE UNIV, 80- *Mem:* Am Crystallog Asn; Am Chem Soc; Sigma Xi. *Res:* Crystal structure studies of inorganic compounds. *Mailing Add:* 2417 Fain St Wichita Falls TX 76308

WILLIAMS, ROBERT ALLEN, b Cleveland, Ohio, Apr 25, 45. NUCLEAR CHEMISTRY. *Educ:* Oberlin Col, BA, 66; Carnegie-Mellon Univ, MS, 69, PhD(nuclear chem), 72. *Prof Exp:* Res assoc nuclear chem, 72-73, presidential intern, 73-74, STAFF MEM NUCLEAR CHEM, LOS ALAMOS SCI LAB, UNIV CALIF, 74- *Mem:* Am Chem Soc; Am Phys Soc; Sigma Xi. *Res:* Pionic nuclear reactions; neutron activation analysis; nuclear spectroscopy; computer applications; data acquisition software systems. *Mailing Add:* 1063 48th St Los Alamos NM 87544

WILLIAMS, ROBERT CALVIN, b Key West, Fla, May 1, 44; m 69; c 1. ANALYTICAL CHEMISTRY, PHYSICAL CHEMISTRY. *Educ:* Univ Kans, BS, 66; Univ Wis-Madison, PhD(phys chem), 72. *Prof Exp:* Instr chem, Univ Nebr-Lincoln, 72-74; sr chemist, Cent Res Labs, 3M Co, 74-79; res specialist, 79, sr res & develop chemist, 79-83, RES & DEVELOP ASSOC, BF GOODRICH CHEM GROUP, 83- *Concurrent Pos:* Res assoc, Univ Nebr-Lincoln, 72-74. *Mem:* Am Chem Soc; Am Phys Soc; Soc Appl Spectros; Coblentz Soc; Am Soc Testing Mat. *Res:* Analytical applications of fourier transform infrared spectroscopy; computer-coupled instrumentation and instrumental methods of analysis; molecular specroscopy, particularly infrared and mass spectroscopy; infrared normal coordinate analysis; polymer characterization. *Mailing Add:* Avon Lake Tech Ctr BF Goodrich Chem Group PO Box 122 Avon Lake OH 44012-1195

WILLIAMS, ROBERT DEE, b Wingate, Ind, June 4, 32; m 60; c 2. VETERINARY MEDICINE, MICROBIOLOGY. *Educ:* Wabash Col, AB, 54; Purdue Univ, DVM, 65, MS, 68, PhD(microbiol), 71. *Prof Exp:* Sanitarian, Ind State Bd Health, 56-60; vet, Food & Drug Admin, 71-75; res vet, Com Solvents Corp, 75-76; DIR ANIMAL SCI RES, INT MINERALS & CHEM CORP, 76- *Concurrent Pos:* Am Vet Med Asn fel, Purdue Univ, 65-66, NIH fel, 67-70; dipl, Am Col Vet Microbiologists, 72. *Mem:* Am Vet Med Asn; Am Asn Indust Vet; Am Soc Microbiol; Int Soc Ecotoxicol & Environ Safety. *Res:* New drugs for animals; toxicology. *Mailing Add:* Rte 25 Box 67 Terre Haute IN 47805

WILLIAMS, ROBERT ELLIS, b Grenada, Miss, Apr 9, 30; m 67; c 3. BIOLOGY. *Educ:* Memphis State Univ, BA, 56; Ohio State Univ, MA, 57, PhD, 61. *Prof Exp:* Res assoc, Ohio State Res Found, 60; entomologist, Army Biol Labs, 61-63, chief insect biol res br, 63-65; asst head microbiol br, Off Naval Res, 65-66, prof officer biol countermeasures, 66-67, head virol dept, Naval Med Res Unit 3, Cairo, Egypt, 68-73; PATRON, ANTIOCH COL, 73- *Mem:* Am Soc Trop Med & Hyg; Entom Soc Am; Am Mosquito Control Asn. *Res:* Entomology; arbovirology; ecology. *Mailing Add:* 437 Fox Ridge Dr Leesburg VA 22075

WILLIAMS, ROBERT FONES, b Bessemer, Ala, July 27, 28; div; c 1. TOPOLOGY. *Educ:* Univ Tex, BA, 48; Univ Va, PhD(math), 54. *Prof Exp:* Asst prof math, Fla State Univ, 54-55; vis lectr, Univ Wis, 55-56; asst prof, Purdue Univ, 56-59; NSF fel & mem, Inst Adv Study, 59-61; asst prof, Univ Chicago, 61-63; from asst prof to prof math, Northwestern Univ, Evanston, 63-87; PROF MATH, UNIV TEX, AUSTIN, 87- *Concurrent Pos:* NSF grant, Univ Geneva, 68-69 & Inst Advan Sci Study, Bures-sur-Yvette, France, 70, 72-73. *Mem:* Am Math Soc; Math Asn Am; AAAS; Coun Am Math Soc. *Res:* Transformation groups; topological dynamics; global analysis; differentiable dynamical systems. *Mailing Add:* Dept Math Univ Tex Austin TX 78712

WILLIAMS, ROBERT GLENN, b Teaneck, NJ, Oct 14, 37; m 65; c 3. OCEANOGRAPHY. *Educ:* NY Univ, BA, 60, MS, 65, PhD(oceanog), 71. *Prof Exp:* Asst res scientist oceanog, NY Univ, 60-61 & 63-65; oceanogr, US Naval Underwater Systs Ctr, New London, Conn, 65-72; prog coordr oceanog, Nat Oceanic & Atmospheric Admin, Rockville, Md, 72-74, dep chief scientist, ship oceanogr, 74, oceanogr, Ctr Environ Assessment Serv, 74-81, phys scientist, Eng Develop Off, 81-83. *Concurrent Pos:* Instr environ sci, Frederick Community Col, Md, 81. *Mem:* Am Geophys Union; Am Soc Limnol & Oceanog. *Res:* Air-sea interaction and dynamic studies of the tropical oceans, and continental shelves, with emphasis on exchange of heat and moisture with the atmosphere on time scales of several weeks; surface wind wave measurement system analysis; underwater acoustics; development of acoustic ocean current measurement systems. *Mailing Add:* 4321 Moxley Valley Dr Mt Airy MD 21771

WILLIAMS, ROBERT HACKNEY, b Providence, RI, Jan 3, 15; m 42; c 2. ORGANIC CHEMISTRY. *Educ:* Univ NC, AB, 35, MA, 37; Temple Univ, PhD(chem), 53. *Prof Exp:* Res chemist, Mobil Res & Develop Lab, 38-42, sr res chemist, Cent Res Div Lab, 46-72, asst to admin mgr, Cent Res Div, 72-75; RETIRED. *Concurrent Pos:* VChmn, Punta Gorda-Charlotte Water & Sewer Bd, 78-81; mem water adv bd, Southwest Fla Regional Planning Coun, 78-81. *Mem:* Am Chem Soc; Am Inst Chem; Sigma Xi. *Res:* Petroleum additives; hydrocracking; lube oil manufacture and composition; radiation chemistry of hydrocarbons; application of nuclear radiation to petroleum processing; radiation and photochemical induced organic chemical reactions; oxidation of hydrocarbons. *Mailing Add:* 23033 Westchester Blvd Apt F-515 Port Charlotte FL 33980-8468

WILLIAMS, ROBERT HAWORTH, b Seattle, Wash, Mar 31, 14; m 39, 54; c 3. PLANT PHYSIOLOGY. *Educ:* Univ Wash, Seattle, BS, 35, MS, 40; Cornell Univ, PhD(plant physiol), 41. *Prof Exp:* Asst bot, Cornell Univ, 36-38, instr, 38-41; from asst prof, Univ Miami, Fla, 41-54, prof marine biol, 50-54, asst dir marine lab, 42-51; prof marine sci, Fla State Univ, 54-55; prof, 55-80, EMER PROF BIOL, UNIV MIAMI, 80- *Concurrent Pos:* Asst, US Frozen Pack Lab, Wash, 35. *Mem:* Am Soc Limnol & Oceanog; Phycol Soc Am; Sigma Xi. *Res:* Carbon dioxide absorption of roots; carbohydrates of the large brown algae; physiology of fern spore germination; ecology of marine algae of Florida. *Mailing Add:* 6790 SW 71st Ct Miami FL 33143

WILLIAMS, ROBERT J(AMES), b Iron Mountain, Mich, Sept 12, 23; m 47; c 4. INDUSTRIAL ENGINEERING. *Educ:* Mich State Univ, BSME, 47, MSME, 53. *Prof Exp:* Instr mech eng, Univ Colo, 48-51; sr indust engr, Boeing Airplane Co, 51-54; from asst prof to assoc prof mech eng, Univ Colo, 52-61, chmn dept, 56-61; indust engr, Fry & Assocs, 61-63; assoc prof, 63-77, PROF ENG DESIGN & ECON EVAL, UNIV COLO, BOULDER, 77- *Concurrent Pos:* Consult, Vet Admin Hosp, Seattle, 53, Boeing Airplane Co, 58, Mountain States Tel & Tel Co, 59-, Babcock & Wilcox Co, Ohio, 60 & Govt of India, 61-63. *Mem:* Am Soc Mech Engrs; Am Soc Eng Educ; Am Inst Indust Engrs; Opers Res Soc Am. *Res:* Process engineering; engineering economics. *Mailing Add:* Dept Mech Eng Univ Colo Boulder CO 80309

WILLIAMS, ROBERT JACKSON, b Iowa City, Iowa, June 13, 31; m 74; c 3. PHYSIOLOGY, BIOPHYSICS. *Educ:* Univ Wis-Madison, BA, 53, MA, 61; Univ Md, PhD(zool), 68. *Prof Exp:* Cryobiologist, Am Found Biol Res, Wis, 60-62; biologist, Naval Med Res Inst, Md, 63-67; res scientist cryobiol, Blood Res Lab, 67-78, SR RES SCIENTIST, TRANSPLANTATION LAB, AM NAT RED CROSS, 78- *Mem:* AAAS; Am Physiol Soc; Soc Cryobiol; Am Soc Plant Physiol; Biophys Soc; NAm Thermal Anal Soc. *Res:* Mechanisms of cell freezing injury; effects of low temperature on geographical distribution; instrumentation. *Mailing Add:* Holland Res & Develop Lab Am Red Cross 15601 Crabbs Branch Way Rockville MD 20855

WILLIAMS, ROBERT K, b Ft Worth, Tex, Jan 6, 28; m 52; c 3. ENTOMOLOGY, CELL BIOLOGY. *Educ:* Agr & Mech Col Tex, BS, 48, MS, 56, PhD(entom), 59. *Prof Exp:* Asst county agent in training, Agr Exten Serv, Agr & Mech Col, Tex, 48, asst county agent, Agr Exten Serv & Wood County, 48, Agr exten Serv & Bowie County, 50 & Agr Exten Serv & Eastland County, 52-54, res asst entom, Col, 54-58; PROF BIOL, ETEX STATE UNIV, 58- *Concurrent Pos:* NSF Col Sci Improv Prog grant, Dept Physiol Chem, Univ Wis, 69-70. *Res:* Physiology; cell physiology; Rotateria. *Mailing Add:* Dept Biol ETex State Univ ETex Station Commerce TX 75428

WILLIAMS, ROBERT L, b Buffalo, NY, July 22, 22; m 49; c 1. PSYCHIATRY, NEUROLOGY. *Educ:* Alfred Univ, BA, 44; Albany Med Col, Union Univ, NY, MD, 46. *Prof Exp:* Chief, Air Force Neurol Ctr, Lackland AFB Hosp, 52-55, chief neuropsychiat serv, 53-55, chief consult, Off Surgeon Gen, USAF, 55-58; from assoc prof to prof psychiat & neurol, Col Med, Univ Fla, 58-72, chmn dept psychiat, 64-72; actg chmn, Neural Dept, 76-77, chmn, Dept Psychiat, 72-90, PROF PSYCHIAT, BAYLOR COL MED, 72-, PROF NEUROL, 76- *Concurrent Pos:* Mem, Nat Adv Ment Health Coun & Nat Adv Neurol Dis & Blindness Coun, 55-58; Fla rep, Comn Ment Illness, Southern Regional Educ Bd, 64-72, chmn, 71-72; psychiat consult, Indust Security Prog, Dept Defense & consult psychiat & neurol, Surgeon Gen, USAF, 66-; pres, Benjamin Rush Soc, 86-88; mem, Accreditation Coun Grad Med Educ, Residency Rev Comt, Psychiatry, 88. *Honors & Awards:* E B Bowis Award, Am Col Psychiatrists. *Mem:* Am Asn Chmn Depts Psychiat (pres, 84-85); Am Med Asn; fel Am Psychiat Asn; fel Am Acad Neurol; Am Electroencephalog Soc; fel Am Col Psychiatrists (pres, 82-83). *Res:* Psychophysiology of sleep; medical education. *Mailing Add:* Dept Psychiat Baylor Col Med 6560 Fannin Suite 832 Houston TX 77030

WILLIAMS, ROBERT LEROY, b St Thomas, Ont, July 9, 28; m 51; c 3. PHYSICS. *Educ:* Univ Western Ont, BSc, 51; Univ BC, MA, 52, PhD(physics), 56. *Prof Exp:* Sci officer, Defense Res Bd Can, 55-59; group leader physics, RCA Victor Co, 59-63; chief physicist, Simtec Ltd, 63-65, gen mgr, 65-66; mem sci staff physics, 66-69, br mgr detector develop, 69-80, SR MEM TECH STAFF, INFRARED DETECTOR DEVELOP, TEX INSTRUMENTS INC, 80- *Mem:* Am Phys Soc; sr mem Inst Elec & Electronics Eng. *Res:* Physics of infrared and nuclear particle detectors; development of detectors; diode, transistor and bulk structures. *Mailing Add:* MS 28 13532 N Central Expressway Dallas TX 75265

WILLIAMS, ROBERT LLOYD, b Coshocton, Ohio, Mar 16, 33; m 59; c 4. INDUSTRIAL ENGINEERING, OPERATIONS RESEARCH. *Educ:* Ohio State Univ, BS & MS, 60, PhD(indust eng), 64. *Prof Exp:* Res assoc & instr indust eng, Ohio State Univ, 60-64; PROF INDUST & SYSTS ENG & CHMN DEPT, OHIO UNIV, 67- *Concurrent Pos:* Chmn, Z-94 comt indust eng terminology, Am Nat Standards Inst, 66-; Hwy & Econ Growth res grant, 66-69; res contract, effects Hwy Warning Signs, 78-82. *Mem:* Am Inst Indust Engrs; Opers Res Soc Am. *Res:* Transportation systems. *Mailing Add:* Dept Indust Eng Ohio Univ Main Campus Athens OH 45701

WILLIAMS, ROBERT MACK, food science & technology, for more information see previous edition

WILLIAMS, ROBERT PIERCE, b Chicago, Ill, Oct 27, 20; m 44; c 2. MICROBIOLOGY. *Educ:* Dartmouth Col, AB, 42; Univ Chicago, SM, 46, PhD(bact, parasitol), 49; Am Bd Med Microbiol, cert. *Prof Exp:* Asst bact, Univ Chicago, 46-47, cur, 47-49; instr, Univ Southern Calif, 49-51; from asst prof to assoc prof, 51-63, actg chmn dept, 61-66, PROF MICROBIOL, BAYLOR COL MED, 63- *Concurrent Pos:* Consult, Vet Admin Hosp & M D Anderson Hosp, 57-75; lectr, Univ Houston, 64; vis mem grad fac, Col Vet Med, Tex A&M Univ, 65-85; chmn bd gov, Am Acad Microbiol, 76-82; vis prof, Univ Wash, 79, Rice Univ, 85-89; mem bd dirs, Nat Found Infectious Dis, 83-92; ed, Appl Environ Biol, 85-90; mem, Coun Sci Soc Pres, 83-87, chmn, 85; educ lectr, Am Soc Microbiol, 91. *Honors & Awards:* I M Lewis Lectr, Am Soc Microbiol, 87. *Mem:* AAAS; hon mem Am Soc Microbiol (pres-elect, 82, pres, 83); Am Chem Soc; fel Am Acad Microbiol; Brit Soc Gen Microbiol; fel Infectious Dis Soc Am; Nat Found Infectious Dis. *Res:* Bacterial pigments, particularly prodigiosin; virulence versus avirulence of microorganisms; host-parasite relationships in bacterial infections; pathogenesis of anthrax and gonorrhea. *Mailing Add:* Dept Microbiol & Immunol Baylor Col Med One Baylor Plaza Houston TX 77030

WILLIAMS, ROBERT SANDERS, b Athens, Ga, Oct 4, 48; m 73; c 3. MOLECULAR BIOLOGY, CARDIOVASCULAR PHYSIOLOGY. *Educ:* Princeton Univ, AB; Duke Univ, MD, 74. *Prof Exp:* Asst prof med, Duke Univ, 80-86, asst prof physiol, 82-90, assoc prof med, 86-90; PROF MED, SOUTHWESTERN MED CTR, UNIV TEX, 90- *Concurrent Pos:* Vis prof biochem, Oxford Univ, Eng, 84-85; mem, Coun Basic Sci, Am Heart Asn; Fogarty Int fel, 84-85. *Honors & Awards:* Young Investr Award, Am Col Cardiol, 79, 81. *Mem:* Am Fed Clin Res; Am Soc Clin Invest; Am Physiol Soc; Am Col Cardiol. *Res:* Mitochondial biogenesis in striated muscles; regulation of gene expression by contractile activity in striated muscles. *Mailing Add:* Univ Tex Southwestern Med Ctr 5323 Harry Hines Blvd Dallas TX 75235-8573

WILLIAMS, ROBERT WALTER, b Palo Alto, Calif, June 3, 20; m 46, 58, 69; c 3. EXPERIMENTAL HIGH-ENERGY PHYSICS. *Educ:* Stanford Univ, AB, 41; Princeton Univ, MA, 43; Mass Inst Technol, PhD(physics), 48. *Prof Exp:* Lab asst, Princeton Univ, 41-42; from jr physicist to assoc physicist, Manhattan Proj, Princeton Univ & Los Alamos Sci Lab, 42-46; res assoc, Mass Inst Technol, 46-48, from asst prof to assoc prof, 48-59; prof, 59-90, EMER PROF PHYSICS, UNIV WASH, 90- *Concurrent Pos:* Sci assoc, Europ Orgn Nuclear Res, 67-68, 74-75, 81-82 & 88-; trustee, Univ Res Asn, 78-84. *Mem:* Fel Am Phys Soc; fel Am Acad Arts & Sci. *Res:* Elementary particle physics using high-energy accelerators; cosmic rays; elementary particles, especially properties of the moon. *Mailing Add:* Physics FM-15 Univ Wash Seattle WA 98195

WILLIAMS, ROBIN, b Southampton, Eng, July 10, 41; US citizen; m 65; c 2. COMPUTER SCIENCE, DISTRIBUTED PROCESSING. *Educ:* Univ London, BSc, 62; NY Univ, MS, 68, PhD(comput sci), 71. *Prof Exp:* Staff mem optical character recognition, Mullard Res Labs, Philips Co, Eng, 62-64, staff mem comput memories, Philips Res Labs, 64-67; asst prof & instr comput sci, NY Univ, 67-72; mgr distrib comput, 72-80, mgr, Database & Distrib Systs Dept, 80-84, MGR, OFF SYSTS DEPT, IBM RES, 84- *Mem:* Asn Comput Mach; Inst Elec & Electronics Engrs. *Res:* Computer graphics; distributed processing; computer aided publishing; image systems. *Mailing Add:* Dept K52/803 IBM Almaden Res Ctr 650 Harry Rd San Jose CA 95120-6099

WILLIAMS, ROBIN O('DARE), b Greensboro, NC, Dec 26, 27; m 53; c 4. METALLURGY, THERMODYNAMICS. *Educ:* Univ Tenn, BS, 48, MS, 50; Carnegie-Mellon Univ, PhD(metall), 55. *Prof Exp:* Metallurgist, Oak Ridge Nat Lab, 48-51, consult, 51-54; res assoc, Res Lab, Gen Elec Co, 54-56; sr res supvr, Cincinnati Milacron, 56-59; metallurgist, Oak Ridge Nat Lab, 59-88; RETIRED. *Mem:* Am Soc Metals Int. *Res:* Preferred orientation of deformed metals; precipitation hardening of alloys; stored energy of deformation; structure and thermodynamics of solid solutions; x-ray diffraction; order-disorder; software development and data reduction; spinodal decomposition; coherent phase equilibria. *Mailing Add:* 906 W Outer Dr Oak Ridge TN 37830

WILLIAMS, ROBLEY COOK, b Santa Rosa, Calif, Oct 13, 08; m 31; c 2. BIOPHYSICS, ELECTRON MICROSCOPY. *Educ:* Cornell Univ, AB, 31, PhD(physics), 35. *Prof Exp:* Asst physics, Cornell Univ, 29-35; from instr to asst prof astron, Observ, Univ Mich, 35-45, from assoc prof to prof physics, 45-50; assoc dir, Virus Lab & res biophysicist, Univ Calif, 50-76, prof biophys, 50-59, prof virol, 59-64, chmn, Dept Molecular Biol, 64-76, prof molecular biol, 64-76, EMER PROF MICROBIOL & BIOPHYS, UNIV CALIF, BERKELEY, 76- *Concurrent Pos:* Vpres & consult, Evaporated Metal Films Corp, 35-56; res assoc, Off Sci Res & Develop, 41-42; coun mem, Int Union Pure & Appl Biophys, 61-69, pres, Comn Molecular Biophys, 61-69; Nat Acad Sci rep, UN Educ Sci & Cultural Orgn, 63-69; chmn, Deep Springs Col, 71- *Honors & Awards:* Longstreth Medal, Franklin Inst, 39, Scott Award, 54. *Mem:* Nat Acad Sci; Biophys Soc (pres, 58 & 59); Electron Micros Soc Am (pres, 51). *Res:* Electron microscopy of biological objects; development of techniques for electron microscopy of virus particles. *Mailing Add:* One Arlington Ct Berkeley CA 94707

WILLIAMS, ROBLEY COOK, JR, b Ann Arbor, Mich, Oct 15, 40; m 68; c 2. PHYSICAL BIOCHEMISTRY, CYTOSKELETON BIOCHEMISTRY. *Educ:* Cornell Univ, BA, 62; Rockefeller Univ, PhD(phys biochem), 68. *Prof Exp:* Nat Inst Arthritis & Metab Dis fel, State Univ NY, Buffalo, 67-68; from asst prof to assoc prof biol, Yale Univ, 69-76; assoc prof, 76-85, PROF MOLECULAR BIOL, VANDERBILT UNIV, 85- *Concurrent Pos:* Mem study sect biophys & biophys chem, NIH, 77-81, chmn, 79-81; fac assoc, US Antarctic Res Prog, McMurdo Sta, 79 & 81, Palmer Sta, 85 & 91. *Mem:* Am Soc Cell Biol; Am Chem Soc; Biophys Soc; Am Soc Biol Chemists. *Res:* Protein-protein association, structure-function relationships in proteins, assembly of microtubules and intermediate filaments. *Mailing Add:* Dept Molecular Biol Vanderbilt Univ Nashville TN 37235

WILLIAMS, ROGER LEA, b Hamilton, Ohio, Jan 5, 41; c 2. CLINICAL PHARMACOLOGY, INTERNAL MEDICINE. *Educ:* Oberlin Col, BA, 63; Univ Chicago, MD, 67. *Prof Exp:* Intern med, Univ Chicago Hosps & Clins, 67-68, resident, 67-71; fel clin pharmacol, 74-77, ASST PROF MED & PHARM, UNIV CALIF, SAN FRANCISCO, 77- *Concurrent Pos:* Consult, Rev Panel New Drug Regulation, HEW, 76-77 & Task Force, Calif Citizen Action Group, 77-78. *Mem:* Am Fedn Clin; Am Soc Clin Pharmacol & Therapeut. *Res:* Clinical research drug risk and efficacy for new and established drug products; academic clinical drug investigation. *Mailing Add:* Univ Calif Drug Study 926 Med Sci Bldg San Francisco CA 94143

WILLIAMS, ROGER NEAL, b Amityville, NY, Apr 3, 35; m 59; c 3. ENTOMOLOGY. *Educ:* Tex Tech Univ, BS, 57; La State Univ, MS, 64, PhD(entom), 66. *Prof Exp:* Res trainee entom, United Brands Co, Honduras, 58-62; res asst, La State Univ, 62-66; res entomologist, IRI Res Inst, Brazil, 66-68; asst prof, Ohio State Univ, Brazil, 68-73; assoc prof, 74-82, PROF ENTOM, OHIO AGR RES & DEVELOP CTR & OHIO STATE UNIV, 82- *Mem:* Entomol Soc Am; Int Orgn Biol Control; Soc Entom Brazil. *Res:*

Biology and control of small fruit insect pests with chemicals and natural enemies; biological control of insect tropical pastures and ranges; natural enemies, chemical control and attractants of sap beetles; pest management tropical crops; cocoa, bananas and other horticultural crops. *Mailing Add:* Dept Entom Ohio Agr Res & Develop Ctr 1680 Madison Ave Wooster OH 44691-4096

WILLIAMS, ROGER RICHARD, b Ogden, Utah, Aug 11, 44; m 68; c 7. CARDIOVASCULAR GENETICS, PREVENTIVE MEDICINE. *Educ:* Weber State Col, BS; Univ Utah Col Med, MD, 71. *Prof Exp:* Med resident internal med, Duke Univ Med Ctr, 71-73; res assoc cancer epidemiol, Nat Cancer Inst, 73-74 & cardiovasc epidemiol, Nat Heart-Lung Inst, 74-75; investr, CV epidemiol, Nat Heart, Lung & Blood Inst, 75-76; from asst prof to assoc prof, 76-85, PROF INTERNAL MED, UNIV UTAH MED SCH, 85- *Concurrent Pos:* Mem & chmn, Epidemiol & Dis Control Sect, NIH, 78-84; prin investr, CV Genetics Res Clin, Univ Utah Sch Med, 77-, dir, Cardiovasc Genetics Res Clin, 80-; chmn, Pharm & Therapeut Comt, Univ Utah Hosp, 80-; chmn, Coord Coun for Diabetes Control Prog, Utah State Dept Health, 84- *Res:* Genetic and environmental determinants of early heart attacks and high blood pressure. *Mailing Add:* 50 N Medical Dr Salt Lake City UT 84132

WILLIAMS, ROGER STEWART, b San Diego, Calif, Feb 15, 41; m 74; c 4. CLINICAL NEUROLOGY, NEUROPATHOLOGY. *Educ:* Emory Univ, MD, 66. *Prof Exp:* Med intern, Grady Mem Hosp, Atlanta, Ga, 66-67, med resident, 67-68; med officer, US Navy, 68-70; neurol resident, Mass Gen Hosp, Boston, Mass, 70-73; instr, 73-78, ASST PROF NEUROL, HARVARD MED SCH, 78- *Concurrent Pos:* Res fel neurol, Mass Gen Hosp, 73-76; res fel neurosci, Joseph P Kennedy, Jr Mem Found, 74-76; assoc neurologist, McLean Hosp, Belmont, Mass, 75-; clin assoc neurol, Mass Gen Hosp, 75-; investr, Schizophrenia Res Found of the Scottish Rite, 78-80; asst neurol, Mass Gen Hosp, 78-87; assoc prof med, Univ Wash Med Sch, 87- *Mem:* Am Acad Neurol; Epilepsy Found Am; Soc Neurosci. *Res:* Experimental neuropathology of the developing nervous system. *Mailing Add:* 2825 Eighth Ave N Billings MT 59101

WILLIAMS, ROGER TERRY, b Covina, Calif, June 15, 36; m 64; c 4. DYNAMIC METEOROLOGY. *Educ:* Univ Calif, Los Angeles, AB, 59, MS, 61, PhD(meteorol), 63. *Prof Exp:* Ford Found fel, Univ Calif, Los Angeles, 63-64; res assoc meteorol, Mass Inst Technol, 64-66; asst prof, Univ Utah, 66-68; assoc prof, 68-74, PROF METEOROL, NAVAL POSTGRAD SCH, 74- *Concurrent Pos:* Mem, FGGE Adv Panel, Nat Res Coun, 74-80. *Mem:* AAAS; fel Am Meteorol Soc. *Res:* Numerical weather prediction; dynamics of the atmosphere and other geophysical systems; application of numerical methods; dynamics of atmospheric waves and fronts. *Mailing Add:* Dept Meteorol Naval Postgrad Sch Monterey CA 93940

WILLIAMS, ROGER WRIGHT, b Great Falls, Mont, Jan 24, 18; m 43; c 2. MEDICAL ENTOMOLOGY, PARASITOLOGY. *Educ:* Univ Ill, BS, 39, MS, 41; Columbia Univ, PhD(med entom), 47; London Sch Hyg & Trop Med, cert appl parasitol & entom, 57. *Prof Exp:* Jr sci asst, Div Cereal Crops, Bur Entom & Plant Quarantine, USDA, 41; asst biol, Cornell Univ, 42; from res asst to res assoc parasitol, Sch Pub Health, Columbia Univ, 44-48, from asst prof to assoc prof med entom, 48-66, actg head div trop med, 70, prof pub health, Med Entom, 66-83; RETIRED. *Concurrent Pos:* NSF sr fel, 56; fel trop med, La State Univ, 57; spec consult, USPHS, Alaska, 49, consult, Ga, 52; corp mem, Bermuda Biol Sta Res, 58-; consult, US Nat Park Serv, VI Govt & Jackson Hole Preserve, Inc, 59 & 61; consult, Rockefeller Found, WI, 63, mem field staff, Nigeria, 64-65; consult, WHO, Burma, 66. *Mem:* Fel AAAS; Entom Soc Am; Am Soc Parasitol; Am Mosquito Control Asn; Am Soc Trop Med & Hyg. *Res:* Biology, physiology control and taxonomy of arthropods of medical importance; biology and chemotherapy of helminths. *Mailing Add:* 732 B Chatam Lane Lakehurst NJ 08733

WILLIAMS, RONALD LEE, b Koleen, Ind, June 26, 36; m 57; c 4. PHARMACOLOGY, MEDICINE. *Educ:* Butler Univ, BS, 59, MS, 61; Tulane Univ, PhD(pharmacol), 64. *Prof Exp:* Asst prof, La State Univ Med Ctr, New Orleans, 66-70, assoc prof pharmacol, 71-84, co-joint assoc prof med, 78-84; consult & pharm, 84-86; ASST DIR HOSP PHARM DEPT CORRECTIONS, STATE COLO, CANON CITY, 86- *Concurrent Pos:* La Heart Asn grants-in-aid, 65-66, 67-68 & 74-75; USPHS res contract, 68; mem, expert adv panel, Renal Drugs & Electrolytes, US Prof Develop Inst, 80-86. *Mem:* AAAS; Am Soc Pharmacol & Exp Therapeut; Soc Exp Biol & Med; NY Acad Sci. *Res:* Effect of autonomic drugs and neurotransmitters upon renal function and the relationships between hemodynamics and tubular changes; treatment of human immunovirus and opportunistic diseases. *Mailing Add:* 1004 Greenwood Ave Canon City CO 81212-3440

WILLIAMS, RONALD LLOYDE, b Northfield, Minn, May 7, 44; m 64; c 2. PHYSICAL CHEMISTRY, ENVIRONMENTAL CHEMISTRY. *Educ:* St Olaf Col, BA, 66; Iowa State Univ, PhD(phys chem), 70. *Prof Exp:* Fel, Univ Calif, Irvine, 70-72; RES CHEMIST, GEN MOTORS RES LABS, 72- *Mem:* Am Chem Soc; Soc Automotive Engrs; Air & Waste Mgt Asn. *Res:* Reaction kinetics; hot atom reactions; unregulated emissions from tires, brakes, refrigeration systems and diesel automobiles; wastewater treatment. *Mailing Add:* 4018 Hillside Royal Oak MI 48073

WILLIAMS, RONALD WENDELL, b Atlanta, Ga, Nov 9, 39; m 63; c 3. SOLID STATE PHYSICS. *Educ:* Christian Bros Col, BSc, 62; Iowa State Univ, PhD(physics), 66. *Prof Exp:* Instr physics, Iowa State Univ, 66-67; staff scientist, Oak Ridge Nat Lab, 67-70; from asst prof to assoc prof, 70-79, PROF ELEC ENG, UNIV VT, 79- *Res:* Band structure and transport properties of metals; digital systems; VLSI circuit design. *Mailing Add:* Dept Comput Sci & Elec Eng Univ Vt Burlington VT 05405

WILLIAMS, ROSS EDWARD, b Carlinville, Ill, June 28, 22; m 58; c 3. PHYSICS. *Educ:* Bowdoin Col, BS, 43; Columbia Univ, MA, 47, PhD(physics), 55. *Prof Exp:* Sr res engr, Sperry Prod Inc, 47-49; consult physicist, Paul Rosenberg Assocs, 53-60; sr res assoc, Oceanog Acoust & Signal Processing, 60-65, asst dir, 65-66, ASSOC DIR, OCEANOG ACOUST & SIGNAL PROCESSING, HUDSON LABS, COLUMBIA UNIV, 66-, PROF OCEAN ENG, 68-, LECTR ELEC ENG, 66- *Concurrent Pos:* Mem, Comt Undersea Warfare, Nat Res Coun-Nat Acad Sci; consult, Naval Res Lab & Nat Acad Sci; chmn bd dirs, Ocean & Atmospheric Sci, Inc, 68- *Mem:* Acoust Soc Am; Am Phys Soc; Am Soc Photogram; sr mem Inst Elec & Electronics Eng. *Res:* Oceanography; acoustic propagation; surveillance system design; signal processing techniques; aerial reconnaissance; automatic mapping; optical data processing; electronic design. *Mailing Add:* 23 Alta Pl Centuck PO Yonkers NY 10710

WILLIAMS, ROY EDWARD, b Cookeville, Tenn, Feb 12, 38; m 59. HYDROGEOLOGY. *Educ:* Ind Univ, Bloomington, BSc, 61, MA, 62; Univ Ill, Urbana, PhD(hydrogeol), 66. *Prof Exp:* Teaching asst phys geol, Ind Univ, Bloomington, 63-64; teaching asst eng geol, Univ Ill, Urbana, 64-66; asst prof, 66-70, PROF HYDROGEOL & HYDROGEOLOGIST, UNIV IDAHO, 70- *Concurrent Pos:* Res asst, Ill State Geol Surv, 64-66; grants, Idaho Water Resources Res Inst & Univ Idaho Res Comt, 66- & Idaho Short Term Appl Res Fund, 68- *Mem:* Am Geophys Union; Sigma Xi. *Res:* Studies of pollution of ground and surface water and the relation between ground water flow systems and certain engineering problems. *Mailing Add:* Col Mines Univ Idaho Moscow ID 83844

WILLIAMS, ROY LEE, b Portsmouth, Va, Feb 20, 37; m 57; c 1. ORGANIC CHEMISTRY. *Educ:* Col William & Mary, BS, 60; Univ Del, PhD(org chem), 65. *Prof Exp:* Res chemist, Am Cyanamid Co, NJ, 64-65; from asst prof to assoc prof, 65-73, PROF CHEM, OLD DOM UNIV, 73-; ASST PROF PHARMACOL, EASTERN VA MED SCH, 77- *Concurrent Pos:* Res grant, Army Med Res Inst, Walter Reed Hosp, Washington, DC, 66-68; consult, Chem & Physics Br, Langley Res Ctr, NASA, Va, 65- *Mem:* Am Chem Soc; fel The Chem Soc; Int Soc Heterocyclic Chem. *Res:* Heterocyclic, organic synthesis, including heterocyclic polymers; medicinals; synthetics. *Mailing Add:* Dept Chem Old Dom Univ Norfolk VA 23508

WILLIAMS, RUSSELL RAYMOND, b Lost Creek, WVa, Apr 11, 26; m 53; c 3. PARASITOLOGY, INVERTEBRATE ZOOLOGY. *Educ:* Ohio State Univ, BSc, 55, MSc, 57, PhD(zool), 63. *Prof Exp:* Lab unit operator, B F Goodrich Chem Co, 45-50; from asst instr to instr zool, Ohio State Univ, 57-63; from asst prof to assoc prof, 63-67, PROF BIOL, WAYNESBURG COL, 67-, CHMN DEPT, 70-, PREMED ADV, 73- *Concurrent Pos:* Res Corp grant, 64-66; vis prof, Univ Northern Colo, 68-; partic, Res Corp Conf for New Sci Chairmen, 71. *Mem:* AAAS; Am Soc Parasitol; Wildlife Dis Asn; Am Micros Soc; Am Inst Biol Sci. *Res:* Life history and taxonomic studies on trematodes. *Mailing Add:* 511 Ross St Waynesburg PA 15370

WILLIAMS, SAM B, RESEARCH ADMINISTRATION. *Prof Exp:* PRES & CHIEF EXEC OFFICER, WILLIAMS INT CORP. *Mem:* Nat Acad Eng. *Mailing Add:* Williams Int Corp 2280 W Maple RD PO Box 200 Walled Lake MI 48390-0500

WILLIAMS, SCOTT WARNER, b Staten Island, NY, Apr 22, 43; c 3. MATHEMATICS. *Educ:* Morgan State Col, BS, 64; Lehigh Univ, MS, 67, PhD(math), 69. *Prof Exp:* Instr, Pa State Univ, 68-69; res assoc, Pa State Univ, 69-71; from asst prof to assoc prof, 71-85, PROF MATH, STATE UNIV NY, BUFFALO, 85- *Concurrent Pos:* Ford Found sr res fel, 80-81; NSF res grant, 83-87; Fulbright lectr, Czech, 86-87; adj prof math, Beijing Teachers, China, 88- *Mem:* Am Math Soc; Nat Asn Mathematicians. *Res:* General topology, completeness, paracompactness, linearly ordered spaces and Baire spaces; algebra, groups and categories; logic; set theory; independence results; topological dynamics. *Mailing Add:* Dept Math State Univ NY Buffalo NY 14214

WILLIAMS, SIDNEY ARTHUR, b Ann Arbor, Mich, Dec 26, 33; m 57; c 1. MINERALOGY. *Educ:* Mich Technol Univ, BS, MS, 57; Univ Ariz, PhD(mineral), 62. *Prof Exp:* Instr mineral, Mich Technol Univ, 60-61, asst prof, 61-63; mineralogist, Silver King Mines, Inc, 63-65; dir res explor geol, Phelps Dodge Corp, 65-82; CONSULT, 82- *Concurrent Pos:* Mineralogist, Brit Mus Natural Hist, 71. *Mem:* Fel Mineral Soc Am; Mineral Asn Can; Brit Mineral Soc; Soc Econ Geol; Mineral Soc Japan. *Res:* Descriptive mineralogy and crystallography; petrology of altered rocks related to ore deposits. *Mailing Add:* PO Box 872 Douglas AZ 85607

WILLIAMS, STANLEY A, b Lawrence, Kans, May 14, 32; m 58; c 2. THEORETICAL PHYSICS. *Educ:* Nebr Wesleyan Univ, BA, 54; Rensselaer Polytech Inst, PhD(physics), 62. *Prof Exp:* NSF fel, Univ Birmingham, 62-63; from asst prof to assoc prof, 63-76, PROF PHYSICS & ASST CHMN DEPT, IOWA STATE UNIV, 76- *Concurrent Pos:* Assoc scientist, Ames Lab, 63-67, scientist, 67- *Mem:* Am Phys Soc; Am Asn Physics Teachers. *Res:* Mathematical physics, principally the application of group theoretic techniques to nuclear and elementary particle physics; structure of fission fragment nuclei. *Mailing Add:* Dept Physics Iowa State Univ Ames IA 50010

WILLIAMS, STANLEY CLARK, b Long Beach, Calif, Aug 24, 39; m 65; c 3. ECOLOGY, MEDICAL ENTOMOLOGY. *Educ:* San Diego State Col, AB, 61, MA, 63; Ariz State Univ, PhD(zool), 68. *Prof Exp:* From asst prof to assoc prof, 67-74, PROF BIOL, SAN FRANCISCO STATE UNIV, 74- *Concurrent Pos:* Res assoc, Calif Acad Sci, 67-; NSF grants, Mexico, 68-72; lectr, Moss Landing Marine Sta, 72-73; mem Int Ctr Arachnological Documentation; ecol consult, Mill Valley, Calif; dir, W Point Acad Sci. *Mem:* Am Arachnology Soc; Ecol Soc Am; Sigma Xi; Brit Arachnology Soc; Soc Syst Zoologists; Asn Biologist in Comput (pres, 85-). *Res:* Invertebrate ecology; scorpion systematics; urban ecology; medical entomology; biostatistics and data analysis. *Mailing Add:* Dept Biol San Francisco State Univ San Francisco CA 94132

WILLIAMS, STEPHEN EARL, b Borger, Tex, Apr 27, 48; m 75. SOIL MICROBIOLOGY, SOIL BIOCHEMISTRY. *Educ:* NMex State Univ, BS, 70, MS, 72; NC State Univ, PhD(soil sci), 77. *Prof Exp:* Plant physiologist soils, Rocky Mountain Forest & Range Exp Sta, Forest Serv, USDA, Albuquerque, NMex, 72; res & teaching assoc, NC State Univ, 75-76; asst prof, 76-82, ASSOC PROF SOILS, UNIV WYO, 82- *Concurrent Pos:* Prin investr, US Dept Energy, Laramie Energy Technol Ctr, Wyo, 77- *Mem:* Am Soc Agron; Soil Sci Soc Am. *Res:* Symbiotic associations between plants and microorganisms such as mycorrhizae and symbiotic nitrogen fixation; revegetation of devastated lands; microbial degradations of organic constituents in waste waters produced from fossil fuel processing. *Mailing Add:* Univ Wyo Div Plant Sci PO Box 3354 Laramie WY 82071

WILLIAMS, STEPHEN EDWARD, b St Louis, Mo, Oct 9, 42; m 68; c 1. MEMBRANE PHYSIOLOGY, MOLECULAR BIOLOGY. *Educ:* Cent Col, Mo, BA, 64; Univ Tenn, Knoxville, MS, 66; Wash Univ, PhD(biol), 71. *Prof Exp:* Lectr plant physiol, Cornell Univ, 70-73; from asst prof to assoc prof, 73-85, PROF BIOL, LEBANON VALLEY COL, 85- *Mem:* AAAS; Am Soc Plant Physiol; Bot Soc Am. *Res:* Chloroplast DNA phylogenetics of droseraceae; electrophysiology, plant sensory physiology, excitable plant cells; carnivorous plants, especially Droseraceae, electron microscopy. *Mailing Add:* Dept Biol Lebanon Valley Col Annville PA 17003

WILLIAMS, STEVEN FRANK, b Tacoma, Wash, May 8, 44; m 66; c 1. FISHERIES, ICHTHYOLOGY. *Educ:* Univ Wash, BS, 66; Univ Calif, Los Angeles, MA, 68; Ore State Univ, PhD(fisheries), 74. *Prof Exp:* Res biologist fisheries, US Peace Corps, Chile, 68-70; ASST PROF BIOL, ST CLOUD STATE UNIV, 74- *Concurrent Pos:* Res asst fisheries, Ore State Univ, 70- *Mem:* Am Fisheries Soc. *Res:* Natural distribution and abundance of fishes and factors which affect them; fish culture, especially optimum culture conditions; ichthyo plankton. *Mailing Add:* Dept Biol St Cloud State Univ St Cloud MN 56301

WILLIAMS, TERENCE HEATON, b Oldham, Eng, Jan 5, 29; m 56; c 3. NEUROANATOMY, ELECTRON MICROSCOPY. *Educ:* Univ Manchester, MB, ChB, 53; Univ Wales, PhD(anat), 60. *Hon Degrees:* DSc, Univ Manchester, 77. *Prof Exp:* House surgeon, Manchester Univ & Royal Infirmary, 53-54; jr registr surg, London Hosp, 55-56; asst lectr anat, Univ Col, Dublin, 57-58; lectr, Univ Wales, 58-61; lectr & sr lectr exp neurol, Univ Manchester, 61-68; vis lectr electron microscopy of nerv syst, Harvard Med Sch, 65-66; prof neuroanat, Sch Med, Tulane Univ, 68-73; PROF ANAT & HEAD DEPT, COL MED, UNIV IOWA, 73- *Concurrent Pos:* Brit Med Res Coun traveling fel, Harvard Med Sch, 64-65; Peck Sci Res award; NIH res awards, 69- *Mem:* Soc Neurosci; Am Asn Anat; Anat Soc Gt Brit & Ireland. *Res:* Neuropeptidergic systems; plasticity of nervous system; small intensely fluorescent cells of sympathetic ganglia; electromicroscopy of the nervous system. *Mailing Add:* Dept Anat Univ Iowa Col Med Iowa City IA 52242

WILLIAMS, TERRY WAYNE, b Los Angeles, Calif, Dec 24, 45. HUMAN TUMOR IMMUNOLOGY, HYBRIDOMA TECHNOLOGY. *Educ:* Univ Calif, San Diego, BA, 67; Univ Calif, Irvine, PhD(molecular biol), 70. *Prof Exp:* Staff fel, Div Infections Dis, Univ Calif, San Diego, 72-79; staff fel, Dept Pediat, Georgetown Univ, Washington, DC, 81-83; res specialist, Biochem Dept, Monsanto Co, 84-86; dir immunol, Div Brunswick Corp, Brunswick Biotechetics, 86-88; ASSOC DIR, IMMUNOTHER DEPT, ST VINCENT MED CTR, 89- *Mem:* Am Asn Immunologists; Tissue Cult Asn; Am Soc Cell Biol. *Res:* Development of immunotherapy for human cancer using cancer cells treated with inteferon and other biological response modifiers as vaccines. *Mailing Add:* Dept Immunother St Vincent Med Ctr 2131 W Third St Los Angeles CA 90057

WILLIAMS, THEODORE BURTON, b Youngstown, Ohio, Sept 9, 49; m 71; c 2. ASTRONOMY. *Educ:* Purdue Univ, BS, 71; Calif Inst Technol, PhD(astron), 75. *Prof Exp:* Res assoc astron, Princeton Univ, 75-77, mem res staff, 77-79; ASST PROF, DEPT PHYSICS & ASTRON, RUTGERS UNIV, 79- *Mem:* Am Astron Soc; Int Astron Union. *Res:* Structure and dynamics of individual galaxies; development of astronomical detector systems. *Mailing Add:* Dept Physics & Astron Rutgers Univ New Brunswick NJ 08903

WILLIAMS, THEODORE J(OSEPH), b Black Lick, Pa, Sept 2, 23; m 46; c 4. AUTOMATIC CONTROL, CHEMICAL ENGINEERING. *Educ:* Pa State Univ, BS, 49, MS, 50, PhD(chem eng), 55; Ohio State Univ, MSEE, 56. *Prof Exp:* Asst prof chem eng, USAF Inst Technol, 53-56; sr eng supvr, Monsanto Co, 56-65; PROF ENG & DIR LAB APPL INDUST CONTROL, PURDUE UNIV, 65- *Concurrent Pos:* Vpres, Simulation Coun, 60-62; vis prof automatic control, Wash Univ, 62-65; vpres, Am Automatic Control Coun, 63-65, pres, 65-67; surv lectr, Int Fedn Automatic Control, Basel, Switz, 63 & Warsaw, Poland, 69; mem bd gov, Am Fedn Info Processing Socs, 65-80, pres, 76-80; plenary lectr, Conf Chem Eng Frankfort, Ger, 70. *Honors & Awards:* Sir Harold Hartley Medal, Inst Measure & Control, London, 75; Silver Core Award, Int Fedn Info Processing, 77; Albert F Sperry Founder Medal, ISA, 90. *Mem:* Fel Instrument Soc (vpres, 65-67, pres, 69); Am Chem Soc; Am Soc Eng Educ; fel Am Inst Chem Engrs; Inst Elec & Electronics Engrs; fel AAAS; fel Inst Measurement & Control; fel Am Inst Chemists. *Res:* Industrial process dynamics and automatic control; application of digital computers to industrial process control and management; theory of separation processes, particularly distillation. *Mailing Add:* Lab Appl Indust Control Purdue Univ 334 Potter Ctr West Lafayette IN 47907-1293

WILLIAMS, THEODORE L, b Denver, Colo, Oct 9, 39; c 2. ELECTRICAL ENGINEERING. *Educ:* Mass Inst Technol, BS, 60; Drexel Inst, MS, 64; Pa State Univ, PhD(control systs), 66. *Prof Exp:* Staff engr, Gen Atronics Corp, 60-64; instr elec eng, Drexel Inst, 64; teaching asst & instr, Pa State Univ, 64-67; from asst prof to assoc prof, 67-87, EMER PROF ELEC ENG, UNIV ARIZ, 87- *Concurrent Pos:* Consult, Burr-Brown Res, 78. *Mem:* AAAS; Inst Elec & Electronics Engrs; Am Soc Eng Educ. *Res:* Design of computer control systems; instrumentation and computational aspects of control systems. *Mailing Add:* 534 N Stone Ave Tucson AZ 85705

WILLIAMS, THEODORE P, b Marianna, Pa, May 24, 33; m 56; c 5. BIOPHYSICS, NEUROSCIENCE. *Educ:* Muskingum Col, BS, 55; Princeton Univ, MA, 57, PhD(phys chem), 59. *Prof Exp:* Res assoc chem, Brown Univ, 59-61, fel psychol, 61-63, asst prof biol & med sci, 63-66; assoc prof biol sci, 66-73, actg chmn dept, 70-71, PROF BIOL SCI, FLA STATE UNIV, 73-, co-dir psychobiol prog, 71-74, DIR INST MOLECULAR BIOPHYSICS, 85- *Concurrent Pos:* Prin investr of numerous NIH, NSF & Dept Energy grants. *Honors & Awards:* Alexander von Humboldt Sr Scientist Award. *Mem:* Asn Res Vision Ophthalmol. *Res:* Visual processes; sensory mechanisms; fast chemical reactions. *Mailing Add:* Dept Biol Sci Fla State Univ Tallahassee FL 32306

WILLIAMS, THEODORE ROOSEVELT, b Washington, DC, Oct 23, 30; m 54; c 4. ANALYTICAL CHEMISTRY. *Educ:* Howard Univ, BS, 52; Pa State Univ, MS, 54; Univ Conn, PhD, 60. *Prof Exp:* Asst instr chem, Univ Conn, 56-59; from instr to assoc prof, 59-66, PROF CHEM, COL WOOSTER, 66-; Case Western Reserve Univ Med Sch, 87-88. *Concurrent Pos:* Res assoc, Harvard Univ, 67-68, Sloan vis prof chem, 69-70; vis prof, Univ Conn, 72-73, Case Western Reserve Univ, 77-78 & Carnegie-Mellon Univ, 82-83; chmn analytical div grad fel comt, Am Chem Soc, 73-; res award, Asn Independent Col & Univ Ohio, 89. *Honors & Awards:* Mfg Chemists Asn Award, 78; Martha Holden Jennings Found Award, 79; Percy L Julian Award, Nat Orgn Prof Advan Black Chemists & Chem Engrs, 90. *Mem:* Am Chem Soc. *Res:* Electroanalytical chemistry; analysis of biological tissues. *Mailing Add:* Dept Chem Col Wooster Wooster OH 44691

WILLIAMS, THEODORE SHIELDS, b Kansas City, Kans, June 2, 11; m 36; c 2. VETERINARY MEDICINE. *Educ:* Kans State Univ, DVM, 35; Iowa State Univ, MS, 46. *Prof Exp:* Col vet, Prairie View State Col, 36; vet inspector, Meat Inspection Div, USDA, 36-45; head dept, Tuskegee Inst, 45-51, dean sch, 47-72, prof Sch Vet & Med, 45-81, emer prof path & parasitol, emer dean, 81; RETIRED. *Concurrent Pos:* Consult, Vet Admin Hosp, Tuskegee, Ala; vis prof path, NY State Vet Col, Cornell Univ, 73. *Mem:* Am Asn Vet Med Cols (pres, 69-70); Am Vet Med Asn. *Res:* Pathological lesions associated with tissue invading migratory parasites in animals. *Mailing Add:* Sch Vet Med Tuskegee Inst Tuskegee Inst AL 36088-2907

WILLIAMS, THOMAS ALAN, b Salem, Ore, Sept 28, 47; m 71; c 1. HERPETOLOGY, EVOLUTIONARY BIOLOGY. *Educ:* Willamette Univ, BA, 71; Drake Univ, MA, 75; Wash State Univ, PhD(zool), 84. *Prof Exp:* From instr to asst prof biol, Ricker Col, 74-77; from instr to asst prof, 81-88, ASSOC PROF BIOL, NCENT COL, 88- *Mem:* Am Soc Ichthyologists & Herpetologists; Soc Study Amphibians & Reptiles; Herpetologists League. *Res:* Biology of salamander granular glands; their structure, function, development and relevance to salamander systematics. *Mailing Add:* Dept Biol NCent Col Naperville IL 60566

WILLIAMS, THOMAS FFRANCON, b Denbighshire, Wales, Jan 30, 28; m 59; c 2. PHYSICAL CHEMISTRY. *Educ:* Univ London, BSc, 49, PhD(chem), 60. *Prof Exp:* From sci officer to prin sci officer, Chem Div, Atomic Energy Res Estab, Harwell, Eng, 49-61; from asst prof to prof, 61-74, DISTINGUISHED SERV PROF CHEM, UNIV TENN, KNOXVILLE, 74- *Concurrent Pos:* Assoc, Northwestern Univ, 57-59; NSF vis scientist, Kyoto Univ, 65-66; Guggenheim fel, Royal Inst Technol, Sweden, 72-73. *Mem:* Am Chem Soc; The Chem Soc. *Res:* Electron spin resonance studies of trapped radicals; free radical reactions at low temperature; electronic structure of radicals; radiation chemistry; radiation-induced ionic polymerization. *Mailing Add:* Dept Chem Univ Tenn Knoxville TN 37996-1600

WILLIAMS, THOMAS FRANKLIN, b Belmont, NC, Nov 26, 21; m 51; c 2. GERIATRIC MEDICINE, DIABETES. *Educ:* Univ NC, BS, 42; Columbia Univ, MA, 43; Harvard Univ, MD, 50. *Hon Degrees:* DSc, Med Col Ohio, 87. *Prof Exp:* Asst chem, Columbia Univ, 42-43; asst med, Johns Hopkins Univ, 51-53 & Boston Univ, 53-54; from instr to prof, Sch Med, Univ NC, 56-68; prof med, Univ Rochester, 68-83; med dir, Monroe Community Hosp, 68-83; DIR, NAT INST AGING, NIH, 83- *Concurrent Pos:* Res fel, Sch Med, Univ NC, 54-56, Markle scholar, 57-61; fel physiol, Vanderbilt Univ, 66-67. *Mem:* Inst Med-Nat Acad Sci; fel AAAS; Am Geriat Soc; fel Am Pub Health Asn; fel Am Col Physicians; fel Geront Soc Am; Asn Am Physicians. *Res:* Diseases of metabolism, especially for chronic illness and aging; metabolic and renal physiology. *Mailing Add:* NIH Bldg 31 Rm 2C02 Bethesda MD 20892

WILLIAMS, THOMAS HENRY, b Jamaica, WI, Apr 21, 34; m 67; c 2. STRUCTURAL CHEMISTRY. *Educ:* Univ WI, BSc, 56; Yale Univ, MS, 60, PhD(chem), 61. *Prof Exp:* Fel Univ Notre Dame, NJ, 61-62; res chemist, 63-74, res fel, 75-84, NMR RES INVESTR, HOFFMANN-LA ROCHE INC, NUTLEY, 85- *Mem:* Am Chem Soc; Sigma Xi. *Res:* Structural elucidation of natural products; applications of nuclear magnetic resonance spectroscopy in organic chemistry; diastereomeric solute-solute interaction of enantiomers; drug and Vitamin D3 metabolites. *Mailing Add:* Seven Patten Terr Cedar Grove NJ 07009

WILLIAMS, THOMAS HENRY LEE, b Deganwy, Wales, UK, May 31, 51; m 73; c 3. REMOTE SENSING, GEOGRAPHIC INFORMATION SYSTEMS. *Educ:* Univ Bristol, Eng, BSc, 72, PhD(geog), 77. *Prof Exp:* From asst prof to assoc prof geog, Univ Kans, 77-86; ASSOC PROF GEOG & DIR GEOSCI REMOTE SENSING GROUP, COL GEOSCI, UNIV OKLA, 86-, DIR, COOP INST APPL REMOTE SENSING, 87-, ASSOC DEAN GEOSCI, 89- *Concurrent Pos:* Consult, UNESCO, Haryana Agr Univ, India, 78-79 & Nat Park Serv, 79; intra-univ prof, dept elec eng, Univ Kans, 83-84, researcher, Radar Systs & Remote Sensing Lab, 83-86. *Mem:* Assoc fel Inst Math & Applns UK; Am Soc Photogram & Remote Sensing. *Res:* Remote sensing of renewable resources with emphasis on digital analysis and interpretation of visible and microwave satellite images; applications of expert systems techniques to image understanding. *Mailing Add:* Col Geosci Univ Okla Norman OK 73019

WILLIAMS, THOMAS R(AY), b Price, Utah, Feb 23, 26; m 68; c 2. ELECTRICAL ENGINEERING. *Educ:* Univ Okla, BSEE, 46; Princeton Univ, MSE, 49, PhD, 54. *Prof Exp:* Instr elec eng, 48-49, lectr, 52-54, asst prof, 54-60, asst chem dept, 70-75, ASSOC PROF ELEC ENG, PRINCETON UNIV, 60- *Res:* Communication theory; detection of signals in noise; stochastic processes; noise mechanisms and properties of noise in electronic devices. *Mailing Add:* 651 Rosedale Rd Princeton Univ Princeton NJ 08543

WILLIAMS, TIMOTHY C, b New York, NY, May 7, 42; m 64; c 2. BEHAVIOR, ETHOLOGY. *Educ:* Swarthmore Col, BA, 64; Harvard Univ, AM, 66; Rockefeller Univ, PhD(animal behav), 68. *Prof Exp:* Asst prof biol, State Univ NY, Buffalo, 69-75; PROF BIOL, SWARTHMORE COL, 76-, DEPT CHMN, 85- *Concurrent Pos:* Scientist, Woods Hole Oceanog Inst, 68-74; invest, Marine Biol Labs, 75-78. *Mem:* Sigma Xi. *Res:* Radar studies of bird orientation and migration, especially over major oceans. *Mailing Add:* Dept Biol Swarthmore Col Swarthmore PA 19081

WILLIAMS, TODD ROBERTSON, b Washington, DC, Nov 3, 45; m 67; c 1. POLYMER CHEMISTRY. *Educ:* Cornell Univ, AB, 67; Univ Calif, Los Angeles, PhD(org chem), 71. *Prof Exp:* Fel, Syntex Res Co, 71-72; sr chemist med chem, 72-77, supvr biomat res, 77-79, res specialist, 79-86, SR RES SPECIALIST, 3M CO, 86- *Mem:* Am Chem Soc. *Res:* Optical polymers. *Mailing Add:* 3025 Lake Elmo Ave N Lake Elmo MN 55042

WILLIAMS, TOM VARE, b Philadelphia, Pa, Dec 27, 38; m 63; c 4. PLANT BREEDING, PLANT PATHOLOGY. *Educ:* Univ Conn, BS, 60; Rutgers Univ, MS, 63, PhD(plant breeding), 66. *Prof Exp:* Agronomist, Soil Conserv Serv, USDA, 66-67; res horticulturist, Birds Eye Div, Gen Foods Corp, 67-70; res dir seed dept, Agr Chem Div, FMC Corp, 70-77; PROJ LEADER & VEG BREEDER, NORTHRUP KING CO, 77- *Concurrent Pos:* Vine Crops Adv Comt, Fla Hort Soc. *Mem:* Am Soc Hort Sci. *Res:* Vegetable variety development. *Mailing Add:* 2329 Pinewoods Circle Naples FL 33942

WILLIAMS, VERNON, b Augusta, Ga, Nov 10, 26; m 69; c 1. MATHEMATICS, EDUCATION. *Educ:* Paine Col, BA, 49; Univ Mich, Ann Arbor, MA, 54; Okla State Univ, EdD, 69. *Prof Exp:* Instr, Paine Col, 49-54 & Fla A&M Univ, 54-56; from asst prof to assoc prof, Southern Univ, Baton Rouge, 56-69, prof math, 69-80; PROF MATH, SETON HALL UNIV, 80- *Concurrent Pos:* Adv, Math Sect, La Educ Asn, 71- *Mem:* Am Math Asn. *Res:* Number theory; educational technology as devoted to higher education as well as secondary education. *Mailing Add:* 309 Heywood Ave Orange NJ 07050

WILLIAMS, VICK FRANKLIN, b Pittsburg, Tex, Apr 30, 36; m 62; c 2. ANATOMY. *Educ:* Austin Col, BA, 58; Univ Tex, MD & PhD(anat), 64. *Prof Exp:* Intern path, Charity Hosp La, New Orleans, 64-65; from instr to asst prof anat, Univ Tex Southwestern Med Sch Dallas, 65-70; assoc prof anat, Dent Sch, Univ Tex, San Antonio & Univ Tex Med Sch, San Antonio, 70-73; assoc prof, 73-79, PROF ANAT, UNIV TEX HEALTH SCI CTR, SAN ANTONIO, 79- *Concurrent Pos:* Consult, Dept Surg, Brooke Army Med Ctr, Ft Sam, Houston, TX, 72-78; mem anat bd, State Tex, 81-; mem fac, Head & Neck Anat Oral Surg Residents, Wilford Hall, US Asn Former Mem Cong, Lackland AFB, Tex, 87- *Honors & Awards:* Borden Award, 64. *Mem:* World Med Asn; Am Asn Anat; Sigma Xi. *Res:* Ultrastructure of the central nervous system of mammals. *Mailing Add:* Dept Cellular & Struct Biol Univ Tex Health Sci Ctr San Antonio TX 78284-7762

WILLIAMS, WALLACE TERRY, b Vincent, Ala, July 7, 42; c 1. CLINICAL NUTRITION. *Educ:* Southern Univ, BS, 64; NDak State Univ, MS, 66; Univ Maine, PhD(nutrit biochem), 69. *Prof Exp:* Asst prof nutrit sci, State Univ NY Buffalo, 69-71, assoc prof, 72-75, prof & chmn, 75-76; sci adminr, Div Assoc Health, NIH, 76-77; prof & chmn family & consumer resources, 77-81, INTERIM DEAN, COL LIB ARTS, WAYNE STATE UNIV, 81- *Concurrent Pos:* Consult, Div Assoc Health Professors, NIH, 69-81, Nat Cancer Inst, 80- & Health Resources Admin, 81-; actg chief clin nutrit, Wayne State Univ Med Ctr, 79-80; prof nutrit, Dean Col Fine & Appl Arts, Univ Akron, Ohio, 87- *Mem:* Am Dietetic Asn; AAAS; Soc Nutrit Educ. *Res:* Effects of diet on lipid metabolism; bio-nutritional status of children and adult populations; the effects of diet on elevation of blood lipids. *Mailing Add:* Dean-Fine & Appl Arts Univ Akron Gazzetta Hall Akron OH 44325

WILLIAMS, WALTER FORD, b Yazoo City, Miss, Oct 7, 27; m 50, 81; c 3. PHYSIOLOGY. *Educ:* Univ Mo, BS, 50, MS, 51, PhD(dairy physiol), 55. *Prof Exp:* Asst, Univ Mo, 53-55, res assoc, 55-57; from asst prof to prof diary physiol, 57-59, PROF DAIRY SCI, UNIV MD, COLLEGE PARK, 76- *Concurrent Pos:* USPHS fel, Nat Cancer Inst, 55-57. *Mem:* AAAS; Am Dairy Sci Asn; Soc Study Reprod; Am Soc Animal Sci. *Res:* Reproductive processes; endocrine regulation of metabolism in relation to reproduction, growth and lactation in domestic and wild mammals. *Mailing Add:* Rte 2 Box 274 Univ Md Beaverdam Rd Laurel MD 20708

WILLIAMS, WALTER JACKSON, JR, b Elkhart, Ind, Jan 17, 25; m 44, 86; c 3. ELECTRICAL ENGINEERING. *Educ:* Purdue Univ, Lafayette, BS, 48, MS, 50, PhD(elec eng), 54. *Prof Exp:* Instr, Purdue Univ, 48-54; from engr to prin engr, Fed Lab, Int Tel & Tel Corp, 54-60; chmn dept elec eng, Ind Inst Technol, 61-64, dean eng, 63-67, vpres & acad dean, 67-75, interim pres, 70-71; sr tech adv, 75-80, dir eng, 80-81, TECH DIR, AEROSPACE-OPTICAL DIV, ITT CORP, FT WAYNE, 81- *Concurrent Pos:* Consult, Int Tel & Tel Corp, Magnavox Co & Bowmar Instrument Corp. *Mem:* Inst Elec & Electronics Engrs; Nat Soc Prof Engrs; Am Soc Eng Educ. *Res:* Systems analysis with emphasis on feedback control. *Mailing Add:* 6518 Dumont Dr Ft Wayne IN 46815

WILLIAMS, WALTER MICHAEL, b Birmingham, Ala, Aug 14, 43; m 72; c 2. CLINICAL PHARMACOLOGY, INTERNAL MEDICINE. *Educ:* Univ Louisville, BS, 65, PhD(pharmacol), 70, MD, 74; Am Bd Internal Med, dipl, 77. *Prof Exp:* Resident internal med, Univ Pittsburgh, 74-77; teaching fel clin pharm, Univ Chicago, 77-79; asst prof, 79-82, ASSOC PROF PHARMACOL, UNIV LOUISVILLE, 82- *Mem:* Sigma Xi; Am Soc Pharmacol & Exp Therapeut; Soc Exp Biol & Med; Am Physiol Soc; NY Acad Sci. *Res:* Clinical pharmacology; basic and clinical studies of drug elimination (metabolism and excretion). *Mailing Add:* Dept Pharmacol & Toxicol Univ Louisville Louisville KY 40292

WILLIAMS, WAYNE WATSON, b Powersville, Mo, Dec 14, 22; m 44; c 7. ENGINEERING, GEOLOGY. *Educ:* Iowa State Univ, BS, 51, MS, 53. *Prof Exp:* Res asst soils, Eng Exp Sta, Iowa State Univ, 51-53; found engr, Des Moines, Iowa, 53-65; assoc prof, 65-76, PROF CIVIL ENG, KANS STATE UNIV, 76- *Concurrent Pos:* NSF res asst, Tex A&M Univ, 67; vis prof, Univ Houston, 77-78. *Mem:* Am Soc Civil Engrs. *Res:* Foundations of structures; shearing resistance; swelling clays; deep foundation; building and construction failures; construction litigation. *Mailing Add:* Dept Civil Eng Kans State Univ Seaton Hall Manhattan KS 66506

WILLIAMS, WELLS ELDON, b Cadillac, Mich, July 8, 19; m 44. FISHERIES. *Educ:* Mich State Univ, BS, 53, MS, 54, PhD(fisheries, wildlife), 58. *Prof Exp:* Asst fisheries & wildlife, 54-56; from instr to asst prof, 56-70, ASSOC PROF NATURAL SCI, MICH STATE UNIV, 70- *Res:* Freshwater ecology; habitat improvement; food conversion and growth rates of fishes; pond culture; physiology of freshwater fishes. *Mailing Add:* Nat Sci Dept Michigan State Univ East Lansing MI 48823

WILLIAMS, WENDELL STERLING, b Lake Forest, Ill, Oct 27, 28; m 52; c 2. SOLID STATE PHYSICS, MATERIALS SCIENCE. *Educ:* Swarthmore Col, BA, 51; Cornell Univ, PhD(physics), 56. *Prof Exp:* Physicist, Leeds & Northrup Co, 51; asst, Cornell Univ, 52-55; physicist, Union Carbide Corp, 56-65, 66-67; sr res vis, Dept Metall, Cambridge Univ, 65-66; from assoc prof to prof physics, ceramic eng & bioeng, Univ Ill, Urbana, 67-87, co-chmn, Bioeng Comt, 71, prin investr, Mat Res Lab, 67-87, dir, Prog Ancient Technol & Archaeol Mat, 80-87; PROF & CHMN, DEPT MAT SCI & ENG, PROF PHYSICS & BIOMED ENG, CASE WESTERN RESERVE UNIV, CLEVELAND, OHIO, 87- *Concurrent Pos:* NSF lectr, Univ PR, 61; mem, Mat Sci Comt, Argonne Ctr Educ Affairs, Argonne Nat Lab, 71-; mem adv comt, Metals & Ceramics Div, Oak Ridge Nat Lab, 72-74, mem & chmn adv comt, High Temperature Mat Lab, 75-79; task coordr energy res, Div Mat Res, NSF, 74-75; consult, Kennametal Inc, sect head metall & mat, Div Mat Res, NSF, 77-78; study dir, Study of Nat Educ Policies Sci, Nat Res Coun, 80; sabbatical vis, Dept Mat Sci & Eng, Imp Col, London, 82. *Mem:* AAAS; fel Am Phys Soc; fel Am Ceramic Soc; Bioelec Repair & Growth Soc (pres, 88-89); Mat Res Soc (pres, 88-89). *Res:* Electrical, thermal and mechanical properties of refractory hard metals; high-strength fibers; defects in solids; low and high temperature thermal conductivity; electrical properties of bone; electrical modification of osteoporosis; implant materials; archaeological materials. *Mailing Add:* Dept Mat Sci & Eng Case Western Reserve Univ Cleveland OH 44106

WILLIAMS, WILLIAM ARNOLD, b Johnson City, NY, Aug 2, 22; m 43; c 3. AGRONOMY, BIOMATHEMATICS. *Educ:* Cornell Univ, BS, 47, MS, 48, PhD(agron), 51. *Prof Exp:* Instr, 51-53, from asst prof to assoc prof, 54-64, PROF AGRON, UNIV CALIF, DAVIS, 65- *Concurrent Pos:* Fulbright scholar, Australia, 60; Rockefeller fel, Cent & SAm, 66; mem staff, NSF Res Vessel Alpha Helix, Amazon Exped, 67. *Mem:* Fel Am Soc Agron; fel Crop Sci Soc Am; Soil Sci Soc Am; Am Soc Plant Physiol; Ecol Soc Am; Am Statist Asn; Brit Ecol Soc; Math Asn Amer; Amer Math Soc; Soc Range Mgmt. *Res:* Systems analysis of annual-type range growth and utilization; competition for nutrients and nutrient cycling, especially sulfur and nitrogen; range and agronomic applications of multivariate data analysis techniques. *Mailing Add:* Dept Agron & Range Sci Univ Calif Davis CA 95616

WILLIAMS, WILLIAM DONALD, b Macon, Ga, Apr 22, 28; m 52; c 4. PHYSICAL CHEMISTRY. *Educ:* Harding Col, BS, 50; Univ Ky, MS, 52, PhD(chem), 54. *Prof Exp:* Assoc prof chem, 54-63, PROF CHEM & CHMN DEPT PHYS SCI, HARDING COL, 63- *Mem:* AAAS; Am Chem Soc. *Res:* Chemistry and kinetics of flames. *Mailing Add:* Dept Chem Harding Col Searcy AR 72143-6845

WILLIAMS, WILLIAM JAMES, b Rio Grande, Ohio, May 9, 35; m 61; c 2. ELECTRICAL ENGINEERING, PHYSIOLOGY. *Educ:* Ohio State Univ, BEE, 58; Univ Iowa, MS, 61, PhD(elec eng), 63; Univ Mich, MS, 66. *Prof Exp:* Res engr, Battelle Mem Inst, 58-60; teaching asst elec eng, Univ Iowa, 60-63; sr res specialist commun theory, Emerson Elec Mfg Co, 63-64; lectr elec eng, 64-65, asst prof elec eng & bioeng, 65-69, assoc prof elec & comput eng & bioeng, 69-73, dir bioelec sci lab, 68-83, PROF ELEC & COMPUT ENG & BIOENG, UNIV MICH, ANN ARBOR, 73- *Concurrent Pos:* Rackham fel & award, 66-67; vis scientist, Sch Med, Johns Hopkins Univ, 74. *Mem:* Inst Elec & Electronics Engrs; Soc Neurosci. *Res:* Electrical and communication biophysics; neurocybernetics; application of signal processing and systems analysis techniques to biological problems, particularly the nervous system; computer applications. *Mailing Add:* 5650 Warren Rd Ann Arbor MI 48105

WILLIAMS, WILLIAM JOSEPH, b Bridgeton, NJ, Dec 8, 26; m; c 3. MEDICINE, BIOCHEMISTRY. *Educ:* Univ Pa, MD, 49. *Prof Exp:* Intern, Hosp Univ Pa, 49-50; sr instr microbiol, Sch Med, Western Reserve Univ, 52-54; resident med, Hosp Univ Pa, 54-55; assoc med, Sch Med, Univ Pa, 55-56, from asst prof to prof, 56-69; PROF MED & CHMN DEPT, STATE UNIV NY HEALTH SCI CTR, 69- *Concurrent Pos:* Am Cancer Soc fel physiol chem, Sch Med, Univ Pa, 50-52, Am Philos Soc Daland fel res clin med, 55-57; Markle scholar, 57-62; USPHS res career develop award, 63-68; asst prof, Sch Med, Wash Univ, 59-60; mem hemat training comt, Nat Inst Arthritis & Metab Dis, 64-68, res career prog comt, 68-72 & thrombosis adv comt, 69-73, chmn, 71-73; mem adv coun, Nat Arthritis Metab & Digestive Dis, NIH, 75-79; vis prof, Med Dept, Monash Univ, Melbourne, Australia, 80; vis scientist, Walter & Eliza Hall Med Res, Melbourne, Australia, 80. *Mem:* Am Soc Hemat; Am Soc Clin Invest; Am Fedn Clin Res; Am Soc Biol Chem; Asn Am Physicians. *Res:* Internal medicine; hematology; blood coagulation; blood cell metabolism. *Mailing Add:* Dept Med State Univ NY Health Sci Ctr Syracuse NY 13210

WILLIAMS, WILLIAM LANE, b Rock Hill, SC, Dec 23, 14. ANATOMY. *Educ:* Wofford Col, BS, 35; Duke Univ, MA, 39; Yale Univ, PhD(anat), 41. *Prof Exp:* Asst anat, Sch Med, Yale Univ, 39-40; instr, Sch Med & Dent, Univ Rochester, 41-42; instr, Sch Med, Yale Univ, 42-43; asst prof, Sch Med, La State Univ, 43-45; from asst prof to assoc prof, Univ Minn, Minneapolis, 45-58; prof & chmn dept, 58-80, EMER PROF ANAT, MED CTR, UNIV MISS, 80- *Concurrent Pos:* Donner Found fel anat, Sch Med, Yale Univ, 42-43; asst vchancellor, Med Ctr, Univ Miss, 75-80. *Mem:* Am Soc Exp Path; Am Physiol Soc; Soc Exp Biol & Med; Am Asn Anat; Am Inst Nutrit. *Res:* Endocrinology, experimental pathology; nutrition; cardiovascular disease; hepatic liposis. *Mailing Add:* Cndm J2 3975 I 55 N Jackson MS 39216

WILLIAMS, WILLIAM LAWRENCE, b St Cloud, Minn, June 14, 19; div; c 2. BIOCHEMISTRY, ANIMAL PHYSIOLOGY. *Educ:* Univ Minn, BS, 42; Univ Wis, MS, 47, PhD(biochem), 49. *Prof Exp:* Asst prof biochem, NC State Univ, 49-50; res biochemist, Lederle Lab, 50-59; from res assoc prof to res prof, 60-76, mem fac biochem, Univ Ga, 76-84; CONSULT, HUMAN INFERTILITY, 84- *Concurrent Pos:* NIH career develop award, 62-, res grant, 64-67, training grant, 65-70; indust consult; dir, Reprod Res Labs. *Mem:* Am Soc Biol Chem; Brit Soc Study Fertil; Soc Study Reprod; Am Fertil Soc; Am Physiol Soc. *Res:* Animal reproduction. *Mailing Add:* 3755 Barnett Shoals Athens GA 30605

WILLIAMS, WILLIAM LEE, atomic physics; deceased, see previous edition for last biography

WILLIAMS, WILLIAM ORVILLE, b Carlsbad, NMex, Oct 19, 40; m 60. MATHEMATICS, MECHANICS. *Educ:* Rice Univ, BA, 62, MS, 63; Brown Univ, PhD(appl math), 67. *Prof Exp:* Assoc res engr, Houston Res Lab, Humble Oil & Ref Co, 64; asst prof, 66-70, assoc prof, 70-76, PROF MATH, CARNEGIE-MELLON UNIV, 76- *Mem:* Soc Natural Philos; Am Math Soc; Math Asn Am. *Res:* Foundations of continuum mechanics; thermodynamics. *Mailing Add:* Dept Math Carnegie-Mellon Univ 5000 Forbes Ave Pittsburgh PA 15213

WILLIAMS, WILLIAM THOMAS, b San Marcos, Tex, Dec 22, 24; m 51; c 1. BIOCHEMISTRY, CELL PHYSIOLOGY. *Educ:* Southwest Tex State Univ, BS, 47; Tex A&M Univ, MS, 51, PhD(biochem), 65. *Prof Exp:* Instr chem, Tex A&M, 47-48; res asst biochem, Tex Agr Exp Sta, 58-61, res chemist, Marine Lab, Tex A&M Univ, 61-62; res biologist, USAF Sch Aerospace Med, Brooks AFB, 63-64, res chemist, 64-72, ASSOC FOUND SCI, SOUTHWEST FOUND RES EDUC, 73- *Mem:* Fel Am Inst Chemists; Am Inst Biol Sci; fel AAAS; Am Chem Soc; Am Soc Microbiol. *Res:* Regulation of cell metabolism and growth; nitrogen nutrition and metabolism; analytical biochemistry. *Mailing Add:* 116 Elm Spring Lane San Antonio TX 78231

WILLIAMS, WILLIE, JR, b Independence, La, Mar 24, 47. THERMAL PHYSICS. *Educ:* Southern Univ, BS, 70; Iowa State Univ, Ms, 72, PhD(physics), 74. *Prof Exp:* Assoc prof, 79-84, CHAIR DEPT, PHYSICS, LINCOLN UNIV, 76-, PROF, 84- *Concurrent Pos:* Mem fac physics, Lincoln Univ, 74-, chmn, Math & Sci Div, 78-; vis prof, Ctr Teaching Innovation, Drexel Univ, 75; founder, dir & prin investr, Laser Progs, Lincoln Advan Sci, Eng & Reinforcement Prog, 80-; bd mem, Women's Tech Prog Lincoln Univ Urban Ctr, 80- & Prime, Inc, 87-; dir, Lincoln Univ Nuclear Energy Training Fel Prog, 85-; liaison officer, Nat Asn Equal Opportunity Higher Educ, 87- *Mem:* Am Asn Physics Teachers; NY Acad Sci; AAAS; Nat Geog Soc; Sigma Xi. *Res:* Experimental studies of low temperature transport phenomena in metals and metal alloys. *Mailing Add:* 1454 Church Hill Pl Reston VA 22094

WILLIAMS, WILLIE ELBERT, b Jacksonville, Tex, June 6, 27; m 51; c 2. MATHEMATIC STATISTICS, COMPUTER SCIENCE. *Educ:* Huston-Tillotson Col, BS, 52; Tex Southern Univ, MS, 53; Mich State Univ, PhD(math educ), 72. *Prof Exp:* Teacher math, Lufkin Independent Schs, 53-59 & Case Western Reserve Univ, 64-73; dept chmn, Cleveland Bd Educ, 60-73; ASSOC PROF MATH, FLA INT UNIV, 73- *Mem:* Math Asn Am; Nat Coun Teachers Math. *Res:* Teacher effectiveness in mathematics and how children learn mathematics. *Mailing Add:* Dept Math Fla Int Univ Tamiami Trail Miami FL 33199

WILLIAMS-ASHMAN, HOWARD GUY, b London, Eng, Sept 3, 25; nat US; m 59; c 4. BIOCHEMISTRY. *Educ:* Cambridge Univ, BA, 46; Univ London, PhD, 59. *Prof Exp:* Biochemist, Chester Beatty Res Inst, Eng, 49-50; from asst prof to prof biochem, Univ Chicago, 53-64; prof pharmacol & exp therapeut & prof reprod biol, Sch Med, Johns Hopkins Univ, 64-69; prof biochem & physiol, 69-73, MAURICE GOLDBLATT PROF BIOL SCI, PRITZKER SCH MED, UNIV CHICAGO, 73- *Concurrent Pos:* Am Cancer Soc scholar, 53-57; USPHS res career award, 62-64. *Honors & Awards:* Amory Prize, Am Acad Arts & Sci, 75. *Mem:* Am Soc Biol Chem; fel Am Acad Arts & Sci. *Res:* Mechanism of hormone action; reproductive physiology; chemical pathology. *Mailing Add:* Ben May Inst Pritzker Sch Med Univ Chicago Chicago IL 60637

WILLIAMSON, ARTHUR ELRIDGE, JR, b Montgomery, Ala, July 6, 26; m 51; c 2. ELECTROOPTICS. *Educ:* Auburn Univ, BEP, 50, MS, 51. *Prof Exp:* Res engr, NAm Aviation, Inc, 51-52; instr physics, Univ Richmond, 52-53; res physicist, Southern Res Inst, 53-55; asst prof physics & res proj dir, Ga Inst Technol, 55-59; chief, Electrooptics Lab, Martin Marietta Corp, 59-73; head, Electro Optics Sect, Southern Res Inst, 73-88; SECY & TREAS, SOUTHERN RES TECH, INC, 88- *Mem:* Am Phys Soc; Optical Soc Am. *Res:* Optics. *Mailing Add:* 501 Bimbo Dr Birmingham AL 35202

WILLIAMSON, ASHLEY DEAS, b Columbus, Ga, June 16, 47; m 73; c 2. AEROSOL MECHANICS & CHEMISTRY. *Educ:* Emory Univ, BS, 68; Calif Inst Technol, PhD(chem), 76. *Prof Exp:* Sci & eng asst, US Army Chem Ctr, Edgewood Arsenal, Md, 69-71; res & teaching asst, Calif Inst Technol, 71-75; scientist, Oak Ridge Nat Lab, 75-79; sr chemist, 79-80, sect head, 81-84, DIV HEAD, SOUTHERN RES INST, 85- *Mem:* Am Chem Soc; Air Pollution Control Asn; Am Asn Aerosol Res. *Res:* Air pollution control and measurement; particulate sampling, characterization and measurement; chemical and physical transformation of condensible vapors in process emissions; laser spectroscopy; multiphoton excitation and ionization; mass spectrometry; gaseous ion chemistry; vacuum ultraviolet spectroscopy; radon measurement and mitigation. *Mailing Add:* Southern Res Inst Box 55305 2000 Ninth Ave S Birmingham AL 35255

WILLIAMSON, CHARLES ELVIN, b Portsmouth, Va, Dec 5, 26; m 52; c 6. BIO-ORGANIC CHEMISTRY, ONCOLOGY. *Educ:* Col William & Mary, BS, 50; Johns Hopkins Univ, PhD(bio-org chem), 70. *Prof Exp:* res chemist, Res Labs, Edgewood Arsenal, 52-79; PRES, JUNGLE-GEMS, INC, 79- *Concurrent Pos:* Res assoc, Sinai Hosp Baltimore, 55-80 & Johns Hopkins Univ Sch Med, 72- *Res:* Microenvironmental forces at biologic binding sites; reactions at cell surfaces; hydrophobic and electrostatic catalyses; neoplastic changes and cancer chemotherapy; tissue culture and micropropagation of plants. *Mailing Add:* 210 E Ring Factory Rd Bel Air MD 21014

WILLIAMSON, CLARENCE KELLY, b McKeesport, Pa, Jan 19, 24; m 51; c 2. MICROBIOLOGY. *Educ:* Univ Pittsburgh, BS, 49, MS, 51, PhD, 55. *Prof Exp:* Instr bact, Sch Pharm, Univ Pittsburgh, 51-55; from asst prof to assoc prof, 55-63, chmn dept microbiol, 62-72, dean, Col Arts & Sci, 71-82, exec vpres Acad Affairs & provost, 82-85, PROF MICROBIOL, MIAMI UNIV, 63-,. *Concurrent Pos:* Consult, Warren-Teed Prod Co, 54-64; consult ed, World Publ Co, 65-68; mem, Com Arts & Sci, Nat Asn State Univ & Land-Grant Col, 75- *Mem:* AAAS; Am Soc Microbiol; fel Am Acad Microbiol; Coun Cols of Arts & Sci (pres, 77-78). *Res:* Microbic dissociation, Pseudomonas aeruginosa; classification and polysaccharides of viridans streptococci; post-streptococcal nephritis. *Mailing Add:* Dept Microbiol Miami Univ Oxford OH 45056

WILLIAMSON, CLAUDE F, b Henderson, Tex, Mar 29, 33; m 59. NUCLEAR PHYSICS. *Educ:* Univ Tex, BS, 55, MA, 56, PhD(physics), 59. *Prof Exp:* Physicist, Saclay Nuclear Res Ctr, France, 60-62; res asst prof nuclear physics, Nuclear Physics Lab, Univ Wash, 62-66; res physicist, Lab Nuclear Sci, 66-75, SR RES SCI, DEPT PHYSICS, MASS INST TECHNOL, 75- *Mem:* AAAS; Am Phys Soc. *Res:* Fast neutron physics; nuclear reaction gamma rays; nuclear structure by electron scattering. *Mailing Add:* Rm 26-431 Mass Inst Technol 77 Massachusetts Ave Cambridge MA 02139

WILLIAMSON, CRAIG EDWARD, b Boston, Mass, July 20, 53; m 74; c 2. AQUATIC ECOLOGY, POPULATION BIOLOGY. *Educ:* Dartmouth Col, AB, 75, PhD(biol), 81; Mt Holyoke Col, MA, 77. *Prof Exp:* Asst prof, 81-87, ASSOC PROF BIOL, LEHIGH UNIV, 87- *Mem:* AAAS; Ecol Soc Am; Am Soc Limnol & Oceanog; Sigma Xi; Int Asn Theoret Appl Limnol. *Res:* Ecology of freshwater zooplankton; comparative limnology; role of selective predation in structuring freshwater zooplankton communities. *Mailing Add:* Dept Biol Williams Hall 31 Lehigh Univ Bethlehem PA 18015

WILLIAMSON, DAVID GADSBY, b Honolulu, Hawaii, June 12, 41; m 63; c 2. CHEMICAL KINETICS. *Educ:* Univ Colo, Boulder, BA, 63; Univ Calif, Los Angeles, PhD(phys chem), 66. *Prof Exp:* Chemist, Nat Bur Standards, Colo, 63; teaching & res asst chem, Univ Calif, Los Angeles, 63-66; fel, Nat Res Coun Can, 67-68; from asst prof to assoc prof, 68-76, PROF CHEM, CALIF POLYTECH STATE UNIV, SAN LUIS OBISPO, 76- *Concurrent Pos:* Res grant, Environ Protection Agency, 72. *Mem:* Am Chem Soc. *Res:* Ozone chemistry and the chemistry of free radicals of importance to atmospheric chemistry; development of energy sources alternate to petroleum products. *Mailing Add:* Dept Chem Calif Polytech State Col San Luis Obispo CA 93401

WILLIAMSON, DAVID LEE, b Humboldt, Nebr, July 17, 30; m 68; c 3. GENETICS. *Educ:* Nebr State Teachers Col, Peru, AB, 52; Univ Nebr, MS, 55, PhD(zool), 59. *Prof Exp:* Instr biol, Dana Col, 55-56; Fulbright scholar, Lab Genetics, Gif-sur-Yvette, France, 59-60; asst prof genetics, Univ Utah, 60-61; NIH fel, Yale Univ, 61-64; res fel, Med Col Pa, 64-66, asst prof, 66-71; assoc prof, 71-81, PROF ANAT SCI, STATE UNIV NY STONY BROOK, 81- *Mem:* AAAS; Am Soc Microbiol; Genetics Soc Am; Int Orgn Mycoplasmology. *Res:* Maternally inherited traits in Drosophila; biology of spiroplasmas. *Mailing Add:* Dept Anat Sci State Univ NY Health Sci Sch Med Stony Brook NY 11794

WILLIAMSON, DENIS GEORGE, b Trail, BC, June 9, 41; m 62; c 3. BIOCHEMISTRY. *Educ:* Univ BC, BSc, 63, PhD(biochem), 68. *Prof Exp:* Med Res Coun fel, 68-71, lectr, 71-72, from asst prof to assoc prof, 72-81, PROF BIOCHEM, UNIV OTTAWA, 81- *Mem:* Can Biochem Soc. *Res:* Metabolism of steroid hormones; purification and characterization of steroid dehydrogenases; estrogen receptors. *Mailing Add:* Dept Biochem Univ Ottawa Ottawa ON K1H 8M5 Can

WILLIAMSON, DONALD ELWIN, b Lansing, Mich, Oct 24, 13; m 40; c 3. INSTRUMENTATION, BIOMEDICAL ENGINEERING. *Educ:* Carleton Col, AB, 35; Univ Mich, MS, 36. *Prof Exp:* Res physicist, Dept Eng Res, Univ Mich, 36, res engr, 44-45; mgr, Profilometer Div, Physicists Res Co, 36-44; res engr, Lincoln Park Industs, 45-47; chief engr & assoc dir res, Baird Assocs, Inc, Mass, 47-52; pres & treas, Williamson Develop Co, 53-60; sci adv to pres, Cordis Corp, 60-79; RETIRED. *Concurrent Pos:* Chmn, Gordon Res Conf Instrumentation, 56. *Mem:* Am Soc Mech Eng; Optical Soc Am. *Res:* Roughness measurement; optics; infrared instruments; physiological and cardiovascular instrumentation. *Mailing Add:* 13001 Old Cutler Rd Miami FL 33156

WILLIAMSON, DOUGLAS HARRIS, b Croydon, Eng, Mar 3, 24; wid; c 2. GEOLOGY. *Educ:* Aberdeen Univ, BSc, 50, PhD, 52. *Prof Exp:* Lectr geol, Aberdeen Univ, 50-53; asst prof, Mt Allison Univ, 53-54, Sr James Dunn prof & head dept, 54-66; head dept, Laurentian Univ, 66-74, assoc dean sci, 69-72, prof geol, 66-80, dean sci, 72-80; Dean Sci, St Mary's Univ, 80-87, prof geol, 80-87; RETIRED. *Concurrent Pos:* Nat Res Coun Can major equip grant, 65, operating grants, 65 & 66; Geol Surv Can grant, Univ NB, 62, NB Res & Productivity Coun grant, 62-66; mem subcomt, Nat Adv Comt Res Geol Sci in Can, 56-57; mem, Comt Deans Arts & Sci, 69-; mem, Coun Ont Univs, 70-72; Univ res grants officer, 70-; mem, Atl Prov Coun Sci, 80- *Mem:* AAAS; Nat Asn Geol Teachers; Can Inst Mining & Metall; Geol Asn Can. *Res:* X-ray crystallography; petrogenesis and synthesis of metamorphic mineral assemblages; mineralogy of fluorspar; exploration of mineral resources of Ontario; geology of Pre-Carboniferous rocks of Southern New Brunswick. *Mailing Add:* Dean's Off Fac Sci St Marys Univ Halifax NS B3H 3C3 Can

WILLIAMSON, EDWARD P, b Lee Co, Fla, July 1, 33; m 58; c 3. ELECTRICAL ENGINEERING, COMMUNICATIONS. *Educ:* Univ Fla, BEE, 55, MSE, 60, PhD(elec eng), 65. *Prof Exp:* Engr, Bendix Radio Div, 55-56; asst prof, 65-70, ASSOC PROF ELEC ENG, TULANE UNIV, 70- *Concurrent Pos:* Res engr, Gulf South Res Inst, 66; res scientist, Kaman Nuclear Div, Colo, 67; NASA fac fel, 74. *Mem:* Inst Elec & Electronics Engrs. *Res:* Digital communications systems; statistical communication theory; communication system signal design and modulation techniques; optimization, estimation and decision theory as applied to communication and radar systems; satellite communication systems; communication channel modelling. *Mailing Add:* Dept Elec Eng Tulane Univ New Orleans LA 70118

WILLIAMSON, FRANCIS SIDNEY LANIER, b Little Rock, Ark, Feb 6, 27; m 78; c 3. ZOOLOGY, PUBLIC HEALTH ADMINISTRATION. *Educ:* San Diego State Col, BS, 50; Univ Calif, MA, 55; Johns Hopkins Univ, ScD, 69. *Prof Exp:* Mus technician vert zool, Univ Calif, 53-55; med biologist, Arctic Health Res Ctr, USPHS, 55-64; dir, Chesapeake Bay Ctr Environ Studies, Smithsonian Inst, 68-75; NIH fel & res assoc, 64-68, assoc pathobiol, Sch Hyg & Pub Health, Johns Hopkins Univ, 65-77; comnr, Dept Health & Social Serv, Alaska, 75-78; prog mgr polar biol & med, NSF, 78-79, chief scientist, Polar Sci Sect, 79-86; DIR, INST ARCTIC BIOL, UNIV ALASKA, FAIRBANKS, 86- *Concurrent Pos:* Consult, Avian ecol, AEC, 59-61; instr, Anchorage Community Col, Univ Alaska, 61-64; consult ecol, Battelle Mem Inst, 67-72; adv, Chesapeake Bay Study, US Army Corps Engrs, 69-75; ed, Biosci, Am Inst Biol Sci, 70 & Condor, Cooper Ornith Soc, 71-74; consult, Md Dept Nat Resources, 72-73; adv environ qual comn, Gov Sci Adv Coun, Md, 72-73. *Honors & Awards:* Exceptional Serv Award, Smithsonian Inst, 75; Dept Interior Commendation, 75. *Mem:* Fel AAAS; Ecol Soc Am; Cooper Ornith Soc; fel Am Ornith Union; Wilson Ornith Soc; Wildlife Dis Asn. *Res:* Taxonomy, behavior, distribution, parasites and viral diseases of birds; taxonomy, distribution, and life histories of helminths; medical helminthology; estuarine ecology; environmental planning; university administration. *Mailing Add:* Inst Arctic Biol Univ Alaska Fairbanks AK 99775

WILLIAMSON, FREDERICK DALE, paint chemistry, for more information see previous edition

WILLIAMSON, HANDY, JR, b Louin, Miss, Oct 24, 45; m 68; c 1. ECONOMIC DEVELOPMENT, MANPOWER PLANNING & TRAINING. *Educ:* Tenn State Univ, MS, 69; Univ Mo, MS, 71, PhD(agr econ), 74; Alcorn State Univ, BS, 77. *Prof Exp:* Assoc prof & assoc regional dir, Agr Econ & Rural Res, Tuskegee Univ, 74-77; assoc prof & res dir, Agr Res, Tenn State Univ, 77-85; dep dir, Res & Univ Rel, Agency Int Develop, 85-88; PROF AGR ECON & HEAD DEPT, UNIV TENN, 88- *Concurrent Pos:* Consult, numerous govt agencies, 74-90; mem, Nat Rural Develop Comn, 75-77 & US Joint Coun Food & Agr Sci, 82-84; agency liaison, White House Comn Hist Black Cols, 85-90. *Mem:* Am Agr Econ Asn; Southern Agr Econ Asn. *Res:* Regional economic development manpower economics and the small firm sector; demographic, historical and empirical analysis. *Mailing Add:* 12108 E Ashton Ct Knoxville TN 37922

WILLIAMSON, HAROLD E, b Racine, Wis, Aug 8, 30; m 57; c 3. PHARMACOLOGY. *Educ:* Univ Wis, BS, 53, PhD(pharmacol, toxicol), 59. *Prof Exp:* Res asst pharmacol, Univ Wis, 55-59, proj assoc, 59-60; from instr to assoc prof, 60-70, PROF PHARMACOL, COL MED, UNIV IOWA, 70- *Mem:* AAAS; Am Soc Pharmacol & Exp Therapeut; Int Soc Nephrology; Soc Exp Biol & Med; Am Soc Nephrology; fel Am Col Clin Pharmacol. *Res:* Renal pharmacology and physiology, especially the effect of diuretics and hormones on electrolyte and water transport. *Mailing Add:* Dept Pharmacol Univ Iowa Col Med Iowa City IA 52242-1194

WILLIAMSON, HUGH A, b Kemp, Tex, Aug 11, 32; m 56; c 4. PHYSICS. *Educ:* North Tex State Univ, BA, 54; Univ Tex, PhD(physics, math), 62. *Prof Exp:* Res scientist, Molecular Physics Res Lab & Mil Physics Res Lab, Univ Tex, 60-63, res fel, 62-63; res scientist, Res Lab, United Aircraft Corp, 63-65, sr res scientist, 65-67; from asst prof to assoc prof, 67-74, PROF PHYSICS, CALIF STATE UNIV, FRESNO, 74- *Res:* Atomic and molecular structure; gaseous electronics; electron scattering processes off neutral atoms including elastic, inelastic and free-free scattering processes. *Mailing Add:* Dept Physics Calif State Univ Fresno CA 93740

WILLIAMSON, HUGH A, b Williamsport, Pa, Apr 20, 27; m 60; c 2. ORGANIC CHEMISTRY, SCIENCE EDUCATION. *Educ:* Bucknell Univ, AB, 60, MA, 61; Cornell Univ, EdD(sci ed, chem), 66. *Prof Exp:* Teacher high sch, Pa, 50-55; prof chem, Lock Haven State Col, 55-90, assoc dean arts & sci, 73-90; RETIRED. *Mem:* Nat Asn Res Sci Teaching. *Res:* Synthesis of cyclooctatetraene compounds; reductions of epoxides. *Mailing Add:* PO Box 209 Woolrich PA 17779

WILLIAMSON, JAMES LAWRENCE, b Rebecca, Ga, Feb 28, 29; m 49; c 3. ANIMAL SCIENCE, ANIMAL NUTRITION. *Educ:* Univ Ga, BSA, 51; Univ Ill, MS, 52, PhD, 57. *Prof Exp:* Mgr, Beef Cattle & Sheep Res, Chow Div, 57-63, mgr, Livestock Res, 63-64, dir res, 64-65, vpres & dir res, Chow Div, 65-83, sr vpres technol develop, Purina Mills, Inc, Ralston Purina Co, 83-86, sr vpres, dir corp res, 86-88; RETIRED. *Concurrent Pos:* Mem, adv comt, USDA, 83- & vet med adv comt, Food & Drug Admin, 85- *Mem:* Am Soc Animal Sci; Am Dairy Sci Asn; Poultry Sci Asn; Agr Res Inst; Indust Res Inst. *Res:* Animal nutrition and management; research management, product acquisition, and business development. *Mailing Add:* Six Fox Meadows St Louis MO 63127-1401

WILLIAMSON, JERRY ROBERT, b Danville, Ill, Feb 14, 38; m 65; c 1. ORGANIC CHEMISTRY, POLYMER CHEMISTRY. *Educ:* Univ Ill, Urbana, BA, 60; Univ Iowa, MS, 63, PhD(org polymer chem), 64. *Prof Exp:* Petrol Res Fund res fel polymer res, Univ Iowa, 61-62, teaching asst gen & org chem, 62-64; asst prof chem & actg chmn div sci, Jarvis Christian Col, 64-66; res fel, Tex Christian Univ, 66-67; asst prof, 67-70, ASSOC PROF CHEM, EASTERN MICH UNIV, 70- *Concurrent Pos:* Partic, State Tech Serv Prog, Mich, 67-69; consult hazardous chem safety, off training serv, JT Baker Chem Co, Phillipsburg, NJ. *Mem:* AAAS; Am Chem Soc. *Res:* Organic polymer chemistry, thermally stable materials; polymer analysis via gel permeation chromatography. *Mailing Add:* Dept Chem Eastern Mich Univ Ypsilanti MI 48197

WILLIAMSON, JOHN HYBERT, b Clarkton, NC, Apr 28, 38; m 63; c 3. GENETICS. *Educ:* NC State Col, BS, 60; Cornell Univ, MS, 63; Univ Ga, PhD(zool), 66. *Prof Exp:* Fel biol, Oak Ridge Nat Lab, 66-67; postdoctoral fel life sci, Univ Calif, Riverside, 67-69; from asst prof to prof zool, Univ Calgary, 69-81, acad admin officer biol dept, 74-76, head dept, 76-78; PROF & CHMN DEPT BIOL, DAVIDSON COL. *Mem:* Genetics Soc Can; Genetics Soc Am. *Res:* Chromosome mechanics; radiation biology; developmental genetics; enzymology. *Mailing Add:* Dept Biol Davidson Col Davidson NC 28036

WILLIAMSON, JOHN RICHARD, b Coventry, Eng, Sept 18, 33; m 61; c 3. BIOCHEMISTRY, BIOPHYSICS. *Educ:* Oxford Univ, BA, 56, MA, 59, PhD(biochem), 60. *Prof Exp:* Dept demonstr biochem, Oxford Univ, 60-61; independent investr, Baker Clin Res Lab, Harvard Univ, 61-63; assoc phys biochem, 63-65, from asst prof to prof phys biochem, Johnson Res Found, 65-76; PROF BIOCHEM, SCH MED, UNIV PA, 75- *Concurrent Pos:* USPHS fel, 61-63; Am Heart Asn grant, 66-72; NIH res grants & contract, 71-; estab investr, Am Heart Asn, 67-72, mem coun basic sci. *Mem:* Am Soc Biol Chem; Am Diabetes Asn; Brit Biochem Soc; NY Acad Sci; Am Physiol Soc. *Res:* Mode of action of hormones and drugs; control of metabolic pathways; effect of hormones on cells and hormone interactions in animal cells; role of anion transport across mitochondrial membranes; myocardial ischemia. *Mailing Add:* Dept Biochem & Biophysics Rm 601 Univ Pa 37th & Hamilton Walk Philadelphia PA 19104

WILLIAMSON, JOHN S, b Jackson, Miss, Oct 24, 58; m 83; c 3. HYBRID ANTIBIOTIC PRODUCTION, DESIGN OF ANTICANCER CHEMOTHERAPEUTICS. *Educ:* Univ Miss, BS, 82; Univ Iowa, PhD(med chem & natural prod), 87. *Prof Exp:* Postdoctoral fel biol, Yale Univ, 87-89; ASST PROF MED CHEM, UNIV MISS, 89- *Concurrent Pos:* Asst res prof, Res Inst Pharmaceut Sci, 90-; Am Soc Pharmacog young investr award, 90; Am Asn Cols Pharm young investr award, 91. *Mem:* Am Chem Soc; Am Soc Microbiol; AAAS; Soc Indust Microbiologists; Am Soc Pharmacog. *Res:* Exploitation of microbiol biosynthetic and catabolic enzymatic systems for use as alternative synthetic reagents. *Mailing Add:* Dept Med Chem Sch Pharm Univ Miss University MS 38677

WILLIAMSON, JOHN W, b Tulsa, Okla, Oct 26, 33; m 61; c 3. MECHANICAL ENGINEERING. *Educ:* Univ Okla, BSc, 55; Ohio State Univ, MSc, 59, PhD(mech eng), 65. *Prof Exp:* Test engr, NAm Aviation, Inc, 56-57; instr mech eng, Ohio State Univ, 57-60, 61-64, res asst, 60-61; from asst prof to assoc prof mech eng, 64-77, PROF MECH ENG & MAT SCI, VANDERBILT UNIV, 77- *Concurrent Pos:* Consult, Aerospace Struct Div, Avco Corp, 66- & E I du Pont de Nemours & Co, Inc, 69. *Mem:* Am Soc Eng Educ; Am Soc Mech Engrs. *Res:* Fluid mechanics, specifically aspects of turbulent fluid flow; energy utilization studies. *Mailing Add:* Dept Mech Eng Vanderbilt Univ Nashville TN 37240

WILLIAMSON, KENNETH DALE, b Drumright, Okla, Sept 4, 20; m 46; c 3. PHYSICAL CHEMISTRY. *Educ:* Univ Okla, BS, 47, MS, 48; Univ Tex, PhD(chem), 54. *Prof Exp:* Chemist, Petrol Exp Sta, US Bur Mines, Okla, 48-50; spec instr chem, Univ Tex, 53; res chemist, Union Carbide Corp, 53-66, group leader res, 66-71, res scientist, 71-85; RETIRED. *Concurrent Pos:* Asst prof, Morris Harvey Col, 56-59 & WVa State Col, 62-64; adj prof, Col Grad Studies, WVa, 71. *Mem:* AAAS; Am Chem Soc; Sigma Xi; N Am Catalyst Soc. *Res:* Physical properties of gas hydrates; physical properties of pure compounds and mixtures; thermodynamics; calorimetry; kinetics of pyrolysis of hydrocarbons; catalysis. *Mailing Add:* 1022 Sand Hill Dr St Albans WV 25177

WILLIAMSON, KENNETH DONALD, JR, systems analysis, cryogenic engineering; deceased, see previous edition for last biography

WILLIAMSON, KENNETH LEE, b Tarentum, Pa, Apr 13, 34; m 56; c 3. ORGANIC CHEMISTRY, STRUCTURAL CHEMISTRY. *Educ:* Harvard Univ, BA, 56; Univ Wis, PhD(org chem), 60. *Prof Exp:* NIH fel, Stanford Univ, 60-61; from asst prof to assoc prof, 61-69, dept chmn, 78-81, PROF CHEM, MT HOLYOKE COL, 69-, MARY E WOOLLEY PROF, 84- *Concurrent Pos:* Mem grad faculty, Univ Mass, 62-; vis prof, Cornell Univ, 66; NSF sci fac fel & fel Univ Liverpool, 68-69; vis assoc, Calif Inst Technol, 75 & 82; secy, Exp Nuclear Magnetic Resonance Spectroscopy Confs, 73-78, chmn, 79; fel, John Simon Guggenheim Found, 75-76; vis prof, Univ Utah, 76, Oxford Univ, Eng, 76 & 83, Dartmouth Col, 86-87 & Harvard Univ, 89-90; mem, Comt Hazardous Substances in the Labs, Nat Res Coun, 81-82. *Mem:* Am Chem Soc; Sigma Xi; AAAS. *Res:* Conformational analysis by means of nuclear magnetic resonance spectroscopy; Xenon nuclear magnetic resonance; carbon relaxation time studies; microscale organic experiments. *Mailing Add:* Dept Chem Mt Holyoke Col South Hadley MA 01075

WILLIAMSON, LUTHER HOWARD, b Osyka, Miss, Oct 9, 36; m 57; c 3. PHYSICAL CHEMISTRY, OIL FIELD CHEMISTRY. *Educ:* La State Univ, BS, 59, MS, 62, PhD(phys chem), 65. *Prof Exp:* Res chemist corrosion, Mobil Res & Develop Corp, 65-70, sr res chemist corrosion, Water Chem, 70-78; staff engr corrosion, Mat Eng, Superior Oil Co, 78-85; TECH SERV CONSULT, TEX SERV, 85- *Mem:* Am Chem Soc; Nat Asn Corrosion Engrs; Soc Petrol Engrs. *Res:* Surface chemistry; corrosion, corrosion inhibition, electrochemistry of corrosion; hydrogen embrittlement; sulfide stress corrosion cracking; water chemistry, oilfield chemistry, chemistry of scale formation; water pollution, air pollution; environmental science. *Mailing Add:* Rte 25 Box 964A Tyler TX 75707

WILLIAMSON, PATRICK LESLIE, b Dickson, NDak, Apr 15, 48; m 70; c 1. CYTOLOGY, BIOCHEMISTRY. *Educ:* Beloit Col, BA, 70; Harvard Univ, MS, 71, PhD(molecular biol), 74. *Prof Exp:* Staff fel, NIH, 74-77; asst prof, 77-83, ASSOC PROF BIOL, AMHERST COL, 83- *Mem:* Am Soc Cell Biol. *Res:* Cell biology; structure and function of the plasma membrane of eukaryote cells; alterations of same in leukemia and sickle cell disease. *Mailing Add:* Dept Biol Amherst Col A103 Webster Amherst MA 01002

WILLIAMSON, PETER GEORGE, b Hemel, UK, May 5, 52. EVOLUTIONARY BIOLOGY, BIOSTRATIGRAPHY. *Educ:* Univ Bristol, BSc, 74, PhD(geol), 79. *Prof Exp:* Res fel, NATO, 79-81; ASST PROF GEOL, HARVARD UNIV, 82- *Mem:* Soc Syst Zool; Geologists Asn; Conchological Soc; Palaeont Asn. *Res:* Quantification of evolutionary mode and tempo in certain well-preserved mollusc sequences in East Africa and Germany; biostratigraphy of the East African hominid deposits; theoretical geometry of shell-coiling. *Mailing Add:* 125 Brooks Ave Arlington MA 02174

WILLIAMSON, RALPH EDWARD, b Wilson, NC, Dec 28, 23; m 46; c 2. PLANT PHYSIOLOGY. *Educ:* NC State Univ, BS, 48; Univ Wis, MS, 50, PhD(bot), 58. *Prof Exp:* Botanist, Chem Corps, Dept Army, 48-49, plant physiologist, 50-51, 53-57 & Soil & Water Conserv Res Div, 57-74, PLANT PHYSIOLOGIST, TOBACCO RES LAB, USDA, OXFORD, NC, 74- *Mem:* Am Soc Plant Physiol; Am Soc Agron; Soil Sci Soc Am. *Res:* Determine major organic constituents of leaf tobacco that, if eliminated or reduced through breeding, cultural, processing, or curing techniques, would result in tobacco varieties less harmful to the consumer. *Mailing Add:* 716 Currituck Dr Raleigh NC 27609

WILLIAMSON, RICHARD CARDINAL, b Minocqua, Wis, Sept 10, 39; m 61; c 4. ELECTROOPTICAL DEVICES, SURFACE ACOUSTIC WAVES. *Educ:* Mass Inst Technol, BS, 61, PhD(physics), 66. *Prof Exp:* Staff scientist, Electronics Res Ctr, NASA, 65-70; mem tech staff, 70-74, assoc group leader, 74-80, GROUP LEADER, MASS INST TECHNOL LINCOLN LAB, 80- *Honors & Awards:* Centennial Award, Inst Elec & Electronics Engrs, 84, Career Achievement Award, 85. *Mem:* Fel Inst Elec & Electronics Engrs; Am Phys Soc; Optical Soc Am. *Res:* Electrooptical devices and applications; optical signal processing; surface acoustic waves and their use in signal processing; ultrasonic investigations; phase transitions; low-temperature physics. *Mailing Add:* Lincoln Lab Mass Inst Technol PO Box 73 Lexington MA 02173

WILLIAMSON, RICHARD EDMUND, b Chicago, Ill, May 23, 27; m 50. MATHEMATICS. *Educ:* Dartmouth Col, AB, 50; Univ Pa, AM, 51, PhD, 55. *Prof Exp:* Res asst, Univ Pa, 55-56; instr, 56-58, from asst prof to assoc prof, 58-66, PROF MATH, DARTMOUTH COL, 66- *Concurrent Pos:* Res fel, Harvard Univ, 60-61. *Mem:* Am Math Soc. *Res:* Analysis. *Mailing Add:* Dept Math Dartmouth Col Hanover NH 03755

WILLIAMSON, ROBERT BRADY, b New Rochelle, NY, Nov 19, 33; m 59; c 3. MATERIAL SCIENCE ENGINEERING, FIRE PROTECTION ENGINEERING. *Educ:* Harvard Univ, AB, 56, SB, 59, PhD(appl physics), 65. *Prof Exp:* Asst prof civil eng, Mass Inst Technol, 65-68; PROF CIVIL ENG, UNIV CALIF, BERKELEY, 68- *Mem:* Am Soc Civil Engrs; Am Soc Testing & Mat; Soc Fire Protection Engrs; Mat Res Soc; Combustion Inst. *Res:* Morphology of solidified materials; fracture of materials; fire research and testing; theory of learning and programmed instruction. *Mailing Add:* Dept Civil Eng Univ Calif Berkeley CA 94720

WILLIAMSON, ROBERT ELMORE, b York Co, SC, Nov 8, 37; m 64; c 2. AGRICULTURAL MECHANIZATION, MACHINERY MANAGEMENT. *Educ:* Clemson Univ, BS, 59, MS, 64; Miss State Univ, PhD (eng), 72. *Prof Exp:* Res assoc, Miss State Univ, 66-71; asst prof, Univ Ga, 71-78; assoc prof, 78-81, PROF AGR ENG, CLEMSON UNIV, 81- *Mem:* Am Soc Agr Engrs; Am Soc Eng Educ; Sigma Xi. *Res:* Fruit and vegetable mechanization; equipment and techniques for chemical application; machinery management and machine systems for agriculture production. *Mailing Add:* 303 Princess Grace Ave Clemson SC 29631

WILLIAMSON, ROBERT EMMETT, b Ashland, Kans, June 9, 37; m 82; c 3. GEOMETRY, TOPOLOGY. *Educ:* Univ Ariz, BS, 59; Univ Calif, Berkeley, PhD(math), 63. *Prof Exp:* Mem, Inst Adv Study, 63-65; vis prof, Univ Warwick, 65-66; asst prof, Yale Univ, 66-69; ASSOC PROF MATH, CLAREMONT GRAD SCH, 69- *Concurrent Pos:* Nat Acad Sci-Air Force Off Sci Res fel, 63-64; mem, Inst Advan Study, Princeton, NJ, 63-65. *Mem:* Am Math Soc. *Res:* Algebraic model of surgery; singularities of smooth maps. *Mailing Add:* Dept Math Claremont Grad Sch Claremont CA 91711

WILLIAMSON, ROBERT MARSHALL, b Madison, Wis, Feb 2, 23; m 50; c 2. PHYSICS. *Educ:* Univ Fla, BS, 43, PhD(physics), 51. *Prof Exp:* Res assoc, Duke Univ, 51-53, from asst prof to assoc prof physics, 53-62; PROF PHYSICS, OAKLAND UNIV, 62- *Concurrent Pos:* Fulbright lectr, Univ Catania, 59-60. *Mem:* Am Phys Soc; Am Asn Phys Teachers; Acoust Soc Am. *Res:* Nuclear spectroscopy; musical acoustics. *Mailing Add:* Dept Physics Oakland Univ Rochester MI 48063

WILLIAMSON, ROBERT SAMUEL, b Cincinnati, Ohio, June 18, 22. PHYSICS. *Educ:* Queens Col, BS, 45; NY Univ, MS, 48; Polytech Inst Brooklyn, PhD(physics), 57. *Prof Exp:* Tutor, 52-56, instr, 56-60, from asst prof to assoc prof, 60-68, from asst dean to assoc dean admin, 68-72, PROF PHYSICS, QUEENS COL, NY, 68- *Mem:* Am Phys Soc; Am Asn Physics Teachers; Am Crystallog Asn; Inst Elec & Electronics Engrs; Sigma Xi. *Res:* X-ray crystallography; electronics. *Mailing Add:* 4726 196th Pl Flushing NY 11358

WILLIAMSON, SAMUEL JOHNS, b West Reading, Pa, Nov 6, 39; m 66. NEUROMAGNETISM, MAGNETIC SENSOR TECHNOLOGY. *Educ:* Mass Inst Technol, SB, 61, ScD(physics), 65. *Hon Degrees:* ScD, NJ Inst Technol, 85. *Prof Exp:* Staff mem, Francis Bitter Nat Magnet Lab, Mass Inst Technol, 65-66; Nat Acad Sci-Nat Res Coun fel, Dept Physics of Solids, Fac Sci, Univ Paris, Orsay, France, 66-67; mem tech staff, NAm Aviation Sci Ctr, 67-70; lectr physics, Univ Calif, Santa Barbara, 70-71; assoc prof, 71-77, PROF PHYSICS, NY UNIV, 77-, PROF PHYSIOL & BIOPHYS, SCH MED, 84-, UNIV PROF, 90- *Concurrent Pos:* Fulbright sr res scholar, 79-80; chmn, 7th Int Conf Biomagnetism, 89. *Mem:* Fel Am Phys Soc; Sigma Xi; AAAS; Soc Neurosci; Inst Elec & Electronics Engrs; Biophys Soc. *Res:* Studies of the magnetic field produced by neural activity of the human brain to elucidate sensory, motor, and cognitive processes and where they take place. *Mailing Add:* Physics Dept NY Univ Two Washington Pl New York NY 10003

WILLIAMSON, STANLEY GILL, b Manhattan, Kans, Aug 28, 38; m 65. MATHEMATICS. *Educ:* Calif Inst Technol, BS, 60; Stanford Univ, MS, 62; Univ Calif, Santa Barbara,PhD(math), 65. *Prof Exp:* From asst prof to assoc prof, 65-75, PROF MATH, UNIV CALIF, SAN DIEGO, 75- *Mem:* Soc Indust Appl Math. *Res:* Combinatorial analysis; computation. *Mailing Add:* Dept Math Univ Calif San Diego La Jolla CA 92093

WILLIAMSON, STANLEY MORRIS, b Chattanooga, Ten, Mar 18, 36; m 66. CHEMISTRY. *Educ:* Univ NC, BS, 58; Univ Wash, PhD(chem), 61. *Prof Exp:* Asst prof chem, Univ Calif, Berkeley, 61-65, from asst prof to assoc prof, 65-74, PROF CHEM, UNIV CALIF, SANTA CRUZ, 74-, DEAN GRAD DIV, 72- *Mem:* AAAS; Am Chem Soc; Royal Soc Chem. *Res:* Fluorine chemistry of compounds of sulfur, nitrogen, oxygen and xenon including preparations and properties. *Mailing Add:* Dept Chem Univ Calif Santa Cruz CA 95064

WILLIAMSON, SUSAN, b Boston, Mass, Dec 29, 36. MATHEMATICS. *Educ:* Radcliffe Col, AB, 58; Brandeis Univ, AM, 61, PhD(math), 63. *Prof Exp:* Instr, Cardinal Cushing Col, 62-63; asst prof math, Boston Col, 63-64; scholar hist sci, Harvard Univ, 64-65; from asst prof to assoc prof, 65-71, dean col, Regis Col, 73-75, PROF MATH, REGIS COL, MASS, 71- *Mem:* Am Math Soc; Math Asn Am; Asn Women Math. *Res:* Associative algebras; commutative rings. *Mailing Add:* 37 Hagen Rd Newton Centre MA 02159

WILLIAMSON, THOMAS GARNETT, b Quincey, Mass, Jan 27, 34; m 61; c 3. NUCLEAR ENGINEERING. *Educ:* Va Mil Inst, BS, 55; Rensselaer Polytech Inst, MS, 57; Univ Va, PhD(physics), 60. *Prof Exp:* Nuclear engr, Alco Prod, Inc, NY, 57-58; dir, Reactor Facil, 77-79, chmn nuclear eng & eng physics dept, 77-90, PROF NUCLEAR ENG, UNIV VA, 60-; SR SCIENTIST, WESTINGHOUSE SAVANNAH RIVER LABS, 90- *Concurrent Pos:* Consult, Gen Atomic Div, Gen Dynamics Corp, 66 & Combustion Eng, Conn, 71, 72; mem, Syst Nuclear Safety & Operating Comt, Va Elec & Power Co, 75-90; mem reactor safety comt, Babcock & Wilcox Co, 76-90; sabbatical leave, Nat Bur Standards, Md, 84-85. *Mem:* Fel Am Nuclear Soc. *Res:* Radioactive isotope usage; neutron activation analysis; radiation shielding; reactor physics. *Mailing Add:* Westinghouse Savnnah River Labs Aiken SC 29802

WILLIAMSON, WALTON E, JR, b Corpus Christi, Tex, May 21, 44; m 65. FLIGHT MECHANICS, OPTIMIZATION. *Educ:* Stanford Univ, BS, 66; Univ Tex, Austin, MS, 67, PhD(aerospace eng), 70. *Prof Exp:* Asst prof optimal control, Univ Tex, Austin, 70-73; mem tech staff, 74-84, SUPVR, ADV SYSTS DIV, SANDIA LABS, ALBUQUERQUE, 84- *Mem:* Am Inst Aeronaut & Astronaut. *Res:* Numerical methods for optimal control; shuttle reentry optimization; hypersonics; ballistic and maneuvering reentry vehicles. *Mailing Add:* Sandia Labs Orgn 9144 PO Box 5800 Bldg 807 Rm 1086 Albuquerque NM 87185

WILLIAMSON, WILLIAM, JR, b Newport, RI, Jan 20, 34; m 57; c 2. ATOMIC PHYSICS, TRANSPORT THEORY. *Educ:* San Francisco State Col, BA, 55; Univ Calif, Berkeley, MA, 58; Univ Colo, PhD(physics), 63. *Prof Exp:* Physicist, US Naval Radiol Defense Lab, Calif, 56-58; Fulbright fel, Frascati Labs, Italy, 61-62; asst physics, Univ Colo, 62-63; fel, Inst Sci & Tech, Univ Mich, 63-64, instr, 64-65; asst prof physics & astron, 65-69, assoc prof, 69-75, PROF PHYSICS, UNIV TOLEDO, 75- *Concurrent Pos:* Vis prof, Univ Adelaide, 71-72; consult, Sandia Nat Labs, Livermore, Calif. *Mem:* Am Phys Soc; Am Asn Physics Teachers; AAAS. *Res:* Nonlinear differential equations (applied). *Mailing Add:* Physics-Astron Univ Toledo Toledo OH 43606

WILLIAMSON, WILLIAM BURTON, Frederick, Okla, Apr 29, 46; m 67; c 3. HETEROGENEOUS CATALYSIS. *Educ:* Tex Tech Univ, BS, 69; Tex A&M Univ, PhD(phys chem), 76. *Prof Exp:* SR RES SCIENTIST, SCI RES, FORD MOTOR CO, 76- *Mem:* Am Chem Soc; Sigma Xi. *Res:* Basic and applied laboratory research on potential automotive emission catalysts (heterogeneous catalysis) for the reduction of nitric oxide and oxidation of carbon monoxide and hydrocarbons. *Mailing Add:* 2521 W Elgin Broken Arrow OK 74012-2212

WILLIAMSON, WILLIAM O(WEN), b Luton, Eng, Jan 30, 11; m 58; c 1. CERAMICS, PETROLOGY. *Educ:* Univ London, BSc, 30 & 31, PhD(geol), 33, DSc(geol, indust chem), 58. *Prof Exp:* Chief asst ceramics, North Staffordshire Col Technol, Eng, 34-42; res chemist, Univ Birmingham, 42-45;

prof officer ceramics, Univ Witwatersrand, 45-47; prin res officer, Chem Res Labs, Commonwealth Sci & Indust Res Orgn, Australia, 47-59; assoc prof ceramic tech, 59-65, prof ceramic sci, 65-76, EMER PROF CERAMIC SCI, PA STATE UNIV, UNIVERSITY PARK, 76- *Mem:* Fel Am Ceramic Soc; Royal Inst Chem. *Res:* Solid state technology; rheology and surface chemistry; high temperature reactions; materials in ancient and modern cultures. *Mailing Add:* 116 Steidle Pa State Univ University Park PA 16802

WILLIARD, PAUL GREGORY, b Mt Carmel, Pa, Dec 18, 50; m 79; c 2. ORGANIC CHEMISTRY, STRUCTURAL CHEMISTRY. *Educ:* Bucknell Univ, BS & MS, 72; Columbia Univ, MPhil, 74, PhD(chem), 76. *Prof Exp:* NIH trainee chem, Mass Inst Technol, 76-78, fel, 78-79; asst prof, 79-85, ASSOC PROF CHEM, BROWN UNIV, 86- *Concurrent Pos:* NIH, Res Career Develop Awards, 88-93. *Mem:* Am Chem Soc; Am Crystal Asn. *Res:* Total synthesis of natural products; synthetic methods; x-ray crystallography. *Mailing Add:* Dept Chem Box H Brown Univ Providence RI 02912

WILLIFORD, WILLIAM OLIN, b San Pedro, Calif, July 18, 33; m 71; c 2. MATHEMATICAL STATISTICS. *Educ:* Pepperdine Col, BA, 57; Fla State Univ, MS, 59; Va Polytech Inst & State Univ, PhD(statist), 67. *Prof Exp:* Instr math, Fla State Univ, 59-62; asst prof, Roanoke Col, 62-63; NIH fel, 64-67; asst prof statist, Univ Ga, 67-76, asst prof computer sci, 74-76; SR BIOSTATISTICIAN, COOP STUDIES PROG COORD CTR, VET ADMIN HOSP, MD, 77- *Mem:* Am Statist Asn; Inst Math Statist; Biomet Soc; Math Asn Am; Soc Clin Trials; Sigma Xi. *Res:* Bayesian estimation; discrete and modified discrete distributions; cooperative clinical trials. *Mailing Add:* Coop Studies Prog Support Ctr Vet Hosp Perry Point MD 21902

WILLIG, MICHAEL ROBERT, b Pittsburgh, Pa, June 7, 52. QUANTITATIVE BIOLOGY. *Educ:* Univ Pittsburgh, BS, 74, PhD(biol), 82. *Prof Exp:* Res fel ecol, Brazilian Nat Acad Sci, 76-78; vis prof terrestrial ecol, LaRoch Col, 79; asst prof ecol & biomet, Loyola Univ, 81-83; ASST PROF BIO, TEX TECH, 83- *Concurrent Pos:* Res fel, Dept Energy, PR, 81, 82 & 84. *Mem:* Ecol Soc Am; Am Soc Naturalists; Soc Study Evolution; Am Soc Mammalogists; AAAS. *Res:* Application of statistical techniques to answer questions of interest in population biology, ecology, behavior and evolution including population estimation, biogeography, conservation and mammalian systematics. *Mailing Add:* Dept Bio Sci Tex Tech Univ Lubbock TX 79409-3131

WILLIGER, ERVIN JOHN, b Szeged, Hungary, June 18, 27; US citizen; m 51; c 4. PLASTICS CHEMISTRY, RUBBER CHEMISTRY. *Educ:* Budapest Tech Univ, dipl chem eng, 50. *Prof Exp:* Res assoc polymers, Inst Plastics Res, Budapest Tech Univ, 50-56; res & develop chemist, Naugatuck Chem, US Rubber Co, 57-63; polymer chemist, Lucidol Div, Wallace & Tiernan, Inc, 63-65; sr res chemist, Res Ctr, Gen Tire & Rubber Co, 65-69; mgr tech serv & mkt, Union Process Co, Akron, 69-73; SR RES CHEMIST, BRECKSVILLE RES & DEVELOP CTR, B F GOODRICH CO, 73- *Mem:* Nat Soc Prof Eng; Am Chem Soc; Soc Plastics Eng. *Res:* Preparative polymer chemistry of thermoplastics and thermosets; structure behavior study of reinforced plastics; preparation and application of unsaturated polyesters and epoxy resins; study of peroxides and other free radical sources; toughening and fatigue of reinforced composites; technology of liquid elastomers; rubber compounding; adhesives and coatings; vinyl technology; compounding and processing testing. *Mailing Add:* 665 Fairwood Dr Tallmadge OH 44278

WILLIGES, GEORGE GOUDIE, b Sioux City, Iowa, May 18, 24; m 47; c 1. PLANT PATHOLOGY, PLANT TAXONOMY. *Educ:* Univ Corpus Christi, BA, 55; Tex A&I Univ, MA, 59; Tex A&M Univ, PhD(plant path), 69. *Prof Exp:* Teacher biol, Sinton Independent Sch Dist, 55-60; assoc prof, 61-77, chmn dept, 77-,80, PROF BIOL, TEX A&I UNIV, 77- CUR HERBARIUM, 71- *Concurrent Pos:* Mem bd dir, Tex Systs Natural Labs, 77. *Mem:* AAAS; Am Phytopath Soc. *Res:* Pathogenic variability, physiological and environmental effects on growth and reproduction of Sclerotium rolfsii; Fusarium diseases of cacti. *Mailing Add:* Dept Biol Tex A&I Univ Santa Gertrudis Kingsville TX 78363

WILLIGES, ROBERT CARL, b Richmond, Va, Mar 19, 42; m 65; c 1. HUMAN FACTORS ENGINEERING. *Educ:* Wittenberg Univ, AB, 64; Ohio State Univ, MA, 66, PhD(eng psychol), 68. *Prof Exp:* Asst human factors, Ohio State Univ, 64-68; from asst prof to assoc prof aviation & psychol, Univ Ill, 68-76; PROF INDUST ENG & PSYCHOL, VA POLYTECH INST & STATE UNIV, 76- *Concurrent Pos:* Asst dir human factors, Hwy Traffic Safety Ctr, Univ Ill, 68-70, assoc head human factors, Aviation Res Lab, 72-76; mem hwy res bd, Nat Acad Sci, 68-70; ed, Human Factors, 76-79. *Honors & Awards:* Jerome H Ely Award, Human Factors Soc, 74. *Mem:* Human Factors Soc (pres, 82-83); Am Psychol Asn; Inst Indust Engrs. *Res:* Research methodology; human performance research; computer-augmented training; human-computer interface; design of computer-generated displays. *Mailing Add:* Va Tech JEOR 0118 Blacksburg VA 24061-0118

WILLINGHAM, ALLAN KING, b Washington, DC, July 11, 41; m 67. BIOCHEMISTRY. *Educ:* George Washington Univ, BS, 63; St Louis Univ, PhD(biochem), 70. *Prof Exp:* Fel biochem, Res Inst Hosp Joint Dis, New York, 70-71; instr, 71-75, asst prof biochem, Col Med, Univ Nebr, Omaha, 75-76; asst prof, 76-78, ASSOC PROF BIOCHEM, KIRKSVILLE COL OSTEOP MED, 78- *Concurrent Pos:* USPHS res grant, Nat Heart, Lung & Blood Inst, 74-80; res grant, Am Osteop Asn, 81-82. *Mem:* AAAS; Sigma Xi. *Res:* Interconversion of phylloquinone and its 2, 3- epoxide and its relationship to the vitamin K-dependent carboxylation of glutamic acid residues to form active clotting proteins; post-translational synthesis and secretion of vitamin K-dependent clotting proteins. *Mailing Add:* Dept Biochem Kirksville Col Osteop Med Kirksville MD 63501

WILLINGHAM, FRANCIS FRIES, JR, b Winston-Salem, NC, June 3, 42; m 70. TAXONOMIC BOTANY, ORNAMENTAL HORTICULTURE. *Educ:* Univ NC, Chapel Hill, AB, 65; Wake Forest Univ, MA, 67, PhD(biol), 73. *Prof Exp:* Instr biol, Pine Crest Prep Sch, 67-69 & Salem Col, 69-71; plant taxonomist, Dept Hort, Callaway Gardens, Ga, 73-74, dir greenhouse opers, 74-78; gen mgr, Res Farms, Houston, Tex, 79-80; pres, Phytotech, Inc, 81-89; MANAGING PARTNER, TURTLE POND NURSERIES, 89- *Concurrent Pos:* Instr philos sci, Gov's Sch NC, 69-72; Hort Res Inst grant, 77-78. *Mem:* Bot Soc Am; Am Fern Soc; Am Soc Plant Taxonomists; Am Soc Hort Sci; Sigma Xi. *Res:* Taxonomy of ornamental horticulture plant materials and tissue culture of ornamentals. *Mailing Add:* 20519 Rhodes Rd Spring TX 77388

WILLINGHAM, MARK C, b Charleston, SC, Jan 16, 46; m 71; c 3. CELL BIOLOGY, PATHOLOGY. *Educ:* Col Charleston, SC, BS, 65; Med Univ SC, MD, 69. *Prof Exp:* Res fel, Dept Path, Med Univ SC, 69-70, resident anatomic path, 70-71; res assoc, 71-75, sr investr,75-80, CHIEF ULTRASTRUCT CYTOCHEM SECT, LAB MOLECULAR BIOL, NAT CANCER INST, NIH, 80- *Mem:* AAAS; Electron Microscopy Soc Am; Am Soc Cell Biol. *Res:* Conducts basic research in the cell and molecular biology of malignancy, with an emphasis on morphological technique in the pathology of neoplasia. *Mailing Add:* Lab Molecular Biol Nat Cancer Inst NIH Bldg 37 Rm 4B15 Bethesda MD 20892

WILLIS, CARL BERTRAM, b Charlottetown, PEI, Nov 27, 37; m 62; c 2. PLANT PATHOLOGY. *Educ:* McGill Univ, BSc, 59; Univ Wis-Madison, PhD(plant path), 62. *Prof Exp:* RES SCIENTIST PLANT PATH, AGR CAN, 62- *Mem:* Can Phytopath Soc; Agr Inst Can; Soc Nematol. *Res:* Forage crops diseases; factors affecting root rots of forage legumes. *Mailing Add:* Res Br Agr Can PO Box 1210 Charlottetown PE C1A 7M8 Can

WILLIS, CARL RAEBURN, JR, b Madison, Wis, Apr 5, 39; m 60; c 4. PHARMACY, PHARMACEUTICAL CHEMISTRY. *Educ:* Purdue Univ, BS, 61, MS, 64, PhD(indust pharm), 66. *Prof Exp:* Teaching asst bionucleonics & mfg pharm, Purdue Univ, 61-62, teaching assoc pharmaceut chem, 62-63; sr pharmaceut chemist, Warren-Teed Pharmaceut, Inc, Rohm & Haas Co, 66-69; supvr prod develop,Pharm Res & Develop Div, Ciba Pharmaceut Co, 69, mgr process & mat technol, 70-71; mgr prod develop, Pharmaceut Div, Ciba-Geigy Corp, 71-72; assoc dir, Drug Regulatory Affairs, Sterling Drug Inc, 72-76, dep dir, 76-78; dir, Drug Regulatory Affairs, Cooper Labs, Inc, 78-79; dir, Drug Regulatory Affairs, 79-83, sr dir, res & develop, 83-84, VPRES OPERS, BERLEX LABS, INC, 84- *Mem:* AAAS; Am Pharmaceut Asn; Acad Pharmaceut Sci; Am Chem Soc; Sigma Xi; Am Asn Pharmaceut Scientists. *Res:* Industrial pharmacy; pharmacokinetics; biopharmaceutics; pharmaceutical dosage form research and development; package materials research and development; drug regulatory affairs. *Mailing Add:* Berlex Labs Inc 300 Fairfield Rd Wayne NJ 07470-7358

WILLIS, CHARLES RICHARD, b Watertown, NY, July 7, 28; m 54; c 2. PHYSICS. *Educ:* Syracuse Univ, BA, 51, PhD, 58. *Prof Exp:* From asst prof to assoc prof, 57-65, PROF PHYSICS, BOSTON UNIV, 68- *Mem:* Am Phys Soc; Am Asn Univ Physicists. *Res:* Statistical mechanics; laser physics; classical many-body problems; quantum optics; solid state physics. *Mailing Add:* 30 Solon St Newton Highland MA 02161

WILLIS, CHRISTOPHER JOHN, b Sutton, Eng, June 6, 34; m 60; c 3. INORGANIC CHEMISTRY. *Educ:* Cambridge Univ, BA, 55, PhD(chem), 58, MA, 59. *Prof Exp:* Fel chem, Univ BC, 58-60, lectr, 60-61; lectr, 61-62, asst prof, 62-66, assoc prof, 66-82, PROF CHEM, UNIV WESTERN ONT, 82- *Concurrent Pos:* Nat Res Coun Can res grant. *Mem:* Chem Inst Can. *Res:* Synthesis and study of fluorinated alchols, alkoxides and related fluorinated ligands. *Mailing Add:* Dept Chem Univ Western Ont London ON N6A 5B7 Can

WILLIS, CLIFFORD LEON, b Chanute, Kans, Feb 20, 13; m 47; c 1. GEOLOGY. *Educ:* Univ Kans, BS, 39; Univ Wash, PhD(geol), 50. *Prof Exp:* Geophysicist, Carter Oil Co, 39-42, geologist, 46-47; instr geol, Univ Wash, 50-52, asst prof, 53-54; chief geologist, Harza Eng Co, Chicago, 54-68, vpres, 68-81; RETIRED. *Concurrent Pos:* Consult geologist, US, Turkey, Greece, Iraq, Jordon, Iran, Pakistan, Belgian Congo, Iceland, Colombia, Venezuela, Ethiopia, Arg, El Salvador & Honduras, 51- *Mem:* Geol Soc Am: Am Asn Petrol Geol; Am Inst Mining, Metall & Petrol Eng; Brit Geol Soc. *Res:* Engineering geology; structural geology; geophysics. *Mailing Add:* 4795 E Quail Creek Dr Tucson AZ 85718

WILLIS, CLIVE, b London, Eng, July 31, 39; Can citizen; m 62; c 2. PHYSICAL CHEMISTRY. *Educ:* Univ Liverpool, BSc Hons, 61, PhD(phys chem), 64. *Prof Exp:* Fel, Univ Calif, Los Angeles, 64-65; res assoc chem physics, Comn Atomic Energy, 65-66; res officer phys chem, Atomic Energy Can Ltd, 66-71; lectr chem, Univ West Indies, 71-73; RES OFFICER PHYS CHEM, NAT RES COUN CAN, 73- *Mem:* Chem Inst Can. *Res:* Laser chemistry; gas phase kinetics; photochemistry; radiation chemistry; discharge phenomena. *Mailing Add:* 62 Queen Elizabeth Driveway Ottawa ON K2P 1E3 Can

WILLIS, D(ONALD) ROGER, b Sutton Coldfield, Eng, Feb 12, 33; m 57; c 2. AERONAUTICAL SCIENCE. *Educ:* Oxford Univ, BA, 53, MA, 57; Princeton Univ, MSE, 57, PhD(aeronaut eng), 59. *Prof Exp:* Res assoc aeronaut sci, Royal Inst Technol, Sweden, 59-61 & Princeton Univ, 61-63; from asst prof to assoc prof, 63-70, PROF ENG SCI, UNIV CALIF, BERKELEY, 70- *Concurrent Pos:* Guggenheim fel, 66. *Mem:* Soc Indust & Appl Math. *Res:* Mathematical theory of rarefied gas dynamics; numerical fluid dynamics. *Mailing Add:* Dept Mech Eng Univ Calif Berkeley CA 94720

WILLIS, DAVID EDWIN, b Cleveland, Ohio, Mar 13, 26; m 48; c 4. GEOPHYSICS, GEOLOGY. *Educ:* Western Reserve Univ, BS, 50; Univ Mich, Ann Arbor, MS, 57, PhD, 68. *Prof Exp:* Computer seismic explor, Keystone Explor Co, 50-52, party chief, 52-54, asst supvr, 54-55; res assoc,

Univ Mich, Ann Arbor, 55-60, assoc res geophysicist, 60-63, res geophysicist, 63-68, actg head geophys lab, 65-67, assoc prof geol & head geophys lab, 68-70; assoc prof geol, 70-73, chmn dept geol sci, 72-76, prof geol, Univ Wis-Milwaukee, 73-; AT OIL & GAS DIV, UNOCAL CORP. *Concurrent Pos:* Lectr, Univ Mich, Ann Arbor, 63-64; geophysicist, Union Oil Co, Int, 80-81. *Mem:* Soc Explor Geophys; Am Geophys Union; Seismol Soc Am; fel Geol Soc Am. *Res:* Seismic and acoustic wave propagation; earthquake seismology; ground vibration studies; geophysical exploration. *Mailing Add:* Unocal Corp Oil & Gas Div 1201 W 55th St PO Box 7600 Los Angeles CA 90051

WILLIS, DAVID LEE, b Pasadena, Calif, Mar 15, 27; m 50; c 3. BIOLOGY, RADIATION BIOLOGY. *Educ:* Biola Univ, BTh, 49, BA, 51; Wheaton Col, Ill, BS, 52; Calif State Univ Long Beach, MA, 54; Ore State Univ, PhD(radiation biol), 63. *Prof Exp:* Teacher high sch, Calif, 52-57; instr biol, Fullerton Col, 57-61; from asst prof to assoc prof, 62-71, chmn dept, 69-85, prof, 71-87, EMER PROF RADIATION BIOL, ORE STATE UNIV, 87- *Concurrent Pos:* Consult, Comn Undergrad Educ Biol Sci, NSF, 66-70; vis investr, Oak Ridge Nat Lab, 68-69; consult, Portland Gen Elec Co, 70-72 & Life Systs, Inc, 84-85; mem, Hanford Health Effects Panel, Richland, WA, 1986. *Mem:* Health Physics Soc; fel Am Sci Affiliation (pres, 75). *Res:* Freshwater radioecology; radionuclide metabolism; radiation effects on reptiles and amphibians; general applications of radiotracer techniques to biology. *Mailing Add:* Dept Gen Sci Ore State Univ Weniger Hall 355 Corvallis OR 97331-6505

WILLIS, FRANK MARSDEN, b Philadelphia, Pa, Sept 5, 26; m 48; c 3. BIOMEDICAL ENGINEERING. *Educ:* Drexel Univ, BS, 51. *Prof Exp:* Asst develop engr, Res & Develop, Atlantic Refining Co, 51-55; area engr oil refining, Mobil Oil Co, 55-57; chief engr, Clark Cooper Co, Inc, 57-59; res engr explosives, Res & Develop, Eastern Lab, 59-62, supt eng & design, 62-65, sect head eng, 65-70, res supv, Polymer Res Develop Lab, 70-78, SR ENGR ASSOC BIOMED, RES & DEVELOP, ENG DEVELOP LAB, E I DU PONT DE NEMOURS & CO, INC, DEL, 78- *Concurrent Pos:* Consult, Lawrence Pump & Engine Co, 59-60; pub rel dir, Safety Div, Am Defense Preparedness Asn, 69-78. *Mem:* Am Soc Mech Engrs. *Res:* Clinical and therapeutic biomedical process; equipment and device research; development and mechanical design. *Mailing Add:* Eng Develop Lab E I du Pont de Memours & Co 101 Beech St Wilmington DE 19898

WILLIS, GROVER C, JR, b Kansas City, Mo, May 25, 21; m 41; c 3. PHYSICAL CHEMISTRY, ELECTROCHEMISTRY. *Educ:* Whittier Col, BA, 52; Univ Ore, MA, 55, PhD(chem), 57. *Prof Exp:* Petrol inspector, Gen Petrol Corp, Calif, 41-43; chemist, W C Hardesty Co, 46-51; assoc, Univ Ore, 53-55, res assoc, 55-57; from asst prof to assoc prof, 57-67, PROF PHYS & ANALYTICAL CHEM, CALIF STATE UNIV, CHICO, 67- *Mem:* Am Chem Soc. *Res:* Anodic oxide formation kinetics; mechanism; reaction rates, thermodynamics, adsorption phenomena and diffusion processes at dropping mercury electrodes; electrochemical instrumentation. *Mailing Add:* Dept Chem Calif St Univ Chico CA 95929

WILLIS, GUYE HENRY, b Los Angeles, Calif, July 1, 37; m 60; c 2. SOIL CHEMISTRY. *Educ:* Okla State Univ, BS, 61; Auburn Univ, MS, 63, PhD(soil chem), 65. *Prof Exp:* RES SOIL SCIENTIST, SOIL & WATER POLLUTION RES DIV, AGR RES SERV, USDA, 65- *Mem:* Am Soc Agron; Soil Sci Soc Am; Am Chem Soc; Am Soc Agr Engrs. *Res:* Fate of agricultural chemicals, including pesticides and fertilizers, in the environment; soil chemistry-plant nutrition relationships. *Mailing Add:* PO Box 25071 University Sta Baton Rouge LA 70894-5071

WILLIS, HAROLD LESTER, b McPherson, Kans, Oct 20, 40. SCIENCE EDUCATION. *Educ:* Emporia Kans State Univ, BA, 62; Univ Kans, PhD(entom), 66. *Prof Exp:* From asst prof to prof biol, Univ Wis-Platteville, 78-81; agr environ consult, 81-; ed, 83-86; WORKSHOP SUPVR, RESOURCES RECOVERY, 86- *Mem:* AAAS. *Res:* Bionomics, taxonomy and zoogeography of Nearctic tiger beetles; author of books and articles on sustainable agriculture. *Mailing Add:* 623 Vine St Wisconsin Dells WI 53965

WILLIS, ISAAC, b Albany, Ga, July 13, 40; m 65; c 2. DERMATOLOGY. *Educ:* Morehouse Col, BS, 61; Howard Univ, MD, 65. *Prof Exp:* Assoc dermat, Sch Med, Univ Pa, 69-70; head internal med res team & dermatologist, Letterman Army Inst, US Army Med Corps, 70-72; res assoc & clin instr dermat, Sch Med, Univ Calif, 70-72; asst prof med dermat, Sch Med, Johns Hopkins Univ, 72-73; asst prof, Vet Admin Hosp-Atlanta, 73-75, assoc prof med dermat, Sch Med, Emory Univ & chief dermat, Vet Admin Med Ctr-Atlanta, 75-81; PROF MED DERMAT, MOREHOUSE SCH MED, 81- *Concurrent Pos:* Asst attend physician, Philadelphia Gen Hosp, 69-70; Dermat Found res award, Univ Pa, 70; attend physician, Univ Calif Med Ctr, 70-72; attend physician, Johns Hopkins Hosp, Baltimore City Hosps & Good Samaritan Hosp, 72-73; consult & lectr dermat, Bur Med & Surg, US Dept Navy, 72-75; consult asst to Prof Dermat, Howard Univ Col Med, 72-; mem Formulary Task Force, Nat Prog Dermat, Am Fedn Clin Res; mem, Gen Med & Study Sect, NIH, 85-89; med dir, McWill Res Lab; mem, Sci Rev Panel Health Res, US Environ Protection Agency, 86- & Inst Rev Bd, West Paces Ferry Hosp, 88- *Honors & Awards:* Frontiers Int Award, 84. *Mem:* Am Med Asn; Nat Med Asn; Am Acad Dermat; Am Fedn Clin Res; Am Dermat Asn; Soc Inst Dermat; Am Photobiol Asn; NY Acad Sci. *Res:* Phototherapy, photochemotherapy; acute and chronic effects of ultraviolet light, including carcinogenesis; effects of light on bacteria and fungi, and effects of heat and humidity on skin. *Mailing Add:* 1141 Regency Rd NW Atlanta GA 30327

WILLIS, JACALYN GIACALONE, b New York, NY, Feb 13, 47; m 80. BEHAVIORAL ECOLOGY. *Educ:* Queens Col, BA, 69, MA, 74; City Univ New York, PhD(biol), 76. *Prof Exp:* Lectr, biol, Queens Col, 68-76; asst prof biol, State Univ NY, Purchase, 76-77; ASSOC PROF BIOL, UPSALA COL, 77- *Concurrent Pos:* Smithsonian Trop Mammal Census, 82-; prin investr, Earthwatch Exped, 84. *Mem:* AAAS; Am Soc Mammalogists; Asn Women Sci; Asn Trop Biol. *Res:* Comparative field studies of the behavioral ecology of New World tree squirrels and flying squirrels; population dynamics of tropical mammals. *Mailing Add:* 531 W Mountain Rd Sparta NJ 07871

WILLIS, JAMES BYRON, b Marietta, Ohio, Jan 24, 18; m 40; c 2. CERAMIC ENGINEERING. *Educ:* Ohio State Univ, BCerE, 39. *Prof Exp:* Ceramic res engr ceramic coatings, Pemco Corp, 40-45, serv mgr, 45-59, dir eng serv, 59-62, sales mgr, Pemco Div, Glidden Co, 62-65, dir develop labs, 65-68, dir res & develop, Pemco Prod Group, SCM Corp, 68-82; RETIRED. *Mem:* Fel Am Ceramic Soc; Sigma Xi. *Res:* Ceramic coating materials and processes; ceramic colorants; casting floxes for heavy metals. *Mailing Add:* Clidden Durkee Dv SCM Co 5601 Eastern Ave Baltimore MD 21224

WILLIS, JAMES STEWART, JR, b West Point, NY, Feb 9, 35; m 59; c 2. PHYSICS. *Educ:* US Mil Acad, BS, 58; Rensselaer Polytech Inst, MS, 64, PhD(physics), 66. *Prof Exp:* US Army, 58-, from instr to prof physics, US Mil Acad, 70-88; RETIRED. *Mem:* AAAS; Am Asn Physics Teachers; Am Soc Eng Educ; Am Phys Soc. *Res:* Type II superconductivity; electron spin resonance studies on color centers and other defects in crystals. *Mailing Add:* PO Box 327 Washington DC 22747

WILLIS, JEFFREY OWEN, b Long Branch, NJ, June 1, 48; m 69. HIGH PRESSURE PHYSICS, SUPERCONDUCTIVITY & MAGNETISM. *Educ:* Univ Ill, Urbana-Champaign, BS, 70, MS, 71, PhD(physics), 76. *Prof Exp:* Res assoc, Naval Res Lab, Nat Res Coun, 75-77; fel, Los Alamos Sci Lab, 78-79, MEM STAFF, LOS ALAMOS NAT LAB, 80- *Mem:* Am Phys Soc; Sigma Xi. *Res:* Magnetic and superconductive properties of metals, under conditions of ultralow temperatures and high pressures; low temperature, specific heat of metals. *Mailing Add:* 1917 40th St Los Alamos NM 87544

WILLIS, JOHN STEELE, b Long Beach, Calif, Jan 19, 35; m 58. PHYSIOLOGY, ZOOLOGY. *Educ:* Univ Calif, Berkeley, AB, 56; Harvard Univ, AM, 58, PhD(biol), 61. *Prof Exp:* Nat Heart Inst fel biochem, Oxford Univ, 61-62; from asst prof to assoc prof physiol, Univ Ill, Urbana, 62-72, prof, 72-91, prof nutrit sci, 81-91; PROF ZOOL & ECOL, UNIV GA, ATHENS, 91- *Concurrent Pos:* Nat Inst Gen Med Sci res grants, 63-; mem, physiol study sect, NIH, 69-73; consult, Basic Sci Rev Bd, Vet Admin, 75-78; assoc ed, Am J Physiol Cell, 76-81. *Mem:* Soc Gen Physiol; Am Soc Zool; Am Physiol Soc; Sigma Xi; Am Inst Nutrit. *Res:* Cold resistance of tissues of hibernating mammals; cation transport; cell physiology of nutrition. *Mailing Add:* Dept Zool Univ Ga Athens GA 30602

WILLIS, JUDITH HORWITZ, b Detroit, Mich, Jan 2, 35; m 58. DEVELOPMENTAL BIOLOGY, INSECT PHYSIOLOGY. *Educ:* Cornell Univ, AB, 56; Harvard Univ, AM, 57, PhD(biol), 61. *Prof Exp:* USPHS res fel, Harvard Univ, 60-61 & Oxford Univ, 61-62; from instr to prof entom, Univ Ill, Urbana, 77-91; PROF & HEAD DEPT ZOOL, UNIV GA, 90- *Concurrent Pos:* Mem, Aging Review Comt, NIH, 76-80; mem, Coun Comt Affairs, AAAS, 77-80 & 79-80, chairperson, Sect G, 87-88; prog dir cellular physiol, Nat Acad Sci, 83; mem, Coun & Exec Bd, Tissue Culture Asn, 86-90. *Mem:* AAAS; Entom Soc Am; Soc Develop Biol; Am Soc Zool; Tissue Culture Asn. *Res:* Gene action in insect metamorphosis, cuticular proteins, insect-tissue culture. *Mailing Add:* Dept Zool Univ Ga Athens GA 30602

WILLIS, LLOYD L, II, b Frederick, Okla, June 10, 43; m 67; c 2. BIOLOGY. *Educ:* Phillips Univ, BS, 65; Univ Va, Med, 67. *Prof Exp:* Teacher gen sci, Roanoke City Pub Sch Syst, 65-68; sci teacher, Pickens Co Pub Sch, 68-69; asst prof biol, Va Western Community Col, 69-70; res asst, bot dept, Univ NC, 70-73; asst prof, 74-77, ASSOC PROF BIOL, PIEDMONT VA COMMUNITY COL, 77- *Mem:* Am Inst Biol Sci; Ecol Soc Am; Nat Sci Teachers Asn; Soc Col Sci Teachers; Am Rhododendron Soc; Nat Asn Biol Teachers. *Res:* Science education biology; innovative teaching techniques; development of required out of class activities; development of independent study courses; development of support-services for part-time faculty in biology. *Mailing Add:* Piedmont Va Community Col Rte 6 Box 1 Charlottesville VA 22901

WILLIS, LYNN ROGER, b Oct 2, 42; m 66; c 3. RENAL PHARMACOLOGY. *Educ:* Univ Iowa, PhD(pharmacol), 70. *Prof Exp:* PROF PHARMACOL & MED, SCH MED, UNIV IND, 83- *Mem:* Am Soc Pharmacol & Exp Therapeut; Am Soc Nephrol; Int Soc Nephrol. *Res:* Hypertension and renal function. *Mailing Add:* Dept Pharmacol MS 346-A Ind Univ Sch Med 635 Barnhill Dr Indianapolis IN 46223

WILLIS, PARK WEED, III, b Seattle, Wash, Nov 18, 25; m 48; c 6. CARDIOVASCULAR DISEASES, INTERNAL MEDICINE. *Educ:* Univ Pa, MD, 48. *Prof Exp:* Intern, Pa Hosp, 48-50; resident internal med, Univ Hosp & Med Sch, Univ Mich, Ann Arbor, 52-53, jr clin instr, Med Sch, 53-54, from instr to assoc prof, 54-65, asst prof postgrad med, 57-59, dir, Div Cardiol, 69-77, prof internal med, 65-79; PROF INTERNAL MED & DIR SECT CARDIOL, COL HUMAN MED, MICH STATE UNIV, 79- *Concurrent Pos:* Attend physician, Vet Admin Hosp, Ann Arbor, 54-59, consult, 59-79; consult health serv, Univ Mich, 56-79; fel, Coun Epidemiol, Am Heart Asn, Coun Clin Cardiol. *Honors & Awards:* Jacob Ehrenzeller Award, 82. *Mem:* Asn Univ Cardiologists (pres, 79-80); fel Am Col Cardiol; fel Am Col Physicians; Am Fed Clin Res; fel Coun Clin Cardiol & Epidemiol, Am Heart Asn. *Res:* Clinical cardiology. *Mailing Add:* B220 Life Sci Bldg Mich State Univ East Lansing MI 48824-1317

WILLIS, PHYLLIDA MAVE, b Wallington, Eng, Mar ll, 18; nat US. PHYSICAL CHEMISTRY. *Educ:* Mt Holyoke Col, AB, 38; Smith Col, AM, 40; Columbia Univ, PhD(phys chem), 46. *Prof Exp:* Teacher sci, Knox Sch, NY, 40-42; asst chem, Columbia Univ, 42-44; from instr to asst prof, Wellesley Col, 46-54; assoc prof & head dept, Newcomb Col, Tulane Univ, 54-60, chmn, dept chem, physics & astron, 60-79, Whitaker prof, 60-83, WHITAKER EMER PROF CHEM, HOOD COL, 83- *Concurrent Pos:* Am Asn Univ Women fel, Oxford Univ, 51-52; NSF fac fel, Univ Minn, 58. *Mem:* AAAS; Am Chem Soc; Soc Appl Spectros; Am Asn Physics Teachers; Sigma Xi. *Res:* Molecular spectroscopy. *Mailing Add:* 120 W Church St Frederick MD 21701

WILLIS, ROBERT D, b Independence, Mo, Apr 23, 48. ATOMIC PHYSICS. *Educ:* Denison Univ, BS, 70; Duke Univ PhD(physics), 76. *Prof Exp:* Fel, Naval Res Lab, Nat Res Coun, 77-79; asst geophysicist, Scripps Inst Oceanog, 79-84; from scientist to sr scientist, Atom Sci, Inc, 84-91; CONSULT, 91- *Mem:* Am Geophys Union; Am Phys Soc. *Res:* Alter sensities tracing analysis; laser spectroscopes, math spectroscopy and isotopic geochemistry. *Mailing Add:* 356 East Dr Oak Ridge TN 37830

WILLIS, RONALD PORTER, b Cowley, Wyo, Sept 20, 26; m 53; c 5. GEOLOGY. *Educ:* Univ Wyo, BS, 52, MA, 53; Univ Ill, PhD(geol), 58. *Prof Exp:* Geologist, Richfield Oil Corp, 53-55; asst, Univ Ill, 55-57; geologist, Richmond Explor Co, 58-61; chief geologist, Bahrain Petrol Co, 61-65; regional geologist, Amoseas, Tripoli, Libya, 65; geologist, Chevron Oil Co, Okla, 65-67; PROF GEOL, UNIV WIS-EAU CLAIRE, 67- *Concurrent Pos:* Fulbright lectr, Univ Benin, Nigeria, 76-77. *Mem:* Geol Soc Am; Am Asn Petrol Geol. *Res:* Stratigraphy; sedimentation; petroleum geology. *Mailing Add:* Geol Dept Univ Wis Eau Claire WI 54701

WILLIS, SUZANNE EILEEN, b New Brunswick, NJ, May 25, 51. ELEMENTARY PARTICLE PHYSICS. *Educ:* Mount Holyoke Col, BA, 72; Yale Univ, MPhil, 74, PhD(physics), 79. *Prof Exp:* Res assoc, Fermi Nat Accelerator Lab, 79-82; from asst prof to assoc prof, dept physics, Univ Okla, 82-88; ASSOC PROF, NORTHERN ILL UNIV, 88- *Mem:* Am Phys Soc; AAAS; Sigma Xi. *Res:* Elementary particle physics; production and decay of charm and heavy quarks; rare decay modes of the muon; neutrino oscillations and decay. *Mailing Add:* Dept Physics Univ Northern Ill Dekalb IL 60115

WILLIS, WAYNE O, b Paonia, Colo, Jan 21, 28; m 51; c 4. SOIL PHYSICS. *Educ:* Colo State Univ, BS, 52; Iowa State Univ, MS, 53, PhD(soil physics), 56. *Prof Exp:* Asst soils res, Iowa State Univ, 52-53; agent soils res, Iowa, 53-56, res soil scientist, Wash, 56-57 & Salinity Lab, Calif, 57-58, res soil scientist, Northern Great Plains Res Ctr, NDak, 58-68, supvry soil scientist and res leader, 68-76, supvry soil scientist & tech adv, 76-81, SUPVRY SOIL SCIENTIST & RES LEADER, AGR RES SERV, USDA, 81- *Mem:* Fel Am Soc Agron; Am Geophys Union; Can Soc Soil Sci; Int Soil Sci Soc; fel Soil Conserv Soc Am; fel Soil Sci Soc Am; Sigma Xi. *Res:* Interrelationships of soil water, soil temperature, plant growth and frozen soils; water conservation; dryland agriculture; crop yield prediction. *Mailing Add:* 721 Cherokee Dr Ft Collins CO 80525-1516

WILLIS, WILLIAM DARRELL, JR, b Dallas, Tex, July 19, 34; m 60; c 1. NEUROPHYSIOLOGY, NEUROANATOMY. *Educ:* Tex A&M Univ, BS & BA, 56; Univ Tex, MD, 60; Australian Nat Univ, PhD(physiol), 63. *Prof Exp:* From asst prof to prof anat, Southwestern Med Sch, Univ Tex, Dallas, 63-70, chmn dept, 64-70; PROF ANAT & PHYSIOL & CHIEF COMP NEUROBIOL, MARINE BIOMED INST, UNIV TEX MED BR GALVESTON, 70-, DIR INST, 78-, ASHBEL SMITH PROF & CHMN DEPT ANAT & NEUROSCI, 86- *Concurrent Pos:* NIH res fel, Australian Nat Univ, 60-62 & Univ Pisa, 62-63; Nat Inst Neurol Dis & Blindness res grant, 63-; mem neurol B study sect, NIH, 68-72, chmn, 70-72, mem neurol dis prog, proj rev comt, 72-76; chief ed, J Neurophysiol, 78-83; Florence & Marie Hall fel, 84-85; Alexander von Humboldt Sr US Scientist Award, 84-85; Jacob K Javits Award, Nat Inst Neurol & Commun Dis & Stroke, 85-; mem Nat Adv Neurol Dis & Stroke Coun, NIH, 88-91; field ed, J Neurosci Lett, 76-, sect ed, J Exp Brain Res, 90- *Honors & Awards:* F W L Kerr Award, Am Pain Soc. *Mem:* AAAS; Am Asn Anat; Am Physiol Soc; Am Pain Soc (treas, 78-81, pres, 82-83); Soc Neurosci (pres, 84-85); Int Asn Study Pain. *Res:* Electrophysiology of the vertebrate spinal cord; somatic sensory pathways; pain mechanisms and descending control of pain transmission. *Mailing Add:* Off Dir 200 University Blvd Galveston TX 77550

WILLIS, WILLIAM HILLMAN, b Trenton, Tenn, Oct 28, 08; m 34; c 2. SOILS, BACTERIOLOGY. *Educ:* Union Univ, Tenn, AB, 30; Iowa State Col, MS, 31, PhD(soil bact), 33. *Prof Exp:* Asst soil technologist, Soil Conserv Serv, USDA, 34-38, assoc soil scientist flood control surv, 38-42; from assoc agronomist to agronomist, 42-66, assoc prof, 42-48, prof agron, 48-79, head dept, 66-79, EMER PROF, LA STATE UNIV, BATON ROUGE, 79- *Concurrent Pos:* Chmn, Southern Regional Soil Res Comt, 73-75. *Mem:* Am Chem Soc; Soil Sci Soc Am; Am Soc Agron; Sigma Xi. *Res:* Soil microbiology; nitrogen fertilization and nutrition of rice; biological nitrogen fixation; microbial nitrate reduction. *Mailing Add:* 256 Mt Vernon Church Rd West Monroe LA 71292

WILLIS, WILLIAM J, b Ft Smith, Ark, Sept 15, 32; m 58; c 5. PHYSICS. *Educ:* Yale Univ, BS, 54, PhD, 58. *Prof Exp:* Physicist, Brookhaven Nat Lab, 58-65; prof physics, Yale Univ, 65-73; PHYSICIST EUROP COUN NUCLEAR RES, GENEVA, SWITZ, 73- *Concurrent Pos:* Physicist, Europ Orgn Nuclear Res, 61-62. *Mem:* Am Phys Soc. *Res:* Elementary particle physics; weak interactions of strange particles, resonances and high energy collisions; hard parton processes; direct photon and lepton production; ultra relativistic heavy ion collisions. *Mailing Add:* NP Div CERN Geneva 1211 Switzerland

WILLIS, WILLIAM RUSSELL, b Moundsville, WVa, Feb 14, 26; m 46; c 4. PHYSICS. *Educ:* WVa Wesleyan Col, BS, 48; Okla State Univ, MS, 50, PhD(chem physics), 54. *Prof Exp:* Chemist, Oak Ridge Nat Lab, 52-55; assoc prof physics, W Liberty State Col, 55-56; prof, WVa Wesleyan Col, 56-65; asst prog dir, NSF, 65-66; sci fac fel, Univ Colo, 66-67; PROF PHYSICS & CHMN DEPT, NORTHERN ARIZ UNIV, 67- *Concurrent Pos:* Consult, Oak Ridge Nat Lab, 55-56, 58-65 & NSF, 67- *Mem:* AAAS; Am Asn Physics Teachers; Sigma Xi. *Res:* Diffusion in solids; molecular physics; ellipsometry. *Mailing Add:* 1700 E Linda Vista Flagstaff AZ 86001

WILLIS, WILLIAM VAN, b Morganton, NC, Oct 15, 37. INORGANIC CHEMISTRY, ANALYTICAL CHEMISTRY. *Educ:* Ga Inst Technol, BS, 60; Univ Tenn, MS, 63, PhD(chem), 66. *Prof Exp:* Res assoc radiation & radiochem, Eng Exp Sta, Ga Inst Technol, 59-61; sci writer nuclear decontamination, Univ Tenn, 63-64; USAEC res fel, 66-67; ASSOC PROF CHEM, CALIF STATE UNIV, FULLERTON, 67- *Concurrent Pos:* Mem, State Regional Water Qual Control Bd. *Mem:* Am Chem Soc. *Res:* Neutron activation analysis; radiochemical tracer analysis; transition metal transport in biological systems. *Mailing Add:* Dept Chem Calif State Univ 800 N State College Blvd Fullerton CA 92631-3547

WILLISTON, JOHN STODDARD, b Ft Madison, Iowa, July 23, 34; m 61. NEUROSCIENCES. *Educ:* Univ Wis-Madison, BS, 61; Calif State Univ, San Francisco, MA, 65; Univ Southern Calif, PhD(physiol, psychol), 68. *Prof Exp:* NIMH fel, Univ Calif, San Francisco, 68-70; from asst prof to assoc prof, 70-82, PROF BIOL, SAN FRANCISCO STATE UNIV, 82- *Concurrent Pos:* Res physiologist, Univ Calif, San Francisco, 76-78. *Mem:* AAAS; Int Brain Res Orgn; Soc Neurosci. *Res:* Neurological substrates of behavioral plasticity; neuroelectrical activity; psychotrophic drugs; distribution of volume conducted event related electrical potentials their generators and clinical applications; natural and artificial learning and associational systems. *Mailing Add:* Dept Biol San Francisco State Univ San Francisco CA 94132

WILLITS, CHARLES HAINES, b Camden, NJ, June 25, 23; m 45; c 2. ORGANIC CHEMISTRY. *Educ:* Wheaton Col, Ill, BS, 44; Ohio State Univ, MS, 48; Ore State Univ, PhD(org chem), 55. *Prof Exp:* Res engr, Battelle Mem Inst, 48-51; from asst prof to assoc prof, Rutgers Univ, Camden, 55-70, actg chmn dept, 60-69, chmn dept, 70-87, PROF CHEM, RUTGERS UNIV, CAMDEN, 70-87. *Mem:* Am Chem Soc; Am Sci Affil. *Res:* Synthesis of purine derivatives; mechanism of Hofmann degradation of amides; Fries rearrangement of higher esters. *Mailing Add:* Dept Chem Rutgers Univ Camden NJ 08102

WILLKE, THOMAS ALOYS, b Rome City, Ind, Apr 22, 32; m 54; c 6. MATHEMATICAL STATISTICS. *Educ:* Xavier Univ, Ohio, AB, 54; Ohio State Univ, MS, 56, PhD(math), 60. *Prof Exp:* Res mathematician, Nat Bur Stand, 61-63; asst prof math, Univ Md, 63-66; assoc prof, 66-72, dir statist lab, 71-73, actg dean, 83-84, PROF MATH, OHIO STATE UNIV, 72-, VPROVOST, COL ARTS & SCI, 73- *Concurrent Pos:* Lectr, Univ Md, 61-63. *Mem:* Math Asn Am; Inst Math Statist; Am Statist Asn. *Res:* Design and analysis of experiments. *Mailing Add:* 4375 Mumford Dr Columbus OH 43220

WILLMAN, JOSEPH F(RANK), b Brownsville, Tex, Dec 3, 31; m 56; c 3. ELECTRICAL ENGINEERING, UNDERWATER ACOUSTICS. *Educ:* Univ Tex, BS, 57, MS, 58, PhD(elec eng), 62. *Prof Exp:* Aerophys engr, Gen Dynamics/Ft Worth, 58-61; res engr, Defense Res Lab, Univ Tex, 61-62; sr res engr, Southwest Res Inst, 62-69; spec res assoc, 69-80, asst dir, 80-89, PROJ MGR, APPL RES LABS, UNIV TEX, AUSTIN, 89- *Mem:* Sigma Xi; Inst Elec & Electronics Engrs; Acoust Soc Am. *Res:* Systems analysis; signal processing; underwater acoustics; sonar systems; radar systems; radio wave propagation; the ionosphere; electromagnetic compatibility. *Mailing Add:* Appl Res Labs Univ Tex PO Box 8029 Austin TX 78713-8029

WILLMAN, VALLEE L, b Greenville, Ill, May 4, 25; m 52; c 9. SURGERY. *Educ:* Univ Ill, BS, 47; St Louis Univ, MD, 51; Am Bd Surg, dipl, 57; Bd Thoracic Surg, dipl, 61. *Prof Exp:* Sr instr, 57-58, from asst prof to assoc prof, 58-64, PROF SURG, SCH MED, ST LOUIS UNIV, 64-, CHMN DEPT, 69- *Concurrent Pos:* McBride fel cancer, Sch Med, St Louis Univ, 56-57; attend physician, Vet Admin Hosp, St Louis Univ Hosp & St Marys Hosp, 57- *Mem:* Soc Univ Surg; Am Surg Asn; Am Physiol Soc; Int Cardiovasc Soc; Am Asn Thoracic Surg. *Res:* Cardiovascular surgery and extracorporeal circulation; author or coauthor of over 210 publications. *Mailing Add:* St Louis Univ Med Ctr 3635 Vista Ave at Grand Blvd PO Box 15250 St Louis MO 63110-0250

WILLMAN, WARREN WALTON, b Chicago, Ill, May 24, 43. APPLIED MATHEMATICS. *Educ:* Univ Mich, BA, 65, BSE, 65; Harvard Univ, PhD(appl math), 69. *Prof Exp:* Mathematician, Shell Develop Co, Shell Oil Co, 69-70; opers res analyst, US Naval Res Lab, 71-82; MATHEMATICIAN, NAVAL WEAPONS CTR, 82- *Res:* Stochastic optimal control theory, estimation theory. *Mailing Add:* Naval Weapons Ctr Code 3807 China Lake CA 93555-6001

WILLMANN, ROBERT B, b Seguin, Tex, May 7, 31; m 62. ASTROPHYSICS. *Educ:* Tex A&M Univ, BS, 54; Univ Wis, PhD(physics), 60. *Prof Exp:* Res assoc physics, Univ Wis, 60-61; from asst prof to assoc prof, 61-71, PROF PHYSICS, PURDUE UNIV, LAFAYETTE, 71- *Mem:* Am Phys Soc. *Res:* Weak and strong interactions in particle physics. *Mailing Add:* Dept Physics Purdue Univ West Lafayette IN 47907

WILLMARTH, WILLIAM W(ALTER), b Highland Park, Ill, Mar 25, 24; m 59; c 4. FLUID MECHANICS, FLUIDS. *Educ:* Purdue Univ, BS, 49; Calif Inst Technol, MS, 50, PhD(aeronaut eng), 54. *Prof Exp:* From res fel to sr res fel aeronaut eng, Calif Inst Technol, 54-58; from assoc prof to prof aeronaut eng, Univ Mich, Ann Arbor, 58-90; RETIRED. *Concurrent Pos:* Consult, Rand Corp, 54-66, Gen Motors Res Lab, 70-74, Bendix Aerospace Systs Div, 73-76, Bendix Res Labs, 75-77, Lear Siegler, 81-85 & Spalding, 83-85; vis fel, Joint Inst Lab Astrophys, Boulder, 63-64. *Honors & Awards:* Off Naval Res Fluid Dynamics Prize, Am Phys Soc, 89. *Mem:* Fel Am Inst Aeronaut & Astronaut; fel Am Phys Soc. *Res:* Condensation of gases; transonic flow; turbulent boundary layer; unsteady aerodynamics; aerodynamic sound; scientific instruments for fluid mechanical measurements; structure of turbulence; turbulent drag reduction; vorticity interaction with free surface. *Mailing Add:* 765 Country Club Rd Ann Arbor MI 48105

WILLMERT, KENNETH DALE, b Kossuth Co, Iowa, Oct 25, 42; m 68. MECHANICAL ENGINEERING. *Educ:* Iowa State Univ, BS, 64; Case Inst Technol, MS, 66; Case Western Reserve Univ, PhD(mech eng), 70. *Prof Exp:* From asst prof to assoc prof, 70-85, PROF MECH ENG, CLARKSON UNIV, 85- *Mem:* Am Soc Mech Engrs; Am Inst Aeronaut & Astronautics; Asn Comput Mach. *Res:* Mechanical design; optimization applied to mechanical and structural systems; kinematic analysis and synthesis; finite element techniques applied to vibration problems. *Mailing Add:* Dept Mech Eng Clarkson Univ Potsdam NY 13676

WILLMES, HENRY, b Bocholt, Ger, Aug 30, 39; US citizen; m 66; c 3. NUCLEAR PHYSICS. *Educ:* Univ Calif, Los Angeles, BS, 61, MA, 62, PhD(physics), 66. *Prof Exp:* Res physicist, Aerospace Res Labs, Wright-Patterson AFB, 65-68; from asst prof to assoc prof, 69-80, PROF PHYSICS, UNIV IDAHO, 80-, CHMN DEPT, 75- *Mem:* Am Phys Soc; Sigma Xi. *Res:* Few nucleon systems; nuclear structure; applications of nuclear technology. *Mailing Add:* 2152 Arbor Crest Rd Moscow ID 83843

WILLMOTT, CORT JAMES, b Oakland, Calif, Dec 18, 46; m 68; c 2. CLIMATOLOGY, QUANTITATIVE METHODS. *Educ:* Calif State Univ, Hayward, BA, 69, MA, 72; Univ Calif, Los Angeles, PhD(geog), 77. *Prof Exp:* Lectr geog & climat, Dept Geog, Univ Del, 76-77, from asst prof to assoc prof, 77-87, PROF GEOG & CLIMAT, DEPT GEOG & OCEANOG PROG, UNIV DEL, 87-, DEPT CHAIR GEOG, 89- *Concurrent Pos:* Prin investr, NASA grants, 80- & NSF grants, 83-90; vis scholar geog, Dept Geog, Univ Victoria, 83 & climat, Dept Meteorol, Univ Md, 84; vis scientist climat, Lab Atmospheres, Goddard Space Flight Ctr, NASA, Greenbelt, Md, 85; chair, Asn Am Geographers, Climate Specialty Group, 86-88; chair prog comt, 1989 Ann Meeting Am Asn Geographers, 88-89; mem, Asn Am Geographers Hons Comt, 89-90 & NSF's Geog, Regional Sci Panel, 88-90; mem prog comt, 27th Cong Int Geog Union, 90-92. *Mem:* Asn Am Geographers; Am Meteorol Soc; Sigma Xi. *Res:* Land-surface influences on climate and climatic change at the continental and global scales; statistics and computational methods. *Mailing Add:* Dept Geog Univ Del Newark DE 19716

WILLMS, CHARLES RONALD, b Rupert, Idaho, June 26, 33; m 55; c 4. BIOCHEMISTRY. *Educ:* Univ Tex, BA, 55; Southwest Tex State Col, MA, 56; Tex A&M Univ, PhD(biochem), 59. *Prof Exp:* Asst prof chem, Southwest Tex State Col, 59-62; res scientist assoc, Clayton Found Biochem Inst, Univ Tex, 62-64; assoc prof, 64-68, chmn dept, 68-75, PROF CHEM, SOUTHWEST TEX STATE UNIV, 68- *Mem:* AAAS; Am Chem Soc; Sigma Xi. *Res:* Enzyme and protein chemistry; carbohydrate metabolism; characterization and isolation of proteolytic enzymes. *Mailing Add:* Dept Chem Southwest Tex State Univ San Marcos TX 78666

WILLNER, DAVID, b Vienna, Austria, July 2, 30; m 54; c 2. ORGANIC CHEMISTRY. *Educ:* Hebrew Univ, Israel, MSc, 56, PhD(org chem), 59. *Prof Exp:* From res asst to res assoc org chem, Weizmann Inst, 59-64; scientist, New Eng Inst Med Res, Conn, 64-66; sr res scientist, Bristol Labs, 66-82, SR RES SCIENTIST II, PHARMACEUT RES & DEVELOP DIV, BRISTOL-MYERS-SQUIBB CO, 82- *Concurrent Pos:* Asst org chem, Bar-Ilan Univ, Israel, 57-59, lectr org chem & reaction mechanism, 62-63; fel, Dept Chem, Univ Southern Calif, 59-61 & Calif Inst Technol, 61-62. *Mem:* Am Chem Soc; AAAS. *Res:* Medicinal chemistry; structure elucidation and synthesis of natural products and physiological active compounds, antibiotics and gastrointestinal pharmacodynamic agents; reaction mechanisms; antitumor agents; drug targeting. *Mailing Add:* Bristol Myers Co Five Research Pkwy Wallingford CT 06492

WILLNER, STEVEN P, b Louisville, Ky. INFRARED OBSERVATION, INSTRUMENTATION. *Educ:* Harvard Col, AB, 71; Calif Inst Technol, PhD(astron), 76. *Prof Exp:* Physicist, Univ Calif, San Diego, 76-81, ASTRONOMER, SMITHSONIAN ASTROPHY OBSERV, 81- *Mem:* Int Astron Union; Am Astron Soc. *Res:* Star formation and activity in galactic nuclei; infrared observation; instrument design, development and construction. *Mailing Add:* 60 Garden St Cambridge MA 02138

WILLOUGHBY, ANNE D, PEDIATRICS. *Educ:* Cornell Univ, MD; Univ Calif, Berkeley, MPH. *Prof Exp:* Epidemiol fel & spec asst pediat, Pregnancy & Perinatology Br, BR CHIEF, PEDIAT, ADOLESCENT & MATERNAL AIDS BR, NAT INST CHILD HEALTH & HUMAN DEVELOP. *Res:* Epidemiology, natural history, pathogenesis, clinical manifestation, treatment and prevention of HIV infection and disease in pregnant women, mothers, women of reproductive age, infants, children, adolescents and families. *Mailing Add:* NIH Nat Inst Child Health & Human Develop Pediat Adolescent & Maternal Aids Br Exec Plaza N Rm 450 6120 Executive Blvd Rockville MD 20892

WILLOUGHBY, DONALD S, medical bacteriology, immunology, for more information see previous edition

WILLOUGHBY, RALPH ARTHUR, b Santa Rosa, Calif, Aug 15, 23; m 47; c 2. MATHEMATICS. *Educ:* Univ Calif, AB, 47, PhD(math), 51. *Prof Exp:* From asst prof to assoc prof math, Ga Inst Technol, 51-55; mem staff, Atomic Energy Div, Babcock & Wilcox Co, 55-57; MEM STAFF, MATH SCI DEPT, THOMAS J WATSON RES CTR, IBM CORP, 57- *Concurrent Pos:* Res partic, Math Panel, Oak Ridge Nat Lab, 54, consult, 54-55. *Mem:* Soc Indust & Appl Math. *Res:* Numerical analysis. *Mailing Add:* 14 Garey Dr Chappaqua NY 10514

WILLOUGHBY, RUSSELL A, b Tilston, Man, July 7, 33; m 54; c 3. VETERINARY MEDICINE. *Educ:* Univ Toronto, DVM, 57; Cornell Univ, PhD(vet path), 65. *Prof Exp:* Pvt pract vet med, Grenfell, Sask, 57-61; asst prof clin vet med, Ont Vet Col, Toronto, 61-62; res asst vet path, Cornell Univ, 62-65; assoc prof, 65-67, assoc dean res, 79-83, PROF CLIN VET MED, ONT VET COL, UNIV GUELPH, 67-, chmn, Dept Clin Studies, 83-86, DIR, EQUINE RES CTR, 86- *Mem:* Am Vet Med Asn; Am Asn Vet Clinicians; Can Vet Med Asn; Am Col Vet Internal Med (secy, 72-81). *Res:* Environmental effects on animals, including heavy metal toxicity, the effects of intensification and the interaction between pollutants and infectious agents. *Mailing Add:* RR No 2 Alora ON N0B 1S0 Can

WILLOUGHBY, SARAH MARGARET C(LAYPOOL), b Bowling Green, Ky, Oct 15, 17; div; c 2. CHEMICAL ENGINEERING, POLYMER CHEMISTRY. *Educ:* Univ Western Ky, BS, 38; Purdue Univ, PhD(chem eng), 50. *Prof Exp:* Chemist, Devoe-Raynolds Co, Inc, 40-42; jr engr, Curtiss-Wright Corp, 42-44; res chemist, Monsanto Chem Co, 50-52; from asst prof to assoc prof chem, 54-83, assoc dir, Ctr Microcrystal Polymer Sci, 75-79, EMER PROF, DEPT CHEM, UNIV TEX, ARLINGTON, 84- *Concurrent Pos:* Pres, Sigma Xi, Univ Tex, Arlington, 66-68; lectr & educator mem, Dallas Soc, Fedn Socs for Paint Technol, 74-83; consult, protective coatings, Albert Halff Eng Co, 80-86; sci book rev, J Appl hem, 82. *Mem:* Sigma Xi; fel Am Inst Chem; Am Chem Soc; NY Acad Sci. *Res:* Protective coatings; organic polymer chemistry; education in chemical engineering; professional registration standards. *Mailing Add:* 1630 Pecan Park Dr Arlington TX 76012

WILLOUGHBY, STEPHEN SCHUYLER, b Madison, Wis, Sept 27, 32; m 54; c 2. MATHEMATICS EDUCATION. *Educ:* Harvard Univ, AB, 53, AMT, 55; Columbia Univ, EdD(math educ), 61. *Prof Exp:* Teacher math & sci, Newton Pub Schs, Mass, 54-57; teacher math, Greenwich Pub Schs, Conn, 57-59; instr educ & math, Univ Wis-Madison, 60-61, asst prof, 61-65; PROF EDUC & MATH, NY UNIV, 65-, CHMN DEPT MATH EDUC, 67- *Concurrent Pos:* Consult, NSF, 73-77. *Mem:* Math Asn Am; Nat Coun Teachers Math (pres, 82-84). *Res:* Learning and teaching mathematics. *Mailing Add:* Dept Math Univ Ariz Tucson AZ 85721

WILLOUGHBY, WILLIAM FRANKLIN, b Washington, DC, Feb 4, 36; m 75; c 5. IMMUNOPATHOLOGY, AEROSPACE MEDICINE. *Educ:* Johns Hopkins Univ, AB, 57, MD, 65, PhD(microbiol), 65. *Prof Exp:* Res fel immunopath, Scripps Clin Res Found, 67-69; asst prof path, Case Western Reserve Univ, 69-72; dir, Va Mason Res Ctr, 72-75; assoc prof path, Johns Hopkins Univ Sch Med, 75-87; PROF & CHMN, PATH, UNIV SC SCH MED, 87- *Concurrent Pos:* Ad hoc consult, Vet Admin, 77-85, NIH, 86-; mem, Path A Study Sect & Spec Study Sect, NIH, 83-86, ad hoc mem, pulmonary dis adv comt, Nat Heart, Lung & Blood Inst, 85; consult, Cotton, Inc, 80-83; mem, Comt Byssinosis, Nat Acad Sci, 81-82, & comt Irritants & Vesicants, 83-84; USAFR, Brig Gen & mobilization asst to Dep Surgeon Gen, HQ, USAF, 90-; fel, Arthritis Found, 67-69. *Honors & Awards:* Edward E Osgood Prize Med Res, E E Osgood Found, 73. *Mem:* Am Asn Immunologists; Am Asn Pathologists; Int Acad Path; Am Thoracic Soc; AAAS; Soc USAF Flight Surgeons; Am Soc Cell Biologists; Reticuloendotheliol Soc; Asn Path Chmn; Soc Med Consults Armed Forces. *Res:* Immunopathology; mechanisms of inflammation, particularly as they affect the lung; macrophage function, including their release of soluble mediators of inflammation. *Mailing Add:* Dept Pathol Univ SC Sch Med Columbia SC 29208

WILLOWS, ARTHUR OWEN DENNIS, b Winnipeg, Man, Mar 26, 41; m 63; c 3. NEUROPHYSIOLOGY. *Educ:* Yale Univ, BS, 63; Univ Ore, PhD(biol), 67. *Prof Exp:* Asst prof, Univ Ore, 67-68, res assoc neurophysiol, 68-69; from asst prof to assoc prof, Univ Wash, 69-75, prof zool, 75-80, dir, Friday Harbor Labs, 73-80; MEM STAFF NEUROBIOL PROG, NSF, 80- *Mem:* Soc Gen Physiol; Soc Neurosci; Am Physiol Soc. *Res:* Neuroethology; neurophysiological basis of behavior. *Mailing Add:* 3140 Mineral Heights Dr Friday Harbor WA 98250

WILLS, CHRISTOPHER J, b London, Eng, Mar 23, 38; m 65; c 1. GENETICS, BIOLOGY. *Educ:* Univ BC, BA, 60, MSc, 62; Univ Calif, Berkeley, PhD(genetics), 65. *Prof Exp:* NIH fel genetics, Univ Calif, Berkeley, 65-66; asst prof biol, Wesleyan Univ, 66-72; assoc prof, 72-78, PROF BIOL, UNIV CALIF, SAN DIEGO, 78- *Concurrent Pos:* NIH res grant, 67; Guggenheim fel, 77-78; var grants, NSF, NIH & Dept Energy. *Mem:* AAAS; Genetics Soc Am; Am Soc Naturalists. *Res:* Maintenance of genetic variability in natural populations; production, through selection in the laboratory, and characterization of isoenzymes in yeast; regulation of yeast isoenzymes; biochemistry. *Mailing Add:* Dept Biol Univ Calif San Diego La Jolla CA 92093

WILLS, DONALD L, b Peoria, Ill, May 12, 24; m 46; c 2. GEOLOGY. *Educ:* Univ Ill, BS, 49, MS, 51; Univ Iowa, PhD(geol), 71. *Prof Exp:* Assoc dir, Ill Dept Conserv, 77-79; DIR FINANCE & BUS, MONMOUTH COL, 79- *Concurrent Pos:* Environ consult, 71- *Mem:* Am Inst Prof Geol; Sigma Xi; Nat Asn Geol Teachers; Geol Soc Am. *Res:* Biostratigraphic studies of Mississippian Chesterian series. *Mailing Add:* 1048 E Euclid Ave Monmouth IL 61462

WILLS, GENE DAVID, b Birmingham, Ala, Apr 11, 34; m 66; c 2. PLANT PHYSIOLOGY, BIOCHEMISTRY. *Educ:* Auburn Univ, BS, 57, MS, 62; Okla State Univ, PhD(bot), 67. *Prof Exp:* Asst bot, Auburn Univ, 59-62 & Okla State Univ, 63-66; PLANT PHYSIOLOGIST WEED CONTROL, DELTA BR EXP STA, 67- *Mem:* Weed Sci Soc Am; Sigma Xi. *Res:* Chemical weed control including studies on ecology and anatomy of weeds and effects of environment on translocation and toxicity of radiolabeled and non-radiolabeled herbicides in weeds. *Mailing Add:* Delta Br Exp Sta PO Box 197 Stoneville MS 38776

WILLS, GEORGE B(AILEY), b Canton, Mo, Nov 24, 28; m 54; c 3. CHEMICAL ENGINEERING. *Educ:* Mass Inst Technol, BS, 54; Univ Wis, MS, 55, PhD(chem eng), 62. *Prof Exp:* Engr, Mallinckrodt Chem Works, 55-57 & Bjorksten Res Labs, 57-61; asst res engr, Phillips Petrol Co, 61-64; PROF CHEM ENG, VA POLYTECH INST & STATE UNIV, 64- *Concurrent Pos:* Consult, A O Smith Corp, 60 & Electrotech Corp, 66- *Mem:* AAAS; Am Inst Chem Engrs; Am Chem Soc; Sigma Xi. *Res:* Mass transfer; electrochemistry; catalysis. *Mailing Add:* Dept Chem Eng Va Polytech Inst & State Univ Blacksburg VA 24061

WILLS, JAMES E, JR, b Tucumcari, NMex, Mar 20, 16. PHYSICS. *Educ:* Miss Col, BA, 36; Univ Va, MA, 38; Univ Tex, PhD(physics), 56. *Prof Exp:* Instr physics, Ga Sch Technol, 38-39; asst prof, Baylor Univ, 46-51; from assoc prof to prof, Stetson Univ, 56-64; chmn dept, 64-72, prof, 64-79, EMER PROF PHYSICS, UNIV NC, ASHEVILLE, 80- *Mem:* Am Phys Soc; Sigma Xi. *Res:* Fast neutron spectroscopy. *Mailing Add:* 36 Mockingbird Rd Swannanoa NC 28778

WILLS, JOHN G, b Greeley, Colo, Feb 4, 31; m 54; c 5. THEORETICAL PHYSICS. *Educ:* San Diego State Col, AB, 53; Univ Wash, MS, 56, PhD(physics), 63. *Prof Exp:* Staff mem, Los Alamos Sci Lab, 56-60; from asst prof to assoc prof, 64-76, PROF PHYSICS, IND UNIV, BLOOMINGTON, 76- *Mem:* Am Phys Soc. *Res:* Nuclear theory; elementary particle theory; scattering theory. *Mailing Add:* Dept Physics Ind Univ Bloomington IN 47405

WILLS, NANCY KAY, b Wytheville, Va, Aug 27, 49. PHYSIOLOGY. *Educ:* Ohio State Univ, BS, 71; Univ Va, MA, 73, PhD(physiol psychol), 77. *Prof Exp:* Res asst neurophysiol, Dept Physiol & Brain Res Inst, Univ Calif, Los Angeles, 74-76; fel physiol, Med Br, Univ Tex, 76-77; fel physiol, Med Sch, 77-80, res assoc fac physiol, Yale Univ, 80-87; ASSOC PROF, DEPT PHYSIOL & BIOPHYS, UNIV TEX MED BR, 87- *Concurrent Pos:* NIH fel, 78-80. *Mem:* Biophys Soc; Soc Gen Physiologists; NY Acad Sci; Am Physiol Soc; Asn Women Sci. *Res:* Electrophysiology and cell biology of ion transport across epithelia. *Mailing Add:* Dept Physiol & Biophysics Univ Tex Med Br Galveston TX 77550-2781

WILLS, WIRT HENRY, b Petersburg, Va, Feb 12, 24; m 54; c 4. DISEASES OF ORNAMENTALS. *Educ:* Univ Richmond, BA, 50; Duke Univ, MA, 52, PhD(bot), 54. *Prof Exp:* From asst prof to prof, 54-68, EMER PROF PLANT PATH, VA POLYTECH INST & STATE UNIV, 90- *Mem:* Am Phytopath Soc. *Res:* Ecology of root diseases; biological control of fungal pathogens in soil-less media. *Mailing Add:* Dept Plant Path & Physiol Va Polytech Inst & State Univ Blacksburg VA 24061

WILLSON, ALAN NEIL, JR, b Baltimore, Md, Oct 16, 39; m 62; c 3. ELECTRICAL ENGINEERING, APPLIED MATHEMATICS. *Educ:* Ga Inst Technol, BEE, 61; Syracuse Univ, MSEE, 65, PhD(elec eng), 67. *Prof Exp:* Instr, Syracuse Univ, 65-67; assoc engr, IBM Corp, 61-64; mem tech staff, Math & Statist Res Ctr, Bell Labs, 67-72; PROF ENG & APPL SCI, UNIV CALIF, LOS ANGELES, 72-, ASST DEAN GRAD STUDIES, SCH ENG & APPL STUDIES, 77- *Concurrent Pos:* Ed, Inst Elec & Electronics Engrs Transactions on Circuits & Systs, 77-79. *Mem:* Inst Elec & Electronics Engrs; Soc Indust & Appl Math; Am Soc Eng Educ; Sigma Xi. *Res:* Theory of nonlinear transistor networks; stability and instability theory for nonlinear distributed networks; studies of nonlinear effects in digital filters. *Mailing Add:* Univ Calif 400 Boelter Hall Los Angeles CA 90024

WILLSON, CARLTON GRANT, b Vallejo, Calif, Mar 30, 39; m 75; c 2. RADIATION CHEMISTRY, POLYMER SYNTHESIS. *Educ:* Univ Calif, Berkeley, BS, 62, PhD(chem), 73; San Diego State Univ, MS, 69. *Prof Exp:* Chemist, Aerojet Gen Corp, Sacramento, 62-64; instr chem & math, Fairfax High Sch, Los Angeles, 65-66; asst prof chem, Long Beach State Univ, 74-75, Univ Calif, San Diego, 76-78; MGR, RES LAB, IBM CORP, 78- *Concurrent Pos:* IBM fel, 85. *Honors & Awards:* Humboldt Sr Scientist Award, 88; Chem Mat Award, Am Chem Soc, 90. *Mem:* Am Chem Soc; Sigma Xi; Am Inst Mech Eng; AAAS; Am Phys Soc; Soc Plastic Engrs. *Res:* Synthetic and mechanistic studies associated with radiation sensitive organic materials, monomers and polymers and their application to resist materials. *Mailing Add:* IBM Almaden Res Ctr K91/801 650 Harry Rd San Jose CA 95120-6099

WILLSON, CLYDE D, b Omaha, Nebr, May 7, 35; m 54; c 4. ORGANIC CHEMISTRY, MOLECULAR BIOLOGY. *Educ:* Univ Calif, Berkeley, BA, 56, PhD(chem), 60. *Prof Exp:* NIH fel bact genetics & protein synthesis, Pasteur Inst, Paris, 60-62; asst prof biochem, Univ Calif, Berkeley, 62-68, res fel entom, Miller Inst, 68-69; INSTR LIFE SCI, LANEY COL, 69- *Concurrent Pos:* Vis prof biol, Brandeis Univ, 74-75. *Res:* Heterocyclic organic chemistry; bacterial enzyme regulation and genetic control; characterization of messenger RNA; biochemistry of communication substances in insects. *Mailing Add:* Dept Life Sci Laney Col 900 Fallon St Oakland CA 94607

WILLSON, DONALD BRUCE, b Bloomington, Ind, Oct 25, 41; m 65; c 4. PHYSICAL INORGANIC CHEMISTRY, EXTRACTIVE METALLURGY. *Educ:* Geneva Col, BA, 63; Tufts Univ, PhD(chem), 69. *Prof Exp:* Res assoc Air Force Off Sci Res, Geneva Col, 69; chemist, Kawecki-Berylco Industs, Inc, Pa, 70-73, proj leader-group leader, 74; tech mgr, M&R Refractory Metals Inc, Winslow, 75-76, tech dir, 76-81; staff res eng, Anaconda Minerals Co, 81-83, sr res scientist, 83-85; tech advisor, Ceralox, Arco Chem Co, 85-86; plant mgr, 86-87, MGR RES & DEVELOP, CERALOX CORP, 87- *Concurrent Pos:* Extractive metallurgy consult. *Mem:* Am Inst Mining, Metall & Petrol Engrs; Am Chem Soc; Am Ceramic Soc. *Res:* Development and production of advanced ceramics and materials; extractive metallurgy and physical, inorganic and analytical chemistry of the refractory, transition, rare earth and noble metals and their compounds; advanced ceramics materials. *Mailing Add:* 3732 N Tres Lomas Pl Tucson AZ 85749-9416

WILLSON, JOHN ELLIS, b Scranton, Pa, May 4, 29; m 55; c 3. VETERINARY MEDICINE, TOXICOLOGY. *Educ:* Pa State Univ, BS, 50; NY State Vet Col, DVM, 54; Am Bd Toxicol, dipl, 81; Acad Toxicol Sci, dipl, 82. *Prof Exp:* Intern, Angell Mem Animal Hosp, Boston, Mass, 57-58, mem staff, 58-61; head dept pharmacol, John L Smith Mem Cancer Res, Chas Pfizer & Co, Inc, NJ, 61-63; sr pathologist, 63-66, asst dir, 66-82, MGR, JOHNSON & JOHNSON RES FOUND, NEW BRUNSWICK, 82- *Concurrent Pos:* Mem coun accreditation, Am Asn Accreditation of Lab Animal Care, 72-76; mem adv coun, Inst Lab Animal Resources, Div Biol Sci, Assembly Life Sci, Nat Res Coun-Nat Acad Sci, 74-77; mem adv coun, NY State Col Vet Med, Cornell Univ, 80-86; adv, Vet Med Educ Comt, NJ, 87- *Honors & Awards:* Philip B Hoffman Res Scientist Award, Johnson & Johnson, 72. *Mem:* Am Vet Med Asn; Soc Toxicol; Am Asn Indust Vet; Am Asn Lab Animal Sci. *Res:* Toxicology of foods, drugs, medical devices and cosmetics; ethylene oxide sterilant residues; laboratory animal husbandry and medicine. *Mailing Add:* 42 Addison Dr Basking Ridge NJ 07920

WILLSON, JOHN TUCKER, b Bismarck, NDak, Aug 26, 24; m 47; c 3. ANATOMY. *Educ:* George Washington Univ, BS, 48, MS, 49; Univ Colo, PhD(anat), 53. *Prof Exp:* Asst, 51-53, from instr to assoc prof, 53-75, PROF ANAT, SCH MED, UNIV COLO, DENVER, 75- *Mem:* AAAS; Am Asn Anat; Am Soc Cell Biol; Microcirc Soc; Sigma Xi. *Res:* Microcirculation, intravascular erythrocyte agglutination, reproduction, fertility and sterility and fine structure. *Mailing Add:* Dept Anat Univ Colo Sch Med 4200 E Ninth Ave Denver CO 80262

WILLSON, KARL STUART, electrochemistry, corrosion, for more information see previous edition

WILLSON, LEE ANNE MORDY, b Honolulu, Hawaii, Mar 14, 47; m 69; c 2. ASTRONOMY, ASTROPHYSICS. *Educ:* Harvard Univ, AB, 68; Univ Mich, MS, 70, PhD(astron), 73. *Prof Exp:* From instr to assoc prof, 73-88, PROF ASTROPHYS, DEPT PHYSICS, IOWA STATE UNIV, 88- *Concurrent Pos:* Annie J Cannon Award, Am Astron Soc, 80-81; mem, Steering Comt, Sect D, AAAS, 87-; chmn, Comt Status Women in Astron, Am Astron Soc, 87-89; dir-at-large, Assoc Univ Res Astron, 90- *Mem:* Int Astron Union; Am Astron Soc; AAAS. *Res:* Problems of stellar atmospheres, particularly theories of mass loss, extended atmospheres, and variable stars. *Mailing Add:* Astron Prog Dept Physics & Astron Iowa State Univ Ames IA 50011

WILLSON, MARY FRANCES, b Madison, Wis, July 28, 38; wid. ECOLOGY, EVOLUTION. *Educ:* Grinnell Col, BA, 60; Univ Wash, PhD(zool), 64. *Prof Exp:* From asst prof to assoc prof zool, Univ Ill, Urbana-Champaign, 65-77, prof ecol, ethology & evolution, 77-90; RES ECOLOGIST, US FOREST SERV, 89-; ADJ PROF ZOOL & BOT, WASH STATE UNIV, PULLMAN, 90- *Concurrent Pos:* Res grants, Chapman Fund, Am Mus Natural Hist, 64 & 79 & Univ Ill Res Bd, 66-70, 77 & 81-85, NSF, 74-78, 86; vis prof, Univ Minn, 78, 80, 82 & 84, Monash Univ, 86. *Mem:* Ecol Soc Am; Am Ornith Union; Brit Ecol Soc; Cooper Ornith Soc; Am Soc Naturalists; Soc Study Evolution; Bot Soc Am; Asn Trop Biol. *Res:* Evolutionary ecology. *Mailing Add:* Forestry Sci Lab PO Box 20909 Juneau AK 99802

WILLSON, PHILIP JAMES, b Detroit, Mich, Apr 23, 26; m 48, 76; c 5. HIGH TEMPERATURE CHEMISTRY, CERAMICS. *Educ:* Wayne State Univ, BA, 51. *Prof Exp:* Technician electronics, US Navy, 45-46 & Gen Motors Corp, 48-50; chemist, Chrysler Corp, 51-55, res supvr chem, 56-88; OWNER, WILLSON ENTERPRISE, INC, 88- *Mem:* Soc Automotive Engrs. *Res:* High temperature structural ceramic materials for turbine engine applications; automotive catalysts for emission control; friction studies. *Mailing Add:* Willson Enterprise Inc 709 Mt Vernon Royal Oak MI 48073

WILLSON, RICHARD ATWOOD, b Minneapolis, Minn. GASTROENTEROLOGY. *Educ:* Univ Minn, BA, 58, BS, 59, MD, 62, MS, 69, Am Bd Internal Med, cert, 70. *Prof Exp:* Intern, Mary Fletcher Hosp, 62-63; resident internal med, Univ Vt, 63-64; resident, Mayo Clin, 66-68, NIH res fel gastroenterol, 68-71; res fel, Liver Unit, Dept Med, King's Col Hosp Med Sch, London, 72-73; asst prof, 73-77, ASSOC PROF MED, UNIV WASH, 77-; HEAD DIV GASTROENTEROL, HARBORVIEW MED CTR, SEATTLE, 73- *Mem:* Am Gastroenterol Asn; Am Asn Study Liver Dis; Am Fedn Clin Res. *Res:* Treatment of acute fulminant hepatic failure and the study of hepatic injury secondary to drugs and drug metabolism. *Mailing Add:* Dept Med Univ Wash Seattle WA 98195

WILLSON, WARRACK GRANT, b San Francisco, Calif, July 15, 43; m 63; c 2. FUEL ENGINEERING, PHYSICAL CHEMISTRY. *Educ:* Univ Northern Colo, BA, 65; Univ Wyo, PhD(chem, physics), 71. *Prof Exp:* Asst prof phys chem, Upper Iowa Col, 70-71; res scientist coal gasification, Univ Wyo, 71-73; res engr reactor eng, E I du Pont de Nemours & Co, Inc, 73-76; group leader coal chems, Occidental Res Corp, Calif, 76-78; proj mgr coal liquefaction, 78-79, mgr gasification & liquefacation, Dept Energy, Grand Forks Energy Technol Ctr, NDak, 79-83; MGR, FUELS PROCESS CHEM DIV, ENERGY RES CTR, UNIV NDAK, 83- *Concurrent Pos:* Adj prof chem eng, Univ NDak, 80- *Mem:* Am Chem Soc; Sigma Xi. *Res:* Conversion of abundant domestic fossil resources and carbonaceous wastes into economically and environmentally acceptable alternate energy sources through combustion, gasification and liquefraction; high pressure/temperature reaction engineering and process development. *Mailing Add:* 1602 Baron Blvd Grand Forks ND 58201-8409

WILLWERTH, LAWRENCE JAMES, b Melrose, Mass, Oct 3, 32; m 56; c 3. PLASTICS CHEMISTRY. *Educ:* Lowell Technol Inst, BS, 72, MS, 75. *Prof Exp:* Jr chemist plastics, Nat Polychem Inc, Mass, 55-60; chemist, Avco Corp, Mass, 60-66; TECH MGR CHEM-PLASTICS, K J QUINN & CO INC, MALDEN, MASS, 66- *Concurrent Pos:* Instr polymer characterization, Eve Div, Lowell Technol Inst, 72- *Mem:* Am Chem Soc; Am Inst Chemists; Soc Plastics Engrs. *Res:* Research and development of polyurethane plastics; attainment of specific properties through rearrangement and addition of various species to the polymer backbone. *Mailing Add:* 160 Flamingo Rd Edgewater FL 32141-7206

WILMER, HARRY A, b New Orleans, La, Mar 5, 17; m 45; c 5. PSYCHIATRY. *Educ:* Univ Minn, BS, 38, MS, 40, MD, 41, PhD(path), 44. *Prof Exp:* Chief psychiat, Palo Alto Clin, Calif, 49-51; prof psychiat, Sch Med, Univ Calif, 64-69; sr psychiatrist, Scott & White Clin, Temple, Tex, 69-72; PROF PSYCHIAT, UNIV TEX HEALTH SCI CTR, 72- *Concurrent Pos:* Consult, Mayo Clin, Mayo Found, 57-58 & Dept Corrections, State Calif, 61-65; ed, Hosp & Community Psychiat, 65-68; Guggenheim fel, Jung Inst, Zurich Switz, 69-70; dir, Int Film Festival-Symp Cult & Psychiat, Univ Tex Health Sci Ctr, 71-78; pres & dir, Inst Humanities, Salado, Tex, 80- *Mem:* Am Psychiat Asn; Int Asn Anal Psychologists; Am Acad Psychoanalysis; Am Col Psychiatrists; Inter-regional Soc Anal Psychologists. *Res:* Therapeutic community; dreams and Jungian psychology. *Mailing Add:* 506 S Ridge Rd Salado TX 76571

WILMER, MICHAEL EMORY, b Washington, DC, Oct 11, 41. INFORMATION SCIENCE. *Educ:* Cath Univ Am, BSEE, 63, MSEE, 67, PhD(elec eng), 68. *Prof Exp:* Prin scientist image processing, Palo Alto Res Ctr, Xerox Corp, 67-; VPRES, PEARSON ELECTRONICS. *Mem:* Sigma Xi; Inst Elec & Electronics Engrs. *Res:* Digital processing of images and speech for enhancement, compression and recognition. *Mailing Add:* Varian Assoc 611 Hansen Way MSC-076 Palo Alto CA 94303

WILMOT, GEORGE BARWICK, b Waterbury, Conn, Oct 27, 28; m 53; c 7. PHYSICAL CHEMISTRY. *Educ:* Rensselaer Polytech Inst, BS, 51; Mass Inst Technol, PhD(phys chem), 54. *Prof Exp:* Res chemist, Naval Ord Sta, 54-73, RES CHEMIST, NAVAL SURFACE WEAPONS CTR, 73- *Mem:* AAAS; Am Chem Soc; Am Phys Soc; Sigma Xi. *Res:* Infrared and Raman spectroscopy; propellants, explosives, combustion; thermodynamics; lasers. *Mailing Add:* 401 Amherst Rd RR Bryans Road MD 20616

WILMOTH, BENTON M, b Big Stone Gap, Va, July 4, 25; m 48; c 3. HYDROLOGY, WATER RESOURCES. *Prof Exp:* Geologist, US Geol Surv, 52-66; HYDROGEOLOGIST, US ENVIRON PROTECTION AGENCY, 66- *Mem:* Geol Soc Am; Am Inst Prof Geologists. *Res:* Ground water availability and quality control in the central Appalachian regions of US; stabilization and removal of unnatural hazardous chemical releases threatening the quality of the natural hydrogeologic environment. *Mailing Add:* 104 Martha Dr St Clairsville OH 43950

WILMOTTE, RAYMOND M, b Paris, France, Aug 13, 08. TELE-COMMUNICATION. *Educ:* Univ Cambridge, Eng, BA, 21, MA, 23, ScD(mech sci), 58. *Prof Exp:* Consult engr, self-employed, 32-83; TECH ANALYST, FED COMMUN COMN, 83- *Concurrent Pos:* Fel Wash Acad Sci. *Mem:* fel Inst Elec & Electronics Engrs. *Res:* Introduction of telecommunication technology; expected application in private and public use. *Mailing Add:* Fed Commun Comn 2025 M St NW Washington DC 20016

WILMS, ERNEST VICTOR, b Winnipeg, Man, Apr 21, 36. ENGINEERING MECHANICS. *Educ:* Univ Man, BSc, 58; Univ Ill, MS, 60, PhD(theoret & appl mech), 63. *Prof Exp:* Res engr, Can Armament Res & Develop Estab, 58-59; from res asst to res assoc theoret & appl mech, Univ Ill, Urbana, 59-62; asst prof mech eng, Univ Sask, 62-64; res engr, Babcock & Wilcox Res Ctr, Ohio, 64-65; from asst prof to assoc prof eng mech, Univ Ala, 65-68; ASSOC PROF CIVIL ENG, UNIV MAN, 68- *Concurrent Pos:* Consult, Army Res Off, NC, 67-68. *Mem:* Am Inst Aeronaut & Astronaut. *Res:* Solid and fluid mechanics; dynamics and vibrations. *Mailing Add:* Dept Civil Eng Univ Manitoba Winnipeg MB R3T 2N2 Can

WILMS, HUGO JOHN, JR, b Gunskirchen, Austria, June 23, 22; US citizen; m 50; c 3. ELECTRICAL ENGINEERING, UNDERSEA WARFARE. *Educ:* Marquette Univ, BEE, 50; Univ Mo, MS, 52. *Prof Exp:* Proj leader underwater acoust, US Navy Underwater Systs Ctr, 51-54, develop supvr sonar systs, 54-61, sr mem syst analytical staff, 61-68, consult tech dir staff, 68-71, head plans & goals div, 71-74, sci adv to comdr, US 6th Fleet, 74-75, antisubmarine warfare adv to comdr, 75-76, head prog & mgt planning staff, 76-79; CONSULT, 80- *Concurrent Pos:* Tech consult; comdr, Oper Test & Eval Force, Norfolk, Va, 61; consult, Adv Sea Based Deterrance Summer Study, US Navy, 64; chmn Captor tech rev comt, Undersea Warfare Res & Develop Planning Coun, 66; consult, Dept Higher Educ, State Conn, 82-85. *Mem:* AAAS; Inst Elec & Electronics Engrs; NY Acad Sci. *Res:* Acoustic propagation in the sea; underwater acoustic sound sources, calibrations and measurements. *Mailing Add:* RR 1 Box 82 Ft Kent ME 04743

WILMSEN, CARL WILLIAM, b Galveston, Tex, Nov 20, 34; m 60; c 2. ELECTRICAL ENGINEERING. *Educ:* Tex A&M Univ, BS, 56; Univ Tex, Austin, BS, 60, MS, 62, PhD(elec eng), 67. *Prof Exp:* Test engr, Gen Dynamics Corp, 56-59; res engr, Tracor, Inc, 60-62; from asst prof to assoc prof, 66-77, PROF ELEC ENG, COLO STATE UNIV, 77- *Mem:* Am Phys Soc; Am Vacuum Soc. *Res:* Optoelectronic integrated circuits; heterojunctiar optoelectronic devices; Auger and XPS analysis; oxide growth on semiconductor. *Mailing Add:* Dept Elec Eng Colo State Univ Ft Collins CO 80523

WILNER, GEORGE DUBAR, b New York, NY, Dec 7, 40. HEMATOLOGY, PATHOLOGY. *Educ:* Northwestern Univ, BS, 62, MD, 65. *Prof Exp:* From instr to prof path, Col Physicians & Surgeons, Columbia Univ, 69-78; ASSOC PROF PATH & ASST PROF MED, DIV CELL BIOL, SCH MED, WASH UNIV & JEWISH HOSP, 78- *Mem:* Am Heart Asn. *Res:* Hemostasis and thrombosis. *Mailing Add:* Albany Med Col Dept Med & Physiol Am Red Cross Hackett Blvd at Clara Barton Dr Albany NY 12208

WILPIZESKI, CHESTER ROBERT, auditory physiology, deafness, for more information see previous edition

WILSDORF, DORIS KUHLMANN, b Bremen, Ger, Feb 15, 22; US citizen; m 50; c 2. MATERIALS SCIENCE. *Educ:* Univ Gottingen, Dipl(physics), 46, Dr rer nat, 47; Univ Witwatersand, DSc (phys metall), 55. *Prof Exp:* Res assoc, Dept Metall, Univ Göttingen, 48 & HH Wills Phys Lab, Bristol Univ, 49-50; lectr physics, Univ Witwatersrand, 50-56; from assoc prof to prof metall eng, Univ Pa, 57-63; prof eng physics, 63-66, UNIV PROF APPL SCI, UNIV VA, 66- *Honors & Awards:* Res Medal, Am Soc Eng Educ, 65 & 66; J Shelton Horsley Award, Va Acad Sci, 65; Heyn Medal, Ger Soc Mat Sci, 88; Achievement Award, Soc Women Engrs, 89; Ragnar Holm Sci Achievement Award, Inst Elec & Electronics Engrs, 91. *Mem:* Fel Am Phys Soc; Am Soc Eng Educ; Ger Metall Soc; Am Inst Mining & Metall Engrs; fel Am Soc Metals. *Res:* Theory of crystal defects, crystal plasticity, tribology; electric contacts. *Mailing Add:* 304 Dept Physics Univ Va Charlottesville VA 22901

WILSDORF, HEINZ G(ERHARD) F(RIEDRICH), b Pennekow, Ger, June 25, 17; nat US; m 50. MATERIALS SCIENCE, PHYSICS. *Educ:* Univ Berlin, Dipl, 44; Univ Göttingen, Dr rer nat, 47; Univ Witwatersrand, DSc, 54. *Prof Exp:* Res asst physics, Univ G-ttingen, 47-49; prin res officer, Nat Phys Lab, SAfrica, 50-56; tech dir solid state sci labs, Franklin Inst, Pa, 56-63; prof mat sci, 63-66, chmn dept, 63-76, Wills Johnson prof, 66-87, WILLIAM G REYNOLDS PROF MAT SCI, UNIV VA, 87- *Concurrent Pos:* Mem, space processing ad hoc adv subcomt, NASA, 75-79; dir, Light Metals Ctr, Univ Va, 84- *Mem:* Fel Am Phys Soc; Electron Micros Soc Am; fel Am Soc Metals; Am Inst Mining, Metall & Petrol Engrs; Sigma Xi. *Res:* Metallurgy; x-ray crystallography; electron diffraction; electron microscopy; metal physics, especially plastic deformation and fracture; thin films. *Mailing Add:* Dept Mat Sci Univ Va Charlottesville VA 22901

WILSEY, NEAL DAVID, b Tunkhannock, Pa, July 27, 37; wid; c 4. MAGNETIC RESONANCE, SEMICONDUCTOR MATERIALS. *Educ:* Hartwick Col, BA, 61; Colo State Univ, MS, 64, PhD(physics), 67. *Prof Exp:* Res physicist solid state physics, US Naval Res Lab, 67-74, head electronic mat sect, Radiation Effects Br, 74-79, head, Radiation Interactions Sect, Semiconductors Br, 80-88, mem strategic planning staff, 88-90, head, Semiconductors Br, 90, HEAD, ELECTRONIC MAT BR, US NAVAL RES LAB, 90- *Concurrent Pos:* Instr, Univ Md, 71-72; vis scientist, Inst Study Defects in Solids, State Univ NY Albany, 78-79. *Mem:* Am Phys Soc; Mat Res Soc; Minerals Metals & Mat Soc. *Res:* Radiation effects, defects in solids; electronic materials. *Mailing Add:* US Naval Res Lab Code 6870 Washington DC 20375-5000

WILSHIRE, HOWARD GORDON, b Shawnee, Okla, Aug 19, 26; m 84; c 3. GEOMORPHOLOGY, SURFACE PROCESSES. *Educ:* Univ Okla, BA, 52; Univ Calif, Berkeley, PhD(geol), 56. *Prof Exp:* Lectr geol, Univ Sydney, 56-60; res fel, Australian Nat Univ, 61; GEOLOGIST, US GEOL SURV, 61- *Mem:* Geol Soc Am; AAAS; Am Geophys Union. *Res:* Structure and petrology of igneous rocks; petrology and processes of the upper mantle; effects of human uses of arid lands and rates of recovery. *Mailing Add:* US Geol Surv 345 Middlefield Rd Menlo Park CA 94025

WILSKA, ALVAR P, physics; deceased, see previous edition for last biography

WILSON, ALAN C, b Kampala, Uganda, July 30, 45; m 76; c 3. HEART DISEASE, LIPID METABOLISM. *Educ:* Univ Aberdeen, Scotland, PhD(biochem), 72. *Prof Exp:* Adj prof, 79-88; adj prof med, 88-89, ASSOC PROF CLIN MED, UNIV MED & DENT NJ, ROBERT W JOHNSON MED SCH, 89- *Mem:* Am Inst Nutrit; Biochem Soc Eng; Biochem Soc Can; Am Soc Biochem & Molecular Biol. *Mailing Add:* Dept Med MEB 589 Univ Med & Dent NJ New Brunswick NJ 08903-0019

WILSON, ALBERT E, b Glenwood Springs, Colo, Jan 17, 27; m 52; c 5. NUCLEAR ENGINEERING. *Educ:* Univ Colo, BS, 50; Univ NMex, MS, 59; Univ Okla, PhD(eng sci), 64. *Prof Exp:* Physicist, Nat Bur Standards, 51-55; staff mem, Los Alamos Sci Lab, 55-59; from instr to asst prof nuclear eng, Univ Okla, 59-66; prof nuclear eng & chmn dept, 66-76, dean sch eng, 76-83, PROF ENG, IDAHO STATE UNIV, 76- *Concurrent Pos:* Mem Sci & Technol Adv Comt, Argonne Nat Lab, 81-88. *Mem:* Am Nuclear Soc; Am Soc Eng Educ; Nat Soc Prof Engrs. *Res:* Nuclear engineering education, especially instrumentation, kinetics and control of nuclear reactor systems. *Mailing Add:* Col Eng Idaho State Univ Pocatello ID 83209

WILSON, ALEXANDER D, b Corning, NY, Sept 12, 45. MECHANICAL ENGINEERING. *Educ:* Mass Inst Technol, BS, 67, MS, 89. *Prof Exp:* Supt prod develop, 86-90, SUPT TECH SERV, LUKENS STEEL CO, 90- *Mem:* Fel Am Soc Metals Int; Am Inst Mining Metall & Petrol Engrs; Am Iron & Steel Inst; Am Soc Civil Engrs. *Mailing Add:* Res Bldg Lukens Steel Co Coatesville PA 19320

WILSON, ALLAN CHARLES, biochemistry, molecular evolution; deceased, see previous edition for last biography

WILSON, ALMA MCDONALD, plant physiology; deceased, see previous edition for last biography

WILSON, ANDREW ROBERT, b Dublin, Ireland, Sept 13, 41; US citizen; m 68. PLASMA PHYSICS. *Educ:* Trinity Col, Univ Dublin, BA, 62, MA, 65; Oxford Univ, DPhil(physics), 68. *Prof Exp:* Fr Govt boursier nuclear physics, Inst Fourier, Grenoble, 62-63; fel solid state physics, Lincoln Lab, Mass Inst Technol, 68-70; systs engr, Elec Supply Bd, Dublin, Ireland, 70-71; prog mgr plasma physics, 71-83, VPRES, MAXWELL LABS, SYSTS, SCI & SOFTWARE, 83- *Mem:* Am Phys Soc; Inst Elec & Electronics Engrs. *Res:* Hydromagnetic theory; system generated electromagnetic theory; radiation transport. *Mailing Add:* Maxwell Labs PO Box 1620 La Jolla CA 92038

WILSON, ANDREW STEPHEN, b Doncaster, Eng, Mar 26, 47; m 75; c 2. ASTRONOMY. *Educ:* Univ Cambridge, BA, 69, MA & PhD(radio astron), 73. *Prof Exp:* Res fel radio astron, State Univ Leiden, 73-75; res fel astron, Univ Sussex, 75-78; from asst prof to assoc prof, 81-86, PROF ASTRON, UNIV MD, COLLLEGE PARK, 86- *Mem:* Fel Royal Astron Soc; fel Am Astron Soc; Int Astron Union; Int Union Radio Sci. *Res:* Crab nebula; radio sources; active galactic nuclei. *Mailing Add:* Dept Astron Univ Md Colllege Park MD 20742

WILSON, ANGUS, b Mexico, Maine, Aug 13, 20; m; c 2. RUBBER CHEMISTRY. *Educ:* Georgetown Univ, BS, 41. *Prof Exp:* Control & anal chemist, E I du Pont de Nemours & Co, Inc, 41-43; prod supvr, Joseph E Seagrams & Sons, Inc, 47-48; control lab supvr, Govt Lab, Univ Akron, 49-52; rubber chemist, US Army Natick Res & Develop Command, 52-68, head rubber technol group, 68-74, head, Rubber & Plastics Group, 74-80; RETIRED. *Mem:* Am Chem Soc. *Res:* Development of methods of testing rubber and elastomeric materials; compounding of phosphazene elastomers; evaluation of experimental elastomers for possible end item applications. *Mailing Add:* Eight Bowdoin Rd Ipswich MA 01938

WILSON, ARCHIE FREDRIC, b Los Angeles, Calif, May 7, 31; m 66; c 2. PULMONARY DISEASES, PULMONARY PHYSIOLOGY. *Educ:* Univ Calif, Los Angeles, BA, 53; Univ Calif, San Francisco, MD, 57, PhD(physiol), 67. *Prof Exp:* Asst prof internal med, Univ Calif, Los Angeles, 67-70; from asst prof to assoc prof, 70-79, vchmn dept, 78-82, PROF INTERNAL MED, UNIV CALIF, IRVINE, 79- *Mem:* Am Col Chest Physicians; Am Fedn Clin Res; Am Col Physicians. *Res:* Asthma; airway physiology and pathophysiology; bronchodilator aerosols; physiology of transcendental meditation; exercise. *Mailing Add:* Dept Med Univ Calif Med Ctr Orange CA 92668

WILSON, ARCHIE SPENCER, b Tekoa, Wash, Jan 19, 21; m 44; c 3. CHEMISTRY. *Educ:* Iowa State Univ, BS, 46; Univ Chicago, MS, 50, PhD, 51. *Prof Exp:* Asst chem, Iowa State Univ, 43-46; res assoc, Gallium Proj, US Dept Navy, 48-49; asst, Univ Chicago, 49-50; instr chem, Univ Nebr, 50-51; sr scientist, Gen Elec Co, 51-65; sr res scientist, Pac Northwest Labs, Battelle Mem Inst, 64-71; assoc chmn dept, 71-78, prof chem, 71-89, EMER PROF CHEM, UNIV MINN, MINNEAPOLIS, 89- *Concurrent Pos:* US sci adv, Int Conf Peaceful Uses Atomic Energy, Geneva, 58; mem, US-UK Ruthenium Conf, 58. *Mem:* Am Chem Soc; fel AAAS; Nat Sci Teachers Asn. *Res:* Crystal structure of compounds of uranium; ruthenium chemistry; solvent extraction of the actinide elements. *Mailing Add:* 14833 58th Pl W Edmonds WA 98026

WILSON, ARMIN GUSCHEL, b Sapulpa, Okla, Dec 13, 16; m 43; c 2. ORGANIC CHEMISTRY, MEDICINAL CHEMISTRY. *Educ:* Rice Inst, BA, 39, MA, 41; Harvard Univ, PhD(org chem), 45. 45. *Prof Exp:* Org chemist, Off Sci Res & Develop, Harvard Univ, 45-47 & Merck & Co, Inc, NJ, 47-52; dept head res div, Bristol-Myers Co, 52-68; chmn math & sci, Mercer County Community Col, 68-69; teacher & counr urban univ prog, Grad Sch Educ, 69-72, PROF ACAD FOUND, LIVINGSTON COL, RUTGERS UNIV, NEW BRUNSWICK, 72- *Concurrent Pos:* Instr, Union Jr Col; chmn, Gordon Conf Med Chem, 67. *Mem:* Am Chem Soc; fel NY Acad Sci. *Res:* Synthetic organic chemistry; structure-action relationships of drugs; reaction mechanisms; photochemistry; nature of science; relationship of science and poetry. *Mailing Add:* 249 Harrison Ave Highland Park NJ 08904

WILSON, BARBARA ANN, solid state physics, for more information see previous edition

WILSON, BARBARA ANN, b Lafayette, Ind, Sept 7, 48; m 91. DETECTORS, PHOTONICS. *Educ:* Mt Holyoke Col, BA, 68; Univ Wis-Madison, PhD(physics), 78. *Prof Exp:* Postdoctoral assoc physics, Univ Wis-Madison, 78; mem tech staff, AT&T Bell Labs, 78-86, tech supvr, 86-88; TECH SUPVR, JET PROPULSION LAB, 88- *Mem:* Fel Am Phys Soc; sr mem Inst Elec & Electronics Engrs; Mat Res Soc. *Res:* Optical characterization of epitaxial semiconductor structures; current emphasis on III-V and Si-based approaches to LWIR detectors; lasers and integrated optoelectronics for laser sensing; optical communication and signal processing. *Mailing Add:* Jet Propulsion Lab MS 302-306 4800 Oak Grove Dr Pasadena CA 91109

WILSON, BARRY WILLIAM, b Brooklyn, NY, Aug 20, 31; m 56; c 1. CELL BIOLOGY, NEUROBIOLOGY. *Educ:* Univ Chicago, BA, 50; Ill Inst Technol, BS & MS, 57; Univ Calif, Los Angeles, PhD(zool), 62. *Prof Exp:* Asst zool, Ill Inst Technol, 56-57; asst, Univ Calif, Los Angeles, 57-58, USPHS cardiovasc trainee, 58, fel, 59-61, jr res zoologist, 62; asst prof poultry husb & animal physiol & asst biologist, Exp Sta, 62-68, assoc prof avian sci & animal physiol & assoc biologist, 68-72, PROF AVIAN SCI, ANIMAL PHYSIOL & ENVIRON TOXICOL, UNIV CALIF, DAVIS, 72- *Mem:* Am Soc Cell Biol; Soc Neurosci; Soc Develop Biol; Soc Toxicol; Tissue Cult Asn; Sigma Xi. *Res:* Cell growth and development; emphasis on muscle and nerve using cell culture and intact animals; regulation of acetylcholinesterase and other molecules of nerve, muscle; muscle disorders, dystrophy, pesticide action and ecotoxicology. *Mailing Add:* Dept Avian Sci Univ Calif Davis CA 95616

WILSON, BASIL W(RIGLEY), b Cape Town, SAfrica, June 16, 09; nat US; m 41; c 4. COASTAL ENGINEERING, PHYSICAL OCEANOGRAPHY. *Educ:* Cape Town Univ, BSc, 31, DSc, 53; Univ Ill, MS, 39, CE, 40. *Prof Exp:* Jr engr, SAfrican Rwys & Harbors, 32-33, asst to res engr, Chief Civil Engrs Dept, 33-41, asst res engr, 41-52; from assoc prof to prof eng oceanog, Tex A&M Univ, 53-61; mem sr staff, Nat Eng Sci Co, Calif, 61-63, assoc dir, 64; dir eng oceanog, Sci Eng Assocs, 64-68; CONSULT OCEANOG ENGR, 68- *Concurrent Pos:* Commonwealth Fund Serv fel, Harkness Found, Univ Ill, 38-39; mem, Permanent Int Asn Navig Congs, London, 57 & Baltimore, 61; coastal eng seminar, US-Japan Coop Sci Prog, Japan, 64; US lectr, NATO Adv Study Inst Berthing & Mooring Ships, Lisbon, Portugal, 65 & Wallingford, Eng, 73; invited lectr, Tech Univ Norway, Trondheim & Shell Develop Corp, Houston, 67; invited lectr, Ocean Eng Prog, Univ Hawaii, Honolulu, 68. *Honors & Awards:* Wellington Prize, Am Soc Civil Engrs, 52, Norman Medal, 69, Moffat-Nichol Harbor & Coastal Eng Award, 83; Inst Award, SAfrican Inst Civil Engrs, 59, Award of Meritorious Res, 84; Overseas Premium Award, Brit Inst Civil Engrs, 68. *Mem:* Emer mem Nat Acad Eng; fel AAAS; hon mem Am Soc Civil Engrs; fel Brit Inst Civil Engrs; fel SAfrican Inst Civil Engrs; Sigma Xi. *Res:* Track stresses; stability of railway vehicles on track; rail steel; economics of railway location; coastal engineering; ocean wave prediction; wave forces on structures; hurricane storm tide prediction; coastal seiches and oscillation in harbors; model engineering; ship motion and mooring problems; tsunami waves and hazards; submarine pipeline stability. *Mailing Add:* 529 Winston Ave Pasadena CA 91107

WILSON, BENJAMIN JAMES, b Pennsboro, WVa, Jan 7, 23; m 78; c 4. MICROBIOLOGY. *Educ:* Univ WVa, AB, 43, MS, 47; George Washington Univ, PhD(microbiol), 55. *Prof Exp:* Med bacteriologist, Biol Warfare Labs, Ft Detrick, 49-51, chief microbiol br, 51-59; assoc prof biol, David Lipscomb Col, 59-65; from asst prof to assoc prof, 63-75, PROF BIOCHEM, SCH MED, VANDERBILT UNIV, 75- *Concurrent Pos:* Consult nutrit sect, Off Int Res, NIH, 65-67 & Food & Drug Admin, 72-; mem subcomt toxicants occurring naturally in foods, Nat Acad Sci, 68-74; contrib ed, Nutrit Reviews, 73-; mem adj fac, Col Vet Med, Univ Tenn, Knoxville, 79- *Mem:* Am Soc Microbiol; Am Chem Soc; Soc Toxicol; NY Acad Sci. *Res:* Mycotoxins; natural toxicants; microbial toxins. *Mailing Add:* RR 1 No 241C White Bluff TN 37187

WILSON, BRAYTON F, b Cambridge, Mass, May 27, 34; m 60; c 2. BOTANY, FORESTRY. *Educ:* Harvard Univ, AB, 55, MF, 57; Australian Forestry Sch, dipl forestry, 59; Univ Calif, Berkeley, PhD(bot), 61. *Prof Exp:* Forest botanist, Harvard Univ, 61-67; asst prof, 67-72, ASSOC PROF FORESTRY, UNIV MASS, AMHERST, 72- *Mem:* AAAS; Bot Soc Am; Am Soc Plant Physiol. *Res:* Tree growth. *Mailing Add:* Dept Forestry Univ Mass Amherst Campus Amherst MA 01003

WILSON, BURTON DAVID, b Los Angeles, Calif, Oct 20, 32; m 58; c 3. ORGANIC CHEMISTRY. *Educ:* Univ Calif, Los Angeles, BS, 54; Univ Ill, PhD, 58. *Prof Exp:* Chemist, Eastman Kodak Co, 57-89; RETIRED. *Mem:* Am Chem Soc. *Res:* Development and production problem solving on chemicals for photographic end uses. *Mailing Add:* 430 Woodland Lane Webster NY 14580

WILSON, BYRON J, b Jackson, Wyo, Feb 2, 31; m 58; c 7. INORGANIC CHEMISTRY. *Educ:* Idaho State Univ, BS, 56; Southern Ill Univ, MA, 58; Univ Wash, PhD(chem), 61. *Prof Exp:* Asst prof chem, Vanderbilt Univ, 61-65; from asst prof to assoc prof, 65-72, PROF CHEM, BRIGHAM YOUNG UNIV, 72- *Mem:* Am Chem Soc. *Res:* Inorganic free-radical research; paper deterioration research. *Mailing Add:* 545 East 3050 N Provo UT 84604

WILSON, CARL C, b Halfway, Ore, Mar 17, 15; m 42; c 2. FOREST FIRE MANAGEMENT & RESEARCH. *Educ:* Univ Idaho, BS, 39; Univ Calif, Berkeley, MS, 41. *Prof Exp:* Fire control asst, Lassen Nat Forest, US Forest Serv, Calif, 46-49 & Plumas Nat Forest, 49, forester, Angeles Nat Forest, 49-55, res forester & proj leader forest fire res, Calif Forest Exp Sta, 56-57, chief forest fire res div, 57-62, asst dir forest fire & eng res & chief forest fire lab, Pac Southwest Forest & Range Exp Sta, 62-73, asst dir & nat fire specialist, Coop Fire Control, State & Pvt Forestry, US Forest Serv, 73-78; FOREST FIRE CONSULT, 78- *Concurrent Pos:* Consult fire mgt, Food & Agr Orgn-UN Environ Prog, UN, Rome, Italy, 75, Calif Dept Forestry, 78-80 & Ministry Natural Resources, Ont Can, 80. *Mem:* Soc Am Foresters; Am Soc Range Mgt; Am Forestry Asn. *Res:* Forest fire science; detection and control of forest fires for the protection of the human environment; author or coauthor of more than 40 technical papers in forestry and forest fire management. *Mailing Add:* Three Maybeck Twin Dr Berkeley CA 94708

WILSON, CARL LOUIS, botany, for more information see previous edition

WILSON, CAROL MAGGART, b Burley, Idaho, Oct 26, 36; m 59; c 2. PHARMACOGENETICS, BIOCHEMICAL GENETICS. *Educ:* Northwestern Univ, BA, 58; Wayne State Univ, PhD(biochem), 63. *Prof Exp:* Fel microbiol, Emory Univ, 63-64, biochem, 64-66; asst prof chem, Wheaton Col, 67-68; clin chemist, RI Hosp, 69-70; res assoc molecular biol, Univ Tex, Dallas, 71-73; res assoc, 74-76, instr, 76-78, ASST PROF PHARMACOL & INTERNAL MED, UNIV TEX HEALTH SCI CTR, DALLAS SOUTHWESTERN MED SCH, 78- *Mem:* Sigma Xi; Genetics Soc. *Res:* Control of eukaryotic gene expression; genetic and hormonal regulation of renin activity and structure; biochemistry and genetic variation of 5 alpha-reductase. *Mailing Add:* 608 Laguna Dr Richardson TX 75080

WILSON, CARROLL KLEPPER, b Denton, Tex, Aug 24, 17; m 41; c 2. MATHEMATICS. *Educ:* NTex State Col, BA, 37; Univ Tex, MA, 40. *Prof Exp:* Teacher pub schs, Tex, 37-41; instr radar, US Civil Serv, 41-43; from assoc prof to prof math, 46-83, actg chmn div math & natural sci, 56-60, head dept, 60-74, EMER PROF MATH, EASTERN NMEX UNIV, 83-; Retired. *Mem:* Math Asn Am. *Res:* Analysis. *Mailing Add:* 1321 S Ave A Portales NM 88130

WILSON, CHARLES B, b Neosho, Mo, Aug 31, 29; m 56; c 3. MEDICINE. *Educ:* Tulane Univ, BS, 51, MD, 54. *Prof Exp:* Resident path, Tulane Univ, 55-56, instr neurosurg, 60-61; resident, Ochsner Clin, 56-60; instr, La State Univ, 61-63; from asst prof to prof, Univ Ky, 63-68; PROF NEUROSURG, UNIV CALIF, SAN FRANCISCO, 68- *Mem:* Am Asn Neurol Surg; Am Asn Neuropath; Soc Neurol Surg. *Res:* Brain tumor chemotherapy. *Mailing Add:* Dept Neurosurg Univ Calif Med Ctr San Francisco CA 94143

WILSON, CHARLES ELMER, b Passaic, NJ, Aug 2, 31; m 58; c 3. NOISE CONTROL, MACHINE DESIGN. *Educ:* NJ Inst Technol, BS, 53, MS, 58; NY UNiv, MS, 62; Polytech Inst NY, PhD(mech eng, 70. *Prof Exp:* Engr, Otis Elevator Co, 53-54; armament & electron officer, USAF, 54-56; engr, Bendix Corp, 56; PROF MECH ENG, MACH DESIGN, MECHANISMS, NOISE CONTROL, NJ INST TECHNOL, 56- *Concurrent Pos:* Prin investr, NSF, NASA & other foundations & industs, 64-81; legal expert & consult mach design, var indust firms, 65-; legal expert & consult noise control, Indust Firms & Govt Agencies, 72-; vis fel, Inst Sound & Vibration Res, Univ Southampton, Eng, 77. *Mem:* Acoust Soc Am; Am Inst Physics. *Res:* Community noise; industrial noise control; hearing protection; transportation noise prediction and control; adhesives and vibration; author of texts in noise control, kinematics, and dynamics of machinery, machine design. *Mailing Add:* 19 Highview Terr Cedar Grove NJ 07009

WILSON, CHARLES MAYE, b Mt Olive, NC, Oct 16, 16. MYCOLOGY. *Educ:* Univ Va, BSc, 41, MA, 42; Harvard Univ, PhD(biol), 50. *Prof Exp:* Nat Res Coun fel, Univ Calif, 50-51; instr biol, Harvard Univ, 51-53; from asst prof to prof bot, McGill Univ, 53-82, chmn, dept biol, 62-70; RETIRED. *Concurrent Pos:* Vis prof, Univ Calif, 57. *Mem:* Bot Soc Am; Mycol Soc Am. *Res:* Cytology and life cycles of the lower fungi. *Mailing Add:* 700 Match Point Dr Apt 202 Virginia Beach VA 23462

WILSON, CHARLES NORMAN, b Seattle, Wash, Mar 28, 47. MATERIALS SCIENCE, PHYSICAL CHEMISTRY. *Educ:* Univ Wash, BS, 69, MS, 70, PhD(ceramic eng), 74. *Prof Exp:* Ceramic engr glass technol, Penberthy Electromelt Co, Seattle, Wash, 68-69; SR SCIENTIST CERAMIC NUCLEAR MAT, RES & DEVELOP, WESTINGHOUSE HANFORD CO, 74- *Mem:* Am Ceramic Soc. *Res:* Fabrication and characterization of ceramic materials including oxide nuclear fuels, boron carbide, ceramic nuclear waste forms, and lithium ceramics for fusion reactor tritium breeding; solar cell materials characterization; biomaterials and bone physical chemistry; testing of spent nuclear fuel as a wasteform for disposal in geologic repositories. *Mailing Add:* 362 Driftwood Ct Richland WA 99352

WILSON, CHARLES OREN, b Salt Lake City, Utah, May 9, 26; m 83; c 4. CLINICAL CHEMISTRY. *Educ:* Stanford Univ, BS, 49; Univ Southern Calif, MS, 59. *Prof Exp:* Prod chemist, Transandino Co, Calif, 49-51; res chemist, Nat Bur Stand, 51-52; sr res chemist, Olin Mathieson Chem Corp, NY, 52-54 & Calif, 54-59; sr res chemist, Nat Eng Sci Corp, 59-61 & Am Potash & Chem Corp, 61-68; chemist, Xerox MDO-Electro-Optical Systs, Inc, 68-69 & Int Chem & Nuclear Corp, 69-70; mgr qual control, Reagents Qual Control, 70-77, tech specialist, 77-83, BIOCHEMIST, CLIN CHEM PROD, ANALYTICAL SYSTS, ABBOTT DIAGNOSTICS CHEM & INSTRUMENTATION, 83- *Concurrent Pos:* Consult, 67- *Mem:* Am Chem Soc. *Res:* Organoboron and organophosphorus chemistry; exotic fuels and polymers; infrared spectroscopy; mass spectrometry; rare earth research; solvent cal extraction; C-14 radioactive organic synthesis; quality control and research of clinical diagnostic reagents; 25 publications in professional chemistry journals. *Mailing Add:* Abbott Diag Div Analytical Systs Reagents Res & Develop 820 Mission St South Pasadena CA 91030

WILSON, CHARLES R, b Baltimore, Md, Jan 25, 29; m 59; c 3. GEOPHYSICS. *Educ:* Case Inst Technol, BS, 51; Univ NMex, MS, 56; Univ Alaska, PhD(geophys), 63. *Prof Exp:* PROF PHYSICS, GEOPHYS INST, UNIV ALASKA, 59- *Concurrent Pos:* Fulbright grant, Paris, 63-64; vis scientist, Nat Ctr Atmospheric Res, 68-69. *Mem:* Am Geophys Union. *Res:* Magnetic storms; geomagnetic micropulsations; auroral infrasonics. *Mailing Add:* 1812 Musk Ox Trail Fairbanks AK 99709

WILSON, CHARLES WOODSON, III, b Columbus, Ohio, Nov 20, 24; m 48; c 4. PHYSICS. *Educ:* Univ Mich, BSE, 47, MS, 48; Wash Univ, PhD(physics), 52. *Prof Exp:* Res assoc, Stanford Univ, 52; res physicist, Prod Dept, Res Div, Texaco Inc, 52-56; physicist, Res & Develop Dept, Chem Div, Union Carbide Corp, 56-62, res scientist, 62-64, group leader, 64-65; PROF PHYSICS & POLYMER SCI, HEAD PHYSICS DEPT & RES ASSOC, INST POLYMER SCI, UNIV AKRON, 65- *Mem:* AAAS; Am Phys Soc; Am Asn Physics Teachers; Sigma Xi. *Res:* Nuclear and electron spin resonance; high polymer physics; energy utilization and conservation. *Mailing Add:* Dept Physics Univ Akron 309 S Rose Blvd Akron OH 44313

WILSON, CHRISTINE SHEARER, b Orleans, Mass. ANTHROPOLOGY. *Educ:* Brown Univ, BA, 50; Univ Calif, Berkeley, PhD(nutrit, anthrop), 70. *Prof Exp:* Asst ed, Nutrit Rev, Sch Pub Health, Harvard Univ, 51-56; nutrit analyst, USDA, 57-58; asst res nutritionist, Univ Calif, San Francisco, 70-71, res assoc nutrit anthrop, 71-74, asst res nutritionist, 74, asst res nutritionist, Dept Epidemiol & Int Health, 75-86, lectr, 74-86, lectr, prog med anthrop, 77-86. *Concurrent Pos:* Consult, Soc Nutrit Educ, 71-72, 79 & 81; USPHS spec res fel, 72-73; lectr, Dept Anthrop, Univ Calif, Riverside, 73; vis prof, Dept Family Studies, Univ Guelph, 78; vis asst prof, Home Econ Dept, San Francisco State Univ, 79; mem, Behav Factors Panel, USDA Competitive Grants Prog, 79; contrib ed, Nutrit Rev, 79-87. *Mem:* Fel Am Anthrop Asn; Am Inst Nutrit; Soc Med Anthrop. *Res:* Food in the culture; social influences on nutritional status; ethnographic field research on diet and nutritional health; dietary methodology; diet and cancer. *Mailing Add:* PO Box 3178 Annapolis MD 21403-0178

WILSON, CHRISTOPHER, INFECTIOUS DISEASE, PEDIATRICS. *Educ:* Univ Calif, MD. *Prof Exp:* ASST PROF MED, ORTHOP HOSP, UNIV WASH, 79- *Mailing Add:* Dept Pediat & Immunol Univ Wash Rd 20 Seattle WA 98195

WILSON, CLAUDE E, analytical chemistry, for more information see previous edition

WILSON, CLAUDE LEONARD, b Ottawa, Kans, Nov 30, 05; m 27; c 1. MECHANICAL ENGINEERING. *Educ:* Kans State Univ, BS, 25, ME, 29, MS, 33. *Hon Degrees:* DSc, Kans State Univ, 82. *Prof Exp:* From asst prof to assoc prof mech eng, Prairie View Agr & Mech Univ, 25-33, supt bldgs & utilities, 33-41, dean eng, 41-66, dean, Col, 66-70, vpres, 70-74; ENG CONSULT, BOVAY ENGRS, INC, 75- *Concurrent Pos:* Pvt consult eng pract, 80- *Mem:* Am Soc Mech Engrs; Am Soc Eng Educ; Am Soc Testing & Mat; Nat Soc Prof Engrs. *Res:* Thermodynamics and heat power; heating and air conditioning; structural analysis and design. *Mailing Add:* PO Box 2848 Prairie View TX 77446

WILSON, CLYDE LIVINGSTON, b Ohio, July 29, 22; m 51; c 2. SOIL PHYSICS. *Educ:* Ohio State Univ, BS, 47, BAE, 48, PhD(agron), 52. *Prof Exp:* Asst soils, Ohio State Univ, 48-52; from res agronomist to res specialist, 52-72, sr res specialist, 72-79, agron systs mgr, 79-82, agr equip eng mgr, Monsanto Co, 82-86, Litigation Consult, 86-87; RETIRED. *Mem:* Am Soc Agron; Am Soc Agr Engrs; Soil Sci Soc Am; Weed Sci Soc Am; Coun Agr Sci & Technol. *Res:* Saturated water flow in tiled lands; chemical soil conditioners; herbicide investigations. *Mailing Add:* 1530 Lynkirk Lane Kirkwood MO 63122

WILSON, COLON HAYES, JR, b Marshallberg, NC, Apr 10, 32; m 56; c 3. RHEUMATOLOGY, INTERNAL MEDICINE. *Educ:* Duke Univ, BA, 52, MD, 56. *Prof Exp:* From intern to asst resident med, Univ Va Hosp, 56-58; resident, Edward J Meyer Mem Hosp, Buffalo, NY, 61-63; res instr, State Univ NY, Buffalo, 64-66; from asst prof to assoc prof rheumatology, 66-73, asst prof phys med, 66-74, PROF MED, MED SCH, EMORY UNIV, 74-, DIR, DIV RHEUMATOLOGY & IMMUNOL, 66- *Concurrent Pos:* Fel, Buffalo Gen Hosp, State Univ NY, Buffalo, 63-66; Nat Inst Arthritis & Metab Dis prog grant, Edward J Meyer Mem Hosp, Buffalo; actg med dir, Arthritis Found, 77-79. *Mem:* Am Fedn Clin Res; Am Rheumatism Asn; Reticuloendothelial Soc; fel Am Col Physicians. *Res:* Significance of various patterns of antinuclear antibody fluorescence with respect to specific diagnosis and prognosis on various collagen vascular diseases; the efficacy of early synovectomy in rheumatoid arthritis in prevention of late deformity and preservation of function. *Mailing Add:* 69 Butler St SE Atlanta GA 30303

WILSON, COYT TAYLOR, b Fulton, Miss, July 27, 13; m 36; c 2. PLANT PATHOLOGY. *Educ:* Ala Polytech Inst, BS, 38, MS, 41; Univ Minn, PhD(plant path), 46. *Prof Exp:* Instr bot, Ala Polytech Inst, 40-41; instr plant path, Univ Minn, 41-43; asst plant pathologist, Exp Sta, Auburn Univ, 44-47, prof plant path & plant pathologist, 47-51, asst dean, Col Agr & assoc dir, Agr Exp Sta, 51-64; from assoc dir to dir, Va Agr Exp Sta, 64-66; assoc dean, Res Div, Va Polytech Inst & State Univ, 66-71, exec assoc dean, 71-78, dir agr & life sci res, 66-78; RETIRED. *Concurrent Pos:* AID short term res consult, Ministry Agr, Iran, 60, Turkey, 66 & EPakistan, 68. *Mailing Add:* 2010 Linwood Lane NW Blacksburg VA 24060

WILSON, CURTIS MARSHALL, b Stillwater, Minn, Sept 11, 26; m 52; c 4. PLANT BIOCHEMISTRY. *Educ:* Univ Minn, BS, 48, MS, 51; Univ Wis, PhD(bot), 54. *Prof Exp:* Asst, Univ Minn, 48-51; from asst prof to assoc prof plant physiol, Rutgers Univ, 54-59; from assoc prof to prof, 66-87, ADJ PROF PLANT PHYSIOL, UNIV ILL, URBANA, 87-; res chemist, Agr Res Serv, USDA, Urbana Ill, 59-87, Peoria, Ill, 87-92. *Mem:* AAAS; Am Soc Plant Physiol; Am Asn Cereal Chemists. *Res:* Plant biochemistry; plant nucleases; seed development and storage proteins. *Mailing Add:* USDA ARS Nat Ctr Agr Utilization Res 1815 N University St Peoria IL 61604

WILSON, CYNTHIA, b Gillingham, Eng, Aug 31, 26; Can citizen. CLIMATOLOGY. *Educ:* Univ London, BA, 47, teachers dipl, 48; McGill Univ, MSc, 58; Laval Univ, PhD, 72. *Prof Exp:* Teacher high sch, Eng, 48-54; res asst meteorol & climat, Meteorol Res Group, McGill Univ, 54-62; asst prof climat, Inst Geog, 64-67, assoc prof climat & researcher, Ctr Nordic Studies, Laval Univ, 67-75; CONSULT CLIMAT, 76- *Concurrent Pos:* Expert, Cold Regions Res & Eng Lab, US Army, 62-66; under contract to Meteorol Serv Can, 66-68. *Honors & Awards:* Darton Prize, Royal Meteorol Soc, 63. *Mem:* Am Meteorol Soc; Can Meteorol Soc; Arctic Inst NAm; Royal Meteorol Soc; Am Quaternary Asn. *Res:* Meteorology and climatology of cold regions; historical climatology. *Mailing Add:* 90 Shamside Gillingham Kent ME7 4BC England

WILSON, DANA E, b Chicago, Ill, Oct 5, 37; m 60; c 3. MEDICINE, METABOLISM. *Educ:* Oberlin Col, AB, 57; Western Reserve Univ, MD, 62. *Prof Exp:* Intern & asst resident internal med, Boston City Hosp, Mass, 62-64, sr resident, 66-67; clin assoc allergy & infectious dis, NIH, 64-66; asst prof clin nutrit & human metab, Mass Inst Technol, 69-71; from asst prof to assoc prof, 71-85, PROF MED, COL MED, UNIV UTAH, 85- *Concurrent Pos:* Fel diabetes & metab, Thorndike Mem Lab, Harvard Med Sch, 67-69. *Mem:* AAAS; Am Fedn Clin Res; Am Diabetes Asn. *Res:* Diabetes and metabolism; lipoproteins and lipid transport. *Mailing Add:* Dept Med Univ Utah Salt Lake City UT 84112

WILSON, DARCY BENOIT, b Rhinebeck, NY, May 14, 36; m 57; c 3. IMMUNOLOGY, IMMUNOBIOLOGY. *Educ:* Harvard Univ, AB, 58; Univ Pa, PhD(zool), 62. *Prof Exp:* Assoc, 63-65, from asst prof to assoc prof path & med genet, 66-74, PROF PATH & HUMAN GENETICS, SCH MED, UNIV PA, 74- *Concurrent Pos:* Res fel transplantation immunol, Wistar Inst, Univ Pa, 62-63, res fel med genet, Sch Med, 65-66; Helen Hay Whitney Found fel, 64-67; USPHS career develop award, 67-72. *Mem:* Am Asn Immunol. *Res:* Immunology of tissue transplantation, particularly immunologic behavior of lymphoid cells in vitro and in vivo. *Mailing Add:* Exp Medicine 11099 N Torrey Pines Rd La Jolla CA 92037

WILSON, DAVID BUCKINGHAM, b Cambridge, Mass, Jan 15, 40; m 63; c 3. BIOCHEMISTRY. *Educ:* Harvard Univ, BA, 61; Stanford Univ, PhD(biochem), 65. *Prof Exp:* Jane Coffin Childs fel biochem, Sch Med, Johns Hopkins Univ, 65-67; from asst prof to assoc prof, 67-83, PROF BIOCHEM, CORNELL UNIV, 83- *Mem:* Am Soc Biol Chemists; Am Soc Microbiol; AAAS. *Res:* Regulation of gene expression in microorganisms; bacterial membrane biochemistry and genetic engineering; mechanisms of host cell lysis by bacteriophage lambda; cloning of thermophillic cellulase genes; regulation of galactose enzyme synthesis in Esherichia coli and yeast and the mechanisms of binding protein transport systems in Esherichia coli; protein excretion by Ecoli; mechanism of cellulose degredation by bacterial cellulases. *Mailing Add:* Dept Biochem Cornell Univ Biotechnol Bldg Rm 458 Ithaca NY 14853

WILSON, DAVID E, b Meade, Kans, Aug 5, 29; m 56; c 2. MATHEMATICS. *Educ:* Kans State Col, Pittsburg, AB, 51, MS, 54; Univ Kans, PhD(math), 67. *Prof Exp:* Asst prof math, Univ Hawaii, 61-65; asst prof, 66-70, ASSOC PROF MATH, WABASH COL, 70-, PROF MATH, 88- *Mem:* Am Math Soc; Math Asn Am. *Res:* Quasicon-formal mappings in n-space. *Mailing Add:* Dept Math Wabash Col Crawfordsville IN 47933

WILSON, DAVID F, b Wray, Colo, Mar 28, 38; m 62, 87. BIOCHEMISTRY. *Educ:* Colo State Univ, BS, 59; Ore State Univ, PhD(biochem), 64. *Prof Exp:* USPHS fel phys biochem, Johnson Res Found, 64-67, Pa Plan scholar, Sch Med, 67-69, res assoc, 67-68, from asst prof to assoc prof, 68-80, PROF BIOCHEM & BIOPHYS, MED SCH, UNIV PA, 80- *Honors & Awards:* Eli Lilly Award, Am Chem Soc, 71. *Mem:* Am Soc Biol Chemists; Biophys Soc. *Res:* Mitochondrial electron transport and energy conservation; cellular energy metabolism; amino acid neurotransmitter function. *Mailing Add:* Dept Biochem & Biophys Univ Pa Med Sch Philadelphia PA 19104-6059

WILSON, DAVID FRANKLIN, b Queens Village, NY, Feb 23, 41; m 65; c 3. NEUROPHYSIOLOGY. *Educ:* Hofstra Univ, BA, 63; Univ Del, MA, 66, PhD(biol sci), 68. *Prof Exp:* USPHS fel, Northwestern Univ, 68-69; from asst prof to assoc prof, 69-78, PROF ZOOL, MIAMI UNIV, 78- *Concurrent Pos:* USPHS grants, 70-72, 73-75 & 87-88 & 89-91. *Mem:* Am Physiol Soc; Soc Neurosci. *Res:* Examination of neuromuscular transmission using intracellular recording techniques. *Mailing Add:* Dept Zool Miami Univ Oxford OH 45056

WILSON, DAVID GEORGE, b Spokane, Wash, Dec 15, 19; m 42; c 3. RANGE CONSERVATION, RESEARCH ADMINISTRATION. *Educ:* Univ Idaho, BS, 47; Agr & Mech Col, Tex, MS, 50, PhD(range mgt), 61. *Prof Exp:* Dep fire warden, Clearwater Timber Protective Asn, 47; instr range mgt, Agr & Mech Col, Tex, 47-48, asst prof, 49-50; from instr to assoc prof range mgt & plants, Univ Ariz, 53-64; range consult, 64-65; res coordr, 65-76, ECOLOGIST & BOTANIST, US BUR LAND MGT, 76- *Mem:* Soc Range Mgt. *Res:* Natural resources. *Mailing Add:* 1429 Valentia St Denver CO 80220

WILSON, DAVID GEORGE, b St Catharines, Ont, Dec 3, 21; m 44; c 2. PLANT BIOCHEMISTRY. *Educ:* Univ Toronto, BA, 44; Queen's Univ, Ont, BA, 49, MA, 50; Univ Wis, PhD(biochem), 53. *Prof Exp:* Agr scientist, Conn Agr Exp Sta, 53-56; PROF BOT, UNIV WESTERN ONT, 56- *Mem:* AAAS; Am Soc Plant Physiol; Can Soc Plant Physiol; Sigma Xi. *Res:* Organic acids and amino acid metabolism in plants; biochemistry of cellulose and lignin degradation. *Mailing Add:* Dept Plant Sci Univ Western Ont London ON N6A 3B7 Can

WILSON, DAVID GORDON, b Sutton Coldfield, UK, Feb 11, 28; m 88; c 2. MECHANICAL ENGINEERING. *Educ:* Univ Birmingham, BSc, 48; Univ Nottingham, PhD(heat transfer), 53. *Prof Exp:* Sr res asst fluid mech & heat transfer, Univ Nottingham, 52-53; sr gas turbine engr, Brush Elec Eng Co, Ltd, UK, 53-55; Commonwealth Fund fel, Mass Inst Technol & Harvard Univ, 55-57; sr gas turbine designer, Ruston & Hornsby, Ltd, UK, 57-58; sr lectr thermodyn, fluid mech & mach design, Univ Ibadan, 58-60; tech dir, 60-61, tech dir & vpres, Northern Res & Eng Corp, Mass, 61-66; assoc prof thermodyn, dynamics & design, 66-71, PROF MECH ENG, MASS INST TECHNOL, 71- *Concurrent Pos:* Ed, Human Power. *Honors & Awards:* Hall Prize, Brit Inst Mech Engrs, 54, Weir Prize, 55. *Mem:* Am Soc Mech Engrs; Brit Inst Mech Engrs; Int Human-Powered-Vehicle Asn. *Res:* Heat transfer and fluid dynamics, especially with regard to turbomachinery; economic and design studies in solid-waste treatment, legislation and transportation and highway safety; solid-waste management; science of human-powered vehicles; turbomachinery design. *Mailing Add:* Mass Inst Technol Rm 3-445 Cambridge MA 02139

WILSON, DAVID J, b Ames, Iowa, June 25, 30; m 52; c 5. PHYSICAL CHEMISTRY. *Educ:* Stanford Univ, BS, 52; Calif Inst Technol, PhD(chem), 58. *Prof Exp:* From instr to prof chem, Univ Rochester, 57-69; PROF CHEM, VANDERBILT UNIV, 69-, PROF ENVIRON ENG, 77- *Concurrent Pos:* Alfred P Sloan fel, 64-66. *Mem:* Am Chem Soc; AAAS. *Res:* Energy transfer in gases; homogeneous gas reactions; pesticide and heavy metal residues; foam flotation; math modelling of unit operations in sanitary engineering; hazardous waste site remediation. *Mailing Add:* Dept Chem Vanderbilt Univ Nashville TN 37235

WILSON, DAVID LOUIS, b Washington, DC, Jan 11, 43; m 67; c 1. NEUROSCIENCES, MOLECULAR BIOLOGY. *Educ:* Univ Md, College Park, BS, 64; Univ Chicago, PhD(biophys), 69. *Prof Exp:* From asst prof to prof physiol & biophys, 72-81, dep dean acad affairs, 82-85, Med Sch, assoc provost res & dean, Grad Sch, 83-85, PROF PHYSIOL & BIOPHYS, MED SCH, UNIV MIAMI, 81-, PROF BIOL, 85-, DEAN, COL ARTS & SCI, 85- *Concurrent Pos:* Helen Hay Whitney fel, Calif Inst Technol, 69-72; NIH grant, 72-74, 75-78, 78-81 & 80-86, NSF grant, 82-85. *Mem:* AAAS; Am Physiol Soc; Soc Neurosci; Am Soc Neurochem. *Res:* Protein synthesis in neurons; axonal transport; theoretical neuroscience; nerve regeneration. *Mailing Add:* Univ Miami Col Arts & Sci PO Box 248004 Coral Gables FL 33124

WILSON, DAVID MERL, b Mosca, Colo, May 25, 41; m 66; c 2. MYCOTOXIN CONTAMINATION, FUNGAL PHYSIOLOGY. *Educ:* Colo State Univ, BS, 64, MS, 66, PhD(plant path), 68. *Prof Exp:* Fel plant path, NC State Univ, 68-69; asst prof bot, Univ Vt, 69-73; from asst prof to assoc prof, 73-82, PROF PLANT PATH, COASTAL PLAIN STA, UNIV GA, 82- *Concurrent Pos:* Assoc referee, Asn Analytical Chemists, 74-; assoc ed, Phytopath J, Am Phytopath Soc, 85-87 & Plant Dis, 90-92. *Honors & Awards:* Res Award, Sigma Xi, 77. *Mem:* Am Chem Soc; Am Peanut Res & Educ Asn; Am Phytopath Soc; Asn Off Anal Chemists; Sigma Xi. *Res:* Mycotoxin contamination of foods and feeds; definition of ways fungi grow in and contaminate commodities with toxic fungal metabolities; analytical chemistry and mycology; control and elimination of toxic fungal metabolites from foodstuffs. *Mailing Add:* Dept Plant Path Coastal Plain Exp Sta Univ Ga PO Box 748 Tifton GA 31793-0748

WILSON, DAVID ORIN, b Portland, Ore, Apr 19, 37; m 56; c 3. SOIL MICROBIOLOGY, PLANT NUTRITION. *Educ:* Wash State Univ, BS, 59, MS, 63; Univ Calif, Davis, PhD(bact nutrit), 69. *Prof Exp:* Scientist, Soil Conserv Serv, USDA, Wash, 59-60 & Biol Lab, Hanford Labs, Gen Elec Co, 62-64; res assoc agron, Cornell Univ, 69-71; asst prof, 71-79, ASSOC PROF, GA EXP STA, UNIV GA, 79- *Mem:* Am Soc Agron; Soil Sci Soc Am. *Res:* Microelement nutrition of plants and bacteria; symbiotic nitrogen fixation; rhizobial legume inoculants; soil microbial nitrogen transformations; mycorrhizae. *Mailing Add:* Dept Agron Univ Ga Griffin GA 30223-1797

WILSON, DELANO D, b Great Falls, Mont, Apr 15, 34; m 59; c 3. AC TRANSMISSION. *Educ:* Mont State Univ, BS, 59. *Prof Exp:* Anal engr, Gen Elec, 59-61, engr-in-charge, TNA, 61-64, sr engr, 64-69, mgr AC transmission, 69-72, mgr, engr studies & proj, Gen Elec, 72-74; prin engr, 74-86, PRES, POWER TECHNOL INC, 86- *Concurrent Pos:* vpres-treas, Technol Assessment Group, 80-85. *Mem:* Fel Inst Elec & Electronics Engrs; Am Nat Standards Inst; Int Conf Large High Voltage Systs. *Mailing Add:* Power Technol Inc PO Box 1058 Schenectady NY 12301

WILSON, DON ELLIS, b Davis, Okla, Apr 30, 44; m 62; c 2. MAMMALOGY, SYSTEMATICS. *Educ:* Univ Ariz, BS, 65; Univ NMex, MS, 67, PhD(biol), 70. *Prof Exp:* Fel ecol, Univ Chicago, 70-71; zoologist, Nat Fish & Wildlife Lab, 71-73; chief mammal sect, US Fish & Wildlife Serv, 73-79; chief, Biol Surv, Nat Mus Natural Hist, 79-90, DIR, BIODIVERSITY PROGS, SMITHSONIAN INST, 90- *Concurrent Pos:* Vis prof, Univ Md, 75, 83 & 84; ed, J Mammal, 76-81. *Mem:* Am Soc Mammalogists (vpres, 84-86, pres, 86-88); AAAS; Soc Syst Zool. *Res:* Systematics and ecology of new world mammals. *Mailing Add:* Nat Mus Natural Hist Washington DC 20560

WILSON, DONALD ALAN, b San Francisco, Calif, Sept 16, 30; m 57; c 3. BIOCHEMISTRY, MICROBIAL PHYSIOLOGY. *Educ:* San Jose State Col, BA, 59; Western Reserve Univ, PhD(microbiol), 65. *Prof Exp:* Fel microbiol, Pioneering Res Div, US Army Natick Labs, 65-66; assoc, 66-67, asst prof, 67-70, ASSOC PROF MICROBIOL, CHICAGO MED SCH, 70- *Res:* Microbial enzymology. *Mailing Add:* Dept of Microbiol Chicago Med Sch 3333 Greenbay Rd N Chicago IL 60064

WILSON, DONALD ALFRED, b Greenville, Maine, Mar 31, 41; m 67; c 2. DENDROCHRONOLOGY, ENTOMOLOGY. *Educ:* Univ Maine, BS, 65; Univ NH, MS, 67. *Prof Exp:* Instr forestry, Dept Forestry Resources, Univ NH, 67-68; instr forestry surv, Sch Forest Resources & Dept Civil Eng, Univ Maine, 68-74; pres, Border Land Consults, 74-78; PRES, LAND & BOUNDARY CONSULTS, INC, 78- *Concurrent Pos:* Lectr, Am Cong Surv & Mapping, 75-87, NH Land Surveyors Asn, 75-, Maine Soc Land Surveyors, 75- & Univ NH, 77- *Honors & Awards:* Surv Excellence Award, Nat Soc Prof Surveyors, Am Cong Surv & Mapping, 84. *Mem:* Am Forestry Asn; fel Am Cong Surv & Mapping. *Res:* Collection and study of Coleoptera, primarily Cicindelidae, Cerambycidae, Cleridae, Scarabaeidae, Carabidae (Elaphrus) and some aquatic families; taxonomy; ecology; taxonomy and parasitology of nymphal cicindelidae. *Mailing Add:* 69 Main St PO Box 322 Newfields NH 03856

WILSON, DONALD BENJAMIN, b Rowley, Alta, June 12, 25; m 48; c 4. FORAGE CROPS. *Educ:* Univ Alta, BSc, 50; Utah State Univ, MS, 54; Ore State Univ, PhD(farm crops), 60. *Prof Exp:* Res officer, Agr Can Res Sta, Lethbridge, 50-68, head, Plant Sci Sect, 68-88; RETIRED. *Concurrent Pos:* Coordr, Forage Crops, chmn, Expert Comt Forage Crops; actg dir, Res Sta, Lethbridge. *Mem:* Can Soc Agron; Agr Inst Can. *Res:* Forage crop production and animal grazing; management of irrigated pastures and administration of agricultural research. *Mailing Add:* 1714 21st St S Lethbridge AB T1K 2G3 Can

WILSON, DONALD BRUCE, b Monticello, Iowa, Nov 4, 33; m 56; c 4. CHEMICAL ENGINEERING. *Educ:* Univ NMex, BS, 56; Princeton Univ, MA, 63, PhD(chem eng), 65. *Prof Exp:* Engr, Phillips Petrol Co, 59-61; assoc prof, NMex State Univ, 64-75, prof chem eng, 75-; RETIRED. *Res:* Macroscopic crystallization; design and optimization; thermodynamics of phase transformations. *Mailing Add:* Dept Chem Eng NMex State Univ Las Cruces NM 88003

WILSON, DONALD LAURENCE, b Hamilton, Ont, Oct 2, 21; m 45; c 6. MEDICINE. *Educ:* Queen's Univ, Ont, MD, CM, 44; Univ Toronto, MA, 48; FRCP(C), 51. *Prof Exp:* From asst prof to assoc prof, 52-67, head dept med, 76-82, PROF MED, QUEEN'S UNIV, 67-, DEAN FAC MED, 82- *Concurrent Pos:* Consult physician, Kingston Gen Hosp, 52-; pres, Coun Col Physicians & Surgeons, Ont, 65-66; pres, Ont Med Asn, 73-74. *Mem:* Endocrine Soc; Am Diabetes Asn; NY Acad Sci; Can Soc Clin Invest; Can Med Asn (pres, 79-80); fel Am Col Physicians. *Res:* Endocrinology and diseases of metabolism. *Mailing Add:* 1601-185 Ontario Kingston ON K7L 2Y7 Can

WILSON, DONALD RICHARD, b Plaistow, NH, Feb 8, 36; m 56; c 3. ORGANIC CHEMISTRY, POLYMER CHEMISTRY. *Educ:* Univ Wash, BS, 58; Univ Calif, Los Angeles, PhD(org chem), 62. *Prof Exp:* Mgr control lab, Am Marietta Co, Wash, 57-58; from res chemist to sr res chemist, E I du Pont de Nemours & Co, Inc, Del, 61-67; sr scientist, Xerox Corp, Webster, 67, res mgr org & polymer chem, 68, develop mgr org & polymer mat, 68-70, technol prog mgr advan xerography, 70-71, prin scientist xerographic mat, 71-72, mgr explor graphic sci, 72-75; res dir chem & catalysis, Celanese Res Co, NJ, 75-82; dir corp res, Pennwalt Corp, Pa, 83; vpres res & secy, Chem Systs Res, Inc, 84; PRES, ADVAN POLYMER TECHNOL, INC, NJ, 85- *Mem:* Am Chem Soc; NAm Membrane Soc; Asn Res Dir (pres, 84-85). *Res:* Organic and polymer synthesis-reaction mechanisms; stereochemistry; organometallics; textile fiber chemistry; high temperature fibers; xerography; xerographic imaging materials; membrane systems; conductive polymers; liquid crystal polymers. *Mailing Add:* Adv Polymer Technol Inc PO Box 6 Glasser NJ 07837-0006

WILSON, DORIS BURDA, b Cleveland, Ohio, July 1, 37; m 68. ANATOMY, EMBRYOLOGY. *Educ:* Ohio Wesleyan Univ, BA, 59; Radcliffe Col, MA, 60; Harvard Univ, PhD(biol), 63. *Prof Exp:* Teaching fel biol, Harvard Univ, 60-62; asst prof zool, San Diego State Col, 63-65; asst prof anat, Sch Med, Stanford Univ, 65-69; res anatomist & lectr, Sch Med, Univ Calif, San Diego, 69-73; assoc prof anat, Sch Med, Univ Calif, Davis, 73-75; assoc prof, 75-77, PROF SURG & ANAT, SCH MED, UNIV CALIF, SAN DIEGO, 77- *Concurrent Pos:* Fulbright vis prof, Taiwan, 70-71. *Mem:* AAAS; Histochem Soc; Teratology Soc; Am Soc Zool; Am Asn Anat. *Res:* Developmental biology; neuroembryology; teratology. *Mailing Add:* Dept Surg/Anat 0604 Univ Calif San Diego La Jolla CA 92093-0604

WILSON, DWIGHT ELLIOTT, JR, b Greensburg, Pa, June 7, 32; m 53; c 4. GENETICS. *Educ:* Yale Univ, BS, 53, PhD(biophys), 56. *Prof Exp:* From asst prof to assoc prof, 56-68, PROF BIOL, 68-, CHMN, DEPT BIOL, RENSSELAER POLYTECH INST, 87- *Concurrent Pos:* NIH fel, 66-67. *Mem:* Am Soc Human Genetics; AAAS. *Res:* Somatic cell genetics. *Mailing Add:* Dept Biol Rensselaer Polytech Inst Troy NY 12181

WILSON, EDGAR BRIGHT, b Gallatin, Tenn, Dec 18, 08; m 35, 55; c 6. CHEMICAL PHYSICS. *Educ:* Princeton Univ, BS, 30, AM, 31; Calif Inst Technol, PhD(phys chem), 33; Harvard Univ, MA, 36. *Hon Degrees:* Dr, from numerous US & foreign univs, 75-82. *Prof Exp:* Fel, Calif Inst Technol, 33-34; jr fel, 34-36, from asst prof to prof, 36-48, Richards prof chem, 48-79, EMER PROF CHEM, HARVARD UNIV, 79- *Concurrent Pos:* Res dir, Underwater Explosives Res Lab, Woods Hole, 42-44; res dir, Weapons Systs Eval Group, 52-53; hon trustee, Woods Hole Oceanog Inst, 79-81. *Honors & Awards:* Prize, Am Chem Soc, 37; Debye Award, Am Chem Soc, 62, Norris Award, 66, G N Lewis Award, Calif Sect, Pauling Award, 72, Richards Medal, 78; Rumford Medal, Am Acad Arts & Sci, 73; Nat Medal of Sci, 75; Feltrinelli Award, Rome, 76; Ferst Award, Sigma Xi, 77; Plyler Award, Am Phys Soc, 78; Welch Found Award, 78; Lippencott Medal, 79; Elliot Cresson Medal, Franklin Inst, 82. *Mem:* Nat Acad Sci; Am Chem Soc; fel Am Phys Soc; Am Acad Arts & Sci; Int Acad Quantum Molecular Sci; Opers Res Soc. *Res:* Quantum mechanics in chemistry; molecular dynamics; microwave spectroscopy. *Mailing Add:* Harvard Chem Lab 12 Oxford St Cambridge MA 02138

WILSON, EDMOND WOODROW, JR, b Selma, Ala, Jan 18, 40; m 65; c 1. PHYSICAL CHEMISTRY. *Educ:* Auburn Univ, BS, 62; Univ Ala, Tuscaloosa, MS, 65, PhD(phys chem), 68. *Prof Exp:* Temporary instr gen chem, Univ Ala, Tuscaloosa, 66-68; res assoc biophys chem, Univ Va, 68-70; asst prof, 70-73, assoc prof, 73-79, PROF PHYS CHEM, HARDING UNIV, 79- *Concurrent Pos:* Vis assoc prof, Okla State Univ, 77. *Mem:* Am Chem Soc; Sigma Xi. *Res:* Calorimetry and spectroscopy of biological molecules containing transition metal ions. *Mailing Add:* 100 Red Oak Lane Searcy AR 72143-4515

WILSON, EDWARD CARL, b Daytona Beach, Fla, Jan 25, 29. INVERTEBRATE PALEONTOLOGY. *Educ:* Univ Calif, Berkeley, BA, 58, MA, 60, PhD(paleont), 67. *Prof Exp:* Cur paleont & chmn div invert, Natural Hist Mus San Diego, 64-67; CUR INVERT PALEONT, LOS ANGELES COUNTY MUS NATURAL HIST, 67- *Mem:* Paleont Soc; Brit Paleont Asn. *Res:* Late Paleozoic corals. *Mailing Add:* Los Angeles Count Mus Natural Hist 900 Exposition Blvd Los Angeles CA 90007

WILSON, EDWARD L, b Ferndale, Calif, Sept 5, 31. FINITE ELEMENTS, COMPUTERS. *Educ:* Univ Calif, Berkeley, BS, 54, MS, 59, DEng(civil eng), 63. *Prof Exp:* Sr res engr, Aerojet Gen Corp, 63-65; from asst prof to assoc prof, 65-71, PROF CIVIL ENG, UNIV CALIF, BERKELEY, 71- *Mem:* Nat Acad Eng; Am Soc Civil Engrs. *Mailing Add:* Dept Civil Eng Univ Calif Berkeley CA 94720

WILSON, EDWARD MATTHEW, b Content, Jamaica, Dec 19, 37; m 62; c 2. DAIRY SCIENCE. *Educ:* McGill Univ, BScAgr, 64, MSc, 66; Ohio State Univ, PhD(dairy sci), 69. *Prof Exp:* Res asst animal genetics, McGill Univ, 64-66; res assoc dairy sci, Ohio State Univ, 66-69; asst prof animal sci, Tuskegee Inst, 69-73; prin physiologist, Coop State Res Serv, USDA, 73-74; DEAN, COOP RES & HEAD, DEPT AGR & NATURAL RESOURCES, LINCOLN UNIV, 74-, DEAN, COL APPL SCI & TECHNOL, 78- *Concurrent Pos:* Mem, Task Force to Repub SAfrica, 74; mem, US Agr Educ Team, Peoples Repub China, 80. *Mem:* Am Dairy Sci Asn. *Res:* Genetic polymorphisms of bovine blood and milk proteins; immunological properties of seminal proteins and their implications in the reproductive process. *Mailing Add:* 11207 E 49th St Kansas City MO 64133

WILSON, EDWARD NATHAN, b Warsaw, NY, Dec 2, 41; m 41; c 2. EDUCATION ADMINISTRATION, SCIENCE POLICY. *Educ:* Cornel Univ, BA, 63; Stanford Univ, MS, 65; Wash Univ, PhD(math), 71. *Prof Exp:* instr, Woodrow Wilson intern prog, Fort Valley State Col, 65-67, Univ Calif, 70-71, Brandeis Univ, 71-73; from instr to assoc prof, 68-87, dean, Univ Col, 86-88, PROF WASH UNIV, 87- *Concurrent Pos:* Dean, Grad Sch Art & Sci, Wash Univ, 83. *Mem:* Am Math Soc; Math Asn Am. *Res:* Lie theory specifically in differential geometry and harmonic analysis. *Mailing Add:* Grad Sch Arts & Sci Wash Univ One Brooking Dr St Louis MO 63130

WILSON, EDWARD OSBORNE, b Birmingham, Ala, June 10, 29; m 55; c 1. BEHAVIORAL BIOLOGY, EVOLUTIONARY BIOLOGY. *Educ:* Univ Ala, BS, 49, MS, 50; Harvard Univ, PhD(biol), 55. *Hon Degrees:* DSc, Duke Univ, Grinnell Col, Lawrence Univ, Univ WFla, Macalaster Col; LHD, Univ Ala, Hofstra Univ; LLD, Simon Fraser Univ; DPhil, Uppsala Univ. *Prof Exp:* Biologist, State Dept Conserv, Ala, 49; Soc Fels jr fel, 53-56, from asst prof to prof zool, 56-76, FRANK B BAIRD JR, PROF SCI, HARVARD UNIV, 76-, MELLON PROF SCI, 90- *Concurrent Pos:* Mem expeds, WIndies & Mex, 53, New Caledonia, 54, Australia & New Guinea, 55, Ceylon, 55 & Surinam, 61; Charles and Martha Hitchcock Prof, Univ Calif, Berkeley, 72; John Simon Guggenheim fel, 77, Comt Selection, 82-; bd dir, World Wildlife Fund, 84-90. *Honors & Awards:* Cleveland Prize, AAAS, 67; Mercer Award, Ecol Soc Am, 71; Founders' Mem Award, Entom Soc Am, 72; Nat Med of Sci, 76; Leidy Medal, 78; Carr Medal, 78; Pulitzer Prize, Gen Nonfiction, 79; L O Howard Award, Entom Soc Am, 85; Tyler Prize, Environ Achievement, 84; Nat Zool Park Medal, 87; Ecol Inst Prize, Germany, 87; Crafoord Prize, Royal Swed Acad, 90; Gold Medal World Wide Fund Nature, 90. *Mem:* Nat Acad Sci; fel Am Acad Arts & Sci; fel Am Philos Soc; Soc Study Evolution (pres, 73); Am Genetic Asn; German Acad Sci; hon mem Brit Ecol Soc; Acad Humanism; Royal Soc London. *Res:* Classification, ecology and behavior of ants; speciation; general sociobiology; chemical communication in animals; biogeography. *Mailing Add:* Mus Comp Zool Harvard Univ Cambridge MA 02138

WILSON, ELIZABETH ALLEN, b Bridgeport, Conn, Jan 6, 50. PETROLEUM GEOLOGY, COASTAL GEOLOGY. *Educ:* Mt Holyoke Col, AB, 72; Univ Del, MS, 74, PhD(geol), 78. *Prof Exp:* Geologist petrol, Shell Oil Co, 77-80; OWNER, METHANE RESOURCES GROUP, LTD, 80- *Mem:* Am Asn Petrol Geologists; Soc Econ Paleontologists & Mineralogists; Int Asn Sedimentologists; Sigma Xi. *Res:* Carbonate diagenesis; seismic evaluation of stratigraphic traps in carbonate environments; coastal marsh deposits and ancient analogues; coal bed methane. *Mailing Add:* 151 Blue Flax Trail Evergreen CO 80439

WILSON, ELIZABETH MARY, MOLECULAR BIOLOGY. *Educ:* Vanderbilt Univ, PhD(biochem), 74. *Prof Exp:* ASSOC PROF ENDOCRINOL, UNIV NC, 83- *Res:* Endocrinology; gene regulation. *Mailing Add:* Dept Pediat & Biochem Univ NC MacNider Bldg CB 7500 Chapel Hill NC 27599

WILSON, ELWOOD JUSTIN, JR, b New York, NY, Nov 28, 17; m 41; c 4. ORGANIC CHEMISTRY. *Educ:* Princeton Univ, AB, 38, MA, 40, PhD(chem), 41. *Prof Exp:* Corn Industs Res Found fel, NIH, 41-42; proj engr, Sperry Gyroscope Co, NY, 42-44; sr chemist, Exp, Inc, 46-47, secy, 47-49, vpres, 49-54; res dir, Detroit Controls Co, 54-59; pres, Adv Tech Labs Div, Am-Standard, 59-64; PRES, E J WILSON ASSOCS, INC, 64-; pres, Epoxon Prods, Inc, 71-82; PRES, HERBTEC, INC, 82- *Concurrent Pos:* Vpres, Flight Res, Inc, 53-54; chmn & dir, Data Cartridge, Inc, 65-67. *Mem:* AAAS; Am Chem Soc. *Res:* Proteins; carbohydrates; fuels; combustion; synthetic organic chemistry; interior ballistics; rockets; petrochemicals; aerospace instruments; nuclear reactors; general and technical management; building specialties. *Mailing Add:* 1125 Westridge Dr Portola Valley CA 94028

WILSON, ERIC LEROY, b Sharon, Pa, Mar 17, 35; m 59; c 2. MATHEMATICS. *Educ:* Westminster Col, Pa, BS, 57; Vanderbilt Univ, PhD(math), 66. *Prof Exp:* From instr to assoc prof, 62-87, chmn dept, 73-76 & 78-81, PROF MATH, WITTENBERG UNIV, 87- *Concurrent Pos:* Asst prof, Univ South, 67-68; prof, Univ Essex, Eng, 81-82; vis scholar, Vanderbilt Univ, 84, 85 & 86. *Mem:* Math Asn Am; Nat Speleol Soc. *Res:* Loop isotopy; graph theory. *Mailing Add:* Dept Math Wittenberg Univ Springfield OH 45501

WILSON, EUGENE M, b Buckhannon, WVa, May 4, 28; m 51; c 2. PLANT PATHOLOGY. *Educ:* Univ WVa, BS, 51, MS, 54; Univ Calif, PhD(plant path), 58. *Prof Exp:* Asst plant path, Univ WVa, 51 & Univ Calif, 54-58; plant pathologist, Cent Res Lab, United Fruit Co, Mass, 58-60; technologist, Shell Chem Co, 60-72; plant pathologist pesticide regulation, 72-73, chief plant path sect, Off Pesticide Progs, 73-74, PROD MGR, US ENVIRON PROTECTION AGENCY, 74-; PROF GEOG, UNIV SOUTHERN ALA. *Mem:* Am Phytopath Soc. *Res:* Physiology of fungi; plant disease control; host parasite relationship; biological control of plant pests. *Mailing Add:* Dept Geog & Geol Univ S Ala 307 University Dr Mobile AL 36688

WILSON, EUGENE MADISON, b Cheyenne, Wyo, July 25, 43; m 65; c 3. CIVIL ENGINEERING. *Educ:* Univ Wyo, BS, 65, MS, 66; Ariz State Univ, PhD(civil eng, transp), 72. *Prof Exp:* Petrol engr, Texaco, Inc, 66-67; instr civil eng & staff asst transp, Ariz State Univ, 68-69; transp planner, Ariz Hwy Dept, 69-70; asst prof civil eng & res assoc transp, Univ Iowa, 70-74; assoc prof, 74-79, PROF CIVIL ENG, UNIV WYO, 79- *Concurrent Pos:* Mem, Transp Res Bd, Nat Acad Sci-Nat Res Coun. *Mem:* Am Soc Civil Engrs; Am Inst Planners; Inst Traffic Engrs. *Res:* traffic assignment; public transportation planning; behavioral and sensitivity studies in transportation; traffic engineering operations. *Mailing Add:* Dept Civil Eng Univ Wyo Laramie WY 82071

WILSON, EVELYN H, b Philadelphia, Pa, Oct 8, 21; m 43; c 2. ORGANIC CHEMISTRY. *Educ:* Bryn Mawr Col, AB, 42; Radcliffe Col, AM, 44, PhD(org chem), 46. *Prof Exp:* Res chemist, Merck & Co, Inc, 46-53; sr scientist, Johnson & Johnson, 53-59; lectr chem, Westfield Sr High Sch, NJ, 49-65; sci supvr, New Brunswick Pub Schs, 65-67; assoc prof sci educ, 67-72, prof sci educ & assoc vpres prog develop & budgeting, 72-88, PROF SCI EDUC, RUTGERS UNIV, NEW BRUNSWICK, 88- *Res:* Educational planning and administration; philosophy of science; science education. *Mailing Add:* Off Prog Develop & Budgeting Rutgers Univ Old Queens Campus New Brunswick NJ 08903

WILSON, EVERETT D, b Covington, Ind, July 13, 28; div; c 4. PHYSIOLOGY, ENDOCRINOLOGY. *Educ:* Ind State Teachers Col, BS, 50, MS, 51; Purdue Univ, PhD(physiol, endocrinol), 60. *Prof Exp:* Teacher pub sch, Ind, 48-50 & 55-56, supvr, 56-57; asst biol, Purdue Univ, 57-58, res asst, 58-60; asst prof zool, Southern Ill Univ, 60-61; assoc prof, 62-64, dean Col Sci, 65-79, PROF, SAM HOUSTON STATE UNIV, 64- *Concurrent Pos:* Lalor res fel, 61; NATO fel, Nat Med Res Inst, 62; chief, Grants Br Pop & Reproduction Ctr, Nat Inst Child Health, NIH, 71-72; adv, Oak Ridge Pop Res Inst, 72-74. *Mem:* Am Soc Zoologists; Endocrine Soc; Am Soc Animal Sci; Brit Soc Study Fertil. *Res:* Factors affecting mammalian reproduction. *Mailing Add:* Dept Sci Sam Houston State Univ Huntsville TX 77341

WILSON, F(RANK) DOUGLAS, b Salt Lake City, Utah, Dec 17, 28; m 50; c 8. HOST-PLANT RESISTANCE. *Educ:* Univ Utah, BS, 50, MS, 53; Wash State Univ, PhD(bot), 57. *Prof Exp:* Actg animal scientist, Animal Sci Dept, Wash State Univ, 56-57; RES GENETICIST, AGR RES SERV, USDA, 57- *Concurrent Pos:* Consult, USAID, 59; assoc ed, Am Soc Agron, 77-80, chmn, Cotton Regist, 89- *Mem:* AAAS; Am Soc Agron. *Res:* Identifying and using natural resistance of cotton plants to insect pests; breeding and genetics of Kenaf, Hibiscus cannabinus, Malvaceae; taxonomy of Hibiscus section Furcaria. *Mailing Add:* USDA Western Cotton Res Lab Cotton & Insect Unit 4135 E Broadway Rd Phoenix AZ 85040

WILSON, F WESLEY, JR, b Washington, DC, Apr 22, 39. MATHEMATICS, ATMOSPHERIC SCIENCES. *Educ:* Univ Md, Coll Park, BS, 61, PhD(Math), 64. *Prof Exp:* Res assoc math, Div Appl Math, Brown Univ, 64-66; asst prof, Univ Mich, Ann Arbor, 66-67; from asst prof to prof math, univ Colo, Boulder, 67; quant analyst, Nat Ctr Atmospheric Res, Boulder, 87-90; TECH STAFF, LINCOLN LAB, MASS INST TECHNOL, 90- *Concurrent Pos:* Vis assoc prof, Univ Md, Coll Park, 70-71. *Mem:* Soc Indust & Appl Math; Am Meteorol Soc. *Res:* Applications of differential topology to problems of nonlinear ordinary differential equations; numerical methods for solving differential equations, multidimensional interpolation and applications in meterology and geophysical data analysis. *Mailing Add:* Lincoln Lab Mass Inst Technol 244 Wood St Lexington MA 02173-9108

WILSON, FOREST RAY, II, b Wichita Falls, Tex, Aug 1, 41. HUMAN PHYSIOLOGY, GERONTOLOGY. *Educ:* Tex Wesleyan Col, BA, 66; Tex Christian Univ, MS, 69; Univ Ill, PhD(physiol), 73. *Prof Exp:* From asst prof to assoc prof, 73-89, PROF PHYSIOL, BAYLOR UNIV, 89- *Concurrent Pos:* Prin investr, Monsanto Found Cancer Res grant, 77-; dir, Masters Clin Geront Prog, Baylor Univ, prof, Inst Biomed Studies. *Mem:* AAAS; Am Zool Soc; Sigma Xi; Isozyme Soc. *Res:* Effects of cations on atherogenesis; many physiological problems faced by man including prostaglandins, extra-renal sites of erythropoietin production, lethality of megadoses of vitamines. *Mailing Add:* PO Box 97388 Baylor Univ Waco TX 76798-7388

WILSON, FRANK B, b Detroit, Mich, Jan 8, 29; m 50; c 7. SPEECH, AUDIOLOGY. *Educ:* Bowling Green Univ, BS, 50; Northwestern Univ, PhD, 56. *Prof Exp:* Speech & hearing clinician, Cerebral Palsy Ctr, Ohio, 51-53, actg dir, 52-53; res asst, Lang Inst, Northwestern Univ, 55-57; asst prof speech, St Louis Univ, 57-59; coordr speech & hearing, Spec Dist for Educ & Training Handicapped Children, St Louis County, Mo, 59-65; dir div speech path, Dept Otolaryngol, Jewish Hosp St Louis, Mo, 66-72; dir res, Spec Sch Dist St Louis County, 73-77; PROF & DEAN REHAB MED, UNIV ALTA, 77- *Concurrent Pos:* Consult, US Off Res, 63-; assoc prof, Wash Univ, 68. *Mem:* Fel Am Speech & Hearing Asn; Am Cleft Palate Asn. *Res:* Articulatory behavior of the retarded child and the efficacy of speech therapy; hearing deviation among orthopedically handicapped children; basis of nonorganic articulation disorders in children; voice disorders in school-age children. *Mailing Add:* Dept Speech Path & Audiol Univ Alta Edmonton AB T6G 2M7 Can

WILSON, FRANK CHARLES, b Ironwood, Mich, June 29, 27; m 50; c 3. POLYMER CHEMISTRY, X-RAY CRYSTALLOGRAPHY. *Educ:* Ripon Col, AB, 52; Mass Inst Technol, BS, 52, PhD(phys chem), 57. *Prof Exp:* RES ASSOC, POLYMER PROD DEPT, E I DU PONT DE NEMOURS & CO, INC, 57- *Mem:* Am Chem Soc; Am Crystallog Asn. *Res:* Structure and morphology of polymers and polymer blends by wide-angle and small-angle x-ray diffraction techniques. *Mailing Add:* Polymers Du Pont Exp Sta PO Box 80323 Wilmington DE 19880-0323

WILSON, FRANK CRANE, b Rome, Ga, Dec 29, 29; m 51; c 3. MEDICINE, ORTHOPEDIC SURGERY. *Educ:* Vanderbilt Univ, AB, 50; Med Col Ga, MD, 54; Am Bd Orthop Surg, dipl, 67. *Prof Exp:* Instr orthop surg, Columbia Univ, 63; from instr to assoc prof, 64-71, PROF ORTHOP SURG, SCH MED, UNIV NC, CHAPEL HILL, 71- *Concurrent Pos:* Markle scholar, 66-71; consult, Watts Hosp, Durham, NC, 65-; chief div orthop surg, NC Mem Hosp, 67- *Honors & Awards:* Nicholar Andry Award, 72. *Mem:* AAAS; Am Col Surg; Asn Am Med Col; Am Acad Orthop Surgeons; Am Orthop Asn. *Res:* Trauma; infections of bones and joints; rheumatoid arthritis. *Mailing Add:* Univ NC Sch Med 236 Burnett Womack Bldg Chapel Hill NC 27514

WILSON, FRANK DOUGLAS, b Salt Lake City, Utah, Dec 17, 28; m 50; c 8. PLANT GENETICS. *Educ:* Univ Utah, BS, 50, MS, 53; Wash State Univ, PhD(bot), 57. *Prof Exp:* Asst biol & genetics, Univ Utah, 51-53; asst bot, Wash State Univ, 53-56, asst agron, 56, jr animal scientist, 56-57; PLANT GENETICIST, WESTERN COTTON RES LAB, AGR RES SERV, USDA, 57- *Concurrent Pos:* Consult, Agron Exp Sta, Int Coop Admin, Cuba, 59-; plant explor, Puerto Rico, 74, E Africa, 75 & Galapagos Islands, 85. *Mem:* AAAS; Crop Sci Soc Am; Asn Taxon Study Trop African Flora. *Res:* Insect resistance in cotton; taxonomy of Hibiscus section Furcaria. *Mailing Add:* Western Cotton Res Lab 4135 E Broadway Phoenix AZ 85040

WILSON, FRANK JOSEPH, b Pittsburgh, Pa; c 2. ANATOMY, CELL BIOLOGY. *Educ:* St Vincent Col, BA, 64; Univ Pittsburgh, PhD(anat, cell biol), 69. *Prof Exp:* Muscular Dystrophy Asn fel, Univ Birmingham, 70-72; from asst prof to assoc prof anat, 72-87, PROF NEUROSCI & CELL BIOL, ROBERT WOOD JOHNSON MED SCH, UNIV MED & DENT NJ, 87- *Mem:* Am Asn Anatomists; AAAS; Am Soc Cell Biol; Soc Neurosci. *Res:* Immunochemistry and biochemistry of the contractile proteins. *Mailing Add:* Dept Neurosci & Cell Biol Robert Wood Johnson Med Sch Piscataway NJ 08854

WILSON, FRED E, b Lenexa, Kans, Dec 23, 37; m 61; c 1. ENDOCRINOLOGY. *Educ:* Univ Kans, BA, 58, MA, 60; Wash State Univ, PhD(zoophysiol), 65. *Prof Exp:* Instr biol, Lewis & Clark Col, 60-61; asst prof zool, 65-71, assoc prof biol, 71-90, PROF BIOL, KANSAS STATE UNIV, 90- *Concurrent Pos:* Physiologist, Agr Exp Sta, Kans State Univ, 65-79; consult, Oak Ridge Grad Sch Biomed Sci, Univ Tenn, 78. *Mem:* Fel AAAS; Am Soc Zoologists; Int Soc Neuroendrocinol; Endocrine Soc; Soc Study Reproduction. *Res:* Avian reproductive physiology; neuroendocrine control of annual reproductive cycles; photoperiodism. *Mailing Add:* Div Biol Kans State Univ Manhattan KS 66506-4901

WILSON, FRED LEE, b Detroit, Mich, Sept 5, 38; m 67; c 2. SCIENCE POLICY, THEORETICAL PHYSICS. *Educ:* Murray State Univ, BA, 59; Univ Kans, PhD(physics), 64. *Prof Exp:* Analyst, US Army, Sci & Technol Ctr, 65-66; sr res, Exxon Prod Res, 66-69; PROF PHYSICS & SCI POLICY, ROCHESTER INST TECHNOL, 69- *Concurrent Pos:* Foreign expert, Shanghai Univ Technol, 87 & 89. *Mem:* Am Phys Soc; Sigma Xi. *Res:* Science and technology policy. *Mailing Add:* 496 Thornell Rd Pittsford NY 14534

WILSON, FREDERICK ALBERT, b Boston, Mass, May 5, 28; c 2. FIELD GEOLOGY, GEOPHYSICS. *Educ:* Brooklyn Col, BA, 66, MA, 70; George Wash Univ, PhD(geol), 81. *Prof Exp:* Lectr earth sci, Brooklyn Col, 68 & mineral field geol, York Col, City Univ New York, 70-71; geologist, US Geol Surv, 72-81; ASST PROF MINERAL, PETROL & FIELD GEOL, HOWARD UNIV, 81- *Concurrent Pos:* Lectr, Howard Univ, 72, Washington Tech Inst, 76 & Howard Univ, 80- *Mem:* Asn Black Geoscientists (secy, 84); fel Geol Soc Am; Am Geophys Union; Asn Geoscientists Int Develop; Nat Asn Black Geologists & Geophysicists. *Res:* Exploration geophysical methods of aeroradioactivity, aeromagnetics and gravity to interpret geology in deeply weathered complex terrains; Caribbean geology. *Mailing Add:* Dept Geol & Geog Howard Univ Washington DC 20059

WILSON, FREDERICK ALLEN, b Winchester, Mass, Aug 22, 37; m 62; c 2. GASTROENTEROLOGY. *Educ:* Colgate Univ, Hamilton, NY, 59; Albany Med Col, NY, MD, 63. *Prof Exp:* Intern med, Hartford Hosp, Conn, 63-64, residency, 64-66; fel gastroenterol, Albany Med Col, NY, 66-67; chief gastroenterol, US Army Hosp, Ft Jackson, SC, 67-69; fel gastroenterol, Southwestern Med Sch, Univ Tex, Dallas, 69-72; asst prof med, Vanderbilt Univ, Nashville, 72-76, assoc prof, 76-82; PROF MED, MILTON S HERSHEY MED CTR, PA STATE UNIV, HERSHEY, 82- *Mem:* Am Gastroenterol Asn; Am Fedn Clin Res; Am Asn Study Liver Dis; Am Soc Clin Invest. *Res:* Gastroenterology involving the cellular and subcellular aspects of bile acid intestinal transport. *Mailing Add:* Dept Med Gastroent Pa State Univ Col Med PO Box 850 Hershey PA 17033

WILSON, FREDERICK SUTPHEN, b Trenton, NJ, Feb 12, 27; m 50; c 3. FAMILY MEDICINE. *Educ:* Dickinson Col, ScB, 48; Thomas Jefferson Univ, MD, 53. *Prof Exp:* Dir clin invest, McNeil Labs, 69-71, dir med serv, 71-74; ASST PROF COMMUNITY MED, MED CTR, TEMPLE UNIV, 74- *Concurrent Pos:* Physician-in-chief & dir family pract ctr/family pract residency prog, Abington Mem Hosp, 74-; med dir, William H Rorea, Inc, 76- *Mailing Add:* 1338 Jercho Rd Abington PA 19001

WILSON, G DENNIS, b Jackson, Tenn, Sept 20, 46; m 67; c 1. EXERCISE PHYSIOLOGY, SPORTS MEDICINE. *Educ:* Union Univ, Jackson, Tenn, BS, 68; Univ Tenn, Knoxville, MS, 70, EdD, 73. *Prof Exp:* Teacher-coach biol & phys educ, Oak Ridge High Sch, Orlando, Fla, 68-69 & teacher-coach, phys educ West High Sch, Knoxville, Tenn, 69-70; instr, Univ Tenn, Knoxville, 70-73; PROF & HEAD, DEPT HEALTH & HUMAN PERFORMANCE, AUBURN UNIV, 73-, PROF INTERDEPT PHYSIOL, 73-; DIR, AUBURN STRENGTH RES CTR, AUBURN ALA, 81- *Concurrent Pos:* Pres, Southeastern Am Col Sports Med, 77-78; chmn res coun, Southern Dist, Am Alliance Health Phys Educ, Recreation & Dance, 79-80 & Position Stands Rev Panel, Nat Fitness Test Rev Comt, Res Consortium, 83-; chmn continuing educ coun, Am Col Sports Med, 80-83; lectr, US Sports Acad, Saudi Arabia & Bahran, 83 & Am Col Sports Med, People to People Prog, People's Repub China, 85; mem exec coun, Int Coun Sports Sci & Phys Educ; mem, Comt Comparative Safeguards & Med Aspects of Sports, Nat Col Athletic Asn. *Mem:* Fel Am Col Sports Med; fel Am Alliance Health, Phys Educ & Dance. *Res:* Use of exercise and training to minimize disease processes such as CHD and obesity, and maximize performance with special emphasis on cardiovascular and metabolic responses to resistive forms of training. *Mailing Add:* Dept Health & Human Performance MC 2050 Auburn Univ Auburn AL 36849

WILSON, GARY AUGUST, b Chicago, Ill, Dec 13, 42; m 65; c 1. MOLECULAR GENETICS, MICROBIOLOGY. *Educ:* Ill State Univ, Normal, BS, 65; Univ Chicago, MS, 68, PhD(microbiol), 70. *Prof Exp:* Fel, Sch Med, Univ Rochester, 70-72, from instr to asst prof microbiol, 72-78, assoc prof, 78-80; dir microbiol res, 80-87, CORP DIR SCI & TECHNOL, MILES INC, 87- *Concurrent Pos:* Am Cancer Soc res grant, 72-, fac res award, 79; vis asst prof, Inst Gulbenkian de Ciencia, Portugal, 74, vis prof, 76; mem biotechnol adv comt, Pharmaceut Mfg Asn, 89- *Mem:* Am Soc Microbiol; AAAS; Sigma Xi. *Res:* DNA mediated transformation in Bacillus subtilis; restriction endonucleases; gene cloning. *Mailing Add:* 55886 Dana Dr Bristol IN 46507

WILSON, GEOFFREY LEONARD, b London, Eng, Oct 26, 24; nat US; m 55; c 2. ACOUSTICS. *Educ:* Oxford Univ, BA, 45, MSc, & MA, 49; Loughborough Univ, PhD, 75. *Prof Exp:* Sci off, Royal Naval Sci Serv, Brit Admiralty, Clarendon Lab, Oxford Univ, 44-48, H M Underwater Detection Estab, 48-51 & Torpedo Exp Estab, 52-53; design engr, Can Westinghouse Co, Ont, 53-59; from asst prof to assoc prof eng res, Pa State Univ, 59-85; BP AM DISTINGUISHED PROF MATH & PHYSICS, JOHNSON C SMITH UNIV, 85- *Concurrent Pos:* Vis res fel, Loughborough Univ Technol, Eng, 68-69; mem, Grad Fac Acoust, Pa State Univ, 75-85; vis scientist, Inst Nat Sci Appliquees, Lyon, France, 82. *Honors & Awards:* Centennial Medal, Inst Elec & Electronics Engrs, 84. *Mem:* Acoust Soc Am; fel Audio Eng Soc; sr mem Inst Elec & Electronics Engrs; fel Brit Inst Elec Engrs; fel Brit Inst Acoust; Sigma Xi. *Res:* Acoustics, especially underwater acoustics and transducer and array design. *Mailing Add:* Johnson C Smith Univ 100 Beatties Ford Rd Charlotte NC 28216-5398

WILSON, GEORGE DONALD, b Chatham, Ont, Can, Jan 17, 25; US citizen; m 50; c 3. MEAT SCIENCE & TECHNOLOGY, MEAT INDUSTRY REGULATORY AFFAIRS. *Educ:* Mich State Univ, BS, 49; Univ Wis, MS, 50, PhD(biochem & meat sci), 53. *Prof Exp:* Instr animal sci, Univ Wis, 52-53; div chief food technol, Am Meat Inst Found, 53-61; tech dir, Klarer of Ky, 61-69, vpres, 69-70; tech dir, Hygrade Food Prod, 70-81; VPRES SCI AFFAIRS, AM MEAT INST, 81- *Concurrent Pos:* Mem adv comm Salmonella, USDA Agr, 75-78, adv comm meat inspection, 81-83; indust adv, Nat Acad Sci, 82- *Honors & Awards:* Co-recipient Indust Achievement Award, Inst Food Technologists, 60. *Mem:* Inst Food Technologists; Am Meat Sci Asn; Agr Res Inst; NY Acad Sci. *Res:* Meat processing; science policy; energy and environmental matters. *Mailing Add:* 1218 U St Washington DC 20020

WILSON, GEORGE PETER, b Truro, NS, Dec 8, 39; m 67; c 4. INDUSTRIAL & MANUFACTURING ENGINEERING. *Educ:* Tech Univ, NS, BEng, 62; Univ Birmingham, Eng, MSc, 64. *Prof Exp:* Indust engr, Can Nat Railways, 65-67; asst prof, 67-75, ASSOC PROF INDUST ENG, TECH UNIV NS, 75-, DEPT HEAD, 76- *Concurrent Pos:* Asst dir, Atlantic Indust Res Inst, 65-76, dir, 76-; pres, Wilson Fuel Co Ltd, 74- & Appl Comput Systs Ltd, 75-, Wentworth Valley Develop Ltd. *Mem:* Inst Indust Engrs; Can Opers Res Inst; fel Eng Inst Can. *Res:* Development of industrial engineering technology for application in smaller sized industry. *Mailing Add:* PO Box 1000 Halifax NS B3J 2X4 Can

WILSON, GEORGE PORTER, III, b Flint, Mich, Nov 10, 27; c 4. VETERINARY SURGERY, ANATOMY. *Educ:* Univ Ill, BS, 51; Univ Pa, VMD, 55; Ohio State Univ, MSc, 59. *Prof Exp:* Intern vet med, Angell Mem Animal Hosp, 55-56; from instr to prof vet clin sci, Col Vet Med, Ohio State Univ, 55-88, prof microbiol, 67-87; RETIRED. *Concurrent Pos:* Ohio State Univ Develop Fund grant, 67-; Mark Morris Found grant, 68; Am Cancer Soc instnl grant, 68-69; vis scientist, Nat Cancer Inst, Environ Epidemiol Br, 78 & Armed Forces Inst Path, 79. *Mem:* AAAS; Am Vet Med Asn; Am Col Vet Surg; NY Acad Sci. *Res:* Veterinary surgery, especially cancer research as related to immunology and epidemiology. *Mailing Add:* 378 Fenway Rd Columbus OH 43214

WILSON, GEORGE RODGER, b Commercial Point, Ohio, July 10, 23; m 46; c 4. AGRICULTURE, ANIMAL SCIENCE. *Educ:* Ohio State Univ, BS, 48, MS, 56, PhD(animal sci), 63. *Prof Exp:* County agent, Butler County, Ohio, 48-54; from asst prof to assoc prof, 54-72, PROF ANIMAL SCI, OHIO STATE UNIV, 72- *Mem:* Am Soc Animal Sci. *Res:* Animal breeding, physiology and production. *Mailing Add:* Dept Animal Sci Ohio State Univ Columbus OH 43210

WILSON, GEORGE SPENCER, b Bronxville, NY, May 23, 39; m 64. BIOCHEMISTRY. *Educ:* Princeton Univ, BA, 61; Univ Ill, MS, 63, PhD, 65. *Prof Exp:* NIH fel, Univ Ill, 65-66, instr, 66-67; asst prof, Univ Ariz, 67-72, assoc prof, 72-80, prof chem, 80-87; PROF CHEM & PHARMACEUT CHEM, UNIV KANS, 87- *Mem:* fel AAAS; Sigma Xi; Electrochem Soc. *Res:* Analytical applications of biochemical reactions; electrochemical synthesis of unusual or unstable products; flow injection analysis; immunology. *Mailing Add:* Dept Chem Univ Kans Lawrence KS 66045

WILSON, GERALD GENE, b Blue Hill, Nebr, Mar 18, 28; m 52; c 4. CHEMICAL ENGINEERING, GAS TECHNOLOGY. *Educ:* Univ Kans, BS, 49; Ill Inst Technol, MS & MGas Tech, 51. *Prof Exp:* Engr, North Shore Gas Co, 51-55; assoc chem engr, Inst Gas Technol, 55-56, supvr dist res, 56-60, sr dist engr, 60-63, coordr indust educ, 63-65, mgr indust educ, 65-69, asst dir educ, 69-77, DIR INDUST EDUC, ILL INST TECHNOL, 77- *Concurrent Pos:* Mem subcomt on distribution design & develop, Oper Sect, Am Gas Asn, 62- *Honors & Awards:* Award of Merit, Am Gas Asn, 66. *Mem:* Am Soc Eng Educ; Nat Asn Corrosion Engrs. *Res:* Internal pipeline coatings; sealants for cast iron bell joints; gas industry leak control technology; techniques for solution of network flow problems; pipeline flow behavior; engineering economics. *Mailing Add:* 3519 S Federal Chicago IL 60609

WILSON, GERALD LOOMIS, b Springfield, Mass, Apr 29, 39; m 58; c 3. ELECTROMAGNETICS, ELECTROMECHANICS. *Educ:* Mass Inst Technol, SB, 61, SM, 63, ScD(mech eng & magnetohydrodynamics), 65. *Prof Exp:* From asst prof to assoc prof elec eng, Mass Inst Technol, 65-75, dir, Elec Power Systs Eng Lab, 75-80, head, Dept Elec Eng & Computers, 78-81, dean eng, 81-91, PROF ENG, MASS INST TECHNOL, 91- *Concurrent Pos:* Ford Found fel, 65-66; consult, Dynatech Corp, 65-66 & Am Elec Power Serv Corp, Mass, 66-67. *Mem:* Nat Acad Eng; fel Inst Elec & Electronics Engrs. *Res:* Magnetohydrodynamic energy conversion; electrohydrodynamics; electromechanics; electric power engineering. *Mailing Add:* Sch Eng Bldg 1 Rm 206 Mass Inst Technol 77 Massachusettes Ave Cambridge MA 02139

WILSON, GOLDER NORTH, b Frederick, Okla, Oct 29, 44; m 67; c 2. BIOCHEMISTRY, GENETICS. *Educ:* Univ Ill, BS, 66; Univ Chicago, PhD(biochem), 70, MD, 72. *Prof Exp:* Intern pediat, New Eng Med Ctr, 72-73; res assoc hematol, NIH, 73-75; resident pediat, 75-76, fel, 76-77, ASST PROF PEDIAT, UNIV MICH, ANN ARBOR, 77- *Res:* Human DNA, its structure and transcription. *Mailing Add:* K2015 Holden Univ Hosp Ann Arbor MI 48109

WILSON, GORDON, JR, b Bowling Green, Ky, Sept 13, 25; m 53; c 2. POLYMER CHEMISTRY. *Educ:* Western Ky State Col, BS, 47; Univ Ky, MS, 49; Purdue Univ, PhD(org chem), 58. *Prof Exp:* Instr chem, Univ Minn, Duluth, 49-54; res chemist, Dow Chem Co, 57-61; assoc prof, 61-65, head, Dept Chem, 65-78, PROF CHEM, WESTERN KY UNIV, 65- *Mem:* AAAS; Am Chem Soc; Sigma Xi. *Res:* Organic fluorine compounds; polymer chemistry, especially polymer bound catalysts. *Mailing Add:* Dept Chem Western Ky Univ Bowling Green KY 42101

WILSON, GREGORY BRUCE, b Columbus, Ohio, Oct 15, 48. CELLULAR IMMUNOLOGY, INBORN ERRORS OF METABOLISM. *Educ:* Univ Calif Los Angeles, BA, 71, PhD(biol-cell biol & immunol), 74. *Prof Exp:* Res asst, Dept Biol, Univ Calif Los Angeles, 72, Dept Microbiol & Immunol, 72-73, Dept Biol, 73-74, res fel, 74; res fel, dept med, Univ Calif San Francisco, 74-75; assoc, Med Univ SC, 75-76, from asst prof to assoc prof immunol, Dept Basic & Clin Immunol & Microbiol, 79-87, assoc prof pediat, 82-87; vpres res & develop, Amtron Inc, 87-89; var teaching positions, 89-91; ASSOC PROF CHEM, FRANCIS MARION COL, 91- *Concurrent Pos:* Mem, Grad Educ Comt, Med Univ SC, 78-, Molecular, Cellular Biol & Pathobiol Fac, 79-, Univ Grad Fac, 76-; mem, Combined Fed Campaign, Nat Health Agencies, 80-, Subcomt Pub Info, 81; chmn, Med & Sci Adv Comt, Cystic Fibrosis Found, 79-80, vchmn, 80-82, comt mem, 82-, bd dirs, 79-, Five Year Plan Comt, 80, Patient Serv Comt, 80, 81 & Speaker Bur Comt, 80, 81; Basil O'Connor grant award, Nat Found March Dimes, 76-79; res fel, Nat Cystic Fibrosis Found, 74-76. *Mem:* AAAS; Sigma Xi; Soc Exp Biol & Med; Reticuloendothelial Soc; NY Acad Sci. *Res:* Role of soluble mediators of cellular immunity in lymphocyte maturation, regulation of the inflammatory response and the development of cell-mediated immune responsiveness; cystic fibrosis as a primary host defense abnormality or immune deficiency disease; structure and mechanism of action of transfer factor. *Mailing Add:* 17 Croft St Greenville SC 29609

WILSON, GUSTAVUS EDWIN, b Philadelphia, Pa, Oct 6, 39; m 61; c 4. BIOCHEMISTRY. *Educ:* Mass Inst Technol, SB, 61; Univ Ill, PhD(org chem), 64. *Prof Exp:* From inst to asst prof chem, Polytech Inst Brooklyn, 64-69; from assoc prof to prof, Polytech Inst NY, 69-80; prof chem, Clarkson Univ, 80-84; PROF & DEPT HEAD CHEM, UNIV AKRON, 84- *Concurrent Pos:* From adj assoc prof to adj prof, Rockefeller Univ, 72-80; vis assoc prof, Sch Med, Univ Pa, 74-76; vis prof, Wash Univ, 78-80. *Mem:* Am Clin Soc; Royal Soc Chem; AAAS; Am Soc Biochem & Molecular Biol; Sigma Xi; NY Acad Sci. *Res:* Magnetic resonance studies of bacterial cell wall metabolism and protein structure. *Mailing Add:* Dept Chem Univ Akron Akron OH 44325-3601

WILSON, HAROLD ALBERT, b Tilton, Ill, Oct 10, 05; m 37; c 2. MICROBIOLOGY. *Educ:* La State Univ, BS, 32, MS, 33; Iowa State Col, PhD(soil microbiol), 37. *Prof Exp:* Prof agr, Panhandle Agr & Mech Col, 35-36; instr & asst, Iowa State Col, 37-38; prof agron, Southwestern La Inst, 38-44; soil conservationist, Soil Conserv Serv, USDA, 44-47; assoc prof bact, Univ, & assoc bacteriologist, Exp Sta, 47-57, prof & bacteriologist, 57-72, EMER PROF BACT, WVA UNIV, 72- *Mem:* Am Soc Agron; Am Soc Microbiol; Soil Sci Soc Am. *Res:* Microbiology of sanitary landfills and sewage decomposition in acid mine water; microbiology of strip mine (coal) spoil. *Mailing Add:* 1011 Boathouse Ct Raleigh NC 27615-5801

WILSON, HAROLD FREDERICK, b Columbiana, Ohio, Aug 15, 22; m 49; c 4. ORGANIC CHEMISTRY. *Educ:* Oberlin Col, AB, 47; Univ Rochester, PhD(chem), 50. *Prof Exp:* Res chemist, Rohm and Haas Co, 50-57, lab head, 57-63, res supvr, 63-68, from asst dir res to dir res, 68-72, vpres, 72-83, chief scientific officer, 81-83; CONSULT, 84- *Concurrent Pos:* Mem bd dirs, Indust Res Inst, 79-82 & Coun Chem Res, 81; chmn, US Nat Comt, Int Union Pure & Appl Chem, 82-84, Finance Comt, 81-89. *Res:* Insecticides; herbicides; growth regulators. *Mailing Add:* PO Box 2152 Cape May NJ 08204

WILSON, HARRY DAVID BRUCE, b Winnipeg, Man, Nov 10, 16; m 41; c 3. ECONOMIC GEOLOGY. *Educ:* Univ Man, BSc, 36; Calif Inst Technol, MS, 39, PhD, 42. *Prof Exp:* Res & explor geologist, Int Nickel Co, Can, 41-47; asst prof geol, Univ Man, 47-49; geologist, Africa & Europe, 49-51; assoc prof geol, Univ Man, 51-57, head dept, 65-72, prof geol, 57-81; RETIRED. *Concurrent Pos:* Mem, Nat Res Coun Can, 69-72; consult, Falconbridge Nickel Mines Ltd & Selco Mining Co Ltd; dir, Selco Mining Co Ltd & Man Mineral Resources Ltd. *Honors & Awards:* Barlow Medal, Can Inst Mining & Metall. *Mem:* Soc Econ Geologists (pres, 76); Geol Soc Am; Geol Asn Can (pres, 65-66); Can Inst Mining & Metall; Royal Soc Can. *Res:* Geology and geochemistry of ore deposits; structure and origin of continental crust. *Mailing Add:* 602-255 Wellington Crescent Winnipeg MB R3M 3V4 Can

WILSON, HARRY W(ALTON), JR, b Homestead, Pa, Nov 15, 24; m 47; c 2. CHEMICAL ENGINEERING. *Educ:* Univ Pittsburgh, BSChE, 48. *Prof Exp:* Develop engr, Goodyear Tire & Rubber Co, 48-50; process engr, Koppers Co, Inc, 50-56; sect head process develop, Callery Chem Co Div, Mine Safety Appliances Co, 56-64, mgr process eng, 64-68, mgr eng, 68-88; RETIRED. *Mem:* Am Inst Chem Engrs. *Res:* Chemical process development; new process design. *Mailing Add:* RR 2 Valencia PA 16059

WILSON, HENRY R, b Webbville, Ky, Mar 6, 36; m 59; c 2. REPRODUCTIVE PHYSIOLOGY, ENVIRONMENTAL PHYSIOLOGY. *Educ:* Univ Ky, BS, 57, MS, 59; Univ Md, PhD(poultry physiol), 62. *Prof Exp:* Asst prof poultry, 62-67, asst poultry physiologist, 62-67, assoc prof & assoc poultry physiologist, 67-74, PROF & POULTRY PHYSIOLOGIST, UNIV FLA, 74- *Mem:* Soc Exp Biol & Med; Soc Study Reproduction; Wildlife Soc; Poultry Sci Asn; World Poultry Sci Asn; Sigma Xi. *Res:* Reproduction in male chickens; heat tolerance in chickens; delaying sexual maturity in chickens; management techniques, fertility and hatchability; game birds. *Mailing Add:* Dept Poultry Sci Univ Fla 6 Mehr Bldg Gainesville FL 32611

WILSON, HERBERT ALEXANDER, JR, b Inverness, Miss, Jan 14, 14; m 41; c 2. AEROSPACE ENGINEERING. *Educ:* Ga Inst Technol, BS, 34. *Prof Exp:* Aeronaut res engr, Langley Mem Aeronaut Lab, Nat Adv Comt Aeronaut, Va, 37-43, head, full scale wind tunnel, 43-54, chief unitary plan wind tunnel div, 54-61, mgr, Proj Fire, Langley Res Ctr, NASA, 61-64, chief appl mat & physics div, 64-70, acting chief, Environ & Space Sci Div, 70-71, asst dir for space, 70-72; exec secy, res & develop incentives study, Nat Acad Eng, Wash, DC, 72-74, consult, 74-75; consult, Argonne Nat Lab, 77-78; consult, Booz-Allen & Hamliton, Inc, Bethesda, 78-79; RETIRED. *Concurrent Pos:* Mem subcomt helicopters, Nat Adv Comt Aeronaut, 47-48; consult, 79- *Honors & Awards:* Except Serv Award, NASA & Spec Serv Award, Langley Res Ctr, 66. *Mem:* Assoc fel Am Inst Aeronaut & Astronaut. *Res:* Space science studies of near earth space and upper atmosphere environmental factors; applications of aerospace knowledge to solution of public sector problems, facilities for research; technical management planning. *Mailing Add:* Three Holly Dr Newport News VA 23601

WILSON, HOWARD LE ROY, b Salem, Ore, Dec 8, 32; m 60; c 3. MATHEMATICS. *Educ:* Willamette Univ, BA, 54; Univ Ill, MS, 60, PhD(educ), 66. *Prof Exp:* Teacher high sch, Ore, 55-56; from instr to asst prof math, Eastern Ore Col, 56-64; from asst prof to assoc prof math & sci educ, 64-80, PROF MATH & MATH EDUC, ORE STATE UNIV, 80- *Concurrent Pos:* Expert in math, Field Sta, Papua, New Guinea, UN Educ Sci & Cult Orgn, 71-73. *Mem:* Nat Coun Teachers Math; Math Asn Am; Am Asn Univ Professors. *Res:* Mathematics education and teacher training. *Mailing Add:* Dept Math Ore State Univ Corvallis OR 97331

WILSON, HOWELL KENNETH, b Savannah, Ga, Aug 28, 37. MATHEMATICS. *Educ:* Ga Inst Technol, BS, 60; Univ Minn, PhD(math), 64. *Prof Exp:* From asst prof to assoc prof math, Ga Inst Technol, 64-69; assoc prof, 69-73, PROF MATH, SOUTHERN ILL UNIV, EDWARDSVILLE, 73- *Mem:* Soc Indust Appl Math. *Res:* Triadic logic. *Mailing Add:* Dept Math & Stat Southern Ill Univ EDWS Edwardsville IL 62026

WILSON, HUGH DANIEL, b Alliance, Ohio, Aug 15, 43; m 70; c 1. SYSTEMATIC BOTANY, ETHNOBOTANY. *Educ:* Kent State Univ, BA, 70, MA, 72; Ind Univ, Bloomington, PhD(bot), 76. *Prof Exp:* Asst prof bot, Univ Wyo, 76-77; from asst prof to assoc prof, 77-90, PROF BIOL, TEX A&M UNIV, 90- *Honors & Awards:* W H Fulling Award, Soc Econ Bot, 81. *Mem:* Am Soc Plant Taxonomists; Soc Econ Bot; fel AAAS; Bot Soc Am. *Res:* Angiosperm biosystematics with emphasis on species complexes that include domesticated taxa. *Mailing Add:* Dept Biol Tex A&M Univ College Station TX 77843

WILSON, HUGH REID, b Ft Monmouth, NJ, Apr 20, 43. VISUAL PSYCHOPHYSICS. *Educ:* Wesleyan Univ, Conn, BA, 65; Univ Chicago, MA, 68, PhD(chem physics), 69. *Prof Exp:* Fel, 69-72, from instr to asst prof, 72-80, PROF OPHTHAL & VISUAL SCI, UNIV CHICAGO, 85- *Concurrent Pos:* NIH grant, 78-95; NSF grants, 82-85. *Honors & Awards:* Res Career Develop Award, NIH. *Mem:* Asn Res Vision & Opthal; AAAS; Optical Soc Am. *Res:* Processing of spatial and motion information by the human visual system; stereopsis and binocular vision; mathematical models of human visual functions. *Mailing Add:* Dept Ophthal Univ Chicago 939 E 57th St Chicago IL 60637

WILSON, IRWIN B, b Yonkers, NY, May 8, 21; m 52; c 2. BIOCHEMISTRY. *Educ:* City Col New York, BS, 41; Columbia Univ, AM, 47, PhD(phys chem), 48. *Prof Exp:* Jr chemist, Picatinny Arsenal, NJ, 42; res assoc, Columbia Univ, 42-45; chemist, Union Carbide & Carbon Corp, 45; instr chem, City Col New York, 46-49; assoc, Col Physicians & Surgeons, Columbia Univ, 48-49, from asst prof to prof biochem, 49-66; prof chem, Univ Colo, Boulder, 66-88; RETIRED. *Honors & Awards:* Asn Res Nerv & Ment Dis Award, 58. *Mem:* Am Chem Soc; Am Soc Biol Chem; Am Acad Neurol. *Res:* Enzymology; protein chemistry; nerve function; peptide hormones. *Mailing Add:* Dept Chem Univ Colo Boulder CO 80309

WILSON, J(AMES) W(OODROW), b Commerce, Tex, Apr 5, 16; m 43; c 3. CHEMICAL ENGINEERING, MATHEMATICS. *Educ:* Stephen F Austin State Col, BA, 35; Univ Tex, BS, 39; Agr & Mech Col, Tex, MS, 41. *Prof Exp:* Instr chem, Agr & Mech Col, Tex, 42; instr chem eng, Columbia Univ, 46-49; from asst prof to prof, 49-77, EMER PROF CHEM ENG, NAVAL POSTGRAD SCH, 77- *Mem:* Am Chem Soc; Am Inst Chem Engrs. *Res:* Explosives; rocket propellants; heat transfer; thermodynamics; fuels and lubricants. *Mailing Add:* 3435 Oak Cluster Rd San Antonio TX 78253-9227

WILSON, JACK BELMONT, b Morgantown, WVa, Dec 1, 21; m 43; c 6. PLANT PATHOLOGY. *Educ:* WVa Univ, BS, 53, MS, 54, PhD(plant path), 57. *Prof Exp:* Instr plant path, Univ Md, 56-57, asst prof, 57-62; res plant pathologist, Potato Handling Res Ctr, USDA, Maine, 62-67; assoc prof plant path & exten plant pathologist & entomologist, WVa Univ, 67-69; asst br chief, Hort Crops Res Br, Mkt Qual Res Div, USDA, 69-73, asst area res dir, NE Region, Agr Res Serv, 73-74, area dir, N Atlantic Area, NE Region, Fed Res, Sci & Educ Agency, 74-85; RETIRED. *Concurrent Pos:* Assoc prof, Univ Maine, 66-67. *Mem:* Am Phytopath Soc; Potato Asn Am; Europ Asn Potato Res. *Res:* Diseases of potatoes and ornamental plants; plant disease and insect diagnosis. *Mailing Add:* 115 E Main St Trumansburg NY 14886

WILSON, JACK CHARLES, b Waterloo, Iowa, Dec 17, 28; m 48; c 4. ALGEBRA. *Educ:* Iowa State Teachers Col, BA, 51; Univ Iowa, MS, 54; Case Western Reserve Univ, PhD(math), 60. *Prof Exp:* Instr math, Cent Col, Iowa, 53-56; asst prof, Fenn Col, 56-59; assoc prof, Cent Col, Iowa, 59-65 & Earlham Col, 65-70; PROF MATH, UNIV NC, ASHEVILLE, 70- *Mem:* Am Math Soc; Math Asn Am. *Res:* Pure mathemtics; functions on algebras. *Mailing Add:* Dept Math Univ NC Asheville NC 28804

WILSON, JACK HAROLD, biochemistry, microbiology, for more information see previous edition

WILSON, JACK LOWERY, b Looxahoma, Miss, June 24, 43; m 66; c 3. ANATOMY. *Educ:* Univ Southern Miss, BS, 64; Univ Miss, MS, 67, PhD(anat), 68. *Prof Exp:* From instr to asst prof, 68-74, ASSOC PROF ANAT, MED UNITS, UNIV TENN, MEMPHIS, 74- *Mem:* Am Asn Anat. *Res:* Relations of age and sex hormones to the dietary induction, high fat and low protein components of cardiac and hepatic lesions in mice, also study of the fine structure of these lesions; cerebrovascular spasm. *Mailing Add:* Dept Anat Univ Tenn Col Med 800 Madison Ave Memphis TN 38163

WILSON, JACK MARTIN, b Camp Atterbury, Ind, June 29, 45; c 2. PHYSICS. *Educ:* Thiel Col, AB, 67; Kent State Univ, MA, 70, PhD(physics, math), 72. *Prof Exp:* From asst prof to assoc prof physics, Sam Houston State Univ, 72-79, chair dept, 79-82; guest scientist, State Univ NY, Stony Brook, 82-84; prof physics, Univ Md, College Park, 84-90; DIR & PROF PHYSICS, CTR INNOVATION UNDERGRAD EDUC, RENSSELAER POLYTECH INST, 90- *Concurrent Pos:* Exec officer, Am Asn Physics Teachers, 82-90; mem educ comt, Am Phys Soc, 82-89; ed, Am Asn Physics Teachers Announcer, 82-90; dir, Md Univ Proj Physics & Educ Technol, 84-89; dir, Cuple Nat Consortium, 90-; dir site comt, Statewide Syst Initiative, NSF, 91. *Mem:* Am Phys Soc; Am Asn Physics Teachers; AAAS; Sigma Xi. *Res:* Computing in physics and education. *Mailing Add:* Ctr Innovation Undergrad Educ Rensselaer Polytech Inc Troy NY 12180

WILSON, JAMES ALBERT, b Boston, Mass, Jan 28, 29. PHYSIOLOGY, BIOCHEMISTRY. *Educ:* Northeastern Univ, BS, 53; Univ Mich, MS, 55, PhD(zool), 59. *Prof Exp:* Res assoc zool, Univ Mich, 58-59; asst prof, 59-72, ASSOC PROF PHYSIOL, OHIO UNIV, 72- *Mem:* AAAS; Am Inst Biol Sci; Am Soc Zoologists; Sigma Xi. *Res:* Pressure-temperature pH effects on the activity of adenosine-triphosphatases from rabbit tissues; contractility of glycerol-extracted muscle fibers and related enzyme activity; factors affecting muscle relaxation. *Mailing Add:* Dept Zool Ohio Univ Athens OH 45701

WILSON, JAMES ALEXANDER, plant breeding, for more information see previous edition

WILSON, JAMES BLAKE, b Albion, Mich, Feb 9, 24; m 49; c 4. APPLIED MATHEMATICS. *Educ:* Univ Fla, BS, 48, PhD(appl math), 57; Cornell Univ, MS, 51. *Prof Exp:* Instr mech, Cornell Univ, 50; instr, Dept Ord, US Mil Acad, 51-54; asst math, Univ Fla, 54-56; from asst prof to prof math, NC State Univ, 57-87, asst head, 78-87; RETIRED. *Mem:* Am Math Soc; Math Asn Am. *Res:* Mechanics; numerical analysis. *Mailing Add:* 1311 Greenwood Circle Cary NC 27511

WILSON, JAMES BRUCE, computer science, biomedical engineering, for more information see previous edition

WILSON, JAMES DENNIS, b The Dalles, Ore, Feb 28, 40; m 62; c 2. ORGANIC CHEMISTRY. *Educ:* Harvard Univ, AB, 62; Univ Wash, PhD(chem), 66. *Prof Exp:* Mem staff, Corp Res Dept, 66-77, mgr res & develop, Detergents & Phosphate Div, 77-82, planning & info dir, 82-87, REGULATORY MGT, HEALTH & SAFETY STAFF, MONSANTO CO, 87- *Concurrent Pos:* Sci comt, Am Indust Health Coun. *Mem:* AAAS. *Res:* Correlation of molecular structure with physical properties, especially solid-state electrical, magnetic and olfactory properties; synthesis of unsaturated nitrogen- and sulfur-compounds; sources and environmental fate of chlorodibenzodioxins; methodologies for human health risk assessment. *Mailing Add:* Monsanto Co 800 N Lindbergh Blvd St Louis MO 63167

WILSON, JAMES FRANKLIN, b Christopher, Ill, Oct 27, 20; m 43; c 2. MICROBIOLOGY, GENETICS. *Educ:* Southern Ill Univ, BS, 44; Iowa State Col, MS, 46; Stanford Univ, PhD, 59. *Prof Exp:* Instr biol, Hartnell Col, 46-64; PROF BIOL, UNIV NC, GREENSBORO, 64- *Mem:* AAAS; Genetics Soc Am; Sigma Xi. *Res:* Application of microsurgical techniques to the study of heterocaryosis and cytoplasmic heredity in Neurospora crassa. *Mailing Add:* Dept Biol Univ NC Greensboro NC 27412

WILSON, JAMES LARRY, b Jackson, Tenn, Feb 9, 42; m 85; c 2. FISHERIES MANAGEMENT, AQUACULTURE. *Educ:* Union Univ, BS, 64; Univ Fla, MS, 67; Univ Tenn, PhD(zool), 70. *Prof Exp:* From asst prof to assoc prof, 70-83, PROF FISHERIES, DEPT FORESTRY, WILDLIFE & FISHERIES, UNIV TENN, 83-, ASSOC DIR, GRAD PROG ECOL, 88- *Concurrent Pos:* Lectr, Tech Aqua Biol Sta, Career Awareness Inst, 78-79; vis scientist prog, Tenn Acad Sci, 83-; chmn regional res proj S-168, USDA/Coop State Res Serv, 84-86; co-chair, Tech Comt, Southern Reg Agr Ctr, 88- *Honors & Awards:* Outstanding Serv to Fisheries Award, Am Fisheries Soc, 85. *Mem:* Am Fisheries Soc; Sigma Xi. *Res:* Management strategies for improvement of game fish yields in Tennessee waters; develop methods for economic rearing and marketing of aquatic animals with economic potential (aquaculture); maximizing production of commercially raised fishes. *Mailing Add:* Dept Forestry Wildlife & Fisheries Univ Tenn PO Box 1071 Knoxville TN 37901-1071

WILSON, JAMES LEE, b Waxahachie, Tex, Dec 1, 20; m 44; c 3. GEOLOGY. *Educ:* Univ Tex, BA, 42, MA, 44; Yale Univ, PhD(geol), 49. *Prof Exp:* Jr geologist, Carter Oil Co, 43-44; assoc prof geol, Univ Tex, 49-52; res geologist, Shell Develop Co & Shell Int Res, 52-66; Wiess prof geol, Rice Univ, 66-78; prof geol, Univ Mich, 79-86; CONSULT, 86- *Concurrent Pos:* Vis prof, Mex, WGer, Turkey, Italy & Can. *Mem:* Fel Geol Soc Am; Paleont Soc; hon Soc Econ Paleont & Mineral (pres, 75-76); hon Am Asn Petrol Geol. *Res:* Cambrian paleontology; Paleozoic biostratigraphy; carbonate petrography and petrology; sedimentology of carbonate strata; tectonics and sedimentation. *Mailing Add:* 1316 Patio Dr New Braunfels TX 78130

WILSON, JAMES LESTER, b Nashville, Tenn, July 18, 25; m 48; c 2. ZOOLOGY. *Educ:* George Peabody Col, BS, 51, MA, 52; Vanderbilt Univ, PhD, 59. *Prof Exp:* Instr biol & phys sci, Ark State Col, 52-54; asst, Vanderbilt Univ, 55-56; from assoc prof to prof biol & chmn div sci & math, Belmont Col, 56-67; assoc prof, 67-70, PROF ZOOL, TENN STATE UNIV, NASHVILLE, 70- *Honors & Awards:* Sullivan Award, George Peabody Col, 51. *Mem:* Fel AAAS; Acarological Soc Am. *Res:* Taxonomy; life history and ecology of water mites; hydracarology. *Mailing Add:* Dept Zool Tenn State Univ Nashville TN 37203

WILSON, JAMES M, b Cuba, NY, June 10, 50; m 72; c 3. CERAMIC POWDER SYNTHESIS & CHARACTERIZATION, DIELECTRIC MATERIALS & PROPERTIES. *Educ:* Syracuse Univ, BA, 88. *Prof Exp:* Jr engr, AVX Inc, 72-76; eng supvr, Mepco Electra Inc, 76-81. *Mem:* Am Ceramics Soc; Nat Inst Ceramic Engrs. *Res:* Dielectric materials for uses in the ceramic capacitor industry. *Mailing Add:* Transelco Div Ferro Corp PO Box 217 Penn Yan NY 14527

WILSON, JAMES R, b Berkeley, Calif, Oct 21, 22; m 49; c 5. THEORETICAL ASTROPHYSICS. *Educ:* Univ Calif, BS, 43, PhD(physics), 52. *Prof Exp:* Physicist, Sandia Corp, 52-53; physicist, Lawrence Livermore Lab, Univ Calif, 53-88; RETIRED. *Mem:* Fel Am Phys Soc; AAAS; Murdock fel; Int Astron Union. *Res:* Astrophysics; gravitational radiation; relativity; nuclear physics. *Mailing Add:* 737 S M St Livermore CA 94550

WILSON, JAMES RUSSELL, b Pittsburgh, Pa, Jan 10, 33. BEHAVIORAL BIOLOGY. *Educ:* Univ Calif, Berkeley, AB, 59, PhD(psychol), 68. *Prof Exp:* Res psychologist, Univ Calif, Berkeley, 63-66; res assoc psychol, Univ Colo, Boulder, 66-68; instr psychol, Univ Calif, Santa Cruz, 69; from asst prof to assoc prof, 69-84, PROF PSYCHOL, UNIV COLO, BOULDER, 84- *Concurrent Pos:* Assoc researcher, Univ Hawaii, 74- *Mem:* Behav Genetics Asn; Res Soc Alcoholism. *Res:* Genetic analysis of behavioral phenotypes, including aggression, sexual behavior, alcohol use and cognitive abilities. *Mailing Add:* Dept Phychol Univ Colo Box 345 Boulder CO 80309-0345

WILSON, JAMES WILLIAM, b Rice, Va, Oct 21, 19; m 50; c 4. MEDICINAL CHEMISTRY. *Educ:* Hampden-Sydney Col, BS, 41; Univ Va, MS, 44, PhD(org chem), 46. *Prof Exp:* Asst chem, Univ Va, 41-43; res chemist, Smith Kline & French Labs, 46-53, asst sect head org chem, 53-54, head med chem sect, 54-66, staff dir, 66-67, assoc dir chem, 67-85; RETIRED. *Mem:* AAAS; Am Chem Soc; NY Acad Sci; Am Inst Chem. *Res:* Medicinal chemistry; analgesics; cardiovascular drugs; psychopharmacological, diuretic and anti-inflammatory agents. *Mailing Add:* 15 Kinterra Rd Wayne PA 19087-4717

WILSON, JAMES WILLIAM, b Linwood, Kans, Sept 12, 36; m 60; c 3. MATHEMATICS EDUCATION, TEACHER EDUCATION IN MATHEMATICS. *Educ:* Kans State Teachers Col, BS, 58 & MA, 60; Stanford Univ, MS, 64 & PhD(math educ), 67; Notre Dame, MS, 65. *Prof Exp:* Teacher math & biol, Marion City Sch, 58-60; instr math, Roosevelt High Lab Sch, 61-62; from res asst to res assoc, Stanford Univ, 62-68; from asst prof to assoc prof math educ, Univ Ga, 68-74; prog mgr, NSF, 74-75; assoc prof, 74-78, PROF MATH EDUC, UNIV GA, 78- *Concurrent Pos:* Chair, Spec Int Theory Math Educ, Am Educ Res Asn, 68-70; ed, J Res Math Educ, 76-82; bd dirs, Nat Coun Teachers Math, 78-81, Sch Sci & Math Asn, 74-76; consult, Nat Assessment Educ Progress, 68-84; vis prof, St Michaels Col, Winvoski, Vt, 67, Univ Tex San Antonio, 81, Univ Hawaii, 82, E Tenn State Univ, 86-88 & Southern Ill Univ, 87-88. *Mem:* Am Educ Res Asn; Nat Coun Teachers Math; Math Asn Am; Sch Sci & Math Asn. *Res:* Development of abilities in mathematical problem-solving. *Mailing Add:* Univ Ga 105 Aderhold Hall Athens GA 30602

WILSON, JAMES WILLIAM ALEXANDER, b Glasgow, Scotland, Dec 24, 44; m 70; c 3. ELECTRICAL ENGINEERING. *Educ:* Heriot-Watt Univ, BSc, 67; Univ Edinburgh, PhD(elec eng), 71. *Prof Exp:* Fel elec eng, Univ Toronto, 71-73; sr engr static power conversion, Reliance Elec Co, Cleveland, 73-76; elec engr, Gen Elec Co, 76-78, unit mgr power circuits & drives, 78-81, br mgr power circuits & systs, 81-84, consult, Res & Develop Strategic Anal, 84-85, mgr, Planning Oper, 86-88, mgr, Power Controls Prog, 88-89, MGR, CONTROL SYSTEMS LAB, GEN ELEC CO, 89- *Mem:* Inst Elec & Electronics Engrs; assoc mem Brit Inst Elec Engrs. *Res:* Development of advanced solid-state power conversion techniques and applications. *Mailing Add:* Gen Elec Co R&D Bldg K1 PO Box 8 Schenectady NY 12301

WILSON, JEAN DONALD, b Wellington, Tex, Aug 26, 32. INTERNAL MEDICINE. *Educ:* Univ Tex, BA, 51, MD, 55; Am Bd Internal Med, dipl, 64. *Prof Exp:* From intern to asst resident internal med, Parkland Mem Hosp, Dallas, 55-58; clin assoc clin biochem, Nat Heart Inst, 58-60; from instr to assoc prof, 60-68, PROF INTERNAL MED, UNIV TEX SOUTHWESTERN MED CTR, DALLAS, 68- *Concurrent Pos:* Estab investr, Am Heart Asn, 60-65; ed, J Clin Invest, 72-77. *Honors & Awards:* Oppenheimer Award, Endocrine Soc, 72; Amory Prize, Am Acad Arts & Sci, 77; Eugene Fuller Award, Am Urol Asn, 83; Lita Annenberg Hazen Award, 86; Dale Medal, Soc Endocrinol, 91. *Mem:* Nat Acad Sci; Am Fedn Clin Res; Am Soc Biol Chem; Endocrine Soc; Asn Am Physicians; Am Soc Clin Invest. *Res:* Mechanism of action of steroid hormones; sexual differentation; androgen physiology. *Mailing Add:* Dept Internal Med Univ Tex Southwestern Med Ctr Dallas TX 75235-8857

WILSON, JERRY D(ICK), b Coshocton, Ohio, May 6, 37; div; c 2. SCIENCE WRITING, PHYSICS. *Educ:* Ohio Univ, BS, 62, PhD(physics), 70; Union Col, MS, 65. *Prof Exp:* Mat behav physicist dielectrics, Gen Elec Co, 63-66; lectr physics, Ohio Univ, 75-78; assoc prof, 79-85, PROF PHYSICS, LANDER COL, 80- *Concurrent Pos:* Fel, Ohio Acad Sci, 72. *Mem:* Am Asn Physics Teachers; Nat Sci Teachers Asn; Am Med Technologists. *Res:* Science writing and science teaching, particularly for non-science students. *Mailing Add:* Dept Sci & Math Lander Col Greenwood SC 29649

WILSON, JERRY LEE, b Heavener, Okla, Jan 30, 38; m 61; c 2. BIOCHEMISTRY. *Educ:* Okla State Univ, BS, 61; Univ Okla, PhD(chem), 67. *Prof Exp:* USPHS res fel, Univ Calif, Davis, 67-69; from asst prof to assoc prof, 69-84, PROF CHEM, CALIF STATE UNIV, SACRAMENTO, 84- *Mem:* Am Chem Soc; Sigma Xi. *Res:* Plant biochemistry; enzymology; protein chemistry. *Mailing Add:* Dept Chem Calif State Univ Sacramento CA 95819

WILSON, JOE BRANSFORD, b Dallas, Tex, June 29, 14; m 44; c 2. BACTERIOLOGY. *Educ:* Univ Tex, BA, 39; Univ Wis, MS, 41, PhD(bact), 47; Am Bd Microbiol, dipl. *Prof Exp:* Instr bact, Univ Tex, 39; asst, Univ Wis-Madison, 39-42, from instr to assoc prof, 46-55, prof bact, 55-81, assoc dean grad sch, 65-69, chmn dept, 68-73, prof med microbiol, 77-81; RETIRED. *Concurrent Pos:* Mem tech adv panel, Off Asst Secy Defense, 52-63. *Mem:* Fel AAAS; Am Soc Microbiol; Soc Exp Biol & Med; Am Asn Immunol; fel Am Acad Microbiol; Sigma Xi. *Res:* Metabolism and pathogenesis of Brucella, Cocci, Vibrio and Leptospira. *Mailing Add:* 36 Pebblebrook Lane Wimberley TX 78676

WILSON, JOE ROBERT, b Colfax, La, Apr 5, 23; m 44; c 3. CIVIL & CONSTRUCTION ENGINEERING. *Educ:* US Naval Acad, BS, 44; Rensselaer Polytech Inst, MCE, 48; Univ Tex, PhD(civil eng), 67. *Prof Exp:* Co commander, Naval Construct Battalion 105 & US Naval Civil Eng Corps, 48-50, proj engr, Pub Works Ctr, Va, 50-52, dir design div, Potomac River Naval Command Pub Works, 52-53, mgr construct div, Off in Charge Construct, Spain, 53-55, dir pub works, Naval Air Sta, Ala, 56-58, staff engr, Off Chief of Naval Opers, 58-60, dir pub works, Naval Sta, CZ, 60-62, dist civil engr, 15th Naval Dist, 62-64; prof civil eng, La Tech Univ, 66-86, head dept, 75-86; RETIRED. *Mem:* Fel Am Soc Civil Engrs; Water Pollution Control Fedn; Am Soc Eng Educ. *Res:* Hydraulics; water pollution control systems; mixing and dispersion phenomena. *Mailing Add:* 1621 Hodges Rd Ruston LA 71270

WILSON, JOHN CLELAND, b Galt, Ont, June 8, 35; m 64; c 4. SMALL COMPUTERS. *Educ:* Univ Toronto, BASc, 58; Univ Waterloo, MSc, 62, PhD(math), 66. *Prof Exp:* Analyst, KCS Data Control Ltd, Ont, 58-62; asst prof math, Univ Waterloo, 66-70, assoc dir comput ctr, 68-70; dir, Univ Toronto, 70-76, assoc prof comput sci, Comput Ctr, 70-76, dir student record serv, 76-79; COMPUT SCIENTIST, COMPUT SYSTS GROUP, UNIV WATERLOO, 80- *Concurrent Pos:* Nat Res Coun fels, Univ Waterloo, 67 & Univ Toronto, 72; Dept Univ Affairs fel, Univ Waterloo, 68. *Mem:* Asn Comput Mach; Inst Elec & Electronics Engrs. *Res:* Microcomputer hardware and software; computer language implementation. *Mailing Add:* RR 3 Wallenstein ON N0B 2S0 Can

WILSON, JOHN COE, b Manhattan, KS, Jul 16, 31; m 55; c 2. GEOLOGY APPLIED ECONOMIC & ENGINEERING. *Educ:* Calif Inst Technol, BS, 53, PhD(geol), 61; Univ Kans, MS, 55. *Prof Exp:* Explor geologist, Bear Creek, Mining Co, 61-64; res geologist, Explor Serv Dept, Kenecott Copper Corp, 64-68, chief geol res, 68-71, dir explor res, 71-77; mgr explor, Anaconda Minerals Co, 77-86; GEN MGR INFO MGT, APPL RESOURCES LTD, 86- *Concurrent Pos:* Prog mgt develop, Harvard Bus Sch, 73. *Mem:* Soc Econ Geologist; Soc Mining Engrs; fel Geol Soc Am. *Res:* Origin and discovery on metalliferous ore deposits; efficient intergration of ore deposit exploration concepts with exploration databases to yield mineral discoveries. *Mailing Add:* 2800 S University Blvd No 3 Denver CO 80210

WILSON, JOHN D(OUGLAS), b Edinburgh, Scotland, Aug 21, 35; m 57; c 3. ELECTRICAL ENGINEERING, COMPUTER ENGINEERING. *Educ:* Univ Edinburgh, BSc, 56; Univ London (UCWI), PhD(elec eng), 67. *Prof Exp:* Electronic engr, Bristol Aircraft Ltd, Eng, 56-59; lectr elec eng, Univ WI, 62-67; assoc prof, 67-76, head dept, 78-84, PROF ELEC ENG, ROYAL MIL COL CAN, 76- *Concurrent Pos:* Can Defence Res Bd grant, 68-; vis prof, Univ Maine, Orono, 85-86; mem, Can Accreditation Vis Team, 82-83 & 85-86. *Mem:* Sr mem Inst Elec & Electronics Engrs. *Res:* Microcomputer applications; acoustic emission studies in aircraft structures; computer aided diagnosties; expert systems. *Mailing Add:* Dept Elec & Computer Eng Royal Mil Col Can Kingston ON K7K 5L0 Can

WILSON, JOHN DRENNAN, b Peoria, Ill, Mar 29, 38; m 67; c 1. PHYSIOLOGY, RADIOBIOLOGY. *Educ:* Carleton Col, BA, 60; Univ Ill, Urbana, MS, 63, PhD(physiol), 66. *Prof Exp:* Res assoc radiobiol, Univ Tex, Austin, 66-72; asst prof, 72-76, ASSOC PROF RADIOBIOL, MED COL VA, 76- *Mem:* AAAS; Radiation Res Soc. *Res:* Lethal and mutagenic effects of radiation on microorganisms; effects of accelerated particles on mammalian systems. *Mailing Add:* Dept Radiol Va Commonwealth Univ Sch Med MCV Sta Box 565 Richmond VA 23298

WILSON, JOHN EDWARD, b Ft Wayne, Ind, Apr 27, 39; m 64; c 3. NEUROCHEMISTRY, ENZYMOLOGY. *Educ:* Univ Notre Dame, BS, 61; Univ Ill, MS, 62, PhD(biochem), 64. *Prof Exp:* From asst prof to assoc prof, 67-75, PROF BIOCHEM, MICH STATE UNIV, 75- *Concurrent Pos:* NSF fel, Univ Ill, Urbana, 64-65. *Honors & Awards:* Javits Nuerosci Investr Award, 85. *Mem:* Am Soc Biol Chem; Am Soc Neurochem; Int Soc Neurochem; Am Chem Soc. *Res:* Brain hexokinase; brain mitochondria; regulation of energy metabolism in brain. *Mailing Add:* Dept Biochem Mich State Univ Wilson Rd Rm 301 East Lansing MI 48824-1319

WILSON, JOHN ERIC, b Champaign, Ill, Dec 13, 19; m 47; c 3. BIOCHEMISTRY. *Educ:* Univ Chicago, SB, 41; Univ Ill, MS, 44; Cornell Univ, PhD(biochem), 48. *Prof Exp:* Asst chem, Univ Ill, 41-44; asst biochem, Med Col, Cornell Univ, 44-48, res assoc, 48-50; from asst prof to assoc prof, 50-60, dir, Neurobiol Prog, 72-73, prof biochem, 65-89, EMER PROF, SCH MED, UNIV NC, CHAPEL HILL, 90- *Concurrent Pos:* Consult, Oak Ridge Nat Lab, 54-57; Kenan prof, Univ Utrecht, Neth, 78, external examr, Univ Malaya, 88. *Mem:* Fel AAAS; Am Chem Soc; Am Soc Biol Chem & Molecular Biol; Am Soc Neurochem; Soc Neurosci; Int Soc Neurochem. *Res:* Effects of experience and behavior on brain metabolism; neurochemistry. *Mailing Add:* Dept Biochem Sch Med Univ NC Chapel Hill NC 27599-7260

WILSON, JOHN F, b Niagara Falls, NY, Dec 23, 22; m 50; c 8. MEDICINE, PATHOLOGY. *Educ:* Univ Cincinnati, MD, 52. *Prof Exp:* Intern pediat, Univ Ark Hosp, 52-53; resident, Children's Hosp, Cincinnati, Ohio, 55-57; instr, Univ Cincinnati, 57-58; from instr to asst prof pediat, 58-68, from instr to asst prof path, 67-87, assoc prof pediat, Univ Utah, 73-87; Pathologist & dir labs, 69-87, CONSULT, PRIMARY CHILDRENS HOSP, 88-; ELECTRON MICROSCOPIST, LDS HOSP, SALT LAKE CITY, 88- *Concurrent Pos:* Smith Kline & French fel hemat, 57-59; from co-prin investr to prin investr gastrointestinal tract in iron deficiency anemia NIH grants, 61-66, prin investr, copper metab in acute leukemia, 63-64; mem comn child nutrit, Food & Nutrit Bd, Nat Res Coun, 64-66; resident path, Univ Utah, 66-69; assoc prog dir, Children's Cancer Study Group A, NIH, Univ Utah, 70-77, path of record, Non-Hodgkins Lymphoma Study, 77-83; mem, Lymphoma Panel, NIH, 78-83. *Mem:* AMA; Am Soc Hemat; Am Fedn Clin Res; Am Soc Clin Path; Col Am Path; Path Soc Gt Brit & Ireland. *Res:* Iron and copper metabolism in iron deficiency; childhood malignancies; childhood non-Hodgkins lymphoma. *Mailing Add:* 1761 Countryside Dr Salt Lake City UT 84106

WILSON, JOHN H, b July 27, 44; US citizen; m 65; c 1. BIOCHEMISTRY, GENETICS. *Educ:* Wabash Col, AB, 66; Calif Inst Technol, PhD(biochem, genet), 72. *Prof Exp:* Asst prof, 73-79, assoc prof, 79-88, PROF BIOCHEM, BAYLOR COL MED, 88- *Concurrent Pos:* Damon Runyon fel biochem, Med Ctr, Stanford Univ, 71-73. *Res:* Genetic recombination in somatic cells. *Mailing Add:* Dept Biochem Baylor Col Med One Baylor Plaza Houston TX 77030

WILSON, JOHN HUMAN, b Wills Point, Tex, Feb 24, 00; m 24; c 2. EXPLORATION GEOPHYSICS. *Educ:* Colo Sch Mines, EM, 23. *Prof Exp:* Geologist, Midwest Ref Co, Colo, 23-26; geologist & geophysicist, Huasteca Petrol Co, Mex, 26-27; asst prof geophys, Colo Sch Mines, 28-29; consult geologist & geophysicist, 29-34; pres, Colo Geophys Corp, 34-37; vpres, Independent Explor Co, Piper Petrol Co, Woodson Oil Co & Wilson Explor Co, 38-61; pres, Piper Petrol Co & Wilson Explor Co, 61-88; RETIRED. *Concurrent Pos:* Explor consult, 50-88. *Mem:* Soc Explor Geophys (secy-treas, 36); Am Asn Petrol Geol; assoc Am Inst Mining, Metall & Petrol Eng. *Res:* Design of geophysical equipment; incipient metamorphism of sediments; seismic velocities; exploration techniques. *Mailing Add:* 1212 W El Paso St Ft Worth TX 76102

WILSON, JOHN NEVILLE, b Portland, Maine, June 13, 18; m 44; c 3. PHYSICS, RESOURCE MANAGEMENT. *Educ:* Rice Univ, BA, 40; Harvard Univ, AM, 41. *Prof Exp:* Res chemist, E I du Pont de Nemours & Co, Va, 41, res physicist, 42-43, tech specialist, Manhattan Dist, Del, 43-44, sr supvr, Wash, 44-45, res physicist, Va, 45-50, tech specialist, Del, 50-51, SC, 51-53, res supvr, Appl Physics Div, Savannah River Plant, 53, res mgr, 54-70, supt Planning & Anal Dept, Savannah River Plant, 70-80; RETIRED. *Concurrent Pos:* Asst physicist, Nat Defense Res Comt, Radio & Sound Lab, Univ Calif, 41-42; assoc physicist, Clinton Labs, Tenn, 44. *Mem:* Soc Rheology. *Res:* Rayon spinning, yarn structure and physical testing; health physics; oceanography; radiation and chemical process instrumentation; electronics; non-destructive testing; computer models for finance and control of large industrial plant and laboratory. *Mailing Add:* 1208 Abbeville Ave NW Aiken SC 29801-3155

WILSON, JOHN PHILLIPS, b Stamford, Conn, June 5, 16; m 40; c 1. MATHEMATICS. *Educ:* Univ Southern Miss, BA, 57; Johns Hopkins Univ, MEd, 60, MS, 70. *Prof Exp:* Res staff asst, Ballistic Anal Lab, Inst Coop Res, Johns Hopkins Univ, 57-62, res assoc, 62-67, res scientist, 67-69; sr res analyst, Thor Div, Falcon Res & Develop Co, Baltimore, 69-81; RETIRED. *Concurrent Pos:* Consult mil opers anal, 81-85. *Mem:* Am Defense Preparedness Asn. *Res:* Military operations analysis: target vulnerability, weapon lethality, weapons systems evaluation, terminal ballistic evaluation of large caliber weapons; geo and celestial navigation for surface vessels; merchant marine industry-shipboard operations. *Mailing Add:* 102 Kenilworth Park Dr Apt 3C Towson MD 21204-2262

WILSON, JOHN RANDALL, b Miami, Fla, June 12, 34; m 59; c 3. PHYSICAL CHEMISTRY. *Educ:* Univ Fla, BS, 56; Univ Wis, MS, 59, PhD(chem), 65. *Prof Exp:* Asst prof chem, Franklin Col, 59-62, Miami Univ, 64-67 & Asheville-Biltmore Col, 67-68; assoc prof, 68-72, chmn dept, 77-81, PROF CHEM, SHIPPENSBURG UNIV, 72-, CHMN DEPT, 85- *Mem:* Am Chem Soc. *Res:* Radiation chemistry; mechanisms of exchange reactions; photochemistry; scientific education in Latin America. *Mailing Add:* Dept Chem Shippensburg Univ Shippensburg PA 17257-2299

WILSON, JOHN SHERIDAN, b Morgantown, WVa, June 17, 44; m 65; c 4. COAL PROCESSES. *Educ:* WVa Univ, BS, 66, MS, 68, PhD(chem eng), 75. *Prof Exp:* Proj leader & engr, Bur Mines, US Dept Interior, 68-76; res supvr, Combustion Res & Develop Br, Res & Develop Admin, 76-77, asst dir, Energy Conversion & Utility Div, 77-79, DIR, COAL PROJ MGT DIV, MORGANTOWN ENERGY TECH CTR, US DEPT ENERGY, 79- *Concurrent Pos:* Proj mgr, Combustion Res & Develop Br, Res & Develop Admin, Dept Energy, 76. *Mem:* Am Chem Soc; Am Soc Mech Engrs; Sigma Xi. *Res:* Coal combustion and furnace analysis; design and operation of fluidized-bed coal combustion; coal gasification; environmental effluent control from coal conversion and utilization processes. *Mailing Add:* 1004 Grand St Morgantown WV 26505

WILSON, JOHN T, b Gainesville, Tex, Apr 27, 38; m 62; c 3. PEDIATRICS, PHARMACOLOGY. *Educ:* Tulane Univ La, BS, 60, MS & MD, 63. *Prof Exp:* From intern to resident clin pediat, Palo Alto-Stanford Med Ctr, Palo Alto, Calif, 63-65; res assoc biochem pharmacol, Univ Iowa, 65-66; res assoc biochem pharmacol & endocrinol, Nat Inst Child Health & Human Develop, Bethesda, Md, 66-68; attend pediatrician & dir lab perinatal med pharmacol, Children's Hosp, San Francisco, 69-70; assoc prof, Med Sch, Vanderbilt Univ, 70-77; PROF PEDIAT & PHARMACOL & CHIEF, SECT CLIN PHARMACOL, SCH MED, LA STATE UNIV, 78- *Concurrent Pos:* Fel neonatal med & dir lab develop pharmacol, Children's Hosp, San Francisco, 68-69; NIH res career develop award, 69 & 72; lectr, Med Ctr, Univ Calif, San Francisco, 69-70; res assoc, J F Kennedy Ctr, 70-; mem, World Health Orgn, Task Force Drugs Breast Milk. *Mem:* AAAS; Soc Pediat Res; Am Soc Pharmacol & Exp Therapeut; Am Soc Clin Pharmacol & Therapeut; Am Acad Pediat. *Res:* Pediatric clinical pharmacology, drug metabolism. *Mailing Add:* Dept Pharmacol PO Box 33932 Shreveport LA 71130-2125

WILSON, JOHN THOMAS, b Tucson, Ariz, Oct 30, 44; m 69. MOLECULAR BIOLOGY, BIOCHEMISTRY. *Educ:* Univ Ariz, BS, 69, PhD(genetics), 74. *Prof Exp:* Fel human genetics, Med Sch, Yale Univ, 74-76, res assoc, 76-78; ASSOC PROF CELL & MOLECULAR BIOL, MED COL GA, 78- *Concurrent Pos:* NIH fel, 75-77. *Mem:* Am Soc Microbiologists; AAAS; Am Chem Soc; Genetics Soc Am; Biophys Soc. *Res:* Structure and organization of human hemoglobin genes through molecular cloning. *Mailing Add:* 112 Harbor Pkwy Clinton CT 06413-2609

WILSON, JOHN THOMAS, JR, b Birmingham, Ala, June 2, 24; m 68. ENVIRONMENTAL MEDICINE, PREVENTIVE MEDICINE. *Educ:* Howard Univ, BS, 46; Columbia Univ, MD, 50; Univ Cincinnati, ScD(indust med), 56. *Prof Exp:* Physician, Div Indust Hyg, NY State Dept Labor, 55-56 & Sidney Hillman Health Ctr, New York, 56-57; chief bur occup health, Santa Clara County Health Dept, Calif, 57-61; life sci adv, Lockheed Aircraft Corp, 61-67, head biol sci res labs, Lockheed Missiles & Space Co, 67-69; asst prof community & prev med, Sch Med, Stanford Univ, 69-71; prof community health pract & chmn dept, Col Med, Howard Univ, 71-74; prof environ health & chmn dept, Sch Pub Health & Community Med Univ Wash, 74-80, prof & dir, 80-87; CONSULT, 87- *Concurrent Pos:* Nat Med Fel fel, 53-55; fel indust med, Univ Cincinnati, 53-56; lectr, Sch Pub Health, Univ Calif, Berkeley, 59-61. *Mem:* Fel Indust Med Asn; fel Am Col Physicians; Am Acad Occup Med; Am Indust Hyg Asn. *Res:* Occupational and environmental medicine; toxicology; industrial hygiene. *Mailing Add:* Dept Pharmacol & Pediat LSU Sch Med PO Box 33932 Shreveport LA 71130

WILSON, JOHN THOMAS, b El Paso, Tex, Apr 30, 47; m 81; c 4. GROUND WATER MICROBIOLOGY. *Educ:* Baylor Univ, BS, 69; Univ Calif, Berkeley, MA, 71; Cornell Univ, PhD(microbiol), 78. *Prof Exp:* Res assoc, agron dept, Cornell Univ, 78, IPA to RSKERL res, 78-79; MICROBIOLOGIST, RES LAB, R S KERR ENVIRON, US ENVIRON PROTECTION AGENCY, 79- *Concurrent Pos:* Adj assoc prof, Rice Univ, 81-83 & 83-; mem, hydrol sect, Am Geophys Union, 83-; adv, Water Sci & Technol Libr Reidel, 84-; assoc ed, Environ Toxicol Chem, J Indust Microbiol & J Contaminant Hydrol, 85- *Mem:* Am Soc Microbiol; Soc Indust Microbiol. *Res:* Biological processes that control behavior of organic contaminants in the subsurface environment, including ground water; activities that promote biological reclamation of polluted subsurface environments; hazardous waste microbiology. *Mailing Add:* Dept Pharmacol La State Univ Med Ctr PO Box 33932 Shreveport LA 71130

WILSON, JOHN TUZO, b Ottawa, Ont, Oct 24, 08; m 38; c 2. GEOPHYSICS, TECTONICS. *Educ:* Univ Toronto, BA, 30; Univ Cambridge, MA, 32, ScD, 58; Princeton Univ, PhD(geol), 36. *Hon Degrees:* LLD, Carleton Univ, 58, Simon Fraser Univ, 78; DSc, Univ Western Ont, 58, Acadia Univ, Mem Univ Newf, 68, McGill Univ, 74, Univ Toronto, 77, Laurentian Univ, 78, Middlebury Col, 81, Yale Univ, 82, McMaster Univ, 82, Royal Mil Col, 83, Lakehead Univ, 84; ScD, Franklin & Marshall Col, 69; DUniv, Univ Calgary, 74. *Prof Exp:* Asst geologist, Geol Surv Can, 36-39; prof geophys, Univ Toronto, 46-74; prin, Erindale Col, 67-74; dir-gen, Ont Sci Ctr, 74-85; RETIRED. *Concurrent Pos:* Vis prof, Australian Nat Univ, 50 & 65; pres, Int Union Geol & Geophys, 57-60; mem, Nat Res Coun Can, 58-64; mem, Defence Res Bd Can, 60-66; distinguished lectr, Univ Toronto, 74-77. *Honors & Awards:* Miller Medal, Royal Soc Can, 56; Blaylock Medal, Can Inst Mining & Metal, 59; Bucher Medal, Am Geophys Union, 68; Penrose Medal, Geol Soc Am, 69; Companion Award, Order Can, 74; J J Carty Medal, Nat Acad Sci, 75; Wollaston Medal, Geol Soc London, 78; M Ewing Medal, Am Geophys Union, 80 & Soc Explor Geophysics, 81; A G Huntsman Award, Bedford Inst Oceanog, 81. *Mem:* Foreign assoc Nat Acad Sci; fel Geol Soc Am; fel Royal Soc Can (pres, 72-73); fel Royal Soc; foreign mem Am Philos Soc; foreign mem Royal Swedish Acad Sci. *Res:* Physics of the earth; continental structure. *Mailing Add:* 27 Pricefield Rd Toronto ON M4W 1Z8 Can

WILSON, JOHN WILLIAM, b Arkansas City, Kans, Aug 6, 40; m 62; c 1. THEORETICAL HIGH ENERGY PHYSICS, HEALTH PHYSICS. *Educ:* Kans State Univ, BS, 62; Col William & Mary, MS, 69, PhD(physics), 75. *Prof Exp:* Aerospace technologist simulation, 63-70, space scientist space physics, 70-76, SR RES SCIENTIST ENERGY SYSTS, LANGLEY RES CTR, NASA, 76- *Concurrent Pos:* Adj asst prof physics, Old Dom Univ, 75-80, adj assoc prof, 80-85, adj prof physics, 85- *Mem:* NY Acad Sci. *Res:* High-energy heavy reaction theory; high-energy transport theory; health physics aspects of high-altitude aircraft and space operations and dosimetry; nuclear induced plasmas and radiolysis; nuclear pumped and electral pumped laser kinetics; solar pumped laser kinetics. *Mailing Add:* NASA Langley Res Ctr Mail Stop 493 Hampton VA 23665

WILSON, JOHN WILLIAM, III, b New York, NY, May 10, 43; m 66. ZOOGEOGRAPHY, PALEONTOLOGY. *Educ:* Amherst Col, BA, 66; Univ Chicago, PhD(evolutionary biol), 72. *Prof Exp:* Asst prof, 72-78, ASSOC PROF BIOL, GEORGE MASON UNIV, 78- *Mem:* AAAS; Soc Study Evolution; Soc Vert Paleont; Am Soc Mammal; Ecol Soc Am. *Res:* Zoogeography of mammals; latitudinal gradients; paleoecology of mammals; changes in resource utilization of mammals during late Cretaceous and Cenozoic; Pleistocene extinctions. *Mailing Add:* Dept Biol George Mason Univ Fairfax VA 22030

WILSON, JOSEPH EDWARD, b Hannibal, Mo, Jan 8, 20; m 45; c 4. POLYMER CHEMISTRY. *Educ:* Univ Chicago, BS, 39; Univ Rochester, PhD(phys chem), 42. *Prof Exp:* Res chemist, Goodyear Aircraft Corp, 42-46, Argonne Nat Lab, 46-47 & Firestone Tire & Rubber Co, 47-50; sr chemist, Bakelite Co Div, Union Carbide & Carbon Corp, 50-57 & J T Baker Chem Co, 57; develop supvr, Atlas Powder Co, 58-61; proj mgr, Kordite Co, 61-64; res dir, Pollock Paper Div, St Regis Paper Co, 64-67; from assoc prof to prof phys chem, Bishop Col, 67-87; ADJ PROF PHYSIOL, TEX COL OSTEOP MED, 87- *Concurrent Pos:* Plastics consult, 67- *Mem:* Am Chem Soc; Soc Plastics Eng. *Res:* Photochemistry; polymerization; stability of polymers; radiation chemistry of plastics; synthesis of blood-compatible plastics for use in artificial organs. *Mailing Add:* Dept Physiol Tex Col Osteop Med Ft Worth TX 76107

WILSON, JOSEPH WILLIAM, b Massena, NY, Apr 11, 34; m 73; c 3. ORGANIC CHEMISTRY. *Educ:* Mass Inst Technol, BS, 56; Ind Univ, PhD(chem), 61. *Prof Exp:* Res assoc, Univ Wis, 61-63; asst prof, 63-70, ASSOC PROF CHEM, UNIV KY, 70- *Mem:* Am Chem Soc. *Res:* Organic photochemistry. *Mailing Add:* Dept Chem Univ KY Lexington KY 40506

WILSON, KARL A, b Buffalo, NY, Jan 19, 47; m 74; c 1. BIOCHEMISTRY. *Educ:* State Univ NY Buffalo, BA, 69, PhD(biochem), 73. *Prof Exp:* Res assoc biochem, Roswell Park Mem Inst, 73-74; res assoc biochem, Purdue Univ, West Lafayette, 74-76; asst prof & assoc fel, Ctr Biochem Res, State Univ NY, 76-80, asst prof & assoc fel, Semantic Cell Genetics & Biochem, 80-83, co-dir biochem prog, 84-88, ASSOC PROF BIOCHEM, STATE UNIV NY, BINGHAMTON, 83-, DIR BIOCHEM PROG, 89- *Concurrent Pos:* NSF grad fel, 69-72. *Mem:* AAAS; Sigma Xi; Am Soc Plant Physiol; Am Chem Soc. *Res:* Mechanism, molecular biology and physiology of proteases and their protein inhibitors; molecular evolution of proteins and protein sequencing; physiology of seed germination; nitrogen metabolism in plants. *Mailing Add:* Dept Biol Sci State Univ NY PO Box 6000 Binghamton NY 13902-6000

WILSON, KATHERINE WOODS, b Los Angeles, Calif, Feb 8, 23. AIR POLLUTION. *Educ:* Univ Calif, BS, 44, MS, 45; Univ Calif, Los Angeles, PhD(chem), 48. *Prof Exp:* Asst prof chem, WVa Univ, 48-53 & Pepperdine Col, 53-54; mem staff, Los Angeles County Air Pollution Control Dist, 54-55; lectr chem & assoc res chemist, Univ Calif, Los Angeles, 55-62; phys chemist, Stanford Res Inst, 62-71; staff officer, Univ Calif, Riverside, 71-72; supv air pollution chemist, Air Pollution Control Serv, San Diego County, 73-74; dir air qual studies, Copley Int Corp, 74-76; dir, Chem & Meteorolgy Dept, Environ Serv, 76-80; consult, 80-88; RETIRED. *Mem:* Am Chem Soc; Air Pollution Control Asn. *Res:* Air pollution; cotton chemistry; chemical analysis of agricultural products. *Mailing Add:* 1011 Hayes Ave San Diego CA 92103-2308

WILSON, KATHRYN JAY, b Virginia, Minn, June 21, 48; m 69. PLANT CELL BIOLOGY, PLANT MORPHOLOGY. *Educ:* Univ Wis-Madison, BA, 71; Ind Univ, Bloomington, MA, 76, PhD(plant sci), 76. *Prof Exp:* asst prof, 76-82, ASSOC PROF BIOL, IND UNIV-PURDUE UNIV, INDIANAPOLIS, 82- *Concurrent Pos:* Res grants, Ind Univ-Purdue Univ, NSF, Whitehall. *Mem:* AAAS; Am Soc Plant Physiologists; Am Bot Soc; Sigma Xi. *Res:* Embryogenesis in plant tissue culture; development of non-articulated laticifer system of the Ascelpiadaceae, with special emphasis on cytodifferentiation of the laticifer as a unique cell type in whole plants and tissue culture. *Mailing Add:* Ind Purdue Univ 723 W Michigan St Indianapolis IN 46202-5132

WILSON, KENNETH ALLEN, b Rio de Janeiro, Brazil, Apr 15, 28; US citizen. PLANT MORPHOLOGY, SYSTEMATIC BOTANY. *Educ:* Miami Univ, BA, 51; Univ Hawaii, MS, 53; Univ Mich, PhD(bot), 58. *Prof Exp:* Botanist, Gray Herbarium & Arnold Arboretum, Harvard Univ, 57-60; from asst prof to assoc prof, 60-67, assoc dean, Sch Letters & Sci, 66-73, PROF BOT, CALIF STATE UNIV, NORTHRIDGE, 67- *Concurrent Pos:* Res assoc, Natural Hist Mus Los Angeles County. *Mem:* Bot Soc Am; Am Soc Plant Taxon; Am Fern Soc; Int Asn Plant Taxon. *Res:* Taxonomy; pteridophytes. *Mailing Add:* Dept Biol Calif State Univ Northridge CA 91330

WILSON, KENNETH CHARLES, b Vancouver, BC, Feb 9, 37; m 62; c 2. CIVIL ENGINEERING, FLUID MECHANICS. *Educ:* Univ BC, BASc, 59; Univ London, MSc & DIC, 61; Queen's Univ, Ont, PhD(civil eng), 65. *Prof Exp:* Hydraul engr, Ingledow, Kidd & Assocs, BC, 61-63 & Int Power & Eng, 65-66; consult hydraul engr, Dept External Affairs, Govt Can, 66-68; sr hydraul engr, T Ingledow & Assocs, Consult Engrs, 68-70; assoc prof, 71-80, PROF CIVIL ENG, QUEEN'S UNIV, ONT, 80- *Mem:* Eng Inst Can. *Res:* Sediment transport in rivers, canals and pipelines; blockage, plug flow and sliding beds in pipelines. *Mailing Add:* Dept Civil Eng Queen's Univ Kingston ON K7L 3N6 Can

WILSON, KENNETH GEDDES, b Waltham, Mass, June 8, 36. PHYSICS. *Educ:* Harvard Univ, BS, 56; Calif Inst Technol, PhD(physics), 61. *Prof Exp:* Jr fel, Harvard Univ, 59-62; Ford Found fel, Europ Orgn Nuclear Res, Geneva, Switz, 62-63; from asst prof to prof physics, Cornell Univ, 63-74, James A Weeks prof, 74-88, dir, Ctr Theory & Simulation in Sci & Eng, 85-88; HAZEL C YOUNGBERG TRUSTEES DISTINGUISHED PROF, OHIO STATE UNIV, 88- *Concurrent Pos:* Mem staff, Stanford Linear Accelerator Ctr, 69-70; mem, Comt Phys Sci, Math & Appln, Nat Acad Sci, 90-, Comt Fed Role in Educ Res, 90-; co-prin investr syst change, NSF. *Honors & Awards:* Nobel Prize for Physics, 82; Dannie Heinemann Prize, 73; Boltzmann Medal, 75; Wolf Prize, 80; Franklin Medal, 82; A C Eringen Medal, 84. *Mem:* Nat Acad Sci; Am Phys Soc; Am Philos Soc; Am Acad Arts & Sci. *Res:* Elementary particle theory. *Mailing Add:* Dept Physics Ohio State Univ 174 W 18th Ave Columbus OH 43210

WILSON, KENNETH GLADE, b Payson, Utah, May 18, 40; m 59; c 3. BOTANY. *Educ:* Univ Utah, BS, 62, PhD(molecular biol), 68. *Prof Exp:* Reliability engr, Hercules Powder, 62-63; PROF BOT, MIAMI UNIV, 67- *Concurrent Pos:* Pres, Modular Genes Miami. *Mem:* Int Soc Plant Molecular Biol; Am Soc Plant Physiologists; Sigma Xi; Tissue Cult Asn. *Res:* The molecular biology of organelles with special interest in the analysis of chloroplast mutations and function of the genes controlling mitochondrial cytochrome oxidase function; investigating the cpDNA of the genus Hosta which is well recognized for its mutant chloroplasts; the sequence of the cox II gene is being studied in carrot and related dicots; tissue culture is being used to produce mutant plants for the plastid investigations and as a source of biologicals; production of specialty compounds is limited to members of the family Apocynaceae. *Mailing Add:* Dept Bot Miami Univ Oxford OH 45056

WILSON, KENNETH SHERIDAN, b Waterloo, Iowa, May 1, 24; m 48, 62; c 1. MYCOLOGY, PLANT PATHOLOGY. *Educ:* Colo Col, BS, 49; Univ Wyo, MS, 50; Purdue Univ, PhD(mycol), 54. *Prof Exp:* Asst, Colo Col, 46-49 & Univ Wyo, 49-50; asst, 52-53, from instr to assoc prof, 54-71, PROF BIOL SCI, PURDUE UNIV, CALUMET CAMPUS, 71- *Mem:* AAAS; Mycol Soc Am; Bot Soc Am; Soc Indust Microbiol; Am Soc Microbiol; Sigma Xi. *Res:* Mycological taxonomy; plant taxonomy; plant morphology; microbiological ecology. *Mailing Add:* 189 W Horn Rd Valparaiso IN 46383

WILSON, KENT RAYMOND, b Philadelphia, Pa, Jan 14, 37; m 67; c 2. CHEMICAL PHYSICS. *Educ:* Harvard Col, AB, 58; Univ Strasbourg, dipl, 59; Univ Calif, Berkeley, PhD(chem), 64. *Prof Exp:* Res fel chem, Harvard Univ, 64-65; res chemist, Nat Bur Stand, 65; asst prof phys chem, 65-71, assoc prof, 71-77, PROF PHYS CHEM, UNIV CALIF, SAN DIEGO, 77- *Concurrent Pos:* Sloan res fel, 70-72; mem comt comput in chem, Div Chem & Chem Technol, Nat Res Coun, 70-72; mem comput & biomath sci study sect, NIH, 71-74; mem panel photochem oxidants, ozone & hydrocarbons, Nat Res Coun, 73-74. *Mem:* Am Phys Soc; Am Chem Soc. *Res:* Molecular dynamics of chemical reactions, particularly in solution; specialized computer systems for solution of scientific problems; computer animation; archaeological chemistry. *Mailing Add:* Dept Chem 0339 Univ Calif San Diego 9500 Gilman Dr La Jolla CA 92093-0339

WILSON, L BRITT, b Amarillo, Tex, Sept 10, 60; m; c 1. NEURAL CONTROL OF THE CIRCULATION, SPINAL CORD NEUROTRANSMITTERS. *Educ:* WTex State Univ, BS, 83; La State Univ Med Ctr, PhD(physiol), 88. *Prof Exp:* Grad asst physiol, La State Univ Med Ctr, 84-88; postdoctoral fel, 88-91, ASST INSTR PHYSIOL, UNIV TEX SOUTHWESTERN MED CTR, 91- *Mem:* Am Physiol Soc; Soc Neurosci; Sigma Xi. *Res:* Neural control of the circulation, with emphasis on spinal cord neurochemistry. *Mailing Add:* Dept Physiol Univ Tex Southwestern Med Ctr 5323 Harry Hines Blvd Dallas TX 75235-9040

WILSON, L KENNETH, b Sacramento, Calif, Sept 22, 10; m 35; c 1. EXPLORATION GEOLOGY. *Educ:* Stanford Univ, AB, 32. *Prof Exp:* Geologist gold mining, Mother Lode Mining Dist, Calif, 32-35; geologist, Calumet & Hecla Copper Co, 35-38; mgr, Auburn Chicago Mine, Calif, 38-39; geologist, Cord Mining Interests, 39-43 & Am Smelting, 43-60; CONSULT GEOLOGIST, 60- *Concurrent Pos:* Gov app, Western Gov Mining Adv Coun, 67- *Mem:* Fel Geol Soc Am; Am Inst Prof Geologists; Am Inst Mining, Metall & Petrol Engrs; Am Asn Petrol Geologists. *Res:* Exploration geology, domestic mineral resources; assistance to legal counsel in mining law, litigation, utilization and acquisition of mineral property. *Mailing Add:* PO Box 7123 Menlo Park CA 94026

WILSON, LARRY EUGENE, b Wapakoneta, Ohio, Nov 17, 35; m 58; c 2. ANALYTICAL CHEMISTRY. *Educ:* Ohio State Univ, BSc, 57, PhD(analytical chem), 62. *Prof Exp:* Analytical chemist, Dow Chem Co, Mich, 63-64; asst prof chem, Mich State Univ, 64-65; analytical chemist, Dow Chem Co, Mich, 65-66, supvr control lab, 66-67, coordr lab technician training, 67-69; asst prof, Lancaster Br, Ohio Univ, 69-72, assoc prof chem, 72-; MEM STAFF, DEPT BIOL, KINGS COL. *Mem:* Am Chem Soc. *Res:* Acid-base equilibria; methods of teaching. *Mailing Add:* Ohio Univ Lancaster OH 43130-1097

WILSON, LAUREN R, b Yates Center, Kans, May 4, 36; m 59; c 2. INORGANIC CHEMISTRY, ENVIRONMENTAL CHEMISTRY. *Educ:* Baker Univ, BS, 58; Univ Kans, PhD(inorg chem), 63. *Prof Exp:* Asst prof chem, Ohio Wesleyan Univ, 63-70, chmn dept, 70-77, dean acad affairs, 77-85, actg provost, 85-86, exec asst to pres, 86-87, prof chem, 70-87; PROF CHEM & VCHANCELLOR ACAD AFFAIRS, UNIV NC ASHEVILLE, 87- *Concurrent Pos:* Mem staff, Oak Ridge Nat Lab, 72-73; vis prof, Ohio State Univ, 68 & 76-77. *Mem:* Am Chem Soc; Royal Soc Chem; AAAS; Coun Undergrad Res; Am Asn Higher Educ. *Res:* Synthesis of transition metal compounds; electrocatalysis of chemically modified electrodes; metal ions in natural and biological systems. *Mailing Add:* 22 Maywood Rd Asheville NC 28804

WILSON, LAURENCE EDWARD, b Aberdeen SDak, June 29, 30; m 57; c 3. INORGANIC CHEMISTRY. *Educ:* Western Wash Col Educ, BA, 52; Univ Wash, PhD(chem), 57. *Prof Exp:* Instr chem, Amherst Col, 56-59; from asst prof to assoc prof, San Jose State Col, 59-63; assoc prof, 63-80, PROF CHEM, KALAMAZOO COL, 80-, CHMN DEPT, 63- *Mem:* AAAS; Am Chem Soc. *Mailing Add:* Dept Chem Kalamazoo Col Kalamazoo MI 49007-3295

WILSON, LAWRENCE ALBERT, JR, b Mt Hope, WVa, Mar 19, 25; m 49; c 3. CHEMICAL & PETROLEUM ENGINEERING. *Educ:* Purdue Univ, BS, 49, PhD(chem eng), 52. *Prof Exp:* Engr, Staple Develop Plant, Am Viscose Corp, 52-53, develop supvr, 53-54; res engr, Gulf Res & Develop Co, 54-56, group leader petrol res, 56-62, sr res engr, 62-66, sect supvr, 66-72, staff engr, 72-76, mgr, prod res dept, 76-81, coordr processing appln, 81-85; RETIRED. *Concurrent Pos:* Lectr, Univ Pittsburgh, 64-81. *Mem:* Am Chem Soc; Am Inst Mining, Metall & Petrol Engrs. *Res:* Petroleum production. *Mailing Add:* 118 Woodshire Rd Greenville PA 16125

WILSON, LEE, b Wichita Falls, Tex, Apr 15, 42; m 69; c 1. ENVIRONMENTAL RESOURCES PROTECTION. *Educ:* Yale Univ, BA, 64; Columbia Univ, PhD(geol), 71. *Prof Exp:* PRES, LESS WILSON & ASSOCS INC, 73- *Mem:* Geol Soc Am; Am Geophys Union; Am Water Resources Asn. *Res:* Applied research related to development and protection of water and other environmental resources. *Mailing Add:* PO Box 931 Sante Fe NM 87504

WILSON, LENNOX NORWOOD, b Quebec City, Que, Feb 15, 32; US citizen; m 58; c 3. FLUID DYNAMICS, ACOUSTICS. *Educ:* Univ Toronto, BASc, 53, MASc, 54, PhD(aerophys), 59. *Prof Exp:* Res asst, Inst Aerophys, Univ Toronto, 53-59; res engr, Armour Res Found, Ill, 59-62; head aerochem, Defense Res Labs, Gen Motors Corp, Calif, 62-66; mgr fluid dynamics & acoust, IIT Res Inst, Ill, 66-71; prof mech & aerospace eng, Univ Mo-Columbia, 71-73; actg head aerospace eng, 78-79, chmn, 81-87, PROF AEROSPACE ENG, IOWA STATE UNIV, 73- *Mem:* Assoc fel Am Inst Aeronaut & Astronaut. *Res:* Turbulence; combustion; shock tubes; chemical kinetics; aerodynamic noise; noise control; engineering acoustics; aerodynamics. *Mailing Add:* Dept Aerospace Eng Iowa State Univ Ames IA 50010

WILSON, LEONARD GILCHRIST, b Orillia, Ont, June 11, 28; m 69; c 1. HISTORY OF MEDICINE. *Educ:* Univ Toronto, BA, 49; Univ London, MSc, 55; Univ Wis, PhD(hist sci), 58. *Prof Exp:* Lectr biol, Mt Allison Univ, 50-53; vis instr hist sci, Univ Calif, 58-59; vis asst prof, Cornell Univ, 59-60; from asst prof to assoc prof hist med, Sch Med, Yale Univ, 60-67; PROF HIST MED & HEAD DEPT, UNIV MINN, MINNEAPOLIS, 67- *Concurrent Pos:* Mem, US Nat Comt Hist Geol, 85-88. *Mem:* AAAS; Am Asn Hist Med; Am Hist Asn; Hist Sci Soc; Int Acad Hist Sci. *Res:* History of biology; history of physiology in the seventeenth century; history of fever; history of geology. *Mailing Add:* Dept Hist Med Univ Minn Minneapolis MN 55455

WILSON, LEONARD RICHARD, b Superior, Wis, July 23, 06; m 30; c 2. GEOLOGY, PALYNOLOGY. *Educ:* Univ Wis, PhB, 30, PhM, 32, PhD(bot), 36. *Prof Exp:* Res assoc asst, Wis Geol & Natural Hist Surv, 31-35; from instr to prof geol & bot, Coe Col, 35-46; prof geol & head dept geol & mineral, Univ Mass, 46-56; prof geol, NY Univ, 56-57; prof, 57-62, res prof, 62-67, George Lynn Gross res prof geol & geophysics, 68-78, EMER PROF GEOL & GEOPHYS, UNIV OKLA, 68-; CUR MICROPALEONT & PALEOBOT, OKLA MUS NATURAL HIST, 71- *Concurrent Pos:* Melhaup scholar, Ohio State Univ, 39-40; consult, Carter Oil Co, 46-56; leader, Am Geog Soc Greenland Ice Cap Exped, 53; res assoc, Am Mus Natural Hist, 57-78; geologist, Okla Geol Surv, 57-78, emer geologist, 78-; adj prof, Univ Tulsa; consult, Jersey Prod Res Corp, 56-62, Humble Oil Co, 63-64 & Sinclair Oil Co, 63-69; consult geologist, 78- *Honors & Awards:* VI Gunnar Erdtman Int Medal, Palynology Soc India, 73. *Mem:* Fel Geol Soc Am; Am Bot Soc; Am Asn Petrol Geol; hon mem Nat Asn Geol Teachers; hon mem Am Asn Stratig Palynologists. *Res:* Stratigraphic and paleoecologic palynology. *Mailing Add:* Okla Mus Natural Hist Univ Okla Norman OK 73069

WILSON, LESLIE, b Boston, Mass, June 29, 41; m 89; c 2. PHARMACOLOGY, BIOCHEMISTRY. *Educ:* Mass Col Pharm, BS, 63; Tufts Univ, PhD(pharmacol), 67. *Prof Exp:* Asst prof pharmacol, Sch Med, Stanford Univ, 69-75; from assoc prof to prof biochem, Univ Calif, Santa Barbara, 75-91, chair, dept biol sci, 87-91. *Concurrent Pos:* USPHS fel, Univ Calif, Berkeley, 67-69; Nat Inst Neurol Dis & Stroke res grant, 70-; mem, Molecular Biol Study Sect, NIH, 78-82; mem, Am Cancer Soc Cell & Develop Biol Sci Adv Comt, 84-, res grant, 87- *Mem:* AAAS; Am Soc Pharmacol & Exp Therapeut; Am Soc Cell Biol; Am Soc Biochem & Molecular Biol. *Res:* Mechanism and regulation of microtubule assembly and function; mechanism of action of antimitotic chemical agents. *Mailing Add:* Dept Biol Sci Univ Calif Santa Barbara CA 93106

WILSON, LESTER A, JR, b Charleston, SC, Apr 30, 17; m 45; c 4. MEDICINE. *Educ:* Col William & Mary, BS, 38; Med Col SC, MD, 42. *Prof Exp:* From asst prof to assoc prof, 51-65, PROF OBSTET & GYNEC, SCH MED, UNIV VA, 65- *Mailing Add:* 1612 Keith Valley Rd Charlottesville VA 22901

WILSON, LINDA S (WHATLEY), b Washington, DC, Nov 10, 36; m 57, 70; c 2. CHEMISTRY, RESEARCH ADMINISTRATION. *Educ:* Tulane Univ, BA, 57; Univ Wis, PhD(inorg chem), 62. *Prof Exp:* Res assoc inorg chem, Univ Wis, 62; Nat Inst Dent Res trainee phys chem & res assoc molecular spectros, Univ Md, 62-64, res asst prof, Molecular Spectros, 64-67; vis res fel, Univ Southampton, 67; vis asst prof, Univ Mo-St Louis, 67-68; asst to vchancellor res, Wash Univ, 68-69, asst vchancellor res, 69-74, assoc vchancellor res, 74-75; assoc vchancellor res, Univ Ill, Urbana, 75-85, assoc dean, Grad Col, 78-85; vpres res, Univ Mich, 85-89; PRES, RADCLIFFE COL, 89- *Concurrent Pos:* Mem, Gen Res Support Adv Comt, NIH, 71-75, chmn, 74-75; mem, Comt Govt Relations, Nat Asn Cols & Univ Bus Officers, 71-77, chmn, 73-75; co-chmn, Panel Eval Div Res Resources, NIH, 75-76; mem, Procurement Policy Adv Comt, Energy Res & Develop Admin, 76-77, Bd-Coun Comt Chem & Pub Affairs, Am Chem Soc, 78-82, Nat Adv Coun Res Resources, NIH, 78-82 & Nat Comn Res, 78-80; mem dirs adv coun, NSF, 80-; mem comt gov & asn affairs, Coun Grad Schs, 81-84; mem comt govt univ rel in support of sci, Nat Acad Sci, 81-83; mem, task force meeting res costs, NIH, 81-83 & adv bd, Nat Coalition Sci & Technol, 83-87; mem bd dirs, Asn Biomed Res, 83-86 & AAAS, 84-88; vis fel, Sci Policy Res Unit, Univ Sussex, Eng, 84; acad vis, London Sch Econ & Polit Sci, 84; coun mem, govt-univ-indust res round table, Nat Acad Sci, 84-; mem, Coun, Inst Med, 86-; mem, Energy Res Adv Bd, DOE, 87-; mem bd dir, Mich Biotechnol Inst, 86-; mem bd dir, Mich Mat Processing Inst, 85-; mem, Task Force 1990's, NSF, 87-88, Comt Govt-Indust Res Collab Biomed Res & Educ, 88-89, Ad Hoc Comt NIH Directorship, 89-90 & Adv Comt Educ & Human Resources, NSF, 90-92; chair, Off Sci & Eng Personnel, Nat Res Coun, 90-93. *Mem:* Inst Med-Nat Acad Sci; fel AAAS; Nat Coun Res Adminr; Soc Res Adminr; Am Chem Soc. *Res:* Molecular spectroscopy; spectroscopic studies of molecular interactions; charge transfer complexes; coordination compounds and hydrogen bonded species; optical studies at high pressures; science policy; research policy; author of numerous technical publications. *Mailing Add:* Radcliffe Col Ten Garden St Cambridge MA 02138

WILSON, LON JAMES, b Mojave, Calif, Sept 4, 44. INORGANIC CHEMISTRY, BIOINORGANIC CHEMISTRY. *Educ:* Iowa State Univ, BS, 66; Univ Wash, PhD(inorg chem), 71. *Prof Exp:* Teaching asst chem, Univ Wash, 66-68; vis asst prof, Univ Ill, 71-73; from asst prof to assoc prof, 73-84, PROF CHEM, RICE UNIV, 85- *Concurrent Pos:* NIH fel, Univ Ill, 71-73. *Mem:* Chem Soc; Am Chem Soc. *Res:* Magnetic and redox properties of transition metal compounds; Mossbauer spectroscopy; iron, copper and nickel containing metalloproteins and their synthetic analogs; lanthanide coordination chemistry. *Mailing Add:* Dept Chem Rice Univ PO Box 1892 Houston TX 77251

WILSON, LORENZO GEORGE, b Appleton, NY, July 25, 38; m 62; c 3. HORTICULTURE, VEGETABLE CROPS. *Educ:* Cornell Univ, BS, 61; Wash State Univ, MS, 64; Mich State Univ, PhD(hort), 69. *Prof Exp:* Res assoc postharvest physiol, United Fruit Co, 63-66; postharvest physiologist, United Fruit Co, 69-75; PROF HORT & EXTEN HORT SPECIALIST VEG CROPS, HORT DEPT, NC STATE UNIV, 75- *Concurrent Pos:* Consult, banana and root crop prod, & postharvest produce handling, Cent & SAm; Gov's Task Force, Farm Econ; mem, Sci & Technol Comt, United Fresh Fruit & Veg Asn; Cong Sci fel, Am Soc Hort Sci, Wash, DC, 90-91. *Mem:* Am Soc Hort Sci; Sigma Xi; Potato Asn Am. *Res:* Investigations to determine optimum cultural practices for potato and sweet potato production in North Carolina, including the use of fertilizers, pesticides and harvesting, handling and storage techniques for enhanced quality maintenance; bananas. *Mailing Add:* Dept Hort Sci NC State Univ Raleigh NC 27695-7609

WILSON, LORNE GRAHAM, b Saskatoon, Sask, Oct 23, 29; US citizen; m 57; c 2. SOIL PHYSICS, HYDROLOGY. *Educ:* Univ BC, BS, 51; Univ Calif, MS, 57, PhD(soil sci), 62. *Prof Exp:* Asst specialist irrig drainage, Univ Calif, 56-58; from asst hydrologist, to assoc hydrologist, 62-67, HYDROLOGIST, WATER RESOURCES RES CTR, UNIV ARIZ, 67- *Res:* Survey of drainage problems in San Joaquin Valley, California; simultaneous flow of air and water during infiltration in soils; subsurface flow characteristics during natural and artificial recharge in stratified sediments. *Mailing Add:* Dept Hydrol Univ Ariz Tucson AZ 85721

WILSON, LOUIS FREDERICK, b Milwaukee, Wis, Nov 22, 32; m 56; c 4. ENTOMOLOGY. *Educ:* Marquette Univ, BS, 55, MS, 57; Univ Minn, PhD(entom), 62. *Prof Exp:* Instr cytol & parasitol, Marquette Univ, 54-57; state entomologist, Minn, 58; from asst prof to assoc prof forestry & entom, Mich State Univ, 67-85; PRIN INSECT ECOLOGIST, N CENT FOREST EXP STA, US FOREST SERV, 58-; PROF FORESTRY & ENTOM, MICH STATE UNIV, 85- *Mem:* Entom Soc Am; Entom Soc Can. *Res:* Insect ecology and behavior; population dynamics; insect impact. *Mailing Add:* Mich State Univ 126 Nat Resources Forestry East Lansing MI 48824

WILSON, LOWELL D, b Pampa, Tex, May 11, 33; m 60; c 2. ENDOCRINOLOGY, BIOLOGICAL CHEMISTRY. *Educ:* Johns Educ: Univ Calif, Berkeley, AB, 55; Univ Chicago, MD, 60; Univ Southern Calif, PhD(biochem), 68. *Prof Exp:* From instr to asst prof med, Sch Med, Univ Southern Calif, 66-68; from asst prof to assoc prof, 68-77, PROF MED & BIOL CHEM, SCH MED, UNIV CALIF, DAVIS, 77- *Mem:* Endocrine Soc; Am Fedn Clin Res; Am Chem Soc; Am Soc Biol Chem. *Res:* Biochemistry; metabolic control processes; hormone action. *Mailing Add:* Dept Internal Med & Biol Chem Univ Calif Sch Med Davis CA 95616

WILSON, LOWELL L, b Egan, Ill, Jan 3, 36; m 55; c 3. ANIMAL SCIENCES. *Educ:* Wis State Univ, BS, 60; SDak State Univ, MS, 62, PhD(animal sci), 64. *Prof Exp:* Res asst animal genetics, SDak State Univ, 60-64; livestock specialist, Purdue Univ, 64-66; assoc prof animal prod, 66-71, PROF ANIMAL SCI, PA STATE UNIV, UNIVERSITY PARK, 71- *Concurrent Pos:* Consult var firms US & Foreign. *Honors & Awards:* Meat Animal Mgt Award, Am Soc Animal Sci, 73. *Mem:* AAAS; Am Genetic Asn; Am Soc Animal Sci; Am Meat Sci Asn. *Res:* Selection indices for beef cattle and estimations of genetic parameters from use of selected sires; ranch x sire and sex x sire interactions; beef cattle and sheep behavior; forage utilization with ruminants; recycling of waste materials through ruminants. *Mailing Add:* Dept Animal Sci Pa State Univ Univ Park PA 16802

WILSON, LYNN O, b Wilmington, Del, July 9, 44. APPLIED MATHEMATICS. *Educ:* Oberlin Col, AB, 65; Univ Wis, PhD(appl math), 70. *Prof Exp:* DISTINGUISHED MEM TECH STAFF, BELL LABS, 70- *Mem:* Am Phys Soc; Soc Indust & Appl Math; Sigma Xi; Inst Mgt Sci. *Res:* Mathematical physics; marketing science. *Mailing Add:* AT&T Bell Labs Rm 7C-517 Murray Hill NJ 07974

WILSON, MABEL F, b Omaha, Nebr, Mar 17, 06; m 29; c 2. ANALYTICAL CHEMISTRY. *Educ:* Buena Vista Col, BS, 27; Mich State Col, MS, 30, PhD(phys chem), 37. *Prof Exp:* Teacher high schs, Iowa, 28-29; lab asst, Mich State Col, 35-36; spec analyst, Burgess Labs, 37-38; spectroscopist, Diamond Alkali Co. 38-44, sr res chemist, 44-52; sr spectroscopist, Res Lab, Air Reduction Co, Inc, 52-56, head instrumental analysis sect, 56-61; group leader plastics div res, Allied Chem Corp, 61-67, res assoc, 67-70; forensic spectroscopist, Lab State Med Examr, Newark, NJ, 70-76; RETIRED. *Mem:* Am Chem Soc; emer mem Soc Appl Spectros; fel Am Inst Chem; fel Am Soc Testing & Mat. *Res:* Trace metals in nonmetallic materials, particularly body tissues and fluids; plastics by means of instrumental analysis including optical and x-ray spectroscopy, x-ray diffraction, atomic, infrared and ultraviolet absorption. *Mailing Add:* c/o Glenside Nursing Ctr 144 Gales Dr New Providence NJ 07974-1926

WILSON, MCCLURE, b Ogden, Utah, July 30, 24; m 50; c 2. RADIOLOGY. *Educ:* Univ Ark, BS, 47, MD, 48. *Prof Exp:* From asst prof to assoc prof radiol, Univ Tex Med Br, 55-63; radiologist, Scott & White Clin, Temple, Tex, 63-64; assoc prof radiol, 64-69, PROF RADIOL, UNIV TEX MED BR GALVESTON, 69- *Mem:* AMA; Radiol Soc NAm; Am Roentgen Ray Soc. *Mailing Add:* Dept Radiol Univ Tex Med Br Galveston TX 77550

WILSON, MARCIA HAMMERQUIST, b Pierre, SDak, Apr 16, 52; m 72. ENDANGERED SPECIES, ORNITHOLOGY. *Educ:* SDak State Univ, BS, 74; Ore State Univ, MS, 78, PhD(wildlife ecol), 84. *Prof Exp:* Lectr avian biol, Ore State Univ, 84; head, Terrestrial Ecol Dept, Charles Darwin Res Sta, Isla Santa Cruz, Galapagos, Ecuador, 85-87; PROG LEADER, PR RES STA, PATUXENT WILDLIFE RES CTR, US FISH & WILDLIFE SERV, 87- *Concurrent Pos:* Mem int affairs comt, Wildlife Soc, 80-82; adj prof, Univ PR, 88-91; chairperson, Caribbean Parrot Working Group, 89-90; partic round table, Conserv Crisis Neotrop Parrots, Am Ornithologists Union, 90. *Mem:* Wildlife Soc; Soc Conserv Biol; Am Ornithologists Union; Int Union Conserv Nature & Natural Resources. *Res:* Habitat requirements of avian game species and general ecology of endangered species; US, Caribbean and Latin American wildlife conservation issues. *Mailing Add:* PR Res Sta US Fish & Wildlife Serv PO Box N Palmer PR 00721

WILSON, MARJORIE PRICE, b Pittsburgh, Pa, Sept 25, 24; m 51; c 2. MEDICINE. *Educ:* Univ Pittsburgh, MD, 49. *Prof Exp:* Intern, Med Ctr, Univ Pittsburgh Hosp, 49-50, resident, Children's Hosp, 50-51; resident, Jackson Mem Hosp, Sch Med, Univ Miami, 54-56; chief, Contractual Res Sect, Res & Educ Serv, Vet Admin, 52-53, chief, Residency & Internship Div, Educ Serv, 56, chief prof training div, 56-60, asst dir, Educ Serv, 60; chief training br, Extramural Prog, Nat Inst Arthritis & Metab Dis, NIH, 60-63, asst to assoc dir training, Off of Dir, 63-64; assoc dir extramural prog, Nat Libr Med, 64-67; assoc dir prog develop, Off Prog Planning & Eval, Off of Dir, NIH, Bethesda, 67-69, asst dir prog planning & eval, 69-70; dir, Dept Inst Develop, Asn Am Med Col, 70-81; sr assoc dean, 81-86, vdean, 86-88, PRES

& CHIEF EXEC OFFICER, EDUC COMT MED GRAD, SCH MED, UNIV MD, BALTIMORE, 88- *Concurrent Pos:* Grants & contracts var agencies, 72-; mem adv bd, Fogarty Int Ctr, NIH, 72-73; mem bd vis, Sch Med, Univ Pittsburgh, 74-; mem, Robert Wood Johnson Health Policy Fels Bd, 75-; mem bd trustees, Analytical Serv, Inc, Arlington, 76-; mem, Nat Bd Med Examiners, 80-87, 89- *Mem:* Inst Med-Nat Acad Sci; Am Fedn Clin Res; fel AAAS; Inst Elec & Electronics Engrs; Asn Am Med Col; AMA. *Res:* Management of medical systems, including education, research and health care; leadership roles and decision making in academic medicine. *Mailing Add:* Educ Comn Foreign Med Grad 3624 Market St Philadelphia PA 19104

WILSON, MARK ALLAN, b Berkeley, Calif, Nov 26, 56; m 76; c 2. INVERTEBRATE PALEONTOLOGY, EVOLUTION. *Educ:* Col Wooster, Ohio, BA, 78; Univ Calif, Berkeley, PhD(paleont), 82. *Prof Exp:* Res geologist, Chevron Oil Field Res Co, 78; teaching asst paleont, Univ Calif, Berkeley, 79-81; from instr to asst prof, 81-85, ASSOC PROF GEOL, COL WOOSTER, OHIO, 85- *Concurrent Pos:* vis prof earth sci, Oxford Univ; tech ed, J Paleontol, 86- *Mem:* Paleont Soc; Sigma Xi; AAAS; Palaeont Asn. *Res:* Paleozoic invertebrate fossil communities in relation to sedimentary environments, paleocology of encrusting and coelobite communities; taxonomy of adherent foraminiferal and bryozoans, application of paleoecology to biostratigraphy. *Mailing Add:* Dept Geol Col Wooster Wooster OH 44691

WILSON, MARK CURTIS, b Ware, Mass, Sept 19, 21. ENTOMOLOGY. *Educ:* Univ Mass, BS, 44; Ohio State Univ, MS, 46. *Prof Exp:* Field Aide, Div Truck Crops Invests, Bur Entom & Plant Quarantine, USDA, 45; asst zool, Ohio State Univ, 44-47; from asst prof to assoc prof, 47-69, PROF ENTOM, PURDUE UNIV, W LAFAYETTE, 69- *Concurrent Pos:* Consult, Adv Comt, Alfalfa Seed Coun, Food & Agr Orgn, Rumania, 71. *Mem:* Entom Soc Am. *Res:* Insect pest management; economic insect thresholds; host plant resistance. *Mailing Add:* Dept Entom Purdue Univ West Lafayette IN 47906

WILSON, MARK VINCENT HARDMAN, b Toronto, Ont, Feb 11, 46; m 70; c 4. VERTEBRATE PALEONTOLOGY. *Educ:* Univ Toronto, BSc, 68, MSc, 70, PhD(geol), 74. *Prof Exp:* Asst prof biol, Queen's Univ, Kingston, Ont, 74-75; from asst prof zool to assoc prof 75-88, PROF ZOOL, UNIV ALTA, 88- *Concurrent Pos:* Res assoc, Dept Vert Palaeont, Royal Ont Mus, 74-; mem, Alta Paleontol Adv Comt, 78-81, 88-90; assoc ed, Paleontographica Canadiana, 81-, J Vert Paleont, 89-91; adj assoc prof geol, Univ Alta, 81-88, adj prof, 88- *Mem:* Soc Vert Paleont; Can Soc Zoologists; Am Soc Ichthyologists & Herpetologists; Soc Syst Zool; Geol Asn Can; Paleont Soc; Paleont Asn. *Res:* Fossil fishes, especially faunal and phlogenetic studies of Tertiary freshwater teleosts; zoological systematics; paleoecology of lacustrine sediments, taphonomy of freshwater organisms; Tertiary insects; anatomy and phylogeny of recent fishes. *Mailing Add:* Dept Zool Univ Alta Edmonton AB T6G 2E9 Can

WILSON, MARLENE MOORE, b Austin, Tex, July 4, 47; m 70; c 1. HUMAN ANATOMY, NEUROENDOCRINOLOGY. *Educ:* Univ St Thomas, BA, 69; Baylor Col Med, PhD(anat), 75. *Prof Exp:* Res asst neuropharmacol, Med Sch, Univ Tex, San Antonio, 69-70; res assoc neuroendocrinol, Health Sci Ctr, Univ Ore, 74-75, res fel endocrinol, 75-76; from asst prof to assoc prof, 76-90, PROF BIOL, UNIV PORTLAND, 90- *Concurrent Pos:* Teaching asst, Baylor Col Med, 70-72; instr anat, Health Sci Ctr, Univ Ore, 74-75; clin res asst prof med, Ore Health Sci Univ, 79-; consult, Providence Med Ctr, 79-84. *Mem:* Am Asn Anatomists; Endocrine Soc; Am Physiol Soc; Soc Neurosci; Geront Soc Am. *Res:* Regulation of pituitary secretion of ACTH by the nervous system; circadian rhythmicity and the role of the hippocampus in regulating the pituitary-adrenal axis; stress and health. *Mailing Add:* Univ Portland 5000 N Willamette Blvd Portland OR 97203-5798

WILSON, MARTIN, b Berlin, Ger, June 12, 13; US citizen; m 47; c 2. CHEMISTRY. *Educ:* Univ Geneva, DSc(constitution of starch), 39. *Prof Exp:* Res chemist, Palestine Potash Co, 43-48, Bonneville Ltd, Utah, 48-55, Kennecott Copper Corp, Utah, 55-56 & Nat Potash Co, NMex, 56; from res chemist to sr res chemist, US Borax Res Corp, 56-62, sr scientist, 62-79; consult, 79; RETIRED. *Mem:* Am Chem Soc. *Res:* Potash refining; phase equilibrium; beneficiation of fluorspar ores and molybdenite ores. *Mailing Add:* 3172 Via Vista No A Laguna Hills CA 92653-2742

WILSON, MARVIN CRACRAFT, b Wheeling, WVa, Aug 7, 43; m 66; c 2. PHARMACY, PHARMACOLOGY. *Educ:* WVa Univ, BS, 66; Univ Mich, Ann Arbor, PhD(pharmacol), 70. *Prof Exp:* Res assoc, Dept Psychiat, Univ Chicago, 70; asst prof, 70-73, assoc prof, 73-80, PROF PHARMACOL, SCH PHARM, UNIV MISS, 80- *Concurrent Pos:* Mem, Int Study Group Invest Drugs as Reinforcers. *Mem:* Sigma Xi; Am Asn Cols Pharm; Am Soc Pharmacol & Exp Therapeut; Soc Stimulus Properties Drugs. *Res:* Neurochemical, neurophysiological and neuropharmacological factors which mediate psychomotor stimulant self-administration behavior; pharmacokinetics of stimulant self-administration; effects of central nervous system drugs on positively and negatively reinforced behavior and group behavior of non-human primates. *Mailing Add:* Dept Pharmacol Univ Miss Sch Pharm University MS 38677

WILSON, MASON P, JR, b Albany, NY, Jan 15, 33; wid; c 2. FLUID MECHANICS, HEAT TRANSFER. *Educ:* State Univ NY Albany, BS, 57; Univ Conn, MS, 60, PhD(mech eng), 68. *Prof Exp:* Mathematician, Res & Develop Lab, Elec Boat Div, Gen Dynamics Corp, 57-58, engr, 58-62; sr analytical engr advan propulsion group, Pratt & Whitney Aircraft, United Aircraft Corp, 62-64, admin supvr heat transfer, 64; res engr, Neptune Res Lab, Neptune Meter Co, 64-68; assoc prof fluid mech & heat transfer, 68-76, PROF MECH ENG & APPL MECH & DIR, UNIV RI, 76- *Concurrent Pos:* Consult, Neptune Res Lab, Neptune Meter Co, 68- *Mem:* Am Soc Mech Engrs. *Res:* Fluidics; thermodynamics; thermophysical & environmental properties; flow instrumentation. *Mailing Add:* Dept Mech Eng & Appl Mech Univ RI Kingston RI 02881

WILSON, MATHEW KENT, b Salt Lake City, Utah, Dec 22, 20; m 44; c 1. PHYSICAL CHEMISTRY. *Educ:* Univ Utah, BS, 43; Calif Inst Technol, PhD(phys chem), 48. *Prof Exp:* Asst, Off Sci Res & Develop, Calif Inst Technol, 43-46; instr chem, Harvard Univ, 48-51, asst prof, 51-56; prof & chmn dept, Tufts Univ, 56-66; head chem sect, 66-74, head, Off Energy-Related Gen Res, 74; dep asst div planning & eval, Math & Phys Sci & Eng Directorate, Nat Sci found, 75-77, div off planning & resources mgt, 77-82, dep asst dir, 83-89, actg asst dir, 89-90, EXEC OFFICER, MATH & PHYS SCI DIRECTORATE, NSF, 90- *Concurrent Pos:* Guggenheim fel & Fulbright scholar, King's Col, London, 54-55; mem, Adv Coun Col Chem, 63-66; ed, Spectrochimia Acta, 64-83. *Mem:* AAAS; Am Chem Soc; Optical Soc Am; fel Am Acad Arts & Sci. *Res:* Molecular spectroscopy. *Mailing Add:* NSF 1800 G St NW Washington DC 20550

WILSON, MERLE R(OBERT), b Rochester, NY, July 16, 32; m 56; c 4. MECHANICAL ENGINEERING. *Educ:* Cleveland State Univ, BME, 55. *Prof Exp:* Study engr, Pipe Mach Co, 53-54; asst engr, Bell Aircraft Co, 54; assoc mech engr, Cornell Aeronaut Lab, 55-63, res engr, 63-66; sr reliability engr, Moog Inc, 66-70, sr design engr, 70-88, engr group leader, 88-90, ENG SECT HEAD, MOOG INC, 90- *Res:* Design and evaluation of servo valves, servo actuators and servo actuation systems relating to military, space and commercial applications; stress and fatigue analysis; finite element analysis and fracture mechanics; design and evaluation of test techniques for purposes of product development and design verification. *Mailing Add:* Moog Inc Proner Airport East Aurora NY 14052-0018

WILSON, MICHAEL FRIEND, b Morgantown, WVa, Jan 13, 27; m 54; c 5. CARDIOLOGY, NUCLEAR MEDICINE. *Educ:* WVa Univ, AB, 49; Univ Pa, MD, 53. *Prof Exp:* From intern to resident med, Presby Hosp, Philadelphia, 53-55; resident physician internal med, Med Ctr, Temple Univ, 55-57; from asst prof to assoc prof physiol & biophys, Col Med, Univ Ky, 60-65; prof physiol & biophys & chmn dept, Med Ctr, WVa Univ, 65-76, clin prof med, 73-76; ASSOC CHIEF STAFF RES, VET ADMIN HOSP, OKLAHOMA CITY, 76-, PROF MED & ASSOC PROF NUCLEAR MED, 76-, DIR NUCLEAR CARDIOL, OKLA MEM HOSP & OKLAHOMA CITY VET ADMIN MED CTR, 79- *Concurrent Pos:* Fel physiol & cardiol, Med Ctr, Temple Univ, 57-58; res fel physiol & biophys, Sch Med, Univ Wash, 58-60; NIH fel, 59-60; vis prof, Sch Med, Univ Nottingham, 72-73. *Mem:* Fel Am Col Cardiol; Am Physiol Soc; Shock Soc; Pavlovian Soc NAm; Am Heart Asn; Fedn Am Soc Exp Biol. *Res:* Neurocirculatory control, cardiovascular dynamics and behavior correlates; myocardial contractility and cardiac function; coronary artery disease. *Mailing Add:* Vet Admin Hosp 921 NE 13th St Oklahoma City OK 73104

WILSON, MICHAEL JOHN, b Iowa City, Iowa, June 3, 42; m 69; c 1. CELL BIOLOGY, MALE REPRODUCTIVE BIOLOGY. *Educ:* St Ambrose Col, BA, 64; Univ Iowa, MS, 67, PhD(zool), 71. *Prof Exp:* NIH fel biochem, Harvard Univ, 71-73; res assoc, 73-75, asst prof, 75-82, ASSOC PROF LAB MED & PATH, UNIV MINN, MINNEAPOLIS, 82-; RES BIOCHEMIST, VET ADMIN MED CTR, MINNEAPOLIS, 76- *Mem:* Am Soc Androl; Am Soc Cell Biol; Endocrine Soc. *Res:* Male reproductive biology; the subcellular responses to hormone interactions in induction of specific effects; the role of proteases in normal and pathological changes of tissue organization of the prostate; constituents of prostatic fluid as markers of prostate disease and in semen function. *Mailing Add:* Res Serv Vet Admin Med Ctr One Veterans Dr Minneapolis MN 55417

WILSON, MIRIAM GEISENDORFER, b Yakima, Wash, Dec 3, 22; m 47; c 5. MEDICINE, PEDIATRICS. *Educ:* Univ Wash, BS, 44, MS, 45; Univ Calif, MD, 50; Am Bd Pediat, dipl. *Prof Exp:* Chief genetics div & dir cytogenics lab, Pediat Pavilion, Los Angeles County-Univ Southern Calif Med Ctr, 65-87; PROF PEDIAT, SCH MED, UNIV SOUTHERN CALIF, 69- *Mem:* Am Acad Pediat; Am Pediat Soc; Am Soc Human Genetics. *Res:* Medical genetics and cytogenetics; medical problems of the newborn and premature infant; growth and development of the infant and child; maternal and child health. *Mailing Add:* Dept Pediat Los Angeles County-Univ Southern Calif Med Ctr Los Angeles CA 90033

WILSON, MONTE DALE, b Pomeroy, Wash, Nov 16, 38; m 62; c 2. GEOLOGY. *Educ:* Brigham Young Univ, BS, 62; Univ Idaho, MS, 68, PhD(geol), 70. *Prof Exp:* Geophysicist, Can Magnetic Reduction, Ltd, Alta, 62-63 & US Army Ballistics Res Labs, 63-65; teacher high schs, Idaho, 65-67; instr geol, Univ Idaho, 68-69; dean arts & sci, 85-86, PROF GEOL, BOISE STATE UNIV, 69- *Concurrent Pos:* Mem, Idaho Bd Registration Prof Geologists, 75-80; Fulbright sr fel, Univ Salzburg, 81-82. *Mem:* Int Mountain Soc; Am Quaternary Asn; Nat Asn Geol Teachers; Geol Soc Am. *Res:* Glacial and periglacial geomorphology; environmental geology. *Mailing Add:* Dept Geol Boise State Univ Boise ID 83725

WILSON, MYRON ALLEN, reliability engineering, mathematical modeling, for more information see previous edition

WILSON, NANCY KEELER, b Walton, NY, Apr 20, 37; m 59; c 3. ENVIRONMENTAL SCIENCE. *Educ:* Univ Rochester, BS, 59; Carnegie-Mellon Univ, MS, 62, PhD(chem), 66. *Prof Exp:* Res assoc chem, Ohio State Univ, Columbus, 66-67; res assoc & lectr chem, Univ NC, Chapel Hill, 67-69; sr staff fel chem, Nat Inst Environ Health Sci, 70-74; res chemist, 74-81, CHIEF, ANALYTICAL METHODS RES SECT, US ENVIRON PROTECTION AGENCY, 81- *Concurrent Pos:* Instr, Point Park Jr Col, 62-63; lectr math, Carnegie-Mellon Univ, 63-65; fac affil, Colo State Univ, 76-77. *Honors & Awards:* Sigma Xi. *Mem:* Am Chem Soc; Am Inst Chemists; Sigma Xi. *Res:* Applications of spectroscopic techniques to problems in physical organic chemistry; environmental chemistry, toxicology and metabolism; author or coauthor of over 150 publications. *Mailing Add:* 1109 Archdale Dr Durham NC 27707

WILSON, NIGEL HENRY MOIR, b Kent, Eng, Aug 8, 44; m 70; c 1. TRANSPORTATION. *Educ:* Imp Col, Univ London, BSc, 65; Mass Inst Technol, SM, 67, PhD(civil eng), 69. *Prof Exp:* Res asst, 66-68, from instr to assoc prof, 68-82, PROF CIVIL ENG, MASS INST TECHNOL, 82- *Concurrent Pos:* Consult, Mass Bay Trans Authority, Budget Off, 85-86; mem, Transp Res Bd; vis prof planning, Dept Civil Eng, Stanford Univ, 77-78. *Res:* Transportation systems analysis; urban transportation; public transportation. *Mailing Add:* Dept Civil Eng Mass Inst Technol 77 Massachusetts Ave Cambridge MA 02139

WILSON, NIXON ALBERT, b Litchfield, Ill, May 20, 30; m 63; c 2. ACAROLOGY. *Educ:* Earlham Col, BA, 52; Univ Mich, MWM, 54; Purdue Univ, PhD(entom), 61. *Prof Exp:* Animal ecologist, Plague Res Lab, Hawaii State Dept Health, 61-62; acarologist, B P Bishop Mus, 62-69; from asst prof to assoc prof, 69-75, PROF BIOL, UNIV NORTHERN IOWA, 75- *Concurrent Pos:* USPHS res grants, B P Bishop Mus, 67-69; grants, Univ Northern Iowa, 69-; res assoc, Fla State Collection of Arthropods, Fla Dept Agr & Consumer Servs, 70-; zool ed, J Iowa Acad Sci, 76-; guest investr, Mus Zool, Nat Univ Mexico, 82; Iowa Sci Found grant, 86-87. *Mem:* Am Soc Mammal; Am Soc Parasitol; Acarological Soc Am; Am Asn Zool Nomenclature; Entomol Soc Am. *Res:* Ectoparasites of vertebrates, especially mites and fleas. *Mailing Add:* Dept Biol Univ Northern Iowa Cedar Falls IA 50614-0421

WILSON, OLIN C(HADDOCK), b San Francisco, Calif, Jan 13, 09; m 43; c 2. ASTRONOMY, OPTICS. *Educ:* Univ Calif, Berkeley, AB, 30; Calif Inst Technol, PhD, 34. *Prof Exp:* Asst astronr, 31-36, astronr, 36-74, EMER ASTRONR, MT WILSON OBSERV, CARNEGIE INST WASH, 75- *Honors & Awards:* Russell Lectr, Am Astron Soc, 77; Bruce Medal, Astron Soc Pac, 84. *Mem:* Nat Acad Sci; Am Astron Soc; AAAS. *Mailing Add:* 616 Terry Lane West Lafayette IN 47906

WILSON, OSCAR BRYAN, JR, b Tex, Aug 15, 22; m 45; c 3. PHYSICS. *Educ:* Univ Tex, BS, 44; Univ Calif, Los Angeles, MA, 48, PhD, 51. *Prof Exp:* Mem tech staff, Hughes Aircraft Co, 51-52; physicist, Soundrive Engine Co, 52-57; assoc prof, 57-62, PROF PHYSICS, NAVAL POSTGRAD SCH, 62- *Concurrent Pos:* Vis prof, ISEN, Lille, France, 89-90. *Mem:* AAAS; Am Phys Soc; Acoust Soc Am; Inst Elec & Electronics Engrs; Sigma Xi. *Res:* Engineering acoustics; underwater acoustics. *Mailing Add:* Dept Physics Naval Postgrad Sch Monterey CA 93943-5000

WILSON, P DAVID, b Roswell, NMex, Oct 4, 33; m 65; c 1. STATISTICS, BIOMATHEMATICS. *Educ:* Univ Colo, Boulder, BA, 56; Univ Minn, Minneapolis, MS, 63; Johns Hopkins Univ, PhD(biostatist), 70. *Prof Exp:* Res assoc, Med Sch, Univ Md, Baltimore, 64-66; consult, Dept Surg, Ctr Study Trauma, Med Sch, Univ Md, Baltimore, 67-70; asst prof biomet, Med Col Va, 70-71; math statistician, Bur Drugs, Food & Drug Admin, HEW, 71-72; assoc prof, Dept Epidemiol & Biostatist, Col Pub Health, Univ SFla, Tampa, 87-88; asst prof, Dept Surg, Med Sch, Univ Md Ctr Study Trauma, 72-74, asst prof, Dept Epidemiol & Prev Med, 74-81, ASSOC PROF, DEPT EPIDEMIOL & PREV MED, MED SCH, UNIV MD, BALTIMORE, 81-87 & 89- *Concurrent Pos:* Prin investr res grants in hypertension, 85-86; comput brain metab, 86-88; consult, nuclear med, Johns Hopkins Univ, 86-; longitudinal data analysis, 91-94; Nat Res Serv Awards fel, USPHS, 82-83. *Mem:* Biomet Soc; Am Statist Asn. *Res:* Mathematics, statistics and computing in biomedical research. *Mailing Add:* Dept Epidemiol & Prev Med Univ Md Sch Med 655 W Baltimore Baltimore MD 21201

WILSON, PAUL ROBERT, b Chicago, Ill, Oct 25, 39. MATHEMATICS. *Educ:* Univ Cincinnati, BA, 61, MA, 62; Univ Ill, PhD(math), 67. *Prof Exp:* Asst prof math, Univ Nebr, Lincoln, 67-71; asst prof math, Alma Col, 71-80; ASSOC PROF MATH, ROCHESTER INST TECHNOL, 80- *Mem:* Am Math Soc. *Res:* Algebra, statistics. *Mailing Add:* Dept Math Rochester Inst Technol Rochester NY 14623

WILSON, PEGGY MAYFIELD DUNLAP, b Austin, Tex, Mar 24, 27; m 75. SURFACE CHEMISTRY, OIL FIELD CHEMISTRY. *Educ:* Univ Tex, BS, 48, PhD(chem), 52. *Prof Exp:* Spec instr chem, Univ Tex, 52-53; sr res technologist, Mobil Res & Develop Corp, 53-67, res assoc, 67-84, mgr, Analytical Serv, 84-85 & Processing & Core Analysis, 85-89; RETIRED. *Concurrent Pos:* Dir, Stone Gap Indust Corp, 68-; State Republican Committeewoman, 71-80; dir, JayBee Mfg Co, 80-88; regent, E Tex State Univ, 81-87; mem Adv Coun, Col Natural Sci Found, Univ Tex, Austin, 87-90. *Mem:* Nat Asn Corrosion Engrs; Am Chem Soc; Soc Petrol Engrs. *Res:* Interfacial tension and contact angles; tertiary oil recovery; corrosion; wellstream processing; rock properties. *Mailing Add:* 1819 W Belt Line Rd Cedar Hill TX 75104-5615

WILSON, PERRY BAKER, b Norman, Okla, Feb 24, 27; m 82; c 2. ACCELERATOR PHYSICS. *Educ:* Wash State Univ, BS, 50, MS, 52; Stanford Univ, PhD(physics), 58. *Prof Exp:* Staff physicist, Linfield Res Inst, Ore, 58-59; res assoc accelerator physicist, High Energy Physics Lab, Stanford Univ, 59-64, assoc dir opers, 64-68; vis scientist, Europ Orgn Nuclear Res, Geneva, Switzerland, 68-69, 77-78; sr res assoc, 69-74, RES (EPRCH) PROF, STANFORD LINEAR ACCELERATOR CTR, STANFORD UNIV, 74- *Concurrent Pos:* Consult, Gen Atomic Div, Gen Dynamics Corp, Calif, 63-64, Phys Electronics Labs, 64, Varian Assocs, 74-77, Western Res Corp, 82-83 & Rocketdyne Div, Rockwell Int Corp, 87-89. *Mem:* AAAS; Am Phys Soc; sr mem Inst Elec & Electronics Engrs; Sigma Xi; fel Am Phys Soc. *Res:* Theory and design of linear electron accelerators and storage rings for high energy particle physics. *Mailing Add:* Stanford Linear Accelerator Ctr PO Box 4349 Stanford CA 94305

WILSON, PHILO CALHOUN, b Westfield, Mass, Jan 29, 24; m 47; c 3. STRATIGRAPHY, MARINE GEOLOGY. *Educ:* Williams Col, AB, 48; Cornell Univ, MS, 50; Wash State Univ, PhD(stratig), 54. *Prof Exp:* From geologist to staff geologist, Sohio Petrol Co, 54-60; area geologist, Champlin Oil Co, 60-63; assoc prof, 63-64, chmn dept, 67-73 & 76-84, prof earth sci, 64-78, distinguished teaching prof, 78-85, EMER DISTINGUISHED TEACHING PROF, STATE UNIV NY COL ONEONTA, 85- *Concurrent Pos:* Mem selection panels rev of NSF Proposals, 65-66 & 70. *Mem:* Am Asn Petrol Geol. *Res:* Sedimentation; regional stratigraphic analysis; Pennsylvanian system of Wyoming. *Mailing Add:* Gull Lane Orleans MA 02653

WILSON, RAPHAEL, b Trenton, NJ, Apr 25, 25. MEDICAL BACTERIOLOGY. *Educ:* Univ Notre Dame, BS, 48; Univ Tex, MA, 51, PhD(bact), 54. *Prof Exp:* From instr to assoc prof biol, St Edward's Univ, 48-59, dean col, 51-58, dir testing & guid, 58-59; from asst prof to assoc prof biol, Univ Notre Dame, 59-71; prof pediat, Baylor Col Med, 71-76; prof biol, Univ Portland, 76-78, pres, 78-81; dir spec progs, Univ San Francisco, 81-82; prof biol, King's Col, 82-89; PRES, HOLY CROSS COL, 90- *Concurrent Pos:* Vis prof, Univ Ulm, Ger, 69-70 & Baylor Col Med, 70-71. *Mem:* Transplantation Soc; Radiation Res Soc; Am Soc Microbiol; Soc Exp Biol & Med; Soc Exp Hemat. *Res:* Germfree life; protection against radiation damage; role of the thymus; clinical gnotobiology and immunology; gastrointestinal microflora. *Mailing Add:* Holy Cross Col 1801 N Michigan Notre Dame IN 46556-0308

WILSON, RAY FLOYD, b Lee Co, Tex, Feb 20, 26; m 57; c 2. ANALYTICAL CHEMISTRY, PHYSICAL CHEMISTRY. *Educ:* Houston-Tillotson Col, BS, 50; Tex Southern Univ, MS, 61; Univ Tex, PhD(chem), 53. *Prof Exp:* Asst, Univ Tex, 51-53; assoc prof, 53-57, PROF CHEM, TEX SOUTHERN UNIV, 57- *Concurrent Pos:* Grants, Res Corp, 53-55, NSF, 54, 56 & Welch Found, 57, 59. *Mem:* Am Chem Soc. *Res:* Interaction of platinum elements with certain organic reagents. *Mailing Add:* Dept Chem Tex Southern Univ Box 227 Houston TX 77004

WILSON, RICHARD, b London, Eng, Apr 29, 26; m 52; c 6. PHYSICS. *Educ:* Oxford Univ, BA, 46, MA & DPhil(physics), 49; Harvard Univ, MA, 56. *Prof Exp:* Res lectr physics, Christ Church, Oxford Univ, 48-53, res off, Clarendon Lab, 53-55; from asst prof to assoc prof, 55-61, PROF PHYSICS, HARVARD UNIV, 61- *Concurrent Pos:* Res assoc, Univ Rochester, 50-51 & Stanford Univ, 51-52; Guggenheim fel, 61 & 69; Fulbright fel, 61 & 69; trustee, Univs Res Asn, 68-74; consult, Energy Res Develop Agency, 75-77, Nuclear Regulatory Comn, 75-, Elec Power Res Inst, 75-76, & Energy Eng Bd, Nat Acad Sci, 82-; vis prof, Univ Grenoble, 81; dir, Regional Ctr Global Environ Change, 90. *Mem:* Am Phys Soc; Am Acad Arts & Sci; Explorers Club; Soc Phys Res. *Res:* Elementary particle physics; environmental physics. *Mailing Add:* Lyman 231 Harvard Univ Cambridge MA 02138

WILSON, RICHARD BARR, b Lincoln, Nebr, Apr 21, 21; m 48; c 2. PATHOLOGY. *Educ:* Univ Nebr, AB, 43, MD, 45; Am Bd Path, dipl, 57. *Prof Exp:* Staff physician, Univ Nebr, Lincoln, 47-52; resident path, Univ Hosp, 53-57, assoc, Col Med, 57-62, from asst prof to assoc prof, 62-70, prof path, dept path, microbiol & anat, Col Med & head electron micros sect, Eppley Inst Cancer Res, 70-89, EMER PROF PATH & MICROBIOL, UNIV NEBR MED CTR, 89- *Concurrent Pos:* Attend pathologist, Vet Admin Hosp, Omaha, 60-89. *Mem:* Col Am Path; Am Soc Clin Path; Electron Micros Soc Am; Int Acad Path; Am Soc Nephrology. *Res:* Renal disease and biopsies; electron microscopy of human biopsies; morphological studies; animal carcinogenesis; ultrastructure of human neoplasms. *Mailing Add:* 727 Leewood Dr Omaha NE 68154-2943

WILSON, RICHARD FAIRFIELD, b Pittsburgh, Pa, Nov 26, 30; m 52; c 6. GEOLOGY. *Educ:* Yale Univ, BS, 52; Stanford Univ, MS, 54, PhD(geol), 59. *Prof Exp:* Geologist, US Geol Surv, 55-62; asst prof geol, 62-66, ASSOC PROF GEOSCI, UNIV ARIZ, 66- *Mem:* AAAS; Paleont Soc; Geol Soc Am. *Res:* Stratigraphy; sedimentation; study of sedimentary rocks. *Mailing Add:* 7445 N Northern Tucson AZ 85704

WILSON, RICHARD FERRIN, b Dundurn, Sask, Jan 8, 20; nat US; m 43; c 4. ANIMAL SCIENCE. *Educ:* Iowa State Col, BS, 43; Univ Ill, MS, 47, PhD(animal sci), 49. *Prof Exp:* Asst animal sci, Univ Ill, 46-49; assoc prof animal husb, SDak State Col, 49-52, actg chmn dept, 51-52; from assoc prof to prof animal sci, 52-85, in charge swine, 52-85, EMER PROF, ANIMAL SCI, OHIO STATE UNIV, 85- *Concurrent Pos:* Consult in Brazil; hosted livestock groups to W Europ and E Europ. *Mem:* Am Soc Animal Sci. *Res:* Swine management, nutrition, physiology and breeding; swine production with emphasis on nutrition and physiology involving toxic feeds resulting from molds. *Mailing Add:* Dept Animal Sci Ohio State Univ 2029 Fyffe Rd Columbus OH 43210

WILSON, RICHARD GARTH, b Montreal, Que, July 30, 45; m 67; c 2. CLIMATOLOGY, AIR POLLUTION. *Educ:* McGill Univ, BSc, 66, MSc, 68; McMaster Univ, PhD(climat), 71. *Prof Exp:* Lectr climat, Dept Geog, Queen's Univ, 67-68; lectr, Dept Geog, McGill Univ, 70-71, asst prof, 71-75; supvr climate inventory, Climate & Data Serv, BC Environ & Land Use Comt Secretariat, 75-78; mgr climate div, Resource Anal Br, 78-79, dir, Air Studies Br, 79-82, DIR, AIR MGT PROG & ASST DIR WASTE MGT, BC MINISTRY ENVIRON, 82- *Mem:* Can Meteorol Soc. *Res:* Solar and terrestrial radiation, evaporation, snow melt, snow mapping, topoclimatology; climate network design; acidic precipitation; ozone control. *Mailing Add:* BC Ministry Environ Parliament Bldg Victoria BC V8V 1X5 Can

WILSON, RICHARD HANSEL, b Madison, Wis, Aug 18, 39; m 64; c 2. PLANT PHYSIOLOGY, BIOCHEMISTRY. *Educ:* Carleton Col, BA, 61; NC State Univ, MS, 64; Ore State Univ, PhD(plant physiol), 67. *Prof Exp:* USPHS fel plant physiol, Univ Ill, Urbana, 67-68; asst prof, Univ Tex, Austin, 68-74; res specialist, Monsanto Chem Co, 74-78; FIELD SCIENTIST, FIELD DEVELOP, EASTERN REGION, SANDOZ CROP PROTECTION, 78- *Mem:* Weed Soc Am. *Res:* Herbicides. *Mailing Add:* 6014 Parkland Dr Corpus Christi TX 78413-3816

WILSON, RICHARD HOWARD, b Spearville, Kans, June 24, 42; m 62; c 1. ZOOLOGY, ANIMAL BEHAVIOR. *Educ:* Kans State Univ, BS, 64, MS, 65; Utah State Univ, PhD(zool), 71. *Prof Exp:* Instr, Prince Makonnen Sec Sch, 65-66; from instr to assoc prof, 66-90, PROF BIOL, UNIV WIS, STOUT, 90- *Mem:* AAAS; Animal Behav Soc; Am Ornith Union; Cooper Ornith Soc; Wilson Ornith Soc; Sigma Xi. *Res:* Animal communication; display postures and signaling mechanisms in birds. *Mailing Add:* Dept Biol Univ Wis-Stout Menomonie WI 54751

WILSON, RICHARD LEE, b Marshalltown, Iowa, Sept 18, 39; m 60; c 3. ENTOMOLOGY. *Educ:* Univ Northern Iowa, BA, 61; Tex A&M Univ, MS, 65; Iowa State Univ, PhD(entom), 71. *Prof Exp:* Teacher, Independent Sch Dist, Iowa, 65-68; res entomologist, Agr Res Serv, USDA, Phoenix, Ariz, 71-77 & Okla State Univ, 77-80; RES ENTOMOLOGIST, AGR RES SERV, IOWA STATE UNIV, USDA, AMES, 80- *Mem:* Entom Soc Am; Sigma Xi. *Res:* Host plant resistance; development of resistant plant introductions to several insect pests. *Mailing Add:* Regional Plant Introduction Sta Iowa State Univ Ames IA 50011

WILSON, RICHARD MICHAEL, b Gary, Ind, Nov 23, 45; m 66. MATHEMATICS. *Educ:* Ind Univ, Bloomington, AB, 66; Ohio State Univ, MS, 68, PhD(math), 69. *Prof Exp:* From asst prof to assoc prof, 69-74, PROF MATH, OHIO STATE UNIV, 74- *Concurrent Pos:* Res fel, A P Sloan Found, 75-77. *Mem:* Math Asn Am; Am Math Soc; Soc Indust & Appl Math. *Res:* Combinatorial mathematics, with emphasis on combinatorial designs and related structures. *Mailing Add:* 2371 Summit St Columbus OH 43202

WILSON, ROBERT BURTON, b Salt Lake City, Utah, June 29, 36; m 62; c 2. EXPERIMENTAL PATHOLOGY. *Educ:* Utah State Univ, BS, 58; Wash State Univ, DVM, 61; Univ Toronto, PhD(physiol), 67. *Prof Exp:* Intern vet med, Angell Mem Animal Hosp, 62-63; asst prof animal sci, Brigham Young Univ, 63-64; res investr, Hosp for Sick Children, 64-67, asst scientist, 67-69; assoc prof nutrit & animal path, Mass Inst Technol, 69-73; prof vet path, Univ Mo, 73-76; prof microbiol & path & chmn dept, Wash State Univ, 76-83, dean, Col Vet Med, 83, 88; CONSULT, 88- *Concurrent Pos:* Lectr physiol, Univ Toronto, 67-69; Mary Mitchell res award, 63; mem, Coun on Arteriosclerosis, Am Heart Asn. *Mem:* Am Vet Med Asn; Am Inst Nutrit; Am Heart Asn. *Res:* Nutrition; diabetes; carcinogenesis; cardiovascular diseases; experimental pathology of nutrition and cardiovascular diseases. *Mailing Add:* Col Vet Med Wash State Univ Pullman WA 99164

WILSON, ROBERT D(OWNING), b Portchester, NY, July 9, 21; m 44; c 3. COMMUNICATIONS, ELECTRICAL ENGINEERING. *Educ:* Rensselaer Polytech Inst, BEE, 42; Cornell Univ, MS, 51. *Prof Exp:* Instr electronics, US Naval Acad, 46-47; res assoc elec eng, Cornell Univ, 47-51, asst prof, 51-56; mgr indust components div, Raytheon Co, 56-60, corp officer, 60-61; dir eng, US Sonics, 61-62; mgr systs support, Raytheon Co, 62-65, prog mgr, 65-66, mgr space systs develop, 66-69, prin engr, 69-74; ELEC PROD ENGR, ILL GAS & ELEC CO, 74- *Concurrent Pos:* Vpres, Zialite Corp, 45-58, pres, 58-74; consult, Gen Elec Co, 51-56. *Mem:* Sr mem Inst Elec & Electronics Engrs; Am Inst Aeronaut & Astronaut. *Res:* Electronics; semiconductors; radio astronomy; vacuum tubes. *Mailing Add:* Ill Gas & Elec Co PO Box 4350 Davenport IA 52808

WILSON, ROBERT E, b Norristown, Pa, Jan 16, 37; div; c 2. STELLAR ASTROPHYSICS, BINARY STARS. *Educ:* Univ Pa, AB, 58, MS, 60, PhD(astron), 63. *Prof Exp:* Asst prof astron, Georgetown Univ, 63-66; Nat Res Coun sr res assoc, Inst Space Studies, 72-74; from assoc prof to prof astron, Univ SFla, 66-75; prof physics & astron, 75-79, PROF ASTRON, UNIV FLA, 79- *Concurrent Pos:* Consult, Goddard Space Flight Ctr, NASA, 65-70; res grants, NASA, 66-69, 76-78, NSF, 70-73, 77-79, 80-81, 83-84 & 85-87; Shapley vis lectr, Am Astron Soc, 81-; Sr Scientist Award, Alexander von Humboldt Found, 79. *Mem:* Am Astron Soc; Int Astron Union; Royal Astron Soc; Astron Soc Pac. *Res:* Theory and observation of binary stars; stellar structure and evolution. *Mailing Add:* Dept Astron Univ Fla Gainesville FL 32611

WILSON, ROBERT E(LWOOD), b Decatur, Ill, July 2, 26; m 49. CHEMICAL ENGINEERING. *Educ:* Univ Ill, AB, 48, BS, 49, PhD(chem eng), 52; Univ Minn, MS, 50. *Prof Exp:* Proj leader eng res, Corn Prod Refining Co, 52-54; from assoc prof to prof chem eng, Univ Dayton, 54-61, head dept, 54-61, dir, Col Eng Grad Prog, 58-61; mgr eng & develop, Thomas J Lipton, Inc, 61-62, assoc dir develop, 62-64; dir res & develop, Int Minerals & Chem Corp, Ill, 64-67; sr vpres & mem bd dirs, Heidrick & Struggles Inc, Chicago, 67-; RETIRED. *Concurrent Pos:* Consult, USAF, 55-56, USN, 59-60 & Eng Exp Sta, Univ Wis, 59-60. *Mem:* Am Soc Eng Educ; Inst Nuclear Mgt; Am Inst Chem Engrs. *Res:* Fluid mechanics; high temperature; bioengineering and transport processes. *Mailing Add:* 1926 Wolf Laurel Dr Sun City Ctr FL 33570

WILSON, ROBERT EUGENE, b Denton, Tex, Apr 16, 32; m 56; c 3. ECOLOGY, GENERAL BIOLOGY. *Educ:* North Tex State Univ, BS, 52, MS, 56; Univ Tex, PhD(microbiol), 63. *Prof Exp:* Instr biol, Col Arts & Indust, 56-59; from asst prof to assoc prof, 63-70, PROF BIOL, E TEX STATE UNIV, 70- *Res:* Cytotoxicity of staphylococcal toxins towards mammalian cells in vitro; changes in serum proteins following x-irradiation; effects of electromagnetic fields on plants and animals; ecology of northeastern Texas pine-oak forests. *Mailing Add:* Dept Biol Sci ETex State Univ Commerce TX 75429

WILSON, ROBERT FRANCIS, b Scranton, Pa, Aug 9, 34; m 72; c 7. THORACIC SURGERY, CARDIOVASCULAR SURGERY. *Educ:* Lehigh Univ, BA, 57; Temple Univ, MD, 58. *Prof Exp:* from instr to assoc prof, 63-71, PROF SURG, SCH MED, WAYNE STATE UNIV, 71- *Concurrent Pos:* Markle scholar acad med; pres med staff, Detroit Gen Receiving Hosp, 72-74; dir, affil prog thoracic surg, Sch Med, Wayne State Univ, 71-89, asst dean, Detroit Gen Hosp affairs, Wayne State Univ, 72-76; chief sect thoracic & cardiovasc surg, Harper Hosp, 72-89. *Mem:* Soc Univ Surg; Am Asn Thoracic Surg; Am Asn Surg Trauma; Am Col Surg; Am Col Chest Physicians; Am Surg Asn. *Res:* Shock; respiratory failure; fluid and electrolytes; trauma. *Mailing Add:* Dept Surg Wayne State Univ Sch Med Detroit MI 48202

WILSON, ROBERT G, b Galesburg, Ill, Aug 30, 30; m 53; c 3. CHEMISTRY, BIOCHEMISTRY. *Educ:* Knox Col, AB, 52; Purdue Univ, MS, 56; Okla State Univ, PhD(chem), 61. *Prof Exp:* Res chemist, Nat Cancer Inst, 63-67; from asst prof to assoc prof, 67-81, PROF BIOCHEM, COL MED NJ, 81- *Concurrent Pos:* NIH fel biochem, Brandeis Univ, 61-63. *Mem:* Sigma Xi; Am Asn Univ Professors; Am Soc Biochem & Molecular Biol; Am Asn Cancer Res. *Res:* Effects of carcinogens on methylation and expression of oncogenes; anthracycline induction of cellular differentiation. *Mailing Add:* Dept Biochem Univ Med NJ Newark NJ 07103

WILSON, ROBERT GRAY, b Wooster, Ohio, Apr 7, 34; m 57; c 2. ELECTRONICS, NUCLEAR PHYSICS. *Educ:* Ohio State Univ, BSc, 56, PhD(physics), 61. *Prof Exp:* Prin physicist, Battelle Mem Inst, 54-58; res asst, Res Found, Ohio State Univ, 58-60; sr physicist, NAm Aviation/Rocketdyne, 61-63; SR MEM TECH STAFF, RES LABS, HUGHES AIRCRAFT CO, 63- *Mem:* Am Phys Soc; Inst Elec & Electronics Eng. *Res:* Ion implantation; semiconductor devices and integrated circuits; electron and ion emission from surfaces; scanning acoustic microscopy; experimental low energy nuclear physics. *Mailing Add:* 20513 Gresham St Canoga Park CA 91306

WILSON, ROBERT HALLOWELL, b Baltimore, Md, July 30, 24; m 48; c 2. ENVIRONMENTAL HEALTH. *Educ:* Univ Rochester, BS, 45; Am Bd Indust Hyg, cert, 62. *Prof Exp:* Jr scientist, Atomic Energy Proj, Univ Rochester, 46-51, instr indust hyg & toxicol & asst scientist, 51-56, scientist, 56-62, chief engr, 62-76, asst prof radiation biol & biophys, 56-83, chief environ health & safety & chief safety officer, 75-89, asst prof toxicol, Sch Med & Dent, 83-89; RETIRED. *Mem:* AAAS; Am Nuclear Soc; Health Physics Soc; Am Indust Hyg Asn. *Res:* Generation, sampling and behavior of aerosols; control of radiation hazards; air safety considerations of nuclear weapons transport and storage; environmental impact of mercury. *Mailing Add:* Sch Med & Dent Univ Rochester PO Box 617 Rochester NY 14642

WILSON, ROBERT JAMES, bacteriology, immunology; deceased, see previous edition for last biography

WILSON, ROBERT JOHN, b St Louis, Mo, Apr 23, 35; m 57; c 5. RADIOLOGICAL PHYSICS, NUCLEAR MEDICINE. *Educ:* St Mary's Univ, Tex, BS & BA, 56; Wash Univ, St Louis, PhD(physics), 63. *Prof Exp:* Res assoc physics, Wash Univ, 63-66; res physicist, US Naval Radiol Defense Lab, San Francisco, 66-69; asst prof radiol, 69-72, assoc prof nuclear med, 72-78, PROF RADIOLOGY, UNIV TENN, MEMPHIS, 78- *Mem:* Am Asn Physicists Med; Am Phys Soc; Soc Nuclear Med; Am Col Radiol; Radiol Soc NAm; Sigma Xi. *Mailing Add:* 513 Fleda Memphis TN 38120

WILSON, ROBERT LAKE, b Gallipolis, Ohio, July 2, 24; m 50, 87; c 4. GEOLOGY. *Educ:* Wheaton Col, AB, 48; Univ Iowa, MS, 50; Univ Tenn, PhD, 67. *Prof Exp:* Asst geol, Univ Iowa, 49-50 & Univ Tenn, 50-52; area geologist, Tenn Div Geol, 52-55; from instr to assoc prof geol, 55-66, PROF GEOL & GEOG, UNIV TENN, CHATTANOOGA, 66- *Concurrent Pos:* NSF fac fel, 60-61. *Mem:* Geol Soc Am. *Res:* Paleozoic stratigraphy and sedimentation of Southern Appalachians; economic geology of the eastern United States. *Mailing Add:* Dept Geol Univ Tenn Chattanooga TN 37402

WILSON, ROBERT LEE, b Champaign, Ill, Mar 7, 17; m 40; c 4. MATHEMATICS. *Educ:* Univ Fla, AB, 38; Univ Wis, MA, 40, PhD(math), 47. *Prof Exp:* Asst math, Univ Wis, 39-41, 46-47; from instr to asst prof, Univ Tenn, 47-56; sr aerophys engr, Gen Dynamics/Convair, Tex, 56-58; prof 58-79, EMER PROF, OHIO WESLEYAN UNIV, 79- *Concurrent Pos:* Adj prof, Tex Christian Univ, 56-58; vis prof & dir comput ctr, Univ Ibadan, 66-68, Univ Western Australia, 80 & Washington & Lee Univ, 81; gov, Math Asn Am, 75-77. *Mem:* Am Math Soc; Math Asn Am; Nat Coun Teachers Math. *Res:* Galois theory; computing. *Mailing Add:* Rte 6 Box 86 Lexington VA 24450

WILSON, ROBERT LEE, b Washington, DC, Jan 16, 46; m 67; c 2. LIE ALGEBRAS. *Educ:* Am Univ, BA, 65; Yale Univ, PhD(math), 69. *Prof Exp:* Instr math, Courant Inst Math Sci, NY Univ, 69-71; from asst prof to assoc prof, 71-80, PROF MATH, RUTGERS UNIV, NEW BRUNSWICK, 80-, CHAIR, 90- *Mem:* Am Math Soc; Math Asn Am. *Res:* Lie algebras over fields of prime characteristic; Kac-Moody Lie algebras. *Mailing Add:* Dept Math Rutgers Univ New Brunswick NJ 08903

WILSON, ROBERT LEE, JR, b Auburn, Ala, Jan 3, 42; m 62; c 2. MATHEMATICS. *Educ:* Ohio Wesleyan Univ, BA, 62; Univ Wis-Madison, MA, 63, PhD(math), 69. *Prof Exp:* Asst prof math, Univ Wis-Madison, 69-75; prof math & computer sci, Washington & Lee Univ, 75-84; sr scientist, Zilog Corp, 84-87; prin engr & scientist, Ford Aerospace Corp, 87-90; PROF MATH, UNIV WIS-MADISON, 90-, DIR MATH OUTREACH, 90- *Mem:* Am Math Soc; Math Asn Am. *Res:* Combinatorics; graph theory; universal algebra; microcomputers; generalizations of group theory. *Mailing Add:* Dept Math Univ Wis 480 Lincoln Dr Madison WI 53706

WILSON, ROBERT NORTON, b Walla Walla, Wash, Oct 7, 27; m 56; c 1. MATHEMATICAL PHYSICS, ATMOSPHERIC PHYSICS. *Educ:* Whitman Col, AB, 48; Stanford Univ, MS, 50, PhD(physics, math), 60. *Prof Exp:* Res scientist, Microwaves, Kane Eng Labs, Calif, 60-65 & Atmospheric Physics, Lockheed Res Lab, 65-71; res scientist, Radiative Transfer & Hydrodynamics, Mission Res Corp, 71-75; RES SCIENTIST, NEW MILLENNIUM ASSOCS, 75- *Mem:* Am Phys Soc; Am Inst Physics. *Res:* Electromagnetic theory and experimentation; multiple quantum effect physics; non-equilibrium statistical mechanics; hydrodynamics; magnetohydrodynamics; radiation physics; acoustic gravitation waves; interaction of electromagnetic radiation with relativistically moving plasma fronts; ground water flow; optical fluorescence detection systems. *Mailing Add:* 16 W Mountain Dr Santa Barbara CA 93103

WILSON, ROBERT PAUL, b Revere, Mo, Dec 28, 41; m 62; c 3. BIOCHEMISTRY, FISH NUTRITION. *Educ:* Univ Mo, Columbia, BSEd, 63, MS, 65, PhD(biochem), 68. *Prof Exp:* Instr agr chem, Univ Mo-Columbia, 68-69; from asst prof to assoc prof biochem, 69-77, head biochem dept, 79-89, PROF BIOCHEM, MISS STATE UNIV, 77- *Concurrent Pos:* Vis prof, Inst Marine Biochem, Aberdeen, Scotland, 84. *Honors & Awards:* Res Award, Catfish Farmers Am, 81. *Mem:* Am Inst Nutrit; Catfish Farmers Am. *Res:* Comparative biochemistry and nutrition of channel catfish; fish nutrition in general. *Mailing Add:* Dept Biochem Miss State Univ Drawer BB Mississippi State MS 39762

WILSON, ROBERT RATHBUN, b Frontier, Wyo, Mar 4, 14; m 40; c 3. PHYSICS. *Educ:* Univ Calif, AB, 36, PhD(physics), 40. *Hon Degrees:* MA, Harvard Univ, 46; DSc, Notre Dame, Univ Bonn, WGer, Harvard, 86, Weslayan, 87. *Prof Exp:* From instr to asst prof physics, Princeton Univ, 40-46, in tech chg isotron develop proj, 42-43; physicist, Los Alamos Sci Lab, 43-46, leader cyclotron group, 43-44, head exp res div, 44-46; assoc prof physics, Harvard Univ, 46-47; prof physics & dir lab nuclear studies, Cornell Univ, 47-67; Ritzma prof, Dept Physics & Enrico Fermi Inst Nuclear Studies, Univ Chicago, 67-80, dir, Fermi Nat Accelerator Lab, 67-78; Michael Pupin prof, Columbia Univ, 80-82; CONSULT, 82- *Concurrent Pos:* Mem, Comt Atomic Casualties, Nat Res Coun, 48-51; exchange prof, Univ Paris, 54-55; mem, Steering Comt, Proj Sherwood, US AEC, 58-; Fulbright fel, 61; chmn bd trustees, Aspen Physics Ctr, 85. *Honors & Awards:* Elliot Cresson Medal, Franklin Inst; Nat Medal of Sci, 73; Fermi Award, 84; del Regato Medal Med, 89. *Mem:* Nat Acad Sci; Am Phys Soc (pres, 85); Am Acad Arts & Sci; Am Philos Soc. *Res:* Nuclear and particle physics. *Mailing Add:* 916 Stewart Ave Ithaca NY 14850

WILSON, ROBERT STEVEN, b Hartford, Conn, Dec 26, 39; m 66; c 3. PHYSICAL CHEMISTRY. *Educ:* Brown Univ, BS, 62, PhD(phys chem), 68; Northern Ill Univ, JD, 84. *Prof Exp:* Fel phys chem, Yale Univ, 68-69; asst prof, 69-73, ASSOC PROF PHYS CHEM, NORTHERN ILL UNIV, 73- *Concurrent Pos:* Res prof, Solid State Sci Div, Argonne Nat Lab, 71-; fel theoret physics, Lorentz Inst, Leiden, Holland. *Mem:* Am Phys Soc; Am Chem Soc; Am Bar Asn; Sigma Xi. *Res:* Statistical mechanics of irreversible processes; optical properties of impurity systems; critical transport properties of fluids; molecular dynamics; computer simulation. *Mailing Add:* 104 N Main St Sycamore IL 60178

WILSON, ROBERT WARREN, b Oakland, Calif, July 26, 09; wid; c 2. VERTEBRATE PALEONTOLOGY. *Educ:* Calif Inst Technol, BS, 30, MS, 32, PhD(vert paleontol), 36. *Prof Exp:* Asst geol, Calif Inst Technol, 30-34, fel, 37-39, Sterling res fel, Yale Univ, 36-37; from instr to asst prof geol, Univ Colo, 39-46, Nat Res Coun fel, 46-47; assoc prof zool & assoc cur vert paleont, Univ Kans, 47-61; prof, 61-75, emer prof paleont & dir mus geol, SDak Sch Mines & Technol, 75; vis prof, Tex Tech Univ, 75-77; Rose Morgan vis prof, 77, ASSOC MUS NATURAL HIST, UNIV KANS, 77-, EMER PROF, 80- *Concurrent Pos:* Guggenheim fel, London, 56-57; Fulbright sr res scholar, Univ Vienna, 67-68; sr res fel, Carnegie Museum Natural Hist, 81; corresp, Natural Hist Museum, Wien, 82- *Honors & Awards:* Arnold Guyot Mem Award, Nat Geog Soc, 74. *Mem:* Fel Geol Soc Am; Paleont Soc; Soc Vert Paleont (secy-treas, 54, pres, 55); Am Soc Mammal. *Res:* Tertiary and late cretaceous mammalian faunas. *Mailing Add:* Mus Natural Hist Univ Kans Lawrence KS 66045

WILSON, ROBERT WOODROW, b Houston, Tex, Jan 10, 36; m 58; c 3. RADIO ASTRONOMY, MOLECULAR CLOUDS. *Educ:* Rice Univ, BA, 57; Calif Inst Technol, PhD(physics), 62. *Prof Exp:* Res fel radio astron, Calif Inst Technol, 62-63; mem tech staff, 63-76, HEAD, DEPT RADIO PHYSICS RES, BELL LABS, 76- *Honors & Awards:* Nobel Prize in Physics, 78; Herschel Medal, Royal Astron Soc, London; Henery Draper Medal, Nat Acad Sci. *Mem:* Nat Acad Sci; Am Acad Arts & Sci; Am Phys Soc; Int Sci Radio Union; Am Astron Soc; Int Astron Union. *Res:* Problems related to the galaxy; absolute flux and background temperature measurements; millimeter-wave measurements of interstellar molecules. *Mailing Add:* Bell Labs HOH-L239 PO Box 400 Holmdel NJ 07733

WILSON, RONALD HARVEY, b Belle Fourche, SDak, Oct 2, 32; m 55; c 3. EXPERIMENTAL SOLID STATE PHYSICS. *Educ:* SDak State Univ, BS, 56, MS, 58; Rensselaer Polytech Inst, PhD(physics), 64. *Prof Exp:* Physicist, Flight Propulsion Lab, Gen Elec Co, 58-59, res trainee, Physics Res Lab, 59-61, physicist, Gen Eng Lab, 61-62; teaching asst physics, Rensselaer Polytech Inst, 62-63; PHYSICIST, GEN ELEC RES & DEVELOP CTR, 63- *Mem:* AAAS; Am Phys Soc; Electrochem Soc. *Res:* Physics and properties of thin films; physics of semiconductor devices; semiconductor processing; energy conversion processes; photoelectrochemistry; solar energy. *Mailing Add:* 669 St Marks Lane Schenectady NY 12309

WILSON, RONALD WAYNE, b Iowa Falls, Iowa, Aug 4, 39. BOTANY, MYCOLOGY. *Educ:* Iowa State Univ, BS, 61; Mich State Univ, PhD(bot), 65. *Prof Exp:* USPHS fel, Med Ctr, Ind Univ, 65-67; from asst prof to assoc prof, 67-77, PROF NATURAL SCI, MICH STATE UNIV, 77- *Mem:* AAAS; Am Inst Biol Sci; Sigma Xi. *Res:* General-liberal studies in science; lysine metabolism. *Mailing Add:* Dept Bot & Plant Path Mich State Univ East Lansing MI 48823

WILSON, RUBY LEILA, b Punxsutawney, Pa, May 29, 31. NURSING. *Educ:* Univ Pittsburgh, BS, 54; Western Reserve, MS, 59; Duke Univ, PhD(higher educ admin), 68. *Prof Exp:* Asst prof, 59, 68, dean Sch Nursing, 71-84, PROF NURSING & MED, DUKE UNIV SCH MED, 71-, ASST TO CHANCELLOR, 84- *Concurrent Pos:* Asst prof Rockefeller Found, 68-71. *Mem:* Inst Med-Nat Acad Sci. *Mailing Add:* Off Chancellor Health Box 3243 Duke Univ Med Ctr Durham NC 27710

WILSON, SAMUEL H, b Washington, DC, Aug 5, 39; c 2. BIOCHEMISTRY. *Educ:* Univ Denver, AB, 61; Harvard Univ, MD, 68. *Prof Exp:* Res Assoc, Lab Biochem Genetics, Nat Heart & Lung Inst, 68-70, sr staff fel, 70-73, RES SCIENTIST, LAB BIOCHEM, NAT CANCER INST, NIH, 73- *Mem:* Am Asn Cancer Res; AAAS; Am Chem Soc; Am Soc Cell Biol; Am Soc Biol Chemists. *Res:* Biochemical studies of mammalian DNA replication proteins. *Mailing Add:* Lab Biochem Nucleic Acid Enzymol Sec Nat Cancer Inst-NIH Bldg 37 Rm 4D-23 Bethesda MD 20892

WILSON, SLOAN JACOB, b Dallas, Tex, Jan 22, 10; m 48; c 4. INTERNAL MEDICINE. *Educ:* Wichita State Univ, AB, 31, MS, 32; Univ Kans, BS, 34, MD, 36; Am Bd Internal Med, dipl, 48. *Prof Exp:* From intern to asst resident, Ohio State Univ Hosp, 36-38, resident res med, 38-39, instr path, 39-40; from asst prof to assoc prof med, 46-59, prof, 59-70, EMER PROF INTERNAL MED, UNIV KANS MED CTR, KANSAS CITY, 70- *Mem:* AAAS; Soc Exp Biol & Med; fel AMA; fel Am Col Physicians; Am Soc Hemat. *Res:* Blood hematology. *Mailing Add:* 5618 W62 Kansas City KS 66113

WILSON, STEPHEN ROSS, b Oklahoma City, Okla, Mar 13, 46; m 67; c 2. SYNTHETIC ORGANIC CHEMISTRY. *Educ:* Rice Univ, BA, 69, MA, 72, PhD(org chem), 72. *Prof Exp:* NIH fel org chem, Calif Inst Technol, 72-74; from asst prof to assoc prof org chem, Ind Univ, Bloomington, 74-80; assoc prof, 80-86, PROF ORG CHEM, NY UNIV, 86- *Concurrent Pos:* Sigma Xi res award, Rice Univ, 72. *Mem:* Royal Soc Chem; Am Chem Soc. *Res:* Development of new approaches to the synthesis of naturally occurring compounds of biological significance, and the structure elucidation and total synthesis of such substances. *Mailing Add:* Dept Chem NY Univ Washington Square New York NY 10003

WILSON, STEPHEN W, b Hackensack, NJ, Feb 6, 52; m 73; c 1. TAXONOMY OF PLANTHOPPERS. *Educ:* Rutgers Univ, BS, 73; Southwest Mo State Univ, MA, 75; Southern Ill Univ, PhD(zool), 80. *Prof Exp:* Asst prof biol, Calif State Univ, Chico, 80-82; from asst prof to assoc prof, 82-89, PROF BIOL, CENT MO STATE UNIV, WARRENSBURG, 89- *Mem:* Entom Soc Am. *Res:* Systematics and ecology of planthoppers (Homoptera: Fulgoroidea) with emphasis on Delphacidae. *Mailing Add:* Dept Biol Cent Mo State Univ Warrensburg MO 64093-5053

WILSON, STEVEN PAUL, b New Castle, Pa, Oct 12, 50; m 72; c 1. NEUROCHEMISTRY. *Educ:* Univ Pittsburgh, BS, 72; Duke Univ, PhD(biochem), 76. *Prof Exp:* Guest worker, Nat Heart, Lung & Blood Inst, NIH, 76-78, staff fel, 78-79; vis scientist, Dept Med Biochem, Wellcome Res Labs, Burroughs Wellcome Co, Research Triangle Park, NC, 79-81; ASST MED RES PROF, DEPT PHARM, DUKE UNIV MED CTR, DURHAM, NC, 82- *Concurrent Pos:* Fel neurol, Sch Med & Dent, George Washington Univ, 76-78. *Mem:* Soc Neurosci. *Res:* Biochemical aspects of neurotransmitter secretion and synapse formation; regulation of catecholamine and opioid peptide biosynthesis, neurobiology of adrenal chromaffin cells. *Mailing Add:* Dept Pharmacol Univ SC Sch Med Columbia SC 29208

WILSON, THEODORE A(LEXANDER), b Elgin, Ill, June 20, 35. AERONAUTICAL ENGINEERING, BIOMECHANICS. *Educ:* Cornell Univ, BEngPhysics, 58, PhD(aeronaut eng), 62. *Prof Exp:* Res scientist, Avco-Everett Res Lab, 62-63 & Jet Propulsion Lab, 63-64; from asst prof to assoc prof aeronaut & eng mech, 64-72, PROF AEROSPACE ENG & MECH, UNIV MINN, MINNEAPOLIS, 72- *Mem:* Am Physiol Soc. *Res:* Respiratory mechanics; acoustics; fluid mechanics. *Mailing Add:* Dept Aerospace Eng & Mech Univ Minn Minneapolis MN 55455

WILSON, THOMAS EDWARD, b Chicago, Ill, Feb 20, 42; m 66; c 2. ENVIRONMENTAL & CHEMICAL ENGINEERING. *Educ:* Northwestern Univ, BS, 64, MS, 67; Ill Inst Technol, PhD(environ eng), 69. *Prof Exp:* Asst prof environ eng, Rutgers Univ, New Brunswick, 67-70; CONSULT POLLUTION CONTROL & WATER TREATMENT, GREELEY & HANSEN, 70- *Mem:* Am Inst Chem Engrs; Int Asn Water Pollution Res; Water Pollution Control Fedn; Am Soc Civil Engrs; Am Water Works Asn; Sigma Xi. *Res:* Pollution control; advanced water treatment; physical-chemical treatment processes; solids and sludge disposal; industrial waste treatment; treatment plant operations. *Mailing Add:* 922 Shoreline Rd-LBS Barrington IL 60010-3815

WILSON, THOMAS G(EORGE), b Annapolis, Md, Jan 19, 26; m 49; c 3. ELECTRICAL ENGINEERING. *Educ:* Harvard Univ, AB, 47, SM, 49, ScD(elec eng), 53. *Prof Exp:* Physicist, US Naval Ord Lab, 48, elec engr, US Naval Res Lab, 49-53; mgr res & develop, Magnetics, Inc, 53-59; assoc prof, 59-63, chmn dept, 64-70, PROF ELEC ENG, DUKE UNIV, 63- *Mem:* Inst Elec & Electronics Engrs; Sigma Xi. *Res:* Magnetic devices, materials and amplifiers; nonlinear electromagnetics; energy conversion. *Mailing Add:* Dept Elec Eng Duke Univ Durham NC 27706

WILSON, THOMAS HASTINGS, b Philadelphia, Pa, Jan 31, 25; m 52; c 4. PHYSIOLOGY. *Educ:* Univ Pa, MD, 48; Sheffield Univ, 51-53, PhD(biochem), 53. *Hon Degrees:* MA, Harvard Univ. *Prof Exp:* Instr physiol, Univ Pa, 49-50; instr biochem, Wash Univ, 56-57; assoc physiol, 57-59, from asst prof to assoc prof, 59-68, PROF PHYSIOL, HARVARD MED SCH, 68- *Mem:* Am Soc Biol Chem; Am Physiol Soc; Brit Biochem Soc; Am Acad Arts & Sci. *Res:* Active transport of materials across cell membranes. *Mailing Add:* Dept Physiol Harvard Med Sch Boston MA 02115

WILSON, THOMAS KENDRICK, b Highland Park, Mich, June 2, 31; m 52; c 5. PLANT MORPHOLOGY. *Educ:* Ohio Univ, BS, 53, MS, 55; Ind Univ, PhD(bot), 58. *Prof Exp:* Asst bot, Ohio Univ, 53-55 & Ind Univ, 55-58; from asst prof to assoc prof, Univ Cincinnati, 58-68; assoc prof, 68-74, PROF BOT, MIAMI UNIV, 74- *Mem:* AAAS; Bot Soc Am; Int Soc Plant Morphol; Int Asn Plant Taxon. *Res:* Comparative morphology of angiosperms; origin and phylogeny of vascular plants; evolution. *Mailing Add:* Dept Bot Miami Univ Oxford OH 45056

WILSON, THOMAS LAMONT, b Salt Lake City, Utah, June 4, 14; m 37; c 2. HIGH POWER BROADCAST TRANSMITTERS. *Educ:* Univ Utah, BS, 40; Univ Louisville, MS, 63. *Prof Exp:* Elec engr, Elec Div, Fed Tel & Radio, 41-46; mgr, eng & develop, Chemetron Corp, 46-77; CONSULT, DIELECTRIC HEATING, 77- *Concurrent Pos:* Tech adv, US Nat Comt Int Electrotech Comm, 75-88. *Honors & Awards:* Achievement Award, Inst Elec & Electronics Engrs, 79 & Centennial Medal, 84. *Mem:* Fel Inst Elec & Electronics Engrs. *Res:* Dielectric heating uses; high frequency high power for industrial heating applications. *Mailing Add:* 1407 Ormsby Lane Louisville KY 40222-3827

WILSON, THOMAS LEE, b Wyoming, Ohio, Nov 4, 09; m 37; c 2. PHYSICAL CHEMISTRY, GENERAL MATHEMATICS. *Educ:* Col Wooster, BS, 30; Univ Wash, MS, 34; Univ Chicago, PhD(chem), 35. *Prof Exp:* Res chemist, Gen Labs, US Rubber Co, 35-42, dept head, 42-54, admin asst, Res & Develop Dept, 54-58, mgr res ctr, 58-65; from asst prof to assoc prof chem, Montclair State Col, 66-76, chmn dept, 71-73, dean, Sch Math & Sci, 73-76; CONSULT, 76- *Mem:* AAAS; Am Chem Soc; fel Am Inst Chem. *Res:* Oceanography; reaction rates of gaseous decomposition; rubber and inorganic chemistry. *Mailing Add:* Seven Rockbrook Dr Camden ME 04843

WILSON, THOMAS LEON, b Alpine, Tex, May 21, 42; m 78; c 2. NEUTRINO & GAMMA-RAY ASTROPHYSICS, QUANTUM. *Educ:* Rice Univ, BA, 64, BSc, 65, MA, 74, PhD(theoret physics), 76. *Prof Exp:* Aerospace engr, 65-76, RES PHYSICIST, NASA JOHNSON SPACE CTR, 76- *Concurrent Pos:* Astronaut instr, Apollo Skylab Prog, 65-72; co-investr, Skylab Prog, NASA, 70-73, fel, 69-76; fel, NASA, 69-76. *Honors & Awards:* Hugo Gernsback Award, Inst Elec & Elec & Electronics Engrs. *Mem:* Am Phys Soc; NY Acad Sci; AAAS. *Res:* Theoretical analysis of mechanisms for gamma-ray bursts from accretion of exotic matter onto neutron stars; detection of relic dark matter in the galaxy; neutrino tomography of the planet earth; "chesire cat" effect in quantum mechanics; pathological problems in relativistic wave equations; neutrino astronomy; physics and astrophysics from a lunar base. *Mailing Add:* NASA Johnson Space Ctr-SN Houston TX 77058

WILSON, THOMAS PUTNAM, b New York, NY, Sept 4, 18; m 44, 80; c 1. HETEROGENOUS CATALYSIS. *Educ:* Amherst Col, BA, 39; Harvard Univ, PhD(chem physics), 43. *Prof Exp:* Res chemist, Manhattan Proj, M W Kellogg Co, NJ, 43-44; res chemist, Kellex Corp, 44-45; res chemist, Manhattan Proj & S A M Labs, Chems & Plastics Div, Union Carbide Corp, 45-46 & 46-62, asst dir res, 62-72, res assoc, 72-73, corp res fel, Res & Develop Dept, Ethylene Oxide/Glycol Div, 73-82, RETIRED. *Mem:* Am Chem Soc; Catalysis Soc. *Res:* Kinetics and catalysis; catalytic reaction mechanisms; ethylene polymerization; catalyst development and characterization. *Mailing Add:* 701 Myrtle Rd South Charleston WV 25314-1117

WILSON, THORNTON ARNOLD, b Sikeston, Mo, Feb 8, 21; m 44; c 3. AERONAUTICS. *Educ:* Iowa State Col, BS, 43; Calif Inst Technol, MS, 48. *Prof Exp:* Mem staff, Boeing Co, 43-57, asst chief tech staff & proj eng mgr, 57-58, vpres & mgr Minuteman Br, Aerosace Div, 62-64, vpres opers & planning, 64-66, exec vpres & dir, 66-68, pres, 68-72, chief exec officer, 69-86, chmn bd, 72-87, EMER CHMN & MEM BD DIRS, BOEING CO, 87- *Concurrent Pos:* Sloan fel, Mass Inst Technol, 52-53; mem bd gov, Iowa State Univ Found; bd dirs, PACCAR Inc, Hewlett-Packard Co & Weyerhaeuser Co. *Honors & Awards:* James Forrestal Award, Nat Security Indust Asn, 75; Wright Bros Trophy, 79; Collier Trophy, Fed Aviation Admin, 82; Daniel Guggenheim Medal & Nat Acad Sci Award. *Mem:* Nat Acad Eng; fel Am Inst Aeronaut & Astronaut; Aerospace Indust Asn. *Res:* Mechanics; aeronautical and astronautical engineering. *Mailing Add:* Boeing Co PO Box 3707 Seattle WA 98124-2207

WILSON, TIMOTHY M, b Columbus, Ohio, Aug 3, 38; c 2. SOLID STATE PHYSICS. *Educ:* Univ Fla, BS, 61, PhD(chem physics), 66. *Prof Exp:* Fel solid state physics, Univ Fla, 66-68, asst prof chem, 68-69; from asst prof to assoc prof physics, 69-78, asst dir exten, Col Arts & Sci, 77-79, assoc dir exten, 79-82, PROF PHYSICS, OKLA STATE UNIV, 78- *Concurrent Pos:* Res staff mem, Solid State Div, Oak Ridge Nat Lab, 74-75. *Mem:* Am Phys Soc; Sigma Xi. *Res:* Theoretical studies of the optical and magnetic properties of impurities and defects in crystalline solids. *Mailing Add:* Dept Physics Okla State Univ Stillwater OK 74078

WILSON, VICTOR JOSEPH, b Berlin, Ger, Dec 24, 28; m 53; c 2. NEUROPHYSIOLOGY. *Educ:* Tufts Col, BS, 48; Univ Ill, PhD(physiol), 53. *Prof Exp:* Res assoc, 56-58, from asst prof to prof neurophysiol, 58-69, PROF NEUROPHYSIOL, ROCKEFELLER UNIV, 69- *Mem:* Am Physiol Soc; Soc Neurosci; Int Brain Res Org. *Res:* Organization and synaptic transmission in the central nervous system, particularly the spinal cord and brain stem; vestibular system. *Mailing Add:* Rockefeller Univ 1230 York Ave New York NY 10021-6399

WILSON, VINCENT L, b Kentfield, Calif, Dec 4, 50. CHEMICAL CARCINOGENESIS, GENE REGULATION. *Educ:* Sonoma State Univ, BS, 73; Univ Calif, Davis, MS, 76; Ore State Univ, PhD(pharmacol & toxicol), 80. *Prof Exp:* Chemist, Cent Path Lab, Inc, Santa Rosa, Calif, 72-73; analytical chemist, Environ Protection Agency, Corvallis Environ Res Lab, Corvallis, Ore, 78-80; chem & viral carcinogenesis training fel, Univ Southern Calif Comprehensive Cancer Ctr & Children's Hosp, Los Angeles, Calif, 80-82; sr staff fel, Nat Cancer Inst, NIH, Bethesda, Md, 82-88; DIR MOLECULAR GENETICS & ONCOL, CHILDREN'S HOSP, DENVER, 88-; ASSOC PROF, DEPT PATH, UNIV COLO MED SCH, DENVER, 88- *Concurrent Pos:* Vis scientist, Danish Cancer Soc, Lab Environ Carcinogenesis, Fibiger laboratoriet, Copenhagen, Denmark, 86. *Mem:* Am Chem Soc; Am Asn Cancer Res; Am Soc Cell Biol; Am Soc Human Genetics; AAAS; Sigma Xi. *Res:* Pharmacology and toxicology of cancer chemotherapeutic agents, radiation therapy, and environmental pollutants with emphasis on mechanisms in mutation and carcinogenesis; determination of genetic controls involved in the physiologic and perturbed processes of embryogenesis, differentiation, aging, and carcinogenesis; basic research in the inherent functions and controlling systems of DNA; genetic markers of diagnostic and prognostic value in the clinical care of cancer patients. *Mailing Add:* Dept Path Childrens Hosp 1056 E 19 Ave Denver CO 80218-1088

WILSON, VOLNEY COLVIN, b Evanston, Ill, Feb 27, 10; m 38; c 2. PLASMA PHYSICS, THERMAL PHYSICS. *Educ:* Northwestern Univ, BS, 32; Ohio State Univ, MA, 34; Univ Chicago, PhD(physics), 38. *Prof Exp:* Instr physics, Univ Chicago, 38-41; res assoc radar res, Mass Inst Technol Radiation Lab, 41-42; res assoc nuclear energy, Univ Chicago, 41-44; res assoc atomic bomb, Los Alamos Lab, N Mex, 44-45; res assoc physics, Gen Elec Res Lab, 45-72; RETIRED. *Mem:* Fel Am Phys Soc. *Res:* Cosmic ray research; nuclear instrumentation; nuclear reactor design; magnetic materials; thermionic emission and gas discharge; thermionic converter. *Mailing Add:* 10555 Pebble Beach Rd Sister Bay WI 54234

WILSON, WALTER DAVIS, b Merced, Calif, Oct 20, 35; m 59; c 3. ATOMIC PHYSICS. *Educ:* Univ Calif, Berkeley, BS, 57, PhD(nuclear eng), 66. *Prof Exp:* Chem engr, Aerojet Gen Nucleonics, Calif, 58-59; mem tech staff high-altitude nuclear effects, Aerospace Corp, 65-69; PROF PHYSICS, CALIF POLYTECH STATE UNIV, SAN LUIS OBISPO, 69- *Concurrent Pos:* Tech consult, Sci Applns, Inc, Calif, 71-75. *Mem:* Am Inst Physics. *Res:* High altitude physics, particularly electromagnetic field propagation, chemistry and trapped radiation; plasma physics; nuclear reactor theory. *Mailing Add:* Dept Physics Calif Polytech State Univ San Luis Obispo CA 93407

WILSON, WALTER ERVIN, b Salem, Ore, Apr 1, 34. RADIOLOGICAL PHYSICS. *Educ:* Willamette Univ, BA, 56; Univ Wis, MS, 58, PhD(physics), 61. *Prof Exp:* RADIATION PHYSICIST, PAC NORTHWEST LABS, BATTELLE MEM INST, 64- *Concurrent Pos:* Fel, Basel Univ, 61-62 & Univ Wis, 62-64; coordr, Radiol Sci Prog, Univ Wash, Tri-Cities Univ Ctr, 88-91. *Mem:* Health Physics Soc; Radiation Res Soc; Am Asn Physicists Med; Am Phys Soc. *Res:* Radiation effects and radiological sciences; radiation transport calculations; computer science; programming languages; numerical analysis. *Mailing Add:* Pac Northwest Labs Battelle Mem Inst PO Box 999 Richland WA 99352

WILSON, WALTER LEROY, b Phoenixville, Pa, Sept 1, 18; m 44; c 2. PHYSIOLOGY. *Educ:* Pa State Teachers Col, West Chester, BS, 41; Univ Pa, PhD(zool), 49. *Prof Exp:* Biologist, Off Sci Res & Develop, Univ Pa, 43-44, asst instr zool, 46-47; biologist, Manhattan Proj, Columbia Univ, 44-46; instr physiol & biophys, Col Med, Univ Vt, 49-52, from asst prof to assoc prof, 52-65; prof biol sci, Oakland Univ, 65-83; RETIRED. *Concurrent Pos:* Lectr, Middlebury Col, 52-53; mem corp, Marine Biol Lab, Woods Hole. *Mem:* Am Physiol Soc; Soc Gen Physiol; Am Soc Zoologists. *Res:* Effects of high temperature on living systems; protoplasmic viscosity changes during cell division; the release of anticoagulant substances from living cells and the inhibitory action of these anticoagulants on cell division; role of cellular cortex in stimulation and cell division. *Mailing Add:* 743 Cambridge Dr Rochester Hills NJ 48309

WILSON, WALTER LUCIEN, JR, b Montgomery, Ala, Apr 26, 27; m 47; c 2. MATHEMATICS. *Educ:* Univ Ala, AB, 50, MA, 51; Univ Calif, Los Angeles, PhD(math), 59. *Prof Exp:* Res engr, NAm Aviation, Inc, Calif, 59-60; assoc prof math, Univ Ala, 60-89; RETIRED. *Mem:* Am Math Soc; Math Asn Am. *Res:* Calculus of variations; numerical analysis; linear programming. *Mailing Add:* 15 Highridge Circle Tuscaloosa AL 35405

WILSON, WALTER R, b South Bend, Ind, May 3, 19; m 44; c 3. ENGINEERING PHYSICS. *Educ:* Univ Mich, BS, 41. *Prof Exp:* Test engr, Res Lab, 41-42, develop engr, Transformer & Allied Prod Lab, 42-49, elec sect head, Switchgear & Control Lab, 49-54, mgr eng res, 54-56, mgr eng, High Voltage Switchgear Dept, 56-61, consult engr, 61-68, mgr elec & mech eng res, Power Delivery Div, 68-76, SR CONSULT & STANDARDS ENGR, GEN ELEC CO, 76- *Concurrent Pos:* US deleg, Int Electro-Tech Comn, Madrid, 59-72. *Honors & Awards:* Alfred Noble Prize, Am Soc Civil Engrs, 44. *Mem:* Nat Soc Prof Engrs; Am Mgt Asn; fel Inst Elec & Electronics Engrs; Sigma Xi. *Res:* High voltage electrical power equipment; dielectrics, magnetics, gaseous discharges and electric contact phenomena; electrical and mechanical instrumentation. *Mailing Add:* One Rabbit Run Wallingford PA 19086

WILSON, WILBUR WILLIAM, b Ferriday, La, Jan 10, 48; m 69; c 2. PHYSICAL CHEMISTRY. *Educ:* Northeast La State Col, BS, 69; Univ NC, PhD(phys chem), 73. *Prof Exp:* Asst prof, 74-80, PROF CHEM, MISS STATE UNIV, 80- *Mem:* Am Chem Soc; Sigma Xi. *Res:* Laser light scattering by macromolecules. *Mailing Add:* Miss State Univ Box 3348 Mississippi MS 39762

WILSON, WILFRED J, b Ferndale, Calif, Mar 4, 30; m 68. EMBRYOLOGY. *Educ:* Sacramento State Col, AB, 52; Univ Calif, Davis, MA, 58, PhD(zool), 64. *Prof Exp:* Teaching asst zool, Univ Calif, Davis, 58-61, assoc, 61-63; from asst prof to assoc prof, 63-70, PROF BIOL, SAN DIEGO STATE UNIV, 70- *Concurrent Pos:* Shell merit fel, Stanford Univ, 69; vis lectr, Burma, 71; Fulbright lectr, US Dept of State, Nat Taiwan Univ, 70-71. *Mem:* Am Soc Zool. *Res:* General and invertebrate biology; crustacean water balance; teaching methods in human biology; early animal development. *Mailing Add:* Dept Biol San Diego State Univ San Diego CA 92182

WILSON, WILLIAM AUGUST, b St Louis, Mo, July 3, 24; m 54; c 4. NEUROPSYCHOLOGY. *Educ:* Univ Calif, AB, 43, PhD(psychol), 56; Yale Univ, MD, 53. *Prof Exp:* Vis instr, psychol, Wesleyan Univ, 50-51; res assoc neurophysiol, Inst of Living, 53-56, dir dept exp psychol, 56-58; vis asst prof psychol, Univ Calif, 58; asst prof, Univ Colo, 59-60; assoc prof, Bryn Mawr Col, 60-64; assoc dean grad sch, 71-72, prof biobehav sci, 69-77, assoc dean

Col Lib Arts & Sci, 78-83, PROF PSYCHOL, UNIV CONN, 64- *Concurrent Pos:* Mem exp psychol res review comt, NIMH, 69-73, chmn, 71-73. *Mem:* AAAS; Am Psychol Asn; Inst Math Statist; Animal Behav Soc; Soc Neurosci. *Res:* Neurological determinants of behavior, especially intersensory effects in perception and learning; function of regions of association cortex and the mechanisms of learning. *Mailing Add:* Dept Psychol U-20 Univ Conn 406 Cross Campus Rd Storrs CT 06268

WILSON, WILLIAM CURTIS, b Orlando, Fla, Dec 29, 27; m 52; c 1. PLANT PHYSIOLOGY. *Educ:* Cornell Univ, BS, 49; Univ Fla, MAgr, 58, PhD(fruit crops), 66. *Prof Exp:* Asst mgr agr res, Fla Agr Res Inst, 57-61; adj assoc horticulturist, Citrus Exp Sta, Agr Res & Educ Ctr, Univ Fla, 66-77; res scientist II, Fla Dept Citrus, 77-80, res scientist III, 80-91; RETIRED. *Concurrent Pos:* Merck & Co grant, 69; Julian C Miller award, Asn Southern Agr Workers, Inc, 66; Ciba-Geigy grants, 70-72; adj assoc prof, Agr Res & Educ Ctr, Univ Fla, 74-81, adj horticulturist, 81- *Mem:* Am Soc Hort Sci; Int Soc Citricult; Plant Growth Regulator Soc Am (secy, 80-81). *Res:* Abscission chemicals to facilitate easier removal of citrus fruit to aid mechanical or hand harvesting; acidity reduction and cold hardy chemicals; fresh fruit research. *Mailing Add:* Fla Dept Citrus 700 Exp Sta Rd Lake Alfred FL 33850

WILSON, WILLIAM D, b Pittsburgh, Pa, Nov 8, 25; m 49; c 5. PARASITOLOGY. *Educ:* Dickinson Col, BS, 50; Univ Kans, MA, 53; Mich State Univ, PhD(microbiol, pub health), 57. *Prof Exp:* Instr, Mich State Univ, 56-57; prof biol, State Univ NY Col Oneonta, 57-81; RETIRED. *Mailing Add:* 8522 Bonita Ilse Dr Lake Worth FL 33467-5531

WILSON, WILLIAM DAVID, b Wilmington, NC, June 28, 44; m 66; c 2. NUCLEIC ACID STRUCTURE & INTERACTIONS, ANTI-CANCER DRUG DESIGN. *Educ:* Univ NC, BS, 66; Purdue Univ, PhD(chem), 70. *Prof Exp:* Fel chem, Purdue Univ, 71; from asst prof to prof, 71-84, REGENTS PROF CHEM, GA STATE UNIV, 84- *Concurrent Pos:* Dir, Lab Biol Sci, Ga State Univ, 79-82; vis prof, Univ Fla, 80 & Inst Cancer Res, London, 88. *Honors & Awards:* Fac Develop Award, Am Cancer Soc. *Mem:* Am Chem Soc; Biophys Soc. *Res:* Biophysical chemistry of nucleic acid structure and interactions; interactions of drugs and model compounds which bind to nucleic acids by very different mechanisms. *Mailing Add:* Dept Chem Ga State Univ Atlanta GA 30303

WILSON, WILLIAM DENNIS, b New York, NY, July 20, 40; m 60, 73; c 2. SOLID STATE PHYSICS. *Educ:* Queens Col, NY, BS, 63, MA, 65; City Univ New York, PhD(physics), 67. *Prof Exp:* Res physicist, Queens Col, NY, 67 & 68-69; fel physics, City Univ New York, 67-68; RES PHYSICIST, LIVERMORE NAT LABS & SANDIA NAT LABS, 69-, DIV SUPVR, THEORET DIV, 74- *Concurrent Pos:* Consult, Lawrence Radiation Lab, Calif, 67-69. *Mem:* Am Phys Soc. *Res:* Interatomic potentials; defects in solids; hydrogen and helium in metals; diffusion. *Mailing Add:* 7036 Corte Del Oro Pleasanton CA 94566

WILSON, WILLIAM ENOCH, JR, b El Dorado, Ark, Jan 15, 33; m 63; c 2. ATMOSPHERIC CHEMISTRY. *Educ:* Hendrix Col, BA, 53; Purdue Univ, PhD(phys chem), 57. *Prof Exp:* Instr chem, Wis State Univ-La Crosse, 55-56; Fulbright res fel, Inst Technol, Munich, Ger, 57-58; sr chemist, Appl Physics Lab, Johns Hopkins Univ, 58-67; assoc fel, Battelle Mem Inst, 67-71; chief atmospheric aerosol res sect, 71-75, chief aerosol res br, 75-77, sci dir, Regional Field Studies Off, 77-88, actg dir, Emission Measurements & Characterization Div, 81-87, DIR ATMOSPHERIC CHEM & PHYSICS DIV, ENVIRON PROTECTION AGENCY, 80- *Concurrent Pos:* Adj prof, Environ Sci & Eng Dept, Sch Pub Health, Univ NC, 73-; vis prof, Peking Univ, People's Rep China, 86; mem bd dirs, Am Asn Aerosol Res, 84-87; prog mgr, Coop Res Atmospheric Sci, People's Rep China, 80-, sr sci adv, 88- *Honors & Awards:* Silver Medal, Environ Protection Agency, 78. *Mem:* AAAS; Am Chem Soc; Am Meteorol Soc; Sigma Xi; Air Pollution Control Asn; Int Aerosol Res Assembly (secy, 86-); Am Asn Aerosol Res. *Res:* Sources, formation, dynamics, transport, removal, and effects of atmospheric pollutants; atmospheric chemistry and physics; molecular spectroscopy; chemical kinetics and thermodynamics pertinent to combustion, propulsion, and air pollution. *Mailing Add:* Environ Protection Agency Research Triangle Park NC 27711

WILSON, WILLIAM EWING, b New Orleans, La, Nov 8, 32; m 61; c 3. MOLECULAR PHARMACOLOGY. *Educ:* King Col, AB, 54; Univ Tenn, MS, 56, PhD(biochem), 59. *Prof Exp:* Res assoc biochem, Okla State Univ, 59-60, instr, 61-63; res biochemist, Radio Isotope Serv, Vet Admin Hosp, Little Rock, 63-64, chief, Biochem Sect, Southern Res Support Ctr, 64-68; asst prof biochem, Sch Med, Univ Ark, Little Rock, 63-68; res chemist, Analytical & Synthetic Chem Br, 68-73, res chemist, Environ Toxicol Br, 73-76, RES CHEMIST, LAB MOLECULAR & INTEGRATIVE NEUROSCI, NAT INST ENVIRON HEALTH SCI, 76- *Mem:* Sigma Xi; Am Chem Soc. *Res:* Interaction of chemicals with membranal enzymes; molecular pharmacology of drugs and environmental agents; analytical biochemistry; neurochemistry; solution interactions. *Mailing Add:* Lab Behav & Neurol Toxicol Nat Inst Environ Health Sci Research Triangle Park NC 27709

WILSON, WILLIAM JAMES FITZPATRICK, b Aberdeen, Scotland, Feb 13, 46; Can citizen; m 81; c 2. STELLAR EVOLUTION, MASS LOSS. *Educ:* Univ BC, BSc, 68; Univ Waterloo, MSc, 70; Univ Calgary, PhD(astrophysics), 77. *Prof Exp:* FEL & INSTR PHYSICS & ASTRON, UNIV CALGARY, 77- *Mem:* Can Astron Soc. *Res:* Computation of stellar evolution from the zero-age main sequence to helium exhaustion, with mass loss due to radiation pressure and convective and rotational turbulent pressure; origin of Wolf-Rayet stars. *Mailing Add:* Dept Physics Univ Calgary Calgary AB T2N 1N4 Can

WILSON, WILLIAM JEWELL, radiology; deceased, see previous edition for last biography

WILSON, WILLIAM JOHN, b Spokane, Wash, Dec 16, 39; m 90; c 2. MILLIMETER WAVE RADIOMETERS, MICROWAVE SPECTROMETERS. *Educ:* Univ Wash, BS, 61; Mass Inst Technol, MS, 63, PhD(elec eng), 70. *Prof Exp:* mem staff, Space Div, USAF, 64-67; mem tech staff, Aerospace Corp, 70-80; asst prof elec eng, Univ Tex, Austin, 76-77; GROUP SUPVR, JET PROPULSION LAB, 80-, SR RES ENGR, 87- *Mem:* Sr mem Inst Elec & Electronics Engrs, Microwave Theory & Tech Soc; Union Radio Sci; Int Astron Union. *Res:* Development of low-noise millimeter-wave and submillimeter-wave radiometers and systems for aircraft and spacecraft which have applications for Earth remote sensing and astrophysics; development of analog and digital microwave spectrometers for spaceborne radiometers. *Mailing Add:* Jet Propulsion Lab 4800 Oak Grove Dr Bldg 168-327 Pasadena CA 91109

WILSON, WILLIAM MARK DUNLOP, b Glasgow, Scotland, Jan 23, 49; Brit citizen; m 73. AGRICULTURE, ANIMAL SCIENCE. *Educ:* Univ Glasgow, BSc hons, 71; Univ Ill, Urbana, MS, 73, PhD(animal sci), 75. *Prof Exp:* Asst animal sci & ruminant nutrit, Univ Ill, Urbana, 71-75; AGR PROJS OFFICER, WORLD BANK, 75- *Mem:* Am Soc Animal Sci; Brit Soc Animal Prod. *Res:* Facets of animal science and agronomy which relate to agriculture in developing countries; appraisal and supervision of agricultural projects in developing countries in Latin America, South Asia, and the Middle East. *Mailing Add:* World Bank 1818 H St NW Washington DC 20433

WILSON, WILLIAM PRESTON, b Fayetteville, NC, Nov 6, 22; m 51; c 5. PSYCHIATRY. *Educ:* Duke Univ, BS, 43, MD, 47. *Prof Exp:* Intern, Gorgas Hosp, CZ, 47-48; staff psychiatrist, State Hosp, Raleigh, NC, 48-49; instr psychiat & asst resident, Sch Med, Duke Univ, 49-52, resident neurol, 52, assoc psychiat & chief resident, 52-54, asst prof psychiat, 55-58; assoc prof, dir psychiat res labs & consult, Hogg Found, Med Br, Univ Tex, 58-60; dir psychiat res labs & staff psychiatrist, Vet Admin Hosp, 58-79; from assoc prof to prof, Med Ctr, 58-85, dir neurophysiol labs, 58-82, EMER PROF PSYCHIAT, DUKE UNIV, 85-; DIR INST CHRISTIAN GROWTH, GRAHAM, NC. *Concurrent Pos:* Fel med, Duke Univ, 52-54; NIH fel, Montreal Neurol Inst, McGill Univ, 54-55; mem, Am Bd Qual EEG, 69-, secy-treas, 71-74. *Honors & Awards:* Ephraim McDowell Award, Am Orthopsychiat Asn. *Mem:* AAAS; AMA; Am Psychiat Asn; Am Psychopath Asn; Asn Res Nerv & Ment Dis; Sigma Xi. *Res:* Clinical psychiatry and neurochemistry; clinical and experimental neurophysiology; electroencephalography. *Mailing Add:* 1209 Virginia Ave Durham NC 27705

WILSON, WILLIAM ROBERT DUNWOODY, b Belfast, Northern Ireland, Sept 11, 41; US citizen; m 67; c 2. TRIBOLOGY, PLASTICITY. *Educ:* Belfast Tech Col, HNC, 64; Queens Univ, Belfast, BS, 63, PhD(mech eng), 67. *Prof Exp:* Student apprentice, Harland & Wolff Ltd, Belfast, 59-63; sr res scientist, Colubus Div, Battelle Mem Inst, 67-71; prof, Univ Mass, Amherst, 71-81; PROF MECH ENG, NORTHWESTERN UNIV, 81- *Mem:* Am Soc Mech Engrs; Soc Mfg Engrs. *Res:* Manufacturing processes and tribology in particular lubrication of metal forming processes, computer-aided design for manufacture. *Mailing Add:* Mech Eng Dept Northwestern Univ Evanston IL 60208-3111

WILSON, WILLIAM THOMAS, b Midvale, Utah, July 22, 32; m 58; c 5. INVERTEBRATE PATHOLOGY, ENTOMOLOGY. *Educ:* Colo Agr & Mech Col, BS, 55; Colo State Univ, MS, 57; Ohio State Univ, PhD(entom), 67. *Prof Exp:* Asst entom, Colo State Univ, 55-57, from instr to asst prof, 60-65; asst, Ohio State Univ, 65-67; asst res pathobiologist, Univ Calif, Irvine, 67-68; RES ENTOMOLOGIST, SCI & EDUC ADMIN-AGR RES, USDA & MEM GRAD FAC, UNIV WYO, 68- *Mem:* Bee Res Asn; Soc Invert Path; Entom Soc Am. *Res:* Chemotherapeutic treatment of insect diseases; invertebrate microbiology; impact of pesticides on bees; diseases and toxicology of the honey bee. *Mailing Add:* USDA ARS Honey Bee 509 W 4th St Weslaco TX 78596-3814

WILT, FRED H, b South Bend, Ind, Dec 12, 34; m 57, 87; c 3. DEVELOPMENTAL BIOLOGY. *Educ:* Ind Univ, AB, 56; Johns Hopkins Univ, PhD(biol), 59. *Prof Exp:* Fel, Carnegie Inst Technol, 59-60; assoc prof biol, Purdue Univ, 60-64; assoc prof, 64-71, PROF ZOOL, UNIV CALIF, BERKELEY, 71- *Concurrent Pos:* NIH spec fel, 63-64; fel, Guggenheim Found, 75. *Honors & Awards:* Fel, AAAS; NIH Merit Award. *Mem:* Soc Develop Biol; Am Soc Cell Biol. *Res:* Regulation of gene expression; roll of cell interaction and nucleo-cytoplasmic interactions during embryonic development. *Mailing Add:* Dept Molecular Cell Biol 371 LSA Univ Calif Berkeley CA 94720

WILT, JAMES WILLIAM, b Chicago, Ill, Aug 28, 30; m 53; c 5. ORGANIC CHEMISTRY. *Educ:* Univ Chicago, AB, 49, MSc, 53, PhD(chem), 54. *Prof Exp:* Instr chem, Univ Conn, 55; from instr to assoc prof, 55-66, chmn dept, 70-77, PROF CHEM, LOYOLA UNIV CHICAGO, 66- *Mem:* Am Chem Soc; Sigma Xi. *Res:* Rearrangements in organic chemistry; reaction mechanisms; diazoalkane chemistry; free radical chemistry; chemistry of bicyclic compounds. *Mailing Add:* 2745 Pauline Ave Glenview IL 60025

WILT, JOHN CHARLES, bacteriology; deceased, see previous edition for last biography

WILT, PAXTON MARSHALL, b Louisville, Ky, July 8, 42; m 67. MOLECULAR SPECTROSCOPY. *Educ:* Centre Col Ky, BA, 64; Vanderbilt Univ, PhD(physics), 67. *Prof Exp:* From asst prof to assoc prof, 67-76, PROF CHEM PHYSICS, CENTRE COL KY, 76- *Concurrent Pos:* Consult, Res Corp Am, 67. *Res:* Molecular vibration-rotation spectres of small polyatomic molecules. *Mailing Add:* 1064 Argyl Dr Danville KY 40422

WILTBANK, WILLIAM JOSEPH, b Clifton, Ariz, Jan 1, 27; m 47; c 5. HORTICULTURE, PLANT PHYSIOLOGY. *Educ:* NMex State Univ, BSAgr, 50; Univ Fla, PhD(fruit crops), 67. *Prof Exp:* Instr voc agr, NMex State Dept Voc Educ, 50-53; instr hort, NMex State Univ, 53-54, ext horticulturist, 54-59; hort adv, US Agency Int Develop, Costa Rica, 59-64;

res asst, 64-68, from asst prof to assoc prof, 68-79, actg chmn dept, 77-79 & 84-86, PROF FRUIT CROPS, UNIV FLA, 79-, ASST CHMN DEPT, 86- *Honors & Awards:* Gourley Award, Am Soc Hort Sci, 71. *Mem:* Am Soc Hort Sci; Am Inst Biol Sci; Int Soc Hort Sci; Int Soc Citricult; AAAS. *Res:* Physiology of plant reproduction; plant tolerance to temperature and water stress; mineral nutrition of plants. *Mailing Add:* 3726 SW Fourth Pl Gainesville FL 32607

WILTON, DONALD ROBERT, b Lawton, Okla, Oct 25, 42; m 65; c 3. ELECTRICAL ENGINEERING. *Educ:* Univ Ill, Urbana, BS, 64, MS, 66, PhD(elec eng), 70. *Prof Exp:* Mem tech staff, Ground Systs Group, Hughes Aircraft Co, 65-67; prof elec eng, Univ Miss, 70-83; PROF ELEC ENG, UNIV HOUSTON, 83- *Concurrent Pos:* Vis prof, Syracuse Univ, 77-78. *Mem:* Inst Elec & Electronics Engrs; Electromagnetics Soc; Sigma Xi. *Res:* Electromagnetic theory; numerical methods applied to electromagnetics; antennas. *Mailing Add:* Dept Elec Eng Univ Houston Houston TX 77004

WILTS, CHARLES H(AROLD), b Los Angeles Co, Calif, Jan 30, 20; m 47; c 3. ELECTRONICS. *Educ:* Calif Inst Technol, BS, 40, MS, 41, PhD(elec eng), 48. *Prof Exp:* Res staff mem, 42-45, asst prof appl mech, 47-52, assoc prof elec eng, 52-57, exec officer for elec eng, 72-75, PROF ELEC ENG, CALIF INST TECHNOL, 57- *Mem:* Inst Elec & Electronics Engrs. *Res:* Design and use of analog computers; feedback control systems; ferromagnetism in metals and alloys; anisotropy and ferromagnetic resonance in thin films. *Mailing Add:* 1431 Brixton Pasadena CA 91105

WILTSCHKO, DAVID VILANDER, b Portland, Ore, Feb 5, 49; m 81; c 2. TECTONICS, TECTONOPHYSICS. *Educ:* Univ Rochester, BA, 71; Brown Univ, MSc, 74, PhD(geol), 78. *Prof Exp:* Vis asst prof, 77-79, asst prof structural geol, Univ Mich, 79-84; asst prof, 84-86, ASSOC PROF STRUCTURAL GEOL, TEX A&M UNIV, 86- *Concurrent Pos:* Assoc dir, Ctr Tectonophysics, 86-89, dir, 89- *Mem:* Geol Soc Am; Am Geophys Union; Am Asn Petrol Geol. *Res:* Tectonics of mountain belts, focusing on the mechanics of continental structures; theoretical model of thrust sheet motion; mechanisms of fault zones; flow of fluids in deforming rock. *Mailing Add:* Dept Geol Tex A&M Univ College Station TX 77843-3115

WILTSE, JAMES CORNELIUS, b Tannersville, NY, Mar 16, 26; m 50; c 2. MILLIMETER WAVES, LASERS. *Educ:* Rensselaer Polytech Inst, BEE, 47, MEE, 52; Johns Hopkins Univ, PhD(eng), 59. *Prof Exp:* Engr, Gen Elec Co, 47-48; instr elec eng, Rensselaer Polytech Inst, 48-51; instr, Johns Hopkins Univ, 53-54, res assoc, 54-58; mgr microwaves & antennas, Electronic Commun, Inc, 59-63, dir advan develop, 63-64; prin scientist & mgr microwaves, radar, optics & lasers, Martin Marietta Corp, 64-73, dir res & technol, 73-78, dir electronics eng, 76; PRIN RES ENGR, ENG EXP STA, GA INST TECHNOL, 78- *Concurrent Pos:* Instr, Naval Reserve Off Sch, Fla, 66-69; mem vis comt, Dept Elec Eng, Univ Fla, 69-70, 73 & 78; mem external adv comt, Ga Inst Technol, 76-78. *Mem:* Fel Inst Elec & Electronics Engrs. *Res:* Microwave and millimeter wave technology, antennas, electromagnetic theory, lasers, infrared and electro-optics with applications to radar, communications, guidance and electronic countermeasures; guided wave propagation; microwave and millimeter-wave technology and communications; lasers, infrared, and optics; quantum electronics; atmospheric propagation; radiometry; radar; antennas; signal processing. *Mailing Add:* Ga Technol Res Inst Atlanta GA 30332

WILTSHIRE, CHARLES THOMAS, b Kansas City, Mo, Apr 5, 41; m 62; c 4. AQUATIC ECOLOGY. *Educ:* Culver-Stockton Col, BA, 63; Drake Univ, MA, 65; Univ Mo-Columbia, PhD(zool), 73. *Prof Exp:* From asst prof to assoc prof, 66-76, PROF BIOL & CHMN DIV NATURAL SCI, CULVER-STOCKTON COL, 76- *Mem:* AAAS; Am Soc Zoologists; Nat Sci Teachers Asn. *Res:* Taxonomy and natural history of conchostracans, such as Cyzicus; ecology of the middle Mississippi River. *Mailing Add:* Dept Nat Sci Culver-Stockton Col College Hill Canton MO 63435

WIMAN, FRED HAWKINS, evolutionary ecology, aquatic ecology, for more information see previous edition

WIMBER, DONALD EDWARD, b Greeley, Colo, Jan 2, 30; c 2. BOTANY, CYTOLOGY. *Educ:* San Diego State Col, BA, 52; Claremont Col, MA, 54, PhD(bot), 56. *Prof Exp:* Res assoc, Dos Pueblos Orchid Co, 54-57; res collabr, Brookhaven Nat Lab, 57-60, asst biologist, 61-63; res collabr, Royal Cancer Hosp, 60-61; from assoc prof to prof, 63-90, EMER PROF BIOL, UNIV ORE, 90- *Concurrent Pos:* NIH fel, 58-61, career develop award, 66-71. *Mem:* Bot Soc Am. *Res:* Genetics and cytology of orchids. *Mailing Add:* Dept Biol Univ Ore Eugene OR 97403

WIMBER, R(AY) TED, b Salina, Utah, Feb 27, 35; m 56; c 7. METALLURGY, MATERIALS SCIENCE. *Educ:* Univ Utah, BS, 56, PhD(metall), 59. *Prof Exp:* From instr to asst prof metall, Wash State Univ, 59-61; sr engr, Hercules Powder Co, Utah, 61-62; res engr, Varian Assocs, Calif, 62-64; res staff engr, Solar Div, Int Harvester Co, 64-67; from assoc prof to prof mech eng, Mont State Univ, 67-74; SCIENTIST, DEERE & CO TECH CTR, 74- *Concurrent Pos:* Vis scientist, Argonne Nat Lab, 85. *Honors & Awards:* NASA Award, 69; Sigma Xi Award, 78. *Mem:* Am Soc Metals; Metall Soc; Am Vacuum Soc. *Res:* Reaction kinetics; surface chemistry; high-temperature materials; coatings. *Mailing Add:* Deere & Co Tech Ctr 3300 River Dr Moline IL 61265-1792

WIMBERLEY, STANLEY, b Detroit, Mich, Dec 22, 27. MARINE GEOLOGY, OCEANOGRAPHY. *Educ:* Johns Hopkins Univ, BA, 52; Univ Tex, MA, 54; Univ Southern Calif, PhD(geol), 64. *Prof Exp:* Res technician, Chesapeake Bay Inst, Johns Hopkins Univ, 51-52; sedimentologist, Gulf Res & Develop Co, 54-58; res assoc, Allan Hancock Found, Univ Southern Calif, 59-61; asst prof geol, Univ PR, 62-65 & Univ South Fla, 65-67; ASSOC PROF GEOL, CHAPMAN COL, 67- *Mem:* Geol Soc Am; Soc Econ Paleont & Mineral; Am Asn Petrol Geol. *Res:* Detrital sediments of barrier islands, beaches and continental shelf; sea floor topography. *Mailing Add:* 1311 E Washing Pl No E4 Santa Anna CA 92701

WIMBERLY, C RAY, b Wichita Falls, Tex, Aug 2, 36; m 59; c 2. HEAT TRANSFER, THERMODYNAMICS. *Educ:* Tex A&M Univ, BS, 61, PhD(mech eng), 68, Univ Ala, MS, 65. *Prof Exp:* Sr engr, Boeing Co, 61-64; proj engr, Marshall Space Flight Ctr, NASA, 64-67; sr specialist, LTV Aerospace Corp, 67-70; asst prof mech eng, Va Mil Inst, 70-73; prof mech eng, Univ Miss, 73-79; prof & head dept mech eng, Mont State Univ, 79-82; dean mech eng, La Tech Univ, 82-87; DEAN ENG, MEMPHIS STATE UNIV, 87- *Concurrent Pos:* Fac fel, Manned Space craft Ctr, NASA, 71-72; prin investr, US Army Res Off, 74-76 & 79; consult eng, Battelle Mem Inst-US Army Missile Command, 75 & Coput Sci Corp, 78. *Mem:* Am Soc Mech Engrs; Am Soc Eng Educ; Sigma Xi. *Res:* Engineering and research in flight dynamics, gas dynamics, fluid mechanics, heat transfer, laser technology, combustion, and associated areas with application to numerous flight and non-flight thermal and fluid dynamic systems. *Mailing Add:* Mech Eng Memphis State Univ Memphis TN 38152

WIMBUSH, MARK HOWARD, b Nairobi, Kenya, June 26, 36; US citizen; m 66; c 2. PHYSICAL OCEANOGRAPHY. *Educ:* Oxford Univ, BA, 57, MA, 64; Univ Hawaii, MA, 63; Univ Calif, San Diego, PhD(phys oceanog), 69. *Prof Exp:* Res assoc phys oceanog, Inst Geophys & Planetary Physics, La Jolla, 69-70; NSF fel, Inst Oceanog Sci, Eng, 70-71; from asst prof to assoc prof phys oceanog, Nova Univ, 71-77; FROM ASSOC PROF TO PROF PHYS OCEANOG, GRAD SCH OCEANOG, UNIV RI, 77- *Mem:* Am Geophys Union; Oceanog Soc. *Res:* Oceanic turbulence, tides and waves; dynamics of shelf and slope regions and interaction of bottom boundary layer flow with underlying sediment; equatorial oceanography. *Mailing Add:* Grad Sch Oceanog Univ RI Narragansett RI 02882-1197

WIMENITZ, FRANCIS NATHANIEL, b Philadelphia, Pa, Mar 9, 22; m 44; c 1. PHYSICS, ENGINEERING. *Educ:* Temple Univ, BA, 49, MA, 51. *Prof Exp:* Physicist acoust, Nat Bur Standards, 51-53; physicist mine fuze develop, Diamond Ord Fuze Lab, 53-58; res supvr nuclear weapons effects, Harry Diamond Labs, 58-64, chief, Nuclear Weapon Effects Br, 64-70, chief, Nuclear Weapons Effects Prog Off, 70-80; fac mem, George Washington Univ, 80-81; assoc sr scientist, Kaman Tempo, 81-87; Scientist, 87-90, CONSULT, WEAPONS EFFECTS SURVIVABILITY/VULNERABILTIY, KAMAN SCI, 90- *Mem:* AAAS. *Res:* Nuclear weapons effects; radiation transport; nuclear weapons electromagnetic pulse measurement; operations research; nuclear weapons effects simulation. *Mailing Add:* 1024 Chiswell Lane Silver Spring MD 20901

WIMER, BRUCE MEADE, b Tuckerton, NJ, Aug 31, 22; m 50; c 3. INTERNAL MEDICINE, HEMATOLOGY. *Educ:* Franklin & Marshall Col, BS, 43; Jefferson Med Col, MD, 46; Am Bd Internal Med, dipl, 56; Am Bd Hemat, dipl, 72. *Prof Exp:* Intern, Hosp, Jefferson Med Col, 46-47, resident internal med & hemat, 48-51; asst internal med & hemat, Guthrie Clin, Sayre, Pa, 53-59; pvt pract, Summit, NJ, 59-61; assoc med dir, Squibb Inst Med Res, 61-62; chief hemat & oncol, Lovelace Bataan Med Ctr, 62-81; ASSOC PROF MED & HEMAT, TEX TECH UNIV HEALTH CTR, 82- *Mem:* AMA; fel Am Col Physicians; Am Soc Hemat; Int Soc Hemat. *Res:* Cancer immunotherapy, especially adoptive leukocyte therapy and application of PHA as biological response modifier; therapeutic applications of bone marrow curettage; heparin-induced thrombocytopenie-thrombosis; hemolytic characteristics of the McLeod phenotype. *Mailing Add:* Dept Internal Med Tex Tech Univ Health Sci Ctr Lubbock TX 79430

WIMER, CYNTHIA CROSBY, b Boston, Mass, Oct 23, 33; m 57; c 3. BEHAVIORAL & NEURAL GENETICS. *Educ:* Wellesley Col, BA, 55; McGill Univ, MA, 58; Rutgers Univ, PhD(psychol), 61. *Prof Exp:* Res assoc phychol, Inst Develop Studies, NY Med Col, 61 & Jackson Lab, 63-69; assoc res scientist, div neurosci, Beckman Res Inst, City of Hope, 69-90; STATIST CONSULT, BAYSHORE DATA SERV, 91- *Res:* Behavior genetics; biometrics; neuroanatomical correlates of behavior. *Mailing Add:* 166 Bayshore Dr Morro Bay CA 93442

WIMER, DAVID CARLISLE, b Champaign, Ill, July 20, 26; m 55; c 1. ANALYTICAL CHEMISTRY, ORGANIC CHEMISTRY. *Educ:* Univ Ill, BS, 51. *Prof Exp:* From chemist to sr chemist, 51-65, group leader invest drugs res, 65-70, analytical res chemist, 70-80, sr analytical chemist, Analytical Res Dept, Abbott Labs, N Chicago, 80-87; RETIRED. *Mem:* Sigma Xi. *Res:* Acid-base interactions; non-aqueous solvent chemistry; functional group analysis, particularly organic nitrogen functions; thin layer chromatography; organic and inorganic qualitative analysis; ultraviolet absorption spectra of inorganic and organic compounds; ion chromatography. *Mailing Add:* 2312 11th St Winthrop Harbor IL 60096

WIMER, LARRY THOMAS, b Stuttgart, Ark, Dec 20, 36; m 59; c 3. INSECT PHYSIOLOGY. *Educ:* Phillips Univ, BA, 57; Rice Inst, MA, 59; Univ Va, PhD(physiol), 63. *Prof Exp:* Instr biol, Northwestern Univ, 62-64; asst prof, 64-70, assoc prof, 70-77, PROF BIOL & CHMN DEPT, UNIV SC, 77- *Mem:* AAAS; Am Soc Zoologists. *Res:* Insect developmental physiology; hormonal regulation of developmental metabolic systems. *Mailing Add:* Dept Biol Univ SC Col Main Campus Columbia SC 29208

WIMER, RICHARD E, b Tulare, Calif, Apr 8, 32; m 57; c 3. BEHAVIORAL GENETICS, NEUROGENETICS. *Educ:* San Jose State Col, AB, 52; Ohio Univ, MSc, 53; McGill Univ, PhD(psychol), 59. *Prof Exp:* Instr psychol, Douglass Col, Rutgers Univ, 58-60; sr res assoc psychiat, NY Med Col, 60-61; assoc staff scientist, Jackson Lab, 61-65, staff scientist, 65-69; SR RES SCIENTIST & CHIEF BEHAV & NEURAL GENETICS SECT, DIV NEUROSCI, CITY OF HOPE MED CTR, 69- *Mem:* Fel Am Psychol Asn; Genetics Soc Am; Am Genetic Asn; Soc Neurosci; Behav Genetics Asn; fel Am Psychol Soc. *Res:* Genetic variations in brain structure and correlated behavioral function. *Mailing Add:* City Hope Med Ctr Div Neurosci 1500 E Duarte Rd Duarte CA 91010

WIMMER, DONN BRADEN, b Pittsburg, Kans, Oct 14, 27; m 49; c 3. COMBUSTION CHEMISTRY, FUEL CHEMISTRY. *Educ:* Univ Kans, BS, 50, MS, 51. *Prof Exp:* Res chemist, 51-65, sr res chemist, 65-77, SECT SUPVR COMBUSTION, PHILLIPS PETROL CO, 77- *Concurrent Pos:* Mem, Eng & Sci Adv Comt, Coord Res Coun, Air Qual Comt, Am Petrol Inst, 76- *Honors & Awards:* Arch T Colewell Award, Soc Automotive Engrs, 74. *Mem:* Int Inst Combustion. *Res:* Reciprocating engine combustion and atmospheric photochemistry. *Mailing Add:* 123 NW Ramblewood Rd Bartlesville OK 74003

WIMMER, ECKARD, b Berlin, Germany, May 22, 36; m 65; c 2. MOLECULAR BIOLOGY, VIROLOGY. *Educ:* Univ Gottingen, diplom chemist, 59, PhD(org chem), 62. *Prof Exp:* Asst org chem, Univ Gottingen, 62-64; res fel biochem, Univ BC, 64-66; res assoc molecular biol, Univ Ill, 66-68; from asst prof to assoc prof microbiol, Sch Med, St Louis Univ, 68-74; assoc prof, 74-79, PROF MICROBIOL, SCH MED, STATE UNIV NY, STONYBROOK, 79-, CHMN DEPT, 84- *Concurrent Pos:* Vis prof, Mass Inst Technol, 69. *Mem:* Am Soc Microbiol; AAAS. *Res:* Molecular biology of animal viruses; biochemistry of nucleic acids and proteins; cell biology; the elucidation of the chemical structure of the poliovirion; mechanisms involved in polio polyprotein processing by proteases and in RNA replication; immunogenic properties, mechanism of entry into the cell, and the molecular basis of pathogencity of the poliovirion. *Mailing Add:* Dept Microbiol Sch Med State Univ NY Stony Brook NY 11794

WIMPRESS, GORDON DUNCAN, JR, b Riverside, Calif, Apr 10, 22; m 46; c 3. RESEARCH ADMINISTRATION. *Educ:* Univ Ore, BA, 46, MA, 51; Univ Denver, PhD(gen semantics), 58. *Hon Degrees:* LLD, Monmouth Col, 70; LHD, Tusculum Col, 71. *Prof Exp:* Instr jour, Whittier Col, 46-51; asst to pres, Colo Sch Mines, 51-59; pres, Monticello Col, 59-64, Monmouth Col, 64-70 & Trinity Univ, 70-77; vchmn, bd gov, 77-82, PRES, SOUTHWEST FOUND BIOMED RES, 82- *Concurrent Pos:* Consult, Burlington Northern RR, 67- & Valero Energy Corp, 85-; exec consult, Donald W Reynolds Found, 83-90. *Mailing Add:* 11102 Whisper Ridge San Antonio TX 78230

WIMS, ANDREW MONTGOMERY, b Phila, Pa, Apr 29, 35; m 59; c 4. PHYSICAL CHEMISTRY, POLYMER CHEMISTRY. *Educ:* Howard Univ, BS, 57, MS, 59, PhD, 67. *Prof Exp:* Res chemist, Nat Bur Standards, 60-68; SR STAFF RES SCIENTIST, GEN MOTORS RES LAB, 69- *Mem:* Am Chem Soc; Sigma Xi. *Res:* Characterization of polymeric materials; light scattering; electron microscopy; electron spectroscopy; x-ray diffraction; data analysis using computer methods. *Mailing Add:* Analytical Chem Dept Gen Motors Res Labs 30500 Mound Rd Warren MI 48090-9055

WINANS, RANDALL EDWARD, b Battle Creek, Mich, Jan 11, 49; m 75; c 3. PHYSICAL ORGANIC CHEMISTRY. *Educ:* Mich Technol Univ, BS, 71; Cornell Univ, MS, 73, PhD(chem), 76. *Prof Exp:* Fel, 75-77, asst chemist, 77-80, chemist, 80-89, GROUP LEADER, ARGONNE NAT LAB, 80-, SR CHEMIST, 90- *Concurrent Pos:* Consult, Gas Res Inst, 79-80, Helene Curtis, 83. *Honors & Awards:* Storch Award Fuel Sci, Am Chem Soc, 88. *Mem:* Am Chem Soc; AAAS. *Res:* Organic chemistry of coals and other fossil fuels; applications of mass spectrometry, microwave induced plasma emission spectroscopy, and synchrotron x-ray absorption spectroscopy; preparation of catalytic materials. *Mailing Add:* Chem Div Argonne Nat Lab Argonne IL 60439

WINAWER, SIDNEY J, b New York, NY, July 9, 31. INTERNAL MEDICINE, GASTROENTEROLOGY. *Educ:* NY Univ, BA, 52; State Univ NY, MD, 56. *Prof Exp:* Asst med, Harvard Med Sch, 62-64, instr, 65-66; asst prof, 66-72, CLIN ASSOC PROF MED, MED COL, CORNELL UNIV, 72- *Concurrent Pos:* Fel med, Boston City Hosp, 62-64, assist physician, 65-66; NIH spec fel, 65-67; asst physician, New York Hosp, 66-; dir gastrointestinal lab, 72-; asst clinician, Sloan-Kettering Inst. *Mem:* Fel Am Col Physicians; Am Gastroenterol Asn; Am Soc Gastrointestinal Endoscopy; Am Fedn Clin Res; Am Col Gastroenterol. *Res:* Clinical investigation in gastrointestinal diseases, particularly morphology, physiology, cell proliferation and other aspects of gastritis; malabsorptive studies such as massive bowel resection; clinical and investigative aspects of gastrointestinal and liver cancer. *Mailing Add:* 1275 York Ave New York NY 10021

WINBORN, WILLIAM BURT, b Victoria, Tex, Oct 6, 31; m 53; c 1. ANATOMY. *Educ:* Univ Tex, BS, 56; La State Univ, PhD(anat), 63. *Prof Exp:* From instr to asst prof anat, Med Units, Univ Tenn, Memphis, 63-68; asst prof, 68-69, ASSOC PROF ANAT, UNIV TEX HEALTH SCI CTR SAN ANTONIO, 69- *Mem:* AAAS; Electron Micros Soc Am; Am Soc Cell Biol; Am Asn Anatomists. *Res:* Electron microscopic studies of Islets of Langerhans and of the gastrointestinal tract; cytochemistry of the gastrointestinal tract. *Mailing Add:* Dept Cell & Struct Biol Univ Tex Health Sci Ctr 7703 Floyd Curl Dr San Antonio TX 78284-7762

WINBOW, GRAHAM ARTHUR, b Sedgley, Eng, Oct 31, 43; m 77; c 2. ACOUSTICS, PETROLEUM ENGINEERING. *Educ:* Cambridge Univ, BA, 65, PhD(physics), 68. *Prof Exp:* Sci Res Coun fel, Univ London, 68-69; NATO fel, Europ Orgn Nuclear Res, Geneva, Switz, 69-70; sr res assoc, Daresbury Lab, Warrington, Eng, 70-75; fel physics, Rutgers Univ, 75-78; sr res physicist geophys, 78-80, res specialist, 80-81, res supvr, 81-85, SR RES ASSOC, EXXON PROD RES CO, 85- *Mem:* Am Phys Soc; Soc Explor Geophys; Soc Indust & Appl Math; Acoust Soc Am. *Res:* Exploration geophysics; theoretical physics. *Mailing Add:* Exxon Prod Res Co PO Box 2189 Houston TX 77252-2189

WINBURY, MARTIN M, b New York, NY, Aug 4, 18; m 42; c 2. PHARMACOLOGY, PHYSIOLOGY. *Educ:* Long Island Univ, BS, 40; Univ Md, MS, 42; NY Univ, PhD(physiol), 51. *Prof Exp:* Economist, US Bur Mines, 42-44; mem staff biochem & pharmacol, Merck Inst Therapeut Res, 44-47; pharmacologist, Div Biol Res, G D Searle & Co, 47-55; sr pharmacologist, Schering Corp, 55-58, dir dept pharmacol, 58-61, assoc dir biol res, 61; dir, Dept Pharmacol, Warner Lambert Co, 61-80, dir Sci Develop, 80-86; PRES, INTERPHARM, 86- *Concurrent Pos:* Mem vis fac, Col Physicians & Surgeons, Columbia Univ; mem vis fac, Rutgers Univ; lectr, Univ Mich Med Sch, 86- *Mem:* NY Acad Sci; Am Soc Pharmacol & Exp Therapeut; Am Chem Soc; Am Heart Asn; Am Col Cardiol. *Res:* Pharmacology and physiology of cardiovascular, coronary and autonomic agents; distribution myocardial flow; microcirculation. *Mailing Add:* Pharm-Ex PO Box 8335 Ann Arbor MI 48107

WINCH, FRED EVERETT, JR, b Mass, June 16, 14; m 39; c 4. FORESTRY, SILVICULTURE. *Educ:* Univ Maine, BS, 36; Cornell Univ, MS, 37. *Prof Exp:* Forestry specialist, Soil Conserv Serv, USDA, 38-40, farm planning technician, 40-43; from asst prof to prof forestry, 43-76, actg head, Dept Natural Resources, 72-73, actg assoc dir agr exten, 73-74, EMER PROF FORESTRY, CORNELL UNIV, 76-; CONSULT, FOREST MGT & ENVIRON MGT, 76- *Mem:* Fel Soc Am Foresters. *Res:* Plantation establishment and early growth; maple syrup and Christmas tree production; forest recreation, resource development and conservation; land use inventory analysis and planning; forest tax impacts. *Mailing Add:* Warner Rd Box 312 Bradford NH 03221

WINCHELL, C PAUL, b Ionia, Mich, Oct 14, 21; m 46; c 3. MEDICINE. *Educ:* Univ Mich, MD, 45. *Prof Exp:* From instr to prof, 51-87,EMER PROF MED, SCH MED, UNIV MINN, MINNEAPOLIS, 87- *Concurrent Pos:* Consult cardiol, Vet Hosp, Minneapolis & Off Hearings & Appeals, Social Security Admin. *Mem:* Am Heart Asn; Am Fedn Clin Res; Am Col Physicians. *Res:* Cardiovascular diseases. *Mailing Add:* 420 Delaware St SE Minneapolis MN 55455

WINCHELL, HARRY SAUL, b Coaldale, Pa, Mar 1, 35; m 64; c 4. MEDICAL PHYSICS, NUCLEAR MEDICINE. *Educ:* Bucknell Univ, BA, 54; Hahnemann Med Col, MD, 58; Univ Calif, Berkeley, PhD(biophys), 61. *Hon Degrees:* DSc, Bucknell Univ, 72. *Prof Exp:* Intern, San Francisco Hosp, Univ Calif, 58-59; univ fel, Univ Calif, Berkeley, 59-61; resident, Mt Sinai Hosp, NY, 61-62; assoc res physician, Donner Lab, Univ Calif, Berkeley, 62-73, lectr med physics, Univ, 66-73; exec vpres & dir res & develop, Medi-Physics, Inc, 72-78, consult, 78-79; CONSULT, 79- *Concurrent Pos:* NSF fel, Univ Calif, Berkeley, 59-60, NIH fel, 60-61; spec consult, Sealab II Exp, La Jolla, Calif, 65. *Honors & Awards:* George Von Hevesy Award, Europ Orgn Nuclear Med, 69. *Mem:* Fel Am Col Physicians; Am Fedn Clin Res; Soc Nuclear Med; Soc Exp Biol & Med. *Mailing Add:* No 1 Via Òneg Lafayette CA 94549

WINCHELL, HORACE, b Madison, Wis, Jan 1, 15; m 37. MINERALOGY, GEOLOGY. *Educ:* Univ Wis, BA & MA, 36; Harvard Univ, MA, 37, PhD(mineral, crystallog), 41. *Hon Degrees:* MA, Yale Univ, 84. *Prof Exp:* Asst geologist, City Bd Water Supply, Honolulu, 38-40; res crystallographer, Hamilton Watch Co, 41-45; from instr to prof, 45-85, EMER PROF MINERAL & EMER CUR MINERAL, PEABODY MUS, YALE UNIV, 85- *Concurrent Pos:* Asst, Conn Geol Natural Hist Surv, 46-52. *Mem:* AAAS; fel Geol Soc Am; fel Mineral Soc Am; Geochem Soc; Soc Econ Geol; Mineral Soc London. *Res:* Physical properties of sapphire; sapphire jewel bearing; diamond dies, design and crystallography; grading of diamond powder; mineralogy of Connecticut; petrology of Oahu, Hawaii; systematic mineralogy; optical mineralogy and crystallography. *Mailing Add:* Dept Geol & Geophys Box 6666 New Haven CT 06511

WINCHELL, ROBERT E, b Wichita, Kans, Sept 21, 31; m 58; c 3. MINERALOGY, CRYSTALLOGRAPHY. *Educ:* Stanford Univ, BS, 56; Mich Col Mining & Technol, MS, 59; Ohio State Univ, PhD(mineral), 63. *Prof Exp:* Jr eng geologist, Bridge Dept, Calif Div Hwys, 57-58; res scientist, AC Spark Plug Div, Gen Motors Corp, 63-66; assoc prof, 66-72, PROF GEOL & MINERAL, CALIF STATE UNIV, LONG BEACH, 72- *Mem:* Mineral Soc Am; Brit Mineral Soc. *Res:* Mineral synthesis and characterization; optical, x-ray and morphological crystallography; x-ray diffraction; electron microscopy; phase equilibrium studies; crystal chemistry, growth and structure analysis; solid state and materials science. *Mailing Add:* Dept Geol Sci Calif State Univ Long Beach CA 90840

WINCHESTER, ALBERT MCCOMBS, b Waco, Tex, Apr 20, 08; m 34; c 1. BIOLOGY. *Educ:* Baylor Univ, AB, 29; Univ Tex, MA, 31, PhD(zool), 34. *Prof Exp:* Instr zool, Univ Tex, 32-34; prof, Lamar Col, 34-35; head dept biol, Ouachita Baptist Col, 35-36; head dept, Tenn State Teachers Col, 36-37; head dept biol & chmn div sci, Okla Baptist Univ, 37-43; prof zool, Baylor Univ, 43-46; head biol dept, Stetson Univ, 46-61; consult, Biol Sci Curric Study, Am Inst Biol Sci, Univ Colo, 61-62; prof, 62-78, EMER PROF BIOL, UNIV NORTHERN COLO, 78- *Concurrent Pos:* Rockefeller fel; Carnegie grant, 48; AAAS vis lectr, 64-78; mem, Colo Bd Exam Basic Sci, 66-78. *Mem:* AAAS; Genetics Soc Am; Am Soc Human Genetics; Am Eugenics Soc. *Res:* Genetics; effects of x-ray on Drosophila; induced sex ratio variation in Drosophila; radiation; human genetics. *Mailing Add:* 2316 15th St No 411 Greeley CO 80631

WINCHESTER, JAMES FRANK, b Glasgow, Scotland, Mar 24, 44. MEDICINE. *Educ:* Royal Col Physicians, MRCP, 72, FRCP, 82; Glasgow Univ, MBChB, 76, MD, 80. *Prof Exp:* From asst prof to assoc prof, 78-86, PROF MED, GEORGETOWN UNIV MED CTR, 87-, DIR, NEPHROLOGY DIV, 88- *Mem:* Int Soc Peritoneal Dialysis (secy & treas, 83-); Am Soc Nephrol; Nat Kidney Found; Int Soc Nephrol; Am Found Clin Res. *Res:* Nutrition and renal disease; drug poisoning; artificial organs; hypertension; general nephrology. *Mailing Add:* Georgetown Univ Hosp 3800 Reservoir Rd NW Washington DC 20007

WINCHESTER, JOHN W, b Chicago, Ill, Oct 8, 29; m 58; c 1. OCEANOGRAPHY. *Educ:* Univ Chicago, AB, 50, SM, 52; Mass Inst Technol, PhD(chem), 55. *Prof Exp:* Fulbright grant, Neth, 55-56; from asst prof to assoc prof geochem, Mass Inst Technol, 56-66; assoc prof meteorol & oceanog, Univ Mich, Ann Arbor, 67-69, prof oceanog & asst dir Great Lakes Res Div, 69-70; chmn dept, 70-76, PROF OCEANOG, FLA STATE UNIV, 70- *Mem:* Am Chem Soc; Geochem Soc; Geol Soc Am; Am Geophys Union; Sigma Xi. *Res:* Atmospheric and marine geochemistry. *Mailing Add:* 2405 Delgado Dr Tallahassee FL 32304

WINCHESTER, RICHARD ALBERT, b Denver, Colo, Nov 20, 21; m 47; c 3. AUDIOLOGY, SPEECH PATHOLOGY. *Educ:* Univ Denver, BA, 47, MA, 48; Univ Southern Calif, PhD(audiol, speech path), 57. *Prof Exp:* Resident audiol & speech path, Orthop Hosp, Los Angeles, 48-50; asst prof, Univ Denver, 50-53; res audiologist, Walter Reed Army Med Ctr, 53-54; dir hearing & speech clin, Vet Admin Hosp, San Francisco, 54-55, res audiologist, Vet Admin Regional Off, Los Angeles, 55-58; asst prof audiol, Sch Med, Temple Univ, 59-63; DIR DIV COMMUN DIS & RES AUDIOLOGIST, CHILDREN'S HOSP PHILADELPHIA, 63-; ASST PROF AUDIOL, SCH MED, UNIV PA, 64- *Concurrent Pos:* Dir audiol & speech path, Otologic Group Philadelphia, 59-66; consult, Pa Acad Ophthal & Otolaryngol, 60-; spec lectr, Univ Md, 63-64. *Mem:* AAAS; Am Speech & Hearing Asn; assoc fel Am Acad Ophthal & Otolaryngol; Am Cleft Palate Asn; Am Audiol Soc. *Res:* Speech in deafness; auditory perception in brain injury; nonorganic deafness; deafness in otosclerosis; congenital mixed deafness; hearing patterns in vestibular disorders; central auditory functions; sound spectrography and cleft palate speech; auditory behavior in infancy and early childhood. *Mailing Add:* 726 Clyde Circle Bryn Mawr PA 19010

WINCHESTER, ROBERT J, b Yonkers, NY, Jan 27, 37. IMMUNOLOGY, GENETICS. *Educ:* Cornell Univ, MD, 63. *Prof Exp:* Prof med, Rockefeller Univ, 60-79, Mt Sinai Sch Med, 80-86; PROF MED, NY UNIV, 86-; CHMN, DEPT RHEUMATICS, HOSP JOINT DIS, 86- *Mailing Add:* Dept Med NYU Med Col Hosp Joint Dis 301 E 17th St New York NY 10003

WINCHURCH, RICHARD ALBERT, b Newark, NJ, June 18, 36; m 61; c 2. IMMUNOLOGY, MICROBIOLOGY. *Educ:* Seton Hall Univ, AB, 58, MS, 67; Rutgers Univ, PhD(microbiol), 70. *Prof Exp:* Sr scientist microbiol, Smith, Kline & French Labs, 70-71, assoc sr investr pharmacol, 71-73; staff fel immunol, Baltimore Cancer Res Ctr, Nat Cancer Inst, 73-76, sr staff fel, 76-77; ASSOC PROF SURG IMMUNOL, SCH MED, JOHNS HOPKINS UNIV, 77- *Mem:* Am Asn Immunologists; Reticuloendothelial Soc; NY Acad Sci; AAAS. *Res:* Cellular immunology; zinc metabolism; trauma; gerontology; infectious diseases. *Mailing Add:* John Hopkins Univ Allergy & Asthma Ctr 301 Bayview Blvd Baltimore MD 21224

WINCKLER, JOHN RANDOLPH, b North Plainfield, NJ, Oct 27, 16; m 43; c 5. PHYSICS. *Educ:* Rutgers Univ, BS, 42; Princeton Univ, PhD(physics), 46. *Hon Degrees:* Doctor, Univ Paul Sabatier, France, 72. *Prof Exp:* Res physicist, Johns-Manville Corp, NJ, 37-42; instr physics, Palmer Lab, Princeton Univ, 46-49; from asst prof to assoc prof, 49-58, PROF PHYSICS, UNIV MINN, MINNEAPOLIS, 58- *Concurrent Pos:* Guggenheim fel, Meudon Observ, 65-66; mem math & phys sci div comt, NSF; Fulbright fel, Univ Paul Sabatien, France, 85. *Honors & Awards:* Space Sci Award, Am Inst Aeronaut & Astronaut, 62; Arctowski Medal, Nat Acad Sci, 78. *Mem:* Fel AAAS; fel Am Phys Soc; fel Am Geophys Union; Int Acad Astronaut. *Res:* High speed flow of gases and shock waves; geomagnetic effects and energy spectrum of primary cosmic rays; atmospheric total radiation; solar produced cosmic rays and energetic processes in solar flares; geomagnetic storm influences on energetic particles in the magnetosphere; space plasma physics, active experiments. *Mailing Add:* Sch Physics & Astron Univ Minn 2012 Irving Ave S Minneapolis MN 55405

WINCKLHOFER, ROBERT CHARLES, b Newark, NJ, Dec 14, 26; m 49; c 2. POLYMER PHYSICS. *Educ:* Columbia Univ, BS, 53. *Prof Exp:* Engr, Cent Res Lab, 53-61, group leader, 61-64, res supvr, 64, Fibers Div, Tech Dept, 64-68, mgr res, 68-70, tech dir heavy denier nylon, 70-72, tech dir advan technol, 72-77, res assoc, 77-84, MGR HIGH PERFORMANCE FIBERS, ALLIED FIBERS, 84- *Mem:* AAAS; Fiber Soc; Marine Technol Soc. *Res:* Fiber polymer physics; crystallization kinetics; polymer characterization; differential thermal analysis of polymers. *Mailing Add:* Allied Fibers Tech Ctr PO Box 31 Petersburg VA 23804

WINDEKNECHT, THOMAS GEORGE, b Owosso, Mich, Feb 13, 35; m 58; c 3. COMPUTER SCIENCE, SYSTEMS ENGINEERING. *Educ:* Univ Mich, BSE, 58, MSE, 59; Case Western Reserve Univ, PhD(systs eng), 64. *Prof Exp:* Mem tech staff systs eng, Space Technol Labs, Inc, 59-62; from asst prof to assoc prof, Case Western Reserve Univ, 64-70; prof elec eng, Mich Technol Univ, 70-72; prof info & comput sci, Ga Inst Technol, 72-73; prof math sci, Memphis State Univ, 73-81; assoc dean eng, 84-85, PROF ENG, OAKLAND UNIV, 81- *Concurrent Pos:* NSF grant, Case Western Reserve Univ, 67-69 & Mich Technol Univ, 70-72. *Mem:* Asn Comput Mach; Inst Elec & Electronics Engrs. *Res:* Formal methods in computer programming; microcomputer programming and graphics; dynamic system theory. *Mailing Add:* Comput Sci & Eng Dept Oakland Univ Rochester MI 48309

WINDELL, JOHN THOMAS, b Hammond, Ind, Apr 4, 30; m 59; c 3. AQUATIC BIOLOGY. *Educ:* Ind Cent Col, BS, 53; Ind Univ, MA, 58, PhD(limnol), 65. *Prof Exp:* Teacher, Griffith High Sch, 55-58; asst prof biol, Ind Cent Col, 58-62; assoc zool, Ind Univ, Bloomington, 62-65; asst prof, Ind Univ Northwest, 65-66; assoc prof, 66-70, PROF BIOL, UNIV COLO, BOULDER, 70- *Concurrent Pos:* Ind Univ fac fel, 66; partic, Int Symp Biol Basis Freshwater Fish Prod, Reading, Eng, 66. *Mem:* AAAS; Am Soc Limnol & Oceanog; Wetland Soc Am; Am Fisheries Soc; Sigma Xi. *Res:* Ecological physiology; biological basis of fish production; food consumption in fishes; conversion coefficients and the ecology of fishes; fish physiology, feeding, digestion, nutrition, population ecology, stream and wetland biology and habitat restoration. *Mailing Add:* Dept EPO Biol Univ Colo Campus Box 334 Boulder CO 80309

WINDELS, CAROL ELIZABETH, b Long Prarie, Minn, July 12, 48; m 70. PHYTOPATHOLOGY. *Educ:* St Cloud State Univ, BA, 70; Univ Minn, MS, 72, PhD(plant path), 80. *Prof Exp:* From jr scientist to scientist, Dept Plant path, Univ Minn, 73-84, asst prof plant path, 84-89, ASSOC PROF PLANT PATH, NW EXP STA, UNIV MINN, CROOKSTON, 89- *Concurrent Pos:* counr-at-large, Am Phytopath Soc, 90-93. *Mem:* Am Phytopath Soc (secy-treas, 87-90); Can Phytopath Soc; Int Soc Plant Path; Mycol Soc Am; Sigma Xi; Am Soc Sugar Beet Technologists. *Res:* Ecology and taxonomy of fusarium species; root rot, stalk rot, and other diseases of corn; ecology and control of seedling and root diseases of sugarbeet and wheat. *Mailing Add:* NW Exp Sta Univ Minn Crookston MN 56716

WINDER, CHARLES GORDON, b Ottawa, Ont, June 13, 22; m 48; c 2. GEOLOGY, GENERAL EARTH SCIENCES. *Educ:* Univ Western Ont, BSc, 49; Cornell Univ, MS, 51, PhD(geol), 53. *Prof Exp:* Lectr, 53-56, from asst prof to assoc prof, 56-64, head dept, 65-71, prof, 64-87, EMER PROF GEOL, UNIV WESTERN ONT, 87- *Mem:* Fel Geol Asn Can; fel Geol Soc Am; Can Soc Petrol Geologists; Sigma Xi. *Res:* Relationship between science and religion. *Mailing Add:* Dept Geol Univ Western Ont London ON N6A 5B7 Can

WINDER, DALE RICHARD, b Marion, Ind, Aug 27, 29; m 53. SOLID STATE PHYSICS. *Educ:* DePauw Univ, AB, 51; Univ Nebr, MA, 54; Case Inst Technol, PhD(physics), 57. *Prof Exp:* Lab & teaching asst physics, DePauw Univ, 50-51, Univ Nebr, 51-54 & Case Inst Technol, 54-57; physicist, Nat Carbon Res Labs, Union Carbide Corp, 57-60; asst prof, 60-64, ASSOC PROF PHYSICS, COLO STATE UNIV, 64- *Concurrent Pos:* Physicist, Boulder Lab, Nat Bur Standards, 62-69; Idaho Nuclear Corp-Asn Western Univs fac appointee, 67. *Mem:* Am Phys Soc; Am Asn Physics Teachers; Am Crystallog Asn; Sigma Xi. *Res:* Photoelectric effect; lattic dynamics; crystal growth; radiation damage; nuclear fuels and moderators; transport property measurements. *Mailing Add:* 1430 W Oak St Ft Collins CO 80521

WINDER, ROBERT OWEN, b Boston, Mass, Oct 9, 34; m; c 5. COMPUTER SCIENCE. *Educ:* Univ Chicago, AB, 54; Univ Mich, BS, 56; Princeton Univ, MS, 58, PhD(math), 62. *Prof Exp:* Engr comput, RCA Corp, 57-58, mem tech staff, RCA Labs, 58-69, head group, 69-75, dir microprocessors, Solid State Div, 75-78; mgr work sta develop, Exxon Off Systs, Syntex Comput Systs, 78-85, vpres, 85-88; MGR PROD DEVELOP, INTEL CORP, 88- *Honors & Awards:* David Sarnoff Award, 76. *Mem:* Fel Inst Elec & Electronics Engrs. *Res:* Threshold logic. *Mailing Add:* 24 Deer Path Princeton NJ 08540

WINDER, WILLIAM CHARLES, b Salt Lake City, Utah, Nov 16, 14; m 39. FOOD SCIENCE, DAIRY INDUSTRY. *Educ:* Utah State Univ, BS, 46, MS, 48; Univ Wis, PhD(dairy indust), 49. *Prof Exp:* Plant supt, Winder Dairy, Utah, 34-45; from instr to assoc prof, 49-60, prof, 60-81, EMER PROF FOOD SCI, UNIV WIS-MADISON, 81- *Mem:* Am Dairy Sci Asn; Sigma Xi. *Res:* Effects of ultrasound on food products; physical and chemical effects of freezing and drying food products; analysis of foods. *Mailing Add:* 418 Critchell Terrace Madison WI 53711

WINDER, WILLIAM W, b Vernal, Utah, Sept 12, 42; m 64; c 8. ENDOCRINOLOGY, EXERCISE PHYSIOLOGY. *Educ:* Brigham Young Univ, BS, 66, PhD(zool), 71. *Prof Exp:* Teaching fel, Wash Univ Sch Med, 71-74, asst prof, 73-79; assoc prof physiol, Univ SDak Sch Med, 79-82, head, physiol & pharmacol sect, 81-82; assoc prof, 82-86, PROF ZOOL, BRIGHAM YOUNG UNIV, 86- *Concurrent Pos:* Prin investr liver metab, NIH grant, 78- *Mem:* Endocrine Soc; Am Physiol Soc; Am Col Sports Med. *Res:* Hormonal regulation of liver and muscle glycogenolysis; gluconeogenesis during exercise. *Mailing Add:* Dept Zool Brigham Young Univ Provo UT 84602

WINDHAGER, ERICH E, b Vienna, Austria, Nov 4, 28; US citizen; m 56; c 2. PHYSIOLOGY, BIOPHYSICS. *Educ:* Univ Vienna, MD, 54. *Prof Exp:* Fel biophys, Harvard Med Sch, 56-58; instr physiol, Med Col, Cornell Univ, 58-61; vis scientist, Biochem Inst, Univ Copenhagen, 61-63; from asst prof to prof physiol, 63-78, MAXWELL M UPSON PROF PHYSIOL & BIOPHYS, MED COL, CORNELL UNIV, 78-, CHMN DEPT PHYSIOL 73- *Concurrent Pos:* Career scientist, Res Coun New York, 63-71 & Irma Hirschl Found, 73-78; sect ed, Am J Physiol, 69-74. *Honors & Awards:* Homer W Smith Award. *Mem:* Am Physiol Soc; Biophys Soc; Int Soc Nephrology; Harvey Soc; Am Soc Nephrology. *Res:* Renal tubular transfer of electrolytes; electrophysiology of the nephron; micropuncture techniques and nephron function; kidney, water and electrolytes. *Mailing Add:* Dept Physiol Cornell Univ Med Col New York NY 10021

WINDHAM, CAROL THOMPSON, b Houston, Tex, Mar 14, 48; m 70; c 2. NUTRITIONAL BIOCHEMISTRY, NUTRITION EDUCATION. *Educ:* Rice Univ, BA, 70; Utah State Univ, PhD(nutrit), 82. *Prof Exp:* ASSOC PROF NUTRIT, DEPT NUTRIT & FOOD SCI, UTAH STATE UNIV, 83- *Mem:* Sigma Xi; Soc Nutrit Educ; AAAS; Am Dietetic Asn; assoc Am Inst Nutrit. *Res:* Evaluation of dietary status of national and international populations; establishment of reliable, documented food composition data bases; determination of food supply, human requirement and dietary standards for vitamin A; development of scientifically based nutrition education materials. *Mailing Add:* Dept Nutrit & Food Sci Utah State Univ Logan UT 84322-8700

WINDHAM, MICHAEL PARKS, b Houston, Tex, Sept 23, 44; m 70; c 2. MATHEMATICS. *Educ:* Rice Univ, BA, 66, MA & PhD(math), 70. *Prof Exp:* Instr math, Univ Miami, 70-71; from asst prof to assoc prof, 71-86, PROF MATH, UTAH STATE UNIV, 86- *Mem:* Soc Indust & Appl Math; Math Asn Am; Am Math Soc; Classification Soc NAm. *Res:* Cluster analysis; numerical optimization. *Mailing Add:* Dept Math & Statist Utah State Univ UMC 3900 Logan UT 84322-3900

WINDHAM, RONNIE LYNN, b Jasper, Tex, Mar 2, 43; m 72. ANALYTICAL CHEMISTRY. *Educ:* Pan Am Univ, BA, 65; Eastern NMex Univ, MS, 69; Tex A&M Univ, PhD(anal chem), 71. *Prof Exp:* Proj chemist, Jefferson Chem Co, 71-81; SUPVR ANALYTICAL CHEM, TEXACO CHEM CO, AUSTIN LABS, 81- *Mem:* Am Chem Soc; Am Soc Qual Control. *Res:* Atomic absorption spectrophotometry; ion chromatography; trace analysis; isotachophoresis. *Mailing Add:* 11608 January Dr Austin TX 78753

WINDHAM, STEVE LEE, b Miss, Sept 19, 22; m 48; c 3. HORTICULTURE. *Educ:* Miss State Col, BS, 43, MS, 48; Mich State Col, PhD(hort), 53. *Prof Exp:* Asst, Miss State Col, 47-48; asst horticulturist 48-51, assoc horticulturist, 53-61, HORTICULTURIST, TRUCK CORPS BR, MISS AGR EXP STA, MISS STATE UNIV, 61-, ADJ ASSOC PROF HORT, 74- *Concurrent Pos:* Asst, Mich State Col, 51-53. *Res:* Vegetable crop response and utilization of nutrient elements. *Mailing Add:* Box 139 Lena MS 39094

WINDHAUSER, MARLENE M, b Lincoln, Nebr, Mar 20, 54; m 75. CHOLESTEROL METABOLISM, DIETARY FIBERS. *Educ:* Colo State Univ, BS, 75, MS, 77; La State Univ, PhD(physiol), 88. *Prof Exp:* Nutritionist, Miss State Bd Health, 78-79; clin dietitian, NMiss Retardation Ctr, 79-81, food serv dir, 81-82; asst mgr, Residence Food Serv, La State Univ, 82-83, res assoc, 89-91; RES DIETITIAN, PENNINGTON BIOMED RES CTR, 91-, DIR, METAB KITCHEN, 91- *Mem:* Am Inst Nutrit; Am Dietetic Asn; Am Heart Asn. *Res:* Cholesterol metabolism as affected by dietary fats, rice bran, rice bran oil and other dietary fibers, including the changes in HMG-CoA reductase activity. *Mailing Add:* 2826 Dakin Ave Baton Rouge LA 70820

WINDHOLZ, THOMAS BELA, b Arad, Romania, Jan 10, 23; US citizen; m 48; c 2. ORGANIC CHEMISTRY. *Educ:* Univ Cluj, MS, 47. *Prof Exp:* Res chemist, Chinoin Pharmaceut Co, Budapest, 48-50; sr res chemist, Res Inst Pharmaceut Indust, 51-54, sect head, 55-56; res chemist, Res Labs, Celanese Corp Am, NJ, 57-59; sr res chemist, 60-63, sect head, 64-69, assoc dir, 70-72, dir, Int Regulatory Affairs, 72-75, dir, Proj Planning & Mgt, 75-80, SR DIR HUMAN HEALTH PROD DEVELOP, RES LABS, MERCK & CO, INC, 81- *Mem:* Am Chem Soc. *Res:* Synthetic organic chemistry; steroids and other natural products; medicinal chemistry, metabolism; international relations and management. *Mailing Add:* 20 E Ninth St Apt 18F New York NY 10003-5944

WINDHOLZ, WALTER M, b Gorham, Kans, Apr 25, 33; m 61; c 3. NUMERICAL ANALYSIS. *Educ:* Ft Hays Kans State Col, AB, 53; Kans State Univ, MS, 58. *Prof Exp:* Mathematician, Thiokol Chem Corp, Utah, 61-65; mathematician, Kaman Sci Corp, 65-83; SR MATHEMATICIAN, SCI APPL INT, INC, 83- *Mem:* Math Asn Am. *Res:* Applied mathematics; structural dynamics; computational electromagnetics. *Mailing Add:* 1555 Gatehouse Circle Apt A1 Colorado Springs CO 80904

WINDHORN, THOMAS H, b New Ulm, Minn, Dec 28, 47. SEMICONDUCTOR DEVICES, LASER DIODES. *Educ:* Mankato State Col, BS, 73; Univ Ill, Urbana-Champaign, MS, 77, PhD(elec eng), 82. *Prof Exp:* Res assoc, Elec Eng Res Lab, Univ Ill Urbana-Champaign, 82; res staff scientist, Lincoln Lab, Mass Inst Technol, 82-87; RES STAFF ENGR, AMOCO CORP RES CTR, 87- *Mem:* Inst Elec & Electronics Engrs; Am Phys Soc. *Res:* Gallium arsenide lasers on Si substrates; surface-emitting diode laser arrays in gallium arsenide; development of high-power diode lasers. *Mailing Add:* Amoco Res Ctr PO Box 3011 MS F-4 Naperville IL 60566

WINDISCH, RITA M, b Pittsburgh, Pa. CLINICAL CHEMISTRY. *Educ:* Duquesne Univ, BS, 60, PhD(chem), 64. *Prof Exp:* Chief clin chemist path, 65-80, DEP CHIEF CLIN CHEM & CRITICAL CARE PATH, MERCY HOSP, 80- *Concurrent Pos:* Clin prof, Sch Med Technol, Carlow Col, 65-; clin prof, Sch Med Technol, Duquesne Univ, 69-; med staff affil, Div Clin Chem & Critical Care, Dept Path, Mercy Hosp, 69- *Mem:* Am Asn Clin Chemists; Am Chem Soc. *Res:* Diabetes, clinical chemistry and toxicology. *Mailing Add:* 25 Thorncrest Dr Pittsburgh PA 15235-5215

WINDLER, DONALD RICHARD, b Centralia, Ill, Feb 4, 40; div; c 1. PLANT TAXONOMY. *Educ:* Southern Ill Univ, Carbondale, BS, 63, MA, 65; Univ NC, Chapel Hill, PhD(bot), 70. *Prof Exp:* From asst prof to assoc prof, 69-77, PROF BIOL, TOWSON STATE UNIV, 77-, CUR HERBARIUM, 69-, CHMN DEPT, 84- *Concurrent Pos:* Nat Defense Educ Act fel, 65-68; actg dean, Col Natural & Math Sci, Towson State Univ, 87-90. *Mem:* Int Asn Plant Taxonomists; Am Soc Plant Taxon; Torrey Bot Club; Sigma Xi; Soc Econ Bot. *Res:* Systematics of Leguminosae, Crotalaria, Mucuna and Neptunia; flora of Maryland and Delaware; lichens of Maryland; plant distribution in the Eastern United States. *Mailing Add:* Herbarium Towson State Univ Baltimore MD 21204

WINDOM, HERBERT LYNN, b Macon, Ga, Apr 23, 41; m 63; c 2. OCEANOGRAPHY, GEOCHEMISTRY. *Educ:* Fla State Univ, BS, 63; Univ Calif, San Diego, MS, 65, PhD(earth sci), 68. *Prof Exp:* PROF OCEANOG, SKIDAWAY INST OCEANOG & GA INST TECHNOL, 68- *Concurrent Pos:* Chmn, GESAMP, Mem GEMSI. *Mem:* Am Soc Limnol & Oceanog; Int Coun Explor Sea; Am Geophys Union. *Res:* Marine environmental quality; chemical oceanography; marine biogeochemistry of trace elements; marine sediments; environmental effects of dredging. *Mailing Add:* Skidaway Inst Oceanog Box 13687 Savannah GA 31416

WINDSOR, DONALD ARTHUR, b Chicago, Ill, Mar 22, 34; m 63, 69; c 4. INFORMATION SCIENCE, BIOLOGICAL DIVERSITY. *Educ:* Univ Ill, Urbana, BS, 59, MS, 60; State Univ NY, Binghamton, MS, 82. *Prof Exp:* Unit leader, 66-67, sect chief, 67-74, INFO SCIENTIST, PROD DEVELOP DEPT, NORWICH EATON PHARMACEUT, 74-; RES DIR, SCI AESTHETICS INST, 69- *Concurrent Pos:* Trustee, Cent NY chap, Nature Conservancy. *Mem:* AAAS; Am Soc Info Sci; Nature Conservancy; Soc Conserv Biol. *Res:* Applications of general systems principles to the investigation of real-world phenomena; evolution models; animal taxonomy; preservation of biological diversity; information transfer using bibliometric traits. *Mailing Add:* Sci Aesthetics Inst PO Box 604 Norwich NY 13815-0604

WINDSOR, JOHN GOLAY, JR, b Chester, Pa, Nov 20, 47. ENVIRONMENTAL CHEMISTRY, MARINE SCIENCE. *Educ:* PMC Col, Pa, BS, 69; Col William & Mary, MA, 72, PhD(marine sci), 77. *Prof Exp:* Res asst, Va Inst Marine Sci, 72-74; res assoc, Mass Inst Technol, 76-78; sr proj scientist, Northrop Servs, Inc, 78-82; ASSOC PROF CHEM OCEANOG, FLA INST TECHNOL, 82- *Mem:* Am Chem Soc; AAAS; Marine Technol Soc. *Res:* Application of modern instrumental methods, for example gas chromatography and mass spectrometry to trace organic analysis of environmental mixtures such as air, soil, water and sediments; development of environmental education programs. *Mailing Add:* Dept Oceanog Fla Inst Technol 150 W Univ Blvd Melbourne FL 32901-6988

WINDSOR, MAURICE WILLIAM, b Kent, Eng, Feb 28, 28; m 53; c 3. PHOTOCHEMISTRY, PHOTOSYNTHESIS. *Educ:* Univ Cambridge, BA, 52, PhD(phys chem), 55, MA, 57. *Prof Exp:* Res assoc, Calif Inst Technol, 55-58, Univ Sheffield, Eng, 56-67; mgr, Chem Sci Dept, TRW Systs, Redondo Beach, Calif, 58-71; prof & chmn dept, 71-74, PROF CHEM & MAT SCI, WASH STATE UNIV, 74- *Concurrent Pos:* Sr scholar, Royal Exhib of 1851, 56-58; guest scientist, Nat Bur Standards, 58-59; prin investr, Sch Aerospace Med, Off Naval Res, Army Res Off, Air Force Off Sci Res Nat Sci Found, 60-; consult, Appl Photophysics, London Eng, 74-77; sr vis fel, Royal Inst, London, Eng, 78-79, Tech Univ, Berlin, 85; vis prof, Univ Paris, Orsay, 79, Univ Cambridge, 84-85, Tech Univ, Berlin, 85. *Mem:* AAAS; Am Chem Soc; Royal Soc Chem London; Royal Inst London; Inter-Am Photochem Soc; Sigma Xi. *Res:* Laser photochemistry and photobiophysics; design and development of instrumentation for ultrafast (nanosecond to picosecond) flash photolysis and kinetic spectroscopy; study of primary events, including electron transfer, in photosynthesis and related model systems. *Mailing Add:* Dept Chem Wash State Univ Pullman WA 99164-4630

WINDSOR, RICHARD ANTHONY, b Baltimore, Md, Aug 7, 43; m 78; c 2. BEHAVIORAL SCIENCE. *Educ:* Morgan State Univ, BS, 69; Univ Ill, MS, 70, PhD(educ res & evaluation), 72; Johns Hopkins Univ, MPH, 76. *Prof Exp:* Asst prof, Col Educ, Ohio State Univ, 72-75; asst prof pub health, Sch Hyg & Pub Health, Johns Hopkins Univ, 76-77; PROF, SCH PUB HEALTH, SCI DIABETES RES & TREAT CTR & SR SCIENTIST, COMPREHENSIVE CANCER CTR, UNIV ALA, BIRMINGHAM, 77- *Concurrent Pos:* Prin investr, Nat Ctr Health Svcs Res, Nat Cancer Inst, Nat heart, Lung & Blood Inst, Ctr Dis Control; sr res fel fac med, Univ Edinburgh; vis prof community med, epidemiol & social res, Univ Manchester, 84. *Mem:* Am Pub Health Asn; Soc Pub Health Educ (vpres, 79, pres elect, 80, pres, 81); Soc Behav Med; Eval Res Soc US; Int Union Health Educ (pres elect, 82-85, pres, 85-88). *Res:* Smoking cessation interventions for pregnant women; public health; maternal and child health; evaluation health services research. *Mailing Add:* Sch Pub Health Univ Ala Birmingham AL 35294

WINE, JEFFREY JUSTUS, b Pittsburgh, Pa, Feb 10, 40; m 66; c 2. NEUROSCIENCES. *Educ:* Univ Pittsburgh, BS, 66; Univ Calif, Los Angeles, PhD(psychol), 71. *Prof Exp:* NIH fel, Dept Biol Sci, 71-72, from asst prof to assoc prof, 72-86, PROF PSYCHOL, STANFORD UNIV, 86- *Mem:* Soc Neurosci; Int Brain Res Orgn; Soc Exp Biol; AAAS; Am Physiol Soc. *Res:* Cystic fibrosis; neurophysiological and neuroanatomical analysis of invertebrate behavior. *Mailing Add:* Cystic Fibrosis Res Lab Stanford Univ Stanford CA 94305-2130

WINE, PAUL HARRIS, b Detroit, Mich, Mar 18, 46; m 74; c 2. CHEMICAL PHYSICS. *Educ:* Univ Mich, BS, 68; Fla State Univ, PhD(phys chem), 74. *Prof Exp:* Robert A Welch fel chem, Univ Tex, Dallas, 74-76; from res scientist to sr res scientist, 77-86, head molecular Sci Br, 87-90, PRIN RES SCIENIST, GA TECH RES INST, 86-, HEAD CHEM PHYSICS BR, 90- *Mem:* Am Chem Soc; Am Phys Soc; Am Geophys Union; Inter-Am Photochem Soc; Sigma Xi. *Res:* Gas phase kinetics; photochemistry; reaction dynamics; lasers; spectroscopy; atmospheric chemistry. *Mailing Add:* Phys Sci Lab Ga Tech Res Inst Ga Inst Technol Atlanta GA 30332-0800

WINE, RUSSELL LOWELL, b Indian Springs, Tenn, Aug 17, 18; m 42; c 3. STATISTICS. *Educ:* Bridgewater Col, BA, 41; Univ Va, MA, 45; Va Polytech Inst, PhD(statist), 55. *Prof Exp:* Instr math, Univ Va, 43-45, Amherst Col, 45-46 & Univ Okla, 46-47; asst prof, Washington & Lee Univ, 47-52; assoc prof statist, Va Polytech Inst, 55-57; from assoc prof to prof, 57-85, EMER PROF STATIST, HOLLINS COL, 85- *Mem:* Fel AAAS; Am Statist Asn; Sigma Xi. *Res:* Least squares; design of experiments; sample surveys; author of two books: Statistics for Scientists and Engineers & Beginning Statistics. *Mailing Add:* 4510 Cloverdale Rd Roanoke VA 24019

WINEBURG, ELLIOT N, b Hornell, NY, May 22, 28; m 57; c 3. BEHAVIORAL MEDICINE. *Educ:* Univ Rochester, BA, 48; Fordham Univ, MA, 50; Univ Zurich, MD, 56. *Prof Exp:* Instr psychol, Hunter Col, 49-59; ASSOC PSYCHIAT, MT SINAI SCH MED, 67-, ASST PROF, 74- *Concurrent Pos:* Asst attend psychiat & dir courses hypnother, Mt Sinai Hosp, 62-; attend psychiat, Parkway Hosp, 65-; vis mem, Albert Einstein Col Med, 74-; dir, Asn Biofeedback Med Group, 79- *Mem:* Fel Am Psychiat Asn; Am Soc Clin Hypn; Biofeedback Soc Am. *Res:* Treatment of stress related disorders of psychological and physical origin; specific use of biofeedback and hypnotherapy as applied to stress related diseases. *Mailing Add:* 145 West 58th St New York NY 10019

WINEFORDNER, JAMES D, b Geneseo, Ill, Dec 31, 31; m 57; c 3. ANALYTICAL CHEMISTRY. *Educ:* Univ Ill, BS, 54, MS, 55, PhD(analytical chem), 58. *Prof Exp:* Fel, Univ Ill, 58-59; from asst prof to assoc prof, 59-67, PROF CHEM, UNIV FLA, 67- *Concurrent Pos:* Hon mem, Japan Soc Analytical Chem, 87. *Honors & Awards:* Meggers Award in Spectros, 69; Am Chem Soc Award Analytical Chem, 73; Am Chem Soc Anal Div Award Chem Instrumentation, 78; Theophilus Redwood Award, 81; Turbern Bergman Award, 87; Anal Div Spectrochem Anal Award, Am Chem Soc, 87. *Mem:* Soc Appl Spectros; Am Chem Soc. *Res:* Atomic, ionic and molecular emission; absorption; fluorescence spectroscopy; gas chromatographic detectors; trace analysis. *Mailing Add:* Dept Chem Univ Fla Gainesville FL 32611

WINEGARD, WILLIAM CHARLES, b Hamilton, Ont, Sept 17, 24; m 47; c 3. METALLURGY. *Educ:* Univ Toronto, BASc, 49, MASc, 50, PhD(metall), 52. *Hon Degrees:* LLD, Univ Toronto, 71; DEng, Mem Univ, 76. *Prof Exp:* Spec lectr, Univ Toronto, 50-52, from asst prof to prof metall, 54-67, asst dean sch grad studies, 64-67; pres & vchancellor, Univ Guelph, 67-75; vchmn, Ont Coun Univ Affairs, 76-77, chmn, 77-82; RETIRED. *Concurrent Pos:* Vis prof, Cambridge Univ, 59-60; consult, Ont Fire Marshall, 50-65, A D Lettle Inc, 60-67; ed, Can Metall Quart, 64-67; pres, Can Bur Int Educ, 71-73; gov, Int Develop Res Ctr, 74-80; mem, Govt Ont Res Found, 78-; fel Guelph, 78. *Honors & Awards:* Alcan Award, Can Inst Mining & Metall, 67. *Mem:* Fel Am Soc Metals; Can Inst Mining & Metall; Can Coun Prof Engrs. *Res:* Solidification of pure metals and alloys; grain boundary migration. *Mailing Add:* House of Commons Rm 256 Confederation Bldg Ottawa ON K1A 0A6 Can

WINEGARTNER, EDGAR CARL, b Cleveland, Ohio, Jan 28, 27; m 49; c 2. COMBUSTIBILITY OF SOLIDS. *Educ:* Ohio State Univ, BMetE, 49. *Prof Exp:* Plant engr beryllium prod, Brush Beryllium Co, 49-51; res engr corrosion, Exxon Co, 51-62; res engr wood technol, Exxon Res & Eng Co, 62-65, res assoc coal combustion, 65-85; CONSULT FUEL TECHNOL, 86- *Honors & Awards:* Fel Am Soc Mech Engrs. *Mem:* Am Soc Mech Engrs; Soc Mining Engrs; Am Soc Metals; Combustion Inst; Nat Asn Corrosion Engrs. *Res:* Investigation of combustion related properties of coal and solid by-products from synthetic fuels, including combustibility, fouling and slagging; size preparation of coal for synthetic fuels processes. *Mailing Add:* 408 Rollingwood Rd Baytown TX 77520

WINEGRAD, SAUL, b Philadelphia, Pa, Mar 15, 31; m 63; c 2. PHYSIOLOGY. *Educ:* Univ Pa, BA, 52, MD, 56. *Prof Exp:* Intern, Peter Bent Brigham Hosp, 56-57; sr asst surgeon, NIH, 57-59, surgeon, 60-61; fel, Nat Heart Inst, 59-60; hon res assoc, Univ Col, Univ London, 61-62; from asst prof to assoc prof, 62-69, PROF PHYSIOL, SCH MED, UNIV PA, 69- *Concurrent Pos:* Assoc, Sch Med, George Washington Univ, 58-61; NSF sr fel, Univ Col, Univ London, 71-72. *Mem:* Am Physiol Soc; Soc Gen Physiol; Biophys Soc; Cardiac Muscle Soc. *Res:* Cardiovascular and muscle physiology. *Mailing Add:* Sch Med Univ Pa Richards Bldg G4 Philadelphia PA 19104

WINEHOLT, ROBERT LEESE, b York, Pa, Sept 20, 39; m 62; c 3. ORGANIC CHEMISTRY. *Educ:* Gettysburg Col, AB, 61; Univ Del, PhD(org chem), 66. *Prof Exp:* Fel, Duke Univ, 66; sr chemist, Hoffman LaRoche, Inc, 67-73, Mallinckrodt, Inc, 73-76; RES ASSOC, CROMPTON & KNOWLES, 77- *Mem:* Am Chem Soc. *Res:* Process development for organic chemicals to plant production. *Mailing Add:* 2631 Filbert Ave Reading PA 19606-2143

WINEK, CHARLES L, b Erie, Pa, Jan 13, 36; m 60; c 3. TOXICOLOGY, PHARMACOLOGY. *Educ:* Duquesne Univ, BS, 57, MS, 59; Ohio State Univ, PhD(pharmacol), 62. *Prof Exp:* Res assoc phytochem, Ohio State Univ, 59-62; res toxicologist, Procter & Gamble Col, 62-63; from asst prof pharmacol & toxicol to assoc prof toxicol, 63-69, PROF TOXICOL, DUQUESNE UNIV, 69-; CHIEF TOXICOLOGIST, ALLEGHENY COUNTY CORONER'S OFF, 66- *Concurrent Pos:* Consult, Dept Anesthesiol, St Francis Hosp, 67-; mem panel ther, Poison Control Ctrs, Dept Health; mem adv comt lab act, Pa Dept Health; mem adv bd, Drug Res Proj, Franklin Inst, Philadelphia; fac mem, Bur Narcotics & Dangerous Drugs, Police Educ Prog; adj prof, Sch Med, Univ Pittsburgh; ed at large toxicol, Marcel Dekker, Inc, New York; ed, Toxicol Newslett, Sch Pharm, Duquesne Univ; ed, Toxicol Ann, 74. *Mem:* Soc Toxicol; Am Acad Forensic Sci; Acad Pharmaceut Sci; Am Asn Poison Control Ctrs; Drug Info Asn. *Res:* Toxicity of antifungal agents; safety evaluations; rapid methods of toxicological analyses. *Mailing Add:* Dept Pharmacol Duquesne Univ Pittsburgh PA 15219

WINEMAN, ALAN STUART, b Wyandotte, Mich, Nov 17, 37; m 64; c 2. APPLIED MECHANICS, APPLIED MATHEMATICS. *Educ:* Univ Mich, BSE, 59; Brown Univ, PhD(appl math), 64. *Prof Exp:* From asst prof to assoc prof, 64-75, PROF APPL MECH, UNIV MICH, 75- *Mem:* Am Acad Mech; Soc Rheology; Am Soc Mech Engrs; Soc Natural Philos. *Res:* Viscoelasticity; numerical methods; nonlinear elasticity; continuum mechanics. *Mailing Add:* Dept Mech Eng & Appl Mech Univ Mich Ann Arbor MI 48109

WINEMAN, ROBERT JUDSON, b Chicago, Ill, 1919; m 44; c 5. BIO-ORGANIC CHEMISTRY, BIOMATERIALS. *Educ:* Williams Col, AB, 41; Univ Mich, MS, 42; Harvard Univ, PhD(chem), 49. *Prof Exp:* Chemist, E I du Pont de Nemours & Co, 42-43; res chemist, Monsanto Chem Co, 49-53, res group leader, 54-60, Monsanto Res Corp, Mass, 60-61, dir, Boston Lab, 61-69; dir biomed res labs, Am Hosp Supply Corp, 69-70; assoc chief, Artificial Kidney-Chronic Uremia Prog, Nat Inst Arthritis, Metab & Digestive Dis, 70-78, DIR, CHRONIC RENAL DIS PROG, NAT INST ARTHRITIS, DIABETES, DIGESTIVE & KIDNEY DIS, 79- *Concurrent Pos:* Instr, Northeastern Univ, 52-53. *Mem:* AAAS; Am Chem Soc; Am Soc Artificial Internal Organs. *Res:* Organic synthesis; steroids; amino acids; sulfur compounds; medical devices; artificial organs. *Mailing Add:* Box 306 East Orleans MA 02643-0306

WINER, ALFRED D, b Lynn, Mass, Dec 24, 25; m 55; c 2. BIOCHEMISTRY. *Educ:* Northeastern Univ, BS, 48; Purdue Univ, MS, 50; Duke Univ, PhD(biochem), 57. *Prof Exp:* Instr org chem, Univ Mass, 50-51; USPHS fel, Med Nobel Inst, Sweden, 58-60; ASSOC PROF BIOCHEM, MED CTR, UNIV KY, 65- *Concurrent Pos:* USPHS career develop award, 60-70. *Mem:* Am Chem Soc; Am Soc Biol Chemists. *Res:* Mechanism of action of dehydrogenase-coenzyme complexes; hormonal effects on enzymes in spermatogenesis. *Mailing Add:* Dept Biochem Univ Ky Med Ctr Lexington KY 40506

WINER, ARTHUR MELVYN, b New York, NY, May 5, 42; c 2. ATMOSPHERIC CHEMISTRY, AIR POLLUTION. *Educ:* Univ Calif, Los Angeles, BS, 64; Ohio State Univ, PhD(phys chem), 69. *Prof Exp:* Asst res chemist, Univ Calif, Riverside, 71-75, assoc res chemist, 76-80, asst dir, Air Pollution Res Ctr, 78-86, res chemist, 80-89, PROF & DIR ENVIRON SCI ENG PROG, UNIV CALIF, LOS ANGELES, 89-; CONSULT, 86- *Concurrent Pos:* Fel chem, Univ Calif, Berkeley, 70-71, prin investr, 79-, consult, 79- *Mem:* Air & Waste Mgt Asn; AAAS; Am Chem Soc; Sigma Xi. *Res:* Applications of longpath infrared and optical spectroscopy to atmospheric systems; indoor air pollution; regional human exposure modeling; air pollution policy. *Mailing Add:* Sch Pub Health Univ Calif Los Angeles CA 90024

WINER, BETTE MARCIA TARMEY, b Boston, Mass, Feb 21, 40; m 63; c 2. PHYSICS, POWER SYSTEM ENGINEERING. *Educ:* Univ Maine, BS, 61; Univ Md, PhD(physics), 69. *Prof Exp:* Staff physics, Lincoln Labs, Mass Inst Technol, 59-63, staff, Nat Magnet Labs, 70-72; engr analysis, Bedford Labs, Raytheon, 72-74; sr staff physics & eng, Arthur D Little, Inc, 74-88; SR STAFF PHYSICS & ENG, US DEPT TRANSP VOLPE, NAT TRANSP SYSTS CTR, 88- *Concurrent Pos:* Vis prof, Univ Md, Catonsville, 69- & Univ Lowell, 70-72. *Mailing Add:* US Dept Transp Volpe Nat Transp Syst Kendall Sq Cambridge MA 02140

WINER, HERBERT ISAAC, b New York, NY, Sept 19, 21; m 43, 70; c 4. FORESTRY. *Educ:* Yale Univ, BA, MF, 49, PhD, 56. *Prof Exp:* Instr forestry, Sch Forestry, Yale Univ, 52-56, asst prof lumbering, 56-64; sci consult, Pulp & Paper Res Inst Can, 63-64, forester, 64-65, sr forester, 65-71, dir, Logging Res Div, 71-75, res dir, 75-76, sr forest engr, Eastern Div, Forest Eng Res Inst Can, 76-78; mgr woodlands oper res, 79-84, mgr woodlands resource progs, Mead Corp, 84-87; LECTR, YALE SCH FORESTRY & ENVIRON STUDIES, 88- *Concurrent Pos:* Mem, Int Union Forest Res Orgns. *Mem:* Soc Am Foresters; Can Inst Forestry; Forest Hist Soc (pres, 87-90); Sigma Xi. *Res:* Forest management; forest history. *Mailing Add:* PO Box 2987 New Haven CT 06515-0087

WINER, JEFFERY ALLAN, b Minneapolis, Minn, Nov 16, 45. NEUROANATOMY, NEUROSCIENCES. *Educ:* Univ Ariz, BA, 67; Univ Tenn, PhD(physiol psychol), 74. *Prof Exp:* Fel neuroanat, Dept Psychol, Duke Univ, 74-76; res assoc, Dept Anat, Harvard Med Sch, 76-77; res assoc neuroanat, Health Ctr, Univ Conn, 77-80; MEM FAC, DEPT PHYSIOL & ANAT, UNIV CALIF, 80- *Concurrent Pos:* NIMH fel, USPHS, 74-77. *Mem:* Soc Neurosci. *Res:* Neuroanatomy of the central auditory and visual systems, including Golgi, electron microscopic and axoplasmic transport methods applied to the morphology and development of the auditory thalamus, midbrain and cerebral cortex. *Mailing Add:* Dept Physiol & Anat Univ Calif Berkeley CA 94720

WINER, RICHARD, b Rochester, NY, Sept 16, 16; m 42; c 2. CHEMICAL ENGINEERING. *Educ:* Univ Ill, BS, 39; Univ Del, MFA, 91. *Prof Exp:* Tech asst, Radford Ord Works, Hercules Powder Co, Va, 42-45, chem engr, Hercules Inc, Exp Sta, Del, 45-46, supvr propellant dept, Allegany Ballistics Lab, 46-53, dir develop, propellants & rockets, 53-59, asst plant mgr, 59-63, dir develop, Chem Propulsion Div, 63-66, dir eng & res, Indust Systs Dept, 66-77, dir, Res Ctr, 78-81; RETIRED. *Mem:* AAAS; Am Chem Soc. *Res:* Solid propellants; rockets; chemical research. *Mailing Add:* 211 Churchill Dr Wilmington DE 19803

WINER, WARD OTIS, b Grand Rapids, Mich, June 27, 36; m 57; c 4. MECHANICAL ENGINEERING, PHYSICS. *Educ:* Univ Mich, BSE, 58, MSE, 59, PhD(mech eng), 62; Cambridge Univ, PhD(physics), 64. *Prof Exp:* Demonstr, Univ Cambridge, 61-63; from asst to assoc prof mech eng, Univ Mich, 63-69; from assoc prof to prof, 69-84, REGENTS PROF MECH ENG, GA INST TECHNOL, 84-, DIR G W WOODRUFF SCH MECH ENG, 88- *Concurrent Pos:* Consult, var indust. *Honors & Awards:* Melville Medal, Am Soc Mech Engrs, 75, Tribology Gold Medal, 86, Charles R Richards Award, 88. *Mem:* Nat Acad Eng; Am Soc Lubrication Engrs; fel Am Soc Mech Engrs; fel AAAS. *Res:* Tribology; high pressure lubricant rheology; fluid mechanics; heat transfer. *Mailing Add:* Sch Mech Eng Ga Inst Technol Atlanta GA 30332-0405

WINESTOCK, CLAIRE HUMMEL, b US, July 7, 32; m 56; c 1. ORGANIC CHEMISTRY. *Educ:* Univ Utah, BS, 52; Univ Wis, PhD(org chem), 56. *Prof Exp:* Res assoc biochem, Columbia Univ, 56-59; res fel chem, Univ Utah, 59, res assoc biochem, Col Med, 60-61, res instr, 61-65; grants assoc, NIH, 65-66, health scientist adminr, Nat Inst Arthritis & Metab Dis, 66-69, exec secy & referral officer, virol study sect, Div Res Grants, NIH, 69-86; SCI ADMINR, HOWARD HUGHES MED INST, 86- *Mem:* AAAS; Am Chem Soc; Sigma Xi. *Res:* Organic synthesis; chemistry of natural products; science administration. *Mailing Add:* 5707 Glenwood Rd Bethesda MD 20817

WINET, HOWARD, b Chicago, Ill, Sept 13, 37; m 68. BIORHEOLOGY, BONE WOUND HEALING. *Educ:* Univ Ill, BS, 59; Univ Calif, Los Angeles, MA, 62, PhD(biophys & cell physiol), 69. *Prof Exp:* Res fel eng sci, Calif Inst Technol, 69-73, res biophysicist eng sci, 73-77; assoc prof physiol, Southern Ill Univ, Carbondale, 77-80; ASSOC PROF RES ORTHOP, DEPT ORTHOP, UNIV SOUTHERN CALIF, 80- *Concurrent Pos:* Adv, Nat Sci Comt, Calif State Comn Teacher Preparation & Licensing, 72-77; vis res assoc eng sci, Calif Inst Technol, 78-; mem spec reproduction study sect, NIH, 78-; prin investr res grant, NIH, 78-; NIH Fogarty fel, Gothenburg Univ, Sweden, 85. *Mem:* Microcirculatory Soc; Am Physiol Soc; Orthop Res Soc; Soc Exp Biol Med Eng; Soc Biomat; Int Soc Biorheology. *Res:* Biophysical fluid mechanics of muco-ciliary systems, transit and propulsion of gametes; blood form in bone, vascular role in osteogenesis; wound healing in bone. *Mailing Add:* c/o Orthop Hosp Univ Southern Calif Med 2400 S Flower St Los Angeles CA 90007

WINETT, JOEL M, b Boston, Mass, Mar 1, 38; m 65; c 3. COMPUTER & MANAGEMENT SCIENCE. *Educ:* Mass Inst Technol, BSEE, 60, EE, 65; Columbia Univ, MSEE, 61. *Prof Exp:* Mem comput systs group, Lincoln Lab, Mass Inst Technol, 61-73, mem radar systs group, 73-74; mgr sci comput, Analytical Sci Corp, 74-79; mgr sci appl support, Sanders Assoc, 79-81; prod mgr, 81-82, dir comput opers, 82-88, DIR QUALITY ASSURANCE, BGS SYSTS, 89- *Concurrent Pos:* Instr, Northeastern Univ, 67-69. *Mem:* Inst Elec & Electronics Engrs; Asn Comput Mach. *Res:* Effective use of computers for scientific and commerical problem solving; compatible batch and interactive operating systems emphasizing human engineered control language; user documentation and training aids; system measurement and operations procedures; software testing and quality assurance. *Mailing Add:* Ten Berkeley Rd Framingham MA 01701

WINFIELD, JOHN BUCKNER, b Kentfield, Calif, Mar 19, 42; m 69; c 3. IMMUNOLOGY, RHEUMATOLOGY. *Educ:* Williams Col, BA, 64; Cornell Univ, MD, 68. *Prof Exp:* Intern internal med, New York Hosp, 68-69; staff assoc immunol, NIH, 69-71; resident, 71-73, instr, 74-75, assoc prof internal med, Univ Va, 76-78; assoc prof, 78-81, PROF INTERNAL MED, UNIV NC, CHAPEL HILL, 81-, CHIEF, DIV IMMUNOL & RHEUMATOLOGY, 78- *Concurrent Pos:* Fel immunol, Rockefeller Univ, 73-75; fel, Arthritis Found, 73-76, sr investr, 76-79. *Mem:* Am Fedn Clin Res; fel Am Col Physicians; Am Rheumatology Asn; Am Asn Immunol; Am Soc Clin Invest. *Res:* Clinical immunology; auto immune diseases. *Mailing Add:* Div Immunol & Rheumatology Univ NC Chapel Hill NC 27514

WINFREE, ARTHUR T, b St Petersburg, Fla, May 5, 42; m 83; c 2. BIOMATHEMATICS, PHYSICAL CHEMISTRY. *Educ:* Cornell Univ, BS, 65; Princeton Univ, PhD(biol), 70. *Prof Exp:* Asst prof math biol, Univ Chicago, 69-72; from assoc prof to prof biol, Purdue Univ, 78-85; PROF ECOL & EVOLUTIONARY BIOL, UNIV ARIZ, 85-, REGENTS PROF, 89- *Concurrent Pos:* Res career develop award, NIH, 73-78; assoc ed, Ecol & Ecol Monographs, 73-77, J Theoret Biol, 77-82, Physica D, 86-91; dir res, Inst Natural Philos, 79-88; Jap Soc Promotion Sci fel, Osaka, 81; J S Guggenheim mem fel, 82. *Honors & Awards:* Einthoven Award in Cardiol, 89. *Res:* Dynamics and timing in biological clocks; discovery of phase singularities and their roles in jet-lag, in sudden cardiac death, and in self-organization of biological and chemical excitable media. *Mailing Add:* Dept Ecol & Evolutionary Biol Univ Ariz Tucson AZ 85721

WINFREY, J C, b Post, Tex, Feb 10, 27; m 47; c 1. ORGANIC CHEMISTRY, ANALYTICAL CHEMISTRY. *Educ:* ETex State Teachers Col, BS & MS, 49. *Prof Exp:* Teacher high sch, Tex, 49-51; chemist, Eagle-Picher Lead Co, 51 & Lone Star Gas Co, 51-56; res chemist, Dow Chem Co, 56-62; analyticalchemist, Res Dept, Signal Oil & Gas Co, 62-68; analytical sect supvr res mgt, Signal Chem Co, 69-71; chief chemist, Geneva Industs, Inc, 71-73; gen mgr & corp secy, Analytical Serv, Inc, 73-75; consult, 75-76; lab mgr, Val Verde Corp, 76-77; staff, 77-80, DIR HYDROCARBON SERV, SOUTHERN PETROL LABS, 80- *Concurrent Pos:* Mem, Adv Comt Proj, 44, Am Petrol Inst, 68-71; analytical consult, Haines & Assocs, 71. *Mem:* Am Chem Soc; Am Soc Testing Mat. *Res:* Instrumental analytical chemistry; organic synthesis of amines and epoxides; gas chromatography; thin layer chromatography; liquid chromatography; mass spectrometry; infrared spectrometry; computer applications in analytical chemistry; spectrochemical analysis of used lube oils. *Mailing Add:* 5215 Georgi Lane Houston TX 77092

WINFREY, RICHARD CAMERON, b Albany, Calif, June 18, 35; m 59; c 4. MECHANICAL ENGINEERING, COMPUTER AIDED DESIGN. *Educ:* Univ Calif, Berkeley, BSME, 63; Univ Calif, Los Angeles, MS, 65, PhD(eng), 69. *Prof Exp:* Mem tech staff, Hughes Aircraft Co, Calif, 63-69; asst prof mech eng, Naval Postgrad Sch, 69-71; mech engr, Naval Civil Eng Lab, 71-72; proj engr Burroughs Corps, 72-75; prin engr, 75-80, CONSULT ENGR, DIGITAL EQUIP CORP, 80- *Mem:* Am Soc Mech Engrs; Sci Res Soc Am; Inst Elec & Electronics Engrs. *Res:* Dynamics of elastic machinery. *Mailing Add:* Digital Equip Corp ML 1-3/E58 Maynard MA 01754

WING, BRUCE LARRY, b Coeur d'Alene, Idaho, Aug 7, 38. BIOLOGICAL OCEANOGRAPHY. *Educ:* San Diego State Col, AB, 60; Univ RI, PhD(oceanog), 76. *Prof Exp:* Fishery biologist zooplankton, US Bur Com Fisheries, 62-75; chief investr oceanog, 75-80, task leader oceanog, 81-82, FISHERY RES BIOLOGIST, AUKE BAY LAB, NAT MARINE FISHERIES SERV, 83- *Concurrent Pos:* Lectr, Univ Alaska Southeast, 87. *Honors & Awards:* C Y Conkel Award, Auke Bay Fisheries Lab, 77. *Mem:* Fel Am Inst Fishery Res Biologists; AAAS; Am Fisheries Soc; Am Soc Limnol & Oceanog; Sigma Xi. *Res:* Effect of environmental variation on plankton composition and fishery productivity; taxonomy of Alaskan marine invertebrates; ellobiopsids. *Mailing Add:* Nat Marine Fisheries Serv PO Box 210155 Auke Bay AK 99821-0155

WING, EDWARD JOSEPH, b Mineola, NY, June 19, 45; m 67; c 2. INFECTIOUS DISEASES, IMMUNOLOGY. *Educ:* Williams Col, BA, 67; Harvard Med Sch, MD, 71. *Prof Exp:* Resident med, Peter Bent Brigham Hosp, Boston, Mass, 71-73; asst surgeon, USPHS, 73-75; fel infectious dis, Stanford Univ Sch Med, 75-77; asst prof, Univ Pittsburgh Sch Med, 77-81, assos prof med, 82-88; PROF MED, MONTEFIORE UNIV HOSP, 89- *Concurrent Pos:* Prin investr, NIH, 83-; interim physician in chief, Montefiore Univ Hosp, 90- *Mem:* Infectious Dis Soc Am; fel Am Fedn Clin Res; Am Soc Microbiol; Am Asn Immunologists; Reticuloendothelial Soc. *Res:* Mechanisms of cell-mediated immunity against intracellular pathogens; role of various cytokines, particularly the colony-stimulating factors in the stimulation of host defenses to microbial pathogens. *Mailing Add:* Montefiore Univ Hosp 3459 Fifth Ave New York NY 15213

WING, ELIZABETH S, b Cambridge, Mass, Mar 5, 32; m 57; c 2. ZOOLOGY. *Educ:* Mt Holyoke Col, BA, 55; Univ Fla, MS, 57, PhD(zool), 62. *Prof Exp:* Asst cur zoo-archaeol, 61-74, asst prof anthrop, 70-75, ASSOC PROF ANTHROP, UNIV FLA, 75-, PROF ZOOL, 90- *Concurrent Pos:* NSF grants, 61-, co-investr, 61-64; Caribbean Res Prog grant, 64 & 65; Ctr Latin Am Studies res grant, 66; US rep, Int Congress Archaeozool; pres, Soc Ethnobiology; assoc cur Fla Mus Natural Hist, 74-78, cur, 78- *Mem:* Am Soc Mammal; Soc Am Archaeol; AAAS. *Res:* Identification and analysis of faunal remains excavated from Indian sites in Southeastern United States and Latin America; prehistoric subsistence and animal domestication in the Andes. *Mailing Add:* Fla Mus Natural Hist Univ Fla Gainesville FL 32611

WING, G(EORGE) MILTON, b Rochester, NY, Jan 21, 23. APPLIED MATHEMATICS. *Educ:* Univ Rochester, BA, 44, MS, 47; Cornell Univ, PhD(math), 49. *Prof Exp:* Scientist, Los Alamos Sci Lab, Univ Calif, 45-46, mem staff, 51-58, 81-87; instr math, Univ Rochester, 46-47; instr, Univ Calif, Los Angeles, 49-51, asst prof, 51-52; assoc prof, Univ NMex, 58-59; mem staff, Sandia Corp, 59-64; prof math, Univ Colo, 64-66 & Univ NMex, 66-73; vis prof, Tex Tech Univ, 75-76; prof math, Southern Methodist Univ, 77-81, chmn, 77-78; RETIRED. *Concurrent Pos:* Consult, Los Alamos Sci Lab, 58-59 & 64-81, 87-, Sandia Corp, 58-59, E H Plesset Assocs, 58-59 & 65-69 & Rand Corp, 58-65; mem, Panel Phys Sci & Eng, Comt Undergrad Prog Math, 63-67. *Mem:* AAAS; Am Math Soc; Math Asn Am; Soc Indust & Appl Math. *Res:* Transport theory; integral equations. *Mailing Add:* 302 Calle Estado Santa Fe NM 87501

WING, JAMES, b Highland Park, Mich, July 8, 29; m 57; c 2. NUCLEAR CHEMISTRY. *Educ:* Univ Tenn, BS, 51; Purdue Univ, MS, 53, PhD(chem), 56. *Prof Exp:* Asst chemist, Argonne Nat Lab, 55-65, assoc chemist, 65-69; res chemist, Anal Div, Nat Bur Standards, 69-75; nuclear chemist, US Nuclear Regulatory Comn, 75-78, sr chem engr & proj mgr, 78-87; RELIABILITY & RISK ANALYST, 87- *Concurrent Pos:* Fulbright lectr, Chinese Univ Hong Kong, 64-65; free-lance translr, 75- *Mem:* Am Chem Soc. *Res:* Nucleidic mass systematics; nuclear activation analysis; cross sections and mechanisms of nuclear reactions; radioactivities of new isotopes; radiochemical separation; toxic vapor detection; computer automation of laboratory experiments; chemical safety for nuclear reactor operation. *Mailing Add:* 15212 Red Clover Dr Rockville MD 20853

WING, JAMES MARVIN, b Anniston, Ala, Mar 17, 20; m 46, 76; c 1. ANIMAL NUTRITION. *Educ:* Berea Col, BS, 46; Colo State Univ, MS, 48; Iowa State Univ, PhD(dairy husb), 52. *Prof Exp:* From asst prof to assoc prof, 51-66, PROF DAIRY SCI, UNIV FLA, 66- *Mem:* Am Soc Animal Sci; Am Dairy Sci Asn. *Res:* Nutrition; digestibility of carotenoids; nucleic acids; medicated feeds; ensilability; digestibility and consumption of herbage; climatic adaptation of cattle; optimum levels of carbohydrates for cattle; evaluation of by-product feed stuffs; environmental quality and the animal industries; international development in Colombia, El Salvador, Paraguay, Viet Nam and Dominican Republic. *Mailing Add:* 2016 NW 20th Lane Gainesville FL 32605

WING, JANET E (SWEEDYK) BENDT, b Detroit, Mich, Oct 12, 25; div; c 4. NUCLEAR REACTOR SAFETY ANALYSIS. *Educ:* Wayne State Univ, Detroit, BS, 47; Columbia Univ, MS, 50. *Prof Exp:* Engr, Gen Motors, 44-48; mathematician, Manhattan Proj, Columbia Univ, 50-51; mem res staff, 51-57, asst group leader, 80-84, MEM RES STAFF, LOS ALAMOS NAT LAB, 68-, ASSOC GROUP LEADER, 88- *Concurrent Pos:* Proj leader, Los Alamos Nat Lab, 78-81. *Mem:* Women Sci & Eng; AAAS; Sigma Xi; Am Nuclear Soc. *Res:* Mathematical modeling and computer solutions of physical systems; radiation-hydrodynamics and transport theory; applications to nuclear weapon design and testing; underground containment of nuclear explosions; pulsations of Cepheid stars. *Mailing Add:* Reactor Safety Analysis Group N-9 MS K556 Los Alamos Nat Lab Los Alamos NM 87545

WING, JOHN FAXON, b Lincoln, Nebr, Jan 27, 34; m 56; c 4. NAVAL ARCHITECTURE, MARINE ENGINEERING. *Educ:* Mass Inst Technol, BS, 55; Harvard Univ, MBA, 57. *Prof Exp:* Proj engr, Alcoa Steamship Co, 57-61; engr, Shipbldg Div, Bethlehem Steel Co, 61-64; sr engr, 64-65, proj engr, 65-66, prin engr, 66-67, res dir, 67-70, vpres, 70-72, sr vpres, 72-81, MANAGING OFFICER TRANSP CONSULT, BOOZ-ALLEN & HAMILTON, INC, 81- *Concurrent Pos:* Lectr, Univ Mich, 66. *Mem:* Soc Naval Archit & Marine Engrs; Sigma Xi. *Res:* Management consulting in transportation; research and analysis of maritime operations. *Mailing Add:* Booz-Allen & Hamilton Inc 4330 East West Hwy Bethesda MD 20814

WING, OMAR, b Detroit, Mich, Mar 2, 28; m 53; c 2. ELECTRICAL ENGINEERING, COMPUTER SCIENCE. *Educ:* Univ Tenn, BS, 50; Mass Inst Technol, MS, 52; Columbia Univ, DEng, 59. *Prof Exp:* Asst elec eng, Mass Inst Technol, 50-52; mem tech staff, Bell Tel Labs, NJ, 52-56; from instr to assoc prof, 56-76, PROF ELEC ENG, COLUMBIA UNIV, 76- *Concurrent Pos:* Fulbright vis lectr, Inst Electronics, Chiao Tung Univ, 61; Ford Found eng resident, Thomas J Watson Res Ctr, Int Bus Mach Corp, 65-66; vis prof, Tech Univ Denmark, 73. *Mem:* Inst Elec & Electronics Engrs. *Res:* Network theory; computer design of networks; design automation; computer simulation of systems; distributed parameter networks; digital filters; computer analysis of large networks. *Mailing Add:* Dept Elec Eng Columbia Univ 1306 Mudd Bldg New York NY 10027

WING, ROBERT EDWARD, b Bridgeport, Conn, Nov 29, 41; div; c 2. CHEMISTRY. *Educ:* Millikin Univ, BA, 63; Southern Ill Univ, Carbondale, PhD(biochem), 67. *Prof Exp:* Northern Regional Lab grant, Southern Ill Univ, Carbondale, 67-68; RES CHEMIST, PLANT POLYMER LAB, NORTHERN REGIONAL RES CTR, USDA, 68- *Concurrent Pos:* Res assoc, Peoria Sch Med, Univ Ill Col Med, 71-87; instr, Bradley Univ, 73- *Honors & Awards:* Indust Res-100 Award, 78. *Mem:* Am Chem Soc; Sigma Xi; Am Electroplaters Soc. *Res:* Water pollution of heavy metal ions; reactive carbohydrate polymers; carbohydrate thin-layer chromatography and enzyme interactions; carbohydrate slow release pesticide formulations. *Mailing Add:* 3215 Willow Knolls 86-D Peoria IL 61614

WING, ROBERT FARQUHAR, b New Haven, Conn, Oct 31, 39; m 63; c 3. ASTRONOMY. *Educ:* Yale Univ, BS, 61; Univ Calif, Berkeley, PhD(astron), 67. *Prof Exp:* From asst prof to assoc prof, 67-76, PROF ASTRON, OHIO STATE UNIV, 76- *Concurrent Pos:* Mem bd, Asn Univs for Res in Astron Inc, 81-87; dir, Perkins Observ, 90- *Mem:* Int Astron Union; Am Astron Soc; fel Royal Astron Soc. *Res:* Spectroscopy and photometry of cool stars, including Mira variables; infrared spectra; determination of chemical composition and effective temperature; chromospheres of cool stars; space astronomy; infrared flux calibration; galactic structure. *Mailing Add:* Dept Astron Ohio State Univ Columbus OH 43210

WING, WILLIAM HINSHAW, b Ann Arbor, Mich, Jan 11, 39; c 2. ATOMIC PHYSICS, LASER SPECTROSCOPY. *Educ:* Yale Univ, BA, 60; Rutgers Univ, New Brunswick, MS, 62; Univ Mich, Ann Arbor, PhD(physics), 68. *Prof Exp:* Res staff physicist, Yale Univ, 68-70, res assoc physics, 70-72, asst prof, 72-74; assoc prof physics, 74-78, PROF PHYSICS & OPTICAL SCI, UNIV ARIZ, 78- *Concurrent Pos:* Res Corp Cottrell grant, 71; Nat Bur Standards, US Dept Com Precision Measurement grants, 74 & 80; Am Chem Soc petrol res grant, 77; vis prof & Joint Inst Lab Astrophysics fel, Univ Colo, 79-80 & Mass Inst Technol, 80; assoc prof, Ecole Normale Superieure, 81; J S Guggenheim fel, 80-81; chmn, Comt Fundamental Constants, Nat Data Adv Bd, Nat Res Coun, 83-86; mem, exec comt, Div Atomic Molecular & Optical Physics, Am Phys Soc, 84-87. *Honors & Awards:* Alexander von Humboldt Sr Scientist Award. *Mem:* Fel Am Phys Soc; Inst Elec & Electronics Eng; fel Optical Soc Am; Sigma Xi; NY Acad Sci; Am Chem Soc. *Res:* Fundamental physical constants; simple atomic and molecular physics; lasers; particle beams; chemical physics; computer sciences; molecular physics. *Mailing Add:* Dept Physics Bldg 81 Univ Ariz Tucson AZ 85721

WINGARD, CHRISTOPHER JON, b Canton, Ohio, Nov 30, 62. COMPARATIVE ANIMAL PHYSIOLOGY. *Educ:* Hiram Col, BA, 84; Univ Akron, MS, 86; Wayne State Univ, PhD(biol), 91. *Prof Exp:* Animal Keeper II, Sea World Ohio Inc, Aurora, 86; lab analyst, ERT Testing Serv Inc, Highland Park, Mich, 88-91; POSTDOCTORAL FEL SMOOTH MUSCLE, DEPT PHYSIOL, UNIV VA, CHARLOTTESVILLE, 91- *Concurrent Pos:* Field lab instr, Bermuda Biol Sta Res, Wayne State Univ, 88-91, instr, Dept Biol, 88-90, lab instr, 90. *Mem:* Am Soc Zoologists; Am Physiol Soc; Int Soc Optical Eng. *Res:* Comparative physiology; development and application of thermal and oxygen imaging systems in the eludification of energetics of physiologial processes; contractile system regulation and control; physiology of aquatic organisms and their response to environmental perturbations. *Mailing Add:* Dept Physiol Health Sci Ctr Univ Va Box 449 Charlottesville VA 22908

WINGARD, DEBORAH LEE, b San Diego, Calif, Aug 20, 52; m 76; c 2. EPIDEMIOLOGY. *Educ:* Univ Calif, Berkeley, BA, 74, MS, 76, PhD(epidemiol), 80. *Prof Exp:* Res specialist, Social Res Group, Sch Pub Health, Univ Calif, Berkeley, 76-77; epidemiologist, Human Pop Lab, Calif State Dept Health Serv, 77-78; asst prof, 80-86, ASSOC PROF, DIV EPIDEMIOL, DEPT COMMUNITY & FAMILY MED, UNIV CALIF, SAN DIEGO, 86- *Concurrent Pos:* Epidemiologist, Lipid Res Clin, Calif, 81-84; consult, DES Action USA, 82-, consult fac San Diego State Univ, 85-, Nat Inst Aging, 86-; prin investr, NIA Grant, 84-87, 85-88, 90-93, AARP Grant, 88, co-prin investr, NIADDK Grant, 83-90. *Mem:* Soc Epidemiol Res; Am Pub Health Asn; Am Heart Assoc; Asn Women Sci. *Res:* The interaction of biological and pyschosocial factors in the maintenance of health, through the study of sex differences in morbidity and mortality. *Mailing Add:* Dept Community & Family Med Univ Calif San Diego 9500 Gilman Dr La Jolla CA 92093-0607

WINGARD, LEMUEL BELL, JR, b Pittsburgh, Pa, July 10, 30; m 65; c 2. BIOCHEMISTRY, BIOTECHNOLOGY. *Educ:* Cornell Univ, BChE, 53, PhD(biochem eng), 65. *Prof Exp:* Res engr, Jackson Lab, E I du Pont de Nemours & Co, 56-58 & Seaford Nylon Plant, 58-61; asst prof chem eng, Cornell Univ, 65-66; assoc prof, Univ Denver, 66-67; from asst prof to assoc prof chem eng, 67-75, adj prof chem eng, 76-77, from assoc prof to prof pharmacol, 72-81, acting chmn dept pharmacol, 80-82, PROF PHARMACOL & ANESTHESIOL, SCH MED, UNIV PITTSBURGH, 81- *Concurrent Pos:* NIH spec fel, Sch Pharm, State Univ NY Buffalo, 70-72; chmn confs enzyme eng, Eng Found, 71 & 73; sabbatical leave, vis assoc prof, Dept Pharmacol, Sch Med, Yale Univ, 79-80; ed, Biosensors; adj prof chem eng, Sch Med, Univ Pittsburgh, 87- *Honors & Awards:* Spec Enzyme Eng Award, Eng Found, 85. *Mem:* Fel AAAS; fel Am Inst Chem Engr; Am Chem Soc; Am Soc Pharmacol & Exp Therapeut. *Res:* Pharmacokinetics; immobilized enzymes; cells and drugs; mechanisms of anticancer drugs; gamma-aminobutyric acid receptor; biosensors. *Mailing Add:* 4444 Gateway Dr Monroeville PA 15146-1030

WINGARD, PAUL SIDNEY, b Akron, Ohio, Jan 10, 30; m 53; c 5. PHYSICAL GEOLOGY. *Educ:* Miami Univ, AB, 52, MS, 55; Univ Ill, PhD(geol), 61. *Prof Exp:* Instr geol, Kans State Univ, 57-61, asst prof, 61-66; assoc prof & asst dean lib arts, 66-67, PROF GEOL & ASSOC DEAN COL ARTS & SCI, UNIV AKRON, 67- *Mem:* AAAS; Geol Soc Am. *Res:* Geochronology and geology of south central Maine; geology of central and southern Colorado; post-Pleistocene geology and life of northern Ohio. *Mailing Add:* Univ Akron Dept Geol Akron OH 44325

WINGARD, ROBERT EUGENE, JR, b Montgomery, Ala, Sept 18, 46. ORGANIC CHEMISTRY, POLYMER CHEMISTRY. *Educ:* Auburn Univ, BS, 68; Ohio State Univ, PhD(org chem), 71. *Prof Exp:* Res chemist, 74-75, sr res chemist, 76, SR RES SCIENTIST ORG CHEM, DYNAPOL, 77- *Concurrent Pos:* Fel org chem, Harvard Univ, 72-73, NIH fel, 72-74. *Mem:* Am Chem Soc; AAAS. *Res:* Synthetic and mechanistic organic chemistry; chemistry of functionalized polymers. *Mailing Add:* Noramco Inc P O Box 1652 Athens GA 30603-1652

WINGATE, CATHARINE L, b Boston, Mass, Sept 7, 22. RADIOLOGICAL PHYSICS, MEDICAL PHYSICS. *Educ:* Simmons Col, BS, 43; Harvard Univ, MA, 48; Columbia Univ, PhD(biophys), 61. *Prof Exp:* Res asst radiation physics, Mass Inst Technol, 43-45; sr technician, Woods Hole Oceanog Inst, 45-46; instr physics, Univ Conn, New London, 48-49; res asst med biophys, Sloan-Kettering Inst Cancer Res, 49-51; instr physics, Adelphi Col, 51-54; res scientist radiol physics, Col Physicians & Surgeons, Columbia Univ, 54-63; radiol physicist, Naval Radiol Defense Lab, 63-66; sr res scientist biophys, NY Univ, 66-67; assoc radiol physicist, Brookhaven Nat Lab, 67-70; asst prof radiol physics, Sch Med & asst dean, Sch Basic Health Sci, State Univ NY Stony Brook, 70-74, res asst prof radiol physics, Sch Med, 74-75, res assoc prof radiol, 75-78; MEM STAFF, HEALTH SCI ADMIN, NIH, 78- *Concurrent Pos:* Consult, Vet Admin Hosp, Northport, NY, 71-74 & Radiation Study Sect, NIH, 76-78; physicist, Nassau County Med Ctr, East Meadow, NY, 74-78; collabr, Brookhaven Nat Lab, 74-78; review ed, Med Phys, 79- *Mem:* Radiation Res Soc; NY Acad Sci; Am Asn Physicists Med; Sigma Xi; Soc Nuclear Med; Soc Magnetic Resonance Med. *Res:* Measurement of ionization parameters; measurement of microscopic dose distributions at a bone-soft tissue interface and around charged particle beams; thermo luminescence; neutron dosimetry; calcium uptake in stressed bone; clinical dosimetry; proton Bragg peak localization for therapy. *Mailing Add:* NIH Westwood Bldg 5333 Westbard Ave Rm 357 Bethesda MD 20892

WINGATE, FREDERICK HUSTON, b Provo, Utah, Dec 21, 32; m 71. GEOLOGY, BOTANY. *Educ:* Univ Utah, BS, 56, MS, 61; Univ Okla, PhD(bot), 74. *Prof Exp:* Subsurface geologist oil explor, Chevron Oil Co, 61-66; western region palynologist biostratig, Cities Serv Co, 73-85; PALYNOLOGICAL CONSULT, 85- *Mem:* Am Asn Stratig Palynologists. *Res:* Palynology, involving organic- and siliceous-walled microfossil studies with application to solution of bistratigraphic and hydrocarbon generation problems. *Mailing Add:* 3052 S Ivan Way Denver CO 80227

WINGATE, MARTIN BERNARD, b London, Eng; c 2. OBSTETRICS & GYNECOLOGY. *Educ:* Univ London, MB, BS, 48, MD, 64; FRCS, 53; FRCS(E), 55; FRCS(C), 66. *Prof Exp:* House surgeon orthop, St Mary's Hosp, London, 48-50 & Obstet Unit, Whittington Hosp, 50-51; demonstr anat, St Mary's Hosp Med Sch, 51-52, prosecutor, 52, sr house officer surg, Hosp, 52-53; house surgeon, Cent Middlesex Hosp, 53-54; locum surg registr, Royal Northern Hosp, 54-55; locum surg registr obstet & gynec, Middlesex Hosp & Hosp for Women, 56-58; sr registr, Southampton Gen Hosp, 58-60; first asst, St George's Hosp, 58-62; sr registr obstet & gynec, Queen Charlotte & Chelsea Hosps, 62-63; asst examr obstet & gynec, Univ Bristol, sr lectr & consult, United Bristol Hosps & Univ Bristol & examr, Cent Midwives Bd, 63-66; assoc prof obstet & gynec, Univ Man, 67-71; prof, Temple Univ, 69-71; prof obstet & gynec & pediat, Thomas Jefferson Univ, 71-75; prof obstet & gynec & pediat, Albany Med Col, 75-77; PROF OBSTET & GYNEC & ASST DEAN, STATE UNIV NY, BUFFALO, 77-, ASST DEAN CONTINUING EDUC, 80- *Concurrent Pos:* Res fel obstet & pediat, Guy's Hosp Med Sch, 62-63. *Mem:* Can Med Asn; Soc Obstet & Gynaec Can; fel Am Col Obstet & Gynec. *Res:* Genetics of human aberrations and malignant lesions of the cervix; surgery of infertility and the transplantation of reproductive organs together with means of modification of the rejection phenomena. *Mailing Add:* Dept Obstet & Gynec Mt Auburn Hosp 330 Mt Auburn St Cambridge MA 02238

WINGE, DENNIS R, b Marshall, Minn, Apr 5, 47. PROTEIN CHEMISTRY. *Educ:* Duke Univ, PhD(biochem), 75. *Prof Exp:* HEAD, RES LAB, DIV HEMAT, UNIV UTAH, 79- *Mem:* Am Soc Biol Chemists; AAAS. *Mailing Add:* Dept Biochem & Med Univ Utah Med Ctr 50 N Medical Dr Salt Lake City UT 84132

WINGELETH, DALE CLIFFORD, b Cleveland, Ohio, June 8, 43; c 3. CLINICAL CHEMISTRY, ORGANIC CHEMISTRY. *Educ:* Cleveland State Univ, BES, 66; Univ Colo, Boulder, PhD(inorg chem), 70. *Prof Exp:* Clin biochemist, St Joseph Hosp, Denver, 70-72; clin chemist, St Lawrence Hosp, Mich, 72; forensic chemist, Poisonlab Inc, 72, vpres & tech dir, Poisonlab Div, Chemed Corp, 72-77; FORENSIC CHEMIST & TOXICOLOGIST, CHEMATOX LAB, INC, 77- *Mem:* Am Chem Soc; Royal Soc Chem; Am Asn Clin Chemists; Forensic Sci Soc; Am Indust Hyg Asn. *Res:* Inorganic hydrides; clinical toxicology. *Mailing Add:* 5401 Western Ave Boulder CO 80301

WINGENDER, RONALD JOHN, b Menominee, Mich, Sept 30, 36; div; c 2. ANALYTICAL CHEMISTRY, ENVIRONMENTAL ANALYSIS. *Educ:* Univ Wis, BS, 59, PhD(analytical chem), 69; Univ Iowa, MS, 61. *Prof Exp:* Chemist, Forest Prod Lab, 61-64; chemist, Ansul Co, Wis, 69, mgr anal res, 69-72; sect head chem, Indust Bio-Test Labs, 72-78; lab & res dir, Clin Bio-Tox Labs, 78-79; chemist, Argonne Nat Lab, 79-87, assoc group leader, 87-89; MGR, ANALYTICAL SERV, DEXTER CORP, 89- *Mem:* Sigma Xi; Am Chem Soc. *Res:* Proton nuclear magnetic resonance of cobalt II and nickel II aminopolycarboxylic acid and polyamine complexes; development of pesticide residue analytical procedures; development of analytical procedures for trace organic pollutants; identification of organic pollutants by gas chromatography and mass spectrometry; development of analytical procedures for drugs in biological fluids; analysis of polymeric coating systems. *Mailing Add:* 3547 Cresent Ave Gurnee IL 60031

WINGER, MILTON EUGENE, b Mayville, NDak, Aug 28, 31; m 54; c 2. MATHEMATICS, STATISTICS. *Educ:* Mayville State Col, BS, 53; Univ NDak, MS, 56; Iowa State Univ, PhD(statist), 72. *Prof Exp:* Inspector eng, US Army CEngr, 57; asst prof math, Univ NDak, 60-68; instr statist, Iowa State Univ, 68-70; assoc prof, 71-78, PROF MATH, UNIV NDAK, 78-, ADV DEPT STATIST, 71-, CHMN DEPT MATH, 84- *Concurrent Pos:* Statist consult, Inst Appl Math & Statist, 74-80; vis lectr, Inst Math Statist, 74-82; assoc ed, Col Math J, 82- *Mem:* Math Asn Am; Am Statist Asn; Am Asn Univ Profs. *Mailing Add:* 623 24th Ave S Grand Forks ND 58201

WINGER, PARLEY VERNON, b Driggs, Idaho, Dec 11, 41; m 67; c 2. AQUATIC BIOLOGY, FISHERIES. *Educ:* Idaho State Univ, BS, 66, BS, 66, MS, 68; Brigham Young Univ, PhD(aquatic biol), 72. *Prof Exp:* Res assoc aquatic biol, Brigham Young Univ, 72-73; asst prof, Tenn Technol Univ, 74-78; LEADER FIELD RES STA AQUATIC BIOL, NAT FISHERY CONTAMINANT RES CTR, US FISH & WILDLIFE SERV, 78- *Concurrent Pos:* Prin investr fisheries, Ctr Health & Environ Studies, Brigham Young Univ, 72-74 & Tenn Valley Authority, 74-76; adj asst prof, Univ Ga, 78- *Mem:* Am Fisheries Soc; Freshwater Biol Asn; NAm Benthic Soc; Soc Int Limnol; Soc Environ Toxicol Chem. *Res:* Effects of habitat alteration on aquatic ecosystems; mitigation techniques to enhance aquatic environments; effect of acid mine drainage and chemical contaminants on aquatic populations; population dynamics of benthic macroinvertebrates; toxicity of aquatic sediments. *Mailing Add:* US Fish & Wildlife Serv Sch Forest Resources Univ Ga Athens GA 30602

WINGERT, LOUIS EUGENE, b Kimball, SDak, Feb 3, 24; m 52; c 3. CHEMICAL ENGINEERING. *Educ:* SDak Sch Mines & Technol, BS, 44. *Prof Exp:* Res engr, Deere & Co, Ill, 46-47; chem engr, Duro Tank Co, 47-48; bd plant engr, US Gypsum Co, Ohio, 48-51, qual supvr, 51-53, qual supt, Que, 53-57; sr chem engr, Minn Mining & Mfg Co, 57-68; RES CHEM ENGR, KIMBERLY-CLARK CORP, NEENAH, 68- *Mem:* Am Chem Soc. *Res:* Gypsum products; thermographic copy paper; lime; paper coatings. *Mailing Add:* 2105 N Garys Lane Appleton WI 54911

WINGET, CARL HENRY, b Noranda, Que, Sept 28, 38; m 64; c 2. FOREST ECOLOGY, TREE PHYSIOLOGY. *Educ:* Univ NB, BScF, 60; Univ Wis, MSc, 62, PhD(forestry, bot), 64. *Prof Exp:* Prof forestry & geod, Laval Univ, 67-73; dir, Laurentian Forestry Res Ctr, 78-82, DIR GEN, RES & TECH SERVS, CAN FORESTRY SERV, 78- *Mem:* Can Inst Forestry. *Res:* Forest resource management; silviculture of tolerant northern hardwoods. *Mailing Add:* Can Forestry Serv Pl Vincent Mossey 351 St Joseph Blvd Hall PQ K1A 1G5 Can

WINGET, CHARLES M, b Garden City, Kans, Dec 26, 25; m; c 4. AEROSPACE SCIENCES. *Educ:* San Francisco State Col, BA, 51; Univ Calif, PhD, 57. *Prof Exp:* Chemist poultry husb, Univ Calif, 51-53, res asst, 53-56, jr res poultry physiologist, 56-57, res fel, 57-59; assoc prof avian physiol, Ont Agr Col, Univ Guelph, 59-63; res scientist, 63-67, proj scientist, biosatellite proj, 67-85, res scientist, Neurosci Br, payload proj scientist, Shuttle EOM-1/2 mission & sci coordr for space sta, Adv Progs Off, 85-86 SEC PAYLOAD SCIENTIST & IML-1 PAYLOAD SCIENTIST, NASA-AMES RES CTR, MOFFETT FIELD, CALIF, 86- *Concurrent Pos:* Nat Inst Neurol Dis & Blindness fel, 57-59; lectr, Univ Calif, Davis, 64-; prof pharmacol, Sch Pharm, Fla A&M Univ, 75-; bd regents distinguished vis scholar, Sch Pharm, Fla Agr & Mech Univ, Tallahassee, 78; consult, Sports Med Div, US Olympic Comt, 82-84; mem, Adv Comt for Minority Insts, Res Ctrs in Minority Insts Prog, NIH, 85- *Honors & Awards:* Paul Bert Award Physiol Res, 77; Arnold D Tuttle Mem Award, Aerospace Med Asn, 82. *Mem:* Poultry Sci Asn; Biophys Soc; Aerospace Med Asn; Int Soc Chronobiol; Am Physiol Soc; Sigma Xi. *Res:* Rhythms and social schedule; hypokinesis and drugs in humans; bird and monkey response to change in photoperiod; physiological changes associated with aeronautical environment; biotelemetry; biorhythm data acquisition and reduction. *Mailing Add:* Space Life Sci 240A-3 NASA Ames Res Ctr Moffett Field CA 94035

WINGET, GARY DOUGLAS, b Dayton, Ohio, Mar 27, 39; m 60; c 3. PLANT BIOCHEMISTRY, PLANT PHYSIOLOGY. *Educ:* Miami Univ, AB, 61, MA, 63; Mich State Univ, PhD(bot), 68. *Prof Exp:* Res chemist, Mound Lab, Monsanto Res Corp, 63-64; asst prof, 66-71, ASSOC PROF BIOL SCI, UNIV CINCINNATI, 71- *Concurrent Pos:* Vis assoc prof biochem, molecular & cell biol, Cornell Univ, 75. *Mem:* Am Soc Photobiol; Am Chem Soc; Am Soc Plant Physiologists; Sigma Xi. *Res:* Inhibitors of photosynthesis; mechanism of photophosphorylation; physiological action of phlorizin. *Mailing Add:* Dept Biol Sci Univ of Cincinnati Cincinnati OH 45221

WINGET, ROBERT NEWELL, b Monroe, Utah, July 11, 42; m 64; c 7. AQUATIC ECOLOGY, BIOLOGY. *Educ:* Univ Utah, BS, 67, MS, 68, PhD(biol sci), 70. *Prof Exp:* Res assoc aquatic ecol, 70-85, ASSOC PROF BIOL, BRIGHAM YOUNG UNIV, HAWAII, 86- *Concurrent Pos:* Proj dir insect control, Div Parks & Recreation, State of Utah, 69-71; consult water qual mgt, US Forest Serv, Intermountain Region, 72-; consult, Cent Utah Proj, US Bur Reclamation, 73-, Vaughn Hansen Assocs, Utah, 77-78 & Sandia Proj, Eastern NMex Univ, 77-78; dir thermal study, Utah Power & Light Co, 74-76, consult aquacult, 77-86; consult impact anal, Westinghouse Corp, 74-76; consult water qual, Eyring Res Inst, 75-83, stream reclamation, Coastal States Energy Co, 79-88, water qual, Getty Mineral Resources Co, 80-85 & Homestake Mining Co, 81-85; prog bd, Hawaii Erwin Educ Asn, 89- *Mem:* Am Sci Teachers Asn; Home Econ Educ Asn; NAm Benthol Soc. *Res:* Environmental impact analyses, especially fisheries resources, water quality and water quality standards; aquaculture using thermal effluents; macroinvertebrate community dynamics. *Mailing Add:* Dept Biol Sci Brigham Young Univ 55-220 Kulanui St Laie Oahu HI 96762

WINGFIELD, EDWARD CHRISTIAN, b Charlottesville, Va, Nov 17, 23; m 85; c 3. PHYSICS, SCIENCE ADMINISTRATION. *Educ:* Univ Va, BA, 46, MA, 49; Univ NC, PhD(physics), 54. *Prof Exp:* Asst prof physics, Univ Richmond, 49-51; physicist, Savannah River Lab, E I Du Pont de Nemours & Co, 54-62; chief, Instrumentation Develop & Serv, United Technol Res Ctr, 62-65, asst mgr eng opers, 65-69, sr consult, 79-87; RETIRED. *Res:* Instrumentation; reactor physics; simulation by computers; aerodynamic testing; management of research and development activities. *Mailing Add:* 137 Tall Timbers Rd Glastonburg CT 06033

WINGO, CHARLES S, b Ruston, La, Oct 5, 49. NEPHROLOGY, EPITHELIAL TRANSPORT. *Educ:* La State Univ, MD, 75. *Prof Exp:* Asst prof, 81-86, ASSOC PROF NEPHROLOGY, UNIV FLA, 86- *Mem:* Am Physiol Soc; Am Col Physicians; Am Fedn Clin Res; AAAS; Am Soc Nephrology; NY Acad Sci. *Mailing Add:* Dept Med Div Nephrology & Hyperten 111G Univ Fla Vet Admin Med Ctr Gainesville FL 32602

WINGO, CURTIS W, b Fair Grove, Mo, Aug 30, 15; m 52; c 2. ENTOMOLOGY. *Educ:* Southwestern Mo State Teachers Col, AB, 36; Univ Mo, MA, 39; Iowa State Col, PhD(entom), 51. *Prof Exp:* Assoc prof entom, Univ Mo-Columbia, 51-59, prof, 59-80; RETIRED. *Mem:* Am Entomol Soc. *Res:* Biology and control of insect parasites of man and domestic animals. *Mailing Add:* RR 7 Columbia MO 65202

WINGO, WILLIAM JACOB, b Ladonia, Tex, July 17, 18; m 40; c 2. BIOCHEMISTRY. *Educ:* Univ Tex, BA, 38, MA, 40; Univ Mich, PhD(biochem), 46. *Prof Exp:* Tutor biochem, Med Br, Univ Tex, 38-39, instr, 39-41, 45-48, lectr biochem & nutrit, 48-54, asst prof biochem, Postgrad Sch Med, 50-54, res assoc, M D Anderson Hosp Cancer Res, 48-50, assoc biochemist, 50-54; ASSOC PROF BIOCHEM, MED COL & SCH DENT, UNIV ALA, BIRMINGHAM, 54- *Mem:* AAAS; Am Soc Biol Chemists; Am Chem Soc; Soc Exp Biol & Med; Soc Protozool. *Res:* Chemistry and metabolism of amino acids; growth and metabolism of ciliate Protozoa; histochemistry; apparatus development. *Mailing Add:* 902 Saulter Rd Birmingham AL 35209

WINGROVE, ALAN SMITH, b Hanford, Calif, Mar 4, 39. ORGANIC CHEMISTRY. *Educ:* Univ Calif, Berkeley, BS, 60; Univ Calif, Los Angeles, PhD(chem), 64. *Prof Exp:* NSF fel, 64-65; asst prof chem, Univ Tex, Austin, 65-71; lectr & sci researcher & writer, 71-73; PROF CHEM, TOWSON STATE UNIV, 73- *Mem:* Sigma Xi; Am Chem Soc. *Res:* Chemistry of second row elements and participation in solvolysis and base-catalyzed cleavages; destabilized carbonium ions; carbenophiles; stereochemistry; synthetic methods. *Mailing Add:* Dept Chem Towson State Univ Towson MD 21204

WINHOLD, EDWARD JOHN, b Brantford, Ont, Jan 3, 28; nat US; m 51; c 3. PHYSICS. *Educ:* Univ Toronto, BA, 49; Mass Inst Technol, PhD(physics), 53. *Prof Exp:* Asst physics, Mass Inst Technol, 49-53, res staff mem, Lab Nuclear Sci, 53-54; from instr to asst prof physics, Univ Pa, 54-57; from asst prof to assoc prof, 57-69, PROF PHYSICS, RENSSELAER POLYTECH INST, 69- *Concurrent Pos:* Vis staff mem, Atomic Energy Res Estab, Harwell, Eng, 68-69 & lab nuclear sci, Mass Inst Technol, 82. *Mem:* Am Phys Soc. *Res:* Experimental nuclear and intermediate energy physics. *Mailing Add:* Dept Physics Rensselaer Polytech Inst Troy NY 12181

WINIARZ, MAREK LEON, b Tczew, Poland, Sept 6, 51; m 76; c 2. FACTORY MEASUREMENT SYSTEMS USING MACHINE VISION, ALGORITHM DEVELOPMENT FOR DIGITAL SYSTEMS. *Educ:* Purdue Univ, BS, 74, MS, 76. *Prof Exp:* Grad teaching asst physics, Purdue Univ, 74-75, grad res asst mech eng, Herrick Labs, 75-76; engr, Ford Motor Co, 76-79; engr & shop foreman, Seaco Indust, 80; test engr, Teledyne McCormick-Selph, 81; consult engr, 82; mgr design eng, BDM Int, 82-90; METROL SPECIALIST, QUAL TECHNOL CTR, GEN ELEC, 90- *Concurrent Pos:* Consult, Pol-Mart Int, 89- *Mem:* Instrument Soc Am; Am Soc Mech Engrs; Soc Automotive Engrs. *Res:* Dimensional and pressure metrology; sensor development for ultra-precision measurements; validation and calibration of sensors. *Mailing Add:* 9838 Forest Glen Dr Cincinnati OH 45242-5922

WINICK, HERMAN, b New York, NY, June 27, 32; m 53; c 3. SYNCHROTRON RADIATION. *Educ:* Columbia Univ, AB, 53, PhD(physics), 57. *Prof Exp:* Asst physics, Columbia Univ, 53-54, asst, Nevis Cyclotron Lab, 54-57; res assoc & instr physics, Univ Rochester, 57-59; res fel, Cambridge Electron Accelerator, Harvard Univ, 59-65, sr res assoc & lectr, 65-73, asst dir, 73; DEP DIR, STANFORD SYNCHROTRON RADIATION LAB, STANFORD UNIV, 73- *Honors & Awards:* Alexander von Humboldt Sr Scientist Prize, 85. *Mem:* Fel Am Phys Soc; AAAS. *Res:* Meson scattering; bremstrahlung research; accelerator design and development; colliding beams; synchrotron radiation production and experimentation. *Mailing Add:* 853 Tolman Dr Stanford CA 94305

WINICK, JEREMY ROSS, b Washington, DC, Aug 24, 47; m 73; c 2. UPPER ATMOSPHERE. *Educ:* Univ Rochester, BS, 69; Harvard Univ, PhD(chem physics), 76. *Prof Exp:* Res assoc, Lab Atmospheric & Space Physics, Univ Colo, 76-80; Nat Res Coun resident & res assoc, Aeronomy Lab, Environ Res Lab, Nat Oceanic & Atmospheric Admin, Boulder, 80-81; PHYSICIST, OPTICAL INFRARED TECHNOL DIV, AIR FORCE GEOPHYS LAB, 82- *Mem:* Am Geophys Union. *Res:* Model atomic and molecular processes that lead to infrared radiation in the upper atmosphere; analysis of field measurements to validate models; kinetics and radiation in upper atmosphere; modeling of upper atmosphere emissions. *Mailing Add:* Air Force Geophys Lab OPS Hanscom AFB MA 01731-5000

WINICK, MYRON, b New York, NY, May 4, 29; m 64; c 2. PEDIATRICS, NUTRITION. *Educ:* Columbia Univ, AB, 51; Univ Ill, Urbana, MS, 52; State Univ NY Downstate Med Ctr, MD, 56. *Prof Exp:* From asst resident pediat to chief resident, Med Col, Cornell Univ, 57-60; Bank Am-Giannini Found fel, Stanford Univ, 62-63, attend pediatrician & instr pediat, Med Col, 63-64; asst prof, Med Col, Cornell Univ, 64-68, from assoc prof to prof 68-71, Birth Defects Treat Ctr, 64-71; ROBERT R WILLIAMS PROF NUTRIT, PROF PEDIAT & DIR INST HUMAN NUTRIT, COL PHYSICIANS & SURGEONS, COLUMBIA UNIV, 72-, DIR, CTR NUTRIT, GENETICS & HUMAN DEVELOP, 75- *Concurrent Pos:* NIH spec fel, 63-64; dir, Birth Defects Treatment Ctr, 64-71; vis prof, Univ Chile, 67; USPHS Career Develop Award, 69-71; mem comt Nutrit, Brain Develop & Behav, Nat Acad Sci, 71-79 & Food & Nutrit Bd, 82-; consult, Pan-AM Health Orgn, 66- *Honors & Awards:* E Mead Johnson Award Pediat Res, 70; Osborne & Mendel Award, Am Inst Nutrit, 76; Agnes Higgins Award, March Dimes Birth Defects Found, 83. *Mem:* Soc Pediat Res; Am Pediat Soc; Am Inst Nutrit; Am Soc Clin Nutrit; Am Acad Pediat; Soc Exp Biol & Med. *Res:* Effects of early malnutrition on subsequent growth and development, particularly of the brain; study of brain growth and subsequent behavior. *Mailing Add:* Inst Human Nutrit Columbia Univ Col P&S 701 W 168th St NY NY 10032

WINICOUR, JEFFREY, b Providence, RI, Apr 12, 38; m 64; c 1. THEORETICAL PHYSICS. *Educ:* Mass Inst Technol, BS, 59; Syracuse Univ, PhD(physics), 64. *Prof Exp:* Res asst, Syracuse Univ, 59-64; res physicist, Aerospace Res Labs, 64-72; ASSOC PROF PHYSICS, UNIV PITTSBURGH, 72- *Concurrent Pos:* Res assoc, Ctr Philos Sci, 77- *Mem:* Am Phys Soc. *Res:* General relativity; equations of motion; gravitational radiation. *Mailing Add:* Dept Physics Univ Pittsburgh Pittsburgh PA 15260

WINICOV, HERBERT, b Brooklyn, NY, Mar 14, 35; c 2. ORGANIC CHEMISTRY, WATER PURIFICATION. *Educ:* Univ Pa, BA, 56; Univ Wis, PhD(chem), 61. *Prof Exp:* Sr chemist, Smith, Kline & French Labs, 60-68, sr investr, 68-86; CHEM SPECIALIST, MAINE YANKEE. *Mem:* AAAS; Sigma Xi; Am Chem Soc. *Mailing Add:* PO Box 345 East Boothbay ME 04544-0345

WINICOV, ILGA, b Riga, Latvia, May 16, 35; US citizen; m 79; c 2. NUCLEIC ACID BIOCHEMISTRY, MOLECULAR GENETICS. *Educ:* Univ Pa, Philadelphia, AB, 56, PhD(microbiol), 71; Univ Wis-Madison, MS, 58. *Prof Exp:* Assoc, Inst Cancer Res, Philadelphia, 72-74, res assoc, 74-76; res asst prof biochem, Fels Res Inst & Dept Biochem, Sch Med, Temple Univ, Philadelphia, 76-79; asst prof, 79-85, ASSOC PROF MICROBIOL & BIOCHEM, SCH MED, UNIV NEV, RENO, 85- *Concurrent Pos:* Vis scientist, NIH, 80-81. *Mem:* Am Soc Microbiol; Am Soc Biol Chemists; Am Asn Cancer Res; Int Soc Plant Molecular Biol; Soc Plant Physiol. *Res:* Eucaryotic gene expression at the level of RNA transcription and processing; characterization of processing products and enzymes in cultured mamr :alian cells; rRNA, tRNA and mRNA processing; stress elicited plant gene expression; cloroplast gene regulation. *Mailing Add:* Dept Biochem Sch Med Univ Nev Reno NV 89557

WINICOV, MURRAY WILLIAM, b New York, NY, July 4, 28; m; c 2. CHEMICAL ANTIMICROBIAL SUBSTANCES. *Educ:* Univ Pa, BS, 49; Brooklyn Polytech Inst, MS, 59. *Prof Exp:* Chemist, Sloan-Kettering Inst, 49-50, Indust Toxicol Labs, 50-53 & Barret Div, Allied Chem & Dye, 53-56; chemist, West Chem Prod Inc, 56-75, res dir, 75-80, res dir, West Agrochem Div, 80-83, vpres res & develop, 83-84, VPRES RES & DEVELOP, WEST AGRO INC, DIV ALFA LAVAL, INC, 84- *Mem:* Am Chem Soc; Am Soc Microbiol; AAAS. *Res:* Formulation of chemical antimicrobials, especially iodine, to maximize effectiveness as disinfectants, sanitizers, and sterilizing agents; development of products for use as human topicals, health care personnel handwash, and bovine teat dips. *Mailing Add:* 4609 Charlotte St Kansas City MO 64110

WINICUR, DANIEL HENRY, b New York, NY, May 6, 39; m 60; c 2. CHEMICAL PHYSICS. *Educ:* City Col New York, BME, 61; Univ Conn, MSME, 63; Univ Calif, Los Angeles, PhD(chem dynamics), 68. *Prof Exp:* Res engr, Space Systs Div, Hughes Aircraft Co, 63-64; Shell Oil fel chem dynamics, A A Noyes Lab Chem Physics, Calif Inst Technol, 68-70; asst prof phys chem, 70-76, asst dean, Col Sci, 79-82, ASSOC PROF PHYS CHEM, UNIV NOTRE DAME, 76-, DEAN ADMIN, 84- *Concurrent Pos:* Registr, Univ Notre Dame, 82- *Mem:* AAAS; Am Chem Soc; Am Phys Soc; Sigma Xi. *Res:* Kinetics and spectroscopy of free-radical species in flames; excited atomic and molecular states; energy transfer processes. *Mailing Add:* 215 Admin Bldg Univ Notre Dame Notre Dame IN 46556

WINICUR, SANDRA, b New York, NY, Oct 4, 39; m 60; c 2. CELL PHYSIOLOGY. *Educ:* Hunter Col, BA, 60; Univ Conn, MS, 63; Calif Inst Technol, PhD(biochem), 71. *Prof Exp:* Asst prof, 70-77, ASSOC PROF BIOL, IND UNIV, SOUTH BEND, 77- *Mem:* Nat Asn Biol Teachers. *Res:* Variations in salivary amylase activity; science & literature. *Mailing Add:* Dept Biol Ind Univ South Bend IN 46634

WINIKOFF, BEVERLY, b New York, NY, Aug 26, 45; m 73; c 2. PUBLIC HEALTH, NUTRITION. *Educ:* NY Univ, MD, 71; Harvard Univ, AB, 66, MPH, 73. *Prof Exp:* Intern, Gen Rose Mem Hosp, Denver, 71-72; res fel, Dept Nutrit, Sch Pub Health, Harvard Univ, 73-74; prog assoc & nutrit specialist, Rockefeller Found, 74-75, asst dir health sci, 75-78; med assoc, 78-84, SR MED ASSOC, INT PROGS, POP COUN, 84- *Mem:* Am Pub Health Asn. *Res:* Development and implementation of nutrition policies and programs; lactation; maternal child health. *Mailing Add:* 333 E 30 St New York NY 10016

WINJE, RUSSELL A, b Britton, SDak, Aug 24, 32; m 59; c 2. POWER ELECTRONICS. *Educ:* Univ Minn, BSEE, 61. *Prof Exp:* Staff engr, Argonne Nat Lab, 61-68 & Fermi Nat Accelerator Lab, 68-76; sr tech staff, Plasma Physics Lab, Princeton Univ, 76-88; SCI APPLN INT CORP, 88- *Mem:* Sr mem Inst Elec & Electronics Engrs. *Res:* Particle accelerator and fusion technology. *Mailing Add:* 30 Wall St Bldg E Princeton NJ 08540

WINJUM, JACK KEITH, b Platte, SDak, Feb 5, 33; m 54; c 3. FOREST CULTURE. *Educ:* Ore State Univ, BS, 55; Univ Wash, MS, 61; Univ Mich, PhD(forest ecol), 65. *Prof Exp:* Forester, US Forest Serv, 55; forest technologist, Forestry Res Ctr, Weyerhaeuser Co, 58-63, regeneration ecologist, 63-73, mgr forest regenerator res, 73-77, forest cult res, Technol Ctr, 77-80, Mt St Helen's Res & Develop, Western Forestry Res Ctr, 80-85, air pollution and forest effects, Nat Acid Precipitation Assessment Prog, 86-87; RES SCIENTIST POTENTIAL FOREST MGT MITIGATION MEASURE GLOBAL WARMING, US ENVIRON PROTECTION AGENCY, CORVALLIS, ORE, 87- *Concurrent Pos:* Affil assoc prof, Col Forest Resources, Univ Wash, Seattle, 80- *Mem:* Soc Am Foresters; Sigma Xi. *Res:* Cone and seed yield of Douglas fir; ecology of forest nurseries; stock handling and field out planting of seedlings in the regeneration period of Douglas fir management; forest regeneration ecology; air pollution and forest effects; managing forests to mitigate global warming. *Mailing Add:* US Environ Protection Agency 200 SW 35th St Corvallis OR 97333

WINKEL, CLEVE R, b Logan, Utah, Mar 20, 32; m 55; c 9. BIOCHEMISTRY, ORGANIC CHEMISTRY. *Educ:* Utah State Univ, BS, 54, MS, 55; Brigham Young Univ, PhD, 70. *Prof Exp:* Chmn, Div Nat Sci, 72-77, chmn dept chem, 84-91, PROF CHEM, RICKS COL, 59- *Honors & Awards:* Catalyst Award, Chem Mfg Asn, 78. *Res:* Enzymology; enzyme mechanism; medical biochemistry. *Mailing Add:* Dept Chem Ricks Col Rexburg ID 83460-0500

WINKELHAKE, JEFFREY LEE, b Champaign, Ill, Oct 5, 45. IMMUNOCHEMISTRY, PHARMACOLOGY. *Educ:* Univ Ill, Urbana-Champaign, BS, 67, MS, 69, PhD(immunochem), 74. *Prof Exp:* Res asst immunol, Walter Reed Army Inst Res, 69-72; res assoc fel cell biol, Jane Coffin Childs Mem Fund Med Res, Salk Inst Biol Studies, 74-76; from asst prof to prof microbiol, Med Col Wis, 76-84; dir pharmacol, Cetus Corp, 84-90; DIR PHARMACOL, PROD DEVELOP, CYTEL CORP, 90- *Concurrent Pos:* Instr hemat & serol, US Army Med Training Ctr, San Antonio, 69-70; assoc scientist, Ctr Great Lakes Res, Univ Wis-Milwaukee, 78-84. *Mem:* Am Chem Soc; Am Asn Immunologists; Am Asn Cancer Res; Am Soc Biol Chemists; Biochem Soc London. *Res:* Immunoglobulin effector functions; homeostasis of immune effector systems; evolutionary aspects; protein and carbohydrate pharmacology; antibody and lymphokine metabolism; biological response modifiers - immunopharmacology; biopharmaceutical drug development. *Mailing Add:* Dept Pharmcol/Chem Cytel Corp 11099 N Torrey Pines Rd La Jolla CA 92037

WINKELMAN, JAMES W, b Brooklyn, NY, Oct 29, 35; m 77; c 4. PATHOLOGY, HOSPITAL ADMINISTRATION. *Educ:* Univ Chicago, AB, 55; Johns Hopkins Sch Med, MD, 59. *Hon Degrees:* MA, Harvard Med Sch, 90. *Prof Exp:* Asst prof path, Sch Med, NY Univ, 65-67; asst dir, Bio-Sci Labs, 67-70, vpres & dir, 70-72, pres & dir, 72-77; exec vpres & dir, Nat Health Labs, 77-80; prof & actg chmn, State Univ NY Health Sci, 80-85, dir labs, 85-86; PROF PATH, HARVARD MED SCH, 86-, VPRES & DIR LABS, BRIGHAM & WOMEN'S HOSP, 86- *Concurrent Pos:* Assoc clin prof path, Los Angeles Med, Univ Calif, 69-80. *Mem:* Am Asn Pathologists; fel Col Am Pathologists; Acad Clin Lab Physicians & Scientists; Am Asn Clin Chemists; Am Fedn Clin Res; AMA. *Res:* Porphyrin uptake in tumors and photodynamic therapy; noninvasive measurements in clinical testing; fiscal analysis of policy and procedures in clinical laboratories. *Mailing Add:* Brigham & Women's Hosp Harvard Med Sch 75 Francis St Boston MA 02115

WINKELMANN, FREDERICK CHARLES, b Brooklyn, NY, Apr 11, 41; m 68; c 2. COMPUTER ANALYSIS, BUILDING ENERGY USE. *Educ:* Mass Inst Technol, BS, 62, PhD(physics), 68. *Prof Exp:* Res assoc physics, Lab Nuclear Sci, Mass Inst Technol, 68-69, Stanford Linear Accelerator Ctr, 69-72, Carleton Univ, Ottawa, Can, 75-76; res assoc, 72-75, STAFF SCIENTIST APPL SCI DIV, LAWRENCE BERKELEY LAB, UNIV CALIF, 76- *Concurrent Pos:* Vis scientist, Group RAMSES, Nat Ctr Sci Res, Orsay, France, 82-83. *Mem:* Am Soc Heating, Refrig & Air Conditioning Engrs. *Res:* Development of software for computer analysis of energy use in buildings. *Mailing Add:* Lawrence Berkeley Lab Bldg 90 Rm 3147 Univ Calif Berkeley CA 94720

WINKELMANN, JOHN ROLAND, b Champaign, Ill; m 62; c 2. VERTEBRATE ZOOLOGY, MAMMALOGY. *Educ:* Univ Ill, Urbana, BS, 54; Univ Mich, Ann Arbor, MA, 60, PhD(zool), 71. *Prof Exp:* Asst prof, 63-80, ASSOC PROF BIOL, GETTYSBURG COL, 80- *Concurrent Pos:* Fac fel grant for res in Mex, Gettysburg Col, 72-73. *Mem:* AAAS; Soc Study Evolution; Am Soc Mammal. *Res:* Biology of nectar-feeding bats. *Mailing Add:* Dept Biol Gettysburg Col Gettysburg PA 17325

WINKELMANN, RICHARD KNISELY, b Akron, Ohio, July 12, 24; c 4. DERMATOLOGY, DERMATOPATHOLOGY. *Educ:* Univ Akron, BS, 47; Marquette Univ, MD, 48; Univ Minn, PhD(dermat), 56. *Prof Exp:* Res assoc chem, Wash Univ, 49, res assoc anat, 50; asst pub health officer, USPHS, Ala, 52-54; from instr to assoc prof dermat, 56-65, assoc prof anat, 64-73, Robert H Kieckhefer prof 75-80, PROF DERMAT, MAYO GRAD SCH MED, UNIV MINN, 65-, PROF ANAT, 73- *Concurrent Pos:* Fel dermat, Mayo Grad Sch Med, 51-52 & 54-56; instr & res assoc, Med Col, Univ Ala, 53-54; consult, dept dermat, Mayo Clin, 56-, chmn dept, 70-75; assoc ed, Invest Dermat, 60-63, 69-72, Dermatol Digest, 65-70, Mayo Clin Proceedings, 68-69; mem bd dirs, Am Acad Dermat, 67-79 & Soc Invest Dermat, 69-70; ed, Dermatol Digest, 71-74; mem, Int Comt Dermat, 72-, Health, Educ & Welfare, Food & Drug Admin. *Mem:* Am Acad Dermat; fel Am Col Physicians; Am Soc Dermatopathologists (pres, 77); Am Asn Anatomists; Am Fedn Clin Res; AAAS; Am Asn Phys Anthropologists; Sigma Xi; Soc Invest Dermat (vpres, 69-70, pres, 70-71); AMA; hon mem NAm Clin Dermat Soc. *Res:* Anatomy; pathobiology; biology of skin. *Mailing Add:* Mayo Clin Rochester MN 55901

WINKELSTEIN, ALAN, b New York, NY, May 27, 35; m 59; c 2. HEMATOLOGY, IMMUNOLOGY-CLINICAL. *Educ:* Univ Mich, BS, 57; State Univ NY, MD, 61. *Prof Exp:* Intern & resident internal med, Univ Calif Los Angeles Sch Med, 61-63; resident internal med, Univ Wash, Seattle, 63-64; fel hemat, Univ Calif Los Angeles Med Sch, 64-67, asst prof hemat, 66-67; from asst prof to assoc prof, 69-80, PROF MED & HEMAT, UNIV PITTSBURGH SCH MED, 80- *Concurrent Pos:* Actg chief div hemat, Univ Pittsburgh Sch Med, 90-91; mem, Immunol Sci Study Sect, NIH, 84-88; mem, Biobehav/Clin subcomt, Drug Abuse AIDS Res Rev Comt, NIDA, 84- *Mem:* Am Soc Hemat. *Res:* Lymphocyte biology-lymphoprolifrative responses, flow cytometry, immune deficiencies including those associated with HIV infection activities of immuno-suppressive compounds and lymphocyte stimulatory activities of interleukins in vitro. *Mailing Add:* Montefiore Univ Hosp Pittsburgh PA 15213

WINKELSTEIN, JERRY A, b Syracuse, NY, Sept 5, 40; m 70; c 2. PEDIATRIC IMMUNOLOGY, COMPLEMENT. *Educ:* Albert Einstein Col Med, MD, 65. *Prof Exp:* DIR, DIV IMMUNOL, JOHNS HOPKINS HOSP, 80-, PROF PEDIAT, 83- *Honors & Awards:* Mead Johnson Award,

Am Acad Pediat. *Mem:* Am Asn Immunol; Am Soc Clin Invest; Am Pediat Soc; Soc Pediat Res; Infectious Dis Soc; Am Soc Microbiol. *Res:* Research focuses on the biology of the complement system. *Mailing Add:* Dept Pediat CMSC 1103 John Hopkins Hosp 601 N Wolfe St Baltimore MD 21205

WINKELSTEIN, WARREN, JR, b Syracuse, NY, July 1, 22; m 47; c 3. MEDICINE, EPIDEMIOLOGY. *Educ:* Univ NC, BA, 43; Syracuse Univ, MD, 47; Columbia Univ, MPH, 50; Am Bd Prev Med, dipl. *Prof Exp:* Dist health officer, Erie County Health Dept, NY, 50-51; regional rep pub health div, Tech & Econ Mission, Mutual Security Agency, Cambodia, Laos & Viet Nam, 51-53; dir div commun dis control, Erie County Health Dept, NY, 53-56; from asst prof to prof prev med, Sch Med, State Univ NY Buffalo, 56-69, chief dept epidemiol, Chronic Dis Res Inst, 57-64; assoc dean, Sch Pub Health, Univ Calif, Berkeley, 70-71, actg dean, 71-72, dean, 72-81, PROF EPIDEMIOL, DEPT BIOMED & ENVIRON HEALTH SCI, SCH PUB HEALTH, UNIV CALIF, BERKELEY, 68-, HEAD, EPIDEMIOL PROG, 87- *Concurrent Pos:* Spec res fel, Nat Heart Inst, 56-67, career develop award, 62-68; Buswell res fel, Univ Buffalo, 58-59; dep health comnr, Erie County Health Dept, 59-62; mem, Heart Dis Control Prog Adv Comt & Air Pollution Training Comt, USPHS, 62-65, Subcomt, Nat Comt Health Statist, 65-68, Res Comt, Am Heart Asn, 66-71, Nat Air Qual Criteria Adv Comt, 69-72, Comn Natural Resources, Nat Res Coun & Ad Hoc Working Group Epidemiol, Nat Cancer Inst, 75-77, Bd Sci Adv, Nat Inst Occup Safety & Health, 83-87 & Adv Comt, III Int Conf AIDS, 86-87; chmn, Panel Arsenic Studies, Am Pub Health Asn, 75-76 & Panel Experts Arch Pub Health, 79; consult, WHO, 76, Dept Health, NY, 84, Nat Acad Sci & USPHS, 86 & 87. *Mem:* Inst Med-Nat Acad Sci; fel Am Pub Health Asn; fel Am Col Prev Med; AAAS; Am Heart Asn; Am Epidemiol Asn; fel Infectious Dis Soc Am. *Res:* Epidemiology of cardiovascular diseases; air pollution and cancer; author of 148 technical publications. *Mailing Add:* Dept Epidemiol Univ Calif Sch Pub Health Berkeley CA 94720

WINKER, JAMES A(NTHONY), b Randall, Minn, Dec 16, 28; m 53; c 5. AERONAUTICS, METEOROLOGY. *Educ:* Univ Minn, BAeroE & BBA, 52. *Prof Exp:* Jr engr, Mech Div, Gen Mills, Inc, 51-54; sr engr, Raven Industs, Inc, 56-60, chief engr, 60-66, res mgr, 66-68, vpres appl technol div, 68-91; CONSULT, 91- *Concurrent Pos:* Expert witness, lighter-than-air aircraft. *Mem:* Sr mem Am Inst Aeronaut & Astronaut. *Res:* Scientific ballooning; atmospheric decelerators; aerial recovery systems; earth and space inflatables. *Mailing Add:* 2805 Poplar Dr Sioux Falls SD 57105

WINKLER, BARRY STEVEN, b New York, NY, Apr 17, 45; m 66; c 2. VISION, RETINAL PHYSIOLOGY. *Educ:* Harpur Col, BA, 65; State Univ NY Buffalo, MA, 68, PhD(physiol), 71. *Prof Exp:* Instr physiol, Sch Med, State Univ NY Buffalo, 70-71; asst prof, 71-78, ASSOC PROF BIOL SCI, OAKLAND UNIV, 78-, ASSOC DIR, RES & ACAD DEVELOP, 83- *Mem:* Am Physiol Soc; Soc Neurosci; Asn Res Vision & Ophthal; AAAS; Soc Res Adminr; Nat Coun Univ Res Adminr. *Res:* Physiology of the retina; analysis of ionic and metabolic contributions to photoreceptor potentials. *Mailing Add:* Eye Res Inst Oakland Univ Rochester MI 48309

WINKLER, BRUCE CONRAD, b Milwaukee, Wis, Sept 25, 37; m 59; c 2. BIOCHEMISTRY. *Educ:* Valparaiso Univ, BA, 59; Iowa State Univ, MS, 62; Univ Okla, PhD(biochem), 67. *Prof Exp:* Instr chem, Cent State Univ, 62-64; fel biochem, Univ Alta, 67-69; asst prof biochem, Kansas City Col Osteop Med, 69-73, actg chmn dept, 73-78; asst prof, 78-82, ASSOC PROF CHEM, UNIV TAMPA, 82- *Res:* Muscle phosphorylase; clinical chemistry, especially proteins and enzymes; protein electrophoresis. *Mailing Add:* Box 94F Univ Tampa Tampa FL 33606-1490

WINKLER, DELOSS EMMET, b Atchison, Kans, Feb 4, 14; m 41; c 2. POLYMER CHEMISTRY. *Educ:* Univ Kans, AB, 36, MA, 39, PhD(chem), 41. *Prof Exp:* Teacher high sch, Kans, 36-37; asst instr chem, Univ Kans, 37-39; chemist, Shell Develop Co, 41-70; real estate salesman, 71-72; chemist, Beckman Instruments, 72-79; RETIRED. *Concurrent Pos:* Consult, Polymer Chem & Technol, 79- *Mem:* Am Chem Soc. *Res:* Vapor phase catalysis; plastics; rubber; oxidation of hydrocarbons; chromatographic polymers for separation of amino acids and polymers for solid phase synthesis of peptides. *Mailing Add:* 133 Lombardy Lane Orinda CA 94563

WINKLER, ERHARD MARIO, b Vienna, Austria, Jan 8, 21; nat US; m 53; c 2. ENVIRONMENTAL GEOLOGY. *Educ:* Univ Vienna, PhD, 45. *Prof Exp:* Asst eng geol, Vienna Tech Univ, 40-45; sci asst geol, Vienna Tech Univ, 46-48; from instr to assoc prof, 48-73, PROF GEOL, UNIV NOTRE DAME, 73- *Honors & Awards:* E B Burwell Jr Award, Geol Soc Am, 75. *Mem:* Fel AAAS; fel Geol Soc Am. *Res:* Decay of stone monuments. *Mailing Add:* Dept Earth Sci Univ Notre Dame Notre Dame IN 46556

WINKLER, HERBERT H, b Highland Park, Mich, June 18, 39; m 61; c 1. MICROBIOLOGY, BIOCHEMISTRY. *Educ:* Kenyon Col, BA, 61; Harvard Univ, PhD(physiol), 66. *Prof Exp:* NSF fel physiol chem, Sch Med, Johns Hopkins Univ, 66-68; from asst prof to assoc prof microbiol, Sch Med, Univ Va, 68-77; prof, 78-80, PROF & VCHMN MICROBIOL, COL MED, UNIV S ALA, 81- *Concurrent Pos:* NIH res career develop award. *Mem:* Am Soc Microbiol; Am Soc Biol Chemists; Am Soc Rickettsiology (pres, 83-85); Am Asn Immunol. *Res:* Transport of molecules across biological membranes; biology of rickettsiae. *Mailing Add:* Dept Microbiol Univ SAla Col Med Mobile AL 36688

WINKLER, JAMES DAVID, b Flint, Mich, Oct 24, 54; m 80; c 3. PHARMACY, BIOCHEMISTRY. *Educ:* Princeton Univ, BA, 76; Med Col Pa, PhD(pharmacol), 87. *Prof Exp:* Postdoctoral fel, 87-89, SR SCIENTIST, SMITHKLINE BEECHAM, 89- *Concurrent Pos:* Assoc prof, Med Col Pa, 87- *Mem:* AAAS; Am Soc Pharmacol & Exp Therapeut; Inflammation Res Asn; NY Acad Sci. *Res:* Mechanisms of inflammation; focus on biochemical mediator of inflammation processes. *Mailing Add:* Dept Pharmacol Smithkline Beecham L532 PO Box 1539 King of Prussia PA 19406

WINKLER, LEONARD P, b New York, NY. ELECTRICAL ENGINEERING, COMPUTER SCIENCES. *Educ:* Polytech Inst Brooklyn, BSEE, 65, MSEE, 67, PhD(elec eng), 71. *Prof Exp:* Res fel, Polytech Inst Brooklyn, 69-70; asst prof eng sci, Richmond Col, NY, 70-77; ASSOC PROF ENG SCI, COL STATEN ISLAND, CITY UNIV NEW YORK, 77- *Concurrent Pos:* NSF res grant, 72. *Mem:* Inst Elec & Electronics Engrs; Sigma Xi. *Res:* Traffic control; microcomputers; communications-computer systems; numerical optimization techniques; stochastic processes. *Mailing Add:* Col Staten Island Staten Island NY 10301

WINKLER, LOUIS, b Elizabeth, NJ, Sept 7, 33; m 57; c 1. ASTRONOMY. *Educ:* Rutgers Univ, BS, 55; Adelphi Univ, MS, 59; Univ Pa, PhD(astron), 64. *Prof Exp:* Engr, Am Bosch Arma Corp, 56-59; proj engr, Philco Corp, 59-64; ASST PROF, PA STATE UNIV, 64- *Mem:* Am Astron Soc. *Res:* Archaeoastronomy; astronomy and astrology of early America; United States of America seismic histories. *Mailing Add:* Dept Astron Pa State Univ University Park PA 16802

WINKLER, MATTHEW M, b Boston, Mass, June 22, 52. CELL BIOLOGY. *Educ:* Univ Calif, Berkeley, PhD(zool), 79. *Prof Exp:* Instr embryol, Marine Biol Lab, Woods Hole, 83; ASST PROF ZOOL, UNIV TEX, 83- *Mailing Add:* Dept Zool Univ Tex Austin TX 78712

WINKLER, MAX ALBERT, b San Antonio, Tex, May 19, 31; m 53; c 4. PHYSICS, MATHEMATICS. *Educ:* St Mary's Univ, BS, 57; Univ Tex, Austin, MA, 62. *Prof Exp:* Physicist, Gen Elec Co, 57-59; MEM STAFF PHYSICS, LOS ALAMOS SCI LAB, 62- *Res:* Optical engineering; lens design; nondestructive testing. *Mailing Add:* Los Alamos Sci Lab Box 1663 Los Alamos NM 87545

WINKLER, NORMAN WALTER, b Englewood, NJ, May 28, 35; m 80; c 2. BIOCHEMISTRY, DERMATOLOGY. *Educ:* Univ Rochester, AB, 57; Univ Chicago, MD, 65, PhD(biochem), 70. *Prof Exp:* Res asst fibrinolysis, Sloan-Kettering Inst Cancer Res, 58-59; from intern to resident internal med, Univ Chicago Hosps & Clins, 68-70; USPHS fel dermat, Med Sch, Univ Ore, 70-72; chief dermat serv, Buffalo Vet Admin Hosp, 73-76; CHIEF, DERMAT SERV, SBUFFALO MERCY HOSP, 76- *Concurrent Pos:* Asst prof dermat, Sch Med, State Univ NY, Buffalo, 72-76. *Mem:* AAAS; Am Acad Dermat; Soc Invest Dermat. *Res:* Enzymology; membrane receptors in cutaneous disease; keratinocyte differentiation. *Mailing Add:* 4174 N Buffalo St Orchard Park NY 14127

WINKLER, PAUL FRANK, b Nashville, Tenn, Nov 10, 42; m 83; c 2. ASTROPHYSICS. *Educ:* Calif Inst Technol, BS, 64; Harvard Univ, AM, 65, PhD(physics), 70. *Prof Exp:* From asst prof to assoc prof physics, 69-81, chmn dept, 80-88, PROF PHYSICS, MIDDLEBURY COL, 81-, CHMN DIV NATURAL SCI, 88- *Concurrent Pos:* Vis scientist, Mass Inst Technol, 73-74, res affil, 74-78, vis scientist, 78-80; sr vis fel, Inst Astron, Cambridge Univ, 85-86; Alfred P Sloan Found res fel, 76-80; vis resident astronr, Cerro Tololo Inter Am Observ, 90-91; vis fel, Joint Inst Lab Astrophys, 91. *Honors & Awards:* Alfred P Sloan Found res fel, 76-80. *Mem:* Am Phys Soc; Am Astron Soc; Int Astron Union. *Res:* Supernova remnants; galactic and extragalactic x-ray sources; atomic and molecular physics. *Mailing Add:* Dept Physics Middlebury Col Middlebury VT 05753

WINKLER, PETER MANN, b Pasadena, Calif, Nov, 9, 46; m 73; c 2. COMPUTABILITY. *Educ:* Harvard Univ, BA, 68; Yale Univ, PhD(math), 75. *Prof Exp:* Mathematician, Dept Defense, 68-70; asst prof math, Stanford Univ, 75-77; from asst prof to assoc prof, 77-89, PROF MATH & COMPUT SCI, EMORY UNIV, 89-; RES MGR, MATH & THEORET COMP SCI, BELLCORE, 89- *Concurrent Pos:* Consult math, Navig Sci, Inc, 78-; Humboldt fel, T H Darmstadt, 84-85. *Mem:* Am Math Soc; Math Asn Am; Soc Indust Appl Math. *Res:* Combinatorics; theoretical computer science. *Mailing Add:* Dept Math & Comput Sci Emory Univ Atlanta GA 30322

WINKLER, ROBERT RANDOLPH, b Washington, DC, June 16, 33; m 55; c 3. ORGANIC CHEMISTRY. *Educ:* Univ Md, BS, 55; Univ Mich, MS, 60, PhD(org chem), 62. *Prof Exp:* Phys sci aide plant indust sta, Agr Res Serv, USDA, Md, 55, chemist, 57-58; teaching asst & res fel, Univ Mich, 58-61; from asst prof to assoc prof, 61-85, PROF ORG CHEM, OHIO UNIV, 85- *Mem:* Sigma Xi; AAAS; Am Chem Soc. *Res:* Chemical education; mechanism and stereochemistry of carbonyl condensation reactions. *Mailing Add:* Dept Chem Ohio Univ Athens OH 45701

WINKLER, SHELDON, b New York, NY, Jan 25, 32; m 61; c 2. DENTISTRY. *Educ:* NY Univ, BA, 53, DDS, 56. *Prof Exp:* From instr to asst prof denture prosthesis, Col Dent, NY Univ, 58-68; asst prof removable prosthodont, State Univ NY, Buffalo, 68-70, assoc prof removable prosthodont, 70-79; prof removable prosthodont & chmn dept, 79-86, asst dean advan studies, continuing educ & res, 87-89; PROF PROSTHODONT, TEMPLE UNIV, PHILADELPHIA, 90- *Concurrent Pos:* Dir mat res, CMP Industs, Inc, 63-65, consult, 65-66; lectr, New York Community Col, 67-68; consult, Coe Labs, Inc, Ill, 67-87; consult dent auxiliary training progs, Bd Coop Educ Serv, Cheektowaga, NY, 70-79; consult, Dental Lab Technol, Erie Community Col, Buffalo, NY, 78-; mem, Bd Consults, Quintessence Int, 80-81; consult, Personal Prod Div, Lever Bros Co, NY, 81-, Dent Schs, Asuncion, Paraguay, Valparaiso, Santiago & Chile, 81-, Off Atty Gen, Commonwealth, PA, 83-, Dent Schs, Bangkok, Thailand & Alexandria, Egypt, 82-, Vet Admin Med Ctr, Philadelphia, PA, 88- *Mem:* Fel Am Col Dent; Am Prosthodont Soc; Am Dent Asn; Am Acad Plastics Res Dent. *Res:* Dental resins and alloys; preservation and embedment of specimens in methyl methacrylate; demineralization of bone; geriatric dentistry; laser radiation applications in dentistry; increasing the bond strength of ceramometal restorations. *Mailing Add:* Sch Dent 3223 N Broad St Philadelphia PA 19140

WINKLER, VIRGIL DEAN, b Danvers, Ill, Feb 9, 17; m 43; c 2. GEOLOGY. *Educ:* Univ Ill, AB & BS, 38, MS, 39, PhD(geol), 41. *Prof Exp:* Instr geol, Univ Ill, 38-39; paleontologist, Creole Petrol Corp, 41-45, chief paleontologist, 45-55, paleont coordr, 55-56, eval geologist, 56-61, eval & opers geologist, 61-63, spec studies & eval geologist, 63-76; GEOL ADV, LAGOVEN S A, 76- *Concurrent Pos:* Prof, Cent Univ Venezuela, 58-59 & 66- *Mem:* AAAS; Paleont Soc; Soc Econ Paleont & Mineral; Geol Soc Am; Asn Geol, Mineral & Petrol, Venezuela (vpres, 54-55, secy-treas, 59-60); Am Asn Petrol Geologists. *Res:* Paleontology of Paleozoic rocks; world-wide occurrence of oil; Mesozoic and Cenozoic stratigraphy of Venezuela. *Mailing Add:* Aptdo 80537 Prados del Este Caracas 1080A Venezuela

WINN, ALDEN L(EWIS), b Portsmouth, NH, Jan 26, 16; m 41; c 2. ELECTRONICS. *Educ:* Univ NH, BS, 37; Mass Inst Technol, MS, 48. *Prof Exp:* Engr & acct phys plant eval, New Eng Gas & Elec Syst, 37-40; asst elec eng, Mass Inst Technol, 45-47, instr, 47-48; from asst prof to assoc prof, 48-54, chmn dept, 52-67, PROF ELEC ENG, UNIV NH, 54- *Mem:* Am Soc Eng Educ; Inst Elec & Electronics Engrs. *Res:* Oceanographic instrumentation; semiconductor devices and circuits. *Mailing Add:* Church Hill Apt 1A Durham NH 03824

WINN, C BYRON, b Canton, Mo, Nov 21, 33; m 58; c 3. AERONAUTICAL ENGINEERING, ELECTRICAL ENGINEERING. *Educ:* Univ Ill, Urbana, BS, 58; Stanford Univ, MS, 60, PhD(aeronaut eng), 67. *Prof Exp:* Engr, Lockheed Missiles & Space Co, 58-60; engr, Martin Marietta Co, 60-62; engr Lockheed Missiles & Space Co, 62-63; res asst, Stanford Univ, 63-67; assoc prof, 67-77, assoc dir, Univ comput Ctr, 70-77, PROF MECH ENG, COLO STATE UNIV, 77-, HEAD DEPT, 83- *Concurrent Pos:* NASA res grant satellite geodesy, Colo State Univ, 67-71, remote sensing in hydrol, 70-72, Off Water Resources res grant optimal control of storm sewer syst, 70-72; consult, Space Res Corp, Que, 70-72, MEPPSCO, Inc, Mass, 71 & USAF, Wright-Patterson AFB, 71-72; vis assoc prof, Univ Newcastle, NSW, 72; dir, Energy Analysis & Diag Ctr, Colo State Univ, 84- & Mfg Excellence Ctr, 88- *Honors & Awards:* Tech Paper Award, Am Inst Aeronaut & Astronaut, 67. *Mem:* Am Inst Aeronaut & Astronaut; Am Soc Mech Engrs. *Res:* Optimal control theory and applications; satellite geodesy; simulation; solar energy. *Mailing Add:* Dept Mech Eng Colo State Univ Ft Collins CO 80523

WINN, EDWARD BARRIERE, b Baltimore, Md, Dec 27, 22; m 49; c 4. TECHNICAL MANAGEMENT, CHEMISTRY. *Educ:* Univ SC, BSEE, 46; Univ Va, MS, 47; Univ Minn, PhD(physics), 50. *Prof Exp:* Elec engr, Westinghouse Elec Corp, 46; res asst, Univ Minn, 48-50; res physicist, Textile Fibers Dept, E I du Pont de Nemours & Co, 50-58, res supvr, 58-62, tech mgr, du Pont de Nemours Int, SA, 62-70; independent consult, 70-74 & SNIA Viscosa SpA, 74-77; mgr, Diamond Shamrock France, 78-80; Europ dir, Process Indust Div, SRI Int, Zurich, Switz, 80-88; EUROP DIR, KRI INT, PRANGINS, SWITZ, 89- *Mem:* Am Phys Soc; Am Chem Soc; Sigma Xi; Soc Chem Indust Europ Sect. *Res:* Physics of high polymers; textile fibers; processing and applications technology of synthetic fibers; physics of electrical insulating materials; electrical insulation technology; industrial and technical marketing; new business ventures in textile, polymers and chemicals. *Mailing Add:* 25 ch de Trembley Prangins 1197 Switzerland

WINN, HENRY JOSEPH, b Lowell, Mass, Mar 2, 27; m 53; c 6. IMMUNOLOGY. *Educ:* Ohio State Univ, BA, 48, MS, 50, PhD(bact), 52. *Prof Exp:* Fel med & bact, Ohio State Univ, 52-54; fel chem, Calif Inst Technol, 54-55; res assoc, Jackson Mem Lab, 55-57, staff scientist, 57-65; assoc immunologist, Mass Gen Hosp, 65-73; asst prof bact, 65-70, assoc prof microbiol & molecular genetics, 69-77; IMMUNOLOGIST.MASS GEN HOSP, 73-; SR ASSOC SURG, HARVARD MED SCH, 77- *Mem:* Am Asn Immunologists. *Res:* Immunology of homotransplantation; immunogenetics. *Mailing Add:* Dept Surg Mass Gen Hosp Boston MA 02114

WINN, HOWARD ELLIOTT, b Winthrop, Mass, May 1, 26; div; c 4. BIOLOGICAL OCEANOGRAPHY, BEHAVIOR-ETHOLOGY. *Educ:* Bowdoin Col, AB, 48; Univ Mich, MS, 50, PhD(zool), 55. *Prof Exp:* Specialist, Am Mus Natural Hist, 54-55; from asst to prof zool, Univ Md, 55-65; PROF OCEANOG & ZOOL, UNIV RI, 65- *Concurrent Pos:* Guggenheim fel, 62-63. *Mem:* AAAS; Am Inst Biol Sci; Am Soc Ichthyol & Herpet; Animal Behav Soc; Am Soc Mammal. *Res:* Comparative animal behavior; biology of fishes; sounds in animals; behavior and sounds of whales; ecology. *Mailing Add:* Grad Sch Oceanog Univ RI Narragansett RI 02882-1197

WINN, HUGH, b St Louis, Mo, Apr 7, 18; m 39; c 2. CHEMICAL & MATERIALS ENGINEERING. *Educ:* Mich Col Mining & Technol, BS, 40; Case Inst Technol, MS, 44, PhD(chem eng), 48. *Prof Exp:* Res engr, Saran Develop Lab, Dow Chem Co, 41-42; asst prof chem eng & dir plastics lab, Case Inst Technol, 42-48; group leader, Firestone Tire & Rubber Co, 48-50 & Defense Res Div, 50-55; mgr nose cone design eng, Missile & Space Vehicle Dept, 55-58, mgr res opers & applns, Aerosci Lab, 58-59, mgr data processing & comput, 59-63, mgr corp eng, 63-71, mgr eng, Lamp Glass Dept, 71-78, PROJ MGR GLASS RESOURCE PLANS & PROG, LAMP GLASS PROD DEPT, GEN ELEC CO, 78- *Concurrent Pos:* Consult, Martin Co, Md, 42, Ohio Chem Co, 44-47 & Frankford Arsenal, 55-58. *Mem:* Am Chem Soc; Soc Plastics Engrs; Am Inst Chem Engrs; Sigma Xi. *Res:* Plastics formulation; evaluation and fabrication; rubber oxidation, compounding and evaluation; high explosive effects; space environment; engineering design and space vehicles; scientific computation and test data reduction; materials and processes. *Mailing Add:* 6524 Kingswood Dr Mayfield OH 44124

WINN, MARTIN, b Brooklyn, NY, Jan 25, 40; m 66; c 2. MEDICINAL CHEMISTRY. *Educ:* Cooper Union Univ, BChE, 61; Northwestern Univ, PhD(org chem), 65. *Prof Exp:* SR CHEMIST, ABBOTT LABS, 65- *Mem:* Am Chem Soc. *Res:* Pharmaceuticals; nonclassical aromatic systems; heterocycles; psychotropic drugs; antihypertensive drugs, diuretics. *Mailing Add:* Dept 47C Abbott Labs Research Div North Chicago IL 60064

WINN, WILLIAM PAUL, b Los Angeles, Calif, Apr 24, 39. ATMOSPHERIC PHYSICS & INSTRUMENTATION. *Educ:* Univ Calif, Berkeley, BS, 61, PhD(physics), 66. *Prof Exp:* Fel physics, Nat Ctr Atmospheric Res, 66-70; from asst prof to assoc prof, 70-77, chmn dept, 77-82, PROF PHYSICS, NMEX INST MINING & TECHNOL, 82- *Mem:* Am Geophys Union; Am Asn Physics Teachers; AAAS; Am Meteorol Soc. *Res:* Thunderstorms. *Mailing Add:* Dept Physics NMex Inst Mining & Technol Socorro NM 87801

WINNER, ROBERT WILLIAM, b Columbus, Ohio, Apr 5, 27; m 51; c 2. AQUATIC ECOLOGY, TOXICOLOGY. *Educ:* Ohio State Univ, PhD(wildlife mgt), 57. *Prof Exp:* Instr zool, 57-59, asst prof biol, 59-65, assoc prof 65-69, PROF ZOOL, MIAMI UNIV, 69- *Mem:* Am Soc Limnol & Oceanog; Soc Environ Toxicol & Chem; Int Soc Limnol; Sigma Xi. *Res:* Evaluation of the effects of toxic chemicals, especially heavy metals on freshwater populations, communities, and ecosystems; evaluating factors which control the structure of freshwater planktonic communities. *Mailing Add:* Dept Zool Miami Univ Oxford OH 45056

WINNETT, GEORGE, environmental sciences, analytical chemistry, for more information see previous edition

WINNICK, JACK, b Chicago, Ill, Sept 20, 37. CHEMICAL ENGINEERING. *Educ:* Univ Ill, BS, 58; Univ Okla, MS, 60, PhD(chem eng), 63. *Prof Exp:* From asst prof to assoc prof chem eng, Univ Mo-Columbia, 63-67, prof, 71-79; Cramer Wilson La Pierre prof eng, 79-80, PROF CHEM ENG, GA INST TECHNOL, 79- *Concurrent Pos:* NSF res grants, 64-94; consult, NASA Manned Spacecraft Ctr, 66-76, Life Syst Inc, 75-78, Combustion Eng, Giner Inc, 86- & Grove Eng, 90-; vis prof, Univ Calif, Berkeley, 69-70, 77 & 84-85, Univ Calif, Los Angeles, 76; Petrol Res Fund grant, 70-73; NASA grants, 72-83; Dept Energy grant, 77- *Honors & Awards:* Krupp Prize Energy Res, 83. *Mem:* Am Inst Chem Engrs; Am Chem Soc; AAAS; Sigma Xi. *Res:* Photoelectrochemistry; electrochemical membrane seperation; electrochemical engineering; fuel cells. *Mailing Add:* Dept Chem Eng Ga Inst Technol Atlanta GA 30332-0100

WINNIE, DAYLE DAVID, b Brandon, Wis, July 20, 35; m 57; c 2. MECHANICAL ENGINEERING, ELECTRONICS. *Educ:* Univ Wis, BS, 58. *Prof Exp:* Aircraft maintenance officer, Charleston AFB, SC, 59-60; prod engr electromech design, Centralab Div, Globe Union Inc, Milwaukee, 60-64; develop engr automatic processing equip, Stoelting Bros Co, Kiel, 64-69; sr res engr, 69-81, STAFF ENGR ELECTROMECH, SOUTHWEST RES INST, 81- *Mem:* Sigma Xi; Am Soc Mech Engrs. *Res:* Electromechanical design; spaceflight mass measurement equipment; automatic machinery design and development; automatic direction finding systems; sub-sea hyperbaric and single atmosphere systems; geophysical anomaly detection systems; automated continuous dairy processing equipment. *Mailing Add:* Southwest Res Inst 6220 Culebra Rd San Antonio TX 78228

WINNIFORD, ROBERT STANLEY, b Portland, Ore, Oct 10, 21; m 44; c 4. PHYSICAL CHEMISTRY. *Educ:* Ore State Col, BS, 43; Calif Inst Technol, MS, 48; Univ Tenn, PhD, 51. *Prof Exp:* Instr chem, Univ Tenn, 47-49; res chemist, Calif Res Corp, Standard Oil Co Calif, 51-63; asst prof chem, 63-67, assoc prof, 67-77, chmn dept, 71-80, RPOF CHEM, WHITWORTH COL, WASH, 77- *Mem:* Am Chem Soc; Sigma Xi. *Res:* Colloid and surface chemistry; nonaqueous solutions; asphalt chemistry and rheology. *Mailing Add:* 41096 Nichol Dr Sweet Home OR 97386-9627

WINNIK, FRANÇOISE MARTINE, b Mulhouse, France, Mar 2, 52; m 80. FLUORESCENCE & SPECTROSCOPY. *Educ:* Ecole Nat Supérieure Chim, Ingénieur Chimiste, 74; Univ Toronto, PhD(org chem), 79. *Prof Exp:* Fel carbohydrate chem, Dept Med Genetics, Univ Toronto, 79-81; MEM RES STAFF ORG & POLYMER CHEM, XEROX RES CTR CAN, 81- *Concurrent Pos:* Lectr, Dept Chem, Univ Toronto, 83-84; vis scientist, Tokyo Inst Technol, 85-86. *Mem:* Am Chem Soc; Chem Inst Can. *Res:* Design and properties of materials used in printing technologies; application of luminescence techniques in polymer science; study of structure-properties relationships of hydrophobic polymers in water. *Mailing Add:* Xerox Res Ctr Can 2660 Speakman Dr Mississauga ON L5K 2L1 Can

WINNIK, MITCHELL ALAN, b Milwaukee, Wis, July 17, 43; m 80. ORGANIC CHEMISTRY, PHOTOCHEMISTRY. *Educ:* Yale Univ, BA, 65; Columbia Univ, PhD(org chem), 69. *Prof Exp:* USPHS fel, Calif Inst Technol, 69-70; asst prof, 70-75, assoc prof, 75-80, PROF ORG CHEM, UNIV TORONTO, 80- *Concurrent Pos:* Assoc prof, Univ Bordeaux, 77-78; world trade fel, Int Bus Mach Corp, San Jose, 82; fel, Japan Soc Prom Sci, Tokyo Inst Technol, 85-86. *Honors & Awards:* A A Vernon Mem Lectr, Northeastern Univ, 83; Xerox Lectrs, Victoria Univ, BC, Can, 87. *Mem:* Inter-Am Photochem Soc; Am Chem Soc; Chem Inst Can. *Res:* Polymer conformation and dynamics; luminescence techniques in polymer science; new techniques for the study of interfaces in polymer materials; polymer colloids. *Mailing Add:* Dept Chem Univ Toronto Toronto ON M5S 1A1 Can

WINNINGHAM, JOHN DAVID, b Mexia, Tex, Dec 28, 40; m 63; c 1. MAGNETOSPHERIC PHYSICS. *Educ:* Tex A&M Univ, BS, 63, MS, 65, PhD(physics), 70. *Prof Exp:* From res asst to res sci asst physics, Univ Tex, Dallas, 66-71, res assoc, 71-73, res scientist, 73-80; MGR, EXP SPACE PHYSICS, SOUTHWEST RES INST, SAN ANTONIO, TEX, 80- *Concurrent Pos:* Consult, Los Alamos Sci Lab, Univ Calif, 74- *Mem:* Am Geophys Union. *Res:* Investigation of the source and acceleration mechanisms of corpuscular fluxes that produce the aurora and concomitant physical processes by means of rocket and satellite instruments. *Mailing Add:* Southwest Res Inst San Antonio TX 78228-0510

WINOGRAD, NICHOLAS, b New London, Conn, Dec 27, 45. ANALYTICAL CHEMISTRY. *Educ:* Rensselaer Polytech Inst, BS, 67; Case Western Reserve Univ, PhD(chem), 70. *Prof Exp:* Asst prof, Purdue Univ, West Lafayette, 70-75, assoc prof chem, 75-79; prof, 79-85, EVAN PUGH

PROF CHEM, PENN STATE UNIV, 85- *Concurrent Pos:* Consult, Shell DevelopCo, 77; mem, Adv Bd Analytical Chem, 86-88, NSF Chem Adv Bd, 87-90; Guggenheim fel, 77-78. *Honors & Awards:* Tex Instruments Found Founders Prize, 84. *Mem:* Am Chem Soc; Electrochem Soc; fel AAAS. *Res:* Characterization of solid surfaces; x-ray photoelectron spectroscopy; secondary ion mass spectrometry; theory of ion impact phenomena on solids. *Mailing Add:* Dept Chem Penn State Univ Davey Lab 152 University Park PA 16802

WINOGRAD, SHMUEL, b Tel Aviv, Israel, Jan 4, 36; m 58; c 2. THEORETICAL COMPUTER SCIENCE. *Educ:* Mass Inst Technol, BS & MS, 59; NY Univ, PhD(math), 68. *Hon Degrees:* Dr, Acad Grenoble, Nat Polytech Inst, 87. *Prof Exp:* Res asst, Mass Inst Technol, 59-61; mem res staff, 61-81, DIR, MATH SCI DEPT, T J WATSON RES CTR, Int Bus Mach Corp, 81- *Concurrent Pos:* Adj prof math, Courant Inst Math Sci, NY Univ, 68; Mackay lectr, Univ Calif, Berkeley, 67-68; vis prof, The Technion, Israel, 72-; Int Bus Mach Corp fel, 72; chmn appl math, Comput Sci & Statist Sect, Nat Acad Sci, 87. *Honors & Awards:* W Wallace McDowell Award, Inst Elec & Electronics Engrs, 74, Comput Pioneer Award, 82. *Mem:* Nat Acad Sci; fel Inst Elec & Electronics Engrs; Asn Comput Mach; Am Math Soc; Math Asn Am; Am Acad Arts & Sci; Soc Indust & Appl Math. *Res:* Computer mathematics; reliable computations; complexity of computations. *Mailing Add:* Math Sci Dept T J Watson Res Ctr PO Box 218 Yorktown Heights NY 10598

WINOGRAD, TERRY ALLEN, b Takoma Park, Md, Feb 24, 46; m 68. COMPUTER SCIENCE. *Educ:* Colo Col, BA, 66; Mass Inst Technol, PhD(appl math), 70. *Hon Degrees:* DSc, Colo Col, 86. *Prof Exp:* Instr math & asst prof elec eng, Mass Inst Technol, 70-74; from asst prof to assoc prof, 74-89, PROF COMPUTER SCI DEPT, STANFORD UNIV, 89- *Concurrent Pos:* Consult, Palo Alto Res Ctr, Xerox Corp, 73-, Action Technol, Inc, Alameda, Calif, Hermenet, Inc, San Francisco, Xerox Palo Alto Res Ctr, spec consult to pres, Fuji Xerox, Japan & to French Govt; mem, Comput Sci & Eng Res Study Panel Artificial Intel, NSF, 75; vis asst prof, Computer Sci Dept, Stanford Univ, 73-74; Mellon Jr fac fel, 77; mem, Spec Interest Group on Computers & Soc, Asn Comput Mach & Nat Bd, Computer Prof for Social Responsibility, 84- *Mem:* Asn Comput Ling; Inst Elec & Electronics Engrs; Union Concerned Scientists; Computer Prof Social Responsibility (pres, 87-90); Asn Comp Mach; Am Asn Artificial Intel. *Res:* Artificial intelligence; computational linguistics; cognitive modelling; author of numerous books and articles. *Mailing Add:* Dept Comput Sci Stanford Univ Stanford CA 94305-2140

WINOKUR, GEORGE, b Philadelphia, Pa, Feb 19, 25; m 51; c 3. GENETICS, EPIDEMIOLOGY. *Educ:* Johns Hopkins Univ, BA, 44; Univ Md, MD, 47; Am Bd Psychiat & Neurol, dipl, 53. *Prof Exp:* Intern med, Church Home & Hosp, 47-48; resident asst, Seton Inst, 48-50; resident neuropsychiat, 50-51, instr psychiat, 51-55, from asst prof to prof, 55-71, PROF PSYCHIAT & HEAD DEPT, SCH MED, WASH UNIV, 71-; PROF PSYCHIAT & HEAD DEPT, COL MED, UNIV IOWA & DIR, IOWA PSYCHIAT HOSP, 71- *Concurrent Pos:* From asst psychiatrist to assoc psychiatrist, Barnes Hosp, 55-71; attend, Malcolm Bliss Psychiat Hosp; ed, J Affective Disorders, 78-; co-ed, Europ Archives Psychiat & Neurol Sci, 85- *Honors & Awards:* Hofheimer Prize Psychiat Res, Am Psychiat Asn, 72; Res Affective Dis Prize, Anna-Monika Found, Switz, 73; Samuel W Hamilton Award, Am Psychopath Asn, 77, Paul Hoch Award,81; Leonard Cammer Award, Columbia Col Physicians & Surgeons, 80; Gold Medal Award, Soc Biol Psychiat, 84. *Mem:* Am Acad Clin Psychiatrists; fel Am Psychiat Asn; Am Col Neuropsychopharmacol; Int Group Study Affective Dis; Am Psychopath Asn; Swiss Psychiat Asn. *Res:* Sexual variables in psychiatric patients and controls; genetics and epidemiological studies of psychiatric diseases. *Mailing Add:* Dept Psychiat Univ Iowa 500 Newton Rd Iowa City IA 52242

WINOKUR, ROBERT MICHAEL, b Minneapolis, Minn, July 2, 42; m 88. ZOOLOGY. *Educ:* Macalester Col, BA, 65; Ariz State Univ, MA, 67; Univ Utah, PhD(biol), 72. *Prof Exp:* Teaching fel biol, Univ Utah, 67-72, asst res prof, 72-73; instr zool, Univ New Eng, Australia, 74-78; ASST PROF BIOL, UNIV NEV, LAS VEGAS, 78- *Honors & Awards:* Dwight D Davis Award, Am Soc Zoologists, 73. *Mem:* Am Soc Zoologists; Am Soc Ichthyologists & Herpetologists; Soc Study Amphibians & Reptiles. *Res:* Comparative morphology of lower vertebrates with emphasis on the microscopic anatomy and integumentary specialization of reptiles and amphibians. *Mailing Add:* Dept Biol Sci Univ Nev Las Vegas NV 89154

WINRICH, LONNY B, b Eau Claire, Wis, July 10, 37; c 5. COMPUTER SCIENCE. *Educ:* Wis State Univ, Eau Claire, BS, 60; Univ Wyo, MS, 62; Iowa State Univ, PhD(appl math), 68. *Prof Exp:* Physicist, Boulder Labs, Nat Bur Standards, 60-62; mathematician, Aerospace Div, Honeywell, Inc, 62-64; instr math & comput sci, Iowa State Univ, 64-68; asst prof comput sci, Univ Mo-Rolla, 68-71; from assoc prof to prof, Univ Wis, La Crosse, 71-85, chmn dept, 71-79; chmn dept, 86-90, PROF COMPUT SCI, UNIV NDAK, 85- *Concurrent Pos:* Consult, Int Bus Mach Corp. *Mem:* Inst Elec & Electronics Engrs; Sigma Xi; Asn Comput Mach; Am Asn Artificial Intel. *Res:* Application of intelligent systems to software testing; software maintenance and documentation; intelligent systems; systems level testing. *Mailing Add:* Dept Comput Sci Univ NDak Box 8181 Univ Sta Grand Forks ND 58202-8181

WINSBERG, GWYNNE ROESELER, b Chicago, Ill, Nov 28, 30; m 50; c 2. EPIDEMIOLOGY. *Educ:* Univ Chicago, MS, 62, PhD(biopsychol), 67. *Prof Exp:* Instr biol, Univ Chicago, 65-67; asst prof anat, Med Sch, Northwestern Univ, Chicago, 67-71, asst prof community health & prev med, 71-76; assoc prof community & family med & assoc dean, Stritch Sch Med, Loyola Univ, Chicago, 76-81; assoc prof prev med, Univ Ill, 82-87; ASSOC PROF PSYCHIAT, RUSH PRESBY ST LUKES MED CTR, 89- *Concurrent Pos:* Lectr, Ill Col Optom, 62-67; consult, Ill Dept Ment Health, 69-72; USPHS grant, Fac Inst Med Care Orgn, Univ Mich, 73 & 74; Nat Endowment Humanities grant, Univ Pa, 74; spec asst to regional health adminr, Region V, USPHS, 74-76; secy, bd trustees, North Communities Health Plan, Inc, Evanston, 76-; sr policy analyst, Off Secy, Dept Health Human Serv, Washington, DC, 79-81; pres, GRW Assoc, Inc, Chicago, 81-; vpres, Efficient Health Systs, Inc, Skokie, Ill, 86-87. *Mem:* Am Pub Health Asn; Am Med Care Review Asn; Group Health Asn Am. *Res:* Social and medical epidemiology; medical care organization; health policy and legislation; mental health and long term care benefits; corporate health care benefits and cost containment. *Mailing Add:* 5533 N Glenwood Chicago IL 60640

WINSCHE, WARREN EDGAR, nuclear engineering; deceased, see previous edition for last biography

WINSKE, DAN, b East Chicago, Ind, Feb 24, 46. SPACE PHYSICS, PLASMA SIMULATION. *Educ:* Purdue Univ, BS, 68; Univ Ill, Champaign-Urbana, MS, 69, PhD(physics), 74. *Prof Exp:* Staff mem, Controlled Thermonuclear Res Div, Los Alamos Sci Lab, 74-79; sr res assoc, Plasma & Fusion Studies Lab, Univ Md, 79-83; STAFF MEM, APPL THEORET PHYSICS DIV, LOS ALAMOS NAT LAB, 83- *Concurrent Pos:* Consult, Off Fusion Energy, Dept Energy, 83; vis scientist, Observatoire de Paris Meudon, 83; assoc ed, J Geophys Res, 89-; prin investr, NASA Space Plasma Theory Prog, Los Alamos Nat Lab, 89- *Mem:* Fel Am Phys Soc; Am Geophys Union. *Res:* Numerical modeling and theoretical analysis of space and laboratory plasmas; study of collisionless shock waves and plasma microinstabilities using particle and hybrid plasma simulation methods. *Mailing Add:* F645 X-1 Los Alamos Nat Lab Los Alamos NM 87545

WINSLOW, ALFRED EDWARDS, b Clinton, Mass, Oct 8, 19; m 44; c 2. ORGANIC CHEMISTRY, POLYMERS & LIQUID PHENOLIC RESINS. *Educ:* Worcester Polytech Inst, BS, 41; Mass Inst Technol, PhD(org chem), 47. *Prof Exp:* Jr chemist, Tenn Eastman Corp, 44-45; asst, Sugar Res Found, Mass Inst Technol, 45-47; res chemist, Union Carbide Chem Co, 47-64; sr res chemist, Borden Chem Co, 64-86; RETIRED. *Mem:* Am Chem Soc. *Res:* Water soluble polymers; condensation polymerizations; reactions in aqueous media; binders; resin analyses; manufacturing procedures and quality control. *Mailing Add:* 2500 Glenwood Park New Albany IN 47150

WINSLOW, CHARLES ELLIS, JR, b Norfolk, Va, July 2, 28; m 50; c 2. CHEMICAL ENGINEERING. *Educ:* Va Polytech Inst, BS, 50; NC State Univ, MS, 52, PhD(chem eng), 56. *Prof Exp:* Chem engr, Va Chem Inc, 56-58, group leader, 58-64, mgr process develop, 64-72, assoc dir res, 72-75, dir develop, 75-90; RETIRED. *Mem:* Am Inst Chem Engrs. *Res:* Inorganic and organic process development and equipment design; chemistry of sulfur dioxide bases on or derived from reducing agents. *Mailing Add:* 1308 W Oceanview Ave Portsmouth VA 23503

WINSLOW, DOUGLAS NATHANIEL, b Lakewood, Ohio. CIVIL ENGINEERING. *Educ:* Purdue Univ, BSCE, 64, MSCE, 69, PhD(construct mat), 73. *Prof Exp:* ASSOC PROF CIVIL ENG, PURDUE UNIV, 73- *Mem:* Am Concrete Inst; Am Soc Testing & Mat; Am Ceramic Soc. *Res:* Microstructure, durability and test methods for various construction materials (cement, concrete, bricks and bituminous mixtures); mercury intrusion porosimetry. *Mailing Add:* 2886 Bridgeway Dr Lafayette IN 47906

WINSLOW, FIELD HOWARD, b Proctor, Vt, June 10, 16; m 45; c 3. ORGANIC CHEMISTRY, POLYMER CHEMISTRY. *Educ:* Middlebury Col, BS, 38; RI State Col, MS, 40; Cornell Univ, PhD(org chem), 43. *Prof Exp:* Res chemist, Manhattan Proj, Columbia Univ, 43-45; mem tech staff, Bell Labs, Inc, 45-87; CONSULT, 87- *Concurrent Pos:* Adj prof, Stevens Inst Technol, 64-67; ed, Macromolecules, 67- *Mem:* Fel AAAS; Am Chem Soc. *Res:* Photochemistry; organic semiconductors; polymer morphology and chemical reactivity; deterioration and stabilization of rubbers and plastics; fluorocarbons. *Mailing Add:* AT&T Bell Labs 7C-217 PO Box 261 Murray Hill NJ 07974

WINSLOW, GEORGE HARVEY, b Washington, DC, June 21, 16; m 44; c 2. PHYSICS. *Educ:* Carnegie Inst Technol, BS, 38, MS, 39, DSc, 46. *Prof Exp:* Instr physics, Carnegie Inst Technol, 38, res physicist, 43-46; assoc physicist, Argonne Nat Lab, 46-81; RETIRED. *Mem:* Am Phys Soc. *Res:* Magnetic moments by molecular beams; high speed deformation of metals; shaped charges; solid state; attempts to find requantization of space quantized atoms at collision; alpha decay theory high temperature physical chemistry. *Mailing Add:* 3004 N Ridge Rd No 301 Ellicot City MD 21043

WINSLOW, LEON E, b Centralia, Ill, Nov 17, 34; m 59; c 6. MATHEMATICAL ANALYSIS. *Educ:* Marquette Univ, BS, 56, MS, 60; Duke Univ, PhD(math), 65. *Prof Exp:* Comput Ctr, 58-59; prin physicist, Battelle Mem Inst, 57-58; instr math, Rockhurst Col, 59-60; asst, Duke Univ, 60-64, res assoc spec projs numerical analysis, 64-65; asst prof math, Rockhurst Col, 65-66; asst prof comput sci, Univ Notre Dame, 66-72; assoc prof comput sci, Wright State Univ, 72-81; PROF COMPUT SCI, UNIV DAYTON, 81- *Mem:* Inst Elec & Electronics Engrs; Asn Comput Mach. *Res:* Systems analysis. *Mailing Add:* 2255 Andrew Rd Dayton OH 45440

WINSOR, FREDERICK JAMES, b Ilion, NY, Aug 22, 21; m 42; c 5. METALLURGY. *Educ:* Rensselaer Polytech Inst, BMetE, 42, MMetE, 44, PhD(metall), 46. *Prof Exp:* Asst, Rensselaer Polytech Inst, 42-46; res metallurgist, Armour Res Found, 46-48; supvr welding res, Standard Oil Co, Ind, 48-51; res engr, E I du Pont de Nemours & Co, 51-59; mgr welding lab, Foster Wheeler Corp, 59-69, dir, Welding Develop Lab, 69-85; CONSULT, 85- *Mem:* Am Soc Metals; Am Welding Soc. *Res:* Welding and metallurgical research and development; manufacturing and fabrication engineering. *Mailing Add:* Two Birchwood Terr Fanwood NJ 07023

WINSOR, LAURISTON P(EARCE), b Johnston, RI, Dec 30, 14; m 42; c 3. ELECTRICAL ENGINEERING. *Educ:* Brown Univ, ScB, 36; Harvard Univ, MS, 37, ScD(elec eng), 46. *Prof Exp:* Asst elec eng, Grad Sch Eng, Harvard Univ, 38-40; instr, Case Inst Technol, 40-46; from asst prof to assoc prof, 46-53, prof elec eng, Rensselaer Polytech Inst, 53-, dir spec projs, Off

Continuing Studies, 72-78, emer prof, rensselaer polytech inst, 79-; RETIRED. *Mem:* Am Soc Eng Educ; Inst Elec & Electronics Engrs. *Res:* Electromechanical energy conversion and control; arc reignition. *Mailing Add:* RD3 Box 163 Troy NY 12180-7336

WINSTEAD, JACK ALAN, b Dixon, Ky, June 13, 32; m 56; c 3. TOXICOLOGY, BIOCHEMISTRY. *Educ:* Univ Ky, BS, 54; Okla State Univ, MS, 59; Univ Ill, PhD(chem), 64. *Prof Exp:* Res officer chem, Mat Lab, USAF Acad, 59-62, res chemist, Sch Aerospace Med, 64-68, Frank J Seiler Res Lab, 68-70, dir, Directorate Chem Sci, 70-72, dep dir, Toxic Hazards Div, Aerospace Med Res Lab, Wright-Patterson AFB, 72-75; prof assoc, Nat Acad Sci, 75-78; dir toxicol rev, Cosmetic, Toiletry & Fragrance Asn, 78-80; prin scientist, Health Effects & Bioassay Div, Tracor Jitco, 80-82; CONSULT, 82- *Mem:* Am Soc Biol Chemists; Am Chem Soc. *Res:* Structure and function of proteins; radiation biochemistry; organic synthesis and toxicology. *Mailing Add:* 6971 E Costilla Pl Englewood CO 80112

WINSTEAD, JANET, b Wichita Falls, Tex, Mar 13, 32. MYCOLOGY. *Educ:* Midwestern Univ, BS, 53; Ohio Univ, MS, 55; Univ Tex, Austin, PhD(bot), 70. *Prof Exp:* Instr biol, Ky Wesleyan Col, 56-57; asst prof, Atlantic Christian Col, 57-65; ASSOC PROF BIOL, JAMES MADISON UNIV, VA, 69- *Mem:* Am Inst Biol Sci; Mycol Soc Am; Sigma Xi. *Res:* Monospore culture of myxomycetes. *Mailing Add:* Dept Biol James Madison Univ Harrisonburg VA 22807

WINSTEAD, JOE EVERETT, b Wichita Falls, Tex, Mar 17, 38; m 80; c 2. BOTANY, ECOLOGY. *Educ:* Midwestern Univ, BS, 60; Ohio Univ, MS, 62; Univ Tex, Austin, PhD(bot), 68. *Prof Exp:* Instr biol, Delta Col, 62; from asst prof to assoc prof, 72-78, PROF BIOL, WESTERN KY UNIV, 78- *Concurrent Pos:* Adj prof, Tenn Tech, 74-90. *Mem:* Ecol Soc Am; Bot Soc Am; Sigma Xi. *Res:* Ecotype differentiation of plant species; natural revegetation of stripmines; differentiation of wood cell types and wood anatomy; environmental physiology. *Mailing Add:* Dept Biol Western Ky Univ Bowling Green KY 42101

WINSTEAD, MELDRUM BARNETT, b Lincolnton, NC, Oct 19, 26; m 59; c 3. ORGANIC CHEMISTRY. *Educ:* Davidson Col, BS, 46; Univ NC, Chapel Hill, MA, 49, PhD(chem), 52. *Prof Exp:* Instr chem, Davidson Col, 46-47; asst, Univ NC, 47-50; from asst prof to prof, 69-71, EMER PROF CHEM, BUCKNELL UNIV, 91- *Concurrent Pos:* Res grants, DuPont res fel, 51-52, Res Corp, 54-56, AAAS, 59-60 & Petrol Res Fund, 61-62; consult, Glyco Chem, Inc, 58 & Sadtler Res Labs, 59-70; vis chem assoc, Calif Inst Technol, 67-68; USPHS spec res fel, 67-68; res assoc, Lawrence Berkeley Lab, Univ Calif, Berkeley, 68-69; vis scientist, Medi-Physics, Inc, Calif, 72-78, Sloan-Kettering Cancer Ctr, 81, Israel Resources Corp, Ltd, Haifa, Israel, 81-82; Bucknell fac res fel, 70-71 & 78-79. *Honors & Awards:* USPHS Award, Nat Inst Gen Med Sci, 74 & 77. *Mem:* Sigma Xi; Am Chem Soc. *Res:* Organic medicinals; preparation and scintigraphic study of pharmaceuticals containing short-lived radiocarbon-11. *Mailing Add:* RD 1 Box 395 Lewisburg PA 17837

WINSTEAD, NASH NICKS, b Durham, Co, NC, June 12, 25; m 49; c 1. PLANT PATHOLOGY. *Educ:* NC State Col, BS, 48, MS, 51; Univ Wis, PhD(plant path), 53. *Prof Exp:* From asst prof to assoc prof, NC State Univ, 53-60, actg chancellor, 81-82, provost & vchancellor, 74-90, PROF PLANT PATH, NC STATE UNIV, 60-, EMER PROVOST & VCHANCELLOR, 90- *Concurrent Pos:* Dir, Inst Biol Sci & asst dir res, NC Agr Exp Sta, 65-67, asst provost, NC State Univ, 67-73, assoc provost, 73-74; Phillips Found internship acad admin, Ind Univ, 65-66; mem, Comt Planned Res Basic Bio-sci during manned earth-orbiting missions, Am Inst Biol Sci-NASA, 65-67; mem bd dirs, Consortium Cooperating Raleigh Cols, 68-90, pres, 71-73 & 83-85; mem, Educ Telecommun Comt, Nat Asn State Univ & Land Grant Col, 80-85; trustee, NC Sch Sci & Math, 85-90. *Honors & Awards:* Res Award, Sigma Xi, 60. *Mem:* Fel AAAS; Am Phytopath Soc; Am Inst Biol Sci; Sigma Xi. *Res:* Vegetable diseases; breeding for resistance; physiology of parasitism. *Mailing Add:* NC State Univ PO Box 7111 Raleigh NC 27695-7111

WINSTEN, SEYMOUR, b Jersey City, NJ, June 14, 26; m 49; c 3. BIOCHEMISTRY. *Educ:* Rutgers Univ, AB, 48, PhD(microbiol, physiol), 56; NY Univ, MSc, 50; Am Bd Clin Chem, dipl. *Prof Exp:* Asst, Merck Inst Therapeut Res, 50-56; assoc microbiol, Univ Pa, 56-57; head dept chem, Albert Einstein Med Ctr, 57-86; DIR LABS, MOSS REHAB HOSP, 75- *Concurrent Pos:* Consult, Atlantic City Hosp, Surgeon Gen US & Walson Gen Hosp; assoc prof biochem, Sch Med, Temple Univ, 70-76; consult, Deborah Heart & Lung Ctr, 65-75. *Honors & Awards:* John Gunther Reinhold Award, 68. *Mem:* Fel Am Asn Clin Chem; Am Chem Soc. *Res:* Clinical chemistry; immunochemistry; chemical diagnosis of disease; mycology; endocrine chemistry and its relationship to various disease processes; nuclear magnetic resonance. *Mailing Add:* Clin Lab Moss Rehab Hosp Philadelphia PA 19141

WINSTON, ANTHONY, b Washington, DC, Dec 5, 25; m 52; c 4. POLYMER CHEMISTRY. *Educ:* George Washington Univ, BS, 50; Duke Univ, MA, 52, PhD, 55. *Prof Exp:* Res chemist, Armstrong Cork Co, 54-59; from asst prof to assoc prof, 59-75, chmn dept, 86-90, PROF CHEM, WVA UNIV, 75- *Concurrent Pos:* Res assoc, Water Res Inst, WVa Univ, 75-82. *Mem:* Am Chem Soc. *Res:* Polymer synthesis and reactions; metal complexing polymers; selective chelating ion exchange resins; polymers for medical applications; stereochemistry; polymer-drug combinations. *Mailing Add:* Dept Chem WVa Univ Morgantown WV 26506

WINSTON, ARTHUR WILLIAM, b Toronto, Ont, Feb 11, 30; US citizen; m 49; c 4. PHYSICS, MATHEMATICS. *Educ:* Univ Toronto, BASc, 51; Mass Inst Technol, PhD(physics), 54. *Prof Exp:* Eng physicist, Nat Res Coun Can, 49-51; res asst, Mass Inst Technol, 51-54; sr engr, Schlumberger Well Surv Corp, 54-57; sr engr, Nat Res Corp, 57-59; chief scientist, Allied Res Assocs, Inc, 59-61; pres, Space Sci, Inc, 61-65; pres, 65-75, chmn, Ikor, Inc, 75-79; PRES, WINCOM CORP, 79- *Concurrent Pos:* Lectr, Northeastern Univ, 57-65, adj prof, 65-; mem Int Dept Com First Trade Mission to Europe on Pollution Controls; judge Mass State Sci Fair; chmn, Northeast Elec & Eng, 60. *Mem:* Am Inst Aeronaut & Astronaut; Inst Elec & Electronics Engrs; Am Geophys Union; Am Phys Soc; Am Inst Mining, Metall & Petrol Engrs. *Res:* Electromagnetic propagation and measurements; nuclear physics applied to geophysics; thin film technology; microprocessors; pollution control devices and systems. *Mailing Add:* Seven Wainwright Rd No 15 Winchester MA 01890

WINSTON, DONALD, b Washington, DC, Apr 4, 31; m; c 2. GEOLOGY. *Educ:* Williams Col, BA, 53; Univ Tex, MA, 57, PhD(geol), 63. *Prof Exp:* From instr to assoc prof, 61-76, PROF GEOL, UNIV MONT, 76- *Mem:* AAAS; Soc Econ Paleont Mineral; Geol Soc Am; Geol Asn Can. *Res:* Stratigraphy and sedimentation, particularly Precambrian rocks; Cambrian paleontology; sedimentary petrology, particularly carbonate petrology of Pennsylvanian rocks and modern carbonate areas. *Mailing Add:* Dept Geol Univ Mont Missoula MT 59812

WINSTON, HARVEY, b Newark, NJ, Aug 11, 26; m 49; c 2. PHYSICAL CHEMISTRY. *Educ:* Columbia Univ, AB, 45, MA, 46, PhD(chem), 49. *Prof Exp:* Asst chem, Columbia Univ, 45-49; Jewett fel, Univ Calif, 49-50, instr chem, 50-51; asst prof, Univ Calif, Los Angeles, 51-52; mem tech staff, Hughes Aircraft Co, 52-58, mgr mat res lab, Semiconductor Div, 58-60; assoc dir quantum electronics lab, Quantatron, Inc, 61, vpres, Quantum Tech Labs, Inc, 61-63; mgr, 63-69, SR SCIENTIST, CHEM PHYSICS DEPT, HUGHES RES LABS, 69- *Mem:* Fel Am Phys Soc; sr mem Inst Elec & Electronics Engrs. *Res:* Solid state spectroscopy; lasers and laser systems; semiconductor physics and devices. *Mailing Add:* 1450 San Remo Pacific Palisades CA 90272

WINSTON, HUBERT, b Wash, DC, May 29, 48. PROCESS CONTROL. *Educ:* NC State Univ, BS, 70, MS, 73, PhD(chem eng), 75. *Prof Exp:* Sr res engr & res specialist, Exxon Prod Res Co, 77-83; asst prof chem eng, 75-77, assoc prof, 83-86, ASST DEAN & DIR ACAD AFFAIRS, COL ENG, NC STATE UNIV, 86- *Mem:* Am Inst Chem Engrs; Am Soc Eng Educ; Instrument Soc Am; Nat Orgn Black Chemists & Chem Engrs. *Mailing Add:* 3967 Wendy Lane Raleigh NC 27606

WINSTON, JOSEPH, electrical engineering, for more information see previous edition

WINSTON, JUDITH ELLEN, b Haverhill, Mass, Mar 11, 45; c 1. SYSTEMATICS, MARINE ECOLOGY. *Educ:* Brown Univ, AB, 66; Univ NH, MS, 70; Univ Chicago, PhD(geophys sci), 74. *Prof Exp:* Postdoctoral fel & invert biologist marine ecol & bryozoan systematics, Smithsonian Marine Sta, Linkport, 74-76; asst prof marine ecol & bryozoan systematics, Indian River Community Col, 76-77; assoc res scientist marine ecol & bryozoan systematics, Johns Hopkins Univ, 77-80; asst cur, 80-85, assoc cur, 85-91, CUR & CHMN MARINE ECOL & BRYOZOAN SYSTEMATICS, DEPT INVERTEBRATES, AM MUS NATURAL HIST, 91- *Concurrent Pos:* Adj prof, City Univ New York, 89- *Mem:* Fel AAAS; Am Asn Mus; Am Soc Zoologists; Soc Syst Zool; Soc Conserv Biol. *Res:* Bryozoan systematics, ecology and evolution of colonial organisms oryozoans, taxonomic procedure and nomenclature, and history of science; field work; scuba duty; traveling duty; laboratory; underwater macro-photography and video recording. *Mailing Add:* Dept Invertebrates Am Mus Natural Hist Cent Park W at 79th St New York NY 10024-5192

WINSTON, PAUL WOLF, b Chicago, Ill, Aug 9, 20; m 48; c 1. BIOLOGY. *Educ:* Univ Mass, BS, 48; Northwestern Univ, MS, 50, PhD, 52. *Prof Exp:* Instr, Brown Univ, 51-52; from instr to prof biol, Univ Colo, Boulder, 52-90; RETIRED. *Mem:* AAAS; Soc Environ Geochem & Health. *Res:* Humidity relations and water balance of terrestrial arthropods, especially cuticular control of water exchange with air; physiology of molybdenum and toxicity of chronic exposure to trace metals in mammals. *Mailing Add:* Dept EPO Biol Univ Colo Boulder CO 80302

WINSTON, ROLAND, b Moscow, USSR, Feb 12, 36; US citizen; m 57; c 3. EXPERIMENTAL PHYSICS, PARTICLE PHYSICS. *Educ:* Shimer Col, BA, 53; Univ Chicago, BS, 56, MS, 57, PhD(physics), 63. *Prof Exp:* Asst prof physics, Univ Pa, 63-64; from asst prof to assoc prof, 64-75, PROF PHYSICS, UNIV CHICAGO, 75- *Concurrent Pos:* Sloan Found fel, 67-69; Guggenheim fel, 77-78. *Honors & Awards:* Charles Greeley Abbot Award, Am Solar Energy Soc, 87- *Mem:* Fel Am Phys Soc; Am Solar Energy Soc. *Res:* Elementary particle physics; leptonic decays of hyperons; muon physics, especially hyperfine effects in muon capture by complex nuclei; solar energy concentrators; infra-red detectors; optics of visual receptors. *Mailing Add:* Dept Physics Univ Chicago 5630 S Ellis EF1-381 Chicago IL 60637

WINSTON, VERN, b Gordon, Nebr, Apr 30, 48; m 70; c 2. VIROLOGY, CELL CULTURE. *Educ:* Univ Nebr-Lincoln, BS, 70, PhD(microbiol), 76. *Prof Exp:* Res assoc, Kans State Univ, 76-80; ASST PROF MICROBIOL, IDAHO STATE UNIV, 81- *Mem:* Am Soc Microbiol; Sigma Xi. *Res:* Viral diseases of fish; immunodiagnostic methods of identifying infectious pancreatic necrosis virus; immunodiagnostic methods of identifying infectious hematopoietic necrosis virus. *Mailing Add:* Dept Microbiol & Biochem Idaho State Univ Box 8094 Pocatello ID 83209-0009

WINSTROM, LEON OSCAR, b Holland, Mich, Apr 8, 12; m 38; c 3. PHYSICAL CHEMISTRY, TEXTILE CHEMISTRY. *Educ:* Hope Col, AB, 34; Carnegie Inst Technol, MS, 37, DSc(phys chem), 38. *Prof Exp:* Instr, Carnegie Inst Technol, 37-38; res chemist, Nat Aniline Chem Div, Allied Chem Corp, 38-53, asst supvr, 53-57, sr scientist, 57-58, group leader, 58-64, res supvr, 64-68, sr res assoc, Spec Chem Div, 68-71; mgr res & develop, Flock Div, Malden Mills, Lawrence, 71-80; CONSULT, 80- *Honors & Awards:* Schoellkopf Medal, 66. *Mem:* Am Chem Soc. *Res:* Vapor and liquid phase hydrogenation and oxidation; ammination by reduction; recovery of organic oxidation products. *Mailing Add:* 57 Maple St East Aurora NY 14052

WINTER, ALEXANDER J, b Vienna, Austria, June 21, 31; nat US; m 59; c 3. IMMUNOBIOLOGY, BACTERIOLOGY. *Educ:* Univ Ill, DVM, 55; Univ Wis, PhD(med & vet path), 59. *Prof Exp:* From asst to assoc prof vet sci, Pa State Univ, 59-63; assoc prof, 63-66, PROF VET MICROBIOL, NY STATE VET COL, CORNELL UNIV, 66- *Concurrent Pos:* Mem, Bacteriol & Mycol Sect, NIH, 71-75. *Mem:* Infectious Dis Soc Am; Am Soc Microbiol; Am Asn Immunologists; Am Col Vet Microbiologists. *Res:* Brucellosis. *Mailing Add:* NY State Col Vet Med Cornell Univ Ithaca NY 14853

WINTER, CHARLES GORDON, b Hanover, Pa, Dec 28, 36; m 58; c 3. BIOCHEMISTRY, MEMBRANES. *Educ:* Juniata Col, BS, 58; Univ Mich, MS, 63, PhD(biochem), 64. *Prof Exp:* Technician, Metab Res Unit, Univ Mich, Ann Arbor, 58-60; Childs Mem Fund Med Res fel phys chem, Sch Med, Johns Hopkins Univ, 64-66; asst prof, 66-73, ASSOC PROF & ACTG CHMN BIOCHEM, SCH MED, UNIV ARK, LITTLE ROCK, 73- *Concurrent Pos:* Hon res assoc, Harvard Univ, 78-79. *Mem:* Am Soc Biochem & Molecular Biol; AAAS; Am Chem Soc; Biophys Soc. *Res:* Structure and function of alkali-cation-dependent. *Mailing Add:* Dept Biochem & Molecular Biol Univ Ark Col Med Little Rock AR 72205

WINTER, CHESTER CALDWELL, b Cazenovia, NY, June 2, 22; m 45; c 3. MEDICINE. *Educ:* Univ Iowa, BA, 43, MD, 46; Am Bd Urol, dipl. *Prof Exp:* Asst prof surg, Sch Med, Univ Calif, Los Angeles, 58-61; PROF UROL, COL MED, OHIO STATE UNIV, 61- *Concurrent Pos:* Mem staff, Univ Hosp, 61- & Children's Hosp, 61- *Mem:* Am Urol Asn; Am Col Surg; Soc Univ Urol; Soc Univ Surg; Am Asn Genito-Urinary Surg. *Res:* Urological surgery; renal hypertension; diagnostic isotopes in urology. *Mailing Add:* 6425 Evening St Worthington OH 43085

WINTER, DAVID ARTHUR, b Windsor, Ont, June 16, 30; m 58; c 3. BIOMEDICAL ENGINEERING, ELECTRICAL ENGINEERING. *Educ:* Queen's Univ, Ont, BSc, 53, MSc, 61; Dalhousie Univ, PhD(physiol), 67. *Prof Exp:* From lectr to asst prof elec eng, Royal Mil Col, Ont, 58-63; from asst prof to assoc prof, Tech Univ, Nova Scotia, 63-69; assoc prof surg, Univ Man, 69-74, adj prof elec eng, 70-74; assoc prof, 74-76, PROF KINESIOLOGY, UNIV WATERLOO, 76- *Concurrent Pos:* Can Coun fel med, eng & sci, Dalhousie Univ, 66-68. *Honors & Awards:* Career Investigators Award, Can Soc Biomech. *Mem:* Inst Elec & Electronics Engrs; Can Med & Biol Eng Soc (pres, 70-74); Int Soc Electrophys Kinesiol; Int Soc Biomech; Can Soc Biomech. *Res:* Signal processing of biological signals; medical image processing; electromyography; biomechanics; locomotion studies; assessment pathological gait. *Mailing Add:* Dept Kinesiology Univ Waterloo Waterloo ON N2L 3G1 Can

WINTER, DAVID F(ERDINAND), b St Louis, Mo, Nov 9, 20; m 44; c 2. ELECTRICAL ENGINEERING, SHOCK & INDUCED VOLTAGES & CURRENT IN DAIRY COWS. *Educ:* Wash Univ, BS, 42; Mass Inst Technol, MS, 48. *Prof Exp:* Mem staff, Radiation Lab, Mass Inst Technol, 42-45, res assoc & asst prof elec eng, Wash Univ, 48-51, assoc prof, 51-54; sr engr, Spec Contract, US Naval Ord Plant, Ind, 50-52, proj head, 52-53; consult, Moloney Elec Co, 51-54, sect engr electronics, 54-57, vpres, chief engr & dir res, 55-66, vpres res & develop, 66-73, vpres eng, central Moloney Div, Colt Indust, 73-74; vpres res & develop, ITT Blackburn, 74-86; PROF, WASH UNIV, 54- *Concurrent Pos:* consult, 86- *Mem:* Fel Inst Elec & Electronics Engrs; Am Soc Agr Eng; Nat Soc Prof Engrs; Sigma Xi. *Res:* High voltage magnetic components for power industry; specialized high power radar and communications; electrical connectors for power industry; electronic grounding for animal confinement with stray voltage. *Mailing Add:* 629 Meadowridge Lane St Louis MO 63122

WINTER, DAVID JOHN, b Painesville, Ohio, May 2, 39; m 65; c 1. MATHEMATICS. *Educ:* Antioch Col, BA, 61; Yale Univ, MS, 63, PhD(math), 65. *Prof Exp:* Instr math, Yale Univ, 65-67; NSF fel, Univ Bonn, 67-68; from asst prof to assoc prof, 68-74, PROF MATH, UNIV MICH, ANN ARBOR, 74- *Concurrent Pos:* Vis assoc prof, Calif Inst Technol, 72-73. *Mem:* Am Math Soc. *Res:* Algebra. *Mailing Add:* Dept Math Univ Mich Ann Arbor MI 48104

WINTER, DAVID LEON, b New York, NY, Nov 10, 33; m 73; c 5. RESEARCH ADMINISTRATION. *Educ:* Columbia Col, AB, 55; Wash Univ, MD, 59. *Prof Exp:* Surg intern, Sch Med, Wash Univ, 59-60; Nat Inst Neurol Dis & Blindness fel, Baylor Univ, 60-62; med res officer, Nat Inst Neurol Dis & Blindness, 62-64; neurophysiologist, Walter Reed Army Inst Res, 64-66, chief dept neurophysiol, 66-71; dep dir life sci, Ames Res Ctr, Moffett Field, Calif, 71-74, dir life sci, NASA HQ, DC, 74-79; dir, 79-84, vpres med res, Sandoz Inc, 84-85; vpres clin res & develop, 85-89, VPRES SCI & EXTERNAL AFFAIRS, SANDOZ PHARMACEUT CORP, 89- *Honors & Awards:* Hans Berger Prize, Am Electroencephalog Soc, 64. *Mem:* AAAS; Am Physiol Soc; Soc Neurosci; Aerospace Med Asn; Am Soc Clin Pharmacol & Therapeut. *Res:* Somatosensory systems; visceral reflexes; autonomic nervous system; psychophysiology; aerospace physiology; drug development. *Mailing Add:* Med Res Sandoz Pharmaceut Corp East Hanover NJ 07936

WINTER, DONALD CHARLES, b Brooklyn, NY, June 15, 48; m 69; c 2. OPTICS. *Educ:* Univ Rochester, BS, 69; Univ Mich, MS, 70, PhD(physics), 72. *Prof Exp:* Mem res staff optics, 72-74, sect ha head optics technol, 74-77, MGR OPTICS DEPT, TRW DEFENSE & SPACE SYSTS GROUP, 77- *Mem:* Optical Soc Am. *Res:* High energy laser optics; optical components; laser diagnostics; interferometry and optical testing; optical system design and optimization. *Mailing Add:* 608 14th St Manhattan Beach CA 90266

WINTER, DONALD F, b Buffalo, NY, Oct 6, 31; div; c 3. APPLIED MATHEMATICS. *Educ:* Amherst Col, BA, 54; Harvard Univ, MA, 59, PhD(appl physics), 62. *Prof Exp:* Mathematician, Air Force Cambridge Res Labs, 54-56; engr, Missile Systs Lab, Sylvania Elec Prod, Inc, 56-58, eng specialist, Appl Res Lab, 58-62, sr eng specialist, 62-63; mem staff, Geo-astrophys Lab, Boeing Sci Res Labs, 63-70; assoc prof, Ctr Quantitative Sci & Dept Oceanog, Univ Wash, 70-74, prof oceanog & appl math, 74-86; PROF MATH, ENG & COMPUTER SCI, UNIV REDLANDS, 86- *Concurrent Pos:* Vis lectr, Univ Manchester, 66-67. *Mem:* Am Math Soc; AAAS; Am Geophys Union; Soc Indust & Appl Math. *Res:* Applied and numerical analysis; methods of mathematical physics with applications to general engineering; hydrodynamical and biological processes in oceanography; growth and transport processes in biological systems. *Mailing Add:* Dept Eng Univ Redlands Redlands CA 92373

WINTER, HARRY CLARK, b New Britain, Conn, Feb 26, 41; m 77. BIOCHEMISTRY. *Educ:* Pa State Univ, BS, 62; Univ Wis, MS, 64, PhD(biochem), 67. *Prof Exp:* NSF fel cell physiol, Univ Calif, Berkeley, 67-68; asst prof biochem, Pa State Univ, 68-75; lectr biol chem, 75-87, ASST RES SCIENTIST, UNIV MICH, ANN ARBOR, 87- *Mem:* Am Chem Soc; Am Soc Plant Physiologists; Am Soc Biol Chemists. *Res:* Biological nitrogen fixation; photosynthesis; biosynthetic pathways of plants and bacteria; enzyme mechanisms. *Mailing Add:* Univ Mich Med Sch 1301 Catherine Rd M5312 0606 Ann Arbor MI 48109-0606

WINTER, HENRY FRANK, JR, b Wooster, Ohio, Dec 25, 36. PHYSIOLOGY. *Educ:* Case Inst Technol, BSc, 58; Baylor Univ, MSc, 62, PhD(physiol, biochem, anat), 65. *Prof Exp:* Asst prof, 65-72, assoc prof physiol, 73-80, PROF PHYSIOL & BIOMED SCI, SCH DENT MED, WASH UNIV, 80- *Mem:* AAAS; Am Physiol Soc; Int Asn Dent Res. *Res:* Oral biology, neurophysiology; instrumentation for medical research; growth and development. *Mailing Add:* 4559 Scott Ave Wash Univ Sch Dent Med St Louis MO 63110

WINTER, HERBERT, b Vienna, Austria, July 31, 24; US citizen; m 46; c 2. CONTROL & ELECTRICAL ENGINEERING. *Educ:* City Col New York, BEE, 49; Univ Mich, MSE, 50. *Prof Exp:* Instr elec eng, City Col New York, 49; engr, Bell Aerospace Textron, 50-54, syst engr, 54-57, group chief systs oper & preliminary design, 57-59, dynamic analytical hypersonic glider, 59-60, supvr preliminary design electromech systs, 60-65, prin scientist, 66-87; RETIRED. *Mem:* Inst Elec & Electronics Engrs; Am Inst Navig. *Res:* Application of optimal control and filtering to inertial navigation; optimization of oceanic air traffic; computer simulation of atmospheric propagation of laser beams; design of gravity gradiometer systems. *Mailing Add:* 75 Chateau Terr Snyder NY 14226-3964

WINTER, HORST HENNING, b Stuttgart, Ger, Sept 9, 41; m 69; c 4. POLYMER ENGINEERING, RHEOLOGY. *Educ:* Univ Stuttgart, Dr-Ing, 73. *Prof Exp:* Privatdozent rheology, Univ Stuttgart, 76-79; assoc prof, 79-84, PROF CHEM ENG, UNIV MASS, AMHERST, 84- *Concurrent Pos:* DFG Fel, Univ Wis-Madison, 73-74; vis prof, Univ Minn, 78 & Max Planck Inst, Polymerforschung, Mainz, Ger, 87-88; ed, Rheologica Acta, 89- *Mem:* Soc Rheology; Soc Plastics Eng; Polymer Processing Soc; Deutsche Rheologische Gesellschaft; Am Inst Chem Engrs. *Res:* Measure the rheology and model the processing behavior of complex polymeric materials such as gels, liquid crystalline polymers, block-copolymers, polymer blends and filled polymers; rheo-optical methods; rheological constitutive equations. *Mailing Add:* Chem Eng Dept Univ Mass Amherst MA 01003

WINTER, IRWIN CLINTON, b Clinton, Okla, July 17, 10; m 38; c 4. PHARMACOLOGY. *Educ:* Allegheny Col, BS, 31; Northwestern Univ, MS, 33, PhD, 34; Univ Tenn, MD, 41. *Prof Exp:* Asst physiol chem, Med Sch, Northwestern Univ, 31-34, fel, 34-35; physiologist, Res Dept, Parke, Davis & Co, Mich, 35-36; instr physiol & pharmacol, Col Med, Baylor Univ, 36-39; assoc prof pharmacol, Sch Med, Univ Okla, 39-42; dir clin res, 46-75, med dir, 59-75, vpres med affairs, 62-75, CONSULT, G D SEARLE & CO, 75- *Concurrent Pos:* Mem coun arteriosclerosis, Am Heart Asn; consult, PMA Found, 75- *Mem:* Soc Exp Biol & Med; Am Soc Pharmacol & Exp Therapeut; AMA; Am Rheumatism Asn; Am Fedn Clin Res; Sigma Xi. *Res:* Liver damage and fat metabolism; physiology and pharmacology of micturition; autonomic pharmacology. *Mailing Add:* 108 Beechwood Dr W Columbus NC 28722

WINTER, JEANETTE E, b New York, NY, Dec 19, 17. MICROBIOLOGY. *Educ:* Brooklyn Col, BA, 37; NY Univ, PhD(microbiol), 60. *Prof Exp:* Instr, 64-68, asst prof, 68-71, ASSOC PROF MICROBIOL, MED SCH, NY UNIV, 71- *Mem:* AAAS; Am Soc Microbiol. *Res:* Mechanism of competence for DNA uptake in bacterial transformation; role of nucleases in DNA integration during bacterial transformation of streptococci. *Mailing Add:* 11011 Queens Blvd Flushing NY 11375

WINTER, JEREMY STEPHEN DRUMMOND, b Duncan, BC, Dec 11, 37; m 61; c 2. PEDIATRICS, ENDOCRINE PHYSIOLOGY. *Educ:* Univ BC, MD, 61; Am Bd Pediat, dipl, 67; FRCP(C), 68. *Prof Exp:* Intern & resident, Montreal Gen, Montreal Children's & Royal Victoria Hosp, 61-64; instr pediat, Univ Pa, 64-67; from asst prof to assoc prof, 67-78, PROF PEDIAT, UNIV MAN, 78- ENDOCRINOLOGIST, HEALTH SCI CTR, WINNIPEG, 71- *Concurrent Pos:* NIH fel endocrinol, Children's Hosp Philadelphia, 64-67; Med Res Coun grant, Univ Man, 67-; consult, St Boniface Hosp, 67-; scientist, Queen Elizabeth II Res Found, 72. *Mem:* Endocrine Soc; Soc Pediat Res; Can Soc Clin Invest; Am Fedn Clin Res; Can Pediat Soc. *Res:* Physiology of the pituitary-gonadal axis during fetal life, childhood and puberty. *Mailing Add:* 678 William Ave Winnipeg MB R3E 0W1 Can

WINTER, JERROLD CLYNE, SR, b Erie, Pa, March 25, 37; m 60; c 4. PHARMACOLOGY. *Educ:* Univ Rochester, BS, 59; State Univ NY, PhD(pharmacol), 66. *Prof Exp:* Asst prof, 67-71, assoc prof, 71-76, PROF PHARMACOL, STATE UNIV NY, BUFFALO, 76- *Res:* Behavioral pharmacology. *Mailing Add:* 102 Farber Hall State Univ NY Buffalo NY 14214

WINTER, JOHN HENRY, b New York, NY, Sept 1, 47. GEOARCHAEOLOGY, PHYSICAL ANTHROPOLOGY. *Educ:* Northern Ariz Univ, BA, 69, MA, 73; Columbia Univ, EdD, CTAS, 89. *Prof Exp:* CHMN, DEPT CHEM, PHYSICS, EARTH SCI, MOLLOY COL, ROCKVILLE CTR, NY. *Concurrent Pos:* Field sta archaeologist, Bahamian Field Sta, San Salvador Island, Bahamas, 80; dir, Int Asn Caribbean Archaeologists, 87-89. *Res:* Palaeopathology of the prehistoric peoples of the Bahamas; geoarchaeology of the Bahamas; environmental conditions of the prehistoric period within the Bahamas. *Mailing Add:* Dept Chem Physics & Earth Sc Molloy Col 1000 Hempstead Ave Rockville Center NY 11570

WINTER, JOSEPH, b New York, NY, July 26, 29; c 2. PHYSICAL METALLURGY. *Educ:* NY Univ, BME, 53, MS, 55, EngScD, 58. *Prof Exp:* Assoc res scientist metall, Res Div, NY Univ, 52-60; ASSOC DIR METALL DEPT, METALS RES LABS, OLIN CORP, 60- *Concurrent Pos:* Adj instr, Cooper Union, 56-61; adj assoc prof, New Haven Col, 62-70. *Honors & Awards:* John M Olin Award. *Mem:* Am Soc Metals; Am Inst Mining, Metall & Petrol Engrs; Brit Inst Metals. *Res:* Solid state bonding; nonferrous physical metallurgy. *Mailing Add:* 90 Linden St New Haven CT 06511

WINTER, KARL A, b Yarmouth, NS, Dec 18, 28; m 57; c 3. RUMINANT NUTRITION. *Educ:* McGill Univ, BSc, 53, MSc, 56; Ohio State Univ, PhD(ruminant nutrit), 62. *Prof Exp:* Grain salesman, Toronto Elevators Ltd, 55-57; res officer ruminant nutrit, Agr Can, 57-65; field res mgr, Tuco Prod Co, Upjohn Co, 65-68; res scientist cattle nutrit, 68-88, spec adv-livestock, 88-89, ASST DIR, RES STA, RES BR, AGR CAN, 90- *Concurrent Pos:* Adj prof, Atl Vet Col, Univ Prince Edward Island, Charlottetown. *Mem:* Can Soc Animal Sci (pres, 81-82); Agr Inst Can (vpres, 76-77 & 80-81); Am Soc Animal Sci. *Res:* Nutrition of ruminant animals, especially the young calf; utilization of non-protein nitrogen and agricultural wastes in cattle feeding and trace elements in ruminant nutrition. *Mailing Add:* Res Sta Agr Can PO Box 1210 Charlottetown PE C1A 7M8 Can

WINTER, MARGARET CASTLE, organic chemistry, environmental chemistry, for more information see previous edition

WINTER, NICHOLAS WILHELM, b Birmingham, Ala, Mar 7, 43; m 65; c 2. THEORETICAL CHEMISTRY, THEORETICAL PHYSICS. *Educ:* Northern Ill Univ, BS, 65; Calif Inst Technol, PhD(chem & physics), 70. *Prof Exp:* Atmospheric physicist, Jet Propulsion Lab, 71-73 & Aerospace Corp, 73-76; atomic & molecular physicist, 76-, CONSENSED MATTER PHYSICIST, LAWRENCE LIVERMORE LAB. *Concurrent Pos:* Fel, Battelle Mem Inst, 69-71; consult, Lawrence Livermore Lab, 71-75; res assoc, Calif Inst Technol, 72-75. *Mem:* Am Phys Soc. *Res:* Molecular structure; potential energy surfaces; excited states; reaction dynamics as applied to atmospheric and laser physics. *Mailing Add:* Lawrence Livermore Lab PO Box 808 Univ Calif Livermore CA 94550

WINTER, OLAF HERMANN, b Erfurt, Ger, Dec 1, 33. CHEMICAL ENGINEERING, CHEMISTRY. *Educ:* Brunswick Tech Univ, BS, 57; Hannover Tech Univ, MS, 60, PhD(chem eng), 63; Univ Akron, MBA, 69. *Prof Exp:* Sr res engr, Res Div, Goodyear Tire & Rubber Co, 63-71; asst to vpres res, Hydrocarbon Res Inc, Trenton Lab, Dynalectron Corp, 72-73; prin engr, 73-80, sr prin engr, process technology planning, Lummus Co, Combustion Eng, Inc, 80-84; technol mgr, H-R Int, Inc, 84-85; consult, 85-86; mgr, Chem Process Res Waste Mgt, Waste Mgt, Inc, 86-87; SUPVRY ENGR, BROWN & ROOT BRAUM, HALLIBURTON CO, 88- *Concurrent Pos:* Teacher, Berlitz Sch Lang, 65-71. *Mem:* Am Chem Soc; Am Inst Chem Engrs; Ger Chem Soc. *Res:* Novel processing methods and equipment; pollution control; gasification and liquefaction of coal; hazardous waste treatment systems; economic evaluations; process planning; facilities planning for petroleum, petrochemical and chemical complexes; feasibility and marketing studies; identification of future technological and economic trends; technology assessments; evaluation of competing processes and licensors. *Mailing Add:* 343 Pioneer Dr Apt 1701 E Glendale CA 91203-1786

WINTER, PETER, b Sheffield, Eng, Nov 4, 46; m 82; c 1. BIOENERGY, ENERGY POLICY. *Educ:* Univ London, BSc, 74, MSC, 76. *Prof Exp:* Ed, Morgan Grampian, 72-76; lectr psychol & jour, Bromley Col Technol, 76-78; EXEC DIR, BIOMASS ENERGY INST, INC, 82- *Concurrent Pos:* Ed, Int Environ & Safety, 78-80; consult commun, 78-82. *Mem:* Solar Energy Soc Can; Can Wood Energy Inst; Int Asn Energy Economists. *Res:* Biotechnology; bioenergy; energy policy; potential of biomass for producing organic chemicals. *Mailing Add:* Winter House Sci Pub Inc PO Box 7131 Sta J Ottawa ON K2A 4C5 Can

WINTER, PETER MICHAEL, b Sverdlovsk, Russia, Aug 5, 34; US citizen; m 64; c 2. ANESTHESIOLOGY. *Educ:* Cornell Univ, AB, 58; Univ Rochester, MD, 62; Am Bd Anesthesiol, dipl, 72. *Prof Exp:* USPHS res fel, Harvard Univ, 65-66; res assoc physiol, State Univ NY Buffalo, 66-67, asst res prof anesthesiol, 67-69; assoc prof anesthesiol, Sch Med, Univ Wash, 69-74, prof, 74-79; chief anesthesiol serv, Vet Admin Hosp, Seattle, 78-79; PROF & CHMN DEPT ANESTHESIOL & CRITICAL CARE MED SCH MED, UNIV PITTSBURGH, 79-, ANESTHESIOLOGIST-IN-CHIEF, UNIV HEALTH CTR HOSP, PITTSBURGH, 79- *Concurrent Pos:* Consult, Virginia Mason Res Ctr, Seattle, 69-; Nat Heart & Lung Inst grant, Sch Med, Univ Wash, 71-74, res career develop award, 72-77. *Mem:* Am Col Chest Physicians; AMA; Am Soc Anesthesiol; Asn Univ Anesthetists; NY Acad Sci. *Res:* Respiration therapy; critical care medicine; hyperbaric physiology; oxygen toxicity. *Mailing Add:* Dept Anesthesiol CCM Univ Pittsburgh Sch Med 4200 Fifth Ave Pittsburgh PA 15260

WINTER, ROBERT JOHN, b Toledo, Ohio, Oct 13, 45; m 72; c 1. PEDIATRIC ENDOCRINOLOGY. *Educ:* Amherst Col, BA, 67; Northwestern Univ, MD, 71. *Prof Exp:* Intern pediat, Hartford Hosp, Conn, 71-72; resident pediat, Boston City Hosp, 72-73; fel pediat endocrinol, Johns Hopkins Univ, 73-75; asst prof, 75-81, ASSOC PROF PEDIAT ENDOCRINOL, CHILDREN'S MEM HOSP & NORTHWESTERN UNIV, CHICAGO, 81-, DIR MED EDUC, 85- *Mem:* Endocrine Soc; Am Diabetes Asn; Acad Pediat; Lawson Wilkins Pediat Endocrine Soc; Soc Pediat Res; Asn Am Med Cols. *Res:* Disorders of growth and of glucose homeostasis; primarily clinical research; research in medical education. *Mailing Add:* Children's Mem Hosp 2300 Children's Plaza Chicago IL 60614

WINTER, ROLAND ARTHUR EDWIN, b Reval, Estonia, Aug 29, 35; US citizen; m 59; c 3. ORGANIC CHEMISTRY, POLYMER CHEMISTRY. *Educ:* Stuttgart Tech Univ, Cand chem, 57; Harvard Univ, AM, 61, PhD(org chem), 65. *Prof Exp:* Res chemist, J R Geigy AG, Basel, Switz, 65-66, res assoc, Geigy Chem Corp, NY, 66-69, group leader, 69-70, Ciba-Geigy Corp, 70-72, sr staff scientist, 72-78, res mgr, 79-86, res fel, 86-88, SR RES FEL, CIBA-GEIGY CORP, 88- *Mem:* Am Chem Soc; Sigma Xi. *Res:* Synthetic organic chemistry; heterocyclic chemistry; high temperature polymers, resins, light stabilizing additives and antioxidants for polymers. *Mailing Add:* 23 Banksville Rd Armonk NY 10504

WINTER, ROLF GERHARD, b Düsseldorf, Ger, June 30, 28; nat US; m 51; c 3. NUCLEAR PHYSICS, ELEMENTARY PARTICLE PHYSICS. *Educ:* Carnegie-Mellon Univ, BS, 48, MS, 51, DSc, 52. *Prof Exp:* From instr to asst prof, Case Western Reserve Univ, 52-54; from asst prof to assoc prof, Pa State Univ, 54-64; chmn dept, 66-72, dean grad studies, Arts & Sci, 81-86, prof, 64-87, CHANCELLOR PROF PHYSICS, COL WILLIAM & MARY, 87- *Concurrent Pos:* Vis physicist & lectr, Carnegie Inst Technol, 55-56, Oxford Univ, 61-62, Univ Wis, 63, Univ Sask, 76, Swiss Inst Nuclear Res & Univ Zurich, 79-80, Oxford Univ, 86-87. *Mem:* Fel Am Phys Soc. *Res:* Intermediate energy nuclear and particle physics. *Mailing Add:* Dept Physics Col William & Mary Williamsburg VA 23185

WINTER, RUDOLPH ERNST KARL, b Vienna, Austria, Nov 27, 35; US citizen; m 64; c 3. ORGANIC CHEMISTRY. *Educ:* Columbia Univ, AB, 57; Johns Hopkins Univ, MA, 59, PhD(org chem), 64. *Prof Exp:* NIH fel chem, Karlsruhe Tech Univ, 62-63 & Harvard Univ, 63-64; asst prof org chem, Polytech Inst Brooklyn, 64-69; ASSOC PROF ORG CHEM, UNIV MO-ST LOUIS, 69- *Concurrent Pos:* Vis res prof, Swiss Fed Univ, Zurich, 75-76. *Mem:* Am Chem Soc; Chem Soc. *Res:* Chemistry of naturally occurring substances, especially terpenes and sesquiterpenes; isolation, structure, reactions and synthesis of natural substances; chemical ecology; photochemical and thermal reactions. *Mailing Add:* Dept Chem Univ Mo St Louis 8001 Natural Bridge Rd St Louis MO 63121-4499

WINTER, STEPHEN SAMUEL, b Vienna, Austria, Feb 27, 26; US citizen; m 51; c 3. SCIENCE EDUCATION. *Educ:* Albright Col, BS, 48; Columbia Univ, PhD(phys chem), 53. *Prof Exp:* Res chemist, Atlas Powder Co, 52-53; asst prof chem, Northeastern Univ, 53-58; asst prof chem & educ, Univ Minn, 58-61; from assoc prof to prof educ, State Univ NY Buffalo, 61-71. dir teacher educ, 68-71; chmn dept, 71-78 & 84-90, PROF EDUC, TUFTS UNIV, 71- *Concurrent Pos:* NSF fac fel, Harvard Univ, 57-58, consult, Proj Physics, 64-70; consult & hon assoc prof, Nat Univ Paraguay, 65; consult, Div Sci Teaching, UNESCO, 69-71. *Mem:* Nat Sci Teachers Asn; Nat Asn Res Sci Teaching; Asn Educ Teachers Sci (pres, 66-67). *Res:* Measurements of outcomes of science instruction; science curriculum in elementary and secondary schools. *Mailing Add:* Dept Educ Tufts Univ Medford MA 02155

WINTER, STEVEN RAY, b Belvidere, Ill, Jan 16, 44; m 70. AGRONOMY, PLANT PHYSIOLOGY. *Educ:* Univ Ill, BS, 66, MS, 68; Purdue Univ, PhD(agron), 71. *Prof Exp:* Asst prof, 71-78, ASSOC PROF CROP PROD, TEX A&M UNIV, 78- *Mem:* Am Soc Agron; Am Soc Sugar Beet Technol. *Res:* Production and physiology of sugar beets on the Texas high plains. *Mailing Add:* 4036 Ricardo Dr Amarillo TX 79109

WINTER, WILLIAM KENNETH, b Manitowoc, Wis, Apr 26, 26; m 63; c 2. PHYSICS. *Educ:* Univ Wis, BA, 50; Kans State Col, MS, 52, PhD(physics), 56. *Prof Exp:* res physicist, Phillips Petrol Co, 56-85; RETIRED. *Mem:* AAAS; Soc Petrol Eng. *Res:* Develop mathematical models for petroleum reservoir simulation. *Mailing Add:* 2312 Hill Dr Bartlesville OK 74006

WINTER, WILLIAM PHILLIPS, b Uniontown, Pa, Aug 17, 38; m 60; c 2. PROTEIN CHEMISTRY. *Educ:* Pa State Univ, University Park, BS, 60, MS, 62, PhD(biochem), 65. *Prof Exp:* Instr biochem, Pa State Univ, University Park, 63-65; res assoc, Univ Wash, 65-67, actg asst prof, 67-69; res assoc, Med Sch, Univ Mich, Ann Arbor, 69-73, asst res scientist, 73-75, assoc res scientist human genetics, 75-77; SR BIOCHEMIST, CTR SICKLE CELL DIS, ASSOC PROF, DEPT MED & GRAD DEPT GENETICS, MED SCH, HOWARD UNIV, 77- *Concurrent Pos:* NIH fel, Univ Wash, 65-66, Am Cancer Soc fel, 66-67; investr, Howard Hughes Med Res Inst, 67-69. *Mem:* Am Soc Human Genetics; AAAS; Am Chem Soc; Am Soc Hemat; NY Acad Sci. *Res:* Structure and function of hemoglobin and other human blood proteins; structural abnormalities in proteins in inherited and congenital disease; therapy for sickle cell anemia. *Mailing Add:* Howard Univ Ctr Sickle Cell Dis 2121 Georgia Ave Washington DC 20059

WINTER, WILLIAM THOMAS, b New York, NY, Nov 14, 44; m 69. X-RAY FIBER DIFFRACTION, COMPUTER AIDED MOLECULAR MODELLING. *Educ:* State Univ NY Col Environ Sci & Forestry, BS, 66, PhD(phys chem), 74. *Prof Exp:* Res assoc biol, Purdue Univ, 73-77; from asst prof to assoc prof chem, Polytech Univ, 77-87, assoc head dept, 82-84; ASSOC PROF CHEM, STATE UNIV NY, COL ENVIRON SCI & FORESTRY, 88- *Concurrent Pos:* Vis asst prof biol, Purdue Univ, 75-77; lectr polymer characterization, Ethicon Div, Johnson & Johnson, 79, 87; res scientist, Ctr Res Plant Macromolecules, Grenoble, France, 84-85; consult & vis scientist, Xerox Res Ctr, Can, 85, 86; vis prof, Univ Grenoble, 87, 90; vis scientist, Agr & Food Res Ctr, Norwich, UK, 88; treas, Cellulose, Paper & Textile Div, Am Chem Soc, 90-92. *Mem:* Am Chem Soc; Soc Complex Carbohydrates; Sigma Xi. *Res:* Structural studies of biologically or physically significant macromolecules by X-ray diffraction computer aided modelling methods and solid state NMR. *Mailing Add:* 315 Baker Lab State Univ NY Col Environ Sci & Forestry Syracuse NY 13210-2786

WINTERBERG, FRIEDWARDT, b Berlin, Ger, June 12, 29; c 1. THEORETICAL PHYSICS. *Educ:* Univ Frankfurt, MS, 53; Univ Goettingen, PhD(nuclear physics), 55. *Prof Exp:* Group leader theoret physics, Res Reactor, Hamburg, Ger, 55-59; asst prof plasma physics & relativity, Case Univ, 59-63; assoc prof physics, 63-68, RES PROF PHYSICS, DESERT RES INST, UNIV NEV SYST, RENO, 68- *Honors & Awards:* Gold Medal, Hermann Oberth-Wernher von Braun Int Space Flight Found, 79. *Mem:* Am Phys Soc; hon mem Hermann Oberth Soc; corresp mem Int Acad Astronaut. *Res:* Neutron physics; plasma physics; magnetohydrodynamics; intense relativistic electron and ion beams; thermonuclear microexplosions and inertial confinement fusion; nuclear rocket propulsion; relativity; atmospheric physics; energy research. *Mailing Add:* Desert Res Inst Univ Nev Syst Reno NV 89507

WINTERBOTTOM, RICHARD, b Livingstone, Zambia, Sept 30, 44; Brit & Can citizen; m 71; c 2. BIOGEOGRAPHY, PHYLOGENETICS. *Educ:* Univ Cape Town, SAfrica, BSc, 67; Queen's Univ, Kingston, Ont, PhD(biol), 71. *Prof Exp:* Teaching & res ichthyol, Smithsonian Inst, Wash DC, 71-72 & Nat Mus Can, Ottawa, 72-73; sr lectr ichthyol, Smith Inst, Rhodes Univ, SAfrica, 73-77; from asst cur to assoc cur, 78-84, CUR ICHTHYOL, ROYAL ONT MUS, TORONTO, 84- *Concurrent Pos:* Adj asst prof, Univ Toronto, 79-86, assoc prof, 86-90, prof, 90-; consult, Mako Films Ltd, Toronto, 80. *Honors & Awards:* Stoye Award, Am Soc Ichthyologists & Herpetologists, 71; Jessup Award, Acad Nat Sci, Philadelphia, 69. *Mem:* Can Soc Zoologists; Am Soc Ichthyologists & Herpetologists (gov, 78-83); Soc Syst Zoologists; Zool Soc Southern Africa; Japan Soc Ichthyol; fel Willi Hennig Soc. *Res:* Systematics, anatomy, phylogeny and biogeography of fishes, primarily the coral reef perciforms of the Indo-Pacific ocean; coral reef ecology; larval fish recruitment; the theory of phylogenetic interpretation. *Mailing Add:* Dept Ichthyol & Herpet Royal Ont Mus 100 Queen's Park Toronto ON M5S 2C6 Can

WINTERBOTTOM, W L, b Pittsburgh, Pa, Sept 27, 30; m 51; c 4. METALLURGY. *Educ:* Drexel Inst Technol, BSc, 58; Carnegie Inst Technol, PhD(metall), 62. *Prof Exp:* Staff scientist, 62-80, PRIN RES SCIENTIST, SCI LAB, FORD MOTOR CO, 80-, MGR, MAT SYSTS RELIABILITY DEPT. *Mem:* Am Inst Mining, Metall & Petrol Engrs; Am Soc Mat; Soc Automotive Eng. *Res:* Surface physics; evaporation of solids; vapor-solid interactions; condensation and nucleation; catalysis; gas monitoring devices; metal joining; fluxless vacuum brazing of aluminum, elec packaging design. *Mailing Add:* 26360 Powers Rd Farmington MI 48334

WINTERCORN, ELEANOR STIEGLER, b Morristown, NJ, Jan 15, 35; m 58. AUDIOLOGY, SPEECH PATHOLOGY. *Educ:* Rockford Col, BA, 56; Univ Wis, MS, 58; Univ Md, PhD, 69. *Prof Exp:* Clin instr speech path & phonetics, Rockford Col, 56-57; speech & hearing therapist, El Paso Cerebral Palsy Treatment Ctr, 58-59; audiologist, 60-66, supvr clin audiol, 66-70, asst chief audiol & speech path serv, 71-80, CHIEF, AUDIOLOGY & SPEECH PATH SERV, VET ADMIN MED CTR, DC, 80- *Concurrent Pos:* Mem res comt hearing aid eval processes, Am Speech & Hearing Asn, 66-67; Vet Admin rep comt hearing, bioacoust & biomech, Nat Res Coun-Nat Acad Sci, 68-71; res assoc & Vet Admin rep comt hearing, Bioacoust Lab, Univ Md, 68-72; res asst prof, Univ Md, 73-; dir, Vet Admin Nat Hearing Aid Testing Prog, 75- *Res:* Hearing aids; speech intelligibility. *Mailing Add:* 9112 Cherbourg Dr Potomac MD 20854

WINTERHALDER, KEITH, b Burrington, Eng, Apr 14, 35; Can citizen. REVEGETATION ECOLOGY, SOIL-PLANT RELATIONSHIPS. *Educ:* Univ Wales, BSc Hons, 56; New Eng Univ, Australia, MSc, 70. *Prof Exp:* Demonstr bot, Univ New Eng, 59-60, lectr, 60-62; sr res asst, Univ Liverpool, 62-65; lectr, 65-69, asst prof, 69-80, ASSOC PROF BIOL, LAURENTIAN UNIV, 80- *Concurrent Pos:* Chmn, veg enhancement tech adv comt, Sudbury, Can, 78-86. *Mem:* Can Bot Asn (treas, 84-86, 89); Can Land Reclamation Asn (vpres, 86-87); Ecol Soc Australia; Brit Ecol Soc; Am Soc Surface Mining & Reclamation; Soc Ecol Restoration. *Res:* Plant-soil relationships; soil biology; industrially-disturbed sites; plant distribution in northern Ontario. *Mailing Add:* Dept Biol Laurentian Univ Sudbury ON P3E 2C6 Can

WINTERLIN, WRAY LAVERNE, b Sioux City, Iowa, July 20, 30; m 56; c 3. ENVIRONMENTAL CHEMISTRY. *Educ:* Univ Nebr, Lincoln, BS, 54, MS, 57. *Prof Exp:* Forestry aide, US Forestry, Calif Inst Technol, Pasadena, 56-57; jr chemist, Dept Water Resources, Calif Bryte, 58-59; lab staff res assoc, 59-65, specialist, 65-78, actg dept chmn, 72-73, ENVIRON CHEMIST, DEPT ENVIRON TOXICOL, UNIV CALIF, DAVIS, 78- *Concurrent Pos:* Prin investr, US Environ Protection Agency, NIH, USDA & other projs, 69-; vis scholar, US Environ Protection Agency, Research Triangle Park, NC, 74-75; vis prof, Cairo, Egypt, 82. *Mem:* Am Chem Soc; Sigma Xi; Soc Environ Toxicol & Chem; AAAS. *Res:* Development of analytical methods for trace quantities of environmental agents, primarily pesticides; isolation and confirmation of trace organics in environmental samples; metabolism and transformation of biologically active agents and waste disposal of pesticides. *Mailing Add:* Dept Environ Toxicol Univ Calif Davis CA 95616

WINTERNHEIMER, P LOUIS, b Evansville, Ind, Feb 9, 31; m 51; c 2. BOTANY. *Educ:* Purdue Univ, West Lafayette, BS, 53; Univ Iowa, MS, 55; Ind Univ, Bloomington, PhD(bot), 71. *Prof Exp:* Assoc prof, 57-70, PROF BIOL, UNIV EVANSVILLE, 70- *Mem:* Am Bot Soc. *Res:* Biosystematic studies of Oenothera biennis and other species. *Mailing Add:* Dept Biol Univ Evansville 1800 Lincoln Ave Evansville IN 47722

WINTERNITZ, WILLIAM WELCH, b New Haven, Conn, June 21, 20; m 49, 84; c 3. MEDICINE. *Educ:* Dartmouth Col, AB, 42; Johns Hopkins Univ, MD, 45. *Prof Exp:* From instr to asst prof med & physiol, Yale Univ, 52-59; from assoc prof to prof med, Col Med, Univ Ky, 59-77; PROF & CHMN, DEPT INTERNAL MED, COL COMMUNITY HEALTH SCI, UNIV ALA, 77- *Mem:* Endocrine Soc; AMA; Am Diabetes Asn. *Res:* Endocrine regulation of metabolism. *Mailing Add:* Sch Med, Univ Ala PO Box 870326 Tuscaloosa AL 35487-0326

WINTERS, ALVIN L, b Enumclaw, Wash, Aug 26, 39; m 60; c 2. VIRAL PATHOGENESIS, ANTIVIRAL IMMUNOMODULATION. *Educ:* Kans State Teachers Col, BA, 64; Kans State Univ, MA, 68, PhD(virol), 69. *Prof Exp:* Instr biol, Kans State Univ, 69; asst prof microbiol, Univ Pa, 71-74; asst prof microbiol, Univ SFla, Tampa, 75-80; ASSOC PROF MICROBIOL & BIOCHEM, UNIV ALA, 80- *Concurrent Pos:* Fel, Univ Pa, 69-71. *Mem:* Am Soc Microbiol; AAAS; Sigma Xi; Am Soc Virol; Int Soc Antiviral Res. *Res:* Biochemistry of adenovirus pathogenesis. *Mailing Add:* Dept Biol PO Box 870344 Tuscaloosa AL 65487-0344

WINTERS, C(HARLES) E(RNEST), b Pratt, Kans, July 15, 16; m 41; c 4. NUCLEAR ENGINEERING. *Educ:* Kans State Col, BS, 37. *Hon Degrees:* ScD, Mass Inst Technol, 42. *Prof Exp:* Chem engr, Mallinckrodt Chem Works, Mo, 40-43; prin engr, Manhattan Dist, US AEC, 43-47; sect chief, Technol Div, Oak Ridge Nat Lab, Carbide & Carbon Chem Co, 47-49, dept head, Eng Res & Develop Technol Div, 49-51, dir, Exp Eng Div, 51-53, asst res dir, 53-55, asst lab dir, Union Carbide Nuclear Co, 55-61; res dir, Parma Res Lab, Ohio, Union Carbide Corp, 62-66, gen mgr, Fuel Cell Dept, 63-69, asst to vpres, Wash, DC, 69-78; CONSULT, 78- *Mem:* Am Chem Soc; fel Am Nuclear Soc; Am Inst Chem Engrs; fel Am Inst Chemists. *Res:* Electrochemical and nuclear engineering; energy generation. *Mailing Add:* 8800 Fernwood Rd Bethesda MD 20817-3014

WINTERS, EARL D, b Rio Grande, Ohio, Aug 28, 37; m 60; c 1. ELECTROCHEMISTRY, ELECTRONIC PACKAGING. *Educ:* Ohio Wesleyan Univ, BA, 59; Mass Inst Technol, PhD(phys chem), 65. *Prof Exp:* Mem tech staff, 65-89, TECH CONSULT, BELL LABS, INC, 89- *Mem:* Am Chem Soc; Electrochem Soc; Electroplaters Soc; Inst Elec & Electronics Engrs. *Res:* Electrodeposition, etching and corrosion of metals. *Mailing Add:* 923 W Sawmill Rd Quakertown PA 18951

WINTERS, HARVEY, b Paterson, NJ, Aug 23, 42; m 65; c 1. MICROBIOLOGY, BIOCHEMISTRY. *Educ:* Fairleigh Dickinson Univ, BS, 64, MS, 66; Columbia Univ, PhD(chem biol), 71. *Prof Exp:* From instr to asst prof, 69-75, assoc prof, 75-79, PROF BIOL, FAIRLEIGH DICKINSON UNIV, 79- *Honors & Awards:* Roon Award, Soc Paint Technol, 73. *Mem:* Am Soc Microbiol; Soc Indust Microbiol; Sigma Xi. *Res:* Microbiology of aqueous coatings; microbiofouling of marine surfaces; desalination. *Mailing Add:* Dept Biol Fairleigh Dickinson Univ Teaneck NJ 07666

WINTERS, LAWRENCE JOSEPH, b Chicago, Ill, June 11, 30; m 61; c 3. ORGANIC CHEMISTRY. *Educ:* Wash Univ, AB, 53; Univ Kans, PhD(chem), 59. *Prof Exp:* Asst chem, Univ Kans, 56-58; fel, Fla State Univ, 59-61; from asst prof to prof, Drexel Univ, 61-72, actg chmn dept, 68-69, asst dean grad sch, 69-71; PROF CHEM & CHMN DEPT, VA COMMONWEALTH UNIV, 72- *Mem:* Am Chem Soc. *Res:* Bipyridine chemistry; organic reaction mechanisms; structure-activity relationships; aliphatic nitro-compounds. *Mailing Add:* Dept Chem Va Commonwealth Univ Richmond VA 23298

WINTERS, MARY ANN, b Paterson, NJ, Nov 14, 37. BIOCHEMISTRY. *Educ:* Seton Hill Col, BA, 67; Univ Pittsburgh, PhD(biochem), 72. *Prof Exp:* Teacher elem & high schs, Pa & Ariz, 56-66; from instr to assoc prof chem, 67-81, ADMIN, SISTERS CHARITY SETON HALL, 81- *Mem:* Am Chem Soc; Leadership Conf Women Religious USA. *Res:* Purification of nucleic acid synthesizing enzymes and the isolation and identification of nucleic acids. *Mailing Add:* Seton Hill Col Greensburg PA 15601

WINTERS, RAY WYATT, b Takoma Park, Md, Feb 17, 42; m 67. PHYSIOLOGICAL PSYCHOLOGY. *Educ:* Mich State Univ, BS, 64, MA, 66, PhD(psychol), 69. *Prof Exp:* Asst prof, 69-74, assoc prof, 74-80, PROF PSYCHOL, UNIV MIAMI, 80- *Concurrent Pos:* NIH & NSF instnl grants, 69-, NIH grant, 72-80. *Mem:* Optical Soc Am. *Res:* Human psychophysical research in conjunction with animal neurophysiology, especially sensory systems and vision. *Mailing Add:* Dept Psychol Univ Miami Coral Gables FL 33124

WINTERS, ROBERT WAYNE, b Evansville, Ind, May 23, 26; m 48, 76; c 1. PEDIATRICS, NUTRITION. *Educ:* Ind Univ, AB, 48; Yale Univ, MD, 52. *Prof Exp:* Intern pediat, Univ Calif, 52-53; from asst to chief resident, Univ NC, 54-56; res fel med, Univ NC, 56-58; asst prof physiol, Univ Pa, 58-61; from assoc prof to prof pediat, Col Physicians & Surgeons, Columbia Univ, 61-81; exec vpres, 81-86, chief exec officer, 86-89, MED DIR & VPRES HEALTH DINE, HOME NUTRIT SUPPORT INC, 89- *Concurrent Pos:* Res fel, Univ Calif, 52-53; res fel biochem, Univ Pa, 58. *Honors & Awards:* E Mead Johnson Prize, 66; Borden Award, 74. *Mem:* Soc Pediat Res; Am Soc Clin Invest; Am Pediat Soc; Am Acad Pediat; Am Physiol Soc. *Res:* Renal and acid base physiology; metabolism of water and electrolytes; intravenous nutrition. *Mailing Add:* Home Nutrit Support Inc 600 Lanidex Plaza Parsippany NJ 07054

WINTERS, RONALD HOWARD, b Los Angeles, Calif, Apr 13, 42; m 76; c 2. PHARMACOLOGY. *Educ:* Calif State Univ, Northridge, BA, 63; Ore State Univ, PhD(pharmacol), 69. *Prof Exp:* Biochemist, Riker Labs, Inc, 64-65; asst to dean undergrad studies, Sch Pharm, Ore State Univ, 72-74, from instr to assoc prof, pharmacol, 68-76, asst dean, 74-76; Assoc Dean, Col Health Related Prof, Wichita State Univ, 77-82; DEAN HEALTH RELS, UNIV ARK, 82- *Concurrent Pos:* Res grants, Ore Heart Asn, 70-72 & Ore Educ Coord Coun, 70-72. *Mem:* AAAS; Sigma Xi; NY Acad Sci; Am Soc Allied Health Professions. *Res:* Cardiovascular pharmacology; anesthesia. *Mailing Add:* Col Health Related Prof Univ Ark Med Sch Little Rock AR 72205-7199

WINTERS, RONALD ROSS, b Marion, Va, June 4, 41; m 60; c 2. NUCLEAR PHYSICS, ASTROPHYSICS. *Educ:* King Col, AB, 63; Va Polytech Inst & State Univ, PhD(physics), 67. *Prof Exp:* Assoc prof physics, Denison Univ, 66-80. *Concurrent Pos:* Consult, Oak Ridge Nat Lab, 72-74;

DIR SCI SEMESTER PROG, GREAT LAKES COL ASN, 75- *Mem:* Sigma Xi; Asn Advan Physics Teaching. *Res:* Measurement of neutron capture cross sections; s-process nucleosynthesis; origin of the earth-moon system. *Mailing Add:* 820 W Broadway Granville OH 43023

WINTERS, STEPHEN SAMUEL, b New York, NY, June 29, 20; m 43; c 2. GEOLOGY. *Educ:* Rutgers Univ, BA, 42; Columbia Univ, MA, 48, PhD(geol), 55. *Prof Exp:* Instr geol, Rutgers Univ, 48-49; from asst prof to assoc prof, 49-66, dean, Div Basic Studies, 64-84, dir honors prog, 67-85, PROF GEOL, FLA STATE UNIV, 66- *Mem:* Fel Geol Soc Am; Paleont Soc; Soc Econ Paleont & Mineral; Am Asn Petrol Geol; Sigma Xi. *Res:* Stratigraphy and invertebrate paleontology of late Paleozoic. *Mailing Add:* Dept Geol 105 Dodd Hall Fla State Univ Tallahassee FL 32306

WINTERS, WALLACE DUDLEY, b New York, NY, June 20, 29; m 53; c 4. NEUROPHARMACOLOGY, CLINICAL PHARMACOLOGY. *Educ:* George Washington Univ, AB, 50; Univ Mich, Ann Arbor, MA, 52; Univ Wis-Madison, PhD(pharmacol), 54; Med Col Wis, MD, 58; Am Bd Med Toxicol, dipl, 77. *Prof Exp:* Asst pharmacol, Univ Mich, Ann Arbor, 51-52 & Univ Wis-Madison, 52-54; instr, Med Col Wis, 54-58; intern, Milwaukee Hosp, Wis, 58-59; Ment Health trainee neuropharmacol, Univ Calif, Los Angeles, 59-61, res pharmacologist, 61-63, assoc prof pharmacol, Sch Med, 63-68, prof pharmacol & psychiat, 68-71; PROF FAMILY PRACT, PHARMACOL, PSYCHIAT, & EMERGENCY MED, SCH MED, UNIV CALIF, DAVIS, 71- *Concurrent Pos:* Ment Health Prog rep pharmacol, Univ Calif, Los Angeles, 61-71, mem, Brain Res Inst & chmn ment health training prog, Educ Comt, 65- 71, mem brain res adv comt, 70-71; mem, Preclin Psychopharmacol Res Rev Comt, 65-69. *Honors & Awards:* A E Bennet Award, Soc Biol Psychiat, 66. *Mem:* AAAS; Am Soc Pharmacol & Exp Therapeut; fel Am Asn Clin Toxicol; Am Asn Poison Control Centers. *Res:* Neuropharmacological action of central nervous system acting drugs; models of psychosis; scheme of anesthetic, excitant, hallucinogen and convulsant drug action; circadian and seasonal rhythm and drug actions; melatonin and analgesia. *Mailing Add:* Dept Pharmacol & Int Med Sch Med Univ Calif Davis CA 95616

WINTERS, WENDELL DELOS, b Herrin, Ill; div; c 3. VIROLOGY, IMMUNOLOGY. *Educ:* Univ Ill, Urbana, BS, 62, MS, 66, PhD(med microbiol), 68. *Prof Exp:* Res asst med microbiol, Univ Ill Col Med, 63-65, teaching asst, 65-68; vis scientist, Nat Inst Med Res, London, 68-71; asst prof surg & med microbiol, Sch Med, Univ Calif, Los Angeles, 71-76; ASSOC PROF MICROBIOL, SCH MED, UNIV TEX HEALTH SCI CTR, SAN ANTONIO, 76- *Concurrent Pos:* Biochemist, Chicago Bd Health, 63-68; microbiologist, Presby St Lukes Hosp, Chicago, 65-68; Med Res Coun Eng grant, Nat Inst Med Res, London, 68-69; Damon Runyon Mem Fund Cancer Res fel, 70-71; consult, Vet Admin Hosp, Sepulveda, Calif, 71-76 & Audie Murphy Mem Vet Admin Hosp, 78-; mem, bd dirs, Aloe Res Found, 90; prog chmn & mem bd dirs, Bioelec Growth & Repair Soc; cong chmn, Int Cong Phytopath. *Mem:* Am Asn Immunologists; Am Asn Cancer Res; Am Soc Microbiol; Tissue Cult Asn; Brit Soc Gen Microbiol; Bioelectromagnetics Soc; Bioelectric Growth & Repair Soc; Soc Exp Biol Med; fel Am Acad Microbiol. *Res:* Mechanisms of virus assembly; immune responses to bacterial, viral and cancer antigens plus immunotherapy substances; characterization of bioactive phytotherapeutic substances; electromagnetic field exposure effects on immunomodulation, embryonic-fetal development and neoplastic marker oncogenes. *Mailing Add:* Dept Microbiol Sch Med Univ Tex Health Sci Ctr San Antonio TX 78284

WINTERSCHEID, LOREN COVART, b Manhattan, Kans, Oct 5, 25; m 48; c 6. MEDICAL SCHOOL ADMINISTRATION. *Educ:* Willamette Univ, BA, 48; Univ Pa, PhD(microbiol), 53, MD, 54. *Prof Exp:* Asst surg, Univ Wash, 57-58, from instr to assoc prof, 58-72, asst dean clin affairs, 72-80, prof surg, Sch Med, 72-89, assoc dean, 80-89, ASSOC DEAN ADMIN, SCH MED, UNIV WASH, 89- *Concurrent Pos:* Resident surgeon, Affil Hosps, Univ Wash, 55-62; NIH fel, 57-60; mem bd trustees, Willamette Univ, 60-; attend surgeon, Univ & King County Hosps, Seattle, 63. *Mem:* AMA. *Res:* General surgery. *Mailing Add:* Sch Med SC-61 Univ Wash Seattle WA 98195

WINTERSTEIN, SCOTT RICHARD, b Charleston, SC, Aug 25, 55. WILDLIFE BIOMETRY, POPULATION DYNAMICS. *Educ:* Northern Ariz Univ, BS, 77; NMex State Univ, MS, 80, PhD(biol), 85; NC State Univ, MSt, 86. *Prof Exp:* Vis prof, Dept Exp Statis, La State Univ, 85-86; ASST PROF WILDLIFE ECOL, DEPT FISHERIES & WILDLIFE, MICH STATE UNIV, 86- *Mem:* Am Ornithologists Union; Biomet Soc; Soc Conserv Biol; Wildlife Soc; Cooper Ornith Soc; Wilson Ornith Soc. *Res:* Dynamics of fish and wildlife populations; statistical methods for sampling fish and wildlife populations; experimental design and statistical analysis; mathematical modeling; conservation biology. *Mailing Add:* Dept Fisheries & Wildlife Mich State Univ East Lansing MI 48824

WINTHROP, JOEL ALBERT, b Elizabeth, NJ, Oct 30, 42. APPLIED MATHEMATICS, ELECTRICAL ENGINEERING. *Educ:* Univ Calif, BA, 64, MA, 70, PhD(math), 71, MEE, 78. *Prof Exp:* Asst prof math, Univ Mo, 71-76; mem tech staff, Bell Tel Labs, 77-80; CONSULT, 80- *Mem:* Am Math Soc; Math Asn Am; Soc Indust & Appl Math; Inst Elec & Electronics Engrs; Sigma Xi. *Res:* Digital signal processing. *Mailing Add:* AT&T Bell Labs Rm 1K306 Crawford Corners Rd Holmdel NJ 07733

WINTHROP, JOHN T, b Evanston, Ill, Feb 7, 38; m 59; c 1. HOLOGRAPHY, OPHTHALMIC LENS DESIGN. *Educ:* Univ Ill, BS, 60; Univ Mich, MS, 62, PhD(physics), 66. *Prof Exp:* Postdoctoral, Univ Mich Inst Sci & Technol, 66-67; res physicist, Am Optical, 67-78, chief scientist, 78-86, vpres prod develop, 87-88; CONSULT, 86-87 & 89- *Mem:* Optical Soc Am; Sigma Xi. *Res:* Optical diffraction, coherence theory and design of progressive-addition ophthalmic lenses; research in physics, reality problem and origin of physical law. *Mailing Add:* 15 Howe St Wellesley MA 02181

WINTHROP, STANLEY OSCAR, b Cowansville, Que, June 22, 27; m 56; c 3. ORGANIC CHEMISTRY. *Educ:* McGill Univ, BEng, 48; Ga Inst Technol, MS, 49; Univ Tex, PhD(org chem), 51. *Prof Exp:* Res chemist, Sterling-Winthrop Res Inst, 52-54; head med chem, Ayerst Res Labs, 54-64; dir res & develop, Lever Bros, Can, 64-69; sci adv, Off Sci & Technol, Can Dept Indust, Trade & Com, 69-71; dir gen, air pollution control directorate, 71-77, DIR GEN, ENVIRON IMPACT CONTROL DIRECTORATE, CAN DEPT ENVIRON, 77- *Mem:* Fel Can Inst Chem; Can Res Mgt Asn. *Res:* Pharmaceuticals; nitrogen heterocycles; fats and oils; detergents; research administration. *Mailing Add:* 1510 Riverside Dr Apt 2702 Ottawa ON K1G 4X5 Can

WINTNER, CLAUDE EDWARD, b Princeton, NJ, Apr 8, 38; m 67; c 2. ORGANIC CHEMISTRY. *Educ:* Princeton Univ, AB, 59; Harvard Univ, MA, 60, PhD(chem), 63. *Prof Exp:* From instr to asst prof, Yale Univ, 63-68; asst prof, Swarthmore Col, 68-69; assoc prof, 69-76, PROF CHEM, HAVERFORD COL, 76- *Concurrent Pos:* Acad guest, Swiss Fed Inst Technol, Zürich, 72-73, 76-77 & 89-90; vis prof chem, Harvard Univ, 84-85. *Honors & Awards:* Lindback Award, 82. *Mem:* Am Chem Soc. *Res:* Organic synthesis; chemical education. *Mailing Add:* 25 Railroad Ave Haverford PA 19041

WINTON, CHARLES NEWTON, b Raleigh, NC, Sept 22, 43; m 66; c 2. SOFTWARE ENGINEERING, SIMULATION. *Educ:* NC State Univ, BS, 65; Univ NC, Chapel Hill, MA & PhD(math), 69. *Prof Exp:* Asst prof math, Univ SC, 69-84; assoc prof comput sci, Univ S Fla, 82-83; assoc prof math sci, 74-82, PROF COMPUT SCI, UNIV NFLA, 83-, DEAN CIS, 89- *Mem:* Inst Elec & Electronics Engrs; Asn Comput Mach; Math Asn Am. *Res:* Medical systems simulation; prototype-based system specification; ring structure theory, including quotient objects and various torsion theories. *Mailing Add:* Col Comput Sci Univ NFla Jacksonville FL 32216

WINTON, HENRY J, b Guthrie, Okla, Mar 29, 29; m 55; c 3. ELECTRICAL ENGINEERING. *Educ:* Purdue Univ, Lafayette, BS, 50; Univ Ill, Urbana, MS, 55; Univ Santa Clara, PhD(elec eng), 70. *Prof Exp:* Engr, Sperry Gyroscope Co, NY, 57-60; assoc prof, 61-76, PROF ELEC ENG, ROSE-HULMAN INST TECHNOL, 76- *Mem:* Inst Elec & Electronics Engrs; Am Soc Eng Educ. *Res:* Modeling and computer simulation of biological control systems. *Mailing Add:* Dept Elec Eng Rose-Hulman Inst Technol Terre Haute IN 47803

WINTON, RAYMOND SHERIDAN, b Raleigh, NC, Jan 4, 40; m 73. MOLECULAR SPECTROSCOPY. *Educ:* NC State Univ, BS, 62; Duke Univ, PhD(physics), 72. *Prof Exp:* Physicist electro-optics, US Army Electronics Command, 65-67; ASST PROF MATH & PHYSICS, COL, 72- *Mem:* Sigma Xi; Am Phys Soc; Math Asn Am; Inst Elec & Electronics Engrs. *Res:* High precision measurements in microwave molecular spectroscopy and saturation effects in molecular absorption spectra. *Mailing Add:* Eng Sch Miss State Univ Mississippi State MS 39762

WINTROUB, HERBERT JACK, b Omaha, Nebr, Aug 22, 21; m 48; c 3. PHYSICS, ELECTRONICS. *Educ:* Univ Southern Calif, BS, 50. *Prof Exp:* Mem tech staff, Hughes Aircraft Co, 50-57; sr engr, Litton Indust, 57-58; mem sr tech staff, Space Tech Lab, TRW, Inc, 58-63, mem tech staff, Electronics Res Lab; HEAD, COMMUN SCI DEPT, ELECTRONICS RES LAB, AEROSPACE CORP, 68- *Concurrent Pos:* Consult, Sch Med, Univ Southern Calif, 65-, adj asst prof, 69-; mem adv group on electron devices, Dept Defense. *Mem:* Inst Elec & Electronics Engrs; Air Force Asn; Armed Force Commun & Electronics Asn. *Res:* Electronic systems research and development in radar, communications, command and control; millimeter-wave systems investigations; applications of electronics to medical research; space systems, microwave/millimeter-wave, receiver, signal processor and transmitter subsystems; management of solid-state device applications research. *Mailing Add:* Aerospace Corp PO Box 92957 M1-111 Los Angeles CA 90009

WINTSCH, ROBERT P, b Toronto, Can, Aug 15, 46; m; c 1. METAMORPHIC GEOLOGY, STRUCTURE PETROLOGY. *Educ:* Beloit Col, BA, 69; Brown Univ, PhD(geol), 75. *Prof Exp:* Asst prof, 75-81, ASSOC PROF, GEOL, DEPT GEOL, IND UNIV, 81- *Concurrent Pos:* Vis prof Inst Tech, Zurich, 81, Dept Earth Sci, Univ Leeds, 82, Dept Geol Sci, Univ Mich, 85, US Geol Surv, 88-89. *Mem:* Geol Soc Am; Am Geophys Union; Mineral Soc Am. *Mailing Add:* Dept Geol Ind Univ Bloomington Bloomington IN 47401

WINTTER, JOHN ERNEST, medicinal chemistry, for more information see previous edition

WINTZ, P(AUL) A, b Batesville, Ind, Mar 7, 35; m 56; c 1. ELECTRICAL ENGINEERING. *Educ:* Purdue Univ, BSEE, 59, MSEE, 61, PhD(elec eng), 64. *Prof Exp:* Engr, Duncan Elec Co, 58-61; from instr to assoc prof, Purdue Univ, West Lafayette, 61-71, prof elec eng, 71-76, asst head res, 72-76; CONSULT, 76- *Concurrent Pos:* Pres, Wintek Corp, 76- *Mem:* Am Soc Eng Educ; Inst Elec & Electronics Engrs. *Res:* Statistical communication theory; data and image processing; information handling. *Mailing Add:* 511 Kossuth St 1801 South St Lafayette IN 47905

WINTZ, WILLIAM A, JR, b Carville, La, June 7, 15; m 42; c 8. CIVIL ENGINEERING. *Educ:* La State Univ, BS, 36, MS, 38; Mass Inst Technol, SM, 51. *Prof Exp:* From instr to prof civil eng, La State Univ, Baton Rouge, 41-80; RETIRED. *Mem:* Am Soc Civil Engrs; Am Soc Eng Educ; Water Pollution Control Fedn; Am Water Works Asn. *Res:* Sanitary engineering; advanced surveying. *Mailing Add:* 1991 Hollydale Ave Baton Rouge LA 70808

WINZENREAD, MARVIN RUSSELL, b Indiandapolis, Ind, Nov 22, 37; m 60; c 2. MATHEMATICS, MICROCOMPUTERS EDUCATION. *Educ:* Purdue Univ, BS, 60; Univ Notre Dame, MS, 64; Ind Univ, Bloomington, EdD(math educ), 69. *Prof Exp:* Teacher high sch, Ind, 60-63; from instr to asst prof math, Northwest Mo State Col, 64-67; lectr, Ind Univ Indianapolis, 69; from asst prof to assoc prof, 69-82, PROF MATH & COMPUT SCI, CALIF STATE UNIV, HAYWARD, 82- *Res:* Mathematics in the inner city school. *Mailing Add:* Dept Math Calif State Univ Hayward CA 94542

WINZER, STEPHEN RANDOLPH, b Orlando, Fla, Nov 19, 44; m 67; c 2. PETROLOGY, GEOCHEMISTRY. *Educ:* Antioch Col, BA, 67; Univ Alta, PhD(petrol & geochem), 73. *Prof Exp:* Res assoc planetol, Goddard Space Flight Ctr, NASA, 73-75; res scientist geol, 75-80, sr scientist, 80-81, MGR, AGGREGATES RES, MARTIN MARIETTA LABS, 81-, STAFF SCIENTIST, 84-, MGR, ADVAN CERAMICS, 85- *Concurrent Pos:* Resident res assoc, Nat Acad Sci-Goddard Space Flight Ctr, NASA, 73-75. *Mem:* Geol Soc Am; Geochem Soc; Am Geophys Union; AAAS; Am Ceramic Soc. *Res:* Petrology, trace element chemistry and isotopic systematics of shock metamorphosed lunar and terrestrial rocks; fundamental mechanisms of fragmentation of rock by explosive loading and applications to blasting operations; synthesis of electronic ceramic materials; high strain rate behavior of structural ceramics. *Mailing Add:* Martin Marietta Labs 1450 S Rolling Rd Baltimore MD 21227

WIORKOWSKI, JOHN JAMES, b Chicago, Ill, Sept 30, 43; m 66; c 1. STATISTICS. *Educ:* Univ Chicago, BS, 65, MS, 66, PhD(statist), 72. *Prof Exp:* Asst prof statist, Grad Prog Health Care Admin, US Army-Baylor Univ, 68-71; res assoc, Univ Chicago, 71-73; asst prof, Pa State Univ, 73-75; assoc prof statist, 75-81, PROF UNIV TEX, DALLAS, 81-, ASST VPRES, ACAD AFFAIRS, 85- *Concurrent Pos:* Assoc dir, Statist Consult & Coop Res Ctr, Pa State Univ, 73-74, dir, 74-75; consult, Fed Energy Admin, 75; asst vpres, acad affairs, 79-; prog head, Math Sci, 78-79; fel acad admin, Am Coun Educ, 81-82. *Mem:* Sigma Xi; Am Statist Asn; Biomet Soc; Inst Math Statist. *Res:* Interest in applied statistics, specifically biostatistics, linear models, time series analysis, genetic statistics. *Mailing Add:* 428 Bedford Richardson TX 75080

WIPKE, W TODD, b St Charles, Mo, Dec 16, 40; c 2. CHEMISTRY. *Educ:* Univ Mo-Columbia, BS, 62; Univ Calif, Berkeley, PhD(chem), 65. *Prof Exp:* Res fel chem, Harvard Univ, 67-69; asst prof, Princeton Univ, 69-75; assoc prof, 75-81, PROF CHEM, UNIV CALIF, SANTA CRUZ, 81- *Concurrent Pos:* NIH res fel, Harvard Univ, 68-69; NIH spec res resource grant, Princeton Univ, 70-75 & Merck prof develop grant, 70-75; consult, Merck, Sharp & Dohme, 70-80, Molecular Design Ltd, 77-; mem bd adv, Chem Abstr Serv, 70-73; dir, NATO Advan Study Inst Comput Rep & Manipulation Chem Info, 73; mem, Nat Res Coun Comt Nat Res Comput Chem, 74-77. *Honors & Awards:* Alexander von Humbolt Award, 87; Computers Chem Award, Am Chem Soc, 87, Skolnik Award, 91. *Mem:* Am Chem Soc; Royal Soc Chem; Asn Comput Mach; Am Asn Artificial Intel. *Res:* Organic synthesis; computer assisted design of organic syntheses; computer assisted prediction of metabolism. *Mailing Add:* Dept Chem Univ Calif Santa Cruz CA 95064

WIRSEN, CARL O, JR, b Arlington, Mass, Aug 11, 42; c 2. MARINE MICROBIOLOGY, OCEANOGRAPHY. *Educ:* Univ Mass, BS, 64; Boston Univ, MA, 66. *Prof Exp:* Res assoc microbiol, Harvard Univ, 66-68; RES SPECIALIST MICROBIOL, WOODS HOLE OCEANOG INST, 68- *Mem:* Am Soc Microbiol; Am Soc Limnol & Oceanog; Sigma Xi. *Res:* The role of microorganisms in the deep sea environment and how the environmental parameters of temperature and pressure influence their activities; microbiological studies of deep sea hydrothermal vents. *Mailing Add:* Dept Biol Woods Hole Oceanog Inst Woods Hole MA 02543

WIRSZUP, IZAAK, b Wilno, Poland, Jan 5, 15; US citizen; m 49; c 1. MATHEMATICS. *Educ:* Univ Wilno, Mag Philos, 39; Univ Chicago, PhD(math), 55. *Prof Exp:* Lectr math, Tech Inst, Wilno, Poland, 39-41; dir bur studies & spec statist, Cent Soc Purchase; dir Soc Anonyme des Monoprix, France, 46-49; from instr to prof, 49-85, EMER PROF MATH, UNIV CHICAGO, 85- *Concurrent Pos:* Dir Surv East Europ Math Lit, proj Univ Chicago, under NSF grant, 56-83; consult sch math study group, Yale Univ & Stanford Univ, 60, 61 & 66; Ford Found consult, Univ Math Progs, Colombia, SAm, 65 & 66; mem, US Comn Math Instr, 69-73; adv math, Encyclopaedia Britannica, 71-; dir, NSF Surv Appl Soviet Res Math Educ, 85-; prin investr, Univ Chicago Sch Math Proj, 83-; dir, Int Math Educ Resource Ctr, Univ Chicago, 88- *Honors & Awards:* Quantrell Award, Univ Chicago, 58. *Mem:* Am Math Soc; Math Asn Am; NY Acad Sci; AAAS. *Res:* Mathematical analysis; international mathematics education. *Mailing Add:* Dept Math Eckhart Hall Univ Chicago 1118 E 58th St Chicago IL 60637

WIRTA, ROY W(ILLIAM), b Big Lake, Wash, Mar 27, 21; m 46; c 3. REHABILITATION BIOMEDICAL ENGINEERING, BIOMECHANICS. *Educ:* Univ Wash, BSME, 47. *Prof Exp:* Test engr, Gen Elec Co, 47-48, develop engr, Gen Eng Lab, 48-52, process engr, 52-56, mech engr, 56-57, sr engr, Hanford Atomic Prod Oper, 57-61, electromech systs engr, Ord Dept, 61-64; sr eng specialist, Biocybernetics Lab, Philco-Ford Corp, Pa, 64-67; sr res scientist, Rehab Eng Ctr, Moss Rehab Hosp, Philadelphia, 67-77; biomech engr, Biomechanics, 77-86; RETIRED. *Concurrent Pos:* Res biomed engr, Vet Admin Med Ctr, San Diego, Calif, 78-86. *Mem:* Am Soc Mech Engrs. *Res:* Electromechanical devices; mechanical arts; integrated man-machine systems; human locomotion; measurement of impaired human performance. *Mailing Add:* 5570 Rab St La Mesa CA 91942-2461

WIRTH, JAMES BURNHAM, b Teaneck, NJ, Jan 15, 41; m; c 3. PHYSIOLOGICAL PSYCHOLOGY, PSYCHIATRY. *Educ:* Cornell Univ, AB, 63, MD, 67; Cambridge Univ, PhD(physiol), 75. *Prof Exp:* Intern med, Cornell Univ Hosps, 67-68; resident neurol, Cleveland Metrop Gen Hosp, 68-69; fel physiol psychol, Inst Neurol Sci, Univ Pa, 71-73; res physiol, Cambridge Univ, 73-75; resident, 75-77, ASST PROF PSYCHIAT, JOHNS HOPKINS HOSP, 77-, CLIN DIR INPATIENT SERVS, 85- *Concurrent Pos:* Mem panel neurol behav, Nat Inst Neurol & Cardiovasc Dis & Stroke, 78. *Mem:* Am Physiol Soc; Am Psychiat Asn; Sigma Xi; Eastern Physiol Asn. *Res:* Ingestive behavior, especially thirst, hunger and sodium appetite; Anorexia Nervosa. *Mailing Add:* Dept Psychiat & Behav Sci Meyer 4-181 600 N Wolfe St Baltimore MD 21205

WIRTH, JOSEPH GLENN, b Onawa, Iowa, Nov 19, 34; m 53; c 1. ORGANIC CHEMISTRY. *Educ:* Univ Wash, BS, 59; Univ Mich, Ann Arbor, MS & PhD(chem), 65. *Prof Exp:* Chemist, Boeing Co, 58-62; res chemist, 65-71, MGR RES & DEVELOP PROD DEVELOP, SILICONE PROD DEPT, RES & DEVELOP CTR, GEN ELEC CO, 71- *Mem:* Am Chem Soc. *Res:* Organic synthesis, nitrogen heterocycles; polymer synthesis; fluorescence, organosilicon chemistry. *Mailing Add:* 13 Lori Ct Pittsfield MA 01201-7145

WIRTHLIN, MILTON ROBERT, JR, b Little Rock, Ark, July 13, 32; m 54; c 5. PERIODONTICS, EPIDEMIOLOGY. *Educ:* Univ Calif, DDS, 56, MS, 68; Am Bd Periodont, dipl, 74. *Prof Exp:* US Navy gen dent officer, 56-58; exec officer, First Dent Co, Fleet marine Force, 68-69, Third Dent Co, 73-74; head periodont, Naval Dent Clin, Long Beach, CA, 69-73; chief epidemiol, Naval Dent Res Inst, Great Lakes, Ill, 74-76, cmndg officer, 76-81; cmndg officer, Naval Regional Dent Ctr, San Francisco, 76-83; asst chief staff, naval med Co, Southwestern Region, 83-85; assoc prof, Univ Pac, 85-86; ASSOC CLIN PROF, UNIV CAL, SAN FRANCISCO, 86- *Concurrent Pos:* Spec proj officer, Bur Med & Surg, US Navy, 64-65, Dir fel, Naval Dent Clin, Long Beach, 71-72, consult periodont, Naval Regional Med Ctr, Great Lakes, Ill, 74-76; clin asst prof periodont, Univ Southern Calif, 70-73; Comt Nat Health Legis, Western Soc Periodont, 72-73; clin asst prof periodont, Univ Ill, 77-81. *Honors & Awards:* Milton & Mary Gabbs Prize in Dentistry, 56; Am Acad Dent Med Cert Merit, 56. *Mem:* Am Dent Asn; Int Asn Dent Res; fel Int Col Dentists; Am Acad Periodont. *Res:* Periodontal new attachment therapy through biological treatment of diseased root surfaces; dental care delivery; experimental atherosclerosis effect on supporting tissues of teeth; application of growth factors to wound healing. *Mailing Add:* Sch Dent Univ Calif Box 0762 San Francisco CA 94143

WIRTSCHAFTER, JONATHAN DINE, b Cleveland, Ohio, Apr 9, 35; m 59; c 5. OPHTHALMOLOGY, NEUROLOGY. *Educ:* Reed Col, BA, 56; Harvard Med Sch, MD, 60; Linfield Col, MS, 63. *Prof Exp:* Intern, Philadelphia Gen Hosp, 60-61; resident neurol, Good Samaritan Hosp, Portland, Ore, 61-63; resident ophthal, Johns Hopkins Hosp, 63-66; fel neurol, New York Neurol Inst, Columbia-Presby Med Ctr, New York, 66-67; from asst prof to assoc prof ophthal & neurol, Col Med, Univ Ky, 67-72, dir div ophthal, 67-74, prof ophthal & neurol, 72-77, chmn dept ophthal, 74-77; PROF OPHTHAL, NEUROL & NEUROSURG, COL MED, UNIV MINN, 77- *Concurrent Pos:* Vis prof, Hadassah Hosp & Hebrew Univ Jerusalem, 73-74; consult surgeon, Vet Admin Hosp, Minneapolis, Minn, 77-; sect ed, Survey of Ophthal, 83- *Mem:* Am Acad Neurol; fel Am Acad Ophthal; fel Am Col Surg; Asn Res Vision & Ophthal; Am Ophthal Soc. *Res:* Clinical neuro-ophthalmology; interactive teaching methods in ophthalmology; optic nerve; experimental pathology and pharmacology; facial spasms. *Mailing Add:* Dept Ophthal Univ Minn Hosp Box 493 516 Delaware St SE Minneapolis MN 55455

WIRTZ, GEORGE H, b Kohler, Wis, Apr 29, 31; m 52; c 2. IMMUNOCHEMISTRY. *Educ:* Univ Wis, BS, 53, MS, 56; George Washington Univ, PhD(biochem), 62. *Prof Exp:* Biochemist, Walter Reed Army Inst Res, 58-62; fel biol, Johns Hopkins Univ, 62-63; from asst prof to assoc prof, 63-72, PROF BIOCHEM, MED CTR, WVA UNIV, 72- *Concurrent Pos:* Vis prof, Mainz Univ, 68; Am Cancer Soc Scholar, Nat Inst Allergy & Infectious Dis, NIH, 78-79. *Mem:* Am Soc Biochem & Molecular Biol; Am Asn Immunol. *Res:* Complementology; vitamin A immunochemistry. *Mailing Add:* Dept Biochem WVa Univ Health Sci Ctr Morgantown WV 26506

WIRTZ, GERALD PAUL, b Wisconsin Rapids, Wis, Dec 22, 37; m 61; c 3. SOLID STATE ELECTROCHEMISTRY, CATALYSIS. *Educ:* St Norbert Col, BS, 59; Marquette Univ, BME, 61; Northwestern Univ, PhD(mat sci), 66. *Prof Exp:* Res scientist, Airco Speer Div, Air Reduction Co, Inc, 66-68; sr staff mem, Mat Res Lab, 68-80, ASSOC PROF CERAMIC ENG, UNIV ILL, URBANA, 68- *Concurrent Pos:* vis lectr, Mideast Tech Univ, Ankara Turkey, 80. *Honors & Awards:* Fulbright-Hayes lectr, Univ Aveiro, 80. *Mem:* Am Ceramic Soc; Sigma Xi. *Res:* Phase equilibria and transformations; magnetic oxides; materials for thick-film hybrid microcircuitry; metallic conductivity in oxides; defect chemistry of oxides; solid electrolytes; high temperature fuel cells. *Mailing Add:* Rte Two Box 159 Urbana IL 61801

WIRTZ, JOHN HAROLD, b Sheboygan, Wis, Nov 13, 23; m 50. NATURAL HISTORY, VERTEBRATE ZOOLOGY. *Educ:* Loyola Univ, Ill, BS, 52; Univ Wyo, MS, 54; Ore State Univ, PhD, 61. *Prof Exp:* Asst prof, 57-70, ASSOC PROF NATURAL HIST & GEN BIOL, PORTLAND STATE UNIV, 70- *Mem:* AAAS; Nat Audubon Soc. *Res:* Visual behavior, mobility and orientation in sciurid rodents. *Mailing Add:* 6545 N Commercial Ave Portland OR 97217

WIRTZ, RICHARD ANTHONY, b New Brunswick, NJ, Aug 16, 44; m 66; c 1. MECHANICAL ENGINEERING. *Educ:* Newark Col Eng, BSME, 66; Rutgers Univ, MS, 68, PhD(heat transfer), 71. *Prof Exp:* Asst prof, 70-76, ASSOC PROF & EXEC OFFICER MECH ENG, CLARKSON COL TECHNOL, 76- *Concurrent Pos:* Prin investr grants, 73-80. *Mem:* Assoc Am Soc Mech Engrs; Am Phys Soc; Sigma Xi. *Res:* Numerical and physical experiments on free convection heat transfer. *Mailing Add:* 2105 Parkway Dr Reno NV 89502

WIRTZ, WILLIAM OTIS, II, b Montclair, NJ, Aug 16, 37; m 72. MAMMALOGY. *Educ:* Rutgers Univ, BA, 59; Cornell Univ, PhD(ecol, evolutionary biol), 68. *Prof Exp:* Res cur, Smithsonian Inst, 62-66; from asst prof to assoc prof zool, 68-88, PROF BIOL, POMONA COL, 88- *Mem:* Am Soc Mammalogists; Ecol Soc Am; Wildlife Soc; Am Soc Zoologists; Am Inst Biol Sci. *Res:* Mammalian population ecology and behavior; fire ecology; predator ecology; carnivores and raptors; evolution and systematics of mammals. *Mailing Add:* Dept Biol Pomona Col Claremont CA 91711

WISBY, WARREN JENSEN, b Denmark, Nov 14, 22; US citizen; m 62; c 2. FISHERY BIOLOGY. *Educ:* Univ Wis, BA, 48, MA, 50, PhD(zool), 52. *Prof Exp:* Res assoc zool, Univ Wis, 52-59; assoc prof marine biol, Inst Marine Sci, Univ Miami, 59-65; dir, Nat Fisheries Ctr & Aquarium Dept of Interior, 65-72; adj prof, Inst Marine Sci, 65-72, assoc dean, Sch Marine & Atmospheric Sci, 72-80, interim dean, 80-82, ASSOC DEAN, SCH MARINE & ATMOSPHERIC SCI, UNIV MIAMI, 82- *Mem:* Am Fisheries Soc; Am Soc Zool; Animal Behav Soc; Sigma Xi. *Res:* Behavior and sensory physiology of marine organisms. *Mailing Add:* 6640 SW 129 Terr Miami FL 33156

WISCHMEIER, WALTER HENRY, b Lincoln, Mo, Jan 18, 11; m 47; c 2. SOIL CONSERVATION. *Educ:* Univ Mo, BS, 53; Purdue Univ, MS, 57. *Prof Exp:* Researcher, Soil & Water Conserv Res Div, Agr Res Serv, USDA, 40-61, res invests leader soil erosion, Corn Belt Br, 61-72, tech adv water erosion, North Cent Region, 72-75; from assoc prof to prof, 65-76, EMER PROF AGR ENG, PURDUE UNIV, WEST LAFAYETTE, 76- *Concurrent Pos:* Consult, soil erosion prediction and control, 76- *Honors & Awards:* Superior Serv Awards, USDA, 59 & 73; Hugh Hammond Bennett Award, Soil Conserv Soc Am, 77. *Mem:* Am Soc Agron; fel Soil Conserv Soc Am; Am Soc Agr Eng; Soil Sci Soc Am. *Res:* Soil and water conservation; quantitative relationship of soil erosion to rainfall characteristics, topographic features, management, productivity level and factor interactions; conservation farm planning; runoff and soil-loss prediction equations. *Mailing Add:* 2009 Indian Trail Dr West Lafayette IN 47906

WISCHMEYER, CARL R(IEHLE), b Terre Haute, Ind, Oct 2, 16; m 45; c 3. ELECTRICAL ENGINEERING. *Educ:* Rose Polytech Inst, BS, 37, EE, 42, ScD, 70; Yale Univ, MEng, 39. *Prof Exp:* Lab asst elec eng, Yale Univ, 37-39; instr, Rice Inst Technol, 39-45; mem tech staff, Bell Tel Labs, Inc, 45-47; from asst prof to prof elec eng, Rice Univ, 47-68, master, Baker Col, 56-68, dir continuing studies, 68; dir educ, Bell Tel Labs, 68-84; TECH REP, WERSI ELECTRONIC ORGANS, 84- *Concurrent Pos:* Consult to indust, 43-68; NSF grant, Eindhoven Technol Univ, 62; mem tech staff, Bell Tel Labs, 63-64, 68-84; bd mgrs, Rose Polytech Inst, 63-67. *Mem:* Am Soc Eng Educ; fel Inst Elec & Electronics Engrs; Am Soc Train & Develop Engrs; Sigma Xi. *Res:* Radio; electronics; instruments and equipment. *Mailing Add:* 25 Lurel Dr Fair Haven NJ 07701

WISCOMBE, WARREN JACKMAN, b St Louis, Mo, Feb 4, 43; m 67; c 2. ATMOSPHERIC PHYSICS. *Educ:* Mass Inst Technol, SB, 64; Calif Inst Technol, MS, 66, PhD(appl math), 70. *Prof Exp:* Res scientist, Systs Sci Software, 69-74; res scientist, Nat Ctr Atmospheric Res, 74-83; PHYS SCIENTIST, GODDARD SPACE LAB CTR, NASA, 83- *Res:* Radiative transfer in planetary atmospheres, particularly bearing on climate problems. *Mailing Add:* Climate Radiation Br Code 913 NASA Goddard Space Lab Ctr Greenbelt MD 20771

WISDOM, JACK LEACH, b Lubbock, Tex, Jan 28, 53; m 81; c 4. NON-LINEAR DYNAMICS, PLANETARY DYNAMICS. *Educ:* Rice Univ, BS, 76; Calif Inst Technol, PhD(physics), 81. *Prof Exp:* NATO fel, Observatoire de Nice, France, 81-82; res assoc, Univ Calif, Santa Barbara; res scientist, 84-85, from asst prof to assoc prof, 85-90, PROF, DEPT EARTH, ATMOSPHERIC & PLANET SCI, MASS INST TECHNOL, 90- *Concurrent Pos:* Presidential Young Investr, 88-93. *Honors & Awards:* Harold C Urey Prize, Div Planet Sci, Am Astron Soc, 86; Helen B Warner Prize, Am Astron Soc, 87. *Mem:* Am Astron Soc. *Res:* Application of modern non-linear dynamics to the dynamics of the solar system. *Mailing Add:* 54-414 Mass Inst Technol Cambridge MA 02139

WISDOM, NORVELL EDWIN, JR, physical chemistry, for more information see previous edition

WISE, BURTON LOUIS, b New York, Nov 24, 24; m 59; c 3. NEUROSURGERY. *Educ:* Columbia Univ, AB, 44; New York Med Col, MD, 47. *Prof Exp:* Clin instr neurol surg, Univ Calif San Francisco, 54, from instr to assoc prof, 55-68, vchmn dept, 65-68, assoc clin prof neurosurg, Sch Med, 68-77; chief dept neurosci, Mt Zion Hosp & Mt Zion Neurol Inst, 75-85; CONSULT, 85- *Concurrent Pos:* Attend neurol surgeon, Ft Miley Vet Admin Hosp, San Francisco, 54-68 & San Francisco Gen Hosp, 58-69; consult neurosurgeon, Laguna Honda Home & Langley Porter Neuropsychiat Inst, 57-68 & Letterman Army Hosp, 58-; chief neurosurg, Mt Zion Hosp & Mt Zion Neurol Inst, 70-74. *Mem:* Am Asn Neurol Surg; AMA; Am Col Surg; Am Fedn Clin Res. *Res:* Metabolic responses to central nervous system lesions; brain stem mechanisms in salt and water homeostasis; effects of hypertonic solutions on cerebrospinal fluid pressure; neuroendocrinology; pediatric neurosurgery and hydrocephalus. *Mailing Add:* Mt Zion Hosp & Neurol Inst 1600 Divisadero St San Francisco CA 94115

WISE, CHARLES DAVIDSON, b Huntington, WVa, June 13, 26; m 47; c 1. INVERTEBRATE ZOOLOGY, ORNITHOLOGY. *Educ:* WVa Univ, AB & MS, 50; Univ NMex, PhD(invert zool), 62. *Prof Exp:* Asst zool, Marshall Univ, 50-51; teacher high sch, WVa, 51-53; instr biol, Amarillo Col, 55-57; res scientist, Inst Marine Sci, Univ Tex, 58-60; asst biol, Univ NMex, 60-61; from asst prof to assoc prof, 61-72, PROF BIOL, BALL STATE UNIV, 72- *Concurrent Pos:* Ind State rep, 66-68; Ind State senator, 68-72; mem, Int Comt Recent Ostracoda, 63-73. *Mem:* Nat Audubon Soc; Sigma Xi. *Res:* Ecology; biological oceanography; marine and freshwater ostracods, especially taxonomy and ecology; ornithology. *Mailing Add:* Dept Biol Ball State Univ Muncie IN 47306

WISE, DAVID HAYNES, b Mineral Wells, Tex, Apr 28, 45; m 67; c 2. POPULATION & COMMUNITY ECOLOGY, SOIL ECOLOGY. *Educ:* Swarthmore Col, BA, 67; Univ Mich, MS, 69, PhD(zool), 74. *Prof Exp:* Instr biol, Albion Col, 69-70; lectr zool, Univ Mich, 70-71; asst prof biol, Univ NMex, 74-76; asst prof, 76-81, ASSOC PROF BIOL SCI, UNIV MD, BALTIMORE COUNTY, 81- *Concurrent Pos:* Alexander von Humboldt fel, Univ Göttinger, Ger, 85-86. *Mem:* Ecol Soc Am; Am Arachnol Soc; Soc Study Evolution; Brit Arachnol Soc; Brit Ecol Soc; US Soil Ecol Soc. *Res:* Population dynamics and regulation of population density; experimental field studies of competition and predation; life history evolution. *Mailing Add:* Dept Biol Sci Univ Md Baltimore County Catonsville MD 21228

WISE, DAVID STEPHEN, b Findlay, Ohio, Aug 10, 45; m 71; c 2. FUNCTIONAL PROGRAMMING LANGUAGES, HEAP BASED MULTIPROCESSING. *Educ:* Carnegie Inst Technol, 67; Univ Wis, MS, 69, PhD(computer sci, 71. *Prof Exp:* Lectr computer sci, Univ Edinburgh, 71-72; from asst prof to assoc prof, 77-87, PROF COMPUTER SCI, IND UNIV, 87- *Concurrent Pos:* Vis assoc prof computer sci, Ore State Univ, 83-84; prin scientist, Computer Res Lab, Tektronix Labs, 87; chair, spec interest group prog lang, Asn Comput Mach, 89-91. *Mem:* Asn Comput Mach. *Res:* Functional (or applicative) programming; developing parallel algorithms, architectures, hardware, and programming style, unified to deliver parallel processing to non- specialists. *Mailing Add:* Computer Sci Dept Ind Univ Lindley Hall Bloomington IN 47405-4101

WISE, DONALD L, b Indianapolis, Ind, May 27, 29; m 52; c 4. CELL PHYSIOLOGY. *Educ:* Wabash Col, AB, 51; NY Univ, MS, 54, PhD, 58. *Prof Exp:* Instr natural sci, Univ Chicago, 57-58; from instr to assoc prof, 58-66, chmn dept, 72-87, PROF BIOL, COL WOOSTER, 66- *Concurrent Pos:* Staff biologist, Comn Undergrad Educ Biol, 67-68; vis prof, George Washington Univ, 67-68; vis prof, Case Western Reserve Univ, 76-77 & Univ Miami, 85-86; consult & examr, N Cent Asn Col & Schs, 75- *Mem:* AAAS; Soc Protozool; Sigma Xi. *Res:* Protozoan, bacterial and cellular metabolism and physiology. *Mailing Add:* Dept Biol Col Wooster Wooster OH 44691

WISE, DONALD U, b Reading, Pa, Apr 21, 31; m 65; c 2. GEOLOGY. *Educ:* Franklin & Marshall Col, BS, 53; Calif Inst Technol, MS, 55; Princeton Univ, PhD(geol), 57. *Prof Exp:* From asst prof to prof geol, Franklin & Marshall Col, 57-68; chief scientist & dep dir, NASA Apollo Lunar Explor Off, 68-69; prof geol, Univ Mass, Amherst, 69-80. *Concurrent Pos:* Pa Geol Surv, 65-66, Geotech & Power Cos Seismic Risk, 72-, Nuclear Regulation Comn, 76-80, various oil co, 73-; vis scientist, Max Planck Inst, Heidelberg, 75 & Univ Rome, 76. *Mem:* AAAS; Geol Soc Am; Am Geophys Union. *Res:* Structural geology; structure and basement features of the middle Rocky Mountains; flow mechanics of rocks; structures of the Appalachian Piedmont; regional fracture analysis; lunar and planetary geology. *Mailing Add:* Dept Geol-Geog Univ Mass Amherst Campus Amherst MA 01003

WISE, DWAYNE ALLISON, b Lewisburg, Tenn, Feb 5, 45; m 66; c 1. CYTOGENETICS. *Educ:* David Lipscomb Col, BA, 67; Fla State Univ, MS, 70, PhD(genetics), 72. *Prof Exp:* Res fel cytogenetics, Health Sci Ctr, Univ Tex, 73-74; instr zool, Duke Univ, 74-75, res assoc cell biol, 75-80; MEM FAC, DEPT BIOL SCI, MISS STATE UNIV, 80- *Concurrent Pos:* NIH fel, 75-77. *Mem:* Sigma Xi. *Res:* Investigation of the control of chromosome structure during the cell cycle and of chromosome distribution at meiosis and mitosis. *Mailing Add:* 1200 S Montgomery St Starkville MS 39759-1136

WISE, EDMUND MERRIMAN, JR, b Jersey City, NJ, Aug 10, 30; m 52; c 2. MICROBIAL BIOCHEMISTRY. *Educ:* Oberlin Col, BA, 52; Harvard Univ, PhD(biochem), 63. *Prof Exp:* Jr biologist, Parke, Davis & Co, 54-55; NIH fel, Med Sch, Tufts Univ, 64-66, from instr to asst prof molecular biol & microbiol, 65-73; SR RES MICROBIOLOGIST, BURROUGHS-WELLCOME CO, 73- *Concurrent Pos:* NIH fel, Med Sch, Harvard Univ, 63-64. *Mem:* AAAS; Am Chem Soc; Am Soc Microbiol. *Res:* Control of enzyme activity; bacterial cell wall synthesis and degradation; design of enzyme inhibitors; microbial cofactor biosynthesis. *Mailing Add:* 709 Kenmore Rd Chapel Hill NC 27514

WISE, EDWARD NELSON, b Athens, Ohio, May 30, 15; m 36; c 2. CHEMISTRY, RESEARCH ADMINISTRATION. *Educ:* Ohio Univ, BS, 37, MS, 38; Univ Kans, PhD(chem), 53. *Prof Exp:* Teacher chem & physics, Gallia Acad High Sch, 38-42; qual control chemist, Baker & Adamson Div, Gen Chem Co, 42; supvr, Standards Lab, WVa Ord Works, 42-45; res engr graphic arts, Battelle Mem Inst, 45-47; mem staff anal instrumentation, Los Alamos Sci Lab, 47-50; tech asst, Hercules Powder Co, 51; from asst prof to prof chem, Univ Ariz, 52-81, from assoc coordr to coordr res, 64-72; RETIRED. *Mem:* Am Chem Soc. *Res:* Electrophoretic deposition of natural and synthetic rubbers; halftone and color separation techniques in the graphic arts; electrostatic image formation and development; xerography; analytical instrumentation; automatic titrimetry and coulometric analysis. *Mailing Add:* 6914 Sesame Lane Tucson AZ 85704-1850

WISE, ERNEST GEORGE, b Dunkirk, NY, Sept 6, 20; m 50; c 2. RADIATION BIOLOGY, MICROBIOLOGY. *Educ:* State Univ NY Col Fredonia, BEd, 42; Columbia Univ, MA, 47; Syracuse Univ, PhD(sci educ), 60. *Prof Exp:* Instr sci & math, State Univ NY Col New Paltz, 47-48, prof biol, State Univ NY Col Oswego, 48-77; RETIRED. *Concurrent Pos:* AEC-NSF acad year fel radiation biol, Cornell Univ, 64-65; AEC equip grant, 68. *Mem:* Health Phys Soc; Int Radiation Protection Asn. *Res:* Science education from elementary grades through the college level. *Mailing Add:* 499 S Shore Dr Boiling Spring Lakes Southport NC 28461

WISE, EVAN MICHAEL, b Cleveland, Ohio, May 15, 52; m 75; c 2. PAPERMAKING, PULPING. *Educ:* Miami Univ, BS, 74; Univ NFlorida, 79. *Prof Exp:* Pulp mill support, St Regis Paper Co, 74-83; prod mgr, Jacksonville Kraft Paper Co, 83-85; MANAGER MILL RES, JEFFERSON SMURFIT CORP, CCA, 85- *Concurrent Pos:* Adj prof, Univ NFla, 76. *Mem:* Tech Asn Pulp & Paper Indust. *Res:* Implementing new developments into current operating systems; innovative ways to produce paper and board products; recycled paper production. *Mailing Add:* 1332 Old Dominion Naperville IL 60540

WISE, GARY E, b Yuma, Colo, July 30, 42; m 62; c 2. CELL BIOLOGY. *Educ:* Univ Denver, BA, 64; Univ Calif, Berkeley, PhD(zool), 68. *Prof Exp:* NIH fel cell biol, Univ Colo, Boulder, 69-71; from asst prof to assoc prof biol, Sch Med, Univ Miami, 72-82; prof & chmn, 82-89, PROF, DEPT ANAT & CELL BIOL, TEX COL OSTEOP MED, 89- *Mem:* Am Soc Cell Biol; Am Asn Anatomists; Int Asn Dent Res. *Res:* Cellular basis of tooth eruption, role of sickled erythrocyte membranes in vaso-occlusion. *Mailing Add:* Dept Anat Tex Col Osteop Med 3500 Camp Bowie Ft Worth TX 76107

WISE, GEORGE HERMAN, b Saluda, SC, July 7, 08; m 37; c 4. ANIMAL NUTRITION, ANIMAL PHYSIOLOGY. *Educ:* Clemson Col, BS, 30; Univ Minn, MS, 32, PhD(dairy husb), 37. *Prof Exp:* Asst dairy husb, Univ Minn, 33-36; assoc, Clemson Col, 37-44; from assoc prof to prof, Kans State Col, 44-47; assoc prof, Iowa State Col, 47-49; prof animal indust, 49-51, head nutrit sect, 49-66, William Neal Reynolds Prof, 51-74, EMER PROF ANIMAL SCI, NC STATE UNIV, 75- *Concurrent Pos:* Mem comt animal nutrit, Nat Res Coun, 51-53; consult, State Exp Sta Div, Agr Res Serv, USDA, 55-62; study leave, Univ Calif, Davis, 66-67. *Honors & Awards:* Award, Am Feed Mfrs Asn, 48; Borden Award, 49; Award of Honor, Am Dairy Sci Asn, 66. *Mem:* Fel Am Soc Animal Sci; Am Dairy Sci Asn (vpres, 63-64, pres, 64-65). *Res:* Nutrition and physiology of animals. *Mailing Add:* 229 Woodburn Rd Raleigh NC 27605

WISE, HAROLD B, b Hamilton, Ont, Feb 14, 37. SOCIAL MEDICINE, INTERNAL MEDICINE. *Educ:* Univ Toronto, MD, 61. *Prof Exp:* Physician, Prince Albert Clin, Sask, Can, 62-63; resident, Kaiser Found Hosp, San Francisco, Calif, 63-64, Montefiore Hosp & Med Ctr, Bronx, NY, 64-65; actg dir ambulatory serv & home care, Morrisania City Hosp, 65-66; dir health ctr, Dr Martin Luther King, Jr Health Ctr, 66-71; dir analytical & develop health teams, Montefiore Hosp & Med Ctr, 71-77; ASSOC PROF COMMUNITY HEALTH, ALBERT EINSTEIN COL MED, 70- *Concurrent Pos:* Milbank Mem Fund fel; dir internship & residency prog social med, Montefiore Hosp & Med Ctr, 69-, dir inst health team develop, 72-; dir, Family Ctr Health. *Mem:* Inst Med-Nat Acad Sci. *Res:* Research into the family and the healing processes. *Mailing Add:* 988 Fifth Ave New York NY 10021

WISE, HENRY, b Ciechanow, Poland, Jan 14, 19; nat US; m 43, 60; c 6. STRUCTURAL CHEMISTRY. *Educ:* Univ Chicago, SB, 41, SM, 44, PhD(phys chem), 47. *Prof Exp:* Res assoc, Univ Chicago, 41-46; dir field lab, NY Univ, 46-47; scientist, Nat Adv Comt Aeronaut, Ohio, 47-49; phys chemist, Calif Inst Technol, 49-55; chmn chem dynamics dept, 55-71, dir catalysis lab, 71-84, SCI FEL, SRI INT, 84-; ADJ PROF MAT SCI & ENG, STANFORD, UNIV, 84- *Concurrent Pos:* Lectr, Sch Eng, Stanford Univ, 60-; vis prof, Israel Inst Technol, 65; Fulbright fel, 65; vis prof, Univ Calif, Berkeley, 77-78; mem comt motor vehicle emission, Nat Acad Sci; vis prof chem eng, Univ Calif, Berkeley, 80-81. *Honors & Awards:* Fulbright Award, 65; McBean Award, 83. *Mem:* Am Chem Soc; Am Phys Soc; Catalysis Soc; Chem Soc. *Res:* Heterogeneous catalysis; chemical kinetics; surface chemistry. *Mailing Add:* Mat Sci & Eng Dept Stanford Univ Stanford CA 94305

WISE, HUGH EDWARD, JR, b Lafayette, Ind, Oct 12, 30. ORGANIC CHEMISTRY. *Educ:* Vanderbilt Univ, BA, 52; Univ Fla, PhD(chem), 61. *Prof Exp:* Proj leader, Tech Serv Lab, Union Carbide Corp, NY, 61-66, proj specialist, 66-68; res chemist, Nalco Chem Co, Ill, 68-70; res chemist, 70-71, field serv supvr, res & develop, Waste Treat Div, Clow Corp, 71-77; ENVIRON SCIENTIST, INDUST TECHNOL DIV, US EPA, 78- *Mem:* Am chem Soc; Water Pollution Control Fedn. *Res:* Environmental science and engineering; waste treatment technology; occurrence and predictability of priority pollutants; industrial process chemistry. *Mailing Add:* US Environ Protection Agency WH-552 401 M St SW Washington DC 20460

WISE, JOHN HICE, physical chemistry, for more information see previous edition

WISE, JOHN JAMES, b Cambridge, Mass, Feb 28, 32; m 67; c 2. PHYSICAL INORGANIC CHEMISTRY, CHEMICAL ENGINEERING. *Educ:* Tufts Univ, BS, 53; Mass Inst Technol, PhD(chem), 65. *Prof Exp:* Res engr, Mobil Res & Develop Corp, 53-55, sr res engr, 56-62, group leader appl res, 65-68, asst mgr appl res, 68-69, supvr appl develop, 69-76, mgr process res & develop, 76-77, vpres planning, 77-84, mgr process & prod res & develop, 84-87, VPRES RES, MOBIL RES & DEVELOP CORP, 87- *Mem:* Nat Acad Eng; AAAS; Am Inst Chem Engrs; Am Chem Soc. *Res:* Catalysis related to petroleum and petrochemical processes. *Mailing Add:* Res Dept Mobil Res & Develop Corp PO Box 1031 Princeton NJ 08540

WISE, JOHN P, b Boston, Mass, Feb 9, 24; m 60; c 3. FISHERIES BIOLOGY, MARINE BIOLOGY. *Educ:* Suffolk Univ, AB, 50; Univ NH, MS, 53. *Prof Exp:* Fishery res biologist, US Bur Com Fisheries, Woods Hole, Mass, 53-60; tech asst officer, Food & Agr Orgn, UN, Rome, Italy, 60-65; chief, Atlantic Tuna Fisheries Res, Nat Marine Fisheries Serv, Miami, 65-73, sr res specialist, 73-74, chief, Resource Assessment Div, 75-77, Data Mgt & Statistics Div, 77-78, Prog Eval Staff, Wash, DC, 78-81; biostatistician, Int Comn Conserv Atlantic Tunas, Madrid, Spain, 81-86; CONSULT, CTR MARINE CONSERV, WASH, DC, 87- *Concurrent Pos:* Consult, Food & Agr Orgn UN, Offshore/Sea Develop Corp, US Agency Int Develop, Heritage Found & KCA Res Inc; adv & comt mem, Univ Miami & Am Univ; adv to res fels, Nat Res Coun, Brazil. *Mem:* AAAS; fel Am Inst Fishery Res Biol; Sigma Xi. *Res:* Stocks of marine animals, involving studies of ecology, growth rates, mortality rates, predation and parasitology, directed at eventual exploitation by man for optimum sustainable yield; approximately 100 publications in assessment of living marine resources, marine biology and oceanography. *Mailing Add:* 4545 Connecticut Ave NW Washington DC 20008

WISE, JOHN THOMAS, b Orangeburg, SC, Nov 29, 26; m 55; c 3. CHEMISTRY, PAPER CHEMISTRY. *Educ:* The Citadel, BS, 46; Purdue Univ, MS, 49. *Prof Exp:* Chemist, Thiokol Chem Co, 49-50; chemist, 52-57, sr chemist, 57-65, group leader res & develop, 65-69, mgr tech serv sect, 69-70, mgr lab serv, 70-73, res coordr, 73-79, TECH SERV MGR, RES & DEVELOP, SONOCO PROD CO, 79- *Concurrent Pos:* Instr, Univ SC, 60; chmn tech comt, Composite Can & Tube Inst, 80- *Mem:* Am Chem Soc; Tech Asn Pulp & Paper Indust. *Res:* Physical and chemical testing; pulp and paper; paper products; utilization of waste products; organic synthesis; instrumentation. *Mailing Add:* 109 Lyndale Dr E Hartsville SC 29550-9606

WISE, LAWRENCE DAVID, b Canton, Ohio, Oct 13, 40; m 67; c 3. ORGANIC CHEMISTRY. *Educ:* Manchester Col, BA, 62; Ohio State Univ, MS, 64, PhD(org chem), 67. *Prof Exp:* Res scientist org chem, Goodyear Tire & Rubber Co, 67-69; res scientist org chem, Warner-Lambert Res Inst, 69-77; res assoc, 77-82, sr res assoc, 82-90, SECT DIR, ORG CHEM, WARNER-LAMBERT/PARKE DAVIS RES, 90- *Mem:* Am Chem Soc; Soc Neurosci. *Res:* Synthetic organic chemistry, particularly heterocycles directed toward drug design. *Mailing Add:* Parke Davis Res 2800 Plymouth Rd Ann Arbor MI 48105

WISE, LOUIS NEAL, b Slagle, La, Jan 27, 21; m 44; c 2. AGRONOMY. *Educ:* Northwestern State Col, BS, 42; La State Univ, BS, 46, MS, 47; Purdue Univ, PhD(agron), 50. *Prof Exp:* Asst, Purdue Univ, 47-50; asst prof, 50-53, from assoc agronomist to agronomist, Exp Sta, 50-66, dir regional res lab, 52-66, dean sch agr, 61-66, vpres agr & forestry, 66-74, PROF AGRON, MISS STATE UNIV, 53-, VPRES AGR, FORESTRY & VET MED, 74- *Mem:* Am Soc Agron. *Res:* Pasture production and management; seed research. *Mailing Add:* 903 S Montgomery St Starkville MS 39759

WISE, MATTHEW NORTON, b Tacoma, Wash, Apr 2, 40; m 65. HISTORY OF PHYSICS, HISTORY OF SCIENCE. *Educ:* Pac Lutheran Univ, BS, 62; Wash State Univ, PhD(physics), 68. *Prof Exp:* Asst prof physics, Auburn Univ, 67-69 & Ore State Univ, 69-71; NSF sci fac fel hist of sci, Princeton Univ, 71-72; lectr, 75-78, ASST PROF HIST, UNIV CALIF, LOS ANGELES, 78- *Mem:* Am Phys Soc; Am Asn Physics Teachers; Hist Sci Soc. *Res:* History of nineteenth and twentieth century physical sciences. *Mailing Add:* Dept Hist 405 Hilgard Ave Univ Calif Los Angeles CA 90024

WISE, MILTON BEE, b Newland, NC, July 17, 29; m 51; c 3. ANIMAL NUTRITION. *Educ:* Berea Col, BS, 51; NC State Col, MS, 53; Cornell Univ, PhD(animal nutrit), 57. *Prof Exp:* Lab supvr, Berea Col, 47-51; asst, NC State Col, 52-53, res assoc, 53-54; asst, Cornell Univ, 54-55, instr animal husb, 55-57; from asst prof to prof animal sci, NC State Univ, 57-70; PROF ANIMAL SCI & HEAD DEPT, VA POLYTECH INST & STATE UNIV, 70- *Mem:* Am Soc Animal Sci; Sigma Xi. *Res:* Mineral and nutrient metabolism; forage utilization; physiology of digestion. *Mailing Add:* Agr & Natural Resources Clemson Univ Clemson SC 29634

WISE, RALEIGH WARREN, b Plainfield, NJ, Sept 30, 28; m 57. RESEARCH ADMINISTRATION, RUBBER CHEMISTRY. *Educ:* Univ Va, BS, 51. *Prof Exp:* Analytical chemist, 51-53, analytical res chemist, 54-56, res group leader, 56-65, res sect mgr, 65-71, group mgr, Instrument & Equip Div, 71-74, dir, 74-75, dir technol, Rubber Chem Div, 75-87, CONSULT, MONSANTO CHEM CO, 87- *Honors & Awards:* Banbury Award, Am Chem Soc, 87. *Mem:* Am Chem Soc; Instrument Soc Am; fel Am Inst Chemists. *Res:* Instrumentation; chemical and elastomer research. *Mailing Add:* 12755 Cold Stream Dr Ft Myers FL 33912

WISE, RICHARD MELVIN, b Greentown, Ohio, Sept 27, 24; m 68; c 3. ORGANIC CHEMISTRY. *Educ:* Mt Union Col, BS, 49; Ohio State Univ, PhD(org chem), 55. *Prof Exp:* Res chemist, Gen Corp, 55-58, sr res chemist, 58-72, res scientis, 72-87; RETIRED. *Mem:* Am Chem Soc. *Res:* Organic research; synthesis of monomers; preparation of polymerization catalysts and polymers; tire cord adhesives; emulsion rubbers. *Mailing Add:* PO Box 459 Uniontown OH 44685

WISE, ROBERT IRBY, b Barstow, Tex, May 19, 15; m 40; c 3. BACTERIOLOGY, INTERNAL MEDICINE. *Educ:* Univ Tex, BA, 37, MD, 50; Univ Ill, MS, 38, PhD(bact), 42; Am Bd Internal Med, dipl, 57. *Hon Degrees:* DSc, Thomas Jefferson Univ, 80. *Prof Exp:* Asst, Div Animal Genetics, Exp Sta, Univ Ill, 38-39, asst instr bact, 39-42; dir, Wichita City-County Pub Health Lab, Tex, 42-43; dir, Houston Pub Health Lab, 43; asst prof bact, Sch Med, Univ Tex, 43-46, dir bact & serol labs, Univ Hosp, 46-50; asst surgeon, USPHS Hosp, New Orleans, La, 50-51; fel med, Univ Minn, 51-53, asst prof, 53-54, asst prof med & bact, 54-55; from asst prof to assoc prof med, 55-59, Magee prof med & head dept, 59-75; asst chief med staff, Vet Admin Med & Regional Off Ctr, Togus, 75-77, chief staff, 77-84; EMER MAGEE PROF MED, JEFFERSON MED COL, 75- *Concurrent Pos:* Bacteriologist, Univ Hosp, Univ Minn, 57-59; assoc mem comn streptococcal dis, Armed Forces Epidemiol Bd, 58-66; physician-in-chief, Thomas Jefferson Univ Hosp, 59-75; mem bd trustees, Magee Mem Hosp, Philadelphia, 59-75 & Drexel Univ, 66-75; mem, Greater Philadelphia Comt Med-Pharmaceut Sci, 63-75, chmn, 70-74; mem adv comt, Inter-Soc Comt Heart Dis Resources, 67-70; mem bd-adv comt registry of tissue reaction, Univs Assoc Res & Educ Path, Inc, 70-75; mem exec comt, Int Cong Internal Med, 71; mem, Nat Brucellosis Tech Comn, 76-79. *Mem:* Am Fedn Clin Res; fel Am Col Physicians; Asn Am Physicians; Am Infectious Dis Soc; Am Pharmaceut Asn. *Res:* Infectious diseases; chemotherapy; antibiotics. *Mailing Add:* 5814 Williamsburg Landing Dr Williamsburg VA 23185

WISE, SHERWOOD WILLING, JR, b Jackson, Miss, May 31, 41; m 65. GEOLOGY, PALEONTOLOGY. *Educ:* Washington & Lee Univ, BS, 63; Univ Ill, MS, 65, PhD(geol), 70. *Prof Exp:* NSF fel, Swiss Fed Inst Technol, 70-71; asst prof, 71-75, assoc prof, 75-80, PROF GEOL, FLA STATE UNIV, 80- *Concurrent Pos:* NSF res grant, 72-; res grant, Petrol Res Fund, 73-81; sci ed, Initial Reports Deep Sea Drilling Proj, 77, 83; co-chief, Deep Sea Drilling Proj, 83, 88; mem, ad hoc comt Antarctic Geoscience, Nat Acad Sci,

83-86; pres, N Am Micropaleontol Sect, Soc Econ Paleontologists & Mineralogists, 86-87; mem, Southern Oceans Panel, Ocean Drilling Project, 87-; chair, Res Comt, SEPM, 88-90; mem, USSAC Comt & Info Handling Panel, Joint Oceanog Inst. *Mem:* Fel AAAS; Am Asn Petrol Geologists; Soc Econ Paleontologists & Mineralogists; Geol Soc Am; Swiss Geol Soc. *Res:* Skeletal ultrastructure; taxonomy and biostratigraphy of fossil calcareous nannoplankton, diatoms and silico flagellates; early diagenesis of carbonate and siliceous sediment; circum-Antarctic and Atlantic marine geology. *Mailing Add:* Dept Geol Fla State Univ Tallahassee FL 32306

WISE, WILLIAM CURTIS, b Louisville, Ky, Nov 24, 40; m 63; c 2. PHYSIOLOGY, COMPUTER SCIENCE. *Educ:* Transylvania Univ, AB, 63; Univ Ky, PhD(physiol & biophys), 67. *Prof Exp:* Physiologist, McDonnell-Douglas Corp, 67-68; From asst prof to assoc prof, 68-80, PROF PHYSIOL, MED UNIV SC, 80-, ASST TO ACAD VPRES, ACAD INFO MGT, 85-, CHIEF INFO OFF, MED CTR, 87- *Concurrent Pos:* Koebig Trust grant physiol, Med Univ SC, 72-73; Nat Cancer Inst res career develop award, 74-79. *Mem:* Sigma Xi; Am Physiol Soc; Soc Gen Physiologists; Biophys Soc; Shock Soc. *Res:* Endotoxic shock; septic shock; renal and acid-base physiology; computers in education. *Mailing Add:* Dept Physiol Med Univ SC 171 Ashley Ave Charleston SC 29425-3100

WISE, WILLIAM STEWART, b Carson City, Nev, Aug 18, 33; m 55; c 3. MINERALOGY, PETROLOGY. *Educ:* Stanford Univ, BS, 55, MS, 58; Johns Hopkins Univ, PhD(geol), 61. *Prof Exp:* Instr geol, Stanford Univ, 58 & Johns Hopkins Univ, 60-61; from asst prof to assoc prof, 61-73, assoc dean, Col Letters & Sci, 79-81, PROF GEOL, UNIV CALIF, SANTA BARBARA, 73-, DEAN, ACAD SKILLS, 81- *Concurrent Pos:* Consult, US Geol Surv, 65-67 & Argonne Nat Lab, 79- *Mem:* Fel Geol Soc Am; fel Mineral Soc Am; Mineral Asn Can; Mineral Soc Gt Brit; Sigma Xi. *Res:* Paragenesis of minerals, principally zeolites and associated minerals, barium silicates; petrology of oceanic volcanoes. *Mailing Add:* 1125 Young St No 910 Honolulu HI 96814

WISEMAN, BILLY RAY, b Sudan, Tex, Mar 28, 37; m 63; c 2. ENTOMOLOGY, HORTICULTURE. *Educ:* Tex Tech Col, BS, 59; Kans State Univ, MS, 61, PhD(entom), 67. *Prof Exp:* Res asst host plant resistance, Kans State Univ, 59-61 & 64-66; res entomologist, Southern Grains Invests, Okla, 66-67; res entomologist, Southern Grain Insects Lab, 67-84, INSECT BIOL POP MGR, RES LAB, AGR RES SERV, USDA, 84- *Concurrent Pos:* Mem grad fac, Univ Ga & Univ Fla; courtesy prof, grad courses in plant resistance to insects, Univ Fla. *Honors & Awards:* Bussart Mem Award, 90. *Mem:* Entom Soc Am; Sigma Xi. *Res:* Entomological research in host plant resistance of small grains, corn, sorghum and vegetable crops and the insects attacking these crops, including feeding stimulants, deterrents, food utilization, behavior, biology and mechanisms of resistance. *Mailing Add:* IBPMRL PO Box 748 Tifton GA 31794-0748

WISEMAN, CARL D, b Chicago, Ill, Oct 25, 25; m 49; c 3. PHYSICAL METALLURGY. *Educ:* Southern Methodist Univ, BS, 50; Univ Calif, MS, 55, PhD(metall), 57. *Prof Exp:* Res engr, Calif, 51-57; metallurgist, Gen Elec Co, 57-59 & Tex Instruments, Inc, 59-64; assoc prof 64-69, PROF ENG MECH, UNIV TEX, ARLINGTON, 69- *Mem:* Am Soc Metals; Am Inst Mining, Metall & Petrol Engrs; Sigma Xi. *Res:* Plastic deformation of metals; nuclear reactor metallurgy; metal failure analysis; metallurgy of semiconductors and thermoelectric materials; surfaces of solids to fields of research. *Mailing Add:* Dept Mech Eng Univ Tex Arlington TX 76019

WISEMAN, EDWARD H, b Portsmouth, Eng, Nov 14, 34; m 57; c 4. BIOCHEMISTRY, PHARMACOLOGY. *Educ:* Univ Birmingham, BSc, 56, PhD(org chem), 59. *Prof Exp:* Fel, Ohio State Univ, 60; res chemist, 61-64, supvr biochem pharmacol, 64-67, asst to res vpres, 67, mgr biochem pharmacol, 67-71, dir pharmacol, 71-76, EXEC DIR RES ADMIN, PFIZER INC, 76- *Mem:* Am Rheumatism Asn; Am Soc Pharmacol & Exp Therapeut. *Res:* Non-steroidal anti-inflammatory agents; biochemistry of metabolic diseases. *Mailing Add:* Pfizer Inc Groton CT 06340

WISEMAN, GEORGE EDWARD, b Brooklyn, NY, May 28, 18; m 45; c 3. INORGANIC CHEMISTRY, ORGANIC CHEMISTRY. *Educ:* St Peter's Col, BS, 40; Polytech Inst Brooklyn, PhD(chem), 56. *Prof Exp:* Assoc prof chem, St John's Univ, NY, 46-59; chmn dept chem, 59-66, assoc grad dean, Conolly, 66-71, prof, 59-86, EMER PROF CHEM, LONG ISLAND UNIV, 87- *Mem:* Am Chem Soc; fel Am Inst Chemists. *Res:* Preparation, properties and structures of organoselenium compounds; heterocyclic compounds; metallic derivatives of aromatic hydrocarbons. *Mailing Add:* 106 Conn Ave Massapequa NY 11758

WISEMAN, GORDON G, b Livingston, Wis, Feb 24, 17; m 42; c 2. PHYSICS. *Educ:* SDak State Univ, BS, 38; Univ Kans, MS, 41, AM, 47, PhD(physics), 50. *Prof Exp:* Instr physics, Culver-Stockton Col, 41-43; from instr to assoc prof, 43-64, PROF PHYSICS, UNIV KANS, 64- *Mem:* Am Phys Soc; Am Asn Physics Teachers. *Res:* Dielectrics; absorption microspectrophotometry; ferroelectricity. *Mailing Add:* Dept Physics Univ Kans Lawrence KS 66044

WISEMAN, GORDON MARCY, b Winnipeg, Man, Feb 24, 34; m 56. BACTERIOLOGY. *Educ:* Univ Man, BSc, 56, MSc, 61; Univ Edinburgh, PhD(bact), 63, DSc, 74. *Prof Exp:* Demonstr, 57-59, from asst prof to assoc prof, 65-75, PROF MED MICROBIOL, FAC MED, UNIV MAN, 75- *Concurrent Pos:* Med Res Coun fel bact, Fac Med, Univ Man, 64 & scholar, 65-70. *Mem:* Am Soc Microbiol; Can Soc Microbiol. *Res:* Neisseria gonorrhoeae; adhesion to host cells in relation to pathogenicity and virulence. *Mailing Add:* Dept Med Microbiol Univ Man Fac Med 753 McDermot Ave Winnipeg MB R3E 0W3 Can

WISEMAN, H(ARRY) A(LEXANDER) B(ENJAMIN), b Montreal, Que, Dec 27, 24; m 44; c 1. ENGINEERING MECHANICS. *Educ:* Univ Sask, BSc, 47; Wash State Univ, MS, 49; Pa State Univ, PhD(eng mech), 54. *Prof Exp:* Instr, Wash State Univ, 47-49; asst & res assoc, Pa State Univ, 49-54; res officer, Nat Res Coun Can, 54-55; assoc prof eng, Univ Del, 55-58; prof civil & mech eng, 58-76, prof biomed eng, Sch Med, 70-76, PROF MECH ENG, UNIV MIAMI, 76- *Mem:* Am Soc Eng Educ; Am Soc Mech Engrs; Nat Soc Prof Engrs; Am Soc Metals; Am Soc Artificial Internal Organs; Sigma Xi. *Res:* Biaxial and triaxial stress and strain relations; prestressed concrete; solid state studies of ultra high pressures; electron microscope metallography; high velocity stress-strain phenomena; photo stress and elasticity; biomedical engineering. *Mailing Add:* 508 Caligula Ave Coral Gables FL 33146

WISEMAN, JEFFREY STEWART, b Athens, Ohio, Dec 8, 48; m 74; c 2. ENZYME INHIBITION, LEUKOTRIENE BIOSYNTHESIS. *Educ:* Ohio Univ, BS, 70; Harvard Univ, PhD(chem), 74. *Prof Exp:* Teaching fel, Stanford Univ, 75-76 & Brandeis Univ, 76-78; sr biochemist, Merrell Dow Res Inst, 79-88; DEPT HEAD, BIOCHEM, GLAXO RES INST, 88- *Mem:* Am Soc Biol Chemists. *Res:* Design and characterization of enzyme inhibitors; mechanisms of action of enzymes of therapeutic interest. *Mailing Add:* Glaxo Res Labs Five Moore Dr Research Triangle Park NC 27709

WISEMAN, JOHN R, b Patriot, Ohio, May 4, 36; m 56; c 3. ORGANIC CHEMISTRY, SYNTHETIC ORGANIC & NATURAL PRODUCTS CHEMISTRY. *Educ:* Univ Colo, BS, 57; Stanford Univ, PhD(chem), 65. *Prof Exp:* NSF fel chem, Univ Calif, Berkeley, 64-65, lectr, 65-66, fel, 66; from asst prof to assoc prof, 66-76, PROF CHEM, UNIV MICH, 76- *Concurrent Pos:* Grants, Petrol Res Fund, 68-73; Res Corp grant, 68-69; Am Cancer Soc grant, 72-77, Nat Cancer Inst, 78-80. *Mem:* AAAS; Am Chem Soc; Royal Soc Chem. *Res:* Strain of bicyclic bridgehead alkenes; reaction of carbonium ions; synthesis of natural products. *Mailing Add:* Dept Chem Univ Mich Ann Arbor MI 48109

WISEMAN, LAWRENCE LINDEN, b Galion, Ohio, Apr 27, 44; c 2. DEVELOPMENTAL BIOLOGY. *Educ:* Hiram Col, AB, 66; Princeton Univ, MA, 69, PhD(biol), 70. *Prof Exp:* Nat Cancer Inst fel, Princeton Univ, 70-71; from asst prof to assoc prof biol, 71-77 CHMN DEPT, 81-, PROF BIOL, COL WILLIAM & MARY, 86- *Concurrent Pos:* Vis scientist, Human Leukemia Prog, Ont Cancer Inst, Toronto, 74-75; spec asst to pres, Univ Colo, 87-88; Am Coun Educ fel, 87-88. *Mem:* AAAS; Soc Develop Biol; Am Soc Zool; Am Soc Cell Biol; Int Soc Develop Biologists. *Res:* Cell adhesion, cell movement; vertebrate embryology. *Mailing Add:* Dept Biol Col William & Mary Williamsburg VA 23185

WISEMAN, PARK ALLEN, b Amsden, Ohio, Dec 29, 18; m 42; c 1. ORGANIC CHEMISTRY. *Educ:* DePauw Univ, AB, 40; Purdue Univ, MA, 42, PhD(org chem), 44. *Prof Exp:* Asst org & phys chem, Purdue Univ, 40-42, Monsanto Chem Co res fel org chem, 44-46; res chemist, Firestone Tire & Rubber Co, Ohio, 46-47; from asst prof to assoc prof, 47-56, head dept, 65-69, prof, 56-81, EMER PROF CHEM, BALL STATE UNIV, 81- *Concurrent Pos:* NSF vis prof, Tech Inst Northwestern Univ, 69-70. *Mem:* AAAS; Am Chem Soc. *Res:* High pressure oxidation of hydrocarbons; catalytic vapor-phase oxidation of hydrocarbons; fluorine chemistry; organic synthesis and natural products. *Mailing Add:* 4204 W Univ St Muncie IN 47304

WISEMAN, RALPH FRANKLIN, b Washington, DC, Sept 1, 21; m 51; c 2. MICROBIOLOGY. *Educ:* Univ Md, BS, 49; Univ Hawaii, MS, 53; Univ Wis, PhD(bact), 56. *Prof Exp:* Lab asst, Nat Cancer Inst, 39-42, med bacteriologist, Nat Inst Dent Res, 49-51; asst bact, Univ Hawaii, 51-53; asst, Univ Wis, 53-55, instr, 55-56; from instr to prof microbiol, 56-87; RETIRED. *Concurrent Pos:* Vis prof, Hacettepe Univ, Turkey, 68-69 & Rega Inst Med Res, Cath Univ, Louvain, 69. *Mem:* Am Soc Microbiol; Asn Gnotobiotics; fel Am Acad Microbiol. *Res:* Intestinal microbiology; germ free-like characteristics in antibiotic-treated animals; animal-microbial ecosystems; water quality. *Mailing Add:* Morgan Sch Biol Sci Univ Ky Lexington KY 40506

WISEMAN, ROBERT S, b Robinson, Ill, Feb 27, 24; m 47; c 1. ILLUMINATING ENGINEERING, ELECTRICAL ENGINEERING. *Educ:* Univ Ill, BSEE, 48, MSEE, 50, PhD(elec eng), 54. *Prof Exp:* Instr & asst prof elec eng, Miss State Col, 48-51; chief, Res Sect, US Army Eng Res & Develop Lab, 51-58, chief, Warfare Vision Br, 58-65; dir, Combat Surv, Night Vision & Target Acquisition Labs, US Army Electronics Command, 65-68, dep for labs, 68-71, dir, Res & Develop/Army Electronics Labs, 71-78, tech dir, 78-79; dep sci & technol, Darcom, 79-80; dir electronics lab, 81-83, dep dir res & develop, 83-84, DIR ELECTRO-OPTICS DEPT, MARTIN MARIETTA AEROSPACE-ORLANDO DIV, 84- *Concurrent Pos:* Mem, Nat Res Coun comt on vision, 54-78. *Mem:* Fel Inst Elec & Electronics Engrs; Am Inst Aeronaut & Astronaut; fel Illum Eng Soc. *Res:* Night vision; combat surveillance and target acquisition; electronics/signals warfare; atmospheric sciences; electronic technology and devices; illuminating engineering; electro-optics. *Mailing Add:* 8451 Bay Hill Blvd Orlando FL 32811

WISEMAN, WILLIAM H(OWARD), b Chillicothe, Ohio, May 11, 29; m 50; c 2. CHEMICAL ENGINEERING. *Educ:* Ohio State Univ, BSChE, 53; Lawrence Col, MS, 55, PhD, 58. *Prof Exp:* Asst tech dir, WVa Pulp & Paper Co, 58-59; asst paper mill supt, 59-61; prod mgr bd div, Brunswick Pulp & Paper Co, Ga, 61-65, asst gen prod mgr, 65-69, opers mgr, 69-71; plant mgr, 71-77, vpres & gen mgr bleached opers, Continental Can Co, 77-85. *Concurrent Pos:* Consult, 85- *Mem:* Tech Asn Pulp & Paper Indust. *Res:* Pulp and paper technology. *Mailing Add:* 2203 Terrace Rd Augusta GA 30904

WISEMAN, WILLIAM JOSEPH, JR, b Summit, NJ, June 16, 43; m 65; c 3. OCEANOGRAPHY. *Educ:* Johns Hopkins Univ, BES, 64, MS, 66, MA, 68, PhD(oceanog), 69. *Prof Exp:* Instr geol, Univ NH, 69-70, asst prof earth sci, 70-71; Chmn dept marine sci, 87-90, from asst prof to assoc prof, 71-80, PROF MARINE SCI, COASTAL STUDIES INST, LA STATE UNIV, BATON ROUGE, 80-, PROF OCEANOG & COASTAL SCI, 87-

Concurrent Pos: Chmn dept, La State Univ, 77-80 & 85-87. *Mem:* Am Geophys Union; Inst Elec & Electronics Engrs; Am Meteorol Soc; Oceanog Soc. *Res:* Estuarine and nearshore circulation. *Mailing Add:* Coastal Studies Inst La State Univ Baton Rouge LA 70803

WISER, CYRUS WYMER, b Wartrace, Tenn, Jan 14, 23; m 45; c 3. AQUATIC ECOLOGY, PHYSIOLOGY. *Educ:* Harding Col, BS, 45; George Peabody Col, MA, 46; Vanderbilt Univ, PhD(biol), 56. *Prof Exp:* Instr biol, David Lipscomb Col, 46-49; assoc prof, Jacksonville State Teachers Col, 49-51, 53-54; vis asst prof, Vanderbilt Univ, 54-55; assoc prof, 56-61, PROF BIOL, MID TENN STATE UNIV, 61- *Mem:* Sigma Xi. *Res:* Physiology, aquatic ecology and limnology; population studies of ponds and lakes; accumulation of radioactive isotopes by aquatic organisms. *Mailing Add:* 814 Minerva Dr Murfreesboro TN 37130

WISER, EDWARD H(EMPSTEAD), b Fatehgarh, India, Jan 21, 31; US citizen; m 57; c 2. AGRICULTURAL ENGINEERING. *Educ:* Iowa State Univ, BS, 53; NC State Univ, MS, 58, PhD(agr eng), 64. *Prof Exp:* From instr to assoc prof agr eng, 57-76, PROF BIOL & AGR ENG, NC STATE UNIV, 76- *Mem:* Am Geophys Union; Am Soc Agr Engrs; Am Soc Agron; Soil Conserv Soc Am; Sigma Xi. *Res:* Prediction of water yield from agricultural watersheds; computer simulation of precipitation and streamflow; statistical methods in hydrology. *Mailing Add:* 404 Dixie Trail Raleigh NC 27607

WISER, HORACE CLARE, b Lewiston, Utah, Jan 26, 33; m 53; c 4. MATHEMATICS. *Educ:* Univ Utah, BA, 53, PhD(math), 61; Univ Wash, BS, 54. *Prof Exp:* From asst prof to assoc prof, 61-74, PROF MATH, WASH STATE UNIV, 74- *Mem:* Am Math Soc; Math Asn Am. *Res:* Undergraduate mathematics curriculum; point set topology. *Mailing Add:* Dept Math Wash State Univ Pullman WA 99164-2930

WISER, JAMES ELDRED, b Wartrace, Tenn, Dec 31, 15; m 41; c 1. ANALYTICAL CHEMISTRY. *Educ:* Mid Tenn State Col, BS, 38; Peabody Col, MA, 40, PhD(sci educ), 47. *Prof Exp:* Teacher high sch, Fla, 38-39 & Ala State Teachers Col, 40-41; teacher physics, Vanderbilt Univ, 42; teacher chem & physics, David Lipscomb Col, 42-46; PROF CHEM & HEAD DEPT CHEM & PHYSICS, MID TENN STATE UNIV, 46- *Concurrent Pos:* Instr, Peabody Col, 44; NSF panelist, 64, 68 & 71. *Mem:* AAAS; emer mem Am Chem Soc; Am Inst Chem. *Res:* Food chemistry; educational psychology. *Mailing Add:* Rte One Box 276 Christiana TN 37037-9764

WISER, NATHAN, b Zurich, Switz, 1935; m 61; c 4. THEORETICAL PHYSICS, MATERIAL SCIENCE. *Educ:* Wayne State Univ, BS, 57; Univ Chicago, MS, 59, PhD(physics), 64. *Prof Exp:* Res assoc physics, Univ Ill, 64-65; sr res scientist, IBM Watson Res Ctr, 65-67; assoc prof, 67-74, PROF PHYSICS, BAR-ILAN UNIV, ISRAEL, 74- *Concurrent Pos:* Fel, NSF, 62-64; Res grant, Nat Bur Standards, 67-70, US-Israel Binational Sci Found, 71-81 & Israel Acad Sci, 76-81; vis prof, Univ Cambridge, Eng, 80-81, 86-87. *Mem:* Fel Am Phys Soc; Europ Phys Soc; Israel Phys Soc. *Res:* Theory of metals, including their electrical, thermal and structural properties; liquid metals; condensed matter physics; theory of phase transitions; dielectric theory; surfaces. *Mailing Add:* Dept Physics Bar-Ilan Univ Ramat-Gan 52100 Israel

WISER, THOMAS HENRY, b Minneapolis, Minn, May 17, 46. CLINICAL PHARMACY. *Educ:* Univ Minn, BS, 71, PharmD, 73. *Prof Exp:* ASST PROF CLIN PHARM, SCH PHARM, UNIV MD, 74-, ASSOC DIR PRIMARY CARE PROG, SCH MED, 75-; PHARMACIST, 71-; PROF PHARM PRACT, CAMPBELL UNIV SCH PHARM. *Concurrent Pos:* Clin pharm practitioner, Univ Md Hosp, 73-, clin pharm consult, 73-, co-dir, Anticoagulant clin, 76-, co-dir, Therapeut Probs Clin, 77-; clin pharm consult, Loch Raven Vet Admin Hosp, 74-, Baltimore City Jail, 77-, Critical Care Nurses' Asn, 78- & Am Pharmaceut Asn Policy Comt Prof Affairs, 78-; mem, Md Comn Nursing, 75-76. *Mem:* Am Pharmaceut Asn; Am Asn Col Pharm; Am Asn Hosp Pharmacists. *Res:* Ambulatory care; medical audits; clinical pharmacy services; drug utilization; adverse reactions; patient education. *Mailing Add:* 1528 Kilarney Dr Cary NC 27511

WISER, WENDELL H(ASLAM), b Fairview, Idaho, Dec 16, 22; m 47; c 5. FUEL ENGINEERING, PHYSICAL CHEMISTRY. *Educ:* Univ Utah, BS, 49, PhD(fuel eng), 52. *Prof Exp:* Res assoc explosives res, Univ Utah, 52-53; asst prof chem eng, Brigham Young Univ, 55-58; pres, Church Col, NZ, 60-65; assoc prof, 65-69, chmn dept, 66-70, PROF FUELS ENG, UNIV UTAH, 69- *Concurrent Pos:* Consult, Power Plant Div, Boeing Airplane Co, 56-58; US ed, Fuel, 70-78. *Honors & Awards:* Henry H Storch Award, Am Chem Soc, 78. *Mem:* Am Inst Aeronaut & Astronaut; fel Brit Interplanetary Soc; Am Chem Soc. *Res:* Jet engine fuels; rocket propellants; production of liquid and gaseous fuels from coal. *Mailing Add:* Dept Fuels Eng Univ Utah Salt Lake City UT 84112

WISER, WINFRED LAVERN, b Wartrace, Tenn, June 14, 26; m 74; c 2. OBSTETRICS & GYNECOLOGY. *Educ:* Middle Tenn Univ, BS, 49; Univ Tenn, MD, 52. *Prof Exp:* Intern, John Gaston Hosp, 53; resident obstet & gynec, Sch Med, Univ Miss, 62; asst prof, Med Ctr, Univ Miss, 67-68; assoc prof, Ctr Health Sci, Univ Tenn, 68-73, prof, 73-76; PROF & CHMN OBSTET & GYNEC, SCH MED, UNIV MISS, 76- *Concurrent Pos:* Dir gynec, Ctr Health Sci, Univ Tenn, 68-76, dep chmn obstet & gynec, 71-76, actg chmn, 74-76; Consult staff, Vet Admin Hosp & Methodist Rehab Ctr, 76- *Mem:* AMA; Am Soc Fertil & Steril; Am Col Surgeons; Am Col Obstetricians & Gynecologists; Sigma Xi. *Res:* Gynecology; congenital anomalies of the uterus; infertility. *Mailing Add:* Dept Obstet & Gynec Dept Med Ctr Univ Miss 2500 N State St Jackson MS 39216

WISHINSKY, HENRY, bio-organic chemistry; deceased, see previous edition for last biography

WISHMAN, MARVIN, b New York, NY, Apr 24, 25; m 53; c 1. POLYMER STABILIZATION, SYNTHETIC FIBER PROCESSES. *Educ:* NY Univ, University Heights, BA, 49; NY Univ, Washington Square, PhD(org chem), 54. *Prof Exp:* Res chemist, Indust Rayon Corp, 55-56; group leader & res mgr, Am Cyanamid, 56-68; br mgr, 68-70, RES DIR, PHILLIPS PETROL CORP, 70- *Mem:* Am Chem Soc. *Res:* Synthetic fibers; acrylic fibers; polypropylene fibers; fiber additives and fiber processes to achieve specific properties. *Mailing Add:* Four Whittington Ct Greenville SC 29615

WISHNER, KATHLEEN L, b Modesto, Calif, June 11, 43; m 73; c 2. ENDOCRINOLOGY, PEDIATRICS. *Educ:* San Francisco State Univ, BA, 63; Univ Calif, San Francisco, PhD(nutrit), 68; Univ Southern Calif, MD, 76. *Prof Exp:* Asst prof Div Nutrit, Univ Minn, St Paul, 68-70; asst prof, Dept Biomed Chem, Sch Pharm, 70-72, asst clin prof, Dept Med, 81-84, ASSOC CLIN PROF, UNIV CALIF, 84- *Concurrent Pos:* Prin-investr, NIH supported proj, 69-73 & 79-81, co-investr, 69-72; physician endocrinol, Pasadena Diabetes & Endocrinol Med Group, 81-; mem, Panel Space Sta Oper Med, Am Inst Biol Sci, 83-85; Space Sta Health Maintenance Fac Consult Comt, NASA, 84-85, mem Space Adv Panel, 85; mem, Sci Adv Panel for Simplesse, Nutra-Sweet Co, 87-; bd dirs, Am Diabetes Asn, 88- *Mem:* Fel Am Acad Pediat; Am Diabetes Asn; Am Dietetics Asn; Am Inst Nutrit; Am Soc Clin Nutritionists. *Res:* Lipo-protein metabolism; dietary factors affecting chylo micron metabolism and the effect of heparin disposition on lipo protein metabolism. *Mailing Add:* Pasadena Diabetes & Edocrinol Med Group 675 S Arroyo PKwy No 420 Pasadena CA 91105

WISHNER, LAWRENCE ARNDT, b New York, NY, Sept 7, 32; m 82; c 2. BIOCHEMISTRY, ANIMAL BEHAVIOR-ETHOLOGY. *Educ:* Univ Md, BS, 54, MS, 61, PhD(food chem), 64. *Prof Exp:* Asst, dairy dept, Univ Md, 57-61; from asst prof to assoc prof, 61-68, chmn dept, 67-71, asst dean, 71-77, PROF CHEM, MARY WASH COL, 68- *Mem:* Am Chem Soc; Am Oil Chem Soc; Am Inst Chem; NY Acad Sci; Sigma Xi. *Res:* Light-induced oxidation of milk; thermal oxidation of fats; autoxidation of tissue lipids in vivo; biological antioxidants; behavior and life history of the eastern chipmunk. *Mailing Add:* 1645 Heatherstone Dr Fredericksburg VA 22401-4845

WISHNETSKY, THEODORE, b New York, NY, July 5, 25; m 48; c 2. FOOD TECHNOLOGY, CRYOGENICS. *Educ:* Cornell Univ, BS, 49, MS, 50; Univ Mass, PhD(food technol), 58. *Prof Exp:* Res assoc, NY Agr Exp Sta, Geneva, 50-54; chemist, Eastman Chem Prod, Inc, 58-62; sr scientist, Air Prod & Chem, Inc, Pa, 62-68; ASSOC PROF FOOD SCI, MICH STATE UNIV, 68- *Mem:* Inst Food Technol. *Res:* Fruit and vegetable processing; processing and marketing of frozen foods; mechanism of changes in frozen foods; application of cryogenic technology to processing problems; low-temperature preservation of foods; controlled atmospheres; packaging. *Mailing Add:* Dept Food Sci & Human Nutrit Mich State Univ East Lansing MI 48824

WISHNIA, ARNOLD, b New York, NY, July 1, 31; m 52; c 3. BIOPHYSICAL CHEMISTRY, MOLECULAR BIOLOGY. *Educ:* Cornell Univ, AB, 52; NY Univ, PhD(biochem), 57. *Prof Exp:* Res assoc chem, Yale Univ, 56-59; from asst prof to assoc prof biochem, Dartmouth Med Sch, 59-66; ASSOC PROF CHEM, STATE UNIV NY, STONY BROOK, 66- *Concurrent Pos:* USPHS sr fel, Dept Natural Philos, Univ Edinburgh, 67. *Mem:* Am Soc Biol Chemists; Am Chem Soc; Biophys Soc. *Res:* Ribosome chemistry. *Mailing Add:* Dept Chem State Univ NY Stony Brook NY 11794

WISHNICK, MARCIA M, b New York, NY, Oct 10, 38; m 60; c 1. PEDIATRICS, GENETICS. *Educ:* Barnard Col, BA, 60; NY Univ, PhD(biochem), 70, MD, 74. *Prof Exp:* Chemist, Lederle Labs, Am Cyanamid Co, 60-66; assoc biochem, Pub Health Res Labs, City New York, 70-71; res assoc pharmacol, Sch Med, NY Univ Med Ctr, 71, asst prof, 77-82, clin assoc prof, 83-87, actg dir, div Human Genetics, 81-90, CLIN PROF PEDIAT, NY UNIV MED CTR, 87- *Concurrent Pos:* Resident pediat, NY Univ-Bellevue Med Ctr, 74-77, asst attend pediatrician, 77-82, assoc attend pediatrician, 83-87, attend pediatrician, 87- *Mem:* AAAS; Am Soc Human Genetics; Am Acad Pediat; Am Med Women's Asn. *Res:* Inborn errors in metabolism. *Mailing Add:* 157 E 81st St No 1A New York NY 10028

WISIAN-NEILSON, PATTY JOAN, b Cuero, Tex, Aug 22, 49; m 76; c 2. CHEMISTRY. *Educ:* Tex Lutheran Col, BS, 71; Univ Tex Austin, PhD(inorg chem), 76. *Prof Exp:* Res assoc, Duke Univ, 76-78; res assoc, Univ Tex, Arlington, 78; res assoc, 79-80, res scientist, Tex Christian Univ, 80-84; asst prof, 84-90, ASSOC PROF, SOUTHERN METHODIST UNIV, 90- *Concurrent Pos:* Prin investr, Army Res Off grant, Tex Christian Univ, 80-84, Southern Methodist Univ, 84-; vis asst prof, Mass Inst Technol, 84-85; NSF-VPW grant, 84-85; prin investr, Am Chem Soc-Petro Res Found, 88- *Mem:* Am Chem Soc; Sigma Xi. *Res:* Synthesis and characterization of new polymers derived from poly(alky-arylphosphazenes); synthesis of new diiron complexes with low-coordinate phosphorus ligands. *Mailing Add:* Dept Chem Southern Methodist-Univ Dallas TX 75275

WISLOCKI, PETER G, b Derby, Conn, Jan 21, 47; m 72; c 2. CHEMICAL CARCINOGENESIS, TOXICOLOGY. *Educ:* Fairfield Univ, BS, 68; Univ Wis, PhD(exp oncol), 74. *Prof Exp:* Fel, Hoffmann-La Rouche, Inc, 74-76, vis scientist, 76-77; asst prof, Eppley Inst Res Cancer, 77-78; res fel, 78-83, sr res fel, 83-87, DIR, MERCK, SHARP & DOHME RES LAB, 88- *Mem:* Am Asn Cancer Res; Am Chem Soc; Am Soc Pharmacol & Exp Therapeut. *Res:* Metabolic activation of carcinogens; binding of carcinogens and reactive intermediates to DNA and protein; toxicological significance of protein bound adducts; metabolism, environmental fate and ecological effects of pesticides. *Mailing Add:* Merck & Co Three Bridges NJ 08887

WISMAN, EVERETT LEE, b Woodstock, Va, Oct 1, 22; m 48; c 3. POULTRY SCIENCE. *Educ:* Va Polytech Inst & State Univ, BS, 46; Cornell Univ, MS, 49; Pa State Univ, PhD(biochem, poultry husb), 52. *Prof Exp:* County agr agent, 47-48, PROF POULTRY SCI, VA POLYTECH INST &

STATE UNIV, 52- *Concurrent Pos:* Bd trustees, Sci Mus of Va, 77-80. *Mem:* Am Poultry Sci Asn; Am Inst Nutrit; Asn Acad Sci (pres, 79); AAAS. *Res:* Role of antibiotics, arsenicals and other feed additives in chick growth stimulation; evaluation of animal by-products in poultry rations. *Mailing Add:* Dept Poultry Sci Va Polytech Inst & State Univ Blacksburg VA 24061-0332

WISMAR, BETH LOUISE, b Cleveland, Ohio, Feb 18, 29. ANATOMY, MEDICAL EDUCATION. *Educ:* Western Reserve Univ, BSc, 51, MSc, 57; Ohio State Univ, PhD(anat), 61. *Prof Exp:* Instr embryol & histol, Col Med, 61-63, asst prof anat, Col Med & Col Arts & Sci, 63-69, ASSOC PROF ANAT, COL MED & COL ARTS & SCI, OHIO STATE UNIV, 69- *Concurrent Pos:* Fac, Creative Prob Solving Inst; consult prob solving & decision making; assoc ed, J Creative Behav. *Mem:* Am Asn Anatomists; Sigma Xi. *Res:* Pulmonary morphology in adult respiratory distress syndrome and sepsis with emphasis on the role of intravascular macrophages; relationship of cognitive style to learning and problem solving abilities of medical students. *Mailing Add:* 4063 Fairfax Dr Columbus OH 43220

WISMER, MARCO, b Switz, Dec 26, 21; nat US; m 52; c 2. POLYMER CHEMISTRY. *Educ:* Swiss Fed Inst Technol, PhD, 48. *Prof Exp:* Res chemist, Amercoat Corp, 49-51; tech mgr, Plastics Div, Ciba Prod Corp, 51-56; res assoc, Springdale Res Ctr, PPG Indust, Inc, 56-62, scientist, 62-64, dir, Advan Res Dept, 64-74, vpres res & develop, 74-84, vpres sci & technol, Res Ctr, 84-87; RETIRED. *Honors & Awards:* IR-100, Indust Res, 66 & 74. *Mem:* Am Chem Soc; Am Soc Test & Mat; Fedn Soc Paint Technol; Am Inst Chem Engrs. *Res:* Synthesis of epoxy resins; polyester resins; polyester polyols; chlorinated compounds; urethane technology; epoxidation technology; synthetic organic chemistry; polyolefin chemistry; unsaturated polyesters; radiation technology and electrodeposition. *Mailing Add:* 12621 Acacia Terr Poway CA 92064-3235

WISMER, ROBERT KINGSLEY, b Atlantic City, NJ, June 18, 45; m 70; c 3. X-RAY CRYSTALLOGRAPHY, CHEMICAL EDUCATION. *Educ:* Haverford Col, BS, 67; Iowa State Univ, PhD(phys chem), 72. *Prof Exp:* Instr chem, Iowa State Univ, 72; systs analyst comput sci, Ames Lab, Energy Res & Develop Admin, 73; asst prof chem, Luther Col, Iowa, 73-74; asst prof chem, Denison Univ, 74-76; from asst prof to assoc prof, 76-85, chmn dept, 83-87, PROF CHEM, MILLERSVILLE UNIV, 85- *Mem:* Am Chem Soc; AAAS; Am Crystallog Asn; Sigma Xi. *Res:* Solution of the phase problem through deconvolution of the Patterson function, especially development of techniques adaptable to small computers; general chemistry and qualitative analysis author. *Mailing Add:* Dept Chem Millersville Univ Millersville PA 17551

WISNER, ROBERT JOEL, b Hannibal, Mo, Jan 18, 25; m 47; c 4. ALGEBRA. *Educ:* Univ Ill, BS, 48, MS, 49; Univ Wash, PhD(math), 53. *Prof Exp:* Assoc math, Univ Wash, 51-53, res mathematician, Pub Opinion Lab, 52-53; instr math, Univ BC, 53-54; from asst prof to assoc prof, Haverford Col, 54-60; assoc prof, Mich State Univ, Oakland, 60-63; assoc prof, 63-70, head dept math, 70-77, PROF MATH SCI, NMEX STATE UNIV, 70- *Concurrent Pos:* Consult, Burroughs Corp, 57-58; NSF fel & mem, Inst Advan Study, 59-60; ed, Rev, Soc Indust & Appl Math, 59- *Mem:* Am Math Soc; Soc Indust & Appl Math; Math Asn Am; Am Statist Asn; Can Math Cong. *Res:* Rings; Abelian groups; number theory. *Mailing Add:* Dept Math Sci NMex State Univ Las Cruces NM 88003

WISNIESKI, BERNADINE JOANN, b Baltimore, Md, Feb 26, 45; m 80. PHYSICAL BIOCHEMISTRY. *Educ:* Univ Md, College Park, BS, 67; Univ Calif, Berkeley, PhD(genetics), 71. *Prof Exp:* Damon Runyon Mem Fund fel, Univ Calif, Los Angeles, 71-73, Celeste Durand Rogers Mem Found fel, 73-74, actg asst prof, 74-75, from asst prof to prof microbiol, 75-90, PROF MICROBIOL & MOLECULAR GENETICS, UNIV CALIF, LOS ANGELES, 90- *Concurrent Pos:* Mem, Jonsson Comprehensive Cancer Ctr, Univ Calif, Los Angeles, 75-, assoc mem, Molecular Biol Inst, 75-; res career develop award, USPHS, 76-81. *Mem:* AAAS; Sigma Xi; Biophys Soc. *Res:* Function and physical structure of animal cell membranes; membrane alterations; photoreactive probes, spin labels and protein insertion into membranes; membrane fusion induced by viral proteins; protein toxins; immunology; tumor necrosis factor. *Mailing Add:* Dept Microbiol & Molecular Genetics Univ Calif Los Angeles CA 90024

WISNIEWSKI, HENRYK MIROSLAW, b Luszkowko, Poland, Feb 27, 31; US citizen; m 54; c 2. NEUROPATHOLOGY, PATHOLOGY. *Educ:* Med Acad, Gdansk, physician dipl, 55; Med Acad, Warsaw, Dr Med, 60, docent, 65. *Prof Exp:* Resident res fel, Med Acad, Gdansk, Poland, 55-58; from asst to assoc prof neuropath, head of lab & assoc dir, Inst Neuropath, Polish Acad Sci, Warsaw, 58-66; from res assoc to asst prof path, Albert Einstein Col Med, Yeshiva Univ, 66-69, from assoc prof to prof neuropath, 69-76; PROF PATH, STATE UNIV NY DOWNSTATE MED CTR, 76-; DIR NY STATE INST BASIC RES, STATEN ISLAND, 76- *Concurrent Pos:* Health Res Coun New York career scientist award, 70-72; Nat Multiple Sclerosis Soc fel, 71-74; NIH fel, 72-77; vis neuropathologist, Univ Toronto, 61-62; vis scientist, lab of Neuropath, Nat Inst Neurol Dis & Blindness, 62-63; consult, Merck Labs, Rahway, NJ, 72-74; dir Demyelinating Dis Unit, Med Res Coun, Newcastle-upon-Tyne, Eng, 74-76. *Honors & Awards:* Weil Award, Am Asn Neuropathologists, 69, Moore Award, 72. *Mem:* Am Asn Neuropath (pres, 83); Polish Asn Neuropath; fel AAAS; Soc Exp Neuropath; World Fedn Neurol. *Res:* Light and ultrastructural studies of the pathological brain; experimental neuropathology; synaptic and axonal pathology; developmental neurobiology; mental retardation; pre and senile dementia; multiple sclerosis and other human and experimental demyelinating diseases; neuronal fibrous protein pathology. *Mailing Add:* NY State Inst Basic Res 1050 Forest Hill Rd Staten Island NY 10314

WISOTZKY, JOEL, b Chicago, Ill, Feb 17, 23; m 49; c 3. DENTISTRY, DENTAL RESEARCH. *Educ:* Cent YMCA Col, BS, 45; Loyola Univ, DDS, 47; Univ Rochester, PhD(exp path), 56. *Prof Exp:* Pvt pract, 48-49; sr asst dent surgeon, Fed Correctional Inst, Tex, 49-51; fel, Univ Rochester, 51-56; res assoc exp path, Dent Med Div, Colgate-Palmolive Co, 56-59; hon assoc res specialist, Bur Biol Res, Rutgers Univ, 57-59; assoc prof res dent med, Sch Dent, Case Western Reserve Univ, 59-63, prof med & dir dent res, 63-72, prof oral biol, 63-85, dir grad training & res, 72-78, emer prof, Sch Dent, 86- *Concurrent Pos:* USPHS res fel, 52-56, career develop award, 59-65. *Res:* Cariology; aging changes in oral tissues; phosphorescence of oral structures; electro-physiology; theoretical oral biology; myofibroblast function studies. *Mailing Add:* 301 Andover Pl S H-173 Sun City Center FL 33573

WISSBRUN, KURT FALKE, b Brackwede, Ger, Mar 19, 30; nat US. POLYMER SCIENCE, RHEOLOGY. *Educ:* Univ Pa, BS, 52; Yale Univ, MS, 53, PhD(phys chem), 56. *Prof Exp:* Dreyfus fel, Univ Rochester, 55-57; res chemist, Celanese Res Co, 57-60, group leader, 60-62, res assoc, 62-72, sr res assoc, 72-90; CONSULT, 90- *Concurrent Pos:* Adj prof chem eng, Univ Del, 74- *Mem:* Am Chem Soc; Soc Rheology; Sigma Xi. *Mailing Add:* One Euclid Ave Apt 4E Summit NJ 07901

WISSEMAN, CHARLES LOUIS, JR, b Seguin, Tex, Oct 2, 20; m 41; c 4. MEDICAL MICROBIOLOGY. *Educ:* Southern Methodist Univ, BA, 41; Kans State Col, MS, 43; Southwestern Univ, MD, 46; Am Bd Path, dipl; Am Bd Microbiol, dipl. *Prof Exp:* Chief chemotherapeut res sect, Dept Virus & Rickettsial Dis, Army Med Serv Grad Sch, Walter Reed Army Med Ctr, DC, 48-54, asst chief dept, 52-54; asst prof med, 57-74, PROF MICROBIOL & HEAD DEPT, SCH MED, UNIV MD, BALTIMORE CITY, 54- *Concurrent Pos:* Instr med, Sch Med, Georgetown Univ & actg dir bact & serol labs, Univ Hosp, 50-54; dep dir comn rickettsial dis, Armed Forces Epidemiol Bd, 57-59, dir, 59-72; consult, Surgeon Gen, US Army, NIH, WHO & Pan-Am Health Orgn. *Mem:* Am Soc Microbiol; Infectious Dis Soc Am; Am Soc Trop Med & Hyg; Am Soc Clin Invest; Am Asn Immunol. *Res:* Infectious diseases; viral and rickettsial diseases; pathogenesis and immunity. *Mailing Add:* Dept Microbiol & Immunol Univ Md Sch Med 660 W Redwood St Baltimore MD 21201

WISSEMAN, WILLIAM ROWLAND, b Halletsville, Tex, Nov 2, 32; m 59; c 3. PHYSICS. *Educ:* NC State Univ, BNuclearEng, 54; Duke Univ, PhD(physics), 59. *Prof Exp:* Res assoc & instr physics, Duke Univ, 59-60; mem tech staff, 60-75, br mgr, 75-84, ASSOC LAB DIR, TEX INSTRUMENTS, INC, 84- *Mem:* Am Phys Soc; fel Inst Elect & Electronics Engrs. *Res:* Electromagnetic wave propagation in solids; properties of semiconductors; solid state microwave sources. *Mailing Add:* 5747 Melshire Dr Dallas TX 75230

WISSIG, STEVEN, ELECTRON MICROSCOPY, MICROVASCULAR PERMEABILITY. *Educ:* Yale Univ, PhD(anat), 56. *Prof Exp:* PROF MICROS ANAT, UNIV CALIF, 58- *Mailing Add:* Dept Anat Univ Calif 1334-S Sch Med San Francisco CA 94143-0452

WISSING, THOMAS EDWARD, b Milwaukee, Wis, Aug 15, 40; m 69. FRESH WATER ECOLOGY. *Educ:* Marquette Univ, BS, 62, MS, 64; Univ Wis-Madison, PhD(zool), 69. *Prof Exp:* Asst aquatic ecol, Marquette Univ, 62-63, asst gen biol, 63-64; asst gen zool, Univ Wis, 64-65, Fed Water Pollution Control Admin trainee aquatic ecol, 65-69; from asst prof to assoc prof, 69-78, PROF ZOOL, MIAMI UNIV, 78- *Concurrent Pos:* Res assoc, Tex A&M Univ, 70; sci ed, Fisheries, Bull Am Fisheries Soc, 84-; ed, Ohio J Sci, 85-88; co-ed, Trans Am Fish Soc, 89- *Mem:* Am Fisheries Soc; Sigma Xi; Am Inst Biol Sci. Int Asn Theoret & Appl Limnol; Am Soc Ichthyol & Herpet; NAm Benthonic Soc. *Res:* Fisheries biology; bioenergetics. *Mailing Add:* Dept Zool Miami Univ Oxford OH 45056

WISSLER, EUGENE H(ARLEY), b Cherokee, Iowa, Dec 18, 27; m 51; c 3. BIOMEDICAL ENGINEERING & SIMULATION. *Educ:* Iowa State Univ, BS, 50; Univ Minn, PhD(chem eng), 55. *Prof Exp:* From asst prof to prof chem eng, 57-67, chmn dept, 69-70, assoc dean col eng, 70-76, PROF CHEM ENG, UNIV TEX, AUSTIN, 67-, ASSOC DEAN GRAD STUDIES, 82- *Concurrent Pos:* NSF fac fel, Univ Mich, 61-62; consult. *Mem:* Am Inst Chem Engrs; Undersea Med Soc; Am Soc Mech Engrs; Aerospace Med Asn. *Res:* Heat and mass transfer in the human; development of computer models for prediction of human responses to exercise and environmental stress. *Mailing Add:* Dept Chem Eng Univ Tex Austin TX 78712

WISSLER, ROBERT WILLIAM, b Richmond, Ind, Mar 1, 17; m 40; c 4. PATHOLOGY. *Educ:* Earlham Col, AB, 39; Univ Chicago, MS, 43, PhD(path), 46, MD, 48; Am Bd Path, dipl, 51. *Hon Degrees:* DSc, Earlham Col, 59; MD, Univ Heidelberg, 73; Siena, 82; DSc, NJ, 82, Ohio State, 90. *Prof Exp:* Asst chem, Earlham Col, 38-39; asst path, 41-43, from instr to prof, 43-72, chmn dept, 57-72, Donald N Pritzker, 72-77, DISTINGUISHED SERV PROF PATH, SCH MED, UNIV CHICAGO, 77- *Concurrent Pos:* Intern, Chicago Marine Hosp, 49-50; mem path study sect, USPHS, 57-61, consult, Surgeon Gen Path Training Comt, 63-68; mem comt path, Nat Acad Sci-Nat Res Coun, 58-69, chmn, 62-69; consult, Armed Forces Inst Path, 61-72, chmn sci adv comt, 66-67; secy-treas, Am Asn Chmn Med Sch Dept Path, 63-64, pres, 67-68; chmn coun arteriosclerosis, Am Heart Asn, 65-66; vpres-dir, Univs Assoc Res & Educ Path, Inc, 65, pres, 69-71; chmn ad hoc comt animal models, Artificial Heart-Myocardial Infarction Prog, Nat Heart Inst & mem Vet Admin Eval & Rev Comt, Res in Path & Lab Med, 66; vchmn bd trustees, Am Asn Accreditation on Lab Animal Care, 67, chmn, 72-74; trustee, Am Bd Path, 68-, secy, 74; mem path adv coun, Vet Admin, 70-74; mem adv comt, Life Sci Res Off, 71-; mem nat adv food comt, Food & Drug Admin, 72-74; pres, Am Bd Pathol, 79-80. *Honors & Awards:* H P Smith Award, Am Soc Clin Path, 76; Joseph B Goldberger Award, Am Med Asn, 79; Gold Headed Cane Award, Am Asn Pathol, 83; Distinguished Achievement Award, Soc Cardiovasc Pathol, 87. *Mem:* Soc Exp Biol & Med; Am Soc Exp Path (vpres, 60-61, pres, 61-62); AMA; Am Asn Path & Bact (vpres, 67, pres, 68-69); Am Asn Cancer Res. *Res:* Protein, lipid nutrition and metabolism; cardiovascular disease; experimental induction and regression of atherosclerosis; cellular immunological reactions including tumor immunity; immunohistochemistry of atherosclerosis; lipoprotein arterial wall cell interaction. *Mailing Add:* Dept Path BH P317 Univ Chicago Chicago IL 60637

WISSNER, ALLAN, b New York, NY, Nov 14, 45; m 78; c 3. ORGANIC CHEMISTRY. *Educ:* Long Island Univ, BS, 67; Univ Pa, PhD(org chem), 71. *Prof Exp:* NIH fel chem, Cornell Univ, 72-74; RES CHEMIST, LEDERLE LABS, AM CYANAMID CO, 74- *Mem:* Am Chem Soc. *Res:* Medicinal chemistry. *Mailing Add:* Metab Dis Ther Res Sect Lederle Labs Pearl River NY 10965

WISSOW, LENNARD JAY, b Philadelphia, Pa, May 23, 21; m 46; c 2. CHEMISTRY. *Educ:* Pa State Col, BS, 42; Duke Univ, AM, 43, PhD(org chem), 45. *Prof Exp:* Asst instr chem & asst org chem res, Duke Univ, 43-45; res org chemist, Publicker, Inc, 45; res & develop chemist, Nat Foam Syst, Inc, Pa, 46-47; sr res & develop chemist, Merck & Co, Inc, 47-51; head develop res, Otto B May, Inc, 51-58; treas, 59-60, CHIEF CHEMIST, J & H BERGE, INC, 58-, PRES, 60- *Mem:* Am Chem Soc; Am Inst Chem; AAAS; Sigma Xi; NY Acad Sci. *Res:* Synthetic organic chemistry; fine organic chemicals and processes; pharmaceuticals; vat dyestuffs and intermediates; research and sales administration. *Mailing Add:* 226 W Rittenhouse Sq Dorchester Philadelphia PA 19103-5768

WIST, ABUND OTTOKAR, b Vienna, Austria, May 23, 26; US citizen; m 63; c 2. ELECTRONICS ENGINEERING, CHEMICAL CATALYSTS. *Educ:* Graz Tech Univ, BS, 48; Univ Vienna, MS, 50, PhD(thermodyn), 51. *Prof Exp:* Technician physics, Vienna Tech Univ, 51-52; res & develop engr, Radiowerke Wien, Austria, 52-54 & Siemens & Halske AG, WGer, 54-58; dir res & develop, Brinkmann Instruments & sr scientist, Fisher Sci, Inc, 64-69; res assoc, Grad Sch Pub Health, Univ Pittsburgh, 70-72 & Dept Chem, 72-73; asst prof comput sci & biophys, 73-83, ASST PROF RADIOL, VA COMMONWEALTH UNIV, 83- *Concurrent Pos:* Adj prof chem, Va Commonwealth Univ, 76-; US deleg in biomed eng to China, 87. *Mem:* Sr mem Inst Elec & Electronics Engrs; Am Chem Soc; NY Acad Sci; Am Asn Phsicists Med. *Res:* Physical and analytical chemistry; solid state devices; new computer systems; reaction kinetics; catalysis; precision and automatic instrumentation in chemistry, physics, medicine and radiology; published book "Electronic Design of Microprossor Based Instrumentation; 59 scientific publications; 10 patents. *Mailing Add:* Dept Radiol Va Commonwealth Univ Richmond VA 23298-0072

WISTENDAHL, WARREN ARTHUR, plant ecology, for more information see previous edition

WISTREICH, GEORGE A, b New York, NY, Aug 12, 32; m 57; c 2. MICROBIOLOGY, ELECTRON MICROSCOPY. *Educ:* Univ Calif, Los Angeles, AB, 57, MS, 61; Univ Southern Calif, PhD(bact), 68. *Prof Exp:* Res asst zool, Univ Calif, Los Angeles, 58-60, res virologist, 60-61; from instr to asst prof biol, 61-71, dir Allied Health Sci Progs, 68-, chmn dept, 72-, ASSOC PROF LIFE SCI, E LOS ANGELES COL, 71-, PROF MICROBIOL, 73- *Concurrent Pos:* Aerospace consult, Garrett Corp, Calif, 66-67; lectr, Upward Bound Prog, East Los Angeles Col, 68- *Mem:* Am Soc Microbiol; Am Inst Biol Sci; NY Acad Sci; fel Am Inst Chem; fel Royal Soc Health; fel Am Acad Microbiol; fel Linnean Soc London. *Res:* Insect pathology; virology and tissue culture; cytology and cytochemistry; undergraduate education in biological sciences; electron microscopy. *Mailing Add:* Dept Life Sci East Los Angeles Col Monterey Park CA 91754

WISTREICH, HUGO ERYK, b Jasto, Poland, Aug 8, 30; nat US; m 58; c 3. BUSINESS MANAGEMENT. *Educ:* Inst Agr Tech, France, Ingenieur, 53; Rutgers Univ, MS, 57, PhD(food tech), 59. *Prof Exp:* Dir res, Reliable Packing Co, 58-60, Preservaline Mfg Co, 60-63 & Dubuque Packing Co, 63-64; vpres technol, 64-75, PRES, B HELLER & CO, 75- *Mem:* AAAS; Am Chem Soc; Inst Food Technol; Am Meat Sci Asn. *Res:* Meat curing; electrical anesthesia in animals; food analysis; nutrition; food-meat biochemistry; bacteriology. *Mailing Add:* 10127 S Seeley Ave Chicago IL 60643

WISWALL, RICHARD H, JR, b Peabody, Mass, Mar 7, 16; m 46; c 5. PHYSICAL CHEMISTRY. *Educ:* Harvard Univ, AB, 37; Princeton Univ, PhD(chem), 41. *Prof Exp:* Chemist, Am Cyanamid Co, NJ, 40-43 & Union Carbide & Carbon Chem Corp, 46-49; chemist, Brookhaven Nat Lab, 49-79; RETIRED. *Concurrent Pos:* Consult, 79- *Mem:* Am Chem Soc; fel Am Nuclear Soc. *Res:* Chemistry of nuclear energy production; fluorine chemistry; fused salts; metal hydrides; energy storage. *Mailing Add:* 331 Beaver Dam Rd Brookhaven NY 11619-9673

WIT, ANDREW LEWIS, b Oceanside, NY, Jan 18, 42; m 65; c 3. CARDIOVASCULAR PHYSIOLOGY, PHARMACOLOGY. *Educ:* Bates Col, BS, 63; Columbia Univ, PhD(pharmacol), 68. *Prof Exp:* Res physiologist, USPHS Hosp, Staten Island, NY, 68-70; assoc, 70-71, from asst prof to assoc prof, 71-81, PROF PHARMACOL, COL PHYSICIANS & SURGEONS, COLUMBIA UNIV, 81- *Concurrent Pos:* Res assoc, Rockefeller Univ, 70-71, vis asst prof, 71-74, adj assoc prof, 74-; NIH grants, 70-; NY Heart Asn sr investr, Columbia Univ, 71-75, Am Heart Asn grant in aid, 72-76. *Mem:* Am Heart Asn; Am Fedn Clin Res; Soc Gen Physiol; Am Physiol Soc; Int Soc Res Cardiac Metab. *Res:* Cardiac electrophysiology, pharmacology and arrhythmias. *Mailing Add:* Col Physicians & Surgeons Columbia Univ Dept Pharmacol 630 W 168th St New York NY 10032

WIT, LAWRENCE CARL, b Chicago, Ill, May 12, 44; m 68; c 3. PHYSIOLOGICAL ECOLOGY. *Educ:* Wheaton Col, BS, 66; Western Ill Univ, MS, 68; Univ Mo, PhD(zool), 75. *Prof Exp:* Asst prof,76-82, ASSOC PROF ZOOL, AUBURN UNIV, 82- *Mem:* AAAS. *Res:* Physiological mechanisms regulating mammalian and reptilian hibernation. *Mailing Add:* 101 Cary Hall Auburn Univ Auburn AL 36849

WITCHER, WESLEY, b Chatham, Va, July 9, 23; m 55; c 2. PLANT PATHOLOGY. *Educ:* Va Polytech Inst, BS, 49, MS, 58; NC State Col, PhD(plant path), 60. *Prof Exp:* Instr voc agr, Pittsylvania County Sch Bd, 49-54; asst county agent, Exten Serv, Va Polytech Inst, 54-56; asst, NC State Col, 57-60; prof forest path, 60-78, veg pathologist, 81-88, EMER PROF FOREST PATH, CLEMSON UNIV, 88- *Mem:* Am Phytopath Soc; Soc Nematol. *Res:* Fungus-nematode complex of tobacco; diseases of highbush blueberries; forest diseases; vegetable diseases. *Mailing Add:* Dept Plant Path & Physiol Clemson Univ Clemson SC 29634-0377

WITCOFSKI, RICHARD LOU, b Peiping, China, Mar 29, 35; US citizen; m 56; c 2. MEDICAL BIOPHYSICS, NUCLEAR MEDICINE. *Educ:* Lynchburg Col, BS, 56; Vanderbilt Univ, MS, 60; Wake Forest Univ, PhD(anat), 67. *Prof Exp:* Res asst, 57-61, from instr to prof, 61-73, PROF RADIOL, BOWMAN GRAY SCH MED, 73- *Mem:* Soc Nuclear Med; Health Physics Soc; Radiation Res Soc; Am Asn Physicists Med. *Res:* Low dose levels-radiation biology; clinical applications of magnetic resonance imaging. *Mailing Add:* Dept Radiol Bowman Gray Sch Med Winston-Salem NC 27103-2796

WITCZAK, ZBIGNIEW J, b Zgierz, Poland, July 13, 47; m 74; c 2. CARBOHYDRATE SWEETENERS CHEMISTRY. *Educ:* Med Acad, Lodz, Poland, MS, 72, PhD(org chem & natural prod), 79. *Prof Exp:* Res asst org chem, dept org chem, Med Acad Lodz, 73-75, asst lectr, 75-79, asst prof, 79-81; res assoc carbohydrate chem, dept biochem, 81-83 & dept food sci, 83-86, ASST PROF CARBOHYDRATE CHEM, WHISTLER CTR CARBOHYDRATE RES, FOOD SCI DEPT, PURDUE UNIV, 86- *Concurrent Pos:* NIH res participation fel, 81-82. *Mem:* Am Chem Soc; Sigma Xi. *Res:* Chemistry of carbohydrates, particularly thio and seleno-sugars; synthesis of new carbohydrate sweeteners; synthetic carbohydrate chemistry. *Mailing Add:* 214 Point Bluff Dr Decatur IL 62521-5508

WITELSON, SANDRA FREEDMAN, b Montreal, Que, Feb 24, 40. NEUROPSYCHOLOGY, NEUROANATOMY. *Educ:* McGill Univ, BSc, 60, MSc, 62, PhD(psychol), 66. *Prof Exp:* Lectr psychol, Yeshiva Univ, 66; NIMH res fel, Sch Med, NY Univ, 66-68; instr, NY Med Col, 68-69; from asst prof to assoc prof psychol, 69-77, PROF, DEPT PSYCHIAT, SCH MED, MCMASTER UNIV, 77- *Concurrent Pos:* Ont Ment Health Found res grant, McMaster Univ, 70-83, assoc mem, Dept Psychol, 76- & Dept Biomed Sci, 79; US NIH, Nat Inst Neurol Dis & Stroke contract & grants, 77- *Honors & Awards:* Morton Prince Award, Am Psychopath Asn; John Dewan Award, Ont Ment Health Found, 77; Clarke Inst Res Fund Prize, Univ Toronto, 78. *Mem:* Fel AAAS; fel Can Psychol Asn; fel Am Psychol Asn; Acad Aphasia; Int Neuropsychol Soc. *Res:* Perception; cognition; language; brain function; developmental psychology; neuroanatomy; sex differences; dyslexia; hemispheric specialization. *Mailing Add:* Dept Psychiat McMaster Univ Hamilton ON L8N 3Z5 Can

WITHAM, ABNER CALHOUN, cardiology, for more information see previous edition

WITHAM, CLYDE LESTER, b Los Angeles, Calif, Jan 15, 48; m 71; c 3. POWDER & PARTICLE TECHNOLOGY, AEROSOL SCIENCE. *Educ:* Brigham Young Univ, BS, 73; Stanford Univ, MS, 77. *Prof Exp:* Chem engr, 73-81, sr chem engr, 81-82, PROG MGR, FINE PARTICLE TECHNOL, SRI INT, 82- *Mem:* Am Inst Chem Engrs; Am Chem Soc; Am Asn Aerosol Res; Soc Plastics Engrs. *Res:* Applied research in fine particle and aerosol technology including product development, process development and air pollution; atomizer development; military smoke; protective equipment and aerosol detection; powder handling; particle size determination; fillers in polymers; pharmaceutical development (especially aerosol drug delivery); air pollution control; industrial hygiene. *Mailing Add:* SRI Int 333 Ravenswood Ave Menlo Park CA 94025

WITHAM, FRANCIS H, b Waltham, Mass, Apr 26, 36; m 61; c 3. PLANT PHYSIOLOGY. *Educ:* Univ Mass, BS, 58, MA, 60; Ind Univ, PhD(plant physiol), 64. *Prof Exp:* Lectr plant physiol, Ind Univ, 63-64; from asst prof to assoc prof biol, 66-79, prof & head, Dept Hort, 79-88, PROF PLANT PHYSIOL, DEPT HORT, PA STATE UNIV, 89- *Mem:* Am Soc Plant Physiol. *Res:* Biosynthesis, chemistry and mechanism of action of phytohormones and their interaction with nucleic acids. *Mailing Add:* Dept Hort Pa State Univ University Park PA 16802

WITHAM, P(HILIP) ROSS, b Stuart, Fla, Apr 11, 17; m 45; c 4. MARINE TURTLE RESEARCH, SPINY LOBSTER RESEARCH. *Educ:* Univ SFla, BS, 73; Univ Okla, MS, 76. *Prof Exp:* Proj leader spiny lobster res, Fla Dept Natural Resources, 63-71, marine turtle coordr, 71-87; RES ASSOC, ROSENTIEL SCH MARINE & ATMOSPHERIC SCI, UNIV MIAMI, 87- *Concurrent Pos:* Mem, Sea Turtles-Dredging Task Force, CEngr, 85-87, Restore Coast Task Force, State Fla, 86 & Southeast Region Marine Turtle Recovery Team, Calif, 87. *Honors & Awards:* Commendation, Gov & Cabinet of Fla, 86. *Mem:* Am Soc Zoologists; Am Soc Ichthyologists & Herpetologists; Am Inst Fisheries Res Biologists; fel Explorers Club. *Res:* Management of marine turtles nesting in unbalanced areas; developing methodology for increasing oceanic stocks of marine turtles. *Mailing Add:* 1457 NW Lake Pt Stuart FL 34994

WITHBROE, GEORGE LUND, b Green Bay, Wis, Dec 14, 38; m 64; c 2. ASTROPHYSICS, SOLAR PHYSICS. *Educ:* Mass Inst Technol, BS, 61; Univ Mich, MS, 63, PhD(astron), 65. *Prof Exp:* Res fel astron, 65-69, res assoc astron, 69-76, LECTR, HARVARD UNIV, 70-; ASTROPHYSICIST, SMITHSONIAN ASTROPHYS OBSERV, 73-, ASSOC DIR, CTR ASTROPHYS. *Concurrent Pos:* Mem, sci adv comt, NASA, 75- & Comt Solar & Space Physics, Space Sci Bd, 81-84; chair, Solar Physics Div, Am Astron Soc, 90-92. *Mem:* Int Astron Union; Am Astron Soc; Am Geophys Union. *Res:* Interpretation of solar and stellar visible, radio and EUV radiation; determination of solar chemical abundances; temperature density structure of solar atmosphere and terrestrial atmosphere; development of plasma diagnostic techniques. *Mailing Add:* Ctr Astrophys 60 Garden St Cambridge MA 02138

WITHEE, WALLACE WALTER, b Minneapolis, Minn, Mar 12, 13; m 38; c 3. AERONAUTICAL ENGINEERING. *Educ:* Univ Minn, BS, 34. *Prof Exp:* Engr, Boeing Co, 36-38; layout engr, Consol Vultee, 38-40, group leader, 40-42, asst proj engr, 42-49, design specialist, 49-50, sr design group engr, 50-53, chief flight test engr, Gen Dynamics/Convair, 53-54, sr design group engr, 54-55, chief exp flight test, 55-56, asst chief flight test, 56-57, asst chief engr, Gen Dynamics/Astronaut, 57-60, sr asst chief engr, 60-61, vpres res, develop

& eng, 61-62, vpres eng, 62-65, dir test opers, Gen Dynamics/Convair Aerospace, 65-66, dept prog dir advan intercontinental ballistic missile, 66-67, prog dir manned orbital space systs, 67-68, dir mil space progs, 68, dir advan space systs, 68-71, prog dir, res & applications modules, 71, dir res & applications progs, 72-76, subcontract proj dir, 76-78, dir energy progs & consult, Gen Dynamics/Convair Aerospace, 78-86; RETIRED. *Mem:* Assoc fel Am Inst Aeronaut & Astronaut; Soc Automotive Engrs; Inst Environ Sci; Am Astronaut Soc; Am Nuclear Soc. *Res:* Technical management in the aerospace industry from conception of missile to system development. *Mailing Add:* 4858 Butterfly Lane La Mesa CA 92041

WITHER, ROSS PLUMMER, b Portland, Ore, Dec 29, 22; m 44; c 3. ORGANIC CHEMISTRY, PULP & PAPER TECHNOLOGY. *Educ:* Univ Ore, BS, 47, MA, 49; Stanford Univ, PhD(chem), 56. *Prof Exp:* Sr res chemist, Cent Res Div, Crown Zellerbach Corp, 55-85; RETIRED. *Mem:* Am Chem Soc; Tech Asn Pulp & Paper Indust; Sigma Xi. *Res:* Cellulose chemistry; pulp and paper research; paper coatings research; specialty papers development. *Mailing Add:* 1526 NE 4th Ave Camas WA 98607

WITHERELL, EGILDA DEAMICIS, b Fall River, Mass, Nov 1, 22; m 56. RADIOLOGICAL PHYSICS, NUCLEAR MEDICINE. *Educ:* Mass Inst Technol, SB, 44; Am Bd Radiol, dip, 53; Am Bd Health Physics, dipl, 60. *Prof Exp:* Mem staff physics, Radiation Lab, Mass Inst Technol, 44-45, mem staff math, Dynamic Anal & Control Lab, 46-47; asst instr chem, Northeastern Univ, 45-56; radiol physicist, Cancer Res Inst, New Eng Deaconess Hosp, Boston, Mass, 47-66 & Peter Bent Brigham Hosp, 66-67; radiol physicist, Newton Wellesley Hosp, Newton Lower Falls, 67-86; RETIRED. *Mem:* Am Col Radiol; Am Asn Physicists Med. *Res:* Radiological physics. *Mailing Add:* PO Box 757 Needham MA 02192

WITHERELL, MICHAEL STEWART, b Toledo, Ohio, Sept 22, 49. ELEMENTARY PARTICLE PHYSICS. *Educ:* Univ Mich, BS, 68; Univ Wis, MA, 70, PhD(physics), 73. *Prof Exp:* Instr, Princeton Univ, 73-75, asst prof, 75-81; from asst prof to assoc prof, 81-86, PROF, UNIV CALIF, SANTA BARBARA, 86- *Concurrent Pos:* Guggenheim Fel. *Honors & Awards:* Panofsky Prize, Am Phys Soc. *Mem:* Am Phys Soc. *Res:* Counter and spark chamber experiments in elementary particle physics. *Mailing Add:* Dept Physics Univ Calif Santa Barbara CA 93106

WITHERELL, PETER CHARLES, b Athol, Mass, Sept 23, 43; m 81; c 2. REGULATORY ENTOMOLOGY, APICULTURE. *Educ:* Univ Mass, Amherst, BS, 65; Univ Calif, Davis, MS, 70, PhD(entom), 73. *Prof Exp:* Res entomologist, Univ Calif, Davis, 73-74; asst res dir, Dadant & Sons Inc, Hamilton, Ill, 75-77; plant protection & quarantine officer, Laredo, Tex, 78-81, sta supvr, Methods Develop Sta, Miami, Fla, 81-85, ASST CTR DIR, METHODS DEVELOP CTR, ANIMAL & PLANT HEALTH INSPECTION SERV, SCI & TECHNOL, USDA, HOBOKEN, NJ, 85- *Mem:* Orgn Prof Employees Dept Agr; Entom Soc Am; Sigma Xi; Am Registry Prof Entomologists. *Res:* Quarantine entomology; inspection; fumigation; heat and cold; treatment of commodities in the import export trade; apiculture, honey bee behavior, Africanized bees, varroa mites. *Mailing Add:* Methods Develop Ctr Animal & Plant Health Inspection Serv USDA 209 River St Hoboken NJ 07030

WITHERS, HUBERT RODNEY, b Stanthorpe, Queensland, Australia, Sept 21, 32; m 59; c 1. RADIATION ONCOLOGY. *Educ:* Univ London, PhD(path), 65, DSc, 82. *Prof Exp:* Res fel, Univ Queensland, 63-65; vis res scientist, Nat Cancer Inst, 66-68; from assoc prof to prof radiotherapy, Univ Tex, 68-80; PROF RADIOTHER, UNIV CALIF-LOS ANGELES, 80- *Concurrent Pos:* Gaggin res fel, Univ Queensland, 63-66; hon mem Europ Soc Therapeut Radiation Oncol, 87; fel, Am Col Radiol. *Honors & Awards:* Finzi Prize, Brit Inst Radiol, 74; Hon Mem, Europ Soc Therapeut Radiation Oncol, 87; Failla Award, Radiation Res Soc, 88; Erskine lectr, Radiol Soc NAm, 88; H S Kaplan Distinguished Scientist Award, Int Asn Radiation Res, 91. *Mem:* Am Col Radiol; Am Soc Therapeut Radiol & Oncol; Radiation Res Soc; Am Radium Soc; Brit Inst Radiol; Royal Australasian Col Radiol. *Res:* Precise quantitation of normal tissue responses to irradiation as used for treatment of cancer. *Mailing Add:* Dept Radiation Oncol Univ Calif Ctr Health Sci Los Angeles CA 90024

WITHERS, JAMES C, b Buna, Tex, Nov 5, 34; m 56; c 3. COMPOSITE MATERIALS, REINFORCEMENTS. *Educ:* Am Univ, BS, 58; Clayton Univ, PhD(chem), 80. *Prof Exp:* Pres, Gen Technol Inc, 61-71; chmn, Deposits & Composites Inc, 71-76; pres, Pora Inc, 76-81; mgr, Arco Metals Co, 81-85; CHIEF EXEC OFFICER, MER CORP, 85- *Concurrent Pos:* Grad fac, VPI, 68-71. *Mem:* Mat Res Soc; Am Ceramic Soc; Am Soc Metals. *Res:* Author of over 60 publications; granted over 24 patents. *Mailing Add:* Mer Corp 7960 S Kolb Rd Tucson AZ 85706

WITHERS, PHILIP CAREW, b Adelaide, SAustralia, Dec 23, 51; m 77; c 1. COMPARATIVE PHYSIOLOGY, PHYSIOLOGICAL ECOLOGY. *Educ:* Univ Adelaide, BSc Hons, 72; Univ Calif, Los Angeles, PhD(biol), 76. *Prof Exp:* Fel zool, Univ Cape Town, 76-78; vis fel zool, Duke Univ, 78-79; ASST PROF BIOL, PORTLAND STATE UNIV, 79- *Concurrent Pos:* Sr lectr, Univ Western Australia. *Mem:* Am Soc Zoologists; Australian Mammal Soc. *Res:* Comparative physiology of terrestrial vertebrates; energetics, water relations and ionic balance of amphibians, reptiles, birds and mammals. *Mailing Add:* Zool Dept Univ Western Australia Nedlands 6009 Western Australia

WITHERS, STEPHEN GEORGE, b Britain, 1953; Brit & Can citizen. ENZYMOLOGY, CARBOHYDRATE CHEMISTRY. *Educ:* Bristol Univ, UK, BSc, 74, PhD(chem), 77. *Prof Exp:* Postdoctoral fel, Dept Biochem, Univ Alta, 77-79, prof asst, 79-82; from asst prof to assoc prof, 82-91, PROF, DEPT CHEM, UNIV BC, CAN, 91-, PROF BIOCHEM, DEPT BIOCHEM, 91- *Honors & Awards:* Merck Award, Can Soc Chem, 89; Corday Morgan Medal, Royal Soc Chem UK, 90. *Mem:* Am Chem Soc; Can Soc Chem; Can Biochem Soc. *Res:* Mechanism of enzyme action, particularly of glycosyl transfer reactions; applications of nuclear magnetic resonance to biochemistry; applications of fluorinated sugars. *Mailing Add:* Dept Chem Univ BC Vancouver BC V6T 1Y6

WITHERSPOON, JOHN PINKNEY, JR, b Hamlet, NC, Feb 28, 31; m 52; c 5. RADIATION ECOLOGY, PLANT ECOLOGY. *Educ:* Emory Univ, BS, 52, MS, 53; Univ Tenn, PhD(bot), 62. *Prof Exp:* Res asst biol, Emory Univ, 55-57; health physicist, 62, ECOLOGIST, OAK RIDGE NAT LAB, 62- *Concurrent Pos:* Adj prof, Univ Tenn, 76-; mem, Nat Acad Sci, 81; mem comt, Nat Coun Radiation Protection & Measurements, 83-84; environ effects ed, Nuclear Safety, 83- *Mem:* Ecol Soc Am; Health Physics Soc. *Res:* Radiological impact assessments of nuclear fuel cycle facilities; environmental health physics. *Mailing Add:* 100 Wade Lane Oak Ridge TN 37830

WITHERSPOON, PAUL A(DAMS), JR, b Pittsburgh, Pa, Feb 9, 19; m 46; c 3. GEOLOGICAL & PETROLEUM ENGINEERING. *Educ:* Univ Pittsburgh, BS, 41; Univ Kans, MS, 51; Univ Ill, PhD(geol, phys chem), 57. *Prof Exp:* Petrol prod engr, Phillips Petrol Co, 41-42 & 45-47, chem process engr, 42-45, petrol reservoir engr, 47-49; asst instr petrol eng, Univ Kans, 49-51, head div petrol eng, State Geol Surv, Ill, 51-57; prof petrol eng, 57-65, PROF GEOL ENG, UNIV CALIF, BERKELEY, 65-, HEAD EARTH SCI DIV & ASSOC DIR, LAWRENCE BERKELEY LAB, 77- *Concurrent Pos:* Mem subpanel nuclear waste disposal, panel on rock mech probs, US Nat Comn Rock Mech, Nat Res Coun, 77-78. *Honors & Awards:* Robert E Horton Award, Am Geophys Union, 69; O E Meinzer Award, Geol Soc Am, 76. *Mem:* Am Geophys Union; Geol Soc Am; Am Asn Petrol Geol; Am Inst Mining, Metall & Petrol Engrs; Sigma Xi. *Res:* Flow of fluids in porous and fractured rocks; regional groundwater flow; well hydraulics; underground storage of fluids; radioactive waste isolation; geothermal systems. *Mailing Add:* 1824 Montery Ave Berkeley CA 94707

WITHERSPOON, SAMUEL MCBRIDGE, anesthesiology, for more information see previous edition

WITHINGTON, HOLDEN W, ENGINEERING. *Prof Exp:* RETIRED. *Mem:* Nat Acad Eng. *Mailing Add:* 8000 SE 20th Mercer Island WA 98040

WITHNER, CARL LESLIE, JR, b Indianapolis, Ind, Mar 3, 18; m 41; c 3. PLANT MORPHOGENETICS. *Educ:* Univ Ill, BA, 41; Yale Univ, MS, 43, PhD(bot), 48. *Prof Exp:* Asst instr bot, Yale Univ, 41-43, asst, 46-47; from instr to prof biol, Brooklyn Col, 48-78, dep chmn dept, 60-64, actg chmn dept, 64-65, EMER PROF BIOL, BROOKLYN COL, 78- *Concurrent Pos:* Resident investr orchids, Brooklyn Bot Garden, 49-76; Guggenheim fel, 61-62; accredited judge, Am Orchid Soc, 61-; orchid consult, NY Bot Garden, 76-79; instr hort, Bellingham Voc-Tech Sch, 79-80; res assoc, Western Wash Univ, 82- *Honors & Awards:* Gold Medal for Distinguished Achievement in Sci & Educ, Am Orchid Soc, 90. *Mem:* Bot Soc Am; Am Soc Plant Physiol; hon mem Am Orchid Soc. *Res:* Orchids; physiology of higher plants in relation to growth and development. *Mailing Add:* 2015 Alabama St Bellingham WA 98226

WITHROW, CLARENCE DEAN, b Hutchinson, WVa, Mar 6, 27; m 53; c 3. PHARMACOLOGY. *Educ:* Davis & Elkins Col, BS, 48; Univ Utah, MS, 55, PhD(pharmacol), 59. *Prof Exp:* Res instr, 59-63, from instr to asst prof, 63-70, ASSOC PROF PHARMACOL, COL MED, UNIV UTAH, 70- *Mem:* AAAS; Am Soc Pharmacol & Exp Therapeut; Sigma Xi. *Res:* Acid-base metabolism, particularly intracellular pH regulation; renal pharmacology; mineralocorticoids; polarography. *Mailing Add:* Dept Pharmacol Univ Utah Col Med Salt Lake City UT 84132

WITHSTANDLEY, VICTOR DEWYCKOFF, III, b New York, NY, Sept 1, 21; m 58; c 3. MOLECULAR SPECTROSCOPY. *Educ:* Cornell Univ, BA, 50; Univ Calif, Berkeley, MA, 52; Pa State Univ, DEd(physics), 66, PhD(physics), 72. *Prof Exp:* Asst seismologist, Geotech Corp, Tex, 52-56; res asst underwater acoustics, Ord Res Lab, Pa State Univ, 59-62; instr math, Juniata Col, 66-67; res assoc, Ctr Air Environ Studies, Pa State Univ, 69-73; prin scientist, Scitek, Inc, 74-75; instr phys sci, Pa State Univ, 75-77; staff scientist, Bacharach Instrument Co, Pittsburgh, 78-80; consult, Ctr Consult, State Col Pa, 80-83; SR PHYSICIST, DEPT DEFENSE, 84- *Mem:* Am Phys Soc. *Res:* Seismic wave and underwater sound studies; magnetic anisotropies of single crystals; computer analysis of time series; optical engineering; infrared spectroscopy of molecules; remote sensing for geophysical and environmental studies; optics; co-holder of one US patent. *Mailing Add:* 127 W Whitehall Rd State College PA 16801

WITIAK, DONALD T, b Milwaukee, Wis, Nov 16, 35; m 55; c 2. ORGANIC CHEMISTRY, MEDICINAL CHEMISTRY. *Educ:* Univ Wis, BS, 58, PhD(med chem), 61. *Prof Exp:* From asst prof to assoc prof med chem, Univ Iowa, 61-67; assoc prof, 67-71, chmn dept, 73-82, PROF MED CHEM, COL PHARM, OHIO STATE UNIV, 71-, KIMBERLY PROF PHARMACOL, 85-, ASSOC DIR BASIC RES, COMPREHENSIVE CANCER CTR, 87- *Concurrent Pos:* NIH fel, 59-61 & mem med chem A study sect, 80-83; consult, Diamond-Shamrock Corp, Ohio, 76-79, Schering Corp, NJ, 77-80 & Adria Labs, Inc, 80-87, Marion Labs, 86-88, G D Searl Co, 85-86, Nova Labs, 89-; rep chem teachers sect, house deleg, Am Asn Col Pharm, 75-76 & chmn, 80-81; vchmn, Div Med Chem, Am Chem Soc, 83, chmn, 84, counr, 90-92; distinguished vis lectr, Col Pharm, Univ Houston, 85; res achievement award, Am Pharmaceut Asn Found-Acad Pharmaceut Sci, 85. *Honors & Awards:* Smith Kline & French Lectr, Div Med Chem, Univ Pittsburgh, 81. *Mem:* Am Chem Soc; Am Pharmaceut Asn; fel Acad Pharmaceut Sci; Am Asn Col Pharm; Sigma Xi; fel Am Asn Pharmaceut Scientists. *Res:* Synthesis of biologically active compounds; stereostructure activity relationships; antiatheroslerotic drugs; central nervous system drugs; carcinogenesis and anticancer agents; chemical synthesis of targets in the antineoplastic, antimetastatic, antiatherogenic, antiaggregatory, carcinogenesis, central and peripheral nervous system areas; assess mechanisms of transformation of synthesized polycyclic aromatic hydrocarbons in tissue culture; analogues of various rationally designed compounds which block the abortifacient activity of PGF2, antilipidemic and antiaggregatory aci-reductone redox compounds, bis dioxopiperazine compounds; preparation of bis alkylating agents as solid tumor specific antineoplastic drugs. *Mailing Add:* Dept Med Chem Ohio State Univ Col Pharm Columbus OH 43210

WITKIN, EVELYN MAISEL, b New York, NY, Mar 9, 21; wid; c 2. MICROBIAL GENETICS. *Educ:* NY Univ, AB, 41; Columbia Univ, MA, 43, PhD(zool), 47. *Hon Degrees:* DSc, NY Med Col, 78. *Prof Exp:* Res assoc bact genetics, Carnegie Inst, 46-49, mem staff genetics, 49-55; assoc prof med, Col Med, State Univ NY Downstate Med Ctr, 55-69, prof, 69-71; prof biol sci, Douglass Col,71-79, BARBARA MCCLINTOCK PROF GENETICS, WAKSMAN INST MICROBIOL, RUTGERS UNIV, 79- *Concurrent Pos:* Am Cancer Soc fel, 47-49; res assoc, Carnegie Inst, 55-71, fel, 56; Waksman lectr, 59. *Honors & Awards:* Prix Charles-Leopold Mayer, Inst France Acad Sci, 77. *Mem:* Nat Acad Sci; Am Soc Microbiol; Genetics Soc Am; Am Soc Nat; Radiation Res Soc; Am Acad Arts & Sci. *Res:* Mechanism of spontaneous and induced mutation in bacteria; genetic effects of radiation; enzymatic repair of DNA damage. *Mailing Add:* Waksman Inst Microbiol Rutgers Univ PO Box 759 Piscataway NJ 08854

WITKIN, STEVEN S, b Brooklyn, NY, Oct 19, 43; m 86; c 3. IMMUNOLOGY, REPRODUCTIVE BIOLOGY. *Educ:* Hunter Col, BA, 65; Univ Conn, MS, 67; Univ Calif, Los Angeles, PhD(microbiol), 70. *Prof Exp:* Staff assoc, Inst Cancer Res, Columbia Univ, 72-74; assoc, Sloan-Kettering Inst Cancer Res, 74-81; ASSOC PROF, MED COL, CORNELL UNIV, 81- *Concurrent Pos:* Fel, Roche Inst Molecular Biol, Nutley, NJ, 70-72. *Mem:* AAAS; Am Soc Microbiol; Soc Gyn Invest; Am Asn Immunologists; Am Fertil Soc; Obstet-Gynec Infectious Dis Soc. *Res:* Reproductive immunology; cancer; spermatozoa. *Mailing Add:* Dept Obstet/Gynec Cornell Univ Med Col 525 E 68th St New York NY 10021

WITKIND, IRVING JEROME, b New York, NY, Mar 28, 17; m 42. STRUCTURAL GEOLOGY, STRATIGRAPHY. *Educ:* Brooklyn Col, BA, 39; Columbia Univ, MA, 41; Univ Colo, PhD(geol), 56. *Prof Exp:* GEOLOGIST, US GEOL SURV, 46- *Mem:* Geol Soc Am. *Res:* Pleistocene geology; localization of sodium sulfate; geologic mapping for environmental purposes in Price 1 degree x 2 degree AMS Sheet, Central Utah; localization of uranium minerals; laccolithic mountains of southeastern Utah and central Montana; stratigraphy and structural geology of southwestern Montana and southeastern Idaho; salt diapirism in central Utah. *Mailing Add:* US Geol Surv Fed Ctr 30 Ammons St Lakewood CO 80226

WITKOP, BERNHARD, b Freiburg, Ger, May 9, 17; nat US; m 45; c 3. BIOLOGICAL CHEMISTRY, PHYSIOLOGICAL CHEMISTRY. *Educ:* Univ Munich, PhD(org chem), 40, ScD, 46. *Prof Exp:* Privat docent, Univ Munich, 46-47; Mellon Found fel, Harvard Univ, 47-48, instr, 48-50, USPHS spec fel, 50-51; vis scientist, NIH, 51-53, chemist, 53-55, Chief, Lab Chem, 57-88, CHIEF SECT METABOLITES, NAT INST ARTHRITIS, METAB & DIGESTIVE DIS, 55-, NIH INST SCHOLAR, 88- *Concurrent Pos:* Mem, Nat Acad Sci-Nat Res Coun, 59-62; vis prof, Kyoto Univ, 61 & Univ Freiburg, 62; mem, Bd Int Sci Exchange, Nat Acad Sci, 75-77; adj prof, Med Sch, Univ Md, Baltimore, 78-; Alexander von Humboldt US sr scientist award, Univ Hamburg, 79; Paul Ehrlich Award Comt, Frankfurt, Ger, 80- *Mem:* Nat Acad Sci; hon mem Pharmacol Soc Japan; Am Acad Arts & Sci; Am Chem Soc; Leopoldina Ger Acad Res Natural Sci; hon mem Chem Soc Japan; hon mem Biochem Soc Japan. *Res:* Alkaloids; arrow and mushroom poisons; oxidation mechanisms; peroxides; ozonides; intermediary and labile metabolites; nonenzymatic selective cleavage of proteins and enzymes; photochemistry of amino acids and nucleotides; venoms of amphibians; biochemical mechanisms; dynamics of modified homopolynucleotides; stimulation of interferon. *Mailing Add:* Nat Inst Diabetes Digestive & Kidney Dis Bldg 8-B1A-11 Bethesda MD 20892

WITKOP, CARL JACOB, JR, b East Grand Rapids, Mich, Dec 27, 20; m 66; c 7. HUMAN GENETICS, ORAL PATHOLOGY. *Educ:* Mich State Col, BS, 44; Univ Mich, DDS, 49, MS, 54; Am Bd Oral Path, dipl, 57. *Prof Exp:* Asst dent surgeon & intern, US Marine Hosp, USPHS, Seattle, Wash, 49-50, sr asst dent surgeon, US Coast Guard Yard, Baltimore, 50; oral pathologist, Nat Inst Dent Res, 50-57, chief human genetics sect, 57-63, chief human genetics br, 63-66; prof human & oral genetics & chmn div, Sch Dent & prof dermat, Med Sch, Univ Minn, Minneapolis, 66-91; RETIRED. *Concurrent Pos:* Fel, Univ Mich, 52-54; consult, Children's Hosp, Washington, DC, 56-, Nat Found Congenital Malformation, 63- & Easter Seal Soc, 63-66; lectr, Schs Med & Dent, Howard Univ, 56-, Georgetown Univ & Johns Hopkins Univ; dent dir, Int Comt Nutrit Nat Develop, Chile, 60 & Paraguay, 65; chief dent sect, Inst Nutrit Cent Am, Panama & Guatemala, 64; mem, Am Bd Oral Path, 78-85, pres, 85; vis prof, Inst Ophthal, Univ London, 80. *Honors & Awards:* Brittle Bone Soc Award, 78. *Mem:* AAAS; Am Soc Human Genetics (secy, 68-70); Am Soc Dermat; Am Dent Asn; fel Am Acad Oral Path (vpres, 60, 72, pres elect, 73, pres, 74). *Res:* Albinism and pigment defects; exfoliative cytology; population isolates; congenital malformations; inherited defects of teeth. *Mailing Add:* Dept Dermat Univ Minn Minneapolis MN 55455

WITKOSKI, FRANCIS CLEMENT, organic chemistry; deceased, see previous edition for last biography

WITKOVSKY, PAUL, b Chicago, Ill, May 24, 37; m 64. SENSORY PHYSIOLOGY. *Educ:* Univ Calif, Los Angeles, BA, 58, MA, 60, PhD(physiol), 62. *Prof Exp:* NIH fel neurophysiol, Sci Res Inst, Caracas, Venezuela, 62-63; instr ophthal, Columbia Univ, 64-65, from asst prof to assoc prof physiol, 65-73; prof anat sci, State Univ NY Stony Brook, 75-; MEM STAFF, DEPT OPHTHAL, NEW YORK UNIV MED CTR. *Concurrent Pos:* Res grants, Nat Inst Neurol Dis & Blindness, 64- & Nat Coun Combat Blindness, 66-67. *Mem:* AAAS; Asn Res Vision & Ophthal; Biophys Soc; Soc Neurosci. *Res:* Central nervous system organization of tactile sensation; neurophysiological organization of the retina. *Mailing Add:* Dept Ophthal NY Univ Med Ctr 550 First Ave New York NY 10016

WITKOWSKI, JOHN FREDERICK, b Beatrice, Nebr, Apr 30, 42; m 74; c 1. ENTOMOLOGY. *Educ:* Univ Nebr, BSc, 65, MSc, 70; Iowa State Univ, PhD(entom), 75. *Prof Exp:* Res rep agr chem, Chemagro Corp, 70-72; res assoc entom, Iowa State Univ, 72-75; ENTOMOLOGIST, UNIV NEBR, 75- *Mem:* Entom Soc Am. *Res:* The biology and chemical control of insects damaging corn and soybeans. *Mailing Add:* 6501 Vine St No 108 Lincoln NE 68505

WITKOWSKI, JOSEPH THEODORE, b Ft Worth, Tex, Oct 29, 42; m 65; c 1. ORGANIC CHEMISTRY, MEDICINAL CHEMISTRY. *Educ:* NTex State Univ, BS, 65, MS, 66; Univ Utah, PhD(chem), 70. *Prof Exp:* Res chemist, Nucleic Acid Res Inst, ICN Pharmaceut, Inc, 69-77; RES CHEMIST, SCHERING CORP, 77- *Mem:* Am Chem Soc; Int Soc Heterocyclic Chem. *Res:* Medical chemistry; design and synthesis of cardiovascular agents. *Mailing Add:* Schering Corp 60 Orange St Bloomfield NJ 07003

WITKOWSKI, ROBERT EDWARD, b Glassport, Pa, Jan 9, 41; m 63; c 2. MICROANALYTICAL CHEMISTRY. *Educ:* Univ Pittsburgh, BS, 62, MS, 73, PhD, 88. *Prof Exp:* Res asst phys measurements, Mellon Inst, Pittsburgh, 62-63, jr fel infrared spectros, 63-67; assoc engr mass spectros, Westinghouse Res Labs, 67-71, engr liquid metal technol, 71-75, sr engr liquid metal technol, 75-80, FEL SCIENTIST ADVAN MAT, WESTINGHOUSE SCI & TECHNOL CTR, 80- *Concurrent Pos:* Res assoc, Sect Minerals, Carnegie Mus Natural Hist, Pittsburgh, 74- *Honors & Awards:* Jacquet-Lucas Award & Gold Medal, 78. *Mem:* Am Chem Soc; Sigma Xi. *Res:* Sodium corrosion and mass transport studies via advanced instrumental microanalytical techniques; South Pole atmospheric particle collection and characterization; microwave excited and DC plasma jet reactor chemistry; chemical vapor deposition technology. *Mailing Add:* 633 Shadyside Dr West Mifflin PA 15122

WITKUS, ELEANOR RUTH, b New York, NY, July 11, 18. BIOLOGY. *Educ:* Hunter Col, BA, 40; Boston Univ, MA, 41; Fordham Univ, PhD(cytol), 44. *Prof Exp:* Instr zool, Marymount Col, NY, 43-44; from instr to assoc prof bot & bact, 44-74, chmn dept, 72-78, PROF BIOL SCI, FORDHAM UNIV, 71- *Mem:* Bot Soc Am; Torrey Bot Club (corresp secy, 53-56); Sigma Xi. *Res:* Botanical cytology. *Mailing Add:* Dept Biol Fordham Univ Bronx NY 10458

WITLIN, BERNARD, b Philadelphia, Pa, July 18, 14; m 44; c 2. MICROBIOLOGY. *Educ:* Univ Calif, Los Angeles, AB, 36; Philadelphia Col Pharm, MSc, 38, DSc(bact, pub health), 40; Am Bd Clin Chem, dipl. *Prof Exp:* Res bacteriologist, Sharp & Dohme, Inc, Pa, 38-40; dir, Barlin Labs, 40-41; bacteriologist, USPHS, US Dept Army & USN, 41-46 & Mellon Inst, 46-50; from assoc prof to prof bact, 50-71, EMER PROF BACT & PUB HEALTH, PA COL OPTOM, 71-, EMER PROF PHARM, COL PHARM & SCI, 81- *Concurrent Pos:* Prof, Philadelphia Col Pharm & Sci, 40-81; assoc prof microbiol & pub health, Philadelphia Col Osteopath Med, 50-68; clin pathologist, 52-81, Emer pathologist, Metrop Hosp, Philadelphia, 81- *Mem:* AAAS; Am Soc Microbiol; Am Pub Health Asn; Am Asn Clin Chemists; Sigma Xi (pres, 71-72). *Res:* Antiseptics and disinfectants; water purification; blood banks; serology. *Mailing Add:* Kennedy House Apt No 1820 1901 John F Kennedy Blvd Philadelphia PA 19103

WITMAN, GEORGE BODO, III, b Upland, Calif, July 19, 45; m 69; c 3. CELL BIOLOGY. *Educ:* Univ Calif, Riverside, BA, 67; Yale Univ, PhD(cellular & develop biol), 72. *Prof Exp:* NIH fel cell biol, Whitman Lab, Univ Chicago, 72-73; NIH fel molecular biol, Lab Molecular Biol & Biophys, Univ Wis-Madison, 73-74; asst prof, dept biol, Princeton Univ, 74-81; staff scientist, 81-82, sr scientist, 83-90, PRIN SCIENTIST, WORCESTER FOUND EXP BIOL, SHREWSBURY, MASS, 90-; ASSOC PROF, DEPT ANATOMY, UNIV MASS MED CTR, WORCESTER, 85- *Concurrent Pos:* Mem, Cancer Ctr, Worcester Found Exp Biol, 81-; dir, Male Fertility Prog, Worcester Found Exp Biol, Shrewsbury, Mass, 85- *Mem:* AAAS; Am Soc Biochem Molecular Biol; Am Soc Cell Biol; Electron Micros Soc Am; Genetics Soc Am; Protein Soc; Soc Study Reproduction. *Res:* Structure, composition, function and development of cell organelles; male reproduction; cell motility; cilia and flagella; microtubule-based force production; sperm maturation and motility. *Mailing Add:* Worcester Found Exp Biol 222 Maple Ave Shrewsbury MA 01545

WITMER, EMMETT A(TLEE), b Bellefonte, Pa, Dec 20, 24; m 48; c 2. AERODYNAMICS. *Educ:* Pa State Univ, BS, 44, MS, 48; Mass Inst Technol, ScD(aeronaut eng), 51. *Prof Exp:* Aeronaut engr, Nat Adv Comt Aeronaut, Ohio, 44; instr appl aerodyn, Pa State Univ, 46-48; aeronaut engr, Naval Supersonic Lab, 49 & Aeroelastic & Struct Res Lab, 50, sr engr, 51-52, proj leader, 52-60, exec officer, 60-62, assoc prof, 60-70, prof, Aeronaut & Astronaut Inst, 70-76, DIR, AEROELASTIC & STRUCT RES LAB, MASS INST TECHNOL, 62-, PROF AERONAUT & ASTRONAUT, 76- *Concurrent Pos:* Consult, Am Sci & Eng Co, 61-, White Sands Missile Range, US Dept Army, 61-, Picatinny Arsenal, 63-, Arthur D Little, Inc, 64-, York Astro Inc, 66 & Watertown Arsenal, 66- *Mem:* Am Inst Aeronaut & Astronaut. *Res:* Aeroelasticity; structural dynamics; transient airloads on wing-body configurations; effects of blast and radiation from explosions on aircraft, missile and ground structures; chemical and nuclear explosion characteristics; dynamic behavior of solids; discrete-element and finite difference methods of static and dynamic elastic-plastic-thermal analysis of structures. *Mailing Add:* Five Bluefish Lane Brewster MA 02631

WITMER, HEMAN JOHN, b Bayonne, NJ, Apr 5, 44; m 70; c 2. VIROLOGY, BIOCHEMISTRY. *Educ:* Delaware Valley Col, BS, 65; Ind Univ, Bloomington, PhD(microbiol), 69. *Prof Exp:* NIH fel, McArdle Lab, Univ Wis-Madison, 69 & Ind Univ, Bloomington, 69-71, vis asst prof microbiol, 71, asst prof, Med Ctr, 71-72; asst prof, 72-77, ASSOC PROF BIOL SCI, UNIV ILL, CHICAGO CIRCLE, 77- *Res:* Controls of gene expression of coliphage T4; biosynthesis of unusual bases in DNA. *Mailing Add:* Dept Biol Sci Univ Ill Chicago IL 60680

WITMER, WILLIAM BYRON, b Clarksville, Tex, June 29, 31; m 55; c 3. INORGANIC CHEMISTRY, PHYSICAL CHEMISTRY. *Educ:* Tex A&M Univ, BS, 52, MS, 58, PhD(chem), 60. *Prof Exp:* Res chemist, Chemstrand Res Ctr, Monsanto Co, NC, 59-64, sr res chemist, 64, supvr spec analysis lab, Textiles Div, Ala, 64-65, supt tech lab, 65-68, supt qual control, Decatur Plant, 68-70, qual control mgr, Lingen Plant, Monsanto (Deutschland), GMBH, WGer, 70-73, supt, Tech Dept, Sand Mountain Plant, 73-79; ASSOC TRAINING SPECIALIST, TEX ENG EXTEN SERV, OIL &

HAZARDOUS MAT TNG DIV, TEX A&M UNIV SYST, 87- *Concurrent Pos:* Supt, Environ Health & Safety, Pensacola Plant, Monsanto Textiles, 79-82; mgr, Environ Analytical Sci Ctr & Contracts, Mon Res Corp, 82-83, assoc dir, Dayton Lab, 83, mgr, Nuclear Opers, Mound Facil, 83-85; owner, Witmer Assocs, Environ & Mgt Servs, 85- *Mem:* Am Soc Qual Control; Am Chem Soc. *Res:* Amine-halogen complexes; polymer characterizition techniques; polymer properties related to synthetic fiber production; investigation of cause, course of a hypersensitivity pneumonities like disease among employees in an industrial setting. *Mailing Add:* 1408 Country Club Rd Argyle TX 76226

WITORSCH, RAPHAEL JAY, b New York, NY, Dec 12, 41; m 64; c 2. PHYSIOLOGY. *Educ:* NY Univ, AB, 63; Yale Univ, MS, 65, PhD(physiol), 68. *Prof Exp:* USPHS trainee, Sch Med, Univ Va, 68-69, NIH fel, 69-70; from asst prof to assoc prof, 70-88, PROF PHYSIOL, MED COL VA, VA COMMONWEALTH UNIV, 88- *Concurrent Pos:* Dir first yr med curriculum, Med Col Va, 75-77; co-investr, NIH Grant Breast Cancer, 75-78; prin investr, NIH grant prolactin binding in normal & neoplastic prostate, 78-85. *Mem:* Endocrine Soc; Am Physiol Soc; Soc Exp Biol & Med; Histochem Soc; Am Soc Andrology. *Res:* Endocrinology; immunohistochemistry; hormone receptors in normal and neoplastic tissues; endocrine toxicology; nature of the hormone prolactin and its interaction with its target tissue. *Mailing Add:* Dept Physiol Box 551 Med Col Va Richmond VA 23298

WITRIOL, NORMAN MARTIN, b Brooklyn, NY, June 9, 40; m 66. OPTICS, CHEMICAL PHYSICS. *Educ:* Polytech Inst Brooklyn, BS, 61; Brandeis Univ, MA, 64, PhD(physics), 68. *Prof Exp:* Res physicist, Phys Sci Lab, US Army Missile Command, 68-77; from asst prof to assoc prof, 77-84, PROF PHYSICS, LA TECH UNIV, 85- *Concurrent Pos:* Asst prof, Univ Ala, Huntsville, 70-77; consult, Columbus Labs, Battelle Mem Inst, 78-79, Naval Res Lab, 85-89, Ballistic Res Lab, 89-90, Chem Res, Develop & Eng Ctr, 89, 91- *Mem:* Am Phys Soc; AAAS; Optical Soc Am. *Res:* Light scattering from particulate matter; computer modelling and simulation of the ignition process and the chemical kinetics in reactive gases; multiphoton processes; design of robotic vision systems. *Mailing Add:* La Tech Univ PO Box 3169 Tech Sta Ruston LA 71272

WITSCHARD, GILBERT, b Morehead City, NC, Mar 13, 33; m 60; c 3. ORGANIC CHEMISTRY. *Educ:* Queens Col, BS, 57; Univ Pittsburgh, PhD(org chem), 63. *Prof Exp:* Res chemist, Hooker Chem Corp, Niagara Falls, 63-70, sr res chemist, 70-84; compound develop mgr, Oxy Chem, Burlington, NJ, 84-86; GROUP LEADER, OXY CHEM, BURLINGTON, NJ, 86- *Mem:* Am Chem Soc; Soc Plastic Engr. *Res:* Organo-phosphorus chemistry; fire retardance; polymer synthesis; polymers stabilization; powder coatings; polyvinyl chloride. *Mailing Add:* 85 Richboro Rd Newtown PA 18940-1534

WITSCHI, HANSPETER R, b Berne, Switz, Mar 17, 33; US citizen; m 63; c 2. PATHOLOGY, ENVIRONMENTAL HEALTH. *Educ:* Univ Berne, MD, 60; Am Bd Toxicol, dipl, 80; Acad Toxicol Sci, dipl, 82. *Prof Exp:* Asst path, Inst Forensic Med, Univ Berne, 61-64; res fel, Toxicol Res Unit, Med Res Coun, Eng, 65-66; res fel exp path, Univ Pittsburgh, 67-69; from asst prof to assoc prof toxicol & pharmacol, Fac Med, Univ Montreal, 69-77; sr res staff mem, Biol Div, Oak Ridge Nat Lab, 77-87; PROF & ASSOC DIR, TOXIC SUBSTANCES RES & TEACHING PROG, UNIV CALIF, DAVIS, 87- *Concurrent Pos:* Assoc ed, Toxicol Appl Pharmacol, 78-86; ed USA, Toxicol, 78-; mem, toxicol study sect, Div Res Grants, NIH, 80-84, rev comt, Nat Inst Environ Health Sci, 86-90, chmn, 89-90; sci adv bd, subcomt Health Effects, Rel Risk Reduction Proj, US Environ Protection Agency; sci rev panel, State Calif Air Resources Bd; bd mem, Am Bd Toxicol, 88-92. *Mem:* Am Soc Pharmacol & Exp Therapeut; Soc Exp Biol & Med; Soc Toxicol; Am Col Toxicol. *Res:* Experimental toxicology; biochemical pathology; interaction of drugs and toxic agents with organ function at the cellular level; pulmonary carcinogenesis. *Mailing Add:* Toxics Prog ITEH Univ Calif Davis CA 95616

WITSENHAUSEN, HANS S, b Frankfurt, Ger, May 6, 30; US citizen; m 61; c 3. APPLIED MATHEMATICS. *Educ:* Free Univ Brussels, ICME, 53, lic sc phys, 56; Mass Inst Technol, SM, 64, PhD(elec eng), 66. *Prof Exp:* Asst elec eng, Free Univ Brussels, 53-57; appln engr, European Ctr, Electronic Assoc, Inc, 57-60, sr engr, Res Div, 60-63; res asst control theory & Lincoln Lab assoc, Electronic Systs Lab, Mass Inst Technol, 63-65; mem tech staff, Math Res Ctr, AT&T Bell Labs, 66-90; SR VIS SCIENTIST, LAB INFO & DECISION SCI, MASS INST TECHNOL, 90- *Concurrent Pos:* Vis prof, Mass Inst Technol, 73; Vinton Hayes sr fel, Harvard Univ, 75-76. *Mem:* Inst Elec & Electronics Engrs; Am Math Soc. *Res:* System theory; optimization; geometry; inequalities; information. *Mailing Add:* Mass Inst Technol Rm 35-410B Cambridge MA 02139

WITT, ADOLF NICOLAUS, b Bad Oldesloe, Ger, Oct 17, 40; m 67; c 2. ASTROPHYSICS. *Educ:* Univ Hamburg, Vordiplom, 63; Univ Chicago, PhD(astrophys), 67. *Prof Exp:* From asst prof to prof astron, Univ Toledo, 67-89, assoc dir, Ritter Observ, 72-76, dir, 76-79, chmn dept phys & astron, 79-81, DISTINGUISHED UNIV PROF ASTRON, UNIV TOLEDO, 89- *Concurrent Pos:* Vis fel, Lab Atmospheric & Space Physics, Univ Colo, Boulder, 75-76; vis prof, Max Planck Inst Astron, Heidelberg, Germany, 82. *Honors & Awards:* Sigma Xi Award. *Mem:* AAAS; Am Astron Soc; Int Astron Union. *Res:* Interstellar matter; photometry; astronomical instrumentation; radiative transfer; scattering by dust. *Mailing Add:* Dept Physics & Astron Univ Toledo Toledo OH 43606

WITT, CHRISTOPHER JOHN, b Flushing, NY, Dec 24, 31; m 58; c 3. SIMULATION TECHNOLOGY. *Educ:* Polytech Univ, BEE, 53, MEE, 54. *Prof Exp:* Sr engr, Sperry Gyroscope Co, 56-63; sect head, Guid & Control, Grumman Aerospace Corp, 63-67, eng mgr, 67-72, mgr & dir, Elec Systs Ctr, 72-87, DIR, CORP DEVELOP LABS, GURMMAN CORP, 87- *Concurrent Pos:* Adj prof systs eng, State Univ NY, Stony Brook, 82-86; mem, ad hoc comt Ultra Reliable Elec, AIA, 88-90; mem, Res & Eng Comt, Nat Security Indust Asn, 91. *Mem:* Inst Elec & Electronics Engrs; Sigma Xi. *Res:* Magnetics; mag amp and mag pulse modulators; inertial component design; space telescope and space telescope control system design; application of high temperature superconductivity and diamond thin films to electronics. *Mailing Add:* Eight White Oak Tree Rd Laurel Hollow NY 11791

WITT, DONALD JAMES, b Chicago, Ill, May 13, 49; m 73; c 2. VIROLOGY, TISSUE CULTURE. *Educ:* Loyola Univ, BS, 71; Ohio State Univ, MS, 73, PhD(virol), 76. *Prof Exp:* Res assoc, Dept Entom, Ohio State Univ, 76-77; sr scientist, Crop Protection Div, Sandoz Inc, 77-78; NIH fel, Dept Cellular Virol & Molecular Biol, Med Sch, 78-80, virologist, 80, MGR BIOCHEM & MOLECULAR BIOL, RES INST, UNIV UTAH, 80- *Mem:* Am Soc Microbiol; Soc Invert Path; Sigma Xi. *Res:* Molecular biology of infectious disease agents, especially viruses; development of diagnostic technology for the identification of infectious disease agents; development of artificial vaccines; application of microorganisms to regulate pest insect populations. *Mailing Add:* 107 S Corncrib Ct Cary NC 27513

WITT, DONALD REINHOLD, b LeMars, Iowa, Apr 15, 23; m 50; c 4. CHEMISTRY. *Educ:* Westmar Col, BS, 48; Univ SDak, MA, 50. *Prof Exp:* Res chemist, 50-62, GROUP LEADER CHEM, PHILLIPS PETROL CO, 62- *Mem:* Am Chem Soc. *Res:* Polymerization of olefins and diolefins to solid polymers; catalyst development related to such reactions. *Mailing Add:* 4909 SE Harvard Dr Bartlesville OK 74006

WITT, ENRIQUE ROBERTO, b Buenos Aires, Arg, May 10, 26; nat US; m 55; c 2. ORGANIC CHEMISTRY. *Educ:* Univ Buenos Aires, Lic, 51, DrChem, 53. *Prof Exp:* Technician, E R Squibb & Sons, Arg, 51-52; res chemist, Arg AEC, 53-55; res assoc, Celanese Chem Co, 56-86; CONSULT ENVIRON SCI, 86- *Mem:* AAAS; Am Chem Soc; Soc Econ Bot. *Res:* Synthetic lubricants; phosphorus compounds; polyester technology; environmental maintenance; anaerobic biological treatment of industrial wastes. *Mailing Add:* 1037 Brock Corpus Christi TX 78412-3343

WITT, HOWARD RUSSELL, b Morden, Man, July 12, 29; m 56; c 3. ELECTRICAL ENGINEERING. *Educ:* Univ Toronto, BASc, 53; Princeton Univ, MSE, 59; Cornell Univ, PhD(elec eng), 62. *Prof Exp:* Asst prof elec eng, Cornell Univ, 62-67; assoc prof eng, 67-71, asst dean sch eng, 69-75, PROF ENG, OAKLAND UNIV, 71-, ASSOC DEAN SCH ENG, 75- *Mem:* Inst Elec & Electronics Engrs; Am Soc Eng Educ; Sigma Xi. *Mailing Add:* Sch Eng Oakland Univ Rochester MI 48309

WITT, JOHN, JR, b Muskegon, Mich, Oct 5, 35; m 64; c 2. ORGANIC CHEMISTRY. *Educ:* Mich State Univ, BS, 57; Univ Ill, PhD(org chem), 61; Univ Chicago, MBA, 78. *Prof Exp:* Res chemist, Ethyl Corp, 60-62; asst head chem process develop, 62-69, mgr spec synthesis & process develop, 69-72, mgr, synthesis develop, 72-78, dir chem develop, 78-81, SR DIR, NUTRASWEET RES & DEVELOP, NUTRASWEET CO, 81- *Mem:* Am Chem Soc. *Res:* Organic synthesis; steroids; heterocyclics; amino acids. *Mailing Add:* 601 E Kensington Rd Mt Prospect IL 60056-1363

WITT, PATRICIA L, PHOTORECEPTORS, IMMUNOLOGY. *Educ:* Univ Wis-Madison, PhD(cell biol), 80. *Prof Exp:* Asst scientist, Univ Wis-Madison, 83-89, assoc scientist, 89-90; ASST RES PROF, DEPT MICROBIOL, MED COL WIS, 90- *Mem:* Int Soc Interferon Res. *Res:* Protein induction by interferons and cytokines in cancer and AIDS. *Mailing Add:* Cancer Ctr Med Col Wis 8701 Watertown Plank Rd Milwaukee WI 53226

WITT, PETER NIKOLAUS, b Berlin, Ger, Oct 20, 18; m 49; c 2. PHARMACOLOGY. *Educ:* Univ T bingen, MD, 46. *Prof Exp:* Asst, Univ T bingen, 45-49; sr asst, Univ Berne, 49-56, privat-docent, 56; from asst prof to assoc prof pharmacol, Col Med, State Univ NY Upstate Med Ctr, 56-66; dir div res, NC Dept Ment Health, 66-81; RETIRED. *Concurrent Pos:* Rockefeller fel, Harvard Med Sch, 52-53; Lederle med fac award, 57-59; adj prof, NC State Univ & Univ NC, Chapel Hill, 66-81. *Honors & Awards:* Buergi Award, 56. *Mem:* Am Soc Pharmacol & Exp Therapeut; Ger Pharmacol Soc; Swiss Pharmacol Soc. *Res:* Effect of drugs on web building behavior of spiders; invertebrate behavior; effect of cardioactive drugs on ion movements in heart muscle; objective testing of fine motor behavior in healthy and diseased human subjects under the influence of drugs. *Mailing Add:* 1623 Park Dr Raleigh NC 27605

WITT, ROBERT MICHAEL, b Waukesha, Wis, Oct 3, 42; m 71; c 2. MEDICAL PHYSICS, RADIOLOGICAL SCIENCES. *Educ:* Univ Wis-Milwaukee, BS, 64; Univ Wis-Madison, MS, 66, PhD(radiol sci), 75. *Prof Exp:* Proj assoc med physics, Dept Radiol, Univ Wis-Madison, 75-77; PHYSICIST NUCLEAR MED, VET ADMIN MED CTR, 77-; ASST PROF RADIOL, MED CTR, IND UNIV, INDIANAPOLIS, 77- *Mem:* Am Asn Physicists Med; AAAS; Soc Nuclear Med; Health Physics Soc; Inst Elec & Electronics Engrs; Int Soc Optical Eng. *Res:* Medical image communication, storage and management; quantitative and qualitative medical imaging. *Mailing Add:* Nuclear Med 115 Vet Admin Med Ctr Indianapolis IN 46202

WITT, SAMUEL N(EWTON), JR, b Seminole, Okla, Jan 29, 28; m 47; c 3. ELECTRONICS ENGINEERING. *Educ:* Tenn Polytech Inst, BS, 50; Ga Inst Technol, MS, 54, PhD(elec eng), 62. *Prof Exp:* Instr elec eng, Tenn Polytech Inst, 50-51; res asst, Ga Inst Technol, 51-53, res engr, 53-61, lectr electronics, 56-61; vpres & chief engr, RMS Eng, Inc, 61-70; DIR ENG, DIGITAL PRODS GROUP, 70- *Mem:* Sr mem Inst Elec & Electronics Engrs. *Res:* Electronic circuits and systems design. *Mailing Add:* 3078 Rockingham Dr Atlanta GA 30327

WITTBECKER, EMERSON LAVERNE, b Freeport, Ill, Feb 25, 17; m 40; c 2. POLYMER CHEMISTRY, TEXTILE CHEMISTRY. *Educ:* Univ Ill, AB, 39; Pa State Univ, MS, 41, PhD(org chem), 42. *Prof Exp:* Jr res assoc, Nylon Res Div, Textile Fibers Dept, E I du Pont de Nemours & Co, Inc, 46-52, res assoc, 52-55, res mgr pioneering res, 55-60, Orlon-Lycra res, 60-64, dir, Carothers Lab, 64-77; RETIRED. *Mem:* Am Chem Soc. *Res:* Condensation polymers; fibers. *Mailing Add:* 8071 Glenbrooke Lane Sarasota FL 34243

WITTCOFF, A HAROLD, b Marion, Ind, July 3, 18; m 46; c 2. ORGANIC CHEMISTRY, INDUSTRIAL CHEMICAL EDUCATION. *Educ:* DePauw Univ, AB, 40; Northwestern Univ, PhD(org chem), 43; Harvard Univ, cert mgt, 64. *Prof Exp:* Head chem res dept, Gen Mills Chem, Inc, 43-56, dir chem res, 56-68, vpres chem res & develop, 68-69, vpres & dir corp res, 69-74, spec adv to the pres, 74-78; dir res & develop, Koor Chem Ltd, Beer-Sheva, Israel, 78-81; MEM STAFF, CHEM SYST INT, LTD 81- *Concurrent Pos:* Adj prof chem, Univ Minn, 73-81. *Honors & Awards:* Minn Award, Am Chem Soc, 76. *Mem:* Am Chem Soc; Am Oil Chem Soc; Fedn Socs Paint Technol; Com Develop Asn; Inst Food Technol; Sigma Xi. *Res:* Phosphatides; polymers; protective coatings; resins and plastics; research adminstration. *Mailing Add:* Chem Syst Inc 303 S Broadway Tarrytown NJ 10591

WITTE, JOHN JACOB, b Passaic, NJ, Mar 10, 32; m 68, 79; c 3. PREVENTIVE MEDICINE, PEDIATRICS. *Educ:* Hope Col, AB, 54; Johns Hopkins Univ, MD, 59; Harvard Univ, MPH(microbiol), 66. *Prof Exp:* Intern & resident pediat, Johns Hopkins Univ, 59-62; med epidemologist infectious dis, Bur Health Educ, Dis Control, 62-65, asst chief immunization br, 66-70, chief, 70-74, dir, Immunization Div, 74-77, med dir, 77-82; ASST, STATE HEALTH OFF DIS CONTROL & AIDS PREV, FLA, 82- *Concurrent Pos:* Consult, Adv Comt Immunizing Agents, Can, 69- & Adv Comt Epidemiol, Can, 69- *Mem:* Fel Am Acad Pediat; Infectious Dis Soc; Am Col Preventive Med; Am Pub Health Asn. *Res:* Epidemiology of communicable and chronic diseases; smoking related diseases; development and field testing of vaccines; complications of vaccine administration; AIDS epidemiology and clinical trials. *Mailing Add:* 1317 Winewood Blvd Tallahassee FL 32301

WITTE, LARRY C(LAUDE), b Jonesboro, Tex, Apr 27, 39; m 62; c 2. MECHANICAL ENGINEERING, MATHEMATICS. *Educ:* Arlington State Col, BSME, 63; Okla State Univ, MSME, 65, PhD(mech eng), 67. *Prof Exp:* Assoc aerodyn engr, Ling-Temco-Vought, Inc, 63-65; res assoc, Argonne Nat Lab, 65-66, asst mech engr, 66-67; from asst prof to assoc prof mech eng, 67-73, chmn dept, 72-76, PROF MECH ENG, UNIV HOUSTON, 73-, CHMN DEPT, 88- *Concurrent Pos:* Heat transfer consult. *Honors & Awards:* Herbert Allen Award, Am Soc Mech Engrs, 75. *Mem:* Sr mem Am Inst Astronaut & Aeronaut; Am Nuclear Soc; Am Inst Astronaut & Aeronaut; Am Soc Heating Refrig & Air Conditioning Eng. *Res:* Boiling heat transfer; high flux heat transfer processes; explosive vapor formation; natural convection in irregular enclosures; thermodynamic efficiency of heat exchangers. *Mailing Add:* Dept Mech Eng Univ Houston Houston TX 77204-4792

WITTE, MICHAEL, b Poland, Mar 15, 11; nat US; m 40; c 4. ORGANIC CHEMISTRY, POLLUTION CHEMISTRY. *Educ:* Loyola Univ, Ill, BS, 37; Univ Ill, MS, 38, PhD(org chem), 41. *Prof Exp:* Asst chem, Univ Ill, 38-41; res chemist, Nat Aniline Div, Allied Chem Corp, 41-47; prod supvr, Gen Aniline & Film Corp, 46-54, prod mgr, NJ, 54-56; pres, Simpson Labs, Inc, 57-59; pres, Carnegies Fine Chem Div, Rexall Drug & Chem Corp, 59-60; pres, M Witte Assocs, 60-87; RETIRED. *Concurrent Pos:* Chem consult; Dept Energy grantee, Solar Energy Proj, 80-81. *Mem:* Am Chem Soc. *Res:* Pharmaceutical intermediates; dyestuffs manufacture; biochemistry; chemical management; solar energy. *Mailing Add:* 3674 N Laurelwood Loop Beverly Hills FL 32665

WITTE, OWEN NEIL, m; c 1. LEUKEMIA & MOLECULAR IMMUNOLOGY. *Educ:* Stanford Univ, MD, 76. *Prof Exp:* Postdoctoral fel, Ctr Cancer Res, Mass Inst Technol, 76-80; from asst prof to assoc prof, 80-86, PROF MICROBIOLOGY & MOLECULAR GENETICS, 86- *Concurrent Pos:* Rev, Virol Study Sect, NIH, 82-83, mem, 85-89; postdoctoral fel, Am Cancer Soc, Calif Div, 83-86; Cancer Res Coordinating Comt, Univ Calif, 84-87; Nat Rev Panel, Leukemia Soc Am, 88-92; numerous invited lectureships. *Honors & Awards:* Allison Eberlin Fund Award, 89; Milken Family Med Found Award, 90. *Mem:* Am Asn Immunologists; Am Soc Clin Invest; Am Soc Microbiol; Am Soc Hemat. *Res:* Development of the immune response; growth regulation of hematopoietic stem cells by the abl oncogene and other mechanisms. *Mailing Add:* Dept Microbiol Univ Calif 405 Hilgard Ave Los Angeles CA 90024

WITTEBORN, FRED CARL, b St Louis, Mo, Dec 27, 34; m 57; c 1. PHYSICS, ASTROPHYSICS. *Educ:* Calif Inst Technol, BS, 56; Stanford Univ, MS, 58, PhD(physics), 65. *Prof Exp:* Res assoc physics, Stanford Univ, 65-68; res scientist, 69-70 & 75-89, CHIEF ASTROPHYS BR, NASA AMES RES CTR, 70-74 & 90- *Concurrent Pos:* Vis scholar, Stanford Univ, 70-75. *Mem:* Am Phys Soc; Am Astron Soc; Astron Soc Pac. *Res:* Infrared astronomy; gravity; atomic physics; low temperature physics. *Mailing Add:* 2071 Madelaine Ct Los Altos CA 94024

WITTEBORT, JULES I, b Findlay, Ohio, Jan 26, 17; m 42; c 2. PHYSICS, MATERIALS SCIENCE. *Educ:* Findlay Col, AB, 39. *Prof Exp:* Textile & rubber engr, Wright-Patterson AFB, 39-46, physicist, 46-53, supvry physicist, 53-66, phys sci adminr, USAF Mat Lab, 66-75; RETIRED. *Concurrent Pos:* Adv & contribr, Off Critical Tables, Nat Acad Sci-Nat Res Coun, 56-; mem ad hoc comts, Mat Adv Bd, Nat Res Coun, 57-58 & 64-65. *Mem:* Am Phys Soc; Sigma Xi. *Res:* Development of synthetic rubber and textile materials for aeronautical applications; thermal, optical magnetic and electronic properties of materials; direction of manufacturing technology program for thermionic and solid state devices and electronic materials; magnetic and semiconductor materials; history of the physics of materials. *Mailing Add:* 3762 Storms Rd Dayton OH 45429

WITTEKIND, RAYMOND RICHARD, b Jamaica, NY, May 9, 29; m 60. ORGANIC CHEMISTRY. *Educ:* Polytech Inst Brooklyn, BS, 51; Columbia Univ, AM, 55, PhD, 59; Seton Hall Univ, JD, 77. *Prof Exp:* Res chemist, McNeil Labs, Inc, 58-61; scientist, Warner-Lambert Res Inst, 61-73, sr scientist, 73-74; patent lawyer, Hoffman-LaRoche Inc, Nutley, 74-78; PATENT LAWYER, AM HOECHST CORP, SOMERVILLE, 78- *Mem:* Am Chem Soc. *Res:* Synthesis of fused ring heterocyclic compounds; stereochemistry and mechanism of organic reactions. *Mailing Add:* 30 Valley View Dr Morristown NJ 07960

WITTELS, BENJAMIN, b Minneapolis, Minn, Jan 22, 26; m 55; c 2. PATHOLOGY, BIOCHEMISTRY. *Educ:* Univ Minn, BA, 48, MD, 52; Am Bd Path, dipl, 57; Am Bd Clin Path, dipl, 70. *Prof Exp:* Assoc prof, 60-70, PROF PATH, MED CTR, DUKE UNIV, 70- *Mem:* Am Asn Path & Bact; Am Col Path. *Res:* Cardiac metabolism; hematology. *Mailing Add:* Dept Path Duke Med Ctr Durham NC 27710

WITTELS, MARK C, b Minneapolis, Minn, July 14, 21; m 51; c 3. MINERALOGY, SOLID STATE PHYSICS. *Educ:* Univ Minn, BS, 47; Mass Inst Technol, PhD(geol, ceramic eng), 51. *Prof Exp:* Staff scientist, Oak Ridge Nat Lab, 51-63; sr solid state physicist, US AEC, 63-74; CHIEF, SOLID STATE PHYSICS & MAT CHEM BR, DIV MAT SCI, US DEPT ENERGY, 74- *Concurrent Pos:* Vis prof, Wash Univ, 68-69. *Mem:* Mineral Soc Am; Am Crystallog Asn; Am Phys Soc. *Res:* Radiation effects in crystalline solids and in lunar materials; x-ray diffraction instrumentation; crystal growth techniques; defects in crystals; stored energy in reactor-irradiated graphite. *Mailing Add:* 8309 Still Spring Ct Bethesda MD 20817

WITTEMANN, JOSEPH KLAUS, b Heilbronn, Ger, Nov 13, 41; US citizen; m 65; c 3. DENTISTRY, PSYCHOLOGY. *Educ:* State Univ NY, BS, 64; Ohio State Univ, MA, 67, PhD(coun), 72. *Prof Exp:* Psychologist, Ohio Penitentiary, 67-68; asst prof psychol, Ohio Dominican Col, 69-72, chmn dept, 70-71; PROF GEN DENT, MED COL VA, 72-, DIR OFF EDUC EVAL, PLANNING & RES, 74- *Concurrent Pos:* Consult Curriculum Develop, Fac Develop, Human Servs Mgt, Res Int Prof Women Dent; mem ad hoc comt learning environ, Asn Am Med Schs, 73- *Mem:* Int Asn Dent Res; Am Educ Res Asn; Am Asn Dent Sch. *Res:* Clinical skills teaching; evaluation research program; pain control and analgesia; institutional research and management; management organization development. *Mailing Add:* Nine N Sheppard St Richmond VA 23221

WITTEN, ALAN JOEL, b Malden, Mass, Dec 11, 49; m 75; c 1. FLUIDS. *Educ:* Univ Rochester, BS, 71, MS, 72, PhD(mech eng), 75. *Prof Exp:* Mem staff, Thermal-Hydraulics Group, 75-78, mgr, Coal Gasification Environ Proj, 78-80, LEADER, APPL PHYS SCI GROUP, OAK RIDGE NAT LAB, 80- *Mem:* Am Geophys Union. *Res:* Geophysical fluid dynamics; ekman layers; diffusion equation with random ejection; atmospheric transport in complex terrain. *Mailing Add:* Bldg 2001 Oak Ridge Nat Lab Oak Ridge TN 37830

WITTEN, EDWARD, b Aug, 1951. PHYSICS. *Educ:* Brandeis Univ, BA, 71; Princeton Univ, MA, 74, PhD, 76. *Prof Exp:* Postdoctoral fel, Harvard Univ, 76-77, jr fel, Harvard Soc Fels, 77-80; prof physics, Princeton Univ, 80-87; PROF, SCH NATURAL SCI, INST ADVAN STUDY, 87- *Concurrent Pos:* MacArthur fel, 82. *Honors & Awards:* Einstein Medal, Einstein Soc Berne, Switz, 85; Award for Phys & Math Sci, NY Acad Sci, 85; Dirac Medal, Int Ctr Theoret Physics, 85; Alan T Waterman Award, NSF, 86; Fields Medal in Math, 90. *Mem:* Fel Nat Acad Sci; fel Am Phys Soc; fel Am Acad Arts & Sci. *Res:* Natural sciences. *Mailing Add:* Sch Natural Sci Inst Advan Study Olden Lane Princeton NJ 08540

WITTEN, GERALD LEE, b Davies County, Mo, May 12, 29; m 51; c 3. SCIENCE EDUCATION. *Educ:* Kans State Teachers Col, 56, MS, 58, EdS(phys sci), 62. *Prof Exp:* Teacher high sch, Kans, 56-62; ASSOC PROF PHYS SCI, EMPORIA KANS STATE COL, 62- *Mem:* Nat Sci Teachers Asn; Am Asn Physics Teachers. *Res:* Development of take home laboratory exercises for high school physics and general education physical science classes. *Mailing Add:* Dept Phys Sci Emporia State Univ 1200 Commercial St Emporia KS 66801

WITTEN, LOUIS, b Baltimore, Md, Apr 13, 21; m 48; c 4. PHYSICS. *Educ:* John Hopkins Univ, BE, 41, PhD(physics), 51; NY Univ, BS, 44. *Prof Exp:* Res assoc, Proj Matterhorn, Princeton Univ, 51-53; instr fluid mech, Univ Md, 52-53; staff scientist, Lincoln Lab, Mass Inst Technol, 53-54; prin scientist, Res Inst Advan Study, Martin Marietta Corp, Md, 54-65, assoc dir, 65-68; head dept physics, 68-74, PROF PHYSICS, UNIV CINCINNATI, 68- *Concurrent Pos:* Adj prof, Drexel Univ, 56-68; Fulbright lectr, Weizmann Inst Sci, Israel, 64-65; trustee, Gravity Res Found, 66-, vpres, 72- *Mem:* AAAS; Am Phys Soc; Am Math Soc; Am Asn Physics Teachers. *Res:* General theory of relativity; theory of particles and fields. *Mailing Add:* Dept Physics Univ Cincinnati Cincinnati OH 45221

WITTEN, MAURICE HADEN, b Jamesport, Mo, Dec 5, 31; m 69; c 3. PHYSICS. *Educ:* Emporia State Univ, BA, 56; Univ Nebr, Lincoln, MA, 60; Univ Iowa, PhD(sci educ, physics), 67. *Prof Exp:* Engr, Int Business Mach Corp, 56-57; from instr to assoc prof, 60-69, PROF PHYSICS, FT HAYS STATE UNIV, 69-, CHMN DEPT, 70- *Concurrent Pos:* NSF sci fac fel, 65-66. *Mem:* Am Asn Physics Teachers; Nat Sci Teachers Asn. *Res:* Nuclear emulsion techniques; physics education. *Mailing Add:* Dept Physics Ft Hays State Univ Hays KS 67601

WITTEN, THOMAS ADAMS, JR, b Raleigh, NC, Aug 24, 44. THEORETICAL PHYSICS. *Educ:* Reed Col, AB, 66; Univ Calif, San Diego, PhD(physics), 71. *Prof Exp:* Instr physics, Princeton Univ, 71-74; foreign collabr, Comn Atomic Energy, Saclay, France, 74-75; ASST PROF PHYSICS, UNIV MICH, ANN ARBOR, 75- *Mem:* Am Phys Soc. *Res:* Renormalization scaling symmetry in extended matter; excitations of systems with long-range order. *Mailing Add:* Univ Chicago 5640 S Ellis Ave Chicago IL 60637

WITTENBACH, VERNON ARIE, b Belding, Mich, Dec 13, 45; m 68; c 2. PLANT PHYSIOLOGY, HERBICIDE ACTION. *Educ:* Mich State Univ, BS, 68, MS, 70, PhD(hort), 74. *Prof Exp:* res biologist, Cent Res & Develop Dept, SR RES BIOLOGIST, AGR PROD DEPT, E I DU PONT DE NEMOURS & CO, 83- *Mem:* Am Soc Plant Physiologists; Am Soc Hort Sci. *Res:* Mechanism of action of herbicides. *Mailing Add:* 609 Greenbank Rd Wilmington DE 19808

WITTENBERG, ALBERT M, b Newark, NJ; m 61; c 2. PHYSICAL ELECTRONICS. *Educ:* Union Col, NY, BS; Johns Hopkins Univ, PhD(physics). *Prof Exp:* MEM TECH STAFF PHYSICS, BELL LABS, 55- *Mem:* Am Phys Soc; Soc Info Display. *Res:* Atomic structure of diatomic molecules; gaseous electronics; radiative heat transfer; optical spectroscopy; photoconductivity; interconnection technology; electron beam and solid state display systems. *Mailing Add:* 19 Exeter Rd Short Hills NJ 07078

WITTENBERG, BEATRICE A, b Berlin, Ger, Nov 6, 28; US citizen; m 54; c 3. BIOCHEMISTRY, PHYSIOLOGY. *Educ:* Univ Toronto, BA, 49, MA, 50; Western Reserve Univ, PhD(pharmacol), 54. *Prof Exp:* Res assoc physiol, Western Reserve Univ, 54-55; cancer res, Delafield Hosp, NY Univ, 55-56; from asst prof to assoc prof physiol, 64-85, PROF PHYSIOL & BIOPHYS, ALBERT EINSTEIN COL MED, 85- *Mem:* Am Soc Physiologists; Am Soc Biol Chem. *Res:* Heme proteins; oxygen supply; isolated adult heart cells. *Mailing Add:* Dept Physiol & Biophys Albert Einstein Col Med Bronx NY 10461

WITTENBERG, JONATHAN B, b New York, NY, Sept 19, 23; m 54; c 3. BIOCHEMISTRY, PHYSIOLOGY. *Educ:* Harvard Univ, BS, 45; Columbia Univ, MA, 46, PhD, 50. *Prof Exp:* From instr to asst prof biochem, Western Reserve Univ, 52-55; asst prof, 55, from asst prof to assoc prof physiol, 56-65, PROF PHYSIOL, ALBERT EINSTEIN COL MED, 65- *Concurrent Pos:* NIH res fel, 52. *Res:* Porphyrins; swimbladders; retia mirabilia; oxygen transport; myoglobin; heme proteins; hemoglobin. *Mailing Add:* Dept Physiol Albert Einstein Col Med Bronx NY 10461

WITTENBERGER, CHARLES LOUIS, bacteriology, biochemistry; deceased, see previous edition for last biography

WITTER, JOHN ALLEN, b Jamestown, NY, Sept 2, 43; m 71; c 1. FOREST ENTOMOLOGY, FOREST STRESS. *Educ:* Va Polytech Inst, BS, 65, MS, 67; Univ Minn, St Paul, PhD(entom), 71. *Prof Exp:* Res technician entom, Southeastern Forest Exp Sta, US Forest Serv, 66-67; res asst, Univ Minn, St Paul, 67-71, res fel, 71-72, instr forest entom, 72; from asst prof to assoc prof forest entom, 72-84, PROF FORESTRY, SCH NATURAL RESOURCES, UNIV MICH, ANN ARBOR, 84- *Mem:* Entom Soc Am; Entom Soc Can; Soc Am Foresters; AAAS. *Res:* Insect impact; forest tree stress; risk rating systems; technology transfer; intensive forestry; defoliators; gypsy moth; spruce budworm; tent caterpillars. *Mailing Add:* 1024 Dana Bldg Univ Mich Main Campus Sch Natural Resources Ann Arbor MI 48109-1115

WITTER, LLOYD DAVID, b Chicago, Ill, May 15, 23; m 50; c 3. FOOD MICROBIOLOGY, FOOD SCIENCE. *Educ:* Univ Wash, Seattle, BS, 45, MS, 50, PhD(microbiol), 53. *Prof Exp:* Asst microbiol, Univ Wash, Seattle, 52-53; res chemist, Continental Can Co, Ill, 53-56; from asst prof to assoc prof, 56-67, PROF FOOD MICROBIOL, UNIV ILL, URBANA, 67- *Concurrent Pos:* Vis prof appl biochem, Univ Nottingham, Eng, 72; vis prof food microbiol, Univ Montevideo, Uruguay, 77; vis prof, Polytech South Bank London, 80; vis scientist, Food Res Asn, Eng, 80. *Mem:* Fel Am Acad Microbiol; Am Soc Microbiol; Am Dairy Sci Asn; Inst Food Technol; NY Acad Sci. *Res:* Microbiology of food and dairy products; psychrophilic bacteria; bacterial growth on solid surfaces; heat resistance and injury of bacteria; food fermentations; osmoregution in microorganisms. *Mailing Add:* Dept Food Sci Univ Ill Urbana IL 61801

WITTER, RICHARD L, b Bangor, Maine, Sept 10, 36; m 62; c 2. POULTRY PATHOLOGY. *Educ:* Mich State Univ, BS, 58, DVM, 60; Cornell Univ, MS, 62, PhD, 64. *Prof Exp:* Res vet, Regional Poultry Res Lab, 64-75, DIR, REGIONAL POULTRY RES LAB, AGR RES SERV, USDA, 75-; CLIN PROF, MICH STATE UNIV, 71- *Concurrent Pos:* Asst prof, Mich State Univ, 64-71. *Honors & Awards:* Am Asn Avian Path Award, 67, 81 & 88; Res Award, Poultry Sci Asn, 71; Res Award, Sigma Xi, 75; Res Award, CPC Int, Inc, 76; Res Award, Mfg Asn Am Feed, 79; Bart Rispens Res Award, World Vet Poultry Asn, 83; Distinguished Serv Award, USDA, 85. *Mem:* Am Vet Med Asn; Am Asn Avian Path; Conf Res Workers Animal Dis; Poultry Sci Asn; Sigma Xi. *Res:* Epizootiology and control of poultry diseases, especially virology and pathology; viral-induced neoplasia, especially Marek's disease and lymphoid leukosis of chickens. *Mailing Add:* Regional Poultry Res Lab 3606 E Mt Hope Rd East Lansing MI 48823

WITTERHOLT, EDWARD JOHN, b Osceola Mills, Pa, Nov 12, 35; m 57; c 5. CROSSWELL TOMOGRAPHY, BOREHOLE GEOPHYSICS. *Educ:* Manhattan Col, BS, 57; Brown Univ, ScM, 59, PhD(appl math), 64. *Prof Exp:* Res proj mathematician, Schlumberger Technol Corp, 63-75; supvr anal & calibration, Seismograph Serv Corp, Tulsa, 75-78; res mgr borehole geophysics, Cities Serv Co, 78-83, res mgr appl geophysics, 83; mgr petrophysics, Cities Serv Oil & Gas, 83-84; MGR BOREHOLE GEOPHYS, B P EXPLOR PROD CO, 84- *Concurrent Pos:* Adj prof math, Univ Tulsa, 78-80; lectr, Am Asn Petrol Geologists, 82- *Mem:* Am Phys Soc; Acoust Soc Am; Soc Petrol Engrs; Soc Explor Geophysicists; Soc Prof Well Log Analysts; Soc Petrol Engr; Technol Comm; Soc Prof Log Analysts (pres Tulsa chap, 80). *Res:* Cross well tomography; reservoir delineation and monitoring; borehole geophysics; production geophysics; well logging; production logging; rock physics. *Mailing Add:* 5151 San Felipe Houston TX 77056

WITTERHOLT, VINCENT GERARD, b New York, NY, Sept 24, 32; m 54; c 6. ORGANIC CHEMISTRY. *Educ:* Queens Col, City Univ NY, BS, 53; Purdue Univ, PhD(org chem), 58. *Prof Exp:* Res chemist, Org Chem Dept, E I Du Pont De Nemours & Co Inc, 58-68, res supvr, 69-73, sr supvr sulfur colors area, Chamber Works, 73-74, chief supvr, Azo Lab, 74-75, div head, Chem Dyes & Pigments Dept, Jackson Lab, 75-80, res supvr, biochem dept, 81-84, from res felto sr res fel, 84-89, DU PONT FEL, AGRICHEM PROD DEPT, E I DU PONT DE NEMOURS & CO, INC, 89- *Mem:* Am Chem Soc. *Res:* Dyes product and process development; agricultural product and process development (herbicides, fungicides, and insecticides). *Mailing Add:* Du Pont Agr Prod Exp Sta E I du Pont de Nemours & Co Inc Wilmington DE 19880-0402

WITTERS, ROBERT DALE, b Cheyenne, Wyo, May 2, 29; m 87. PHYSICAL CHEMISTRY. *Educ:* Univ Colo, BA, 51; Mont State Univ, PhD(phys chem), 64. *Prof Exp:* Chemist, E I du Pont de Nemours & Co, 51-53; asst prof chem, State Univ NY Col Plattsburgh, 59-62 & Mont State Univ, 62-63; res fel, Harvey Mudd Col, 64-65; from asst prof to assoc prof, 65-78, PROF CHEM, COLO SCH MINES, 78- *Concurrent Pos:* Brown innovative teaching grant, 75. *Mem:* Am Chem Soc; Am Crystallog Asn; Sigma Xi. *Res:* X-ray crystallography. *Mailing Add:* Dept Chem Colo Sch Mines Golden CO 80401

WITTERS, WELDON L, reproductive physiology, biology; deceased, see previous edition for last biography

WITTICK, JAMES JOHN, b New York, NY, Aug 17, 30; m 56; c 4. ANALYTICAL CHEMISTRY, QUALITY CONTROL. *Educ:* Col Holy Cross, BS, 52; Tufts Univ, MS, 55; Univ Pa, PhD(chem), 66. *Prof Exp:* Chemist, Merck Sharp & Dohme Res Labs, NJ, 55-57, group leader phys & analytical res, 57-60, unit head pharmaceut analysis, Pa, 60-62, sr res chemist, 65-68, res fel, 68-70, assoc dir qual control, 70-80, DIR QUAL CONTROL OPERS, MERCK CHEM DIV, MERCK & CO, INC, 80- *Mem:* Am Chem Soc; AAAS. *Res:* Purity and structure determination of organic compounds; pharmaceutical analysis; electro-analytical chemistry; x-ray diffraction; chromatography. *Mailing Add:* 401 Stoney Brook Dr Bridgewater NJ 08807-1962

WITTIE, LARRY DAWSON, b Bay City, Tex, Mar 9, 43; m 72; c 2. INTELLIGENT SYSTEMS. *Educ:* Calif Inst Technol, BS, 66; Univ Wis-Madison, MS, 67, PhD(comput sci), 73. *Prof Exp:* Systs programmer, Calif Inst Technol, 63-66 & IBM Corp, Sunnyvale, 66; NASA trainee comput sci, Univ Wis-Madison, 66-69, res asst, 69-72; asst prof comput sci, Purdue Univ, 72-73; from asst prof to assoc prof, 73-87, PROF, STATE UNIV NY, BUFFALO, 87-; MEM STAFF, DEPT COMPUT SCI, STATE UNIV NY, STONY BROOK. *Concurrent Pos:* Nat lectr, Asn Comput Machinery, 78-79 & 79-80; prin investr, NSF, 77-, Air Force, 82-, NASA, 82-, Navy, 88-; sci adv, Army, 81; consult industry, 77- *Mem:* Asn Comput Mach; Sigma Xi; Inst Elec & Electronics Engrs; Soc Neurosci. *Res:* Distributed operating systems to control modular computers; parallel information processing in networks, with emphasis on computer architecture interconnection techniques for efficient communications among millions of microcomputers; simulation of large brain models; neural distributed memory mechanisms. *Mailing Add:* Dept Comput Sci State Univ NY 1426 Lab Office Bldg Stony Brook NY 11794-4400

WITTIG, GEORG FRIEDRICH KARL, b Berlin, Ger, June 16, 97. CHEMISTRY. *Educ:* Univ Tubingen, Ger, 16. *Prof Exp:* Prof & inst dir, Tubingen Univ, 44-56; prof, 56-67, EMER PROF CHEM, HEIDELBERG UNIV, 67- *Honors & Awards:* Nobel Prize in Chem, 79; Roger Adams Award, Am Chem Soc, 73. *Mem:* Nat Acad Sci; Swiss Chem Am. *Mailing Add:* Organisch-Chemisches Inst der Univ Heidelberg Im Neuenheimer Feld 7 Heidelberg Germany

WITTIG, GERTRAUDE CHRISTA, b Glauchau, Ger, Oct 4, 28; US citizen. HISTORY OF SCIENCE, INSECT PATHOLOGY. *Educ:* Univ Tuebingen, Dr rer nat(zool, bot, biochem), 55. *Prof Exp:* Teacher, Musterschule Glauchau, Ger, 46-47 & Preuniv Sch Neckarsulm, 55; Ger Res Asn prin investr, Zool Inst, Univ Tuebingen, 56-58; res fel entom, Univ Calif, Berkeley, 58-59; microbiologist, Insect Path Pioneering Lab, Entom Res Div, Agr Res Serv, USDA, Md, 59-62; res microbiologist, Forestry Sci Lab, Pac Northwest Forest & Range Exp Sta, US Forest Serv, Ore, 62-68; assoc prof biol sci, 68-75, PROF BIOL SCI, SOUTHERN ILL UNIV, EDWARDSVILLE, 75- *Concurrent Pos:* Lalor Found fel, 58; Ger Acad Exchange Serv & Ministry Educ Baden-Wuerttemberg res grants, 58; Fulbright travel grant, 58-59; consult, Univ Ariz, 61; adj assoc prof, Ore State Univ, 62-68. *Mem:* AAAS; Hist Sci Soc; Nat Women's Studies Asn; Am Inst Biol Sci; Am Asn Univ Prof. *Res:* Biological ultrastructure; morphology and histology of insects; insect pathology; history of science, especially gendeo studies. *Mailing Add:* Dept Biol Sci Southern Ill Univ Edwardsville IL 62026

WITTIG, KENNETH PAUL, b Pittsburgh, Pa, Aug 19, 46; m 69. COMPARATIVE PHYSIOLOGY. *Educ:* St Vincent Col, BS, 68; Kent State Univ, PhD(animal physiol), 74; State Univ NY, Albany, MBA, 84. *Prof Exp:* Teacher biol & chem, St Thomas Aquinas High Sch, 68-69; from asst prof to assoc prof biol, 74-80, chmn dept, 79-83, chmn, Long Range Planning, 83-84, PROF BIOL & DEAN SCI, SIENA COL, 84- *Mem:* AAAS; Sigma Xi; Am Inst Biol Sci; Am Soc Zoologists; NY Acad Sci. *Res:* Investigation of the neuroendocrine control of calcium metabolism in the crayfish; salt metabolism in amphibians. *Mailing Add:* Dept Biol Siena Col Loudonville NY 12211

WITTING, HARALD LUDWIG, b Duisburg, Ger, Sept 23, 36; US citizen; m 60; c 3. PLASMA PHYSICS. *Educ:* Mass Inst Technol, BS & MS, 59, ScD(plasma), 64. *Prof Exp:* PHYSICIST, GEN ELEC RES & DEVELOP CTR, 64- *Mem:* AAAS; Am Phys Soc; Sigma Xi. *Res:* Gas discharges; electrodes; plasma light sources. *Mailing Add:* 11 Wendy Lane Burnt Hills NY 12027

WITTING, JAMES M, b Chicago, Ill, Jan 14, 38; m 62; c 3. PHYSICAL OCEANOGRAPHY, HYDRODYNAMICS. *Educ:* John Carroll Univ, BS, 59, MS, 60; Mass Inst Technol, PhD(physics), 64. *Prof Exp:* Assoc physicist, IIT Res Inst, 64-66, res physicist, 66; asst prof hydrodyn, Dept Geophys Sci, Univ Chicago, 66-70; phys sci adminr, Off Naval Res, US Naval Res Lab, 70-73, oceanogr, Phys Oceanog Prog, 72-73, head, Phys Oceanog Br, 73-84; DIR, ADVAN TECHNOL OPERS CTR, ARC PROF SERV, ROCKVILLE, MD, 84- *Mem:* Am Phys Soc; Am Geophys Union. *Res:* Waves in dispersive media; water waves; undular bores; numerical modeling of oceanic circulation. *Mailing Add:* 9400 Athens Rd Fairfax VA 22030

WITTING, LLOYD ALLEN, b Chicago, Ill, May 18, 30; m 56; c 3. NUTRITION, LIPID CHEMISTRY. *Educ:* Univ Ill, BS, 52, MS, 53, PhD, 56. *Prof Exp:* Asst food technol, Univ Ill, 52-55; assoc biochemist, Am Meat Inst Found, Chicago, 55-57; proj assoc physiol chem, Univ Wis, 57-59; med res assoc, Mendel Res Lab, Elgin State Hosp, Ill, 59-68, actg dir, 68-70, res scientist, 69-72; assoc prof food sci & technol, Tex Woman's Univ, 72-74; consult, NTex Educ & Training Coop, Inc, 75-76; sr chemist, Supelco Inc, 76-77, tech dir biochem res & mfg, 77-84, DIR REGULATORY COMPLIANCE, SUPELCO INC, 84- *Concurrent Pos:* Asst prof, Col Med, Univ Ill, 62-72. *Honors & Awards:* Merit Award, Am Oil Chem Soc, 85. *Mem:* Am Inst Nutrit; Am Oil Chem Soc; Am Soc Biol Chemists. *Res:* High performance liquid chromatography; polyunsaturated fatty acids; glycolipids; clinical and animal nutrition; chemistry and biochemistry of lipids; tocopherol; gas chromatography. *Mailing Add:* Supelco Inc Supelco Park Bellefonte PA 16823-0048

WITTKE, DAYTON D, b Matador, Tex, Oct 9, 32; m 56; c 5. NUCLEAR & MECHANICAL ENGINEERING. *Educ:* Brigham Young Univ, BES, 56; Univ Ill, MS, 61, PhD(nuclear eng), 66. *Prof Exp:* Mech engr, Atomic Prod Div, Gen Elec Co, Wash, 56; res asst nuclear eng, Univ Ill, Urbana, 63-65; from asst prof to assoc prof mech & nuclear eng, Univ Nebr, Lincoln, 65-74; mgr, Opers Tech Support Serv, 74-77, mgr generating sta eng, 77-80, mgr, Eng Div, 80-85, VPRES, OMAHA PUB POWER DIST, 85- *Concurrent Pos:* Tech consult, Omaha Pub Power Dist, 67-74; consult, Wittke & Assocs, 70- *Mem:* Am Nuclear Soc; Am Soc Mech Engrs; Am Soc Eng Educ; Nat Soc Prof Engrs. *Res:* Heat transfer and fluid flow problems associated with nuclear engineering applications. *Mailing Add:* Omaha Pub Power Dist 444 S 16th St Mall Omaha NE 68102

WITTKE, JAMES PLEISTER, b Westfield, NJ, Apr 2, 28; m 52; c 2. OPTICAL PHYSICS. *Educ:* Stevens Inst Technol, ME, 49; Princeton Univ, MA, 52, PhD(physics), 55. *Prof Exp:* Instr physics, Princeton Univ, 54-55; MEM TECH STAFF, RCA LABS, 55- *Mem:* Fel Am Phys Soc; Inst Elec & Electronics Engrs; Optical Soc Am; Sigma Xi. *Res:* Microwave spectroscopy; masers; lasers; fiber optics; optical instrumentation. *Mailing Add:* 244 Russell Rd Princeton NJ 08540

WITTKE, PAUL H, b Toronto, Ont, Aug 17, 34; m 60; c 2. ELECTRICAL ENGINEERING. *Educ:* Univ Toronto, BASc, 56; Queen's Univ, Ont, MSc, 62, PhD(elec eng), 65. *Prof Exp:* Defense sci serv officer radar, Defence Res Telecommun Estab, Ont, 56-60; res assoc statist commun theory, Queen's Univ, 60-65, asst prof elec eng, 65-67, prof & head dept elec eng, 77-86, ASSOC PROF ELEC ENG, QUEEN'S UNIV, ONT, 67-, HEAD COMMUN GROUP, 68-, ASSOC DEAN RES, 90- *Concurrent Pos:* Acad vis, Imp Col, Univ London, 71-72; chmn, Comn C, Int Union Radio Sci Can, 76-82; vis prof, Space Systs Sect, Commun Res Ctr, Ottawa & Nat Univ Singapore, 82-83; chmn Can Nat Comt, Int Union Radio Sci, 87- *Mem:* Inst Elec & Electronics Engrs. *Res:* Radar signal processing; statistical communication theory; digital communications systems. *Mailing Add:* Dept Elec Eng Queen's Univ Kingston ON K7L 3N6 Can

WITTKOWER, ANDREW BENEDICT, b London, Eng, Nov 7, 34; m 57; c 2. ATOMIC PHYSICS. *Educ:* McGill Univ, BSc, 55; Cambridge Univ, MSc, 59; Univ London, PhD(atomic physics), 67. *Prof Exp:* Proj physicist, High Voltage Eng Corp, 59-64, res physicist, 64-67, assoc dir res, 67-71; sr vpres, Extrion Corp, 71-75, mkt & asst gen mgr, Extrion Div, Varian Assoc, 75-78; vpres & gen mgr, Nova Assoc, 78-80; vpres & gen mgr, Ion Implementation Div, Eaton Corp, 80-83; pres, Zymet Corp, 83-85; BD CHMN, IBIS TECH CORP, 87- *Honors & Awards:* Semmy Award, Semiconductor Equip Mfrs Inst, 86. *Mem:* Fel Am Phys Soc; fel Brit Inst Physics. *Res:* Atomic physics applied to the development of ion accelerators; author or coauthor of over 100 publications. *Mailing Add:* 352 Granite St Rockport MA 01966

WITTLE, JOHN KENNETH, b Lancaster, Pa, July 20, 39; c 1. INORGANIC CHEMISTRY, ANALYTICAL CHEMISTRY. *Educ:* Franklin & Marshall Col, AB, 62; Purdue Univ, Lafayette, PhD(inorg chem), 68. *Prof Exp:* Inorg chemist, 67-68, proj engr, 68-72, mgr polymer technol, 72-75, MGR DIELECTRIC MAT LAB, GEN ELEC CO, 75- *Mem:* AAAS; Am Chem Soc; Royal Soc Chem; Am Inst Mining, Metall & Petrol Eng. *Res:* Insulation systems for electrical systems; decomposition of electrical insulation under electrical stress; materials application. *Mailing Add:* 1740 Conestoga Rd Chester Springs PA 19425-1810

WITTLE, LAWRENCE WAYNE, b Mt Joy, Pa, Nov 20, 41; m 64, 77; c 2. PHYSIOLOGY. *Educ:* Lebanon Valley Col, BS, 63; Univ Va, PhD(biol), 68. *Prof Exp:* NIH fel, Inst Marine Sci, Univ Miami, 68-70; asst prof, 70-77, ASSOC PROF BIOL, ALMA COL, 77-, CHMN BIOL, 77- *Mem:* AAAS; Am Soc Zool; Int Soc Toxinology; Sigma Xi. *Res:* Physiological and pharmacological properties of marine toxins; parathyroid physiology of uradele amphibians. *Mailing Add:* Dept Biol Alma Col Alma MI 48801

WITTLIFF, JAMES LAMAR, b Taft, Tex, June 15, 38; m 62; c 2. MOLECULAR ENDOCRINOLOGY, CLINICAL BIOCHEMISTRY. *Educ:* Univ Tex, Austin, BA, 61, PhD(molecular biol), 67; La State Univ, MS, 63. *Prof Exp:* USPHS fel biochem regulation, Oak Ridge Nat Lab, Tenn, 67-69; asst prof biochem, Sch Med & Dent, 69-74, assoc prof biochem & head sect endocrine biochem, Cancer Ctr, Univ Rochester, 75-76; prof biochem & chmn dept, sch Med & Dent, 76-83, DIR, HORMONE RECEPTOR LAB, JAMES GRAHAM BROWN CANCER CTR, UNIV LOUISVILLE, 76- *Concurrent Pos:* Vis prof, Univ Dusseldorf, Institut fur Physiologische Chemie II, WGer, 74, Univ Innsbruck Frauenklinik, Austria, 76, SAfrican Asn Clin Biochemists, 82, Wolmarans Lab, Univ Pretoria, RSA, 88; vis prof biochem, La State Univ Sch Med, Shreveport, 79; fel, Nat Acad Clin Biochem, 84; spec lectureship, Cath Med Ctr, Seoul, Korea, 85; distinguished vis prof, SAfrican Inst Med Res, Johannesburg, 86; vis lectr, Univ Zimbabwe, Harare, 86; ed-in-chief, Clin Biochem, 89; MDS health lectureship award, Can Soc Clin Chem, 89. *Honors & Awards:* George Grannis Award Excellence in Res, Nat Acad Clin Biochem, 85; Special Award, Cath Med Col, Korea, 85; SKBL Award, Clin Ligand Assay Soc, 88; Pfizer Lectr, Montreal Gen Hosp, Can, 90. *Mem:* AAAS; Am Chem Soc; Am Asn Cancer Res; Am Soc Biol Chemists; Endocrine Soc; fel Nat Acad Clin Biochem; hon mem SAfrican Asn Clin Biochemists. *Res:* Hormonal control of protein and nucleic acid synthesis; role of specific hormone receptors in target cell response. *Mailing Add:* Hormone Receptor Lab James Graham Brown Cancer Ctr Univ Louisville Louisville KY 40292

WITTMAN, JAMES SMYTHE, III, b Ft Bragg, NC, Mar 1, 43; m 86; c 4. MANAGEMENT, NUTRITION. *Educ:* La Col, BA, 64; Tulane Univ, PhD(biochem), 70; Fairleigh Dickinson Univ, MBA, 78. *Prof Exp:* Chemist, Southern Regional Res Lab, USDA, 64-65; org chemist, US Customs Lab, La, 65-66; asst biochemist, Hoffmann-La Roche Inc, 68-70, sr biochemist, 70-77, clin res scientist, 77-79; tech dir, Batter-Lite Foods Inc, 80-81; PROF, ROCK VALLEY COL, 81- *Concurrent Pos:* Consult food indust, 81- *Mem:* Am Inst Nutrit. *Res:* Reduced-calorie food formulations; fructose; biochemical nutrition; regulation of intermediary metabolism and protein synthesis; clinical studies, vitamin E and calorie sweeteners; computers in management. *Mailing Add:* Dept Bus Rock Valley Col 3301 N Mulford Rockford IL 61101

WITTMAN, WILLIAM F, b Pittsburgh, Pa, Oct 10, 37; m 64; c 6. ORGANIC CHEMISTRY. *Educ:* Carnegie Inst Technol, BS, 59; Univ Nebr, PhD(org chem), 65. *Prof Exp:* Sr res chemist, 3M Co, 64-72, patent liaison specialist & patent agent, 72-76, patent & info supvr, 77, patent & res admin mgr, 78-82, staff intellectual property scientist, 85, CORP INTELLECTUAL PROPERTY SCIENTIST, 3M CO, 90. *Res:* Pharmaceutical, agrichemical, biomaterials and biopolymer patents. *Mailing Add:* 260-6B-05 3M Co 3M Ctr St Paul MN 55101

WITTMANN, HORST RICHARD, b Worms, Ger, Jan 31, 36; US citizen; div; c 3. SOLID STATE ELECTRONICS. *Educ:* Graz Uni, Phd(physics),64. *Prof Exp:* Asst exp physics,Graz Univ,63-64; space physicist, Boelkow, Ger, 64-66; res scientist quantum electronics, Phys Sci Lab, Missile Command, Redstone Arsenal, Ala, 66-70; chief, Elec Br, US Army Res Off, 70-75, asst dir, Electronics Div, 75-84; DIR, ELECTRONIC & MAT SCI, AIR FORCE OFF SCI RES, 84- *Concurrent Pos:* Res asst, Duke Univ, 71-73; adj prof, NC State Univ, 75-; Secy Army fel, 77; Fulbright fel, Tech Univ, Vienna, Austria, 77-78. *Mem:* Am Phys Soc; Inst Elec & Electronics Engrs. *Res:* Semiconductor lasers; solid state electronics; 3-5 compounds. *Mailing Add:* Air Force Off Sci Res AFOSR-NE Bldg 410 Bolling AFB Washington DC 20332

WITTMER, MARC F, b Basel, Switz, May 4, 45. SEMICONDUCTOR MATERIALS. *Educ:* Univ Basel, Switz, dipl, 70, PhD(solid state physics), 75. *Prof Exp:* Res scientist, solid state physics, Calif Inst Technol, 75-77; physicist res & develop, Brown Boveri Co, Baden, Switz, 77-81; res asst solid state physics, Swiss Fed Inst Technol, Zurich, 81; RES STAFF MEM SOLID STATE PHYSICS, IBM, INC T J WATSON RES CTR, 81- *Mem:* Inst Elec & Electronics Engrs; Mat Res Soc; Swiss Phys Soc; Europ Phys Soc; Sigma Xi. *Res:* Research in electronic and materials properties of metal-semiconductor interfaces and reactions, in contact and interconnect structures and materials for VLSI, in fabrication and properties of epitaxial layers on semiconductor materials. *Mailing Add:* Box 171 Col Greene Rd Yorktown Heights NY 10598

WITTMUSS, HOWARD D(ALE), b Papillion, Nebr, July 15, 22; m 50; c 3. AGRICULTURAL ENGINEERING. *Educ:* Univ Nebr, BS, 47, MS, 50, PhD(soil physics, agr eng), 56. *Prof Exp:* Construct engr, Diamond Eng Co, Nebr, 47-49; irrig engr, Irrig Res Div, Soil Conserv Serv, USDA, 49-50; asst soil physics, 52-56, asst prof agr eng, 56-58, ASSOC PROF AGR ENG, UNIV NEBR, LINCOLN, 58- *Mem:* Am Soc Agr Engrs; Soil Conserv Soc Am. *Res:* Irrigation efficiency; soil structure; tillplant system of corn production; land shaping; soil and water conservation. *Mailing Add:* 201 Agr Eng Univ Nebr Lincoln NE 68503

WITTNEBERT, FRED R, b Bayonne, NJ, July 8, 11; m 36; c 1. MECHANICAL ENGINEERING. *Educ:* Stevens Inst Technol, ME, 33. *Prof Exp:* From jr engr to works mgr, Sperry Prod, Inc, 33-41; chief engr, War Prod Div, Eversharp, Inc, 41-44; pres, Wittnebert-Jones Corp, 44-53; dir phys lab, Parker Pen Co, 53-54, dir labs, 54-58, tech dir, 58-67, vpres tech develop, 67-76; RETIRED. *Concurrent Pos:* Dir, Panoramic Corp, 67-76 & Omniflight Helicopters, Inc, 72-78; mem res adv comt & spec consult, Agency Int Develop, US State Dept, 74-78. *Res:* Metal flaw detection; engine and generator design; instrument development; control and servo design; capillary and surface phenomenon; research and engineering administration. *Mailing Add:* 3516 Crystal Springs Rd Janesville WI 53545

WITTNER, MURRAY, b New York, NY, Apr 23, 27; m 55; c 2. PHYSIOLOGY, PARASITOLOGY. *Educ:* Univ Ill, ScB, 48, ScM, 49; Harvard Univ, PhD, 55; Yale Univ, MD, 61. *Prof Exp:* Instr path & parasitol & consult path, 56-57, PROF PATH & PARASITOL & DIR PARASITOL LABS, ALBERT EINSTEIN COL MED, 67- *Concurrent Pos:* Career scientist, Health Res Coun, New York, 67-; attend physician, Bronx Munic Hosp Ctr, Lincoln Hosp & Albert Einstein Col Med Hosp; dir, Trop Dis Clin, Lincoln Hosp & Bronx Munic Hosp Ctr. *Mem:* Soc Protozool; Am Soc Cell Biol; Am Asn Path & Bact; Am Soc Zool; Am Soc Parasitol; Sigma Xi. *Res:* Physiology and biochemistry of oxygen poisoning; physiology of parasites; experimental pathology. *Mailing Add:* Dept Path Albert Einstein Col Med Bronx NY 10461

WITTROCK, DARWIN DONALD, b Primghar, Iowa, Oct 20, 49; m 79; c 2. HELMINTHOLOGY, ELECTRON MICROSCOPY. *Educ:* Univ Northern Iowa, BA, 71; Iowa State Univ, MS, 73, PhD(parasitol), 76. *Prof Exp:* PROF BIOL, UNIV WIS-EAU CLAIRE, 76- *Mem:* Am Soc Parasitologists; Am Micros Soc; Sigma Xi. *Res:* Ultrastructural studies on organ systems of digenetic trematodes and helminthological surveys from mammals. *Mailing Add:* Dept Biol Univ Wis Eau Claire WI 54701

WITTRY, DAVID BERYLE, b Mason City, Iowa, Feb 7, 29; m 55; c 5. SEMICONDUCTOR PHYSICS, ELECTRON OPTICS. *Educ:* Univ Wis, BS, 51; Calif Inst Technol, MS, 53, PhD(physics), 57. *Prof Exp:* Res fel, Calif Inst Technol, 57-59; from asst prof to assoc prof, 59-69, PROF MAT SCI & ELEC ENG, UNIV SOUTHERN CALIF, 69- *Concurrent Pos:* Consult, Appl Res Labs Inc, 58-83, Hughes Aircraft, 58-59, Exp Sta, E I du Pont de Nemours & Co, 62-71, NAm Aviation, 61-63, Gen Tel & Electronics Res Labs, 66-72, Electronics Res Div, Rockwell Int, 76-81, Atlanta Richfield Co Corp Technol Lab, 81-, Jet Propulsion Lab, 85- & Microbean Inc, 88-; Guggenheim fel, Univ Cambridge, 67-68; vis prof, Univ Osaha prefecture, 74 & Ariz State Univ, 81; consult, Atlanta Richfield Co Corp Technol Lab, 81-, Jet Propulsion Lab, 85-, Microbean Inc, 88-; coun mem, Microbean Anal Soc, 70-72, pres, 88; dir phys sci, Electron Micros Soc Am, 79-87, pres, 82-84. *Mem:* Am Phys Soc; Electron Micros Soc Am; Microbeam Anal Soc; Sigma Xi. *Res:* Scanning electron microprobe instrumentation; quantitative electron probe microanalysis, electron microprobe applications to solid state electronics, electron spectroscopy in TEM, secondary ion mass spectrometry; x-ray optics. *Mailing Add:* 1036 S Madison Pasadena CA 91106

WITTRY, ESPERANCE, b Marshall, Minn, Jan 13, 20. BIOLOGY. *Educ:* Col St Catherine, BA, 46; Univ Notre Dame, MS, 54, PhD(biol), 60. *Prof Exp:* Assoc prof, 60-71, PROF BIOL, COL ST CATHERINE, 71- *Mem:* Sigma Xi. *Res:* Physiology; radiation biology; neurophysiology. *Mailing Add:* Dept Biol Col St Catherine St Paul MN 55105

WITTRY, JOHN P(ETER), b Aurora, Ill, Sept 6, 29; m 51; c 6. AERONAUTICAL & ASTRONAUTICAL ENGINEERING. *Educ:* St Louis Univ, BS, 51; USAF Inst Technol, MS, 56; Univ Mich, AAE, 62. *Prof Exp:* USAF, 51-, proj off aircraft nuclear propulsion, Res & Develop Command, Wright-Patterson AFB, Ohio, 56-58, adv propulsion technologist, AEC, Hq, Germantown, Md, 58-60, instr astronaut, USAF Acad, 62-72, assoc prof astronaut & head dept astronaut & comput sci, 73-78, prof & head, Dept Astronautics & Comput Sci, 73-78, chmn eng div & vdean fac, USAF Acad, 78-84; ACAD DEAN, CALIF MARITIME ACAD, 84- *Mem:* Am Soc Eng Educ. *Res:* Space nuclear power systems; astrodynamics; inertial guidance and control systems. *Mailing Add:* CMA PO Box 1392 Vallejo CA 94590

WITTSON, CECIL L, psychiatry; deceased, see previous edition for last biography

WITTWER, JOHN WILLIAM, b Columbus, Ohio, Apr 35; m 59; c 3. PERIODONTOLOGY. *Educ:* Ohio State Univ, DDS, 59, MSc, 65. *Prof Exp:* Instr periodont, Ohio State Univ, 61-64; asst prof, Sch Dent, Loyola Univ Chicago, 66-69; assoc prof, 69-75, PROF PERIODONT, SCH DENT, UNIV LOUISVILLE, 75- *Mem:* Int Asn Dent Res; Am Acad Periodont. *Mailing Add:* 323 Bramton Rd Louisville KY 40207

WITTWER, LELAND S, b Belleville, Wis, Apr 26, 19. ANIMAL NUTRITION. *Educ:* Mich State Univ, BS, 52; Cornell Univ, MS, 54, PhD(dairy prod), 56. *Prof Exp:* Asst prof animal sci, Univ Mass, 56-58; PROF ANIMAL SCI, UNIV WIS-RIVER FALLS, 58- *Mem:* Sigma Xi; AAAS; Am Dairy Sci Asn; Am Soc Animal Sci. *Mailing Add:* Glarner Haus 700 Second Ave Apt 5204 New Glarus WI 53574

WITTWER, ROBERT FREDERICK, b Boonville, NY, Sept 18, 40; m 68; c 3. SILVICULTURE. *Educ:* State Univ NY Col Environ Sci & Forestry, BS, 66, PhD(forestry), 74. *Prof Exp:* Forester, NY State Dept Environ Conserv, 66-69; NDEA fel, State Univ NY Col Environ Sci & Forestry, 69-73; from asst prof to assoc prof forestry, Univ Ky, 74-80; ASSOC PROF FORESTRY, OKLA STATE UNIV, 82- *Mem:* Soc Am Foresters. *Res:* Forest soil productivity; forest fertilization; nutrient cycling; nutrition of forest trees. *Mailing Add:* Dept Forestry Okla State Univ Stillwater OK 74078

WITTWER, SYLVAN HAROLD, b Hurricane, Utah, Jan 17, 17; m 38; c 4. HORTICULTURE. *Educ:* Utah State Agr Col, BS, 39; Univ Mo, PhD(hort), 43. *Prof Exp:* Asst, Univ Mo, 40-43, instr hort, 43-46; from asst prof to emer prof hort, Mich State Univ, 46-86, dir, Agr Exp Sta, 65-83; agr mgt consult, USAID-Belize, 87-89; RETIRED. *Concurrent Pos:* Consult, Rockefeller Found, Mex, 68-69, Ford Found, Ceylon, 69- & UN Develop Prog, 71-; mem agr bd, Nat Acad Sci, 71-73; chmn bd agr & renewable resources, Nat Acad Sci-Nat Res Coun, 73-77; climate res bd, Nat Acad Sci, Nat Res Coun, 78-83; mem, US Cong Food Adv Bd, Off Technol Assessment, 77-82, Liaison Comt, food & agr, Int Inst Appl Systs Anal, 77-81, NASA Adv Coun, space & terrestrial appl, 78-82 & V I Lenin All-Union Acad of Agr Sci, USSR, 78-; tech adv agr adv groups, UNDP-Egypt, 86 & 90, China, 80, 81, 83, 85, 87 & 89. *Honors & Awards:* Campbell Award, AAAS, 57; Tanner Lectr Award, Inst Food Technol, 80; Am Farm Bur Fedn Serv Award, 82. *Mem:* Fel AAAS; Soc Develop Biol; Am Soc Hort Sci; Am Soc Plant Physiol; Bot Soc Am. *Res:* Physiology of reproduction in horticulture; plant growth regulators for improving fruit set and control flowering; nutrition of horticultural crops; radioisotopes in mineral nutrition of plants; agricultural communications; minimizing agricultural production; biological limits in agricultural productivity. *Mailing Add:* 1767 Hitching Post East Lansing MI 48823

WITULSKI, ARTHUR FRANK, b Denver, Colo, May 27, 58. RESONANT CONVERTERS, HIGH-POWER-FACTOR RECTIFIERS. *Educ:* Univ Colo, BS, 81, MS, 86, PhD(elec eng), 88. *Prof Exp:* Design engr, Storage Technol Corp, 81-83; teaching asst power electronics, Dept Elec & Computer Eng, Univ Colo, 83-84, res asst, 84-88, res assoc, 88-89; ASST PROF POWER ELECTRONICS, DEPT ELEC & COMPUTER ENG, UNIV ARIZ, 89- *Mem:* Inst Elec & Electronics Engrs. *Res:* Power electronics, especially resonant and quasi-resonant dc-dc switching converters and regulators; high-power-factor ac to dc conversion; distributed electronic power systems; design and analysis of high-frequency ferrite magnetic circuit components. *Mailing Add:* Dept Elec & Computer Eng Univ Ariz Tucson AZ 85721

WITZ, DENNIS FREDRICK, b Milwaukee, Wis, Dec 10, 38; div; c 4. MICROBIOLOGY. *Educ:* Carroll Col, Wis, BS, 61; Univ Wis-Madison, MS, 64, PhD(bact), 67. *Prof Exp:* RES ASSOC MICROBIOL, UPJOHN CO, 67- *Mem:* Am Soc Microbiol; Soc Indust Microbiol. *Res:* Process of biological nitrogen fixation by microorganisms; biosynthesis of antibiotics and secondary metabolites; production of antibiotics by fermentation. *Mailing Add:* 1400-89-1 Upjohn Co Kalamazoo MI 49081

WITZ, GISELA, b Breslau, Ger, Mar 16, 39; US citizen. CANCER. *Educ:* NY Univ, BA, 62, MS, 65, PhD(phys org chem), 69. *Prof Exp:* Fel biochem, Sloan-Kettering Inst Cancer Res, 69-70; assoc res scientist cancer res, NY Univ, Med Ctr, 70-73, res scientist environ med, 73-77, asst prof environ med, 77-80; asst prof commun med, Rutgers Med Sch, 80-86; ASSOC PROF ENVIRON & COMMUN MED, ROBERT WOOD JOHNSON MED SCH, UNIV MED & DENT NJ, 86- *Mem:* NY Acad Sci; Am Asn Cancer Res; Am Chem Soc; Sigma Xi; Soc Toxicol. *Res:* Chemical carcinogenesis; mode of action of tumor promoters; mechanism of benzene hematoxicity and mechanisms of oxidant injury. *Mailing Add:* 240 Lurline Dr Basking Ridge NJ 07920

WITZ, RICHARD L, b New Lisbon, Wis, Jan 12, 16; m 41; c 2. AGRICULTURAL ENGINEERING. *Educ:* Univ Wis, BS, 39; Purdue Univ, MS, 42. *Prof Exp:* Asst, Purdue Univ, 39-42; exten specialist, Mich State Univ, 42-45; from asst prof & asst agr engr to assoc prof agr eng & assoc agr engr, 45-57, prof 57-83, EMER PROF AGR ENG & AGR ENG, EXP STA, NDAK STATE UNIV, 83- *Honors & Awards:* George W Kable Award, Am Soc Agr Engrs. *Mem:* Fel & sr mem Am Soc Agr Engrs; Am Soc Eng Educ. *Res:* Rural electrification; building design; sewage disposal; farm water treatment. *Mailing Add:* 1525 N 28th St Fargo ND 58105

WITZEL, DONALD ANDREW, b Artesian, SDak, Sept 9, 26. VETERINARY PHYSIOLOGY. *Educ:* Univ Minn, BS, 53, DVM, 57; Iowa State Univ, MS, 65, PhD(vet physiol), 70. *Prof Exp:* Res vet physiol, Nat Animal Dis Lab, Agr Res Serv, USDA, 61-72, vet med officer physiol & toxicol, Vet Toxicol & Entom Res Lab, 72-90; RETIRED. *Concurrent Pos:* Consult, Baylor Col Med, 72-74. *Mem:* AAAS; Am Vet Med Asn; NY Acad Sci; Am Soc Vet Physiologists & Pharmacologists. *Res:* Electrophysiological studies of the visual system of domestic animals as related to toxicological problems. *Mailing Add:* 3720 Sweetbriar Dr Bryan TX 77802

WITZELL, O(TTO) W(ILLIAM), b Baltimore, Md, Nov 14, 16; m 42. MECHANICAL ENGINEERING. *Educ:* Johns Hopkins Univ, BE, 37; Purdue Univ, MSME, 49, PhD, 51. *Prof Exp:* Marine engr, US Maritime Comn, 40-46; from instr to prof mech eng, Purdue Univ, 46-64; prof & chmn dept, Univ Calif, Santa Barbara, 64-66; dean, Grad Sch, Drexel Univ, 66-83, prof mech eng, 76-83; RETIRED. *Concurrent Pos:* Consult, US Steel Corp, 57- & Allison Div, Gen Motors Corp, 58-59; prog dir, off inst prog, NSF, 62-63. *Mem:* AAAS; Am Soc Mech Engrs; Am Soc Eng Educ. *Res:* Determination of physical and chemical thermodynamic properties. *Mailing Add:* 9306 Pebble Creek Dr Lutz FL 33549

WITZEMAN, JONATHAN STEWART, b Phoenix, Ariz, June 18, 57; m 80; c 2. POLYMER CHEMISTRY, ORGANIC CHEMISTRY. *Educ:* Northern Ariz Univ, BS, 79; Univ Calif, Santa Barbara, PhD(org chem), 84. *Prof Exp:* Postdoctoral assoc, Dept Chem, Univ Chicago, Chicago, Ill, 84-85; PRIN CHEMIST, EASTMAN KODAK INC, EASTMAN CHEM CO, KINGSPORT, TENN, 85- *Concurrent Pos:* Adj prof, Dept Chem, ETenn State Univ, Johnson City, 91. *Mem:* Am Chem Soc; Sigma Xi. *Res:* Mechanistic study of reactions of industrial importance; development of new monomers, crosslinkers and polymers for coating applications. *Mailing Add:* Eastman Chem Co B-230 Kingsport TN 37662

WITZGALL, CHRISTOPH JOHANN, b Hindelang, Ger, Feb 25, 29; US citizen; m 64; c 3. OPERATIONS RESEARCH, NUMERICAL ANALYSIS. *Educ:* Univ Munich, PhD(math), 58. *Prof Exp:* Res assoc math, Princeton Univ, 59-60, Univ Mainz, 60-62 & Argonne Nat Lab, 62; mathematician, Nat Bur Standards, 62-66 & Boeing Co, 66-73; actg chief, Opers Res Div, 79-82, MATHEMATICIAN, CTR COMPUTER APPL MATH, NAT INST STANDARDS & TECHNOL, 73- *Concurrent Pos:* Vis prof, Univ Tex, Austin, 71 & Univ Würzburg, 72; assoc ed, Math Prog, 73-82. *Mem:* Soc Indust & Appl Math; AAAS; Am Math Soc; Opers Res Soc Am. *Res:* Further development of operations research, numerical analysis and programming languages as needed for planning and systems applications; computational geometry. *Mailing Add:* Ctr Comput Appl Math Nat Inst Standards & Technol Gaithersburg MD 20899

WITZIG, WARREN FRANK, b Detroit, Mich, Mar 26, 21; m 42; c 4. NUCLEAR ENGINEERING, PHYSICS. *Educ:* Rensselaer Polytech Inst, BS, 42; Univ Pittsburgh, MS, 44, PhD(physics), 52. *Prof Exp:* Res engr, Westinghouse Elec Corp, 42-46, from engr & scientist to proj mgr, Bettis Labs, 46-60; sr vpres & dir, NUS Corp, 60-67; prof nuclear eng & head dept, PA State Univ, University Park, 67-86; EMER PROF & DEPT HEAD, PA STATE UNIV, UNIVERSITY PARK, 86- *Concurrent Pos:* Mem nuclear standards bd, US Am Standards Inst, 65; mem comt radioactive waste mgt, Nat Acad Sci; mem, Pa Gov Adv Comt Nuclear Energy & Asn Am Univs Rev Comt on EBR II Fast Breeder Reactor; bd dirs, Gen Pub Utilities Nuclear Corp, 84-; mem, PSE&G, Nuclear Oversight Comt, 83-91, TV Operating Rev Comt, 85-, Tenn Valley Auth Nuclear Safety Rev Bd, 86-91; chmn, Westinghouse Nuclear Safety & Environ Oversight Comt, 88- *Mem:* Inst Elec & Electronics Engrs; fel Am Nuclear Soc; Am Phys Soc; fel,AAAS; Sigma Xi. *Res:* Nuclear reactor engineering and physics; reactor safety; nuclear safeguards; heat transfer; nuclear fuel costs; reactor plant siting; reactor design and operation. *Mailing Add:* 1330 Park Hills Ave E State College PA 16803

WITZLEBEN, CAMILLUS LEO, b Dickinson, NDak, Apr 20, 32; m 56; c 6. PATHOLOGY. *Educ:* Univ Notre Dame, BS, 53; St Louis Univ, MD, 57. *Prof Exp:* Resident path, St Louis Univ, 57-60; NSF fel, Hosp Sick Children, London, Eng, 60-61; fel, Harvard Univ, 61-62; dir labs, Children's Hosp Med

Ctr Northern Calif, 64-66; dir labs, Cardinal Glennon Mem Hosp Children, St Louis, 66-73; PROF PATH & PEDIAT, UNIV PA, 73-; DIR PATH, CHILDREN'S HOSP PHILADELPHIA, 73- *Concurrent Pos:* Captain, USAF, 62-64; consult, San Francisco Gen Hosp, 64-66; asst prof, Univ Calif, 64-66; from asst prof to prof, St Louis Univ, 66-73; NIH res grant, 68-74. *Mem:* Pediat Path Soc. *Res:* Hepatobiliary system; pediatric disease. *Mailing Add:* Path Med Pediat Univ Pa Childrens Hosp 34th St & Civic Center Blvd Philadelphia PA 19104

WITZMANN, FRANK A, b Bucyrus, Ohio, June 6, 54; m 77; c 2. MUSCLE PHYSIOLOGY & BIOCHEMISTRY, EXERCISE PHYSIOLOGY. *Educ:* Defiance Col, BA, 76; Ball State Univ, MS, 78; Marquette Univ, PhD(biol), 81. *Prof Exp:* Asst prof biol, Marquette Univ, 81-82; asst prof, Defiance Col, 82-85; ASST PROF BIOL, IND UNIV-PURDUE UNIV, COLUMBUS, 85- *Mem:* Sigma Xi; Am Col Sports Med. *Res:* Role of humoral mechanisms in muscle atrophy and endocrine influence on muscle; qualitative changes induced in skeletal and cardiac muscle by anabolic steroids and growth hormone. *Mailing Add:* Ind Univ-Purdue Univ Columbus 2080 Bakalar Dr Columbus IN 47203

WIXOM, ROBERT LLEWELLYN, b Philadelphia, Pa, July 6, 24; m 86; c 2. BIOCHEMISTRY, NUTRITION. *Educ:* Earlham Col, AB, 47; Univ Ill, PhD(biochem), 52. *Prof Exp:* Asst biochem, Univ Ill, 48-52; from instr to assoc prof, Sch Med, Univ Ark, 52-64; assoc prof, 64-72, PROF BIOCHEM, SCH MED, UNIV MO-COLUMBIA, 72- *Concurrent Pos:* Fel, Univ Ill, 55; Lalor Found res fel, 58; NIH spec res fel, 70-71; NIH res serv fel, 78-79; Univ Mo res develop fel, 86. *Mem:* AAAS; Am Soc Biochem & Molecular Biol; Am Chem Soc; Soc Exp Biol & Chem; Am Inst Nutrit; Sigma Xi; Protein Soc. *Res:* Requirements of essential amino acids in man; glycine serine in chick nutrition; biosynthesis of amino acids in microorganisms and plants; inborn errors of amino acid metabolism; total parenteral administration of amino acids; role of histidine in man; relation of iron/ferritin/hemosiderin; role of ubiquitin in protein degradation. *Mailing Add:* Dept Biochem Sch Med Univ Mo Columbia MO 65212

WIXSON, BOBBY GUINN, b Abilene, Tex, Mar 19, 31; m 52; c 2. AQUATIC BIOLOGY, SANITARY ENGINEERING. *Educ:* Sul Ross State Col, BS, 60, MA, 61; Tex A&M Univ, PhD(biol oceanog), 67. *Prof Exp:* Asst instr sci, Sul Ross State Col, 58-60, instr, 60-61, dir col planetarium & asst to dean of men, 59-61; asst oceanog, Tex A&M Univ, 61-63, asst aquatic biol, 63-65, asst water pollution res, 65-67; from asst prof to prof environ health, Univ Mo-Rolla, 67-87, dir, Int Ctr, 69-84; actg provost, 83, dean, Int Progs, 85-87, DEAN, COL SCI, CLEMSON UNIV, 87- *Concurrent Pos:* Consult, United Nations Environ Prog, US Environ Protection Agency; Proj Hope, Brazil; secy-treas, US Found Int Econ Policy. *Mem:* Nat Water Pollution Control Fedn; Am Water Resources Asn; Soc Environ Geochem & Health (pres, 78-81); Sigma Xi. *Res:* Aquatic pollution and industrial waste disposal problems; environmental, earth & marine sciences. *Mailing Add:* Kinard Lab Rm 120 Clemson Univ Clemson SC 29634

WIXSON, ELDWIN A, JR, b Winslow, Maine, Nov 30, 31; m 54, 76; c 6. MATHEMATICS OF FINANCE. *Educ:* Univ Maine, BS, 53; Colby Col, MST, 62; Temple Univ, MSEd, 62; Univ Mich, PhD, 69. *Prof Exp:* Teacher, High Sch, Maine, 53-54 & 57-60; TV teacher, Maine Dept Educ, 60-61; teacher, High Sch, Maine, 62-63; assoc prof math, Keene State Col, 63-65; assoc prof, 66-69, chmn dept, 66-83, acad dean, 83-86, PROF MATH, PLYMOUTH STATE COL, 69- *Concurrent Pos:* Pvt financial consult, 66- *Mem:* Math Asn Am; Nat Coun Teachers Math. *Res:* Mathematics education, especially at the undergraduate college level; financial matters, especially annuities of all types and real estate or business financing. *Mailing Add:* Dept Math Plymouth State Col Plymouth NH 03264

WLODEK, STANLEY T, b Haiduki, Poland, Sept 23, 30; US citizen; m 55; c 3. METALLURGY, PHYSICAL METALLURGICAL & MATERIALS SCIENCE ENGINEERING. *Educ:* Queens Univ, Ont, BSc Hons, 52; Mass Inst Technol, Sm, 54, ScD, 56. *Prof Exp:* Unit mgr, Metals Res Lab, Union Carbide Corp, 56-60, AEBG, 60-64 & Graham Lab, Jal Steel, 64-69; res dir, Cabot Corp, Kokomo, Ind, 69-79; unit mgr, GE AEBG, 60-64, mgr proc develop, GE TTL Lab, 79-85, SR STAFF ENGR, GEN ELEC AIRCRAFT ENGINES, 85- *Mem:* Fel Am Soc Metals Inc; Minerals Metals & Mat Soc. *Res:* Physical and process metallurgy of high performance Ni, Co and Fe base alloys: corrosion and coating technology; structural studes of super alloys to elucidate structure- property interactions, particulary of jet engine alloys. *Mailing Add:* Gen Elec Aircraft Engines Bldg 500 M-87 Cincinnati OH 45215

WNUK, MICHAEL PETER, b Katowice, Poland, Sept 12, 36; US citizen; m 64; c 1. ENGINEERING MECHANICS. *Prof Exp:* Asst prof physics, Tech Univ Krakow, 59-64, assoc prof, 64-66; from asst prof to assoc prof mech eng, SDaK State Univ, 66-82; PROF ENG MECH, UNIV WIS, 82- *Concurrent Pos:* Sr res fel, Calif Inst Technol, 67-68; distinguished vis scholar, Cambridge Univ, 68-70; vis scholar, Acad Mining & Metall, Krakow, 74; Nat Acad Sci sponsored vis prof & co-ed of proceedings, Fac Technol & Metall, Univ Belgrade, 80; vis prof, Northwestern Univ, 80-81. *Mem:* Sigma Xi; NY Acad Sci. *Res:* Solid mechanics with particular emphasis on mechanics of fracture; initiation and subsequent propagation of fracture in non-linear range of material behavior (ductile and time-dependent fracture). *Mailing Add:* 3436 Dousman Apt D Milwaukee WI 53212

WOBESER, GARY ARTHUR, b Regina, Sask, Feb 12, 42; m 65; c 2. WILDLIFE PATHOLOGY. *Educ:* Univ Toronto, BSA, 63; Univ Guelph, MSc, 66, DVM, 69; Univ Sask, PhD(vet path), 73. *Prof Exp:* Assoc prof, 73-78, PROF VET PATH, WESTERN COL VET MED, UNIV SASK, SASKATOON, 78- *Concurrent Pos:* Coun mem, Wildlife Dis Asn, 75-78, chmn student awards comt, 76. *Mem:* Wildlife Dis Asn; Can Soc Environ Biologists; Can Asn Vet Pathologists; Am Fisheries Soc; Can Vet Med Asn. *Res:* Diseases of wild life, with particular emphasis on infectious, degenerative and toxic problems. *Mailing Add:* Dept Vet Path Western Col Vet Med Univ Sask Saskatoon SK S7N 0W0 Can

WOBSCHALL, DAROLD C, b Wells, Minn, Feb 24, 32; m 57; c 3. BIOPHYSICS, ELECTRICAL ENGINEERING. *Educ:* St Olaf Col, BA, 53; State Univ NY, Buffalo, MA, 60, PhD(biophys), 66. *Prof Exp:* Res assoc physics, Univ Buffalo, 58-60; assoc physicist, Cornell Aeronaut Lab, 60-62; cancer res scientist, Roswell Park Mem Inst, 66-67; from asst prof to assoc prof eng, State Univ NY, Buffalo, 67-85; pres, Index Electronics Inc, 85-90; VPRES, SENSOR PLUS, 90- *Concurrent Pos:* NIH spec fel, 66-67. *Mem:* Am Phys Soc; Biophys Soc; Inst Elec & Electronics Engrs. *Res:* Sensors; bioengineering; electronics instrumentation. *Mailing Add:* 25 Blossom Heath Williamsville NY 14221

WOBUS, REINHARD ARTHUR, b Norfolk, Va, Jan 11, 41; m 67; c 2. GEOLOGY. *Educ:* Wash Univ, AB, 62; Harvard Univ, MA, 63; Stanford Univ, PhD(geol), 66. *Prof Exp:* From asst prof to assoc prof, 66-78, PROF GEOL, WILLIAMS COL, 78-, DEPT HEAD, 88- *Concurrent Pos:* Geologist, Cent Regional Geol Br, US Geol Surv, 67-86; vis prof geol, Colo Col, 76, 82-83; instr geol field course, Colo State Univ, summers, 77-84; coun, Council Undergrad Res, 87-89. *Mem:* Fel Geol Soc Am; Nat Asn Geol Teachers; Am Geophys Union; Sigma Xi. *Res:* Igneous and metamorphic petrology; Precambrian geology of Southern Rocky Mountains; mid-tertiary volcanism in central Colorado. *Mailing Add:* Dept Geol Williams Col Williamstown MA 01267

WOCHOK, ZACHARY STEPHEN, plant physiology, for more information see previous edition

WODARCZYK, FRANCIS JOHN, b Chicago, Ill, Dec 11, 44. MOLECULAR DYNAMICS, SPECTROSCOPY. *Educ:* Ill Inst Technol, BS, 66; Harvard Univ, AM, 67, PhD(chem), 71. *Prof Exp:* Lectr chem, Harvard Univ, 69; res assoc, Univ Calif, Berkeley, 71-73; res chemist, Cambridge Res Labs, USAF, 73-77, prog mgr, Off Sci Res, 77-78; mem tech staff, Rockwell Int Sci Ctr, 78-85; prog mgr, USAF Off Sci Res, 85-90; PROG DIR, NSF, 90- *Concurrent Pos:* Lectr, Univ Calif, Berkeley, 73. *Mem:* Am Chem Soc; AAAS; Sigma Xi. *Res:* Atomic and molecular spectroscopy; energy transfer and reaction kinetics; laser physics and chemistry; surface analysis; thin film coatings research. *Mailing Add:* Chem Div NSF 1800 G St NW Washington DC 20550

WODARSKI, JOHN STANLEY, b Philadelphia, Pa, Feb 27, 43; m 64; c 1. RESEARCH ADMINSTRATION, HEALTH SCIENCES. *Educ:* Fla State Univ, BS, 65; Univ Tenn, MSSW, 67; Wash Univ, St Louis, Mo, PhD(social work), 70. *Prof Exp:* Instr sociol, Sam Houston State Univ, 67-68; asst prof social work, George Warren Brown Sch Social Work, Wash Univ, 70-74; assoc prof social work, Sch Social Work, Univ Md, 75-78; prof social work & dir res ctr, Sch Social Work, Univ Ga, 78-88; ASSOC VPRES RES & GRAD STUDIES, UNIV AKRON, 88- *Concurrent Pos:* Res dir, Ctr Studies Crime & Delinquency, NIMH, 70-75; adj fac res scientist, Inst Behav Res & Ctr Social Orgn Sch, Johns Hopkins Univ, 75-77; prin investr, spec progs proj grants, Univ Ga, 78-80, grad sch fac award, 79, res found grant, 80-82 & 82-83, social work training br grant, NIMH, 79-83, social work educ br grant, 83-86, Ga Dept Human Resources grant, 82-83, Ford Found grant, 82-83, Edna McConnell-Clark Found, Ga Alliance Children & Trust Fund, Atlanta grant, 83-84, US dept transp grant, 83-86 & Gov emergency fund grant, Ga state, 84-85; co-prin investr, Ga Dept Human Resources grant, 81-84 & Ga Dept Community Affairs grant, 83-84; prin investr grants, US Dept Educ, Univ Affil Prog, 86-88, Edward W Hazen Found, 86-87, US Dept Health & Human Serv, 87-; res assoc, Dept Health & Human Serv, 86-88. *Mem:* Am Psychol Asn; Am Sociol Asn; Asn Advan Behav Ther; Coun Social Work Educ; Nat Asn Social Workers; Nat Coun Crime & Delinquency. *Res:* Mental health programs with emphasis on the provision of services to rural areas and social work training for minorities; services for children and youth, specifically in regard to family violence and prevention of alcohol abuse; aging with emphasis on training social workers to work with older adults; health; school social work; projects on energy conservation for college campuses, comprehensive treatment for college and high school students who abuse alcohol, treatment for families with runaway children and evaluation of different child welfare procedures. *Mailing Add:* Res & Grad Studies Rm 208 Univ Akron 138 Fir Hill Akron OH 44325

WODICKA, VIRGIL O, b St Louis, Mo, Mar 5, 15; m 41; c 2. FOOD TECHNOLOGY. *Educ:* Rutgers Univ, PhD(food sci), 56. *Prof Exp:* Food technol consult, 75-89; RETIRED. *Mem:* Inst Food Technol; Am Soc Qual Control; Am Chem Soc; Am Asn Cereal Chemists; Am Oil Chemists Soc; AAAS; Am Inst Nutrit. *Mailing Add:* 1307 Norman Pl Fullerton CA 92631

WODINSKY, ISIDORE, b New York, NY, Mar 6, 19; m 42; c 2. ONCOLOGY. *Educ:* Brooklyn Col, BA, 39; George Washington Univ, MS, 51. *Prof Exp:* Biologist, Nat Cancer Inst, 46-55, sci adminr, 55-57; HEAD CANCER CHEMOTHER RES, LIFE SCI SECT, ARTHUR D LITTLE, INC, 59- *Mem:* Int Union Against Cancer; Am Asn Cancer Res; Int Soc Chemother; NY Acad Sci. *Res:* Chemotherapy of cancer; carcinogenesis; biology and kinetics of experimental neoplasms; cryobiology. *Mailing Add:* Two Driftwood Rd Jamaica Plain MA 02130

WODZICKI, ANTONI, b Krakow, Poland, July 15, 34; US citizen; m 62; c 4. GEOLOGY. *Educ:* Univ Otago, NZ, BE, 56; Univ Minn, MS, 61; Stanford Univ, PhD(geol), 65. *Prof Exp:* Geologist econ geol, NZ Geol Surv, 54-75, geologist petrol, 58-59; vis asst prof mineral & geochem, Portland State Univ, 75-76; vis asst prof mineral & econ geol, Univ Ore, 76-77; from asst prof to assoc prof, econ geol & mineral, 77-88, PROF ECON GEOL, WESTERN WASH UNIV, 88- *Concurrent Pos:* Actg chief petrologist, NZ Geol Surv, 69-70; mem mineral resources comt, NZ Nat Develop Conf, 69-70; hon lectr, Victoria Univ, Wellington, 74-75; proj geologist, Nat Uranium Resource Eval, 78-80 & Wildreness Study Area Assessment, 82-83; exchange fel, Polish Acad Sci, 85; geol consult, Haitian Govt, 88. *Mem:* Royal Soc NZ; NZ Geol Soc; NZ Geochem Group (chmn, 71-73); Polish Geol Soc. *Res:* Field geology, economic geology, petrology, geochemistry and particularly the application of these disciplines to the finding, evaluation and understanding of the origin of ore deposits. *Mailing Add:* Dept Geol Western Wash Univ 516 High St Bellingham WA 98225

WODZINSKI, RUDY JOSEPH, b Chicago, Ill, June 12, 33; m 56; c 3. MICROBIAL BIOCHEMISTRY. *Educ:* Loyola Univ, Ill, BS, 55; Univ Wis, MS, 57, PhD(bact), 60. *Prof Exp:* Asst bact, Univ Wis, 55-60; sr res scientist, Squibb Inst Med Res, New Brunswick, NJ, 60-62; res microbiologist, Int Minerals & Chem Corp, Ill, 62-64, supvr microbial biochem, 64-68, mgr animal sci, 68-70; prof biol sci, Univ Cent Fla, 70-75 & 77-89, Gordon J Barnett prof environ sci, 75-77, PROF MOLECULAR BIOL & MICROBIOL, UNIV CENT FLA, 89- *Concurrent Pos:* Consult, microbiol of the eye, Frontier Contact Lenses, 76-81, viruses in waste water, WORLDCO, 77-80, methane generation, Anaerobic Energy Systs, Inc, 78-82, fermentation-biotechnology, William Underwood Co, 76-83, Pet Inc, 83-85, biotechnology investments, G K Scott & Co, 82-84, Rooney PAC, Chahill, 89-; nat coun, Fla br, Am Soc Microbiol, 78-81, vchmn, Environ, Gen & Appl Microbiol, 76-77, chmn, 77-78, mem bd, Pub Sci Affairs, 85-, Comt Agr, Food & Ind Microbiol, 79-85, chmn, 85-, counr Environ, Gen & Appl Div, 85-; mem, bd dirs & secy-treas, Convert-EDA, 79- & Energy Develop Assocs, 87-, Gov Energy Res Task Force, State Fla, 80-, adv panel, Fla Solar Energy Res Ctr, 83-, adv coun, Inst Food, Arg Sci Ctr Biomass, Univ Fla, 83. *Mem:* Am Soc Microbiol; fel Am Acad Microbiol; Sigma Xi. *Res:* Enzymology; molecular biology; virology; microbial genetics and physiology; fermentation; available moisture requirements of microorganisms; water and waste microbiology. *Mailing Add:* Dept Molecular Biol & Microbiol Univ Cent Fla Orlando FL 32816

WOEHLER, KARLHEINZ EDGAR, b Berlin, Ger, June 5, 30; m 56; c 1. PHYSICS. *Educ:* Univ Bonn, BS, 53; Aachen Tech Univ, Dipl, 55; Univ Munich, PhD(physics), 62. *Prof Exp:* Physicist commun technol, Siemens & Halske, Ger, 55-59; res assoc plasma physics, Max Planck Inst Physics, 59-62; asst prof physics, US Naval Postgrad Sch, 62-64; sr res assoc, Inst Plasma Physics, 64-65; assoc prof physics, 65-72, chmn dept physics & chem, 74-79, PROF PHYSICS, NAVAL POSTGRAD SCH, 72-, CHMN DEPT PHYSICS, 87- *Concurrent Pos:* Consult physicist, Atomics Int Div, NAm Aviation, Inc, 63-64; Nat Acad Sci res grant, Res Labs, NASA, 66; consult, Naval Electronics Lab Ctr, Calif, 68, 69 & 72. *Mem:* Am Phys Soc; AAAS; Sigma Xi. *Res:* Plasma physics; general relativity and cosmology. *Mailing Add:* Dept Physics Naval Postgrad Sch Monterey CA 93943

WOEHLER, MICHAEL EDWARD, b Appleton, Wis, Feb 16, 45. IMMUNOLOGY, MEDICAL MICROBIOLOGY. *Educ:* Northwestern Univ, BA, 67; Marquette Univ, PhD(microbiol), 71. *Prof Exp:* Fel biochem, Univ Ga, 71-74; mem tech staff immunol, GTE Labs, Inc, 74-80; MEM STAFF, PHARMACIA, INC, 80- *Mem:* Am Soc Microbiol; AAAS; Sigma Xi; NY Acad Sci. *Res:* Immunology, specifically structure function relationships of immunoglobulins G and E antibodies. *Mailing Add:* Pharmacia Inc 800 Centennial Ave Piscataway NJ 08854

WOEHLER, SCOTT EDWIN, b Albany, Ga, Sept 19, 52; m 89; c 1. NUCLEAR MAGNETIC RESONANCE, FOURIER TRANSFORM INFRARED. *Educ:* Ill Benedictine Col, BS, 74; Univ Mich, MS, 77; Univ Louisville, PhD(chem), 86. *Prof Exp:* Sales & serv engr, Tret-O-Lite, St Louis, 77-79; res asst phys chem, Univ Ky, 79-82; fel biochem, Ga State Univ, 87-88; nuclear magnetic resonance specialist, Purdue Univ, 88-90; NUCLEAR MAGNETIC RESONANCE SPECIALIST, NORTHWESTERN UNIV, 90- *Concurrent Pos:* Consult, Molecular Modeling Workshop, Fac Southeastern US Cols, 89. *Res:* Electron self exchange through nuclear magnetic resonance; polymer reactions through Fourier Transform Infrared. *Mailing Add:* Analytical Serv Lab Dept Chem Northwestern Univ Evanston IL 60208-3113

WOELFEL, JULIAN BRADFORD, b Baltimore, Md, Dec 17, 25; m 48; c 3. DENTISTRY. *Educ:* Ohio State Univ, DDS, 48. *Prof Exp:* PROF DENT, COL DENT, OHIO STATE UNIV, 48- *Concurrent Pos:* Consult & Am Dent Asn res assoc, Nat Bur Stands, 57-63; consult, Vet Admin Hosp, Dayton, Ohio, 66-69 & Fed Penitentiary, Chillicothe, 65-66; pres, Carl O Boucher Prosthodontic Conf, 67- *Honors & Awards:* Int Asn Dent Res Award, 67. *Mem:* Am Prosthodont Soc; Am Dent Asn; Int Asn Dent Res; fel Am Col Dent; Acad Denture Prosthetics; Sigma Xi. *Res:* Prosthodontic dentistry; denture base resins; clinical evaluation of complete dentures; electromyography; jaw and denture movement; mandibular motion in three dimensions; accuracy of impression materials; soft and hard tissue and facial dimension changes beneath complete dentures during six years; computer analysis of mandibular resorption. *Mailing Add:* 4345 Brookie Ct Columbus OH 43214

WOELKE, CHARLES EDWARD, b Seattle, Wash, Jan 8, 26; m 47; c 2. FISHERIES, MARINE ECOLOGY. *Educ:* Univ Wash, BS, 50, PhD(fisheries), 68. *Prof Exp:* Aquatic biologist, Ore Fisheries Comn, 50-51; fisheries biologist, 51-64, res scientist fisheries, 68-75; chief res & develop, Wash Dept Fisheries, 75-79; asst to dir, Intergovt Affairs, 79-82; CONSULT BIOLOGIST, 82- *Concurrent Pos:* Affil prof, Univ Wash, 69- *Mem:* Fel Am Inst Fishery Res Biol; Nat Shellfisheries Asn. *Res:* Molluscan commercial shellfish; bioassays with bivalve embryos; development of in situ bioassays with bivalve larvae; development of water quality standards and criteria; biometrics and ecological systems analysis. *Mailing Add:* 2378 Crestline Blvd Olympia WA 98502

WOERNER, DALE EARL, b Oak Hill, Kans, Jan 15, 26; m 50; c 6. ANALYTICAL CHEMISTRY. *Educ:* Kans State Univ, BS, 49; Univ Ill, MS, 51, PhD(analytical chem), 53. *Prof Exp:* Asst chem, Univ Ill, 49-53; assoc prof, Hanover Col, 53-55; from asst instr to asst prof, Kans State Univ, 55-58; from asst prof to assoc prof, 58-66, PROF CHEM, UNIV NORTHERN COLO, 66- *Mem:* Am Chem Soc; Sigma Xi. *Res:* Amperometric titrations; spectroscopy. *Mailing Add:* Dept Chem Univ Northern Colo Greeley CO 80639

WOERNER, ROBERT LEO, b Evanston, Ill, Apr 21, 48. LASER ISOTOPE SEPARATION, FUSION TARGET FABRICATION. *Educ:* Mass Inst Technol, SB & SM, 71, PhD(physics), 74. *Prof Exp:* Fel liquid helium, Dept Physics, Mass Inst Technol, 74-75; mem tech staff physics, Bell Tel Labs, Holmdel, NJ, 75-76; physicist, Lawrence Livermore Lab, 76-84; DIR, STRATEGY & PLANNING, PAC BELL, 91- *Mem:* Am Phys Soc. *Res:* Integrated experiments to demonstrate the atomic vapor laser isotope separation process. *Mailing Add:* 2535 Chateau Way Livermore CA 94550

WOESE, CARL R, b Syracuse, NY, July 15, 28. EVOLUTION, BACTERIA PHYLOGENCY. *Educ:* Yale Univ, PhD(biophysics), 53. *Prof Exp:* PROF PHYSICS, UNIV ILL, 64-, PROF MICRO BIOL, CTR ADVAN STUDY, 64- *Mem:* Nat Acad Sci; Acad Art & Sci. *Mailing Add:* Dept Microbiol Univ Ill 131 Burrill Hall 407 S Goodwin Ave Urbana IL 61801

WOESSNER, DONALD EDWARD, b Milledgeville, Ill, Oct 6, 30; m 58; c 1. NUCLEAR MAGNETIC RESONANCE & IMAGING. *Educ:* Carthage Col, AB, 52; Univ Ill, PhD(chem), 57. *Prof Exp:* Asst phys chem, Univ Ill, 55-57, fel chem, 57-58; sr res technologist, 58-62, res assoc, 62-84, SR RES ASSOC, DALLAS RES LAB, MOBIL RES & DEVELOP CORP, 84- *Concurrent Pos:* assoc ed, J Chem Physics, 71-73 & J Magnetic Resonance, 80- *Honors & Awards:* W T Doherty Award, Am Chem Soc, 75. *Mem:* AAAS; Am Phys Soc; NY Acad Sci; Am Chem Soc; Clay Minerals Soc; Soc Petrol Engrs. *Res:* Use of nuclear magnetic resonance to analyze structure and motions in solids, liquids and heterogenous systems; conventional liquid and solid measurements; magic angle spinning and cross polarization; magic angle spinning of solids and absorbed materials; nuclear magnetic resonance imaging. *Mailing Add:* Mobil Res & Develop Corp Dallas Res Lab 13777 Midway Rd Dallas TX 75244

WOESSNER, JACOB FREDERICK, JR, b Pittsburgh, Pa, May 8, 28; m 53; c 2. BIOCHEMISTRY. *Educ:* Valparaiso Univ, BA, 50; Mass Inst Technol, PhD(biochem), 55. *Prof Exp:* Asst, Mass Inst Technol, 53-55; Lilly fel natural sci, Univ Mich, 55-56; res asst prof biochem, 56-64, assoc prof, 64-72, assoc prof med, 72-80, PROF BIOCHEM, SCH MED, UNIV MIAMI, 72-, PROF MED, 80- *Concurrent Pos:* Investr, Labs Cardiovasc Res, Howard Hughes Med Inst, Fla, 56-71; vis scientist, Max Planck Inst Protein & Leather Res, 61-62; mem gen med B study sect, NIH, 71-75. *Mem:* Am Chem Soc; Geront Soc; Am Rheumatism Asn; Am Soc Biol Chemists; Biochem Soc. *Res:* Formation, metabolism and aging of connective tissues; proteolytic enzymes; lysosomes and tissue resorption; arthritis. *Mailing Add:* Univ Miami Sch Med R-127 PO Box 016960 Miami FL 33101

WOESSNER, RONALD ARTHUR, b Pittsburgh, Pa, Apr 27, 37; m 56; c 3. FORESTRY, GENETICS. *Educ:* WVa Univ, BS, 63; NC State Univ, MS, 66, PhD(forest genetics), 68. *Prof Exp:* Asst prof plant & forest sci, Tex A&M Univ, 68-74, assoc prof forestry, 74, assoc geneticist, Tex Forest Serv, 68-74; supv res & devel, Jari Florestal, Amazon Basin, Nat Bulk Carriers Inc, 74-81; mgr, forest productivity, 81-88, MGR, LANDS & FOREST PRODUCTIVITY, MEAD CORP, 88- *Mem:* Soc Am Foresters; AAAS; Sigma Xi; Commonwealth Forestry Asn; Int Soc Trop Foresters. *Res:* Genetic improvement of pine and hardwood species, temperate and tropical; quantitative genetics of forest trees; provenance and progeny testing; seed orchards, genotype-environment interactions; inter-population crossing; wood density, nurseries and regeneration; herbicides for site preparation and release. *Mailing Add:* Mead Coated Bd Div Woodland Dept PO Box 9908 Columbus GA 31908-9908

WOESTE, FRANK EDWARD, b Alexandria, Ky, Feb 11, 48. WOOD CONSTRUCTION, SYSTEM SIMULATION & RELIABILITY. *Educ:* Univ Ky, BSAE, 70, MSAE, 72; Purdue Univ, PhD(agr eng), 75. *Prof Exp:* Vis asst prof wood eng res, Wood Res Lab, Purdue Univ, West Lafayette, 75-77; from asst prof to assoc prof, 77-90, PROF WOOD ENG RES, VA TECH, BLACKSBURG, 90- *Mem:* Forest Prod Res Soc; Am Soc Testing & Mat; Am Soc Agr Engrs; Truss Plate Inst. *Res:* Engineered wood trusses; lumber properties research; farm structures research; light-frame building fire research. *Mailing Add:* Dept Agr Eng Va Polytech Inst & State Univ Blacksburg VA 24061

WOFFORD, IRVIN MIRLE, b White Co, Ga, Dec 11, 16; m 38. AGRONOMY. *Educ:* Univ Ga, BSA, 48; Univ Fla, MSA, 49; Mich State Col, PhD(farm crops), 53. *Prof Exp:* Instr agron, Univ Fla, 49-51, asst agronomist, Exp Sta, 53-56; asst, Mich State Col, 51-53; dir agron, Southern Nitrogen Co, Inc, 56-64; MGR AGR PUB RELS, KAISER AGR CHEM, 64- *Mem:* Am Soc Agron. *Res:* Crop management and production; fertilizer studies; variety testing; date of planting; rotations; plant population studies. *Mailing Add:* 1325 Lavon Ave Savannah GA 31406

WOFSY, LEON, b Stamford, Conn, Nov 23, 21; m 42; c 2. CHEMISTRY, IMMUNOLOGY. *Educ:* City Col New York, BS, 42; Yale Univ, MS, 60, PhD(chem), 61. *Prof Exp:* PROF IMMUNOL, UNIV CALIF, BERKELEY, 64- *Mem:* Am Chem Soc. *Res:* Study of antibody specificity; mechanisms of cellular differentiation. *Mailing Add:* Dept Microbiol & Cell Biol Univ Calif Berkeley CA 94720

WOFSY, STEVEN CHARLES, b New York, NY, June 24, 46. ATMOSPHERIC CHEMISTRY, AQUATIC CHEMISTRY. *Educ:* Univ Chicago, BS, 66; Harvard Univ, PhD(chem), 71. *Prof Exp:* Res assoc, Smithsonian Astrophys Observ, 71-74; lectr, 74-77, assoc prof, Div Appl Sci, 77-82, SR RES FEL, HARVARD UNIV, 82- *Concurrent Pos:* Res assoc, Nat Acad Sci-Nat Res Coun, 71-73. *Honors & Awards:* MacIlwane Award, Am Geophys-Union, 81. *Mem:* Am Geophys Union; AAAS; Am Soc Limnol & Oceanog. *Res:* Photochemistry and biogeochemistry of atmospheric gases; gases and nutrients in marine and fresh waters; human impact on the global environment. *Mailing Add:* Pierce Hall Harvard Univ Cambridge MA 02138

WOGAN, GERALD NORMAN, b Altoona, Pa, Jan 11, 30; m 57; c 2. PHARMACOLOGY. *Educ:* Juniata Col, BS, 51; Univ Ill, MS, 53, PhD(physiol), 57. *Prof Exp:* Instr physiol, Univ Ill, 56-57; asst prof, Rutgers Univ, 57-61; res assoc food toxicol, 61-62, from asst prof to assoc prof, 62-68, PROF TOXICOL, MASS INST TECHNOL, 68- *Mem:* Nat Acad Sci; AAAS; Am Soc Pharmacol Exp Ther; Soc Toxicol; Am Asn Cancer Res. *Res:* Chemical carcinogenesis; physiological and biochemical responses to toxic substances; mechanisms of action of carcinogens and mutagens; environmental carcinogenesis. *Mailing Add:* Div Toxicol Rm 16-333 Mass Inst Technol Cambridge MA 02139

WOGEN, WARREN RONALD, b Forest City, Iowa, Feb 19, 43; m 69; c 2. MATHEMATICS. *Educ:* Luther Col, Iowa, BA, 65; Ind Univ, Bloomington, MS, 67, PhD(math), 69. *Prof Exp:* Assoc prof, 69-80, PROF MATH, UNIV NC, CHAPEL HILL, 80- *Concurrent Pos:* Vis prof, Ind Univ, 76. *Mem:* Am Math Soc; Math Asn Am. *Res:* Operator theory and operator algebras. *Mailing Add:* Dept Math Univ NC Chapel Hill NC 27599-3250

WOGMAN, NED ALLEN, b Spokane, Wash, Oct 25, 39; m 59; c 3. NUCLEAR CHEMISTRY, PHYSICAL CHEMISTRY. *Educ:* Wash State Univ, BS, 61; Purdue Univ, PhD(phys chem), 66. *Prof Exp:* Sr res scientist, 65-68, mgr radiol chem, 68-72, res assoc, Battelle Pac Northwest Lab, 72-79, mgr radiol chem, 79-86, MGR TECHNOL DEVELOP, 86- *Concurrent Pos:* Lectr, Joint Ctr Grad Study, Wash State Univ, Univ Wash, Ore State Univ, 72-; mem sci comt, Nat Coun Radiation Protection, 73- *Mem:* Am Chem Soc; Am Nuclear Soc. *Res:* Rates and mechanisms of biological, meteorological, oceanographic and ecological processes studied using natural and artificial radionuclides, development of sensitive multidimensional gamma-ray spectrometer systems for trace radionuclide measurements; neutron, alpha, and beta detection systems. *Mailing Add:* 108 Somerset Richland WA 99352

WOGRIN, CONRAD A(NTHONY), b Denver, Colo, Apr 16, 24; m 51; c 3. ELECTRICAL ENGINEERING, COMPUTER SCIENCE. *Educ:* Yale Univ, BE, 49, MEng, 51, DEng, 55. *Prof Exp:* From instr to assoc prof elec eng, Yale Univ, 51-66; dir, Univ Comput Ctr, 67-85, chmn comput sci, 85-86, PROF COMPUT & INFO SCI, UNIV MASS, AMHERST, 67- *Concurrent Pos:* Consult, Mitre Corp, 61-65, Goddard Space Flight Ctr, 65-72 & United Aircraft Corp Syst Ctr, 66-70. *Mem:* AAAS; Inst Elec & Electronics Engrs; Asn Comput Mach. *Res:* Digital computers; computer systems and languages; information and control systems; digital image processing. *Mailing Add:* Dept Comput Sci Univ Mass Amherst MA 01002

WOHL, JOSEPH G, system analysis; deceased, see previous edition for last biography

WOHL, MARTIN H, b New York, NY, Feb 12, 35; m 57; c 2. CHEMICAL ENGINEERING. *Educ:* Cornell Univ, BChE, 57. *Prof Exp:* Res engr, Plastics Div, Monsanto Co, 57-63, res specialist, Plastic Prod & Resins Div, 63-68, tech group leader process develop, 68-70, tech supt film, Fabricated Prod Div, Monsanto Com Prod Co, 70-75, mgr, Res & Technol, Fabricated Prod Div, Monsanto Plastics & Resins Co, 75-78, proj dir, New Prod Develop Dept, 78-83, proj dir, Plastics Div, Monsanto Polymer Prod Co, 83-86; MGR, BUSINESS DEVELOP, MONSANTO CHEM CO, 86- *Mem:* Am Inst Chem Engrs; Soc Plastic Engrs; Soc Automotive Engrs; Soc Adv Mat & Process Eng; Soc Mfg Engrs. *Res:* Process development in the field of high polymers; fluid flow and heat transfer to non-Newtonian fluids; research and engineering administration; advanced composites; engineering plastics. *Mailing Add:* Monsanto Co 800 N Lindbergh Blvd St Louis MO 63167

WOHL, PHILIP R, b 1944; US citizen; m; c 2. ECOLOGICAL MODELING, DYNAMICAL SYSTEMS. *Educ:* Queens Col, NY, BA, 66; Cornell Univ, PhD(appl math), 71. *Prof Exp:* Asst prof, NY Univ, 71-72; res assoc math, Carleton Univ, 72-74; asst prof, 74-80, ASSOC PROF & GRAD PROG DIR, OLD DOM UNIV, 80- *Mem:* Soc Indust & Appl Math; Math Asn Am; Soc Math Biol. *Res:* Classical analysis and methods of ordinary and partial differential equations; perturbation methods in fluid mechanics; low Reynolds number hydrodynamics; blood flow and other flows of suspensions in tubes; biofluid mechanics; mathematical modelling in biology. *Mailing Add:* Dept Math Sci Old Dom Univ Norfolk VA 23529

WOHL, RONALD A, b Basel, Switz, Nov 25, 36. ORGANIC CHEMISTRY, MEDICINAL CHEMISTRY. *Educ:* Univ Basel, PhD(org chem), 65. *Prof Exp:* Lectr org chem, Univ Basel, 65-66; res assoc, Yale Univ, 66-67; asst prof, Rutgers Univ, 67-74; sr org chemist, 74-76, sect head org chem, 76-82, ASST DIR MED CHEM, BERLEX LABS, 82- *Mem:* Am Chem Soc. *Res:* Stereochemistry; cardiovascular drugs. *Mailing Add:* Berlex Labs Inc 110 E Hanover Ave Cedar Knolls NJ 07927-2007

WOHLEBER, DAVID ALAN, b Pittsburgh, Pa, Oct 1, 40; m 61; c 4. CHEMISTRY. *Educ:* Univ Pittsburgh, BS, 62; John Carroll Univ, MS, 67; Kent State Univ, PhD(chem), 70. *Prof Exp:* Chemist anal chem, Develop Lab, Standard Oil Co, Ohio, 61-63, chemist polymer chem, Res Lab, 63-66; sr res scientist phys chem, 70-74, sect head, 74-79, mgr Extractive Metall Div, 79-81, mgr process chem & physics div, 81-84, TECH CONSULT, ALCOA TECH CTR, ALUMINUM CO AM, 84- *Mem:* Am Chem Soc; Sigma Xi. *Res:* Chlorination technology; inorganic and physical chemistry of alumina refining and aluminum smelting; physical adsorption; polymer synthesis and characterization. *Mailing Add:* Alcoa Ctr Pittsburgh PA 15069

WOHLERS, HENRY CARL, b New York, NY, Feb 2, 16; m 49; c 2. AIR POLLUTION, ENVIRONMENTAL AFFAIRS. *Educ:* St Lawrence Univ, BS, 39; Stanford Univ, PhD(phys chem), 49. *Prof Exp:* Asst chemist, Boyce Thompson Inst Plant Res, 36-37 & 39-40; assoc chemist, Gen Chem Co, 40-44; supvr, Ford, Bacon & Davis, 44; sr chemist, Dorr Co, 44-46; sr anal chemist, Stanford Res Inst, 49-61; dir tech serv, Bay Area Air Pollution Control Dist, 61-65; from assoc prof to prof environ sci, Drexel Univ, 65-72; CONSULT ENVIRON AFFAIRS, 72- *Concurrent Pos:* Consult to var UN orgn. *Mem:* Air Pollution Control Asn; Am Chem Soc. *Res:* All phases of air pollution and global environmental problems. *Mailing Add:* 1313 Lincoln Ave No 1107 Eugene OR 97401-3966

WOHLFORD, DUANE DENNIS, b Newcastle, Ind, May 20, 37; m 66; c 3. GEOLOGY, PETROLOGY. *Educ:* Univ Wis, BS, 59; Univ Colo, PhD(geol), 65. *Prof Exp:* Asst prof geol, 64-74, ASSOC PROF EARTH SCI, STATE UNIV NY COL ONEONTA, 74- *Mem:* AAAS; Geol Soc Am. *Res:* Petrology of the Precambrian rocks of the southern Adirondack Mountains of New York. *Mailing Add:* Dept Earth Sci State Univ NY Col Oneonta Oneonta NY 13820

WOHLFORT, SAM WILLIS, b Toledo, Ohio, June 8, 26; m 49; c 3. ANALYTICAL CHEMISTRY, TECHNICAL MANAGEMENT. *Educ:* Univ Toledo, BS, 48; Ohio State Univ, MS, 50. *Prof Exp:* Asst instrumental & spectrog anal, Ohio State Univ, 50-53; chemist, Sharp-Schurtz Co, 53-56; chemist, Libbey-Owens-Ford Glass Co, 56-58, group leader anal res, 58-65; supvr anal methods & mfg processes res & develop, Milchem Inc, Tex, 65-69; supvr phys measurements dept, 69-73, supvr mat technol dept, 73-77, supvr process technol, 77-82, prin engr, 82-85, prin scientist, 85-90, RES STAFF MEM, MARTIN MARIETTA ENERGY SYSTS INC, 90- *Res:* Electron microscopy; x-ray diffraction and fluorescence; ultraviolet, infrared and visible spectrophotometry; differential thermal analysis; thin films; corrosion protection; vacuum deposition; x-ray stress analysis; incipient failure detection; non-destructive testing; metallography; radwaste disposal; effluent treatment, especially gas, liquid and solid; UF6 containment; lab information management systems software development; NQA-1 lead QA auditor; INPO QA surveillance; lab controls and standards; QA/QC technical management. *Mailing Add:* Martin Marietta Energy Systs Inc PO Box 628 MS-2145A Piketon OH 45661

WOHLGELERNTER, DEVORA KASACHKOFF, b Washington, DC, Apr 1, 41; m 67; c 7. MATHEMATICS. *Educ:* Yeshiva Univ, BA, 61, MA, 63, PhD(math), 70. *Prof Exp:* Asst prof, 70-77, ASSOC PROF MATH, BARUCH COL, 77- *Mem:* Am Math Soc. *Res:* Polynomial approximation of functions of a complex variable. *Mailing Add:* 116 W 72nd New York NY 10023

WOHLMAN, ALAN, b New York, NY, May 17, 36; m 58; c 3. BIOCHEMICAL PHYSIOLOGY. *Educ:* NY Univ, BA, 59; Princeton Univ, MS, 65, PhD(biol sci), 66. *Prof Exp:* Res scientist, Bellevue Med Ctr NY Univ, 56-59 & Denver Labs, Stanford, Conn, 59-62; fac biol, State Univ NY, Stony Brook, 66-67; dir res pharmaceut, Denver Labs, Toronto, 67-70; div dir, Connaught Labs, Univ Toronto, 70-74; dir qual control & head, Armour Pharm Co, Kankakee, Ill, 74-77; DIR RES, FRITO-LAY INC, DALLAS, 77- *Concurrent Pos:* NIH fel, Princeton Univ, 63-66. *Mem:* AAAS; Sigma Xi; Food Res Inst. *Res:* Technical management, cell biology, biochemistry, physiology, optics, basic food and drug research, food and drug stability, nutrition and regulatory affairs. *Mailing Add:* 3036 Georgetown St Houston TX 77005

WOHLPART, KENNETH JOSEPH, b New York, NY. FOOD TECHNOLOGY, AGRICULTURAL PRODUCTS INSPECTION. *Educ:* St Bonaventure Univ, BS, 53; Purdue Univ, MS, 58. *Prof Exp:* Res asst anal chem, Union Carbide Corp, 55-56; asst food technologist food prod develop, Gen Foods Corp, 57-60; food pilot plant mgr, Beech Nut Foods, 60-63; sr food technologist, Prod Eng Div, Aluminum Co Am, 63-86; FOOD INSPECTOR, PA DEPT AGR, 86- *Mem:* Inst Food Technologists. *Res:* Product-package interaction; sensory evaluation; package-process development. *Mailing Add:* 4113 Impala Dr Pittsburgh PA 15239

WOHLRAB, HARTMUT, b Berlin, WGer, July 2, 41; m 67; c 2. BIOCHEMISTRY. *Educ:* Rensselaer Polytech Inst, BS, 62; Stanford Univ, PhD(biophys), 68. *Prof Exp:* Asst biochem, Univ Munich, 70-72; Staff scientist, 72-81, SR SCIENTIST, BOSTON BIOMED RES INST, 81-; RES ASSOC, DEPT BIOL CHEM & MOLECULAR PHARMACOL, HAVARD MED SCH,73- *Concurrent Pos:* Estab investr, Am Heart Asn, 73-78; adj prof, Inst Gen Path, Cath Univ, Rome, Italy, 88- *Mem:* Geront Soc; Biophys Soc; Fedn Europ Biochem Socs; Am Aging Asn; AAAS; Am Soc Biol Chem & Molecular Biol. *Res:* Mitochondrial biochemistry; biochemistry of developing and aging, or senescent, tissues; function and molecular structure of membrane transport proteins. *Mailing Add:* Boston Biomed Res Inst 20 Staniford St Boston MA 02114

WOHLSCHLAG, DONALD EUGENE, b Bucyrus, Ohio, Nov 6, 18; m 43; c 3. MARINE ECOLOGY, SUBLETHAL STRESSES ON FISHES. *Educ:* Heidelberg Col, BS, 40; Ind Univ, PhD, 49. *Prof Exp:* Lab asst chem & zool, Heidelberg Col, 40-41; res assoc limnol, Univ Wis, 49; from asst prof to prof biol, Stanford Univ, 49-65; prof zool & dir, 65-70, prof zool & marine studies, 70-86, EMER PROF, MARINE SCI INST, UNIV TEX, 86. *Concurrent Pos:* Prin investr Arctic fisheries, Off Naval Res & Arctic Inst NAm, 52-55; prin investr Antarctic fisheries biol, NSF, 58-65; prin investr Gulf coastal fishery biol res, several funding sources, 65-85; coord comt, Nat Acad Sci, Nat Res Coun, Int Biol Prog, 67-70, consult, Interoceanic Canal, 69-70; consult, NSF Tundra Biome Panel, Off Polar Prog, 71-74, Inst Ecol, 74-; ed, Contrib Marine Sci, Univ Tex Marine Sci Inst, 75-88. *Mem:* AAAS; Am Fisheries Soc; Am Soc Limnol & Oceanog; Am Soc Zoologists; Ecol Soc Am. *Res:* Ecology of fishes; metabolism and growth; population dynamics; toxic stresses. *Mailing Add:* Marine Sci Inst Univ Tex Port Aransas TX 78373

WOHLTMANN, HULDA JUSTINE, b Charleston, SC, Apr 10, 23. PEDIATRIC DIABETES & ENDOCRINOLOGY. *Educ:* Col Charleston, BS, 44; Med Col SC, MD, 49; Am Bd Pediat, dipl, 55. *Prof Exp:* From instr to asst prof pediat, Sch Med, Wash Univ, 53-61, USPHS spec res fel biochem, Univ, 61-63; asst prof pediat, 65-70, PROF PEDIAT, MED UNIV SC, 70- *Mem:* AAAS; fel Am Acad Pediat; Am Diabetes Asn; Endocrine Soc; Am Pediat Soc; Am Fedn Clin Res. *Res:* Pediatric metabolism and endocrinology; tight metabolic control of diabetes mellitus in children and adolescents and assessing the effects in later development. *Mailing Add:* Med Univ SC 171 Ashley Ave Charleston SC 29425

WOISARD, EDWIN LEWIS, b Newark, NJ, Jan 21, 26; m 53, 69; c 7. TECHNICAL MANAGEMENT, OPERATIONS RESEARCH. *Educ:* Drew Univ, BA, 50; Lehigh Univ, MS, 52, PhD(physics), 59. *Prof Exp:* Instr physics, Moravian Col, 56-59; assoc res physicist, Res Lab, Whirlpool Corp, 59-61; proj leader, Weapons Systs Eval Div, Inst Defense Anal, 61-67; exec vpres, John D Kettelle Corp, 67-70; asst, Navy Net Assessment & Midrange Objectives, Off Chief Naval Opers, US Navy, 71-79; prin scientist, Ramcor Inc, 79-89; dir advan technol, Unmanned Aerial Vehicle Prog Off, Gen Dynamics Corp, 90; SR STAFF, STRATEGIC ANALYSIS, INC, 91- *Concurrent Pos:* Consult, 70-71. *Mem:* Res Soc Am; Sigma Xi; AAAS; Opers Res Soc Am; Mil pers Res Soc. *Res:* Thermoelectricity; solid state physics; microwave absorption; systems and defense analysis; planning and management of research and development; technology review and assessment. *Mailing Add:* 5020 King David Blvd Annandale VA 22003

WOJCICKI, ANDREW, b Warsaw, Poland, May 5, 35; nat US; m 68; c 2. INORGANIC CHEMISTRY, ORGANOMETALLIC CHEMISTRY. *Educ:* Brown Univ, BS, 56; Northwestern Univ, PhD(chem), 60. *Prof Exp:* Asst chem, Northwestern Univ, 56-58, assoc, 59; NSF fel inorg chem, Univ Nottingham, 60-61; from asst prof to assoc prof, 61-69, PROF INORG CHEM, OHIO STATE UNIV, 69- *Concurrent Pos:* Vis assoc prof, Case Western Reserve Univ, 67; US sr scientist award, Humboldt Found, Ger, 75-76; Guggenheim Found fel, 76; vis scholar, Univ Calif, Berkeley, 84; vis prof, Univ Bologna, Italy, 88. *Mem:* AAAS; Am Chem Soc; Royal Soc Chem. *Res:* Synthesis and mechanism of reactions of inorganic and organometallic compounds. *Mailing Add:* Dept Chem Ohio State Univ Columbus OH 43210

WOJCICKI, STANLEY G, b Warsaw, Poland, Mar 30, 37; US citizen; m 61. HIGH ENERGY PHYSICS. *Educ:* Harvard Univ, AB, 57; Univ Calif, Berkeley, PhD(physics), 62. *Prof Exp:* Physicist, Lawrence Radiation Lab, Univ Calif, 61-66; from asst prof to assoc prof, 66-74, PROF PHYSICS, STANFORD UNIV, 74- *Concurrent Pos:* NSF fel, 64-65; Alfred P Sloan Found fel, 68-72; Guggenheim fel, 73-74; Alexander von Humnoldt sr scientist award, 80. *Mem:* Fel Am Phys Soc. *Res:* Resonances in high energy physics; candle power violation; electron-positron annihilations; muon production in hedronic interactions. *Mailing Add:* Dept Physics Stanford Univ Stanford CA 94305

WOJCIECHOWSKI, BOHDAN WIESLAW, b Wilno, Poland, Jan 29, 35; Can citizen; m 59; c 2. CHEMICAL ENGINEERING, CATALYSIS. *Educ:* Univ Toronto, BASc, 57, MASc, 58; Univ Ottawa, PhD(phys chem), 60. *Prof Exp:* Fel photolysis, Nat Res Coun, 60-61; fel kinetics, Univ Ottawa, 61-62; sr res chemist, Res & Develop, Socony Mobil Oil Co, Inc, 62-65; from asst prof to assoc prof chem eng, 65-72, PROF CHEM ENG, QUEEN'S UNIV, ONT, 72- *Concurrent Pos:* UN Develop Prog tech expert, Brazil, 72; consult to petrol indust, catalyst mfrs. *Mem:* AAAS; Sigma Xi; fel Chem Inst Can; Am Inst Chem Engrs; Catalysis Soc; Can Soc Chem Eng. *Res:* Kinetics; catalysis; absorption; petroleum refining; process design; oceanography; recovery of oils from sand and shale; mathematical systems for highly coupled reactions; Fischer Tropsch synthesis. *Mailing Add:* Dept Chem Eng Queen's Univ Kingston ON K7L 3N6 Can

WOJCIECHOWSKI, NORBERT JOSEPH, b Evanston, Ill, Aug 4, 27; m 54. DRUG INFORMATION. *Educ:* Univ Ill, BS, 51, MS, 61; Loyola Univ, PhD(pharmacol), 67. *Prof Exp:* Sr anal chemist chem, Abbott Labs, Inc, 51-56; grad asst pharmacol, Stritch Sch Med, Loyola Univ, 61-66; from asst prof to assoc prof pharmaceut, 66-86, PROF CLIN PHARM, COL PHARM, UNIV TENN, 86- *Concurrent Pos:* Instr pharmacol, Col Basic Sci, Univ Tenn, 66-80, asst prof pharmacol, 80-89; assoc ed, Drug & Therap Lett, Drug Info Ctr, Univ Tenn, 82-90; consult, Bur Medicaid-Formulary Adv Comt, Tenn Dept Health & Environ, 86- *Mem:* Sigma Xi; Am Asn Col Pharm; Am Pharm Asn. *Res:* Cardiovascular pharmacology - antiarrhythmic activity of compounds in isolated atrial tissue; antihypertensive studies in hypertensive rats; written reviews and evaluations of new drugs. *Mailing Add:* Drug Info Ctr Univ Tenn 877 Madison Ave Suite 210 Memphis TN 38163

WOJCIK, ANTHONY STEPHEN, b Chicago, Ill, Sept 18, 45; m 69; c 2. DESIGN AUTOMATION, ARTIFICIAL INTELLIGENCE. *Educ:* Univ Ill, Urbana-Champaign, BS, 67, MS, 68, PhD(comput sci), 71. *Prof Exp:* From asst prof to prof comput sci, Ill Inst Technol, 71-86, chmn dept, 78-84; PROF COMPUT SCI & CHMN DEPT, MICH STATE UNIV, 86- *Concurrent Pos:* Mem tech staff, Bell Tel Labs, 74; resident assoc, Argonne Nat Lab, 82-85, vis scientist, 86. *Mem:* Asn Comput Mach; Inst Elec & Electronics Engrs. *Res:* Computer architecture; reliable design of digital systems; design automation of digital systems; logic design; multiple-valued logic; artificial intelligence. *Mailing Add:* Dept Comput Sci Ill Inst Technol Chicago IL 60616

WOJCIK, JOHN F, b Ashley, Pa, Nov 12, 38; m 60; c 6. PHYSICAL CHEMISTRY. *Educ:* King's Col, Pa, BS, 60; Cornell Univ, PhD(phys chem), 65. *Prof Exp:* Asst prof chem, St Francis Col, Pa, 65-66; from asst prof to assoc prof, 77-83, PROF CHEM, VILLANOVA UNIV, 84-, CHMN CHEM, 90- *Concurrent Pos:* Petrol Res Fund res grant, 66-68. *Mem:* Am Chem Soc; Sigma Xi. *Res:* Cyclodextrin chemistry; nature of the hydrophobic bond; fast reactions by temperature jump; classical experimental physical chemistry. *Mailing Add:* Dept Chem Villanova Univ Villanova PA 19085

WOJNAR, ROBERT JOHN, b Thompsonville, Conn, Jan 29, 35; m 58; c 3. IMMUNOBIOLOGY, BIOCHEMISTRY. *Educ:* Univ Conn, BA, 56, MS, 60, PhD(biochem), 64. *Prof Exp:* NIH fel, Yale Univ, 63-65; staff scientist, Worcester Found Exp Biol, 65-68; res investr, 68-69, SR INVESTR BIOCHEM IMMUNOL, SQUIBB INST MED RES, 69- *Mem:* AAAS; NY Acad Sci. *Res:* Cellular immunology; inflammation; nucleic acids; immunopharmacology; allergy and drug hypersensitivity; steroid biology. *Mailing Add:* 147 Oak Creek Rd E Windsor Township NJ 08561

WOJTAL, STEVEN FRANCIS, b Albany, NY, Oct 3, 52; m 85. STRAIN MEASUREMENT DEFORMATION MECHANISMS IN ROCKS. *Educ:* Brown Univ, BS, 74; Johns Hopkins Univ, MA, 76, PhD(geol), 82. *Prof Exp:* From instr to assoc prof, 79-88, PROF GEOL, OBERLIN COL, 88- *Concurrent Pos:* Chair, Dept Geol, Oberlin Col, 89-; assoc ed, J Structural Geol, 91- *Mem:* Geol Soc Am; Am Geophys Union; Sigma Xi. *Res:* Examining naturally-occurring rock structures using theoretical models of and experimental data on the physical properties of crystalline aggregates and continuous media. *Mailing Add:* Dept Geol Oberlin Col Oberlin OH 44074

WOJTKOWSKI, PAUL WALTER, b Buffalo, NY, May 16, 45. ORGANIC CHEMISTRY. *Educ:* State Univ NY, Buffalo, BA, 67; Univ Notre Dame, PhD(org chem), 71. *Prof Exp:* Assoc, Iowa State Univ, 71-73 & Squibb Inst Med Res, Princeton, NJ, 73-74; res chemist, Pioneering Res Lab, Textile Fibers Dept, 74-76, res chemist, Petrochem Dept, 76-79, sr res chemist, Feedstocks Div, Cent Res & Develop Dept, 80-82, RES ASSOC, AGR PROD DEPT, E I DU PONT DE NEMOURS & CO, 82- *Mem:* Am Chem Soc; Sigma Xi. *Res:* Organic chemistry; organoboranes; low temperature organic photochemistry; insect chemistry; synthesis of beta-lactam antibiotics; condensation polymers and their fibers; heterogeneous catalysis; agricultural chemical process development. *Mailing Add:* E I du Pont de Nemours & Co Exp Sta Wilmington DE 19880-0402

WOJTOWICZ, JOHN ALFRED, b Niagara Falls, NY, Oct 12, 26; div; c 3. INDUSTRIAL CHEMISTRY. *Educ:* Univ Buffalo, BA, 54; Niagara Univ, MS, 66. *Prof Exp:* Analyst chem anal, E I du Pont de Nemours & Co, Inc, 45-50, develop analyst chem anal & process develop, 50-54, develop chemist, 54-56; res chemist inorg synthesis & chem kinetics, 56-61, sr res chemist org synthesis & process develop, 61-71, res assoc inorg & org synthesis & process develop, consult scientist, 77-85, SR CONSULT SCIENTIST, INORG & ORG SYNTHESIS PROCESS & PROD DEVELOP, OLIN CORP, 85- *Concurrent Pos:* Consult scientist, inorg & org synthesis & process develop. *Mem:* Am Chem Soc; Sigma Xi. *Res:* Organic and inorganic synthesis; process and product development in the areas of hypochlorites and chloroisocyanurates. *Mailing Add:* Olin Corp PO Box 586 Cheshire CT 06410-3225

WOJTOWICZ, PETER JOSEPH, b Elizabeth, NJ, Sept 22, 31; m 53; c 3. THEORETICAL PHYSICS, MATHEMATICAL MODELING. *Educ:* Rutgers Univ, BSc, 53; Yale Univ, MS, 54, PhD(phys chem), 56. *Prof Exp:* Mem tech staff & theoret chem physicist, RCA Labs, 56-78, head electron optics & deflection res, 78-84; SR MEM TECH STAFF, DAVID SARNOFF RES CTR, 84- *Mem:* Fel Am Phys Soc. *Res:* Statistical mechanics; theory of liquid and solid states; theory of magnetism and properties of magnetic substances; liquid crystals; electron optics and magnetic deflection in display systems. *Mailing Add:* David Sarnoff Res Ctr Princeton NJ 08543-5300

WOLAK, JAN, b Lubasz, Poland, Mar 8, 20; US citizen; m 59; c 3. MECHANICAL ENGINEERING. *Educ:* Univ London, BSc, 50; Wash Univ, MS, 60; Univ Calif, Berkeley, PhD(mech eng), 65. *Prof Exp:* Col apprenticeship, AEI-Gen Elec Co, Eng, 50-52, steam turbine design engr, 52-56; lectr mech eng, Wash Univ, 56-60; assoc, Univ Calif, Berkeley, 60-63, asst res specialist, Inst Eng Res, 63-65; from asst to assoc prof, 65-84, PROF MECH ENG, UNIV WASH, 84- *Concurrent Pos:* Fac fel, NASA, 79-80; res consult, Boeing Com Airplane Co, 85- *Mem:* Am Soc Mech Engrs; Brit Inst Mech Engrs; Am Soc Eng Educ; fel Soc Mfg Engrs. *Res:* Mechanical behavior of materials; metal removal and forming processes; erosion by abrasive particles; evaluation of large plastic strains; friction and wear. *Mailing Add:* Dept Mech Eng FU-10 Univ Wash Seattle WA 98195

WOLAVER, LYNN E(LLSWORTH), b Springfield, Ill, Mar 10, 24; m 49; c 2. ENGINEERING, MATHEMATICS. *Educ:* Univ Ill, BS, 49, MS, 50; Univ Mich, PhD(info & control eng), 64. *Prof Exp:* Instr elec eng, Univ Ill, 49-50; electronic engr, Wright Air Develop Ctr, USAF, 50-51, electronic scientist, 51-56, aero-res engr, Aerospace Res Labs, 56-63, dep dir, Appl Math Lab, 63-67, dir, 67-71, chmn dept syst eng & assoc dean res, 71-79, PROF, AIR FORCE INST TECHNOL, USAF, 63-, DEAN RES, 79- *Concurrent Pos:* Lectr eve sch, Wright State Univ, 64-71; mem, Midwestern Simulation Coun, 53- *Mem:* NY Acad Sci; Inst Elec & Electronics Engrs; Soc Indust & Appl Math; fel Brit Interplanetary Soc; Am Soc Eng Educ; AAAS; Biofeedback Soc Am. *Res:* Navigation; astrodynamics; simulation; nonlinear systems analysis; modeling nonlinear systems driven by random noise; bioengineering; motion sickness research. *Mailing Add:* 1380 Timberwyck Ct Fairborn OH 45324

WOLBARSHT, MYRON LEE, b Baltimore, Md, Sept 18, 24; div; c 3. BIOPHYSICS, BIOMEDICAL ENGINEERING. *Educ:* St Johns Col, AB, 50; Johns Hopkins Univ, PhD (biol biophys), 58; Am Bd Laser Surg, dipl, 85. *Prof Exp:* Chief physicist, Naval Med Res Inst, 58-68; PROF OPHTHALMOL & BIOMED ENG, ASSOC PROF PHYSIOL, PSYCHOL DEPT, DUKE UNIV, ASST DIR INT CARDIAC CATHETERIZATION, DEPT MED, 84- *Concurrent Pos:* Guest scientist, Naval Med Res Inst, 54-58; res assoc, Psychiat Inst, Med Sch, Univ Md, 54-60; res fel biol, Johns Hopkins Univ, 58-63; consult, York Hosp, Pa, 63-68; mem exec panel, Nat Res Coun Armed Forces Comt Vision, 63-; chmn eye hazards subcomt, Laser Safety Comt, Am Nat Standards Inst, 68; lectr, psychol dept, Duke Univ, 68-; US rep, Tech Comt 76 Laser Safety, Int Electro Tech Comn, 76-; mem, US Nat Comt Photobiol, 78-; mem bd, Am Soc Laser Med & Surg, 82- *Honors & Awards:* Mark Award, Am Soc Laser Med & Surg. *Mem:* Am Physiol Soc; Inst Elec & Electronics Engrs; Soc Gen Physiol; Optical Soc Am; Royal Soc Med; Am Soc Laser Med & Surg; Laser Inst Am (pres, 82-83). *Res:* Laser safety; biomedical engineering applications to ophthalmology; structure and function of sense organs, especially vision, chemoreception, mechanoreception; electrophysiology of central nervous system; laser surgery. *Mailing Add:* Dept Psychol Duke Univ Durham NC 27706

WOLBER, WILLIAM GEORGE, b Detroit, Mich, Feb 19, 27; m 50; c 5. INSTRUMENT ENGINEERING. *Educ:* Univ Mich, BS(chem eng) & BS(eng math), 49 & MS, 50. *Prof Exp:* Engr-group leader, Uniroyal Tire Div, 50-54; proj engr-prog mgr, Bendix Res Labs, 54-62, dept mgr, 62-66, sr prin engr, 66-73, sr res planner, 73-76, sr res consult, 76-81; sr tech adv, 81-86, chief tech off, 86-89, EXEC ENGR, CUMMINS ELECTRONICS CO, 90- *Concurrent Pos:* Chmn, Invention & Patent Comt, Bendix Res Labs, 74-78, corp gatekeeper-instr, Bendix Corp, 77-81; reader, Soc Automotive Engrs; NSF peer grant reviewer, 80-; mem, bd dirs, Weed Instrument, 84-; chmn, Invention Rev Comt, Cummins Electronics, 85- *Honors & Awards:* Corp Tech Achievement Award, Bendix Corp, 75. *Mem:* Soc Automotive Engrs. *Res:* Research and development in measuring instruments and sensors; their application to control systems such as automobile engine control; theory and practice of precision instrument calibration. *Mailing Add:* Cummins Electronics Co 2851 State St Columbus IN 47201

WOLBERG, DONALD LESTER, b New York, NY, Dec 18, 45; c 7. VERTEBRATE PALEONTOLOGY. *Educ:* New York Univ, BA, 68; Univ Minn, PhD(geol), 78. *Prof Exp:* Curator & teaching asst geol, New York Univ, 67-69; ed, Encyclopedia Britannica, 69-71; teaching assoc, Univ Minn, 71-75; ed & geologist, Minn Geol Survey, 75-76; geologist, Nat Biocentric, Inc, 76-77; asst prof, Univ Wis-River Falls, 77-78; PALEONTOLOGIST, NMEX BUR MINES & MINERAL RESOURCES, 78- *Concurrent Pos:* Res assoc, Minn Messenia Exped Greece, 71-73; contract writer & ed, Encyclopedia Britannica, 71-73; ed, J Paleont; mem, NMex Coal Surface Mining Comn, 82-; mem, Paleont Collecting Guidelines Comt, Nat Acad Sci; mem, Govt Liason Comt, Soc Vert Paleont; adj assoc prof geol, Geosci Dept, NMex Inst Mining & Technol; bd dirs, Geronimo Springs, Mus, San Juan Basin Regional Coal Team & Environ Legis & Regulations. *Mem:* Soc Vert Paleont; Soc Econ Paleontologists & Mineralogists; Paleont Soc; Asn Geoscientists Int Develop; Sigma Xi; Int Soc Cryptozoology; Paleont Res Inst; Geol Soc Am. *Res:* Late Mesozoic and early Tertiary stratigraphy; paleontolgy and paleoecology of North America. *Mailing Add:* NMex Bur Mines & Mineral Resources Socorro NM 87801

WOLBERG, GERALD, b New York, NY, Aug 18, 37; m 67; c 2. IMMUNOLOGY, MICROBIOLOGY. *Educ:* NY Univ, BA, 58; Univ Ky, MS, 63; Tulane Univ, PhD(immunol), 67. *Prof Exp:* Fel immunol, Pub Health Res Inst of City of New York, 67-68, NIH fel, 68-70; MEM STAFF, WELLCOME RES LABS, BURROUGHS WELLCOME CO, 70- *Mem:* Am Asn Immunol; Am Soc Microbiol; Sigma Xi. *Res:* Immunosuppression; immunoactivation. *Mailing Add:* Burroughs Wellcome Co 3030 Cornwallis Rd Research Triangle Park NC 27709

WOLBERG, WILLIAM HARVEY, b July 10, 31; US citizen; m 55; c 6. SURGERY, ONCOLOGY. *Educ:* Univ Wis-Madison, BS, 53, MD, 56. *Prof Exp:* Intern, Ohio State Univ, 56-57; resident surg, 57-61, chmn gen surg, 72-75, from instr to assoc prof, 61-71, PROF SURG, UNIV WIS-MADISON, 71- *Concurrent Pos:* Consult merit rev bd oncol, Vet Admin, 72. *Mem:* Am Asn Cancer Res; Am Col Surg; Asn Acad Surg; Western Surg Asn. *Res:* Nucleic acid synthesis in human tumors; immune response to human tumors. *Mailing Add:* Dept Surg Univ Wis 600 Highland Ave Madison WI 53792

WOLCOTT, MARK WALTON, b Mansfield, Ohio, Apr 16, 15; m 41; c 2. SURGERY, THORACIC SURGERY. *Educ:* Lehigh Univ, BA, 37; Univ Pa, MD, 41. *Prof Exp:* Instr surg, Grad Sch Med, Univ Pa, 50-52; assoc, Med Col Pa, 52-53; asst clin prof, Med Col Ga, 54-57; chief of surg, Vet Admin Hosp, Coral Gables, Fla, 57-64; chief of res in surg, Vet Admin Cent Off, DC, 64-70; PROF SURG, SCH MED, UNIV UTAH, 70-; CHIEF STAFF, VET ADMIN HOSP, SALT LAKE CITY, 70- *Concurrent Pos:* Assoc prof, Sch Med, Univ Miami, 57-64; asst clin prof, Med Sch, George Washington Univ, 65-70. *Mem:* Am Thoracic Soc; AMA; Soc Thoracic Surg; Am Col Surg; Am Col Chest Physicians. *Res:* Coagulation defects associated with thoracic surgery; deep hypothermia and extra corporeal circulation; tissue transplantation; clostridial infection and hyperbaric medicine. *Mailing Add:* Dept Surg Vet Med Ctr 500 Foothill Blvd Salt Lake City UT 84148

WOLCOTT, THOMAS GORDON, b San Diego, Calif, Dec 22, 44; m 68; c 2. PHYSIOLOGICAL ECOLOGY, BIOTELEMETRY. *Educ:* Univ Calif, Riverside, BA, 66; Univ Calif, Berkeley, PhD(zool), 71. *Prof Exp:* Vis asst prof biol, Univ Calif, Riverside, 71-72; asst prof zool, 72-78, assoc prof, 78-85, PROF MARINE, EARTH & ATMOSPHERIC SCI, NC STATE UNIV, 85- *Concurrent Pos:* NSF, Nat Oceanog & Atmospheric Admin grants, 77-80, 83-86 & 89-; res assoc, Smithsonian Environ Res Ctr, 85- *Mem:* AAAS; Am Soc Zoologists; Am Soc Limnol & Oceanog; Sigma Xi; Crustacean Soc; Int Soc Biotelemetry. *Res:* Physiological ecology of marine invertebrates and terrestrial crabs; behavioral biotelemetry and instrumentation development, transports and recruitment of planktonic larvae using "smart" drifters. *Mailing Add:* Marine Earth & Atmospheric Sci Dept NC State Univ PO Box 8208 Raleigh NC 27695-8208

WOLD, AARON, b NY, May 8, 27; m 57; c 3. INORGANIC CHEMISTRY. *Educ:* Polytech Inst Brooklyn, BS, 46, MS, 48, PhD(chem), 52. *Prof Exp:* Res assoc chem, Univ Conn, 51-52; from instr to asst prof, Hofstra Col, 52-56; mem staff, Lincoln Lab, Mass Inst Technol, 56-63; assoc prof, 63-67, prof eng & chem, 67-80, VERNON K KRIEBLE PROF CHEM, BROWN UNIV, 80- *Concurrent Pos:* Ed, J Solid State Chem, 68-75, mem adv bd, 75-; assoc ed, Inorg Chem, 74-76 & Mat Res Bull, 77-; consult, Exxon Res & Develop Labs, 76-, Dow Chem Co, Midland, Mich, 80-83, Gen Telephone Labs, Waltham, Mass, 80- & Kodak, 85- *Mem:* Am Chem Soc; Sigma Xi. *Res:* Solid state chemistry of rare earths and transition elements; new synthetic techniques for the optimization of transparent far infrared chalcogenides and phosphides; preparation and properties of new superconductors; development of new techniques for the preparation of thin films. *Mailing Add:* Dept Chem & Div Eng Brown Univ Providence RI 02912

WOLD, DONALD C, b Fargo, NDak, Sept 24, 33; m 56; c 3. NEUTRINO AND GAMMA-RAY ASTROPHYSICS, ACOUSTICS OF SPEECH. *Educ:* Univ Wis-Madison, BA, 55, MA, 57; Ind Univ, Bloomington, PhD(physics), 68. *Prof Exp:* Lectr physics, Forman Christian Col, WPakistan, 58-63, head dept, 61-63; asst res physicist, Univ Calif, Los Angeles, 68-69; assoc prof physics, 69-74, head dept, 70-74, chmn dept physics & astron, 74-89, PROF PHYSICS, UNIV ARK, LITTLE ROCK, 74- *Concurrent Pos:* Proj dir energy conserv plan, State Ark, 76-77; HEW res fel, 78. *Honors & Awards:* Donaghey Urban Mission Award, 79. *Mem:* Acoust Soc Am; Am Phys Soc; Am Asn Physics Teachers; Asn Comput Mach; Inst Elec & Electronics Engrs. *Res:* Acoustics and perception of speech to investigate the properties of physiologically significant acoustical features of the voice; the study of astrophysical sources and high-energy particle interactions using the GRANDE research facility. *Mailing Add:* Dept Physics & Astron Univ Ark 2801 S Univ Ave Little Rock AR 72204

WOLD, FINN, b Stavanger, Norway, Feb 3, 28; nat US; m 53; c 2. BIOCHEMISTRY. *Educ:* Okla State Univ, MS, 53; Univ Calif, PhD(biochem), 56. *Prof Exp:* Res assoc biochem, Univ Calif, 56-57; from asst prof to assoc prof, Univ Ill, 57-66; prof biochem, Med Sch, Univ Minn, Minneapolis, 66-74; head dept, Univ Minn, St Paul, 74-79, prof biochem, 74-81; ROBERT A WELCH PROF CHEM, MED SCH, UNIV TEX, HOUSTON, 82- *Concurrent Pos:* Lalor res award, 58; Guggenheim fel immunochem, London, Eng, 60-61; USPHS res career develop award, 61-66; vis prof chem, Nat Taiwan Univ, 71; consult, biochem & molecular biol fel rev comt, Nat Inst Gen Med Sci, 66-70 & biochem training comt, 71-74; consult, biochem & biophys res eval comt, Vet Admin, 69-71 & res serv merit rev basic sci, 72-75; consult, res personnel comt, Am Cancer Soc, 74-77; vis prof biochem, Rice Univ, 74; consult, adv comt on nucleic acids & protein synthesis, Am Cancer Soc, 79-82. *Mem:* AAAS; Am Soc Biol Chemists; Am Chem Soc. *Res:* Protein chemistry; physical, chemical and biological properties of proteins and glycoproteins; relation of protein structure and function; mechanism of enzyme action. *Mailing Add:* Dept Biochem Molecular Biol Med Sch Univ Tex PO Box 20708 Houston TX 77225

WOLD, RICHARD JOHN, b Oshkosh, Wis, Oct 23, 37; m 60; c 2. MARINE GEOPHYSICS. *Educ:* Univ Wis, BS, 60, PhD(geophys), 66. *Prof Exp:* Lectr geophys, Univ Wis, 66-67; asst prof geol & geophys, 67-70, chmn dept geol sci, 70-72, assoc dean res, Grad Sch, 72-73, dir, Gt Lakes Res Facility, 73-75; assoc prof geol sci, Univ Wis-Milwaukee, 70-77; geophysicist, US Geol Surv, 77-80 & Br Electromagnetism & Geomagnetism, 80-82; chief geophysicist, Western Geophys Corp, 82-83 & EG&G Geometrics, 83-86; PRES, TERRASENSE, INC, 86- *Concurrent Pos:* Mem, Nat Adv Coun, Univ-Nat Oceanog Lab Syst, 74-75; assoc br chief, Off Marine Geol, Atlantic-Gulf of Mex Br, US Geol Surv, 75-78. *Mem:* Am Geophys Union; Soc Explor Geophys; Am Asn Petrol Geologists. *Res:* Geophysical studies in Great Lakes; geophysical and geological studies of inland lakes; marine geophysical instrumentation. *Mailing Add:* Terra Sense PO Box 3651 Sunnydale CA 94088

WOLDA, HINDRIK, b Wageningen, Neth, May 24, 31; m 58; c 4. POPULATION ECOLOGY, PHYSIOLOGICAL ECOLOGY. *Educ:* Univ Groningen, Neth, BSc, 55, MSc, 58, PhD(ecol), 63. *Prof Exp:* Teaching asst ecol, Univ Groningen, 58-60, sci officer, 60-63, sr sci officer, 63-68, reader, 68-71; BIOLOGIST, SMITHSONIAN TROP RES INST, 71- *Concurrent Pos:* Res fel ecol, Univ Sydney, Australia, 64-65; vis prof, Univ Wash, 88. *Honors & Awards:* Award, Smithsonian Inst, 84. *Mem:* Fel Royal Dutch Acad Sci; Am Soc Naturalists; AAAS; Ecol Soc Am; Int Asn Ecol; Int Soc Trop Ecol; Asn Trop Biol; Linnean Soc London; Royal Entomol Soc London. *Res:* Temporal and spatial variations in abundance of tropical insect species; diapause in tropical insects; systematics of leafhoppers; population ecology of insects; population genetics of polymorphic landsnails; community studies of tropical forests; seasonality and stability of insect populations. *Mailing Add:* Smithsonian Trop Res Inst APO Miami FL 34002-0011

WOLDEGIORGIS, GEBRETATEOS, b Asmara, Eritrea, Ethiopia. MEMBRANE BIOCHEMISTRY, LIPID METABOLISM. *Educ:* Haile Sellassie Univ, Addis Ababa, Ethiopia, BS, 69; Univ Wis-Madison, MS, 73, PhD(nutrit biochem), 76. *Prof Exp:* Asst chem, Haile Sellassie Univ, Addis Ababa, Ethiopia, 69-70; res asst biochem, 70-76, res assoc med, 76-80, proj assoc, 80-81, ASST SCIENTIST MED, UNIV WIS-MADISON, 81- *Concurrent Pos:* Vis scientist biochem, Inst Med Biochem, Univ Oslo, Norway, 85. *Mem:* Am Soc Biol Chemists; AAAS. *Res:* Membrane biochemistry; ion transport and bioenergetics; receptor binding; protein purification and reconstitution; lipid metabolism and chemistry; intermediary metabolism; production of monoclonal antibody and protein chemistry. *Mailing Add:* Dept Medicine Univ Wis 1415 Linden Dr Nutrit Sci Bldg Rm 431 Madison WI 53706

WOLDE-TINSAE, AMDE M, b Ethiopia, Apr 9, 47; m 84; c 4. ENGINEERING MECHANICS. *Educ:* Johns Hopkins Univ, BES, 70; Univ Calif, Berkeley, MS, 71; State Univ NY, Buffalo, PhD(civil eng), 76. *Prof Exp:* Vis asst prof civil eng & eng mech, McMaster Univ, Hamilton, Ont, 76-77; from asst prof to assoc prof civil eng, Iowa State Univ, 77-83; assoc prof, 83-88, chmn, Struct Tech Group, Dept Civil Eng, 85-90, DIR GRAD STUDIES, DEPT CIVIL ENG, UNIV MD, COLLEGE PARK, 83-, PROF CIVIL ENG, 88- *Concurrent Pos:* Consult, Doefer Eng Consults, Designers & Fabricators, 80, Johnson Machine Works, Inc, 80, WTA Eng, Inc, 84-, Occup Safety & Health Admin, US Dept Labor, 84-85, UN Develop Prog, 86-87 & 88- 89; proj engr, Anal Nuclear Reactor Containment Vessels, Nuclear Regulatory Comn, US Dept Energy, 80-81, Anal Barriers Turbine Missiles Nuclear Reactor Power Plants, 81-83; dir, Ethiopian Sci Soc NAm; vol, Vols Tech Assistance. *Honors & Awards:* Innovation in Civil Eng Award of Merit, Am Soc Civil Engrs, 91. *Mem:* Am Soc Civil Engrs; Nat Soc Prof Engrs; Int Standards Orgn; Masonry Soc; Int Asn Shell & Spatial Structures. *Res:* Non-linear analysis and stability of large space enclosures; soil structure interactions in bridges; behavior of integral and jointless bridge structures; rehabilitation and retrofitting of bridges; limit analysis of nuclear reactor containment vessels; analysis of turbine missile impact in nuclear power plants; masonry structures. *Mailing Add:* Dept Civil Eng Univ Md College Park MD 20742

WOLDSETH, ROLF, b Trondheim, Norway, Jan 18, 30; US citizen; m 55; c 4. ATOMIC PHYSICS, NUCLEAR PHYSICS. *Educ:* Tech Univ Norway, BSc, 55; Wash Univ, St Louis, PhD(physics), 65. *Prof Exp:* Res physicist, Joint Estab for Nuclear Energy Res, Norway, 55-56; asst prof physics, Rensselaer Polytech Inst, 63-67 & Wake Forest Univ, 67-70; DIR APPLN LAB, KEVEX CORP, FOSTER CITY, CALIF, 70- *Mem:* Am Phys Soc. *Res:* Positron annihilation; x-ray spectra; photo-nuclear reactions; fast neutron induced reactions. *Mailing Add:* Kevex Instruments 355 Shoreway Rd San Carlos CA 94070

WOLEN, ROBERT LAWRENCE, b New York, NY, May 20, 28. PHARMACOLOGY. *Educ:* West Chester State Col, BS, 50; Univ Del, MS, 51, PhD(biochem), 60. *Prof Exp:* Lectr atomic energy, Oak Ridge Inst Nuclear Studies, 60-61; lab dir biochem, Res Inst, St Joseph Hosp, 61-62; res scientist, Lilly Res Labs, 62-65, res assoc pharmacol, 65-75, RES ADV PHARMACOL, LILLY LAB CLIN RES, 75-; ASSOC PROF PHARMACOL, SCH MED, IND UNIV, INDIANAPOLIS, 73- *Mem:* Am Chem Soc; Sigma Xi; Am Soc Pharmacol & Exp Therapeut; Am Asn Clin Chemists; Am Soc Chem Pharmacol & Therapeut. *Res:* Phase I clinical pharmacology including drug metabolism, pharmacokinetics, drug interactions and therapeutic drug analysis. *Mailing Add:* Dept LC 789 Lilly Lab Clin Res Wishard Mem Hosp 1001 W Tenth St Indianapolis IN 46202

WOLF, A(LFRED) A(BRAHAM), b Philadelphia, Pa, July 21, 35; m 57; c 2. ELECTRICAL ENGINEERING. *Educ:* Drexel Univ, BSc, 53; Univ Pa, MSc, 54, PhD(elec eng), 58; Univ Juarez, Mex, ScD(biomed), 77, MD, 78. *Prof Exp:* Head anal group, Anti-Submarine Div, US Naval Air Develop Ctr, 49-54; asst prof elec eng, Univ Pa, 54-59; chief scientist & tech asst to vpres, Gen Dynamics-Electronics, 59-61; dir res, Emerson Radio & Phonograph Corp, Md, 61-63; distinguished prof elec eng, chmn grad res & actg dir, Sch Continuing Educ, Drexel Univ, 63-65; coordr & tech dir, Aerospace Systs Div, Radio Corp Am, Mass, 65-67; assoc tech dir res, Naval Ship Res & Develop Ctr, 67-78; PRES, PRIME RES CORP, MD, 78- *Concurrent Pos:* Lectr, Drexel Univ, 58-59, vis prof, 65-66; vis assoc prof, Univ Rochester, 60-61; vis prof, Univ Md, 67-68; vis prof, George Washington Univ, 68-69, adj prof, 72-; asst secy, Int Mil Electronics Conf, DC, 62, secy, 63. *Mem:* Sr mem Inst Elec & Electronics Engrs; Sigma Xi (vpres, 60-61). *Res:* Mathematical theory of nonlinear systems; stochastic processes; theory of physical and human systems; mathematical analysis; superconductivity; high temperature superconducting organic compounds with high critical fields; etiology and pathogenesis of cancer; mechanism of division of biological cells; mathematical biology. *Mailing Add:* 562 Ferry Pt Rd Annapolis MD 21403

WOLF, ALBERT ALLEN, b Nashville, Tenn, Sept 2, 35; m 56; c 4. ELEMENTARY PARTICLE PHYSICS. *Educ:* Vanderbilt Univ, BA, 58, MA, 60; Ga Inst Technol, PhD(physics), 66. *Prof Exp:* Physicist, Aladdin Electronics Div, Aladdin Industs, Inc, 58; engr, Sperry Rand, Inc, 59-61; instr physics, Ga Inst Technol, 61-64; from asst prof to assoc prof, 65-83, chmn dept, 83-89, PROF PHYSICS, DAVIDSON COL, 84- *Concurrent Pos:* Guest prof, Univ Ulm, WGer, 71-72. *Mem:* Am Phys Soc; Am Asn Physics Teachers; Am Chem Soc. *Res:* Quantization of non-linear fields; computer simulation. *Mailing Add:* 6600 Alexander Rd Charlotte NC 28270-2802

WOLF, ALFRED PETER, b New York, NY, Feb 13, 23; m 46; c 1. ORGANIC CHEMISTRY, NUCLEAR CHEMISTRY. *Educ:* Columbia Univ, BA, 44, MA, 48, PhD(chem), 52. *Hon Degrees:* PhD, Univ Uppsala, Sweden, 83. *Prof Exp:* MEM STAFF, CHEM DEPT, BROOKHAVEN NAT LAB, 51-, DIR, CYCOTRON-PET PROG, 76- *Concurrent Pos:* Adj prof, Columbia Univ, 53-; ed, J Labelled Compounds, 65-; consult, Univ Mass, 65-66, Philip Morris, 66-, NIH, 66 & Int Atomic Energy Agency, 69-; adv, Ital Nat Res Coun, 69-; mem eval panel, Nat Bur Standards, 72-; ed, Radiochim Acta, 77- & assoc ed, J Nuclear Med, 78-81; consult, Nat Bur Standards, 72-76 & adv panel chem, Nat Res Coun, 77-80; vis comn, Atomic Res Ctr, Julich, Ger, 80- & Los Alamos Nat Lab, 81-; chmn chem dept, Brookhaven Nat Lab, 82-87; Japan Soc Prom Sci fel, 84. *Honors & Awards:* Award, Am Chem Soc, 71; Aebersold Award, Soc Nuclear Med, 81; GV Hevesy Medal, Hevesy Found, 86; JK Javits Neurosci Investr Award, 86; Esselen Award, 88. *Mem:* Nat Acad Sci; Royal Soc Chem; Soc Ger Chem; Soc Nuclear Med; Am Chem Soc; Am Chem Soc. *Res:* Radiopharmaceutical research and nuclear medicine; organic reaction mechanisms; chemical effects of nuclear transformations; chemistry of carbon-11, nitrogen-13, oxygen-15, fluorine-18 and iodine-123; accelerators for nuclide production and radiopharmaceutical production. *Mailing Add:* PO Box 1043 Setauket NY 11733

WOLF, BARRY, b Chicago, Ill, June 19, 47; m 71; c 2. BIOCHEMICAL GENETICS. *Educ:* Univ Ill, PhD(biol chem) & MD, 74; Am Bd Pediat, dipl, 79; Am Bd Med Genetics, dipl, 82, cert clin genetics & clin biochem genetics. *Prof Exp:* Assoc prof, 82-85, PROF HUMAN GENETICS & PEDIAT, MED COL VA, 85- *Concurrent Pos:* Residency, Children's Mem Hosp, Northwestern Univ, 74-76; fel, Human Genetics, Yale Univ Sch Med, 76-78. *Honors & Awards:* Borden Award Nutrit, Am Inst Nutrit, 87; E Mead Johnson Award Pediat Res, Am Acad Pediat. *Mem:* Am Soc Clin Invest; Soc Pediat Res; Am Soc Human Genetics; Soc Inherited Metab Dis; Soc Study Inborn Errors Metab; Am Soc Clin Nutrit; Am Inst Nutrit. *Res:* Inherited metabolic diseases particularly the vitamin-responsive disorders; biochemical and clinical characterization of the biotin-responsive disorder and biotindase deficiency. *Mailing Add:* Dept Human Genetics & Pediat Med Col Va PO Box 33 MCV Richmond VA 23298

WOLF, BENJAMIN, b Deerfield, NJ, Dec 2, 13; m 40; c 2. SOIL CHEMISTRY, SOIL & PLANT ANALYSIS. *Educ:* Rutgers Univ, BS, 35, MS, 38, PhD(soil chem), 40. *Prof Exp:* Asst instr soil chem, Rutgers Univ, New Brunswick, 40-41; soil chemist, Seabrook Farms Co, 41-49; CONSULT, DR WOLF'S AGR LABS, 49- *Concurrent Pos:* Consult veg & floricult crops, Cent Am, SAm & Caribbean; assoc ed, Commun Soil Sci & Plant Anal, 70- *Mem:* Fel AAAS; Soil Sci Soc Am; Am Soc Hort Sci; Am Chem Soc. *Res:* Soil and plant analysis; herbicides; fluid fertilizer. *Mailing Add:* Dr Wolf's Agr Labs 6851 SW 45th St Ft Lauderdale FL 33314-3239

WOLF, BENJAMIN, b Detroit, Mich, June 27, 26; m 52; c 2. MICROBIOLOGY. *Educ:* Wayne State Univ, BS, 49; Univ Mich, MS, 52; Univ Pa, PhD(microbiol), 59. *Prof Exp:* From asst instr to asst prof, 57-69, assoc prof, 69-80, PROF MICROBIOL, SCH VET MED, UNIV PA, 80- *Mem:* Am Soc Microbiol. *Res:* Medical microbiology; immunology. *Mailing Add:* Dept Pathobiol Univ Pa 3800 Spruce St Philadelphia PA 19104

WOLF, BEVERLY, b Chicago, Ill, Jan 14, 35. MICROBIOLOGY. *Educ:* Univ Colo, BA, 55; Univ Calif, Los Angeles, PhD(biochem), 59. *Prof Exp:* Guest investr, Rockefeller Inst, 59-61; res assoc, Harvard Univ, 61-63; asst res biologist, Univ Calif, Berkeley, 65-72; vis asst prof, Mills Col, 72-73; SR SCIENTIST, CETUS CORP, 73- *Res:* Biochemical genetics of neurospora; genetical transformation of pneumococcus; origin and direction of deoxyribonucleic acid synthesis in Escherichia coli; temperature sensitive DNA synthesis mutants of Escherichia coli; microbial antibiotic production. *Mailing Add:* 5825 Huntington Ave Richmond CA 94804

WOLF, CAROL EUWEMA, b New Castle, Pa, June 11, 36; m 58; c 2. COMPUTER THEORY, COMPUTER SCIENCE EDUCATION. *Educ:* Swarthmore Col, BA, 58; Cornell Univ, MA, 62, PhD(math), 64. *Prof Exp:* Asst prof math, State Univ NY, Brockport, 68-75; asst prof comput sci, Iowa State Univ, 75-86; ASSOC PROF COMPUT SCI, PACE UNIV, 86-, CHAIR, COMPUTER SCI DEPT, 88- *Mem:* Asn Comput Mach; Math Asn Am; Asn Women Math; Inst Elec & Electronics Engrs Computer Soc. *Res:* Study of context-free graph grammers; computer science. *Mailing Add:* Dept Comput Sci Pace Univ One Pace Plaza New York NY 10038

WOLF, CHARLES TROSTLE, b West Reading, Pa, Mar 20, 30; m 53; c 3. MATHEMATICS. *Educ:* Millersville State Col, BS, 53; Univ Del, MS, 59. *Prof Exp:* Teacher, Pequea Valley High Sch, 53-56; asst math, Univ Del, 56-58; asst prof, Shippensburg State Col, 58-61; ASSOC PROF MATH, MILLERSVILLE UNIV, 61- *Mem:* Am Asn Univ Prof; Nat Coun Teachers Math. *Res:* Modern mathematics. *Mailing Add:* Dept Math Millersville Univ Millersville PA 17551

WOLF, CLARENCE J, b St Louis, Mo, Nov 11, 31; m 87; c 6. POLYMERIC STABILITY, EXTRATERRESTIAL STUDIES. *Educ:* Univ Mo, Columbia, BS, 53; Purdue Univ, PhD (phys chem), 57. *Prof Exp:* Sr radiation chemist, Union Carbide Res Labs, 57-61; scientist, 61-70, mgr res, 70-72, chief scientist, 72-78, prin scientist, 78-83, STAFF MGR &RES FEL, MCDONNELL DOUGLAS RES LABS, 83- *Concurrent Pos:* Assoc prof physics, Univ Col, Wash Univ, 65-72, adj prof mats sci, 88-; gen chmn, Gordon Res Conf on composites, 86. *Mem:* Am Chem Soc; Am Soc Composites; Planetary Soc. *Res:* Organic analysis of meteorites and lunar samples; development of paralysis and thermal analysis methods; solvent effects and resin systems. *Mailing Add:* 15939 Wetherburn Rd Chesterfield MO 63017-7340

WOLF, DALE DUANE, b Alma, Nebr, June 16, 32; m 52; c 4. AGRONOMY. *Educ:* Univ Nebr, BSc, 54, MSc, 57; Univ Wis, PhD(agron), 62. *Prof Exp:* Asst prof agron, Univ Conn, 62-67; asst prof, 67-71, ASSOC PROF AGRON, VA POLYTECH INST & STATE UNIV, 71- *Mem:* Am Soc Agron; Crop Sci Soc Am. *Res:* Forage crops; plant physiology. *Mailing Add:* Dept Agron Va Polytech Inst & Sta Univ Blacksburg VA 24061

WOLF, DALE E, b Kearney, Nebr, Sept 6, 24; m 45; c 4. AGRONOMY. *Educ:* Univ Nebr, BSc, 43; Rutgers Univ, PhD(farm crops, weed control), 49. *Prof Exp:* Asst farm crops, Rutgers Univ, 46-49, assoc prof & assoc res specialist, 49-50; agronomist, USDA, 47-50; asst mgr agr chem res, 50-54, mgr, 54-56, asst dist sales mgr, 56-59, dist sales mgr, 60-64, sales mgr biochem, Del, 64-67, mgr, Planning Div, 67-68, asst dir, Agr Div, 68-70, dir, Indust Specialities Div, 71-72, dir mkt agrichem, 72-75, asst gen mgr, Biochem Dept, 75-78, gen mgr, Biochem Dept, 78-79, VPRES BIOCHEMICALS, E I DU PONT DE NEMOURS & CO, INC, 79-, CHMN BD, ENDO LAB, INC, SUBSID E I DU PONT DE NEMOURS & CO, INC, 79- *Mem:* Am Soc Agron; Nat Agr Chemicals Asn. *Res:* Agricultural chemicals; weed control. *Mailing Add:* PO Box 3825 Greenville DE 19807

WOLF, DANIEL STAR, b Indianapolis, Ind, Aug 25, 49; m 77, 84; c 1. OCEAN ENGINEERING, FINITE ELEMENT METHODS. *Educ:* Purdue Univ, BS, 71, MS, 73, PhD(aeronaut eng), 76. *Prof Exp:* Teaching asst aeronaut eng, Purdue Univ, 71-76; asst prof ocean eng, Fla Atlantic Univ, 76-79; PROJ ENGR, GOVT PROD DIV, PRATT & WHITNEY AIRCRAFT, UNITED TECHNOL CORP, 79- *Res:* Fatigue crack propagation; solid mechanics structures; application of the finite element method to thermal and structural analysis; patent in turbine vane assembly. *Mailing Add:* 4049 N Browning Dr West Palm Beach FL 33406

WOLF, DIETER, b Sindelfingen, WGer, Nov 3, 46. THEORETICAL MATERIALS SCIENCE. *Educ:* Univ Stuttgart, Diplom, 70, Dr rer nat(physics), 73. *Prof Exp:* Res assoc mat sci, Max Planck Inst Metal Res, Stuttgart, 72-74; res asst prof physics, Univ Utah, 74-77; asst scientist, Mat Sci Div, 77-79, scientist, 79-86; SR SCIENTIST, ARGONNE NAT LAB, 86-; PVT LECTR, UNIV DORT, WGER, 79- *Mem:* Am Phys Soc; Sigma Xi; Am Cer Soc; Nat Res Soc. *Res:* Diffusion in crystals; theory of magnetic resonance and relaxation; atomic, molecular and defect motions in metals, metal oxides and ionic crystals studied by nuclear magnetic resonance; Mossbauer effect and quasielastic neutron scattering; computer simulation and theory of solid interfaces, grain boundaries and point defects. *Mailing Add:* Div Mat Sci Argonne Nat Lab Argonne IL 60439

WOLF, DON PAUL, b Lansing, Mich, Aug 8, 39; m 67; c 2. BIOCHEMISTRY, REPRODUCTIVE BIOLOGY. *Educ:* Mich State Univ, BS, 61, MS, 62; Univ Wash, PhD(biochem), 67. *Prof Exp:* Fel biochem, Univ Geneva, 67-68 & Univ Calif, Davis, 68-71; asst prof obstet & gynec & biophys, Univ Pa, 71-77, assoc prof, 77-; PROF, DEPT OBSTET & GYNEC, UNIV TEX HEALTH SCI CTR. *Mem:* AAAS; Am Fertil Soc; Soc Study Reproduction; Am Soc Cell Biol; Am Soc Biol Chemists. *Res:*

Characterization of reproductive processes; the role of cortical granules in fertilization and early development; sperm excluding mechanism operative in animal ova; isolation and characterization of cervical and tracheal mucin. *Mailing Add:* Dept Obstet & Gynec Univ Tex Med Sch 6431 Fannin Suite 3270 Houston TX 77030

WOLF, DUANE CARL, b Springfield, Mo, Apr 7, 46; m 68. SOIL MICROBIOLOGY. *Educ:* Univ Mo-Columbia, BS, 68; Univ Calif, Riverside, PhD(soils), 73. *Prof Exp:* Asst prof soils, Univ Md, College Park, 73-78; assoc prof, 79-81, PROF SOILS, UNIV ARK, FAYETTEVILLE, 81- *Honors & Awards:* Fel, Am Soc Agron, 90. *Mem:* Am Soc Agron; Soil Sci Soc Am; Am Soc Microbiol; AAAS; Sigma Xi. *Res:* Renovation of onsite wastewater; degradation of organic chemicals in soil; microbiology and biochemistry of nitrogen transformations in soil. *Mailing Add:* Dept Agron Univ Ark Fayetteville AR 72701

WOLF, EDWARD CHARLES, b Los Angeles, Calif, Jan 8, 54; m 80. ENZYMOLOGY, METABOLIC REGULATION. *Educ:* Calif State Univ, Fresno, BA, 76; Univ Calif, Los Angeles, PhD(biochem), 81. *Prof Exp:* Lectr chem, Calif State Univ, Fresno, 82-83; ASST PROF CHEM & BIOCHEM, UNIV NEV, LAS VEGAS, 83- *Mem:* Sigma Xi; Am Chem Soc; AAAS. *Res:* Identity and regulation of the converging pathways of arginine, ornithine and proline metabolism; Neurospora crassa. *Mailing Add:* Dept Chem Univ Nev Las Vegas NV 89154

WOLF, EDWARD D, b Quinter, Kans, May 30, 35; m 55; c 3. SURFACE PHYSICS. *Educ:* McPherson Col, BS, 57; Iowa State Univ, PhD(phys chem), 61. *Prof Exp:* Res assoc, Princeton Univ, 61-62; sr chemist, Atomics Int Div, NAm Aviation, Inc, 63-64, res specialist, 65, mem tech staff, Sci Ctr Div, 64-65; mem tech staff, Hughes Res Labs, Malibu, 65-67, sr staff chemist, 67-72, sect head electron beam surface physics, 72-74, sr scientist, 74-78; PROF & DIR, NAT RES & RESOURCE FACIL SUBMICRON STRUCT, CORNELL UNIV, 78- *Concurrent Pos:* Res assoc, Univ Calif, Berkeley, 68; vis prof, eng dept, Cambridge Univ, 86-87; guest prof, Tech Univ Vienna, Austria, 87; chmn bd, Boilistics Inc, 86- *Mem:* Am Phys Soc; Electron Micros Soc Am; fel Inst Elec & Electronics Eng. *Res:* Ionic mobilities of high temperature inorganic liquids; magnetohydrodynamic energy conversion; field emission and scanning electron microscopy; scanning electron beam diagnostics and microfabrication. *Mailing Add:* Eight Highgate Circle Ithaca NY 14850

WOLF, EDWARD LINCOLN, b Cocoa, Fla, Nov 22, 36; m 58; c 2. CONDENSED MATTER PHYSICS, TUNNELING SPECTROSCOPY. *Educ:* Swarthmore Col, AB, 58; Cornell Univ, PhD(exp physics), 64. *Prof Exp:* Fel physics, Dept Physics & Coord Sci Lab, Univ Ill, Urbana, 64-66, res assoc, 67; sr physicist, Res Labs, Eastman Kodak Co, Rochester, 68-75; assoc prof, 75-80, prof physics, Ames Lab, Iowa State Univ, 81-85; PROF & HEAD DEPT PHYSICS, POLYTECH UNIV, 86- *Concurrent Pos:* Vis fel, Cavendish Lab, Univ Cambridge, 73-74, Univ Pa, 81, IBM, Corp Watson Res, 82. *Mem:* Fel Am Phys Soc; AAAS. *Res:* Superconductivity; electron tunneling; physics of surfaces and interfaces; ultra high vacuum; superconducting proximity effect; photoemission spectroscopy and electron energy loss spectroscopy; scanning tunneling microscopy; superconductivity; high Tc superconductivity. *Mailing Add:* 34 Plaza St No 607 Brooklyn NY 11238-5038

WOLF, ELIZABETH ANNE, b Leeds, UK; US citizen; m 54; c 2. NUCLEAR PHYSICS. *Educ:* Oxford Univ, BA, 51, PhD(nuclear physics), 55. *Prof Exp:* Dept demonstr physics, Clarendon Lab, Oxford Univ, 52-55, res staff, Lab Archaeol, 55-56; res assoc physics, Biophys Lab, Harvard Med Sch, 56-57; lectr, Univ Conn, Waterbury, 75-79; from asst prof to assoc prof, 79-89, PROF PHYSICS, SOUTHERN CONN STATE UNIV, 89-, CHAIR DEPT, 87- *Concurrent Pos:* Newsletter ed, New Eng Sect, Am Phys Soc, 81-86; hon sr res physicist, Churchill Hosp, Oxford, UK, 84. *Mem:* Am Phys Soc; Am Asn Physics Teachers; Am Asn Physicists Med. *Res:* Experimental nuclear physics; deuteron-deuteron interaction; inelastic neutron scattering; measurement of very low level radiation; pulmonary edema detection by back scattering of gamma radiation. *Mailing Add:* Physics Dept Southern Conn State Univ New Haven CT 06515

WOLF, EMIL, b Prague, Czech, July 30, 22. MATHEMATICAL PHYSICS, OPTICS. *Educ:* Bristol Univ, BSc, 45, PhD(physics), 48; Univ Edinburgh, DSc(physics), 55. *Prof Exp:* Res asst optics, Cambridge Univ, 48-51; res asst & lectr math physics, Univ Edinburgh, 51-54; res fel theoret physics, Univ Manchester, 54-59; assoc prof optics, 59-61, prof physics, 61-87, prof optics, 78-87, WILSON PROF OPTICAL PHYSICS, UNIV ROCHESTER, 87- *Concurrent Pos:* Vis scientist, NY Univ, 57; Guggenheim fel & vis prof, Univ Calif, Berkeley, 66-67; vis prof, Univ Toronto, 74-75; ed, Progress in Optics. *Honors & Awards:* Frederic Ives Medal, Optical Soc Am, 77; Albert A Michelson Medal, Franklin Inst, 80; Max Born Award, Optics Soc Am, 87; Marconi Medal, Ital Nat Res Coun, 87. *Mem:* Fel Optical Soc Am (pres, 78); fel Am Phys Soc; fel Brit Inst Physics; hon mem Optical Soc India; fel Franklin Inst; Optics Soc Am; hon mem Optical Soc Am. *Res:* Theoretical optics; electromagnetic theory; co-author of book. *Mailing Add:* Dept Physics & Astron Univ Rochester Rochester NY 14627

WOLF, ERIC W, b Frankfurt am Main, Ger, Feb 20, 22; US citizen; m 49; c 2. SYSTEMS SCIENCE, LIBRARY AUTOMATION. *Educ:* City Col New York, BEE, 49; Ohio State Univ, MS, 51. *Prof Exp:* Electronic scientist, Wright Air Develop Ctr, 49-53; res engr, Lincoln Lab, Mass Inst Technol, 53-59; tech adv tech ctr, Supreme Hq Allied Powers Europe, Hague, 59-60; dir comput prog, Data & Info Systs Div, Int Tel & Tel Corp, 61-62, tech dir, 62-65; tech dir, Naval Command Systs Support Activity, 65-70; sr scientist, Bolt Beranek & Newman Inc, 70-80, mgr wash opers, Commun Systs Div, 80-87, prog mgr, 87-91; PRIN SCIENTIST, FIAT LUX, 91- *Mem:* Comput Soc; Commun Soc; Inst Elec & Electronics Engrs. *Res:* Design and development of computer-based information systems; computer communications; computer networks; information management; man-machine interaction in natural language. *Mailing Add:* Fiat Lux 6300 Waterway Dr Falls Church VA 22044-1316

WOLF, FRANK JAMES, b Xenia, Ohio, Nov 7, 16; m 42; c 3. ORGANIC CHEMISTRY. *Educ:* Miami Univ, AB, 38; Univ Ill, PhD(org chem), 42. *Prof Exp:* Sr chemist, Merck & Co, Inc, 42-59, asst dir microbiol, Merck Sharp & Dohme Res Labs, 59-70, dir animal drug metab & radiochem, 70-77, SR SCIENTIST, MERCK SHARP & DOHME RES LABS, 78- *Concurrent Pos:* Asst, Nat Defense Res Comt, 41-42. *Mem:* Am Soc Pharmacol & Exp Therapeut; Am Chem Soc. *Res:* Phthalides; substituted sulfaquinoxalines; benzotriazines; antibiotics; vitamin B-12; catalysis; biochemical separations; animal and human drug metabolism; animal tissue residue; mass spectroscopy; biochemical toxicology; pharmacokinetics. *Mailing Add:* 38 Genesee Trail Westfield NJ 07090-2706

WOLF, FRANK LOUIS, b St Louis, Mo, Apr 18, 24; m 47; c 4. MULTIVARIATE ANALYSIS. *Educ:* Wash Univ, BS, 44, MA, 48; Univ Minn, PhD(math), 55. *Prof Exp:* Tech supvr chem eng, Carbon & Carbide Chem Corp, Tenn, 44-46; instr math, St Cloud State Col, 49-51; from instr to prof, 52-90, EMER PROF MATH, CARLETON COL, 90- *Concurrent Pos:* NSF fel, 60-61. *Mem:* AAAS; Math Asn Am; Am Statist Asn. *Res:* Multivariate descriptive statistics; statistical education. *Mailing Add:* 12 Bundy Ct Northfield MN 55057

WOLF, FRANKLIN KREAMER, b Norman, Okla, Sept 30, 35; m 62; c 3. INDUSTRIAL ENGINEERING, OPERATIONS RESEARCH. *Educ:* Iowa State Univ, BS, 57, PhD(eng valuation, statist), 70; Univ Wis, MS, 62. *Prof Exp:* Officer, US Army CEngrs, 57-59; engr, Rheem Semiconductor, 59-60; res asst, Univ Wis, 60; engr, Martin Co, Denver, 61-64; asst prof indust eng, Iowa State Univ, 64-70; assoc prof, 70-77, PROF INDUST ENG & CHMN, WESTERN MICH UNIV, 77- *Mem:* Am Inst Indust Engrs; Am Soc Eng Educ; Inst Mgt Sci; Opers Res Soc Am; Sigma Xi. *Res:* Engineering economics and study of life estimation and methods of depreciation for capital recovery in regulated industries. *Mailing Add:* Dept Indust Eng Western Mich Univ Kalamazoo MI 49008

WOLF, FREDERICK TAYLOR, b Auburn, Ala, July 11, 15; m 45; c 2. BOTANY. *Educ:* Harvard Univ, AB, 35; Univ Wis, MA, 36, PhD(bot), 38. *Prof Exp:* Nat Res Coun fel, Harvard Univ, 38-39; from instr to prof bot, Vanderbilt Univ, 39-81; RETIRED. *Mem:* Mycol Soc Am; Bot Soc Am; Am Soc Plant Physiol; Brit Mycol Soc; Int Soc Human & Animal Mycol. *Res:* Mycology; physiology of fungi; plant physiology. *Mailing Add:* 3611 Saratoga Dr Nashville TN 37205

WOLF, GEORGE, b Vienna, Austria, June 16, 22; nat US; m 48; c 3. NUTRITION, BIOCHEMISTRY. *Educ:* Univ London, BSc, 44; Oxford Univ, DPhil, 47. *Prof Exp:* Res fel, Chester Beatty Res Inst, Royal Cancer Hosp, Eng, 47-48, Harvard Univ, 48-50 & Univ Wis, 50-51; from asst prof to assoc prof animal nutrit, Univ Ill, Urbana, 51-62; from assoc prof to prof physiol chem 62-88, EMER PROF PHYSIOL CHEM, MASS INST TECHNOL, 88-; ADJ PROF NUTRIT SCI, UNIV CALIF, BERKELEY, 88- *Concurrent Pos:* Guggenheim fel, 58-59. *Honors & Awards:* Osborne-Mendel Award, Am Inst Nutrit, 77. *Mem:* Am Soc Biol Chemists; Am Inst Nutrit. *Res:* Metabolism and function of vitamin A and beta-carotene. *Mailing Add:* Dept Nutrit Sci Morgan Hall Univ Calif Berkeley CA 94720

WOLF, GEORGE ANTHONY, JR, b East Orange, NJ, Apr 20, 14; m 39; c 2. MEDICINE. *Educ:* NY Univ, BS, 36; Cornell Univ, MD, 41; Am Bd Internal Med, dipl, 48. *Hon Degrees:* DSc, Union Univ, 65, Univ Vermont, 86. *Prof Exp:* Intern med, New York Hosp, 41-42; asst, Med Col, Cornell Univ, 42-43, instr, 43, fel med res, 44-46, asst prof clin med, 49-52; prof clin med & dean, Col Med, Univ Vt, 52-61; prof med, Sch Med & vpres med & dent affairs, Tufts Univ, 61-66; dean & provost, Univ Kans Med Ctr, Kansas City, 66-70; prof, 70-79, EMER PROF MED, COL MED, UNIV VT, 79- *Concurrent Pos:* Asst resident med, New York Hosp, 42-43, resident physician, 43-44, physician to out-patients, 44, asst attend physician & dir out-patient dept, 49-52; dir, Tufts New Eng Med Ctr, 61-66; attend physician, Med Ctr Hosp, Vt, 70-81, emer attend physician, 80- *Mem:* Harvey Soc; fel Am Col Physicians; fel NY Acad Med; Asn Am Med Cols. *Res:* Internal medicine; medical education; medical school administration. *Mailing Add:* Rte 1 Box 237 Nashville Rd Jericho VT 05465

WOLF, GEORGE WILLIAM, b Newark, NJ, Jan 15, 43; m 80; c 2. ASTRONOMY. *Educ:* Univ Pa, BA, 65, MS, 67, PhD(astron), 70. *Prof Exp:* Res assoc astron, Mt John Observ, Univ Canterbury, NZ, 68-69; PROF ASTRON, SOUTHWEST MO STATE UNIV, 71- *Concurrent Pos:* NSF fel, 77-78. *Mem:* Am Astron Soc. *Res:* Observational astronomy in the areas of linear and circular polarimetry, spectroscopy and photoelectric photometry of stars. *Mailing Add:* Dept Physics & Astron Southwest Mo State Univ Springfield MO 65804-0094

WOLF, GERALD LEE, b Sidney, Nebr, Apr 2, 38; m 64; c 2. RADIOLOGY. *Educ:* Univ Nebr, BS, 62, MS, 64, PhD(physiol & pharmacol), 65; Harvard Univ, MD, 68. *Prof Exp:* Asst prof physiol, Col Med, Univ Nebr, Omaha, 68-69, asst prof physiol & med, 69-71, asst prof pharmacol & radiol & dir radiol res, 71-80; PROF RADIOL, UNIV PA MED CTR, 80- *Concurrent Pos:* Life Ins Med res fel, 64-69; intern, Col Med, Univ Nebr, 68-69; res consult, Omaha Vet Admin Hosp, 69-, clin investr, 70-73. *Mem:* Am Col Radiol; fel Am Soc Clin Pharmacol & Therapeut; Asn Am Med Cols; AMA; Am Soc Pharmacol & Exp Therapeut. *Res:* Renal and endocrine participation in electrolyte homeostasis; physiology of the pump-perfused dog kidney. *Mailing Add:* Dept Radiol Mass Gen Hosp 55 Fruit St Boston MA 02114

WOLF, HAROLD HERBERT, b Quincy, Mass, Dec 19, 34; m 57; c 2. PHARMACOLOGY. *Educ:* Mass Col Pharm, BS, 56; Univ Utah, PhD(pharmacol, exp psychiat), 61. *Prof Exp:* From asst prof to assoc prof, 61-69, Kimberly prof pharmacol & chmn dept, Col Pharm, Ohio State Univ, 69-76; dean, Col Pharm, 76-89, PROF PHARMACOL & TOXICOL, COL PHARM, UNIV UTAH, 76- *Concurrent Pos:* NIH & Alcohol, Drug Abuse & Ment Health Admin res grants, 63-; Am Asn Cols Pharm vis lectr, 64-; mem pharm rev comt, NIH, 69-71; Fulbright-Hays sr scholar, Univ Sains Malaysia, 74; mem, Joint Comn on Prescription Drug Use, 77-80, Biomed Res Support Comt, NIH, 78-79, bd of dir, Am Found for Pharmaceut Educ, 78, comt of pres, Asn Acad Health Ctr, 78; external exam, Univ Sains, Malaysia, 80; Am Pharmaceut Asn Task Force Educ, 81-84; Comn on Goals, Am Soc Hosp Pharm, 82-84; bd dir, Am Coun on Pharm Educ, 85-88; chmn, Comn Implement Change in Pharmaceut Educ, Am Asn Col Pharm, 89. *Honors & Awards:* Distinguished Educ Award, Am Asn Col Pharm, 88. *Mem:* AAAS; Am Asn Col Pharm (pres, 78); Am Pharmaceut Asn; fel AAAS; fel Acad Pharmaceut Sci; Am Soc Pharmacol & Exp Therapeut. *Res:* Investigation of effects of drugs on central nervous system; psychotropics, central nervous system stimulants, anticonvulsants, narcotic analgetics, thermoregulation and animal behavior. *Mailing Add:* Univ Utah Col Pharm Salt Lake City UT 84112

WOLF, HAROLD WILLIAM, b Chicago, Ill, July 13, 21; m 44; c 4. ENVIRONMENTAL HEALTH. *Educ:* Univ Iowa, BS, 49, MS, 50; Univ Calif, Los Angeles, DrPH, 65. *Prof Exp:* From sanit engr to sr sanit engr, Fed Water Pollution Control Admin, USPHS, Calif, 50-68, asst chief res & develop, Water Supply & Sea Resources, Ohio, 68, dir div criteria & standards, Bur Water Hygiene, Md, 68-70; prof civil eng, Tex A&M Univ, 70-84; CONSULT ENGR, 84- *Concurrent Pos:* Res fel, Calif Inst Technol, 60-62; lectr, Univ Calif, Los Angeles, 65-66; dir, Dallas Water Reclamation Res Ctr, 70-75; mem, Nat Drinking Water Adv Coun, 75-79. *Mem:* AAAS; Asn Environ Eng Profs; Am Soc Civil Eng; Am Pub Health Asn. *Res:* Water quality criteria; air-borne pathogens; hazardous waste disposal. *Mailing Add:* 1007 Rose Circle College Station TX 77840

WOLF, HELMUT, b Einöd, Ger, Jan 19, 24; US citizen. MECHANICAL ENGINEERING, HEAT TRANSFER. *Educ:* Case Inst Technol, BSME, 48; Purdue Univ, MSME, 50, PhD(mech eng), 58. *Prof Exp:* Instr mech eng, Purdue Univ, 48-50; develop engr, Eastman Kodak Co, 50-53; res assoc heat transfer, Jet Propulsion Ctr, Purdue Univ, 53-58; prin scientist, NAm Aviation, Inc, 58-61; prof mech eng, Univ Ark, Fayetteville, 61-71, Raymond F. Giffels distinguished prof eng, 71-; RETIRED. *Concurrent Pos:* Resident res assoc, Argonne Nat Lab, 64-66; res grant & consult, Cent Transformer Corp, 66-67; staff adv, Inst Reactor Develop, Nuclear Res Ctr, Karlsruhe, WGer, 70-71. *Mem:* Am Soc Mech Engrs; Am Nuclear Soc; Am Soc Eng Educ; Sigma Xi. *Res:* Heat transfer of forced and free-convection shear flows and conduction transfer. *Mailing Add:* 1913 Quail Run Dr NE Albuquerque NM 87122

WOLF, HENRY, b Munich, Ger, Aug 30, 19; mat US; m 44; c 4. APPLIED MATHEMATICS, CELESTIAL MECHANICS. *Educ:* Univ Toronto, BA, 46, MA, 47; Brown Univ, PhD, 50. *Prof Exp:* Instr math, Univ Toronto, 46-48; res assoc appl math, Brown Univ, 48-50; from asst prof to assoc prof math, Hofstra Col, 50-56; prin engr, Repub Aviation Corp, 56-57, group engr, 57-58, sr group engr, 58-59, develop engr, Appl Math Sect, 59-60, chief numerical methods, 60-63; sr scientist & secy treas, Anal Mech Assocs, Jericho, 63-74, vpres, 74-84, pres, 84-89, CHMN BD, ANALTICAL MECH ASSOCS, JERICHO, 90- *Mem:* Am Math Soc; Am Inst Aeronauts & Astronauts. *Res:* Plastic wave propagation; interior ballistics and aeroelasticity; numerical analysis; celestial mechanics; cosmic dust; air data systems. *Mailing Add:* 110-18 68th Ave Forest Hills NY 11375

WOLF, IRA KENNETH, b New York, NY, Nov 14, 42; m 64; c 1. MATHEMATICS. *Educ:* Tufts Univ, BA, 64; Yale Univ, MA, 66; Rutgers Univ, PhD(math), 71. *Prof Exp:* Asst prof math, Brooklyn Col, 71-; MEM STAFF, DEPT MATH, PACE UNIV. *Mem:* Math Asn Am. *Res:* Category theory. *Mailing Add:* 12 Rugby Rd Roslyn NY 11577

WOLF, IRVING W, b Nashville, Tenn, July 12, 27; m 54; c 3. CHEMICAL ENGINEERING. *Educ:* Vanderbilt Univ, BE, 47, MS, 49; Ill Inst Technol, PhD(chem eng), 51. *Prof Exp:* Instr chem eng, Ill Inst Technol, 50-51; res & develop engr, Gen Elec Co, 51-55, consult heat transfer, 55-58, proj engr, 58-61, mgr functional films sect, 61-63; head mat res, 63-65, mgr mat & devices res, 65-71, corp consult, 71-72, mgr magnetic mat, Magnetic Tape Div, 72-78, MGR PLASTICS & CHEM ENG, AMPEX CORP, 72- *Mem:* AAAS; Inst Elec & Electronics Engrs; Electrochem Soc. *Res:* Electrodeposition of magnetic thin films for information storage; vacuum deposition and sputtering of thin films for electronic application. *Mailing Add:* Box 5481 Redwood City CA 94063

WOLF, JACK KEIL, b Newark, NJ, Mar 14, 35; m 55; c 3. ELECTRICAL ENGINEERING. *Educ:* Univ Pa, BS, 56; Princeton Univ, MSE, 57, MA, 58, PhD(elec eng), 60. *Prof Exp:* Instr, Syracuse Univ, 60-62; assoc prof elec eng, NY Univ, 63-65; from assoc prof to prof, Polytech Inst Brooklyn, 65-73; chmn dept elec & comput eng, Univ Mass, Amherst, 73-75, prof elec eng, 73-84; PROF ELEC ENG & COMPUT SCI, UNIV CALIF, SAN DIEGO, 85- *Concurrent Pos:* Consult, govt & indust, 63-; NSF sr fel, Univ Hawaii, 71-72; Guggenheim fel, 79-80. *Mem:* Fel Inst Elec & Electronics Engrs. *Res:* Statistical communications theory; algebraic coding theory; magnetic recording; computer networks. *Mailing Add:* Dept Elec & Computer Sci Univ Calif San Diego La Jolla CA 92037

WOLF, JAMES S, b Cleveland, Ohio, July 26, 33; m 57. METALLURGY. *Educ:* Case Inst Technol, BS, 54, MS, 60; Univ Fla, PhD(metall), 65. *Prof Exp:* Res scientist, Lewis Lab, NASA, 57-69; assoc prof mat eng, 69-77, PROF MAT ENG, CLEMSON UNIV, 77- *Mem:* Am Soc Metals; Am Inst Mining, Metall & Petrol Engrs; Brit Inst Metals; Nat Asn Corrosion Engrs; Sigma Xi. *Res:* High temperature oxidation and deformation of metals and alloys; biomedical materials; secondary metals. *Mailing Add:* 222 Holly Ave Clemson SC 29631

WOLF, JAMES STUART, b Chicago, Ill, Mar 1, 35; m 58; c 2. MEDICINE, SURGERY. *Educ:* Grinnell Col, AB, 57; Univ Ill, BS, 59, MD, 61. *Prof Exp:* USPHS res fel transplantation, 64-66, from instr to prof surg, Med Col Va, 67-76; chief surg, McGuire Vet Admin Hosp, 68-76; PROF SURG & CHMN DIV TRANSPLANTATION, NORTHWESTERN UNIV MED SCH, 76-, ASSOC DEAN MED EDUC, 90- *Concurrent Pos:* Attend surgeon, McGuire Vet Admin Hosp, 67-76, Northwestern Mem Hosp, Lakeside Vet Admin Hosp, Evanston Hosp, 76- *Mem:* Transplantation Soc; Am Soc Nephrology; Asn Acad Surg; fel Am Col Surg; Soc Univ Surg. *Res:* Transplantation immunology; clinical and experimental organ transplantation. *Mailing Add:* Northwestern Univ Med Sch Dept Surg 303 E Chicago Ave Chicago IL 60611

WOLF, JOSEPH A(LLEN), JR, b Tacoma, Wash, Nov 26, 33; m 60; c 1. APPLIED MECHANICS. *Educ:* Stevens Inst Technol, ME, 55; Univ Calif, Los Angeles, MS, 57; Mass Inst Technol, ScD(mech eng), 67. *Prof Exp:* Mem tech staff, Hughes Aircraft Co, 55-62; asst prof mech & struct, Univ Calif, Los Angeles, 66-71; sr res engr, Eng Mech Dept, Gen Motors Res Labs, 71-80, staff res engr, 80-90, STAFF RES ENGR, GEN MOTORS SYSTS ENG CTR, 90- *Concurrent Pos:* Adj prof, Dept Mech Eng, Wayne State Univ, Detroit, 89- *Mem:* Fel Am Soc Mech Engrs; Soc Automotive Engrs; Am Acad Mech. *Res:* Dynamics; mechanical and structural vibrations; structural-acoustic interaction. *Mailing Add:* Systs Analysis & Info Mgt Dept Gen Motors Systs Eng Ctr 1151 Crooks Rd Troy MI 48084

WOLF, JOSEPH ALBERT, b Chicago, Ill, Oct 18, 36. GEOMETRY. *Educ:* Univ Chicago, BS, 56, MS, 57, PhD(math), 59. *Prof Exp:* From asst prof to assoc prof, 62-66, PROF MATH, UNIV CALIF, BERKELEY, 66- *Concurrent Pos:* Mem, Inst Advan Study, Princeton, 60-62 & 65-66; prin investr, NSF, 64-; Miller res prof, Univ Calif, 72-73 & 83-84; ed, Geometriae Dedicata, Math Reports, J Math Systs, Estimation & Control, Nova J Algebra & Geom, Lett in Math Physics; hon prof, Nat Univ Cordoba, Arg, 89. *Honors & Awards:* Medaille de l'Universite, Univ Liege, 77. *Res:* Use of group theory to study differential geometry, complex manifolds, harmonic analysis, and a few areas in particle physics and control theory. *Mailing Add:* Dept Math Univ Calif Berkeley CA 94720

WOLF, JULIUS, b Boston, Mass, Aug 15, 18; m 45; c 3. MEDICINE. *Educ:* Boston Univ, SB, 40, MD, 43. *Prof Exp:* Chief med serv, 54-72, CHIEF OF STAFF, VET ADMIN HOSP, BRONX, 72-; PROF CLIN MED, MT SINAI SCH MED, 68-, ASSOC DEAN VET ADMIN PROGS, 72- *Concurrent Pos:* Assoc clin prof med, Col Physicians & Surgeons, Columbia Univ, 62-68; chmn, Vet Admin Lung Cancer Study Group, 62- *Res:* Lung cancer chemotherapy. *Mailing Add:* Vet Admin Ctr 130 W Kingsbridge Rd Bronx NY 10468

WOLF, KATHLEEN A, b Dallas, Tex, Sept 10, 46. ENVIRONMENTAL RESEARCH. *Educ:* Univ Washington, BS, 68; San Diego State Univ, MS, 71; Univ Southern Calif, PhD(chem), 80. *Prof Exp:* Phys scientist, Rand Corp, 73-87; source reduction, Source Reduction Res Partnership, 88-90; DIR, INST RES & TECH ASSISTANCE, 90- *Concurrent Pos:* Res asst chem, Univ Southern Calif, 76-80, res assoc, 81-83. *Mem:* Am Chem Soc. *Res:* Quantum mechanics; spectroscopy; ozone layer depletion; hazardous waste disposal; risk analysis. *Mailing Add:* Inst Res & Tech Assistance 3727 W Sixth St Suite 505 Los Angeles CA 90020

WOLF, KENNETH EDWARD, b Chicago, Ill, Oct 22, 21; m 48; c 3. MICROBIOLOGY. *Educ:* Utah State Univ, BS, 51, MS, 52, PhD(fish path), 56. *Prof Exp:* Microbiologist, Eastern Fish Dis Lab, US Fish & Wildlife Serv, 54-72, dir, 72-77; sr res scientist, Nat Fish Health Res Lab, 77-86; RETIRED. *Concurrent Pos:* Ed, Fish Health News. *Mem:* AAAS; Am Soc Microbiol; Tissue Cult Asn; Wildlife Dis Asn; Int Asn Aquatic Animal Med. *Res:* Diseases of fishes, especially of viral etiology; methods of cultivation of fish cells and tissues; life cycle studies on salmonid whirling disease, myxosoma cerebralis; immunization of salmonids against ichthyophthiriasis. *Mailing Add:* Rte 3 Box 79 Kearneyville WV 25430

WOLF, LARRY LOUIS, b Madison, Wis, Oct 21, 38; m 65; c 2. MATING SYSTEMS, COMMUNITY ORGANIZATION. *Educ:* Univ Mich, BS, 61; Univ Calif, Berkeley, PhD(zool), 66. *Prof Exp:* Assoc zool, Univ Calif, Berkeley, 64-66; Elsie Binger Naumberg res fel ornith, Am Mus Natural Hist, 66-67; from asst prof zool to assoc prof biol, 67-76, PROF BIOL, SYRACUSE UNIV, 76- *Concurrent Pos:* Ecol adv panel, NSF, 77-79; ed Ecol, Ecological Monographs, 84-87; vis prof, Univ Miami, 82. *Mem:* Am Ornith Union; Int Soc Behav Ecol; Am Soc Naturalists; Brit Ecol Soc; Ecol Soc Am. *Res:* Ecological determinants of social systems, principally in birds and insects; community organization; coevolution; plant population biology. *Mailing Add:* Dept Biol Syracuse Univ Syracuse NY 13244-1270

WOLF, LESLIE RAYMOND, b East Chicago, Ind, Feb 18, 49; m 72; c 2. ORGANIC CHEMISTRY, POLYMER CHEMISTRY. *Educ:* Lewis Univ, BA, 71; Pa State Univ, PhD(chem), 74. *Prof Exp:* Chemist, Rohm & Haas Co, 73-78; sr res chemist, DeSoto, Inc, 78-81; SR RES CHEMIST, AMOCO OIL CO, 81- *Concurrent Pos:* Lectr chem, Eve Div, LaSalle Col, 74-78, Lewis Univ, 85- *Mem:* Am Chem Soc. *Res:* Fuels research. *Mailing Add:* Amoco Res Ctr PO Box 3011 Naperville IL 60565-7011

WOLF, LOUIS W, b Saginaw, Mich, May 3, 28; m 52; c 4. ENGINEERING MECHANICS, APPLIED MATHEMATICS. *Educ:* Univ Mich, BSE, 52, MSE, 55, PhD(eng mech), 63. *Prof Exp:* Engr, Carrier Corp, 52-54; res assoc eng mech, Univ Mich, 54-61, instr, 55-60; res engr, Systs Div, Bendix Corp, 61-62; asst prof, 62-65, ASSOC PROF MECH ENG, UNIV MICH-DEARBORN, 65- *Mem:* Asn Comput Mach; Soc Indust & Appl Math. *Res:* Fluid mechanics; nonlinear elastic shells; computer technology. *Mailing Add:* Dept Mech Eng Univ Mich Deerborn 4901 Evergreen Rd Dearborn MI 48128

WOLF, MARVIN ABRAHAM, b Syracuse, NY, Dec 26, 25; m 54; c 2. MICROMETEOROLOGY, AIR POLLUTION. *Educ:* NMex Inst Mining & Technol, BS, 51; Univ Wash, MS, 62. *Prof Exp:* Assoc res engr, Boeing Airplane Co, Wash, 58-60; res meteorologist, Meteorol Res Inc, Calif, 60-63; physicist, Pac Missile Range, Calif, 63-66; res assoc, Pac Northwest Labs,

Battelle-Mem Inst, 66-77; res assoc, Air Res Ctr, Ore State Univ, 77-80, assoc prof, Dept Atmospheric Sci, 80-86; CONSULT, 82- *Concurrent Pos:* Consult, US Dept Energy, 84. *Mem:* Am Meteorol Soc; Am Geophys Union; Sigma Xi. *Res:* Micrometeorological and mesometeorological processes which control the suspension, transport, diffusion and deposition of airborne materials from agricultural and industrial sources. *Mailing Add:* 654 NW Stewart Pl Corvallis OR 97330

WOLF, MATTHEW BERNARD, b Los Angeles, Calif, May 5, 35; m 60; c 2. PHYSIOLOGY. *Educ:* Univ Calif, Los Angeles, BSc, 57, MSc, 62, PhD(physiol), 67. *Prof Exp:* Sr engr, Bendix Corp, 58-63; asst prof biomed eng, Univ Southern Calif, 67-72; assoc prof eng, Univ Ala, Birmingham, 72-76; ASSOC PROF PHYSIOL, UNIV SC, 76- *Concurrent Pos:* Consult, Rand Corp, 61-72; USPHS fel, 63-67. *Mem:* Am Physiol Soc; Biomed Eng Soc. *Res:* Cell biophysics; ion transport. *Mailing Add:* Dept Physiol Univ SC Sch Med Columbia SC 29208

WOLF, MERRILL KENNETH, b Cleveland, Ohio, Aug 28, 31; m 58. NEUROBIOLOGY, TISSUE CULTURE. *Educ:* Yale Col, BA, 45; Western Reserve Univ, MD, 56. *Prof Exp:* Intern med, Peter Bent Brigham Hosp, 56-57; res assoc, Nat Inst Neurol Dis & Blindness, 57-59; res fel neurol, Med Sch, Harvard Univ, 59-64, instr anat, 64-69, asst prof neuropath, 69-71, assoc prof, 71-72; PROF ANAT, MED SCH, UNIV MASS, 72-, PROF NEUROL, 78- *Concurrent Pos:* Lectr, Med Sch, Harvard Univ, 72-79; vis prof, Sch Med, Stanford Univ, 77. *Honors & Awards:* Javits Award, NINCDS, 84. *Mem:* Am Asn Anatomists; Soc Neurosci; Tissue Cult Asn. *Res:* Neurological mutant mice with central nervous system disorders; organotypic culture, cytology, genetics. *Mailing Add:* Dept Anat Sch Med Univ Mass 55 Lake Ave N Worcester MA 01655

WOLF, MONTE WILLIAM, b Whittier, Calif, Sept 1, 49. ORGANIC CHEMISTRY. *Educ:* Univ Calif, Santa Barbara, BS, 71; Univ Southern Calif, PhD(org chem), 76. *Prof Exp:* Res org chem, Univ Calif, San Diego, 76-77; asst prof chem, Bethel Col, Minn, 77-78; vis asst prof, Univ Calif, San Diego, 78; ASSOC PROF CHEM, OGLETHORPE UNIV, 78- *Mem:* Sigma Xi; Am Chem Soc. *Res:* Photophysical processes of excited state aromatic ketones. *Mailing Add:* 11220 El Rey Dr Whittier CA 90606

WOLF, NEIL STEPHAN, b Brooklyn, NY, Nov 15, 37; m 59; c 2. PLASMA PHYSICS. *Educ:* Queens Col, NY, BS, 58; Stevens Inst Technol, MS, 60, PhD(physics), 66. *Prof Exp:* Res assoc physics, Space Physics Labs, G C Dewey Corp, 65-67; asst prof, 67-72, chmn, dept physics & astron, 74-77 & 84-88, assoc prof & coordr, dept sci, 72-80, PROF PHYSICS, DICKINSON COL, 80-, DIR, DICKINSON CTR EUROP STUDIES, BOLOGNA, ITALY, 88- *Concurrent Pos:* Res Corp Frederick Cottrell res grant, 67-; NSF fel, Univ Calif, Irvine, 81, vis assoc prof, 81-82, vis prof, 90-91; vis physicist, Univ Innsbruck, Austria, 83 & Univ Calif, Irvine, 84-88. *Honors & Awards:* Merck Lab Develop Award, 89. *Mem:* Am Inst Physics. *Res:* Plasma physics related to control of instabilities in plasma immersed in strong magnetic fields; current drive in tokamaks. *Mailing Add:* Dept Physics Dickinson Col Carlisle PA 17013

WOLF, NORMAN SANFORD, b Kansas City, Mo, July 22, 27; m 67, 76; c 1. EXPERIMENTAL PATHOLOGY, RADIOBIOLOGY. *Educ:* Kans State Univ, BS & DVM, 53; Northwestern Univ, PhD(exp path), 60; Am Col Lab Animal Med, dipl, 55. *Prof Exp:* Dir dept animal care, Med Sch, Northwestern Univ, 53-58; vis scientist & NSF fel, Pasteur Lab, Inst Radium, Paris, 60-61; consult radiation biol, Path & Physiol Sect, Biol Div, Oak Ridge Nat Lab, 61-62; res asst prof exp biol, Baylor Col Med, 62-68; assoc prof path & mem radiol sci group, Sch Med, 68-90, PROF PATH & AFFIL PROF COMP MED, SCH MED, UNIV WASH, 90- *Concurrent Pos:* Consult, Animal Quarters, Vet Admin Res Hosp, Chicago, 54-60 & Vet Admin Hosp, Seattle, 73-79; vis scientist, P MacCallum Cancer Res Inst, Melbourne, Australia, 88. *Mem:* Radiation Res Soc; Am Soc Exp Path; Int Soc Exp Hemat; Am Col Lab Animal Med. *Res:* Hematopoietic regeneration and transplantation following ionizing radiation; immune competence; hemopoietic stem cell identification; control of hematopoiesis by hematopoietic organ stroma; recovery after irradiation by cells and tissues; senescence of hemopoietic stem cells. *Mailing Add:* Dept Path Univ Wash Sch Med Seattle WA 98195

WOLF, P S, PHARMACOLOGY. *Prof Exp:* MEM STAFF, DEPT CORP DEVELOP, STERLING DRUG INC. *Mailing Add:* Dept Corp Develop Sterling Drug Inc 90 Park Ave New York NY 10016

WOLF, PAUL LEON, b Detroit, Mich, Oct 4, 28; m 52; c 3. PATHOLOGY. *Educ:* Wayne State Univ, BA, 48; Univ Mich, MD, 52; Am Bd Path, cert path anat & clin path, 60. *Prof Exp:* Intern, Detroit Receiving Hosp, 52-53; resident path, Wayne State Univ Hosps, 56-60, from asst prof to prof, 60-68; dir, Clin lab, Stanford Univ Med Ctr, 68-74; PROF PATH, UNIV CALIF, SAN DIEGO, 74- *Concurrent Pos:* Assoc, Detroit Receiving Hosp; mem staff, Dearborn Vet Admin Hosp. *Mem:* AMA; Am Soc Clin Path; Am Asn Path & Bact; Col Am Path. *Res:* Anatomic pathology; histochemistry; immunopathology; cancer immunology. *Mailing Add:* Univ Hosp 225 W Dickenson PO Box 3548 San Diego CA 92103

WOLF, PAUL R, b Mazomanie, Wis, June 13, 34; m 59; c 3. PHOTOGRAMMETRY,CIVIL ENGINEERING. *Educ:* Univ Wis-Madison, BSCE, 60, MSCE, 66, PhD(civil eng), 67. *Prof Exp:* Hwy engr, Wis Hwy Comn, 60-63; instr civil eng, Univ Wis-Madison, 63-67; asst prof, Univ Calif, Berkeley, 67-70; assoc prof, 70-74, PROF CIVIL & ENVIRON ENG, UNIV WIS-MADISON, 74- *Concurrent Pos:* Mem, panel geod & cartog, Nat Acad Sci, 72-74. *Honors & Awards:* Bausch & Lomb Photogram Award, 66; Talbert Abrams Award III, Am Soc Photogram, 71; Fennel Award, Am Cong Surv & Mapping, 79. *Mem:* Am Cong Surv & Mapping; Am Soc Photogram; Am Soc Civil Engrs; Soc Automotive Engrs. *Res:* Geodesy; cartography; photogrammetry with concentration on close-range and terrestrial applications; accident reconstruction. *Mailing Add:* Dept Civil & Environ Eng Univ Wis Madison WI 53706

WOLF, PHILIP FRANK, b New York, NY, Apr 12, 38; div; c 3. PHYSICAL ORGANIC CHEMISTRY. *Educ:* NY Univ, BS, 60; Columbia Univ, MA, 61, PhD(org chem), 64. *Prof Exp:* Fel org chem, Yale Univ, 64-65; res chemist, Union Carbide Corp, 65-70, from proj scientist to res scientist, 70-74, group leader org chem, 74-76, sr res scientist, 76-77, assoc dir res & develop, 77-84, DIR RES & DEVELOP, UNION CARBIDE CORP, 84- *Mem:* Am Chem Soc; Sigma Xi. *Res:* Studies on oxidation-oxygen transfer mechanisms, substitution reactions of ethylene oxide, free radical telomerization, heterogeneous gas phase kinetics, Diels-Alder reactions and water soluble polymers. *Mailing Add:* 401 Jaguar Lane Bridgewater NJ 08807

WOLF, RAOUL, b 1946. PEDIATRICS, PULMONARY MEDICINE. *Educ:* Univ Witwatersrand, Johannesburg, SAfrica, MD, 69. *Prof Exp:* DIR ALLERGY & CLIN IMMUNOL, DEPT ALLERGY & IMMUNOL, WYLER CHILDREN'S HOSP, 79- *Mailing Add:* Dept Allergy & Immunol Wyler Children's Hosp Chicago IL 60637

WOLF, RICHARD ALAN, b Pittsburgh, Pa, Nov 10, 39; m 71; c 2. SPACE PHYSICS. *Educ:* Cornell Univ, BEngPhys, 62; Calif Inst Technol, PhD(nuclear astrophys), 66. *Prof Exp:* Res fel physics, Calif Inst Technol, 66; mem tech staff, Bel Tel Labs, 66-67; from asst prof to assoc prof space sci, 67-74, PROF SPACE PHYSICS & ASTRON, RICE UNIV, 74- *Concurrent Pos:* Mem, Inst Advan Study, 69 & 74-75. *Mem:* Am Geophys Union. *Res:* Physics of the solar wind; magnetosphere and ionosphere. *Mailing Add:* Dept Space Physics & Astron Rice Univ Houston TX 77251-1892

WOLF, RICHARD CLARENCE, b Lancaster, Pa, Nov 28, 26; m 52; c 2. PHYSIOLOGY, ENDOCRINOLOGY. *Educ:* Franklin & Marshall Col, BS, 50; Rutgers Univ, PhD(zool), 54. *Prof Exp:* Waksman-Merck fel, Rutgers Univ, New Brunswick, 54-55; Milton fel, Sch Dent Med, Harvard Univ, 55-56, USPHS fel, 56-57; asst prof physiol, Primate Lab, 57-61, assoc prof, 61-66, co-dir, 68-70, MEM ENDOCRINOL-REPRODUCTION PHYSIOL PROG, UNIV WIS-MADISON, 63-, DIR, 70-, PROF PHYSIOL, 66-CHMN DEPT, 71- *Concurrent Pos:* Mem res career award comt, NIH, 70-72, mem contract res & adv comt, 73-76; mem sci adv bd, Yerkes Regional Primate Res Ctr, Emory Univ, 72-78; consult, Ford Found, 72- *Mem:* Soc Study Reprod; Endocrine Soc; Am Physiol Soc; Brit Soc Endocrinol; Brit Soc Study Fertil. *Res:* Endocrinology of pregnancy. *Mailing Add:* Serv Mem Inst Rm 123 1300 University Ave Madison WI 53706

WOLF, RICHARD EDWARD, JR, b Philadelphia, Pa, Mar 22, 41. MOLECULAR BIOLOGY. *Educ:* Univ Cincinnati, BA, 63, MS, 68, PhD(microbiol), 70. *Prof Exp:* Res fel microbiol & molecular genetics, Med Sch, Harvard Univ, 70-73, instr, 73-75; asst prof, 75-80, ASSOC PROF BIOL SCI, UNIV MD BALTIMORE COUNTY, 80- *Res:* Molecular mechanisms of growth rate-dependent regulation of gene expression. *Mailing Add:* 7104 Black Rock Ct Columbia MD 21046-1465

WOLF, RICHARD EUGENE, b Dixon, Ill, June 25, 36; m 60; c 3. ORGANIC CHEMISTRY, POLYMER CHEMISTRY. *Educ:* Northern Ill Univ, BSEd, 57; Univ San Francisco, MS, 65; Univ Calif, Berkeley, PhD(chem), 68. *Prof Exp:* Sect leader org synthesis, Desoto Inc, 68-70, mgr, Long Range Res Dept, 70-74, res scientist, 74-77, contract & explor res mgr, 77-79, dir int licensing, 79-87, sr dir int licensing, Desoto, Inc, 87-90; MANAGING DIR, INT & CENTRES, VALSPAR CORP, 91- *Concurrent Pos:* Chmn patent comt, DeSoto, Inc, 71-76, chmn basic res comt, 74-76. *Mem:* Am Chem Soc; Fedn Socs Paint Technol; Nat Micrographics Asn; fel Am Inst Chem; Licensing Exec Soc. *Res:* Photochemistry; photoconduction in organic molecules; photopolymerization; emulsion polymerization; coatings technology; water treatment. *Mailing Add:* 210 Tully Pl Prospect Heights IL 60070

WOLF, ROBERT E, b Houston, Tex, Jan 20, 42; m 67; c 1. IMMUNOLOGY, RHEUMATOLOGY. *Educ:* Baylor Univ, BA, 64; Univ Tex, Galveston, MD, 69, PhD, 73. *Prof Exp:* PROF MED & CHIEF RHEUMATOLOGY, MED CTR, LA STATE UNIV, 77- *Mem:* Am Rheumatism Asn; Am Asn Immunologists. *Res:* Aspects of cellular immunology and immunopharmacology relating to rheumatic diseases. *Mailing Add:* Dept Med La State Univ Med Ctr PO Box 33932 Shreveport LA 71130

WOLF, ROBERT LAWRENCE, b New York, NY, Aug 7, 28; m 57; c 1. PHYSIOLOGY, BIOCHEMISTRY. *Educ:* Duke Univ, BS, 50, MD, 52; Am Bd Internal Med, dipl, 66. *Prof Exp:* Intern med & surg serv, Mt Sinai Hosp, 52-53, chief resident path, 53-54, asst resident med, 56-57, chief resident, 57-58, clin asst, Mt Sinai Hosp & Mt Sinai Sch Med, 58-65, res asst, 60-66, ASST CLIN PROF MED, MT SINAI SCH MED, 66- *Concurrent Pos:* Arthritis & Rheumatism Found res fel, 58-59; res fel med, Mt Sinai Hosp, 58-61; AEC byprod mat license for res, 59-; var pharmaceut corps res grants, 59-; USPHS res grants, 59-; Lupus Erhthematosus Found, Inc res grants, 61-; mem coun high blood pressure res & coun circulation, Am Heart Asn; asst attend physician, Mt Sinai Hosp, 65-; Syntex Res Found res grants, 65-; Ciba Co res grant, 66; guest lectr, Nat Univ Colombia, 66, Cath Univ Chile, 66 & Inst Med Res, Univ SAm, Buenos Aires, 66; G D Searle & Co res grant, 66-68; Health Res Coun of City New York res grant, 66-69; US partic, Int Atomic Energy Agency Symp Radioimmunoassay & Related Procedures in Clin Med & Res, Istanbul, Turkey, 73; consult ed, AMA Drug Eval, 73-74. *Mem:* Am Nuclear Soc; Am Physiol Soc; fel Am Col Physicians; Am Soc Internal Med; Am Col Chest Physicians. *Res:* Hypertension; circulatory physiology and cardiology. *Mailing Add:* Dept Med Mt Sinai Sch Med 20 E 74th St New York NY 10029

WOLF, ROBERT OLIVER, b Mansfield, Ohio, Mar 14, 25; m 70; c 2. ORAL BIOLOGY. *Educ:* NCent Col, BA, 50; Ohio State Univ, MA, 52, DDS, 58. *Prof Exp:* Res asst physiol genetics, Ohio State Univ, 52-54; res asst oral biol, Col Dent, 57-58; comn officer, USPHS, 58; intern clin dent, USPHS Hosp, New Orleans, La, 58-59; investr salivary physiol & biochem, Human Genetics Br, Nat Inst Dent Res, 59-70, Oral Med & Surg Br, 70-73, Lab Oral Med, 73-77 & Clin Invest Br, 77-81; prin investr, salivary physiol & biochem, Lab Biol Struct, Nat Inst Dent Res, 81-85; RETIRED. *Concurrent Pos:* Clin assoc prof, Sch Dent, Georgetown Univ, 80- *Mem:* Am Dent Asn; Am Soc Human Genetics; Int Asn Dent Res; Am Inst Biol Sci. *Res:* Human and animal salivary physiology, biochemistry, enzymology and genetics, especially isoamylases; human salivary gland disease, diagnosis and treatment. *Mailing Add:* 5515 Johnson Ave Bethesda MD 20817

WOLF, ROBERT PETER, b Long Branch, NJ, Oct 27, 39; m 60; c 2. ENVIRONMENTAL PHYSICS. *Educ:* Mass Inst Technol, BS, 60, PhD(physics), 63. *Prof Exp:* From asst prof to prof physics, 63-88, DIR ACAD COMPUT, HARVEY MUDD COL, 86- *Concurrent Pos:* NSF sci faculty fel, Oxford Univ, 69-70; vis scientist, Mass Inst Technol, 76, Univ Toronto, 79, Stanford, 85-86. *Mem:* Am Phys Soc. *Res:* Energy resources; phase transitions; solar energy development; philosophy of science; chaos; neural networks. *Mailing Add:* Dept Physics Harvey Mudd Col Claremont CA 91711

WOLF, ROBERT STANLEY, b New York, NY, May 14, 46; m 75. MATHEMATICAL LOGIC. *Educ:* Mass Inst Technol, BS, 66; Stanford Univ, MS, 67, PhD(math), 74. *Prof Exp:* Vis asst prof math, Univ Ore, 73-74; vis scholar, Stanford Univ, 74-75; from lectr to assoc prof, 75-85, PROF MATH, CALIF POLYTECH STATE UNIV, SAN LUIS OBISPO, 85- *Mem:* Am Math Soc; Asn Symbolic Logic. *Res:* Continuing study of set theories with intuitionistic logic; theory of infinite games, specifically Almost-Borel games; point-set topology, mathematical biology, artificial intelligence and mathematics education. *Mailing Add:* Dept Math Calif Polytech State Univ San Luis Obispo CA 93407

WOLF, ROBERT V(ALENTIN), b St Louis, Mo, June 5, 29; m 75. METALLURGICAL ENGINEERING. *Educ:* Univ Mo, BS, 51, MS, 52. *Prof Exp:* From instr to assoc prof, 51-67, PROF METALL ENG, SCH MINES, UNIV MO, ROLLA, 67-, ASST DEAN, SCH MINES & METALL, 81- *Concurrent Pos:* Partner, Askeland, Kisslinger & Wolf. *Mem:* Am Soc Metals; Am Foundrymen's Soc; Am Inst Mining & Metall Engrs; Am Soc Nondestructive Testing; Am Welding Soc; Sigma Xi. *Res:* Metals casting; nondestructive testing. *Mailing Add:* Dept Metall Eng Univ Mo Rolla MO 65401

WOLF, STANLEY MYRON, b Washington, DC, July 12, 39; m 63; c 1. METALLURGICAL ENGINEERING, MATERIALS SCIENCE. *Educ:* Va Polytech Inst & State Univ, BS, 60; Cornell Univ, MS, 63; Mass Inst Technol, PhD(metall), 72. *Prof Exp:* Metallurgist res & develop, Mat & Control Div, Tex Instruments, 62-63; researcher, US Army Mat Res Agency, 64-66, chief data processing br, Eighth US Army Hq, 66-67, metallurgist, US Army Mat & Mech Res Ctr, 71-74; metallurgist res, AEC, 74-75; metallurgist, ERDA, 75-77; METALLURGIST, US DEPT ENERGY, 77- *Mem:* Am Inst Metall Engrs; Am Soc Metals; Nat Asn Corrosion Engrs; Am Soc Testing Mat. *Res:* Mechanical and structural stability of metals, ceramics and polymers, including glasses, coatings and weld/braze joints; aqueous and gaseous corrosion; powder metallurgy; structure-property-processing correlations; research administration and program management. *Mailing Add:* Mat Sci Div US Dept Energy Washington DC 20545

WOLF, STEPHEN NOLL, b Biloxi, Miss, Dec 4, 44; m 74; c 1. ACOUSTICS. *Educ:* Lebanon Valley Col, BS, 66; Univ Md, PhD(molecular spectros), 72. *Prof Exp:* RES PHYSICIST UNDERWATER ACOUST, US NAVAL RES LAB, 71- *Concurrent Pos:* Exchange scientist, Defence Res Estab Atlantic Dartmouth, NS, Can, 86-88. *Mem:* Sigma Xi; Acoust Soc Am. *Res:* At-sea experimental studies of propagation of sound in shallow water. *Mailing Add:* Code 5160 US Naval Res Lab Overlook Ave SW Washington DC 20375-5000

WOLF, STEVEN L, b Chicago, Ill, May 15, 44; m 71; c 2. REHABILITATION MEDICINE, PHYSICAL THERAPY. *Educ:* Clark Univ, AB, 63; Boston Univ, MS, 68; Emory Univ, MS, 72, PhD(anat), 73. *Prof Exp:* Staff phys ther, USPHS, Boston, 66-68; instr phys ther, Boston Univ, 69-70; from asst prof to assoc prof, 75-85, PROF ANAT REHAB, EMORY UNIV, 85- *Concurrent Pos:* Assoc ed, F A Davis, Phildelphia & J Head Trauma Rehab, 86-; prin investr, Emory Univ Rehab Res, 74, 82, 83 & 88; Catherine Worthingham fel, Am Phys Ther Asn, 86; bd dirs, Biofeedback Soc Am, 83-86; vis prof, Univ Gothenberg, 87-88; consult, EMPI Inc, 83-87, Verimed Inc & Medronics Inc, 87- *Honors & Awards:* Golden Pen Award, Am Phys Ther Asn, 80. *Mem:* Biofeedback Cert Inst Am (secy, 81-84, vpres, 85-86 & pres, 87-89); Am Phys Ther Asn; Am Cong Rehab Med; NY Acad Sci; Biofeedback Soc Am. *Res:* Use of physiological monitoring to enhace self-control of movement among patients with neuromuscular disorders; new techniques to achieve sensory-motor integration following central nervous system trauma. *Mailing Add:* Ctr Rehab Med 1441 Clifton Rd NE Atlanta GA 30322

WOLF, STEWART GEORGE, JR, b Baltimore, Md, Jan 12, 14; m 42; c 3. INTERNAL MEDICINE, PHYSIOLOGY. *Educ:* Johns Hopkins Univ, AB, 34, MD, 38. *Hon Degrees:* MD, Gothenburg Univ, 68. *Prof Exp:* Intern med, NY Hosp, 38-39, from asst resident to resident, 39-42; from asst prof to assoc prof med, Med Col, Cornell Univ, 46-52; prof med & consult prof psychiat, Sch Med, Univ Okla, 52-67, regents prof med, psychiat, neurol & behav sci, 67-70, head dept med, 52-69, prof physiol, Sch Med, 68-70; prof med, Univ Tex Syst, 70-77; dir, Marine Biomed Inst, 70-77; prof internal med & physiol, Univ Tex Med Br Galveston, 70-77; vpres med affairs & mem staff, St Lukes Hosp, Bethlehem, Pa, 77-82; DIR, TOTTS GAP MED RES LABS, BANGOR, PA, 58-; PROF MED, TEMPLE UNIV, 77- *Concurrent Pos:* Res fel, Bellevue Hosp, 39-42; Nat Res Coun fel, Cornell Univ, 41-42; assoc vis neuropsychiatrist, Bellevue Hosp, 48-52; asst attend physician in charge psychosom clin, Cornell Univ, 46-52; mem comt psychiat, Nat Res Coun, 48-52, mem comt vet med probs, 51-52; head psychosom sect, Okla Med Res Found, 52-, head neurosci sect, 67-70; mem spec study group, Off Res & Develop, Dept Defense, 52-55; mem pharmacol & exp therapeut study sect, 56-57; mem comt prof educ, 56-63, chmn, 57-63; mem gen med study sect, NIH, 57-61, chmn gastroenterol training grant comt, 58-61; mem adv comt, Space Med & Behav Sci, NASA, 60-61 & Nat Adv Heart Coun, 61-65; mem coun ment health, AMA, 60-64; consult, Europ Off, Off Int Res, NIH, 63-64; mem adv comt admis, Nat Formulary, 65-69; mem bd regents, Nat Libr Med, 65-69, chmn, 68-69; mem comt int progs, Am Heart Asn, 65-70, chmn, 65-70; mem educ & supply panel, Nat Adv Comn Health Manpower, 66-67; mem, Nat Adv Environ Health Sci Coun, 78-82; mem bd visitors, Dept Biol, Boston Univ, 78- & Ctr Soc Res, Lehigh Univ, 80-90; chmn adv comt, Francis Clark Wood Inst Hist Med, Col Physicians of Philadelphia, 79-90; mem bd dirs, Inst Advan Studies Immunol & Aging, 84-; mem, Sci Adv Comt, Lehigh Univ, 74-, chmn, 80-89. *Honors & Awards:* Award, Am Gastroentrol Asn, 42; Hofheimer Prize, Am Psychiat Asn, 52; Hans Selye Award, Am Inst Stress, 88. *Mem:* Am Psychosom Soc (pres, 61-62); Am Gastroenterol Asn (pres, 69-70); fel Am Col Physicians; Pavlovian Soc (pres, 66-67); Am Col Clin Pharmacol & Chemother (pres, 66-67); Am Soc Clin Invest; Asn Am Physicians; distinguished fel, Am Psychiat Asn. *Res:* Gastrointestinal, cardiovascular, sensory and neural physiology. *Mailing Add:* RD 1 Box 1120 G Bangor PA 18013

WOLF, STUART ALAN, b Brooklyn, NY, Sept 15, 43; m 65; c 2. SOLID STATE PHYSICS. *Educ:* Columbia Col, AB, 64; Rutgers Univ, MS, 66, PhD(physics), 69. *Prof Exp:* Res assoc physics, Case Western Reserve Univ, 69-72; res physicist, 72-82, supvry res physicist, 82-86, HEAD, MAT PHYSICS BR, US NAVAL RES LAB, 86- *Concurrent Pos:* Assoc prof & lectr, George Washington Univ, 79-; panel mem, Interagency Advan Power Group-Superconductivity Panel, 80-; vis scholar, Univ Calif, Los Angeles, 81-82; mem, Elec, Magnetic & Optical Panel, Metall Soc; organizer, Gordon Conf Superconducting Films; NSF Rev panels. *Mem:* Fel Am Phys Soc; Am Vacuum Soc; Sigma Xi; Metall Soc. *Res:* Cryogenics; superconductivity; transport properties; vacuum system design; thin films; magnetic shielding; Josephison devices. *Mailing Add:* Naval Res Lab Code 6340 Washington DC 20375-5000

WOLF, THOMAS, b Sept 10, 32; US citizen; m 59; c 2. ANALYTICAL CHEMISTRY. *Educ:* Cambridge Univ, BA, 55, MA, 61; Univ RI, PhD(anal chem), 66. *Prof Exp:* Chemist, Coates Bros, Eng, 55-57; lab supvr, Wymat Corp, NJ, 57-58; chemist, Enthone Inc, Conn, 58-59 & Eltex Res Corp, RI, 59-62; sr res chemist, 66-75, RES ASSOC, COLGATE-PALMOLIVE CO, 75- *Concurrent Pos:* Course dir, Ctr Prof Advan, 79- *Mem:* Am Chem Soc; Am Soc Testing & Mat. *Res:* Liquid chromatography; thermal analysis; laboratory automation. *Mailing Add:* 609 S Fifth Ave Highland Park NJ 08904-2628

WOLF, THOMAS MARK, b Cincinnati, Ohio, Dec 25, 44; m 69; c 1. CHILD-CLINICAL PSYCHOLOGY, BEHAVIORAL MEDICINE. *Educ:* Univ Cincinnati, BA, 66; Miami Univ, Ohio, MA, 67; Univ Waterloo, Ont, PhD(psychol), 71. *Prof Exp:* Asst prof psychol, State Univ NY, Cortland, 70-74, assoc prof, 74-75; assoc prof, 75-82, PROF PSYCHOL, LA STATE UNIV MED SCH, 82- *Concurrent Pos:* Postdoctorate psychol, St Louis Univ, 74-76; consult psychologist, Youth Study Ctr, New Orleans, 77, New Orleans pub schs, 79-80, St Bernard Group Home, 79-86, Cent City Ment Health Clin, 80-89, Asn Cath Charities New Orleans, 87-89, Child & Adolescent Ment Health Prog, 89- *Mem:* Am Psychol Asn; Soc Behav Med; Asn Am Med Col. *Res:* Perceived mistreatment and health during professional training; lifestyle characteristics stress, coping and health of medical students; health promotion and disease prevention. *Mailing Add:* Dept Psychiat La State Univ Med Ctr 1542 Tulane Ave New Orleans LA 70112

WOLF, THOMAS MICHAEL, b Highland Park, Mich, Dec 28, 42; c 2. GENETICS EVOLUTION. *Educ:* Western Mich Univ, BS, 65; Wayne State Univ, MS, 67, PhD(genetics), 72. *Prof Exp:* From instr to assoc prof biol, 80-88, ASSOC PROF BIOL, WASHBURN UNIV, TOPEKA, 88- *Mem:* Genetics Soc Am; Entomol Soc Am. *Res:* Evolutionary biology of petridine pigments in dipterans. *Mailing Add:* Dept Biol Washburn Univ Topeka Topeka KS 66621

WOLF, WALTER, b Frankfurt, Germany, May 25, 31; US citizen; m 55; c 2. RADIOCHEMISTRY, RADIOPHARMACY. *Educ:* Univ of the Repub, Uruguay, BSc, 49, MS, 52; Univ Paris, PhD, 56. *Prof Exp:* Asst chem, Univ of the Repub, Uruguay, 51-52; asst chem, Nat Cent Sci Res, France, 55-56, attache, 56; assoc prof org chem & biochem, Concepcion Univ, 56-58; traveling fel, McGill Univ, 58; res assoc, Amherst Col, 58-59; res assoc org chem, 59-62, vis asst prof pharmaceut chem, 61-62, from asst prof to assoc prof, 62-70, chmn dept biomed chem, 70-74, PROF BIOMED CHEM, UNIV SOUTHERN CALIF, 70-, DIR, RADIOPHARM PROG, 68-; DIR RADIOPHARM SERV, LOS ANGELES COUNTY-UNIV SOUTHERN CALIF MED CTR, 70- *Concurrent Pos:* Consult radiopharm, Int Atomic Energy Agency & US Vet Admin. *Honors & Awards:* G Czezniak Prize Nuclear Med, Israel, 80, 86. *Mem:* Am Chem Soc; Soc Nuclear Med; Radiation Res Soc; Acad Pharmaceut Sci. *Res:* Non-invasive monitoring of drug biodistribution targeting and metabolism; chemistry and biochemistry of organic iodo compounds; radioiodination; non-invasive magnetic resonance; radiopharmacokinetics; nuclear magnetic resonance. *Mailing Add:* Dept Biomed Chem Univ Southern Calif Los Angeles CA 90033

WOLF, WALTER ALAN, b New York, NY, Mar 9, 42; m 62; c 2. BIOCHEMISTRY, EXPERT SYSTEMS. *Educ:* Wesleyan Univ, BA, 62; Brandeis Univ, MA, 64, PhD(org chem), 67; Rochester Inst Technol, MS, 86. *Prof Exp:* Fel, Mass Inst Technol, 67-70; asst prof biochem, Colgate Univ, 70-77; from asst prof to assoc prof chem, Eisenhower Univ, 77-84; ASST PROF COMPUT SCI, ROCHESTER INST TECHNOL, 85- *Concurrent Pos:* Vis prof, State Univ NY Agr & Tech Col Morrisville, 71-72, State Agr Exp Sta, Geneva, 80-; ed, Chem Ed Compacts, J Chem Educ; res corp grant, Colgate Univ, 71-72; grants, NSF. *Mem:* AAAS; Am Chem Soc; Royal Soc Chem; Asn Comput Mach. *Res:* Pheromone biosynthesis; expert system user interfaces; expert systems. *Mailing Add:* 1229 Birdsey Rd Waterloo NY 13165

WOLF, WALTER J, b Hague, NDak, May 2, 27; m 49; c 5. BIOCHEMISTRY, PROTEIN CHEMISTRY. *Educ:* Col St Thomas, BS, 50; Univ Minn, PhD(biochem), 56. *Prof Exp:* Assoc chemist, 56-58, chemist, 58-61, prin chemist, 61-68, leader meal prod res, 68-85, LEAD SCIENTIST, NORTHERN REGIONAL RES CTR, USDA, 85- *Concurrent Pos:* Assoc ed, Cereal Chem, 70-73 & Cereal Sci Today, 70-73; consult, Int Nutrit Anemia Consultative Group, Nutrit Found, 81-82; off deleg, US-People's Repub China Soybean Symp, 83; ed, Food Microstructure, 84-; ed bd, J Sci Food Agr, 88-; mem, US Delegation, codex alimentarius comt veg proteins, 86- *Mem:* AAAS; Am Asn Cereal Chemists; Am Chem Soc; Am Oil Chemists Soc; Inst Food Technologists. *Res:* Isolation and characterization of soybean and jojoba proteins; protein interactions, denaturation and structure; food uses of soybean proteins. *Mailing Add:* Nat Ctr Agr Utilization Res 1815 N University St Peoria IL 61604

WOLF, WARREN WALTER, b Oakdale, Pa, Dec 10, 41; m 74; c 2. GLASS SCIENCE, CERAMICS ENGINEERING. *Educ:* Pa State Univ, University Park, BS, 63; Ohio State Univ, PhD(ceramic eng), 68; Xavier Univ, MBA, 77. *Prof Exp:* Res engr, Ferro Corp, 63-65; sr scientist glass technol, Owens-Corning Fiberglass Corp, 68-75, supvr glass res & develop, 75-78, mgr glass res & develop, 78-84, lab dir, anal & mat res, 84-90, LAB DIR, CPG RES & DEVELOP MAT SERV, OWENS-CORNING FIBERGLASS CORP, 90- *Mem:* Am Ceramic Soc; Nat Inst Ceramic Engrs. *Res:* Glass fibers, especially high tensile strength glass fibers; alkali resistant glass for portland cement reinforcement; textile fibers; glass surface chemistry; high temperature glass properties; glass structural investigations; optical glass research. *Mailing Add:* 8056 Eliot Dr Reynoldsburg OH 43068

WOLF, WAYNE ROBERT, b Kenton, Ohio, May 18, 43; m 71; c 1. ANALYTICAL CHEMISTRY, REFERENCE MATERIALS. *Educ:* Kent State Univ, BS, 65, PhD(chem), 69. *Prof Exp:* Res chemist, Aerospace Res Lab, USAF, Wright-Patterson AFB, Ohio, 67-71; Nat Res Coun assoc, 71-72, res chemist, Human Nutrit Res Div, Agr Res Serv, 71-75, RES CHEMIST, NUTRIENT COMPOS LAB, NUTRIT CTR, USDA, 75- *Concurrent Pos:* USDA-NIST res assoc, SRMP, Nat Inst Standards & Technol, 88- *Mem:* AAAS; Am Chem Soc. *Res:* Instrumental analytical methodology for nutrient content of foods, trace element nutrition, biological availability of trace elements; atomic absorption spectrometry; gas-liquid chromatography; mass spectrometry. *Mailing Add:* Nutrient Compos Lab Nutrit Ctr USDA Beltsville MD 20705

WOLF, WERNER PAUL, b Vienna, Austria, Apr 22, 30; nat US; m 54; c 2. PHYSICS, MAGNETIC MATERIALS. *Educ:* Oxford Univ, BA, 51, MA & DPhil(physics), 54. *Hon Degrees:* MA, Yale Univ, 65. *Prof Exp:* Res assoc physics, Clarendon Lab, Oxford Univ, 54-56, Imp Chem Industs res fel, 57-59; res fel appl physics, Harvard Univ, 56-57; univ lectr physics, Oxford Univ, 59-63; from assoc prof to prof physics & appl sci, Yale Univ, 63-76, Becton prof & chmn dept eng & appl sci, 76-81, chmn, Coun Eng, 81-84, RAYMOND J WEAN PROF ENG & APPL SCI & PROF PHYSICS, YALE UNIV, 84- , CHMN, DEPT APPL PHYSICS, 90- *Concurrent Pos:* Consult, Hughes Aircraft Co, E I du Pont de Nemours & Co, Inc, 57 & Gen Elec Co, 60 & 66-, Mullard Res Labs, 61, Watson Res Ctr, IBM, 62-66; res collabr, Brookhaven Nat Lab, 66-80; gen conf chmn, Conf Magnetism & Magnetic Mat, 71, adv comt chmn, 71-72; vis prof, Munich Tech Univ, 69; sr vis fel, Oxford Univ, 80 & 84; Alexander von Humboldt Found sr US scientist award, 83; vis fel, Corpus Christi Col, 84, 87 & vis guest fel, Royal Soc London, 87. *Mem:* Am Asn Crystal Growth; fel Am Phys Soc; fel Inst Elec & Electronics Engrs. *Res:* Magnetism; experimental and theoretical study of magnetic materials, especially at low temperatures; magnetic cooling, relaxation, microwave resonance, optical properties, crystal fields and anisotropy; magnetic thermal properties, critical points and magnetic phase transitions; synthesis of new materials and growth of single crystals. *Mailing Add:* Becton Ctr Yale Univ PO Box 2157 New Haven CT 06520-2157

WOLFARTH, EUGENE F, b Washington, DC, June 24, 32; m 57; c 3. PHYSICAL ORGANIC CHEMISTRY. *Educ:* Univ Md, BS, 54; Ohio State Univ, PhD(chem), 61. *Prof Exp:* Res chemist, Res & Develop Command, Wright Patterson AFB, Ohio, 57; asst instr chem, Ohio State Univ, 58-60; res assoc, Res Labs, 61-76, mgr, Kodak Legal Dept, Litigation Group, 76-82, RES ADMINR, KODAK RES LABS, EASTMAN KODAK CO, 82- *Res:* Kinetics and reaction mechanisms; organic synthesis; color photography, the mechanisms of development and designing new film systems. *Mailing Add:* 19 Rippingale Rd Pittsford NY 14534

WOLFE, ALAN DAVID, b New York, NY, Mar 25, 29; m 69. BIOCHEMISTRY, TOXICOLOGY. *Educ:* Queens Col, NY, BS, 52; Mass Inst Technol, SM, 56, George Washington Univ, JD, 62; Univ Md, College Park, PhD(microbiol), 70. *Prof Exp:* Molecular biologist, 58-76, ASST CHIEF BIOCHEM & PRIN INVESTR BIOCHEM, WALTER REED ARMY INST RES, 76- *Concurrent Pos:* Secy Army fel, US Dept Defense, 72-73. *Mem:* Am Soc Microbiol; AAAS; Am Chem Soc. *Res:* Molecular biology and pharmacology of organophosphates; biochemistry of bacterial toxins; protein and nucleic acid synthesis; the mode of action of growth inhibitors. *Mailing Add:* Dept Appl Biochem Walter Reed Army Inst Res Washington DC 20012

WOLFE, ALLAN FREDERICK, b Olyphant, Pa, Oct 22, 38; m 63; c 4. INVERTEBRATE PHYSIOLOGY, HISTOLOGY. *Educ:* Gettysburg Col, BA, 63; Drake Univ, MA, 65; Univ Vt, PhD(zool), 68. *Prof Exp:* PROF BIOL, LEBANON VALLEY COL, 68- *Mem:* AAAS; Am Soc Zool; Am Inst Biol Sci. *Res:* Histology, cell biology and physiology of Artemia; invertebrate reproductive systems; sensory receptors, behavior. *Mailing Add:* Dept Biol Lebanon Valley Col Annville PA 17003

WOLFE, ALLAN MARVIN, b New York, NY, Nov 4, 37; m 64; c 3. MEDICINE, METABOLISM. *Educ:* Cornell Univ, MB, 58; NY Univ, MD, 62. *Prof Exp:* Asst instr med, State Univ NY Downstate Med Ctr, 64-66; diabetes & arthritis consult & chief educ & info control prog, Nat Ctr Chronic Dis Control, 66-68; clin asst prof med, State Univ NY Downstate Med Ctr, 68-72; med dir, Arthritis Found, NY Chap, 68-70; dir med res infrared eng, Barnes Eng Co, 69-71, dir med mkt, 70-71; dir med res devices & instrumentation, Survival Technol Inc, 71-75; vpres res & develop pharmaceut, McGaw Labs Div, Am Hosp Supply Corp, 75-81; PRES, VOXELL, 81- *Concurrent Pos:* Prin investr, Voc Rehab Serv, 68-69, Arthritis Ctr Prog, Pub Health Serv, 69-72, Mayor's Orgn Task Force Comprehensive Health Planning, 70-72 & Sudden Cardiac Death & Onset Myocardial Infarction, 71-73; gen partner, Utah Venture Partners, Salt Lake City, 90- *Mem:* AMA; Am Soc Parenteral & Enteral Nutrit. *Res:* Human metabolism; amino acid chemistry; medical devices; parenteral pharmacologic agents. *Mailing Add:* Voxell 27075 Cabot Rd Laguna Hills CA 92653

WOLFE, BARBARA BLAIR, b Sharon, Pa, Oct 7, 40. RESEARCH ADMINISTRATION. *Educ:* Wayne State Univ, BS, 63, Med, 67, PhD, 74. *Prof Exp:* Asst dir comput serv, Wayne State Univ, 66-82; assoc vpres, State Univ NY, Albany, 82-85; asst vpres, Univ Minn, 85-88; EXEC DIR, CICNET INC, 88- *Concurrent Pos:* Mem Comput Soc, Inst Elec & Electronics Engrs. *Mem:* Asn Comput Mach; Inst Elec & Electronics Engrs; Am Educ Res Asn; Am Statist Asn. *Res:* Research administration. *Mailing Add:* 2520 Broadway Dr Rm 100 Lauderdale MN 55113

WOLFE, BERNARD MARTIN, b Killdeer, Sask, Dec 31, 34; m 70; c 1. MEDICINE, BIOCHEMISTRY. *Educ:* Univ Sask, BA, 56; Oxford Univ, BM & BCh, 63, MA, 67; McGill Univ, MSc, 67; FRCP (C) , 68; Royal Col Physicians & Surgeons, cert Endocrinol & Metab, 85. *Prof Exp:* House physician & surgeon med, Guy's Hosp, London, Eng, 63-64; from jr asst resident to sr asst resident, Royal Victoria Hosp, Montreal, Que, 64-68; Med Res Coun Can centennial fel, Cardiovasc Res Inst, Med Ctr, Univ Calif, San Francisco, 68-70; from asst prof to assoc prof, 70-80, PROF MED, UNIV WESTERN ONT, 80- *Concurrent Pos:* Consult & chief endocrinol & metab, Univ Hosp, London, Ont, 72-87; mem Coun on Arteriosclerosis, Am Heart Asn; hon lectr biochem, Univ Western Ont, 72-; chmn, Can Lipoprotein Conf, 85-86; chmn, Dept Med Div Endocrinol & Metab, Dept Med, Univ Western Ont, 87-91. *Mem:* Can Soc Clin Invest; Can Soc Endocrinol & Metab; Can Med Asn; Can Atherosclerosis Soc. *Res:* Clinical investigation of the effects of diets, hormones and drugs on lipid, carbohydrate and amino acid metabolism in man and experimental animals. *Mailing Add:* 17 Metamora Crescent London ON N6G 1R2 Can

WOLFE, BERTRAM, b New York, NY, June 26, 27; m 50; c 3. ENERGY TECHNOLOGY, BUSINESS MANAGEMENT. *Educ:* Princeton Univ, BA, 50; Cornell Univ, PhD(nuclear physics), 54. *Prof Exp:* Physicist, Eastman Kodak Co, 54-55; physicist, Nuclear Energy Div, Gen Elec Co, 55-56, mgr develop reactor physics, 57-59, mgr conceptual design & anal, 59-64, mgr plant eng & develop, Advan Prod Oper, 64-69; assoc dir, Pac Northwest Labs, Battelle Mem Inst, 69-70; vpres & tech dir, Wadco Corp, 70; gen mgr, Breeder Reactor Dept, 70-73 & Fuel Recovery & Irradiation Prod Dept, 74-78, vpres & gen mgr, Nuclear Energy Progs Div, 78-84 & Nuclear Technol & Fuel Div, 84-87, VPRES & GEN MGR, GEN ELEC NUCLEAR ENERGY, GEN ELEC CO, 87- *Concurrent Pos:* Mem, Nuclear Power Oversight Comt; mem bd dirs, Am Nuclear Energy Coun; mem bd dirs, Nuclear Mgt & Resources Comt; Am Nuclear Soc (pres). *Honors & Awards:* Walter H Zinn Award, Am Nuclear Soc. *Mem:* Nat Acad Eng; Am Phys Soc; fel Am Nuclear Soc; AAAS; Sigma Xi; Nat Res Coun. *Res:* Nuclear power technology; advanced energy technology. *Mailing Add:* 15453 Via Vaquero Monte Sereno CA 95030

WOLFE, CARVEL STEWART, b Minneapolis, Minn, June 11, 27; m 54; c 3. MATHEMATICS, INTEGER PROGRAMMING. *Educ:* Univ Ariz, BS, 50, MS, 51; Walden Univ, PhD, 83. *Prof Exp:* Asst math, Univ Wash, 51-53; asst prof, Shepherd State Col, 53-54; asst, Univ Md, 54-56; asst prof, 56-63, ASSOC PROF MATH, US NAVAL ACAD, 63- *Mem:* AAAS; Am Math Soc; Math Asn Am. *Res:* Numerical analysis; linear programming; integer programming. *Mailing Add:* Dept Math US Naval Acad Annapolis MD 21402

WOLFE, CHARLES MORGAN, b Morgantown, WVa, Dec 21, 35; div; c 2. ELECTRICAL ENGINEERING, SOLID STATE ELECTRONICS. *Educ:* Univ WVa, BSEE, 61, MSEE, 62; Univ Ill, PhD(elec eng), 65. *Prof Exp:* Mem staff, Lincoln Lab, Mass Inst Technol, 65-75; dir semiconductor res lab, 79-90, Samuel C Sachs prof, 82-90, PROF ELEC ENG, WASHINGTON UNIV, 75- *Honors & Awards:* Electronics Award, Electrochem Soc, 78; Jack A Morton Award, Inst Elec & Electronics Engrs, 90. *Mem:* Nat Acad Eng; Am Asn Univ Prof; fel Inst Elec & Electronics Engrs; Am Phys Soc; Electrochem Soc; AAAS. *Res:* Preparation and characterization of semiconductor materials for solid state devices. *Mailing Add:* Washhington Univ Box 1127 St Louis MO 63130

WOLFE, DAVID M, b Philadelphia, Pa, Oct 27, 38; div; c 3. HIGH ENERGY PHYSICS. *Educ:* Univ Pa, BA, 59, MS, 61, PhD(physics), 66. *Prof Exp:* Res assoc physics, Enrico Fermi Inst, Univ Chicago, 66-69; vis asst prof physics, Univ Wash, 69-71; from asst prof to assoc prof, 71-79, PROF PHYSICS, UNIV NMEX, 79- *Concurrent Pos:* Vis scientist, Brookhaven Nat Lab, Upton, NY, 78-79; sci assoc, CERN, Geneva, Switz, 83-85. *Mem:* Am Phys Soc; Fedn Am Scientists; Sigma Xi. *Res:* Experimental research in the nucleon-nucleon and antinucleon-nucleon interactions. *Mailing Add:* Dept Physics & Astron Univ NMex Albuquerque NM 87131

WOLFE, DOROTHY WEXLER, b Springfield, Ill, Aug 20, 20; m 42; c 1. MATHEMATICS. *Educ:* Univ Ill, BS, 41; Wayne State Univ, MA, 53; Univ Pa, PhD(math), 66. *Prof Exp:* Asst math, Univ Pa, 54-55; instr, Swarthmore Col, 62-64; from asst prof to prof, 65-85, EMER PROF MATH, WIDENER UNIV, 85- *Mem:* Am Math Soc; Sigma Xi; Math Asn Am. *Res:* Combinatorics; metric spaces. *Mailing Add:* 245 Hathaway Lane Wynnewood PA 19096

WOLFE, DOUGLAS ARTHUR, b Dayton, Ohio, July 6, 39; m 59; c 3. MARINE CHEMISTRY, POLLUTION BIOLOGY. *Educ:* Ohio State Univ, BSc, 59, MSc, 61, PhD(physiol chem), 64; Stanford Univ, MSc, 81. *Prof Exp:* Chief biogeochem prog, Radiobiol Lab, US Bur Com Fisheries, 64-70; dir estuarine res, Atlantic Estuarine Fisheries Ctr, Nat Marine Fisheries Serv, 70-75; staff dir ecol, 75-77, dep dir, Outer Continental Shelf Environ Assessment Prog, 77-81, dir oper progs, off marine pollution assessment, 81-83, CHIEF SCIENTIST, OCEAN ASSESSMENT DIV, NAT OCEANIC & ATMOSPHERIC ADMIN, 83- *Concurrent Pos:* Adj asst prof, NC State Univ, 66-70, adj assoc prof, 70-75; chief scientist I marine biol, Nuclear Ctr, PR, 69-70; consult, Panel on Zinc, Nat Acad Sci, 73-75; proj mgr, Nat Oceanic & Atmospheric Admin-Environ Protection Agency, Energy Proj on Fate & Effects of Petrol, 75-80; mem, Nat Oceanic & Atmospheric Admin-Ctr Nat Exploitation of Oceans, Int Comn for Study of the Ecol Effects of the Amoco Cadiz Oil Spill, 78-81; head marine pollution deleg, Nat Ocean & Atmospheric Admin, People's Repub China, 83; Nat Ocean & Atmospheric Admin rep, Sci Group of London Dumping Conv, 83-; affil prof, George Mason Univ, 88- *Mem:* Estuarine Res Fedn; Am Malacol Union; Am Soc Limnol & Oceanog; Sigma Xi. *Res:* Comparative biochemistry of carotenoids and lipids; ecology and biology of molluscs; biogeochemistry of petroleum, radioisotopes and trace metals in marine environment; ocean dumping and marine waste disposal management. *Mailing Add:* Ocean Assessment Div N/OMA3 Nat Oceanic & Atmospheric Admin Rockville MD 20852

WOLFE, EDWARD W, b Brooklyn, NY, Jan 21, 36; m 56; c 5. GEOLOGY. *Educ:* Col Wooster, BA, 57; Ohio State Univ, PhD(geol), 61. *Prof Exp:* Instr geol, Col Wooster, 59-61; GEOLOGIST, US GEOL SURV, 61- *Mem:* AAAS; Geol Soc Am; Sigma Xi; Am Geophys Union. *Res:* Areal geology and volcanology. *Mailing Add:* Cascades Volcano Observ 5400 MacArthur Blvd Vancouver WA 98661

WOLFE, GENE H, b Calumet City, Ill, May 18, 36; m 60; c 6. STATIC ELIMINATION, ELECTROSTATIC PRINTING. *Educ:* Univ Ill, BS, 58, MS, 59. *Prof Exp:* Res physicist, IIT Res Inst, 60-64; RES PHYSICIST, R R DONNELLEY & SONS, 64- *Mem:* Optical Soc Am; Laser Inst Am; Int Soc Optical Eng; Elec Overstress/Electrostatic Discharge Asn. *Res:* Development of equipment utilizing electrostatics for various graphics arts processes, primarily printing; devices and methods to control static electricity. *Mailing Add:* 1309 Buffalo Calumet City IL 60409

WOLFE, GORDON A, b Chicago, Ill, Sept 8, 31; m 56; c 2. SOLID STATE PHYSICS. *Educ:* Ill Inst Technol, BS, 60; Univ Mo, MS, 63, PhD(physics), 67. *Prof Exp:* Instr physics, Univ Mo, 64-67; asst prof, 67-74, ASSOC PROF PHYSICS, SOUTHERN ORE STATE COL, 74- *Mem:* AAAS; Am Asn Physics Teachers. *Res:* Anharmonic effects in crystals. *Mailing Add:* Dept Physics Southern Ore State Col Ashland OR 97520

WOLFE, HARRY BERNARD, b Vancouver, BC, Dec 29, 27; m 52; c 4. OPERATIONS RESEARCH. *Educ:* Univ BC, BA, 49, MA, 51; Columbia Univ, PhD(physics), 56. *Prof Exp:* Sr staff mem oper res, Mass, 56-63, mgr, San Francisco Opers Res Group, Calif, 63-69, SR STAFF MEM, HEALTH CARE SECT, ARTHUR D LITTLE, INC, MASS, 69- *Mem:* Inst Mgt Sci; Opers Res Soc Am; Can Opers Res Soc. *Res:* Health care; hospitals; management science, including inventory and scheduling theory and marketing analysis; nuclear physics. *Mailing Add:* Arthur D Little Inc 35 Acorn Park Cambridge MA 02140

WOLFE, HARVEY, b Baltimore, Md, Apr 14, 38; m 59; c 4. OPERATIONS RESEARCH, STATISTICS. *Educ:* Johns Hopkins Univ, BES, 60, MSE, 62, PhD(opers res), 64. *Prof Exp:* Opers res asst, Johns Hopkins Hosp, 60-64, opers res assoc, 64; from asst prof to assoc prof, 64-72, chmn, 86, PROF INDUST ENG, UNIV PITTSBURGH, 72- *Concurrent Pos:* Res assoc, Grad Sch Pub Health, Univ Pittsburgh, 64-67, adj assoc prof, 67-77, prof, 77-, mem grad fac, 66-; consult, Blue Cross Western Pa, 65-, dir res, 67-69; consult, Dept Med & Surg Study Group, Vet Admin, 66-72, Nat Ctr Health Serv Res, Dept HEW, 67-, Social Security Admin, 73- & Blue Cross/Blue Shield Greater New York, 75-; partner, Wolfe-Shuman Consult, 75-; bd dirs, Actronics, Inc, 81- *Mem:* Opers Res Soc Am; Inst Indust Engrs; Am Soc Eng Educ; Inst Mgt Sci. *Res:* Applications of operations research to the health services; manufacturing systems. *Mailing Add:* Dept Indust Eng Univ Pittsburgh Pittsburgh PA 15261

WOLFE, HERBERT GLENN, b Uniontown, Kans, Mar 14, 28; m 50; c 3. DEVELOPMENTAL GENETICS. *Educ:* Kans State Univ, BS, 49; Univ Kans, PhD(zool), 60. *Prof Exp:* Assoc staff scientist physiol genetics, Jackson Lab, Maine, 60-63; from asst prof to assoc prof, 63-70, PROF PHYSIOL GENETICS, UNIV KANS, 70- *Concurrent Pos:* Mem genetics standards subcomt, Inst Lab Animal Resources, 65-68; NIH spec res fel, Harwell, Eng, 69-70; vis scientist, Worcester Found for Exp Biol, Shrewsbury, Mass, 77. *Mem:* Soc Develop Biol; Genetics Soc Am; Sigma Xi; Soc Study Reproduction. *Res:* Physiological genetics, specifically genetic control of physiological and developmental processes related to reproduction; blood proteins and hematopoiesis; pigmentation in mice; gene regulation of y chromosome length. *Mailing Add:* Dept Physiol & Cell Biol Univ Kans Lawrence KS 66045

WOLFE, HUGH CAMPBELL, physics; deceased, see previous edition for last biography

WOLFE, JAMES F, b York, Pa, Oct 5, 36; m 59; c 2. ORGANIC CHEMISTRY. *Educ:* Lebanon Valley Col, BS, 58; Ind Univ, PhD(chem), 63. *Prof Exp:* Res assoc, Duke Univ, 63-64; from asst prof to assoc prof, 64-74, dept head, 81-89, PROF CHEM, VA POLYTECH INST & STATE UNIV, 74-, VPROVOST ACAD AFFAIRS, 90- *Mem:* Am Chem Soc. *Res:* Use of multiple anions in organic synthesis; mechanisms of heteroaromatic nucleophilic substitution; synthesis of new CNS agents. *Mailing Add:* Dept Chem Va Polytech Inst & State Univ Blacksburg VA 24061

WOLFE, JAMES FREDERICK, b Bell, Calif, Dec 1, 48; m 71; c 1. POLYMERS. *Educ:* Occidental Col, Calif, AB, 70; Univ Iowa, PhD(org chem), 75. *Prof Exp:* Vis scientist polymers, Mat Lab, Polymer Br Wright-Patterson AFB, 76; SR POLYMER CHEMIST, POLYMER SYNTHESIS RES, SRI INT, 76- *Mem:* Am Chem Soc. *Res:* Defining molecular structural requirements and developing synthesis conditions and techniques to give aromatic, heterocyclic polymers capable of liquid crystalline order; polyphenylene benzo bisthiazole. *Mailing Add:* Lockheed 0/9350 B/204 3251 Hanover St Palo Alto CA 94304-1191

WOLFE, JAMES H, b Salt Lake City, Utah, Jan 7, 22; m 56. MATHEMATICS. *Educ:* Univ Utah, BA, 42; Harvard Univ, MA, 43, PhD(math), 48. *Prof Exp:* PROF MATH, UNIV UTAH, 48- *Mem:* Am Math Soc. *Res:* Topology and integration theory; matrices. *Mailing Add:* Dept Math Univ Utah 237 John Widstoe Bldg Salt Lake City UT 84112

WOLFE, JAMES LEONARD, b Milton, Fla, May 5, 40; m 63. VERTEBRATE ZOOLOGY, ANIMAL BEHAVIOR. *Educ:* Univ Fla, BS, 62; Cornell Univ, PhD(vert zool), 66. *Prof Exp:* Asst prof biol, Univ Ala, University, 66-68; from asst to assoc prof zool, Res Ctr, Nat Space Technol Lab, Miss State Univ, 68-77, prof zool & adj assoc prof wildlife & fish, 77-81, dir, 81-85; exec dir, Archbold Biol Sta, 85-88; DEAN GRAD STUDIES & RES, EMPORIA STATE UNIV, 88- *Mem:* Animal Behav Soc; Am Soc Mammal; Ecol Soc Am. *Res:* Ecology and behavior of mammals. *Mailing Add:* Grad Studies & Res Emporia State Univ Emporia KS 66801

WOLFE, JAMES PHILLIP, b Randolph Field, Tex, July 16, 43; m 66; c 3. SOLID STATE PHYSICS. *Educ:* Univ Calif, Berkeley, BA, 65, PhD(physics), 71. *Prof Exp:* Asst res physicist, Univ Calif, Berkeley, 71-76; from asst prof to assoc prof, 76-81, PROF PHYSICS, UNIV ILL, URBANA-CHAMPAIGN, 81- *Concurrent Pos:* Prin investr, Mat Res Lab, Univ Ill, 76-, Cottrell res grant, Res Corp, 77-78, NSF grants, 78- & Air Force Off Sci Res grant, 79-83; prog dir, DOE, Sol St Sci, Univ Ill, 89-91; Humboldt Sr Scientist, 88-89. *Mem:* Fel Am Phys Soc. *Res:* Physics of semiconductors; optical and microwave studies of photo-excited phases; thermal transport in crystals; phonon imaging; electron and nuclear magnetic resonance in solids. *Mailing Add:* Loomis Lab Univ Ill 1110 W Green Urbana IL 61801

WOLFE, JAMES RICHARD, JR, reaction mechanisms, polymer synthesis, for more information see previous edition

WOLFE, JAMES WALLACE, b Ludlowville, NY, Apr 11, 32; m 54; c 2. NEUROPHYSIOLOGY, PSYCHOLOGY. *Educ:* Univ Calif, Riverside, BA, 63; Univ Rochester, PhD(psychol), 66. *Prof Exp:* Res psychologist, Army Med Res Lab, Ft Knox, Ky, 66-68; res neurophysiologist sch aerospace med, USAF, 68-, liason officer, Off Sci Res, Tokyo, Japan, 85-; RETIRED. *Concurrent Pos:* Lectr, Univ Louisville, 67-68 & St Mary's Univ, Tex, 69-81; adj prof, Univ Tex, San Antonio, 85- *Mem:* Int Brain Res Orgn; Aerospace Med Asn; Barany Soc. *Res:* Cerebellar integration of sensory information; effects of drugs on electrophysiological responses; neurophysiological control of oculomotor function. *Mailing Add:* 4110 Spring View San Antonio TX 78222

WOLFE, JOHN A(LLEN), b Riverton, Iowa, June 3, 20; m 48, 82; c 2. GEOLOGICAL & MINING ENGINEERING. *Educ:* Colo Sch Mines, GeolE & EM, 47, MS, 54; Columbia Pac Univ, PhD, 83. *Prof Exp:* Res geologist, Ideal Cement Co, 48-51, chief geologist, 51-63, dir explor, 63-65; pres, Pan-Asian Tech Servs, 74-84, PRES, TAYSAN COPPER, INC, 72-; VPRES & DIR, KENMORE MINERALS, 87- *Concurrent Pos:* Consult porphyry copper & other metals & nonmetallics, Philippines & Southeast Asia; geol consult, 65-; prof lectr, Univ Philippines, 83-86. *Mem:* Geol Soc Am; Am Inst Mining, Metall & Petrol Engrs; Soc Econ Geol; Geol Soc Philippines; Asn Geologists Int Develop. *Res:* Origin and nature of porphyry copper deposits and development of mineral deposits in tectonic environments; geomorphology; quaternary explosive volcanism; Philippine geochronology and mineral resource economics. *Mailing Add:* MCC PO Box 1868 Makati Metro-Manila 1299 Philippines

WOLFE, LAUREN GENE, b Kenton, Ohio, Nov 7, 39; m 66. PATHOLOGY. *Educ:* Ohio State Univ, DVM, 63, MS, 65, PhD(vet path), 68; Am Col Vet Path, dipl, 68. *Prof Exp:* Asst prof path, Univ Ill Med Ctr, 68-71; assoc prof microbiol, Rush-Presby-St Luke's Med Ctr, 71-74, prof, 74-81; prof & head path & parasitol, 81-87, PROF & HEAD PATHOBIOL, AUBURN UNIV, 88- *Concurrent Pos:* Asst microbiologist, Presby-St Luke's Hosp, 68-71. *Mem:* AAAS; Am Soc Microbiol; Am Asn Pathologists; Am Asn Immunol; Am Asn Cancer Res. *Res:* Tumor pathobiology and immunology; oncology. *Mailing Add:* Dept Pathobiol 166 Greene Hall Auburn Univ AL 36849

WOLFE, LEONHARD SCOTT, b Auckland, NZ, Mar 23, 26; Can citizen; m 60; c 2. BIOCHEMISTRY, NEUROCHEMISTRY. *Educ:* Univ NZ, BSc, 47; Cambridge Univ, PhD(insect physiol, biochem), 52, ScD, 76; Univ Western Ont, MD, 58; FRCP(C), 72, FRSC, 73. *Prof Exp:* Jr lectr zool, Univ Canterbury, 49-50; assoc entomologist, Agr Res Inst, Can Dept Agr, 52-54; from asst prof neurochem to assoc prof neurol & neurosurg, 60-70, PROF NEUROL & NEUROSURG, MONTREAL NEUROL INST, MCGILL UNIV, 70-, DIR, DONNER LAB EXP NEUROCHEM, 65- *Concurrent Pos:* Nat Res Coun Can med res fel, 59-60; Sister Elizabeth Kenny Found scholar, 60-61; career investr, Med Res Coun Can, 63-, mem grants comt neurol sci, 70-74, mem priorities selection & rev comt, 72-75; hon lectr biochem, McGill Univ, 60-70, prof biochem, 71-; consult dermat res unit, Royal Victoria Hosp, Montreal, 65-67. *Honors & Awards:* Heinrich Waelsh Lectr, Columbia Univ. *Mem:* Int Brain Res Orgn; Am Soc Biol Chemists; Can Biochem Soc; Can Physiol Soc; Int Soc Neurochem; Soc Neurosci; Am Soc Neurochem. *Res:* Entomology; biology and control of biting flies; insect cholinesterases; metabolism of insecticides; biochemistry and function of complex glycolipids in neurones; membranes; role of lipid anions in excitable tissues; convulsive states; biosynthesis, release and action of prostaglandins and eicosanoids; biochemistry of lipid storage diseases and degenerative neurological diseases. *Mailing Add:* Montreal Neurol Inst 3801 University St Montreal PQ H3A 2B4 Can

WOLFE, PAUL JAY, b Mansfield, Ohio, Oct 2, 38; m 60; c 2. EXPLORATION GEOPHYSICS. *Educ:* Case Inst Technol, BS, 60, MS, 63, PhD(nuclear physics), 66. *Prof Exp:* Design engr, Lamp Div, Gen Elec Co, 60-61; asst prof physics, 66-71, chmn dept, 72-75, ASSOC PROF PHYSICS & GEOL SCI, WRIGHT STATE UNIV, 71. *Concurrent Pos:* NSF geophysicist fel, US Bur Mines, Denver, 79-80. *Mem:* Soc Explor Geophysicists; Am Asn Physics Teachers. *Res:* Seismic exploration techniques related to hydrocarbons and coal; shallow geophysical techniques for hydrology, engineering and archeology; subsurface cavity detection with seismic and gravity methods. *Mailing Add:* Dept Geol Wright State Univ Dayton OH 45435

WOLFE, PETER E, b Hammonton, NJ, Apr 27, 11. ENVIRONMENTAL CONSULTING, HYDROLOGY-GROUND WATER. *Educ:* Rutgers State Univ, BS, 33; Princeton Univ, MA, 40, PhD(geol), 41. *Prof Exp:* Instr geol, Princeton Univ, 38-41, res assoc, 45-46; dir, Nfld Dept Natural Resources, 41-45; prof, 46-81, EMER PROF GEOL, RUTGERS UNIV, 81- *Concurrent Pos:* Vis prof geol, Osmania Univ India, 57-81. *Mem:* Geol Soc Am; Sigma Xi. *Res:* Ground water recycling research; environmental geology. *Mailing Add:* 204 Pleasant Mills Rd Hammonton NJ 08037

WOLFE, PETER NORD, b Lakewood, Ohio, July 24, 29; m 51; c 3. PHYSICS. *Educ:* Ohio Wesleyan Univ, BA, 51; Ohio State Univ, MS, 52, PhD(physics), 55. *Prof Exp:* NSF fel, 54-55; from res physicist to mgr systs physics dept, Res Labs, Westinghouse Elec Corp, 55-72, mgr transformer technol, 72-75, mgr laser fusion activities, 75-78; staff mem, 78-80, PROJ LEADER, LOS ALAMOS NAT LAB, 80- *Mem:* Am Phys Soc; Inst Elec & Electronics Engrs. *Res:* High power lasers; power technology; applied physics; microwave spectra. *Mailing Add:* 155 Tunyo Los Alamos NM 87544

WOLFE, PHILIP, b San Francisco, Calif, Aug 11, 27; m 68; c 1. APPLIED MATHEMATICS. *Educ:* Univ Calif, AB, 48 & PhD(math), 54. *Prof Exp:* Instr math, Princeton Univ, 54-57; mathematician, Rand Corp, 57-66; MEM RES STAFF, IBM RES, 66-, ASST CHMN, MATH SCI DEPT, 86- *Concurrent Pos:* Prof eng math, Columbia Univ, 68-77; chmn, Math Prog Soc, 78-80, vchmn, 80-82. *Mem:* Fel AAAS; Math Prog Soc; fel Econ Soc. *Res:* Mathematics of optimization; linear and nonlinear programming. *Mailing Add:* Math Sci Dept IBM Res PO Box 218 Yorktown Heights NY 10598

WOLFE, RALPH STONER, b New Windsor, Md, July 18, 21; m 50; c 3. MICROBIAL BIOCHEMISTRY. *Educ:* Bridgewater Col, BS, 42; Univ Pa, MS, 49, PhD, 53. *Prof Exp:* Asst instr microbiol, Univ Pa, 47-49, instr, 51-52; asst limnol, Acad Natural Sci, Pa, 49-50; from instr to assoc prof, 53-60, PROF MICROBIOL, UNIV ILL, URBANA, 61- *Concurrent Pos:* NSF fel, 58; Guggenheim fel, 60 & 75. *Mem:* Nat Acad Sci; Am Soc Microbiol; Am Acad Arts & Sci; Am Soc Biol Chemists. *Res:* Metabolism and physiology of bacteria; methanogens; archaebacteria. *Mailing Add:* Dept Microbiol 131 Burrill Hall Univ Ill Urbana IL 61801

WOLFE, RAYMOND, b Hamilton, Ont, Apr 8, 27; m 54; c 3. SOLID STATE PHYSICS. *Educ:* Univ Toronto, BA, 49, MA, 50; Bristol Univ, PhD(physics), 55. *Prof Exp:* Physicist, Eastman Kodak Co, NY, 50-52; physicist, Gen Elec Co, Eng, 54-57; supvr, Magnetic Bubble Mat Group, 67-83, MEM TECH STAFF, BELL TEL LABS, 57-, PHYSICS MAT RES 83- *Concurrent Pos:* Ed, Appl Solid State Sci. *Mem:* Fel Am Phys Soc. *Res:* Theoretical and experimental solid state physics; transport and optical properties of semiconductors and metals; thermoelectric materials and devices; magnetic materials; magnetic bubble devices; magneto-optic materials and devices. *Mailing Add:* Mat Physics Res Lab AT&T Bell Labs 1D 445 Murray Hill NJ 07974

WOLFE, RAYMOND GROVER, JR, b Oakland, Calif, June 1, 20; m 46; c 3. BIOCHEMISTRY. *Educ:* Univ Calif, AB, 42, MA, 48, PhD(biochem), 55. *Prof Exp:* Biochemist, Donner Lab, Univ Calif, 48-55; Nat Found Infantile Paralysis fel chem, Univ Wis, 55-56; Lalor res fel, 56, from asst prof to prof, 56-83, EMER PROF CHEM, UNIV ORE, 83- *Concurrent Pos:* Guggenheim fel, Inst Biochem, Univ Vienna, 63-64; vis prof, Bristol Univ, 70-71, Cornell Univ, 78-79. *Mem:* AAAS. *Res:* Enzyme catalytic mechanism; structure-function relationship of polymeric enzymes; enzyme kinetics and inhibition; protein structure studies, particularly in dehydrogenases. *Mailing Add:* Dept Chem Univ Ore Eugene OR 97403

WOLFE, REUBEN EDWARD, b Schuylkill Haven, Pa, Dec 19, 27; m 60; c 2. PULSE CODED SIGNAL SYSTEM OPERATIONS FOR TARGET LOCATION & COMMUNICATIONS, MILLIMETER RECEIVER DESIGN CHARACTERISTICS FOR HARDWARE COLLECTORS. *Educ:* Pa State Col, BS, 49; Pa State Univ, BS, 59. *Prof Exp:* Res engr, Sylvania Elec Corp, 50-54; sci adv, Haller, Raymond & Brown Inc, 54-57; prin engr, Topp Industs, 57-59; staff engr, HRB-Singer Corp, 59-87, Singer Corp, 87-88, Hadson Corp, 88-90; STAFF ENGR, E SYSTS INC, 90- *Mem:* Nat Soc Prof Engrs (secy-treas, 73-75); Am Inst Elec Engrs (secy, 66-68). *Res:* Development of linear modulated fast wave electronic beam devices for power spectral density values greater than 50 kilowatts and phase stabilities exceeding six degrees over an octave bandwidth. *Mailing Add:* 150 S Science Park Rd State College PA 16801

WOLFE, ROBERT KENNETH, b Chattanooga, Tenn, Sept 5, 29; m 59; c 2. COMPUTER SCIENCE, ENGINEERING. *Educ:* Ga Inst Technol, BChE, 52, PhD(chem eng), 56. *Prof Exp:* Res asst chem eng, Ga Inst Technol, 51-52; chem engr, Mallinckrodt Chem Works, 55-60; systs eng mgr comput, Int Bus Mach, 60-68; opers res mgr planning, Owens Ill, 68-73; PROF INDUST ENG & COMPUT SCI & ENG, UNIV TOLEDO, 73- *Concurrent Pos:* Tenn Eastman fel, Ga Inst Technol, 52-54; res contracts, Ohio Dept Transp & Fed Hwy Admin, 75-, Edison Indust Systs Ctr, 88-; chmn systs eng PhD prog, Univ Toledo, 76-78. *Mem:* Sigma Xi; sr mem Am Inst Indust Engrs; Am Inst Chem Engrs. *Res:* Development and use of design and decision making models using computers, informations systems and simulation. *Mailing Add:* 4930 Spring Mill Ct Toledo OH 43615

WOLFE, ROBERT NORTON, physics; deceased, see previous edition for last biography

WOLFE, ROGER THOMAS, b Mt Vernon, Ill, July 31, 32; m 56; c 2. ORGANIC CHEMISTRY. *Educ:* Bradley Univ, BS, 54; Rensselaer Polytech Inst, PhD(chem), 59; Salmon P Chase Col, JD, 69. *Prof Exp:* Lab asst anal chem, Bradley Univ, 52-54; asst gen & org chem, Rensselaer Polytech Inst, 54-56; res chemist, Sterling-Winthrop Res Inst, NY, 56-60, from assoc patent agent to patent agent, 60-65, patent agent, Hilton-Davis Chem Co Div, Sterling Drug Co, 65-66, asst dir res & develop, 66-69, vpres res & develop, 70-75, patent attorney, Hilton-Davis Chem Co Div, 69-77, vpres res admin & legal affairs, 75-77, ASST TO CORP DIR, SAFETY & ENVIRON AFFAIRS, STERLING DRUG INC, 77- *Mem:* Am Chem Soc. *Mailing Add:* Three Kingsbury Ct New York NY 14618

WOLFE, SETH AUGUST, JR, b Baltimore, Md, July 3, 44; m 76; c 2. NEUROIMMUNOLOGY, PSYCHONEUROIMMUNOLOGY. *Educ:* Dickenson Col, BA, 70; Johns Hopkins Univ, PhD(immunol), 83. *Prof Exp:* Res contractor opthal & immunol, Dept Opthal, Sch Med, Johns Hopkins Univ, 76-81; biologist, Lab Neurochem & Neuroimmunol, Nat Inst Child Health & Human Develop, NIH, 83-84, guest researcher, Clin Immunol Sect, Gerontol Res Ctr, Nat Inst Aging; staff fel, Neurobiol Lab, Neurosci Br, Addiction Res Ctr, Nat Inst Drug Abuse, 86-90; ASST PROF IMMUNOL & NEUROIMMUNOL, DEPT MED MICROBIOL & IMMUNOL, OHIO STATE UNIV, 90- *Mem:* Fedn Am Soc Exp Biol; Am Asn Immunologists; Soc Neurosci; AAAS; NY Acad Sci. *Res:* Receptors & communication molecules shared by the immune, endocrine and central nervous systems; mechanisms of communication and integration of neuroendocrine and immune responses and the involvement of these mechanisms in disease processes and pharmacologic intervention. *Mailing Add:* Dept Med Microbiol & Immunol Graves Hall Ohio State Univ 333 W Tenth Ave Columbus OH 43210-1239

WOLFE, STEPHEN LANDIS, b Sept 23, 32; m 82; c 3. CELL BIOLOGY. *Educ:* Bloomsburg State Col, BS, 54; Ohio State Univ, MS, 59; Johns Hopkins Univ, PhD(biol), 62. *Prof Exp:* NIH fel zool, Univ Minn, 62-63; from asst prof to assoc prof, 63-82, sr lectr zool, Univ Calif, Davis, 82-87; EMER PROF, UNIV CALIF, 87- *Concurrent Pos:* Vis prof, Yale Univ, 73. *Res:* Fine and molecular structure of chromatin. *Mailing Add:* Div Biomed Sci Univ Calif Riverside CA 92521

WOLFE, STEPHEN MITCHELL, b Winter Haven, Fla, Nov 13, 49; m. PLASMA PHYSICS, LASER PHYSICS. *Educ:* Mass Inst Technol, SB, 71, PhD(physics), 77. *Prof Exp:* physicist, Francis Bitter Nat Magnet Lab, 77-80,' PHYSICIST, PLASMA FUSION CTR, MASS INST TECHNOL, 80- *Mem:* Am Optical Soc; Am Phys Soc. *Res:* Fusion research; plasma diagnostics; optically pumped lasers; cyclotron resonance masers. *Mailing Add:* Plasma Fusion Ctr Mass Inst Technol 175 Albany St Cambridge MA 02139

WOLFE, WALTER MCILHANEY, b Baltimore, Md, Aug 15, 21; m 45; c 3. OBSTETRICS & GYNECOLOGY. *Educ:* Univ Md, MD, 46; Am Bd Obstet & Gynec, dipl, 57. *Prof Exp:* Asst prof, 65-68, actg chmn dept, 69-72, ASSOC PROF OBSTET & GYNEC, SCH MED, UNIV LOUISVILLE, 68- *Concurrent Pos:* Proj dir family planning, Dept HEW Grant, Louisville Gen Hosp, 71- *Mem:* Am Col Obstet & Gynec. *Res:* Applications of current technological and educational techniques to community reproductive health. *Mailing Add:* Dept Obstet/Gynec Univ Louisville Sch Med Louisville KY 40208

WOLFE, WILLIAM LOUIS, JR, b Yonkers, NY, Apr 5, 31; m 55; c 3. OPTICS, ELECTRICAL ENGINEERING. *Educ:* Bucknell Univ, BS, 53; Univ Mich, MS, 56, MSE, 66. *Prof Exp:* Asst proj engr, Sperry Gyroscope Co, 52-53; from res asst to res assoc infrared & optics, Univ Mich, 53-57, from assoc res engr to res engr, 57-66, lectr elec eng, 62-66; chief engr, Honeywell Radiation Ctr, 66-68, mgr, lectro-Optics Dept, Honeywell, Inc, 68-69; PROF OPTICAL SCI, UNIV ARIZ, 69- *Concurrent Pos:* Lectr, Northeastern Univ, 68-69; mem panel of comt undersea warfare-assessment electro optics, Nat Acad Sci; mem adv comt, Army Res Off, study panel on Army Countermine Adv Comt & adv comt, Nat Bur Standard & Air Force Systs Command; consult, var orgn; Army sci bd. *Mem:* Fel Optical Soc Am; sr mem Inst Elec & Electronics Engrs; fel Soc Photoelectronic Instrumentation Engrs (pres elect). *Res:* Optical materials for infrared use; radiometry; space navigation using star trackers; electro-optical system design; infrared simulation; infrared reconaissance and surveillance systems; optical scattering. *Mailing Add:* Optical Sci Ctr Univ Ariz Tucson AZ 85721

WOLFE, WILLIAM RAY, JR, b Grafton, WVa, Nov 16, 24; m 52; c 2. PHYSICAL CHEMISTRY. *Educ:* WVa Wesleyan Col, BS, 49; Western Reserve Univ, MS, 50, PhD(phys chem), 53. *Prof Exp:* Asst phys chem, Western Reserve Univ, 50-52; res chemist cent res dept, 52-60, chemist, Explosives Dept, 60-64, chemist develop dept, 65-68, RES ASSOC, POLYMER PROD DEPT, EXP STA, E I DU PONT DE NEMOURS & CO, INC, 68- *Mem:* Am Chem Soc; Sigma Xi. *Res:* Fused salt and aqueous electrochemistry; energy conversion; high temperature chemistry; catalysis; plastics processing; microwave and dielectric film. *Mailing Add:* 1007 Parkside Dr Wilmington DE 19803

WOLFENBARGER, DAN A, b White Plains, NY, Sept 23, 34; m 59; c 3. PLANT BREEDING & GENETICS. *Educ:* Univ Fla, BSA, 56; Iowa State Univ, MS, 57; Ohio State Univ, PhD(entom), 61. *Prof Exp:* Entomologist, Agr Exp Sta, Tex A&M Univ, 61-65; entomologist cotton insect res, Agr Res Serv, USDA, 65-82. *Mem:* Entom Soc Am. *Res:* Activity of and resistance and mode of inheritance to insecticides against tobacco bollworm, beet armyworm, boll weevil and tobacco budworm; sound for detection of feeding; post-harvest treatment tests on subtropical fruits (citrus and mangoes) against fruit flies of Anastrepha; effects of insect growth regulators and biological and chemical insecticides on Heliothis virescens and Heliocoverpa zea and Spodoptera exigua. *Mailing Add:* Agr Res Serv Cotton Insects Res Unit USDA 2413 E Hwy 83 Weslaco TX 78596

WOLFENBERGER, VIRGINIA ANN, b Fort Worth, Tex, Sept 15, 48. ENVIRONMENTAL PHYSIOLOGY, BEHAVIORIAL SCIENCE. *Educ:* Univ Tex, Arlington, BS, 70, MA, 73; Tex A&M Univ, PhD(biol), 81; Univ Houston, Clear Lake, MA, 85. *Prof Exp:* Asst prof biol, Xavier Univ La, 81-82; PROF PHYSIOL & CHEM, TEX CHIROPRACTIC COL, 82- *Res:* Physiological responses of an organism to various environmental stimuli and ecological factors; responses of humans to stresses and other psychological factors. *Mailing Add:* 5912 Spencer Pasadena TX 77505

WOLFENDEN, RICHARD VANCE, b Oxford, Eng, May 17, 35; US citizen; m 65; c 2. BIOCHEMISTRY. *Educ:* Princeton Univ, AB, 56; Oxford Univ, BA & MA, 60; Rockefeller Inst, PhD(biochem), 64. *Prof Exp:* Asst prof biochem, Princeton Univ, 64-70; from assoc prof to prof biochem, Sch Med, 70-83, DISTINGUISHED PROF BIOCHEM & NUTRIT, UNIV NC, CHAPEL HILL, 83- *Concurrent Pos:* Mem, NSF Adv Panel Molecular Biol, 74-77, NIH Study Sect Biorg & Natural Prods, 82-86; vis fel, Exeter Col, Oxford, 69, 76. *Mem:* Fel AAAS; Am Chem Soc; Am Soc Biol Chemists. *Res:* Physical organic chemistry in relation to enzyme-catalyzed reactions. *Mailing Add:* Dept Biochem Univ NC Sch Med Chapel Hill NC 27514

WOLFENSTEIN, LINCOLN, b Cleveland, Ohio, Feb 10, 23; m 43, 57; c 3. THEORETICAL HIGH ENERGY PHYSICS. *Educ:* Univ Chicago, BS, 43, MS, 44, PhD(physics), 49. *Prof Exp:* Physicist, Nat Adv Comt Aeronaut, 44-46; instr physics, 48-49, from asst prof to prof, 49-78, UNIV PROF PHYSICS, CARNEGIE-MELLON UNIV, 78- *Concurrent Pos:* NSF sr fel, Europ Orgn Nuclear Res, Geneva, Switz, 64-65; vis prof, Univ Mich, 70-71; Guggenheim fel, 74-75 & 83-84; mem physics adv comt, NSF, 74-77; Fairchild vis scholar, Calif Tech, 88. *Mem:* Nat Acad Sci; Am Phys Soc; AAAS. *Res:* Nuclear collisions; weak interactions. *Mailing Add:* Dept Physics Carnegie-Mellon Univ Pittsburgh PA 15213

WOLFERSBERGER, MICHAEL GREGG, b Northampton, Pa, June 14, 44; m 65. MEMBRANE BIOLOGY. *Educ:* Lebanon Valley Col, BS, 66; Temple Univ, PhD(biochem), 71. *Prof Exp:* Res assoc biochem, Lab Exp Dermat, Albert Einstein Med Ctr, Philadelphia, 71-73, assoc mem div res, 73-75; asst prof, Div Natural Sci & Math, Rosemont Col, 74-77; res assoc, 77-85, SR RES ASSOC BIOL, TEMPLE UNIV, 85- *Concurrent Pos:* Lectr, Cabrini Col, 76-77 & LaSalle Col, 81-82; consult, Rohm & Haas Co, 79-80 & 83-84; Roche fel Microbiol Inst, Swiss Fed Inst Technol, Zurich, 84-85; vis scientist, Gen Physiol & Biochem, Univ Milan,85; consult, Advanced Technol Ctr, Southeastern Pa, 87; prin investr, USDA Comp res grant, 87- & NIH res grant, 90- *Mem:* AAAS; NY Acad Sci; Am Soc Zool; Soc Invert Path; Am Soc Biochem & Molecular Biol. *Res:* Mechanism of active solute transport in insect epithelial cells; mechanism of action of insecticidal bacterial toxins. *Mailing Add:* Dept Biol Temple Univ Philadelphia PA 19122

WOLFF, ALBERT ELI, biostatistics, epidemiology, for more information see previous edition

WOLFF, ARTHUR HAROLD, b Trenton, NJ, Dec 23, 19; m 46; c 3. ENVIRONMENTAL HEALTH. *Educ:* Mich State Univ, DVM, 42. *Prof Exp:* Res investr, Commun Dis Ctr, USPHS, 46-50, res biologist, Sanit Eng Ctr, 50-58, chief training br, Div Radiol Health, 58-60; consult, UN Food & Agr Orgn, Italy, 60-61; chief radiation bio-effects prog, Nat Ctr Radiol Health, USPHS, 61-68; asst dir res, Consumer Protection & Environ Health, Dept HEW, 68-69, asst dir radiation prog, US Environ Protection Agency, 69-71; assoc dean res, 77-80, prof environ health & chmn dept, Sch Pub Health, 71-82, EMER PROF, UNIV ILL MED CTR, 82-; DOCENT, SMITHSONIAN MUS NATURAL HIST, 82- *Concurrent Pos:* Fel, Duke Univ & Oak Ridge Inst Nuclear Studies, 50-51; exec secy, Nat Adv Comt Radiation, 59-61; consult & mem expert panels, UN Food & Agr Orgn, 60-63 & WHO, 61-; sci adv, Comn Europ Communities, 78-79; sci adv, US Environ Protection Agency, 80-; consult, Nat Res Coun, Nat Acad Sci, 81-82. *Honors & Awards:* K F Meyer Award, 73. *Mem:* AAAS; Am Vet Med Asn; Am Pub Health Asn. *Res:* Radiation biology; comparative oncology; environmental toxicology. *Mailing Add:* 4285 Embassy Park Dr NW Washington DC 20016

WOLFF, DAVID A, b Cleveland, Ohio, Nov 2, 34; m 58, 76; c 4. CELL BIOLOGY, VIROLOGY. *Educ:* Col Wooster, AB, 56; Univ Cincinnati, MS, 60, PhD(microbiol, virol), 65. *Prof Exp:* From asst prof to prof virol & microbiol, Ohio State Univ, 64-78; MEM STAFF, NIH, 78- *Concurrent Pos:* Res grant, 66-69; Am-Swiss Found Sci exchange lectr, Switz, 70; res leave, Univ Uppsala, Sweden, 71. *Mem:* AAAS; Am Soc Microbiol; Sigma Xi. *Res:* Viral-induced cytopathic effects and relation of lysosomal enzymes; purification of virus; electron microscopy of virus infected cells; purification of lysosomes; virus interactions with synchronized cells. *Mailing Add:* Dep Asst Dir Off Prog Activ Nat Inst Gen Med Sci GPA Bldg WB Rm 936 Bethesda MD 20892

WOLFF, DONALD JOHN, b New York, NY, Feb 23, 42. BIOCHEMISTRY, PHARMACOLOGY. *Educ:* Fordham Univ, BS, 63; Univ Wis, PhD(biochem), 69. *Prof Exp:* Asst prof, 72-80, ASSOC PROF PHARMACOL, RUTGERS MED SCH, COL MED & DENT NJ, 80- *Concurrent Pos:* NIH trainee pediat, J P Kennedy, Jr Labs, Med Sch, Univ Wis, 68-72. *Mem:* AAAS. *Res:* Calcium-binding proteins; regulation of cyclic nucleotide metabolism by calcium ion. *Mailing Add:* Dept Pharmacol UMDNJ Robert Wood Johnson Med Sch Box 101 Piscataway NJ 08854

WOLFF, EDWARD A, b Chicago, Ill, Oct 31, 29; m 51; c 3. ELECTRICAL ENGINEERING. *Educ:* Univ Ill, BSEE, 51; Univ Md, MS, 53, PhD, 61. *Prof Exp:* Electronic scientist, US Naval Res Lab, 51-54; proj engr, Md Electronic Mfg Corp, Litton Indust, 56-59 & Electromagnetic Res Corp, 59-61; staff consult & mgr, Space Eng Lab, Aero Geo Astro Corp, Keltec Indust, Inc, Md, 61-65, chief engr, 65-67; vpres, Geotronics, Inc, 67-71; head, Syst Study Off, 71-73, ASSOC CHIEF COMMUN & NAVIG DIV, NASA GODDARD SPACE FLIGHT CTR, 73-; SR STAFF MEM, MRJ, INC, 89- *Concurrent Pos:* Mem, Md Gov Sci Resources Adv Bd, treas, Joint Bd Sci Educ. *Honors & Awards:* Centennial Medal, Inst Elec & Electronics Engrs, 85. *Mem:* Nat Soc Prof Engrs; fel Inst Elec & Electronics Engrs; Am Inst Aeronaut & Astronaut; Antennas & Propagation Soc (pres, 77). *Res:* Antennas; microwave components; electromagnetic waves. *Mailing Add:* 1021 Cresthaven Dr Silver Spring MD 20903

WOLFF, ERNEST N, b St Paul, Minn. GEOLOGY, MINING ENGINEERING. *Educ:* Univ Alaska, BS, 41; Univ Ore, MS, 59, PhD(geol), 65. *Prof Exp:* Field asst, Alaska Territorial Dept Mines, 39-40; observer geophys, Carnegie Inst Dept Terrestrial Magnetism, 41-46; observer in chg, Univ Alaska, 46-48, res assoc & asst prof mining eng, Sch Mines, 51-57; asst prof geol, Colo State Univ, 59-66; assoc prof geol & mining eng, Univ Alaska, 66-67; assoc prof geol, Colo State Univ, 67-69; PROF EXPLOR ENG & ASSOC DIR, MINERAL INDUST RES LAB, UNIV ALASKA, 69- *Mem:* Am Inst Mining, Metall & Petrol Engrs; Sigma Xi. *Res:* Economic geology of Alaska; regional Alaskan economics. *Mailing Add:* PO Box 10705 Fairbanks AK 99710

WOLFF, FREDERICK WILLIAM, b Berlin, Ger, Aug 21, 20; c 3. PHARMACOLOGY, MEDICINE. *Educ:* Univ Durham, MB, BS, 46, MD, 57. *Hon Degrees:* Georgeian USSR Acad Sci, Dipl, 81. *Prof Exp:* House physician, Royal Victoria Infirmary, Univ Durham, 46-47; house physician, med registr & resident med officer, Southend-on-Sea Gen Hosp, Eng, 47-50; med registr, Whittington Hosp, 53-54; clin pharmacologist, Wellcome Res Inst, 55-59; asst prof med, Sch Med & Endocrine Clin, Johns Hopkins Univ, 59-63; PROF MED, SCH MED, GEORGE WASHINGTON UNIV, 65-; PRES, INST DRUG DEVELOP, WASHINGTON, DC, 80- *Concurrent Pos:* Sr res asst, Post-Grad Med Sch, Univ London & Whittington Hosp, 55-59; consult, Food & Drug Admin, DC & Children's Med Ctr, DC, 71-79. *Mem:* Am Diabetes Asn; Am Soc Pharmacol & Exp Therapeut; Am Fedn Clin Res; Am Heart Asn; Royal Soc Med. *Res:* Therapeutics; clinical pharmacology; endocrinology; diabetes; hypertension. *Mailing Add:* 10908 Piney Meeting House Rd Potomac MD 20854

WOLFF, GEORGE LOUIS, b Hamburg, Ger, Aug 24, 28; US citizen; m 53; c 2. GENETICS, TOXICOLOGY. *Educ:* Ohio State Univ, BS, 50; Univ Chicago, PhD(zool), 54. *Prof Exp:* Biologist, Nat Cancer Inst, 56-58; res assoc, Inst Cancer Res, 58-63, supvr animal colony, 58-68, asst mem, 63-72, geneticist, 68-72; chief mammalian genetics br, 72-74, chief div mutagenic res, 74-79, sr sci coordr genetics, 79-88, SR RES SCIENTIST, NAT CTR TOXICOL RES, US FOOD & DRUG ADMIN, 88- *Concurrent Pos:* USPHS fel, Nat Cancer Inst, 54-56; prof assoc, Nat Acad Sci, 56-57; consult, Am Asn Accreditation Lab Animal Care, 70-75; mem, HEW subcomt, Environ Mutagenesis, 73-81; from asst prof to assoc prof biochem, Univ Ark, 73-89, assoc prof interdisciplinary toxicol & biochem, 81-90, prof biochem, 89-, prof interdisciplinary toxicol & biochem, 90-; assoc ed, Lab Animal Sci, 64-76. *Mem:* Soc Toxicol; Am Asn Cancer Res; Environ Mutagen Soc; Genetics Soc Am; Soc Exp Biol & Med. *Res:* Developmental and molecular genetics of carcinogenic response to environmental toxicants in mammals and obesity; genetic aspects of toxicology; regulation of gene expression in mutant mice. *Mailing Add:* Div Comp Toxicol Nat Ctr Toxicol Res Jefferson AR 72079

WOLFF, GEORGE THOMAS, b Irvington, NJ, Nov 27, 47; m 72; c 3. AIR POLLUTION METEOROLOGY & CHEMISTRY. *Educ:* NJ Inst Technol, BSChe, 69; NY Univ, MS, 70; Rutgers Univ, PhD(environ sci), 74. *Prof Exp:* Assoc engr, Interstate Sanitation Comn, 73-77; sr scientist, 77-81, group leader, 78-86, sr staff scientist, 81-86, PRIN SCIENTIST & SECT MGR, RES LABS, GEN MOTORS CORP, 86- *Concurrent Pos:* Prog chmn, Int Conf Carbonaceous Aerosols, 80; adj prof atmospheric & oceanic sci, Univ Mich, 84-88; consult, Clean Air Sci Adv Comt, US Environ Protection Agency, 85-87 & mem, 87- *Honors & Awards:* John Campbell Award, Gen Motors Res Labs, 84, Environmental Achievement Award, 83. *Mem:* Am Meteorol Soc; Air & Waste Mgt Asn (tech prog chmn, 85, dir, 86-89, vpres, 88-89); Sigma Xi. *Res:* Pollutant transport; chemical composition of aerosols; sources of aerosols; effect of aerosol composition on visibility; fate of air pollutants and acid precipitation; effect of pollutants on climate. *Mailing Add:* Environ Sci Dept Gen Motors Res Labs Warren MI 48090

WOLFF, GUNTHER ARTHUR, b Essen, Ger, Mar 31, 18; nat US; m 45; c 2. PHYSICAL CHEMISTRY, SOLID STATE CHEMISTRY. *Educ:* Univ Berlin, BS, 44, MS, 45; Tech Univ, Berlin, ScD(theoret inorg chem), 48. *Prof Exp:* Res assoc, Fritz-Haber Inst, Ger, 44-50, sci asst head & dep chief, 50-53; consult & sr res scientist, Signal Corps Res & Develop Labs, US Dept Army, NJ, 53-60; sr group leader mat res solid state res dept, Harshaw Chem Co, Ohio, 60-63; dir mat res, Erie Tech Prod, Inc, 63-64; prin scientist, Tyco Labs, Inc, 64-70; consult chemist, Lighting Res & Tech Serv Oper, Gen Elec Co, East Cleveland, 70-77; sr scientist, Epidyne, Inc, 77-78; sr engr, Nat Semiconductor Corp, 78-81; G A CONSULT, NPO, 81- *Concurrent Pos:* Mem comn crystal growth, Int Union Crystallog, 66-75, mem, Am Comt Crystal Growth, 67-72; chmn, Gordon Conf Chem & Metall of Semiconductors, 65; chmn, Int Union Crystallog Topical Meeting Crystal Morphol & Its Rel to Crystal Struc & Environ Conditions, 69. *Mem:* Am Asn Crystal Growth; Am Chem Soc; Electrochem Soc; fel Mineral Soc Am; fel Am Inst Chemists; AAAS; Am Crystallog Asn; Am Ceramic Soc; NY Acad Sci. *Res:* Crystal growth and dissolution, including evaporation and etching; crystal imperfections; electroluminescence and luminescence; semiconductors and ceramics; chemical bonding; solid state chemistry and physics. *Mailing Add:* 3776 N Hampton Rd Cleveland Heights OH 44121

WOLFF, HANNS H, electrical engineering, electrophysics; deceased, see previous edition for last biography

WOLFF, IVAN A, b Louisville, Ky, Feb 10, 17; m 41; c 4. ORGANIC CHEMISTRY, RESEARCH ADMINISTRATION. *Educ:* Univ Louisville, BA, 37; Univ Wis, MA, 38, PhD(org chem), 40. *Prof Exp:* Fel biochem, Univ Wis, 40-41; asst chemist northern regional res lab bur agr chem & eng, Sci & Educ Admin Agr Res, USDA, 41-42, from assoc chemist to chemist, 43-48, unit leader, 48-54, asst head cereal crops sect, 54-58, chief indust crops lab northern utilization res & develop div, Agr Res Serv, 58-69, dir, 69-80, EMER

DIR, EAST REGION RES CTR, AGR RES SERV, USDA, 81- *Concurrent Pos:* Mem subcomt natural toxicants food protection comt, Nat Acad Sci-Nat Res Coun, 70-74; consult, chem & mgt, 82- *Mem:* Am Chem Soc; Am Oil Chem Soc; Soc Econ Bot (pres, 64-65); Inst Food Technol. *Res:* Biochemistry, nutrition, processing utilization of farm commodities, and the constituents, components and derivatives from them; authored or coauthored over two hundred professional publications. *Mailing Add:* 124 Weldy Ave Oreland PA 19075

WOLFF, JAN, b Dusseldorf, Ger, Apr 25, 25; US citizen; m 55; c 2. BIOCHEMISTRY, ENDOCRINOLOGY. *Educ:* Univ Calif, BA, 45, PhD(physiol, biochem), 49; Harvard Univ, MD, 53. *Prof Exp:* Teaching asst, Univ Calif, 46 & res asst, 46-49; res asst, Harvard Univ, 54-55; from surgeon to sr surgeon, 55-63, MED DIR, NAT INSTS HEALTH, 63-, ASSOC CHIEF, CLIN ENDOCRINOL BR, 65-, CHIEF, SECT ENDOCRINE BIOCHEM, 75- *Concurrent Pos:* NSF sr fel, London & Paris, 58-59; vis prof, Univ Naples, 68 & Univ Lyon, 83. *Mem:* Am Thyroid Asn; Am Soc Biol Chemists; Endocrine Soc; Protein Soc. *Res:* Chemistry and biochemistry of tubulin, microtubule assembly, microtubule-associated proteins, and ligands that regulate polymerization such as colchicine and its analogues, vinblastine, calcium and calmodulin and fluorescent probes; properties; activation by calmodulin and host-cell penetration of adenylate cyclase from Bordetella pertussis and its role as a virulence factor; properties and function of biological membranes, particularly thyroid membranes, and the relation of adenylate cyclase to receptors and membrane organization, transport, secretory mechanisms, and tubulin-membrane interactions. *Mailing Add:* Rm 8N312 NIH Clin Ctr Bethesda MD 20892

WOLFF, JOHN B, b Ger, May 5, 25; nat US; m 50; c 1. BIOPHYSICS. *Educ:* Hunter Col, AB, 50; Johns Hopkins Univ, MA, 51, PhD(biol), 55. *Prof Exp:* NIH fel, 52-54; biochemist, Smithsonian Inst, 54-58; vis scientist & res assoc, Nat Inst Arthritis & Metab Dis, NIH, 58-60, chemist, Nat Inst Neurol Dis & Blindness, 60-62, health scientist adminr, 62-65, Div Res Grants, 65-90; RETIRED. *Honors & Awards:* Distinguished Serv Award, Biophys Soc, 91. *Mem:* Fel AAAS; Am Soc Biochem & Molecular Biol; Biophys Soc (treas, 71-78). *Res:* Microbial and plant biochemistry; enzymology. *Mailing Add:* 5609 Roosevelt St Bethesda MD 20817-6739

WOLFF, JOHN SHEARER, III, b Rochester, NY, Feb 9, 41; m 63; c 2. BIOCHEMISTRY, VIROLOGY. *Educ:* Wittenberg Univ, AB, 62; Univ Cincinnati, PhD(biochem), 66. *Prof Exp:* Asst prof biol, Fed City Col, 69-70; sr investr, 70-71, LAB HEAD MOLECULAR BIOL, JOHN L SMITH MEM FOR CANCER RES, PFIZER, INC, 71- *Concurrent Pos:* Nat Inst Allergy & Infectious Dis fel immunochem, Col Med, Univ Ill, 68-69. *Mem:* AAAS; Am Chem Soc; Am Soc Microbiol; Reticuloendothelial Soc. *Res:* Relationships between biochemical and biological activities of tumor viruses, especially RNA tumor viruses. *Mailing Add:* Licensing Dev Div 31st FL Pfizer Inc 235 E 42nd St New York NY 10017-5755

WOLFF, MANFRED ERNST, b Berlin, Ger, Feb 14, 30; nat US; div; c 3. MEDICINAL CHEMISTRY. *Educ:* Univ Calif, BS, 51, MS, 53, PhD(pharmaceut chem), 55. *Prof Exp:* Asst, Univ Calif, 52-55; res fel, Univ Va, 55-57; sr med chemist, Smith Kline & French Labs, 57-60; from asst prof to assoc prof pharmaceut chem, Univ Calif, San Francisco, 60-65, prof, 65-, chmn dept, 70-; AT ALERGAN PHARMACEUT. *Concurrent Pos:* Vis prof, Imperial Col Sci, London, 67-68; ed, Burger's Med Chem, 79-81. *Mem:* Am Chem Soc; Am Pharmaceut Asn; fel Am Acad Pharmaceut Sci; Sigma Xi. *Res:* Synthesis of potential anabolic, anti-inflammatory or anti-aldosterone hormone analogs; synthesis of aldosterone, of cardiac glycosides and aglycones; steroid chemistry and biochemistry. *Mailing Add:* 72 Suffolk Ave North Babylon NY 11703

WOLFF, MANFRED PAUL, b New York, NY, Apr 26, 38; m 88; c 2. GEOLOGY, STRATIGRAPHY-SEDIMENTOLOGY. *Educ:* Hofstra Univ, BS, 61; Univ Rochester, MS, 63; Cornell Univ, PhD(geol), 67. *Prof Exp:* Grad instr, Cornell Univ, 64-67, from asst prof to assoc prof geol, 67-74, actg chmn dept, 71-75; assoc prof, 74-81, PROF GEOL, HOFSTRA UNIV, 81- *Concurrent Pos:* Wilson P Foss fel, Cornell Univ, 63-65; Am Penrose Bequest grant, Geol Soc Am, 66; mem adv comt, Nassau-Suffolk Regional Planning Bd, 66-69; NSF grant, 69; Hofstra Univ res awards, 75-90; mem sci adv comt, NY State Eng Mgt Offices. *Mem:* Geol Soc Am; Soc Econ Paleont & Mineral; Nat Asn Geol Teachers. *Res:* Ancient and recent clastic and carbonate depositional environments; hurricane effects on beaches; coastal processes; hurricane effects on coastal stabilization and structures; sedimentology of beaches and barrier islands; nearshore shelf processes. *Mailing Add:* Dept Geol Hofstra Univ Hempstead NY 11550

WOLFF, MARIANNE, b Berlin, Ger; US citizen; m 52; c 2. PATHOLOGY, SURGICAL PATHOLOGY. *Educ:* Hunter Col, BA, 48; Columbia Univ, MD, 52. *Prof Exp:* Intern med, Presby Hosp, New York, 52-53; asst resident lab, Mt Sinai Hosp, New York, 53-54, asst resident path, St Luke's Hosp, 54-56; from instr to asst prof surg path, 56-70, assoc prof clin surg path, 70-82, ASSOC PROF SURG PATH, COL PHYSICIANS & SURGEONS, COLUMBIA UNIV, 70-, PROF CLIN SURG PATH, 82-; ASSOC SURG PATHOLOGIST, PRESBY HOSP, NEW YORK, 71- *Concurrent Pos:* Resident, Presby Hosp, New York, 56-57, asst surg pathologist, 68-71; from asst pathologist to assoc pathologist, Roosevelt Hosp, New York, 57-68; mem, Arthur Purdy Stout Soc Surg Pathologist, 68- *Mem:* Fel Am Soc Clin Path. *Mailing Add:* Dept Path Morristown Mem Hosp 100 Madison Ave Morristown NJ 07960

WOLFF, MILO MITCHELL, b Glen Ridge, NJ, Aug 9, 23; m 54; c 5. PHOTOPOLARIMETRY OF ASTEROIDS, MODELS OF FUNDAMENTAL PARTICLES. *Educ:* Upsala Col, BS, 48; Univ Pa, MS, 53, PhD(physics), 58. *Prof Exp:* Electronic engr, Philco Corp, Pa, 49-51; lectr electronics, Community Col, Temple Univ, 52-53; instr physics, Univ Pa, 58; Univ Ky-Agency Int Develop asst prof, Bandung Tech Inst, Indonesia, 58-61, assoc prof, 62; res physicist, Mass Inst Technol, 63-69; prof physics, Nanyang Univ, Singapore, 70-72; mem tech staff, Aerospace Corp, Los Angeles, 72-75; chief, Sci & Technol Sect, Econ Comn Africa-UN, Addis Ababa, 75-78; MEM STAFF, INT TECHNOL ASN, LONG BEACH, 78- *Concurrent Pos:* Asia Found vis prof physics, Vidyalankara Univ, Ceylon, 66-68; mem, US-Pakistan Sci Surv Team, US NSF, 74; mem methane gas panel, Nat Acad Sci, 74; vis prof, Nanjing Univ, China, 81; vis astronr, Paris Observ, 82. *Honors & Awards:* Apollo Navig Team Award, 69. *Mem:* Inst Elec & Electronics Engrs; Am Phys Soc. *Res:* Electronics; computer science; methods of cultural adaptation to technology in traditional societies; nuclear physics; space physics; upper atmosphere; technical development in Southeast Asia; the analysis of polarized light scattered from asteriods, moons, and particulate surfaces; numerous publications. *Mailing Add:* 1600 Nelson Ave Manhattan Beach CA 90266

WOLFF, NIKOLAUS EMANUEL, b Munich, Ger, July 7, 21; nat US; m 54; c 3. CHEMISTRY, ELECTRONICS. *Educ:* Munich Tech Univ, Cand, 48; Princeton Univ, MA, 51, PhD(chem), 52. *Prof Exp:* Asst instr, Princeton Univ, 50-52, instr chem, 52-53; res chemist, Jackson Lab & Exp Sta, E I du Pont de Nemours & Co, 53-58; mem tech staff, Labs, David Sarnoff Res Ctr, RCA Corp, 59-63, head mat processing res, 63-66; assoc lab dir, Process Res & Develop Lab, 67-68; mgr mat progs, Xerox Corp, 68-69, mgr photoreceptor technol, 69-71, mgr mat info technol group, 71-76; TECHNOL CONSULT TO MGT, 76- *Mem:* Fel AAAS; Am Chem Soc; emer Tech Asn Pulp & Paper Indust; fel Am Inst Chemists; Soc Photog Sci & Eng. *Res:* Steroid, fluorine and polymer chemistry; organometallics; electronic properties of organic materials; chemistry of recording media; electrophotography; solid state technology and integrated circuits. *Mailing Add:* PO Box 1003 Hanover NH 03755-1003

WOLFF, PETER A, b Oakland, Calif, Nov 15, 23. NONLINEAR OPTICS, DILUTED MAGNETIC SEMICONDUCTORS. *Educ:* Univ Calif, Berkeley, BS, 45, PhD(physics), 51. *Prof Exp:* Postdoctoral, Lawrence Radiation Lab, 51-52; prof physics, Univ Calif, San Diego, 65-66; mem staff, Bell Labs, 52-65, mem tech staff, 66-70; prof physics, Mass Inst Technol, 70-88, dir, Res Lab Electronics, 76-81, dir, Nat Magnetics Lab, 81-88; FEL, NEC RES INST, 89- *Mem:* Fel Am Phys Soc. *Res:* Semiconductor physics; magnetism; optics; nonlinear optics of semiconductors and diluted magnetic semiconductors. *Mailing Add:* NEC Res Inst Four Independence Way Princeton NJ 08540

WOLFF, PETER HARTWIG, b Krefeld, Ger, July 8, 26; US citizen; m 62; c 4. MEDICINE, PSYCHOBIOLOGY. *Educ:* Univ Chicago, BS, 47, MD, 50. *Prof Exp:* Asst psychiat, Harvard Med Sch, 56-59, instr, 58-61, assoc, 61-64, asst prof, 64-71; res assoc, 56-61, ASSOC, CHILDREN'S HOSP MED CTR, 61-, DIR RES, 64-; PROF PSYCHIAT, HARVARD MED SCH, 71- *Concurrent Pos:* Fel neurophysiol, Univ Chicago, 51-52; instr, Boston Psychoanal Inst, 67- *Honors & Awards:* Kirby Collier Mem Lectr, Rochester, NY, 63; Helen Sargent Prize, Menninger Found, 66; Felix & Helene Deutsch Prize, Boston Psychoanal Inst, 66; Sandor Rado Lectr, Columbia Univ, 69. *Mem:* Fel Am Psychiat Asn. *Res:* Developmental psychobiology; biological basis of behavior. *Mailing Add:* Children's Hosp Med Ctr 300 Longwood Ave Boston MA 02115

WOLFF, ROBERT JOHN, b Marquette, Mich, Jan 22, 52; m 74. ARACHNOLOGY, INVERTEBRATE ZOOLOGY. *Educ:* Hope Col, BA, 74; Western Mich Univ, MA, 76; Univ Wis-Madison, PhD(biol), 85. *Prof Exp:* ASSOC PROF BIOL & GEOG, TRINITY CHRISTIAN COL, 80- *Concurrent Pos:* Fac aquatic biol, Au Sable Inst Environ Studies, 86 & 87; mem, Educ Div, Argonne Nat Labs, 89-, res fac, 91; mem educ comn, Am Soc Zoologists, 90-; vis asst prof biol sci, Univ Ill, Chicago, 90; sci educ consult, Ill Sch Dist 161, 90-91; field assoc, Field Mus Natural Hist, 91. *Mem:* Sigma Xi; Am Arachnological Soc; Am Soc Zoologists; Am Inst Biol Sci; Soc Conserv Biol. *Res:* Biology, ecology and taxonomy of spiders and other invertebrates; conservation biology of tarantulas and fauna of natural areas. *Mailing Add:* Trinity Christian Col 6601 W College Dr Palos Heights IL 60463

WOLFF, ROBERT L, b Marion, Tex, Dec 12, 39; m 60; c 2. AGRICULTURE. *Educ:* Tex A&I Univ, BS, 66; Tex A&M Univ, MS, 68; La State Univ, Baton Rouge, PhD(agr), 71. *Prof Exp:* Asst prof agr, Tex A&I Univ, 66-72; asst prof, 72-77, PROF & DEPT CHMN AGR, SOUTHERN ILL UNIV, CARBONDALE, 77- *Mem:* Am Soc Agr Eng. *Res:* Agricultural mechanization. *Mailing Add:* Dept of Agr Educ & Mech Southern Ill Univ Carbondale IL 62901

WOLFF, ROGER GLEN, b Eureka, SDak, Sept 7, 32; m 59; c 2. WEATHERING OF ROCK. *Educ:* SDak Sch Mines, BS, 58; Univ Ill, MS, 60, PhD(geol), 61. *Prof Exp:* Hydrologist, 61-79, CHIEF RES, WATER RESOURCES DIV, US GEOL SURV, DEPT INTERIOR, 79- *Mem:* Fel Geol Soc Am; Am Geophys Union; Int Asn Hydrologists. *Res:* Weathering of rock, impact on water quality; role of confining beds on movement of water and dissolved solutes. *Mailing Add:* 11909 Hitching Post Lane Rockville MD 20852

WOLFF, RONALD GILBERT, b Lewiston, Idaho, Jan 17, 42; m; c 1. VERTEBRATE PALEONTOLOGY, VERTERBRATE MORPHOLOGY. *Educ:* Whitman Col, AB, 64; Univ Ore, MA, 66; Univ Calif, Berkeley, PhD(paleont), 71. *Prof Exp:* Fel anat, Univ Chicago, 71-72, res assoc, 72-73; asst prof, 73-78, ASSOC PROF ZOOL, UNIV FLA, 78- *Concurrent Pos:* Prin investr grant, Col Arts & Sci, Univ Fla, 77-79 & Div Environ Biol, NSF,

78-83; co-investr grant, Nat Geog Soc, 78-79. *Mem:* Soc Vert Paleont; Paleont Soc; Sigma Xi; Ecol Soc Am; Am Soc Mammalogists; Am Soc Zoologists. *Res:* Population and community ecology of mammals; environment of human evolution; Tertiary and Quaternary vertebrate communities. *Mailing Add:* Dept Zool Univ Fla Gainesville FL 32611

WOLFF, RONALD KEITH, b Brantford, Ont, July 25, 46; m 72; c 4. AEROSOLS IN BIOMEDICINE, MUCOCILIARY CLEARANCE. *Educ:* Univ Toronto, BSc, 68, MSc, 69, PhD(med biophysics), 72, Dipl, Am Bd Toxicol, 83. *Prof Exp:* Res fel respiratory physiol, McMaster Univ, 73-76; sr scientist, Lovelace Inhalation Toxicol Res Inst, 76-88; RES SCIENTIST, LILLY RES LABS, 88- *Mem:* Soc Toxicol; Am Indust Hyg Asn; Am Thoracic Soc; Am Physiol Soc; Am Asn Physicists in Med; Am Asn Aerosol Res. *Res:* Inhalation toxicology; deposition and clearance of inhaled particles from the lung including effects of toxic agents, inhalation carcinogenesis; drug delivery. *Mailing Add:* Lilly Res Labs PO Box 708 Greenfield IN 46140

WOLFF, SHELDON, b Peabody, Mass, Sept 22, 28; m 54; c 3. CYTOGENETICS, RADIOBIOLOGY. *Educ:* Tufts Col, BS, 50; Harvard Univ, MA, 51, PhD(biol), 53. *Prof Exp:* Biologist, Oak Ridge Nat Lab, 53-66, mem sr res staff, 65-66; PROF CYTOGENETICS, UNIV CALIF, SAN FRANCISCO, 66-, DIR, LAB RADIOBIOL & ENVIRON HEALTH, 83- *Concurrent Pos:* Mem subcomt radiobiol, Nat Acad Sci-Nat Res Coun, 61-77, space sci bd, 74-78, Comt nuclear sci, 73-75; vis prof, Univ Tenn, 62; mem comt 15 environ biol & chmn panel radiation biol of comt 15, Space Sci Bd, Nat Acad Sci, 62, mem comt postdoctoral fels div biol & agr, 62-65; consult spec facil prog, NSF, 62-64; mem exec comt, Nat Acad Sci-Nat Res Coun Space Biol Summer Study, 68, mem exec comt priorities study for NASA, 70, mem subcomt genetics effects adv comt to Environ Protection Agency, Div Med Sci, 70-; mem safe drinking water comt, Nat Acad Sci, 76-81; prog chmn, XIII Int Cong Genetics; mem comt federal res ionizing radiation & mem comt chem environ mutagens, Nat Res Coun-Nat Acad Sci, 80-81; mem, Mammalian Genetics study Sect, NIH, 80-83; chmn, Health & Environ Res Adv Comt, US Dept Educ, 87- *Honors & Awards:* E O Lawrence Award, US AEC, 73; Environ Mutagen Soc Award,82. *Mem:* Radiation Res Soc; Bot Soc Am; Am Soc Cell Biol; Sigma Xi; Environ Mutagen Soc (pres, 80). *Res:* Chromosome structure; radiation genetics and cytology; chromosome structure; genetics and cytology. *Mailing Add:* Lab Radiobiol Univ Calif San Francisco CA 94143-0750

WOLFF, SHELDON MALCOLM, b Newark, NJ, Aug 19, 30; m 56; c 3. INFECTIOUS DISEASES, IMMUNOLOGY. *Educ:* Univ Ga, BS, 52; Vanderbilt Univ, MD, 57; Am Bd Internal Med, dipl. *Hon Degrees:* Dr, Fed Univ Rio de Janeiro, Brazil, 75. *Prof Exp:* Intern med, Sch Med, Vanderbilt Univ, 57-58, asst resident, 58-59; sr resident, Bronx Munic Hosp Ctr & Albert Einstein Col Med, 59-60; clin assoc, Nat Inst Allergy & Infectious Dis, 60-62, clin investr, 62-63, sr investr, 63-65, head physiol sect, Lab Clin Invest, 64-74, clin dir & chief lab clin invest, 68-77; ENDICOTT PROF MED & CHMN DEPT, SCH MED, TUFTS UNIV, 77-; PHYSICIAN-IN-CHIEF, NEW ENG MED CTR HOSP, 77- *Concurrent Pos:* Res asst, Sch Med, Vanderbilt Univ, 56-59; lectr, Sch Med, Georgetown Univ, 62-69, clin prof, 69-77; consult infectious dis, Nat Naval Med Ctr, Md, 70- *Honors & Awards:* Super Serv Award, NIH, 71; Squibb Award, 76; Phillips Award, Am Col Physicians, 87. *Mem:* Inst Med-Nat Acad Sci; Asn Am Physicians; master Am Col Physicians; Am Acad Arts & Sci; Am Soc Clin Investigation. *Res:* Mechanisms of host responses. *Mailing Add:* Dept Med New England Med Ctr, 750 Washington St Boston MA 02111

WOLFF, SIDNEY CARNE, b Sioux City, Iowa, June 6, 41; m 62. ASTROPHYSICS. *Educ:* Carleton Col, BA, 62; Univ Calif, Berkeley, PhD(astron), 66. *Hon Degrees:* DSc, Carleton Col, 85. *Prof Exp:* Res astronr, Lick Observ, Univ Calif, Santa Cruz, 67; from asst astronr to assoc astronr, Inst Astron, Univ Hawaii, 67-75, astronr, 75-, assoc dir, 76-; dir, Kitt Peak Nat Observ, 84-87, DIR NAT OPTICAL ASTRON OBSERV, TUCSON, AZ, 87- *Mem:* Am Astron Soc; Int Astron Union; Astron Soc Pac. *Res:* Stellar spectroscopy; photoelectric photometry; magnetic stars. *Mailing Add:* Nat Optical Astron Observ PO Box 26732 Tucson AZ 85726

WOLFF, STEVEN, b New York, NY, Apr 15, 43; m 84; c 1. ORGANIC CHEMISTRY. *Educ:* Williams Col, BA, 65; Yale Univ, PhD(org chem), 70. *Prof Exp:* Fel org chem, Squibb Inst Med Res, 70-71; res assoc, Rockefeller Univ, 71-73, from asst prof to assoc prof org chem, 71-87; SR SCIENTIST, CHEM RES, HOFFMANN-LA ROCHE UNIV, 87- *Mem:* Am Chem Soc; Royal Soc Chem; Int Am Photochem Soc; Sigma Xi. *Res:* Mechanistic organic photochemistry; organic synthesis. *Mailing Add:* Hoffmann-La Roche Inc 340 Kingsland St Nutley NJ 07110

WOLFF, THEODORE ALBERT, b Philadelphia, Pa, Feb 24, 43; m 62. ENTOMOLOGY, MEDICAL PARASITOLOGY. *Educ:* NMex Highlands Univ, BS, 65; Univ NC, Chapel Hill, MSPH, 69; Univ Utah, PhD(biol), 76. *Prof Exp:* Vol biol, Peace Corps, Malaysia, 65-68; environ scientist entom, NMex Environ Improv Agency, 69-72; teaching fel biol, Univ Utah, 72-74; environ scientist, NMex Environ Improv Agency, 74-76, dir radiation protection prog, 76-81; TECH STAFF MEM, SANDIA NAT LABS, 81- *Mem:* Assoc Sigma Xi; Am Mosquito Control Asn; Am Pub Health Asn; Health Physics Soc. *Res:* Systematic studies of mountain Aedes mosquitoes; subgenus Ochlerotatus of Arizona and New Mexico; the transportation of radioactive materials. *Mailing Add:* Sandia Nat Labs Orgn 3314 Bldg MO-194 Rm 3 Albuquerque NM 87185

WOLFF, WILLIAM FRANCIS, b Newark, NJ, June 17, 21; m 48; c 4. ORGANIC CHEMISTRY. *Educ:* Yale Univ, BS, 47. *Prof Exp:* Res chemist, Standard Oil Co, Ind, 47-53 & Pa Salt Mfg Co, 53-54; res chemist, 54-58, sr proj chemist, 58-61, sr res scientist, 61-82, INDEPENDENT RES, STANDARD OIL CO, IND, 82- *Mem:* AAAS; Am Chem Soc. *Res:* Hydrocarbon polymers; organic sulfur and chlorine compounds; carbons and aromatic complexes; heterogeneous catalysis; electromagnetism. *Mailing Add:* 205 Rich Road Park Forest IL 60466-1611

WOLFFE, ALAN PAUL, b Burton-on-Trent, Staffordshire, UK, June 21, 59; m 82. PROTEIN-NUCLEIC ACID INTERACTIONS, GENE REGULATION. *Educ:* Oxford Univ, BA, 81; Med Res Coun, London, PhD(molecular biol), 84. *Prof Exp:* Postdoctoral molecular biol, Carnegie Inst, Baltimore, Md, 84-86, prin investr, 87; prin investr, Lab Molecular Biol, Nat Inst Diabetes Digestive & Kidney Dis, 87-90, CHIEF, LAB MOLECULAR EMBRYOL, NAT INST CHILD HEALTH & HUMAN DEVELOP, NIH, BETHESDA, MD, 90- *Mem:* Am Soc Microbiol; AAAS; Am Asn Cancer Res; Am Soc Cell Biol. *Res:* Molecular mechanisms responsible for establishing and maintaining stable states of gene activity; integration of replication, transcription and chromatin assembly especially during early vertebrate development; germ-cell specific gene expression. *Mailing Add:* Lab Molecular Embryol Nat Inst NIH Bldg 6 Rm 131 Bethesda MD 20892

WOLFGRAM, FREDERICK JOHN, neurochemistry, for more information see previous edition

WOLFHAGEN, JAMES LANGDON, organic chemistry, for more information see previous edition

WOLFHARD, HANS GEORG, b Basel, Switz, Apr 2, 12; nat US; m 40; c 3. ATOMIC & MOLECULAR PHYSICS. *Educ:* Univ Goettingen, Dr Rer Nat(physics), 38. *Prof Exp:* Scientist, Aeronaut Res Sta, Brunswick, Ger, 39-46; res scientist, Imp Col, London, Royal Aircraft Estab, Eng, 46-56 & Bur Mines, 56-59; head, Physics Dept, Reaction Motors Div, Thiokol Chem Corp, Denville, NJ, 59-63; SR RES STAFF, INST DEFENSE ANALYSIS, ALEXANDRIA, VA, 63- *Mem:* Fel Am Optical Soc; Am Inst Aeronaut & Astronaut; Combustion Inst. *Res:* Combustion research; ballistic missile research. *Mailing Add:* Inst Defense Anal 1801 N Beauregard St Alexandria VA 22311

WOLFLE, DAEL (LEE), b Puyallup, Wash, Mar 5, 06; m 29; c 3. SCIENCE POLICY. *Educ:* Univ Wash, BS, 27, MS, 28; Ohio State Univ, PhD(psychol), 31. *Hon Degrees:* DSc, Drexel Inst Technol, 56, Ohio State Univ, 57 & Western Mich Univ, 60. *Prof Exp:* Instr psychol, Ohio State Univ, 29-32; prof, Univ Miss, 32-36; examr biol sci, Univ Chicago, 36-39, from asst prof to assoc prof psychol, 38-45; exec secy, Am Psychol Asn, 46-50; dir, Comn Human Resources & Advan Training, Assoc Res Couns, 50-54; exec officer, AAAS, 54-70; actg dean archit & urban planning, 72-73, prof pub affairs, 70-76, EMER PROF PUB AFFAIRS, GRAD SCH PUB AFFAIRS, UNIV WASH, 76- *Concurrent Pos:* Civilian training adminr electronics, US Army Sig Corps, 41-43; tech aide, Off Sci Res & Develop, 44-46; mem or vchmn, Bd Trustees, Russell Sage Found, 61-78; mem or chmn, Bd Trustees, James McKeen Cattell Fund, 62-82; trustee, Pac Sci Ctr Found, 62-80; mem ed adv bd, Sci Yearbk, Encycl Brittanica, 67-77; mem res adv comt, Am Coun Educ, 68-73; trustee, Biosci Info Servs, 68-74; mem or chmn, Geophys Inst Adv Comt, Univ Alaska, 69-; mem manpower inst, Nat Indust Conf Bd, 70; chmn rev comt sci resources studies, NSF, 72-73 & 82-83; mem comt grad med educ, Asn Am Med Cols, 72-75; mem, US-USSR Joint Group Experts Sci Policy, 73-82; mem comn human resources, Nat Acad Sci-Nat Res Coun, 74-78; mem bd trustees, Biol Sci Curriculum Study, 80-85; mem coun, AAAS, 49-51, 87-88 & 90-, exec officer, 54-70, pres, Pac Div, 92. *Honors & Awards:* Montgomery Lectr, Univ Nebr, 59; Walter Van Dyke Bingham Lectr, Columbia Univ, 60; Herbert S Langfeld Lectr, Princeton Univ, 69. *Mem:* AAAS; Am Psychol Asn (exec secy 46-50); Am Coun Educ (secy, 66-67). *Res:* Education, utilization, mobility, supply and demand trends of scientific and specialized personnel. *Mailing Add:* Grad Sch Pub Affairs Univ Wash Seattle WA 98195

WOLFLE, THOMAS LEE, b Eugene, Ore, Apr 24, 36; m 86; c 2. LABORATORY ANIMAL SCIENCE, ANIMAL BEHAVIOR. *Educ:* Tex A&M Univ, BS, 59, DVM, 61; Univ Calif, Los Angeles, MA, 67, PhD(physiol psychol), 70; Am Col Lab Animal Med, dipl, 66. *Prof Exp:* Vet primate colony mgt, Sch Aerospace Med, Brooks AFB, Tex, 61-65, chief comp toxicol lab, Aeromed Res Lab, Holloman AFB, 65-66; chief flight environ br, Aerospace Med Res Labs, Wright-Patterson AFB, Ohio, 70-73, asst chief weapons effects br, Sch Aerospace Med, Brooks AFB, Tex, 73-75; vet dir, USPHS, 75-88; DIR INST LAB ANIMAL RESOURCES, NAT ACAD SCI, 88- *Concurrent Pos:* Adj prof psychol, Wright State Univ, 71-73, consult, Lab Animal Med, 72-73; exec dir, Interagency Res Animal Comt, NIH, 82-88; adj prof vet med, Univ Md, 84- *Honors & Awards:* Commendation Medal, USPHS, 83, Outstanding Serv Medal, 88. *Mem:* Am Vet Med Asn; Am Vet Soc Animal Behav; Am Asn Lab Animal Sci; Am Soc Primatology; Int Primatological Soc; Asn Primate Vets; Am Soc Lab Animal Practitioners. *Res:* Animal behavior, canine medicine; impact on research and animals of early non-specific environmental influences; identification of animal models of human disease; biological and behavioral effects of housing on laboratory animals; pain and stress in animals. *Mailing Add:* NAS 2101 Constitution Ave NW Washington DC 20418

WOLFMAN, EARL FRANK, JR, b Buffalo, NY, Sept 14, 26; m 46; c 3. SURGERY. *Educ:* Harvard Univ, BS, 46; Univ Mich, MD, 50; Am Bd Surg, dipl, 58. *Prof Exp:* From instr to assoc prof surg, Med Sch, Univ Mich, 57-66, asst to dean, 60-61, asst dean, 61-64; chmn div & dept & assoc dean, 66-78, PROF SURG, SCH MED, UNIV CALIF, DAVIS, 66- *Concurrent Pos:* Consult, Vet Admin Hosps, Travis AFB & Martinez, Calif, 66 -; chief div surg serv, Sacramento Med Ctr, 66-78. *Mem:* Fel Am Col Surgeons; AMA; Soc Surg Alimentary Tract; Asn Acad Surg; Sigma Xi. *Mailing Add:* Dept Surg Univ Calif Sch Med Prof Bldg 4301 X St Sacramento CA 95817

WOLFNER, MARIANA FEDERICA, b Caracas, Venezuela, Dec 30, 53; US citizen; m 85. DROSOPHILA DEVELOPMENT, GENE REGULATION. *Educ:* Cornell Univ, BA, 74; Stanford Univ, PhD(biochem), 81. *Prof Exp:* Damon Runyon-Walter Winchell Cancer Fund fel, dept biol, Univ Calif, San Diego, 81-82; sr fel, Am Cancer Soc, 83; ASST PROF MOLECULAR DEVELOP, SECT GENETICS & DEVELOP, CORNELL UNIV, 83- *Res:* Determining the molecular basis for the regulation of gene expression during development; isolation of Drosophila DNA sequences which are expressed sex-specifically and examination of their DNA structure, their tissue-specific and developmental expression and the ways in which this expression is affected by major developmental regulatory genes. *Mailing Add:* Dept Genetics Cornell Univ Main Campus Ithaca NY 14853

WOLFORD, JACK ARLINGTON, b Brookville, Pa, Dec 5, 17; m 44; c 2. PSYCHIATRY. *Educ:* Allegheny Col, AB, 40; Univ Pa, MD, 43; Am Bd Psychiat, dipl, 53. *Prof Exp:* Intern med, Allegheny Gen Hosp, Pittsburgh, Pa, 43; resident psychiat, Warren State Hosp, 44-46, sr psychiatrist, 48-51, clin dir, 51-56; dir, Hastings State Hosp, Nebr, 56-58; asst prof psychiat, Sch Med, 58-69, chief social psychiat, Western Psychiat Inst & Clin, 58-72, dir community ment health, Retardation Ctr, 67-74, psychiatrist-in-chief & actg chmn dept psychiat, Sch Med, 72-73, PROF PSYCHIAT, SCH MED, UNIV PITTSBURGH, 69-, DIR ADULT SERV, WESTERN PSYCHIAT INST & CLIN, 72-, DIR COMMUNITY MENT HEALTH/MENT RETARDATION CTRS, 76- *Concurrent Pos:* Resident psychiat, Psychiat Inst & Clin, Pittsburgh, 53; asst prof psychiat, Univ Nebr, 56-58; vis fac sem, Lab Community Psychiat, Harvard Univ; mem, Governor's Adv Comt to Dept Welfare & Comn Ment Health; mem commun adv comt, NIMH; mem med adv comt, Gov Adv Comt Ment Health & Ment Retardation, Dept Pub Welfare & Comn Ment Health, 74 & 75, Comprehensive Comt Ment Health Planning, 75 & Pa Asn Community Ment Health & Ment Retardation Ctrs, 75; pres, Group for Advan Psychiat, 77-79; secy, Pa Asn Community Ment Health & Ment Retardation Providers, 80- *Mem:* Fel Am Psychiat Asn (vpres, 75); fel Am Col Psychiat; AMA. *Res:* Social and community psychiatry; urban mental health and illness. *Mailing Add:* 3811 O'Hara St Pittsburgh PA 15213

WOLFORD, JAMES C, b Fairmont, Nebr, July 20, 20; m 50; c 3. MECHANICS, MECHANICAL ENGINEERING. *Educ:* Univ Nebr, BS, 47, MS, 52; Purdue Univ, PhD(kinematics), 56. *Prof Exp:* Test engr, Gen Elec Co, 47-48, design engr, 48-50; design engr, Cecil W Armstrong & Assocs, 50-51; from instr to asst prof mech, 54-58, assoc prof kinematics & mach design, 58-63, PROF KINEMATICS & MACH DESIGN, UNIV NEBR, LINCOLN, 63- *Mem:* Am Soc Mech Engrs; Am Soc Eng Educ; Sigma Xi. *Res:* Design of machine elements; kinematics of mechanisms. *Mailing Add:* Dept of Mech Eng Univ of Nebr Lincoln NE 68588

WOLFORD, JOHN HENRY, b Osgood, Ind, June 11, 36. AVIAN PHYSIOLOGY, POULTRY SCIENCE. *Educ:* Purdue Univ, BS, 58; Mich State Univ, MS, 60, PhD(avian physiol), 63. *Prof Exp:* From asst prof to assoc prof poultry sci, Mich State Univ, 63-74; prof animal & vet sci & chmn dept, Univ Maine, Orono, 74-80; PROF POULTRY SCI & DEPT HEAD, VA POLYTECH INST & STATE UNIV, 80- *Mem:* Poultry Sci Asn; Sigma Xi; World Poultry Sci Asn. *Res:* Reproductive physiology of the turkey breeder hen; physiological alterations of fatty liver syndrome in laying chickens. *Mailing Add:* Dept Poultry Sci Va Polytech Inst & State Univ Blacksburg VA 24061

WOLFORD, RICHARD KENNETH, b Pa, Jan 3, 32. CHEMISTRY. *Educ:* WVa Wesleyan Col, BS, 53; Univ Ky, MS, 55, PhD(chem), 59. *Prof Exp:* Asst chem, Univ Ky, 53-58; chemist, Nat Bur Standards, 58-66; PHYS SCI ADMINSTR, NAT OCEAN SURV, NAT OCEANIC & ATMOSPHERIC ADMIN, 66- *Mem:* AAAS; Am Chem Soc; Sigma Xi. *Res:* Oceanography. *Mailing Add:* N/CGx11 NOAA NOS 6001 Executive Blvd Rm 1016 Rockville MD 20852

WOLFRAM, LESZEK JANUARY, b Krakow, Poland, Feb 24, 29; US citizen; m 53; c 2. FIBER SCIENCE, PROTEIN CHEMISTRY. *Educ:* Politechnika, Lodz, Poland, BSc, 53, MSc, 55; Univ Leeds, PhD(protein chem), 61. *Prof Exp:* Chemist, Gillette Res Lab, Reading, UK, 61-63, sr chemist, ground leader & prin scientist, Gillette Res Inst, Rockville, Md, 63-72; sci liaison officer, Int Wool Secretariat, Australia, 72-73; asst dir res, Personal Care Div, The Gillette Co, Boston, 74-77; dir res, 77-79, VPRES RES, CLAIROL RES LABS, 79- *Concurrent Pos:* Assoc ed, J Soc Cosmetic Chemists, 78-79, ed, 79-83. *Honors & Awards:* Medal Award, Soc Cosmetic Chemists. *Mem:* Am Chem Soc; NY Acad Sci; Fiber Soc; Soc Cosmetic Chemists; AAAS. *Res:* Physical chemistry of synthetic polymers and proteins; structure of fibers and biological tissues; evaluative techniques for hair and skin; organ biosurfaces and their properties; structure and properties of melanius. *Mailing Add:* Clairol Res Labs Two Blachley Rd Stamford CT 06902

WOLFRAM, THOMAS, b St Louis, Mo, July 27, 36; m; c 5. SOLID STATE PHYSICS. *Educ:* Univ Calif, Riverside, AB, 59, PhD(physics), 63; Univ Calif, Los Angeles, MA, 60. *Prof Exp:* Mem tech staff, Atomics Int, 60-63; mem tech staff, NAm Rockwell Sci Ctr, 63-68, group leader solid state physics, 68-72, dir physics & chem, Rockwell Int Sci Ctr, 72-74; prof, 74-83, CHMN DEPT PHYSICS, UNIV MO, COLUMBIA, 83-; VPRES & GEN MGR, AMOCO LASER CO, 87- *Concurrent Pos:* Adj prof physics, Univ Calif, Riverside, 68-69; dir, PHYS Technol Div, Amoco corp, Napreville, Ill, 83-86. *Mem:* Fel Am Phys Soc. *Res:* Lattice dynamics; spin waves; superconductivity; electronic and optical properties; physics and chemistry of surfaces; catalysis. *Mailing Add:* Amoco Laser Co 1251 Frontenac Rd Naperville IL 60540

WOLFROM, GLEN WALLACE, b Freeport, Ill, Apr 8, 47; m 69; c 2. ANIMAL NUTRITION, BIOSTATISTICS. *Educ:* Western Ill Univ, BS, 69; Southern Ill Univ, MS, 72; Univ Mo-Columbia, PhD(animal nutrit), 76. *Prof Exp:* Mgr nutrit & chem, Contech Lab-Pet Inc, 76-78; RESEARCHER NUTRIT & PROJ LEADER, PITMAN-MOORE INC, 78- *Mem:* AAAS; Am Soc Animal Sci. *Res:* Animal growth and development; ruminant nutrition. *Mailing Add:* Physiol Res PO Box 207 Pitman-Moore Inc Terre Haute IN 47808-0207

WOLFSBERG, KURT, b Hamburg, Ger, Nov 1, 31; nat US; m 55; c 3. RADIOCHEMISTRY, NUCLEAR CHEMISTRY. *Educ:* St Louis Univ, BS, 53; Wash Univ, St Louis, MA, 55, PhD(chem), 59. *Prof Exp:* Assoc group leader, Isotope Geochem Group, 80-88, STAFF MEM, RADIOCHEM GROUP, LOS ALAMOS NAT LAB, 59-, SECT LEADER, ISOTOPE GEOCHEM GROUP, 88- *Concurrent Pos:* Consult aircraft nuclear propulsion comt nuclear measurements & standards, US Air Force, 56-57; mem subcomt radiochem, Nat Acad Sci, 72-76; guest & Fulbright grantee, Univ Mainz, WGer, 74-75, guest scientist, 82, 85; Fulbright award, tech proj officer, Nev Terminal Waste Storage Proj, 78-79. *Mem:* AAAS; Am Chem Soc; fel Am Inst Chemists; Mat Res Soc. *Res:* Emanation techniques; mass and charge distribution in fission; high temperature diffusion of fission products; lanthanide and actinide chemistry; properties of very heavy nuclides; nuclear waste management, sorptive properties of geologic media, migration of radionuclides; geochemistry; solar neutrinos; radiochemical separation. *Mailing Add:* 303 Venado Los Alamos NM 87544

WOLFSBERG, MAX, b Hamburg, Germany, May 28, 28; nat US; m 57; c 1. ISOTOPE EFFECTS, CHEMICAL PHYSICS. *Educ:* Wash Univ, St Louis, AB, 48, PhD(chem), 51. *Prof Exp:* Asst chem, Wash Univ, St Louis, 48-50; assoc chemist, Brookhaven Nat Lab, 51-54, from chemist to sr chemist, 54-69; Regents' lectr, 68, chemn dept, 74-80, PROF CHEM, UNIV CALIF, IRVINE, 69- *Concurrent Pos:* NSF sr fel, 58-59; vis prof chem, Cornell Univ, 63 & Ind Univ, 65; prof, State Univ NY Stony Brook, 66-69; Alexander von Humboldt award, 77; guest prof, Deutsche Forschungspemeinschaft, Univ Ulm, Fed Repub Ger, 86. *Mem:* Am Chem Soc. *Res:* Theoretical chemistry; isotope effects, energy transfer, chemical reactions, electronic structure of molecules and molecular dynamics. *Mailing Add:* Dept of Chem Univ of Calif Irvine CA 92717

WOLFSON, ALFRED M, b New York, NY, May 23, 99; m 26, 58; c 2. PLANT MORPHOLOGY, EVOLUTION. *Educ:* Cornell Univ, BS, 21; Univ Wis, MA, 22, PhD(bot), 24. *Prof Exp:* Res fel, Belgian-Am Educ Found, 24-26 & US Nat Res Coun, Cornell Univ, 26-27; res asst, United Fruit Co, 27-30; chmn dept, 30-69, EMER PROF BIOL, MURRAY STATE UNIV, KY, 69- *Mem:* Fel AAAS; Sigma Xi; Nat Geog Soc; Am Inst Biol; Audubon Soc. *Res:* Cytology and genetics of Spherocarpos Donnellii; spore germination in Pellia. *Mailing Add:* 310 N 14th St Murray KY 42071

WOLFSON, BERNARD T, b Chicago, Ill, Mar 16, 19; m 42; c 2. ENERGY CONVERSION & UTILIZATION, BUSINESS OF PROGRAM DEVELOPMENT & MARKETING. *Educ:* Ill Inst Technol, BS, 40; Ohio State Univ, MS, 50, PhD, 60. *Prof Exp:* Chem & metall engr, Process Control Dept, Carnegie-Ill Steel Corp, Ind, 40-41; chem engr, Eng & Oper Div, Kankakee Ord Works, Ill, 41-42; tech supvr prod & inspection, Atlas Imp Diesel Engine Co, 42; rotary wing aircraft develop engr, Mech Br, Rotary Wing Unit, Propeller Lab, Wright Patterson AFB, Ohio, 42-44; aeronaut flight test engr, Flight Test Div, Ames Res Ctr, Moffett Field, Calif, 45-46; asst chief rotary wing aerodyn develop eng, Aerodyn Br, Rotary Wing Unit, Aircraft Lab, Wright Patterson AFB, 46-51, sr propulsion res scientist, Fluid Dynamics Res Br, Aerospace Res Lab, 51-61; aerospace propulsion scientist, Propulsion Div, Eng Sci Directorate, Air Force Off Sci Res, 61-66, 67-83, actg dir div, 66-67; TECH & MGT CONSULT, 85- *Concurrent Pos:* Mem panels & steering comts, Liquid Rocket Combustion Instability & Solid Rocket Combustion, Interagency Chem Rocket Propellants Group, Dept Defense Advan Res Projs Agency-NASA, 62-; adv panel air-breathing propulsion & power plants, NASA, 65-; ad hoc comt SRAMJET, Supersonic Combustion & Appln, Air Force Systs Command, 65-; Joint Army-Navy-Air Force-NASA Panels Liquid & Solid Rockets & Air-Breathing Combustion; magnetohydrodyn panel & working groups, Interagency Advan Power Group; mem steering group, Joint Army Navy NASA Air Force Interagency Propulsion Comt, 79- *Mem:* Combustion Inst; Am Inst Aeronaut & Astronaut; Sigma Xi. *Res:* Fuel-air deflagrations and explosions; alternative fuels; high speed reacting flow characterization; non-interference instrumentation and diagnostics; plasma propulsion and electric power generation; air-breathing propulsion; gas dynamic, electric, explosive-driven and magnetohydrodynamic lasers; plasma dynamics; combustion instability. *Mailing Add:* 4797 Lake Valencia Blvd W Palm Harbor FL 34684-3924

WOLFSON, EDWARD A, preventive medicine, medical education; deceased, see previous edition for last biography

WOLFSON, JAMES, b Chicago, Ill, Mar 16, 43; m 71. PHYSICS. *Educ:* Grinnell Col, BA, 64; Mass Inst Technol, PhD(physics), 68. *Prof Exp:* Mem staff physics lab nuclear sci, Mass Inst Technol, 68-70, asst prof physics, 70-76; mem staff, Fermi Nat Accelerator Lab, 76-80. *Mem:* AAAS; Am Phys Soc. *Res:* Experimental high energy physics. *Mailing Add:* 694 Sterling Court Naperville IL 60540

WOLFSON, JOSEPH LAURENCE, b Winnipeg, Man, July 22, 17; m 44; c 2. NUCLEAR PHYSICS. *Educ:* Univ Man, BSc, 42, MSc, 43; McGill Univ, PhD(physics), 48. *Prof Exp:* Asst res officer physics, Atomic Energy Can, Ltd, Chalk River, 48-55; physicist, Jewish Gen Hosp, Montreal, Que, 55-58; assoc res officer, Nat Res Coun Can, 58-64; prof physics, Univ Sask, 64-74; dean sci, 74-80, prof physics, 74-82, ADJ PROF, CARLETON UNIV, OTTAWA, 83- *Mem:* Am Phys Soc; Can Asn Physicists. *Res:* Nuclear spectroscopy. *Mailing Add:* 951 Blythdale Rd Ottawa ON K2A 3N9 Can

WOLFSON, KENNETH GRAHAM, b New York, NY, Nov 21, 24; m 47; c 2. MATHEMATICS. *Educ:* Brooklyn Col, BA, 47; Johns Hopkins Univ, MA, 48; Univ Ill, PhD(math), 52. *Prof Exp:* From instr to assoc prof, 52-58, chmn dept, 61-75, PROF MATH, RUTGERS UNIV, 60-, DEAN, GRAD SCH, 75- *Mem:* Am Math Soc; Math Asn Am. *Res:* Spectral theory of differential equations; rings of linear transformations; structure of rings. *Mailing Add:* Dept Math Rutgers Univ New Brunswick NJ 08903

WOLFSON, LEONARD LOUIS, b Wilkes Barre, Pa, Dec 13, 19; m 47; c 1. QUALITY CONTROL, STERILIZATION. *Educ:* Univ Chicago, BS, 50, MS, 51. *Prof Exp:* Microbiologist, Wilson & Co, Chicago, 54-57; group leader microbiol, Nalco Chem Co, Chicago, 57-70; corp dir qual assurance & regulation affairs, Will Ross Inc, Milwaukee, 70-78 & Ipco Corp, White Plains, NY, 78-88; CONSULT, LEONARD L WOLFSON, 88- *Concurrent Pos:* Pres, Wilro Sci Labs, 72-78. *Mem:* NY Acad Sci; Am Soc Microbiol; Soc Indust Microbiol. *Res:* Medical and dental devices; quality control; drugs; cosmetics; disinfectants; product development; sterilization. *Mailing Add:* 36 Maple Wood Rd Hartsdale NY 10530

WOLFSON, RICHARD L T, b San Francisco, Calif, Apr 13, 47; m 70; c 2. SOLAR PHYSICS, SOLAR ENERGY. *Educ:* Swarthmore Col, BA, 69; Univ Mich, MS, 71; Dartmouth Col, PhD(physics), 76. *Prof Exp:* From asst prof to assoc prof, 76-87, PROF, MIDDLEBURY COL, 87-, CHMN, DEPT PHYSICS, 88- *Concurrent Pos:* Vis scientist, High Altitude Observ, Nat Ctr Atmospheric Res, 80-81; prin investr, NSF, Dept Energy, NASA; auth. *Mem:* Am Phys Soc; Am Asn Physics Teachers. *Res:* Theoretical work on magnetohydrodynamics of space plasmas, especially in application to the solar corona and solar wind; experimental work on control strategies for solar energy systems. *Mailing Add:* Dept Physics Middlebury Col Middlebury VT 05753

WOLFSON, ROBERT JOSEPH, b Philadelphia, Pa, Sept 11, 29; m 53; c 2. OTOLARYNGOLOGY. *Educ:* Temple Univ, BA, 52, MS, 61; Hahnemann Med Col, MD, 57. *Prof Exp:* PROF OTOLARYNGOL & BRONCHO-ESOPHAGOLOGY, & HEAD DEPT, MED COL PA, 69- *Mem:* Am Acad Ophthal & Otolaryngol; Am Otol Soc; Am Laryngol, Rhinol & Otol Soc; Royal Soc Med; AMA. *Mailing Add:* Dept Otolaryngol Hahnemann Univ Hosp 1920 Chestnut St Suite 700 Philadelphia PA 19103

WOLFSON, SEYMOUR J, b Detroit, Mich, Feb 13, 37; m 58; c 4. COMPUTER SCIENCE. *Educ:* Wayne State Univ, BS, 59, PhD(physics), 65; Univ Chicago, MS, 60. *Prof Exp:* Res assoc physics, Wayne State Univ, 63-65; sr scientist, Comput Sci Corp, 65-68; asst prof, 68-73, ASSOC PROF COMPUT SCI, WAYNE STATE UNIV, 73- *Concurrent Pos:* Consult, Comput Sci Corp, 68-71, Lincorp Corp, 72- & Dept Housing & Urban Develop, 75-; secy, Comput Sci Bd, 74-76 & New York Carpet World, 80; chmn, Nat Comput Conf Bd, 81-82. *Mem:* Am Phys Soc; Asn Comput Mach; Am Arbit Asn; Inst Elec & Electronics Engrs. *Res:* Computer networks and applications; numerical methods. *Mailing Add:* 18803 Hilton Southfield MI 48075

WOLFSON, SIDNEY KENNETH, JR, b Philadelphia, Pa, June 14, 31; m 58; c 3. NEUROSURGERY, BIOMEDICAL ENGINEERING. *Educ:* Univ Pa, AB, 51; Univ Chicago, MD, 58. *Prof Exp:* Resident surg, Univ Pa Hosp, 59-63, asst prof surg res, Sch Med, Univ Pa, 63-68; assoc prof surg, Univ Chicago, 68-71; dir surg res, Michael Reese Hosp & Med Ctr, 68-71; assoc prof neurosurg, 71-77, dir surg res, 74-78, PROF NEUROSURG, UNIV PITTSBURGH, 78-, DIR SURG RES, MONTEFIORE HOSP, 71- *Concurrent Pos:* Career Develop Award, Nat Heart & Lung Inst, 63; chmn spec study sect, Nat Inst Arthritis & Metab Dis, 74-75; consult, Artificial Heart Assessment Panel, NIH, 74, Med Devices Prog, Nat Heart & Lung Inst, 73-74; Peripheral Vascular Diag Lab, 77- *Mem:* Am Soc Artificial Internal Organs; Am Asn Neurol Surgeons; Soc Acad Surgeons; Soc Neurosci; Am Soc Hypertension; Inst Elec & Electronics Engrs. *Res:* Hypothermia and circulatory arrest; artificial pancreas; cerebral blood flow; computerized patient and data management systems; implantable glucose electrode; epidemiological studies of cerebral ischemia and stroke; relationship of ultrasound imaging and doppler blood flow to risk factors in cardiovascular disease; experimental model of neurogenic hypertension. *Mailing Add:* 205 Buckingham Rd Pittsburgh PA 15215

WOLGA, GEORGE JACOB, b New York, NY, Apr 2, 31; div; c 3. PHYSICS. *Educ:* Cornell Univ, BEngPhys, 53; Mass Inst Technol, PhD(physics), 57. *Prof Exp:* Asst physics, Mass Inst Technol, 53-56, instr, 57-60, asst prof, 60-61; from asst prof to assoc prof, 61-68, PROF ELEC ENG & APPL PHYSICS, CORNELL UNIV, 68- *Concurrent Pos:* Consult, Gen Elec, Sylvania & US Naval Res Lab, vpres-dir res, Lansing Res Corp, 64-; head, Laser Physics Br, US Naval Res Lab, 68-70. *Mem:* Am Phys Soc; Inst Elec & Electronics Engrs; Optical Soc Am; Mat Res soc; Soc Info Display. *Res:* Excited state spectroscopy; molecular physics; quantum electronics; molecular energy transfer and relaxation; modern optical spectroscopic instruments; infrared tunable lasers and spectroscopy; picosecond optoelectronics; laser processing of semiconductors; electroluminescent displays. *Mailing Add:* Sch Elec Eng 237 Phillips Hall Cornell Univ Ithaca NY 14853

WOLGAMOTT, GARY, b Alva, Okla, July 23, 40; m 62; c 2. MICROBIOLOGY, MOLECULAR BIOLOGY. *Educ:* Northwestern State Col, Okla, BS, 63; Okla State Univ, PhD(microbiol), 68. *Prof Exp:* From asst prof to assoc prof, 68-75, PROF MICROBIOL, SOUTHWESTERN OKLA STATE UNIV, 75- CHMN DEPT, DIV ALLIED HEALTH SCI, 77- *Concurrent Pos:* NSF fel col med, Univ Iowa, 71; NASA-Am Soc Eng Educ fel, Johnson Space Ctr, Houston, Tex; mem, Med Technol Rev Comt, Clin Microbiol Prog Rev Comt & Nat Accrediting Agency Clin Lab Sci. *Mem:* AAAS; Am Soc Microbiol. *Res:* Study of the mode of action of specific chemotherapeutic agents on the activities of microorganisms, including their physiology and microstructure; action of microbial hemolysins on tissue culture cells ultrastructure; virulence analysis of space flown microautoflora from astronauts. *Mailing Add:* Div Allied Health Sci Southwestern Okla State Univ Weatherford OK 73096

WOLGEMUTH, CARL HESS, b Bareville, Pa, Apr 18, 34; m 54; c 2. MECHANICAL ENGINEERING. *Educ:* Pa State Univ, BS, 56; Ohio State Univ, MS, 58, PhD(mech eng), 63. *Prof Exp:* Instr mech eng, Ohio State Univ, 56-63; from asst prof to assoc prof, 63-77, actg dept head, 83-84, PROF MECH ENG, PA STATE UNIV, 77-, ASSOC DEAN ENG, 84- *Honors & Awards:* Ralph R Teetor Award, Soc Automotive Engrs, 65. *Mem:* Fel Soc Automotive Engrs; Am Soc Eng Educ; Am Soc Mech Engrs; Sigma Xi. *Res:* Application of thermodynamics and heat transfer to the study of power-producing systems. *Mailing Add:* Pa State Univ 101 Hammond Bldg University Park PA 16802

WOLGEMUTH, DEBRA JOANNE, b Lancaster, Pa, July 28, 47; m 70; c 1. DEVELOPMENTAL BIOLOGY, CELL BIOLOGY. *Educ:* Gettysburg Col, BA, 69; Vanderbilt Univ, MA, 71; Columbia Univ, MPhil, 72, PhD(human genetics), 77. *Prof Exp:* Res assoc reproductive physiol, Vanderbilt Univ, 71-72; fel RNA synthesis, Sloan-Kettering Inst Cancer Res, 77-78, molecular cell biol, Rockefeller Univ, 78-80; ASST PROF HUMAN GENETICS & DEVELOP, COLUMBIA UNIV, 80- *Mem:* Am Soc Cell Biol; Soc Develop Biol; Int Soc Develop Biologists; AAAS; NY Acad Sci. *Res:* Cellular and molecular biology of mammalian gametogenesis, fertilization, and early embryogenesis; elucidating the structure of sperm chromatin and the role of oocyte products in the activation of development. *Mailing Add:* Dept Human Genetics & Develop Columbia Univ Col Physicians & Surgeons 630 W 168th St New York NY 10032

WOLGEMUTH, KENNETH MARK, b Mechanicsburg, Pa, Dec 23, 43; m 70. MARINE GEOCHEMISTRY. *Educ:* Wheaton Col, BS, 65; Columbia Univ, MS, 69, PhD(geochem), 72. *Prof Exp:* Asst prof geol, Dickinson Col, 71-76; vis prof geophys, Fed Univ, Bahia, Brazil, 76-78; SR SCIENTIST, TERRA TEK, INC, 78- *Concurrent Pos:* Vis asst prof geol, World Campus Afloat, Chapman Col, 75; asst prof oceanog, Fla State Univ, 78. *Mem:* Geol Soc Am; Am Geophys Union; AAAS. *Res:* Trace element geochemistry of surface waters; geothermal well stimulation; geochemistry of hydrothermal systems; elastic moduli of tar sand for oil recovery. *Mailing Add:* Terra Tek Inc 420 Wakara Way Salt Lake City UT 84108

WOLGEMUTH, RICHARD LEE, b Lebanon, Pa, June 29, 45; m 68; c 4. PHYSIOLOGY, PHARMACOLOGY. *Educ:* Ashland Col, BSc, 68; Ohio State Univ, MS, 75, PhD(physiol), 75. *Prof Exp:* Jr pharmacologist, Warren-Teed Pharmaceut, Rohm & Haas, 69-74; teaching asst physiol, Ohio State Univ, 73-75; scientist, Rohm & Haas Co, 75-77; sr res scientist drug metab, 77-82, proj leader, 83-84, mgr pharmacol & med chem, 84-85, dir proj coord, 86-87, DIR REGULATORY AFFAIRS NEW DRUGS, ADRIA LABS INC, 88- *Concurrent Pos:* Consult, Nat Inst Occup Safety & Health, 75-76 & Poly Sci Inc, 76-79. *Mem:* Soc Exp Biol & Med; Am Soc Cancer Chemother; NY Acad Sci; AAAS; Regulatory Affairs Profs Soc. *Res:* Gastrointestinal physiology especially pancreatic function, proteolytic enzymes and digestion, enteric bacteria; renal stones especially calcium and phosphorus metabolism, mechanism and treatment of urolithiasis; anthracycline metabolism and toxicity; antiemetics. *Mailing Add:* Adria Labs PO Box 16529 Columbus OH 43216-6529

WOLICKI, ELIGIUS ANTHONY, b Buffalo, NY, May 10, 27; m 54; c 4. NUCLEAR PHYSICS. *Educ:* Canisius Col, BS, 46; Univ Notre Dame, PhD(physics), 50. *Prof Exp:* Asst physics, Univ Notre Dame, 46-47, asst, 47-48; res assoc nuclear physics, Univ Iowa, 50-52; nuclear physicist, Nuclear Sci Div, 52-66, consult & actg assoc, 66-77, assoc supt, Radiation Technol Div, 77-81, assoc supt, Condensed Matter & Radiation Sci Div, US Naval Res Lab, 81-84; PRES, WOLICKI ASSOCS, INC, 84- *Concurrent Pos:* AEC fel, 48-50. *Honors & Awards:* Centennial of Sci Award, Univ Notre Dame, 65. *Mem:* Fel Am Phys Soc; fel Inst Elec & Electronics Engrs. *Res:* Nuclear reactions, and applications; electrostatic accelerators; radiation detectors; applications of nuclear radiation, nuclear techniques and ion beam accelerators; radiation damage, and radiation effects; radiation hardness assurance; single event upsets in very large scale integrated circuits. *Mailing Add:* 1310 Gatewood Dr Alexandria VA 22307

WOLIN, ALAN GEORGE, b New York, NY, Apr 2, 33; m 54; c 4. FOOD SCIENCE. *Educ:* Cornell Univ, BS, 54, MS, 56, PhD, 58. *Prof Exp:* Sr proj leader yeast tech div, Fleischmann Labs, Standard Brands, Inc, 58-60; head dairy tech lab, Vitex Labs Div, Nopco Chem Co, 60-62; prod improv mgr, 62-67, QUAL COORDR, M&M CANDIES DIV, MARS, INC, 67- *Concurrent Pos:* Ed, J Appl Microbiol, 72-; mem res comt, Nat Confectioners Asn, 75-, tech comt, Grocery Mfrs Asn, 73- & NJ Pub Health Adv Comn, 73-75. *Mem:* Am Dairy Sci Asn; Am Soc Microbiol; Inst Food Technol; Soc Consumer Affairs Prof; Am Chem Soc. *Res:* Use of radioisotopes in study of food flavors; dairy starter cultures; enzymes; fermentation; antimicrobial agents in milk; food fortification; phosphatase activity of chocolate milk; natural bacterial inhibitors in raw milk. *Mailing Add:* 44 Stonehenge Rd Morristown NJ 07960

WOLIN, HAROLD LEONARD, b Brooklyn, NY, June 22, 27; m 56; c 3. MICROBIOLOGY, CLINICAL CHEMISTRY. *Educ:* Univ Calif, AB, 50, MA, 52; Cornell Univ, PhD(microbiol/biochem), 56. *Prof Exp:* Res scientist, Pac Yeast Prod Co, 56-57; instr, Hahnemann Med Col, 57-59; asst prof microbiol col med & dent, Seton Hall Univ, 59-63; CLIN ASSOC PROF MICROBIOL, COL MED & DENT, NJ, 63-; ASSOC DIR CLIN LAB, BROOKDALE HOSP CTR, BROOKLYN, NY, 63- *Mem:* Am Soc Microbiol; Brit Soc Gen Microbiol; Am Asn Clin Chem; Nat Acad Clin Biochem. *Res:* Clinical microbiology and chemistry; microbial physiology. *Mailing Add:* 1772 Slocum St Hewlett NY 11557

WOLIN, LEE ROY, b Cleveland, Ohio, Dec 8, 27; m 50; c 3. PSYCHOLOGY, NEUROPHYSIOLOGY. *Educ:* Los Angeles State Col, BS, 50; Cornell Univ, PhD(psychol), 55. *Prof Exp:* Mem fac psychol, Sarah Lawrence Col, 55-59; res assoc develop, Child Study Ctr, Clark Univ, 59-60; res assoc, Cleveland Psychiat Inst, 60-69, dir lab neuropsychol, 69-73; asst prof neurosurg, 70-76, ASST CLIN PROF PHYSIOL, CASE WESTERN RESERVE UNIV, 76-; DIR LAB NEUROPSYCHOL & EEG, OHIO MENT HEALTH & MENT RETARDATION RES CTR, 73-; ADMIN DIR, DRUG ABUSE TREAT UNIT, CLEVELAND PSYCHIAT INST, 75- *Concurrent Pos:* Lectr, Lakewood High Exten, Ohio State Univ, 63-66 & Bedford Exten, Cleveland State Univ, 66; lectr, Kent State Univ, 65. *Mem:* AAAS; Am Psychol Asn; NY Acad Sci. *Res:* Neurophysiology of vision; behavioral and neurophysiological manifestations of brain disfunction; perceptual, cognitive and emotional aspects of neuropsychiatric disorders perception. *Mailing Add:* 12228 Fairview Ct Cleveland OH 44106

WOLIN, MEYER JEROME, b Bronx, NY, Nov 10, 30; m 55; c 2. MICROBIAL ECOLOGY, MICROBIAL PHYSIOLOGY. *Educ:* Cornell Univ, BS, 51; Univ Chicago, PhD(microbiol), 54. *Prof Exp:* NIH fel microbiol, Univ Minn, 54-55; fel, Univ Ill, Urbana, 55-56, from asst prof to assoc prof dairy sci, 56-67, assoc prof microbiol, 65-67, prof dairy sci & microbiol, 67-74; CHIEF RES SCIENTIST DIV LAB & RES, NY STATE HEALTH DEPT, 74-; PROF ENVIRON HEALTH & TOXICOL, GRAD SCH PUB HEALTH SCI, STATE UNIV NY, ALBANY, 85- *Concurrent Pos:* NSF sr fel, Univ Newcastle, 64-65; mem microbial chem study sect, NIH, 71-75; chmn dept environ health & toxicol, Grad Sch Pub Health Sci, State Univ NY, Albany, 85-87. *Mem:* AAAS; Am Chem Soc; Am Soc Microbiol; NY Acad Sci; Am Soc Biol Chem. *Res:* Microbial biochemistry and ecology; fermentations in anaerobic ecosystems; intestinal tract microbiology; interspecies interactions; hydrogen metabolism; methane production. *Mailing Add:* Wadsworth Ctr Labs & Res Empire State Plaza Albany NY 12201-0509

WOLIN, MICHAEL STUART, b Brooklyn, NY, Sept 11, 53; m 87; c 1. VASCULAR REGULATION, OXYGEN PHYSIOLOGY. *Educ:* Harpur Col, BA, 75; Yale Univ, MSc, 76, MPhil, 77 & PhD(chem), 81. *Prof Exp:* Postdoctoral, Tulane Univ Sch Med, 81-83, instr pharmacol, 82-83; asst prof, 83-89, ASSOC PROF PHYSIOL, NY MED COL, 89- *Concurrent Pos:* Biomed Res Scholar Awardee, C H Revson Found, 83-85; prin investr, NIH, 84-, Am Lung Asn Grant, 84-86; prog comt, Am Heart Asn Cardiolpulmonary Council, 85-; prin investr, Am Heart Asn Grant, 86-; nat res serv award, NIH, 81-83; estab investr award, Am Heart Asn, 89-; prog comt, Am Thoracic Soc, 90- *Honors & Awards:* Albert Hyman Res Award, Am Heart Asn, 83. *Mem:* Am Physiol Soc; AAAS; Am Heart Asn; Int Soc Free Radical Res; Microcirculatory Soc; Am Thoraic Soc. *Res:* Reconstruction of the regulation of soluble guanylate cyclase; basic mechanisms of vascular regulation by oxygen tension, reactive oxygen matabolites and the intracellular mediator cyclic GMP, with a focus on those which are unique to the pulmonary circulation versus systemic circulations. *Mailing Add:* Dept Physiol NY Med Col Valhalla NY 10595

WOLIN, SAMUEL, b New York, NY, Feb 12, 09; m 42; c 2. ELECTRONICS, PHYSICS. *Educ:* City Col New York, BS, 30, MS, 31; Columbia Univ, AM, 41. *Prof Exp:* Instr, Pub Schs, NY, 30-42; instr radio eng, Army Air Force Officers Div, Yale Univ, 42-44; instr elec commun, Radar Sch, Mass Inst Technol, 44-45; radio engr, Victor Div, Radio Corp Am, 45-46; res engr, Bartol Res Found, Franklin Inst, 46-47; electronics engr, US Naval Base Sta, Naval Air Mat Ctr, 47; electronic scientist, US Naval Air Develop Ctr, 47-57; sr electronic engr, McDonnell Aircraft Corp, Mo, 56-60; sr elec engr, Adv Systs Res Dept, Lockheed Electronics Co, 60-61; sr design staff engr, Boeing Co, Pa, 61-63; sr res engr, Brown Engr Co, Ala, 63-66; res engr, Boeing Co, Ala, 66-69; electronic engr, US Army Missile Command, Redstone Arsenal, 69-80; CONSULT SCIENTIST, 80- *Honors & Awards:* Belden Medal. *Mem:* AAAS; sr mem Inst Elec & Electronics Engrs; Soc Indust & Appl Math. *Res:* Electronic systems engineering for space vehicles; guided missiles; aircraft and vertical takeoff and landing; applied electromagnetic theory; antennas; radomes; radio propagation; optics; reliability; applied mathematics. *Mailing Add:* 2205 Wimberly Rd NW Huntsville AL 35816-1237

WOLINSKI, LEON EDWARD, b Buffalo, NY, Apr 3, 26; m 54; c 3. ORGANIC POLYMER CHEMISTRY. *Educ:* Univ Buffalo, BA, 49, PhD(chem), 51. *Prof Exp:* Res chemist film dept, E I du Pont de Nemours & Co, 51-57, staff scientist org polymer chem, 57-59, res assoc, 59-70; CORP DIR RES, PRATT & LAMBERT CO, 70- *Concurrent Pos:* Lectr, Canisius Col, 55-59. *Mem:* Am Chem Soc; Am Inst Chemists. *Res:* Grignard reagents; organo silicon chemistry; surface chemistry; adhesives; condensation and addition polymers; coatings; films. *Mailing Add:* 35 Parkview Terr Cheektowaga NY 14225

WOLINSKY, EMANUEL, b New York, NY, Sept 23, 17; m 47; c 2. INFECTIOUS DISEASES. *Educ:* Cornell Univ, BA, 38, MD, 41. *Prof Exp:* Intern med, New York Hosp, 43-44, resident, 44-45; asst dir tuberc res, Trudeau Lab, Trudeau Found, 46-56; asst prof med, Case Western Reserve Univ, 56-62, from asst prof to assoc microbiol, 62-68, prof med, 68-88, prof path, 81-88, EMER PROF MED & PATH, SCH MED, CASE WESTERN RESERVE UNIV, 88- *Concurrent Pos:* Director microbiol, Cleveland Metrop Gen Hosp, 59-; former mem tuberc panel, US-Japan Coop Med Sci Prog; former mem strep & staph comn, Armed Forces Epidemiol Bd; former assoc ed, Am Rev Respiratory Dis. *Honors & Awards:* Trudeau Medal, 86. *Mem:* Am Soc Microbiol; Am Thoracic Soc; Infectious Dis Soc Am. *Res:* Medical microbiology and pulmonary diseases; tuberculosis bacteriology and experimental chemotherapy; infectious diseases. *Mailing Add:* Metrop Gen Hosp Cleveland OH 44109

WOLINSKY, HARVEY, b Cleveland, Ohio, June 3, 39; m; c 2. MEDICINE, PATHOLOGY. *Educ:* Western Reserve Univ, AB, 60; Univ Chicago, MD & MS, 63, PhD(path), 67. *Prof Exp:* Intern, Univ Chicago Hosps, 63-64; asst resident internal med, Mt Sinai Hosp, Cleveland, Ohio, 64-65; resident chest serv, Bronx Municipal Hosp Ctr, NY, 67-68; assoc, Albert Einstein Col Med, 68-70, assoc path, 69-70, asst prof med & path, 70-73, assoc prof med, 73-77, prof med & path, 77-81; CLIN PROF MED, MT SINAI SCH MED, 81- *Concurrent Pos:* USPHS res fel, Univ Chicago Hosps, 65-67; USPHS res career develop award, Nat Heart & Lung Inst, 72-77; assoc attend physician, Bronx Munic Hosp Ctr, NY, 68-71, attend physician, 72-81; mem coun arteriosclerosis, Am Heart Asn, 70- & mem coun high blood pressure res, 72-; attend physician, Mt Sinai Hosp & Med Ctr, 81- *Mem:* Am Thoracic Soc; Am Soc Exp Path; Fedn Am Socs Exp Biol; Am Soc Clin Invest; NY Acad Sci; Am Col Cardiol. *Res:* Comparative pathology; structure, function and biochemistry of blood vessels; effects of hormonal and mechanical factors on blood vessel structure. *Mailing Add:* Dept Med Mt Sinai Sch Med New York NY 10029

WOLINSKY, IRA, b New York, NY, Mar 30, 38; m 65; c 2. NUTRITION, BIOCHEMISTRY. *Educ:* City Col New York, BS, 60; Kans Univ, MS, 65, PhD(biochem), 68. *Prof Exp:* Lectr, Hebrew Univ Hadassah Med Sch, 68-74; vis scientist, Dalton Res Ctr, Univ Mo, 74; assoc prof nutrit, Pa State Univ, 74-79; PROF NUTRIT, UNIV HOUSTON, 79- *Mem:* Am Inst Nutrit; Soc Exp Biol & Med; Sigma Xi; Soc Clin Nutrit. *Res:* Nutritional biochemistry of bone; calcium metabolism; sports nutrition. *Mailing Add:* Dept Human Develop Univ Houston Houston TX 77204-6861

WOLINSKY, JERRY SAUL, b Baltimore, Md, Nov 26, 43; m 69; c 2. NEUROVIROLOGY, NEUROIMMUNOLOGY. *Educ:* Ill Inst Technol, BS; Univ Ill, MD, 69. *Prof Exp:* From instr to asst prof neurol, Univ Calif, San Francisco, 73-78; assoc prof neurol, Sch Med, Johns Hopkins Univ, 78-83, assoc prof immunol & infectious dis, Sch Hyg & Pub Health, 79-83; PROF NEUROL, HEALTH SCI CTR, UNIV TEX, HOUSTON, 83-, GRAD FAC VIROL, GRAD SCH BIOMED SCI, 84- *Concurrent Pos:* David M Olkon scholar, 68-69; Basil O'Connor Starter Res grant, 75-78; res assoc neurovirol, Vet Admin Hosp, San Francisco, 75-78; prin investr, Nat Mult Sclerosis Soc, 86-; mem, NIH Immunol Sci study sect, 85- *Mem:* Fel Am Acad Neurol; AAAS; Am Soc Microbiol; Am Neurol Asn; Am Soc Clin Invest; Am Soc Virol. *Res:* Virus and autoimmune diseases of the nervous system. *Mailing Add:* Rm 222 A PO Box 20708 Houston TX 77225

WOLINSKY, JOSEPH, b Chicago, Ill, Dec 3, 30; m 51; c 5. ORGANIC CHEMISTRY. *Educ:* Univ Ill, BS, 52; Cornell Univ, PhD, 56. *Prof Exp:* Proj assoc, Univ Wis, 56-58; from asst prof to assoc prof 58-67, PROF CHEM, PURDUE UNIV, 67- *Mem:* Am Chem Soc. *Res:* Chemistry of terpenes, alkaloids and related natural products. *Mailing Add:* Dept Chem Purdue Univ West Lafayette IN 47907

WOLK, COLEMAN PETER, b New York, NY, Sept 28, 36; m 65; c 1. DEVELOPMENTAL MICROBIOLOGY. *Educ:* Mass Inst Technol, SB & SM, 58; Rockefeller Inst, PhD(biol), 64. *Prof Exp:* Nat Acad Sci-Nat Res Coun res fel biol, Calif Inst Technol, 64-65; from asst prof to assoc prof, 65-74, PROF BOT, MICH STATE UNIV, 74- *Honors & Awards:* Darbaker Prize, Bot Soc Am. *Mem:* Phycol Soc Am; Am Soc Microbiol; Soc Develop Biol; Am Soc Plant Physiol. *Res:* Physiological, biochemical and genetic bases of development and nitrogen fixation in cyanobacteria. *Mailing Add:* 166 Plant Biol-Bot Mich State Univ East Lansing MI 48824

WOLK, ELLIOT SAMUEL, b Springfield, Mass, Aug 5, 19; m 50; c 3. MATHEMATICS. *Educ:* Clark Univ, AB, 40; Brown Univ, ScM, 47, PhD(math), 54. *Prof Exp:* Instr math, 50-56, chmn dept, 67-73, from asst prof to prof, 56-88, EMER PROF MATH, UNIV CONN, 88- *Concurrent Pos:* Consult elec boat div, Gen Dynamics Corp, 55-58. *Mem:* Am Math Soc; Math Asn Am. *Res:* Partially ordered sets; general topology. *Mailing Add:* Dept Math Univ Conn Storrs CT 06268

WOLK, ROBERT GEORGE, b New York, NY, Mar 10, 31; m 56; c 5. ORNITHOLOGY, EVOLUTION. *Educ:* City Univ NY, BS, 52; Cornell Univ, MS, 54, PhD(vert zool), 59. *Prof Exp:* Asst prof biol, St Lawrence Univ, 57-63; assoc prof vert morphol & behav, Adelphi Univ, 63-67; cur life sci, Nassau County Mus, 67-78; exec dir, Nature Sci Ctr, 78-82; lectr zool, Greensboro Col, 82-83; dir educ, 83-91, PROG DIR, NC STATE MUS NATURAL SCI, 91- *Mem:* Am Ornithologists Union; Brit Ornithologists Union; Am Asn Mus; Sigma Xi; Cooper Ornith Soc; Wilson Ornith Soc. *Res:* Behavioral adaptations and functional morphology of birds; avian vision; reproductive behavior of the black skimmer; evolution, systematics, and distribution of gulls, terns, and skimmers. *Mailing Add:* NC State Mus Nat Sci PO Box 27647 Raleigh NC 27611

WOLKE, RICHARD ELWOOD, b East Orange, NJ, June 2, 33; m 64; c 2. VETERINARY PATHOLOGY. *Educ:* Cornell Univ, BS, 55, DVM, 62; Univ Conn, MS, 66, PhD(vet path), 68. *Prof Exp:* Vet, Am Soc Prev Cruelty Animals, 62-63; pvt practice, 63-64; NIH path trainee, Univ Conn, 64-68, res assoc, 68-69, res assoc ichthyopath, 69-70; asst prof, 70-75, assoc prof ichthyopath, 75-81, PROF AQUACULT & PATH, UNIV RI, 86- *Concurrent Pos:* Conn Res Comn grant, Univ Conn, 69-70; Nat Oceanic & Atmospheric Admin Sea grant, Univ RI, 70-80; vis prof, Unit Aquatic Pathobiol, Stirling Univ, Scotland, 78-79; adj prof comp med, Tufts Univ, 81-86. *Mem:* Wildlife Dis Asn; Int Asn Aquatic Animal Med; World Maricult Soc; Sigma Xi; NY Acad Sci. *Res:* Ichthyopathology; comparative pathology; inflammation. *Mailing Add:* 25 Main St Carolina RI 02812

WOLKE, ROBERT LESLIE, b Brooklyn, NY, Apr 2, 28; m 64; c 1. NUCLEAR CHEMISTRY, UNIVERSITY ADMINISTRATION. *Educ:* Polytech Inst Brooklyn, BS, 49; Cornell Univ, PhD, 53. *Prof Exp:* Res assoc nuclear chem, Enrico Fermi Inst, Univ Chicago, 53-56; nuclear chemist, Gen Atomic Div, Gen Dynamics Corp, 56-57; from asst prof to assoc prof chem, Univ Fla, 57-60; assoc prof, Univ Pittsburgh, 60-67, dir, Wherrett Lab Nuclear Chem, 61-77 & Univ Off Fac Develop, 77-88, prof, 67-89, EMER PROF CHEM, UNIV PITTSBURGH, 89- *Concurrent Pos:* Res partic, Oak Ridge Nat Lab, 58 & 59; vis prof, Univ PR, 70; vis prof, USAID, Univ Oriente, Venezuela, 73; acad dean, Semester at Sea, 82. *Mem:* Am Chem Soc; Am Asn Higher Educ. *Res:* Nuclear reactions; recoil studies, interaction of energetic ions with matter; natural radioactivity; marine radioactivity. *Mailing Add:* Dept Chem Univ Pittsburgh Pittsburgh PA 15260

WOLKEN, GEORGE, JR, b Jersey City, NJ, Nov 11, 44; m 67; c 2. CHEMICAL PHYSICS, THEORETICAL CHEMISTRY. *Educ:* Tufts Univ, BS, 66; Harvard Univ, PhD(chem physics), 71. *Prof Exp:* Fel, Max Planck Inst Aerodyn, Univ Göttingen, 71-72; asst prof chem, Ill Inst Technol, 72-74; mem staff, Battelle Mem Inst, 74-81; PATENT ATTY & CONSULT, 74- *Mem:* Am Chem Soc; Am Phys Soc. *Res:* Theoretical chemical kinetics, both of gas phase reactions and heterogeneous reactions. *Mailing Add:* 6602 Hawthorne St Worthington OH 43085

WOLKEN, JEROME JAY, b Pittsburgh, Pa, Mar 28, 17; m 45, 56; c 4. BIOPHYSICS. *Educ:* Univ Pittsburgh, BS, 46, MS, 48, PhD(biophys), 49. *Prof Exp:* Res fel, Mellon Inst, 43-47 & Rockefeller Inst, 51-52; dir, Biophys Res Lab, Eye & Ear Hosp, Mellon Col Sci, 53-64, head, dept biol sci, 64-67, PROF BIOPHYS, CARNEGIE-MELLON UNIV, 64- *Concurrent Pos:* AEC fel, 49-51; Am Cancer Soc fel, 51-53; asst prof sch med, Univ Pittsburgh, 53-57, from assoc prof to prof, 57-66; Nat Coun Combat Blindness fel, 57; career prof, USPHS, 62-64; guest prof, Pa State Univ, 63; vis prof, Univ Paris, 67-68, Univ Col, Univ London, 71, Pasteur Inst, Paris, 72 & Princeton Univ, 78. *Mem:* Fel AAAS; fel Optical Soc Am; Am Chem Soc; fel Am Inst Chemists; Soc Gen Physiol; Am Soc Photobiol; Am Soc Cell Biol; Biophys Soc. *Res:* Biophysics; photobiology, optics and vision; issued 2 patents. *Mailing Add:* 5817 Elmer St Pittsburgh PA 15232

WOLKO, HOWARD STEPHEN, b Buffalo, NY, Apr 30, 25; m 50; c 4. MECHANICAL ENGINEERING, SOLID MECHANICS. *Educ:* Univ Buffalo, BS, 49, MS, 53; George Washington Univ, ScD(mech), 67. *Prof Exp:* Design engr, Sci Instruments Div, Am Optical Co, 50-52; res assoc exp mech, Cornell Aeronaut Lab, Cornell Univ, 52-55; chief struct res, Bell Aircraft Corp, 55-59; head solid mech, Eng Sci Directorate, Air Force Off Sci Res, 59-63; chief struct mech, Off Advan Res & Technol, NASA, 65-67; prof mech eng, Tex A&M Univ, 67-72; prof & chmn dept, Memphis State Univ, 72-73; asst dir sci & technol dept, 73-80, SPEC ADV AERONAUT DEPT, NAT AIR & SPACE MUS, SMITHSONIAN INST, 80- *Concurrent Pos:* Lectr, Univ Buffalo, 53-59. *Mem:* Soc Eng Sci; Soc Exp Stress Anal. *Res:* Coupled thermomechanics; continuum mechanics; materials science. *Mailing Add:* Nat Air & Space Mus Smithsonian Inst Washington DC 20560

WOLKOFF, AARON WILFRED, b Toronto, Ont, Feb 12, 44; m 66; c 2. ANALYTICAL CHEMISTRY. *Educ:* Univ Toronto, BSc, 65, MSc, 67, PhD(org chem), 71. *Prof Exp:* Fel, Inst Environ Sci & Eng, Univ Toronto, 71-72; teaching master math & chem, Seneca Col Appl Arts & Technol, 72-73; res scientist environ anal, Can Centre Inland Waters, Dept Environ, 73-77; MGR, WATERS CHROMATOGRAPHY DIV, MILLIPORE LTD, CAN, 77- *Mem:* Am Chem Soc; Chem Inst Can. *Res:* Development of new analytical methods on the applications of high pressure liquid chromatography. *Mailing Add:* 3680 Nashua Dr Mississauga ON L4V 1M5 Can

WOLKOFF, HAROLD, b Brooklyn, NY, June 10, 23; m 49; c 2. MECHANICAL ENGINEERING. *Educ:* Polytech Inst Brooklyn, BME, 49; City Col New York, MBA, 56. *Prof Exp:* Prof 51-85, EMER PROF ENG, NEW YORK CITY TECH COL, 85- *Mem:* Am Soc Eng Educ; Soc Mfg Engrs; Am Indust Arts Asn. *Res:* Strength of materials; engineering drawing. *Mailing Add:* Dept Mech Eng Technol NY Tech Col 300 Jay St Brooklyn NY 11201

WOLL, EDWARD, b New York, NY, May 29, 14. MECHANICAL ENGINEERING. *Educ:* Mass Inst Technol, BS, 35; Rensselaer Polytech Inst, ME, 46. *Prof Exp:* Gen mgr, Small Aircraft Dept, Gen Elec Co, 58-63, vpres, Mil Eng Div & Group Eng Div, 68-79, vpres & gen mgr, Group Advan Eng Div, 70-79; RETIRED. *Mem:* Nat Acad Eng; fel Soc Automotive Engrs; fel Am Inst Aeronaut & Astronaut; Am Helicopter Soc. *Mailing Add:* PO Box 702 Boca Grande FL 33921

WOLL, HARRY JEAN, b Farmington, Minn, Aug, 25, 20; m 47; c 2. ELECTRICAL & ELECTRONICS ENGINEERING. *Educ:* NDak State Univ, BS, 40; Univ Pa, PhD(elec eng), 53. *Prof Exp:* Asst, Ill Inst Technol, 40-41; res & develop engr, RCA, 41-43, res & develop supvr, 53-58, mgr appl res, 58-63, chief engr, Aerospace Systs Div, 63-69, div vpres gov eng, Govt & Com Systs, 69-75, div vpres & gen mgr, Automated Systs, 75-81, staff vpres & chief engr, Electronic Prod, Systs & Serv, 81-85; RETIRED. *Concurrent Pos:* Chmn, Trustees Moore Sch Elec Eng, Univ Pa, 76-90; chmn, Aerospace Indust Asn Tech Coun, 77 & chmn, Fel Comt, 79-80. *Mem:* AAAS; fel Inst Elec & Electronics & Engrs. *Res:* Linear circuit theory; solid state circuits; communications; electro-optics; aerospace systems; automatic test systems. *Mailing Add:* PO Box 679 Concord MA 01742

WOLL, JOHN WILLIAM, JR, b Philadelphia, Pa, May 19, 31; m 52, 84; c 2. MATHEMATICS. *Educ:* Haverford Col, BS, 52; Princeton Univ, PhD, 56. *Prof Exp:* Instr math, Princeton Univ, 56-57; asst prof, Lehigh Univ, 57-58, Univ Calif, 58-61 & Univ Wash, 61-68; PROF MATH, WESTERN WASH STATE UNIV, 68- *Mem:* Am Math Soc. *Res:* Functional analysis and stochastic processes. *Mailing Add:* Dept Math Western Wash State Univ Bellingham WA 98225

WOLLA, MAURICE L(EROY), b Minot, NDak, May 13, 33; m 53; c 5. ELECTRICAL & COMPUTER ENGINEERING. *Educ:* NDak State Univ, BS, 56; Mich State Univ, PhD(elec eng), 66. *Prof Exp:* Instr elec eng, NDak State Univ, 56-58 & Mich State Univ, 58-65; assoc prof, Colo State Univ, 65-66; assoc prof, 66-72, PROF ELEC ENG, CLEMSON UNIV, 72- *Concurrent Pos:* US Dept Interior res grants, 67-69. *Mem:* Simulation Coun; Inst Elec & Electronics Engrs; Am Soc Eng Educ. *Res:* Analysis and simulation of physical systems; computer sciences; software engineering; educational computing systems; real-time computing systems. *Mailing Add:* Schlumberger Indust Div Sangamo Elec 180 Technol Pkwy Norcross GA 30092

WOLLAN, DAVID STRAND, b Boston, Mass, Mar 25, 37; m 79; c 2. SCIENCE POLICY, SOLID STATE PHYSICS. *Educ:* Amherst Col, AB, 59; Univ Ill, MS, 61, PhD(physics), 66. *Prof Exp:* Asst prof physics, Va Polytech Inst & State Univ, 66-74; physicist, 74-89, SR EXEC SERV, US ARMS CONTROL & DISARMAMENT AGENCY, 89- *Mem:* Am Phys Soc; Inst Elec & Electronics Engrs; Sigma Xi. *Res:* Electron and nuclear magnetic resonance in solids; strategic arms control. *Mailing Add:* 6026 Grove Dr Alexandria VA 22307

WOLLAN, JOHN JEROME, b Chicago, Ill, July 7, 42; div; c 2. LOW TEMPERATURE PHYSICS. *Educ:* St Olaf Col, BA, 64; Iowa State Univ, PhD(physics), 70. *Prof Exp:* Vis asst prof physics, Univ Ky, 70-73; Nat Res Coun assoc physics, Air Force Mat Lab, 73-74; STAFF MEM PHYSICS, LOS ALAMOS SCI LAB, 74- *Mem:* Am Phys Soc. *Res:* Development and evaluation of superconducting wire for magnetic energy transfer and storage applications for fusion power systems, in particular, static and fast pulse energy losses in superconducting wire. *Mailing Add:* 3225 Lake Shore Dr Florence SC 29501

WOLLENBERG, BRUCE FREDERICK, b Buffalo, NY, June 14, 42; m 65; c 4. POWER SYSTEMS OPERATIONS. *Educ:* Rensselaer Polytech Inst, BEE, 64, MEng, 66; Univ Pa, PhD(systs eng), 74. *Prof Exp:* Engr, Leeds & Northrop Co, 66-70, sr engr, 70-74; SR ENGR, POWER TECHNOLOGIES INC, 74-; PMN CONSULT, ENERGY MGT SYSTS DIV, CONTROL DATA CORP. *Concurrent Pos:* Adj assoc prof, Rensselaer Polytech Inst, 79- *Mem:* Sigma Xi; Inst Elec & Electronics Engrs. *Res:* Methods for secure and optical operation of electric power systems; software implementation; field testing. *Mailing Add:* Dept Elec Eng Univ Minn 200 Union St SE Minneapolis MN 55455

WOLLENSAK, JOHN CHARLES, b Rochester, NY, Dec 16, 32; m 57; c 5. ORGANIC CHEMISTRY, ORGANOMETALLIC CHEMISTRY. *Educ:* Col Holy Cross, BS, 54; Mass Inst Technol, PhD(org chem), 58; Mich State Univ, MBA, 80. *Prof Exp:* Res chemist, 58-66, supvr chem res, 66-81, asst to vpres res, 81-83, DIR CHEM RES & DEVELOP, ETHYL CORP, 83- *Mem:* Am Chem Soc; assoc Int Union Pure & Appl Chem. *Res:* Organic synthesis; organometallics; administering research on pharmaceuticals, agricultural chemical intermediates, bromine-containing organics, organometallics, alkylated phenols and anilines, polymer intermediates, long chain olefins and alcohols, antioxidants, and additives for petroleum products. *Mailing Add:* Ethyl Corp PO Box 14799 Baton Rouge LA 70898

WOLLER, WILLIAM HENRY, b San Antonio, Tex, Feb 28, 33; m 58; c 3. PHYSICAL PHARMACY, COSMETIC CHEMISTRY. *Educ:* Univ Tex, Austin, BS, 55. *Prof Exp:* Mfg pharmacist drug prod, 56-64, res & develop scientist drugs & cosmetics, 65-73, RES & DEVELOP MGR DRUGS & COSMETICS & TECH DIR, DERMAT PROD TEX DIV, ALCON LABS, INC, NESTLE S A, SWITZ. *Mem:* Am Pharm Asn; Soc Cosmetic Chemists. *Res:* Development of vehicles for organic peroxides for use in the treatment of dermatological disorders; screening alpha hydroxy acids for use in treatment of icthyosis; coal tar fractions. *Mailing Add:* 3314 Yorktown San Antonio TX 78230

WOLLIN, GOESTA, b Ystad, Sweden, Oct 4, 22; nat US; m 50; c 1. CLIMATOLOGY, CHEMISTRY. *Educ:* Hermods Col, Sweden, Phil; Columbia Univ, MS, 53. *Prof Exp:* Newspaper reporter, Ystads Allehanda, 39-41; free lance writer, Swedish Newspapers, 45-50; asst, 50-56, Columbia Univ, res scientist, Lamont Geol Observ, Columbia Univ, 56-86; RES CONSULT, HAYWOOD, COMMUNITY COL, CLYDE, NC, 86- *Mem:* AAAS; NY Acad Sci; Glaciol Soc; Explorers Club. *Res:* Holocene and pleistocene climates and marine sedimentation; micropaleontological research of marine sediments; the relationship between climatic changes and variations in the earth's magnetic field; chemical research on origin of life; inventions for firefighting, agricultural spraying, and snow-skiing. *Mailing Add:* 617 Jones Cove Rd Clyde NC 28721

WOLLMAN, HARRY, b Brooklyn, NY, Sept 26, 32; m 57; c 3. ANESTHESIOLOGY. *Educ:* Harvard Univ, AB, 54, MD, 58; Am Bd Anesthesiol, dipl, 64. *Prof Exp:* Intern med & surg, Univ Chicago Clins, 58-59; resident, Hosp Univ Pa, 59-63, assoc in anesthesia, Univ, 63-65, from asst prof to prof, 65-70, ROBERT DUNNING DRIPPS PROF ANESTHESIA & CHMN DEPT, SCH MED, UNIV PA, 72-, PROF PHARMACOL, 71- *Concurrent Pos:* NIH res trainee, 59-63; Pharmaceut Mfrs Asn fel, 60-61; consult, Vet Admin Hosp, Philadelphia, 63-64, 79- & Valley Forge Army Hosp, 65-66; mem pharm & toxicol training grants comt, NIH, 66-68, mem anesthesia training grants comt, 71-73, mem surg A study sect, 74-78; mem anesthesia drug panel drug efficacy study, Comt Anesthesia, Nat Acad Sci-Nat Res Coun, 70-71, mem comt adverse reactions to anesthesia drugs, 71-72; assoc ed, Anesthesiol, 70-75; prin investr, Anesthesia Res Ctr, Univ Pa, 72-78, prog dir anesthesia res training grant, 72-, chmn comt studies involving human beings, 72-76; chmn, Clin Pract Exec Com, 76-80; John Harvard scholar. *Honors & Awards:* Detur Award. *Mem:* Am Soc Anesthesiologists; Soc Acad Anesthesia Chmn (pres, 77-78); Am Physiol Soc; Sigma Xi; Asn Univ Anesthetists. *Res:* Circulatory physiology; cerebral blood flow and metabolism; regional blood flow during anesthesia. *Mailing Add:* Bahnemann Univ Hosp Broad & Vine Sts Philadelphia PA 19102

WOLLMAN, LEO, b New York City, NY, Mar 14, 14; m 85; c 2. TRANSSEXUALISM, HYPNOTHERAPY. *Educ:* Columbia Univ, BS, 34; NY Univ, MS, 38, Royal Col Physicians & Surgeons, Edinburgh, MD, 42; Rochdale Col, PhD, 72. *Hon Degrees:* DSc, Univ Mich, 73. *Prof Exp:* Pres, Royal Med Soc, 40-41; peripatetic psychiat, World Fed Mental Health, 62-88; dir med, World Med Asn, 64-88, secy, Acad Psychosomat Med, 65-66, ed dent & med, J Am Soc Psychol Dent & Med, 67-84, adv ed hypnosis, Japanese J Hypnosis, 68-88, lic private sch teacher, Univ State NY, 73; VPRES HYPNOSIS FIRST WORLD CONG SOPHROL, 70- *Concurrent Pos:* Hon mem, Argentina Soc Hypnotherap, 62-88, life mem, NY State marriage, family, child counselor, 70-88, mem Soc Med Jurisprudence, 74-88, fel Am Col Sexologists, 79-88, exec dir, Inst Human Develop, 80-88. *Mem:* AAAS; Am Soc Abdominal Surgeons; Nat Acupuncture Res Soc; Am Soc Law & Med; Am Soc Clin Hypno; Am Soc Psychoprophylax Obstet; Am Med Writers Asn; Am Soc Psychosomat Dent & Med (pres, 66-69); Soc Sci Study Sex (pres, 79-82). *Res:* Transsexualism and transvestism using hypnosis; books on obesity, sexology, marriage, divorce, hypnosis; acupuncture therapy for medical dysphoria. *Mailing Add:* 3813 Poplar Ave Brooklyn NY 11224-1301

WOLLMAN, SEYMOUR HORACE, b New York, NY, May 17, 15; m 44; c 3. ENDOCRINOLOGY, CYTOLOGY. *Educ:* NY Univ, BS, 35, MS, 36; Duke Univ, PhD(physics), 41. *Hon Degrees:* MD, Univ Goteborg, Sweden, 83. *Prof Exp:* Scientist, 48-85, EMER SCIENTIST THYROID GLAND, PHYSIOL, CELL BIOL & TUMOR, NAT CANCER INST, NIH, 85- *Honors & Awards:* Dunhill Lectr, 5th Int Thyroid Cong, 65. *Mem:* Am Thyroid Asn (vpres, 70); Am Soc Cell Biol; European Thyroid Asn; Am Physiol Soc. *Res:* Aspects of synthesis and secretion of thyroid hormone; growth and involution of thyroid gland; production of thyroid tumors. *Mailing Add:* NIH Bldg 37 Rm 1E20 Bethesda MD 20892

WOLLMER, RICHARD DIETRICH, b Los Angeles, Calif, July 27, 38. OPERATIONS RESEARCH, STATISTICS. *Educ:* Pomona Col, BA, 60; Columbia Univ, MA, 62; Univ Calif, Berkeley, PhD(eng sci), 65. *Prof Exp:* Oper res sci, Rand Corp, 65-70; PROF, CALIF STATE UNIV, LONG BEACH, 70- *Concurrent Pos:* Consult, USAF Adv Bd Int Trans, 68-70; mem, Naval Res Adv Comt, 74-78; lectr, Univ Calif, Los Angeles, 70 & 74; res mathematician, Univ Southern Calif, 73-75; vis assoc prof, Stanford Univ, 76-77; mathematician, Elec Power Res Inst, 77; consult, McDonnell Douglas Corp, 78-80, 85- & Logicon, 79-81. *Mem:* Opers Res Soc Am; Inst Mgt Sci; Math Prog Soc. *Res:* Mathematics programming; network flows; dynamic programming; Markov processes. *Mailing Add:* Dept Info Syst Calif State Univ 1250 Bellflower Blvd Long Beach CA 90840

WOLLNER, THOMAS EDWARD, b Rochester, Minn, Dec 30, 36; m 58; c 2. ORGANIC CHEMISTRY, POLYMER CHEMISTRY. *Educ:* St John's Univ, Minn, BA, 58; Wash State Univ, PhD(chem), 64. *Prof Exp:* Res mgr polymer res, 3M Co, 64-74, lab mgr, 75-77, dir, Chem Res Lab, 77-81 & Indust & Consumer Sector Res Lab, 81-84, res & develop dir, 84-86, res & develop vpres, Indust & Consumer Sector, 86-87, STAFF VPRES, CORP RES LABS, 3M CO, 87- *Mem:* Am Chem Soc; Sigma Xi. *Res:* Adhesives, binders and coatings; optical and electrical properties of organics. *Mailing Add:* 3M Co 3M Ctr Bldg 220-4E-01 St Paul MN 55144-1000

WOLLSCHLAEGER, GERTRAUD, b Muenchen, Ger, Feb 28, 24; US citizen; m 48; c 3. MEDICINE, RADIOLOGY. *Educ:* Univ Munich, physicum, 54, MD & PhD, 57; State Univ NY, MD, 65. *Prof Exp:* Instr radiol, Albert Einstein Col Med, 61-64; from asst prof to assoc prof, Sch Med, Univ Mo, Columbia, 64-71; asst chief sect neuroradiol, William Beaumont Hosp, Royal Oak, Mich, 71-72; PROF RADIOL & NEURORADIOL, SCH MED, WAYNE STATE UNIV, 73- *Concurrent Pos:* Res fel radiol, Albert Einstein Col Med, Yeshiva Univ, 61-62, NIH spec fel neuroradiol, 62-64; res grant, Univ Mo-Columbia, 64-71; co-dir, NIH spec fel training prog, 67-71, consult, Crippled Children's Serv, 69-71. *Mem:* AAAS; Asn Univ Neuroradiol; Radiol Soc NAm; Inst Elec & Electronics Eng; AMA; Sigma Xi. *Res:* Postmortem cerebral angiography; cerebral microangiography; microtumor-circulation. *Mailing Add:* 5885 Wing Lake Rd Birmingham MI 48010

WOLLUM, ARTHUR GEORGE, II, b Chicago, Ill, July 26, 37; m 60; c 2. SOILS, MICROBIOLOGY. *Educ:* Univ Minn, BS, 59; Ore State Univ, MS, 62, PhD(soils), 65. *Prof Exp:* Forester, Gifford Pinchot Nat Forest, USDA, 59-60, res forester, Pac Northwest Forest & Range Exp Sta, 60-61; asst soils, Ore State Univ, 64-65, asst prof, 65-67; asst prof, NMex State Univ, 67-71; assoc prof soils, 71-76, chmn Ecol Prog, 85-89, PROF SOILS, NC STATE UNIV, 76- *Concurrent Pos:* Vis prof, Ohio State Univ, 78-79. *Mem:* Soc Am Foresters; fel Soil Sci Soc Am; Am Soc Microbiol; Sigma Xi; fel Am Soc Agron. *Res:* Microbiology of nodule-formation of nonleguminous plants; ecology of nitrogen fixing plants in forested ecosystems; nitrogen cycle in forested ecosystems; microbiology of environmental pollution; rhizobial ecology of stressed environments. *Mailing Add:* Dept Soil Sci NC State Univ Box 7619 Raleigh NC 27695-7619

WOLLWAGE, JOHN CARL, b Chicago, Ill, Oct 11, 14; wid; c 3. PAPER CHEMISTRY. *Educ:* Northwestern Univ, BS, 34; Lawrence Col, MS, 36, PhD(paper chem), 38. *Prof Exp:* Mem staff, Res Dept, Hammermill Paper Co, Pa, 36; mem staff chem res, Beveridge-Marvellum Co, Mass, 37; res chemist, Kimberly-Clark Corp, 39-40, 42, tech supt, Mill, 41-42, war prod develop, 43-45, mill mgr, 45-49, asst tech dir, 49-52, dir res, 52-55, mgr foreign opers, 55-59, gen mgr creped wadding mfg processes, Consumer Prod Div, 59-62, vpres mfg, 62-68, vpres, C W Mfg, Res & Eng, 68-71, vpres corp res & eng, 71-73; vpres res, Inst Paper Chem, 73-78; dir, Memline, Inc, Suring, 78-88; RETIRED. *Concurrent Pos:* Tech Asn Pulp & Paper Indus fel, 63-; dir, Upson Co, Lockport, NY, 78-84. *Mem:* Tech Asn Pulp & Paper Indust (pres, 63-64); Am Chem Soc; Can Pulp & Paper Asn; AAAS; Sigma Xi. *Res:* Alum and its effect on hydrogen ion concentration of paper; flocculation of paper making fibers. *Mailing Add:* Nine Lawrence Ct Appleton WI 54911

WOLLWAGE, PAUL CARL, b Appleton, Wis, Mar 15, 41; m 65; c 2. PULPING AND BLEACHING CHEMISTRY, ENVIRONMENTAL SCIENCES PULP & PAPER. *Educ:* St Olaf Col, BA, 63; Inst Paper Chem, MS, 66, PhD(chem), 69. *Prof Exp:* Sr res chemist, St Regis Tech Ctr, 69-79; BLEACHING SCIENTIST, WEYERHAUSER TECH CTR, 79- *Mem:* Tech Asn Pulp & Paper Indust. *Res:* Pulping and bleaching processes; pulp characterization; wood chemicals. *Mailing Add:* Weyerhaeuser Co Weyerhauser Tech Ctr 2G25 Tacoma WA 98477

WOLMA, FRED J, b Albuquerque, NMex, Dec 10, 16; m 43; c 2. MEDICINE, SURGERY. *Educ:* Univ Tex, BA, 40, MD, 43; Am Bd Surg, dipl, 53. *Prof Exp:* Intern, Med Col Va, 43-44, resident surg, 47-48; res physician, St Mary's Infirmary, Galveston, Tex, 46-47; resident surg, 48-51, from instr to assoc prof, 51-69, chief div gen surg, 67-70, PROF SURG, UNIV TEX MED BR GALVESTON, 69- *Concurrent Pos:* Consult, St Mary's Hosp & Galveston, Tex. *Mem:* AMA; Am Col Surg; Soc Surg Alimentary Tract; Am Asn Cancer Res; Am Asn Surg Trauma. *Res:* Clinical medicine; peripheral vascular surgery. *Mailing Add:* Dept Surg Univ Tex Med Br Galveston TX 77550

WOLMAN, ABEL, sanitary engineering; deceased, see previous edition for last biography

WOLMAN, ERIC, b New York, NY, Sept 25, 31; m 63; c 2. CANCER CONTROL. *Educ:* Harvard Univ, AB, 53, MA, 54, PhD(appl math), 57. *Prof Exp:* Mem tech staff, AT&T Bell Labs, 57-66, head Traffic Systs Anal Dept, 66-68, head Traffic Res Dept, 68-72, head Network Eng Dept, 72-77, head, Opers Res Tech & Database Res Dept, 77-80, head Advan Comput Systs Dept, 81-82, head, Human Performance Eng Dept, AT&T Bell Labs, 83-87; VPRES COMMUNITY RES, MICH CANCER FOUND, 88- *Concurrent Pos:* Vis lectr, Harvard Univ, 64; mem comt fire res, Nat Acad Sci-Nat Res Coun, 66-70; mem ad hoc eval panel for Fire Prog, 71-74, mem eval panel, Inst Appl Technol, 74-78, Nat Eng Lab, Nat Bur Standards, 78-80; mem, Working Group Info Technol, NSF, 80-81; chmn, bd trustees, Soc Indust & Appl Math, 81-82. *Mem:* Fel AAAS; Am Pub Health Asn; Soc Indust & Appl Math; Opers Res Soc Am; Sigma Xi. *Res:* Cancer detection, care and epidemiology. *Mailing Add:* Mich Cancer Found 110 E Warren Ave Detroit MI 48201-1379

WOLMAN, MARKLEY GORDON, b Baltimore, Md, Aug 16, 24; m 51; c 4. GEOLOGY. *Educ:* Johns Hopkins Univ, BA, 49; Harvard Univ, MA, 51, PhD(geol), 53. *Prof Exp:* GEOLOGIST, US GEOL SURV, 51-; interim provost, 87, 89-90, PROF GEOG & CHMN, DEPT GEOG & ENVIRON ENG, JOHNS HOPKINS UNIV, 58- *Honors & Awards:* Asn Am Geog Award, 72; John Wesley Powell Award, US Geol Surv, 89; Cullum Geog Medal, Am Geog Soc, 89. *Mem:* Nat Acad Sci; Geol Soc Am; Am Geophys Union; Asn Am Geog; Am Acad Arts & Sci; AAAS. *Res:* River morphology; water resources. *Mailing Add:* Dept Geog & Environ Eng Johns Hopkins Univ Baltimore MD 21218

WOLMAN, SANDRA R, b New York, NY, Nov 23, 33; m 63; c 2. PATHOLOGY, CYTOGENETICS. *Educ:* Radcliffe Col, AB, 55; NY Univ, MD, 59. *Prof Exp:* Intern path, Bellevue Hosp, New York, 59-60, resident, 60-63; asst pathologist, Morristown Mem Hosp, 64-66; asst pathologist, Monmouth Med Ctr, 66-67; asst prof clin path, Sch Med, NY Univ, 67-71, from asst prof to prof, 72-88; assoc attend pathologist, Bellevue & Univ Hosps, 76-85, attend pathologist, 85-88; ASSOC MED DIR, MICH CANCER FOUND, 88- *Concurrent Pos:* Teaching fel path, Sch Med, NY Univ, 62-64, Nat Cancer Inst res fel oncol, 63, Children's Cancer Res Found fel, 64; asst pathologist, Bellevue Hosp, 63-64, asst vis pathologist, 67-76, consult pathologist, Morristown Mem Hosp, 66-73; assoc pathologist, French & Polyclin Hosps, 70-71; mem, Path B Study Sect, NIH, 76-80; mem, Nat Large Bowel Cancer Proj, 80-83, VA Merit Rev Bd, Basic Med Sci, 82-84; chair, fac coun, NY Univ, 85-86, Gordon Conf on Cancer, 88; dir, Am Asn Cancer Res, 88- *Mem:* AAAS; Am Soc Human Genetics; Am Asn Cancer Res; Tissue Cult Asn. *Res:* Clinical cytogenetics; tumor cytogenetics; leukemic cell differentiation. *Mailing Add:* Cancer Genetics Mich Cancer Found 110 E Warren Ave Detroit MI 48201-1379

WOLNIAK, STEPHEN M, CELL MOTILITY, SIGNAL TRANSDUCTION. *Educ:* Univ Calif, Berkeley, PhD(bot), 79. *Prof Exp:* ASST PROF PLANT CELL BIOL & OPTICAL PRIN MICROS, UNIV MD, 81- *Mailing Add:* Dept Bot Univ Md College Park MD 20742

WOLNY, FRIEDRICH FRANZ, b Troppau, Czech, Aug 24, 31; US citizen; m 56; c 1. ORGANIC POLYMER CHEMISTRY. *Educ:* Munich Tech Univ, BS, 54, MS, 55. *Prof Exp:* Group leader resin res, Sued-West-Chemie, WGer, 56-58, asst res & develop dir, 59-64; chemist, 65-68, group leader, 69-72, mgr resin res, 73-83, VPRES RES, SCHENECTADY CHEM, INC, NY, 84- *Mem:* Electrochem Soc WGer; Am Chem Soc; Soc Automotive Eng. *Res:* Organic polymer chemistry, poly condensation products, phenol formaldehyde resins, applied in friction materials. *Mailing Add:* 686 Plank Rd Clifton Park NY 12065-2097

WOLOCHOW, HYMAN, b Richdale, Can, Feb 10, 18; nat US; m 43; c 2. MICROBIOLOGY. *Educ:* Univ Alta, BSc, 38, MSc, 40; Univ Calif, PhD(microbiol), 50. *Prof Exp:* Asst dairy bact, Univ Alta, 38-41; teaching asst bact, Univ Calif, 47-48 & plant nutrit, 48-50, asst res bacteriologist, 50-59; assoc bacteriologist, Naval Biol Lab, Naval Supply Ctr, Univ Calif, 59-73, assoc res bacteriologist, 73-80; RETIRED. *Concurrent Pos:* Consult, US Dept Defense. *Mem:* Sigma Xi. *Res:* Dairy bacteriology; enzyme chemistry; medical microbiology; aerobiology. *Mailing Add:* 18313 Pepper St Castro Valley CA 94546

WOLOCK, FRED WALTER, b Whitinsville, Mass, Mar 8, 26. MATHEMATICAL STATISTICS. *Educ:* Holy Cross Col, BS, 47; Cath Univ, MS, 49; Va Polytech Inst, PhD(math statist), 64. *Prof Exp:* Instr math, Lewis Col, 53-54; instr, Iona Col, 54-56; instr, St John's Univ, NY, 56-57; instr, Worcester Polytech Inst, 57-60; asst prof math & statist, Boston Col, 64-65; assoc prof, 65-77, PROF MATH & STATIST, SOUTHEASTERN MASS UNIV, 77- *Concurrent Pos:* Consult, Nat Ctr Air Pollution Control, USPHS, 65- & Berkshire Hathaway Co, 66- *Mem:* Inst Math Statist; Am Statist Asn; Math Asn Am. *Res:* Experimental design and applications of statistics to biological and medical sciences; industrial applications of statistics. *Mailing Add:* Dept Math Southeastern Mass Univ North Dartmouth MA 02747

WOLOCK, IRVIN, b Baltimore, Md, June 21, 23; m 51; c 3. MATERIALS ENGINEERING, COMPOSITES. *Educ:* Johns Hopkins Univ, BS, 43, ME, 49, DrEng, 50. *Prof Exp:* Res asst, SAM Labs, Columbia Univ, 43-44; chemist, Nat Bur Standards, 50-57; supvr mat res engr, Naval Res Lab, DC, 57-73, liaison scientist, Off Naval Res, London, 73-74, SUPVR MAT RES ENGR, NAVAL RES LAB, DC, 74- *Concurrent Pos:* Ed, J Soc Plastics Engrs, 58-64. *Mem:* Am Soc Testing & Mat; Soc Advan Mat & Process Eng; AAAS; Sigma Xi. *Res:* Failure behavior of plastics and composites; transparent plastics; composites applications. *Mailing Add:* 401 Scott Dr Silver Spring MD 20904-1065

WOLOS, JEFFREY ALAN, SOLUBLE MEDIATORS, T-CELLS. *Educ:* NY Upstate Med Ctr, PhD(microbiol), 79. *Prof Exp:* ASST PROF IMMUNOL, THOMAS JEFFERSON UNIV, 81- *Mailing Add:* Cell Biol Merrell Dow Res Inst 2110 E Galbraith Rd Cincinnati OH 45215

WOLOSEWICK, JOHN J, b Chicago, Ill, July 26, 45. CYTOSKELETON ELECTRON MICROSCOPY. *Educ:* Univ Chicago, PhD(zool & cell biol), 74. *Prof Exp:* ASSOC PROF ANAT, UNIV ILL, CHICAGO, 84- *Mailing Add:* Dept Anat & Cell Biol Univ Ill Chicago MC 512 PO Box 6998 Chicago IL 60680

WOLOSHIN, HENRY JACOB, b Philadelphia, Pa, Aug 23, 13; m 48; c 3. RADIOLOGY. *Educ:* Temple Univ, BS, 34, MD, 38; Am Bd Radiol, dipl, 47. *Prof Exp:* Instr, Sch Med, Temple Univ, 47-48, assoc, 48-55, from asst prof to prof radiol, 55-89, assoc, Univ Hosp, 47-89; RETIRED. *Honors & Awards:* Russell P Moses Mem Award. *Mem:* AMA; fel Am Col Radiol; Am Roentgen Ray Soc. *Mailing Add:* Foxcroft Apts No 627 Jenkintown PA 19046

WOLOVICH, WILLIAM ANTHONY, b Hartford, Conn, Oct 15, 37; m 59; c 2. ELECTRICAL ENGINEERING, CONTROL SYSTEMS. *Educ:* Univ Conn, BS, 59; Worcester Polytech Inst, MS, 61; Brown Univ, PhD(elec sci), 70. *Prof Exp:* Systs analyst, Res Lab, United Aircraft Corp, 59-61; control systs engr, Electronics Res Labs, NASA, 64-70; from asst prof to assoc prof eng, 70-77, PROF ENG, BROWN UNIV, 77- *Mem:* Inst Elec & Electronics Engrs. *Res:* Linear multivariable control systems; computational methods for the analysis and synthesis of large scale systems; stability and optimization of dynamical systems. *Mailing Add:* Div Eng Brown Univ Providence RI 02912

WOLOWYK, MICHAEL WALTER, b Rain-Amlech, Ger, Oct 26, 42; Can citizen; m 66; c 2. PHARMACOLOGY, CELL BIOLOGY. *Educ:* Univ Alta, BSc, 65, PhD(pharmacol), 69. *Prof Exp:* Asst prof clin pharm, 71-75, assoc prof pharm, 75-81, PROF PHARM, UNIV ALTA, 81-, HON PROF PHARMACOL, 81- *Concurrent Pos:* Med Res Coun fel, Wellcome Res Labs, Beckenham, Eng, 69-71, physiol lab, Cambridge, Eng, 77-78. *Mem:* Pharmacol Soc Can; Micros Soc Can. *Res:* Cardiovascular pharmacology and cellular mechanisms of heart disease; calcium channel antagonist/agonist drugs; membrane transport of ions, amino acids and drugs; marine pharmacology. *Mailing Add:* Dept Pharm Univ Alta Edmonton AB T6G 2M7 Can

WOLPE, JOSEPH, psychiatry, psychology, for more information see previous edition

WOLPOFF, MILFORD HOWELL, b Chicago, Ill, Oct 28, 42; m 85; c 1. PHYSICAL ANTHROPOLOGY. *Educ:* Univ Ill, Urbana, AB, 64, PhD(anthrop), 69. *Prof Exp:* Asst prof anthrop, Case Western Reserve Univ, 68-70; assoc prof, 70-78, PROF ANTHROP, UNIV MICH, ANN ARBOR, 78- *Concurrent Pos:* NSF res grant, Transvaal Mus, Nat Mus Kenya, 72; NSF grant human evolution, 76-79; NSF, Rackham Found & Nat Acad Sci grants. *Mem:* Fel Am Asn Phys Anthrop; fel Am Anthrop Asn; Sigma Xi. *Res:* Paleoanthropology; human origins and evolution; evolution theory; biomechanics; computer analysis; dental variation; worldwide fossil hominid study. *Mailing Add:* Dept Anthrop Univ Mich Main Campus Ann Arbor MI 48109-1382

WOLSEY, WAYNE C, b Battle Creek, Mich, Nov 12, 36; m 65; c 2. INORGANIC CHEMISTRY. *Educ:* Mich State Univ, BS, 58; Univ Kans, PhD(chem), 62. *Prof Exp:* Sr res chemist chem div, Pittsburgh Plate Glass Co, Ohio, 62-65; asst prof chem, 65-72, assoc prof, 72-80, PROF CHEM, MACALESTER COL, 80- *Concurrent Pos:* Vis asst prof, Ariz State Univ, 71-72; vis sr res fel, Bristol Univ, 78-79; guest scientist, Oak Ridge Nat Lab, 87-88. *Mem:* AAAS; Am Chem Soc; Am Asn Univ Prof. *Res:* Coordination compounds; chemistry of chlorine and nitrogen compounds; laboratory computing; chemistry experiments. *Mailing Add:* Dept Chem Macalester Col St Paul MN 55105-1899

WOLSKY, ALAN MARTIN, b Brooklyn, NY, May 17, 43; m 69; c 1. PHYSICS, RESEARCH ADMINISTRATION. *Educ:* Columbia Col, AB, 64; Univ Pa, MS, 65, PhD(physics), 69. *Prof Exp:* Nat Res Coun fel math physics, Courant Inst Math Sci, NY Univ, 70, vis mem, 71; asst prof physics, Temple Univ, 71-75; dir technol eval, 86-89, MEM STAFF, ARGONNE NAT LAB, 75-, SR ENERGY SYSTS SCIENTIST, EES DIV, 85-, ASSOC DIV DIR, ES DIV, 89- *Concurrent Pos:* Prin investr, Argonne Nat Lab, 78- *Mem:* Am Phys Soc; Int Asn Economists. *Res:* Feasibility of reducing petroleum consumption by using biomass as an alternative source of hydrocarbons and by recycling, recovering or upgrading industrial waste materials; feasibility of recovering carbon dioxide from stationary combustion to provide carbon dioxide for enhanced oil recovery; mathematical economics; theory of social choice and input-output analysis; potential uses of high temperature superconductors. *Mailing Add:* Argonne Nat Lab 9700 Cass Ave Argonne IL 60439

WOLSKY, ALEXANDER, b Budapest, Hungary, Aug 12, 02; US citizen; m 40; c 2. HISTORY & THEORIES OF EVOLUTION. *Educ:* Univ Budapest, DPhil, 28. *Prof Exp:* Researcher & dir biol, Biol Res Inst Hungary Acad Tihany, 29-45; prof gen zool, Univ Budapest, 45-48; prin sci officer gen sci, UNESCO, 48-54; prof exp embryol, Fordham Univ, NY, 54-66; prof biol, Marymount Col, Tarrytown, 66-72; adj prof radiation biol, Med Sch, NY Univ, 72-86; adj prof biol, Concordia Univ, Montreal, 87-91; RETIRED. *Mem:* Hungarian Acad Sci; Sigma Xi. *Res:* Analysis of development and regeneration of arthropods (insects, crustaceans) and amphibians, both morphological and physiological (respiration); functions (optics) of the arthropodan eye; evolution (both progressive and regressive) of eyes and the mechanism of evolution in general. *Mailing Add:* 407 Mount Echo Rd RR 4 Sutton Junction PQ J0E 2K0

WOLSKY, MARIA DE ISSEKUTZ, b Kolozsvar, Romania, June 17, 16; nat US; m 40; c 2. CELL BIOLOGY, HISTORY & PHILOSOPHY OF SCIENCE. *Educ:* Med Univ, Budapest, MD, 43. *Prof Exp:* Asst pharmacol, Univ Budapest, 38-39; res assoc, Hungarian Biol Res Inst, 40-45; adj prof biol, NY Univ, 84-85; from instr to prof biol, 56-81, EMER PROF, MANHATTANVILLE COL, 81- *Concurrent Pos:* Res assoc, Fordham Univ, 58-62. *Mem:* AAAS. *Res:* Cell physiology; theories of evolution; cell differentiation. *Mailing Add:* 4800 Maisonneuve Blvd W/150 Montreal PQ H3Z 1M2

WOLSKY, SUMNER PAUL, b Boston, Mass, Aug 21, 26; m 50; c 2. PHYSICAL CHEMISTRY. *Educ:* Northeastern Univ, BS, 47; Boston Univ, MA, 49, PhD(chem), 52. *Prof Exp:* Mem res staff, Raytheon Co, 52-61; dir lab phys sci, P R Mallory & Co, Inc, 61-74, dir res & develop lab phys sci, 74-76, vpres res & develop, 76-81; PRES, ANSUM ENTERPRISES, INC, 81- *Mem:* AAAS; Am Chem Soc; Am Phys Soc; Inst Elec & Electronics Engrs; Electrochem Soc. *Res:* Batteries; semiconductors; surface physics; physical electronics; sputtering; thin films; vacuum microbalances; environment; occupational health. *Mailing Add:* 1900 Coconut Rd Boca Raton FL 33432

WOLSSON, KENNETH, b Paterson, NJ, Oct 12, 33. MATHEMATICS. *Educ:* Brooklyn Col, BS, 54; Columbia Univ, AM, 55; NY Univ, PhD(math), 62. *Prof Exp:* Prin res mathematician, Repub Aviation Corp, 62-63, sr res mathematician, 63-64; asst prof, 64-77, ASSOC PROF MATH, FAIRLEIGH DICKINSON UNIV, TEANECK, 77- *Mem:* Am Math Soc; Math Asn Am. *Res:* Partial and ordinary differential equations. *Mailing Add:* 697 W End Ave New York NY 10025

WOLSTENHOLME, DAVID ROBERT, b Bury, Eng, Nov 5, 37; m 63. MOLECULAR BIOLOGY, CELL BIOLOGY. *Educ:* Univ Sheffield, BSc, 58, PhD(genetics), 61, DSc(genetics), 73. *Prof Exp:* Fel zool, Univ Wis, 61-62, res assoc, 62-63; vis lectr genetics, Univ Groningen, 63-64; res fel biol, Beermann Div, Max Planck Inst Biol, 64-67; res assoc, Whitman Lab, Univ Chicago, 67-68; assoc prof, Kans State Univ, 68-70; assoc prof, 70-72, chmn, dept biol, 80-83, PROF BIOL, UNIV UTAH, 72- *Concurrent Pos:* Nat Inst Gen Med Sci res career develop award, Univ Utah, 72-76; mem, Molecular Biol Study Sect, NIH, 73-77. *Mem:* Brit Genetical Soc; Am Soc Cell Biol; Genetics Soc Am. *Res:* Structure, replication, and evolution of animal and plant mitochondrial DNA. *Mailing Add:* Dept Biol Univ Utah Salt Lake City UT 84112

WOLSTENHOLME, WAYNE W, b Philadelphia, Pa, Dec 24, 48; m 85; c 3. MUSCARINIC RECEPTORS, OPIOID RECEPTORS. *Educ:* Villanova Univ, BS, 70; Temple Univ, MS, 74, PhD(pharmacol), 78. *Prof Exp:* Asst prof pharmacol, Al-Faateh Univ, Tripol, Libya, 78-79, Univ PR Pharm Sch, 80-83; ASSOC PROF PHARMACOL, MED SCH, UNIV CENT DEL CARIBE, 83- *Concurrent Pos:* Coordr, Computer Facil, Univ Cent del Caribe, 85-91, prin investr, Neurosci Proj, 86-91. *Mem:* Sigma Xi; AAAS. *Res:* Receptor identification of muscarinic receptors in the isolated vasdeferens and neuromodulatory role of opioids and peptides in the same preparation. *Mailing Add:* Dept Pharmacol Univ Cent del Caribe Call Box 60-327 Bayamon PR 00621-6032

WOLSZON, JOHN DONALD, b Chicago, Ill, Jan 27, 29; m 53; c 6. ANALYTICAL CHEMISTRY. *Educ:* Univ Ill, BS, 51; Pa State Univ, PhD(anal chem), 55. *Prof Exp:* Instr chem, Marshall Univ, 55-58; asst prof, Univ Mo, 58-63; ASSOC PROF CHEM, PURDUE UNIV, 63- *Mem:* Am Chem Soc; Water Pollution Control Fedn; Am Water Works Asn. *Res:* Methods of chemical analysis; water and waste water chemistry. *Mailing Add:* Sch Civil Eng Purdue Univ Lafayette IN 47907

WOLT, JEFFREY DUAINE, b Grand Forks, NDak, Oct 18, 51; m 80; c 2. SOIL CHEMISTRY, ENVIRONMENTAL CHEMISTRY. *Educ:* Colo State Univ, BS, 73; Auburn Univ, MS, 76, PhD(soil chem), 79. *Prof Exp:* From asst prof to assoc prof plant & soil sci, Univ Tenn, 84-89; SR SCIENTIST, ENVIRON CHEM LAB, DOWELANCO, 89- *Concurrent Pos:* Consult, 83-; vis scientist award, Am Soc Agron, 89; vis scientist agron & soils, Univ Hawaii, 89-90. *Mem:* Am Soc Agron; Soil Sci Soc Am; AAAS; Sigma Xi. *Res:* Methods development and applications in soil solution chemistry; soil chemistry of pesticides; organic waste management; acid precipitation impacts on plants and soils; solute transport and retention in the vadose zone. *Mailing Add:* Environ Chem Lab DowElanco 9001 Bldg Midland MI 48641

WOLTER, J REIMER, b Halstenbek, Ger, May 9, 24; US citizen; m 52; c 4. OPHTHALMOLOGY, PATHOLOGY. *Educ:* Univ Hamburg, MD, 49. *Prof Exp:* Intern med, Med Sch, Univ Hamburg, 49-50, resident instr ophthal, 50-53; res assoc path, 53-56, from asst prof to assoc prof ophthal, 56-64, PROF OPHTHAL, MED SCH, UNIV MICH, ANN ARBOR, 64- *Concurrent Pos:* Chief ophthal serv, Vet Admin Hosp, Ann Arbor, 62-85; ed, J Pediat Ophthal, 67-81. *Mem:* AMA; Am Ophthal Soc; Am Acad Ophthal & Otolaryngol; Asn Res Ophthal; Ger Ophthal Soc. *Res:* Clinical ophthalmology; opthalmic pathology. *Mailing Add:* Eye Clin Univ Hosp Ann Arbor MI 48105

WOLTER, JAN D(ITHMAR), b Ann Arbor, Mich, Jan 31, 59. ASSEMBLY SEQUENCE PLANNING, COMPUTER-AIDED MECHANICAL DESIGN. *Educ:* Univ Mich, BS, 81, MS, 83, PhD(computer eng), 88. *Prof Exp:* ASST PROF COMPUTER SCI, TEX A&M UNIV, 88- *Concurrent Pos:* NSF presidential young investr award, 90. *Res:* Assembly sequence planning. *Mailing Add:* Dept Computer Sci Texas A&M Univ College Station TX 77843-3112

WOLTER, JANET, b Chicago, Ill, Apr 24, 26; m 73. MEDICAL ONCOLOGY. *Educ:* Cornell Col, AB, 46; Univ Ill, MD, 50. *Prof Exp:* Instr med, Univ Ill Col Med, 55-57, clin asst prof, 57-70, asst prof med, 70-71; from asst prof to assoc prof, 71-81, PROF MED, RUSH MED COL, PRESBY-ST LUKE'S MED CTR, 81- *Concurrent Pos:* Staff physician respiratory, Univ Ill Hosp, 55-57; tumor clin physician oncol, Univ Ill Res & Educ Hosp, 63; res assoc oncol, Presby-St Luke's Hosp, 63-66, consult oncol, 66; consult oncol, WSuburban Hosp, 66; asst attend physician, Presby-St Luke's Med Ctr, 67-70, assoc attend physician, 70-72, sr attend physician, 72, consult oncol, 72-; consult oncol, Copley Hosp, 72- *Mem:* Am Soc Clin Oncol; Am Asn Cancer Res; Eastern Coop Oncol Group; Nat Surg Adj Breast Proj. *Res:* Cancer chemotherapy. *Mailing Add:* Sect Med Oncol 1725 W Harrison St Suite 830 Chicago IL 60612

WOLTER, KARL ERICH, b New York, NY, Nov 8, 30. PLANT PHYSIOLOGY. *Educ:* State Univ NY Col Forestry, Syracuse, BS, 58; Univ Wis, PhD(plant physiol), 64. *Prof Exp:* Proj leader biodegradation & biotechnol res, Forest Prod Lab, Forest Serv, USDA, 63-86; ASST PROF & SR SCIENTIST, DEPT HORT, UNIV WIS, 86- *Concurrent Pos:* Vis prof, Iowa State Univ, 71-72; Japanese Sci & Technol grant; vis scientist, Tsukuba, Japan, 81-82. *Mem:* Am Soc Plant Physiol; Scand Soc Plant Physiol; Int Plant Growth Substances Asn. *Res:* Growth, differentiation and nutrition of plants and secondary cambium, specifically tree species; host pathogen interactions, action, and characterization of plant cell wall degrading enzymes. *Mailing Add:* Dept Hort Univ Wis Madison WI 53703

WOLTER, KIRK MARCUS, b Evanston, Ill; m 81; c 3. SCIENCE ADMINISTRATION, MATHEMATICAL STATISTICS. *Educ:* St Olaf Col, BA, 70; Iowa State Univ, MS, 72, PhD(statist), 74. *Prof Exp:* Chief, Statist Res Div, US Bur Census, 83-88; VPRES, A C NIELSEN, 88- *Concurrent Pos:* Prof lectr, George Washington Univ, 75-; mem coun, Int Asn Survey Statisticians, 85-89. *Honors & Awards:* Bronze Medal & Silver Medal, US Dept Com. *Mem:* Fel Am Statist Asn; Int Statist Inst; Int Asn Surv Statisticians; Inst Math Statist. *Res:* Variance estimation for complex sample surveys; measurement of decennial census undercount; evaluation of statistical programs; statistical policy; statistical administration. *Mailing Add:* A C Nielsen Nielsen Plaza Northbrook IL 60062

WOLTERINK, LESTER FLOYD, b Marion, NY, July 28, 15; m 38; c 2. BIOPHYSICS. *Educ:* Hope Col, AB, 36; Univ Minn, MA, 40, PhD(zool), 43. *Prof Exp:* Lab asst, Hope Col, 34-36; lab asst, Univ Minn, 36-41; from instr to prof physiol, 41-84, asst exp sta, 45-81, EMER PROF PHYSIOL, MICH STATE UNIV, 84- *Concurrent Pos:* Assoc physiologist, Argonne Nat Lab, Ill, 48; proj scientist biosatellite proj, Ames Res Ctr, NASA, 65-66; mem subcomt nitrogen oxides, Nat Acad Sci-Nat Res Coun, 73-75. *Mem:* Am Soc Zool; Biophys Soc; Am Physiol Soc; Radiation Res Soc; Brit Biol Eng Soc. *Res:* Biological rhythms, oscillatory time series; physiological models. *Mailing Add:* Dept Physiol Mich State Univ East Lansing MI 48824

WOLTERS, ROBERT JOHN, b St Louis, Mo, Nov 7, 40; m 70; c 2. PHARMACEUTICAL CHEMISTRY, PHARMACOLOGY. *Educ:* St Louis Col Pharm, BS, 65; NDak State Univ, MS, 68, PhD(pharmaceut chem), 71. *Prof Exp:* SUPVRY CHEMIST, US FOOD & DRUG ADMIN, ROCKVILLE, 71- *Mem:* Am Chem Soc; Am Pharmaceut Asn; Sigma Xi. *Res:* Synthesis of potential pharmaceutically active compounds; mescaline analogs. *Mailing Add:* 18645 Hedgegrove Terr Olney MD 20832

WOLTERSDORF, OTTO WILLIAM, JR, b Philadelphia, Pa, June 19, 35; m 57; c 2. MEDICINAL CHEMISTRY. *Educ:* Gettysburg Col, AB, 56; Pa State Univ, MS, 59. *Prof Exp:* Res assoc org synthesis, 59-65, from res chemist to sr res chemist, 65-73, res fel, 73-83, SR RES FEL, MERCK SHARP & DOHME RES LABS, 83- *Mem:* Am Chem Soc. *Res:* Organic synthesis; radioisotope synthesis. *Mailing Add:* 200 Dorset Way Chalfont PA 18914-2322

WOLTHUIS, ROGER A, b Champaign, Ill, Mar 30, 37; m 65; c 2. CARDIOVASCULAR PHYSIOLOGY. *Educ:* Univ Mich, BA, 63; Mich State Univ, MS, 65, PhD(physiol), 68. *Prof Exp:* Prin res scientist, Technol Inc, Cardiovasc Res Lab, NASA Johnson Space Ctr, 68-74; chief cardiovasc res internal med, Sch Aerospace Med, Brooks AFB, 74-79; sr staff scientist, Medtronic Inc, 79-81; VPRES MED PROD, METRICOR INC, 82- *Concurrent Pos:* Partic, Apollo & Skylab Med Experiments. *Mem:* Am Col Cardiol; Am Heart Asn; Aerospace Med Asn; Am Physiol Soc; Inst Elec & Electronics Engrs. *Res:* Studies on man's physiological adaptation to normal and zero gravity environments; studies on exercise stress testing for detection of coronary artery disease; studies on hypertension and its treatment; development of fiber optic sensors & systems. *Mailing Add:* Metricor Inc 18800 142nd Ave NW Woodinville WA 98072

WOLTZ, FRANK EARL, b Bethlehem, Pa, Nov 29, 16; m 47; c 2. PHYSICAL CHEMISTRY. *Educ:* Bethany Col, WVa, BS, 38; Univ WVa, MS, 40, PhD(chem), 43; Ohio Univ, MS, 70. *Prof Exp:* Mat engr, Westinghouse Elec Corp, Pa, 42-44; lab mgr, Goodyear Synthetic Rubber Corp, Ohio, 44-47, res chemist, Goodyear Tire & Rubber Co, 47-50, rubber compounder, 50-53, supvr opers anal, Goodyear Atomic Corp, Piketon, 53-67, supt eng develop, 67-83, supt nuclear criticality safety, 83-85; RETIRED. *Concurrent Pos:* Consult radiation adv bd, Ohio Dept Health, 74- *Mem:* Am Nuclear Soc. *Res:* Electrical insulating varnishes; analytical test methods; rubber manufacture; inert electrode systems; rubber compounding for use in tire manufacturing; fluid flow; material and energy optimization of gaseous diffusion processes; computer control of chemical processes; nuclear criticality safety. *Mailing Add:* 400 E Third St Waverly OH 45690

WOLTZ, SHREVE SIMPSON, b Clifton, Va, Apr 9, 24; m 47; c 2. HORTICULTURE. *Educ:* Va Polytech Inst, BS, 43; Rutgers Univ, PhD(soils, plant physiol), 51. *Prof Exp:* Dir fertilizer res, Baugh & Sons Co, 51-53; asst horticulturist, 53-62, assoc plant physiologist, 62-68, plant physiologist, Agr Res & Educ Ctr, Bradenton, 53-62, PROF PLANT PHYSIOL, UNIV FLA, 68- *Mem:* Am Soc Hort Sci; Am Soc Plant Physiol; Am Phytopath Soc; Scand Soc Plant Physiol. *Res:* Plant nutrition; gladiolus and chrysanthemum culture; soil fertility; physiology of disease. *Mailing Add:* Res & Educ Ctr 9611 Oakrun Dr Bradenton FL 34202

WOLVEN-GARRETT, ANNE M, b New York, NY, Feb 2, 25; m 78. REGULATORY AFFAIRS, RISK ASSESSMENT. *Educ:* Hunter Col, BA, 45. *Prof Exp:* Group leader pharmacol, Schering Corp, NJ, 45-47; group leader, Leberco Labs, NJ, 47-52, from asst dir to assoc dir labs, 52-72; mgr toxicol, Alza Res Corp, 72-75; sr toxicologist, Shell Chem Co, 75-76; environ toxicologist, Syntex Corp, 76-78; CONSULT TOXICOL & REGULATORY AFFAIRS, A M WOLVEN, INC, 78- *Concurrent Pos:* Round Table discussant, Gordon Conf Toxicol & Safety Eval, 70, chairperson, 76 & 84; lectr, Ctr Prof Advan, 72 & 88; consult toxicol, Nat Inst Drug Abuse, 74-75 & Ministry Health, Mex, 75; mem hazardous mat adv comt, Environ Protection Agency, 75-77, mem sci adv bd, 75-79; lectr, Howard Univ, 77, Int Chem Consult, 90. *Mem:* AAAS; Soc Toxicol; Soc Cosmetic Chem; Am Soc Microbiol; NY Acad Sci; Soc Comp Ophthal. *Res:* Eye and skin irritation and absorption phenomena; pesticide toxicology. *Mailing Add:* A M Wolven Inc 175 W Wieuca Rd Suite 118 Atlanta GA 30342

WOLVERTON, BILLY CHARLES, b Scott Co, Miss, Oct 13, 32; m 55; c 1. CHEMISTRY. *Educ:* Miss Col, BS, 60; Occidental Univ, PhD(environ eng), 78. *Prof Exp:* Res asst, Med Ctr, Univ Miss, 60-63; res chemist, US Naval Weapons Lab, 63-65; br chief res chemist, Air Force Armament Lab, 65-71; ENVIRON & SR SCIENTIST, NAT SPACE TECHNOL LAB, BAY ST LOUIS, MISS, 71- *Concurrent Pos:* Mem Panel Unconventional Approaches to Aquatic Weed Control & Utilization, Nat Acad Sci, 75- *Honors & Awards:* Super Sci Achievement Award, Dept Navy, 65; Sci Technol Utilization Award, Am Inst Aeronaut & Astronaut, 70; Except Sci Serv Medal, NASA, 75. *Mem:* Water Pollution Control Fedn; AAAS; Econ Botany Soc. *Res:* Vascular, aquatic plants as biological filtration systems for removing domestic and industrial pollutants from wastewater; utilization of harvested plant material as renewable sources of feed, fertilizer and methane; converting raw sewage to potable water; use of plants for purifying and revitalizing air inside energy-efficient homes and other closed facilities; converting raw sewage to potable water. *Mailing Add:* Wolverton Environ Serv 726 Pine Grove Rd Picayune MS 39466

WOLYNES, PETER GUY, b Chicago, Ill, Apr 21, 53. CONDENSED MATTER CHEMICAL PHYSICS. *Educ:* Ind Univ, AB, 71; Harvard Univ, PhD(chem physics), 76. *Prof Exp:* Fel, Mass Inst Technol, 75-76; asst prof, Harvard Univ, 76-79, assoc prof, 79-80; assoc prof, 80-83, PROF CHEM, UNIV ILL, URBANA-CHAMPAIGN, 83-, PROF PHYSICS, 85- *Concurrent Pos:* Vis scientist, Max Planck Inst Biophys Chem, 77; vis prof, Inst Molecular Sci, Okazaki, Japan, 82. *Honors & Awards:* Award in Pure Chem, Am Chem Soc, 86. *Mem:* Nat Acad Sci; Am Chem Soc; NY Acad Sci; AAAS; Am Phys Soc. *Res:* Theory of chemical dynamic phenomena in condensed phases, especially kinetics, tunneling and electronic structure in liquids; theory of the glassy state; biophysical applications. *Mailing Add:* Dept Chem Noyes Lab Box 32-1 505 S Mathews Ave Urbana IL 61801

WOLYNETZ, MARK STANLEY, b Kitchener, Ont, Nov 11, 45. STATISTICS. *Educ:* Univ Waterloo, BMath, 69, MMath, 70, PhD(statist), 74. *Prof Exp:* Statistician rd safety, Ministry Transp Can, 74-75; STATISTICIAN AGR RES, DEPT AGR, CAN, 75- *Res:* Application of statistical techniques in agriculture research; specifically in problems of categorical data, data screening, censored observations. *Mailing Add:* 1514 Edgediffe Ottawa ON K1Z 8G1 Can

WOMACK, EDGAR ALLEN, JR, b Humboldt, Tenn, Oct 29, 42; m 63; c 2. ENERGY RESEARCH & DEVELOP. *Educ:* Mass Inst Technol, BS, 63, MS, 65, PhD(physics), 69. *Prof Exp:* Br engr reactor develop, br chief & asst div dir, US AEC Comn, Washington DC, 68-75; sr tech consult reactor plant eng, eng mgr, mgr proj mgt, vpres sales, 75-85, VPRES RES & DEVELOP, MCDERMOTT INC, BABCOCK & WILCOX CO, 85- *Concurrent Pos:* Lectr, Cath Univ Am, 72-73; mem Indust Adv Bd, Am Soc Mech Engrs; guest lectr, Mass Inst Technol, 80-82; chmn, fed sci comt, Indust Res Inst, 88-89. *Mem:* Sigma Xi; AAAS; Am Asn Artificial Intelligence; Am Soc Mech Engrs; Indust Res Inst. *Res:* Energy systems, instruments materials and mechanics. *Mailing Add:* 1562 Beeson St Alliance OH 44601

WOMACK, FRANCES C, b Owensboro, Ky, Mar 23, 31; m 53; c 2. GENETICS, ENZYMOLOGY. *Educ:* Vanderbilt Univ, BA, 52, MA, 55, PhD(biol), 62. *Prof Exp:* Asst prof genetics, 62-63 & 64-65, res assoc, 63-64, res assoc enzymol, 65-72, ASST PROF ENZYMOL, SCH MED, VANDERBILT UNIV, 72- *Res:* Protein structure, function and their relation to genetic information; bacteriophage genetics. *Mailing Add:* 418 Lynwood Blvd Nashville TN 37205

WOMACK, JAMES E, b Anson, Tex, Mar 30, 41; m 63; c 2. GENETICS. *Educ:* Abilene Christian Col, BS, 64; Ore State Univ, PhD(genetics), 68. *Prof Exp:* From asst prof to assoc prof biol, Abilene Christian Col, 68-73; vis scientist, Jackson Lab, 73-75, staff scientist, 75-77; PROF VET PATH & GENETICS, TEX A&M UNIV, 77- *Mem:* AAAS; Genetics Soc Am; Am Genetics Asn (pres, 85); Am Soc Human Genetics. *Res:* Comparative gene mapping; somatic cell genetics of cattle; mammalian developmental genetics; gene transfer in animals; mapping the bovine genome. *Mailing Add:* Dept Vet Path Tex A&M Univ College Station TX 77843

WOMBLE, DAVID DALE, b Coffeyville, Kans, Oct 10, 49; m 72. MOLECULAR BIOLOGY, BIOCHEMISTRY. *Educ:* Ohio Univ, BS, 71; Univ Wis-Madison, PhD(biochem), 76. *Prof Exp:* Trainee biochem, Univ Wis-Madison, 71-75, res asst, 75-76, res assoc molecular biol, 76-77, trainee pathobiol, 77-79, asst scientist molecular biol, 79-81; sr res assoc, 81-86, RES ASST PROF, MED SCH, NORTHWESTERN UNIV, 86- *Mem:* Am Soc Microbiol; AAAS; Sigma Xi. *Res:* Nuclei acid structure and organization; regulation of replication and genetic expression. *Mailing Add:* Dept Cell Molecular & Struct Biol Northwestern Univ Med Sch 303 E Chicago Ave Chicago IL 60611

WOMBLE, EUGENE WILSON, b High Point, NC, June 27, 31; m 59; c 4. MATHEMATICS. *Educ:* Wofford Col, BS, 52; Univ NC, Chapel Hill, MA, 59; Univ Okla, PhD(math), 70. *Prof Exp:* Teacher, Kernersville High Sch, 56-58; instr math, Wake Forest Col, 59-61; asst prof, Pfeiffer Col, 61-66; spec instr, Univ Okla, 69-70; prof, 70-72, CHARLES A DANA PROF MATH, PRESBY COL, 72- *Mem:* Math Asn Am; Nat Coun Teachers Math. *Res:* Foundations of convexity; convexity structures. *Mailing Add:* Two Shell Creek Dr Clinton SC 29325

WOMMACK, JOEL BENJAMIN, JR, b Benton, Ky, Dec 5, 42; m 67; c 2. AGRICULTURAL CHEMISTRY. *Educ:* David Lipscomb Col, BS, 64; Vanderbilt Univ, PhD(org chem), 68. *Prof Exp:* Res chemist, 68-74, res supvr, 74-76, res mgr, 76-78, gen supvr, 78-79, asst mgr, 80-81, site mgr, 81-84, PLANNING MGR, BIOCHEM DEPT, E I DU PONT DE NEMOURS & CO, INC, 85- *Mem:* Am Chem Soc. *Mailing Add:* 12 Wineberry Dr Hockessin DE 19707-2126

WONDERGEM, ROBERT, b Sheboygan, Wis, Jan 17, 50; m 72; c 2. BIOLOGICAL TRANSPORT, ELECTROPHYSIOLOGY. *Educ:* Calvin Col, BS, 72; Med Col Wis, Milwaukee, PhD(physiol), 77. *Prof Exp:* Nat Cancer Inst fel cell biol, McArdle Lab Cancer Res, Univ Wis, Madison, 77-78; from asst prof to assoc prof, Quillen-Dishner Col Med, E Tenn State Univ, 78-89, PROF PHYSIOL, JAMES H QUILLEN COL MED, ETENN STATE UNIV, 89- *Concurrent Pos:* Prin investr, Mt Desert Island Biol Lab, Maine, 85- *Mem:* Am Physiol Soc; Am Soc Cell Biol; Soc Gen Physiologists; Am Asn Study Liver Dis; Biophys Soc. *Res:* Regulation of biological transport and cell volume regulation in liver cells. *Mailing Add:* Dept Physiol James H Quillen Col Med ETenn State Univ PO Box 19 780A Johnson City TN 37614

WONDERLING, THOMAS FRANKLIN, b Utica, Ohio, Feb 4, 15; m 41; c 2. AGRICULTURE. *Educ:* Ohio State Univ, BS, 39. *Prof Exp:* Teacher bd educ, Ohio, 39-45; farm mgr, Tiffin State Hosp, State Dept Pub Welfare, Ohio, 45-46, farm mgr, Lima State Hosp, 46-48; supt outlying farms, Ohio Agr Exp Sta, 48-63; coordr res opers, Ohio Agr Res & Develop Ctr, 63-69, coordr res opers & phys plant, 69-74; coordr res opers & supvr, North Appalachian Water Exp Sta, 74-79; RETIRED. *Mailing Add:* 4353 S Honeytown Rd Apple Creek OH 44606

WONG, ALAN YAU KUEN, b Hong Kong, Feb 6, 37; Can citizen; m 67; c 3. PHYSICS, BIOPHYSICS. *Educ:* Dalhousie Univ, BSc, 62, MSc, 63, PhD(biophys), 67. *Prof Exp:* Res assoc comput sci, 66-68, lectr biophys, 68-71, from asst prof to assoc prof biophys, 71-84, PROF, DALHOUSIE UNIV, 84- *Concurrent Pos:* Can Heart Found fel biophys & bioeng res lab, Dalhousie Univ, 68-71, Med Res Coun Can res scholar, 71-76; fel circulation, Am Heart Asn. *Mem:* Biophys Soc; Soc Math Biol; NY Acad Sci; Biophys Soc Can; Can Physiol Soc. *Res:* Excitation-contraction coupling of cardiac muscle; coronary flow; sodium-calcium exchange in excitable tissue; ventricular dynamics; intrathoracic cardiovascular neurons. *Mailing Add:* Dept Physiol & Biophys Dalhousie Univ Halifax NS B3H 4H7 Can

WONG, ALFRED YIU-FAI, b Macao, Portugal, Feb 4, 37; m 65. PLASMA PHYSICS. *Educ:* Univ Toronto, BASc, 58, MA, 59; Univ Ill, MSc, 61; Princeton Univ, PhD(plasma physics), 63. *Prof Exp:* Res assoc plasma physics lab, Princeton Univ, 62-64; from asst prof to assoc prof physics, 64-72, PROF PHYSICS, UNIV CALIF, LOS ANGELES, 72- *Concurrent Pos:* Sloan res fel, 66-68. *Honors & Awards:* Plasma Physics Res Award, Am Phys Soc, 85. *Mem:* Fel Am Phys Soc; Am Geophys Union. *Res:* Waves and radiation from plasmas; nonlinear phenomena; confinement system and space plasmas. *Mailing Add:* Dept Physics Univ Calif Los Angeles 405 Hilgard Ave Los Angeles CA 90024

WONG, ANTHONY SAI-HUNG, b Hong Kong, Apr 5, 51; US citizen; m 75; c 2. TECHNOLOGY INCUBATOR, SEMICONDUCTOR SENSORS. *Educ:* San Jose State Univ, BS, 74; Case Western Reserve Univ, MS, 76, PhD(biomed eng), 85. *Prof Exp:* Clin engr med instruments, Vet Admin Hosp, 75-76; process engr & res asst semiconductor sensors, Case Western Reserve Univ, 76-85, res assoc, 85-86; asst prof microelectronics, Cleveland State Univ, 86-87; staff scientist electronic mat, 87, proj mgr, 87, MGR SENSORS, ELECTRONIC MAT, GOULD ELECTRONICS & TECHNOL CTR, 88- *Concurrent Pos:* Sensor consult numerous clients, 81-87; integrated circuit lab mgr semiconductor mat, processes, sensors, Case Western Res Univ, 83-84; chmn, Microsensor & Fabrication, Sensors Expo 86, 86. *Mem:* Inst Elec & Electronics Engrs; Am Vacuum Soc; Am Chem Soc; Soc Photo-Optical Instrumentation Engrs; Mat Res Soc. *Res:* Technology management-develop and/or acquire technology for present and future electronic industry; develop semiconductor sensors for industrial and medical applications; technology gate-keeper-keep abreast in new development and chart out strategy for company growth. *Mailing Add:* 35129 Curtis Blvd Eastlake OH 44095-4001

WONG, BING KUEN, b Shanghai, China, Oct 4, 38; m 66; c 2. MATHEMATICAL ANALYSIS. *Educ:* Pittsburg State Univ, Kansas, AB, 61; Univ Ill, MA, 63, PhD(math), 66. *Prof Exp:* Asst prof math, Univ Western Ill, 65-66; asst prof, Rochester Inst Technol, 66-68; PROF MATH & COMPUT SCI, WILKES COL, 68-, CHMN DEPT, 68-84, 90- *Mem:* Am Math Soc; Math Asn Am; Asn Comput Mach. *Res:* Analysis. *Mailing Add:* Dept Math & Comput Sci Wilkes Univ Wilkes-Barre PA 18766

WONG, BRENDAN SO, b Hong Kong, Feb 25, 47; US citizen; m 72; c 1. BIOPHYSICS, PHYSIOLOGY. *Educ:* Univ San Francisco, BS, 71; Southern Ill Univ, MA, 74, PhD(biophys), 78. *Prof Exp:* FEL BIOPHYS, NAT INST NEUROL & COMMUN DIS & STROKE, 78-; ASSOC PROF, DEPT PHYSIOL, BAYLOR COL DENT, DALLAS, 83- *Mem:* Biophys Soc; Am Physiol Soc; Soc Neurosci; Soc Exp Biol Med; Sigma Xi. *Res:* Single-channel patch-clamp studies on tissue-cultured and dissociated primary cells; voltage clamp studies on the marine worm myxicola; electrophysiology of acanthocephalans; study of water dynamics in biological systems using pulsed nuclear magnetic resonance. *Mailing Add:* Dept Physiol Baylor Col Dent 3302 Gatson Ave Dallas TX 75246

WONG, CHAK-KUEN, b Macao, China; m 70; c 2. DESIGN & ANALYSIS OF ALGORITHMS, COMPUTER-AIDED DESIGN. *Educ:* Univ Hong Kong, BA, 65; Columbia Univ, MA, 66, PhD(math), 70. *Prof Exp:* RES STAFF MEM COMPUT SCI, T J WATSON RES CTR, IBM CORP, 69-, MGR DESIGN ALGORITHMS, 85- *Concurrent Pos:* Vis assoc prof, Univ Ill, Urbana, 72-73; vis prof, Columbia Univ, 78-79. *Honors & Awards:* Outstanding Invention Award, IBM Corp, 71. *Mem:* Fel Inst Elec & Electronics Engrs; NY Acad Sci; Asn Comput Mach. *Res:* Computer-aided design algorithms; very large scale integration algorithms; analysis of optimum and near-optimum algorithms in computing; application of matheematics to computers and computing. *Mailing Add:* T J Watson Res Ctr IBM Corp Yorktown Heights NY 10598

WONG, CHEUK-YIN, b Kwangtung, China, Apr 28, 41; m 66; c 3. PHYSICS. *Educ:* Princeton Univ, AB, 61, MA, 63, PhD(physics), 66. *Prof Exp:* Physicist, Oak Ridge Nat Lab, 66-68; res fel physics, Niels Bohr Inst, Copenhagen, Denmark, 68-69; PHYSICIST, OAK RIDGE NAT LAB, 69- *Concurrent Pos:* Vis scientist, Mass Inst Technol, 82-83; vis prof, Inst Nuclear Study, Univ Tokyo, Japan, 88. *Mem:* Fel Am Phys Soc. *Res:* Theoretical studies of nuclear properties, nuclear reactions, high energy nucleus-nucleus collisions, particle production and physics of strong fields; dynamics of nuclear fluid; high energy nucleus; nucleus collisions. *Mailing Add:* Oak Ridge Nat Lab Oak Ridge TN 37830

WONG, CHI SONG, b Cheng Tak, Hunan, China, May 26, 38; m 66; c 3. MATHEMATICAL STATISTICS, OPERATOR THEORY. *Educ:* Nat Taiwan Univ, BS, 62; Univ Ore, MS, 66; Univ Ill, Urbana, MS, 67, PhD(functional anal), 69. *Prof Exp:* Tutor, Chinese Univ, Hong Kong, 62-65; asst prof math, Southern Ill Univ, Carbondale, 69-71; asst prof, 71-73, assoc prof, 73-76, PROF MATH, UNIV WINDSOR, 76- *Concurrent Pos:* Can Nat Res Coun fels, 72-76. *Mem:* Can Math Cong; Can Math Soc. *Res:* Using algebra, functional analysis, geometry and topology to characterize certain classes of self maps which have fixed points; mathematical analysis. *Mailing Add:* Dept Math & Statist Univ Windsor Windsor ON N9B 3P4 Can

WONG, CHI-HUEY, b Taiwan, Aug 3, 48; m 76; c 2. BIOORGANIC CHEMISTRY, ENZYME TECHNOLOGY. *Educ:* Nat Taiwan Univ, BS, 70, MS, 77; Mass Inst Technol, PhD(chem), 82. *Prof Exp:* Asst res fel & lectr biochem, Acad Sci & Nat Taiwan Univ, 74-79; fel, Harvard Univ, 82-83; asst prof org chem, 83-86, assoc prof, 86-87, PROF CHEM, TEX A&M UNIV, 87- *Concurrent Pos:* Consult, Bioinfo, Boston, 83-85, Miles Lab, 85-86. *Honors & Awards:* Searle Scholar Award, 85; Presidential Young Investr Award, NSF, 86. *Mem:* Am Chem Soc; NY Acad Sci; AAAS. *Res:* Enzymes in organic synthesis; protein chemistry and biochemistry; synthesis of bioactive molecules; enzyme stabilization and functionalization; enzyme design by chemical modification and directed mutagenesis. *Mailing Add:* Dept Chem Res Inst Scripps Clin 10666 N Torrey Pines Rd La Jolla CA 92037-1027

WONG, CHIU MING, b Canton, China, July 8, 35; m 60; c 2. ORGANIC CHEMISTRY. *Educ:* Nat Taiwan Univ, BSc, 59; Univ NB, PhD(chem), 64. *Prof Exp:* Asst, Univ NB, 64, res fels, 64-65; res fels, Harvard Univ, 65-66; from asst prof to assoc prof, 66-75, PROF ORG CHEM, UNIV MAN, 75- *Mem:* Am Chem Soc. *Res:* Synthesis and structure-reactivity studies of anthracycline antitumor antibiotics. *Mailing Add:* 31 Folkestone Blvd Winnipeg MB R3P 0B4 Can

WONG, CHUEN, US citizen. PHYSICS. *Educ:* Chung Chi Col, Hong Kong, dipl sci, 60; Case Western Reserve Univ, PhD(physics), 67. *Prof Exp:* Demonstr physics, Chung Chi Col, Hong Kong, 60-63; from instr to asst prof, 67-83, ASSOC PROF PHYSICS, UNIV LOWELL, 83- *Mem:* Am Phys Soc; Optical Soc Am. *Res:* Experimental solid state physics; elastic constants; semiconductors. *Mailing Add:* Dept Physics Univ Lowell One University Ave Lowell MA 01854

WONG, CHUN WA, b Hong Kong, China, Jan 22, 38; US citizen; m 67; c 2. NUCLEAR FORCES, NUCLEON STRUCTURE. *Educ:* Univ Calif, Los Angeles, BS, 59; Harvard Univ, AM, 60, PhD(physics), 65. *Prof Exp:* Res assoc physics, Princeton Univ, 65-66, Oxford Univ, 66-67 & Saclay, 67-68; actg assoc prof, 70-71, FROM ASST PROF TO PROF PHYSICS, UNIV CALIF, LOS ANGELES, 69- *Concurrent Pos:* Alfred P Sloan Found fel, 70-72. *Mem:* Fel Am Phys Soc. *Res:* Description of the properties of atomic nuclei in terms of the fundamental forces acting between their constituents. *Mailing Add:* Dept Physics Univ Calif Los Angeles CA 90024-1547

WONG, CHUN-MING, b Hong Kong, Brit Crown Colony, Nov 12, 40; m 71. ORGANIC POLYMER CHEMISTRY. *Educ:* Univ Calif, Berkeley, BS, 65; Wayne State Univ, MS, 66; NDak State Univ, PhD(chem), 73. *Prof Exp:* Chemist, Inmont Corp, 66-70; res chemist, E I du Pont de Nemours & Co, Inc, 74-77; Chrysler Corp, 77-80; Boeing Comm Airplane Co, 80-89. *Mem:* Am Chem Soc. *Res:* Low volatile organic compound coatings aimed at reducing pollution. *Mailing Add:* 5745 NE 63rd St No 104 Seattle WA 98115

WONG, DAVID TAIWAI, b Hong Kong, Nov 6, 35; US citizen; m 63; c 3. BIOCHEMISTRY. *Educ:* Seattle Pac Col, BS, 60; Ore State Univ, MS, 64; Univ Ore, PhD(biochem), 66. *Prof Exp:* Fel biophys chem, Univ Pa, 66-68; sr biochemist, Lilly Res Labs, Eli Lilly & Co, 68-72, res biochemist, 73-77, sr res scientist, 78-89, RES ADV, LILLY RES LABS, ELI LILLY & CO, 90- *Concurrent Pos:* Adj prof biochem, 86-, neurobiol, 90- *Mem:* NY Acad Sci; Int Soc Neurochem; Am Soc Neurochem; Am Soc Pharmacol & Exp Therapeut; Sigma Xi; Soc Neurosci. *Res:* Biochemistry of neurotransmission; synthetic chemicals which block the uptake of specific neurotransmitters investigated as potentially useful therapeutic agents for mental disorders; discoverer of fluoxetine (prozac) and tomoxetine for treatment of mental depression; biochemical mechanism of ionophorous agents. *Mailing Add:* 1640 Ridge Hill Lane Indianapolis IN 46217

WONG, DAVID YUE, b Swatow, China, Apr 16, 34; US citizen; m 60; c 2. PHYSICS. *Educ:* Hardin-Simmons Univ, BA, 54; Univ Md, College Park, PhD(physics), 58. *Prof Exp:* Theoretical physicist, Univ Calif, Berkeley, 58-60, from asst prof to assoc prof, 60-67, PROF PHYSICS, UNIV CALIF, SAN DIEGO, 67-; PROVOST, WARREN COL, 86- *Concurrent Pos:* Alfred P Sloan fel, Univ Calif, San Diego, 63-66. *Mem:* Am Inst Physics; Am Phys Soc. *Res:* Theoretical high energy physics. *Mailing Add:* Dept Physics Univ Calif San Diego La Jolla CA 92093-0319

WONG, DENNIS MUN, b San Francisco, Calif, Dec 14, 44; m 69. CELLULAR IMMUNOLOGY. *Educ:* Calif State Univ, San Francisco, BA, 69, MA, 74; Georgetown Univ, PhD(microbiol), 77. *Prof Exp:* Fel, Naval Med Res Inst, 77-79; STAFF FEL, BUR BIOLOGICS, DIV BLOOD & BLOOD PROD, NIH, 80- *Mem:* AAAS; Am Soc Microbiol; NY Acad Sci; Am Asn Clin Histocompatibility Testing; Sigma Xi. *Res:* Cellular immunology; human histocompatibility antigens; immunogenetic system; development of hybridoma antibodies against cell surface antigens. *Mailing Add:* 1428 Cutstone Way Silver Spring MD 20905

WONG, DEREK, b Shanghai, China, Jan 22, 46; m 72; c 1. OPERATIONS RESEARCH, MATHEMATICAL PROGRAMMING. *Educ:* Hong Kong Baptist Col, BSc, 68; Fla State Univ, MS, 75, PhD(statist), 77. *Prof Exp:* Asst prof math prog, Northern Ill Univ, 77-85. *Mem:* Inst Mgt Sci; Am Statist Asn; Inst Math Statist. *Res:* Optimization theory. *Mailing Add:* Naval Systs Ctr Code 811 San Diego CA 92152

WONG, DONALD TAI ON, b Honolulu, Hawaii, Nov 1, 26; m 54; c 2. IMMUNOLOGY. *Educ:* St Louis Univ, BS, 49; Wash Univ, PhD(microbiol), 53. *Prof Exp:* Res chemist, Dept Bact Microbial Chem Sect, Walter Reed Army Inst Res, Army Med Ctr, Washington, DC, 52-61; res chemist, Blood Antigen Lab, Div Animal Husb, Agr Res Ctr, Md, 61-65; res chemist, dept immunochem, Div Commun Dis & Immunol, Walter Reed Army Inst Res, Army Med Ctr, 65-81, RETIRED. *Mem:* AAAS; Am Chem Soc. *Res:* Oxidative metabolism in microorganisms; alternate pathways and carbon-2-carbon-2 condensation mechanisms; immunoglobulin specifity and structure; mechanisms involved with immediate type hypersensitivity reactions. *Mailing Add:* 5918 Mustang Dr Riverdale MD 20737

WONG, DOROTHY PAN, b Nanking, China, July 8, 37; US citizen; m 68. PHYSICAL CHEMISTRY. *Educ:* Univ Okla, BS, 57; Univ Minn, MS, 59; Case Inst Technol, PhD(phys chem), 64. *Prof Exp:* Res chemist, Continental Oil Co, 57; assoc chemist, Airforce Midway Lab, Univ Chicago, 59-60; asst prof phys chem, Calif State Col Fullerton, 64-65; res assoc quantum chem, Princeton Univ, 65-66; asst prof, 66-67, assoc prof, 68-73, PROF PHYS CHEM, CALIF STATE UNIV, FULLERTON, 70- *Mem:* Am Chem Soc; Am Phys Soc. *Res:* Non-empirical quantum mechanical calculations for geometry of molecules; molecular properties and rotation barriers of nitrogen compounds and for other molecules of current chemical interest. *Mailing Add:* Dept Chem Calif State Univ 800 N State College Blvd Fullerton CA 92634

WONG, E(UGENE), b Nanking, China, Dec 24, 34; nat US; m 56; c 3. ELECTRICAL ENGINEERING, COMPUTER SCIENCE. *Educ:* Princeton Univ, BSE, 55, AM, 58, PhD(elec eng), 59. *Prof Exp:* Mem res staff, Int Bus Mach Corp, 55-56 & 60-62; NSF fel, Cambridge Univ, 59-60; from asst prof to prof elec eng, Elec Eng & Computer Sci Dept, Univ Calif, Berkeley, 62-90, chmn, 85-89; ASSOC DIR PHYS SCI & ENG, OFF SCI & TECHNOL POLICIES, WHITE HOUSE, WASHINGTON, DC, 90- *Concurrent Pos:* Guggenheim fel, Cambridge Univ, 68-69; consult, Ampex Corp; Vinton Hayes sr fel, Harvard Univ, 76-77; consult, Honeywell, Inc, 78- *Mem:* Nat Acad Eng; Asn Comput Mach; fel Inst Elec & Electronics Engrs. *Res:* Stochastic processes; data base systems. *Mailing Add:* Off Sci & Technol Policies The White House 1600 Pensylvania Ave NW Washington DC 20506

WONG, EDWARD CHOR-CHEUNG, b Hong Kong, China, Jan 16, 52; US citizen; c 3. RAS, SYSTEM ARCHITECTURE. *Educ:* Fordham Univ, BA, 73; Columbia Univ, BS & MS, 74. *Prof Exp:* Jr engr, IBM, 74-75, assoc engr, 75-76, sr assoc engr, 76-79, staff engr, 79-82, develop eng mgr, 82-89, SR ENGR, IBM, 89- *Concurrent Pos:* Adj lectr, Columbia Univ, 82; Adj prof, Computer Software Develop Grad Prog, Marist Col, 83- *Mem:* Asn Comput Mach; sr mem Inst Elec & Electronics Engrs. *Res:* Cache design; large system architecture; reliability, availability, serviceability architecture/design, software design/development; algorithms. *Mailing Add:* PO Box 1909 Poughkeepsie NY 12601

WONG, EDWARD HOU, b Hankow, China, Oct 5, 46; c 2. INORGANIC CHEMISTRY. *Educ:* Univ Calif, Berkeley, BS, 68; Harvard Univ, PhD(inorg chem), 74. *Prof Exp:* Res assoc boron chem, Univ Calif, Los Angeles, 74-76; asst prof, Fordham Univ, 76-78; from asst prof to assoc prof, 78-89, PROF CHEM, UNIV NH, 89- *Concurrent Pos:* St Mary's fel, Durham Univ, UK, 86. *Mem:* Am Chem Soc. *Res:* Polyhedral borane and carborane chemistry; transition metal complexes in catalysis; poly-phospline hetrocycles. *Mailing Add:* Dept Chem Parsons Hall Univ NH Durham NH 03824

WONG, FULTON, b Kwangtung, China, Nov 9, 48; m 76; c 2. VISUAL SCIENCES. *Educ:* Univ Redlands, Calif, BS, 72; Rockefeller Univ, NY, PhD(biophysics, neurophysiol), 77. *Prof Exp:* Res assoc vision, Dept Biol Sci, Purdue Univ, 77-78, vis asst prof biol, 78-79; asst prof, Dept Physiol & Biophys, Med Br, Univ Tex, 79-86; assoc prof, Dept Ophthal & Anat & Cell Biol, Univ Ill, Chicago, 86-89; ASSOC PROF, DEPT OPHTHAL & NEUROBIOL, DUKE UNIV, 89- *Concurrent Pos:* Adj res assoc, Rockefeller Univ, 77-78; mem, Marine Biomed Inst, 79-86; vis res scientist neurobiol, Yale Univ, 85-86. *Mem:* AAAS; Soc Gen Physiologists; NY Acad Sci; Sigma Xi; Asn Res Vision & Ophthal. *Res:* Hereditary eye diseases caused by mutations in photorecptor-specific genes. *Mailing Add:* Dept Ophthalmol Duke Univ Med Ctr Box 3802 DUMC Durham NC 27710

WONG, GEORGE SHOUNG-KOON, b Hong Kong, July 21, 35; Can citizen. ACOUSTICAL MEASUREMENTS, ACCOUSTICAL STANDARDS. *Educ:* Manchester Univ, MSc, 63, PhD, 65. *Prof Exp:* From asst res officer to assoc res officer, 66-79, SR RES OFFICER, NAT RES COUN, CAN, 79- *Concurrent Pos:* Chmn, Int Electrotech Comn Comt TC 29, Can Standards Coun, 85-, Acoust Soc Am Standards Comt S1-Acoust, 90- *Mem:* Fel Inst Elec Engrs UK; fel Acoust Soc Am; Inst Mech Engrs UK; Inst Prod Engrs UK. *Res:* Precision acoustical measurements developed precision primary acoustical standards calibration system; theoretical prediction and experimental confirmation on the variation of the ratio of specific heats in air with humidity and temperature; calculated a new value for the speed of sound and other physical properties in air. *Mailing Add:* Inst Nat Measurement Standards Nat Res Coun Can Ottawa ON K1A 0R6 Can

WONG, GEORGE TIN FUK, b Hong Kong, Nov 29, 49; m 74. HYDROGEOCHEMISTRY, ENVIRONMENTAL CHEMISTRY. *Educ:* Calif State Univ, Los Angeles, BS, 71; Mass Inst Technol, MS, 73; PhD(chem oceanog), 76. *Prof Exp:* Asst prof, 76-82, ASSOC PROF OCEANOG, OLD DOMINION UNIV, 82-, DOCTORAL PROG DIR, 85- *Mem:* Am Soc Limnol & Oceanog; AAAS; Am Geophys Union; Geochem Soc. *Res:* Trace elements; radiogeochemistry; physical and analytical chemistry of natural waters; environmental chemistry; redox chemistry of ground water; hydrogeochemistry. *Mailing Add:* Inst Oceanog Old Dominion Univ Hampton Blvd Norfolk VA 23508

WONG, HANS KUOMIN, b Canton, Kwangtung, China, Apr 30, 36; US citizen; m 67; c 2. PHYSICAL & INORGANIC CHEMISTRY. *Educ:* NDak State Univ, BS, 59; Univ Minn, PhD(phys chem), 65. *Prof Exp:* Sr chemist, Itek Corp, 65-70; sr chemist phys chem, Olivetti Corp Am, 71-80; sr chemist phys chem, Bacharach Inst Co, 80-82; asst prof, Cumberland Col, 83-84; sr chemist, Olin Hunt, 84-87; CONSULT, 87- *Mem:* Am Chem Soc. *Res:* Electrophotography; photoconductivity; dye sensitization mechanism; photochemistry. *Mailing Add:* 1269 Ayala Dr No 4 Sunnyvale CA 94086-5541

WONG, HARRY YUEN CHEE, b Kapaa, Hawaii, Oct 23, 17; m 43; c 3. PHYSIOLOGY, ENDOCRINOLOGY. *Educ:* Okla State Univ, BS, 42; Univ Southern Calif, MS, 47, PhD(endocrinol, physiol), 50. *Hon Degrees:* Vis prof, Shanghai Inst Cardiovasc Dis, China, 88. *Prof Exp:* Asst physiol, Univ Southern Calif, 46-48, lab assoc anat, 48-49; assoc prof biol, Andrews Univ, 49-51; dir basic endocrine res, Freedmen's Hosp, 52-60; from instr to assoc prof, Sch Med, 51-66, PROF PHYSIOL, COL MED, HOWARD UNIV, 66-, DIR ENDOCRINOL & METAB, 53-, PROF PHYSIOL & BIOPHYSICS, GRAD SCH ARTS & SCI, 75. *Concurrent Pos:* Consult, Off Surgeon Gen, US Air Force, 63-; vis prof hormone lab, II Med Clin, Univ Hamburg, 69; vis scientist, Armed Forces Inst Path, DC, 70 & 82; fel, coun arteriosclerosis, Am Heart Asn; mem, Int Cong Physiol Sci, Int Cong Pharmacol, Int Cong Hormonal Steroids & Int Cong Endocrinol; consult to chief dept med & clin sci, Brooks AFB; fel coun arteriosclerosis, Am Heart Asn; vis prof, People's Repub China, 82 & 88, Nanking Med Univ, 88; vis prof, Shanghai Med Univ & Shanghai Inst Cardiovascular Disease, China, 88. *Honors & Awards:* Distinguished Fac Award for Outstanding Res, Howard Univ, 82. *Mem:* Am Physiol Soc; Endocrine Soc; NY Acad Sci; AMA; fel Coun Arteriosclerosis; Am Heart Asn. *Res:* Lipid mobilizing factor; factor of stress and exercise, drugs and sex hormones in atherosclerosis; enzymes, catecholamines and hormones of heart and adrenal glands; lipid & lipoprotein changes in chickens, man, miniature pig, rabbits & gerbils; chemical determination of androgens; toxicology of benzodiazepines in lipid metabolism and atherosclerosis. *Mailing Add:* Dept Physiol & Biophys Howard Univ Col Med Washington DC 20059

WONG, HORNE RICHARD, b Hong Kong, Jan 9, 23; Can citizen; m 58; c 3. SAWFLY SYSTEMATICS, FOREST ENTOMOLOGY. *Educ:* Univ Man, BSA, 47; Mich State Univ, MS, 50; Univ Ill, PhD(entom), 60. *Prof Exp:* Sr agr asst, Can Dept Agr, 47-48, officer-in-charge forest insect surv, 49-52, res officer, 53-60; res officer, 61-65, res scientist, Can Dept Agr, 66-88, EMER RES SCIENTIST, NORTHERN FOREST RES CTR, CAN DEPT FORESTRY, 89- *Concurrent Pos:* Mem, Can Comt Common Names Insects, 58-70; pres, Entom Soc Alta, 79. *Mem:* Entom Soc Am; Sigma Xi; Entom Soc Can. *Res:* Systematics, biology and phylogeny of sawflies (Hymenopter, Symphyta); life history and habits of forest insects. *Mailing Add:* Northern Forestry Ctr Can Dept Forestry 5320 122 St Edmonton AB T6H 3S5 Can

WONG, JAMES B(OK), b Canton, China; nat US; m 46; c 3. CHEMICAL ENGINEERING, ECONOMICS. *Educ:* Univ Md, BS, 49, BChE, 50; Univ Ill, MS, 51, PhD(chem eng), 54. *Prof Exp:* Asst chem eng, Univ Ill, 50-53; chem engr, Standard Oil Co, Ind, 53-55; engr, Shell Develop Co, Calif, 55-61; sr planning engr, Chem Group, Dart Indust Inc, 61-64, prin planning engr, 64-66, supvr planning & econ, 66-67, mgr long range planning & econ, 67, chief economist, 67-72, dir econ & opers analysis, 72-78, dir, Int Technol, 78-81; PRES, JAMES B WONG ASSOC, INC, 81- *Mem:* Am Chem Soc; Am Inst Chem Engrs. *Res:* Filtration of aerosols; fluid mechanics; process design; economics and planning; technologies on polyethylene, polypropylene, polystryrene and other polymers; general operations analysis; international licensing; technology transfer; technology transfer and product distribution, Asian and Pacific basin regions. *Mailing Add:* 2460 Venus Dr Los Angeles CA 90046

WONG, JAMES CHIN-SZE, b Hong Kong, Dec 5, 40; Can citizen. MATHEMATICS. *Educ:* Univ Hong Kong, BA, 63; Univ BC, PhD(math), 69. *Prof Exp:* Nat Res Coun Can fel, McMaster Univ, 69-71; from asst prof to assoc prof, 71-83, PROF MATH, UNIV CALGARY, 84- *Mem:* Can Math Soc; London Math Soc; Am Math Soc. *Res:* Functional analysis. *Mailing Add:* Dept Math & Stat Univ Calgary Calgary AB T2N 1N4 Can

WONG, JEFFREY TZE-FEI, b Hong Kong, Aug 5, 37; Can citizen; m 61; c 3. BIOCHEMISTRY. *Educ:* Univ Toronto, BA, 59, PhD(biochem), 62. *Prof Exp:* From asst prof to assoc prof, 65-76, PROF BIOCHEM, FAC MED, UNIV TORONTO, 76- *Concurrent Pos:* Med Res Coun Can grant, Univ Toronto, 65- *Mem:* Can Soc Biochem; Am Soc Biochem. *Res:* Enzyme kinetics and mechanism; biochemical evolution; blood substitutes. *Mailing Add:* Dept Biochem Univ Toronto Toronto ON M5S 1A8 Can

WONG, JO YUNG, b Canton, China; Can citizen; m 62; c 1. TRANSPORT TECHNOLOGY, TRANSPORT SYSTEMS ANALYSIS. *Educ:* Tsinghua Univ, China, BSc Hons, 55; Univ Newcastle-upon-Tyne, Eng, PhD(eng), 67, DSc, 86. *Prof Exp:* Res assoc eng, Univ Newcastle-upon-Tyne, Eng, 65-67; res engr, Logging Develop Corp, 67-68; from asst prof to assoc prof, 68-78, PROF ENG, DEPT MECH & AEROSPACE ENG, & DIR, TRANSP TECHNOL RES LAB, CARLETON UNIV, 78-; PRES, VEHICLE SYSTS DEVELOP CORP, 77- *Concurrent Pos:* Assoc ed, J Terramech, 73-; vis prof, Univ Warwick, Eng, 74-75 & Cemoter, Italian Nat Res Coun, 85, 87-; dir,

Ottawa-Carleton Inst Mech & Aeronautical Eng, 87-90; consult, govt agencies & industs, NAm, Europ & Asia. *Honors & Awards:* Starley Premium Award, Inst Mech Engrs, Eng, 84 & 88. *Mem:* Fel Inst Mech Engrs; fel Am Soc Mech Engrs; Soc Automotive Engrs; Int Soc Terrain-Vehicle Systs. *Res:* Ground transport technology; off-road transport technology; road vehicle dynamics and safety; air cushion technology; advanced guided ground transport systems. *Mailing Add:* Carleton Univ Colonel By Dr Ottawa ON K1S 5B6 Can

WONG, JOE, b Hong Kong, Aug 8, 42; m 69; c 3. PHYSICAL CHEMISTRY, SOLID STATE SCIENCE. *Educ:* Univ Tasmania, BSc, 65, BSc Hons, 66; Purdue Univ, Lafayette, PhD(phys chem), 70. *Hon Degrees:* DSc, Univ Tasmania, 85. *Prof Exp:* Analytical chemist, Australian Titan Prod, Tasmania, 62-63; res asst, Electrolytic Zinc Co Australasia, 63-64 & Dept Chem, Univ Tasmania, 64-65; res chemist, Electrolytic Zinc Co Australasia, Tasmania, 66; res asst chem, Walker Lab, Rensselaer Polytech Inst, 66-67 & Purdue Univ, West Lafayette, 67-70; phys chemist, Corp Res & Develop, Gen Elec Co, 70-86; CONSULT, 86- *Concurrent Pos:* Adj lectr chem, Royal Hobart Col, 66; adj prof chem, State Univ NY, Albany, 80-85; sci & technol agency fel, 91. *Honors & Awards:* 1st Prize Optical Micros, Am Ceramic Soc, 75 & 77; Dushman Award, General Elec, 84; Humboldt Award, 91. *Mem:* Am Chem Soc; Royal Australian Chem Inst; Am Phys Soc; AAAS; fel Am Inst Chemists. *Res:* Molten salt chemistry; thermodynamic and spectroscopic studies; spectroscopy of simple inorganic glasses; thin films; deposition and structure; impurity diffusion in semiconductors; microstructure of non-ohmic Zn0 ceramics; metallic glasses; EXAFS spectroscopy; coal science; synchrotron radiation research; near edge structure spectroscopy, high resolution electron microscopy, electron energy loss spectroscopy. *Mailing Add:* Lawrence Livermore Nat Lab PO Box 808 Chem Mat Sci L369 Livermore CA 94551

WONG, JOHN LUI, b Macau, June 12, 40; US citizen; m 64; c 2. MOLECULAR EPIDEMIOLOGY, COAL CHEMISTRY. *Educ:* Cheng Kung Univ, BS, 62; Univ Calif, Berkeley, PhD(org chem), 66. *Prof Exp:* From asst prof to assoc prof chem, 66-74, vchmn dept, 75-76, chmn, 87-88, PROF CHEM, UNIV LOUISVILLE, 74-, ASSOC ONCOL, CANCER CTR, 75-, ASSOC PHARMACOL & TOXICOL. *Concurrent Pos:* Consult, indust & govt, 79- *Mem:* Sigma Xi; Am Chem Soc; Am Asn Cancer Res; fel Am Inst Chemists. *Res:* Chemical carcinogenesis and molecular epidemiology; synthesis of polycyclic amines and natural product analogs; modern coal and environmental studies. *Mailing Add:* Dept Chem Univ Louisville Louisville KY 40292

WONG, JOHNNY WAI-NANG, b Hong Kong, Nov 22, 47. COMPUTER SYSTEMS. *Educ:* Univ Calif, BS, 70, MS, 71, PhD(comput sci), 75. *Prof Exp:* From asst prof to assoc prof, 75-85, PROF, DEPT COMPUT SCI, UNIV WATERLOO, 85- *Res:* Modeling and analysis of computer systems and networks. *Mailing Add:* Dept Comput Sci Univ Waterloo Waterloo ON N2L 3G1 Can

WONG, K(WEE) C, b Burma; US citizen. WATER TREATMENT, CHEMICAL ETCHING. *Educ:* Univ Rangoon, Burma, BSc, 65. *Prof Exp:* Chemist, Chamberlain Mfg Co, 71-74; chief chemist, Bonewitz Chem Serv, 74-79; prin chemist, Dart Industs, 79-84; res chemist, Inland Specialty, Great Lakes Chem, 84-87; PROJ LEADER, DEXTER ELECTRONIC MAT, 87- *Mem:* Am Chem Soc; Royal Soc Chem; fel Am Inst Chemists. *Res:* Improving and finding new processes and products for the printed circuit board industry and wastewater treatment; granted several patents. *Mailing Add:* 5118 W Davit Ave Santa Ana CA 92704

WONG, KAI-WAI, b Aug 7, 38; nat US citizen; m; c 3. PHYSICS. *Educ:* Duke Univ, BS, 59; Northwestern Univ, MS, 60, PhD(physics), 63. *Prof Exp:* Res assoc physics, Northwestern Univ, 63; res assoc, Univ Iowa, 63-64; from asst prof to assoc prof, 64-72, PROF PHYSICS, UNIV KANS, 72-, DIR HIGH TC LAB, 89- *Concurrent Pos:* Vis assoc prof, Univ Southern Calif, 69-71; vis prof, Univ Calif, Los Angeles, 72-73 & 79-80; hon prof, Univ Hong Kong, 72. *Mem:* Am Phys Soc. *Res:* Theoretical physics; many-body problems; statistical mechanics; high Tc theory. *Mailing Add:* Dept Physics Univ Kans Lawrence KS 66045

WONG, KAM WU, b Hong Kong, Mar 8, 40; m 70; c 2. PHOTOGRAMMETRY, GEODETIC SURVEYING. *Educ:* Univ NB, BSc, 64; Cornell Univ, MSc, 66, PhD(photogram), 68. *Prof Exp:* From asst prof to assoc prof, 67-76, PROF CIVIL ENG, UNIV ILL, URBANA-CHAMPAIGN, 76- *Honors & Awards:* Talbert Abrams Award, Am Soc Photogram, 70; Walter L Huber Res Prize, Am Soc Civil Engrs, 71. *Mem:* Am Soc Photogram; Am Soc Civil Engrs; Can Inst Surv. *Res:* Metric vision; digital photogrammetry; geodetic surveying. *Mailing Add:* Dept Civil Eng Univ Ill 205 N Mathews Ave Urbana IL 61801

WONG, KEITH KAM-KIN, b Hong Kong, Feb 11, 29; US citizen; m 61; c 2. BIOCHEMISTRY, PHYSIOLOGY. *Educ:* Southwestern at Memphis, BA, 55; Univ Tenn, MSc, 57; NY Univ, PhD(biol), 69; Monmouth Col, MBA, 85. *Prof Exp:* Res asst chemother, Sloan-Kettering Inst Cancer Res, 57-58; mem staff biochem & drug metab, Worcester Found Exp Biol, 60-63; res assoc biochem pharmacol, Schering Corp, 63-66; res assoc exp hemat, NY Univ, 66-69; sr res investr drug metab, Squibb Inst Med Res, 69-79; SECT LEADER BIOCHEM, WALLACE LABS, CARTER-WALLACE INC, 79- *Mem:* AAAS; Am Chem Soc; NY Acad Sci. *Res:* Drug metabolism; biochemical pharmacology; biogenesis of erythropoietin; metabolism of biogenic amines; amino acid activation and transfer; protein synthesis; transformation of nucleic acid. *Mailing Add:* 16 Desmet Milltown NJ 08850

WONG, KIN FAI, b Kwangtung, China, Nov 6, 44; m 71; c 2. UNIT OPERATIONS, CHEMICAL SAFETY. *Educ:* Ariz State Univ, BSE, 65; Univ Ill, Urbana, MS, 67, PhD(chem engr), 70. *Prof Exp:* Res engr process develop, Western Res Ctr, Stauffer Chem Co, 70-76, sr res engr, De Guigne Tech Ctr, 76-79; PROJ OFFICER, OFF TOXIC SUBSTANCES, US ENVIRON PROTECTION AGENCY, 79- *Mem:* Am Inst Chem Engrs. *Res:* Thermodynamics and kinetics; chemical protective clothing selection; occupational exposure and controls, work place exposure modelling; process economics; waste product recovery and pollution control; industrial hygiene; asbestos abatement. *Mailing Add:* 12716 Serpentine Way Silver Spring MD 20904

WONG, KING-LAP, b Canton, China. PLASMA HEATING & CONFINEMENT, TOKAMAK PHYSICS. *Educ:* Chinese Univ, Hong Kong, BSc, 68; Univ Del, MS, 70; Univ Wis, PhD(physics), 75. *Prof Exp:* Res assoc, Columbia Univ, 75-76; res assoc, 76-78, res staff, 78-80, res physicist, 80-86, PRIN RES PHYSICIST, PLASMA PHYSICS LAB, PRINCETON UNIV, 86- *Concurrent Pos:* Prin investr L3/ACT-1 proj, USDOE, 77-84. *Mem:* Am Phys Soc; Inst Elec & Electronics Engrs. *Res:* Plasma heating and confinement; adiabatic compression; toroidal plasma confinement; thermonuclear fusion; linear and nonlinear plasma wave phenomena; radio-frequency wave heating and current drive; impurity transport. *Mailing Add:* Plasma Physics Lab James Forrestal Campus Princeton Univ Princeton NJ 08540

WONG, KIN-PING, b China, Aug 14, 41; m 68; c 2. BIOTECHNOLOGY, PHYSICAL CHEMISTRY. *Educ:* Univ Calif, Berkeley, BS, 64; Purdue Univ, PhD(phys & biol chem), 68. *Prof Exp:* Res fel phys biochem, Med Ctr, Duke Univ, 68-70; from asst prof to assoc prof chem, Univ SFla, Tampa, 70-75; from assoc prof to prof biochem, 75-83, dean grad studies, Univ Kans Med Ctr, 80-83; DEAN & PROF CHEM, SCH NATURAL SCI, CALIF STATE UNIV, 83- *Concurrent Pos:* Am Cancer Soc res grant, Univ SFla, Tampa, 70-71, NIH biomed res grant, 71-72, Cottrell res grant, 71-, Damon Runyon cancer res grant, 72-74; USPHS res career develop award, Nat Inst Gen Med Sci, 73; vis scientist, Max Planck Inst Molecular Genetics, 73; European Molecular Biol Orgn sr fel, Wallenberg Lab, Univ Uppsala, Sweden, 75; res grants, Nat Inst Gen Med Sci & Nat Heart, Lung & Blood Inst, 74-; vis prof, Univ Tokyo, Hongo, 79; prog dir biophys, NSF, Wash, DC, 82-83; US Dept Com sea grant, 87-, Milheim Found Cancer Res grant, 86-87; adj prof biochem & biophys, Univ Calif, San Francisco, adj prof med, Fresno Med Prog, 86-; Nat Col Sea grant, 87-90; chmn, chief exec officer & pres, RiboGene, Inc, 90-91. *Mem:* Am Soc Biol Chemists; Biophys Soc; Sigma Xi; AAAS; Am Chem Soc; Protein Soc; fel Royal Soc Chemists; fel Am Inst Chemists. *Res:* Physical biochemistry of protein and nucleic acids; mechanism of protein folding; ribosome structure; physicochemical studies of ribosomal proteins and RNAs; the molecular mechanism on the assembly of ribosome; mechanism of RNA folding; development of a potential anti-tumor drug: angiogenesis inhibitor from cartilage; protein aging; chemical carcinogensis. *Mailing Add:* Sch Natural Sci Calif State Univ Fresno CA 93740

WONG, KWAN Y, b Hong Kong, June 12, 37; US citizen; m 66; c 2. ELECTRICAL & SYSTEMS ENGINEERING. *Educ:* Univ New South Wales, BS, 60, ME, 63; Univ Calif, Berkeley, PhD(elec eng), 66. *Prof Exp:* Mem res staff systs, 66-77, MGR, INT BUS MACH CORP RES LAB, SAN JOSE, CALIF, 77- *Mem:* Sr mem Inst Elec & Electronics Engrs; Sigma Xi. *Res:* Image processing; data compression; pattern recognition; process control. *Mailing Add:* IBM Almaden Res Ctr Dept K51-801 650 Harry Rd San Jose CA 95120-6099

WONG, LAURENCE CHENG KONG, b Shanghai, China, Oct 2, 33; US citizen; m 85; c 1. RESEARCH ADMINISTRATION, SCIENCE POLICY. *Educ:* Nat Defense Med Ctr, BS, 55; Univ Rochester, MS, 64, PhD(pharmacol & toxicol), 67. *Prof Exp:* Asst prof pharmacol, Univ Calif Sch Med, San Francisco, 67-70; sr scientist, Bionetic Res Lab, 70-72; sr scientist, Nornich Pharmacol Co, 72-74; group leader, Ortho Pharmaceut Co, 74-78; assoc dir biol sci, Midwest Inst, 78-79; dir safety, Revlon Health Care, 79-86; DIR BIOL SCI, ADRIA LABS, 86- *Concurrent Pos:* Adj prof pharmacol, Sch Med, Howard Univ, 81-; vis prof, Shanghai Inst Pharmaceut Indust, 86-; consult, Am Health Found, 86- *Mem:* Soc Toxicol; Soc Clin Toxicol; Soc Xenobiotics; Am Chinese Toxicol Soc; Regulatory Affairs Prof Soc. *Res:* New drug discoveries and assessing the pharmacological and toxicological properties for efficacy and safety of the drugs in therapeutical usages in patients. *Mailing Add:* 7681 Seminary Ridge Columbus OH 43235

WONG, MAURICE KING FAN, b Shanghai, China, Apr 9, 32. MATHEMATICAL PHYSICS. *Educ:* Univ Hong Kong, BSc, 54; Berhmans Col, AB, 58, MA, 61; Univ Birmingham, PhD(math physics), 64. *Prof Exp:* Res assoc, Inst Advan Studies, Dublin, 64-65; res fel res inst nat sci, Woodstock Col, Md, 65-68; from asst prof to assoc prof, 69-79, PROF MATH, FAIRFIELD UNIV, 79- *Concurrent Pos:* Fel, St Louis Univ, 69. *Mem:* AAAS; Am Phys Soc; Am Math Soc; Math Asn Am. *Res:* Lie groups; superconductivity; M-ssbauer effect; elementary particles; nuclear physics; quantum theory. *Mailing Add:* Dept Math & Comput Sci Fairfield Univ Fairfield CT 06430

WONG, MING MING, b Singapore, Jan 3, 28; US citizen. PARASITOLOGY. *Educ:* Wilmington Col, Ohio, BS, 52; Ohio State Univ, MS, 53; Tulane Univ La, PhD(med parasitol), 63. *Prof Exp:* Med technologist, Good Samaritan Hosp, Zanesville, Ohio, 54-55; teacher, Diocesan Girls' Sch, Hong Kong, 55-56; demonstr parasitol & bact fac med, Univ Hong Kong, 56-59; teaching asst med parasitol med sch, Tulane Univ La, 59-63, NIH res fel trop med, 63-64, res assoc, 64-65; res assoc fac med, Univ Malaya, 64-65, lectr parasitol, 65-67; asst res parasitologist, 67-73, ASSOC RES PARASITOLOGIST, PRIMATE RES CTR, UNIV CALIF, DAVIS, 73- *Concurrent Pos:* WHO res grant, Univ Malaya, 66-67; NIH grants primate res ctr, Univ Calif, Davis, 70-75. *Mem:* Am Soc Parasitol; Am Soc Trop Med & Hyg; Royal Soc Trop Med & Hyg; Am Heartworm Soc; Am Soc Clin Pathologists. *Res:* Filariasis; primate parasitology; immunology of parasitic diseases. *Mailing Add:* Dept Vet Microbiol & Immunol Univ Calif Davis CA 95616

WONG, MORTON MIN, b Canton, China, Oct 2, 24; US citizen; m 56; c 4. ELECTROCHEMISTRY, HYDROMETALLURGY. *Educ:* Univ Calif, BS, 51. *Prof Exp:* Trainee, Am Potash & Chem Corp, 51-53; researcher, US Bur Mines, 53-54, asst proj leader, 54-56, res proj leader, 56-60, res group leader, 60-62, res proj coordr, 62-71, res supvr, 71-82; SR ENG ASSOC, UNION SCI & TECHNOL DIV, UNOCAL, 82- *Concurrent Pos:* Vis lectr, People's Repub China, 81. *Honors & Awards:* Distinguished Serv Award, US Dept Interior, 81; Hofman Spec Prize, Int Lead Consortium, 82; Fred L Hartley Award, 88. *Mem:* Am Inst Mining, Metall & Petrol Eng; Am Inst Chem. *Res:* Fused-salt electrolysis of rare earths; hydrometallurgical and electrolytical processing of copper-lead-zinc sulfide ores; metallurgical treatment of titanium ore; benefication and extraction of platinum-group metals from ores; recovery of minerals from geothermal brines; detoxification of solid wastes. *Mailing Add:* 413 Lavender Lane Placentia CA 92670

WONG, NOBLE POWELL, b Baltimore, Md, Apr 30, 31; m 61; c 4. FOOD CHEMISTRY, NUTRITION. *Educ:* Univ Md, BS, 53; Pa State Univ, MS, 58, PhD(dairy sci), 61. *Prof Exp:* Res chem, Food & Drug Admin, 61-66; RES CHEMIST, USDA, 66- *Mem:* Am Chem Soc; Am Dairy Sci Asn; Inst Food Technol. *Res:* Nutrition and composition of dairy products; bioavailability of minerals. *Mailing Add:* 8716 Camille Dr Rockville MD 20854

WONG, NORMAN L M, b Hong Kong, Jan 14, 45; Can citizen; c 2. NEPHROLOGY, PATHOPHYSIOLOGY. *Educ:* Sir George Williams Univ, BS, 66; McGill Univ, MS, 69 & PhD(physiol), 72. *Prof Exp:* Post doc fel, Renal & Electrolyte Div, McGill Univ, 72-74, lectr, dept exp med & prof asst, Renal & Electrolyte Div, 74-76; res assoc, Dept Med, Univ Hosp, 76-83, res med staff, 83-89, asst prof med, 83-89, ASSOC PROF, UNIV BC, 89- *Concurrent Pos:* Res med staff, Dept Med, Vancouver Gen, 83- *Mem:* Am Soc Nephrol; Am Physiol Soc; Can Physiol Soc; Can Soc Clin Invest; NY Acad Sci; Int Soc Nephrol. *Res:* Renal micropuncture of mammalian nephron with special emphasis on electrolyte transport, particularly directed at understanding of kidney tubule handling of calcium, magnesium and phosphate in various physiological and pathophysiological states; mechanism of atrial peptide release in viro and vitro; in normal and abnormal physiological conditions. *Mailing Add:* Dept Med Univ BC 2211 Wesbrook Mall Vancouver BC V6T 2B5 Can

WONG, PATRICK YUI-KWONG, b Kiangsi, China, Nov 25, 44. BIOCHEMISTRY, BIOCHEMICAL PHARMACOLOGY. *Educ:* Nat Taiwan Norm Univ, BSc, 67; Univ Vt, PhD(biochem), 75. *Prof Exp:* Fel, Med Col Wis, 74-75; instr pharmacol, Col Basic Med Sci, Univ Tenn, Memphis, 75-79; assoc prof, 79-85, PROF, DEPT PHARMACOL, NY MED COL, VALHALLA, 85- *Concurrent Pos:* Young investr award, NIH, 78-80 & career res develop award, 81-86; spec dental res award, Nat Inst Dental Res, 78-82; vis scientist, Karolinska Inst, Stockholm, Sweden, 82-83. *Mem:* Am Chem Soc; Am Soc Biol Chemists; Am Soc Pharm & Exp Therapeut. *Res:* Control and regulation of leukotrienes and prostaglandin synthesis and metabolism in cardiovascular disorders and inflammation process in arthritis. *Mailing Add:* Dept Pharmacol NY Med Col Valhalla NY 10595-1691

WONG, PAUL WING-KON, US citizen. PEDIATRICS, GENETICS. *Educ:* Univ Hong Kong, MD, 58; Univ Man, MSc, 67; Am Bd Pediat, dipl, 64. *Prof Exp:* Instr pediat, Children's Mem Hosp, Northwestern Univ, Chicago, 63-64; from asst prof to prof pediat, Chicago Med Sch, 67-73; prof pediat & dir metab unit, Abraham Lincoln Sch Med, Univ Ill Med Ctr, 73-76; PROF PEDIAT & DIR GENETIC SECT, RUSH MED SCH & PRESBY-ST LUKE MED CTR, CHICAGO, 76- *Concurrent Pos:* USPHS res fel biochem & med genetics, Northwestern Univ, Chicago, 62-64; Children's Res Fund fel, Ment Retardation Res Unit, Royal Manchester Children's Hosp, Eng, 65-67; attend physician, Cook County Hosp, Chicago, Ill, 65-72, consult, 72-; dir infant's aid perinatal res labs & premature & newborn nurseries, Mt Sinai Hosp, 67-73; attend physician, Univ Ill Hosp & Presby-St Luke Med Ctr, 73- *Mem:* Am Pediat Soc; Am Fedn Clin Res; Soc Pediat Res; Am Soc Human Genetics; Cent Soc Clin Res. *Res:* Metabolic diseases; human genetics; atherosclerosis. *Mailing Add:* Dept Pediat Rush Univ 600 S Paulina St Chicago IL 60612

WONG, PETER ALEXANDER, b Honan, China, Apr 9, 41; US citizen; m 66; c 1. CHEMICAL INSTRUMENTATION. *Educ:* Pac Union Col, BS, 62; Rensselaer Polytech Inst, PhD(chem), 69. *Prof Exp:* US AEC grant, Purdue Univ, 67-69; from asst prof to assoc prof, 69-88, PROF CHEM, ANDREWS UNIV, 88- *Mem:* Am Chem Soc; Sigma Xi; Fedn Am Scientists. *Res:* Use of microcomputers with chemical instruments. *Mailing Add:* Dept Chem Andrews Univ Berrien Springs MI 49104

WONG, PETER P, b Shanghai, China, Dec 12, 41; m 64; c 2. NITROGEN FIXATION, CELL-CELL RECOGNITION. *Educ:* San Francisco State Col, BS, 66; Ore State Univ, BA, 67, PhD(plant physiol), 71. *Prof Exp:* Res assoc biochem, Univ Wis, 70-72; res assoc agr chem, Wash State Univ, 72-74, instr bot, 74-76; from asst prof to assoc prof, 76-85, PROF BIOL, KANS STATE UNIV, 85- *Concurrent Pos:* prin investr, NSF, USDA, 75- *Mem:* Am Soc Plant Physiologists; Am Soc Microbiol; AAAS; Int Soc Plant Molecular Biol; Am Acad Microbiol. *Res:* Physiology and biochemistry of legume root nodule development; mechanism of recognition between rhizobia and legumes. *Mailing Add:* Div Biol Kans State Univ Manhattan KS 66506

WONG, PO KEE, b Canton City, China, May 5, 34; US citizen; m 65; c 2. RESEARCH ADMINISTRATION, SCIENCE EDUCATION. *Educ:* Cheng-Kung Univ, Taiwan, BSc, 56; Univ Utah, MSc, 61; Calif Inst Technol, Degree Eng, 66; Stanford Univ, PhD(aeronaut & astronaut), 70. *Prof Exp:* Teaching asst thermodynamics elasticity, Cheng-Kung Univ, 58-59; res & teaching asst appl math & mech eng, Univ Utah, 59-61; res & teaching asst appl math, Calif Inst Technol, 61-65; sr scientist appl mech, Lockheed Missles & Space Co, 66-68; res asst aeronaut & astronaut eng, Stanford Univ, 68-70; lectr & researcher, Univ Santa Clara, 70 & 71 & Ames Ctr, NASA, 70; engr I, Breeder Reactor Dept, Gen Elec, Sunnyvale, 72-73; specialist engr, Nuclear Serv Co, Campbell, Calif, 73; engr, Stone & Webster Eng Co, Boston, Mass, 74; PRES, SYST RES CO, BROOKLINE, MASS, 76- *Concurrent Pos:* Teacher math & sci, Hong Kong YMCA English Col, 59; consult, pressure transducer, Consolidated Electrodynamics Co, 62-65; consult, Flanco Serv, Inc & Air Res Co, Phoenix, 72-73; teacher math & sci, Boston Pub Sch, Mass, 79- *Mem:* Am Soc Mech Engrs; AAAS; Int Asn Stru Mech Reactor Technol. *Res:* Trajectory solid angle, generalized stream functions, magneto-viscoelasto dynamics and visco-elasto dynamics; formulation and solution of multi-reservoir transient problem; physical economic model by means of the solution of a system of indeterminate structures which provide impacts in science; mathematics and engineering. *Mailing Add:* 50 Bradley St Somerville MA 02145-2924

WONG, PUI KEI, b Canton, China, Nov 7, 35; m 67; c 1. MATHEMATICS. *Educ:* Pac Union Col, BS, 56; Carnegie Inst Technol, MS, 58, PhD(math), 62. *Prof Exp:* Instr math, Carnegie Inst Technol, 60-62; asst prof, Lehigh Univ, 62-64; from asst prof to assoc prof, 64-72, PROF MATH, MICH STATE UNIV, 72-, ASSOC DEAN COL NATURAL SCI. *Mem:* Math Asn Am; Soc Indust & Appl Math. *Res:* Stability and oscillation theory of differential equation; function-theoretic differential equations; non-linear boundary value problems. *Mailing Add:* Dept Math Wells Hall Mich State Univ East Lansing MI 48824

WONG, RODERICK SUE-CHEUN, b Shanghai, China, Oct 2, 44; c 2. MATHEMATICAL ANALYSIS. *Educ:* San Diego State Col, AB, 65; Univ Alta, PhD(math), 69. *Prof Exp:* From asst prof to assoc prof, 69-79, PROF MATH, UNIV MAN, 79-, HEAD, DEPT APPL MATH, 86- *Concurrent Pos:* Nat Res Coun Can grant, Univ Man, 69-; Killam res fel, Can Coun, 82-84. *Mem:* Can Math Soc; Can Appl Math Soc (pres, 89-90). *Res:* Asymptotic expansions; special functions. *Mailing Add:* Dept Appl Math Univ Man Winnipeg MB R3T 2N2 Can

WONG, ROMAN WOON-CHING, b Canton, China, June 17, 48; m 73; c 2. ALGEBRA, MATHEMATICS. *Educ:* Chinese Univ, Hong Kong, BS, 70; Sam Houston State Univ, MA, 72; Rutgers Univ, PhD(math), 77. *Prof Exp:* Asst prof, Syracuse Univ, 77-78; ASSOC PROF MATH, WASHINGTON & JEFFERSON COL, 78- *Concurrent Pos:* NATO travel grant math conf, Univ Antwerp, 78; res grant, Am Philos Soc, 78-79. *Mem:* Am Math Soc; Math Asn Am. *Res:* Category theory and homological algebra; group rings and free algebras. *Mailing Add:* Dept Math Washington & Jefferson Col Washington PA 15301

WONG, ROSIE BICK-HAR, b Shanghai, China; US citizen; m; c 2. BIOCHEMISTRY, IMMUNOLOGY. *Educ:* Mt Mary Col, BSc, 65; Med Col Wis, PhD(biochem), 69. *Prof Exp:* Res assoc biochem, Rockefeller Univ, 69-71; res assoc virol & immunol, Rutgers State Univ, 71-77; SCIENTIST IMMUNOL, AGR DIV, AM CYANAMID CO, 77- *Honors & Awards:* Am Cyanamid Sci Achievement Award, 83. *Mem:* Am Chem Soc; Sigma Xi. *Res:* Protein structure and function relationship; radioimmunoassay, transplantation antigen system in mouse; hybridoma; chicken immunology; immunoassays for environmental analysis. *Mailing Add:* Agr Div Am Cyanamid Co Princeton NJ 08543-0400

WONG, RUTH (LAU), b Hong Kong, Nov 25, 25; m 52; c 4. PATHOLOGY. *Educ:* Lingnan Univ, MD, 48. *Prof Exp:* Resident path, Children's Hosp, Wash, DC, 51-52; resident, Duke Univ Hosp, 52-53; resident, Michael Reese Hosp, 53-54; res asst, La Rabida Sanitarium, 55-56; res asst, Univ Chicago, 56-57; from asst prof to assoc prof, 57-69, PROF PATH, UNIV ILL COL MED, 69- *Concurrent Pos:* Fel, Michael Reese Hosp, Chicago, Ill, 54-55. *Mem:* Int Acad Path. *Res:* Surgical pathology; serotonin content of mast and enterochromaffin cells; relationship of mast cells to tissue response; medical information science; data retrieval of pathology records. *Mailing Add:* Dept Path Univ Ill Med Sch 1853 W Polk Chicago IL 60612

WONG, S(OON) Y(UCK), b San Antonio, Tex, Mar 4, 20; m 49; c 3. CHEMICAL ENGINEERING. *Educ:* Univ Tex, BS & MS, 43, PhD(chem eng), 49. *Prof Exp:* Sr res engr, Skelly Oil Co, 43-46; res assoc, Jefferson Chem Co, 50-53, staff engr, 53-56; tech asst to dir, 56-61, supt process lab, Petrol Prod Res & Develop Div, 61-63, supvr process develop sect, Petrochem Res Div, 64-65, proj coordr, Petrochem Dept, 66-68, asst to mgr, Petrol Prods Res Div, 68-72, tech asst to mgr proj develop, Res & Develop Dept, 72-76, PROJ COORDR, MINING RES DIV, CONTINENTAL OIL CO, 76- *Res:* Minerals extraction and coal processing projects. *Mailing Add:* 7014 Desilow Dr San Antonio TX 78240-2428

WONG, SAMUEL SHAW MING, b Beijing, China, May 10, 37; m 67; c 2. PHYSICS, THEORETICAL PHYSICS. *Educ:* Int Christian Univ, Tokyo, BA, 59; Purdue Univ, MS, 61; Univ Rochester, PhD(theoret physics), 65. *Prof Exp:* From asst prof to assoc prof, 69-78, PROF PHYSICS, UNIV TORONTO, 78- *Mem:* Am Phys Soc; Can Asn Physics. *Res:* Nuclear structure theory; intermediate energy nuclear theory, high energy nuclear physics. *Mailing Add:* Dept Physics Univ Toronto Toronto ON M5S 1A1 Can

WONG, SHAN SHEKYUK, b Mankassar, Indonesia, May 10, 45; US citizen; m 73; c 2. BIOCHEMISTRY, CLINICAL CHEMISTRY. *Educ:* Ore State Univ, BS, 70; Ohio State Univ, PhD(biochem), 74. *Prof Exp:* Fel chem, Temple Univ, 74-76; instr, Ohio State Univ, 78, res assoc, 76-78; from asst prof to assoc prof, 78-84, PROF BIOCHEM, UNIV LOWELL, 84- *Concurrent Pos:* Clin chemist, Univ Tex, Houston, 85-87. *Mem:* Am Soc Biochem & Molecular Biol; Am Chem Soc; NY Acad Sci; Sigma Xi. *Res:* Regulation of enzyme activities; diagnostic enzymology. *Mailing Add:* Dept Chem Univ Lowell Lowell MA 01854

WONG, SHEK-FU, b Canton, China, Dec 5, 43; m 73. ATOMIC PHYSICS, MOLECULAR PHYSICS. *Educ:* Chung Chi Col, Chinese Univ Hong Kong, BSc, 65; Univ Del, PhD(physics), 72. *Prof Exp:* Instr physics, Chung Chi Col, Chinese Univ Hong Kong, 65-67; fel, 72-74, res staff, 74-75, asst prof, 75-78, ASSOC PROF, ENG & APPL SCI, YALE UNIV, 78- *Mem:* Am Phys Soc. *Res:* Crossed-beam electron impact experiments, with particular interest in the resonant vibrational and rotational excitation in small molecules. *Mailing Add:* 96 Spring Garden Hamden CT 06514

WONG, SHI-YIN, b Hoiping, Canton, China, Apr 27, 41; US citizen; m 67; c 1. ORGANIC CHEMISTRY. *Educ:* Univ Calif, Los Angeles, BS, 64; Univ Southern Calif, PhD(chem), 68. *Prof Exp:* MEM TECH STAFF, HUGHES RES LAB, HUGHES AIRCRAFT CO, MALIBU, CALIF, 69- *Concurrent Pos:* Air Force Off Sci Res fel, Hughes Res Lab, 72-73. *Res:* Electrohydrodynamics of liquid crystal; molecular correlation of liquid crystal. *Mailing Add:* 407 W College Los Angeles CA 90012

WONG, SIU GUM, b San Francisco, Calif, Feb 21, 47; m 79. OPTOMETRY, PUBLIC HEALTH. *Educ:* Univ Calif, Berkeley, BS, 68, OD, 70, MPH, 72. *Prof Exp:* Res assoc community med, Sch Med, St Louis Univ, 72-73; asst prof optom & pub health, Col Optom, Univ Houston, 73-78; optom consult, 84-91, CHIEF, AREA OPTOM SERV BR, INDIAN HEALTH SERV, 78- *Concurrent Pos:* Consult, Health Power Assoc, Inc, New Orleans, 74. *Honors & Awards:* Pub Health Serv Commendation Award, 82; Outstanding Serv Medal, USPHS, 85, Unit Commendation, 85, 88 & 90, Meritorious Serv Medal, 89. *Mem:* Fel Am Acad Optom; Am Optom Asn; Am Pub Health Asn. *Res:* Community optometry; quality assurance. *Mailing Add:* Indian Health Serv 505 Marquette Ave NW Suite 1502 Albuquerque NM 87102-2162

WONG, STEWART, pharmacology, for more information see previous edition

WONG, TANG-FONG FRANK, b Canton, China, Jan 21, 44; m 69; c 2. THEORETICAL HIGH ENERGY PHYSICS. *Educ:* Chinese Univ Hong Kong, BSc, 65; Brown Univ, PhD(physics), 70. *Prof Exp:* Res assoc physics, Brookhaven Nat Lab, 69-71; res fel physics, Rutgers Univ, New Brunswick, 71-73, vis asst prof, 73-74, asst prof, 74-78; MEM STAFF, BELL LABS, 79- *Mem:* Am Phys Soc. *Res:* High energy behavior of renormalizable field theories; symmetry and symmetry breaking in field theories; strong interaction phenomenology. *Mailing Add:* 16 Ross Hall Blvd Piscataway NJ 08854

WONG, TING-WA, ENDOCRINE PATHOLOGY, CELL PATHOLOGY. *Educ:* Univ Chicago, MD, 57, PhD(org chem), 70. *Prof Exp:* ASSOC PROF PATH, UNIV CHICAGO, 58- *Mailing Add:* Dept Path Univ Chicago 5841 S Maryland Ave Chicago IL 60637

WONG, TUCK CHUEN, b Canton, Kwangtong, China, Mar 28, 46; US citizen; m 72; c 3. NUCLEAR MAGNETIC RESONANCE OF SULFACTANT SOLUTION & LIQUID CRYSTALS, NUCLEAR MAGNETIC RESONANCE STUDY OF BIOMOLECULAR STRUCTURE. *Educ:* Chinese Univ Hong Kong, BSc, 69; Univ Mich, MS, 71, PhD(chem), 74. *Prof Exp:* Fel res, Dept Chem, Ind Univ, 74-75; Killam fel res, Univ BC, 75-76; asst prof teaching, Dept Chem, Tufts Univ, 76-81; assoc prof teaching & res, 81-89, DIR NUCLEAR MAGNETIC RESONANCE FACIL, UNIV MO, 81-, PROF TEACHING & RES, 89- *Concurrent Pos:* Vis prof, Chem Ctr, Univ Lund, Sweden, 87-88; consult, Unilever Res Lab, NJ, 90- *Mem:* Am Chem Soc; Int Soc Magnetic Resonance. *Res:* Nuclear magnetic resonance investigation of micelles, liquid crystals and surfactant/polymer systems; structure and dynamics of biological molecules in solution by multi-dimensional nuclear magnetic resonance, magnetic relaxation and molecular modeling. *Mailing Add:* Dept Chem Univ Mo Columbia MO 65211

WONG, VICTOR KENNETH, b San Francisco, Calif, Nov 1, 38; m 64; c 3. PHYSICS. *Educ:* Univ Calif, Berkeley, BS, 60, PhD(physics), 66. *Prof Exp:* Fel physics, Ohio State Univ, 66-67, res assoc, 67-68; lectr, 68-69, asst prof physics, Univ Mich, Ann Arbor, 69-76; assoc prof, 76-82, chmn physics discipline, 79-80, chmn, Dept Natural Sci, 80-83, PROF PHYSICS, UNIV MICH-DEARBORN, 82-, DEAN, COL ARTS, SCI & LETT, 83- *Mem:* Am Phys Soc. *Res:* Many-body theory; equilibrium properties of interacting bosons at low temperatures; two-band superconductors; liquid helium; superfluidity; quantum fluids; critical phenomena; surface phenomena; and dielectric formulation of Bose liquids. *Mailing Add:* Acad Affair Univ Mich Flint Flint MI 48503

WONG, WAI-MAI TSANG, b Hong Kong, Apr 14, 41; Brit citizen; m 69; c 1. POLYMER SCIENCE. *Educ:* Nat Taiwan Normal Univ, BSc, 65; Univ Guelph, MSc, 69; Case Western Reserve Univ, PhD(macromolecular sci), 74. *Prof Exp:* Teaching asst physics, Nat Taiwan Normal Univ, 64-67; res assoc polymer sci, Case Western Reserve Univ, 74-76; RES SCIENTIST POLYMER SCI, COLUMBUS DIV, BATTELLE MEM INST, 76- *Mem:* Am Phys Soc. *Res:* Structure-property relationship of synthetic and bio-polymers. *Mailing Add:* 1390 Darcann Dr Columbus OH 43220

WONG, WANG MO, b Canton, Kwangtung, China; US citizen; m 49; c 2. CHEMICAL ENGINEERING. *Educ:* Sun Yat-Sen Univ, BS, 39; Univ Iowa, MS, 49, PhD(chem eng), 54. *Prof Exp:* Chief chem engr, US Rubber Co, Joliet Arsenal, Ill, 55-60; res engr, US Borax & Chem Corp, Calif, 60-61; sr engr, Armour Pharmaceut Co, Kankakee, 61-77; CHEM ENGR, ARTHUR G MC KEE & CO, 77- *Mem:* Am Chem Soc; Am Inst Chem Engrs. *Res:* Research and process development in bench scale and pilot plant of organic and inorganic chemical process; specialization in continuous process development. *Mailing Add:* 4233 Oakwood Lane Matteson IL 60443

WONG, WARREN JAMES, b Masterton, NZ, Oct 16, 34; m 62; c 3. ALGEBRA. *Educ:* Univ Otago, NZ, BSc, 55, MSc, 56; Harvard Univ, PhD(math), 59. *Prof Exp:* Lectr math, Univ Otago, NZ, 60-63, sr lectr, 64; assoc prof, 64-68, PROF MATH, UNIV NOTRE DAME, 68- *Concurrent Pos:* Vis fel, Univ Auckland, 69; ed, Proceedings Am Math Soc, 88-90. *Mem:* Am Math Soc; Math Asn Am; Australian Math Soc; Sigma Xi. *Res:* Finite group theory; groups of Lietype. *Mailing Add:* Dept Math Univ Notre Dame PO Box 398 Notre Dame IN 46556-0398

WONG, YIU-HUEN, b Hong Kong, May 9, 46; m 72. PHYSICS, MATERIALS SCIENCE. *Educ:* Mass Inst Technol, BS, 67; Univ Wis-Madison, MS, 69, PhD(physics), 73. *Prof Exp:* Res fel, Rutgers Univ, 73-76; asst prof physics, Wayne State Univ, 76-79; MEM TECH STAFF, BELL LABS, 79- *Concurrent Pos:* Co-prin investr grant, Army Res Off, 78-79. *Mem:* Am Phys Soc; Inst Elec & Electronics Engrs; Electrochem Soc. *Res:* Laser spectroscopy of matter; surface and subsurface characterization of solids; transport properties of condensed matter; microelectronics; telecommunication systems. *Mailing Add:* Bell Labs MS7F-305 600 Mountain Ave Murray Hill NJ 07974

WONG, YUEN-FAT, b Kwangtung, China, Sept 22, 35; US citizen; m 62; c 3. MATHEMATICS. *Educ:* Cornell Univ, PhD(math), 64. *Prof Exp:* ASSOC PROF MATH, DEPAUL UNIV, 64- *Concurrent Pos:* NSF fel, DePaul Univ, 65-67. *Mem:* Am Math Soc. *Res:* Algebraic topology. *Mailing Add:* Dept Math DePaul Univ Chicago IL 60614

WONG-RILEY, MARGARET TZE TUNG, b Shanghai, China, Oct 20, 41; US citizen; m 70; c 2. NEUROANATOMY, ANATOMY. *Educ:* Columbia Univ, BS, 65, MA, 66; Stanford Univ, PhD(anat), 70. *Prof Exp:* Asst prof anat & neuroanat, Univ Calif, San Francisco, 73-80, assoc prof anat, 80-81; assoc prof anat, 81-84, PROF ANAT & CELLULAR BIOL, MED COL WIS, 84- *Concurrent Pos:* Fight for Sight fel, Univ Wis, 70-71; NIH fel, Lab Neurophysiol, Nat Inst Neurol Dis & Stroke, 72-73; Alexander Ryan Endowment Fund fel, Univ Calif, San Francisco, 74-75; visual sci B study sect, NIH, 80-84; mem, Behav & Neurosci Study Sect 1, NIH, 86-90 & Reviewers Reserve, 90-94; pres, Milwaukee chap, Soc Neurosci, 88. *Mem:* AAAS; Soc Neurosci; Asn Am Anat. *Res:* Structural and functional organization of the mammalian visual system; functionally related metabolic adjustments in neurons as revealed by cytochrome oxidase histo-and cytochemistry. *Mailing Add:* Dept Anat & Cellular Biol Med Col Wis PO Box 26509 Milwaukee WI 53226

WONG-STAAL, FLOSSIE, b China; US citizen. RETROVIRUSES, AIDS RESEARCH. *Educ:* Univ Calif, LA, BA, & PhD(molecular biol), 72. *Prof Exp:* Teaching asst, Univ Calif, LA, 68-69, res asst, 70-72; postdoctoral fel, Univ Calif, San Diego, 72-73; Fogerty fel, Nat Cancer Inst, NIH, 73-75, vis assoc, 75-76, Cancer expert, 76-78, sr investr, 78-81; chief, Molecular Genetics Hematopoietic Cells Sect, Lab Tumor Cell Biol, Nat Cancer Inst, 81-90; PROF MED & BIOL, UNIV CALIF, SAN DIEGO, 90- *Concurrent Pos:* Ed, J AIDS res & human retroviruses, Microbial Pathogenesis, DNA, Cancer Res, Leukemia, 87- *Mem:* Am Soc Virol; Am Asn Cancer Res. *Mailing Add:* Dept Med 0613-M Univ Calif San Diego La Jolla CA 92093-0613

WONHAM, W MURRAY, b Montreal, Que, Nov 1, 34. APPLIED MATHEMATICS, SYSTEMS ENGINEERING. *Educ:* McGill Univ, BEng, 56; Cambridge Univ, PhD(eng), 61. *Prof Exp:* Asst prof elec eng, Purdue Univ, 61-62; res mathematician, Res Inst Advan Study, 62-64; assoc prof appl math & eng, Brown Univ, 64-70; assoc prof, 70-72, PROF ELEC ENG, UNIV TORONTO, 72 - *Concurrent Pos:* Consult, Electronics Res Ctr, NASA, Mass, 65-70. *Mem:* Fel Inst Elec & Electronics Engrs; Soc Indust & Appl Math. *Res:* Systems theory. *Mailing Add:* Dept Elec Eng Univ Toronto Toronto ON M5S 1A4 Can

WONNACOTT, THOMAS HERBERT, b London, Ont, Nov 29, 35; m 80; c 7. MATHEMATICAL & APPLIED STATISTICS. *Educ:* Univ Western Ont, BA, 57; Princeton Univ, PhD(math statist), 63. *Prof Exp:* Asst prof, Wesleyan Univ, 61-66; assoc prof math, 66-80, ASSOC PROF STATIST, UNIV WESTERN ONT, 80- *Mem:* Am Statist Asn. *Res:* Statistics in social science; textbook writing. *Mailing Add:* Dept Statist Univ Western Ont London ON N6A 5B9 Can

WONSIEWICZ, BUD CAESAR, b Buffalo, NY, Aug 23, 41; m 63; c 3. METALLURGY. *Educ:* Mass Inst Technol, SB, 63, PhD(metall), 66. *Prof Exp:* Asst prof metall, Mass Inst Technol, 66-67; mem tech staff, Bell Tel Labs, 67-88; VPRES SCI & TECHNOL, US W ADVAN TECHNOL, 88- *Mem:* AAAS; Am Inst Mining, Metall & Petrol Engrs; Am Soc Metals; Inst Elec & Electronics Engrs; Asn Comput Mach. *Res:* Software engineering; software quality and productivity; software methodology; software project management. *Mailing Add:* US W Advan Technol 6200 S Quebec St Suite 270 Englewood CO 80111

WOO, CHIA-WEI, b Shanghai, China, Nov 13, 37; m 60; c 4. THEORETICAL PHYSICS. *Educ:* Georgetown Col, BS, 56; Washington Univ, MA, 61, PhD(physics), 66. *Prof Exp:* Appl mathematician, Monsanto Co, Mo, 59-63; res assoc physics, Washington Univ, 66; asst res physicist, Univ Calif, San Diego, 66-68; from asst prof to prof physics, Northwestern Univ, Evanston, 68-79, chmn dept, 74-79; provost, Revelle Col & prof physics, Univ Calif, San Diego, 79-83; PRES & PROF PHYSICS, SAN FRANCISCO STATE UNIV, 83- *Concurrent Pos:* Lectr, Wash Univ, 59-61; consult, Argonne Nat Lab, 68-81, Exxon Res & Eng Co, 81-; vis assoc prof, Univ Ill, Urbana-Champaign, 70-71; Alfred P Sloan res fel, 71-73. *Mem:* Fel Am Phys Soc. *Res:* Quantum many body theory; low temperature physics; surface physics; liquid crystals. *Mailing Add:* Custom Pass House Eight Fei Ngo Shan Rd Kowloon Hong Kong

WOO, CHUNG-HO, b Sept 7, 43; Can citizen; m 68; c 1. MATERIALS SCIENCE, SOLID STATE PHYSICS. *Educ:* Univ Hong Kong, BSc, 67; Univ Calgary, MSc, 69; Univ Waterloo, PhD(physics), 73. *Prof Exp:* Asst res officer, 77-80, assoc res officer, 81-82, res officer, 83-85, sr res officer, 86-87, SR SCIENTIST, ADV MAT, WHITESHELL NUCLEAR RES ESTAB, ATOMIC ENERGY CAN LTD, PINAWA, 88- *Concurrent Pos:* Adj prof, Dept Physics, Univ Man, 78- *Mem:* Am Soc Testing & Mats. *Res:* Radiation damage; defects and mechanical properties; dislocations and point-defect dislocation interaction; color centers; electronic states in metals; computer simulation and numerical analysis; molecular quantum chemistry; diffusion mechanisms in metals. *Mailing Add:* Mat Sci Br Atomic Energy Can Ltd Pinawa MB R0E 1L0 Can

WOO, DAH-CHENG, b Shanghai, China, Dec 18, 21; US citizen. HYDRAULICS, HYDROLOGY. *Educ:* Hangchow Christian Col, BS, 44; Univ Mich, Ann Arbor, MA, 48, PhD(hydraul eng), 56. *Prof Exp:* Res asst struct & hydraul res, Univ Mich, Ann Arbor, 48-50; design engr, Ayres, Lewis, Norris & May, Consult Engrs, 51-56, hydraul engr, 57-62; proj engr, Univ Mich, Ann Arbor, 56-57; HYDRAUL ENGR, US FED HWY ADMIN, 62- *Mem:* Am Geophys Union; Am Soc Civil Engrs; Int Asn Sci Hydrol; Int Asn Hydraul Res; Int Water Resources Asn. *Res:* Urban water resources; small watershed hydrology. *Mailing Add:* US Fed Hwy Admin Washington DC 20590

WOO, GAR LOK, b Canton, China, Jan 14, 35; US citizen; m 64; c 2. ORGANIC CHEMISTRY. *Educ:* Univ Calif, Berkeley, BS, 59; Mass Inst Technol, PhD(org chem), 62. *Prof Exp:* Res chemist, 62-69, sr res chemist, 69-75, SR RES ASSOC, CHEVRON RES CO, RICHMOND, 75- *Mem:* Am Chem Soc. *Res:* Physical organic chemistry, mechanism, stereochemistry and synthesis; exploratory petrochemicals and surfactants; sulfur and organosulfur chemistry. *Mailing Add:* 200 Blackfield Dr Tiburon CA 94920-2074

WOO, GEORGE CHI SHING, b Shanghai, China, Feb 15, 41; Can citizen; m 67; c 3. OPTOMETRY. *Educ:* Col Optom Ont, OD, 64; Ind Univ, Bloomington, MS, 68, PhD(physiol optics), 70; Melbourne Univ, LOSc, 79. *Hon Degrees:* OD, Univ Waterloo, 87. *Prof Exp:* Optometrist, Can Red Cross, 64-66; clin instr optom, Ind Univ, 66-67; dir, Ctr Sight Enhancement, 84-87; asst prof, 70-74, assoc prof, 74-80, PROF OPTOM, UNIV WATERLOO, 80- *Concurrent Pos:* Res assoc & assoc instr, Ind Univ, 66-70; vis prof, Pa Col Optom, 83; vis scholar, Cambridge Univ, UK, 84; head & prof, Dept Diag Sci, Hong Kong Polytech, 87-89. *Honors & Awards:* Herbert Moss Mem lectr, Int Optom & Optical League, 90. *Mem:* Am Acad Optom; Can Asn Optom. *Res:* Low vision; photometry; contrast sensitivity function and optics of the eye; refraction; myopia. *Mailing Add:* Sch Optom Univ Waterloo Waterloo ON N2L 3G1 Can

WOO, JAMES T K, b Shanghai, China, June 7, 38. ORGANIC CHEMISTRY. *Educ:* Wabash Col, BA, 61; Univ Md, PhD(chem), 67. *Prof Exp:* Asst, Univ Md, 61-63, res asst, 63-66, fel, 66-67; res chemist, Dow Chem Co, Mich, 67-71; mem staff, Horizon Res Corp, 71-72; sr chemist, 72-74, scientist, 74-83, SR SCIENTIST, GLIDDEN CO, 83- *Mem:* Am Chem Soc; AAAS; Sigma Xi; Am Inst Chemists. *Res:* Polymer chemistry; graft copolymer; coatings. *Mailing Add:* 6551 Speita Rd Medina OH 44256

WOO, KWANG BANG, b Kyoto, Japan, Jan 25, 34; m 63; c 2. ROBOTICS & AUTOMATION, INTELLIGENT SYSTEMS. *Educ:* Yonsei Univ, Korea, BE, 57, ME, 59; Ore State Univ, MS, 62, PhD(elec eng), 64. *Prof Exp:* Instr elec eng, Yonsei Univ, Korea, 59-60; res mem, Sci Res Inst, Ministry Defense, Korea, 57-60; res assoc, Ore State Univ, 64; asst prof elec eng & biomed eng, Wash Univ & res assoc, Ctr Biol Natural Systs, 66-71; sr staff fel, Off Dept Dir, Div Cancer Treatment, Nat Cancer Inst, 71-76; scientist, Biol Markers Lab, Frederick Cancer Res Ctr, 76-79 & John Hopkins Oncol Ctr, 79-82; assoc dean, Grad Sch, 83-87, PROF ELEC ENG, YONSEI UNIV, 82- *Concurrent Pos:* Fel biophys, Inst Sci Technol, Univ Mich, 65-66. *Honors & Awards:* Medal of Honor, Dong Baek Jang, Repub Korea. *Mem:* AAAS; Inst Elec & Electronics Engrs; Korean Sci Eng Asn Am (pres, 80-81); Biophys Soc; Korean Inst Elec Eng. *Res:* Robotics and automation; intelligent systems; tumor-marker interactions; untrasonic imaging. *Mailing Add:* Dept Elec Eng Yonsei Univ Seoul Republic of Korea

WOO, LECON, b Chung Kin, Sept 9, 45; US citizen; m 78; c 2. POLYMER CHEMISTRY, MEDICAL & HEALTH SCIENCES. *Educ:* Kansas State Univ, BS, 67; Univ Chicago, MS, 73, PhD(chem physics), 73. *Prof Exp:* Sr chemist, E I du Pont de Nemours & Co, Inc, 73-78; sr res scientist, Atlantic & Richfield, 78-82; sr scientist, 82-84, TECH DIR, BAXTER HEALTHCARE, 84- *Concurrent Pos:* Consult polymer physics, 78- *Honors & Awards:* I-R 100 Award, 77. *Mem:* AAAS; Sigma Xi; Am Chem Soc; Am Inst Physics; Soc Plastics Engrs; NAm Thermal Analysis Soc. *Res:* Structure property of synthetic and natural polymers; biomaterial development; dynamic mechanical & rheological properties; fracture properties of polymers. *Mailing Add:* Baxter Healthcare Rte 120 & Wilson Rd Round Lake IL 60073-0490

WOO, NAM-SUNG, Korean citizen. COMPUTER-AIDED VLSI DESIGN, PARALLEL PROCESSING. *Educ:* Seoul Nat Univ, BS, 75; Korea Advan Inst Sci, MS, 77; Univ Md, PhD(computer sci), 83. *Prof Exp:* Design engr, SamSung, GTE Telecommun, Inc, 77-79; lectr computer systs, Dept Computer Sci, Dong-Guk Univ, 79-80; res asst, Univ Md, College Park, 80-83; MEM TECH STAFF, AT&T BELL LABS, 83- *Concurrent Pos:* Lectr, Univ Md, College Park, 83. *Mem:* Inst Elec & Electronics Engrs. *Res:* Computer-aided design of very large scale integration and parallel processing; developing a synthesis environment for field programmable gate array. *Mailing Add:* AT&T Bell Labs Rm 3D-439 600 Mountain Ave Murray Hill NJ 07974-2070

WOO, NORMAN TZU TEH, b Shanghai, China, Sept 28, 39. MATHEMATICS. *Educ:* Wabash Col, BA, 62; Southern Methodist Univ, MS, 64; Wash State Univ, PhD(math), 68. *Prof Exp:* PROF MATH, CALIF STATE UNIV, FRESNO, 68- *Mem:* Am Math Soc. *Res:* Number theory of mathematics. *Mailing Add:* Dept Math Calif State Univ Fresno CA 93740

WOO, P(ETER) W(ING) K(EE), b Canton, China, June 22, 34; m 66; c 3. ORGANIC CHEMISTRY, RADIOCHEMISTRY. *Educ:* Stanford Univ, BS, 55; Univ Ill, PhD(chem), 58. *Prof Exp:* From assoc res chemist to sr res chemist, Parke, Davis & Co, 58-71, res scientist, 71-77, RES ASSOC, PHARMACEUT RES DIV, WARNER-LAMBERT/PARKE-DAVIS, 77- *Mem:* Am Chem Soc. *Res:* Radioabeling, synthesis, isolation and structural elucidation of medicinal agents and related metabolites: antibiotics, enzyme inhibitors, anti-inflammatory and anti-hypertensive agents; chemistry of peptides, nucleosides, beta-lactams, aminoglycosides, carbohydrates, and macrolides. *Mailing Add:* Park Davis Pharmaceut Res Div Warner Lambert 2800 Plymouth Rd Ann Arbor MI 48105

WOO, SAVIO L C, b Shanghai, China, Dec 20, 44. PHENYLKETONURIA, HEPATIC GENE THERAPY. *Educ:* Loyola Col, BSc, 66; Univ Wash, Seattle, PhD(biochem), 71. *Prof Exp:* Postdoc fel neurol sci, dept psychiat, Univ BC, 71-73; res assoc, 73-74, from instr to assoc prof, 74-83, PROF, DEPT CELL BIOL, BAYLOR COL MED, 84-, PROF, INST MOLECULAR GENETICS, 85- *Concurrent Pos:* Res assoc, Howard Hughes Med Inst, 76-77, assoc investr, 77-79, investr, 79-; mem, ad hoc grant review comt, Cystic Fibrosis Found, 81-86, ad hoc NIH Grant Study Sect Physiol Chem, 83, Grant Study Sect Molecular Biol, 83-85, sci adv meeting, Metab Dis Res Prog, Nat Inst Diabetes, Digestive & Kidney Dis, 87, bd sci counr, Nat Inst Child Health & Human Develop, 88-; organizer & chmn, Gordon Res Conf, Colby, Sawgu Col, 85; co-organizer, Serle UCLA Symp, Keystone, Colo, 86; mem, US del human genetics, Sect US-Peoples Rep China Coop Med Health Protocol, Chinese Ministry Pub Health & US Dept Health & Human Serv, 83, US-Japan Coop Prog Recombinant DNA Res, Japanese Ministry Educ & US Dept Health & Human Serv, 87; mem, internal adv comt, Baylor Col Med, 86-, internal expert adv comt, 87-, inst biosafety comt recombinant DNA res, 78-; mem, external adv comt, Albert Einstein Col, 86, Univ Houston, 87-; cell biol res fel prog, 79-, med genetics res fel prog, 85-; dir & organizer, Interdept Grad Training Prog, 87- *Mem:* Inst Med-Nat Acad Sci; NY Acad Sci; Soc Study Inborn Errors Metab; Am Soc Biol Chemists; Am Soc Human Genetics; Am Soc Cell Biol; Soc Inherited Metab Disorders. *Res:* Molecular basis of genetic disorders in man, to develop analytical methods for prenatal diagnosis and carrier screening; exploration of possibility of correcting such genetic disorders by somatic gene therapy. *Mailing Add:* Dept Cell Biol Baylor Col Med Tex Med Ctr Houston TX 77030

WOO, SHIEN-BIAU, b Shanghai, China, Aug 13, 37; m 63; c 2. ATOMIC PHYSICS, MOLECULAR PHYSICS. *Educ:* Georgetown Col, BS, 57; Wash Univ, MA, 61, PhD(physics), 64. *Prof Exp:* Instr math, Univ Mo, St Louis, 61-62; res assoc physics, Joint Inst Lab Astrophys, Univ Colo, 64-66; from asst prof to assoc prof, 66-82, trustee, 76-82, PROF PHYSICS, UNIV DEL, 82- *Concurrent Pos:* Consult, Control Data Corp & Mid-Atlantic Consortium Energy Res, 81 & various law firms, 79-81; prin investr grant, NSF, 78-81 & Asn Res Opthamal, 81-83; lt gov, Del, 85-89; inst fel, Kennedy Sch Govt, Harvard Univ, 89. *Mem:* Am Phys Soc; Am Asn Physics Teachers; AAAS. *Res:* Inference of ion-molecule reaction cross sections from rate constants; ion velocity distributions; photodetachment of molecular negative ions; photodetachment theory: zero-core-contribution model. *Mailing Add:* Dept Physics Univ Del Newark DE 19711

WOO, TSE-CHIEN, b Nanking, China, Mar 6, 24, US citizen; m 57; c 2. APPLIED MECHANICS, APPLIED MATHEMATICS. *Educ:* Ord Eng Col, China, BS, 46; Univ Wash, MS, 54; Brown Univ, PhD(appl math), 60. *Prof Exp:* Res asst interior ballistics, Ballistics Res Inst, China, 46-48; res assoc weapon eng, Navy Res Inst, 48-51; sr develop engr, Taiwan Fertilizer Co, 51-52; res asst mech eng, Univ Wash, 53-54; res asst appl math, Brown Univ, 54-60; sr researcher basic eng, Glass Res Ctr, PPG Indust, Inc, 60-62, eng assoc, 62-64, sr eng assoc, 64-67; assoc prof mech eng, 68-71, prof math, 77-80, PROF MECH ENG, UNIV PITTSBURGH, 71- *Concurrent Pos:* Lectr, Univ Pittsburgh, 60-67; consult, Glass Res Ctr, PPG Indust, Inc, 68-; bd dirs, Mid-West Mech Conf, 71-75, co-chmn, 13th Mid-West Mech Conf, 73. *Mem:* Soc Rheology; Math Asn Am; fel Am Soc Mech Engrs; Sigma Xi; Am Acad Mech; Am Asn Univ Prof. *Res:* Finite elasticity; theory of viscoelasticity and its applications to the thermal stress problems; viscoelastic properties of glass at elevated temperatures. *Mailing Add:* Dept Mech Eng Univ Pittsburgh Pittsburgh PA 15261

WOO, YIN-TAK, b Shanghai, China, Oct 23, 47; US citizen; m 77; c 2. TOXICOLOGY, CANCER RESEARCH. *Educ:* McGill Univ, Bsc, 70; Univ Toronto, MSc, 71, PhD(biochem & pharmacol), 75; Am Bd Toxicol, dipl, 81. *Prof Exp:* Instr biochem & oncol, dept med, Tulane Univ, 75-79, guest lectr, dept pharmacol, 78-79; sr toxicologist, Sci Applns Int Corp, 79-88; TOXICOLOGIST, US ENVIRON PROTECTION AGENCY, 88- *Concurrent Pos:* Mem, peer rev comt, Ambient Water Qual Criteria Doc, US Environ Protection Agency, Cincinnati, 78-79, consult, struct-activ team, Wash, DC, 82-; co-ed, J Environ Sci Health, Part C, Environ Carcinogenesis Revs, 83-; sect ed, J Am Col Toxicol, Chem Carcinogenesis, 85-; vis prof, Med Col Wis, 85; NCI chem selection working group, 88-; mem workgroup, Nat Acad Sci, 90- *Honors & Awards:* Bronze Medal, US Environ Protection Agency, 87; J Seifter Mem Award, US Environ Protection Agency, 90. *Mem:* Am Asn Cancer Res; Europ Asn Cancer Res; Am Soc Pharmacol & Exp Therapeut; Am Col Toxicol; Soc Toxicol. *Res:* Environmental and industrial chemical carcinogenesis and toxicology; structure-activity relationships analysis; risk assessment; metabolism and reaction mechanisms; chemical induction of cancer; environmental and occupational carcinogenesis; cancer prevention. *Mailing Add:* PO Box 2429 Springfield VA 22152

WOOD, ALASTAIR JAMES JOHNSTON, b Edingburgh, Scotland, UK, Oct 13, 46; US citizen; m 72; c 2. PHARMACOLOGY, MEDICINE. *Educ:* Univ St Andrews, MBChB, 70; Royal Col Physicians, UK, MRCP, 76. *Prof Exp:* House physician, Maryfield Hosp, 71; res fel in drug monitoring, Dundee Univ, 71-72, med registr, 72-75, lectr, therapeut & pharmacol, 75-76; res fel, clin pharmacol, 76-78, asst prof med & pharmacol, 78-82, assoc prof med & pharmacol, 82-85, PROF MED & PHARMACOL, VANDERBUILT UNIV, SCH MED, 85- *Concurrent Pos:* Mem, Geront & Geriat Study Sect, 84-88; chmn, Geront & Geriat Rev Comt, NIH, 88-89. *Mem:* Am Soc Clin Invest; fel Am Col Physicians; Am Fedn Clin Res; Am Soc Pharmacol & Exp Therapeut; Am Soc Clin Pharmacol & Therapeut; British Pharmacol Soc. *Res:* Control of factors responsible for interindividual variability in the response to drugs. *Mailing Add:* Dept Med & Pharmacol Vanderbilt Univ Sch Med Nashville TN 37232

WOOD, ALBERT D(OUGLAS), b Providence, RI, June 8, 30; m 75; c 5. GAS DYNAMICS, ENERGY CONVERSION. *Educ:* Brown Univ, ScB, 51, PhD(eng), 59; Harvard Univ, MS, 53. *Prof Exp:* Instr eng, Brown Univ, 56-58; staff scientist, Avco Syst Div, Avco Corp, 58-64, chief exp gas dynamics sect,

64-66, prin res scientist, Avco Everett Res Lab, 66-69; fluid dynamicist, Off Naval Res, Boston, 69-81, dir sci, 81, head, Mech Div, Arlington, 81-85, dep dir, Technol Progs, 85-89, DIR, APPL RES & TECHNOL, OFF NAVAL RES, ARLINGTON, VA, 89- *Mem:* Sigma Xi; Am Inst Aeronaut & Astronaut. *Res:* Theoretical and experimental aerophysics; high temperature gas dynamics; chemical kinetics; laser technology; aerospace propulsion. *Mailing Add:* 1130 Secretariat Ct Great Falls VA 22066

WOOD, ALBERT E(LMER), b Cape May Court House, NJ, Sept 22, 10; wid; c 3. VERTEBRATE PALEONTOLOGY. *Educ:* Princeton Univ, BS, 30; Columbia Univ, MA, 32, PhD(geol), 35. *Hon Degrees:* MA, Amherst Col, 54. *Prof Exp:* Asst biol, Long Island Univ, 30-33, tutor, 33-34; from asst geologist to geologist, US Army Engrs, 36-41 & geologist, 46; from asst prof to prof, 46-70, chmn dept, 62-66, EMER PROF BIOL, AMHERST COL, 70- *Concurrent Pos:* Partic, Paleont Expeds, Western US, 28, 31-32 & 35; dir, Amherst Col Exped, 48, 57, 60, 63, 65 & 68; assoc cur vert paleont, Pratt Mus, Amherst Col, 48-70; NSF sr fel, Naturhistorisches Mus, Basel, Switz, 66-67; vis prof, Dept Paleont, Univ Calif, Berkeley, 72. *Mem:* Fel AAAS; Am Soc Mammalogists; Paleont Soc; fel Geol Soc Am; Soc Study Evolution; hon mem Soc Vert Paleont (secy-treas, 58-59, vpres, 60, pres, 61). *Res:* Rodent and lagomorph classification, paleontology and evolution. *Mailing Add:* 20 E Mechanic St Cape May Court House NJ 08210

WOOD, ALEXANDER W, b Newburgh, NY, Nov 2, 44; m; c 2. ONCOLOGY. *Educ:* NY Univ, PhD(basic med sci), 71. *Prof Exp:* Res group chief, Hoffmann LaRoche Inc, 81-85; assoc mem, Roche Inst Molecular Biol, 85-87; DISTINGUISHED RES LEADER, ROCHE RES CTR, 87- *Concurrent Pos:* Adj prof, NY Univ Med Sch. *Mem:* Am Soc Biochem & Molecular Biol; Am Asn Cancer Res; Am Soc Pharmacol & Exp Therapeut. *Res:* Oncogenes and tumor suppressor genes. *Mailing Add:* Dept Oncol Bldg 86 Roche Res Ctr Nutley NJ 07110

WOOD, ALLEN D(OANE), b Englewood, NJ, Aug 16, 35; m 57; c 4. MECHANICAL ENGINEERING. *Educ:* Purdue Univ, BSME, 57, MSME, 61, PhD(mech eng), 63. *Prof Exp:* RES SCIENTIST, LOCKHEED RES LAB, 63- *Mem:* Am Inst Aeronaut & Astronaut. *Res:* Instrumentation for remote measurement of automobile exhaust emissions; environmental pollution studies; electrically-powered vehicles; infrared systems; spectroscopy; lasers; shock tubes; gas dynamics. *Mailing Add:* 3212 Cowper St Palo Alto CA 94306

WOOD, ALLEN JOHN, b Milwaukee, Wis, Oct 1, 25; m 49; c 2. ELECTRICAL ENGINEERING. *Educ:* Marquette Univ, BEE, 49; Ill Inst Technol, MSEE, 51; Rensselaer Polytech Inst, PhD(elec eng), 59. *Prof Exp:* Engr, Allis Chalmers Mfg Co, Wis, 49-50; asst, Ill Inst Technol, 50-51; eng analyst, Gen Elec Co, 51-59; mem tech staff, Hughes Aircraft Co, 59-60; sr engr, Elec Utility Eng Oper, Gen Elec Co, 60-69; PRIN CONSULT, CFO & MEM BD DIR, POWER TECHNOL, INC, 69- *Concurrent Pos:* Adj prof, Rensselaer Polytech Inst, 68- *Mem:* AAAS; Nat Soc Prof Engrs; Inst Elec & Electronics Engrs; Am Nuclear Soc. *Res:* Engineering analysis of complex technical or economic systems; business and system planning simulations and studies. *Mailing Add:* 901 Vrooman Ave Schenectady NY 12309

WOOD, ANNE MICHELLE, b Cleveland, Ohio, Aug 4, 51; m. MICROBIAL ECOLOGY, PHYTOPLANKTON ECOLOGY. *Educ:* Univ Corpus Christi, BA, 73; Univ Ga, PhD(zool), 80. *Prof Exp:* Res asst, Cent Power & Light, 73; engr tech II, Tex Water Qual Bd, 73-74; teaching asst, Univ Ga, 74-78, instr, 78, res asst, 76-80; fel, Univ Chicago, 81-84, res asst prof, 84-89, res assoc prof, 89-90; ASST PROF, UNIV ORE, 90- *Concurrent Pos:* Consult, Cent Power & Light, 73-75 & LGL, Inc, 78; NIH trainee, 81-83; fel, Am Asn Univ Women, 83-84. *Honors & Awards:* Provasoli Prize, 88. *Mem:* Am Soc Limnol & Oceanog; Phycol Soc Am; Ecol Soc Am; Oceanog Soc Diatom Res; Sigma Xi. *Res:* Plankton ecology and evolution, particularly the effect of environmental parameters on production, distribution and physiology; the relationship of metallo-organic interactions with trace metal bioavailability; environmental impact assessment in marine systems. *Mailing Add:* Dept Biol Univ Ore Eugene OR 97403

WOOD, BENJAMIN W, b Cardston, Alta, Nov 13, 38; US citizen; m 59; c 6. BOTANY, RANGE SCIENCE. *Educ:* Brigham Young Univ, BS, 63, MS, 67; Ore State Univ, PhD(range sci), 71. *Prof Exp:* Asst prof biol, Boise Col, 67-68; instr range mgt, Ore State Univ, 69-70, asst prof, 70-71; ASST PROF BOT & RANGE SCI, BRIGHAM YOUNG UNIV, 71- *Concurrent Pos:* Navajo-Kaiparowits Power Generating Sta fel, Ctr Health & Environ Studies, Brigham Young Univ, 71- *Mem:* Soc Range Mgt. *Res:* Desert ecology; wildlife habitat improvement. *Mailing Add:* Dept Bot & Range Sci Brigham Young Univ Provo UT 84602

WOOD, BETTY J, b Oklahoma City, Okla, Oct 24, 55. MOLECULAR BIOLOGY. *Educ:* NC State Univ, BS, 78; Univ Ga, PhD(molecular genetics), 85. *Prof Exp:* PLANT PHYSIOLOGIST, AGR RES SERV, USDA, 85- *Res:* Restriction fragment analysis of members of the genus Sacchrum; molecular cytological analysis of drought adaption. *Mailing Add:* 99-1440 Aiea Heights Dr No 52 Aiea HI 96701

WOOD, BOBBY EUGENE, b Martinsville, Va, May 9, 39; m 65; c 2. PHYSICS, MATHEMATICS. *Educ:* Berea Col, BA, 61; Vanderbilt Univ, MA, 64. *Prof Exp:* Engr aerospace, Aro, Inc, 64-80; ENGR AEROSPACE, ARVIN/CALSPAN, 80- *Mem:* Optical Soc Am; Am Inst Aeronaut & Astronaut; Inst Environ Sci; Int Soc Optical Eng. *Res:* Optical properties of materials, especially radiative properties of condensed gases on cryogenically cooled optical components; contamination and contaminant optical effects; contaminant monitoring device testing. *Mailing Add:* Arvin/Calspan MS 650 Arnold AFB TN 37389

WOOD, BRUCE, b Kintnersville, Pa, Apr 9, 38; m 63; c 2. MATHEMATICS, MECHANICAL ENGINEERING. *Educ:* Pa State Univ, BS, 60; Univ Wyo, MS, 64; Lehigh Univ, PhD(math), 67. *Prof Exp:* Asst air pollution control engr, Bethlehem Steel Corp, 60-62; asst math, Univ Wyo, 63-64 & Lehigh Univ, 64-67; asst prof, 67-71, ASSOC PROF MATH, UNIV ARIZ, 71- *Mem:* Am Math Soc. *Res:* Linear approximation theory; summability theory; theory of complex variables; linear positive operators. *Mailing Add:* Dept Math Univ Ariz Tucson AZ 85721

WOOD, BRUCE WADE, b Morganfield, Ky, Oct 22, 51; m 75; c 4. TREE NUT PHYSIOLOGY, BREEDING & HORTICULTURE. *Educ:* Univ Ky, BS, 73, MS, 75; Mich State Univ, PhD(forestry), 79. *Prof Exp:* HORTICULTURIST, SCI EDUC, AGR RES SERV, USDA, 79- *Mem:* Am Soc Hort Sci; Int Soc Hort Sci. *Res:* Maximizing nut production efficient of nut trees by regulating growth and developmental processes via bioregulating chemicals and genetic systems. *Mailing Add:* USDA Southeastern Fruit & Tree Nut Res Lab Byron GA 31008

WOOD, BYARD DEAN, b Cardston, Alta, Can, Mar 30, 40; US citizen; m 65; c 3. MECHANICAL ENGINEERING, HEAT TRANSFER & SOLAR ENERGY. *Educ:* Utah State Univ, BS, 63, MS, 66; Univ Minn, Minneapolis, PhD(mech eng), 70. *Prof Exp:* Assoc prof heat transfer & exp methods, 70-80, PROF ENG, MECH ENG, ARIZ STATE UNIV, 80- *Concurrent Pos:* Consult, Solar Energy Res & Educ Found, 77-, Solar Rating & Cert Corp, Nat Bur Standards & ERG, Inc. *Mem:* Fel Am Soc Mech Engrs; Am Soc Heating, Refrig & Air-Conditioning Engrs; Combust Inst; Int Solar Energy Soc. *Res:* Solar energy research, especially heat transfer and experimental methods in thermal and photovoltaic solar heating and cooling systems; fire research, especially heat and mass transfer associated with large turbulent flames. *Mailing Add:* Dept Mech Eng Ariz State Univ Tempe AZ 85287

WOOD, CALVIN DALE, b Salt Lake City, Utah, July 13, 33; m 55; c 5. NUCLEAR PHYSICS, SHOCK HYDRODYNAMICS. *Educ:* Univ Calif, Berkeley, AB, 57, PhD(high energy physics), 61. *Prof Exp:* Res asst high energy physics, Lawrence Radiation Lab, Univ Calif, Berkeley, 58-61, physicist, 61-62; asst prof physics, Univ Utah, 62-64; SR DESIGN PHYSICIST, LAWRENCE LIVERMORE LAB, 64- *Concurrent Pos:* Expert witness, accident reconstruction; Crown Zellerbach Scholar, 57. *Mem:* Am Phys Soc; Ital Phys Soc; Sigma Xi. *Res:* Shock hydrodynamics; neutron cross sections; high speed digital computers; nuclear processes; chemical high explosives; material properties under high strain rates; high speed optics. *Mailing Add:* Lawrence Livermore Lab PO Box 808 L-35 Livermore CA 94550

WOOD, CARL EUGENE, b Alice, Tex, Aug 28, 40; m 73; c 3. MARINE ECOLOGY, INVERTEBRATE ZOOLOGY. *Educ:* Tex A&M Univ, BS, 62, MS, 65; PhD(fisheries), 69. *Prof Exp:* Fishery technician, Nat Marine Fisheries Serv, 63-64; res asst limnol, Tex A&M Univ, 65-67; limnologist, Tenn Valley Authority, 67-69; ASSOC PROF INVERT & MARINE BIOL, TEX A&I UNIV, 69- *Concurrent Pos:* Maricult consult, Flato Corp, 73-74. *Mem:* Am Soc Limnol & Oceanog; Fedn Estuarine Res. *Res:* Marine invertebrate ecology; shrimp of the suborder Natantia systematics; primary productivity of estuaries. *Mailing Add:* Box 158 Dept Biol Tex A&I Univ Kingsville TX 78363

WOOD, CARLOS C, b Turloc, Calif, June 19, 13. AERONAUTICAL ENGINEERING. *Educ:* Col Pac, BA, 33; Calif Inst Technol, MSME, 34, MSAE, 35. *Prof Exp:* Mem eng staff, Douglas Aircraft Corp, 37-42, chief preliminary design, Douglas Aircraft Eng, Santa Monica, 42-55, chief engr, Long Beach Div, 55-59, dir advan eng planning, 59-60; eng mgr & vpres, Sikorsky Aircraft Div, United Aircraft Corp, 60-70, consult, 70-75; RETIRED. *Concurrent Pos:* Dir & vpres, Claude C Wood & Co, Lodi, Calif, 46-81; mem sci adv bd, USAF; mem aerospace eng bd, Nat Acad Eng. *Mem:* Emer mem Nat Acad Eng; emer fel Am Inst Aeronaut & Astronaut. *Mailing Add:* 145 Bonniebrook Dr Napa CA 94558

WOOD, CAROL SAUNDERS, b Pennington Gap, Va, Feb 9, 45. MATHEMATICAL LOGIC. *Educ:* Randolph-Macon Woman's Col, AB, 66; Yale Univ, PhD(math), 71. *Prof Exp:* Gast dozent math, Univ Erlangen-N rnberg, WGer, 71-72; lectr, Yale Univ, 72-73; res assoc prof, Dept Math, Rutgers Univ, New Brunswick, NJ, 85-86; vis instr math, 70-71, asst prof, 73-80, assoc prof, 80-86, PROF, WESLEYAN UNIV, 86-, CHAIR, 90- *Concurrent Pos:* Vis, Inst Advan Study, Princeton, 82, 85-86; vis mem, MSRI, Berkeley, 89-90. *Mem:* Am Math Soc; Math Asn Am; Asn Symbolic Logic; Asn Women Math (pres, 91-93). *Res:* Application of model theory to algebra. *Mailing Add:* Dept Math Wesleyan Univ Middletown CT 06457

WOOD, CARROLL E, JR, b Roanoke, Va, Jan 13, 21. BOTANY. *Educ:* Roanoke Col, BS, 41; Univ Pa, MS, 43; Harvard Univ, AM, 47, PhD(biol), 49. *Prof Exp:* Instr biol, Harvard Univ, 49-51; from asst prof to assoc prof bot, Univ NC, 51-54; assoc cur, 54-70, CUR, ARNOLD ARBORETUM, HARVARD UNIV, 70-, PROF BIOL, 72-, MEM FAC ARTS & SCI, 63- *Concurrent Pos:* Lectr biol, Harvard Univ, 64-72. *Mem:* Am Soc Plant Taxon; Bot Soc Am; Int Asn Plant Taxon. *Res:* Flora of southeastern United States; biosystematics and taxonomy of flowering plants. *Mailing Add:* 64 W Rutland Sq Boston MA 02118

WOOD, CHARLES, b London, Eng, Nov 6, 24; US citizen; m 50; c 3. EXPERIMENTAL SOLID STATE PHYSICS. *Educ:* Univ London, BSc, 51, MSc, 55, PhD(physics), 62. *Prof Exp:* Physicist, Gen Elec Res Labs, Eng, 51-53 & Electronic Tubes Ltd, 53-54; dep group leader, Caswell Res Lab, Plessey Co, 54-56; group supvr, Res Div, Philco Corp, Pa, 56-60; sect head solid state prod group, Kearfott Div, Gen Precision Instruments, NJ, 60-61; dir thermoelec, Intermetallic Prod, Inc, 61-63; mgr, Mat Res Dept, Xerox Corp Res Labs, NY, 63-67; head dept, 67-70, PROF PHYSICS, NORTHERN ILL UNIV, 67- *Mem:* Am Phys Soc; Sigma Xi. *Res:* Semiconductors; photoconductors; thin films; crystal growth; Hall-effect devices; thermoelectricity; electrophotography. *Mailing Add:* 3795 Berwick Dr Flintridge CA 91011

WOOD, CHARLES D, US citizen. AUTOMOTIVE ENGINEERING. *Educ:* Rice Univ, BS, 56; Southern Methodist Univ, MS, 62. *Prof Exp:* Test engr, Ling-Temco-Vought Inc, 58-61, propulsion engr, 61-62; sr res engr, Dept Automotive Res, 62-67, sect mgr, 67-74, DIR DEPT ENGINE & VEHICLE RES, AUTOMOTIVE RES DIV, SOUTHWEST RES INST, 74- *Mem:* Soc Automotive Engrs. *Res:* Engine development; development of off-highway equipment; fuel-air explosive devices for earthmoving, rock ripping, hard-facing deposition, ship propulsion, undersea seismic shock generation, dredging and pumping, minefield clearance and marine icebreaking. *Mailing Add:* Dept Engine & Vehicle Res Southwest Res Inst PO Drawer 28510 San Antonio TX 78284

WOOD, CHARLES DONALD, b Ravena, Ky, Feb 4, 25; m 49; c 4. PHARMACOLOGY. *Educ:* Univ Ky, BS, 49, MS, 50; Univ NC, PhD(physiol), 57. *Prof Exp:* Neurophysiologist, Nat Inst Neurol Dis & Blindness, 53-55; instr physiol & pharmacol, Med Sch, Univ Ark, 57-59, from asst prof to assoc prof pharmacol, Med Ctr, 59-67; prof physiol & pharmacol, 67-77, head med commun, 71-78, actg head pharmacol, 86-88, PROF PHARMACOL & THERAPEUT, MED SCH, LA STATE UNIV, 77- *Concurrent Pos:* Consult, US Naval Aerospace Med Inst, 62- *Mem:* Aerospace Med Asn; Am Phys Soc. *Res:* Temporal lobe epilepsy; influence of temporal lobe structures on behavior; neuropharmacology; pulmonary edema; oxygen toxicity; neurogenic hypertension; antimotion sickness drugs; aerospace pharmacology; drugs and athletic performance. *Mailing Add:* PO Box 422 Greenwood LA 71033

WOOD, CHARLES EVANS, b San Francisco, Calif, May 14, 52; m 79; c 3. FETAL PHYSIOLOGY & ENDOCRINOLOGY. *Educ:* Univ Calif, Berkeley, AB, 74, San Francisco, PhD(endocrinol), 80. *Prof Exp:* Fel, Cardiovasc Res Inst, Univ Calif, San Francisco, 80-83; asst prof, 83-88, ASSOC PROF PHYSIOL, UNIV FLA, 88- *Mem:* Am Physiol Soc; Endocrine Soc. *Res:* Cortisol negative feedback control of fetal and adult adrenocorticotrophic hormone secretion; cardiovascular mechanoreceptor and chemoreceptor control of adrencorticotrophic hormone, cortisol, renin and vasopressin secretion. *Mailing Add:* Dept Physiol Univ Fla Box J-274 Gainesville FL 32610

WOOD, CHRISTOPHER MICHAEL, b Manchester, Eng, Feb 21, 47; Brit & Can citizen; m 71; c 2. COMPARATIVE PHYSIOLOGY. *Educ:* Univ BC, BSc, 68, MSc, 71; Univ E Anglia, PhD(comp physiol), 74. *Prof Exp:* Fel, Univ Calgary, 74-76; from asst prof to assoc prof, 76-85, PROF BIOL, MCMASTER UNIV, 85- *Mem:* Soc Exp Biol; Am Soc Zool; Can Soc Zool; AAAS. *Res:* Circulation, respiration, gas exchange, acid base regulation, osmoregulation and homeostasis in teleost fish and decapod crustaceans; environmental acid and heavy metal toxicology of fishes. *Mailing Add:* Dept Biol McMaster Univ Hamilton ON L8S 4K1 Can

WOOD, CORINNE SHEAR, b Baltimore, Md; m 46; c 4. PHYSICAL ANTHROPOLOGY, MEDICAL ANTHROPOLOGY. *Educ:* Univ Calif, Riverside, BA, 68, PhD(anthrop), 73. *Prof Exp:* Res asst med, Johns Hopkins Univ, 50-55, Sinai Hosp, Baltimore, 55-58 & Johns Hopkins Univ, 59-61; med technologist, Riverside Community Hosp, Calif, 62-71; teaching asst anthrop, Univ Calif, Riverside, 68-70; from asst prof to assoc prof, 73-85, PROF ANTHROP, CALIF STATE UNIV, FULLERTON, 85- *Concurrent Pos:* Res mem, Univ Calif Med Ctr, San Francisco, 73; lectr, Univ Calif, Riverside, 74-75, lab consult, 75. *Mem:* Fel Am Anthrop Asn; Am Asn Phys Anthropologists; Med Anthrop Soc; fel AAAS. *Res:* Human physiological variables related to disease and disease vectors; health conditions of American Indian populations and Western Samoa; women, nutrition and health; interrelationship of human culture and disease; primary health care in Maori populations of New Zealand; leprosy research in Pakistan. *Mailing Add:* Dept Anthrop Calif State Univ Fullerton CA 92634

WOOD, CRAIG ADAMS, b Rochester, NY, Jan 31, 41; m 64; c 2. COMPUTER SCIENCE, MATHEMATICS. *Educ:* Col Wooster, BA, 62; Fla State Univ, MS, 63, PhD(math), 67. *Prof Exp:* Instr math, Fla State Univ, 67-68; from asst prof to assoc prof, Okla State Univ, 68-73, NASA res grant, 69-70; assoc prof math sci & head div, Univ Houston, Victoria Campus, 73-76, assoc prof comput sci & math & dir, Comput Ctr, 77-78, prof, 78-79; PROF COMPUT SCI & CHMN DEPT, STEPHEN F AUSTIN STATE UNIV, 79- *Concurrent Pos:* Am Coun Educ fel, Acad Admin, 76-77. *Mem:* Asn Comput Mach. *Res:* Commutative ring theory in algebra; algebraic equations in numerical analysis; computer graphics and applications. *Mailing Add:* Comput Sci Dept Stephen F Austin State Univ Nacogdoches TX 75962-3063

WOOD, DARWIN LEWIS, b East Orange, NJ, July 21, 21; m 45; c 6. PHYSICS, CHEMISTRY. *Educ:* Princeton Univ, AB, 42; Ohio State Univ, PhD(physics, physiol), 50. *Prof Exp:* Physicist, Rohm and Haas Co, 42-46; fel, Univ Mich, 50-52, asst prof physics, 53-56; MEM TECH STAFF, CHEM DEPT, BELL LABS, INC, 56- *Mem:* Fel Optical Soc Am. *Res:* Polymers; proteins; optics; spectroscopy; crystal spectra; ions in crystals. *Mailing Add:* Chem Dept Bell Labs Inc AT&T Bell Labs 1A-346 Murray Hill NJ 07974

WOOD, DAVID, b Woodlawn, Ill, Oct 10, 28; m 58; c 3. CHROMATOGRAPHY, DOCUMENTATION. *Educ:* Univ Ill, BS, 50; Univ Wis, PhD(chem), 56. *Prof Exp:* Chemist, Velsicol Chem Corp, 50-52; asst chem, Univ Wis, 52-56; chemist, Spencer Kellogg & Sons, Inc, 56-59; assoc res chemist, Sterling-Winthrop Res Inst, 59-75, group leader, 75-88, sect head, 88-90, ASST RES DIR, ANALYTICAL SCI, STERLING RES GROUP, 90- *Mem:* Am Chem Soc. *Res:* Synthetic organic chemistry; synthesis of pharmaceuticals. *Mailing Add:* Sterling Res Group Rensselaer NY 12144-3493

WOOD, DAVID ALVRA, b Flora Vista, NMex, Dec 21, 04; m 37; c 5. PATHOLOGY. *Educ:* Stanford Univ, AB, 26, MD, 30. *Prof Exp:* EMER PROF PATH, SCH MED, & EMER DIR, CANCER RES INST, UNIV CALIF, SAN FRANCISCO, 72- *Concurrent Pos:* Historian, Am Asn Cancer Educ, 81-, historian, Am Cancer Soc, Calif div, 86-; CONSULT, 77. *Honors & Awards:* Am Cancer Soc Award, 50; Col Am Pathologists Award, 58; Lucy Wortham James Award, James Ewing Soc, 70. *Mem:* Fedn Am Socs Exp Biol; Am Cancer Soc (pres, 56-57); Am Asn Cancer Res; Col Am Pathologists (pres, 52-55); Asn Am Cancer Insts (pres, 70-72); Sigma Xi. *Res:* Neoplastic diseases; dual pulmonary circulation; exfoliative cytology; evaluation of cancer education in medical and dental schools; oral contraceptives and tumors of the breast, epidemiological and morphological correlations. *Mailing Add:* 54 Commonwealth Ave San Francisco CA 94118

WOOD, DAVID BELDEN, b Glendale, Calif, Nov 15, 35; m 56; c 3. TECHNICAL MANAGEMENT, ASTRONOMY. *Educ:* Univ Calif, Berkeley, AB, 57, PhD(astron), 63. *Prof Exp:* Mem tech staff astron, Bellcomm, Inc, 67-69, supvr astrophys, 69-71; mem adv plans staff, Goddard Space Flight Ctr, NASA, 71-76, opers res analyst, Appln Systs Analysis Off, 76-80; MEM TECH STAFF, THEATER C 3 SYSTS, MITRE CORP, 80- *Concurrent Pos:* Pres, Sirius Ware. *Mem:* AAAS; Am Astron Soc. *Res:* Eclipsing binary stars; extragalactic research; photoelectric photometry; space astronomy; operations research; computer modeling. *Mailing Add:* Mitre Corp MS J-111 Burlington Rd Bedford MA 01730

WOOD, DAVID COLLIER, b Jacksonville, Fla, Sept 7, 54; m 88. PROTEIN PURIFICATION, RECOMBINANT PROTEIN TECHNOLOGY. *Educ:* Duke Univ, BS, 76; Univ NC, PhD(protein chem), 81. *Prof Exp:* Postdoctoral fel, Dept Microbiol, Univ Ala, Birmingham, 81-84; STAFF SCIENTIST, MONSANTO CORP RES, MONSANTO CO, ST LOUIS, MO, 85- *Mem:* Am Chem Soc; assoc mem Am Soc Biochem & Molecular Biol; AAAS. *Res:* Protein biochemistry; purification and characterization of proteins from heterologous expression systems; structure-function analysis via mutagenesis and active-site modification. *Mailing Add:* Monsanto Co AA4I 700 Chesterfield Village Pkwy St Louis MO 63198

WOOD, DAVID DUDLEY, b Wilmington, Del, May 3, 43; m 64; c 2. LYMPHOKINES, IMMUNOREGULATION. *Educ:* Harvard Univ, BA, 65; Rockefeller Univ, PhD(cell biol), 70. *Prof Exp:* Sr res biologist, Merck Sharp & Dohme Res Labs, 72-75, res fel, 75-81, sr res fel, 81-82, assoc dir, 82-83; dir Immunol, Ayerst Labs Res Inc, 83-87; DIR INST ARTHRITIS & AUTOIMMUNITY, MILES INC, 87- *Mem:* Am Asn Immunologists; NY Acad Sci; Int Soc Immunopharmacol. *Res:* Interleukin-1, immunoregulation, cartilage and bone metabolism. *Mailing Add:* Miles Inc 400 Morgan Lane W Haven CT 06516

WOOD, DAVID LEE, b St Louis, Mo, Jan 8, 31; m 60; c 2. FOREST ENTOMOLOGY. *Educ:* State Univ NY Col Envrion Sci & Forestry, Syracuse Univ, BS, 52; Univ Calif, Berkeley, PhD(entom), 60. *Prof Exp:* Asst entomologist, Boyce Thompson Inst Plant Res, 59-60; from lectr entom & asst entomologist to assoc prof entom & assoc entomologist, Univ Calif, 60-70, assoc dean, Grad Div, 83-85, chmn dept entom sci, 85-90, PROF ENTOM & ENTOMOLOGIST, UNIV CALIF, BERKELEY, 70- *Honors & Awards:* Founders Award Mem Lectr, Entom Soc Am, 86. *Mem:* AAAS; Sigma Xi; Entom Soc Am; Soc Am Foresters; fel Entom Soc Can; fel Entom Soc Can. *Res:* Forest insect behavior and pest management; insect-host relationships, especially host selection behavior, insect pheromones and host resistance with special emphasis on bark beetles. *Mailing Add:* Dept Entom Sci Univ Calif Berkeley CA 94720

WOOD, DAVID OLIVER, b Rome, Ga. BACTERIAL GENETICS. *Educ:* Berry Col, BA, 72; Med Col Ga, MS, 75, PhD(microbiol), 78. *Prof Exp:* Res fel microbiol, Med Col, Va Commonwealth Univ, 77-79; ASST PROF MICROBIOL, UNIV SOUTH ALA, 79- *Mem:* Am Soc Microbiol. *Res:* Molecular basis of bacterial pathogenicity; pseudomonas visulence factors. *Mailing Add:* Dept Microbiol & Immunol Med Col Univ SAla 307 University Blvd Mobile AL 36688

WOOD, DAVID ROY, b Mar 3, 35; US citizen; m 67. SPECTROSCOPY. *Educ:* Friends Univ, AB, 56; Univ Mich, MS, 58; Purdue Univ, PhD(physics), 67. *Prof Exp:* Instr physics, Friends Univ, 58-59; instr math & physics, Scattergood Sch, 59-61; res assoc physics, Purdue Univ, 67; asst prof, 67-74, ASSOC PROF PHYSICS, WRIGHT STATE UNIV, 74- *Mem:* Optical Soc Am. *Res:* Experimental atomic spectroscopy; analysis of the energy level structure of the lead atom and ion; Fabry-Perut interferometry and Zeeman effect analysis; spectral line profiles in the vacuum ultraviolet; spectra of multiply-ionized atoms. *Mailing Add:* 1813 Shady Lane Dayton OH 45432

WOOD, DAVID S(HOTWELL), b Akron, Ohio, May 21, 20; m 45; c 1. MATERIALS SCIENCE. *Educ:* Calif Inst Technol, BS, 41, MS, 46, PhD(mech eng), 49. *Prof Exp:* Asst, Calif Inst Technol, 42-44; staff engr, Univ Calif, 44-46; asst mach design, 46-49, lectr, 49-50, from asst prof to assoc prof mech eng, 50-61, assoc dean students, 68-74, PROF MAT SCI, CALIF INST TECHNOL, 61- *Mem:* AAAS; Am Soc Mech Engrs; Am Soc Metals; Am Inst Mining, Metall & Petrol Engrs. *Res:* Plastic strain waves in metals; mechanical properties of metals subjected to dynamic loading; dislocations in crystals. *Mailing Add:* 590 Elm Ave Sierra Madre CA 91024

WOOD, DAVID WELLS, b Amesville, Ohio, May 22, 38; m 70. MATERIALS SCIENCE, PHYSICAL CHEMISTRY. *Educ:* Univ Pac, BS, 60, MS, 61; Univ Utah, PhD(ceramic eng), 65. *Prof Exp:* Petrol Res Fund grant calorimetry, Univ Pac, 60-61; Army res grant high pressure chem, 63-65; res chemist, E I du Pont de Nemours & Co, Inc, 65-68; proj leader composites, Burlington Industs, 68-70, sr res chemist, Corp Res & Develop, 70-77, mem tech staff, Indust Div, 77-78, mgr qual control, Indust Div, 78-79; mgr new prod, 79-85, RES ASSOC, HEXCEL CORP, DUBLIN, CALIF, 85- *Mem:* AAAS; Am Chem Soc; Am Ceramic Soc; NAm Thermal Analysis Soc; Soc Advan Mat & Process Eng. *Res:* Solid state chemistry and physics relating to polymers in the form of plastics and fibers or in conjunction with each other in composites. *Mailing Add:* Hexcel Corp 11711 Dublin Blvd Dublin CA 94568-0705

WOOD, DENNIS STEPHENSON, structural geology, tectonics, for more information see previous edition

WOOD, DERICK, b Bolton, Eng, July 19, 40. COMPUTER SCIENCE. *Educ:* Univ Leeds, BS, 63, dipl electronic comput, 64, PhD(math), 68. *Prof Exp:* Comput asst, Univ Leeds, 64-68; asst res scientist, Courant Inst Math Sci, NY Univ, 68-70; from asst prof to assoc prof comput sci, McMaster Univ, 70-78, prof, 78-82; PROF COMPUT SCI, UNIV WATERLOO, 82- *Mem:* Asn Comput Mach; Am Math Soc; Can Info Processing Soc; Inst Elec & Electronics Engrs; Europ Asn Comput Sci. *Res:* Formal language theory; data structure theory; analysis of algorithms; computational geometry; very large scale integration theory. *Mailing Add:* Dept Comput Sci Univ Waterloo Waterloo ON N2L 3G1 Can

WOOD, DON JAMES, b Northeast, Pa, July 28, 36; m 59; c 2. FLUID MECHANICS. *Educ:* Carnegie Inst Technol, BS, 58, MS, 59, PhD, 61. *Prof Exp:* Asst prof eng mech, Clemson Univ, 61-62; asst prof civil eng, Duke Univ, 62-66; PROF CIVIL ENG, UNIV KY, 66- *Concurrent Pos:* Res engr & consult, NASA, 63-66. *Honors & Awards:* Huber Res Prize, Am Soc Civil Engrs, 75; Western Elec Fund, Am Soc Eng Educ, 76. *Mem:* Am Soc Civil Engrs; Am Soc Mech Engrs; Soc Eng Sci; Am Inst Aeronaut & Astronaut; Am Soc Eng Educ. *Res:* Fluid transients; water hammer problems; water distribution systems; hydrotransport. *Mailing Add:* Dept Civil Eng Univ Ky Lexington KY 40506

WOOD, DONALD EUGENE, b St Paul, Minn, Dec 26, 30; m 54; c 3. NUCLEAR PHYSICS. *Educ:* Univ Nev, BS, 51; Northwestern Univ, MS, 53, PhD(physics), 56. *Prof Exp:* Asst physics, Northwestern Univ, 51-55; physicist, Hanford Labs, Gen Elec Co, 55-58, sr physicist, 58-63; res scientist, Kaman Sci Corp, 63-68, mgr nuclear prod, 68-74, sr scientist, Nuclear Serv Prog, 74-78; staff scientist, Rockwell Hanford Opers, 78-83, prin scientist, 83-87; FEL SCIENTIST, WESTINGHOUSE HANFORD CO, 87- *Concurrent Pos:* Lectr, Hanford Grad Ctr, Univ Wash, 58-63. *Honors & Awards:* George Wash Signature Award, 89. *Mem:* Fel Am Nuclear Soc; Soc Risk Assessment. *Res:* Nuclear fuel cycle safety; risk assessment probabilistic analysis; reliability prediction; activation analysis; regulatory analysis. *Mailing Add:* MSIN B2-19 Westinghouse Hanford Co PO Box 1970 Richland WA 99352

WOOD, DONALD ROY, b Keats, Kans, Apr 17, 21; m 43; c 2. AGRONOMY. *Educ:* Kans State Col, BS, 43; Colo State Univ, MS, 49; Univ Wis, PhD, 56. *Prof Exp:* From asst prof agron & asst agronomist to assoc prof agron & assoc agronomist, 47-63, PROF AGRON & AGRONOMIST, COLO STATE UNIV, 63- *Concurrent Pos:* Asst, Univ Wis, 50-51; res assoc, Univ Calif, 60-61; res assoc, Inst Nutrit Cent Am, Panama, 74. *Mem:* Fel AAAS; Genetics Soc Am; Genetics Soc Can; Am Soc Agron; Am Phytopath Soc. *Res:* Dry field bean breeding; dry bean disease resistance; breeding for improved nutritional value. *Mailing Add:* 1920 Sheeley Dr Ft Collins CO 80526

WOOD, EARL HOWARD, b Mankato, Minn, Jan 1, 12; m 36; c 4. PHYSIOLOGY. *Educ:* Macalester Col, BA, 34; Univ Minn, BS, 36, MS, 39, PhD(physiol), 40; MD, 41. *Hon Degrees:* DSc, Macalester Col, 50; DMed, Univ Bern, 82. *Prof Exp:* Instr physiol, Univ Minn, 39-40; Nat Res Coun fel pharmacol, Univ Pa, 41-42; instr pharmacol, Harvard Med Sch, 42; assoc prof physiol, 42-60, PROF PHYSIOL & MED, MAYO GRAD SCH MED, UNIV MINN, 42-; CONSULT, MAYO CLIN, 42- *Concurrent Pos:* Sci consult to Surgeon Gen, Aeromedical Ctr, USAF, Heidelberg, 46; career investr, Am Heart Asn, 61-; vis scientist, Univ Bern, 65-66 & Univ Col, Univ London, 72-73, Univ Kiel, 73 & 75 & McGill Univ, 86; distinguished lectr, Am Col Chest Physicians, 74; Am Physiol Soc travel award, Int Physiol Cong, Oxford Univ; mem, first med teaching deleg China, Am Col Physicians; Humboldt sr scientist award, 83. *Honors & Awards:* Mayo Distinguished Lectr Med Sci, 78; Alza Lectr, Biomed Eng Soc, 78; John Phillips Mem Award, Am Col Physicians, 83; Res Award, Aerospace Med Asn, 83; Carl Wiggers Award, Am Physiol Soc; Res Achievement & Gold Heart Awards, Am Heart Asn; Stewart Mem Lectr, Royal Aeronaut Soc, London, 88; C Ludwig Ehremuinz Award, Ger, 89. *Mem:* AAAS; Am Physiol Soc (pres, 80-81); Soc Exp Biol & Med; Am Heart Asn; hon fel Am Col Cardiol; Am Soc Clin Invest; Royal Netherlands Acad Arts & Sci; Biomed Eng Soc (pres elect); Cent Soc Clin Invest; Aerospace Med Asn. *Res:* Electrolyte metabolism of cardiac and voluntary muscle; glucose reabsorption in amphibian kidney; effect of cardiac glycoside on electrolyte metabolism; cardiopulmonary effects of gravitational and inertial forces, aerospace medicine; computer based quantitative imaging techniques; cardiovascular and respiratory physiology of man. *Mailing Add:* Mayo Rochester MN 55901

WOOD, EDWARD C(HALMERS), b Tucson, Ariz, Apr 30, 23; m 43; c 4. ENGINEERING. *Educ:* Univ Ariz, BA, 46, BSEE, 48. *Prof Exp:* Nuclear engr & mgr, Gen Elec Co, Wash, 49-62; mgr geophysics, Stanford Res Inst, 62-67, dir appl progs, 67-69, exec dir, Irvine, 70-72, staff scientist, 72-76; MEM STAFF, GEN ELEC CO, CALIF, 76- *Mem:* Am Nuclear Soc; Inst Elec & Electronics Engrs. *Res:* Institutional and environmental studies; management. *Mailing Add:* 385 Cervantes Portola Valley CA 94025

WOOD, ELWYN DEVERE, b Everett, Wash, Sept 15, 34; m 66; c 2. MARINE GEOCHEMISTRY. *Educ:* Western Wash State Col, BA(chem) & BA(educ), 64; Univ Wash, MS, 66; Univ Alaska, PhD(chem oceanog), 71. *Prof Exp:* Oceanogr, Univ Wash, 66-67 & PR Nuclear Ctr, 70-75; oceanogr, Outer Continental Shelf Off, Bur Land Mgt, Dept Interior, 75-81; CONSULT, 81- *Res:* Trace element chemistry in the marine environment; determination and evaluation of natural and induced radioactivity in the environment; design and administration of environmental studies on the outer continental shelf; evaluation of near-shore currents for the disposal of wastes. *Mailing Add:* 133 Crapemyrtle Rd Covington LA 70433

WOOD, ERIC F, b Vancouver, BC; c 2. ENVIRONMENTAL DATA ANALYSIS, GROUNDWATER ANALYSIS. *Educ:* Univ BC, BASc Hons, 70; Mass Inst Technol, SM, 72, ScD(civil eng), 74. *Prof Exp:* Res scholar, Int Inst Appl Systs Analysis, Austria, 74-76; from asst prof to assoc prof, 76-86, PROF WATER RESOURCES, PRINCETON UNIV, 86- *Concurrent Pos:* Assoc ed, Water Resources Res, 77-82, Rev Geophys, J Forecasting & Appl Math & Computation; mem, numerous comts, Am Geophys Union, NASA, Am Meteorol Asn, Nat Acad Sci, NSF & Environ Protection Agency, 77-; dir, Water Resources Prog, Princeton Univ, 80-, actg chmn, Dept Civil Eng, 86-87; vis sr scientist, Inst Hydrol, UK, 83-84; vis prof, Politecnico de Milano, Italy, 88-89. *Honors & Awards:* Robert E Horton Award, Am Geophys Union, 77. *Mem:* Am Geophys Union; Opers Res Soc Am. *Res:* Hydroclimatology with emphasis on land-atmospheric interactions and determining the hydrologic impacts from climate change. *Mailing Add:* Dept Civil Eng Princeton Univ Princeton NJ 08540

WOOD, EUNICE MARJORIE, b Venice, Calif, Sept 5, 27. CELL BIOLOGY, ELECTRON MICROSCOPY. *Educ:* Rutgers Univ, New Brunswick, BS, 48; Mt Holyoke Col, MA, 50; Harvard Med Sch, PhD(anat), 68. *Prof Exp:* Instr zool, Wellesley Col, 50-52; lectr, Barnard Col, Columbia Univ, 53-55; res technician hemat, City Hope Med Ctr, Calif, 57-59; res assoc, Inst Cancer & Blood Res, 59-62; instr biol, Mt Holyoke Col, 62-63; USPHS fel, Med Sch, Univ Southern Calif, 67-68; from asst prof to assoc prof, 68-77, PROF BIOL, CALIF STATE UNIV, LONG BEACH, 77- *Concurrent Pos:* Electron micros consult oncol unit, Med Sch, Univ Southern Calif, 68-72. *Mem:* Am Soc Cell Biologists; Am Asn Anat. *Res:* Electron microscopy and cytochemistry of invertebrate glands. *Mailing Add:* NOUA Scotia Inst Tech 5685 Lee St Halifax NS B3J 3C4 Can

WOOD, F(REDERICK) B(ERNARD), b Sacramento, Calif, Dec 17, 17; m 42; c 2. ELECTRICAL ENGINEERING. *Educ:* Univ Calif, BS, 41, MS, 48, PhD(elec eng), 53. *Prof Exp:* Asst elec eng, Univ Calif, 40-41, asst resonator coupling, 49-52; mem staff radar, Mass Inst Technol, 41-46; staff engr, Res & Develop Lab, 52-58, staff engr, Advan Systs Develop Div Lab, 59-61, proj engr, 61-64, staff engr, 65-70, adv engr, 70-72, adv engr, Systs Develop Div, 72-75, adv engr, Systs Commun Div, 75-76, ADV ENGR, GEN PROD DIV, INT BUS MACH CORP, 76- *Mem:* AAAS; Inst Elec & Electronics Engrs; Soc Social Responsibilities Sci; Soc Gen Systs Res; NY Acad Sci; Sigma Xi. *Res:* Development of computer-communication systems; simulation; computer programming; information theory; data communication; cybernetics; philosophy of science. *Mailing Add:* 2346 Lansford Ave San Jose CA 95125

WOOD, FERGUS JAMES, b London, Ont, May 13, 17; nat US; m 46; c 2. TIDAL DYNAMICS, COASTAL FLOODING. *Educ:* Univ Calif, AB, 38. *Prof Exp:* Asst astron, Univ Mich, 40-42; instr physics & astron, Pasadena City Col, 46-48 & John Muir Col, 48-49; asst prof physics, Univ Md, 49-50; assoc physicist, Appl Physics Lab, Johns Hopkins Univ, 50-55; sci ed, Encycl Americana, 55-60; aeronaut & space res scientist & sci asst to dir off space flight progs, NASA, 60-61; prog dir foreign sci info, NSF, 61-62; phys scientist, Off Dir, US Coast & Geod Surv, 62-70; phys scientist, Off Dir, Nat Ocean Surv, Nat Oceanic & Atmospheric Admin, 70-73, res assoc, Off Dir, Nat Ocean Surv, Rockville, Md, 73-77; GEOPHYS CONSULT TIDAL DYNAMICS, 78- *Res:* Environmental geoscience; wind-profile studies over navy ships at sea; perigean and proxigean spring tide analysis and potential for coastal flooding; gravitational-geophysical correlations; science education, history and film documentation. *Mailing Add:* 3103 Casa Bonita Dr Bonita CA 91902

WOOD, FORREST GLENN, b South Bend, Ind, Sept 13, 18; m 46, 79. MARINE ZOOLOGY. *Educ:* Earlham Col, AB, 40; Yale Univ, MS, 50. *Prof Exp:* Resident biologist, Lerner Marine Lab, BWI, 50-51; cur, Marine Studios & Res Lab, Marineland, Fla, 51-63; head marine biosci facil, Naval Missile Ctr, Calif, 63-70; sr scientist & consult, Ocean Sci Dept, 70, head marine biosci prog off, Under Sea Sci Dept, Naval Undersea Ctr, 72-77, STAFF SCIENTIST, BIOSCI DIV, NAVAL OCEAN SYSTS CTR, 77- *Concurrent Pos:* Mem, comt sci adv, Marine Mammal Comn, 84-88. *Mem:* Soc Vert Paleontol; Am Soc Mammalogy; Sigma Xi; Soc Marine Mammalogy; Am Elasmobranch Soc. *Res:* Behavior of toothed whales, sharks and octopuses. *Mailing Add:* 2651-201 Front St San Diego CA 92103-6540

WOOD, FRANCIS C, JR, b Philadelphia, Pa, Oct 20, 28; m 58, 91; c 2. DIABETES, NUTRITION. *Educ:* Princeton Univ, AB, 50; Harvard Med Sch, MD, 54; Am Bd Internal Med, dipl, 63 & 74. *Prof Exp:* Intern, King County Hosp, Seattle, Wash, 54-55; resident, Vet Admin Hosp, 55-56; resident, Univ Wash Hosp, 60-61; from instr to asst prof, Sch Med, 61-68, from asst prog dir to prog dir, Clin Res Ctr, 62-70, assoc dean, 70-76, ASSOC PROF MED, SCH MED, UNIV WASH, 68- *Concurrent Pos:* Res fel, Harvard Med Sch & Peter Bent Brigham Hosp, 58-60; chief staff, Seattle Vet Admin Hosp, 70-76; dir physician educ, Providence Med Ctr, Seattle, 76-87. *Mem:* Endocrine Soc; Am Diabetes Asn; fel Am Col Physicians. *Mailing Add:* Dept Med RG-26 Univ Wash Seattle WA 98195

WOOD, FRANCIS CLARK, medicine; deceased, see previous edition for last biography

WOOD, FRANCIS EUGENE, b Kirksville, Mo, Sept 19, 32; m 55; c 4. ENTOMOLOGY. *Educ:* Univ Mo-Columbia, BS, 58, MS, 62; Univ Md, PhD(entom), 70. *Prof Exp:* Exten entomologist, Univ Mo-Columbia, 60-64; exten entomologist, Univ Md, 64-76, from asst prof to assoc prof entom, 71-89; RETIRED. *Mem:* Entom Soc Am. *Res:* Taxonomy of Coleoptera; household insects; youth entomology. *Mailing Add:* Dept Entom Univ Md College Park MD 20742

WOOD, FRANCIS PATRICK, b Seattle, Wash, June 18, 17. ELECTRICAL ENGINEERING. *Educ:* Gonzaga Univ, AB, 40; Alma Col, Calif, STL, 48; Stanford Univ, MS, 51. *Prof Exp:* Instr elec eng, Gonzaga Univ, 42-44; instr elec eng, High Sch, Wash, 50; from asst prof to emer prof elec eng, Seattle Univ, 52-87, head dept, 59-63; RETIRED. *Mem:* Am Soc Eng Educ; Inst Elec & Electronics Engrs. *Res:* Power system analysis; stability; control systems. *Mailing Add:* Dept Elec Eng Seattle Univ Seattle WA 98122

WOOD, FRANK BRADSHAW, b Jackson, Tenn, Dec 21, 15; m 45; c 4. ASTRONOMY. *Educ:* Univ Fla, BS, 36; Princeton Univ, MA, 40, PhD(astron), 41. *Prof Exp:* Res assoc, Princeton Univ, 46; Nat Res Coun fel, Steward Observ, Univ Ariz & Lick Observ, Univ Calif, 46-47; asst prof astron & asst astronr, Univ Ariz, 47-50; from assoc prof to prof astron, Univ Pa, 50-54, Flower prof & chmn dept, 54-68; assoc chmn dept, 71-76, dir, optical astron observ, 68-80, PROF ASTRON, UNIV FLA, 80- *Concurrent Pos:* Mem comts, Int Astron Union, 38 & 42, mem orgn comt & chmn comt int progs, 42, pres comt, 42, 68-71; Fulbright fel, Mt Stromlo Observ, Australian Nat Univ, 57-58, pres comt, 38 & 82-85; exec dir, Flower & Cook Observ, Univ Pa, 50-54, dir, 54-68; NATO sr fel sci, 73; Am Astron Soc vis lectr astron, 73-; Fulbright fel, Inst Astron & Space Physics, Buenos Aires, Argentina, 77. *Mem:* Can Astron Soc; Am Astron Soc; Royal Astron Soc; hon mem Royal Astron Soc NZ; hon mem Int Amateur & Prof Photoelec Photom. *Res:* Photoelectric photometry eclipsing binary stars; 051019311of spectrophotometric data in the far ultraviolet as taken from the Copernicus satellite; emphasis on close double stars. *Mailing Add:* Dept Astron SSRB Univ Fla Gainesville FL 32611

WOOD, FREDERICK STARR, b Redding, Conn, Nov 21, 21; m 47; c 2. CHEMICAL ENGINEERING. *Educ:* Univ Va, BChE, 43. *Prof Exp:* Sect leader, Res Dept, Am Oil Co, 44-64, sr opers analyst, 64-72, sr opers analyst, Standard Oil Co, Ind, 72-77, sr oper researcher, 77-80; consult statist, 80-88. *Mem:* Soc Automotive Engrs; Am Inst Chem Engrs; Am Statist Asn; Royal Statist Soc; Am Chem Soc. *Res:* Fuels and lubrication process and development; design and analysis of experiments with computers; Author of two books and 18 patents. *Mailing Add:* 1414 Del Vista Dr Valparaiso IN 46383-3320

WOOD, GALEN THEODORE, b Philadelphia, Pa, Feb 7, 29; m 55; c 3. NUCLEAR PHYSICS. *Educ:* Wash Univ, BS, 51, PhD(physics), 56. *Prof Exp:* Physicist, Argonne Cancer Res Hosp, Univ Chicago, 55-57; NSF res fel nuclear spectros, Inst Theoret Physics, Univ Copenhagen, 57-59; asst prof physics & res nuclear spectros, Univ Pa, 59-65, NSF res grant radioactive nuclei, 63-65; assoc physicist, Argonne Nat Lab, 65-69; ASSOC PROF PHYSICS, CLEVELAND STATE UNIV, 69- *Mem:* Am Phys Soc. *Res:* Nuclear spectroscopy of radiations from radioactive decay and nuclear reactions, decay scheme, gamma-gamma directional and polarization correlations, magnetic moments, lifetimes and nuclear magnetic hyperfine fields; electron accelerator developments; nuclear spectroscopy. *Mailing Add:* Dept Physics Cleveland State Univ Euclid Ave & 24th St Cleveland OH 44115

WOOD, GARNETT ELMER, b Gloucester, Va, Feb 14, 29; m 53; c 2. BIOCHEMISTRY. *Educ:* Va State Col, BS, 51, MS, 56; Georgetown Univ, PhD(chem), 66. *Prof Exp:* Res microbiologist, Div Vet Med, Walter Reed Army Med Ctr, 56-64; RES CHEMIST, DIV CHEM & PHYSICS, FOOD & DRUG ADMIN, 65- *Concurrent Pos:* Lectr chem, Univ DC, 73- *Mem:* AAAS; NY Acad Sci; Sigma Xi; Am Chem Soc; Am Oil Chemists Soc. *Res:* Chemistry of toxic and deleterious compounds that may arise in certain foods as a result of handling, storage and/or processing. *Mailing Add:* 1717 Verbena St NW Washington DC 20012

WOOD, GARY WARREN, b Rochester, NY, Sept 9, 41. IMMUNOLOGY. *Educ:* Kalamazoo Col, BA, 63; Univ Mich, MS, 65; State Univ NY Buffalo, PhD(microbiol), 71. *Prof Exp:* Fel immunol, 71-73, instr path, 73-74, asst prof, 74-79, ASSOC PROF PATH, UNIV KANS MED CTR, 79-, DIR, DIAG IMMUNOL LAB, 73- *Mem:* Am Asn Immunologists; Sigma Xi; Reticuloendothelial Soc. *Res:* Immunobiology of the maternal/fetal interrelationship; in situ immune response to tumors. *Mailing Add:* Dept Path Univ Kans Med Ctr 39th St & Rainbow Blvd Kansas City KS 66103

WOOD, GENE WAYNE, b Bedford, Va, Oct 23, 40; m 65; c 2. WILDLIFE ECOLOGY. *Educ:* Va Polytech Inst & State Univ, BS, 63; Pa State Univ, MS, 66, PhD(agron), 71. *Prof Exp:* Instr wildlife mgt, Pa State Univ, University Park, 67-71, asst prof wildlife ecol, 71-74; from asst prof forestry to assoc prof, 74-81, PROF FORESTRY, BELLE W BARUCH RES INST, CLEMSON UNIV, 81- *Mem:* Ecol Soc Am; Wildlife Soc; Soc Am Foresters. *Res:* Effects of silvicultural practices on animal population and habitat; nutrient distribution in forest ecosystems. *Mailing Add:* Dept Wildlife Mgt Clemson Univ Main Campus Clemson SC 29634

WOOD, GEORGE MARSHALL, b Fairfield, Conn, Jan 20, 33; m 52; c 3. ENGINEERING SCIENCE, PHYSICS. *Educ:* Univ Ga, BS, 59; Rensselaer Polytech Inst, MS, 68, PhD(eng sci), 74. *Prof Exp:* SR RES SCIENTIST INSTRUMENTATION, NASA/LANGLEY RES CTR, 59- *Concurrent Pos:* Adj assoc prof nuclear eng, Rensselaer Polytech Inst, 75-83; adj assoc prof physics, Univ New Orleans, 82; mgr tethered satellite syst oper, flight exp, NASA Hq, 89-90. *Mem:* Am Inst Aeronaut & Astronaut; Am Soc Mass Spectrometry; Am Chem Soc. *Res:* Mass spectrometry; analytical instrumentation; boundary layer analysis; ionization processes and charged particle behavior; surface chemistry; ion physics; metal oxide catalysis; gas-gas interactions; data enhancement through deconvolution methods. *Mailing Add:* Instrument Res Div NASA/Langley Res Ctr Hampton VA 23665

WOOD, GERRY ODELL, b Oklahoma City, Okla, Nov 19, 43; m 65; c 2. INDUSTRIAL HYGIENE, CHEMISTRY. *Educ:* Univ Okla, BSCh, 65; Univ Tex, Austin, PhD(phys chem), 69; Am Bd Indust Hyg, cert, 76. *Prof Exp:* Res asst phys chem, Kerr-McGee Res Ctr, Okla, 65; fel, 69-71, STAFF MEM CHEM, LOS ALAMOS NAT LAB, 72- *Mem:* Am Indust Hyg Asn; Am Chem Soc. *Res:* Air sampling techniques; analytical methods development; chemical kinetics; photochemistry; vapor adsorption in sorbent beds; dynamics of gas phase reactions. *Mailing Add:* 2233 37 St Los Alamos Nat Lab MS K486 Los Alamos NM 87545

WOOD, GLEN MEREDITH, b Dallas, Tex, Apr 17, 20; m 50; c 7. AGRONOMY. *Educ:* RI State Col, BS, 47; Rutgers Univ, MS, 48, PhD(agr), 50. *Prof Exp:* Asst, Rutgers Univ, 47-50; from assoc prof & assoc agronomist to prof & agronomist, 50-85, actg chmn, Dept Agron, 53-55, EMER PROF PLANT & SOIL SCI, UNIV VT, 85- *Concurrent Pos:* Golf Course Supts Asn Am res grants, 67, 68 & 71-74; assoc prof & assoc agronomist, Wash State Univ, 69-70; tech adv, Vol Tech Asst, 76-; vis prof, Cornell Univ, 80. *Mem:* Am Soc Agron; Crop Sci Soc Am; Nat Audubon Soc; Sierra Club; Nature Conservancy. *Res:* Cold hardiness in ladino clover; physiological and environmental studies with birdsfoot trefoil, perennial ryegrass and other forage crops; forage utilization by poultry; turfgrass management; shade and drouth studies with turfgrasses; application of infrared photography to turfgrass research; cold hardiness studies with forage and turfgrasses; marginal land pasture renovation studies using sheep, goats and cattle. *Mailing Add:* RR 2 Box 5550 Jericho VT 05465

WOOD, GORDON HARVEY, b Trail, BC, Jan 30, 40; m 64; c 2. SCIENTIFIC NUMERIC DATABASES, PHYSICS. *Educ:* Univ BC, BASc, 63, MASc, 65, PhD(physics), 69. *Prof Exp:* Res officer, 69-79, RES COUN OFFICER, NAT RES COUN CAN, 80- *Concurrent Pos:* Chmn, Can Nat Comt, Comt Data Sci & Technol; mem, NRC Assoc Comt Tribology; secy gen, Comt Data Sci & Technol. *Mem:* Can Asn Physicists; Chem Inst Can; Am Sci Affil. *Res:* Collections of scientific-technical numeric data which can be searched, retrieved and manipulated by computer. *Mailing Add:* Nat Res Coun Can Montreal Rd Ottawa ON K1A 0S1 Can

WOOD, GORDON WALTER, b NS, Can, Apr 6, 33; m 56; c 2. MASS SPECTROMETRY, SCIENCE EDUCATION. *Educ:* Mt Allison Univ, BSc, 55, MSc, 56; Syracuse Univ, PhD(org chem), 62. *Prof Exp:* Elem sch teacher, NS, 51-52; chemist, Paints Div, Can Industs Ltd, 56-58; fel with A C Cope, Mass Inst Technol, 62-63; from asst prof to assoc prof, 63-75, PROF CHEM, UNIV WINDSOR, 75- *Concurrent Pos:* Vis assoc res scientist, Space Sci Lab, Univ Calif, Berkeley, 69-70; sr researcher, Dept Med Biochem, Fac Med, Univ Dijon, 76-77; assoc dean, Fac Grad Studies & Res, Univ Windsor, 79-82, dean, 82-85, vpres acad, 85- *Mem:* Am Chem Soc; Asn Educ Teachers Sci; fel Chem Inst Can. *Res:* Applications of field desorption mass spectrometry and fast atom bombardment to problems in organic and biochemistry; science education improvements through revised teacher training curricula. *Mailing Add:* VP Academic Univ Windsor Windsor ON N9B 3P4 Can

WOOD, HARLAND G, b Delavan, Minn, Sept 2, 07; m 29; c 3. BIOCHEMISTRY, MICROBIOLOGY. *Educ:* Macalester Col, BA, 31; Iowa State Col, PhD, 35. *Hon Degrees:* ScD, Macalester Col, 46, Northwestern Univ, 72 & Univ Cinn, 82. *Prof Exp:* From instr to asst prof bact, Iowa State Col, 36-43; assoc prof physiol chem, Univ Minn, 43-46; prof biochem & dir, Case Western Reserve Univ, 46-65, dean sci, 67-69, univ prof, 70-78, PROF BIOCHEM, CASE WESTERN RESERVE UNIV, 65-, EMER UNIV PROF, 78- *Concurrent Pos:* Nat Res Coun fel, Univ Wis-Madison, 35-36; Fulbright fel, Univ Dunedin, 55; Commonwealth fel, Max Planck Inst, Munich, Ger, 62; mem adv coun, Life Ins Med Res Fund, 57-62; mem, training grant comt, NIH, 65-69; mem adv bd, Am Cancer Soc, 65-69; mem, Pres' Adv Comt, 67-71; mem coun, Int Union Biochem, 67-76; NIH res fel, Univ Ga, 69; mem, Phys Study Sect, NIH, 73-77; Fulbright sr scholar, Australia, 76; Humboldt Found Sr US scientist award, Munich, Ger, 79; Wellcome vis prof, Basic Med Sci, St Louis Univ, 89. *Honors & Awards:* Eli Lilly Award, 42; Carl Neuberg Medal, 52; Modern Med Award for Distinguished Achievement, 68; Lynen Medal & lectr, 72; Lynen Mem lect, 13th Int Cong Biochem, 85; Selman A Waksman Award, Nat Acad Sci, 86; Rosentiel Med Res Award, 87; Nat Medal Sci, 89; William C Rose Award in Biochem & Nutrit, 90; Humboldt Found Sr US Scientist Award, Munich, Ger, 79. *Mem:* Nat Acad Sci; Am Acad Arts & Sci; Am Soc Biol Chemists (pres, 59); Am Chem Soc; Soc Am Microbiol; Int Union Biochem (secy gen, 70-73, pres, 79-85). *Res:* Polyphosphate metabolism, autotrophic growth; structure of enzymes; mechanism of enzyme action; role of biotin, B-12 and metals. *Mailing Add:* Dept Biochem Case Western Reserve Univ Cleveland OH 44106

WOOD, HAROLD SINCLAIR, b Washington, DC, Aug 4, 22. CHEMICAL ENGINEERING. *Educ:* Cornell Univ, BChE, 44. *Prof Exp:* Asst, Geol Lab, Univ Kans, 41; phys chemist, Univ Okla, 42; chem engr, Standard Oil Co, Ind, 44-50; head supvr eng lab, Res & Develop Dept, Mid-Continent Petrol Corp, 50-55; owner, Remwood Chem Co, Okla, 55-58; process engr, Refining & Chem Div, Bechtel Corp, Calif, 58-61; head process design gas plant construct, 61-77, CHIEF ENGR, MALONEY-CRAWFORD TANK CORP, 77-; MGR APPLN ENG, ARROW ENG, INC, TULSA. *Mem:* Am Chem Soc; Am Inst Chem Engrs. *Res:* Commercial desalting unit employing fiberglass as contacting agent; petroleum, natural gas and their products. *Mailing Add:* 217 E 24th St Tulsa OK 74114-1217

WOOD, HARRY ALAN, b Albany, NY, Apr 24, 41; div; c 2. VIROLOGY. *Educ:* Middlebury Col, AB, 63; Purdue Univ, MS, 65, PhD(plant virol), 68. *Prof Exp:* RES SCIENTIST, BOYCE THOMPSON INST PLANT RES, 68- *Mem:* Soc Invert Pathol; Am Soc Microbiol; Am Soc Virol; AAAS. *Res:* Physical and biological properties of insect and plant viruses with special interest in insect virus genetics. *Mailing Add:* Boyce Thompson Inst Cornell Univ-Tower Rd Ithaca NY 14853

WOOD, HENDERSON KINGSBERRY, b Huntington, WVa, Feb 24, 13; m 38; c 1. GENETICS, PHYSIOLOGY. *Educ:* Ohio Wesleyan Univ, BA, 37; Fisk Univ, MA, 40; Ind Univ, PhD(zool), 53. *Prof Exp:* Instr biol, Ala State Teachers Col, 40-44; from instr to asst prof, Fisk Univ, 44-48; prof biol, Tenn State Univ, 48-80, head, Dept Biol Sci, 56-80. *Concurrent Pos:* Grad consult, Tenn State Univ, 52-62. *Mem:* Am Genetic Asn; Nat Inst Sci; Sigma Xi. *Res:* Protozoan genetics; physiology of Protozoa. *Mailing Add:* 3949 Drakes Branch Rd Nashville TN 37218

WOOD, HENRY NELSON, biochemistry, for more information see previous edition

WOOD, HOUSTON GILLEYLEN, III, b Tupelo, Miss, Oct 4, 44; m 65, 91; c 2. FLUID DYNAMICS, MECHANICAL ENGINEERING. *Educ:* Miss State Univ, BA, 65, MS, 67; Univ Va, PhD(appl math), 78. *Prof Exp:* Engr isotope separation theory, Oak Ridge Gaseous Diffusion Plant, Union Carbide Corp, 67-73; res engr gas centrifuge theory res lab eng sci, Univ Va, 73-77; engr gas centrifuge theory & fluid dynamics, Oak Ridge Gaseous Diffusion Plant, Nuclear Div, Union Carbide Corp, 77-81; ASSOC PROF MECH & AEROSPACE ENG, SCH ENG & APPL SCI, UNIV VA, CHARLOTTESVILLE, VA, 81- *Mem:* Soc Indust & Appl Math; Am Phys Soc; Am Inst Aeronaut & Astronaut; Am Soc Mech Engrs. *Res:* Fluid dynamics; computational methods; centrifugation. *Mailing Add:* Sch Eng & Appl Sci Univ Va Thornton Hall Charlottesville VA 22901

WOOD, HOWARD JOHN, III, b Baltimore, Md, July 19, 38; m 61, 77; c 3. ASTRONOMY, PHYSICS. *Educ:* Swarthmore Col, BA, 60; Ind Univ, MA, 62, PhD(astron), 65. *Prof Exp:* Res asst, Sproul Observ, 57-59, Goethe Link Observ, 58-62 & Lowell Observ, 62-63; from instr to assoc prof astron, Univ Va, 64-70; staff astronr, Europ Southern Observ, Santiago, Chile, 70-75; Fulbright vis prof, Univ Observ, Vienna, 75-77; vis asst prof dept astron, Ind Univ, Bloomington, 78-81; asst to dir, Cerro Tololo, Interam Observ, La Serena, Chile, 82-84; sr scientist, Northrop Serv Inc, NASA-Goddard Space Flight Ctr, 84-85; PHYSICIST-ASTRON, NASA GODDARD SPACE FLIGHT CTR, 85-, MGR OPTICAL TELESCOPE ASSEMBLY, HUBBLE SPACE TELESCOPE, 90. *Concurrent Pos:* Guest investr, McDonald Observ, 59-60, Lowell Observ, 62-65 & Kitt Peak Nat Observ, 63-69; NSF grants, 66-70, 79-82; guest prof, Univ Observ Vienna, 80-81. *Mem:* Am Astron Soc; Sigma Xi; Int Astron Union; Space Studies Inst. *Res:* Photoelectric and spectrophotometric studies of the Balmer lines in the spectra of the magnetic and related stars; Zeeman spectroscopy of magnetic stars; photometric studies of asteroids; photography of Mars; optical alignment and testing of aerospace optics; optics lead for DIRBE instrument on the cosmic Background Explorer. *Mailing Add:* 15806 Pinecroft Ln Bowie MD 20716-1737

WOOD, IRWIN BOYDEN, b Concord, NH, Apr 27, 26; m 84; c 3. PARASITOLOGY. *Educ:* Univ NH, BS, 49, MS, 51; Kans State Univ, PhD(parasitol), 58. *Prof Exp:* Asst zoologist, Univ NH, 50; chemist, Cyanamid Agr Res Div, 52-54, parasitologist, 54-56, res parasitologist & group leader, Agr Div, 58-64, mgr animal res & develop, Cyanamid Int, 64-74, dir Animal Prod Res & Develop, Cyanamid Int, Wayne, 74-77, prin scientist, Parasitic Chemotherapy & Immunol Cyanamid Agr Res Div, Princeton, 77-84, assoc res fel, 84-86, RES FEL, CHEMOTHER & IMMUNOL CYANAMID AGR RES DIV, PRINCETON, 87-, RES FEL, ANIMAL INDUST PROD DEVELOP, 91- *Concurrent Pos:* Chmn, Anthelminthic Guidelines Comt, 83- *Honors & Awards:* Sci Achievement Award, Am Cyanamid Co, 86. *Mem:* Am Soc Parasitologists; World Asn Advan Vet Parasitologists; Am Asn Vet Parasitologists. *Res:* Chemotherapy and physiology of helminths; host-parasite relations; bacterial chemotherapy; acaricides; animal health and feed product development. *Mailing Add:* Am Cyanamid Co PO Box 400 Princeton NJ 08540

WOOD, JACK SHEEHAN, b St Albans, Vt, Oct 31, 31; m 58; c 2. ENVIRONMENTAL PHYSIOLOGY. *Educ:* Univ Maine, Orono, BS, 54; Mich State Univ, MA, 60, PhD(ecol, animal physiol), 63. *Prof Exp:* From asst prof to assoc prof, 63-75, PROF BIOMED SCI, WESTERN MICH UNIV, 75-, DIR PUB SERV, 81- *Concurrent Pos:* Water qual dir, Mich SCent Planning & Develop Region, 75- *Mem:* AAAS; Wildlife Soc; Am Inst Biol Sci; Soc Exp Biol & Med. *Res:* Physiological response to adverse environmental conditions, including general systematic stress responses, reproductive inhibition and related phenomena in vertebrates; water quality management. *Mailing Add:* Dept Biol & Health Sci Western Mich Univ Kalamazoo MI 49008

WOOD, JACKIE DALE, b Picher, Okla, Feb 16, 37; m 56; c 2. PHYSIOLOGY, NEUROBIOLOGY. *Educ:* Kans State Univ Pittsburg, BS, 64, MS, 66; Univ Ill, PhD(physiol), 69. *Prof Exp:* Asst prof biol, Williams Col, 69-71; from asst prof physiol to assoc prof, Univ Kans Med Ctr, Kansas City, 71-78, prof, 78-79; prof & chmn, physiol dept, Sch Med, Univ Nev, Reno, 79-85; PROF & CHMN, PHYSIOL DEPT, COL MED, OHIO STATE UNIV, COLUMBUS, 85- *Concurrent Pos:* Alexander von Humboldt fel; NIH career develop award, 74. *Honors & Awards:* Hoffman LaRoche Prize Gastrointestinal Res, 86. *Mem:* AAAS; Am Gastroenterol Asn; Am Soc Zool; Soc Neurosci; Am Physiol Soc. *Res:* Electrical and synapatic behavior of gastrointestinal nerve cells in relation to control and coordination of secretion; absorption and motor function of the gut and disease. *Mailing Add:* Dept Physiol 4084 Graves Hall Ohio Stae Univ 333 W 10th Ave Columbus OH 43210-1239

WOOD, JAMES ALAN, b Richmond, Va, Sept 16, 39; m 65; c 1. MATHEMATICAL ANALYSIS. *Educ:* Georgetown Univ, BS, 61; Univ Va, MA, 63, PhD(math), 66. *Prof Exp:* From instr to asst prof math, Georgetown Univ, 65-69; asst prof, 69-72, ASSOC PROF MATH, VA COMMONWEALTH UNIV, 72- *Concurrent Pos:* NSF res grant, 68-70 & NSF-Nat Inst Educ grant, 80-82. *Mem:* Am Math Soc. *Res:* Operational calculus and dynamical systems; multiplier theory. *Mailing Add:* Dept Math Sci Va Commonwealth Univ Richmond VA 23284

WOOD, JAMES BRENT, III, b Oct 25, 42; US citizen; m 69; c 2. ORGANIC CHEMISTRY, GENERAL CHEMISTRY. *Educ:* Univ Denver, BS, 65; Univ Ariz, MS, 70, PhD(chem), 71. *Prof Exp:* Asst prof chem, Mobile Col, 71-73; assoc prof chem & head dept, Palm Beach Atlantic Col, 73-83, chmn, div natural sci & math, 76-83; ASSOC PROF, PALM BEACH COMMUNITY COL, 83- *Concurrent Pos:* Instr physics, Palm Beach Community Col, 74-83; instr chem, Fla Atlantic Univ. *Mem:* Am Chem Soc. *Res:* Chemical education. *Mailing Add:* Palm Beach Community Col Cent 4200 Congress Ave Lake Worth FL 33461

WOOD, JAMES C, JR, b Spartanburg, SC, Aug 21, 39; m 64; c 2. SOLID STATE PHYSICS. *Educ:* Clemson Univ, BS, 61, MS, 63; Univ Va, PhD(physics), 66. *Prof Exp:* Res physicist, Cent Res Div, Am Cyanamid Co, Conn, 66-71; sr res scientist, TRW Eastern Res Lab, 71-73; head, dept sci teaching physics & phys sci, 73-84, actg head, nuclear eng dept, 78-80, CHMN INDUST & ENG TECHNOL DIV, TRI- COUNTY TECH COL, 84- *Mem:* Am Phys Soc; Am Asn Physics Teachers; Sigma Xi; Am Soc Eng Educators. *Mailing Add:* Tri-County Tech Col Pendleton SC 29670

WOOD, JAMES DOUGLAS, b Aberdeen, Scotland, Jan 25, 30; m 56; c 3. NEUROCHEMISTRY. *Educ:* Aberdeen Univ, BSc, 51, PhD(biochem), 54; Univ Saskatchewan, DSc, 85. *Prof Exp:* Res officer, Can Dept Agr, 54-57; assoc scientist, Fisheries Res Bd, Can, 57-61; head biochem group, Defence Res Med Labs, Can, 61-63, head physiol chem sect, 63-68; head dept, 68-87, PROF BIOCHEM, UNIV SASK, 68- *Concurrent Pos:* Mem, Med Res Coun, Can, 76-82. *Mem:* Am Soc Neurochem; Int Soc Neurochem; Can Biochem Soc. *Res:* Gamma-aminobutyric acid metabolism and function. *Mailing Add:* Dept Biochem Univ Sask Saskatoon SK S7N 0W0 Can

WOOD, JAMES KENNETH, b Boulder, Colo, Jan 29, 42; m 66; c 2. SYNTHETIC ORGANIC CHEMISTRY. *Educ:* Colo State Univ, BS, 64; Kans State Col, MS, 65; Ohio State Univ, PhD(chem), 69. *Prof Exp:* From asst prof to assoc prof, 69-82, PROF CHEM, UNIV NEBR, OMAHA, 82- *Mem:* Am Chem Soc; Sigma Xi; Nat Sci Teachers Asn. *Res:* Synthesis of novel and biologically active compounds; development of new synthetic techniques and methods; development of new methods for the resolution of racemates. *Mailing Add:* 1622 Hillside Dr Omaha NE 68114

WOOD, JAMES LEE, b Cordele, Ga, Sept 5, 40; m 60; c 4. INORGANIC CHEMISTRY, THERMOCHEMISTRY. *Educ:* Vanderbilt Univ, BA, 62, PhD(inorg chem), 66. *Prof Exp:* Res fel chem, Rice Univ, 65-66; asst prof, 66-69, ASSOC PROF CHEM, DAVID LIPSCOMB COL, 69- *Concurrent Pos:* Sr fel, Rice Univ, 71-73; Indust Res 100 Award, Indust Res Mag; actg dir, Hazardous Mat Training Inst, State Tenn, 78-, consult hazardous mat, Off Civil Defense. *Mem:* Am Chem Soc. *Res:* Thermodynamics and reaction calorimetry; fluorine chemistry; coordination compounds. *Mailing Add:* 4024 General Bates Dr Nashville TN 37204-4020

WOOD, JAMES MANLEY, JR, b Birmingham, Ala, July 5, 27; m 53; c 3. PHYSICAL CHEMISTRY. *Educ:* Howard Col, BA, 47; Univ Wis, PhD, 52. *Prof Exp:* Res chemist, La, 52-69; RES ADVISOR, ETHYL CORP, 52- *Mem:* Am Chem Soc; Electrochem Soc. *Res:* Molecular spectra; electrochemistry of fused salts; high energy batteries; decomposition of organometallic compounds, vapor plating; zeolite chemistry; high temperature chemistry; semiconductor materials; lubricant testing. *Mailing Add:* Ethyl Corp PO Box 14799 Baton Rouge LA 70898

WOOD, JAMES W, b Seattle, Wash, Jan 22, 25; m 53; c 2. BIOLOGY. *Educ:* Univ Wash, BS, 50, MS, 58. *Prof Exp:* Aquatic biologist, Fish Comn Ore, 50-55, fish pathologist, 55-60; fish pathologist, Wash State Dept Fisheries, 60-70, supvr fish cult res, 70-77, fish qual control supvr, 77-86; RETIRED. *Concurrent Pos:* Consult, Int Pac Salmon Fisheries Comn, 64-65, Can Dept Fisheries, 66 & Repub of Chile Dept Fisheries, 70-71. *Mem:* Am Fisheries Soc; Am Inst Fishery Res Biol; Wildlife Dis Asn. *Res:* Infectious and nutritional diseases of salmonid fishes. *Mailing Add:* 8124 NE 157th Bothell WA 98011

WOOD, JOE GEORGE, b Victoria, Tex, Dec 8, 28; c 1. ANATOMY. *Educ:* Univ Houston, BS, 53, MS, 58; Univ Tex, Galveston, PhD(anat), 62. *Prof Exp:* Asst biol, Univ Houston, 56-58; instr anat, Dent Br, Univ Tex, 61 & Sch Med, Yale Univ, 62-63; asst prof, Sch Med, Univ Ark, 63-66; assoc prof, 66-70, asst dean acad develop, Univ Tex Med Sch, San Antonio, 67-69; prof neurobiol & anat, 70-88, chmn, dept, Univ Tex Med Sch, Houston, 70-84. *Concurrent Pos:* USPHS trainee, Univ Tex, Galveston & Sch Med, Yale Univ, 62-63; mem neuroanat vis scientist prog, USPHS, 65-66; consult to vchancellor health affairs, Univ Tex, 68-70. *Honors & Awards:* Res Award, Sigma Xi, 62. *Mem:* Am Asn Anat; Soc Exp Biol & Med; Electron Micros Soc Am; Soc Neurosci; Histochem Soc; Am Soc Cell Biol; Am Soc Neurochem. *Res:* Histochemistry and cytochemistry of neurons; histochemical and electron microscopic localization of biogenic amines and their relation to nerve function in animals under stress and drug administration. *Mailing Add:* Dept Anat Sci Okla Univ Health Sci Ctr Col Med PO Box 26901 Oklahoma City OK 73190

WOOD, JOHN ARMSTEAD, b Roanoke, Va, July 28, 32; div; c 2. METEORITICS, PLANETARY SCIENCE. *Educ:* Va Polytech Inst, BS, 54; Mass Inst Technol, PhD(geol), 58. *Prof Exp:* Geologist, Smithsonian Astrophys Observ, 59; Am Chem Soc-Petrol Res Fund fel, Cambridge Univ, 59-60; geologist, Smithsonian Astrophys Observ, 60-62; res assoc, Enrico Fermi Inst Nuclear Studies, Univ Chicago, 62-65; assoc dir, Harvard-Smithsonian Ctr Astrophys, 81-86, GEOLOGIST, SMITHSONIAN ASTROPHYS OBSERV, 65- *Concurrent Pos:* Res assoc, Harvard Col Observ, 60-; vchmn, Lunar Sample Anal Planning Team, 71-73; prof pract geol, Dept Geol Sci, Harvard Univ, 76- *Honors & Awards:* NASA Medal, 73; J Lawrence Smith Award, Nat Acad Sci, 76; Frederick C Leonard Award, Meteoritical Soc, 78. *Mem:* Nat Acad Sci; fel Am Geophys Union; Meteoritical Soc (pres, 70-72); Am Astron Soc; Int Astron Union; fel AAAS. *Res:* Study of meteorites as samples of primordial planetary material; lunar petrology and geophysics; origin of the planets. *Mailing Add:* Harvard-Smithsonian Ctr Astrophys 60 Garden St Cambridge MA 02138

WOOD, JOHN D(UDLEY), b Brooklyn, NY, Dec 5, 30; m 56, 81; c 6. PHYSICAL METALLURGY, MATERIALS SELECTION. *Educ:* Case Inst Technol, BS, 53; Lehigh Univ, MS, 59, PhD(metall eng), 62. *Prof Exp:* Student engr, Long Lines Dept, Am Tel & Tel Co, 53-54; prod metallurgist, Kaiser Aluminum & Chem Corp, 56-58; instr phys metall, Lehigh Univ, 60-61; res metallurgist, Alcoa Res Labs, Aluminum Co Am, 61-62; asst prof metall eng, 62-65, assoc prof metall, 65-78, PROF METALL, LEHIGH

UNIV, 78- *Concurrent Pos:* Consult, Aluminum Div, Howmet Corp, 65-82; mat engr, Allied Chem Corp, 76-77; course dir & lectr, Ctr Prof Adv, 78- *Mem:* Am Inst Mining, Metall & Petrol Engrs; Am Soc Metals; Am Soc Testing & Mat; Brit Inst Metals. *Res:* Physical metallurgy of aluminum alloys; corrosion and stress corrosion; selection of materials; failure analysis; non-destructive evaluation. *Mailing Add:* Dept Metall & Mat Eng Lehigh Univ Whataker Lab Five Bethlehem PA 18015

WOOD, JOHN GRADY, b Atlanta, Ga, Aug 1, 42; m 78; c 2. NEUROBIOLOGY. *Educ:* Ga State Univ, BS, 67; Emory Univ, PhD(anat), 71. *Prof Exp:* Fel neurobiol, Inst Animal Physiol, Cambridge, Eng, 71-73 & City Hope Med Ctr, Duarte, Calif, 73-74; asst prof anat, Univ Tenn, 74-76, assoc prof, 76-80, adj assoc prof anat, Ctr Health Sci, 80-; MEM STAFF, DEPT ANAT & CELL BIOL, SCH MED, EMORY UNIV, GA. *Concurrent Pos:* Fel, Nat Mult Sclerosis Soc, 71-72 & Mult Sclerosis Soc Gt Brit & Northern Ireland, 72-73; independent res fel neurobiol, Friday Harbor Marine Labs, Friday Harbor, Wash, 74; Alfred P Sloan Found res fel, 76. *Mem:* Am Asn Anat; Soc Neurosci; Am Soc Cell Biol; Am Soc Neurochem; Am Soc Zoologists. *Res:* Synthesis, posttranslational modification and transport of glycoconjugates in neurons; fundamental mechanics of axonal transport; cytochemical and biochemical studies of synapse formation; role of protein phosphorylation systems in neurite outgrowth and synapse formation, particularly in regard to phosphoxylation of cytoskeletal proteins; studies of the segration of cytoskeletal proteins in neurons and mechanisms of cytoskeletal clysfunction in neurodegenerative diesases such as Alzheimer's disease. *Mailing Add:* Dept Anat & Cell Biol Emory Univ Sch Med Atlanta GA 30322

WOOD, JOHN HENRY, b Calgary, Alta, Nov 18, 24; nat US; m 50; c 2. PHARMACOKINETICS, BIOPHARMACEUTICS. *Educ:* Univ Man, BSc, 46, MSc, 47; Ohio State Univ, PhD(phys chem), 50. *Prof Exp:* Proj chemist, Colgate-Palmolive Co, 50-53; sr res assoc chem, Rensselaer Polytech Inst, 53-54; proj chemist, Colgate-Palmolive Co, 54-56, group leader, 56-57; head phys chem sect, Prod Div, Bristol-Myers Co, NJ, 57-61, head phys chem dept, 61-65, asst dir res & develop labs, 65-67, dir chem res, 67-69; prof, 69-89, EMER PROF PHARM, MED COL VA, VA COMMONWEALTH UNIV, 90- *Concurrent Pos:* Mem comt rev, US Pharmacopoeia, 70-75. *Mem:* Am Chem Soc; fel Am Asn Pharmaceut Sci. *Res:* Drug micellar phenomena; rheology; physical pharmacy; biopharmaceutics and pharmacokinetics; pharmacogenetics; saturation and competitive metabolism. *Mailing Add:* 1504 Cedarbluff Dr Richmond VA 23233-4804

WOOD, JOHN HERBERT, b Michigan City, Ind, Oct 12, 29; wid; c 3. SOFTWARE SYSTEMS. *Educ:* Purdue Univ, BS, 51; Mass Inst Technol, PhD(solid state physics), 58. *Prof Exp:* Res assoc solid state physics, Mass Inst Technol, 58-62, asst prof, 62-66, dir coop comput lab, 64-65; consult, 65-66, STAFF MEM, LOS ALAMOS SCI LAB, 66- *Concurrent Pos:* Res assoc, Atomic Energy Res Estab, Harwell, Eng, 74-75. *Res:* Calculation of atomic wave functions and energy levels; calculation of energy band structures; calculation of molecular structure via complete neglect of differential overlap and scattered wave methods; in-house computer consultant. *Mailing Add:* Group C-10 MS B296 Los Alamos Sci Lab Los Alamos NM 87545

WOOD, JOHN KARL, b Logan, Utah, July 8, 19; m 47; c 4. PHYSICS. *Educ:* Utah State Agr Col, BS, 41; Pa State Col, MS, 42, PhD(physics), 46. *Prof Exp:* Asst petrol refining, Pa State Univ, 44-46; optical engr, Bausch & Lomb Optical Co, NY, 46-48; from asst prof to assoc prof physics, Univ Wyo, 48-56; PROF PHYSICS, UTAH STATE UNIV, 56- *Concurrent Pos:* NSF sci fac fel, Sweden, 66. *Mem:* Am Phys Soc; Am Soc Metals; Optical Soc Am. *Res:* Crystal orientation in metals studies by means of x-rays; Raman spectroscopy; pole figures of the effect of some cold rolling mill variables on low carbon steel; light; molecular and atomic physics; general mathematics; sound. *Mailing Add:* 1359 Juniper Dr Logan UT 84321

WOOD, JOHN LEWIS, b Homer, Ill, Aug 7, 12; m 41; c 2. BIOCHEMISTRY. *Educ:* Univ Ill, BS, 34; Univ Va, PhD(org chem), 37. *Hon Degrees:* DSc, Blackburn Univ, 55. *Prof Exp:* Asst biochem, Med Sch, George Washington Univ, 37-38; asst, Med Col, Cornell Univ, 38-39; assoc chemist, Eastern Regional Res Lab, Bur Agr Chem & Eng, USDA, 41-42; asst biochem, Med Col, Cornell Univ, 42-44, asst prof, 44-46; assoc prof, Col Med, Univ Tenn, Memphis, 46-50, prof biochem, Med Units, 50-71, head dept biochem, 52-55, chmn dept, 55-67, alumni distinguished serv prof biochem, Ctr Health Sci, 71-80, assoc dean, Grad Sch-Med Sci, 78-79, actg dean, 79-80; sr staff scientist, Life Sci Res Off, Fedn Am Soc Exp Biol, 80-81; CONSULT LIFE SCI, 81- *Concurrent Pos:* Finney-Howell fel, Harvard Univ, 39-41; Guggenheim fel, 54; USPHS spec res fel, 65; Nat Acad Sci-Polish Acad Sci exchange visitor, 70; vis prof, Rhodes Col, 83. *Mem:* AAAS; Am Chem Soc; Am Soc Biol Chemists; Soc Exp Biol & Med; Am Asn Cancer Res; Am Asn Univ Prof. *Res:* Biochemistry of amino acids; proteins; carcinogenesis; thiocyano derivatives; sulfur compounds. *Mailing Add:* 49 Sevier St Memphis TN 38111

WOOD, JOHN MARTIN, b Huddersfield, Eng, Mar 22, 38; m 62; c 2. BIOCHEMISTRY, INORGANIC CHEMISTRY. *Educ:* Univ Leeds, BSc, 61, PhD(biochem), 64. *Prof Exp:* Lectr org chem, Leeds Col Technol, Eng, 61-63; res assoc microbiol, Univ Ill, Urbana, 64-66, from asst prof to prof biochem, 66-74; PROF BIOCHEM & ECOL & DIR, FRESHWATER BIOL INST, UNIV MINN, 74-; MEM STAFF, GRAY FRESHWATER BIOL INST, NAVARRE, MN. *Honors & Awards:* Gold Medal Chem, Synthetic Org Chem Mfrs Asn US, 72. *Mem:* Am Chem Soc; AAAS; Freshwater Soc. *Res:* The environmental chemistry of trace metals in water and the mechanism of action of metallo-enzymes. *Mailing Add:* 1729 Dove Lane Mound MN 55364

WOOD, JOHN STANLEY, b Stoke-on-Trent, Eng, Oct 9, 36; m 62; c 2. CHEMISTRY. *Educ:* Univ Keele, BA, 58; Univ Manchester, PhD(chem), 62. *Prof Exp:* Res assoc chem, Mass Inst Technol, 62-64; lectr, Univ Southampton, 64-70; PROF CHEM, UNIV MASS, AMHERST, 70- *Mem:* Royal Soc Chem; Am Crystallog Asn. *Res:* Inorganic chemistry; x-ray crystallography; studies of stereochemistries and electron structures of inorganic compounds. *Mailing Add:* Dept Chem Univ Mass Amherst Campus Amherst MA 01003

WOOD, JOHN WILLIAM, b Tacoma Park, Md, Nov 14, 41; m 65; c 2. DIFFERENTIAL TOPOLOGY, TOPOLOGY OF VARIETIES. *Educ:* Harvard Univ, BA, 63; Univ Calif, Berkeley, PhD(math), 68. *Prof Exp:* Instr math, Princeton Univ, 68-69, lectr, 70-71, mem, Inst Advan Study, 69-70 & 71; Ritt asst prof, Columbia Univ, 72-75; assoc prof, 72-81, PROF MATH, UNIV ILL, CHICAGO, 81- *Concurrent Pos:* Vis res scholar, Univ Geneva, 72. *Mem:* Am Math Soc. *Res:* Algebraic and differential topology; topology of algebraic varieties. *Mailing Add:* Math Dept Univ Ill Chicago M/C 249 Chicago IL 60680

WOOD, JOSEPH M, b Richmond, Ind, May 2, 21; m 57; c 2. BOTANY, PALEOBOTANY. *Educ:* Ind Univ, BA, 53, PhD(plant morphol, paleobot), 60; Univ Mich, MSc, 56. *Prof Exp:* From instr to assoc prof bot & paleobot, 57-73, asst dir div biol sci, 71-75, PROF BIOL SCI, UNIV MO-COLUMBIA, 73- *Mem:* Paleont Soc; Paleont Asn; Int Asn Angiosperm Paleobot; Int Orgn Palaeobotanists; Am Asn Stratig Palynologists; Sigma Xi. *Res:* Paleozoic and Mesozoic plant macro/micro fossil morphology, anatomy, evolutionary, ecological, and stratigraphic studies. *Mailing Add:* Div Biol Sci Tucker Hall Univ Mo Columbia MO 65211

WOOD, KENNETH GEORGE, b Niagara Falls, Ont, Jan 11, 24; nat US; m 48; c 3. LIMNOLOGY. *Educ:* Univ Toronto, BA, 47, MA, 49; Ohio State Univ, PhD(hydrobiol), 53. *Prof Exp:* Asst prof biol, Buena Vista Col, 53-55 & RI Col, 55-56; prof, Thiel Col, 56-65; from assoc prof to prof biol, State Univ NY, Col Fredonia, 65-87; RETIRED. *Concurrent Pos:* Sabbatical leave, Calspan Corp, NY, 72-73; Fulbright scholar, Madurai-Kamaraj Univ, Madurai, India, 80-81. *Mem:* NAm Benthological Soc; Int Asn Theoret & Appl Limnol. *Res:* Ecology of aquatic animals; primary productivity; inorganic carbon dioxide. *Mailing Add:* Rte 1 Box 1322 Whigham GA 31797

WOOD, KRISTIN LEE, b Burlington, Colo, Dec 27, 62; m 86; c 1. ENGINEERING DESIGN, DESIGN FOR MANUFACTURING. *Educ:* Colo State Univ, BS, 85; Calif Inst Technol, MS, 86, PhD(mech eng), 89. *Prof Exp:* Engr, tech support, Digital Equip Corp, 85; grad teaching asst, Calif Inst Technol, 85-88, grad res asst, 86-89; ASST PROF, UNIV TEX, AUSTIN, 89. *Concurrent Pos:* Eng aid A roadway design, Colo Dept Highways, 82-83, eng aide B, 84; Lab tutor & monitor, Ctr Computer Asst Eng Colo State Univ, 84-85. *Mem:* Am Soc Mech Engrs; Am Soc Eng Educ; Inst Elec & Electronics Engrs Computer Soc. *Res:* Design theory; computer-integrated engineering; design for manufacture; applied mechanics in the design of mechanical components, features and assemblies. *Mailing Add:* Dept Mech Eng Univ Tex ETC 5-160 Austin TX 78712-1063

WOOD, KURT ARTHUR, b Springfield, Minn, July 2, 56; m 79. VIBRATIONAL SPECTROSCOPY, CONDENSED PHASES. *Educ:* Univ Calif, Davis, BS, 77, Berkeley, PhD(chem), 81. *Prof Exp:* ASST PROF CHEM, ST OLAF COL, 81- *Mem:* Am Phys Soc; Am Chem Soc; Am Sci Affil. *Res:* Applications of spectroscopy, particularly vibrational spectroscopy, to intra-and intermolecular interactions of molecules in condensed phases, to energy transfer and to the structure of matter. *Mailing Add:* Ind Coatings Rohm & Haas 727 Norristown Rd Spring House PA 19477

WOOD, LAWRENCE ARNELL, b Peekskill, NY, Jan 15, 04; m 51; c 2. PHYSICS. *Educ:* Hamilton Col, AB, 25; Cornell Univ, PhD(physics), 32. *Prof Exp:* From asst to instr physics, Cornell Univ, 27-35; res physicist, 35-43, chief rubber sect, 43-62, CONSULT RUBBER, NAT BUR STANDARDS, 62- *Concurrent Pos:* Deleg, Int Rubber Technol Conf, London, 38, 48 & 62, Kuala Lumpur, Malaysia, 68 & Rio de Janeiro, Brazil, 74. *Mem:* Fel Am Phys Soc; Am Chem Soc; Sigma Xi. *Res:* Semiconductors; Hall effect; blocking layer photocells; physics and technology of polymers, especially synthetic rubbers and natural rubber. *Mailing Add:* Nat Bur Standards Washington DC 20234

WOOD, LEONARD ALTON, b Gratiot Co, Mich, Aug 22, 22; m 42, 77; c 3. GEOLOGY, GROUND WATER HYDROLOGY. *Educ:* Mich State Univ, BS, 46. *Prof Exp:* Geologist, Water Resources Div, US Geol Surv, Mich, 46-51, Tex, 52-63, Colo, 63-67, Wash, DC, 67-74 & Reston, Va, 74-80, coordr subsurface waste disposal studies, 71-78, staff hydrologist, Water Resources Div, 67-80; CONSULT HYDROGEOLOGIST, S S PAPADOPULOS & ASSOCS, 80- *Concurrent Pos:* Chmn, Hydrogeology Div, Geol Soc Am, 81-82. *Mem:* Geol Soc Am; Am Asn Petrol Geologists; Am Geophys Union; Asn Eng Geologists; Am Inst Prof Geologists. *Res:* Occurrence of ground water; relation of ground water to surface water; contamination of ground water; protection of ground water resources. *Mailing Add:* 10406 Hunter Ridge Dr Oakton VA 22124

WOOD, LEONARD E(UGENE), b Burr Oak, Kans, June 10, 23; m 47. CIVIL ENGINEERING. *Educ:* Kans State Univ, BS & MS, 49; Purdue Univ, PhD(civil eng), 56. *Prof Exp:* Instr appl mech, Kans State Univ, 49-53; from instr to assoc prof, 55-80, PROF ENG MAT, PURDUE UNIV, 80- *Concurrent Pos:* Consult, Tech Studies Adv Comt, Bldg Res Adv Bd, Nat Acad Sci-Nat Res Coun, 57-61, Conoco, 82-88, Mobile, 85-87, US CEngr, 85, Nat High Inst, 89-91. *Mem:* Hon mem Am Soc Testing & Mat; Asn Asphalt Paving Technol; Am Soc Eng Educ; Transp Res Bd. *Res:* Engineering materials; rock mechanics; corrosion studies; viscoelastic behavior of bituminous mixes; cold and hot recycling of asphalt pavements; compositional studies of asphalts. *Mailing Add:* Sch Civil Eng Purdue Univ West Lafayette IN 47907

WOOD, LEONARD E(UGENE), b Elwood, Ind, Nov 2, 27; m 58; c 4. ENVIRONMENTAL SCIENCES, GEOLOGY. *Educ:* Univ Ky, BS, 52, MS, 57; Mich State Univ, PhD(geol), 58. *Prof Exp:* Instr geol, Mich State Univ, 55-58; staff geologist, Mobil Oil Co, 58-61, US Geol Surv, 62-63 &

Army Res Off, 63-65; prog mgr environ sci, Adv Res Projs Agency, Thailand, 65-72; chief, environ control group, Off Res, 72-83, CHIEF, ENG & HWY, OPER TECHNOL TRANSFERS, FED HWY ADMIN, 83- *Concurrent Pos:* Instr exten, Univ Va, 64-65; adv ed, J Develop Areas, 68- *Res:* Management in environmental sciences; geology, soils, water resources, vegetation, fauna and meteorology in United States, Europe and Southeast Asia; geologic considerations in excavation; environmental research on high elevations; highways and the bio-environment; highway environmental interface, especially air, noise, water, ecology. *Mailing Add:* 7109 Murray Lane McLean VA 22101

WOOD, LOUIS L, b Washington, DC, July 26, 31; m 58, 68; c 3. ORGANIC CHEMISTRY. *Educ:* Univ Del, BS, 53; Ohio State Univ, PhD(org chem), 59. *Prof Exp:* Res chemist, W R Grace & Co, Wash Res Ctr, Clarkville, 58-81; chem adv, Purification Eng Inc, Columbia, Md, 81-86; CHEM ADV, RHONE-POULENC, SAVAGE, MD, 86- *Mem:* Am Chem Soc; AAAS. *Res:* Polymers; organic chemical synthesis; textile applications; foam technology; immobilzation of cells; enzymes; monoclonal antibodies. *Mailing Add:* 11760 Gainsborough Rd Potomac MD 20854-3246

WOOD, LOWELL THOMAS, b Ada, Okla, Sept 8, 42; m 66; c 2. HIGH TEMPERATURE SUPERCONDUCTORS, FIBER OPTIC SENSORS. *Educ:* Univ Kans, BS, 64; Univ Tex, Austin, PhD(physics), 68. *Prof Exp:* Asst prof physics, Univ Tex, Austin, 68-69; asst prof, 69-73, chmn dept, 75-80, ASSOC PROF PHYSICS, UNIV HOUSTON, 73- *Concurrent Pos:* NASA grant, Univ Houston, 70-78 & NSF grant, 72-77; mem Col Bd Comt, Vector/Schlumberger, 78-82, sr res scientist, 80-81; assoc dean, NSM, 84-89. *Mem:* Am Phys Soc; Optical Soc Am. *Res:* Optical properties of high temperature superconductors; fiber optic microdistortion sensors; undergraduate laboratory development using computers. *Mailing Add:* Dept Physics Univ Houston 4800 Calhoun Houston TX 77204-5504

WOOD, MARGARET GRAY, b Jamaica, NY, May 23, 18; wid; c 3. MEDICINE, DERMATOLOGY. *Educ:* Univ Ala, BA, 41; Woman's Med Col Pa, MD, 48. *Hon Degrees:* Dr Med Sci, Med Col Pa, 90. *Prof Exp:* Assoc, 53-68, from asst prof to assoc prof, 68-77, PROF DERMAT, SCH MED, UNIV PA, 78- *Concurrent Pos:* Assoc, Grad Div Med Dermat, Univ Pa, 53-66, assoc prof, 66-71; asst prof, Woman's Col Pa, 58-66, vis asst prof, 66-; asst vis physician, Philadelphia Gen Hosp, 54-70, consult, 70-77; consult, Philadelphia Vet Hosp & Dent Sch, Univ Pa, 66-,& Vet Sch Med, 70-; bd corp, Med Col Pa, 79-84, vchmn, 84-; mem & chmn, Library Comt, Univ Pa, 84-87; peer rev comt, Am Soc Dermatopathol, 78-81; bd dirs, Alleghany Health Syst, Inc, 89-, United Hosp, Inc, 91. *Honors & Awards:* Commonwealth Bd Award, 81. *Mem:* Histochem Soc: Soc Invest Dermat; AMA; Am Med Women's Asn; Am Acad Dermat; Int Soc Dermatol. *Res:* Histochemistry; dermatopathology. *Mailing Add:* 6386 Church Rd Philadelphia PA 19151

WOOD, MAURICE, b Durham Co, Eng, June 28, 22; US citizen. FAMILY MEDICINE. *Educ:* Newcastle upon Tyne, MB & BS, 45; Kings Col Med Sch, Univ Durham, MRC, 66; FRCP(G), 75; Am Bd Family Pract, dipl, 79. *Prof Exp:* House physician, Newcastle Gen Hosp, 45; house surgeon, Sunderland Royal Infirmary, 46 & Obstet & Gynecol, Queen Elizabeth Hosp Gateshead, 50; maj & sr med officer, Port Said, Mid E Oper, Royal Army Med Corps, 46-49; pvt med pract, S Shields, Eng, 50-71; dir res, 71-87, from assoc prof to prof, 71-88, EMER PROF, DEPT FAMILY PRACT, MED COL VA, 88- *Concurrent Pos:* Surg registr, Sunderland Royal Infirmary, 46; lectr pract nurse training, N Eastern Fac, Royal Col Gen Practr, 65-71; clin asst, Dept Psychol Med, S Shields Gen Hosp, 66-71; gen pract teaching group, Univ Newcastle, 69-71; mem res comt, N Eastern Fac, Royal Col Gen Practr, 60-70, chmn, Pract Orgn Comt, 69-70, recorder, Nat Collab Study, Med Res Coun, 69-73 & vis patterns gen pract, NE Fac, 70; sr recorder collabor study, Dept Health & Human Serv, UK, 64-67; A D Williams distinguished scholar, 71-72; mem, US Nat Comt Vital & Health Statist, 77-80; MS McCleod vis prof, Univ Adelaide, S Australia, 82; adv, data & commun comt, Dept Family Pract, Med Col Va, curric & eval comt, pract orgn & mgt comt & behav sci training, proj dir, Predoc Med Educ Grant Award, 72-80 & Grad Med Educ Grant Award, 76-86, prin investr, Epidemiol Diabetes Study, 80-83, mem, steering comt, Family Pract Fac Develop Prog, 78-81 & Chest Pain Study Group, 80-83, prin investr & prog dir family med, 72-86; mem, res comt, Soc Teachers Family Med, 71-77, comt terminol, 72-79 & adv comn prof & hosp activ, 72-79, Mead Johnson Award Comt, Am Acad Family Physicians, 77-79, bd dirs, Ambulatory Sentinel Pract Network & Nat Cancer Inst Task Force, 86-; chmn, steering comt, NAm Primary Care Res Group, 72-79, pres, 79-83, exec dir, 83; data rec subcomt chmn, Soc Teachers Family Prac, 72-77; consult, Div Educ, Am Acad Family Physicians, 75-; chmn, Community Oriented Primary Care Study Group, Inst Med, NAS, 83-85; treas bd dirs, Ambulatory Sentinel Pract Network, 86-; reviewer, J Am Med Asn, 80; consult, Nat Comt Vital & Health Statist, 76-77, chmn tech consult panel for Ambulatory Med Care Minimum Data Set, 78-80, NIH, 79-, Nat Ctr Health Statist, 80-; mem, Int Primary Care Res Network, 81-; secy res comt, World Orgn Nat Col, Acad & Acad Asn Gen Practr & Family Physicians, 76-83 & mem, 75-, comt voc, 75-80 & mem classification comt, 83-; mem, Int Primary Care Res Network, 81- *Mem:* Inst Med-Nat Acad Sci; Brit Med Asn (pres, 69-); fel Royal Col Gen Practr; fel Am Acad Family Physicians; Soc Teachers Family Med; Int Epidemiol Asn. *Res:* Author of over 64 articles. *Mailing Add:* MCV Sta Box 251 Richmond VA 23298-0001

WOOD, MICHAEL BRUCE, b Glasgow, Mont, Oct 7, 43. BONE BLOOD FLOW. *Educ:* Franklin & Marshall Col, AB, 65; McGill Univ, MD, CM, 69. *Prof Exp:* Asst prof, orthod surg, Med Col Ohio, 77-79; from asst prof to assoc prof orthod surg, Mayo Grad Sch Med, 79-89, DIR, MAYO MICROSURG TRAINING CTR, MAYO CLINIC, 86-, PROF ORTHOP SURG, 89- *Concurrent Pos:* Consult, Sect Hand Surg, Mayo Clin, 78- *Honors & Awards:* Henry W Meyending Essay Award, Am Fracture Asn, 74. *Mem:* Am Acad Orthop Surgeons; Am Soc Surg Hand; Am Soc Reconstructive Microsurg; Int Soc Reconstructive Microsurg. *Res:* Bone blood flow; nerve graft revascularization; microsurgical free tissue transfer. *Mailing Add:* Dept Orthop Surg Mayo Clin 200 SW First St Rochester MN 55905

WOOD, NANCY ELIZABETH, b Martins Ferry, Ohio. SPEECH PATHOLOGY. *Educ:* Ohio Univ, BS, 43, MS, 47; Northwestern Univ, PhD(speech path), 52. *Prof Exp:* Assoc prof lang path, Case Western Reserve Univ, 52-60; consult specialist, Off Educ, Dept Health, Educ & Welfare, 60-62, chief neurol & sensory dis res, USPHS, 62-64; prof commun dis, 65-74, PROF SURG, SCH MED, UNIV SOUTHERN CALIF, 65-, DIR COMMUN DIS, 71-, PROF & RES DIR, SCH JOUR, 75- *Concurrent Pos:* Asst dir, Cleveland Hearing & Speech Ctr, 52-56, coordr clin serv, 56-59, dir lang dis, 59-60. *Mem:* AAAS; Soc Res Child Develop; fel Am Speech & Hearing Asn; fel Am Psychol Asn. *Res:* Language development, disorders and pathology; differential diagnosis of young children; aphasia; mental retardation; hearing loss; test design; communication science research; memory, perception and auditory processing. *Mailing Add:* Dept Jour 1695 Univ Southern Calif University Park Los Angeles CA 90089

WOOD, NORMAN KENYON, b Perth, Ont, Dec 1, 35; m 69; c 4. ORAL PATHOLOGY, ORAL SURGERY. *Educ:* Univ Toronto, DDS, 58; Cook County Hosp, Chicago, dipl oral surg, 65; Northwestern Univ, MS, 66, PhD(oral path), 68; Am Bd Oral Surg, dipl, 70; Am Bd Oral Path, dipl, 71; Am Bd Oral Med, dipl, 86. *Prof Exp:* Pvt pract, 58-62; res assoc biol mat, Dent Sch, Northwestern Univ, 67-68; asst prof oral path, Loyola Univ, Chicago, 68-70, from assoc prof to prof oral diag, Dent Sch, 70-89, chmn dept oral diag, radiol & path, 70-86; DEAN, FAC DENT, UNIV ALTA, 89- *Concurrent Pos:* Consult, Hines Vet Admin Hosp, Ill, 71-89. *Mem:* Am Acad Oral Path; Sigma Xi; Am Asn Dent Schs. *Res:* Induction of hamster cheek pouch carcinomas; seeding during incisional biopsy. *Mailing Add:* Fac Dent Univ Alta Edmonton AB T8B 1J9 Can

WOOD, NORRIS PHILIP, b Binghamton, NY, July 8, 24; m 55; c 2. BACTERIOLOGY. *Educ:* Hartwick Col, BS, 49; Cornell Univ, MNS, 51; Univ Pa, PhD(microbiol), 55. *Prof Exp:* From asst prof to assoc prof microbiol, Agr & Mech Col Tex, 55-63; from asst prof to assoc prof bact, 63-72, PROF MICROBIOL, UNIV RI, 72-, CHMN DEPT, 70- *Concurrent Pos:* Res partic, Oak Ridge Nat Lab, 58 & 62; mem, State Adv Comt Regional Med Prog, 66-76. *Mem:* AAAS; Am Chem Soc; Am Soc Microbiol; Sigma Xi. *Res:* Bacterial physiology; intermediary metabolism; chemistry of microorganisms; microbial ecology. *Mailing Add:* Dept Microbiol Univ RI Kingston RI 02881

WOOD, O LEW, b Hurricane, Utah, Apr 26, 36; m 57; c 4. QUARTZ RESONATOR TRANSDUCERS. *Educ:* Brigham Young Univ, BS, 58; Univ Calif, Los Angeles, MS, 62; Univ Utah, PhD(radiation biophys), 68. *Prof Exp:* Sr engr, Sperry Utah Co, 62-65; res scientist, Fluidonics Res Lab, ITE-Imperial Corp, 65-67, dir res biomed eng, BioLogics Inc, 67-69; assoc prof & collab environ pollution, Utah State Univ, 69-71; prog dir radiologic technol, Weber State Col, 70-75; CHMN & CHIEF EXEC OFFICER, QUARTEX & QUARTZTRONICS, INC, 79- *Concurrent Pos:* Consult, O Lew Wood Assocs, 69- *Mem:* Sr mem Inst Elec & Electronics Engrs; sr mem Instrument Soc Am. *Res:* Development of biomedical and air pollution instrumentation; development of quartz resonator transducers to measure force, pressure, temperature, weight and acceleration; management of technology transfer. *Mailing Add:* 811 E Woodshire Circle Murray UT 84107

WOOD, OBERT REEVES, II, b Sacramento, Calif, Jan 18, 43; m; c 2. ELECTRICAL ENGINEERING, QUANTUM ELECTRONICS. *Educ:* Univ Calif, Berkeley, BS, 64, MS, 65, PhD(elec eng), 69. *Prof Exp:* MEM TECH STAFF LASER RES, AT&T BELL LABS, 69- *Mem:* Am Phys Soc; Optical Soc Am; Inst Elec & Electronics Engrs; AAAS; Sigma Xi. *Res:* Quantum electronics; nonlinear optics; plasma physics. *Mailing Add:* Rm 4C-434 AT&T Bell Labs Holmdel NJ 07733

WOOD, PAUL MIX, b Champion, NY, Dec 28, 30; m 55; c 2. COMPUTER-BASED SYSTEMS ANALYSES. *Educ:* Syracuse Univ, BS, 56, MS, 58; Rensselaer Polytech Inst, MS, 71, MBA, 75. *Prof Exp:* Opers res analyst, Gen Dynamics Pomona, 62-63, Hamilton Standard, United Aircraft, 63-64 & Tech Opers Res, Inc, 64-65; res scientist, Travelers Res Ctr, Inc, 65-68; res scientist, NY State DOH, 68-89, RES SCIENTIST, NY STATE DSAS, 90- *Concurrent Pos:* Mem, Nat Cost Effectiveness Consult Comt, Opers Res Soc Am, 69-72 & Nat Data Comt, Coun Regional Networks Genetics, 86-89; vpres, RPI-MBA Asn, 88-90. *Mem:* Opers Res Soc Am; Am Statist Asn. *Res:* Computer-based systems analyses for decision support in health and in health systems. *Mailing Add:* 24 N Hill Dr Ballston Lake NY 12019

WOOD, PAULINE J, b Springdale, Pa, Nov 7, 22. DEVELOPMENTAL BIOLOGY, HISTOLOGY. *Educ:* Adrian Col, BS, 51; Univ Mich, MS, 54, PhD(zool), 60. *Prof Exp:* Instr zool, Univ Mich, 58-59; res instr embryol, Dent Sch, Univ Wash, 59-61; asst prof zool, Knox Col, 61-62; from asst prof to assoc prof, 62-73, PROF ZOOL, UNIV DETROIT, 73- *Mem:* AAAS; Am Soc Zoologists; Am Inst Biol Sci; Reticuloendothelial Soc. *Res:* Phylogeny of mesenchymal and hemopoietic cells. *Mailing Add:* Dept Biol Univ Detroit 4001 W McNichols Rd Detroit MI 48221

WOOD, PETER DOUGLAS, b London, Eng, Aug 25, 29; m 53; c 1. BIOCHEMISTRY, CHEMISTRY. *Educ:* Univ London, BSc, 52, MSc, 56, PhD(chem), 62, DSc, 72. *Prof Exp:* Chemist, Weston Res Labs, Eng, 52-55; res chemist, Imp Chem Industs Australia & NZ, 56-59; res asst chem, Univ Sask, 59; res assoc, Inst Metab Res, Oakland, Calif, 62-68; ADJ PROF MED, MED CTR, STANFORD UNIV, 69- *Concurrent Pos:* Fel coun arteriosclerosis, Am Heart Asn, 68; dep dir, Stanford Heart-Dis Prev Prog, 71- *Mem:* Fel Royal Soc Chem; Am Inst Nutrit; Am Soc Clin Nutrit; Am Heart Asn; Am Oil Chem Soc; Sigma Xi. *Res:* Lipid chemistry, metabolism and methodology; exercise. *Mailing Add:* Dept Med Standford Ctr Res Stanford Univ 730 Welch Rd Suite B Dis Prevention Palo Alto CA 94304

WOOD, PETER JOHN, b Barnard Castle, Durham, Eng, Aug 20, 43. CARBOHYDRATE CHEMISTRY, FOOD SCIENCE. *Educ:* Univ Birmingham, BS, 65, PhD(carbohydrate chem), 69. *Prof Exp:* RES SCIENTIST CARBOHYDRATES, FOOD RESEARCH INST, 69- *Mem:* Am Asn Cereal Chemists. *Res:* Connective tissue glycosaminoglycans;

polysaccharides of potato cell wall; rapeseed carbohydrates; oat endosperm cell wall polysaccharides; analytical methodology; use of boronate esters for glc analysis and methanol analysis; dye-binding by polysaccharides; dietary fiber. *Mailing Add:* Res Br Food Res Inst Agr Can Ottawa ON K1A 0C6 Can

WOOD, RANDALL DUDLEY, b Palmer, Ky, Aug 3, 36; m 59; c 1. BIOCHEMISTRY, ORGANIC CHEMISTRY. *Educ:* Univ Ky, BS, 59, MS, 61; Tex A&M Univ, PhD(biochem), 65. *Prof Exp:* Scientist, Oak Ridge Assoc Univs, 66-70; assoc prof, Stritch Sch Med, Loyola Univ Chicago & Hines Vet Admin Hosp, 70-71; assoc prof med & biochem, Sch Med, Univ Mo-Columbia, 71-76; PROF BIOCHEM, TEX A&M UNIV, COLLEGE STATION, 76- *Concurrent Pos:* AEC fel, 65-66; Consult, Planning & Design of Sci & Biotechnol Facil. *Mem:* Am Inst Nutrit; Am Asn Cancer Res; Am Chem Soc; Am Oil Chem Soc; Am Soc Biol Chemists. *Res:* Lipid biochemistry and metabolism of normal, tumor and embryonic tissues; biosynthesis, metabolism and occurrence of alkyl glyceryl ethers and plasmalogens; structural and metabolic relationships between molecular species of various classes; the metabolic fate of unnatural dietary fatty acids in processed foods; lipid metabolism in plants; nutrition; effect of dietary fat on serum cholesterol and serum lipids. *Mailing Add:* Dept Biochem & Biophys Tex A&M Univ College Station TX 77843

WOOD, RAYMOND ARTHUR, b Middletown, NY, Nov 28, 24. ZOOLOGY, PARASITOLOGY. *Educ:* Mt St Mary's Col, Md, BS, 50; Univ Notre Dame, MS, 53, PhD, 55. *Prof Exp:* Instr & spec lectr anat & physiol, Ind Univ, 54-55; asst prof, Pan Am Col, 55-56; PROF ZOOL & CHMN DIV, ORANGE COUNTY COMMUNITY COL, 56- *Concurrent Pos:* NSF grants, Bermuda Biol Sta, 56, Comp Anat Inst, Harvard Univ, 63, Col Biol Inst, Williams Col, 66, Marine Lab, Duke Univ, 65, 67 & Marine Labs, Naples, Italy, 70; Sigma Xi grant, 57. *Mem:* Fel AAAS; Am Soc Parasitologists; Am Soc Zoologists; Soc Syst Zool; fel Royal Soc Trop Med & Hyg. *Res:* Systematics of monogenea. *Mailing Add:* Div Biol Orange County Community Col Middletown NY 10940

WOOD, REGINALD KENNETH, b Unity, Sask, July 12, 37; m 58; c 3. COMPUTER PROCESS CONTROL, MODELING & SIMULATION. *Educ:* Univ Sask, BE, 58; Univ Alta, MSc, 60; Northwestern Univ, PhD(chem eng), 63. *Prof Exp:* Res engr, Can Chem Co Ltd, 59-60; res assoc foam separation, Northwestern Univ, 62-63; asst prof chem eng, Univ Ottawa, 63-66; from asst prof to assoc prof, 66-72, PROF CHEM ENG, UNIV ALTA, 72- *Concurrent Pos:* Guest worker, Control Eng Div, Warren Spring Lab, Stevenage, Eng, 72-; vis res fel, Comput Aided Design Ctr, Cambridge, Eng, 80-81. *Mem:* Am Inst Chem Engrs; Can Soc Chem Engrs; Instrument Soc Am; Soc Comput Simulation; Am Soc Eng Educ. *Res:* Dynamics and control of chemical and mineral process systems. *Mailing Add:* Dept Chem Eng Univ Alta Edmonton AB T6G 2E1 Can

WOOD, RICHARD ELLET, b Farmington, Utah, Mar 3, 28; m 48; c 5. NUCLEAR PHYSICS, ENERGY SYSTEMS. *Educ:* Univ Utah, BS, 52, PhD(physics), 55. *Prof Exp:* Res assoc neutron cross sects, Brookhaven Nat Lab, 53-54; nuclear engr, Gen Elec Co, 55-56, supvr low power test opers, 56-59, supvr initial eng test opers, 59-60, supvr anal, Idaho Test Sta, Air Craft Nuclear Propulsion Dept, 60-61, physicist, Atomic Power Equip Dept, 61-62, mgr, Idaho Eng, Nuclear Mat & Propulsion Oper, Idaho Test Sta, 62-68; chief nuclear eng br, Idaho Opers Off, AEC, 68-74; dir, Reactor Support Div, US Energy Res & Develop Admin, 74-76, dir, Energy & Technol Div, 76-84, ASST MGR, PROJS & ENERGY PROGS, US DEPT ENERGY, 84- *Mem:* Am Nuclear Soc. *Res:* Nuclear engineering and neutron physics. *Mailing Add:* 783 E 1500 North Shelley ID 83274

WOOD, RICHARD FROST, b Lebanon, Tenn, June 6, 31; m; c 3. PHYSICS. *Educ:* Fordham Univ, BS, 53; Ohio State Univ, MS, 56, PhD(physics), 59. *Prof Exp:* Res specialist, Opers Res, NAm Aviation, Inc, 58-60; asst prof physics, Univ Fla, 60-62; RES SCIENTIST & HEAD THEORY SECT, SOLID STATE DIV, OAK RIDGE NAT LAB, 62- *Concurrent Pos:* Vis prof, Univ Uppsala, 60-61, NSF fel, 61-62. *Mem:* AAAS; fel Am Phys Soc. *Res:* Theoretical solid state physics; lattice defects; lattice dynamics; optical properties and electronic structure of solids. *Mailing Add:* Solid State Div Oak Ridge Nat Lab PO Box 2008 Bldg 2000 MS 6056 Oak Ridge TN 37830

WOOD, RICHARD LEE, b Ft Dodge, Iowa, Sept 8, 30; m 55; c 3. VETERINARY MICROBIOLOGY. *Educ:* Univ Mo-Columbia, DVM, 61; Iowa State Univ, MS, 66, PhD(vet microbiol), 70. *Prof Exp:* VET MED OFFICER, NAT ANIMAL DIS CTR, AGR RES SERV, USDA, 61- *Mem:* Am Vet Med Asn; Conf Res Workers Animal Dis; US Animal Health Asn. *Res:* Epizootiology, pathogenesis and immunology of bacterial diseases of swine. *Mailing Add:* 620 River Oak Dr Ames IA 50010

WOOD, RICHARD LYMAN, b Allamore, Tex, Jan 2, 29; m 51; c 2. CYTOLOGY. *Educ:* Linfield Col, BA, 50; Univ Wash, PhD(zool), 57. *Prof Exp:* From instr to asst prof anat, Univ Wash, 59-64; assoc prof, Univ Minn, Minneapolis, 64-70; prof biol struct, Sch Med, Univ Miami, 70-74; PROF ANAT, SCH MED, UNIV SOUTHERN CALIF, 74- *Concurrent Pos:* NIH fel, Univ Wash, 57-59; NIH res grant, Dept Biol Struct, Univ Wash, 60-64 & Dept Anat, Univ Minn, 64-70; NSF res grant, Dept Anat, Univ Southern Calif, 78-82. *Mem:* Am Soc Cell Biol; Am Asn Anat; Electron Micros Soc Am; Soc Develop Biol. *Res:* Cell junctions; cellular biology; animal cytology and histology. *Mailing Add:* Dept Anat Univ Southern Calif Sch Med 2025 Zonal Ave Los Angeles CA 90033

WOOD, ROBERT CHARLES, b Lakewood, Ohio, May 7, 29; div. ONCOGENESIS. *Educ:* Lehigh Univ, BA, 51, MS, 52; Univ Md, PhD, 55. *Prof Exp:* Sr res microbiologist, Wellcome Res Labs, Burroughs Wellcome & Co, 56-60; asst prof microbiol, Sch Med, George Washington Univ, 60-64; asst prof, 64-65, ASSOC PROF MICROBIOL, UNIV TEX MED BR GALVESTON, 65- *Concurrent Pos:* Res fel bact physiol, Univ Pa, 55-56. *Mem:* AAAS. *Res:* Folic acid metabolism; glycolipid metabolism; enzymic basis of oncogenesis. *Mailing Add:* Dept Microbiol Univ Tex Med Br Galveston TX 77550

WOOD, ROBERT E, b Philadelphia, Pa, May 16, 38; m 62; c 2. INDUSTRIAL ENGINEERING. *Educ:* Ga Inst Technol, BS, 60, MS, 62, PhD(physics), 65. *Prof Exp:* From asst prof to assoc prof physics, Emory Univ, 64-75; sr opers res analyst, 75-76, mgr indust eng, Southern Rwy Syst, 76-82; MGR INDUST ENG, NORFOLK SOUTHERN CORP, 82- *Concurrent Pos:* Consult, Ga Inst Technol, 65, fel chem, 66; consult, Allied Gen Nuclear Serv, 73-75 & Aston Co, 75-76. *Mem:* AAAS; Am Phys Soc. *Res:* Application of models to transportation industry. *Mailing Add:* Norfolk Southern Corp 125 Spring St Box 110 Atlanta GA 30303

WOOD, ROBERT HEMSLEY, b Brooklyn, NY, May 8, 32; div; c 2. PHYSICAL CHEMISTRY. *Educ:* Calif Inst Technol, BS, 53; Univ Calif, PhD(chem), 57. *Prof Exp:* From instr to assoc prof, 57-70, chmn dept, 69-71, PROF CHEM, UNIV DEL, 70- *Mem:* Am Chem Soc. *Res:* Solution thermodynamics; electrolytes, non-electrolytes, non-aqueous, and high temperature. *Mailing Add:* Dept Chem Univ Del Newark DE 19716

WOOD, ROBERT M(CLANE), b Ithaca, NY, Apr 4, 28; m 51; c 2. TECHNICAL MANAGEMENT. *Educ:* Univ Colo, BS, 49; Cornell Univ, PhD(exp physics), 53. *Prof Exp:* Asst elem physics, Cornell Univ, 49-50, asst solid state physics, 51-53; res engr automatic control systs, Douglas Aircraft Co, Santa Monica, 53-54 & Missile & Space Systs Div, 56-63, dep chief engr advan space technol, 63-64, asst dir sci & future systs res & develop, 64-66, dep dir res & develop, 66-71, asst dir detection, designation & discrimination, 71-76, dir res & develop, 76-84, DIR ADVAN SYSTS & TECHNOL, SPACE STA DIV, MCDONNELL DOUGLAS ASTRONAUT CO, 84- *Concurrent Pos:* Consult, Flight Refueling, Inc, 55. *Mem:* AAAS; Am Inst Aeronaut & Astronaut; Am Phys Soc. *Res:* Heat transfer; missile and space system design; radar discrimination; research and development management. *Mailing Add:* 1727 Candlestick Lane Newport Beach CA 92660

WOOD, ROBERT MANNING, b Bronxville, NY, May 13, 38; c 2. ATOMIC PHYSICS, MOLECULAR PHYSICS. *Educ:* Princeton Univ, AB, 60; Univ Wis, PhD(physics), 64. *Prof Exp:* Res assoc physics, Univ Wis, 64-66; asst prof, 66-80, ASSOC PROF PHYSICS, UNIV GA, 80- *Mem:* Am Phys Soc. *Res:* Ion-atom and ion-molecule collisions. *Mailing Add:* Dept Physics Univ Ga Athens GA 30601

WOOD, ROBERT WINFIELD, b Detroit, Mich, Dec 29, 31; m 59; c 3. RADIATION BIOPHYSICS. *Educ:* Univ Detroit, BS, 53; Vanderbilt Univ, MA, 55; Cornell Univ, PhD(biophys), 61. *Prof Exp:* Radiol physicist, AEC, 62-73; DIR PHYS & TECHNOL RES DIV, OFF HEALTH & ENVIRON RES, US DEPT ENERGY, 73- *Concurrent Pos:* Nat Inst Gen Med Sci fel, 61-62. *Mem:* AAAS; Health Physics Soc; Radiation Res Soc; Sigma Xi. *Res:* Electron spin resonance, aromatic hydrocarbon negative ions and irradiated biological compounds; radiological physics; dosimetry; biomedical instrumentation. *Mailing Add:* Off Health & Environ Res US Dept Energy Washington DC 20545

WOOD, ROBERTSON HARRIS LANGLEY, b Lynchburg, Va, Aug 22, 24; m 51; c 5. ENVIRONMENTAL PHYSIOLOGY. *Educ:* Col William & Mary, BS, 49; Columbia Univ, AM, 50; Cornell Univ, PhD(biol), 65. *Prof Exp:* Instr sociol, Winthrop Col, 50-51; pvt bus, 52-56; researcher, Bur Commercial Fisheries, US Fish & Wildlife Serv, 56-57; res asst biol, Woods Hole Oceanog Inst, 57-58, Inst Fish Res, Univ NC, 59 & Lerner Lab, Am Mus Natural Hist, 59-60; assoc marine scientist & head dept, Va Inst Marine Sci, 61-67, sr marine scientist & head dept environ physiol, 67-69; prof zool & chmn dept, Univ NH, 69-72; prof environ studies & dir prog, Sweet Briar Col, 72-82; RETIRED. *Concurrent Pos:* From asst prof to assoc prof, Col William & Mary, 61-69; asst prof, Univ Va, 63-69. *Mem:* AAAS; Am Soc Limnol & Oceanog; Am Soc Zoologists; Animal Behav Soc; Estuarine Res Soc. *Res:* Physiological and behavioral effects upon marine organisms of changes in sensory and biochemical characteristics of environment. *Mailing Add:* 104 Monacan Pl Elon VA 24572-3411

WOOD, RODNEY DAVID, b Lansing, Mich, Aug 19, 32; m 61; c 4. CHEMICAL ENGINEERING. *Educ:* Yale Univ, BE, 54; Mich State Univ, MS, 59; Northwestern Univ, PhD(chem eng), 63. *Prof Exp:* Instr mech eng, Mich State Univ, 56-59; asst prof, Univ Nebr, 62-64; res engr, Tex Instruments Inc, 64-70; sr engr, Sherwin Williams Chem, 70-72, tech dir, 73-77; dir process eng, 77-80, MGR STERILE DRUGS, MEAD JOHNSON & CO, 80- *Honors & Awards:* Award, Am Inst Chem Engrs, 64. *Mem:* Am Chem Soc; Am Inst Chem Engrs; Sigma Xi. *Res:* Thermodynamics, food, and pharmaceutical processing. *Mailing Add:* 524 S Plaza Dr Evansville IN 47715

WOOD, ROGER CHARLES, b Minneapolis, Minn, Jan 17, 32; m 54; c 4. COMPUTER SCIENCE, APPLIED MATHEMATICS. *Educ:* Univ Minn, BS, 54, MS, 56; Univ Calif, Los Angeles, PhD(eng), 66. *Prof Exp:* Dynamics engr, Ryan Aeronaut Co, 54-57, sr dynamics engr, 57-58, sr opers res engr, 58-59; res engr, Univ Calif, Los Angeles, 64-65; opers res analyst, Syst Develop Corp, 59-60, sect head, 60-61, opers res scientist, 61-65; from asst prof to assoc prof elec eng, 65-72, chmn dept, 71-75, PROF ELEC ENG & COMPUT SCI, UNIV CALIF, SANTA BARBARA, 72- *Concurrent Pos:* Consult, GE Tempo, 66-68, Gen Motors Res Lab, 68-73, Human Factors Res Corp, 68-74, Gen Res Corp, 77-80 & USN, 76-; mem bd dir, Concordia Historical Inst Systs, 78- *Mem:* Inst Elec & Electronics Engrs; Opers Res Soc Am; Asn Comput Mach; Simulation Coun. *Res:* Analog, hybrid and digital computer applications; quantization theory; time sharing system scheduling; operations research; communications theory; simulation; computer architecture. *Mailing Add:* Dept Elec Eng & Comput Sci Univ Calif Santa Barbara CA 93106

WOOD, SCOTT EMERSON, b Ft Collins, Colo, Apr 9, 10; m 36; c 1. PHYSICAL CHEMISTRY. *Educ:* Univ Denver, BS, 30, MS, 31; Univ Calif, PhD(chem), 35. *Prof Exp:* Res assoc chem, Mass Inst Technol, 35-40; from instr to asst prof, Yale Univ, 40-48; from assoc prof to prof, Ill Inst Technol, 48-75, admin officer & vchmn dept, 60-62, actg assoc dean for res, 62-64, EMER PROF CHEM, ILL INST TECHNOL, 75- *Concurrent Pos:* Consult,

Argonne Nat Lab, 60-80, Exxon Nuclear, 74-75 & 78-79; assoc ed, J Chem Phys & sect ed, Chem Abstr, 61-63; vis prof, Univ Col, Dublin, 66-67. *Mem:* AAAS; Am Chem Soc; Am Phys Soc. *Res:* Vapor pressures of nonaqueous solutions; density and coefficient of expansion of solutions; index of refraction; theory and thermodynamics of nonaqueous solutions; chromatography; paper electrochromatography; fused salts. *Mailing Add:* 1575 Belvidere St Apt 116 El Paso TX 79912-2635

WOOD, SPENCER HOFFMAN, b Portland, Ore, Nov 18, 38; m 61; c 3. GEOLOGY. *Educ:* Colo Sch Mines, GE, 64; Calif Inst Technol, MS, 70, PhD(geol), 75. *Prof Exp:* Seismol engr, Geophys Serv, Mobil Oil Corp, 64-65, geophysicist, Mobil Oil Libya, Ltd, 65-68; instr geol, Occidental Col, 74-76; geologist, US Geol Surv, Nat Ctr Earthquake Res, 76-77; asst prof, 78-80, PROF GEOL, BOISE STATE UNIV, 80- *Concurrent Pos:* Vis asst prof geol, Univ Ore, 76. *Mem:* Geol Soc Am; Am Geophys Union; Sigma Xi; Am Asn Petrol Geologists. *Res:* Hydrogeology, geomorphology, neotectonics, geology of geothermal areas, tephrochronology and obsidian dating; geophysics. *Mailing Add:* Dept Geol & Geophys Boise State Univ 1910 Univ Dr Boise ID 83725

WOOD, STEPHEN CRAIG, b Cleveland, Ohio, Sept 28, 42; m 67; c 4. PHYSIOLOGY. *Educ:* Kent State Univ, BS, 64, MA, 66; Univ Ore, PhD(physiol), 70. *Prof Exp:* Nat Res Coun res assoc physiol, Submarine Med Res Lab, Groton, Conn, 70-71; asst prof zoophysiol, Aarhus Univ, 71-72; asst prof physiol, Southern Ill Univ, Edwardsville, 72-74; from asst prof to prof physiol, Sch Med, 74-87, SR SCIENTIST LOVELACE MED FOUND, NMEX, 87- *Concurrent Pos:* Sr vis res fel, Marine Biol Lab, Plymouth, Eng, 72; Danish Natural Sci Res Coun grant, Comoro Islands Coelacanth Exped, 72. *Mem:* Am Physiol Soc; Scand Physiol Soc. *Res:* Diving physiology; comparative physiology of respiration and blood gas transport; metabolism and function of red blood cells; environmental physiology. *Mailing Add:* Lovelace Medical Foundation Albuquerque NM 87108

WOOD, STEPHEN LANE, b Logan, Utah, July 2, 24; m 47; c 3. ENTOMOLOGY. *Educ:* Utah State Univ, BS, 46, MS, 48; Univ Kans, PhD(entom), 53. *Prof Exp:* High sch instr, Utah, 48-50; asst instr biol, Univ Kans, 50-53; syst entomologist, Can Dept Agr, 53-56; from asst prof to prof, 56-89, EMER PROP ZOOL & ENTOM, BRIGHAM YOUNG UNIV, 89- *Concurrent Pos:* Ed, Great Basin Naturalist, 70-89; vis res prof entom, Univ de Costa Rica, 63-64, Univ de los Andes, Merida, Venezuela, 69-70. *Mem:* AAAS; Entom Soc Am; Coleopterists' Soc. *Res:* Systematics of Scolytidae and Platypodidae. *Mailing Add:* Life Sci Mus Brigham Young Univ Provo UT 84602

WOOD, SUSAN, b Snainton, Eng, Jan 13, 48; US citizen. MATERIALS SCIENCE ENGINEERING. *Educ:* Victoria Univ, BSc Hon, 69; Univ Pittsburgh, MS, 73, PhD(mat eng), 76. *Prof Exp:* Sr engr, 76-80, MGR, RES & DEVELOP CTR, WESTINGHOUSE, 80- *Mem:* Mat Res Soc; Asn Women Sci; Am Inst Mining, Metall & Petrol Engrs. *Res:* Microstructural evaluation of metals and semiconductors, particularly by transmission electron microscope; surface and subsurface (buried layer) structure changes by ion implatation; analysis of ion beam effects on materials. *Mailing Add:* 2374 Colins Rd Pittsburgh PA 19013

WOOD, THOMAS H, b Mineola, NY, Apr 17, 53; m 87. MULTIPLE QUANTUM WELLS. *Educ:* Brown Univ, ScB, 75; Univ Ill, Urbana, MS, 76, PhD(physics), 80. *Prof Exp:* DISTINGUISHED MEM TECH STAFF, AT&T BELL LABS, 80- *Mem:* Am Phys Soc; Optical Soc Am; Inst Elec & Electronics Engrs. *Res:* Optoelectronic properties of multiple quantum wells, especially their use as optical modulators. *Mailing Add:* AT&T Bell Labs Rm L115 Crawford Hill Holmdel-Keyport Rd Holmdel NJ 07733-0400

WOOD, THOMAS HAMIL, b Atlanta, Ga, June 22, 23; m 51; c 3. BIOPHYSICS. *Educ:* Univ Fla, BS, 46; Univ Chicago, PhD(biophys), 53. *Prof Exp:* Res assoc, Univ Chicago, 53; from asst prof to assoc prof, 53-63, PROF PHYSICS, UNIV PA, 63- *Concurrent Pos:* NSF sr fel, Inst Radium, Paris, 61-62; vis prof, Univ Leicester, 67-68 & Ein Shams Univ, Cairo, Egypt, 80-81. *Mem:* Biophys Soc; Radiation Res Soc; Genetics Soc Am. *Res:* Effects of radiations on microorganisms; influence of temperature and protective agents; cellular freezing; bacterial conjugation; genetic recombination. *Mailing Add:* Dept Physics Univ Pa Philadelphia PA 19104

WOOD, THOMAS KENNETH, b Cleveland, Ohio, June 12, 43; m 65; c 2. BEHAVIORAL ECOLOGY, INSECT SYSTEMATICS. *Educ:* Wilmington Col, AB, 64; Cornell Univ, PhD(entom), 68. *Prof Exp:* Prof biol, Wilmington Col, 68-80; MEM FAC, DEPT ENTOM & APPL ECOL, UNIV DEL, 80- *Concurrent Pos:* NSF grant, 75-80. *Mem:* Entom Soc Am; Entom Soc Can; AAAS. *Res:* Parental care in insects, host-plant interaction; membracid-ant associations, membracid systematics. *Mailing Add:* Dept Entom & Appl Ecol Univ Del Newark DE 19711

WOOD, TIMOTHY E, b San Francisco, Calif, Aug 27, 48. ROOT SYMBIOSES, SOIL BIOGEOCHEMISTRY. *Educ:* Univ Calif, Santa Barbara, BA, 70; Yale Univ, MFS, 72, PhD(forest ecol), 80. *Prof Exp:* Res asst forest ecol, Yale Univ, 72-74; res assoc, Wood Hole Marine Biol Lab, 75-76; SR ECOLOGIST SOILS ECOL, NATIVE PLANTS, INC, 81- *Res:* Mineral nutrition of natural plant communities, including root systems, soil chemistry, and soil microbiology; symbiotic root associations involving mycorrhizal fungi, Frankia actomycenes, and Rhizobium bacteria. *Mailing Add:* Native Plants Inc 417 Wakara Way Salt Lake City UT 84108

WOOD, TIMOTHY SMEDLEY, b Port Washington, NY, Dec 4, 42; m 64; c 2. INVERTEBRATE ZOOLOGY. *Educ:* Earlham Col, AB, 64; Univ Colo, PhD(zool), 71. *Prof Exp:* Asst prof, 71-76, ASSOC PROF BIOL SCI, WRIGHT STATE UNIV, 76-, DIR, ENVIRON STUDIES PROG, 75- *Mem:* AAAS; Am Soc Zoologists; Am Micros Soc; Int Asn. *Res:* Aquatic community ecology; structure and function of animal colonies, with emphasis on the Ectoprocta. *Mailing Add:* Dept Biol Sci Wright State Univ Dayton OH 45435

WOOD, VAN EARL, b New York, NY, May 25, 33; m 58; c 2. PHYSICS. *Educ:* Union Col, BS, 55; Case Inst Technol, MS, 59, PhD(physics), 61. *Prof Exp:* Fel, 60-73, prin physicist, 73-89, SR PHYSICIST, COLUMBUS LABS, BATTELLE MEM INST, 89- *Mem:* Am Phys Soc; Inst Elec & Electronics Engrs; Sigma Xi. *Res:* Theoretical solid-state physics, materials science and optics; related applied mathematics. *Mailing Add:* 7332 S Section Line Rd Delaware OH 43015

WOOD, WALLACE D(EAN), b Owego, NY, Dec 28, 10; m 37; c 2. ENGINEERING. *Educ:* Rochester Inst Technol, EE, 32; Cornell Univ, AB, 36. *Prof Exp:* Res engr, Taylor Instrument Co, 36-40, dir, 40-70, dir, Taylor Instrument Process Control Div, Standards Lab, Sybron Corp, 70-74. *Mem:* Instrument Soc Am; Nat Soc Prof Engrs; Am Soc Mech Engrs. *Res:* Automatic control analysis; design and maintenance of calibrating standards; test and process equipment. *Mailing Add:* 154 Clover Hills Dr Rochester NY 14618

WOOD, WARREN WILBUR, b Pontiac, Mich, Apr 9, 37; m 61; c 1. GEOCHEMISTRY, GROUND WATER HYDROLOGY. *Educ:* Mich State Univ, BS, 59, MS, 62, PhD(geol), 69. *Prof Exp:* Geologist II, Mich Hwy Dept, 62-63; res hydrologist, US Geol Surv, 64-77, asst chief off radiohydrol, 77-78; assoc prof, dept geosci, Tex Tech Univ, 78-81; RES HYDROLOGIST, US GEOL SURV, 81- *Concurrent Pos:* Consult hydrologist, 78-81. *Mem:* Am Geophys Union; fel Geol Soc Am; Nat Water Well Asn. *Res:* Geochemistry of ground water; artifical recharge; solute transport; hydrology of radioactive waste isolation. *Mailing Add:* US Geol Surv Nat Ctr MS431 Reston VA 22092

WOOD, WILLIAM BAINSTER, b Surry Co, NC, Feb 7, 31; m 52; c 4. MEDICINE. *Educ:* Univ NC, BS, 53, MD, 56. *Prof Exp:* Intern, NC Mem Hosp, Chapel Hill, 56-57, resident internal med, 57-59; assoc physiol, George Washington Univ, 61-63; from instr to assoc prof, 63-68, PROF MED & ASSOC DEAN, SCH MED, UNIV NC, CHAPEL HILL, 84- *Concurrent Pos:* Fel chest dis, Sch Med, Univ NC, 59-60; attend physician, consult & dir pulmonary lab, NC Mem Hosp, Chapel Hill, 63-; attend physician, Gravely Sanatorium, 63-; sr scientist, Wrightsville Marine-Biomed Lab, 66-; hon sr lectr, Cardio Thoracic Inst, London, Eng, 73-74; dir med educ & res, Eastern NC Hosp, 74-, med dir, Wilson, NC, 75-77. *Mem:* Am Thoracic Soc; AMA; Am Fedn Clin Res; Am Soc Internal Med; Soc Med Col. *Res:* Pulmonary diseases; morphology of lung in disease; mechanics and control of respiration; effects of hyperbaric atmospheres on respiration; respiratory physiology in deep sea diving; immunological and hypersensitivity lung diseases. *Mailing Add:* Sch Med Pulmonary Dis Univ NC Chapel Hill Sch Med Chapel Hill NC 27514

WOOD, WILLIAM BARRY, III, b Baltimore, Md, Feb 19, 38; m 61; c 2. GENETICS, MOLECULAR BIOLOGY. *Educ:* Harvard Univ, AB, 59; Stanford Univ, PhD(biochem), 64. *Prof Exp:* Air Force Off Sci Res, Nat Acad Sci-Nat Res Coun fel molecular biol, Univ Geneva, 63-64; from asst prof to prof, Calif Inst Technol, 65-77; chmn, 77-83, PROF, DEPT MOLECULAR, CELLULAR & DEVELOP BIOL, UNIV COLO, 77- *Concurrent Pos:* Guggenheim fel, Dept Molecular, Cellular & Develop Biol, Univ Colo, Boulder, 75-76. *Honors & Awards:* US Steel Award, Nat Acad Sci, 69. *Mem:* Nat Acad Sci; AAAS; Am Acad Arts & Sci; Am Soc Biol Chemists; Soc Develop Biol; Genetics Soc Am. *Res:* Genetic control of bacteriophage; developmental genetics of the neuratode Caenorhabditis elegans; determination of sex and cell fates in the early embryo. *Mailing Add:* Univ Colo Campus Box 347 Boulder CO 80309

WOOD, WILLIAM BOOTH, b New York, NY, Sept 9, 22. PHYSIOLOGY. *Educ:* Ft Hays Kans State Col, MS, 51; Univ Kans, PhD(physiol, anat), 59. *Prof Exp:* Lab asst physiol, Univ Kans, 55-56, asst, 56-58, res assoc, 58, asst instr, 58-59; from instr to assoc prof, 59-73, PROF PHARMACOL, COL MED, UNIV TENN, MEMPHIS, 73- *Mem:* AAAS; assoc Am Physiol Soc; Am Soc Pharmacol & Exp Therapeut. *Res:* Respiratory physiology and pharmacology of cardiovascular system. *Mailing Add:* Dept Pharmacol Univ Tenn Ctr Health Sci Memphis TN 38163

WOOD, WILLIAM C, b Fairbury, Ill, May 3, 40. ONCOLOGY, SURGICAL ONCOLOGY. *Educ:* Harvard Univ, MD, 66. *Prof Exp:* ASSOC PROF SURG, SCH MED, HARVARD UNIV, 82-; DIR, CANCER CTR, MASS GEN HOSP, 82-, CHIEF, SURG ONCOL, 85- *Mem:* Soc Surg Oncol; Am Soc Clin Oncol; Am Asn Immunol. *Mailing Add:* Dept Med Mass Gen Hosp Cox Ctr Boston MA 02114

WOOD, WILLIAM EDWIN, b Plainfield, NJ, Mar 25, 47; m 71; c 3. MATERIALS SCIENCE. *Educ:* Univ Notre Dame, BS, 69; Univ Calif, Berkeley, MS, 70, DEngr, 73. *Prof Exp:* Asst prof, 73-77, assoc prof, 77-81, PROF MAT SCI, ORE GRAD CTR, 81- *Mem:* Am Soc Metals; Am Welding Soc. *Res:* Physical mechanical metallurgy; alloy developmental design; ferrous and non ferrous materials. *Mailing Add:* Dept Mat Sci Ore Grad Ctr 19600 NW Von Newman Beaverton OR 97006

WOOD, WILLIAM IRWIN, b Bloomington, Ind, Nov 8, 47; m 73. BIOCHEMISTRY, MOLECULAR BIOLOGY. *Educ:* Cornell Univ, BA, 70; Harvard Univ, MA, 71, PhD(biochem), 77. *Prof Exp:* Comn officer comput sci, USPHS, NIH, 71-73, staff fel biochem, 78-81; scientist, 82-84, sr scientist, 85-89, STAFF SCIENTIST, GENENTECH, INC, 90- *Mem:* AAAS. *Res:* Control of gene expression in eucaryotes; cloning and expression of genes for animal science; cloning and expression of Factor VIII and the molecular basis of hemophilia. *Mailing Add:* Genentech Inc 460 Pt San Bruno Blvd South San Francisco CA 94080

WOOD, WILLIAM OTTO, b Oklahoma City, Okla, Apr 7, 25; m 50; c 1. CELL BIOLOGY, BIOPHYSICS. *Educ:* Southern Methodist, BS, 50;. *Prof Exp:* Indust bacteriologist, Joseph E Seagrams & Sons, Inc, Ky, 50-53; phys chem scientist, Bur Mines, US Dept Interior, Ore, 53-54; serologist, USDA, 59-61; tissue culture res cytologist, Nat Animal Dis Lab, Iowa, 61-64; res biologist, tissue culture res labs, biophys lab, US Dept Army, 64-69, physical sci adminr, 69-74; RETIRED. *Concurrent Pos:* Independent grant, 66-69; chem res coordr, Dept Army, 69-74. *Mem:* AAAS; Tissue Culture Asn; Soc Cell Biol; AAAS; Soc Indust Microbiol. *Res:* Comparative studies in tissue culture preservation; in vitro studies of regenerating liver cells wounded by ballistic missles; wounding and hyperbaric oxygenation; biochemistry of "Choisonne" Goat kidney cells in vitro; primary cell culture metabolism in vitro; lymphoid tissue culture for mucosal disease studies; serum specificity for vesicular stomatitis on field samples; serum neutralization method for vesicular stomatitis; the culture of cells from a "cloissone" goat kidney. *Mailing Add:* 384 Foxwood Lake Estates Lakeland FL 33809

WOOD, WILLIAM WAYNE, b Terry, Mont, Nov 1, 24; m 46; c 5. STATISTICAL MECHANICS. *Educ:* Mont State Col, BS, 47; Calif Inst Technol, PhD(chem), 51. *Prof Exp:* Staff mem, Los Alamos Sci Lab, 50-58, group leader, 58-71, staff mem, 71-81; prof, dept physics, Carroll Col, Helena, Mont, 81-90; CONSULT, 90- *Concurrent Pos:* Vis prof, Univ Colo, 69-70. *Mem:* AAAS; Sigma Xi; fel Am Phys Soc. *Res:* Theory of optical activity; detonation; Monte Carlo and molecular dynamics methods in statistical mechanics. *Mailing Add:* 1615 Cleveland Helena MT 59601

WOOD, WILLIS AVERY, b Johnson City, NY, Aug 6, 21; m 47; c 3. ENZYMOLOGY, BIOINSTRUMENTATION. *Educ:* Cornell Univ, BS, 47; Ind Univ, PhD(microbiol), 50. *Prof Exp:* From asst prof to assoc prof dairy bact, Univ Ill, 50-58; prof agr chem, Mich State Univ, 58-61, prof biochem, 61-68, chmn dept biochem, 68-74; DIR MICROBIOL & BIOCHEM, SALK INST BIOTECHNOL-INDUST ASSOC, 82- *Concurrent Pos:* Dir, Gilford Instrument Labs, 67-81; mem & chmn adv comt, Nat Inst Dent Res, NIH, 71-78, mem, Nat Caries Adv Comt, 79-80, Nat Adv Dent Res Coun, 86-; pres, Neogen Corp, 81-82. *Honors & Awards:* Lilly Award, Am Soc Microbiol, 55. *Mem:* Am Chem Soc; Am Soc Microbiol (pres, 79); Am Soc Biol Chemists. *Res:* Chemical activities of microorganisms; amino acid metabolism; protein structure; instrumentation; biotechnology-bioconversion. *Mailing Add:* Salk Inst Biotechnol-Indust Assoc PO Box 85200 San Diego CA 92138

WOODALL, DAVID MONROE, b Perryville, Ark, Aug 2, 45; m 66. NUCLEAR ENGINEERING, ENGINEERING PHYSICS. *Educ:* Hendrix Col, BA, 67; Columbia Univ, MS, 68; Cornell Univ, PhD(appl physics), 76. *Prof Exp:* Nuclear engr, Westinghouse Nuclear Energy Systs, 68-70; asst prof mech eng, Univ Rochester, 74-77; from asst prof to assoc prof nuclear eng, 77-79, assoc prof & chmn chem & nuclear eng, 80-83, PROF CHEM & NUCLEAR ENG, UNIV NMEX, 84- *Concurrent Pos:* Vis, Ctr Nuclear Studies, Nat Univ Mex, 83-84. *Mem:* Am Phys Soc; Am Nuclear Soc; Nat Soc Prof Engrs; Inst Elec & Electronics Engrs. *Res:* Plasma physics; nuclear engineering; reactor physics. *Mailing Add:* Idaho Nat Eng Lab PO Box 1625 Idaho Falls ID 83415-2211

WOODALL, JERRY M, RESEARCH ADMINISTRATION. *Prof Exp:* IBM FEL, INT BUS MACH, 85- *Mem:* Nat Acad Eng. *Mailing Add:* IBM T J Watson Res Ctr PO Box 218 19-115 Yorktown Heights NY 10598

WOODALL, WILLIAM ROBERT, JR, b Augusta, Ga, May 29, 45; m 66; c 2. FUEL CHEMISTRY, ENVIRONMENTAL SCIENCE. *Educ:* Univ Ga, BS, 67, MS, 69, PhD(entom), 72. *Prof Exp:* Ecol consult, Ga Power Co, 72, environ specialist, 72-78, supvr, Environ Ctr, 78-80, mgr, Power Supply Labs, 80-84; mgr, 84-91, GEN MGR, ENVIRON AFFAIRS DEPT, 91- *Mem:* Ecol Soc Am; Soc Power Indust Biologists (pres, 77-78). *Res:* Thermal effects; ecology of large rivers, their floodplains, oxbow lakes and swamps; invasion of exotic bivalves; entrainment and impingement of aquatic organisms; larval fish and invertebrate drift; nutrient cycling; coal chemistry. *Mailing Add:* 15 Clarendon Ave Avandale Estates GA 30002

WOODARD, DAVID W, b Philadelphia, Pa, Sept 28, 38; m 62; c 2. MICROWAVE & OPTICAL SEMICONDUCTORS, SEMICONDUCTOR MICROFABRICATION. *Educ:* Princeton Univ, BS, 62; Rutgers Univ, MS, 64; Cornell Univ, PhD(appl physics), 79. *Prof Exp:* Mem tech staff, RCA Labs, 62-65; prin engr, Cayuga Assocs, Inc, 68-77; res asst, 65-68 & 77-79, SR RES ASSOC, CORNELL UNIV, 79- *Concurrent Pos:* Consult, Narda Microwave Corp, 77-87 & Simmonds Precision Corp, 87-89, Mission Res, Inc, 91- *Mem:* Inst Elec & Electronics Engrs. *Res:* Gallium-arsenic materials and devices for microwave and high speed optical applications; non-linear optical effects in quantum wells. *Mailing Add:* 300 Iroquois Rd Ithaca NY 14850

WOODARD, GEOFFREY DEAN LEROY, b Big Rapids, Mich, Nov 16, 15; m 48; c 5. TOXICOLOGY. *Educ:* George Washington Univ, BS, 39, PhD(pharmacol), 51. *Prof Exp:* Lab asst, Div Pharmacol, USDA, 36-39, assoc pharmacologist, 39-51, pharmacologist, 51-57; pres, Woodard Res Corp, 57-76; CONSULT TOXICOL, 76- *Concurrent Pos:* Instr chem & biochem, USDA Grad Sch, 43-57. *Mem:* AAAS. *Res:* Bioassay methods: eye irritation, pyrogen assay, posterior pituitary; toxicology: protocol design acute-chronic studies; drug metabolism: hydroquinone, carbamate pesticides, ddt and analogues. *Mailing Add:* PO Box 445 Isle of Palms SC 29451

WOODARD, HELEN QUINCY, b Detroit, Mich, Aug 8, 00. BIOCHEMISTRY, RADIOBIOLOGY. *Educ:* Stetson Univ, BS, 20; Columbia Univ, AM, 21, PhD(chem), 25. *Prof Exp:* Res chemist, 25-56, assoc biochemist, 56-66, EMER ASSOC BIOCHEMIST, MEM CTR CANCER & ALLIED DIS, 66-; CONSULT, SLOAN-KETTERING INST, 68- *Concurrent Pos:* Asst mem, Sloan-Kettering Inst, 48-60, assoc mem, 60-68; asst prof, Med Col, Cornell Univ, 52-64, assoc prof, 64-68; adj prof environ med, NY Univ, 84. *Mem:* AAAS; Am Chem Soc; Health Physics Soc; Sigma Xi. *Res:* Chemical effects of gamma and x-rays; serum and tissue phosphatases; clinical biochemistry in bone disease; effects of radiation on bone; metabolism of bone-seeking isotopes; biological effects of radiation; chemical composition of human tissues for use in calculating absorbed radiation dose. *Mailing Add:* Mem Cancer Ctr 1275 York Ave New York NY 10021

WOODARD, HENRY HERMAN, JR, b Salisbury, Mass, Dec 18, 25; m 49; c 2. GEOLOGY. *Educ:* Dartmouth Col, AB, 47, AM, 49; Univ Chicago, PhD, 55. *Prof Exp:* Geologist, US Geol Surv, 47-49 & State Develop Comn, Maine, 50-51; assoc prof geol, Beloit Col, 53-66, chair dept, 55-87, chmn, Div Natural Sci & Math, PROF GEOL, BELOIT COL, 66- *Concurrent Pos:* Geologist, Geol Surv Nfld, 55 & US Geol Surv, 57. *Honors & Awards:* Neil Miner Award, Nat Asn Geol Teachers. *Mem:* Geol Soc Am; Geochem Soc; Nat Asn Geol Teachers; Am Geophys Union. *Res:* Diffusion in naturally occurring silicates; geology of Newfoundland and Boulder batholith; structure and petrology of southwestern Maine; contact alteration associated with Tertiary stocks in central Colorado; sanidines from Ordovician of Wisconsin; structural geology and petrology of the eastern contact zone vermilion batholith, Minnesota. *Mailing Add:* Dept Geol Beloit Col Beloit WI 53511

WOODARD, JAMES CARROLL, b Birmingham, Ala, Nov 19, 33. NUTRITION, COMPARATIVE PATHOLOGY. *Educ:* Auburn Univ, DVM, 58; Mass Inst Technol, PhD(nutrit, path), 65. *Prof Exp:* Parasitologist, Fla Vet Diag Lab, Fla Livestock Bd, 58-59; instr vet path, Auburn Univ, 61-62; res assoc, Mass Inst Technol, 63-65; NIH fel, 65-66, asst prof, 66-70, head dir comp path, 74-80, assoc prof, 78, PROF PATH, COL MED & VET MED, UNIV FLA, 78- *Mem:* Am Vet Med Asn; Int Acad Path; Am Inst Nutrit; Am Soc Exp Path. *Res:* Pathology, biochemical relationships of disease to microscopic pathology. *Mailing Add:* Div Comp & Exp Path Univ Fla Col Vet Med Box J145 JHMHC Gainesville FL 32610

WOODARD, KENNETH EUGENE, JR, b Middletown, Ohio, Oct 7, 42; m 69; c 3. ELECTROCHEMICAL ENGINEERING. *Educ:* Univ Cincinnati, BS, 65. *Prof Exp:* Prod & develop engr chloralkali, Olin Mathieson Chem Corp, 65-69; prod supvr chloralkali & fluorocarbons, Vulcan Mat Co, 69-74; sect mgr chloralkali, Olin Corp, 74-75, mgr electrochem develop chloralakali, Chem Group, 75-79, mgr electrochem technol, 79-91, PROG MGR HAN DEVELOP, OLIN CORP, 91- *Mem:* Electrochem Soc; Int Electrochem Soc; Am Inst Chem Engrs. *Mailing Add:* 128 Hummingbird Dr NW Cleveland TN 37312-9702

WOODARD, RALPH EMERSON, b Nelsonville, Ohio, May 27, 21; m 43; c 3. RESEARCH ADMINISTRATION, REACTOR PHYSICS. *Educ:* Wittenberg Col, BA, 53; Oak Ridge Sch Reactor Technol, dipl, 57; Indust Col Armed Forces, dipl, 70; George Washington Univ, MSBA, 71. *Prof Exp:* Res physicist, Wright Air Develop Ctr, US Dept Air Force, 53-56, gen physicist, Oak Ridge Nat Lab, 56-57, nuclear physicist, Wright Air Develop Ctr, 57-61, gen engr res & tech plans, Aeronaut Systs Div, 61-63, supvry physicist & dep dir gen physics res lab, Aerospace Res Labs, 63-71, phys sci adminr, 71-75, phys sci adminr, Air Force Wright Aeronaut Labs, 75-79, phys sci adminr, Air Force Mat Lab, 80-82; RETIRED. *Concurrent Pos:* Instr physics, Wittenberg Col, 57-58; lectr, Res Mgt, Air Force Inst Tech, 72-75. *Mem:* Am Phys Soc. *Res:* Laboratory management; plasma physics; solid state physics; mathematics; chemistry; fluid mechanics; metallurgy and ceramics; energy conversion; research management. *Mailing Add:* 305 Gordon Rd Springfield OH 45504

WOODARD, RICHARD P, b Kansas City, Kans, Aug 10, 55. QUANTUM GRAVITY, QUANTUM FIELD THEORY. *Educ:* Case Western Reserve Univ, BS, 77; Harvard Univ, AM, 79, PhD(physics), 84. *Prof Exp:* Postdoctoral physics, Univ Tex, Austin, 83-85 & Univ Calif, Santa Barbara, 85-87; res asst prof, Brown Univ, 87-89; ASST PROF PHYSICS, UNIV FLA, GAINESVILLE, 89- *Res:* Quantum gravity in the larger context of Laguargian field theory and elementary particle physics; string field theory and methods of regularization. *Mailing Add:* Physics Dept Univ Fla Gainesville FL 32611

WOODARD, ROBERT LOUIS, science education, astronomy; deceased, see previous edition for last biography

WOODBREY, JAMES C, b Sebago Lake, Maine, Oct 16, 34; m 56; c 3. POLYMER SCIENCE, MATERIALS SCIENCE. *Educ:* Univ Maine, BS, 56; Mich State Univ, PhD(phys chem, physics, org chem), 60. *Prof Exp:* Sr phys res chemist, Res Div, W R Grace & Co, Md, 60-61; sr res chemist, Explor & Fund Res, Plastics Div, Monsanto Co, Mass, 61-65, res projs leader, Explor & Fund Mat Res, Cent Res Dept, Mo, 65-67, Mat & Composites Res & Develop, New Enterprises Div, 67-69, res projs mgr & sci fel, 69-77, New Prod Develop Dept, Polymer Prod Co, 77-79, res projs mgr & sr sci fel, 79-83, res dir & sr fel, Photoresist Prod, Technol Div, dir res & mfg, Electronic Mat Co, 86-89; TECHNOL CONSULT, 89- *Mem:* Am Chem Soc; AAAS; Am Phys Soc; Soc Plasics Engrs. *Res:* Chemical, physical and mechanical properties, molecular and solid-state structures and preparations of macromolecular and polymer, ceramic and metal composite materials, photopolymers and photoresists; magnetic resonance and vibrational spectros; solution thermodynamics; surface physical chemistry. *Mailing Add:* 339 Ridge Trail Chesterfield MO 63017-3066

WOODBRIDGE, JOSEPH ELIOT, b Philadelphia, Pa, July 15, 21; m 49; c 6. CLINICAL CHEMISTRY, PHYSICAL CHEMISTRY. *Educ:* Princeton Univ, PhD(chem), 48. *Prof Exp:* Chemist, Manhattan Proj, 44; from res chemist to group leader, Atlantic Refining Co, 46-60; dir res, Hartman-Leddon Co, 60-66; res dir, Sadtler Res Labs, Inc, Pa, 66-68; dir clin res, Worthington Biochem Corp, NJ, 68-71; vpres diag prod, Princeton Biomedix Inc, 71-81; PRES, ALLADIN DIAGNOSTICS INC, 81- *Mem:* Am Chem Soc; Am Asn Clin Chemists; Sigma Xi; Am Soc Clin Pathol. *Res:* Synthetic detergents; petrochemicals; clinical reagents; mass spectrometry. *Mailing Add:* 84 Bayard Lane Princeton NJ 08540

WOODBURN, MARGY JEANETTE, b Pontiac, Ill, Sept 5, 28. FOOD SCIENCE, MICROBIOLOGY. *Educ:* Univ Ill, Urbana, BS, 50; Univ Wis, Madison, MS, 56, PhD(exp foods), 59. *Prof Exp:* Instr foods & nutrit, Univ Wis, Madison, 56-57; from assoc prof to prof, Purdue Univ, 59-69; assoc dean res, 80-87, PROF & HEAD, DEPT FOOD & NUTRIT, COL HOME ECON, ORE STATE UNIV, 69-, ASSOC DIR, ORE AGR EXP STA, 82-

Concurrent Pos: Nat Res Coun res assoc, Ft Detrick, Md, 68. *Honors & Awards:* Borden Award, 76. *Mem:* Am Dietetic Asn; Am Home Econ Asn; Am Soc Microbiol; Inst Food Technol; Sigma Xi; Am Inst Nutrition. *Res:* Food microbiology; staphylococcal enterotoxins; foodborne pathogenic bacteria; consumer practices influencing food quality and safety. *Mailing Add:* Dept Foods & Nutrit Ore State Univ Milan Hall 108 Corvallis OR 97331-5103

WOODBURN, RUSSELL, b Central City, Ky, Feb 16, 07; m 33; c 2. CIVIL ENGINEERING. *Educ:* Univ Ky, BS, 29. *Prof Exp:* Field engr, Erie RR Co, 29-30 & Mo State Hwy Dept, 30-33; engr soil & water res & field opers, USDA, 33-56, dir sedimentation lab, Agr Res Serv, Miss, 56-61, chief southern br, Soil & Water Conserv Res Div Ga, 61-64; assoc coordr res & dir eng exp sta, 64-72, from assoc prof to prof civil eng, 64-72, EMER PROF CIVIL ENG, UNIV MISS, 72- *Concurrent Pos:* Alderman & commissioner, Elec Power, Water & Sewer Depts, Oxford, Miss, 73-81. *Mem:* Fel Am Soc Agr Engrs. *Res:* Soil and water conservation and research; engineering education and structures; research administration. *Mailing Add:* Box 283 University MS 38677

WOODBURN, WILTON A, b Pittsburgh, Pa, Nov 2, 26; m 56; c 3. MECHANICAL ENGINEERING. *Educ:* Carnegie Inst Technol, BS, 47, MS, 56. *Prof Exp:* Res engr, Eng Design Div, Alcoa Res Labs, Aluminum Co Am, 47-65 & Fabricating Metall Div, 65-72, eng assoc, 72-80, sr tech specialist, Fabricating Technol Div, Alcoa Tech Ctr, 80-88; RETIRED. *Mem:* Am Soc Mech Engrs. *Res:* Fabricating processes, particularly hot and cold rolling; computerized control for preset, guage and flatness. *Mailing Add:* 69 Indian Fields Trail Lower Burrell PA 15068

WOODBURNE, MICHAEL O, b Ann Arbor, Mich, Mar 8, 37; m 60; c 2. VERTEBRATE PALEONTOLOGY, STRATIGRAPHY. *Educ:* Univ Mich, BS, 58, MS, 60; Univ Calif, Berkeley, PhD(paleont), 66. *Prof Exp:* Mus technician vert paleont, Univ Calif, Berkeley, 62-65, mus scientist, 65-66; res assoc, Princeton Univ, 66; from lectr to assoc prof, 66-77, PROF GEOL, UNIV CALIF, RIVERSIDE, 78- *Mem:* Soc Vert Paleont; Paleont Soc; Soc Study Evolution; fel Geol Soc Am. *Res:* Mammalian paleontology, including Australian marsupials; biostratigraphy and paleontology of the Mojave Desert. *Mailing Add:* Dept Earth Sci Univ Calif Riverside CA 92521

WOODBURNE, RUSSELL THOMAS, b London, Ont, Nov 2, 04; US citizen; m 34; c 3. ANATOMY. *Educ:* Univ Mich, AB, 32, MA, 33, PhD(anat), 35. *Prof Exp:* From instr to prof, 36-74, chmn dept, 58-73, EMER PROF ANAT, MED SCH, UNIV MICH, ANN ARBOR, 75- *Mem:* Am Asn Anat (secy-treas, 64-72, pres, 74-75); Can Asn Anat. *Res:* Structure of mammalian midbrain; pleura; blood vessels of pancreas, liver, urinary bladder, ureter and urethra. *Mailing Add:* 2520 Hawthorne Ann Arbor MI 48104

WOODBURY, DIXON MILES, b St George, Utah, Aug 6, 21; m 45; c 3. PHARMACOLOGY. *Educ:* Univ Utah, BS, 42, MS, 45; Univ Calif, PhD(zool), 48. *Prof Exp:* Asst zool, Univ Calif, 44-47; res instr physiol & pharmacol, 47-50, asst res prof, 50-53, assoc res prof pharmacol, 53-59, assoc prof, 59-61, PROF PHARMACOL, COL MED, UNIV UTAH, 61-, CHMN DEPT, 72- *Concurrent Pos:* USPHS grant, Col Med, Univ Utah, 47-50; NIH res career award, 62-; Nat Neurol Res Found scientist, Univ Utah, 58; mem pharmacol training comt, NIH, 61-65; mem adv panel on epilepsies, Surgeon Gen, USPHS, 66-69; mem prog proj comt A, Nat Inst Neurol Dis & Stroke, 67-71, comt anticonvulsants, 68- & vis scientist, Epilepsy Sect, 78; prof, State Univ Leiden, 68; mem neurobiol merit rev bd, Vet Admin, 72-74; mem prof adv bd & nat bd dirs, Epilepsy Found Am, 73-; distinguished res prof, Univ Utah, 73-74. *Mem:* Am Soc Pharmacol & Exp Therapeut; Endocrine Soc; Soc Neurosci; Am Soc Neurochem; Int Soc Neurochem. *Res:* Brain electrolyte metabolism; convulsive disorders; endocrinology; effects of hormones on the nervous system; neurochemistry; development of the nervous system; blood-brain barrier; metabolism of drugs. *Mailing Add:* 3118 Crestview Circle Bountiful UT 84010

WOODBURY, ERIC JOHN, b Washington, DC, Feb 9, 25; m 46; c 3. LASER SYSTEMS. *Educ:* Calif Inst Technol, BS, 47, PhD(physics), 51. *Prof Exp:* Asst, Calif Inst Technol, 47-51; mem tech staff, 51-53, group head & sr staff engr, Electronics Dept, Guided Missile Lab, 53-60, sr staff physicist, Radar & Missile Electronics Lab, 61-62, sr scientist, 62-63, asst dept mgr, Laser Develop Dept, 63-66, mgr laser dept, 66-69, mgr laser develop dept, 69-72, mgr tactical laser systs lab, 72-76, chief scientist, Laser Systs Div, 76-79, chief scientist, Electro-Optical Engr Div, Hughes Aircraft Co, 79-81; ASST PROF, CALIF STATE UNIV, NORTHRIDGE, 82- *Concurrent Pos:* Mem indust adv comt, Tech Educ Res Ctr, 72-75; mem local adv comt, Caltech TV proj, The Mechanical Universe. *Mem:* Am Phys Soc; fel Inst Elec & Electronics Engrs; Sigma Xi. *Res:* Noise in electronic systems; application of solid state devices to electronic devices; electronic systems for use in missiles and satellites; experimental nuclear physics; laser systems and components. *Mailing Add:* 18621 Tarzana Dr Tarzana CA 91356

WOODBURY, GEORGE WALLIS, JR, b Oct 13, 37; US citizen; m 60; c 2. PHYSICAL CHEMISTRY. *Educ:* Univ Idaho, BS, 59; Univ Minn, PhD(phys chem), 64. *Prof Exp:* Res assoc chem, Univ Minn, 64-65 & Cornell Univ, 65-66; from asst prof to assoc prof, 66-74, PROF CHEM, UNIV MONT, 74- *Mem:* Am Chem Soc. *Res:* Statistical mechanics of cooperative phenomena and adsorption. *Mailing Add:* Dept Chem Univ Mont Missoula MT 59812

WOODBURY, JOHN F L, b London, Eng, Dec 22, 18; Can citizen; m 43; c 3. MEDICINE. *Educ:* Dalhousie Univ, BSc, 39, MD, CM, 43; Royal Col Physicians Can, cert, 52; FRCP(C), 72. *Prof Exp:* Assoc prof, 46-49, PROF MED, DALHOUSIE UNIV, 69- *Mem:* Fel Am Col Physicians; Am Rheumatism Asn; Can Rheumatism Asn; Can Med Asn. *Res:* Etiology and immunology of arthritis. *Mailing Add:* Dept Med Rheumat Dalhousie Univ Fac Med Sir Chas Tupper Bldg Halifax NS B3H 4H7 Can

WOODBURY, JOHN WALTER, b St George, Utah, Aug 7, 23; m 49; c 4. PHYSIOLOGY, BIOPHYSICS. *Educ:* Univ Utah, BS, 43, MS, 47, PhD(physiol), 50. *Prof Exp:* Lab asst physics, Univ Utah, 42-43; staff mem, Radiation Lab, Mass Inst Technol, 43-45; res asst physiol, Univ Utah, 45-47; from instr to asst prof physiol, Sch Med, Univ Wash, 50-57, from assoc prof to prof physiol & biophys, 57-73; PROF PHYSIOL, SCH MED, UNIV UTAH, 73- *Concurrent Pos:* Mem, NIH Physiol Study Sect, Dept Health & Human Serv, 78-82. *Mem:* AAAS; Am Physiol Soc; Biophys Soc; Inst Elec & Electronics Eng; Soc Neurosci. *Res:* Electrophysiology of excitable tissues; ion transport through membranes; characteristics of anion channels; anion permeability; human voluntary motor performance. *Mailing Add:* Dept Physiol Univ Utah Col Med 410 Chipeta Way Salt Lake City UT 84108

WOODBURY, MAX ATKIN, b St George, Utah, Apr 30, 17; m 47; c 4. BIOMATHEMATICS, COMPUTER SCIENCE. *Educ:* Univ Utah, BS, 39; Univ Mich, MS, 40, PhD(math), 48; Univ NC, Chapel Hill, MPH, 77. *Prof Exp:* Instr math, Univ Mich, 47-49; Off Naval Res grant & mem, Inst Advan Study, 49-50; res assoc math & econ, Princeton Univ, 50-52; assoc prof statist, Univ Pa, 52-54; prin investr, Logistics Res Proj, George Washington Univ, 54-56; res prof math, Col Eng, NY Univ, 56-62, prof exp neurol, Med Ctr, 63-65; EMER PROF, MED CTR, DUKE UNIV, 66-, PROF COMPUT SCI, 71- *Concurrent Pos:* Gov & indust consult, 51-; mem opers res adv coun, New York, 64-68; mem diag radiol adv group, Nat Cancer Inst, 74-77; sr fel, Ctr for Demographic Studies, 75-; mem several nat comts weather modification & NIH & Food & Drug Admin study sects. *Mem:* Fel AAAS; fel Inst Math Statist; fel Am Statist Asn; Sigma Xi. *Res:* Computing; statistics; models in biology and medicine; quantitative models of information about biomedical systems require mathematics, probability and basic sciences for formulation, statistics of estimation of parametrics and testing, computing and numerical analysis for calculation. *Mailing Add:* Biol-Math Dept Duke Univ Med Ctr Box 3200 Durham NC 27710

WOODBURY, RICHARD C, b Salt Lake City, Utah, Apr 19, 31; m 54; c 6. SOLID STATE ELECTRONICS. *Educ:* Univ Utah, BS, 56; Stanford Univ, MS, 58, PhD(elec eng), 65. *Prof Exp:* Engr, Oscilloscope Circuit Design, Hewlett-Packard Co, Calif, 56-59; asst prof elec eng, Brigham Young Univ, 59-62; res assoc, Stanford Electronics Labs, 62-65; from asst prof to assoc prof, 65-72, PROF ELEC ENG, BRIGHAM YOUNG UNIV, 72- *Concurrent Pos:* NSF grants, 68-72; res consult to numerous firms. *Mem:* Inst Elec & Electronics Engrs. *Res:* Magnetic devices; silicon monolithic integrated circuits; silicon and cadmium sulfide photovoltaic cells; radiation hardening of silicon electron voltaic and photovoltaic cells; vacuum microelectronics. *Mailing Add:* Dept Elec Eng Brigham Young Univ 458 CB Provo UT 84602

WOODBURY, RICHARD PAUL, b Kenosha, Wis, Aug 16, 50; m 71; c 2. POLYMER STABILIZATION. *Educ:* Univ Wis-Whitewater, BS, 72; Mich State Univ, PhD(chem), 76. *Prof Exp:* Sr chemist, Borg Warner Chem, 76-83; sr res chemist, 83-89, MGR ORG CHEM RES, W R GRACE & CO, NASHUA, NH, 89- *Mem:* Am Chem Soc; Int Union Pure & Appl Chem. *Res:* Synthesis of biologically important peptides and amino acids; nitroparaffins; heterocyclic compounds. *Mailing Add:* Nine Woodland Dr Amherst NH 03031

WOODBURY, ROBERT A, b Sep 1, 04; US citizen; m; c 3. PHARMACOLOGY. *Educ:* Univ Kans, BS, 24, MS, 28, PhD(physiol), 31; Univ Chicago, MD, 34. *Prof Exp:* From asst prof to prof pharmacol, Sch Med, Univ Ga, 34-47, chmn, 41-47; prof & chmn pharmacol, med units, Univ Tenn, 47-75; RETIRED. *Concurrent Pos:* Mem, study sect & rev comt, NIH, 65-67. *Mem:* Sigma Xi; Am Soc Physiol; Am Soc Pharmacol & Exp Therapeut. *Res:* Cardiovascular system; autonomic system; uterine pharmacology and physiology. *Mailing Add:* 532 W Clover Dr Memphis TN 38119

WOODCOCK, ALFRED HERBERT, b Atlanta, Ga, Sept 7, 05; m 41; c 3. OCEANOGRAPHY. *Hon Degrees:* DSc, Long Island Univ, 61. *Prof Exp:* Technician, Woods Hole Oceanog Inst, 31-42, res assoc, 42-46, oceanogr, 46-63; res assoc geophys, Hawaii Inst Geophys, 63-72, RES AFFIL, DEPT OCEANOG, UNIV HAWAII, 72-; EMER SCIENTIST, WOODS HOLE OCEANOG INST, 88- *Mem:* AAAS; fel Am Meteorol Soc; assoc Am Geophys Union. *Res:* Marine meteorology; air-sea interaction; sea-salt nuclei in marine air and their role in cloud, rain and fog formation; Hawaii Alpine Lake, permafrost and mountain breathing studies. *Mailing Add:* 45-090 Namoku St Apt 912 Kaneohe HI 96744-5320

WOODCOCK, CHRISTOPHER LEONARD FRANK, b Essex, Eng, July 9, 42; m 64; c 3. CELL BIOLOGY, BIOCHEMISTRY. *Educ:* Univ Col, Univ London, BSc, 63, PhD(bot), 66. *Prof Exp:* Res fel biophys, Univ Chicago, 66-67; res fel bot, Harvard Univ, 67-69, lectr biol, 69-72; from asst prof to assoc prof, 72-78, PROF ZOOL, UNIV MASS, AMHERST, 78- *Mem:* Am Soc Cell Biol; Micros Soc Am. *Res:* Cell ultrastructure and function; information processing in cells; chromatin structure and function. *Mailing Add:* Dept Zool Univ Mass Amherst MA 01003

WOODCOCK-MITCHELL, JANET LOUISE, b Barre, Vt, Sept 6, 49; m 75. CELL BIOLOGY. *Educ:* Univ Conn, BA, 71, MS, 74, PhD(biochem), 79. *Prof Exp:* Postdoctoral fel, Johns Hopkins Univ, 79-81; postdoctoral fel, 81-86, RES ASST PROF, UNIV VT, 86- *Mem:* Am Soc Cell Biol; Am Thoracic Soc. *Res:* Differentiation of lung epithelial cells during development and in repair after lung injury; growth factors and signals inducing differentiation of epithelial cells. *Mailing Add:* Dept Physiol Univ Vt Burlington VT 05405

WOODFIELD, F(RANK) W(ILLIAM), JR, b Astoria, Ore, Mar 1, 18; m 44; c 3. CHEMICAL ENGINEERING. *Educ:* Ore State Col, BS, 39; Columbia Univ, MS, 40. *Prof Exp:* Chem engr process eng & eng res, E I du Pont de Nemours & Co, Inc, 40-47; increasing responsibility group leader to mgr eng develop, Gen Elec Co, 47-55, mgr reactor mat develop, 55-62, specialist contract admin, 62-63, mgr programming, 64; dep staff mgr programming & tech develop, Pac Northwest Lab, Battelle Mem Inst, 65-66, mgr develop,

Fast Flux Test Facility, 67-68, asst lab dir tech serv, 68-70, mgr facilities planning & eng, 70-73, prog mgr plenum fill exp, 74; mgr res & technol ctr, Exxon Nuclear Co Inc, 75-81, mgr logisitics, 81-83; RETIRED. *Concurrent Pos:* Chmn, Nuclear Eng Div, Am INst Chem Engrs, 67. *Mem:* Am Chem Soc; Am Nuclear Soc; Am Inst Chem Engrs. *Res:* Chemical engineering diffusional operations; technical and administrative management of research and development in a broad-spectrum laboratory. *Mailing Add:* 81 McMurray St Richland WA 99352

WOODFILL, MARVIN CARL, b Los Angeles, Calif, June 13, 38; m 80, 88; c 5. COMPUTER SYSTEM ENGINEERING. *Educ:* Iowa State Univ, BS, 59, MS, 61, PhD(elec eng), 64. *Prof Exp:* Asst prof elec eng, Iowa State Univ, 59-64; from asst prof to prof elec eng, 66-80, PROF COMPUT SCI, ARIZ STATE UNIV, 80- *Concurrent Pos:* Engr consult, Sperry Flight Systs, Ariz, 66-67, semiconductor prod div, Motorola, Inc, 68-70, govt electronics div, 71, St Joseph's Hosp & Med Ctr, Phoenix, 70-81 & City Prescott, 79- *Mem:* Inst Elec & Electronics Engrs. *Res:* Development of mini-micro computer systems for engineering, medical and governmental applications. *Mailing Add:* Comput Sci Dept Ariz State Univ Tempe AZ 85287

WOODFIN, BEULAH MARIE, b Chicago, Ill, June 22, 36. BIOCHEMISTRY. *Educ:* Vanderbilt Univ, BA, 58; Univ Ill, Urbana, MS, 60, PhD(biochem), 63. *Prof Exp:* Res assoc biochem, Univ Mich, 63-66, instr, 66-67; asst prof, 67-78, ASSOC PROF BIOCHEM, SCH MED, UNIV NMEX, 78- *Concurrent Pos:* USPHS fel, Univ Mich, 63-65. *Mem:* AAAS; Am Chem Soc; NY Acad Sci; Sigma Xi; Am Soc Biochem Molecular Biol; Protein Soc. *Res:* Interaction of viral proteins with mitochondria; experimental mouse model of Reye's syndrome; ornithine carbamoyl transferase deficiency. *Mailing Add:* Dept Biochem Univ NMex Sch Med Albuquerque NM 87131-5221

WOODFORD, DAVID A(UBREY), b Cleethorpes, Eng, Sept 17, 37; m 61; c 3. PHYSICAL METALLURGY. *Educ:* Univ Birmingham, BSc, 60, PhD(metall), 63. *Hon Degrees:* DSc, Univ Birmingham, 81. *Prof Exp:* Res fel, Univ Birmingham, 63-64; res metallurgist, Mat & Processes Lab, 64-73, STAFF METALLURGIST, CORP RES & DEVELOP, GEN ELEC CO, 73- *Concurrent Pos:* Assoc ed, J Eng Mat & Tech, 74- & Fatigue Eng Mat & Struct, 78- *Honors & Awards:* A H Geisler Award, 72. *Mem:* Am Soc Metals; Am Soc Testing & Mat; Am Soc Mech Engrs. *Res:* High temperature mechanical properties; creep and radiation damage; strain aging superplasticity; temper embrittlement; thermal fatigue; cavitation erosion; environmental embrittlement; author or coauthor of 60 publications. *Mailing Add:* Rm 215 Materials Res Ctr Troy NY 12180

WOODFORD, JAMES, b Roanoke, Va, Mar 18, 46. FORENSIC COMPARISON ANALYSIS, MEDICINAL CHEMISTRY. *Educ:* Emory Univ, MS, 70, PhD(chem), 73. *Prof Exp:* Postdoctoral, Univ Kans, 73-74; Res assoc chem, Dept Chem, Emory Univ, 74-75; DIR, WEB RES, ATLANTA, GA, 75- *Concurrent Pos:* Res award, Nat Cancer Inst, 73; vis scientist & guest seminar speaker, Metro Police Forensic Sci Lab, Scotland Yard; lectr chem, Chem Dept, Emory Univ, 76-77; instr clin chem, Sch Pharm, Mercer Univ, 77-78; bd dir, Metrop Atlanta Coun Alcohol & Drugs, 79- *Mem:* Am Chem Soc; AAAS. *Res:* Testing for court purposes; forensic comparison analyses of evidence. *Mailing Add:* PO Box 941156 Atlanta GA 30341

WOODGATE, BRUCE EDWARD, b Eastbourne, Sussex, Eng, Feb 19, 39; m 65; c 2. ASTROPHYSICS. *Educ:* Univ London, BSc, 61, PhD(astron), 65. *Prof Exp:* From res asst to res assoc physics, Univ Col, Univ London, 65-71; sr res assoc, Columbia Univ, 71-74, assoc dir Astrophys Lab, 72-74; sci systs analyst, 74-75, ASTROPHYSICIST, GODDARD SPACE FLIGHT CTR, 75-, PROJ SCIENTIST, SOLAR MAXIMUM MISSION, 83- *Concurrent Pos:* Mem working group, NASA Outlook for Space Study, 74. *Mem:* Am Astron Soc. *Res:* X-ray astronomy; astronomy of supernova remnants and the interstellar medium; solar physics and solar-terrestrial relations, solar flares; remote sensing of earth resources; solar flare physics using UV and x-ray spectroscopy; supernova remnant and interstellar medium physics using optical narrow bend imaging; space instrumentation, UV, x-ray, optical. *Mailing Add:* 1708 Truro Rd Goddard Space Flight Ctr NASA Crofton MD 21114

WOODHAM, DONALD W, b Jacksonville, Ala, July 9, 29; m 54; c 1. ANALYTICAL CHEMISTRY, INSECT REARING. *Educ:* Jacksonville State Univ, BS, 54; Am Inst Chemists, cert, 73. *Prof Exp:* Chem aide, Phosphate Develop Works, Tenn Valley Authority, 54-55, control chemist, 55-57; res chemist, Pesticides Chem Br, Entom Res Div, 57-65, res chemist, Plant Pest Control Div, Miss, 65-67, supvry chemist, 67-70, chemist in-chg plant pest control prog, Animal & Plant Health Inspection Serv, 70-75, lab supvr, Pink Bollworm Rearing Facil, Animal & Plant Health Inspection Serv, Agr Res Serv, 75-76, new pest detection & surv staff, USDA Aphis PPQ, 76-80, emergency progs, 80-82, TECHNOL ANALYSIS & DEVELOP STAFF, USDA, 82- *Mem:* AAAS; Am Chem Soc; Entom Soc Am; Am Inst Chemists; Sigma Xi. *Res:* Analysis of phosphorous intermediates used in the synthesis of nerve gas; pesticide residues in crops; animal products and miscellaneous materials; analysis of environmental samples for pesticide and herbicide residues; development of new methods for the analysis of pesticides and pesticide residues; confirmation methods for identification purposes; methods utilized in the rearing, irradiation and shipment of pink bollworm moths. *Mailing Add:* Rte 1 Rural Box 145 Guntersville AL 35976-0145

WOODHOUR, ALLEN F, b Newark, NJ, Feb 21, 30; m 55; c 1. VIROLOGY, BACTERIOLOGY. *Educ:* St Vincent Col, AB, 52; Cath Univ Am, MS, 54, PhD(bact), 56. *Prof Exp:* Bacteriologist, Walter Reed Army Inst Res, 56-57; res assoc virol, Charles Pfizer & Co, Inc, 57-60; res assoc, 60-64, asst dir, Dept Virus Dis, 64-66, dir viral vaccine res, 66-73, asst dir virus & cell biol res, 68-73, sr dir & asst area head virus cell biol res, 74-78, EXEC DIR BACT VACCINES & ADMIN AFFAIRS, VIRUS & CELL BIOL RES, MERCK INST THERAPEUT RES, 78- *Mem:* AAAS; Am Asn Immunol; Sigma Xi; Int Asn Biol Stand; Am Soc Microbiol. *Res:* Use of adjuvants in immunology; metabolizable vegetable oil water-in-oil adjuvant; respiratory viruses for vaccine development; development of bacterial vaccines; antivirals. *Mailing Add:* 8003 River Pl Carmel CA 93923

WOODHOUSE, BERNARD LAWRENCE, b Norfolk, Va, Aug 14, 36; m 64; c 3. PHARMACOLOGY. *Educ:* Howard Univ, BS, 58, MS, 63, PhD(pharmacol), 73. *Prof Exp:* Instr zool, A&T Col NC, 63-64; from instr to asst prof, 64-73, assoc prof, 73-79, PROF BIOL, SAVANNAH STATE COL, 79- *Concurrent Pos:* Hoffmann-La Roche res grant & NIH res grant, 75- *Res:* Study of the mechanism of the antihypertensive effects of beta adrenergic blocking drugs on various species of animals. *Mailing Add:* Dept Biol Savannah State Col Savannah GA 31404

WOODHOUSE, EDWARD JOHN, b Norwich, Eng, Apr 22, 39. CHEMISTRY, TOXICOLOGY. *Educ:* Univ Nottingham, BSc, 61, PhD(chem), 64. *Prof Exp:* Res assoc inorg chem, Ore State Univ, 64-67; sr chemist, 67-74, prin chemist, 74-80, HEAD BIOANALYTICAL CHEM SECT, MIDWEST RES INST, 80- *Mem:* AAAS; Am Chem Soc; NY Acad Sci; Int Asn Forensic Toxicol. *Res:* Radiochemistry; analytical chemistry; solvent extraction; analysis of drugs in body fluids and dosage form; marijuana chemistry and metabolism; analysis metabolism and distribution of xenobiotics in animal systems; analytical chemistry support for toxicology studies. *Mailing Add:* Midwest Res Inst 425 Volker Blvd Kansas City MO 64110

WOODHOUSE, JOHN CRAWFORD, chemistry; deceased, see previous edition for last biography

WOODHOUSE, WILLIAM WALTON, JR, soils; deceased, see previous edition for last biography

WOODHULL-MCNEAL, ANN P, b Orange, NJ, Oct 20, 42. BIOPHYSICS, HUMAN POSTURE. *Educ:* Swarthmore Col, BA, 64; Univ Wash, PhD(physiol, biophys), 72. *Prof Exp:* NIH fel physiol & biophys, Univ Wash, 72; from asst prof to assoc prof biol, 72-88, PROF PHYSIOL, DIV NATURAL SCI, HAMPSHIRE COL, 88- *Concurrent Pos:* Lectr neurobiol, Harvard Univ, 75. *Mem:* Am Col Sports Med; Nat Womens Studies Asn. *Res:* Electromyographic and photographic studies of human posture; women and science. *Mailing Add:* Div Natural Sci Hampshire Col Amherst MA 01002

WOODIN, SARAH ANN, b New York, NY, Dec 27, 45; m 80; c 1. MARINE ECOLOGY. *Educ:* Goucher Col, BA, 67; Univ Wash, PhD(marine ecol), 72. *Prof Exp:* Asst prof ecol, Univ Md, 72-75; asst prof ecol, Johns Hopkins Univ, 75-80; from res assoc prof to res prof, 80-87, PROF, DEPT BIOL, UNIV SC, 87- *Concurrent Pos:* Res grant, div biol oceanog, NSF, 74- & div systs, 85-88, div ecol, 89-; marine ed, Ecol & Ecological Monographs, 78-81; mem, animal resources comt, NIH, 83-87. *Mem:* Ecol Soc Am; Am Soc Zool; fel AAAS; Am Soc Limnol & Oceanog; Sigma Xi; Soc Study Evolution. *Res:* Benthic ecology; life history strategies of organisms, particularly in fauna; functional morphology of polychaetes; functional groups of infauna. *Mailing Add:* Dept Biol Univ SC Columbia SC 29208

WOODIN, TERRY STERN, b New York, NY, Dec 25, 33; m 54; c 5. BIOCHEMISTRY. *Educ:* Alfred Univ, BA, 54; Univ Calif, Davis, MA, 65, PhD(biochem), 67. *Prof Exp:* Res assoc biochem, Univ Calif, Davis, 67-68; adj asst prof, Univ Nev, Reno, 68-69, asst prof, 69-72; asst prof biochem, Humboldt State Univ, 72-77; ASSOC PROF, UNIV NEV, RENO, 77- *Mem:* AAAS; Am Chem Soc; Sigma Xi; Am Soc Plant Physiologists. *Res:* Sulfate metabolism in fungi; thermophilic fungi; temperature effects on membrane structure and function. *Mailing Add:* Biochem Dept Univ Nev Reno NV 89557

WOODIN, WILLIAM GRAVES, b Dunkirk, NY, July 22, 14; m 40; c 2. ALLERGY. *Educ:* Cornell Univ, AB, 36, MD, 39. *Prof Exp:* Intern & asst resident med, Univ Hosps Cleveland, 39-41; asst, Med Col, Cornell Univ, 46, instr, 47-48; from instr to asst prof, 48-54, clin assoc prof, 54-63, CLIN PROF MED, STATE UNIV NY UPSTATE MED CTR, 63-, DIR ALLERGY CLIN, 48- *Concurrent Pos:* Fel, Roosevelt Hosp, New York, 47-48; consult hosps, 50- *Mem:* AMA; fel Am Acad Allergy; NY Acad Sci. *Mailing Add:* 109 S Warren St Syracuse NY 13202

WOODIN, WILLIAM HARTMAN, III, b New York, NY, Dec 16, 25; m 48, 77; c 4. ZOOLOGY. *Educ:* Univ Ariz, BA, 50; Univ Calif, MA, 56. *Prof Exp:* Dir, 54-71, EMER DIR, ARIZ-SONORA DESERT MUS, 72-; PRES, WOODIN LAB, 81- *Mem:* Fel AAAS. *Res:* Herpetology; taxonomy; desert ecology. *Mailing Add:* 3600 N Larrea Lane Tucson AZ 85715

WOODING, FRANK JAMES, b Pontiac, Ill, Feb 1, 41; m 64; c 3. AGRONOMY. *Educ:* Univ Ill, Urbana, BS, 63; Kans State Univ, MS, 66, PhD(agron), 70. *Prof Exp:* Res assoc plant physiol, Pa State Univ, 69-70; from asst prof to assoc prof, 70-82, PROF AGRON, AGR EXP STA, 82-, HEAD, DIV PLANT & ANIMAL SCI, UNIV ALASKA, 88- *Mem:* Am Soc Agron; Crop Sci Soc Am; Soil Sci Soc Am. *Res:* Plant nutrition; plant growth under arctic and subarctic conditions; crop physiology; crop production; introduction, multiplication, maintenance, evaluation and cataloging of plant germ plasm; introducing plants with origins in northern latitudes or high elevation to Alaska to test their suitablitity. *Mailing Add:* Div Plant & Animal Sci Univ Alaska Fairbanks AK 99775-0080

WOODING, WILLIAM MINOR, b Waterbury, Conn, Aug 24, 17; m 40; c 2. EXPERIMENTAL STATISTICS, CHEMISTRY. *Educ:* Polytech Inst Brooklyn, BChemE, 53. *Prof Exp:* Analyst inorg chem, Scovill Mfg Co, Conn, 36-40; technician, Am Cyanamid Co, 41-46, chemist, 46-51, res chemist, 51-56, coordr personnel admin serv, 56-57; asst chief chemist, Revlon, Inc, NY, 57-61, assoc res dir, 61-65; assoc res dir, Carter-Wallace, Inc, 65-67, dir tech serv, Carter Prod Div, 67-75, corp dir statist serv, 75-82; CONSULT, 82-; WRITER, 85- *Mem:* AAAS; Am Chem Soc; fel Am Inst Chemists; fel Am Soc Qual Control; fel Soc Cosmetic Chem; Am Statist Asn; Am Soc Test Mat; Biomet Soc; Soc Clin Trials. *Res:* Experimental design and applied statistics, principally in medical and biological fields; teaching statistical techniques to biomedical personnel. *Mailing Add:* 4690 Maquam Shore Rd Swanton VT 05488

WOODLAND, BERTRAM GEORGE, b Mountain Ash, Wales, Apr 4, 22; m 52; c 2. STRUCTURAL GEOLOGY, MICRO-FABRICS. *Educ:* Univ Wales, BSc, 42; Univ Chicago, PhD(geol), 62. *Prof Exp:* Exp officer, Ministry Home Security, Gt Brit Air Ministry, 43-46; res asst mineral surv, Ministry Town & Country Planning, 46-49; asst res officer, Ministry Housing & Local Govt, 49-54; from instr to asst prof geol, Univ Mass, 54-56; asst prof, Mt Holyoke Col, 56-58; assoc cur, Chicago Natural Hist Mus, 58-62; cur, 63-87, EMER CUR, FIELD MUS NATURAL HIST, 87- *Concurrent Pos:* Consult, Petrol Brasileiro Depex, Rio de Janeiro, Brazil, 55-56, Coronet Instrnl Films, 70-80, ITT Res Inst, 75-79, Denoyer-Cappert, 79-80 & Harza Eng Co, 83-84. *Mem:* Fel Geol Soc London; Geol Asn London. *Res:* Metamorphism; igneous rocks; cone-in-cone structure; tectonics and microstructures. *Mailing Add:* Dept Geol Field Mus Natural Hist Chicago IL 60605

WOODLAND, DOROTHY JANE, b Warren, Ohio, Sept 20, 08. PHYSICAL CHEMISTRY. *Educ:* Col Wooster, BS, 29; Ohio State Univ, MSc, 30, PhD(chem), 32. *Prof Exp:* Asst, Col Wooster, 28-29 & Ohio State Univ, 29-32; from instr to asst prof chem, Wellesley Col, 32-38; assoc prof & head dept, Western Col, 38-42, prof, 42-44; prof & head dept, 44-74, EMER PROF CHEM, JOHN BROWN UNIV, 74- *Mem:* Emer mem Am Chem Soc; Sigma Xi. *Res:* Surface energy; relation between radius of curvature of droplets and surface energy. *Mailing Add:* Box 3048 John Brown Univ Siloam Springs AR 72761

WOODLAND, WILLIAM CHARLES, b Highland Park, Mich, Nov 22, 19; m 44; c 3. PHYSICAL CHEMISTRY. *Educ:* Col Wooster, BA, 41; Carnegie Inst Technol, MS, 49, DSc(phys chem), 50. *Prof Exp:* Instr org microanal, NY Univ, 49-51; chemist, Jackson Lab, 51, Chambers Works, NJ, 51-64, color specialist, Washington Works, Parkersburg, WVa, 64-72, SR CHEMIST, WASH WORKS, E I DU PONT DE NEMOURS & CO, INC, PARKERSBURG, 72- *Res:* Thermoplastic resins; color technology; specialized analytical chemistry. *Mailing Add:* Nine Ashwood Dr Vienna WV 26101

WOODLEY, CHARLES LAMAR, b Atlanta, Ga, Sept 12, 41; m 64; c 1. BACTERIAL GENETICS. *Educ:* Ga State Univ, BS, 72, MS, 75; Univ Ga, PhD(microbiol), 80. *Prof Exp:* Lab technician, US Army Nutrit Labs, 64-66; lab technician, 66-72, MICROBIOLOGIST, CTR DIS CONTROL, 72- *Mem:* Am Soc Microbiol. *Res:* Genetic aspects of genus mycobacterium; organisms that cause tuberculosis. *Mailing Add:* Chamblee 7 FOB 1600 Clifton Rd Atlanta GA 30333

WOODLEY, CHARLES LEON, b Montgomery, Ala, Jan 22, 44; m 66; c 1. BIOCHEMISTRY. *Educ:* Univ Ala, BS, 66, MS, 68; Univ Nebr, PhD(chem), 72. *Prof Exp:* Res assoc biochem, Univ Nebr, Lincoln, 71-74; ASST PROF BIOCHEM, UNIV MISS MED CTR, 74- *Mem:* Sigma Xi. *Res:* Control of protein synthesis initiation and translation in eukaryotic and viral systems. *Mailing Add:* Dept Biochem Univ Miss Sch Med 2500 N State St Jackson MS 39216

WOODMAN, DANIEL RALPH, b Portland, Maine, Apr 20, 42; m 67. VIROLOGY, CLINICAL MICROBIOLOGY. *Educ:* Univ Maine, Orono, BS, 64; Univ Md, College Park, MS, 66, PhD(microbiol), 72. *Prof Exp:* Asst microbiol, Univ Md, College Park, 64-67; naval officer virol, Deseret Test Ctr, Salt Lake City, 67-69; instr, Univ Col, Univ Md, 70; exec officer, US Naval Unit, Ft Detrick, Md, 71-74, head virol div, 74-78, head microbiol br, Naval Med Res Inst, 78-80, HEAD MICROBIOL SECT, NAT NAVAL MED CTR, BETHESDA, 80- *Mem:* Am Soc Microbiol; Asn Mil Surgeons US; Am Soc Trop Med & Hyg; AAAS. *Res:* Animal virology with a special interest in viral immunology and chemotherapeutics; arbovirus replication and pathogenicity. *Mailing Add:* Dept Path George Washington Univ 2121 Eye St NW Washington DC 20052

WOODMAN, PETER WILLIAM, b Gloucester, Eng; m; c 2. TOXICOLOGY, CARCINOGENESIS. *Educ:* Univ Bath, B Pharm, Hons, 70, PhD(molecular pharmacol & med chem), 74. *Prof Exp:* Systs analyst pharmaceut mfg, Eli Lilly, Eng, 69; teaching asst pharm & pharmacol, Univ Bath, 70-74; proj assoc oncol, McArdle Lab Cancer Res, 74-76; asst mem biochem & pharmacol, St Jude Childrens Res Hosp, 76-80; tech dir & dept mgr, Health Sci, Dynamac Corp, 80-90; SR ASSOC, PRECEPTS INC, 90- *Concurrent Pos:* Mem ad hoc, Nat Large Bowel Cancer Proj, Nat Cancer Inst, 76-80; mem SBIR rev bd, Nat Inst Environ Health Sci, 85-88. *Res:* Medicinal, chemical, and environmental risk assessment programs involving evaluation of the metabolism, pharmacology, toxicology, pharmacokinetics, and carcinogenicity of chemical compounds. *Mailing Add:* 20624 Dubois Ct Montgomery Village MD 20879-4048

WOODMANSEE, DONALD ERNEST, b Lexington, Ky, May 23, 41; m 64; c 2. CHEMICAL ENGINEERING, GEOCHEMISTRY. *Educ:* Univ Del, BChE, 63; Univ Ill, MS, 65, PhD(chem eng), 68. *Prof Exp:* Mgr, Geosci Br, 80-82, CHEM ENGR FUELS RES, GEN ELEC CORP RES & DEVELOP CTR, 67- *Concurrent Pos:* Mgr planning & resources, Mat Labs, Gen Elec Res & Develop Ctr, 83-85. *Honors & Awards:* Industrial Res 100, 77. *Mem:* Am Inst Chem Engrs. *Res:* Conversion of fossil fuels to clean energy forms; research on a one ton per hour, twenty atmosphere coal gasifier. *Mailing Add:* 1470 Dean St Schenectady NY 12309

WOODMANSEE, ROBERT ASBURY, b Cincinnati, Ohio, Mar 6, 26; m 49; c 3. MARINE ECOLOGY. *Educ:* Univ Miami, BS, 48, MS, 49; Western Reserve Univ, PhD(zool), 52. *Prof Exp:* Asst zool, Western Reserve Univ, 49-52, technician biochem, 52; oceanogr, US Naval Hydrog Off, 52-54; from asst prof to prof biol, Miss Southern Col, 54-64; assoc marine scientist, Va Inst Marine Sci, 64-65; from assoc prof to prof biol, Univ SAla, 65-72; HEAD ECOL SECT, GULF COAST RES LAB, 72- *Mem:* Am Soc Limnol & Oceanog; Ecol Soc Am. *Res:* Ecology of plankton; seasonal distribution and daily vertical migrations of zooplankton; distribution of planktonic diatoms in estuary; primary productivity. *Mailing Add:* RR 2 No 70 Laurel MS 38440

WOODMANSEE, ROBERT GEORGE, b Albuquerque, NMex, Sept 11, 41; m 63; c 2. RANGE ECOLOGY, SYSTEMS ECOLOGY. *Educ:* Univ NMex, BS, 67, MS, 69; Colo State Univ, PhD(forest ecol & soils), 72. *Prof Exp:* Fel grassland ecol, 72-74, sr res ecologist, Natural Resource Ecol Lab, 74-78, asst dir grassland biome, 75-76, assoc prof, dept range sci, Colo State Univ, 78-82; prog dir, Ecosyst Studies, NSF, 82-84; PROF, DEPT RANGE SCI, COLO STATE UNIV, 82-, DIR, NATURAL RESOURCE LABS, 84- *Concurrent Pos:* Consult interactions biochem cycles, Steering Comt, Swedish Univ Agr; prin investr, NSF; mem, US Nat Comt-Sci Comt Prob Environ, Med Adv Bd, Exec Comt Asn Ecosyst Res Ctr, Nat Astrological Soc, Comt Planetary Biol & Chem Evol. *Mem:* Ecol Soc Am; Soc Range Mgt; AAAS; Am Inst Biol Sci; Sigma Xi. *Res:* Field experimentation and simulation modeling of nutrient cycling in grassland and agricultural ecosystems; long-term ecological research. *Mailing Add:* 703 E Co Rd 68 Ft Collins CO 80524

WOODRIFF, ROGER L, b Bozeman, Mont. MATHEMATICAL & MACHINE MODELS OF THE BRAIN. *Educ:* Mont State Univ, BS, 64; Univ Wis-Milwaukee, MS, 65. *Prof Exp:* Asst math, Mont State Univ, 63-64; instructing asst, Univ Wis-Milwaukee, 64-65; vis instr, Mid East Tech Univ, Ankara, 65-67; asst prof, Humboldt State Col, 67-70; PROF MATH, MENLO COL, 70-; PRES, MENLO RES ASSOC, 85- *Concurrent Pos:* Comput & educ consult, Menlo Res Assoc, 81-; ed & rev, Math Texts, Harper & Row. *Mem:* Am Math Soc; Math Asn Am; Am Asn Univ Profs; Am Col Math. *Res:* Mathematical models of cognition and their implementation as computer programs or computer hardware architecture, particularly cellular and hierarchial paradigms. *Mailing Add:* Menlo Res Assocs 1000 El Camino Atherton CA 94025

WOODRING, J PORTER, b Philipsburg, Pa, Sept 29, 32; m 55; c 2. ANIMAL PHYSIOLOGY, ENTOMOLOGY. *Educ:* Pa State Univ, BS, 54; Univ Minn, MS, 58, PhD(entom & zool), 60. *Prof Exp:* From instr to assoc prof, 60-70, PROF ZOOL & PHYSIOL, 71-, PROF ENTOM, LA State Univ, Baton Rouge, 81- *Concurrent Pos:* Humboldt scholar, 67-68 & 88. *Mem:* Am Soc Zool; Entomol Soc Am; AAAS. *Res:* Regulation of insect growth and development; hormonal regulation of metabolism. *Mailing Add:* Dept Zool & Physiol La State Univ Baton Rouge LA 70803-1725

WOODROW, DONALD L, b Washington, Pa, Nov 25, 35; m 60; c 2. STRATIGRAPHY, SEDIMENTOLOGY. *Educ:* Pa State Univ, BS, 57; Univ Rochester, MS, 50, PhD(geol), 65. *Prof Exp:* From asst prof to assoc prof, 65-75, PROF GEOL, HOBART & WILLIAM SMITH COLS, 75- *Concurrent Pos:* NSF res grant, 68-78; Res Corp res grant, 69-75; vis res geologist, Univ Reading, 71-72 & 79-80; consult, var indust orgns, 74-; vis res fel, Univ Rochester, 80-82; adj prof, State Univ NY, Binghamton. *Mem:* Geol Soc Am; Am Asn Petrol Geologists; Soc Econ Paleontologists & Mineralogists; Int Asn Sedimentol. *Res:* Sedimentology of Paralic and lake sediments; Upper Devonian stratigraphy and sedimentology of the northern hemisphere; paleomagnetism of lake sediments; paleomagnetism. *Mailing Add:* Dept Geosci Hobart & William Cols Geneva NY 14456

WOODRUFF, CALVIN WATTS, medicine, pediatrics; deceased, see previous edition for last biography

WOODRUFF, CHARLES MARSH, JR, b Columbia, Tenn, Aug 26, 44; m 73, 87. ENVIRONMENTAL GEOLOGY & HYDROLOGY. *Educ:* Vanderbilt Univ, BA, 66, MS, 68; Univ Tex, Austin, PhD(geol), 73. *Prof Exp:* Geologist, Tenn Div Geol, 69-70; res scientist geol, Bur Econ Geol, Univ Tex, Austin, 72-83; CONSULT GEOLOGIST & PRIN, CHARLES WOODRUFF & ASSOCS, 83- *Concurrent Pos:* Geol consult, Coastal Mgt Prog, Tex Gen Land Off, 74-76; lectr, Dept Geol Sci, Univ Tex, Austin, 76-83; lectr, Sch Archit, Dept Community & Regional Planning, Univ Tex, Austin, 87. *Mem:* Am Asn Petrol Geologists; AAAS; Sigma Xi; Geol Soc Am. *Res:* Assessment of land, water, mineral and energy resources; terrain analysis; assessment of environmental impacts. *Mailing Add:* 711 W 15th St Austin TX 78701-1707

WOODRUFF, CLARENCE MERRILL, b Kansas City, Mo, Apr 8, 10; m 37; c 2. AGRONOMY, SOIL SCIENCE. *Educ:* Univ Mo, BS, 32, MA, 39, PhD, 53. *Prof Exp:* Lab technician, Exp Sta, Univ Mo, 32-34; supt in chg soil conserv exp sta, USDA, 34-38; from instr to prof soils, 38-76, chmn dept, 67-70, EMER PROF AGRON, EXP STA, COL AGR, UNIV MO-COLUMBIA, 76- *Mem:* Fel Am Soc Agron; fel Soil Sci Soc Am; assoc Am Geophys Union; Sigma Xi; fel AAAS. *Mailing Add:* 135 Mumford Hall Univ Mo Columbia MO 65211

WOODRUFF, DAVID SCOTT, b Penrith, Eng, June 12, 43; Australian citizen; m 72; c 2. EVOLUTION AND CONSERVATION OF ANIMAL SPECIES. *Educ:* Univ Melbourne, BSc, 65, PhD(zool), 73. *Prof Exp:* Tutor biol, Trinity Col, Univ Melbourne, 66-69; Frank Knox fel biol, Harvard Univ, 69-71, Alexander Agassiz lectr biogeog, 72, res fel biol, Mus Comp Zool, 73-74; asst prof biol, Purdue Univ, 74-80; assoc prof, 80-86, PROF BIOL, UNIV CALIF, SAN DIEGO, 86- *Concurrent Pos:* Lectr ecol, Comn Exten Courses, Harvard Univ, 72-74; dir, Res Initiative & Support Prog Develop Ecol, Purdue Univ, 77-79; coordr, Proj in Conserv Sci, Univ Calif, San Diego, 87- *Mem:* Soc Study Evolution; fel AAAS; Soc Syst Zool; Ecol Soc Am; fel Linnean Soc London; Soc Conserv Biol. *Res:* Genetics, ecology and evolution of land snails; evolution of animal species including gibbons and chimpanzees; host-parasite coevolution; schistosomiasis; conservation biology with field projects in Thailand. *Mailing Add:* Dept Biol Univ Calif San Diego La Jolla CA 92093-0116

WOODRUFF, EDYTHE PARKER, b Bellwood, Ill, Jan 15, 28; m 50; c 2. TOPOLOGY. *Educ:* Univ Rochester, BA, 48, MS, 52; Rutgers Univ, New Brunswick, MS, 67; State Univ NY Binghamton, PhD(math), 71. *Prof Exp:* From asst prof to assoc prof math, Trenton State Col, 71-90; RETIRED. *Concurrent Pos:* Vis, Inst Advan Study, 79-80 & 81. *Mem:* Am Math Soc; Math Asn Am; Asn Women Mathematicians; Sigma Xi. *Res:* Topology of Euclidean 3-space; monotone decompositions, P-lifting; crumpled cubes; shrinkable decompositions. *Mailing Add:* 11 Fairview Ave East Brunswick NJ 08816

WOODRUFF, GENE L(OWRY), b Conway, Ark, May 6, 34; m 61; c 2. NUCLEAR ENGINEERING. *Educ:* US Naval Acad, BS, 56; Mass Inst Technol, SM, 63, PhD(nuclear eng), 65. *Prof Exp:* Reactor supvr, Mass Inst Technol, 63-65; from asst prof to assoc prof, Univ Wash, 65-76, from asst dir to dir, Nuclear Reactor Labs, 65-76, chmn dept, 81-84, PROF NUCLEAR ENG, UNIV WASH, 76-, DEAN, GRAD SCH & VICE PROVOST, 84- *Concurrent Pos:* Consult, Pac Northwest Labs, Battelle Mem Inst, 67-70, Math Sci Northwest, 70- & Los Alamos Nat Lab, 77- *Mem:* Am Nuclear Soc. *Res:* Nuclear reactor physics, especially neutron spectra; fusion reactor engineering. *Mailing Add:* Grad Sch Univ Wash Seattle WA 98195

WOODRUFF, HAROLD BOYD, b Bridgeton, NJ, July 22, 17; m 42; c 2. MICROBIOLOGY. *Educ:* Rutgers Univ, BS, 39, PhD(microbiol), 42. *Prof Exp:* Asst soil microbiol, Rutgers Univ, 38-42; res microbiologist, Merck Sharp & Dohme Res Labs, 42-46, head res sect, Microbiol Dept, 47-49, from asst dir to dir, 49-57, dir microbiol & natural prod res dept, 57-69, exec dir biol sci, Merck Inst Therapeut Res, 69-73, exec adminr, Merck Sharp & Dohme Res Labs, MSD (Japan) Co, LTD, 73-82; PRES, SOIL MICROBIOL ASSOCS, INC, 82 - *Concurrent Pos:* Lectr, US Off Educ; ed, Appl Microbiol, 53-62; mem bd dirs, Am Soc Microbiol Found, 72-74, (pres,74); mem bd trustees, Biol Abstracts, 72-77, (treas 74-77); mem sci adv comt, Charles F Kettering Res Lab, 72-75; mem exec bd, US Fedn Cult Collections, 73-76; mem bd trustees, Am Type Cult Collection, 81-87. *Honors & Awards:* Charles Thom Award, Soc Indust Microbiol, 73. *Mem:* Am Soc Microbiol (treas, 64-70); Soc Indust Microbiol (pres, 54-56); Am Chem Soc; Am Acad Microbiol; Brit Soc Gen Microbiol; Sigma Xi; AAAS; hon mem Am Soc Microbiol; hon mem Kitasato Inst, Japan. *Res:* Antibiotics; physiology of microorganisms; production of chemicals by microorganisms; analytical procedures using microorganisms; isolation of natural products. *Mailing Add:* 797 Valley Rd Watchung NJ 07060

WOODRUFF, HUGH BOYD, b Plainfield, NJ, Mar 3, 49; m 73; c 1. ANALYTICAL & COMPUTER CHEMISTRY. *Educ:* Trinity Col, BS, 71; Univ NC, PhD(analytical chem), 75. *Prof Exp:* Res assoc chem, Ariz State Univ, 75-77; sr res chemist analytical chem, Merck Sharp & Dohme Res Labs, 77-82, res fel, 82-85, mgr, 85-88, DIR, DEPT COMPUTER RESOURCES, MERCK SHARP & DOHME RES LABS, 88- *Concurrent Pos:* Chmn, Comput Div, Am Chem Soc, 89. *Mem:* Am Chem Soc; Coblentz Soc; Soc Appl Spectros; Drug Info Asn. *Res:* Computer applications in chemistry; pattern recognition; intelligent computer systems for infrared spectral interpretation; computer resource planning. *Mailing Add:* Merck Sharp & Dohme Res Labs PO Box 2000 R86-200 Rahway NJ 07065-0900

WOODRUFF, JAMES DONALD, b Baltimore, Md, June 20, 12; m 39; c 3. OBSTETRICS & GYNECOLOGY. *Educ:* Dickinson Col, BS, 33; Johns Hopkins Univ, MD, 37. *Prof Exp:* From instr to assoc prof gynec, 42-60, from assoc prof to prof gynec & obstet, 60-75, assoc prof path, 63-75, RICHARD W TELINDE PROF GYNEC & PATH, SCH MED, JOHNS HOPKINS UNIV, 75-, HEAD GYNEC PATH LAB, JOHNS HOPKINS HOSP, 51- *Concurrent Pos:* Chief gynecologist, Md Gen Hosp, 51-58 & Hosp for Women of Med, 58-62. *Mem:* Int Soc Study Vulvar Dis (pres, 73-75); Am Asn Obstet & Gynec (pres, 77); Am Gynec Soc; fel Am Col Obstet & Gynec. *Res:* Gynecologic pathology; study of functional activity of ovarian neoplasms and vulvar disease. *Mailing Add:* Johns Hopkins Hosp Baltimore MD 21205

WOODRUFF, JOHN H, JR, b Barre, Vt, Dec 14, 11; m 50; c 1. RADIOLOGY. *Educ:* Univ Vt, BS, 35, MD, 38. *Prof Exp:* Intern, US Marine Hosp, 38-39; resident radiol, Mary Fletcher Hosp, 39-40 & Royal Victoria Hosp, 40-41; asst prof, Univ Vt, 42-44; pvt pract, Calif, 46-50; clin instr, 50-52, asst prof, 52-54, assoc clin prof, 54-80, ADJ ASSOC PROF RADIOL, UNIV CALIF, LOS ANGELES, 80-; CHIEF RADIOL, US VET ADMIN HOSP, SEPULVEDA, 71- *Concurrent Pos:* Chief radiologist, Los Angeles, County Harbor Gen Hosp, 52-65; consult, Terminal Island Fed Prison, 57-65 & Long Beach Vet Admin Hosp, 59-68; radiologist, Univ Calif Med Ctr, 65-67; clin prof radiologic serv, Univ Calif, Irvine, 67-68, clin prof radiologic sci, 68-70; chief radiol, San Fernando Vet Admin Hosp, 68-71. *Mem:* Am Roentgen Ray Soc; Radiol Soc NAm; AMA; Am Col Radiol. *Res:* Radiologic aspects of diseases of kidneys, lungs, gastro-intestinal tract and of trauma to the abdomen and its contents. *Mailing Add:* 2618 Palos Verdes Dr W Palos Verdes Estates CA 90274

WOODRUFF, JUDITH J, b Boston, Mass, Feb 17, 39. IMMUNOLOGY. *Educ:* Cornell Univ, MD, 64. *Prof Exp:* PROF PATH, HEALTH SCI CTR, STATE UNIV NY, 80- *Mailing Add:* 319 Hampton Ave Rensselaer NY 12144

WOODRUFF, KENNETH LEE, b Phoenixville, Pa, Oct 10, 50; m 76; c 2. MINERAL ENGINEERING, METALLURGY. *Educ:* Pa State Univ, BS, 72. *Prof Exp:* Eng asst, Bethlehem Mines Corp, 70-71; staff consult & res engr, Nat Ctr Resource Recovery Inc, 72-75; proj mgr, Wehran Eng Corp, 75-76; pres, Resource Recovery Serv Inc, 76-81; mgr, Bio-Tech Energy Corp, 82-83; CONSULT, RESOURCE RECOVERY, KENNETH L WOODRUFF & ASSOC, 83- *Concurrent Pos:* Tech prog chmn, Ninth Nat Waste Processing Conf, Am Soc Mech Engrs, 78-80; chmn, Solid Waste Proc Div, Am Soc Mech Engrs, 85-86; assoc ed, Solid Waste Handbk, J Wiley & Sons, 86. *Mem:* Am Inst Mining, Metall & Petrol Engrs; Am Inst Chem Engrs; Am Inst Mech Engrs; Am Soc Testing & Mat; Nat Asn Environ Prof; Air & Waste Mgt Asn; Soc Plastics Engrs. *Res:* Energy and materials recovery from municipal and industrial solid waste; unit operations and processes; hazardous waste reclamation and recovery systems; trommel for pre-processing of solid waste for resource recovery. *Mailing Add:* 182 Walton Dr Morrisville PA 19067-0042

WOODRUFF, LAURENCE CLARK, insect physiology; deceased, see previous edition for last biography

WOODRUFF, RICHARD IRA, b Glen Ridge, NJ, Aug 19, 40; m 62; c 3. DEVELOPMENTAL BIOLOGY, REPRODUCTIVE BIOLOGY. *Educ:* Ursinus Col, BS, 62; West Chester State Col, MEd, 65; Univ Pa, PhD(biol), 72. *Prof Exp:* Teacher high sch, 62-66; from instr to assoc prof, 66-72, PROF BIOL, WEST CHESTER STATE COL, 72- *Concurrent Pos:* Res fel, Univ Pa, 72- *Mem:* AAAS; Am Soc Zoologists. *Res:* Developmental biology; electrophysiological events during egg formation. *Mailing Add:* Dept Biol West Chester Univ West Chester PA 19383

WOODRUFF, ROBERT EUGENE, b Kennard, Ohio, July 20, 33; m 54; c 2. ENTOMOLOGY. *Educ:* Ohio State Univ, BSc, 56; Univ Fla, PhD(entom), 67. *Prof Exp:* Entomologist, Ky State Health Dept, 56-58 & Plant Indust Div, Fla Dept Agr, 58-88; RETIRED. *Concurrent Pos:* Mem, Orgn Trop Studies, NSF, Costa Rica, 64; ed, Coleopterists Bulletin, 71-75; mem bd dirs, NAm Beetle Fauna Proj, 75-81; adj cur Natural Sci, Fla State Mus, 76-; NSF fossil amber insects, Dominican Republic, 77-78; consult, FAD/UN, 87-91, Ill Nat Hist Serv, 89-90, Tex A&M Univ, 89-90; assoc ed, Fla Entom Soc, 70-82. *Mem:* Entom Soc Am; Soc Syst Zool; Asn Trop Biol; Coleopterists Soc (pres, 78); Sigma Xi. *Res:* Systematic entomology; taxonomy, ethology, ecology of beetles of the family Scarabaeidae, especially myrmecophilous and termitophilous species; fossil amber insects of the Dominican Republic; mineralogy (research on Larimar from Dominican Republic); insects of Grenada (FAO). *Mailing Add:* 3517 NW Tenth Ave Gainesville FL 32605

WOODRUFF, RONNY CLIFFORD, b Greenville, Tex, Mar 12, 43; m; c 1. GENETICS. *Educ:* ETex State Univ, BS, 66, MS, 67; Utah State Univ, PhD(zool), 72. *Prof Exp:* NIH reproduction & develop training grant, Univ Tex, Austin, 71-73, asst prof zool, 73-74; sr res asst genetics, Univ Cambridge, 74-76; res assoc zool, Univ Okla, 76-77; from asst prof to assoc prof, 77-85, PROF BIOL, BOWLING GREEN STATE UNIV, 85- *Concurrent Pos:* Mem, Environ Protection Agency Gene-Tox Comt, 80-83; mem, Comt Standards Drosophilia Assays, Am Soc Testing & Mat, 84-85; Fulbright fel, Kenya, 87; vis res fel, Univ New Eng, Armidale, Australia, 91. *Honors & Awards:* Outstanding Young Scientist Award, Sigma Xi, 81. *Mem:* Genetics Soc Am; Environ Mutagen Soc; Sigma Xi; Soc Study Evolution. *Res:* Structure and function of transposable DNA elements in drossphilia melanogaster; mutagenesis. *Mailing Add:* Dept Biol Sci Bowling Green State Univ Bowling Green OH 43402

WOODRUFF, SUSAN BEATTY, b Wakefield, RI, Aug 18, 40; m 63; c 2. THEORETICAL CHEMISTRY. *Educ:* Oberlin Col, AB, 62; Johns Hopkins Univ, MAT, 63; Univ Calif, Irvine, PhD(chem), 77. *Prof Exp:* MEM STAFF CHEM & FEL, LOS ALAMOS SCI LAB, 77- *Mem:* Am Chem Soc. *Res:* Chemical dynamics; chemical kinetics; classical trajectory methodology; quantum chemistry; potential energy surfaces; surface chemistry; computer capabilities. *Mailing Add:* 120 Dos Brazos Los Alamos NM 87544-2431

WOODRUFF, TRUMAN OWEN, b Salt Lake City, Utah, May 26, 25; m 48. THEORETICAL SOLID STATE PHYSICS. *Educ:* Harvard Univ, AB, 47; Oxford Univ, BA, 50; Calif Inst Technol, PhD(physics), 55. *Prof Exp:* Res assoc physics, Univ Ill, 54-55; physicist, Res Lab, Gen Elec Co, 55-62; sr scientist, Hughes Aircraft Co Res Labs, 86-87; prof, 62-85, EMER PROF PHYSICS, MICH STATE UNIV, 85- *Concurrent Pos:* Vis prof, Univ Ariz, 67; Fulbright fel, Univ Pisa, 68-69; consult, solid state physics; consult physicist, 87- *Mem:* Fel Am Phys Soc. *Res:* Exploitation of quantum and statistical mechanics to explain static and time-dependent characteristics of solids and to a lesser extent liquids. *Mailing Add:* Suite 305 514 S Barrington Ave Los Angeles CA 90049

WOODRUFF, WILLIAM LEE, b Seward, Nebr, Oct 21, 38; m 63; c 3. REACTOR PHYSICS. *Educ:* Nebr Wesleyan Univ, BA, 60; Univ Nebr, MS, 64; Tex A&M Univ, PhD(nuclear eng), 70. *Prof Exp:* Lab asst & technician physics, Nebr Wesleyan Univ, 60-63, vis lectr, 63-64; instr, Univ Omaha, 64-66; NUCLEAR ENGR, ARGONNE NAT LAB, 68- *Concurrent Pos:* Inst Atomic Energy, Brazil, 75- *Mem:* Am Nuclear Soc; Sigma Xi. *Res:* Methods and computer code development and physics analysis for design and safety of liquid metal and gas-cooled fast breeder reactors and for reduced enrichment research reactors. *Mailing Add:* 2905 Hickory Ct Woodridge IL 60517

WOODS, ALAN CHURCHILL, JR, b Baltimore, Md, July 1, 18; m 44; c 4. SURGERY. *Educ:* Princeton Univ, AB, 40; Johns Hopkins Univ, MD, 43; Am Bd Surg, dipl, 51. *Prof Exp:* Intern & asst resident surgeon, Johns Hopkins Hosp, 44-45; resident surg, Henry Ford Hosp, 45-56; from asst resident surgeon to resident surgeon, Johns Hopkins Hosp, 48-49, surgeon & surgeon chg outpatient dept, 50; asst, 49, from instr to asst prof, 49-67, ASSOC PROF SURG, JOHNS HOPKINS UNIV, 67- *Concurrent Pos:* William Stewart Halsted fel surg, Johns Hopkins Univ, 49-50. *Mem:* Fel Am Col Surgeons; Soc Head & Neck Surg. *Res:* Abdominal, head and neck surgery. *Mailing Add:* 207 Wendover Rd Baltimore MD 21218

WOODS, ALEXANDER HAMILTON, b Tuxedo, NY, July 26, 22; m 56; c 2. IMMUNOLOGY. *Educ:* Harvard Univ, BS, 44; Johns Hopkins Univ, MD, 52; Am Bd Internal Med, dipl, 60. *Prof Exp:* Instr med, Sch Med, Duke Univ, 55-56; asst prof med & microbiol, Med Ctr, Univ Okla, 58-64; prof immunol, 64-77, ASSOC PROF INTERNAL MED, COL MED, UNIV ARIZ, 64- *Concurrent Pos:* Res fel biochem, Duke Univ, 56-58; clin investr, Vet Admin Hosp, Oklahoma City, 59-61; dir res, Vet Admin Hosp, Tucson, 64-70; assoc chief staff, Vet Admin Hosp, Tuscon, 70-77. *Mem:* AAAS; Am Chem Soc; NY Acad Sci; Brit Biochem Soc. *Res:* Immunochemistry; hematology; cancer chemotherapy. *Mailing Add:* 5805 Camino Escalante Tucson AZ 85718

WOODS, ALFRED DAVID BRAINE, b St John's, Nfld, July 16, 32; m 54; c 3. SOLID STATE PHYSICS, LOW TEMPERATURE PHYSICS. *Educ:* Dalhousie Univ, BSc, 53, MSc, 55; Univ Toronto, PhD(low temperature physics), 57. *Prof Exp:* Res fel low temperature physics, Univ Toronto, 57-58; res officer solid state physics, 58-78, head, Neutron & Solid State Physics Br, 71-78, sr adv to the exec vpres, Atomic Energy Can Ltd, Res Co, 79-86; SR

ADV STRATEGIC TECH MGT , ATOMIC ENERGY CAN LTD, RES CO, 86. *Mem:* Am Phys Soc; Can Asn Physicists; Royal Soc Can. *Res:* Dynamics of condensed matter using inelastic neutron scattering. *Mailing Add:* Atomic Energy Can Ltd Res Co 275 Slater St Ottawa ON K1A 1E5 Can

WOODS, ALVIN EDWIN, b Murfreesboro, Tenn, Mar 17, 34; m 59; c 2. FOOD CHEMISTRY, BIOCHEMISTRY. *Educ:* Mid Tenn State Univ, BS, 56; NC State Univ, MS, 58, PhD(food flavor), 62. *Prof Exp:* From asst prof to assoc prof, 61-69, PROF CHEM, MID TENN STATE UNIV, 69- *Concurrent Pos:* Consult, Off Int Res, NIH, 65-67. *Mem:* Am Chem Soc; fel Am Inst Chemists; Sigma Xi. *Res:* Enzyme stereochemistry and kinetics; food flavors, metal ions in biological systems; organo phosphorous compounds; bromine in biological systems. *Mailing Add:* Dept Chem Mid Tenn State Univ Murfreesboro TN 37132

WOODS, CHARLES ARTHUR, b Sherman, Tex, Dec 23, 40; m 63; c 3. PALEONTOLOGY. *Educ:* Univ Denver, BA, 64; Univ Mass, PhD(zool), 70. *Prof Exp:* Asst prof zool, Univ Denver, 70-71; assoc prof, Univ Vt, 71-79; PROF ZOOL, UNIV FLA, 79-; CURATOR MAMMAL, FLA MUS NATURAL HIST, 79-, CHMN, NATURAL SCI, 79- *Concurrent Pos:* Consult, govt Haiti, 83-, Mill Pond Press, 85-, Nat Acad Sci, 88-90. *Mem:* Am Soc Mammalogists; Am Soc Zool; Sigma Xi; Soc Study Evolution; Soc Vert Paleont; Soc Syst Zool. *Res:* Biogeography of new and old world mammals; island biology; biological conservation; birds and mammals of the West Indies; small mammal biology. *Mailing Add:* Fla Mus Natural Hist Gainesville FL 32611

WOODS, CHARLES WILLIAM, b Akron, Ohio, June 1, 28; m 61; c 2. ORGANIC CHEMISTRY. *Educ:* Ohio State Univ, BS, 51; Univ Md, PhD(org chem), 58. *Prof Exp:* Chemist, E I du Pont de Nemours & Co, Ky, 57-58; RES CHEMIST, ENTOM RES DIV, USDA, 59- *Mem:* AAAS; Am Chem Soc; Sigma Xi. *Res:* Synthesis of chemosterilants for insects; synthesis of radioactive organic compounds. *Mailing Add:* 4320 Underwood St University Park Hyattsville MD 20782

WOODS, CLIFTON, III, b Mecklenburg Co, NC, Aug 28, 44. PHYSICAL INORGANIC CHEMISTRY. *Educ:* NC Cent Univ, BS, 66; NC State Univ, MS, 69, PhD(chem), 71. *Prof Exp:* Asst prof chem, Univ Fla, 71-73, Bowling Green State Univ, 73-74; asst prof, 74-82, ASSOC PROF CHEM, UNIV TENN, 83- *Mem:* Am Chem Soc; Sigma Xi. *Res:* Synthesis, characterization, electrochemistry and catalysis of organorhodium complexes. *Mailing Add:* Dept Chem Univ Tenn Knoxville TN 37996-1600

WOODS, DONALD LESLIE, b London, Eng, Mar 4, 44; Can citizen; m 68; c 2. PLANT BREEDING. *Educ:* Univ London, BPharm, 65, MPhil, 68; Univ Man, PhD(plant sci), 71. *Prof Exp:* Res assoc plant breeding, Dept Plant Sci, Univ Man, 71-76; res specialist plant chem, Dept Agron, Univ Minn, 76; RES SCIENTIST PLANT BREEDING, RES STA, AGR CAN, 77- *Concurrent Pos:* Overseas aid prog, Kenya, 83 & 84. *Res:* Secondary plant metabolites and their biosynthesis, function, utilization, and environmental and genetic control. *Mailing Add:* Res Sta Agr Can Box 29 Beaverlodge AB T0H 0C0 Can

WOODS, DONALD ROBERT, b Sarnia, Ont, Apr 17, 35; m 61; c 3. COLLOID & SURFACE CHEMISTRY. *Educ:* Queen's Univ, Ont, BSc, 57; Univ Wis, MS, 58, PhD(chem eng), 61. *Prof Exp:* Instr chem eng, Univ Wis, 59-60, Athlone fel, 61-63; asst prof chem eng, 64-68, assoc prof, 68-74, chmn, 79-82, PROF CHEM ENG, MCMASTER UNIV, 74-, DIR, ENG & MGT PROG, 86- *Concurrent Pos:* C D Howe mem fel, van't Hoff Lab, State Univ Utrecht, 70-71. *Honors & Awards:* Ben Dasher Award. *Mem:* Am Inst Chem Engrs; Am Soc Eng Educ; Am Chem Soc; Am Asn Cost Engrs; Can Soc Chem Engrs; fel Chem Inst Can. *Res:* Surface behavior; separation of immiscible liquid and liquid-solid systems for waste water treatment and chemical processing; properties of emulsions; stability and separation of emulsions; physical separations; cost estimation; problem solving. *Mailing Add:* Dept Chem Eng McMaster Univ Hamilton ON L8S 4L7 Can

WOODS, EDWARD JAMES, b Timmins, ON, Nov 9, 36; m 58, 82; c 3. MATHEMATICS. *Educ:* Queen's Univ, BSc, 57; Princeton Univ, PhD(physics), 62. *Prof Exp:* Fel physics, Univ Alberta, 62-63; res assoc, Univ Md, 63-64, asst prof, 64-68; from asst prof to assoc prof, 68-76, PROF MATH, QUEEN'S UNIV, 76- *Concurrent Pos:* Vis prof, Heidelberg Univ, 76 & Kyoto Univ, 77. *Mem:* Can Math Soc. *Res:* Operator algebras; ergodic theory. *Mailing Add:* Dept Math Queen's Univ Kingston ON K7L 3N6 Can

WOODS, FRANK ROBERT, b Mt Vernon, NY, June 20, 16; m 42; c 3. HYDRODYNAMICS, GAS DYNAMICS. *Educ:* NY Univ, BA, 41, MS, 47, PhD(physics), 55. *Prof Exp:* Instr physics, Univ NH, 48-53, from asst prof to assoc prof, 53-57; physicist, Boeing Airplane Co, 57-58; assoc prof physics, Mont State Col, 58-63; lectr aerospace eng & eng physics, Univ Va, 63-73, sr scientist, 63-69, prin scientist, Dept Aerospace Eng & Eng Physics, 69-73; master, Hill Sch, Pottstown, Pa, 73-81; assoc prof physics, Lock Haven State Col, 81-82; RETIRED. *Concurrent Pos:* Adj prof, Millersville Univ Pa, 82-86. *Mem:* Am Phys Soc; Sigma Xi. *Res:* Hydrodynamics; scattering; theoretical hydrodynamics. *Mailing Add:* 301 N Rutland Ave PO Box 374 Brooklyn WI 53521

WOODS, FRANK WILSON, b Covington, Va, Apr 1, 24; m 72; c 8. FOREST ECOLOGY, ENVIRONMENTAL SCIENCES. *Educ:* NC State Col, BS, 49; Univ Tenn, MS, 51, PhD, 57. *Prof Exp:* Instr, Univ Tenn, 52; res forester, Southern Forest Exp Sta, US Forest Serv, 53-58; asst prof silvicult, Duke Univ, 58-64, assoc prof forest ecol, 64-69; PROF FORESTRY, UNIV TENN, KNOXVILLE, 69- *Concurrent Pos:* Vis res partic, Oak Ridge Inst Nuclear Studies, 64; vis res assoc, Univ Calif, Los Angeles, 65; NSF-Soc Am Foresters vis lectr, 66; consult, Orgn Econ Develop, Portugal, 69-, Oak Ridge Nat Lab, 71-, Venezuela, 78-79, Peru, 83, Costa Rica, 84, Ecuador, 62-63. *Mem:* Soc Am Foresters; Int Asn Trop Biologists; fel AAAS. *Res:* Dendroecology of air pollution; effects of environment on tree growth; agroforestry; strip mine revegetation. *Mailing Add:* Univ Tenn Dept Forestry PO Box 1071 Knoxville TN 37996

WOODS, GEORGE THEODORE, b Tyro, Kans, Aug 21, 24; m 48; c 3. VETERINARY MEDICINE, PUBLIC HEALTH. *Educ:* Kans State Univ, DVM, 46; Univ Calif, MPH, 59; Purdue Univ, MS, 60. *Prof Exp:* Inspector animal dis eradication, Ill State Dept Agr, 46-47; supvr lab animal med, Med Sch, Northwestern Univ, 47-48; vet, 48-49; asst prof vet path & hyg, 49-59, assoc prof vet microbiol, pub health & res, 59-66, prof, 66-86, EMER PROF VET MICROBIOL & PUB HEALTH, RES COL VET MED, UNIV ILL, URBANA, 86-; ADMINR DEWITT-PIATT BI-COUNTY HEALTH DEPT CI, 88- *Concurrent Pos:* Trainee, USPHS, 58-59. *Mem:* Am Vet Med Asn; Am Pub Health Asn; Asn Teachers Vet Pub Health & Prev Med (pres); Am Asn Food Hyg Veterinarians. *Res:* Preventive veterinary medicine; epidemiology; viral respiratory diseases of cattle and swine; bovine myxovirus para-influenza 3. *Mailing Add:* Dept Vet Pathobiol Col Vet Med 2001 S Lincoln Univ Ill Urbana IL 61801

WOODS, GERALDINE PITTMAN, b West Palm Beach, Fla; m 45; c 3. NEUROEMBRYOLOGY. *Educ:* Howard Univ, BS, 42; Radcliffe Col, MA, 43; Harvard Univ, PhD(neuroembryol), 45. *Hon Degrees:* DSc, Benedict Col, 77 & Talladega Col, 80; LHD, Meharry Med Col, 88 & Howard Univ, 89. *Prof Exp:* Instr biol, Howard Univ, 45-46; spec consult, Nat Inst Gen Med Sci, NIH, 69-87; EMER CHMN BD DIRS, HOWARD UNIV, 88- *Concurrent Pos:* Mem, Nat Adv Coun, Gen Med Sci Inst, NIH, 64-68; chmn, Defense Adv Comt Women in Serv, 68; mem, Gen Res Support Prog Adv Comt, Div Res Resources, NIH, 70-73 & 77-78; mem bd trustees, Calif Mus Found Calif Mus Sci & Indust, 71-; mem air pollution manpower develop adv comt, Environ Protection Agency, 73-75; mem bd dirs, Robert Wood Johnson Health Policy Fels, Inst Med-Nat Acad Sci, 73-78, Nat Comn Cert Physicians Assts, 74-78; mem bd trustees, Atlanta Univ, 74- & mem Calif Post Sec Educ Comn, 74-78; chmn bd trustees, Howard Univ, 75-88, mem bd trustees, 89; chmn bd dirs, Howard Univ Found, 84-88; mem bd dirs, Charles R Drew Univ Med & Sci, 91; hon chmn, Nat Inst Sci, 85. *Honors & Awards:* Kellogg Lectr, Univ Ark; Scroll of Merit, Nat Med Asn, 79. *Mem:* Inst Med-Nat Acad Sci; AAAS; Fedn Am Scientists; Nat Inst Sci; Am Asn Univ Professors; NY Acad Sci. *Res:* Encouraging the participation of minorities in the regular programs at the National Institutes of Health and developing two programs, Minority Biomedical Research Support Program and Minority Access to Research Careers, that would further assist colleges with minorities move into research and research training to project more participation in health and scientific careers. *Mailing Add:* 12065 Rose Marie Lane Los Angeles CA 90049

WOODS, J(OHN) M(ELVILLE), b Denver, Colo, May 14, 22; m 49; c 3. CHEMICAL ENGINEERING. *Educ:* Univ Kans, BS, 43; Univ Wis, PhD(chem eng), 53. *Prof Exp:* Instr chem eng, Univ Wis, 48-50; asst prof, Univ RI, 50-52; asst prof chem eng, Purdue Univ, 52-57, assoc prof, 57-77; MEM STAFF, DEPT CHEM ENG, PRAIRIE VIEW A&M UNIV. *Concurrent Pos:* Consult chem eng, 77- *Mem:* Sigma Xi; Am Chem Soc; Am Inst Chem Engrs. *Res:* Applied reaction kinetics; process simulation and optimization. *Mailing Add:* PO Box 397 Prairie View TX 77446-0897

WOODS, JAMES E, histochemistry, for more information see previous edition

WOODS, JAMES STERRETT, b Lewistown, Pa, Feb 26, 40; m 69; c 1. TOXICOLOGY, ENVIRONMENTAL EPIDEMIOLOGY. *Educ:* Princeton Univ, AB, 62; Univ Wash, PhD(pharmacol), 70; Univ NC, MPH, 78; Am Bd Toxicol, dipl. *Prof Exp:* Res assoc pharmacol, Sch Med, Yale Univ, 70-72; head biochem toxicol sect, Lab Environ Toxicol, Nat Inst Environ Health Sci, NIH, 72-78; HEAD, EPIDEMIOL & ENVIRON HEALTH RES PROG, HEALTH & POP STUDY CTR, BATTELLE HUMAN AFFAIRS RES CTRS, 78-; RES PROF, DEPT ENVIRON HEALTH, UNIV WASH, 79- *Concurrent Pos:* Founding pres, Pac Northwest Asn Toxicol, 84-85; trustee, Seattle Biomed Res Inst, 89-; counr, Soc Toxicol, 90-; chmn sci coun & mem bd dir, Pac Sci Ctr, 81-87, adv comt herbicides, Vet Admin, 87- *Mem:* Am Asn Cancer Res; Am Soc Pharmacol & Exp Therapeut; Soc Epidemiol Res; NY Acad Sci; Am Pub Health Asn. *Res:* Biochemical toxicology; environmental epidemiology; biomarkers of toxicant exposures. *Mailing Add:* Battelle Human Affairs Res Ctrs 4000 NE 41st St Seattle WA 98105

WOODS, JAMES WATSON, b Lewisburg, Tenn, Feb 20, 18; m 44; c 3. MEDICINE. *Educ:* Univ Tenn, BA, 39; Vanderbilt Univ, MD, 43; Am Bd Internal Med, dipl. *Prof Exp:* Instr, Sch Med, Univ Pa, 47-48; from asst prof to prof med, Sch Med, Univ Nc, Chapel Hill, 64-87; RETIRED. *Concurrent Pos:* NIH spec fel, Dept Med, Vanderbilt Univ, 67-68; mem med adv bd, Coun High Blood Pressure Res, Am Heart Asn. *Mem:* Int Soc Hypertension; Am Fedn Clin Res; Am Clin & Climat Asn; fel Am Col Physicians. *Res:* Cardiology; pathogenesis of hypertension. *Mailing Add:* Dept Med Univ NC Sch Med Chapel Hill NC 27514

WOODS, JIMMIE DALE, b Albuquerque, NMex, Oct 8, 33; m 56; c 3. MATHEMATICS, STATISTICS. *Educ:* US Coast Guard Acad, BS, 55; Trinity Col, Conn, MS, 63; Univ Conn, PhD(math statist), 68. *Prof Exp:* US Coast Guard, 51-81, from instr to assoc prof math, US Coast Guard Acad, 60-68, asst head dept, 64-68, prof math & head dept, 68-81; PROF MGT, HARTFORD GRAD CTR, 81-, DEAN SCH MGT, 90- *Mem:* Am Statist Asn; Am Math Soc; Math Asn Am; Am Soc Qual Control. *Res:* Application of matrix techniques to statistical distribution theory; mathematical modeling in management science. *Mailing Add:* RD 2 Box 136B Wheeler Rd Stonington CT 06378

WOODS, JOE DARST, b Knoxville, Iowa, Jan 2, 23; m 45; c 2. INORGANIC CHEMISTRY. *Educ:* Cent Col, Iowa, BS, 44; Iowa State Univ, MS, 50, PhD(chem), 54. *Prof Exp:* From instr to assoc prof, 52-64, chmn dept, 69-78 & 82-85, prof, 64-85, EMER PROF CHEM, DRAKE UNIV, 85- *Concurrent Pos:* Fulbright-Hays lectr, St Louis Univ, Philippines, 67-68. *Mem:* AAAS; Am Chem Soc. *Res:* Mechanism of decomposition of chlorates; radioactive tracers; oxidation of metals; dissolution of metals in acids. *Mailing Add:* 4107 Ardmore Rd Des Moines IA 50310

WOODS, JOHN WHITCOMB, b Winchester, Mass, Dec 11, 52; m 75; c 1. MEMBRANE TRAFFIC THROUGH GOLGI APPARATUS. *Educ:* Univ Md, PhD(biol), 82. *Prof Exp:* NIH res fel, 83-86, LEUKEMIA SOC SPEC FEL CELL BIOL, SCH MED, YALE UNIV, 86- *Honors & Awards:* Argall L & Anna G Hull Cancer Res Award, 86. *Mem:* Am Soc Cell Biol. *Res:* Defining the pathways traversed and the mechanisms by which the cell regulates the intracellular movements of the transferrin receptor and other membrane bound receptors. *Mailing Add:* Dept Cell Biol Yale Univ Sch Med 333 Cedar St PO Box 3333 New Haven CT 06510

WOODS, JOHN WILLIAM, b Wash, DC, Dec 5, 43; m 72; c 2. IMAGE PROCESSING, IMAGE CODING. *Educ:* Mass Inst Technol, BS, 65, MS, 67, PhD (commun), 70. *Prof Exp:* Sr engr, Electronics Res Dept, Lawrence Livermore Lab, 73-76; prof elec & comput eng, Rensselaer Polytech Inst, 76-88; prog dir, Circuits Signal Processing, NSF, Wash, 87-88; CONSULT, 88- *Concurrent Pos:* Admin Comt, Inst Elec & Electronics Engrs, ASSP Soc, 86-89; vis prof, Delft Univ Technol, Netherlands, 85-86. *Mem:* Inst Elec & Electronics Engrs. *Res:* Image processing including image restoration and image coding; parallel architectures to realize algorithms developed in the imaging area. *Mailing Add:* 43 Longview Dr Clifton Park NY 12065-0213

WOODS, JOSEPH, b Rochester, NY, Nov 13, 56. ELECTRICAL ENGINEERING, MATERIALS SCIENCE. *Educ:* Univ Rochester, BS, 79; Univ Tex, Austin, PhD(physics), 86. *Prof Exp:* Postdoctoral assoc mat sci, Mass Inst Technol, 86-90; ASST PROF PHYSICS, UNIV NEBR, 90- *Mem:* Am Phys Soc; Am Vacuum Soc. *Res:* Condensed matter experimental physics; surface magnetism of ferromagnetic materials measured with polarized light scattering and spin resolved electron spectroscopies; amorphous and crystalline bulk alloys and thin film and multilayered structures. *Mailing Add:* 260 Behlen Lab Univ Nebr Lincoln NE 68588-0111

WOODS, JOSEPH JAMES, b Camden, NJ, June 21, 43; m 68; c 2. PHYSIOLOGY. *Educ:* St Jospeh's Col, Pa, BS, 65; Rutgers Univ, PhD(physiol), 71. *Prof Exp:* From asst prof to assoc prof, 70-78, PROF BIOL, QUINNIPIAC COL, 78-, CHMN DEPT, 81- *Concurrent Pos:* NIH co-investr, 71-81. *Mem:* Sigma Xi. *Res:* Application of the principles of muscle biophysics to muscular exercise; muscle activity and oxygen consumption during varying rates of positive and negative work; muscle fatigue mechanisms using electromyogram indices. *Mailing Add:* Dept Sci Quinnipiac Col Mt Carmel Ave Hamden CT 06518

WOODS, KEITH NEWELL, b Wilmington, Del, Jan 28, 41; m 68; c 2. MATERIALS SCIENCE, CERAMICS. *Educ:* Stanford Univ, BS, 62; Univ Mich, MSE, 63; Northwestern Univ, PhD(mat sci), 68. *Prof Exp:* Adv res engr, Appl Res Lab, Sylvania Elec Prods, Inc, 68-69; staff ceramist, Res & Develop Ctr, Gen Elec Co, 69-72, tech mgr, Nuclear Fuels Dept, 72-77; RES MGR, EXXON NUCLEAR CO, INC, 77- *Mem:* Am Ceramic Soc; Am Nuclear Soc. *Res:* Physical properties of crystalline ceramics and glasses; fabrication and irradiation behavior of nuclear fuels. *Mailing Add:* 2020 Harris Rd Richland WA 99352

WOODS, KENNETH R, b Chicago, Ill, Apr 14, 25; m 46; c 2. ENERGY, ENVIRONMENT. *Educ:* Northwestern Univ, Evanston, BSME, 48. *Prof Exp:* Res engr, Ingersoll Utility Unit Div, Borg Warner Corp, 48-49; test engr, Underwriter's Labs, Inc, 49-52; responsible charge, Indust Hyg & Radiation Safety Div, Argonne Nat Lab, 52-55; monitoring partner, Thulin, Woods & Isensee, 55-60; pres, Woods & Assocs, Inc, 60-86; PRES, ENERGY-ENVIRON, INC, 86- *Concurrent Pos:* Prin investr, Am Soc Mech Engrs, Am Inst Architects, Am Soc Heating Refrig & Air Conditioning Eng, Am Soc Bldg & Construct Inspectors & Am Acad Environ Med, 82-87. *Mem:* Am Inst Architects; Am Cong Surv & Mapping; Nat Soc Prof Engrs; Sigma Xi; Int Solar Energy Soc; AAAS. *Res:* Technologies that do more with less in the energy field and promulgate living in harmony with nature in the environmental field. *Mailing Add:* 1264 Harvest Court Naperville IL 60564-8956

WOODS, LAUREN ALBERT, b Aurora Co, SDak, Spet 10, 19; m 44; c 3. PHARMACOLOGY. *Educ:* Dakota Wesleyan Univ, BA, 39; Iowa State Col, PhD(org chem), 43; Univ Mich, MD, 49. *Prof Exp:* Asst, Nat Defense Res Comt, Iowa State Col, 43-44; from instr to prof pharmacol, Univ Mich, 46-60, actg chmn dept, 56; prof & head dept, Col Med, Univ Iowa, 60-70; prof, 81-84, vpres health sci, 70-76 & 81-84, assoc provost, Acad Med Col Va, 77-78, actg vpres health sci, 78-81, EMER PROF PHARMACOL, VA COMMONWEALTH UNIV, 84- *Mem:* AAAS; Am Chem Soc; Am Soc Pharmacol & Exp Therapeut; NY Acad Sci. *Res:* Metabolism of drugs; chemical structure; biological activity relationships; compounds affecting the central nervous system; mechanisms of development of tolerance and physical dependence to narcotics; radioactive tracer studies; histochemical distribution of drugs. *Mailing Add:* Box 613 Va State Med Col Richmond VA 23298-0001

WOODS, MARIBELLE, b Albany, Ga, Aug 16, 19. MICROBIOLOGY. *Educ:* Univ Chattanooga, BS, 42; Yale Univ, MS, 48. *Prof Exp:* Bioassayer, Chattanooga Med Co, 42-48, pharmacologist, 48-66; pharmacologist, Chattem Drug & Chem Co, 66-78, MICROBIOLOGIST, CHATTEM, INC, 78- *Concurrent Pos:* Pharmacologist, Brayten Pharmaceut Co, 48-73. *Mem:* NY Acad Sci. *Res:* Laxative action of senna; premenstrual tension; uterine physiology; antacids. *Mailing Add:* 311 Guild Dr Chattanooga TN 37421-3920

WOODS, MARY, b Webster Groves, Mo, Dec 22, 23. INORGANIC CHEMISTRY. *Educ:* Rosary Col, AB, 45; Univ Ill, MA, 47; Univ Wis, PhD(chem), 61. *Prof Exp:* Teacher, Trinity High Sch, Ill, 49-53; from instr to assoc prof chem, 53-73, PROF CHEM, ROSARY COL, 73- *Concurrent Pos:* Consult, Argonne Nat Lab, 75- *Mem:* Am Chem Soc; Sigma Xi. *Res:* Complex ion equilibria and kinetics; kinetics of redox reactions of actinide ions in solution. *Mailing Add:* Dept Natural Sci Rosary Col River Forest IL 60305-1099

WOODS, PHILIP SARGENT, b Concord, NH, Nov 25, 21; div; c 2. CELL BIOLOGY, MOLECULAR BIOLOGY. *Educ:* Mich State Univ, BS, 47; Univ Wis, PhD(cytol), 52. *Prof Exp:* Head cellular res sect, Dept Radiobiol, US Army Med Res Lab, Ky, 52-53; USPHS fel, Columbia Univ, 53-55; assoc cytochemist biol dept, Brookhaven Nat Lab, 55-61; assoc prof biol sci, Univ Del, 61-67; prof biol, Queen's Col, NY, 67-85; GUEST CYTOCHEMIST, BROOKHAVEN NAT LAB, 86- *Mem:* Am Soc Cell Biol; Sigma Xi; Electron Micros Soc Am. *Res:* Biosynthesis of macromolecules; nucleic acid and protein metabolism within cells; autoradiography with tritium-labeled precursors; fine structure of cells and chromosomes; electron microscopy. *Mailing Add:* Dept Biol Brookhaven Nat Lab Upton NY 11973

WOODS, RALPH ARTHUR, b Norwich, Eng, Mar 26, 41; m 66; c 2. METALLURGY, WELDING ENGINEERING. *Educ:* Birmingham Univ, BSc, 63, PhD(metall), 66. *Prof Exp:* Mat engr, Sikorsky Helicopters Div, United Technologies Corp, 66-67; WELDING ENGR, KAISER ALUMINUM & CHEM CORP, 67- *Honors & Awards:* C H Jennings Mem Award, Am Welding Soc, 72. *Mem:* Am Soc Metals. *Res:* Aluminum welding research; aluminum alloy metallurgy; aluminum vacuum brazing. *Mailing Add:* 2292 Camino Brazos Pleasanton CA 94566

WOODS, RAYMOND D, b Evangeline, La, Sept 14, 10. GEOLOGY. *Educ:* Univ Tex, BS, 31, MS, 34. *Prof Exp:* Res geologist, Exxon Corp, asst chief geologist, sr petrol geologist, 32-72; RETIRED. *Concurrent Pos:* Consult, Exxon, 72. *Mem:* Geol Soc Am; Am Asn Petrol Geologists; Soc Econ Paleontologists & Mineralogists. *Mailing Add:* 4100 Jackson Ave Apt 484 Austin TX 78731

WOODS, RICHARD DAVID, b Lansing, Mich, Sept 4, 35; m 57; c 3. CIVIL ENGINEERING. *Educ:* Univ Notre Dame, BSCE, 57, MSCE, 62; Univ Mich, Ann Arbor, PhD(civil eng), 67. *Prof Exp:* Proj engr, Air Force Weapons Lab, NMex, 62-63; instr civil eng, Mich Technol Univ, 63-64; from asst prof to assoc prof, 67-77, PROF CIVIL ENG, UNIV MICH, ANN ARBOR, 77- *Concurrent Pos:* Consult, 67-; Fed Ger Res Coun vis prof, Inst Soil & Rock Mech, Univ Karlsruhe, 71-72; found eng consult, Consumers Power Co, Mich, 72- *Honors & Awards:* Collingwood Prize, Am Soc Civil Engrs, 68. *Mem:* Am Soc Civil Engrs; Am Soc Eng Educ; Am Soc Testing & Mat. *Res:* Vibrations of soils; foundations and structures; seismology; earthquake engineering. *Mailing Add:* 700 Mt Pleasant Ann Arbor MI 48103

WOODS, ROBERT CLAUDE, b Atlanta, Ga, Mar 24, 40; m 63; c 2. CHEMICAL PHYSICS, INTERSTELLAR CHEMISTRY. *Educ:* Ga Inst Technol, BS, 61; Harvard Univ, AM, 62, PhD(phys chem), 65. *Prof Exp:* Instr chem, US Naval Acad, 65-67; from asst prof to assoc prof, 67-77, PROF CHEM, UNIV WIS-MADISON, 77- *Concurrent Pos:* Mem, Comt Atomic & Molecular Physics, Nat Res Coun, 84-; vis prof, Justus-Liebig Univ, Giessen, WGer, 85. *Honors & Awards:* Nobel Laureate Signature Award, Am Chem Soc, 84. *Mem:* Am Chem Soc; Am Phys Soc; Sigma Xi; fel Am Inst Chemists; AAAS. *Res:* Microwave spectroscopy of molecular ions and of transient species present in electrical discharges and related theoretical problems; radioastronomy and the chemistry of the interstellar medium. *Mailing Add:* Dept Chem Univ Wis 1101 University Ave Madison WI 53706

WOODS, ROBERT JAMES, b London, Eng, Feb 8, 28; m 58; c 4. RADIATION CHEMISTRY, SCIENCE COMMUNICATIONS. *Educ:* Univ London, BSc, 49, PhD(org chem), 51; Imp Col London, dipl, 51. *Prof Exp:* Nat Res Coun Can fel, Prairie Regional Lab, Sask, 51-53; Univ NZ res fel, Victoria, NA, 53-54; res assoc, Univ Sask, 55-62; sr res fel, Royal Mil Col Sci, Eng, 62-63; from asst prof to assoc prof, 63-76, head dept chem, 84-88 PROF CHEM, UNIV SASK, 76- *Concurrent Pos:* Vis prof, Saclay Nuclear Res Ctr, France, 72-73. *Mem:* Chem Inst Can; fel Royal Soc Chem; Can Nuclear Soc. *Res:* Radiation chemistry of organic compounds, both pure and in aqueous solution. *Mailing Add:* Dept Chem Univ Sask Saskatoon SK S7N 0W0 Can

WOODS, ROBERT OCTAVIUS, b Evanston, Ill, Feb 17, 33; m 65; c 2. VACUUM SCIENCE, MASS SPECTROMETRY. *Educ:* Princeton Univ, BSE, 62, MSE, 64, MA, 65, PhD(aerospace & mech sci), 67. *Prof Exp:* MEM TECH STAFF, SANDIA NAT LABS, 67- *Concurrent Pos:* Cong fel, Am Soc Mech Engrs, 91-92. *Mem:* Am Phys Soc; Sigma Xi; Am Soc Mech Engrs; fel Brit Interplanetary Soc. *Res:* Design of instrument systems for research from flight vehicles: small rockets, aircraft, stratospheric balloons and satellites. *Mailing Add:* 7513 Harwood Ave NE Albuquerque NM 87110

WOODS, ROGER DAVID, b Los Angeles, Calif, Mar 28, 24; m 52; c 3. THEORETICAL PHYSICS, COMPUTER SCIENCE. *Educ:* Univ Redlands, AB, 45; Univ Calif, Los Angeles, MA, 49, PhD, 54. *Prof Exp:* Res physicist, Univ Calif, Los Angeles, 54; asst prof physics, Univ Miami, 54-61; assoc prof, Univ Redlands, 61-65; assoc prof, 65-72, PROF PHYSICS, SAN BERNARDINO VALLEY COL, 72- *Mem:* AAAS; Am Phys Soc. *Res:* Nucleon-nuclei scattering; molecular structure; electron structure of atoms; energy bands in solids; generalized theory of gravitation; visible, ultraviolet and infrared spectroscopy; scientific computer application. *Mailing Add:* Dept Physics San Bernardino Valley Col San Bernardino CA 92410

WOODS, ROY ALEXANDER, b Columbia, Mo, Oct 31, 13; m 34; c 1. PHYSICS, ELECTRONICS. *Educ:* Lincoln Univ, Mo, AB, 34; Boston Univ, AM, 46 & 48, EdD, 60. *Prof Exp:* Teacher high schs, Mo, 34-41; instr electronics & radar technician, US Navy, 42-45; prof physics & chmn div natural sci, Va State Col, 48-68; prof, 68-80, EMER PROF PHYSICS, NORFOLK STATE COL, 80- *Concurrent Pos:* Electronic engr, Lab Electronics Res & Develop, Mass, 52-53, 59-60. *Mem:* Am Asn Physics Teachers. *Res:* Instrumentation for naval radar. *Mailing Add:* 6028 Wesleyan Dr Norfolk VA 23502

WOODS, SHERWYN MARTIN, b Des Moines, Iowa, June 25, 32; m 71. PSYCHIATRY, PSYCHOANALYSIS. *Educ:* Univ Wis, BS, 54, MD, 57, PhD(psychoanal), 77; Am Bd Psychiat & Neurol, dipl, 65, cert psychoanal, 77. *Prof Exp:* Intern, Philadelphia Gen Hosp, 57-58; resident, Univ Hosps, Univ Wis Med Sch, 58-61, instr psychiat, Med Sch, 61; dir grad educ, 63-82, from asst prof to assoc prof, 63-74, PROF PSYCHIAT, SCH MED, UNIV SOUTHERN CALIF, 74-, DIR STUDENT PSYCHIAT SERV, 66-, DIR PSYCHOANALYSIS EDUC, 82- *Concurrent Pos:* NIMH career teacher award, 64-66; clin assoc, Southern Calif Psychoanalysis Inst, 64-68, mem, 68-, instr, 69-, supv & training analyst, 78-; examr, Am Bd Psychiat & Neurol, 67-; consult, Calif Dept Ment Hyg, 70-; pres, Am Asn Dir Psychiat Residency Training, 74-76; chmn, Residency Rev Comt Psychiat, 87-89, Am Psychiat Asn Comt Grad Educ, 85- *Mem:* Fel Am Psychiat Asn; Asn Advan Psychother; fel Am Col Psychiat. *Res:* Medical education; human sexuality; psychotherapy. *Mailing Add:* Univ Southern Calif 1934 Hosp Place Suite 1A21-E Los Angeles CA 90033

WOODS, STEPHEN CHARLES, b Pasadena, Calif, Feb 17, 42; m 88; c 2. PHYSIOLOGICAL PSYCHOLOGY. *Educ:* Univ Wash, BS, 65, BS, 66, PhD(physiol & psychol), 70. *Prof Exp:* Asst to assoc prof psychol, Columbia Univ, 70-72; from asst prof to assoc prof, 72-77, chmn, Dept Psychol, 81-88, PROF PSYCHOL, UNIV WASH, 77- *Concurrent Pos:* Mem psychobiol Study Sect, NSF, 83-86; councilor, NAm Asn Study Obesity, 84-86; mem biopsychol Study Sect, NIH, 86-; pres, Int Congress Physiol Food & Fluid Intake, 86; pres, Soc Study Ingestive Behav, 88- *Mem:* Am Psychol Asn; Am Diabetes Asn; Soc Neurosci; Soc Study Ingestive Behav(pres 88-89). *Res:* Neural and hormonal control appetite and the regulation of body weight; influence of learning on metabolic processes; numerous articles published in various journals. *Mailing Add:* Dept Psychol Univ Wash Seattle WA 98195

WOODS, STUART B, b Pathlow, Sask, Apr 26, 24; m 49; c 2. SOLID STATE PHYSICS. *Educ:* Univ Sask, BA, 44, MA, 48; Univ BC, PhD, 52. *Prof Exp:* Res officer physics, Nat Res Coun Can, 52-59; from assoc prof to prof physics, 59-89, assoc dean Fac Grad Studies & Res, 74-83, EMER PROF PHYSICS, UNIV ALTA, 89- *Concurrent Pos:* Vis prof, Univ Bristol, 66-67 & Univ St Andrews, Scotland, 73-74. *Mem:* Am Phys Soc; Can Asn Physicists. *Res:* Low temperature and solid state physics, chiefly experimental transport properties in solids and high Tc superconductivity. *Mailing Add:* Dept Physics Univ Alta Edmonton AB T6G 2J1 Can

WOODS, THOMAS STEPHEN, b Florence, Ala, Dec 13, 44; m 66; c 2. ORGANIC CHEMISTRY. *Educ:* Auburn Univ, BS, 67; Univ Ill, PhD(org chem), 71. *Prof Exp:* Res chemist, Div Med Chem, Walter Reed Army Inst Res, 72-74; res chemist, Biochem Dept, Du Pont Co, 74-79, res supvr, synthesis, 79-80, lab adminr, 80-82, res supvr, formulations, 82-84, licensing mgr, Agr Prod Dept, 85-87, mgr res & develop Europ, Mideast & Africa, Agr Prod Dept, du Pont de Nemours, Paris, 87-90, RES MGR, FORMULATIONS, DU PONT AGR PRODS, 90- *Mem:* Am Chem Soc. *Res:* Synthesis of novel organic compounds of biological utility; heterocyclic chemistry; organosulfur and selenium chemistry; theories of tautomerism and resonance; organic photochemistry; agricultural product formulations. *Mailing Add:* Du Pont Agr Prod Exp Sta Bldg 402 Wilmington DE 19880-0402

WOODS, W(ALLACE) KELLY, b Claremore, Okla, Dec 10, 12; m 37; c 4. NUCLEAR ENGINEERING. *Educ:* Stanford Univ, AB, 34; Mass Inst Technol, MS, 36, DSc(chem eng), 40. *Prof Exp:* Instr chem eng, Mass Inst Technol, 36-40; chem engr, Tech Div, Eng Dept, E I du Pont de Nemours & Co, 40-43, tech specialist, Explosives Dept, 43-46; tech & managerial assignments with plutonium prod reactors, Hanford Labs, Gen Elec Co, 46-55, mgr nuclear mat, Vallecitos Atomic Lab, 55-61 & prog, Hanford Labs, 61-63, consult engr, 63-65; consult engr, Douglas United Nuclear, Inc, Wash, 65-71; coordr, Ore Nuclear & Thermal Energy Coun, Ore State Govt, 71-75, energy site coordr, 75-78; prof nuclear eng, Ore State Univ, 78-88; RETIRED. *Mem:* Am Chem Soc; fel Am Nuclear Soc (treas, 59-61); Am Inst Chem Engrs. *Res:* Heat transfer to boiling liquids; fluid flow; irradiation effects in solids. *Mailing Add:* 714 Tillman Ave SE Salem OR 97302

WOODS, WALTER RALPH, b Grant, Va, Dec 2, 31; m 53; c 2. ANIMAL NUTRITION. *Educ:* Murray State Univ, BS, 54; Univ Ky, MS, 55; Okla State Univ, PhD, 57. *Prof Exp:* Instr, Okla State Univ, 56-57; from asst prof to assoc prof animal husb, Iowa State Univ, 57-62; from assoc prof to prof, Univ Nebr-Lincoln, 62-71; head dept animal sci, Purdue Univ, West Lafayette, 71-85; DEAN AGR & DIR AGR EXP STA, 85-, DIR, COOP EXTEN SERV, KAN STATE UNIV, 87- *Mem:* Am Soc Animal Sci; Am Dairy Sci Asn; Sigma Xi. *Res:* Protein and nonprotein nitrogen utilization in beef cattle and sheep; energy utilization as influenced by diet composition and processing. *Mailing Add:* Dean Agr & Dir Agr Exp Sta & Coop Exten Serv Kans State Univ Manhattan KS 66506

WOODS, WENDELL DAVID, b Liberal, Kans, Dec 5, 32; m 64; c 2. OPHTHALMOLOGY, ECOSANOIDS. *Educ:* Univ Mo, BS, 58, MS, 59, PhD(biochem), 65. *Prof Exp:* Res asst, 63-66, from instr to asst prof, 66-73, ASSOC PROF OPHTHAL, SCH MED, EMORY UNIV, 73- *Concurrent Pos:* NIH grant vision, Emory Univ, 72- *Mem:* AAAS; Asn Res Ophthal; Am Chem Soc; Sigma Xi. *Res:* Ecosanoid metabolism in the cornea and in the outflow area of the eye; oxidative effects in glaucoma. *Mailing Add:* Lab Ophthal Res Emory Univ Atlanta GA 30322-0856

WOODS, WILLIAM A, KNOWLEDGE REPRESENTATION, NATURAL LANGUAGE PROCESSING. *Educ:* Ohio Wesleyan Univ, BA, 64; Harvard Univ, AM, 65, PhD(appl math), 67. *Prof Exp:* Asst prof comput sci, Harvard Univ, 67-70; sr scientist, Bolt Beranek & Newman Inc, 70-76, prin scientist, 76-83; chief scientist, Appl Expert Systs Inc, 83-88; PRIN TECHNOLOGIST, ON TECHNOL INC, 88- *Concurrent Pos:* Lectr, 70-75, prof pract comput sci, Harvard Univ, 85-; lectr, Mass Inst Technol, 74-75; Fulbright scholar, Univ Sussex, 78. *Mem:* Asn Comput Ling (pres, 73-74); Am Asn Artificial Intel; Asn Comput Mach; Am Soc Info Sci. *Res:* Knowledge representation; machine reasoning; natural language processing; human/computer communication; complex evolutionary systems; development of ATN grammars, the lunar question-answering system and contributions to speech understanding and semantic network theory. *Mailing Add:* On Technol Inc One Cambridge Ctr Cambridge MA 02142

WOODS, WILLIAM FRED, b Ala. AGRICULTURAL ECONOMICS. *Educ:* Auburn Univ, BS, 60, MS, 61. *Prof Exp:* Asst dep admin, 80-85, PUB POLICY SPECIALIST, EXTEN SERV, USDA, 85- *Concurrent Pos:* Mem, Nat Pub Policy Educ Comt. *Mem:* Am Agr Econ Asn. *Res:* Agricultural policy; public policy education; agricultural extension education. *Mailing Add:* 902 Lauren Lane SE Vienna VA 22180

WOODS, WILLIAM GEORGE, b Superior, Wis, Dec 21, 31; m 52; c 3. PHYSICAL ORGANIC CHEMISTRY. *Educ:* Univ Calif, Los Angeles, BS, 53; Calif Inst Technol, PhD, 57. *Prof Exp:* Asst, Calif Inst Technol, 53-55; res chemist, Res Lab, Gen Elec Co, 56-58; sr res chemist, 58-74, mgr agr res & develop, 74-76, SR SCIENTIST, US BORAX RES CORP, 76- *Concurrent Pos:* Res officer, Commonwealth Sci & Indust Res Orgn, Melbourne, Australia, 62-63. *Mem:* Am Inst Mining, Metall & Petrol Engrs; Am Chem Soc; Sigma Xi. *Res:* Carbonium ion and pyrolysis mechanism; semi-inorganic polymer systems; semi-empirical molecular orbital calculations; nuclear magnetic resonance spectroscopy; organometallic and organic synthesis; herbicide synthesis and metabolism; direct agricultural research programs; fire retardants and polymer additives; corrosion electro-chemistry; precious metal extractive metallurgy; boron chemistry. *Mailing Add:* 1011 N Lemon Fullerton CA 92632-6794

WOODS, WILNA ANN, b Apr 20, 29. CANCER RESEARCH. *Educ:* Stanford Univ, BA, 51, PhD, 55. *Prof Exp:* Res assoc, Case Western Reserve Univ, Cleveland, 54-68; CHIEF, CONTRACT REV BR, NAT CANCER INST, NIH, 89- *Res:* Cancer; microbiology; chemistry; anatomy; numerous technical publications. *Mailing Add:* NIH Nat Cancer Inst Contracts Rev Br Westwood Bldg Rm 803 Bethesda MD 20892

WOODSIDE, DONALD G, orthodontics, for more information see previous edition

WOODSIDE, JOHN MOFFATT, b Toronto, Ont, Jan 23, 41; m 67; c 2. MARINE GEOPHYSICS, GEOLOGY. *Educ:* Queen's Univ, BSc, 64; Mass Inst Technol, MSc, 68; Cambridge Univ, PhD(geophys), 76. *Prof Exp:* Tech officer geophys, Geol Surv Can, 64; res asst marine geophys, Woods Hole Oceanog Lab, 66-68; sci officer, Bedford Inst Oceanog, 68-71; res asst, Dept Geod & Geophys, Cambridge Univ, 71-76; res scientist marine geophys, Atlantic Geosci Ctr, Geol Surv Can, 76-89; ASSOC PROF, FREE UNIV AMSTERDAM, 89- *Concurrent Pos:* Int Bus Mach Corp fel, Dept Geod & Geophys, Cambridge Univ, 75-76; mem bd dirs, Int Gravity Bur, 78-; sr geophysicist, Proj for Regional Offshore Prospecting in East Asia & Pac, UN, 79-80. *Mem:* Fel Geol Asn Can; fel Royal Astron Soc; Am Geophys Union; Soc Explor Geophysicists; Am Asn Petrol Geologists; Can Geophys Union. *Res:* Structural and tectonic aspects of evolution of ocean basins with particular emphasis on neotectonics of collision zones, nature of passive continental margins, and subduction zones; areas of primary interest are Mediterranean Sea, Banda Sea and the Southwestern Pacific; potential fields and seismic methods. *Mailing Add:* De Boelelaan 1085 Amsterdam 1081 HV Netherlands

WOODSIDE, KENNETH HALL, b Northampton, Mass, June 18, 38; m 60; c 2. PHYSIOLOGY, BIOCHEMISTRY. *Educ:* Oberlin Col, AB, 59; Univ Rochester, PhD(biochem), 69. *Prof Exp:* Res assoc physiol, Col Med, Pa State Univ, 68-70, asst prof physiol & head multidiscipline labs, 70-76; asst prof med, Sch Med, Univ Miami, 76-79; res biochemist, Mt Sinai Med Ctr, Miami, 79-81; PROF BIOCHEM & CHMN DEPT, SOUTHEASTERN COL OSTEOP MED, NORTH MIAMI BEACH, 81- *Mem:* Am Physiol Soc; AAAS; Sigma Xi. *Res:* Hormonal and non-hormonal regulation of protein biosynthesis and degradation; pulmonary macrophage function. *Mailing Add:* Southeastern Col Osteopathic Med 1750 NE 168 St North Miami Beach FL 33162

WOODSIDE, WILLIAM, b Ft William, Ont, July 5, 31; m 53; c 3. PHYSICS, MATHEMATICS. *Educ:* Queen's Univ, Belfast, BSc, 51, MSc, 59, DSc, 62. *Prof Exp:* Asst Royal Naval Sci Serv, Baldock, Eng, 51 & Toronto, 53; master physics, Ridley Col, Can, 52-55; res officer, Nat Res Coun, Can, 54-58; res physicist, Gulf Res & Develop Co, 58-60; master math, Ridley Col, 60-66; asst prof, 66-70, ASSOC PROF MATH, QUEEN'S UNIV, ONT, 70-, CHMN ENG MATH, 80- *Mem:* Can Math Soc; Math Asn Am. *Res:* Optimization theory; operations research; mathematics; education; heat transfer in porous media. *Mailing Add:* Dept Math & Statist Queen's Univ Kingston ON K7L 3N6 Can

WOODSON, HERBERT H(ORACE), b Stamford, Tex, Apr 5, 25; m 51; c 3. ELECTRICAL ENGINEERING. *Educ:* Mass Inst Technol, SB & SM, 52, ScD, 56. *Prof Exp:* Elec engr, US Naval Ord Lab, Md, 52-54; asst elec eng, Mass Inst Technol, 54-55, from instr to prof, 55-71; Alcoa Found prof, Univ Tex, Austin, 72-75, chmn dept, 71-81, prof elec eng, 71-83, Tex Atomic Energy Found prof, 80-83, DIR, CTR ENERGY STUDIES, UNIV TEX, AUSTIN, 77-, ERNEST H COCKRELL CENTENNIAL CHAIR ENG, 87- *Concurrent Pos:* Actg dean eng, Univ Tex, Austin, 87- *Honors & Awards:* Nikola Tesla Award, Inst Elec & Electronics Engrs, 84. *Mem:* Nat Acad Eng; Am Soc Eng Educ; fel Inst Elec & Electronics Engrs. *Res:* Electrical energy conversion and control; power system engineering. *Mailing Add:* Ernest Cockrell Jr Hall 10310 Univ Tex Austin TX 78712

WOODSON, JOHN HODGES, b Hartford, Conn, May 25, 33; m 60; c 3. PHYSICAL CHEMISTRY, THEORETICAL CHEMISTRY. *Educ:* Wesleyan Univ, BA, 55; Northwestern Univ, PhD(phys chem), 59. *Prof Exp:* Asst prof chem, Wesleyan Univ, 59-61; from asst prof to assoc prof, 61-70, PROF CHEM, SAN DIEGO STATE UNIV, 70- *Mem:* Am Chem Soc; Sigma Xi. *Res:* Valence theory; kinetics of oscillating reactions; ion-selective electrodes. *Mailing Add:* Dept Chem San Diego State Univ San Diego CA 92182-0328

WOODSON, PAUL BERNARD, neurobiology, for more information see previous edition

WOODSON, ROBERT D, m 69; c 3. HEMATOLOGY, PHYSIOLOGY. *Educ:* Univ Chicago, MD, 63. *Prof Exp:* PROF MED, UNIV WIS, 82- *Concurrent Pos:* Vis prof, Univ Zurich, Switz, 84-85. *Mem:* Am Physiol Soc; Am Fedn Clin Res; Am Soc Hemat; Int Soc Hemat. *Res:* Control of oxygen delivery; role of oxygen dissociation wave in oxygen transport. *Mailing Add:* Dept Med Univ Wis H4/540 CSC Madison WI 53792

WOODSTOCK, LOWELL WILLARD, b Harvey, Ill, July 6, 31; m 61; c 2. SEED PHYSIOLOGY, SEED VIGOR & DETERIORATION. *Educ:* Univ Ill, BS, 54; Univ Wis, PhD(bot), 59. *Prof Exp:* Nat Cancer Inst fel, Royal Bot Garden, Scotland, 59-61 & Univ Wis, 61-62; res assoc, Argonne Nat Lab, 62-63; res plant physiologist, USDA, 63-67, leader seed qual invests, 67-73, res plant physiologist, Agr Res Serv, 73-76; assoc tech dir, Agr & Water Res Ctr, Riyadh, Saudi Arabia, 76-78; RES PLANT PHYSIOLOGIST, AGR RES SERV, USDA, 78- *Concurrent Pos:* Am Seed Res Found grant, 69-72; intergovt employee exchange act trainee, Ore State Univ, 73-74; ad hoc prof, hort dept, Univ Md, 82-; consult, Food & Agr Orgn, China, 85. *Mem:* Crop Sci Soc Am; Am Soc Hort Sci; Am Soc Plant Physiol; Asn Off Seed Analysis; Japanese Soc Plant Physiologist. *Res:* Molecular basis of seed germination, vigor, deterioration and storage; biochemical and physiological basis of seed response to environmental stress; freeze-drying seeds for longer storage. *Mailing Add:* 10503 Royal Rd Silver Spring MD 20903

WOODWARD, ARTHUR EUGENE, b Los Angeles, Calif, Oct 16, 25; m 52; c 1. PHYSICAL CHEMISTRY OF POLYMERS, POLYMER PHYSICS. *Educ:* Occidental Col, BA, 49, MA, 50; Polytech Inst Brooklyn, PhD(phys chem), 53. *Prof Exp:* Asst Occidental Col, 50; US Govt grantee chem, Cath Univ Louvain, 53-54; res fel, Harvard Univ, 54-55; asst prof, Pa State Univ, 55-59, from asst prof to assoc prof physics, 59-64; assoc prof chem, 64-66, PROF CHEM, CITY COL NEW YORK, 67- *Concurrent Pos:* Guggenheim fel, Queen Mary Col, London, 62-63. *Mem:* Am Chem Soc; Am Phys Soc. *Res:* Physical and chemical properties of polymer crystals; nuclear magnetic resonance of high polymers; polymer morphology and infrared spectroscopy. *Mailing Add:* Dept Chem City Col New York New York NY 10031

WOODWARD, CLARE K, b Houston, Tex, Dec 10, 41; m 67. BIOCHEMISTRY. *Educ:* Smith Col, BA, 63; Rice Univ, PhD(biol), 67. *Prof Exp:* fel phys chem, 68-70, asst prof lab med, Med Sch, Univ Minn, Minneapolis, 70-72; from asst prof to assoc prof biochem, 72-81, PROF BIOCHEM & BIOL SCI, UNIV MINN, ST PAUL, 81- *Concurrent Pos:* Fel genetics, Univ Minn, St Paul, 67-68. *Mem:* Biophys Soc; Am Soc Biol Chemists. *Res:* Protein chemistry; protein dynamics, hydrogen exchange, NMR; protein folding. *Mailing Add:* Dept Biochem Univ Minn St Paul MN 55108

WOODWARD, DAVID WILLCOX, b Oxford, NY, July 24, 13; m 38; c 1. ORGANIC CHEMISTRY. *Educ:* Amherst Col, AB, 34; Harvard Univ, PhD(org chem), 37. *Prof Exp:* Res chemist, Chem Dept, E I du Pont de Nemours & Co, Inc, 37-51, res supvr, Photo Prod Dept, 51-54, res mgr, 45-68, lab dir, Photo Prod Dept, 68-76; RETIRED. *Mem:* Am Chem Soc; Soc Photog Sci & Eng. *Res:* Photography; furane and polymer chemistry; cyanides; nitriles; photochemistry; photopolymerization; dyes. *Mailing Add:* 103 Taylor Lane Kennett Square PA 19348

WOODWARD, DONALD JAY, b Detroit, Mich, Aug 1, 40; m 63; c 3. PHYSIOLOGY. *Educ:* Univ Mich, BS, 62, PhD(physiol), 66. *Prof Exp:* From instr to assoc prof physiol, Dept Physiol & Ctr Brain Res, Sch Med & Dent, Univ Rochester, 66-75; PROF CELL BIOL & PHYSIOL, UNIV TEX HEALTH SCI CTR, DALLAS, 75- *Mem:* Am Physiol Soc; Neurosci Soc. *Res:* Developmental neurobiology; cerebellar neurophysiology and neuropharmacology. *Mailing Add:* Dept Cell Biol 5323 Harry Hines Blvd Dallas TX 75235

WOODWARD, DOW OWEN, b Logan, Utah, Dec 1, 31; m 56; c 4. MOLECULAR BIOLOGY. *Educ:* Utah State Univ, BS, 56; Yale Univ, MS, 57, PhD(bot), 59. *Prof Exp:* Mem res staff radiobiol, Aerospace Med Ctr, USAF, 59-62; assoc prof, 62-74, PROF BIOL, STANFORD UNIV, 74- *Mem:* Genetics Soc Am; Am Soc Biol Chemists. *Res:* Biochemical genetics in microorganisms; membrane structure and function; cytoplasmic inheritance; biological rhythms; enzymology. *Mailing Add:* Dept Biol Sci Stanford Univ Stanford CA 94305

WOODWARD, EDWARD ROY, b Chicago, Ill, Sept 6, 16; m 39; c 2. SURGERY. *Educ:* Grinnell Col, BA, 38; Univ Chicago, MD, 42; Am Bd Surg, dipl, 51. *Prof Exp:* Asst resident, Univ Clins, Univ Chicago, 47-49, instr & resident, Grad Sch Med, 49-50 & 51-52, instr, 52-53; from asst prof to assoc prof, Med Ctr, Univ Calif, Los Angeles, 53-57; head dept, 57-82, PROF SURG, COL MED, UNIV FLA, 57- *Concurrent Pos:* Douglas Smith fel surg, Univ Chicago, 46-47; Markle scholar, 52-57. *Mem:* Am Physiol Soc; Soc Univ Surg; Am Surg Asn; Am Gastroenterol Asn; fel Am Col Surgeons. *Res:* Physiology of gastrointestinal tract. *Mailing Add:* Dept Surg Box J-286 Univ Fla Med Col Gainesville FL 32610-0286

WOODWARD, ERVIN CHAPMAN, JR, b Long Beach, Calif, Apr 8, 23; m 49; c 2. RADIATION PHYSICS. *Educ:* Univ Calif, PhD(physics), 52. *Prof Exp:* PHYSICIST, LAWRENCE LIVERMORE LAB, UNIV CALIF, 52- *Mem:* Am Phys Soc. *Res:* Spectroscopy; hyperfine structure; nuclear spin and isotope shift; high speed optics. *Mailing Add:* 3876 Stanford Way Livermore CA 94550

WOODWARD, FRED ERSKINE, b Boston, Mass, Aug 20, 21; m 44; c 4. SURFACE CHEMISTRY, ORGANIC POLYMER CHEMISTRY. *Educ:* Dartmouth Col, AB, 43; Univ Ill, MA, 44, PhD(org chem), 46. *Prof Exp:* Chemist, Gen Aniline & Film Corp, Pa, 46-48, sr chemist, Dyestuff & Chem Div, 48-54, mgr indust sect, Antara Tech Dept, 54-58, asst prog mgr appln res, 58-62; dir res & develop, Nopco Chem Co, NJ, 62-69; PRES, SURFACE CHEMISTS FLA, INC, 69- *Mem:* Am Chem Soc; Am Oil Chem Soc; Soc Lubrication Eng; Am Inst Mining & Metall Engrs; Am Asn Textile Chemists & Colorists. *Res:* Surface active agents; lubrication; defoaming; detergency; synthesis and applications research; surface chemistry of antimicrobial agents; mechanisms of clay dewatering; iodophors. *Mailing Add:* Surface Chemists 328 W 11th St Riviera Beach FL 33404-4399

WOODWARD, J GUY, b Carleton, Mich, Nov 19, 14; m 45; c 3. PHYSICS. *Educ:* NCent Col, BA, 36; Mich State Col, MS, 38; Ohio State Univ, PhD(physics), 42. *Prof Exp:* Asst physics, Mich State Col, 36-39; asst physics, Ohio State Univ, 39-42; res physicist, Mfg Co, RCA Corp, 42, res engr, 42-72, res fel, RCA Labs, 72-83; RETIRED. *Honors & Awards:* Emile Berliner Award, Audio Eng Soc, 68. *Mem:* Fel AAAS; Acoust Soc Am; Hon Mem Audio Eng Soc (pres, 71-72); fel Inst Elec & Electronics Engrs; Sigma Xi. *Res:* Physical optics; room and music acoustics; radio interference from motor vehicles; underwater sound; electroacoustic transducers; high fidelity phonograph systems; viscometry; video digital and audio magnetic tape recording. *Mailing Add:* 208 Laurel Circle Princeton NJ 08540

WOODWARD, JAMES FRANKLIN, b Boston, Mass, Dec 22, 41. EXPERIMENTAL GRAVITATION, PULSAR ASTROPHYSICS. *Educ:* Middlebury Col, AB, 64; NY Univ, MS, 69; Univ Denver, PhD(hist sci), 72. *Prof Exp:* PROF HIST SCI, CALIF STATE UNIV, 72- *Concurrent Pos:* Adj prof physics, Calif State Univ, 80- *Mem:* Int Soc Gen Relativity & Gravitation; NY Acad Sci; Hist Sci Soc; Astron Soc Pac. *Res:* Experimental investigation of relativistic gravitation; Mach's principle; interrelation of gravity with the other forces of nature. *Mailing Add:* Dept Hist & Physics Calif State Univ Fullerton CA 92634

WOODWARD, JAMES KENNETH, b Anderson, Mo, Feb 5, 38; m 60; c 2. PHARMACOLOGY. *Educ:* Southwest Mo State Col, BS & AB, 60; Univ Pa, PhD(pharmacol), 67. *Prof Exp:* Pharmacologist, Stine Lab, E I du Pont de Nemours & Co, Inc, Del, 63-67, res pharmacologist, 67-71, head cardiovasc-autonomic dis res, MERRELL DOW PHARMACEUT INC, 71-72, head cardiovasc-respiratory dis res, 72-74, pharmacol dept, Merrell-Nat Labs, 74-78, preclin pharmacol dept, Merrell Res Ctr, 78-81 & Merrell Dow Res Ctr, 81-83, assoc dir res admin, 83-88, dir biol develop, Merrell Dow Res Inst, Merrell Dow Pharmaceut Inc, 88-91, DIR REGULATORY AFFAIRS, MARION MERRELL DOW INC, 91- *Concurrent Pos:* Lectr, Sch Med, Univ Pa, 67-71. *Mem:* NY Acad Sci; AAAS. *Res:* Cardiovascular-autonomic pharmacology, especially etiology, pathology and treatment of hypertension; pharmacology of autacoids, especially non-sedating antihistamines; pharmacokinetic/pharmacodynamic relationships. *Mailing Add:* Int Regulatory Affairs Marion Merrell Dow Inc 2110 E Galbraith Rd Cincinnati OH 45215

WOODWARD, JOE WILLIAM, b Teral, Okla, Oct 31, 37; m 58; c 3. CHEMICAL ENGINEERING. *Educ:* Tex A&M Col, BS, 60; Calif Inst Technol, MS, 61, PhD(chem eng), 65. *Prof Exp:* Sr res engr, US Army assigned to Jet Propulsion Lab, NASA, 65-66; CONSULT, E I DU PONT DE NEMOURS & CO, INC, 66- *Concurrent Pos:* Fel, Calif Inst Technol, 65-66. *Mem:* Am Inst Chem Engrs; Nat Soc Prof Engrs. *Res:* Polymer processing; rheology; mathematical models; computer process control. *Mailing Add:* 2726 Elizabeth Port Neches TX 77651

WOODWARD, KENT THOMAS, b Cleveland, Ohio, Dec 11, 23; m 49; c 3. RADIOTHERAPY. *Educ:* Clemson Univ, BS, 47; Univ SC, MD, 47; Univ Rochester, PhD(radiation biol), 66; Univ Pa, MS, 73. *Prof Exp:* Intern, Boston City Hosp, Boston, Mass, 47-48; staff mem, med res, Georgetown Univ Hosp, DC, 48-49; staff mem chest dis, Fitzsimons Army Hosp, Denver, Colo, 49, 50-51; intern med, Walter Reed Army Hosp, DC, 51-52; staff mem biomed res, Los Alamos Sci Lab, 52-56; chief biophys dept, Walter Reed Army Inst Res, 56-60, dir div nuclear med, 62-68; prog dir radiation, Nat Cancer Inst, 68-69; fel radiation ther, Univ Tex, M D Anderson Hosp & Tumor Inst Houston, 69-72; assoc prof radiol, Univ Pa, 73-77; assoc prof radiation ther, Duke Univ Med Ctr, 78-; MYRTLE BEACH RADIOTHER CTR, SC. *Mem:* Sigma Xi; Am Med Asn; Am Soc Therapeut Radiol & Oncol; Am Col Radiol; fel Am Col Physicians. *Res:* Biological effects ionizing radiation on normal tissues and tumors. *Mailing Add:* 704 Dogwood Ave Myrtle Beach SC 29577

WOODWARD, LEE ALBERT, b Omaha, Nebr, Apr 22, 31; m 52; c 4. STRUCTURAL GEOLOGY. *Educ:* Univ Mont, BA, 58, MS, 59; Univ Wash, PhD(geol), 62. *Prof Exp:* Geologist, US Bur Reclamation, 58 & Pan Am Petrol Corp, 62-63; instr geol, Olympic Col, 63-65; from asst prof to assoc prof, 65-73, chmn dept, 70-76, PROF GEOL, UNIV NMEX, 73- *Concurrent Pos:* NSF fel, 58-61; Monsanto fel, 61-62; assoc ed, Geol Soc Am Bull, 77-82; NATO fel, 73. *Mem:* Geol Soc Am; Am Asn Petrol Geologists; Soc Mining Engrs; Soc Econ Geologists. *Res:* Regional tectonics of western United States; mineral exploration. *Mailing Add:* Dept Geol Univ NMex Albuquerque NM 87106

WOODWARD, LEROY ALBERT, b Hartford, Conn, Nov 22, 16; m 42; c 2. PHYSICS. *Educ:* Ga Inst Technol, BS, 43; Univ Mich, MS, 47. *Prof Exp:* Mem sci staff, US Navy Underwater Sound Lab, Columbia Univ, 44; contract physicist, David Taylor Model Basin, US Dept Navy, DC, 45; from instr to asst prof physics, Ga Inst Technol, 47-51, assoc prof physics & res physicist, Eng Exp Sta, 51-55; res physicist, Scripto, Inc, 55-58, dir res, 58-60; res asst prof physics, Eng Exp Sta, 60-63, from asst prof to assoc prof, 63-82, EMER ASSOC PROF PHYSICS, GA INST TECHNOL, 82- *Mem:* Sigma Xi. *Res:* Optics and optical microscopy. *Mailing Add:* 834 Oakdale Rd NE Atlanta GA 30307

WOODWARD, STEPHEN COTTER, b Atlanta, Ga, July 19, 35; m 57; c 2. PATHOLOGY. *Educ:* Emory Univ, MD, 59. *Prof Exp:* Pathologist, Georgetown Univ Hosp, 64-68; asst prof, 64-68, ASSOC PROF PATH, SCH MED, GEORGETOWN UNIV, 68-; PATHOLOGIST, HUNTER LAB, SIBLEY HOSP, 68- *Concurrent Pos:* Consult, Children's Hosp, 64-, Vet

Admin Hosp, 65- & comt skeletal syst, Div Med Sci, Nat Res Coun, 68-; attend pathologist, DC Gen Hosp, 67- *Mem:* Col Am Path; Am Soc Clin Path; Am Soc Exp Path; AMA. *Res:* Fibroplasia; collagen elaboration; effects of endocrine and vulnerary agents upon wound repair; quality control methods for clinical laboratories. *Mailing Add:* Dept Path Vanderbilt Univ Nashville TN 37232

WOODWARD, TED K, b Peoria, Ill, Nov 11, 60; m 88. SEMICONDUCTOR DEVICE PHYSICS, OPTOELECTRONIC DEVICES. *Educ:* Univ Tex, Austin, BS, 83; Calif Inst Technol, MS, 85, PhD(appl physics), 88. *Prof Exp:* MEM TECH STAFF, AT&T BELL LABS, 88- *Mem:* Am Phys Soc; Inst Elec & Electronics Engrs; Optical Soc Am; Sigma Xi. *Res:* Semiconductors; optical, electro-optical and electrical investigations of electronics and optoelectronic materials and devices; quantum electronics and bandgap engineering; photonic switching device research; basic physics of semiconductor materials and devices; fabrication. *Mailing Add:* AT&T Bell Labs Rm 4B-525 Crawfords Corner Rd Holmdel NJ 07733-1988

WOODWARD, THEODORE ENGLAR, b Westminster, Md, Mar 22, 14; m 38; c 3. MICROBIOLOGY, MEDICINE. *Educ:* Franklin & Marshall Col, BS, 34; Univ Md, MD, 38; Am Bd Internal Med, dipl. *Hon Degrees:* DSc, Western Med Col, 50 & Franklin & Marshall Col, 54. *Prof Exp:* Asst prof med, Sch Med & Univ Hosp, Univ Md, 46-48, assoc prof med & dir sect infectious dis, Sch Med, 48-54, prof med & head dept, Sch Med, 54-81, distinguished physician, Vet Admin Hosp, 81-86; RETIRED. *Concurrent Pos:* Instr, Sch Med, Johns Hopkins Univ, 46-48, lectr, 48-54; attend physician, Vet Admin Hosp, 46-48, consult, 48-; consult, State Health Dept, Md, 50; mem comt Int Ctrs Med Res & Training, USPHS, 61-62; mem US Adv Comt, US-Japan Coop Med Sci Prog, 65- *Mem:* Am Soc Clin Invest; AMA; Am Clin & Climat Asn; Am Asn Physicians; master Am Col Physicians. *Res:* Infectious and rickettsial diseases; enteric diseases including typhoid fever; internal medicine. *Mailing Add:* Dept Med Rm N3W40 Univ Md Hosp Baltimore MD 21201

WOODWARD, VAL WADDOUPS, b Preston, Idaho, July 26, 27; m 47, 67; c 3. GENETICS. *Educ:* Utah State Univ, BS, 50; Kans State Univ, MS, 50; Cornell Univ, PhD(genetics), 53. *Prof Exp:* NIH fel & guest assoc biologist, Brookhaven Nat Lab, 53-55; assoc prof genetics, Kans State Univ, 55-58; prof biol & chmn dept, Univ Wichita, 58-61; assoc prof, Rice Univ, 61-67; PROF GENETICS, UNIV MINN, ST PAUL, 67- *Concurrent Pos:* Fel, Birmingham Univ, 62-63, Open Univ, 86-87. *Mem:* AAAS; Am Soc Cell Biol; Genetics Soc Am; Sigma Xi. *Res:* Gene-enzyme transport; cell wall; neurospora; self-assembly of membrane and other organelle proteins; behavior genetics. *Mailing Add:* Genetics Dept Univ Minn 742 Biosci St Paul MN 55108

WOODWELL, GEORGE MASTERS, b Cambridge, Mass, Oct 23, 28; m 55; c 4. ECOLOGY, BOTANY. *Educ:* Dartmouth Col, AB, 50; Duke Univ, AM, 56, PhD(bot), 58. *Hon Degrees:* DSc, Williams Col, 77; Miami Univ, 84; Carleton Col, 88; Muhlenberg Col, 90. *Prof Exp:* From asst prof to assoc prof bot, Univ Maine, 57-61; sr ecologist, Brookhaven Nat Lab, 61-75; founder & dir, Ecosysts Ctr, Marine Biol Lab, 75-85; FOUNDER & DIR, WOODS HOLE RES CTR, 85- *Concurrent Pos:* Assoc, Conserv Found, 58-61, mem bd trustees, 75-77; lectr, Sch Forestry, Yale Univ, 67-; founding mem bd trustees, Environ Defense Fund, 67-68 & 73-; founding mem, bd trustees, Natural Resources Res Coun, 70-, vchmn 74-, World Wildlife Fund, 70, chmn bd, 80-84; chmn, Suffolk County NY Coun Environ Qual, 72; trustee, Conserv Found, 75-77; bd trustees, Sea Educ Asn, 80-84; founding mem bd, World Resources Inst, 82-, bd trustees, Ruth Mott Fund, 84-, chmn, 89-; chmn, Conf Long-term biol consequences nuclear war, 82-83; mem, Conn River Watershed Coun, 84. *Honors & Awards:* NY Bot Garden Sci Award, 75; Distinguished Serv Award, Am Inst Biol Sci, 82. *Mem:* Nat Acad Sci; Ecol Soc Am (vpres, 67, pres, 77-78); Am Inst Biol Sci; Brit Ecol Soc; fel Am Acad Arts & Sci; Sigma Xi; fel AAAS. *Res:* Structure, function and development of terrestrial and aquatic ecosystems; biotic impoverishment, especially effects of ionizing radiation and other toxins such as pesticides; biotic contributions to the global carbon cycle and climatic change; biotic contributions to the global carbon cycle. *Mailing Add:* Woods Hole Res Ctr Box 296 Woods Hole MA 02543

WOODWICK, KEITH HARRIS, b Tappen, NDak, Jan 4, 27; m 51; c 3. INVERTEBRATE ZOOLOGY. *Educ:* Jamestown Col, BS, 49; Univ Wash, MS, 51; Univ Southern Calif, PhD(zool), 55. *Prof Exp:* Asst zool, Univ Wash, 49-51; instr, Univ Southern Calif, 54, asst, 54-55; from instr to assoc prof zool, 55-66, chmn dept, 65-69, PROF ZOOL, CALIF STATE UNIV, FRESNO, 66-, COORDR MARINE SCI, 68- *Mem:* Am Inst Biol Sci; Soc Syst Zool; Sigma Xi. *Res:* Systematics and larval development of polychaetes; Enteropneusta. *Mailing Add:* Dept Biol Calif State Univ Fresno CA 93710

WOODWORTH, CURTIS WILMER, b Reading, Pa, Aug 30, 42; m 64; c 2. ORGANIC CHEMISTRY. *Educ:* Albright Col, BS, 64; Princeton Univ, PhD(chem), 69. *Prof Exp:* Res chemist, Am Cyanamid Co, Bound Brook, 68-73, group leader, 73-75, dept head, Lederle Labs, 75-79, SECT DIR, MED RES DIV, AM CYANAMID CO, PEARL RIVER, 79- *Mem:* Am Chem Soc. *Res:* Structure-reactivity relationships; process research and development on pharmaceuticals and fine chemicals; fermentation; proces development; biotechnology. *Mailing Add:* 16 Westminster Pl Old Tappan NJ 07675

WOODWORTH, JOHN GEORGE, b Lockport, NY, Apr 16, 48; m 78; c 2. NUCLEAR & PLASMA PHYSICS. *Educ:* Trent Univ, BSc, 72, MSc, 74; Univ Toronto, PhD(physics), 78. *Prof Exp:* Physicist nuclear physics, 78-85, proj leader, ICF Prog, 86-88, SR SCIENTIST, LAWRENCE LIVERMORE LAB, 88- *Mem:* Am Phys Soc. *Res:* Intertial confinement fusion. *Mailing Add:* Lawrence Livermore Lab L-490 PO Box 808 Livermore CA 94550

WOODWORTH, MARY ESTHER, b Grand Rapids, Mich. ANIMAL VIROLOGY, REGULATORY MECHANISMS IN EUKARYOTIC CELLS. *Educ:* Univ Mich, Ann Arbor, BS, 57; Temple Univ, MS, 65, PhD(biochem), 68. *Prof Exp:* Fel bacteriophage genetics, Univ Mich, 68-72; res assoc DNA topology, Syracuse Univ, NY, 72-73; res scientist tumor virol, Sch Med, Johns Hopkins Univ, 73-76; cancer res scientist, Roswell Park Mem Inst, 77-89, actg chair tumor & cell biol, 83-84; asst prof viral oncol, Roswell Park Div, State Univ NY, Buffalo, 77-86, assoc prof molecular & cell biol, 86-89; CHAIR & PROF MICROBIOL, MIAMI UNIV, OXFORD, OHIO, 89- *Concurrent Pos:* Prin investr, Nat Cancer Inst, NIH, 77-80 & 80-88; Nat Cancer Inst res career develop award, NIH, 78-83; mem bd dirs, Health Res, Inc, Buffalo, NY, 84-87; Nat Found lectr, Am Soc Microbiol, 89-90, comt mem, Genetic & Molecular Microbiol, 90- *Mem:* Am Soc Microbiol; Am Soc Biochem & Molecular Biol; Am Soc Virol; Sigma Xi; Am Asn Univ Women; Am Women Sci. *Res:* Molecular biology of tumor viruses utilizing SV40 as a model system to study the regulation of eukaryotic DNA replication and gene expression. *Mailing Add:* Dept Microbiol Miami Univ Oxford OH 45056

WOODWORTH, ROBERT CUMMINGS, b Cambridge, Mass, Nov 11, 30; m 52; c 3. PROTEIN CHEMISTRY, BIOINORGANIC CHEMISTRY. *Educ:* Univ Vt, BS, 53; Pa State Univ, PhD(chem), 57. *Prof Exp:* Res chemist, Nat Inst Allergy & Infectious Dis, 56-60; from instr to assoc prof, 61-75, PROF BIOCHEM, COL MED, UNIV VT, 75- *Concurrent Pos:* USPHS fel, Clin Chem Lab, Malmo Gen Hosp, Sweden, 60-61; USPHS spec fel & vis prof, Inorg Chem Lab, Oxford Univ, 68-69; Fogarty Int sr fel, Oxford Univ, 76-77; MRC Can vis scientist, Univ BC, 87-88. *Mem:* Am Chem Soc; Am Soc Biochem & Molecular Biol; Sigma Xi; Am Heart Asn; AAAS; Soc Values Higher Educ. *Res:* Protein structure-function; nature of iron-binding proteins; mechanisms of metal binding and release; role of bound anions, physiological function and structure; protein structure and function. *Mailing Add:* Dept Biochem Univ Vt Given Med Bldg Burlington VT 05405

WOODWORTH, ROBERT HUGO, biology; deceased, see previous edition for last biography

WOODY, A-YOUNG MOON, b Pyungyang, Korea, Mar 7, 34; US citizen; m 65; c 2. BIOCHEMISTRY, CHEMISTRY. *Educ:* Univ Calif, Berkeley, BS, 59; Cornell Univ, PhD(biochem), 64. *Prof Exp:* Res assoc chem, Cornell Univ, 64-65; res assoc microbiol, Univ Ill, Urbana, 65-66; res assoc biochem, 67-69; res assoc develop biol, Dept Zool, Ariz State Univ, 72-74; res assoc biochem, 76-88, RES SCIENTIST BIOCHEM, DEPT BIOCHEM, COLO STATE UNIV, FT COLLINS, 88- *Mem:* Am Chem Soc; Sigma Xi. *Res:* Structure and function of proteins and enzymes; mechanism of RNA polymerase. *Mailing Add:* Dept Biochem Colo State Univ Ft Collins CO 80523

WOODY, CHARLES DILLON, b Brooklyn, NY, Feb 6, 37; m 59; c 2. NEUROPHYSIOLOGY. *Educ:* Princeton Univ, AB, 57; Harvard Med Sch, MD, 62. *Prof Exp:* Intern med, Strong Mem Hosp, Univ Rochester, 62-63; resident, Boston City Hosp, Mass, 63-64; res assoc, Lab Neurophysiol, NIH, 64-67, res officer, Lab Neural Control, 68-71; assoc prof anat, physiol & psychiat, 71-76, PROF ANAT & PSYCHIAT, MENT RETARDATION CTR, NEUROPSYCHIAT INST, UNIV CALIF, LOS ANGELES, 77- *Concurrent Pos:* Harvard Moseley fel & Nat Acad Sci exchange fel neurophysiol, Inst Physiol, Czech Acad Sci, 67-68; res fel neurol, Harvard Med Sch, 63-64. *Honors & Awards:* Nightingale Prize, Brit Biol Eng Soc & Int Fedn Med Electronics & Biol Eng, 69. *Mem:* AAAS; Am Physiol Soc; Soc Neurosci; Biomed Eng Soc. *Res:* Neurophysiology of learning and memory; neurophysiology of learned motor performance; electrophysiologic data analysis by linear filter techniques employing digital computers. *Mailing Add:* Dept Anat & Psychiat Univ Calif Los Angeles Med Ctr Los Angeles CA 90024

WOODY, CHARLES OWEN, JR, b Somerville, Tenn, Oct 28, 30; m 60; c 4. REPRODUCTIVE PHYSIOLOGY. *Educ:* Miss State Univ, BS, 57, MS, 59; NC State Univ, PhD(animal sci), 63. *Prof Exp:* Trainee endocrinol, Univ Wis-Madison, 63, proj assoc reproductive physiol, 64-68; assoc prof, 68-81, PROF ANIMAL SCI, UNIV CONN, 81- *Mem:* Am Soc Animal Sci; Brit Soc Study Fert; Soc Study Reproduction; NY Acad Sci. *Res:* Gamete physiology and in vitro fertilization; testis growth and function. *Mailing Add:* Dept Animal Sci Univ Conn U-40 Storrs CT 06268

WOODY, CRAIG L, b Baltimore, Md, Mar 26, 51; m 73. EXPERIMENTAL PARTICLE PHYSICS. *Educ:* Johns Hopkins Univ, BA, 73, MA, 74, PhD(physics), 79. *Prof Exp:* Res asst physics, Johns Hopkins Univ, 74-78; res assoc physics, Stanford Linear Accelerator Ctr, 78-79; MEM STAFF, BROOKHAVEN NAT LAB, 79- *Mem:* Am Phys Soc. *Res:* Detectors and instrumentation for high energy particle physics, especially calorimetry and electromagnetic radiation detectors. *Mailing Add:* Brookhaven Nat Lab Physics Dept Bldg 510A Upton NY 11973

WOODY, ROBERT WAYNE, b Newton, Iowa, Dec 5, 35; m 65; c 2. BIOPHYSICAL CHEMISTRY, SPECTROSCOPY. *Educ:* Iowa State Col, BS, 58; Univ Calif, Berkeley, PhD(chem), 62. *Prof Exp:* Res assoc phys chem, Cornell Univ, 62-64, Nat Inst Gen Med Sci fel, 63-64; asst prof, Univ Ill, Urbana, 64-70; from assoc prof to prof, Ariz State Univ, 70-75; PROF BIOCHEM, COLO STATE UNIV, 75- *Concurrent Pos:* John Simon Guggenheim Mem Fel, 81-82; mem, Molecular & Cellular Biophys Study Sect, NIH, 83-87; Fogarty Sr Int fel, 89-90. *Mem:* AAAS; Am Chem Soc; Am Soc Biochem & Molecular Biol; Biophys Soc. *Res:* Optical properties of molecules; structure of proteins; protein-nucleic acid interactions; interaction of small molecules with proteins. *Mailing Add:* Dept Biochem Colo State Univ Ft Collins CO 80523

WOODYARD, JAMES DOUGLAS, b San Antonio, Tex, Oct 8, 38. ORGANIC CHEMISTRY. *Educ:* Tex Christian Univ, BA, 61, MA, 63, PhD(chem), 67. *Prof Exp:* NSF fel Univ Ill, Chicago, 66-67; from asst prof to assoc prof, 67-74, PROF CHEM, WTEX STATE UNIV, 78- *Mem:* Am Chem Soc; Royal Soc Chem. *Res:* Reaction of carbenes and stereochemistry of carbene reactions; triplet state of organic molecules. *Mailing Add:* Dept Chem WTex State Univ Canyon TX 79016

WOODYARD, JAMES ROBERT, b Pittsburgh, Pa, July 18, 36; m 60; c 4. PHYSICS. *Educ:* Duquesne Univ, BEd, 60; Univ Del, MS, 62, PhD(physics), 66. *Prof Exp:* asst prof physics, Univ Ky, 67-69; mem tech staff, Gen Tel & Electronics Labs, 69-71; asst prof physics, Univ Hartford, 71-72; asst prof, Trenton State Col, 72-75; assoc prof, Div Sci & Technol, Col Lifelong Learning, 75-82, ASSOC PROF, DEPT ELEC & COMPUTER ENG, WAYNE STATE UNIV, 83- *Concurrent Pos:* AEC postdoctoral fel, 65-67; Danforth Found assoc, 77-86. *Mem:* AAAS; Am Phys Soc; Am Vacuum Soc; Sigma Xi; Am Asn Physics Teachers; Am Asn Univ Professors; Inst Elec & Electronics Engrs; Mat Res Soc. *Res:* Particle surface interactions; instructional methods; solid state device materials. *Mailing Add:* Dept Elec & Computer Eng Wayne State Univ Detroit MI 48202

WOOFTER, HARVEY DARRELL, b Glenville, WVa, Jan 31, 23; m 44. WEED SCIENCE, ENTOMOLOGY. *Educ:* WVa Univ, BS, 43; Ohio State Univ, MS, 49, PhD(agron), 53. *Prof Exp:* County agr agt, WVa, 43-44 & 46-48; asst agron, Ohio State Univ, 48-49; res agronomist, Chem Corps Biol Warfare Labs, US Dept Army, Md, 49-54; field res rep, Pittsburgh Coke & Chem Co, 54-56; field res rep, Chemagro Corp, 56-62, asst supvr field res, 62-64; mgr field res, Diamond Alkali Co, 64-66; mgr prod develop, Ciba Agrochem Co, Fla, 66-68; mgr, Fla Res Sta, Velsicol Chem Corp, Hoffman-La Roche, Inc, 68-72, mgr agrochem sta, 72-76, dir res & develop, Maag Agrochem, 76-86; PRES, WOOFTER CONSULTS, INC, 86- *Mem:* Am Soc Agron; Entom Soc Am; Weed Sci Soc Am; Aquatic Plant Mgt Soc; Plant Growth Regulator Soc Am. *Res:* Weed control; development of new herbicides, insecticides, fungicides, nematocides, bactericides, defoliants, desiccants and plant growth regulators. *Mailing Add:* 3040 Nassau Dr Vero Beach FL 32960

WOOL, IRA GOODWIN, b Newark, NJ, Aug 22, 28; m 50; c 2. BIOCHEMISTRY, MOLECULAR BIOLOGY. *Educ:* Syracuse Univ, AB, 49; Univ Chicago, MD, 53, PhD(physiol), 54. *Prof Exp:* Intern med, Beth Israel Hosp, Boston, Mass, 54-55, asst resident, 55-56; from asst prof to assoc prof physiol, 57-65, assoc prof biochem, 64-65, PROF BIOCHEM, UNIV CHICAGO, 65-, A J CARLSON PROF BIOL SCI, 73- *Concurrent Pos:* Fel physiol, Harvard Univ, 56; Commonwealth Fund fel, Univ Chicago, 56-57; vis scientist, Dept Biochem, Cambridge Univ, 60-61; vis prof, Wayne State Univ, 64, 66 & Fla State Univ, 66; Lederle sci lectr, Lederle Labs, 66; vis fac mem, Mayo Grad Sch Med, 66; vis prof, Rutgers Univ, 67; ed, Vitamins & Hormones & J Biol Chem; vis res scientist, Max-Planck Inst Molecular Genetics, Berlin, Ger, 73-74; mem, Molecular Biol Study Sect, NIH, 74-78. *Honors & Awards:* Ginsburg Award, Univ Chicago, 52; Bernstein lectr, Beth Israel Hosp, Harvard Med Sch, 64; Alexander von Humboldt spec fel, Fed Repub Ger, 73-74. *Mem:* AAAS; Brit Biochem Soc; Am Soc Biol Chemists; Am Soc Cell Biol. *Res:* Structure and function of eukaryotic ribosomes; nucleic acid-protein interactions. *Mailing Add:* Dept Biochem & Molecular Biol 920 E 58th St Chicago IL 60637

WOOLARD, HENRY W(ALDO), b Clarksburg, WVa, June 2, 17; m 41; c 2. FLUID DYNAMICS, AERODYNAMICS. *Educ:* Univ Mich, BS, 41; Univ Buffalo, MS, 54. *Prof Exp:* Aeronaut engr, Nat Adv Comt Aeronaut, 41-46; from asst prof to assoc prof aeronaut eng, Univ WVa, 46-48, actg head dept, 46-48; res aerodynamicist, Cornell Aeronaut Lab, Inc, 48-57; sr staff engr, Appl Physics Lab, Johns Hopkins Univ, 57-63; sr res specialist, Lockheed-Calif Co, 63-67; mem tech staff, TRW Systs Group, Calif, 67-70; pres, Beta Technol Co, Calif, 70-71; aerospace engr, Flight Dynamics Lab, Wright Patterson AFB, 71-85; ENG CONSULT, 85- *Concurrent Pos:* Lectr aerodynamics, Univ Buffalo, 55-57. *Mem:* Assoc fel Am Inst Aeronaut & Astronaut; Sigma Xi. *Res:* Numerous investigations in fluid dynamics and aerodynamics published in various journals and government, industry and university reports. *Mailing Add:* 1249 W Magill Ave Fresno CA 93711-1428

WOOLCOTT, WILLIAM STARNOLD, b Coffeyville, Kans, Apr 14, 22; m 46; c 2. VERTEBRATE ZOOLOGY. *Educ:* Austin Peay State Col, BS, 47; Peabody Col, MA, 48; Cornell Univ, PhD(vert zool), 55. *Prof Exp:* Asst prof biol, Carson-Newman Col, 49-53; assoc prof, 55-67, PROF BIOL, UNIV RICHMOND, 67- *Mem:* Sigma Xi; Am Soc Ichthyol & Herpet. *Res:* Morphological and ecological aspects of fishes. *Mailing Add:* Dept Biol Univ Richmond Richmond VA 23173

WOOLDRIDGE, DAVID DILLEY, b Seattle, Wash, Mar 12, 27; m 48, 70; c 5. FOREST SOILS, FOREST HYDROLOGY. *Educ:* Univ Wash, BS, 50, PhD(forestry), 61. *Prof Exp:* Forester, Rayonier Inc, 50-52, res forester, 53-56; res forester, Forest Hydrol Lab, US Forest Serv, Wash, 56-68; assoc prof forest hydrol, Col Forest Resources, Univ Wash, 68-83; RETIRED. *Concurrent Pos:* Asst prof, Univ Wash, 65-68; consult, King County Flood Control Div, 66-; consult local eng firms & US CEngr; res grants, Off Water Resources & Res & US Forest Serv, Wash State Dept Ecol, Environ Protection Agency & Wash State Dept Natural Resources; consult, 83- *Mem:* Sigma Xi; Soc Am Foresters; Soil Sci Soc Am; Am Geophys Union. *Res:* Watershed classification, Thailand; turbidity-suspended sediment relations in forest streams; disposal of stabilized municipal-industrial sewage sludge; impacts of clear-cutting on soil nutrient balance; effects of slash burial on surface water quality; soil properties under pristine western hemlock. *Mailing Add:* Forest Resources Univ Wash Seattle WA 98195

WOOLDRIDGE, DAVID PAUL, b Terre Haute, Ind, Dec 25, 31; m 56. ENTOMOLOGY. *Educ:* Ind Univ, BS, 56, PhD(zool), 62. *Prof Exp:* Asst prof biol, Wilkes Col, 62-63; asst prof zool, Southern Ill Univ, 63-67; assoc ed, Biol Abstr, 67-68; from asst prof to assoc prof, 68-78, PROF BIOL, PA STATE UNIV, 78- *Mem:* Coleopterist's Soc. *Res:* Taxonomy of aquatic Coleoptera; systematics of world Limnichidae. *Mailing Add:* Dept Biol Pa tate Univ Ogontz Campus Abington PA 19001

WOOLDRIDGE, DEAN E, b Chickasha, Okla; c 3. PHYSICS, ENGINEERING. *Educ:* Univ Okla, BA, MS; Calif Inst Technol, PhD(physics), 36. *Prof Exp:* Mem tech staff, Bell Tel Lab, 36-46; dir electronic res & develop, Hughes Aircraft Co, 46-53; pres, Thompson Ramo Wooldridge Inc, 53-62; res assoc eng, Calif Inst Technol, 62-72. *Concurrent Pos:* Chmn, NIH Study Comt, 64-65; mem bd trustees, Calif Inst Technol, 75-78. *Honors & Awards:* Raymond E Hackett Award, 55; Westinghouse Award, AAAS, 63. *Mem:* Nat Acad Sci; Nat Acad Eng; Am Phys Soc; Inst Elec & Electronics Engrs; Am Inst Aeronaut & Astronaut. *Mailing Add:* 4545 Via Esperanza Santa Barbara CA 93110

WOOLDRIDGE, GENE LYSLE, b Randalia, Iowa, Apr 16, 24; m 45; c 5. ATMOSPHERIC SCIENCES, PHYSICS. *Educ:* Upper Iowa Col, BS, 44; Mankato State Col, MS, 61; Colo State Univ, PhD(atmospheric sci), 70. *Prof Exp:* Instr physics, Rochester State Jr Col, Minn, 61-62; instr, Mankato State Col, 62-64, asst prof atmospheric sci, 65-67; from assoc prof to prof atmospheric sci, Utah State Univ, 77-87; RETIRED. *Mem:* Am Meteorol Soc. *Res:* Mesoscale circulations and transport processes; mesoscale-macroscale and mesoscale-microscale energy interactions and mechanisms. *Mailing Add:* 1908 Wallenberg Dr Ft Collins CO 80526

WOOLDRIDGE, KENT ERNEST, b Waukegan, Ill, Apr 23, 42; m 68; c 2. MATHEMATICS. *Educ:* Univ Chicago, BS, 64; Univ Ill, Urbana, PhD(math), 75. *Prof Exp:* Asst prof math, 74-81, ASSOC PROF MATH, CALIF STATE COL, 81- *Mem:* Am Math Soc; Math Asn Am. *Res:* Number theory. *Mailing Add:* Dept Comput Sci Calif State Univ Chico CA 95929

WOOLES, WALLACE RALPH, b Lawrence, Mass, Mar 8, 31; m 51; c 5. PHYSIOLOGY, PHARMACOLOGY. *Educ:* Boston Col, BS, 58, MS, 61; Univ Tenn, PhD(physiol), 63. *Prof Exp:* From instr to assoc prof pharmacol, Med Col Va, 63-70; PROF PHARMACOL & ASSOC VCHANCELLOR HEALTH AFFAIRS, E CAROLINA UNIV, 70- *Mem:* AAAS; Soc Toxicol; Int Soc Res Reticuloendothelial Systs; Am Soc Pharmacol & Exp Therapeut; Sigma Xi. *Res:* Radiation injury and lipid metabolism; alcohol and lipid metabolism; reticuloendothelial system and drug metabolism. *Mailing Add:* Sch Med ECarolina Univ Greenville NC 27834

WOOLF, C R, b Cape Town, SAfrica, Jan 26, 25; Can citizen; m 52; c 3. PULMONARY DISEASES, INTERNAL MEDICINE. *Educ:* Univ Cape Town, BSc, 44, MB, ChB, 47, MD, 51; FRCP(C), 57; FRCP, 78. *Prof Exp:* Jr asst path, Univ Cape Town, 49, jr asst med, 50-51; house physician, Brompton Hosp, London, Eng & London Chest Hosp, 52-53; resident chest serv, Bellevue Hosp, New York, 53-54; res assoc, Ont Heart Found, 55-63; from asst prof to assoc prof, 63-73, dir Tri-Hosp Respiratory Serv, 63-86, PROF MED, UNIV TORONTO, 73-, ASST DEAN, CONTINUING MED EDUC, 85- *Concurrent Pos:* Francis Esther res fel cardio-respiratory, Univ Toronto, 54-55; AMA, Med Res Coun Can, Ont Heart Found, Ont Tuberc Asn & Can Tuberc Asn grants; sr physician, Toronto Gen Hosp, 58-; div postgrad training prog, Dept Med, Univ Toronto, Ont, Can. *Mem:* Am Thoracic Soc; Am Col Chest Physicians; Can Thoracic Soc; Soc Med Col; Asn Can Med Col. *Res:* Respiratory physiology; cardiac dyspnea; role of oxygen in the regulation of respiration; respiratory rehabilitation; effects of smoking; surgical treatment of emphysema; personalized continuing medical education. *Mailing Add:* Continuing Educ Fac Med Univ Toronto One Kings Col Circle Toronto ON M5S 1A8 Can

WOOLF, CHARLES MARTIN, b Salt Lake City, Utah, Aug 23, 25; m 50; c 4. GENETICS. *Educ:* Univ Utah, BS, 48, MS, 49; Univ Calif, PhD(genetics), 54. *Prof Exp:* Asst geneticist, Lab Human Genetics, Univ Utah, 50-51, dir, 57-61, res instr genetics, Univ, 53-55, asst prof, 55-59, assoc prof, 59-61; dean col lib arts, 73-75, assoc prof genetics, 61-64, vpres grad studies & res, 76-79, dean grad col, 76-86, PROF ZOOL ARIZ STATE UNIV, 64- *Mem:* Am Soc Human Genetics (treas, 61-63); Genetics Soc Am. *Res:* Genetics of congenital malformation; consanguinity and genetic effects; drosophila behavior and developmental genetics. *Mailing Add:* Dept Zool Ariz State Univ Tempe AZ 85287-1501

WOOLF, HARRY, b New York, NY, Aug 12, 23; c 4. SCIENCE POLICY. *Educ:* Univ Chicago, BS, 48, MA, 49; Cornell Univ, PhD(hist sci), 55. *Hon Degrees:* DSc, Whitman Col, 79, Am Univ, 82; LHD, Johns Hopkins Univ, 83 & St Lawrence Univ, 86. *Prof Exp:* Instr physics, Boston Univ, 53-55; from asst prof to prof hist, Univ Wash, 55-61; prof hist sci, Johns Hopkins Univ, 61-76, chmn, 6-72, provost, 72-76; dir, 76-87, PROF, INST ADVAN STUDY, 87- *Concurrent Pos:* Instr hist, Brandeis Univ, 54-55; trustee, Assoc Univ Inc, Brookhaven Nat Lab & Nat Radio Astron Observ, 72-; chmn bd, Univ Res Asn, Inc, 79-89; mem, Corp Vis Comt, dept physics, Mass Inst Technol, 79-85, Adv Coun, dept philos, Princeton Univ, 80-84, adv bd, Stanford Humanities Ctr, 81-87, Wissensch-Kol Beriat, Berlin, 81-87 & Sci Adv Bd, Alexander von Humboldt Found, 85- *Mem:* Sigma Xi; fel Am Philos Soc; AAAS; Royal Astron Soc; Hist Sci Soc; fel Am Acad Arts & Sci. *Res:* History of science, with emphasis on physics and astronomy; issues involving the intersection of basic science and education with modern technology. *Mailing Add:* Inst Advan Study Olden Lane Princeton NJ 08540

WOOLF, J(ACK) R(OYCE), b Trinidad, Tex, June 10, 24; m 48; c 3. MECHANICAL ENGINEERING. *Educ:* Agr & Mech Col, Tex, BS & MS, 48; Purdue Univ, PhD, 51. *Prof Exp:* Asst instr, Agr & Mech Col, Tex, 47-48, prof mech eng & res engr, Eng Exp Sta, 56-57, asst to dean eng, 57; instr, Purdue Univ, 48-51; res engr, Consol Aircraft Co, 51-56; dean, 57-58, actg pres, 58-59, pres, 59-68, UNIV PROF MECH ENG, UNIV TEX, ARLINGTON, 68- *Concurrent Pos:* Lectr, Southern Methodist Univ, 53-56. *Mem:* Am Soc Mech Engrs; Am Soc Eng Educ. *Res:* Heat transfer; thermodynamics and transfer properties of fluids; aerothermodynamics. *Mailing Add:* 3115 Woodford Arlington TX 76010

WOOLF, NEVILLE JOHN, b London, Eng, Sept 15, 32; US citizen; m 72; c 2. ASTROPHYSICS. *Educ:* Manchester Univ, BSc, 56, PhD(astrophys), 59. *Prof Exp:* Res assoc astron, Lick Observ, Univ Calif, 59-61; res assoc, Princeton Univ, 61-65; assoc prof, Univ Tex, 65-67; prof, Univ Minn, 67-74; PROF ASTRON, STEWARD OBSERV, UNIV ARIZ, 74- *Concurrent Pos:* Dir, Minn Observ, Univ Minn, 67-74; Nat Acad Sci/Nat Res Coun sr fel, NASA Goddard Inst Space Studies, 65-67; actg dir, Flandrau Planetarium,

Univ Ariz, 77; actg dir, Mult Mirror Telescope Observ, 78-79. *Mem:* Am Astron Soc; Int Astron Union. *Res:* Astrophysics, observational, theoretical and instrumental. *Mailing Add:* Dept Astron Steward Observ Univ Ariz Tucson AZ 85721

WOOLF, WILLIAM BLAUVELT, b New Rochelle, NY, Sept 18, 32. MATHEMATICS. *Educ:* Pomona Col, BA, 53; Claremont Col, MA, 55; Univ Mich, PhD(math), 60. *Prof Exp:* Instr math, Mt San Antonio Jr Col, 54 & Univ Mich, 54-59; from instr to assoc prof, Univ Wash, 59-68; staff assoc, Am Asn Univ Profs, 68-69, assoc secy & dir admin, 69-79; actg exec ed, 84-85, MANAGING ED MATH REVIEWS, 79- *Concurrent Pos:* Fulbright res scholar, Univ Helsinki, 63-64; invité prof, Univ Louis Pasteur, Strasbourg, France, 87-88. *Mem:* Am Math Soc; AAAS; Math Asn Am. *Res:* Functions of a complex variable; mathematics education. *Mailing Add:* Am Math Soc PO Box 6248 Providence RI 02940-6248

WOOLFENDEN, GLEN EVERETT, b Elizabeth, NJ, Jan 23, 30; m 54; c 3. ORNITHOLOGY, BEHAVIORAL ECOLOGY. *Educ:* Cornell Univ, BS, 53; Univ Kans, MA, 56; Univ Fla, PhD(zool), 60. *Prof Exp:* Instr biol, Univ Fla, 59-60; from instr to assoc prof zool, 60-70, PROF BIOL, UNIV SFLA, TAMPA, 70- *Concurrent Pos:* Res Soc Am res grant, 61; consult, Encephalitis Res Ctr, Tampa, 61-64; res assoc, Archbold Biol Sta, 70-, Field Mus Natural Hist, 82-; vis prof, Univ Kans, 85, Int Ornithol Cong, 78- *Honors & Awards:* Brewster Award, Am Ornithol Union, 85. *Mem:* Am Ornith Union; Wilson Ornith Soc; Cooper Ornith Soc; Brit Ornith Union; Animal Behav Soc. *Res:* Behavioral ecology; communal breeding; long-term study of individually marked birds investigates the demographic and habitat features of cooperative breeding. *Mailing Add:* Dept Biol Univ SFla 4202 Fowler Ave Tampa FL 33620

WOOLFOLK, CLIFFORD ALLEN, b Riverside, Calif, June 5, 35; m 57; c 2. MICROBIOLOGY. *Educ:* Univ Calif, Riverside, BA, 57; Univ Wash, Seattle, MSc, 59, PhD(microbiol), 63. *Prof Exp:* Res asst microbiol, Univ Wash, 57-59 & 62-63; USPHS fel enzymol lab biochem, Nat Heart Inst, 63-65; asst prof, 65-68, ASSOC PROF MICROBIOL, UNIV CALIF, IRVINE, 68- *Mem:* AAAS; Am Soc Microbiol; Brit Soc Gen Microbiol; Am Chem Soc; Am Soc Biol Chemists. *Res:* Microbial physiology; hydrogenase and hydrogenase mediated reduction of inorganic compounds; cumulative feedback inhibition of glutamine synthetase from Escherichia coli; bacterial purine oxidizing enzymes. *Mailing Add:* Dept Microbiol & Biochem Univ Calif Irvine CA 92717

WOOLFOLK, ROBERT WILLIAM, b Riverside, Calif, Feb 9, 37; m 68; c 4. PHYSICAL CHEMISTRY. *Educ:* Univ Calif, Riverside, BA, 58; Univ Calif, Berkeley, PhD(phys chem), 64. *Prof Exp:* Staff scientist, Chem Syst Div, United Technol Corp, 63-65; PHYS CHEMIST & DIR BUS DEVELOP, SRI INT, 65- *Mem:* Am Chem Soc; Am Phys Soc; Am Inst Aeronaut & Astronaut; AAAS; Sigma Xi. *Res:* Shock wave phenomenon; gas phase reaction kinetics; photochemistry; fluorine chemistry; explosive sensitivity; nonideal explosions; combustion and marketing research and development. *Mailing Add:* SRI Int 1235 Jefferson Davis Hwy Suite 1104 Arlington VA 22202

WOOLFORD, ROBERT GRAHAM, b London, Ont, Apr 14, 33; m 55; c 2. ORGANIC CHEMISTRY. *Educ:* Univ Western Ont, BSc, 55, MSc, 56; Univ Ill, PhD(org chem), 59. *Prof Exp:* From asst prof to assoc prof chem, 59-68, assoc dean sci, 67-75, PROF CHEM, UNIV WATERLOO, 68-, ASSOC CHMN DEPT, 77- *Concurrent Pos:* Res fel chem, Univ Ill, 56-58. *Mem:* Fel Chem Inst Can. *Res:* Electroorganic chemistry of halogenated carboxylic acids; synthesis of polymers. *Mailing Add:* Dept Chem Univ Waterloo Waterloo ON N2L 3G1 Can

WOOLHISER, DAVID A(RTHUR), b La Crosse, Wis, Jan 21, 32; m 57; c 3. CIVIL ENGINEERING, AGRICULTURAL ENGINEERING. *Educ:* Univ Wis, BS(agr) & BS(civil eng), 55, PhD(civil eng), 62; Univ Ariz, MS, 59. *Prof Exp:* Asst agr engr, Univ Ariz, 55-58; hydraul engr, Agr Res Serv, USDA, 58-63; asst prof civil eng, Cornell Univ, 63-67; RES HYDRAUL ENGR, AGR RES SERV, USDA, 67- *Concurrent Pos:* Mem assoc fac, Dept Civil Eng, Colo State Univ, 70-84; vis prof, Imperial Col, London, 77-78; vis scientist, Inst Hydrol, Wallingford, UK, 77-78; adj prof, Dept Hydrol & Water Resources, Univ Ariz, 81- *Honors & Awards:* Robert E Horton Award, Am Geophys Union, 83; Arid Lands Hydrol Award, Am Soc Civil Engrs, 88 & Nat Acad Eng, 90. *Mem:* Am Soc Civil Engrs; Am Soc Agr Engrs; fel Am Geophys Union. *Res:* Simulation of hydrologic systems, including numerical solutions of unsteady, spatially varied flow; stochastic models of climate. *Mailing Add:* USDA-ARS 2000 E Allen Rd Tucson AZ 85719

WOOLLAM, JOHN ARTHUR, b Kalamazoo, Mich, Aug 10, 39; c 2. PHYSICS, ELECTRICAL ENGINEERING. *Educ:* Kenyon Col, AB, 61; Mich State Univ, MS, 63, PhD(physics), 67; Case Western Reserve Univ, MS, 78. *Prof Exp:* Res physicist mat, Lewis Res Ctr, NASA, 67-74, head, Cryophysics Sect, 74-78, res physicist photovoltaics, 78-80; GEORGE HOLMES REGENTS PROF, DEPT ELEC ENG, UNIV NEBR, 80- *Concurrent Pos:* Asst, Mich State Univ, 61-62, res asst, 62-67; vis scientist, Mat Sci Ctr, Mass Inst Technol, 72 & Francis Bitter Nat Magnet Lab, 77-; adj scientist, Oberlin Col, 78-; consult, Wright Patterson AFB, 78-; mem, Interagency Adv Power Group. *Honors & Awards:* Achievement Awards, NASA/Lewis Res Ctr, 72 & 77. *Mem:* Am Phys Soc; Am Carbon Soc; Inst Elec & Electronics Engrs; Mat Res Soc. *Res:* Physics of solids: electronic properties; superconductivity, magnetism; photovoltaics: energy conversion in Ga As; Ga As integrated circuits, electro-optics and semiconductors. *Mailing Add:* Dept Elec Eng 209N WSEC Univ Nebr Lincoln NE 68588-0511

WOOLLETT, ALBERT HAINES, b Oxford, Miss, Jan 23, 30; m 64; c 3. PHYSICS. *Educ:* Univ Miss, BA, 49, MS, 50; Univ Okla, PhD(physics), 56. *Prof Exp:* Asst prof physics, Fisk Univ, 56-57; instr, Reed Col, 57-59; assoc prof, High Point Col, 59-61; asst prof, Va Polytech Inst & State Univ, 61-63; asst prof, 63-72, ASSOC PROF PHYSICS, MEMPHIS STATE UNIV, 72- *Mem:* Am Asn Physics Teachers; Optical Soc Am; assoc mem Am Astron Soc; Sigma Xi. *Res:* Infrared and Raman spectroscopy; physical optics. *Mailing Add:* 5409 Pecan Groove Lane Memphis TN 38120

WOOLLEY, DONALD GRANT, b Magrath, Alta, Dec 12, 25; m 47; c 5. AGRONOMY, PHYSIOLOGY. *Educ:* Utah State Univ, BSc, 51, MSc, 56; Iowa State Univ, PhD(crop physiol), 59. *Prof Exp:* Asst prof agron, Iowa State Univ, 59-60; res officer, Can Dept Agr, 60-63; assoc prof agron, 63-67, head, Farm Mgt Dept & Classification Officer, 76-80, PROF AGRON, COL AGR, IOWA STATE UNIV, 67- *Concurrent Pos:* Consult, World Bank, SE Asia, 74-75. *Mem:* Am Soc Agron; Am Soc Crop Sci. *Res:* Crop physiology, climatology and crambe production. *Mailing Add:* 1816 Bel Air Dr Ames IA 50010

WOOLLEY, DOROTHY ELIZABETH SCHUMANN, b Wapakoneta, Ohio, Feb 2, 29; m 50; c 3. PHYSIOLOGY, PHARMACOLOGY. *Educ:* Bowling Green State Univ, BS, 50; Ohio State Univ, MS, 56; Univ Calif, Berkeley, PhD(physiol), 61. *Prof Exp:* NSF fel, Univ Calif, Berkeley, 61-62, asst res physiologist, 62-65, lectr physiol, 63-64; from asst prof to assoc prof physiol & environ toxicol, 65-74, PROF ANIMAL PHYSIOL, UNIV CALIF, DAVIS, 74- *Concurrent Pos:* Lectr, Sch Med, Univ Calif, San Francisco, 60-65. *Mem:* AAAS; Am Physiol Soc; Endocrine Soc; Am Soc Pharmacol & Exp Therapeut. *Res:* Effects of hormones, drugs and neurotoxins on brain electrical activity, neurochemistry and behavior in rats and monkeys. *Mailing Add:* Dept Animal Physiol Univ Calif Davis CA 95616

WOOLLEY, EARL MADSEN, b Richfield, Utah, Apr 10, 42; m 66; c 8. PHYSICAL CHEMISTRY, ANALYTICAL CHEMISTRY. *Educ:* Brigham Young Univ, BS, 66, PhD(phys chem), 69. *Prof Exp:* Nat Res Coun Can fel, Univ Lethbridge, 69-70; assoc prof, 70-77, PROF ANALYTICAL & PHYS CHEM, BRIGHAM YOUNG UNIV, 77-, CHAIR, 89- *Mem:* Am Chem Soc; Calorimetry Conf. *Res:* Thermodynamics of solutions containing reacting and/or non-reacting components including surfactants, strong electrolytes, acids, bases, and hydrogen bonding species. *Mailing Add:* Dept Chem Brigham Young Univ Provo UT 84602-1022

WOOLLEY, JOSEPH TARBET, b Denver, Colo, Jan 25, 23; m 52; c 3. PLANT PHYSIOLOGY. *Educ:* Utah State Agr Col, BS, 48, MS, 52; Univ Calif, PhD(plant physiol), 56. *Prof Exp:* Asst prof, 56-65, ASSOC PROF PLANT PHYSIOL, UNIV ILL, URBANA, 65-; PLANT PHYSIOLOGIST, AGR RES SERV, USDA, 56- *Mem:* Am Soc Plant Physiol; Am Soc Agron; Soc Exp Biol. *Res:* Physics of water in plants and soils; reflection of light by leaves; soil-plant-atmosphere interactions. *Mailing Add:* 1701 W Bradley Ave Champaign IL 61821

WOOLLEY, LEGRAND H, b Salt Lake City, Utah, Apr 22, 31; m 53; c 4. ORAL PATHOLOGY, DENTISTRY. *Educ:* Univ Mo-Kansas City, DDS, 58; Univ Ore, MS, 66; Am Bd Oral Path, dipl, 73. *Prof Exp:* Asst prof path, Dent Sch, 66-70, sr clin instr dent & oral med, 67-71, assoc prof dent, Med Sch, 70-71, assoc prof path, Dent Sch, 70-75, chmn dept path, 76-80, PROF PATH, SCH DENT, ORE HEALTH SCI UNIV, 78-, DIR ORAL PATH GRAD PROG, 81- *Concurrent Pos:* Am Cancer Soc fel, 66-68; consult, Fairview Hosp, Salem, Ore, 66-68. *Mem:* Fel Am Acad Oral Path. *Res:* Effects of therapeutic radiation on dental structures. *Mailing Add:* 2864 Forest Lodge Rd Pebble Beach CA 93953

WOOLLEY, TYLER ANDERSON, b Los Angeles, Calif, Apr 3, 18; m 80; c 3. ZOOLOGY. *Educ:* Univ Utah, BS, 39, MS, 41; Ohio State Univ, PhD(entom), 48. *Prof Exp:* Sr asst comp anat, Univ Utah, 38-39; asst zool, Ohio State Univ, 46-47, asst instr, 47-48; from asst prof to assoc prof, 48-58, PROF ZOOL, COLO STATE UNIV, 58- *Res:* Acarology; Oribatei; taxonomy; biology; invertebrate zoology. *Mailing Add:* Dept Zool & Entom Colo State Univ Ft Collins CO 80523

WOOLRIDGE, EDWARD DANIEL, b Jackson, Miss, June 7, 32; m 63; c 2. DENTISTRY, FORENSIC DENTISTRY. *Educ:* Lynchburg Col, BS, 53; Med Col Va, DDS, 57; George Washington Univ, MEd, 81; Am Bd Forensic Odontol, dipl, 71. *Prof Exp:* Private pract dent, Farmville, Va, 59-62; liaison officer, USCG, NY, 65-66, sr dent officer, 69-74; sr dent officer, USCG Acad, 74-77; CHIEF DENT OFFICER, USCG HQ, 77- *Concurrent Pos:* Consult, Off Chief Med Examrs, Rockland, NY, 71-, vis asst prof, Med Univ SC, 75-; clin instr, Sch Dent Med, Tufts Univ, 76-; clin field instr, Baltimore Col Dent Surg, Univ Md, 78- *Honors & Awards:* Commendation Medal, USPHS, 73. *Mem:* Fel Am Col Dent; fel Am Acad Forensic Sci; Soc Med Jurisprudence. *Mailing Add:* Commandant G-KOM USCG Washington DC 20590

WOOLRIDGE, ROBERT LEONARD, b Garretson, SDak, Oct 13, 19; m 46; c 1. MEDICAL BACTERIOLOGY, IMMUNOLOGY. *Educ:* Univ SDak, BA, 41; Univ Chicago, MSc, 43; Keio Univ, Japan, DSc(microbiol), 60. *Prof Exp:* Res asst path, Sch Med, Univ Chicago, 43-48; microbiologist, US Naval Med Res Unit 4, 48-59 & Unit 2, 59-66; health scientist adminr, US-Japan Coop Med Sci Prog, NIH, 66-68, chief, Pac Off, Japan, 68-71, Dept HEW health scientist liaison officer to Dept State, 71-72, chief prev br, Cancer Control Prog, 72-76, prog dir, Diag Res & Prev, Div Div Cancer Res Resources & Ctrs, 76-78, exec scientist, Univ Asn Res & Educ Pathol, 78-85; consult, 85-89; RETIRED. *Concurrent Pos:* Res assoc, Univ Chicago, 49-52; asst clin prof, Univ Wash, 61-65; tech consult, WHO, Geneva, 62-66. *Honors & Awards:* Provincial Health Dept Award, Rep China, 66. *Mem:* Fel AAAS; Am Soc Microbiol; Am Sci Prev Oncol; Asia-Pac Acad Ophthal; NY Acad Sci; Am Pharmaceut Asn; Fed Am Soc Exp Biol. *Res:* Vaccine development; immunological prophylaxis; laboratory diagnosis of respiratory viruses, arboviruses and the trachomainclusion conjunctivitis agents; epidemiology and chemoprophylaxis studies in trachomatous children in the Far East. *Mailing Add:* 2318 Lost Rd Martinsburg WV 25401

WOOLSEY, CLINTON NATHAN, b Brooklyn, NY, Nov 30, 04; m 42; c 3. NEUROPHYSIOLOGY. *Educ:* Union Col, NY, AB, 28; Johns Hopkins Univ, MD, 33. *Hon Degrees:* ScD, Union Col, 68. *Prof Exp:* Asst physiol, Sch Med, Johns Hopkins Univ, 33-34, instr, 34-39, from assoc to assoc prof physiol, 41-48; Charles Sumner Slichter res prof neurophysiol, 48-75, dir lab neurophysiol, 60-73, biomed unit coordr, Waisman Ctr, 73-78; EMER PROF NEUROPHYSIOL, UNIV WIS-MADISON, 75- *Concurrent Pos:*

Rockefeller Found fel, Johnson Found, Univ Pa, 38-39; fel neuromuscular physiol, Johns Hopkins Univ, 39-41; hon lectr, Albany Med Col, 66; hon mem fac med, Univ Chile. Spec consult, Ment Health Study Sect, NIH, 52-53 & neurol study sect, 53-56 & 57-58; mem nat adv coun, Nat Inst Neurol Dis & Blindness, 58-62, spec consult neurol prog proj comt, 63-67, mem bd sci counr, 65-69; mem div med sci, Nat Res Coun, 52-58 & Inter-Coun Adv Comt Career Res Professorships, 60-61; mem med exchange mission pharmacol & physiol of nerv syst, USSR, 58; US organizer, Int Brain Res Orgn-UNESCO vis sem, Chile, 66, workshop, Shanghai, China, 80 & workshop, Algiers, Algeria, 81; mem, Nat Acad Sci-Colciencias Panel Grad Educ & Res Biol, Colombian Univs, 72. *Honors & Awards:* Medalist, Univ Brussels, 68; James Arthur lectr, Am Mus Nat Hist, New York, 52; Hines lectr, Emory Univ, 61; Bishop lectr, Wash Univ, 61; Von Monakow lectr, Univ Zurich, 68; J Hughlings Jackson Mem lectr, Montreal Neurol Inst, 71; Donald D Matson lectr, Harvard Med Sch, 75; J L O'Leary Mem lectr, 76; Ralph Gerard Award, Soc Neuroscience. *Mem:* Nat Acad Sci; Am Physiol Soc; hon mem Am Neurol Asn; assoc Am Asn Neurol Surg; Soc Neurosci; Int Brain Res Orgn. *Res:* Cerebral localization; sensory and motor. *Mailing Add:* Dept Neurophysiol 627 Waisman Ctr Univ Wis Madison WI 53705-2280

WOOLSEY, GERALD BRUCE, b Brooks, Ga, Aug 16, 37; m 60; c 3. PHYSICAL CHEMISTRY. *Educ:* Univ SC, BS, 60, PhD(phys chem), 67. *Prof Exp:* Chemist, Tenn Corp, Cities Serv Co, 60-63; res chemist, 67-69, develop supvr, 69-71, res supvr, 71-74, mfg supvr, 74-76, bus specialist, 76-78, STAFF RES CHEMIST, SAVANNAH RIVER PLANT, E I DU PONT DE NEMOURS & CO, INC, 79- *Mem:* Am Chem Soc; Sigma Xi. *Res:* Thermodynamics of solutions; hydration in solutions of concentrated electrolytes; amide solutions; polyester films; nuclear waste disposal; nuclear propulsion. *Mailing Add:* 114 Lynwood Dr Aiken SC 29803

WOOLSEY, MARION ELMER, b Croft, Kans, July 27, 19; m 40; c 3. MICROBIOLOGY. *Educ:* Univ Tex, Austin, BA, 64, MA, 66, PhD(microbiol), 68. *Prof Exp:* Res assoc immunol, Univ Tex, Austin, 68, asst prof microbiol, 68-70; asst prof, 70-76, ASSOC PROF MICROBIOL, UNIV TULSA, 76- *Concurrent Pos:* NIH fel, Univ Tex, Austin, 68. *Mem:* Am Soc Microbiol. *Res:* Basic and applied research in medical microbiology, immunology and immunochemistry. *Mailing Add:* 5702 E Seventh St Tulsa OK 74112

WOOLSEY, NEIL FRANKLIN, b Tieton, Wash, Apr 30, 35; m 56; c 5. ORGANIC CHEMISTRY OF COAL, ORGANOMETALLIC ARENE CHEMISTRY. *Educ:* Univ Portland, BS, 57; Univ Wis, PhD(org chem), 62. *Prof Exp:* NSF fel, Imp Col, London, 61-63; res assoc org chem, Iowa State Univ, 63-65; from asst prof to assoc prof, 65-77, PROF CHEM, UNIV NDAK, 77- *Mem:* Am Chem Soc; Sigma Xi. *Res:* Reductive cleavage of aryl ethers; organometallic arene chemistry; structure, reactions and analytical chemistry of coal and coal derived materials. *Mailing Add:* 1819 N Fourth St Grand Forks ND 58201

WOOLSEY, ROBERT M, b Chicago, Ill, May 30, 31; m 67; c 1. MEDICINE, NEUROLOGY. *Educ:* St Louis Univ, BS, 53, MD, 57. *Prof Exp:* Intern med, St Louis Univ Hosp, 57-58; resident neurol, Univ Mich Hosp, 58-61; from instr to assoc prof, 62-75, PROF NEUROL, ST LOUIS UNIV, 75- *Concurrent Pos:* Fel neuropath, Col Physicians & Surgeons, Columbia Univ, 61-62. *Mem:* AMA; Asn Res Nerv & Ment Dis; Am Acad Neurol; Am Electroencephalog Soc; Am Paraplegia Soc. *Res:* Electroencephalography. *Mailing Add:* Dept Neurol 1325 S Grand Ave St Louis MO 63104

WOOLSEY, THOMAS ALLEN, b Baltimore, Md, Apr 17, 43; m 69; c 2. NEUROANATOMY. *Educ:* Univ Wis-Madison, BS, 65; Johns Hopkins Univ, MD, 69. *Prof Exp:* Intern surg, Barnes Hosp St Louis, Mo, 69-70; from asst prof to assoc prof, 71-83, coordr neural sci prog, 80-84, SR MCDONNELL NEUROSCIENTIST, MCDONNEL CTR STUDIES HIGHER BRAIN FUNCTION, WASH UNIV MED SCH, 82-, PROF NEUROL, NEUROSURG, ANAT, NEUROBIOL, CELL BIOL & PHYSIOL, 84-, GEORGE H & ETHEL R BISHOP SCHOLAR NEUROSCI, 84- *Concurrent Pos:* NIH fel anat, Med Sch, Wash Univ, 70-71, Nat Inst Neurol Dis & Stroke grant, 72-, consult NIH, NSF; reviewer Brain Res, J Comp Neurol Sci; dir, James L O'Leary Div Exper Neurol & Neurobiol Surg, Med Sch, Wash Univ, 84- *Honors & Awards:* Pickney J Harman Lectr, Am Asn Anatomists; McKnight Neurosci Develop Award. *Mem:* Am Asn Anat; Soc Neurosci; Am Neurol Asn; Am Acad Neurol; Microvascular Soc; Develop Neurobiol. *Res:* Structure and function of somatosensory areas of the cerebral cortex; computer applications to morphology. *Mailing Add:* Dept Exp Neurol & Neurosurg Wash Univ Sch Med Box 8057 St Louis MO 63110

WOOLSON, EDWIN ALBERT, b Takoma Park, Md, Oct 2, 41; m 61; c 4. PESTICIDE CHEMISTRY. *Educ:* Univ Md, BS, 63, MS, 66, PhD(soil chem), 69. *Prof Exp:* Analyst soil testing, Univ Md, 62; phys sci aide, USDA, 62-63, chemist, 63-65, anal chemist, 65-67, res chemist, 67-85; mgr chem, Am Biogenics Corp, 85-86; PRES & DIR CHEM, EPL-BIO-ANALYTICAL SERV, 86- *Mem:* Am Chem Soc; Am Soc Agron; Soil Sci Soc Am; Weed Sci Soc Am; Coun Agr Sci Technol. *Res:* Behavior and fate of arsenic, herbicides and insecticides in soil and water; method development for pesticides in soils, plants and water; toxicity of pesticides to plants; impurities in pesticides; bioaccumulation; residue analysis in soil, plants and water. *Mailing Add:* EPL Bio-Analytical Servs Inc PO Box 1708 Decatur IL 62525-1708

WOOLSTON, DANIEL D, b Churchville, NY, Oct 18, 26; m 50; c 2. ELECTRONICS. *Educ:* Iowa State Col, BS, 51. *Prof Exp:* Physicist & electronic scientist, US Naval Ord Lab, 51-63; vpres, 63-75, PRES, UNDERWATER SYSTS, INC, 75- *Mem:* Inst Elec & Electronics Engrs. *Res:* Underwater telemetry; hydroacoustic mine research and development; underwater explosive parameter data gathering and analysis for underwater mine research programs; sediment velocimeter research, design and development of the instrumentation; underwater sound propagation studies. *Mailing Add:* 11810 Pittson Rd Wheaton MD 20906

WOOLVERTON, WILLIAM L, b Birmingham, Ala, Nov 28, 50. BEHAVIORAL PHARMACOLOGY. *Educ:* Univ Chicago, PhD(pharm), 77. *Prof Exp:* Asst prof, 80-84, ASSOC PROF PHARM, UNIV CHICAGO, 84- *Concurrent Pos:* Pres, Int Study Group Investingating Drugs as Reinforcers, 85- *Mem:* Am Soc Pharm & Exp Therapeut; Soc Neurosis; Behav Pharm Soc. *Mailing Add:* Dept Pharmacol & Physiol Sci Univ Chicago 947 E 58 St Chicago IL 60637

WOOSLEY, RAYMOND LEON, b Roundhill, Ky, Oct 2, 42; m 84; c 2. CLINICAL PHARMACOLOGY. *Educ:* Western Ky State Univ, BS, 64; Univ Louisville, PhD(pharmacol), 67; Univ Miami, MD, 73. *Prof Exp:* Sr pharmacologist, Meyer Labs, Inc, 68-69, dir res pharmaceut, 69-71; from intern to resident med, Vanderbilt Univ Hosp, Vanderbilt Univ, 73-74, fel clin pharmacol, 74-75, from asst prof to prof med & pharmacol, 75-88; PROF & CHMN, PHARMACOL DEPT, GEORGETOWN UNIV, 88- *Concurrent Pos:* NIH fel pharmacol, Med Sch, Univ Louisville, 67-68, lectr, 69-71; instr med & clin pharmacol, Vanderbilt Univ, 75. *Mem:* Soc Exp Biol & Med; Am Soc Pharmacol & Exp Therapeut; Am Col Clin Pharmacol; fel Am Col Physicians; fel Am Col Cardiologists. *Res:* Drug induced lupus erythematosus; and clinical pharmacology of new antiarrhythmic drugs. *Mailing Add:* Dept Pharmacol Georgetown Univ 3900 Reservoir Rd NW Washington DC 20007

WOOSLEY, ROYCE STANLEY, b Caneyville, Ky, June 17, 34; m 59; c 2. ORGANIC CHEMISTRY. *Educ:* Western Ky Univ, BS, 56; Univ Conn, MS, 59; Ohio Univ, PhD(org chem), 67. *Prof Exp:* Sect leader tech serv, Olin Mathieson Chem Corp, 58-62; from asst prof to assoc prof chem, 66-74, chmn dept, 69-75 & 88-89, PROF CHEM, WESTERN CAROLINA UNIV, 74- *Mem:* Am Chem Soc; Sigma Xi; Int Union Pure & Appl Chem. *Res:* Organic free-radicals; Friedel-Crafts reactions; terpenes. *Mailing Add:* PO Box 1164 Cullowhee NC 28723

WOOSLEY, STANFORD EARL, b Texarkana, Tex, Dec 8, 44; div. ASTROPHYSICS. *Educ:* Rice Univ, BA, 66, MS, 69, PhD(astrophys), 71. *Prof Exp:* Res assoc, Rice Univ, 71-73; res fel, Calif Inst Technol, 73-75; from asst prof to assoc prof, 75-83, PROF ASTRON, LICK OBSERV, UNIV CALIF, SANTA CRUZ, 83-, CHMN DEPT, 83- *Concurrent Pos:* Consult, Lawrence Livermore Lab, 74-, Los Alamos Lab, 88-; NSF & NASA grants, 77-83; mem coun, Am Astron Soc. *Honors & Awards:* Fel Am Phys Soc. *Mem:* Am Phys Soc; Am Astron Soc; Int Astron Union. *Res:* Nuclear astrophysics, nucleosynthesis supernovae, gamma-ray bursts. *Mailing Add:* UCO/Lick Observ Univ Calif Santa Cruz CA 95064

WOOSTER, HAROLD ABBOTT, b Hartford, Conn, Jan 3, 19; m 41, 68; c 4. INFORMATION SCIENCE. *Educ:* Syracuse Univ, AB, 39; Univ Wis, MS, 41, PhD(physiol chem), 43. *Prof Exp:* Asst, Toxicity Lab, Univ Chicago, 43-46; res assoc, Pepper Lab, Univ Pa, 46-47; sr fel, Mellon Inst, 47-56; dir res commun, Air Force Off Sci Res, 56-59, chief info sci div, 59-62, dir info sci, 62-70; chief res & develop br, Nat Libr Med, Lister Hill Nat Ctr Biomed Commun, 70-74, sr info scientist, 74-85 spec asst prog develop, 74-85; RETIRED. *Concurrent Pos:* Exec secy panel info sci & tech, Comt Sci & Tech Info, Off Dir Defense Res & Eng, 65-66; adj instr, Grad Sch Libr Sci, Drexel Inst Technol, 67. *Mem:* AAAS; Am Soc Info Sci; Sigma Xi. *Res:* Biomedical communications; computer-aided instruction; medical television; information storage and retrieval. *Mailing Add:* 8807 Mead St Bethesda MD 20817-3223

WOOSTER, WARREN SCRIVER, b Westfield, Mass, Feb 20, 21; m 48; c 3. OCEANOGRAPHY. *Educ:* Brown Univ, BSc, 43; Calif Inst Technol, MS, 47; Univ Calif, PhD, 53. *Prof Exp:* Asst res oceanogr, Scripps Inst, Univ Calif, 51-58, assoc res oceanogr, 58-61; dir off oceanog, UNESCO, 61-63; prof oceanog, Scripps Inst Oceanog, Univ Calif, 63-73, chmn grad dept, 67-69; prof oceanog & dean, Rosenstiel Sch Marine & Atmospheric Sci, Univ Miami, 73-76; dir, 79-81, PROF MARINE STUDIES & FISHERIES, INST MARINE STUDIES, UNIV WASH, 76- *Concurrent Pos:* Dir invests, Coun Hydrobiol Invests, Peru, 57-58; pres, Sci Comm Oceanic Res, 68-72, Int Explor Sea, 82-85. *Mem:* Fel Am Geophys Union; fel Am Meteorol Soc. *Res:* Descriptive oceanography of the Pacific Ocean; physical, chemical and fishery oceanography; ocean affairs. *Mailing Add:* Sch Marine Affairs HF-05 Univ Wash Seattle WA 98195

WOOTEN, FRANK THOMAS, b Fayetteville, NC, Sept 24, 35; m 62; c 3. TECHNICAL MANAGEMENT. *Educ:* Duke Univ, BS, 57, PhD(elec eng), 64. *Prof Exp:* Sr engr, Electronic Res Lab, Corning Glass Works, 64-67; res engr, Res Triangle Inst, 67-71, mgr biomed eng, 71-75, exec asst to pres, 75-80, vpres, 80-89, PRES, RES TRIANGLE INST, 89- *Mem:* Asn Advan Med Instrumentation; Inst Elec & Electronics Engrs; Sigma Xi. *Res:* Medical instrumentation; analysis of pulmonary sound; technology transfer; electronics. *Mailing Add:* Res Triangle Inst PO Box 12194 Research Triangle Park NC 27709

WOOTEN, FREDERICK (OLIVER), b Linwood, Pa, May 16, 28; m 52; c 2. SOLID STATE PHYSICS. *Educ:* Mass Inst Technol, BS, 50; Univ Del, PhD(chem), 55. *Prof Exp:* Staff physicist, All Am Eng Co, 54-57; res chemist, Lawrence Livermore Lab, 57-72; vchmn dept appl sci, 72-73, PROF APPL SCI, UNIV CALIF, DAVIS, 72-, CHMN DEPT, 73- *Concurrent Pos:* Vis prof, Drexel Univ, 64; lectr, Univ Calif, Davis, 65-72; vis prof, Chalmers Univ Technol, Sweden, 67-68; consult, Lawrence Livermore Nat Lab, 72-; vis prof, Heriot-Watt Univ, Scotland, 79, Trinity Col, Ireland, 86. *Mem:* AAAS; Am Phys Soc; NY Acad Sci; Mat Res Soc. *Res:* The structure of amorphous solids. *Mailing Add:* Dept Appl Sci Univ Calif Davis CA 95616

WOOTEN, JEAN W, b Douglasville, Ga, Jan 11, 29; m 52; c 1. BOTANY. *Educ:* NGa Col, BS, 46; Fla State Univ, MS, 64, PhD(bot), 68. *Prof Exp:* Asst prof bot, Iowa State Univ, 68-74; asst prof, 74-77, ASSOC PROF BIOL, UNIV SOUTHERN MISS, 78- *Mem:* Am Soc Plant Taxonomists; Soc Study Evolution; Int Asn Plant Taxon; Bot Soc Am; Sigma Xi. *Res:* Evolution and systematics in aquatic vascular plants. *Mailing Add:* Southern Sta Box 5018 Hattiesburg MS 39406

WOOTEN, MARIE W, CANCER RESEARCH. *Educ:* Tex Women's Univ, PhD(biochem), 83. *Prof Exp:* RES ASSOC, COLD SPRING HARBOR LABS, 85- & UNIV ALA, BIRMINGHAM, 86- *Res:* Protein kinase; oncogenesis. *Mailing Add:* Dept Zool Auburn Univ 331 Funchess Hall Auburn AL 36849-5414

WOOTEN, WILLIS CARL, JR, b Homerville, Ga, Mar 9, 22; m 49; c 3. POLYMER CHEMISTRY. *Educ:* Univ NC, PhD(chem), 50. *Prof Exp:* Sr res chemist, 51-60, admin asst, 60-65, div head, res lab, 65-75, DIR, POLYMER RES DIV, TENN EASTMAN CO, 76- *Mem:* Am Chem Soc; Sigma Xi. *Res:* Synthetic fibers and plastics; organic chemistry. *Mailing Add:* 105 Pine Lane Chapel Hill NC 27514

WOOTTEN, HENRY ALWYN, b Salisbury, Md, May 3, 48; m 80; c 2. ASTRONOMY, ASTROPHYSICS. *Educ:* Univ Md, BS, 70; Univ Tex, MS, 76, PhD(astron), 78. *Prof Exp:* Res fel astron, Calif Inst Technol, 78-81; res assoc astron, Rensselaer Polytech Inst, 82; SCIENTIST, NAT RADIO ASTRON OBSERV, 82- *Mem:* Am Astron Soc; Int Astron Union. *Mailing Add:* Nat Radio Astron Observ Edgemont Rd Charlottesville VA 22903-2475

WOOTTEN, MICHAEL JOHN, b Somersham, Eng, Mar 2, 44; m 66; c 2. PHYSICAL CHEMISTRY, WATER CHEMISTRY. *Educ:* Univ Leicester, BS, 66, PhD(phys chem), 69. *Prof Exp:* Res fel chem, Carnegie-Mellon Univ, 69-71; sr res fel phys chem, Univ Southampton, Eng, 71-73; sr engr, Westinghouse Europe, Belg, 73-75, sr engr, 75-77, mgr, Westinghouse Res & Develop Ctr, Pittsburgh, 77-80, MGR, NUCLEAR TECHNOL DIV, PITTSBURGH, 80-, MGR DIAGNOSTICS & MONITORING, POWERS SYSTS DIV, WESTINGHOUSE, 80- *Mem:* Nat Asn Corrosion Engrs; Royal Inst Chem, Eng; Electrochem Soc Eng. *Res:* The physical chemistry of aqueous solutions at high temperatures and pressures with particular reference to power generation. *Mailing Add:* Westinghouse Powers Systs Div Res & Develop Ctr Bldg 701 PO Box 2728 Pittsburgh PA 15230

WOOTTERS, WILLIAM KENT, b Houston, Tex, July 7, 51; m 82; c 2. QUANTUM INFORMATION THEORY, QUANTUM CHAOS. *Educ:* Stanford Univ, BS, 73; Univ Tex Austin, PhD(physics), 80. *Prof Exp:* Instr & postdoctoral res assoc physics, Univ Tex, Austin, 80-82; asst prof, 82-89, ASSOC PROF PHYSICS, WILLIAMS COL, 89- *Concurrent Pos:* Vis res assoc, Univ Tex, Austin, 85-86; vis assoc prof, Santa Fe Inst, 89-90. *Mem:* Am Phys Soc; Sigma Xi; Coun Undergrad Res. *Res:* Foundations of quantum mechanics, with special emphasis on optimal measurement strategies, quantum information theory, relation between quantum and classical mechanics, and quantum chaos. *Mailing Add:* Dept Physics Williams Col Williamstown MA 01267

WOOTTON, DONALD MERCHANT, b Paonia, Colo, Apr 13, 16; m 41; c 1. PARASITOLOGY, ZOOLOGY. *Educ:* Santa Barbara State Col, BA, 41; Univ Wash, MS, 43; Stanford Univ, PhD(biol), 49. *Prof Exp:* Actg instr zool & bot, Santa Barbara State Col, 41-42, instr zool, 49-52; asst prof, Univ Calif, Santa Barbara, 52-56; prof, 57-80, dept chmn, 76-80, EMER PROF BIOL, CALIF STATE UNIV, CHICO, 80- *Concurrent Pos:* Vis prof, Univ Mich Biol Sta, Pellston, Mich, 57-72. *Mem:* Am Soc Parasitol. *Res:* Helminths; branchipods; limnology. *Mailing Add:* 811 Terry Dr Emmett ID 83617

WOOTTON, JOHN FRANCIS, b Penn Yan, NY, May 31, 29; m 59; c 4. BIOCHEMISTRY. *Educ:* Cornell Univ, BS, 51, MS, 53, PhD(biochem), 60. *Prof Exp:* Clin chemist, Clifton Springs Sanitarium & Clin, NY, 56; Nat Found fel chem, Univ Col London, 60-62; from asst prof to prof physiol chem, 62-79, assoc dean, Grad Sch, 80-83, PROF BIOCHEM, NY STATE COL VET MED, CORNELL UNIV, 79- *Concurrent Pos:* Hon res asst, Univ London, 62; vis scientist, Lab Molecular Biol, Univ Postgrad Med Sch, Cambridge, Eng, 69-70, Nat Inst Med Res, Mill Hill, London, 85-86; temp sr res assoc, Stanford Univ Med Sch, 77-78. *Mem:* Am Chem Soc; Am Soc Biol Chemists; AAAS; Sigma Xi. *Res:* Enzymology; proteolytic enzymes; regulatory & transport proteins; relationship of protein structure to function; synthesis and applications of photolabile compounds. *Mailing Add:* Dept/Sect Physiol 817 VRT Cornell Univ Ithaca NY 14853

WOOTTON, PETER, b Peterborough, Eng, Apr 30, 24; m 47; c 3. MEDICAL PHYSICS, RADIOLOGY. *Educ:* Univ Birmingham, BSc, 44. *Prof Exp:* Physicist, Res & Develop Labs, Farrow's Br, Reckitt & Coleman Ltd, 44-48; radiation physicist, Royal Infirmary, Glasgow, Scotland, 48-51; instr radiol physics, Univ Tex M D Anderson Hosp, 51-53; radiation physicist, Tumor Inst, Swedish Hosp, Seattle, Wash, 53-64; from asst prof to assoc prof radiol, 64-72, PROF RADIOL, UNIV WASH, 72- *Concurrent Pos:* Instr, Med Sch, Univ Ore, 54-60, Univ Seattle, 56 & Penrose Cancer Hosp, Colorado Springs, Colo, 56-57; clin asst prof & radiation physicist, Univ Wash, 59-64; mem tech adv bd radiation control, Dept Health, Wash, 62-68, comt radiation oncol studies, Nat Cancer Inst, 77-81 & sci comt, 25, Nat Coun Radiation Protection & Measurements, 67-83. *Mem:* Am Asn Physicists Med (pres, 78); Soc Nuclear Med; fel Am Col Radiol; Am Soc Therapeut Radiol; Brit Inst Physics; Am Col Med Physics. *Res:* Applications of radiation physics in medicine, especially dosimetry of all types of ionizing radiations and the effects of physical parameters such as high pressure oxygen and pulsed radiation in radiobiology; fast neutron therapy. *Mailing Add:* NN136 Univ Hosp RC-08 Univ Wash Seattle WA 98195

WOPSCHALL, ROBERT HAROLD, b Glendale, Calif, May 1, 40; m 62; c 2. PHYSICAL CHEMISTRY. *Educ:* Harvey Mudd Col, BS, 62; Univ Wis, PhD(phys chem), 67. *Prof Exp:* RES CHEMIST, E I DU PONT DE NEMOURS & CO, INC, 66- *Mem:* Am Chem Soc; Sigma Xi. *Res:* Stationary electrode polarography; adsorption of electroactive species on electrodes and coupled chemical reactions; photopolymerization; electroless deposition; photoresists; printed circuits manufacturing; photographic science. *Mailing Add:* 1118 S Dolton Ct Wilmington DE 19810

WORDEN, DAVID GILBERT, b Minneapolis, Minn, Mar 9, 24; m 47; c 2. SOLID STATE PHYSICS, PLASMA PHYSICS. *Educ:* Earlham Col, AB, 50; Iowa State Univ, PhD(physics), 56. *Prof Exp:* Physicist, Gen Elec Res Lab, 56-61; mgr surface physics sect, Electro-Optical Systs, Inc, 61-65, electron device res sect, 65-66, electron & image device dept, 66-67; chmn, Dept Physics, NDak State Univ, 67-68, prof physics & acad vpres, 68-79; mgr, Univ Relations, Gen Elec Corp Res & Develop, 79-85; RETIRED. *Concurrent Pos:* Assoc part-time prof, Calif State Col, Los Angeles, 65-67. *Res:* Properties of thin films; physical and chemical adsorption; electron emission; ion production on surfaces; low energy gaseous electrical discharges in magnetic fields; university industry relationships; manpower; science and engineering policy. *Mailing Add:* 4028 Chaucer Place Slingerlands NY 12159

WORDEN, EARL FREEMONT, JR, b Portsmouth, NH, Nov 30, 31; m 60; c 1. PHYSICAL CHEMISTRY. *Educ:* Univ NH, BS, 53, MS, 55; Univ Calif, PhD(chem), 59. *Prof Exp:* CHEMIST, LAWRENCE LIVERMORE NAT LAB, UNIV CALIF, 58- *Honors & Awards:* Louis A Stroit Award, Soc Appl Spectros, 85. *Mem:* Am Chem Soc; Sigma Xi; Optical Soc Am; Soc Appl Spectros. *Res:* Atomic and molecular optical spectroscopy; laser isotope separation; laser spectroscopy of atoms. *Mailing Add:* Lawrence Livermore Nat Lab Univ Calif Livermore CA 94550

WORDEN, PATRICIA BARRON, SCIENCE EDUCATION. *Educ:* Harvard Univ, PhD(biol), 64. *Prof Exp:* MEM, ARLINGTON SCH COMT, ARLINGTON TWP, 79- *Mailing Add:* 27 Jason St Arlington MA 02174-6446

WORDEN, PAUL WELLMAN, JR, b San Angelo, Tex, Mar 1, 45. PHYSICS. *Educ:* Rice Univ, BA, 67; Stanford Univ, MS, 69, PhD(physics), 76. *Prof Exp:* Res asst, 68-76, scholar, 76-78, res assoc physics, 78-80, SR RES ASSOC, STANFORD UNIV, 80- *Concurrent Pos:* Consult, Jet Propulsion Lab, Pasadena, Calif, 83-84. *Mem:* AAAS; Am Phys Soc; Sigma Xi. *Res:* Experimental gravitation; equivalence principle; cryogenic applications to basic research. *Mailing Add:* W W Hansen Lab Physics Stanford Univ Stanford CA 94305

WORDEN, SIMON PETER, b Mt Clemens, Mich, Oct 21, 49. SOLAR PHYSICS. *Educ:* Univ Mich, BS, 71; Univ Ariz, PhD(astron), 75. *Prof Exp:* Res asst astron, Kitt Peak Nat Observ, 71-75; astrophysicist, Sacramento Peak Observ, Air Force Geophys Lab, 75-80; MEM FAC, DEPT ASTRON, UNIV CALIF, LOS ANGELES, 80- *Mem:* Am Astron Soc; Royal Astron Soc. *Res:* The study of large-scale convective motions on the sun and observation and interpretation of stellar phenomena related to solar surface activity; development of techniques for high resolution imaging through the Earth's atmosphere. *Mailing Add:* 6757 N 27th St Arlington VA 22213

WORDINGER, ROBERT JAMES, b Philadelphia, Pa, Feb 5, 45; m 74; c 4. ANATOMY, ELECTRON MICROSCOPY. *Educ:* Pa State Univ, BS, 67; Clemson Univ, MS, 69, PhD(animal physiol), 72. *Prof Exp:* Fel physiol, Sch Vet Med, Univ Pa, 72-73; asst prof biol, St Bonaventure Univ, 73-76; asst prof path, Sch Med Sci, Univ Ark, 76-77; asst prof, 78-80, ASSOC PROF ANAT, TEX COL OSTEOP MED, UNIV NTEX, 80- *Concurrent Pos:* Head histol, Nat Ctr Toxicol Res, 76-77; vchmn, Dept Anat, Tex Col Osteop Med, NTex State Univ, 80- *Mem:* Soc Study Reproduction; Histochem Soc; Am Soc Cell Biol. *Res:* Estrogenic changes in the mammalian reproductive system using histochemistry and electron microscopy; factors influencing in vitro embryo development and differentiation within mammals; implantation. *Mailing Add:* Dept Anat Univ NTex Tex Col Osteop Med Ft Worth TX 76107

WOREK, WILLIAM MARTIN, b Joliet, Ill, May 7, 54; m 85; c 1. HEAT & MASS TRANSFER, ENERGY SYSTEMS. *Educ:* Ill Inst Technol, BS, 76, MS, 77 & PhD(mech eng), 80. *Prof Exp:* Instr, Ill Inst Technol, 77-80, vis asst prof mech eng, 80-83, asst prof mech & aero eng, 83-86; ASSOC PROF, UNIV ILL, CHICAGO, 86- *Mem:* Am Soc Mech Engrs; Am Soc Heating, Refrig & Air Conditioning Engrs; Sigma Xi. *Res:* Fundamental heat and mass transfer problems which are related to specific problems encountered in advanced energy systems; design and simulation of hybrid heat transfer components and advanced energy systems. *Mailing Add:* Dept Mech Eng M-C 251 Univ Ill Box 4348 Chicago IL 60680

WORF, GAYLE L, b Garden City, Kans, Nov 17, 29; m 52; c 2. PLANT PATHOLOGY. *Educ:* Kans State Univ, BS, 51, MS, 53; Univ Wis, PhD(plant path), 61. *Prof Exp:* County agt, Kans State Univ, 55-58; plant pathologist, Iowa State Univ, 61-63; PLANT PATHOLOGIST, UNIV WIS-MADISON, 63- *Mem:* Am Phytopath Soc. *Res:* Interpreting current research in plant pathology and analyzing its application to field situations; diagnostic procedures; economical appraisal of disease outbreaks; effective control programs. *Mailing Add:* Dept Plant Path Univ Wis Madison 284 Russell Labs Madison WI 53706

WORGUL, BASIL VLADIMIR, b New York, NY, June 30, 47; m 69; c 2. CELL BIOLOGY, RADIATION BIOLOGY. *Educ:* Univ Miami, BS, 69; Univ Vt, PhD(zool), 74. *Prof Exp:* Staff assoc, Columbia Univ, 74-75, NIH fel & res assoc cell biol, 75-78, from asst prof to assoc prof, Dept Ophthal, Col Physicians & Surgeons, 78-90, PROF RADIATION BIOL, 90-, DIR, EYE RADIATION & ENVIRON RES LAB, 85- *Concurrent Pos:* Consult, Cataract Panel Nat Adv Eye Coun, 80-83; adj prof, Dept Biol Sci, Fordham Univ, 82-; adv, NRCP, 83-; mem, Non-ionizing Radiation Effects Panel, US Dept Labor, 84- *Honors & Awards:* Robert E McCormick Res Scholar. *Mem:* Soc Gen Physiol; Radiation Res Soc; AAAS; Asn Res Vision & Ophthal; Am Soc Cell Biol. *Res:* The cytopathomechanism of radiation cataractogenesis; control of growth and differentiation in ocular epithelia. *Mailing Add:* Col Physicians & Surgeons Eye Radiation Res Lab Columbia Univ 630 W 168th St New York NY 10032

WORK, CLYDE E(VERETTE), b Bridgeport, Nebr, Jan 31, 24; m 48; c 5. MATERIALS SCIENCE & MECHANICAL ENGINEERING. *Educ:* Univ Ill, BS, 45, MS, 48, PhD, 52. *Prof Exp:* Res asst, Univ Ill, 46-47, instr, 47-51, res assoc, 51-52, asst prof, 52-53; assoc prof, Rensselaer Polytech Inst, 53-57; head dept, Mich Technol Univ, 57-69, prof eng mech, 57-68, assoc dean eng, 69-84; DEAN ENG, WESTERN NEW ENG COL, 84- *Concurrent Pos:* UNESCO expert eng mat, Maulana Azad Col Technol, Bhopal, India, 68-69; vis prof mech eng, Univ Ilorin, Nigeria, 78-79. *Honors & Awards:* Dudley Medal, Am Soc Testing & Mat, 54; Western Elec Award, Am Soc Eng Educ, 68, Distinguished Educator Award, Mech Div, 84; Tatnall Award, Soc Exp Mech, 78. *Mem:* Am Soc Testing & Mat; fel Soc Exp Stress Analysis (pres, 72-73); Am Soc Eng Educ; Am Acad Mechs. *Res:* Mechanical behavior of engineering materials; fatigue of metals; effects of fluctuation in stress amplitude; temperature time effects; torsional loading. *Mailing Add:* Sch Eng Western New Eng Col Springfield MA 01119-2684

WORK, HENRY HARCUS, b Buffalo, NY, Nov 11, 11; m 45; c 4. PEDIATRICS, PSYCHIATRY. *Educ:* Hamilton Col, AB, 33; Harvard Univ, MD, 37; Am Bd Pediat, dipl, 47; Am Bd Psychiat & Neurol, dipl, 50, cert child psychiat, 60. *Prof Exp:* Psychiat serv adv, US Children's Bur, 47-49; assoc prof pediat & psychiat, Univ Louisville, 49-55; from assoc prof to prof psychiat, Univ Calif, Los Angeles, 55-72; dep med dir, Am Psychiat Asn, 72-83; RETIRED. *Mem:* Am Psychiat Asn; fel Am Orthopsychiat Asn; Am Pub Health Asn; Am Acad Pediat; Am Col Psychiat. *Res:* Identification of child with mother and other relatives. *Mailing Add:* 7001 Glenbrook Rd Bethesda MD 20814

WORK, JAMES LEROY, b Lancaster, Pa, Feb 6, 35; m 55; c 2. POLYMER CHEMISTRY, PHYSICAL CHEMISTRY. *Educ:* Franklin & Marshall Col, BA, 62; Univ Del, PhD(phys chem), 70. *Prof Exp:* Res scientist, 52-77, RES ASSOC, ARMSTRONG CORK CO, LANCASTER, 77- *Mem:* Am Chem Soc. *Res:* Structure-property relationships in polymers and polymeric composites. *Mailing Add:* 913 Sherry Lane Lancaster PA 17601-2023

WORK, ROBERT WYLLIE, b Chicago, Ill, July 10, 07; wid. SPIDER SILK. *Educ:* Univ Ill, BS, 29; Cornell Univ, PhD, 32. *Prof Exp:* Chemist, Swift & Co, Ill, 29; asst chem, Cornell Univ, 29-32; res engr, Gen Elec Co, 33-41; chief chemist, Celanese Corp Am, 41-43, dir phys res, Res Lab, 43-52, asst mgr res, 52-54, asst mgr opers, Textile Div, 54-55, asst tech dir, 55-56; mgr tech res, Chemstrand Res Ctr, Inc, 57-64; dir textile res & prof textiles, 64-73, EMER PROF TEXTILES, NC STATE UNIV, 73- *Mem:* Am Micros Soc; Am Chem Soc; hon mem Fiber Soc; fel Brit Textile Inst; Am Arachnological Soc. *Res:* Man-made and spider fibers; physico-chemical properties of major and minor ampullate silk fibers of orb-web-building spiders (Araneae), their relations to taxa and to spider behavioral patterns. *Mailing Add:* Centennial Campus, Research Dr NC State Univ PO Box 8301 Raleigh NC 27695-8301

WORK, STEWART D, b Chicago, Ill, Oct 17, 37; m 59; c 4. ORGANIC CHEMISTRY. *Educ:* Oberlin Col, AB, 59; Duke Univ, PhD(org chem), 63. *Prof Exp:* Fel, Duke Univ, 63 & Purdue Univ, 63-64; from asst prof to assoc prof, 64-73, PROF CHEM, EASTERN MICH UNIV, 73- *Concurrent Pos:* Spec asst, Provost, 85- *Mem:* Am Chem Soc; Sigma Xi. *Res:* Base-catalyzed condensation reactions; organo-silicon chemistry. *Mailing Add:* Dept Chem Eastern Mich Univ Ypsilanti MI 48197

WORK, TELFORD HINDLEY, b Selma, Calif, July 11, 21; c 3. BIOLOGY, EPIDEMIOLOGY. *Educ:* Stanford Univ, AB, 42, MD, 46; Univ London, dipl, 49; Johns Hopkins Univ, MPH, 52. *Prof Exp:* Res assoc filariasis, Fiji Islands Colonial Med Serv, 49-51; staff mem virus labs, Rockefeller Found, NY, 52; mem, US Naval Med Res Unit 3, Egypt, 53-54; staff mem virus res ctr, Rockefeller Found, Poona, India, 54-55, dir, 55-58; staff mem virus labs, Rockefeller Found, NY, 58-60; chief virus & rickettsia sect, Commun Dis Ctr, USPHS, 61-67; prof infectious & trop dis, 66-69, prof immunol & microbiol, 69-71; vchmn sch pub health, 66-69, head div infectious & trop dis, 69-73, PROF INFECTIOUS & TROP DIS, MED MICROBIOL & IMMUNOL, MED CTR, UNIV CALIF, LOS ANGELES, 71- *Concurrent Pos:* Mem expert comt virus dis, WHO, 60-; dir, WHO Regional Arbovirus Lab, 61-66; assoc mem comn viral infections, Armed Forces Epidemiol Bd, 61-69; Pan-Am Health Orgn consult, Arg, 62, Venezuela, 64, Jamaica, 66; regents lectr, Univ Calif, Los Angeles, 63; consult, US Army Engrs Medico-Ecol Study, Atlantic-Pac Interoceanic Canal Studies, 67-, Fogarty sr int fel, Australia, 78-79; prof epidemiol & virol, Pahlavi Univ, Shiroz, Iran, 78; vis prof microbiol, Univ Western Australia, NZealand, 78; Arbouiduses, Dept Environ, Papua, New Guinea, 78; epidemologist, Dept Health, Western Australia, 79; consult epidemial, Clark County Health Dist, 80. *Honors & Awards:* Richard Moreland Taylor Award, 81. *Mem:* AAAS; Cooper Ornith Soc; Am Ornith Union; Am Soc Trop Med & Hyg; Am Pub Health Asn. *Res:* Medical ecology, epidemiology and medicine of virus infections, especially arthropod-borne viruses; tropical medicine. *Mailing Add:* Dept Pub Health Univ Calif Los Angeles 405 Hilgard Ave Los Angeles CA 90024

WORK, WILLIAM JAMES, b Carmel, Calif, Feb 23, 48; m 76. POLYMER CHEMISTRY, ORGANIC CHEMISTRY. *Educ:* Univ Santa Clara, BS, 70; Univ Ill, PhD(org chem), 76. *Prof Exp:* CHEMIST POLYMER CHEM, ROHM & HAAS CO, 76- *Mem:* Am Chem Soc; AAAS. *Res:* Morphology of polymer blends and composites; synthesis of grafted polymers. *Mailing Add:* 1288 Burnett Rd Huntington Valley PA 19006-2706

WORKER, GEORGE F, JR, b Ordway, Colo, June 1, 23; m 47; c 4. AGRONOMY, BOTANY. *Educ:* Colo State Univ, BS, 49; Univ Nebr, MS, 53. *Prof Exp:* Asst county agent, Nebr, 49-51, asst in agron, 53; specialist in agron & supt, IMP Valley Agr Ctr, Univ Calif, Davis, 53-87; RETIRED. *Concurrent Pos:* Consult, Kufra Agr Pro, Libya, 67-70, Hawaiian Agron, Iran, 75, N Yeman, 81 & Sahel Area Western Africa, 75. *Mem:* Am Soc Agron. *Res:* Grain sorghum production, plant function, breeding and adaption to desert climate; adaption of other field crops such as barley, sugar beets and flax to southwestern desert areas. *Mailing Add:* PO Box 546 Julian CA 92036

WORKMAN, GARY LEE, b Birmingham, Ala, Apr 21, 40; m 67; c 1. PHYSICAL CHEMISTRY. *Educ:* Col William & Mary, BS, 64; Univ Rochester, PhD(phys chem), 69. *Prof Exp:* Res fel chem, Ohio State Univ, 69-70; Nat Acad Sci res assoc, Marshall Space Flight Ctr, NASA, 70-72; dir, PBR Electronics, Inc, 72-76; prof sci technol, Athens State Col, 76-80; DIR & RES ASSOC, UNIV ALA, HUNTSVILLE, 80- *Concurrent Pos:* Dir, Mat Proc Lab, 89- *Mem:* Am Phys Soc; Soc Advan Mat & Process Eng; Optical Soc Am; Sigma Xi; sr mem Inst Elec & Electronics Engrs. *Res:* Systems interfacing; electrooptics applications; laser welding in space; microgravity materials processing; nondestructive evaluation; materials processing; robotics and industrial automation. *Mailing Add:* Univ Ala RI-A6 Huntsville AL 35899

WORKMAN, JOHN PAUL, b Salem, Ore, Feb 18, 43; m 64; c 2. RANGE MANAGEMENT, AGRICULTURAL ECONOMICS. *Educ:* Univ Wyo, BS, 65; Utah State Univ, MS, 67, PhD(range econ), 70. *Prof Exp:* From asst prof to assoc prof, 70-80, PROF RANGE ECON, UTAH STATE UNIV, 81- *Concurrent Pos:* Mem, Rangeland Mgt Comt, Nat Acad Sci; mem, Range Sci Panel, NSF; range mgt consult, 80- *Mem:* AAAS; Soc Range Mgt; Western Agr Econ Asn. *Res:* Economics of range utilization, range improvement and range livestock production. *Mailing Add:* Dept Range Sci Utah State Univ Logan UT 84322-5230

WORKMAN, MARCUS ORRIN, b Canton, Ohio, Sept 20, 40. INORGANIC CHEMISTRY. *Educ:* Manchester Col, BA, 62; Ohio State Univ, PhD(inorg chem), 66. *Prof Exp:* Teaching assoc chem, Ohio State Univ, 65-66; res assoc, Northwestern Univ, 66-67; asst prof chem, Univ Va, 67-74; ASSOC PROF CHEM, THOMAS NELSON COMMUNITY COL, 75- *Concurrent Pos:* Adv Res Projs Agency fel, 66-67. *Mem:* Am Chem Soc. *Res:* Coordination complexes of transition metals with polydentate ligands; complexes with oxide and sulfoxide donor atoms; complexes of lanthanides and actinides. *Mailing Add:* Thomas Nelson Community Col Box 9407 Hampton VA 23670-0407

WORKMAN, MILTON, b Chicago Heights, Ill, Oct 1, 20; m 49; c 6. PLANT PHYSIOLOGY. *Educ:* Colo Agr & Mech Col, BS, 50; Univ Calif, PhD(plant physiol), 54. *Prof Exp:* Asst veg crops, Univ Calif, 50-54; from instr to assoc prof hort, Purdue Univ, 54-63; prof hort, Colo State Univ, 63-85; RETIRED. *Mem:* Fel Am Soc Hort Sci. *Res:* Pre and post harvest physiology. *Mailing Add:* 3033 Moore Lane Ft Collins CO 80526

WORKMAN, RALPH BURNS, b Omaha, Nebr, June 25, 24; m 51; c 4. ECONOMIC ENTOMOLOGY, VEGETABLE CROPS. *Educ:* Colo State Univ, BS, 51, MS, 52; Ore State Univ, PhD(entom), 58. *Prof Exp:* Res asst entom, Colo State Univ, 52-55; from asst entomologist to assoc entomologist, Agr Res Ctr, Univ Fla, 58-85; RETIRED. *Mem:* Entom Soc Am. *Res:* Economic entomology; biology and control of cruciferous and potato insects. *Mailing Add:* 30 Coquina Ave St Augustine FL 32084

WORKMAN, WESLEY RAY, b Mich, Feb 1, 26; m 48; c 4. ORGANIC CHEMISTRY. *Educ:* Mich State Univ, BS, 49, MS, 50; Univ Minn, PhD(org chem), 54. *Prof Exp:* Sr chemist, 3M Co, 54-68, mgr photog sci & photo chem, 68-72, dir imaging, Res Div, 72-74, dir, Systs Res Lab, 74-78, DIR UNIV RELATIONS, CENT RES LABS, 78- *Mem:* Am Chem Soc; Soc Photog Scientists & Engrs. *Res:* Photochemistry. *Mailing Add:* 3403 Garth Rd No 513 Batown TX 77521

WORKMAN, WILLIAM EDWARD, b Richmond, Va, May 13, 41; m 59; c 2. ENVIRONMENTAL GEOLOGY. *Educ:* Univ Va, BS, 62, MS, 64; Univ Tex, Austin, PhD(geol), 68. *Prof Exp:* Asst prof geol, Albion Col, 68-73; geoscientist, Palmer & Baker Engrs, Inc, 73-75; supvry environ geologist, US CEngr, Mobile, 75-79; CONSULT GEOLOGIST, 79- *Concurrent Pos:* Sigma Xi-Sci Res Soc Am grants in aid of res, 63 & 66; geologist C, Va Div Mineral Resources, Charlottesville, 63-; consult, Palmer & Baker Engrs, Inc, 75-; environ consult to var pvt firms, 75- *Mem:* Sigma Xi. *Res:* Regional metamorphism in Llano Uplift, Texas; coastal erosion; engineering geology relative to coastal processes. *Mailing Add:* PO Box 745 Lillian AL 36549

WORKMAN, WILLIAM GLENN, b Sheridan, Wyo, Mar 19, 47; m 72; c 2. NATURAL RESOURCE ECONOMICS, ENVIRONMENTAL ECONOMICS. *Educ:* Univ Wyo, BS, 69; Utah State Univ, MA, 72, PhD(resource econ), 78. *Prof Exp:* Res asst econ, Utah State Univ, 69-72; from asst prof to assoc prof econ, Univ Alaska, 79-90; ECON CONSULT & ADJ PROF, EASTERN ORE STATE COL, 90- *Concurrent Pos:* Assoc economist, Frank Orth & Assocs, Seattle, 80-82; vis assoc prof, Dept Agr & Resource Econ, Univ Md, 85-86. *Mem:* Asn Environ & Resource Econ. *Res:* Economic analysis of land use and public land use policy. *Mailing Add:* 2202 Linda Lane LaGrande OR 97850

WORLEY, DAVID EUGENE, b Cadiz, Ohio, Aug 6, 29; m 68; c 2. PARASITOLOGY, PUBLIC HEALTH & EPIDEMIOLOGY. *Educ:* Col Wooster, AB, 51; Kans State Univ, MS, 55, PhD(parasitol), 58. *Prof Exp:* Assoc res parasitologist, Parke, Davis & Co, 58-62; from asst prof to assoc prof, 62-72, PROF VET RES LAB, MONT STATE UNIV, 72- *Concurrent Pos:* Consult, Brazil & Czech; pres, Rocky Mountain Conf Parasitol, 74-75. *Mem:* Am Soc Parasitol; Wildlife Dis Asn; Am Asn Vet Parasitol; World Asn Advan Vet Parasitol. *Res:* Zoology; helminthology, including chemotherapy of parasitic infections and helminth life cycles and host-parasite relationships; parasitic zoonoses, especially trichinosis and echinococcosis. *Mailing Add:* Vet Molecular Biol Lab Mont State Univ Bozeman MT 59717

WORLEY, FRANK L, JR, b Kansas City, Mo, Oct 9, 29; m 52; c 1. CHEMICAL ENGINEERING. *Educ:* Univ Houston, BS, 52, MS, 59, PhD(chem eng), 65. *Prof Exp:* Chem engr, Nyotex Chem, Inc, 52-55 & Stauffer Chem Co, 55-57; instr chem eng, Univ Houston, 57-59; Fulbright lectr, Univ Guayaquil & Cent Univ Ecuador, 59-61; from instr to assoc prof, 61-72, PROF CHEM ENG, UNIV HOUSTON, 72-, ASSOC DEAN ENG, 85- *Concurrent Pos:* Vis scientist, Div Meteorol, Environ Protection Agency,

71. *Mem:* Fel Am Inst Chem Engrs. *Res:* Modelling of atmospheric dispersion and reactions; air pollution control; computer aided design and analysis; process control. *Mailing Add:* Dept Chem Eng Univ Houston 4800 Calhoun University Park Houston TX 77204-4814

WORLEY, JIMMY WELDON, b Bowie, Tex, May 2, 44; m 66; c 3. TRACE ORGANIC ANALYSIS, INDUSTRIAL HYGIENE. *Educ:* Midwestern State Univ, BS, 66; Univ Ill, Urbana-Champaign, PhD(org chem), 71. *Prof Exp:* Res assoc, Wesleyan Univ, 71; sr res chemist, 71-75, res specialist, 75-78, res group leader, 78-80, sr res group leader, 80-82, res mgr, Pesticide Residues, 82-84, mgr, Res Process Environ, 84-87, mgr, Chem & Microbiol, 87-89, MGR, PROCESS TECHNOL, MONSANTO CO, ST LOUIS, 89- *Mem:* Am Chem Soc. *Res:* Process chemistry research and development to support manufacture of agricultural chemicals. *Mailing Add:* Monsanto Co 800 N Lindbergh Blvd St Louis MO 63167

WORLEY, JOHN DAVID, b Texarkana, Tex, Dec 10, 38; m 63; c 4. BIOPHYSICAL CHEMISTRY. *Educ:* Hendrix Col, BA, 60; Univ Okla, PhD(phys chem), 64. *Prof Exp:* NIH fel biophys chem, Northwestern Univ, 64-66; asst prof chem, Univ Cincinnati, 66-70; asst prof, 70-77, ASSOC PROF CHEM, ST NORBERT COL, 78- *Mem:* Am Chem Soc; Sigma Xi. *Res:* Protein structure and denaturation in solution; solutions of nonelectrolytes; hydrogen bonding. *Mailing Add:* Dept Chem St Norbert Col De Pere WI 54115

WORLEY, RAY EDWARD, b Robbinsville, NC, May 4, 32; m 55; c 3. HORTICULTURE, AGRONOMY. *Educ:* NC State Col, BS, 54, MS, 58; Va Polytech Inst, PhD(agron), 61. *Prof Exp:* Asst field crops, NC State Col, 56-58; instr agron, Va Polytech Inst, 58-61; asst horticulturist, 61-72, assoc horticulturist, 72-80, prof hort, 80-85, prof & actg dept head, 85-87, PROF HORT, GA COASTAL PLAIN EXP STA, 87- *Honors & Awards:* L M Ware Award, 81; Distinguished Res Award, Sigma Xi, 90. *Mem:* Am Soc Hort Sci. *Res:* Pecan tree nutrition, management and physiology; vegetable and forage nutrition and physiology. *Mailing Add:* Hort Dept Ga Coastal Plain Exp Sta Tifton GA 31794

WORLEY, RICHARD DIXON, b Little Rock, Ark, Dec 24, 26; m 51; c 2. PHYSICS. *Educ:* Hendrix Col, BS, 49; Univ Ark, MA, 51; Univ Chicago, MS, 60; Univ Calif, Berkeley, PhD(physics), 63. *Prof Exp:* Nuclear physicist, Wright-Patterson AFB, Ohio, 51-54; nuclear engr, Douglas Air Craft Co, Calif, 57-59; sr physicist, Lawrence Radiation Lab, Univ Calif, 63-70; Physicist Mason & Hanger, Pantex Plant, Silas Maspm Co, 70-81; PROF PHYSICS, DYERSBURG STATE COMMUNITY COL, 81- *Concurrent Pos:* Teacher, Memphis State Univ. *Mem:* Am Phys Soc. *Res:* Characteristic x-ray production by ion bombardment of both polycrystalline and single-crystal targets; atomic beams and the hyperfine interaction; high explosive research and development. *Mailing Add:* 920 Dianne Dr Dyersburg TN 38024

WORLEY, ROBERT DUNKLE, b Trenton, NJ, Jan 24, 25; m 50; c 3. PHYSICS. *Educ:* Williams Col, AB, 49; Columbia Univ, AM, 51, PhD, 55. *Prof Exp:* From asst to lectr, Columbia Univ, 50-54; mem staff, Bell Labs, Inc, 54-86, supvr, 60-86; RETIRED. *Mem:* Am Phys Soc; Acoust Soc Am. *Res:* Heat capacities of superconductors; sound transmission in the ocean; operations research; sonar systems. *Mailing Add:* 71 Galway Dr Mendham NJ 07945

WORLEY, S D, b Russellville, Ala, Jan 31, 42; m 64; c 2. ORGANIC CHEMISTRY, SOLID STATE PHYSICS. *Educ:* Auburn Univ, BS, 64; Univ Tex, PhD(chem), 69. *Prof Exp:* Res chemist, Johnson Manned Spacecraft Ctr, 69-72; asst prof phys chem, Cleveland State Univ, 72-73; prog officer, Off Naval Res, 73-74; from asst to assoc prof, 74-82, PROF ORG CHEM, AUBURN UNIV, 82- *Concurrent Pos:* Vis prof chem, Univ Wis, 75, Calif Inst Technol, 78; vis scientist, Nat Bur Standards, 76. *Mem:* Am Chem Soc; AAAS. *Res:* Spectroscopic studies of surfaces and catalysts; synthesis of new water disinfectants; molecular orbital calculations; photoelectron spectroscopy. *Mailing Add:* Dept Chem Auburn Univ Auburn AL 36849

WORLEY, WILL J, b Gibson City, Ill, Aug 2, 19; m 54; c 3. THEORETICAL MECHANICS, APPLIED MECHANICS. *Educ:* Univ Ill, BS, 43, MS, 45, PhD, 52. *Prof Exp:* From instr to prof, 43-89, EMER PROF THEORET & APPL MECH, UNIV ILL, URBANA, 89- *Concurrent Pos:* Consult, Magnavox Co, Ind, 56-58, Ill, 58-65, Chris-Kaye Mfg Co, Ill, 57, A O Smith Corp, Wis, 59 & attorneys, Ill, 59-; proj dir, Wright Air Develop Ctr Nonlinear Mech Proj, 59-61, NASA Proj, 63-67; consult, Int Bus Mach Corp, NY, 60-61 & 63-64, Calif, 61; chmn, Shock & Vibration Comt, Appl Mech Div, Am Soc Mech Engrs, 62-63, chmn, awards comt, Mach Design Div, 64-65; pres, Worley Systs Inc, 77-; vis fel, Inst Sound & Vibration Res, Univ Southampton, Eng, 79. *Mem:* Am Soc Mech Engrs. *Res:* Acoustical noise reduction; mechanical vibrations and nonlinear mechanics; static and dynamic behavior of plates and shells; optimum structural design; mechanical properties of materials; failure investigation of components and systems; systems approach to prevention and analysis of system failure. *Mailing Add:* 2106 Zuppke A3 Urbana IL 61801-6706

WORLOCK, JOHN M, b Kearney, Nebr, Feb 15, 31; div; c 2. SOLID STATE PHYSICS. *Educ:* Swarthmore Col, BA, 53; Cornell Univ, PhD(physics), 62. *Prof Exp:* NSF fel, 62-63; actg asst prof physics, Univ Calif, Berkeley, 63-64; mem tech staff, Bell Labs, Inc, 64-84; DISTINGUISHED MEM PROF STAFF, BELLCORE, 85-91. *Mem:* AAAS; Fedn Am Scientists; fel Am Phys Soc. *Res:* Lattice dynamics; transport and optical properties of non-metallic crystals; light scattering; phase transitions; electron-mole droplets; optical properties of semiconductors; heterostructures; superlattices; microstructures. *Mailing Add:* Navesink Ctr Rm 3X283 Bellcore Red Bank NJ 07701

WORMAN, JAMES JOHN, b Allentown, Pa, Feb 17, 40; m 61; c 4. ORGANIC CHEMISTRY. *Educ:* Moravian Col, BS, 61; NMex Highlands Univ, MS, 64; Univ Wyo, PhD, 68. *Prof Exp:* Instr chem, Moravian Col, 62-63; res asst, Univ Wyo, 65-67; from asst prof chem to assoc prof, SDak State Univ, 67-76, prof, 77-88; VIS PROF, CHEM DEPT, DARTMOUTH COL, 88- *Mem:* Am Chem Soc; Royal Soc Chem. *Res:* Theoretical and experimental organic photochemistry, including syntheses and reaction of large ring nitrogen heterocycles. *Mailing Add:* Chem Dept Dartmouth Col Hanover NH 03755-1894

WORMSER, ERIC M, b Ger, Apr 30, 21; nat US; m 47; c 2. ENGINEERING PHYSICS. *Educ:* Mass Inst Technol, BS, 42. *Prof Exp:* Test engr, Universal Camera Co, 42-45; physicist, Hillyer Instrument Co, 47-49 & Servo Corp Am, 49-52; chief engr, Barnes Eng Co, 52-58, vpres, 58-68, exec vpres, 68-76; PRES, WORMSER SCI CORP, 76- *Mem:* Optical Soc Am; Inst Elec & Electronics Engrs; Instrument Soc Am; NY Acad Sci; fel Optical Soc Am; fel Can Aeronaut & Space Inst. *Res:* Infrared instruments, systems and detector development. *Mailing Add:* Wormser Sci Corp 66 Doral Farm Rd Stamford CT 06902

WORMSER, GARY PAUL, b Wilmington, Del, Jan 17, 47. INFECTIOUS DISEASES, INTERNAL MEDICINE. *Educ:* Univ Pa, BA, 68; Johns Hopkins Univ, MD, 72. *Prof Exp:* Chief infectious dis, Bronx Vet Admin Hosp, 77-81; CHIEF INFECTIOUS DIS, NY MED COL, 81-, PROF MED, 85- *Concurrent Pos:* Dir, Ctr Study & Treat AIDS & Lyme Dis Res Ctr, NY Med Col, 85- *Mem:* Am Fedn Clin Res. *Res:* AIDS-HIV infection; lyme disease; infection control. *Mailing Add:* 301 E 79th St New York NY 10021

WORMSER, HENRY C, b Strasbourg, France, Sept 10, 36; US citizen; m 63; c 2. PHARMACEUTICAL CHEMISTRY. *Educ:* Temple Univ, BSc, 59, MSc, 61; Univ Wis, PhD(pharmaceut chem), 65. *Prof Exp:* From asst prof to assoc prof, 65-76, PROF PHARMACEUT CHEM, WAYNE STATE UNIV, 76- *Mem:* Am Chem Soc; Am Pharmaceut Asn. *Res:* Synthesis of model compounds to be used in study of drug-enzyme or drug-receptor site interactions in an effort to determine specific mechanisms of drug activity. *Mailing Add:* 5420 Hammersmith Dr West Bloomfield MI 48033

WORMUTH, JOHN HAZEN, b Cobleskill, NY, Dec 9, 44; m 67; c 2. BIOLOGICAL OCEANOGRAPHY. *Educ:* Hope Col, BA, 66; Scripps Inst Oceanog, PhD(oceanog), 71. *Prof Exp:* Biol oceanogr, Intersea Res Corp, 71-72; asst prof, 72-77, ASSOC PROF, TEX A&M UNIV, 77- *Mem:* Am Soc Limnol & Oceanog; Am Geophys Union. *Res:* Ecology and sampling problems associated with cephalopods; neuston communities; zooplankton ecology, particularly pteropods. *Mailing Add:* Dept Oceanog Tex A&M Univ College Station TX 77843

WORNICK, ROBERT C(HARLES), b Bellows Falls, Vt, June 13, 24; m 53; c 2. CHEMICAL ENGINEERING, ANIMAL FEED TECHNOLOGY. *Educ:* Columbia Univ, BChE, 50. *Prof Exp:* Analytical lab asst, Chas Pfizer & Co, Inc, 42 & 46-48, analytical chemist, 49-50, head food & beverage sect, Tech Serv Dept, 51-53, head chem labs, Agr Res Dept, 53-59, prof engr, Agr Develop Dept, 60-65, dir develop res labs, Agr Div, 66-70, dir agr licensing, Pfizer Int, 70-75, sci dir, Latin Am, 75-84; PRES, ROBERT C WARNICK PE & ASSOCS, 84-, CONSULT. *Concurrent Pos:* Russian abstractor, Chem Abstr; prof engr licenses, Ind, Ill, & Fla. *Honors & Awards:* Tech Service Award, Nat Feed Ingredients Asn. *Mem:* Am Inst Chem Engrs; Nat Soc Prof Engrs; Am Chem Soc; Am Asn Feed Micros; Animal Nutrit Res Coun. *Res:* Vitamin enrichment, processing and preservation of foods; leavening agents, acidulants and antioxidants; micro-ingredient supplementation of animal feeds; feed processing and formulation; equipment evaluation; food and feed quality control; product development. *Mailing Add:* 13855 SW 78th Ct Miami FL 33158-1109

WOROCH, EUGENE LEO, b Kenosha, Wis, Mar 18, 22; m 49; c 4. MEDICINAL CHEMISTRY. *Educ:* Univ Wis, BS, 44, MS, 45, PhD(org chem), 48; Univ Chicago, MBA, 71. *Prof Exp:* Proj dir natural prod, Wis Alumni Res Found, 48-49; group leader & consult, Bjorksten Labs, 49; res assoc natural prod, Mayo Clin, 49-51; group leader, Glidden Co, 51-58; head dept org chem res, Abbott Labs, 58-75, dir, Div Antibiotic Res, 75-77, Div Sci Serv, 77-84, Div Chem & Analytical Serv, 84-86; RETIRED. *Concurrent Pos:* Vpres & treas, Ill Sci Lect Asn, 80-; vpres & dir, Clara Abbott Found, 88- *Mem:* Am Chem Soc; AAAS. *Res:* Natural products; steroids; antibiotics; peptides and structural chemistry. *Mailing Add:* 485 Greenvale Rd Lake Forest IL 60045

WORONICK, CHARLES LOUIS, b Meriden, Conn, Dec 4, 30. BIOCHEMISTRY, CLINICAL CHEMISTRY. *Educ:* Univ Conn, BS, 53; Univ Calif, Berkeley, MS, 55; Univ Wis, PhD(biochem), 59. *Prof Exp:* Asst prof chem, Brown Univ, 62-66; assoc non-clin investr path, Pa Hosp, 66-68; BIOCHEMIST, MED RES LABS & DEPT MED, HARTFORD HOSP, 68- *Concurrent Pos:* NSF fel, 59-61; fel enzyme chem, Nobel Med Inst, Stockholm, 59-62; Nat Cancer Inst fel, 61-62; assoc path, Med Sch, Univ Pa, 66-68; consult staff biochem, Hartford Hosp, 68-; consult staff clin chem, John Dempsey Hosp, Conn, 75-; asst prof labs med, Sch Med, Univ Conn, Farmington, 73-; mem comn toxicol, Clin Chem Div, Subcomt Cholinesterases, Int Union Pure & Appl Chem, 76-81; assoc non-clinical investr, Pa Hosp, Philadelphia, 66-68. *Mem:* Am Asn Clin Chem; Am Chem Soc; NY Acad Sci; fel Nat Acad Clin Biochem; AAAS; fel Asn Clin Scientists. *Res:* Enzyme kinetics; equilibria and mechanisms; immunochemistry; leukocyte function; biostatistics. *Mailing Add:* Med Res Lab Hartford Hosp Hartford CT 06115-0729

WORRALL, JOHN GATLAND, b Cleethorpes, Eng, May 22, 38; Can citizen. DENDROLOGY. *Educ:* Univ Durham, BSc, 59; Yale Univ, MF, 64, PhD(forestry), 69. *Prof Exp:* ASSOC PROF FORESTRY, UNIV BC, 68- *Mem:* Sigma Xi. *Res:* Environmental control of cambial activity; breakage of dormancy seeds and plants. *Mailing Add:* Fac Forestry Univ BC Vancouver BC V6T 1W5 Can

WORRALL, RICHARD D, b Waterloo, Eng, May 31, 38; m 60; c 2. CIVIL & TRANSPORTATION ENGINEERING. *Educ:* Durham Univ, BSc, 60; Northwestern Univ, MS, 61, PhD(civil eng), 66. *Prof Exp:* Asst traffic engr, City Planning Dept, Newcastle, Eng, 61-62; sr res fel city & transp planning, Lower Swansea Valley Proj, Univ Wales, 62-63; res assoc transp planning & traffic flow theory, Northwestern Univ, Evanston, 63-69; MGR, PEAT MARWICK MITCHELL & CO, 69- *Concurrent Pos:* Mem, Hwy Res Bd, Nat Acad Sci-Nat Res Coun, 63-, mem comts land use eval, freeway opers & origin-destination, 66- *Mem:* Am Soc Civil Engrs. *Res:* Transportation and city planning; traffic flow theory; transportation systems control. *Mailing Add:* 206 Carrwood Rd Great Falls VA 22066-3722

WORRALL, WINFIELD SCOTT, b Cheltenham, Pa, Jan 12, 21; m 49; c 2. ORGANIC CHEMISTRY. *Educ:* Haverford Col, BS, 42; Harvard Univ, MA, 47, PhD(chem), 49. *Prof Exp:* Res chemist, Monsanto Chem Co, 50-54; from instr to assoc prof chem, Trinity Col, Conn, 54-64; assoc prof chem, 64-68, assoc prof gen sci studies, 68-77, PROF GEN SCI STUDIES, STATE UNIV NY COL PLATTSBURGH, 78- *Res:* Steroids; heterocyclics. *Mailing Add:* 51 Prospect Ave Plattsburgh NY 12901

WORRELL, FRANCIS TOUSSAINT, b Hartford, Conn, Apr 19, 15; m 48, 72; c 2. PHYSICS. *Educ:* Univ Mich, BSE, 36; Univ Pittsburgh, MS, 40, PhD(physics), 41. *Prof Exp:* Mem staff, Physicists Res Co, Mich, 36-37; asst, Univ Pittsburgh, 37-41; instr, Univ Tenn, 41-42; staff mem, Radiation Lab, Mass Inst Technol, 42-46; res assoc, Inst Metals, Univ Chicago, 46-47; asst prof physics, Rensselaer Polytech Inst, 47-55; assoc prof, DePauw Univ, 55-58; assoc prof, Beloit Col, 59-60; staff mem, Lincoln Lab, Mass Inst Technol, 60-63; prof physics, Univ Lowell, 63-80; RETIRED. *Concurrent Pos:* Fulbright lectr, Al-Hikma Univ Baghdad, 58-59; vis lectr, Univ Bristol, 72-73 & Mass Maritime Acad, 81. *Mem:* AAAS; Am Asn Physics Teachers. *Res:* Design of experiments; structure of materials; thermionic emission; atmospheric optics. *Mailing Add:* 11 Old Salt Lane Yarmouth Port MA 02675-1234

WORRELL, JAY H, b Manchester, NH, July 14, 38; m 59; c 3. PHYSICAL INORGANIC CHEMISTRY, CHEMICAL ENGINEERING. *Educ:* Univ NH, BS, 61, MS, 63; Ohio State Univ, PhD(inorg chem), 66. *Prof Exp:* Res assoc chem, State Univ NY Stony Brook, 66-67; from asst to assoc prof, 67-78, PROF CHEM, UNIV SFLA, TAMPA, 78-, DIR, CTR GEN CHEM, 89- *Concurrent Pos:* Indust consult; mem bd dirs & ed secy, Inorganic Syntheses Inc; Summer Inst for high sch teachers; steering comt excellence in math, sci, computer & technol, Univ SFla. *Mem:* Am Chem Soc; Am Inst Chem Engrs. *Res:* Preparation, properties and theory of coordination compounds; stereochemistry and inorganic reaction kinetics for ligand substitution and oxidation reduction processes; applied industrial chemistry; chemical engineering. *Mailing Add:* Dept Chem Univ SFla Tampa FL 33620

WORRELL, JOHN MAYS, JR, b El Paso, Tex, Oct 3, 33; m 66; c 1. MATHEMATICS, MEDICINE. *Educ:* Univ Tex, BA, 54, MD, 57, PhD(math), 61. *Prof Exp:* Intern med, Denver Gen Hosp, Colo, 57-58; instr math, Univ Tex, 58-59; NSF fel, 61-62; mem tech staff math res, Sandia Corp, NMex, 62-72; PROF MATH, OHIO UNIV, 72-, DIR INST MED & MATH, 75- *Concurrent Pos:* Consult clin med & biomed sci. *Mem:* Am Math Soc; Am Med Asn. *Res:* Problems having topological character; clinical medicine; biological processes. *Mailing Add:* Inst Med & Math Ohio Univ Athens OH 45701

WORRELL, WAYNE L, b Rock Island, Ill, Oct 25, 37; m 68; c 2. ELECTRICAL CERAMICS, SENSOR MATERIALS. *Educ:* Mass Inst Technol, BS, 59, PhD(metall), 63. *Hon Degrees:* MA, Univ Pa, 71. *Prof Exp:* Fel metall, Univ Calif, Berkeley, 63-64, lectr, 64-65; from asst prof to assoc prof, 65-74, PROF MAT SCI, UNIV PA, 74-, ASSOC DEAN, 86- *Concurrent Pos:* Chmn comt high temp sci & technol, Nat Acad Sci-Nat Res Coun, 74-77; consult to indust & govt orgns; vis prof, Dept Chem, Univ Calif, Berkeley, 75-76; ed, Prog Solid State Chem, 77-86; mem chem eng, Div Rev Comt, Argonne Nat Lab, 79-86; Japan Soc Prom Sci lect fel, 82; Max Planck Soc fel, Stuttgart, Fed Repub Ger, 82-83. *Honors & Awards:* Outstanding Achievement Award High Temperature Mat, Electrochem Soc, 88, Carl Wagner Mem Award, 89. *Mem:* Electrochem Soc; fel Am Soc metals Int; fel Am Ceramic Soc; Am Inst Mining, Metall & Petrol Engrs; Sigma Xi. *Res:* Ceramic sensors; ceramic coatings; corrosion at elevated temperatures; high-temperature materials chemistry; solid state electrochemistry; electrical ceramics. *Mailing Add:* Dept Mat Sci Univ Pa 3231 Walnut St Philadelphia PA 19104-6272

WORREST, ROBERT CHARLES, b Hartford, Conn, July 6, 35; m 57; c 1. RADIATION BIOLOGY, MARINE ECOLOGY. *Educ:* Williams Col, BA, 57; Wesleyan Univ, MA, 64; Ore State Univ, PhD(radiation biol, physiol & zool), 75. *Prof Exp:* From asst prof to prof radiation biol, sr res, Ore State Univ, 77-86; MGR STRATOSPHERE OZONE RES PROG, US ENVIRON PROTECTION AGENCY, 87- *Concurrent Pos:* Res grants, NASA, 75 & 77, US Environ Protection Agency, 79, 82-85 & 85-86, & Agency Int Develop, 82; proj leader photobiol prog, US Environ Protection Agency, 80-82; mem, pub comt, Estuarine Res Fedn, 80-84 & gov bd, 82-84. *Mem:* AAAS; Am Soc Photobiol; Ecol Soc Am; Am Soc Limnol & Oceanog; NY Acad Sci; Pac Estuarine Res Soc (pres, 82-84); Soc Risk Analysis. *Res:* Assessment of the impact of increased solar ultraviolet radiation and other environmental stress factors upon the biosphere; effects of gamma (cobalt-60) and mixed gamma/thermal neutron (Triga Reaction) radiation on mammalian reproduction. *Mailing Add:* Environ US EPA 200 SW 35th St Corvallis OR 97333-4996

WORSHAM, ARCH DOUGLAS, b Culloden, Ga, Feb 22, 33; m 56, 82; c 5. WEED SCIENCE, ALLELOPATHY. *Educ:* Univ Ga, BSA, 55, MS, 57; NC State Univ, PhD(crop sci), 61. *Prof Exp:* From exten asst prof to exten assoc prof, 60-67, assoc prof, 67-69, PROF CROP SCI, NC STATE UNIV, 69- *Concurrent Pos:* Res grants, 66-; consult, 67-; ed, S Weed Sci Soc, 69-71, 87-89, vpres, 91. *Honors & Awards:* Outstanding Publ Award, Weed Sci Soc Am, 75. *Mem:* Weed Sci Soc Am; Am Soc Agron; Am Soc Plant Physiologists; Sigma Xi; Int Soc Chem Ecol. *Res:* Pesticides; crop science; weed science; weed biology basic and applied weed science research in agronomic crops and non-tillage crop production and biology and control of specific weeds; weed suppression through allelopathy. *Mailing Add:* Dept Crop Sci Campus Box 7620 NC State Univ Raleigh NC 27695-7620

WORSHAM, JAMES ESSEX, JR, physical chemistry, biomedical engineering; deceased, see previous edition for last biography

WORSHAM, LESA MARIE SPACEK, b Sellersville, Pa, Dec 4, 50; m 73. PROTEIN CHARACTERIZATION. *Educ:* Ursinus Col, BS, 72; Univ NC Chapel Hill, PhD(chem), 77. *Prof Exp:* Consult atmospheric chem, Res Triangle Inst, 77-78; RES ASSOC BIOCHEM, QUILLEN-DISHNER COL MED E TENN STATE UNIV, 79- *Mem:* Am Chem Soc; Sigma Xi; AAAS. *Res:* Isolation fatty acid synthetase from Euglena gracilis to determine its molecular weight and study its properties and to break down this complex to study its submits. *Mailing Add:* 525 Ranier Dr Kingsport TN 37663

WORSHAM, WALTER CASTINE, b Turbeville, SC, Aug 17, 38; m 59; c 2. TEXTILE CHEMISTRY. *Educ:* Col Charleston, BS, 61; Univ NC, PhD(phys chem), 66. *Prof Exp:* Chemist, Fiber Industs, Inc, 66-67; chemist, Emery Industs, Inc, Mauldin, 67-75, tech mgr, 75-79; VPRES & GEN MGR, ETHOX CHEM, INC, 79- *Mem:* Am Chem Soc; Am Asn Textile Chemists & Colorists. *Res:* Kinetics of photochemical reactions; chemistry of textile processing. *Mailing Add:* PO Box 5094 Sta B Ethox Chem Inc Greenville SC 29606

WORSLEY, THOMAS R, b Brooklyn, NY, June 1, 42. MARINE GEOLOGY, BIOSTRATIGRAPHY. *Educ:* City Col New York, BS, 65; Univ Tenn, MS, 67; Univ Ill, PhD(geol), 70. *Prof Exp:* Res prof oceanog, Univ Wash, 70-77, PROF GEOL, OHIO UNIV, 76- *Concurrent Pos:* NSF grant, 73-75, 75-77 & 77- *Mem:* Soc Econ Paleont & Mineralogy. *Res:* Computer application in marine micropaleontology; erosion-sedimentation of the globe. *Mailing Add:* Dept Geol Ohio Univ Main Campus Athens OH 45701

WORSTELL, HAIRSTON G(EORGE), b West Plains, Mo, Jan 5, 20; m 44; c 1. MECHANICAL ENGINEERING. *Educ:* Univ Okla, BSME, 50. *Prof Exp:* Design engr, Boeing Aircraft Co, Kans, 50-51; staff engr, Sandia Corp, NMex, 51-53; proj engr, Phillips Petrol Co, Okla, 53-59; staff engr, Los Alamos Sci Lab, 59-66, asst group leader mech eng, 66-67, assoc group leader, 67-72, alt group leader, 72-77, consult, 77-86; RETIRED. *Mem:* Am Vacuum Soc. *Res:* Mechanical design and supervision of fellow engineers in development of large ultra high vacuum linear accelerator systems employing unusual fabrication and metallurgical techniques. *Mailing Add:* 207 McDougal Belen NM 87002

WORT, ARTHUR JOHN, b Fordingbridge, Eng, Mar 26, 24; m 57; c 2. MICROBIOLOGY, MEDICINE. *Educ:* Univ Durham, MB, BS, 56; FRCP(C), 70, FRCPath, 77. *Prof Exp:* Demonstrator path, Univ Durham, 57-60; sr registrar microbiol, Royal Victoria Infirmary, 60-65; consult pathologist, N Tees Hosp, 65-68; asst prof, Univ Toronto, 68-70; asst prof, Dalhousie Univ, 70-; head, mmicrobiol div, IWK Hosp, 70-89; RETIRED. *Concurrent Pos:* Consult microbiol, Grace Hosp, 72-; mem, Conjoint Comt, Med Lab Training, Can Med Asn (Nat Comt). *Mem:* Am Soc Microbiol; Path Soc Gt Brit & Ireland; Can Soc Microbiol; NY Acad Sci. *Res:* Epidemiology and immunology of pertussis; streptococci. *Mailing Add:* PO Box 112 Waverley NS B0N 2S0 Can

WORTH, DONALD CALHOUN, b Brooklyn, NY, Oct 20, 23; m 46; c 4. SOLAR ENERGY APPLICATIONS, SCIENCE EDUCATION. *Educ:* Carnegie Inst Technol, BS, 44; Yale Univ, MS, 48, PhD(physics), 49. *Prof Exp:* Instr physics, Berea Col, 50-52, assoc prof, 52-53; from asst prof to assoc prof, Int Christian Univ, Tokyo, 54-60, dean, Col Liberal Arts, 70-74 & 76-80, prof physics, 60-89; RETIRED. *Concurrent Pos:* Vis asst prof, Ala Polytech Inst, 51; vis assoc prof, Univ Chicago, 53-54; NSF fel, Univ Va, 58-59; vis prof, Univ Wis, 63-64 & State Univ NY Stony Brook, 68-69; vis researcher, Lawrence Berkeley Nat Lab, Univ Calif, 81-82, 86-87. *Mem:* Int Solar Energy Soc; Am Phys Soc; Am Asn Physics Teachers. *Res:* Low energy nuclear polarization studies, especially in nucleon-nucleon scattering; physics education; solar energy utilization. *Mailing Add:* 2918-H Regent St Berkeley CA 94705

WORTH, ROBERT MCALPINE, b Kiangsi Prov, China, Aug 27, 24; US citizen; m 51; c 2. EPIDEMIOLOGY. *Educ:* Univ Calif, Berkeley, BA, 50, PhD(epidemiol), 62; Univ Calif, San Francisco, MD, 54; Harvard Univ, MPH, 58. *Prof Exp:* Intern med, Southern Pac Hosp, San Francisco, 54-55; resident family pract, San Mateo Community Hosp, 55-56; physician leprosy, Hawaii Dept Health, 56-57, asst chief chronic dis, 57, health officer, 58-60; prof epidemiol, Sch Pub Health, Univ Hawaii, Manoa, 63-85; chief scientist, Agent Orange Progs, Ctr Dis Control, Atlanta, 86-88; CHIEF, DIV COMMUN DIS CONTROL, DEPT HEALTH, HAWAII, 88- *Concurrent Pos:* Hooper Found res fel, Univ Hong Kong, 61-63. *Mem:* Am Pub Health Asn. *Res:* Leprosy epidemiology and control; disease survey methodology; automation medical record systems for epidemiologic and quality control purposes; public health in modern China. *Mailing Add:* Dept Health PO Box 3378 Honolulu HI 96801

WORTH, ROY EUGENE, b Broxton, Ga, Mar 24, 38; m 65; c 1. FUNCTIONAL ANALYSIS, COMPUTER PROGRAMMING. *Educ:* Univ Ga, BS, 60, MA, 62, PhD(math), 68. *Prof Exp:* Asst prof math, WGa Col, 63-68; asst prof, 68-71, ASSOC PROF MATH, GA STATE UNIV, 71- *Mem:* Math Asn Am. *Res:* Henstock integral. *Mailing Add:* Dept Math & Comput Sci Ga State Univ Atlanta GA 30303-3083

WORTHEN, HOWARD GEORGE, b Provo, Utah, Dec 21, 25; m 50; c 5. PEDIATRICS, BIOCHEMISTRY. *Educ:* Brigham Young Univ, AB, 47; Northwestern Univ, MD, 51; Univ Minn, PhD, 61. *Prof Exp:* From instr to asst prof pediat, Univ Minn, 56-62; assoc prof, Med Col, Cornell Univ, 62-65;

PROF PEDIAT, UNIV TEX HEALTH SCI CTR DALLAS, 65- *Mem:* Am Soc Exp Path; Electron Micros Soc Am; Harvey Soc; Soc Pediat Res; Am Soc Cell Biologists. *Res:* Renal disease; histochemistry; electron microscopy. *Mailing Add:* 12008 Browning Lane Dallas TX 75230

WORTHEN, LEONARD ROBERT, b Woburn, Mass, Dec 28, 25; m 55; c 3. MICROBIOLOGY. *Educ:* Mass Col Pharm, BS, 50; Temple Univ, MS, 52; Univ Mass, PhD(bact), 57. *Prof Exp:* Instr pharmacol, Sch Nursing, Holyoke Hosp, 55-57; asst prof pharm, 57-63, assoc prof pharmacog, 63-70, PROF PHARMACOG, UNIV RI, 70-, DIR ENVIRON HEALTH SCI PROG, 72- *Concurrent Pos:* assoc dean, Univ RI, 84- *Mem:* Am Pharmaceut Asn; Am Soc Pharmacog; Sigma Xi. *Res:* Fungal metabolites, particularly antibiotics and other metabolites of medicinal importance; natural products from marine sources. *Mailing Add:* Col Pharm Univ RI Kingston RI 02881

WORTHING, JURGEN, b Brooklyn, NY, Aug 11, 24; m 49; c 2. ENERGY CONSERVATION BY PUBLIC UTILITIES. *Educ:* Polytech Inst Brooklyn, BEE, 48. *Prof Exp:* Staff engr, Servomechanisms, Inc, 48-55; staff engr, Repub Aviation Corp, 55; dir eng & sr vpres, Trio Labs, Inc, NY, 55-68; tech dir, NY State Legis Comn Energy Systs, 75-78; tech coun, NY State Energy Comt, 78-84; PRES, WORTHING ENTERPRISES, 68- *Mem:* Sr mem Inst Elec & Electronics Engrs. *Res:* Cable television; medical instrumentation; energy legislation, its effects on production, distribution and use of energy, and socio-economic effects; microcomputer applications. *Mailing Add:* 75 Weaving Lane Wantagh NY 11793

WORTHINGTON, CHARLES ROY, b Penola, Australia, May 17, 25; m 59; c 3. BIOPHYSICS. *Educ:* Univ Adelaide, PhD(physics), 55. *Prof Exp:* Res assoc crystallog, Polytech Inst Brooklyn, 55-57; staff mem biophys, Biophys Res Unit, Med Res Coun, King's Col, Univ London, 58-61; from asst prof to assoc prof physics, Univ Mich, Ann Arbor, 61-69; prof chem & physics, 69-72, PROF PHYSICS & BIOL, MELLON COL SCI, CARNEGIE-MELLON UNIV, 72- *Concurrent Pos:* Consult, Nat Bur Standards, DC, 57-58. *Mem:* Biophys Soc; Am Crystallog Asn. *Res:* Membrane structure; molecular organization of biological systems and theories of biological mechanisms; x-ray biophysics and microscopy. *Mailing Add:* Mellon Col Sci Carnegie-Mellon Univ Pittsburgh PA 15213

WORTHINGTON, JAMES BRIAN, b Sandwich, Ill, Nov 29, 43; m 65; c 1. ANALYTICAL CHEMISTRY. *Educ:* Augustana Col, BA, 65; Purdue Univ, Lafayette, PhD(analytical chem) 70. *Prof Exp:* Res chemist, 70-72, group leader environ anal, 72-75, mgr environ analysis & comput serv, 75-76, mgr sci servs, 76-77, dir environ affairs,77-84, MGR TECHNOL, DIAMOND SHAMROCK CORP, 84- *Honors & Awards:* Am Inst Chemists Award. *Mem:* Am Chem Soc. *Res:* Kinetic methods of analysis; design of chemical instrumentation; real time computer automation; environmental related analyses. *Mailing Add:* 2723 Colony Park Dr Sugar Land TX 77479-2629

WORTHINGTON, JOHN WILBUR, b Lexington, Ky, June 13, 18; m 43; c 2. ELECTRICAL ENGINEERING. *Educ:* Univ Ky, BSEE, 49. *Prof Exp:* Tech adv, Tech Training Command, Scott AFB, US Dept Air Force, Ill, 49-51, engr, Wright-Patterson AFB, 51-52, elec engr, Griffis AFB, 52-58, supv elec engr, Ground Elec Eng Installation Agency, 58-60, dep div chief, 60, proj elec engr, Rome Air Develop Ctr, 60-62; elec engr, opers res off, Defense Commun Agency, 62-74, elec engr, Defense Commun Eng Ctr, 74-85; RETIRED. *Mem:* Sr mem Inst Elec & Electronics Engrs. *Res:* Operations research; war gaming; simulation; systems engineering; systems survivability/endurability. *Mailing Add:* 8416 Doyle Dr Alexandria VA 22308

WORTHINGTON, LAWRENCE VALENTINE, oceanography, for more information see previous edition

WORTHINGTON, RALPH ERIC, b Coventry, Eng, July 28, 26; m 59; c 3. CHEMISTRY, FERTILIZER TECHNOLOGY. *Educ:* Univ Birmingham, Eng, BSc hons, 47, PhD(chem), 52. *Prof Exp:* Sci officer U metal prod, UK Atomic Energy Authority, 47-49, sr sci officer isotope separation, 52-55; chem res mgr clays, English Clays, Lovering & Pochin Ltd, Eng, 55-58; process develop mgr fertilizers, Assoc Chem Co, Eng, 58-63; tech dir fertilizers, Goulding Chem Ltd, Ireland, 63-76; VPRES TECH URANIUM EXTRACTION & FERTILIZER PROCESS DEVELOP, UNC RECOVERY CORP, 77- *Concurrent Pos:* Res scholar, Birmingham Univ, Royal Aircraft Estab, 49-52; mem bd, Inst Indust Res & Standards, Ireland, 69-72. *Honors & Awards:* Award Technol Univ Ireland, Govt Ireland, 72. *Mem:* Soc Chem Indust, UK; Am Chem Soc; Fertilizer Soc; Inst Chem Ireland. *Res:* Technology of recovery of secondary elements particularly related to phosphoric acid manufacture; general fertilizer technology. *Mailing Add:* UNC Recovery Corp c/o 1201 Jadwin Ave Suite 202 Richland WA 99352

WORTHINGTON, RICHARD DANE, b Houston, Tex, Sept 20, 41; m 64. MORPHOLOGY, HERPETOLOGY. *Educ:* Univ Tex, Austin, BA, 63; Univ Md, MS, 66, PhD(zool), 68. *Prof Exp:* USPHS trainee, Univ Chicago, 68-69; asst prof biol, 69-74, ASSOC PROF BIOL SCI, UNIV TEX, EL PASO, 74- *Mem:* Soc Syst Zool; Soc Study Evolution; Ecol Soc Am; Am Soc Ichthyologists & Herpetologists; Soc Study Amphibians & Reptiles; Am Soc Plant Taxonomists. *Res:* Evolutionary biology of caudate amphibians; mountain floras of the southwest United States; evolutionary morphology; lizard ecology; ecology; floras. *Mailing Add:* Dept Biol Sci Univ Tex El Paso TX 79968

WORTHINGTON, ROBERT EARL, b Kingston, Ga, Jan 2, 29; m 60; c 3. LIPID CHEMISTRY, FOOD SCIENCE. *Educ:* Berry Col, BSA, 52; NC State Col, MS, 55; Iowa State Univ, PhD(biochem), 62. *Prof Exp:* Res instr agr & biol chem, NC State Col, 55-56; assoc biochem & biophys, Iowa State Univ, 56-61, asst prof animal sci, 61-64; asst prof, 64-76, ASSOC PROF FOOD SCI, GA STA, UNIV GA, 76- *Mem:* AAAS; Am Chem Soc; Am Oil Chem Soc. *Res:* Lipid chemistry of foods. *Mailing Add:* 979 McDonough Rd Hampton GA 30228-1527

WORTHINGTON, THOMAS KIMBER, b New York, NY, Aug 4, 47. LOW TEMPERATURE PHYSICS. *Educ:* Franklin & Marshall Col, BA, 69; Wesleyan Univ, PhD(physics), 75. *Prof Exp:* FEL PHYSICS, RUTGERS UNIV, 75- *Mem:* Am Inst Physics; Am Phys Soc. *Res:* Specific heat and thermal conductivity of granular aluminum films; magnetics. *Mailing Add:* IBM Thomas J Watson Res Ctr PO Box 218 Yorktown Heights NY 10598

WORTHINGTON, WARD CURTIS, JR, b Savannah, Ga, Aug 8, 25; m 47; c 2. MEDICAL SCHOOL ADMINISTRATION, ANATOMY. *Educ:* The Citadel, BS, 52; Med Col SC, MD, 52. *Prof Exp:* Intern surg, Boston City Hosp, Mass, 52-53; instr anat, Sch Med, Johns Hopkins Univ, 53-56; asst prof, Col Med, Univ Ill, 56-57; from asst prof to assoc prof, 57-66, asst dean, 66-69, assoc dean,70-77 & 82, actg vpres, 75-77, vpres acad affairs, 77-82, PROF ANAT, MED UNIV SC, 66- *Concurrent Pos:* NIH spec fel, Dept Human Anat, Oxford Univ, 64-65; chmn, Asn Anat, 69-77, secy-treas & pres, 71-76. *Mem:* AAAS; Am Asn Anat; Endocrine Soc; Am Physiol Soc. *Res:* Medical history, anatomy and physiology of pituitary circulation; histology; neuroendocrinology. *Mailing Add:* Dept Anat Med Univ SC 171 Ashley Ave Charleston SC 29425-2201

WORTHMAN, ROBERT PAUL, b Chicago, Ill, Aug 3, 19; m 46; c 2. VETERINARY ANATOMY. *Educ:* Kans State Univ, DVM, 43; Iowa State Univ, MS, 53. *Prof Exp:* Asst prof vet anat, Wash State Univ, 46-49 & Iowa State Univ, 49-53; from asst prof to assoc prof, PROF VET ANAT, WASH STATE UNIV, 65- *Mem:* Am Vet Med Asn; Am Asn Anat; Am Asn Vet Anatomists (pres, 75); World Asn Vet Anatomists (vpres, 75); Sigma Xi. *Res:* Veterinary surgical and functional anatomy; techniques of teaching museum specimen preparation. *Mailing Add:* Wash State Univ McCoy 210 Pullman WA 99164

WORTHY, GRAHAM ANTHONY JAMES, b Bournemouth, Eng, Sept 13, 56; Can citizen; m 77; c 1. MARINE MAMMALS. *Educ:* Univ Guelph, BSc, 79, MSc, 82, PhD(environ physiol), 85. *Prof Exp:* Postdoctoral researcher, Univ Calif, Santa Cruz, 87-90; ASST PROF NATURAL HIST PHYSIOL, TEX A&M UNIV, GALVESTON, 90- *Concurrent Pos:* Lectr, Univ Calif, Santa Cruz, 87-90. *Mem:* Soc Marine Mammal; Am Cetacean Soc; Am Soc Mammalogists; Int Asn Aquatic Animal Med; Am Soc Zoologists. *Res:* Physiological ecology of marine mammals through the study of their energetics growth and nutrition; defining the appropriate criteria for proper care of marine mammals in captivity. *Mailing Add:* Tex A&M Univ 4700 Ave U Galveston TX 77551

WORTHY, THOMAS E, laboratory medicine, for more information see previous edition

WORTIS, MICHAEL, b New York, NY, Sept 28, 36; m 64; c 2. THEORETICAL PHYSICS, SOLID STATE PHYSICS. *Educ:* Harvard Univ, BA, 58, MA, 59, PhD(physics), 63. *Prof Exp:* Miller fel physics, Univ Calif, Berkeley, 62-64; NSF fel, Fac Sci, Univ Paris, 64-65; vis prof, Pakistan AEC, WPakistan, 65; res asst prof, Univ Ill, Urbana, 66, from asst prof to prof physics, 66-89; PROF PHYSICS, SIMON FRASER UNIV, 87- *Concurrent Pos:* A P Sloan Found fel, Univ Ill, Urbana, 67-69; Ford Found consult, Univ Islamabad, WPakistan, 71. *Honors & Awards:* Fulbright-Hays lectr, Pakistan, 78, Morocco 86,. *Mem:* Am Phys Soc. *Res:* Statistical physics; magnetic phenomena; phase transitions; phase transitions and related phenomena in strongly coupled and complex systems; surface and interface phenomena; biomembranes. *Mailing Add:* Dept Physics Simon Fraser Univ Burnaby BC V5A 1S6 CAN

WORTMAN, BERNARD, b Brooklyn, NY, Apr 23, 24; m 52; c 3. BIOCHEMISTRY. *Educ:* Syracuse Univ, AB, 48; Univ Tex, MA, 51; Ohio State Univ, PhD(physiol), 55. *Prof Exp:* From res asst to res asst prof ophthal, Sch Med, Wash Univ, 55-65; res assoc prof, Albany Med Col, 65-66; res chemist, Food & Drug Admin, Wash, DC, 66-69; scientist adminr, Nat Eye Inst, 69-80; sci adminr, Nat Inst Aging, 80-82; vis prof, Col Optom, Univ Calif, 82-83; spec asst, Nat Inst Arthritis, Metabolic & Kidney Dis, 83-84; sr staff consult, Fedn Am Socs Exp Biol, 84-87; CONSULT, TECHREVIEW, 84- *Concurrent Pos:* NIH spec fel, 60-61; estab investr, Am Heart Asn, 61-65; vis scientist, Am Physiol Soc, 63; consult to lab serv, Vet Admin Hosp, Albany, 65-66; res assoc prof, Med Sch, George Washington Univ, 67-70. *Mem:* Am Physiol Soc; Soc Gen Physiol. *Res:* Cell physiology; general metabolism of cornea; biosynthesis of sulfated mucopolysaccharides in cornea; biochemistry of connective tissue; macromolecular biochemistry. *Mailing Add:* 6407 Tone Dr Bethesda MD 20817-5815

WORTMAN, JIMMIE J(ACK), b NC, Feb 23, 36; m 54, 84; c 2. SOLID STATE PHYSICS, ELECTRONICS. *Educ:* NC State Univ, BS, 60; Duke Univ, MS, 62, PhD(elec eng), 65. *Prof Exp:* Mem tech staff, Bell Tel Labs, 60-62; res engr, Res Triangle Inst, 62-68, mgr, Eng Physics Dept, 68-72, dir, Eng Div, 72-75, dir, Energy & Environ Res Div, 75-80. *Concurrent Pos:* Asst prof, exten div, NC State Univ, 62-63, adj prof, 74, prof, 80-85; lectr, Duke Univ, 64-70. *Mem:* Inst Elec & Electronics Engrs; Mat Res Soc; Electrochem Soc. *Res:* Microelectronics, semiconductor device theory and fabrication; thin film phenomena; integrated circuits. *Mailing Add:* Dept Elec & Computer Eng NC State Univ Box 7911 Raleigh NC 27695

WORTMAN, ROGER MATTHEW, physics, for more information see previous edition

WORTS, GEORGE FRANK, JR, b Toledo, Ohio, Apr 24, 16; m 50; c 1. GEOLOGY, HYDROGEOLOGY. *Educ:* Stanford Univ, BS, 39. *Prof Exp:* Geologist, Ground Water Br, US Geol Surv, 40-50, geologist in charge, Long Beach, 52-56, dist geologist, Sacramento, 56-58, br area chief, Pac Coast Area, 58-62, dist chief, Nev, 62-74, part-time, 74-89; RETIRED. *Concurrent Pos:* Water res eval, Azores, Guyana, Philippines, Guam, Israel, Egypt, SKorea & Somalia. *Mem:* Fel Geol Soc Am; Am Geophys Union; Asn Eng Geol. *Res:* National and international hydrology, especially of arid regions; ground-water resources, particularly quantitative analysis, water quality, coastal hydrology and water management; direction of complex hydrologic studies and applied research. *Mailing Add:* 163 Tahoe Dr Carson City NV 89703

WORZALA, F(RANK) JOHN, b Milwaukee, Wis, Nov 13, 33; m 54; c 9. NUCLEAR METALLURGY. *Educ:* Univ Wis, BS, 56, MS, 58; Carnegie Inst Technol, MS, 62, PhD(metall eng), 65. *Prof Exp:* Engr, Hanford Labs, Gen Elec Co, Wash, 56-57; engr, Bettis Atomic Power Lab, Westinghouse Elec Corp, 58-65, res scientist, 66-67; NSF fel & res scientist, Grenoble Nuclear Res Ctr, France, 65-66; assoc prof nuclear mat, 67-77, PROF ENG METALL & MINING ENG, UNIV WIS-MADISON, 77- *Mem:* Am Nuclear Soc; Am Inst Mining, Metall & Petrol Engrs; Am Soc Metals. *Res:* Nuclear materials; irradiation damage in metals; fracture mechanics; applied superconductivity and superconducting materials. *Mailing Add:* Dept Metall 222 Minerals & Metals Univ Wis Madison 1505 University Ave Madison WI 53706

WORZEL, JOHN LAMAR, b W Brighton, NY, Feb 21, 19; m 41; c 4. GEOPHYSICS. *Educ:* Lehigh Univ, BS, 40; Columbia Univ, MA, 48, PhD, 49. *Prof Exp:* Res assoc, Woods Hole Oceanog Inst, 40-46; geodesist, Columbia Univ, 47-49, res assoc geol, 48-49, instr, 49-51, asst prof, 51-52, assoc prof geophys, 52-57, prof, 57-72, asst dir, Lamont Geol Observ, 57-64, actg dir, 64-65, assoc dir, 65-72; prof geophys & dep dir earth & planetary sci div, Marine Biomed Inst, Univ Tex Med Br, Galveston, 72-74; actg dir, Geophys Lab, Marine Sci Inst, Univ Tex, Austin, 74-75, prof geol sci, 72-79, prof marine studies, 75-79, dir, Geophhys Lab, Marine Sci Inst, 75-79; RETIRED. *Concurrent Pos:* Geophys consult, Off Naval Res, 50; mem, USN Deep Water Propagation Comt, 50-60; Guggenheim Found fel, 63; chmn spec study group, Int Union Geod & Geophys; prof geosci, Univ Tex, Dallas, 72-79; adj prof geol sci, Rice Univ, 74-77; pres, Palisades Geophys Inst, 74-; mem, US Nat Comt Int Geol Correlation Prog, 78-79. *Mem:* AAAS; Seismol Soc Am; Soc Explor Geophysicists (vpres, 78-79); Am Phys Soc; Geol Soc Am. *Res:* Gravity at sea; seismic refraction and reflection at sea; underwater photography; tectonophysics; marine geophysics; marine instrumentation. *Mailing Add:* 1091 S Shore Dr Southport NC 28461

WOSILAIT, WALTER DANIEL, b Racine, Wis, Feb 4, 24; m 48; c 1. PHARMACOLOGY, BIOCHEMISTRY. *Educ:* Wabash Col, BA, 49; Johns Hopkins Univ, PhD(biol), 53. *Prof Exp:* Jr instr pharmacol, Western Reserve Univ, 53-56; asst prof, State Univ NY Downstate Med Ctr, 56-63, assoc prof, 63-65; from assoc prof to prof, 65-89, EMER PROF PHARMACOL, SCH MED, UNIV MO-COLUMBIA, 89- *Concurrent Pos:* USPHS res grant, State Univ NY Downstate Med Ctr, 56-65; USPHS res grant, Sch Med, Univ Mo-Columbia, 65- *Mem:* Am Soc Biol Chemists; Am Soc Pharmacol & Exp Therapeut; Harvey Soc. *Res:* Anticoagulant drug interactions; pharmacokinetics, computer simulations of drug interactions; mathematical modeling of human fetal growth. *Mailing Add:* Dept Pharmacol Univ Mo Sch Med Columbia MO 65212

WOSINSKI, JOHN FRANCIS, b North Tonawanda, NY, Dec 30, 30; m 56; c 2. MATERIAL SCIENCE, MINERALOGY. *Educ:* Denison Univ, BS, 53; Brown Univ, MS, 58. *Prof Exp:* Geologist mapping, US Geol Surv, 56-57; mineralogist glass, 58-60, supvr petrog, 61-64, supvr refractories, 65-67, mgr refractories, 67-82, PROJ MGR REFRACTORIES, CORNING GLASS WORKS, 82- *Concurrent Pos:* Prin investr Apollo 14-15 mission samples, NASA, 71-72; chmn refractories div, Am Ceramic Soc, 78-79; chmn, C-8 div, Am Soc Testing & Mat, 82-87; consult glass refractory appln, Corning Eng. *Mem:* Geol Soc Am; Sigma Xi; fel Am Ceramic Soc; Am Soc Testing & Mat; Soc Glass Technol. *Res:* Develop, test, evaluate and recommend refractory materials used in corporate glass melting endeavors; glass-refractory manufacturing problems and archaeological excavations. *Mailing Add:* Corning Glass Works HP ME E-5 Corning NY 14831

WOSKE, HARRY MAX, b Reading, Pa, Feb 26, 24; m 72; c 7. INTERNAL MEDICINE, CARDIOLOGY. *Educ:* Columbia Univ, AB, 45; Long Island Col Med, MD, 48. *Prof Exp:* Instr med, Sch Med, Univ Pa, 55-57, assoc, 57-64, from adj asst prof to adj assoc prof, 64-69, assoc prof med & assoc dean, 69-73; assoc dean, 74, prof med & chief med serv, 73-76, CLIN PROF MED, NJ-RUTGERS MED SCH, 77-, ACTG DEAN, 75-; CHIEF CARDIOL, HUNTERDON MED CTR, NJ, 77- *Concurrent Pos:* NIH fel cardiol, Sch Med, Univ Pa, 55-57. *Mem:* Fel Am Col Physicians; fel Am Col Cardiol; fel NY Acad Sci. *Res:* Exercise projects. *Mailing Add:* Township Line Rd Belle Mead NJ 08502

WOSKOV, PAUL PETER, b Montour Falls, NY, Apr 23, 50; m 76; c 2. PLASMA DIAGNOSTICS IR & MILLIMETER-WAVES. *Educ:* Rensselaer Polytech Inst, BS, 72, MS, 74, PhD(electrophysics), 76. *Prof Exp:* Res asst lasers, Rensselaer Polytech Inst, 72-75, asst electromagnetics, 75-76; staff scientist lasers & plasma diagnostics, Francis Bitter Nat Magnet Lab, 76-80, PROJ LEADER IR & MILLIMETER-WAVE PLASMA DIAG DEVELOP, PLASMA FUSION CTR, MASS INST TECHNOL, 80- *Concurrent Pos:* Consult, Xsirius Superconductivity, Inc, 89-; chmn, 8th Topical Cont High Temperature Plasma Diagnostics, 90. *Mem:* Am Phys Soc; AAAS; Inst Elec & Electronics Engrs. *Res:* High power far-infrared laser and gyrotron development and research; plasma diagnostics of controlled fusion plasmas; development of infrared to millimeter-wave technologies; application of millimeter/submillimeter-wave technologies to high temperature superconductor development. *Mailing Add:* Bldg NW16-134 Mass Inst Technol Cambridge MA 02139

WOSTER, PATRICK MICHAEL, b Omaha, Nebr, Jan 5, 55; m 84; c 1. COMPUTER ASSISTED DRUG DESIGN. *Educ:* Univ Nebr, BS, 78, PhD(med chem), 86. *Prof Exp:* NIH postdoctoral fel chem, Rensselaer Polytech Inst, 86; NIH postdoctoral fel med chem, Univ Mich, 87; lectr, 88-89, ASST PROF MED CHEM, WAYNE STATE UNIV, 89- *Mem:* Am Chem Soc; Am Asn Cols Pharm. *Res:* Synthesis and biological evaluation of novel, rationally designed enzyme inhibitors as potential therapeutic agents. *Mailing Add:* 528 Shapero Hall Wayne State Univ Detroit MI 48202

WOSTMANN, BERNARD STEPHAN, b Amsterdam, Neth, Nov 6, 18; US citizen; m 46; c 5. BIOCHEMISTRY, NUTRITION. *Educ:* Univ Amsterdam, BS, 40, MS, 45, DSc, 48. *Prof Exp:* Instr org chem, Univ Amsterdam, 43, instr biochem, 45-48, lectr, 48-50, sci off, 48-55; from asst prof to assoc prof, 55-65, PROF, 65-, EMER PROF BIOL SCI, UNIV NOTRE DAME, 87- *Concurrent Pos:* Asst dir, Neth Inst Nutrit, 48-55; Rockefeller res fel, 50-51. *Mem:* AAAS; Am Inst Nutrit; Soc Exp Biol & Med; Asn Gnotobiotics (pres, 66-68); NY Acad Sci. *Res:* Biochemical background of host-contaminant relationship; role of intestinal flora in physiology and nutrition. *Mailing Add:* Dept Biol Sci Lobund Lab Univ Notre Dame Notre Dame IN 46556

WOTHERSPOON, NEIL, b New York, NY, Oct 24, 30; m 54; c 1. PHYSICAL CHEMISTRY, INSTRUMENTATION. *Educ:* Polytech Inst Brooklyn, BS, 52, PhD, 57; NY Univ, MBA, 82. *Prof Exp:* Res scientist, Radiation & Solid State Lab, NY Univ, 57-68; asst prof biophys & bioeng, Mt Sinai Sch Med, 68-76; sr scientist, Technicon Instruments Corp, 76-85; ASST PROF ELECTRO-MECH TECH, CITY TECH COL, CITY UNIV NY, 86-; SR PHYSICIST, DEPT HEALTH, BUR RADIATION CONTROL, CITY NY, 87- *Concurrent Pos:* Prog dir, Conf Radiation Control. *Mem:* Am Chem Soc; Asn Comput Mach; Inst Elec & Electronics Engrs. *Res:* Instrumentation for physics, chemistry and biomedical sciences, including optical electronic and computer techniques for automated analysis; analog and digital data acquisition, interfacing and processing. *Mailing Add:* 163 DeKalb Ave Brooklyn NY 11217

WOTIZ, HERBERT HENRY, b Vienna, Austria, Oct 8, 22; nat US; m 47; c 3. ENDOCRINE BIOCHEMISTRY. *Educ:* Providence Col, ScB, 44; Yale Univ, PhD(org chem), 51. *Prof Exp:* Instr, 51-53, asst prof, 53-55, assoc res prof, 55-61, assoc prof, 61-63, res assoc, R D Evans Mem Hosp, 55-68, PROF BIOCHEM, SCH MED, BOSTON UNIV, 63-, RES PROF UROL, 79-, DIR, H H HUMPHREY CANCER RES CTR, 84- *Concurrent Pos:* Res fel, R D Evans Mem Hosp, 50-55; USPHS sr res fel, 60-64 & res career develop fel, 65-69; assoc ed, Steroids, 85- *Mem:* Am Chem Soc; Am Soc Biol Chemists; Endocrine Soc; Am Asn Cancer Res; AAAS. *Res:* Steroid metabolism and analysis; mechanism of hormone action. *Mailing Add:* Dept Biochem Boston Univ Sch Med Boston MA 02118

WOTIZ, JOHN HENRY, b Moravska Ostrava, Czech, Apr 12, 19; nat US; m 45; c 3. HISTORY OF CHEMISTRY. *Educ:* Furman Univ, BS, 41; Univ Richmond, MS, 43; Ohio State Univ, PhD(org chem), 48. *Prof Exp:* Asst chem, Univ Richmond, 41-43 & Ohio State Univ, 43-44 & 46-47; from instr to assoc prof, Univ Pittsburgh, 48-57; prof chem & chmn dept, Marshall Univ, 62-67; chmn dept, 67-69, prof, 67-90, EMER PROF CHEM, SOUTHERN ILL UNIV, 90- *Concurrent Pos:* At Fed Security Agency, 44; Nat Acad Sci exchange prof, numerous Far East and Europ Socialist countries, 69-74; mem int activ comt, Am Chem Soc, 75-; vis prof & lectr, Japan, 84. *Honors & Awards:* Dexter Award, 82. *Mem:* Hist Sci Soc; Am Chem Soc. *Res:* Propargylic rearrangement; radical ions from vicinal diamines; institutional research in eastern socialist southeast Asian and Pacific Ocean countries; history of chemistry. *Mailing Add:* Dept Chem Southern Ill Univ Carbondale IL 62901

WOTT, JOHN ARTHUR, b Fremont, Ohio, Apr 10, 39; m 59; c 3. HORTICULTURE. *Educ:* Ohio State Univ, BS, 61; Cornell Univ, MS, 66, PhD(hort), 68. *Prof Exp:* Instr, Coop Exten Serv, Ohio State Univ, 61-64; res asst hort, Cornell Univ, 64-68; from asst prof to assoc prof hort, Purdue Univ, West Lafayette, 68-78, prof 78-81; PROF URBAN HORT, UNIV WASH, SEATTLE, 81-, ASSOC DIR, CTR URBAN HORT, 90- *Concurrent Pos:* County exten agent 4-H, Wood County, Bowling Green, Ohio, 61-64; mdm, Home Hort Working Group, Am Soc Hort Sci, 69-85, chmn, 82, chmn, Ornamental & Turf Working Group, 84; mem, Int Plant Propagation, Eastern Region, 70-79, Int pres, 84, int secy treas, 86-; mem, Prod Rev Comt, Am Hort Soc, 79-80. *Mem:* Am Soc Hort Sci; Am Hort Soc; Int Plant Propagators Soc; Am Asn Bot Gardens & Arboreta. *Res:* Nutrition of cuttings during propagation; application of nutrient mist; horticultural problems of homeowners, such as foliage plants, annuals and perennials; herbicides in annual flowers; morphological and physiological response of woody plants to flooding; adaptation and use of plants in the urban environment. *Mailing Add:* Ctr Urban Hort GF-15 Univ Wash Seattle WA 98195

WOTZAK, GREGORY PAUL, b New York, NY, Dec 12, 44; m 68; c 1. CHEMICAL ENGINEERING. *Educ:* Rensselaer Polytech Inst, BChE, 65; Princeton Univ, MS, 67, PhD(chem eng), 68. *Prof Exp:* From asst prof to assoc prof chem eng, Rensselaer Polytech Inst, 68-77, prof, 77-80; MEM FAC CHEM ENG, CLEVELAND STATE UNIV, 80- *Mem:* Am Inst Chem Engrs. *Res:* Chemical reaction engineering; stochastic simulation; transport phenomena; chemical physics. *Mailing Add:* Dept Chem Eng Cleveland State Univ Euclid Ave at E 24th Cleveland OH 44115

WOUK, ARTHUR, b New York, NY, Mar 25, 24; m 44; c 2. MATHEMATICS. *Educ:* City Col New York, BS, 43; Johns Hopkins Univ, MA, 47, PhD(math), 51. *Prof Exp:* Instr math, Johns Hopkins Univ, 47-50 & Queens Col, NY, 50-52; mathematician, Proj Cyclone, Reeves Instrument Corp, 52-54; supvr, Math Analysis Sect, Missile Systs Lab, Sylvania Elec Prod, Inc, 54-56, sr eng specialist, Appl Res Lab, Sylvania Electronic Systs Div, Gen Tel & Electronics Corp, 56-62; vis prof, Math Res Ctr, Univ Wis, 62-63; from assoc prof to prof appl math, Northwestern Univ, Evanston, 63-72; chmn dept, Univ Alta, 72-77, prof comput sci, 72-83; mathematician, Army Res Off, Triangle Park, NC, 83-89; RETIRED. *Concurrent Pos:* Ed, Commun, Asn Comput Mach, 58-64; consult, Appl Res Lab, Sylvania Electronic Systs Div, Gen Tel & Electronics Corp & Argonne Nat Lab, 63-69; ed, Soc Indust & Appl Math Rev, 63-83. *Mem:* Am Math Soc; Soc Indust & Appl Math; Asn Comput Mach. *Res:* Numerical and functional analysis. *Mailing Add:* 3849 Birchwood Dr Boulder CO 80304-1428

WOUK, VICTOR, b New York, NY, Apr 27, 19; m 41; c 2. HIGH POWER ELECTRONICS, ELECTRIC HYBRID VEHICLES. *Educ:* Columbia Univ, BA, 39; Calif Inst Technol, MS, 40, PhD(elec eng), 42. *Prof Exp:* Res engr, Res Labs, Westinghouse Elec Corp, 42-45; circuit engr, NAm Philips Co, NY, 45-47; pres & chief engr, Beta Elec Corp, 47-57; vpres eng & res, Sorenson & Co, Inc, 57-60; pres, Electronic Energy Conversion Corp, 60-63; gen mgr electronic energy conversion dept, Gulton Industs, Inc, NY, 63-68, dir electronics res, 68-70; pres, Petro-elec Motors, Ltd, 70-76; PRES, VICTOR WOUK ASSOCS, 76- *Concurrent Pos:* US rep tech comt elec rd vehicles, Int Electrotech Comn, 69; consult, Dept Energy, 76-81; vpres bd gov, NY Acad Sci; mem gov bd, Metrol Sect, Soc Automotive Engrs, 81- *Mem:* Fel AAAS; Soc Automotive Engrs; Inst Elec & Electronics Engrs; Fedn Am Sci; fel NY Acad Sci (vpres); Sigma Xi. *Res:* Static electricity generation by gasoline; television amplifier and sweep circuits; high voltage power supplies; regulated alternating and direct current power supplies; electronic controls for electric vehicles; electric and hybrid vehicles. *Mailing Add:* 1225 Park Ave New York NY 10128-1707

WOURMS, JOHN P, b New York, NY, Apr 30, 37; m 72; c 1. CELL BIOLOGY, DEVELOPMENTAL BIOLOGY. *Educ:* Fordham Univ, BS, 58, MS, 60; Stanford Univ, PhD(biol), 66. *Prof Exp:* Am Cancer Soc fel, Harvard Univ, 66-68; Nat Res Coun Can grant & asst prof cell develop biol, Dept Biol, McGill Univ, 68-71, res assoc, Dept Path, 71-72; assoc res scientist, NY Ocean Sci Lab, 72-76; assoc prof, 77-82, PROF BIOL SCI, CLEMSON UNIV, 82- *Concurrent Pos:* Vis lectr, Biol Labs, Harvard Univ, 71-72; assoc ed, J Exp Zool, 79-83; prin investr, NSF Grant, 82-84 & 87-89; Guggenheim fel, 84-85; vis prof, Univ Colo, 84-85 & Univ Calif, Santa Cruz, 85; chief scientist, Nat Oceanic & Atmospheric Admin Deep Submersible Dives, 87, 88 & 90; assoc ed, Environ Biol Fishes, 84-, J Morphol, 91- *Mem:* Am Soc Cell Biol; Am Soc Ichthyologists & Herpetologists; Soc Develop Biol; Int Soc Develop Biologists; Marine Biol Asn UK; Am Soc Zoologists; Explorer's Club. *Res:* Cell differentiation; cell ultrastructure; reproduction and development of fishes and marine invertebrates; oogenesis; ultrastructure and chemistry of extra-cellular matrices; biology of annual fishes; evolution of development; elasmobranch biology; maternal-fetal relationship in viviparous fishes. *Mailing Add:* Dept Biol Sci Clemson Univ Clemson SC 29634-1903

WOVCHA, MERLE G, b Virginia, Minn, Dec 19, 38. BIOCHEMISTRY. *Educ:* Univ Minn, BA, 62, BS, 64, MS, 67, PhD(biochem), 71. *Prof Exp:* NIH fel, Dept Biol Chem, Univ Mich, 71-74; res scientist, 74-77, sr res scientist, 77-81, SR SCIENTIST, FERMENTATION RES & DEVELOP, UPJOHN CO, 81- *Mem:* Am Soc Microbiol. *Res:* Microbial sterol bioconversions; streptomycete metabolism and replication; bacteriophage biochemistry. *Mailing Add:* 103 S Prairie Ave Kalamazoo MI 49007

WOYCHIK, JOHN HENRY, b Scranton, Pa, Mar 30, 30; m 53; c 2. BIOCHEMISTRY, PHYSIOLOGY. *Educ:* Univ Scranton, BS, 53; Univ Tenn, MS, 55, PhD(biochem), 57. *Prof Exp:* Chemist, Northern Regional Lab, 57-63, prin chemist, Eastern Mkt & Res Div, 63-74, CHIEF, DAIRY LAB, EASTERN REGIONAL RES CTR, USDA, 74- *Mem:* Am Chem Soc; Am Dairy Sci Asn; Am Soc Biol Chemists; Inst Food Technologists. *Res:* Isolation and characterization of milk and cereal proteins; glycoproteins; basic research on milk components and dairy product development. *Mailing Add:* Eastern Regional Res Ctr USDA 600 E Mermaid Lane Philadelphia PA 19118

WOYCHIK, RICHARD P, b Arcadia, Wis, Oct 28, 52. MOLECULAR BIOLOGY, BACTERIOLOGY. *Educ:* Univ Wis-Madison, BS, 77; Case Western Reserve Univ, PhD(molecular biol), 84. *Prof Exp:* Postdoctoral, Harvard Med Sch, 84-87; RES SCIENTIST, OAK RIDGE NAT LAB, 87- *Mem:* AAAS; Am Soc Microbiol; Fedn Am Socs Exp Biol. *Res:* Molecular analysis of developmental genes in mice utilizing DNA probes derived from transgenic mice. *Mailing Add:* Biol Div Oak Ridge Nat Lab PO Box 2009 Oak Ridge TN 37831

WOYSKI, MARGARET SKILLMAN, b W Chester, Pa, July 26, 21; m 48; c 4. GEOLOGY. *Educ:* Wellesley Col, BA, 43; Univ Minn, MS, 45, PhD(geol), 46. *Prof Exp:* Instr geol, Univ Minn, 46; geologist, Mo Geol Surv, 46-48; instr geol, Univ Wis, 48-52; lectr, Calif State Col, Long Beach, 63-67; from asst prof to assoc prof, 67-74, chmn earth sci, 73-76, PROF GEOL, CALIF STATE UNIV, FULLERTON, 74-, ASSOC DEAN NATURAL SCI & MATH, 81- *Mem:* Mineral Soc Am; Geol Soc Am. *Res:* Intrusive rocks of central Minnesota; Precambrian sediments of Missouri; laboratory manuals for historical and physical geology; geologic guidebooks of Southern California; source locations of artifacts; petrology of peninsular range batholith. *Mailing Add:* Dept Geol Calif State Univ 800 N State Col Blvd Fullerton CA 92634

WOZAB, DAVID HYRUM, b Los Angeles, Calif, April 17, 23; m 56; c 1. LIMESTONE HYDROLOGY, IRRIGATION DEVELOPMENT. *Educ:* Univ Southern Calif, BA, 50, MA, 52. *Prof Exp:* Proj mgr, Food & Agr Orgn, UN, Rome, Italy, 61-80, rep, Tanzania, 80-83, rep, Jamaica & Bahamas, 83-86. *Concurrent Pos:* Consult, World Bank & Jamaican Govt, 87-88, Food & Agr Orgn, UN, 86. *Mem:* fel Geol Soc Am; Am Geophy Union; AAAS; Am Water Works Asn; Nat Water Well Asn; Asn Eng Geol. *Res:* Flow of water in karatified limestone. *Mailing Add:* 10132 SW 79th Ave Miami FL 33156

WOZENCRAFT, JOHN MCREYNOLDS, b Dallas, Tex, Sept 30, 25; m 63; c 2. ELECTRICAL ENGINEERING, COMMUNICATIONS. *Educ:* US Mil Acad, BS, 46; Mass Inst Technol, SM & EE, 51, ScD, 57. *Prof Exp:* Asst elec eng, Mass Inst Technol, 55-57, from asst prof to prof, 57-72, head commun div, Lincoln Lab, 69-72; dean res, 72-76, PROF ELEC ENG, NAVAL POSTGRAD SCH, 76- *Mem:* Fel Inst Elec & Electronics Engrs. *Res:* Application of information theory to practical communication problems; algorithmic languages for digital computation. *Mailing Add:* 6560 Brookdale Dr Carmel CA 93923

WOZNIAK, WAYNE THEODORE, b Chicago, Ill, Oct 13, 45; m 71; c 3. COLOR ANALYSIS. *Educ:* Ill Benedictine Col, BS, 67; Fla State Univ, PhD(phys inorg chem), 71. *Prof Exp:* Res assoc phys chem, Princeton Univ, 71-73; res assoc chem physics, Univ Ill, 73-75; res assoc biophys, Am Dent Asn, 75-78, head, Physics Lab, 78-79, lab dir, 79-80, asst secy, 80-89, ASST DIR, COUN DENT MAT, INSTRUMENTS & EQUIP, AM DENT ASN, 89- *Concurrent Pos:* Instr, Univ Ill, 74. *Mem:* Am Chem Soc; Int Asn Dent Res. *Res:* Spectroscopy of dental materials and calcified tissue; color analysis; applications of vibrational spectroscopy to biological systems. *Mailing Add:* Am Dent Asn 211 E Chicago Ave Chicago IL 60611

WRAIGHT, COLIN ALLEN, b London, Eng, Nov 27, 45. BIOPHYSICS. *Educ:* Univ Bristol, BSc, 67, PhD(biochem), 71. *Prof Exp:* Fel biophys, State Univ Leiden, 71-72; assoc, Cornell Univ, 72-74; asst prof biol, Univ Calif, Santa Barbara, 74-75; asst prof, 75-81, ASSOC PROF BIOPHYS & BOT, UNIV ILL, URBANA-CHAMPAIGN, 81- *Mem:* Biophys Soc; AAAS; Am Soc Photobiol. *Res:* Membrane functions and mechanisms of electron and ion transport in biological energy conservation. *Mailing Add:* Dept Plant Biol Univ Ill Urbana Campus 505 S Goodwin Ave Urbana IL 61801

WRASIDLO, WOLFGANG JOHANN, b Beuthen, Ger, Oct 31, 38; US citizen; c 2. POLYMER CHEMISTRY. *Educ:* San Diego State Univ, BA, 63; Univ Nurnberg, MS, 67. *Prof Exp:* Sr res chemist, Whittaker Corp, 63-65; res chemist, US Naval Ord Lab, 67-69; staff mem polymers, Sci Res Labs, Boeing Co, 69-71; STAFF CHEMIST, UOP, 71- *Mem:* Am Chem Soc. *Res:* Solid state polymers; synthesis and characterization of high temperature polymers; mechanical, thermal and morphological behavior of polymers; reverse osmosis membranes. *Mailing Add:* 307 Prospect St La Jolla CA 92037-4652

WRATHALL, DONALD PRIOR, b Pittsburgh, Pa, Mar 9, 36; m 62; c 4. PHYSICAL CHEMISTRY. *Educ:* Brigham Young Univ, BA, 64, PhD(phys chem), 68. *Prof Exp:* NIH fel, Yale Univ, 67-68; sr res chemist, Eastman Kodak Co, 68-87; EASTMAN GELATINE CORP, 87- *Res:* Thermodynamics of reactions at solid-liquid interfaces and with macromolecules in solution. *Mailing Add:* Eastman Gelatine Corp 227 Washington St Peabody MA 01960

WRATHALL, JAY W, b Salt Lake City, Utah, May 12, 33; m 69; c 7. INORGANIC CHEMISTRY. *Educ:* Brigham Young Univ, BS, 57, MS, 59; Ohio State Univ, PhD(inorg chem), 62. *Prof Exp:* Asst prof chem, Univ Calif, Berkeley, 62-64 & Univ Hawaii, 64-69; assoc prof, Church Col Hawaii, 69-73, prof & div chmn, 74-85, PROF PHYS SCI, BRIGHAM YOUNG UNIV, HAWAII CAMPUS, 77-, CHMN 74- *Mem:* Am Chem Soc. *Res:* Coordination chemistry; reactions of coordinated ligands; biological activity of transition metal complexes. *Mailing Add:* Brigham Young Univ Hawaii Campus Dept Phys Sci Laie HI 96762

WRATHALL, JEAN REW, b Brooklyn, NY, Dec 3, 42; m 60; c 2. GENETICS, CELL BIOLOGY. *Educ:* Univ Utah, BS, 64, PhD(genetics, molecular biol), 69. *Prof Exp:* Asst prof biol, State Univ NY Col Geneseo, 69-70; from instr to asst prof genetics, Med Col, Cornell Univ, 70-74; res assoc, 74-75, ASST PROF ANAT, MED COL, GEORGETOWN UNIV, 75- *Concurrent Pos:* Damon Runyon Mem Fund fel, Cornell Univ, 71- *Mem:* AAAS; Soc Develop Biol; Soc Cell Biologists; Tissue Cult Asn; Soc Neurosci. *Res:* Control of differentiated function in eukaryotic cells in culture; abnormal functions in malignant cells in culture; 5-bromodeoxyuridine suppression of melanin synthesis and tumorigenicity of melanoma cells. *Mailing Add:* Dept Anat Georgetown Univ Sch Med 3900 Reservoir NW Washington DC 20007

WRAY, GRANVILLE WAYNE, b Elk City, Okla, Dec 16, 41; m 65; c 1. CELL BIOLOGY, BIOCHEMISTRY. *Educ:* Phillips Univ, BS, 63; Okla State Univ, MS, 65; Univ Tex, PhD(cell biol), 70. *Prof Exp:* Res asst biochem, Okla State Univ, 63-65; res asst, Univ Tex M D Anderson Hosp & Tumor Inst, 65-66; asst prof cell biol, Baylor Col Med, 72-84; grants assoc, NIH, 84-85, health scientist adminr, Nat Heart Lung & Blood Inst/NIH, 86-88, DEP DIR EP, NIDR/NIH, 88- *Concurrent Pos:* Damon Runyon Cancer res fel, McArdle Lab Cancer Res, Univ Wis-Madison, 71-72. *Honors & Awards:* Mike Hogg Award, 68. *Mem:* AAAS; Am Soc Cell Biologists. *Res:* Isolation, morphology and biochemistry of the mammalian metaphase chromosome. *Mailing Add:* NIH Westwood Bldg Rm 502 5333 Westbard Ave Bethesda MD 20892

WRAY, H LINTON, b Charlotte, NC, Apr 2, 40; m; c 3. ENDOCRINOLOGY. *Educ:* Univ Pa, MD, 66. *Prof Exp:* Asst chief, 77-84, Chief Endocrine & Metab Serv, 84-85, CHIEF DEPT CLIN INVEST, WALTER REED ARMY MED CTR, 85- *Mem:* Endocrine Soc; Am Physiol Soc; Am Col Physicians; Am Soc Bone & Mineral Res. *Res:* Hormonal control of calcium metabolism. *Mailing Add:* Dept Clin Invest Walter Reed Army Med Ctr Washington DC 20307-5001

WRAY, JAMES DAVID, b Norton, Kans, Oct 3, 36; c 1. ASTRONOMY. *Educ:* Univ MNex, BS, 59; Univ Cincinnati, MS, 62; Northwestern Univ, PhD(astron), 66. *Prof Exp:* Res assoc meteoritics, Univ NMex, 62-64, dir, Inst Meteoritics, 66-67; asst prof astron, Northwestern Univ, 67-72; res scientist astron, Univ Tex, Austin, 72-85; PRES, SCI TECH ASTRON RES, 85- *Concurrent Pos:* Consult, Boller & Chivens Div, Perkin-Elmer Corp, 73- *Mem:* Int Astron Union; Am Astron Soc. *Res:* Space astronomy, ultraviolet stellar spectroscopy; extra-galactic research, surface distribution of color in galaxies; digital image processing; numerical data base management and applications. *Mailing Add:* 21200 Todd Valley Rd No 54 Forest Hill CA 95631-9511

WRAY, JOE D, b Conway, Ark, Sept 30, 26; m 51; c 5. MATERNAL & CHILD HEALTH, NUTRITION. *Educ:* Stanford Univ, BA, 47, MD, 52; Univ NC, MPH, 67; Am Bd Pediat, dipl. *Hon Degrees:* Dr, Hacettepe Univ, Ankara, 83. *Prof Exp:* Intern, Charity Hosp La, New Orleans, 51-52; intern & resident pediat, Grace-New Haven Community Hosp, Conn, 54-56; chief resident, Hacettepe Children's Hosp, Ankara, Turkey, 56-58, assoc pediatrician, 58-61; vis prof pediat, Fac Med, Univ Valle, Cali, Colombia, 61-66; vis prof community med & pediat, Ramathibodi Hosp Med Sch, Mahidol Univ, Bangkok, Thailand, 67-74; fel, Ctr Advan Study Behav Sci, 74-75; vis prof maternal & child health & int health, Harvard Sch Pub Health, 75-78, sr lectr, 78-81, head, dept pop sci, 78-80, dir, Off Int Health Prog, 79-81; dep dir Ctr Pop & Family Health, 81-87, PROF CLIN PUB HEALTH,

SCH PUB HEALTH, COLUMBIA UNIV, 81- *Concurrent Pos:* Mem field staff health & pop, Rockefeller Found, NY, 60-76; chairperson, gov bd, Nat Coun Int Health, 84-86. *Mem:* Fel Am Acad Pediat; Am Pub Health Asn. *Res:* Infant and preschool child nutrition; growth and development; nutrition and infection; delivery of health services to children in developing countries; family planning. *Mailing Add:* Ctr Pop & Family Health Columbia Univ 60 Haven Ave New York NY 10032

WRAY, JOHN L, b Maryville, Mo, June 17, 35; m 58; c 3. ELECTRONICS ENGINEERING, TECHNICAL MANAGEMENT. *Educ:* Univ Mo, Columbia, BS, 57; Stanford Univ, MS, 58; Univ Santa Clara, MBA, 66. *Prof Exp:* Test planning officer, USAF, 58-62; mgr mkt res, Gen Elec Co, 62-78; vpres, Quadrex Corp, 78-84; pres, Systrol, Inc, 84-86; pres, Gooselake Lumber Co, 87-89; vpres, Renewable Resources, 89; SR CONSULT, VPRES & CHIEF FINANCIAL OFFICER, THOMAS RES CORP, 90- *Mem:* Am Soc Mech Engrs. *Res:* Effects on eyes of thermal energy from nuclear detonations; water jet pumps for boiling water reactors; instrumentation for nuclear reactors. *Mailing Add:* 19992 Buckhaven Lane Saratoga CA 95070

WRAY, JOHN LEE, b Charleston, WVa, July 10, 25; m 52; c 1. GEOLOGY. *Educ:* Univ WVa, BS, 50, MS, 51; Univ Wis, PhD(geol), 56. *Prof Exp:* Teaching asst geol, Univ WVa, 48-51; geologist, WVa State Hwy Dept, 51-53; res asst chem, Univ Wis, 53-56; res assoc, Res Ctr, Marathon Oil Co, 56-86; RETIRED. *Concurrent Pos:* Adj prof geol, Colo Sch Mines, 70-83. *Mem:* Geol Soc Am; Paleont Soc; Am Asn Petrol Geologists; Soc Econ Paleont & Mineral. *Res:* Paleontology; fossil algae; carbonate sedimentology; biostratigraphy. *Mailing Add:* 3755 Chataway Ct Colorado Springs CO 80906-4388

WRAY, PORTER R, b Chester, Pa, Jan 21, 13; m 39; c 4. ENGINEERING. *Educ:* Swarthmore Col, BS, 34. *Prof Exp:* Qual observer, Carnegie Steel Co, 34-36; res asst, Res Lab, US Steel, 36-37; metall serv engr, Carnegie-Ill Steel Corp, 37-41, mgr stainless, 41-43, asst chief metallurgist, Duquesne Works, 43-44, metall engr alloy, US Steel, 44-55, gen mgr metall, 55-69, dir metall eng qual control & serv, 69-78; RETIRED. *Concurrent Pos:* Consult, US State Dept, NATO, 59; mem, Adv Coun, Mat Res Agency, US Army, 66-68; dir, Am Nat Standards Inst, 70-73 & Metal Properties Coun, 74-79; chair, Steel Indust, Am Nat Metric Coun, 75-78. *Mem:* Fel Am Soc Metals; Am Iron & Steel Inst; Soc Automotive Engrs; Am Nat Standards Inst; Am Nat Metric Coun. *Mailing Add:* 1290 Boyce Rd Pittsburgh PA 15241

WRAY, STEPHEN DONALD, b Leeds, Yorkshire, Eng, Sept 23, 47. ORDINARY DIFFERENTIAL EQUATIONS. *Educ:* Univ Adelaide, BSc, 67; Flinders Univ, BSc, 68, MSc, 69, PhD(math), 74. *Prof Exp:* Asst prof math, Carleton Univ, 77-78 & Mt Allison Univ, 78-81; asst prof, 81-83, ASSOC PROF MATH, ROYAL ROADS COL, VICTORIA, BC, 83- *Res:* Ordinary differential equations and Dirichlet integral inequalities associated with formally symmetric ordinary differential expressions of Sturm-Liouville type. *Mailing Add:* Dept Math Royal Roads Col Victoria BC V0S 1B0 Can

WRAY, VIRGINIA LEE POLLAN, b Grove, Okla, Mar 20, 40; m 65. BIOCHEMISTRY. *Educ:* Okla State Univ, BS, 62, MS, 66; Univ Tex Grad Sch Biomed Sci Houston, PhD(biochem), 70. *Prof Exp:* Res asst biochem virol, Col Med, Baylor Univ, 66; RES ASST PROF CELL BIOL, BAYLOR COL MED, 73- *Concurrent Pos:* Fel, McArdle Lab Cancer Res, Univ Wis-Madison, 70-72. *Mem:* Am Soc Cell Biologists; Am Chem Soc. *Res:* Membrane biochemistry; composition and function of nuclear and plasma membranes. *Mailing Add:* 7505 Water Lily Way Columbia MD 21046

WREDE, DON EDWARD, medical physics, biophysics, for more information see previous edition

WREDE, ROBERT C, JR, b Cincinnati, Ohio, Oct 19, 26; m 48; c 3. MATHEMATICS. *Educ:* Miami Univ, Ohio, BS, 49, MA, 50; Ind Univ, PhD(math), 56. *Prof Exp:* Instr math, Miami Univ, Ohio, 50-51; from instr to assoc prof, 55-63, PROF MATH, CALIF STATE UNIV, SAN JOSE, 63- *Concurrent Pos:* Consult phys & res lab, Int Bus Mach Corp, Calif, 56-58; Hunter's Point Radiation Lab, 60. *Mem:* Am Math Soc; Math Asn Am; Tensor Soc. *Res:* Relativity theory; differential geometry; vector and tensor analysis. *Mailing Add:* Dept Math McQuarry Hall Calif State Univ San Jose CA 95114

WREFORD, STANLEY S, b Detroit, Mich, Mar 18, 49; m 67; c 1. CATALYTIC CHEMISTRY, CARBOXYLATION CHEMISTRY. *Educ:* Univ Mich, BS, 70; Mass Inst Technol, PhD(inorg chem), 74. *Prof Exp:* Assoc prof inorg chem, Harvard Univ, 74-78, Univ Toronto, 78-80; TECH MGR SPECIALTY CHEMICALS, E I DU PONT DE NEMOURS & CO, INC, 80- *Mem:* Am Chem Soc. *Res:* Homogeneous catalysis of carboxylation reactions. *Mailing Add:* RD 1 Box 11 Bassett Rd Salem NJ 08079

WRENN, MCDONALD EDWARD, b New York, NY, Apr 16, 36; m 67; c 2. ENVIRONMENTAL HEALTH, RADIOLOGICAL HEALTH. *Educ:* Princeton Univ, AB, 58; NY Univ, MS, 62, PhD(nuclear eng, environ health), 67. *Prof Exp:* Res scientist, US Energy Res & Develop Admin, 62-67, from instr to asst prof, 67-72, biomed scientist radiobiol, Div Biomed & Environ Res, 73-75; assoc prof environ med, Med Ctr, NY Univ, 72-79; prof pharm & dir, Radiobiol Div, 79-86, DIR, ENVIRON RADIATION & TOXICOL LAB, UNIV UTAH SCH MED, 86- *Concurrent Pos:* Mem, Nat Coun Radiation Protection & Measurements, 71-; mem & former chmn N-13 comt, Am Nat Standards Inst, 72- *Mem:* Radiation Res Soc; Health Physics Soc; fel Am Pub Health Asn; Am Inst Biol Sci; Am Indust Hygiene Asn. *Res:* Biological effects of environmental agents on man and animals, particularly radiations and radioactive materials; environmental cycling and transport of trace and radioactive elements; mammalian metabolism of actinides and development of environmental radiation detection instruments. *Mailing Add:* Environ Radiation & Toxicol Lab Univ Utah 1771 S 900 W No 10 Salt Lake City UT 84104

WRENN, WILLIAM J, b Los Angeles, Calif, July 18, 35. SYSTEMATICS & TAXONOMY OF CHIGGERS, ECTOPARASITIC MITES ON TERRESTRIAL VERTEBRATES. *Educ:* Calif State Univ, BS, 62, MA, 64; Univ Kans, PhD(entom), 72. *Prof Exp:* Teaching asst biol & zool, Dept Biol, Calif State Univ, 62-64; teaching asst biol & biomet, Entom Dept, Univ Kans, 64-66; teaching asst zool & entom, Univ Mich, Ann Arbor, 66-68; from asst prof to assoc prof, 69-87, PROF BIOL & ENTOM, DEPT BIOL, UNIV NDAK, 87- *Concurrent Pos:* Res assoc entom, Natural Hist Mus, Los Angeles County, 80-; adj prof, Calif State Univ, Long Beach, 81. *Mem:* Acarol Soc Am; Entom Soc Am; Soc Syst Zool; Soc Study Evolution; Soc Vector Ecol. *Res:* Systematics of parasitic acarines, especially chiggers; chigger-host relationships, ecology, biology, immune responsiveness and taxonomy, especially of western hemisphere species. *Mailing Add:* Dept Biol Univ NDak Grand Forks ND 58202-8238

WRENSCH, DANA LOUISE, b Greenwood, Miss, Oct 3, 46; m 74; c 2. ACAROLOGY, GENETICS. *Educ:* Ohio State Univ, BSc, 68, MS, 70, PhD(genetics), 72. *Prof Exp:* NIH trainee, 72-75, ADJ ASST PROF ENTOM, OHIO STATE UNIV, 78-; ASST PROF BOT & BACT, OHIO WESLEYAN UNIV, 81- *Concurrent Pos:* Consult, IBP, 72 & 73, Ministry Educ & Cult, Brazil, 78; lectr genetics, Ohio State Univ, 82. *Mem:* AAAS; Acarol Soc Am; Am Genetic Asn; Entom Soc Am. *Res:* Population biology and genetics of spider mites. *Mailing Add:* 3819 Granden Columbus OH 43214

WRIDE, W(ILLIAM) JAMES, b Garfield, Wash, Dec 3, 21; m 48; c 3. CHEMICAL ENGINEERING. *Educ:* State Col Wash, BS, 43, MS, 46; Iowa State Col, PhD(chem eng), 48. *Prof Exp:* Asst chem & chem eng, State Col Wash, 41-42; co-owner, mgr & consult engr, Ames Eng & Testing Serv, 48-51; chem & process engr, Philips Petrol Co, 52-66, mgr chem eng fundamentals br, Res & Develop Dept, 66-72, mgr eng res br, 72-74, process technol consult, 74-77, mgr, eng & technol, Energy Minerals Div, 78-84; RETIRED. *Mem:* Am Inst Chem Engrs. *Res:* Mercaptan formation in petroleum distillates; polyolefin resin process development; petrochemical process optimization; kinetics and mass transfer; process design and economic evaluation; research and development of alternate energy sources. *Mailing Add:* 633 SE Oakridge Dr Bartlesville OK 74006

WRIEDT, HENRY ANDERSON, b Melbourne, Australia, Feb 6, 28; nat US; m 60; c 3. PHYSICAL CHEMISTRY, METALS. *Educ:* Univ Melbourne, BMetE, 49; Mass Inst Technol, ScD(metall), 54. *Prof Exp:* Asst metall, Mass Inst Technol, 49-53; technologist, Appl Res Lab, US Steel Corp, 53-55, scientist, Edgar C Bain Lab Fund Res, 55-66, sr scientist, 66-72, sr scientist, Res Lab, 72-76, assoc res consult, Res Lab, 76-83; CONSULT, 83- *Concurrent Pos:* Ed screener, tech writer, Int Ctr Diffraction Data, 90- *Mem:* Am Inst Mining, Metall & Petrol Eng. *Res:* Physical chemistry of metals; phase equlibria and gas-metal reactions in metallic systems; thermodynamics of metallic systems. *Mailing Add:* 148 Washington St Pittsburgh PA 15218-1352

WRIGHT, ALAN CARL, b Bangor, Maine, Aug 16, 39; m 64; c 2. CHEMISTRY. *Educ:* Univ Maine, Orono, BS, 61; Univ Fla, PhD(chem), 66. *Prof Exp:* Res chemist, Cent Res Div, Am Cyanamid Co, 66-69; from asst prof to assoc prof, 70-82, chmn dept earth & phys sci, 75-78, PROF CHEM, EASTERN CONN STATE UNIV, 82- *Mem:* Am Chem Soc; Sigma Xi; AAAS. *Res:* Development of new laboratory experiments for organic chemistry involving compounds of biological interest. *Mailing Add:* Dept Phys Sci Eastern Conn State Univ Willimantic CT 06226

WRIGHT, ALDEN HALBERT, b Missoula, Mont, Apr 23, 42; m 67; c 2. ALGORITHMS. *Educ:* Dartmouth Col, BA, 64; Univ Wis-Madison, PhD(math), 69. *Prof Exp:* Vis asst prof math, Univ Utah, 69-70; from asst prof to prof math, Western Mich Univ,70-83; assoc prof, 83-86, PROF COMPUT SCI, UNIV MONT, 86- *Mem:* Asn Comput Mach; Inst Elec & Electronics Engrs Comput Soc. *Res:* Finding all solutions to a system of polynomial equations; genetics algorithms; string search algorithms. *Mailing Add:* Dept Comput Sci Univ Mont Missoula MT 59812

WRIGHT, ANDREW, b Edinburgh, Scotland, Jan 28, 35; m 57; c 2. MOLECULAR BIOLOGY. *Educ:* Univ Edinburgh, BSc, 57, PhD(biochem), 60. *Prof Exp:* Fel biochem, Univ Minn, 60-62; fel biol, Mass Inst Technol, 63-67; PROF MOLECULAR BIOL, MED SCH, TUFTS UNIV, 67- *Mem:* Fedn Am Socs Exp Biol; Sigma Xi. *Res:* Structure and function of bacterial cell membranes and mechanisms of DNA replication. *Mailing Add:* Dept Molecular Biol Tufts Univ Med Sch 136 Harrison Ave Boston MA 02111

WRIGHT, ANN ELIZABETH, b Mooringsport, La, Feb 20, 22; m 40; c 1. RADIOLOGICAL PHYSICS, BIOPHYSICS. *Educ:* Univ Houston, BA, 65; Univ Tex Grad Sch Biomed Sci, Houston, MS, 67, PhD(radiol physics), 70. *Prof Exp:* Instr radiol, Baylor Col Med, 68-70, asst prof radiol physics, 70-75; assoc prof, 75-82, PROF RADIOL, UNIV TEX MED BR GALVESTON, 82- *Concurrent Pos:* Consult physicist radiother. *Honors & Awards:* R J Shalek Award. *Mem:* Am Inst Physics; Am Asn Physicist Med (pres, 82); Sigma Xi; fel Am Col Med Physicist; Am Soc Therapeut Radio & Oncol; Asn Univ Radiol. *Res:* Radiological physics measurements using solid state devices; nuclear magnetic resonance; spectroscopy. *Mailing Add:* Res & Develop Oncol D84 Univ Tex Med Br Galveston TX 77550

WRIGHT, ANTHONY AUNE, b Los Angeles, Calif, Jan 4, 43; m 65. ANIMAL BEHAVIOR, PSYCHOPHYSICS. *Educ:* Stanford Univ, BA, 65; Columbia Univ, MA, 70, PhD(psychol), 71. *Prof Exp:* Instr psychol, Columbia Univ, 69-71; asst prof, Univ Tex, Austin, 71-72, from asst prof to assoc prof, 72-81, PROF NEURAL SCI, SENSORY SCI CTR, HOUSTON, 81- *Concurrent Pos:* Fogarty sr int fel NZ, 83-84. *Mem:* Psychonomic Soc; Asn Res Vision & Ophthal; Sigma Xi. *Res:* Animal sensory processes; discrimination learning; theoretical psychophysics; color vision; memory cognition. *Mailing Add:* Sensory Sci Ctr UT Health Sci Ctr PO Box 20036 Houston TX 77225

WRIGHT, ARCHIBALD NELSON, b Toronto, Ont, May 22, 32; m 55, 87; c 5. PHYSICAL CHEMISTRY. *Educ:* McGill Univ, BSc, 53, PhD, 57. *Prof Exp:* Grace Chem fel with Prof F S Dainton, Univ Leeds, 57-59; res fel with Prof C A Winkler, McGill Univ, 59-63; phys chemist, 63-68, mgr photochem br, Chem Lab, 68-72, mgr reactions & processes br, 72-73, mgr planning & resources, Mat Sci & Eng, Res & Develop Ctr, Gen Elec Co, 73-78; VPRES & DIR RES & DEVELOP, SYNERGISTICS INDUST, MONTREAL, 78- *Concurrent Pos:* Chmn, Macromolecular & Eng Div, Chem Inst Can, 83-85, Vinyl Div, 88-89, gen chmn, Viny Retec, 88, tech prog chmn, Antec, 91; mem bd, Can Soc Chem, 88-; comt mem, Can Nat, Int Union Pure & Appl Chem, 91- *Mem:* Am Chem Soc; Am Phys Soc; Royal Soc Chem; NY Acad Sci; Chem Inst Can; Soc Plastics Engrs. *Res:* Gas phase kinetics, especially reactions of hydrogen and nitrogen atoms and excited nitrogen molecules; anionic polymerization; reactions at clean metal surfaces; photolysis; photopolymerization; polymer chemistry; author of 20 referred technical publications; awarded 24 patents. *Mailing Add:* Synergistics Industs Ltd 177 St Andre St St Remi PQ J0L 2L0 Can

WRIGHT, BARBARA EVELYN, b Pasadena, Calif, Apr 6, 26; m 51; c 3. BIOCHEMISTRY, DIFFERENTIATION. *Educ:* Stanford Univ, PhD(microbiol), 51. *Prof Exp:* Biologist, Nat Heart Inst, 53-61; res assoc, Huntington Labs, Mass Gen Hosp, 61-67; res dir, Boston Biomed Res Inst, 67-82, RES PROF, DEPT MICROBIOL, UNIV MONT, 82- *Concurrent Pos:* Nat Res Coun fel, Carlsberg Lab, Copenhagen, Denmark, 50-51, Childs Mem Fund fel, 51-52; tutor, Harvard Univ & assoc prof, Harvard Med Sch; ed ann rev microbiol, Exp Mycol & J Bact; Found for Microbiol lectr, 70-71; Consult, Miles Lab, 80-83; dir, Stella Duncan Mem Res Inst, 82-; mem, Govenor's Coun Sci & Technol, 83-85; pres, Pac Div, AAAS, 84-85; consult, Advancing Sci Excellence NDak, 86-88. *Mem:* Am Soc Biol Chem; Am Soc Microbiol. *Res:* Biochemical basis of differentiation in the slime mold; kinetic modelling of metabolic networks in steady state and while undergoing transitions, as in aging and development. *Mailing Add:* Div Biol Sci Univ Mont Missoula MT 59812

WRIGHT, BILL C, b Waterford, Miss, June 15, 30; m 59; c 3. SOIL SCIENCE. *Educ:* Miss State Univ, BS, 52, MS, 56; Cornell Univ, PhD(soil sci), 59. *Prof Exp:* From asst prof to assoc prof soil sci, Miss State Univ, 59-64; assoc soil scientist to assoc dir, Indian Agr Prog Rockefeller Found, Int Agr Develop Serv, NY, 64-70, agr proj leader, Turkey, 70-77, prog officer Africa & Middle East, 77-82; sabbatical study, NC State Univ, 82-83; asst dean agr int progs, Okla State Univ, 83-86; CHIEF, PARTY MGT, AGR RES & TECHNOL PROJ, PAKISTAN, WINROCK INT, 86- *Concurrent Pos:* Mem, Bd Sci & Technol for Int Develop, Nat Acad Sci, 80-82, bd dirs, MidAm Int Agr Consortium. *Honors & Awards:* Farmers Terai Award, 71. *Mem:* Int Soc Soil Sci; Am Soc Agron; Soil Sci Soc Am; Indian Soc Agron; Indian Soil Sci Soc. *Res:* Soil-phosphorus reactions products; phosphate components of fertilizers; evaluation techniques for fertilizers in field and laboratory; fertilizer use and cultural management of cereal crops in India; wheat production in areas of low rainfall; organization and management of agricultural research and production in developing countries. *Mailing Add:* Aid/Islamabad Washington DC 20523

WRIGHT, BRADFORD LAWRENCE, physics, for more information see previous edition

WRIGHT, BYRON TERRY, b Waco Tex, Oct 19, 17; wid; c 3. NUCLEAR PHYSICS. *Educ:* Rice Univ, BA, 38; Univ Calif, Berkeley, PhD(physics), 41. *Prof Exp:* Physicist, Navy Radio & Sound Lab, Calif, 41-42 & Manhattan Dist, Calif, Tenn & NMex, 42-46; from asst prof to assoc prof, 46-56, PROF PHYSICS, UNIV CALIF, LOS ANGELES, 56- *Concurrent Pos:* Fulbright res scholar, 56-57; Guggenheim fel, 63-64; Ford Found fel, Europ Orgn Nuclear Res, 63-64. *Mem:* Am Phys Soc. *Res:* Accelerators; nuclear structure. *Mailing Add:* 1225 Chickory Lane Los Angeles CA 90049

WRIGHT, CHARLES CATHBERT, b Hanford, Calif, Dec 18, 19; m 41; c 3. CHEMISTRY. *Educ:* Univ Calif, Los Angeles, BA, 41. *Prof Exp:* Chemist, A R Maas Chem Co, 46-47, res chemist, 47-48; independent consult, 48-49; gen mgr, Oilwell Res, Inc, 49-50, pres, 50-72, chmn bd, 72-74; tech advr, Eng Dept, 74-77, SR ENG CONSULT, ARABIAN AM OIL CO, 77- *Concurrent Pos:* Lectr, Univ Southern Calif, 57-; pres, Am Coun Independent Labs, 64-66; dir water technol, Rhodes Corp, 66-68. *Honors & Awards:* Max Hecht Award, 77. *Mem:* Am Chem Soc; Am Water Works Asn; Am Inst Chemists; Am Soc Testing & Mat; Soc Petrol Eng. *Res:* Treatment of drilling fluids and the chemistry of oilwell production; oilfield corrosion control; water for subsurface injection. *Mailing Add:* 443B Tecate Rd Suite 105 Tecate CA 92080

WRIGHT, CHARLES DEAN, b Yankton, SDak, June 25, 30; m 52; c 4. POLYMER CHEMISTRY. *Educ:* Augustana Col, SDak, BA, 52; Univ Minn, PhD(org chem), 56. *Prof Exp:* Instr org chem, Univ Minn, 55-56; res chemist, 56-64, mgr res & new prod groups, 64-77, sr res specialist, 77-84, SCIENTIST, ADHESIVES, COATINGS & SEALERS DIV, 3M CO, 84- *Mem:* Am Chem Soc. *Res:* Stereospecific polymers; oxidizers for rocket fuels; general polymer chemistry; adhesive compounding and testing; adhesion; new business development; relationships of science and Christianity; epoxy chemistry; cyanoacrylate chemistry. *Mailing Add:* 14 Oakridge Dr White Bear Lake MN 55110-1839

WRIGHT, CHARLES GERALD, b Boynton, Pa, June 12, 30; m 53; c 1. ENTOMOLOGY. *Educ:* Univ Md, BS, 51, MS, 53; NC State Univ, PhD(entom), 58. *Prof Exp:* Entomologist, Wilson Pest Control, 58-63; from asst prof to assoc prof, 63-75, PROF ENTOM, NC STATE UNIV, 75- *Concurrent Pos:* Vchmn, NC Struct Pest Control Comt, 67-80, mem, 83-89. *Mem:* Entom Soc Am; Nat Asn Cols & Teachers Agr; Sigma Xi. *Res:* Urban and industrial entomology; cockroaches; insecticide movement in structures. *Mailing Add:* Box 7613 NC State Univ Raleigh NC 27695

WRIGHT, CHARLES HUBERT, b Appleton City, Mo, Oct 30, 22; m 52; c 5. ANALYTICAL CHEMISTRY. *Educ:* Univ Mo, PhD(chem), 52. *Prof Exp:* Instr analytical chem, Univ Mo, 52-54; chemist, US Radium Corp, 54-58; analytical group leader, Spencer Chem Co, 58-66; supvr analytical sect, Gulf Res & Develop Co, 66-71; sr res chemist, 71-80, RES ASSOC, PITTSBURG & MIDWAY COAL MINING CO, 80- *Mem:* AAAS; Am Chem Soc. *Res:* Analytical chemistry of fertilizers, herbicides, polymers and fuels. *Mailing Add:* 6115 W 85th Terr Overland Park KS 66207

WRIGHT, CHARLES JOSEPH, b Montour Falls, NY, May 27, 38. SYNTHETIC ORGANIC & NATURAL PRODUCTS CHEMISTRY. *Educ:* Univ Rochester, BS, 60; Mass Inst Technol, MS, 62. *Prof Exp:* From res chemist to sr res chemist, 64-82, res assoc, Res Labs, 82-86, TECH ASSOC SECT SUPVR MASS SPECTROMETRY, ANALYTICAL TECHNOL DIV, EASTMAN KODAK CO, 86- *Mem:* Am Chem Soc; Am Soc Mass Spectrometry. *Res:* Mass spectrometry of organic compounds. *Mailing Add:* 1210 Majestic Way Webster NY 14580

WRIGHT, CHARLES R B, b Lincoln, Nebr, Jan 11, 37; m 90; c 2. ALGEBRA, DISCRETE MATHEMATICS. *Educ:* Univ Nebr, BA, 56, MA, 57; Univ Wis, PhD(math), 59. *Prof Exp:* Res fel math, Calif Inst Technol, 59-60, instr, 60-61; from asst prof to assoc prof, 61-72, assoc dean, Col Liberal Arts, 73-77, PROF MATH, UNIV ORE, 72- *Concurrent Pos:* Vis prof math, Mich State Univ, 77. *Mem:* Am Math Soc; Math Asn Am. *Res:* Finite groups and computational group theory. *Mailing Add:* Dept Math Univ Ore Eugene OR 97403

WRIGHT, CLARENCE PAUL, b Cliffside, NC, Apr 15, 39. GENETICS. *Educ:* Lenoir-Rhyne Col, BS, 62; Univ Utah, MS, 65, PhD(genetics), 68. *Prof Exp:* ASSOC PROF BIOL, WESTERN CAROLINA UNIV, 68- *Mem:* Genetics Soc Am. *Res:* Developmental genetics. *Mailing Add:* Dept Biol Western Carolina Univ Cullowhee NC 28723

WRIGHT, DANIEL CRAIG, b Rockford, Ill, Nov 15, 54; m 76; c 2. PLANT GROWTH REGULATION, TISSUE CULTURE. *Educ:* WVa Univ, BA, 75; Univ Md, MS, 78 & PhD(bot), 79. *Prof Exp:* Teach asst, bot, Univ Md, 75-78; lab asst, USDA Weed Res Lab, 78-79; res assoc Hort Dept, Purdue Univ, 79-81; asst prof bot, Alfred Univ, 81-82; res scientist, Brooklyn Bot Garden, 82-87, res mgr, 87-90; RES COMPUTER SPECIALIST, DIAL CORP, 90- *Concurrent Pos:* Adj asst prof bot, Pace Univ, 86-90. *Mem:* Am Soc Plant Physiol; Am Soc Hort Sci; Sigma Xi. *Res:* Biochemistry and physiology of plant growth and development; the effects of various herbicides and other growth regulation on woody plant growth. *Mailing Add:* 14815 N Deerskin Dr Fountain Hills AZ 85268

WRIGHT, DANIEL GODWIN, b St Louis, Mo, July 27, 45; m 67; c 1. HEMATOLOGY, LEUKOCYTE BIOLOGY. *Educ:* Yale Univ, BA, 67, MD, 71. *Prof Exp:* Assoc, Nat Inst Allergy & Infectious Dis, NIH, 73-77, sr investr, Nat Cancer Inst, 78-80; CHIEF DEPT HEMAT, WALTER REED ARMY INST RES, WASH, DC, 80-; PROF MED, SCH MED, UNIFORMED SERV, UNIV HEALTH SCI, 88- *Concurrent Pos:* Adj assoc prof med, Sch Med, George Washington Univ, 85-88; assoc prof med, Uniformed Serv, Univ Health Sci, 80-87. *Mem:* Am Soc Hemat; Am Soc Cell Biol; Am Asn Immunologists; Int Soc Exp Hemat; Am Fedn Clin Res; Am Soc Clin Invest. *Res:* Secretory functions of human neutrophils; regulation of myelopoiesis and neutrophil production. *Mailing Add:* Dept Hemat Walter Reed Army Inst Res Washington DC 20307-5100

WRIGHT, DAVID ANTHONY, b Baltimore, Md, Aug 19, 41; m 62; c 3. DEVELOPMENTAL BIOLOGY, GENETICS. *Educ:* Univ Md, College Park, BS, 63; Univ Ill, Urbana, MS, 65; Wash Univ, PhD(biol), 68. *Prof Exp:* NIH fel med genetics, Univ Tex M D Anderson Hosp & Tumor Inst Houston, 68-70; asst prof biol, Univ Tex Grad Sch Biomed Sci Houston, 70-75; assoc biologist & asst prof biol, 75-80, ASSOC PROF MOLECULAR GENETICS, UNIV TEX CANCER CTR, M D ANDERSON HOSP & TUMOR INST, 80- *Mem:* AAAS; Am Soc Zoologists; Soc Develop Biol; Genetics Soc Am. *Res:* Patterns and control of gene expression during embryogenesis, especially of enzyme phenotypes in nuclear-cytoplasmic hybrids in amphibians. *Mailing Add:* Dept Molecular Genetics 1515 Holcombe Box 45 Houston TX 77030

WRIGHT, DAVID FRANKLIN, b Quincy, Mass, Feb 19, 29; m 52; c 1. PHYSICAL CHEMISTRY. *Educ:* Tufts Univ, BS, 51; Ohio State Univ, PhD(chem), 57. *Prof Exp:* PROF BASIC SCI & CHMN DEPT, MASS MARITIME ACAD, 69- *Mem:* Am Chem Soc. *Res:* Low temperature thermodynamics; clathrates of hydroquinone. *Mailing Add:* Mass Maritime Acad Buzzards Bay MA 02532

WRIGHT, DAVID GRANT, b Am Fork, Utah, Aug 21, 46; m 70; c 6. TOPOLOGY. *Educ:* Brigham Young Univ, BS, 70; Univ Wis-Madison, MA, 72 & PhD(math), 73. *Prof Exp:* Lectr math, Univ Wis-Madison, 74; fel math, Mich State Univ, 74-76; from asst prof to assoc prof math, Utah State Univ, 76-83; assoc prof, 83-88, PROF MATH, BRIGHAM YOUNG UNIV, 88- *Mem:* Am Math Soc. *Res:* Geometric topology including piecewise linear topology and topological embeddings in manifolds. *Mailing Add:* Dept Math Brigham Young Univ Provo UT 84602

WRIGHT, DAVID LEE, b Mattoon, Ill, Dec 1, 49; m 69; c 1. MATHEMATICS. *Educ:* David Lipscomb Col, BA, 71; Columbia Univ, MS, 73, PhD(math), 75. *Prof Exp:* Asst prof math, 75-81, ASSOC PROF MATH, WASH UNIV, 81- *Mem:* Sigma Xi; Am Math Soc. *Res:* Behavior of polynomial algebras, their automorphisms, their stable structure. *Mailing Add:* Dept Math Wash Univ St Louis MO 63108

WRIGHT, DAVID PATRICK, b Pocahontas, Ark, Apr 11, 43; m 75. NUCLEAR PHYSICS. *Educ:* St Edward's Univ, BS, 65; Univ Tex, Austin, PhD(nuclear physics), 74. *Prof Exp:* Instr math, Univ Tex, Austin, 75-76; vis asst prof physics, Univ Southwestern La, 76-77; ASST PROF PHYSICS & MATH, ST EDWARD'S UNIV, 77- *Mem:* Am Phys Soc. *Res:* Nuclear structure; medical physics; energy studies. *Mailing Add:* Dept Phys & Biol Sci St Edwards Univ Austin TX 78704

WRIGHT, DENNIS CHARLES, b Flint, Mich, Oct 14, 39. BEHAVIORAL NEUROSCIENCE. *Educ:* Univ Mich, Ann Arbor, BA, 60; Univ Calif, Berkeley, PhD(psychol), 69. *Prof Exp:* Asst prof, 68-73, assoc chair, 81-85, ASSOC PROF PSYCHOL, UNIV MO-COLUMBIA, 73- *Mem:* Behav Teratology Soc; Psychonomic Soc; Am Psychol Soc. *Res:* Psychopharmacology; behavioral toxicology/teratology; biochemistry and electrophysiology of learning and memory. *Mailing Add:* Dept Psychol Univ Mo Columbia MO 65211

WRIGHT, DEXTER V(AIL), b Milford, Conn, Sept 12, 23; m 44; c 2. MECHANICAL ENGINEERING. *Educ:* Univ Conn, BSEE, 44; Univ Pittsburgh, MSEE, 47. *Prof Exp:* Res engr, Westinghouse Res & Develop Ctr, 44-61, fel engr, 61-66, engr in charge mech vibrations, 66-67, mgr dynamics, Westinghouse Sci & Technol Ctr, 67-90, CONSULT ENGR, WESTINGHOUSE SCI & TECHNOL CTR, 90- *Concurrent Pos:* Mem, Am Nat Standards S2 Comt. *Mem:* Am Soc Mech Engrs; Acoust Soc Am. *Res:* Vibration of machines and structures; noise control and acoustics; vibration and acoustic instrumentation; structural vibration due to fluid flow. *Mailing Add:* 104 Kings Dale Rd Pittsburgh PA 15221

WRIGHT, DONALD C, b Altus, Okla, Sept 27, 51; m 74; c 1. ORGANIC BIOCHEMISTRY, ANALYTICAL BIOCHEMISTRY. *Educ:* Northeast La Univ, BS, 72; Univ Mo, Columbia, MS, 74; Kans State Univ, PhD(biochem), 78. *Prof Exp:* Res assoc biochem, Miss State Univ, 77-78; assoc prof chem, Eastern Ill Univ, 78-79; chemist, Ill Environ Protection Agency, Springfield, 79-83; independent consult, 83-85; chemist, Southwest Labs, Tulsa, Okla, 85-86; GAS CHROMATOGRAPHY/MASS SPECTROMETRY MGR, LANGSTON LABS, LEAWOOD, KANS, 86- *Mem:* Am Chem Soc. *Res:* Investigator in environmental gas chromatography mass spectrometry (GC/MS); mass spectral techniques combined with capillary gas chromatography; artificial intelligence techniques for processing data streams of organic compound classes by computer. *Mailing Add:* 13301 W 80 Terr Lenexa KS 66215-2504

WRIGHT, DONALD N, b Provo, Utah, Dec 6, 35; m 57; c 4. BACTERIOLOGY, BIOCHEMISTRY. *Educ:* Univ Utah, BS, 58; Iowa State Univ, PhD(bact), 64. *Prof Exp:* Chief bacteriologist & head serol div, Philadelphia Naval Hosp, Pa, 60-62; res bacteriologist, Naval Biol Lab, Calif, 64-69; assoc prof, 69-74, PROF MICROBIOL, BRIGHAM YOUNG UNIV, 74- *Mem:* AAAS; Am Soc Microbiol; Brit Soc Gen Microbiol; Am Oil Chem Soc. *Res:* Physiology of microorganisms, particularly growth, inhibition, taxonomy and aerosol behavior of the Mycoplasma; biochemical responses and growth rate control of microorganisms as a function of their environment. *Mailing Add:* Dept Microbiol Brigham Young Univ Provo UT 84602

WRIGHT, DOUGLAS TYNDALL, b Toronto, Ont, Oct 4, 27; m 55; c 5. STRUCTURAL ENGINEERING. *Educ:* Univ Toronto, BASc, 49; Univ Ill, MSc, 52; Cambridge Univ, PhD(eng), 54. *Hon Degrees:* DEng, Carleton Univ, 67; LLD, Brock Univ, 67, Concordia Univ, 82; DSc, Mem Univ Nfld, 69; LHD, Northeastern Univ, 85; DU, Strathelyde Univ, 89. *Prof Exp:* Struct designer, Morrison Hershfield Millman & Huggins, 49-52; lectr assoc prof, Queen's Univ, 54-58; prof civil eng, Univ Waterloo, 58-67, chmn dept, 58-63, dean eng, 59-66; chmn, Ont Comt Univ Affairs, 67-72, chmn, Comn Post Secondary Educ Ont, 69-72, dep prov secy social develop, 72-78, dep minister cult & recreation, 79-80; PRES & VCHANCELLOR, DEPT SYSTS DESIGN ENG, UNIV WATERLOO, 81- *Concurrent Pos:* Athlone fel, Trinity Col, 52-54; fel, Australia, 74; consult, Dutch & Mex Pavilions, Expo 67, 65-67, Palacio de Los Deportes, Olympic Games, Mex, 66-68 & Cinesphere Dome Theatre & Forum, Ont Place, 69-70; dir, Bell Can, London Life, Westinghouse Can, Electrohome Ltd. *Honors & Awards:* Gold Medal, Ont Asn Prof Engrs, 90. *Mem:* Fel Eng Inst Can; fel Am Soc Civil Engrs; Inst Pub Admin Can; Int Asn Bridge & Struct Engrs; fel Can Acad Eng. *Res:* Public policy; space frame structures; aseismic design; author of over 60 publications on structural engineering, education and manpower issues. *Mailing Add:* Univ Waterloo Waterloo ON N2L 3G1 Can

WRIGHT, EDWARD KENNETH, b Bluefields, Nicaragua, Nov 9, 30; US citizen; div; c 2. OPERATIONS ANALYSIS, INDUSTRIAL MANAGEMENT. *Educ:* Mass Inst Technol, BS, 52; Carnegie Inst Technol, MS, 61. *Prof Exp:* Dept asst, Missile Div, NAm Aviation, Inc, 55-57; mech engr, Bettis Atomic Power Lab, Westinghouse Elec Corp, 57-59; consult, Westinghouse Elec Corp & Rolls Royce, Ltd, 59-61; mech engr, Analytical Serv, Inc, 61-64; engr, Sperry Rand Systs Group, Sperry Gyroscope Co, 64-65; mech engr, Analytical Serv, Inc, 65-69; asst dir systs analysis, 69-70, assoc dir syst definition, 70-71, staff engr strategic analysis, 71-73, dir minuteman/site defense integration, Develop Planning Div, 73-75, prog mgr planning, Environ Energy Conserv Div, 76-81, proj engr strategic petrol res dir, 81-87, proj engr, Strategic Defense Weapons Dir, 87-89, PROJ ENGR, STRATEGIC DEFENSE INITIATIVE ARCHITECTURES, PHASE ONE ENG TEAM, AEROSPACE CORP, 89- *Concurrent Pos:* Fel, Ctr Advan Eng Study, Mass Inst Technol, 75-76. *Mem:* AAAS; Inst Mgt Sci; Opers Res Soc Am; Am Nuclear Soc; Am Inst Aeronaut & Astronaut. *Res:* Nuclear reactors for submarines; operations analysis, operations research, cost-effectiveness analysis and weapon systems evaluation for present and proposed strategic aerospace systems, strategic planning, and space weapons architecture development. *Mailing Add:* Aerospace Corp 2350 E El Segundo Blvd El Segundo CA 90245-4691

WRIGHT, ELISABETH MURIEL JANE, b Ottawa, Ont, July 28, 26; m 58; c 3. MATHEMATICS. *Educ:* Univ Toronto, BA, 49, MA, 50, cert, 51; Wash Univ, PhD(educ), 57. *Prof Exp:* Specialist schs, Ont, 51-55; asst prof educ, Wash Univ, 57-61, asst prof math, 59-61; from asst prof to prof math, Calif State Univ, Northridge, 63-89; RETIRED. *Concurrent Pos:* Consult, Santa Barbara Schs, 61-62 & Minn Nat Lab, State Dept Educ, 62-67. *Mem:* AAAS; Math Asn Am; Am Educ Res Asn. *Res:* Psychological problems in mathematics education; curriculum development. *Mailing Add:* Dept Math Calif State Univ Northridge CA 91324

WRIGHT, ERNEST MARSHALL, b Belfast, N Ireland, June 8, 40; Brit citizen. MEMBRANE BIOLOGY, EPITHELIAL TRANSPORT. *Educ:* Univ London, BSc, 61, DSc, 78; Univ Sheffield, England, Ph D(physiol), 64. *Prof Exp:* Res asst physiol, Univ Sheffield, 63-65; res fel biophysics, Harvard Univ, 65-66; res fel physiol, 66-67, from asst prof to assoc prof, 67-74, PROF PHYSIOL, 74-, CHMN DEPT PHYSIOL, UNIV CALIF LOS ANGELES, 87- *Concurrent Pos:* Mem, physiol study sect, NIH, 82-86, chmn, 83-86, Jacob Javits Neurosci investr, 85- *Mem:* Am Physiol Soc; Am Biophys Soc; Am Nephrology Soc; Am Soc Gen Physiologists; Brit Physiol Soc. *Res:* Transport of solutes and water across epithelial tissues from the kidney, intestine and brain; transport of ions and nutrients (sugars, amino acids, carboxylic acids) across plasma membranes of epithelial cells. *Mailing Add:* Dept Physiol Univ Calif Med Sch Los Angeles CA 90024-1751

WRIGHT, EVERETT JAMES, b Meriden, Conn, Sept 20, 29; m 51; c 1. ORGANIC CHEMISTRY. *Educ:* Hobart Col, BS, 51; Univ Del, MS, 54, PhD(chem), 57. *Prof Exp:* Chemist, Olin Industs, 51-52; RES CHEMIST, E I DU PONT DE NEMOURS & CO, 57- *Mem:* Am Chem Soc. *Res:* Organic nitrogen heterocycles; aliphatic nitrogen compounds; polymerization; textile chemicals; application techniques; fibers and fabrics; personnel and industrial relations. *Mailing Add:* 216 N Star Rd Newark DE 19711-2935

WRIGHT, FARRIN SCOTT, b Fallston, NC, Dec 3, 36; m 57; c 3. AGRICULTURAL ENGINEERING, PLANT SCIENCE. *Educ:* Clemson Univ, BS, 59, MS, 61; NC State Univ, PhD(agr eng), 66. *Prof Exp:* Res asst agr eng, Clemson Univ, 57-59; res instr, NC State Univ, 60-66; AGR ENGR, USDA, 66- *Concurrent Pos:* Fel prog, Nat Cotton Coun, 59-60; mem, Suffolk City Sch Bd, 77-87. *Mem:* Am Soc Agr Engrs; Sigma Xi; Am Peanut Res & Educ Soc. *Res:* Agricultural research, development of concepts and improvement of peanut production equipment and peanut harvesting machinery; soil and water management-irrigation and conservation tillage. *Mailing Add:* Tidewater Res Ctr Holland Sta PO Box 7099 Suffolk VA 23437

WRIGHT, FARROLL TIM, b Hume, Mo, June 24, 41; m 65; c 4. STATISTICS, MATHEMATICS. *Educ:* Univ Mo, AB, 63, AM, 64, PhD(statist), 68. *Prof Exp:* Asst prof math, Univ Mo-Rolla, 67-68; asst prof statist, Univ Iowa, 68-71, assoc prof, 71-75; from assoc prof to prof math, Univ Mo, Rolla, 75-89, PROF STATIST, UNIV MO, COLUMBIA, 89- *Mem:* Fel Am Statist Asn; fel Inst Math Statist; Int Statist Inst. *Res:* Order restricted inference; behavior of sums of independent variables; reliability and life testing. *Mailing Add:* Dept Statist Univ Mo Columbia MO 65211

WRIGHT, FRANCIS HOWELL, b New York, NY, Jan 30, 08; m 69; c 2. PEDIATRICS. *Educ:* Haverford Col, BS, 29; Johns Hopkins Univ, MD, 33; Am Bd Pediat, dipl, 41. *Hon Degrees:* DMSc, Haverford Col, 80. *Prof Exp:* Asst pediat, Johns Hopkins Univ, 34-35; asst, Columbia Univ, 35-36, instr, 36-38; from asst prof to prof, Sch Med, Univ Chicago, 40-73, chmn dept, 46-62; exec secy, Am Bd Pediat, 69-77; RETIRED. *Concurrent Pos:* Fel bact & path, Rockefeller Inst, 38-40; mem, Am Bd Pediat, 62-67, from vpres to pres, 64-67. *Honors & Awards:* Jacoby Award, 80. *Mem:* AAAS; Am Pediat Soc; Soc Pediat Res (vpres, 52); Am Acad Pediat. *Res:* Virology; infant care; psychologic adjustment of infants and children. *Mailing Add:* 105 Kendal at Longwood Kennett Square PA 19348

WRIGHT, FRANCIS STUART, b Pittsfield, Mass, Feb 24, 29; m 58; c 6. PEDIATRIC NEUROLOGY. *Educ:* Univ Mass, BS, 51; Univ Rochester, MD, 55; Am Bd Pediat, dipl, 64; Am Bd Psychiat & Neurol, dipl & cert neurol, 66, cert child neurol, 72. *Prof Exp:* From asst prof to assoc prof, 68-76, prof pediat & neurol, Med Ctr, Univ Minn, Minneapolis, 76-81; PROF PEDIAT & NEUROL, OHIO STATE UNIV, 81- *Concurrent Pos:* Fel pediat, Med Ctr, Univ Minn, Minneapolis, 56-58, Nat Inst Neurol Dis & Blindness spec fel pediat neurol, 60-63; consult, Minneapolis Pub Sch Syst, 65-; assoc, Grad Fac, Univ Minn, 67- *Res:* Developmental electrophysiology. *Mailing Add:* Dept Pediat Children's Hosp 700 Children's Dr Columbus OH 43205

WRIGHT, FRED BOYER, b Roanoke, Va, Dec 14, 25; m 48; c 2. MATHEMATICS. *Educ:* Univ NC, BA, 47, MA, 48; Univ Chicago, PhD(math), 53. *Prof Exp:* Instr math, Univ NC, 48-49; sr mathematician adv bd simulation, Univ Chicago, 53-54, consult systs res; from instr to prof math, Tulane Univ, 54-68; chmn dept, 68-76, PROF MATH, UNIV NC, CHAPEL HILL, 68- *Concurrent Pos:* Vis prof, Cambridge Univ, 58-59 & Northwestern Univ, 63; Sloan Found fel, 58-62; chmn comt regional develop, Nat Acad Sci-Nat Res Coun, 69-71, pregrad fel panel, 71-72; mem math adv panel, NSF, 71-; pregrad fel panel (chmn, 73-74), 71-74; actg chmn opers res curric, Univ NC, Chapel Hill, 72. *Mem:* Am Math Soc; Math Asn Am; London Math Soc; Sigma Xi. *Res:* Algebra; functional analysis; history of math. *Mailing Add:* Buttons Lane Chapel Hill NC 27514

WRIGHT, FRED(ERICK) D(UNSTAN), b Genoa, Italy, Mar 31, 16; US citizen; m 44; c 5. MINING ENGINEERING. *Educ:* Harvard Univ, BS, 38; Columbia Univ, MS, 42. *Prof Exp:* Shift boss, New Consol Goldfields, SAfrica, 38-39; supt, Johns-Manville Corp, NY, 39-41; mining engr, US Bur Mines, 46-53; tech dir, San Martin Mining Co, Inc, Mex, 53-54; prof mining eng, Univ Ill, Urbana, 54-67; prof, 67-81, EMER PROF MINING ENG, UNIV KY, 81- *Concurrent Pos:* Mining consult, 54- *Honors & Awards:* Rock Mech Award, Am Inst Mining, Metall & Petrol Engrs, 86. *Mem:* Am Inst Mining, Metall & Petrol Engrs; Am Soc Eng Educ. *Res:* Rock mechanics; strata control; design of underground structures; drilling, blasting, loading and transportation in mines; operations research. *Mailing Add:* 1059 Mackey Pike Univ Ky Nicholasville KY 40356

WRIGHT, FRED MARION, b Aurora, Ill, Sept 29, 23; m 47. MATHEMATICS. *Educ:* Denison Univ, BA, 44; Northwestern Univ, MS, 49, PhD(math), 53. *Prof Exp:* Instr math, Denison Univ, 47; asst, Northwestern Univ, 47-51; from instr to assoc prof, 53-64, PROF MATH, IOWA STATE UNIV, 64- *Concurrent Pos:* Vis asst prof, Univ Mich, 57-58. *Mem:* Am Math Soc. *Res:* Continued fractions and function theory. *Mailing Add:* Dept Math 400 Carver Iowa State Univ Ames IA 50011

WRIGHT, FREDERICK FENNING, b Princeton, NJ, Mar 16, 34; c 2. MARINE GEOLOGY, ESTUARINE OCEANOGRAPHY. *Educ:* Columbia Univ, BS, 59, AM, 61; Univ Southern Calif, PhD(geol), 67. *Prof Exp:* Teaching asst geol, Columbia Univ, 59-61; res asst, Univ Southern Calif, 61-65; eng geologist, Div Water Resources, State of Calif, 65-66; asst prof marine sci, Inst Marine Sci, Univ Alaska, Fairbanks, 66-72, asst prof oceanog & exten oceanogr, Marine Adv Prog, Anchorage, 72-74; oceanog consult, 74-75; dir, Alaska Coastal Mgt Prog, Off Gov, State of Alaska, 75; Alaska OCS res mgt officer, Outer Continental Shelf Environ Assessment Prog. Off Gov, State of Alaska & Nat Oceanic & Atmospheric Agency, 75-81; CONSULT, 81- *Concurrent Pos:* Asst dir, NSF Inst Oceanog, Marine Lab, Tex A&M Univ, 63; Geol Soc Am Penrose res grant, 64-66; dir, NSF Inst Geol Oceanog, Douglas Marine Sta, Univ Alaska, 68. *Mem:* Am Geophys Union; Am Soc Limnol & Oceanog; Arctic Inst NAm. *Res:* Inshore oceanography and sedimentation in subarctic; fisheries oceanography; coastal resource management and planning. *Mailing Add:* Box 240537 Douglas AK 99824-0537

WRIGHT, FREDERICK HAMILTON, b Washington, DC, Dec 2, 12; m 70. PHYSICS. *Educ:* Haverford Col, BA, 34; Calif Inst Technol, PhD(physics), 48. *Prof Exp:* Aerodynamicist, Douglas Aircraft Co, Inc, 40-46; res engr, Jet Propulsion Lab, Calif Inst Technol, 46-59; div mgr & prog mgr, Space Gen Corp, 59-69; mem staff, Aerojet Electrosysts Co, Azusa, 69-78, CONSULT, 78- *Mem:* Am Inst Aeronaut & Astronaut; Am Phys Soc; Inst Elec & Electronics Engrs. *Res:* Sensor systems; fluids; geophysics. *Mailing Add:* 515 Palmetto Dr Pasadena CA 91105

WRIGHT, GEORGE CARLIN, synthetic organic chemistry, for more information see previous edition

WRIGHT, GEORGE EDWARD, b Milwaukee, Wis, Oct 21, 41; m 65; c 3. MEDICINAL CHEMISTRY. *Educ:* Univ Ill, Chicago, BS, 63, PhD(chem), 67. *Prof Exp:* Sr res asst chem, Univ Durham, 66-68; asst prof med chem, Sch Pharm, Univ Md, Baltimore, 68-74; assoc prof, 74-78, actg assoc dean, 78-80, dean grad studies, 80-84, PROF PHARMACOL, MED SCH, UNIV MASS, 78- *Concurrent Pos:* Affil assoc prof chem, Clark Univ, 76-78, affil prof, 78-; vis prof, Inst Exp Physics, Univ Warsaw, 80-81. *Mem:* Am Chem Soc; The Chem Soc; Int Soc Heterocyclic Chem; Int Soc Antiviral Res. *Res:* Synthesis and structure of nucleosides and nucleotides; nuclear magnetic resonance spectroscopy; DNA polymerase inhibitors; oncogene proteins. *Mailing Add:* Dept Pharmacol Univ Mass Med Sch Worcester MA 01605

WRIGHT, GEORGE GREEN, b Ann Arbor, Mich, Aug 17, 16; m 57; c 3. IMMUNOLOGY, MICROBIOLOGY. *Educ:* Olivet Col, BA, 36; Univ Chicago, PhD(bact), 41. *Prof Exp:* Instr immunol, Univ Chicago, 41-42; vis investr, NIH, 46-48; med microbiologist, Ft Detrick, Md, 48-71; asst dir, Biol Labs, Mass State Dept Pub Health, 71-77, dir, 77-81; res asst prof internal med, Univ Va, 83-87; consult, 87-90; RETIRED. *Concurrent Pos:* Logan fel Univ Chicago, 41-42; Nat Res Coun fel, Calif Inst Technol, 42-43, fel immunol, 43-46; US Secy Army res fel, Oxford Univ, 57-58; vis lectr appl microbiol, Harvard Sch Pub Health, 72-83; lectr med, Sch Med, Tufts Univ, 72-83. *Mem:* Am Soc Microbiol; Am Asn Immunol. *Res:* Anti-phagocytic effects of bacterial toxins; elaboration , isolation and characterization of microbial antigens and toxins; production of biologics; immunity in anthrax. *Mailing Add:* 1707 Yorktown Dr Charlottesville VA 22901

WRIGHT, GEORGE JOSEPH, b Allendale, Ill, June 4, 31; m 53; c 2. DRUG METABOLISM, BIOCHEMISTRY. *Educ:* Univ Ill, Urbana, BS, 52, PhD(animal nutrit & biochem), 60. *Prof Exp:* Med researcher human biochem, L B Mendel Res Lab, Elgin State Hosp, Ill, 60-62; biochemist toxicol & indust hyg, Biochem Res Lab, Dow Chem Co, Midland, Mich, 62-67; head, Drug Metab Dept, Merrell-Nat Labs, Richardson-Merrell, Inc, 67-79; dir, Drug Metab, A H Robins Co, 79-; G WRIGHT CONSULT INC. *Concurrent Pos:* Lectr chem, Eve Col, Univ Cincinnati, 70-72; consult clin and preclin pharmacol, Walter Reed Army Inst Res & Nat Cancer Inst. *Mem:* AAAS; Soc Toxicol; Am Soc Pharmacol & Exp Therapeut; Am Soc Clin Pharmacol & Therapeut; Sigma Xi. *Res:* Drug disposition; bioavailability; biopharmaceutics; pharmacokinetics; enzymology; toxicology. *Mailing Add:* G Wright Consult Inc 11634 Almahurst Ct Cincinnati OH 45249

WRIGHT, GEORGE LEONARD, JR, b Ludington, Mich, Feb 8, 37; m 64; c 2. IMMUNOCHEMISTRY, TUMOR IMMUNOLOGY. *Educ:* Albion Col, BA, 59; Mich State Univ, MS, 62, PhD(microbiol, path, biochem), 66. *Prof Exp:* Fel immunol & immunochem, Sch Med, George Washington Univ, 66-67, asst prof microbiol & immunochem, 67-73; assoc prof, 73-76, DIR IMMUNOL PROG, EASTERN VA MED SCH, 75-, DEP CHMN, DEPT MICRO/IMMUNOL, 86-, PROF DEPT MICROBIOL & IMMUNOL, 76- *Concurrent Pos:* Sigma Xi award sci res, 66; spec consult, Vet Admin Hosp, Wilmington, Del, 66-73; deleg & rapporteur, Tuberc Panel, US-Japan Coop Med Sci Prog, Tokyo, 68 & 70; Hartford Found grant, 68-73; NIH grant, 72-74; Nat Cancer Inst contract, 75-78; Am Cancer Soc grant, 77-79; spec consult, Beckman Instruments; consult, Washington, DC Vet Admin & Children's Hosps; mem subcomt, US-Japan Med Sci Prog Stand Mycobact Antigens. *Honors & Awards:* NIH Career Develop Award, 75-80. *Mem:* AAAS; Am Asn Cancer Res; Am Soc Microbiol; Am Asn Immunologists; Sigma Xi; Am Urol Asn; Soc Basic Urol Res; Soc Urologic Oncol. *Res:* Separation, isolation and immunobiological characterization of specific mycobacterial antigens; identification and isolation of human prostate tumor-associated antigens; monoclonial antibodies to urogenital cancer cells for diagnosis and patient management. *Mailing Add:* Dept Microbiol & Immunol Eastern Va Med Norfolk VA 23507

WRIGHT, HAROLD E(UGENE), b Hillsdale, Mich, Dec 28, 20; m 49; c 2. MECHANICS. *Educ:* Univ Dayton, BME, 49; Univ Cincinnati, MS, 60; Mich State Univ, PhD, 66. *Prof Exp:* From instr to assoc prof mech eng, Univ Dayton, 49-62; assoc prof, 62-77, prof mech eng, US Air Force Inst Technol, 77-; AT ANTIOCH COL. *Concurrent Pos:* Mech engr, Nat Cash Register Co, 51-52; combustion engr, City of Dayton, 52-53; maintenance dir, Miami Valley Hosp, 53-57. *Mem:* Am Soc Mech Engrs; Am Soc Eng Educ; Nat Soc Prof Engrs. *Res:* Gas dynamics; applied mechanics; heat transfer; shock tubes utilization, both gas and liquid drivers. *Mailing Add:* Dept Humanity Antioch Col Yellow Spring OH 45387

WRIGHT, HARRY TUCKER, JR, b Louisville, Ky, July 21, 29. PEDIATRICS, EPIDEMIOLOGY. *Educ:* Wake Forest Col, BS, 51, MD, 55; Univ Calif, Berkeley, MPH, 75. *Prof Exp:* From instr to assoc prof, 63-73, PROF PEDIAT, UNIV SOUTHERN CALIF, 73-, HEAD, DIV INFECTIOUS DIS & VIROLOGY, CHILDREN'S HOSP, LOS ANGELES, 75- *Concurrent Pos:* Teaching fel pediat, Case Western Reserve Univ, 58-59; NIH res training grant, 61-64; mem infectious dis comt, Am Acad Pediat. *Mem:* Am Acad Pediat; Am Fedn Clin Res; Soc Pediat Res; Am Pediat Soc; Am Soc Microbiol; fel Infectious Dis Soc. *Res:* Clinical and laboratory studies of cytomegalovirus and other herpes viruses; central nervous system syndromes of viral etiology and newborn infections; virology; hospital acquired infections; infectious disease. *Mailing Add:* Childrens Hosp Los Angeles PO Box 54700 Terminal Annex Los Angeles CA 90054

WRIGHT, HARVEL AMOS, b Mayflower, Ark, July 6, 33; m 54; c 2. MATHEMATICS. *Educ:* Ark State Teachers Col, BS, 54; Univ Ark, MA, 56; Univ Tenn, PhD, 67. *Prof Exp:* Teacher high sch, Ark, 55-56; instr math & physics, Ark State Teachers Col, 56-58; instr math, Univ Tenn, 58-62; physicist, 62-74, HEAD, BIOL & RADIATION PHYSICS SECT, OAK RIDGE NAT LAB, 74- *Concurrent Pos:* Vis scientist, Europ Ctr Nuclear Res, Switz, 71-72. *Mem:* Health Physics Soc; Radiation Res Soc. *Res:* Interaction of radiation with matter; dosimetry of ionizing radiation; plasma physics; theory of real variable. *Mailing Add:* Oak Ridge Nat Lab Box 2008 Oak Ridge TN 37831

WRIGHT, HASTINGS KEMPER, b Boston, Mass, Aug 22, 28; m 54; c 4. MEDICINE, PHYSIOLOGY. *Educ:* Harvard Univ, AB, 50, MD, 54. *Prof Exp:* Asst prof surg, Western Reserve Univ, 62-66; assoc prof, 66-72, PROF SURG, SCH MED, YALE UNIV, 72- *Concurrent Pos:* Crile fel surg, Western Reserve Univ, 61-62. *Mem:* AAAS; Soc Univ Surg; Am Fedn Clin Res; Am Gastroenterol Asn; Am Surg Asn. *Res:* Gastrointestinal fluid and electrolyte absorption. *Mailing Add:* Dept Surg Yale-New Haven Hosp 333 Cedar St New Haven CT 06510

WRIGHT, HELEN S, b Kirkland Lake, Ont, Mar 29, 36. APPLIED NUTRITION, NUTRITION ASSESSMENT. *Educ:* Pa State Univ, PhD(nutrit), 69. *Prof Exp:* Assoc prof, 75-85, PROF NUTRIT, PA STATE UNIV, 85- *Mem:* Soc Nutrit Educ; Am Inst Nutrit; Am Dietetic Asn. *Mailing Add:* S-125A Human Dev University Park PA 16802

WRIGHT, HENRY ALBERT, b Modesto, Calif, June 1, 35; m 61; c 4. FIRE ECOLOGY, RANGE MANAGEMENT. *Educ:* Univ Calif, Davis, BS, 57; Utah State Univ, MS, 62, PhD(range mgt), 64. *Prof Exp:* Exten specialist range mgt, Univ Calif, Davis, 57-58; range aid, Intermountain Forest & Range Exp Sta, Boise, Idaho, 60-64, res assoc, 64-67; from asst prof to assoc prof, 67-74, prof, 74-78, HORN PROF RANGE MGT, TEX TECH UNIV, 78-, CHAIRPERSON RANGE & WILDLIFE MGT, 81- *Concurrent Pos:* Mem grad coun, Tex Tech Univ, 71-74; educ comt, Southwest Interagency Fire Coun, 71-, mem tenure & privilege comt, 79-81. *Mem:* Soc Range Mgt. *Res:* Fire ecology particularly developing prescription techniques for burning rangeland communities and studying the effect of fire on their total ecosystem. *Mailing Add:* 3704 69th Dr Lubbock TX 79413

WRIGHT, HERBERT EDGAR, JR, b Malden, Mass, Sept 13, 17; m 43; c 4. PALEOECOLOGY. *Educ:* Harvard Univ, AB, 39, AM, 41, PhD(geol), 43. *Hon Degrees:* DSc, Trinity Col, Dublin, 66; PhD, Lund Univ, Sweden, 87. *Prof Exp:* Instr geol, Brown Univ, 46-47; from asst prof to prof, 47-74, regents' prof geol, ecol & bot, 74-88, DIR LIMNOL RES CTR, UNIV MINN, MINNEAPOLIS, 63- *Concurrent Pos:* Geologist, US Geol Surv, DC, 46-47 & 52-53 & Minn Geol Surv, DC, 52-53; mem Boston Col-Fordham archaeol exped, 47; Wenner-Gren fel, 51; mem, archaeol exped, Oriental Inst, 51, 54-55, 60, 63, 64 & 70; Guggenheim Fel, 54-55. *Honors & Awards:* Pomerance Award, Archeol Inst Am, 84. *Mem:* Nat Acad Sci; Geol Soc Am; Ecol Soc Am; Am Soc Limnol & Oceonog; Am Quaternary Asn (pres, 71-72). *Res:* Pleistocene geology and paleoecology; vegetation history; paleolimnology. *Mailing Add:* Dept Geol Univ Minn Minneapolis MN 55455

WRIGHT, HERBERT FESSENDEN, b Worcester, Mass, July 19, 17; m 41; c 1. MEDICINAL CHEMISTRY, BIOLOGY & MOLECULAR BIOLOGY. *Educ:* Oberlin Col, AB, 40; Cornell Univ, MS, 42, PhD(org chem), 44; Am Inst Chem, cert. *Prof Exp:* Asst chem, Cornell Univ, 42-43, Off Sci Res & Develop antimalarial proj, 43-44; res chemist, Lever Bros Co, Mass, 44-45; res assoc, Mass Inst Technol, 45-46; instr chem, Tufts Col, 46-48; res chemist, Arthur D Little, Inc, 48-49; instr chem, Yale Univ, 49-52; sr res chemist, Olin Industs, 52-54; pres & res dir, W Elsworth Co, Inc, 59-64; from asst prof to assoc prof, Univ New Haven, 64-66, chmn dept, 67-79, dir environ studies, 70-79, prof biol & phys sci, 66-87; RETIRED. *Concurrent Pos:* Consult chem, 54-; dir, Univ Res Insts Conn, 67-77. *Mem:* AAAS; Am Chem Soc; NY Acad Sci; fel Am Inst Chem; Am Inst Biol Sci. *Res:* Synthetic drugs and vitamin A, organic ortho silicate esters; polymers; fungicides; antimetabolites; organic electrode reactions; organic synthetic methods; molecular biology; forensic science; biochemical genetics; nutrition; science education. *Mailing Add:* 1041 Rodge Rd Hamden CT 06517-1620

WRIGHT, HERBERT N, b Berwyn, Ill, May 23, 28; m 52, 86; c 3. PSYCHOPHYSICS, AUDIOLOGY. *Educ:* Grinnell Col, BA, 50; Ind Univ, MA, 53; Northwestern Univ, PhD(audiol), 58. *Prof Exp:* Res assoc otolaryngol, Northwestern, 56-58; res assoc blind mobility, Auditory Res Ctr, 60-61; from res instr to res assoc prof otolaryngol, 61-63, assoc prof, 63-88, PROF OTOLARYNGOL, STATE UNIV NY HEALTH SCI CTR, SYRACUSE, 88- *Concurrent Pos:* Res fel psychoacoustics, Cent Inst Deaf, St Louis, Mo, 58-60 & Syracuse Univ, 61-63. *Mem:* AAAS; fel Am Speech & Hearing Asn; Acoust Soc Am; Psychonomic Soc; Int Soc Audiol; Sigma Xi. *Res:* Psychophysics of individual differences, especially with reference to disorders of the olfactory system; dysfunctions of the sense of smell. *Mailing Add:* State Univ NY Health Sci Ctr 750 E Adams St Syracuse NY 13210

WRIGHT, IAN GLAISBY, b Fredericton, NB, Sept 15, 35; m 58; c 3. SYNTHETIC ORGANIC CHEMISTRY. *Educ:* Univ BC, BA, 57; Univ NB, Fredericton, MSc, 59; Univ Wis-Madison, PhD(org chem), 65. *Prof Exp:* Fel with A I Scott, Univ BC, 63-65; sr org chemist, 65-73, RES SCIENTIST PROCESS RES & DEVELOP DIV, LILLY RES LABS, ELI LILLY & CO, 73- *Mem:* AAAS; Am Chem Soc; Royal Soc Chem. *Res:* Chemistry and biological activity of cephalosporin antibiotics; chemistry of natural products from plant and microbiological sources; synthesis of pharmaceutical and agricultural product candidates; manufacturing process design and development. *Mailing Add:* Eli Lilly & Co Dept IC 742 Bldg 110 Indianapolis IN 46285-0002

WRIGHT, JAMES ARTHUR, b Toronto, Ont, Dec 29, 41; m 70. GEOPHYSICS. *Educ:* Univ Toronto, BASc, 64, MSc, 65, PhD(physics), 68. *Prof Exp:* Nat Res Coun Can fel, Brunswick Tech Univ, 68-69; from asst prof to assoc prof physics, 69-82, PROF EARTH SCI, 90- *Concurrent Pos:* Mem geomagnetism subcomt, Assoc Comt Geod & Geophys, Nat Res Coun Can, 71-73, Lithprobe Sub-Comt, 88-92. *Mem:* Can Soc Explor Geophysicists; Geol Asn Can; Am Geophys Union; Soc Explor Geophysicists; Asn Prof Engrs & Geoscientists. *Res:* Geomagnetism and the earth's interior; exploration geophysics; marine heat flow. *Mailing Add:* Dept Earth Sci Mem Univ Nfld St Johns NF A1B 3X5 Can

WRIGHT, JAMES EDWARD, b Little Rock, Ark, Sept 6, 46; m 66; c 2. EXERCISE PHYSIOLOGY. *Educ:* Fairleigh Dickinson Univ, BS, 69; Miss State Univ, PhD(zool), 73. *Prof Exp:* Asst prof biol sci, Simon's Rock Early Col, 73-75; NIH fel & asst res physiologist exercise & environ physiol, Inst Environment Stress, Univ Calif, Santa Barbara, 75-77; staff, Exercise Physiol Div, US Army Res Inst Environ Med, 77-82; TECH ED & SPEC ASST PUBL, WEIDER HEALTH & FITNESS CORP, WOODLAND HILLS, CALIF, 82- *Concurrent Pos:* Athletic training & nutritional consult, Sports Science Consult, Natick, Mass, 78; assoc ed, Nat Strength Conditioning Asn J, 81-; mem, Int Coun Phys Fitness Res, 80- *Mem:* AAAS; Am Col Sports Med. *Res:* Longitudinal and cross sectional investigation of functional physical and physiological changes produced by weight/strength training and body building; anabolic steroid and other ergogenic aid effects on muscle size, strength, endurance and performance; comparative responses of males and females to strength training and bodybuilding; occupational fitness requirements. *Mailing Add:* 9001 Eames Ave Northridge CA 91324

WRIGHT, JAMES ELBERT, b Kerrville, Tex, Oct 7, 40; m 62; c 2. MEDICAL ENTOMOLOGY, BACTERIOLOGY. *Educ:* Tex A&M Univ, BS, 63; Ohio State Univ, PhD(med entom, bact), 66; Am Registry Cert Entom, cert. *Prof Exp:* Res asst mosquitoes entom, Ohio State Univ, 63-66, res assoc, 66; res entomologist & res leader, Boll Weevil Res Lab, Sci & Educ Admin-Agr Res, 66-, RES ENTOMOLOGIST, SUBTROP COTTON INSECT RES LAB, AGR RES SERV, USDA. *Mem:* Fel AAAS; Entom Soc Am; Am Chem Soc; Soc Invert Path; Am Mosquito Control Asn. *Res:* Insect juvenile and molting hormones interrelationships; insect biochemistry; mosquito biology; diapause mechanisms; livestock arthropods; biting flies; stable fly and horn fly biology; studies on temperature and photoperiod; biology of livestock ticks; insect growth regulators; boll weevil and cotton insects; sterility; nutrition. *Mailing Add:* SubTrop Cotton Insect Res Lab Agr Res Serv USDA 2413 E Hwy 83 Weslaco TX 78596

WRIGHT, JAMES EVERETT, JR, b Deepstep, Ga, Apr 28, 23; m 48; c 3. GENETICS. *Educ:* Univ Ga, BS, 46; Cornell Univ, PhD(genetics), 50. *Prof Exp:* From instr to prof, 49-83, EMER PROF GENETICS, PA STATE UNIV, 83- *Mem:* Genetics Soc Am; Am Fisheries Soc; Am Genetic Asn. *Res:* Genetics and breeding of fishes. *Mailing Add:* 02 Buckhout Lab Pa State Univ University Park PA 16802

WRIGHT, JAMES FOLEY, b Tulsa, Okla, Feb 9, 43; m 64; c 2. ENVIRONMENTAL SCIENCES. *Educ:* Cent State Univ, Okla, BS, 69; Iowa State Univ, PhD(nuclear chem), 74. *Prof Exp:* Fel physics, Ames Lab, Iowa, 74; scientist, Battelle Northwest Labs, 74-75; mem sci staff planning, Los Alamos Sci Lab, 75-80; PRES, TECH SYSTS, 80- *Concurrent Pos:* Dir, Nat Particle-Beam Study Group, 78-79. *Mem:* Am Phys Soc; Inst Elec & Electronics Engrs. *Res:* Directed energy weapons, nuclear weapons, plasma physics, systems analysis; Monte Carlo risk analysis; hydrodynamics. *Mailing Add:* 45 Lafayette Pl Midland TX 79705-5322

WRIGHT, JAMES FRANCIS, b Philadelphia, Pa, Jan 18, 24; m 55; c 3. COMPARATIVE PATHOLOGY, TOXICOLOGY. *Educ:* Univ Pa, VMD, 51; Univ Calif, Davis, PhD(comp path), 69. *Prof Exp:* Vet in-chg animal quarantine, Plum Island Animal Dis Lab, USDA, NY, 54-57; vet, Nat Zool Park, Smithsonian Inst, 57-62; res scientist, USPHS, US Air Force Radiobiol Lab, Univ Tex, Austin, 62-64, Yerkes Regional Primate Res Ctr, Emory Univ, 64-65 & Radiobiol Lab, Univ Calif, Davis, 65-69; chief toxicol studies sect, Twinbrook Res Lab, Environ Protection Agency, 69-73, chief path studies sect, Health Effects Res Lab, 73-82; VET & PATHOLOGIST, NC ZOOL PARK, ASHEBORO, NC, 88- *Concurrent Pos:* Adj assoc prof, Sch Vet Med, NC State Univ, 79-82, vis assoc prof path, 82-; consult pathologist, NC Zool Park, 82-; vis assoc prof, Vet Med, 83- *Mem:* Am Vet Med Asn; Am Asn Zoo Vets; Am Asn Lab Animal Sci; Wildlife Dis Asn; Radiation Res Soc. *Res:* pathology and toxicology; wild & exotic animals; diseases of wild animals; laboratory animal medicine. *Mailing Add:* 8508 E Lake Ct Raleigh NC 27612

WRIGHT, JAMES LOUIS, b Pleasant Grove, Utah, Aug 28, 34; m 58; c 7. AGRONOMY, AGRICULTURAL METEOROLOGY. *Educ:* Utah State Univ, BS, 59, MS, 61; Cornell Univ, PhD(soil physics), 65. *Prof Exp:* SOIL SCIENTIST, AGR RES SERV, USDA, 65-; ASSOC PROF RES CLIMATOL & AFFIL PROF SOILS, UNIV IDAHO, 70- *Mem:* Am Soc Agron; Soil Sci Soc Am; Sigma Xi. *Res:* Microclimate investigations to determine rate of evaporation of water from agricultural crops using energy balance and micrometeorological approaches. *Mailing Add:* Rt No 1 Box 161 Kimberly ID 83341

WRIGHT, JAMES P, b St Petersburg, Fla, Apr 10, 34; m 56; c 4. ASTROPHYSICS. *Educ:* Univ Fla, BS, 56; Univ Chicago, PhD(chem), 61. *Prof Exp:* NSF-Nat Res Coun assoc, Inst Space Studies, 61-63; vis asst prof math, Math Res Ctr, Univ Wis, 63-64; astrophysicist, Smithsonian Astrophys Observ, 64-70; PROG DIR, NSF, 70- *Concurrent Pos:* Lectr astron, Harvard Univ, 64-70. *Mem:* Am Phys Soc; Am Astron Soc; Int Astron Union. *Res:* Properties of high temperature gases in electromagnetic fields; theoretical cosmological world models; observable effects of use of general relativity theory; high energy astrophysics. *Mailing Add:* 4119 W Woodbine St Chevy Chase MD 20815-5043

WRIGHT, JAMES R, b Riversdale, NS, July 19, 16; m 42; c 2. SOIL CHEMISTRY. *Educ:* McGill Univ, BSc, 40; Mich State Univ, MS, 48, PhD(soil chem), 53. *Hon Degrees:* DSc, Acadia Univ, 80. *Prof Exp:* Asst chemist, NS Dept Agr, 40-41, asst prov chemist, 45-50; head soil genesis sect, Soil Res Inst, Can Dept Agr, 51-61, dir, Kentville Res Sta, 61-78; CONSULT SOILS & AGR, 79- *Concurrent Pos:* Lectr, NS Agr Col, 45-50; prof, Acadia Univ, 65-75; asst dir gen, Res Br, Agr Can, 67-68; reg-treas, NS Inst Agrology, 81-85. *Mem:* Am Soc Agron; Soil Sci Soc Am; fel Chem Inst Can; Can Soc Soil Sci (pres, 61-62); fel Agr Inst Can. *Res:* Chemical nature of soil organic matter; soil genesis, fertility and plant nutrition. *Mailing Add:* PO Box 668 Chester NS B0J 1J0 Can

WRIGHT, JAMES ROSCOE, b White Hall, Md, July 7, 22; m 50; c 2. ORGANIC CHEMISTRY, TECHNICAL MANAGEMENT. *Educ:* Md State Col Salisbury, BSEd, 46; Wash Col, BS, 48; Univ Del, MS, 49, PhD(chem), 51; Harvard Bus Sch, Prog Mgt Dev, Cert, 67. *Prof Exp:* Res chemist, Southwest Res Inst, 51-52 & Chevron Res Corp, Chevron Oil Co, 52-60; from chemist to sr chemist, Nat Bur Standards, 60-66, phys sci adminstr, 66-72, dir, Ctr for Bldg Technol, 72-74, dep dir, Inst Appl Technol, 74-76, actg dir, 76-78, dep dir, nat eng lab, 78-85; consult, 85-88; RETIRED. *Concurrent Pos:* Asst prof, Trinity Univ, Tex, 51-52; com Sci & Tech fel, US Patent Off, 64-65; del, Int Union Testing & Res Labs Mat & Struct, 68-88, pres, 71-72 & 82-85; mem, USA/Egypt Working Group, Technol, Res & Develop, 76-78; dir & treas, St Mark Elderly Housing Corp, 87- *Honors & Awards:* Gold Medal Award, Dept Com, 75; hon mem, Int Union Testing & Res Labs, Mat & Structures, 85. *Mem:* Emer mem Am Chem Soc; Am Soc Testing & Mat; Int Union Testing & Res Labs Mat & Struct (vice-pres, 79-82, pres, 82-85, outgoing pres, 85-88. *Res:* Organosilicon compounds; radiation effects on organic lubricants; emulsification of petroleum products; photochemical stability of organic building materials; application of performance concept in building research, codes and standards; managerial organization development. *Mailing Add:* 6204 Lone Oak Dr Bethesda MD 20817

WRIGHT, JAMES SHERMAN, b Seattle, Wash, Aug 27, 40; m 63; c 1. THEORETICAL CHEMISTRY. *Educ:* Stanford Univ, BS, 62; Univ Calif, Berkeley, PhD(chem), 68. *Prof Exp:* Nat Ctr Sci Res, France grant, Fac Sci, Orsay, France, 69-70; asst prof, 70-77, ASSOC PROF CHEM, CARLETON UNIV, 77- *Mem:* Am Chem Soc. *Res:* Chemistry. *Mailing Add:* Dept Chem Carleton Univ Ottawa ON K1S 5B6 Can

WRIGHT, JEFFERY REGAN, b Spokane, Wash, Nov 25, 50; m 74; c 2. INFORMATION SYSTEMS, PUBLIC DECISION. *Educ:* Univ Wash, BA, 75, BSE, 75, MSCE, 77; Johns Hopkins Univ, PhD(environ eng), 82. *Prof Exp:* Res asst prof, Dept Civil Eng, Univ Wash, 76-78; ASSOC PROF CIVIL ENG, SCH CIVIL ENG, PURDUE UNIV, 87-; DIR, IND WATER RESOURCES RES CTR. *Concurrent Pos:* Vpres, Omtek Eng, Inc, 83-; dir, Scheduling Systs Res Inst, 85- *Mem:* Am Soc Civil Eng; Opers Res Soc Am; Inst Mgt Sci; Am Soc Eng Educ; Soc Comput Simulation; Am Geophys Union. *Res:* Public systems modeling such as facilities location, project planning and management, scheduling systems; decision support and information systems design and resource acquisition-allocation; development of computer-support systems. *Mailing Add:* Sch Civil Eng Purdue Univ West Lafayette IN 47907

WRIGHT, JOE CARROL, b Benton, Ark, Feb 6, 33; m 53; c 4. ORGANIC CHEMISTRY. *Educ:* Ouachita Baptist Univ, BS, 54; Univ Ark, MS, 62, PhD(org mech), 66. *Prof Exp:* Asst prof chem, Mobile Col, 64-66; assoc prof, 66-72, PROF CHEM, HENDERSON STATE UNIV, 72-, DEAN SCH SCI, 69- *Mem:* Am Chem Soc; Royal Soc Chem. *Res:* Isotope effect studies in organic reaction mechanism determinations. *Mailing Add:* Off Dean Sch Sci Henderson State Univ Arkadelphia AR 71923

WRIGHT, JOHN CLIFFORD, b Livingston, Mont, Jan 29, 19; m 44; c 2. LIMNOLOGY. *Educ:* Mont State Univ, BS, 41; Ohio State Univ, PhD, 50. *Prof Exp:* From instr to assoc prof, Mont State Univ, 49-61, assoc coordr, Ctr Environ Studies, 71-73, prof bot, 61-; RETIRED. *Concurrent Pos:* NSF sr fel, 59-60; exec secy, XV Int Cong Limnol, 61-62; dir, Ctr Environ Studies, Mont State Univ, 66-71. *Mem:* AAAS; Ecol Soc Am; Am Soc Limnol & Oceanog; Int Asn Theoret & Appl Limnol. *Res:* Oceanography; ecology. *Mailing Add:* 1716 Willoway Bozeman MT 59715

WRIGHT, JOHN CURTIS, b Lubbock, Tex, Sept 17, 43; m 68; c 2. ANALYTICAL & PHYSICAL CHEMISTRY, SOLID STATE PHYSICS. *Educ:* Union Col, BS, 65; Johns Hopkins Univ, PhD(physics), 70. *Prof Exp:* Student fel, Purdue Univ, 70-72; from asst prof to assoc prof, 72-80, PROF CHEM, UNIV WIS, 80- *Honors & Awards:* William F Meggars Award, Optical Soc Am, 80. *Mem:* Sigma Xi; Am Chem Soc; Am Phys Soc. *Res:* Lasers in analytical chemistry; fluorescence methods; solid state chemistry; coherent Raman methods; energy transfer and relaxation processes. *Mailing Add:* Dept Chem Univ Wis Madison WI 53706

WRIGHT, JOHN FOWLER, b London, Ont, July 28, 21; US citizen; m 49; c 2. PHYSICAL CHEMISTRY. *Educ:* Univ Western Ont, BSc, 45, MSc, 46. *Prof Exp:* Res assoc, Res Labs, Eastman Kodak Co, 46-83; RETIRED. *Mem:* Am Chem Soc. *Res:* Polymer synthesis; surface treatments of polymers; colloid chemistry; surface active agents. *Mailing Add:* 85 Greatwood Circle Fairport NY 14450

WRIGHT, JOHN JAY, b Torrington, Conn, July 10, 43; m 65; c 2. ATOMIC PHYSICS. *Educ:* Worcester Polytech Inst, BS, 65; Univ NH, PhD(physics), 70. *Prof Exp:* Fel lasers, Joint Inst Lab Astrophys, Colo, 69-70; asst prof, 70-77, PROF PHYSICS, UNIV NH, 77- *Mem:* Am Inst Physics; Am Phys Soc; Am Asn Physics Teachers. *Res:* Optical pumping; liquid crystals. *Mailing Add:* Dept Physics Dermitt Hall Univ NH Durham NH 03824

WRIGHT, JOHN MARLIN, b Long Beach, Calif, Feb 2, 37; m 58; c 2. CHEMICAL INSTRUMENTATION, MAGNETIC RESONANCE. *Educ:* Calif Inst Technol, BS, 60; Harvard Univ, PhD(chem), 67. *Prof Exp:* Res fel chem, Harvard Univ, 67-70; assoc specialist, 70-73, specialist & lectr chem, 73-79, SR DEVELOP ENGR CHEM, UNIV CALIF, SAN DIEGO, 79- *Concurrent Pos:* Lectr, Univ Calif, San Diego, 73-76. *Mem:* Am Chem Soc; AAAS. *Res:* Application of nuclear magnetic resonance and mass spectrometry to chemical problems; development of new instrumental techniques. *Mailing Add:* Dept Chem B-014 Univ Calif San Diego La Jolla CA 92093-0314

WRIGHT, JOHN RICKEN, b Batesville, Ark, Jan 3, 39; m 64; c 1. BIOINORGANIC CHEMISTRY. *Educ:* Ark State Univ, BS, 60; Univ Miss, MS, 67, PhD(chem), 71. *Prof Exp:* Res assoc, Dept Bot, Wash Univ, 67-68; NIH fel, Fla State Univ, 72-73; PROF CHEM, DEPT PHYS SCI, SOUTHEASTERN OKLA STATE UNIV, 73- *Mem:* Sigma Xi. *Res:* The metabolism and transport of biological forms of copper; bismuth, boron and copper antibody labels for biomedical applications. *Mailing Add:* Dept Phys Sci Southeastern Okla State Univ Durant OK 74701

WRIGHT, JOHNIE ALGIE, b Atlas, Ala, Feb 5, 24; m 49; c 2. HORTICULTURE. *Educ:* Tenn Polytech Inst, BS, 47; Iowa State Col, MS, 49; La State Univ, PhD, 60. *Prof Exp:* Asst prof hort, Tenn Polytech Inst, 49-53; assoc prof, 53-60, PROF HORT, LA TECH UNIV, 60-, ASSOC DEAN, COL LIFE SCI, 72-, PROF AGRON, 77-, HEAD DEPT AGRON & HORT, 77- *Concurrent Pos:* Asst, La State Univ, 60. *Mem:* Am Soc Hort Sci. *Res:* Nutrient culture of roses; methods of watering greenhouse roses; container nursery stock studies. *Mailing Add:* 610 Durden Ave Ruston LA 71270

WRIGHT, JON ALAN, b Tacoma, Wash, Jan 18, 38; m 62; c 3. THEORETICAL PHYSICS, PARTICLE PHYSICS. *Educ:* Calif Inst Technol, BS, 59; Univ Calif, Berkeley, PhD(physics), 65. *Prof Exp:* Physicist, Aerospace Corp, 65; res assoc physics, Univ Calif, San Diego, 65-67; assoc prof, 67-73, PROF PHYSICS, UNIV ILL, URBANA, 73- *Mem:* Am Inst Physics. *Res:* Theoretical elementary particle physics; non-linear physics. *Mailing Add:* La Jolla Inst 7855 Foy Ave Suite 320 La Jolla CA 92037

WRIGHT, JOSEPH D, b Trochu, Alta, Nov 11, 41; m 69; c 4. CHEMICAL ENGINEERING. *Educ:* Univ Alta, BSc, 63; Cambridge Univ, PhD(chem eng), 67. *Prof Exp:* Res engr, Shawinigan Chem Div, Gulf Oil Can, 67-69; asst prof 69-74, assoc prof, 74-78, PROF CHEM ENG, MCMASTER UNIV, 78-; VPRES & CTR MGR, TECHNOL & ENG SYSTS LAB, XEROX RES CTR CAN, 87- *Concurrent Pos:* Trustee CACHE Corp, chmn, CACHE Task Force on Real-time Comput; exec dir, Sheridan Res Park Asn, 79-; mgr, Mat Processing Lab, Xerox Res Centre Can, 77-85, Technol Strategy Off, Xerox Corp, 85-86 & Technol & Eng Systs Lab, 86-87; vchmn, Comput & Systs Technol Div, Am Inst Chem Eng, 88-89, chmn, 89-90; chmn, Ont Centre Mat Res, 88-; exec bd, Precarn Corp, 88- *Mem:* Chem Inst Can; Am Inst Chem Engrs; Am Chem Soc; Can Soc Chem Eng. *Res:* Chemical plant simulation and control extending to direct digital control and optimization of chemical plants and simulations; computer process control; specialty materials; management of technology. *Mailing Add:* Xerox Res Ctr Can 2660 Speakman Dr Mississauga ON L5K 2L1 Can

WRIGHT, JOSEPH WILLIAM, JR, b Indianapolis, Ind, Oct 30, 16; m 42, 65; c 5. MEDICINE. *Educ:* Univ Mich, AB, 38, MD, 42. *Prof Exp:* Intern, Univ Mich Hosp, 42-43; resident, Ind Univ Hosp, 46-48; ASSOC PROF OTOLARYNGOL, SCH MED, IND UNIV, INDIANAPOLIS, 58- *Concurrent Pos:* Fel surg, Northwestern Univ, 55-56; consult, Off Surgeon Gen, US Army, 49-; chmn otolaryngol serv, Community Hosp, 57-62. mem bd trustees; mem, Ind State Hearing Comn; pres, Wright Inst Otol; consult, Crossroad Rehab Ctr, Indianapolis, Ind, & Indianapolis Speech & Hearing Ctr; consult mem, St Vincent's Hosp; pres, Am Coun Otolaryngol, 79-80; mem, Otosclerosis Study Group, 75- *Mem:* AMA; Am Acad Ophthal & Otolaryngol (vpres,86); Am Laryngol, Rhinol & Otol Soc; Pan-Am Med Asn; Int Col Surg; Am Otol Soc; Triological Soc. *Res:* Bioceramics as applied to otology; endolymphatic ducts in tomography. *Mailing Add:* 7474 Holliday East Dr Indianapolis IN 46260

WRIGHT, JUDITH, b Akron, Ohio, Oct 25, 41; c 2. CHEMOSTRATIGRAPHY, PALEOCEANOGRAPHY. *Educ:* Calif State Univ, Northridge, BS, 79; Univ Ore, Eugene, MS, 82, PhD(geol), 85. *Prof Exp:* Res asst conodont biostratig, dept geol, Calif State Univ, Northridge, 76-77; head soil analyst & eng geologist, R T Frankian & Assocs, Burbank, Calif, 78; explor geologist petrol, Gulf Oil Corp, Wyo, 80; res assoc geochemist stratig, dept geol, Univ Ore, 79-85, res fel chemostratig, 85-86; res asst stable isotopes, Dept Geol, Ariz State Univ, 86-87. *Mem:* Geol Soc Am; Geochem Soc; Am Geophys Union; Soc Econ Paleontologists & Mineralogists; corresp mem Int Union Geol Sci. *Res:* Use of trace and rare earth element geochemistry of recent and fossil minerals, mainly apatite, as paleochemical indicators for ancient oceanic environments and for chemostratigraphic correlation; stable isotope studies of fluid inclusions in salts. *Mailing Add:* 802 45th St No 3-207 Auburn WA 98002

WRIGHT, KENNETH A, zoology, parasitology, for more information see previous edition

WRIGHT, KENNETH C, INTERVENTIONAL RADIOLOGY, IMAGING. *Educ:* Tex A&M Univ, PhD(vet physiol), 77. *Prof Exp:* ASST PROF PHYSIOL, SYST CANCER CTR, ANDERSON HOSP, UNIV TEX, 80- *Mailing Add:* Dept Exp Diagnostic Radiol Cancer Ctr MD Anderson Hosp Univ Tex 1515 Holcombe Blvd Box 057 Houston TX 77030

WRIGHT, KENNETH JAMES, b Pittsburgh, Pa, Aug 26, 39; m 66; c 2. SCIENCE EDUCATION METHODOLOGY, ENVIRONMENTAL SCIENCES. *Educ:* Portland State Univ, BS, 62; Univ Idaho, PhD(inorg chem), 72. *Prof Exp:* Anal & res chemist, Harvey Aluminum Corp, Ore, 63-66; chmn div phys sci, 72-77, INSTR CHEM & ENVIRON SCI, N IDAHO COL, 71- *Concurrent Pos:* NSF trainee chem, Univ Idaho, 67, Nat Defense Educ Act fel, 68-70, consult, 78-, Coeur d'Alene Hazardous Mat Comt, 86- *Mem:* Am Chem Soc; Sigma Xi; Nat Sci Teachers Asn. *Res:* Environmental effects of air and water pollutants; energy resources and conservation; science education methodologies; analytical chemistry of environmental pollutants. *Mailing Add:* Div Phys Sci NIdaho Col 1000 W Garden Ave Coeur d'Alene ID 83814

WRIGHT, KENNETH OSBORNE, b Ft George, BC, Nov 1, 11; c 1. ASTROPHYSICS. *Educ:* Univ Toronto, BA, 33, MA, 34; Univ Mich, PhD(astron), 40. *Hon Degrees:* DSc, Nicolas Copernicus Univ, Torun, Poland, 73. *Prof Exp:* Asst astron, Univ Toronto, 33-34; asst, Dom Astrophys Observ, 36-40, astromr, 40-60, asst dir, 60-66, dir, 66-76, guest worker, 76-84; RETIRED. *Concurrent Pos:* Lectr, Univ BC, 43-44; spec lectr, Univ Toronto, 60-61; res asst, Mt Wilson & Palomar Observ, 62; chmn assoc comt astron, Nat Res Coun Can, 71-74. *Honors & Awards:* Gold Medal, Royal Astron Soc Can, 33. *Mem:* Int Astron Union; Am Astron Soc; Royal Astron Soc Can (pres, 64-66); fel Royal Soc Can; Can Astron Soc. *Res:* Stellar radial velocities; observations of stellar line intensities; curves of growth; stellar atmospheres, peculiar A-type stars; observation and analysis of atmospheres of giant eclipsing systems. *Mailing Add:* 202-1375 Newport Victoria BC V8S 5E8 Can

WRIGHT, LAUREN ALBERT, b New York, NY, July 9, 18. GEOLOGY. *Educ:* Univ Southern Calif, AB, 40, MS, 43; Calif Inst Technol, PhD(geol), 51. *Prof Exp:* From jr geologist to asst geologist, US Geol Surv, 42-46; assoc geologist, State Div Mines, Calif, 47-51, sr mining geologist, 51-54, supv mining geologist, 54-61; prof geol, 61-85, EMER PROF GEOL, PA STATE UNIV, 85- *Mem:* Fel Geol Soc Am; Am Soc Econ Geologists; fel AAAS; Sigma Xi. *Res:* Geologic occurrence, origin and economics of industrial minerals; stratigraphic-tectonic evolution of southwestern Great Basin; structural geology; extensional tectonics. *Mailing Add:* 500 E Marylyn Ave Apt E69 State Col University Park PA 16801

WRIGHT, LEMUEL DARY, b Nashua, NH, Mar 1, 13; m 41; c 5. BIOCHEMISTRY. *Educ:* Univ NH, BS, 35, MS, 36; Ore State Col, PhD(biochem), 40. *Prof Exp:* Asst biochem, Pa State Univ, 36-37; asst nutrit chemist, Ore State Col, 37-40; fel, Univ Tex, 40-41; instr biochem, Sch Med, Univ WVa, 41-42; res biochemist, Med Res Div, Merck Sharp & Dohme Div, 42-47, dir nutrit res, 47-50, dir microbiol chem, 50-56; PROF NUTRIT, CORNELL UNIV, 56- *Mem:* Am Chem Soc; Am Soc Biol Chemists; Soc Exp Biol & Med; Am Inst Nutrit. *Res:* Vitamin B complex; microbiological methods of assay; bacterial growth factors; biogenesis and metabolism of pyrimidines. *Mailing Add:* 308 Savage Hall Cornell Univ Ithaca NY 14850

WRIGHT, LEON WENDELL, b Los Angeles, Calif, July 16, 23; m 50; c 4. ORGANIC CHEMISTRY, FOOD SCIENCE & TECHNOLOGY. *Educ:* Univ Calif, Los Angeles, BS, 46, MS, 47; Univ Del, PhD(chem), 51. *Prof Exp:* Instr chem, Mont State Col, 47-49 & Univ Del, 49-51; res chemist, Houdry Process Corp, 51-56; from res chemist to sr res chemist, Atlas Chem Indust, Inc, 56-67, prin chemist, Chem Res Dept, 67-70, supvr, Org & Process Res Group, 70-77, prin chemist, ICI US, Inc, 77-86; RETIRED. *Mem:* Am Chem Soc; Sigma Xi. *Res:* Catalytic hydrogenation and hydrogenolysis; hetero and homogeneous activation of hydrogen; isomerization of polyhydric alcohols; carbohydrate conversion processes; sunscreen processes; kinetics, catalysis, process development of carbohydrate conversion processes, monnitol, sorbitol, food and cosmetic chemicals including sun blockers; catalyst development. *Mailing Add:* 215 Oakwood R Fairfax Wilmington DE 19803

WRIGHT, LOUIS E, b Buras, La, Oct 18, 40; m 69; c 2. THEORETICAL PHYSICS, NUCLEAR PHYSICS. *Educ:* La State Univ, BS, 61; Duke Univ, PhD(physics), 66. *Prof Exp:* Res assoc physics, Duke Univ, 66; prog mgr theoret physics, US Army Res Off-Durham, 66-69; asst prof physics, Duke Univ, 69-70; from asst prof to assoc prof, 70-80, PROF PHYSICS, OHIO UNIV, 80- *Concurrent Pos:* Vis physicist, Inst Theoret Physics, Univ Frankfurt, 68-69; Alexander von Humboldt fel, Inst Nuclear Physics, Main Univ, 81-82. *Mem:* Am Phys Soc; Sigma Xi; AAAS. *Res:* Theoretical nuclear physics; electron scattering, pion production, virtual photon spectra, bremsstrahlung and pair production. *Mailing Add:* 98 Mulligan Rd Athens OH 45701

WRIGHT, MADISON JOHNSTON, b Washington, DC, Apr 9, 24; m 54; c 4. AGRONOMY. *Educ:* Univ NC, BA, 47; Univ Wis, MS, 50, PhD(agron, bot), 52. *Prof Exp:* Asst prof agron, Univ Wis, 52-59; assoc prof, Cornell Univ, 59-68, chmn dept, 70-75, prof agron, 68-89; RETIRED. *Mem:* Am Soc Agron; Crop Sci Soc Am; AAAS; Am Inst Biol Sci. *Res:* Crop management. *Mailing Add:* Dept Agron Cornell Univ Ithaca NY 14853

WRIGHT, MARGARET HAGEN, b San Francisco, Calif; m 65; c 1. NONLINEAR PROGRAMMING, NUMERICAL ANALYSIS. *Educ:* Stanford Univ, BS, 64, MS, 65, PhD(comput sci), 76. *Prof Exp:* Develop engr, Sylvania Electron Syst, 65-71; sr res assoc, Stanford Univ, 76-88; MEM TECH STAFF, AT&T BELL LABS, MURRAY HILLS, NJ, 88- *Concurrent Pos:* Mem, Bd Dir, Special Interest Group Numerical Anal, Asn Comput Mach, 79-82; assoc ed, J Sci Statist Comput, Soc Indust & Appl Math, 81- *Mem:* Asn Comput Mach; Soc Indust & Appl Math; Math Prog Soc. *Res:* Methods for nonlinear programming, particularly unconstrained, linearly constrained and nonlinearly constrained optimization; mathematical software; numerical linear algebra; software library development. *Mailing Add:* AT&T Bell Labs 600 Mountain Ave Murray Hill NJ 07974

WRIGHT, MARGARET RUTH, b Rochester, NY, Mar 24, 13. ZOOLOGY. *Educ:* Univ Rochester, AB, 34, MS, 38; Yale Univ, PhD(zool), 46. *Prof Exp:* Asst, Univ Rochester, 36-38; histol technician, Med Sch, Yale Univ, 38-41, asst, Osborn Zool Lab, 42-43; instr biol, Middlebury Col, 43-46; from instr to prof, 46-78, EMER PROF ZOOL, VASSAR COL, 78- *Concurrent Pos:* Mem exped, Alaska, 36; grant, Nat Inst Neurol Dis & Blindness, 54-64; Vassar Col fels, 54-55, 63-64 & 68-69; cur, Natural Hist Mus, Vassar Col; vis scientist, Marine Biol Asn UK, Plymouth, Devon, 68-69, 76 & 81. *Mem:* Fel AAAS; Am Soc Zool; Sigma Xi; NY Acad Sci; Am Inst Biol Sci. *Res:* Limnology and biogeography; experimental morphology; trophic action in sensory systems; ecology of cladocera. *Mailing Add:* Dept Biol Vassar Col Poughkeepsie NY 12601

WRIGHT, MARY LOU, b Milford, Mass, Dec 4, 34. CYTOLOGY, DEVELOPMENTAL BIOLOGY. *Educ:* Col Our Lady Elms, BS, 57; Univ Detroit, MS, 66; Univ Mass, PhD(develop biol), 72. *Prof Exp:* TEACHING & RES BIOL, COL OUR LADY ELMS, 58- *Concurrent Pos:* Sigma Xi Res grant, 73; Res Corp grants, 76-88; consult, NSF, 81-84 & grants; NIH grant, 88-91. *Mem:* Int Soc Chronobiol; Am Soc Zoologists; Tissue Cult Asn; AAAS; Sigma Xi; Cell Kinetics Soc. *Res:* Hormonal control of amphibian metamorphosis; regulation of cell proliferation rhythms in tadpole hindlimb epidermis; influence of photoperiod and daylength on tadpole growth and development. *Mailing Add:* Col Our Lady of Elms Chicopee MA 01013

WRIGHT, MAURICE ARTHUR, b Coventry, Eng, Aug 2, 35; US citizen; m 60; c 3. COMPOSITE MATERIALS, FRACTURE MECHANICS. *Educ:* Univ Wales, Swansea, UK, BSc, 59, PhD(metall), 62. *Prof Exp:* Staff scientist, Tyco Labs, 62-65; sr scientist, Nat Norton Res Corp, 65-68; from assoc prof to prof, Univ Tenn, 68-84; PROF & DIR, MAT TECHNOL CTR, SOUTHERN ILL UNIV, CARBONDALE, 84- *Concurrent Pos:* Vis prof, Technischen Hochschule, Aachen, Ger, 73 & USAF Acad, 83-84; Alcoa Found res award, 75-76; dir, Mat Div, Univ Tenn, 76-84, head, Aviations Systs Div, 79-84; consult, Univ Ala, Huntsville, 79, 81 & 84, Sverdrup Technol, Arvin-Calspan, Am Cyanamid & Nat Ctr Mfg Sci, 79, 84 & Los Alamos Weapons Lab, NMex, 80. *Mem:* Am Carbon Soc; Soc Advan Mat & Process Eng. *Res:* Relationships that exist between processing, microstructure and the reality physical and mechanical properties of materials, primarily fiber reinforced composites; basic mechanisms of fatigue and fracture of materials. *Mailing Add:* Mat Technol Ctr Southern Ill Univ Carbondale IL 62901-4033

WRIGHT, MAURICE MORGAN, b Assiniboia, Sask, July 29, 16; m 45; c 3. ELECTROCHEMISTRY, HETEROGENEOUS CATALYSIS. *Educ:* Univ BC, BA & BASc, 38; Princeton Univ, MA, 48, PhD(chem), 52. *Prof Exp:* Res chemist, Cominco, Ltd, 38-45 & 49-81; RETIRED. *Concurrent Pos:* Indust fel, Nat Res Coun Can, 49-52. *Mem:* Sigma Xi; Chem Inst Can. *Res:* Electrowinning and refining of metals; electrolytic hydrogen; ammonia synthesis; physical methods of analysis; deuterium separation, analysis and exchange reactions; reactive metals; anodic films; corrosion; protective coatings; lead-acid battery chemistry. *Mailing Add:* 1454 Willowdown Rd Oakville ON L6L 1X3 Can

WRIGHT, MELVYN CHARLES HARMAN, b Leicester, Eng, Sept 1, 44; Brit citizen. RADIO ASTRONOMY. *Educ:* Cambridge Univ, BA, 66, MA, 69, PhD(astron), 70. *Prof Exp:* Res assoc astron, Nat Radio Astron Observ, 70-72 & Calif Inst Technol, 72-77; RES ASTRONR, RADIO ASTRON LAB, 72- *Mem:* Fel Royal Astron Soc; Am Astron Soc. *Res:* Galactic and extragalactic radio astronomy; design and implementation of radio interferometers. *Mailing Add:* Radio Astron Lab Univ Calif Berkeley CA 94720

WRIGHT, NORMAN SAMUEL, b BC, Dec 8, 20; m 49. PLANT PATHOLOGY. *Educ:* Univ BC, BSA, 44, MSA, 46; Univ Calif, PhD(plant path), 51. *Prof Exp:* Plant pathologist, Sci Serv, Plant Path Lab, Vancouver Res Sta, Can Dept Agr, 46-60, head plant path sect, 60-81, sr plant pathologist, 81-85; RETIRED. *Mem:* Potato Asn Am; Can Phytopath Soc. *Res:* Potato viruses. *Mailing Add:* 10480 No Three Rd Richmond BC V7A 1W9 Can

WRIGHT, OSCAR LEWIS, b Murphysboro, Ill, Apr 21, 17; m 39; c 3. ORGANIC CHEMISTRY, PHYSICAL CHEMISTRY. *Educ:* Southern Ill Univ, BEd, 38; Univ Mo, BS, 47, PhD(org & phys chem), 49. *Prof Exp:* Teacher pub sch, Ill, 39-40; anal chemist, Wis Steel Co, 40-41 & E I du Pont de Nemours & Co, 41-43; assoc prof, Southwest Mo State Col, 49-50; head dept chem, Col Emporia, 50-52; org chemist, Continental Oil Co, 52-55, Pittsburgh Coke & Chem Co, 55-58 & MSA Res Corp Div, Mine Safety Appliance Co, 58-61; prof chem & chmn div sci & math, Rockhurst Col, 61-67; PROF CHEM & CHMN DEPT, NORTHEAST LA UNIV, 67- *Mem:* Sigma Xi. *Res:* Organic reactions of ion-exchange agents; steric factors in aromatic electrophilic substitution; organometallics; gas-solid reaction systems; effect of solvents in aromatic electrophilic substitution reactions. *Mailing Add:* 11 Kathy Lane Monroe LA 71203

WRIGHT, PAUL ALBERT, b Nashua, NH, June 15, 20; m 43; c 3. REPRODUCTIVE ENDOCRINOLOGY. *Educ:* Bates Col, SB, 41; Harvard Univ, AM, 42, PhD(endocrinol), 44. *Prof Exp:* Asst biol, Harvard Univ, 42-43, instr, 44-45; instr zool & physiol, Univ Wash, 45-46; instr biol sci, Boston Univ, 46-47; from instr to assoc prof zool, Univ Mich, 47-58; assoc prof, Univ NH, 58-62, endocrinologist, Agr Exp Sta, 58-68, chmn dept, 63-69, prof zool, 62-85; RETIRED. *Mem:* AAAS; Am Soc Zoologists (treas, 65-68); Soc Exp Biol & Med; Soc Study Reproduction. *Res:* Ovulation in the frog; physiology of melanophores of Amphibia; blood sugar studies in lower vertebrates; control of corpus luteum life by the uterus. *Mailing Add:* 20979 Cornell Ave Port Charlotte FL 33952

WRIGHT, PAUL MCCOY, b Alfalfa Co, Okla, Sept 11, 04; m 30; c 3. PHYSICAL CHEMISTRY. *Educ:* Wheaton Col, Ill, BS, 26; Ohio State Univ, MS, 28, PhD(phys chem), 30. *Prof Exp:* Asst chem, Wheaton Col, Ill, 25-26; asst gen chem, Ohio State Univ, 26-28, asst phys chem, 28-29; from asst prof to prof chem, 29-70, actg chmn dept chem & geol, 39-40, chmn dept geol, 40-59, chmn dept chem, 40-69, dir field camp, SDak, 46-52 & 60, EMER PROF CHEM, WHEATON COL, ILL, 70- *Mem:* AAAS; Am Chem Soc; fel Am Inst Chemists. *Res:* Dimensions of vapor particles; eutectics of explosive mixtures; equilibria of glycerol esters; radiochemistry. *Mailing Add:* GoYe Village Apt 416 Tahlequah OK 74464

WRIGHT, PETER MURRELL, b Toronto, Ont, Sept 26, 32; m; c 4. CIVIL ENGINEERING. *Educ:* Univ Sask, BS, 54, MSc, 61; Univ Colo, Boulder, PhD(struct), 68. *Prof Exp:* Engr, Dorman-Long Ltd, Eng, 54-57; asst prof struct, Univ Sask, 57-66; assoc dean eng, 81-85, actg dean archit, 84-88, PROF STRUCT, UNIV TORONTO, 68- *Honors & Awards:* Queen's Can Silver Jubilee Medal, 77. *Mem:* Can Soc Civil Engrs (pres, 81-82). *Res:* Automatic design of steel building frames including member selection. *Mailing Add:* Dept Civil Eng Univ Toronto Toronto ON M5S 1A4 Can

WRIGHT, PHILIP LINCOLN, b Nashua, NH, July 9, 14; m 39; c 3. ZOOLOGY. *Educ:* Univ NH, BS, 35, MS, 37; Univ Wis, PhD(zool), 40. *Prof Exp:* From instr to prof zool, 39-85, chmn dept, 56-69 & 70-71, EMER PROF, UNIV MONT, 85- *Concurrent Pos:* Ed, Gen Notes & Rev, J Mammal, 66-67; sabbatical leave, Africa, 70; vis Maytag prof zool, Ariz State Univ, 80. *Mem:* Am Soc Mammalogists; Am Soc Zoologists; Wildlife Soc; Am Ornitholgists Union; fel AAAS. *Res:* Reproductive cycles of birds and mammals; especially Mustelidae. *Mailing Add:* Dept Zool Univ Mont Missoula MT 59812

WRIGHT, RAMIL CARTER, b Hastings, Nebr, May 16, 39; m 63. MICROPALEONTOLOGY, BIOSTRATIGRAPHY. *Educ:* Rice Univ, BA, 60; Univ Ill, MS, 62, PhD(geol), 64. *Prof Exp:* Asst dir, Waterways Exp Sta, Corps Engrs, 64-66; NSF fel, Museo Argentino de Ciencias Naturales, Argentina, 66-67; vis prof, Hamline Univ, St Paul, Minn, 68; from asst prof to assoc prof, Beloit Col, Wis, 68-76; from assoc prof to prof geol, Fla State Univ, 76-81, ASSOC CHAIRPERSON, DEPT GEOL, EXXON PROD RES CO, 81- *Concurrent Pos:* Int Working Group, Int Geol Correlation Proj, 75-78. *Mem:* Geol Soc Am; Paleont Soc; AAAS; Sigma Xi; Soc Econ Paleontologist & Mineralogists. *Res:* Foraminiferal ecology and biostratigraphy; Miocene paleoenvironments. *Mailing Add:* Exxon Prod Res Co Box 2189 Rm St-4151A Houston TX 77252

WRIGHT, RICHARD KENNETH, b Richmond, Ind, Sept 22, 39. COMPARATIVE IMMUNOLOGY. *Educ:* San Diego State Univ, BS, 67, MS, 70; Univ Calif, Santa Barbara, PhD(immunol), 73. *Prof Exp:* Teaching asst, San Diego State Univ, 67-69; assoc biol sci, Univ Calif, Santa Barbara, 71-73, res fel, Sch Med, Los Angeles, 73-75, asst res anatomist, 75-81, assoc res, 81-84; dir res & develop, Physicians Labs Inc, 84-89; PROPERTY MASTER, MOTION PICTURE, 89- *Mem:* Int Soc Develop & Comp Immunol (secy & treas, 80-86); Am Soc Zoologists, Div Comp Immunol (secy, 76-79); Am Asn Immunologists; Sigma Xi. *Res:* Development and production of immunological assays for the detection of immediate and delayed type hypersensitivities to dietary proteins and environmental allergens. *Mailing Add:* 1315 Stanford St-1 Santa Monica CA 90404

WRIGHT, RICHARD N(EWPORT), b Syracuse, NY, May 17, 32; m 59; c 4. CIVIL ENGINEERING. *Educ:* Syracuse Univ, BS, 53, MS, 55; Univ Ill, PhD(civil eng), 62. *Prof Exp:* Jr engr, Pa RR Co, 53-54; from instr to assoc prof, 57-70, prof civil eng, Univ Ill, Urbana, 70-72 & 73-74; chief struct sect, 71-72, dep dir-tech, 72-73, dir, Ctr for Bldg Technol, 74-90, DIR, BLDG & FIRE RES LAB, NAT INST STANDARDS & TECHNOL, 91- *Concurrent Pos:* Chmn mgt group A, Am Soc Civil Engrs, 78-79; chmn, Lighting Res Inst, 82-83 (mem bd 82-); mem bd, Int Coun Bldg Res, 80- (pres, 83-86); US chmn, US Japan Panel Wind & Seismic Effects, 83- *Honors & Awards:* Gold Medal, US Dept Com, 82; Presidential Award, Illum Eng Soc, 83; Fed Eng Yr, Nat Soc Prof Engrs, 88. *Mem:* Am Soc Eng Educ; fel Am Soc Civil Engrs; Int Union Testing & Res Labs Mat & Struct; Earthquake Eng Res Inst; Sigma Xi; Nat Soc Prof Engrs; fel AAAS. *Res:* Analysis, behavior and design of structures; technologies for the formulation and expression of standards; performance criteria and measurement technology for buildings. *Mailing Add:* Bldg & Fire Res Lab Nat Inst Standards & Technol Gaithersburg MD 20899

WRIGHT, RICHARD T, b Haddonfield, NJ, June 28, 33; m 61; c 3. AQUATIC ECOLOGY. *Educ:* Rutgers Univ, AB, 59; Harvard Univ, PhD(biol), 63. *Prof Exp:* NSF fel, Inst Limnol, Univ Uppsala, 63-65; PROF BIOL, GORDON COL, 65- *Concurrent Pos:* NSF res grants, 66-71, 73-79, 80-83, 84-88, 88-91; NSF sci fac fel, Ore State Univ, 69-70; acad chair, Ausable Inst, 82 - *Mem:* AAAS; Am Soc Limnol & Oceanog; Am Sci Affil; Am Soc Microbiologists. *Res:* Factors controlling the density and productivity of planktonic bacteria in estuaries and coastal waters; relationships between biological knowledge and christian thought. *Mailing Add:* Dept Biol Gordon Col Wenham MA 01984

WRIGHT, ROBERT ANDERSON, b El Paso, Tex, July 22, 33; m 55; c 4. BOTANY. *Educ:* NMex State Univ, BS, 55, MS, 60; Univ Ariz, PhD(bot), 65. *Prof Exp:* PROF BIOL, WTEX STATE UNIV, 64- *Mem:* AAAS; Biomet Soc; Ecol Soc Am; Soc Study Evolution; Brit Ecol Soc; Sigma Xi. *Res:* Vegetation changes; statistical ecology; net primary productivity. *Mailing Add:* WTex State Univ Box 237 Canyon TX 79015

WRIGHT, ROBERT L, b Buckhannon, WVa, Sept 7, 30; m 52; c 1. ORGANIC CHEMISTRY, RUBBER CHEMISTRY. *Educ:* WVa Wesleyan Col, BS, 52. *Prof Exp:* Anal trainee, Monsanto Co, 52, res chemist, 55-64, sr res chemist, 64-75, res specialist, 75-80, sr res specialist, 80-89; RES CONSULT TECHNOL, 90- *Mem:* Am Chem Soc. *Res:* Process development; exploratory synthesis in the field of rubber chemicals. *Mailing Add:* 95 Blue Hill Lane Akron OH 44333-3417

WRIGHT, ROBERT RAYMOND, science policy; deceased, see previous edition for last biography

WRIGHT, ROBERT W, b Auburn, NY, Aug 2, 32; m 55. PHYSICAL ORGANIC CHEMISTRY, POLYMER CHEMISTRY. *Educ:* NY State Col Forestry, BS, 59, PhD(chem), 64. *Prof Exp:* Chemist, Owens-Ill, Inc, Ohio, 64-65, res chemist, 65-66, sect leader polymer chem, 66-69; group dir synthetic mat, 69-76, MGR ANALYTICAL SCI, CORP RES CTR, INT PAPER CO, 77- *Mem:* Am Chem Soc; Tech Asn Pulp & Paper Indust; Soc Plastics Engrs; Coun Agr & Chemurgic Res. *Res:* Plastics processing; polymer coatings and adhesives; polymer characterization; analytical chemistry. *Mailing Add:* Int Paper Co Long Meadow Rd Tuxedo Park NY 10987

WRIGHT, ROGER M, b Long Beach, Calif, Feb 19, 35; m 58; c 2. CHEMICAL ENGINEERING, HEAT TRANSFER. *Educ:* Univ Calif, Berkeley, BS, 56, PhD(chem eng), 61. *Prof Exp:* Res engr, Lawrence Radiation Lab, Calif, 57-61; SR ENG SPECIALIST, AIRES MFG CO, 61- *Mem:* Am Inst Chem Engrs; Am Inst Aeronaut & Astronaut. *Res:* Heat transfer, including research in forced-convection boiling, compact heat exchangers and spacecraft thermal radiators; spacecraft environmental control systems; mass-transfer systems; thermal systems design and analysis. *Mailing Add:* 6235 Monita St Long Beach CA 90803

WRIGHT, ROGER NEAL, b Springfield, Ill, Nov 14, 42; m 67; c 2. MECHANICAL BEHAVIOR, WIRE PROCESSING. *Educ:* Mass Inst Technol, BS, 65, DSc(metall), 69. *Prof Exp:* Sr res metallurgist, Allegheny Ludlum Res Labs, 68-71; sr engr, Westinghouse Res Labs, 71-74; from asst prof to assoc prof mat eng, Rensselaer Polytech Inst, 74-83, exec officer, 83-90, actg chmn, 84-86, PROF MAT ENG, RENSSELAER POLYTECH INST, 83-, DIR, HIGH TEMPERATURE TECHNOL PROG, 90- *Concurrent Pos:* Consult metall & metals processing, var co & govt agencies, 74-; prin investr, NSF, Gen Motors, Int Bus Mach, Am Iron & Steel Inst, Int Copper Asn & NY State Energy Res & Develop Authority, 76-; chmn, Eng Mat Coun, Soc Mfg Engrs & trustee, Fedn Mat Soc, 80-83; mem, bd dirs, Metall Soc, 83-85. *Mem:* Metall Soc; Soc Metall Engrs; Am Soc Metals Int; Am Soc Mech Engrs. *Res:* Mechanical properties of materials; thermal processing of materials. *Mailing Add:* Dept Mat Eng Rensselaer Polytech Inst Troy NY 12180

WRIGHT, RUSSELL EMERY, b Muscatine, Iowa, June 19, 39; m 63; c 3. MEDICAL ENTOMOLOGY. *Educ:* Iowa State Univ, BSc, 63, MS, 66; Univ Wis-Madison, PhD(entom), 69. *Prof Exp:* From asst prof to assoc prof entom, Univ Guelph, 69-76; assoc prof, 76-82, PROF, OKLA STATE UNIV, 82- *Mem:* Entom Soc Am; Am Mosquito Control Asn. *Res:* Behavior, biology and control of insect pests of livestock; arthropod borne viruses. *Mailing Add:* Dept Entom Okla State Univ Main Campus Stillwater OK 74078

WRIGHT, SAMUEL D, b Hibbing, Minn, Sept 23, 52; m 80; c 1. PHAGOCYTOSIS, RECEPTOR BIOLOGY. *Educ:* Carletor Col, BA, 74; Harvard Univ, PhD(biol), 79. *Prof Exp:* Asst prof, 84-89, ASSOC PROF, ROCKEFELLER UNIV, 89- *Mem:* Am Soc Cell Biol; Am Soc Immunologists. *Res:* Study of receptors on leukocytes that mediate adhesion; migration and phagocytosis. *Mailing Add:* Dept Cellular Physiol & Immunol Rockefeller Univ 1230 York Ave New York NY 10021

WRIGHT, STEPHEN E, b Searcy, Ark, Mar 20, 42; m 65; c 2. ONCOLOGY, VIROLOGY. *Educ:* Hendrix Col, BA, 63; Univ Ark, MD, 67. *Prof Exp:* Instr med hematol/oncol, Sch Med, Univ Wash, 73-74, asst prof, 74-75; chief, oncol, Vet Admin Med Ctr, Salt Lake City, Utah, 75-82; ASST PROF MED MOLECULAR BIOL, COL MED, UNIV UTAH, 75-; RES ASSOC CANCER RES, VET ADMIN MED CTR, SALT LAKE CITY, UTAH, 82- *Concurrent Pos:* Prin investr, Southwest Oncol Group, 75-85, Vet Admin Grant, 76-79 & 79-82. *Res:* Mechanisms of oncogenic transformation; tumor vaccine production; tumor antibody production. *Mailing Add:* Vet Admin Ctr Hemat Oncol Sect 111 6010 Amarillo Blvd W Amarillo TX 79106-1797

WRIGHT, STEPHEN GAILORD, b San Diego, Calif, Aug 13, 43; m 70; c 2. CIVIL ENGINEERING. *Educ:* Univ Calif, Berkeley, BS, 66, MS, 67, PhD(civil eng), 69. *Prof Exp:* From asst prof to assoc prof, 69-84, PROF CIVIL ENG, UNIV TEX, AUSTIN, 84- *Mem:* Am Soc Civil Engrs. *Res:* Soil mechanics and foundations; slope stability; foundations for offshore structures. *Mailing Add:* Dept Civil Eng Univ Tex Austin TX 78712

WRIGHT, STEVEN MARTIN, b Oak Park, Ill, Aug 25, 53; m 75; c 2. FLUORINE CHEMISTRY. *Educ:* Elmhurst Col, BA, 75; Marquette Univ, PhD(inorg chem), 80. *Prof Exp:* Asst prof chem, Lakeland Col, 80-82; asst prof, 82-87, ASSOC PROF, DEPT CHEM, UNIV WIS, STEVENS POINT, 87- *Concurrent Pos:* Vis asst prof, Univ Wis-Milwaukee, 81-82; instr, Marquette Univ, 80-82. *Mem:* Am Chem Soc; AAAS; Sigma Xi; NY Acad Sci. *Res:* New synthetic methods to form new and existing graphite interalation compounds; interalation of fluorides and graphite. *Mailing Add:* Chem Dept Univ Wis Stevens Point WI 54481

WRIGHT, STUART JOSEPH, b Arvida, Que, Jan 17, 52; US citizen. TROPICAL ECOLOGY. *Educ:* Princeton Univ, BA, 74; Univ Calif, Los Angeles, PhD(biol), 80. *Prof Exp:* Fel trop biol, 80-81, Res Assoc, Smithsonian Trop Res Inst, 81-; AT DEPT EDUC, WAKE FOREST UNIV, SALEM, NC. *Concurrent Pos:* Vis asst prof ecol, Univ Mont, 81-82. *Mem:* Ecol Soc Am; Asn Trop Biol; Am Ornithologists Union. *Res:* Theoretical and empirical investigations of the relative abundances of species with emphasis on tropical communities of birds, lizards and trees. *Mailing Add:* Dept Educ Wake Forest Univ Box 27524 Winston Salem NC 27109

WRIGHT, STUART R(EDMOND), b Calgary, Alta, Aug 13, 23; m 49; c 3. CHEMICAL ENGINEERING, CHEMISTRY. *Educ:* Univ Alta, BSc, 46, MSc, 47; Northwestern Univ, PhD, 50. *Prof Exp:* Engr, Aluminum Co Can, Ltd, 50-54; engr, 54-66, sr res engr, 66-69, res assoc chem, 69-74, res assoc, Div Eng, 74-85, CONSULT, E I DU PONT DE NEMOURS & CO, INC, 85- *Mem:* Am Inst Chem Engrs; Sigma Xi. *Res:* Industrial chemicals; dye intermediates. *Mailing Add:* 15 Marlton Rd Woodstown NJ 08098

WRIGHT, SYDNEY COURTENAY, b Vancouver, BC, Oct 16, 23; m 48; c 3. PHYSICS. *Educ:* Univ BC, BA, 43; Univ Calif, PhD(physics), 49. *Prof Exp:* NSF fel, 49-50, res assoc, 50-55, from asst prof to assoc prof, 55-68, PROF PHYSICS, ENRICO FERMI INST, UNIV CHICAGO, 69- *Concurrent Pos:* Consult, Brookhaven Nat Lab, 53 & Argonne Nat Lab, 57-60. *Mem:* Am Phys Soc. *Res:* Experimental particle physics; particle accelerator design. *Mailing Add:* 5831 S Blackstone Chicago IL 60637

WRIGHT, TERRY L, b Houston, Tex, July 9, 52. SYNTHETIC HETEROCYCLIC CHEMISTRY. *Educ:* Tex Christian Univ, BS, 74; Univ Chicago, MS, 75, PhD(org chem), 79. *Prof Exp:* Sr res chemist, 79-81, proj leader, Western Div, 81-86, RES LEADER, DOW CHEM USA, 86-, RES SAFETY DIR, 88- *Mem:* Am Chem Soc. *Res:* Development of new areas of heterocyclic chemistry particularly the synthesis of new heterocyclic ring systems and for the design, synthesis and development of new pharmaceuticals; process development for agricultural chemicals. *Mailing Add:* 349 Alcatraz Ave Oakland CA 94618-1303

WRIGHT, THEODORE ROBERT FAIRBANK, b Kodaikanal, India, Apr 10, 28; US citizen; m 51. DEVELOPMENTAL GENETICS. *Educ:* Princeton Univ, AB, 49; Wesleyan Univ, MA, 54, Yale Univ, PhD(zool), 59. *Prof Exp:* Asst prof biol, Johns Hopkins Univ, 59-65; assoc prof, 65-75, PROF BIOL, UNIV VA, 75- *Concurrent Pos:* Mem genetics study sect, NIH, 72-74; fel, Max Planck Inst Biol, 75-76. *Mem:* Fel AAAS; Genetics Soc Am; Soc Develop Biol. *Res:* Developmental genetics of embryonic mutants in Drosophila; genetic and molecular analysis of the functional organizaion of the genome of Drosophila. *Mailing Add:* Dept Biol Univ Va Charlottesville VA 22901

WRIGHT, THOMAS CARR, JR, PATHOLOGY, CELL GROWTH CONTROL. *Educ:* Harvard Univ, MD, 77. *Prof Exp:* ASSOC PROF PATH, SCH MED, HARVARD UNIV, 83- *Mailing Add:* Dept Path Sch Med Harvard Univ 25 Shattuck St Boston MA 02115

WRIGHT, THOMAS L, b Chicago, Ill, July 26, 35; div; c 2. PETROLOGY, VOLCANOLOGY. *Educ:* Pomona Col, AB, 57; Johns Hopkins Univ, PhD(geol), 61. *Prof Exp:* Geologist, Washington, DC, 61-64, staff geologist, Hawaiian Volcano Observ, 64-69, geologist, Geol Div, Washington, DC, 69-74, geologist, Reston, Va, 74-84, SCIENTIST-IN-CHARGE, HAWAIIAN VOLCANO OBSERV, HAWAII NAT PARK, US GEOL SURV, 84- *Mem:* Mineral Soc Am; Am Geophys Union. *Res:* Igneous petrology; petrology and mineralogy of Hawaiian basalt; study of crystallization of basalt in the lava lakes of Kilauea volcano; chemical and stratigraphic study of the basalts of the Columbia River plateau. *Mailing Add:* US Geol Surv 959 Nat Ctr Reston VA 22092

WRIGHT, THOMAS OSCAR, b Jasper, Ala, July 9, 40; m 65. STRATIGRAPHY, STRUCTURE. *Educ:* Auburn Univ, BS, 65; George Washington Univ, MS, 71, PhD(geochem), 74. *Prof Exp:* Oceanogr, Nat Oceanog Data Ctr, 65-69; lectr geol, Bryn Mawr Col, 73-75; asst prof geol, Allegheny Col, 75-78; prog assoc geol, 78-80, PROG DIR STRUCT & TECTONICS, NSF, 80- *Concurrent Pos:* Lectr, George Washington Univ, 73-75 & Carbon Lehigh Intermediate Sch Unit, 76-78; proj geologist explor, Cyprus Explor Ltd, 70- *Mem:* Geol Soc Am. *Res:* Geology and geochemistry of sedimentary rocks and modern analogs; teaching of geology; Ordovician clastic rocks of central Appalachians; sedimentology and structure of pre-cambrian and paleozoic rocks; northern Victoria Land; Antarctica. *Mailing Add:* 3531 T St Washington DC 20007

WRIGHT, THOMAS PAYNE, b Ft Worth, Tex, Dec 23, 43; m 66; c 3. PHYSICS, PLASMA PHYSICS. *Educ:* St Bonaventure Univ, BS, 66; NMex State Univ, MS, 68, PhD(physics), 69. *Prof Exp:* Staff mem physics, 69-80, SUPVR THEORET DIV, SANDIA NAT LABS, 80- *Mem:* Am Phys Soc. *Res:* Plasma waves and instabilities; electromagnetic theory; theory of kinetic equations; laser-plasma interaction; statistical mechanics; relativistic electron beams; magnetohydrodynamics; intense charged particle beam sources and transport. *Mailing Add:* 341 Big Horn Ridge NE Albuquerque NM 87122

WRIGHT, THOMAS PERRIN, JR, b Great Falls, SC, June 23, 39; m 61; c 2. TOPOLOGY. *Educ:* Davidson Col, AB, 60; Univ Wis, MA, 63, PhD(math), 67. *Prof Exp:* ASSOC PROF MATH, FLA STATE UNIV, 67- *Concurrent Pos:* NSF grant, Fla State Univ, 67-72. *Mem:* Am Math Soc. *Res:* Topology of manifolds. *Mailing Add:* Math Dept Fla State Univ Tallahassee FL 32306-3027

WRIGHT, THOMAS WILSON, b Fergus Falls, Minn, Oct 23, 33; m 55; c 3. CONTINUUM MECHANICS. *Educ:* Cornell Univ, BCE, 56, MCE, 57, PhD(mech), 64. *Prof Exp:* Sr engr struct, AAI Corp, Md, 59-61; asst prof mech, Johns Hopkins Univ, 64-67; MECH ENGR, US ARMY BALLISTIC RES LABS, ABERDEEN PROVING GROUND, 67- *Concurrent Pos:* Prof mech eng, Johns Hopkins Univ. *Mem:* Soc Natural Philos; Am Soc Mech Engrs; Soc Eng Sci. *Res:* Nonlinear wave propagation and the mechanics of deformable media. *Mailing Add:* 515 Cornell Ave Swarthmore PA 19081-2401

WRIGHT, VERNON LEE, b Muscatine, Iowa, Feb 20, 41; c 3. BIOMETRICS, POPULATION DYNAMICS. *Educ:* Iowa State Univ, BS, 64; Purdue Univ, MS, 66; Wash State Univ, PhD(zool), 71. *Prof Exp:* Fel biometrics, Cornell Univ, 70-72; wildlife biometrician, Iowa Conserv Comn, 72-76; conserv res analyst, Ill Dept Conserv, 76-78; PROF EXP STATIST, SCH FORESTRY, WILDLIFE & FISHERIES, LA STATE UNIV, BATON ROUGE, 78- *Mem:* Wildlife Soc; Biometric Soc; Ecol Soc Am; Sigma Xi. *Res:* Sampling from and modeling of naturally occurring populations; statistical aspects of wildlife management and administration; measuring peoples' attitudes toward conservation issues. *Mailing Add:* Sch Forestry Wildlife & Fisheries La State Univ Baton Rouge LA 70803

WRIGHT, WALTER EUGENE, b Terre Haute, Ind, July 16, 24; m 51; c 5. BIOPHARMACEUTICS. *Educ:* Purdue Univ, BS, 48, MS, 50, PhD(pharmaceut chem), 53. *Prof Exp:* Sr biochemist, Eli Lilly & Co, 53-65, res scientist, 65-69, res assoc, 69-86; RETIRED. *Mem:* Am Chem Soc; NY Acad Sci; Am Soc Microbiol; Sigma Xi. *Res:* Intestinal and drug absorption; active transport; study of the absorption, metabolism and excretion of new medicinal agents in experimental animals. *Mailing Add:* 7553 N Audubon Rd Indianapolis IN 46250

WRIGHT, WAYNE MITCHELL, b Sanford, Maine, July 12, 34; m 59; c 4. ACOUSTICS. *Educ:* Bowdoin Col, AB, 56; Harvard Univ, MS, 57, PhD(appl physics), 61. *Prof Exp:* Res fel, Harvard Univ, 61-62; from asst prof to assoc prof, 62-75, PROF PHYSICS & CHMN DEPT, KALAMAZOO COL, 75- *Concurrent Pos:* Vis prof physics, Naval Postgrad Sch, 69-70; vis prof atmospheric sci, Univ Mich, 79-80; res fel mech eng, Univ Tex Austin, 88-89. *Mem:* Fel Acoust Soc Am; Am Asn Physics Teachers. *Res:* Physical acoustics; experimental studies of finite-amplitude sound phenomena in air; optoacoustics. *Mailing Add:* Dept Physics Kalamazoo Col Kalamazoo MI 49007

WRIGHT, WILBUR HERBERT, physics, for more information see previous edition

WRIGHT, WILLIAM BLYTHE, JR, b Washington, DC, Sept 29, 18; m 48; c 2. PHARMACEUTICAL CHEMISTRY. *Educ:* Univ Va, BS, 39; Univ Mich, PhD(chem), 42. *Prof Exp:* Resin res chemist, Rohm and Haas Co, Pa, 42-47; pharmaceut res chemist, Bound Brook Labs, Am Cyanamid Co, 47-55 & Lederle Labs, 55-87; RETIRED. *Mem:* Am Chem Soc. *Res:* Coatings; plywood adhesives; ion exchange resins; pharmaceuticals. *Mailing Add:* 18 Clinton Pl Woodcliff Lake NJ 97675-8299

WRIGHT, WILLIAM F(RED), b Salt Lake City, Utah, Nov 12, 18; m 55; c 2. MECHANICAL ENGINEERING. *Educ:* Vanderbilt Univ, BE, 40, MA, 41; Harvard Univ, MS, 47. *Prof Exp:* Instr appl math, 46-47, from instr to assoc prof, 47-70, PROF MECH ENG, VANDERBILT UNIV, 70- *Concurrent Pos:* Instr, US Naval Acad, 59-60. *Mem:* Am Soc Eng Educ; Am Soc Mech Engrs. *Res:* Mechanical design; heat transfer; thermodynamics. *Mailing Add:* 2709 Wortham Ave Nashville TN 37215

WRIGHT, WILLIAM HERBERT, III, b Newton, Mass, Feb 13, 43; c 2. STRUCTURAL GEOLOGY. *Educ:* Middlebury Col, BA, 65; Ind Univ, Bloomington, MA, 67; Univ Ill, Urbana, PhD(geol), 70. *Prof Exp:* Explor geologist, Chevron Oil Co, Colo, 66; from asst prof to assoc prof, 69-78, PROF GEOL, SONOMA STATE UNIV, 78- *Mem:* Geol Soc Am; Am Geophys Union. *Res:* Folding, metamorphic structures; geology of geothermal energy resources; structural evolution of mountain belts; structure and geologic history of Calaveras Formation and Sierra Foothills Melange, Sierra Nevada, California. *Mailing Add:* Dept Geol Sonoma State Univ 1801 E Cotati Ave Rohnert Park CA 94928

WRIGHT, WILLIAM LELAND, b Darbyville, Ohio, Aug 14, 30; m 52; c 3. WEED SCIENCE. *Educ:* Ohio Univ, BS, 53, MS, 57; Purdue Univ, PhD(plant physiol), 64. *Prof Exp:* Plant physiologist, Dow Chem Co, Tex, 57-58; from plant physiologist to sr plant physiologist, Eli Lilly & Co, Washington, DC, 58-65, head plant sci res, 65-72, prod plans adv, Elanco Prod Co, 72-73, regulatory serv adv, 73-74, mgr agr regulatory serv, 74-76, dir, Agr Regulatory Serv, 76-80, dir, Govt Affairs, 80-82, dir, Int Plant Sci Res, Indianapolis, Ind, 82-85, dir, Discovery Res Plant Sci, 85-87, dir, Qual Assurance Res & Develop, 87-89; RETIRED. *Mem:* Sigma Xi; Am Hort Soc; Am Soc Plant Physiologists; Weed Sci Soc Am. *Res:* Chemical weed control; plant growth regulation, insecticides and aquatic weed control; fate of herbicides in environment; product planning; regulatory affairs; government affairs. *Mailing Add:* 1410 Bowman Dr Greenfield IN 46140

WRIGHT, WILLIAM RAY, b Iola, Wis, Aug 16, 41; m 63. SOIL MORPHOLOGY. *Educ:* Wis State Univ-River Falls, BS, 66; Univ Md, College Park, MS, 69, PhD(soils), 72. *Prof Exp:* From asst prof to assoc prof, 72-86, PROF & CHMN, SOIL SCI, UNIV RI, 86- *Mem:* Am Soc Agron; Soil Sci Soc Am; Soil Conserv Soc Am. *Res:* Soil genesis, classification and land use. *Mailing Add:* Dept Natural Resources Univ RI Kingston RI 02881

WRIGHT, WILLIAM REDWOOD, b Philadelphia, Pa, Sept 17, 27; m 56; c 3. PHYSICAL OCEANOGRAPHY. *Educ:* Princeton Univ, BA, 50; Univ RI, MS, 65, PhD(oceanog), 70. *Prof Exp:* Teacher, RI Pvt Sch, 50-52; pub info officer, Woods Hole Oceanog Inst, 60-62, res asst phys oceanog, 62-70, asst scientist, 70-75; supvry oceanogr, Northeast Fisheries Ctr, Nat Marine Fisheries Serv, 76-81; pres, Bermuda Biol Sta Res, Inc, 77-86; OCEANOGR, ASSOC SCIENTISTS AT WOODS HOLE, INC, 79- *Concurrent Pos:* Reporter, Auburn Citizen-Advertiser, NY, 52-54 & Providence J, RI, 54-60. *Res:* Deep circulation of the world oceans; coastal circulation. *Mailing Add:* Box 54 Woods Hole MA 02543

WRIGHT, WILLIAM ROBERT, b Cleveland, Ohio, Nov 24, 28. THEORETICAL PHYSICS. *Educ:* Harvard Univ, AB, 51, MA, 52, PhD(physics), 57. *Prof Exp:* Asst prof physics, Univ Kans, 57-61; from assoc prof to prof, 61-86, head dept, 62-66, EMER PROF PHYSICS, UNIV CINCINNATI, 86- *Concurrent Pos:* Vis mem dept physics, Oxford Univ, 67-68. *Mem:* Am Phys Soc; Am Asn Physics Teachers. *Res:* Solid state effects in nuclear orientation; Green's function techniques in magnetism; variational methods in statistical mechanics. *Mailing Add:* Dept Physics ML 11 Univ Cincinnati Cincinnati OH 45221

WRIGHT, WILLIAM V(AUGHN), b Winston-Salem, NC, Sept 15, 31; m 55; c 3. COMPUTER SYSTEMS FOR MOLECULAR BIOLOGY, VIRTUAL REALITY. *Educ:* Duke Univ, BSEE, 53; Harvard Univ, SM, 54; Univ NC, PhD(comput sci), 72. *Prof Exp:* Engr, IBM Corp, 58-67, sr engr, Research Triangle Park, 67-90; VIS RES ASSOC PROF, UNIV NC, CHAPEL HILL, 90- *Concurrent Pos:* Adj assoc prof comput sci, Univ NC, Chapel Hill, 72-90. *Mem:* Sr mem Inst Elec & Electronics Engrs; Asn Comput Mach; Sigma Xi. *Res:* Computer applications in molecular biology; interactive computer graphics; computing systems architecture and implementation. *Mailing Add:* 104 Campbell Lane Chapel Hill NC 27514-7802

WRIGHT, WILLIAM VALE, b Long Beach, Calif, Dec 4, 29; m 71; c 4. ENGINEERING, AERONAUTICS. *Educ:* Calif Inst Technol, BS, 51, PhD(mech eng, physics), 55. *Prof Exp:* Res asst, Calif Inst Technol, 50-52; mem tech staff, Hughes Aircraft Co, 53-54; mgr semiconductor mats, TRW Semiconductors, Inc, 54-57, mgr, Solid State Div, Electro Optical Systs, Inc, 57-60, vpres, 60-65; dir sci & eng, Environ Sci Serv Admin, 66-68; pres, Flight Test Res, Inc, 68-72; prog develop mgr autonetics, Rockwell Int Corp, 72-75; vpres, COR, Inc, 75-78; dir eng & develop, Ball Corp, 78-80; tech dir, Food Mach Group, FMC Corp, 80-81; assoc prof mech eng technol, Univ NC, Charlotte, 84-89; PRES, ENTEG SYSTS INC, 81- *Concurrent Pos:* Lectr, Univ Calif, Los Angeles, 55-60; lectr eng, Univ NC, Charlotte, 74-75; NSF fel, 52-53. *Mem:* AAAS; Inst Elec & Electronics Engrs; assoc fel Am Inst Aeronaut & Astronaut. *Res:* Research and engineering management; aeronautics; solid state materials and devices; computer sciences. *Mailing Add:* 2412 Cloister Dr Charlotte NC 28211

WRIGHT, WILLIAM W, b Feb 17, 49. REPRODUCTIVE BIOLOGY, ENDOCRINOLOGY. *Educ:* State Univ NY, Binghamton, PhD(biol sci), 78. *Prof Exp:* Asst prof, 82-88, ASSOC PROF POP DYNAMICS, SCH HYG & PUB HEALTH, JOHNS HOPKINS UNIV, 88- *Mem:* Am Soc Cell Biol; Soc Study Reproduction. *Res:* Male reproductive biology. *Mailing Add:* Dept Pop Dynamics Johns Hopkins Univ 615 N Wolfe St Rm 3606 Baltimore MD 21205

WRIGHT, WILLIAM WYNN, b Baltimore, Md, Aug 13, 23; m 45; c 5. PHARMACEUTICAL CHEMISTRY. *Educ:* Loyola Col, Md, BS, 44; Georgetown Univ, MS, 46, PhD(biochem), 48. *Prof Exp:* Chemist, Nat Bur Stand, Washington, DC, 45; chief antibiotic chem br, US Food & Drug Admin, 45-55, dir antibiotic control labs, 55-57, dir antibiotic res, 57-64, dept dir div antibiotics & insulin cert, 64-69, dep dir pharmaceut res & testing, 69-71, dir div drug biol, 71-75, dep assoc dir pharmaceut res & testing, Bur Drugs, 75-79; SR SCIENTIST, US PHARMACOPEIA, ROCKVILLE, MD, 79- *Concurrent Pos:* Mem, WHO Expert Panels on Antibiotics, 60-75 & Biol Stand, 75- *Mem:* Fel AAAS; Am Chem Soc; fel Asn Off Anal Chemists (pres, 76-77); NY Acad Sci; fel Acad Pharmaceut Sci; Sigma Xi. *Res:* Antibiotic testing by chemical, physical, microbial and biological methods; absorption, excretion, distribution and tissue residues of antibiotics; bacterial susceptibility; pharmaceutical analysis. *Mailing Add:* 1301 Dilston Pl Silver Spring MD 20903

WRIGHT, WOODRING ERIK, b San Francisco, Calif, June 21, 49; m 71; c 2. CELL BIOLOGY, SOMATIC CELL GENETICS. *Educ:* Harvard Univ, BA, 70; Stanford Univ, PhD(med microbiol), 74, MD, 75. *Prof Exp:* Fel, Pasteur Inst, Paris, 75-78; asst prof, Univ Tex Health Sci Ctr, Dallas, 78-85; ASSOC PROF CELL BIOL, UNIV TEX SOUTHWESTERN MED CTR, DALLAS, 85- *Concurrent Pos:* Prin investr, Am Heart Asn, 78-79 & 80-82, res grants, NIH, 78-, & Muscular Dystrophy Asn, 80-; Nat Inst Aging, Res Career Develop Award, 78. *Honors & Awards:* Lyndon Baines Johnson Award, Am Heart Asn, 78; Merit Award, Nat Inst Aging, 86- *Mem:* Geront Soc; Am Soc Cell Biol; AAAS. *Res:* Using somatic cell genetic and molecular genetic approaches to probe the mechanisms regulating cell differentiation and aging. *Mailing Add:* Dept Cell Biol Univ Tex Southwestern Med Ctr 5323 Harry Hines Blvd Dallas TX 75235

WRIGHTON, MARK STEPHEN, b Jacksonville, Fla, June 11, 49; m 68. PHOTOCHEMISTRY, INORGANIC CHEMISTRY. *Educ:* Fla State Univ, BS, 69; Calif Inst Technol, PhD(chem), 72. *Hon Degrees:* DSc, Univ WFla, 83. *Prof Exp:* From asst prof to prof chem, Mass Inst Technol, 72-81, Frederick G Keyes prof, 81-89, dept head, 87-90, CIBA-GEIGY PROF CHEM, MASS INST TECHNOL, 89-, PROVOST, 90- *Concurrent Pos:* Alfred P Sloan fel, 74-76 & Dreyfus grant, 75-80; mem, Chem Res Eval Panel for Air Force Off Sci Res, 76-80; Standard Oil Co Calif vis energy prof, Calif Inst Technol, 77; distinguished vis lectr, Univ Tex, Austin, 77; div ed, J Electrochem Soc, 80-82; mem, Mat Res Coun, Defense Adv Res Proj Agency, 81-; MacArthur Prize fel, 83-88; mem, chem adv comt, NSF, 84-87; bd, Chem Scis & Technol, NRC, 86- *Honors & Awards:* Herbert Newby McCoy Award, Calif Inst Technol, 72; Am Chem Soc Award, 80 & 88; E O Lawrence Award, US Dept Energy, 83. *Mem:* Fel AAAS; Am Chem Soc; Electrochem Soc; Am Acad Arts & Sci. *Res:* Excited state processes in transition metal containing molecules; photoelectrochemistry; surface chemistry; catalysis; energy conversion; molecular electronics. *Mailing Add:* Dept Chem Mass Inst Technol Cambridge MA 02139

WRISTON, JOHN CLARENCE, JR, b Boston, Mass, Aug 12, 25; m 45; c 4. BIOCHEMISTRY. *Educ:* Univ Vt, BS, 48; Columbia Univ, PhD(biochem), 53. *Prof Exp:* Nat Found Infantile Paralysis fel & instr biochem, Sch Med, Univ Colo, 53-55; from asst prof to assoc prof, 55-69, PROF CHEM, UNIV DEL, 69- *Mem:* Am Soc Biol Chemists. *Res:* Protein chemistry; structure and mechanism of action of L-asparaginase. *Mailing Add:* Dept Chem Univ Del Newark DE 19711

WROBEL, JOSEPH JUDE, b Chicago, Ill, Mar 18, 47; m 70; c 2. CHEMICAL PHYSICS. *Educ:* Loyola Univ, Chicago, BS, 68; Univ Fla, PhD(chem physics), 76. *Prof Exp:* RES PHYSICIST, EASTMAN KODAK CO RES LABS, 75- *Res:* Image storage and display systems. *Mailing Add:* 29 Red Cedar Dr Rochester NY 14616

WROBEL, JOSEPH STEPHEN, b Syracuse, NY, Aug 15, 39. SOLID STATE PHYSICS. *Educ:* Syracuse Univ, BS, 61, MS, 64, PhD(physics), 67. *Prof Exp:* Res asst physics, Syracuse Univ, 61-66; mem tech staff, Tex Instruments Inc, 66-86; ASSOC PROF COMPUT SCI ENG, UNIV ARK, 86- *Mem:* Asn Comput Mach; Am Phys Soc. *Res:* Semiconductor materials; computer automation; photoconductivity; infrared physics. *Mailing Add:* Dept Comput Sci Eng 309 Eng Bldg Univ Ark Fayetteville AR 72701

WROBEL, RAYMOND JOSEPH, food science & technology; deceased, see previous edition for last biography

WROBEL, WILLIAM EUGENE, b Syracuse, NY; m. ENVIRONMENTAL ASSESSMENT, WATER QUALITY. *Educ:* Syracuse Univ, BA, 68; New York Univ, MS, 73. *Prof Exp:* Biologist, NY State Atomic & Space Develop Authority, 69-72; sr ecologist, Dames & Moore, 72-79; projs dir, VTN Ore, Inc, Subsid VTN Corp, 79-84; VPRES OPERS, ECOTECH, INC, 89- *Mem:* Am Soc Civil Eng. *Res:* Performance and management of interdisciplinary environmental studies for mining and energy facilities, pipelines, chemical plants, refineries, dredging activities, shoreline and land developments, and pollution control facilities; specializes in soil and groundwater studies for solid and hazardous waste facilities; hazardous waste studies and remediation. *Mailing Add:* 5355 Bull Run Dr Baton Rouge LA 70817-2756

WROBLEWSKI, JOSEPH S, b Chicago, Ill, June 8, 48; m 82; c 2. BIOLOGICAL OCEANOGRAPHY, FISHERIES OCEANOGRAPHY. *Educ:* Univ Ill, BSc, 70; Fla State Univ, MSc, 72, PhD(oceanog), 76. *Prof Exp:* Asst prof biol oceanog, Dalhousie Univ, 76-84; res scientist, Bigelow Lab Ocean Sci, 84-89; PROF FISHERIES OCEANOG, MEM UNIV NFLD, 89- *Concurrent Pos:* Prin investr, Natural Sci & Eng Res Coun Can, 76-, NSF, 80-, Off Naval Res, 84-89; indust res chair, Nat Sci & Eng Res Coun Can, 89- *Mem:* Am Soc Limnol & Oceanog; Sigma Xi; Am Geophys Union. *Res:* Numerical modeling of marine ecosystems; theory of plankton patchiness; ocean circulation-marine biomass interactions; numerical analysis; applied statistics; computer-generated movie productions of global ocean productivity; operational fisheries oceanography. *Mailing Add:* Ocean Sci Ctr Mem Univ Nfld St John's NF A1B 3X7 Can

WROGEMANN, KLAUS, b Berlin, Ger, Dec 8, 40; m 67; c 3. BIOCHEMISTRY, GENETICS & MOLECULAR BIOLOGY. *Educ:* Univ Marburg, MD, 66; Univ Man, PhD(biochem), 69. *Prof Exp:* Intern surg & med, Med Hosp, Hanover, Ger, 70; from asst prof to prof, 70-79, PROF BIOCHEM, FAC MED, UNIV MAN, 79-, PROF HUMAN GENETICS, 85- *Concurrent Pos:* Vis prof, Max-Planck Inst Immunol, Freiburg, Ger, 77-78; vis prof, Inst Molecular Biol, Strasbourg, France, 84-85. *Honors & Awards:* Rh Inst Award, 82. *Mem:* Can Biochem Soc; Am Soc Human Genetics; Int Study Group Heart Res; Am Soc Biochem & Molecular Biol. *Res:* Metabolism of normal and dystrophic heart and skeletal muscle; molecular basis of genetic diseases; basic and applied studies of steroid hormone receptors. *Mailing Add:* Dept Biochem Univ Man Winnipeg MB R3E 0W3 Can

WROLSTAD, RONALD EARL, b Oregon City, Ore, Feb 5, 39; m 77; c 2. FOOD SCIENCE, AGRICULTURAL CHEMISTRY. *Educ:* Ore State Univ, BS, 60; Univ Calif, Davis, PhD(agr chem), 64. *Prof Exp:* Grad scientist, Unilever Res Lab, Eng, 64-65; res assoc, 65-66, asst prof, 66-71 assoc prof, 71-80, PROF FOOD SCI, ORE STATE UNIV, 80- *Concurrent Pos:* Sabbatical leave, Plant Dis Div, Dept Sci & Indust Res, Auckland, NZ, 72-73; vis prof, Cornell Univ, 79-80. *Mem:* Inst Food Technol; Am Chem Soc; Sigma Xi. *Res:* Composition of foods as indices of authenticity and quality; sugars; anthocyanin pigments; flavonoids; acids; color degradation; adulteration. *Mailing Add:* Dept Food Sci Ore State Univ Corvallis OR 97331

WROTENBERY, PAUL TAYLOR, b Pollok, Tex, Apr 24, 34; m 54; c 2. INFORMATION SYSTEMS. *Educ:* Univ Tex, BS, 58, MA, 62, PhD(physics, chem), 64. *Prof Exp:* Res scientist, Defense Res Labs, Tex, 58; sr scientist & proj dir, Tracor Inc, 58-64; sci consult & mgr, Int Bus Mach Corp, Dallas & Washington, DC, 64-68; dir comput serv dept, Tracor Inc, 68, sr vpres & dir, Tracor Comput Corp, 68-70; pres & chmn bd, United Systs Int, 70-74; pres, Equimatics Co, 74-76; dir, Informatics Inc, 74-76, group vpres & dir, 76-79; sr staff to Gov & Dir Budget & Planning, State of Tex, 79-82; chmn Tex State Bd Ins, 89-90; PVT INVESTR & EXEC CONSULT, 79- *Res:* Surfaces and solid-liquid interfaces; semiconductor electrolyte interface properties; signal processing; management information systems; systems analysis; strategic planning, management development. *Mailing Add:* 3411 Monte Vista Austin TX 78731

WRUCKE, CHESTER THEODORE, JR, b Portland, Ore, Oct 24, 27; m 54; c 3. GEOLOGY. *Educ:* Stanford Univ, BS, 51, MS, 52, PhD, 66. *Prof Exp:* GEOLOGIST, US GEOL SURV, 52- *Mem:* Geol Soc Am. *Res:* Mineral resource assessment and general geologic studies of areas in the western United States having complex structural settings and igneous rocks. *Mailing Add:* US Geol Surv MS-901 345 Middlefield Rd Menlo Park CA 94025

WU, ADAM YU, b Nanking, Aug 2, 46. PHYSICS, CHEMICAL PHYSICS. *Educ:* Tunghai Univ, BS, 68; Univ Chicago, MS, 71, PhD(physics), 76. *Prof Exp:* RES ASSOC CHEM, NAT RES COUN CAN, 75- *Res:* High pressure neutron scattering experiments; high pressure ultrasonic experiments. *Mailing Add:* Ctr High Tech Mat EECE Bldg Rm 125 Univ NMex Albuquerque NM 87131

WU, ALAN SEEMING, b Hong Kong, China, June 4, 40; Can citizen. PUBLIC HEALTH. *Educ:* Univ Alta, DDS, 66; Univ Toronto, DDPH, 69. *Prof Exp:* DIR DENT SERVS, OTTAWA-CARLETON REGIONAL HEALTH DEPT, 69- *Concurrent Pos:* Guest lectr, Univ Ottawa, 71- & lectr, Univ Toronto, 84-; chief surveyor, Ont Coun Community Health Accreditation, 81- *Mem:* Can Dent Asn; Can Pub Health Asn. *Mailing Add:* Dent Div Ottawa-Carleton Regional Health Dept 495 Richmond Rd Ottawa ON K2A 4A4 Can

WU, ALBERT M, b Tainan, Taiwan, Mar 28, 40; m 72. GLYCO-IMMUNOCHEMISTRY, CARBOHYDRATE CHEMIST. *Educ:* Nat Taiwan Univ, BS, 65; NY Med Col, PhD(biochem), 75. *Prof Exp:* Staff assoc glyco-immunochem, Dr E A Kabat Lab, Dept Microbiol, Col Physicians & Surgeons, Columbia Univ, New York, NY, 76-79, sr staff assoc, 79-82; assoc prof glyco-immunochem, Dept Vet Path, Tex A&M Univ, College Sta, Tex, 82-89; PROF GLYCO-IMMUNOCHEM, CHANG-GUNG MED COL, TAU-YUAN, TAIWAN, 89- *Honors & Awards:* Units Award Super Serv, USDA, 89. *Mem:* Am Soc Biochem & Molecular Biol; Soc Complex Carbohydrates; Sigma Xi. *Res:* Isolation, purification and characterization of glycoproteins; immunochemical studies on the binding properties of antibodies and lectins structural and immunochemical studies on epitopes or antigenic determinants. *Mailing Add:* Dept Molecular & Cell Biol Immunochem Lab Chang-Gung Med Col Kwei-San Tau-Yuan Taiwan

WU, ALFRED CHI-TAI, b Chekiang, China, Jan 24, 33; m 67; c 2. THEORETICAL PHYSICS. *Educ:* Wheaton Col, BS, 55; Univ Md, PhD(physics), 60. *Prof Exp:* Mem, Inst Advan Study, 60-62; from asst prof to assoc prof, 62-80, PROF PHYSICS, UNIV MICH, ANN ARBOR, 80- *Concurrent Pos:* John Simon Guggenheim Mem Found fel, 68-69. *Mem:* Am Phys Soc. *Res:* Quantum field theory; particle physics. *Mailing Add:* Dept Physics & Astron Bldg Univ Mich Rm 709 Ann Arbor MI 48109

WU, ANNA FANG, b Chengtu, China, Mar 25, 40; US citizen; m 66; c 1. MEDICINE, BIOCHEMISTRY. *Educ:* Cornell Univ, BA, 62; Mass Inst Technol, MS, 65, PhD(chem), 67; Univ Chicago, MD, 74; Am Bd Internal Med, cert, 77. *Prof Exp:* Res assoc biochem, Muscle Inst, 67-68; res assoc, Col Physicians & Surgeons, Columbia Univ, 68-71; resident, 75-77, instr, 77-79, assoc, 79-80, ASST PROF MED, MED SCH, NORTHWESTERN UNIV, 80- *Concurrent Pos:* Med dir employee health, Northwestern Mem Hosp, 79- *Mem:* Sigma Xi; Am Col Physicians; Am Occup Med Asn. *Res:* Synthetic peptides; synthetic nucleotides; fractionation of erythrocytes. *Mailing Add:* Dept Med Med Sch Northwestern Univ Chicago IL 60611

WU, CHANGSHENG, b Liaoning, Manchuria, China, Oct 3, 23; nat US; m 55; c 2. EXPLORATION GEOPHYSICS. *Educ:* Nat Southwestern Assoc Univs, China, 44; Univ Tex, BS, 47; Rice Univ, MA, 63, PhD(geophys), 66. *Prof Exp:* Seismologist, United Geophys Co, 47-51; party chief, Tex Seismog Co, 52-53; rev geophysicist, Precision Explor Co, 53-55 & Ralph E Fair, Inc, 55-56; seismic prospecting expert, Tech Assistance Admin, UN, 57-60; asst geophys, Rice Univ, 62-64; res geophysicist, Western Geophys Co, 66-69, sr res geophysicist, 69-74, sr staff scientist, 74-84, mgr interpretation & spec processing, 79-84, CHIEF SCIENTIST & MGR INTERPRETATION, CHINA, WESTERN GEOPHYS CO, LITTON INDUSTS, INC, 85- *Mem:* Soc Explor Geophysicists. *Res:* Elastic wave propagation and seismic data interpretation. *Mailing Add:* 2275 Woodland Springs Dr Houston TX 77077

WU, CHAU HSIUNG, b Taipei, Taiwan, Feb 10, 41; US citizen; m 66; c 2. PHARMACOLOGY, ELECTROPHYSIOLOGY. *Educ:* Nat Taiwan Univ, BS, 63; Univ Miami, MS, 68, PhD(pharmacol), 71. *Prof Exp:* Res assoc pharmacol, Med Ctr, Duke Univ, 74-75, asst adj prof, 75-76, asst med res prof, 76-77; asst prof, 77-82, ASSOC PROF PHARMACOL, MED SCH, NORTHWESTERN UNIV, 82- *Concurrent Pos:* Fel, Med Ctr, Duke Univ, 70-72, Muscular Dystrophy Asn fel, 72-74; NIH res career develop award, 81-86. *Mem:* Am Chem Soc; Biophys Soc; Soc Gen Physiologists; Soc Neurosci; Am Soc Pharmacol & Exp Therapeut; Int Soc Toxinology. *Res:* Mechanisms of action of neurotoxins on membrane ionic channels in nerves and muscles. *Mailing Add:* Dept Pharmacol/Searle 8-477 Northwestern Univ Med Sch 303 E Chicago Ave Chicago IL 60611

WU, CHENG-HSIAO, b Lukang, Taiwan, Feb 5, 43; US citizen; m 72; c 2. SEMICONDUCTOR DEVICES, SOLID STATE PHYSICS. *Educ:* Nat Taiwan Univ, BS, 65; Univ Rochester, MS, 67, PhD(physics), 72. *Prof Exp:* Fel physics, New York Univ, 72-74, City Col, City Univ NY, 74-75; vis scientist solid state physics, Max-Planck Inst, Stuttgart, Germany, 75-77; res assoc, Univ Rochester, 78-80; mem tech staff, RCA Lab, Princeton, NJ, 80-83; ASSOC PROF ELEC ENG, UNIV MO-ROLLA, 83- *Concurrent Pos:* Consult, Xerox Weber Res Ctr, 79-80; vis prof, Inst Theoret Physics, Univ Stattgart, Ger, 90. *Mem:* Am Phys Soc; Inst Elec & Electronics Engrs; Sigma Xi; NY Acad Sci. *Res:* Random walk theory and transport in disordered materials; semiconductor device modeling. *Mailing Add:* Dept Elec Eng Univ Mo Rolla MO 65401

WU, CHENG-WEN, b Taipei, Taiwan, June 19, 38; m 63; c 3. BIOPHYSICS, BIOCHEMISTRY. *Educ:* Nat Taiwan Univ, MD, 64; Case Western Reserve Univ, PhD(biochem), 69. *Prof Exp:* Res assoc phys biochem, Cornell Univ, 69-71; from asst prof to assoc prof biophys, Albert Einstein Col Med , 72-78, prof biochem, 78-80; prof pharmacol sci, State Univ NY Stony Brook, 80-90; DIR, INST BIOMED SCI, ACADEMIA SINICA, REPUB CHINA, 88- *Concurrent Pos:* NIH spec fel biophys, Yale Univ, 71-72; Am Cancer Soc res grant, Albert Einstein Col Med, 72-75, NIH res grant, 72-79 & res career develop award, 72-77; Irma T Hirschl Sci Award, 77-82. *Honors & Awards:* Catacosinos Prof Award Cancer Res. *Mem:* Am Soc Pharmacol & Exp Therapeut; Am Chem Soc; Am Soc Biochem & Molecular Biol; Biophys Soc; Am Asn Cancer Res. *Res:* Regulation and mechanism of gene expression; carcinogenesis; optical studies of nucleic acid and protein interaction; fast reactions in biological systems; absorption and emission spectroscopy. *Mailing Add:* Inst Biomed Sci Academia Sinica Taipei 11529 Taiwan

WU, CHIEN-SHIUNG, b Shanghai, China, May 31, 12; nat US; m 42; c 1. PHYSICS. *Educ:* Nat Cent Univ, China, BS, 34; Univ Calif, PhD, 40. *Hon Degrees:* DSc, Princeton Univ, 58, Smith Col, 59, Goucher Col, 60, Rutgers Univ, 61, Yale Univ, 67, Russell Sage Col, 71, Harvard Univ, Bard Col & Adelphi Univ, 74, Dickinson Col, 75; LLD, Chinese Univ Hong Kong, 69. *Prof Exp:* Res fel & lectr, Univ Calif, 40-42; asst prof, Smith Col, 42-43; instr, Princeton Univ, 43-44; sr scientist, Columbia Univ, 44-47, assoc, 47-52, from assoc prof to prof, 52-72, Pupin prof physics, 72-81; RETIRED. *Concurrent Pos:* Mem, Adv Comt to Dir, NIH, 75-82; hon prof, Nanking Univ, Sci & Technol Univ, Beijing Univ, Tsao Hwa Univ, Nan Kai Univ, People's Repub China & Padua Univ, Italy. *Honors & Awards:* Res Award, Res Corp, 59; Award, Am Asn Univ Women, 60; Comstock Award, Nat Acad Sci, 64; Achievement Award, Chi-Tsin Cult Found, 65; Scientist of Year, Indust Res Magazine, 74; Tom Bonner Prize, Am Phys Soc, 75; Nat Sci Medal, 75; Wolf Prize in Physics, 78; Nishina Mem Lectr, Univs Tokyo, Osaka & Kyoto, 83; Asteroid named in honor, 90. *Mem:* Nat Acad Sci; hon fel Royal Soc Edinburgh; Am Phys Soc (pres, 75); Chinese Acad Sci; fel AAAS; Am Acad Arts & Sci. *Res:* Nuclear physics; non-conservation of parity in beta decay. *Mailing Add:* 15 Clairmont Ave Apt 73 New York NY 10027

WU, CHIH, b Changsha, China, Apr 13, 36; m 66; c 4. THERMODYNAMICS. *Educ:* Cheng Kung Univ, Taiwan, BS, 57, MS, 61; Univ Ill, PhD(mech eng), 66. *Prof Exp:* Instr eng, Cheng Kung Univ, 59-61; instr, Univ Ill, 61-66; from asst to assoc prof, 66-78, PROF ENG, US NAVAL ACAD, 78- *Concurrent Pos:* Prof physics, Loyola Col, Md, 66-69; prof, Johns Hopkins Univ, 68- *Mem:* Am Soc Mech Engrs; Am Soc Eng Educ; Chinese Soc Mech Engrs. *Res:* Transport properties; ionized gas; magnetohydrodynamics; multi-phase fluid; energy conversion; computer-assisted education. *Mailing Add:* 1528 Wild Cranberry Dr Crownsville MD 21032

WU, CHING KUEI, biology, genetics, for more information see previous edition

WU, CHING-HSONG GEORGE, b Taipei, Taiwan, Apr 22, 39; m 67; c 2. CHEMICAL KINETICS & CHEMICAL VAPOR DEPOSITION, DIAMOND THIN FILMS. *Educ:* Nat Taiwan Univ, BS, 62; NMex Highland Univ, MS, 65; Univ Calif, Berkeley, PhD(chem), 69. *Prof Exp:* SR RES CHEMIST, SCI RES LAB, FORD MOTOR CO, DEARBORN, 69- *Mem:* Am Chem Soc; Mat Res Soc. *Res:* Gas phase kinetics; free radical reactions; mechanism of smog formation; energy transfer; mechanism of chemical vapor deposition; three-way catalyst research, characterization, and modeling; cvd diamond films. *Mailing Add:* 30691 Turtle Creek Dr Farmington Hills MI 48331

WU, CHING-SHENG, b Nanjing, China, Nov 11, 29; US citizen; m 61; c 1. THEORETICAL PLASMA PHYSICS, PLASMA ASTROPHYSICS. *Educ:* Nat Taiwan Univ, BSE, 54; Va Polytech Inst, MS, 56; Princeton Univ, PhD(plasma physics), 59. *Prof Exp:* Sr scientist, physics sect, Jet Propulsion Lab, Calif Inst Technol, 59-68; res prof space plasma-physics, Inst Fluid Dynamics & Appl Math, 68-76, RES PROF SPACE PLASMA PHYSICS, INST PHYS SCI & TECHNOL, UNIV MD, 76- *Concurrent Pos:* Vis mem, Atomic Energy Res Estab, UK Atomic Energy Authority, 61-62; vis assoc prof, dept math, Mass Inst Technol, 66-67 & Inst Fluid & Appl Math, Univ Md, 67; vis prof, dept physics, Nat Taiwan Univ, 68-69 & Fed Univ Rio Grande do Sul, Porto Alegre, Brazil, 77; consult, Nat Aeronaut & Space Admin, 77-84; hon prof, Chinese Acad Sci, 79. *Mem:* Am Geophys Union; NY Acad Sci; Int Union Radio Sci; Sigma Xi; fel Am Phys Soc; Fel, NY Acad Sci, 87. *Res:* Study of radio emission and major plasma processes in solar-terrestrial environment; investigation of basic plasma instabilities relevant to laboratory and space plasmas. *Mailing Add:* Inst Phys Sci & Technol Univ Md College Park MD 20742

WU, CHING-YONG, US citizen. INDUSTRIAL ORGANIC CHEMISTRY. *Educ:* Nat Taiwan Univ, BS, 55; Univ Pittsburgh, PhD(chem), 61. *Prof Exp:* Nat Res Coun Can fel chem, 63-65; fel, Mellon Inst Sci, 65-67; res chemist, Gulf Res & Develop Co, 67-71, sr res chemist, 71-83; SR SCIENTIST, KOPPERS CO, INC, 83- *Mem:* Am Chem Soc; Catalysis Soc. *Res:* Physical organic chemistry and petrochemical research; homogeneous and heterogeneous catalysis, polymer synthesis and new polymerization. *Mailing Add:* 104 Wilmer Dr Pittsburgh PA 15238-1608

WU, CHISUNG, organic chemistry, for more information see previous edition

WU, CHUEN-SHANG C, b Taiwan, July 10, 32; US citizen; m 67; c 1. PROTEIN CONFIRMATION, ENZYMOLOGY. *Educ:* Univ Calif, Berkeley, PhD(agr chem) 67. *Prof Exp:* ASSOC RES BIOCHEMIST, CARDIOVASC RES INST, UNIV CALIF, SAN FRANCISCO, 82- *Mem:* Am Soc Biochem & Molecular Biol; Protein Soc; Soc Chinese Bioscientists Am. *Res:* Conformation of peptides and proteins in surfactant solutions; structure-function relationship of acetyleholinesterase. *Mailing Add:* Cardiovasc Res Inst Univ Calif Box 0524 San Francisco CA 94143

WU, CHUN-FANG, b Fujien, China, Feb 4, 47; m 71; c 3. NEUROPHYSIOLOGY, NEUROGENETICS. *Educ:* Tunghai Univ, Taiwan, BS, 69; Purdue Univ, PhD(neurobiol), 76. *Prof Exp:* Res fel neurobiol, Cal Inst Technol, 76-79; from asst prof to assoc prof, 79-89, PROF BIOL, UNIV IOWA, 89- *Concurrent Pos:* Vis prof pharmacol, Nat Taiwan Univ, 87 & vis prof physics, Univ Tokyo, 88; assoc ed, J Neurogenetics. *Mem:* Soc Neurosci; Biophys Soc; AAAS; Sigma Xi. *Res:* Genetic analysis of neurophysiological processes; function and development of Drosophila nervous system. *Mailing Add:* Dept Biol Univ Iowa Iowa City IA 52242

WU, CHUNG, b Foochow, China, Dec 13, 19; nat US; m 50; c 4. BIOCHEMISTRY. *Educ:* Fukien Christian Univ, China, BS, 41; Univ Mich, MS, 48, PhD(biol chem), 52. *Prof Exp:* Res asst, Mayo Clin, 52-56; from instr to assoc prof biol chem, Med Sch, Univ Mich, Ann Arbor, 56-86; RETIRED. *Concurrent Pos:* Lectr, summer sem, Academea Sinica, Taiwan, 68. *Mem:* Am Soc Biol Chem; Am Chem Soc; Am Asn Cancer Res. *Res:* Mechanisms of enzyme action; enzymology of cancer; metabolic controls; drug allergy. *Mailing Add:* 731 Hillmont St Santa Rosa CA 95409

WU, CHUNG PAO, b Kwantung, China, May 15, 42; US citizen; m 71; c 3. SOLID STATE ELECTRONICS. *Educ:* Yale Univ, BS, 65, MS, 66, PhD(physics), 68. *Prof Exp:* Res physicist, Electron Accelerator Lab, Yale Univ, 68-70; asst prof physics, Nanyang Univ, 70-72; MEM TECH STAFF INTEGRATED CIRCUIT TECHNOL, SARNOFF RES CTR, RCA CORP, 73- *Mem:* Am Phys Soc; Inst Elec & Electronics Engrs. *Res:* Development of methods which accurately control the generation and implantation of ions in solids; study and characterization of ion implantation and laser annealing techniques for semiconductor device fabrication. *Mailing Add:* David Sarnoff Res Ctr Princeton NJ 08540

WU, DAISY YEN, b Shanghai, China, June 12, 02; US citizen; m 24; c 5. NUTRITION, BIOCHEMISTRY. *Educ:* Ginling Col, China, BA, 21; Columbia Univ, MA, 23; Chinese-French Acad, China, dipl, 44; UN Lang Training Course, dipl, 63. *Prof Exp:* Asst biochem, Peking Union Med Col, 23-24; res assoc biochem, Med Col Ala, 49-53; tech assoc nutrit, Food Conserv Div, UNICEF, 60-64; assoc pub health nutrit, Inst Human Nutrit, Columbia Univ, 64-71; res assoc nutrit, St Luke's Hosp Ctr, 71-86; RETIRED. *Concurrent Pos:* Consult, 86- *Mem:* Am Inst Nutrit; fel Am Pub Health Asn; fel Royal Soc Health; Sigma Xi; NY Acad Sci. *Res:* Proteins and amino acids; metabolic studies in man; vegetarian diets and dietaries; development, design and administration of an information retrieval system on obesity; author of supplement to Hsien Wu's Principles of Nutrition and Hsien Wu. *Mailing Add:* 449 E 14th St New York NY 10009

WU, DAO-TSING, b China, Nov 6, 33; m; c 1. CHEMICAL ENGINEERING. *Educ:* Univ Calif, Berkeley, BS, 54; Princeton Univ, MSE, 56; Mass Inst Technol, DSc(chem eng), 62. *Prof Exp:* Res engr, E I du Pont de Nemours & Co, Inc, 61-66, staff engr, 66-69, res supvr, 69-76, res assoc, 76-79, res fel, 79-86, sr res fel, 86-89, DU PONT FEL, E I DU PONT DE NEMOURS & CO, INC, 89- *Mem:* Am Chem Soc; Am Inst Chem Engrs; Sigma Xi. *Res:* Polymer solution thermodynamics; non-aqueous colloidal dispersion; paint research; computer modeling and simulation. *Mailing Add:* E I du Pont de Nemours & Co Inc 3500 Grays Ferry Ave Philadelphia PA 19146

WU, ELLEN LEM, b Shanghai, China, Dec 6, 30; m 54; c 2. PHYSICAL CHEMISTRY. *Educ:* Carleton Col, BA, 54; Univ Minn, PhD(phys chem), 62. *Prof Exp:* Sr res chemist, Appl Res Div, 62-71, res assoc, 71-79, SR RES ASSOC, PROCESS RES & TECH SERV DIV, MOBIL RES & DEVELOP CORP, 79- *Mem:* Am Chem Soc; Am Crystallogrs Asn. *Res:* Catalysis research; physicochemical methods employed to elucidate nature of catalysts used in hydrocarbon conversion processes. *Mailing Add:* 1203 Woodruff Rd Glassboro NJ 08028

WU, EN SHINN, b Kwangtung, China, Apr 20, 43. CHEMICAL PHYSICS. *Educ:* Nat Taiwan Univ, BS, 65; Cornell Univ, PhD(appl physics), 72. *Prof Exp:* Res assoc chem, Syracuse Univ, 72-74; asst prof, 74-79, ASSOC PROF PHYSICS, UNIV MD, BALTIMORE COUNTY, 79- *Mem:* Am Phys Soc; Biophys Soc; Sigma Xi; Optical Soc Am. *Res:* General properties of simple fluids and fluid mixtures, in particular, their thermodynamic behaviors near the critical points; the experimental techniques employed are primarily light-scattering and small angle X-ray scattering. *Mailing Add:* Dept Physics Univ Md Baltimore County Catonsville MD 21228

WU, FA YUEH, b China, Jan 5, 32; m 63; c 3. THEORETICAL PHYSICS. *Educ:* Chinese Naval Col, BS, 54; Nat Tsing Hua Univ, MS, 59; Wash Univ, PhD(physics), 63. *Prof Exp:* Res assoc physics, Wash Univ, 63; asst prof, Va Polytech Inst, 63-67; from asst prof to prof, 67-89, DISTINGUISHED PROF PHYSICS, NORTHEASTERN UNIV, 89- *Concurrent Pos:* Sr Fulbright res fel, Australian Nat Univ, 73; staff assoc, Nat Sci Found, 83-84; vis prof, Inst Lorentz, 80, Nat Tsing Hua Univ, 74 & 88, Univ Wash, 87, Ecole Polytech Fed, Laussane, 75, 78, 85, 91, Nat Taiwan Univ, 84. *Mem:* Fel Am Phys Soc. *Res:* Many body theory; theory of quantum liquids; statistical mechanics; solid state theory. *Mailing Add:* Dept Physics Northeastern Univ Boston MA 02115

WU, FELICIA YING-HSIUEH, b Taipei, Taiwan, Feb 27, 39; US citizen; m 63; c 3. BIOCHEMISTRY, BIOPHYSICS. *Educ:* Nat Taiwan Univ, BS, 61; Univ Minn, MS, 63; Case Western Reserve Univ, PhD(org chem), 69. *Prof Exp:* Med technician biochem, US Naval Med Res Unit 2, Taipei, Taiwan, 63-65; res asst, Dept Chem, Case Western Reserve Univ, 65-69; res assoc, Sect Biochem & Molecular Biol, Cornell Univ, 69-71; res assoc pharmacol, Yale Univ, 71-72; instr/assoc biophysics, Albert Einstein Col Med, 72-78, asst prof biochem, 78-79; from assoc prof to prof pharmacol sci, State Univ NY, Stony Brook, 80-90; RES FEL, INST BIOMED SCI, ACADEMIA SINICA, TAIPEI, TAIWAN, 88-, SPEC MED RES CHAIR PROF, 89- *Concurrent Pos:* Vis prof, Pasteur Inst & Inst Gustave Roussy, Paris, 79-80; Catacosinos Prof Cancer Res, Stony Brook Res Found, 80-90; adj prof pharmacol, Nat Taiwan Univ, 88-90, adj prof toxicol, 90-; grants, NSF & NIH, 72-93, NIEHS, 82-83; ad hoc reviewer, NIH & NSF. *Mem:* Am Soc Biochem & Molecular Biol; Am Biophys Soc; Am Chem Soc; Am Soc Pharmacol & Exp Therapeut; Soc Chinese Biosci Am; Am Asn Cancer Res; fel Am Inst Chemists; Sigma Xi; Chinese Biochem Soc; Chinese Soc Cell & Molecular Biol. *Res:* Role of metal ions in gene expression; mechanism of action of anti-tumor drugs; oncosuppressive mechanisms of adenoassociated virus; author or co-author of 72 publications. *Mailing Add:* Inst Biomed Sci Academia Sinica Taipei 11529 Taiwan

WU, FELIX F, b China, Dec 1, 43; US citizen. POWER SYSTEMS. *Educ:* Nat Taiwan Univ, BS, 65; Univ Pittsburgh, MS, 68; Univ Calif, Berkeley, PhD(elec eng & comput sci), 72. *Prof Exp:* Asst prof elec eng, Univ Pittsburgh, 72-74, asst prof elec eng & comput sci, Univ Calif, Berkeley, 74-78; engr, Pac Gas & Elec Co, 76-77; ASSOC PROF ELEC ENG & COMPUT SCI, UNIV CALIF, BERKELEY, 78- *Concurrent Pos:* Consult, Pac Gas & Elec Co, 77-, Solar Energy Res Inst, 80-81, Elec Power Res Inst, 81-82; vis prof, Shanghai Jiao Tung Univ, 80, Swiss Fed Inst Technol, 82. *Mem:* Inst Elec & Electronics Engrs; AAAS. *Res:* Analysis methods for electric power systems planning and operations. *Mailing Add:* Dept Elec Eng & Comput Sci Univ Calif Berkeley CA 94720

WU, FRANCIS TAMING, b Shanghai, China, May 27, 36; m 66; c 1. GEOPHYSICS. *Educ:* Nat Taiwan Univ, BS, 59; Calif Inst Technol, PhD(geophys), 66. *Prof Exp:* Asst prof geophys, Boston Col, 68-69; from asst prof to assoc prof, 70-76, PROF GEOPHYS, STATE UNIV NY, BINGHAMTON, 76-, CHMN, DEPT GEOL SCI & ENVIRON STUDIES, 89- *Concurrent Pos:* Bd dirs, Seismol Soc Am, 86- *Mem:* AAAS; Am Geophys Union; Seismol Soc Am. *Res:* Faulting as a dynamic phenomenon, its seismic radiation, rate of growth and driving mechanism; near source strong ground motion; fault gauge and mechanics of faulting; crustal structures. *Mailing Add:* Dept Geol State Univ NY Binghamton NY 13902-6000

WU, HAI, b Tunghai, Kiangsu, China, Aug 22, 36; US citizen; m 66; c 3. MECHANICAL ENGINEERING, APPLIED MECHANICS. *Educ:* Nat Cheng Kung Univ, Taiwan, BS, 54; Univ Iowa, MS, 63; Case Inst Technol, PhD(fluid mech), 69. *Prof Exp:* Mech engr HVAC design, Syska & Hennessy, Engrs, 65; sr res engr fluid mech heat transfer, Mech Res Dept, Sci Res Lab, 69-75, prin res assoc engr engine analysis, 76-80, prin staff engr, comput modeling & simulation, Control Systs Dept, 80-83, PRIN STAFF ENGR, FLUID & STRUCT DYNAMICS, COMPUT AIDED ENG DEPT, SYSTS RES LAB, ENG & RES STAFF, FORD MOTOR CO, 84- *Concurrent Pos:* Vis lectr, Detroit Inst Technol, 74. *Mem:* Sigma Xi; Am Soc Mech Engrs. *Res:* Energy management, engine research, automotive safety, lubrication, squeeze films, computer modeling, system optimization, and structural dynamics. *Mailing Add:* Eng Computer Ctr Mail Drop 4 PO Box 2053 20000 Rotunda Dr Dearborn MI 48121

WU, HENRY CHI-PING, b Fenghua, Chekiang, China, May 21, 35; m 65; c 3. BIOCHEMISTRY. *Educ:* Nat Taiwan Univ, MD, 60; Harvard Univ, PhD(biochem), 66. *Prof Exp:* Jane Coffin Childs Fund fel, Mass Inst Technol, 66-67, Med Found fel, 67-69; asst prof to prof microbiol, Univ Conn Health Ctr, 69-80; PROF MICROBIOL, UNIFORMED SERV UNIV HEALTH SCI CTR, 80- *Mem:* AAAS; Am Soc Microbiol; Am Soc Biol Chemists. *Res:* Biogenesis of membrane proteins in bacteria; biochemical and genetic studies of cell surface. *Mailing Add:* Dept Microbiol Uniformed Serv Univ Health Sci Ctr 4301 Jones Bridge Rd Bethesda MD 20814-4799

WU, HOFU, b Taipei, Taiwan, Mar 28, 49; c 2. SOLAR ENERGY. *Educ:* Tamkang Univ, BArch, 71; Univ Ill, MArch, 75. *Prof Exp:* Lectr archit, Ill Ctr Col, Peoria, 76-77; lectr, Univ Mich, 78-81, asst prof archit, 81-84; ASST PROF SOLAR ARCHIT, ARIZ STATE UNIV, 84- *Concurrent Pos:* Researcher, archit & planning res lab, Univ Mich, 80-84; dir, environ test lab, Col Archit & Environ Design, 84- *Mem:* Am Soc Heating, Refrigerating & Air Conditioning; Am Inst Architects; Asn Comput Mach; Int Solar Energy Soc. *Res:* Computer aided energy design for architecture and passive solar heating and cooling strategies for hot and arid region; energy management of large institutional buildings and automatic controls of intelligent buildings. *Mailing Add:* Dept Archit Ariz State Univ Tempe AZ 85287

WU, HSIN-I, b Tokyo, Japan, May 25, 37; m 64; c 1. ECOLOGICAL PHYSICS, BIOSYSTEMS MODELING. *Educ:* Tunghai Univ, Taiwan, BS, 60; Univ Mo, MS, 64, PhD(physics), 67; Tex A&M Univ, MS, 77. *Prof Exp:* From asst prof to assoc prof physics, Southeast Mo State Univ, 67-76; res assoc, Tex A&M Univ, 76-77, res scientist, 77-78, sr res scientist biosysts, 78-80; from asst prof to assoc prof bioeng, 80-87, PROF BIOENG, TEX A&M UNIV, 87- *Concurrent Pos:* Adj sr res scientist, Gansu Grassland Ecol Res Inst, Lanzhou, China, 88- *Mem:* Sigma Xi; Int Soc Ecol Modeling. *Res:* Ecological physics; application of saturated rate kinetics in ecological modeling; applied nonlinear dynamics in ecological modelling. *Mailing Add:* Ctr Biosysts Modelling Tex A&M Univ College Station TX 77843-3131

WU, HUNG-HSI, b Hong Kong, May 25, 40; US citizen; m 76; c 1. MATHEMATICS. *Educ:* Columbia Col, AB, 61; Mass Inst Technol, PhD(math), 63. *Prof Exp:* Res assoc math, Mass Inst Technol, 63-64; mem, Inst Advan Study, 64-65; from asst prof to assoc prof, 65-73, PROF MATH, UNIV CALIF, BERKELEY, 73- *Concurrent Pos:* Alfred P Sloan fel, 71-73. *Mem:* Am Math Soc. *Res:* Differential geometry; complex manifolds. *Mailing Add:* Dept Math Univ Calif Berkeley CA 94720

WU, I-PAI, b Chingkaing Kaingsu, China, June 23, 33; m 63; c 2. HYDROLOGY, HYDRAULICS. *Educ:* Nat Taiwan Univ, BS, 55; Purdue Univ, MS, 60, PhD(civil eng), 63. *Prof Exp:* Hydraul engr, Ind Flood Control & Water Resources Comn, 61-62 & 63-64; asst prof civil eng, Chico State Col, 64-66; from asst prof to assoc prof agr eng, 66-76, assoc agr engr, 71-80, PROF AGR ENG, UNIV HAWAII, 76- *Concurrent Pos:* Fulbright prof, Dept Agr Eng, Univ Khartoum, Sudan, 79-80, Fac Agr Eng, Israel Inst Technol, Technion, Haifa, Israel, 86. *Mem:* Am Soc Agr Engrs; Am Soc Civil Engrs; Am Water Resources Asn. *Res:* Small watershed hydrology; hydraulics of surface irrigation; sprinkler irrigation; drip irrigation system design. *Mailing Add:* Dept Agr Eng Univ Hawaii Manoa Honolulu HI 96822

WU, JAMES CHEN-YUAN, b Nanking, China, Oct 5, 31; nat US; m 57; c 2. MECHANICAL ENGINEERING. *Educ:* Gonzaga Univ, BS, 54; Univ Ill, MS, 55, PhD(mech eng, appl math), 57. *Prof Exp:* Mech engr, Wah Chang Corp, NY, 54; mem res staff, Mass Inst Technol, 57; asst prof, Gonzaga Univ, 57-59; chief res br, Douglas Aircraft Co, 59-65; PROF AEROSPACE ENG, GA INST TECHNOL, 65- *Concurrent Pos:* European Atomic Energy Comn sr vis fel, Ispra Res Ctr, Italy, 62-63; consult, Lockheed-Georgia Co, 76- *Mem:* Am Soc Eng Educ; Am Inst Aeronaut & Astronaut; Am Astronaut Soc; Sigma Xi. *Res:* Gas dynamics; thermodynamics; boundary layer theory; viscous flows; numerical analysis. *Mailing Add:* Sch Aerospace Eng Ga Inst Technol Atlanta GA 30332

WU, JIA-HSI, b Formosa, July 6, 26; m 56; c 1. PLANT PHYSIOLOGY, VIROLOGY. *Educ:* Univ Taiwan, BA, 50; Cornell Univ, MS, 52; Wash Univ, PhD(bot), 58. *Prof Exp:* Instr plant physiol, Univ Taiwan, 52-55; fel, Univ Wis, 58-59; asst botanist, Univ Calif, Los Angeles, 59-63; asst prof plant physiol, Tex Tech Col, 63-65; asst biologist, Univ Calif, San Diego, 65-66; asst prof cell physiol, 66-72, assoc prof cell physiol, 72-80, PROF BIOL SCI, CALIF STATE POLYTECH UNIV, POMONA, 80- *Concurrent Pos:* NSF grants, 64-68 & 70-72. *Res:* Cell physiology. *Mailing Add:* Dept Biol Calif State Polytech Univ Pomona 3801 W Temple Ave Pomona CA 91768

WU, JIANN-LONG, b Chang-hua, Taiwan. RADIOCHEMISTRY, RADIOPHARMACEUTICALS. *Educ:* Fu-Jen Cath Univ, Taiwan, BS, 68; Northeast La Univ, MS, 72; Va Polytech Inst & State Univ, PhD(chem), 77. *Prof Exp:* Instr anal chem, Chem Dept, Northeast La Univ, 72; fel, Nuclear Med Dept, Med Ctr, Univ Mich, 77-79; res scientist, 79-84, MGR RES & DEVELOP, MEDI-PHYSICS INC, SUBSID HOFFMANN-LA ROCHE, 84- *Mem:* Soc Nuclear Med; Am Chem Soc. *Res:* Design, synthesis and development of radiolabeled compounds for organ imaging and biological function studies. *Mailing Add:* 2570 Royal Oaks Dr Alamo CA 94507

WU, JIN, b Nanking, China, Apr 9, 34; m 61; c 3. FLUID MECHANICS, HYDRAULIC ENGINEERING. *Educ:* Nat Cheng Kung Univ, Taiwan, BSc, 56; Univ Iowa, MSc, 61, PhD(mech, hydraul), 64. *Prof Exp:* Res scientist, Hydronautics, Inc, 63-66, sr res scientist, 66-69, head fluid motions div, 66-72, prin res scientist, 69-74, head geophys fluid dynamics div, 72-74; from assoc prof to prof, 74-80, H FLETCHER BROWN PROF MARINE STUDIES & CIVIL ENG, UNIV DEL, 80-, DIR, AIR-SEA INTERACTION LAB, 80- *Concurrent Pos:* Consult, Hydronautics, Inc, 74- & US Naval Res Lab, 80-; adv, Taiwan Hydraulics Lab, Nat Cheng-Kung Univ, Taiwan, 75-; hon prof, Shandong Col Oceanog, China, 81- *Mem:* Am Geophys Union; Am Soc Civil Engrs; Academia Sinicia. *Res:* Geophysical and environmental fluid dynamics; air-sea interaction; coastal and ocean engineering. *Mailing Add:* Air-Sea Interaction Lab Col Marine Studies Univ Del Lewes DE 19958

WU, JIN ZHONG, b Chengdu, China, Jan 5, 45; m 76; c 1. DENSITY FUNCTIONAL THEORY, COMPUTATIONAL PHYSICS. *Educ:* Beijing Univ, China, BS, 67; Univ Sci & Technol China, MS, 81; Univ Cincinnati, MS, 83, PhD(physics), 88. *Prof Exp:* Assoc engr, Beijing Semiconductor Devices Inst, 70-79; asst researcher, Inst Semiconductors, Chinese Acad Sci, 81; postdoctoral assoc, 88-91, ASST RES SCIENTIST, QUANTUM THEORY PROJ, UNIV FLA, 91- *Mem:* Am Phys Soc. *Res:* Electronic properties and structures in films; energy deposition of massive swift particles in ultra-thin films; renopmalization of anisotropic Fermi surface of two-dimensional electron systems; semiconductor devices physics and microelectronics; develop films stopping fortran code. *Mailing Add:* Quantum Theory Proj, Williamson Hall Univ Fla Gainesville FL 32611-2085

WU, JOHN NAICHI, b Soochow, China, Sept 10, 32; US citizen; m 61; c 2. MATERIALS SCIENCE ENGINEERING. *Educ:* Nat Taiwan Univ, BSME, 55; Univ Fla, MSEM, 61, PhD(eng sci), 65. *Prof Exp:* Sr res specialist appl mech, Babcock & Wilcox Co Res Ctr, 62-66, group supvr, 67-77; MGR, MAT & PROCESSES LAB, GEN ELEC CO, 77- *Concurrent Pos:* Chmn, Div Cable & Wire Comt, Tech Paper Comt, chief, Energy Design Rev, TSBO, Gen Elec Co; bd dirs, Erie Int Inst. *Mem:* Am Acad Mech; Am Soc Mech Engrs; Am Soc Acoust; Am Soc Metals. *Res:* Theoretical and experimental mechanics; mechanics of sandwich structures; finite element analysis; nonlinear mechanics and dynamics; earthquake engineering in terms of structural design; nuclear steam generating systems. *Mailing Add:* 139 Putnam Dr Erie PA 16509

WU, JONATHAN TZONG, b Taipei, Taiwan, Dec 7, 51; US citizen; m 75; c 1. GEOSYNTHETICS, NUMERICAL METHODS. *Educ:* Nat Taiwan Univ, BS, 74; Va Polytech Inst & State Univ, MS, 76; Purdue Univ, PhD(civil eng), 80. *Prof Exp:* Asst prof, 80-86, ASSOC PROF GEOTECH ENG, UNIV COLO, DENVER, 86- *Concurrent Pos:* Consult, Colo Dept Hwy, 84-, sr engr, 85; prin investr, Geosynthetic Res Projs, 84-; mem, Soil Placement & Improv Comt, Am Soc Civil Engrs, 87- & Geosynthetic Comt, Nat Res Coun, 91-; vis prof geotech eng, Univ Tokyo, 89. *Mem:* Am Soc Civil Engrs; Int Soc Soil Mech & Found Eng; Int Geotextile Soc; NAm Geotextile Soc; Am Drill Shaft Soc. *Res:* Geosynthetics in earth reinforcement and drainage applications; application of finite element methods in geotechnical engineering; seepage and ground water flow and their effects on earth structures. *Mailing Add:* 15916 E Mercer Circle Aurora CO 80013

WU, JOSEPH M, b China, Aug 1, 47; US citizen; m 75; c 2. GENE REGULATION, DEVELOPMENTAL CONTROL. *Educ:* McGill Univ, BS, 70; Fla State Univ, MS, 72, PhD(biol sci), 75. *Prof Exp:* Fel, Temple Univ, 76-77, res instr, 77-78; from asst prof to assoc prof, 78-87, PROF BIOCHEM, NY MED COL, 87- *Concurrent Pos:* Prin investr, NIH grants, 79- *Res:* Mechanism of eukaryotic protein synthesis regulation; molecular mechanism of interferon action. *Mailing Add:* Dept Biochem Basic Sci Bldg NY Med Col Rm 127 Valhalla NY 10595

WU, JOSEPH WOO-TIEN, b Taiwan; US citizen; c 2. ORGANIC CHEMISTRY, ENZYMOLOGY. *Educ:* Tainan Cheng-Kung Univ, BS, 65; Worcester Polytech Inst, MS, 69; Univ Pa, PhD(chem), 72. *Prof Exp:* Fel enzymol dept biochem & human genetics, Univ Pa, 72-74; sr chemist bio-org chem, New Eng Nuclear Co, 74-76; SR SCIENTIST IMMUNOCHEM, INSTRUMENTATION LAB INC, 76- *Mem:* Am Chem Soc. *Res:* Modification of protein enzyme surface; organic and enzymatic reaction mechanism and analysis of biochemical compounds by antibody, enzyme and fluorometer. *Mailing Add:* Hoffmann-La Roche Inc 340 Kingsland St Nutley NJ 07110-1199

WU, JULIAN JUH-REN, b Shanghai, China, June 15, 35; US citizen; m. NUMERICAL METHODS, FINITE ELEMENT METHODS. *Educ:* Taiwan Univ, BS, 58; Rice Univ, MS, 62; Columbia Univ, ME, 66; Rensselaer Polytech Inst, PhD(mech), 70. *Prof Exp:* Sr scientist composites, Advan Technol Div, Avco Co, 66-67; proj engr eng anal, Teledyne Advan Mat Co, 67-68; res assoc mech, Rensselaer Polytech Inst, 68-71; res engr struct dynamics, US Army Watervliet Arsenal, 71-76, res mathematician numerical methods, 76-83; PROG MGR APPL MATH & MATH SCI, US ARMY RES OFF, 83- *Concurrent Pos:* Adj prof lasticity & vibrations, Rensselaer Polytech Inst, 76-83; adj prof, Dept Mech Eng, Duke Univ, 84- *Mem:* Soc Appl & Indust Math; Am Soc Mech Engrs; NY Acad Sci; Am Acad Mechanics. *Res:* Analytical and numerical research on nonlinear structural dynamics and stability; nonlinear continuum mechanics; perturbation methods. *Mailing Add:* US Army Res Off PO Box 12211 Research Triangle Park NC 27709-2211

WU, JUNG-TSUNG, b Taiwan, China, Feb 17, 36; US citizen; m 65; c 2. REPRODUCTIVE BIOLOGY, EMBRYOLOGY. *Educ:* Nat Taiwan Univ, BS, 58; Univ Wis-Madison, MS, 66, PhD(endocrinol & reprod biol), 69. *Prof Exp:* Teaching asst zool, Nat Taiwan Univ, 60-62; fel reprod biol, Med Ctr, Univ Kans, 69-71; res assoc, Worcester Found Exp Biol, 71-72, staff scientist reprod biol, 72-85; ASST PROF OBSTET & GYNEC, MED SCH, UNIV MASS, 85- *Mem:* Soc Study Reprod; Am Fertil Soc. *Res:* Implantation; embryo development and transport; hybridization; pathenogenesis; fertilization. *Mailing Add:* 47 Floral St Shrewsbury MA 01545

WU, KENNETH KUN-YU, b Kaohsiung, Taiwan, July 6, 41; US citizen; m 69; c 2. MEDICAL SCIENCE, BIOLOGY. *Educ:* Yale Univ, MS, 69; Nat Taiwan Univ, MD, 66. *Prof Exp:* From instr to asst prof, Univ Iowa, 73-76; assoc prof & chief coagulation & thrombosis unit, 76-81, PROF MED, RUSH MED COL, RUSH PRESBY ST LUKE'S MED CTR, 81- *Concurrent Pos:* Adv consult, NIH, 76-, mem prog proj rev comt, 76- *Mem:* Am Asn Immunol; Am Soc Hemat; Am Fedn Clin Res; Int Soc Thrombosis & Hemostasis. *Res:* Thrombosis and hemostasis; platelet physiology; pathophysiology and biochemistry. *Mailing Add:* Div Hemat & Oncol Med Dept Univ Texas Med Sch PO Box 20708 Houston TX 77225

WU, KONRAD T, b Canton, China, July 1, 48; US citizen; m 74; c 2. GAS-PHASE SPECTROSCOPY, MOLECULAR BEAM CHEMISTRY. *Educ:* Fu-Jen Catholic Univ, Taiwan, BS, 70; State Univ NY, Albany, PhD(chem), 76. *Prof Exp:* Postdoctoral res, Univ Tex, Austin, 76-77 & Columbia Univ, 77-78; chemist, Mt Sinai Sch Med, 78-80; asst prof, 80-87, ASSOC PROF CHEM, STATE UNIV NY, OLD WESTBURY, 87- *Concurrent Pos:* Vis scholar, Dept Chem, Wesleyan Univ, 83-85; vis prof, State Univ NY, Stony Brook, 87-88. *Mem:* Fel Int Biog Soc; Am Chem Soc; Am Phys Soc; Sigma Xi. *Res:* Gas-phase spectroscopy; molecular reaction dynamics; studies of flowing-afterglow and chemiluminescence. *Mailing Add:* Dept Chem State Univ NY Old Westbury NY 11568

WU, KUANG MING, b Taiwan, Nov 30, 49; m 74. STRUCTURAL MECHANICS, CONTINUUM MECHANICS. *Educ:* Nat Taiwan Univ, BS, 71; Princeton Univ, MS, 74, PhD(civil eng), 77. *Prof Exp:* Assoc sr res engr, 77-, STAFF RES ENGR, GEN MOTORS RES LABS. *Res:* Structural dynamics; elasticity; composite materials. *Mailing Add:* Thermosci Gen Motors Res Labs 30500 Mound Rd PO Box 9055 Warren MI 48090

WU, LANCELOT T L, b Taiwan, China. DIGITAL SIGNAL PROCESSING, INFORMATION THEORY. *Educ:* Nat Taiwan Univ, BS, 74; State Univ NY, Albany, MS, 78; Columbia Univ, PhD(statist), 82. *Prof Exp:* Mem tech staff, Bell Labs, 82-83, MEM TECH STAFF RES, BELL COMMUN RES, 83- *Mem:* Inst Elec & Electronics Engrs; Sigma Xi. *Res:* Statistical communications; digital communications; digital signal processing. *Mailing Add:* 19 Alpine Dr Morris Township NJ 07960

WU, LEI, b Beijing, People's Repub China, Jan 14, 63. DYNAMICS PROPERTIES, PHASE TRANSITION. *Educ:* Peking Univ, BS, 83; Univ Chicago, PhD(physics), 90. *Prof Exp:* POSTDOCTORAL FEL CHEM & CTR MAT SCI & ENG, MASS INST TECHNOL, 90- *Mem:* Am Phys Soc. *Res:* Dynamical properties in glassy material and complex systems; phase transition in liquid crystal, surfactant, polymer, gel and protein systems; computer simulation of sandpile to show universality and scaling of self-organized criticality; dielectric, quadrupolar and specific heat measurements. *Mailing Add:* Mass Inst Technol Rm 13-2062 77 Massachusetts Ave Cambridge MA 02139

WU, LILIAN SHIAO-YEN, b Peiking, China, July 6, 47. APPLIED MATHEMATICS. *Educ:* Univ Md, BS, 68; Cornell Univ, MS, 72, PhD(appl math), 74. *Prof Exp:* RES STAFF APPL MATH, THOMAS J WATSON RES CTR, IBM CORP, 73- *Concurrent Pos:* Vis scientist, Marine Biol Lab, 75-77; adj prof, NY Univ Grad Sch Bus, 88- *Mem:* Am Statist Asn; Int Inst Forecasters. *Res:* Business planning and forecasting. *Mailing Add:* BB5 Bedford Mews 208 Harris Rd Bedford Hills NY 10507

WU, MING-CHI, b Nantou, Taiwan, Nov 13, 40; US citizen; m 68; c 2. BIOCHEMISTRY. *Educ:* Nat Taiwan Univ, BS, 63; Univ Wis-Madison, MS, 68, PhD(biochem), 70. *Prof Exp:* Res fel physiol chem, Sch Med, Johns Hopkins Univ, 69-71; res assoc biochem, Sch Med, Univ Pittsburgh, 71-73; res assoc hemat, Howard Hughes Med Inst, 74-80, asst prof med, Sch Med, Univ Miami, 75-80, assoc prof, 81-82; ASSOC PROF BIOCHEM, TEX COL OSTEOPATH MED, 82-; ASSOC PROF & VCHMN, DEPT BIOCHEM, UNIV NTEX, 82- *Mem:* Am Chem Soc; Am Fedn Clin Res; Sigma Xi; Am Soc Biol Chemists; Int Soc Exp Ment Hematol. *Res:* Control of granulopoiesis; proteases from cultured cancer cells. *Mailing Add:* Dept Biochem Univ NTex Tcom Denton TX 76203

WU, MIN-YEN, b Taiwan, China, Oct 1, 40; m 68; c 1. ELECTRICAL ENGINEERING, COMPUTER SCIENCE. *Educ:* Nat Taiwan Univ, BS, 62; Univ Ottawa, MSc, 65; Univ Calif, Berkeley, PhD(elec eng), 68. *Prof Exp:* Actg asst prof elec eng, Univ Calif, Berkeley, 68-69; asst prof, 69-74, ASSOC PROF ELEC ENG, UNIV COLO, BOULDER, 74- *Concurrent Pos:* Independent prof, IBM Corp, Boulder, 72-74, consult, 74-75. *Mem:* AAAS; Inst Elec & Electronics Engrs. *Res:* Control and system theory; social and economical systems; mathematical ecology. *Mailing Add:* Dept Elec & Engr Ecee 0-02 Univ Colo Boulder CO 80309

WU, MU TSU, b Changhwa, Taiwan, Oct 25, 29; US citizen; m 57; c 4. ORGANIC CHEMISTRY, MEDICINAL CHEMISTRY. *Educ:* Nat Taiwan Univ, BS, 51; Univ Md, PhD(pharmaceut chem), 61; Tohoku Univ, Japan, DSc(chem), 61. *Prof Exp:* Res chemist, Ord Res Inst, 51-58; res assoc pharmaceut chem, Univ Md, 58-62, assoc res prof, 64-65; res assoc chem, Univ NH, 62-64; sr res chemist, 65-72, res fel, 72-78, SR RES FEL, MERCK & CO, INC, 78- *Mem:* AAAS; Am Chem Soc; Am Inst Chemists. *Res:* Synthetic organic and medicinal chemistry. *Mailing Add:* 35 Lance Dr Clark NJ 07066-2717

WU, PEI-RIN, b Taoyuan, Taiwan, Feb 16, 35; US citizen; m 58; c 3. ELECTRICAL ENGINEERING, ELECTROMAGNETISM. *Educ:* Taipei Inst Technol, dipl, 55; Univ Tenn, MS, 60; Univ Mich, PhD(elec eng), 67. *Prof Exp:* Engr elec, Sintong Chem Works Inc, 56-59; asst prof elec eng, SDak Sch Mines & Technol, 60-64; res asst, Radiation Lab, Univ Mich, 66-67; SR STAFF ELECTROMAGNETICS, LINCOLN LAB, MASS INST TECHNOL, 67- *Mem:* Inst Elec & Electronics Engrs. *Res:* Antennas; electromagnetics; scattering; radar data analysis; radar decoy designs; reentry physics and signatures; satellite signatures; identification; signature discrimination techniques; near-field measurement techniques. *Mailing Add:* Hiddenwood Path Lincoln MA 01773

WU, RAY J, b Peking, China, Aug 14, 28; nat US; m 56; c 2. BIOCHEMISTRY. *Educ:* Univ Ala, BS, 50; Univ Pa, PhD, 55. *Prof Exp:* Asst instr biochem, Univ Pa, 51-55, Damon Runyon fel cancer res, 55-57; from asst to assoc, Pub Health Res Inst New York, 57-61, assoc mem, 61-66; assoc prof, 66-72, assoc chmn sect, 75-77, chmn sect, 77-79, PROF BIOCHEM, MOLECULAR & CELL BIOL, CORNELL UNIV, 72- *Concurrent Pos:* NSF sr fel, MRC Lab, Cambridge, England, 71; vis assoc prof, Mass Inst Technol, 72; hon res prof, Academia Sinica, China, 82-; hon prof, Peking Med Col, 83-, Fudan Univ, 83- *Mem:* Am Chem Soc; Sigma Xi; Am Soc Biol Chem; fel Chinese Acad Sci. *Res:* DNA sequence analysis; cancer research; recombinant DNA research; gene synthesis; enzymology. *Mailing Add:* Dept Biochem 316 Biotechnol Bldg Cornell Univ Ithaca NY 14853

WU, RICHARD LI-CHUAN, b Tainan, Taiwan, Aug 21, 40; m 68; c 1. MATERIALS SCIENCE ENGINEERING. *Educ:* Nat Cheng Kung Univ, Taiwan, BS, 63; Univ Kans, PhD(chem), 71. *Prof Exp:* Res chemist, Aerospace Res Labs, Wright-Patterson Air Force Base, 71-75; adj res prof eng, Chem & Brehm Lab, Wright State Univ, 75-86; SR SCIENTIST, UES, INC, 87- *Mem:* Am Chem Soc; Am Mat Res Soc; Sigma Xi. *Res:* Diamond and diamond-like carbon films; thin film coating; mass spectrometry; high temperature chemistry; vaporization processes; ion-molecule reactions; thermodynamics. *Mailing Add:* 384 Merrick Dr Xenia OH 45385

WU, ROBERT CHUNG-YUNG, b Kao-Hsiung, Taiwan, Oct 16, 43; US citizen; m 69; c 2. EXTREME ULTRAVIOLET SPECTROSCOPY, LASER NONLINEAR SPECTROSCOPY. *Educ:* Nat Taiwan Norm Univ, BS, 68; Univ Ill, MS, 70, PhD(chem), 73. *Prof Exp:* Postdoctoral fel, Chem Dept, Univ Iowa, 76-77; postdoctoral fel, Chem Dept, Univ Southern Calif, 73-75, res scientist, Physics Dept, 77-78, asst res prof, 78-88, ASSOC RES PROF, SPACE SCI CTR, UNIV SOUTHERN CALIF, 88- *Concurrent Pos:* Consult, Res Div, Nat Tech Systs, 80-82, Rocketdyne Div, Rockwell Int Co, 84-85 & Plasma Lab, Univ Calif, Los Angeles, 88. *Mem:* Optical Soc Am; Am Geophys Union; Soc Appl Spectros. *Res:* Molecular autoionization; molecular photodissociation; nonlinear laser spectroscopy; ultraviolet laser generation; photochemical processes in planetary environments; aeronomy. *Mailing Add:* Space Sci Ctr & Dept Physics Univ Southern Calif Los Angeles CA 90089-1341

WU, ROY SHIH-SHYONG, b Shanghai, China, Nov 15, 44; US citizen. CELL BIOLOGY, BIOCHEMISTRY. *Educ:* Univ Calif, Berkeley, AB, 67; Albert Einstein Col Med, Bronx, NY, PhD(biochem), 72. *Prof Exp:* NIH fel develop biol, Dept Zool, Univ Calif, Berkeley, 72-74; fel cell biol, Children's Hosp, Oakland, Calif, 74-75; scientist cell biol, Biotech Res Lab Inc, Rockville, Md, 75-80; MEM STAFF, NIH, 80- *Concurrent Pos:* Lectr, Dept Zool, Univ Calif, Berkeley; peer reviewer grants, NSF & NIH. *Mem:* AAAS; Am Soc Cell Biol; Am Chem Soc; Am Soc Biol Chemists; Am Asn Cancer Res. *Res:* Studying the molecular mechanism of coordination of protein and DNA synthesis during cell proliferation; development of new assays for DNA repair enzymes, the methytransferases. *Mailing Add:* CTEP/DCT/NCI/NIH EPN Rm 734 Bethesda MD 20892

WU, SAU LAN YU, b Hong Kong, China; US citizen; m 67. EXPERIMENTAL HIGH ENERGY PHYSICS. *Educ:* Vassar Col, BA, 63; Harvard Univ, MA, 64, PhD(physics), 70. *Prof Exp:* Res assoc, Mass Inst Technol, 70-72, res physicist, 72-77; from asst prof to assoc prof, 77-83, PROF, PHYSICS DEPT, UNIV WIS-MADISON, 83- *Concurrent Pos:* Vis scientist, Deutsches Elektronen-Synchrotron, 70-72, 77-86, Brookhaven Nat Lab, 72-75, Europ Orgn Nuclear Res, 75-77, 81-; prin investr, US Dept Energy, 77-, investr award, 80; Romnes fel, Univ Wis, 81; Enrico Fermi prof physics, Univ Wis, 90, Hilldale prof, 91. *Mem:* Am Phys Soc. *Res:* Electron-positron colliding beam physics at high energies using the detector TASSO and the machine PETRA to obtain the first evidence of three jet events which signified the existence of gluons; preparation and construction of the new detector ALEPH to study the production and decay of the neutral intermediate boson and the production of the charged intermediate boson. *Mailing Add:* Dept Physics Univ Wis 1150 University Ave Madison WI 53706

WU, SHERMAN H, b Hupeh, China, Aug 21, 38; US citizen. ELECTRICAL & SYSTEMS ENGINEERING. *Educ:* Northwestern Univ, BSEE, 61, MS, 63, PhD(elec eng), 65. *Prof Exp:* Asst engr, Ill Bell Tel Co, 61-62; lectr systs eng, Univ Ill, Chicago, 64; staff engr, TRW Systs Group, Calif, 65; from asst prof to assoc prof elec eng, 65-76, PROF ELEC ENG, MARQUETTE UNIV, 76- *Concurrent Pos:* NSF grant, 66-68. *Mem:* Inst Elec & Electronics Engrs. *Res:* Nonlinear pulse-modulation in aerospace and physiological systems. *Mailing Add:* Dept Elec Eng & Comput Sci Marquette Univ Milwaukee WI 53233

WU, SHI TSAN, b Nanchang, China, July 31, 34; m 64; c 3. AEROSPACE ENGINEERING. *Educ:* Nat Taiwan Univ, BS, 56; Ill Inst Technol, MS, 59; Univ Colo, PhD(aerospace eng sci), 67. *Prof Exp:* Res asst eng sci, Harvard Univ, 59-62; res fel aeronaut, NY Univ, 62-63; asst aerospace eng sci, Univ Colo, 63-64; res asst solar physics, High Altitude Observ, Nat Ctr Atmospheric Res, 64-67; from asst prof to assoc prof, 67-72, PROF MECH ENG, UNIV ALA, HUNTSVILLE, 72-, ADJ PROF PHYSICS, 73- *Concurrent Pos:* Consult, Wyle Labs, 68-; sr Fulbright-Hays Scholar, 75-76; Australia-Am Educ Found prof space physics, La Trobe Univ, Australia, 75-76; solar physics coordr, Study of Interplanetary Phenomena & Spec Comt on Solar Terrestrial Physics; dir, Ctr Space Plasma & Aeronomic Res, 86-; distinguished prof, Mech/Aeronaut, 90- *Honors & Awards:* Flag Award, Am Inst Aeronaut & Astronaut. *Mem:* AAAS; Am Phys Soc; Am Geophys Union; fel Am Inst Aeronaut & Astronaut. *Res:* Plasmadynamics; magnetohydrodynamics and its astrogeophysical applications; boundary layer type flows; kinetic theory; radiative gas dynamics and other fluid mechanics problems; numerical methods, computational field mechanics. *Mailing Add:* Ctr Space Plasma & Aeronomic Res Univ Ala Huntsville AL 35899

WU, SHIEN-MING, b Chekiang, China, Oct 28, 24; m 59; c 2. MECHANICAL ENGINEERING. *Educ:* Univ Wis, PhD(mech eng), 62. *Prof Exp:* Asst prof mech eng, 62-65, assoc prof, 65-68, prof mech eng & statist, 68-80, RES PROF, UNIV WIS-MADISON, 81- *Honors & Awards:* C R Richards Award, Am Soc Mech Engrs, 81. *Mem:* Am Soc Mech Engrs; Soc Mfg Engrs. *Res:* Drilling and metal cutting; engineering statistics; stochastic processes; system analysis. *Mailing Add:* Dept MEAM Brown 3424C GG Univ Mich Ann Arbor MI 48109-2125

WU, SING-CHOU, b China, June 2, 36; m 64; c 2. STATISTICS, ECONOMETRICS. *Educ:* Nat Taiwan Univ, BA, 59; Utah State Univ, MS, 66; Colo State Univ, PhD(statist), 70. *Prof Exp:* Economist, Bank of China, 61-63; programmer, Comput Ctr, Utah State Univ, 65-66; asst statist, Colo State Univ, 66-69; from asst prof to assoc prof, 69-75, PROF STATIST, CALIF STATE POLYTECH UNIV, SAN LUIS OBISPO, 75- *Mem:* Am Statist Asn; Chinese Statist Soc; Japan Statist Soc. *Res:* Design of experiment; statistical computation. *Mailing Add:* Dept Statist Calif Polytech State Univ San Luis Obispo San Luis Obispo CA 93407

WU, SING-YUNG, b Cheng-tu, China, July 5, 39. ENDOCRINOLOGY, NUCLEAR MEDICINE. *Educ:* Univ Wash, PhD(exp path), 69; Johns Hopkins Univ, MD, 72. *Prof Exp:* Intern med, Univ Chicago, 71-72; resident, Univ Calif, Irvine, 72-73; instr, Univ Wash, 73-75; fel, Univ Calif, Los Angeles, 75-77; from asst prof to assoc prof, 77-90, PROF MED, UNIV CALIF, IRVINE, 90- *Concurrent Pos:* Staff physician, Vet Admin Med Ctr, Long Beach, 77- *Mem:* Am Thyroid Asn; Soc Nuclear Med; Soc Clin Res; Endocrine Soc. *Res:* Biochemical study of endocrine physiology, the thyroid in particular; clinical application of radioisotopes: radioimmunoassay and radio-immune-detection of pathological foci including cancer. *Mailing Add:* 5901 E Seventh St Long Beach CA 90822

WU, SOUHENG, b Taiwan, China, Jan 16, 36; m 65; c 1. PHYSICAL CHEMISTRY. *Educ:* Nat Cheng Kung Univ, BS, 58; Univ Kans, PhD(chem), 65. *Prof Exp:* Researcher, Taiwan Sugar Corp, 58-61; res chemist, 65-69, staff chemist, 69-71, res assoc, 71-85, RES FEL, E I DU PONT DE NEMOURS & CO, INC, 85- *Mem:* Am Phys Soc; Am Chem Soc; Soc Rheology. *Res:* Polymer blends; interfacial sciences; adhesion. *Mailing Add:* Exp Sta E I du Pont de Nemours & Co Inc Wilmington DE 19880-0356

WU, TAI TE, b Shanghai, China, Aug 2, 35; m 66; c 1. BIOCHEMISTRY, MOLECULAR BIOLOGY. *Educ:* Univ Hong Kong, MB & BS, 56; Univ Ill, Urbana, BS, 58; Harvard Univ, SM, 59, PhD(eng), 61. *Prof Exp:* Res fel struct mech, Harvard Univ, 61-63; asst prof eng, Brown Univ, 63-65; res assoc biol chem, Harvard Med Sch, 65-66; from asst prof to assoc prof biomath, Med Col, Cornell Univ, 67-70; from assoc prof to prof physics & eng sci, 70-74, prof biochem, molecular & cell biol, Eng Sci & Appl Math, 74-85, PROF BIOCHEM, MOLECULAR BIOL & CELL BIOL, BIOMED ENG, ENG SCI & APPL MATH, NORTHWESTERN UNIV, 85- *Concurrent Pos:* Gordon McKay fel, Harvard Univ, 58; res scientist, Hydronaut, Md, 62; res fel biol chem, Harvard Med Sch, 64; chmn comt biophys, & mem comt biomed eng, Northwestern Univ, 72-80; res career develop award, NIH, 74; C T Loo scholar, China Inst, 59; vis prof biol, Mass Inst Technol. *Mem:* Am Soc Biochem & Molecular Biol; Am Soc Microbiol; Biophys Soc. *Res:* Structure and functions of macromolecules, especially those of antibodies and related proteins. *Mailing Add:* Dept Biochem Molecular Biol & Cell Biol Northwestern Univ Evanston IL 60208

WU, TAI TSUN, b Shanghai, China, Dec 1, 33; m 67. PHYSICS. *Educ:* Univ Minn, BS, 53; Harvard Univ, SM, 54, PhD(appl physics), 56. *Prof Exp:* Jr fel, Soc Fels, 56-59, from asst prof to assoc prof, 59-66, GORDON MCKAY PROF APPL PHYSICS, HARVARD UNIV, 66- *Concurrent Pos:* Mem, Inst Advan Study, 58-59, 60-61 & 62-63; vis prof & NSF sr fel, Rockefeller Univ, 66-67; Guggenheim Mem Found fel, Deutsches Elektronen-Synchotron, Hamburg, Ger, 70-71; Kramers prof, Inst Theoret Physics, Univ Utrecht, Neth, 77-78. *Mem:* Am Acad Arts & Sci. *Res:* Electromagnetic theory; statistical mechanics; elementary particles. *Mailing Add:* Appl Sci Pierce Hall Harvard Univ Cambridge MA 02138

WU, TAI WING, b Hong Kong; Can citizen; m 74; c 1. CLINICAL BIOCHEMISTRY, CLINICAL DIAGNOSTICS & THERAPEUTICS. *Educ:* Chinese Univ Hong Kong, BSc, 66; Univ Toronto, MSc, 68, PhD(biochem), 71. *Prof Exp:* Med Res Coun Can molecular biol, Univ BC, 71-73; sr res chemist, Res Labs, Eastman Kodak Co, Ny, 73-79, sr res assoc, 79-86; PROF CLIN BIOCHEM, PROF SURG, FAC MED, UNIV TORONTO, 86- *Concurrent Pos:* Referee publ, Can J Biochem, Nat Res Coun Can, 73-; referee papers, Biochem, J Biol Chem, 73-, Clin Chem, 81-; sr vis fel biotechnol, Cornell Univ, Ithaca, Ny, 84-86, vis prof, Dept Chem, 84-85 & Sch Appl Phys & Eng, 85-86. *Honors & Awards:* Meuser Prize. *Mem:* Biophys Soc; Can Biol Socs; AAAS; Am Asn Clin Chem. *Res:* Discovery and identification of novel and clinically useful enzymes and antioxidants; mechanistic studies of organ transplant rejection; exploration of multiple molecular diagnoses of diseases, especially of the liver, heart, lung and nervous system; protein chemistry and enzymology. *Mailing Add:* Toronto Gen Hosp ES 3-404B 200 Elizabeth St Toronto ON M5G 2C4 Can

WU, TE-KAO, b Feng-Shan, Taiwan, Oct 12, 48; m 76; c 1. ELECTRICAL ENGINEERING. *Educ:* Nat Taiwan Univ, BEE, 70; Univ Miss, MS, 73, PhD(elec eng), 76. *Prof Exp:* Res assoc elec eng, Univ Miss, 76-78; scientist assoc res antenna eng, Lockheed Missiles & Space Co Inc, 78-80; SR STAFF ENGR, SPERRY MICROWAVE ELECTRONICS, 80- *Concurrent Pos:* Res asst, Univ Miss, 71-76; res fels, Army Res Orgn, 76-78 & Rome Air Develop Ctr, 76-78. *Mem:* Sigma Xi; Inst Elec & Electronics Engrs. *Res:* Electromagnetics; scattering; antennas; microwave biological effects; electromagnetic compatibiligy; electromagnetic pulses; microwave circuits and devices. *Mailing Add:* Huges Aircraft Co Radar Systs Group R/A102 PO Box 92426 Los Angeles CA 90009

WU, THEODORE YAO-TSU, b Changchow, China, Mar 20, 24; US citizen; m 50; c 2. FLUID MECHANICS, ENGINEERING SCIENCE. *Educ:* Chiao Tung Univ, BS, 46; Iowa State Col, MS, 48; Calif Inst Technol, PhD(aeronaut), 52. *Prof Exp:* Res fel appl mech, 52-55, from asst prof to assoc prof, 55-61, PROF ENG SCI, CALIF INST TECHNOL, 61- *Concurrent Pos:* Mem, Fluid Mech Comt, Am Inst Aeronaut & Astronaut, 49- & Hydrodyn H-5 & H-8 Panels, Soc Naval Architects & Marine Engrs, 63; fac mem, Div Eng & Appl Sci, Calif Inst Technol, Pasedena, 52-; mem, Am Towing Tank Conf, 56-, chmn, 72-74; Guggenheim fel & vis prof, Univ Hamburg, 64-65; mem, Int Towing Tank Conf, Wave Resistance Comt, 67-78, chmn, 77-78; mem, Comt Recommendation US Army Basic Sci Res, Nat Res Coun, 75-78; vchmn, Div Fluid Dynamics, Am Phys Soc; adj prof, Shanghai Jiao Tong Univ, China, 79-; hon prof, Northwestern Polytech Univ, Xian, China, 79- & Harbin Shipbldg Eng Inst, China, 87-; Russell Severance Springer vis prof, Univ Calif, Berkeley, 80; vis fel, Japan soc Promotion Sci, 82; co-ed, Advan Appl Mech, 82-; consult var co & corp. *Mem:* Nat Acad Eng; fel Am Phys Soc; assoc fel Am Inst Aeronaut & Astronaut; Ger Soc Appl Math & Mech; hon fel Acad Sinica; Sigma Xi; Soc Naval Architects & Marine Engrs; Soc Indust & Appl Math. *Res:* Fluid mechanics of compressible, viscous, heat-conducting fluids, water waves, jets, cavity, wake, boundary layer and stratified flows; biophysical and geophysical fluid mechanics; author of various publications. *Mailing Add:* Dept Eng & Sci 104-44 Calif Inst Technol Pasadena CA 91125

WU, TIEN HSING, b Shanghai, China, Mar 2, 23; nat US; m 52; c 2. CIVIL ENGINEERING. *Educ:* St John's Univ, BS, 47; Univ Ill, MS, 48, PhD(civil eng), 51. *Prof Exp:* Civil engr, Deleuw, Cather & Co, Ill, 51-52; State Hwy Div, Ill, 52-53; from asst prof to prof civil eng, Mich State Univ, 53-65; PROF CIVIL ENG, OHIO STATE UNIV, 65- *Concurrent Pos:* Consult geotech eng. *Honors & Awards:* US Antartica Serv Medal, 67; State of Art Award, Am Soc Civil Engrs, 90. *Mem:* Am Soc Civil Engrs; Soc Soil Mech Found Eng. *Res:* Soil mechanics; geotechnical engineering. *Mailing Add:* Dept Civil Eng Ohio State Univ 2070 Neil Ave Columbus OH 43210

WU, TING KAI, b Nanking, China, Aug 5, 37; m 66; c 2. CHEMISTRY. *Educ:* Mt St Mary's Col, Md, BS, 60; Columbia Univ, MA, 61, PhD(phys chem), 65. *Prof Exp:* Res assoc, Columbia Univ, 65; res chemist, 65-72, sr res chemist, 72-78, SR RES SUPVR, E I DU PONT DE NEMOURS & CO, INC, 78- *Concurrent Pos:* Vis assoc prof, Nat Sci Coun China, Nat Taiwan Univ, 70-71. *Mem:* AAAS; Am Chem Soc; NY Acad Sci; Am Inst Physics; Soc Plastics Engrs. *Res:* Structure and properties of molecules and macromolecules; spectroscopic analyses of molecular structure; polymer characterization. *Mailing Add:* Polymer Prod Dept Exp Sta Du Pont Co Wilmington DE 19898

WU, TSE CHENG, b Hong Kong, Aug 21, 23; nat US; m 63; c 3. ORGANIC CHEMISTRY, POLYMER CHEMISTRY. *Educ:* Yenching Univ, China, BS, 46; Univ Ill, MS, 48; Iowa State Univ, PhD(org chem), 52. *Prof Exp:* From res asst to res assoc, Iowa State Univ, 48-53; res chemist, Textile Fibers Dept, E I du Pont de Nemours & Co, Inc, 53-60; res chemist, Silicone Prod Dept, Gen Elec Co, 60-71; sr res chemist, Abcor, Inc, 71-77; res assoc, Corp Res Ctr, Allied-Signal Inc, 77-88; RETIRED. *Concurrent Pos:* Consult. *Mem:* Am Chem Soc; Sigma Xi. *Res:* Monomer and polymer syntheses; acrylic, vinyl, polyester, polycarbonate, polyamide, silicone, and bioabsorbable polymers; rubber, fiber, plastic, membrane, and coating; biomaterials and biomedical implantation devices; organometallic, organosilicon, and fluorine chemistry; interpenetrating polymer networks. *Mailing Add:* 14-E Dorado Dr Morristown NJ 07960

WU, TSU MING, b Taipei, Taiwan, Dec 18, 36; c 2. PHYSICS. *Educ:* Univ Taiwan, BS, 59; Univ Pa, PhD(physics), 66. *Prof Exp:* Fel physics, Case Western Reserve Univ, 66-68; asst prof, 68-71, assoc prof, 71-80, PROF PHYSICS, STATE UNIV NY, BINGHAMTON, 80- *Mem:* Am Phys Soc. *Res:* Many-body problems in solid state physics, especially superconductivity and magnetism; biophysics. *Mailing Add:* Dept Physics State Univ NY Binghamton NY 13901

WU, WEN-HSIEN, PHARMACOLOGY, ALGOLOGY. *Educ:* Nat Taiwan Univ, MD, 58. *Prof Exp:* PROF & CHMN ANESTHESIOL & PHARMACOL, NJ SCH MED, 79-, DIR, PAIN MGT CTR, 80- *Mailing Add:* Dept Anesthesiol UMDNJ NJ Med Sch 185 S Orange Ave Newark NJ 07103

WU, WILLIAM CHI-LIANG, b Shanghai, China, Dec 24, 32; US citizen; m 57; c 3. CHEMICAL ENGINEERING, POLYMER SCIENCE. *Educ:* Mass Inst Technol, SB, 54, SM, 55, ChE, 58. *Prof Exp:* Res asst, Fuels Res Lab, Mass Inst Technol, 55-57; res engr, US Indust Chem Co Div, Nat Distillers & Chem Corp, 57-59, proj leader, 59-61, res supvr, 61; develop engr, Mobil Chem Co Div, Mobil Oil Corp, 62-63, sr develop engr, 63, group leader polymer process develop, 63-67, sect leader, 67-68; res mgr polymer develop, Chemplex Co, 68-77; MGR RES & DEVELOP, MOBIL CHEM CO, 77- *Mem:* Am Inst Chem Engrs; Am Chem Soc; Soc Plastics Engrs. *Res:* Polymerization of ethylene; stereospecific polymerization of propylene and butene-1; polyesterification of terephthalic acid; chemical modifications of polyolefins; polymer additives; polymer blends. *Mailing Add:* 11 Tall Acres Dr Pitts Ford NY 14539

WU, WILLIAM GAY, b Portland, Ore, Feb 5, 31; m 57; c 3. MEDICAL MICROBIOLOGY, IMMUNOLOGY. *Educ:* Ore State Univ, BS, 49, MS, 61; Univ Utah, PhD(immunol, cell biol), 62. *Prof Exp:* Lab asst soil microbiol, Ore State Univ, 57-58, res asst vet microbiol, 58-59; res asst immunol, Univ Utah, 59-62; from asst prof to assoc prof, 62-70, chmn dept microbiol, 67-72, chmn dept biol, 76-81, PROF MICROBIOL, SAN FRANCISCO STATE UNIV, 70- *Concurrent Pos:* Res Corp grants, 64-65; NSF grants, 65-67, NIH grant, 87-; vis prof, Tulane Univ, 70-71; acad vis, Stanford Res Inst, 82. *Mem:* AAAS; Am Soc Microbiol; Am Asn Immunol. *Res:* Study of surface antigens of bacteria; acute disease mechanisms. *Mailing Add:* Dept Biol San Francisco State Univ 1600 Holloway Ave San Francisco CA 94132

WU, WU-NAN, b Kaohsiung, Taiwan, Mar 16, 38; US citizen; m 71; c 2. PHARMACY. *Educ:* Kaohsiung Med Col, BS, 61; Ohio State Univ, PhD(pharm), 72. *Prof Exp:* Technologist drug anal, Taiwan Prov Hyg Labs, 62-63; asst pharm, Taipei Med Col, 63-65; specialist, Bristol Res Inst Taiwan, 65-67; res scientist, 78-79, sr scientist, 79-84, PRIN SCIENTIST DRUG METAB, MCNEIL PHARMACEUT INC, 85- *Concurrent Pos:* Res assoc, Col Pharm, Ohio State Univ, 67-72, postdoctoral researcher, 74-77; res fel, Col Pharm, Univ Fla, 73-74. *Mem:* Am Chem Soc; Am Pharmacog Soc; Am Asn Pharm Scientist; Int Soc Study Xenobiotics. *Res:* Isolation and structural elucidation of biologically active natural products; drug metabolism and disposition; pharmacokinetics; biotransformation. *Mailing Add:* R W Johnson Pharmaceut Res Inst McNeil Pharmaceut Inc McKean & Welsh Rd Spring House PA 19477

WU, XIZENG, b Guiyang, China, Aug 22, 44; div; c 2. DIAGNOSTIC IMAGING, THEORETICAL PHYSICS. *Educ:* Nanjing Normal Univ, BS, 65; City Univ NY, PhD(physics), 83. *Prof Exp:* Fel theoret physics, Mass Inst Technol, 83-85, Univ Cinninat, 85-88; fel med physcis, 88-89, instr med physics, 89-90, ASST PROF MED PHYSICS, UNIV ALA, BIRMINGHAM, 90- *Mem:* Am Phys Soc; Am Asn Med Physicists. *Res:* Theoretical high energy physics; unification theories in high energy physics; diagnostic imaging author of 15 articles and a book. *Mailing Add:* Dept Radiol Univ Ala Birmingham 619 S 19th St Birmingham AL 35233

WU, YAO HUA, b Soochow, China, July 16, 20; nat US; m 50; c 2. ORGANIC CHEMISTRY. *Educ:* Chiao Tung Univ, BS, 43; Univ Nebr, MS, 48, PhD(org chem), 51. *Prof Exp:* Pharmaceut chemist, Int Chem Works, Shanghai, 43-47; res chemist, Smith-Dorsey Co, 51-53; res chemist, Mead Johnson & Co, 53-60, sr res fel, 60-70, dir chem res, 70-76, dir clin publ, 77-80, dir res planning, 80-83; sr staff mem, Res & Develop Div, Bristol-Myers Pharmaceut, 84-90; RETIRED. *Mem:* Am Chem Soc. *Res:* Synthetic pharmaceuticals. *Mailing Add:* 9200 Farmington Dr Evansville IN 47712

WU, YING VICTOR, b Peking, China, Nov 1, 31; nat US; m 60; c 1. PHYSICAL CHEMISTRY, BIOCHEMISTRY. *Educ:* Univ Ala, BS, 53; Mass Inst Technol, PhD(phys chem), 58. *Prof Exp:* Asst phys chem, Mass Inst Technol, 53-57; res assoc chem, Cornell Univ, 58-61; RES CHEMIST, NAT CTR AGR UTILIZATION RES, USDA, 61- *Concurrent Pos:* Fel, Mass Inst Technol, 57-58. *Mem:* AAAS; Am Chem Soc; Am Asn Cereal Chemists; Am Soc Biochem Molecular Biol; Inst Food Technologists; Sigma Xi; Protein Soc. *Res:* Physical chemistry of protein, protein structure, cereal protein concentrates and isolates; optical rotatory dispersion; circular dichroism; hydrogen ion equilibria; distillers' grains and solubles; reverse osmosis and ultrafiltration; food science and technology. *Mailing Add:* Nat Ctr Agr Utilization Res 1815 N University St Peoria IL 61604

WU, YING-CHU LIN (SUSAN), b Peking, China, June 23, 32; m 59; c 3. ENERGY, FLUIDS. *Educ:* Nat Taiwan Univ, BS, 55; Ohio State Univ, MS, 59; Calif Inst Technol, PhD(aeronaut), 63. *Prof Exp:* Engr, Taiwan Hwy Bur, 55-56; sr engr, Electro-Optical Syst, Pasadena, Calif, 63-65; from asst prof to assoc prof, 65-73, prof aerospace, Space Inst, Univ Tenn, 73-88; PRES & CHIEF EXEC OFFICER, ENG RES CONSULT INC, 88- *Concurrent Pos:* Amelia Earhart fel, 58, 59 & 62; lab mgr, Res & Develop Lab, 77-81, adminr, Energy Conversion Res & Develop Prog, 81-88. *Honors & Awards:* Achievment Award, Soc Women Engrs, 85. *Mem:* Assoc fel Am Inst Aeronaut & Astronaut; fel Am Soc Mech Engrs; Sigma Xi; Soc Women Engrs. *Res:* Magnetohydrodynamic power generation. *Mailing Add:* Eng Res & Consult Inc PO Box 417 Tullahoma TN 37388

WU, YONG-SHI, Hubei, China, Aug 4, 42; m 70; c 2. QUANTUM FIELD THEORY, TOPOLOGICAL INVESTIGATIONS IN PHYSICS. *Educ:* Peking Univ, MS, 65; Academia Sinica, PhD(equivalent), 70. *Prof Exp:* Res fel, Instit Physics, 65-78, asst prof, Instit Theoret Physics, 78-80, assoc prof, physics, Academia Sinica, 80-82; res assoc, physics, Dept Physics, Univ Washington, 82-84; assoc prof, 84-87, PROF PHYSICS, DEPT PHYSICS, UNIV UTAH, 87- *Concurrent Pos:* Vis scientist, Instit Adv Sci Study, Bures sur Yvette, Paris, 79, Inst Theoret Physics, State Univ NY, Stony Brook, 81, Inst Theoret Physics, Univ Calif, Santa Barbara, 86, Inst Solid State Physics, Univ Tokyo, 90; mem Instit Adv Study, Princeton, 81-82. *Mem:* Am Phys Soc. *Res:* Unity of forces and matter by use of quantum field theory and string theory and topological aspects of physical phenomena in various branches of physics from cosmology down to the deepest structure in matter. *Mailing Add:* Dept Physics Univ Utah Salt Lake City UT 84112

WU, YUNG-CHI, b Canton, China, Oct 3, 23; US citizen; m 45; c 2. THERMODYNAMICS. *Educ:* Sun Yat-Sen Univ, BS, 47; Univ Houston, MS, 52; Univ Chicago, PhD(chem), 57. *Prof Exp:* Chemist, Res & Develop Lab, Portland Cement Asn, 57-62, Watson Res Ctr, Int Bus Mach Corp, 63-66 & Oak Ridge Nat Lab, 66-67; CHEMIST, NAT INST STANDARDS & TECHNOL, 67- *Mem:* Electrochem Soc; Am Chem Soc; Sigma Xi. *Res:* Electrolyte solutions; thermodynamics; electrolytic conductivity. *Mailing Add:* Nat Inst Standards & Technol Gaithersburg MD 20899

WU, YUNG-KUANG, b Chung-li, Taiwan, Dec 15, 33; m 63; c 1. ELECTRICAL ENGINEERING. *Educ:* Nat Taiwan Univ, BS, 56; Kans State Univ, MS, 60; Univ Mich, PhD(elec eng), 65. *Prof Exp:* Asst res engr, Radiation Lab, Univ Mich, 64-65; from asst prof to prof to prof elec eng, Southeastern Mass Univ, 65-72; mem staff, Lincoln Lab, Mass Inst Technol, 72-74; mem staff, Charles Stark Draper Lab, 74-75; RES ENGR, NAT HWY TRAFFIC SAFETY ADMIN, DEPT TRANSP, 75-; MITRE CORP, MCLEAN, VA. *Mem:* Sr mem Inst Elec & Electronics Engrs. *Res:* Electromagnetic compatibility; radar brake; electronic engine control systems; application of electromagnetic theory such as reentry blackout problems; magnetohydrodynamic boundary-layer control; strategic communications. *Mailing Add:* Usaotea 4501 Ford Ave Alexandria VA 22302

WU, ZHEN, b Sozhou, China. SURFACE SCIENCE. *Educ:* Columbia Univ, MA, 80, MPhil, 81, PhD(physics), 84. *Prof Exp:* Fel physics, Princeton Univ, 84-85, res physicist, 86-87; RES SCIENTIST PHYSICS, MICROELECTRONICS SCI LAB, COLUMBIA UNIV, 87- *Mem:* Am Phys Soc; Optical Soc Am; AAAS; Am Vacuum Soc. *Res:* Interactions between radiation and matter (such as atoms, molecules, and surfaces of various solids). *Mailing Add:* Dept Elec Eng Columbia Univ 1312 SW Mudd Bldg New York NY 10027

WUBBELS, GENE GERALD, b Preston, Minn, Sept 21, 42; m 67; c 3. CHEMISTRY, BIOCHEMISTRY. *Educ:* Hamline Univ, BS, 64; Northwestern Univ, PhD(chem), 68. *Prof Exp:* From asst prof to prof, 68-79, chmn dept, 75-78 & 83-85, PROF ORG & BIOL CHEM, GRINNELL COL, 79-, DACK PROF, 86- *Concurrent Pos:* Res assoc, State Univ NY Buffalo, 74-75; vis prof, Univ Leiden, Neth, 81-82; res grant, Am Chem Soc-Petrol Res Fund, 71-73, 74-77, 78-80, 81-83 & 84-86 & NSF, 86-; charter counr, Coun Undergrad Res; ed, Surv Progress Chem, 80-86; prog dir, NSF, 90-92. *Honors & Awards:* Catalyst Awardee, Chem Mfg Asn, 89. *Mem:* Am Chem Soc; InterAm Photochem Soc; AAAS. *Res:* Catalytic mechanisms for photosubstitution, photoreduction and photoaddition reactions of aromatic compounds. *Mailing Add:* Dept Chem Grinnell Col Grinnell IA 50112

WUCHTER, RICHARD B, b Wadsworth, Ohio, July 21, 37; m 72. ORGANIC CHEMISTRY. *Educ:* Western Reserve Univ, AB, 59; Cornell Univ, PhD(org chem), 63. *Prof Exp:* Asst org chem, Cornell Univ, 60-62; group leader process res, 63-70, pollution control chem synthesis, Fluid Process Lab, 70-89, SEPARATIONS RESEARCHER, ROHM & HAAS CO, 90- *Mem:* Am Chem Soc. *Res:* Monomer synthesis and process development; plastics; modifiers for plastics; fibers; pollution control and ion exchange syntheses. *Mailing Add:* 1521 Old Welsh Rd Huntington Valley PA 19006

WUDL, FRED, b Cochabamba, Bolivia, Jan 8, 41; US citizen; m 67. ORGANIC CHEMISTRY, MATERIALS SCIENCE. *Educ:* Univ Calif, Los Angeles, BS, 64, PhD(chem), 67. *Prof Exp:* Fel org chem, Harvard Univ, 67-68; asst prof, State Univ NY, Buffalo, 68-73; MEM TECH STAFF, BELL LABS, 73-; PROF, UNIV CALIF, SANTA BARBARA, 73- *Concurrent Pos:* Assoc dir, Inst Polymers & Organic Solids, 82- *Mem:* Fel AAAS; Am Chem Soc; Sigma Xi. *Res:* Organic conductors; organic synthesis; heterocycles; organometallic compounds and complexes; polymer science. *Mailing Add:* Dept Chem Univ Calif Santa Barbara CA 93106

WUEBBLES, DONALD J, b Breese, Ill, Jan 28, 48; m 70; c 3. ATMOSPHERIC CHEMISTRY & PHYSICS. *Educ:* Univ Ill, BS, 70, MS, 72; Univ Calif, PHD, 83. *Prof Exp:* Res scientist, Nat Oceanic & Atmospheric Admin, 72-73, Univ Colo, Boulder, 73; RES SCIENTIST, LAWRENCE LIVERMORE NAT LAB, 73- *Concurrent Pos:* Prin investr, Environ Protection Agency, 79-, NASA, 84-, Dept Energy, 84-; sci adv, US deleg, Orgn Econ Coop & Develop, 81; mem, adv panel, NASA High Speed Res Prog, 89; mem, Nat Res Coun Working Group on Solar Influences, 90-91, Int Comn Meteorol Upper Atmosphere Working Group on Modeling Middle Atmosphere. *Mem:* Am Geophys Union; Am Meteorol Soc. *Res:* Computational modelling of the physical, chemical, and radiative processes in the atmosphere; atmospheric ozone and anthropogenic effects on it; greenhouse gases and their effects on climate. *Mailing Add:* Lawrence Livermore Nat Lab L-262 7000 East Ave Livermore CA 94550

WUENSCH, BERNHARDT J(OHN), b Paterson, NJ, Sept 17, 33; m 60; c 2. CRYSTALLOGRAPHY, CERAMICS. *Educ:* Mass Inst Technol, SB, 55, SM, 57, PhD(crystallog), 63. *Prof Exp:* Res fel crystallog, Inst Mineral Petrog, Univ Berne, 63-64; from asst prof to prof ceramics, 64-85, TDK prof mat sci & eng, 85-90, DIR, CTR MATS SCI & ENG, MASS INST TECHNOL, 88- *Concurrent Pos:* Ford fel eng, 64-66; vis prof crystallog, Univ Saarland, Saarbrucken, Germany, 73; adv ed, Physics & Chem Minerals, 76-85; assoc ed, Can Mineralogist, 78-80, ed, Zeit Kristallogr, 81-88; vis scientist, Max Planck Inst fur Festkorperforschung, Ger, 81; USA Nat Comt Crystallography, Nat Acad Sci, 80-82, 89- *Mem:* Mineral Asn Can; Am Crystallog Asn; fel Am Ceramic Soc; fel Mineral Soc Am; Electrochem Soc; Mats Res Soc. *Res:* X-ray and neutron diffraction; crystal structure determination; relation between crystal structure and crystal properties; diffusion and point defects. *Mailing Add:* Dept Mat Sci & Eng Mass Inst Technol Cambridge MA 02139

WUENSCHEL, PAUL CLARENCE, b Erie, Pa, May 13, 21; m 50; c 6. GEOPHYSICS. *Educ:* Colo Sch Mines, GeolEngr, 44; Columbia Univ, PhD(geol), 55. *Prof Exp:* Res assoc, Columbia Univ, 46-52; dir geol & geophys res, Res, Inc, 52-55; res assoc, 55-74, geophysicist & sr scientist, 74-83, GEOPHYS CONSULT, GULF RES & DEVELOP CO, 83- *Mem:* Hon mem Soc Explor Geophys; Seismol Soc Am; Acoust Soc Am; Am Geophys Union; Europ Asn Explor Geophys. *Res:* Seismology; potential, electrical and seismic methods of geophysical exploration. *Mailing Add:* 128 Marian Ave Glenshaw PA 15116

WUEPPER, KIRK DEAN, b Bay City, Mich, Mar 18, 38; c 3. DERMATOLOGY, EXPERIMENTAL DERMATOPATHOLOGY. *Educ:* Univ Mich, MD, 63; Am Bd Dermat, dipl, 70. *Prof Exp:* USPHS fel, Scripps Clin & Res Found, 68-70, assoc dermat & path, 70-72; assoc prof, Med Sch, Ore Health Sci Ctr, 72-76, PROF DERMAT, ORE HEALTH SCI UNIV, 76- *Concurrent Pos:* Res career develop award, 71-76. *Honors & Awards:* Charles W Burr Award for Res, 64. *Mem:* Am Asn Immunol; Am Soc Clin Invest; Soc Invest Dermat (secy-treas, 79-84, pres, 86-87); Am Fedn Clin Res; Am Acad Dermat; Asn Am Phys. *Res:* Molecular basis of keratinization, keratinocyte adhesiveness, and pathogenesis of immunologic skin diseases. *Mailing Add:* Dept Dermat Ore Health Sci Univ 3181 SW Sam Jackson Park Rd Portland OR 97201

WUERKER, RALPH FREDERICK, b Los Angeles, Calif, Jan 18, 29; div; c 3. PHYSICS. *Educ:* Occidental Col, BA, 51; Stanford Univ, PhD(physics), 60. *Prof Exp:* Engr, AiRes, Inc, 56-58; mem tech staff, Res Lab, Ramo-Wooldridge, Inc, 58-59; mem sr staff, Res Lab, Space Technol Labs, Inc, 60-61; mem sr staff, Quantatron Corp, 61-62; MEM PROF STAFF, TRW SYSTS GROUP, TRW INC, REDONDO BEACH, 63- *Concurrent Pos:* Consult, Lawrence Livermore Labs, 73-; assoc dir, HIPAS Observ, Fairbanks, Alaska. *Honors & Awards:* Res Soc Award, TRW Systs, 66. *Mem:* Am Asn Physics Teachers; Am Phys Soc; Optical Soc Am. *Res:* Holography and coherent optics; plasma particle resonances; superconductivity; electrooptics; lasers; physical optics; plasma physics; electron and general experimental physics. *Mailing Add:* 15123 Cordary Ave Lawndale CA 90260

WUEST, PAUL J, b Philadelphia, Pa, Feb 26, 37; m 61; c 4. PLANT PATHOLOGY, MYCOLOGY. *Educ:* Pa State Univ, BS, 58, PhD(plant path), 63. *Prof Exp:* Asst, 58-63, asst prof, 64-68, assoc prof, 68-74, PROF PLANT PATH, PA STATE UNIV, 74- *Concurrent Pos:* Fel, Univ Guelph, 70-71; agr consult, 70- *Mem:* Am Phytopath Soc; Can Phytopath Soc; Soc Nematol; Mycol Soc Am; Am Mushroom Inst; Sigma Xi. *Res:* Diseases of the commercial mushroom; soil treatment and disease occurrence; fungicide tolerance; epidemiology; pest management; worker exposure to pesticides. *Mailing Add:* Dept Plant Path 211 Buckhout Lab University Park PA 16802

WUJEK, DANIEL EVERETT, b Bay City, Mich, Oct 26, 39; m 66. BOTANY. *Educ:* Cent Mich Univ, BS, 61, MA, 62; Univ Kans, PhD(bot), 66. *Prof Exp:* From asst prof to assoc prof bot, Wis State Univ, La Crosse, 66-68; assoc prof, 68-73, PROF BOT, CENT MICH UNIV, 73-, RES PROF, 78- *Concurrent Pos:* Wis State Univ fac grant, 66-68; NSF grant, 71-72, 76-77 & 87; Cent Mich Univ fac res grant, 71-85; vis prof, Univ Minn, Univ Mont & Univ S Fla; pres, Mich Electron Microscopy Forum. *Honors & Awards:* Dimond Award, Bot Soc Am, 75. *Mem:* AAAS; Bot Soc Am; Am Micros Soc; Phycol Soc Am; Int Phycol Soc. *Res:* Algal life history studies and electron microscopy, including ecology and relations to water quality. *Mailing Add:* Dept Biol Cent Mich Univ Mt Pleasant MI 48859

WUKELIC, GEORGE EDWARD, b Steubenville, Ohio, Sept 17, 29; m 55; c 3. REMOTE SENSING, SPACE PHYSICS. *Educ:* WVa Univ, AB, 52. *Prof Exp:* Prin physicist, Columbus Div, 52-60, sr physicist, 60-72, assoc sect mgr, Space Systs & Appln Sect, 72-77, staff scientist, Water & Land Resources Dept, 77-85, PROG MGR, EARTH SCI DEPT, PAC NORTHWEST DIV, BATTELLE MEM INST, 85- *Mem:* Am Soc Photogram; Am Geophys Union. *Res:* Space; geophysics; remote sensing applications; satellite earth resource surveys; image processing-multispectral data. *Mailing Add:* Pac Northwest Div Battelle Blvd Richland WA 99352

WULBERT, DANIEL ELIOT, b Chicago, Ill, Dec 17, 41; c 2. MATHEMATICS. *Educ:* Knox Col, Ill, BA, 63; Univ Tex, MA, 64, PhD(math), 66. *Prof Exp:* Vis asst prof math, Univ Lund, 66-67; asst prof, Univ Wash, 67-74; assoc prof, 74-80, PROF MATH, UNIV CALIF, SAN DIEGO, 80- *Concurrent Pos:* Fel, Univ Lund, 66-67; NSF res grant, 68-; vis prof, Northwestern Univ, 77-80. *Mem:* Am Math Soc. *Res:* Approximation theory; functional analysis. *Mailing Add:* Dept Math Univ Calif at San Diego La Jolla CA 92093

WULF, RONALD JAMES, b Davenport, Iowa, July 24, 28; m 59; c 3. TOXICOLOGY, PHARMACOLOGY. *Educ:* Univ Iowa, BS, 50, MS, 57; Purdue Univ, PhD(biochem), 64. *Prof Exp:* Res chemist, John Deere & Co, 50-52; asst pharmacol, Univ Iowa, 54-57; res pharmacologist, Lederle Labs, Am Cyanamid Co, NY, 57-61; teaching asst biochem, Purdue Univ, 61-64; assoc prof pharmacol, Univ Conn, 64-70; DIR BIOL RES, CARTER WALLACE INC, 70- *Mem:* AAAS; Am Chem Soc; Soc Toxicol. *Res:* Drug safety evaluations; chemically induced fibrinolysis; inhalation toxicology. *Mailing Add:* Carter Wallace Inc PO Box 1 Cranbury NJ 08512-0001

WULF, WILLIAM ALLAN, b Chicago, Ill, Dec 8, 39; m 61; c 2. COMPUTER SCIENCE. *Educ:* Univ Ill, BSc, 61, MSc, 63; Univ Va, DSc(comput sci), 68. *Prof Exp:* Instr comput sci, Univ Va, 63-68; from asst prof to assoc prof, 68-75, actg dept head, 78-79, prof comput sci, Carnegie-Mellon Univ, 75-81; founder & chmn, Tartan Labs, 81-87; AT&T PROF ENG, UNIV VA, 88-; ASST DIR, NSF, 88- *Concurrent Pos:* Chmn working group on syst implementation lang, Int Fedn Info Processing, 72-78; assoc ed, Transactions Prog Languages & Systs, 79-; asst dir, NSF, 88-90. *Mem:* Asn Comput Mach; fel Inst Elec & Electronics Engrs; Int Fedn Info Processing; fel AAAS. *Res:* Computer languages and their translators, operating systems, methodology and computer architecture. *Mailing Add:* Dept Comput Sci Univ Va Charlottesville VA 22903

WULFERS, THOMAS FREDERICK, b Cape Girardeau, Mo, Oct 4, 39; m 63; c 2. ORGANIC CHEMISTRY. *Educ:* St Louis Univ, BS, 61; Wash Univ, MA, 63; Univ Chicago, PhD(chem), 65. *Prof Exp:* Res chemist, Shell Oil Co, Mo, 65-72; sr res chemist, 72-80, mgr indust prod res, 72-85, MGR LUBRICANTS RES & DEVELOP & VPRES LICENSING, ATLANTIC RICHFIELD CO, ILL, 85- *Mem:* Am Chem Soc. *Res:* Petroleum products research; lubricants formulation and processing. *Mailing Add:* 907 Woodland St Houston TX 77009-6543

WULFF, BARRY LEE, b Mt Kisco, NY, Feb 17, 40; m 66; c 2. ECOLOGY, BOTANY. *Educ:* State Univ NY, Cortland, BS, 65; Col William & Mary, MA, 68; Ore State Univ, PhD(bot), 70. *Prof Exp:* From asst prof to assoc prof, 70-82, PROF BIOL, EASTERN CONN STATE UNIV, 82- *Concurrent Pos:* Pres, Natural Resources Coun Conn, 76-87. *Mem:* AAAS; Am Bryological & Lichenological Soc; Mycol Soc Am; Sigma Xi. *Res:* Lichens; fungi; marine algae; ecology of cryptogamic plants. *Mailing Add:* Dept Biol Eastern Conn State Univ Willimantic CT 06226

WULFF, DANIEL LEWIS, b Santa Barbara, Calif, Mar 29, 37; m 57; c 3. MOLECULAR GENETICS. *Educ:* Calif Inst Technol, BS, 58, PhD(chem), 62. *Prof Exp:* Fel, Inst Genetics, Univ Koln, 62-63, & biol dept, Harvard Univ, 63-65; from asst prof to prof biol, dept molecular biol & biochem, Univ Calif, Irvine, 65-79, assoc dean biol sci, 75-79; PROF BIOL, DEPT BIOL SCI & DEAN COL SCI & MATH, STATE UNIV NY, ALBANY, 80- *Mem:* AAAS; Genetics Soc Am; Am Soc Microbiol; Am Soc Biochem & Molecular Biol. *Res:* Transcriptional and translational control mechanisms in the regulation of gene expression in bacteriophage lambda. *Mailing Add:* Col Sci & Math Chem B27 State Univ NY Albany NY 12222

WULFF, JOHN LELAND, b Oakland, Calif, Mar 19, 32; m 68; c 4. MATHEMATICS. *Educ:* Sacramento State Col, AB, 54; Univ Calif, Davis, MA, 57, PhD(math), 66. *Prof Exp:* From instr to assoc prof, 55-68, chmn dept, 68-71, PROF MATH, CALIF STATE UNIV, SACRAMENTO, 69- *Res:* Measure theory and integration. *Mailing Add:* Dept Math & Statist Calif State Univ Sacramento 6000 J St Sacramento CA 95819

WULFF, VERNER JOHN, b Essen, Ger, Aug 16, 16; m 42; c 3. NEUROPHYSIOLOGY. *Educ:* Wayne Univ, AB, 38; Northwestern Univ, MA, 40; Univ Iowa, PhD(zool), 42. *Prof Exp:* Asst zool & comp vert anat, Wayne Univ, 36-38; asst zool, embryol & endocrinol, Northwestern Univ, 38-40; asst, Univ Iowa, 40-42, instr & assoc mammalian physiol, 46-47; asst prof mammalian neurophysiol, Univ Ill, 47-51; prof zool & chmn dept, Syracuse Univ, 51-60; assoc dir res, 61-, EMER SR SCIENTIST AT MASONIC MED RES LAB. *Mem:* AAAS; Am Soc Zool; Soc Exp Biol & Med. *Res:* Electrophysiology of retinae; relation between photochemical events and retinal action potential. *Mailing Add:* Six Slaytonbush Lane Utica NY 13501

WULFF, WOLFGANG, b Darmstadt, Ger, June 16, 33, US citizen; m 64; c 2. MECHANICAL & AEROSPACE ENGINEERING, COMPUTER SIMULATION. *Educ:* Winterthur Inst Technol, Switz, BSME, 58; Ill Inst Technol, MSME, 62, PhD(mech & aerospace eng), 68. *Prof Exp:* Proj engr, Escher-Wyss Ltd, Zurich, 58-60 & Eppi Precision Prod, Ill, 60-63; res engr, IIT Res Inst, 63-68; from asst prof to assoc prof mech eng, Ga Inst Technol, 68-74; TENURED SCIENTIST, BROOKHAVEN NAT LAB, 74- *Concurrent Pos:* Adj prof, NY Inst Technol, 79-80. *Honors & Awards:* Achievement Award, US Nuclear Regulatory Comn, 89. *Mem:* Am Soc Mech Engrs; Combustion Inst; NY Acad Sci; Am Nuclear Soc; Sigma Xi; Am Asn Univ Prof. *Res:* Fabric flammability; thermal control of space vehicles; atmospheric vortices; thermohydraulics of reactor systems; nuclear power reactor safety; nuclear power plant simulation. *Mailing Add:* 11 Hamilton Rd Setauket NY 11733

WULFMAN, CARL E, b Detroit, Mich, Nov 29, 30; m 52; c 4. THEORETICAL PHYSICS, THEORETICAL CHEMISTRY. *Educ:* Univ Mich, BS, 53; Univ London, PhD(org chem), 57. *Prof Exp:* Instr chem, Univ Tex, 56-57; assoc prof, Defiance Col, 57-61; chmn dept, 61-74, PROF PHYSICS, UNIV PAC, 61- *Concurrent Pos:* Vis mem, Ctr Theoret Studies, Coral Gables, Fla, 67; NSF sci fac fel, Oxford Univ, 67-68; vis prof, Japan Soc Promotion Sci, 74-75, Max Born vis prof, Hebrew Univ, 88; coop res, NZ, Can, Mex & Israel. *Mem:* Am Phys Soc; Am Asn Physics Teachers. *Res:* Transformation properties of dynamical equations; continuous groups; applications to chemical kinetics, atomic and molecular quantum mechanics. *Mailing Add:* Dept Physics Univ Pac Stockton CA 95211

WULFMAN, DAVID SWINTON, b Detroit, Mich, Sept 1, 34; m 61; c 3. SYNTHETIC ORGANIC CHEMISTRY, EXPLOSIVES ENGINEERING. *Educ:* Univ Mich, Ann Arbor, BS, 56; Dartmouth Col, AM, 58; Stanford Univ, PhD(chem), 62; Alliance Francaise, Paris, France, IVe, French, 74. *Prof Exp:* Res asst chem, Univ Mich, 54-56; sr develop engr, Hercules Inc, Utah, 61-63; from asst prof to assoc prof, 63-77, PROF CHEM, UNIV MO-ROLLA , 77-, SR INVESTR, ROCK MECH & EXPLOSIVES RES CTR, 88-; CHIEF SCIENTIST, TDI, 88- *Concurrent Pos:* Consult, Dept Chem, Stanford Univ, 69; lectr, Wash Univ, 70 & Chem Soc France, 75; res assoc, Ctr Nat Res Sci, Ecole Normale Superieure, Paris, 74-75; vis prof, St Marys Univ, Halifax NS Can, 81-82 & Dalhousie Univ, 84; consult, Metalbright, Burns & Roe & IRECO. *Mem:* Am Chem Soc; Royal Soc Chem; Chem Inst Can; Asn Can Studies US; Soc Explosive Engrs. *Res:* Synthesis and study of theoretically important molecules; homogeneous catalysis; kinetics of photochemical processes; biological action of diazo compounds; physical organic chemistry; demilitarization of explosive and propellants. *Mailing Add:* Dept Chem Univ Mo Rolla MO 65401

WULLERT, JOHN R, II, b Abington, Pa, Feb 27, 62; m 86. DISPLAYS FOR GRAPHICS & VIDEO, OPTICAL SIGNAL PROCESSING. *Educ:* Lafayette Col, BS, 84; Carnegie Mellon Univ, MS, 87. *Prof Exp:* MEM TECH STAFF, ELECTRONIC DISPLAYS & OPTICAL SIGNAL PROCESSING, BELLCORE, 86- *Mem:* Soc Info Display; Inst Elec & Electronics Engrs; Optical Soc Am. *Res:* Investigating electrical-optical conversions for the purposes of generating images for video displays as well as for generation of inputs to optical signal processing systems. *Mailing Add:* 331 Newman Spring Rd Red Bank NJ 07701

WULLSTEIN, LEROY HUGH, b Nampa, Idaho, Nov 23, 31; m 56; c 1. BIOGEOGRAPHY, SOIL MICROBIOLOGY. *Educ:* Univ Utah, BS, 57; Ore State Univ, MS, 61, PhD(microbiol), 64. *Prof Exp:* Asst prof soil sci, Univ BC, 64-66; from asst prof to assoc prof, 66-75, PROF BIOGEOG, UNIV UTAH, 75-, ASSOC PROF BIOL, 80-, PROF GEOG, 85- *Concurrent Pos:* Sr Fulbright res fel nitrogen fixation, Ireland, 72-73; consult, Brookhaven Labs, 74-; mem, Utah Statewater Pollution Control Comn. *Mem:* Am Soc Microbiol; Am Chem Soc; Soil Sci Soc Am. *Res:* Nitrogen transformations; oak biogeography; endemism. *Mailing Add:* Dept Geog Univ Utah Salt Lake City UT 84112

WULPI, DONALD JAMES, metallurgical engineering, mechanics; deceased, see previous edition for last biography

WUN, CHUN KWUN, b Canton, China, Feb 15, 40; US citizen; m 64; c 2. BACTERIOLOGY, ENVIRONMENTAL POLLUTION CONTROL. *Educ:* Chung Chi Col, Chinese Univ Hong Kong, BS, 64; Springfield Col, MS, 69; Univ Mass, Amherst, MS, 71, PhD(lipid chem), 74. *Prof Exp:* Asst educ officer sci, Educ Dept Hong Kong, 65-66; from asst instr to instr biol, Springfield Col, 68-70; res asst lipid chem, Univ Mass, Amherst, 70-73; asst prof biol, Springfield Col, 73-74, assoc res prof, 81-82; FEL, DEPT ENVIRON SCI, UNIV MASS, AMHERST, 75- *Concurrent Pos:* Adj asst prof biol, Springfield Col, 75- *Mem:* Am Soc Microbiol; Sigma Xi; Soc Appl Bacteriol. *Res:* Lipid metabolism, particularly triglyceride synthesis and its control in mycobacterium smegmatis; water pollution; the use of fecal sterols as an indicator of fecal pollution of water; column method for rapid extraction and gas-liquid chromatography quantitation of algal chlorophylls; new media for the isolation/identification of E coli and other enteric organisms; rapid procedure for the isolation and drug susceptibility determination of mycobacteria; recycling of agricultural and domestic wastes. *Mailing Add:* 1177 Boston Rd Springfield MA 01119-1309

WUNDER, BRUCE ARNOLD, b Monterey Park, Calif, Feb 10, 42; m 63; c 2. PHYSIOLOGICAL ECOLOGY, VERTEBRATE ZOOLOGY. *Educ:* Whittier Col, BA, 63; Univ Calif, Los Angeles, PhD(vert zool), 68. *Prof Exp:* NIH fel, Inst Arctic Biol, Univ Alaska, 68-69; from asst prof to assoc prof, Colo State Univ,69-84, asst chmn dept, 78-79, 83-84, interim chmn, 84-85, chmn prog ecol studies, 85-87 PROF ZOOL, COLO STATE UNIV, 84-, CHMN DEPT, 85- *Concurrent Pos:* Small mammal ecologist, Biol Res Assocs, Inc & consult, Thorne Ecol Inst, 72-; assoc prof zool, Univ Mich Biol Sta, 76 & 78; prof zool, Univ Mont Biol Sta, 81, 83 & 85; Alexander von Humboldt fel, Frankfurt, Ger, 79-80; prof biol, Rocky Mt Biol Sta, 87 & 90. *Mem:* AAAS; Am Soc Zoologists; Am Soc Mammalogists; Ecol Soc Am; Sigma Xi. *Res:* Temperature regulation and energetics; water balance and mechanisms of evaporative water loss, particularly in vertebrates; feeding strategies and distribution patterns in vertebrates; digestive physiology of vertebrates. *Mailing Add:* Dept Biol Colo State Univ Ft Collins CO 80523

WUNDER, CHARLES C(OOPER), b Pittsburgh, Pa, Oct 2, 28; m 62; c 3. PHYSIOLOGY, BIOPHYSICS. *Educ:* Washington & Jefferson Col, AB, 49; Univ Pittsburgh, MS, 52, PhD(biophys), 54. *Prof Exp:* Asst biophys, Univ Pittsburgh, 49-51; assoc physiol, 54-56, asst prof, 56-63, assoc prof physiol & biophys, 63-71, PROF PHYSIOL & BIOPHYS, UNIV IOWA, 71- *Concurrent Pos:* NIH res career develop award, 61-66; vis scientist & NIH spec fel, Mayo Clin, 66-67. *Mem:* Soc Exp Biol & Med; Am Soc Zoologists; Soc Develop Biol; Biophys Soc; Am Physiol Soc; Am Soc Gravitational & Space Biol; Am Soc Biomech; Aerospace Med Asn. *Res:* Environmental biophysics of growth and function; gravitational biology. *Mailing Add:* Physiol Res Bldg Oakdale Campus Univ Iowa Oakdale IA 52319

WUNDER, WILLIAM W, b Lake Park, Iowa, June 4, 30; m 60; c 3. POPULATION GENETICS, ANIMAL SCIENCE. *Educ:* Iowa State Univ, BS, 58; Mich State Univ, MS, 64, PhD(dairy cattle breeding), 67. *Prof Exp:* Asst prof exten dairy sci, Univ Ky, 67-68; asst prof, 68-74, assoc prof, 74-79, PROF ANIMAL SCI, IOWA STATE UNIV, 79- *Mem:* Am Dairy Sci Asn; Am Soc Animal Sci. *Res:* Influence of corrective versus random mating on net income and type in Holstein dairy cattle. *Mailing Add:* Dept Animal Sci 101 Kildee Iowa State Univ Ames IA 50011

WUNDERLICH, BERNHARD, b Brandenburg, Ger, May 28, 31; nat US; m 53; c 2. POLYMER CHEMISTRY. *Educ:* Univ Frankfurt, BSc, 54; Northwestern Univ, PhD, 57. *Prof Exp:* Instr chem, Northwestern Univ, 57-58; from instr to asst prof, Cornell Univ, 58-63; from assoc prof to prof, 63-88, EMER PROF CHEM, RENSSELAER POLYTECH INST, 88-; PROF CHEM & DISTINGUISHED SCIENTIST, UNIV TENN, 88- *Concurrent Pos:* Consult, E I du Pont de Nemours & Co, 63-88; Humboldt fel, 87; adv prof, Fudan Univ, Shanghai, China, 88- *Honors & Awards:* Mettler Award for Thermal Anal, 71. *Mem:* Am Chem Soc; fel NAm Thermal Anal Soc; Int Confedn Thermal Anal; fel Am Phys Soc. *Res:* Physical chemistry of the solid state of high polymers; transitions of high polymers at elevated temperatures and high pressures; thermal analysis. *Mailing Add:* 200 Baltusrol Rd Knoxville TN 37922

WUNDERLICH, FRANCIS J, b Philadelphia, Pa, Mar 9, 38; m 62; c 5. PHYSICAL CHEMISTRY. *Educ:* Villanova Univ, BS, 59; Georgetown Univ, PhD(chem), 64. *Prof Exp:* Res asst chem, Villanova Univ, 57-58, instr, 59; instr, Georgetown Univ, 59-61, res assoc, 61-63; fel molecular physics, 63-65; from asst prof to assoc prof physics & chem, Col Virgin Islands, 65-69; ASST PROF PHYSICS, VILLANOVA UNIV, 69- *Concurrent Pos:* Dir, NSF Grant, Undergrad Sci Equip Prog, Col Virgin Islands, 65-67, dir, Etelman Astron Observ, 66-67. *Honors & Awards:* Award, Am Inst Chemists, 59. *Mem:* Am Chem Soc; Royal Soc Chem. *Res:* Theoretical molecular physics; gas phase free radicals; laser-induced gas phase reactions. *Mailing Add:* Dept Physics Villanova Univ Villanova PA 19085

WUNDERLICH, JOHN R, TUMOR IMMUNOLOGY, CELLULAR IMMUNOLOGY. *Educ:* Stanford Univ, MD, 64. *Prof Exp:* SR INVESTR, NIH, 70- *Mailing Add:* Immunol Br NCI NIH Bldg 10 Rm 4B17 Bethesda MD 20892

WUNDERLICH, MARVIN C, b Decatur, Ill, May 8, 37; m 60; c 2. MATHEMATICS. *Educ:* Concordia Teachers Col, Ill, BS, 59; Univ Colo, PhD(math), 64. *Prof Exp:* Asst prof math, State Univ NY, Buffalo, 64-67; assoc prof math, Northern Ill Univ, 67-72, prof, 72-88; DIR MATH SCI PROG, NAT SECURITY AGENCY, US DEPT DEFENSE, FT MEADE, MD, 88- *Concurrent Pos:* NSF res grant, 66-; vis, Univ Nottingham, 72-73. *Mem:* Am Math Soc; Asn Comput Mach. *Res:* Number theory; computing mathematics. *Mailing Add:* Nat Security Agency US Dept Defense Ft Meade MD 20455

WUNDERLY, STEPHEN WALKER, b Cleveland, Ohio, May 24, 45; m 69; c 2. MEASUREMENT OF LOW ENERGY RADIOACTIVITY, SYNTHETIC ORGANIC CHEMISTRY. *Educ:* Col Wooster, BA, 63; Univ Cincinnati, MS, 71, PhD(chem), 75. *Prof Exp:* Res assoc chem, Univ BC, 74-76, Univ Calif, San Francisco, 76-77 & Univ Southern Calif, 77-78; STAFF SCIENTIST, BECKMAN INSTRUMENTS INC, 78- *Mem:* Am Chem Soc; Am Inst Chem. *Res:* Photochemistry; alkaloid synthesis and biosynthesis; synthesis of polycyclic aromatics; emulsion chemistry and emulsifer properties; chemistry with solid phase supported reagents; detection and measurement of low energy radionuclides; nuclear chemistry. *Mailing Add:* Beckman Instruments NSO 2500 Harbor Blvd PO Box 3100 Fullerton CA 92635-1080

WUNDERMAN, IRWIN, b New York, NY, Apr 24, 31; m 51; c 3. ELECTROOPTICS, ELECTRONIC INSTRUMENTATION. *Educ:* City Col New York, BSEE, 52; Univ Southern Calif, MSEE, 56; Stanford Univ, EEE, 61, PhD(elec eng), 64. *Prof Exp:* Jr engr draftsman, Lockheed-Calif Co, 52, jr engr, 52-53, res engr, 53-56; lab sect leader, Hewlett Packard Co, 56-61, co-founder, Hewlett Packard Assocs, 61-65, lab mgr, Hewlett Packard Corp Labs, 65-67; pres & gen mgr, Cintra Inc, Cintra Physics Int, 67-71. *Concurrent Pos:* Researcher, scientist & author, 71- *Mem:* AAAS; sr mem Inst Elec & Electronics Engrs; Optical Soc Am; Am Inst Physics; Sigma Xi. *Res:* Electrooptics instrumentation; radiometry; photometry; optoelectronic solid state devices and circuits; computer architecture and systems; modeling of nonlinear physical systems; physics of optics and quanta; wave/particle dilemma; photons, granular mechanics and the classical origin of quanta theory in mathematical physics. *Mailing Add:* 655 Eunice Ave Mountain View CA 94040

WUNDERMAN, RICHARD LLOYD, b New York, NY, Jan 14, 52. VOLCANOLOGY, MAGNETOTELLURICS. *Educ:* Univ Calif, Berkeley, AB, 76; Mich Technol Univ, MA, 83, PhD, 88. *Prof Exp:* Volcanologist, US Geol Surv, 81-82; Res Volcanologist, Mich Technol Univ, 80-87; CONSULT, 88- *Mem:* Soc Explor Geophysicists; Am Geophys Union; Sigma Xi; Geol Soc Am. *Res:* Recognition and description of volcanic hazards; crustal studies of volcanic terraines. *Mailing Add:* PO Box 355 Aurora CO 80040-0355

WUNSCH, ABRAHAM DAVID, b Brooklyn, NY, Dec 15, 39; c 2. APPLIED MATHEMATICS. *Educ:* Cornell Univ, BEE, 61; Harvard Univ, SM, 62, PhD(eng & appl physics). *Prof Exp:* Asst prof, 69-82, ASSOC PROF ELEC ENG, UNIV LOWELL, 82- *Mem:* Sigma Xi. *Res:* Electromagnetic theory; antennas; functions of a complex variable. *Mailing Add:* 111 Louise Rd Belmont MA 02178

WUNSCH, CARL ISAAC, b New York, NY, May 5, 41; m 70; c 2. PHYSICAL OCEANOGRAPHY. *Educ:* Mass Inst Technol, SB, 62, PhD(geophys), 66. *Prof Exp:* Lectr oceanog, Mass Inst Technol, 66-67, from asst prof to prof, 67-76, head, Dept Earth & Planetary Sci, 77-81, CECIL & IDA GREEN PROF PHYS OCEANOG, MASS INST TECHNOL, 76- *Concurrent Pos:* Vis sr investr, Cambridge Univ, 69, 74-75 & 81-82; vis prof, Univ Wash, 80 & Harvard Univ, 80; ed, Monographs in Mech & Appl Math, Cambridge Univ Press, 81-89; Fulbright fel, 81-82; Guggenheim Found fel, 81-82; secy, Navy Res Prof, 85-89. *Honors & Awards:* James B Macelwane Award, Am Geophys Union, 71, Maurice Ewing Medal, 90; Founders Prize, Tex Instruments Found, 75, A G Huntsman Prize, 88. *Mem:* Nat Acad Sci; Am Geophys Union; Royal Astron Soc; Am Acad Arts & Sci; Soc Indust & Appl Math. *Res:* Sea level, general circulation; tides; ocean acoustics; climate; author of several scientific papers. *Mailing Add:* Dept Earth Atmospheric & Planetary Sci Rm 54-1524 Mass Inst Technol Cambridge MA 02139

WUNZ, PAUL RICHARD, JR, b Erie, Pa, Oct 18, 23; m 48; c 3. ORGANIC CHEMISTRY, TEACHING. *Educ:* Pa State Col, BS, 44, MS, 47; Univ Del, PhD(chem), 50. *Prof Exp:* Instr chem, Univ Del, 50; asst prof chem & head dept, Augsburg Col, 50-51; res chemist, Nopco Chem Co, 51-53; res chemist & group leader, Callery Chem Co, 53-57; from asst prof to assoc prof chem, Geneva Col, 57-65; chmn dept, 65-73, prof chem, Indiana Univ Pa, 65-81; RETIRED. *Concurrent Pos:* Vis prof, Univ Ariz, 82. *Mem:* Am Chem Soc; Sigma Xi. *Res:* Synthetic organic chemistry; organometallic compounds; pharmaceuticals, steroids; heterocyclic compounds. *Mailing Add:* 219 Oriole Ave Indiana PA 15701

WUONOLA, MARK ARVID, b Astoria, Ore, Oct 9, 47. ORGANIC CHEMISTRY, MEDICINAL CHEMISTRY. *Educ:* Mass Inst Technol, SB, 69; Harvard Univ, AM, 70, PhD(org chem), 73. *Prof Exp:* Res fel, Harvard Univ, 73-76; res chemist org chem, Cent Res & Develop Dept, E I du Pont de Nemours & Co, Inc, 76-82, res chemist, biochem dept, 82-; AT TOXICOL DIV, CHEM PHYSIOL BR, CHEM RES & DEVELOP CTR. *Mem:* Am Chem Soc. *Res:* Chemistry of natural products; synthetic organic chemistry; heterocyclic chemistry. *Mailing Add:* Dupont MPD Exp Sta Bldg 353 Rm 302 PO Box 80353 Wilmington DE 19880-0353

WUORINEN, JOHN H, JR, b New York, NY, Aug 9, 31; m 56; c 4. ELECTRICAL ENGINEERING. *Educ:* Columbia Univ, AB, 53, MS, 56, PhD(elec eng), 63. *Prof Exp:* Res asst elec eng, Electronics Res Lab, Columbia Univ, 54-56, instr, Univ, 56-62; mem tech staff, 62-64, supvr digital device integration, 64-68, head, Electronic Subsyst Design Dept, 68-74, head, Memory & Call Progress Syst Dept, 74-80, HEAD, SPECIAL SYSTS DEPT, BELL LABS, 81- *Concurrent Pos:* Chmn solid state circuits conf, Inst Elec & Electronics Engrs, 76-77. *Mem:* Inst Elec & Electronics Engrs; Sigma Xi. *Res:* Digital integrated circuits; solid state electronics; digital memory systems; telephone switching. *Mailing Add:* PO Box 304 Castine ME 04421

WURDACK, JOHN J, b Pittsburgh, Pa, Apr 28, 21; m 59; c 2. BOTANY. *Educ:* Univ Pittsburgh, BS, 42; Univ Ill, BS, 49; Columbia Univ, PhD, 52. *Prof Exp:* Asst bot, Univ Pittsburgh, 42; tech asst bot, NY Bot Garden, 49-52; from asst cur to assoc cur, Nat Mus Natural Hist, Smithsonian Inst, 52-60, assoc cur, Div Phanerogams, 60-63, cur bot, 63-90, EMER CUR, NAT MUS NATURAL HIST, SMITHSONIAN INST, 91- *Concurrent Pos:* Mem exped, Venezuela, 50-59 & 72 & Peru, 62. *Mem:* Am Soc Plant Taxonomists; Torrey Bot Club; Int Asn Plant Taxon. *Res:* Taxonomy of Melastomataceae and flowering plants of northern South America. *Mailing Add:* Dept Bot Nat Mus Natural Hist Smithsonian Inst Washington DC 20560

WURM, JAROSLAV, b Czech, Apr 12, 35; US citizen; m 64; c 2. COMPRESSOR & PRIME MOVERS, AIR CONDITIONING & REFRIGERATION. *Educ:* Czech Tech Univ, MS, 58. *Prof Exp:* Dep head, Compressor Div, Dept Compressor Develop, Czech, 58-66; ASST DIR, SPACE CONDITIONING RES, INST GAS TECHNOL, 68- *Concurrent Pos:* Pres, Air Conditioning Comn, Int Inst Refrig, 83-91. *Mem:* Am Soc Heating Refrig & Air Conditioning Engrs; Int Inst Refrig; Am Solar Energy Soc; Ger Cold & Climate Soc. *Res:* Energy conversion systems including conceptual, component, and systems evaluation; compressors, refrigeration, air conditioning and air quality, prime movers, and heat transfer. *Mailing Add:* Inst Gas Technol 4201 W 36th St Chicago IL 60632

WURMSER, LEON, b Zurich, Switz, Jan 31, 31; US & Swiss citizen; m 58; c 3. PSYCHOANALYSIS, DRUG ABUSE. *Educ:* Univ Basel, MD, 58. *Prof Exp:* Staff psychiatrist, Sheppard Pratt Hosp, Baltimore, 62-65; dir out patient, Sinai, 66-69; clin asst prof psychol & dir, Drug Abuse Ctr, Johns Hopkins Univ, 69-71; from assoc prof to prof psychiat, Univ Md, 71-83, dir alcohol & drug abuse prog, 77-83. *Concurrent Pos:* Teacher, Switz, Ger & Sweden. *Honors & Awards:* Lewis B Hill Award, Inst Psychoanal, 75. *Mem:* Am Psychiat Asn; Am Psychoanal Asn; Swiss Med Soc; Swiss Psychiat Soc; AAAS. *Res:* Psychodynamics of compulsive substance abuse; shame, shame conflicts, and defense against shame; the defense against superego; author of over 250 scientific articles. *Mailing Add:* 200 E Joppa Rd Baltimore MD 21204

WURSIG, BERND GERHARD, b WGermany, Nov 9, 48; m 69; c 2. BEHAVIORAL BIOLOGY, MARINE MAMMALOGY. *Educ:* Ohio State Univ, BSc, 71; State Univ NY, PhD(behav biol), 78. *Prof Exp:* Fel, NIH & Univ Calif, Santa Cruz, 78-81; staff mem, Moss Landing Marine Labs, 81-89; PROF & DIR, MARINE MAMMAL RES PROG, TEX, 89- *Concurrent Pos:* Researcher, Nat Geog Soc, 74-77, biomed res fel, 77. *Mem:* Am Behav Soc; Explorer's Club; Am Soc Mammalogists; Natural History Soc; Int Soc Cryptozool; Int Soc Marine Mammalogy (pres, 90-). *Res:* Behavior and ecology of cetaceans and pinnipeds; movement and migration patterns of dolphins and whales; field research techniques for cetaceans; comparison of wild and captive dolphin behavior; sociobiology and ecology of mammals. *Mailing Add:* Marine Mammal Res Prog Tex A&M Univ Galveston TX 77553

WURST, GLEN GILBERT, b Mt Holly, NJ, Apr 17, 45. GENETICS. *Educ:* Juniata Col, BS, 67; Univ Pittsburgh, PhD(biol), 75. *Prof Exp:* Teaching asst biol, Univ Pittsburgh, 67-71, teaching fel, 71-75; asst prof, 75-84, ASSOC PROF BIOL, ALLEGHENY COL, 84- *Concurrent Pos:* Vis assoc res scientist, Johns Hopkins Univ, 80, 81 & 83-84. *Mem:* Genetics Soc Am; Soc Develop Biol; Sigma Xi; AAAS. *Res:* Developmental genetics of Drosophila Melanogaster. *Mailing Add:* Dept Biol Allegheny Col Meadville PA 16335

WURST, GLORIA ZETTLE, b Steelton, Pa, Jan 13, 46. ANIMAL PHYSIOLOGY, EMBRYOLOGY. *Educ:* Juniata Col, BS, 66; Univ Pittsburgh, MS, 70, PhD(biol), 74. *Prof Exp:* Res assoc zool, Univ Calif, Berkeley, 75-78; from asst prof to assoc prof zool, Weber State Col, 78-88, PROF ZOOL, WEBER STATE UNIV, 88- *Mem:* AAAS; Am Soc Zoologists; Am Soc Ichthyologists & Herpetologists; Herpetologists' League; Sigma Xi. *Res:* Developmental morphology of pineal complex and pituitary gland in amphibians and reptiles; population genetics and evolution of salamanders. *Mailing Add:* Dept Zool Weber State Univ Ogden UT 84408-2505

WURST, JOHN CHARLES, b Defiance, Ohio, Jan 11, 36; m 58; c 3. CERAMIC ENGINEERING. *Educ:* Univ Dayton, BME, 57, MEngSc, 68; Univ Ill, Urbana, PhD(ceramic eng), 71. *Prof Exp:* Sr res ceramist, 57-75, assoc prof, 73-83, ASSOC DIR RES INST, UNIV DAYTON, 75-, PROF, GRAD FAC, 83- *Concurrent Pos:* Consult mem comt coatings, Mat Adv Bd, Nat Acad Eng, 67-71. *Mem:* Fel Am Ceramic Soc; Sigma Xi; Nat Coun Univ Adminr; Nat Inst Ceramic Engrs; Ceramic Educ Coun. *Res:* High temperature protective coatings; corrosion; infrared transmitting materials; ceramics; sintering. *Mailing Add:* Univ Dayton Res Inst Dayton OH 45469

WURSTER, CHARLES F, b Philadelphia, Pa, Aug 1, 30; c 3. ENVIRONMENTAL SCIENCES & TOXICOLOGY, ORNITHOLOGY. *Educ:* Haverford Col, SB, 52; Univ Del, MS, 54; Stanford Univ, PhD(org chem), 57. *Prof Exp:* Asst, Univ Del, 52-54 & Stanford Univ, 54-57; Fulbright fel, Innsbruck Univ, 57-58; res chemist, Monsanto Res Corp, 59-62; res assoc, Dartmouth Col, 62-65; asst prof biol sci, 65-71, ASSOC PROF ENVIRON TOXICOL, MARINE SCI RES CTR, STATE UNIV NY, STONY BROOK, 71- *Concurrent Pos:* Founder & mem bd trustees, Environ Defense Fund, 67-; vis scientist & spec asst to dir, Nat Cancer Inst, Bethesda, Md, 78-79; vis prof, Dept Chem, Microbiol & Plant Physiol, Univ Bergen, Norway, 81; mem bd dir, Defenders Wildlife, 75-84 & 87- *Mem:* AAAS. *Res:* Ecological and physiological effects of stable chemical pollutants; effects of chlorinated hydrocarbons on marine plankton communities; chemical pollutants and avian reproduction; environmental protection via legal action; public policy and environmental quality; public interest science. *Mailing Add:* Marine Sci Res Ctr State Univ NY Stony Brook NY 11794-5000

WURSTER, DALE E, b Sparta, Wis, Apr 10, 18; m 44; c 2. PHARMACY. *Educ:* Univ Wis, BS, 42, PhD, 47. *Prof Exp:* From instr to prof pharm, Univ Wis, 47-71; prof pharm & pharmaceut chem & dean col pharm, NDak State Univ, 71-72; prof pharm & dean Col Pharm, 72-84, EMER PROF & DEAN, UNIV IOWA, 84-, INTERIM DEAN, 91- *Concurrent Pos:* Am Asn Cols Pharm-NSF vis scientist, 63-66; consult, USPHS, 66-72; mem revision comt, US Pharmacopoeia, 61-72; sci adv, Wis Alumni Res Found, 68-72; phys sci adminr, US Navy, 60-63. *Honors & Awards:* Res Achievement Award, Am Pharmaceut Asn, 65; George B Kaufman mem lectr, Ohio State Univ, 68; Indust Pharm Technol Award, Am Pharmaceut Asn Acad Pharmaceut Sci, 80; Hancher-Finkbine Medal, Univ Iowa, 84. *Mem:* Acad Pharmaceut Sci (pres, 75); Am Pharmaceut Asn; hon mem Rumanian Soc Med Sci; Soc Invest Dermat; Am Asn Cols Pharm; Sigma Xi; Am Asn Hosp Pharmacists; Fed Int Pharmaceut; Controlled Release Soc. *Res:* Physical factors influencing dissolution kinetics; diffusion kinetics in biological membranes, drug release mechanisms from pharmaceutical systems, percutaneous absorption, air-suspension microencapsulation coating and granulating technique. *Mailing Add:* 16 Brickwood Knoll RR 6 Iowa City IA 52240

WURSTER, DALE ERIC, JR, b Madison, Wis, Jan 19, 51; m 75; c 2. SURFACE CHEMISTRY, THERMAL ANALYSIS. *Educ:* Univ Wis, BS, 74; Purdue Univ, PhD(phys pharm), 79. *Prof Exp:* Asst prof pharm, Sch Pharm, Univ NC, 79-82; asst prof, 83-86, ASSOC PROF, COL PHARM, UNIV IOWA, 87- *Mem:* Acad Pharmaceut Sci; Mat Res Soc; Am Chem Soc; Sigma Xi; Am Asn Col Pharm; Am Asn Pharmaceut Scientists. *Res:* Surface phenomena, especially adsorption-desorption thermodynamics; dissolution mechanisms; heats of solution and of interaction; heat evolution upon tablet compression; solution and differential scanning calorimetry; physics of tablet compression; analytical aspects of FTIR. *Mailing Add:* Col Pharm Univ Iowa Iowa City IA 52242

WURSTER-HILL, DORIS HADLEY, b Washington, DC, Sept 9, 32; div; c 1. CYTOGENETICS. *Educ:* George Washington Univ, BS, 54; Stanford Univ, MA, 56, PhD(biol), 58; Am Bd Med Genetics, dipl, 84. *Prof Exp:* Res assoc endocrinol, 62-65, res assoc cytogenetics, 67-77, from instr to assoc prof cytogenetics, 67-80, res assoc prof path, 80-83, RES PROF PATH, DARTMOUTH MED SCH, 83- *Concurrent Pos:* USPHS trainee cytogenetics, Dartmouth Med Sch, 65-67. *Mem:* AAAS; Am Soc Human Genetics; Am Soc Mammalogists; Am Asn Cancer Res; Soc Anal Cytol; AAAS. *Res:* Clinical cytogenetics; comparative mammalian cytogenetics; cancer cytogenetics. *Mailing Add:* Dept Path Dartmouth Med Sch Hanover NH 03756

WURTELE, MORTON GAITHER, b Harrodsburg, Ky, July 25, 19; m 42; c 2. DYNAMIC METEOROLOGY. *Educ:* Harvard Univ, BS, 40; Univ Calif, Los Angeles, MA, 44, PhD, 53. *Prof Exp:* Asst prof meteorol, Mass Inst Technol, 53-58; assoc prof, 58-64, vchmn dept, 69-72, chmn dept, 72-76, PROF METEOROL, UNIV CALIF, LOS ANGELES, 64- *Concurrent Pos:* Fulbright grants, Univ Sorbonne, 49 & Hebrew Univ, Israel, 65; NATO sr fel, 62; consult, Atmospheric Sci Lab, White Sands Missile Range, 65-, Jet Propulsion Lab, 78- *Mem:* Fel Am Meteorol Soc; Am Geophysics Union. *Res:* Small- and medium-scale atmospheric motions; sound propagation; atmospheric-ocean interaction; remote sensing; air pollution. *Mailing Add:* Dept Atmospheric Sci Univ Calif Los Angeles CA 90024-1565

WURTELE, ZIVIA SYRKIN, statistics, econometrics, for more information see previous edition

WURTH, MICHAEL JOHN, b Highland Park, Ill, Mar 31, 37; m 65; c 1. ORGANIC CHEMISTRY, PHOTOGRAPHY. *Educ:* Lake Forest Col, BA, 64; Northwestern Univ, Evanston, PhD(org chem), 69. *Prof Exp:* ASSOC PROF CHEM, OKLAHOMA CITY UNIV, 68- *Mem:* Am Chem Soc. *Res:* Synthesis of new bicyclic heterocyclic compounds; photographic chemistry, especially emulsions and developing agents. *Mailing Add:* 2409 NW 17th St Oklahoma City OK 73107

WURTH, THOMAS JOSEPH, b St Louis, Mo, June 13, 28; m 58; c 3. CHEMICAL ENGINEERING. *Educ:* Washington Univ, St Louis, BS, 52; Univ Mo-Kansas City, MBA, 67. *Prof Exp:* Develop engr, Mathieson Chem Corp, NY, 52-53, proj engr, Olin Mathieson Chem Corp, Ill, 53-55, pilot plant supvr, 55-59; process engr, R W Booker & Assoc, Mo, 59-61; process engr, Austin Co, Ill, 61-63; res engr, Gulf Res & Develop Co, 63-66, sr proj engr, 66-69, planning specialist, Chem Dept, Develop Div, 69-71, dir planning, Develop Div, 71-75, mgr planning & bus anal, Spec Chem, 75-80, COORDR TECHNOL DEVELOP, SPEC CHEM, GULF OIL CHEM CO, 80- *Mem:* Am Asn Cost Engrs. *Res:* Process development of chemical processes; process design of chemical plants. *Mailing Add:* 5054 Bayou Vista Dr Houston TX 77091

WURTMAN, JUDITH JOY, b Brooklyn, NY, Aug 4, 37; m 59; c 2. OBESITY, PREMENSTRUAL SYNDROME. *Educ:* Wellesley Col, BA, 59; Harvard Univ, MAT, 60; George Washington Univ, PhD(cell biol), 71. *Prof Exp:* Asst prof biol & nutrit, Newton Col, 72-74; RES SCIENTIST, MASS INST TECHNOL, 74-, FEL, DEPT BRAIN & COGNITIVE SCI, 74- *Concurrent Pos:* Consult nutrit educ, Newton Pub Schs, 72-74; NIH fel, 74-76; instr, Radcliffe Seminars-Harvard Exten Courses, 76-; counselor, Obesity Clin, 81-; commentator, Boston TV Sta, 81- *Mem:* Soc Nutrit Educ; Am Dietetic Asn; Nutrit Today; Inst Food Technol; Sigma Xi. *Res:* Regulation of food intake in laboratory animal and human, especially in obesity and other eating disorders. *Mailing Add:* Dept Brain & Cognitive Sci Mass Inst Technol Cambridge MA 02139

WURTMAN, RICHARD JAY, b Philadelphia, Pa, Mar 9, 36; m 59; c 2. NEUROSCIENCE, METABOLISM. *Educ:* Univ Pa, AB, 56; Harvard Med Sch, MD, 60. *Prof Exp:* Intern & asst resident med, Mass Gen Hosp, 60-62; res assoc, Lab Clin Sci, NIMH, 62-64, med res officer, 65-67; assoc prof, 67-70, prof endocrinol & metab, 70-80, PROF NEUROENDOCRINE REGULATION, MASS INST TECHNOL, 80-, PROF NEUROPHARMACOL, WHITAKER COL, 84-, DIR, CLIN RES CTR, 85- *Concurrent Pos:* Josiah Macy, Jr Found fel, Mass Gen Hosp, 60-62; res fel endocrinol, Mass Gen Hosp, 64-65; clin assoc med, Mass Gen Hosp, 65-; vis lectr, Am Chem Soc, 66; lectr, Harvard Med Sch, 69-; mem, Preclin Psychopharmacol Study Sect, NIMH, 71-75; Am Inst Biol Sci Adv panel, Biosci Prog, NASA; res adv bd, Parkinson's Dis, Am Parkinson's Dis Asn, Tourette Syndrome Asn; assoc, Neurosci Res Prog; chmn, Life Sci Adv Comt, NASA, 79-; Sterling vis prof, Boston Univ, 81; invited prof, Univ Geneva, 81;

chmn, Air Force Life Sci adv bd, 85- *Honors & Awards:* Am Therapeut Soc Prize, 66; Soc Biol Psychiat Prize, 66; John Jacob Abel Award, Am Soc Pharmacol Exp Therapeut, 68; Alvarenga Prize & Lect, Col Physicians Philadelphia, 70; Ernst Oppenheimer Prize, Endocrine Soc, 73; Foster Elting Bennett Lectr, Am Neurol Asn, 74; Louis B Flexner Lectr, 75; Pfizer Lectr, State Univ NY Buffalo, 80; Zale Lectr, Univ Tex, 80; McCallum Lectr, Univ Toronto, 81; Osborne & Mendel Award, 82; Ciba-Geigy Drew Award, 82; Rufus Cole Lectr, Rockefeller Univ, 85. *Mem:* Am Physiol Soc; Am Soc Pharmacol & Exp Therapeut; Am Soc Neurochem; Am Soc Biol Chem; Am Soc Clin Invest. *Res:* Neuroendocrinology; neuropharmacology; biological rhythms; pineal gland; catecholamines; amino acid metabolism; effects of nutrition on brain; biological effects of light; acetylcholine; membrane phosphatides. *Mailing Add:* Dept Brain & Cognitive Sci Mass Inst Technol E 25-604 Cambridge MA 02139

WURTZ, ROBERT HENRY, b St Louis, Mo, Mar 28, 36; m; c 2. NEUROPHYSIOLOGY, NEUROPSYCHOLOGY. *Educ:* Oberlin Col, AB, 58; Univ Mich, Ann Arbor, PhD(physiol psychol), 62. *Prof Exp:* Res assoc neurophysiol, Wash Univ, 62-65; res fel, Lab Neurophysiol, NIH, 65-66, res scientist, Lab Neurobiol, NIMH, 66-78, CHIEF, LAB SENSORIMOTOR RES, NAT EYE INST, 78- *Honors & Awards:* W Alden Spencer Award, Columbia Univ, 87. *Mem:* Nat Acad Sci; Int Brain Res Orgn; Soc Neurosci (pres, 90-91); Asn Res Vision & Opthal; Am Physiol Soc; Am Acad Arts & Sci. *Res:* Neurophysiological basis of behavior, specifically the physiology of vision and movement. *Mailing Add:* Lab Sensorimotor Res Bldg 10 Rm 10C101 Nat Eye Inst Bethesda MD 20892

WURZBURG, OTTO BERNARD, b Grand Rapids, Mich, Aug 1, 15; m 40; c 6. STARCH CARBOHYDRATE CHEMISTRY. *Educ:* Univ Mich, BS, 38, MS, 39. *Prof Exp:* Chemist & supvr cent control, Nat Starch Prod, Nat Starch & Chem Corp, 39-44, res chemist & supvr starch res, 45-55, assoc res dir, 56-68, vpres res, Starch Div, 68-73, sr vpres, 73-80; CONSULT, 80- *Concurrent Pos:* Mem bd dirs, Customaize Inc, 74-; ed, Modified Starches: Properties & Uses, Food Regulatory. *Honors & Awards:* Alsberg Schock Mem Award, 86, Corn Refiners Asn, Inc & Am Asn Cereal Chemists, Inc. *Mem:* Am Chem Soc; Inst Food Technologists; Am Asn Cereal Chemists. *Res:* Starch; carbohydrates; industrial applications; author. *Mailing Add:* RFD 1 Box 138E St Johnsbury VT 05819-9801

WUSKELL, JOSEPH P, b New York, NY, Nov 14, 38. CHEMISTRY, PHARMACEUTICAL CHEMISTRY. *Educ:* Univ Conn, BA, 60; Univ Minn, PhD(chem), 67. *Prof Exp:* Chemist, Merck Sharp & Dohme Res Labs, 60-62; sr chemist, Ott Chem Co, Corn Prod Co, 67-68; group leader, Quaker Oats Co, 68-83; RES ASSOC, UCONN HEALTH CTR, 86- *Mem:* Am Chem Soc; Royal Soc Chem. *Res:* Organic synthesis and reaction mechaisms. *Mailing Add:* 11 Lowell Rd West Hartford CT 06119-1584

WUSSOW, GEORGE C, b Milwaukee, Wis, Mar 10, 23; m 49; c 3. ORAL SURGERY, ORAL PATHOLOGY. *Educ:* Marquette Univ, DDS, 49. *Prof Exp:* From instr to assoc prof, 53-67, PROF ORAL SURG, MARQUETTE UNIV, 67-, CHMN DEPT, 61-, LECTR, SCH DENT HYG, 70- *Concurrent Pos:* Attend oral surg, Vet Admin Ctr, Wood, Wis & mem consult staff, Milwaukee County Gen Hosp, 57-; consult, Great Lakes Naval Hosp, 70-; mem adv bd, Milwaukee Area Tech Col, 70- *Mem:* Am Soc Oral Surg; Am Acad Oral Path; fel Am Col Dent; Int Asn Oral Surg; fel Royal Soc Health. *Res:* Clinical evaluation of proteolytic enzymes in the management of impacted mandibular third molars. *Mailing Add:* 1730 N 118th St Milwaukee WI 53226

WUST, CARL JOHN, b Providence, RI, July 2, 28; m 51; c 5. IMMUNOLOGY. *Educ:* Providence Col, BS, 50; Brown Univ, MSc, 53; Ind Univ, PhD(microbiol), 57. *Prof Exp:* Electron microscopist, Ind Univ, 53-55; NIH fel, Yale Univ, 57-59; biochemist, Biol Div, Oak Ridge Nat Lab, 59-70; assoc prof, 70-74, PROF MICROBIOL, UNIV TENN, KNOXVILLE, 74-, PROF MED BIOL, 83- *Mem:* AAAS; Am Soc Microbiol; Am Asn Immunologists; Sigma Xi; Soc Exp Biol & Med. *Res:* Leukemia antigens/ induced differentiation and induced cell growth; immunity to viral infections. *Mailing Add:* Dept Microbiol Univ Tenn Knoxville TN 37996-0845

WUTHIER, ROY EDWARD, b Rushville, Nebr, Nov 11, 32; m 56; c 2. BIOCHEMISTRY. *Educ:* Univ Wyo, BS, 54; Univ Wis, MS, 58, PhD, 60. *Prof Exp:* Asst biochem, Univ Wis, 55-60; res fel, Forsyth Dent Ctr, Harvard Med Sch, 60-63, asst mem staff & assoc biol chem, 63-69; assoc prof biochem, Depts Orthop Surg & Biochem, Col Med, Univ Vt, 69-75; PROF CHEM, COL ARTS & SCI & COORDR BIOCHEM, COL MED, UNIV SC, 75- *Concurrent Pos:* Chmn, Gordon Res Conf, Calcium Phosphates, 89. *Honors & Awards:* Biol Mineralization Basic Res Award, Int Asn Dent Res, 82. *Mem:* AAAS; Am Soc Biol Chem; Am Chem Soc; Am Soc Bone & Mineral Res; Int Asn Dent Res. *Res:* Mechanism of calcification; lipid and membrane involvement in calcification; role of matrix resicles in calcification; fatty acid metabolism in cartilage. *Mailing Add:* Dept Chem Phys Sci Ctr Univ SC Thomas Jones Bldg Columbia SC 29208

WUTS, PETER G M, b Schiedamm, Holland, July 24, 50; US citizen; m 78; c 1. SYNTHETIC ORGANIC CHEMISTRY. *Educ:* Univ Washington, BS, 73; Northwestern Univ, PhD(chem), 78. *Prof Exp:* Fel chem, Calif Inst Technol, 78-80; asst prof chem, Univ Mich, 80-85; AT UPJOHN CO, 85- *Concurrent Pos:* NIH fel, 80. *Mem:* Am Chem Soc. *Res:* Development of methodology and process development for drug candidates. *Mailing Add:* UpJohn Co 1500-91-2 Kalamazoo MI 49001

WUU, TING-CHI, b Salt County, China, Sept 27, 34; US citizen; m 62; c 3. BIOCHEMISTRY. *Educ:* Nat Taiwan Univ, BSc, 58, MSc, 60; McGill Univ, PhD(biochem), 67. *Prof Exp:* Prof asst biochem res, McGill Univ, 67-69; asst prof, 69-74, ASSOC PROF BIOCHEM, MED COL OHIO, 75- *Mem:* AAAS. *Res:* Structure and function of peptides and proteins; brain, peptides and proteins; isolation and characterization of hormone-binding proteins of neurohypophysis; biological, chemical and physical properties of neurophysins and neurosecretory granules; biosynthesis of brain peptides and proteins. *Mailing Add:* Dept Biochem Med Col Ohio Caller Serv No 10008 Toledo OH 43699

WU-WONG, JINSHYUN RUTH, b Hong Kong, Brit citizen. PROTEIN CHEMISTRY. *Educ:* Nat Taiwan Univ, Repub of China, BSc, 77; Ohio State Univ, PhD(biochem), 81. *Prof Exp:* Res assoc, Ohio State Univ, 81 & Chinese Univ, Hong Kong, 82-85; res fel, Univ Tex Med Sch, 85-88; SR RES SCIENTIST, ABBOTT LABS, 88- *Concurrent Pos:* Travel fel, Int Union Biochem, 85. *Mem:* Am Soc Biochem & Molecular Biol; Biochem Soc UK; AAAS. *Res:* Characterization and purification of endothelin converting enzymecs from tissues/cells. *Mailing Add:* Pharmaceut Prod Div D47V AP10 Abbott Lab Abbott Park IL 60064-3500

WYANT, GORDON MICHAEL, b Frankfurt, Ger, Mar 28, 14; c 5. ANESTHESIOLOGY. *Educ:* Univ Bologna, MD, 38; Royal Col Physicians & Surgeons Eng, dipl, 45; Am Bd Anesthesiol, dipl, 53; Royal Col Physicians & Surgeons Can, dipl anesthesiol, 52, FRCP(C), 63. *Prof Exp:* Asst prof anesthesia, Col Med, Univ Ill, 50-53; asst prof surg & head, Div Anesthesia, Stritch Sch Med, Loyola Univ, Ill, 53-54; prof, 54-71, EMER PROF ANESTHESIA, UNIV SASK, 71- *Honors & Awards:* Gold Medal, Can Anesthesiol Soc. *Mem:* Am Soc Anesthesiol; fel Am Geriat Soc; fel Am Col Anesthesiol; Can Anaesthetists Soc. *Res:* Related clinical and basic sciences of anesthesia; principle management. *Mailing Add:* No 7 Oasis Manor Estates 3415 Calder Cresent Saskatoon SK S7J 4Z9 Can

WYANT, JAMES CLAIR, b Morenci, Mich, July 31, 43; m 71; c 1. OPTICS. *Educ:* Case Inst Technol, BS, 65; Univ Rochester, MS, 67, PhD(optics), 68. *Prof Exp:* Optical engr & head, Optical Eng Sect, Itek Corp, 68-74; FROM ASST PROF TO PROF OPTICS, OPTICAL SCI CTR, UNIV ARIZ, 74-; PRES, WYKO CORP, 84- *Concurrent Pos:* Assoc ed, J Optical Soc Am & Optical Eng; mem, Int Comn Optics, US Nat Comt, Am Inst Physics; mem bd dir, Optical Soc Am, 79-81, mem exec comt, 80-81; vis prof, Univ Rochester, 83; assoc ed, Appl Optics, 83; chmn, Gordon Conf on Holography & Optical Info Processing, 84. *Mem:* Optical Soc Am; Soc Photo-Optical Instrumentation Engrs (pres, 86). *Res:* Interferometry; holography; optical testing; optical processing; optical properties of the atmosphere; active optics; application of microcomputer to optics. *Mailing Add:* Optical Sci Ctr Univ Ariz Tucson AZ 85721

WYATT, BENJAMIN WOODROW, b Farrar, Ga, Dec 24, 16; m 48; c 4. CHEMISTRY. *Educ:* Southwestern Univ, Tex, BS, 37; Univ Tex, MA, 40, PhD(org chem), 43. *Prof Exp:* Tutor chem, Univ Tex, 38-43; assoc mem & asst to patent agent, Sterling-Winthrop Res Inst, 43-50, patent agent, 50-61, from asst dir to assoc dir, 61-74, dir, Patent Div, 74-78, vpres patents, 78-87; RETIRED. *Mem:* Am Chem Soc. *Res:* Organic chemistry. *Mailing Add:* 317 Loudonville Rd Loudonville NY 12211

WYATT, CAROLYN JANE, food science, for more information see previous edition

WYATT, COLEN CHARLES, b Geneva, NY, Dec 10, 27; m 48; c 5. HORTICULTURE. *Educ:* Cornell Univ, BS, 53. *Prof Exp:* Sr res horticulturist, H J Heinz Co, 54-64; univ horticulturist & asst dir maintenance, Bowling Green State Univ, 65-66; plant breeder, Libby McNeill & Libby, Ohio, 67-72; PLANT BREEDER, PETO SEED CO, 72- *Res:* Breeding processing and fresh market tomatoes for worldwide production, including the development of disease resistance for tobacco mosaic virus, spotted wilt virus, nematodes bacterial wilt and other major diseases; breeding squash for high vitamin content and disease resistance. *Mailing Add:* PO Box 1255 Rte 4 Woodland CA 95695

WYATT, ELLIS JUNIOR, b Norton, Kans, Oct 30, 30; m 53; c 3. PARASITOLOGY, INVERTEBRATE ZOOLOGY. *Educ:* Lewis & Clark Col, BS, 57; Ore State Univ, MS, 61, PhD(zool), 71. *Prof Exp:* Aquatic biologist, Ore State Fish Comn, 57, 58-60 & 61; asst prof biol, Cent Ore Col, 61-65 & 67-68; aquatic biologist, Ore State Fish Comn, 68-71; from asst prof to assoc prof, 71-77, chmn dept, 72-82, PROF BIOL, HAMLINE UNIV, 77- *Concurrent Pos:* Lab teaching asst, Ore State Univ, 57. *Mem:* AAAS; Am Soc Parasitologists; Am Soc Zoologists; Am Fisheries Soc. *Res:* Parasitic protozoa of fresh water fishes; bacteriology, helminthology, mycology, therapeutics and toxicology of fresh water fishes; ecology of fish parasitism. *Mailing Add:* Dept Biol Hamline Univ St Paul MN 55104

WYATT, GERARD ROBERT, b Palo Alto, Calif, Sept 3, 25; m 85; c 3. INSECT BIOCHEMISTRY, MOLECULAR BIOLOGY. *Educ:* Univ BC, BA, 45; Cambridge Univ, PhD(natural sci), 50. *Prof Exp:* Sci officer, Insect Path Res Inst, Can Dept Agr, Ont, 50-54; asst prof biochem, Yale Univ, 54-60, from assoc prof to prof biol, 60-73; head dept, 73-75, PROF BIOL, QUEEN'S UNIV, ONT, 73- *Concurrent Pos:* Guggenheim Mem fel, 54; Killam res fel, 85. *Mem:* Am Soc Biochem & Molecular Biol; fel Royal Soc Can. *Res:* Composition of nucleic acids; biochemistry and physiology of insects; composition of insect hemolymph; carbohydrate metabolism and its regulation; physiology of development; actions of insect hormones; insect vitellogenins; insect molecular biology. *Mailing Add:* Dept Biol Queen's Univ Kingston ON K7L 3N6 Can

WYATT, JAMES L(UTHER), b Williamsburg, Ky, May 13, 24; m 46; c 2. METALLURGICAL ENGINEERING. *Educ:* Univ Ky, BS, 47, MS, 48; Mass Inst Technol, ScD(metall), 53. *Prof Exp:* Instr metall, Univ Ky, 47-48; proj engr, Titanium Div, Nat Lead Co, 48-50; head dept metall eng, Horizons, Inc, 53-57; consult & assoc, Booz-Allen & Hamilton, 57-61; vpres prog develop, Armour Res Found, 61-63; vpres new prod develop, Joy Mfg Co, Pa, 63-67; vpres corp develop, Nat Gypsum Co, NY, 67-69; Max Factor & Co, Calif, 69-71; pres, Wyatt & Co, 71-72; PRES & OWNER, AMBASSADOR INDUSTS INC, 72- *Concurrent Pos:* US rep, World Metall Cong Steelmaking, 51. *Mem:* Am Inst Mining, Metall & Petrol Engrs; Am Soc Metals; Marine Technol Soc. *Res:* Extractive and physical metallurgy of titanium, zinc, zirconium, uranium; development of heavy industrial machinery; mining and construction; chemical engineering. *Mailing Add:* 510 NE Golden Harbour Dr Boca Raton FL 33432

WYATT, JEFFREY RENNER, b Hampton, Va, Jan 1, 46; m 69; c 3. ANALYTICAL CHEMISTRY. *Educ:* Univ Calif, Riverside, AB, 67; Northwestern Univ, Evanston, PhD(chem), 71. *Prof Exp:* Teaching assoc chem, Univ Kans, 71-72; Nat Res Coun res fel, 72-73, RES CHEMIST, NAVAL RES LAB, 73- *Mem:* Am Chem Soc; Am Soc Mass Spectrometry. *Res:* Mass spectrometry; chemical dynamics; trace analysis. *Mailing Add:* Naval Res Lab Code 6110 Washington DC 20375

WYATT, PHILIP JOSEPH, b Los Angeles, Calif, Apr 16, 32; m 57; c 3. LIGHT SCATTERING, ELECTROOPTICAL INSTRUMENTS. *Educ:* Univ Chicago, BA, 52, BS, 54; Univ Ill, MS, 56; Fla State Univ, PhD, 59. *Prof Exp:* Staff mem, Los Alamos Sci Lab, 59; prin scientist, Aeronutronic Div, Ford Motor Co, 59-62; dir adv planning, Plasmadyne Corp, 62-63; mem tech staff, DRC Inc, 63-66; sr sci specialist, EG&G, Inc, 66-68; pres & chmn, Sci Spectrum, Inc, 68-82; PRES & CHMN, WYATT TECHNOL CORP, 82- *Concurrent Pos:* Vis lectr physics, Univ Calif, Santa Barbara, 81-82. *Mem:* Fel Am Phys Soc; fel Optical Soc Am; Am Soc Microbiol; Sigma Xi. *Res:* Light scattering studies of microparticles; development of new assays and identification techniques using resonance light scattering; quality control for beverages and foods using electrooptical techniques; development of light scattering instrumentation; bioassays for antimicrobials in serum, antineoplastic drugs, drug residues in food, and toxicants in water; antibiotic susceptibility testing. *Mailing Add:* PO Box 3003 Santa Barbara CA 93130-3003

WYATT, RAYMOND L, b Salisbury, NC, Nov 23, 26; m 52; c 1. PLANT MORPHOLOGY, PLANT TAXONOMY. *Educ:* Wake Forest Col, BS, 48; Univ NC, MA, 54, PhD, 56. *Prof Exp:* Instr biol, Mars Hill Col, 48-52 & Univ NC, 55-56; from asst prof to assoc prof, 56-75, lectr, NSF Insts Sci Teachers, 59-60, PROF BIOL, WAKE FOREST UNIV, 75- *Mem:* AAAS. *Res:* Embryology of Asarum; floral morphology and phylogeny of Aristolochiaceae and Annonaceae; survival of American chestnut in North Carolina. *Mailing Add:* Dept Biol Wake Forest Univ Reynolds Sta Winston-Salem NC 27109

WYATT, RICHARD JED, b Los Angeles, Calif, June 5, 39. PSYCHOPHARMACOLOGY. *Educ:* Johns Hopkins Univ, BA, 61, MD, 64; Am Bd Psychiat & Neurol, dipl, 71. *Hon Degrees:* Dr, Univ Centrale Venezuela, 77. *Prof Exp:* Intern pediat, Western Reserve Univ Hosp, 64-65; resident psychiat, Mass Ment Health Ctr, Boston, 65-67; clin assoc, Lab Clin Psychobiol, 67-69, res psychiatrist, Lab Clin Psychopharmacol, 67-69, dir, Div Spec Mental Health Res, 77-87, assoc dir intramural res, 77-87, CHIEF NEUROPSYCHIAT BR, NIMH NEUROSCI CTR, 77- *Concurrent Pos:* Teaching asst, Harvard Univ, 65-67; consult, Sch Syst, Falls Church, Va, 69-70, Head Start, Washington, DC, 70-72 & Chestnut Lodge Hosp, 84-; instr, Found Advan Educ Sci, NIH & Washington Sch Psychiat, 71-; assoc prof psychiat, Med Ctr, Stanford Univ, 73-74; clin prof, Med Ctr, Duke Univ, 75-; adj prof psychiat, Uniformed Servs Univ Sch Med, 80-; ed, Foundations of Modern Psychiat, 82- & Psychiat Briefs, 85-; assoc ed, Psychiat, 85-, Am J Psychiat, 88- & Schizophrenia Bull, 88-; vis prof psychiat, Columbia Univ, 87- *Honors & Awards:* Harry Solomon Res Award, Mass Ment Health Ctr, Boston, 68; A E Bennett Award Clin Psychiat Res, Soc Biol Psychiat, 71; Psychopharmacol Award, Am Psychol Asn, 71; Stanley R Dean Res Award, Am Col Psychiatrists, 82; Daniel Efron Award, Am Col Neuropsychopharmacol, 83; Arthur P Noyes Award Schizophrenia Res, 86. *Mem:* Fel Am Psychiat Asn; Soc Psychophysiol Study Sleep; Psychiat Res Soc; Soc Biol Psychiat; fel Am Col Neuropsychopharmacol; Am Asn Geriat Psychiat; Asn Clin Psychosocial Res; Int Psychogeriat Asn; Soc Neurosci. *Res:* Etiology and treatment of major psychiatric disorders; neural plasticity; sleep and memory; author of numerous scientific publications. *Mailing Add:* Rm 536 NIMH Neurosci Ctr Neuropsychiat Br St Elizabeth's Hosp NIMH Washington DC 20032

WYATT, ROBERT EDWARD, b Charleston, SC, July 15, 50; m 78; c 1. POPULATION BIOLOGY, PLANT SYSTEMATICS. *Educ:* Univ NC, AB, 72; Duke Univ, PhD(bot), 77. *Prof Exp:* Asst prof biol, Tex A&M Univ, 77-79; from asst prof to assoc prof, 79-83, PROF BOT, UNIV GA, 88- *Concurrent Pos:* Vis asst prof biol, Univ Va, 80; mem bd sci adv, Highlands Biol Sta, Highlands, NC; prin investr on grants, Nat Sci Foun & Whitehall Found; Sarah Moss fel, Guggenheim fel. *Honors & Awards:* George R Cooley Award, Am Soc Plant Taxonomists. *Mem:* Am Soc Plant Taxonomists; Bot Soc Am; Soc Study Evolution; Ecol Soc Am; Am Soc Naturalists; Am Bryological & Lichenological Soc; Sigma Xi. *Res:* Reproductive biology of plants, especially Asclepias, Arenaria, and Aesculus; population dynamics, breeding systems, pollination and fruit-set; plant biosystematics and evolutionary ecology; population ecology of bryophytes. *Mailing Add:* Dept Bot Univ Ga Athens GA 30602

WYATT, ROBERT EUGENE, b Chicago, Ill, Nov 11, 38; m 64; c 1. THEORETICAL CHEMISTRY. *Educ:* Ill Inst Technol, 61; Johns Hopkins Univ, MA, 63, PhD(chem), 65. *Prof Exp:* NSF fel chem, Keele Univ, 65-66 & Harvard Univ, 66-67; from asst prof to assoc prof, 67-76, PROF CHEM, UNIV TEX, AUSTIN, 76- *Honors & Awards:* Prize Medal, Int Acad Quantum Molecular Sci, 80. *Mem:* AAAS; Am Chem Soc; Am Phys Soc. *Res:* Scattering theory; theoretical chemical dynamics; molecular multiphoton processes. *Mailing Add:* Dept Chem Univ Tex Austin TX 78712-1104

WYATT, ROGER DALE, b Albemarle, NC, Apr 16, 48; m 68; c 2. MICROBIOLOGY, POULTRY SCIENCE. *Educ:* NC State Univ, BS, 70, MS, 72, PhD(microbiol), 74. *Prof Exp:* Asst prof, 74-77, ASSOC PROF, DEPT POULTRY SCI, UNIV GA, 77- *Mem:* Poultry Sci Asn; Am Soc Microbiol. *Res:* Biological effects of dietary mycotoxins on poultry and evaluation of antifungal compounds for use in grain and poultry feeds. *Mailing Add:* Dept Animal Sci Univ Ga Athens GA 30602

WYBAN, JAMES A, b Cleveland, Ohio, Jan 23, 51; m 80; c 2. AQUACULTURE PRODUCTION TECHNOLOGY, INDUSTRY DEVELOPMENT. *Educ:* Northwestern Univ, BS, 73; Univ Hawaii, PhD(zool), 81. *Prof Exp:* Res assoc, Inst Cancer Res & fel, Marine Biol Lab, 78; res asst, Hawaii Inst Marine Biol, 76-81; pres-owner, Northshore Fish & Produce, 81-85; co-mgr, US Aid Coop Res Prog & prin investr, Hawaii Shrimp Prog, 84-85; PRIN INVESTR, US MARINE SHRIMP FARMING PROG, 85- & MARINE INSTRUMENTATION PROG, 87- *Concurrent Pos:* Adj prof, dept animal sci, Univ Hawaii, 87- *Mem:* Am Soc Zool; Fedn Am Scientists; Nat Wildlife Fedn; World Aquaculture Soc. *Res:* Marine shrimp production technology for US commercial application combining economic evaluation, system design and biological performance targeting economic optimization. *Mailing Add:* Oceanic Inst Makapuu Point PO Box 25280 Honolulu HI 96825

WYCH, ROBERT DALE, b Kingsley, Iowa, Mar 3, 48; c 2. CROP PHYSIOLOGY, AGRONOMY. *Educ:* Iowa State Univ, BS, 70; Univ Calif, Davis, MS, 74, PhD(plant physiol), 77. *Prof Exp:* Asst prof small grain physiol, Dept Agron & Plant Genetics, Univ Minn, 77-82; res agronomist, 82-86, PROD RES MGR, PIONEER HI-BRED INT INC, 86- *Mem:* Am Soc Plant Physiol; Am Soc Agron; Crop Sci Soc Am; Plant Growth Regulator Soc Am; Weed Sci Soc Am. *Res:* Influence of genotype and environment on seed corn production. *Mailing Add:* NAS Prod Dept Pioneer Hi-Bred Int Inc 7000 Pioneer Pkwy PO Box 256 Johnston IA 50131

WYCKOFF, DELAPHINE GRACE ROSA, b Beloit, Wis, Sept 11, 06; m 42. MICROBIOLOGY, BACTERIOLOGY. *Educ:* Univ Wis, PhB, 27, PhM, 28, PhD(bact), 38. *Prof Exp:* From instr to asst prof bact, NDak Agr Col, 28-37; from instr to assoc prof bact & bot, 38-57, prof bact, 57-72, EMER PROF BACT, WELLESLEY COL, 72- *Concurrent Pos:* Consult, Traveling Sci Teachers Prog, Oak Ridge Inst Nuclear Studies & Biol Sci Curric Study, Am Inst Biol Sci. *Mem:* Fel AAAS; Am Soc Microbiol; Am Acad Microbiol. *Res:* Physiological variation; induced mutations in actinomycetes; bactericidal agents; antibiotics from actinomycetes; biochemical activities of yeasts; marine halophilic bacteria; soil microbiology. *Mailing Add:* 78 Cedar St Newington CT 06111

WYCKOFF, HAROLD ORVILLE, b Traverse City, Mich, Apr 26, 10; m 40; c 2. PHYSICS. *Educ:* Univ Wash, BS, 34, PhD(physics), 40. *Prof Exp:* Jr physicist, Nat Bur Stand, 41-42, from asst physicist to assoc physicist, 42-43; expert consult, Ninth US Army Air Force, Europe, 43-45; physicist & asst chief, X-ray Sect, Nat Bur Stand, 45-49, chief, 49-53, chief radiation physics lab, 53-66; dep sci dir, Armed Forces Radiobiol Res Inst, 66-71; consult, Bur Radiol Health, 71-74; CONSULT RADIATION PHYSICS, 74- *Concurrent Pos:* Secy & mem, Int Comn Radiation Units & Measurements, 53-69, chmn, 69-85, prin sci counr, 85; mem comt, Int Comn Radiol Protection, 53-69; mem adv comt, Health Physics Div, Oak Ridge Nat Lab, 62-69; mem bd, Nat Coun Radiation Protection & Measurements, 64-70; consult, Fed Aviation Agency, 71-74; mem US Delegation, Sci Comt Effects of Atomic Radiation, UN, 76-, mem Sec Standards Dosimetry Lab, Sci Comt Int Atomic Energy Agency. *Honors & Awards:* Silver Medal, Dept Com, 52, Gold Medal, 60; Gold Medal, Radiol Soc NAm, 63; Gold Medal, XIII Int Cong Radiol, 73; Gold Medal, Am Col Radiol, 76. *Mem:* Health Physics Soc; fel Am Phys Soc; Radiation Res Soc; fel Am Col Radiol; Radiol Soc NAm (treas, 70-76). *Res:* Radiation physics; radiation protection and measurement. *Mailing Add:* 4108 Montpelier Rd Rockville MD 20853

WYCKOFF, HAROLD WINFIELD, b Niagara Falls, NY, Dec 3, 26; m 55; c 3. MOLECULAR BIOPHYSICS. *Educ:* Antioch Col, BS, 49; Mass Inst Technol, PhD(biophys), 55. *Prof Exp:* Res assoc biol, Mass Inst Technol, 55; NIH fel, Cambridge Univ, 56; res physicist, Am Viscose Corp, Pa, 57-63; ASSOC PROF MOLECULAR BIOPHYS, YALE UNIV, 63- *Mem:* Am Crystallog Asn; Biophys Soc. *Res:* Structure and function of biological macromolecules, especially enzymes, as determined by x-ray diffraction analysis. *Mailing Add:* Dept Biophys & Biochem Yale Univ Sch Med 333 Cedar St New Haven CT 06510

WYCKOFF, JAMES M, b Niagara Falls, NY, July 3, 24; m 47, 68; c 5. NUCLEAR PHYSICS. *Educ:* Antioch Col, BS, 48; Univ Rochester, MS, 52. *Prof Exp:* Electronics technician, Airborne Instruments Lab, 46-47; asst physics, Antioch Col, 47-48; res asst, Univ Rochester, 48-51; physicist, Nat Bur Stand, 51-67; health physicist, Stanford Linear Accelerator Ctr, 67-68; physicist, 68-71, coordr, Radiation Safety Prog, 71-76, LIAISON OFFICER, STATE & LOCAL GOVT AFFAIRS, NAT BUR STANDARDS, 76- *Concurrent Pos:* Mem comt 22, Nat Coun Radiation Protection & Measurement, 66-72, mem off telecommun policy side effects subcomt, 72-75; mem, Interagency Comt Fed Guid Occup Exposures to Ionizing Radiation, 74-75; rep, Nat Conf State Legislatures' Model Interstate Tech Info Clearinghouse, 76-; rep, Fed Lab Consortium Technol Transfer, 77-; exec secy comt, Fed Labs Off Sci & Technol Policy, 76-77. *Honors & Awards:* Metcalf Award, Fed Lab Consortium Technol Transfer, 82; Presidents Award, Technol Transfer Soc, 84. *Mem:* Am Phys Soc; Sigma Xi; Fedn Am Scientists; Technol Transfer Soc (vpres, 83-). *Res:* Detection of high energy x-rays; measurement of attenuation coefficients; induced radioactivity and development of on-line computer system; application of measurements to safe and effective use of x-rays; radioactivity, ultraviolet light, lasers, electromagnetic and ultrasonic radiation sources; development of measurement systems adequate to the protection of those near such radiation sources; review of mechanisms of technology transfer. *Mailing Add:* 24300 Hanson Ct Gaithersburg MD 20879

WYCKOFF, RALPH WALTER GRAYSTONE, b Geneva, NY, Aug 9, 97; m 27; c 3. PHYSICAL CHEMISTRY, BIOPHYSICS. *Educ:* Hobart Col, BS, 16; Cornell Univ PhD(chem), 19. *Hon Degrees:* MD, Masaryk Univ, Brno, 47; ScD, Univ Strasbourg, 52, Hobart Col, 75. *Prof Exp:* Instr anal chem, Cornell Univ, 17-19; phys chemist, Carnegie Inst Geophys Lab, 19-27; assoc mem subdiv biophys, Rockefeller Inst, 27-38; with Lederle Labs, 38-40, assoc res dir, 40-42; tech dir, Reichel Labs, 42-43; lectr, Univ Mich, 43-45; sr scientist, NIH, 45, scientist dir, 46-52, sci attache, Am Embassy, Eng, 52-54, biophysicist, 54-59; prof, 59-84, EMER PROF PHYSICS & BACTERIOL, UNIV ARIZ, 84- *Concurrent Pos:* Res assoc, Calif Inst Technol, 21-22; dir res, Nat Ctr Sci Res, France, 58-62 & 69; consult, Duval Corp, 61-84. *Honors & Awards:* Medal, Pasteur Inst. *Mem:* Nat Acad Sci; fel AAAS; fel Am Chem Soc; fel Am Phys Soc; fel Am Acad Arts & Sci; foreign mem Royal Soc London. *Res:* Structure of crystals; effect of radiation on cells; development of air-driven ultracentrifugation of proteins and viruses; electron microscopy; purification of viruses and macromolecules; ultra-soft x-rays; ultrastructure of fossils; application of x-ray and electron optical techniques to ore minerals. *Mailing Add:* 4741 E Cherry Hills Dr Tucson AZ 85718

WYCKOFF, SUSAN, b Santa Cruz, Calif, Mar 18, 41; m 67. ASTRONOMY. *Educ:* Mt Holyoke Col, AB, 62; Case Western Reserve, PhD(astrophys), 67. *Prof Exp:* Post doctoral fel, Univ Mich, 67-68; asst prof physics, Albion Col, 68-70; res assoc astron, Univ Kans, 70-72; Smithsonian res fel, Wise Observ, Tel Aviv Univ, 72-75; prin res fel, Royal Greenwich Observ, 75-78 & 83; vis prof, Dept Astron, Ohio State Univ, 78-79; assoc prof, 79-82, PROF PHYSICS, ARIZ STATE UNIV, 82- *Concurrent Pos:* Hon lectr, Women Astron, Smith Col, 78; vis prof, Univ Heidelberg, 80; Millan scholar, Denison Univ, 82; mem, comt comet spectroscopy, Int Astron Union, coun, Am Astron Soc, 85-; discipline specialist spectroscopy & spectrophotometry, Int Halley Watch, NASA, 82-, Space Telescope Working Group Key Projs, Extra Galactic Astron, 84-88; chmn, Comt Status Women, Am Astron Soc, 84-85; space sci adv comt, Nat Acad Sci, 86-87; adv comt, NSF, 90. *Honors & Awards:* Shapley Lectr, Am Astron Soc, 85. *Mem:* Int Astron Union; Royal Astron Soc. *Res:* Cool stars, comets and supernovae; normal and active galaxies; quasars and cosmology. *Mailing Add:* Dept Physics Ariz State Univ Tempe AZ 85287

WYCOFF, SAMUEL JOHN, b Berry, Ala, Feb 25, 29. PUBLIC HEALTH, DENTISTRY. *Educ:* Univ Ala, Tuscaloosa, BS, 50; Univ Ala, Birmingham, DMD, 54; Univ Mich, Ann Arbor, MPH, 59; Am Bd Dent Pub Health, dipl, 65. *Prof Exp:* Staff dent officer, Div Dent Health, USPHS, Washington, DC, 59-63, regional prog dir, Boston, Mass, 63-65; assoc prof prev dent & community health & chmn dept, Sch Dent Loyola Univ Chicago, 67-69; PROF GEN DENT & CHMN DEPT, SCH DENT, UNIV SAN FRANCISCO, 69- *Concurrent Pos:* Fels, Univ Calif, San Francisco, 69-; consult, Coun Dent Educ, Am Dent Asn, 68- & Calif State Health Manpower Coun, 70-; mem comt acad affairs, Am Asn Dent Schs, 71-; mem tech adv comt, Calif State Dept Pub Health, 72-; mem, Pierre Fouchard Acad. *Honors & Awards:* H Trendley Dean Citation, 65. *Mem:* Am Dent Asn; fel Am Pub Health Asn; Am Asn Hosp Dent; fel Am Col Dent; Int Dent Fedn. *Res:* Epidemiology of oral disease; studies on health care delivery system and health manpower needs. *Mailing Add:* Dept Dent Univ Calif San Francisco CA 94143

WYDEVEN, THEODORE, b Wausau, Wis, Jan 18, 36; m 63; c 2. PHYSICAL CHEMISTRY. *Educ:* Marquette Univ, BS, 58; Univ Wash, PhD(phys chem), 64. *Prof Exp:* RES SCIENTIST ENVIRON CONTROL SYSTS, NASA AMES RES CTR, 64- *Mem:* Am Chem Soc. *Res:* Advanced environmental control systems for purifying water, recycling oxygen and controlling atmospheric trace contaminants; development of plasma polymerized thin film technology. *Mailing Add:* 1130 Revere Dr Sunnyvale CA 94087

WYETH, NEWELL CONVERS, b Wilmington, Del, Oct 13, 46; m 72; c 2. PHYSICS. *Educ:* Princeton Univ, AB, 67; Univ Calif, Berkeley, PhD(physics), 74. *Prof Exp:* Physicist appl physics, Eng Physics Lab, E I du Pont de Nemours & Co, 68-69; assoc scientist solid state physics, Inst Energy Conversion, Univ Del, 74-79; SR SCIENTIST, SCI APPLN INT CORP, 79- *Mem:* Am Phys Soc; AAAS; Math Asn Am. *Res:* Underwater acoustics; biomembrane modeling. *Mailing Add:* Marine Technol Div Sci Appln Int Corp McLean VA 22102

WYGANT, J(AMES) F(REDERIC), b Hornell, NY, Aug 4, 19; m 44; c 7. CERAMIC ENGINEERING, PETROLEUM TECHNOLOGY. *Educ:* Alfred Univ, BS, 41, MS, 48; Mass Inst Technol, ScD(ceramics), 50. *Prof Exp:* Chemist, Mo Portland Cement Co, 41-43; proj engr, Eng Res Dept, Standard Oil Co, Ind, 50-55, sect head res & develop dept, Am Oil Co, 55-62, asst dir res dept, 62-68, dir eng & explor res, Res & Develop Dept, 68-71, dir prod & explor res, 71-73; dir corp res, Standard Oil Co, Ind, 73-75, mgr corp serv, 75-79; RETIRED. *Concurrent Pos:* Consult, 79- *Honors & Awards:* Purdy Award, Am Ceramic Soc, 52; Pace Award, Inst Ceramic Engrs, 59. *Mem:* Fel Am Ceramic Soc; Inst Ceramic Engrs. *Res:* Non-petroleum energy resources and processing; process industry refractories and insulations; industrial research evaluation and management; fireproofing; organic materials of construction; cements and concretes; corrosion protection; carbon; petroleum processing; advanced energy resources; alternative fossil fuels. *Mailing Add:* 1306 Francisco Dr Tallahassee FL 32304

WYGNANSKI, ISRAEL JERZY, b Warsaw, Poland, June 3, 35; US citizen; m 62; c 2. TURBULENCE. *Educ:* McGill Univ, Montreal, BEng, 61, MEng, 62, PhD(aerodyn), 64. *Prof Exp:* Asst prof heat transfer, Univ BC, 64-65; sr res scientist, Boeing Sci Res Labs, 65-72; PROF FLUID DYNAMICS, TEL-AVIV UNIV, 72- & AEROSPACE ENG, UNIV ARIZ, 85- *Concurrent Pos:* Chmn, Eng Dept, Tel-Aviv Univ, 72-76, dean fac eng, 77-80; consult, Israel Aircraft Industs, 72-76; vis prof, Univ Southern Calif, 76-77 & Univ Ariz, 81-83; fel, Inst Advan Study, Berlin, Ger, 90-91. *Mem:* Nat Acad Eng; fel Am Phys Soc; Am Inst Aeronaut & Astronaut. *Res:* Low speed aerodynamics, high life devices, delay of flow separation and drag reduction; turbulent shear flow control and enhancement of mixing. *Mailing Add:* Dept Aerospace & Mech Eng Univ Ariz Tucson AZ 85721

WYKES, ARTHUR ALBERT, b Boston, Mass, May 21, 23; m 56; c 2. BIOCHEMICAL PHARMACOLOGY & TOXICOLOGY. *Educ:* Univ Ill, BS, 45; Univ Wis, MS, 49; Purdue Univ, PhD(pharmaceut chem, pharmacol, biochem), 57. *Prof Exp:* Res asst biochem, Res Div, Armour & Co, 45-46; res asst biochem, Univ Wis, 47-49; teaching asst pharmaceut chem, Purdue Univ, 49-51; res biochemist, Armour & Co, 51-52; res biochemist, Res Div, Int Minerals & Chem Corp, 52-53; res biochemist, Res Dept, Baxter Labs, Inc, 53-55; teaching asst pharmaceut chem, Purdue Univ, 55-57; sr res biochem pharmacologist, Chem Pharmacol Sect, Abbott Labs, 57-61; chief & supvry res pharmacologist, chem pharmacol sect, US Air Force Sch Aerospace Med, 61-67; sr pharmacologist, Chem & Life Sci Labs, Res Triangle Inst, 67-68; PHARMACOLOGIST & SR DRUG & TOXICOL INFO SPECIALIST, NAT LIBR MED, NIH, 68- *Concurrent Pos:* Assoc clin prof pharmacol, Div Pharmacol & Physiol, Med Ctr, Duke Univ, 67-69; vis assoc prof pharmacol, Milton S Hershey Med Ctr, 71-77; consult, Int Life Sci Inst, 84. *Mem:* Fel AAAS; Am Chem Soc; Int Soc Biochem Pharmacol; Soc Toxicol; Drug Info Asn; Am Soc Pharmacol & Exp Therapeut; Fedn Am Socs Exp Biol; Asn Govt Toxicologists (pres-elect); Sigma Xi; Int Asn Study Fatty Acids & Lipids. *Res:* Drug and toxic agent effects on biological, enzyme and metabolic systems at cellular and subcellular levels, especially as influenced by environmental agents, nutritional factors, other drugs and toxic chemicals; drug interactions; neurochemistry; basic pharmacology and toxicology; enzymology; central nervous system and cardiovascular drugs; drugs of abuse; biogenic amines; psychopharmacology drugs and central nervous system agents; actions of prescription and experimental drugs, nutritional supplements and factors, biotoxins and hazardous chemicals; pharmacology, toxicology, biotechnology products, nutritional agents and their uses; computerized biomedical information systems for query/response activities; specialized information services relevent to toxicology, pharmacology, nutrition, biomedicine and chemistry. *Mailing Add:* 18900 Diary Rd Gaithersburg MD 20879

WYKLE, ROBERT LEE, b Belmont, NC, Mar 17, 40; m 71. BIOCHEMISTRY. *Educ:* Western Carolina Univ, BS, 63; Univ Tenn, Memphis, PhD(biochem), 70. *Prof Exp:* Teacher, Waynesville High Sch, 63-65; teaching asst biochem, Med Sch, Univ Tenn, 65-67; fel, Oak Ridge Assoc Univs, 68-70, assoc scientist, 70-71, scientist, 76-80; ASSOC PROF, DEPT BIOCHEM, BOWMAN GRAY SCH MED, 80- *Mem:* Am Soc Biol Chemists. *Res:* Biochemistry and function of lipids in normal and neoplastic cells with special interest in ether-linked lipids. *Mailing Add:* Dept Biochem Bowman Gray Sch Med 300 Hawthorne Rd Winston-Salem NC 27103

WYLD, HENRY WILLIAM, JR, b Portland, Ore, Oct 16, 28; m 55; c 3. PHYSICS. *Educ:* Reed Col, BA, 49; Univ Chicago, MS, 52, PhD(physics), 54. *Prof Exp:* Instr physics, Princeton Univ, 54-57; from asst prof to assoc prof, 57-63, PROF PHYSICS, UNIV ILL, URBANA, 63- *Concurrent Pos:* Consult, Space Tech Labs, Inc, Calif, 57-63; NSF sr fel, Oxford Univ, 63-64; Guggenheim fel, Europ Coun Nuclear Res, 71. *Mem:* Am Phys Soc. *Res:* Theoretical high energy and plasma physics. *Mailing Add:* Dept Physics Loomis Lab Physics Univ Ill 1110 W Green St Urbana IL 61801

WYLDE, RONALD JAMES, b St Louis, Mo, Feb 7, 21; m 47; c 3. INSTRUMENTATION, PHYSICAL MEASUREMENTS. *Educ:* Washington Univ, St Louis, BA, 42; Univ Md, MS, 52. *Prof Exp:* Physicist, Underwater Acoust Div, Naval Ord Lab, USB, 42-48, head physics br, Naval Gun Factory, 48-54, supt elec lab, Naval Eng Exp Sta, 54-57, head, Instrumentation Div, 57-63 & Elec Div, US Navy Marine Eng Lab, 63-68, head elec dept, Naval Ship Res & Develop Ctr, 68-71; CONSULT, 71- *Mem:* Sr mem Inst Elec & Electronics Engrs. *Res:* Electronic circuitry of measuring systems; electromechanical transducers; amplitude-modulating transducers; electronic devices for physical measurement. *Mailing Add:* 5320 18th St-Blvd E Bradenton FL 34203

WYLER, ALLEN RAYMER, b Seattle, Wash, Sept 3, 43; m 83. EPILEPSY, CLINICAL NEUROSURGERY. *Educ:* Univ Wash, BA, 65, MD, 69. *Prof Exp:* Actg instr neurosurg, Univ Wash, 74-75, from asst prof to assoc prof, Dept Neurol Surg, 75-84; chief neurosurg, Harborview Med Ctr, Seattle, Wash, 77-84; PROF NEUROSURG, UNIV TENN, MEMPHIS, 84- *Concurrent Pos:* Prof, Univ Wash Hosp, Seattle, 74-84; teacher investr award, NIH, 77-83; mem staff, Baptist Hosp, Memphis, Tenn, 84- *Mem:* Am League Against Epilepsy; Am Asn Neurol Surgeons; Res Soc Neurol Surgeons. *Res:* Active clinical and basic neurophysiological investigations into the basic mechanisms of focal epilepsy. *Mailing Add:* Dept Neurosurg Coleman Bldg A202 Univ Tenn 956 Court Ave Memphis TN 38163

WYLER, DAVID J, b Dec 21, 44; m; c 2. IMMUNOPARASITOLOGY, TROPICAL MEDICINE. *Educ:* Harvard Univ, MD, 70. *Prof Exp:* PROF MED, SCH MED, TUFTS UNIV, 85- *Concurrent Pos:* Physician, New Eng Med Ctr, 83- *Mem:* AAAS; Am Col Physicians; Am Soc Microbiol; Am Soc Trop Med Hyg; Am Assoc Immunol; Am Soc Clin Invest; Infectious Dis Soc Am. *Res:* Infectious diseases. *Mailing Add:* New Eng Med Ctr 750 Washington St NEMCH 041 Boston MA 02111

WYLER, JOHN STEPHEN, mechanical engineering, fluid mechanics, for more information see previous edition

WYLER, OSWALD, b Scuol, Grisons, Switz, Apr 2, 22; m 60; c 3. MATHEMATICS. *Educ:* Swiss Fed Inst Technol, dipl, 47, Dr sc math, 50. *Prof Exp:* Asst inst geophys, Swiss Fed Inst Technol, 46-50; lectr math, Northwestern Univ, 51-53; from asst prof to assoc prof, Univ NMex, 53-65; PROF MATH, CARNEGIE-MELLON UNIV, 65- *Mem:* Am Math Soc; Math Asn Am; Swiss Math Soc. *Res:* Categorical algebra; categorical topology; theory of convergence spaces; continuous lattices. *Mailing Add:* Dept Math Carnegie-Mellon Univ Pittsburgh PA 15213

WYLIE, C J, b Morganton, NC, June 20, 26. NUCLEAR ENGINEERING. *Educ:* Univ SC, BSEE, 50. *Prof Exp:* Chief elec engr, Duke Power Co, NC, 70-84; MEM ADV COMT REACTOR SAFEGUARDS, US NUCLEAR REG COMN, 84- *Concurrent Pos:* Elec Utility Engr, 84- *Mem:* Fel Inst Elec & Electronics Engrs. *Res:* Electrics and design of power generator plants. *Mailing Add:* 9610 Lawyers Rd Charlotte NC 28227

WYLIE, CLARENCE RAYMOND, JR, b Cincinnati, Ohio, Sept 9, 11; m 35, 58; c 2. GEOMETRY. *Educ:* Wayne State Univ, BA & BS, 31; Cornell Univ, MS, 32, PhD(math), 34. *Prof Exp:* Instr & asst prof math, Ohio State Univ, 34-46; prof & head dept & actg dean col eng, USAF Inst Technol, 46-48; prof, Univ Utah, 48-69, head dept, 48-67; prof, 69-71, chmn dept, 70-76, Kenan prof math, 71-78, WILLIAM R KENAN, JR PROF EMER, FURMAN UNIV, 78- *Concurrent Pos:* Consult, Gen Elec Co, NY, 37 & Briggs Mfg Co, Mich, 41; mech engr, Wright Field Propellor Lab, Ohio, 43-46 & Aero Prod Div, Gen Motors Corp, 45-47; lectr, Educ Prog, Union Carbide Corp, 65-70. *Mem:* Fel AAAS; Am Math Soc; Math Asn Am; Am Soc Eng Educ; Sigma Xi. *Res:* Projective geometry, especially line geometry; applied mathematics, especially mechanical vibrations. *Mailing Add:* Seven Merritt View Terr Greenville SC 29609

WYLIE, DOUGLAS WILSON, b Saskatoon, Sask, Nov 12, 26; nat US; m 51; c 4. PHYSICS. *Educ:* Univ NB, BSc, 47; Dalhousie Univ, MSc, 49; Univ Conn, PhD(solid state physics), 62. *Prof Exp:* Asst, Brown Univ, 49-50; instr, Univ NB, 50-51; from instr to prof, Univ Maine, 51-68; chmn dept, 68-73, PROF PHYSICS, WESTERN ILL UNIV, 68- *Mem:* AAAS; Am Phys Soc; Am Asn Physics Teachers. *Res:* Solid state physics; radiation damage; electron spin resonance. *Mailing Add:* Dept Physics Western Ill Univ Macomb IL 61455

WYLIE, EVAN BENJAMIN, b Sask, Can, Jan 14, 31; US citizen; m 55; c 3. HYDRAULICS, CIVIL ENGINEERING. *Educ:* Univ Denver, BS, 53; Univ Colo, MS, 55; Univ Mich, PhD(hydraul), 64. *Prof Exp:* Asst engr, City Engr Off, Englewood, Colo, 54-55; jr res off hydraul sect, Div Mech Eng, Nat Res Coun Can, 55-56; struct design engr, Ford Motor Co Can, Ltd, 56-59; asst prof civil eng, Univ Denver, 59-62; from asst prof to assoc prof, 65-70, PROF CIVIL ENG, UNIV MICH, ANN ARBOR, 70-, CHMN DEPT, 84- *Mem:* Fel Am Soc Civil Engrs; fel Am Soc Mech Engrs; Am Soc Eng Educ; Int Asn Hydraul Res; Earthquake Eng Res Inst. *Res:* Basic and applied research in fluid transients in closed and open conduits, and fluid flow in porous media. *Mailing Add:* Dept Civil Eng Univ Mich 2340 G G Brown Ann Arbor MI 48109-2125

WYLIE, HAROLD GLENN, b Wingham, Ont, Oct 15, 27; m 53; c 2. ENTOMOLOGY. *Educ:* Univ Toronto, BA, 49; Univ Oxford, PhD(entom), 53. *Prof Exp:* Res scientist, Belleville, Ont, 49-50 & 51-72, res scientist entom, Can Dept Agr, Winnipeg, 72-87; RETIRED. *Concurrent Pos:* Res scientist, Commonwealth Inst Biol Control, Zurich, 50-51; adj prof entom, Univ Man, 82-88. *Mem:* hon mem Entom Soc Can. *Res:* Parasitic hymenoptera and diptera; manipulation of insect parasites for pest control; host selection behavior of insect parasites. *Mailing Add:* 643 Silverstone Ave Winnipeg MB R3T 2V8 Can

WYLIE, KYRAL FRANCIS, b Akron, Ohio, Feb 24, 31; m 56. ENGINEERING, PHYSICS. *Educ:* Kent State Univ, BS, 58, MA, 60; Ohio State Univ, PhD(eng), 70. *Prof Exp:* Res physicist, Mound Lab, Ohio, 60-65, sr res physicist, 65-66; res fel, Ohio State Univ, 70; from asst prof to assoc prof mech eng, Univ Miss, 70-78; STAFF MEM, LOS ALAMOS NAT LAB, 78- *Concurrent Pos:* Consult physicist, Nuclear Med Lab, Grandview Hosp, Dayton, Ohio, 60-66. *Mem:* Am Soc Mech Engrs; Am Phys Soc; Health Physics Soc; Sigma Xi. *Mailing Add:* PO Box 254 Santa Fe NM 87504

WYLIE, RICHARD MICHAEL, b Louisville, Ky, June 17, 34; m 69. BIOLOGY, NEUROPHYSIOLOGY. *Educ:* Harvard Univ, BA, 56, MA, 58, PhD(biol), 62. *Prof Exp:* Fel neurophysiol, Univ Utah, 62-65, res assoc, 65-66; res assoc, Rockefeller Univ, 66-69; RES PHYSIOLOGIST, DEPT MED NEUROSCI, WALTER REED ARMY INST RES, 69- *Mem:* Soc Neurosci; assoc mem Am Physiol Soc. *Res:* Biophysics of sensory mechanisms; mechanisms of sensory discrimination in central nervous systems; integration in sensory and motor systems. *Mailing Add:* Dept Med Neurosci Walter Reed Army Inst Res Washington DC 20307

WYLLIE, GILBERT ALEXANDER, b Saltcoats, Scotland, Jan 11, 28; US citizen; m 57; c 3. BIOLOGY, ECOLOGY. *Educ:* Col Idaho, BS, 58; Sacramento State Col, MA, 60; Purdue Univ, PhD(ecol), 63. *Prof Exp:* Assoc prof biol, WTex State Univ, 63-65; asst prof, 65-66, ASSOC PROF BIOL, BOISE STATE UNIV, 66- *Mem:* Ecol Soc Am. *Res:* Effects of environment on morphology, distribution, behavior of invertebrates and lower vertebrates. *Mailing Add:* Dept Biol Boise State Univ 1910 University Dr Boise ID 83725

WYLLIE, LORING A, ENGINEERING. *Educ:* Univ Calif, BS, 60, MS, 62. *Prof Exp:* CHMN BD, H J DEGENKOLB ASSOC, 86- *Concurrent Pos:* Consult, var agencies; chmn, Comt Concrete & Masonry Struct, Am Soc Civil Engrs, 81-84; dir, Am Concrete Inst, 85-88, mem, Tech Activ Comt, 82-88 & Comt 318, Standard Bldg Code, 72-; dir, Earthquake Eng Res Inst, 86-89. *Mem:* Nat Acad Eng; fel Am Soc Civil Engrs; fel Am Concrete Inst; Earthquake Eng Res Inst. *Res:* Analysis of existing buildings; design for correction of structural deficiencies; strengthening of buildings for improved seismic performance; author of various publications. *Mailing Add:* H J Degenkolb Assoc 350 Sansome St Suite 900 San Francisco CA 94104

WYLLIE, PETER JOHN, b London, Eng, Feb 8, 30; m 56; c 3. GEOLOGY, GEOCHEMISTRY. *Educ:* Univ St Andrews, BSc, 52 & 55, PhD, 58. *Hon Degrees:* DSc, Univ St Andrews, 74. *Prof Exp:* Geologist, Brit NGreenland Exped, 52-54; asst lectr geol, Univ St Andrews, 55-56; asst lectr geochem, Pa State Univ, 56-58, asst prof, 58-59, assoc prof petrol, 61-65, actg head, Dept Geochem & Mineral, 62-63; res fel chem, Leeds Univ, Eng, 59-60, lectr exp petrol, 60-61; prof petrol & geochem, Univ Chicago, 65-83, master phys sci, Col Div & assoc dean col & Phys Sci Div, 72-73, Homer J Livingston Prof, 78-83, chmn geophys sci, 79-82; chmn & prof geol, Div Geol & Planetary Sci, 83-87, PROF GEOL, CALIF INST TECHNOL, PASADENA, 87- *Concurrent Pos:* Managing ed, J Petrol, 65-67, ed, J Geol, 67-83, ed-in-chief, Springer-Verlan Monograph Series, 67-; mem, Comn Exp Petrol, Int Union Geophys, 70-; chmn, Award Comt, Mineral Soc Am, 70-71, counr, 70-72; mem subcomt, Penrose Medal Award, Geol Soc Am, 71-73; subcomt, Nat Medal Sci, 80-82, counr, 82-85; mem vis comt, Dept Geol & Geophys, Woodshole Oceanog Inst, 72; Goldschmidt Medal Comt, Geochem Soc, 72-74, counr, 82-85; mem, Adv Panel Earth Sci, NSF, 75-78, chmn, Earth Sci Div Adv Comt, 79-82; team leader, Basatic Volcanism Study Proj, Lunar & Planetary Inst, 76-79; vpres, Int Mineral Asn, 78-86; mem ed bd, Tectonophys, 78-; mem, Macelwane Award Comt, Am Geophys Union, 78-80; mem, US Nat Comt Geol, 78-80; mem, US Nat Comt, Int Union Geodesy & Geophys, 80-84; vchmn coordn comt, Int Union Comn Litosphere, 81-84; mem, Comt Ocean Drilling, Nat Acad Sci, 81-82; mem, Arthur Holmes Medal Comt, Europ Geophys Union, 83; mem adv bd, Sch Earth Sci, Stanford Univ, 83-88; pres, Int Mineral Asn. 86-; Louis Murray vis fel, Univ Cape Town, 87- *Honors & Awards:* Polar Medal, Her Majesty Queen Elizabeth, 54; Res Award, Mineral Soc Am, 65; Quantrell Award, 79; Wollaston Medal, Geol Soc London, 82; Abraham-Gottlob-Werner Medal, Ger Mineral Soc, 87. *Mem:* Foreign assoc Nat Acad Sci; foreign fel Indian Geophys Union; fel Am Geophys Union; fel Geol Soc Am; hon mem Mineral Soc London; fel Mineral Soc Am; fel Am Acad Arts & Sci; fel Royal Soc London. *Res:* Igneous and metamorphic petrology; experimental petrology; high pressure studies on hydrothermal systems; application of phase equilibrium studies to batholiths, andesites, kimberlites, and carbonatites; author of two textbooks and one book on geosciences. *Mailing Add:* Calif Inst Technol 1201 E California Blvd Pasadena CA 91125

WYLLIE, THOMAS DEAN, b Hinsdale, Ill, Dec 4, 28; m 50; c 3. PLANT PATHOLOGY. *Educ:* San Diego State Col, AB, 52; Univ Minn, MS, 57, PhD(plant path), 60. *Prof Exp:* PROF PLANT PATH, UNIV MO-COLUMBIA, 60- *Mem:* Am Phytopath Soc. *Res:* Mycotoxin and mycotoxicoses research; computer modeling; ecological relationships of non-specific soil borne pathogenic fungi on the soybean. *Mailing Add:* Dept Plant Path 108 Waters Hall Univ Mo Columbia MO 65211

WYLLY, ALEXANDER, b New York, NY, Sept 2, 20; m 43; c 2. OPERATIONS RESEARCH, COMPUTER SCIENCE. *Educ:* Ga Inst Technol, BS, 41; Mass Inst Technol, SM, 46; Calif Inst Technol, PhD(aerodyn eng), 51. *Prof Exp:* Aerodynamicist, Rand Corp, 46-51, chief missile preliminary designs, 51-54; founder & vpres comput sci, Planning Res Corp, 54-69, vpres pac opers, 69-73; CONSULT, 73- *Concurrent Pos:* Bd dir, Planning Res Corp, 54-74; trustee, Oceanic Found. *Mem:* Am Soc Info Sci. *Res:* Systems analysis; computer systems software design and applications. *Mailing Add:* 1015 Wilder Ave Apt 601 Honolulu HI 96822

WYLLYS, RONALD EUGENE, b Phoenix, Ariz, May 14, 30; m 57; c 4. INFORMATION SCIENCE. *Educ:* Ariz State Univ, BA, 50; Univ Wis-Madison, PhD(info sci), 74. *Prof Exp:* Mathematician, Dept Defense, Washington, DC, 54-59; assoc, Planning Res Corp, Los Angeles, 59-61; comput systs specialist, Syst Develop Corp, Santa Monica, 61-66; chief systs analyst, Univ Librs, Univ Wis-Madison, 66-69, lectr comput sci, 66-72; Dean, 82-90, PROF LIBR & INFO SCI, GRAD SCH LIBR & INFO SCI, UNIV TEX, AUSTIN, 72- *Concurrent Pos:* Assoc math, George Washington Univ, 54-59. *Mem:* AAAS; Am Soc Info Sci; Am Statist Asn; Asn Records Mgrs & Adminrs. *Res:* Information storage and retrieval; statistical linguistics and other programmatic techniques for the analysis of information content. *Mailing Add:* Grad Sch Libr & Info Sci Univ Tex Austin TX 78712-1276

WYLY, LEMUEL DAVID, JR, b Seneca, SC, Aug 9, 16; m 38; c 2. NUCLEAR PHYSICS. *Educ:* The Citadel, BS, 38; Univ NC, MA, 39; Yale Univ, PhD(physics), 49. *Prof Exp:* Instr physics, Ga Sch Technol, 39-41, asst prof, 46; asst instr, Yale Univ, 46-48; from assoc prof to prof, 49-58, REGENTS' PROF PHYSICS, GA INST TECHNOL, 58- *Concurrent Pos:* Consult, Oak Ridge Nat Lab, 52- *Mem:* Fel Am Phys Soc; Sigma Xi. *Res:* Nuclear energy levels; proportional and scintillation and solid state detectors; decay schemes of radioactive isotopes and from neutron capture. *Mailing Add:* Dept Physics Ga Inst Technol Atlanta GA 30332

WYMA, RICHARD J, b Grand Rapids, Mich, June 25, 36; m 64; c 3. PHYSICAL CHEMISTRY, INORGANIC CHEMISTRY. *Educ:* Hope Col, AB, 58; Univ Mich, MS, 60, PhD(phys chem), 64. *Prof Exp:* Asst prof chem, Geneva Col, 64-69; ASSOC PROF CHEM, IND-PURDUE UNIV, INDIANAPOLIS, 70- *Concurrent Pos:* Consult, City Indianapolis & Inst Adv Res, Ind-Purdue Univ, 76-80. *Mem:* Am Chem Soc; Soc Appl Spectros; Coblentz Soc. *Res:* Application of molecular spectroscopy to structure determination and to bounding theories of inorganic systems; chemistry of sulfides, phosphines, boranes and transition metal complexes; chemical education research; methods for solving chemical problems. *Mailing Add:* IUPUI Dept Chem 1125 E 38th St Indianapolis IN 46205-2820

WYMAN, BOSTWICK FRAMPTON, b Aiken, SC, Aug 22, 41; m 75; c 2. MATHEMATICS. *Educ:* Mass Inst Technol, SB, 62; Univ Calif, Berkeley, MA, 64, PhD(math), 66. *Prof Exp:* Instr math, Princeton Univ, 66-68; asst prof, Stanford Univ, 68-72; assoc prof, 72-82, PROF MATH, OHIO STATE UNIV, 82- *Concurrent Pos:* Vis asst prof, Univ Oslo, 70-71; vis assoc prof, Univ Notre Dame, 78-79; vis prof, Univ Genoa, Italy, 88-90. *Mem:* Am Math Soc; Math Asn Am; Inst Elec & Electronics Engrs. *Res:* Algebraic system theory; linear control system design; algebraic number theory. *Mailing Add:* Dept Math Ohio State Univ Columbus OH 43210

WYMAN, DONALD, b Templeton, Calif, Sept 18, 03; m 27; c 4. HORTICULTURE. *Educ:* Pa State Col, BSA, 26; Cornell Univ, MSA, 31, PhD(hort), 35. *Prof Exp:* Instr, Pa State Univ, 27-29; investr, Cornell Univ, 29-31, instr, 31-35; horticulturist, 36-70, EMER HORTICULTURIST, ARNOLD ARBORETUM, HARVARD UNIV, 70- *Honors & Awards:* Coleman Award, Am Asn Nurserymen, 49 & 51; NY Hort Soc Distinguished Serv Award, 60; Garden Club Fedn Am Medal Honor, 65; Veitch Medal, Royal Hort Soc, 69; George Robert White Medal Honor, Mass Hort Soc, 70; Arthur Hoyt Scott Gold Medal, Swarthmore Col, 71; L H Bailey Medal, Am Hort Soc, 71. *Mem:* Am Soc Hort Sci (vpres, 52-53); Am Hort Soc (pres, 61-62); Am Asn Bot Gardens & Arboretums; fel Nat Recreation & Park Asn. *Res:* Ornamental horticulture; plant propagation; winter hardiness; selection of best varieties of woody plants for landscape use. *Mailing Add:* 59 Jericho Rd Weston MA 02193

WYMAN, DONALD PAUL, b Cleveland, Ohio, Feb 4, 31; m 54; c 2. TECHNICAL MANAGEMENT, ORGANIC CHEMISTRY. *Educ:* Ohio Univ, BS, 53; Mich State Univ, PhD(org chem), 57. *Prof Exp:* Group leader, Lubrizol Corp, 57-61; fel, Mellon Inst, 61-64; res mgr, Borg Warner Chem, 64-68, Scott Graphics, 68-74; tech dir, GAF, 74-79; vpres, Rochester Midland Corp, 79-87; PRES, UNYTEX RES CORP, 87- *Concurrent Pos:* Chmn, T-3A comt, Nat Asn Corrosion Engrs. *Mem:* AAAS; Am Asn Metals; Am Chem Soc; Nat Asn Corrosion Engrs. *Res:* Polymers; plastics; corrosion and scale inhibitors; disinfectants; organic chemicals. *Mailing Add:* Unytex Res Corp 333 Hollenbeck St PO Box 818 Rochester NY 14603-0818

WYMAN, GEORGE MARTIN, b Budapest, Hungary, Oct 13, 21; nat US; m 51; c 1. CHEMICAL DYNAMICS, PHOTOCHEMISTRY. *Educ:* Cornell Univ, AB, 41, MS, 43, PhD(org chem), 44. *Prof Exp:* Res chemist, Gen Chem Co, 44-45, Gen Aniline & Film Corp, 45-49 & Nat Bur Standards, 49-54; chief spectros sect, Qm Res & Develop Ctr, US Dept Army, 54-57, sci adv, European Res Off, 57-60, dir chem div, 60-77, chief Chem Br, 77-83, dir, Chem & Biol Sci Div, US Army Res Off, 83-85; CONSULT, 85- *Concurrent Pos:* Adj prof chem, Univ NC, Chapel Hill, 73-77, 86- *Mem:* Am Chem Soc; Int Am Photochem Soc; Europ Photochem Asn. *Res:* Spectrophotometry; cis-trans isomerization of conjugated compounds; organic photochemistry; excited state chemistry of dyes. *Mailing Add:* 2231 Cranford Rd Durham NC 27706

WYMAN, HAROLD ROBERTSON, chemistry; deceased, see previous edition for last biography

WYMAN, JEFFRIES, b West Newton, Mass, June 21, 01; m 54; c 2. MOLECULAR BIOLOGY. *Educ:* Harvard Univ, AB, 23; Univ London, PhD, 27. *Prof Exp:* From instr to assoc prof zool, Harvard Univ, 28-51; sci adv, US Embassy, Paris, 51-54; dir, UNESCO Sci Off MidE, 55-58; guest prof, Biochem Inst, Univ Rome & Inst Regina Elena, 60-84; RETIRED. *Concurrent Pos:* Past secy gen, Europ Molecular Biol Orgn. *Mem:* Nat Acad Sci; Am Acad Arts & Sci. *Mailing Add:* 29 Ave De La Motte Picquet Paris 75007 France

WYMAN, JOHN E, b Amsterdam, NY, Feb 20, 31; m 52; c 4. CHEMISTRY. *Educ:* Univ Mich, BS, 52, Purdue Univ, MS, 55, PhD, 56. *Prof Exp:* Res chemist, Linde Co, Union Carbide Corp, 56-58, res chemist, Union Carbide Chem Co, 58-59; res chemist, Spec Proj Dept, Monsanto Chem Co, 59-60, res group leader, Monsanto Res Corp, 60-65; MEM SCI STAFF, ITEK CORP, 65- *Mem:* Am Chem Soc. *Res:* Photochemistry; complex transition element organometallic chemistry; metal carbonyls; propellant, explosive and inorganic chemistry; graphic arts, film and paper coatings. *Mailing Add:* 17 Monadnock Dr Westford MA 01886-3018

WYMAN, MAX, mathematics; deceased, see previous edition for last biography

WYMAN, MILTON, b Cleveland, Ohio, Oct 11, 30; m 56; c 2. VETERINARY MEDICINE, OPHTHALMOLOGY. *Educ:* Ohio State Univ, DVM, 63, MS, 64; Am Col Vet Ophthal, dipl. *Prof Exp:* Res assoc ophthal, Cols Med & Vet Med, 62-64, instr vet ophthal, Col Vet Med, 64-66, from asst prof to prof vet ophthal & med, 66-75, CHIEF COMP VET OPHTHAL & MED, COL VET MED, OHIO STATE UNIV, 75-, CHIEF SMALL ANIMAL SERV, 72-, ASSOC PROF OPHTHAL, COL MED, 72- *Mem:* Am Soc Vet Ophthal; Am Col Vet Ophthalmologists (past pres); Am Vet Med Asn; Am Asn Vet Clin. *Res:* Congenital ocular defects in dogs and their relationship to man; glaucoma in the basset hound; ocular fundus anomaly in collies; medical application of soft contact lenses in animals and man. *Mailing Add:* 427 E Lorain St Oberlin OH 44074

WYMAN, RICHARD VAUGHN, b Painesville, Ohio, Feb 22, 27; m 47; c 1. GEOLOGICAL ENGINEERING, ECONOMIC GEOLOGY. *Educ:* Case Western Reserve Univ, BS, 48; Univ Mich, MS, 49; Univ Ariz, PhD(geol eng), 74. *Prof Exp:* Geologist econ geol, NJ Zinc Co, 49; geologist, Cerro de Pasco Corp, 50-52; geologist, NJ Zinc Co, 52-53; geologist uranium, Western Gold & Uranium Inc, 53-54, chief geologist, 54-55, gen mgr, 55-57, vpres, 57-59; tunnel supt tunnel construct, Reynolds Elec & Eng Co, 61-63; construct supt, Sunshine Mining Co, 63-65; engr, Reynolds Elec & Eng Co, 65-67, asst mgr, 67-69; lectr, Univ Nev, 69-74, assoc prof, 74-80, chmn dept, 76-80, chmn, civil & mech eng, 84-90, PROF ENG, UNIV NEV, 80-, CHMN, CIVIL & ENVIRON ENG, 90- *Concurrent Pos:* Pres explor geol, Intermountain Explor Co, 59-; consult, C K Geoenergy Corp, Latin Am, 77-; Univ Nev, Las Vegas Senate Res Grant Geothermal, 77-; mem peer rev comt, Nevada Nuclear Waste Isolation Proj, Dept Energy, 78-82. *Mem:* Am Inst Mining Metall & Petrol Engrs; Am Soc Civil Engrs; Soc Econ Geol; Geol Soc Am; Asn Eng Geol; Sigma Xi; distinguished mem Soc Mining Engrs; Arctic Inst NAm. *Res:* Geothermal energy utilization; ore genesis; tunnel construction and design. *Mailing Add:* 610 Bryant Ct Boulder City NV 89005

WYMAN, ROBERT J, b Syracuse, NY, June 8, 40. DEVELOPMENTAL NEUROBIOLOGY, DEVELOPMENTAL GENETICS. *Educ:* Harvard Univ, AB, 60; Univ Calif, Berkeley, MA, 63, PhD(biophys), 65. *Hon Degrees:* MA, Yale Univ. *Prof Exp:* Math analyst, Tech Res Group, Inc, 59; NSF res fel appl sci, Calif Inst Technol, 66; from asst prof to assoc prof, 66-80, PROF BIOL, YALE UNIV, 80- *Concurrent Pos:* Vis scientist, Nobel Inst, Stockholm, 70-71, Med Res Coun, Cambridge, Eng, 74 & Univ Basel, Switz, 77; mem physiol study sect, NIH, 76-80; mem bd sci adv, Microgenesis Corp. *Mem:* Soc Neurosci; Int Brain Res Orgn; Sigma Xi; Brit Soc Exp Biol. *Res:* Genetics of Drosophila nervous system; neural generation of motor output in insects and vertebrates; insect physiology; molecular cloning of genes necessary for the specificity of neural connectivity. *Mailing Add:* Dept Biol Yale Univ New Haven CT 06511

WYMAN, STANLEY M, b Cambridge, Mass, Aug 3, 13; c 4. MEDICINE, RADIOLOGY. *Educ:* Harvard Univ, AB, 35, MD, 39. *Prof Exp:* Radiologist, Mass Gen Hosp, 47-68; asst clin prof radiol, Harvard Med Sch, 54-75; vis radiologist, 68-77, radiologist, 77-81, SR RADIOLOGIST, MASS GEN HOSP, 81-; EMER PROF RADIOL, HARVARD MED SCH, 85- *Concurrent Pos:* Consult, USN, 57-73; clin prof radiol, Harvard Med Sch, 77-85. *Honors & Awards:* Silver Medal, Roentgen Ray Soc, 52; Gold Medal, Am Col Radiol, 72; Gold Medal, Radiol Soc NAm, 74. *Mem:* Radiol Soc NAm (pres, 68); Roentgen Ray Soc; AMA; Am Col Radiol (pres, 71). *Res:* Cardiovascular radiology. *Mailing Add:* Zero Emerson Pl Boston MA 02114

WYMER, JOSEPH PETER, b Marion, Va, July 31, 23; m 55; c 3. INDUSTRIAL ENGINEERING. *Educ:* Va Polytech Inst & State Univ, BSIE, 47; Univ Southern Calif, MSIE, 69; Golden State Univ, PhD(indust eng), 83. *Prof Exp:* Methods & standards engr, Brunswick, Balke-Collender, Va, 47-49; RCA Victor, Va, 49-51; mfg engr, Convair Div, Gen Dynamics Corp, Tex, 51-55; chief indust engr, O'Sullivan Rubber Corp, Va, 55-59; chief indust engr, Gen Instruments Corp, Va, 59; indust engr, Boeing Airplane Co, Wash, 59-61; prof & chmn dept, 61-83, EMER PROF INDUST ENG, CALIF STATE POLYTECH UNIV, POMONA, 83- *Concurrent Pos:* Indust eng consult, Lawry Foods, Inc, McDonnell Douglas Corp, Electronic Specialties, Taylor Corp & others, Southern Calif, 61- *Mem:* Inst Indust Engrs; Nat Soc Prof Engrs; Am Soc Eng Educ. *Mailing Add:* Dept Indust Eng Calif State Polytech Univ Pomona 3801 W Temple Ave Pomona CA 91768

WYMER, RAYMOND GEORGE, b Colton, Ohio, Oct 1, 27; m 48; c 4. NUCLEAR CHEMISTRY. *Educ:* Memphis State Col, BS, 50; Vanderbilt Univ, MA & PhD, 53. *Prof Exp:* Mem staff, Oak Ridge Nat Lab, 53-56; assoc prof, Ga Inst Technol, 56-58; chief nuclear chem, Indust Reactor Labs, 58-59; res chemist, 59-62, sect chief, 62-73, assoc dir, 73-82, DIR, CHEM TECHNOL DIV, OAK RIDGE NAT LAB, 73- *Concurrent Pos:* Ed, Radiochimica Acta. *Mem:* Am Inst Chem Engrs; Sigma Xi; fel Am Nuclear Soc; Am Chem Soc. *Res:* Colloid, radiation, transuranium element and complex ion chemistry; kinetics; nuclear fuel cycle. *Mailing Add:* 188A Outer Dr Oak Ridge TN 37830

WYMORE, ALBERT WAYNE, b New Sharon, Iowa, Feb 1, 27; m 49; c 4. SYSTEM THEORY, SYSTEMS ENGINEERING. *Educ:* Iowa State Univ, BS, 49, MS, 50; Univ Wis, PhD(math), 55. *Prof Exp:* Consult, Pure Oil Co, 55-57; dir, Comput Ctr, Univ Ariz, 57-67, head dept systs eng, 59-74, prof systs eng, 59-87; OWNER & PRIN SYSTS ENGR, SANDS SYSTS ANALYSIS & DESIGN SYSTS, 87- *Concurrent Pos:* Consult, RCA, 58 & 73, IBM, San Jose, 60-61, Lockheed, Ga, 69, Centro Agronomotrop de Invest y Ensenanza, Costa Rica, 75-76, Gen Elec Tempo Ctr Advan Studies, 77-78, IBM Fed Syst Div, 82-85, Siemens WGer, 86-88 & 90, Bellcore, 85-88, Unisys, 88, City Tucson, Dept Energy & Environ, 89-91, Agr Res Serv, USDA, 89 & Bechtel, 91. *Mem:* AAAS; Am Inst Indust Engrs; Am Math Soc. *Res:* Mathematical system theory; system design methodology; information systems; software engineering; agricultural systems; communication systems. *Mailing Add:* Sands Systs Analysis & Design Systs 4301 Camino Kino Tucson AZ 85718

WYMORE, CHARLES ELMER, inorganic chemistry; deceased, see previous edition for last biography

WYNBLATT, PAUL P, b Alexandria, Egypt, June 30, 35; m 59; c 2. MATERIALS SCIENCE, SURFACE SCIENCE. *Educ:* Univ Manchester, BScTech, 56; Israel Inst Technol, MS, 58; Univ Calif, Berkeley, PhD(metall), 66. *Hon Degrees:* Dr, Univ Aix-Marseille, 87. *Prof Exp:* Res metallurgist, Israel Atomic Energy Comn Labs, 58-62; staff scientist res staff, Ford Motor Co, 66-81; dir, Ctr Study Mat, 85-88, PROF METALL ENG & MAT SCI, CARNEGIE-MELLON UNIV, 81- *Concurrent Pos:* Bd dir, Am Inst Mining Metallurg & Petrol Engrs, Metallurg Soc, 77-80. *Mem:* Fel Am Soc Metals; Am Vacuum Soc; Am Inst Mining Metall & Petrol Engrs-Metall Soc; Mat Res Soc. *Res:* Equilibrium surface composition of materials; structure and composition of solid-solid interfaces; thermal stability of thin films; particle migration phenomena; catalytic materials. *Mailing Add:* Dept Metall Eng & Mat Sci Carnegie-Mellon Univ Pittsburgh PA 15213

WYNDER, ERNST LUDWIG, b Ger, Apr 30, 22; nat US. PREVENTIVE MEDICINE, EPIDEMIOLOGY. *Educ:* NY Univ, BA, 43; Wash Univ, BS & MD, 50. *Prof Exp:* Intern, Georgetown Univ Hosp, 50; asst prof prev med, Grad Sch Med Sci, Med Col, Cornell Univ, 54-56, assoc prof, 56-69; PRES & MED DIR, AM HEALTH FOUND, 69- *Concurrent Pos:* Asst, Sloan-Kettering Inst Cancer Res, 52-54, assoc, 54-60, assoc mem, 60-69, assoc scientist, 69-71; jr asst resident, Mem Hosp for Cancer & Allied Dis, 51-52, sr asst resident, 52-54, clin asst physician, 54-64, asst attend physician, 64-69, consult epidemiologist, 69-; clin vis asst, James Ewing Hosp, 54-64, asst vis physician, 64-68; mem, Task Force Lung Cancer, Tobacco Working Group, 67-; mem, Nat Cancer Plan, 71; ed, Prev Med J, 72. *Mem:* AMA; Am Asn Cancer Res; Am Pub Health Asn; NY Acad Sci. *Res:* Environmental factors affecting major chronic disease development, preventive medical aspects. *Mailing Add:* Am Health Found 860 United Nations Plaza New York NY 10017

WYNDRUM, RALPH WILLIAM, JR, b Brooklyn, NY, Apr 20, 37; m 60; c 4. ELECTRICAL ENGINEERING, TELECOMMUNICATIONS. *Educ:* Columbia Univ, BS, 59, MS, 60, MS(bus admin), 78; NY Univ, EngScD(elec eng), 63. *Prof Exp:* Mem tech staff, 63-65, supvr explor circuit appln, 65-69, dept head loop transmission systs, 69-79, head adv transmission systs dept, 79-87, dir, systs anal, 87-90, DIR, QUAL PROCESSES, BELL LABS, INC, 90- *Concurrent Pos:* Asst prof, NY Univ, 63-64; mem, Nat Basic Sci Comt, 66-70; adj prof, Newark Col Eng, 67-70; prof elec eng & computer sci, Stevens Inst Technol, 81-89; mem bd dirs, Inst Elec & Electronics Engrs, 88-90, vpres publ, 90. *Mem:* Fel Inst Elec & Electronics Engrs. *Res:* Network synthesis; integrated circuit design; telephone transmission design; quality systems. *Mailing Add:* Bell Labs Ho 31-324 Crawfords Corner Rd Holmdel NJ 07733-1988

WYNEKEN, JEANETTE, b Bloomington, Ill, July 14, 56. ZOOLOGY. *Educ:* Ill Wesleyan Univ, BA, 78; Univ Ill, PhD(biol), 88. *Prof Exp:* Res assoc, Univ Ill, 88-89; RES ASSOC, FLA ATLANTIC UNIV, 90. *Mem:* Am Soc Zoologist; Sigma Xi; AAAS; Soc Study Amphibians & Reptiles; Soc Study Evolution; Herpetologists League. *Res:* Understanding how organisms contend with their environments; the behavior, functional morphology and physiology of sea turtles; understanding the adapture significance of characteristics unique to marine turtles or unique to migratory species. *Mailing Add:* Dept Biol Sci Fla Atlantic Univ Boca Raton FL 33431-0991

WYNER, AARON D, b New York, NY, Mar 17, 39; m 63; c 4. ELECTRICAL ENGINEERING, MATHEMATICS. *Educ:* Queens Col, NY, BS, 60; Columbia Univ, BS, 60, MS, 61, PhD(elec eng), 63. *Prof Exp:* Asst prof elec eng, Columbia Univ, 63; mem tech staff, Math Res Ctr, 63-74, HEAD, COMMUN ANALYSIS RES DEPT, BELL LABS, 74- *Concurrent Pos:* Consult, T J Watson Res Ctr & IBM Corp, 63; adj prof, Columbia Univ, 64-; Guggenheim fel, 66-67; vis scientist, Weizmann Inst Sci & Israel Inst Technol, 69-70; vis prof, Polytech Inst Brooklyn, 71, Princeton Univ, 83 & 90. *Honors & Awards:* Centennial Award, Inst Elec & Electronics Engrs. *Mem:* AAAS; Inst Elec & Electronics Engrs. *Res:* Communication theory; information theory; probability. *Mailing Add:* Bell Labs Mountain Ave New Providence NJ 07974

WYNGAARD, JOHN C, b Madison, Wis, Dec 4, 38; m 65; c 2. FLUID DYNAMICS. *Educ:* Univ Wis-Madison, BSc, 61, MSc, 62; Pa State Univ, PhD(mech eng), 67. *Prof Exp:* Res physicist, Air Force Cambridge Res Labs, 67-75; physicist, Wave Propagation Lab, Nat Oceanic & Atmospheric Admin, Boulder, 75-79; PHYSICIST, MESOSCALE RES SECT, NAT CTR ATMOSPHERIC RES, 79-, HEAD, 81-; PROF METEOROL, PA STATE UNIV, 80- *Concurrent Pos:* Vis assoc prof atmospheric sci, Univ Wash, 73; fel, Coop Inst Res Environ Sci, Boulder, 75-79. *Mem:* Am Meteorol Soc; Am Phys Soc; Sigma Xi. *Res:* The structure and dynamics of turbulent flows, particularly in the lower atmosphere. *Mailing Add:* Dept Meteorol Pa State Univ 508 Walker Bldg University Park PA 16802

WYNGAARDEN, JAMES BARNES, b East Grand Rapids, Mich, Oct 19, 24; div; c 5. BIOCHEMISTRY, METABOLISM. *Educ:* Univ Mich, MD, 48. *Hon Degrees:* DSc, Univ Mich, 80, Med Col Ohio, 84, Univ Ill, Chicago, 85, George Wash Univ, 86, Univ SC, 89 & Western Mich Univ, 89; PhD, Tel Aviv Univ, 87. *Prof Exp:* Asst pharmacol, Med Sch, Univ Mich, 46-48; mem med house staff, Mass Gen Hosp, 48-52; vis investr, Pub Health Res Inst New York, 53; investr, Nat Heart Inst, 53-54 & Nat Inst Arthritis & Metab Dis, 54-56; assoc prof med & biochem, Sch Med, Duke Univ, 56-61, prof med & assoc prof biochem, 61-65; prof med & chmn dept, Univ Pa, 65-67; Frederic M Hanes prof med & chmn dept, Duke Univ, 67-82; dir NIH, 82-89; assoc dir, Life Sci Off, Sci & Tech Policy, The White House, 89-90; FOREIGN SECY, NAT ACAD SCI & INST MED, 90-; DIR, HUGO, 90- *Concurrent Pos:* Dalton scholar med res, Mass Gen Hosp, 51; consult, Vet Admin Hosp, Durham, NC; consult, Off Sci & Technol, Exec Off President, 66-72; mem adv comt biol & med, AEC, 67-69; mem adv bd, Howard Hughes Med Inst, 69-82; mem bd sci counrs, Nat Inst Arthritis, Metab & Digestive Dis, 71-74; mem, President's Sci Adv Comt, 72-73; mem exec comn, Assembly Life Sci, 72-77; mem, Coun Govt-Univ-Indust Res Roundtable, Nat Acad Sci, 84-90. *Mem:* Nat Acad Sci; Inst Med Nat Acad Sci; Am Soc Clin Invest; Am Acad Arts & Sci; Asn Am Physicians (pres, 78-79); fel Royal Col Physicians London; Am Soc Biochem & Molecular Biol. *Res:* Control of purine synthesis; purine metabolism in normal and gouty man; metabolism of iodine and steroids; oxalate synthesis; inborn errors of metabolism/human genetics. *Mailing Add:* Foreign Secy US Nat Acad Sci 2101 Constitution Ave Washington DC 20418

WYNN, CHARLES MARTIN, SR, b New York, NY, May 8, 39; m 66; c 4. ORGANIC CHEMISTRY. *Educ:* City Col NY, BChE, 60; Univ Mich, MS, 63, PhD(chem), 65. *Prof Exp:* Instr gen chem, Univ Mich, 65-67; US Peace Corps lectr chem, Malayan Teachers' Col, 67-69; from asst prof to prof chem, 79-83, gen chem coordr, Eastern Conn State Univ; from asst prof to assoc prof phys sci, 69-74, asst to provost, 74-75, CHMN DEPT, OAKLAND COMMUNITY COL, 64-, PROF PHYS SCI, 74- *Concurrent Pos:* Vis scholar, Wesleyan Univ, 87. *Mem:* Am Educ Sci Asn. *Res:* Structural directivity in diene synthesis; author of laboratory manuals and natural science textbook. *Mailing Add:* Dept Phys Sci Eastern Conn State Univ Willimantic CT 06226-2295

WYNN, CLAYTON S(COTT), chemical engineering; deceased, see previous edition for last biography

WYNN, JAMES ELKANAH, b Pennington Gap, Va, Feb 7, 42; m 64; c 1. MEDICINAL CHEMISTRY, ANALYTICAL CHEMISTRY. *Educ:* Va Commonwealth Univ, BS, 64, PhD(med chem), 69. *Prof Exp:* Res fel med & anal chem, Med Col Va, 69; from asst prof to assoc prof med chem, 69-77, PROF PHARM, COL PHARM, UNIV SC, 77- *Concurrent Pos:* Comn prod scholar grant, Col Pharm, Univ SC, 70-71, lectr, Proj Upward Bound, 70- *Mem:* Am Chem Soc; Am Pharmaceut Asn; Am Asn Cols Pharm; Sigma Xi. *Res:* Organic chemistry; cancer chemotherapeutic agents of the alkylating type; synthesis, testing and correlation of activity with physical parameters; mechanism of dimenthyl sulfoxide interaction with isolated enzyme systems; synthesis of agents for urolithiasis treatment. *Mailing Add:* 306 Ayers Circle Summerville SC 29483

WYNN, WILLARD KENDALL, JR, b Raleigh, NC, Mar 28, 32; m 63; c 2. PLANT PATHOLOGY. *Educ:* NC State Univ, BS, 55; Univ Fla, PhD(plant path), 63. *Prof Exp:* Assoc plant pathologist, Boyce Thompson Inst Plant Res, 63-68; ASSOC PROF PLANT PATH, UNIV GA, 68- *Mem:* Am Phytopath Soc. *Res:* Physiology of uredospore germination and rust infection; rust resistance. *Mailing Add:* Dept Plant Path Univ Ga Athens GA 30602

WYNNE, BAYARD EDMUND, b Pittsburgh, Pa, Sept 6, 30; div; c 3. GROUP DESIGN MAKING & COMMUNICATION SUPPORT SYSTEMS, ORGANIZATIONAL MANAGEMENT & LEADERSHIP. *Educ:* Carnegie-Mellon Univ, BSCE, 53, BSIM, 55, MSIA, 56; Univ Minn, PhD(info systs & psychol), 72. *Prof Exp:* Assoc, Booz Allen & Hamilton, Chicago, Ill, 56-61; reg admin serv mgr opers res, Arthur Andersen & Co, Detroit, Mich, 61-62; mgr opers res, Weyerhaeuser Co, Tacoma, Wash, 62-67; corp group vpres, Super Valu Stores, Inc, Hopkins, Minn, 67-70; mem bds educ, Orono & Twin Cities Metro Area, Minn, 70-72; sr lectr grad fac, Bus Sch, Univ Wis, Milwaukee Campus, 72-76; prin, Arthur Andersen & Co, Chicago, Ill, 76-86; PROF DECISION & INFO SYSTS, SCH BUS, IND UNIV, 86- *Concurrent Pos:* Mem bd dirs, Houston-Starr Co, Pa, 65-88; pres & consult, Wynne Affils, 69-; dir, Inst Res on Mgt Info Systs, Ind Univ, 86-89, ref fac mem, Inst Study of Develop Disabilities, 89-; res coordr, Soc Info Mgt, 88-90; vis res scientist, FAW Res Inst Appl Knowledge Processin, Ulm Univ, Ger, 91-; assoc ed, J Int Info Systs, 91- *Mem:* Acad Mgt; Soc Info Mgt; Inst Mgt Sci; Am Psychol Soc; Decision Sci Inst; Inst Indust Engrs; Asn Comput Mach; Am Psychol Asn; AAAS. *Res:* Interaction of organization and information systems in practice; design, development, and application of interactive, collaborative, "anytime-anyplace" group support systems; management, decision making, and leadership. *Mailing Add:* 2238 E Cape Cod Dr Bloomington IN 47401-6124

WYNNE, ELMER STATEN, b El Paso, Tex, Oct 23, 17; m 38, 78; c 2. MEDICAL MICROBIOLOGY. *Educ:* Univ Tex, BA, 38, MA, 44, PhD(bact), 48; Am Bd Microbiol, cert microbiol & bact. *Prof Exp:* Asst bact, Univ Tex, 38-39, tutor, 39-42, instr, 46, res assoc, 46-48, res bacteriologist, M D Anderson Hosp & Tumor Inst, 50-58, assoc prof microbiol, Dent Br, Univ, 58-59; asst prof, Univ Okla, 48-50; bacteriologist, USAF Sch Aerospace Med, 59-60, res prof bact & chief microbiol, 60-67, sr microbiologist, 68-69; from assoc prof to prof, St Phillips Col, 70-82, Med Lab Technol, 70-76, prog dir, 70-75; RETIRED. *Mem:* AAAS; Am Soc Microbiol; fel Am Acad Microbiol; Sigma Xi. *Res:* Enteric bacteriology; bacterial antagonism; physiology of Clostridium spore germination; microbiological aspects of cancer research; aerospace microbiology; hand disinfection; live hybrid vaccine for bacillary dysentery. *Mailing Add:* 10514 Auldine San Antonio TX 78220

WYNNE, JOHNNY CALVIN, b Williamston, NC, May 17, 43; m 61; c 3. PLANT BREEDING, PLANT GENETICS. *Educ:* NC State Univ, BS, 65, MS, 68, PhD(crop sci), 74. *Prof Exp:* From instr to assoc prof, 68-82, PROF & HEAD CROP SCI, NC STATE UNIV, 89- *Honors & Awards:* Bailey Award, Am Peanut Res & Educ Asn. *Mem:* Am Soc Agron; Am Peanut Res & Educ Asn; fel AAAS. *Res:* Improvement of cultivated peanuts through breeding for higher productivity, disease resistance, insect resistance, nitrogen fixation efficacy, and better quality; collaborative research on peanuts in Thailand and the Philippines. *Mailing Add:* Dept Crop Sci NC State Univ Box 7620 Raleigh NC 27695-7620

WYNNE, KENNETH JOSEPH, b Rumford, RI, Jan 17, 40; m 67; c 2. POLYMER SCIENCE, INORGANIC CHEMISTRY. *Educ:* Providence Col, BS, 61; Univ Mass, Amherst, MS & PhD(chem), 65. *Prof Exp:* Fel inorg chem, Univ Calif, Berkeley, 65-67; asst prof, Univ Ga, 67-73; sci officer, Off Naval Res, 73-83; PROG MGR, ORG & POLYMERIC MAT, 84- *Concurrent Pos:* Vis scientist, IBM, San Jose, 83-84; vis scholar, Stanford Univ, 83-84; prog mgr, Solid State Chem, NSF, 88-89. *Mem:* AAAS; Sigma Xi; Am Chem Soc. *Res:* Polymer chemistry; inorganic polymers; electroactive polymers; synthetic inorganic and organometallic chemistry. *Mailing Add:* Off Naval Res Chem Div 800 N Quincy St Arlington VA 22217

WYNNE, LYMAN CARROLL, b Lake Benton, Minn, Sept 17, 23; m 47; c 5. PSYCHIATRY, PSYCHOLOGY. *Educ:* Harvard Med Sch, MD, 47; Harvard Univ, PhD(soc psychol), 58. *Prof Exp:* Intern med, Peter Bent Brigham Hosp, Boston, 47-48; USPHS res fel, Harvard Univ, 48-49; Moseley traveling fel, London, Eng, 49-50; Rantoul fel psychol, Harvard Univ, 50; resident, Mass Gen Hosp, 51; resident, NIMH & St Elizabeth Hosp, Washington, DC, 52-53; psychiatrist, Lab Socio-Environ Studies, NIMH, Md, 54, Adult Psychiat, 54-71, chief family studies sect, 57-67, chief adult psychiat br, 61-71; chmn dept, 71-77, PROF PSYCHIAT SCH MED & DENT, UNIV ROCHESTER, 71- *Concurrent Pos:* Mem fac, Wash Sch Psychiat, 56-71; mem fac, Wash Psychoanal Inst, 60-71, teaching analyst, 66-71; consult & collab investr, WHO, 65-; mem-at-large, Div Behav Sci, Nat Res Coun, 69-72; psychiatrist-in-chief, Strong Mem Hosp, 71-77; mem rev comt career develop awards, NIMH, 72-76; vis lectr, Am Univ Beirut, 63-64. *Honors & Awards:* Commendation medal, USPHS, 65, Meritorious Serv Medal, 66; Fromm-Reichmann Award, Am Acad Psychoanal, 66; Hofheimer Prize, Am Psychiat Asn, 66; Salmon Lectr, 73; Stanley R Dean Res Award, Am Col Psychiatrists, 76; McAlpin Res Achievement Award, 77; Distinguished Contributions to Family Ther Res Award, Am Asn Marriage & Family Ther, 82, Distinguished Achievement in Family Ther Res Award, 81, Distinguished Prof Contributions to Family Ther Award, 86. *Mem:* Am Psychiat Asn; Psychiat Res Soc; Soc Life Hist Res Psychopath; Am Family Ther Asn (pres, 86-87); Asn Clin Psychosocial Res. *Res:* Family research and therapy; schizophrenia; cross-cultural studies; child development. *Mailing Add:* Dept Psychiat Univ Rochester Sch Med & Dent Rochester NY 14642

WYNNE, MICHAEL JAMES, b St Louis, Mo, Feb 4, 40. BIOLOGY. *Educ:* Wash Univ, St Louis, BA, 62; Univ Calif, Berkeley, PhD(botany), 67. *Prof Exp:* From asst prof biol to assoc prof biol, Univ Texas, Austin, 69-76; PROF BIOL, UNIV MICH, ANN ARBOR, 76- *Concurrent Pos:* Instr, Hopkins Marine Sta, Stanford Univ, 68, Friday Harbor Labs, Univ Wash, 70 & 78, Marine Biol Lab, Woods Hole, 71 & 72; vis res assoc, Univ Melbourne Parkville, 81. *Honors & Awards:* Darbaker Prize, Bot Soc Am, 71; G W Prescott Award, Phycol Soc Am, 87. *Mem:* Int Phycol Soc (pres, 88-89); Phycol Soc Am (secy, 73-75, vpres, 76, pres, 77); Bot Soc Am; Brit Phycol Soc; Am Soc Plant Syst. *Res:* Marine phycology, systematics of marine algae, morphology, and biogeography; the algae floor of Alaska, the Caribbean, and various other regions, particular interest on the red algae order Ceramiales. *Mailing Add:* Dept Biol & Herbarium Univ Mich Ann Arbor MI 48109

WYNNE-EDWARDS, HUGH ROBERT, b Montreal, Que, Jan 19, 34; m 56, 72, 85; c 4. GEOLOGY. *Educ:* Aberdeen Univ, BSc, 55; Queen's Univ, Ont, MA, 57, PhD(geol), 59. *Hon Degrees:* DSc, Mem Univ, 75. *Prof Exp:* Tech officer geol, Geol Surv Can, 58-59; lectr, Queen's Univ, Ont, 59-61, from asst prof to assoc prof, 61-68, prof geol & head dept geol sci, 68-72; Cominco prof geol & head, Dept Geol Sci, Univ BC, 72-77; asst secy, Ministry State for Sci & Technol, Govt Can, Ottawa, 77-79; sci dir, 79-80, VPRES & CHIEF SCI OFFICER, ALCAN INT LTD, MONTREAL, 80- *Concurrent Pos:* Vis fel, Aberdeen Univ, 65-66 & Univ Witwatersrand, 72; pres, Can Geosci Coun, 74; UN consult, India, 76; mem sci adv comt, Can Broadcasting Comn, 78-84; mem, Conseil de la Politique Sci, Quebec, 81-; dir, Soquip, 83-; dir & trustee, Sci Coun Can, 83- & Royal Victorial Hosp, 84- *Honors & Awards:* Spendiarov Prize, 72. *Mem:* AAAS; Asn Sci & Technol Community Can (pres, 77); Royal Soc Can. *Res:* Resources and physical environment; science policies. *Mailing Add:* Tech Corp 1199 W Hastings St Vancouver BC V6E 2K5 Can

WYNNYCKYJ, JOHN ROSTYSLAV, b Ukraine, Nov 4, 32; Can citizen; m 56; c 3. EXTRACTIVE METALLURGY, HIGH TEMPERATURE PROCESSES. *Educ:* McGill Univ, BEng, 56; Univ Toronto, MASc, 65, PhD(metall), 68. *Prof Exp:* Res engr, Int Nickel Co Can, 56-59; develop supvr, DuPont Can Ltd, 59-64; Nat Res Coun Can fel, Max Planck Inst Phys Chem, 68-69; assoc prof, 69-82, PROF CHEM ENG & EXTRACTIVE METALL, UNIV WATERLOO, 82- *Concurrent Pos:* Consult, metal extraction; vis scientist, Nat Res Coun, Ottawa, 79; div mineral eng, Commonwealth Sci & Indust Res Orgn, Australia, 84. *Honors & Awards:* Plummer Gold Medal, 81. *Mem:* Can Inst Mining & Metall; Metall Soc Am Inst Mining Metall & Petrol Engrs. *Res:* Mechanisms of high temperature heterogeneous reactions significant in metals extraction; coal combustion and ash fouling; sintering, selective reduction; iron ore pelletizing; silicothermic production of magnesium metal; agglomeration. *Mailing Add:* Dept Chem Eng Univ Waterloo Waterloo ON N2L 3G1 Can

WYNSTON, LESLIE K, b San Diego, Calif, Jan 5, 34; m 85; c 1. BIOCHEMISTRY. *Educ:* San Diego State Univ, BS, 55; Univ Calif, Los Angeles, MS, 58, PhD(physiol chem), 60. *Prof Exp:* Instr biochem, Med Sch, Northwestern Univ, 60-61; lectr, Med Sch, Univ Calif, San Francisco, 61-63; USPHS fel, Max Planck Inst Protein & Leather Res, 63-65; from asst prof to assoc prof chem, 65-75, PROF CHEM, CALIF STATE UNIV, LONG BEACH, 75- *Concurrent Pos:* Supvr, Metab Res Lab, Chicago Wesley Mem Hosp, 60-61; consult, NAm Aviation, Inc, 65-67; vis prof, Univ Zurich, 71-72; exchange prof, Nat Chung Hsing Univ Taiwan, 75-76. *Mem:* AAAS; Am Chem Soc; Soc Wine Educr; NY Acad Sci. *Res:* Protein purification and characterization; chemical isolation procedures; chromatographic and electrophoretic methods. *Mailing Add:* Dept Chem & Biochem Calif State Univ Long Beach CA 90840-3903

WYNVEEN, ROBERT ALLEN, b Baldwin, Wis, July 24, 39; m 64; c 4. HEALTH PHYSICS, MEDICAL PHYSICS. *Educ:* Univ Wis-River Falls, BS, 61; Rutgers Univ, MS, 63, PhD(radiation biophys), 72. *Prof Exp:* Health physicist, Argonne Nat Lab, 63-65; radiol health physicist, Rutgers Med Sch, Rutgers Univ, 65-76; health physics mgr, Argonne Nat Lab, 76-87, assoc mgr, Occup Health & Safety, 82-87, mgr, Environ Health & Safety Dept, 87-91, DIR, ENVIRON SAFETY & HEALTH DIV, ARGONNE NAT LAB, 91- *Concurrent Pos:* Radiol health physics consult, Colgate-Palmolive Res Ctr, 68-76, Warner-Lambert Res Ctr, 72-76, Ortho Diag & Pharmaceut, Inc, 74-76 & Fusion Energy Corp, 75-76. *Mem:* Am Asn Physicists Med; Nat Health Physics Soc. *Res:* Immediate and transient effects of radiation, especially ionizing, microwave and laser, on biological systems' functions with emphasis on cellular energy production and active transport across membranes; environmental hazard assessment. *Mailing Add:* Environ Safety & Health Div Argonne Nat Lab 9700 S Cass Ave Argonne IL 60439

WYON, JOHN BENJAMIN, b London, Eng, May 3, 18; m 46; c 2. EPIDEMIOLOGY, DEMOGRAPHY. *Educ:* Cambridge Univ, BA, 40, MB, BCh, 42; Harvard Univ, MPH, 53. *Prof Exp:* Med officer, Friends Ambulance Unit, Ethiopia, 43-45; med missionary to India from Church Missionary Soc, London, 47-52; res assoc epidemiol, Sch Pub Health, Harvard Univ, 53-58, instr, 58-60, res fel, 60-61, res assoc pop studies, 61-62, asst prof, 62-66, lectr pop studies & sr res assoc, Ctr Pop Studies, 66-71, sr lectr pop studies, 71-88; RETIRED. *Concurrent Pos:* Field dir, India-Harvard-Ludhiana Pop Study & asst prof, Christian Med Col, Ludhiana, India, 53-60. *Mem:* Am Pub Health Asn. *Res:* Community-oriented approaches to primary health care and population control; community epidemiology; field research on births, deaths and migrations in developing countries; development of local education units to demonstrate implications of population changes. *Mailing Add:* Dept Pop Sci 665 Huntington Ave Boston MA 02115

WYRICK, PRISCILLA BLAKENEY, b Greensboro, NC, Apr 28, 40. BACTERIOLOGY. *Educ:* Univ NC, Chapel Hill, BS, 62, MS, 67, PhD(bact), 71. *Prof Exp:* Technician clin microbiol, NC Mem Hosp, 62-64, asst supvr, 64-65, supvr in chg mycol & mycobact, 65-66; from asst prof to assoc prof, 73-88, PROF SCH MED, UNIV NC, CHAPEL HILL, 88- *Concurrent Pos:* Med Res Coun fel, Nat Inst Med Res, London, Eng, 71-73; consult, Dept Hosp Labs, NC Mem Hosp, 73- *Mem:* Am Soc Microbiol; Brit Soc Gen Microbiol; Am Acad Microbiol. *Res:* Bacterial L-forms; bacterial ultrastructure; pathogenesis of Chlamydia; medical microbiology. *Mailing Add:* Dept Microbiol CB 7290 FLOB Univ NC Sch Med Chapel Hill NC 27599

WYRICK, RONALD EARL, b Kansas City, Mo, Nov 4, 44; m 66; c 4. BIOCHEMISTRY, ALLERGY. *Educ:* Calif State Col, Stanislaus, BA, 68; Univ Calif, Davis, PhD(biochem), 74. *Prof Exp:* Indust res allergist, 74-81, vpres sci affairs, 81-84, VPRES OPERS, HOLLISTER-STIER, SUBSID MILES INC, WASH, 84- *Res:* Elucidation of allergy mechanisms to provide research directions for potential new treatments for the allergic condition. *Mailing Add:* Hollister-Stier Box 3145 Spokane WA 99220-3153

WYRICK, STEVEN DALE, b Greensboro, NC, Oct 23, 51; m 87. MOLECULAR MODELING, TISSUE CULTURE. *Educ:* Univ NC, Chapel Hill, BS, 74, PhD(med chem), 77. *Prof Exp:* Res assoc med chem, Research Triangle Inst, 78-82; ASST PROF, UNIV NC, CHAPEL HILL, 82- *Mem:* Am Chem Soc; Sigma Xi; AAAS. *Res:* Radiosynthesis, molecular modeling and drug design and synthesis of novel dopamine antagonists and anti-cerebral edema agents. *Mailing Add:* 2901 Bainbridge Dr No L Durham NC 27713-1446

WYRTKI, KLAUS, b Tarnowitz, Ger, Feb 7, 25; m 53; c 2. PHYSICAL OCEANOGRAPHY. *Educ:* Kiel Univ, PhD(phys oceanog), 50. *Prof Exp:* Scientist, Ger Hydrographic Inst, Hamburg, 50-51; res fel oceanog, Kiel Univ, 51-54; scientist, Inst Marine Res, Djakarta, 54-57 & Int Hydrographic Bur, Monaco,58; res officer, Commonwealth Sci & Indust Res Orgn, Australia, 58-61; res oceanogr, Scripps Inst, Univ Calif, 61-64; PROF OCEANOG, UNIV HAWAII, 64- *Concurrent Pos:* Ed, Atlas Phys Oceanog for Int Indian Ocean Exped, 65-72; chmn, NPac Exp, 74-82. *Honors & Awards:* Rossenstiel Award, Univ Miami, 81; Maurice Ewing Medal, Am Geophys Union, 89; Sverdrup Gold Medal, Am Meterol Soc, 91. *Mem:* AAAS; fel Am Geophys Union; Am Meteorol Soc; Oceanography Soc. *Res:* General circulation of the oceans; water masses; equatorial circulation; air-sea energy exchange; ocean-atmosphere interaction; El Niño, Sea Level. *Mailing Add:* Dept Oceanog Univ Hawaii Honolulu HI 96822-2336

WYRWICKA, WANDA, b Poland, Sept 8, 12, US Citizen; m 46; c 1. FEEDING, CONDITIONING. *Educ:* Univ Poznan, Poland, PhD(zool), 47. *Prof Exp:* Res anat, Univ Calif, Los Angeles, 66-78; key investr, Ctr Ulcer Res & Educ, Wadsworth Vet Admn Med Ctr, Los Angeles, 74-78. *Concurrent Pos:* Consult, Lab Neuropsychol, Vet Admn Med Ctr, Sepulveda, 66-71. *Honors & Awards:* Pavlovian Soc Award, 77. *Mem:* Am Physiolog Soc; Soc Neurosci; Pavlovian Soc Am. *Res:* Feeding responses and their representations in the brain; initiation and inhibition of eating; social influences on feeding and food preferences; feeding disorders; articles on conditioning and feeding; books on problems of conditioning, food preferences, and brain control of food intake. *Mailing Add:* Dept Anat Univ Calif Los Angeles CA 90024

WYSE, B O, b Columbia, SC, July 20, 27; m 52; c 2. ORGANIC CHEMISTRY. *Educ:* Erskine Col, AB, 48; Vanderbilt Univ, MA, 51; Univ SC, PhD(chem), 57. *Prof Exp:* Instr math, Erskine Col, 49-51; chemist, Celanese Corp, SC, 51-53; chemist, Tech Sect, 56-61, SR RES CHEMIST, TECH SECT, E I DU PONT DE NEMOURS & CO, INC, 61- *Mem:* Am Chem Soc. *Res:* Acrylic polymerization processes and reaction mechanisms; isocyanate chemistry; solution and melt spinning processes of elastomers and polyamides. *Mailing Add:* 4813 Shorewood Dr Chattanooga TN 37416

WYSE, BONITA W, b Lorain, Ohio, Oct 2, 45. DIETETICS, MAMMALIAN PHYSIOLOGY. *Educ:* Colorado State Univ, PhD(nutrit), 77. *Prof Exp:* PROF NUTRIT, UTAH STATE UNIV, 81-, DEAN, COL FAMILY LIFE, 86- *Concurrent Pos:* Mem, Arteriosclerosis-Hypertension Lipid Metab Adv Comt, Nat Heart Lung & Blood Inst, NIH, 84-87. *Mem:* Am Dietetic Asn; Exp Sta Comn Orgn & Policy. *Mailing Add:* Dean Off Col Family Life Utah State Univ Logan UT 84322-2900

WYSE, DAVID GEORGE, b Kamloops, BC, Aug 15, 41. HYPERTENSION RESEARCH, CARDIOELECTROPHYSIOLOGY. *Educ:* McGill Univ, PhD(pharmacol), 69; Univ Calgary, MD, 74. *Prof Exp:* PROF PHARMACOL & THERAPEUT, UNIV CALGARY, 78-, PROF CARDIOL, 85-; HEAD CARDIOL, FOOTHILLS HOSP, 86- *Concurrent Pos:* Scholar, Alta Heritage Found Med Res. *Mem:* Can Cardiovasc Soc; Can Soc Clin Invest; Am Soc Pharmacol & Exp Thera; Can Soc Clin Pharmacol. *Res:* Pathaphysiology of hypertension; management of ventricular arrhythmics; antiarrhythmic drugs. *Mailing Add:* Div Cardiol Univ Calgary Foothills Hosp 1403 29th St NW Calgary AB T2N 2T9 Can

WYSE, FRANK OLIVER, b Milwaukee, Wis, Apr 22, 30; m 67; c 3. MATHEMATICS. *Educ:* Harvard Univ, AB, 52; Princeton Univ, AM, 55; Ore State Univ, PhD(math), 64. *Prof Exp:* Instr math, Lehigh Univ, 55-58; from instr to asst prof, Ore State Univ, 58-70; prof & chmn dept, Talladega Col, 70-73; ASSOC PROF MATH, CLARK COL, 73- *Concurrent Pos:* Asst prof, Cleveland State Univ, 66-70. *Mem:* Am Math Soc. *Res:* Algebra; topology. *Mailing Add:* 1611 E Nancy Creek Dr NE Atlanta GA 30319

WYSE, GORDON ARTHUR, b San Jose, Calif, July 12, 40; m 63; c 3. NEUROBIOLOGY. *Educ:* Swarthmore Col, BA, 61; Univ Mich, MA, 63, PhD(zool), 67. *Prof Exp:* from asst prof to assoc prof, 66-87, actg dir, Neurosci & Behav Prog, 88-89, PROF ZOOL, UNIV MASS, AMHERST, 87- *Concurrent Pos:* Nat Inst Neurol Dis & Stroke res grant, 69-78; vis scholar, Stanford Univ, 72-73; vis prof, neurobiol, Harvard Med Sch, 82-83. *Mem:* AAAS; Am Soc Mammal; Soc Neurosci. *Res:* Comparative neurobiology; neural integration of central and sensory information to control rhythmic and other behavior patterns; neurotransmitters and neuromodulators. *Mailing Add:* Dept Zool Univ Mass Amherst MA 01003

WYSE, JOHN PATRICK HENRY, b Kamloops, BC, July 28, 48; m 72; c 4. MEDICAL SCIENCE, ANATOMY. *Educ:* Univ BC, BSc, 71, MD, 75; Univ Calgary, PhD(med sci anat), 78. *Prof Exp:* Asst prof, 78-81, ASSOC PROF ANAT, FAC MED, UNIV CALGARY, 81- *Concurrent Pos:* MRC grant, Med Res Coun Can, 78-80; Nat Retinitis Pigmentosa Found Can grant, 78-82; Alberta Mental Health grant & Alberta Heritage Found med res grant, 81; fel, Med Res Coun, 75-78. *Mem:* Asn Res Vision & Ophthal; Soc Neurosci; Can Asn Anatomists. *Res:* Histological, functional and genetic investigation of inherited ophthalmic defects in the BW rat; dopamine neurochemisty of retina, hypothalamus and nigrostriatal system of BW rat; morphometric investigations of mechanisms regulating rod outer segment renewal in vertebrate retinae. *Mailing Add:* 1632 14th Ave NW No 261B Calgary AB T2N 1M7 Can

WYSHAK, GRACE, b Boston, Mass. BIOSTATISTICS. *Educ:* Smith Col, BA, 49; Harvard Univ, MSHyg, 56; Yale Univ, PhD(biomet), 64. *Prof Exp:* Res assoc epidemiol, Harvard Univ, 56-60; instr math, Albertus Magnus Col, 64-65; assoc prof biomet, Yale Univ, 65-75; LECTR, HARVARD SCH PUB HEALTH & MED SCH, 75- *Concurrent Pos:* NIH res career develop award, 68-72; consult, NIH, 70-, Vet Admin Coop Studies Ctr, 72- & Radcliffe Inst Prog Health Care, 75-77; statist consult, Mass Gen Hosp; lectr, Preventive Med, Harvard Med Sch, 78-82, lectr on med, 82- *Mem:* Sigma Xi; Am Statist Asn; Biomet Soc; Am Epidemiol Asn; Int Epidemiol Asn. *Res:* Inheritance of twinning; biometric and epidemiologic methods; statistical methods in virology; statistical applications in psychiatry; cancer epidemiology; statistical applications in reproductive health and in preventive medicine. *Mailing Add:* 32 Commonwealth Ave Brookline MA 02167

WYSKIDA, RICHARD MARTIN, b Perrysburg, NY, Sept 2, 35; m 62; c 2. INDUSTRIAL ENGINEERING, OPERATIONS RESEARCH. *Educ:* Tri-State Col, BS, 60; Univ Ala, Tuscaloosa, MS, 64; Okla State Univ, PhD(indust eng), 68. *Prof Exp:* Elec engr, Philco Corp, 60-62; aerospace technol, Marshall Space Flight Ctr, NASA, 62-68; from asst prof to assoc prof, 68-78, PROF INDUST ENG, UNIV ALA, HUNTSVILLE, 78- *Concurrent Pos:* Consult, Gen Res Corp, Revere, Missile Command, Battelle Mem Inst & Mantech; part-time assoc prof, Ala A&M Univ, 71-76. *Mem:* Am Inst Indust Engrs; Opers Res Soc Am; Am Soc Eng Educ; Inst Mgt Sci. *Res:* Temperature sensitive cushioning systems; routing models; cost estimation models. *Mailing Add:* Dept Indust Eng Univ Ala Huntsville 4701 University Dr Huntsville AL 35899

WYSOCKI, ALLEN JOHN, b Chicago, Ill, Dec 22, 36; m 63; c 2. INDUSTRIAL ORGANIC CHEMISTRY. *Educ:* Loyola Univ Chicago, BS, 58; Northwestern Univ, Evanston, PhD(org chem), 63. *Prof Exp:* Res chemist, IIT-Res Inst, 62-64; div res mgr, Soap & Household Prod Div, Armour-Dial, Inc, Phoenix, 64-77; tech mgr, 77-84, TECH DIR, CHEM PROD DIV, DE SOTO, INC, 84- *Mem:* Am Chem Soc. *Res:* Development of detergents and other household products. *Mailing Add:* 715 E Cherry Lane Arlington Heights IL 60004-3217

WYSOCKI, ANNETTE BERNADETTE, b Raleigh, NC, Dec 31, 54; m 87. CELL-EXTRACELLULAR MATRIX INTERACTIONS. *Educ:* E Carolina Univ, BS, 78, MS, 80; Univ Tex, Austin, PhD(nursing), 86. *Prof Exp:* Staff nurse, Univ Va Med Ctr, 78-79, Seton Med Ctr, Austin, Tex, 81-86; res & teaching asst nursing, Univ Tex, Austin, 82-84, sr res assoc, Univ Southwestern Med Ctr, 86-88, NIH postdoctoral, 88-89; NIH postdoctoral, Cornell Univ Med Col, 89-91; DIR NURSING RES, NY UNIV MED CTR, 91-, RES ASST PROF, 91- *Concurrent Pos:* Mem, res comt, Parkland Mem Hosp, Dallas, 87-89, NY Hosp, New York, 90-91. *Mem:* AAAS; Asn Operating Rm Nurses; Am Soc Cell Biol. *Res:* Clinical and laboratory studies of acute and chronic wound healing; factors leading to the development of chronic wounds; prevention of chronic wounds; cell and extracellular matrix interactions in normal and chronic wound healing; adhesion, integrin receptors and proteolysis. *Mailing Add:* Dept Nursing & Inst Reconst Plastic Surg NY Univ Med Ctr 560 First Ave New York NY 10016

WYSOCKI, CHARLES JOSEPH, b Utica, NY, May 4, 47; m 68; c 3. REPRODUCTION, CHEMORECEPTION. *Educ:* Col Oswego, State Univ NY, BA, 73; Fla State Univ, Tallahassee, MS, 76, PhD(psychobiol), 78. *Prof Exp:* Fel vomeronasal organ & reproduction, Univ Pa, 78-79, asst mem genes & reproduction, 80-83, assoc mem, chem commun & genetics, Monell Chem Senses Ctr, 83-90, RES ASST PROF, DEPT ANIMAL BIOL, SCH VET MED, UNIV PA, 85-, MEM, GENETICS & CHEM COMMUN, 90- *Concurrent Pos:* Lectr, Dept Biol, Univ Pa, 78-; asst prof, Dept Psychol, Rutgers Univ, 79-80; ed, Focus on Fragrance, 84-; prin investr, Individual Differences in Human Olfaction, NIH, 85-; co-author, co-prin investr, Nat Geograph Smell Surv. *Honors & Awards:* Kenji Nakanishi Award in Olfaction, 88- *Mem:* Asn Chemoreception Sci; NY Acad Sci; Sigma Xi; Soc Neurosci. *Res:* Role of genetics in individual differences in odor perception; regulation of reproductive physiology and behavior by odors; analysis of 1.5 million returns of Nat Geograph Smell Survey. *Mailing Add:* Monell Chem Senses Ctr Univ Pa 3500 Market St Philadelphia PA 19104-3308

WYSOCKI, JOSEPH J(OHN), b Cohoes, NY, Aug 6, 28; m 57; c 3. LIQUID CRYSTAL DEVICES, THIN FILM DEVICES. *Educ:* Mass Inst Technol, BS & MS, 54. *Prof Exp:* Trainee, Naval Ord Lab, 51-54; res engr, RCA Labs, 54-67; sr res scientist, 67-86, PRIN SCIENTIST, XEROX CORP, 86- *Concurrent Pos:* Instr, Trenton Jr Col, 60-62. *Mem:* Inst Elec & Electronics Engrs. *Res:* Semiconductor and solar cell devices; transistors; diodes; radiation damage to solar cells; lithium-doped, radiation-resistant solar cells; liquid crystals; display devices; thin-film transistors; fabrication; testing and evaluation of solid-state devices of various types; ferroelectric liquid crystals. *Mailing Add:* 544 Crest Circle Webster NY 14580

WYSOLMERSKI, THERESA, b West Rutland, Vt, Oct 25, 32. ZOOLOGY, DEVELOPMENTAL BIOLOGY. *Educ:* Col St Rose, BS, 59; Univ Notre Dame, MS, 61; Rutgers Univ, New Brunswick, PhD(zool, ecol), 73. *Prof Exp:* Teacher, St John's Cath Acad, 56-59; instr chem, 59-60, from asst prof to assoc prof biol, 61-73, PROF BIOL, COL ST ROSE, 73- *Concurrent Pos:* Col rep, Hudson Valley Mohawk League Consortium, 69-70; NSF Faculty Develop Grant. *Mem:* Am Soc Zool; Am Inst Biol Sci; Sigma Xi. *Res:* Distribution of animal populations and their energy impact on forest floors; cytoskeletal proteins. *Mailing Add:* Col St Rose Albany NY 12203

WYSONG, DAVID SERGE, b Glasgow, Ky, Apr 20, 34; m 58; c 5. PLANT PATHOLOGY. *Educ:* Colo State Univ, BS, 58, MS, 61; Univ Ill, PhD(plant path), 64. *Prof Exp:* ASSOC PROF PLANT PATH & EXTEN PLANT PATHOLOGIST, UNIV NEBR-LINCOLN, 64- *Mem:* Am Phytopath Soc. *Res:* Practical application of plant pathology. *Mailing Add:* 820 E Sanborn Dr Lincoln NE 68505

WYSS, JAMES MICHAEL, b Ft Wayne, Ind, Mar 11, 48; m 73; c 3. NEUROBIOLOGY, NEUROANATOMY. *Educ:* Concordia Col, Ft Wayne, Ind, BS, 70; Lutheran Sch Theol, Chicago, MDiv, 74; Wash Univ, PhD(psychobiol), 76. *Prof Exp:* Instr biol, Wash Univ, 74-77, teaching fel anat, Wash Univ, 76-79; asst prof anat & psychol, 79-84, assoc prof cell biol & anat, 84-88, PROF CELL BIOL, UNIV ALA, BIRMINGHAM, 88- *Concurrent Pos:* Assoc prof med, Univ Ala, Birmingham, 85, mem, Neuropsychiat Prog, 79-, Hypertension Res Prog 81- *Mem:* AAAS; Soc Neurosci; Am Asn Anatomists; Am Physiol Soc; Am Heart Asn. *Res:* Understanding the mechanisms by which the central and peripheral nervous system participate in the development and maintenance of hypertension; role that the renal nerves and hypothalmus play in hypertension; mechanisms of antonomic control by limbic cortex. *Mailing Add:* Dept Cell Biol Univ Ala Birmingham AL 35294

WYSS, MAX, b Zürich, Switz, Sept 10, 39; US citizen; m 70; c 2. SEISMOLOGY. *Educ:* Swiss Fed Inst Technol, dipl, 64; Calif Inst Technol, MS, 67, PhD(geophysics), 70. *Prof Exp:* Res scientist geophysics, Univ Calif, San Diego, 70; res scientist seismology, Lamont-Doherty Geol Observ, Columbia Univ, 70-71, res assoc, 71; from asst prof to assoc prof, 72-83, PROF GEOL, UNIV COLO, BOULDER, 83-, CIRES FEL, GEOL SCI, 72- *Concurrent Pos:* Ed, Pure & Appl Geophys, 74; vis prof geophys, Univ Karlsruhe, Ger, 75, Fed Inst Technol, Zürich, 79, 80 & Nat Univ Mex, 72; explor geophys, Geotest AG, Bern, Switz, 64; res scientist, Univ Mainz, Mainz, Ger, 65; res asst, Calif Inst Technol, 65-70; vis scientist, Graefenberg Array, Erlangen, Ger, 85-86; Sr US Scientist Award, Alexander von Humboldt, 85-86; mem, various comts, Geol Sci Dept, Univ Colo, 73-89, Earth Dynamics Adv Subcomt, NASA, 77, Proposal Rev Panel, NSF, LOCI, 77, Adv Comt, Asian Inst Technol, 85-; chmn, sub-comn earthquake prediction, IASPEI, 87-; mem eval comt, Can Earthquake Prediction, 89- *Mem:* Am Geophys Union; Seismol Soc Am; Geol Soc Am. *Res:* Earthquake source mechanism; earthquake predictions; seismic risk; editor one book. *Mailing Add:* Dept Geol Univ Colo Boulder Boulder CO 80309

WYSS, ORVILLE, b Medford, Wis, Sept 10, 12; m 41; c 3. MICROBIOLOGY. *Educ:* Univ Wis, BS, 37, MS, 38, PhD(bact), 41. *Prof Exp:* Asst bact, Univ Wis, 37-41; res bacteriologist, Wallace & Teirnan Prod, 41-45; assoc prof bact, 45-48, chmn dept microbiol, 59-69, 75-76, prof, 48-82, EMER PROF MICROBIOL, UNIV TEX, AUSTIN, 82- *Concurrent Pos:* Fulbright grant, Univ Sydney, 71; Tribhuvan Univ, Nepal, 78. *Mem:* AAAS; Am Chem Soc; hon mem Am Soc Microbiol (pres, 65); Am Soc Biol Chemists; Am Acad Microbiol. *Res:* Bacterial physiology and genetics; microbial survival. *Mailing Add:* 6550 Yank Way Arvada CO 80004-2293

WYSS, WALTER, b Matzendorf, Switz, Mar 26, 38; m 61; c 4. MATHEMATICS, PHYSICS. *Educ:* Swiss Fed Inst Technol, dipl phys, 61, Dr Sc Nat(math physics), 65. *Prof Exp:* Instr physics, Swiss Fed Inst Technol, 61-66; instr, Princeton Univ, 66-68; from asst prof to assoc prof math & physics, 68-77, PROF PHYSICS, UNIV COLO, BOULDER, 77- *Concurrent Pos:* Swiss Nat Found stipend, Princeton Univ, 66-68; res fel, Univ Colo, Boulder, 69-70, NSF res grant, 70-72. *Mem:* Am Math Soc; Int Asn Math Phys. *Res:* Axiomatic theory of quantized fields; general relativity; functional analysis; infinite parameter lie groups; stochastic processes and combinatorics. *Mailing Add:* 2810 Iliff Boulder CO 80303

WYSSBROD, HERMAN ROBERT, b Louisville, Ky, Oct 17, 41; m 63. PHYSIOLOGY, BIOPHYSICS. *Educ:* Univ Louisville, BEE, 63, PhD(physiol), 68. *Prof Exp:* Asst prof physiol, 72-73, ASSOC PROF PHYSIOL & BIOPHYS, MT SINAI SCH MED, 74- *Concurrent Pos:* NIH res career develop award, Mt Sinai Sch Med, 72-77; asst prof biophys chem, Mt Sinai Grad Sch Biol Sci, City Univ New York, 68-73; vis asst prof, Rockefeller Univ, 71-73, vis assoc prof, 74-78; investr, NY Heart Asn, 77-81. *Mem:* AAAS; Am Physiol Soc; Am Chem Soc; Soc Exp Biol & Med; NY Acad Sci. *Res:* Conformation-function relationships of biologically active peptides; transmembrane transport. *Mailing Add:* Chem Dept Univ Louisville Louisville KY 40292

WYSTRACH, VERNON PAUL, b St Paul, Minn, May 8, 19; m 49; c 4. ORGANIC CHEMISTRY. *Educ:* Univ Minn, BCh, 41; Univ Rochester, PhD(org chem), 44. *Prof Exp:* Res chemist, Am Cyanamid Co, 44-52, group leader, Res Div, 52-54, group leader, Basic Res Dept, 54-59, mgr chem synthesis sect, Contract Res Dept, 59-66, mgr appl res sect, 66-67, mgr prod res sect, Chem Dept, Cent Res Div, 67-72, prod mgr, Prod Develop Dept, Chem Res Div, 72-74, employment supvr, 74-81; RETIRED. *Concurrent Pos:* Adj assoc prof chem, Sacred Heart Univ, 81-85, adj prof, 85- *Honors & Awards:* Am Cyanamid Co Award, Univ Cambridge, 61-62. *Mem:* Emer mem Am Chem Soc. *Res:* Cyanamide derivatives and nitrogen heterocycles; organic phosphorus compounds; rocket propellants and explosives; chemistry of adhesion; fire retardants. *Mailing Add:* 20 Westfield Rd Wilton CT 06897-2325

WYTTENBACH, CHARLES RICHARD, b South Bend, Ind, Jan 28, 33; m 59; c 3. DEVELOPMENTAL BIOLOGY. *Educ:* Ind Univ, AB, 54, MA, 56; Johns Hopkins Univ, PhD(biol), 59. *Prof Exp:* From instr to asst prof anat, Univ Chicago, 59-66; asst prof zool, 66-70, assoc prof, 70-75, chmn dept, 76-83, PROF PHYSIOL & CELL BIOL, UNIV KANS, 75- *Concurrent Pos:* Managing ed, Univ Kans Sci Bull, 68-74; mem corp, Marine Biol Lab, Woods Hole, Mass. *Mem:* Soc Develop Biol; Am Soc Zoologists. *Res:* Developmental biology, particularly mechanisms of insecticide-induced teratogenesis in chicks. *Mailing Add:* Dept Physiol & Cell Biol Univ Kans Haworth Hall Lawrence KS 66045-2106

WYZGA, RONALD EDWARD, b New Bedford, Mass, Aug 10, 42; m 69; c 2. PUBLIC HEALTH. *Educ:* Harvard Col, AB, 64; Fla State Univ, MS, 66; Harvard Univ, ScD, 71. *Prof Exp:* Tech staff, Orgn Econ Coop & Develop, 71-75; tech mgr, 77-83, PROJ MGR, ELEC POWER RES INST, 75-, SR PROJ MGR, 84- *Concurrent Pos:* Lectr, Am Col, Paris, 73-75. *Mem:* Am Statist Asn; Biomet Soc; Soc Risk Assessment. *Res:* Environmental risk assessment; health effects of air pollution; environmental cost-benefit analyses. *Mailing Add:* 4690 Smith Grade Santa Cruz CA 95060

WYZGOSKI, MICHAEL GARY, b Pontiac, Mich, Dec 25, 43; m 67; c 3. POLYMER SCIENCE, MATERIALS ENGINEERING. *Educ:* Oakland Univ, BS, 64; Univ Mich, MS, 70, PhD(mat eng), 73. *Prof Exp:* Plastics engr, Dow Chem Co, 66-68; RES ENGR, GEN MOTORS RES LAB, 73- *Mem:* Am Phys Soc; Am Chem Soc. *Res:* Durability of plastic and elastomeric materials, including oxidative and physical aging, stress cracking and ozone cracking; fatigue fracture of composites and unfilled thermoplastics; impact fracture. *Mailing Add:* Polymers Dept 12 Mile & Mound Rds Warren MI 48090-9055

X

XIE, YA-HONG, b Beijing, China, June 21, 56; US citizen; m; c 2. NOVEL PROPERTIES OF SI & GE, MOLECULAR BEAM EPITAXY. *Educ:* Purdue Univ, BS, 81; Univ Calif, Los Angeles, MS, 84, PhD(elec eng), 86. *Prof Exp:* MEM TECH STAFF, AT&T BELL LABS, MURRAY HILL, NJ, 86- *Mem:* Mat Res Soc. *Res:* Increased V functionality of Si based materials; optical and electronic crystal growth of GeSi alloys by molecular beam epitaxy; epitaxy of dissimilar materials: stability and relaxation. *Mailing Add:* AT&T Bell Labs MH1E-342 600 Mountain Ave Murray Hill NJ 07974-2070

XINTARAS, CHARLES, b New Bedford, Mass, Sept 5, 28; m 57; c 2. INDUSTRIAL HEALTH. *Educ:* Harvard Univ, AB, 52; Univ Cincinnati, ScD(indust health), 64. *Prof Exp:* Res chemist, Filtrol Corp, 52-55; eng inspector, Los Angeles County, 55-59; pub health adv, USPHS, 59-62, pharmacologist brain res, 62-74, asst chief, Behav Motivation Factors Br, 71-76, chief, Support Serv Br, Nat Inst Occup Safety & Health, 76-80; SCIENTIST OCCUP MED, WHO, GENEVA, 81- *Concurrent Pos:* Consult behav toxicol, WHO; mem, Permanent Comn & Int Asn Occup Health, Soc Neurosci. *Res:* Behavioral toxicology; behavioral and neurophysiological indicators for the monitoring and early detection of potential industrial health and safety problems. *Mailing Add:* Occup Health/WHO Geneva 27 1211 Switzerland

Y

YAAKOBI, BARUKH, b Petach-Tikva, Israel, Dec 10, 36; US citizen; m 68; c 2. ATOMIC & MOLECULAR PHYSICS. *Educ:* Hebrew Univ, MS, 62, PhD(physics), 67. *Prof Exp:* Scientist magnetic fusion, AEC, France, 70-74; SR SCIENTIST, LAB LASER ENERGETICS, UNIV ROCHESTER, 74- *Concurrent Pos:* Consult, Hampshire Instruments, Rochester, 84- *Mem:* Fel Am Phys Soc. *Res:* Plasma diagnosis using primarily x-ray physics and x-ray spectroscopy as applied to inertially confined fusion research. *Mailing Add:* Univ Rochester 250 E River Rd Rochester NY 14623-1299

YABLON, ISADORE GERALD, b Montreal, Que, May 30, 33; m 62; c 2. ORTHOPEDIC SURGERY. *Educ:* McGill Univ, BSc, 54; Univ Toronto, MD, 58. *Prof Exp:* Instr orthop, McGill Univ, 67-71; from asst prof to assoc prof, 71-78, PROF ORTHOP SURG, BOSTON UNIV SCH MED, 78- *Concurrent Pos:* Vis surgeon orthop, Univ Hosp, 71- & Boston City Hosp, 71- *Mem:* Am Acad Orthop Surg; Orthop Res Soc; Can Orthop Soc; Can Orthop Res Soc; fel Royal Col Physicians & Surgeons Can. *Res:* Developing technique of joint homografting for clinical application; methods to prevent graft rejection. *Mailing Add:* 75 E Newton St Boston MA 02118

YABLON, MARVIN, b New York, NY, Oct 30, 35; m 68; c 2. APPLIED STATISTICS, PATTERN RECOGNITION. *Educ:* New York Univ, BEE, 57, MEE, 60, MS, 64, PhD(operatons res), 77. *Prof Exp:* Res engr, Grumman Corp, 62-69; operations res analyst, Roosevelt Hosp, 73-76; statistician, Med Ctr, New York Univ, 76-78; from asst prof to assoc prof, 78-90, PROF MATH, JOHN JAY COL, 91- *Concurrent Pos:* Consult, Med Ctr, New York Univ, 78-82; prof, PhD prog criminal justice, Grad Sch & Univ Ctr, City Univ NY, 88- *Mem:* Sigma Xi; Am Statistical Asn; Inst Elec & Electronics Engrs; Operations Res Soc Am; Soc Indust & Appl Math; NY Acad Sci. *Res:* Pattern recognition; applied statistics; stochastic processes; applications of operations research. *Mailing Add:* 35-17 172nd St Flushing NY 11358

YABLONOVITCH, ELI, b Puch, Austria, Dec 15, 46; Can citizen. OPTICS, SOLID STATE PHYSICS. *Educ:* McGill Univ, BSc, 67; Harvard Univ, AM, 69, PhD(appl physics), 72. *Prof Exp:* Mem tech staff, Bell Labs, 72-74; asst prof appl physics, Harvard Univ, 74-76, assoc, 76-79; res assoc & group head, Exxon Res Ctr, 79-84; mem tech staff, 84-90, DISTINGUISHED MEM STAFF, BELLCORE, 90- *Concurrent Pos:* A P Sloan fel, 78-79. *Honors & Awards:* Adolph Lomb Medal, Optical Soc Am, 78. *Mem:* Fel Am Phys Soc; fel Optical Soc Am; sr mem Inst Elec & Electronics Engrs. *Res:* Nonlinear optics; laser-plasma interaction; laser induced chemistry; solar cells; semiconductor surfaces; photonic band structure. *Mailing Add:* 11 Blueberry Lane Leonardo NJ 07737-1804

YABLONSKI, MICHAEL EUGENE, b Minneapolis, Minn, July 11, 40; m 62; c 4. MEDICINE, PHYSIOLOGY. *Educ:* Univ Minn, BS, 65, MD, 67, PhD(physiol), 73. *Prof Exp:* Intern surg, Albert Einstein Col Med, 68; resident physician, Naval Air Develop Ctr, 71-73; resident ophthal, Univ Minn Hosps, 73-76; res ophthalmologist, Mt Sinai Sch Med, 77-; STAFF MEM, CORNELL UNIV. *Concurrent Pos:* Glaucoma fel, Wash Univ, St Louis, 76-77. *Res:* Transport physiology; transcapillary exchange; physiology of the eye and vision. *Mailing Add:* Dept Ophthal Mt Sinai Hosp 520 E 70th St New York NY 10021

YABLONSKY, HARVEY ALLEN, b New York, NY, Nov 24, 33; m 64; c 2. PHYSICAL CHEMISTRY. *Educ:* Brooklyn Col, BS, 54, MA, 58; Stevens Inst Technol, MS, 57, PhD(phys chem), 64; Am Inst Chem, cert. *Prof Exp:* Res chemist, NY State Dept Health, 55; lectr chem, Brooklyn Col, 55-56; teaching asst, Stevens Inst Technol, 56-59; lectr, Hunter Col, 60-63; asst prof, US Merchant Marine Acad, 63-64; head dept phys chem, Prod Div, Bristol Myers, 64-69; PROF CHEM, DEPT PHYS SCI, KINGSBOROUGH COL, 69- *Concurrent Pos:* Res biochemist, Messinger Res Found, 56-58; lectr, Hunter Col, 63-64 & Rutgers Univ, 67-69; independent consult. *Mem:* Am Chem Soc; fel Am Inst Chem; Sigma Xi. *Res:* Kinetics of redox systems; structure of solutions; sorption at surfaces; complex ion chemistry; fiber and powder rheology; piezoelectricity of biological materials; nonuniform surface photometry; pharmacokinetics. *Mailing Add:* Dept Phys Sci Kingsborough Col Brooklyn NY 11235

YACHNIN, STANLEY, b New York, NY, June 28, 30; m 60; c 2. INTERNAL MEDICINE, HEMATOLOGY. *Educ:* NY Univ, MD, 54; Am Bd Internal Med, dipl, 62. *Prof Exp:* House officer, Peter Bent Brigham Hosp, Boston, Mass, 54-55, jr asst resident med, 55-56, sr resident, 60-61; from asst prof to assoc prof, 61-69, head, Sect Hemat-Oncol, 66-81, PROF MED, SCH MED, UNIV CHICAGO, 69- *Concurrent Pos:* USPHS res fel, Peter Bent Brigham Hosp & Harvard Med Sch, 58-60; Markle scholar acad med, 63-68. *Mem:* AAAS; Am Soc Clin Invest; Am Fedn Clin Res; Am Soc Hemat; Asn Am Physicians. *Res:* Hemolytic anemias; complement; lymphocyte transformation; mitogenic proteins; paroxysmal nocturnal hemoglobinuria; alpha feto protein; cholesterol metabolism and cell growth. *Mailing Add:* Dept Med Univ Chicago Sch Med Chicago IL 60637

YACHNIS, MICHAEL, b Athens, Greece, Mar 22, 22. OCEAN ENGINEERING. *Educ:* Mil Acad , Greece, BS, 43; Mil Tech Col, Greece, BSCE, 51; George Washington Univ, MS, 56, MA, 62, DSc(struct eng), 68. *Prof Exp:* From lieutenant to major, Greek Army Corps Engrs, 43-55; struct eng specialist, Adams Fabricated Steel, 55-56; struct br mgr, Naval Facil Eng Command, 56-63, chief struct engr & ocean eng consult, 63-72, chief engr, 72-88; PROF, SCH ENG, GEORGE WASHINGTON UNIV, 70- *Concurrent Pos:* Chmn, Task Comt E, Am Soc Civil Engrs, mem, Coun Ocean Eng; mem, Welding Res Coun, Eng Found. *Honors & Awards:* Presidential Rank Award, 82. *Mem:* Nat Acad Eng; fel Am Soc Civil Engrs; Soc Mil Engrs; Nat Soc Prof Engrs; Marine Technol Soc; Sigma Xi; Am Welding Soc; Undersea Med Soc. *Res:* Fleet readiness; author of various publications; granted one patent. *Mailing Add:* 4200 Military Rd NW Washington DC 20015

YACKEL, JAMES W, b Sanborn, Minn, Mar 6, 36; m 60; c 3. MATHEMATICS. *Educ:* Univ Minn, BA, 58, MA, 60, PhD(math), 64. *Prof Exp:* John Wesley Young res instr math, Dartmouth Col, 64-66; from asst prof to assoc prof, 66-76, prof & assoc dean sci, 76-87, VCHANCELLOR, ACAD AFFAIRS, PURDUE UNIV, 87- *Mem:* Am Math Soc; fel AAAS; Inst Math Statist; Math Asn Am. *Res:* Stochastics processes and graph theory; probability theory; combinatorial theory. *Mailing Add:* Purdue Univ Calumet Hammond IN 46323-2094

YACKEL, WALTER CARL, b Evanston, Ill, Nov 20, 42; m 73. FOOD SCIENCE & TECHNOLOGY. *Educ:* Univ Ill, BS, 65, MS, 68, PhD(food sci), 71. *Prof Exp:* Food technologist consumer prod, 70-75, food scientist, Corp Res, 76-86, APPLNS MGR FOOD & INDUST PROD, A E STALEY MFG CO, 86- *Mem:* Inst Food Technologists; Sigma Xi. *Res:* Product and process development of food ingredients. *Mailing Add:* 19 Timber Bluff Trail Decatur IL 62521

YACOUB, KAMAL, b Nazareth, Palestine, Nov 11, 32; US citizen; m 64; c 3. ELECTRICAL ENGINEERING. *Educ:* Univ Pa, MSEE, 58, PhD(elec eng), 61. *Prof Exp:* Assoc elec eng, Moore Sch Elec Eng, Univ Pa, 61-63, asst prof, 63-66; from asst prof to assoc prof, 66-73, chmn dept, 72-87, PROF ELEC ENG, UNIV MIAMI, 73- *Honors & Awards:* Res Award, Am Soc Eng Educ. *Mem:* Inst Elec & Electronics Engrs; Am Soc Eng Educ; Sigma Xi. *Res:* Acoustic transmission fluctuations in Florida straits; tidal modulation of environmental and acoustic parameters; spectral analysis and modelling of relationships; relationship between partial and multiple coherence functions; estimation through order statistics; time series analysis. *Mailing Add:* 1525 Ancona Ave Coral Gables FL 33146

YACOWITZ, HAROLD, b New York, NY, Feb 17, 22; m 41; c 3. NUTRITION. *Educ:* Cornell Univ, BS, 47, MNS, 48, PhD(animal nutrit), 50. *Prof Exp:* Assoc res biochemist, Parke, Davis & Co, 50-51; from asst prof to assoc prof poultry nutrit, Ohio State Univ, 51-55, assoc prof, Agr Exp Sta, 51-55; head nutrit res dept, Squibb Inst Med Res, 55-59; dir appl res, Nopco Chem Co, 59-61; res assoc, Fairleigh Dickinson Univ, 61-82; PRES, AIMS INC, 80- *Concurrent Pos:* Nutrit consult, 61-; owner, H Yacowitz & Co, nutrit consult lab. *Mem:* Am Chem Soc; NY Acad Sci; Poultry Sci Asn; Am Soc Animal Sci; Am Inst Nutrit; Am Asn Lab Animal Sci. *Res:* Vitamin B-12 microbiological assays; vitamin interrelationships; antibiotic absorption and effects of dietary antibiotics on chicks and hens; vitamin requirements; antifungal agents; calcium and fat metabolism in man and animals; atherosclerosis; nutritional value of human and animal foods; marine sources of animal protein; developed new animal identification system. *Mailing Add:* 221 Second Ave Piscataway NJ 08854

YADAV, KAMALESHWARI PRASAD, b Burhiatikar, India, Jan 5, 37; m 57; c 2. ANALYTICAL BIOCHEMISTRY. *Educ:* Univ Bihar, BSc, 59; Univ Mo, MS, 61, PhD(biochem), 66. *Prof Exp:* Instr animal husb, Ranchi Agr Col, Bihar, 59-61; res asst agr chem, Univ Mo, 62-66; res biochemist, Falstaff Brewing Corp, 66-70, sr biochemist, 70-74; PRES, CHEMCO INDUSTS, INC, 74- *Mem:* Am Soc Brewing Chemists. *Res:* Brewing and fermentation. *Mailing Add:* 1133 Clayton Pl Dr St Louis MO 63131

YADAV, RAGHUNATH P, b Kanpur, India, Jan 2, 35; m 64; c 3. ENTOMOLOGY. *Educ:* Agra Univ, BS, 56, MS, 58; La State Univ, PhD(entom), 64. *Prof Exp:* From asst prof to assoc prof entom, 64-74, assoc prof entom, 74-88, PROF BIOL, SOUTHERN UNIV, BATON ROUGE, 88- *Mem:* Entom Soc Am. *Res:* Artificial diet media for rearing of sugarcane borer; laboratory techniques for the detection of sugarcane borer resistance to insecticides; response of sweet corn hybrids to corn earworm damage in southern central Louisiana; possible antibiosis in sweet corn hybrids to corn earworm. *Mailing Add:* Dept Biol Sci Southern Univ & A&M Col Baton Rouge LA 70813

YADAVALLI, SRIRAMAMURTI VENKATA, b Secunderabad, India, May 12, 24; US citizen; m 52. PHYSICS, ELECTRICAL ENGINEERING. *Educ:* Andhra Univ, India, BS, 42, MS, 45; Univ Calif, MS, 49, PhD(elec eng), 53. *Prof Exp:* Officer in charge, Physics & Chem Labs, Eng Res Dept, Hyderabad, India, 46-48; asst elec eng, Univ Calif, 49-52; mem tech staff, Gen Elec, 53-59; physicist, Stanford Res Inst, 59-67,staff scientist, 67-68, staff scientist eng sci, 68-77; DIR RES, SHASTRA INC, 77- *Concurrent Pos:* Lectr, Univ Calif, 52 & 57-58, res engr, 59; consult engr, Gen Elec, 59; consult, Raytheon Co, 59-62, Litton Indust, 62-64, McGraw Hill Book Co, 62-70 & Rand Corp, 71-78 & 85-87, Misc Orgn, 77- *Mem:* AAAS; Inst Elec & Electronics Engrs; Am Phys Soc; NY Acad Sci; Sigma Xi. *Res:* Stochastic processes; electron and plasma physics; statistical and mathematical physics; electrohydrodynamics; optics; biomathematics. *Mailing Add:* Shastra Inc PO Box 1231 Palo Alto CA 94302

YADEN, SENKALONG, b Nagaland, India, Apr 21, 35. ZOOLOGY. *Educ:* Wilson Col, Bombay, India, BSc, 56; Univ Bombay, MSc, 58; Univ Minn, PhD(zool), 65. *Prof Exp:* From assoc prof to prof biol, Jarvis Christian Col, 67-81; STAFF MEM, WILEY COL, TEX, 88- *Mem:* AAAS; NY Acad Sci; Am Inst Biol Sci. *Res:* Ecology of freshwater organisms. *Mailing Add:* 3712 Lexington Ave Tyler TX 75701

YADVISH, ROBERT D, MATERIAL QUALIFICATION OF IN GAAS & INP SUBSTRAITS LASER FABRICATION OF INP LASERS. *Educ:* Jersey City State Col, BA, 77, BS, 89. *Prof Exp:* Electronic device processor, 81-82, sr tech asst, electronic device fabrication, 82-86, MEM TECH STAFF, LASER DEVELOP, AT&T BELL LABS, 86- *Res:* Much research was completed on the effects of silicon nitride layers on graphite strip heaters and the recrystallization of the thin film Si on Sio2; the effects of high temperature operation of lattice mached and strained in Ga As/INP quantum well lasers. *Mailing Add:* 142 North St Bayonne NJ 07002

YAEGER, JAMES AMOS, b Chicago, Ill, Aug 10, 28; m 52; c 4. HISTOLOGY. *Educ:* Ind Univ, DDS, 52, MS, 55; Univ Ill, PhD(anat), 59. *Prof Exp:* Instr anat & clin dent, 55-57; from asst prof to prof histol & head dept, Col Dent, Univ Ill, Chicago, 59-68; head dept, 68-78, PROF BIOSTRUCT & FUNCTION, SCH DENT MED, UNIV CONN, 68- *Concurrent Pos:* Resident assoc, Argonne Nat Lab, 63-64; vis prof, Univ Sidney, 77-; exec dir, Annual Midwest Seminar Dent Med, 64-66; regional ed, Archives Oral Biol, 68-76; mem, Dent Training Comt, Nat Inst Dent Res, 68-72. *Mem:* Am Asn Anat; Int Asn Dent Res. *Res:* Physiology of mastication. *Mailing Add:* Dept of Biostruct & Function Univ Conn Health Ctr Farmington CT 06030

YAEGER, ROBERT GEORGE, b Rochester, NY, Oct 25, 17; m 53; c 3. PARASITOLOGY. *Educ:* Univ Rochester, AB, 50; Univ Tex, MA, 52; Tulane Univ, PhD(parasitol, biochem), 55. *Prof Exp:* Instr bact & parasitol, Med Sch, Univ Tex, 52-54; from instr to prof parasitol & med, Sch Med, Tulane Univ, 55-88; RETIRED. *Concurrent Pos:* Vis prof, Sch Med, St George Univ, 79-85; consult, AID, 87. *Mem:* Am Soc Parasitol; Am Soc Trop Med & Hyg; Soc Exp Biol & Med; Am Inst Nutrit; Soc Protozool. *Res:* Nutritional and immunological relationships of host and parasite; culture methods; Chagas' disease; other parasitic protozoa which infect man; venomous animals. *Mailing Add:* 6052 Shetland Dr New Orleans LA 70131-3940

YAES, ROBERT JOEL, b New York, NY, July 11, 42; m 86. RADIATION ONCOLOGY, THEORETICAL RADIATION BIOLOGY. *Educ:* Mass Inst Technol, Cambridge, SB, 63, ScD(physics), 67; Mem Univ Nfld, St Johns, BS, 78, MD, 80. *Prof Exp:* Fac assoc physics, Univ Tex, Austin, 68-70; Humboldt Found fel, Johannes Gutenberg Univ, Mainz, Ger, 71-72; asst prof physics, Mem Univ Nfld, 72-76; resident radiation ther, Mem Sloan-Kettering Cancer Ctr, 80-82 & radiation oncol, State Univ NY Downstate Med Ctr, 82-85; fel radiation med, State Univ NY, Stoney Brook, 85-86; ASST PROF RADIATION MED, UNIV KY MED CTR, 86- *Concurrent Pos:* Alexander von Humboldt Stiftung fel, 71; attend radiotherapist, Cent Baptist Hosp, Lexington, Ky, 86-89 & St Claires Med Ctr, Morehead, Ky, 89- *Mem:* Am Soc Therapeut Radiol & Oncol; Radiol Soc NAm; Radiation Res Soc; Am Soc Physicists Med; Am Phys Soc; AMA. *Res:* Mathematical modeling of radiation effects on tumors and on normal tissues; treatment of malignant gliomas with California-252 neutron brachytherapy and hyperfractionation. *Mailing Add:* Dept Radiation Med Univ Ky Med Ctr Lexington KY 40536-0084

YAFET, YAKO, b Istanbul, Turkey, Jan 2, 23; nat US; m 49; c 3. SOLID STATE PHYSICS. *Educ:* Tech Univ Istanbul, ME, 45; Univ Calif, PhD(physics), 52. *Prof Exp:* Res assoc physics, Univ Ill, 52-54; physicist, Westinghouse Elec Corp Res Labs, 54-60; MEM TECH STAFF, BELL TEL LABS, INC, 60- *Mem:* Fel Am Phys Soc. *Res:* Electronic properties of semiconductors and metals; theoretical solid state physics. *Mailing Add:* 47 Curtiss Pl Maplewood NJ 07040

YAFFE, LEO, b US, July 6, 16; Can citizen; m 45; c 2. RADIOCHEMISTRY. *Educ:* Univ Man, BSc, 40, MSc, 41; McGill Univ, PhD, 43. *Hon Degrees:* DLett, Trent Univ; DSc, Univ Man. *Prof Exp:* Proj leader nuclear chem & tracer res, Atomic Energy Can, Ltd, 43-52; spec lectr radiochem, 52-54, assoc prof, 54-58, chmn dept chem, 65-72, vprin admin & prof fac, 74-81, MacDonald prof chem, 58-84, EMER PROF, MCGILL UNIV, 84- *Concurrent Pos:* Dir res, Int Atomic Energy Agency, Vienna, Austria, 63-65; res collabr, Brookhaven Nat Lab. *Honors & Awards:* Nuclear Chem Award, Am Chem Soc, 82; Seaborg Medal, Am Nuclear Soc, 88; Officer, Order of Can, 88. *Mem:* Fel AAAS; fel Chem Inst Can; fel Royal Soc Can; fel Am Inst Physics. *Res:* Nuclear chemistry; fission studies; chemistry applied to archaeology. *Mailing Add:* Dept Chem McGill Univ Montreal PQ H3A 2K6 Can

YAFFE, ROBERTA, b Chelsea, Mass, Jan 2, 44. ORGANIC CHEMISTRY. *Educ:* Bryn Mawr Col, AB, 65; Mass Inst Technol, PhD(chem), 70. *Prof Exp:* Teaching asst org chem, Mass Inst Technol, 65-66; sr chemist, Beacon Res Labs, Texaco, Inc, 69-75, res chemist, 75-83, sr res chemist, Petroleum Prod Res, 83-85; TECH REP, ADDITIVES FOR LUBRICANTS, CIBA-GEIGY CORP, 85- *Mem:* Am Chem Soc; Am Soc Lubrication Engrs; Nat Lubricating Grease Inst. *Res:* Non-crankcase automotive lubricants; aircraft lubricants; tractor transmission-differential-hydraulic lubricants; lubricant base rocks. *Mailing Add:* PO Box K Glenham NY 12527

YAFFE, RUTH POWERS, b Duluth, Minn, June 4, 27; m 76; c 2. RADIOCHEMISTRY. *Educ:* Macalester Col, BA, 48, PhD(phys chem), 51. *Prof Exp:* AEC fel radiochem, Ames Lab, 51-52; chemist, Oak Ridge Nat Lab, 52-53; instr chem, Univ Tenn, 55-56; from asst prof to assoc prof, 57-66, PROF CHEM, SAN JOSE STATE UNIV, 66-, COORDR NUCLEAR SCI FACIL, 72- *Concurrent Pos:* Consult radiation safety & chem safety training, 80-; outstanding prof, San Jose State Univ, 83; outstanding prof, Calif State Univ, 84. *Mem:* Am Nuclear Soc; Am Chem Soc; Health Physics Soc; Sigma Xi. *Res:* Chemistry of ruthenium; environmental soil and water analysis for radionuclides; fast radiochemistry, nuclear spectroscopy. *Mailing Add:* Dept Chem San Jose State Univ San Jose CA 95192

YAFFE, SUMNER J, b Boston, Mass, May 9, 23; c 5. PEDIATRICS, PHARMACOLOGY. *Educ:* Harvard Univ, AB, 45, MA, 50; Univ Vt, MD, 54; Am Bd Pediat, dipl, 60. *Prof Exp:* From intern to sr asst resident, Children's Hosp, Boston, 54-56; exchange resident, St Mary's Hosp, London, Eng, 56-57; from instr to asst prof pediat, Stanford Univ, 59-63; assoc prof, State Univ NY Buffalo, 63-66, prof pediat, 66-75, assoc chmn dept, 69-75; prof pediat & pharmacol, Univ Pa, 75-81; head div clin pharmacol, Children's Hosp Philadelphia, 75-81; DIR, CTR RES MOTHERS & CHILDREN, NAT INST CHILD HEALTH & HUMAN DEVELOP, NIH, BETHESDA, MD, 81- *Concurrent Pos:* Res fel metab, Children's Hosp, Boston, 57-59; teaching fel pediat, Harvard Med Sch, 56; Fulbright scholar, St Mary's Hosp, London, Eng, 56-57; Am Heart Asn adv res fel, 60; Lederle med fac award, 62; attend pediatrician, Palo Alto-Stanford Hosp, 59, dir newborn nursery serv, 60, prof dir clin res ctr premature infants, 62; dir pediat renal clin, Stanford Med Ctr, 60; attend pediatrician, Children's Hosp, Buffalo, 63-75; prog dir, Clin Res Ctr Children, 63; prog consult, Nat Inst Child Health & Human Develop, 63, mem training grant comt, 63, mem reprod biol comt, 65; on leave, Dept Pharmacol, Karolinska Inst, Stockholm, Sweden, 69-70; Jose Albert mem lectr, 81; fel, Japan Soc Prom Sci. *Mem:* Soc Pediat Res; Am Acad Pediat; Am Soc Clin Pharmacol & Therapeut; Am Pediat Soc; Am Soc Pharmacol & Exp Therapeut; Perinatal Res Soc. *Res:* Pediatric clinical pharmacology; neonatal, perinatal, fetal and pediatric pharmacology; developmental pharmacology; drug metabolism; drug disposition in sick infants and children; bilirubin metabolism and binding albumin; drug effects upon the mother, fetus and infant. *Mailing Add:* Nat Inst Child Health & Human Develop 7910 Woodmont Ave Bethesda MD 20205

YAGER, BILLY JOE, b Cameron, Tex, Dec 16, 32; m 53; c 4. PHYSICAL ORGANIC CHEMISTRY. *Educ:* Southwest Tex State Col, BS, 53; Tex A&M Univ, MS, 60, PhD(chem), 62. *Prof Exp:* Instr chem, Tex A&M Univ, 61-62; from asst prof to assoc prof, 62-71, chmn dept, 75-87, PROF CHEM, SOUTHWEST TEX STATE UNIV, 71- *Mem:* Am Chem Soc; Sigma Xi. *Res:* Solvent effects upon saponification rate constants; effect of solvent composition upon activity of reactants. *Mailing Add:* Dept Chem Southwest Tex State Univ San Marcos TX 78666

YAGER, JAMES DONALD, JR, b Milwaukee, Wis, Dec 29, 43; m 68, 81; c 3. CELL BIOLOGY, GENETICS. *Educ:* Marquette Univ, BS, 65; Univ Conn, PhD(cell/develop biol), 71. *Prof Exp:* Postdoctoral fel, McAndes Lab Cancer Res, Univ Wis, 71-74; asst prof biol sci, Dartmouth Col, 74-77, asst prof path, Dartmouth Med Sch, 77-80, assoc prof anat & path, 80-81, prof anat, 86-89; assoc prof environ med, NY Univ Med Ctr, 81-83, assoc prof anat, 83-86; PROF & DIR, DIV TOXICOL SCI, JOHNS HOPKINS SCH HYG & PUB HEALTH, 89- *Concurrent Pos:* Assoc dir, Norris Cotton Cancer Ctr, Dartmouth/Hitchcock Med Ctr, 83-89; actg chair, Dept Pharmacol & Toxicol, Dartmouth Med Sch, 87- 89. *Mem:* AAAS; Am Asn Cancer Res; Am Soc Cell Biologists; Sigma Xi; Am Asn Pathologists; Soc Toxicol. *Res:* Carcinogenesis. *Mailing Add:* Div Toxicol Sci Dept Environ Health Sci Johns Hopkins Sch Hyg & Pub Health 615 N Wolfe St Baltimore MD 21205

YAGER, JANICE L WINTER, b Green Bay, Wis, June 17, 40; m 66; c 2. GENETIC TOXICOLOGY, TOXICOLOGY. *Educ:* Univ Wis-Madison, BS, 62; Univ Calif, Berkeley, MPH, 72, PhD(toxicol), 82. *Prof Exp:* Med microbiologist, Stanford Med Ctr, Palo Alto, Calif, 62-63; asst scientist, dept physiol chem, Univ Wis-Madison, 64-65; assoc scientist, dept biochem, Col Med Sci, Univ Minn, Minneapolis, 66-68; res assoc, dept med, Univ Calif, San Francisco, 69-72; med microbiologist, Pac Med Ctr, San Francisco, 73-74; assoc res specialist, 75-82, RES TOXICOLOGIST, DEPT BIOMED ENVIRON HEALTH SCI, UNIV CALIF, BERKELEY, 82- *Concurrent Pos:* Lectr toxicol, dept biomed environ health sci, Univ Calif, San Francisco, 82-85; consult, Tetra-Tech, Inc, Lafayette, Calif, 83, Occup Safety & Health Admin, Washington, DC, 83-84 & Genen-Cor, Inc, S San Francisco, Calif, 84-; res fel, NIH, Fogarty Int Fel Award, Acad Finland, Helsinki, 86-87. *Honors & Awards:* Tebbens Award, Am Indust Hyg Asn, 81. *Mem:* AAAS; Am Conf Govt Indust Hygienists; Am Indust Hyg Asn; Environ Mutagen Soc. *Res:* Development of biological monitoring methods for early assessment of occupational exposures utilizing principles of genetic toxicology; inhalation toxicology; effect of toxicants on immune status. *Mailing Add:* Elec Power Res Inst 3412 Hillview Ave PO Box 10412 Palo Alto CA 94303

YAGER, PHILIP MARVIN, b Los Angeles, Calif, Aug 5, 38; m 73; c 2. EXPERIMENTAL HIGH ENERGY PHYSICS. *Educ:* Univ Calif, Berkeley, BA, 61; Univ Calif, San Diego, MS, 64, PhD(physics), 71. *Prof Exp:* From lectr to assoc prof, 68-75, PROF PHYSICS, UNIV CALIF, DAVIS, 81- *Mem:* AAAS; Am Phys Soc; Sigma Xi. *Res:* Hadronic interactions at high energy. *Mailing Add:* Dept Physics Univ Calif Davis CA 95616

YAGER, ROBERT EUGENE, b Carroll, Iowa, Apr 13, 30; m 55; c 2. SCIENCE EDUCATION, PLANT PHYSIOLOGY. *Educ:* State Col Iowa, BS, 50; Univ Iowa, MS, 53, PhD, 57. *Prof Exp:* Res asst plant physiol, 55-56, from instr to assoc prof sci educ, 56-67, PROF SCI EDUC, UNIV IOWA, 67- *Concurrent Pos:* Dir sec sci training prog, NSF, 59-, dir, In-Serv Inst, 61-, dir, Summer Inst, 63-, dir acad yr prog for sci supv, summer inst in-serv teachers, coop col sch sci prog & undergrad pre-serv teacher educ prog; ed, J Nat Asn Res Sci Teaching, 64-, scope, sequence & coord dir, 90; pres, Nat Asn Sci Technol Soc. *Mem:* Nat Asn Res Sci Teaching; Nat Sci Teachers Asn; Nat Asn Biol Teachers. *Res:* Chemical control of abscission processes. *Mailing Add:* Dept Sci Educ Univ Iowa Iowa City IA 52242-1478

YAGER, ROBERT H, MICROBIOLOGY, LABORATORY ANIMAL MEDICINE. *Educ:* Univ Pa, MD, 39. *Prof Exp:* Exec secy, Inst Lab Animal Resources, Nat Acad Sci, 75-86; RETIRED. *Mailing Add:* 318 Forrestland Ct West Columbia SC 29169

YAGGY, PAUL FRANCIS, b Detroit, Mich, Aug 4, 23; m 45; c 4. AERONAUTICAL & ELECTRICAL ENGINEERING. *Educ:* San Jose State Col, BSEE, 63. *Prof Exp:* Aeronaut engr, Ames Res Ctr, NASA, 46-50 & 51-58, aeronaut engr large scale aerodyn br, 58-61, asst br chief, 61-65, tech dir, US Army Aeronaut Res Lab, 65-68, dir lab, 68-72, dir lab, US Army Air Mobility Res & Develop Lab, 72-74; RETIRED. *Concurrent Pos:* Mem subcomt aircraft aerodyn, NASA, 67-70; mem fluid dynamics panel, Adv Group Aerospace Res & Develop, NATO, 67- *Mem:* Am Helicopter Soc; Soc Automotive Engrs; assoc fel Am Inst Aeronaut & Astronaut. *Res:* Aeronautical sciences; aerodynamics; flight vehicles for air mobile systems development. *Mailing Add:* 1381 E Sheffield Ave Campbell CA 95008

YAGHJIAN, ARTHUR DAVID, b Jan 1, 43; US citizen; m 73. ELECTROMAGNETIC THEORY. *Educ:* Brown Univ, ScB, 64, ScM, 66, PhD(elec eng), 69. *Prof Exp:* Asst prof physics & elec eng, Hampton Inst, 69-70; Nat Res Coun res assoc, Nat Bur Standards, 71-73, electronics engr, Electromagnetic Fields Div, 73-82; ELECTRONICS ENGR, ELECTROMAGNETICS DIRECTORATE, HANSCOM AFB, 82- *Concurrent Pos:* Vis scientist, Rome Lab, 80; assoc ed, Inst Elec & Electronics Engrs Trans Antennas & Propagation, 83-86, Radio Sci, 88-89; guest prof, Tech Univ Denmark, 89. *Mem:* Inst Elec & Electronics Engrs; Sigma Xi; Int Union Radio Scientists; Electromagnetic Soc. *Res:* Near-field antenna measurements; electromagnetic scattering and radiation. *Mailing Add:* 115 Wright Rd Concord MA 01742

YAGI, FUMIO, b Seattle, Wash, July 14, 17; m 54. MATHEMATICS. *Educ:* Univ Wash, BS, 38, MS, 41; Mass Inst Technol, PhD(math), 43. *Prof Exp:* Fel, Inst Advan Study, 43; instr, Univ Wash, 46-49, asst prof, 49-53; mathematician, Ballistic Res Labs, Aberdeen Proving Ground, 53-56; sr res engr, Jet Propulsion Lab, Calif Inst Technol, 56-58, res specialist, 58-63; appl mathematician, Grumman Aircraft Eng Corp, Bethpage, NY, 63-66, mem systs anal staff, 66-67, group supvr systs anal, 67-77; RETIRED. *Concurrent Pos:* Lectr, Univ Md, 56, Univ Calif, Los Angeles, 57-61 & Adelphi Univ, 63-64 & 66; adj prof, C W Post Col, Long Island Univ, 66-77. *Mem:* Am Math Soc; Sigma Xi. *Res:* Analysis; space trajectory and guidance studies; systems performance and error analysis. *Mailing Add:* 2914 Sahalee Dr E Redmond WA 98053

YAGI, HARUHIKO, b Sendai, Japan, June 27, 39; m 69; c 2. ORGANIC CHEMISTRY, DRUG METABOLISM. *Educ:* Tohoku Univ, Japan, MS, 65, PhD(systhesis of isoquinoline alkaloid), 68. *Prof Exp:* Asst org synthesis, Pharmaceut Inst, Tohoku Univ, Japan, 68-69; res asst electrochem, Univ Conn, 69-70; res assoc chem kinetics, Johns Hopkins Univ, 70-71; VIS SCIENTIST DRUG METABOLISM & CARCINOGENESIS, LAB BIOORG CHEM, NAT INST ARTHRITIS, DIGESTIVE & KIDNEY DIS, NIH, 71- *Concurrent Pos:* Fel, Univ Conn, 69-70 & Johns Hopkins Univ, 70-71. *Mem:* Pharmaceut Soc Japan; Am Chem Soc. *Res:* Organic synthesis; natural product chemistry; carcinogenesis. *Mailing Add:* 11 Candlelight Ct Potomac MD 20854

YAGIELA, JOHN ALLEN, b Washington, DC, July 23, 47; m 70; c 2. DENTAL ANESTHESIA, DENTAL PHARMACOLOGY. *Educ:* Univ Calif, Los Angeles, DDS, 71; Univ Utah, PhD(pharmacol), 75. *Prof Exp:* From asst prof to assoc prof oral biol, Sch Dent, Emory Univ, 75-82; assoc prof pain control, Sch Dent, 82-83, resident anesthesiol, Sch Med, 82-83, assoc dean acad affairs, 85-86, assoc dean, Acad & Admin Affairs, 86-89, PROF PAIN CONTROL, SCH DENT, UNIV CALIF, LOS ANGELES, 83-, PROF ANESTHESIOL, SCH MED, 83- *Concurrent Pos:* Consult physician, Dent Serv, Vet Admin, Wadsworth Med Ctr, 83-; chmn, Sect Pharmacol & Therapeut, Am Asn Dent Schs, 83; consult, Coun Dent Therapeut, Am Dental Asn, 87; chmn, fellowship Comt, Am Dent Soc Anesthesiol, 87-90; chmn, Pharmacol, Toxicol, Therapeut Group, Int Asn Dent Res, 90-; ed, Anesthesia Progress, 90- *Mem:* Int Asn Dent Res; Am Asn Dent Schs; fel Am Dent Soc Anesthesiol; Am Dent Asn; AAAS. *Res:* Pharmacology and toxicology of local anesthetic agents and drugs used in dental anesthesia and sedation. *Mailing Add:* Ctr Health Sci Sch Dent Univ Calif Los Angeles CA 90024

YAGLE, RAYMOND A(RTHUR), b Aspinwall, Pa, Dec 29, 23; m 54; c 3. NAVAL ARCHITECTURE, OCEAN ENGINEERING. *Educ:* Univ Mich, BSE, 44, MSE, 47. *Prof Exp:* Jr engr, Cornell Aeronaut Lab, Inc, 46; res engr, Frederic Flader, Inc, 47; res asst to res engr, 50-54, asst prof eng mech, 55-60, assoc prof naval archit & marine eng, 60-64, prof, 64-76, PROF MARINE ENG, ENG RES INST, UNIV MICH, ANN ARBOR, 76- *Concurrent Pos:* Consult, Rand Corp; chmn ship res comt, Nat Acad Sci-Nat Res Coun. *Mem:* Marine Technol Soc; Am Soc Naval Engrs; Soc Naval Archit & Marine Engrs. *Res:* Fuel sprays, atomization and combustion; fluid mechanics generally, resistance studies and naval architecture applications in particular; ship structures; computer applications. *Mailing Add:* Dept Naval Ardn Marine Eng Univ Mich Ann Arbor MI 48109

YAGUCHI, MAKOTO, b Yokohama, Japan, Oct 19, 30; m 60; c 1. PROTEIN CHEMISTRY, AGRICULTURAL CHEMISTRY. *Educ:* Tokyo Univ Agr, BAgr, 53; Univ Calif, Davis, MS, 57, PhD(agr chem), 63. *Prof Exp:* Res assoc protein chem, Purdue Univ, 63-64; res food technologist, Univ Calif, Davis, 64-65; from asst res officer to assoc res officer, 65-77, SR RES OFFICER, DIV BIOL SCI, NAT RES COUN CAN, 77- *Concurrent Pos:* Vis scientist, Max-Planck-Inst for Molecular Genetics, Berlin, 74-75; vis prof, Japan Soc Promotion Sci, 78. *Honors & Awards:* Jap Govt Res Award, Foreign Specialists, 87. *Mem:* AAAS; Protein Soc; Inst Food Technol; NY Acad Sci; Can Biochem Soc; Am Chem Soc. *Res:* Ribosomal proteins and histones; proteins and enzymes from thermophilic and halophilic bacteria; structure and function of cellulases and xylanases. *Mailing Add:* Inst Biol Sci Nat Res Coun Ottawa ON K1A 0R6 Can

YAHIKU, PAUL Y, statistics, for more information see previous edition

YAHIL, AMOS, b Tel Aviv, Israel, Nov 28, 43; m 72; c 3. ASTROPHYSICS, ASTRONOMY. *Educ:* Hebrew Univ, Jerusalem, BSc, 66; Calif Inst Technol, PhD(physics), 70. *Prof Exp:* Lectr, Tel Aviv Univ, 70-71; mem, Inst Advan Study, 71-73; lectr, Tel Aviv Univ, 73-75, sr lectr, 75-77; from asst prof to assoc prof, 77-83, PROF ASTROPHYS, STATE UNIV NY, STONY BROOK, 83- *Concurrent Pos:* Chercheur Invite, Univ Montreal, 75 & 76-77; vis assoc prof, Calif Inst Technol, 77 & 78; vis prof, Nordita, Copenhagen, 81; vis fel, Inst Astron, Cambridge, 72, 81 & 84-85; vis prof, Res Inst Fundamental Physics, Kyoto Univ, 86. *Honors & Awards:* Fullam Award, 82; Guggenheim Award, 84. *Mem:* Am Astron Soc; Int Astron Union. *Res:* Physical cosmology; dynamics and evolution of clusters of galaxies; galactic structure; stellar collapse; supernovae; molecular clouds. *Mailing Add:* Astron Prog ESS Bldg State Univ NY Stony Brook NY 11794

YAHNER, JOSEPH EDWARD, b Chicago, Ill, June 16, 31; m 58; c 3. SOIL CLASSIFICATION. *Educ:* Purdue Univ, BS, 54; Ore State Univ, MS, 61, PhD(soils), 63. *Prof Exp:* Agronomist, Purdue Univ-Brazil Proj, US AID Contract, Vicosa, Brazil, 63-67, assoc prof, 70-78, EXTEN AGRONOMIST, PURDUE UNIV, WEST LAFAYETTE, 67-, PROF AGRON, 78- *Mem:* Am Soc Agron; Soil Conserv Soc Am. *Res:* Use of soil maps and soil information in land use planning; on-site waste disposal for homes or small commercial establishments; soil map use for land appraisal. *Mailing Add:* Dept Agron Purdue Univ West Lafayette IN 47907

YAHNER, RICHARD HOWARD, b McKees Rocks, Pa, June 21, 49; m 72; c 2. WILDLIFE SCIENCES. *Educ:* Pa State Univ, BS, 71; Univ Tenn, MS, 73; Ohio Univ, PhD(zool), 77. *Prof Exp:* Fel, Smithsonian Inst, 77-78; asst prof, Univ Minn, 78-81; assoc prof, 81-89, PROF, PA STATE UNIV, 89- *Concurrent Pos:* Wildlife consult. *Mem:* Am Soc Mammalogists; Wildlife Soc. *Res:* Effects of habitat alteration, manipulation, and fragmentation on ecology of vertebrates; habitat management; conservation biology. *Mailing Add:* Forest Resources Sch & Lab Pa State Univ University Park PA 16802

YAHR, MELVIN DAVID, b New York, NY, Nov 18, 17; m 48; c 4. NEUROLOGY. *Educ:* NY Univ, AB, 39, MD, 43; Am Bd Psychiat & Neurol, dipl, 49. *Prof Exp:* Res asst, Col Physicians & Surgeons, Columbia Univ, 48-50, instr, 50-51, assoc, 51-53, from asst prof to assoc prof neurol, 53-70, from asst dean to assoc dean, 59-73, Merritt prof neurol, 70-73; HENRY P & GEORGETTE GOLDSCHMIDT PROF NEUROL & CHMN DEPT, MT SINAI SCH MED, 73- *Concurrent Pos:* Nat Res Coun res assoc & asst neurologist, Neurol Inst, Presby Hosp, 48-50, asst attend neurologist, 50-53, assoc attend, 53-61, attend, 61-; asst adj neurol serv, Lenox Hill Hosp, 48-49, adj, 49-54, assoc, 54-60; asst neurologist, Montefiore Hosp, 49-51; mem neurol study sect, NIH, 50-; consult, USPHS, 52-55 & Neuro-Psychiat Inst, NJ, 52-59; med dir, Parkinson's Dis Found, 58-73; mem comt drug ther in neurol, Nat Inst Neurol Dis & Blindness, 59-; dir & neurologist-in-chief, Mt Sinai Hosp, NY, 73- *Honors & Awards:* Golden Plate Award, Am Acad Achievement, 69; Lucy Moses Award, Col Physicians & Surgeons, Columbia Univ, 72; William Hammond Award, Col Med, NY Univ, 79, Solomon A Berson Award, 85. *Mem:* Am Epilepsy Soc; Am Neurol Asn; Asn Am Med Cols; Asn Res Nerv & Ment Dis; fel Am Acad Neurol; fel NY Acad Sci; fel NY Acad Med. *Res:* Cause and treatment of epilepsy; cerebro-vascular diseases; Parkinsonism and multiple sclerosis; mechanisms, causes, methods, and treatment of Parkinson's and allied diseases. *Mailing Add:* Dept Neurol Mt Sinai Sch Med New York NY 10029

YAKAITIS, RONALD WILLIAM, b Baltimore, Md, Oct 13, 41; m 68; c 2. ANESTHESIOLOGY. *Educ:* Loyola Col, BS, 63; Univ Md, MD, 67. *Prof Exp:* Staff instr anesthesia, US Naval Hosp, Oakland, Calif, 71-73; asst prof, Med Univ SC, 73-75; asst prof anesthesia, Univ Ariz Med Ctr, 75-77; STAFF ANESTHESIOLOGIST, KINO COMMUNITY HOSP, TUCSON, 77- *Concurrent Pos:* Consult anesthesia, Univ Calif, San Francisco, 72-73, Vet Admin Hosp, Charleston, SC, 73-75 & Vet Admin Hosp, Tucson, Ariz, 75-; clin assoc, Univ Ariz Med Ctr, 77-; consult, Tucson Hosps Med Educ Prog, 78- *Mem:* Am Soc Anesthesiologists; Int Anesthesia Res Soc; Am Col Physicians; AMA. *Res:* Pulmonary ultramicroscopic and biochemical changes due to oxygen toxicity; cardiovascular drug pharmacokinetics during acid-base imbalance; new techniques for intraoperative anesthetic management. *Mailing Add:* 3390 N Campbell Suite 110 Tuscon AZ 85719

YAKAITIS-SURBIS, ALBINA ANN, b Harvey, Ill, Feb 16, 23; m 67. ANATOMY. *Educ:* Univ Chicago, BS, 45, MS, 49; Univ Minn, PhD(anat), 55. *Prof Exp:* Instr biol, Univ Akron, 46-47; instr, Univ Minn, 47-49, asst anat, 49-54; res assoc path, Med Sch, Univ Mich, 59-61; ASST PROF ANAT, SCH MED, UNIV MIAMI, 61- *Concurrent Pos:* George fel, Detroit Inst Cancer Res, 56-58; res assoc, Vet Admin Hosp, Miami, 64- *Mem:* Am Soc Hemat; Am Asn Anat. *Res:* Oncology; electron microscopy; pathology. *Mailing Add:* Dept Anat & Cell Biol Univ Miami Sch Med Miami FL 33101

YAKATAN, GERALD JOSEPH, b Philadelphia, Pa, May 20, 42; m 64; c 2. PHARMACY, DRUG DISCOVERY & DEVELOPMENT. *Educ:* Temple Univ, BS, 63, MS, 65; Univ Fla, PhD(pharmaceut sci), 71. *Prof Exp:* Asst prof pharm, Univ Tex, Austin, 72-76, assoc prof, 76-80, asst dir, Drug Dynamics Inst, 75-80; dir, Pharmacokinetics/Drug Metab, Warner-Lambert Co, 80-83, vpres, prod develop, 83-87; vpres pharmaceut develop, Immunetech Pharmaceuts, 87-90; PRES/CHIEF EXEC OFF, TANABE RES LABS, USA, 90- *Concurrent Pos:* Adj prof pharm, Univ Mich, 81-83; vis prof, Rutgers Univ, 83-87; predoctoral fel, NIH & NSF. *Mem:* Am Pharmaceut Asn; fel Am Asn Pharmaceut Sci; fel Am Col Clin Pharmacol; Drug Info Asn; NY Acad Sci. *Res:* Pharmacokinetics; biopharmaceutics; analysis of drugs in biological fluids; drug stability; drug discovery and development; drug delivery. *Mailing Add:* Tanabe Res Labs USA 11045 Roselle St San Diego CA 92121

YAKEL, HARRY L, b Brooklyn, NY, July 24, 29. X-RAY CRYSTALLOGRAPHY. *Educ:* Polytech Inst Brooklyn, BS, 49; Calif Inst Technol, PhD(chem), 52. *Prof Exp:* Fel chem, Calif Inst Technol, 52-53; GROUP LEADER METALS & CERAMICS, OAK RIDGE NAT LAB, 53- *Mem:* AAAS; Am Chem Soc; Am Crystallog Asn; Mineral Soc Am; Sigma Xi. *Res:* Structural studies of solids using x-ray diffraction methods. *Mailing Add:* 129 Westlook Circle Oak Ridge TN 37830

YAKIN, MUSTAFA ZAFER, b Isparta, Turkey, May 8, 52; m 78; c 1. MATHEMATICAL PROGRAMMING. *Educ:* Istanbul Tech Univ, BS, 73; Univ Mich, MSE, 75, PhD(indust eng), 81. *Prof Exp:* Teaching asst indust eng, Univ Mich, 77-80; ASST PROF OPERS RES, UNIV HOUSTON, 80- *Concurrent Pos:* Lectr opers res, Eastern Mich Univ, 79-80. *Mem:* Opers Res Soc Am; Inst Mgt Sci; Inst Indust Engrs; Math Prog Soc. *Res:* Developing mathematical models for capacity planning in electric power systems; design and operation of computer integrated manufacturing systems. *Mailing Add:* Quantitative Mgt Sci Dept Univ Houston Univ Park 4800 Calhoun Rd Suite 270-MH Houston TX 77204-6282

YAKOWITZ, HARVEY, b Baltimore, Md, Feb 1, 39; m 80; c 2. HAZARDOUS WASTE MANAGEMENT. *Educ:* Univ Md, College Park, BS, 59, MS, 62, PhD(mat sci), 70. *Prof Exp:* METALLURGIST, NAT BUR STANDARDS, 59- *Concurrent Pos:* Panel mem interagency bd, US Civil Serv Comn, 66-77; US Dept Com vis res worker, Dept Eng, Cambridge Univ, 70-71; assigned, Orgn Econ Coop & Develop, Paris, 83- *Mem:* Sigma Xi. *Res:* Electron probe microanalysis; scanning electron microscopy; optical microscopy; divergent beam x-ray microdiffraction; recycling methodology; risk assessment; economics in environmental policy selection; international environmental policy; transfrontier movements of hazardous wastes; waste reduction. *Mailing Add:* Nat Bur Standards Physics Bldg Rm B326 Gaithersburg MD 20899

YAKOWITZ, SIDNEY J, b San Francisco, Calif, Mar 8, 37; m 63; c 2. STATISTICS, COMPUTER SCIENCE. *Educ:* Stanford univ, BS, 60; Ariz state Univ, MS, 65, MA, 66, PhD(elec eng), 67. *Prof Exp:* Fac assoc elec eng, Ariz State Univ, 65-66; asst prof systs eng, 66-68, assoc prof math, 68-71, assoc prof systs eng, 68-, assoc prof, 71-76, PROF INDUST ENG, UNIV ARIZ, 76- *Concurrent Pos:* Nat Res Coun fel, Naval Postgrad Sch, 70-71. *Mem:* Inst Math Statist; Inst Elec & Electronics Engrs. *Res:* Sequential design of statistical experiments; statistical decision theory; applications to adaptive control theory; pattern recognition. *Mailing Add:* Dept Systs Eng Univ Ariz Tucson AZ 85721

YAKSH, TONY LEE, b San Angelo, Tex, June 14, 44; m 74; c 1. NEUROPHARMACOLOGY, NEUROPHYSIOLOGY. *Educ:* Ga Inst Technol, BS, 66; Univ Ga, MS, 68; Purdue Univ, PhD(neurobiol), 71. *Prof Exp:* Res asst, Purdue Univ, 67-71; mem res staff, Biomed Lab, US Army, Edgewood Arsenal, Md, 71-73; asst scientist, Sch Pharm, Univ Wis-Madison, 73-76; vis scientist, Dept Physiol, Univ Col, Univ London, 76-77; assoc consult pharmacol & neurosurg, Mayo Clin, 77-88; PROF ANESTHESIOL & PHARMACOL, UNIV CALIF, SAN DIEGO, 88- *Mem:* Am Physiol Soc; Int Asn Study Pain; Soc Neurosci; Sigma Xi. *Res:* Pharmacology of opiate action; physiology and pharmacology of pain transmission; role of neuropeptides in behavior. *Mailing Add:* Dept Anesthesiol 0818 Univ Calif San Diego 9500 Gilman Dr La Jolla CA 92093-0818

YAKUBIK, JOHN, b Fords, NJ, Sept 23, 28; m 52; c 2. PHARMACEUTICAL CHEMISTRY. *Educ:* Rutgers Univ, BS, 49; Purdue Univ, MS, 50, PhD(pharmaceut chem), 52. *Prof Exp:* Res assoc, Squibb Inst Med Res, 52-55; from scientist to sr scientist, Schering Corp, 55-61, mgr, Pharmaceut Develop Dept, 61-65, dir sci liaison, Schering Labs, 65-71, dir new prod planning, 71-74, dir corp prod develop, Schering Corp, Bloomfield, 74-79, dir bus develop, 79-80, dir int regulatory affairs, 80-81, DIR NEW PROD DEVELOP, PHARMACEUT PROD DIV, SCHERING PLOUGH INT, KENILWORTH, 82- *Mem:* AAAS; Soc Cosmetic Chem; Am Pharmaceut Asn; NY Acad Sci. *Res:* Pharmaceutical product development; pharmaceutical and medicinal chemistry; pharmacology. *Mailing Add:* 65 Stratford Dr Colonia NJ 07067

YAKURA, HIDETAKA, b Sapporo, Japan. IMMUNOLOGY, PATHOLOGY. *Educ:* Hokkaido Univ, MD, 72, PhD(path), 78. *Prof Exp:* Res fel immunol, Farber Cancer Ctr, Harvard Med Sch, 76-78; res assoc immunol, Mem Sloan-Kettering Cancer Ctr, 78-83; from asst prof to assoc prof path, Asahikawa Med Col, 83-89; DIR, DEPT MICROBIOL & IMMUNOL, TOKYO METROP INST NEUROSCI, 89- *Concurrent Pos:* Vis investr, Mem Sloan-Kettering Cancer Ctr, 88; assoc scientist, Nat Ctr Neurol & Psychiat, Nat Inst Neurosci, Tokyo, Japan, 91- *Mem:* Am Asn Immunologists. *Res:* Regulatory mechanisms of B lymphocyte differentiation; function of receptor-type protein tyrosine phosphatases. *Mailing Add:* Tokyo Metropolitan Inst Neurosci Fuchu Toyko 183 Japan

YAKURA, JAMES K, b Los Angeles, Calif, Nov 1, 33; m 61; c 3. AERONAUTICAL ENGINEERING. *Educ:* Univ Calif, Los Angeles, BS, 55, MS, 57; Stanford Univ, PhD(aeronaut), 62. *Prof Exp:* Mem tech staff, Hughes Aircraft Co, 55-58, sr staff engr, 62-65; mem sr staff, Nat Eng Sci Co, 65-66; staff engr, 66-69, asst dir, 69-71, assoc systs planning dir, 71-72, dir, Vehicle Systs Off, 72-74, dir, Shuttle Interface Off, 75-77, ASSOC DIR, PAYLOAD INTEGRATION OFF, AEROSPACE CORP, EL SEGUNDO, 77- *Mem:* Am Inst Aeronaut & Astronaut. *Res:* Hypersonic aerodynamics. *Mailing Add:* 7732 Gonzaga Pl Westminster CA 92683

YALCINTAS, M GÜVEN, b Milas-Mugla, Turkey, Apr 11, 45; Turkish & US citizen; m 70; c 1. HEALTH & MEDICAL PHYSICS, WASTE MANAGEMENT. *Educ:* Univ Ankara, Turkey, BS, 65; Univ Rochester, MS, 71, PhD(radiation biol-med physics), 74. *Prof Exp:* Fel, Argonne Nat Lab, 74-75; dir med physics & asst prof radiol physics, Izmir Med Sch, 75-76; corp health physicist, EMI Med Inc, 76-77; PROJ MGR, OAK RIDGE NAT LAB, 77- *Concurrent Pos:* Turkish Atomic Energy grad fel; consult, UN Develop Prog, 79-83; adj prof radiation biol, Tenn Technol Univ, 86- *Mem:* Am Nuclear Soc; Health Physics Soc. *Res:* Comparison of risk from radiation in diagnostic radiology; development of a telemetry system that provides real life images from a radiation survey on the field; radiation biology; dosimetry. *Mailing Add:* 1120 W Outer Dr Oak Ridge TN 37830

YALE, CHARLES E, b Aurora, Ill, Mar 21, 25; m 48; c 4. MEDICINE, SURGERY. *Educ:* Univ Ill, Urbana, BS, 49; Case Western Reserve Univ, MD, 55; Univ Cincinnati, DSc(surg), 61. *Prof Exp:* Resident surg, Univ Cincinnati, 56-62, instr, 62-64; from asst prof to assoc prof, 64-72, PROF SURG, MED SCH, UNIV WIS-MADISON, 72-, VCHMN DEPT, 73- *Concurrent Pos:* NIH grant, Univ Wis-Madison, 66-71; clin investr surg, Cincinnati Vet Admin Hosp, 62-64; attend surgeon, Univ Wis Hosps, 64-, dir gnotobiotic lab, Univ Wis-Madison, 65-81. *Mem:* Soc Surg Alimentary Tract; Asn Gnotobiotics (pres, 70-71); Asn Acad Surg; Am Col Surg; AMA; Sigma Xi. *Res:* Gastrointestinal surgery; intestinal obstruction and strangulation; wound healing; surgical infections and septic shock; gnotobiotics; surgery for morbid obesity. *Mailing Add:* Dept Surg Sch Med Univ Wis 600 Highland Ave Madison WI 53792

YALE, HARRY LOUIS, b Chicago, Ill, Dec 18, 13; m 43; c 3. ORGANIC CHEMISTRY. *Educ:* Univ Ill, BSc, 37; Iowa State Col, PhD(org chem), 40. *Prof Exp:* Res chemist, Nat Defense Res Comt, 40-41 & Shell Develop Co, 41-45; res chemist, Squibb Inst Med Res, 46-67, sr res fel, 67-79; consult, 79-88; RETIRED. *Honors & Awards:* Lasker Award. *Mem:* Am Chem Soc; NY Acad Sci; Swiss Chem Soc. *Res:* Organometallic compounds; quinoline derivatives; furan; high temperature oxidation and chlorination of olefins; chelate compounds; explosives; antituberculous drugs; diuretics; ataractic agents; natural products. *Mailing Add:* Four New York Ave New Brunswick NJ 08901

YALE, IRL KEITH, b Billings, Mont, Mar 13, 39. MATHEMATICS. *Educ:* Univ Mont, BA, 60; Univ Calif, Berkeley, PhD(math), 66. *Prof Exp:* Instr math, Univ Mont, 64-65; asst prof, Morehouse Col, 66-67; asst prof, 67-73, assoc prof, 73-80, PROF MATH, UNIV MONT, 80- *Mem:* Am Math Soc; Math Asn Am. *Res:* Functional and harmonic analysis. *Mailing Add:* Dept Math Univ Mont Missoula MT 59812-1032

YALE, PAUL B, b Geneva, NY, Apr 28, 32; m 71; c 5. MATHEMATICS. *Educ:* Univ Calif, Berkeley, BA, 53; Harvard Univ, MA & PhD(math), 59. *Prof Exp:* Asst prof math, Oberlin Col, 59-61; from asst prof to assoc prof, 61-74, PROF MATH, POMONA COL, 74- *Concurrent Pos:* NSF sci fac fel, 67-68. *Honors & Awards:* Lester R Ford Award, 67. *Mem:* Am Math Soc; Math Asn Am; Asn Comput Mach. *Res:* Geometry; symmetry; group theory; computer graphics. *Mailing Add:* Dept Math Pomona Col 810 N College Ave Claremont CA 91711

YALE, SEYMOUR HERSHEL, b Chicago, Ill, Nov 27, 20; m 43; c 2. RADIOLOGY. *Educ:* Univ Ill, BS, 44, DDS, 45. *Prof Exp:* Asst clin dent, Col Dent, Univ Ill, 48-49, from instr to asst prof, 49-54, assoc prof Dept Radiol, 56, prof & head dept, 57-61, admin asst dean, 61-63, from asst dean to actg dean, 63-65, dean, 65-87, EMER DEAN, COL DENT, UNIV ILL, 87- *Concurrent Pos:* Res consult, Hines Vet Admin Hosp, Ill, 59; consult, West Side Vet Admin Hosp, Chicago, 61-, dent proj sect, Nat Inst Radiol Health, 61-, div radiol health, Bur State Serv, 63- & Vet Admin Res Hosp, Chicago, 63-; mem sect comt dent film specifications, US Am Stand Inst, 59-, subcomt 16, Nat Comt Radiation Protection, 63-, gen res support adv comt, Dept Health, Educ & Welfare, 65-, Mayor's Comt Heart, Cancer & Stroke, Chicago & comt dent care ment ill, Ill State Dept Ment Health; mem, Grad Fac Dept Radiol, Col Med, Univ Ill, Chicago, Hillel Bd; founder, Ctr Res Periodont Dis & Oral Molecular Biol, 77; organizer & chmn, Nat Conf Hepatitis-B Dent, 82; organizer & dir, Univ Taskforce Primary Health Care Proj, Univ Ill, Chicago; chmn, Univ Ill, Univ Stockholm & Univ Gothenberg Conf Geriatrics, 85; prof dent & health sources mgt, Sch Pub Health, Univ Ill, Chicago, 87- *Honors & Awards:* Harry Sicher Meml Lect Award, Am Col Stomologic Surgeons, 83. *Mem:* AAAS; Am Dent Asn; Am Acad Dent Radiol; fel Am Col Dent; Int Asn Dent Res; hon fel, Acad Gen Dent; Am Acad Oral Roentgenology; NY Acad Sci; Am Pub Health Asn; Sigma Xi. *Res:* Morphology; radiographic anatomy; radiation biology and control. *Mailing Add:* 155 N Harbor Dr Chicago IL 60601

YALKOVSKY, RAFAEL, b Chicago, Ill, Oct 11, 17. OCEANOGRAPHY. *Educ:* Univ Chicago, BS, 46, MS, 55, PhD(geol), 56. *Prof Exp:* Geologist, US Corps Engrs, Wash, 49-50; jr engr, State Div Hwys, Calif, 50-52; jr engr geophys comput, Western Geophys Co, 52-54; assoc geol engr, Crane Co, Ill, 55-56; asst prof geol, Mont State Univ, 56-61 & State Univ NY Col New Paltz, 61-62; from asst prof to prof geol, 62-83, EMER PROF GEOL & OCEANOG, BUFFALO STATE UNIV COL, 84-; FREELANCE SCI WRITER, 84- *Concurrent Pos:* Res grant, Mont State Univ, 59-60 & 60-61; vis investr, Archives of Indies, Naval Mus Madrid, 59-60 & 68, Spanish Inst Oceanog, Spain, 59-60 & Royal Spanish Acad Hist, 68; vis scientist, US Coast

Geodetic Surv ship Discoverer, 69; sci corresp, Nat Public Radio, WBFO, 75-78, UN press corresp, Geneva, 75, New York, 73, 76, 77 & 79-81; vis scholar, Univ London Law Sch, 75; rep, KPBX, Radio, Spokane, 79-81; mem, Malaspina Expedition; vis investr, Mus Naval, Madrid, 81-82; lectr, Kings Col London, 87. *Honors & Awards:* Silver Pin Award, Am Geophys Union. *Mem:* Fel AAAS; Nat Asn Geol Teachers; Am Geophys Union; fel NY Acad Sci; Nat Sci Teachers Asn; Sigma Xi; Nat Asn Sci Writers; Int Sci Writers Asn. *Res:* Marine geology and geochemistry of marine sediments; water resources; history of science; international law of the sea; public policy. *Mailing Add:* Snug Harbor PO Box 398 Grand Island NY 14072-0398

YALKOWSKY, SAMUEL HYMAN, b New York, NY, Dec 5, 42; m 62; c 2. PHARMACEUTICAL CHEMISTRY. *Educ:* Columbia Univ, BS, 65; Univ Mich, Ann Arbor, MS, 68, PhD(pharm chem), 69. *Prof Exp:* Sr res scientist pharm res, Upjohn Co, 69-; AT COL PHARM, UNIV ARIZ. *Mem:* Am Asn Pharmaceut Sci; Am Chem Soc. *Res:* Physical chemistry of surfaces and micelles; solubility and related phenomena; drug product formulation. *Mailing Add:* Dept Pharmaceut Sci Univ Ariz Tucson AZ 85721

YALL, IRVING, b Chicago, Ill, Jan 31, 23. MICROBIAL PHYSIOLOGY. *Educ:* Brooklyn Col, BA, 48; Univ Mo, MA, 51; Purdue Univ, PhD(bact), 55. *Prof Exp:* Asst bact, Univ Mo, 49-51; res fel, Purdue Univ, 54-56, resident res assoc biochem, Argonne Nat Lab, 56-57; from asst prof to assoc prof microbiol, 57-68, prof, 68-83, EMER PROF MICROBIOL & IMMUNOL, UNIV ARIZ, 84- *Mem:* Fel AAAS; Am Soc Microbiol; Am Chem Soc. *Res:* Phosphorus metabolism in wastewaters; intermediary metabolism of microorganisms; control of nucleic acids. *Mailing Add:* 6124 E Rosewood Tucson AZ 85711-1637

YALMAN, RICHARD GEORGE, b Indianapolis, Ind, Apr 16, 23; m 44; c 2. INORGANIC CHEMISTRY. *Educ:* Harvard Univ, BS, 43, MA, 47, PhD(chem), 49. *Prof Exp:* Jr chemist, Monsanto Chem Co, 44; res chemist & sr group leader, Mound Lab, 49-50; from asst prof to prof chem, Antioch Col, 50-83, chmn dept, 58 & 61-66, chmn dept, 72-83; RETIRED. *Concurrent Pos:* Consult, Signal Corps, Air Force Off Sci Res Projs, Antioch, 50-58, Monsanto Co, 55 & Kettering Res Lab, 67-71; pres, Mad River Chem Co, 67-; consult, Yellow Springs Instrument Co, 71-72; vis scientist, Electronics Br, Avionics Lab, Wright-Patterson AFB, 79- *Mem:* AAAS; Am Chem Soc; Royal Soc Chem; Indian Chem Soc; Soc Chem Indust. *Res:* Kinetics; metal complexes; high temperature synthesis; metalloporphines; hydrothermal properties of oil shale. *Mailing Add:* 440 San Pasquale Santa Fe NM 87501

YALOVSKY, MORTY A, b Montreal, Que, May 14, 44; m 70; c 2. STATISTICS. *Educ:* McGill Univ, BSc, 65, MSc, 68, PhD(statist), 76. *Prof Exp:* Lectr math, 70-74, asst prof, 74-78, assoc mgt sci, 79-82, ASSOC DEAN ACAD, MCGILL UNIV, 83- *Mem:* Am Statist Asn; Statist Soc Can. *Res:* Goodness-of-fit testing; regression models and time series analysis; applications of quantitative techniques to administrative sciences; biostatistics. *Mailing Add:* Fac Mgt McGill Univ Montreal PQ H3A 1G5 Can

YALOW, A(BRAHAM) AARON, b Syracuse, NY, Sept 18, 19; m 43; c 2. PHYSICS, MEDICAL BIOPHYSICS. *Educ:* Syracuse Univ, AB, 39; Univ Ill, MS, 42, PhD(physics), 45. *Prof Exp:* Asst physics, Syracuse Univ, 39-41 & Univ Ill, 41-42; asst physicist, Nat Defense Res Comt, 43; asst physics, Univ Ill, 44-45; asst engr, Fed Telecommun Labs, NY, 45-46; asst prof physics, NY State Maritime Col, 47-48; from instr to assoc prof, 48-66, prof, 66-83, EMER PROF PHYSICS, COOPER UNION SCH ENG, 83- *Concurrent Pos:* Consult physicist, Montefiore Hosp, 46-80. *Mem:* AAAS; Am Phys Soc; Am Asn Physics Teachers; assoc fel Am Col Radiol. *Res:* Neutron resonance absorption and scattering; microwave transmission; medical applications of radioactive isotopes; Mössbauer effect. *Mailing Add:* 3242 Tibbett Ave Bronx NY 10463-3801

YALOW, ROSALYN SUSSMAN, b New York, NY, July 19, 21; m 43; c 2. MEDICAL PHYSICS. *Educ:* Hunter Col, AB, 41; Univ Ill, MS, 42, PhD(physics), 45; Am Bd Radiol, dipl, 51. *Hon Degrees:* Various from US Univs, 74-89. *Prof Exp:* Instr physics, Univ Ill, 44-45; lectr & temp asst prof, Hunter Col, 46-50; physicist & asst chief, Radioisotope Serv, Vet Admin Hosp, Bronx, 50-70, actg chief, Radioisotope Serv, 68-70; res prof, Dept Med, Mt Sinai Sch Med, 68-74, distinguished serv prof, 74-79; chmn, dept clin sci, Montefiore Hosp & Med Ctr, 80-85; distinguished prof at large, 79-85, EMER PROF, ALBERT EINSTEIN COL MED, YESHIVA UNIV, 85-; SR MED INVESTR, VET ADMIN, 72- *Concurrent Pos:* Consult, Radioisotope Unit, Vet Admin Hosp, Bronx, 47-50 & Lenox Hill Hosp, 52-62; secy, US Nat Comt Med Physics, 63-67; chief, Radioimmunassay Ref Lab, Vet Admin Med Ctr, 69-, Nuclear Med Serv, 70-80, sr med investr, 72-; mem, Med Adv Bd, Nat Pituitary Agency, 68-71; Int Atomic Energy Agency expert, Inst Atomic Energy, Sao Paulo, 70; mem, Endocrinol Study Sect, NIH, 69-72, Bd Sci Coun, Nat Inst Arthritis, Metab & Digestive Dis, 72-75 & 78-81, Task Force Immunol Dis, Nat Inst Allergy & Infectious Dis, 72-73; consult, New York City Dept Health, 72-; co-ed, Hormone & Metab Res, 73-; mem, Comt Eval of Nat Pituitary Agency, Nat Res Coun, 73-74; Albert Lasker Basic Med Res Award, 76; WHO consult, Radiation Med Ctr, Bombay, 78; dir, Solomon A Berson Res Lab, Vet Admin Hosp, Bronx, 73-, Solomon A Berson distinguished prof-at-large, Mt Sinai Sch Med, City Univ, NY, 86- *Honors & Awards:* Nobel Prize, 77; Lilly Award, Am Diabetes Asn, 61, Banting Medal, 78, Rosalyn S Yalow Res & Develop Award, 78; Fed Woman's Award, 61; Van Slyke Award, Am Asn Clin Chem, 68; Gairdner Found Int Award, 71; Am Col Physicians Award, 71; Koch Award, Endocrine Soc, 72; A Cressy Morrison Award, Natural Sci, NY Acad Sci 75; Boehringer-Mannheim Award, Am Asn Clin Chemists, 75; Sci Achievement Award, AMA, 75; Gratum Genus Humanum Gold Medal, World Fedn Nuclear Med & Biol, 78; G von Heresy medal, 78; Theobold Smith Award, 82; Nat Medal of Sci, 88. *Mem:* Nat Acad Sci; Am Acad Arts & Sci; Endocrine Soc (pres elect, 77-78, pres, 78-79); fel NY Acad Sci; hon mem Am Gastroenterol Asn; foreign assoc French Acad Med; Radiation Res Soc; Am Asn Physicists Med; Am Col Radiol; Biophys Soc; Am Diabetes Asn; Am Physiol Soc; Soc Nuclear Med. *Res:* Medical use of radioisotopes, radioimmunoassay and radiation chemistry. *Mailing Add:* Vet Admin Hosp 130 W Kingsbridge Rd Bronx NY 10468

YAM, LUNG TSIONG, b Canton, China, Apr 10, 36; US citizen; m 64; c 2. INTERNAL MEDICINE. *Educ:* Nat Taiwan Univ, MD, 60. *Prof Exp:* From instr to asst prof med, Sch Med, Tufts Univ, 67-72; assoc hemat, Scripps Clin & Res Found, 72-74; assoc prof, 74-80, PROF MED, UNIV LOUISVILLE, 80-; CHIEF HEMAT-ONCOL, VET ADMIN HOSP, 74- *Concurrent Pos:* Res assoc, New Eng Med Ctr Hosp, 68-72, head cytol & histochem, 70-72. *Mem:* Am Soc Hemat; Am Soc Histochem & Cytochem. *Res:* Use of morphologic approach to study problems related to hematology-oncology; use of cytochemistry, immunochemistry and electrophoresis for isoenzymes to identify the origin of normal and neoplastic cells. *Mailing Add:* Dept Med Univ Louisville Box 35260 Louisville KY 40292-0001

YAMADA, EICHI, b Fukuoka, Japan, May 17, 22; m 47; c 2. ANATOMY, CYTOLOGY. *Educ:* Kyushu Imp Univ, MD, 45; DSc, Kyushu Univ, 50. *Prof Exp:* From assoc prof to prof anat, Fac Med, Kyushu Univ, 49-69; prof anat, Sch Med, Kurume Univ, 56-60; prof, 69-83, EMER PROF ANAT, FAC MED, TOKYO UNIV, 83-; PROF ANAT, SCH MED, FUKUOKA UNIV, 83- *Concurrent Pos:* Postdoctoral fel, Sch Med, Univ Wash, 54-55; res assoc, Rockefeller Inst Med Res, 58-59; vis prof, Jules Stein Eye Inst, Univ Calif, Los Angeles, 67-68; sr res sci, Sch Med, Yale Univ, 86-; hon mem, Am Asn Anatomists, 83. *Honors & Awards:* Setoh Award, Japan Soc Electron Micros, 58; Yamaji Sci Award, Yamaji Sci Found, 67; Brasilian Garibaldi Order, 91. *Mem:* Am Soc Cell Biol. *Res:* Electron microscopy of cells and tissues; gall bladder epithelium; renal glomerulus; megakaryocyte; centriole; ocular tissue. *Mailing Add:* Hirao 3-9-15 Chuo-ku Fukuoka 810 Japan

YAMADA, ESTHER V, b London, Ont, July 19, 23; m 53; c 1. BIOCHEMISTRY. *Educ:* Univ Western Ont, BSc, 45, PhD(biochem), 51; McGill Univ, MSc, 47. *Prof Exp:* Nat Cancer Inst fel, Univ Western Ont, 52-55; res assoc, Karolinska Inst, Sweden, 55-57; res fel, NIH, 57-59; lectr, 59-60, asst prof, 60-66, assoc prof, 66-75, PROF BIOCHEM, FAC MED, UNIV MAN, 66- *Mem:* AAAS; Can Biochem Soc; Am Soc Biol Chemists; NY Acad Sci. *Res:* Transcription and translation; skeletal muscle metabolism; bioenergetics; membrane; bound enzymes; heart metabolism. *Mailing Add:* Dept Biochem Med Sci Bldg Univ Man Winnipeg MB R3E 0W3 Can

YAMADA, KENNETH MANAO, Minneapolis, Minnesota, 44; m. EXTRA-CELLULAR MATRIX, CELL ADHESION. *Educ:* Stanford Univ, PhD(biol sci), 71, MD, 72. *Prof Exp:* CHIEF, MEMBRANE & BIOCHEM SECT, NAT CANCER INST, NIH, 80- *Mailing Add:* 9305 Wadsworth Dr Bethesda MD 20817

YAMADA, MASAAKI, b Japan, Aug 9, 42; m 71; c 2. PLASMA PHYSICS. *Educ:* Univ Tokyo, Japan, BS, 66, MS, 68; Univ Ill, PhD(physics), 73. *Prof Exp:* Res asst physics, Univ Ill, 69-73; res assoc, Princeton Univ, 73-75, res staff, 75-78, res physicist, plasma physics, Plasma Physics Lab, 78-82, PRIN RES PHYSICIST, PRINCETON UNIV, 82- *Mem:* Fel Am Phys Soc; Phys Soc Japan. *Res:* Experimental studies of plasma physics, including spheromak physics, general magnetohydrodynamics and micro-instabilities and transport properties of plasmas; magnetic reconnection. *Mailing Add:* Plasma Physics Lab PO Box 451 Princeton Univ Forrestal Campus Princeton NJ 08544

YAMADA, RYUJI, b Hiroshima City, Japan, Jan 3, 32; m 60; c 2. SUPERCONDUCTING MAGNETS, ACCELERATOR. *Educ:* Hiroshima Univ, Japan, BS, 54; Univ Tokyo, Japan, MS, 56, PhD(physics), 62. *Prof Exp:* Res assoc high energy physics, Inst Nuclear Study, Univ Tokyo, Japan, 56-68; res assoc, high energy physics, Brookhaven Nat Lab, 63-65; res assoc accelerator physics, Cornell Univ, 65-66; accelerator consult, 67, PHYSICIST HIGH ENERGY PHYSICS, FERMI NAT ACCELERATOR LAB, 68- *Mem:* Am Phys Soc. *Res:* Development of super conducting magnets for Energy Doubler Project and Colliding Detectors; high energy experiments using existing 400 GeV Proton Synchrotron; construction of accelerators, including 16 inch cyclotron, 1 GeV INS electron synchrotron, 10 GeV Cornell Electron Synchrotron, and NAL 500 GeV Proton Synchrotron. *Mailing Add:* 1651 White Pines Ct Naperville IL 60563

YAMADA, SYLVIA BEHRENS, b Hamburg, Ger, May 7, 46; m 75; c 2. POPULATION ECOLOGY. *Educ:* Univ BC, BSc, 68, MSc, 71; Univ Ore, PhD(marine ecol), 74. *Prof Exp:* Biologist fisheries, Pac Biol Sta Fisheries & Marine Serv, Dept Environ Can, 74-79; asst prof, Wellesley Col, Mass, 80-81; adj prof, Sch Oceanog, 81-83, res assoc fisheries, wildlife & zool, 83-84, RES ASSOC ZOOL, ORE STATE UNIV, 84- *Concurrent Pos:* Course coordr, Marine Ecol, Marine Biol Lab, Woods Hole, Mass, 80; lectr, Biol Dept, Univ Manitoba, 73, dept zool & fisheries & wildlife, Ore State Univ, 82- *Mem:* Ecol Soc Am; Western Soc Naturalists; Sigma Xi. *Res:* Chemical marking of salmon; incorporation and retention of trace elements in salmon tissue; ecology of intertidal mollusks, (Littorina sitkana, L scutulata, L planaxis, L littorea, L saxatilis, Batillaria attramentatia, Mytilus californianus and M edulis); molluskan aquaculture. *Mailing Add:* Zool Dept Ore State Univ Corvallis OR 97331

YAMADA, TETSUJI, b Osaka, Japan, May 9, 42; US citizen; m 67; c 2. METEOROLOGY, ENVIRONMENTAL SCIENCES. *Educ:* Osaka Univ, BE, 65, ME, 67; Colo State Univ, PhD(civil eng), 71. *Prof Exp:* Fel civil eng, Colo State Univ, 71-72; mem res staff meteorol, Princeton Univ, 72-76; meteorologist, Argonne Nat Lab, 76-81; mem staff, Los Alamos Nat Lab, 81-90; PRES, YAMADA SCI & ART CORP, 90- *Honors & Awards:* Soc Award, Meteorol Soc Japan, 84. *Mem:* Am Meteorol Soc; Meteorol Soc Japan; Sigma Xi. *Res:* Turbulence theory; numerical modelling of mountain flows, sea breeze and air pollution. *Mailing Add:* Yamada Sci & Art Corp 147 Monte Rey Dr S Los Alamos NM 87544

YAMADA, YOSHIKAZU, b Honokaa, Hawaii, May 20, 15; m 50; c 4. ORGANIC CHEMISTRY. *Educ:* Univ Hawaii, BS, 37; Univ Mich, MS, 38; Purdue Univ, PhD(chem), 50. *Prof Exp:* Res chemist, Davidson Corp, Ill, 50-53; from sr proj engr to res engr, Mergenthaler Linotype Co, NY, 53-59; sr res chemist, Bell & Howell Co, Ill, 59-60, prin res chemist, Res Labs, Calif, 60-72; PRES, YAMADA-GRAPHICS CORP, IRVINE, 72- *Mem:* AAAS; Am Chem Soc; Sigma Xi. *Res:* Photosensitive systems; graphic media. *Mailing Add:* 6151 Sierra Bravo Rd Irvine CA 92715

YAMAGISHI, FREDERICK GEORGE, b Reno, Nev, Sept 14, 43; m 68; c 2. ORGANIC CHEMISTRY. *Educ:* Univ Calif, Los Angeles, BS, 65, PhD(org chem), 72; Calif State Col, MS, 67. *Prof Exp:* Fel, Dept Physics, Univ Pa, 72-73, res assoc org chem, 73-74; MEM TECH STAFF ORG CHEM, HUGHES RES LAB, HUGHES AIRCRAFT CO, 74-, STAFF CHEMIST, 85- *Mem:* Am Chem Soc. *Res:* Organic conductors; one-dimensional materials; conducting polymers; liquid crystals; plasma polymerized thin films; nonlinear optical polymers. *Mailing Add:* Exp Study Dept 3011 Malibu Canyon Rd Malibu CA 90265

YAMAGUCHI, MASATOSHI, b San Leandro, Calif, Mar 12, 18; m 42; c 3. PLANT PHYSIOLOGY, BIOCHEMISTRY. *Educ:* Univ Calif, BS, 40, PhD(agr chem), 50. *Prof Exp:* Prin lab tchnician, 41 & 46-50, instr, 50-52, from asst olericulturist to assoc olericulturist, 50-64, lectr, 64-73, prof veg crops, 73-81, olericulturist, 64-81, EMER PROF, UNIV CALIF, DAVIS, 81- *Concurrent Pos:* Fulbright res scholar, 59-60; vis prof, Univ Man, 67-68; consult veg crops, Sao Paulo & Parana, Brazil, 75-76, Beijing Veg Res Ctr, Beijing, China, 88. *Honors & Awards:* Japan Nat Sci Found Award, 79. *Mem:* Am Chem Soc; Am Soc Hort. *Res:* Chemical constituents and quality of vegetables; physiological disorders of vegetable crops; biochemistry and physiology of vegetable fruit development and ripening; author of one book. *Mailing Add:* Dept Veg Crops Univ Calif Davis CA 95616

YAMAGUCHI, TADANORI, b Miyazaki, Japan, Jan 17, 49. METAL OXIDE SEMICONDUCTOR. *Educ:* Miyakonojyo Tech Col, BSEE, 69. *Prof Exp:* Res engr, Semiconnductor Div, Sony Corp, Japan, 69-77; PROF PROJ MGR, TEKTRONIX, INC, 78- *Res:* Device and technology development for advanced metaloxide semiconductor-large scale integration-very large scale integrations, including state-of-the-art device structure, also new device structure in n-channel metal oxide semiconductor, complementary metal oxide semiconductor, high-voltage metal oxide semiconductor, and advanced bipolar semiconductor integrated circuits; investigation of metal oxide semiconductor device physics and modeling. *Mailing Add:* 12757 NW Hartford St Portland OR 97229

YAMAKAWA, KAZUO ALAN, b San Jose, Calif, June 18, 18; m 55. SOLID STATE PHYSICS. *Educ:* Stanford Univ, AB, 40; Princeton Univ, MA, 47, PhD(physics), 49. *Prof Exp:* Br chief, Ballistic Res Labs, Aberdeen, Md, 49-55; group head semiconductor develop, Hughes Res Labs, Malibu, Calif, 59-61; dept mgr laser mat, Electro Optical Syst, Pasadena, Calif, 61-63; group scientist semi conductor mat, Autometics, Amaheim, Calif, 63-68; SR MEM TECH STAFF PHOTOVOLTAICS, JET PROPULSION LABS, 70- *Mem:* Fel Am Phys Soc; Sigma Xi. *Mailing Add:* 1459 Bonita Terr Monterey Park CA 91754

YAMAMOTO, DIANE M, b Madison, Wis, Nov 3, 46. MEDICINAL CHEMISTRY, ORGANIC CHEMISTRY. *Educ:* Univ Wis-Madison, BS, 68; Univ Calif, Berkeley, PhD(org chem), 73. *Prof Exp:* Res asst, Univ Calif, Berkeley, 69-73; res assoc, Univ Ariz, 73-75; sr chemist, Wyeth Labs Div, Am Home Prods Corp, 75-77; scientist, Ortho Pharmaceut Corp, Johnson & Johnson, 78-82; res investr, Dupont Crit Care, 82-83, sr res investr, 83-89; PATENT LIAISON, ABBOTT LABS, 89- *Mem:* Am Chem Soc; Sigma Xi. *Res:* Chemistry of biologically active compounds, including design, synthesis, structure-activity relationships and pertinent mechanistic studies. *Mailing Add:* 36143 N Grand Oaks Ct Gurnee IL 60031-1409

YAMAMOTO, HARRY Y, b Honolulu, Hawaii, Nov 26, 33; m 57; c 2. BIOCHEMISTRY. *Educ:* Univ Hawaii, BS, 55; Univ Ill, MS, 58; Univ Calif, Davis, PhD(biochem), 62. *Prof Exp:* From asst prof to assoc prof, plant physiol, Univ Hawaii, Manoa, 61-70, actg assoc dean res, 80-82, dept chmn, 82-88, PROF PLANT PHYSIOL, UNIV HAWAII, MANOA, 70- *Concurrent Pos:* USPHS spec fel, Charles F Kettering Res Lab, 68-69; vis scientist, Inst Animal Physiol, Babraham, Cambridge, Eng, 76. *Honors & Awards:* Samuel Cate Prescott Award, Inst Food Technologists, 69. *Mem:* Am Chem Soc; Am Soc Plant Physiol; Inst Food Technologists; AAAS. *Res:* Carotenoid function; photosynthesis; food biotechnology. *Mailing Add:* Dept Plant Molecular Physiol Univ Hawaii Honolulu HI 96822

YAMAMOTO, HIROSHI, b Osaka, Japan, Aug 23, 46; m 76; c 1. IMMUNOLOGY. *Educ:* Osaka Univ, BS, 70, MS, 73, PhD(immunol) & DMS, 77. *Prof Exp:* Postdoctoral immunol, Scripps Clin Res Found, 77-78, res assoc, 78-79; assoc prof, Kochi Med Sch, Japan, 80-90; HEAD IMMUNOL, NAT INST NEUROSCI, NAT CTR NEUROL & PSYCHIAT, TOKYO, 90- *Concurrent Pos:* Lectr, Kochi Med Sch, Japan, 90- *Mem:* Am Asn Immunologists. *Res:* Cell surface molecules that govern communications in immunology and neurology. *Mailing Add:* Div Neuroimmunol Nat Inst Neurosci Nat Ctr Neurol & Psychiat Ogawahigashi Kodaira Tokyo 187 Japan

YAMAMOTO, JOE, b Los Angeles, Calif, Apr 18, 24; m 47; c 2. PSYCHIATRY. *Educ:* Univ Minn, BS, 46, MB, 48, MD, 49. *Prof Exp:* Asst prof psychiat, Sch Med, Univ Okla, 55-58, asst prof, 58-61; from asst prof to assoc prof psychiat, Sch Med, 58-66, mem fac, Psychoanal Inst, 66-69, prof psychiat, Sch Med, Univ Southern Calif, 69-77; dir, Adult Ambulatory Serv, 77-78, PROF PSYCHIAT, UNIV CALIF, LOS ANGELES, 77- *Concurrent Pos:* Clin dir, Adult Outpatient Psychiat Clin, Los Angeles County-Univ Southern Calif Med Ctr, 58-77; mem comt psychopath, NIH, 84-86. *Mem:* AAAS; fel Am Acad Psychoanal (pres, 78); fel Am Psychiat Asn; fel Am Col Psychiatrists; fel Am Asn Soc Psychiat (vpres 84-86). *Res:* Clinical and preventive psychiatry; social class factors, Asian/Pacific factors. *Mailing Add:* 760 Westwood Plaza Box 18 760 Westwood Plaza Los Angeles CA 90024-1759

YAMAMOTO, KEITH ROBERT, b Des Moines, Iowa, Feb 4, 46. MOLECULAR BIOLOGY. *Educ:* Iowa State Univ, BSc, 68; Princeton Univ, PhD(biochem sci), 73. *Prof Exp:* Res asst biochem, Dept Biochem & Biophys, Iowa State Univ, 67-68; NIH trainee biochem sci, Princeton Univ, 68-73; fel biochem, Lab Gordon M Tomkins, 73-75, from asst prof to assoc prof, 76-83, PROF BIOCHEM, UNIV CALIF, SAN FRANCISCO, 83-, VCHMN, DEPT BIOCHEM & BIOPHYS, 85- *Concurrent Pos:* Fel, Helen Hay Whitney Found, 73-76; fel consult, Found Res Hereditary Dis, 74; mem sci adv comt, Dam Runyon-Walt Winchell Cancer Found, 77-81; NIH res career develop award, 77-82, mem, Molecular Biol Study Sect, 86-90 & chmn, 87-90; Dreyfus teacher scholar award, 82-86; mem, Genetics Biol Rev Panel, NSF, 84-87; mem, Molecular Biol Study Sect, NIH, 86-90, chair, 87-90, mem, Dir Subcomt Intramural Res Prog Rev Process, 89 & Nat Adv Coun Human Genome Res, 90-92; assoc ed, J Molecular Biol, 88- & Cell Regulation, 88-; mem, Panel Sci Responsibility/Conduct Res, 90-91. *Honors & Awards:* Swerling Lectr, Harvard Med Sch, 82; Rudin Lectr, Columbia Univ, 83; Weissberger Lectr, Univ Rochester, 83; Bernard Axelrod Lectr, Purdue Univ, 88; Markey Lectr, Harvard Med Sch, 90; Lindner Lectr, Weizmann Inst, 90; Greg Pincus Medal, 90. *Mem:* Nat Acad Sci; Am Soc Biol Chemists; Am Soc Develop Biol; Am Soc Microbiol; AAAS; Am Soc Biochem & Molecular Biol; Am Soc Cell Biol; fel Am Acad Arts & Sci. *Res:* Mechanisms of steroid receptor action; regulation of eukaryotic transcription; regulation and maintenance of chromosome structure. *Mailing Add:* Dept Biochem & Biophys Univ Calif San Francisco CA 94143

YAMAMOTO, MITSUYOSHI, b Tokyo, Japan, Jan, 1923. SUPERCONDUCTIVE APPLICATION TO ELECTRICAL MACHINES. *Educ:* Tokyo Univ, Bachelor, 46, Dr Eng, 61. *Prof Exp:* Dir, Heavy Apparatus Eng Lab, Toshiba Co, Ltd, 78-83; prof elec mach, Saitemca Univ, 84-88; PROF ELEC MACH, TOKUSHOKU UNIV, 88- *Mem:* Fel Inst Elec & Electronics Engrs. *Res:* Superconductive application to electrical machines; control technique electrical machines. *Mailing Add:* Mechanical System Eng Takushoku Univ 815-1 Tate-Machi Hachioji-Shi 193 Japan

YAMAMOTO, NOBUTO, b Tagawa City, Japan, Apr 25, 25; m 54; c 3. MICROBIOLOGY, BIOPHYSICS. *Educ:* Kurume Inst Technol, BS, 47; Kyushi Univ, MS, 53; Nagoya Univ, PhD(bact), 58. *Prof Exp:* Asst prof, Sch Med, Gifu Univ, 58-62; vis scientist molecular biol, NIH, 62-63; from asst prof to assoc prof microbiol, Fels Res Inst & Sch Med, Temple Univ, 63-71, prof, 71-80; MEM FAC, DEPT MICROBIOL, HAHNEMANN MED CCL, 80- *Concurrent Pos:* Vis researcher virol, Inst Cancer Res, Philadelphia, 59-61; vis scientist, Dept Bact, Ind Univ, 61-62. *Mem:* Am Soc Microbiol; Am Asn Cancer Res. *Res:* Virology; genetics; cancer research; molecular biology. *Mailing Add:* Dept Microbiol & Immunol Hahnemann Univ Sch Med Broad & Vine Sts Philadelphia PA 19102

YAMAMOTO, RICHARD, b Wapato, Wash, May 27, 27; m 50; c 7. VETERINARY MICROBIOLOGY, MYCOPLASMOLOGY. *Educ:* Univ Wash, BS, 52; Univ Calif, MA, 55, PhD(microbiol), 57. *Prof Exp:* Asst specialist vet pub health, Univ Calif, 57-58, asst res microbiologist, 59; asst prof vet serol & asst vet serologist, Ore State Univ, 59-61; asst microbiologist, Univ Calif, Davis, 62-64, lectr, 64-67, assoc prof, 67-70, assoc microbiologist, 64-70, PROF & MICROBIOLOGIST, UNIV CALIF, DAVIS, 70-, DIR, MASTER PREV VET MED PROG, SCH VET MED, 90- *Concurrent Pos:* Assoc ed, Poultry Sci, 67-72 & 80-86 & Avian Dis J, 75-88. *Honors & Awards:* Tom Newman Int Award, 67; Nat Turkey Fedn Res Award, 70. *Mem:* Am Soc Microbiol; Poultry Sci Asn; Am Asn Avian Path; US Animal Health Asn; World Poultry Sci Asn; Int Orgn Mycoplasmology; Am Asn Vet Lab Diag; hon dipl Am Col Vet Microbiologists, 90. *Res:* Host parasite interactions (infection, immunity, epidemiology) with reference to avian species; mycoplasmas and viruses; development of rapid diagnostic terts for avian diseases. *Mailing Add:* Dept Epidemiol & Prev Med Univ Calif Davis CA 95616-8735

YAMAMOTO, RICHARD KUMEO, b Honolulu, Hawaii, June 29, 35; m 61; c 3. PHYSICS. *Educ:* Mass Inst Technol, SB, 57, PhD(physics), 63. *Prof Exp:* Res staff physics, 63-64, from instr to assoc prof, 64-75, PROF PHYSICS, MASS INST TECHNOL, 75- *Mem:* Fel Am Phys Soc; AAAS; NY Acad Sci. *Res:* High energy nuclear physics; spectrometer techniques; study of lepton and quark interactions; electronic detectors. *Mailing Add:* Dept Physics Rm 24-043 Mass Inst Technol Cambridge MA 02139

YAMAMOTO, RICHARD SUSUMU, b Honolulu, Hawaii, May 15, 20; m 46; c 1. BIOCHEMISTRY. *Educ:* Univ Hawaii, AB, 46; George Washington Univ, AM, 49; Johns Hopkins Univ, ScD, 54. *Prof Exp:* Res assoc biochem, Johns Hopkins Univ, 54-55; biochemist, Lab Nutrit & Endocrinol, Nat Inst Arthritis & Metab Dis, 55-62; biochemist biol, 62-70, biochemist etiology, Exp Path Br, 70-76, biochemist, Carcinogen Metabolism & Toxicol Br, Nat Cancer Inst, 76-84; SR SCI ADV, TECH RESOURCES, INC, 84- *Concurrent Pos:* Vis scientist, Dept Molecular Oncol, Univ Tokyo & Div Biochem, Japanese Nat Cancer Ctr. *Mem:* AAAS; Am Inst Nutrit; Soc Exp Biol & Med; Am Asn Cancer Res; Am Asn Clin Chemists; Sigma Xi; Soc Toxicol. *Res:* Vitamin B12; nutritional obesity; lipid metabolism; chemical carcinogenesis; nutrition and endocrines in carcinogenesis. *Mailing Add:* 11109 Jolly Way Kensington MD 20895

YAMAMOTO, ROBERT TAKAICHI, b Hawaii, May 26, 27. ENTOMOLOGY. *Educ:* Univ Hawaii, BA, 53; Univ Ill, MS, 55, PhD, 57. *Prof Exp:* Res assoc entom, Univ Ill, 57-60; mem staff, Entom Res Div, USDA, 60-65; ASSOC PROF ENTOM, NC STATE UNIV, 65- *Mem:* Entom Soc Am. *Res:* Insect physiology and behavior. *Mailing Add:* 1116 Terrell Lane Raleigh NC 23666

YAMAMOTO, SACHIO, b Petaluma, Calif, Dec 12, 32; m 58; c 3. MARINE CHEMISTRY, PHYSICAL CHEMISTRY. *Educ:* Univ Calif, Berkeley, BS, 55; Iowa State Univ, PhD(phys chem), 59. *Prof Exp:* Res chemist, Calif Res Corp, 59-63; res chemist, US Naval Radiol Defense Lab, 63-69, Naval Undersea Ctr, 69-71; res & develop mgr, Naval Ocean Systs Ctr, 71-90; DIR, ASIAN OFF, OFF NAVAL RES, 82-84, 90- *Mem:* Am Chem Soc; Royal Soc Chem. *Res:* Environmental sciences; trace metal analysis; gas solubility; x-ray fluorescence analysis. *Mailing Add:* Asian Off Off Naval Res APO San Francisco CA 96503-0007

YAMAMOTO, TATSUZO, b Hardieville, Alta, Feb 8, 28. VIROLOGY. *Educ:* Univ Alta, BSc, 52, MSc, 55; Yale Univ, PhD(virol), 61. *Prof Exp:* Fel microbiol, Univ Toronto, 61-62; from asst prof to assoc prof virol, 62-74, PROF MICROBIOL, UNIV ALTA, 74- *Concurrent Pos:* Consult, Govt Can, 72-; sabbatical res assoc award, Can Int Develop Res Ctr, 79-80. *Mem:* Am Soc Microbiol; Am Fisheries Soc; Nat Geog Soc; Can Nat Geog Soc; Wildlife Dis Asn. *Res:* Replication and structure of animal viruses; viral diseases of fish; infectious pancreatic necrosis; dermal hyperplastics. *Mailing Add:* Dept Microbiol M330 Biol Sci Bldg Univ Alta Edmonton AB T6G 2E9 Can

YAMAMOTO, WILLIAM SHIGERU, b Cleveland, Ohio, Sept 22, 24; m 65; c 3. PHYSIOLOGY, COMPUTER SCIENCE. *Educ:* Park Col, AB, 45; Univ Pa, MD, 49. *Hon Degrees:* MS, Univ Pa, 71. *Prof Exp:* Instr physiol, Sch Med, Univ Pa, 52-53, assoc, 55-57, instr biostatist & asst prof physiol, 57-66, prof physiol, 66-70; prof physiol & biomath, Sch Med, Univ Calif, 70-71; PROF COMPUT MED & CHMN DEPT, SCH MED, GEORGE WASHINGTON UNIV, 71- *Concurrent Pos:* Mem study sect, NIH, 63-65, mem Nat Adv Coun Res Resources, 71-75; consult, Health Care Tech Div, Nat Ctr Health Serv Res & Develop, 68-80 & Am Col Preventive Med, 80- *Mem:* AAAS; Am Physiol Soc; Asn Comput Mach; Biomed Eng Soc; Am Col Prev Med; Am Col Med Informatics. *Res:* Computer applications in health services; physiology of respiratory regulation by carbon dioxide homeostasis. *Mailing Add:* Dept Comp Med George Washington Univ Med Ctr 2300 K St NW Washington DC 20037

YAMAMOTO, Y LUCAS, b Hokkaido, Japan, Jan 19, 28; Can citizen; m 58; c 3. NEUROSURGERY, NUCLEAR MEDICINE. *Educ:* Hokkaido Univ, BSc, 48, MD, 52; Yokohama Nat Univ, PhD(radiobiol), 61. *Prof Exp:* Intern med, Int Cath Hosp, Tokyo, Japan, 53; resident neurosurg, Med Ctr, Georgetown Univ, 54-58; res assoc nuclear med & radiobiol, Med Dept, Brookhaven Nat Lab, 58-61; res assoc neurosurg res, 61-68, from asst prof to assoc prof, 68-80, PROF NEUROL & NEUROSURG, MONTREAL NEUROL INST, MCGILL UNIV, 80- *Concurrent Pos:* Mem, Am Bd Nuclear Med, 73- *Mem:* Soc Nuclear Med; Can Asn Nuclear Med; Am Col Nuclear Physicians; Can Neurosurg Soc. *Res:* Neurological science; cerebral circulation. *Mailing Add:* Montreal Neurol Inst 3801 University St Montreal PQ H3A 2B4 Can

YAMAMOTO, YASUSHI STEPHEN, b Topaz, Utah, Aug 6, 43. ORGANIC CHEMISTRY, PHOTOGRAPHIC CHEMISTRY. *Educ:* Univ Wis, BS, 65; Pa State Univ, PhD(org chem), 71. *Prof Exp:* Sr res chemist, Eastman Kodak Co Res Labs, 71-78; assoc prof chem, Rochester Inst Technol, 78-83; sr ed, Ullmann's Encycl Indust Chem, 83-88; info specialist, 88-90, PATENT LIAISON, DU PONT CENT RES & DEVELOP EXP STA, 90- *Concurrent Pos:* Coordr experiential learning, Rochester Inst Technol, 78-81. *Mem:* Am Chem Soc. *Mailing Add:* Du Pont Cent Res & Develop Exp Sta 301/310 Wilmington DE 19880-0301

YAMAMURA, HENRY ICHIRO, b Seattle, Wash, June 25, 40; m 64; c 1. NEUROCHEMISTRY, NEUROPHARMACOLOGY. *Educ:* Univ Wash, BS, 64, MS, 68, PhD(pharmacol), 69. *Prof Exp:* Pharm intern, Seattle, Wash, 60-64, staff pharmacist, 64-66; spec lectr pharmacol, Seattle Pac Col, 66-67; asst prof, 75-77, ASSOC PROF, DEPT PHARMACOL, COL MED, UNIV ARIZ, 77- *Concurrent Pos:* NIMH spec fel pharmacol, Sch Med, Johns Hopkins Univ, 72-75; NIMH Res Scientist Develop Awardee, 75- *Mem:* AAAS; Soc Neurosci; Am Soc Neurochem; Am Soc Pharmacol & Exp Therapeut. *Res:* Release and uptake of brain neurohumoral transmitters; demonstration of brain neurotransmitter receptors and their alterations in neuropsychiatric disorders. *Mailing Add:* Dept Pharmacol & Biochem-Psych Univ Ariz Col Med 1501 N Campbell Tucson AZ 85724

YAMANAKA, WILLIAM KIYOSHI, b Kauai, Hawaii, Mar 19, 31; m 58; c 3. NUTRITION, COMMUNITY HEALTH. *Educ:* Univ Hawaii, BS, 55; Univ Calif, PhD(nutrit), 69. *Prof Exp:* Res assoc nutrit, Children's Hosp of East Bay, Oakland, Calif, 59-64; res asst, Univ Calif, Berkeley, 64-69; assoc prof nutrit, Univ Mo-Columbia, 69-75; MEM FAC, UNIV WASH, 75- *Concurrent Pos:* Nutrit adv, Delta Area Head Start, Mo, 69-74. *Mem:* Am Dietetic Asn; Soc Nutrit Educ; Am Soc Clin Path; Am Home Econ Asn; Am Pub Health Asn. *Res:* Role of lipids in cardiovascular disease; effect of protein malnutrition of growing mammalian organisms; applied nutrition programs in developing countries; nutrition for cancer patients. *Mailing Add:* Dept Epidemiol Univ Wash Seattle WA 98195

YAMANE, GEORGE M, b Honolulu, Hawaii, Aug 9, 24; m 51; c 3. ORAL MEDICINE, ORAL PATHOLOGY. *Educ:* Haverford Col, AB, 46; Univ Minn, Minneapolis, DDS, 50, PhD(oral path), 63. *Prof Exp:* Asst chem & zool, Univ Hawaii, 43-44; asst oral path & diag, Univ Minn, Minneapolis, 51-53; asst prof oral path, Univ Ill, Chicago, 57-59; asst prof oral path, Univ Wash, 59-63, dir tissue lab, 60-63; prof oral diag, med & roentgenol & chmn div, Univ Minn, Minneapolis, 63-70; asst dean res & postdoc prog, 77-80, prof oral med & radiol & chmn dept, 70-83, PROF BIODENT SCI, NJ DENT SCH, UNIV MED & DENT NJ, 83- *Concurrent Pos:* Consult, Children Orthop Hosp & Med Ctr, Seattle, 60-63, Vet Admin Hosps, American Lake, Wash, 62-63 & Minneapolis, 64-70 & div dent health, Wyo State Bd Health, 66-70; vis scientist, Armed Forces Inst Pathol, 82; vis prof, NY Univ Col Dent, 82, adj prof, 84- *Mem:* Fel AAAS; fel Am Acad Oral Path; fel Int Col Dent; Int Asn Dent Res; fel Am Col Dent; Sigma Xi. *Res:* Bone tissue formation; radiobiology; magnesium metabolism; psychosomatic etiology of oral lesions. *Mailing Add:* Dept Biodent Sci Univ Med & Dent NJ Newark NJ 07103

YAMANOUCHI, TAIJI, b Tokyo, Japan, Aug 16, 31; m 61; c 2. PHYSICS. *Educ:* Tokyo Univ Ed, BS, 53, MS, 55; Univ Rochester, PhD(physics), 60. *Prof Exp:* Res assoc physics, Univ Rochester, 60-65, sr res assoc, 65-69; PHYSICIST, FERMI NAT ACCELERATOR LAB, 69- *Mem:* Am Phys Soc. *Res:* Experimental particle physics. *Mailing Add:* Fermi Nat Accelerator Lab PO Box 500 Batavia IL 60510

YAMARTINO, ROBERT J, b Waltham, Mass, Nov 17, 44. COMPLEX TERRAIN-GEOMETRY FLOW TURBULENCE, POLLUTANT MODELING OXIDANT & ACID RAIN MODEL DEVELOPMENT. *Educ:* Tufts Univ, BS, 66; Stanford Univ, PhD(physics), 74. *Prof Exp:* Res asst-assoc, Stanford Linear Accelerator Ctr Nat Lab, 67-72; postdoctoral fel, Purdue Univ, 72-74; scientist, Argonne Nat Lab, 74-80; sr scientist, Geomet GmbH, West Berlin, 80-81; sr scientist, Environ Res & Technol Inc, 81-85; PRIN SCIENTIST, SIGMA RES CORP, 85- *Concurrent Pos:* Consult, Geos GmbH & Geomet GmbH, 79-; adv, Environ Protection Agency Ger, 79-; reviewer & contribr, Var World Meteorol Orgns, Europ Monitoring Eval Prog, 83- & Dutch-Ger Phoxa Prog Rev & Workshops; mem, Comt on Meteorol Aspects of Air Pollution, Am Meteorol Soc, 91- *Mem:* Sigma Xi; Am Meteorol Soc; Am Phys Soc; Air & Waste Mgt Asn; Union Concerned Scientists. *Mailing Add:* 36 Orchard Rd Bedford MA 01730

YAMASHIRO, STANLEY MOTOHIRO, b Honolulu, Hawaii, Nov 26, 41; m 64; c 2. BIOMEDICAL ENGINEERING. *Educ:* Univ Southern Calif, BS, 64, MS, 66, PhD(elec eng), 70. *Prof Exp:* Mem tech staff elec eng, Hughes Aircraft Co, 64-70; res assoc, 70-71, asst prof, 71-74, ASSOC PROF BIOMED ENG, UNIV SOUTHERN CALIF, 74- *Mem:* Fedn Am Soc Exp Biol. *Res:* Cardiopulmonary physiology; application of control theory and computer technology to biological systems. *Mailing Add:* Dept Biomed Eng 1451 Univ Southern Calif University Park Los Angeles CA 90089

YAMASHIROYA, HERBERT MITSUGI, b Honolulu, Hawaii, Sept 14, 30; m 47; c 4. MICROBIOLOGY, VIROLOGY. *Educ:* Univ Hawaii, BA, 53; Univ Ill, Chicago, MS, 62, PhD(microbiol), 65; Registry Technologists, cert. *Prof Exp:* Supvr clin lab, Atomic Bomb Casualty Comn, Nat Acad Sci-Nat Res Coun, Hiroshima, Japan, 56-58; assoc scientist, IIT Res Inst, 64-65, res scientist, 65-68, sr scientist, 68-71; asst prof path, 71-76; asst dir hosp lab, 71-76; ASSOC PROF PATH, UNIV ILL COL MED, 76-, HEAD VIROL LAB, UNIV ILL HOSP, 71-, DIR GRAD STUDIES PATHOL, 81- *Concurrent Pos:* Mem adv bd, Clin Lab & Blood Bank, State Ill, 79-; lectr microbiol, Cook County Grad Sch Med, 77-; lectr, Ill Col Podiatric Med, 81-; lectr & consult med virol, VA West Side Hosp, 82- *Honors & Awards:* Tanner Shaughnessy Award, Microbiol, 82. *Mem:* AAAS; Am Soc Microbiol; Am Soc Clin Path; Sigma Xi; Am Asn Pathol. *Res:* Tissue culture and its application to viruses and rickettsiae; viral pathogenesis; immunologic techniques and the immune mechanism; diagnostic procedures in microbiology and immunology. *Mailing Add:* 808 S Wood Hosp Lab Pathol Univ Ill Col Med Chicago IL 60612

YAMAUCHI, HIROSHI, b Honolulu, Hawaii, Mar 26, 23; m 60; c 3. THEORETICAL PHYSICS. *Educ:* Univ Hawaii, BS, 47; Harvard Univ, MA, 48, PhD(physics), 50. *Prof Exp:* From instr to asst prof physics, Colby Col, 50-54; theoret physicist, Nuclear Develop Assoc, Inc, NY, 54-55; from asst prof to assoc prof math, Univ Hawaii, 55-65; assoc prof math & physics, 66-69, PROF MATH & PHYSICS, CHAMINADE UNIV, HONOLULU, 69- *Mem:* Am Phys Soc. *Res:* Foundations of quantum mechanics; quantum field theory; mathematical physics; theoretical nuclear physics. *Mailing Add:* Dept Math-Phys Chaminade Univ 3140 Waialae Ave Honolulu HI 96816

YAMAUCHI, MASANOBU, b Maui, Hawaii, Mar 3, 31; m 58; c 2. INORGANIC CHEMISTRY. *Educ:* Univ Hawaii, BA, 53; Univ Mich, MS, 58, PhD(chem), 61. *Prof Exp:* Asst prof chem, Univ NMex, 60-65; from asst prof to assoc prof, 65-74, PROF CHEM, EASTERN MICH UNIV, 74- *Mem:* Am Chem Soc. *Res:* Chemistry of boron hydrides and related compounds. *Mailing Add:* Dept Chem Eastern Mich Univ Ypsilanti MI 48197

YAMAUCHI, TOSHIO, b Newell, Calif, Feb 13, 45; m; c 2. MEDICAL GENETICS, PEDIATRICS. *Educ:* Northwestern Univ, BA, 66, PhD(biol sci), 72; Univ Tex Med Sch, MD, 79. *Prof Exp:* Fel med genetics, Univ Tex Syst Cancer Ctr, M D Anderson Hosp & Tumor Inst, 72-74, res assoc med genetics, 74-77, mem staff, Univ Tex Med Sch, 76-79. *Concurrent Pos:* Resident pediat, Hermann Hosp, 79-82. *Mem:* AMA; Am Acad Pediat. *Res:* Study of enzymes which are involved in the metabolism and activation of chemical carcinogens and the development of systems to study these enzymes in the human population. *Mailing Add:* 3425 Hwy Six Suite 107 Sugar Land TX 77478

YAMAZAKI, HIROSHI, b Hokkaido, Japan, Sept 5, 31; m 61; c 3. BIOCHEMISTRY. *Educ:* Hokkaido Univ, BS, 54, MS, 56; Univ Wis, PhD(biochem), 60. *Prof Exp:* Proj assoc biochem, Univ Wis, 61-63, res assoc, 63-65; res officer biol, Atomic Energy Can Ltd, 65-67; from asst prof to assoc prof, 67-77, PROF BIOL, CARLETON UNIV, 77- *Mailing Add:* Dept Biol Carleton Univ Ottawa ON K1S 5B6 Can

YAMAZAKI, RUSSELL KAZUO, b Topaz, Utah, Nov 23, 42; m 66. BIOCHEMICAL PHARMACOLOGY. *Educ:* Col Wooster, BA, 64; Mich State Univ, PhD(biochem), 69. *Prof Exp:* Res assoc biochem, Case Western Reserve Univ, 69-71; vis asst prof pharmacol, Univ Va, 71-73, asst prof, 73-78; ASSOC PROF PHARMACOL, WAYNE STATE UNIV, 78- *Mem:* Am Soc Biol Chemists; Am Chem Soc; Am Soc Plant Physiologists; AAAS. *Res:* Actions of hormones on the metabolism of mitochondria, peroxisomes and lysosomes. *Mailing Add:* Dept Pharmacol Sch Med Wayne State Univ Detroit MI 48201-3421

YAMAZAKI, WILLIAM TOSHI, b San Francisco, Calif, May 10, 17; m 42; c 3. CEREAL CHEMISTRY. *Educ:* Univ Calif, BS, 39, MS, 41; Ohio State Univ, PhD(agr chem), 50. *Prof Exp:* CHEMIST IN CHG SOFT WHEAT QUAL LAB, AGR RES SERV, USDA, OHIO AGR RES & DEVELOP CTR, 63- *Concurrent Pos:* Adj prof agron, Ohio State Univ & Ohio Agr Res & Develop Ctr, 57- *Mem:* AAAS; Am Asn Cereal Chem; Am Chem Soc. *Res:* Chemical and physical basis for processing quality in soft wheat and soft wheat flour. *Mailing Add:* 748 Hamilton Ave Wooster OH 44691

YAMBERT, PAUL ABT, b Toledo, Ohio, May 15, 28; m 50; c 5. CONSERVATION. *Educ:* Univ Mich, BS, 50, MS, 51, MA, 55, PhD(conserv), 60. *Prof Exp:* Field scout exec, Boy Scouts of Am, 51-52; teacher, Ann Arbor High Sch, Mich, 53-57; prof natural resources, Wis State Univ, Stevens Point, 57-69; dean outdoor labs, 69-74, PROF FORESTRY DEPT, SOUTHERN ILL UNIV, CARBONDALE, 74- *Concurrent Pos:* Vis prof, Univ Mich, 63. *Res:* Environmental attitude and knowledge; interpretation for the handicapped; outdoor recreation. *Mailing Add:* Forestry Dept Southern Ill Univ Bldg 0180 Carbondale IL 62901

YAMDAGNI, RAGHAVENDRA, b Aligarh, India, June 30, 41; m; c 2. PHYSICAL CHEMISTRY, MASS SPECTROMETRY. *Educ:* Allahabad Univ, BSc, 57, MSc, 59, PhD, 65. *Prof Exp:* Res assoc, Cornell Univ, 65-68; res assoc mass spectrom, Univ Alta, 68-75; prof assoc, 75-78, SR INSTR, DEPT CHEM, UNIV CALGARY, 78- *Mem:* Am Soc Mass Spectrom; Chem Inst Can. *Res:* Ion molecule reactions; thermodynamic and reaction kinetics studies; nuclear magnetic resonance spectroscopy. *Mailing Add:* Dept Chem Univ Calgary Calgary AB T2N 1N4 Can

YAMIN, MICHAEL, computer sciences, physical chemistry, for more information see previous edition

YAMIN, SAMUEL PETER, b New York, NY, July 26, 38; div; c 2. PARTICLE PHYSICS. *Educ:* Mass Inst Technol, SB, 60; Univ Pa, MS, 61, PhD(physics), 66. *Prof Exp:* Res assoc physics, Brookhaven Nat Lab, 66-69; asst prof physics, Rutgers Univ, New Brunswick, 69-75; assoc physicist, 75-78, PHYSICIST, BROOKHAVEN NAT LAB, 78- *Mem:* Am Phys Soc. *Res:* Experimental elementary particle physics; particle beam optics; musical acoustics. *Mailing Add:* Bldg 911B Brookhaven Nat Lab Upton NY 11973

YAMINI, SOHRAB, b Iran, Aug 11, 53; m 86. GASTROENTEROLOGY. *Educ:* Nat Univ Iran, MD, 78. *Prof Exp:* Resident surg, 82-85, fel gastroenterol, 86-87, ASST PROF GASTROENTEROL, KING-DREW MED CTR, UNIV CALIF, LOS ANGELES, 87-, ASSOC INVESTR, DIGESTIVE DIS RES CTR, 88- *Concurrent Pos:* Clin investr, Digestive Dis Res Ctr, 87; dir endoscopy, King-Drew Med Ctr, 87- *Res:* Multiple studies regarding effect of alcohol on gastrointestinal system; treatment of peptic ulcer disease and prevention of recurrence of ulcer. *Mailing Add:* 11600 Wilshire Blvd LL18 Los Angeles CA 90025

YAMINS, J(ACOB) L(OUIS), b Fall River, Mass, Jan 8, 14; m 48; c 1. FOOD SCIENCE. *Prof Exp:* Res assoc, Res Lab Org Chem, Mass Inst Technol, 39-42; sr chemist, Nat Fireworks, Inc, Mass, 43-46; sr chemist, Biochem Div, Interchem Corp, NJ, 47-48; res chemist & proj leader, Res Labs Div, Nat Dairy Prod Corp, 48-49, asst to vpres & dir res, 50-54, asst to pres, 54-58; dept head fundamental studies, Res & Develop Div, Am Sugar Co, NY, 58-65, dir sci develop, 65-67; VIS PROF FOOD SCI, RUTGERS UNIV, NEW BRUNSWICK, 67-; CONSULT, 69- *Concurrent Pos:* Adv chem dept, Adelphi Col, 50-; abstractor, Chem Abstrs. *Mem:* Fel AAAS; Am Chem Soc; Sigma Xi; Soc Chem Indust; fel NY Acad Sci; Inst Food Technologists. *Res:* Protein hydrolysis and isolation of amino acids; vitamin syntheses; preparation of primary explosives; tall oil; sterol isolation and syntheses; tocopherols; flavors; syntheses of long chain surface active agents and bactericides; antioxidants; baked products; carbohydrates; food product development; nutrition; single cell proteins; packaging. *Mailing Add:* PO Box 456 Sunderland MA 01375

YAN, JOHNSON FAA, b Amoy, China, May 21, 34; m 70. PHYSICAL CHEMISTRY. *Educ:* Nat Taiwan Univ, BS, 59; Kent State Univ, MS, 65, PhD(chem), 67. *Prof Exp:* Chief lab, Hwa Ming Pulp & Paper Manufactory, Chu-nan, Taiwan, 59-62; res assoc & fel, Cornell Univ, 67-69; develop assoc pulp & paper, Bowaters Carolina Corp, 69-77; SCI SPECIALIST, WEYERHAEUSER CORP, 77- *Mem:* AAAS; Am Chem Soc. *Res:* Physical chemistry, surface and colloid; polymers; biopolymers; pulp and paper. *Mailing Add:* 3801 SW 326th St Federal Way WA 98023-2611

YAN, MAN FEI, b Kwang Tung, China, Dec 26, 48; m 77; c 4. CERAMIC SCIENCE, PHYSICAL METALLURGY. *Educ:* Mass Inst Technol, BS, 70, ScD(mat sci & eng), 76; Univ Calif, Berkeley, MS, 71. *Prof Exp:* Mem Tech Staff, 76-87, DISTINGUISHED MEM TECH STAFF, AT&T BELL LABS, 87-, SUPVR, 87- *Honors & Awards:* Ross Coffin Purdy Award, Am Ceramic Soc, 80. *Mem:* Am Ceramics Soc; Am Soc Metals; Am Optical Soc. *Res:* Sintering of solids; fiber optics; grain boundaries; grain boundary migration; glass fiber processing; electronic ceramics; magnetic ferrites; varistors; ceramic capacitor; metal-ceramic interfaces. *Mailing Add:* AT&T Bell Labs Rm 6C-308 600 Mountain Ave Murray Hill NJ 07974-2070

YAN, MAXWELL MENUHIN, b Russia, May 20, 19; Can citizen; m 44; c 5. RECONSTITUTED WOOD PRODUCTS, WOOD SCIENCE. *Educ:* Univ Man, BSc, 40, MSc, 42; McGill Col, PhD(wood chem), 47. *Prof Exp:* Inspector asphalts, Can Nat Testing, 42; chemist oil refining, Imp Oil Regina, 42-45; mgr panelbd res, Wood Compos Bd, Abitibi Paper Co Ltd, 47-80, assoc dir res, 81-84; CONSULT, MAXWELL M YAN & ASSOCS, 84- *Concurrent Pos:* Mem comt fibrebd, Can Gov Specif Bd, 69-84; mem, Nat Adv Comt Forest Prod Res, Can Forestry Serv, Ottawa, 71-78; chmn, Panelbd Res Prog Adv Comt, Fed Forest Prod Lab, Ottawa, 71-78; mem, US Joint Coatings, Forest Prod Indust Steering Comt, 73-80; adj prof forest prod, Fac Forestry, Univ Toronto, 75-90. *Honors & Awards:* Pioneer Wood Award, Forest Prod Res Soc, 76. *Mem:* Forest Prod Res Soc; Soc Wood Sci & Technol; AAAS. *Res:* Wood composition board; wood fiber utilization; hardboard; particle board; wood and lignin chemistry; exterior siding and molding; educational programs and videos. *Mailing Add:* 61 Stonedene Blvd Willowdale ON M2R 3C8 Can

YAN, SAU-CHI BETTY, b Hong Kong, Nov 25, 54; US citizen; c 1. BIOTECHNOLOGY, CARBOHYDRATE BIOCHEMISTRY. *Educ:* Cent Mo State Univ, BS, 75; Iowa State Univ, PhD(biochem), 80. *Prof Exp:* Postdoctoral fel biochem, St Paul Ramsey Med Ctr, 80-82; postdoctoral fel biochem, Univ Tex Med Sch, 82-85; sr biochemist biotechnol, 85-88, SR SCIENTIST CARDIOVASC, ELI LILLY & CO, 89- *Concurrent Pos:* Grant reviewer, NSF, 90; chmn travel awards, Protein Soc, 90. *Mem:* Am Soc Biochem & Molecular Biol; Protein Soc; AAAS; Soc Chinese Bioscientists Am. *Res:* Using expertise in protein chemistry and carbohydrate analytical chemistry to discover new therapeutics for cardiovascular diseases. *Mailing Add:* Biochem Res Dept Eli Lilly & Co Bldg 88-4 307 E McCarty St Indianapolis IN 46285

YAN, TSOUNG-YUAN, b Tainan, Taiwan, Sept 17, 33; m 61; c 2. CHEMICAL ENGINEERING, FUEL TECHNOLOGY. *Educ:* Nat Taiwan Univ, BS, 59; Purdue Univ, MS, 62, PhD(chem eng), 63. *Prof Exp:* Teacher, Tainan Sch, 52-55; from res engr to sr engr, Mobil Oil Corp, 62-70, from res assoc to sr res assoc, Mobil Res & Develop Corp, 78-82, RES SCIENTIST, MOBIL RES & DEVELOP CORP, 82- *Mem:* Am Chem Soc; Sigma Xi. *Res:* Catalysis, petroleum refining; coal upgrading and utilization; uranium in-situ leaching and refining processes; biomass conversion; lube oil processing; environmental control; process evaluation; fuel combustion; petrochemicals; natural gas processing; waste minimization. *Mailing Add:* Mobil Res & Develop Corp PO Box 1025 Princeton NJ 08540

YAN, TUNG-MOW, b Keelung, Taiwan, Nov 27, 36; m 64; c 2. HIGH ENERGY PHYSICS. *Educ:* Nat Taiwan Univ, BS, 60; Nat Tsinghua Univ, Taiwan, MS, 62; Harvard Univ, PhD(physics), 68. *Prof Exp:* Res assoc physics, Stanford Linear Accelerator Ctr, 68-70, vis scientist, 73-74; from asst to assoc prof, 70-81, PROF PHYSICS, CORNELL UNIV, 81- *Concurrent Pos:* Alfred P Sloan Found fel, 74-78. *Mem:* Am Phys Soc. *Res:* Structure of elementary particles; properties of quantum field theories. *Mailing Add:* Newman Lab Nuclear Studies Cornell Univ Ithaca NY 14853

YANABU, SATORU, b Japan, July 15, 41; m 65; c 3. ELECTRICAL ENGINEERING, VACUUM CIRCUIT BREAKER. *Educ:* Univ Tokyo, BEng, 64, Dr Eng, 90; Univ Liverpool, PhD(elec eng), 71. *Prof Exp:* Mgr, High Power Lab, 76-85, sr mgr, High Voltage & High Power Lab, 85-90, SR MGR, POWER TRANSMISSION & SUBSTA ENG, TOSHIBA COOP, 90- *Concurrent Pos:* Mem, Int Conf Large High Voltage Elec Systs, 81- *Mem:* Fel Inst Elec & Electronics Engrs. *Res:* Gas circuit breakers and vacuum circuit breakers. *Mailing Add:* Toshiba Hibiya Off Power Transmission & Substa 1-6 Uchisaiwai-Cho Chiyoda-Ku Tokyo 100 Japan

YANAGIHARA, RICHARD, b Honolulu, Hawaii, Aug 3, 46; m; c 2. VIRAL PATHOGENESIS, NEUROVIROLOGY. *Educ:* Univ Hawaii, Honolulu, BA, 68; Univ Cincinnati, MD, 72; Johns Hopkins Univ, MPH, 85. *Prof Exp:* Intern pediat, Health Sci Ctr, Univ Colo, Denver, 72-73, resident, 73-74, teaching fel infectious dis, 77-79; resident pediat, Univ Calif, San Francisco, 76-77; res assoc, 74-76, res physician, Guam, 79-82, spec expert, 82-84, MED OFFICER, NAT INST NEUROL & COMMUN DIS & STROKE, NIH, BETHESDA, MD, 85- *Mem:* Infectious Dis Soc Am; Am Soc Virol; Am Soc Trop Med & Hyg; Soc Pediat Res. *Res:* Epidemiology and epizootiology of vector-borne viral zoonoses, particularly those involving nonarthropod vectors; pathogenesis of persistent viral infections using animal models; role of heavy metals and essential minerals in neurodegenerative diseases occurring in high incidence in the western Pacific region; human retroviruses. *Mailing Add:* Nat Inst Neurol & Commun Dis & Stroke NIH Bldg 36 Rm 5B-21 Bethesda MD 20892

YANAGIMACHI, RYUZO, b Sapporo, Japan, Aug 27, 28; m 60. REPRODUCTIVE BIOLOGY. *Educ:* Hokkaido Univ, BSc, 53, DSc(biol), 60. *Prof Exp:* Res scientist, Worcester Found Exp Biol, 60-64; lectr biol, Hokkaido Univ, 64-66; from asst prof to assoc prof, 66-73, PROF ANAT, SCH MED, UNIV HAWAII, 73- *Mem:* Soc Study Reproduction; Am Asn Anat; Am Soc Cell Biol; Brit Soc Study Fertil. *Res:* Biology of reproduction, particularly biology and physiology of gametes and early development of mammals. *Mailing Add:* Dept Anat Univ Hawaii Burns Med Sch 1960 East-West Rd Honolulu HI 96822

YANAGISAWA, SAMUEL T, b Berkeley, Calif, Feb 18, 22; m 52; c 3. ELECTRICAL ENGINEERING. *Educ:* Univ Calif, Berkeley, BSEE Hons, 42. *Prof Exp:* Qual control engr, Machlett Labs, Inc, 43-45, develop engr, 46-48, sr develop engr, 48-55, sect head spec prod eng, 55-57, chief engr display & storage tubes, 57-60, prod line mgr, 60-63; vpres eng & mfg, Warnecke Electron Tubes, Ill, 63-67; mgr, Imaging Sensors Div, Varo Inc, 67-68, gen mgr, Electron Devices Div, 68-72, vpres, 71-72, exec vpres & dir, 72-74, pres & chief exec officer, 74-76, chmn bd & chief exec officer, 76-78, chmn bd, pres & chief exec officer, 78-87, pres & dir, Electron Devices Div, 72-87; RETIRED. *Concurrent Pos:* Mem, eng adv coun, Univ Tex, Dallas, SMU; mem sci bd, US Army, 86-90; mem, chmn bd visitor, Univ Tex, McDonald Observ. *Mem:* Inst Elec & Electronics Engrs; Am Vacuum Soc; Sigma Xi. *Res:* Management; design, development and production engineering of high power transmitting and microwave tubes; circuitry; vacuum switches; storage and display tubes; supervoltage electron accelerators; image converters and intensifiers; crossed field tubes; design development and production of image intensified and night vision systems. *Mailing Add:* 7708 Chalkstone Dr Dallas TX 75248

YANAI, HIDEYASU STEVE, b Tokyo, Japan, Feb 26, 28; nat US; m 56; c 2. ANALYTICAL CHEMISTRY, PHYSICAL CHEMISTRY. *Educ:* Tokyo Agr Col, BS, 47; Calif State Polytech Col, BS, 53; Univ Minn, PhD(phys chem), 58. *Prof Exp:* Sr scientist, Rohm & Haas, 58-69, anal lab head, Res Div, 69-73, anal res proj leader, Res Div, 73-79, proj leader, Plastics Res Dept, 79-81, res sect mgr plastics, 81-88, plastics bus mgr, 89-91, PAC REGION MGR, ROHM & HAAS CO, JAPAN, 91- *Res:* Plastics, polymer chemistry or polymers. *Mailing Add:* Pac Region Rohm & Haas Co Philadelphia PA 19105

YANARI, SAM SATOMI, b Gilcrest, Colo, May 27, 23; m 51; c 3. IMMUNOLOGY, BIOCHEMISTRY. *Educ:* Univ Chicago, BS, 48, PhD, 52. *Prof Exp:* Res assoc, Univ Chicago, 52-53; res assoc, Armour & Co, 53-56; res assoc, Minn Mining & Mfg Co, 56-64; head, Biochem Dept, Armour Pharmaceut Co, Ill, 65-69; res dir & vpres, Wilson Labs, 69-72; dir res lab, Div Allergy & Clin Immunol, Henry Ford Hosp, Detroit, Mich, 72-89; chief res officer, Covalent Tech Corp; RETIRED. *Mem:* Am Chem Soc; Am Fedn Clin Res; Am Asn Clin Chem; Am Acad Allergy. *Res:* Mechanism of enzyme action; protein chemistry and structure; proteolytic enzymes; absorption and fluorescence spectroscopy; hormones; immunology. *Mailing Add:* 13328 Wales Huntington Woods MI 48070

YANCEY, ASA G, b Atlanta, Ga, 1916; m; c 4. SURGERY. *Educ:* Mich Univ, MD, 41. *Prof Exp:* Intern, City Hosp, Cleveland, 41-42; DIR, GRADY MEM HOSP, ATLANTA, 72-; PROF SURG, EMORY UNIV, 72- *Mem:* Inst Med-Nat Acad Sci; Am Surg Asn; Nat Med Asn; Am Col Surgeons; AMA. *Mailing Add:* 2845 Engle Rd NW Atlanta GA 30318

YANCEY, PAUL HERBERT, b Whittier, Calif, July 4, 51; m 78; c 1. ENVIRONMENTAL PHYSIOLOGY, COMPARATIVE BIOCHEMISTRY. *Educ:* Calif Inst Technol, BS, 73; Scripps Inst Oceanog, Univ Calif, PhD(marine biol), 78. *Prof Exp:* NATO fel physiol, Univ St Andrews, Scotland, 78-80; asst prof, 81-86, chair, 88-91, ASSOC PROF BIOL, WHITMAN COL, 86- *Concurrent Pos:* Res Corp grant, 86; vis scientist, Nat Inst Health, 87-88; PEW fel, Mt Desert I Biol Lab, 90. *Mem:* AAAS; Soc Develop Biol; Am Inst Biol Sci. *Res:* Biochemical and physiological adaptations in muscle proteins of fishes to temperature, pressure and salinity; properties of osmoregulatory compounds in laboratory mammals and desert mammals. *Mailing Add:* Biol Dept Whitman Col 345 Boyer Walla Walla WA 99362

YANCEY, ROBERT JOHN, JR, b Austin, Tex, Sept 17, 48; m 69; c 1. VETERINARY MICROBIOLOGY, MICROBIAL PATHOGENESIS. *Educ:* Univ Tex, Austin, BA, 71, PhD(microbiol), 77. *Prof Exp:* Instr med & gen microbiol, Univ Tex, Austin, 76-77, res assoc, Dept Microbiol, 77-78; fel, Dept Microbiol & Immunol, Univ Tex Health Sci Ctr Dallas, 78-79 & Sch Med, Univ Mo, 79-80; RES SCIENTIST VET THERAPEUT, UPJOHN CO, 80- *Concurrent Pos:* Coordr microbiol teaching labs, Univ Tex, Austin, 76-77; conf group leader med microbiol, Univ Tex Health Sci Ctr Dallas & bacteriol lab instr, 78-79; bacteriol lab lectr, Sch Med, Univ Mo, 79-80. *Mem:* Sigma Xi; Am Soc Microbiol; AAAS; NY Acad Sci. *Res:* Molecular mechanisms of bacterial pathogenesis including the colonization of the gut by obligate anaerobes; mechanisms of exotoxin production; role of bacterial motility; disease pathogenesis of slaphylococcal mastitis; antibiotic killing of intracellular pathogens. *Mailing Add:* 7922-19-MR Upjohn Co Kalamazoo MI 49001

YANCEY, THOMAS ERWIN, b Lowville, NY, July 24, 41; m 71; c 3. PALEONTOLOGY, SEDIMENTARY PETROLOGY. *Educ:* Univ Calif, Berkeley, BA, 66, MA, 69, PhD(paleont), 71. *Prof Exp:* Lectr, Univ Malaya, 71-75; prof geol, Idaho State Univ, 75-80; MEM STAFF, DEPT GEOL, TEX A&M UNIV, 80- *Concurrent Pos:* Consult. *Mem:* Geol Soc Am; Soc Econ Paleontologists & Mineralogists; Am Asn Petrol Geologists; Paleont Soc; Paleont Asn. *Res:* Paleoecology and systematics of Upper Paleozoic invertebrates, primarily molluscs and brachipods; studies of molluscs of all ages. *Mailing Add:* Dept Geol Tex A&M Univ College Station TX 77843-3115

YANCHICK, VICTOR A, b Joliet, Ill, Dec 3, 40; m 63; c 3. PHARMACY. *Educ:* Univ Iowa, BS, 62, MS, 66; Purdue Univ, PhD(pharm), 68. *Prof Exp:* Instr pharm, Purdue Univ, 66-68; from asst prof to assoc prof, 68-79, actg asst dean, 72-74, asst dean acad affairs, Col Pharm, 74-81, assoc dean, 81, PROF PHARM, UNIV TEX, AUSTIN, 79- *Honors & Awards:* Parenteral Drug Asn Res Awards, 66 & 68. *Mem:* Am Pharmaceut Asn; Am Soc Hosp Pharmacists; Acad Pharmaceut Sci; Am Asn Cols Pharm. *Res:* Parenteral drugs, primarily inactivation by other agents; drug interactions and incompatibilities; drug-nutrient interactions; geriatric pharmacotherapeutics. *Mailing Add:* Dean Pharm Univ Okla Main Campus Norman OK 73019

YANCIK, JOSEPH J, b Mt Olive, Ill, Dec 1, 30; m 55; c 2. MINING ENGINEERING. *Educ:* Univ Ill, Urbana-Champaign, BS, 54; Mo Sch Mines, MS, 56; Univ Mo-Rolla, PhD(mining eng), 60. *Prof Exp:* Mining engr, St Joseph Lead Co, 55-58, res assoc mining, Mo Sch Mines, 58-60; mgr res & develop explosives, Monsanto Co, 60-70; asst dir mining, 72-77, CHIEF DIV MINING RES, US BUR MINES, US DEPT INTERIOR, WASHINGTON, DC, 70- *Concurrent Pos:* Mem, Emergency Minerals Admin, US Dept Interior, 71-; vpres res & tech servs, Nat Coal Asn, 77-; chmn bd, Bituminous Coal Res, Inc, 77-79, vchmn bd & pres, 79-81; dir, Off Energy, Int Trade Admin, US Dept Com. *Mem:* Am Inst Mining, Metall & Petrol Engrs. *Res:* Mining research in areas of operations research in areas of operations, health and safety; explosives in areas of utilization, thermochemical and hydrodynamic properties; coal utilization. *Mailing Add:* Off Energy Rm 4413 US Dept Com 14th & Constitution Aves NW Washington DC 20230

YANDERS, ARMON FREDERICK, b Lincoln, Nebr, Apr 12, 28; m 48; c 2. ENVIRONMENTAL TOXICOLOGY. *Educ:* Nebr State Col, AB, 48; Univ Nebr, PhD(zool), 53. *Prof Exp:* Res assoc genetics, Oak Ridge Nat Lab & Northwestern Univ, 53-54; biophysicist, US Naval Radiol Defense Lab, 55-58; assoc geneticist, Argonne Nat Lab, 58-59; assoc prof zool, Mich State Univ, 59-65, prof & asst dean, Col Natural Sci, 65-69; dean, Col Arts & Sci, 69-82, DIR, ENVIRON TRACE SUBSTANCES RES CTR & SINCLAIR COMP MED RES FARM, UNIV MO, 83- *Concurrent Pos:* Vis scientist, Commonwealth Sci & Indust Res Orgn, Canberra, Australia, 66; mem bd dir & consult, Assoc Midwestern Univs, 66-68; chmn bd & pres, Argonne Univs Asn, 70-77; mem, Mo Dioxin Adv Comt, 84- & adv comt environ hazards, US Vet Admin, 85- *Mem:* Fel AAAS; Environ Mutagen Soc; Soc Environ Toxicol & Chem; Radiation Res Soc; Sigma Xi; Genetics Soc Am. *Res:* Environmental toxicology. *Mailing Add:* Environ Trace Substances Res Ctr & Sinclair Comp Med Res Farm Univ Mo Columbia MO 65203-9497

YANEY, PERRY PAPPAS, b Columbus, Ohio, July 28, 31; m 61; c 3. ELECTROOPTICS. *Educ:* Univ Cincinnati, EE, 54, MS, 57, PhD(physics), 63. *Prof Exp:* Design engr, Baldwin Piano Co, Ohio, 54-55; res physicist, St Eloi Corp, 55-59; physicist, Electronics & Ord Div, Avco Corp, 59-62; Univ Cincinnati res fel, Wright-Patterson AFB, 62-63; assoc res physicist, Res Inst, 63-65, from asst prof to assoc prof physics, 65-77, PROF PHYSICS, UNIV DAYTON, 77- *Concurrent Pos:* Consult, Optical Spectros & Reheis Chem Co, 74 & Aero-Propulsion Lab, Wright-Patterson AFB, 81-; consult & vis scientist, Univ Southern Calif, 75. *Mem:* Am Inst Aeronaut & Astronaut; Am Phys Soc; Optical Soc Am; Inst Elec & Electronics Engrs; Am Asn Physics Teachers; Sigma Xi. *Res:* Laser probe measurements in combustion and flowing gases; coherent anti-Stokes Raman spectroscopy in gas discharges and plasmas; Raman spectroscopy in solids and gases; lasers and their applications; electrooptical instrumentation techniques; optical and electrical properties of ions in crystals; electroluminescence in solids. *Mailing Add:* 4424 Renwood Dr Dayton OH 45429

YANG, AN TZU, b Shanghai, China, Oct 5, 23; US citizen; m 52; c 2. DYNAMICS. *Educ:* Northwestern Col Eng, China, BS, 46; Ohio State Univ, MS, 50; Columbia Univ, DEngSc, 63. *Prof Exp:* Res engr, Columbia Univ, 55-56; vis prof, Inst Math, Univ Rio Grande do Sul, Brazil, 59-60; sr engr, Res & Develop Div, Am Mach & Foundry Co, 60-64; from asst prof to assoc prof eng, 64-71, NSF res grant, 66-69, PROF ENG, UNIV CALIF, DAVIS, 71- *Concurrent Pos:* NSF res grant, Stanford Univ, 70-71. *Mem:* Am Soc Mech Engrs. *Res:* Kinematic analysis and synthesis of mechanisms; dynamics of mechanical systems. *Mailing Add:* Dept Mech Eng Univ Calif Davis CA 95616

YANG, ARTHUR JING-MIN, b Nanjing, Jiang-Hsu, Nov 28, 47; m; c 2. STRUCTURE PROPERTY RELATIONSHIP OF A COMPOSITE, RUBBER ELASTICITY. *Educ:* Fu Jen Univ, Taiwan, BS, 70; Brown Univ, PhD(chem), 75. *Prof Exp:* Assoc prof chem, Chem Dept, Fu Jen Univ, 75-80; sr res scientist, Amherst Col, Mass, 80-84; SR RES SCIENTIST CHEM & PHYSICS, RES & DEVELOP, ARMSTRONG WORLD INDUSTS, INC, 84- *Concurrent Pos:* Vis prof, Brown Univ, 77-78. *Res:* Structure property relationship in a composite; macroscopic material properties from the microscopic interaction; viscoelasticity of a polymer material. *Mailing Add:* Res & Develop Armstrong World Industs Inc PO Box 3511 Lancaster PA 17604

YANG, C(HENG) Y(I), b Tientsin, China, Dec 17, 30; m 61. STRUCTURAL DYNAMICS. *Educ:* Nat Taiwan Univ, BS, 53; Purdue Univ, MS, 58; Mass Inst Technol, DSc(civil eng), 62. *Prof Exp:* Asst civil eng, Nat Taiwan Univ, 54-56, Purdue Univ, 56-58 & Mass Inst Technol, 58-61, res engr, 61-62; asst prof, Univ Ill, 62-66; ASSOC PROF, UNIV DEL, 66- *Mem:* Assoc mem Am Soc Civil Engrs; Am Astronaut Soc; Am Soc Eng Educ; Soc Exp Stress Anal. *Res:* Wave propagation in solids investigated by method of characteristics; behavior of structural systems under deterministic and probabilistic loads; structural safety studies. *Mailing Add:* Dept Civil Eng Univ Del Newark DE 19716

YANG, CHAO-CHIH, b Changsha, China, Dec 17, 28; m 57; c 1. COMPUTER SCIENCE. *Educ:* Chinese Naval Col Technol, BS, 53; Nat Chiao Tung Univ, MS, 62; Northwestern Univ, MS, 64, PhD(elec eng), 66. *Prof Exp:* Asst prof elec eng & info sci, Wash State Univ, 66-67; prof comput sci, Nat Chiao Tung Univ, 67-71; scientist, Int Bus Mach, San Jose Res Lab, 71-72; assoc prof info sci, 72-79, PROF COMPUT & INFO SCI, UNIV ALA, BIRMINGHAM, 79- *Mem:* Inst Elec & Electronics Engrs; Asn Comput Mach; Sigma Xi. *Res:* Design, implementation and analysis of computer algorithms related to theory of computation, operating systems principles. *Mailing Add:* Dept Comput & Info Sci Bldg 4 Rm 242 Univ Ala Univ Sta Birmingham AL 35294

YANG, CHAO-HUI, b Taichung, Taiwan, Aug 20, 28; m 63; c 2. MATHEMATICS. *Educ:* Nat Taiwan Univ, BS, 51; Univ Mich, MA, 55; Univ Cincinnati, PhD(math), 58. *Prof Exp:* Res fel, Inst Math Sci, NY Univ, 58-59, assoc res scientist, 59-61; lectr math, Rutgers Univ, 61-64; PROF MATH, STATE UNIV NY COL ONEONTA. *Mem:* Am Math Soc; Math Asn Am. *Res:* Maximal binary matrices; integral equations; integrability of trigonometric series; combinatorial and functional analyses. *Mailing Add:* Dept Math State Univ NY Col Oneonta NY 13820

YANG, CHARLES CHIN-TZE, b Shanghai, China, Aug 20, 22; US citizen; m 57; c 3. MECHANICS, METALLURGY. *Educ:* Henry Lester Inst Technol, BSc, 42; Univ Mich, Ann Arbor, MSE, 47; Univ Calif, Berkeley, PhD(mech eng, metall), 51. *Prof Exp:* Asst prof mech eng, Mass Inst Technol, 51-54; sr res engr, Ford Res Ctr, Ford Motor Co, 54-57; assoc prof, Univ Mich, Ann Arbor, 57-60; adv engr, Int Bus Mach Corp, 60-63; sr staff scientist, San Diego Div, Gen Dynamics Corp, 63-68, staff scientist, Pomona Div, 68-77; SR ENGR, GEN ELEC CO, CALIF, 77- *Concurrent Pos:* Sr staff specialist, Tech Resources Inc, 71- *Mem:* Am Soc Metals; Am Soc Mech Engrs. *Res:* Powder metallurgy, boron fibrous-aluminum powder composites; nitrocellulose tape metal processing and dinitrocellulose; structural design analysis and evaluation of engineered components, machineries, large steam turbine-generators, gas turbines, and nuclear reactors; statistics; dynamics; heat transfer. *Mailing Add:* 1309 Miller Ave San Jose CA 95129

YANG, CHEN NING, b Hefei, Anhwei, China, Sept 22, 22; m 50; c 3. PHYSICS. *Educ:* Southwest Assoc Univ, China, BSc, 42; Univ Chicago, PhD(physics), 48. *Hon Degrees:* DSc, Princeton Univ, 58, Polytechnic Inst Brooklyn, 65, Univ Wroclaw, 74, Gustavus Adolphus Col, Minn, 75, Univ Md, 79, Univ Durham, Eng, 79 & Fundan Univ, China, 84, Eidg Technische Hochschule, Switz, 87. *Prof Exp:* Instr physics, Univ Chicago, 48-49; mem, Inst Advan Study, Princeton, 49-66, prof, 55-66; ALBERT EINSTEIN PROF PHYSICS & DIR, INST THEORET PHYSICS, STATE UNIV NY, STONY BROOK, 66- *Concurrent Pos:* Bd trustees, Woods Hole Oceanog Inst, 62-78, Rockefeller Univ, 70-76, Salk Inst, 78-79 & Ben Gurion Univ, 81-; mem, Governing Coun, Courant Inst Math Sci, 63- & Sci Adv Comt, IBM,

66-71; chmn, Panel Theoret Physics, Physics Surv Comt, Nat Acad Sci, 65, Div Particles & Fields, Int Union Pure & Appl Physics, 72-76 & Fachbeirat, Max Planck Inst Physics, Munich, 80-83; bd dirs, AAAS, 75-79, Neurosci Inst, 83-88 & Sci Am, Inc, 83-90; hon prof, numerous univs, China; distinguished prof-at-large, Chinese Univ Hong Kong. *Honors & Awards:* Nobel Prize in Physics, 57; Albert Einstein Commemorative Award, 57; Rumford Prize, 80; Nat Medal Sci, 86; Liberty Award, 86. *Mem:* Nat Acad Sci; Am Philos Soc; Am Phys Soc; Brazilian Acad Sci; Venezuelan Acad Sci; Royal Spanish Acad Sci; Polish Acad Sci; Sigma Xi. *Res:* Theoretical physics. *Mailing Add:* Inst Theoret Physics State Univ NY Stony Brook NY 11794-3840

YANG, CHIA HSIUNG, b Peikang, Taiwan, Sept 24, 40; US citizen; m 69; c 3. NUCLEAR PHYSICS. *Educ:* Tunghai Univ, Taiwan, BSc, 62; Tsing Hua Univ, Taiwan, MSc, 65; Washington Univ, MA, 67, PhD(physics), 71. *Prof Exp:* From asst prof to assoc prof, 71-80, PROF PHYSICS, SOUTHERN UNIV, 80-, CHMN, 87- *Concurrent Pos:* NASA res grant, 72-79. *Mem:* Sigma Xi; Am Phys Soc. *Res:* Microscopic study of isotropic superfluidity of neutron star matter by Yang and Clark method which combines Bardeen-Cooper-Schrieffer and correlated basis function theories. *Mailing Add:* 1062 Stoneliegh Dr Baton Rouge LA 70808

YANG, CHIH TED, b Chung King, China, Jan 23, 40; m 68; c 2. MORPHOLOGY, HYDROLOGY. *Educ:* Cheng Kung Univ, Taiwan, BS, 62; Colo State Univ, MS, 65, PhD, 68. *Prof Exp:* Assoc prof scientist, Ill State Water Surv, 68-74; hydraul engr, US Army Corp Engrs, 74-78; tech rev staff, 79-88, INT & SPECIAL PROJS COORDR, US BUR RECLAMATION, 88- *Concurrent Pos:* Adj prof, Univ Colo, Denver, 82- *Honors & Awards:* Robert E Horton Award, Am Geophys Union, 72; Walter L Hurber Res Prize, Am Soc Civil Engrs, 73, J C Stevens Award, 80. *Mem:* Am Geophys Union; Am Soc Civil Engrs. *Res:* Hydraulics, hydrology, morphology, sedimentation, hydraulic structure design. *Mailing Add:* US Bur Reclamation Code 3240 PO Box 25007 Denver CO 80225

YANG, CHING HUAN, b Hunan, China, Sept 7, 20; m 52; c 3. MECHANICS. *Educ:* Nat Cent Univ, China, BS, 43; Lehigh Univ, PhD(appl mech), 51. *Prof Exp:* Res assoc plasticity, Fritz Eng Lab, Lehigh Univ, 47-51; engr, Anaconda Copper Mining Co, 52-55; sr engr, Convair Div, Gen Dynamics Corp, 56-57, staff scientist, Sci Res Lab, 57-62; mem tech staff, Defense Res Corp, Calif, 62-68; PROF ENG, STATE UNIV NY, STONY BROOK, 68- *Mem:* Combustion Inst. *Res:* Applied mechanics; combustion theories; explosion phenomena; nonequilibrium thermodynamics; gaseous kinetics. *Mailing Add:* Dept Mech Eng State Univ NY Stony Brook NY 11794

YANG, CHUI-HSU (TRACY), b Hunan, China, Nov 19, 38; US citizen; m 64; c 2. RADIATION BIOPHYSICS. *Educ:* Tunghai Univ, BS, 59; North Tex State Univ, MS, 64; Univ Ill, PhD(biophys), 67. *Prof Exp:* Appointee biol, Argonne Nat Lab, 67-69; biophysicist, Lawrence Berkeley Lab, Univ Calif, 69-90; RADIATION HEALTH OFFICER, NASA JOHNSON SPACE CTR, 90- *Mem:* Cancer Res Asn; Biophys Soc; Radiation Res Soc; Evniron Mutagenesis Soc; Am Soc Therapeut Radiol & Oncol; Tissue Cult Asn. *Res:* Effects of radiation on membrane, development and longevity; mechanisms and kinetics of recovery; space biology; responses of cultured mammalian cells to heavy ions and other environmental factors; mechanisms of cell transformation in vitro by radiation; somatic mutation by radiation. *Mailing Add:* Bldg 74B Rm 106 Univ Calif Lawrence Berkeley Lab Berkeley CA 94720

YANG, CHUN CHUAN, b Taichung, Taiwan, Jan 25, 36; m 66; c 2. MECHANICAL ENGINEERING, FLUID DYNAMICS. *Educ:* Cheng-Kung Univ, BS, Taiwan, 59; Univ RI, MS, 65; Yale Univ, PhD(eng), 69. *Prof Exp:* Jr engr, Taiwan Mach Mfg Corp, 61-63; res asst heat transfer, Univ RI, 63-65; mem res staff geophys fluid dynamics, Yale Univ, 69-72; res scientist, Xonics, Inc, 72-74; res engr, Heat Transfer Res, Inc, 74-88; ROCKWELL INT, ROCKETDYNE DIV, 88- *Mem:* Am Soc Mech Engrs; Sigma Xi. *Res:* Heat transfer and fluid mechanics in two-phase flow; computational fluid dynamics. *Mailing Add:* 10211 Melvin Ave Northridge CA 91324

YANG, CHUNG CHING, b Peking, China, Sept 13, 38; US citizen; m 62; c 2. SOLID STATE PHYSICS, QUANTUM ELECTRONICS. *Educ:* Univ Calif, Los Angeles, BS, 66; Harvard Univ, MS, 67, PhD(solid state physics), 72. *Prof Exp:* Mem sci staff, 72-78, res mgr, 78-80, TECH RES MGR, XEROX CORP, 80- *Mem:* Am Phys Soc. *Res:* Quantum optics; phase transition; transport across materials interfaces. *Mailing Add:* 19550 Braemar Ct Saratoga CA 95070

YANG, CHUNG SHU, b Peking, China, Aug 8, 41; m 66; c 2. BIOCHEMISTRY. *Educ:* Nat Taiwan Univ, BS, 62; Cornell Univ, PhD(biochem), 67. *Prof Exp:* Fel biochem, Scripps Clin & Res Found, 67-69; res assoc, Yale Univ, 69-71; asst prof, 71-75, assoc prof, 75-79, PROF BIOCHEM, COL MED NJ, 79- *Honors & Awards:* Fac Res Award, Am Cancer Soc. *Mem:* Am Soc Biochemists; Am Chem Soc; Am Soc Pharmacol; Am Inst Nutrit; Am Asn Cancer Res. *Res:* Mechanisms of biological oxygenation and carcinogen activation, etiology and modification of carcinogenesis. *Mailing Add:* Dept Biochem Univ Med & Dent NJ 185 S Orange Ave Newark NJ 07103

YANG, CHUNG-CHUN, b Kiang-su, China, Nov 21, 42; m 67; c 3. PURE MATHEMATICS, APPLIED MATHEMATICS. *Educ:* Nat Taiwan Univ, BS, 64; Univ Wis-Madison, MS, 66, PhD(math), 69. *Prof Exp:* Res assoc math, Mich State Univ, 69-70; RES MATHEMATICIAN, NAVAL RES LAB, 70- *Concurrent Pos:* Comt mem Chinese translation comt, Am Math Soc; vis prof, Elect Engr & Comput Sci Dept, IIT, 83-84. *Mem:* Am Math Soc; Japanese Math Soc. *Res:* Factorization theory in the function theory of one complex variable; applications of theory of meromorphic functions to some physical and engineering problems; pattern recognition and image processing in computer science area. *Mailing Add:* Hong Kong Univ Sci & Tech Hong Kong Hong Kong

YANG, CHUNG-TAO, b Pingyang, China, May 4, 23; m 57; c 3. MATHEMATICS. *Educ:* Chekiang Univ, BS, 46; Tulane Univ, PhD, 52. *Prof Exp:* Asst math, Chekiang Univ, 46-48; asst, Nat Acad Sci, China, 48-49; instr, Nat Taiwan Univ, 49-50; res assoc, Univ Ill, 52-54; mem staff, Inst Advan Study, 54-56; from asst prof to assoc prof, 56-61, chmn dept, 78-83, PROF MATH, UNIV PA, 61- *Mem:* Am Math Soc; Math Asn Am; Sigma Xi. *Res:* General and algebraic topology; topological and differential transformation groups. *Mailing Add:* 311 Hidden River Rd Narberth PA 19072

YANG, DA-PING, b Peiping, China, Oct 5, 33; US citizen; m 67; c 2. GENETICS, CELL BIOLOGY. *Educ:* Nat Taiwan Univ, BSc, 56; Ottawa Univ, Can, MSc, 64, PhD(genetics), 69. *Prof Exp:* Agronomist, Rice Res Inst, Taiwan; instr, Ottawa Univ, Can, 64-68; fel biol, Nat Res Coun, Can, 68-70; sr res scientist & supvr, Cytol Lab, 71-82, supvr, Monoclonal Antibody Unit, 83-85; SR RES SCIENTIST EXP BIOL & IMMUNOL SECT, WYETH-AYERST LABS, INC, 85- *Mem:* AAAS; Tissue Cult Asn; Environ Mutagen Soc; NY Acad Sci. *Res:* Mammalian cell culture and cytogenetics; genetic toxicology; mutagenesis; carcinogenesis; somatic cell genetics; monoclonal antibody and hybridoma research. *Mailing Add:* Div Res & Develop Wyeth-Ayerst Labs Inc PO Box 8299 Philadelphia PA 19101-1245

YANG, DARCHUN BILLY, b Taipei, Taiwan, July 17, 45; m 71; c 2. POLYMER CHEMISTRY, HETEROGENEOUS & HOMOGENEOUS CATALYSIS. *Educ:* Tamkang Univ, Taiwan, BS, 69; Furman Univ, MS, 73; Univ Ga, PhD(inorg chem), 77. *Prof Exp:* Analytic technician, Alpha Lab, Can, 74-75; fel, Univ Ga, 78-80; sr chemist, 80-86, staff chemist, Exxon Res & Eng Co, 87-88; STAFF CONSULT, STANLEY WORKS, 88-; SR SCIENTIST, LOCTITE CORP. *Concurrent Pos:* Chem teacher, Mil Acad Taiwan & Shulin High Sch, 69-71. *Mem:* Am Chem Soc; Sigma Xi. *Res:* Expert in adhesive and sealant research, especially in structure and instant adhesives; anaerobic, photo and thermal cure chemistries; thermal stability; polymerization kinetic by real time FTIR; surface chemistries; primers for low energy plastics and metal surfaces; structure-property relationships; coating, adhesive, and medical applications of polymers; thermal analysis of polymers; homogeneous and heterogeneous catalysis; new and improved catalysts; catalyst performance evaluation; physical characterization of catalyst and carrier; waste treatments. *Mailing Add:* 48 Lostbrook Rd West Hartford CT 06117

YANG, DAVID CHIH-HSIN, b Hsinchiang, China, Jan 8, 47; m 71. BIOCHEMISTRY. *Educ:* Nat Taiwan Univ, BSc, 68; Yale Univ, PhD(biochem), 73. *Prof Exp:* Res assoc, Rockefeller Univ, 73-75; from asst prof to assoc prof, 75-90, PROF CHEM, GEORGETOWN UNIV, 90- *Mem:* Am Chem Soc; AAAS; Am Asn Univ Professors; Am Soc Biol Chemists; NY Acad Sci. *Res:* Structure of amino acyl-tRNA synthetases; conformational analysis of protein and nucleic acid; fluorescence spectroscopy; chemical modification of nucleic acid; animal lectins. *Mailing Add:* Dept Chem Georgetown Univ 654 Reiss 34th & O St NW Washington DC 20057

YANG, DOMINIC TSUNG-CHE, b Tainan, Taiwan, Oct 9, 33; US citizen; m 62; c 2. ORGANIC CHEMISTRY. *Educ:* St Benedict's Col, Kans, BS, 59; Univ Ga, PhD(org chem), 69. *Prof Exp:* Chemist, Nalco Chem Co, Ill, 59-64; instr toxicol & NIH res grant, Vanderbilt Univ, 68-70; from asst prof to assoc prof, 70-79, PROF ORG CHEM, UNIV ARK, LITTLE ROCK, 79- *Mem:* Am Chem Soc; Sigma Xi. *Res:* Organic synthesis of polycyclic aromatic hydrocarbons. *Mailing Add:* Dept Chem Univ Ark Little Rock AR 72204

YANG, DOROTHY CHUAN-YING, b Shanghai, China, May 27, 18; US citizen; div; c 2. PEDIATRICS, NEUROLOGY. *Educ:* St Johns Univ, China, MD, 45; New York Med Col, MMSc, 50; Am Bd Pediat, dipl, 52. *Prof Exp:* Intern, St Luke's Hosp, Shanghai, China, 44-45; med resident, Govt Hosp, Free China, 45-46; resident pediat, Children's Ctr, New York, 47-48; asst resident, New Eng Hosp Women & Children, Boston, Mass, 48-49; asst resident, Syracuse Med Ctr, 49-50; instr pediat, New York Med Col, 52-54, assoc, 54-55, asst clin prof, 55-56, asst prof, 56-60; res neurol, Children's Hosp Philadelphia, 61-62; assoc prof pediat, New York Med Col, 64-69, asst prof neurol, 66-69; CLIN ASSOC PROF PEDIAT, STATE UNIV NY DOWNSTATE MED CTR, 69-; ASSOC DIR STANLEY S LAMM INST DEVELOP DIS, LONG ISLAND COL HOSP, 69- *Concurrent Pos:* Teaching & res fel, New York Med Col, 50-52; NIH spec fel pediat neurol, 60-63; asst pediatrician, Flower & Fifth Ave Hosp, New York, 54-60; pediatrician, Collab Study Neurol Dis & Blindness, NIH, 57-60, neurologist, 64-69; asst vis pediatrician, Metrop Hosp, New York, 58-60; spec training electroencephalog, Grad Hosp, Univ Pa, 62-63; assoc attend pediatrician, Flower & Fifth Ave & Metrop Hosps, 64-69, asst attend neurologist, 66-69; assoc vis pediatrician, Kings County Hosp Ctr, 71-; consult pediat neurol, St John's Episcopal Hosp, Brooklyn, NY, 72-80. *Mem:* Fel Am Acad Pediat; Child Neurol Soc; Am Acad Neurol; Am Med EEG Asn. *Mailing Add:* 110 Amity St Brooklyn NY 11201

YANG, EDWARD S, b Nanking, China, Oct 16, 37; US citizen; m 61; c 2. ELECTRONICS. *Educ:* Cheng Kung Univ, Taiwan, BS, 57; Okla State Univ, MS, 61; Yale Univ, PhD(eng & appl sci), 66. *Prof Exp:* Jr engr, IBM Corp, 61-62, as soc engr, 62-63; asst electronics, Yale Univ, 63-65; from asst prof to assoc prof elec eng, 65-75, PROF ELEC ENG, COLUMBIA UNIV, 75-, CHMN DEPT, 87- *Concurrent Pos:* Dir, NSF res grant & prin investr, 67-81, NASA res grant, 67-69; contracts, Dept Energy, 77-82 & Dept Defense, 77-, IBM Corp, 85- *Mem:* Inst Elec & Electronics Engrs; Am Phys Soc. *Res:* Schottky Barriers, p-n junction, bipolar transistor; optoelectronics, high Tc superconductor devices. *Mailing Add:* Dept Elec Eng Columbia Univ New York NY 10027

YANG, FUNNEI, MOLECULAR BIOLOGY, HUMAN GENETICS. *Educ:* Univ Wash, PhD(microbiol & immunol), 80. *Prof Exp:* ASST PROF HUMAN GENETICS, UNIV TEX, 82- *Mailing Add:* Dept Cellular & Struct Biol Univ Tex Health Sci Ctr 7703 Floyd Curl Dr San Antonio TX 78284

YANG, GENE CHING-HUA, b Kaohsuing, Taiwan, Feb 26, 38; US citizen; m 70; c 1. MICROBIOLOGY, INFECTIOUS DISEASE. *Educ:* Tunghai Univ, BS, 60; Northwestern Univ, MS, 66; Mich State Univ, PhD(microbiol), 70. *Prof Exp:* Asst prof, Med Ctr, Univ Tenn, 71-76; ASSOC PROF MICROBIOL & DIR MED TECHNOL PROG, SAGINAW VALLEY STATE COL, 76- *Concurrent Pos:* Mem bd, Dr Sun Yet Sen Inst, 77-; consult, St Luke's Hosp, Saginaw, 77- *Mem:* AAAS; Am Soc Med Technol; Am Soc Microbiol; Sigma Xi. *Res:* Pathogenic mechanism of enteric infection. *Mailing Add:* Dept Eng & Sci Saginaw Valley State Col 2250 Pierce Rd University Center MI 48710

YANG, GRACE L, b Queichow, China; m 64; c 2. STATISTICS. *Educ:* Univ Calif, Berkeley, MA, 63, PhD(statist), 66. *Prof Exp:* From asst prof to assoc prof, 73-78, PROF STATIST, UNIV MD, COLLEGE PARK, 78- *Mem:* Am Statist Asn; Inst Math Statist. *Res:* Mathematical statistics; biostatistics. *Mailing Add:* 6106 Neilwood Dr Rockville MD 20852

YANG, H(SUN) T(IAO), b Hangchow, China, May 19, 24; m 65; c 3. AEROSPACE ENGINEERING. *Educ:* Univ Wash, MS, 50; Calif Inst Technol, PhD(aeronaut, math), 55. *Prof Exp:* Res fel, Calif Inst Technol, 55-56; res assoc, Inst Fluid Dynamics, Univ Md, 56-58; res scientist, Eng Ctr, 58-63, ASSOC PROF AERONAUT ENG, UNIV SOUTHERN CALIF, 63- *Concurrent Pos:* Fulbright-Hays lectr, 64-65; consult various indust concerns. *Mem:* Am Phys Soc; Am Inst Aeronaut & Astronaut. *Res:* Fluid dynamics; aerodynamics. *Mailing Add:* Dept Aerospace Eng 0651 Univ Southern Calif Los Angeles CA 90089

YANG, HENRY T Y, b Chungking, China, Nov 29, 40; US citizen; m 67; c 2. ENGINEERING EDUCATION, COMPUTATIONAL METHODS. *Educ:* Nat Taiwan Univ, BS, 62; WVa Univ, MS, 65; Cornell Univ, PhD, 68. *Prof Exp:* Struct engr, Gilbert Assocs, 68-69; from asst prof to assoc prof aeronaut & astronaut, 69-76, sch head, 79-84, PROF AERONAUT & ASTRONAUT, PURDUE UNIV, 76-, DEAN ENG, 84- *Concurrent Pos:* Vis scientist, Air Force Flight Dynamics Lab, 76; Neil A Armstrong distinguished prof, Purdue Univ, 88. *Mem:* Nat Acad Eng; fel Am Inst Aeronaut & Astronaut; Am Soc Eng Educ; Am Soc Mech Engrs; Am Soc Civil Engrs. *Res:* Aeroelasticity; finite elements; structural dynamics; manufacturing; engineering education. *Mailing Add:* 864 Rose St West Lafayette IN 47906

YANG, HO SEUNG, b Korea, Dec 13, 47; m 74; c 2. FOOD SCIENCE & TECHNOLOGY, BIOCHEMICAL ENGINEERING. *Educ:* Seoul Nat Univ, BS, 69; Univ Minn, MS, 71; Mass Inst Technol, PhD(food sci), 76; Sangamon State Univ, MBA, 82. *Prof Exp:* Res assoc, Mass Inst Technol, 76; res scientist, Watson Res Ctr, IBM, 76-78; biochem eng, A E Staley Mfg Co, 78-82, mgr bioprocessing res, 83-87; PROJ MGR, UNIVERSAL FOODS CORP, 87- *Mem:* Inst Food Technologists; Am Chem Soc; Soc Indust Microbiol. *Res:* Enzyme technology; carbohydrate and protein processing; fermentation; applied biochemistry; food process development; biotechnical scale up. *Mailing Add:* Sunkyung Ind 500 Juneia Dong Changan-Ku Kyungki-Do Republic of Korea

YANG, HONG-YI, b Tientsin, China, Mar 26, 33; US citizen; m 65; c 3. IMMUNOLOGY. *Educ:* Nat Taiwan Univ, MD, 61; Univ Chicago, PhD(path), 67; Am Bd Path, cert, 69. *Prof Exp:* Res & teaching asst path, Univ Chicago, 63-67; from asst prof to assoc prof, 68-83, PROF PATH, JOHN A BURNS SCH MED, UNIV HAWAII, MANOA, 83- *Concurrent Pos:* Assoc path, St Francis Hosp, 73-; consult pathologist, Castle Mem Hosp, 79- *Mem:* AMA; Am Asn Pathologists; Am Soc Exp Path; Am Soc Clin Pathologists. *Res:* Cellular immunity and macrophage functions; ultrastructural study of soft tissue tumors; immunopathology of renal disease. *Mailing Add:* Dept Path/Biol T510 Univ Hawaii 1960 East-West Rd Manoa HI 96822

YANG, HOYA Y, b Amoy, China, June 3, 12; nat US; m 46; c 2. FOOD TECHNOLOGY. *Educ:* Nanking Univ, BS, 36; Ore State Col, MS, 40, PhD(food technol), 44. *Prof Exp:* From asst prof to prof, 43-77, EMER PROF FOOD SCI, ORE STATE UNIV, 77- *Concurrent Pos:* Consult, Food Industries. *Mem:* Am Chem Soc; Am Soc Enol; Inst Food Technologists; Am Soc Microbiol; Sigma Xi. *Res:* Food fermentation; food additives; food enzymes; food and food product analysis. *Mailing Add:* 1020 NW 30th Corvallis OR 97330-4442

YANG, IN CHE, b Taiwan, Feb 7, 34; US citizen; m 66; c 3. HYDROCHEMISTRY, INORGANIC CHEMISTRY. *Educ:* Nat Taiwan Univ, BSc, 56; Carleton Univ, MSc, 66; Univ Wash, PhD(nuclear & inorg chem), 71. *Prof Exp:* Radiochemist radiation ecol, Univ Wash, 72, res assoc isotope geol, 72-75, res asst prof, 75-78; radiochemist isotope hydrol, 78-80, radiochem sect chief, Denver Cent Lab, Water Resource Div, 80-84, PROJ CHIEF HYDROCHEM, NUCLEAR HYDROL PROG, US GEOL SURV, 84- *Concurrent Pos:* Task mem, Standard Method Comt, Standard Method, 80- *Mem:* Am Geophys Union. *Res:* Applications of radioactive and stable isotopes to geology and hydrology; geochronology using carbon-14; tritium and potassium-argon dating; stable isotope ratios in hydrology using oxygen-18/oxygen-16, carbon-13/carbon-12 and deuterium/hydrogen ratios. *Mailing Add:* US Geol Surv MS 421 Denver Fed Ctr Box 25046 Lakewood CO 80225

YANG, JEN TSI, b Shanghai, China, Mar 18, 22; m 49; c 2. BIOPHYSICAL CHEMISTRY. *Educ:* Nat Cent Univ, China, BS, 44; Iowa State Univ, PhD(biophys chem), 52. *Prof Exp:* Asst anal chem, Nat Cent Univ, China, 46-47; res assoc protein chem, Iowa State Univ, 52-54; res fel polypeptide & protein chem, Harvard Univ, 54-56; res chemist, Am Viscose Corp, 56-59; assoc prof biochem, Dartmouth Med Sch, 59-60; assoc prof, 60-64, PROF BIOCHEM, UNIV CALIF, SAN FRANCISCO, 64- *Concurrent Pos:* Fel, John Simon Guggenheim, 59-60, Commonwealth, 67, Japan Soc Prom Sci, 75; vis prof, Nat Sci Coun, Taiwan, 67, Univ Tokyo, Japan, 67, Shanghia Inst Biochem, Academia Sinica, China, 78 & 79, Inst Biol Chem, Academia Sinica, Taiwan, 82, Nanjing Univ, China, 86, Inst Protein Res, Acad Sci USSR, 90; vis scholar Exchange Prog Partic, Comt Scholarly Commun with People's Repub China, Nat Acad Sci, 89. *Mem:* Am Chem Soc; Am Soc Biochem & Molecular Biol; Protein Soc. *Res:* Physical chemistry of biopolymers; chiroptical properties of proteins. *Mailing Add:* Cardiovasc Res Inst Box 0524 Univ Calif San Francisco CA 94143-0524

YANG, JENN-MING, b Hsin-Chu, Taiwan, Dec 20, 57; m 84; c 1. HIGH TEMPERATURE COMPOSITE MATERIALS, PROCESSING & MECHANICAL BEHAVIOR. *Educ:* Nat Tsing-Hua Univ, Taiwan, BS, 79; Univ Del, PhD(mat sci eng), 86. *Prof Exp:* Engr, Tex Instruments, Taiwan, 81-82; ASST PROF MAT SCI ENG, UNIV CALIF, LOS ANGELES, 86- *Concurrent Pos:* NSF presidential young investr award, 90. *Mem:* Mat Res Soc; Am Ceramic Soc; Metall Soc. *Res:* Processing and mechanical behavior of metal; intermetallic, ceramic and carbon matrix composites; aerospace structural and propulsional materials; high temperature deformation and fracture. *Mailing Add:* Dept Mat Sci & Eng Univ Calif Los Angeles CA 90024

YANG, JEONG SHENG, b Taiwan, July 11, 34; m 60; c 3. TOPOLOGY. *Educ:* Taiwan Normal Univ, BS, 58; Univ Ala, Tuscaloosa, MA, 63; Univ Miami, PhD(math), 67. *Prof Exp:* Asst math, Taiwan Normal Univ, 57-62; asst prof, La State Univ, New Orleans, 66-68; vis asst prof, Univ Miami, 68-69; from asst prof to assoc prof, 69-83, PROF MATH, UNIV SC, 83- *Mem:* Am Math Soc; Math Asn Am. *Res:* Function spaces; topological groups; transformation groups. *Mailing Add:* Dept Math Univ SC Columbia SC 29208

YANG, JIH HSIN (JASON), b Formosa, 1937; m 71; c 2. CHEMICAL ENGINEERING, FOOD SCIENCE. *Educ:* Univ Ill, Urbana, BS, 63, MS, 66, PhD(food sci), 68. *Prof Exp:* Res proj engr foods, Procter & Gamble Co, 68-70; sr prod technologist, Kroger Co, 70-74; res & develop mgr, Beatrice Co, 74-78; DIR RES, SANDOZ NUTRIT CORP, 78- *Mem:* Am Inst Chem Engrs; Inst Food Technol; Am Mgt Asn; Am Asn Cereal Chemists; Res & Develop Assocs. *Res:* Development of new foods and processes for medical and health care applications. *Mailing Add:* Sandoz Nutrit Corp 1541 Vernon Ave S Minneapolis MN 55416

YANG, JOHN YUN-WEN, b Changsha, China, May 19, 30; US citizen; m 58; c 3. ENVIRONMENTAL CHEMISTRY. *Educ:* St Benedict's Col, Kans, BS, 52; Univ Kans, PhD(chem), 57. *Prof Exp:* Res assoc recoil carbon-14, Brookhaven Nat Lab, 57-59; sr chemist, US Naval Radiol Defense Lab, Calif, 59-60; res specialist, Atomics Int Div, NAm Aviation, Inc, 60-62, mem tech staff, Sci Ctr, 62-67; sr res scientist, Western NY Nuclear Res Ctr, 67-72; prin chemist, Environ Systs Dept, Calspan Corp, 72-76; SR CHEMIST, LINDE DIV, UNION CARBIDE CORP, 76- *Concurrent Pos:* Adj assoc prof, State Univ NY, Buffalo, 70-76. *Mem:* Am Chem Soc; Am Nuclear Soc; NY Acad Sci. *Res:* Environmental air and water chemistry; physiocochemical methods of wastewater treatment; air pollution technology; radiation and isotope applications; water treatment and corrosion protection; radiation photochemistry; electrochemical processes. *Mailing Add:* 121 Ranch Trail Williamsville NY 14221

YANG, JULIE CHI-SUN, b Beijing, China, June 10, 28. ANALYTICAL INORGANIC CHEMISTRY. *Educ:* Tsing Hua Univ, China, BS, 49; Ind Univ, MA, 52; Univ Ill, PhD, 55. *Prof Exp:* Asst chem, Ind Univ, 50-52; res chemist, Res Ctr, Johns-Manville Corp, 55-59, sr res chemist, 59-67, res assoc, 67-72; sr group leader, 72-74, RES MGR, CONSTRUCT PROD DIV, W R GRACE & CO, 75- *Concurrent Pos:* Mem, Hwy Res Bd, Nat Acad Sci-Nat Res Coun. *Mem:* Am Chem Soc; Am Ceramic Soc; Clay Minerals Soc. *Res:* Inorganic silicate and cement chemistry; synthesis; properties; structures; material research; analytical instrumentation. *Mailing Add:* W R Grace & Co 62 Whittemore Ave Cambridge MA 02140

YANG, KEI-HSIUNG, b Taiwan, Dec 10, 40; US citizen; m 71; c 2. QUANTUM OPTICS, MEDICAL PHYSICS. *Educ:* Nat Taiwan Univ, BS, 64; Univ Notre Dame, MS, 67; Univ Calif, Berkeley, PhD(physics), 74. *Prof Exp:* Teaching asst, Univ Notre Dame, 65-67; teaching asst, Univ Calif, Berkeley, 67-69; mem tech staff res, Bell Tel Labs, Murray Hill, NJ, 69; res asst, Lawrence Berkeley Lab, 69-73; mem staff, Res & Develop Ctr, Gen Elec Corp, 73-79; MEM RES STAFF, T J WATSON RES CTR, IBM CORP, YORKTOWN HEIGHTS, NY, 79- *Mem:* Am Phys Soc; Soc Info Display. *Res:* Continuous tunable coherent vacuum ultraviolet source; medical x-ray devices; liquid crystal displays. *Mailing Add:* T J Watson Res Ctr IBM Corp PO Box 218 Yorktown Heights NY 10598

YANG, KWANG-TZU, b China, Nov 12, 28; nat US; m 53; c 5. MECHANICAL ENGINEERING. *Educ:* Ill Inst Technol, BS, 51, MS, 52, PhD(heat transfer), 55. *Prof Exp:* From asst prof to assoc prof, 56-62, chmn dept, 68-69, chmn dept aerospace & mech eng, 69-78, PROF MECH ENG, UNIV NOTRE DAME, 62-, VIOLA D HANK PROF ENG, 85- *Concurrent Pos:* Consult, Dodge Mfg Corp, 56-; NSF res grants, 56-; Off Naval Res grant, 61-64, Navy, 85- *Mem:* AAAS; fel Am Soc Mech Engrs; Am Soc Eng Educ. *Res:* Boundary-layer theory; nonlinear methods; forced and free convection; hydrodynamic stability; thermal radiation. *Mailing Add:* Dept Aerospace & Mech Eng Univ Notre Dame Notre Dame IN 46556

YANG, MAN-CHIU, b Hankow, China, Aug 16, 46; m 71; c 1. BIOCHEMISTRY. *Educ:* Chinese Univ Hong Kong, BSc, 70; Univ Nebr, PhD(chem), 74. *Prof Exp:* RES ASSOC, STATE UNIV NY, BUFFALO, 74- *Mem:* Biophys Soc; Am Chem Soc. *Res:* Study of the intermediates and enzymic systems in oxidative phosphorylation and photophosphorylation. *Mailing Add:* 25 Tanbark Circle Don Mills ON M3B 1N7 Can

YANG, MARK CHAO-KUEN, b Tsuchuan, China, Dec 14, 42; m 68; c 1. STATISTICS. *Educ:* Nat Taiwan Univ, BS, 64; Univ Wis, MS, 67, PhD(statist), 70. *Prof Exp:* Asst prof, 70-75, ASSOC PROF STATIST, UNIV FLA, 75- *Concurrent Pos:* Consult, Redstone Arsenal, US Army Command, Ala, 72-73 & Offshore Power Co, Fla, 73-74. *Mem:* Am Statist Asn; Inst Math Statist; AAAS. *Res:* Applied probability; stochastic processes; time series analysis. *Mailing Add:* Dept Math Western Ill Univ Adams St Macomb IL 61455

YANG, MEILING T, b Taiwan, Apr 21, 31; US citizen; m 60. INSTRUMENT MODIFICATION & FABRICATION. *Educ:* Chung Kung Univ, BS, 54; Tex Tech Univ, MS, 58; Purdue Univ, PhD(chem), 62. *Prof Exp:* Teaching res fel, Purdue Univ, 62; vis asst prof res, La State Univ, 63-64; res chemist, Ethyl Corp, 64-70; asst soil chemist lab supvr, Agr Ext Serv, Tex A&M Univ, 71-81; res scientist, Anal Procedure Develop, Tex Tech Univ, 85-87; RETIRED. *Concurrent Pos:* Consult chemist, anal procedure develop, 81-85. *Mem:* Am Chem Soc. *Res:* Analytical methodology development; evaluation and comparison of analytical methods for trace mineral determination and enzyme assays; synthesis of specialty chemicals; spectroscopy, chromatography and polarography. *Mailing Add:* PO Box 768 Nipamo CA 93444

YANG, NIEN-CHU, b Shanghai, China, May 1, 28; nat US; m 54; c 3. CHEMISTRY. *Educ:* St John's Univ, China, BS, 48; Univ Chicago, PhD, 52. *Prof Exp:* Res assoc, Mass Inst Technol, 52-55; res fel, Harvard Univ, 55-56; from asst prof to assoc prof, 56-63, PROF CHEM, UNIV CHICAGO, 63- *Concurrent Pos:* Alfred P Sloan fel, 60-64; Nat Cancer Inst fel, 73-74; Guggenheim fel, 74-75. *Honors & Awards:* Gregory & Freda Jalpern Award, NY Acad Sci, 81. *Mem:* AAAS; Am Chem Soc; Royal Soc Chem; Sigma Xi. *Res:* Photochemistry; chemistry of nucleic acids; organic synthesis. *Mailing Add:* Dept Chem Univ Chicago SCL-429 Chicago IL 60637

YANG, OVID Y H, b Korea; US citizen. PATHOLOGY, IMMUNOLOGY. *Educ:* Yonsei Univ, Korea, MD, 50; Univ Ottawa, PhD, 62. *Prof Exp:* Dir clin lab, Park Place Hosp, Port Arthur, Tex, 69-76; DIR CLIN LAB, TITUS COUNTY MEM HOSP, MT PLEASANT, TEX, 76- *Mem:* Am Soc Exp Path; Int Acad Path; Am Soc Clin Path; Col Am Path; AMA. *Res:* Chemical carcinogenesis; cancer immunology. *Mailing Add:* 3107 Fareway Mt Pleasant TX 75455

YANG, PAUL WANG, b Taichung, Taiwan, Oct 14, 53; m 83; c 1. SPECTROSCOPIC DATA PROCESSING. *Educ:* Nat Chung Hsing Univ, BS, 75; Ohio Univ, PhD(anal chem), 83. *Prof Exp:* Res asst org anal, Union Indust Lab, Indust Res Inst, 77-79; res asst chem, Univ Calif, Riverside, 82-83; res assoc bioanal, Div Chem, Nat Res Coun, 83-88; SCIENTIST, LAB SERVICES BR, ONT MINISTRY ENVIRON, 88- *Concurrent Pos:* Consult, Bomem Inc, 86- *Mem:* Chem Inst Can. *Res:* Development of spectroscopic techniques and signal processing algorithms to facilitate the observation, interpretation of fundamental properties of lipid surfactant molecules and spectroscopic chromatographic data. *Mailing Add:* Trace Organics Min Environ 125 Resources Rd Rexdale ON M9W 5L1 Can

YANG, PHILIP YUNG-CHIN, b Pin-ton, Taiwan, Nov 20, 54; US citizen; m 86; c 1. POLYMERS DEVELOPMENT, STRUCTURE PROPERTY RELATIONSHIPS. *Educ:* Fen Chia Univ, BE, 76; Cleveland State Univ, MS, 79, Univ Ill, PhD(polymer sci & eng), 82. *Prof Exp:* Advan chemist, B F Goodrich Co, 82-85, sr chemist, geon vinyl, 85-86, sr chemist, Div Elastomers & Latex, 87-89, RES & DEVELOP ASSOC, SPECIALTY POLYMERS & CHEMICALS DIV, B F GOODRICH CO, 89- *Mem:* Am Chem Soc; Am Phys Soc; Am Inst Chem Engr; Soc Plastic Engr. *Res:* Research on polymeric materials, polymerization process and products development, structure property relationship; synthetic latexex development. *Mailing Add:* 401 Hurst Dr Bay Village OH 44140

YANG, RALPH TZU-BOW, b Chungking, China, Sept 18, 42; US citizen; m 72; c 2. CHEMICAL ENGINEERING, PHYSICAL CHEMISTRY. *Educ:* Nat Taiwan Univ, BS, 64; Yale Univ, MS, 68, PhD(chem eng), 71. *Prof Exp:* Res assoc, NY Univ, 71-72 & Argonne Nat Lab, 72-73; scientist, Alcoa Lab, Aluminum Co Am, 73-74; group leader, Brookhaven Nat Lab, 74-78; assoc prof, 78-81, PROF CHEM ENG, STATE UNIV NY, BUFFALO, 82-, DEPT CHMN, 90- *Concurrent Pos:* Mem adv bd, Am Carbon Soc, 85-; program dir, NSF, 87-88; mem, adv bd, Ind Eng Chem Res, 91-93, Adsorption Sci Tech, 90- *Mem:* Am Inst Chem Engrs; Am Chem Soc; Am Carbon Soc. *Res:* Heterogeneous kinetics; surface chemistry; diffusion; carbon; gas separation; adsorption processes. *Mailing Add:* Dept Chem Eng State Univ NY Buffalo Amherst NY 14260

YANG, SEN-LIAN, b Taipei, Taiwan, Jan 10, 38; Chinese citizen; m 66; c 2. OBSTETRICS & GYNECOLOGY. *Educ:* Nat Taiwan Univ, MD, 63. *Prof Exp:* Resident obstet & gynec, Nat Taiwan Univ Hosp, 64-68; res fel immunol, US Naval Res Unit 2, 68-70; Ford Found fel reproductive biol & immunobiol, Dept Obstet & Gynec, Univ Pa, 70-71; resident obstet & gynec, Chicago Lying-in Hosp, Univ Chicago, 72-73; from instr to asst prof, 73-80, ASSOC PROF OBSTET & GYNEC, UNIV CHICAGO, 80- *Concurrent Pos:* Vis staff obstet & gynec, Nat Taiwan Univ Hosp, 68-70; fel reproductive biol & immunobiol, Chicago Lying-in Hosp, Univ Chicago, 73-74. *Mem:* Int Fedn Gynec & Obstet; Am Fertil Soc; Soc Study Reproduction; Am Col Obstetricians & Gynecologists; Asn Obstet & Gynec Repub China. *Res:* Immunobiology of reproductive medicine. *Mailing Add:* Chicago Lying-In Hosp-OG 5841 S Maryland Ave Chicago IL 60637

YANG, SHANG FA, b Tainan, Taiwan, Nov 10, 32; m 65; c 2. PLANT PHYSIOLOGY. *Educ:* Nat Taiwan Univ, BS, 56, MS, 58; Utah State Univ, PhD(plant biochem), 62. *Prof Exp:* Res fel, Univ Calif, Davis, 62-63; res assoc biochem, NY Univ Med Ctr, 63-64 & Univ Calif, San Diego, 64-66; from asst biochemist to assoc biochemist, 66-74, PROF VEG CROPS & BIOCHEMIST, UNIV CALIF, DAVIS, 74- *Concurrent Pos:* NSF res grant, 67-; vis prof, Univ Konstanz, 74, Nat Taiwan Univ, 83, Univ Cambridge, 83; NIH res grant, 75- *Honors & Awards:* Campbell Award, Am Inst Biol Sci, 69; Guggenheim Fel, 82; Res Award, Int Plant Growth Substances Asn, 85. *Mem:* Nat Acad Sci; AAAS; Am Soc Biol Chemists; Am Soc Plant Physiol; Am Soc Hort Sci. *Res:* Biosynthesis and hormonal action of ethylene; postharvest biochemistry of fruits and vegetables; biochemical effects of sulfur dioxide on vegetation. *Mailing Add:* Dept Veg Crops Univ Calif Davis CA 95616

YANG, SHEN KWEI, b Chung-King, China, May 4, 41; US citizen; m 68; c 1. PHYSICAL & PHARMACEUTICAL CHEMISTRY, BIOCHEMISTRY. *Educ:* Nat Taiwan Univ, BS, 64; Wesleyan Univ, MA, 69; Yale Univ, MPh, 70, PhD(biophys chem), 72. *Prof Exp:* Fel biochem, Yale Univ, 72-73; res fel chem, Calif Inst Technol, 73-75; sr staff fel, Nat Cancer Inst, NIH, 75-77; assoc prof, 77-81, PROF PHARMACOL, SCH MED, UNIFORMED SERV UNIV HEALTH SCI, 81- *Concurrent Pos:* USPHS traineeship, Yale Univ, 72. *Honors & Awards:* Sci Achievement Award, Chinese Med & Health Asn, 78. *Mem:* Am Soc Pharmacol & Exp Therapeut; Am Asn Cancer Res; Am Soc Biol Chemists; Int Soc Study Xenobiotics. *Res:* Environmental toxicology, drug metabolism and chemical carcinogenesis. *Mailing Add:* Dept Pharmacol Uniformed Serv Univ Health 4301 Jones Bridge Rd Bethesda MD 20814

YANG, SHIANG-PING, b Hankow, China, Mar 5, 19; nat US; m 60. NUTRITION. *Educ:* Nat Cent Univ, China, BS, 42; Iowa State Univ, MS, 49, PhD(nutrit), 56. *Prof Exp:* Animal husbandman, Nan-An Dairy Farms, China, 42-45; tech trainee, USDA, 45-46; sr animal husbandman, Chinese Nat Relief & Rehab Admin, 46-47; res assoc nutrit, Iowa State Univ, 49-56; chemist, Mead Johnson Res Ctr, 56-57; asst prof food & nutrit, Purdue Univ, 57-62; assoc prof, Va Polytech Inst, 62; prof, La State Univ, Baton Rouge, 63-69; chmn dept, 69-76, PROF FOOD & NUTRIT, TEX TECH UNIV, 69- *Concurrent Pos:* Fulbright lectr & vis prof, Nat Taiwan Univ, 59-60. *Mem:* Int Asn Bioinorg Scientists; Soc Environ Geochem & Health; Am Inst Nutrit. *Res:* Diets and carcinogenesis; protein metabolism; nutritional improvement of dietary proteins; factors influencing the qualities of food. *Mailing Add:* Dept Food & Nutrit Tex Tech Univ PO Box 4170 Lubbock TX 79409

YANG, SHI-TIEN, nuclear reactor physics, nuclear reactor safety, for more information see previous edition

YANG, SHUNG-JUN, b Tientsin, China, Jan 13, 34; US citizen; m 64; c 3. CELL BIOLOGY, RADIATION BIOLOGY. *Educ:* Taiwan Univ, BS, 55; Univ Toronto, MS, 58; NC State Col, PhD(genetics), 62. *Prof Exp:* Fel cytogenetics, Baylor Col Med, 62-63, res instr, 63-64; res assoc radiation biol, Sch Med, Stanford Univ, 64-68; res assoc cell biol, Albert Einstein Col Med, 73-74; RADIATION BIOLOGIST, METHODIST HOSP, BROOKLYN, NY, 74- *Concurrent Pos:* Clin asst prof radiation biol, State Univ NY Downstate Med Ctr, 75- *Mem:* NY Acad Sci; Tissue Cult Asn; AAAS; Radiation Res Soc; Int Soc Anal Cytol. *Res:* Proliferation kinetics of mammalian cells in culture; cellular effects of anti-tumor agents; immune response in cancer patients; flow cytometry. *Mailing Add:* Dept Radiation Oncol Methodist Hosp 506 Sixth St Brooklyn NY 11215

YANG, SONG-YU, b Wu-Xi, China, Oct 27, 38; m 65; c 2. GENE CLONING & DNA SEQUENCING, ENZYMOLOGY. *Educ:* Peking Med Col, China, MD, 60; City Univ NY, MA, 83, PhD(biochem), 84. *Prof Exp:* Instr biochem, Peking Med Col, 60-75; asst prof cancer immunol, Shanghai Inst Cell Biol, Academia Sinica, 75-81; res asst biochem, City Col, City Univ NY, 81-84, res assoc, 84-88; RES SCIENTIST BIOCHEM & MOLECULAR BIOL, INST BASIC RES DEVELOP DISABILITIES, 88- *Concurrent Pos:* Investr, Am Heart Asn, 91- *Honors & Awards:* L J Curtman Prize, City Col NY, 84. *Mem:* Am Soc Biochem & Molecular Biol; Am Chem Soc; AAAS; Sigma Xi. *Res:* Structure and expression of genes of fatty acid degradation; metabolism of saturated and unsaturated fatty acids; kinetics of coupled enzyme reactions; structure and function of multienzyme complex. *Mailing Add:* Dept Pharm Inst Basic Res Develop Disabilities 1050 Forest Hill Rd Staten Island NY 10314

YANG, TAH TEH, b Shanghai, China, Aug 15, 27; m 63; c 2. FLUID MECHANICS. *Educ:* Shanghai Inst Technol, BS, 48; Okla State Univ, MS, 57; Cornell Univ, PhD(mech eng), 61. *Prof Exp:* Engr, Kaohsiung Harbor Bur, Taiwan, 49-55; proj engr, Wright Aeronaut Div, Curtiss Wright Corp, 60-62; from asst prof to assoc prof fluids, 62-69, PROF FLUIDS, CLEMSON UNIV, 69- *Concurrent Pos:* Prin investr, NSF grant, 63-64; US Army res grant, 63-66, res contract, 70-73; NASA grant, 69-72 & 84-85; consult, Singer Co, 76- & Avco Corp, 77-; res grants, US Air Force, 75-76, Singer Co, 77-79, US Navy, 80-82 & SC Energy Res & Develop Ctr, 85-86, United Technol, 87-89 & Morgantown Energy Technol Ctr, Dept Energy, 89-92. *Mem:* Fel Am Soc Mech Engrs; Am Inst Aeronaut & Astronaut. *Res:* Thermal engineering. *Mailing Add:* Box 1254 Clemson SC 29631

YANG, TA-LUN, b Tien Tsin, China, Sept 20, 37; Taiwan citizen; m 64; c 2. MEASUREMENT & TESTING, REAL-TIME ANALYSIS SYSTEMS. *Educ:* Nat Taiwan Univ, BS, 60; Univ Calif, Berkeley, MS, 63, PhD(appl mech), 67. *Prof Exp:* Mem tech staff, Bellcomm Inc, 68-72; CHIEF ENGR, ENSCO INC, 72- *Mem:* Am Soc Mech Engrs; Inst Elec & Electronics Engrs; Am Railway Eng Asn. *Res:* Development of measurement technology and data processing techniques for dynamic testing and evaluation of rail, highway equipment and guideways. *Mailing Add:* Ensco Inc 5400 Port Royal Rd Springfield VA 22151

YANG, TSANYEN, b Ping-tung, Taiwan, Oct 22, 49; m 80; c 1. MICROBIAL PHYSIOLOGY & ENZYMOLOGY, PROTEIN CHEMISTRY. *Educ:* Tunghai Univ, Taiwan, BS, 70; McNeese State Univ, La, MS, 74; Univ Houston, PhD(microbiol), 78. *Prof Exp:* Res fel biochem, Sch Med, Univ Pa, 78-79; res assoc molecular biol, Sloan-Kettering Inst, NY, 79-81; ASST PROF BIOL & BIOL SCI, BOWLING GREEN STATE UNIV, 81- *Mem:* AAAS; Am Soc Microbiol; NY Acad Sci. *Res:* Bacterial cytochrome oxidases, dehydrogenases and mutagenesis; properties, purification and characterization, and kinetic studies of enzymes; genetic alterations leading to changes in enzymes and proteins proper. *Mailing Add:* 3435 Kenwood Blvd Toledo OH 43606

YANG, TSU-JU (THOMAS), b Fengshang, Taiwan, Repub China, Aug 14, 32; m 61; c 3. PATHOBIOLOGY, IMMUNOBIOLOGY. *Educ:* Nat Taiwan Univ, BVM, 55; Ministry Exam, Taipei, Taiwan, DVM, 59; McGill Univ, PhD(immunol), 71. *Prof Exp:* Assoc mem immunol, Academia Sinica Inst

Zool, 61-64; res assoc cytogenetics, Dept Animal Biol, Univ Pa, 64-66; res fel immunol, Dept Microbiol, Univ Minn, 66-67; demonstr immunol, McGill Univ, 68-71; from asst prof to assoc prof immunol, Univ Tenn Mem Res Ctr, 71-75; assoc prof, 75-78, PROF PATHOBIOL, UNIV CONN, STORRS, 78-, ASST HEAD, 88-, JOINT APPOINTMENT, CELL & MOLECULAR BIOL, 89- *Concurrent Pos:* Vis fel, Walter & Eliza Hall Inst Med Res, Melbourne, Australia, 83; vis prof, Nat Taiwan Univ, Taipei, 90. *Honors & Awards:* Ralston Purina Small Animal Res Award, Am Vet Med Asn, 88. *Mem:* Am Asn Cancer Res; Am Asn Immunologists; Am Soc Microbiol; AAAS; Am Asn Vet Immunologists. *Res:* Spontaneous regression of tumors; canine immunology; bovine immunology; mode of action of membrane reactive agents: antibodies and lectins; monoclonal antibodies. *Mailing Add:* Dept Pathobiol U-89 Univ Conn Storrs CT 06269-3089

YANG, TSUTE, b Yangchow, China, Nov 16, 16; nat US; m 42; c 2. ELECTRICAL ENGINEERING, COMPUTER SCIENCE. *Educ:* Chiao Tung Univ, China, BS, 38; Harvard Univ, MS, 46; Univ Pa, PhD, 60. *Prof Exp:* Assoc prof elec eng, Univ Toledo, 46-53 & Univ Pa 57-60; sr engr, Remington Rand Univac, Pa, 53-57; res engr, Radio Corp Am, NJ, 60-62; prof elec eng, Villanova Univ, 62-80; vis prof, Dept Elec & Comput Eng, Univ Clemson, 80-82; RETIRED. *Concurrent Pos:* Consult, Tracor Corp, Pa, Naval Air Develop Ctr, Pa, NASA, Va & Burroughs Corp, Pa. *Mem:* Am Asn Univ Prof; Inst Elec & Electronics Engrs; Am Soc Eng Educ. *Res:* Network theory; computer architecture and design; communication systems. *Mailing Add:* 116 Cambridge Dr Clemson SC 29631

YANG, WEI-HSUIN, b Apr 1, 36; m 64; c 2. MECHANICS, APPLIED MATHEMATICS. *Educ:* Cheng Kung Univ, Taiwan, BS, 58; Univ Wash, MS, 62; Stanford Univ, PhD(mech), 65. *Prof Exp:* Res engr, Boeing Co, 62; res fel, Calif Inst Technol, 64-65; from asst prof to assoc prof eng mech, 65-77, PROF APPL MECH, UNIV MICH, ANN ARBOR, 77- *Concurrent Pos:* Ed consult, Math Rev, 66-67; consult, Gen Motors Corp, 67- *Res:* Solid mechanics including elasticity, plasticity and viscoelasticity; large deformation and numerical analysis. *Mailing Add:* Dept Appl Mech Eng Univ Mich Ann Arbor MI 48109-2125

YANG, WEITAO, b Chaozhou, Guangdong, Mar 31, 61. ELECTRONIC STRUCTURE THEORY, CHEMICAL DYNAMICS. *Educ:* Beijing Univ, BS, 82; Univ NC, Chapel Hill, PhD(chem), 86. *Prof Exp:* Res assoc, Univ NC, Chapel Hill, 86-87; res assoc, Univ Calif, Berkeley, 88-89; ASST PROF, DUKE UNIV, DURHAM, 89- *Mem:* Am Phys Soc; Am Chem Soc; Sigma Xi. *Res:* Development and application of the density-functional theory of atoms and molecules; ab initio theory for large molecules; structure and reactivity of large molecules; chemical dynamics and molecular spectroscopy. *Mailing Add:* Dept Chem Duke Univ Durham NC 27706

YANG, WEN JEI, b Taiwan, China, Oct 14, 31; m 60; c 3. HEAT TRANSFER, BIOENGINEERING. *Educ:* Nat Taiwan Univ, BS, 54; Univ Mich, MS, 56, PhD(mech eng), 60. *Prof Exp:* Res engr, Sci Lab, Ford Motor Co, 57-58; res engr, Inst Indust Sci, Tokyo Univ, 60-61; lectr mech eng, 61-62, from asst prof to assoc prof, 62-70, PROF MECH ENG, UNIV MICH, ANN ARBOR, 70- *Concurrent Pos:* Consult, Tamano Works, Mitsui Shipbldg Co, Japan, 60-61, Tecumseh Prod Res Lab, 66-67, Atomic Power Div, Westinghouse Elec Corp, 67-68, Borg-Warner Mach, 68-71 & Environ Protection Agency, 70-72; ed-in-chief, Int J Biomed Eng, 71-73; invited vis prof, Inst Space & Aeronaut Sci, Univ Tokyo, 75; guest prof, Inst Tech Chem Eng, Tech Univ, Berlin, 75-76; consult, Ex-Cell-O, 76-80, Energy Develop Assocs, 79-80, Panasonic, 79-, Bendix Corp, 79-80, KMMCO, 81-, Sarns, 82 & Japan Life Int, 84-; fel, Japan Soc Prom Sci, 83; Italian Nat Res Coun fel, 86; hon vis prof, Tsinghua Univ, Beijing, China, 88; Deutsche Forschungsgemeinschaft fel, 88. *Honors & Awards:* Heat Transfer Mem Award, Am Soc Mech Engrs, 84; C Strouhal Award in Fluid Visualization, 89. *Mem:* Japan Soc Flow Visualization; fel Am Soc Mech Engrs; Sigma Xi; AAAS; Nat Soc Prof Engrs; NY Acad Sci; hon mem Chinese Soc Theoret & Appl Mech; corresp mem Ital Acad Med & Surg; assoc mem Am Inst Aeronaut & Astronaut. *Res:* Heat exchangers; heat transfer enhancement; flow visualization; thermograph and digital image processing; thermal physiology; cardiovascular fluid mechanics; hyperthermia; natural convection; applied optics; transport phenomena in materials processing; gas turbine blades cooling. *Mailing Add:* Dept Mech Eng Univ Mich Ann Arbor MI 48109-2125

YANG, WEN-CHING, b Taipei, China, Nov 11, 39; US citizen; m 68; c 2. CHEMICAL ENGINEERING. *Educ:* Nat Taiwan Univ, BS, 62; Univ Calif, Berkeley, MS, 65; Carnegie-Mellon Univ, PhD(chem eng), 68. *Prof Exp:* Sr engr, 68-76, FEL ENGR, WESTINGHOUSE ELEC CORP, PITTSBURGH, 76- *Concurrent Pos:* Lectr, Univ Pittsburgh, 80 & 83; mem, PhD Adv Comt, Lehigh Univ, 82-84; vchmn, Prog Comt, Group 3b, Am Inst Chem Engrs, 83-85 & chmn, 86-87, vchmn, Prog Comt, Group 3c, 91- *Mem:* Am Inst Chem Engrs; Am Chem Soc. *Res:* Mass transfer; thermodynamics; chemical vapor deposition and high temperature technology; fluidized bed technology; fluidized bed coal gasification and combustion; pneumatic transport; wastes disposal. *Mailing Add:* 236 Mt Vernon Ave Export PA 15632

YANG, WEN-KUANG, b Taiwan, China, Oct 19, 36; m 63; c 2. BIOCHEMISTRY, MEDICINE. *Educ:* Nat Taiwan Univ, MD, 62; Tulane Univ, PhD(biochem), 66; Educ Coun Foreign Med Grad, cert, 62. *Prof Exp:* Vis investr enzym, 66-68, staff biochemist, 68-73, GROUP LEADER, BIOL DIV, OAK RIDGE NAT LAB, 73- *Concurrent Pos:* Res fel nutrit & metab, Sch Med, Tulane Univ, 63-66; Damon Runyon Mem res fel cancer, 66-68; lectr biomed sci, 69-74, adj prof biomed sci, Oak Ridge Biomed Grad Sch, Univ Tenn, 74-; ad hoc mem, Review Comt & Study Sect, NIH, 77-; mem, NIH Exp Virol Study Sect, 82-86, Nat Cancer Inst, Can Biol Immunol Contract Review Comt, 88-91. *Mem:* Am Soc Biochem & Molecular Biol; Am Asn Cancer Res; Environ Mutagen Soc; Formosan Med Asn; Sigma Xi. *Res:* Chromosomal transposable genes and environmental carcinogenesis; transgenic mouse and retroviral vecters; genetic expression in cancer tissues; host cell-RNA oncogenic virus interaction; enzymology of DNA synthesis in mammalian normal and tumor cells; isoaccepting transfer RNA of mammalian tissues; molecular biology; genetics. *Mailing Add:* Biol Div Oak Ridge Nat Lab PO Box 2009 Oak Ridge TN 37831-8077

YANG, WILLIAM C T, ANTI-MICROBIAL DRUGS, ANTI-ARRHYTHMIC DRUGS. *Educ:* Univ Southern Calif, PhD(zool), 56. *Prof Exp:* EMER PROF PHARMACOL, SCH MED, UNIV SOUTHERN CALIF, 89- *Mailing Add:* Dept Pharmacol & Nutrit Univ Southern Calif 2025 Zonal Ave Los Angeles CA 90033

YANG, XIAOWEI, b Shanghai, China, July 19, 54; m 83; c 2. BIOMEDICAL SIGNAL PROCESSING, SPEECH RECOGNITION & DATA COMPRESSION. *Educ:* Wuhan Univ, China, BS, 78; EChina Normal Univ, MS, 82; Univ Md, MS, 88, PhD(elec eng), 89. *Prof Exp:* Asst researcher electrophysics, Wuhan Univ, 78-79; lectr electronics, EChina Normal Univ, 82-83; teaching asst elec eng, 84-86, res asst, 86-89, POSTDOCTORAL RES ASSOC BIOMED ENG, UNIV MD, 89-; PRES BIOMED ENG, MULTICHANNEL CONCEPTS, INC, 91- *Concurrent Pos:* Vis engr biomed eng, Micro Probe, Inc, 89-90; prin investr, Nat Inst Neurol Disorders & Stroke, NIH, 91- *Mem:* Inst Elec & Electronice Engrs; Inst Elec & Electronics Engrs Biomed Eng Soc; Inst Elec & Electronics Engrs Signal Processing Soc; Inst Elec & Electronics Engrs Pattern Analysis & Mach Intel; Inst Elec & Electronics Engrs Info Theory Soc; Inst Elec & Electronics Engrs Commun Soc. *Res:* Auditory representation of acoustic signals; speech recognition; speech data compression; neural signal discrimination; synaptic connectivity identification in neural networks; mathematical modeling of biological systems; biomedical instrumentation. *Mailing Add:* 13904 Grey Colt Dr Gaithersburg MD 20878

YANICK, NICHOLAS SAMUEL, b Oakburn, Man, Dec 4, 07; nat US; m 53. PHYSICAL CHEMISTRY. *Educ:* Univ Man, BSc, 30, MSc, 32; NY Univ, PhD(chem), 35. *Prof Exp:* Chemist, Olin-Mathieson Corp, Va, 35-36; res chemist, US Gypsum Co, Ill, 36-38; chief chemist, Wahl-Henius Inst, Ill, 38-45; scientist, Metall Labs, Univ Chicago, 45; chief chemist, Chapman & Smith Co, 45-48; res chemist, John F Jelke Co, 49-51; sr scientist, Res & Develop Div, Kraft, Inc, 51-70; TECH CONSULT, 70- *Mem:* Am Chem Soc; Inst Food Technologists. *Res:* Colloid-chemical aspects of food products; solubilities; physico-chemical properties of organic compounds; winterization of vegetable oils, separation of fatty acids and fats; evaluation of edible proteins; vanilla; evaluation and formulation of retail and institutional, nutritive and dietetic food products; food ingredients application; food technology; development of industrial food products. *Mailing Add:* 1643 David Dr Escondido CA 92026

YANIV, SHLOMO STEFAN, b Poznan, Poland, Sept 11, 31; US citizen; div; c 2. HEALTH PHYSICS, RADIOLOGICAL RISK ASSESSMENT. *Educ:* Israel Inst Technol, BS, 54, Ingenieur, 55; Univ Pittsburgh, MS, 65, DSc(radiation health), 69; Am Bd Health Physics, Cert, 66, recert, 81, 85 & 89. *Prof Exp:* Radiation protection engr, Israel AEC, 58-62; univ health physicist, Grad Sch Pub Health, Univ Pittsburgh, 63-67, asst prof health physics & asst prof radiol, Sch Med, 69-72; sr health physicist, Prod Standards Br, Directorate Regulatory Standards, US Atomic Energy Comn, 72-75, tech asst to dir, Div Safeguards Fuel Cycle & Environ Res, 75-79, SR HEALTH PHYSICIST/RADIATION PHYSICS & HEALTH EFFECTS SPECIALIST, RADIATION PROTECTION & HEALTH EFFECTS BR, OFF NUCLEAR REGULATORY COMN, WASH, DC, 79- *Concurrent Pos:* Am Cancer Soc grant, Univ Pittsburgh, 71-72; radiol physicist, Montefiore Hosp, 67-72; consult, radiol & health physics, several hosps, insts, & com co, Pittsburgh, 64-72 & Interagency Sci Review Group, 79-80; mem subcomt, Nuclear Measurements Standards, Am Nat Standards Inst, 76-82, sci work group, Interagency Task Force Health Effects Ionizing Radiation, 78-79, adv group, Harvard Univ Sch Pub Health, sci subpanel Inoizing Radiation Risk Assessment, 82-, Comt Interagency Radiation Res & Policy Coord, 87-; adv, World Health Orgn, 81. *Mem:* Health Physics Soc; Am Asn Physicists in Med. *Res:* Radiological risk assessment; environmental aspects of radionuclides use; radiation dosimetry; health effects of population exposure to ionizing radiation. *Mailing Add:* 18 Cedarwood Ct Rockville MD 20852

YANIV, SIMONE LILIANE, b France, May 17, 38; US citizen; c 2. BUILDING ACOUSTICS, PSYCHOACOUSTICS. *Educ:* Univ Pittsburgh, BS, 66, MS, 68, PhD(noise control psychoacoust), 72. *Prof Exp:* Noise pollution, load cell metrology & force measurements consult, Allegheny County Health Dept, 72-73; bioacoust scientist effects noise on people, US Environ Protection Agency, Noise Abatement & Control Off, 73-74; GROUP LEADER FORCE GROUP, NAT INST STANDARDS & TECHNOL, 86- *Mem:* Am Soc Testing & Mat. *Res:* Force metrology; sound absorption, propagation and isolation; acoustic measurements. *Mailing Add:* Bldg 202 Rm 220 Nat Inst Standards & Technol Washington DC 20234

YANKAUER, ALFRED, b New York, NY, Oct 12, 13; m 48; c 2. PUBLIC HEALTH. *Educ:* Dartmouth Col, BA, 34; Harvard Univ, MD, 38; Columbia Univ, MPH, 47. *Prof Exp:* Dist health officer, New York City Dept Health, 48-50; asst prof prev med & pub health, Med Col, Cornell Univ, 48-50 & Sch Med, Univ Rochester, 50-52; dir, Bur Maternal & Child Health, NY State Dept Health, 52-61; regional adv maternal & child health, Pan-Am Health Orgn, WHO, 61-66; sr lectr maternal & child health, Sch Pub Health, Harvard Univ, 66-70, sr lectr health serv admin, 70-73; PROF COMMUNITY MED & PEDIAT, MED SCH, UNIV MASS, WORCESTER, 73- *Concurrent Pos:* Dir, Maternal & Child Health Serv, Health Bur, Rochester, NY, 50-52; lectr, Albany Med Col, 52-61; WHO vis prof, Madras Med Col, India, 57-59; ed, Am J Pub Health, 75-90. *Honors & Awards:* Job Lewis Smith Award, Am Acad Pediat, 79; Excellence Award, Am PUb Health Asn, 90. *Mem:* Fel Am Pub Health Asn; fel Am Acad Pediat. *Res:* Maternal and child health; school health; social medicine; health care. *Mailing Add:* Dept Community & Family Med Univ Mass Med Sch Worcester MA 01655

YANKEE, ERNEST WARREN, b Hayward, Calif, Nov 18, 43; m 65; c 2. ORGANIC CHEMISTRY, MEDICINAL CHEMISTRY. *Educ:* La Sierra Col, BA, 65; Univ Calif, Los Angeles, PhD(org chem), 70. *Prof Exp:* RES ASSOC ORG CHEM, UPJOHN CO, 70- *Mem:* Am Chem Soc. *Res:* Synthesis and structure-activity relationships of prostaglandins. *Mailing Add:* Upjohn Co Kalamazoo MI 49001-0199

YANKEE, RONALD AUGUST, b Franklin, Mass, May 24, 34. MEDICINE. *Educ:* Tufts Univ, BS, 56; Yale Univ, MD, 60. *Prof Exp:* Intern med, Univ Va, 60-61; resident, Univ Mich, 62-63; sr investr, Nat Cancer Inst, 63-75; assoc prof, Sidney Farber Cancer Inst, Harvard Med Sch, 73-79; PROF, BROWN UNIV, 79-; DIR, RHODE ISLAND BLOOD CTR, 79- *Mem:* Transplantation Soc; Am Soc Hemat. *Res:* Bone marrow transplantation; histocompatibility; platelet transfusion therapy; cancer chemotherapy. *Mailing Add:* Dept Med Brown Univ Brown Sta Providence RI 02912

YANKELL, SAMUEL L, b Bridgeton, NJ, July 4, 35; m 58; c 3. DENTISTRY, TOXICOLOGY. *Educ:* Ursinus Col, BS, 56; Rutgers Univ, MS, 57, PhD, 60; Univ Pa, Sch Dent Med, RDH, 81. *Prof Exp:* Instr, Georgian Court Col, 60; sr res biochemist, Colgate-Palmolive Co, 60-63; head dept biochem & pharmacol, Smith, Miller & Patch, 63-65; sr pharmacologist, Menley & James Res Labs, 66-67; sect head biol labs, Smith Kline & French Inter-Am Corp, 67-69; head dept biol sci, Menley & James Labs, 69-74; RES PROF, SCH DENT MED, UNIV PA, 74- *Mem:* Am Chem Soc; NY Acad Sci; Int Asn Dent Res; Am Soc Pharmacol & Exp Therapeut; Sigma Xi. *Res:* Nutrition; dermatology; dental research. *Mailing Add:* Univ Pa Sch Dent Med 4001 Spruce St Philadelphia PA 19104

YANKO, WILLIAM HARRY, b Monessen, Pa, Jan 6, 19; m 42; c 2. RADIOCHEMISTRY, ORGANIC CHEMISTRY. *Educ:* Geneva Col, BS, 40; Pa State Univ, PhD(org chem), 44. *Prof Exp:* Org res chemist, Cent Res Lab, Monsanto Chem Co, 43-46; org res chemist, Clinton Labs, Tenn, 46-47; radiochem res group leader, Cent Res Lab, Monsanto Chem Co, 47-60, group leader, 61-78, sr chemist, 78-82; RETIRED. *Mem:* Am Chem Soc; AAAS. *Res:* Synthetic antimalarials; anticancer drugs; radioisotopic synthesis. *Mailing Add:* 5612 Royalwood Dr Centerville OH 45429

YANKWICH, PETER EWALD, b Los Angeles, Calif, Oct 20, 23; m 45; c 3. PHYSICAL CHEMISTRY, SCIENCE EDUCATION. *Educ:* Univ Calif, BS, 43, PhD(chem), 45. *Prof Exp:* Res chemist, Radiation Lab, Univ Calif, 44-46, instr chem, 47-48; from asst prof to assoc prof, Univ Ill, Urbana, 48-57, prof chem, 57-88; EXEC OFFICER, DIRECTORATE EDUC & HUMAN RESOURCES, NSF, 85- *Concurrent Pos:* NSF fel, Calif Inst Technol & Brookhaven Nat Lab, 60-61; vpres acad affairs, 77-82. *Mem:* Fel AAAS; Sigma Xi; Am Chem Soc; fel Am Phys Soc. *Res:* Chemical kinetics; isotope effects. *Mailing Add:* EHR Rm 516 NSF Washington DC 20550

YANNAS, IOANNIS VASSILIOS, b Athens, Greece, Apr 14, 35; div; c 2. BIOMATERIALS, TISSUE REGENERATION. *Educ:* Harvard Col, AB, 57; Mass Inst Technol, MS, 59; Princeton Univ, MA, 65, PhD(phys chem), 66. *Prof Exp:* Res phys chemist, W R Grace & Co, 59-63; asst prof mech eng, 66-68, Du Pont asst prof, 68-69, assoc prof, 69-78, PROF POLYMER SCI & ENG, MASS INST TECHNOL, 78- *Concurrent Pos:* Polymer consult. *Honors & Awards:* Fred O Conley Award, Soc Plastics Engrs, 82; Founders Award, Soc Biomat, 82; Doolittle Award, Am Chem Soc, 88. *Mem:* Inst Med-Nat Acad Sci; Am Chem Soc; NY Acad Sci; fel Am Inst Chemists; Am Soc Cell Biol; Soc Neurosci. *Res:* Natural and synthetic polymers; analogs of extracellular matrix have been synthesized and used to induce regeneration of skin (dermis and epidermis) in humans and rodents; also peripheral nerve has been regenerated in rodents. *Mailing Add:* Dept Mech Eng Rm 3-334 Mass Inst Technol Cambridge MA 02139

YANNITELL, DANIEL W, b Johnson City, NY, Sept 26, 41; m 70; c 2. MECHANICS, APPLIED MATHEMATICS. *Educ:* Webb Inst Technol, BS, 62; Cornell Univ, PhD(mech), 67. *Prof Exp:* Res assoc & instr mech, Cornell Univ, 67; asst prof eng mech, 67-72, assoc prof eng sci, 72-77, ASSOC PROF MECH ENG, LA STATE UNIV, BATON ROUGE, 77- *Concurrent Pos:* Rocket Propulsion Lab, USAF, 81-82 & 86. *Mem:* Am Acad Mech. *Res:* Fluid mechanics; analytical and computational fluid dynamics. *Mailing Add:* Dept Mech Eng 2505 CEBA Bldg La State Univ Baton Rouge LA 70803

YANNONI, COSTANTINO SHELDON, b Boston, Mass, May 20, 35; div; c 5. STRUCTURAL CHEMISTRY. *Educ:* Harvard Col, AB, 57; Columbia Univ, MA, 60, PhD(chem), 67. *Prof Exp:* Res chemist, Union Carbide Res Inst, 66-67; res staff mem, Watson Res Ctr, 67-71, RES STAFF MEM, RES LAB, IBM CORP, 71- *Concurrent Pos:* Chmn, Exp Nuclear Magnetic Resonance Conf, 81-82 & Gordon Conf on Magnetic Resonance, 87. *Mem:* Am Chem Soc. *Res:* Structure and dynamics of reactive intermediates; structure of molecules at interfaces. *Mailing Add:* IBM Almaden Res Ctr Res Div IBM 650 Harry Rd San Jose CA 95120-6099

YANNONI, NICHOLAS, b Boston, Mass, Aug 3, 27; m 55; c 4. PHYSICAL CHEMISTRY. *Educ:* Boston Univ, BA, 54, PhD(chem), 61; Boston Col, MBA, 80. *Prof Exp:* Res fel chem, Mellon Inst, 54-55; staff scientist, Device Develop Corp, 60-61; physicist, 61-64, chief energetics br, 64-73, chief, opto-electronic physics br, Air Force Cambridge Res Labs, 74-79, CHIEF SIGNAL PROCESSING & TIMING DEVICES, ROME AIR DEVELOP CTR, 80- *Honors & Awards:* Am Inst Chemists Medal, 54. *Mem:* Sigma Xi; Am Crystallog Asn; Am Chem Soc; Am Phys Soc. *Res:* Crystal structure analysis; optics; atomic frequency/time standards; quartz oscillators. *Mailing Add:* 31 Lafayette Rd Newton Lower Falls MA 02162

YANO, FLEUR BELLE, US citizen. THEORETICAL PHYSICS. *Educ:* Columbia Univ, BS, 54; Univ Southern Calif, MA, 58; Univ Rochester, PhD(physics), 66. *Prof Exp:* From asst prof to assoc prof, 64-73, assoc dean instruction, Sch Letters & Sci, 78-79, resident dir int prog Sweden & Denmark, 84-85, PROF PHYSICS, CALIF STATE UNIV, LOS ANGELES, 73- *Concurrent Pos:* Res Corp grant, Calif State Univ, Los Angeles & State Univ Groningen, 72-73, Calif Inst Technol, 77-78 & Uppsala Univ, 84-85. *Mem:* Am Phys Soc. *Res:* Theoretical nuclear physics; intermediato energy physics; radiative muon capture by complex nuclei; pion correlations (pion interferometry Brown-Hanbury-Twiss Effect) in heavy ion collision. *Mailing Add:* Dept Physics Calif State Univ Los Angeles CA 90032

YANOFSKY, CHARLES, b New York, NY, Apr 17, 25; m 49; c 3. MOLECULAR BIOLOGY. *Educ:* City Col New York, BS, 48; Yale Univ, MS, 50, PhD(microbiol), 51. *Hon Degrees:* DSc, Univ Chicago, 80 & Yale Univ, 81. *Prof Exp:* Res asst microbiol, Yale Univ, 51-54; asst prof, Sch Med, Western Reserve Univ, 54-58; assoc prof biol, 58-61, PROF BIOL, STANFORD UNIV, 61- *Concurrent Pos:* Lederle med fac award, 55-57; Am Heart Asn career investr, 69- *Honors & Awards:* Eli Lilly Award, 59; US Steel Found Award, 64; Howard Taylor Ricketts Award, 66; Lasker Med Res Award, 71; Waksman Award, Nat Acad Sci, 72; Louisa Gross Horwitz Prize, 76; Genetics Soc Am Medal, 83; Gairdner Found Int Award, 85. *Mem:* Nat Acad Sci; AAAS; Am Acad Arts & Sci; Am Soc Microbiol; Genetics Soc Am; foreign mem Royal Soc; Leopold Ger Acad Naturalists. *Mailing Add:* Dept Biol Sci Stanford Univ Stanford CA 94305-5020

YANOW, GILBERT, b Los Angeles, Calif, Oct 15, 35; m 63; c 3. PHYSICS, SOLAR ENERGY. *Educ:* Univ Calif, Los Angeles, BA, 59; Univ Queensland, MS, 65; Australian Nat Univ, PhD(physics), 71. *Prof Exp:* Res specialist space sci, Douglas Aircraft Co, 56-63; lab instr physics, Univ Queensland, 64-65; sr lab instr, Australian Nat Univ, 65-71; sr engr, Martin Marietta, 71-72; res specialist, McDonnell Douglas Aircraft Co, 72-74; staff mem physics, HEAD EDUC OUTREACH, JET PROPULSION LAB, CALIF INST TECHNOL, 74- *Mem:* Am Radio Relay League; Sigma Xi. *Res:* Solar energy utilization; high speed gas dynamics; teaching methods; science education. *Mailing Add:* Wagontrail Rd Diamond Bar CA 91765

YANOWITCH, MICHAEL, b Minsk, Russia, Feb 1, 23; nat US; wid; c 2. APPLIED MATHEMATICS. *Educ:* Cooper Union, BEE, 43; NY Univ, MS, 50, PhD(math), 53. *Prof Exp:* Elec engr, Philco Corp, 43-46; instr elec eng, Polytech Inst Brooklyn, 48-49; asst, Inst Math Sci, NY Univ, 50-52, assoc res scientist, 57-58; sr mathematician, Reeves Instrument Corp, 52-57; assoc prof math, 58-62, PROF MATH, ADELPHI UNIV, 62- *Concurrent Pos:* Consult, Surv Bur Corp, 59-60 & Grumman Aircraft Eng Corp, 61-63; vis scientist, Nat Ctr Atmospheric Res, 65-66. *Mem:* AAAS; Am Math Soc; Soc Indust & Appl Math; Math Asn Am. *Res:* Asymptotics; wave motion; atmospheric waves. *Mailing Add:* Dept Math & Comput Sci Adelphi Univ Garden City NY 11530

YANTIS, PHILLIP ALEXANDER, b Portland, Ore, Mar 30, 28; m 54; c 3. AUDIOLOGY. *Educ:* Univ Wash, BA, 50; Univ Mich, MA, 52, PhD(audiol), 55. *Prof Exp:* Res assoc physiol acoust, Univ Mich, 55-58, instr audiol, 57-60; from asst prof to assoc prof, Case Western Reserve Univ, 60-65; prog dir speech path & audiol, 68-74, PROF AUDIOL, UNIV WASH, 65- *Concurrent Pos:* Dir, Dept Audiol, Cleveland Hearing & Speech Ctr, 60-65; mem rev panel, Neurol & Sensory Dis Proj, USPHS, 67-69; field reader, Bur Educ Handicapped, HEW, 72-73, mem rev panel, Speech & Hearing Training, Rehab Serv Admin, 77-78; chmn, Coun Prof Standards Speech-Lang Path & Audiol, 82-84. *Mem:* Acad Rehab Audiol; Acoust Soc Am; hon fel, Am Speech, Lang & Hearing Asn (vpres, 69-71, pres, 75); Am Auditory Soc. *Res:* Detection of aural harmonics; bone conduction audiometry; speech audiometry; middle ear immittance. *Mailing Add:* Dept Speech & Hearing Sci JG-15 Univ Wash Seattle WA 98195

YANTIS, RICHARD P, b Westerville, Ohio, July 1, 32; m 59; c 2. MATHEMATICS, OPERATIONS RESEARCH. *Educ:* US Naval Acad, BS, 54; Univ NC, MA, 62; Ohio State Univ, PhD(indust eng), 66. *Prof Exp:* USAF, 54-74, intel officer, 54-56, instr navig, 56-60, from instr to assoc prof math, USAF Acad, 62-70, assoc prof opers res, USAF Inst Technol, 71-74; teacher, Columbus Acad, 75-76; PROF MATH, OTTERBEIN COL, 76- *Concurrent Pos:* Proj leader underground coal mining proj, Battelle Mem Inst, 74-75. *Mem:* Sigma Xi; Nat Coun Teachers Math. *Res:* Linear programming; integer linear programming. *Mailing Add:* 265 Storington Rd Westerville OH 43081

YANUSHKA, ARTHUR, b NY, Dec 2, 48. MATHEMATICS. *Educ:* Fordham Univ, BA, 70; Univ Ill, MS & PhD(math), 74. *Prof Exp:* Asst prof math, Univ Mich, 74-76 & Kans State Univ, 76-77; asst prof, 77-80, ASSOC PROF MATH & COMPUT SCI, CHRISTIAN BROS COL, 80- *Concurrent Pos:* Prin investr, Math Sci Sect, NSF, 78, 79. *Mem:* Math Asn Am. *Res:* Finite geometry and combinatorics. *Mailing Add:* 242 W Edwin Memphis TN 38104

YAO, ALICE C, b Philippines; US citizen. NEONATAL-PRENATAL MEDICINE, PEDIATRIC CARDIOLOGY. *Educ:* Far Eastern Univ, Manila, AA, 53, MD, 58. *Prof Exp:* Internship & residency pediat, Bellevue New York Univ Med Ctr, 59-61; fel pediat cardiol, NY Heart Asn, State Univ NY Downstate Med Ctr, 61-62, Johns Hopkins Hosp, 62-63; res fel neonatal physiol, Karolinska Hosp & Inst, Sweden, 65-68; from asst prof to assoc prof, 68-77, PROF PEDIAT, STATE UNIV NY, DOWNSTATE MED CTR, BROOKLYN, 77- *Concurrent Pos:* Attend physician, Univ Hosp Brooklyn, Childrens Med Ctr, Brooklyn, 69-; vis pediat, Kings County Hosp, Brooklyn, 69-; vis res scientist, Dept Pediat & Clin Pharmacol, Karolinska Hosp, Stockholm, Sweden, 74-75 & 82-83, Univ Hosp, Trondheim, Norway, 90-; vis res fel, Brown Univ Prog Med, Women & Infant Hosp, RI, 75 & 82. *Mem:* Soc Pediat Res; Am Pediat Soc; Am Physiol Soc; Am Acad Pediat; Sigma Xi; Am Col Cardiol. *Res:* Factors influencing postnatal maturation of the neonatal circulation; placental transfusion; blood volume; hemorrhage; hypocia; special interest on the splanchnic, peripheral circulations in relation to the systemic circulation. *Mailing Add:* State Univ NY Health Sci Ctr 450 Clarkson Ave Brooklyn NY 11203

YAO, DAVID D, b Shanghai, China, July 14, 50; m 79; c 2. QUEUEING NETWORKS, OPTIMIZATION CONTROL. *Educ:* Univ Toronto, MASc, 81, PhD, 83. *Prof Exp:* Assoc prof, Harvard Univ, 86-88; asst prof, 83-86, PROF, COLUMBIA UNIV, 88- *Honors & Awards:* Presidential Young Investr Award, NSF. *Mem:* Inst Elec & Electronics Engrs; Opers Res Soc Am. *Res:* Optimization and control of discrete event stochastic systems focusing on queueing networks and their applications in communication systems and computer integrated manufacturing systems; stochastic optimization techniques in such systems. *Mailing Add:* IEOR Dept Columbia Univ 312 Mudd New York NY 10027-6699

YAO, JAMES T-P, b Shanghai, China, July 7, 33; m 58; c 3. STRUCTURAL ENGINEERING, ENGINEERING MECHANICS. *Educ:* Univ Ill, Urbana, BS, 57, MS, 58, PhD(civil eng), 61. *Prof Exp:* From asst prof to prof civil eng, Univ NMex, Albuquerque, 61-71; postdoctoral preceptor eng mech, Columbia Univ, NY, 64-65; prof civil eng, Purdue Univ, West Lafayette, Ind, 71-88, asst head, 83-88; PROF & HEAD CIVIL ENG, TEX A&M UNIV, 88- *Concurrent Pos:* Asst dean grad sch, Purdue Univ, WLafayette, Ind, 84-87; ed, J Struct Eng, Am Soc Civil Engrs, 90-, vchmn, Struct Div Exec Comt, 90-91, chmn, 91-92. *Honors & Awards:* Am Freudenthal Medal, Am Soc Civil Engrs, 90; Max Planck Res Award, Alexander von Humboldt Found & Max Planck Soc, 90. *Mem:* Am Soc Civil Engrs; Am Concrete Inst; Am Soc Eng Educ; Am Asn Univ Professors; Earthquake Eng Res Inst; NAm Fuzzy Info Processing Soc (pres, 85-88). *Res:* Structural fatigue; earthquake engineering; structural safety, control, and reliability; civil engineering education. *Mailing Add:* Dept Civil Eng Tex A&M Univ College Station TX 77843-3136

YAO, JERRY SHI KUANG, b Peiping, China, Oct 12, 25; m 46; c 2. PHOTOGRAPHIC CHEMISTRY. *Educ:* Peking Univ, BS, 46; Mont State Col, PhD, 60. *Prof Exp:* Fel biochem, Mont State Col, 60-61; asst prof org & gen chem, Wis State Col, Stevens Point, 61-62; assoc prof, Dubuque, 62-63; res chemist, Gaf Corp, 63-66, tech specialist, Photo & Reproduction Div, 66-76, group leader res & develop, 76-82; PRES, EMPIRE STATE SCI, INC, 82- *Mem:* Soc Photog Sci & Eng; Am Chem Soc. *Res:* Heterocyclic chemistry in relation to photography; silver halide emulsions. *Mailing Add:* PO Box 300 Mendon NY 14506

YAO, JOE, b Antung, China, Feb 11, 30; m 68. WOOD SCIENCE, WOOD CHEMISTRY. *Educ:* Chung Hsing Univ, Taiwan, BS, 54; Mont State Univ, MS, 58; NC State Univ, PhD(wood sci, wood technol), 65. *Prof Exp:* Asst prof, 63-70, ASSOC PROF WOOD SCI & TECHNOL, MISS STATE UNIV, 70-; assoc wood technologist, Miss Forest Prod Utilization Lab, 70-79. *Concurrent Pos:* Asst forester, Miss Agr Exp Sta, 63-67; Forest Prod Utilization Lab, 67-70, assoc wood technologist, 70-79; vis prof, Nat Chung Hsing Univ, Taiwan, 73-74. *Mem:* AAAS; Soc Wood Sci & Technol; Forest Prod Res Soc; Am Inst Chemists; NY Acad Sci. *Res:* Wood particleboard properties; wood capillary structure; water diffusion in wood; shrinkage and related properties; low grade hardwood utilization; utilization of recycled wood fiber material. *Mailing Add:* 4115 Baynard Dr Houston TX 77072

YAO, KUAN MU, b Chekiang, China, Sept 25, 23; US citizen; m 57; c 3. CIVIL ENGINEERING. *Educ:* Pei-Yang Univ, BS, 47; Univ NC, MS, 55, PhD(sanit eng), 68. *Prof Exp:* Sr engr, Prov Govt Taiwan, 48-58; teaching fel water resources, Univ NSW, Australia, 59; sr lectr civil eng, Canterbury Univ, NZ, 60-65; res assoc sanit eng, Univ NC, 68; sr specialist, Camp, Dresser & McKee, Boston, 68-71; Sanit engr, 72-78, water qual mgt adv WHO, 79-85; CONSULT, PROV GOVT, TAIWAN, 86- *Concurrent Pos:* Consult water supply eng, Mil Construct Comn, Taiwan, 57-58; prof sanit eng, Univ Eng & Technol, Lahore, Pakistan, 72-75. *Mem:* Water Pollution Control Fedn. *Res:* Fundamental and practical studies of water filtration, tube settling and sewer hydraulics; mathematical modeling and computer application in engineering. *Mailing Add:* THUDB PO Box 81-465 Taipei Taiwan

YAO, KUNG, b Hong Kong, Nov 24, 38; US citizen; m 69. COMMUNICATIONS, SYSTEMS ENGINEERING. *Educ:* Princeton Univ, BSE, 61, MA, 63, PhD(elec eng), 65. *Prof Exp:* Res asst commun eng, Princeton Univ, 63-65; Nat Acad Sci-Nat Res Coun res fel syst eng, Univ Calif, Berkeley, 65-66; asst prof commun eng, 66-72, assoc prof, 72-78, PROF COMMUN ENG, UNIV CALIF, LOS ANGELES, 78- *Concurrent Pos:* Prin investr, NSF initiation grant, 68-71 & Off Naval Res, 72- *Mem:* AAAS; Inst Elec & Electronics Engrs; Am Math Soc. *Res:* Research, teaching and consulting in communications theory and signal processing. *Mailing Add:* Dept Elec Eng Univ Calif 405 Hilgard Ave Los Angeles CA 90024

YAO, MENG-CHAO, b Taipei, Taiwan, Mar 21, 49; US citizen; m 74; c 1. CHROMOSOME STRUCTURE, GENE REARRANGEMENT. *Educ:* Nat Taiwan Univ, BS, 71; Univ Rochester, MS, 74, PhD(biol), 75. *Prof Exp:* Postdoctoral fel cell biol, Yale Univ, 75-78; from asst prof to assoc prof develop biol, Wash Univ, St Louis, 78-86; MEM STAFF GENETICS, FRED HUTCHINSON CANCER RES CTR, 86- *Concurrent Pos:* NIH res career develop award, 84; vis scholar, Univ Wash, 85-86, affil prof, Dept Zool, 88-; mem, Genetics Study Sect, NIH, 87-89. *Mem:* Am Soc Cell Biol; AAAS. *Res:* Various types of DNA rearrangements during development and growth of cells, primarily in the ciliated protozoan tetrahymena; gene amplification; telomere formation; chromosome breakage; DNA deletion; palindrome formation and DNA elimination. *Mailing Add:* 3724 Cascadia Ave S Seattle WA 98144

YAO, NENG-PING, b Shanghai, China, Oct 13, 38; m 67; c 1. ADVANCED BATTERIES & FUEL CELLS, ELECTROCHEMISTRY. *Educ:* Univ Calif, Los Angeles, BS, 63, MS, 65, PhD(chem eng), 69; Univ Chicago, MBA, 79. *Prof Exp:* Res engr, Univ Calif, Los Angeles, 63-69; mem tech staff chem technol, Atomics Int Div, N Am Rockwell Corp, 69-71; sect head electrochem technol, Heliotek Div, Textron Inc, 71-72; mem staff, Chem Eng Div, 72-76, assoc dir energy storage prog, 76-77, dir, off electrochem projs, Argonne Nat Lab, 77-85; PRES, YAO INT INC, 85- *Concurrent Pos:* Adj prof, Purdue Univ, 80-; vis prof, Helsinki Univ Technol, Finland, 86-90. *Mem:* AAAS; Electrochem Soc; Am Chem Soc; Sigma Xi. *Res:* Electrochemical energy conversion systems; batteries and fuel cells; air and water pollution control systems. *Mailing Add:* 350 Meadow Ct Clarendon Hills IL 60514

YAO, SHANG JEONG, b Canton, China, June 6, 34; US citizen; c 1. CHEMICAL PHYSICS, BIOMEDICAL SCIENCES. *Educ:* Univ Ore, MA, 61; Univ Minn, Minneapolis, PhD chem, 66. *Prof Exp:* Asst prof phys sci, Wilbur Wright Col, Chicago City Col, 68-69; assoc surg res, Michael Reese Hosp & Med Ctr, Chicago, 69-71; asst prof neurosurg & chem, 71-79, res assoc prof, 79-87, RES PROF NEUROL SURG & RES CHEMIST, SCH MED, UNIV PITTSBURGH, 87- *Concurrent Pos:* Robert A Welch Found fel theoret chem, Tex A&M Univ, 66-67; fel, Northwestern Univ, Evanston, 67-68; instr, Univ Chicago, 70-71; sr res scientist, Montefiore Hosp, Pittsburgh, Pa, 71-; prin investr, John A Hartford Found grant, 77-80 & NIH, 79-; mem, Spec Rev Surg & Bioeng Study Sect, NIH, 87; distinguished lect & Lingan prof chem, Sun Yatsen Univ, China, 88. *Mem:* Am Phys Soc; Am Soc Artificial Internal Organs; Soc Neurosci; Sigma Xi. *Res:* Irreversible thermodynamics; bioenergetics; quantum theory of enzyme specificity and theory of catalyst facilitated tunneling; implantable energy sources; neuroscience; electrochemical urea removal for hemodialysis; electrochemical sensors. *Mailing Add:* 1695 Hastings Mill Rd Pittsburgh PA 15241

YAO, SHI CHUNE, b Taiwan, Dec 31, 46. HEAT TRANSFER, NUCLEAR ENGINEERING. *Educ:* Nat Tsing Univ, Taiwan, BS, 68; Univ Calif, Berkeley, MS, 71, PhD(nuclear eng), 74. *Prof Exp:* Engr, Argonne Nat Lab, 74-77; from asst prof to assoc prof, 77-83, PROF MECH ENG, CARNEGIE-MELLON UNIV, 83- *Concurrent Pos:* Consult, Nuclear Ctr & Steam Turbine Div, Westinghouse Elec Co, 78-, Cooling of Electronic Equip, Int Tel & Tel, 85-86 & USX, 87-88. *Mem:* Am Soc Mech Engrs; Combustion Inst; Am Inst Chem Engrs. *Res:* Two phase flow and heat transfer; droplet flow; particle flow; cooling of electronic equipment; nuclear reactor thermal hydraulics; rod bundles; spray combustion; continuous castings. *Mailing Add:* Dept Mech Eng Carnegie Mellon Univ Pittsburgh PA 15213

YAO, YORK-PENG EDWARD, b Canton, China, Sept 11, 37; m 65; c 4. THEORETICAL HIGH ENERGY PHYSICS. *Educ:* Univ Calif, Berkeley, BS, 60; Harvard Univ, MA, 63, PhD(physics), 64. *Prof Exp:* Assoc mem natural sci, Inst Advan Study, 64-66; asst prof physics, 66-72, assoc prof, 72-78, PROF PHYSICS, UNIV MICH, ANN ARBOR, 78- *Mem:* Am Phys Soc. *Res:* Quantum field theory; elementary particle physics. *Mailing Add:* Dept Physics Univ Mich Ann Arbor MI 48109

YAP, FUNG YEN, b Jamaica, WI, Oct 12, 33. ATMOSPHERIC SCIENCE, MARINE SCIENCE. *Educ:* Brandeis Univ, BA, 58; Johns Hopkins Univ, PhD(physics), 67. *Prof Exp:* Asst prof physics, Wilson Col, 66-74, chmn dept, 72-74; sr digital programmer, Comput Sci Technicolor Assocs, 74-76; mem tech staff, 76-77, sr mem, 77-78, tech mgr, 78-80, SR SCIENTIST, COMPUT SCI CORP, 80- *Concurrent Pos:* Proj dir, NSF Award for Purchase of Instnl Sci Equip for Physics Dept, Wilson Col, 68-70. *Mem:* Sigma Xi; Am Phys Soc; Am Asn Physics Teachers. *Res:* Satellite navigation and remote sensing; atmospheric modelling; ozone modelling; solar ultraviolet flux determination by satellites; determination of temperature and humidity for storm prediction from satellite sounding measurements; determination of nuclear decay schemes; environmental pollution; artificial satellite image navigation; satellite remote sensing of ocean color. *Mailing Add:* 11032 Firethorn Dr Cupertino CA 95014

YAP, WILLIAM TAN, b Amoy, China, Aug 10, 34; US citizen; m 69; c 2. PHYSICAL CHEMISTRY. *Educ:* Mass Inst Technol, BS, 56, MS, 58, PhD(phys chem), 64. *Prof Exp:* Res chemist, Res Ctr, Hercules, Inc, 64-69; vis assoc biophys chem, NIH, 69-71; RES CHEMIST, NAT BUR STANDARD, 72- *Res:* Biophysical chemistry; solution properties of proteins and other macromolecules; electroanalytical chemistry. *Mailing Add:* 6204 Mori St McLean VA 22101

YAPEL, ANTHONY FRANCIS, JR, b Soudan, Minn, Aug 14, 37; m 60; c 3. DRUG DELIVERY SYSTEMS, BIOMATERIALS. *Educ:* St John's Univ, Minn, BA, 59; Univ Minn, Minneapolis, PhD(phys chem), 67. *Prof Exp:* Sr res chemist, Minn Mining & Mfg Co, 66-71, res specialist, 71-75, sr res specialist, 75-78, mgr biokinetics res, 78-81, mgr, 81-86, lab mgr, biomat res, 86-88, LAB MGR, DRUG DELIVERY/EXPLORATORY RES, MINN MINING & MFG CO, 88- *Concurrent Pos:* Mem bd dir, Univ Minn Inst Technol Alumni Asn, 75-, vpres, 77-78, pres, 78-79; mem bd dir, St John's Univ Alumni Asn, 78-81, vpres, 80-81. *Mem:* Am Chem Soc; AAAS. *Res:* Fast reaction, temperature-jump relaxation and enzyme kinetics; physical chemistry of membranes; reverse osmosis phenomena; structure-activity correlations on biological systems; column chromatography; drug delivery systems; wound management studies; biomaterials; transdermal drug delivery; pulmonary-aerosol drug delivery. *Mailing Add:* 1935 Hythe St St Paul MN 55113

YAPHE, WILFRED, bacteriology; deceased, see previous edition for last biography

YAQUB, ADIL MOHAMED, b Jordan, Jan 19, 28; nat US; m 51; c 2. ALGEBRA. *Educ:* Univ Calif, AB, 50, MA, 51, PhD(math), 55. *Prof Exp:* Asst, Univ Calif, 50-55; from instr to asst prof math, Purdue Univ, 55-60; assoc prof, 60-67, PROF MATH, UNIV CALIF, SANTA BARBARA, 67- *Mem:* Am Math Soc; Math Asn Am. *Res:* Algebraic structures; ring theory; number theory. *Mailing Add:* Dept Math Univ Calif Santa Barbara CA 93106

YAQUB, JILL COURTANEY DONALDSON SPENCER, b Almondsbury, Eng, Dec 17, 31; m 59. GEOMETRY. *Educ:* Oxford Univ, BA, 53, MA, 57, PhD(math), 60. *Prof Exp:* Asst lectr math, Royal Holloway Col, Univ London, 56-59; instr, Wash Univ, 60-61; asst prof, Tufts Univ, 61-63; from asst to prof math, Ohio State Univ, 75-83; RETIRED. *Concurrent Pos:* Alexander von Humboldt fel, 70-71 & 80-81. *Mem:* Am Math Soc. *Res:* Non-Desarguesian planes, inversive planes and their automorphism groups. *Mailing Add:* 4922 Olentangy Blvd Columbus OH 43214

YARAMANOGLU, MELIH, b Istanbul, Turkey, July 20, 47; m 73. MATHEMATICAL WATERSHED MODELING. *Educ:* Middle East Tech Univ, BS, 71, MS, 73; Univ Md, PhD(agr eng), 78. *Prof Exp:* Res assoc, Dept Agr Eng, Univ Md, 78-80, asst prof hydrol, 80-84; DEPT HEAD & PROJ ENGR, TRANSP SYSTS DIV, ENG & ECON RES INC, 84- *Concurrent Pos:* Consult, Photovoltaic Systs Simulation & Design. *Mem:* Am Soc Agr Eng; Soc Comput Simulation. *Res:* Mathematical modeling of watersheds; hydrology; water quality; structured systems analysis; air traffic control software design. *Mailing Add:* 9200 Edwards Way No 703 Adelphi MD 20783

YARAR, BAKI, b Adana, Turkey, Feb 28, 41; US citizen; m 71; c 2. HYDROMETALLURGY, PLASTICS RECYCLING. *Educ:* Mid E Tech Univ, Ankara, BSc, 65, MSc, 66; Univ London, Eng, PhD(phys chem) & DIC, 69. *Prof Exp:* From instr to assoc prof chem, Mid E Tech Univ, Ankara, 70-79; vis prof mineral technol, Univ BC, Vancouver, 79-80; assoc prof, 80-86, PROF METALL ENG, COLO SCH MINES, 86- *Concurrent Pos:* Consult, nat & int govt & indust, 80-; vis prof, Concepcion Univ & La Serena Univ, Chile, 85-; chmn, MPD Fundamentals Comt, Soc Mining Engrs, 89-90. *Honors & Awards:* Cert of Recognition, Metall Soc, 86. *Mem:* Am Chem Soc; Soc Mining Engrs; Mat Res Soc; Sigma Xi. *Res:* Applied surface chemistry; extractive metallurgy; flotation; floculation and gold processing; author of over 120 technical publications; awarded one patent; high temperature superconductor processing. *Mailing Add:* Dept Metall Eng Colo Sch Mines Golden CO 80401

YARBOROUGH, LYMAN, b Cushing, Okla, Feb 13, 37; m 61; c 1. PHASE EQUILBRIA, NATURAL GAS PROCESSING. *Educ:* Okla State Univ, BS, 59, MS, 61, PhD(chem eng), 64. *Prof Exp:* Staff res engr, Amoco Prod Co, Chicago, 64-76, res supvr, Tulsa, 76-78, plant eng supvr, 78-81, mgr process eng, Chicago, 81-82 & Houston, 82-88; MGR PROCESS DEVELOP, AMOCO CAN PETROL CO, CALGARY, 88- *Mem:* Am Inst Chem Engrs; Am Chem Soc; Soc Petrol Engrs. *Res:* Thermodynamics and phase behavior of fluids; hydrocarbons and mixtures of hydrocarbons; natural gas processing plant design. *Mailing Add:* Amoco Can Petrol Co PO Box 200 Sta M Calgary AB T2P 2H8 Can

YARBOROUGH, WILLIAM WALTER, JR, b Tylertown, Miss, Jan 6, 45; m 68. PLASMA PHYSICS. *Educ:* Univ Chattanooga, AB, 67; Vanderbilt Univ, PhD(physics), 74. *Prof Exp:* From asst to assoc prof, 74-88, PROF PHYSICS, PRESBY COL, 88-, CHMN, DEPT PHYSICS, 79- *Mem:* Am Asn Physics Teachers; Am Inst Physics. *Res:* Low energy theta pinch devices, particularly losses from such devices. *Mailing Add:* Dept Physics Presby Col PO Box 975 Clinton SC 29325

YARBRO, CLAUDE LEE, JR, b Jackson, Tenn, Sept 26, 22; m 51; c 3. ECOLOGY. *Educ:* Lambuth Col, BA, 43; Univ NC, PhD(biochem), 54. *Prof Exp:* Actg prof math & physics, Lambuth Col, 46-47; instr physics, Union Col, Tenn, 48; instr biochem, Vanderbilt Univ, 49-51; asst, Univ NC, 51-54, res assoc, 54-57, instr, 54-60; biologist, Biol Br, Res & Develop Div, US AEC, US Dept Energy, 60-67, biol scientist, Res Contracts Br, Lab & Univ Div, 67-72, biol scientist, Res & Develop Admin, 72-76, life scientist, Oak Ridge Opers, 76-84, SCIENTIST II, OAK RIDGE ASSOC UNIVS, 87- *Concurrent Pos:* Consult, Environ Sci, Waste Mgt, Educ & Training, 84- *Mem:* Ecol Soc Am; AAAS; fel Am Inst Chemists; Sigma Xi; NY Acad Sci; Am Forestry Asn. *Res:* Phospholipid chemistry and metabolism; mechanism of renal calculus; formation and physical chemistry of calcium phosphate; ecological succession on sandstone bluffs. *Mailing Add:* 147 Alger Rd Oak Ridge TN 37830

YARBRO, JOHN WILLIAMSON, b Chattanooga, Tenn, Sept 15, 31; m 79. INTERNAL MEDICINE, MEDICAL ONCOLOGY. *Educ:* Univ Louisville, BA, 52, MD, 56; Univ Minn, PhD(biochem), 65. *Prof Exp:* Asst prof med, Univ Minn, 65-68; assoc prof, Univ Ky, 68-70, Univ Pa, 70-73; dir cancer ctrs prog, Nat Cancer Inst, 73-75; PROF MED, UNIV MO, 75- *Mem:* Am Col Physicians; Am Soc Hemat; Asn Community Cancer Ctrs (pres, 85-86); Am Soc Clin Oncol (secy-treas, 88). *Res:* Efficacy and mechanisms of drugs used in the treatment of malignant diseases. *Mailing Add:* 2604 Luan Ct Columbia MO 65203

YARBROUGH, ARTHUR C, JR, b Mitchell, Ga, June 12, 28. ORGANIC CHEMISTRY, SCIENCE EDUCATION. *Educ:* Ga Southern Col, BS, 49; George Peabody Col, MA, 52. *Prof Exp:* Pub sch instr, Ga, 49-51 & 52-53; asst prof chem, Oxford Col, Emory Univ, 53-55; instr, George Peabody Col, Vanderbilt Univ, 56-57; ASSOC PROF CHEM, TOWSON STATE UNIV, 57- *Concurrent Pos:* Consult var high & jr high schs, Md, 57-, Univ Md, 63 & Baltimore County Fire Dept, 67- *Mem:* Am Chem Soc. *Res:* Reactions of transition elements. *Mailing Add:* Dept Chem Towson State Col Baltimore MD 21204

YARBROUGH, CHARLES GERALD, b Lumberton, NC, Oct 13, 39; m 60; c 2. ZOOLOGY. *Educ:* Wake Forest Univ, BS, 61, MA, 63; Univ Fla, PhD(zool), 70. *Prof Exp:* From instr to asst prof biol, Campbell Col, 64-75, assoc prof, 75-80; prof biol & chmn, 80-85, ASST DEAN & PROF BIOL, DIV SCI & MATH, WINGATE COL, 85- *Concurrent Pos:* Chapman res grant, Am Mus Natural Hist, 64; AEC proj ecol researcher, Battelle Mem Inst, 68; prin investr, NC Bd Sci & Technol, 71-72 & US Dept Interior Endangered Species Prog, 78-82; vis grad prof, Univ Va, 75. *Mem:* Am Ornith Union; Cooper Ornith Soc; Sigma Xi; Ecol Soc Am. *Res:* Influence of physical factors, nutrients and heavy metals on biotic communities; metabolism and temperature regulation in vertebrates; ecological implications of energetics in animals. *Mailing Add:* Div Sci & Math Wingate Col Wingate NC 28174

YARBROUGH, DAVID WYLIE, b Long Beach, Calif, May, 7, 37. THERMODYNAMICS & HEAT TRANSFER, THERMAL INSULATION. *Educ:* Ga Inst Technol, BChE, 60, MS, 61, PhD(chem eng), 66. *Prof Exp:* From asst prof to assoc prof, 68-76, assoc dean, Grad Studies & Res, 76-79, prof chem eng, 76-87, PROF CHEM ENG & CHMN, TENN TECHNOL UNIV, 87- *Concurrent Pos:* Mem res staff, Oak Ridge Nat Lab, 79-80, adj res participant, 80-; gov bd, Int Thermal Conductivity Conf; fel, Int Thermal Conductivity Conf, 87, chmn 19th Conf, 86. *Honors & Awards:* Univ Res Award, 86. *Mem:* Am Inst Chem Engrs; Am Soc Eng Educ; Sigma Xi; Am Soc Testing & Mats; NY Acad Sci. *Res:* Thermodynamics; physical properties; applied mathematics; energy conservation. *Mailing Add:* Tenn Technol Univ Box 5013 Cookeville TN 38505

YARBROUGH, GEORGE GIBBS, b Houston, Tex, Jan 20, 43; m 64; c 2. NEUROPHARMACOLOGY, PHARMACOLOGY. *Educ:* Univ Houston, BS, 68; Vanderbilt Univ, PhD(pharmacol), 72. *Prof Exp:* Fel physiol, Univ Man, 72-73; from lectr to asst prof, Univ Sask, 73-75; res fel, Merck Inst Therapeut Res, 75-80, sr res fel neuropsychopharmacol, 80-83, sr res fel, 83-84; vpres, 84-86, SR VPRES, PHARMACOL DIV, PANLABS, INC, 86- *Concurrent Pos:* Med Res Coun Can scholar, Univ Sask, 74-75. *Mem:* Am Soc Pharmacol & Exp Therapeut; Soc Neurosci; Am Chem Soc. *Res:* Physiology and pharmacology of synaptic transmission in the mammalian central nervous system. *Mailing Add:* 11804 N Creek Pkwy S Bothell WA 98011-8805

YARBROUGH, JAMES DAVID, b Stockton, Mo, Jan 15, 33; m 54; c 3. ANIMAL PHYSIOLOGY, TOXICOLOGY. *Educ:* Huntingdon Col, AB, 60; Univ Ala, MS, 61, PhD(biol), 64. *Prof Exp:* Asst prof biol, Univ Ala, 64-68; from assoc prof to prof zool, Miss State Univ, 68-78, head, Dept Biol Sci, 78-84, distinguished prof biol sci, 78-87; prof & head physiol, Univ Mo-Kansas City 87-89; DEAN, COL ARTS & SCI, UNIV ALA, 89- *Mem:* Am Physiol Soc; Am Soc Pharmacol & Exp Therapeut; Am Soc Cell Biol; Sigma Xi; Am Soc Biochem & Molecular Biol. *Res:* Cytotoxcity of insecticides, mechanisms of vetebrate insecticide resistance, adaptive liver growth as induced by xenobiotics. *Mailing Add:* Col Arts & Sci Univ Ala Tuscaloosa AL 35487-0268

YARBROUGH, KAREN MARGUERITE, b Memphis, Tenn, Mar 4, 38. GENETICS, MICROBIOLOGY. *Educ:* Miss State Univ, BS, 61, MS, 63; NC State Univ, PhD(genetics), 67. *Prof Exp:* Asst prof biol, 67-70, assoc prof microbiol, 70-81, prof & actg vpres, Acad Affairs, 81-82, dir, Inst Genetics, 72-82, vpres Res & Extended Serv, 82-90, PROF BIOL SCI, UNIV SOUTHERN MISS, 76-, VPRES RES & PLANNING, 90- *Concurrent Pos:* Assoc prof biol, Univ Southern Miss, 70-71, asst dean & res coordr, Col Sci & Technol, 76-81. *Mem:* Genetics Soc Am; Am Genetic Asn; Sigma Xi; NY Acad Sci; Am Soc Human Genetics; Am Dermatoglyphic Asn. *Res:* Population genetics; human genetics; dermatoglyphics; zoology. *Mailing Add:* PO Box 5116 Univ Southern Miss Hattiesburg MS 39406

YARBROUGH, LYNN DOUGLAS, b Ft Worth, Tex, July 17, 30; m 61; c 3. COMPUTER SCIENCE. *Educ:* Rice Inst, BA, 53; Univ Ill, MS, 55. *Prof Exp:* Comput prog analyst, MacDonnell Aircraft Corp, 55; from sr engr to prin scientist, Space & Info Systs Div, NAm Aviation Inc, 55-65; asst dir, Comput Ctr, Harvard Univ, 65-67; sr scientist, Arcon Corp, 67-75; prin software engr, 75-86, PRIN OPERS ANALYST, DIGITAL EQUIPMENT CORP, 86- *Mem:* Asn Comput Mach. *Res:* Design and development of computer operating systems, computer languages and computer graphic systems; statistical methods of project management and control. *Mailing Add:* 128 Simonds Rd Lexington MA 02173

YARBROUGH, LYNWOOD R, b Cherokee, Ala, July 26, 40; m 75. BIOCHEMISTRY, BIOPHYSICS. *Educ:* Univ Northwestern Ala, BS, 65; Purdue Univ, PhD(biochem, molecular biol), 71. *Prof Exp:* Res assoc molecular biol, Albert Einstein Col Med, 71-73, res assoc biophys, 73-75; from asst prof to assoc prof, 75-89, PROF BIOCHEM, MED CTR, UNIV KANS, 90- *Concurrent Pos:* Damon Runyon Found res fel, 72-74. *Mem:* Biophys Soc; Am Chem Soc; Sigma Xi; Am Soc Cell Biol; Am Soc Biochem & Molecular Biol; Am Soc Microbiol. *Res:* Structure of tubulin genes and regtlation of tubulin gene expression; assembly of subunit proteins; cardiac stress responses and aging. *Mailing Add:* Dept Biochem & Molecular Biol Univ Kans Med Ctr Kansas City KS 66103

YARCHOAN, ROBERT, b New York, NY, July 21, 50; m 81; c 1. CLINICAL PHARMACOLOGY, RETROVIRUSES. *Educ:* Amherst Col, BA, 71; Univ Pa, MD, 75. *Prof Exp:* Clin assoc, Metab Br, Nat Cancer Inst, 78-81, investr, 81-84, Investr, Clin Oncol Prog, 84-88, sr investr, 88-89, CHIEF, RETROVIRAL DIS SECT, CLIN ONCOL PROG, NAT CANCER INST, NIH, 90- *Concurrent Pos:* Assoc ed, J Immunol, 86-89, AIDS Res & Human Retroviruses, 87-, AIDS, 90- *Honors & Awards:* Inventor's Award, US Dept Com, 87. *Mem:* Am Asn Immunol; Am Fedn Clin Res; AAAS; Sigma Xi; Clin Immunol Soc. *Res:* Immunodeficiency diseases, particularly AIDS; interactions between viruses & the immune system; developing therapics for AIDS. *Mailing Add:* Bldg 10 Rm 13N248 NIH Bethesda MD 20892

YARD, ALLAN STANLEY, b Rocktown, NJ, Nov 18, 27; m 57; c 2. PHARMACOLOGY. *Educ:* Rutgers Univ, BS, 52; Med Col Va, PhD(pharmacol, biochem), 56. *Prof Exp:* Asst prof pharmacol, Rutgers Univ, 55-56; asst prof, Med Col Va, 56-60; sr scientist, Ortho Res Found, 60-63, Ortho res fel, 63-67; group chief, Hoffmann-LaRoche, 67-73, dir acad & govt liaison, 73-75, asst dir, Drug Regulatory Affairs, 75-84; DIR, DRUG REGULATORY AFFAIRS, BERLEX LABS INC, 84- *Concurrent Pos:* Mem co-adj fac, Rutgers Univ, 63- *Mem:* AAAS; Am Soc Pharmacol & Exp Therapeut; Soc Study Reproduction; Am Chem Soc; NY Acad Sci. *Res:* Drug metabolism; effect of drugs on liver metabolism; synthesis of hydrazino compounds; biochemistry and pharmacology of the oviduct, uterus and fertility control agents. *Mailing Add:* Berlex Lab Inc 300 Fairfield Rd Wayne NJ 07470

YARDLEY, DARRELL GENE, b Gorman, Tex, Apr 15, 48; m 72; c 3. POPULATION GENETICS. *Educ:* Univ Tex, Austin, BA, 71, MA, 72; Univ Ga, PhD(zool), 75. *Prof Exp:* Asst prof, 75-80, ASSOC PROF ZOOL, CLEMSON UNIV, 80- *Concurrent Pos:* Vis assoc prof cell biol, Baylor Col Med, Houston, Tex. *Mem:* Genetics Soc Am; Soc Study Evolution. *Res:* Molecular and biochemical evolutionary genetics of fishes. *Mailing Add:* Dept Biol Sci 201 Sikes Hall Clemson Univ Clemson SC 29634

YARDLEY, DONALD H, b Estevan, Sask, Sept 9, 17; m 42. MINING ENGINEERING. *Educ:* Queen's Univ, Ont, BSc, 41 & 46, MSc, 47; Univ Minn, PhD(geol), 51. *Prof Exp:* Asst to chief engr, Hardrock Gold Mines, 41-42; field engr, Andowan Mines, 45-46; party chief, Can Geol Surv, 47-48; asst mining, Univ Minn, St Paul, 47-49, from instr to assoc prof, 49-77; prof civil & mining eng, Univ Minn, Minneapolis, 77-; RETIRED. *Res:* Geochemical exploration; heavy metals; geological engineering; economic geology. *Mailing Add:* 2107 Fairways Lane St Paul MN 55113

YARDLEY, JAMES THOMAS, III, b Taft, Calif, May 15, 42; m 66, 76; c 3. PHYSICAL CHEMISTRY. *Educ:* Rice Univ, BA, 64; Univ Calif, Berkeley, PhD(chem), 67. *Prof Exp:* From asst prof to assoc prof chem, Univ Ill, Urbana, 67-77; sr chem physicist, 77-80, sr res assoc, 80-82, res mgr, 82-85, ASSOC DIR, ALLIED CORP, 85- *Concurrent Pos:* Dreyfus Found teacher-scholar award, 70-75; Alfred P Sloan fel, 72-73. *Mem:* Am Phys Soc; Am Chem Soc. *Res:* Molecular spectroscopy; vibrational energy transfer; molecular lasers; molecular dynamics. *Mailing Add:* Corp Res Ctr Allied Corp PO Box 1021R Morristown NJ 07960

YARDLEY, JOHN FINLEY, b St Louis, Mo, Feb 1, 25; m 46; c 5. AERONAUTICAL ENGINEERING. *Educ:* Iowa State Univ, BS, 44; Wash Univ, MS, 50. *Prof Exp:* Stress analyst aircraft design, McDonnell Aircraft Corp, 46-48, strength group engr, 48-51, proj stress engr, 51-54, chief strength engr, 54-58, proj engr Mercury capsule design, 58-60, base mgr spacecraft flight testing, Cape Canaveral, 60-64, Gemini Prog tech dir, 64-66, tech dir, 66-68, vpres & dep gen mgr, Eastern Div, McDonnell-Douglas Astronaut Co, 68-72, gen mgr, Eastern Div, 73-74; assoc adminr, Manned Space Flight, NASA, 74-76, Space Flight, 76-78, Space Transport Syst, 78-81; pres, McDonnell-Douglas Astronaut, 81-88; RETIRED. *Honors & Awards:* Inst Aerospace Sci Achievement Award, 61; John J Montgomery Award, 63; Spirit of St Louis Award, Am Soc Mech Engrs, 73; Am Astronaut Space Flight Award, 78; Goddard Award Trophy, 83; Elmer A Sperry Award, 86; Von Karman Astronautics Lectureship Award, 88. *Mem:* Nat Acad Eng; Am Inst Aeronaut & Astronaut; Am Astronaut Soc; Int Acad Astronaut. *Res:* Applied mechanics; integration of advanced systems into space vehicles. *Mailing Add:* 14319 Cross Timbers Ct Chesterfield MO 63107

YARDLEY, JOHN HOWARD, b Columbia, SC, June 7, 26; m 52; c 3. PATHOLOGY. *Educ:* Birmingham-Southern Col, AB, 49; Johns Hopkins Univ, MD, 53; Am Bd Path, dipl, 59. *Prof Exp:* Intern internal med, Vanderbilt Univ Hosp, 53-54; from instr to assoc prof, 54-72, assoc dean acad affairs, 77-84, PROF PATH, SCH MED, JOHNS HOPKINS UNIV, 72- *Concurrent Pos:* From asst resident to resident, Johns Hopkins Hosp, 54-58. *Mem:* Am Asn Path & Bact; Int Acad Path; Am Soc Exp Path; Am Gastroenterol Asn. *Res:* Gastrointestinal diseases; electron microscopy. *Mailing Add:* Dept Path Johns Hopkins Hosp 600 N Wolfe St Baltimore MD 21205

YARGER, DOUGLAS NEAL, b Omaha, Nebr, July 13, 37; m 60; c 4. ATMOSPHERIC PHYSICS, METEOROLOGY. *Educ:* Iowa State Univ, BS, 59; Univ Ariz, MS, 62, PhD(meteorol), 67. *Prof Exp:* Asst prof, 67-71, assoc prof meteorol & climatol, 71-80, PROF METEOROL & AGRON, IOWA STATE UNIV, 80- *Mem:* Am Meteorol Soc; Optical Soc Am. *Res:* Radiative transfer in the atmosphere; physical meteorology. *Mailing Add:* Dept Agron 253 Sci Iowa State Univ Ames IA 50011

YARGER, FREDERICK LYNN, b Lindsey, Ohio, Mar 8, 25; m 48; c 2. PHYSICS OF HIGH PRESSURES, LASER ELECTRO OPTICS. *Educ:* Capital Univ, BSc, 50; Ohio State Univ, MSc, 53, PhD(physics), 60. *Prof Exp:* Res asst physics, Los Alamos Sci Lab, 52, mem staff, 53-55 & 56; sr engr, Columbus Div, NAm Aviation, Inc, Ohio, 56-58; res asst physics, Ohio State Univ, 58-60; supvry physicist, Nat Bur Standards, Colo, 60-64; sci specialist, Edgerton, Germeshausen & Grier, Inc, Nev, 64-65; sr res physicist, Falcon Res & Develop Co, Colo, 65-66; from assoc prof to prof physics, 66-87, EMER PROF, NMEX HIGHLANDS UNIV, 87- *Concurrent Pos:* Vis staff mem, Los Alamos Sci Lab, 74-; vis prof, Nat Univ Mex, 68, 72, 77, 81, 84, 85, 87, 88, 89 & 90. *Mem:* Am Phys Soc; Optical Soc Am; Sigma Xi; Mex Phys Soc. *Res:* High pressure equations of state of liquids and solids using X-ray diffraction and spectroscopic techniques; biomedical instrumentation development. *Mailing Add:* Dept Physics NMex Highlands Univ Las Vegas NM 87701

YARGER, HAROLD LEE, b Ypsilanti, Mich, Mar 15, 40; m 65; c 3. EXPLORATION GEOPHYSICS. *Educ:* Antioch Col, BS, 62; State Univ NY Stony Brook, MA, 65, PhD(physics), 68. *Prof Exp:* Instr physics, State Univ NY Stony Brook, 67-68; res assoc, Northwestern Univ, 68-69; res assoc, Kans Geol Surv, Univ Kans, 69-70, res assoc geophys, 70-77, assoc scientist, 77-84, adj assoc prof physics & geol, 78-86, sr geophys scientist, 84-86; STAFF GEOPHYS, CHEVRON, 86- *Concurrent Pos:* Res contracts, NASA, 72-74 & 73-75, US Geol Surv, 74-76, 78-79 & 83-85, Ark Geol Comn, 79-81 & Gravity, Inc, 83-84. *Mem:* Geol Soc Am; Am Geophys Union; Am Phys Soc; Soc Explor Geophys. *Res:* Gravity and magnetics; remote sensing; airplane and satellite imagery for exploration and management of earth resources; high energy physics; bubble chamber work. *Mailing Add:* Chevron 1301 McKinney Houston TX 77010

YARGER, JAMES G, b Waverly, Iowa, Sept 15, 51; m; c 2. METABOLIC ENGINEERING, FOOD & DRUG ADMINISTRATION PETITIONS. *Educ:* Univ Iowa, BA, 74; Brandeis Univ, PhD(biol), 81. *Prof Exp:* Fel, Harvard Univ, 81-83; res scientist, Miles Inc, 83-85, sr res scientist, 85-87; STAFF SCIENTIST, AMOCO TECHNOL CO, 87-, TASK FORCE LEADER, 91- *Mem:* Am Soc Microbiol; Sigma Xi; AAAS. *Res:* Eukaryotic metabolic engineering; cloning and homologous-heterologous expression of metabolis genes in E coli and fungi; classical yeast genetics; industrial strain development; commercial product development; quality control and Food and Drug Administration product petition filings. *Mailing Add:* Amoco Technol Co PO Box 3011 MCF2 Naperville IL 60566

YARGER, WILLIAM E, b Houston, Tex, Apr 1, 37; m 63; c 3. NEPHROLOGY. *Educ:* Tex Christian Univ, BA, 59; Baylor Univ, MD, 63. *Prof Exp:* Intern internal med, St Luke's Hosp, New York, 63-64; resident, Baylor Affil Hosps, Houston, Tex, 64-67; lieutenant commander environ physiol, Naval Med Field Res Lab, NC, 67-69; NIH spec fel renal physiol, New York Univ, 69-71; from asst prof to assoc prof med, Duke Univ, 71-78; chief nephrology, 78-82, CHIEF MED, DURHAM VET ADMIN MED CTR, NC, 82- *Concurrent Pos:* Vchmn, dept med, Duke Univ, 85- *Mem:* Am Soc Clin Invest; Am Soc Nephrology; Am Phys Soc; Int Soc Nephrology. *Res:* Renal pathophysiology with particular interest in renal prostaglandins and thromboxan. *Mailing Add:* C8017 Durham Vet Admin Med Ctr 508 Fulton St Durham NC 27705

YARIAN, DEAN ROBERT, b Warsaw, Ind, Oct 12, 33; m 54; c 7. ORGANIC CHEMISTRY. *Educ:* DePauw Univ, BA, 55; Univ Wash, PhD(chem), 60. *Prof Exp:* Asst lectr, Univ Wash, 58-59; sr chemist, Cent Res Lab, 3M Co, St Paul, 60-65, Paper Prod Div, 65-70, chemist specialist, 70-78, mgr carbonless paper develop, Paper Prod Lab, 78-81, prod develop mgr, Telecommun Prod Div, 81-85, TECH MGR, ELEC SPECIALTIES DIV, 3M CO, AUSTIN, 85- *Mem:* Am Chem Soc. *Res:* Organic synthesis; imaging chemistry; colloids; paper chemistry; statistics. *Mailing Add:* 3M Co Bldg 147-45-02 PO Box 2963 Austin TX 78769-2963

YARINGTON, CHARLES THOMAS, JR, b Sayre, Pa, Apr 26, 34; m 63; c 3. OTORHINOLARYNGOLOGY. *Educ:* Princeton Univ, AB, 56; Hahnemann Med Col, MD, 60; Am Bd Otolaryngol, dipl, 65. *Prof Exp:* Rotating intern, Rochester Gen Hosp, 60-61; res gen surg, Dartmouth Med Sch Affil Hosps, 61-62; asst otolaryngol, Sch Med, Univ Rochester, 62-63, instr, 63-65; chief, Eye, Ear, Nose & Throat Serv, US Army Hosp, Ft Carson, Colo, 65-67; asst prof, Sch Med WVa Univ, 67-68; from assoc prof to prof otorhinolaryngol & chmn dept, Col Med, Univ Nebr, Omaha, 68-74; CLIN PROF OTOLARYNGOL, UNIV WASH, 74-; HEAD & NECK SURGEON & CHIEF, DEPT SURG, MASON CLIN, 89- *Concurrent Pos:* Res otolaryngol, Sch Med, Univ Rochester, 62-65; consult, Colo State Hosp, 65-67 & Vet Admin Hosp, WVa, 67-68; chief, Vet Admin Hosp, Omaha & Univ Nebr Hosp, Omaha, 68-74; chief ear, nose & throat, Mason Clin & Virginia Mason Hosp, 74-89; consult, Surgeon Gen, US Air Force, 76-; bd gov, Am Acad Otolaryngol & Head & Neck Surg; clin prof surg, Uniformed Serv Univ, Bethesda, 86- *Honors & Awards:* Prof Doctor Ignacio Barraquer Mem Award, 68; Honor Award, Am Acad Ophthal & Otolaryngol, 74; Sir Henry Wellcome Prize, 84. *Mem:* Fel Am Col Surgeons; Soc Head & Neck Surgeons; Am Laryngol, Rhinol & Otol Soc; Am Laryngol Asn; Am Acad Otolaryngol; Am Broncho-Esophagolog Asn (pres, 87-88); Ophthal Soc (pres, 87-88). *Res:* studies in the pathology and therapy of congenital and neoplastic defects in the head and neck; histopathology of salivary gland disease. *Mailing Add:* Mason Clin 1100 Ninth Ave Seattle WA 98101

YARINSKY, ALLEN, b Brooklyn, NY, May 6, 29; m 52; c 3. PARASITOLOGY, CHEMOTHERAPY. *Educ:* City Col New York, BS, 51; Columbia Univ, MS, 53; Univ NC, MS, 57, PhD, 61. *Prof Exp:* Med bacteriologist, Ft Detrick, Md, 54-56; res scientist, New York Dept Health, 61-65; res biologist, Sterling-Winthrop Res Inst, 66-69, head, Parasitol Dept, 69-79, dir Qual Assurance Dept, 79-89; CONSULT, 89- *Mem:* Am Soc Parasitol; Am Soc Trop Med & Hyg; Royal Soc Trop Med & Hyg; Sigma Xi. *Res:* Research on Clostridium botulinum type E toxin; immunological relationships of experimental Trichinella spirals infections; laboratory diagnosis of Protozoan and helminth infections; chemotherapy of parasitic infections; quality assurance; good clinical practices. *Mailing Add:* 53 Paxwood Rd Delmar NY 12054

YARIS, ROBERT, b New York, NY, Oct 16, 35; m 64. PHYSICAL CHEMISTRY. *Educ:* Univ Calif, Los Angeles, BS, 58; Univ Wash, PhD(phys chem), 62. *Prof Exp:* Res assoc phys chem, Univ Minn, 62-64; from asst prof to assoc prof chem, 64-70, PROF CHEM, WASH UNIV, 70- *Concurrent Pos:* Alfred P Sloan fel, 66- *Mem:* Am Phys Soc. *Res:* Theoretical and quantum chemistry; time-dependent perturbation theory; many body theory. *Mailing Add:* Dept Chem Wash Univ-Lindell-Skinker Blvd St Louis MO 63130

YARIV, A(MNON), b Tel Aviv, Israel, Apr 13, 30; nat US; m 54; c 3. SEMICONDUCTOR LASERS, INTEGRATED OPTICS. *Educ:* Univ Calif, Berkeley, BS, 54, MS, 56, PhD(elec eng), 59. *Prof Exp:* Asst, Univ Calif, Berkeley, 55-58, res assoc, 58-59; mem tech staff, Bell Tel Labs, 59-66; prof elect eng, 66-82, THOMAS G MYERS PROF ELEC ENG & APPL PHYSICS, CALIF INST TECHNOL, 82- *Concurrent Pos:* Indust consult; chmn bd & cofounder, Ortel Corp, Alhambra, Calif. *Honors & Awards:* Quantum Electronics Award, Inst Elec & Electronics Engrs, 80; Ives Medal, Optical Soc Am, 86; Pender Award, 85. *Mem:* Fel Nat Acad Eng; Nat Acad Sci; fel Am Optical Soc; Am Phys Soc; fel Inst Elec & Electronics Engrs. *Res:* Lasers; nonlinear optics; optical communication. *Mailing Add:* Dept Elec Eng Calif Inst Technol 1201 E Calif Blvd Pasadena CA 91125

YARKONY, DAVID R, b Bronx, NY, Jan 28, 49. CHEMICAL PHYSICS. *Educ:* State Univ NY, Stony Brook, BA, 71; Univ Calif, Berkeley, PhD(chem), 75. *Prof Exp:* Res asst, Mass Inst Technol, 75-77; ASST PROF CHEM, JOHNS HOPKINS UNIV, 77- *Concurrent Pos:* Consult, Inst Comput Appln Sci & Eng, 78-79. *Mem:* Am Chem Soc. *Res:* Electronic structure theory; energy transport in solids. *Mailing Add:* Dept Chem Johns Hopkins Univ 3400 N Charles St Baltimore MD 21218

YARLAGADDA, RADHA KRISHNA RAO, b Velpucherla, India, Apr 1, 38; m 66; c 3. ELECTRICAL ENGINEERING. *Educ:* Univ Mysore, BE, 59; SDak State Univ, MS, 61; Mich State Univ, PhD(elec eng), 64. *Prof Exp:* From asst prof to assoc prof, 66-78, PROF ELEC ENG, OKLA STATE UNIV, 78- *Concurrent Pos:* NSF res grant, 67-69; Air Force res grant, 81- *Honors & Awards:* Premium Award, Brit Inst Elec Engrs, 66. *Mem:* Inst Elec & Electronics Engrs; Sigma Xi. *Res:* Digital signal processing; communication theory. *Mailing Add:* Sch Elec Eng Okla State Univ Stillwater OK 74075

YARMOLINSKY, ADAM, b New York, NY, Nov 17, 22; div; c 4. SCIENCE POLICY. *Educ:* Harvard Univ, BA, 43. *Hon Degrees:* LLB, Yale Univ, 48. *Prof Exp:* Prin dep asst secy defense, Dept Defense, 65-66; Ralph Waldo Emerson Univ prof, Univ Mass, 72-77; pvt pract law, Kominers, Fort, Schlefer & Boyer, Washington, DC, 79-85; prof policy sci, Health Care Policy, Nat Security, 85-86, PROVOST & VPRES ACAD AFFAIRS, UNIV MD, BALTIMORE, 86- *Concurrent Pos:* Coun US Amrs Control & Disarmament 77-79; consult, Off Technol Assessment, 74-77; chmn comt sci & law, Asn the Bar, NY, 84-87; Comdr, Foreign Relations, Am Law Inst. *Mem:* Nat Acad Sci; AAAS; Int Inst Strategic Studies. *Res:* Large organization functionalism. *Mailing Add:* Provost Off Univ Md Baltimore MD 21228

YARMOLINSKY, MICHAEL BEZALEL, b New York, NY, Jan 18, 29; m 62; c 1. MOLECULAR BIOLOGY. *Educ:* Harvard Univ, AB, 50; Johns Hopkins Univ, PhD(biol), 54. *Prof Exp:* Instr pharmacol, Col Med, NY Univ, 54-55; res assoc, McCollum-Pratt Inst, Johns Hopkins Univ, 58-61, asst prof, 61-63; res chemist, NIH, 64-70; dir res, CNRS Rech Biol Molec, Paris, 70-76; dir molecular genetics sect, Cancer Biol Prog, Frederick Cancer Res Fac, Md, 76-84; chief develop biochem & genetics sect, 84-88, CHIEF MICROBIAL GENETICS & BIOCHEM SECT, LAB BIOCHEM, NAT CANCER INST, NIH, BETHESDA, MD, 88- *Concurrent Pos:* NSF fel, Pasteur Inst, Paris, 63-64. *Mem:* Am Soc Biol Chemists; Am Soc Microbiol; fel AAAS. *Res:* Protein biosynthesis and its regulation; interactions between temperate bacteriophage and its host; replication control and partition of plasmids in bacteria. *Mailing Add:* Lab Biochem Nat Cancer Inst NIH Bldg 37 4D-15 Bethesda MD 20892

YARMUSH, DAVID LEON, b New York, NY, June 10, 28. APPLIED MATHEMATICS. *Educ:* Harvard Univ, BA, 49; Princeton Univ, PhD(math), 59. *Prof Exp:* Mathematician Chem & Radiation Labs, Army Chem Ctr, Md, 52-54; asst math, Princeton Univ, 54-56; mathematician, Tech Res Group, Inc, 56-67; res scientist, Courant Inst, NY Univ, 65-76; res assoc, Columbia Univ, 76-83, assoc res scientist, Dept Biol, 84-90; RETIRED. *Mem:* Am Math Soc. *Res:* Dynamic theory of games; radiation transport theory; structural vibrations and sound radiation; computer programming of deduction procedures; computer study of conformation of proteins. *Mailing Add:* 333 E 14th St New York NY 10003-4208

YARMUSH, MARTIN LEON, b Brooklyn, NY, Oct 8, 52; m 78; c 2. APPLIED IMMUNOLOGY, BIOSEPARATIONS. *Educ:* Yeshiva Univ, BA, 75; Rockefeller Univ, PhD(biochem & immunol), 79; Yale Univ, MD, 84. *Prof Exp:* Res chemist biochem, NIH, 78-79; prin res assoc chem eng, Mass Inst Technol, 84-88; ASSOC PROF SURG, HARVARD MED SCH & MASS GEN HOSP, 86-; PROF BIOCHEM ENG & BIOCHEM, CTR ADV BIOTECHNOL & MED, RUTGERS UNIV, 88- *Concurrent Pos:* Consult, NIH, Sepracor, Inc, Cabot Corp, Delta Biotechnol Ltd, Valio Finnish Dairy Cooperatives, E I Dupont de Nemours & Co, Inc, Ortho Diag, Union Carbide; adj asst prof, Sch Vet Med, Univ Pa, 80-83, adj assoc prof, 83-86. *Honors & Awards:* Pres Young Investr Award; Lucille D Markey Scholar Award; NIH Nat Res Secy Award. *Mem:* Am Chem Soc; Am Inst Chem Engrs; Am Asn Immunologists; fel Am Soc Artificial Internal Organs; AMA; Am Col Sportsmed. *Res:* Applied immunology; molecular and cellular bioengineering; bioseparations; artificial organs. *Mailing Add:* 35 Cheryl Dr Sharon MA 02067

YARNALL, JOHN LEE, b Elizabeth, NJ, Jan 27, 32; m 53; c 3. INVERTEBRATE ZOOLOGY, BIOLOGY. *Educ:* Univ Mont, BS, 53, MA, 62; Stanford Univ, PhD(biol), 72. *Prof Exp:* Asst prof, 69-72, assoc prof, 72-78, PROF BIOL, HUMBOLDT STATE UNIV, 78- *Mem:* AAAS; Am Soc Zoologists. *Res:* Invertebrate functional morphology and behavior, especially locomotion and feeding. *Mailing Add:* Dept Biol Humboldt State Univ Arcata CA 95521

YARNELL, JOHN LEONARD, b Topeka, Kans, Mar 1, 22; m 52; c 4. SOLID STATE PHYSICS, NUCLEAR ENGINEERING. *Educ:* Univ Kans, AB, 47, AM, 49; Univ Minn, PhD(physics), 52. *Prof Exp:* Asst instr physics & math, Univ Kans, 47-49; asst, Univ Minn, 49-51; staff mem, Physics Div, Los Alamos Sci Lab, 52-65, group leader, Physics Div, 65-81; CONSULT, 81- *Mem:* Fel Am Phys Soc. *Res:* Lattice dynamics; neutron diffraction; reactors; cryogenics; solid state physics. *Mailing Add:* 205 El Conejo Los Alamos NM 87544

YARNELL, RICHARD ASA, b Boston, Mass, May 11, 29; c 4. ANTHROPOLOGY, ETHNOBOTANY. *Educ:* Duke Univ, BS, 50; Univ NMex, MA, 58; Univ Mich, PhD(anthrop), 63. *Prof Exp:* From instr to assoc prof anthrop, Emory Univ, 62-71; assoc prof anthrop, 71-75, assoc chmn dept, 73-75, PROF ANTHROP, UNIV NC, CHAPEL HILL, 75- *Mem:* Fel AAAS; Soc Econ Bot; Soc Ethnobiol. *Res:* Analysis of archaeological plant remains; evolution of plant domestication; aboriginal plant utilization; cultural ecology; economic botany. *Mailing Add:* Dept Anthrop Univ NC CB3115 Chapel Hill NC 27599

YARNS, DALE A, b Jackson, Minn, July 9, 30; m 54; c 2. ANIMAL PHYSIOLOGY. *Educ:* Univ Minn, BS, 56; SDak State Col, MS, 58; Univ Md, PhD(animal sci), 64. *Prof Exp:* Res asst dairy husb, SDak State Univ, 56-58; lab technician animal sci, Univ Calif, Davis, 58-61; animal husbandryman, Beef Cattle Br, USDA, 61-64; assoc res physiol, Animal Med Ctr, 64-66; asst prof, 66-72, ASSOC PROF PHYSIOL & CHMN DEPT BIOL, WAGNER COL, 72- *Res:* Comparative cardiac and ruminant physiology. *Mailing Add:* Dept Biol Wagner Col 631 Howard Ave Staten Island NY 10301

YAROSEWICK, STANLEY J, b Epping, NH, Sept 10, 39; m 64; c 2. ATOMIC PHYSICS, SPECTROSCOPY. *Educ:* Univ NH, BS, 61; Clarkson Col Technol, MS, 63, PhD(physics), 66. *Prof Exp:* Asst prof physics, Clarkson Col Technol, 66-69; assoc prof, 69-74, PROF PHYSICS, WEST CHESTER STATE COL, 74- *Mem:* Am Asn Physics Teachers. *Res:* Atomic emission spectra. *Mailing Add:* Provost Dir West Chester State Col West Chester PA 19383

YARRINGTON, ROBERT M, b Peoria, Ohio, Sept 26, 28; m 66; c 2. CHEMICAL ENGINEERING. *Educ:* Ohio State Univ, BSc, 51, MSc, 56, PhD(chem eng), 58. *Prof Exp:* Group leader res & develop, Am Cyanamid Co, 58-74; SECT HEAD RES & DEVELOP, ENGELHARD MINERAL & CHEM, 74- *Mem:* Am Inst Chem Engrs; Sigma Xi. *Res:* Development of catalytic processes related to the petroleum and energy fields. *Mailing Add:* 327 Parsonage Rd Edison NJ 08837-2108

YARUS, MICHAEL J, b Pikeville, Ky, Mar 2, 40; m 62, 85; c 3. MOLECULAR BIOLOGY, BIOCHEMISTRY. *Educ:* Johns Hopkins Univ, BA, 60; Calif Inst Technol, PhD(biophys), 66. *Prof Exp:* USPHS & NIH fels biochem, Stanford Univ, 65-67; from asst prof to assoc prof, 67-79, PROF MOLECULAR, CELLULAR & DEVELOP BIOL, UNIV COLO, BOULDER, 79- *Concurrent Pos:* USPHS & NIH grant, 68- *Mem:* AAAS; Fedn Am Soc Exp Biol. *Res:* Minute viruses; transfer RNA; mammalian embryogeny; control of translation, origin of the code. *Mailing Add:* Dept Molecular Cellular & Develop Biol Campus Box 347 Univ Colo Boulder CO 80309

YASAR, TUGRUL, b Ankara, Turkey, Sept 23, 41; US citizen; m 85; c 3. MAGNETIC RECORDING, THIN FILM TECHNOLOGY. *Educ:* Robert Col, Istanbul, BS, 63; Princeton Univ, MA, 67, PhD(elec eng in solid state sci), 68. *Prof Exp:* Res engr photoelectronics, Bendix Res Labs, 67-73; chief eng, Gen Instruments, 73-76; dir eng, Nat Micronetics, Inc, 76-79, opers mgr, 79-81, vpres, 81-90; VPRES & GEN MGR, MAGNETO-OPTICS DIV, MRC, 90- *Concurrent Pos:* Mem affil fac, Col Eng, Wayne State Univ, 68-69; mem adv coun, Ulster Community Col, 79- *Mem:* Inst Elec & Electronics Engrs; Int Soc Hybrid Microelectronics; Am Vacuum Soc; Sigma Xi; Am Ceramics Soc. *Res:* Photoemission from semiconductors; photoelectronic devices; imaging devices; amorphous semiconductors; ultrasonics; thin film technology; magnetic devices; semiconductor manufacturing technology; magneto-optics. *Mailing Add:* PO Box 94 West Hurley NY 12491

YASBIN, RONALD ELIOTT, b Brooklyn, NY, Apr 27, 47; m 72; c 2. INDUSTRIAL MICROBIOLOGY. *Educ:* Pa State Univ, BS, 68; Cornell Univ, MS, 70; Univ Rochester, PhD(microbiol), 74. *Prof Exp:* Asst prof microbiol & molecular genetics, Pa State Univ, 76-81; ASST PROF MICROBIOL, SCH MED, UNIV ROCHESTER, 81- *Concurrent Pos:* Vis scientist, Brookhaven Nat Lab, 79-82. *Mem:* Am Soc Microbiol; Environ Mutagen Soc; Genetics Soc; Sigma Xi. *Res:* The role of DNA repair systems in the production of mutations and carcinogenesis events; use of recombinant DNA technology to explore pathogenesis and secondary metabolites. *Mailing Add:* Dept Biol Sci UMBC 5401 Wilkens Ave Baltimore MD 21228

YASHON, DAVID, b Chicago, Ill, May 13, 35; c 3. NEUROSURGERY. *Educ:* Univ Ill, BSM, 58, MD, 60; FRCS(C), 69. *Prof Exp:* Instr neurosurg, Univ Chicago, 65-66; asst prof, Case Western Reserve Univ, 66-69; assoc prof, 69-74, PROF NEUROSURG, OHIO STATE UNIV, 74- *Mem:* Cong Neurol Surg; Am Asn Neurol Surg; Soc Univ Surg; Am Acad Neurol; Asn Acad Surg. *Res:* Cerebral physiology and metabolism during circulatory deficiency; spinal cord injury and metabolic effects. *Mailing Add:* 1492 E Broad St Apt 1100 Columbus OH 43215

YASKO, RICHARD N, b Conemaugh, Pa, Aug 29, 35; m 64; c 2. NUCLEAR PHYSICS. *Educ:* Pa State Univ, BS, 57, MS, 61, PhD(physics), 63. *Prof Exp:* Fel nuclear physics, Argonne Nat Lab, 63-64; asst prof physics, Villanova Univ, 64-67; adv physicist, IBM Corp, Endicott, 68-77; prod assurance mgr, Avco Syst Div, Avco Corp, 78-83; sr engr, Norden Systs, United Technol, Merrimack, NH, 83-86; ASSOC PROF PHYSICS, WINONA STATE UNIV, 89- *Mem:* Am Phys Soc; Soc Advan Mat & Process Eng. *Res:* Surface physics; semiconductor device physics; ion implantation, spreading resistance diffusion profiling and capacitance; voltage testing of MO5 devices. *Mailing Add:* 1317 W Wincrest Dr Winona MN 55987

YASMINEH, WALID GABRIEL, b Amman, Jordan, Jan 21, 31; US citizen; m 60; c 3. BIOCHEMISTRY. *Educ:* Am Univ Cairo, BSc, 53; Univ Minn, Minneapolis, MSc, 63, PhD(biochem), 66. *Prof Exp:* From jr scientist to assoc scientist pediat, 59-65, asst prof lab med, 67-72, ASSOC PROF LAB MED, SCH MED, UNIV MINN, MINNEAPOLIS, 72- *Concurrent Pos:* Grad Sch grant, Sch Med, Univ Minn, Minneapolis, 68- *Mem:* Am Chem Soc. *Res:* Mammalian constitutive heterochromatin and repetitive DNA, nature, origin, function and relation to disease. *Mailing Add:* 2735 Mackubin No 10 St Paul MN 55113-2362

YASSO, WARREN E, b New York, NY, Oct 19, 30; m 57; c 2. GEOLOGY, GEOMORPHOLOGY. *Educ:* Brooklyn Col, BS, 57; Columbia Univ, MA, 61, PhD(geomorphol), 64. *Prof Exp:* Instr earth sci, Adelphi Univ, 61-64; asst prof geol, Va Polytech Inst, 64-66; ASSOC PROF SCI EDUC, TEACHERS COL, COLUMBIA UNIV, 66- *Concurrent Pos:* Prog dir educ, Nat Sea Grant Prog, Nat Oceanic & Atmospheric Admin, 78-79; dir, Microcomput Software Sci Proj, NY Power Authority, 84-86. *Mem:* AAAS; fel Geol Soc Am; Int Asn Sedimentol; Nat Asn Geol Teachers; Nat Sci Teachers Asn. *Res:* Coastal and continental shelf geological processes; curriculum research in earth sciences. *Mailing Add:* 528 Franklin Turnpike Ridgewood NJ 07450

YASUDA, HIROTSUGU, b Kyoto, Japan, Mar 24, 30; m 68; c 3. POLYMER CHEMISTRY, PHYSICAL CHEMISTRY. *Educ:* Kyoto Univ, BS, 53; State Univ NY Col Environ Sci & Forestry, MS, 59, PhD(polymer & phys chem), 61. *Prof Exp:* Fel, State Univ NY Col Environ Sci & Forestry, Syracuse, 61; chemist, Camille Dreyfus Lab, Res Triangle Inst, 61-63; res assoc, Ophthalmic Plastic Lab, Mass Eye & Ear Infirmary, 63-64 & Cedars-Sinai Med Ctr, 64-65; guest scientist, Royal Inst Technol, Sweden, 65-66; head membrane & med polymer sect, Camille Dreyfus Lab, Res Triangle Inst, 66-75, mgr polymer dept, 75-77; prof chem eng, Univ Mo-Rolla, 78-88, sr investr, Mat Res Ctr, 78-88, dir, Thin Film Inst, 85-88, chmn, chem eng, Univ Mo, Columbia, 88-90, DIR, CTR SURFACE SCI & PLASMIC TECHNOL, UNIV MO, COLUMBIA, 90- *Mem:* Am Chem Soc; AAAS. *Res:* Preparation and characterization of polymers; transport phenomena through polymer membrane; biomedical application of polymers, membrane technology, plasma polymerization and surface modifications; vacuum deposition of polymers. *Mailing Add:* Ctr Surface Sci & Plasmic Technol Eng Complex Univ Mo Columbia MO 65211

YASUDA, STANLEY K, b Pahoa, Hawaii, Jan 7, 31; m 55; c 3. ANALYTICAL CHEMISTRY. *Educ:* Park Col, BA, 53; Kans State Univ, MS, 55, PhD(anal chem), 57. *Prof Exp:* MEM STAFF, LOS ALAMOS NAT LAB, 57- *Concurrent Pos:* Sr analyst, Chemagro Corp, 64. *Mem:* Fel Am Inst Chemists; Am Chem Soc; Sigma Xi. *Res:* Microanalytical methods for analysis of explosive and non-explosive materials, utilizing wet and instrumental techniques. *Mailing Add:* 75 San Juan Los Alamos NM 87544

YASUI, GEORGE, b Olympia, Wash, May 7, 22; m 50; c 4. CHEMICAL ENGINEERING. *Educ:* Univ Denver, BS, 44; Univ Mich, MS, 48; Univ Wash, PhD(chem eng), 57. *Prof Exp:* Chem engr, Varnish Lab, Sherwin-Williams Co, 44-45 & Chem Eng Div, Argonne Nat Lab, 48-54; res scientist, Lockheed Missiles & Space Co, 56-60, staff scientist, 60-62, mgr res & develop staff, 62-63, asst mgr nuclear eng, 63-64, sr staff engr, 64-71; proj engr, Environ Qual Eng, Inc, 71-72; staff engr, Lockheed Missiles & Space Co, 72-90; RETIRED. *Mem:* Am Inst Chem Engrs; Sigma Xi. *Res:* Development of nosetips and heating shields for reentry vehicles; Analyses and testing of materials exposed to nuclear weapons; lasers; space environments. *Mailing Add:* 3491 Janice Way Palo Alto CA 94303

YASUKAWA, KEN, b New York, NY, Sept 7, 49; m 72; c 2. ECOLOGY, ANIMAL BEHAVIOR. *Educ:* State Univ NY, Stony Brook, BS, 71; Ind Univ, MS, 75, PhD(zool), 77. *Prof Exp:* Fel animal behav, Field Res Ctr, Rockefeller Univ, 77-80; from asst prof to assoc prof, 80-90, PROF BIOL, DEPT BIOL, BELOIT COL, 90- *Concurrent Pos:* Prin Investr, NSF, 84, 87-89 & 90-93; vis prof, Dept Zool, Univ Wis-Madison; ed, Asn Field Ornithologists. *Mem:* Soc Am Naturalists; Animal Behav Soc; Ecol Soc Am; fel Am Ornithologists Union; Sigma Xi. *Res:* Avian behavioral ecology; evolution of mating systems; function and causation of aggression; territoriality; population regulation; dominance hierarchies; communication behavior; function and development of avian vocal behavior. *Mailing Add:* Dept Biol Beloit Col Beloit WI 53511

YASUMURA, SEIICHI, b New York, NY, Sept 28, 32; m 63; c 2. ENDOCRINOLOGY. *Educ:* Occidental Col, AB, 58; Univ Cincinnati, PhD(anat), 62. *Prof Exp:* Instr, 64-66, asst prof, 66-77, ASSOC PROF PHYSIOL, STATE UNIV NY DOWNSTATE MED CTR, 77- *Concurrent Pos:* Fel, State Univ Groningen, 62-63; NSF fel, State Univ NY Downstate Med Ctr, 63-64; res collabr, Brookhaven Nat Lab, 75-; consult, Nat Inst Environ Health Sci, 80-82. *Mem:* Endocrine Soc; Am Physiol Soc. *Mailing Add:* Dept Physiol State Univ NY Downstate Med Ctr Brooklyn NY 11203

YASUNOBU, KERRY T, b Seattle, Wash, Nov 21, 25; m 52; c 1. BIOCHEMISTRY. *Educ:* Univ Wash, PhD(biochem), 54. *Prof Exp:* Res scientist, Univ Tex, 54-55; res assoc, Med Sch, Univ Ore, 55-58; asst prof chem, 58-62, assoc prof biochem, 62-64, PROF BIOCHEM, UNIV HAWAII, MANOA, 64- *Concurrent Pos:* NSF sr fel, 63-64; NIH sr fel, 71-72. *Mem:* Am Chem Soc; Am Soc Biol Chemists; Sigma Xi. *Res:* Enzymology, especially oxidative, heme-enzymes and proteolytic enzymes. *Mailing Add:* 3270 Melemele Pl Honolulu HI 96822

YATES, ALBERT CARL, b Memphis, Tenn, Sept 29, 41; m 62; c 2. THEORETICAL PHYSICAL CHEMISTRY. *Educ:* Memphis State Univ, BS, 65; Ind Univ, Bloomington, PhD(chem physics), 68. *Prof Exp:* Res assoc chem, Univ Southern Calif, 68-69; from asst prof to assoc prof, Ind Univ, Bloomington, 69-74; assoc prof chem & assoc univ dean grad educ & res, Univ Cincinnati, 74-76, prof chem & univ dean grad educ & res, 76-; PROVOST, WASH STATE UNIV. *Mem:* Am Phys Soc; Am Chem Soc. *Res:* Collisions of charged particles with atomic and molecular systems; heavy-particle collisions; photo-absorption processes. *Mailing Add:* Rte 2 Box 698 Sand Rd Pullman WA 99163-9648

YATES, ALDEN P, engineering construction; deceased, see previous edition for last biography

YATES, ALLAN JAMES, b Calgary, Alta, May 23, 43; m 68; c 2. NEUROPATHOLOGY, NEUROCHEMISTRY. *Educ:* Univ Alta, MD, 67; Univ Toronto, PhD(neurochem), 72; FRCP(C). *Prof Exp:* From asst prof to assoc prof, 75-84, PROF NEUROPATH, OHIO STATE UNIV, 84-, HEAD, NEUROPATH, 82- *Mem:* Am Asn Neuropath; Am Soc Neurochem; Am Soc Path; Int Soc Neurochem; Int Soc Neuropath. *Res:* Physiological and biochemical aspects of glycolipids and their roles in diseases of the nervous system, with special reference to nerve degeneration, regeneration and gliomas. *Mailing Add:* Dept Path Upham Hall Rm 111 Ohio State Univ Col Med 473 W 12th Ave Columbus OH 43210

YATES, ANN MARIE, b Ogdensburg, NY, Sept 29, 40. X-RAY ANALYSIS, METEORITES. *Educ:* St Lawrence Univ, BS, 62; Ariz State Univ, PhD(chem), 66. *Prof Exp:* Asst prof anal chem, Ariz State Univ, 66-67; NIH res assoc inorg chem, Univ Pittsburgh, 67-68; dir labs, Chemalytics Inc, 68-74; mem chem fac, Maricopa County Community Col Dist, 76-77; mem chem fac, 77-78, RES SPECIALIST, ARIZ STATE UNIV, 78- *Mem:* Am Chem Soc; Sigma Xi; Meteoritical Soc. *Res:* Microanalytical techniques; x-ray fluorescence; x-ray diffraction; meteorite analysis. *Mailing Add:* 1627 E Wesleyan Dr Tempe AZ 85282

YATES, CHARLIE LEE, b Harrellsville, NC, Apr 8, 36; c 3. FLUID MECHANICS, AEROCHEMISTRY. *Educ:* Va Polytech Inst & State Univ, BS, 58; Calif Inst Technol, MS, 59; Johns Hopkins Univ, PhD, 78. *Prof Exp:* Assoc engr, Westinghouse Elec Corp, 59-60; assoc engr, Appl Physics Lab, Johns Hopkins Univ, 60-64, sr engr, 64-79; Dept Chem Eng, Hampton Univ, 83-86; assoc prof, Mech Eng Dept, 79-83, ASSOC PROF, AEROSPACE & OCEAN ENG DEPT, VA POLYTECH INST & STATE UNIV, 87- *Mem:* Am Inst Aeronaut & Astronaut; Am Soc Eng Educ; Sigma Xi. *Res:* Surface gravity wave physics; wave-wave interactions. *Mailing Add:* Dept Aerospace & Ocean Eng Va Polytech Inst State Univ Blacksburg VA 24061

YATES, CLAIRE HILLIARD, b Cornwall, Ont, Mar 27, 20; m 42; c 3. ANALYTICAL CHEMISTRY, PHARMACEUTICAL CHEMISTRY. *Educ:* Sir George Williams Col, BSc, 46; McGill Univ, PhD(biochem), 51. *Prof Exp:* Prod chemist, Charles E Frosst & Co, Montreal, 41-47, res chemist, 48-69, anal unit head, Pharmaceut Res Dept, Merck Frosst Can, 69-85, res fel, 80-85; RETIRED. *Concurrent Pos:* Lectr, Sir George Williams Col, 48-61. *Mem:* Chem Inst Can. *Res:* Steroids; synthesis of radioactive organic substances; analyses of drug formulations; stability of drugs in dosage forms. *Mailing Add:* 994 Second Ave Verdun PQ H4G 2W8 Can

YATES, EDWARD CARSON, JR, b Raleigh, NC, Nov 3, 26; m 52; c 1. AEROELASTICITY, UNSTEADY AERODYNAMICS. *Educ:* NC State Univ, BS, 48, MS, 49; Univ Va, MS, 53; Va Polytech Inst & State Univ, PhD(eng mech), 59. *Prof Exp:* Aerospace engr, 49-87, chief scientist, Loads & Aeroelasticity Div, 82-86, SR AEROSPACE ENGR, INTERDISCIPLINARY RES OFF, LANGLEY RES CTR, NASA, 87-; PROF, GEORGE WASHINGTON UNIV, 68- *Concurrent Pos:* Lectr physics, Va Polytech Inst & State Univ, 59-67; adj assoc prof, NC State Univ, 64-75; assoc ed, J Aircraft, 72-78; vis prof, Univ Rome, La Sapienza, 90; mem, Sci Comt, Int Asn Boundary Element Methods, 90- *Mem:* Assoc fel Am Inst Aeronaut & Astronaut. *Res:* Subsonic, transonic and supersonic aerodynamics; unsteady aerodynamics; aeroelasticity; structural dynamics. *Mailing Add:* NASA Langley Res Ctr Hampton VA 23665-5225

YATES, FRANCIS EUGENE, b Pasadena, Calif, Feb 26, 27; m 49; c 5. PHYSIOLOGY. *Educ:* Stanford Univ, BA, 47, MD, 51. *Prof Exp:* Intern, Philadelphia Gen Hosp, 50-51; instr physiol, Harvard Med Sch, 55-57, assoc, 57-59, asst prof, 59-60; from assoc prof to prof, Stanford Univ, 60-69, actg exec head dept physiol, 64-69; PROF BIOMED ENG, DEPT BIOMED ENG, UNIV SOUTHERN CALIF, 69-, DIR, BIOMED ENG CTR, 77-; PROF CHEM ENG, MED & PHYSIOL, RALPH & MARJORIE CRUMP PROF MED ENG & DIR, CRUMP INST MED ENG, UNIV CALIF, LOS ANGELES, 80- *Concurrent Pos:* Res fel physiol, Harvard Univ, 53-55; Markle scholar med sci, Harvard Med Sch, 59; mem, physiol training comt, Nat Inst Gen Med Sci, 64-70, mem med scientist training prog comt, 71-73; vis prof, Stanford Univ, 69-; sect ed endocrinol & metab, Am J Physiol, 69-74; managing ed, Annals Biomed Eng, 71-74; consult prin scientist, Alza Corp; mem sci info prog adv comt, Nat Inst Neurol & Communicative Dis & Stroke, 76-; managing ed, Am J Physiol, Regulatory Integrative & Comp Physiol, 76-; mem, Space Biol Panel, NASA, 79-80; mem, Panel Basic Biomed Sci, Human Resources Comn, Nat Res Coun, 79-; mem adv bd biol, Harvey Mudd Col, 79- *Honors & Awards:* Upjohn Award, Endocrine Soc, 62. *Mem:* AAAS; Biomed Eng Soc (pres, 74-75); Am Physiol Soc; Endocrine Soc; NY Acad Sci; Sigma Xi. *Res:* Metabolism and inactivation of adrenal cortical hormones; analysis of endocrine feedback systems. *Mailing Add:* Univ Calif-Los Angeles 405 Hilgard Ave Los Angeles CA 90024-1654

YATES, GEORGE KENNETH, b Chicago, Ill, Sept 24, 25; m 84; c 4. SPACE PHYSICS. *Educ:* Harvard Univ, AB, 50; Univ Chicago, MS, 55, PhD(physics), 64; Suffolk Univ, JD, 80. *Prof Exp:* RES PHYSICIST, AIR FORCE GEOPHYS LAB, 64- *Mem:* Am Phys Soc; Am Geophys Union. *Res:* Physics of the near space environment. *Mailing Add:* 86 Tadmuck Rd Westford MA 01886

YATES, HAROLD W(ILLIAM), b Hagerstown, Md, Oct 17, 23; m 47; c 3. CHEMICAL ENGINEERING. *Educ:* Johns Hopkins Univ, BE, 44, MA, 50. *Prof Exp:* Jr instr physics, Johns Hopkins Univ, 44-47, asst optics, 47-50; physicist, Optics Div, US Naval Res Lab, 50-57; chief engr, Field Eng Dept, Barnes Eng Co, 57-67; DIR, OFF RES, NAT ENVIRON SATELLITE SERV, NAT OCEANIC & ATMOSPHERIC ADMIN, 67- *Mem:* Optical Soc Am. *Res:* Optical properties of atmosphere; infrared transmission; refraction; scintillation military applications of infrared; target and background measurements; detection systems; tracking; optical instrument design for field research; photography; radiometry; spectroscopy. *Mailing Add:* Nat Environ Satellite Serv Nat Oceanic & Atmospheric Admin Suitland MD 20689

YATES, HARRIS OLIVER, b Paducah, Ky, Apr 14, 34; m 54; c 3. BIOLOGY. *Educ:* David Lipscomb Col, BA, 56; George Peabody Col, MA, 57; Vanderbilt Univ, PhD(biol), 65. *Prof Exp:* Instr, 57-59, from asst prof to assoc prof, 63-68, PROF BIOL & CHMN DEPT, DAVID LIPSCOMB COL, 68- *Res:* Experimental plant taxonomy. *Mailing Add:* Dept Biol David Lipscomb Univ Nashville TN 37204

YATES, JAMES T, b Forney, Tex, June 8, 40; m 61; c 2. GERIATRICS, DIAGNOSTIC AUDIOLOGY. *Educ:* Tex Tech Univ, BA(psychol) & BA(audiol), 65, MA, 66; Univ Denver, PhD(audiol), 70. *Prof Exp:* Teaching asst commun, Tex Tech Univ, 65-66; lab instr anat, Univ Denver, 66-69; dir audiol, Tex Tech Univ, 70-78; PROF & CHMN AUDIOL, SCH MED, UNIV HAWAII, 78- *Concurrent Pos:* Consult, various govt orgn & co, 70-78; prin investr grants & contracts, 70-; consult audiologist, Lubbock State Sch Ment Retardation, 72-75, WTex Rehab Ctr, 75-78; consult mgr, Kuakini Hosp, Honolulu, 90- *Mem:* Fel Am Speech & Hearing Asn; Acoust Soc Am; Acad Rehab Audiol. *Res:* Geriatric audiology; central auditory processing: development, deteriation and diagnosis of disorders of processing in children and adults. *Mailing Add:* SPA-Med 1410 Lower Campus Dr Univ Hawaii Honolulu HI 96822

YATES, JEROME DOUGLAS, b Center Point, Ark, Jan 5, 35; m 69; c 4. POULTRY NUTRITION. *Educ:* Univ Ark, BSA, 58, MS, 59; Mich State Univ, PhD(poultry nutrit), 64. *Prof Exp:* Nutrit technician, Mich State Univ, 59-63; res assoc nutrit & food sci, Soup Co, 64-68, res scientist, 69-85, prin scientist, 86-90, SR RES PROG MGR, CAMPBELL SOUP CO, 90- *Mem:* Poultry Sci Asn; World Poultry Sci Asn. *Res:* Poultry nutrition emphasizing the influence of nutrients and other dietary components on quality of poultry meat; mineral and amino acid nutrition. *Mailing Add:* Campbell Soup Co PO Box 179 Farmington AR 72730

YATES, JEROME WILLIAM, b Rockford, Ill, Nov 9, 36; m 79; c 2. GERONTOLOGY, EPIDEMIOLOGY. *Educ:* Lawrence Univ, AB, 61; Univ Ill, MD, 65; Harvard Sch Pub Health, MPH, 81. *Prof Exp:* Prof med, Univ Vt Col Med, 78-82; assoc dir, Nat Cancer Inst, 82-87; PROF MED ONCOL, STATE UNIV NY MED SCH, BUFFALO, 87-; ASSOC DIR CLIN AFFAIRS, ROSWELL PARK MEM INST, 87- *Concurrent Pos:* Ed, J Psychol Oncol, Nat Cancer Inst, 83-, J Nat Cancer Inst, 82-87; consult, Health Care Financing Admn, 81-82, World Health Orgn, 84, Nat Cancer Inst, 87- *Mem:* Am Asn Cancer Res; Am Soc Clin Oncol; Am Pub Health Asn; Am Soc Prev Oncol; Am Asn Cancer Inst. *Res:* Leukemia management; cancer in elderlies; health care delivery issues and supportive care. *Mailing Add:* Assoc Dir Clin Affairs Roswell Park Mem Inst Elm St Buffalo NY 14263

YATES, JOHN THOMAS, JR, b Winchester, Va, Aug 3, 35; m 58; c 2. PHYSICAL CHEMISTRY, SURFACE CHEMISTRY. *Educ:* Juniata Col, BS, 56; Mass Inst Technol, PhD, 60. *Prof Exp:* Res assoc chem, Mass Inst Technol, 60; instr & asst prof, Antioch Col, 60-63; Nat Res Coun-Nat Bur Standards res assoc, 63-65; staff mem, Phys Chem Div, Nat Bur Standards, 65-74, chief, Surface Processes & Catalysis Sect, 74-78; R K MELLON PROF CHEM, UNIV PITTSBURGH, 81-, DIR, PITTSBURGH SURFACE SCI CTR, 81- *Concurrent Pos:* Sr vis fel, Univ EAnglia, 70-71 & 72; trustee, Am Vacuum Soc, 74; mem bd dirs, Catalysis Soc, 76-81; Sherman Fairchild scholar, Calif Inst Technol, 77-78; chmn, Div Colloid & Surface Chem, Am Chem Soc, 80; distinguished vis lectr, Univ Tex, Austin, 78; vchmn & chmn, Gordon Conf on Molecular Dynamics of Surfaces, 79 & 82; mem vis comt, Div Chem & Chem Eng, Caltech, 86, comt, US Army Basic Sci Res, NRC, 87; chmn rev panel, Chem & Laser Sci Div, Los Alamos Nat Lab, 87; rev panel, Sci & Technol Ctrs, NSF, 88; adv bd, Petrol Res Found, 88; chmn, div chem, Am Phys Soc, 89. *Honors & Awards:* Silver Medal, US Dept Com, 73, Gold Medal, 82; Samuel Wesley Stratton Award, Nat Bur Standards, 78; Kendall Award, Am Chem Soc; E W Morley Medal, Am Chem Soc, 90. *Mem:* Am Chem Soc; Am Phys Soc; Am Vacuum Soc. *Res:* Spectra of adsorbed molecules; heterogeneous catalysis; kinetics of adsorption and desorption; electron impact studies of adsorbed species; electronic properties of the chemisorbed layer; semiconductor surface chemistry; tribology and surface chemistry. *Mailing Add:* Dept Chem Univ Pittsburgh Pittsburgh PA 15260

YATES, JON ARTHUR, b Independence, Mo, Dec 14, 47. FILARIASIS, PARASITE IMMUNOLOGY. *Educ:* Univ Mo, BS, 73; Tulane Univ, MS, 76, PhD(parasitol), 81. *Prof Exp:* Res assoc, Tulane Univ, Colciencias Int Ctr Med Res, Cali & El Porvenir, Columbia, 77-80; postdoctoral res scholar immunol, Dept Epidemiol, Sch Pub Health, Univ Mich, 81-84, sr res fel epidemiol, 84-86; ASST PROF BIOL PARASITOL IMMUNOL, OAKLAND UNIV, ROCHESTER HILLS, 86- *Concurrent Pos:* Dir, Field Parasitol Lab, El Porvenir, Columbia, 78-79; scientist, Indo-US Res Initiative, Rajamundry, India, 85; guest lectr, Nat Inst Commun Dis, Dehli, India, 85; adj asst prof epidemiol, Sch Pub Health, Univ Mich, 86-, vis lectr, Grad Summer Session Epidemiol, 89; mem, Sci Working Group Filariasis, WHO, 87 & Trop Med Deleg Peoples Repub China, 88. *Mem:* Am Soc Trop Med & Hyg; Am Soc Parasitologists; Royal Soc Trop Med & Hyg; Am Soc Trop Vet Med. *Res:* Biology and immunology of parasitic nematode infections; vaccine models; chemotherapy; immunopathologic mechanisms in parasitic diseases; vector biology host-parasite interactions. *Mailing Add:* Dept Biol Sci Oakland Univ Rochester Hills MI 48309-4401

YATES, KEITH, b Preston, Eng, Oct, 22, 28; Can citizen; m 53; c 3. PHYSICAL ORGANIC CHEMISTRY. *Educ:* Univ BC, BA, 56, MSc, 57, PhD(org chem), 59; Oxford Univ, DPhil(phys chem), 61. *Prof Exp:* From asst prof to assoc prof chem, 61-67, asst dean, Sch Grad Studies, 67-70, chmn dept, 74-85, PROF CHEM, UNIV TORONTO, 68- *Honors & Awards:* Syntex Award, Chem Inst Can. *Mem:* Chem Inst Can; fel Royal Soc Can. *Res:* Physical and theoretical organic chemistry; acidity functions and reaction mechanisms. *Mailing Add:* Dept Chem Univ Toronto 15 King Col Circle Toronto ON M5S 1A1 Can

YATES, LELAND MARSHALL, b Stevensville, Mont, Feb 11, 15; c 4. PHYSICAL CHEMISTRY. *Educ:* Mont State Univ, BA, 38, MA, 40; Wash State Univ, PhD(chem), 55. *Prof Exp:* Instr chem & physics, Custer County Jr Col, 40-42 & 45-47; instr chem, Univ Mont, 47-49; asst, Wash State Univ, 49-51; from instr to assoc prof, 51-71, prof chem, 71-78, EMER PROF CHEM, UNIV MONT, 78- *Mem:* Am Chem Soc. *Res:* Equilibrium constants and thermodynamics of complex ions; analysis for small concentration of ions. *Mailing Add:* 610 Hastings Ave Missoula MT 59801

YATES, MARY ANNE, b Circleville, Ohio, Apr 26, 49. NUCLEAR CHEMISTRY. *Educ:* Univ Rochester, NY, BA, 71; Carnegie-Mellon Univ, MS, 73, PhD(nuclear chem), 76. *Prof Exp:* Postdoctoral fel, Los Alamos Sci Lab, 76-79; STAFF MEM, LOS ALAMOS NAT LAB, 79- *Honors & Awards:* Award of Excellence, Dept Energy, 87. *Mem:* Sigma Xi; Am Phys Soc. *Res:* Research and documentation of a variety of defense related projects. *Mailing Add:* Los Alamos Nat Lab MS J514 PO Box 1663 Los Alamos NM 87545

YATES, PETER, b Wanstead, Eng, Aug 26, 24; m 50; c 3. ORGANIC CHEMISTRY, ORGANIC PHOTO CHEMISTRY. *Educ:* Univ London, BSc, 46; Dalhousie Univ, MSc, 48; Yale Univ, PhD(chem), 51. *Prof Exp:* Instr chem, Yale Univ, 51-52; from instr to asst prof, Harvard Univ, 52-60; prof chem, 60-90, univ prof, 86-90, EMER UNIV PROF CHEM, UNIV TORONTO, 90- *Concurrent Pos:* Sloan Found fel, 57-60; vis prof, Yale Univ, 66; vis prof, Princeton Univ, 77. *Honors & Awards:* Centennial Medal, Govt Can, 67; Merck Sharp & Dohme lectr, Chem Inst Can, 63; Chem Inst Can Medal, 84. *Mem:* Am Chem Soc; Chem Inst Can; Royal Soc Can. *Res:* Structural, synthetic and mechanistic organic chemistry, including natural and photochemical products, aliphatic diazo and heterocyclic compounds. *Mailing Add:* Dept Chem Univ Toronto Toronto ON M5S 1A1 Can

YATES, RICHARD ALAN, b Oakland, Calif, July 14, 30; m 58; c 1. BIOCHEMISTRY, MICROBIOLOGY. *Educ:* Univ Calif, Berkeley, BA, 52, PhD(biochem), 56. *Prof Exp:* RES BIOCHEMIST, CENT RES & DEVELOP DEPT, E I DU PONT DE NEMOURS & CO, INC, 56- *Mem:* Am Soc Microbiol; Inst Food Technologists; AAAS; NY Acad Sci. *Res:* Biochemical control mechanisms; microbial genetic alterations; production of feedstock chemicals from renewable sources; single cell protein production and adaptation for human foods. *Mailing Add:* 233 Prospect Dr Wilmington DE 19803

YATES, RICHARD LEE, b Red Oak, Iowa, June 15, 31; m 61; c 2. MATHEMATICS. *Educ:* Fla Southern Col, BS, 52; Univ Fla, MS, 54, PhD(math), 57. *Prof Exp:* Asst, Univ Fla, 52-56; asst prof, Univ Houston, 57-60; from asst prof to assoc prof, Kans State Univ, 60-67; assoc prof & chmn math sect, 67-70, prof & acad dean, 70-75, exec asst to chancellor, 75-80, PROF MATH, PURDUE UNIV, 80- *Mem:* Math Asn Am. *Res:* Classical number theory; modern algebra; lattice theory. *Mailing Add:* Dept Math Purdue Univ Calumet Campus 2233 171st St Hammond IN 46323

YATES, ROBERT DOYLE, b Birmingham, Ala, Feb 28, 31; m 55; c 2. CYTOLOGY. *Educ:* Univ Ala, BS, 54, MS, 56, PhD(anat), 60. *Prof Exp:* Instr gross anat & neuroanat, Univ Tex Med Br, Galveston, 61-64, from asst prof to prof microanat, 64-70; PROF ANAT & CHMN DEPT, SCH MED, TULANE UNIV, LA, 72- *Concurrent Pos:* Fel, Med Ctr, Univ Ala, 61-62; NIH career res develop award, 64. *Honors & Awards:* Golden Apple Award, SAMA. *Mem:* AAAS; Am Soc Cell Biol; Am Asn Anat; Am Soc Neuropath; Fedn Am Soc Exp Biol. *Res:* Electron microscopic studies of reversible alterations in the organelles and inclusions of cells subjected to experimentally induced stresses. *Mailing Add:* Dept Anat Sch Med Tulane Univ New Orleans LA 70112

YATES, ROBERT EDMUNDS, b Bisbee, Ariz, Aug 15, 26; m 47; c 4. PHYSICAL CHEMISTRY. *Educ:* Univ Ariz, BS, 48, MS 49; Mich State Univ, PhD(phys chem), 52. *Prof Exp:* Asst chem, Mich State Univ, 49-51; res engr, Dow Chem Co, 52-58; res chemist, Aerojet-Gen Corp, 58-61 & Rocket Power, Inc, 61-65; res chemist, Aerojet-Gen Corp, Calif, 66-67, chem specialist, 67-71; PHYSICIST, MCCLELLAN AFB, SACRAMENTO, 71- *Mem:* Am Chem Soc. *Res:* Boron, fluorine and high temperature chemistry; thermodynamics and spectroscopy. *Mailing Add:* 7313 Pine Grove Way Folsom CA 95630-1923

YATES, SCOTT RAYMOND, b Long Beach, Calif, Dec 1, 54. SOIL PHYSICS, CONTAMINANT TRANSPORT. *Educ:* Univ Wis-Madison, BS, 80; NMex Inst Mining & Technol, MS, 82; Univ Ariz, PhD(soil physics), 85. *Prof Exp:* Sr sci specialist simulation, Computer Sci Corp, 85-86; soil scientist soil physics, US Environ Protection Agency, 86-87; SOIL SCIENTIST SOIL PHYSICS, AGR RES SERV, USDA, 88- *Concurrent Pos:* Mem task force comt, Am Soc Civil Engrs, 87-89; prin investr grant, US Environ Protection Agency, 87-91, res grant, US Golf Asn, 91-; mem comt, W-82, Western Regional Res Proj, 88-; adj asst prof, Dept Soil & Environ Sci, Univ Calif, Riverside, 89-; mem Emil Truog Award comt, Am Soc Agron, 89-91, monographs comt, 90-; mem Water Sci Technol Bd, Nat Res Coun, 90- *Mem:* Sigma Xi; Am Soc Agron; Am Geophys Union; Int Geostatist Asn; Int Soc Soil Sci; Soil Sci Soc Am. *Res:* Transport of pesticides and other toxic substances in the environment; developing analytical and numerical solutions to soil-physical and hydrological models; application of models for simulating the transport of water contaminants and energy in the subsurface; volatilization processes; geostatistics. *Mailing Add:* Dept Soil & Environ Sci UCR Pesticides & Water Qual Res Agr Res Serv USDA Riverside CA 92521

YATES, SHELLY GENE, b Altus, Okla, Feb 29, 32; m 54; c 4. CHEMISTRY, NATURAL PRODUCTS. *Educ:* Southwestern State Col, Okla, BS, 56; Okla State Univ, MS, 58. *Prof Exp:* Org chemist, Northern Regional Res Ctr, Agr Res Serv, USDA, 58-88; RETIRED. *Mem:* Am Chem Soc. *Res:* Natural products; isolation, characterization and analysis. *Mailing Add:* 5619 N Plaza Dr Peoria IL 61614

YATES, STEVEN WINFIELD, b Memphis, Mo, Apr 19, 46; m 75; c 1. RADIOCHEMISTRY. *Educ:* Univ Mo-Columbia, BS, 68; Purdue Univ, Lafayette, PhD(chem), 73. *Prof Exp:* Res asst chem, Purdue Univ, Lafayette, 71-73; fel, Argonne Nat Lab, 73-75; from asst prof to assoc prof, 75-85, PROF CHEM, UNIV KY, 85- *Concurrent Pos:* Proctor & Gamble fel, 72; dir gen chem, 85-87. *Honors & Awards:* Res Found Award, Univ Ky, 81. *Mem:* Am Chem Soc; Am Phys Soc; Sigma Xi. *Res:* Level structures of transitional and deformed nuclei; heavy-ion reactions; inelastic scattering and transfer reactions; nuclear isomerism and high-spin phenomena; neutron induced reactions; neutron scattering; nuclear spectroscopy. *Mailing Add:* Dept Chem Univ Ky Lexington KY 40506-0055

YATES, VANCE JOSEPH, b Smithville, Ohio, Oct 25, 17; m 42; c 4. VETERINARY VIROLOGY. *Educ:* Ohio State Univ, BSc, 40, DVM, 49; Univ Wis, PhD, 60. *Prof Exp:* Instr high sch, Ohio, 40-41; asst prof animal path, Univ RI, 49-50, assoc prof & assoc res prof, 51-55, head dept, 51-79, prof animal path & res prof, 55-83; RETIRED. *Concurrent Pos:* Mem temp staff, Rockefeller Found, 63-64; vis prof dept exp biol, Baylor Col Med, 71-72. *Mem:* Am Vet Med Asn; Am Asn Avian Path. *Res:* Avian virology and pathology; oncogenicity of avian adenoviruses. *Mailing Add:* 40 Spring Hill Rd Kingston RI 02881

YATES, WILLARD F, JR, b Findlay, Ohio, June 20, 30; m 65; c 2. PLANT TAXONOMY, CYTOGENETICS. *Educ:* Eastern Ill Univ, BS, 58; Ind Univ, MA, 60, PhD(bot), 67. *Prof Exp:* Instr biol, Cumberland Col, 60-62; asst prof, Ball State Univ, 65-67; assoc prof, 67-78, PROF BOT, BUTLER UNIV, 78- *Mem:* Tissue Cult Asn; Bot Soc Am; Am Soc Plant Taxon; Plant Molecular Biol Asn. *Res:* Plant cytotaxonomy; phytochemistry; plant tissue culture. *Mailing Add:* Dept Bot Butler Univ 4600 Sunset Ave Indianapolis IN 46208

YATES-PARKER, NANCY L, b Jackson, Miss, Nov 14, 56; m 84; c 1. PESTICIDE REGISTRATION. *Educ:* Miss State Univ, BS, 78; Univ Mo, MS, 80; NC State Univ, PhD(hort), 84. *Prof Exp:* Intern, Prod Develop Div, Monsanto Agr Co, St Louis, Mo, 80- 81, prod develop rep, Omaha, Nebr, 84-85, Kansas City, Mo, 85-86, prod develop assoc, St Paul, Minn, 86-88, sr registration specialist, St Louis, Mo, 88-89, MGR REGISTRATION, ENVIRON & PUB AFFAIRS, MONSANTO AGR CO, ST LOUIS, MO, 89- *Mem:* Weed Sci Soc Am. *Res:* Development of pesticides; residue analysis of food commodities for pesticide residue. *Mailing Add:* Monsanto Agr Co 800 N Lindbergh Blvd St Louis MO 63167

YATSU, FRANK MICHIO, b Los Angeles, Calif, Nov 28, 32; m 55; c 1. NEUROLOGY. *Educ:* Brown Univ, AB, 55; Case Western Reserve Univ, MD, 59. *Prof Exp:* From asst prof to assoc prof neurol, Univ Calif Med Ctr, San Francisco, 67-75, vchmn dept, 73-75; prof & chmn dept, Univ Ore Health Sci Ctr, 75-82; PROF NEUROL & CHMN DEPT, UNIV TEX HEALTH SCI CTR, HOUSTON, 82- *Concurrent Pos:* Chief neurol serv, San Francisco Gen Hosp, 69-75; mem cardiovasc A res study comt, Am Heart Asn, 74-77; mem neurol disorders prog, Proj A Rev Comt, Nat Inst Neurol & Commun Disorders & Stroke, NIH, 75-79; mem adv coun, Epilepsy Ctr of Ore, 75- *Mem:* Am Acad Neurol; Am Neurol Asn; Am Soc Neurochem; Int Soc Neurochem. *Res:* Brain ischemia and atherosclerosis. *Mailing Add:* Dept Neurol Univ Tex Health Sci Ctr 6431 Fannin St Suite 7044-MSMB Houston TX 77030

YATSU, LAWRENCE Y, b Pasadena, Calif, Aug 2, 25; m 54; c 2. PLANT PHYSIOLOGY. *Educ:* Mich State Univ, BS, 49; Univ Calif, MS, 50; Cornell Univ, PhD, 60. *Prof Exp:* Chemist, Strong, Cobb & Co, 54-55; res assoc, Cornell Univ, 55-60; plant physiologist, Field Lab Tung Invest, USDA, 60-61, RES CHEMIST, SOUTHERN REGIONAL RES CTR, USDA, 61- *Concurrent Pos:* Adj assoc prof biol, Tulane Univ, 73- *Mem:* AAAS; Am Chem Soc; Bot Soc Am; Am Soc Plant Physiol; Am Inst Biol Scientists; Sigma Xi. *Res:* Cell biology; biochemistry. *Mailing Add:* 7611 Dalewood Rd New Orleans LA 70126

YATVIN, MILTON B, b New Brunswick, NJ, Nov 12, 30; m 52; c 3. PHYSIOLOGY, RADIOBIOLOGY. *Educ:* Rutgers Univ, BS, 52, MS, 54, PhD(endocrinol, reproductive physiol), 62. *Prof Exp:* Instr dairy sci, Rutgers Univ, 55-56; lectr reproductive physiol, Univ PR, 57-59; from instr to assoc prof, Univ Wis-Madison, 63-71, prof radiobiol, Med Sch, 71-88, prof human oncol, 77-88; PROF RADIATION ONCOL, DEPT BIOSCI & MOLECULAR BIOL, SCH MED, ORE HEALTH SCI UNIV, 88- *Concurrent Pos:* NIH fel endocrinol, Rutgers Univ, 62-63. *Mem:* Biophys Soc; Radiation Res Soc; Soc Exp Biol & Med; Am Physiol Soc; Sigma Xi. *Res:* Cell damage and repair after exposure to ionizing radiation; nucleic acid-membrane relationships in cells. *Mailing Add:* 5226 SW Northwood Ave Portland OR 97201

YAU, CHEUK CHUNG, b Hong Kong, Apr 21, 50; US citizen; m 77; c 2. POLYESTERS. *Educ:* Univ Hawaii, BS, 74; Ga Inst Technol, PhD(chem), 79. *Prof Exp:* Develop chemist res & develop, Tenn Eastman Co, 79-83, res chemist, Eastman Chem Div, 83-84, sr chemist Res & Develop, 84-89, PRIN RES CHEMIST, EASTMAN CHEM DIV, EASTMAN KODAK CO, 89- *Concurrent Pos:* Reviewer, J Org Chem, Am Chem Soc, 83- *Mem:* Am Chem Soc. *Res:* Understanding of polyester systems to improve their performance; predicting formulations of polyesters to meet property specifications. *Mailing Add:* 313 Highridge Rd Kingsport TN 37660

YAU, CHIOU CHING, b Taiwan, Dec 31, 34; US citizen; m 66; c 2. FIBER & POLYMER SCIENCE. *Educ:* Nat Cheng Kung Univ, Taiwan, BS, 58; Ga Inst Technol, MS, 66; NC State Univ, PhD(fiber & polymer sci), 72. *Prof Exp:* Chem engr, Chinese Petrol Corp, 60-65; res chemist, Am Enka Corp, Akzona, Inc, 66-69; mem tech staff, Emery Indust Inc, 72-77; res scientist, Kimberly-Clark Corp, 77-79; sr res chemist, Kendall Co, 79-81; PROG CHEMIST, GILLETTE CO, 81- *Mem:* Am Chem Soc; Soc Plastics Engrs; Fedn Soc Coatings Technol. *Res:* Polymer synthesis; polymer characterization by thermal analysis; structure and property relationships of polymers; fiber chemistry; flame retardants for fibers and plastics; nonwovens, polymer emulsions and coatings. *Mailing Add:* Gillette Co Gillette Park Boston MA 02106

YAU, KING-WAI, b China, Oct 27, 48; m 75; c 2. NEUROPHYSIOLOGY, VISION. *Educ:* Princeton Univ, AB, 71; Harvard Univ, PhD(neurobiol), 76. *Prof Exp:* Res fel neurobiol, Sch Med, Stanford Univ, 76-79, physiol, Univ Cambridge, Eng, 79-80; from asst prof to assoc prof, 80-85, PROF PHYSIOL & BIOPHYS, MED BR, UNIV TEX, 85- *Concurrent Pos:* Vis fel, Trinity Col, Univ Cambridge, 80-81. *Honors & Awards:* Rank Prize, Eng, 80. *Mem:* Biophys Soc; Soc Neurosci; Asn Res Vision & Ophthal. *Res:* Retinal physiology. *Mailing Add:* Dept Neurosci Johns Hopkins Univ Med Sch 720 Rutland Ave Baltimore MD 21205

YAU, LEOPOLDO D, b Surigao, Philippines, Aug 15, 40; US citizen; m 71; c 1. SEMICONDUCTOR PROCESS TECHNOLOGY, MOS DEVICE PHYSICS. *Educ:* Univ San Carlos, BS, 62; Univ Minn, MS, 65; Univ Ill, PhD(elec eng), 69. *Prof Exp:* Asst prof elec eng, Univ Phillipines, 70-71, Univ Ill, 71-73; staff, Bell Telephone Labs, 73-78; sr scientist, 78-84, prin engr, 84-86, FEL, INTEL, 86- *Concurrent Pos:* Fel, Univ Ill, 65-67. *Mem:* Inst Elec & Electronics Engrs. *Res:* Deep level impurities in silicon; short-channel device model ofmos-transistors; applications of electron-beam lithography; measurements of process-induced distortion in silicon; dynamic random access memory technology; ultra-thin dielectrics; low pressure CVD; plasma enhanced CVD. *Mailing Add:* 3539 NW Brunson Crest Loop Portland OR 97229

YAU, SHING-TUNG, b Kwuntung, China, Apr 4, 49; m; c 2. MATHEMATICS. *Educ:* Univ Calif, Berkeley, PhD(math), 71; Harvard Univ, PhD, 87. *Prof Exp:* Fel, Inst Advan Study, Princeton Univ, 71-72; asst prof math, State Univ NY, Stony Brook, 72-73; vis asst prof, Stanford Univ, 73-74, from asst prof to prof, 74-80; prof, Inst Advan Study, Princeton, 80-83; prof math, Univ Calif, San Diego, 83-87; PROF MATH, HARVARD UNIV, 87- *Concurrent Pos:* Distinguished vis prof, State Univ New York, Stony Brook, 90; Fairchild Distinguished Scholar, Calif Tech, 90; MacArthur fel, 85; mem bd, Math Sci, Nat Acad Sci. *Honors & Awards:* Veblen Prize, Am Mat Soc, 81; Certy Prize, Nat Acad Sci, 81; Field's Medal, 82; Humboldt Found Sr Scientist Award, 82. *Mem:* Am Math Soc; NY Acad Sci; Acad Arts & Sci; Am Phys Soc; Soc Indust & Appl Math; AAAS. *Res:* Differential geometry. *Mailing Add:* Dept Math Harvard Univ Cambridge MA 02138

YAU, STEPHEN SHING-TOUNG, b Hong Kong, China, Apr 12, 52; m 85; c 1. KOHN-ROSSI COHOMOLOGY, ELECTRICAL ENGINEERING. *Educ:* Staten Univ NY, Stony Brook MA, 74, PhD(math, 76). *Prof Exp:* Mem res, Inst Advan Study, 76-77 & 81-82; asst prof math, Harvard Univ, 77-80; assoc prof, 80-84, Univ Senate, admin, 86-89, PROF MATH, UNIV ILL, CHICAGO, 84- *Concurrent Pos:* Sloan Res Fel, Alfred Sloan Res Found, 80-84; co-prin investr, NSF, 80-84, prin investr, 84-88 & 89-91; vis prof, math, Princeton Univ, 81; vis assoc prof, Univ Southern Calif, 83-84; vis prof, Yale Univ, 84-85; vis prof, Inst Mittag-Leffler, Royal Swedish Acad of Sci, 87; vis prof, Johns Hopkins Univ, 89-90; prin investr, Army Res Off, 89-92; ed, J Algebraie Geom & Singularities, 90- *Mem:* Am Math Soc; Inst Elec & Electronics Engrs; Soc Indust & Appl Math. *Res:* Development of structure theory for weakly elliptic singularities and classify them; solving of real co-dimension three complex plateau problems for strongly psuedoconvex manifolds; to completely understand complex structures of hupersurface singularities; classify finite dimensional estimation algebras in nonlinear filtering theory. *Mailing Add:* Dept Math Statist & Comput Sci Univ Ill Chicago Box 4348 Chicago IL 60680

YAU, STEPHEN SIK-SANG, b Wusei, Kiangsu, China, Aug 6, 35; m 64; c 2. COMPUTER SCIENCE, ELECTRICAL ENGINEERING. *Educ:* Nat Taiwan Univ, BS, 58; Univ Ill, Urbana, MS, 59, PhD(elec eng), 61. *Prof Exp:* From asst prof to prof elec eng, 61-70, chmn Dept Comput Sci, 72-77, PROF ELEC ENG & COMPUT SCI, NORTHWESTERN UNIV, EVANSTON, 70-, CHMN DEPT, 77- *Concurrent Pos:* Consult, Battelle Columbus Lab, 76-77 & Syst Develop Corp, 77. *Honors & Awards:* Levy Medal, Franklin Inst, 63; Golden Plate Award, Am Acad Achievement, 64. *Mem:* AAAS; Am Soc Eng Educ; Inst Elec & Electronics Engrs; Asn Comput Mach; Soc Indust & Appl Math; Sigma Xi. *Res:* Reliability and maintainability of computing systems; software engineering; computer architecture; fault tolerant computing; pattern recognition. *Mailing Add:* Dept Elec Eng & Comput Sci Northwestern Univ 633 Clark St Evanston IL 60201

YAU, WALLACE WEN-CHUAN, b Shianghai, China, Feb 20, 37; US citizen; m 61; c 4. ANALYTICAL CHEMISTRY, POLYMER PHYSICS. *Educ:* Nat Taiwan Univ, BS, 59; Univ Mass, PhD(phys chem), 66. *Prof Exp:* RES CHEMIST ANALYTICAL CHEM, CENT RES & DEVELOP DEPT, E I DU PONT DE NEMOURS & CO, INC, 65- *Honors & Awards:* IR-100 Award. *Mem:* Am Chem Soc. *Res:* Studies of polymer structures and properties using viscometry, osmometry, chromatographic, optical and mechanical characterization techniques; separation science; sec; viscometry; light scattering. *Mailing Add:* DuPont Exp Sta PO Box 80228 Wilmington DE 19880-0228

YAU, WEN-FOO, b Shanghai, China, June 11, 35; US citizen; m 62. ENGINEERING MECHANICS, APPLIED MATHEMATICS. *Educ:* Nat Taiwan Univ, BS, 58; Univ Mass, MS, 61; Princeton Univ, PhD, 65. *Prof Exp:* Instr civil eng, Univ Mass, 61-63; res assoc eng mech, Univ Ky, 65-66, asst prof, 66-69; res engr, Eastern Lab, 69-72, staff engr, 72-78, res staff engr, 78-81, RES ASSOC, SAVANNAH RIVER LAB, E I DU PONT DE NEMOURS & CO, INC, 81- *Mem:* Am Soc Mech Engrs. *Res:* Operation and safety considerations of the Savannah River Plant for production of nuclear materials. *Mailing Add:* 2103 Maple Dr N Augusta SC 29801

YAU-YOUNG, ANNIE O, BIOCHEMISTRY & IMMUNOLOGY, CELL BIOLOGY. *Educ:* Boston Univ, PhD(biochem), 78. *Prof Exp:* Sr scientist, Liposome Technol Inc, 84-89; LICENSING ASSOC, OFF TECHNOL LICENSING, UNIV CALIF, BERKELEY, 89- *Mailing Add:* Off Technol Licensing 2150 Shattuck Ave Suite 510 Berkeley CA 94704

YAVERBAUM, SIDNEY, b New York, NY, Jan 28, 23; m 67; c 2. MEDICAL MICROBIOLOGY, IMMUNOCHEMISTRY. *Educ:* Univ Pa, PhD(med microbiol), 52. *Prof Exp:* Asst, Univ Pa, 51-52, res assoc, 52-53; fel microbiol, Boyce Thompson Inst Plant Res, 54-55; res med bacteriologist, Bio-Detection Br, Phys Defense Div, US Dept Army, Ft Detrick, 55-70; sr res biologist, Corning Glass Works, 70-76; mem staff, Wampole Div, Carter Wallace Inc, 76-80; MEM STAFF, TECHNICON, 80- *Mem:* AAAS; Am Soc Microbiol; Am Chem Soc; Sigma Xi; NY Acad Sci. *Res:* Cytology of yeasts, fungi and bacteria; genetics and nutrition of bacteria; assay of fungicides; physiology of aerobic sporeforming bacteria; biochemical composition of microorganisms; radioactive antibodies; solid-phase radioimmunoassay; immobilized enzyme research; automated instrumentation for microbiology. *Mailing Add:* 703 Monticello Ct Edgewood MO 21040

YAVORSKY, JOHN MICHAEL, b Renovo, Pa, June 11, 19. WOOD TECHNOLOGY. *Educ:* State Univ NY, BS, 42, MS, 47, PhD(wood eng), 55. *Prof Exp:* Res assoc & assoc prof wood utilization, State Univ NY Col Forestry, Syracuse, 48-56; chief wood utilization sect, Forestry Div, Food & Agr Orgn, UN, Rome, Italy, 57-63, proj mgr forestry proj, Lima, Peru, 63-67, sr proj officer, UN Spec Fund, NY, 67; prof forestry, 67-84, dean, Sch Continuing Educ, 73-84, EMER DEAN SCH CONTINUING EDUC, STATE UNIV NY COL ENVIRON SCI & FORESTRY, SYRACUSE, 84- *Mem:* Forest Prod Res Soc; Soc Am Foresters; Soc Wood Sci & Technol; Int Soc Trop Foresters. *Res:* Planning and supervision of continuing education activities in forestry and forest products technology; world forestry aspects of forest industries development. *Mailing Add:* 1216 Meadowbrook Dr Syracuse NY 13224

YAVORSKY, PAUL M(ICHAEL), b Donora, Pa, Jan 6, 25; m 50; c 3. CHEMICAL ENGINEERING, PHYSICAL CHEMISTRY. *Educ:* Univ Pittsburgh, BS, 47, MS, 49, PhD, 56. *Prof Exp:* Asst eng & nuclear physics, Univ Pittsburgh, 47-49; chemist, Res Div, Consol Coal Co, Pa, 56-60, group leader, 60-71; res supvr explor eng, US Bur Mines, 71-73; asst dir coal liquefaction, US Off Coal Res, 73-74; div mgr, US Energy Res & Develop Admin, US Dept Energy, 74-79; mgr, Mine Equip Test Facil, US Bur Mines, 79-87; RETIRED. *Honors & Awards:* Bituminous Coal Res Awards, 57 & 64. *Mem:* Am Chem Soc; Am Inst Chem Engrs. *Res:* Coal energy conversion; engineering development of liquefaction and gasification. *Mailing Add:* 714 Birch Ave Monongahela PA 15063

YAWS, CARL LLOYD, b Yoakum, Tex, Oct 1, 38; m 58; c 4. CHEMICAL ENGINEERING, CHEMISTRY. *Educ:* Tex A&I Col, BS, 60; Univ Houston, MS, 63, PhD(chem eng), 65. *Prof Exp:* Res chem engr, Baytown Labs, Esso Res & Eng Co, Tex, 63-65; process develop engr, Res & Develop Dept, Ethyl Corp, La, 65-68; sr engr & mem tech staff, Chem Mat Div, Tex Instruments, Inc, 68-77; ASSOC PROF CHEM ENG, LAMAR UNIV, 77- *Mem:* Am Inst Chem Engrs; Am Chem Soc. *Res:* Chemicals; petrochemicals; petroleum refining; pollution control; solvent extractions; distillation; chlorination; fluorination; hydrogenation; desulfurization; catalytic reactions; chemical vapor deposition; fluorocarbons; chlorocarbons; specialty hydrocarbons; lubricants; gasolines; ultra purity silicon; carbides; nitrides. *Mailing Add:* 685 Birchwood Port Neches TX 77651

YAYANOS, A ARISTIDES, b Buffalo, NY, Jan 31, 40; m 66; c 1. BIOPHYSICS, DEEP-SEA BIOLOGY. *Educ:* Univ Buffalo, BA, 61; Pa State Univ, MS, 65, PhD(biophysics), 67. *Prof Exp:* Fel, Univ Calif, 67-68, asst res physiologist, 68-78, assoc res biophysicist, 78-83, RES BIOPHYSICIST & SR LECTR, SCRIPPS INST OCEANOG, UNIV CALIF, SAN DIEGO, 83- *Concurrent Pos:* NIH res career develop award, 70-75. *Mem:* Am Chem Soc; Math Asn Am; Am Soc Microbiol; AAAS. *Res:* Physiology of deep-sea invertebrates; deep-sea microbiology; physics and biology at high pressures; radiation biology. *Mailing Add:* PO Box 3412 Rancho Santa Fe CA 92067

YAZ, ENGIN, b Istanbul, Turkey, May 11, 54; m 80. CONTROL SYSTEMS, SIGNAL PROCESSING. *Educ:* Bosphorus Univ, BS, 76, MS, 79, PhD(elec eng), 82. *Prof Exp:* Teaching asst elec eng, Bosphorus Univ, Turkey, 76-79; instr elec eng, Yildiz Univ, Turkey, 79-82; researcher control systs, Marmara Res Inst & Res Inst Basic Sci, 82-84; asst prof, 85-87, ASSOC PROF ELEC ENG, UNIV ARK, 87- *Concurrent Pos:* Consult, Sun Electronics Co, Turkey, 78; prin investr, Univ Ark, 86-87; res fel, Coord Sci Lab, Univ Ill, Urbana-Champaign, 84. *Honors & Awards:* Halliburton Found Award, 87. *Mem:* Inst Elec & Electronics Engrs; Am Soc Eng Educ. *Res:* Non-linear and stochastic control and estimation; stability theory with applications to robotics; neural networks and adaptive controllers; intelligent control and information processing. *Mailing Add:* Elec Eng Dept BEC 3217 Univ Ark Fayetteville AR 72701

YAZICIGIL, HASAN, b Konya, Turkey, Nov 24, 52; m 77; c 1. GROUND-WATER MODELLING, SURFACE-WATER FORECASTING. *Educ:* Middle East Tech Univ, BS, 74; Iowa State Univ, MS, 77; Purdue Univ, PhD(systs eng), 80. *Prof Exp:* Soil surv proj engr, Gen Directorate Hwy Dept, Turkey, 74-75; proj asst, Iowa State Univ, 75-77; instr, Purdue Univ Sch Civil Eng, 77-80; res assoc, 80-81, ASST PROF HYDROGEOL ENG GEOL, SOUTHERN ILL UNIV, CARBONDALE COAL RES CTR, 81- *Concurrent Pos:* Consult engr, Willow Creek Consults, 81-, Sargent Limestone Quarry, 76-77; res assoc, Coal Res Ctr, Southern Ill Univ, 80- *Mem:* Am Geophys Union; Oper Res Soc Am; Nat Water Well Asn; Int Asn Hydrogeologists; Chambers Turkish Geol Engrs. *Res:* Ground water and surface water modelling; environmental modelling; optimal operations of surface and underground reservoir systems using forecasts. *Mailing Add:* Dhahran Int Airport Box 144 OPM No 771 Dhahram 583 Saudia Arabia

YAZULLA, STEPHEN, b Jersey City, NJ, Sept 3, 45; m 67, 83; c 2. NEUROSCIENCE, RETINA. *Educ:* Univ Scranton, BS, 67; Univ Del, MA, 69, PhD(psychol), 71. *Prof Exp:* Postdoctoral fel biol, Univ Del, 71-72, neurobiol, Harvard Univ, 72-74; from asst prof to assoc prof, 74-86, PROF NEUROBIOL, STATE UNIV NY, STONY BROOK, 86- *Concurrent Pos:* NIH res grant, prin investr, State Univ NY, Stony Brook, 76-; adj mem, NIH-Nat Eye Inst Vis A Study Sect, Div Res Grants, 85, mem, 86-90, Reviewers Reserve, 90- *Mem:* Sigma Xi; Asn Res Vision & Ophthal; Soc Neurosci; Int Soc Eye Res; Int Brain Res Orgn. *Res:* Cellular mechanisms and anatomical organization underlying information processing in the vertebrate retina; light and electron microscopic autoradiography; immunocytochemistry; electrophysiology; pharmacology; biochemistry; cellular basis of visual processing. *Mailing Add:* Dept Neurobiol & Behav State Univ NY Stony Brook NY 11794-5230

YCAS, MARTYNAS, b Voronezh, Russia, Dec 10, 17; nat US; m 45; c 3. BIOLOGY, COMPUTER APPLICATIONS. *Educ:* Univ Wis, BA, 47; Calif Inst Technol, PhD(embryol), 50. *Prof Exp:* Instr, Univ Wash, 50-51; biologist pioneering res labs, Corps, US Dept Army, 51-56; from asst prof to prof microbiol, 56-88, EMER PROF MICROBIOL, SCH MED, STATE UNIV NY UPSTATE MED CTR, 88- *Res:* Biochemical evolution; computer applications to protein structure. *Mailing Add:* 109 Croyden Rd Syracuse NY 13224

YEADON, DAVID ALLOU, b New Orleans, La, Nov 10, 20; m 49; c 2. CHEMISTRY. *Educ:* Loyola Univ, La, BS, 40; Univ Detroit, MS, 42. *Prof Exp:* Asst & lab instr, Univ Detroit, 40-42; chemist, Gelatin Prod Corp, Mich, 42-43; chemist, Esso Standard Oil Co, La, res chemist, Esso Labs, 43-50; res chemist, Southern Regional Res Lab, Naval Store Div, USDA, 50; res chemist, Alpine Corp, Miss, 50-53; res chemist, Oilseed Crops Lab, 53-65, RES CHEMIST, COTTON FINISHES LAB, SOUTHERN REGIONAL RES LAB, USDA, 65- *Mem:* Am Chem Soc; Sigma Xi; Am Asn Textile Chemists & Colorists. *Res:* Synthetic rubber; hydrocarbons; polymers, resins and coatings; chemistry, synthesis and applications of fats and oils; modifications to improve utilization of cotton; fire retardant cotton textiles. *Mailing Add:* 1460 Pressburg St New Orleans LA 70122

YEAGER, ERNEST BILL, b Orange, NJ, Sept 26, 24. ELECTROCHEMISTRY. *Educ:* Montclair State Col, BA, 45; Western Reserve Univ, MS, 46, PhD(phys chem), 48. *Hon Degrees:* Dr, Montclair State Col, 83. *Prof Exp:* Asst physics & phys chem, Case Western Reserve Univ, 45-47, from instr to assoc prof chem, 48-58, actg chmn dept, 64-65, chmn dept, 69-72, chmn fac senate, 72-73, PROF CHEM, CASE WESTERN RESERVE UNIV, 58-, DIR CASE CTR ELECTROCHEM SCI, 76-, FRANK HOVORKA CHAIR CHEM, 83-, PROF CHEM ENG, 84-, FRANK HOVORKA EMER PROF CHEM, 90- *Concurrent Pos:* Consult, Union Carbide Corp, 55- & Gen Motors Corp, 70; vis prof & NATO fel, Univ Southampton, 68; mem comt undersea warfare, Nat Acad Sci-Nat Res Coun, 63-73; rep mem, phys sci div, Nat Res Coun, 69-73, mem, comt Battery Mat & Fuel Cells for Vehicular Appln, Nat Mat Adv Bd, Nat Res Coun, 79-81 & comt Electrochem Aspects Energy Conserv & Prod, Nat Mat Adv Bd, 85-87; mem comn electrochem, Int Union Pure & Appl Chem, 69-75; mem underwater sound adv group, Off Naval Res, 72-74; mem, vis comt, Brookhaven Nat Lab & Montclair State Col, NJ, 78-80; mem, adv comt on USSR & Eastern Europe, Comn Int Relations, Nat Acad Sci, 80-; mem, electrolysis technol adv comt, US Dept Energy, 82-83 & advan fuel cell working group, 84-85; mem, rev comt chem technol div, Argonne Nat Lab, 84- *Honors & Awards:* Acoust Soc Am Award, 56; Cert of Commendation, US Navy, 73; Acheson Medal, Electrochem Soc, 80. *Mem:* Fel AAAS; fel Acoust Soc Am (vpres, 67-68); Am Chem Soc; Int Soc Electrochem (vpres, 67-68, pres, 70-71); Electrochem Soc (vpres, 62-64, pres, 64-65). *Res:* Ultrasonics; electrode kinetics; electrolytes; relaxation spectroscopy; electrocatalysis. *Mailing Add:* Dept Chem Case Western Reserve Univ Cleveland OH 44106

YEAGER, HENRY, JR, b Dallas, Tex, Sept 13, 33; m 76; c 1. PULMONARY DISEASE. *Educ:* Johns Hopkins Univ, MD, 57. *Prof Exp:* Assoc prof, 77-90, PROF MED, DIV PULMONARY & CRITICAL CARE MED, MED CTR, GEORGETOWN UNIV, 90- *Mem:* Am Col Physicians; Am Thoracic Soc; Am Asn Immunologists; Am Fedn Clin Res; AMA; Am Col Chest Physicians; Am Physiol Soc; Soc Leukocyte Biol. *Res:* Pulmonary macrophage physiology; mycobacterial disease. *Mailing Add:* Med Ctr Georgetown Univ 3800 Reservoir Rd Washington DC 20007

YEAGER, HOWARD LANE, b Pittsburgh, Pa, Dec 24, 43. MEMBRANE SCIENCE, ION EXCHANGE. *Educ:* Univ Pittsburgh, BS, 65; Univ Wis-Madison, MS, 67; Univ Alta, PhD(chem), 69. *Prof Exp:* Lectr anal chem, Univ Wis, 69-70; from asst prof to assoc prof, 70-83, PROF CHEM, UNIV CALGARY, 83-, DEPT HEAD, 88- *Concurrent Pos:* Vis scientist, Nat Res Coun Can, 78-79. *Mem:* Am Chem Soc; Electrochem Soc; Chem Inst Can; AAAS; Sigma Xi. *Res:* Transport in ion exchange membranes, ion exchange thermodynamics and synthetic membrane technology. *Mailing Add:* Dept Chem Univ Calgary Calgary AB T2N 1N4 Can

YEAGER, JOHN FREDERICK, b Orange, NJ, Jan 3, 27; m 57; c 4. CHEMISTRY. *Educ:* NJ State Teachers Col, Montclair, BA, 49; Western Reserve Univ, MS, 51, PhD(phys chem), 53. *Prof Exp:* Res chemist, Nat Carbon Co, Union Carbide Corp, 53-60, res group leader, Consumer Prod Div, 60-65, res tech mgr primary batteries, 65-67, tech mgr Leclanche cells, 67-72, dir battery develop lab, Battery Prod Div, 72-75, dir technol, 76-84, vpres Technol, Battery Div, 84-86; VPRES TECHNOL, EVEREADY BATTERY CO INC, 86- *Mem:* Electrochem Soc. *Res:* Electrochemistry; plating; batteries; fuel cells. *Mailing Add:* Checkerboard Sq St Louis MO 63164

YEAGER, PAUL RAY, b Sherman, Tex, Feb 7, 31; m 56; c 2. PHYSICS, INSTRUMENT ENGINEERING. *Educ:* Austin Col, BA, 52; George Washington Univ, MS, 70. *Prof Exp:* Engr instrumentation, Nat Adv Comt Aeronaut, 52-55, sr proj engr, 55-59; leader vacuum measurements group, 59-63, head vacuum measurements sect, 63-69, head environ measurements sect, 69-74, HEAD GEN RES INSTRUMENTATION BR, NASA, 74- *Mem:* Am Vacuum Soc. *Res:* Mass spectrometry; vacuum instrumentation; thermal measurements. *Mailing Add:* 115 Paradise Point Rd Grafton VA 23692

YEAGER, SANDRA ANN, b Philadelphia, Pa, Jan 4, 39. ORGANIC CHEMISTRY, BIOCHEMISTRY. *Educ:* Thiel Col, AB, 60; Univ NH, MS, 63, PhD(org chem), 68. *Prof Exp:* Asst prof chem, Hudson Valley Community Col, NY, 62-64; res asst, Children's Cancer Res Found, Boston, Mass, 68-69; asst prof chem, Pa State Univ, Mont Alto, 69-73; asst prof org & biochem, Wilson Col, 73-74; ASSOC PROF ORG & BIOCHEM, MILLERSVILLE UNIV, 74- *Concurrent Pos:* Res assoc, Southwest Res Inst, San Antonio, Tex, 81; vis fac, Lehigh Univ, 85-86, Univ NH, 87. *Mem:* Am Chem Soc; World Future Soc; Sigma Xi; Asn Women Sci. *Res:* Analysis of biochemically important substances using varied chromatographic and electrophoretic techniques; synthesis of possible cancer chemotherapeutics; nuclear magnetic resonance in vitro and in vivo of metalloporphyrins. *Mailing Add:* Millersville Univ Millersville PA 17551

YEAGER, VERNON LEROY, b Williston, NDak, Nov 20, 26; m 47; c 4. ANATOMY. *Educ:* Minot State Col, BS, 49; Univ NDak, PhD(anat), 55. *Prof Exp:* Teacher sci, Garrison High Sch, 49-51; prof anat, Univ NDak, 55-67; assoc prof, St Louis Univ, 67-68; Rockefeller Found vis prof, Mahidol Univ, Thailand, 68-71; PROF ANAT, ST LOUIS UNIV, 71- *Concurrent Pos:* NSF fel, Northwestern Univ, Chicago, 60. *Honors & Awards:* Hektoen Gold Medal, AMA, 78. *Mem:* Am Asn Anat; Sigma Xi; Asn Clin Anatomists. *Res:* Pathology of connective tissues; cancer of the larynx. *Mailing Add:* Dept Anat St Louis Univ Sch Med St Louis MO 63104

YEAGLE, PHILIP L, b 1949; US citizen; m. BIOLOGICAL MEMBRANES. *Educ:* St Olaf Col, BA, 71; Duke Univ, PhD(chem), 74. *Prof Exp:* Fel, Univ Va, 74-78; asst prof, 78-83, ASSOC PROF BIOCHEM, STATE UNIV NY, BUFFALO, 83- *Mem:* Am Chem Soc; Am Soc Biochem & Molecular Biol; Biophys Soc; Asn Res Vision & Opthal. *Res:* Structure of cell membranes and relations of structure to function, including membrane fusion, photoreceptor function, and cholesterol protein interactions in membranes using biochemical and biophysical techniques. *Mailing Add:* Dept Biochem 140 Farber Hall State Univ NY Health Sci Ctr 3435 Main St Buffalo NY 14214

YEAKEY, ERNEST LEON, b Sikeston, Mo, Aug 5, 34; m 61; c 2. ORGANIC CHEMISTRY. *Educ:* Southeast Mo State Col, BS, 56; State Univ, Iowa, PhD(org chem), 60. *Prof Exp:* Res chemist, 60-67, supvr explor res, Jefferson Chem Co, 67-76, mgr res, 76-80, MGR NEW PROD DEVELOP, TEXACO CHEM CO, 80- *Mem:* Am Chem Soc. *Res:* Hydrogenation of nitriles to amines; reductive amination of alcohols to amines; synthetic routes to alpha olefins; catalytic synthesis of ethyleneamines; synthesis of heterocyclic amines. *Mailing Add:* Texaco Chem Co PO Box 15730 Austin TX 78761-5730

YEARGAN, JERRY REESE, b Kirby, Ark, Jan 31, 40; m 59; c 2. ELECTRICAL ENGINEERING. *Educ:* Univ Ark, BSEE, 61, MSEE, 65; Univ Tex, Austin, PhD(elec eng), 67. *Prof Exp:* From asst engr to assoc engr, Tex Instruments, Inc, 61-63; res engr, Univ Tex, 66-67; from asst prof to assoc prof elec eng, 67-77, PROF ELEC ENG, UNIV ARK, FAYETTEVILLE, 77-, HEAD, DEPT ELEC ENG, 80- *Mem:* Inst Elec & Electronics Engrs; Am Phys Soc; Am Soc Eng Educ. *Res:* Metal-oxide-semiconductor devices, particularly charge-coupled and optoelectronic metal-oxide-semiconductor devices. *Mailing Add:* Dept Elec Eng BEC 3217 Univ Ark Fayetteville AR 72701

YEARGAN, KENNETH VERNON, b Clanton, Ala, Feb 12, 47; m 72; c 1. ENTOMOLOGY, ECOLOGY. *Educ:* Auburn Univ, BS, 69; Univ Calif, Davis, PhD(entom), 74. *Prof Exp:* From asst prof to assoc prof, 74-84, PROF ENTOMOL, UNIV KY, 84- *Mem:* Entom Soc Am; Int Orgn Biol Control; Sigma Xi. *Res:* Biological control of insect pests of soybeans and forage crops; theory and practice of population sampling; integrated pest management; ecology and behavior of insects and spiders. *Mailing Add:* Dept Entom Univ Ky Lexington KY 40546

YEARGERS, EDWARD KLINGENSMITH, b Houma, La, Apr 27, 38; m 75; c 2. BIOPHYSICS. *Educ:* Ga Inst Technol, BS, 60; Emory Univ, MS, 62; Mich State Univ, PhD(biophys), 66. *Prof Exp:* US AEC, res fel radiation physics, Oak Ridge Nat Lab, 66-67; NIH res fel theoret chem, Czech Acad Sci, 67; asst prof, 68-70, ASSOC PROF BIOL, GA INST TECHNOL, 70- *Res:* Molecular biophysics; protein structure; molecular aspects of aging. *Mailing Add:* Sch Biol Ga Inst Technol 225 N Ave NW Atlanta GA 30332

YEARIAN, MASON RUSSELL, b Lafayette, Ind, July 5, 32; m 56, 65; c 3. NUCLEAR PHYSICS, HIGH ENERGY PHYSICS. *Educ:* Purdue Univ, BS, 54; Stanford Univ, MS, 56, PhD(physics), 61. *Prof Exp:* Res assoc physics, Univ Pa, 59-61; from asst prof to assoc prof, 61-71, PROF PHYSICS, STANFORD UNIV, 71-, DIR HIGH ENERGY PHYSICS LAB, 73- *Res:* Electron scattering from nuclei and nucleons; nucleon form factors; charge distribution in nuclei; high energy particle physics. *Mailing Add:* Dept Physics-Varian Bldg Rm 214 Stanford Univ Stanford CA 94305-1684

YEARIAN, WILLIAM C, b Lake Village, Ark, May 20, 37; m 60; c 1. ENTOMOLOGY. *Educ:* Univ Ark, Fayetteville, BS, 60, MS, 61; Univ Fla, PhD(entom), 66. *Prof Exp:* From asst prof to assoc prof, 65-74, PROF ENTOM, UNIV ARK, FAYETTEVILLE, 74- *Mem:* Soc Invert Path; Entom Soc Am. *Res:* Forest entomology; applied insect pathology. *Mailing Add:* Dept Entom Univ Ark Fayetteville AR 72701

YEARICK, ELISABETH STELLE, b Spokane, Wash, July 1, 13. NUTRITION, BIOCHEMISTRY. *Educ:* Univ Wis, BS, 34, MS, 35; Univ Iowa, PhD(nutrit), 60; Am Bd Nutrit, cert human nutrit, 65. *Prof Exp:* Asst dir dietetics, Duke Hosp, 48-53, asst prof, Duke Univ, 53; assoc prof nutrit, Univ Iowa, 53-57; assoc prof, WVa Univ, 60-66; prof, 66-79, EMER PROF NUTRIT & FOODS, ORE STATE UNIV, 79- *Mem:* Am Inst Nutrit; Am Dietetic Asn. *Mailing Add:* 145 NW 16th No 406 Corvallis OR 97330

YEAROUT, PAUL HARMON, JR, b Coeur d'Alene, Idaho, Sept 15, 24; m 49. MATHEMATICS. *Educ:* Reed Col, BA, 49; Univ Wash, MS, 58, PhD, 61. *Prof Exp:* Instr math, Reed Col, 51-52; instr exten ctr, Portland State Col, 52-55; asst prof, Knox Col, 59-62; assoc prof 62-67, PROF MATH, BRIGHAM YOUNG UNIV, 67- *Concurrent Pos:* Consult sch med, Univ Wash, 59-64. *Mem:* Math Asn Am. *Res:* Abstract algebra; theory and structure of groups and rings. *Mailing Add:* Paul Harmon Yearout Jr 2174 N 250 E Provo UT 84604

YEARY, ROGER A, b Cleveland, Ohio, Apr 26, 32; div; c 4. TOXICOLOGY, PHARMACOLOGY. *Educ:* Ohio State Univ, DVM, 56; Am Bd Vet Toxicol, dipl; Am Bd Toxicol, dipl. *Prof Exp:* Staff sr toxicologist, Charles Pfizer & Co, Inc, Conn, 60-61; chief toxicol sect, Lakeside Labs Div, Colgate-Palmolive Co, Wis, 61-65; exten vet toxicologist, Coop Exten Serv, 65-67, from assoc prof to prof vet physiol & pharmacol, 67-84, ADJ PROF VET PHYSIOL & PHARMACOL, COL VET MED, OHIO STATE UNIV, 84- *Concurrent Pos:* Toxicol consult, Lakeside Labs Div, Colgate-Palmolive Co, 65-67 & Minn Mining & Mfg Co, 67-77; NIH res grants, toxicol study sect, 73-77; consult, Chemlawn Corp, 77-81 & O M Scott & Sons, 78-81; dir, Employee & Environ Health, Chemlawn Corp, 81-86, vpres health & safety, 86-; corp safety officer, Ecolab, Inc, 88-90; vpres Health, Safety & Environ Affairs, Ecolab, Inc & Chemlawn Serv Corp, 90- *Mem:* AAAS; Am Soc Pharmacol & Exp Therapeut; Am Vet Med Asn; Soc Toxicol; Am Asn Clin Chemists. *Res:* Environmental and occupational health. *Mailing Add:* Chemlawn Serv Corp 8275 N High St Columbus OH 43235

YEATMAN, CHRISTOPHER WILLIAM, b Port Pirie, Australia, Aug 6, 27; m 54; c 4. FOREST GENETIC DIVERSITY. *Educ:* Univ Adelaide, BSc, 51; Australian Sch Forestry, Canberra, dipl, 51; Yale Univ, MF, 57, PhD(forest genetics), 66. *Prof Exp:* Forester, Dept Woods & Forests, SAustralia, 51; forest officer, Forestry Comn Gt Brit, 51-53; asst, Petawawa Forest Exp Sta, 53-54, res forest officer, 54-66, res scientist, 66-89; Can proj mgr, Can Forest Tree Seed Ctr, Asn Southeast Asian Nations, Thailand, 89-91; CONSULT, 91- *Concurrent Pos:* Mem & secy, Comt Forest Tree Breeding Can, 55-66, exec secy, 66-80; forest comt expert, Gen Resources, 78-88; comt, Experts Plant Gene Resources, Can, 82-89. *Mem:* Biomet Soc; Can Inst Forestry; Genetics Soc Can; Sigma Xi. *Res:* Silviculture; plantation establishment; tree breeding and forest genetics; genecology. *Mailing Add:* PO Box 721 Deep River ON K0J 1P0 Can

YEATMAN, HARRY CLAY, b Ashwood, Tenn, June 22, 16; m 49; c 2. ZOOLOGY. *Educ:* Univ NC, AB, 39, MA, 42, PhD(zool), 53. *Prof Exp:* Asst zool, Univ NC, 39-42, instr, 47-50; from asst prof to assoc prof biol, 50-59, chmn dept, 71-76, prof, 59-80, KENAN PROF BIOL, UNIV SOUTH 80-, ELDER HOSTEL PROF, 87- *Concurrent Pos:* Consult, US Nat Mus, 48-, Woods Hole Oceanog Inst, 60- & SEATO, US Army, Thailand, 66-; vis prof marine biol, VA Inst Marine Sci, Gloucester Point, VA, 67; consult, Univ Tehran, Iran, 72- & WHO, 76-; Brown Found fel, Univ South, 84. *Mem:* AAAS; Soc Syst Zool; Am Soc Limnol & Oceanog; Am Soc Ichthyologists & Herpetologists; Am Ornith Union. *Res:* Limnology; taxonomy and ecology of freshwater and marine copepods. *Mailing Add:* Dept Biol Univ South Sewanee TN 37375

YEATMAN, JOHN NEWTON, b Washington, DC, Apr 30, 20; m 54; c 2. FOOD SCIENCE. *Educ:* Univ Md, BS, 44; Univ Calif, Los Angeles, MS, 48. *Prof Exp:* Plant physiologist, USDA, 44-47; plant physiologist, Chem Corps, US Dept Army, Ft Detrick, 48-53; res food technol & leader, Qual Eval Invest, Agr Res Serv, USDA, 54-68, dir color res lab, Mkt Qual Res Div, 68-71; res food technol, Bur Foods, Div Food Technol, Food & Drug Admin, 71-75; CONSULT FOOD STANDS-QUAL EVAL, 75- *Mem:* Inst Food Technol; Inter-Soc Color Coun; Sigma Xi. *Res:* Research and development of methods for standards improvement by objectively measuring by physical and chemical means identity and quality factors in processed fruits and vegetables and their products; food standards and quality evaluation; instruments and inspection lighting. *Mailing Add:* 11106 Cherry Hill Rd Adelphi MD 20783

YEATS, FREDERICK TINSLEY, b Gadsden, Ala, Apr 4, 42; m 69; c 2. BOTANY. *Educ:* Miss Col, BS, 64; Univ Miss, MS, 67; Univ SC, PhD(biol), 71. *Prof Exp:* PROF BIOL, HIGH POINT COL, 69- *Mem:* Sigma Xi; Am Inst Biol Sci. *Res:* Developmental morphology in fern gametophytes; embryology in the genus Smilax. *Mailing Add:* Dept Biol High Point Col High Point NC 27262

YEATS, ROBERT SHEPPARD, b Miami, Fla, Mar 30, 31; m 52; c 5. GEOLOGY. *Educ:* Univ Fla, AB, 52; Univ Wash, MS & PhD(geol), 58. *Prof Exp:* Exploitation engr, Shell Oil Co, 58-62, sr prod geologist, 62-64, sr geologist, 64-67; from assoc prof to prof geol, Ohio Univ, 67-77; chmn dept, 77-85, PROF GEOL, ORE STATE UNIV, 77- *Concurrent Pos:* Consult, F Beach Leighton & Assocs, Calif & Energy Resources Br, US Geol Surv, 75; co-chief scientist, Deep Sea Drilling Proj, 73-74, 78; chmn, Structural Geol & Tectonics Div, Geol Soc Am, 84-85, Cordilleran sect, Geol Soc Am, 88-89, working group 1, Int Lithosphere Prog, 87-90, task group, Holocene earthquakes, 90- *Mem:* AAAS; fel Geol Soc Am; Am Geophys Union; Seismol Soc Am; Am Asn Petrol Geol. *Res:* Structural evolution of Pacific continental margin of Americas; application of plate tectonics to petroleum accumulation; active folds and faults, particularly in contractile continental areas. *Mailing Add:* Dept Geosci Ore State Univ Corvallis OR 97331-5506

YEATS, RONALD BRADSHAW, b Newcastle-upon-Tyne, Eng, Mar 17, 41; Can citizen; m 62; c 2. MICROCOMPUTER APPLICATIONS, CHEMICAL INFORMATION RETRIEVAL. *Educ:* Univ Durham, BSc, 62, PhD(org chem), 65. *Prof Exp:* Mayo fel, Univ Western Ont, 65-67, lectr org chem, 67-68; asst researcher, Univ Hosp Ctr, Univ Sherbrooke, 70-71; asst prof, 68-74, assoc prof, 74-80, PROF ORG CHEM, BISHOPS UNIV, LENNOXVILLE, QUE, CAN, 80- *Concurrent Pos:* Hon prof chem, Univ Western Ont, 87- *Mem:* Royal Soc Chem; Chem Inst Can. *Res:* Natural products and synthesis; synthetic methods of organic chemistry; mechanism of solvolysis reactions. *Mailing Add:* Dept Chem Bishop's Univ Lennoxville PQ J1M 1Z7 Can

YEATTS, FRANK RICHARD, b Altoona, Pa, Mar 5, 36; m 60; c 2. THEORETICAL PHYSICS. *Educ:* Pa State Univ, BS, 58; Univ Ariz, MS, 63, PhD(physics), 64. *Prof Exp:* From asst prof to assoc prof, 64-81, PROF PHYSICS, COLO SCH MINES, 82- *Mem:* Am Geophys Union; Am Phys Soc; Am Asn Physics Teachers. *Res:* Mathematical physics. *Mailing Add:* Dept Physics Colo Sch Mines Golden CO 80401

YECK, ROBERT GILBERT, b La Valle, Wis, Dec 6, 20; m 44; c 2. AGRICULTURAL ENGINEERING. *Educ:* Univ Wis, BS, 48; Univ Mo, MS, 53, PhD(agr eng), 60. *Prof Exp:* Proj leader environ & animals, Agr Eng Div, USDA, 48-51, lab leader bioclimatic studies, 51-58, invests leader animal environ, 58-60, br chief livestock eng & rural housing, 70-72, staff scientist, waste mgt & microbiol, 72-75, staff scientist farmstead eng & rural housing, agr res, USDA, 75-80; vis prof, 80-89, CONSULT, AGR ENG, UNIV MD, 89- *Concurrent Pos:* Mem Agr bd, Nat Acad Sci, 71-74; consult, Food & Agr Orgn, UN, Italy, 64-67 & Agr Eng Grad Prog, Agrarian Univ, Peru, UN develop proj, 67-70; chmn, Int Symp Livestock Wastes, 71 & USDA Task Force Recycled Animal Wastes, 74; mem, President's Solar Domestic Pol Rev Group, 78; consult, Grain Storage World Bank, 84-86. *Mem:* Fel Am Soc Agr Engrs; Sigma Xi. *Res:* Farmstead engineering; rural housing; waste management; solar energy; anaerobic fermentation processes; livestock shelters; environmental stress; bioclimatic studies. *Mailing Add:* 14301 Northwyn Dr Silver Spring MD 20904

YEDINAK, PETER DEMERTON, b Bath, NY, Aug 20, 39; m 63, 77; c 3. THEORETICAL PHYSICS, COMPUTER SCIENCE THEORY. *Educ:* Union Col, BS, 62; Clark Univ, MA, 67, PhD(chem physics), 68. *Prof Exp:* from asst prof to assoc prof, 67-72, PROF PHYSICS, WESTERN MD COL, 72- *Concurrent Pos:* Consult to Army, 68-73. *Mem:* AAAS; Am Inst Physics; Am Asn Physics Teachers; Am Phys Soc; Am Asn Univ Prof. *Res:* Theoretical lattice dynamics; determination of transition rates for acoustical energy absorption by molecules in a solid containing isotopic defects. *Mailing Add:* Dept Physics Western Md Col Westminster MD 21157

YEE, ALBERT FAN, b Canton, China, Nov 1, 45; US Citizen; m 87; c 1. POLYMER MATERIALS, MECHANICAL BEHAVIOR. *Educ:* Univ Calif, Berkeley, BS, 67, PhD(chem), 71. *Prof Exp:* Staff, Gen Elec Corp R&D, 71-85; PROF, MAT SCI, UNIV MICH, ANN ARBOR, 85- *Mem:* Fel Am Phys Soc; Am Chem Soc; Plastics & Rubber Instit, London; Soc Rheology; Mat Res Soc. *Res:* Mechanical behavior of polymeric materials & composites; physics of polymers; nature and origins of molecular relaxations in polymer glasses; toughening of polymer alloys and composites; nonliner viscoelasticity. *Mailing Add:* Dept Mat Sci & Eng Dow Bldg Univ Mich Ann Arbor MI 48109-2136

YEE, ALFRED A, b Aug 5, 25; US citizen; m; c 8. ENGINEERING, STRUCTURAL ENGINEERING. *Educ:* Rose-Hulman Inst Technol, BSCE, 48; Yale Univ, ME, 49. *Hon Degrees:* DE, Rose-Hulman Inst Technol, 76. *Prof Exp:* Pres, Alfred A Yee & Assocs, Inc, 53-82, tech vpres, Alfred A Yee Div, Leo A Daly, 82-89; PRES, APPL TECHNOL CORP, 84- *Concurrent Pos:* Mem, Comt Connection Details & Comt Seismic Design, Prestressed Concrete Inst, dir, 69-72; hon struct consult, Singapore Housing Develop Bd, 80-; mem, Spec Comt Offshore Installations, Am Bur Shipping, Comt 512 Precast Struct Concrete, Am Concrete Inst-Am Soc Civil Engrs; consult mem, Comt 357 Offshore Concrete Struct, Am Concrete Inst. *Honors & Awards:* Martin P Korn Award, Prestressed Concrete Inst, 65, Robert J Lyman Award, 84. *Mem:* Nat Acad Eng; hon mem & fel Am Concrete Inst; hon mem & fel Am Soc Civil Engrs; Pre-Stressed Concrete Inst; Soc Naval Architects & Marine Engrs; Earthquake Eng Res Inst; Int Asn Bridge & Struct Eng; Nat Soc Prof Engrs. *Res:* Offshore and oceangoing vessels; reinforcing steel bar splices and concrete framing systems for high rise buildings; precast prestressed concrete construction; granted several patents; author of various publications. *Mailing Add:* Appl Technol Corp 1441 Kapiolani Blvd Suite 810 Honolulu HI 96814

YEE, JOHN ALAN, b Salt Lake City, Utah, Feb 11, 47; m 68; c 3. BONE BIOLOGY, CELL BIOLOGY. *Prof Exp:* Asst prof, 74-80, ASSOC PROF ANAT, TEX TECH UNIV HEALTH SCI CTR, 80- *Mem:* Am Asn Anatomists; Am Soc Bone & Mineral Res; AAAS. *Res:* Elucidating the factors which regulate bone cell functions and how cell function changes with age. *Mailing Add:* Dept Anat Tex Tech Univ Health Sci Ctr Lubbock TX 79430

YEE, KANE SHEE-GONG, b Kwangtung, China, Mar 26, 34; US citizen; m 62; c 2. APPLIED MATHEMATICS. *Educ:* Univ Calif, Berkeley, BS, 57, MS, 58, PhD(appl math), 63. *Prof Exp:* Assoc prof math, Univ Fla, 66-68; from assoc prof to prof math, Kans State Univ, 68-84; with Lawrence Livermore Nat Lab, 84-87; WITH LOCKHEED MISSILE & SPACE CO, SUNNYVALE, 87- *Concurrent Pos:* Consult, Lawrence Livermore Lab, Univ Calif, 66-, grant, 70- *Mem:* Soc Indust & Appl Math. *Res:* Mathematical physics; numerical solution to partial differential equations. *Mailing Add:* 23350 Toyonita Rd Los Altos Hills CA 94022

YEE, RENA, b Hong Kong, June 1,36; US citizen; m 68; c 1. POLYMER-PHYSICAL CHEMISTRY. *Educ:* Chung Chi Col, Hong Kong, BS, 59; Univ NDak, Grand Fork, MS, 61; Univ Mass, Amherst, PhD(polymer phys chem), 67. *Prof Exp:* Res chemist polymer chem, Hercules Res Ctr, Wilmington, Del, 67-71; res assoc biochem, Univ Pa, 71-72; RES CHEMIST POLYMER PHYS CHEM, RES DEPT, NAVAL WEAPONS CTR, CHINA LAKE, CALIF, 73- *Mem:* Am Chem Soc; Sigma Xi. *Res:* Polymer structural characterizations of polyolefin and collagen with x-ray diffraction, small light scattering refractive index-birefringence measurements and tritium exchange methods; surface chemistry of nitramines. *Mailing Add:* 908 Sylvia Ave Ridgecrest CA 93555-3155

YEE, SINCLAIR SHEE-SING, b China, Jan 20, 37; US citizen; m 61; c 2. BIOENGINEERING, MICROELECTRONICS. *Educ:* Univ Calif, Berkeley, BS, 59, MS, 61, PhD(elec eng), 65. *Prof Exp:* Res engr, Lawrence Livermore Lab, 64-66; from asst prof to prof elec eng, Univ Wash, 66-77, dir microtechnol lab, 74-77 CONSULT, 77- *Concurrent Pos:* Consult, Lawrence Livermore Lab, 66-75, Beckman Instruments, 77- & Tektronics Inc, 77-78; NIH spec res fels, Case Western Reserve Univ, 72-73 & Bioeng Ctr, Univ Wash, 73-74. *Mem:* Am Phys Soc; fel Inst Elec & Electronics Engrs. *Res:* Semiconductor physics and devices; bioinstrumentation; microelectronic devices; microsensors. *Mailing Add:* Dept Elec Eng Univ Wash Seattle WA 98195

YEE, TIN BOO, b Canton, China, Feb 25, 15; US citizen; m 74. CHEMISTRY, CERAMICS. *Educ:* Ark State Col, BS, 38; Univ Ark, MS, 40; Univ Ill, AM, 50, PhD(chem), 54. *Prof Exp:* Chemist, Chem Warfare Serv, Huntsville Arsenal, Ala, 42-45; asst chemist, State Geol Surv, Ill, 45-55; RES CHEMIST, REDSTONE ARSENAL, 55- *Concurrent Pos:* Instr, Exten, Univ Ala, 60; vis res scientist, Union Indust Res Inst, Taiwan, 70-71. *Mem:* Am Chem Soc; Am Ceramic Soc; AAAS. *Res:* Sol-gel process for making optical ceramics; develop durable rubberized material for repairing worn out sole and heel of shoes; mutations in flowers and plants by radiations; material research in microelectronics. *Mailing Add:* 719 Erskine St NW Huntsville AL 35805

YEE, TUCKER TEW, b Toyshun, Canton, China, Mar 9, 36; US citizen; m 68; c 1. ORGANIC CHEMISTRY, PHYSICAL ORGANIC CHEMISTRY. *Educ:* Knox Col, Ill, BA, 60; Univ Mass, PhD(org chem), 64. *Prof Exp:* Res assoc, Princeton Univ, 64-65; res chemist, Eastern Lab, E I du Pont de Nemours & Co, Inc, 65-67; sr res chemist, Arco Chem Co Div, Atlantic Richfield Co, Pa, 67-72; RES CHEMIST, NAVAL WEAPONS CTR, CHINA LAKE, CALIF, 72- *Mem:* Sigma Xi; Am Chem Soc. *Res:* Chemistry of nitrogen containing heterocycles; radiochemical tracer technique; general organic syntheses; organic polymer syntheses and polymer applications. *Mailing Add:* 908 Sylvia Ave Ridgecrest CA 93555

YEE, WILLIAM C, b Boston, Mass, Sept 23, 28; m 54; c 3. CHEMICAL & ENVIRONMENTAL ENGINEERING. *Educ:* Tufts Univ, BS, 48; Univ Tenn, MS, 59. *Prof Exp:* Chemist, Allied Chem & Dye Corp, 48-52; chemist, 52-59, chem engr, 59-72, GROUP LEADER, OAK RIDGE NAT LAB, 72- *Concurrent Pos:* Session chmn, Gordon Conf on Water, 66. *Mem:* Sigma Xi; Am Inst Chem Engrs; Am Chem Soc. *Res:* Environmental impact statement preparation; aquaculture; waste heat utilization; nutrition economics; nuclear desalination; agro-industrial complexes; waste and water treatment; ion exchange; radiation effects on metals; corrosion of nuclear materials; synthetic detergent technology. *Mailing Add:* 113 Westover Dr Oak Ridge TN 37830

YEEND, WARREN ERNEST, b Colfax, Wash, May 14, 36; m 64, 85; c 1. GEOLOGY. *Educ:* Wash State Univ, BS, 58; Univ Colo, MS, 61; Univ Wis, PhD(geol), 65. *Prof Exp:* GEOLOGIST, US GEOL SURV, 65- *Honors & Awards:* Spec Achievement Award, US Geol Surv, 75. *Res:* Surficial geology in an area of oil shade interest in western Colorado; geomorphology; gold bearing gravels of the Sierra Nevada; economic geology; engineering geology along the Trans-Alaska pipeline; mapping and copper resource evaluation in southern Arizona; placer gold in Alaska; surficial geology in Alaska. *Mailing Add:* US Geol Surv 345 Middlefield Rd Menlo Park CA 94025

YEGULALP, TUNCEL MUSTAFA, b Konya, Turkey, Nov 5, 37; m 63; c 2. MINING ENGINEERING, OPERATIONS RESEARCH. *Educ:* Tech Univ Istanbul, MS, 61; Columbia Univ, DES(mining), 68. *Prof Exp:* Mining engr, Mineral Res & Explor Inst, Turkey, 61-63; res engr, Mobil Res & Develop Corp, 67-69; prog mgr, Mineral Res & Explor Inst, Turkey, 71-72; from asst prof to assoc prof, 72-85, PROF MINING, COLUMBIA UNIV, 85- *Concurrent Pos:* Dir, NY Mining & Mineral Resources Res Inst, 87- *Mem:* Am Inst Mining, Metall & Petrol Engrs; Inst Mgt Sci; Opers Res Soc Am. *Res:* Mineral economics; hydraulic transport of coal in underground mines; earthquake forecasting; statistical methods in geomechanics; queueing models for open-pit mining. *Mailing Add:* Henry Krumb Sch Mines Columbia Univ New York NY 10027

YEH, BILLY KUO-JIUN, b Foochow, China, Aug 28, 37; m 65; c 3. INTERNAL MEDICINE, CARDIOVASCULAR DISEASES. *Educ:* Nat Taiwan Univ, MD, 61; Univ Okla, MS, 63; Columbia Univ, PhD(pharmacol), 67; Am Bd Internal Med, dipl & cert cardiovasc dis. *Prof Exp:* Intern med, Nat Taiwan Univ Hosp, 60-61; resident path, Med Ctr, Univ Okla, 63; teaching asst, Col Physicians & Surgeons, Columbia Univ, 64-67; asst resident, Emory Univ Affil Hosps, 68-69; staff cardiologist & chief sect clin pharmacol, Div Cardiol, Mt Sinai Med Ctr, Miami Beach, 69-71; co-actg chief, Heart Sta, Jackson Mem Hosp, 72-73; assoc dir, Div Clin Investr, Miami Heart Inst, Miami Beach, 73-76; asst prof med, 70-73, asst prof pharmacol, 72-73, clin asst prof, 73-76, CLIN ASSOC PROF MED, SCH MED, UNIV MIAMI, 76- *Concurrent Pos:* Fel, Univ Okla, 62-63; fel pharmacol, Col Physicians & Surgeons, Columbia Univ, 63-64; fel med, Sch Med, Emory Univ & Grady Mem Hosp, 67-68; fel Coun Clin Cardiol, Am Heart Asn. *Mem:* Fel Am Col Physicians; fel Am Col Cardiol; Am Soc Pharmacol & Exp Therapeut; Am Physiol Soc; Am Fedn Clin Res; fel Col Chest Physicians. *Res:* Clinical pharmacology; clinical cardiovascular disease. *Mailing Add:* Dept Med Univ Miami Sch Med 315 Palermo Ave Coral Gables FL 33134

YEH, CAVOUR W, b Nanking, China, Aug 11, 36; US citizen; m 60; c 2. MICROWAVES, FIBER OPTICS. *Educ:* Calif Inst Technol, BS, 57, MS, 58, PhD(elec eng), 62. *Prof Exp:* Res asst elec eng, Calif Inst Technol, 58-62; from asst prof to assoc prof, Univ Southern Calif, 62-67; assoc prof, 67-72, PROF ELEC ENG, UNIV CALIF, LOS ANGELES, 72- *Concurrent Pos:* Consult, Jet Propulsion Lab, 62-, Hughes Res Lab, 74-80. *Mem:* Fel Inst Elec & Electronics Engrs; fel Optical Soc Am; Int Union Radio Sci. *Res:* Theoretical & experimental aspect of electromagnetic waves: low-loss mm/sub-mm waveguides, ultra high speed fiber optics local network area, intergrated fiber optics structures; scattering & diffraction of wave by arbitrarily shape dielectric objects, moving medium, microwave antenna, multiple scattering effects & moving particle. *Mailing Add:* Elec Eng Dept Univ Calif Westwood Los Angeles CA 90024

YEH, CHAI, b Hangchow, China, Sept 21, 11; nat US; m 36; c 2. ELECTRONICS. *Educ:* Chekiang Univ, BS, 31; Harvard Univ, MS, 34, ScD(commun), 36. *Prof Exp:* Asst, Harvard Univ, 35-36; prof elec eng, Peiyong Univ, China, 36-37; prof, Tsinghua Univ, 37-48; vis prof, Univ Kans, 48-56; res engr, Inst Sci & Technol, 56-61, from assoc prof to prof, 61-81, EMER PROF ELEC ENG, UNIV MICH, ANN ARBOR, 81- *Mem:* Sr mem Inst Elec & Electronics Engrs. *Res:* Microwave electronics and engineering; linear and nonlinear circuit analysis. *Mailing Add:* 1821 Alhambra Dr Ann Arbor MI 48103

YEH, EDWARD H Y, b Hsin-Chu, Taiwan, Jan 1, 30; US citizen; m 67; c 2. ELEMENTARY PARTICLE PHYSICS, FLUIDS. *Educ:* Nat Taiwan Univ, BS, 52; Kyushu Univ, Fukuoka, Japan, MS, 57; Univ NC, Chapel Hill, PhD(physics), 60. *Prof Exp:* Lectr physics, Nat Taiwan Univ, Taipei, 53-55; res staff Dept Nuclear Physics, Col Gen Educ, Tokyo Univ, Meguroku, Japan, 57-58; res physicist, Nuclear Data Group, Nat Acad Sci-Nat Res Coun, Wash, DC, 60-61; sr physicist, Res Electro Magnetic Waves, Germeshausen & Grier, Inc, Boston, 61-63; res scholar theoretical physics, Dublin Inst Adv Scis, 63-66; prof physics & astron, Moorhead State Univ, Minn, 66-87; PHYSICIST, RES INFRARED MICROWAVES, NAVAL WEAPONS CTR, CHINA LAKE, 84- *Concurrent Pos:* Reviewer, Am Math Soc, 73-; vis prof, Fairfield Univ, Conn, 73-74. *Mem:* NY Acad Sci; Am Physical Soc; Int Soc Gen Relativity & Gravitation; Soc Photo-Optical Instrumentation Engrs. *Res:* Infrared electrooptics; computer simulation; microwaves and hydrodynamics; elememtary particle physics; group theory; atomic and molecular physics; theoretical physics; electromagnetism; author of over thiry publications on the above research areas. *Mailing Add:* 628 S Gemstone St Ridgecrest CA 93555

YEH, FRANCIS CHO-HAO, b Hankow, China, Dec 20, 45; m 69; c 1. QUANTITATIVE GENETICS, POPULATION GENETICS. *Educ:* Univ Calgary, BSc, 70, PhD(genetics), 74. *Prof Exp:* Geneticist quantitative genetics, 74-80, TECH ADV GENETICS, BC MINISTRY FORESTS, VICTORIA, 80- *Concurrent Pos:* Adj assoc prof, dept forest sci, Univ Alberta, 80- *Mem:* NAm Quantitative Forest Genetics Group; Int Union Forestry Res Orgn; Can Tree Improvement Asn; Genetics Soc Can; Genetics Soc Am. *Res:* Genetic structure of forest trees and breeding theories. *Mailing Add:* Dept Forests Sci Univ Alta Edmonton AB T6G 2M7 Can

YEH, GENE HORNG C, b Chia-Yi, Taiwan, May 10, 48; nat US; m 77; c 1. POLYMER CHEMISTRY. *Educ:* Cheng Kung Univ, Taiwan, BS, 70; Wake Forest Univ, Winston-Salem, NC, MA, 74; Univ Wash, Seattle, PhD, 80. *Prof Exp:* Res chemist, Phillips Petrol Co, 80-86; sr res chemist, Crown Zellerbach Corp, 86-87; SR PROJ LEADER, JAMES RIVER CORP, 87- *Mem:* Am Chem Soc. *Res:* Ziegler-Natta and stereospecific polymerizations; emulsion polymerization; water soluble polymers; thermoplastic elastomers; polymer blend; polymer characterization; plastics packaging materials. *Mailing Add:* James River Corp 2199 Williams St San Leandro CA 94577

YEH, GEORGE CHIAYOU, b Kagi, Taiwan, Oct 3, 26; US citizen; m 57; c 4. PHYSICAL CHEMISTRY, CHEMICAL PHYSICS. *Educ:* Taiwan Univ, BSc, 50; Univ Tokyo, DEng, 53; Univ Toronto, MSc, 55, PhD(chem engr), 57. *Prof Exp:* Lectr, Japanese Engrs Union, Tokyo, 51-53; assoc prof chem eng, Auburn Univ, 57-61; assoc prof, 61-63, dir res & patent affairs, 73-79, PROF CHEM ENG, VILLANOVA UNIV, 63- *Concurrent Pos:* Prin investr res grants, Petrol Res Fund, 63-65 & NASA, 65-69; independent res, Thermodyne Co, 70- *Honors & Awards:* Am Inst Chem Award. *Mem:* Fel Am Inst Chemists; AAAS; Am Inst Chem Engrs; Am Chem Soc; Japanese Soc Chem Eng. *Res:* Reactions at interfaces; interfacial phenomena; catalysis; quantum mechanics; solid state chemistry; electrochemical analysis; liquids separation; gas separation; energy conversion; vapor engine; biomedical engineering; US and foreign patents. *Mailing Add:* Dept Chem Eng Villanova Univ Villanova PA 19085

YEH, GOUR-TSYH, b Lunhoutsun, Taiwan, Dec 5, 40; US citizen; m 71; c 2. HYDROSCIENCES, ATMOSPHERIC SCIENCES. *Educ:* Nat Taiwan Univ, BS, 64; Syracuse Univ, MS, 67; Cornell Univ, PhD(hydrol), 69. *Prof Exp:* Res assoc atmospheric diffusion, Cornell Univ, 69-71; vis res scientist air-sea interaction, NASA, Houston, 71-72; sr environ engr thermal hydraul, Ebasio Serv Inc, 72-73; sr hydraul-environ engr fluid mech, Stone & Webster Eng Corp, 73-77; res scientist hydrol, Oak Ridge Nat Lab, 77-82, sr res scientist, 82-89; PROF CIVIL ENG, PA STATE UNIV, 89- *Concurrent Pos:* Adj prof, Northeastern Univ, 74-75. *Mem:* AAAS; Am Soc Civil Engrs; Am Geophys Union; Am Meteorol Soc; Nat Soc Prof Engrs. *Res:* Groundwater hydrology; computational fluid dynamics; environmental transport; numerical modeling; geochemical modeling. *Mailing Add:* 402 Candlewood Dr State College PA 16803

YEH, GREGORY SOH-YU, b Shanghai, China, Apr 11, 33; US citizen; m 59; c 2. POLYMER PHYSICS, ELECTRON MICROSCOPY. *Educ:* Holy Cross Col, BS, 57; Cornell Univ, MS, 60; Case Inst Technol, PhD(polymer physics), 66. *Prof Exp:* Res physicist, Goodyear Res Ctr, 60-61; sr res physicist, Gen Res Ctr, 61-64; res fel, Case Inst Technol, 66-67; asst prof chem & metall eng, 67-69, assoc prof, 69-72, PROF MAT & METALL & CHEM ENG, UNIV MICH, ANN ARBOR, 72- *Concurrent Pos:* Fulbright fel, 73 & 83; Humboldt fel, 74, 76 & 77. *Mem:* Am Phys Soc; Electron Micros Soc Am; Am Chem Soc. *Res:* Polymer structure and properties; morphology and crystal structure; mechanical properties of polymers; electron diffraction. *Mailing Add:* Chem Eng Univ Mich Dow Bldg 3042 Ann Arbor MI 48109

YEH, HEN-GEUL, b Taiwan, China; US citizen. DIGITAL SIGNAL PROCESSING, SATELLITE COMMUNICATION SYSTEMS. *Educ:* Univ Calif, Irvine, PhD(control & digital signal processing), 82. *Prof Exp:* Mem tech staff, Parker-Hannifin Corp, 79-81; PROF DIGITAL SIGNAL PROCESSING, CALIF STATE UNIV, LONG BEACH, 83-; SCIENTIST, MAGNAVOX GOVT & INDUST ELECTRONICS CORP, 87- *Mem:* Sr mem Inst Elec & Electronics Engrs. *Res:* Digital signal processing; satellite communication; digital demodulator; digital modem; systolic arrays. *Mailing Add:* 24306 Carlene Lane Lomita CA 90717

YEH, HERMAN JIA-CHAIN, b Taipei, Taiwan, Nov 15, 39. PHYSICAL CHEMISTRY. *Educ:* Cheng-Kung Univ, Taiwan, BS, 63; Univ Mass, Amherst, PhD(chem), 68. *Prof Exp:* Fel chem, Univ Mass, 68-69; vis fel, Lab Chem, 70-71, staff fel, 72-74, sr staff fel, 74-76, RES CHEMIST, NAT INST ARTHRITIS, METABOLISM & DIGESTIVE DIS, 76- *Mem:* AAAS; Am Chem Soc. *Res:* Nuclear magnetic resonance spectroscopy. *Mailing Add:* 6000 Marquette Terr Bethesda MD 20817-1752

YEH, HSIANG-YUEH, b Tainan Hsien, Taiwan, Apr 1, 40; m 69; c 2. MECHANICS. *Educ:* Cheng King Univ, Taiwan, BSE, 62; Univ NMex, MSCE, 67, PhD(civil eng), 69. *Prof Exp:* Jr engr, Taiwan Pub Works Bur, 63-65; ASSOC PROF CIVIL ENG, PRAIRIE VIEW AGR & MECH COL, 69- *Mem:* Am Soc Civil Engrs; Am Soc Eng Educ. *Res:* Reliability analysis, random vibration and fatigue damage of structural systems. *Mailing Add:* PO Box 2244 Prairie View A&M Univ Prairie View TX 77446

YEH, HSI-HAN, b Shanghai, China, Nov 10, 35; m 66; c 2. ELECTRICAL ENGINEERING, CONTROL SYSTEMS. *Educ:* Nat Taiwan Univ, BSc, 56; Chiao Tung Univ, MSc, 61; Univ NB, MSc, 63; Ohio State Univ, PhD(elec eng), 67. *Prof Exp:* Asst engr, Taiwan Power Co, 58-60; res assoc elec eng, Ohio State Univ, 66-67; from asst prof to assoc prof elec eng, Univ Ky, 67-85; ELEC ENGR, FLIGHT-DYNAMICS LAB, WRIGHT-PATTERSON AFB, 85- *Concurrent Pos:* Fac res, NASA, 73-74 & Air Force Off Sci Res, 81; vis prof, Nat Chiao Tung Univ, 75-76. *Mem:* Inst Elec & Electronics Engrs; Sigma Xi. *Res:* Distributed-parameter systems; computer-controlled systems; learning and adaptive control systems; robust control theory; US patents. *Mailing Add:* 1181 Mint Springs Dr Fairborn OH 45324-5728

YEH, HSUAN, b Shanghai, China, Dec 1, 16. MECHANICAL ENGINEERING, AERONAUTICAL ENGINEERING. *Educ:* Chiao-Tung Univ, BS, 36; Mass Inst Technol, SM, 44, ScD(mech eng), 50. *Prof Exp:* Instr mech eng, Mass Inst Technol, 48-50; res assoc, Johns Hopkins Univ, 50-52, from asst prof to assoc prof, 52-56; PROF MECH ENG, UNIV PA, 56-, DIR TOWNE SCH CIVIL & MECH ENG, 60- *Concurrent Pos:* Mem, Franklin Inst. *Mem:* Am Inst Aeronaut & Astronaut; Am Soc Eng Educ; Am Phys Soc; fel Am Soc Mech Engrs. *Res:* Turbomachinery; fluid and analytical mechanics; aerodynamics; magnetohydrodynamics. *Mailing Add:* 4431 Osage Ave Philadelphia PA 19104

YEH, HSU-CHI, b Taipei, Taiwan, Sept 30, 40; US citizen; m 66; c 2. AEROSOL SCIENCE, MECHANICAL ENGINEERING. *Educ:* Nat Taiwan Univ, BS, 63; Univ Minn, MS, 67, PhD(mech eng), 72. *Prof Exp:* Teaching asst mech eng, Univ Minn, 64-65, res asst mech eng & aerosol physics, 65-72; res assoc, 72-73, RES SCIENTIST AEROSOL PHYSICS, INHALATION TOXICOL RES INST, LOVELACE BIOMED & ENVIRON RES INST, 73-, AEROSOL SCI GROUP SUPVR, 82- *Concurrent Pos:* Mem, Sci Comt, 57, Task Group 2 Lung Dosimetry Model, Nat Coun Radiation Protection & Measurements, 84, bd dirs, Am Asn Aerosol Res, 89- *Mem:* Health Physics Soc; Am Indust Hyg Asn; Air Pollution Control Asn; Am Asn Aerosol Res. *Res:* Aerosol science and technology; inhalation toxicity associated with inhaled aerosols and the particle deposition in mammalian lungs including mammalian airway morphometry. *Mailing Add:* Lovelace Inhalation Toxicol Res Inst PO Box 5890 Albuquerque NM 87185

YEH, HUN CHIANG, b Nanking, China, June 27, 35; m 65; c 2. METALLURGY, CERAMICS. *Educ:* Nat Taiwan Univ, BS, 58; Brown Univ, MS, 63; Ill Inst Technol, PhD, 66. *Prof Exp:* Sr metallurgist, Res & Develop Lab, Corning Glass Works, 66-69; asst prof metall eng, 69-75, actg chmn dept, 76-77, assoc prof metall eng, 75-80, PROF MAT SCI, CHMN ENG DEPT, CLEVELAND STATE UNIV, 80- *Mem:* Metall Soc; Am Ceramic Soc. *Res:* X-ray studies of metals and ceramics; mechanical behaviors of materials; grinding of metals; processing of refractory oxides and silicon nitrides. *Mailing Add:* Garrett Ceramics Components 19800 S Van Nest Ave Torrance CA 90509

YEH, J J, electronics, for more information see previous edition

YEH, JAMES JUI-TIN, b Tainan, Formosa, Apr 26, 27; US citizen. MATHEMATICS. *Educ:* Taiwan Univ, BS, 50; Univ Minn, MA, 54, PhD(math), 57. *Prof Exp:* Instr math inst technol, Univ Minn, 57-58 & Mass Inst Technol, 58-60; asst prof, Univ Rochester, 60-64; vis mem, Courant Inst Math Sci, NY Univ, 64-65; assoc prof, 65-68, PROF MATH, UNIV CALIF, IRVINE, 68-, CHMN DEPT. *Mem:* Am Math Soc. *Res:* Integration in function spaces; functional analysis; stochastic processes. *Mailing Add:* Dept Math Univ Calif Irvine CA 92717

YEH, K(UNG) C(HIE), b Hangchow, China, Aug 4, 30; m 57; c 4. ELECTRICAL ENGINEERING. *Educ:* Univ Ill, BS, 53; Stanford Univ, MS, 54, PhD(elec eng), 58. *Prof Exp:* Asst, Electronics Lab, Stanford Univ, 54-58; res assoc elec eng, 58-59, from asst prof to assoc prof, 59-67, assoc, Ctr Advan Study, 73-74, PROF ELEC ENG, UNIV ILL, URBANA, 67- *Concurrent Pos:* Mem, US Comn G, Int Sci Radio Union, vice-chmn, 81-84, chmn, 85-87; assoc ed, Radio Sci J, 79-81, ed, 83-86; US Panel mem Electromagnetic Wave-Propogation Panel, Adv Group Aerospace Res & Develop, 84; prog chmn of Adv Group for Aerospace Res & Develop Panel Symposium on "Scattering & Propogation in Random Media", 87. *Mem:* Fel Inst Elec & Electronics Engrs; Am Geophys Union; Am Phys Soc; Am Meteorol Soc. *Res:* Propagation, ionosphere and plasma; radio propagation; ionospheric dynamics. *Mailing Add:* Dept Elec & Comput Eng Univ Ill 1406 W Green St Urbana IL 61801-2991

YEH, KWAN-NAN, b Taichung, Taiwan, Feb 27, 38; m 65; c 2. TEXTILE CHEMISTRY. *Educ:* Nat Taiwan Univ, BS, 61; Tulane Univ, La, MS, 65; Univ Ga, PhD(chem), 70. *Prof Exp:* Cotton Found res assoc fel, Nat Bur Standards, 70-72, res chemist, 72-73; from asst prof to assoc prof, 73-85, PROF TEXTILES & CONSUMER ECON, UNIV MD, COLLEGE PARK, 85- *Mem:* Am Chem Soc; Am Asn Textile Chemists & Colorists; Textile Inst; Fiber Soc. *Res:* Thermodynamics; thermochemistry; textile and polymer flammability; textile chemistry; dyeing and finishing. *Mailing Add:* Dept Textiles & Consumer Econ Univ Md College Park MD 20742

YEH, KWO-YIH, b Tao-yuan, Taiwan, Jan 20, 42; m 71. CELL CULTURE. *Educ:* Taiwan Normal Univ, BS, 65; Washington Univ, St Louis, PhD(biol), 75. *Prof Exp:* Instr biol, Chung-Li High Sch, Taiwan, 65-66; teaching asst develop biol, Taiwan Normal Univ, 67-71; res asst, Washington Univ, St Louis, 71-75, res assoc, 75-77, res asst prof, 77-78; sr res scientist cell biol, Southern Res Inst, 78-83; AT ST LUKES-ROOSEVELT HOSP, 83-; ASSOC RES SCIENTIST, COLUMBIA UNIV, 83- *Mem:* AAAS; Am Soc Cell Biol; Am Soc Zoologists; Soc Develop Biol; Am Gastroenterol Asn. *Res:* Developmental biology in vertebrates; structural and enzymic development of intestine; developmental endocrinology; cell proliferation and differentiation. *Mailing Add:* St Lukes-Roosevelt Hosp Amsterdam Ave 114th St New York NY 10025

YEH, LAI-SU LEE, b Hunan, China, June 24, 42, US citizen; m 72; c 2. ENZYMOLOGY. *Educ:* Taiwan Normal Univ, BS, 65; Sacramento State Univ, MS, 69; Univ Calif, Davis, PhD(agr chem), 79. *Prof Exp:* Res assoc enzym, Dept Biochem, Univ Utah, 75-77, res specialist clin biochem, Dept Pediat, 77-78; RES SCIENTIST BIOCHEM, NAT BIOMED RES FOUND, 80- *Mem:* Am Chem Soc. *Mailing Add:* 11317 Stryver Ct Gaithersburg MD 20878

YEH, LEE-CHUAN CAROLINE, b Jan 3, 54; US citizen. BIOCHEMISTRY. *Educ:* Fu-Jen Cath Univ, BS, 76; Univ Ga, MS, 78; Ore State Univ, PhD(nutrit), 84. *Prof Exp:* Postdoctoral fel pharmacol, 84-85 & biochem, 85-90, RES INSTR BIOCHEM, UNIV TEX HEALTH SCI CTR, 90- *Mem:* Am Soc Biochem & Molecular Biol. *Res:* Higher-order structure, function, and interactions of RNAs and proteins. *Mailing Add:* Dept Biochem Univ Tex Health Sci Ctr San Antonio TX 78284-7760

YEH, NAI-CHANG, b Taiwan, Repub China, Dec 15, 61. CONDENSED MATTER PHYSICS. *Educ:* Nat Taiwan Univ, Bachelor, 83; Mass Inst Technol, PhD(physics), 88. *Prof Exp:* ASST PROF PHYSICS, CALIF INST TECHNOL, 89- *Concurrent Pos:* Vis scientist, Thomas J Watson Res Ctr, Int Bus Mach Corp, 89-; Alfred P Sloan Found res fel, 90. *Mem:* Am Phys Soc; Mat Res Soc. *Res:* Physical properties of conventional and high-temperature superconductors; electrical transport; magnetic and microwave experimental studies of vortex phases; vortex dynamics; phase transitions; dissipation in type-II superconductors. *Mailing Add:* Dept Physics 114-36 Calif Inst Technol Pasadena CA 91125

YEH, NOEL KUEI-ENG, b Malacca, Malaysia, Dec 15, 37; m 65; c 2. PARTICLE PHYSICS, MEDICAL PHYSICS. *Educ:* Williams Col, BA, 61; Yale Univ, MS, 62, PhD(physics), 66. *Prof Exp:* Res assoc physics, Nevis Labs, Columbia Univ, 66-68, instr univ, 68-69; from asst prof to assoc prof, 69-80, chmn dept physics, appl physics & astron, 81-87, PROF PHYSICS, STATE UNIV NY, BINGHAMTON, 80- *Concurrent Pos:* Vis prof, Max-Planck Inst Physics & Astrophys, Munich, WGer, 76-77; vis physicist, Brookhaven Nat Lab, 63-74, Stanford Linear Accelerator Ctr, 70-73. *Mem:* Am Phys Soc. *Res:* Elementary particles; medical physics; radiation physics; particle interaction in solid state physics; radiation in environment. *Mailing Add:* Dept Physics State Univ NY PO Box 6000 Binghamton NY 13902-6000

YEH, PAI-T(AO), b Canton, China, Feb 5, 20; nat US; m 51; c 2. CIVIL ENGINEERING. *Educ:* Chekiang Univ, BS, 43; Purdue Univ, MSCE, 49, PhD, 53. *Prof Exp:* Asst engr, Chekiang Waterway Off, 43-44; engr, Kwak Wha Eng Develop Co, 44-48; res engr, 53-58, from asst prof to assoc prof hwy eng, 59-82, EMER PROF CIVIL ENG, PURDUE UNIV, WEST LAFAYETTE, 82- *Mem:* AAAS; Am Soc Photogram; Sigma Xi; Am Soc Prof Engrs; Am Mil Eng. *Res:* Aerial photograph interpretation of engineering soils; hydrological problems. *Mailing Add:* Dept Civil Eng Purdue Univ West Lafayette IN 47907

YEH, PAUL PAO, b Sun-Yang, China, Mar 25, 27; US citizen; m 53; c 4. ELECTRICAL & ELECTRONICS ENGINEERING. *Educ:* Univ Toronto, BASc, 51; Univ Pa, MS, 60, PhD(elec eng), 66. *Prof Exp:* Design engr, James R Kearny Corp Can Ltd, 51 & Can Gen Elec Co, 51-56; asst prof elec technol, Broome Technol Community Col-State Univ NY, Binghamton, 56-57; sr design engr, H K Porter Co, Inc, Pa, 57; transformer engr, I-T-E Circuit Breaker Co, 57-58 & Kuhlman Elec Co, Mich, 58-59; sr design engr, Fed Pac Elec Co, NJ, 59-61; chief engr, Eisler Transformer Co, 61; assoc prof elec eng, Newark Col Eng, 61-67; supvr performance anal, Autonetics Div, N Am Rockwell Corp, 67-70; advan systs engr, S3-A Avionics, Lockheed-Calif Co, 70-72; supvr module commun systs, Site Defense Prog, McDonnell Douglas Corp, Huntington Beach, 72-73; mem tech staff, Aerospace Corp, 73-78, SR STAFF ENGR, ADVAN DEVELOP PROJ, LOCKHEED-CALIF CO, 78- *Concurrent Pos:* Consult, Eisler Transformer Co, NJ, 61-62, Standard Transformer Co, Ohio, 62, H K Porter Co, Inc, Va, 62-65, Consol Edison Co NY, Inc & Niagara Transformer Co, NY, 63-64 & Pub Serv Elec & Gas Co, NJ, 66; lectr, Calif State Univ, Long Beach, 67-73. *Honors & Awards:* Achievement Award, Lockheed-Calif Co. *Mem:* AAAS; sr mem Inst Elec & Electronics Engrs; Am Inst Aeronaut & Astronaut; Am Asn Univ Prof; Nat Mgt Asn. *Res:* Weapon delivery computer mechanization; air-to-air missile launch control problems; magnetic anomaly detection in antisubmarine warfare systems; communication and systems readiness verification as applied to antiballistic missile programs; Cruise missile launch and control. *Mailing Add:* Dept 7251 Bldg 310 Plan B-6 Lockheed Calif Co Box 551 Burbank CA 91520

YEH, POCHI ALBERT, b 1948; m; c 2. OPTICS OF LAYERED MEDIA. *Educ:* Nat Taiwan Univ, BS, 71; Caltech, MS, 75, PhD(physics), 77. *Prof Exp:* Mem tech staff, Rockwell Sci Ctr, 77-85, principle scientist, optics, 85-90; PROF, UNIV CALIF, SANTA BARBARA, 89- *Concurrent Pos:* Vis prof, Nat Taiwan Univ, 83-87. *Honors & Awards:* Rudolf Kingslake Medal & Prize, 89. *Mem:* Fel Inst Elec & Electronics Engrs; fel Optical Soc Am. *Res:* Nonlinear optics and optical phase conjugation. *Mailing Add:* PO Box 1085 Thousand Oaks CA 91360

YEH, PU-SEN, b Hualien, Taiwan, July 7, 35; m 64; c 3. AEROTHERMODYNAMICS, HEAT TRANSFER. *Educ:* Nat Taiwan Univ, BSME, 58; Univ Ill, MS, 62; Rutgers Univ, PhD(mech eng), 67. *Prof Exp:* PROF ENG, JACKSONVILLE STATE UNIV, 67-, HEAD DEPT, 83- *Mem:* Am Soc Eng Educ; Am Soc Mech Engrs. *Res:* Ignition and combustion of liquid fuel droplets; heat transfer problem of underground residential transformer; acoustic sootblower; computer-aided drafting and design. *Mailing Add:* Dept Eng Jacksonville State Univ Jacksonville AL 36265

YEH, RAYMOND T, b Hunan, China, Nov 5, 37; m 66; c 1. COMPUTER SCIENCE. *Educ:* Univ Ill, BS, 61, MA, 63, PhD(math), 66. *Prof Exp:* Asst prof comput sci, Pa State Univ, 66-69; assoc prof comput sci & elec eng, Univ Tex, Austin, 69-74, prof, 74-, chmn dept comput sci, 75-; AT INT SOFTWARE SYSTS. *Concurrent Pos:* Ed transactions software eng, Inst Elec & Electronics Engrs. *Mem:* Am Math Soc; Asn Comput Mach; Inst Elec & Electronics Eng. *Res:* Software engineering; data base management. *Mailing Add:* Syscorp Int 9420 Research Blvd Echelon III Suite 200 Austin TX 78759

YEH, SAMUEL D J, b Kunming, China, Apr 23, 26; US citizen; m 59; c 2. BIOCHEMISTRY. *Educ:* Nat Defense Med Ctr, Shanghai, MD, 48; Johns Hopkins Univ, ScD(biochem), 60. *Prof Exp:* Instr med, Nat Defense Med Ctr, 48-53; asst resident, Lutheran Hosp, Md, 53-54; fel med, 54-57, from instr to asst prof biochem, Sch Hyg & Pub Health, Johns Hopkins Univ, 60-63; from instr to asst prof med, Med Col, Cornell Univ, 63-76, assoc prof clin med, 76-84; asst attend physician, 69-70 & 72-75, assoc attend physician, 75-83, ATTEND PHYSICIAN, MEM HOSP, NEW YORK CITY, 83-; ASSOC PROF MED, MED COL, CORNELL UNIV, 84- *Concurrent Pos:* Assoc prof clin med, Med Col Cornell Univ, 76-84, assoc prof med, 84- *Mem:* AAAS; Am Chem Soc; Am Inst Nutrit; Soc Nuclear Med; Am Col Nuclear Physicians. *Res:* Interrelationship between nutrients; intestinal absorption and marginal deficiencies; role of protein synthesis on metabolic functions; tumor localizing radionuclides. *Mailing Add:* Dept Med Imaging Sloan Kettering Cancer Ctr 1275 York Ave New York NY 10021

YEH, SHU-YUAN, b Kwangtung, China, June 26, 26; US citizen; m 57; c 2. PHARMACEUTICAL CHEMISTRY, PHARMACOLOGY. *Educ:* Nat Defense Med Ctr, Taiwan, BS, 51; Univ Iowa, MS, 57, PhD(pharmaceut chem), 59. *Prof Exp:* Pharmacist, Army, Navy & Air Force Hosp, Taiwan, 51-52; clin, Off of President, Taiwan, 53-55; res asst pharm, Univ Iowa, 55-59, res assoc med, 59-63, from res instr to res asst prof drug metab, Col Med, 63-70; asst prof, Univ Ky, 71; PHARMACOLOGIST, ADDICTION RES CTR, NAT INST DRUG ABUSE, 72- *Mem:* AAAS; Am Pharmaceut Asn; Acad Pharmaceut Sci; Am Soc Pharmacol & Exp Med; NY Acad Sci. *Res:* Development of methods for detection, isolation and identification of drug and its metabolites in the biological fluids. *Mailing Add:* Addiction Res Ctr Nat Inst Drug Abuse PO Box 5180 Baltimore MD 21224-0180

YEH, TSYH TYAN, b Taiwan, Sept 11, 41; m; c 2. FLUID DYNAMICS. *Educ:* Nat Taiwan Univ, BS, 54; Univ Calif, San Diego, MS, 68, PhD(eng sci), 71. *Prof Exp:* Res engr, Univ Calif, 71-74; mech engr, Argonne Nat Lab, 74-80; MECH ENGR, FLOW METERING, NAT INST STANDARDS & TECHNOL, 80- *Mem:* Am Phys Soc. *Mailing Add:* 7701 Rydal Terr Rockville MD 20855

YEH, WILLIAM WEN-GONG, b Szechwan, China, Dec 5, 38; m 67; c 2. WATER RESOURCES, HYDROLOGY. *Educ:* Cheng-Kung Univ, BS, 61; NMex State Univ, MS, 64; Stanford Univ, PhD(civil eng), 67. *Prof Exp:* Actg asst prof civil eng, Stanford Univ, 67; asst res engr eng syst, 67-69, from asst prof to prof eng syst, 69-83, PROF CIVIL ENG, UNIV CALIF, LOS ANGELES, 83-, CHMN CIVIL ENG, 85- *Concurrent Pos:* Asst res engr, Dry-Lands Res Inst, Univ Calif, Riverside, 72-73; lectureship, Nat Coun Sci & Technol, Mexico, 73, Nat Polytech Inst, Mexico 81; expert water resources, Div Hydrol, Unesco, 74; Eng Found Fel Award, Eng Found, 81. *Honors & Awards:* Robert E Horton Award, Am Geophys Union, 89. *Mem:* Am Soc Civil Engrs; Am Geophys Union; Am Water Resources Asn. *Res:* Hydrology; water resources; optimization of large-scale water resources systems. *Mailing Add:* 4531C Boelter Hall Univ Calif Los Angeles CA 90024

YEH, YEA-CHUAN MILTON, b Szu-Chuan, China, Apr 16, 43. PHOTOVOLTAIC DEVICE, SOLAR CELL. *Educ:* Nat Taiwan Univ, BS, 65; UCLA, MS, 69, PhD(elec eng), 73. *Prof Exp:* Head, Radio Multiplex Relay Sta, Pen-Ho, Taiwan, 65-66; asst instr, Nat Taiwan Univ, 66-67; teaching assoc, UCLA, 67-72; mem tech staff, Jet Propulsion Lab, 72-82; DIR, APPL SOLAR ENERGY CORP, 82-; PRES, MYK TECHNOL, INC, 83- *Mem:* Inst Elec & Electronics Engrs; Sigma Xi; Int Solar Energy Soc. *Res:* Initiated and developed the technology of the world's first large-scale production-line in manufacturing high-efficiency gallium arsenide solar cell on gallium arsenide or germanium substrates using Metal-Organic Chemical-Vapor-Deposition technique; granted two patents. *Mailing Add:* Appl Solar Energy Corp 1454 Princeton St Santa Monica CA 90404

YEH, YIN, b Chungking, China, Nov 1, 38; US citizen; m 61; c 2. QUANTUM ELECTRONICS, CHEMICAL PHYSICS. *Educ:* Mass Inst Technol, BS, 60; Columbia Univ, PhD(physics), 65. *Prof Exp:* Res assoc physics, Columbia Radiation Lab, Columbia Univ, 65-66; Lawrence Radiation Lab fel, Lawrence Livermore Lab, 66-67, sr physicist, 67-72; assoc prof, 72-78, PROF APPL SCI, UNIV CALIF, DAVIS, 78- *Concurrent Pos:* Lectr, St Mary's Col, Calif, 67-68; lectr, Univ Calif, Davis, 71-72; consult, Lawrence Livermore Lab, 72- *Mem:* AAAS; Sigma Xi; Biophys Soc; Am Phys Soc. *Res:* Application of laser spectroscopy to study dynamic phenomena in chemical physics, biophysics and solid state physics. *Mailing Add:* Dept Appl Sci & Eng Univ Calif Davis CA 95616

YEH, YUNCHI, b Oct 16, 30; m; c 4. GROWTH FACTORS & CANCER, GENETICS. *Educ:* Univ Calif, San Francisco, PhD(biochem & biophys), 64. *Prof Exp:* PROF BIOCHEM, UNIV ARK MED SCI, 79- *Mem:* Am Soc Biochem & Molecular Biol. *Res:* Virus inhibitors; growth factors and cancer. *Mailing Add:* Dept Biochem & Molecular Biol Univ Ark Med Sch 4301 W Markham St Little Rock AR 72205

YEILDING, LERENA WADE, deoxyribonucleic acid, for more information see previous edition

YEISER, JIMMIE LYNN, b Owensboro, Ky, Feb 12, 52; m 79; c 2. VEGETATION MANAGEMENT, FOREST REGENERATION. *Educ:* Univ Ky, BS, 74, MS, 76; Tex A&M Univ, PhD(forestry), 80. *Prof Exp:* From asst prof to assoc prof, 80-90, PROF FORESTRY, UNIV ARK, MONTICELLO, 91- *Concurrent Pos:* DuPont grant biol sci, 89, 90 & 91. *Mem:* Soc Am Foresters; Weed Sci Soc Am. *Res:* Impact of competition from undesired herbaceous and woody species on crop forest species; management and manipulation of the undesired species with herbicides and forest litter for increased survival and growth. *Mailing Add:* Dept Forest Resources Univ Ark PO Box 2414 Monticello AR 71655

YELENOSKY, GEORGE, b Vintondale, Pa, July 20, 29; m 63; c 3. PLANT PHYSIOLOGY. *Educ:* Pa State Univ, BS, 55, MS, 58; Duke Univ, DF, 63. *Prof Exp:* Res forester, Northeastern Forest Exp Sta, US Forest Serv, 55-56 & 58-61; Int Shade Tree Conf res fel, 63-64; SUPVRY PLANT PHYSIOLOGIST, CITRUS INVESTS, USDA, 64- *Concurrent Pos:* Mem, Coun Agr Sci & Technol. *Mem:* Am Soc Plant Physiol; Am Soc Hort Sci. *Res:* Forestry; soil aeration and tree growth; cold hardiness of citrus trees; cryobiology membership; improve survival, growth and production of citrus; low temperature stress; identifying, characterizing and manipulating stress avoidance; tolerance systems that have evolved in citrus germplasm. *Mailing Add:* USDA Agr Res Serv 2120 Camden Rd Orlando FL 32803

YELICH, MICHAEL RALPH, b Chicago Ill, June 26, 48; m 76. CIRCULATORY SHOCK, METABOLISM. *Educ:* Northwestern Univ, BA, 70, Univ Ark, MS, 73, PhD(physiol), 75. *Prof Exp:* Res assoc, Dept Pediat, 75-77, res assoc & instr, Dept Physiol, 77-81, adj instr, 81-82, asst prof, Dept Physiol, Stritch Sch Med, 83-90, CLIN ASST PROF, DEPT PHYSIOL & PHARMACOL, SCH DENT, LOYOLA UNIV, CHICAGO, 90- *Concurrent Pos:* Dir res, dept Anesthesiol, Mount Sinai Hosp, Chicago, 81-83; asst prof, depts Physiol & Anesthesiol, Rush Med Col, Chicago, 81-83; NIH prin investr, 86-88. *Mem:* Am Physiol Soc; Sigma Xi; AAAS; Shock Soc; Int Endotoxin Soc. *Res:* Study of pancreatic islet dysfunction during endotoxic and septic shock in relationship to the role of cytokines (IL1, TNF etc) on the glucose dyshomeostasis which occurs during circulatory shock. *Mailing Add:* Dept Physiol & Pharmacol Sch Dent Loyola Univ Med Ctr 2160 S First Ave Maywood IL 60153-5594

YELLE, ROGER V, b Taunton, Mass, July 19, 57. PLANETARY ATMOSPHERES, RADIATIVE TRANSFER. *Educ:* Worcester Polytech Inst, BS, 78; Univ Wis-Madison, MS, 80, PhD(physics), 84. *Prof Exp:* Teaching asst, Univ Wis-Madison, 78-79, res asst, 79-80 & 82-84; mem prof staff, Plasma Physics Lab, Princeton Univ, 80-82; res assoc, 84-88, sr res assoc, 88-89, ASST RES SCIENTIST, LUNAR & PLANETARY LAB, UNIV ARIZ, 89- *Concurrent Pos:* Prin investr, NSF, 85-87 & NASA, 87- *Mem:* Am Geophys Union; Am Astron Soc. *Res:* Physical and chemical processes in planetary atmospheres; role of radiation in the initiation of chemical cycles and in the thermal balance in planetary atmospheres. *Mailing Add:* 9th Floor Gould-Simpson Bldg Univ Ariz Tucson AZ 85721

YELLEN, JAY, b New York, NY, Dec 17, 48; m 71; c 1. ALGEBRA. *Educ:* Polytech Inst Brooklyn, BS, 69, MS, 71; Colo State Univ, PhD(math), 75. *Prof Exp:* Teaching asst math, Polytech Inst Brooklyn, 69-71; teaching asst, Colo State Univ, 71-75; asst prof, Allegheny Col, 75-76; ASST PROF MATH, STATE UNIV NY COL, FREDONIA, 76- *Mem:* Am Math Soc. *Res:* Group representation theory, particularly groups of central type; algebraic coding theory. *Mailing Add:* Dept Math Sci Fla Inst Technol 150 W University Ave Melbourne FL 32904

YELLIN, ABSALOM MOSES, b Tel-Aviv, Israel, July 25, 36; US citizen; m 65; c 2. PSYCHOPHYSIOLOGY, CHILD PSYCHIATRY. *Educ:* Univ Del, BA, 65, MA, 68, PhD(psychol), 70. *Prof Exp:* Scholar, Neuropsychiat Inst, Univ Calif, Los Angeles Sch Med, 69-71; asst prof res, Neuropsychiat Inst, Univ Calif, Los Angeles, 71-72; asst prof, Dept Psychiat, Univ Calif, Davis, 72-74; asst prof, Univ Minn, 74-80, dir, Clin Studies Unit, 74-83, dir res, Div Child Adolescent Psychiat, 74-87, assoc prof, Dept Psychiat, 80-87, dir, Lab Neurosci, 83-87, assoc dir, Child Psychiat Inpatient Unit, 85-87; CHILD PSYCHIATRIST, PARK NICOLLET MED CTR, MINNEAPOLIS, MINN, 88- *Concurrent Pos:* Consult pychologist, Minn. *Mem:* Am Psychol Asn; fel Int Soc Pychophysiol; NY Acad Sci; Soc Psychophysiol Res; Int Soc Develop Psychobiol; Sigma Xi. *Res:* Attention and information processing; attention deficit disorders and other disorders involving attentional deficits, chronobiology, mental retardation, psychophysiology; child psychology and psychiatry; psychopharmacology; voluntary control of autonomic functions. *Mailing Add:* 1437 Wisconsin Ave N Golden Valley MN 55427-3958

YELLIN, EDWARD L, b Brooklyn, NY, July 2, 27; m 48; c 3. CARDIAC PHYSIOLOGY, BIOENGINEERING. *Educ:* Colo State Univ, BS, 59; Univ Ill, MS, 61, PhD(mech eng), 64. *Prof Exp:* Res assoc surg, 65-66, assoc surg & physiol, 66-68, asst prof, 68-72, assoc prof surg & physiol, 72-79, PROF SURG & PHYSIOL & BIOPHYS, ALBERT EINSTEIN COL MED, 79- *Concurrent Pos:* NIH sr fel cardiovasc tech, Univ Wash, 64-65; fel, Coun on Circulation, Am Heart Asn. *Mem:* Am Physiol Soc; Biomed Eng Soc; Am Soc Mech Eng; Cardiovasc Syst Dynamics Soc; Am Soc Artificial Internal Organs; Inst Elec & Electronics Engrs Eng Med & Biol Soc. *Res:* Cardiac dynamics, particularly left ventricular filling and mitral valve dynamics; diastolic function of the heart; natural and artificial heart valves. *Mailing Add:* 38 Lakeside Dr New Rochelle NY 10801

YELLIN, HERBERT, b New York, NY, May 27, 35; m 63; c 5. HEALTH SCIENCES, EXPERIMENTAL NEUROLOGY. *Educ:* City Col New York, BA, 56; Univ Calif, Los Angeles, PhD(anat & neurophysiol), 66. *Prof Exp:* Cytologist, Div Labs, Cedars of Lebanon Hosp, Los Angeles, 57-58; cytologist radiation path, Armed Forces Inst Path, US Army, 58-60; staff fel exp neurol, Nat Inst Neurol Dis & Blindness, NIH, 66-68; res physiologist trophic nerve functions, Nat Inst Neurol Dis & Stroke, NIH, 68-73; res physiologist neuronal develop & regeneration, Nat Inst Neurol & Communicative Dis & Stroke, NIH, 73-76; grants assoc sci admin, div res grants, 76-77, prog dir, Nat Eye Inst, 77-78, health scientist admin, Nat Inst Neurol & Commun Dis & Stroke, 78-, EXEC SECY, RESPIRATORY & APPL PHYSIOL STUDY SECT, NIH. *Concurrent Pos:* Lectr anat, Sch Med & Dent, Georgetown Univ, 67-75; councilman assembly scientists, Nat Inst Neurolog & Communicative Dis & Stroke, NIH, 73-75. *Mem:* Am Asn Anatomists; Am Physiol Soc; Soc Neurosci; NY Acad Sci. *Res:* Sensorimotor characteristics of posture and locomotion; interrelationships of neurons and skeletal muscle. *Mailing Add:* NINDS NIH Fed Bldg Rm 9C14 7550 Wisconsin Ave Bethesda MD 20892

YELLIN, JOSEPH, b Tel Aviv, Israel, Apr 21, 38; US citizen; m 68; c 3. PHYSICS, ARCHAEOLOGY. *Educ:* Univ Del, BS & BA, 60; Univ Calif, Berkeley, PhD(physics), 65. *Prof Exp:* Teaching asst physics, Univ Calif, Berkeley, 60-62, res asst, 62-65, res physicist, Lawrence Berkeley Lab, 65-73; co-dir, Archaeometry Lab, 73-86, ASSOC PROF PHYSICS & ARCHAEOL, HEBREW UNIV, JERUSALEM, ISRAEL, 73-, DIR, ARCHAEOMETRY UNIT, 86- *Concurrent Pos:* Vis sr lectr, Hebrew Univ, 70-71; vis assoc prof, Univ Calif, Los Angeles, 78-79; sr res assoc, Nat Res Coun, Jet Propulsion Lab, Pasadena, Calif, 80-82. *Mem:* Am Phys Soc; NY Acad Sci. *Res:* Gamma-ray spectroscopy with application to element analysis; provenance of ancient ceramics obsidian and flint through chemical fingerprinting employing gamma-ray spectroscopy; neutron activation analysis. *Mailing Add:* Inst Archaeol Hebrew Univ Jerusalem Israel

YELLIN, STEVEN JOSEPH, b San Francisco, Caiif, Dec 27, 41. EXPERIMENTAL HIGH ENERGY PHYSICS. *Educ:* Calif Inst Technol, BS, 63, PhD(physics), 71. *Prof Exp:* Physicist, German Electron Synchrotron, 71-73; asst prof & asst res physicist, 73-80, assoc prof residence & assoc res physicist, 80-88, PROF RESIDENCE & RES PHYS, UNIV CALIF, SANTA BARBARA, 88- *Res:* Electromagnetic interactions in elementary particle physics. *Mailing Add:* Dept Physics Univ Calif Santa Barbara CA 93106

YELLIN, TOBIAS O, b Tel Aviv, Israel, Aug 22, 34; US citizen; m 66; c 3. BIOCHEMISTRY, PHARMACOLOGY. *Educ:* Philadelphia Col Pharm, BS, 59, MS, 62; Univ Del, PhD(chem), 66. *Prof Exp:* Sr scientist, Abbott Labs, Ill, 66-71; sr pharmacologist, Smith Kline & French Labs, Pa, 71-73; head gastroenterol sect, Rorer Res Labs, Pa, 73-75; mgr gastroentestinal pharmacol, ICI Americas Inc, Wilmington, Del, 75-81; RES DIR BIOPROD OPER, BECKMAN INSTRUMENTS, INC, PALO ALTO, CALIF, 81- *Concurrent Pos:* Vis instr pharmacol, Med Col Pa, 73-81. *Mem:* AAAS; Am Chem Soc; Am Soc Pharmacol & Exp Therapeut. *Res:* Gastrointestinal pharmacology and biochemistry; histamine and histamine antagonists; biology and chemistry of peptides. *Mailing Add:* Smith Kline & French 709 Swedeland Rd Swedeland PA 19479

YELLIN, WILBUR, b Passaic, NJ, Nov 1, 32; m 77; c 3. PHYSICAL MEASUREMENTS, SPECTROSCOPY. *Educ:* Rutgers Univ, BSc, 54; Cornell Univ, PhD, 62. *Prof Exp:* Chemist, Nat Starch Corp, 54; asst chem, Cornell Univ, 54-58, res asst, 58-59; sr res chemist, Ozalid Div, Gen Aniline & Film Corp, 59-62; res chemist, Procter & Gamble Co, 62-70, sect head Physical Sci, 70-82, sect head Anal Chem, 82-90; CONSULT, 90- *Mem:* AAAS; Soc Appl Spectros; Coblentz Soc; Am Chem Soc. *Res:* Application of spectroscopic and physical measurement techniques to chemical problems and structure determinations; structure and physical properties of stratum corneum; metal ion complexes, photochemistry; Raman, infrared, uvvis and mass spectrometry. *Mailing Add:* 373 Compton Rd Cincinnati OH 45215-4145

YELON, ARTHUR MICHAEL, b New York, NY, Apr 15, 34; m 58; c 2. SOLID STATE PHYSICS. *Educ:* Cornell Univ, BA, 55; Case Inst Technol, MS, 59, PhD(physics), 61. *Prof Exp:* Asst physics, Case Inst Technol, 55-57, instr, 57-61; assoc mem res staff, Res Ctr, Int Bus Mach Corp, 61-62, mem res staff, 62-63; mem res staff, Lab Electrostatics & Metal Phys, Grenoble, France, 63-66; assoc prof appl sci, Yale Univ, 66-72; assoc prof, 72-74, dir, thin film res group, 84-90, PROF ENG PHYSICS, ECOLE POLYTECH, UNIV MONTREAL, 74- *Mem:* Am Phys Soc; Am Vacuum Soc; Am Asn Physics Teachers; Inst Elec & Electronics Engrs; Can Asn Physicists; Mat Res Soc. *Res:* Surface physics; electron tunneling; structure, electronic properties of polymers; amorphous silicon; photovoltaics; high energy ion implantation. *Mailing Add:* Dept Eng Physics Post Box 6079 Sta A Montreal PQ H3C 3A7 Can

YELON, WILLIAM B, b Brooklyn, NY, Aug 23, 44; m 66; c 3. EXPERIMENTAL SOLID STATE PHYSICS, NEUTRON & GAMMA RAY SCATTERING. *Educ:* Haverford Col, BA, 65; Carnegie Mellon Univ, MS, 67, PhD(physics), 70. *Prof Exp:* Fel physics, Brookhaven Nat Lab, 70-72; res physicist, Inst Laue-Langevin, 72-75; assoc prof, 75-84, PROF PHYSICS, UNIV MO-COLUMBIA, 84-, SR RES SCIENTIST, LEADER NEUTRON SCATTERING, UNIV RES REACTOR, 75- *Concurrent Pos:* Mem, Neutron Diffraction Comn, Int Union Crystalography, 77-; ed, Neutron Diffraction Newsletter, 77-82; consult neutron scattering, 85- *Honors & Awards:* Chancellors Award Outstanding Res, 86- *Mem:* Am Phys Soc; Am Crystallog Asn; Mat Res Soc. *Res:* Neutron scattering; studies of phase transitions; dynamics and statics of nearly one and two dimensional systems; gamma ray diffraction; crystallography; Mossbauer scattering; neutron powder diffraction; magnetism. *Mailing Add:* Univ Mo Res Reactor Res Park Columbia MO 65211

YELTON, DAVID BAETZ, b Cincinnati, Ohio, Jan 16, 45. MICROBIOLOGY. *Educ:* Mass Inst Technol, BS, 66; Univ Mass, Amherst, MS, 69, PhD(microbiol), 71. *Prof Exp:* Instr cell biol, Med Sch, Univ Md, 72-73; asst prof, 73-80, ASSOC PROF MICROBIOL, WVA UNIV MED CTR, 80- *Mem:* Am Soc Microbiol; AAAS. *Res:* Genetics; molecular biology. *Mailing Add:* Dept Microbiol & Immunol Med Ctr WVa Univ 2095 HSN Morgantown WV 26506

YEMMA, JOHN JOSEPH, b Youngstown, Ohio, July 14, 33; m 78; c 7. CYTOCHEMISTRY, CELL BIOLOGY. *Educ:* Youngstown State Univ, BS, 61; George Peabody Col, MA, 65; Pa State Univ, PhD(cytochem), 71. *Prof Exp:* Instr biol, Pa State Univ, 65-71; assoc prof, 71-80, PROF CELL BIOL, YOUNGSTOWN STATE UNIV, 80- *Concurrent Pos:* Youngstown State Univ res grant, 72-79, 81-86; NIH trainee, 75. *Mem:* AAAS; Am Bot Soc; Am Inst Biol Sci; Sigma Xi. *Res:* Developmental biochemistry of biomembranes during stages of the cell cycle. *Mailing Add:* Dept Biol Youngstown State Univ 410 Wick Ave Youngstown OH 44555

YEN, ANDREW, b New York, NY, Mar 4, 48; m 75; c 3. CANCER BIOLOGY, REGULATORY BIOLOGY. *Educ:* Haverford Col, BA, 69; Univ Wash, MS, 70; Cornell Univ, PhD(biophys), 76. *Prof Exp:* Res fel, Harvard Univ, 76-78; mem staff, Sloan-Kettering Inst, 78-81; fac dir, assoc prof flow cytometry & internal med, Sch Med, Univ Iowa, 81-88; AT DEPT PATH, CORNELL UNIV, 88- *Concurrent Pos:* Woodrow Wilson fel; Leukemia Soc Am fel. *Mem:* Am Asn Cancer Res; Am Soc Cell Biol; Soc Anal Cytol; Biophys Soc. *Res:* Cellular and molecular mechanisms regulating cell growth and differentiation in normal and neoplastic cells. *Mailing Add:* Dept Path Vet Col Cornell Univ Ithaca NY 14853

YEN, BELINDA R S, b Szechuen, China, Sept 2, 42; US citizen; m; c 2. IMMUNOLOGY. *Educ:* Southern Ill Univ, Carbondale, BS, 62; Univ Ark, Fayetteville, MS, 66, PhD(immunol), 71. *Prof Exp:* Am Heart Asn fel, 71-72, Arthritis Found res fel, 72-73; res assoc clin immunol, Sch Med, Case Western Reserve Univ, 73-74; res fel, 74-77, PROJ SCIENTIST, CLEVELAND CLIN FOUND, 77- *Concurrent Pos:* Staff microbiologist, Cleveland Clin Found. *Mem:* AAAS; NY Acad Sci; Am Soc Microbiol; Sigma Xi. *Res:* Clinical immunology; regulation of immune response; tumor immunology; clinical virology. *Mailing Add:* Cleveland Clin Found 9500 Euclid Ave Cleveland OH 44195-5140

YEN, BEN CHIE, b Canton, China, Apr 14, 35. HYDRAULICS, FLUID MECHANICS. *Educ:* Nat Taiwan Univ, BS, 56; Univ Iowa, MS, 59, PhD(mech & hydraul), 65. *Prof Exp:* Civil engr, Army Engrs Br, Taiwan, 57-58, jr engr, Water Resources Planning & Develop Comn, Taiwan, 58; res assoc, Inst Hydraul Res, Univ Iowa, 60-64 & Princeton Univ, 64-66; prof civil eng & mem Ctr Advan Studies, Univ Va, 88-91; from asst prof to assoc prof hydraul, 67-76, PROF CIVIL ENG, UNIV ILL, URBANA, 76- *Concurrent Pos:* Vis prof, Univ Karlsruhe, Germany, 74-75, Fed Sch Polytech, Lausanne, Switz, 82, Vrije Univ, Brussels, Belg, 83, Nat Taiwan Univ, Taipei, 83, Univ Stuttgart, Germany, 83 & 84 & Univ New SWales, 85; res fel, Hudraul Res Sta, Dept Environ, UK, 75; dir, Second Int Conf Urban Storm Drainage, 81, Fourth Int Symp Stochastic Hydraul, 84, Int Conf Channel Flow & Catchment Runoff, 89; pres, Joint Comt Urban Drainage, Int Asn Hydraul Res & Int Asn Water Pollution Res & Control, 82-86; ed, Advan Hydrosci, 85-; UK Dept Environ Res Fel; Fulbright Distinguished Sr Lectr Award, 88; assoc ed, J Hydraul Eng, 88-, J Hydraul Res, 85-; Nat Sci Coun distinguished lectr, 89. *Mem:* Am Geophys Union; Int Asn Hydraul Res; Am Soc Civil Engrs; Int Water Resources Asn; Sigma Xi; Int Asn Hydrol Scientist. *Res:* Open-channel hydraulics; sediment transport; surface water hydrology; probabilistics hydrology; risk and reliability analysis; floods; urban drainage; infiltration. *Mailing Add:* Dept Civil Eng Univ Ill Urbana IL 61801

YEN, BEN-TSENG, b Canton, China, Jan 19, 32; m 65; c 2. CIVIL ENGINEERING. *Educ:* Nat Taiwan Univ, BS, 55; Lehigh Univ, MS, 59, PhD(struct eng), 63. *Prof Exp:* Res asst civil eng, Fritz Lab, 57-60, res assoc, 60-64, from asst prof to assoc prof, 64-77, PROF CIVIL ENG, LEHIGH UNIV, 77- *Concurrent Pos:* Consult industs, 61- *Mem:* Am Soc Civil Engrs; Struct Stability Res Coun (secy, 66-69); Sigma Xi. *Res:* Mechanical properties of materials; behavior and strength of civil engineering structures; fatigue and fracture of materials and structures. *Mailing Add:* Dept Civil Eng Fritz Lab 13 Lehigh Univ Bethlehem PA 18015

YEN, BING CHENG, b Shantung Province, China, June 13, 34; m 64; c 3. CIVIL ENGINEERING, SOIL MECHANICS. *Educ:* Nat Taiwan Univ, BS, 56; Univ Utah, PhD(civil eng), 63. *Prof Exp:* Found engr, Utah State Hwy Dept, 62-63; asst prof civil eng, Univ Utah, 63-64; from asst prof to assoc prof, 64-78, PROF CIVIL ENG, CALIF STATE UNIV, LONG BEACH, 78- *Concurrent Pos:* Consult, Dept Engrs, Los Angeles County, 65-70 & Woodward-Clyde Consults, 72-78; Geotech consult. *Mem:* Am Soc Civil Engrs. *Res:* Soil mechanics and foundation engineering; i slope stability; earthquake effects on soil engineering; marine geomechanics; rock mechanics. *Mailing Add:* Dept Civil Eng Calif State Univ 1250 Bellflower Blvd Long Beach CA 90840

YEN, CHEN-WAN LIU, b Tainan, Taiwan, Jan 26, 32; US citizen; m 58; c 2. AEROSPACE SCIENCE. *Educ:* Nat Taiwan Univ, BS, 54; Mass Inst Technol, PhD(physics), 64. *Prof Exp:* Tech staff space sci, KMS Technol, 68-72; MEM TECH STAFF SPACE SCI, JET PROPULSION LAB, CALIF INST TECHNOL, 72- *Mem:* Am Inst Aeronaut & Astronaut. *Mailing Add:* 867 Marymount Lane Claremont CA 91711-1513

YEN, DAVID HSIEN-YAO, b Tsingtao, Shantung, Apr 18, 34; m 64; c 3. APPLIED MATHEMATICS, ENGINEERING MECHANICS. *Educ:* Nat Taiwan Univ, BS, 56; Mich State Univ, MS, 61; NY Univ, PhD(math), 66. *Prof Exp:* Asst prof civil eng & mech, Bradley Univ, 61-62; from asst prof to assoc prof, 65-72, PROF MATH, MICH STATE UNIV, 75- *Concurrent Pos:* Vis scientist, Inst Comp Appln Sci & Eng, NASA Langley Res Ctr, 74. *Mem:* Am Math Soc; Am Soc Mech Engrs; Soc Indust Appl Math; Am Acad Mech. *Res:* Mathematical mechanics; methods of applied mathematics. *Mailing Add:* Dept Math Mich State Univ East Lansing MI 48824-1027

YEN, I-KUEN, b Singapore, May 29, 30; US citizen; m 58; c 2. CHEMICAL ENGINEERING. *Educ:* Nat Taiwan Univ, BS, 54; Mass Inst Technol, SM, 56, ScD(chem eng), 60. *Prof Exp:* Chem engr, Crucible Steel Co, 59-61; res scientist pollution control, Am Standard Corp, 61-64; group leader eng, C F Braun & Co, 64-68; dir contract res, Occidental Res Corp, 68-78, dir health, safety & environ, 78-83; PRIN, IKE YEN ASSOCS, 83- *Concurrent Pos:* Adj prof, Newark Col Eng, 61-62, Stevens Inst Technol, 63-64, Harvey Mudd Col, 85-86 & USC, 87-88. *Mem:* Am Inst Chem Engrs; Am Chem Soc; NY Acad Sci. *Res:* Inorganic chemicals processing; health and safety; pollution control; process development; contract research; research management. *Mailing Add:* 867 Marymount Lane Claremont CA 91711-1513

YEN, JONG-TSENG, b Tainan, Taiwan, Feb 12, 42; US citizen; m 70; c 2. ANIMAL SCIENCES & NUTRITION. *Educ:* Nat Taiwan Univ, BS, 64; Univ Ill, MS, 70, PhD(animal nutrit), 75. *Prof Exp:* Teaching asst animal sci, animal sci dept, Nat Taiwan Univ, 65-68; res asst swine nutrit, animal sci dept, Univ Ill, 68-74, res assoc, 75; res assoc biochem dept, Univ Mo, 75-77, res assoc animal sci dept, 77-78; RES ANIMAL SCI, AGR RES SERV, USDA, 78- *Concurrent Pos:* Swine consult, US Feed Grains Coun, 82. *Mem:* Am Soc Animal Sci; Am Inst Nutrit. *Res:* Mechanism of growth stimulation by antibiotics; hepatic portal absorption of nutrients and gut metabolites; vitamin C in swine nutrition; post weaning stress in relation to nutrition; growth repartitioning agents for pigs. *Mailing Add:* US Meat Animal Res Ctr PO Box 166 Clay Center NE 68933

YEN, NAI-CHYUAN, b Shaoxing, China, Apr 12, 36; US citizen; m 67; c 3. SIGNAL & INFORMATION PROCESSING. *Educ:* Cheng Kung Univ, BS, 57; Univ RI, MS, 62; Harvard Univ, PhD(appl physics), 71. *Prof Exp:* Elec engr electronics, Bendix Corp, 62-66; res physicist acoust, Naval Underwater Systs Ctr, 72-79; ocean engr, US Coast Guard Res & Develop Ctr, 79-82; RES PHYSICIST, NAVAL RES LAB, 82- *Honors & Awards:* Alan Berman Res Publ Award, 89. *Mem:* Acoust Soc Am; Inst Elec & Electronics Engrs; AAAS; Asn Comput Mach. *Res:* Radiation, transmission and reception of underwater acoustic waves; noise generation mechanisms and control; non-linear phenomena and oscillations; signal processing; electronic and electro-optics instrumentation; image processing and pattern recognition. *Mailing Add:* Naval Res Lab 4555 Overlook Ave SW Washington DC 20375-5000

YEN, PETER KAI JEN, b China, Feb 10, 22. DENTISTRY, BIOPHYSICS. *Educ:* West China Union Univ, DDS, 47; Harvard Univ, DMD, 54. *Prof Exp:* Intern children's dent, Forsyth Dent Infirmary, 49-50; asst oral path, 54-57, instr dent med, 57-61, clin assoc, 61-67, asst prof orthod, 67-70, ASSOC PROF ORTHOD, SCH DENT MED, HARVARD UNIV, 70- *Concurrent Pos:* Fel orthod, Forsyth Dent Infirmary, 50-52 & 55-57; assoc orthop, Med Ctr, Boston Children's Hosp, 65-74, sr assoc, 74- *Mem:* Fel AAAS; Am Dent Asn; Am Asn Orthod; NY Acad Sci; Int Asn Dent Res. *Res:* Cranio-facial growth and development; bone dynamics. *Mailing Add:* 52 Saxton St Boston MA 02125

YEN, S C YEN, b Peking, China, Feb 22, 27; m 58; c 3. REPRODUCTIVE ENDOCRINOLOGY. *Educ:* Chee-Loo Univ, China, BS, 49; Univ Hong Kong, MD, 54, DSc, 80; Am Bd Obstet & Gynec, dipl, 66; Am Bd Reproductive Endocrinol, cert, 73. *Prof Exp:* Intern med, Queen Mary Hosp, Hong Kong, 54-55; resident obstet & gynec, Johns Hopkins Hosp, 56-60; instr, Johns Hopkins Univ, 58-60; chief dept, Guam Mem Hosp, 60-62; from asst prof to assoc prof reprod biol, Sch Med, Case Western Reserve Univ, 62-70; assoc dir obstet, Univ Hosps of Cleveland, 68-70; chmn dept, 72-87, PROF OBSTET & GYNEC, UNIV CALIF 70-, PROF REPRODUCTIVE MED & W R PERSONS CHAIR, 87- *Concurrent Pos:* Teaching & res fels, Harvard Med Sch, 62. *Mem:* Endocrine Soc; Am Diabetes Asn; Am Soc Gynec Invest (pres, 81); fel Am Col Obstet & Gynec; Asn Am Physicians. *Res:* Obstetrics and gynecology. *Mailing Add:* Dept Reprod Med Univ Calif Sch Med La Jolla CA 92093

YEN, TEH FU, b Kunming, China, Jan 9, 27; US citizen; m 59. ENVIRONMENTAL SCIENCE, BIOCHEMICAL ENGINEERING. *Educ:* Cent China Univ, BS, 47; WVa Univ, MS, 53; Va Polytech, PhD(org chem), 56. *Hon Degrees:* Dr, Pepperdine Univ, 82. *Prof Exp:* Asst, Cent China Univ, 47-48, Yunnan Univ, 48-49 & WVa Univ, 50-53; sr res chemist, Res Div, Goodyear Tire & Rubber Co, 55-59; fel petrol chem, Mellon Inst, 59-65, sr fel, Carnegie-Mellon Univ, 65-68; assoc prof, Calif State Univ, Los Angeles, 68-69; assoc prof, Dept Med, Dept Chem Environ Eng, 69-80, PROF, DEPT CIVIL & ENVIRON ENG, UNIV SOUTHERN CALIF, 80- *Concurrent Pos:* Hon prof, Univ Petrol, Beijing, China, E China Univ Chem Eng. *Mem:* Sr mem Am Chem Soc; Am Inst Physics; Am Inst Chem Engrs; Am Soc Petrol Engrs; fel Royal Soc Chem; fel, Inst Petrol UK. *Res:* Alternative processes or materials leading to clean environment; biogeoorganic chemistry; geomicrobiology, fossil fuels extraction, conversion control technology, asphaltenes; biochemical energy conversion, energy and environment; biomaterials; heavy metals in environments; hazardous waste science. *Mailing Add:* Dept Civil & Environ Eng Univ Southern Calif 224A KAP Los Angeles CA 90089-2531

YEN, TERENCE TSIN TSU, b Shanghai, China, May 2, 37; m 64; c 3. BIOCHEMISTRY, GENETICS. *Educ:* Nat Taiwan Univ, BS, 58; Univ NC, PhD(biochem, genetics), 66. *Prof Exp:* RES SCIENTIST, RES LABS, ELI LILLY & CO, 65- *Mem:* AAAS; Sigma Xi; Am Soc Biochem & Molecular Biol; Am Diabetes Asn; NAm Asn Study Obesity. *Res:* Biochemical defects and treatment of metabolic diseases in higher organisms; obesity; diabetes; functions of growth factors. *Mailing Add:* Lilly Res Labs Eli Lilly & Co Lilly Corp Ctr Indianapolis IN 46285

YEN, WILLIAM MAOSHUNG, b Nanking, China, Apr 5, 35; US citizen; m 78; c 1. LASER SPECTROSCOPY. *Educ:* Univ Redlands, Calif, BS, 56; Wash Univ, St Louis, PhD(physics), 62. *Prof Exp:* Res assoc physics, Wash Univ, St Louis, 62 & Stanford Univ, 62-65; from asst prof to prof, Univ Wis-Madison, 65-90; GRAHAM PURDUE PROF PHYSICS, UNIV GA, ATHENS, 86- *Concurrent Pos:* Vis prof, Univ Tokyo, 72, Univ Paris, Orsay, France, 76 & J Goethe Univ, Frankfurt, Ger, 85; consult, Lawrence Livermore Nat Lab, 75-86, Argonne Nat Lab, 77-82, Fla State Univ Bd Regents, 88- & Rosemont Corp, St Paul, Minn, 89-; hon prof, Univ San Antonia de Abad, Cusco, Peru, 81; Alexander von Humboldt Found sr US scientist award, 85 & 90; vchair, Luminescence & Display Sect, Electrochem Soc, 90- *Mem:* Fel Am Phys Soc; fel Optical Soc Am; fel AAAS; Electrochem Soc; Sigma Xi. *Res:* Laser spectroscopic techniques for study of optical properties of solids specially the dynamical properties, transfer and relaxation, of optically excited states of light emitting materials. *Mailing Add:* Dept Physics Univ Ga Athens GA 30602-9986

YEN, YIN-CHAO, b China, 27; US citizen; m 60; c 3. ENGINEERING PHYSICS, EARTH SCIENCES. *Educ:* Nat Taiwan Univ, BS, 51; Kans State Col, MS, 56; Northwestern Univ, PhD(chem eng), 60. *Prof Exp:* Chem engr, 60-62, res chem engr, 62-67, res phys scientist, 67-68, supvry res phys scientist, 68-82, RES PHYS SCIENTIST, US ARMY COLD REGIONS RES & ENG LAB, 82- *Concurrent Pos:* From adj asst prof to adj assoc prof, Univ NH, 65-71 & vis prof, 71-74; invited lectr, Academia Sinica, China, 85. *Honors & Awards:* Civilian Qual Increase Award, US Army Cold Regions Res & Eng Lab, 64, 65 & 71, Civilian Spec Act Award, 65, Meritorious Performance Award, 65, Outstanding Performance Award, 71. *Mem:* AAAS; Sigma Xi; Am Inst Chem Engrs. *Res:* Thermophysical properties of snow and ice; effect of density inversion and thermal instability phenomena associated with the ice-water systems and heat transfer characteristics in snow and ice involving phase change; atmospheric stability and turbulent heat exchange. *Mailing Add:* Four Willow Spring Lane & Eng Lab Hanover NH 03755

YEN, YOU-HSIN EUGENE, fields & waves, communication theory, for more information see previous edition

YENCHA, ANDREW JOSEPH, b Pa, July 3, 38; m 67. PHYSICAL CHEMISTRY, CHEMICAL PHYSICS. *Educ:* Univ Calif, Berkeley, BS, 63, Univ Calif, Los Angeles, PhD(chem), 68. *Prof Exp:* Res asst chem, Dow Chem Co, Calif, 60-62; res asst, Univ Calif, Berkeley, 62-63, teaching asst, Los Angeles, 64-65, res asst, 65-68; fel, Yale Univ, 68-70; from asst prof to assoc prof, 70-89, PROF, STATE UNIV NY, ALBANY, 89- *Mem:* Am Soc Mass Spectros; Am Inst Physics; Am Chem Soc; Int Am Photochem Soc. *Res:* Molecular beam, molecular spectroscopy and Penning ionization electron spectroscopy. *Mailing Add:* Dept Chem & Physics State Univ NY Albany NY 12222

YENDOL, WILLIAM G, b Pomona, Calif, Feb 22, 31; m 59; c 2. ENTOMOLOGY. *Educ:* Calif State Polytech Univ, BS, 53; Purdue Univ, MS, 57, PhD(entom), 64. *Prof Exp:* Tech rep agr chem, L H Butcher Chem Co, 58-59; entomologist, Lake States Forest Exp Sta, 63-65; assoc prof, 65-71, PROF ENTOM RES, PESTICIDE LAB, PA STATE UNIV, UNIVERSITY PARK, 71- *Concurrent Pos:* Consult, US/USSR, 78 & AID, 80; pres, EBESA, 82. *Mem:* Entom Soc Am; Soc Invert Path; Int Orgn Biol Control; Am Inst Biol Sci; Sigma Xi. *Res:* Insect pathology; biological control; insect pest management; biochemistry; aerial application technology. *Mailing Add:* Pesticide Lab Dept Entom Pa State Univ University Park PA 16802

YENER, YAMAN, b Ankara, Turkey, Oct 18, 46; m 76; c 1. THERMOFLUIDS ENGINEERING, RADIATIVE TRANSFER. *Educ:* Mid East Tech Univ, Ankara, BS, 68, MS, 70; NC State Univ, Raleigh, PhD(mech eng), 73. *Prof Exp:* From asst prof to assoc prof mech eng, Mid East Tech Univ, 74-82, chmn dept, 78-80; assoc prof mech eng, 82-89, actg chmn dept, 89-91, PROF MECH ENG, NORTHEASTERN UNIV, 89- *Concurrent Pos:* Actg chmn, Heat Tech & Energy Res Unit, Sci & Tech Res Coun Turkey, 78-80. *Mem:* Assoc mem Sigma Xi; Am Soc Mech Engrs. *Res:* Simultaneous radiation and convection in radiatively participating media; radiating aerosols; stability of natural convection in enclosed spaces; modeling of reacting coal particle behavior under simultaneous devolatilization and combustion; integrated design for productivity growth. *Mailing Add:* Mech Eng Dept Northeastern Univ Boston MA 02115

YENISCAVICH, WILLIAM, b Girardville, Pa, June 30, 34; m 54; c 2. NUCLEAR ENGINEERING. *Educ:* Drexel Inst Technol, BS, 57; Carnegie-Mellon Univ, MS, 62, PhD(metall eng), 63. *Prof Exp:* Engr, Bettis Atomic Power Lab, Westinghouse Elec Corp, 57-66; group leader welding eng, Mat Systs Div, Union Carbide Corp 66-67, mgr eng sci, 67-69, mgr process eng, 69-71; US Navy Rep, Subcontractors in Mfg Nuclear Reactor Cores, Naval Ship Propulsion, 84-87; mgr welding eng, 71-76, mgr mat control, 76-80, mgr welding eng, 80-84, SPEC ASSIGNMENT TO THE BETTIS GEN MGR, BETTIS ATOMIC POWER LAB, WESTINGHOUSE ELEC CORP, 87- *Concurrent Pos:* Vis asst prof mat sci & metall eng, Purdue Univ, 67-68; chmn basic res subcomt, high alloys comt, Welding Res Coun, 68-71; mem bd dirs, Metal Properties Coun, 69-72. *Mem:* Am Welding Soc; Am Soc Metals; Am Inst Mining, Metall & Petrol Engrs. *Res:* Irradiation damage in reactor fuels, control rods and structural materials; causes of cracking during welding and effects of small cracks on service performance; solidification of metals; design testing and manufacture of nuclear power reactors. *Mailing Add:* Bettis Atomic Power Lab PO Box 79 West Mifflin PA 15122

YEN-KOO, HELEN C, b Shanghai, China, May 8, 25; US citizen; m 58; c 1. DRUG SIDE EFFECTS, MECHANISM OF DRUG ADDICTION. *Educ:* Franco-Chinese Univ, BS, 46; Ohio State Univ, MS, 51; George Washington Univ, PhD(pharmacol), 55. *Prof Exp:* Lab asst pharmacog, Franco-Chinese Univ, 42-48; mgr manufacture, Nat Biol Prod, 46-49; teaching asst pharm, Ohio State Univ, 50-51 & pharmacol, George Washington Univ, 51-55; res assoc cancer, Med Lab, Roswell Park Res Inst, 55-56; res assoc pharmacol, Ortho Res Inst, Johnson & Johnson, 56-60; res assoc psychopharmacol, Geigy Pharmaceut Co, 60-64; group leader, Endo Lab, Du Pont Chem Co, 64-66; GROUP LEADER & ACTG CHIEF PSYCHOPHARMACOL & TOXICOL, DIV DRUG BIOL, FOOD & DRUG ADMIN, 66- *Concurrent Pos:* Vis scientist, Naking Med Sch, People's Repub China, 86. *Mem:* NY Acad Sci; Am Soc Pharmacol & Therapeut; Am Soc Toxicol; Sigma Xi; Chinese Health & Med Asn (vpres, 78-80, pres, 80). *Res:* Drug effects on animal behavior, memory and learning activities. *Mailing Add:* Dept Drug HIN-176 Food & Drug Admin 200 C St Washington DC 20204

YENNIE, DONALD ROBERT, b Paterson, NJ, Mar 4, 24; m 50; c 2. THEORETICAL HIGH ENERGY PHYSICS. *Educ:* Stevens Inst Technol, ME, 45; Columbia Univ, PhD(physics), 51. *Prof Exp:* Instr physics, Stevens Inst Technol, 46-47; mem, Inst Advan Study, 51-52; from instr to asst prof, Stanford, Univ, 52-57; from assoc prof to prof, Univ Minn, Minneapolis, 57-64; PROF PHYSICS, CORNELL UNIV, 64-, MEM STAFF, LAB NUCLEAR STUDIES, 64- *Concurrent Pos:* NSF sr fel, 60-61; mem prog adv comt, Stanford Linear Accelerator Ctr, 65-68, vis scientist, 70-71 & 85; vis prof, Univ Paris VI, 78-79; Guggenheim fel, 78-79; vis scientist, Fermi Nat Accelerator Lab, 84 & 88; vis prof, Inst Theoret Physics, 85 & 88. *Honors & Awards:* Alexander von Humboldt Res Award, 90. *Mem:* Am Phys Soc; AAAS; Am Asn Univ Prof. *Res:* Quantum field theory; theory of high energy electromagnetic interactions; theory of Lamb shift and hyperfine splitting; renormalization theory; infrared divergence in quantum electrodynamics; theory of integer quantum Hall effect. *Mailing Add:* Newman Lab Cornell Univ Ithaca NY 14853

YENSEN, ARTHUR ERIC, b Nampa, Idaho, Oct 13, 44; m 66, 86; c 1. ECOLOGY. *Educ:* Col Idaho, BS, 66; Ore State Univ, MA, 71; Univ Ariz, PhD(zool), 73. *Prof Exp:* asst prof biol, Millsaps Col, 73-78; asst prof biol, Boise State Univ, 78-; AT DEPT BIOL, COL IDAHO. *Concurrent Pos:* Ed, Murrelet, 83-87. *Mem:* Ecol Soc Am; Am Soc Mammalogists; Br Ecol Soc; Sigma Xi. *Res:* Ecology and biosystematics of mammals. *Mailing Add:* Dept Biol Col Idaho Caldwell ID 83605

YENTSCH, CHARLES SAMUEL, b Louisville, Ky, Sept 13, 27. MARINE BIOLOGY. *Educ:* Univ Louisville, BS, 50; Fla State Univ, MS, 53. *Prof Exp:* Asst marine biol, Fla State Univ, 52-53; asst biol oceanog, Univ Wash, 53-55; res assoc marine ecol, Woods Hole Oceanog Inst, 55-67; assoc prof oceanog, Nova Univ, 67-69, assoc prof marine biol, 69-71; prof marine sci & dir marine sta, Univ Mass, Amherst, 71-77; EXEC DIR, BIGELOW LAB OCEAN SCI, BOOTHBAY HARBOR, 77- *Mem:* Am Soc Limnol & Oceanog; Phycol Soc Am. *Res:* Marine phytoplankton ecology. *Mailing Add:* Samoset Rd Boothbay Harbor ME 04537

YEO, YUNG KEE, b Kyungbuk, Korea, Apr 24, 38; US citizen; m 64; c 2. SEMI CONDUCTOR PHYSICS, LOW-TEMPERATURE PHYSICS. *Educ:* Seoul Nat Univ, BS, 61; Univ Southern Calif, PhD(physics), 72. *Prof Exp:* Res asst, 69-72, res assoc, Univ Southern Calif, 72; physicist, Develco Inc, 73-74; res assoc, Univ Ore, 74-77; resident scientist, Avionics Lab, Wright-Patterson AFB, 77-78; sr physicist, Systs Res Lab, 78-80; res scientist, Universal Energy Systs, 80-84; from asst prof to assoc prof, 87-90, PROF, AIR FORCE INST TECHNOL, 90- *Concurrent Pos:* Adj lectr, Air Force Inst Technol, 81. *Mem:* Am Phys Soc; Korean Scientists & Engrs Asn Am; Asn Korean Physicist Am. *Res:* Ion-implantation techniques used for fabrication of electronic devices and/or opto-electronic applications; electrical and optical properties of various compound semiconductors such as gallium arsenide and alluminum gallium arsenide. *Mailing Add:* Dept Eng Physics Sch Eng AFIT/ENP Air Force Inst Technol Wright-Patterson AFB Dayton OH 45433

YEOMAN, LYNN CHALMERS, b Evanston, Ill, May 17, 43; m 66; c 3. SCIENCE EDUCATION. *Educ:* DePauw Univ, BA, 65; Univ Ill, Champaign, PhD(biochem), 70. *Prof Exp:* From instr to assoc prof, 72-84, PROF PHARMACOL, BAYLOR COL MED, 84- *Concurrent Pos:* NIH training grant & univ fel, Baylor Col Med, 70-72; prin investr, Cancer Prog Proj grant, 73-; consult, Colon Cancer Working Group, M D Anderson Hosp & Tumor Inst, 85, Bristol Labs, 86. *Mem:* Am Asn Cancer Res; Am Soc Biol Chemists; Am Chem Soc; Am Asn Immunologists; Am Soc Cell Biol; Soc Exp Biol & Med. *Res:* Studies on colon cell autocrine receptors, redifferentiating drugs and peptides, and anticancer drug resistance; tumor markers, tumor antigens and monoclonal antibodies. *Mailing Add:* Dept Pharmacol Baylor Col Med Houston TX 77030

YEOMANS, DONALD KEITH, b Rochester, NY, May 3, 42; m 70; c 2. CELESTIAL MECHANICS. *Educ:* Middlebury Col, BS, 64; Univ Md, MS, 67, PhD(astron), 70. *Prof Exp:* Supvr, Comput Sci Corp, 73-75; ASTRON, JET PROPULSION LAB, 76- *Concurrent Pos:* Deleg, Interagency Working Group, NASA, 82-86, discipline specialist, Int Halley Watch, 82-89, prin investr comet rendezvous mission, 86-; mem, NASA Comets & Asteroids Sci Working Group, 74-85, Discovery Prog Working Group, 90- *Honors & Awards:* Asteroid 2956 Yeomans Award; Except Serv Medal, NASA, 86. *Mem:* Int Astron Union; Am Astron Soc; Hist Sci Soc. *Res:* Comet and asteroid motions; future spacecraft mission studies; history of science. *Mailing Add:* 833 Chehalem Rd La Canada CA 91011

YEON-SOOK, YUN, b Korea, July 25, 50; m 76; c 2. TUMOR IMMUNOLOGY, CANCER CHEMOTHERAPY. *Educ:* Ewha Woman Univ, Korea, BS, 72, MS, 74, PhD(biochem), 79. *Prof Exp:* Res asst biochem, Col Pharm, Ewha Woman Univ, Korea, 72-75; researcher, Lab Molecular Biol, Kaeri, Korea, 76-78; chief, Lab Biochem, 79-84, CHIEF, LAB IMMUNOL, KOREA CANCER CTR HOSP, 84- *Concurrent Pos:* Guest researcher, Exp Immunol Br, Nat Cancer Inst, NIH, 86-88; consult, Col Pharm, Ewha Woman Univ, 90- *Mem:* Am Asn Immunologists; Am Asn Cancer Res. *Res:* Screening of biological response modifiers and anti-cancer drug from oriental medicinal plants; LAK cell phenomenon; radiation damage on immune system. *Mailing Add:* Mido Apt 110-807 Daechi-dong Kangnamku Seoul 135-282 Republic of Korea

YEOWELL, DAVID ARTHUR, b London, Eng, Jan 3, 37; m 64; c 3. DRUGS. *Educ:* Bristol Univ, BSc, 58, PhD, 61. *Prof Exp:* Swiss Nat Fund fel org chem, Univ Zurich, 61-62; Imp Chem Indust res fel biogenetics, Univ Liverpool, 62-64; sr org chemist, Wellcome Res Labs, Burroughs Wellcome Co, 64-68, sr develop chemist, Chem Develop Labs, 68-71, head develop res, 71-73, mgr chem develop labs, 73-81, dir develop labs, 81-86, VPRES TECH DEVELOP, BURROUGHS WELLCOME CO, 86- *Mem:* Am Chem Soc; Royal Soc Chem. *Res:* Application of organic, physical and analytical chemical knowledge to devising economic and practical chemical and dosage form production scale processes for pharmaceuticals and their quality assurance. *Mailing Add:* Vpres Tech Develop Burroughs Wellcome Co 3030 Cornwallis Rd Research Triangle Park NC 27709

YERANSIAN, JAMES A, analytical chemistry; deceased, see previous edition for last biography

YERAZUNIS, STEPHEN, b Pittsfield, Mass, Aug 21, 22; m 54; c 2. CHEMICAL ENGINEERING. *Educ:* Rensselaer Polytech Inst, BChE, 47, MChE, 48, DChE, 52. *Prof Exp:* Engr, Manhattan Proj, Oak Ridge, Tenn, 43-46 & Gen Elec Co, Mass, 47; from instr to assoc prof chem eng, Rensselaer Polytech Inst, 48-63, assoc dean, Sch Eng, 66-79, prof chem eng, 63-; RETIRED. *Concurrent Pos:* Consult, Knolls Atomic Power Lab, NY, 56-72, NY State Dept Ment Hyg, 68-72 & Jet Propulsion Lab, 77-78. *Mem:* Fel Am Inst Chem Engrs; Am Soc Eng Educ. *Res:* Mass and heat transfer; vapor-liquid equilibrium; activation of radio-nuclides and their transport; appraisal of electrical energy alternatives; guidance of an autonomous planetary rover. *Mailing Add:* 107 Carriage Dr Pittsburgh PA 15237

YERBY, ALONZO SMYTHE, b Augusta, Ga, Oct 14, 21; m 43; c 3. PREVENTIVE MEDICINE ADMINISTRATION. *Educ:* Univ Chicago, BS, 41; Meharry Med Col, MD, 46; Harvard Univ, MPH, 48; Am Bd Prev Med, dipl, 53. *Hon Degrees:* DSc, Meharry Med Col, 77. *Prof Exp:* Asst to dir food res inst, Univ Chicago, 42-43; intern, Coney Island Hosp, Brooklyn, NY, 46-47; health officer-in-training, New York City Dept Health, 47-48; field med officer, UN Int Refugee Orgn, US Zone, Ger, 48-49, dep chief med affairs, Off US High Comnr Ger, 49-50; assoc med dir, Health Ins Plan Greater New York, 50-54, consult, 54; regional med consult, Off Voc Rehab, 54-57; dep comnr med affairs, NY State Dept Soc Welfare, 57-60; exec dir med servs, New York City Dept Health & med welfare adminr, Dept Welfare, 60-65 & coordr welfare servs, Dept Hosps, 64-65, comnr, Dept Hosps, 65-66; prof health serv admin, Sch Pub Health, Harvard Univ, 66-82, head dept, 66-75; prof & dir, Div Health Serv Admin, Sch Med, Uniformed Serv Univ Health Sci, 82-88; RETIRED. *Concurrent Pos:* Staff physician, Sidney Hillman Health Ctr, New York, 51-53; adj asst prof admin med, Sch Pub Health, Columbia Univ, 60-66; lectr, Yale Univ, 65-66; mem, Surg Gen Adv Comt Urban Health Affairs, USPHS & task force on orgn community health servs, Nat Comn Health Servs, 63-66; mem summer study sci & urban develop, Dept Housing & Urban Develop & President's Off Sci & Tech, 66; adv comt on rels with state health agencies, Dept Health, Educ & Welfare, 66; mem, President's Nat Adv Comn Health Manpower, 66-67; mem nat pub health training coun, NIH, 69-71, consult, 71-, mem bd adv, John E Fogarty Int Ctr, 72-76, vis prof, 75; WHO consult health manpower training, Govt Sierra Leone, WAfrica, 70; found mem, Inst Med of Nat Acad Sci, mem Coun, 70-73; adv, Ministry Health Kuwait, 76; mem, Nat Prof Standards Rev Coun, OASH, HEW, 77-; dep asst, Sect Health for Intergovt Affairs & asst surg gen, US Dept Health & Human Serv/USPHS, 80-81. *Mem:* Inst Med-Nat Acad Sci; fel Am Pub Health Asn; fel Am Col Prev Med; Am Public Welfare Asn; NY Acad Med. *Res:* Health service administration; public health practice. *Mailing Add:* 4104 Byforde Ct Kensington MD 20895

YERG, DONALD G, b Lewistown, Pa, Mar 4, 25; m 48; c 3. METEOROLOGY. *Educ:* Pa State Univ, BS, 46, MS, 47, PhD(meteorol), 53. *Prof Exp:* Asst agr eng, Univ Calif, 47-48; instr math, Univ Alaska, 48-50; consult, Tech Info Div, Library of Cong, 51-52; asst ionosphere res, Pa State Univ, 52-53; lectr physics, Univ PR, 53-55; assoc prof, 55-60, PROF PHYSICS & DEAN GRAD SCH, MICH TECHNOL UNIV, 60- *Mem:* AAAS; Am Meteorol Soc; Int Soc Biometeorol; Am Geophys Union; Sigma Xi. *Res:* Meteorology of ionospheric regions; micrometeorology; biometeorology. *Mailing Add:* 112 E Seventh Ave Mich Technol Univ Houghton MI 49931

YERG, RAYMOND A, b Jersey City, NJ, Apr 4, 17; m 46; c 3. AEROSPACE MEDICINE, OCCUPATIONAL MEDICINE. *Educ:* Seton Hall Col, BS, 38; Georgetown Univ, MD, 42; Harvard Univ, MPH, 55. *Prof Exp:* Comdr, 1st Missile Div, Vandenberg AFB, US Air Force, Calif, 59-61, dep bioastronaut Air Force Eastern Test Range, 61-65, comdr, Aerospace Med Res Lab, Wright-Patterson AFB, Ohio, 65-68, chief sci & tech div, Hq US Air Force, The Pentagon, 68-72; corp med dir, Perlcin-Elmer Corp, 78-80 & 82-87 corp med dir, Am Can Co, 72; RETIRED. *Concurrent Pos:* Consult Occup Med. *Honors & Awards:* AMA Spec Aerospace Med Citation, 62; Air Force Asn Meritorious Award Support Mgt, 64. *Mem:* Fel Aerospace Med Asn; fel Am Col Prev Med; NY Acad Sci; fel Am Acad Occup Med; fel Am Occup Med Asn; fel Am Col Physicians. *Mailing Add:* 11808 Beekman Pl Potomac MD 20854

YERGANIAN, GEORGE, b New York, NY, June 14, 23; m 50; c 2. BIOLOGY. *Educ:* Mich State Univ, BS, 47; Harvard Univ, PhD(biol), 50. *Prof Exp:* Instr bot, Univ Minn, 50-51; AEC fel, Brookhaven Nat Lab, 51-52 & Boston Univ, 52-53; USPHS res fel, Boston Univ & Children's Cancer Res Found, 53-54; res assoc, Children's Cancer Res Found, Children's Hosp Med Ctr, 54-78; res assoc path, Harvard Univ, 54-78; chief, Lab Cytogenetics & Exp Carcinogenesis, Sidney Farber Cancer Inst, 56-78; PROF, INST ENVIRON BIOMED SCI, NORTHEASTERN UNIV, 78- *Honors & Awards:* Prof Dikran H Kabakjian Sci Award, Armenian Students Asn, 77. *Mem:* Genetics Soc Am; Tissue Culture Asn; Am Soc Human Genetics; Am Asn Cancer Res; Am Soc Cell Biologists. *Res:* Mammalian cytogenetics with emphasis on dwarf species of hamsters. *Mailing Add:* Dept Pop Studies Harvard Sch Pub Health 677 Huntington Ave Boston MA 02115

YERGER, RALPH WILLIAM, b Reading, Pa, July 31, 22; m 54; c 4. SYSTEMATIC ICHTHYOLOGY. *Educ:* Pa State Univ, BS, 43, MS, 47; Cornell Univ, PhD(zool), 50. *Prof Exp:* Teacher high sch, Pa, 43; instr biol, Pa State Univ, Altoona Ctr, 47-48; instr nature study, Reading Mus, 48; from asst prof to assoc prof zool, Fla State Univ, 50-61, actg head dept, 52-54, assoc chmn undergrad studies, 75-77, assoc dean, Col Arts & Sci, 77-78 & 79-83, actg dean, Col Arts & Sci, 78-79, prof biol, 61- 82, univ serv prof, 83-88; RETIRED. *Mem:* Am Soc Ichthyologists & Herpetologists; Am Inst Biol Scientist; Am Fisheries Soc; Sigma Xi. *Res:* Taxonomy, ecology and distribution of fresh and salt water fishes of the southeastern United States, Central America and the Caribbean. *Mailing Add:* 2917 Woodside Dr Tallahassee FL 32312-2830

YERGEY, ALFRED L, III, b Philadelphia, Pa, Sept 17, 41; m 63, 91; c 3. ANALYTICAL CHEMISTRY, CHEMICAL KINETICS. *Educ:* Muhlenberg Col, BS, 63; Pa State Univ, PhD(chem), 67. *Prof Exp:* Res fel chem, Rice Univ, 67-69; chemist, Esso Res & Eng Co, NJ, 69-71; sr scientist, Sci Res Instruments Corp, 71-77; res chemist, 77-85, SECT CHIEF, NAT INST CHILD HEALTH & HUMAN DEVELOP, NIH, 85- *Mem:* Am Chem Soc; Am Soc Bone Mineral Res; Am Inst Nutrit. *Res:* Calcium kinetics of skeleton; applications of mass spectrometry; stable isotope applications to clinical problems; quadrupole mass spectrometry. *Mailing Add:* NIH Bldg 10 Rm 6C101 Bethesda MD 20892

YERGIN, PAUL FLOHR, b New York, NY, Apr 21, 23; m 47; c 2. NUCLEAR PHYSICS. *Educ:* Union Univ, NY, BS, 44; Columbia Univ, MA, 49, PhD, 53. *Prof Exp:* Asst electronics res, Gen Elec Co, NY, 42; asst, Gen Physics Lab, Union Univ, NY, 42-43; physicist radiation lab, Columbia Univ, 44-45, mem sci staff, 45-52; from instr to asst prof, Univ Pa, 52-56; from asst prof to assoc prof, 56-74, PROF PHYSICS, RENSSELAER POLYTECH INST, 74- *Concurrent Pos:* Res affil, Lab Nuclear Sci, Mass Inst Technol, 80-82. *Mem:* Am Phys Soc; Am Asn Physics Teachers; Sigma Xi; Am Asn Univ Professors. *Res:* Photonuclear reactions; photopion reactions. *Mailing Add:* 92 Willett St Apt 4A Albany NY 12210

YERICK, ROGER EUGENE, b Kingsville, Tex, July 6, 32. ANALYTICAL CHEMISTRY. *Educ:* Tex Col Arts & Indust, BS, 53. *Prof Exp:* Res asst anal chem, Iowa State Univ, 53-57; asst prof chem, Tex Col Arts & Indust, 57-58; from asst prof to assoc prof, 58-65, PROF CHEM, LAMAR UNIV, 65-, DEAN COL SCI, 74- *Concurrent Pos:* Educ consult, Spec Training Div, Oak Ridge Assoc Univs, 62- *Mem:* AAAS; Am Chem Soc; Sigma Xi. *Res:* Analytical chemistry of chelates; analytical radiochemistry; analytical applications of liquid scintillation counting techniques. *Mailing Add:* Box 173 Kingsville TX 78363

YERKES, WILLIAM D(ILWORTH), JR, b Wilkes Barre, Pa, May 29, 22; m 47; c 3. ENVIRONMENTAL SCIENCES. *Educ:* State Col Wash, BS, 48, PhD(plant path), 52. *Prof Exp:* Asst, State Col Wash, 48-52; asst plant pathologist, Mex Agr Prog, Rockefeller Found, 52-56, assoc plant pathologist, 56-60; microbiologist, Pioneering Res Lab, Kimberly-Clark Corp, 60-72; chmn, Dept Environ Sci, Grand Valley State Col, 72-79, mem staff, Sch Health Sci, 80-87; RETIRED. *Mem:* Am Soc Civil Engrs; Water Pollution Control Fedn; Nat Environ Health Asn; Sigma Xi. *Res:* Ecology of water pollution; effects of pollutants on aquatic biota; solid and hazardous waste management; food sanitation; biochemistry and biophysics of photosynthesis. *Mailing Add:* 3679 Blackfoot Ct Grandville MI 49418

YESAIR, DAVID WAYNE, b Newbury, Mass, Sept 9, 32; m 54; c 3. BIOCHEMISTRY. *Educ:* Univ Mass, BS, 54; Cornell Univ, PhD(biochem), 58. *Prof Exp:* Asst biochem, Cornell Univ, 55-57, res assoc, 58; res biochemist, Lederle Labs Div, Am Cyanimid, 59-61; NSF fel, Reading Univ, Eng, 61-62; sr scientist, biochem group, Arthur D. Little, Inc, 62-66, head, biochem and pharmacol group, 66-71 & 72-77, mgr, biomed sci sect, 77-82, vpres, 78-84; PRES & CHIEF EXEC OFFICER, BIOMOLECULAR PROD INC, 84- *Concurrent Pos:* Nat Cancer Inst spec res award, Inst Org Chem, Paris, France, 71-72; lectr, Mass Inst Technol, 72-82; biotechnol adv bd, Univ Conn, 86- *Mem:* Am Chem Soc; NY Acad Sci; Am Asn Cancer Res; Am Soc Pharmacol & Exp Therapeut; fel Am Inst Chemists; AAAS; Am Soc Toxicol; Sigma Xi; Int Soc Studying Xenobiotics (secy, 90-91); Am Asn Pharmaceut Scientist; fel NSF; fel, Leukemia Soc Am. *Res:* Lipid biochemistry; cancer, obesity and diabetes; metabolism and mode of action of drugs affecting lipid metabolism; isolation, characterization and metabolic action of biologically active agents; development of drug delivery systems. *Mailing Add:* BioMolecular Prod Inc PO Box 347 Byfield MA 01922

YESINOWSKI, JAMES PAUL, b LaSalle, Ill, Mar 22, 50. PHYSICAL CHEMISTRY, INORGANIC CHEMISTRY. *Educ:* Univ Ill, Urbana, BS, 71; Univ Cambridge, PhD(chem), 74. *Prof Exp:* Res assoc phys chem, Mass Inst Technol, 74-76; Staff Scientist, Miami Valley Labs, Procter & Gamble Co, 76-84; mem prof staff, Calif Inst Technol, 84-88; ADJ ASSOC PROF CHEM & DIR, MAX T ROGERS, NUCLEAR MAGNETIC RESONANCE FAC, MICH STATE UNIV, 88- *Concurrent Pos:* Am Cancer Soc fel, 76. *Mem:* Am Chem Soc; Am Geophys Union. *Res:* Nuclear magnetic resonance spectroscopy of solids, theory & applications; nuclear magnetic resonance applied to biological mineralization, surfaces, solid state chemistry and homogeneous nucleation; nuclear magnetic resonance spectroscopy of inorganic complexes, surfactant, phospholipid and liquid crystalline systems. *Mailing Add:* Dept Chem Mich State Univ East Lansing MI 48824-1322

YESKE, RONALD A, b Wisconsin Rapids, Wis, Oct 28, 46; m 68; c 1. MATERIALS SCIENCE ENGINEERING. *Educ:* Marquette Univ, BME, 68; Northwestern Univ, PhD(mat sci), 73. *Prof Exp:* Asst prof metall eng, Univ Ill, Urbana, 74-78; mgr corrosion res, Res & Develop Ctr, Westinghouse, 78-82; sr res assoc & assoc prof corrosion, 82-87, VPRES, INST PAPER CHEM, 87- *Mem:* Nat Asn Corrosion Engrs; Am Soc Metals; Tech Assoc Pulp & Paper Indust; Paper Indust Mgr Asn. *Res:* Mechanisms of corrosion, corrosion-assisted cracking and fatigue crack propagation; materials performance in power plant materials; materials for the pulp and paper industry. *Mailing Add:* Inst Paper Sci & Technol 575 14th St NW Atlanta GA 30318

YESNER, RAYMOND, b Columbus, Ga, Apr 18, 14; m 47; c 3. PATHOLOGY. *Educ:* Harvard Univ, AB, 35; Tufts Col, MD, 41; Yale Univ, MA, 72. *Prof Exp:* Intern & res, Beth Israel Hosp, Boston, Mass, 41-44; pathologist & chief lab serv, Vet Admin Hosp, Newington, 47-53; from asst clin prof to assoc clin prof, Med Sch, Yale Univ, 49-64, assoc prof, 64-72; prof, 72-84, EMER PROF PATH, MED SCH, YALE UNIV, 82-; DIR AUTOPSY SERV, YALE MED SCH, 87- *Concurrent Pos:* Consult pathologist, Coop Study of Prostate, Vet Admin, chmn path panel, Lung Cancer Chemother Study Group, 58-, mem path res eval comt, sr physician, 71-74, radiation ther oncol group, 82-; pathologist & chief lab serv, Vet Admin

Hosp, West Haven, 53-74, chief staff, 69-74, chief pathologist, 74-77, dir anat path, 77-87; chmn, WHO Lung Cancer Comt, Geneva, 77; path panel, Int Asn Study Lung Cancer, 82- *Honors & Awards:* Health Memorial Award, Univ Tex, 84. *Mem:* Emer mem AMA; emer mem Am Asn Path & Bact; emer mem Col Am Path; emer mem Int Acad Path. *Res:* Changes in blood viscosity; liver, lung and gastrointestinal disease; carcinoma of lung and prostate and bladder. *Mailing Add:* Path Dept Yale Med Sch 310 Cedar St New Haven CT 06510

YESSIK, MICHAEL JOHN, b Webster, Mass, Nov 22, 41; m 70. SOLID STATE PHYSICS. *Educ:* Williams Col, BA, 62; Syracuse Univ, PhD(solid state sci), 66; Univ Cambridge, MA, 67. *Prof Exp:* NATO fel, Cavendish Lab, Univ Cambridge, 66-67, NSF fel, 67-68; sr res scientist physics, Sci Res Staff, Ford Motor Co, 68-75; mgr process develop, Photon Sources, Inc, 75-77, mgr new prod develop, 77-80, dir res & develop, 80-83; vpres res & develop, Bausch & Lomb Ophthal Inst, 83-85; eng mgr, Spectrophysics Inc, 85-89; PRES, SURFACE SCI LABS, 89- *Mem:* Optical Soc Am; Sigma Xi; AAAS; Am Phys Soc; Inst Elec & Electronics Engrs. *Res:* Electronic and magnetic properties of metals and alloys; physical properties of high temperature ceramics; high-power gas laser development; materials processing using high-power lasers. *Mailing Add:* 1206 Charleston Rd Mountain View CA 94043

YETT, FOWLER REDFORD, b Johnson City, Tex, Oct 18, 19; m 45; c 3. APPLIED MATHEMATICS. *Educ:* Univ Tex, BS, 43, MA, 52; Iowa State Univ, PhD(appl math), 55. *Prof Exp:* Res chemist & chem engr, Manhattan Proj, 43-45; owner, camera supply co, 46-49; teaching fel, Univ Tex, 49-52; instr math, Iowa State Univ, 52-55; asst prof, Long Beach State Col, 55-56 & Univ Tex, 56-65; chmn dept math, 65-68, PROF MATH, UNIV S ALA, 65- *Concurrent Pos:* Sr res engr, NAm Aviation, Inc, Downey, CA, 56-57 & 59; fac res assoc, Boring Co, Seattle, Wash. *Mem:* Am Math Soc; Math Asn Am. *Res:* Nonlinear differential equations. *Mailing Add:* Dept Math Univ SAla 307 University Dr Mobile AL 36688

YEUNG, DAVID LAWRENCE, b Hong Kong, Dec 18, 39; Can citizen; m 66; c 2. INFANT NUTRITION, GERIATRIC NUTRITION. *Educ:* Univ Toronto, BA, 63, MA, 66, PhD(nutrit), 70. *Prof Exp:* From asst prof to assoc prof nutrit, Univ Guelph, 70-76; infant nutritionist, 76-82, mgr, Dept Nutrit Res, 82-84, COORDR CORP NUTRIT, H J HEINZ CO CAN LTD, 82- & MGR NUTRIT RES, 84-; ASSOC PROF NUTRIT, DEPT NUTRIT SCI, FAC MED, UNIV TORONTO, 82- *Concurrent Pos:* Coordr food & nutrit sci, Inst Study & Appln Integrated Develop, 76-; consult, Can Pediat Soc, 78- & consult ed nutrit res, 82-85; ed, Infant Nutrit Inst, 82-; vis assoc prof, Sun Yet Sen Univ Med Sci, Guangzhou, China, 85-, W China Univ Med Sci, Chengdu, China, 87-, Zhejiang Med Univ, 90; adj prof, Ryerson Polytechnic Inst, Toronto, 90-91. *Mem:* Can Soc Nutrit; Can Pub Health Asn; Am Inst Nutrit; Can Inst Food Sci & Technol; Can Pediat Soc. *Res:* Infant nutrition, feeding and growth; food habits, nutritional status and requirement of the elderly; maternal and infant nutrition in China, the Far East & Africa; international nutrition. *Mailing Add:* Dept R & D H J Heinz Co Can Ltd 5650 Yonge St 16th Fl North York ON M2M 4G3 Can

YEUNG, EDWARD SZESHING, b Hong Kong, Feb 17, 48; m 71; c 2. ANALYTICAL CHEMISTRY, PHYSICAL CHEMISTRY. *Educ:* Cornell Univ, AB, 68; Univ Calif, Berkeley, PhD(chem), 72. *Prof Exp:* From instr to assoc prof, 72-81, PROF CHEM, IOWA STATE UNIV, 81- *Concurrent Pos:* Sloan Found fel, 74; ed, Progress Anal Spectros & Anal Chem. *Honors & Awards:* Chem Instrumentation Award, Am Chem Soc, 87. *Mem:* Am Chem Soc; NAm Photochem Soc. *Res:* Pollution monitoring; high resolution spectroscopy; photochemistry. *Mailing Add:* Dept Chem Iowa State Univ Ames IA 50011

YEUNG, KATHERINE LU, b Shanghai, China, July 28, 43; US citizen; m 68; c 2. CANCER, PHARMACOLOGY. *Educ:* Univ Houston, BS, 65, MS, 68. *Prof Exp:* Res asst, 68-75, res assoc, 75-80, ASST PHARMACOLOGIST, CANCER CHEMOTHER, UNIV TEX SYST CANCER CTR, 80- *Mem:* Sigma Xi; NY Acad Sci; Am Soc Pharm & Exp Therapeut; Am Asn Cancer Res. *Res:* Cancer chemotherapy; metabolism and distribution of antitumor agents in patients and experimental animals. *Mailing Add:* 9850 Pagewood 206 Houston TX 77042-5525

YEUNG, PATRICK PUI-HANG, b Kweiyang, China, Sept 26, 42; US citizen; m 70; c 1. ORGANIC CHEMISTRY, TEXTILE CHEMISTRY. *Educ:* Auburn Univ, BSc, 67; Ga Inst Technol, PhD(org chem), 75. *Prof Exp:* Res chemist dyes, Toms River Chem Corp, Ciba-Geigy Corp, 75-80; MEM STAFF, AM COLOR & CHEM CORP, 80- *Mem:* Am Chem Soc; Am Asn Textile Chemists & Colorists; AAAS. *Res:* Dyes and chemicals. *Mailing Add:* PO Box 8261 Holland MI 49422-8261

YEUNG, REGINALD SZE-CHIT, b Hong Kong, Apr 30, 32; US citizen; m 66; c 1. CHEMICAL ENGINEERING, CHEMISTRY. *Educ:* Univ London, BSc, 54; Mass Inst Technol, SM, 59, ScD(chem eng), 61. *Prof Exp:* Engr, Cabot Corp, 61-63; ENGR, SHELL OIL CO, HOUSTON, 64- *Mem:* Am Chem Soc; Am Inst Chem Engrs. *Res:* Process development and design; reactor design; flame reactor design; insecticide production. *Mailing Add:* 10607 Creektree Dr Houston TX 77070

YEUNG, RONALD WAI-CHUN, b Hong Kong, July 19, 45; US citizen; m 70; c 1. FREE-SURFACE MECHANICS, SHIP & OFFSHORE DYNAMICS. *Educ:* Univ Calif, Berkeley, BS, 68, MS, 70, PhD(eng), 73. *Prof Exp:* Naval architect, Advan Marine Technol, Litton Ship Systs, 70-71; res assoc hydrodyn, Dept Ocean Eng, Mass Inst Technol, 73-74, from asst prof to assoc prof naval archit, 74-82; PROF HYDROMECH, DEPT NAVAL ARCHIT & OFFSHORE ENG, UNIV CALIF, BERKELEY, 82-, DEPT CHAIR, 89- *Concurrent Pos:* Instr, Long Beach Naval Shipyard, Univ Calif, Los Angeles, 71; prin investr, NSF, 75-83, fluid dynamics, Off Naval Res, 77-; consult, Maritech Inc, 74-77, Hydronautics Inc, 79-80, Ecodynamics, 86-88, Chevron Oil, 88-90 & USCG, 88; rep, US-Japan Sci exchange shallow-water probs, 79-80; assoc ed, J Ship Res, 80-, Computers & Fluids, 84- & J Eng Math, 86-; Fulbright-Hayes sr fes, Dept Appl Math, Univ Adelaide, S Australia, 81; vis distinguished scientist, Alexander von Humboldt Found, Univ Hamburg, WGer, 88-89. *Honors & Awards:* US Distinguished Scientist, Alexander von Humboldt Found, 88. *Mem:* Soc Naval Architects & Marine Engrs; Am Soc Eng Educ. *Res:* Theory of surface gravity waves; wave-ship or wave-structure interaction; theory of ship-ship interaction hydrodynamics; numerical methods in free-surface flows; non-linear breaking wave modeling; wave-viscosity interaction physics. *Mailing Add:* Dept Naval Archit & Offshore Eng Univ Calif Berkeley CA 94720

YEVICH, JOSEPH PAUL, b McKees Rocks, Pa, Sept 20, 40; m 64; c 2. NEUROSCIENCES. *Educ:* Carnegie-Mellon Univ, BS, 62, MS, 67, PhD(org chem), 69. *Prof Exp:* Chemist, Gulf Res & Develop Corp, 62-65; sr scientist, 69-74, sr investr, 74-76, sr res assoc, 76-80, SR RES SCIENTIST, MEAD JOHNSON & CO, 80- *Concurrent Pos:* Assoc dir, 82-86, dir, Bristol-Myers Squibb, 86- *Mem:* AAAS; Am Chem Soc; Sigma Xi; Soc Neurosci; Int Cong Heterocyclic Chem; NY Acad Sci. *Res:* Design and synthesis of potential medicinal agents; design and synthesis of anti-inflammatory agents, bronchodilators and medistor-release inhibitors; CNS area on anxiolytics, antipsychotics, antidepressants and cognition enhancers, neuroprotective agents. *Mailing Add:* 115 Crest Rd Southington CT 06489-2807

YEVICH, PAUL PETER, b Berwick, Pa, June 16, 24; m. HISTOPATHOLOGY. *Educ:* Pa State Univ, BA, 49. *Prof Exp:* Histologist chem warfare lab med directorate, Army Chem Ctr, US Dept Defense, 50-54, histopathologist, 54-60; res histopathologist physiol sect, Div Occup Health, US Dept Health, Educ & Welfare, 61-66; res biologist, Invert Sect, Environ Protection Agency, 67-71, res team leader, Invert Sect, Engiron Res Lab, 71-86; CONSULT, UNIV RI, SAIC, ENVIRON PROTECTION AGENCY, UNIV SFLA, 88- *Mem:* Soc Invert Path. *Res:* Comparative histology and pathology; toxic effects of various compounds on cells and tissues of various species of animals; invertebrate histology and effects of toxic compounds on invertebrates; histopathologic effects of oil pollutants, metals, organics and sewage sludge on marine life; immune responses of molusks. *Mailing Add:* 12 Old Mountain Rd Peace Dale RI 02882

YEVICK, DAVID OWEN, b New York City, May 3, 54; c 2. PHYSICS OPTICAL COMMUNICATIONS, OPTO-ELECTRONIC DEVICE PHYSICS. *Educ:* Harvard Univ, AB, 73; Princeton Univ, MS, 75, PhD(physics), 77. *Prof Exp:* Res asst physics, State Univ NY, Stony Brook, 77-79; researcher optics, Inst Optical Res, Stockholm 79-83; RESEARCH SOLID STATE PHYSICS, UNIV LUND, SWEDEN, 83-; ASSOC PROF ELEC ENG, QUEEN'S UNIV, KINGSTON, 89- *Concurrent Pos:* Consult, Xerox Parc, 83-84, Bell Commun Res, 87, Amoco Technol Co & Corning Glass, 89; assoc prof elec eng, Pa State univ, 86-89. *Mem:* Inst Elec & Electronics Engrs; fel Optical Soc Am; Am Phys Soc; Int Soc Optical Eng. *Res:* Numerical simulations of guided-wave and semiconductor opto-electronic devices, basic physics of opto-electronic devices; optical and electron transport processes in semiconductors. *Mailing Add:* Dept Elec Eng Queen's Univ Kingston ON K7L 3N6 Can

YEVICK, GEORGE JOHANNUS, b Berwick, Pa, Apr 24, 22; m 45. PHYSICS. *Educ:* Mass Inst Technol, BSc, 42, DSc(physics), 47. *Hon Degrees:* MEng, Stevens Inst Technol, 58. *Prof Exp:* Staff mem, Radiation Lab, Mass Inst Technol, 44-46; from asst prof to assoc prof, 48-57, PROF PHYSICS, STEVENS INST TECHNOL, 57-, PROF ENG PHYSICS, 77- *Mem:* Am Phys Soc; Soc Photo-Optical Instrument Engr; NY Acad Sci. *Res:* Theory of elementary particles; dynamical theory of many interacting particles; causal theory of quantum mechanics; control of thermonuclear fusion. *Mailing Add:* Dept Physics/Eng Physics Stevens Inst Technol Castle Point Hoboken NJ 07030

YEZ, MARTIN S(IMON), b Chicago, Ill, Apr 25, 20; m 46; c 1. MECHANICAL ENGINEERING, AERONAUTICAL ENGINEERING. *Educ:* Ill Inst Technol, BSME, 46; DePaul Univ, MA, 51. *Prof Exp:* Eng draftsman, Pullman Standard Car Mfg Co, 46; chief power engr, Commonwealth Edison Co, 47-49; prof eng, Wright Br, Chicago City Jr Col, 50-51; eng designer, Northrop Aircraft Co, 51-54; PROF ENG, EL CAMINO COL, 54- *Mem:* Am Cong Surv & Mapping. *Res:* Drafting room practices; surveying; geology; plant propagation. *Mailing Add:* 4029 Via Larqavista Palos Verdes Estate CA 90274

YGUERABIDE, JUAN, b Laredo, Tex, Oct 9, 35; m 56; c 4. BIOCHEMISTRY, BIOPHYSICS. *Educ:* St Mary's Univ, Tex, BS, 57; Univ Notre Dame, PhD(phys chem), 62. *Prof Exp:* Res assoc, Radiation Lab, Univ Notre Dame, 61-63; mem res staff, Sandia Corp, NMex, 63-68; res assoc biochem, Stanford Univ, 68-69; lectr & res assoc biochem & biophys, Yale Univ, 69-72; ASSOC PROF BIOL, UNIV CALIF, SAN DIEGO, 72- *Concurrent Pos:* Fel, Radiation Lab, Univ Notre Dame, 62-63; consult prof chem, Univ NMex, 66-68; sci consult, Sandia Corp, NMex, 68-69. *Mem:* AAAS; Biophys Soc; Sigma Xi. *Res:* Structure, conformation and function of proteins and biological membranes; nanosecond fluorescence spectroscopy; mathematical physics. *Mailing Add:* Dept Biol C-O16 Univ Calif San Diego La Jolla CA 92093

YI, CHO KWANG, plant biochemistry, plant pathology, for more information see previous edition

YIAMOUYIANNIS, JOHN ANDREW, b Hartford, Conn, Sept 25, 42; div; c 5. BIOCHEMISTRY, STATISTICS. *Educ:* Univ Chicago, SB, 63; Univ RI, PhD(biochem), 67. *Prof Exp:* Fel develop biol, Western Reserve Univ Sch Med, 67-68; assoc ed, Chem Abst Serv, 68-72; sci dir, Nat Health Fedn, 74-80; exec dir, Health Action, 80-84; CONSULT, 75- *Concurrent Pos:* Co-ed, Fluoride, 72-81; pres, Safe Water Found, 79- *Mem:* Am Chem Soc; Int Soc Fluoride Res. *Res:* Ganglioside biosynthesis; biochemical differentiation; subcellular particles; RNA synthesis in isolated nuclei; biological effects of inorganic fluoride; epidemiological ramifications of toxic substances; science and politics; high performance health. *Mailing Add:* 6439 Taggart Rd Delaware OH 43015

YIANNOS, PETER N, b Olympia, Greece, Nov 27, 32; US citizen; m 62; c 3. PHYSICAL CHEMISTRY, ENGINEERING. *Educ:* Univ Mo-Rolla, BS, 56; Lawrence Univ, MS, 58, PhD(phys chem), 60. *Prof Exp:* From res group leader to sr res group leader, Scott Paper Co, 60-65, sect head, 65-66, mgr pioneering res, 66-67; assoc prof eng, Widener Col, 67-69; mgr paper res, 69-73, dir paper res, 73-76, vpres prod develop, 76-79, vpres consumer res & develop, 79-81, vpres int res & develop, 81-83, VPRES FIBER TECHNOL, SCOTT PAPER CO, 83- *Concurrent Pos:* Instr, Tech Develop Prog, Scott Paper Training Course, 64-69; lectr eng, Eve Div, PMC Cols, 66-67; tech consult, Scott Paper Co, 67-69; adj prof, Widener Col, 72-74. *Honors & Awards:* Albert Award, Tech Asn Pulp & Paper Indust. *Mem:* Am Chem Soc; Am Inst Chem Eng; Tech Asn Pulp & Paper Indust. *Res:* Molecular forces and surface phenomena; fibers and fiber bonding; wood technology and pulping; mechanical properties of fibers and sheet assemblies; technical management; materials science. *Mailing Add:* 2304 Empire Dr Wilmington DE 19810-2707

YIELDING, K LEMONE, b Auburn, Ala, Mar 25, 31; m 73; c 5. MOLECULAR BIOLOGY, MEDICINE. *Educ:* Ala Polytech Inst, BS, 49; Univ Ala, MS, 52, MD, 54. *Prof Exp:* Intern, Med Ctr, Univ Ala, 54-55; clin assoc, NIH, 55-57; resident, USPHS Hosp, 57-58; sr investr, Nat Inst Arthritis & Metab Dis, 58-64; prof biochem, assoc prof med & chief lab molecular biol, Med Ctr, Univ Ala, Birmingham, 64-80; CHMN & PROF ANAT MED, UNIV SALA, MOBILE, 80- *Concurrent Pos:* Asst prof med, Georgetown Univ, 58-64; consult, USPHS, 64- *Mem:* AAAS; Am Soc Pharmacol & Exp Therapeut; Soc Exp Biol & Med; Am Soc Biol Chem; Am Asn Anatomists; Am Asn Pathologists. *Res:* Molecular basis for biological regulation, including both genetic mechanisms and control of enzyme activity; elucidation of disease mechanisms and drug action in molecular terms. *Mailing Add:* Univ Tex Grad Sch Biomed Sci Galveston TX 77550-2764

YIH, CHIA-SHUN, b Kweiyang, China, July 25, 18; nat US; m 49; c 3. MECHANICS, APPLIED MATHEMATICS. *Educ:* Nat Cent Univ, China, BS, 42; Univ Iowa, MS, 47, PhD(fluid mech), 48. *Prof Exp:* Asst, Nat Bur Hydraul Res, China, 42-43; jr bridge engr, Nat Bur Bridge Design, 43-45; instr, Nat Kweichow Univ, 45; instr math, Univ Wis, 48-49; lectr, Univ BC, 49-50; assoc prof civil eng, Colo State Univ, 50-52; asst prof fluid mech & res engr, Univ Iowa, 52-54, assoc prof, 54-56; from assoc prof to prof eng mech, 56-68, Stephan P Timoshenko Univ prof, 68-88, EMER PROF FLUID MECH, UNIV MICH, ANN ARBOR, 88-; EMER PROF ENG SCI, UNIV FAL, 90- *Concurrent Pos:* Res assoc, Nat Ctr Sci Res, Univ Nancy, France, 51-52; NSF sr fel, Cambridge Univ, 59-60; consult, Huyck Felt Co, 60-64; Guggenheim fel, 64; vis prof, Univ Paris & Univ Grenoble, 70-71, Univ Karlsruhe, 77-78; von Humboldt award, 77-78; mem academia sinica, US Nat Acad Eng; ed, Adv Appl Mech, 71-82; grad res prof, Univ Fla, 86-90. *Honors & Awards:* Henry Russel Lectr, Univ Mich, 74; Theodore Von Kármán Medal, Am Soc Civil Engrs, 81; Fluid Dynamics Prize, Am Phys Soc, 85, Otto Laporte Award, 89. *Mem:* Nat Acad Eng; fel Am Phys Soc; Academia Sinica. *Res:* Fluid mechanics, especially flows of nonhomogeneous fluids, geophysical fluid mechanics, waves and hydrodynamic stability; author 3 books and over 120 articles on fluid mechanics and applied mathematics. *Mailing Add:* Dept Mech Eng & Appl Mech Univ Mich Ann Arbor MI 48109

YIH, ROY YANGMING, b Changsha, China, Oct 5, 31; nat US; m 60; c 3. AGRICULTURAL CHEMISTRY. *Educ:* Nat Taiwan Normal Univ, BS, 56; Univ SC, MS, 59; Rutgers Univ, PhD(plant physiol, biochem), 63. *Prof Exp:* Sr scientist, 62-71, lab head, 72-73, proj leader, 73-81, RES DEPT MGR, ROHM & HAAS CO, SPRINGHOUSE, 82- *Mem:* Am Chem Soc; Weed Sci Soc Am. *Res:* Agricultural products (herbicides, fungicides, and insecticides) development worldwide. *Mailing Add:* 94 Windover Lane Doylestown PA 18901

YII, ROLAND, b Chengtu, China, Aug 11, 19; m 54; c 3. ELECTRICAL ENGINEERING. *Educ:* Nanking Univ, BS, 45; Brown Univ, MS, 53; Univ Pa, PhD(elec eng), 65. *Prof Exp:* Mgr printing dept, Sprague Elec Co, 52-54; prog & dept mgr, Burroughs Corp, 55-65; prof elec eng, Villanova Univ, 65-68; prof, 68-77, RES SCIENTIST, UNIV FLA, 77- *Res:* Solid state electronic and magnetic circuits; computer logic and variable radix computer. *Mailing Add:* Eglin Grad Eng Ctr Univ Fla Box 1918 Eglin AFB FL 32542

YILDIZ, ALAETTIN, b Surmene, Turkey, Jan 5, 22; US citizen. AERO ASTRONAUTICS, CARDIOLOGY. *Educ:* Tech Univ West Berlin, MS, 56, PhD(fluid mech & heat transfer), 65. *Prof Exp:* Res eng aerodynamics, Res Lab, Siemens, Berlin, 60-62; asst prof & res assoc mech eng, Tech Univ West Berlin, 62-65; res scientist aeronaut & astronaut, Edcliff Instruments contractor NASA, 67-69; med scientist cardiol, City Hope Nat Med Ctr, Calif, 67-69, res, Univ Hosp, Vakifgureba, Turkey, 76-79, dir biomed eng, Nat Med Enterprises, Los Angeles, Calif, 79-80, scientist comn team, Fac Med, King Faisal Univ, 80-81, scientist med sci, Royal Al Hada Hosp, Saudi Arabia, 81-82; FLUID MECH & HEAT TRANSFER CONSULT, LA JOLLA, CALIF, 83- *Concurrent Pos:* Fel aeronaut sci, Calif Inst Technol, Pasadena, 69; sci advisor, Prog Data Co, Calif & Univ Calif, Los Angeles, 71-76. *Mem:* Am Inst Aeronaut & Astronaut; Ger Engrs Asn; Turkish Engrs Asn; Am Soc Mech Engrs. *Res:* Resistance coefficient of intimal surface on normal and pathological human aortas with special emphasis on development of atherosclerosis and stenosis of vessels; potential role of hydro-dynamics of the structures of blood vessels; implantation of electronic systems in human encephalon for self cerebellum stimulation; ballistics of spinning projectiles; establishing the dimensionless Yildiz-number related to aero-astronautics. *Mailing Add:* PO Box 2605 La Jolla CA 92038

YIM, GEORGE KWOCK WAH, b Honolulu, Hawaii, Jan 7, 30; m 52; c 5. PHARMACOLOGY. *Educ:* Univ Iowa, BS, 52, MS, 54, PhD(pharmacol), 56. *Prof Exp:* Instr pharmacol, Univ Iowa, 55-56; from asst prof to assoc prof, 56-70, PROF PHARMACOL DEPT HEAD, PURDUE UNIV, WEST LAFAYETTE, 83- *Concurrent Pos:* USPHS career develop award, 61-66; NIH spec fel, 66-67. *Mem:* AAAS; Am Asn Col Pharm; Am Soc Pharmacol & Exp Therapeut; Soc Toxicol; Soc Neurosci. *Res:* Pharmacological control of appetite; stress and endorphins; cardiovascular and central actions of neuro active agents; biochemical mechanisms of cancer cachexia and anorexia. *Mailing Add:* Dept Pharmacol Purdue Univ Sch Pharm Lafayette IN 47907

YIM, W(OONGSOON) MICHAEL, materials science; deceased, see previous edition for last biography

YIN, CHIH-MING, b Szechwan, China, July 2, 43; m 68; c 3. INSECT PHYSIOLOGY, INVERTEBRATE ENDOCRINOLOGY. *Educ:* Taiwan Nat Univ, BSc, 66; Univ Sask, PhD(biol), 72. *Prof Exp:* Fel entomol, Univ Mo-Columbia, 72-74, res assoc, 74-78; assoc, Cornell Univ, 78; asst prof, 78-82, ASSOC PROF ENTOM, UNIV MASS, 82- *Mem:* Entom Soc Am; Sigma Xi; AAAS. *Res:* Hormonal control of growth, development and diapause in insects. *Mailing Add:* Dept Entom Fernald Hall Univ Mass Amherst MA 01003

YIN, FAY HOH, b Peking, China, Mar 10, 32; US citizen; m 59; c 2. VIROLOGY. *Educ:* Univ Wis-Madison, BA, 54, MS, 55, PhD(biochem), 60. *Prof Exp:* Res asst biochem, Univ Wis-Madison, 60; res assoc virol, Dept Path, Univ Pa, 63-65; RES CHEMIST, CENT RES DEPT, E I DUPONT DE NEMOURS & CO, INC, 66- *Mem:* Am Soc Microbiol; Am Soc Virol. *Res:* Biochemical studies of arbovirus and picornavirus replication. *Mailing Add:* 1804 Bellewood Rd Wilmington DE 19803

YIN, FRANK CHI-PONG, b Kunming, Yunnan, China, June 21, 43; US citizen; m 75; c 2. BIOMECHANICS, HEMODYNAMICS. *Educ:* Mass Inst Technol, BS, 65, MS, 67; Univ Calif, San Diego, PhD(bioeng), 70, MD, 73. *Prof Exp:* Res asst bioeng, dept appl mech, Univ Calif, San Diego, 67-70, intern med, Univ Hosp, 73-74, asst resident, 74-75; clin assoc cardiovasc, Nat Inst Aging, NIH, 75-77; fel cardiol, 77-78, asst prof med & physiol, 78-83, assoc prof, 83-88, PROF MED & BIOMED ENG, JOHNS HOPKINS MED INST, 88- *Concurrent Pos:* Mem cardiovasc pulmonary study sect, Nat Heart, Lung & Blood Inst, NIH, 83-; Estab investr, Am Heart Asn, 83-88. *Mem:* Am Heart Asn; fel Am Physiol Soc; Biomed Eng Soc; Am Soc Clin Invest; Biophys Soc; Am Soc Mech Eng. *Res:* Application of engineering principles and techniques to ventricular function, muscle mechanics and hemodynamics; determination of mechanical properties of biological tissue under multiaxial loading conditions. *Mailing Add:* Cardiol Div 538 Carnegie Bldg Johns Hopkins Hosp Baltimore MD 21205

YIN, HELEN LU, CELL MOTILITY. *Educ:* Harvard Univ, PhD(physiol), 76. *Prof Exp:* ASSOC PROF MED, SCH MED, HARVARD UNIV & MEM STAFF, MASS GEN HOSP, 78- *Mailing Add:* Dept Physiol Univ Tex Southwest Med Ctr 5323 Harry Hines Blvd Dallas TX 75235

YIN, LO I, b Wuchang, China, Apr 19, 30; US citizen; m 58; c 2. PHYSICS. *Educ:* Cent China Univ, BA, 49; Carleton Col, BA, 51; Univ Rochester, MA, 52, BS, 56; Univ Mich, MS, 59, PhD(physics), 63. *Prof Exp:* Res physicist, Bendix Res Lab, Bendix Corp, 64-67; AEROSPACE TECHNOLOGIST, NASA GODDARD SPACE FLIGHT CTR, 67- *Concurrent Pos:* Vis prof chem, Univ Md, 72-76, adj prof, 77- *Honors & Awards:* IR-100 Award, 79; NASA Inventor of the Year, 80. *Mem:* Am Phys Soc; AAAS; Am Nuclear Soc; Soc Photo-Optical Instrumentation Engrs. *Res:* Atomic and nuclear physics; x-ray spectroscopy; x-ray and gamma-ray spectroscopy; x-ray and gamma-ray imaging; lixiscope, low intensity x-ray imaging scope. *Mailing Add:* 1207 Downs Dr Silver Spring MD 20904

YIN, TOM CHI TIEN, b Kunming, China, Jan 7, 45; US citizen; m 72. NEUROPHYSIOLOGY, BIOENGINEERING. *Educ:* Princeton Univ, BSE, 66; Univ Mich, PhD(elec eng), 73. *Prof Exp:* Fel neurophysiol, State Univ NY, Buffalo, 74; fel physiol, Johns Hopkins Univ, 74-77; asst prof, 77-83, ASSOC PROF NEUROPHYSIOL, UNIV WIS-MADISON, 83- *Mem:* Soc Neurosci; Sigma Xi; Inst Elec & Electronics Engrs. *Res:* Neurophysiology of sensory and motor systems. *Mailing Add:* Dept Neurophys 283 B Med Sci Univ Wis Madison WI 53706

YING, KUANG LIN, b Kiangsu, China, June 12, 27; Can citizen; m 55; c 3. GENETICS, CYTOGENETICS. *Educ:* Nat Taiwan Univ, BSc, 52; Univ Sask, PhD(cytol, genetics), 61. *Prof Exp:* Sr specialist plant breeding & genetics, Sino-Am Joint Comn Rural Reconstruct, 61-64; Med Res Coun Can res assoc, Univ Sask, 64-67, from asst prof human cytogenetics to assoc prof, dept pediat, 73-78; dir cytogenetics & prenatal detection lab, Valley Children's Hosp & Guidance Clin, Fresno, Calif, 78-81; HEAD SECT CYTOGENETICS, DIV MED GENETICS, CHILDREN'S HOSP, LOS ANGELES, CALIF, 81-; ASSOC CLIN PROF, DEPT PEDIAT, SCH MED, UNIV SOUTHERN CALIF, LOS ANGELES, 81- *Concurrent Pos:* Assoc prof, Grad Sch, Nat Taiwan Univ, 62-63. *Mem:* Genetics Soc Can; Am Soc Human Genetics; Tissue Cult Asn. *Res:* Human cytogenetics; prenatal detection of genetic disorders. *Mailing Add:* Div Med Genetics Children's Hosp 4650 Sunset Blvd Los Angeles CA 90027

YING, RAMONA YUN-CHING, b Taiwan, July 2, 57; US citizen; m 84; c 1. ELECTROCHEMISTRY, BATTERY. *Educ:* Univ Calif, Los Angeles, BS, 79; Univ Mich, Ann Arbor, PhD(chem eng), 85. *Prof Exp:* STAFF RES SCIENTIST PHYS CHEM, GEN MOTORS RES LABS, 83- *Mem:* Electrochem Soc. *Res:* Electroless plating, alloy plating; electrocatalysis, fuel cell; lead acid battery, corrosion. *Mailing Add:* 2944 Reese Sterling Heights MI 48310

YING, SEE CHEN, b Shanghai, China, Apr 4, 41; m 82. SOLID STATE PHYSICS. *Educ:* Univ Hong Kong, BSc, 63 & 64; Brown Univ, PhD(physics), 68. *Prof Exp:* Res assoc physics, Brown Univ, 68-69; asst res scientist, Univ Calif, San Diego, 69-71; from asst prof to assoc prof, 71-80, PROF PHYSICS, BROWN UNIV, 80- *Concurrent Pos:* Res fel, A P Sloan Found, 72; Alexander von Humboldt Found US sr scientist award, 76. *Res:* Theoretical solid state physics; electronic properties of surfaces and interfaces; phase transitions on surfaces and low dimensional systems. *Mailing Add:* Dept Physics Brown Univ Providence RI 02912

YING, WILLIAM H, b Shanghai, China, July 1, 35; m 64. CIVIL ENGINEERING. *Educ:* Cheng Kung Univ, Taiwan, BS, 57; Univ Mo-Rolla, MS, 61; Okla State Univ, PhD(civil eng), 65. *Prof Exp:* Engr, Chau & Lee Archit & Civil Engrs, Hong Kong, 57-58 & Pan-Ocean, Ltd, Okinawa, 58-60; asst prof mech math, 64-67, assoc prof civil eng, 67-72, PROF CIVIL ENG, CALIF STATE UNIV, LONG BEACH, 72- *Mem:* Am Soc Civil Engrs; Am Soc Eng Educ; Am Inst Aeronaut & Astronaut. *Res:* Solid mechanics, including plates and shells, elasticity and dynamics of structures. *Mailing Add:* Dept Civil Eng Calif State Univ 1250 Bellflower Blvd Long Beach CA 90840

YINGST, RALPH EARL, b Lebanon, Pa, Aug 5, 29; m 64. INORGANIC CHEMISTRY. *Educ:* Univ Chicago, AB, 50; Lebanon Valley Col, BS, 55; Univ Pittsburgh, PhD(chem), 64. *Prof Exp:* Instr chem, Johnstown Col, 61-63; asst prof, 64-70, ASSOC PROF CHEM, YOUNGSTOWN STATE UNIV, 70- *Mem:* AAAS; Am Chem Soc; Sigma Xi. *Res:* Coordination compounds of metals with pyridine and substituted pyridines, especially those containing olefinic linkages, such as 2-vinylpyridine and 2-allylpyridine; optically active metal complexes, especially cobalt. *Mailing Add:* Dept Chem Youngstown State Univ Youngstown OH 44555

YIP, CECIL CHEUNG-CHING, b Hong Kong, June 11, 37; m 60; c 2. BIOCHEMISTRY, ENDOCRINOLOGY. *Educ:* McMaster Univ, BSc, 59; Rockefeller Univ, PhD(biochem, endocrinol), 63. *Prof Exp:* Res assoc endocrinol, Rockefeller Univ, 63-64; from asst prof to assoc prof, 64-74, PROF, BANTING & BEST DEPT MED RES, CH BEST INST, UNIV TORONTO, 74-, CHARLES H BEST PROF, 87-, CHMN, 90- *Concurrent Pos:* Med Res Coun Can med res scholar, 67-71. *Honors & Awards:* Charles H Best Prize, 72. *Mem:* AAAS; Am Soc Biol Chem; Can Biochem Soc; Am Chem Soc. *Res:* Insulin receptor structure and function; hormone-receptor interaction. *Mailing Add:* C H Best Inst Univ Toronto Toronto ON M5G 1L6 Can

YIP, GEORGE, b Oakland, Calif, Nov 14, 26; m 53; c 1. BIOCHEMISTRY, FOOD TECHNOLOGY. *Educ:* Univ Calif, Berkeley, BS, 51; Georgetown Univ, MS, 59. *Prof Exp:* Chemist, Nat Canners Asn, Calif, 51-52; chemist, Div Food Chem, 55-56, res chemist pesticides, 56-63, sect chief herbicides & plant growth regulators, 63-71, chief, Biochem Technol Br, 71-72, CHIEF, INDUST CHEM CONTAMINANT BR, DIV CHEM TECHNOL, FOOD & DRUG ADMIN, 72- *Mem:* Am Chem Soc; Asn Off Anal Chem. *Res:* Methods of analysis for industrial chemical contaminants in foods; identification of unknown contaminants including degradation products. *Mailing Add:* 5211 Kipling St Springfield VA 22151-2928

YIP, JOSEPH W, b Hong Kong, Sept 17, 48; m 77; c 2. DEVELOPMENTAL NEUROBIOLOGY. *Educ:* Wash State Univ, BS, 71; Univ Calif, San Francisco, PhD(physiol), 77. *Prof Exp:* Fel neurobiol, Wash Univ, St Louis, 77-80; ASSOC PROF PHYSIOL, SCH MED, UNIV PITTSBURGH, 81- *Mem:* Soc Neurosci. *Res:* Specificity of synapse formation. *Mailing Add:* Dept Physiol Sch Med Univ Pittsburgh Pittsburgh PA 15261

YIP, KWOK LEUNG, b Canton, China, Sept 23, 44; m 72; c 2. MEDICAL IMAGING, DIGITAL RADIOGRAPHY. *Educ:* Chung Chi Col, Chinese Univ Hong Kong, BS, 65, dipl educ, 66; Providence Col, MS, 70; Lehigh Univ, PhD(physics), 73. *Prof Exp:* Teacher physics, King's Col, Hong Kong, 66-68; postdoctoral res fel physics, Univ Ill, Urbana, 73-75; sr tech specialist physics, Wilson Ctr Technol, Xerox Corp, 75-84; SR STAFF RES SCIENTIST PHYSICS, RES LABS, EASTMAN KODAK CO, 84- *Mem:* Am Phys Soc; Int Soc Optical Eng. *Res:* Electronic printer, display and digitizer technologies; laser beam scanning and recording systems; medical imaging; digital radiography; systems modeling and analysis. *Mailing Add:* Eastman Kodak Co Res Labs Kodak Park Bldg 82A Rochester NY 14650-2123

YIP, RICK KA SUN, b 1952. CELL BIOLOGY, NEUROANATOMY. *Educ:* Med Col Wis, PhD(anat), 85. *Prof Exp:* FEL CELL BIOL, UNIV TEX, 84- *Mailing Add:* Dept Physiol Univ Md Med Sch 660 W Redwood St Baltimore MD 21201

YIP, SIDNEY, b Peking, China, Jan 28, 36; US citizen; m 58. NUCLEAR ENGINEERING. *Educ:* Univ Mich, BS, 58, MS, 59, PhD(nuclear eng), 62. *Prof Exp:* Res fel, Inst Sci Technol, Univ Mich, 62-63; res assoc eng physics, Cornell Univ, 63-65; from asst prof to assoc prof nuclear eng, 65-73, PROF NUCLEAR ENG, MASS INST TECHNOL, 73- *Concurrent Pos:* John Simon Guggenheim fel; Alexander von Humboldt found sr scientist award, Ger. *Mem:* Fel Am Phys Soc; Am Nuclear Soc. *Res:* Atomistic simulation of materials properties and behavior; statistical mechanics of dense fluids; molecular dynamics studies of atomic and polymeric glasses, interfacial phenomena and fracture. *Mailing Add:* Dept Nuclear Eng 24-208 Mass Inst Technol Cambridge MA 02139

YIP, YUM KEUNG, PROTEIN CHEMISTRY, HEMOGLOBIN. *Educ:* Ind Univ, PhD(pharmacol), 69. *Prof Exp:* ASSOC PROF MED MICROBIOL, MED SCH, NY UNIV, 77- *Res:* Cellular immunology; interferon gamma. *Mailing Add:* Dept Microbiol Sch Med NY Univ 550 First Ave New York NY 10016

YIRAK, JACK J(UNIOR), b Omaha, Nebr, Oct 10, 18; m 45; c 2. CHEMICAL ENGINEERING. *Educ:* Iowa State Col, BS, 40; Lawrence Col, MS, 42, PhD, 44. *Prof Exp:* Chem engr, Union Bag & Paper Corp, 44-48, group leader, 48-51, proj engr, Semi-Chem Pulp Mill, 51-54, asst pulp mill supt, 54-55; pulp mill proj engr, Bleached Div, Union Camp Corp, 55-60, construct proj engr, 60-64, asst construct engr, 64-66, construct proj mgr, 66-81, asst to proj dir, 81-89; RETIRED. *Mem:* Tech Asn Pulp & Paper Indust. *Res:* Pulp, paper and related products; tall oil. *Mailing Add:* 27 Sulgrave Rd Savannah GA 31406

YI-YAN, ALFREDO, b Lima, Peru, Aug 23, 49. INTEGRATED OPTICS, OPTOELECTRONICS. *Educ:* Univ Glasgow, Scotland, BSc, 74, PhD(elec eng), 79. *Prof Exp:* Fel, Univ Glasgow, Scotland, 78-81; sr res engr, Nat Telecommunications Study Ctr, France, 81-87; MEM TECH STAFF, BELL COMMUN RES, NJ, 87- *Concurrent Pos:* Consult, Linn Prod Ltd, Glasgow, 81. *Mem:* Optical Soc Am. *Res:* Research work in optoelectronics integrated circuits with emphasis on semiconductor-grafted guided-wave devices fabricated using the epitaxial lift-off technique. *Mailing Add:* 124 Wedgewood Circle Eatontown NJ 07724

YNGVESSON, K SIGFRID, b Lidkõ, Sweden, Mar 23, 36; m 65; c 2. ELECTRICAL ENGINEERING. *Educ:* Chalmers Univ Technol, Sweden, Civ ing, 58, Tekn Lic, 65, Tekn Dr, 68. *Prof Exp:* Res asst electron physics, Chalmers Univ Technol, 59-64, res assoc, 66-68; res scholar physics, Univ Calif, Berkeley, 64-66, postdoctoral res asst, 68-70; assoc prof, 70-78, PROF ELEC ENG, UNIV MASS, 78- *Concurrent Pos:* Vis prof, Chalmers Univ Technol, 77 & 85. *Honors & Awards:* Wallin Prize, Royal Acad Sci, 69. *Mem:* Inst Elec & Electronics Engrs; Am Phys Soc. *Res:* Low noise receivers, especially for millimeter waves; focal plane arrays for millimeter wave imaging. *Mailing Add:* Dept Elec & Computer Eng Univ Mass Amherst MA 01003

YNTEMA, JAN LAMBERTUS, b Neth, Oct 5, 20; m 48; c 4. PHYSICS. *Educ:* Free Univ, Amsterdam, NatPhilDrs, 48, DrPhysics, 52. *Prof Exp:* Res assoc physics, Princeton Univ, 49-52; asst prof, Univ Pittsburgh, 52-55; assoc physicist, 55-68, SR PHYSICIST, ARGONNE NAT LAB, 68- *Mem:* Am Phys Soc. *Res:* Radioactivity; gases at high temperatures; nuclear physics. *Mailing Add:* 5125 Grand Ave Western Springs IL 60558

YNTEMA, MARY KATHERINE, b Urbana, Ill, Jan 20, 28. MATHEMATICS. *Educ:* Swarthmore Col, BA, 50; Univ Ill, AM, 61, PhD(math), 65. *Prof Exp:* Teacher, Am Col Girls, Istanbul, 50-54 & Columbus Sch Girls, Ohio, 54-57; programmer, Lincoln Lab, Mass Inst Technol, 57-58; teacher high sch, Mont, 59-60; asst prof math, Univ Ill, Chicago, 65-67; asst prof comput sci, Pa State Univ, 67-71; assoc prof, 71-81, coordr math systs prog, 75-77, 80-81 & 84-86, chmn fac senate, 77-78, PROF MATH, SANGAMON STATE UNIV, 81-, COORDR MATH SYST PROG, 88-, DIV MATH & COMPUTER SECT, 89- *Mem:* Am Math Soc; Asn Comput Mach; Math Asn Am; Sigma Xi; Am Asn Univ Prof. *Res:* Automata; context-free languages. *Mailing Add:* Dept Math Sci Sangamon State Univ Springfield IL 62794-9243

YOAKUM, ANNA MARGARET, b Loudon, Tenn, Jan 13, 33. ANALYTICAL CHEMISTRY, PHYSICAL METALLURGY. *Educ:* Maryville Col, AB, 54; Univ Fla, MS, 56, PhD(anal chem), 60. *Prof Exp:* Supvr control lab, Greenback Indust, Inc, 56-59; sr res chemist, Chemstrand Res Ctr, Inc, 60-64; mem res staff, Oak Ridge Nat Lab, 64-69; EXEC VPRES & LAB DIR, STEWART LABS, INC, 67- *Mem:* Am Chem Soc; Soc Appl Spectros; NY Acad Sci; Am Soc Test & Mat; fel Am Inst Chem; Sigma Xi. *Res:* Analytical chemistry and trace analysis; research and method development in emission, flame, atomic absorption, x-ray fluorescence and infrared spectroscopy. *Mailing Add:* Rte 4 Twin Coves Box 418 Lenoir City TN 37771

YOBURN, BYRON CROCKER, b Danbury, Conn, Nov 26, 50; m 76; c 2. PSYCHOPHARMACOLOGY. *Educ:* Boston Univ, BA, 73; Hollins Col, MA, 76; Northeastern Univ, PhD(exp psychol), 79. *Prof Exp:* Fel neurobehav sci, Col Physicians & Surgeons, Columbia Univ, 79-81,; res scientist, NY State Psychiat Inst, 81-82; res assoc, 82-83, instr pharmacol, Med Col, Cornell Univ, 83-87; asst prof, 87-90, ASSOC PROF, COL PHARM, ST JOHN'S UNIV, 90- *Mem:* Soc Neurosci; NY Acad Sci; Am Soc Pharmacol & Exp Therapeut; Sigma Xi. *Res:* Basic mechanisms in opioid pharmacology, especially tolerance and dependence; opioid receptor plasticity. *Mailing Add:* Dept Pharmaceut Sci Col Pharm St John's Univ Grand Central & Utopia Pkwys Queens NY 11439

YOCCA, FRANK D, b Brooklyn, NY, July 22, 55. RECEPTOR PHARMACOLOGY, SEROTONIN. *Educ:* Manhattan Col, BS, 77; St John's Univ, MS, 81; NY Univ, PhD(pharmacol), 83. *Prof Exp:* Postdoctoral fel, Mt Sinai Sch Med, 83-84; postdoctoral fel, Bristol-Myers Co, 84-85, res scientist, 85-87, sr res scientist, 87-89, assoc dir, 89-91, GROUP LEADER, BRISTOL-MYERS SQUIBB CO, 91- *Concurrent Pos:* Adj asst prof, Dept Anesthesiol, Mt Sinai Sch Med, 89- *Mem:* Soc Neurosci; Am Soc Pharmacol & Exp Therapeut; NY Acad Sci; Am Chem Soc; Sigma Xi. *Res:* Delineation and classification of 5-HT, -like receptors in brain and periphery; 5-HTip receptors, specifically their regulation and possible heterogeniety; neuropsychopharmacology. *Mailing Add:* Dept 404 CNS Neuropharmacol Five Research Pkwy Wallingford CT 06492-7660

YOCH, DUANE CHARLES, b Parkston, SDak, Nov 4, 40; m 64; c 2. MICROBIOLOGY, BIOCHEMISTRY. *Educ:* SDak State Univ, BS, 63, MS, 65; Pa State Univ, PhD(microbiol), 68. *Prof Exp:* Asst res microbiologist, Agr Exp Sta, Dept Cell Physiol, Univ Calif, Berkeley, 68-69, res specialist, 70-78; from asst prof to assoc prof biol, 78-85, PROF BIOL, UNIV SC, COLUMBIA, 85- *Mem:* Am Soc Microbiol. *Res:* Biochemistry and molecular biology of nitrogen; iron-sulfur proteins; methylotrophic N2 fixation. *Mailing Add:* Dept Biol Univ SC Columbia SC 29208

YOCHELSON, ELLIS LEON, b Washington, DC, Nov 14, 28; m 50; c 3. INVERTEBRATE PALEONTOLOGY. *Educ:* Univ Kans, BS, 49, MS, 50; Columbia Univ, PhD, 55. *Prof Exp:* Asst, Univ Kans & Columbia Univ, 50-52; paleontologist-geologist, US Geol Surv, 52-85; RES ASSOC, SMITHSONIAN INST, 64- *Concurrent Pos:* Mem, Nat Res Coun, 57-71; treas, Int Paleont Asn, 71-75; organizer, NAm Paleontol Conv, 69, ed proceedings, 70-71; secy-gen, Ninth Int Cong Carboniferous Stratig & Geol, 79. *Honors & Awards:* Erasmus Haworth Award, Dept Geol, Univ Kans. *Mem:* Sigma Xi; Paleont Soc (pres, 76); Soc Syst Zool (secy, 62-65); Hist Earth Sci Soc (secy-treas, 81-84, secy, 85-88, pres, 89). *Res:* Systematics and evolution of Paleozoic gastropods; phylogeny of Mollusca, especially early Paleozoic major taxa; enigmatic Paleozoic fossils; paleontology; stratigraphy-sedimentation; history of science; taxonomy; evolution. *Mailing Add:* Dept Paleobiol Nat Mus Natural Hist Washington DC 20560

YOCHIM, JEROME M, b Chicago, Ill, Feb 23, 33; m 57; c 2. ENDOCRINOLOGY, PHYSIOLOGY. *Educ:* Univ Ill, BS, 55, MS, 57; Purdue Univ, PhD(biol sci), 60. *Prof Exp:* NIH fel anat, Col Med, Univ Ill, 60-62; from asst prof to assoc prof physiol, 62-70, PROF PHYSIOL, UNIV KANS, 71- *Concurrent Pos:* NIH career develop award, 71-76. *Mem:* AAAS; Am Asn Anatomists; Am Physiol Soc; Soc Study Reprod; Endocrine Soc; Sigma Xi. *Res:* Physiology of reproduction. *Mailing Add:* Dept Physiol & Cell Biol Univ Kans Lawrence KS 66045

YOCKEY, HUBERT PALMER, b Alexandria, Minn, Apr 15, 16; m 46; c 3. THEORETICAL BIOLOGY, MOLECULAR BIOLOGY. *Educ:* Univ Calif, Berkeley, AB, 38, PhD(physics), 42. *Prof Exp:* Jr physicist, Nat Defense Res Comt, Calif, 41-42; physicist radiation lab, Univ Calif, 42-44; sr physicist, Tenn Eastman Corp, 44-46; group leader irradiation physics, NAm Aviation, Inc, 46-52; chief nuclear physics, Convair Div, Gen Dynamics Corp, Tex, 52-53; asst dir health & physics div, Oak Ridge Nat Lab, Tenn, 53-59; mgr res & develop div, Aerojet-Gen Nucleonics Corp, 59-62 & Hughes Res Labs, 63-64; chief Reactor Br, 64-80, chief Army Pulse Radiation Div, 80-85, DIR, NUCLEAR EFFECTS DIRECTORATE, COMBAT SYSTS TEST ACTIV ABERDEEN PROVING GROUND, 85- *Concurrent Pos:* Consult, Oak Ridge Nat Lab, 60. *Mem:* Am Phys Soc; Am Nuclear Soc; Radiation Res Soc; Health Phys Soc. *Res:* Application of information theory to origin of life, genetic code, calculated information content of cytochrome c aging and radiation effects; established fast pulsed reactor facility; radiation effects in ferroelectrics; solid state physics; pulsed reactors. *Mailing Add:* 1507 Balmoral Dr Bel Air MD 21014

YOCOM, PERRY NIEL, b Auburn, Maine, Sept 27, 30; m 62; c 3. INORGANIC CHEMISTRY. *Educ:* Pa State Univ, BS, 54; Univ Ill, PhD, 58. *Prof Exp:* Mem tech staff, 57-70, RES GROUP HEAD, DAVID SARNOFF RES CTR, SUBSID SRI INT, 70- *Concurrent Pos:* Lectr, Advan Study Inst, NATO, 72. *Honors & Awards:* David Sarnoff Award, Sci, 70. *Mem:* AAAS; Am Chem Soc; Electrochem Soc; Sigma Xi; Mineral Soc Am. *Res:* Chemistry of fused salts; crystal growth; defects in solids; rare earth phase chemistry; luminescence. *Mailing Add:* David Sarnoff Res Ctr CN 5300 Subsid SRI Int Princeton NJ 08543-5300

YOCUM, CHARLES FREDRICK, b Storm Lake, Iowa, Oct 31, 41; m 82; c 1. BIOCHEMISTRY. *Educ:* Iowa State Univ, BS, 63; Ind Univ, PhD(biochem), 71. *Prof Exp:* Biochemist protein chem, ITT Res Inst, 63-68; NIH Fel biochem, Cornell Univ, 71-73; from asst prof to assoc prof biol, 73-81, PROF BIOL & CHEM, UNIV MICH, 82-, CHMN BIOL, 85- *Concurrent Pos:* Mem, adv panel on metabolic biol, NSF, 82-85, adv panel chem life processes, 86, adv panel plant sci ctrs, 88-, adv panel USDA Photosynthesis Prog, 89; res grants, NSF & USDA. *Mem:* Am Chem Soc; Am Soc Plant Physiologists; AAAS; Biophys Soc; Am Soc Biol Chemists. *Res:* Mechanisms of photosynthetic oxygen evolution and energy transduction in chloroplasts and blue-green algae. *Mailing Add:* Dept Biol Univ Mich Ann Arbor MI 48109-1048

YOCUM, CONRAD SCHATTE, b Swarthmore, Pa, Mar 29, 19; m 46; c 3. PLANT PHYSIOLOGY. *Educ:* Col William & Mary, BS, 40; Univ Md, MS, 47; Stanford Univ, PhD, 52. *Prof Exp:* Asst marine biol, Va Fisheries Lab, 46-47; asst plant physiol, Hopkins Marine Sta, 47-48; instr plant physiol, Harvard Univ, 52-55; asst prof, Cornell Univ, 55-61; from assoc prof to prof plant physiol, Univ Mich, Ann Arbor, 61-88; RETIRED. *Mem:* Am Soc Plant Physiol; Sigma Xi. *Res:* Photosynthesis; respiration; tropisms; nitrogen fixation. *Mailing Add:* 2080 Newport Rd Ann Arbor MI 48103

YOCUM, RONALD HARRIS, b Darby, Pa, June 2, 39. CHEMISTRY. *Educ:* Gettysburg Col, BA, 61; Univ Pa, PhD(org chem), 65. *Prof Exp:* Dir res, Latin Am, 73-77, Designed Prod Dept, 77-78, dir prod res, 78-80, dir res, Mich Div, Dow Chem USA, 80-; AT NORCHEM, INC. *Mem:* Am Chem Soc. *Mailing Add:* 571 Chaswil Ct 11500 Northlake Dr Cincinnati OH 45255-5602

YODAIKEN, RALPH EMILE, b Johannesburg, SAfrica, Aug 25, 28; US citizen. PATHOLOGY, OCCUPATIONAL HEALTH. *Educ:* Univ Witwatersrand, MB & BCh, 56; Johns Hopkins Univ, MPH, 76. *Prof Exp:* Lectr path, Univ Witwatersrand, 58-63; assoc pathologist, Buffalo Gen Hosp, NY, 63-67; assoc prof path, Sch Med, Univ Cincinnati, 68-71; prof path & assoc prof med, Sch Med, Emory Univ, 71-77; asst researcher, Johns Hopkins Univ, 76-76; SR MED OFFICER & CHMN SR ADV STAFF, NAT INST OCCUP SAFETY & HEALTH, 77- *Concurrent Pos:* Chief ultra structure res, Vet Admin Hosp, 71-76. *Mem:* Fel Am Col Preventive Med; Soc Occup & Environ Health; fel Am Col Path; fel Collegium Ramazzini; Am Occup Med Assoc; Acad Occup Med; Acad Prev Med; Am Pub Health Assoc. *Res:* Vascular pathology with special reference to diabetes. *Mailing Add:* 7100 Oak Forest Lane Bethesda MD 20817

YODER, CHARLES FINNEY, b Cincinnati, Ohio, July 18, 43; m 70; c 2. CELESTIAL MECHANICS, PLANETOLOGY. *Educ:* Univ Calif, Santa Barbara, BA, 68, PhD(physics), 73. *Prof Exp:* Fel, Dept Earth & Space Sci, Univ Calif, Los Angeles, 73-76; MEM TECH STAFF, JET PROPULSION LAB, 76- *Mem:* Am Geophys Union; Am Astron Soc. *Res:* Effect of tidal friction and gravitational resonances on planetary satellites; rotational dynamics; core mantle coupling mechanisms. *Mailing Add:* Jet Propulsion Lab MS 183-501 4800 Oak Grove Dr Pasadena CA 91109

YODER, CLAUDE H, b West Reading, Pa, Mar 16, 40; m 66; c 2. INORGANIC CHEMISTRY. *Educ:* Franklin & Marshall Col, BA, 62; Cornell Univ, PhD(chem), 66. *Prof Exp:* Asst prof, 66-74, assoc prof, 74-80, PROF CHEM, FRANKLIN & MARSHALL COL, 80-, CHMN DEPT, 74- *Concurrent Pos:* Dreyfus Found teacher scholar, 71. *Mem:* AAAS; Am Chem Soc. *Res:* Bonding in organometallic compounds. *Mailing Add:* 2946 Kings Lane Lancaster PA 17601-1617

YODER, DAVID LEE, b Bellefontaine, Ohio, June 23, 36; m 61; c 3. PLANT PATHOLOGY, SOIL MICROBIOLOGY. *Educ:* Goshen Col, BA, 60; Mich State Univ, MS, 68, PhD(plant path), 71. *Prof Exp:* Plant pathologist, Hunt-Wesson Foods, Inc, 71-78; PLANT BREEDER, GILROY FOODS, INC, 78- *Mem:* AAAS; Am Soc Hort Sci; Am Plant Selections. *Res:* Soil-borne plant diseases; disease of tomatoes; soil fungistasis; onion genetics, pepper genetics and diseases of onion, garlic and pepper; tissue culture. *Mailing Add:* Gilroy Foods Inc 1350 Pachenko Pass Gilroy CA 95020

YODER, DONALD MAURICE, b Elkhart Co, Ind, Jan 3, 20; m 45, 85; c 3. AGRICULTURAL CHEMISTRY. *Educ:* Wabash Col, BA, 42; Cornell Univ, PhD(plant path), 50. *Prof Exp:* Sr fel biol res div, Union Carbide Chem Co, 50-54, head div, 54-61, mem staff tech develop, Agr Chem Div, 61-67, sr analyst mkt res & technol deleg UAR, Union Carbide Tech Serv Co, 67-68; mgr prod develop agr chem, Basf-Wyandotte Corp, 68-74, mgr regist & toxicol, 74-84, res fel 85-87; RETIRED. *Concurrent Pos:* Fel, Boyce Thompson Inst Plant Res, 50. *Mem:* Am Chem Soc; Am Phytopath Soc; Am Inst Biol Sci; Agr Res Inst. *Res:* Evaluation of organic chemicals for agricultural uses; agricultural chemicals and food; pesticide regulations. *Mailing Add:* Nine Meriden Rd RD 2 Box 165 Boonton NJ 07005

YODER, ELDON J, civil engineering; deceased, see previous edition for last biography

YODER, ELMON EUGENE, b Wolford, NDak, Oct 10, 21; m 47; c 4. AGRICULTURAL ENGINEERING, CIVIL ENGINEERING. *Educ:* Ore State Univ, BS, 47 & 54, MS, 61. *Prof Exp:* Civil engr design, Consumers Power Inc, 48-54; agr engr, Khon Trup Agr Univ, Thailand, 54-56; civil engr, Ore State Univ, 57-61; AGR ENGR RES, USDA, LEXINGTON, 62- *Concurrent Pos:* Agr engr, Univ Ky, 65-66 & 76-78. *Mem:* Am Soc Agr Engrs; Tobacco Workers; Sigma Xi. *Res:* Tobacco mechanization and processing; improvement of tobacco combustion in health-related research. *Mailing Add:* Agr Eng Univ Ky Lexington KY 40506

YODER, HATTEN SCHUYLER, JR, b Cleveland, Ohio, Mar 20, 21; m 59; c 2. EXPERIMENTAL PETROLOGY. *Educ:* Univ Chicago, SB, 41, cert(meteorol), 42 & 46; Mass Inst Technol, PhD(petrol), 48. *Hon Degrees:* Dr, Univ Paris, 81. *Prof Exp:* Petrologist, 48-71, dir, 71-86, DIR EMER, CARNEGIE INST WASH GEOPHYS LAB, 86- *Concurrent Pos:* Vis prof, Calif Inst Technol, 58, Univ Tex, 64, Univ Colo, 66 & Univ Cape Town, 67. *Honors & Awards:* Mineral Soc Am Award, 54; Day Medal, Geol Soc Am, 62; Arthur L Day Prize & lectureship, Nat Acad Sci, 72; A G Werner Medal, Ger Mineral Soc, 72; Wollaston Medal, Geol Soc London, 79- *Mem:* Nat Acad Sci; Mineral Soc Am (pres, 72); Geol Soc Am; Am Geophys Union; Geochem Soc; Am Chem Soc. *Res:* Experimental petrology; phase equilibria in mineral systems; properties of minerals at high pressure and high temperature; hydrothermal mineral synthesis; experimental heat transfer in silicates. *Mailing Add:* Carnegie Inst Wash Geophys Lab 5251 Broad Branch Rd NW Washington DC 20015-1305

YODER, JOHN MENLY, b Ft Wayne, Ind, Oct 4, 31; m 60; c 2. ENDOCRINOLOGY, IMMUNOCHEMISTRY. *Educ:* Purdue Univ, BS, 53, PhD(animal physiol), 61. *Prof Exp:* Res biochemist, Ames Co Div, 61-72, SR RES SCIENTIST, AMES CO DIV, MILES LABS, INC, 72- *Mem:* Am Chem Soc. *Res:* Plant and animal physiology; silage fermentation; protein purification; characterization of proteins and polysaccharides by immunochemistry; gonadotropins; hepatitis antigen; antibody production; factor VIII; thyroxine assay; hemoglobin AIC. *Mailing Add:* Diag Div Miles Inc PO Box 70 Elkhart IN 46515

YODER, LEVON LEE, b Middlebury, Ind, June 22, 36; m 60; c 1. MUSICAL ACOUSTICS. *Educ:* Goshen Col, BA, 58; Univ Mich, MA, 61, PhD, 63. *Prof Exp:* Asst prof physics & chmn dept, Millikin Univ, 63-65; assoc prof, 65-71, chmn dept, 65-76, PROF PHYSICS, ADRIAN COL, 71-, CHMN DEPT, 79- *Mem:* Catgut Acoust Soc; Am Asn Physics Teachers; Acoust Soc Am. *Res:* Acoustical properties of violins. *Mailing Add:* 2499 Sword Hwy Adrian MI 49221

YODER, NEIL RICHARD, b Wichita, Kans, Mar 27, 37; m 68. PARTICLE PHYSICS. *Educ:* Kans State Teachers Col, BA, 59; Pa State Univ, PhD(physics), 69. *Prof Exp:* Instr physics, Mich State Univ, 65-67; sr res assoc, Univ Md, College Park, 67-; AT CYCLOTRON FACIL, IND UNIV. *Mem:* Am Phys Soc; Sigma Xi. *Res:* Phenomenological analysis of moderate energy nucleon-nucleon data; application of computers to on-line analysis of nuclear physics experimental data. *Mailing Add:* Cyclotron Facil Ind Univ 2401 Milo B Sampson Lane Bloomington IN 47401

YODER, OLEN CURTIS, b Fairview, Mich, Jan 26, 42; m 67. PLANT PATHOLOGY. *Educ:* Goshen Col, BA, 64; Mich State Univ, MS, 68, PhD(plant path), 71. *Prof Exp:* From asst prof to assoc prof, 71-83, PROF PLANT PATH, CORNELL UNIV, 83- *Concurrent Pos:* USDA res grant, 72-75 & 78-; Rockefeller Found res grant, 74 & 77-; sabbatical, Stanford Univ, 78-79; study leave, Mass Int Technol, 83-84; prog mgr, USDA/Competitive Res Grants, 85-86. *Mem:* AAAS; Am Phytopath Soc; Am Soc Microbiol; Sigma Xi. *Res:* Genetics and molecular biology of fungal pathogenicity to plants. *Mailing Add:* Dept Plant Path Cornell Univ Ithaca NY 14853

YODER, PAUL RUFUS, JR, b Huntingdon, Pa, Feb 6, 27; m 48; c 4. OPTICS. *Educ:* Juniata Col, BS, 47; Pa State Univ, MS, 50. *Prof Exp:* Assoc prof physics & math, Bridgewater Col, 50-51; optical physicist, US Army Frankford Arsenal, 51-61; proj engr, 61-67, eng dept mgr, 67-82, asst to dir res, Perkin-Elmer Corp, 82-86; sr scientist, Taunton Technologies Inc, 86-91; CONSULT OPTICAL ENG, 85- *Concurrent Pos:* Lectr geometric optics, Univ Conn Exten, 77-86. *Mem:* Fel Optical Soc Am; fel Int Soc Optical Engr; German Soc Appl Optics; Sigma Xi. *Res:* Design, development, fabrication and test of specialized optical instrumentation; author of two books and 50 articles in field of optics. *Mailing Add:* 1220 Foxboro Dr Norwalk CT 06851

YODER, ROBERT E, b Richmond, Va, Jan 1, 30; m 52; c 3. HEALTH PHYSICS, INDUSTRIAL HYGIENE. *Educ:* Appalachian State Teachers Col, BS, 51. *Hon Degrees:* ScD, Harvard Univ, 63. *Prof Exp:* Jr health physicist, Oak Ridge Nat Lab, 54-57; from instr to asst prof health physics, Sch Pub Health, Harvard Univ, 57-66; group leader & hazards control dept head, Lawrence Livermore Lab, Univ Calif, 65-72; asst dir nuclear facil, Div Oper Safety, US Atomic Energy Comn, 72-75; dir health safety & environ, Rockwell Int, Rocky Flats Plant, Colo, 75-86; sr scientist, Sci Applications Int Corp, 86-90; CONSULT RADIATION SAFETY, 58-64, 90- *Mem:* AAAS; Health Physics Soc; Sigma Xi; NY Acad Sci. *Res:* Aerosol technology; basic properties of aerosols; physics and chemistry of small particles and their influence on health; radiation safety and environment protection program administration in nuclear research and production institutions. *Mailing Add:* 1526 Villisca Terr Rockvile MD 20879-2437

YODER, WAYNE ALVA, b Grantsville, Md, July 6, 43; m 67; c 2. INVERTEBRATE ZOOLOGY, ENTOMOLOGY. *Educ:* Goshen Col, BA, 65; Mich State Univ, MS, 71, PhD(zool), 72. *Prof Exp:* From asst prof to assoc prof, 72-85, PROF BIOL, FROSTBURG STATE UNIV, MD, 85- *Mem:* Am Inst Biol Sci; Am Soc Zoologists; Entom Soc Am; Sigma Xi. *Res:* Systematic and ecological studies of the mites associated with silphid beetles. *Mailing Add:* Dept Biol Frostburg State Univ Frostburg MD 21532

YODER WISE, PATRICIA S, b Wadsworth, Ohio, July 2, 41; m 73; c 2. MANAGEMENT & MARKETING, GERONTICS. *Educ:* Ohio State Univ, BSN, 63; Wayne State Univ, MSN, 68; Tex Tech Univ, EdD, 84. *Prof Exp:* Spec assignment nurse, Univ Hosp, 63-64; instr, Med Surg Nursing, Aultman Hosp, 65-66; res asst, Nursing Process, Wayne State Univ, 68; educ dir, Ohio Nurses' Asn, 68-72; asst dir nursing, Mount Clemons Hosp, 72-73; clin instr nursing, West Shore Community Col, 74-75; assoc prof & head nursing, Ferris State Col, 75-77; from asst prof to assoc prof & dir, Univ Colo, 77-79; PROF & EXEC ASSOC DEAN, TEX TECH UNIV, HEALTH SCI CTR, 79- *Concurrent Pos:* Vchair & chair, Mich Nurses' Asn, 72-76; consult, self employed, 73-; vchair & mem, Mich Bd Nursing, 77; coun continuing educ, Am Nurses' Asn, 78-82 & 82-85; acad adv panel, Hosp Satellite Network, 83- *Mem:* Am Nurses' Asn Coun (secy, 78-82). *Res:* Management, continuing education, marketing and teaching as they relate to nursing; gerontics nursing, especially involving pets. *Mailing Add:* Sch Nursing Tex Tech Univ Health Sci Ctr Lubbock TX 79430

YODH, GAURANG BHASKAR, b Ahmedabad, India, Nov 24, 28; nat US; m 54; c 3. PHYSICS. *Educ:* Univ Bombay, BSc, 48; Univ Chicago, MS, 51, PhD(physics), 55. *Prof Exp:* Instr physics, Stanford Univ, 54-56; res fel, Tata Inst Fundamental Res, India, 57-58; res physicist, Carnegie Inst Technol, 58-59, asst prof physics, 59-61; prof physics, Univ Md, College Park, 65-88; PROF PHYSICS, UNIV CALIF, IRVINE, 88- *Concurrent Pos:* Consult, US Naval Res Lab, DC, 65-69 & Argonne Nat Lab, 66-70; vis prof, Univ Ariz, 66-67; vis scientist, Goddard Space Flight Ctr, NASA, 76-77; prog officer elementary particle physics, NSF, 78-80. *Mem:* Fel Am Phys Soc. *Res:* Experimental and phenomenological study of high energy interactions of elementary particles and cosmic rays; gamma ray astronomy. *Mailing Add:* Dept Physics Univ Calif Irvine CA 92717

YOERGER, ROGER R, b LeMars, Iowa, Feb 17, 29; m 71; c 4. AGRICULTURAL ENGINEERING. *Educ:* Iowa State Univ, BS, 49, MS, 51, PhD(agr eng), 57. *Prof Exp:* Instr & asst prof agr eng, Iowa State Univ, 49-56; assoc prof, Pa State Univ, 56-58; prof, 58-78, head dept agr eng, 78-85, EMER PROF AGR ENG, UNIV ILL, URBANA, 85- *Concurrent Pos:* Consulting engr, MESAC Enterprises, 56- *Honors & Awards:* Massey Ferguson Medal, Am Soc Agr Engrs, 89. *Mem:* Fel Am Soc Agr Engrs; Am Soc Eng Educ; AAAS; Sigma Xi. *Res:* Off-road vehicles; noise reduction; vibration and operator comfort; field crop production equipment. *Mailing Add:* Dept Agr Eng Univ Ill 1304 W Pennsylvania Ave Urbana IL 61801

YOESTING, CLARENCE C, b Apr 5, 12; US citizen; m 40; c 2. PHYSICS, SCIENCE EDUCATION. *Educ:* Cent State Col, Okla, BS, 36; Univ Okla, MEd, 47, EdD(sci educ), 65. *Prof Exp:* Teacher sci, Lacy Schs, Okla, 36-38, Loyal Schs, 38-40, Newkirk, Okla, 41-42 & Ponca Mil Acad, 45-47; prin & teacher, Tonkawa, Okla, 47-51; counr & teacher, Northeast High Sch, Oklahoma City, 51-61; prof physics, Cent State Univ Okla, 61-77; RETIRED. *Mem:* Nat Sci Teachers Asn; Am Inst Physics; Am Asn Physics Teachers. *Mailing Add:* 1716 S Rankin Edmond OK 73013

YOFFA, ELLEN JUNE, b Boston, Mass, Aug 18, 51. VERY LARGE SCALE INTEGRATION DESIGN. *Educ:* Mass Inst Technol, BS, 73, PhD(physics), 78. *Prof Exp:* Fel, 78-80, RES STAFF MEM, THOMAS J WATSON RES CTR, IBM, 80- *Mem:* Am Phys Soc; Sigma Xi. *Res:* Creating tools for very large scale integration design automation, and construction of interactive systems for large scale circuit design, wiring, and data interchange standards. *Mailing Add:* IBM T J Watson Res Ctr PO Box 218 Yorktown Heights NY 10598

YOGANATHAN, AJIT PRITHIVIRAJ, b Colombo, Sri Lanka, Dec 6, 51. CHEMICAL ENGINEERING. *Educ:* Univ Co, Univ London, England, BSc, 73; Calif Inst Technol, Pasadena, PhD(chem eng), 78. *Prof Exp:* Postdoctoral fel chem eng, Calif Inst Technol, 77-79; from asst prof chem eng to assoc prof chem eng, 79-88, PROF CHEM ENG, GA INST TECHNOL, 88-, PROF MECH ENG, 89- *Concurrent Pos:* Alexander von Humbolt fel, Humboldt Foundation, W Ger, 85; mem, Int Standards Orgn sub comt prosthetic heartvalves, 85-; adj prof, div cardiovascular med, Univ Ala Med Sch, 87-; co-dir, Bioengineering Ctr, Ga Inst Technol, 89-; mem, Surg & Bioeng Study Sect, NIH, 88-92. *Honors & Awards:* Goldsmid Medal, Univ Col, Univ London, 73; Edwin Walker Award, Brit Inst Mech Eng, 89. *Mem:* Int Standards Orgn; Am Inst Chem Engrs; Am Soc Echocardiography; Sigma Xi; Am Soc Eng Educ; Am Heart Asn; Am Soc Mech Eng. *Res:* Physics of blood flow in the human heart and its large vessels; non-invasive Doppler techniques; magnetic resonance imaging; computational fluid dynamics. *Mailing Add:* Sch Chem Eng Ga Inst Tech Atlanta GA 30332-0100

YOGORE, MARIANO G, JR, b Iloilo City, Philippines, Dec 29, 21; m 45; c 7. PARASITOLOGY, PUBLIC HEALTH. *Educ:* Univ Philippines, MD, 45; Johns Hopkins Univ, MPH, 48, DrPH, 57; Philippine Bd Prev Med & Pub Health, dipl, 56. *Prof Exp:* From instr to prof parasitol, Univ Philippines, 45-67; res assoc & assoc prof, Univ Chicago, 67-69, res assoc & prof parasitol, 69-; RETIRED. *Concurrent Pos:* USPHS res fel, Dept Microbiol, Univ Chicago, 59-61; mem, Nat Res Coun Philippines, 57- *Mem:* Am Soc Trop Med & Hyg. *Res:* Immunity to parasitic diseases with special interest in schistosomiasis. *Mailing Add:* 7620 S Crandon Chicago IL 60649

YOH, JOHN K, b Shanghai, China, Oct 9, 44; US citizen. PHYSICS. *Educ:* Cornell Univ, BA, 64; Calif Inst Technol, MS, 66, PhD(physics), 70. *Prof Exp:* Res asst high energy physics, Calif Inst Technol, 67-70; res assoc, Rutgers Univ, 70; NATO fel, Cern Europ Orgn Nuclear Res, 70-71, vis scientist, 71-73; res assoc, Columbia Univ, 73-77, asst prof, 77-80; SCIENTIST RES & ADMIN, FERMI NAT ACCELERATOR LAB, 80- *Res:* Colliding anti-proton proton project; experimental dileptons and high energy physics. *Mailing Add:* MS 223 Fermi Nat Accelarator Lab PO Box 500 Batavia IL 60510

YOHE, CLEON RUSSELL, b New York, NY, July 8, 41; m 66; c 1. ALGEBRA. *Educ:* Univ Pa, AB, 62; Univ Chicago, MS, 63, PhD(math), 66. *Prof Exp:* Asst prof, 66-71, ASSOC PROF MATH, WASH UNIV, 71- *Mem:* AAAS; Am Math Soc. *Res:* Structure theory of rings. *Mailing Add:* Dept Math Wash Univ St Louis MO 63130

YOHE, JAMES MICHAEL, b Delaware, Ohio, June 8, 36; m 61; c 3. COMPUTER SCIENCE, MATHEMATICS. *Educ:* DePauw Univ, BA, 57; Univ Wis-Madison, MS, 62, PhD(math), 67. *Prof Exp:* Asst prof math, Math Res Ctr, Univ Wis-Madison, 67-68; asst prof, Pa State Univ, 68-69; proj assoc, Math Res Ctr, Univ Wis-Madison, 69-71, asst dir, 71-75, assoc dir, 75-78; dir acad comput, Univ Wis-Eau Claire, 79-86; dir, computing servs, Bradley Univ, 86-89; DIR, INFO SYSTS & COMPUT SERVS, UNIV NORTHERN IOWA, 89- *Concurrent Pos:* Lectr, Univ Wis, 71-72, asst prof comput sci, 73-74. *Mem:* Am Math Soc; Math Asn Am; Asn Comput Mach. *Res:* Computer arithmetic; interval arithmetic; computer systems programming; graph theory. *Mailing Add:* Info Systs & Comput Servs Univ Northern Iowa Cedar Falls IA 50614-0007

YOHE, THOMAS LESTER, b Bryn Mawr, Pa, Oct 30, 47; m 83; c 1. PROCESS DESIGN, ANALYTIC CHEMISTRY. *Educ:* Slippery Rock State Col, BA, 74; Drexel Univ, MS, 76, PhD(environ sci), 82. *Prof Exp:* Teaching asst environ eng, Drexel Univ, 76-77, res asst, 77- 78, res assoc, 78-79; consult water quality, self employed, 79-81; dir res, 81-84, DIR RES & ENVIRON AFFAIRS, PHILADELPHIA SUBURBAN WATER CO, 84- *Concurrent Pos:* Mem, Res Adv Coun, Am Water Works Assoc, 88- *Mem:* Am Chem Soc; Am Water Works Asn; Sigma Xi. *Res:* Optimize existing methods of drinking water treatment and analyses; develop new treatment and analysis methods. *Mailing Add:* 1406 Morstein Rd West Chester PA 19380

YOHEM, KARIN HUMMELL, b Warren, Ohio, Dec 4, 55; m 79. CANCER RESEARCH, HUMAN MALIGNANT MELANOMA. *Educ:* Kent State Univ, BS, 77, MS, 80; Univ Ariz, PhD(plant path), 82. *Prof Exp:* Postdoctoral fel, Cancer Ctr, 83-87, res assoc, 88-90, RES ASST PROF CANCER BIOL, DEPT ANAT, COL MED, UNIV ARIZ, 91- *Concurrent Pos:* Consult, Upjohn Corp, 88, Glaxo, Inc, 89-90. *Mem:* Am Asn Cancer Res; Am Soc Cell Biol; AAAS; PanAm Soc Pigment Cell Res. *Res:* Invasion and metastasis; second messengers; intracellular calcium; growth factors; cytokines; multidrug resistance; experimental therapeutics; biological response modifiers. *Mailing Add:* Dept Anat Col Med Univ Ariz 1501 N Campbell Ave Tucson AZ 85724

YOHN, DAVID STEWART, b Shelby, Ohio, June 7, 29; m 50; c 5. MICROBIOLOGY. *Educ:* Otterbein Col, BS, 51; Ohio State Univ, MS, 53, PhD, 57; Univ Pittsburgh, MPH, 60. *Prof Exp:* Res assoc, Univ Pittsburgh, 56-60, asst res prof microbiol, Grad Sch Pub Health, 60-62; res prof, State Univ NY Buffalo, 62-71; assoc cancer res scientist, Roswell Park Mem Inst, 62-69; dir, Comprehensive Cancer Ctr, 73-88, PROF VIROL, OHIO STATE UNIV, 69-, DEP DIR, COMPREHENSIVE CANCER CTR, 88- *Concurrent Pos:* Consult, Nat Cancer Inst, 70-; mem med & sci adv bd & bd trustees, Leukemia Soc Am, 71-; secy gen, Int Asn Comp Res Leukemia & Related Dis, 74- *Honors & Awards:* Leadership Award, Leukemia Soc, 85. *Mem:* Am Soc Virol; Am Soc Microbiol; Am Asn Cancer Res; Am Asn Immunol. *Res:* Mammalian and oncogenic viruses; virus host-cell relationships; tumor immunology. *Mailing Add:* Ohio State Univ Comprehensive Cancer Ctr Suite 1132 300 W Tenth Ave Columbus OH 43210

YOHO, CLAYTON W, b Glen Dale, WVa, Dec 4, 24; m 49, 88; c 3. ORGANIC CHEMISTRY. *Educ:* W Liberty State Col, BSc, 49; Univ Pittsburgh, MSc, 51, PhD(org chem), 57. *Prof Exp:* Jr fel org res, Mellon Inst Indust Res, 51-54; process develop chemist, Merck & Co, Inc, 57-60; res supvr org res, 65-71, STAFF RES SUPVR, JOHNSON WAX, 71-, SR CHEMIST, 60-, TECH INVESTR, 76-, SCI ASSOC, 82-, MGR NEW TECHNOL, 83- *Mem:* AAAS; Am Chem Soc. *Res:* Process development work involving vitamins B-1, B-12, and gibrel; organic synthesis work in the areas of adhesives, insect repellents, insect attractants and insecticides; product development of oral hygiene products. *Mailing Add:* 1232 Pleasant Valley Dr Baltimore MD 21228-2649

YOHO, ROBERT OSCAR, public health education; deceased, see previous edition for last biography

YOHO, TIMOTHY PRICE, b Nov 8, 41. DEVELOPMENTAL BIOLOGY, ENTOMOLOGY. *Educ:* West Liberty State Col, BS, 67; WVa Univ, PhD(develop biol & entomol), 72. *Prof Exp:* Fel res, WVa Univ, 72-74; asst prof, 74-77, ASSOC PROF BIOL, LOCK HAVEN COL, 77-, PROF BIOL

SCI. *Concurrent Pos:* Liaison dir, Pa Comt Correspon Creation Evolution Controversy, 80- *Mem:* Sigma Xi; Entomol Soc Am; Am Inst Biol Sci. *Res:* Electron microscopy; biochemistry; electrophysiology to study the photodynamic effect of light on dye-fed insects; death in visible light-exposed insects caused by food, drug and cosmetic dyes; danger to consumer and new insecticide development. *Mailing Add:* Dept Biol Lock Haven Col Lock Haven PA 17745

YOKE, JOHN THOMAS, b New York, NY, Feb 27, 28; m 56; c 3. SYNTHETIC INORGANIC & ORGANOMETALLIC CHEMISTRY. *Educ:* Yale Univ, BS, 48; Univ Mich, MS, 50, PhD(chem), 54. *Prof Exp:* Res chemist, US Army Chem Corp, 54-56, Univ Chicago, 55-56, Procter & Gamble Co, 56-58; instr chem, Univ NC, 58-59; asst prof, Univ Ariz, 59-64; assoc prof, 64-70, PROF CHEM, ORE STATE UNIV, 70- *Concurrent Pos:* Minn Mining & Mfg Co fel, Univ Mich. *Mem:* Am Chem Soc. *Res:* Inorganic synthesis; coordination chemistry; group V compounds; fused salt electrochemistry; oxidation of ligands; catalysis. *Mailing Add:* 13 NW Edgewood Dr Corvallis OR 97330

YOKEL, FELIX Y, b Vienna, Austria, July 13, 22; US citizen; m 46; c 3. GEOTECHNICAL ENGINEERING, STRUCTURAL ENGINEERING. *Educ:* Univ Conn, BS, 59, MS, 61, PhD(civil eng), 63. *Prof Exp:* Tech dir, Hazbani River Diversion Proj, 50-56; design engr, Griswold Eng, 56-60; chief found engr, John Clarkeson, Consult Engr, 60-63; sr partner, Clarkeson, Clough & Yokel, Consult Engrs, 63-68; res engr, 68-78, chief geotech eng, 78-85, SR RES ENG, NAT BUR STANDARDS, 85- *Concurrent Pos:* Chmn masonry comt, Am Nat Standards Inst; chmn comt found & excavation standards, Am Soc Civil Engrs. *Honors & Awards:* Silver Medal, Dept Com, 76. *Mem:* Am Soc Civil Engrs. *Mailing Add:* Nat Inst Standards & Technol Gaithersburg MD 20099

YOKEL, ROBERT ALLEN, b Rockford, Ill, June 22, 45; m 69; c 2. NEUROBEHAVIORAL TOXICOLOGY. *Educ:* Univ Wis-Madison, BS, 68; Univ Minn, Minneapolis, PhD(pharmacol), 73. *Prof Exp:* Res fel drug abuse, Ctr Res Drug Dependence, dept psychol, Concordia Univ, 73-75; asst prof pharmacol, Drug & Poison Info Ctr, dept pharmacol, Univ Cincinnati, 75-78, vis asst prof, Col Pharm, 78-79; asst prof, 79-84, ASSOC PROF PHARMACOL & TOXICOL, COL PHARM, UNIV KY, 84- *Concurrent Pos:* Prin investr, NIH res grants, 80-; NIH res career develop award. *Mem:* Behav Pharmacol Soc; Soc Neurosci; AAAS; Am Soc Pharmacol & Exp Therapeut; Am Asn Col Pharm; Soc Toxicol. *Res:* Toxicology of aluminum; neurobehavioral, pharmacokinetic and other factors defining and contributing to aluminum toxicity; identification of drugs useful in the treatment of aluminum-induced toxicity. *Mailing Add:* Pharm Bldg Univ Ky Med Ctr Rose St Lexington KY 40536-0082

YOKELSON, BERNARD J(ULIUS), b Brooklyn, NY, Sept 14, 24; m 46; c 2. ELECTRICAL ENGINEERING, TELEPHONE COMMUNICATIONS. *Educ:* Columbia Univ, BS, 48; Polytech Inst Brooklyn, MEE, 54. *Prof Exp:* Mem tech staff transmission & switching, AT&T Bell Labs 48-54, supvr switching syst & circuit develop, 54-59, head network & syst develop, 59-66, dir, Oper Systs Lab, 66-74 & Electronic Power Systs Lab, 74-76, dir, Local Switching Systs Lab, 76-80, dir, Toll Digital Switching Lab, 80-88; CONSULT, 88- *Mem:* Fel Inst Elec & Electronics Engrs; Sigma Xi. *Res:* Telephone communications; electronic switching systems; automation of telephone operator services; digital computers. *Mailing Add:* 2535 The Fifth Fairway Roswell GA 30076

YOKELSON, HOWARD BRUCE, b Elizabeth, NJ, July 29, 56. SYNTHETIC INORGANIC & ORGANOMETALLIC CHEMISTRY, POLYMER CHEMISTRY. *Educ:* Univ Del, BS, 78; Univ Wis-Madison, MS, 80, PhD(chem), 87. *Prof Exp:* Teaching asst, dept chem, Univ Wis-Madison, 78-80, rest asst, 83-87; res chemist, Teva Pharmaceut Co, Jerusalem, 81-82; RES CHEMIST, AMOCO CHEM CO, CHICAGO, 87- *Mem:* Am Chem Soc; Sigma Xi. *Res:* Organosilicon, physical organic and polymer chemistry. *Mailing Add:* Amoco Chem Co PO Box 3011 Naperville IL 60566-7011

YOKELSON, M(ARSHALL) V(ICTOR), b Brooklyn, NY, Nov 18, 18; m 53; c 2. METALLURGICAL ENGINEERING. *Educ:* City Col New York, BChE, 38; Polytech Inst Brooklyn, MMetE, 51. *Prof Exp:* Metallurgist, Chance Vought Corp, 47-48; res metallurgist, Gen Cable Corp, 48-56, res supvr, 56-57, chief metallurgist, 57-67, chief metall engr, 67-85; METALLURGICAL CONSULT, 85- *Concurrent Pos:* Metall consult, 85- *Mem:* Am Soc Metals Int; Am Inst Mining, Metall & Petrol Engrs; Wire Asn Int; fel Am Soc Testing & Mat. *Res:* Work-hardening and annealing characteristics of copper; fabricating methods; laboratory evaluation and service behavior of metallic components of power and communications cables. *Mailing Add:* 65 Sandy Hill Rd Westfield NJ 07090-2826

YOKLEY, PAUL, JR, b Mitchellville, Tenn, Aug 3, 23; m 52; c 2. ZOOLOGY. *Educ:* George Peabody Col, BS, 49, MA, 50; Ohio State Univ, PhD(zool), 68. *Prof Exp:* Instr biol, 50-53, asst prof, 53-68, PROF ZOOL, UNIV NALA, 68- *Concurrent Pos:* Sci consult, Colbert Co Schs, 67-68; Tenn Game & Fish Comn res grant, 69-72; fisheries scientist, Am Fisheries Soc; Am Fisheries res scientist fel; consult, Tenn Valley Authority, 71-; res grants, USDA, 73, 77, 78 & Corps of Engrs, 76, 78, 80, 81, 84, Tenn Wildlife Resources Agency, 87, & numerous others. *Honors & Awards:* Res Award, Asn Southeastern Biologists, 70; State Conserv Educr Year, Ala Wildlife Fedn, 72. *Mem:* Soc Syst Zool; Am Malacol Union. *Res:* Life history and ecology of freshwater mussels; ecology of the freshwater mussels in the Tennessee River. *Mailing Add:* Dept Biol Univ NAla Box 5048 Florence AL 35632-0001

YOKOSAWA, AKIHIKO, b Kofu, Japan, Nov 19, 27; US citizen; m 57; c 3. HIGH ENERGY PHYSICS. *Educ:* Tohoku Univ, Japan, BS, 51; Univ Cincinnati, MS, 53; Ohio State Univ, PhD(physics), 57. *Prof Exp:* Assoc prof physics, Ill State Univ, 57-59; physicist, 59-70, SR PHYSICIST, ARGONNE NAT LAB, 70- *Mem:* Fel Am Phys Soc. *Res:* Elementary particle physics. *Mailing Add:* Argonne Nat Lab 9700 S Cass Ave Argonne IL 60439

YOKOYAMA, MELVIN T, b Honolulu, Hawaii, Jan 22, 43. NUTRITION, MICROBIOLOGY. *Educ:* Univ Hawaii, BS, 66; Univ Ill, MS, 69, PhD(nutrit sci), 71. *Prof Exp:* Res asst dairy nutrit, Univ Ill, 66-69 & Nutrit Sci Prog, 66-71; res assoc nutrit biochem, Wash State Univ, 71-75; asst prof, 75-80, ASSOC PROF ANIMAL NUTRIT, MICH STATE UNIV, 80- *Mem:* Am Soc Animal Sci; Am Soc Microbiol; AAAS; Am Inst Nutrit; Sigma Xi. *Res:* Nutrition-microbiology. *Mailing Add:* Dept Animal Husb Mich State Univ East Lansing MI 48824

YOKOYAMA, MITSUO, immunology, immunohematology, for more information see previous edition

YOKOYAMA, SHOZO, b Miyazaki, Japan, Jan 15, 46; m 77; c 1. POPULATION GENETICS, MOLECULAR EVOLUTION. *Educ:* Miyazaki Univ, BS, 68; Kyushu Univ, MS, 71; Univ Washington, PhD(biomath), 77. *Prof Exp:* Res assoc genetics, Univ Tex, Houston, 77-78; instr psychiat, 78-79, ASST PROF PSYCHIAT & GENETICS, WASH UNIV, ST LOUIS, 80-, ASST PROF BIOL, 82- *Mem:* Genetics Soc Am; AAAS; Am Soc Human Genetics; Genetics Soc Japan; Am Soc Naturalists; Soc Study Evolution. *Res:* Theoretical and experimental population genetics, which study the mechanisms of evolutionary processes. *Mailing Add:* Ecol & Evolution Univ Ill Urbana Campus 505 S Goodwin Ave Urbana IL 61801

YOLDAS, BULENT ERTURK, b Turkey, Feb 19, 38; US citizen; m; c 2. CERAMICS, GLASS TECHNOLOGY. *Educ:* Ohio State Univ, BCerE, 63, MS, 64, PhD(glass & refractory), 66. *Prof Exp:* Sr engr mat sci, Owens-Ill Tech Ctr, 66-74; fel scientist mat sci, Westinghouse Res Labs, 74-84; FEL SCIENTIST MAT SCI, GLASS RES CTR, PPG INDUST, INC, 84- *Honors & Awards:* George Westinghouse Award, 84. *Mem:* Fel Am Ceramic Soc; Sigma Xi. *Res:* Coating technology for electronic and consumer products; formation of glass and ceramic materials by chemical polymerization; high surface area; catalytic materials; porous ceramic, metal-organic compounds; optics. *Mailing Add:* 1605 Jamestown Pl Pittsburgh PA 15235

YOLE, RAYMOND WILLIAM, b Middlesbrough, Eng, Feb 21, 27; Can citizen; m 57; c 4. PETROLEUM GEOLOGY, CORDILLERAN GEOLOGY. *Educ:* Univ New Brunswick, BSc, 47; Johns Hopkins Univ, MA, 58; Univ BC, PhD(geol), 65. *Prof Exp:* Geologist, Calif Standard Co, Alta, 47-51, asst to vpres explor, 51-53, dist stratigr, 53-56; from asst prof to assoc prof geol, 63-82, chmn dept, 67-70, PROF GEOL, CARLETON UNIV, 82- *Concurrent Pos:* Vis res geologist, Univ Reading, 70-71; assoc ed, Can Soc Petrol Geol, 74-; vis prof, Fed Univ Pernambuco, Recife, Brazil, 76. *Mem:* Can Soc Petrol Geol; Petrol Soc; Can Inst Mining & Metall; fel Geol Asn Can; Am Asn Petrol Geol; Int Asn Sedimentologists. *Res:* Petroleum geology of Canadian offshore basins; stratigraphy and tectonic history of Canadian cordillera; Paleozoic stratigraphy and sedimentology. *Mailing Add:* Dept Earth Sci Carleton Univ Colonel By Dr Ottawa ON K1S 5B6 Can

YOLKEN, HOWARD THOMAS, b Birmingham, Ala, Jan 29, 38; m 67; c 1. MATERIALS PROCESSING, NONDESTRUCTIVE EVALUATION. *Educ:* Univ Md, BS, 60, PhD(mat sci), 70. *Prof Exp:* Res metallurgist, 60-70, dep chief, Off Standard Ref Mat, 71-75, chief, Off Measurements Nuclear Technol, 76-81, CHIEF OFF NONDESTRUCTIVE EVAL, NAT INST STANDARDS & TECHNOL, 82- *Concurrent Pos:* Mem, Nat Mat Adv Bd Comts, Life Cycle Eng, Nat Res Coun, 68-88 & Indust Energy Conserv, 87-; mem tech adv comt nuclear safeguards, Int Atomic Energy Asn, 78 & 79; founding chmn, Gordon Res Conf Nondestructive Eval, 83; sci fel, US Cong, 84-85; chmn, Working Group Intel Processing Mat, White House Off Sci & Technol Policy, 85-; founding ed-in-chief Res Nondestructive Eval, J Am Soc Nondestructive Testing, 88- *Honors & Awards:* Bronze Medal, US Dept of Com, 81. *Mem:* Am Phys Soc; Am Soc Nondestructive Evaluation. *Res:* Fostered, at the national level, the concept and pursuit of intelligent processing of materials; directed major research effort on nondestructive evaluation for real-time materials process control; pioneered research in the application of ellipsometry to clean surfaces in ultrahigh vacuum. *Mailing Add:* Nat Inst Standards & Technol Mat Bldg Rm B344 Gaithersburg MD 20899

YOLLES, SEYMOUR, chemistry; deceased, see previous edition for last biography

YOLLES, STANLEY FAUSST, b New York, NY, Apr 19, 19; m 42; c 2. MEDICINE, PSYCHIATRY. *Educ:* Brooklyn Col, AB, 39; Harvard Univ, AM, 40; NY Univ, MD, 50; Johns Hopkins Univ, MPH, 57. *Prof Exp:* Parasitologist, Sector Malaria Lab, US Dept Army, 41-42, assoc dir, 42-44; intern, USPHS Hosp, Staten Island, NY, 50-51, resident psychiat, Lexington, Ky, 51-54; staff psychiatrist, Ment Health Study Ctr, NIMH, 54-55, from assoc dir to dir, 55-60, from assoc dir to dep dir extramural progs, 60-63, from dep dir to dir, NIMH, 63-70, assoc adminr for ment health, US Dept Health, Educ & Welfare, 68-70; prof & chmn dept, 71-81, psychiatrist in chief, Univ Hosp, 80-81, EMER PROF PSYCHIAT & BEHAV SCI, STATE UNIV NY, STONYBROOK, 81- *Concurrent Pos:* Clin prof psychiat, George Washington Univ, 67-71; spec consult, NY City Bd Educ; mem, Prof Adv Bd, Int Comt Against Ment Illness, 68-, Nat Adv Panel, Am Jewish Comt, 69-72, Expert Adv Panel Ment Health, WHO, 69-81, Med Adv Comt, Am Joint Distribution Comt, 70-; trustee, NY Sch Psychiat; sr consult, Southside Hosp, Bay Shore, South Oaks Hosp, Amityville & Nassau County Med Ctr, NY. *Mem:* AAAS; fel Am Psychiat Asn; fel Am Pub Health Asn; fel Am Col Psychiat; fel NY Acad Sci. *Res:* Community mental health; mental health administration; epidemiology of mental health. *Mailing Add:* Two Soundview Ct Stony Brook NY 11790

YOLLES, TAMARATH KNIGIN, b New York, NY, Feb 27, 19; m 42; c 2. PUBLIC HEALTH. *Educ:* Bellevue Med Ctr, New York Univ, MD, 51. *Prof Exp:* Assoc dir, Malaria Lab, Sector, Trinadad, 43-45; dep dir, USPHS Outpatients Clinic, Washington, DC, 53, health maintenance officer, Off Surgeon Gen, 61-64, dir, Off Personnal Technol, HHS, 64-70, asst surgeon gen, 70-71; PROF COMMUN MED, SCH MED, HEALTH SCI CTR,

STATE UNIV NY, STONY BROOK, 71-, ASSOC DEAN CONTINUING MED EDUC, 74- *Concurrent Pos:* Consult, Health Serv & Mental Health Advan Lab, Dept HHS, 71-73; mem, Emergency Med Serv Comt, Nat Acad Sci, 75-; vchmn, Regional Emergency Med Serv Coun, 80- *Mem:* Am Col Emergency Physicians; AMA. *Res:* Malaria vectors in Guyana and French Guiana; sehistosomiasis. *Mailing Add:* Two Soundview Ct Stony Brook NY 11790

YOLLICK, BERNARD LAWRENCE, b Toronto, Ont, Mar 24, 22; nat US; m 47; c 2. ANATOMY, SURGERY. *Educ:* Univ Toronto, MD, 45; Am Bd Surg, dipl, 57; Am Bd Otolaryngol, dipl, 67. *Prof Exp:* Instr anat, Col Med, Univ Sask, 47-49; clin asst prof, Col Med, Baylor Univ, Houston, 54-60; asst prof, Univ Tex Dent Sch, Houston, 54-60; lectr surg, Univ Tex Postgrad Sch Med, 57-67; ASST PROF OTOLARYNGOL, UNIV TEX HEALTH SCI CTR DALLAS, 67- *Concurrent Pos:* Fel surg, Am Cancer Soc, 53-54; consult, Houston Pulmonary Cytol Proj, 59 & Vet Admin Hosp, Dallas; adj assoc prof anat, Baylor Col Dent, Dallas, 89- *Mem:* Soc Human Genetics; Am Asn Anat; AMA; fel Am Col Surg; Soc Head & Neck Surgeons. *Res:* Induction of bone tumors in animals using heavy metals; experimental surgery in animals. *Mailing Add:* 4229 Bobbitt Dr Dallas TX 75229-4136

YON, E(UGENE) T, b Mt Hope, WVa, Dec 29, 36; m 60; c 2. ELECTRICAL ENGINEERING, SOLID STATE ELECTRONICS. *Educ:* Univ Cincinnati, EE, 60; Case Western Reserve Univ, MS, 62, PhD, 65. *Prof Exp:* Sr staff engr, Electronics Div, Avco Corp, 65-67; asst prof elec eng, Case Western Reserve Univ, 67-70, assoc dir solid state electronics labs, 67-77, assoc prof, 70-77; vpres, Combustion Eng, Inc, 84-88; vpres, 78-83, VPRES, BOOZ, ALLEN, & HAMILTON, INC, 88- *Concurrent Pos:* Consult, Avco Corp, 64-65 & 67-68, Babcock & Wilcox Corp, 67-, Am Radiation Res, Inc, 67-, Keithley Instruments, Inc, 68- & Solon Assocs, Inc, 68- *Mem:* Inst Elec & Electronics Engrs. *Res:* Semiconductor devices; biomedical engineering; technology, assessment and forecasting. *Mailing Add:* 316 Green Oak Ct Longwood FL 32779

YONAN, EDWARD E, b Beirut, Lebanon, Apr 15, 43; US citizen; m 72; c 2. ORGANIC SYNTHESIS. *Educ:* Univ Wis, BS, 70. *Prof Exp:* Chemist, Hodag Chem Corp, 72-73 & Velsicol Chem Corp, 73-76; sr chemist, PPG Industs, Inc, 76-80; consult, 80-86; RES ASSOC, G D SEARLE & CO, 86- *Mem:* Am Chem Soc; Fire Retardant Chem Asn. *Res:* Organic synthesis and process development of agricultural chemicals such as herbicides, pesticides, phosgene, and related chemistry, carbonates, chloroformates, isocyanutes, and fire retardant additives. *Mailing Add:* G D Searle & Co 49010 Pkwy Skokie IL 60077

YONAS, GEROLD, b Cleveland, Ohio, Dec 8, 39; m 61; c 2. PULSED POWER, INERTIAL FUSION. *Educ:* Cornell Univ, BS, 62; Calif Inst Technol, PhD(eng sci), 66. *Prof Exp:* Physicist & mgr electron beam res dept, Physics Int Co, 67-72; div supvr, Sandia Labs, 72-73, dept mgr fusion res, 73-78, dir pulsed energy progs, 78-89, DIR LAB DEVELOP, SANDIA LABS, 89- *Concurrent Pos:* Mem adv bd, Am for Energy Independence; assoc ed, J Fusion Energy. *Mem:* Am Phys Soc; Am Nuclear Soc; Sigma Xi. *Res:* Fusion; intense electron and ion beams; pulsed high voltage technology; inertial confinement fusion; laser technology; plasma physics; high pressure physics. *Mailing Add:* Orgn 400 Sandia Nat Labs PO Box 5800 Albuquerque NM 87185

YONCE, LLOYD ROBERT, b Roscoe, Mont, Sept 27, 24; m 48; c 2. PHYSIOLOGY. *Educ:* Mont State Col, BS, 49; Oregon State Univ, MS, 52; Univ Mich, PhD(physiol), 55. *Prof Exp:* Instr physiol, Ore State Univ, 51-52; instr, Med Sch, Univ Mich, 55-56; asst prof, 57-61, ASSOC PROF PHYSIOL, SCH MED, UNIV NC, CHAPEL HILL, 61- *Concurrent Pos:* Fel, Med Col Ga, 56-57; USPHS spec fel physiol, Univ Gothenburg. *Mem:* Fedn Am Soc Exp Biol; Am Microcirculation Soc. *Res:* Cardiovascular physiology; neurophysiology; physiology of diving animals. *Mailing Add:* Dept Physiol CB No 7545 Univ NC Med Sch Chapel Hill NC 27599

YONDA, ALFRED WILLIAM, b Cambridge, Mass, Aug 10, 19; m 49, 75; c 4. MATHEMATICS. *Educ:* Univ Ala, BS, 52, MA, 54. *Prof Exp:* Mathematician, Rocket Res, Redstone Arsenal, Ala, 53 & US Army Ballistic Res Labs, Aberdeen Proving Ground, Md, 54-56; instr math, Temple Univ, 56-57; assoc scientist, Res & Adv Develop Div, Avco Corp, 57-59; sr proj mem tech staff, Radio Corp Am, 59-66; mgr comput anal & prog dept, Raytheon Corp, Bedford, 66-70, prin engr, Missile Systs Div, 70-73; mgr systs anal & prog dept, Eastern Div, Needham, Mass, 73-76, software eng, Atlantic Oper, 77-82, sr mem tech staff, Commun Systs Div, 83-87, SR MEM TECH STAFF, C3 SECTOR, GOVT SYSTS CORP, GTE SYLVANIA, NEEDHAM, MASS, 87- *Concurrent Pos:* Hon fel, Advan Level Telecom Training Ctr, Fac, New Delhi, India. *Mem:* AAAS; Inst Elec & Electronics Engrs; Math Asn Am; NY Acad Sci. *Res:* Computer systems analysis; simulation; communications systems analysis; numerical analysis; telecommunications systems. *Mailing Add:* 12 Sunset Dr Medway MA 02053

YONEDA, KOKICHI, ELECTROMICROSCOPY, CELL BIOLOGY. *Educ:* Mara Med Col, Japan, MD, 68. *Prof Exp:* PROF PATH, UNIV KY, 86- *Mailing Add:* Univ Ky Med Sch 800 Rose St Univ Ky Med Sch Lexington KY 40536

YONETANI, TAKASHI, b Kagawa-ken, Japan, Aug 6, 30; US citizen; m 58; c 1. BIOCHEMISTRY, BIOPHYSICS. *Educ:* Osaka Univ, BA, 53, PhD(biochem), 60. *Hon Degrees:* MD, Univ Umea, Sweden, 84. *Prof Exp:* Res fel biochem, Johnson Found, Univ Pa, 58-61; Swedish Med Res Coun res fel, Nobel Med Inst, Stockholm, 62-64; from asst prof to assoc prof phys biochem, 64-68, PROF PHYS BIOCHEM, DEPT BIOCHEM & BIOPHYS, UNIV PA, 68- *Concurrent Pos:* USPHS career develop award, 67-72. *Mem:* AAAS; Am Soc Biol Chemists; Am Chem Soc; Biophys Soc. *Res:* Purification, crystallization and characterization of cytochrome oxidase, alcohol dehydrogenase and cytochrome c peroxidase; determination of structure and function of these enzymes by spectrophotometry, electron spin resonance and x-ray diffraction techniques; heart disease; artificial hemoglobin; metalloporphyrin synthesis. *Mailing Add:* C601 Richards Bldg 37th & Hamilton Walk Philadelphia PA 19104-6089

YONG, R(AYMOND), b Singapore, Apr 10, 29; Can citizen; m 61; c 2. ENVIRONMENT & GEOTECHNICAL ENGINEERING, WASTE MANAGEMENT. *Educ:* Washington & Jefferson Col, BA, 50; Mass Inst Technol, BSc, 52; Purdue Univ, MSc, 54; McGill Univ, MEng, 58, PhD(soil mech), 60. *Prof Exp:* From asst prof to assoc prof civil eng, 59-65, prof civil eng & appl mech, 65-73, WILLIAM SCOTT PROF CIVIL ENG & APPL MECH, MCGILL UNIV, 73-, DIR, GEOTECH RES CTR, 76- *Honors & Awards:* Izaak Walton Killam Mem Prize, Can Coun, 85; Chevalier de l'ordre nat du Quebec, 85; Charles B Dudley Prize, Am Soc Testing & Mat. *Mem:* Am Soc Civil Engrs; Brit Inst Civil Engrs; Am Soc Testing & Mat; Soc Rheology; Int Soc Terrain-Vehicle Systs; fel Engr Inst Can; Fel Royal Soc Can. *Res:* Soil mechanics; soil physics; nonlinear mechanics; plasticity; geo-environmental research technology. *Mailing Add:* McGill Univ Geotech Res Ctr 817 Sherbrooke St W Montreal PQ H3A 2K6 Can

YONG, YAN, b Wuhan, China, July 23, 55; m 87. STOCHASTIC EARTHQUAKE MODELING, WAVE PROPAGATION IN STRUCTURAL NETWORKS. *Educ:* Wuhan Inst Bldg Mat, dipl eng mech, 82; Univ Ill, Urbana- Champaign, MS, 83, PhD (aerospace eng), 87. *Prof Exp:* Res asst, Univ Ill, 82-84; res assoc, 87-88, ASST PROF, FLA ATLANTIC UNIV, 88- *Concurrent Pos:* NSF presidential young investr award, 90; vis scientist, Wright-Patterson AFB, 90. *Mem:* Am Soc Civil Eng; Am Inst Aeronaut & Astronaut; Acoust Soc Am. *Res:* Dynamics and control of large-scale complex structures; stochastic modeling of earthquake ground motions; dynamic response and control of civil structures under earthquake and wind excitations; random vibration; earthquake eng; author of nine publications. *Mailing Add:* Dept Ocean Eng Fla Atlantic Univ Boca Raton FL 33431

YONGE, KEITH A, b London, Eng, June 22, 10; m 47; c 4. PSYCHIATRY. *Educ:* McGill Univ, MD, CM, 48; Univ London, dipl psychol med, 52. *Prof Exp:* Mem staff psychiat, Med Res Coun, Eng, 51-52; dir, Ment Health Clin, Can, 52-54; from asst prof to assoc prof psychiat, Univ Sask, 55-57, clin dir, Univ Hosp, 55-57; prof & head dept, 55-75, EMER PROF PSYCHIAT, UNIV ALTA, 75- *Concurrent Pos:* Can Coun leave fel, 71-72. *Mem:* Psychiat Asn; Can Med Asn; Can Psychiat Asn; Can Ment Health Asn. *Res:* Basic and clinical psychiatry; phenomenology of depression; cognitive effects of cannabis; nature of human aggression. *Mailing Add:* 4345 Kingscote Rd RR 3 Cobble Hill BC V0R 1L0 Can

YONGUE, WILLIAM HENRY, b Charlotte, NC, Aug 21, 26; m 48; c 1. PROTOZOOLOGY, AQUATIC ECOLOGY. *Educ:* Johnson C Smith Univ, BS, 49; Univ Mich, MS, 62; Va Polytech Inst & State Univ, PhD(zool), 72. *Prof Exp:* Head dept sci, West Charlotte High Sch, 59-70; from instr to asst prof, 70-73, assoc prof, 73-85, ADJ PROF ZOOL, VA POLYTECH INST & STATE UNIV, 86- *Concurrent Pos:* Res assoc & proj scientist, Univ Mich Biol Sta, 69-75; regional sci coordr, NC State Dept Pub Inst. *Mem:* Soc Protozoologists; Nat Sci Teachers Asn; Sigma Xi; fel AAAS; Am Micros Soc. *Res:* The ecology of freshwater protozoans using polyurethane foam substrates for sampling and as microhabitats; lentic plankton dynamics; effects of heat and toxicants on protists; intracellular parasitism effects in blood of water snakes; science education. *Mailing Add:* 2400 Hildebrand St Charlotte NC 28216

YONKE, THOMAS RICHARD, b Kankakee, Ill, Nov 30, 39; m 63; c 3. ENTOMOLOGY, SYSTEMATICS. *Educ:* Loras Col, BS, 62; Univ Wis, Madison, MS, 64, PhD(entom), 67. *Prof Exp:* Instr entom, Univ Wis, Madison, 66-67; from asst prof to assoc prof, 67-77, PROF ENTOM, UNIV MO-COLUMBIA, 77-, DIR, ENTOM RES MUS, 78-, CHMN DEPT, 80- *Concurrent Pos:* Pres grad fac senate, Univ Mo-Columbia, 80-81. *Mem:* Entom Soc Am; Soc Syst Zool; AAAS; Sigma Xi. *Res:* Biology and systematics of Hemiptera; taxonomy of immature Heteroptera; biology of Cicadellidae; taxonomy of Coreidae. *Mailing Add:* Dept Entom 1-87 AGR Univ Mo Columbia MO 65211

YONKERS, ANTHONY J, b Muskegon, Mich, June 17, 38; c 4. SURGERY, OTORHINOLARYNGOLOGY. *Educ:* Univ Mich, MD, 63. *Prof Exp:* From instr to assoc prof otorhinolaryngol, 68-72, vchmn dept, 72-74, 72-76, PROF OTOLARYNGOL & MAXILLOFACIAL SURG, UNIV NEBR MED CTR, 76-, CHMN DEPT, 74- *Concurrent Pos:* Guest ed, Ear, Nose & Throat J, 81; prin investr, Univ Nebr Med Ctr, 83-84 & 85-; comdr, Mil & Hosp Order St Lazarus, Jerusalem, 84-; mil consult to surg gen, 84-, colonel flight surgeon, Med Corps, US Air Force & Dept Nebr Reserve Officers Asn US, 85. *Mem:* Am Acad Otolaryngol; fel Am Col Surg; Am Acad Facial Plastic & Reconstructive Surg; Am Acad Otolaryngic Allergy; Am Broncho-Esophagological Asn; Am Asn Cosmetic Surgeons; Triological Soc; Aerospace Med Asn; AMA. *Res:* Study of interferon in the treatment of juvenile papillomatosis; evaluation of new methods of management of children with clefts of the lip and palate; recognition and management of sleep apnea. *Mailing Add:* Dept Otolaryngol Univ Nebr Med Ctr 42nd & Dewey Ave Omaha NE 68105-1065

YONUSCHOT, GENE R, b Brooklyn, NY, Oct 29, 36; m 66; c 3. NUTRITION. *Educ:* Calif Polytech Inst, BS, 63; Univ Mo-Columbia, PhD(biochem), 69. *Prof Exp:* Instr biochem, Univ NC, 69-71; asst prof biochem, George Mason Univ, 71-76; chmn, Biochem Dept, WVa Sch Osteop Med, 76-78; ASSOC DEAN BASIC SCI, NEW ENG COL OSTEOP MED, 78- *Res:* Acidic chromosomal proteins and t-RNA in relation to the control of cell division and differentiation; medical school curriculum; lanthanide series elements in biology. *Mailing Add:* Dept Biochem & Nutrit New Eng Col Osteop Med 605 Pool Rd Biddeford ME 04005

YOO, BONG YUL, b Pusan, Korea, June 30, 35; Can citizen. PLANT PHYSIOLOGY, CELL BIOLOGY. *Educ:* Seoul Nat Univ, BSc, 58; Okla State Univ, MSc, 61; Univ Calif, Berkeley, PhD(bot), 65. *Prof Exp:* Nat Res Coun Can fel, 65-66; from asst prof to assoc prof, 66-77, PROF BIOL, UNIV NB, FREDERICTON, 77- *Mem:* Am Soc Cell Biol; Am Soc Plant Physiol; Bot Soc Am; Can Soc Cell Biol; Sigma Xi. *Res:* Biogenesis of plant cell organelles. *Mailing Add:* Dept Biol Univ NB Col Hill Box 4400 Fredericton NB E3B 5A3 Can

YOO, MAN HYONG, b Seoul, Korea, June 27, 35; m 60; c 3. SOLID STATE PHYSICS, APPLIED MATHEMATICS. *Educ:* Mich State Univ, BS, 60, MS, 62, PhD(phys metall), 66. *Prof Exp:* Spec res grad asst, Mich State Univ, 61-65, res assoc, 65-67, res assoc mech metall, 66-67; res staff scientist, Metals & Ceramics Div, 67-79, SR RES STAFF & TASK LEADER, OAK RIDGE NAT LAB, 79- *Concurrent Pos:* Guest scientist, Nuclear Res Ctr, Jülich, W Ger, 84-85; vis prof, Tohoku Univ, Sendai, Japan, 88-89. *Mem:* Mineral, Metals & Mat Soc; Am Soc Metals Int; Mat Res Soc; Sigma Xi; Am Asn Advan Sci. *Res:* Mechanical properties of materials; dislocation theory of plastic deformation and fracture; kinetic theory of lattice defects and irradiation effects; small angle neutron scattering; high-temperature intermetallic ordered alloys. *Mailing Add:* Oak Ridge Nat Lab PO Box 2008 Oak Ridge TN 37831-6115

YOO, TAI-JUNE, b Seoul, Korea, Mar 7, 35; US citizen; m 63; c 3. IMMUNOLOGY, INTERNAL MEDICINE. *Educ:* Seoul Nat Univ, MD, 59; Univ Calif, Berkeley, PhD(med physics), 63. *Prof Exp:* Teaching asst biophys, Div Med Physics, Univ Calif, Berkeley, 60-61, res asst, Lawrence Radiation Lab, 61-63; asst in med, Sch Med, Wash Univ, 63-66; sr cancer res scientist, Roswell Park Mem Inst, 66-68; res prof biol, Niagara Univ, 68-69; asst prof med, Col Med, Univ Iowa, 72-75, assoc prof, 75-80; AT DEPT MED, DIV ALLERGY & IMMUNOL, UNIV TENN HEALTH SCI CTR. *Concurrent Pos:* NIH fel immunol, Wash Univ, 65-66; from intern to asst resident, Barnes Hosp, St Louis, Mo, 63-65; assoc resident, Sch Med, NY Univ, 68-69; clin investr, Vet Admin Hosp, Iowa City, Iowa, 72-75. *Mem:* AAAS; Biophys Soc; Am Asn Immunol; NY Acad Sci; Am Fedn Clin Res. *Res:* Molecular and cellular biology of immune phenomena; immunologic and allergic disorders; mechanism of autoimmunity of cochlea and hearing loss. *Mailing Add:* Dept Med Div Allergy & Immunol Univ Tenn Health Sci Ctr 956 Court Rm H300 Memphis TN 38163

YOOD, BERTRAM, b Bayonne, NJ, Jan 6, 17; m 44; c 3. MATHEMATICAL ANALYSIS. *Educ:* Yale Univ, BS, 38, PhD(math), 47; Calif Inst Technol, MS, 39. *Prof Exp:* From instr to asst prof math, Cornell Univ, 47-53; from asst prof to prof, Univ Ore, 53-72; prof, 72-82, EMER PROF MATH, PA STATE UNIV, 82- *Concurrent Pos:* Vis assoc prof, Univ Calif, 56-57; vis res assoc, Yale Univ, 58-59; mem, Inst Adv Study, 61-62; vis prof, Univ Edinburgh, 69-70, Weizmann Inst, Israel, 78-79, Kans State Univ, 84-85, Reed Col, 85-86, Univ Ottawa, 87. *Mem:* Am Math Soc; Sigma Xi. *Res:* Banach algebra; Banach spaces; analysis. *Mailing Add:* Dept Math Pa State Univ University Park PA 16802

YOON, DO YEUNG, b Inchon, Korea, Jan 22, 47; m 71; c 3. POLYMER CHEMISTRY. *Educ:* Seoul Nat Univ, BS, 69; Univ Mass, MS, 71, PhD(polymer sci), 73. *Prof Exp:* Res assoc chem, Stanford Univ, 73-75; RES SCIENTIST POLYMER MAT, RES LAB, IBM CORP, 75- *Mem:* Am Chem Soc; Am Phys Soc. *Res:* Conformational statistics and conformation-dependent properties of polymers; morphology and properties of polymers. *Mailing Add:* IBM Almaden Res Lab K93 801 650 Harry Rd San Jose CA 95120-6099

YOON, EUIJOON, b Seoul, Korea, May 4, 60; m 88; c 1. THIN FILM GROWTH, CHARACTERIZATION OF STRUCTURAL & ELECTRONIC PROPERTIES OF SEMICONDUCTORS. *Educ:* Seoul Nat Univ, BS, 83, MS, 85; Mass Inst Technol, PhD(electronic mat), 90. *Prof Exp:* Prin Mem Tech Staff, AT&T BELL LABS, 90- *Mem:* Mat Res Soc; Minerals, Metals & Mat Soc; Am Vacuum Soc. *Res:* Heteroepitaxy of gallium arsenic on silicon by metal-organic chemical vapor deposition (MOCVD); hydrogen passivation of gallium arsenic surfaces. *Mailing Add:* AT&T Bell Labs 600 Mountain Ave Rm 7C-313 Murray Hill NJ 07974

YOON, HYO SUB, b Kyungbook, Korea, Apr 17, 35; US citizen; m 72. BIOENGINEERING, BIOMEDICAL ENGINEERING. *Educ:* Seoul Nat Univ, BS, 59; Univ Cincinnati, MS, 65; Pa State Univ, PhD(solid state sci), 71. *Prof Exp:* Res metallurgist, Sci Res Inst, Ministry Nat Defense, Korea, 59-62; res fel metall eng, Univ Cincinnati, 62-66; res asst solid state sci, Pa State Univ, 66-71; NIH trainee biophysics, 71-74, SR RES ASSOC, DEPT BIOMED ENG, RENSSELAER POLYTECH INST, 74- *Mem:* Am Soc Metals; Am Ceramic Soc; Inst Elec & Electronics Engrs; Am Crystallog Asn; Am Phys Soc; Sigma Xi. *Res:* Biomedical ultrasonics; bone biomechanics; crystal physics; biomaterials; dental materials; nondestructive testing; elasticity; calcified tissues; animal ultrasound. *Mailing Add:* Six Tower Heights Loudonville KY 12211

YOON, JI-WON, b Kangjin, Korea, Mar 28, 39; US citizen; m 68; c 2. DIABETES MELLITUS, VIRUS-INDUCED DISEASES. *Educ:* Chosun Univ, Korea, BS, 59, MS, 61; Univ Conn, Storrs, MS, 71, PhD(path), 73. *Prof Exp:* Asst prof cell biol, Chosun Univ, 65-67, assoc prof microbiol, Med Sch, 67-69; teaching asst, Univ Conn, Storrs, 69-73; res fel pathobiol, Sloan-Kettering Cancer Inst, 73-74; staff fel, 74-76, sr staff fel, 76-78, SR INVESTR VIROL, NIH, MD, 78- *Concurrent Pos:* Adj fac, Med Sch, Howard Univ, 79-82. *Mem:* Am Soc Microbiol; Genetics Soc Am; Tissue Culture Asn Am; NY Acad Sci; AAAS. *Res:* Role of viruses and autoantibodies in the pathogenesis of insulin dependent diabetes mellitus; cultivation and characterization of human, non-human primate and murine pancreatic beta cell cultures in microculture system; virus and cell interaction in the disease process in human and animal. *Mailing Add:* Dept Microbiol Univ Health Sci Chicago Med Sch 333 Greenbay Rd North Chicago IL 60064

YOON, JONG SIK, b Suwon, Korea, Jan 25, 37; US citizen; m 62; c 3. GENETICS, EVOLUTION. *Educ:* Yonsei Univ, Korea, BS, 61; Univ Tex, Austin, MA, 64, PhD(genetics), 65. *Prof Exp:* Res scientist assoc IV, Univ Tex, Austin, 62-65; res assoc oncol, M D Anderson Hosp, Univ Tex, 66-68; asst prof genetics & cytol, Yonsei Univ, Korea, 68-71; res scientist IV & V genetics, 71-74, instr cell biol, 74-75, res scientist genetics, Univ Tex, Austin, 75-78; assoc prof, 78-82, PROF BIOL SCI, BOWLING GREEN STATE UNIV, 82- *Concurrent Pos:* Dir, Nat Drosophila Species Resource Ctr, 82- *Honors & Awards:* Young Scientist Award, Int Union Against Cancer, 70. *Mem:* Genetics Soc Am; Soc Study Evolution; AAAS; Sigma Xi; Am Genetic Asn; Environ Mutagen Soc; Am Soc Naturalists. *Res:* Cytogenetics; mutation; oncogenetics; radiation genetics; genome organization, speciation and evolution of Drosophila and other species; genic balance between euchromatin and heterochromatin of chromosomes; sister chromatid exchange of mammalian and human chromosomes. *Mailing Add:* Dept Biol Sci Bowling Green State Univ Bowling Green OH 43403-0212

YOON, PETER HAESUNG, b Seoul, Korea, April 3, 58; m 87; c 2. SPACE PHYSICS. *Educ:* Yonsei Univ, Seoul, Korea, BS, 80; Mass Inst Technol, PhD(plasma physics), 87. *Prof Exp:* Postdoctoral assoc space plasma physics, Ctr Space Res, Mass Inst Technol, Cambridge, Ma, 87-89; ASST RES SCIENTIST SPACE PLASMA PHYSICS, INST PHYS SCI & TECHNOL, UNIV MD, 89- *Mem:* Sigma Xi; Am Phys Soc; Am Geophys Union. *Res:* Space and astrophysical plasma; radio emission process; plasma turbulence; nonlinear processes and kinetic and transport theory. *Mailing Add:* Inst Phys Sci & Technol Univ Md College Park MD 20742

YOON, RICK J, b Korea, Oct 5, 43; US citizen; m 71; c 2. GLASS-TO-METAL BONDING. *Educ:* Han Yang Univ, Seoul, Korea, BS, 69; Iowa State Univ, Ames, MS, 72, PhD(ceramic eng), 77. *Prof Exp:* Mat scientist, Anchor Hocking Corp, 78-79; res scientist, Kerr Div, Sybron Corp, 79-81; chief scientist, Rauland Div, Zenith Corp, 81-82; eng mgr, SPTC, Inc, 82-84; proj mgr, Bourns Instruments Inc, 84-88; PRES, IJ RES, 88- *Concurrent Pos:* Consult, Smar Equip Industs, Brazil, 90, Daeha Co, Korea, 91 & Teltek Serv. *Mem:* Am Ceramic Soc; Nat Inst Ceramic Engrs; Am Soc Metals; Int Soc Hybrid Microelectronics; Int Electronics Packaging Soc; Sigma Xi. *Res:* Glass-to-metal seal/bonding for commercial applications; ceramic-to-metal bonding including titanium and molybdenum seal to glass or ceramic materials. *Mailing Add:* Seven Shooting Star Irvine CA 92714

YOPP, JOHN HERMAN, b Paducah, Ky, Nov 13, 40; m 65; c 3. PLANT GROWTH REGULATORS & PHYCOLOGY. *Educ:* Georgetown Univ, BS, 62; Univ Louisville, PhD(biol), 69. *Prof Exp:* Lectr human physiol, Catherine Spalding Col, 64-65; asst biol, NASA, 65-67; lectr, Univ Louisville, 68, researcher radioisotopes, Atomic Energy Comn, 68-69; res fel plant physiol, Nat Res Coun, NASA, 69-70; assoc prof, 74-79, assoc dean, Col Sci, 84-86, PROF BOT, SOUTHERN ILL UNIV, 79-, DEAN GRAD SCH, 86- *Concurrent Pos:* Prin investr, Environ Protection Agency, Ill, 73-74, NASA-NGR, 73-77, Ill Soybean Prog Operating Bd, 75-86, Nat Marine Fisheries Bd, 76-78, NASA, 77-79 & 82-86, Fisheries Bd, 76-78, NASA, 77-79 & 82-86, NIH Biomed Res Prog; lectr, Sigma Xi, 82-85; consult, Int Corn Prod Corp & Abbott Labs, 85-88; mem, Bd Nat Resources & Conserv, Ill, 87- , Exec Comt of Grad Deans, African-Am Inst; chmn, Coun Res Policy & Grad Educ, NASULGC. *Honors & Awards:* Kaplan Award, Sigma Xi. *Mem:* Am Soc Plant Physiologists; Soc Study Origin Life; Plant Growth Regulator Soc Am; Sigma Xi. *Res:* Mechanisms of adaptation of plants to extreme environments, particularly hypersalinity and drought; soybean, cyanobacteria and aquatic plants; newly discovered plant hormones, the brassinosteroids; physiological mechanisms of adaptation to stress in plants. *Mailing Add:* Plant Biol Dept Southern Ill Univ Carbondale IL 62901

YORDY, JOHN DAVID, b St John's, Mich, Sept 17, 42; m 66; c 3. ORGANIC CHEMISTRY. *Educ:* Goshen Col, BA, 67; Mich State Univ, PhD(org chem), 74. *Prof Exp:* Teacher chem & math, Wesley High Sch, Oturkpo, Nigeria, 67-70; res scientist, Lubrizol Corp, 74-77; chair, Div Natural Sci, 81-87, PROF CHEM, GOSHEN COL, 77-, CHAIR, DEPT CHEM, 90- *Concurrent Pos:* Vis prof chem, Univ Nairobi, 87-88. *Mem:* Am Chem Soc. *Res:* Lubrication chemistry; synthesis and transformations of cyclopropanol derivatives; natural products isolation, identification and syntheses; heterocyclic chemistry. *Mailing Add:* Dept Chem Goshen Col Goshen IN 46526

YORE, EUGENE ELLIOTT, b Columbus, Ohio, Mar 6, 39; c 2. CONTROL SYSTEMS, MECHANICAL ENGINEERING. *Educ:* Ohio State Univ, BSME, 62; Univ Calif, Berkeley, MS, 63, PhD(mech eng), 66. *Prof Exp:* Sr prin res scientist, Honeywell Inc, 66-70, sect chief, 70-73, mgr, 73-76, dir res, 76-78; dep sci & technol, asst secy of Army, 78-81; corp dir, Eng & Mfg Prod, 81-83, VPRES PRECISION WEAPONS OPERS, HONEYWELL INC, 84- *Concurrent Pos:* NASA trainee, Univ Calif, Berkeley, 64-66; proj mgr, res & develop team, President's Pvt Sector Surv on Govt Cost Control, Grove Comn, 83-84. *Mem:* Inst Elec & Electronics Engrs; Am Soc Mech Engrs; Am Inst Aeronaut & Astronaut; Am Defense Preparedness Asn. *Res:* Identification of dynamic systems and component parameter identification. *Mailing Add:* Honeywell Inc WA34-4e02 6500 Harbour Heights Pkwy Everett WA 98204

YORIO, THOMAS, b New York, NY, Aug 27, 48; m 70; c 2. MEMBRANE PHYSIOLOGY, RENAL PHARMACOLOGY. *Educ:* Herbert H Lehman Col, BA, 71; City Univ NY, PhD(pharmacol), 75. *Prof Exp:* Res asst, Dept Ophthal, Mt Sinai Sch Med, NY, 74-75, fel, 75-77; asst prof, 77-82, ASSOC PROF, DEPT PHARMACOL, TEX COL OSTEOP MED, 82- *Concurrent Pos:* Prin investr, Nat Kidney Found, 77-78, Am Heart Asn, 78-81, Nat Inst Arthritis, Metab & Digestive Dis, NIH, 79-82, minority hypertension res develop grant, Nat Heart, Lung & Blood Inst, 80-90, 87-92 & Tex Heart Asn, 88-90. *Mem:* Sigma Xi; NY Acad Sci; Asn Res Vision & Ophthal; Am Soc Pharmacol & Exp Therapeut; Europ Membrane Club; Am Heart Asn, Kidney Cardiovasc Dis; Soc Exp Biol & Med. *Res:* Hormonal regulation of epithelial transport of ions and water; role of lipid metabolism and prostaglandins in the response to aldosterone and antidiuretic hormone. *Mailing Add:* Dept Pharmacol Camp Bowie Montgomery Tex Col Osteop Med 3500 Camp Bowie Blvd Ft Worth TX 76107

YORK, ALAN CLARENCE, b Asheboro, NC, Jan 24, 52; m 73; c 2. WEED SCIENCE, COTTON PRODUCTION. *Educ:* NC State Univ, BS, 74, MS, 76, PhD(agron), 79. *Prof Exp:* Cotton specialist, 79-83, WEED SPECIALIST, NC STATE UNIV, 83- *Concurrent Pos:* Vpres, Precision Agri-Res Serv, Inc, 90- *Mem:* Weed Sci Soc Am; Am Peanut Res & Educ Soc;

Coun Agr Sci Technol. *Res:* Development of weed management systems for agronomic crops, emphases on product performance, application technology, pesticide interactions, rotational crop responses, and integration of weed management programs into total production systems; extension and PGR evaluations. *Mailing Add:* Crop Sci Dept NC State Univ Box 7620 Raleigh NC 27695-7620

YORK, CARL MONROE, JR, b Macon, Ga, July 2, 25; m 52; c 3. PHYSICS. *Educ:* Univ Calif, Berkeley, AB, 46, MA, 50, PhD(physics), 51. *Prof Exp:* Fulbright fel physics, Univ Manchester, 51-52; res fel, Calif Inst Technol, 52-54; asst prof, Univ Chicago, 54-59; Ford Found & Guggenheim Found fel, Europ Orgn Nuclear Res, Geneva, 59-60; from assoc prof to prof, Univ Calif, Los Angeles, 60-69, asst chancellor res, 65-69, assoc dean grad div, 63-65; tech asst basic sci, Off Sci & Technol, Exec Off of the President, 69-72; vchancellor acad affairs, Univ Denver, 72-74; consult, 74-77; staff scientist, Lawrence Berkeley Lab, 77-81; consult, 81-84; dir, Syst Develop Found, 84-88; CONSULT, 88- *Concurrent Pos:* Consult, Argonne Nat Lab, 57-61, TRW Systs, Calif, 61-, Film Assocs, Calif, 62-, Lawrence Radiation Lab, 66, NSF, 72-76, Fedn Rocky Mountain States, 74-75, Colo Energy Res Inst, 75-76, Calif Energy Resource, Conserv & Develop Comn, 75-76, Pac Gas & Elec Co, 77-84 & Lawrence Livermore Nat Lab, 81-83. *Mem:* AAAS; Am Phys Soc. *Res:* Cosmic rays; elementary particles; positive sigma hyperon; high energy accelerator design; pion-nucleon scattering and production; muon decay and interactions; photo-production of pions; regional and state energy policies and plans; federal budgets for basic science and national science policy; validation of energy data bases and models; basic research in information science. *Mailing Add:* 26 Schooner Hill Oakland CA 94618

YORK, CHARLES JAMES, b Calif, Sept 28, 19; m 44; c 4. VIROLOGY, BACTERIOLOGY. *Educ:* Univ Calif, AB, 43; Ohio State Univ, DVM, 48; Cornell Univ, PhD(virol, bact), 50. *Prof Exp:* Asst bact, Univ Calif, 41-43, bacteriologist, Vet Sci Dept, 43-44; sr bacteriologist, Med Res Dept, Ohio State Univ, 44-46, bacteriologist, Vet Col, 46; res assoc, Vet Virus Inst, Cornell Univ, 48-52; dir virus res lab, Pitman-Moore Co, Ind, 52-63; prof vet sci & head dept vet res lab, Mont State Univ, 63-65; dir inst comp biol, Zool Soc San Diego, 65-70; ASSOC PROF COMP PATH, MED SCH, UNIV CALIF, SAN DIEGO, 67- *Concurrent Pos:* Mem WHO. *Mem:* Soc Exp Biol & Med; Am Vet Med Asn; Tissue Cult Asn; US Animal Health Asn; Am Pub Health Asn. *Res:* Virus diseases of man and animals; comparative medicine for animal research models. *Mailing Add:* 5451 Yerba Anita Dr San Diego CA 92115

YORK, DAVID ANTHONY, b Birmingham, Eng, Mar 29, 45; m 67; c 2. ANIMAL PHYSIOLOGY, NUTRITION. *Educ:* Southampton Univ, UK, BSc, 66, PhD(physiol), 69. *Prof Exp:* Res assoc obesity, New Eng Med Ctr Hosp, 69-70; res assoc, Sch Med, Univ Calif, Los Angeles, 70-71; lectr nutrit, Dept Nutrit, Univ Southampton, 71-82, sr lectr, Dept Human Nutrit, 82-88, reader, fac med, 88-89; PROF & CHIEF, PHYSIOL & OBESITY RES, PENNINGTON BIOMED RES CTR, LA STATE UNIV, 90- *Concurrent Pos:* Mem, UK Nat Comt Nutrit & Food Sci, 87-89; prof, Dept Physiol, Med Ctr, La State Univ, 90- *Mem:* Nutrit Soc UK; Am Inst Nutrit; Biochem Soc UK; British Asn Study Obesity; Am Asn Study Obesity. *Res:* Control of food intake and energy expenditure; metabolic and endorcrine basis for development of obesity in experimental models; role of the autonomic nervous system in control of energy balance. *Mailing Add:* Pennington Biomed Ctr La State Univ 6400 Perkins Rd Baton Rouge LA 70808

YORK, DEREK H, b Yorkshire, Eng, Aug 12, 36; m 61; c 1. GEOPHYSICS. *Educ:* Oxford Univ, BA, 57, DPhil(physics), 60. *Prof Exp:* Lectr, 60-62, from asst prof to assoc prof, 62-74, PROF PHYSICS, UNIV TORONTO, 74-; AT LAB GEOL & GEOCHEM, UNIV NICE, FRANCE. *Concurrent Pos:* Chmn subcomt isotope geophys & mem comt geod & geophys, Nat Res Coun Can, 67. *Mem:* Am Geophys Union; Can Asn Physicists. *Res:* Isotopic geophysics; temporal evolution of continents; reversals of earth's magnetic field. *Mailing Add:* Physics Dept Univ Toronto 215 Huron St Toronto ON M5S 1A1 Can

YORK, DONALD GILBERT, b Shelbyville, Ill, Oct 28, 44; m 66; c 4. INTESTELLAR MATTER, OBSERVATIONAL COSMOLOGY. *Educ:* Mass Inst Technol, BA, 66; Univ Chicago, PhD(astrophysics), 70. *Prof Exp:* From res asst to res assoc, 70-72, res staff astrophysics, Princeton Univ, 72-82; AT ASTRON & ASTROPHYS CTR, UNIV CHICAGO. *Honors & Awards:* Distinguished Serv Award, NASA, 75. *Mem:* Int Astron Union; Am Astron Soc. *Res:* Determination physical properties of interstellar gas and dust, using ultraviolet and visual spectroscopic techniques, in our own galaxy as well as distant galactic systems. *Mailing Add:* Astron & Astrophys Ctr Univ Chicago 5640 S Ellis Ave Chicago IL 60637

YORK, DONALD HAROLD, b Moose Jaw, Sask, Jan 30, 44; m 66; c 3. NEUROPHYSIOLOGY, NEUROSCIENCE. *Educ:* Univ BC, BSc, 65, MSc, 66; Monash Univ, Australia, PhD(neurophysiol), 69. *Prof Exp:* Asst prof physiol, Queen's Univ, Ont, 68-75; assoc prof, 75-81, PROF PHYSIOL, SCH MED, UNIV MO-COLUMBIA, 82- *Concurrent Pos:* Med Res Coun Can grant, Queen's Univ, Ont, 69-72; scholar, Med Res Coun Can, 70-75; vis fel, Australian Nat Univ, 83-84, Adv Bd Nat Hydrocephalus Found, 87-; exec coun, Am Soc Neuromonitoring, 88- *Mem:* Sigma Xi; Can Physiol Soc; Am Physiol Soc; Soc Neurosci; Pharmacol Soc Can; Inst Elect & Electronics Engrs. *Res:* Motor control; evoked potentials and movement; synaptic transmission in central nervous system. *Mailing Add:* Dept Physiol Univ Mo Sch Med Columbia MO 65212

YORK, GEORGE KENNETH, II, b Tucson, Ariz, July 1, 25; m 47; c 5. MICROBIOLOGY. *Educ:* Stanford Univ, AB, 50; Univ Calif, PhD(microbiol), 60. *Prof Exp:* Asst bacteriologist, Nat Canners Asn, 51-53; actg asst prof, 58-60, asst prof food sci & technol, 60-66, EXTEN MICROBIOLOGIST, UNIV CALIF, DAVIS, 66- *Mem:* NY Acad Sci; Am Soc Microbiol; Inst Food Technol. *Res:* Food microbiology; food-borne infections and intoxications; thermomicrobiology; modes of inhibition of microbes by chemicals; treatment and disposal of waste. *Mailing Add:* Dept Food Sci & Technol Univ Calif Davis CA 95616

YORK, GEORGE WILLIAM, b St Louis, Mo, Sept 26, 45; m 68; c 2. LASERS. *Educ:* St Louis Univ, BS, 67; Univ Mo, Rolla, MS, 69, PhD(physics), 71. *Prof Exp:* Res assoc physics, Joint Inst Lab Astrophys, 72-74; MEM STAFF PHYSICS, LOS ALAMOS SCI LAB, 74- *Mem:* Am Phys Soc; Laser Inst Am. *Res:* Development of high power gas lasers utilizing preionized electrical discharges in metal vapor system. *Mailing Add:* Los Alamos Nat Lab CLS5 Mail Stop E543 Los Alamos NM 87545

YORK, HERBERT FRANK, b Rochester, NY, Nov 24, 21; m 47; c 3. PHYSICS, SCIENCE POLICY. *Educ:* Univ Rochester, AB, 42, MS, 43; Univ Calif, PhD, 49. *Hon Degrees:* DSc, Case Western Reserve Univ, 60; LLD, Univ San Diego, 64; DrHumL, Claremont Grad Sch, 74. *Prof Exp:* Asst physics, Univ Rochester, 42-43; physicist, Radiation Lab, Univ Calif, 43-54, assoc dir, 54-58, dir, Lawrence Livermore Lab, 52-58, asst prof physics, 51-54; dir, Advan Res Projs Div, Inst Defense Anal & chief scientist, Advan Res Projs Agency, US Dept Defense, 58, dir defense res & eng, Off Secy Defense, 58-61; chancellor, 61-64 & 70-72, grad dean, 69-70, PROF PHYSICS, UNIV CALIF, SAN DIEGO, 65-; DIR, INST GLOBAL CONFLICT & COOP, 83- *Concurrent Pos:* Mem,Sci Adv Bd, US Air Force, 53-57, ballistic missile adv comt, Secy Defense, 55-58 & Sci Adv Panel, US Army, 56-58; mem, President's Sci Adv Comt, 57-58 & 64-67, vchmn, 65-67; mem gen adv comt, US Arms Control & Disarmament Agency, 61-69; Guggenheim fel, 72-73; mem continuing comt, Conf Sci & World Affairs, 73-76; sr consult, Off Secy Defense, 77-81; mem, Defense Sci Bd, 78-81, US Ambassador, Comp Test Ban Talks, Geneva, 79-81. *Honors & Awards:* E O Lawrence Award, 84. *Mem:* Am Phys Soc; Am Acad Arts & Sci; Am Inst Aeronaut & Astronaut; Inst Elec & Electronics Engrs; Int Acad Astronaut. *Res:* Science and public affairs; disarmament problems. *Mailing Add:* 6110 Camino de la Costa La Jolla CA 92037

YORK, J(ESSE) LOUIS, b Plains, Tex, May 1, 18; m 45, 75; c 2. AIR POLLUTION CONTROL, ENVIRONMENTAL PERMITS. *Educ:* Univ NMex, BSE, 38; Univ Mich, MS, 40, PhD(chem eng), 50. *Prof Exp:* From instr to prof chem & metall eng, Univ Mich, 42-70, proj dir, Res Inst, 42-70; environ scientist, Stearns-Roger Corp, 70-72, chief environ scientist, 72-83; CONSULT, 83- *Concurrent Pos:* Mem bd, Colo Sch Mines Found, 73- *Mem:* Am Chem Soc; Am Soc Mech Engrs; Air Pollution Control Asn; Am Inst Chem Engrs; Nat Soc Prof Engrs. *Res:* Suspensions of fine particles; sprays; nozzles; combustion; air and water pollution control; entrainment; desalination; filtration; environmental studies and analyses. *Mailing Add:* 3557 S Ivanhoe St Denver CO 80237

YORK, JAMES LESTER, b Peoria, Ill, Nov 12, 42; m 70; c 2. GERONTOLOGY, EXPERIMENTAL PSYCHOLOGY. *Educ:* Bradley Univ, AB, 65; Univ Ill, PhD(pharmacol), 72. *Prof Exp:* Postdoctoral pharmacol, State Univ NY-Buffalo, 72-74; RES SCIENTIST, NY STATE RES INST ALCOHOLISM, BUFFALO, 74- *Concurrent Pos:* Prin investr, NIH res grants, 76-78 & 87-93; referee-prof journals, Psychopharmacol, 78-90, Pharmacol, Biochem, Behav, 79-90, Alcohol, 84-90, Physiol & Behav, 85-89, Alcoholism, 89-90; res assoc prof, Dept Psychol, State Univ NY-Buffalo, 81- *Mem:* Fedn Am Soc Exp Biol; Res Soc Alcoholism; Int Soc Biomed Res Alcoholism; Soc Stimulus Properties Drugs. *Res:* Age-related changes in psychomotor performance and physiological function; tolerance to alcohol and other drugs; after-effects of alcohol intoxication; chronic alcohol toxicity on motor/muscle systems. *Mailing Add:* 783 Chestnuthill Rd East Aurora NY 14052

YORK, JAMES WESLEY, JR, b Raleigh, NC, July 3, 39; m 61; c 2. THEORETICAL PHYSICS, GRAVITATION. *Educ:* NC State Univ, BS, 62, PhD(physics), 66. *Prof Exp:* Asst prof physics, NC State Univ, Raleigh, 65-68; res assoc physics, Princeton Univ, 68-69, lectr, 69-70, asst prof, 70-73; from assoc prof to prof physics, 73-88, dir, inst field physics, 84-90, AGNEW H BAHNSON JR PROF PHYSICS, 89- *Concurrent Pos:* Vis asst prof, Univ Maryland, 72; vis prof, Univ Paris, 76 & Univ Tex, 79, 87; vis scientist, Harvard-Smithsonian Ctr Astrophys, 77; mem, Int Soc Gen Relativity & Gravitation, 80- *Honors & Awards:* Third Prize, Gravity Res Found, Mass, 75; Alfred Schild Mem Lectr, Univ Tex, 79. *Mem:* Fel Am Phys Soc; AAAS. *Res:* Gravitation and relativity; mathematical, astrophysical and quantum theoretic aspects. *Mailing Add:* Dept Physics & Astron Univ NC Chapel Hill NC 27599-3255

YORK, JOHN LYNDAL, b Morton, Tex, Aug 14, 36; m 58; c 2. BIOCHEMISTRY, PHYSICAL ORGANIC CHEMISTRY. *Educ:* Harding Col, BS, 58; Johns Hopkins Univ, PhD(biochem), 62. *Prof Exp:* NIH trainee, 62-64; biochemist, Stanford Res Inst, 64-65; asst prof biochem, Med Units, Univ Tenn, Memphis, 65-68; assoc prof, 68-77, PROF BIOCHEM, SCH MED, UNIV ARK, LITTLE ROCK, 77- *Concurrent Pos:* Vis prof, Karolinska Inst, Stockholm, 74-75; fel, Swedish Med Res Coun, 75. *Mem:* Am Soc Biol Chem; Am Chem Soc; Sigma Xi; AAAS. *Res:* Nature of the active site and mechanisms of action of non-heme iron proteins; Mossbauer effect; mechanisms of oxidation and oxygenation; structure and function of fibrinogen; mechanism of detoxicification by glutathione transferases. *Mailing Add:* Dept Biochem No 516 Univ Ark Med Col Little Rock AR 72205

YORK, JOHN OWEN, b Parkin, Ark, July 11, 23; m 49; c 2. PLANT BREEDING, PLANT GENETICS. *Educ:* Miss State Univ, BS, 48, MS, 50; Tex A&M Univ, PhD(plant breeding), 62. *Prof Exp:* From instr to assoc prof, 52-68, PROF AGRON, UNIV ARK, FAYETTEVILLE, 68- *Mem:* Am Soc Agron; Am Genetic Asn; Sigma Xi. *Res:* Hybrid corn and grain sorghum breeding and production. *Mailing Add:* Dept Agron PS 115 Univ Ark Fayetteville AR 72703

YORK, OWEN, JR, b Evansville, Ind, Oct 18, 27; m 48; c 3. ORGANIC CHEMISTRY. *Educ:* Evansville Col, BA, 48; Univ Ill, MA, 50, PhD(org chem), 52. *Prof Exp:* Asst, Univ Ill, 48-52; res chemist, Hercules Powder Co, 52-56; head dept chem, Ill Wesleyan Univ, 56-60; res supvr, W R Grace & Co, 60-61; assoc prof, 61-64, chmn dept, 64-73, reader & table leader, 64-72, chief reader, Advan Placement Chem, 72-76, assoc dir, Reading-Advan Placement Prog Educ Testing Serv, 79-84, PROF CHEM, KENYON COL, 64- *Concurrent Pos:* NSF fel, Stanford Univ, 68-69. *Mem:* AAAS; Am Chem Soc; NY Acad Sci. *Res:* Grignard reaction; oxidation; electrophilic substitution; isomerization of aromatic acids; synthetic photochemistry; catalysis; biomass; macromolecules. *Mailing Add:* Dept Chem Kenyon Col Gambier OH 43022

YORK, RAYMOND A, b Baldwin, Kans, May 15, 17; c 3. INTERNATIONAL TECHNOLOGY TRANSFER. *Educ:* Univ Kans, BSEE, 41. *Prof Exp:* Mgr, Gen Elec, Syracuse, NY, 50-69, gen mgr, International Lic Div, New York City, 69-82; consult, 82-88; DEVELOPER REAL ESTATE SUBDIV, 88- *Concurrent Pos:* Indust Standard Domestic & Intl. *Mem:* Fel Inst Elec & Electronics Engrs. *Res:* Working with equipment for the measurement of the amount of nuclear exposure and for metal detection. *Mailing Add:* 1214 James St Syracuse NY 13203

YORK, SHELDON STAFFORD, b New Haven, Conn, Oct 29, 43; m 68; c 2. BIOPHYSICAL CHEMISTRY. *Educ:* Bates Col, BS, 65; Stanford Univ, PhD(biochem), 71. *Prof Exp:* Am Cancer Soc res fel chem, Calif Inst Technol, 70-72; ASST PROF CHEM, UNIV DENVER, 72- *Mem:* Am Chem Soc. *Res:* Protein-DNA recognition processes, specifically the conformational changes within the lac repressor protein which affect its ability to bind to the lac operator. *Mailing Add:* Dept Chem Univ Denver Denver CO 80208-0001

YORKE, JAMES ALAN, b Peking, China, Aug 3, 41; US citizen; m 63; c 8. APPLIED MATHEMATICS, BIOMATHEMATICS. *Educ:* Columbia Univ, AB, 63; Univ Md, College Park, PhD(math), 66. *Prof Exp:* From res assoc to res assoc prof, 66-72, RES PROF MATH, INST PHYS SCI & TECHNOL, UNIV MD, COLLEGE PARK, 72-, DIR, 88- *Concurrent Pos:* Guggenheim fel, 80-81. *Mem:* AAAS; Am Math Soc; Math Asn Am; Soc Indust & Appl Math. *Res:* Qualitative ordinary differential equations; applications in epidemiology. *Mailing Add:* Inst Phys Sci & Technol Univ Md College Park MD 20742-2431

YORKS, TERENCE PRESTON, b Syracuse, NY, Apr 29, 47; m 87. SYSTEM MODELLING. *Educ:* Colo State Univ, BS, 69; NMex Highlands Univ, MS, 71; Tex A&M Univ, PhD(food sci), 76. *Prof Exp:* Res assoc animal sci, Colo State Univ, 76-78, res assoc modelling, natural resource ecol lab, 80-82; opers res analyst, Forest Serv, USDA, 79; consult modelling, Poudre Valley Res Inst, 82-83; sr tech info specialist chem, Environ Protection Agency, 83-84; staff officer admin, Nat Acad Sci, 84-85; systs scientist & asst prof range mgt, Univ Wyo, 85-87; INDEPENDENT CONSULT, 87- *Concurrent Pos:* Postdoctorate animal sci, Colo State Univ, 76-78; vis asst prof, Range Sci Dept, Utah State Univ, 82-83, postdoctorate ecol, 89-90. *Mem:* Soc Range Mgt; Wildlife Soc. *Res:* Elucidation and monitoring of ecosystem-level interactions that can affect long-term human food and fiber supplies; computer-based natural resource, energy-flow and risk analyses; microbiological and chemical toxicology. *Mailing Add:* 45 E 500 N Logan UT 84321

YORQUE, RALF RICHARD, biology, resource management, for more information see previous edition

YOS, DAVID ALBERT, b Trenton, NJ, Apr 24, 23; m 46; c 2. BOTANY, INFORMATION SCIENCE. *Educ:* NY Univ, AB, 48; Univ Mo, MA, 52, MLS, 81; Univ Iowa, PhD(bot, plant anat), 60. *Prof Exp:* Instr biol, Burlington Col, 52-59; asst prof bot, Univ Wis-Green Bay, 60-62; teacher, High Sch, NJ, 63; from asst prof to prof biol, Eastern NMex Univ, 63-81, adj prof & sci librn, 81-82; SCI LIBRN, ILL STATE UNIV, NORMAL, 82- *Mem:* Sigma Xi; AAAS. *Res:* Fluorescence microscopy of plant tissues; development of periderm; microtechnique; history of microscopy and photomicrography. *Mailing Add:* Milner Library Ill State Univ Normal IL 61761

YOS, JERROLD MOORE, b Clinton, Iowa, Jan 1, 30; m 60; c 3. RE-ENTRY PHYSICS. *Educ:* Univ Nebr, AB, 52, MS, 54, PhD(physics), 56. *Prof Exp:* SR SCIENTIST, SYSTS DIV, AVCO CORP, 57- *Mem:* Am Phys Soc. *Res:* Electromagnetics; gas physics; kinetic theory of gases; electrical discharges. *Mailing Add:* 1001 Main St 34 Woburn MA 01801

YOSHIDA, AKIRA, b Okayama, Japan, May 10, 24; m 54. BIOCHEMISTRY, GENETICS. *Educ:* Univ Tokyo, MS, 47, DSc, 54. *Prof Exp:* From instr to asst prof chem, Univ Tokyo, 51-54, assoc prof chem & biochem, 54-60; res assoc biochem, Univ Pa, 61-63; res chemist, NIH, 63-65; res prof med genetics, Univ Wash, 64-72; DIR, DEPT BIOCHEM GENETICS, CITY HOPE MED CTR, 72- *Concurrent Pos:* Rockefeller Found Int scholar, 55-56. *Honors & Awards:* Japanese Human Genetics Am Soc Award, 81. *Mem:* AAAS; Am Soc Biol Chem; Soc Hemat; Soc Human Genetics. *Res:* Study on the changes of protein structure and properties of the enzymes due to mutation; regulatory mechanism of gene action. *Mailing Add:* Dept Biochem Genetics City Hope Med Ctr 1500 Duarte Rd Duarte CA 91010

YOSHIDA, FUMITAKE, b Saitama, Japan, Mar 20, 13; m 41; c 2. BIOENGINEERING & BIOMEDICAL ENGINEERING. *Educ:* Kyoto Univ, BEng, 37, DEng, 51. *Prof Exp:* Engr, Chem Eng Plant Design, Hitachi Ltd, 37-45; asst prof, 46-51, prof, 51-76, EMER PROF CHEM ENG, KYOTO UNIV, 76- *Concurrent Pos:* Lectr, Kyoto Univ, 40-45; res assoc, Univ Wis-Madison, 59; vis prof, Univ Calif, Berkeley, 63, Univ Pa, 70; assoc ed, Chem Eng J, 72-; guest prof, Univ Dortmund, Ger, 87; ed, Chem Eng Sci, 87- *Mem:* Nat Acad Eng; Soc Chem Engrs Japan (vpres, 67-69); Am Inst Chem Engrs; Am Chem Soc; Japanese Soc Artificial Organs; Am Soc Artificial Interal Organs. *Res:* Mass transfer operations in chemical engineering, such as gas absorption, distillation; applications of chemical engineering in medicine and bio process industries. *Mailing Add:* Kyoto Univ 2 Matsugasaki-Yobikaeshicho Sakyoku Kyoto 600 Japan

YOSHIDA, TAKESHI, b Fukuoka, Japan, July 24, 38; m 64; c 2. IMMUNOPATHOLOGY. *Educ:* Univ Tokyo, MD, 63, Dr Med Sci, 70. *Prof Exp:* Res mem immunol, Dept Tuberc, NIH, Tokyo, 64-71; asst prof, Dept Path, State Univ NY Buffalo, 71-74; from asst prof to assoc prof path, Univ Conn Health Ctr, 74-82, prof, 82-84; CLIN PROF, TOKYTO INST IMMUNOPHARMACOL, 84-; DIR, 87- *Concurrent Pos:* Res assoc, Dept Path, NY Univ Med Ctr, 67-68; vis assoc, Lab Immunol, Nat Inst Allergy & Infectious Dis, Bethesda, 68-69, vis scientist, 71; Buswell fel, State Univ NY Buffalo, Buffalo, 71-74, asst dir, Ctr Immunol, 73-74; Nat Inst Allergy & Infectious Dis res career develop award, 75-80. *Mem:* Am Asn Immunologists; Am Asn Pathologists; Am Asn Univ Pathologists; Reticuloendothelial Soc; NY Acad Sci. *Res:* Mechanisms of cell-mediated immunity, biological and physicochemical characterizations of effector molecules produced in vitro by stimulated lymphocytes, and in vivo activities of lymphokines. *Mailing Add:* Tokyo Inst Immunopharmacol 3-41-8 Takada Toshimaku Tokyo 171 Japan

YOSHIKAMI, DOJU, b Heart Mountain, Wyo. NEUROBIOLOGY, NEUROSCIENCE. *Educ:* Reed Col, BA, 65; Cornell Univ, PhD(biochem & molecular biol), 70. *Prof Exp:* NIH fel appl physics, Cornell Univ, 70-71; NSF fel neurobiol, Harvard Univ Sch Med, 71-73, instr, 73-76, prin res assoc, 76-78; from asst prof to assoc prof, 78-86, PROF BIOL, UNIV UTAH, 86- *Res:* Cellular and molecular neurobiology; structure and function of synapses. *Mailing Add:* Dept Biol Univ Utah Salt Lake City UT 84112

YOSHIKAWA, HERBERT HIROSHI, b South Dos Palos, Calif, May 13, 29; m 60. NUCLEAR ENGINEERING, SOLID STATE PHYSICS. *Educ:* Univ Chicago, PhB, 48, MS, 51; Univ Pa, PhD(physics), 58. *Prof Exp:* Sr engr physics, Hanford Labs, Gen Elec Co, 58-65; mgr graphite res & develop radiation effects, Pac Northwest Lab, Battelle-Northwest Lab, 65-70; mgr mat eng, 70-78, mgr technol, Hanford Eng Develop Lab, 78-87, MGR STRATEGIC PLANNING, WESTINGHOUSE HANFORD CO, 88- *Mem:* Am Nuclear Soc; Am Phys Soc; AAAS; Sigma Xi. *Res:* Technology of fast breeder, light water and fusion reactors; materials development, irradiation testing and reactor environment characterization. *Mailing Add:* 2712 W Klamath Kennewick WA 99336

YOSHIMOTO, CARL MASARU, b Honolulu, Hawaii, Apr 27, 22; m 57; c 2. ENTOMOLOGY. *Educ:* Iowa Wesleyan Col, BA, 50; Kans State Univ, MS, 52; Cornell Univ, PhD(entom), 55. *Prof Exp:* Entomologist, Entom Res Div, USDA, 55-57 & BP Bishop Mus, 58-69; sr res scientist, Dept Agr, Can Forestry Serv, 69-87; RETIRED. *Concurrent Pos:* Affil fac, Grad Sch, Univ Hawaii, 64-69; NSF grant, Brit Mus, London, Eng, 67-68. *Mem:* Entom Soc Am; Entom Soc Can; Sigma Xi. *Res:* Taxonomy of Hymenoptera; insect behavior and dispersal; zoogeography; biological control. *Mailing Add:* Six Charing Rd Nepean ON K2G 0Z5 Can

YOSHINAGA, KOJI, b Yokohama, Japan, Mar 20, 32; m 61. ENDOCRINOLOGY. *Educ:* Univ Tokyo, BSc, 55, MSc, 57, PhD(agr), 60. *Prof Exp:* Trainee physiol reproduction, Worcester Found Exp Biol, 61-64; vis scientist, Agr Res Coun Unit Reproduction Physiol & Biochem, Cambridge, 64-66; staff scientist, Worcester Found Exp Biol, 66-69; res assoc anat, Harvard Med Sch, 69, asst prof, 69-72, assoc prof, 72-79; HEALTH SCIENTIST ADMINR, REPRODUCTION SCI BR, POP RES, NAT INST CHILD HEALTH & HUMAN DEVELOP, NIH, 78- *Concurrent Pos:* Pop Coun fel, Worcester Found Exp Biol, 62-63 & Agr Res Coun Unit Reproduction Physiol & Biochem, Cambridge, 64-65, Lalor Found fel, 65-66; adj prof, Georgetown Univ Sch Med, 84- *Mem:* Am Asn Anat; Endocrine Soc; Soc Study Reproduction; Am Physiol Soc; Soc Study Fertil. *Res:* Endocrinology and physiology of reproduction in female animals, especially the mechanisms involved in ovo-implantation, ovarian function and relationship between the egg development and hormone action. *Mailing Add:* Reproduction Sci Br NIH EPN Rm 603 Bethesda MD 20892

YOSHINO, KOUICHI, b Matsuyama, Japan, Jan 1, 31. CHEMICAL PHYSICS, MOLECULAR SPECTROSCOPY. *Educ:* Tokyo Univ Educ, BS, 53, MS, 55, PhD(physics), 72. *Prof Exp:* Physicist, Indust Res Inst Kanagawa, Japan, 54-58 & Govt Indust Res Inst, Tokyo, 58-65; res physicist, Air Force Cambridge Res Lab, 65-76; RES ASSOC, HARVARD-SMITHSONIAN CTR ASTROPHYS, 76- *Concurrent Pos:* Res physicist, Wentworth Inst, 61-63. *Mem:* Phys Soc Japan; Spectros Soc Japan; Optical Soc Am. *Res:* Determination of properties of ground and excited electronic states of molecules or atoms by high resolution vacuum ultraviolet spectroscopy such as nitrogen, oxygen, carbon monoxide and rare gases. *Mailing Add:* Harvard Col Observ 60 Garden St Cambridge MA 02138

YOSHINO, TIMOTHY PHILLIP, b Turlock, Calif, Apr 5, 48; m 75; c 2. PARASITOLOGY, IMMUNOBIOLOGY. *Educ:* Univ Calif, Santa Barbara, BA, 70, MA, 71, PhD(biol), 75. *Prof Exp:* Res assoc parasitol, Univ Calif, Santa Barbara, 71-73; res assoc invert immunol, Lehigh Univ, 75-77; from asst prof to assoc prof zool, Univ Okla, 78-88; ASSOC PROF PATHOBIOL, UNIV WIS-MADISON, 88- *Concurrent Pos:* USPHS res fel, Nat Inst Allergy & Infectious Dis, 75. *Honors & Awards:* Res Career Develop Award, Nat Inst Allergy & Infectious Dis, 84. *Mem:* Am Soc Parasitologists; Soc Invert Path; Am Soc Zoologists; AAAS; Am Soc Trop Med Hyg. *Res:* Humoral and cellular mechanisms of internal defense in bivalve and gastropod molluscs; immunobiology of schistosome-mollusc interactions; parasitic castration in molluscs. *Mailing Add:* Dept Pathobiol Sci Univ Wis-Madison 2015 Linden Dr W Madison WI 53706

YOSS, KENNETH M, b Hudson, Iowa, Jan 13, 26; m 55; c 3. ASTRONOMY. *Educ:* Univ Mich, BS, 48, MS, 50, PhD(astron), 53. *Prof Exp:* Asst prof astron & physics, Wilson Col, 52-53; from asst prof to assoc prof, La State Univ, 53-59; assoc prof astron, Mt Holyoke Col, 59-64; PROF ASTRON, UNIV ILL, URBANA, 64- *Mem:* Am Astron Soc; Int Astron Union. *Res:* Spectrophotometry of objective prism and slit spectra; spectral and luminosity classification. *Mailing Add:* Univ Ill 1002 W Green St Rm 103 Urbana IL 61801

YOSS, ROBERT EUGENE, b Spooner, Wis, Nov 28, 24; m 47; c 3. NEUROANATOMY, NEUROLOGY. *Educ:* Univ Tenn, MD, 48; Univ Mich, MS & PhD(neuroanat), 52. *Prof Exp:* From instr to asst prof anat, Med Sch, Univ Mich, 49-54; from asst prof to assoc prof, 57-70, PROF NEUROL, MAYO GRAD SCH MED, UNIV MINN, 70-, CONSULT, MAYO CLIN, 57- *Concurrent Pos:* Fel neruol, Mayo Grad Sch Med, Univ Minn, 55-57. *Mem:* Am Asn Anat. *Res:* Anatomy of spinal cord; narcolepsy. *Mailing Add:* 920 SW Seventh Ave Rochester MN 55902

YOST, FRANCIS LORRAINE, b Punxsutawney, Pa, Dec 18, 08; m 54. THEORETICAL PHYSICS. *Educ:* Univ Ky, BS, 29, MS, 31; Univ Wis, PhD(physics), 36. *Prof Exp:* Asst physics, Univ Wis, 31-36 & Purdue Univ, 36-37; physicist, US Rubber Co, Mich, 37-42; sr tech aide, Div 17, Nat Defense Res Comt, 43-44; Off Sci Res & Develop, Eng & France, 44-45, Washington, 45-46; chief math anal subdiv, US Naval Ord Lab, 46-47 & weapons anal div, 50-54; assoc prof physics, Ill Inst Technol, 47-51; prof physics, Univ Ky, 54-74, head dept, 54-65; RETIRED. *Concurrent Pos:* Vis prof, Univ Indonesia, 56-58. *Mem:* Am Phys Soc; Am Asn Physics Teachers; Sigma Xi. *Res:* Nuclear physics; physical properties of rubber; naval ordnance; guided missile lethalities. *Mailing Add:* 1320 Cooper Rd Lexington KY 40502

YOST, FREDERICK GORDON, b Norwalk, Conn, Aug 29, 40; m 63; c 2. METALLURGY. *Educ:* Polytech Inst Brooklyn, BSMetE, 66; Iowa State Univ, PhD(metall), 72. *Prof Exp:* MEM TECH STAFF, SANDIA LABS, 72- *Mem:* Am Soc Metals; Am Inst Mining, Metall & Petrol Engrs. *Res:* Kinetics of solid state phase transformations in metals and alloys, especially those in microelectronics. *Mailing Add:* Orgn 2134 Bldg 858 Rm 2256 Sandia Labs Albuquerque NM 87185

YOST, GAROLD STEVEN, b Denver, Colo, Sept 8, 49; m 78. DRUG METABOLISM, MEDICINAL CHEMISTRY. *Educ:* Bethel Col, BS, 71; Univ Hawaii, MS, 74; Colo State Univ, PhD(chem), 77. *Prof Exp:* Fel pharmaceut chem, Univ Calif, San Francisco, 77-78; instr chem, Towson State Univ, 78-81; ASST PROF, COL PHARM, WASH STATE UNIV, 81- *Concurrent Pos:* Vis lectr, Dept Pharmacol, Johns Hopkins Univ, 79-81. *Mem:* Am Chem Soc; Int Soc Study Xenobiotics; Sigma Xi. *Res:* Marine and terrestrial natural products including poisonous feedstocks; suicidal inhibitors of cytochrome P-450 and other enzymes; stereoselective drug metabolism involving cytochrome P-450 and glucuronyltransferase. *Mailing Add:* Dept Pharm Wash State Univ Pullman WA 99164

YOST, JOHN FRANKLIN, b Brodbecks, Pa, Mar 21, 19; m 43; c 2. ORGANIC CHEMISTRY. *Educ:* Western Md Col, BS, 43; Johns Hopkins Univ, AM, 48, PhD(org chem), 50. *Hon Degrees:* DS, Western Md Col, 64. *Prof Exp:* Jr instr, Johns Hopkins Univ, 46-50; res chemist, Synthetic Rubber & Plastics Lab, US Rubber Co, 43-44 & 46; agr chemist & group leader, Agr Chem Labs, 50-57, mgr tech dept, Phosphate & Nitrate Div, 57, dir plant indust develop, Agr Div, 58-62, dir prod develop & govt registrn, 62-66, DIR AGR RES & DEVELOP, INT DEPT, AM CYANAMID CO, 66- *Mem:* Am Chem Soc. *Res:* Synthetic organic chemistry; agricultural product analysis, formulation and registration; pesticides, fertilizers, feed additives and veterinary products; research administration. *Mailing Add:* Box 2049 RD No 2 Spring Grove PA 17362-1198

YOST, JOHN R(OBARTS), JR, b Phoenixville, Pa, July 10, 23; m 47; c 2. CHEMICAL ENGINEERING. *Educ:* Ursinus Col, BS, 44; Univ Pa, MS, 49. *Prof Exp:* Process develop engr, Res Labs, Sharples Corp, 49-51; process develop engr, Merck & Co, 51-54; sect head, Process Eng, E R Squibb & Sons Div, Olin Mathieson Chem Corp, 54-61, head chem pilot plant, 61-67; vpres mfg, Ott Chem Co, 67-70, exec vpres, 70-73; PRES, MUSKEGON CHEM CO, 74- *Mem:* Am Chem Soc; Am Inst Chem Engrs. *Res:* Pharmaceutical compounds; unit processes leading to synthetic organic compounds; distillation and solvent extraction. *Mailing Add:* Koch Chem Co 1725 Warner St Whitehall MI 49461-1829

YOST, RICHARD A, b Martins Ferry, Ohio, May 31, 53; m 79; c 2. MASS SPECTROMETRY, INSTRUMENTATION. *Educ:* Univ Ariz, BS, 74; Mich State Univ, PhD(anal chem), 79. *Prof Exp:* Fel, NSF, 75-79 & Am Chem Soc Anal Div, 77-78; from asst prof to assoc prof, 79-89, PROF CHEM, UNIV FLA, 89- *Concurrent Pos:* Vis scientist, LaTrobe Univ, Victoria, Australia, 77; consult, Lawrence Livermore Nat Lab, 80-84, Finnigan MAT, 80-, Bristol-Myers Squibb, 89- *Mem:* Am Chem Soc; Am Soc Mass Spectrometry. *Res:* Development of new analytical chemistry techniques using modern instrumentation and digital computers; application of new techniques in areas such as environmental, clinical and forensic chemistry. *Mailing Add:* Dept Chem Univ Fla Gainesville FL 32611

YOST, ROBERT STANLEY, b Pottsville, Pa, Jan 24, 21; m 43; c 2. ORGANIC CHEMISTRY. *Educ:* Pa State Univ, BS, 42; Duke Univ, PhD(org chem), 48. *Prof Exp:* Chemist, Hercules Powder Co, 42-43; lab instr chem, Duke Univ, 43-44; chemist, 44-46 & 47-59, head process res group, Redstone Res Labs, 59-68, CHEMIST, RES LABS, ROHM & HAAS CO, 68- *Mem:* Am Chem Soc. *Res:* Organic synthesis including synthetic resins and explosives; process research. *Mailing Add:* 4542 Holiday Heights Dr Oakwood GA 30566

YOST, WILLIAM A, b Dallas, Tex, Sept 21, 44; m 69; c 2. PSYCHOACOUSTICS, PSYCHOPHYSICS. *Educ:* Colo Col, BA, 66; Ind Univ, Bloomington, PhD(psychol), 70. *Prof Exp:* NSF fel, Univ Calif, San Diego, 70-71; asst prof speech psychol, Univ Fla, 71-74, assoc prof psychol, 74-77; assoc prof, 77-78, PROF HEARING SCI, PSYCHOL & OTOLARYNGOL, LOYOLA UNIV, CHICAGO, 79- *Concurrent Pos:* Mem & chmn, Comt Acoust Standards, 72-75; prog officer, Nat Sci Found, 82-84; adv panel, Nat Sci Found, 80-86; assoc ed, J Acoust Soc Am, 87- *Mem:* Fel Acoust Soc Am; Am Psychol Asn; Sigma Xi; Int Audiol Soc; AAAS; Asn Res Otolaryngol (pres, 89, secy-treas, 85-87); Am Psychol Soc. *Res:* Binaural hearing; pitch perception; speech perception; auditory sensitivity and discrimination; noise pollution. *Mailing Add:* Parmly Hearing Inst Loyola Univ Chicago 6525 N Sheridan Rd Chicago IL 60626

YOST, WILLIAM LASSITER, b Washington, DC, Mar 14, 23; m 51; c 4. ORGANIC CHEMISTRY. *Educ:* Univ Va, BS, 44, MS, 47, PhD(chem), 49. *Prof Exp:* AEC fel biol Sci, Calif Inst Technol, 49-50, NIH fel, 50-51; res chemist, US Naval Ord Test Sta, 51-52; sr chemist, Ciba Pharmaceut Co, 52-65, head appl math, 65-67, dir chem develop & appl math, 67-69; pres, Union Data Corp, 69-85; vpres, Action Biomed Comput, 78-85; RETIRED. *Mem:* AAAS; Am Chem Soc; Asn Comput Mach; NY Acad Sci. *Res:* Application of mathematics and computer techniques to the solution of problems arising in medical and pharmaceutical research. *Mailing Add:* Rte 1 Box 66 Esmont VA 22937-9708

YOTIS, WILLIAM WILLIAM, b Almyros, Greece, Jan 17, 30; US citizen; m 57; c 3. MICROBIOLOGY. *Educ:* Wayne State Univ, BS, 54, MS, 56; Northwestern Univ, PhD(microbiol), 60. *Prof Exp:* Asst microbiol, Wayne State Univ, 54-56; clin bacteriologist, Univ Hosp, Univ Mich, 56-57; from instr to assoc prof, 60-72, interim chmn, 64-65, PROF MICROBIOL, MED SCH, LOYOLA UNIV CHICAGO, 73- *Concurrent Pos:* Res grants, NIH, 60-89, Eli Lilly Res Labs, 65-66, Syntex Res Labs, 66-68, Upjohn Co, 66-67, Off Naval Res grant, 89-92; consult, Am Type Cult Collection, 66-; vis scientist, Argonne Nat Lab, 77-88. *Mem:* AAAS; Am Med Asn; Am Soc Microbiol; Am Asn Dent Res; NY Acad Sci. *Res:* Investigations on mechanisms of microbial pathogenicity; nonspecific host defense mechanisms; bactericidal activity of body fluids; staphylococcal host-parasite relationship; hormonal influence on infection; bacterial physiology; isotachophoresis and scanning isoelectric focusing of medically important proteins; cariostasis by fluoride; oral microbiology with emphasis on potential periodonto pathogens. *Mailing Add:* Loyola Univ Med Sch 2160 S First Ave Maywood IL 60153

YOU, KWAN-SA, b Seoul, Korea, Oct 26, 41; m 73; c 2. ENZYMOLOGY, CELL BIOLOGY. *Educ:* Seoul Univ, BS, 66; Brandeis Univ, PhD(biochem), 71. *Prof Exp:* Fel biochem, Scripps Clin & Res Found, 71-73; res biochemist, Univ Calif, San Diego, 73-77; asst prof metab, dept pediat, Duke Univ Med Sch, 77-83; VIS SCIENTIST, W ALTON JONES CELL SCI CTR, LAKE PLACID, NY, 84- *Concurrent Pos:* Grant reviewer, div physiol, cellular & molecular biol, NSF, 84-; manuscript reviewer, J Am Chem Soc, 86- *Mem:* Am Soc Biol Chemists; Am Chem Soc; AAAS; Am Asn Clin Chemists; Tissue Cult Asn; Am Soc Microbiol. *Res:* Structure, function and mechanism of action of enzyme; stereospecificity of enzyme-catalyzed reactions; mechanism of action of hormone and growth factors. *Mailing Add:* Five Westwood Townhouse 40 Valley Rd Lake Placid NY 12946-1042

YOUD, THOMAS LESLIE, b Spanish Fork, Utah, Apr 2, 38; m 62; c 5. SOIL MECHANICS, EARTHQUAKE ENGINEERING. *Educ:* Brigham Young Univ, BES, 64; Iowa State Univ, PhD(civil eng), 67. *Prof Exp:* Res & civil engr, US Geol Surv, Menlo Park, 67-84; PROF CIVIL ENG, BRIGHAM YOUNG UNIV, PROVO, UTAH, 84- *Mem:* Am Soc Civil Engrs; Int Soc Soil Mech & Found Engrs; Earthquake Eng Res Inst; Seismol Soc Am. *Res:* Liquefaction of soils caused by earthquakes. *Mailing Add:* 1132 E 1010 N Orem UT 84057

YOUDELIS, W(ILLIAM) V(INCENT), b Edmonton, Alta, Aug 1, 31; m 56; c 2. PHYSICAL METALLURGY. *Educ:* Univ Alta, BSc, 52; McGill Univ, MEng, 56, PhD(metall eng), 58. *Prof Exp:* Mine engr, Steep Rock Iron Mines, Ont, 52-54; from asst prof to assoc prof phys metall, Univ Alta, 58-65; assoc prof, 65-68, PROF PHYS METALL, UNIV WINDSOR, 68-, HEAD DEPT ENG MAT, 72-; PRES, YOUDELIS ASSOCS INC, 77- *Mem:* AAAS; Am Soc Metals; Am Inst Mining, Metall & Petrol Engrs; Can Inst Mining & Metall; Metals Soc. *Res:* Solidification; kinetics of phase transformations; dental alloys; electronic materials. *Mailing Add:* Dept Eng Mat Univ Windsor Windsor ON N9B 3P4 Can

YOUKER, JAMES EDWARD, b Cooperstown, NY, Nov 13, 28; div; c 2. RADIOLOGY. *Educ:* Colgate Univ, AB, 50; Univ Buffalo, MD, 54; Am Bd Radiol, dipl, 60. *Prof Exp:* Asst prof radiol, Med Col Va, 61-63; from asst prof to assoc prof, Univ Calif, San Francisco, 64-68; PROF RADIOL & CHMN DEPT, MILWAUKEE COUNTY GEN HOSP, WIS, 68- *Concurrent Pos:* NIH res fel, Allmanna Sjukhuset, Malmo, Sweden, 62 & 63; attend radiologist, Proj HOPE, Indonesia, 58; USPHS grant dir, Training Radiologist & Technician Teams in Mammography, & co-dir, Training Prog Cardiovasc Radiol, 65-68. *Mem:* Radiol Soc NAm; Am Col Radiol; Asn Univ Radiol; Int Soc Lymphology. *Res:* Pulmonary function changes with lymphographic contrast media; pathology of congenital heart disease; chylous ascites and the spectrum of the disease. *Mailing Add:* Dept Radiol Med Col Wis 8701 W Watertown Plank Rd Milwaukee WI 53226

YOUKER, JOHN, b Auburn, NY, Sept 7, 43; m 67; c 4. PHYSICAL CHEMISTRY, ANALYTICAL CHEMISTRY. *Educ:* Rensselaer Polytech Inst, BS, 65, PhD(phys chem), 69. *Prof Exp:* Chemist, Coated Abrasive & Tape Div, Norton Co, 68-69; from asst prof to assoc prof, 69-81, PROF CHEM, HUDSON VALLEY COMMUNITY COL, 81-, DEPT CHAIR, CHEM, 89- *Concurrent Pos:* Dep dir, Rensselaer County Off Disaster Preparedness, 81-82. *Mem:* Am Chem Soc; Sigma Xi. *Res:* Spectroscopy; analytical chemistry; environmental science. *Mailing Add:* 203 Winter St Extension Troy NY 12180

YOULA, DANTE C, b Brooklyn, NY, Oct 17, 25. ELECTRICAL ENGINEERING, MICROWAVE ENGINEERING. *Educ:* City Col New York, BEE, 47; NY Univ, MS, 50. *Prof Exp:* Engr, Jet Propulsion Labs, Calif Inst Technol, Pasadena, 51-54; PROF ELEC ENG & COMPUTER SCI, POLYTECH UNIV, 55- *Concurrent Pos:* Ed, Trans on Circuit Theory, Inst Elec & Electronic Engrs, 65-67. *Honors & Awards:* Baker Award, Inst Elec & Electronics Engrs, 64, 100th Yr Commemorative Medal for Contrib to Circuit Theory & Systs, 84, Field Award in Systs Sci & Eng, 88; Guillemin-Cauer Circuit Theory Award, 73; Tech Achievement Award, Circuit & Systs Soc, Inst Elec & Electronics Engrs, 91. *Mem:* Nat Acad Eng; fel Inst Elec & Electronics Engrs; Sigma Xi; NY Acad Sci. *Res:* Network theory and synthesis-linear and nonlinear; stability problem of time-varying structures;

statistical communication theory; noise in linear and nonlinear systems; radar detection estimation and coding; feedback system theory. *Mailing Add:* Dept Elec Eng & Comput Sci Polytech Univ Long Island Ctr Farmingdale NY 11735

YOULE, RICHARD JAMES, b Concord, Calif, Sept 20, 52; m 81; c 1. PROTEIN TOXINS, CELL MEMBRANES. *Educ:* Albion Col, AB, 74; Univ SC, PhD(biol), 77. *Prof Exp:* Staff fel, NIMH, 78-80, sr staff fel, lab neurochem, 81-84, sr investr, 84-88, SECT CHIEF, SURG NEUROL BR, NAT INST NEUROL & COMMUN DIS & STROKE, NIH, 88- *Res:* The mechanism of protein toxins (ricin and diptheria), inhibition of protein synthesis and how they can be linked to cell surface binding species, such as monoclonal antibodies, to create cell-type-specific toxins; physiological role of ricin and lectins. *Mailing Add:* Nat Inst Neurol & Commun Dis & Stroke NIH Bldg 10A Rm 3E-68 Bethesda MD 20205

YOUM, YOUNGIL, b Seoul, Korea, Jan 2, 42; m 68; c 3. KINEMATICS OF MECHANISMS, BIOMECHANICS. *Educ:* Utah State Univ, Logan, BS, 68; Univ Wis-Madison, MS, 70 & 73, PhD(eng mech), 76. *Prof Exp:* Res asst biomech, Univ Wis-Madison, 68-73; assoc res scientist bioeng, Univ Iowa, Iowa City, 74-78; asst prof, 78-80, ASSOC PROF MECH ENG, CATH UNIV AM, WASHINGTON, DC, 80- *Concurrent Pos:* Consult, Nat Bur Standards, 79-81 & Walter Reed Army Med Ctr, 79- *Mem:* Am Soc Mech Engrs; Am Soc Biomech; Sigma Xi; Korean Soc Engrs & Scientists. *Res:* Three dimemsional (spatial) analysis of the mechanisms, kinematics, kinetics, vibrations and biomechanics of the human upper extremity (the hand, the elbow, the wrist joint); design of prosthetic limbs and total replacement of human articular joint; robotic research. *Mailing Add:* Dept Mech Eng Cath Univ Am 620 Michigan Ave N Washington DC 20064

YOUMANS, HUBERT LAFAY, b Lexsy, Ga, Aug 2, 25; m 51; c 1. ANALYTICAL CHEMISTRY. *Educ:* Emory Univ, AB, 49, MS, 50; La State Univ, PhD(anal chem), 61. *Prof Exp:* Chemist, Savannah River Plant, E I du Pont de Nemours & Co, SC, 52-57; develop chemist, Sucrochem Div, Colonial Sugars Co, Lab, 57-58; asst prof chem, Ft Hays Kans State Col, 61-64; res chemist, Atlas Chem Indust, Inc, Del, 64-67; from assoc prof to prof chem, Western Carolina Univ, 67-89; RETIRED. *Mem:* Am Chem Soc. *Res:* Absorptiometry; analytical separations; communications for chemistry students. *Mailing Add:* PO Box 1450 Cullowhee NC 28723

YOUMANS, JULIAN RAY, b Baxley, Ga, Jan 2, 28; m 54; c 3. NEUROSURGERY. *Educ:* Emory Univ, BS, 49, MD, 52; Univ Mich, MS, 55, PhD(neuroanat), 57; Am Bd Neurol Surg, dipl, 60. *Prof Exp:* From asst prof to assoc prof neurosurg, Sch Med, Univ Miss, 59-63; from assoc prof to prof, Med Col SC, 63-67, chief div, 63-67; PROF NEUROSURG & CHIEF DEPT, SCH MED, UNIV CALIF, DAVIS, 67- *Mem:* AMA; Am Acad Neurol; Am Asn Surg of Trauma; Am Col Surg; Am Asn Automotive Med. *Res:* Physiology of cerebral blood flow. *Mailing Add:* Dept Neurol Surg Univ Calif Sch Med 4301 X St Sacramento CA 95817

YOUMANS, WILLIAM BARTON, b Cincinnati, Ohio, Feb 3, 10; m 32; c 3. PHYSIOLOGY. *Educ:* Western Ky State Col, BS, 32, MA, 33; Univ Wis, PhD(med physiol), 38; Univ Ore, MD, 44. *Prof Exp:* Instr biol, Western Ky State Col, 32-35; from asst to instr physiol, Univ Wis, 35-38; from instr to prof, Med Sch, Univ Ore, 38-52, head dept, 45-52; chmn dept, 52-71, prof, 52-76, EMER PROF PHYSIOL, SCH MED, UNIV WIS-MADISON, 76- *Concurrent Pos:* USPHS spec fel, 61-62; intern, Henry Ford Hosp, 44-45; mem physiol study sect, USPHS, 52-56, mem physiol training comt, 58-62. *Mem:* Am Physiol Soc; Am Soc Pharmacol & Exp Therapeut. *Res:* Innervation of intestine; gastrointestinal motility; visceral reflexes; cardiac innervation, neurohormones; angiotensin. *Mailing Add:* 162 Benson Rd Port Angeles WA 98362

YOUNATHAN, EZZAT SAAD, b Deirut, Egypt, Aug 25, 22; nat US; m 58; c 2. BIOCHEMISTRY. *Educ:* Univ Cairo, BSc, 44; Fla State Univ, MA, 53, PhD, 55. *Prof Exp:* Chemist, Govt Labs, Egypt, 44-50; Seagrams' Int Training Prog fel, 50-51; res asst, Fla State Univ, 51-55, asst prof, 58-59; res assoc, Col Med, Univ Ill, 55-57; from asst prof to assoc prof biochem, Sch Med, Univ Ark, Little Rock, 59-66, actg head dept, 63-66; NIH spec fel & vis prof, Inst Enzyme Res, Univ Wis, 66-67; PROF BIOCHEM, LA STATE UNIV, BATON ROUGE, 68- *Concurrent Pos:* Vis prof, Dept Biochem, Univ Calif, Berkeley, 78. *Mem:* Am Soc Biochem & Molecular Biol; Am Chem Soc; Soc Exp Biol & Med; AMA. *Res:* Mechanism of enzyme action; protein chemistry; control of carbohydrate metabolism; experimental diabetes. *Mailing Add:* Dept Biochem La State Univ Baton Rouge LA 70803-1806

YOUNATHAN, MARGARET TIMS, b Clinton, Miss, Apr 25, 26; m 58; c 2. FOOD SCIENCE, NUTRITION. *Educ:* Univ Southern Miss, BA, 46, BS, 50; Univ Tenn, MS, 51; Fla State Univ, PhD(food & nutrit), 58. *Prof Exp:* Instr food & nutrit, Ore State Univ, 51-55; Qm Food & Container Inst res fel, Fla State Univ, 58-59; instr pediat, Sch Med, Univ Ark, Little Rock, 62-65, asst prof, 65-68; assoc prof, 71-79, PROF HUMAN NUTRIT & FOOD, LA STATE UNIV, BATON ROUGE, 79- *Concurrent Pos:* Consult, Ark State Health Dept, 62-68; consult, USAID, Sierra Leone, WAfrica, 84 & Jamaica, WI, 87. *Mem:* Inst Food Technol; Am Home Econ Asn; Am Dietetic Asn; Am Meat Sci Asn; Am Inst Nutrit. *Res:* Heme pigments; antioxidants; lipid oxidation; infant and child nutrition; effects of products of lipid oxidation on biological tissue. *Mailing Add:* Human Nutrit & Food Sch Human Ecol La State Univ Baton Rouge LA 70803-4300

YOUNES, MAGDY K, b Damanhur, Egypt, July 13, 39. EXERCISE PHYSIOLOGY, RESPIRATION. *Educ:* Alexandria Univ, Egypt, MD, 62; McGill Univ, Montreal, PhD(pulmonary physiol), 73; FRCP(C). *Prof Exp:* PROF MED, HEALTH SCI CTR, MAN UNIV, 82- *Mem:* Am Thoracic Soc; Am Physiol Soc; Am Col Chest Physicians. *Mailing Add:* Dept Med Univ Man 700 Williams Ave Winnipeg MB R3E 0Z3 Can

YOUNES, MUAZAZ A, endocrinology, clinical chemistry, for more information see previous edition

YOUNES, USAMA E, b Haifa, Israel, Sept 6, 49; US citizen; m 86. POLYMER COMPOSITES, THERMOSETTING POLYMER SYNTHESES. *Educ:* Warren Wilson Col, BA, 71; Western Carolina Univ, MS, 73; Univ New Orleans, PhD(org chem), 78. *Prof Exp:* Postdoctoral fel polymer chem, Carnegie Mellon Univ, 78-80; PRIN RES SCIENTIST, ARCO CHEM CO, 80- *Res:* Syntheses and manufacturing of advanced composite materials through structural reaction injection moldings; syntheses of polyurethanes and polystyrenes as well as inherently flame retardant polymers; 25 patents; author of eleven publications. *Mailing Add:* C27 Hollybrook Newtown Square PA 19073

YOUNG, AINSLIE THOMAS, JR, b Norman, Okla, Mar 3, 43; m 68. POLYMER CHEMISTRY. *Educ:* Memphis State Univ, BS, 66, MS, 68; Univ Ky, PhD(phys chem), 71. *Prof Exp:* Fel, Univ Calif, Berkeley, 71-73; fel, La State Univ, Baton Rouge, 73-74; res chemist polymer chem, Ctr Res Lab, Mead Corp, 74-80; res chemist polymer chem, Los Alamos Nat Lab, 80-82, sect leader polymer chem, 82-85, proj mgr fusion prog, 85; CONSULT, 85- *Mem:* Am Chem Soc; Sigma Xi; AAAS. *Res:* Applied polymer science. *Mailing Add:* 360 Cheryl Ave Los Alamos NM 87544

YOUNG, ALLEN MARCUS, b Ossining, NY, Feb 23, 42; div. ECOLOGY. *Educ:* State Univ NY New Platz, BA, 64; Univ Chicago, PhD(zool), 68. *Prof Exp:* Orgn Trop Studies, Inc fel for study in Costa Rica, Univ Chicago, 68-70; asst prof biol, Lawrence Univ, Appleton, Wis, 70-75; cur & head Dept Invertebrate Zool, 75-90, CUR ZOOL, MILWAUKEE PUB MUS, 91- *Concurrent Pos:* NSF fel for study in Costa Rica, Lawrence Univ, 71-74; NSF res grant, 75-; res assoc, Nat Mus Costa Rica, 75-; Am Cocoa Res Inst grant, 78- *Mem:* AAAS; Ecol Soc Am; Lepidop Soc; Asn Trop Biol. *Res:* Population biology and behavior of neotropical Lepidoptera and Cicadidae; ecology of laboratory populations of Tribolium; insect pollination of cocoa. *Mailing Add:* 234 N Broadway No 207 Milwaukee WI 53202

YOUNG, ALVIN L, b Laramie, Wyo, Aug 3, 42; m 66; c 2. PESTICIDE CHEMISTRY & TOXICOLOGY. *Educ:* Univ Wyo, BS, 64, MS, 65; Kans State Univ, PhD(agron), 68. *Prof Exp:* Proj scientist environ res, Armament Lab, Air Force Base, US Airforce, 68-71, environ sci consult, Occup & Environ Health Lab, Brooks Air Force Base, San Antonio, 77-79, Sch Aerospace Med Epidemiol Div, 79-81; asst prof life sci, dept life sci, US Air Force Acad, Colo, 71-74, assoc prof biol sci, dept chem & biol sci, 74-77, dir res, 75-77; spec asst environ sci, dept med & surg, Vet Admin, Washington, DC, 81-83; sr policy analyst life sci, Off Sci & Technol Policy, Exec Off of the Pres, Washington, DC, 84-87; DIR OFF AGR BIOTECHNOL, USDA, WASHINGTON, DC, 87- *Concurrent Pos:* Vis assoc prof, Univ Colo, Colorado Springs, 75-77; adv environ toxicol, Seveso Authority, Milan, Italy, 77-85, Australian Dept Vet Affairs, Woden Act, 83-85; mem, Whitehouse Agent Orange Working Groups, Washington, DC, 81-86, Joint Coun Food & Agr Sci, USDA, 84-87. *Honors & Awards:* Woodroof Lectr, 91. *Mem:* Sigma Xi; Am Soc Agron; Weed Sci Soc Am; Am Chem Soc; AAAS; Soc Environ Toxicol & Chem. *Res:* Environmental fate, toxicology and human risks of the tetrachlorodibenzo-p-dioxins; ecological impact and toxicity of the pesticides used in Vietnam. *Mailing Add:* 708 Seven Pines Lane Waldorf MD 20601

YOUNG, ANDREW TIPTON, b Canton, Ohio, Apr 4, 35; m 54, 63, 68; c 2. ASTRONOMICAL PHOTOMETRY, PLANETARY ASTRONOMY. *Educ:* Oberlin Col, BA, 55; Harvard Univ, MA, 57, PhD(astron), 62. *Prof Exp:* Res fel, Observ, Harvard Univ, 60-65, lectr astron, 62-65; asst prof, Univ Tex, 65-67; mem tech staff, Aerospace Corp, 67-68 & Jet Propulsion Lab, 68-73; res scientist, Tex A&M Univ, 73-75, vis asst prof physics, 75-78, res scientist, 78-80; ADJ ASSOC PROF ASTRON, SAN DIEGO STATE UNIV, 81- *Concurrent Pos:* Co-investr TV experiments, Mariner Mars, 69 & 71; consult, NASA; assoc ed, ICARUS, 77- *Mem:* Fel AAAS; Am Astron Soc; fel Royal Astron Soc; fel Optical Soc Am; Int Astron Union. *Res:* Observational astronomy; astronomical photometry and instrumentation; photomultipliers; scintillation; planetary physics. *Mailing Add:* Dept Astron San Diego State Univ San Diego CA 92182-0334

YOUNG, ARTHUR, b New York, NY, Jan 4, 40; m 60; c 2. ASTRONOMY. *Educ:* Allegheny Col, BS, 60; Ind Univ, MA, 65, PhD(astron), 67. *Prof Exp:* From asst prof to assoc prof, 67-74, PROF ASTRON, SAN DIEGO STATE UNIV, 74- *Concurrent Pos:* Acad consult, Spitz Labs, 63-; vis staff scientist, High Altitude Observ, 81-82. *Mem:* Int Astron Union; Am Astron Soc. *Res:* Spectroscopy of cool stars; chromospheres and stellar magnetic activity; binary stars; hot pulsating stars. *Mailing Add:* Dept Astron San Diego State Univ San Diego CA 92115

YOUNG, ARTHUR WESLEY, b Shenandoah, Iowa, May 14, 04; m 29; c 3. AGRONOMY, SOILS. *Educ:* Iowa State Col, BS, 29, MS, 30, PhD(soil microbiol), 32. *Prof Exp:* Instr agr bact, Univ Tenn, 32-34; prof agron, Panhandle Agr & Mech Col, 34-35; from assoc prof to emer prof agron, Tex Tech Univ, 35-69, from actg head dept to head dept, 37-69; CONSULT, 69- *Concurrent Pos:* Consult, Bunge y Born, Argentina, 68-71; independent agr consult, 69-; chmn, State Seed & Plant Bd, Tex; res consult, Dept of Agr, Lubbock Christian Col, 75- *Honors & Awards:* Agr Chem Award, W Tex Agr Chem Inst, 69; Gerold W Thomas Outstanding Agriculturalist Award, Tex Tech Univ, 70. *Mem:* AAAS; Sigma Xi; Am Soc Agron; Soil Sci Soc Am; hon mem Int Crop Improve Asn (vpres, 58-59, pres, 60-61). *Res:* Soil bacteriology, chemistry and fertility; direction of microbial studies on influence of organic matter and fertilizer to selected soils. *Mailing Add:* 3305 45th St Lubbock TX 79413

YOUNG, AUSTIN HARRY, b Brighton, Mass, Oct 25, 28; m 58; c 2. RADIATION CHEMISTRY, PHYSICAL TESTING OF PLASTICS. *Educ:* Tufts Univ, BS, 50; Univ Wis, PhD(chem), 59. *Prof Exp:* Develop chemist, Fabrics & Finishes Dept, EI du Pont de Nemours & Co, 50-51, supvr, Explosives Dept, 51-52; res asst, Army Chem Ctr, Edgewood, Md, 53-54; from res chemist to sr res chemist, A E Staley Mfg Co, 58-69, res assoc, 69-78, sr scientist, 78-83, mgr, Technol Group Polymerizable Prod Dept, 84-85, sr res scientist, 85-87; RETIRED. *Concurrent Pos:* Consult, A E Staley Mfg Co, 87; consult, Archer Daniels Midland Co, 87-90, A E Staley Mfg Co, 91- *Honors & Awards:* IR-100 Award, 78. *Mem:* AAAS; Am Chem Soc; Am Asn Cereal Chem; Sigma Xi; Soc Plastics Engrs. *Res:* Starch, colloid, polymer and radiation chemistry; kinetics; polymer synthesis and characterization; coatings; chromatography; thermal analysis; rheology; radiotracer techniques; graphic arts; health physics; microscopy; computer science; degradable plastics; environmental science; plastics technology; elementary particle physics; physics. *Mailing Add:* 151 Point Bluff Decatur IL 62521

YOUNG, BERNARD THEODORE, b Tarentum, Pa, Apr 13, 30; m 55; c 5. PHYSICS. *Educ:* Slippery Rock State Univ, BS, 52; Tex A&M Univ, MS, 61, PhD(physics), 64. *Prof Exp:* Res & develop engr, Tex Div, Dow Chem Co, 56-59; res asst physics, Tex A&M Univ, 60-63; from assoc prof to prof, Sam Houston State Univ, 63-68, dir, 65-68; assoc dean, 68-70, prof physics & dean grad sch, 70-88, VPRES ACAD AFFAIRS, ANGELO STATE UNIV, 88- *Concurrent Pos:* Mem elem transparency proj, Tex Educ Agency, 66-67. *Mem:* Am Phys Soc. *Res:* Molecular spectroscopy; atomic structure. *Mailing Add:* Off Acad Affairs Angelo State Univ 2601 W Ave N San Angelo TX 76909

YOUNG, BING-LIN, b Honan, China, Feb 3, 37; m 64; c 1. HIGH ENERGY PHYSICS. *Educ:* Nat Taiwan Univ, BS, 59; Univ Minn, PhD(physics), 66. *Prof Exp:* Res assoc physics, Ind Univ, 66-68 & Brookhaven Nat Lab, 68-70; from asst prof to assoc prof , 70-74, assoc prof, 74-79, PROF PHYSICS, IOWA STATE UNIV, 79- *Concurrent Pos:* Assoc physicist, Ames Lab, US AEC, 70-74, physicist, 74-79, sr physicist, 79-; hon prof, Zhengzhon Univ, & Henan Univ, China. *Mem:* Am Phys Soc. *Res:* Theoretical physics of the elementary particles. *Mailing Add:* Dept Physics Iowa State Univ Ames IA 50011

YOUNG, BOBBY GENE, microbiology, genetics; deceased, see previous edition for last biography

YOUNG, BRUCE ARTHUR, b Sydney, Australia, Jan 16, 39; m 65; c 2. AGRICULTURE, ANIMAL PHYSIOLOGY. *Educ:* Univ New Eng, Australia, BRurSc, 62 MRurSc, 65, PhD(physiol), 69. *Prof Exp:* From asst prof to assoc prof, 68-78, PROF ANIMAL PHYSIOL, UNIV ALTA, 78- *Concurrent Pos:* Pres, World Conf Animal Prod, 88-93; int consult, animal prod. *Honors & Awards:* Excellence Res Award, Can Soc Animal Sci; Award Excellence Physiol, Can Asn Animal Breeders. *Mem:* Am Soc Animal Sci; Can Soc Animal Sci; Can Physiol Soc; Agr Inst Can. *Res:* Environmental physiology of animals; adaptation to physical environment; livestock production and energy metabolism in cold climates. *Mailing Add:* Dept Animal Sci Univ Alta Edmonton AB T6G 2E2 Can

YOUNG, C(HARLES), JR, b Washington, Pa, June 3, 23; m 47; c 2. ELECTRONICS, MECHANICAL ENGINEERING. *Educ:* Ill Inst Technol, BSME, 45; Mo Sch Mines, MSME, 50. *Prof Exp:* Instr mech eng, Univ RI, 46-47; from instr to asst prof, Mo Sch Mines, 47-51; design engr, 51-57, supvry electronics engr, 57-66, SR PROJ MGR SWIMMER SYSTS, US NAVAL ORD LAB, US NAVAL SURFACE WEAPONS CTR, 66- *Concurrent Pos:* Subj specialist electronics, Nat Home Study Coun, 62- *Mem:* Am Soc Mech Engrs; Am Ord Asn; Marine Technol Soc; Am Defense Preparedness Asn. *Res:* Application of electronic, mechanical and radioactive materials and principles to the design, development and production of combat swimmers' and divers' weapons. *Mailing Add:* 12205 Cedar Hill Dr Silver Spring MD 20904

YOUNG, C(LARENCE) B(ERNARD) F(EHRLER), b Birmingham, Ala, May 13, 08; m 34; c 3. CHEMISTRY, ELECTROMETALLURGY. *Educ:* Howard Col, BS, 30; Columbia Univ, MS, 32, PhD, 34. *Hon Degrees:* PhD(law), Freeman-Hardeman Univ, 82. *Prof Exp:* Asst chem, Howard Col, 27-30; lectr, Columbia Univ, 30-32; asst cur, Chandler Chem Mus, 32-34; tech dir, US Res Corp, NY, 34-37; dir, Nat Southern Prod Corp, 43-52, vpres, 44, exec vpres, 44-52; pres, Aply N Austin Co, 52-55 & Cracker Asphalt Corp, 55-69; PRES, YOUNG REFINING CORP, 71-, CHEM ENGR, 76- *Concurrent Pos:* Pres, Warrior Asphalt Corp, 49-55, Ala Southern Warehouses, 49-91, Cytho Corp, 49-, M'Lady, Inc, 52-60, Auromet Corp, NY, 53-56 & Abaca Chem Corp, 54-56; vpres, Findan Corp, 55-57. *Mem:* AAAS; Electrochem Soc; fel Inst Chemists. *Res:* Surface acting agents; chemistry for electroplating; metal finishing; production of fatty acids; resin and fatty acids from the Southern pine tree; refineries for production of petroleum light oils, asphalts and its products. *Mailing Add:* PO Box 796 Douglasville GA 30133

YOUNG, C(ECIL) G(EORGE), JR, electronics, for more information see previous edition

YOUNG, CHARLES ALBERT, b Dodge Center, Minn, Oct 11, 11; m 50; c 2. INDUSTRIAL ORGANIC CHEMISTRY. *Educ:* Purdue Univ, BS, 33; Univ Notre Dame, MS, 34, PhD(org chem), 36. *Prof Exp:* Asst chem, Univ Notre Dame, 33-36; staff chemist, Jackson Lab, E I du Pont de Nemours & Co, Inc, 36-60, staff chemist, Exp Sta, 60-76; RETIRED. *Mem:* Am Chem Soc. *Res:* Synthetic rubber; organic water repellants; adhesives; dye synthesis and application; industrial and automotive finishes. *Mailing Add:* 111 Beech Lane Wilmington DE 19804-2309

YOUNG, CHARLES EDWARD, b Petrolia, Ont, June 1, 41. LASER MASS-SPECTROMETRY, SURFACE SCIENCE. *Educ:* Univ Toronto, BSc, 63; Univ Calif, Berkeley, PhD(chem), 66. *Prof Exp:* Res assoc chem, Mass Inst Technol, 66-68; mem tech staff, Bell Labs, 68-70; CHEMIST, ARGONNE NAT LAB, 70- *Mem:* Am Chem Soc; Am Phys Soc. *Res:* Laser photoionization detection of neutrals in (1) trace analysis, and (2) in studies of the dynamics of desorption from surfaces via photon- and electron-stimulated desorption and sputtering; synchrotron radiation surface research. *Mailing Add:* CHM 200 Argonne Nat Lab 9700 S Cass Ave Argonne IL 60439

YOUNG, CHARLES GILBERT, b Pa, Feb 25, 30; m 59; c 3. ELECTROOPTICS, SYSTEMS ENGINEERING. *Educ:* Elizabethtown Col, BA, 52; Univ Conn, MS, 56, PhD(physics), 61. *Prof Exp:* Physicist, Brookhaven Nat Lab, 56 & Navy Electronics Lab, Calif, 57; instr physics, Conn Col, 57-59; from res asst to instr, Univ Conn, 59-62; consult, Am Optical Co, 61-62, sr physicist, 62-65, mgr systs res dept, 65-70, gen mgr laser prod dept, 70-73, dir prod develop, 73-77; dir eng, Kollmorgen Corp, 77-79; asst tech mgr, New Eng Res Appln Ctr, Univ Conn, 79-80, dir tech develop combustion eng, 80-86; pres, Instaread Corp, 86-87; tech dir, infra red search & track, 87-88, mgr electro-optical eng, 88-90, MGR AIR DEFENSE INIATIVES, MARTIN MARIETTA, 90- *Concurrent Pos:* Adj instr, Sch Bus, Univ Conn, 80. *Mem:* Sr mem Inst Elec & Electronics Engrs; Optical Soc Am; Am Phys Soc; Int Soc Optical Eng. *Res:* Development and design of electro-optic sensor and tracker systems; numerous patents and publications. *Mailing Add:* 1331 College Pt Winter Park FL 32789

YOUNG, CHARLES STUART HAMISH, b Plymouth, UK, Feb 2, 44; m 70; c 1. ANIMAL VIRUS GENETICS. *Educ:* Oxford Univ, BA, 66, DPhil(genetics), 69. *Prof Exp:* Lectr genetics, Biol Dept, Princeton Univ, 70-71; sci officer, Virol Inst Med Res Coun, 71-74; asst prof, 74-82, res scientist, 82-85, ASSOC PROF VIROL & GENETICS, DEPT MICROBIOL, COLUMBIA UNIV, 85- *Mem:* Am Soc Microbiol; Am Soc Virol. *Res:* Mechanisms of genetic recombination and gene expression in eukaryotic cells using DNA- containing animal viruses as model systems. *Mailing Add:* Dept Microbiol Columbia Univ Health Sci Ctr New York NY 10032

YOUNG, CHARLES WESLEY, b Enid, Okla, Dec 2, 29; m 52; c 3. DAIRY BREEDING & MANAGEMENT. *Educ:* Okla State Univ, BS, 56; NC State Univ, MS, 58, PhD(animal indust), 61. *Prof Exp:* From asst prof to assoc prof, 60-69, PROF ANIMAL SCI, UNIV MINN, ST PAUL, 69- *Concurrent Pos:* Columnist, Hoard's Dairyman, 76-82. *Mem:* Am Dairy Sci Asn. *Res:* Systems of breeding in dairy cattle; relationship between size and production in dairy cattle; economics of selection for milk yield in dairy cattle; milk pricing and component differentials. *Mailing Add:* Dept Animal Sci Univ Minn 1404 Gortner Ave St Paul MN 55108

YOUNG, CHARLES WILLIAM, b Denver, Colo, Nov 19, 30; m 63; c 3. INTERNAL MEDICINE, CANCER. *Educ:* Columbia Univ, AB, 52; Harvard Univ, MD, 56. *Prof Exp:* Intern med, Second Med Div, Bellevue Hosp, New York, 56-57, resident, Second Med Div, Bellevue Hosp & Mem Hosp, 57-59; asst prof med, Med Col, Cornell Univ, 66-76; from res assoc to assoc, 62-71, ASSOC MEM, SLOAN-KETTERING INST CANCER RES, 71-; ASSOC PROF MED, MED COL, CORNELL UNIV, 76- *Concurrent Pos:* Attend physician, Mem Hosp, 78- *Mem:* AAAS; Am Asn Cancer Res; Am Soc Clin Oncol; Am Fedn Clin Res; NY Acad Sci. *Res:* Oncology; biochemical pharmacology; embryology. *Mailing Add:* Chemother Serv Mem Sloan-Kettering Cancer Ctr 1275 York Ave New York NY 10021

YOUNG, CLIFTON A, b 1943; US citizen. INORGANIC CHEMISTRY. *Educ:* Haverford Col, BA, 65; Dartmouth Col, AM, 68; Tufts Univ, PhD(chem), 77. *Prof Exp:* Mem res staff, P R Mallory Co Inc, 77-78; mem tech staff, GTE Labs Inc, Gen Tel & Electronics Corp, 78-80; asst prof, Tufts Univ, 80-81; asst prof, Univ Dallas, 81-85; asst prof, Drury Col, 85-90; ASST PROF, DICKINSON STATE UNIV, 90- *Mem:* Am Chem Soc; Electrochem Soc; Int Soc Electrochem; Royal Soc Chem. *Res:* Battery research; chemistry of metal-ammonia and other metal solutions; non-aqueous solvents; battery chemistry. *Mailing Add:* Dickinson State Univ Dickinson ND 58601

YOUNG, CLYDE THOMAS, b Durham, NC, Aug 22, 30; m 55; c 5. FOOD SCIENCE. *Educ:* NC State Univ, BS, 52, MS, 55; Okla State Univ, PhD(food sci), 70. *Prof Exp:* Instr chem & math, NC State Univ, 54-56; asst head claims dept, Anderson-Clayton & Co, Ga, 58-59; res chemist, Ga Inst Technol, 59-60; asst res chemist, Ga Sta, Univ Ga, 60-76; assoc prof, 76-80, PROF, NC STATE UNIV, 80- *Mem:* Am Peanut Res & Educ Soc; Inst Food Technologists. *Res:* Major investigations in changes and variations in biochemical constituents of peanuts-factors affecting aroma, flavor, color, maturation and protein during development, harvesting, curing, storage, roasting and processing. *Mailing Add:* Dept Food Sci NC State Univ Raleigh NC 27695-7624

YOUNG, DANA, b Washington, DC, Oct 6, 04; m 30; c 2. MECHANICAL ENGINEERING. *Educ:* Yale Univ, BS, 26, MS, 30; Univ Mich, PhD(eng mech), 40. *Prof Exp:* Field engr, Marland Oil Co, 26-27; asst dist engr, Shell Petrol Corp, 27-28; instr eng mech, Yale Univ, 28-30; struct engr, United Engrs & Constructors, Inc, 30-34; prof civil eng & appl mech, Univ Conn, 34-42; prof eng mech, Univ Tex, 42-50; prof mech eng, Univ Minn, 50-53; chmn dept civil eng, Yale Univ, 53-55, Sterling prof civil eng, 53-62, dean sch eng, 55-59; sr vpres, 62-71, consult, Southwest Res Inst, 71-83; CONSULT, 83- *Concurrent Pos:* Consult, Space Technol Labs, Inc, 54-62; mem comt ship struct design, Nat Res Coun, 55-63, adv comt to Nat Bur Standards, 56-61, div eng & indust res, 59-64 & spec comt appl res, US Air Force Sci Adv Bd, 61; comnr, Nat Comn Prod Safety, 68-70. *Honors & Awards:* Pressure Vessel & Piping Medal, Am Soc Mech Engrs, 80. *Mem:* Am Soc Civil Engrs; hon mem Am Soc Mech Engrs; Soc Exp Stress Anal; Am Inst Aeronaut & Astronaut. *Res:* Vibrations; theory of elasticity; structural mechanics; dynamics of rotating machines; pressure vessels and piping. *Mailing Add:* 8700 Post Oak Lane Apt 228 San Antonio TX 78217-5165

YOUNG, DAVID A, b Carmel, Calif, Sept 26, 42. PHYSICAL CHEMISTRY. *Educ:* Pomona Col, BA, 64; Univ Chicago, PhD(chem), 67. *Prof Exp:* PHYSICIST, LAWRENCE LIVERMORE LAB, 67- *Mem:* Am Phys Soc. *Res:* Statistical mechanics of gases, liquids and solids. *Mailing Add:* 1874 Peary Way Livermore CA 94550

YOUNG, DAVID A(NTHONY), b Pittsburgh, Pa, Jan 6, 21; m 45; c 5. CHEMICAL ENGINEERING, POLYMER CHEMISTRY. *Educ:* Pa State Univ, BS, 42. *Prof Exp:* Chem engr, Sun Oil Co, 42-47 & Warwick Wax Co, 47-48; group leader, Chem Eng Process Develop Dept, Spencer Chem Co, 48-59; chief engr, Eng Sci Div, Am Metal Prod Co, 59-62; sect head solution polymerization, Gen Tire & Rubber Co, Ohio, 62-70; sr process engr, J F Pritchard & Co, Kansas City, 70-76; staff engr, Kansas City Div, Allied Signal Aerospace, 76-84; RETIRED. *Concurrent Pos:* Instr, Kans State Teachers Col, 59. *Mem:* Am Inst Chem Engrs. *Res:* Design, construction and start-up of new chemical plants; hydrogen sulfide and sulphur dioxide recovery from stack gases; evaporation and burning of pulp mill waste liquors; equipment design and procurement for new polymer facilities; pilot plant production of new polymers including polyethylenes, nylon, specialized poly BD's, propyleneoxide rubber and polyimides; treatment to detoxify chemical wastewater streams. *Mailing Add:* 9501 Meadow Ln Leawood KS 66206

YOUNG, DAVID ALLAN, b Wilkinsburg, Pa, May 26, 15; m 34; c 1. INSECT TAXONOMY. *Educ:* Louisville Univ, AB, 39; Cornell Univ, MS, 42; Univ Kans, PhD(entom), 51. *Prof Exp:* Instr, Louisville Univ, 46-48; asst, Univ Kans, 48-49; entomologist, Insect Identification & Parasite Introd Sect, Entom Res Br, Agr Res Serv, USDA, 50-57; from assoc prof to emer prof entom, Nc State Univ, 80-85; RETIRED. *Mem:* Entom Soc Am; Sigma Xi. *Res:* Insect taxonomy; taxonomy of the Auchenorrhynchous Homoptera; reclassification of Cicadellinae (Homoptera: Auchenorrhyncha). *Mailing Add:* 612 Buck Jones Rd Raleigh NC 27606

YOUNG, DAVID ALLEN, b Pomona, Calif, July 31, 46; c 2. BOTANY, BIOCHEMICAL SYSTEMATICS. *Educ:* Calif State Univ, Fullerton, BA, 71, MA, 72; Claremont Grad Sch & Univ Ctr, PhD(bot), 75. *Prof Exp:* Lab instr & instr bot, Calif State Univ, Fullerton, 70-72; res asst, Rancho Santa Ana Bot Garden, Claremont, Calif, 72-75; asst prof, Union Col, 75-76; asst prof, 76-80, prof, Univ Ill, Urbana, 80-; AT DEPT BOT, CORNELL UNIV. *Mem:* Bot Soc Am; Am Asn Plant Taxonomists; Int Asn Plant Taxonomists; Soc Study Evolution; Am Soc Naturalists. *Res:* Biochemical systematics of the angiosperms; angiosperm phylogeny; cladistics. *Mailing Add:* Univ Okla Col Arts & Sci 601 Elm Rm 221 Norman OK 73019-0315

YOUNG, DAVID BRUCE, b Pittsburgh, Pa, Mar 13, 45; m 65; c 3. PHYSIOLOGY. *Educ:* Univ Colo, BA, 67; Ind Univ, PhD(physiol), 72. *Prof Exp:* From instr to asst prof, 72-77, assoc prof, 77-81, PROF PHYSIOL, SCH MED, UNIV MISS, 81- *Concurrent Pos:* NIH trainee, Sch Med, Univ Miss, 72-74. *Mem:* Fedn Am Soc Exp Biol. *Res:* Fluid and electrolyte balance control mechanisms; hypertension. *Mailing Add:* Dept Physiol Univ Miss Med Ctr 2500 N State St Jackson MS 39216-4505

YOUNG, DAVID CALDWELL, b Memphis, Tenn, June 18, 24; m 55; c 3. ORGANIC CHEMISTRY. *Educ:* Davidson Col, BS, 46; Univ Fla, MS, 48, PhD(chem), 50. *Prof Exp:* Chemist, Edgar C Britton Res Lab, Dow Chem Co, 50-53, group leader, 53-62, patents coordr, 62-65, asst to dir, 65-70, asst to dir chem biol res, 70-74, asst to dir pharmaceut res & develop, 74-76, sr res specialist, Hydrocarbons & Energy Res, 77-80, sr environ specialist, Environ Serv Dept, 81-86; RETIRED. *Concurrent Pos:* Mem bd dirs, Am Chem Soc, 84-89, exec comt, 87-89. *Mem:* Am Chem Soc. *Res:* Leuckart reaction; phthalaldehydic acid; organic research and process development; chemical patents; environmental regulations. *Mailing Add:* 1223 Holyrood St Midland MI 48640

YOUNG, DAVID GRIER, entomology, for more information see previous edition

YOUNG, DAVID MARSHALL, b Minot, ND, Aug 26, 42; m 63; c 4. PATHOLOGY. *Educ:* Colo State Univ, DVM, 66; Ohio State Univ, MS, 67, PhD(comp path), 70. *Prof Exp:* Comp pathologist, Lab Toxicol, Nat Cancer Inst, 70-77, head, Comp Path Sect, DCT, NIH, 73-77; prof vet sci & head dept, 77-85, PROF COMP PATH, DEPT VET SCI, MONT STATE UNIV, 77-, COORDR BIOMED RES PROG, 86- *Concurrent Pos:* Sr staff fel, NIH, 70-73; consult pathologist, Statutory Adv Comt, Food & Drug Admin, Dept Health, Educ & Welfare, 71; mem fac, Found Advan Educ in the Sci, NIH, 72-77; prof, Wash, Alaska, Mont, Idaho Med Prog, Mont State Univ, 78-, dir, Am Indian res opportunities prog, 84- *Honors & Awards:* C L Davis Jour Award, C L Davis Found Advan Vet Path, 73; Maurice Shahan Res Award, 66. *Mem:* AAAS; Int Acad Path; Am Asn Path; Am Vet Med Asn; Soc Pharmacol & Environ Path; Am Asn Cancer Res; Sigma Xi; Orthop Res Soc. *Res:* Comparative pathology; endocrine pathology; cancer; mineral metabolism; orthopedic pathology; animal models of human disease; toxicology. *Mailing Add:* Dept Vet Sci Mont State Univ Bozeman MT 59715

YOUNG, DAVID MATHESON, b London, Eng, Apr 19, 28; m 53; c 6. ENVIRONMENTAL CHEMISTRY. *Educ:* Univ London, BSc, 48, PhD(chem), 49. *Prof Exp:* Asst lectr phys chem, St Andrews Univ, 49-51; res assoc, Amherst Col, 51-52; res fel, Nat Res Coun Can, 52-53; Royal Mil Col, Can, 53-54 & 55-56; Humboldt scholar, Phys Chem Inst, Munich, 54-55; res chemist, 56-58, supvr res & develop lab, 58-65, asst res mgr, 65-72, mgr chem res & develop, 72-75, mgr environ affairs, 75-82, MGR SCI AFFAIRS, DOW CHEM CAN, INC, 82- *Mem:* Chem Inst Can. *Res:* Physical adsorption of gases; boron chemistry; gas chromatography; heterogeneous catalysis; photochemistry. *Mailing Add:* 1619 Lancaster Ave Sarnia ON N7V 3S7 Can

YOUNG, DAVID MICHAEL, b Bluffton, Ind, Oct 11, 35; m 72; c 4. MEDICINE. *Educ:* Duke Univ, BS, 57, MD, 59. *Prof Exp:* Staff scientist, Lab Cellular Physics & Metab, Nat Heart Inst, NIH, 60-62; asst prof biol, dept biol, McCollum-Pratt Inst, NIH, 60-62; from asst prof to assoc prof biol chem, Sch Med, Harvard Univ, 65-79, chmn PhD prog cell biol 72-76; chief, Lab Phys Biochem, Mass Gen Hosp, 65-79; chmn, dept biochem & molecular biol, 79-81, PROF BIOCHEM, DEPT BIOCHEM & MOLECULAR BIOL, COL MED, UNIV FLA, 79-, PROF MED, 79-, PROF PEDIAT, 85- *Concurrent Pos:* Vis scientist, McCollum-Pratt Inst, Johns Hopkins Univ, 62-63; mem res allocations comt, Am Heart Asn, 76-79; cell biol study sect, NIH, 77-81 & cell physiol study sect, 78-82, chmn, 80-; ed-in-chief, J Molecular & Cellular Biochem, 83- *Mem:* Am Soc Biol Chem; AAAS; Biophys Soc; Am Soc Clin Invest; Am Soc Cell Biol; NY Acad Sci. *Res:* Studies of the structure and function of cellular growth factors, particularly as they relate to cancer research and abnormal growth of cells; studies on nerve growth factor as it relates both to the development of the nervous system and it's potential role in promoting wound healing; biochemical and cell biology approaches aimed at understanding the control of cell growth. *Mailing Add:* Dept Biochem Univ Fla Box J-245 Gainesville FL 32610

YOUNG, DAVID MONAGHAN, JR, b Boston, Mass, Oct 20, 23; m 49; c 3. MATHEMATICS. *Educ:* Webb Inst Naval Archit, BS, 44; Harvard Univ, MA, 47, PhD(math), 50. *Prof Exp:* Instr & res assoc math, Harvard Univ, 50-51; mathematician, Aberdeen Proving Ground, 51-52; assoc prof math, Univ Md, 52-55; mgr math anal dept, Ramo-Wooldridge Corp, 55-58; prof math & dir comput ctr, 58-70, PROF MATH & COMPUT SCI & DIR CTR NUMERICAL ANALYSIS, UNIV TEX, AUSTIN, 70- *Mem:* Am Math Soc; Soc Indust & Appl Math; Math Asn Am; Asn Comput Mach. *Res:* Numerical analysis, especially the numerical solution of partial differential equations by finite difference methods; high-speed computing. *Mailing Add:* 3406 Monte Vista Dr Austin TX 78731

YOUNG, DAVID THAD, b Port of Spain, Trinidad, Apr 18, 43; US citizen; m 66, 90; c 6. SPACE PHYSICS. *Educ:* Univ Southwestern La, BS, 64; Rice Univ, MS, 67, PhD(space physics), 70. *Hon Degrees:* Venia Docendi, Univ Bern, 80. *Prof Exp:* Grad fel, Dept Space Physics & Astron, Rice Univ, 64-70, res assoc, 70-71; sr res assoc, Physikalisches Inst, Univ Bern, 72-81; mem staff, Space Plasma Physics Group, Los Alamos Nat Lab, 81-87; INST SCIENTIST, SOUTHWEST RES INST, SAN ANTONIO, TEX, 88- *Concurrent Pos:* Consult, Europ Space Agency, NASA; vis scientist, Div Plasma Physics, Royal Inst Technol, Stockholm, 72- *Mem:* Am Geophys Union; Europ Geophys Soc. *Res:* Magnetospheric plasma physics, particularly the development of ion mass spectrometers for space flight and the analysis and application of this data. *Mailing Add:* Southwest Res Inst PO Drawer 28510 San Antonio TX 78284

YOUNG, DAVID W, b Oblong, Ill, Aug 16, 09. HYDROCARBON BLENDING TO FORM IMPROVED PRODUCTS. *Educ:* Univ Ky, BS, 31, MS, 35. *Hon Degrees:* DSc, Transylvania Univ, 35 & Queens Col, 68. *Prof Exp:* Res chemist, Ky Agr Experiment Sta, 31-35, Laurel Hill Res Lab, Gen Chem Co, 36-40; sr res chemist, Esso Res & Eng Co, 40-55 & Sinclair Res Inc, 55-65; res assoc, Atlantic Richfield Co, 65-71; vpres, Compino Labs, 71-72; partner, RB MacMullin Assoc, 72-73; pres, David W Young & Assoc, 75-87; CONSULT, 87- *Concurrent Pos:* Consult, US Army, US Armed Forces, 59-62; lectr, Sigma Xi, 63-67, Purdue Univ, 68-70 & Am Chem Soc Speakers Bur, 70-88; mem bd dirs, Asn Consult Chemists & Chem Engrs. *Honors & Awards:* Pioneering Award, Am Inst Chem, 67. *Mem:* Nat Acad Sci; Am Inst Chemists (pres, 71-73); Nat Asn Corrosion Eng; Am Chem Soc; AAAS. *Res:* Consulting on use of polymerization catalysts to form high polymers from hydrocarbon olefins. *Mailing Add:* 18508 Clyde Ave Homewood IL 60430

YOUNG, DAVIS ALAN, b Abington, Pa, Mar 5, 41; m 65; c 3. PETROLOGY, MINERALOGY. *Educ:* Princeton Univ, BSE, 62; Pa State Univ, MS, 65; Brown Univ, PhD(geol), 69. *Prof Exp:* Asst prof geol, NY Univ, 68-73; assoc prof geol, Univ NC, Wilmington, 73-78; assoc prof, 78-83, PROF GEOL, CALVIN COL, 83- *Concurrent Pos:* NSF instnl grant, NY Univ, 69-70. *Mem:* Mineral Soc Am; Geol Soc Am; Sigma Xi; Am Sci Affil; Nat Asn Geol Teachers. *Res:* Precambrian igneous and metamorphic geology of New Jersey and southeastern Pennsylvania; petrology of syenites; history of geology in relation to Christianity. *Mailing Add:* Dept Geol Geog & Environ Studies Calvin Col Grand Rapids MI 49506

YOUNG, DELANO VICTOR, b Honolulu, Hawaii, Nov 17, 45; m 70; c 1. MAMMALIAN CELL MANUFACTURING, CELL GROWTH CONTROL. *Educ:* Stanford Univ, BS, 67; Columbia Univ, PhD(biochem), 73. *Prof Exp:* Fel cell biol, Salk Inst Biol Studies, 73-75; asst prof chem, Boston Univ, 75-84; Bioassay Syst Corp, 84-86; sr scientist, Damon Biotech, 86-88, dir, 88-90, HEAD, CELL BIOL DEPT, ABBOTT BIOTECH, 90- *Concurrent Pos:* Vis scientist, Dept Biochem, Harvard Univ, 82-83; prin investr, NIH grants, 76-79, 81-84. *Mem:* AAAS; Sigma Xi; Am Soc Microbiol; Am Soc Biochem & Molecular Biol. *Res:* Large scale mammalian cell manufacturing, including characterization of cells and maximization of genetic expressions; cell growth control and signal transduction, especially as related to lymphokine, cytokine and growth factor action. *Mailing Add:* Abbott Biotech 119 Fourth Ave Needham Heights MA 02194

YOUNG, DENNIS LEE, b St Louis, Mo, Jan 22, 44; m 78; c 2. MATHEMATICAL STATISTICS, APPLIED STATISTICS. *Educ:* St Louis Univ, BS, 65; Purdue Univ, MS, 67, PhD(statist), 70. *Prof Exp:* Asst prof statist, NMex State Univ, 70-75; assoc prof, 75-84, PROF STATIST, ARIZ STATE UNIV, 84- *Concurrent Pos:* Consult, Motorola Inc, 83-; expert witness, Ariz Atty Gen Off, 86. *Mem:* AAAS; Inst Math Statist; Am Statist Asn. *Res:* Multivariate statistical analysis; applications of statistical methods. *Mailing Add:* Dept Math Ariz State Univ Tempe AZ 85287-1804

YOUNG, DONALD ALCOE, b Fredericton, NB, Oct 21, 29; m 55; c 3. GENETICS, PLANT BREEDING. *Educ:* McGill Univ, BSc, 52; Univ Wis, MS, 54, PhD(genetics), 57. *Prof Exp:* Res officer, Can Dept Agr, 57-66, sect head potato breeding, 66-73, prog mgr, 73-86. *Concurrent Pos:* Mem, Work Planning Comt Potato Breeding, 59-; chmn, Work Planning Comt Potato Texture, 64- *Mem:* Potato Asn Am; Can Soc Hort Sci; Genetics Soc Can. *Res:* Potato breeding; sample selection as related to potato quality; disease resistance. *Mailing Add:* 1289 Lincoln Rd RR 1 Fredericton NB E3B 4X2 Can

YOUNG, DONALD C, b Paducah, Ky, Feb 25, 33; m 51; c 4. INORGANIC CHEMISTRY, AGRICULTURAL CHEMISTRY. *Educ:* Univ Calif, Riverside, BA, 61, PhD(inorg chem), 66. *Prof Exp:* From res asst petrochem to sr res chemist, 53-69, from res assoc to sr res assoc fertilizer chem, 69-78, staff consult, Union Oil Co, 78-89, SR STAFF CONSULT, SCI & TECHNOL DIV, UNOCAL, 89- *Concurrent Pos:* Res scholar, Univ Calif, Riverside, 66-68. *Mem:* Am Chem Soc. *Res:* Chemistry and technology of polyphosphoric acid; transition-metal complexes in homogeneous catalysis; carborane chemistry; transition metal complexes of dicarbollide ion; fertilizer and soil chemistry; herbicide and plant growth regulator chemistry. *Mailing Add:* PO Box 76 Brea CA 92621

YOUNG, DONALD CHARLES, b Fremont, Ohio, June 29, 44; div; c 3. ANALYTICAL CHEMISTRY. *Educ:* Harvard Univ, AB, 66; Univ NC, Chapel Hill, PhD(anal chem), 71. *Prof Exp:* Res assoc chem, Purdue Univ, 71-72; asst prof chem, Oakland Univ, 72-78; proj chemist, Gulf Oil Corp, 78-84, sr res chemist, 84; sr res chemist, 85-89, RES ASSOC, CHEVRON RES CO, 89- *Mem:* Am Chem Soc; Sigma Xi; Soc Appl Spectros. *Res:* Nuclear magnetic resonance methods of analysis; infrared spectroscopy; process analysis; polymer analysis. *Mailing Add:* Chevron Res Co PO Box 1627 Richmond CA 94802-0627

YOUNG, DONALD EDWARD, b Lake Zurich, Ill, June 13, 22; m 47; c 3. HIGH ENERGY PHYSICS. *Educ:* Ripon Col, BA, 46; Univ Minn, MS, 51, PhD(nucleon scattering), 59. *Prof Exp:* Asst physics, Univ Minn, 49-53; physicist, Labs, Gen Mills Co, 53-60; head, Physics Div, Midwestern Univs Res Asn, 60-67; prof physics, Univ Wis, 67; physicist, 67-90, EMER SCIENTIST, FERMI NAT ACCELERATOR LAB, 90- *Concurrent Pos:* Consult, G H Gillespie Assoc, 89-, Super Conducting Super Collider Lab, 90-; pres, Particle Accelerator Corp, 90- *Mem:* Fel Am Phys Soc; Sigma Xi; Am Asn Advan Sci; Inst Elec & Electronics Engrs. *Res:* Proton linear accelerator design; high energy particle accelerators design and operation; magnetic field measurements; nuclear physics and radioactivity; proton-proton scattering; dosimetry; radiation damage; antiproton production and collection. *Mailing Add:* Fermi Nat Accelerator Lab PO Box 500 Batavia IL 60510

YOUNG, DONALD F(REDRICK), b Joplin, Mo, Apr 27, 28; m 50; c 5. ENGINEERING MECHANICS, BIOMEDICAL FLUID ENGINEERING. *Educ:* Iowa State Univ, BS, 51, MS, 52, PhD(eng mech), 56. *Prof Exp:* Res assoc, 52-53, from instr to assoc prof eng mech, 52-61, prof eng sci & mech, 61-74, DISTINGUISHED PROF ENG, IOWA STATE UNIV, 74- *Mem:* Am Soc Eng Educ; fel Am Soc Mech Engrs. *Res:* Fluid mechanics; biomechanics. *Mailing Add:* Dept Aerospace Eng & Eng Mech Iowa State Univ Ames IA 50011

YOUNG, DONALD RAYMOND, b Beaver Falls, Pa, May 11, 54; m 76; c 2. PHYSIOLOGICAL PLANT ECOLOGY, COMPUTER MODELING. *Educ:* Clarion Univ Pa, BS, 75; Univ Wyo, MS, 79, PhD(bot), 82. *Prof Exp:* Asst prof bot, Univ Wyo, 82-83; res scholar phys ecol, Univ Calif, Los Angeles, 83-84; ASST PROF BIOL, VA COMMONWEALTH UNIV, 84- *Mem:* Ecol Soc Am; Bot Soc Am; Sigma Xi. *Res:* Effects of microclimate, crown architecture and leaf display on the water relations, photosynthesis and small-scale distribution patterns of understory plants; physiological ecology of Paulownia tomentosa as related to use in reclamation. *Mailing Add:* Lab Biomed & Environ Sci Univ Calif 900 Veterans Ave Los Angeles CA 90024

YOUNG, DONALD REEDER, b Logan, Utah, July 21, 21; m 46; c 4. PHYSICS. *Educ:* Utah State Agr Col, BA, 42; Mass Inst Technol, PhD(physics), 49. *Prof Exp:* Mem staff, Radiation Lab, Mass Inst Technol, 42-45, asst, Insulation Lab, 45-49; tech engr, 49-52, proj engr, 52-61, sr engr & mgr device & mat characterization, 61-71, res staff mem, IBM Corp, 71-86; PROF PHYSICS, LEHIGH UNIV, 86- *Concurrent Pos:* Alexander von Humboldt US sr scientist award, 80-81; mgr interface physics, IBM Corp, 77-86. *Mem:* Fel Am Phys Soc; fel Inst Elec & Electronics Engrs. *Res:* Electrical breakdown; ferroelectric materials; superconductors; semiconductors. *Mailing Add:* Fairchild Lab Lehigh Univ Bethlehem PA 18015

YOUNG, DONALD STIRLING, b Belfast, Northern Ireland, Dec 17, 33; c 3. CLINICAL PATHOLOGY. *Educ:* Aberdeen Univ, MB & ChB, 57; Univ London, PhD(chem path), 62. *Prof Exp:* Lectr mat med, Aberdeen Univ, 58-59; resident chem path, Royal Postgrad Med Sch London, 62-64; vis scientist, NIH, 65-67, chief clin chem, 67-77; head clin chem, Mayo Clin, 77-84; PROF, DEPT PATH & LAB MED, UNIV PA, 84- *Concurrent Pos:* Chmn bd ed, Clin Chem, Am Asn Clin Chemists, 73-78 & 85-87, pres, 80. *Honors & Awards:* J H Roe Award, Am Asn Clin Chemists, 73, Ames Award, 77; Bernard Gerulat Award, Am Asn Clin Chemists, 77, Van Slyke Award, 85; NIH Dir Award, 77; Gerard B Lambert Award, Gerard B Lambert Awards Orgn, 75; W Roman Lectr, Australian Asn Clin Biochem, 79; Gerald T Evans Award, Acad Clin Lab Physicians & Scientists, 81; Corning Lectr, Asn Clin Biochem, 85. *Mem:* Brit Asn Clin Biochem; Am Asn Clin Chemists (pres, 80); Acad Clin Lab Physicians & Scientists; Int Fedn Clin Chem (vpres, 82-85, pres, 85-90). *Res:* Clinical chemistry; development and application of high resolution analytical techniques in the clinical laboratory; optimized use of the clinical laboratory. *Mailing Add:* Dept Path & Lab Med Hosp Univ Pa 3400 Spruce St Philadelphia PA 19104-4283

YOUNG, EARLE F(RANCIS), JR, b Pittsburgh, Pa, Aug 27; wid; c 3. ENVIRONMENTAL CONTROL, PROCESS METALLURGY. *Educ:* Carnegie Inst Technol, BS, 49. *Prof Exp:* Res engr chem eng, Babcock & Wilcox Co, 49-51; res engr, Olin Mathieson Chem Corp, 51-54, proj group leader, 54-56; process engr, Jones & Laughlin Steel Corp, 56-57, sr develop engr, 57-58, res supvr, 58-61, supvr ore res, 61-65, asst dir chem & raw mat serv, Tech Serv Div, 65-69, dir environ control, 69-72, gen mgr, 73-76; dir environ affairs, Am Iron & Steel Inst, 76-78, asst vpres, 78-80, vpres energy & environ, 80-85, environ, 80-87, environ & energy, 87-89; RETIRED. *Concurrent Pos:* Instr, Carnegie Inst Technol, 57-58. *Mem:* Am Chem Soc; Am Inst Chem Engrs; Soc Mining Engrs; Am Iron & Steel Inst. *Res:* Upgrading of iron ores; fluid dynamics of steelmaking; air and stream pollution. *Mailing Add:* 1600 N Oak St N0720 Arlington VA 22209-2764

YOUNG, EDMOND GROVE, b Govans, Md, Oct 29, 17; m 46; c 3. FLUORINE CHEMISTRY. *Educ:* Univ Md, BS, 38, PhD(org chem), 43. *Prof Exp:* Asst chem, Univ Md, 38-43; chemist, Sharples Chem, Inc, Mich, 43-44; res chemist, E I du Pont de Nemours & Co, Inc, 44-48; tech sales, Kinetics Chem, Inc, 48-49, sales mgr aerosol propellants, 49-50; sales develop, Kinetic Chem Div, E I Du Pont de Nemours & Co, Inc, 50-52, mgr kinetic chem div, 52-57, mgr develop conf, 57-68, mgr develop conf & govt liaison, Develop Dept, 68-73, mgr bus develop, Cent Res & Develop Dept, 73-82; CONSULT, 82- *Concurrent Pos:* Mem, Franklin Inst; consult, NASA, 82-83, Jet Propulsion Lab, 84-88. *Mem:* AAAS; Am Chem Soc; Com Develop Asn. *Res:* Reaction of metallo-organics; chemistry of fluorinated compounds; commercial chemical development and marketing. *Mailing Add:* PO Box 67 Mickleton NJ 08056-0067

YOUNG, EDWARD JOSEPH, b Roselle, NJ, Feb 18, 23; m 55; c 2. GEOCHEMISTRY, MINERALOGY. *Educ:* Rutgers Univ, BS, 48; Mass Inst Technol, MS, 50, PhD(geol), 54. *Prof Exp:* GEOLOGIST, US GEOL SURV, 52- *Mem:* Geol Soc Am; Mineral Soc Am; Soc Econ Geol; Mineral Asn Can. *Res:* Petrology; petrology of granitic rocks. *Mailing Add:* US Geol Surv Fed Ctr M5973 Denver CO 80225

YOUNG, EDWIN H(AROLD), b Detroit, Mich, Nov 4, 18; m 44; c 2. CHEMICAL ENGINEERING, METALLURGICAL ENGINEERING. *Educ:* Univ Detroit, BChE, 42; Univ Mich, MSE, 49 & 51. *Prof Exp:* Jr chem engr, Wright Air Develop Ctr, 42-43; instr chem eng, Univ Detroit, 46-47; from instr to prof, 47-89, EMER PROF CHEM & METALL ENG, UNIV MICH, ANN ARBOR, 89-; CONSULT ENGR, 47- *Honors & Awards:* Award, Nat Soc Prof Engrs, 77; Donald Q Kern Award, Am Inst Chem Engrs, 79. *Mem:* Am Chem Soc; fel Am Soc Mech Engrs; fel Am Inst Chem Engrs; fel Am Inst Chemists; Nat Soc Prof Engr (pres, 68-69); fel Am Soc Heating Refrig & Air Conditioning Engrs. *Res:* Heat transfer; process design; design of process equipment; forensic engineer; technical expert on explosions and fires. *Mailing Add:* Dept Chem & Metall Eng Univ Mich Ann Arbor MI 48109

YOUNG, ELEANOR ANNE, b Houston, Tex, Oct 8, 25. NUTRITION. *Educ:* Incarnate World Col, BA, 47; St Louis Univ, MEd, 55; Univ Wis, PhD(nutrit), 68. *Prof Exp:* Asst prof foods & Nutrit, Incarnate World Col, 53-63 & 68-72; sr res assoc, Univ Tex Health Sci Ctr, Dept Med, 68-72, asst prof med, 72-77; assoc prof med, 77-87, PROF MED, DEPT MED, UNIV TEX HEALTH SCI CTR, SAN ANTONIO, 87-; ASSOC PROF FOODS & NUTRIT, INCARNATE WORD COL, 72- *Concurrent Pos:* Consult, Audie Murphy Vet Admin Hosp, San Antonio, 73-, Med Ctr Hosp, 73- *Mem:* Am Inst Nutrit; Am Bd Human Nutrit; Am Soc Clin Nutrit; Am Dietetic Asn; Am Pub Health Asn. *Res:* Metabolic response and feeding-fasting intervals in man; metabolism of intravenously infused maltose; lactose intolerance; nutritional adaptations after partial small bowel resections; effect of semistarvation diets on the gastrointestinal tract and heart; effect of gastric stapling on the gastrointestinal tract; nutrition, health and aging. *Mailing Add:* Univ Tex Health Sci Ctr Dept Med 7703 Floyd Curl San Antonio TX 78284-7878

YOUNG, ELIZABETH BELL, b Franklinton, NC, July 2, 29. SPEECH PATHOLOGY, AUDIOLOGY. *Educ:* NC Col Durham, AB, 48, MA, 50; Ohio State Univ, PhD(speech sci, speech & hearing ther), 59. *Prof Exp:* Chmn dept English, Barber-Scotia Col, 48-53; chmn dept speech, Talladega Col, 53-55; asst prof, Va State Col, 56-57; prof speech correction & dir speech clin, Fla A&M Univ, 59; chmn dept English & speech, Fayetteville State Col, 59-63; asst prof speech path, Col Dent, Howard Univ, 63-64; chmn dept English & lang, Md State Col, 65-66; prof speech path, undergrad & grad prog & supvr speech & audiol training clin prog, Cath Univ Am, 66-79, supvr speech clin, 69-79; staff aide, US House Rep, 80-82; prof commun, Univ DC, 82-83; CONSULT & LECTR, LOCAL NAT AGENCIES & ORGN, 83- *Concurrent Pos:* Consult, nat orgns, mgt firms, etc, 83- *Mem:* Fel Am Speech & Hearing Asn. *Res:* Pathology of speech and hearing mechanism and speech science; observations in the field of speech and hearing pathology and science. *Mailing Add:* 8104 W Beach Dr NW Washington DC 20012

YOUNG, ELTON THEODORE, b Brush, Colo, May 1, 40; m 62; c 3. MOLECULAR BIOLOGY. *Educ:* Univ Colo, BA, 62; Calif Inst Technol, PhD(biophys), 67. *Prof Exp:* Fel molecular biol, Univ Geneva, 67-69; from asst prof to assoc prof, 69-81, PROF BIOCHEM & GENETICS, UNIV WASH, 81- *Mem:* Fedn Am Soc Exp Biol. *Res:* Regulation of gene expression and protein transport in yeast. *Mailing Add:* Dept Biochem Univ Wash SJ-70 Seattle WA 98195

YOUNG, ERIC D, b Elko, Nev, July 9, 45; m; c 2. BIOMEDICAL ENGINEERING. *Educ:* Calif Inst Technol, BS, 67; Johns Hopkins Univ, PhD(biomed eng), 72. *Prof Exp:* Postdoctoral fel, Dept Pharmacol & Physiol Sci, Univ Chicago, 72-74, res assoc, 74-75; from asst prof to assoc prof, Dept Biomed Eng, Sch Med, Johns Hopkins Univ, 75-87, assoc prof, Dept Neurosci, 81-87 & Dept Otolaryngol, 85-87, PROF, DEPTS BIOMED ENG, NEUROSCI & OTOLARYNGOL, SCH MED, JOHNS HOPKINS UNIV, 87- *Concurrent Pos:* NIH postdoctoral fel, 72-74, res career develop award, 79-83; coordr, Undergrad Prog Biomed Eng, Whiting Sch Eng, Johns Hopkins Univ, 81-; mem, tech comt Physiol & Psychol Acoust, Acoust Soc Am, 77-80, Commun Dis Rev Comt, Nat Inst Neurol Commun Dis & Stroke; ad hoc mem, Commun Sci Study Sect, NIH, 80, Hearing Sci Study Sect, 83; ad hoc grant reviewer, Whittaker Found, 87-90; ad hoc manuscript reviewer, J Neurophysiol, Hearing Res, J Acoust Soc Am; consult, Res Triangle Inst, 90. *Mem:* Sr mem Biomed Eng Soc; fel Acoust Soc Am; Asn Res Otolaryngol; Soc Neurosci; AAAS; Sigma Xi. *Res:* Biomedical engineering; neuroscience; otolaryngology/head and neck surgery. *Mailing Add:* Dept Biomed Eng Johns Hopkins Sch Med 720 Rutland Dr Baltimore MD 21205

YOUNG, EUTIQUIO CHUA, b Del Gallego, Philippines, July 17, 32; m 61; c 3. MATHEMATICAL ANALYSIS. *Educ:* Far Eastern Univ, Manila, BS, 54; Univ Md, MA, 60, PhD(math), 62. *Prof Exp:* Asst prof math, Univ Conn, 61-62; head dept math, Far Eastern Univ, Manila, 62-64; assoc prof, De la Salle Col, Manila, 64-65; from asst prof math to assoc prof, 65-74, PROF MATH, FLA STATE UNIV, 74- *Mem:* Math Asn Am; Am Math Soc. *Res:* Cauchy problems, uniqueness of solutions of boundary value problems and comparison theorems for partial differential equations. *Mailing Add:* Dept Math Fla State Univ Tallahassee FL 32306

YOUNG, FRANCIS ALLAN, b Utica, NY, Dec 29, 18; m 45; c 2. EYE VISION & PRIMATOLOGY. *Educ:* Tampa Univ, Fla, BS, 41; Case Western Reserve Univ, MA, 45; Ohio State Univ, PhD(psychol), 49. *Prof Exp:* From instr to prof psychol, 48-88, DIR, PRIMATE RES CTR, WASH STATE UNIV, 57-, EMER PROF, 88- *Concurrent Pos:* Consult, Ment Health Res Inst, Wash, 54-67, Vet Admin Blind Rehab Ctr, Palo Alto, Calif, 74-84, Nat Inst Neurol, Commun Dis & Stroke, 76-84; vis res prof, Sch Med, Univ Ore, 63-64, Med Sch, Univ Uppsala, Sweden, 71 & Optom Sch, Univ Houston, 79-81; actg asst dir, Regional Primate Res Ctr, Univ Wash, 66-68; chmn, Gov's Comt Sexual Psychopath, Wash, 66-68; mem, res adv comt, Dept Insts, Wash, 69-70. *Honors & Awards:* Apollo Award, Am Optom Asn, 80. *Mem:* AAAS; fel Am Psychol Asn; Am Primatol Soc; Asn Res Vision & Ophthal; Int Soc Myopia Res (secy-treas, 78-); fel NY Acad Sci. *Res:* Role of genetics and environment in the development of visual refractive errors including myopia in humans and sub-human primates, with emphasis on the role of accommodation; mechanisms of accommodation; video display terminal syndrome; role of vision in motion sickness. *Mailing Add:* Primate Res Ctr Wash State Univ Pullman WA 99164-1170

YOUNG, FRANK, digital electronics, analog servomechanisms; deceased, see previous edition for last biography

YOUNG, FRANK COLEMAN, b Roanoke, Va, June 10, 35; m 59; c 3. EXPERIMENTAL PLASMA PHYSICS, NUCLEAR DIAGNOSTICS. *Educ:* Johns Hopkins Univ, BA, 57; Univ Md, PhD(nuclear physics), 62. *Prof Exp:* NSF res fel, US Naval Res Lab, 62-63; from asst prof to assoc prof physics, Univ Md, 63-72; RES PHYSICIST, US NAVAL RES LAB, 72- *Concurrent Pos:* Vis scientist, Sandia Nat Lab, 86-87; chmn, Int Conf Plasma Sci, Inst Elec & Electronic Engrs, 87, exec comt Plasma Sci & Applns comt, 91- *Mem:* Am Phys Soc; Inst Elec & Electronics Engrs; Nuclear & Plasma Sci Soc. *Res:* Experimental nuclear physics; application of nuclear techniques to studies of hot dense plasmas; plasmas for x-ray lasers. *Mailing Add:* 100 Mel Mara Dr Oxon Hill MD 20745-1019

YOUNG, FRANK E, b Mineola, NY, Sept 1, 31; m 56; c 5. BIOTECHNOLOGY, MICROBIOLOGY. *Educ:* State Univ NY, MD, 56; Western Reserve Univ, PhD(microbiol), 62. *Hon Degrees:* DSc, Roberts Wesleyan Col, Rochester, 83 & Houghton Col, Houghton, 84. *Prof Exp:* From intern to resident path, Univ Hosps, Cleveland, Ohio, 56-60; from instr to asst prof, Western Reserve Univ, 62-65; from assoc mem to mem, Depts Microbiol & Exp Path, Scripps Clin & Res Found, 65-70; prof microbiol, path, radiation biol & biophys & chmn dept microbiol, Sch Med & Dent, Univ Rochester, 70-79, dean, 79-84, vpres health affairs, 81-84; comnr, Food & Drug Admin, 84-89; DEP ASST SECY HEALTH, SCI & ENVIRON, DEPT HEALTH & HUMAN SERV, 89- *Concurrent Pos:* Am Cancer Soc res grant, 62-; NIH res grants, 65-, training grant, 70-; NSF res grant, 70-72; fac res assoc, Am Cancer Soc, 62-70; assoc prof, Univ Calif, San Diego, 67-70; dir clin microbiol labs, Strong Mem Hosp, 70-79, microbiologist-in-chief, 76-79; dir, Health Dept Labs, Monroe County, 70-79, mem bd, Am Asn Med Cols. *Honors & Awards:* Edward Mott Moore Award, 85. *Mem:* Inst Med-Nat Acad Sci; Am Soc Microbiol; AAAS; Am Acad Microbiol; Sigma Xi. *Res:* Mechanism of deoxyribonucleic and mediated transformation of bacterial and animal cells; regulation of bacterial cell surface; pathobiology of Neisseria gonorrhoeae. *Mailing Add:* Hubert H Humphrey Bldg Rm 701H 200 Independence Ave SW Washington DC 20201

YOUNG, FRANK GLYNN, b New York, NY, Dec 29, 16; m 41. CHEMISTRY, CATALYSIS. *Educ:* Dartmouth Col, AB, 37; Columbia Univ, PhD(org chem), 41. *Prof Exp:* Asst chem, Columbia Univ, 37-41; res chemist, Union Carbide Corp, 41-55, group leader radiation & isotope chem, 55-60, chem physics, Parma Res Labs, 60-63, sr res scientist, 63-83; CONSULT & PRES, CATALYSIS, INC, 85- *Concurrent Pos:* Lectr Catalysis, Ctr Prof Advan, 75-84 & McGraw-Hill, 77-81. *Mem:* Emer mem Am Chem Soc; Int Cong Catalysis. *Res:* Mechanisms of catalytic processes; isotopic tracer studies; heterogeneous catalysis and surfaces. *Mailing Add:* 3913 Butternut Ct Brandon FL 33511-7961

YOUNG, FRANK HOOD, b Baltimore, Md, Dec 31, 39; m 61; c 4. COMPUTER SCIENCE EDUCATION, SOFTWARE ENGINEERING. *Educ:* Haverford Col, BA, 61; Univ Pa, MA, 63, PhD(math), 68. *Prof Exp:* Instr math, Temple Univ, 65-68; from asst prof to prof, Knox Col, Ill, 68-87, chmn dept, 76-81; PROF COMPUT SCI & CHMN DEPT, ROSE-HULMAN INST TECHNOL, IND, 87- *Concurrent Pos:* Vis asst prof, Dept Comput Sci, Univ Ill, 73; Fulbright vis sr lectr, Dept Comput Sci, Univ Lagos, Nigeria, 75-76; vis assoc prof, Dept Comput Sci, Univ Iowa, 81-82. *Mem:* Asn Comput Mach; Math Asn Am. *Res:* Software engineering; reusable design schemes; analysis of algorithms; computer science education. *Mailing Add:* Dept Comput Sci Rose-Hulman Inst Technol 5500 Wabash Ave Terre Haute IN 47803

YOUNG, FRANK NELSON, JR, b Oneonta, Ala, Nov 2, 15; m 43; c 2. BIOLOGY. *Educ:* Univ Fla, BS, 38, MS, 40, PhD(biol), 42. *Prof Exp:* Asst prof biol, Univ Fla, 46-49; from asst prof to prof, 49-85, EMER PROF ZOOL, IND UNIV, BLOOMINGTON, 85- *Concurrent Pos:* Guggenheim fel, 60-61; fel, La State Univ, 63; res assoc, Fla Dept Agr, 72, US Nat Mus Natural Hist, 90. *Mem:* Am Soc Zool; Soc Study Evolution. *Res:* Taxonomy and ecology of aquatic Coleoptera; medical entomology; speciation and extinction of animals; land snails of genus Liguus. *Mailing Add:* Dept Biol Ind Univ Bloomington IN 47405-6801

YOUNG, FRANKLIN, b Beijing, China, Feb 1, 28; nat US; m 82. NUTRITION, BIOCHEMISTRY. *Educ:* Mercer Univ, AB, 51; Univ Fla, BSA, 52, MAgr, 54, PhD(nutrit), 60. *Prof Exp:* Asst vet sci, Univ Fla, 54-60, fel biochem, 60-61; res assoc, Bowman Gray Sch Med, 61-65, res instr prev med & assoc biochem, 65-66; assoc prof food & nutrit sci, Univ Hawaii, 66-83; prof & chair foods & nutrit, Univ Utah, 83-85; prof & chair health, 85-87, PROF, WEST CHESTER UNIV PA, 88- *Mem:* Am Inst Nutrit; NY Acad Sci; Am Chem Soc. *Res:* Atherosclerosis, lipid metabolism, hypertension and human nutrition. *Mailing Add:* Dept Health Chester Univ West Chester PA 19383

YOUNG, FRANKLIN ALDEN, JR, b Harrisburg, Pa, Mar 14, 38; m 59; c 3. MATERIALS SCIENCE. *Educ:* Univ Fla, BIE, 60, MSE, 63; Univ Va, DSc(mat sci), 68. *Prof Exp:* Instr metall, Clemson Univ, 63-65, asst prof mat eng, 68-70; assoc prof, 70-75, PROF DENT MAT & CHMN DEPT, COL DENT MED, MED UNIV SC, 75- *Concurrent Pos:* pres, Implant Res Group, Int Asn Dental Rec, 87-88; J ed, Dent Mat, 83- *Mem:* Am Soc Metals; Int Asn Dent Res; Sigma Xi; fel Acad Dent Mat; Mats Res Soc; Int Asn Dent Res. *Res:* Materials of medicine and dentistry; titanium alloy surfaces and their effects on dental and orthopaedic implants; CAD-CAM applications to dental and orthopaedic prosthesis. *Mailing Add:* Dept Mat Sci Med Univ SC 171 Ashley Ave Charleston SC 29425-2641

YOUNG, FRED M(ICHAEL), b Dallas, Tex, Aug 29, 40; m 63; c 2. HEAT TRANSFER, FLUID MECHANICS. *Educ:* Southern Methodist Univ, BSME, 63, MSME, 65, PhD(mech eng), 67. *Prof Exp:* Propulsion engr, Gen Dynamics/Ft Worth, 63-64; from asst prof to assoc prof mech eng, Lamar Univ, 67-74, dir grad eng studies, 72-74; head eng & appl sci, Portland State Univ, 74-79; DEAN ENG, LAMAR UNIV, 79- *Concurrent Pos:* Consult, Mobil Oil, Bethlehem Steel & Beaumont Yard. *Mem:* Am Soc Mech Engrs; Am Soc Eng Educ. *Res:* Two phase heat transfer; unsteady fluid flow; transient boiling; compter-aided design. *Mailing Add:* Col Eng Box 10057 Lamar Univ Beaumont TX 77710

YOUNG, FREDERICK GRIFFIN, b Niagara Falls, Ont, Nov 7, 40; m 63; c 2. PETROLEUM GEOLOGY. *Educ:* Queen's Univ, Ont, BSc, 63; McGill Univ, MSc, 64, PhD(geol), 70. *Prof Exp:* Geologist, Hudson's Bay Oil & Gas Co, 64-66; res scientist stratig, Inst Sedimentary & Petrol Geol, Geol Surv Can, 69-78; sr staff geologist, 78-80, chief geologist, 80-83, mgr, Frontiers Explor, 83-87, SR EXPLOR, HOME OIL CO, 87- *Mem:* Am Asn Petrol Geologists; Soc Econ Paleont & Mineral; Can Soc Petrol Geologists; Sigma Xi. *Res:* Physical stratigraphy; lithofacies analyses; geology of Upper Precambrian and Cambrian; clastic sedimentation; trace fossils; Mesozoic and Cenozoic geology of Mackenzie Delta area; petroleum geology. *Mailing Add:* Home Oil Co Ltd 324 Eighth Ave SW Calgary AB T2P 2Z5 Can

YOUNG, FREDERICK J(OHN), b Buffalo, NY, May 19, 31; m 54; c 3. ELECTRICAL ENGINEERING. *Educ:* Carnegie Inst Technol, BS, 53, MS, 54, PhD(elec eng), 56. *Prof Exp:* From instr to prof elec eng, Carnegie-Mellon Univ, 55-71; res engr, Labs, Westinghouse Elec Corp, 57-63, consult engr, 63-71, adv engr, 71-73; LSI CONSULT, 74- *Concurrent Pos:* Consult, Westinghouse Air Brake Co, 56, Union Switch & Signal Div, 57 & 60, Cornell Aeronaut Lab, Inc & Concrete Accessories Corp, 59, Oak Ridge Nat Labs, 74-, Westinghouse Res, 74-, Allied Chem Corp, 76-, TRW, Inc, 77-, ERDA, 77- & Mech Res, Inc, 77-; ed, Proceedings of Inst Elec & Electronics Engrs, 68; reviewer, Zent Br Math, E Ger Acad Sci, 70-; secy bd, Erebus Ltd, Eng, 75-; pres, Frontier Timber Co, 75- *Honors & Awards:* Young Elect Engr Award, 63. *Mem:* Inst Elec & Electronics Engrs; Sigma Xi. *Res:* Acoustical horns; electromagnetic field theory in ferrous media; magnetodydrodynamics; plasmadynamics; magneto-oceanography; magneto-mechanical devices; couple transmission lines in VLSI. *Mailing Add:* 800 Minard Run Rd Frontier Timber Co Bradford PA 16701

YOUNG, FREDERICK WALTER, JR, b Hebron, Va, Sept 13, 24; m 50; c 2. PHYSICAL CHEMISTRY. *Educ:* Hampden-Sydney Col, BS, 44; Univ Va, PhD(chem), 50. *Prof Exp:* Instr, Hampden-Sydney Col, 44-46; res assoc, 50-51, chemist, Solid State Div, 56-69, ASSOC DIR SOLID STATE DIV, OAK RIDGE NAT LAB, 69- *Concurrent Pos:* Res assoc, Univ Va, 51-56. *Mem:* Fel AAAS; Am Crystallog Asn; fel Am Phys Soc; Am Asn Crystal Growth; Mat Res Soc. *Res:* Chemical properties of metal surfaces; observations of dislocations in metals; radiation damage in metals. *Mailing Add:* 2900 Gallaher Ferry Rd Knoxville TN 37932

YOUNG, GALE, b Baroda, Mich, Mar 5, 12; m 49; c 2. NUCLEAR PHYSICS. *Educ:* Milwaukee Sch Eng, BS, 33; Univ Chicago, BS & MS, 36. *Prof Exp:* Asst math biophys res, Univ Chicago, 36-40; head dept math & physics, Olivet Col, 40-42; physicist, Manhattan Dist Proj, Univ Chicago, 42-46; physicist, Clinton Labs, Tenn, 46-48; tech dir, Nuclear Develop Assocs, 48-55, vpres, Nuclear Develop Corp Am, 55-61; div vpres, United Nuclear Corp, 61-62; asst dir, Oak Ridge Nat Lab, 67-71, consult, 71-88; RETIRED. *Concurrent Pos:* Mem sci adv bd, US Dept Air Force, 54-58. *Mem:* Soc Indust & Appl Math; Math Asn Am. *Res:* Nuclear reactors; applied mathematics. *Mailing Add:* 4534 High Vista Ln Knoxville TN 37931

YOUNG, GEORGE ANTHONY, b New York, NY, Nov 8, 19; m 49; c 4. METEOROLOGY, EXPLOSION PHENOMENA. *Educ:* NY Univ, BS, 48, MS, 49, PhD(meteorol), 65. *Prof Exp:* Res asst meteorol, NY Univ, 49-50; RES ASSOC, NAVAL SURFACE WARFARE CTR, 50- *Mem:* Am Meteorol Soc. *Res:* Micrometeorology; hydrodynamics; turbulence; underwater explosions; environmental effects; effects of conventional and nuclear explosions in air and water, including military damage and environmental impact. *Mailing Add:* 3611 Janet Rd Silver Spring MD 20906-4353

YOUNG, GEORGE JAMISON, b Hornell, NY, Aug 31, 25; m 46; c 3. PHYSICAL CHEMISTRY. *Educ:* Rensselaer Polytech Inst, BS, 50; Lehigh Univ, MS, 52, PhD(phys chem), 54. *Prof Exp:* Instr phys chem, Lehigh Univ, 54-55; fel, Mellon Inst, 55-56; from asst prof to assoc prof, Pa State Univ,

56-58; prof chem, Alfred Univ, 58-61; pres, Surface Processes Corp, 61-69; DIR CORP ENG, PITNEY-BOWES, INC, 69- *Mem:* Royal Soc Chem. *Res:* Engineering science; operations analysis; research administration. *Mailing Add:* 125 Burlington Rd Harwinton CT 06791

YOUNG, GEORGE ROBERT, b Monmouth, Ill, Mar 9, 25; m 46; c 4. BIOLOGICAL CHEMISTRY. *Educ:* Univ Ind, BS, 49, PhD(biol chem), 56; Northwestern Univ, MS, 52. *Prof Exp:* Fel biol chem, Univ Ind, 55-56; instr, Dent Br, Univ Tex, 56-57, asst prof, 57-62; from assoc prof to prof biochem & nutrit, Sch Dent, 62-84, coord grad studies, 67-72, prof med, Sch Med, 73-84, PROF BIOCHEM & NUTRIT, SCH BASIC LIFE, UNIV MO-KANSAS CITY, 85-, CHMN DEPT, SCH DENT, 62- *Mem:* Am Chem Soc; Int Asn Dent Res. *Res:* Collagen and collagenase; invasiveness of cells; solubility of tooth enamel; nutrition and periodontal metabolism. *Mailing Add:* 66210-1121 Overland Park KS 66210

YOUNG, GERALD A, b Plainwell, Mich, Jan 9, 36; m 76; c 3. NEUROPSYCHOPHARMACOLOGY. *Educ:* Colo Col, BA, 58; Western Mich Univ, MA, 68; McMaster Univ, PhD(psychol), 73. *Prof Exp:* Res psychol, Kalamazoo State Hosp, 72-73; asst prof, Western Mich Univ, 73-74; res assoc, 74-81, SR RES ASSOC, UNIV MD, BALTIMORE, 81- *Concurrent Pos:* Adj asst prof pharmacol, Univ Md, Baltimore, 78-79, adj assoc prof, 79- *Mem:* Am Soc Pharmacol & Exp Therapeut; Soc Neurosci. *Res:* Neuropsychopharmacology of psychoactive drugs; electroencephalogram spectral analysis; self-administration of drugs of abuse by experimental animals; behavioral pharmacology. *Mailing Add:* Dept Pharmacol & Toxicol Sch Pharm Univ Md 20 N Pine St Baltimore MD 21201

YOUNG, GILBERT FLOWERS, JR, b Mayesville, SC, Sept 23, 22. NEUROLOGY, PEDIATRICS. *Educ:* Col Charleston, BA, 42; Univ NC, MA, 46, Med Col SC, MD, 47; Am Bd Pediat, dipl, 58; Am Bd Psychiat & Neurol, dipl, 66, cert neurol with spec competence child neurol, 69. *Prof Exp:* Intern, Med Col Va Hosp, 47-48; instr pharmacol, Univ NC, 48-49; resident pediat, Roper Hosp, Charleston, SC, 51-53; pvt pract, 53-56; resident child develop, Children's Hosp, Columbus, Ohio, 56-57; from asst prof to assoc prof pediat, 57-71, from asst prof to assoc prof neurol, 60-71, PROF NEUROL & PEDIAT, MED UNIV SC, 71- *Concurrent Pos:* Resident, Mass Gen Hosp, 60-63; consult, Med Univ SC Hosp & Roper Hosp, Charleston. *Mem:* AMA; Am Acad Pediat; Am Acad Neurol; Sigma Xi. *Res:* Child neurology and development; congenital encephalopathies. *Mailing Add:* 1057 S Shem Dr Mt Pleasant SC 29464

YOUNG, GLENN REID, b Kingsport, Tenn, Aug 22, 51; m 80; c 2. NUCLEAR PHYSICS. *Educ:* Univ Tenn, BA, 73; Mass Inst Technol, PhD(physics), 77. *Prof Exp:* Fel Mass Inst Technol, 77-78; fel nuclear physics, 78-80, mem res staff, 80-87, GROUP LEADER, OAK RIDGE NAT LAB, 87- *Mem:* Sigma Xi; Am Physical Soc. *Res:* Heavy ion macrophysics and reactions; ultra relativistic heavy ion reactions. *Mailing Add:* Physics Div Bldg 6003 PO Box 2008 Oak Ridge Nat Lab Oak Ridge TN 37831-6373

YOUNG, GRANT MCADAM, b Glasgow, Scotland, Aug 23, 37; m 60; c 3. STRATIGRAPHY, SEDIMENTOLOGY. *Educ:* Glasgow Univ, BSc, 60, PhD(geol), 67. *Prof Exp:* Res demonstr geol, Univ Wales, 62-63; from lectr to assoc prof, 63-78, PROF GEOL, UNIV WESTERN ONT, 78- *Concurrent Pos:* Nat Res Coun Can & Geol Surv Can grants, Univ Western Ont, 65-; co-leader Int Geol Correlation Prog Projs, 72- *Mem:* Soc Econ Paleont & Mineral; Geol Soc Am; Geol Asn Can. *Res:* Paleoclimatology glacigenic rocks in global correlation; Huronian rocks of Ontario; Upper Precambrian rocks of Arctic Canada, Canadian Cordillera and South Australia; geochemistry of sedimentary rocks. *Mailing Add:* Dept Geol Univ Western Ont London ON N6A 5B7 Can

YOUNG, H(ENRY) BEN, JR, b Rockdale, Tex, Oct 13, 13; m 41; c 2. TECHNICAL MANAGEMENT, MECHANICAL ENGINEERING. *Educ:* Rice Univ, BSME, 37; Harvard Univ, AMP, 58. *Prof Exp:* From mech engr to chief engr & dir overseas projs, Mission Mfg Co, Houston, Tex, 37-56; dir eng & res, W-K-M valve div, ACF Industs, Inc, Houston, Tex, 56-59, corp vpres eng & res, New York, 59-63, vpres mgr & res, 63-65, vpres & gen mgr, W-K-M Valve Div, Houston, Tex, 65-72; dir capital expenditures, Marathon Mfg Co, Houston, 72-73; mgr & eng consult, Houston, 73-75 & 77-78; chmn bd dir & pres, CEO Chronister Valve Co, Houston, 75-77; vpres admin, Lanzagorta Int Inc, 78-84; RETIRED. *Concurrent Pos:* Consult mgt & eng, 84- *Mem:* Am Soc Mech Engrs; Nat Soc Prof Engrs. *Mailing Add:* 10122 Del Monte Houston TX 77042

YOUNG, HARLAND HARRY, b Portland, Ore, July 29, 08; m 32; c 2. ORGANIC CHEMISTRY. *Educ:* Reed Col, BA, 29; Mass Inst Technol, PhD(org chem), 32. *Prof Exp:* From anal chemist to res chemist, Swift & Co, 24-36, in charge new prod develop div, 36-39, asst to chief chemist, 39-41, asst chief chemist, 41-46, from asst dir res to dir res, 46-70; pres, Res Adv Serv, Inc, 71-82; RETIRED. *Mem:* Am Chem Soc; Am Leather Chem Asn; Am Inst Chem Engrs. *Res:* Adhesives; foods; fats; industrial oils; colloids; emulsions; detergents; proteins; in-plant pollution control; byproduct utilization. *Mailing Add:* 4724 Wolf Rd Western Springs IL 60558

YOUNG, HAROLD EDLE, b Arlington, Mass, Sept 4, 17; m 43; c 4. FORESTRY. *Educ:* Univ Maine, BS, 37; Duke Univ, MF, 46, PhD(tree physiol), 48. *Prof Exp:* Field asst, US Forest Serv, 37-40; from instr to prof forestry, Univ Maine, 48-82; BIOMASS MGR, JAMES W SEWALL CO, 82- *Concurrent Pos:* Fulbright res scholar, Norway, 63-64; vis appointment, Dept Forestry, Australian Nat Univ, 68-69; consult, Off Opers Anal, US Air Force, 51-59. *Honors & Awards:* Hitchcock Award, Forest Prod Res Soc Am, 74; Burckhart-Medaille, Fac Forestery, Univ Gottingen, 80. *Mem:* Fel AAAS; Soc Am Foresters; corresp mem Soc Forestry Finland. *Res:* Soils; biomass studies within complete forest concept. *Mailing Add:* 77 or 71 Forest Ave Orono ME 04473

YOUNG, HAROLD HENRY, b Malone, NY, Sept 11, 27; m 51; c 5. ANALYTICAL CHEMISTRY. *Educ:* St Michael's Col, Vt, BS, 51. *Prof Exp:* Chem tech, Works Lab, Gen Elec Co, 51-52; shift supvr, Ind Ord Works, E I du Pont de Nemours & Co, Inc, Ind, 52-54, lab supvr, Savannah River Plant, SC, 54-57; tech reviewer, Div Civilian Appln, US Dept Energy, Oak Ridge, Tenn, 57-58, non-destructive testing specialist, Div Isotopes Develop, Washington, DC, 58-59, isotopes training specialist, 59-62, nuclear educ & training specialist, 62-73, educ & training specialist, Div Biomed & Environ Res, 73-75, sr training coordr, Div Univ & Manpower Develop Progs, US AEC, 75- 77, chief, Instnl Prog Br, Educ Progs Div, 77-85, chief, Lab Prog Br, Univ & Ind Prog Div, Off Energy Res, US Dept Energy, Washington, DC, 82-99; RETIRED. *Mem:* Am Nuclear Soc; AAAS. *Res:* Quality control of nitrocellulose, plutonium and special nuclear materials; radioisotope and radiation applications. *Mailing Add:* Ten Holly Gaithersburg MD 20545

YOUNG, HARRISON HURST, JR, b Drumright, Okla, Sept 24, 19; m 42; c 2. PHYSICAL CHEMISTRY. *Educ:* Princeton Univ, AB, 40; Columbia Univ, PhD(chem), 50; Fordham Univ, JD, 74. *Prof Exp:* Asst, Nat Defense Res Comt, 41-45; instr chem, Williams Col, Mass, 47-50; res group leader, Wesvaco Div, Food Mach & Chem Corp, FMC Corp, 50-52, res sect mgr, 52-56, asst to dir res, Westvaco Mineral Prod Div, 56-58, mgr detergent applns res, Inorg Res & Develop Dept, 59-62, tech recruitment mgr, Chem Div, 62-70, mem staff chem group, Patent & Licensing Dept, 70-85, patent attorney, 75-85; RETIRED. *Mem:* Am Chem Soc. *Res:* Reaction kinetics; industrial inorganic chemicals; detergent applications; agricultrual pesticides. *Mailing Add:* 529 Custis Rd Glenside PA 19038-2011

YOUNG, HENRY H(ANS), b Des Moines, Iowa, June 28, 20; m 45. CHEMICAL ENGINEERING. *Educ:* Iowa State Univ, BS, 42. *Prof Exp:* Asst engr, Midwest Res Inst, 46-48, assoc engr, 48-55, sr engr, 55-65; res coordr, Res Found Kans, 65-67; INDUST SPECIALIST, UNIV MO-KANSAS CITY, 67- *Concurrent Pos:* Chief chem sect, Joint US-Brazil Mil Comn, Rio de Janeiro; US Army Officer, Chem Corps, 42-46. *Mem:* Am Inst chem Engrs; Nat Soc Prof Engrs; Nat Asn Mgt & Tech Assistance Ctrs. *Res:* Combustion; heat transfer; filtration; extraction; evaporation; fire protection engineering; information transfer; research proposal evaluation. *Mailing Add:* 1118 E 41st St Kansas City MO 64110

YOUNG, HEWITT H, b Willoughby, Ohio, May 16, 23; m 45; c 6. INDUSTRIAL & HUMAN FACTORS ENGINEERING. *Educ:* Case Inst Technol, BSME, 44, MSIE, 50;Ariz State Univ, PhD, 66. *Prof Exp:* Engr, Lamp Develop Lab, Gen Elec Co, Ohio, 46-48; instr graphics, Case Inst Technol, 48-50; indust engr jet engines, Tapco Div, Thompson Prod, Inc, Ohio, 50-53; prof indust eng, Purdue Univ, 53-67; chmn indust eng fac, 67-76, prof, 67-88, EMER PROF ENG, ARIZ STATE UNIV, 88- *Concurrent Pos:* Chmn, Col-Indust Comt Mat Handling Educ, 66-68 & Nat Coun Indust Eng Acad Dept Heads, 71-72; NASA/Am Soc Eng Educ fel, 67 & US Air Force/Am Soc Eng Educ fel, 70. *Mem:* Am Soc Eng Educ; fel Am Inst Indust Engrs (vpres, 72-75, 77-79). *Res:* Large-scale systems analysis; performance analysis; human engineering. *Mailing Add:* Dept Indust Eng Ariz State Univ Tempe AZ 85287-5906

YOUNG, HO LEE, b Canton, China, July 15, 20; nat US; m 49; c 1. ENVIRONMENT SCIENCES. *Educ:* Lingnan Univ, BS, 43; Univ Calif, PhD, 54. *Prof Exp:* Asst gen biol & comp anat, Nat Med Col Shanghai, China, 44-49; student res physiologist, Univ Calif, 52-54, jr res physiologist, 54-56; res assoc radiol, Stanford Univ, 56-57; asst res physiologist, Med Ctr, Univ Calif, San Francisco, 57-63; res scientist, NASA Ames Res Ctr, 63-72; chief chem lab, 72-74, chief air lab, 74-76, chief lab br, 74, chief water sect, 76-79, Region 9 Qual Assurance Officer, US Environ Protection Agency, 79-85. *Concurrent Pos:* Sr res fels, San Francisco Heart Asn, 60-62; Abraham Rosenberg res fel, 50-52. *Mem:* Am Chem Soc; Sigma Xi; Chinese Am Chem Soc. *Res:* Cell physiology; cell particulates; cellular and fat metabolism; enzyme systems; protein synthesis; radiation effect on cells; electrolyte transport in skeletal muscle and isolated cells; effect of toxic materials and environmental stresses on biochemical and physiological process of living systems. *Mailing Add:* 5978 Greenridge Rd Castro Valley CA 94546

YOUNG, HOBART PEYTON, b Evanston, Ill, Mar 9, 45. APPLIED MATHEMATICS, ECONOMICS. *Educ:* Harvard Univ, BA, 66; Univ Mich, PhD(math), 70. *Prof Exp:* Economist, Nat Water Comn, 71; assoc prof math, Grad Sch, City Univ New York, 71-75; res scholar & dept chmn, Int Inst Appl Systs Anal, 76-81; PROF PUB POLICY, UNIV MD, 81- *Concurrent Pos:* Vis prof, Univ Bonn, Univ Paris & Yale Univ; consult, numerous govt agencies; dir grants, US Army Res Off, 73-74 & NSF, 75-82. *Honors & Awards:* Lester R Ford Award, Math Asn Am, 76. *Mem:* Economet Soc; Am Polit Sci Asn; Opers Res Soc Am. *Res:* Public economics; mathemattical models of voting and representation; environmental policy. *Mailing Add:* Ctr Int Sec Studies Sch Pub Affairs Univ Md College Park MD 20742

YOUNG, HONG YIP, b Wailuku, Hawaii, Nov 27, 10; m 37; c 3. AGRICULTURAL CHEMISTRY. *Educ:* Univ Hawaii, BS, 32, MS, 33. *Prof Exp:* Sci aide, Pineapple Res Inst, 33-40, from jr chemist to chemist, 40-67; from assoc agronomist to agronomist, 67-75, EMER AGRONOMIST, AGR EXP STA, COL TROP AGR, UNIV HAWAII, 75- *Concurrent Pos:* Vis scientist, Int Rice Res Inst, 65-66; consult, Indian Agr Res Inst, 68. *Mem:* Am Chem Soc; Sigma Xi. *Res:* Plant, soil, hormone and pesticide residue analysis; mineral nutrition of plants. *Mailing Add:* 676 Hakaka St Honolulu HI 96816

YOUNG, HOWARD ALAN, b Ossining, NY, Jan 29, 48. MOLECULAR IMMUNOLOGY. *Educ:* Univ Mass, BS, 69; Univ Wash, MS, 72, PhD(microbiol), 74. *Prof Exp:* Postdoctoral fel, Nat Cancer Inst, 74-76, staff fel, 76-79; head, Tech Serv, Bethesda Res Labs, 81-83; staff scientist, 81-83 & 83-88, HEAD, CELLULAR & MOLECULAR IMMUNOL SECT, NAT CANCER INST, FREDERICK CANCER RES & DEVELOP CTR, 88- *Mem:* Am Soc Microbiol; Int Interferon Soc; AAAS; Am Asn Immunologists. *Res:* Investigation of gene regulation in the immune system with special emphasis on interferon gamma; gene regulation and large granular lymphocyte specific gene expression. *Mailing Add:* Cellular & Molecular Immunol Sect Nat Cancer Inst FCRDC Lab Exp Immunol Bldg 560 Rm 31-23 PO Box B Frederick MD 21702-1201

YOUNG, HOWARD FREDERICK, b Fond du Lac, Wis, Nov 13, 18; m 42; c 3. ZOOLOGY. *Educ:* Univ Wis, BA, 46, MA, 47, PhD, 50. *Prof Exp:* Instr, Univ Ark, 50, asst prof, 50-53; assoc prof biol, Western Ill State Col, 53-55; from asst prof to assoc prof, 55-63, PROF BIOL, UNIV WIS, LA CROSSE, 63- *Concurrent Pos:* NSF grant, 59-60. *Mem:* Sigma Xi; Northeastern Bird Banding Asn; Wilson Ornith Soc; Am Ornith Union. *Res:* Ornithology; ecology; behavior; population. *Mailing Add:* W5903 Apple Orchard Lane La Crosse WI 54601

YOUNG, HOWARD SETH, b Birmingham, Ala, July 7, 24; m 45; c 7. PHYSICAL CHEMISTRY. *Educ:* Birmingham Southern Col, BS, 42; Brown Univ, PhD(chem), 48. *Prof Exp:* Chemist, Eastman Kodak Co, 44-46 & 48-51, sr res chemist, 51-62, from res assoc to sr res assoc, 63-70, head, Eng Res Div, 70-75, head, Phys & Anal Chem Res Div, 75-84, asst dir, 84-85, dir res labs, Eastman Chem Div, 85-89; RETIRED. *Mem:* Sigma Xi; Am Chem Soc. *Res:* Inorganic chemistry; catalysis. *Mailing Add:* 1909 E Sevier Ave Kingsport TN 37664

YOUNG, HUGH DAVID, b Ames, Iowa, Nov 3, 30; m 60; c 2. SCIENCE EDUCATION. *Educ:* Carnegie Inst Technol, BS, 52, MS, 53, PhD(physics), 59, Carnegie-Mellon Univ, BFA, 72. *Prof Exp:* From instr to assoc prof physics, 56-76, head dept natural sci, 62-74, PROF PHYSICS, CARNEGIE-MELLON UNIV, 76- *Mem:* Am Phys Soc; Am Asn Physics Teachers. *Res:* Physics textbook writing; new teaching materials for introductory college physics courses. *Mailing Add:* Dept Physics Carnegie-Mellon Univ Pittsburgh PA 15213

YOUNG, IAN THEODORE, b Chicago, Ill, Dec 15, 43; m 77; c 3. ELECTRICAL ENGINEERING, ANALYTICAL CYTOLOGY. *Educ:* Mass Inst Technol, SB & SM, 66, PhD(elec eng), 69. *Prof Exp:* From instr to assoc prof elec eng, Mass Inst Technol, 67-79; group leader, Lawrence Livermore Lab, 78-81; PROF APPL PHYSICS, TECH UNIV DELFT, NETH, 81- *Concurrent Pos:* Vincent Hayes fel, Mass Inst Technol, 69-71, consult, Lincoln Labs, 72-78 & Coulter Biomed Res Corp, 75-78; fel, Neth Orgn Res, 75-77; guest prof, Tech Univ Neth, 75-76, Tech Univ Sweden, 76 & Tech Univ Lausanne, Switz, 80; mem cytol automation comn, Nat Cancer Inst, NIH, 77-81. *Honors & Awards:* Res Prize, Schlumberger Found, 90. *Mem:* Inst Elec & Electronics Engrs; Med & Biol Soc; Soc Anal Cytol; fel Royal Microscopal Soc. *Res:* Quantitative microscopy; image processing; pattern recognition; signal processing. *Mailing Add:* Dept Appl Physics Loreutzweg 1 Tech Univ Delft Delft 2600 GA Netherlands

YOUNG, IN MIN, b Seoul, Korea, July 13, 26; US citizen; m 53; c 3. AUDIOLOGY, PSYCHOACOUSTICS. *Educ:* Yonsei Univ, MD, 48; Jefferson Med Col, MSc, 66. *Prof Exp:* Instr otolaryngol, Sch Med, Yonsei Univ, 54-59; resident, Newark Eye & Ear Infirmary, NJ, 59-60; res audiologist, 60-65, from asst prof to assoc prof, 65-72, PROF AUDIOL, JEFFERSON MED COL, 72- *Concurrent Pos:* Dir, Hearing & Speech Ctr, Thomas Jefferson Univ Hosp, Philadelphia, 72-, audiologist, Affil Staff, 73-; consult otolaryngol, US Naval Hosp, Philadelphia; United Fund grant, 73-76. *Mem:* Acoust Soc Am; Am Speech & Hearing Asn; Am Neurotol Soc; Am Audiol Soc; Sigma Xi. *Res:* Auditory threshold and suprathreshold adaptation; Bekesy audiometry and marking. *Mailing Add:* Jefferson Med Col Thomas Jefferson Univ Philadelphia PA 19107

YOUNG, IRVING, b New York, NY, Aug 15, 22; m 48; c 3. MEDICINE, PATHOLOGY. *Educ:* Johns Hopkins Univ, AB, 43, MD, 46. *Prof Exp:* Asst pathologist, Kings County Hosp, Brooklyn, NY, 49-51; assoc dir labs, Div Path, 52-71, CHMN DIV LABS, ALBERT EINSTEIN MED CTR, 71-; CLIN PROF PATH, SCH MED, TEMPLE UNIV, 75- *Mem:* AMA; Fedn Am Soc Exp Biol. *Res:* Immunomorphologic correlation; histochemistry of chromosomes; surgical pathology. *Mailing Add:* Labs EM 7615 Mountain Ave Elkins Park PA 19117

YOUNG, IRVING GUSTAV, b Brooklyn, NY, Dec 10, 19; m 41; c 2. ANALYTICAL CHEMISTRY, PHYSICAL CHEMISTRY. *Educ:* City Col New York, BS, 39; Polytech Inst Brooklyn, MS, 50; Temple Univ, PhD, 67. *Prof Exp:* Res asst, Bellevue Hosp, Columbia Univ, 39-42; asst chemist, Picatinny Arsenal, 42-44; battery technologist, US Elec Mfg Corp, 44-51; sr res chemist, Int Resistance Co, 51-56, sr res scientist, 59-64; chief chemist, Transition Metals & Chem Co, 56-57; asst res scientist, Leeds & Northrup Co, 57-59; res fel, Temple Univ, 64-65; chemist, Advan Technol Staff, Indust Div, Honeywell, Inc, 65-74, develop supvr, Honeywell Power Sources Ctr, 74-76; mgr, Energy Progs, Am Nat Standards Inst, 76-83; LECTR CHEM, OGONTZ CAMPUS, PA STATE, 83- *Mem:* Am Chem Soc; Instrument Soc Am; Air Pollution Control Asn; Water Pollution Control Fedn; Am Soc Test & Mat; Inst Elec & Electronics Engrs. *Res:* Electrochemistry; process analyzers; air and water pollution; standards coordination. *Mailing Add:* 22 Four Leaf Rd Levittown PA 19056

YOUNG, J LOWELL, b Perry, Utah, Dec 13, 25; m 50; c 4. SOIL BIOCHEMISTRY & BIOLOGY, SOIL CLAY MINERALOGY. *Educ:* Brigham Young Univ, BS, 53; Ohio State Univ, PhD(soils), 56. *Prof Exp:* Asst agron, Agr Exp Sta, Ohio State Univ, 53-56, fel agr biochem, Univ, 56-57; chemist, Agr Res Serv, USDA, 57-60; asst prof, 57-63, assoc prof, 63-78, PROF SOIL SCI, ORE STATE UNIV, 78-; RES CHEMIST, AGR RES SERV, USDA, 60- *Concurrent Pos:* Assoc ed, Soil Sci Soc Am J, 75-80. *Mem:* AAAS; Am Soc Agron; Soil Sci Soc Am; Int Soc Soil Sci; Int Humic Substances Soc; Sigma Xi; Inst Alternative Agri. *Res:* Enzymes in cotyledons of germinating seeds; amino acids of soils, humic substances, root exudates; D-amino acids uptake and metabolism by higher plants; nitrogen and soil particulates as nonpoint-source pollutants eroded from agricultural lands; soil properties affecting beneficial root-fungus symbioses (endomycorrhiza); forms of nitrogen in soil; anhydrous NH3 reactions with organic clay-size and colloidal organic-minerals as primary feeder surfaces for soil; clay size and colloidal organic-minerals as primary feeder surfaces for soil microbes and plant roots. *Mailing Add:* Dept Crops & Soil Sci Ore State Univ Corvallis OR 97331

YOUNG, JACK PHILLIP, b Huntington, Ind, Oct 28, 29; m 55; c 5. ANALYTICAL CHEMISTRY. *Educ:* Ball State Teachers Col, BS, 50; Univ Ind, PhD(chem), 55. *Prof Exp:* Chemist, 55-82, SR STAFF MEM, OAK RIDGE NAT LAB, 82- *Mem:* Fel AAAS; Am Chem Soc; Sigma Xi. *Res:* Actinide chemistry; spectroscopy of solutions, molten salts, solid state compounds; laser spectroscopy; photoionization studies; applications of lasers to chemical analysis. *Mailing Add:* 100 Westlook Circle Oak Ridge TN 37830

YOUNG, JAMES ARTHUR, JR, b Tacoma, Wash, Feb 12, 21; m 43; c 2. PHYSICS, ELECTRICAL ENGINEERING. *Educ:* Calif Inst Technol, BS, 43; Univ Wash, PhD(physics), 53. *Prof Exp:* Radar officer, US Army Signal Corps, 43-46; res engr rocket instrumentation, Jet Propulsion Lab, Calif Inst Technol, 46-47; teaching fel physics, Univ Wash, 47-53; co-dir, Radio Res Lab, Bell Tel Labs, Holmdel, 53-76; proj mgr UNDP Brazil, Int Telecommunications Union, 75-78, affil prof elec engr, Univ Wash, 78-, adj prof, 82-87, dean eng sci, 85-87, EMER PROF, COGSWELL COL NORTH, 88- *Mem:* Inst Elec & Electronics Engrs; Am Phys Soc. *Res:* Communications research; encoding, modulation transmission and switching of information signals, particularly for high radio frequency and optical media. *Mailing Add:* 13921 Silven Ave NE Bainbridge Island WA 98110

YOUNG, JAMES CHRISTOPHER F, b Charlottetown, PEI, Apr 1, 40; m 66; c 3. HEALTH & SAFETY, ANALYTICAL CHEMISTRY. *Educ:* Mt Allison Univ, BSc, 60; McMaster Univ, MSc, 62; Mass Inst Technol, PhD(org chem), 71. *Prof Exp:* Teacher sci, Kitchener-Waterloo Col & Voc Sch, 62-64; lectr chem, Sir Wilfred Laurier Univ, 64-66; RES SCIENTIST, PLANT RES CENTRE, AGR CAN, 72- *Concurrent Pos:* Rockefeller Found fel, NY Col Forestry, Syracuse Univ, 71-72. *Mem:* AAAS; Chem Inst Can; Am Chem Soc. *Res:* Natural product chemistry; organic chemistry analytical methodology; mycotoxins; ergot alkaloids; chemicals used in animal communication; insect pheromones; mass spectrometry. *Mailing Add:* Agr Can Plant Res Centre Ottawa ON K1A 0C6 Can

YOUNG, JAMES EDWARD, b Wheeling, WVa, Jan 18, 26; m 48; c 1. DIFFERENTIAL GEOMETRY IN PARTICLE PHYSICS. *Educ:* Howard Univ, BS, 46, MS, 49; Mass Inst Technol, MS, 51, PhD(physics), 53. *Prof Exp:* Instr physics, Hampton Inst, 46-49; fel acoustics, Mass Inst Technol, 53-55; staff mem physics, Los Alamos Sci Lab, 57-70; PROF PHYSICS, INST THEORET PHYSICS, MASS INST TECHNOL, 70- *Concurrent Pos:* Shell B P fel, Univ Southampton, 56; consult, Gen Atomics, Calif, 57-58; Nat Acad Sci-Nat Res Coun & Ford fel, Bohr Inst, Copenhagen, 61-62; vis assoc prof, Univ Minn, 64; res asst, Oxford Univ, 65-66; visitor, Inst Theoret Physics, Mass Inst Technol, 68-69; adv study med sci, Harvard Uninv, 83-85; fel, Dept Anat, Tufts Med Sch, 85-86. *Mem:* Am Phys Soc. *Res:* Spin manifolds and cohomology of field theories; non-linear statistical mechanisms of N vortex system; investigation of neural peptides with superfusion of brain tissues; real time electrophoresis in serological material substances. *Mailing Add:* Dept Physics Mass Inst Technol N Massachusetts Ave Cambridge MA 02139

YOUNG, JAMES FORREST, b Meadville, Pa, June 22, 43; m 71. LASER PHYSICS, ENGINEERING. *Educ:* Mass Inst Technol, BS, 65, MS, 66; Stanford Univ, PhD(elec eng), 70. *Prof Exp:* Res assoc, E L Ginzton Lab, 70-75, RES PROF ELEC ENG, STANFORD UNIV, 75- *Concurrent Pos:* Consult, Coherent, Inc, 70-75, Spectra-Physics Inc, 76- & Bell Tel Labs, 77- *Mem:* Am Inst Physics; fel Optical Soc Am; Inst Elec & Electronics Engrs. *Res:* Quantum electronics; nonlinear optics; experimental techniques and instruction. *Mailing Add:* Edward L Ginzton Lab Stanford Univ Stanford CA 94306

YOUNG, JAMES GEORGE, b Milwaukee, Wis, July 18, 26; m 54; c 3. INDUSTRIAL PHARMACY. *Educ:* Univ Wis, BS, 48, MS, 49; Univ NC, PhD(pharm), 52. *Prof Exp:* Asst prof pharmaceut chem, Med Col Va, 51-54 & Univ Tenn, 56-58; sr chemist, Riker Labs, Inc, 58-60, dir prod develop, 60-72; DIR DEVELOP, G D SEARLE & CO, 72- *Mem:* AAAS; Am Chem Soc; Am Pharmaceut Asn; NY Acad Sci. *Res:* Pharmaceutical aerosol formulation; drug stabilization; general pharmaceutical development. *Mailing Add:* Pharmaquest Corp PO Box 10608 San Raphael CA 94912-0608

YOUNG, JAMES H, b LaFayette, Ky, Mar 19, 41; m 63; c 2. AGRICULTURAL ENGINEERING. *Educ:* Univ Ky, BSAE, 62, MSAE, 64; Okla State Univ, PhD(eng), 66. *Prof Exp:* From asst prof to assoc prof, 66-76, PROF BIOL & AGR ENG, NC STATE UNIV, 76- *Concurrent Pos:* Bd dirs, Am Soc Agr Engrs, 82-84, 88-90. *Mem:* Fel Am Soc Agr Engrs; Am Soc Eng Educ; Am Peanut Res Educ Soc; Am Soc Heating, Refrigerating & Air-Conditioning Engrs. *Res:* Heat and mass transfer in biological materials including tobacco, wheat and peanuts; growth simulation of peanuts as affected by environment; automatic monitoring of weather parameters affecting crop production. *Mailing Add:* NC State Univ Campus Box 7625 Raleigh NC 27695-7625

YOUNG, JAMES HOWARD, b Norfolk, Va, May 9, 24; m 50; c 1. MATHEMATICS. *Educ:* Univ Va, BA, 48; Duke Univ, MA, 51. *Prof Exp:* Instr physics, Norfolk Div, Col William & Mary, 47-51; instr math, Johns Hopkins Univ, 55-74, res scientist, Inst Coop Res, 51-69; sr res analyst, Falcon Res & Develop, 69-81; sr res analyst, Ketron, Inc, 81-84; RETIRED. *Mem:* Am Math Soc. *Res:* Weapons systems analysis; military operations research. *Mailing Add:* 106 Regester Ave Baltimore MD 21212

YOUNG, JAMES R(ALPH), b Rushville, Ill, June 22, 30; m 52; c 1. ELECTRICAL ENGINEERING, MATHEMATICS. *Educ:* Univ Ill, BS, 56, MS, 57, PhD(elec eng), 60. *Prof Exp:* Instr elec eng, Univ Ill, 56-60; assoc prof, NMex State Univ, 60-65; SR RES ENGR, SRI INT, 65- *Concurrent Pos:* Elec engr, Barber Colman Co, Ill, 56-59. *Mem:* Inst Elec & Electronics Engrs. *Res:* Network theory; applications of topology to network synthesis; applications of dynamic programming techniques; analysis of linear systems. *Mailing Add:* SRI Int 333 Ravenswood Ave Menlo Park CA 94025

YOUNG, JAMES ROGER, b Fordland, Mo, June 14, 23; m 45; c 3. PHYSICS. *Educ:* Park Col, AB, 46; Univ Mo, BA, 49, PhD(physics), 52. *Prof Exp:* Res physicist, Res Lab, Gen Elec Co, Schenectady, 51-63, mgr advan develop vacuum prod, 63-64, mgr eng, 64-66, mgr Plasma Light Sources, Res & Develop Ctr, 74-85; RETIRED. *Mem:* Am Vacuum Soc (treas, 73-83); Am Phys Soc; Am Inst Physics. *Res:* Vacuum physics; physical electronics. *Mailing Add:* 34 Nott St Rexford NY 12148

YOUNG, JANIS DILLAHA, b Little Rock, Ark, July 12, 27; m 56; c 2. BIOCHEMISTRY. *Educ:* Hendrix Col, BA, 49; Univ Okla, MS, 51; Univ Calif, Berkeley, PhD(biochem), 60. *Prof Exp:* Biochemist, Armour Labs, Chicago, Ill, 51-54; instr chem, Colby Col, 54-55; NIH trainee virol, Univ Calif, Berkeley, 59-61; assoc res scientist, Lab Med Entom, Kaiser Found Res Inst, 61-70; assoc res biochemist, Space Sci Lab & Adj Assoc Prof Immunol, Dept Bact & Immunol, Univ Calif, Berkeley, 71-79; res assoc, dept biol chem, Harvard Med Sch, 79-85; assoc biochemist, Lowell Labs, McLean Hosp, Belmont, Mass, 79-82, RES ASSOC, DIV INFECTIOUS DIS CHILDREN'S HOSP MED CTR, 82-; DIR, PEPTIDE SYNTHESIS FACIL, DEPT BIOL CHEM, UNIV CALIF, LOS ANGELES, 85- *Honors & Awards:* Alexander von Humboldt Sr Scientist Award. *Mem:* Am Chem Soc; Am Soc Biol Chemists. *Res:* Solid phase peptide synthesis. *Mailing Add:* Dept Biol Chem Univ Calif CHS 33-257 Los Angeles CA 90024-1737

YOUNG, JAY ALFRED, b Huntington, Ind, Sept 8, 20; m 42, 62; c 18. CHEMICAL SAFETY, HAZARDOUS CHEMICALS LABELING. *Educ:* Univ Ind, BS, 39; Oberlin Col, AM, 40; Univ Notre Dame, PhD, 50. *Prof Exp:* Chief chemist, Asbestos Mfg Co, Ind, 40-42; ord engr, US War Dept, DC, 42-44; from instr to prof chem, King's Col, Pa, 49-69; vis prof, Carleton Univ, 69-70; Hudson Prof Chem, Auburn Univ, 70-75; vis prof chem, Fla State Univ, 75-77; mgr tech publ, Chem Mfrs Asn, DC, 77-80; CHEM SAFETY CONSULT, 80- *Concurrent Pos:* Ed, Int J Chem Health & Safety, 80-83; expert witness & tech consult in labeling hazardous chem & prod liability, 79- *Mem:* AAAS; Am Chem Soc. *Res:* Safe use, handling, precautionary labeling and disposal of chemicals; application of chemical reactions to manufacturing processes and to consumer uses and concomitant prevention of injury and property damage and loss. *Mailing Add:* 12916 Allerton Lane Silver Spring MD 20904

YOUNG, JAY MAITLAND, b Louisville, Ky, Nov 26, 44. IMMUNOASSAYS, PHYSIOLOGICAL DIAGNOSTICS. *Educ:* Vanderbilt Univ, BA, 66; Yale Univ, MS, 67, MPh, 68, PhD(chem), 71. *Prof Exp:* Asst prof chem, Bryn Mawr Col, 70-76; res biochemist & proj mgr, 77-82, int clin specialist, 82-84, CLIN PROJ MGR, ABBOTT LABS, 84- *Concurrent Pos:* NIH fel, Oxford Univ, 71-72; vis scientist, Inst Cancer Res, Philadelphia, Pa, 75-76. *Mem:* AAAS; Am Chem Soc. *Res:* Radio-immuno assays; enzyme-immuno assays; antibody production and characterization; clinical studies of assays for cancer markers and thyroid hormones for indications of pregnancy; cardiovascular status and metabolic function. *Mailing Add:* Abbott Diag Div D-9YG AP8B One Abbott Park Rd Abbott Park IL 60064-9075

YOUNG, JERRY H, b Fitzhugh, Okla, Aug 4, 31; m 52; c 1. ENTOMOLOGY. *Educ:* Okla State Univ, BS, 55, MS, 56; Univ Calif, Berkeley, PhD(parasitol), 59. *Prof Exp:* Prof entom, Okla State Univ, 59-; RETIRED. *Mem:* Entom Soc Am; Entom Soc Can. *Res:* Cotton insect control; mite morphology; Hymenoptera taxonomy; integrated and biological control of cotton insects. *Mailing Add:* 2421 S 78th St Lincoln NE 68506

YOUNG, JERRY WESLEY, b Mulberry, Tenn, Aug 19, 34; m 59; c 2. ANIMAL NUTRITION. *Educ:* Berry Col, BS, 57; NC State Univ, MS, 59, PhD(animal nutrit), 63. *Prof Exp:* Res asst animal nutrit, NC State Univ, 57-63; USPHS fel biochem, Inst Enzyme Res, Univ Wis, 63-65; asst prof animal nutrit, 65-70, assoc prof animal sci & biochem, 70-74, PROF ANIMAL SCI & BIOCHEM, IOWA STATE UNIV, 74- *Honors & Awards:* Am Feed Indust Award, Outstanding Res Dairy Nutrit, Am Dairy Sci Asn, 87. *Mem:* Am Chem Soc; Am Dairy Sci Asn; Am Inst Nutrit; Am Soc Animal Sci; Sigma Xi. *Res:* Ketosis and fatty liver lactating dairy cows; volatile fatty acid metabolism in ruminants; gluconeogenesis and glucose metabolism in ruminants, in vivo kinetics of glucose; effects of nutrition upon reproduction; effects of animal products on cholesterol and lipoprotein metabolism; effects of niacin in dairy nutrition. *Mailing Add:* Dept Animal Sci Iowa State Univ Ames IA 50010

YOUNG, JOHN A, b Washington, DC, July 4, 39; m 88; c 2. METEOROLOGY. *Educ:* Miami Univ, BA, 61; Mass Inst Technol, PhD(meteorol), 66. *Prof Exp:* NSF fel, Univ Oslo, 66; from asst prof to assoc prof, 66-69, chmn, 84-87, PROF METEOROL, UNIV WIS-MADISON, 75- *Concurrent Pos:* NSF res grant, 71-; vis assoc prof, Mass Inst Technol, 73-74; mem Global Atmospheric Res Prog, Nat Acad Sci, 73-78, 80-86; vis scientist, Oxford Univ, 87, Nat Meteorol Ctr, 88. *Mem:* AAAS; fel Am Meteorol Soc. *Res:* Dynamic meteorology and oceanography; monsoons; numerical modeling; planetary boundary layers. *Mailing Add:* Dept Meteorol Univ Wis Madison WI 53706

YOUNG, JOHN A, b Nampa, Idaho, Apr 24, 32; m 54. ELECTRONICS ADMINISTRATION. *Educ:* Ore State Univ, Corvallis, BSEE, 53; Stanford Univ, Calif, MBA, 58. *Hon Degrees:* DPhil, Col Idaho, 86, Purdue Univ, 87 & Ore State Univ, 89; DSc, Univ Idaho, 91. *Prof Exp:* Mem, Mkt Planning Staff, Hewlett-Packard Co, Palo Alto, Calif, 58-61, regional sales mgr, 61-62, mem, Corp Finance Staff, 62-63, mkt mgr, Microwave Div, 63, gen mgr, 63-68, vpres & group mgr, Electronic Prod Group, 68-74, exec vpres, Instrument, Computer Systs & Components Group, 74-77, chief oper officer, 77-78, PRES, HEWLETT-PACKARD CO, PALO ALTO, CALIF, 77-, CHIEF EXEC OFFICER, 78- *Concurrent Pos:* Dir, Wells Fargo Bank, Wells Fargo & Co & Chevron Corp; mem, Am Electronics Asn, 68-86; dir, Kaspes Instruments Inc, 70-75, Am Express Mutual Funds, 70-75, Dillingham Corp, 74-81, SRI Int, 79-85; chmn, Pres Comn Indust Competitiveness, 83-85; dir & trustee, Found Malcolm Baldridge Nat Qual Award, 88-, pres, 90-91. *Honors & Awards:* Distinguished Pub Serv, NSF, 90; Medal of Achievement, Am Electronics Asn, 90; Prime Minister's Trade Award, Prime Minister Toshiki Kaifu of Japan, 90. *Mem:* Nat Acad Eng; fel Am Acad Arts & Sci. *Mailing Add:* Off Pres Hewlett-Packard Co PO Box 10301 Palo Alto CA 94303

YOUNG, JOHN ALBION, JR, b Newport, RI, Aug 29, 09; m 37. PETROLEUM GEOLOGY. *Educ:* Brown Univ, PhB, 32, ScM, 34; Harvard Univ, PhD(geol), 46. *Prof Exp:* Asst geol, Brown Univ, 32-37; asst paleont, Harvard Univ, 37-39; instr geol, Mich State Col, 39-44; geologist, Sun Oil Co, 44-46; asst prof geol, Syracuse Univ, 46-47; geologist, Sun Oil Co, 47-50, staff geologist, 50-70; vpres, secy, dir & dir oil & gas div, Tax Shelter Adv Serv, Inc, Pa, 70-73; vpres, World Resources, 73-81; sr vpres explor & prod, 74-81, SR EXPLOR CONSULT, OMNI-EXPLOR INC, RADNOR, 81- *Mem:* Fel Geol Soc Am; Am Asn Petrol Geologists. *Res:* Petroleum geology; stratigraphy; subsurface geology. *Mailing Add:* 142 Hunters Lane Devon PA 19333

YOUNG, JOHN CHANCELLOR, nuclear physics, environmental sciences, for more information see previous edition

YOUNG, JOHN COLEMAN, b Leesville, La, July 13, 42; m 64; c 1. STATISTICS. *Educ:* Northwestern State Univ, BA, 64, MS, 65; Southern Methodist Univ, PhD(statist), 71. *Prof Exp:* Asst instr math, Northwestern State Univ, 64-65; asst prof, McNeese State Univ, 67-69; instr statist, Southern Methodist Univ, 69-71; from asst prof to assoc prof, 71-77, PROF MATH, MCNEESE STATE UNIV, 77- *Mem:* Am Statist Asn. *Res:* Multivariate analysis; discrimination, analysis of variance and goodness of Fit test for multivariate populations. *Mailing Add:* Dept Statist McNeese State Univ 4100 Ryan St Lake Charles LA 70609

YOUNG, JOHN DING-E, b Taipei, Taiwan, Dec 27, 58; Brazil; m 85; c 2. TUMOR IMMUNOLOGY, MEMBRANE BIOCHEMISTRY. *Educ:* Nat Univ Brasilia, Brazil, MD, 80; Rockefeller Univ, PhD(immunol), 83. *Prof Exp:* Intern, med resident, Nat Univ Brasilia, 79-80; post doctorat fel immunol, Jane Coffin Childs, Rockefeller Univ, 83-85; ASSOC PROF IMMUNOL, ROCKEFELLER UNIV, 85- *Concurrent Pos:* Vis scholar bibl, Nat Inst Basic Biol Japan, Nat Sci Found, 84; scholar, Lucille P Markey Scholar, Rockefeller Univ, 85-; investigator, Cancer Res Inst, 87-; prin invest, Rockefeller Univ, 88; consult, Bristol Myer Pharmaceut Co, 88. *Honors & Awards:* New Initiatives Award, Irvington Inst Med Res, 88. *Res:* Discovered several molecules from lymphocytes that are used to kill tumor cells and other foreign targets, pore-forming protein, serine esterases, a DNA fragmenting protein called leukalexin; isolation and cloning of these molecules to apply them in immunotherapy.. *Mailing Add:* Cellular Physiol & Immunol Lab Rockfeller Univ 1230 York Ave New York NY 10021

YOUNG, JOHN FALKNER, b Tyler, Tex, Apr 3, 40; m 77; c 2. PHARMACODYNAMICS, RISK ASSESSMENT. *Educ:* NTex State Univ, BA, 63; Univ Houston, BS, 66; Univ Fla, MS, 69, PhD(pharmaceut res), 73. *Prof Exp:* SUPVRY RES BIOLOGIST, NAT CTR TOXICOL RES, 73- *Concurrent Pos:* Exec secy, Agent Orange Working Group Sci Panel, 86-89; adj appointments, Col Pharm, Dept Pharmacol & Interdisciplinary Toxicol, Univ Ark Med Sci, Dept Biol, Univ Ark, Pine Bluff. *Mem:* Am Pharmaceut Asn; Soc Appl Spectros; Sigma Xi; Teratology Soc; Soc Comput Simulation. *Res:* Application of the principles of pharmacokinetics to teratogenic research; development of analytical procedures used to quantitate chemicals from biological fluids and tissues; simulation of data on hybrid computer; investigation of the utility of pharmacokinetics and pharmacodynamics in risk assessment. *Mailing Add:* Div Reprod & Develop Toxicol Nat Ctr Toxicol Res Jefferson AR 72079

YOUNG, JOHN H, b Shamokin, Pa, Aug 16, 40. PHYSICS. *Educ:* Gettysburg Col, BA, 62; Univ NH, MS, 64; Clark Univ, PhD, 69. *Prof Exp:* Asst physics, Clark Univ, 64-66; asst prof, 70-77, ASSOC PROF PHYSICS, UNIV ALA, BIRMINGHAM, 77- *Mem:* Am Phys Soc. *Res:* Description of the gravitational field of a rotating mass in the general theory by exact means; two body problem in general relativity; three body nucleon problem. *Mailing Add:* Dept Physics Univ Ala Birmingham AL 35294

YOUNG, JOHN KARL, b Minneapolis, Minn, Aug 15, 51; m 77; c 2. NEUROENDOCRINOLOGY. *Educ:* Cornell Univ, BS, 72; Univ Calif, Los Angeles, PhD(anat), 77. *Prof Exp:* Teaching fel anat, dept anat, Univ Minn, 77-79; asst prof anat, Howard Univ, 79-86; CONSULT, 86- *Concurrent Pos:* Prin investr, NIH res grant, Howard Univ, 80-83. *Mem:* Am Asn Anatomists; Am Physiol Soc. *Res:* Possible neuroendocrine basis of anorexia nervosa; specialized glial cells that may mediate effects of glucose upon hypothalamic regulation of feeding. *Mailing Add:* Dept Anat Howard Univ 520 W St NW Washington DC 20059

YOUNG, JOHN W(ATTS), b San Francisco, Calif, Sept 24, 30; c 2. ASTRONAUTICS, AERONAUTICAL ENGINEERING. *Educ:* Ga Inst Technol, BSAE, 52. *Hon Degrees:* LLD, Western State Univ Col Law, 69; DrApplSci, Fla Technol Univ, 70. *Prof Exp:* Pilot on first manned Gemini flight, 65, backup pilot for Gemini 6, command pilot, Gemini 10 mission, 66, backup command module pilot for Apollo 7, command module pilot for Apollo 10, 69, backup spacecraft commander for Apollo, 13, spacecraft commander of Apollo 16, 72, backup spacecraft commander for Apollo 17, 72, ASTRONAUT, NASA, 62- *Honors & Awards:* Ivan C Kinchloe Award, Soc Exp Test Pilots, 72. *Mem:* Fel Am Astronaut Soc; assoc fel Soc Exp Test Pilots; Am Inst Aeronaut & Astronaut. *Mailing Add:* NASA-Johnson Space Ctr Mail Code ACS Houston TX 77058

YOUNG, JOHN W(ESLEY), JR, b Baltimore, Md, June 16, 24; m 50; c 1. COMPUTER SCIENCE. *Educ:* Albright Col, AB, 45; Harvard Univ, MS, 55. *Prof Exp:* Supvry analyst data processing, US Dept Defense, Washington, DC, 49-56; SR CONSULT ANALYST, SYSTS ENG, SCRIPPS RANCH, NCR CORP, 57- *Concurrent Pos:* Guest lectr, Univ Calif, Los Angeles, 66-

Mem: Asn Comput Mach; Inst Elec & Electronics Engrs; Asn Comput Ling; Sigma Xi. *Res:* Design, application and programming of digital computers, especially problem formulation languages, simulation, artificial intelligence and operating systems; data base systems; query languages. *Mailing Add:* 460 Ferne Ave Palo Alto CA 94306

YOUNG, JOHN WILLIAM, b Toronto, Ont, Nov 16, 12; nat US; m 39; c 1. MATHEMATICS, COMPUTER SCIENCE. *Educ:* Univ Fla, AB, 34, BS, 37, MA, 40, PhD(math), 52. *Prof Exp:* Prin & sch teacher, Fla, 34-42; instr math, Univ Fla, 46-54; appl sci rep, Int Bus Mach Corp, Ga, 54-55; mathematician & comput prog consult, IBM Res Comput Ctr, NY, 55-57; mathematician & comput prog consult, Missile Test Ctr, Radio Corp Am, 57-58; head anal & info processing, Res Div, Radiation, Inc, 59-60, head eng comput serv, 61-68, mem serv staff, 68-71; mem sr specialist staff, Data Systs Div, Martin Marietta Corp, Fla, 72-75; vis assoc prof comput sci, Fla Technol Univ, 75-76; mem sr tech staff, Educ Comput Corp, Orlando, 76-81; RETIRED. *Concurrent Pos:* Consult, Sci Systs Serv, Melbourne, 76; comput consult, 81- *Mem:* Nat Coun Teachers Math; Math Asn Am; Asn Comput Mach. *Res:* Applications of electronic digital computers; computer performance measurement and evaluation; simulation of computer systems. *Mailing Add:* 2873 A-5 S Osceola Ave Orlando FL 32806-5440

YOUNG, JOSEPH HARDIE, b Salt Lake City, Utah, Aug 11, 18; m 66; c 1. BIOLOGY. *Educ:* Stanford Univ, PhD(biol), 54. *Prof Exp:* From instr to asst prof zool, Tulane Univ, 54-59; from asst prof to assoc prof biol, 59-65, chmn dept, 66-80, PROF BIOL, SAN JOSE STATE UNIV, 65- *Mem:* AAAS; Entom Soc Am; Am Soc Zool; Sigma Xi. *Res:* Insect embryology; arthropod morphology; marine biology. *Mailing Add:* 6888 Goldpine Ct San Jose CA 95120

YOUNG, JOSEPH MARVIN, b Marshall, Tex, Oct 16, 19; m 42; c 4. PATHOLOGY, ANATOMY. *Educ:* Harvard Univ, BS, 43; Johns Hopkins Univ, MD, 45. *Prof Exp:* Intern surg, Johns Hopkins Hosp, 45-46; resident surg, 46-51, resident path, 51-54, CHIEF LAB SERV, VET ADMIN HOSP, 54-; PROF PATH, UNIV TENN MED UNITS, MEMPHIS, 62- *Mem:* AAAS; Col Am Path; Am Asn Path & Bact; Am Soc Exp Path; Sigma Xi. *Res:* Joint reactions and spread of cancer. *Mailing Add:* 1145 Oak Ridge Dr Memphis TN 38111-8117

YOUNG, KEITH PRESTON, b Buffalo, Wyo, Aug 18, 18; m 49; c 3. PALEONTOLOGY, STRATIGRAPHY. *Educ:* Univ Wyo, BA, 40, MA, 42; Univ Wis, PhD(stratig), 48. *Prof Exp:* From asst prof to assoc prof, 48-58, PROF GEOL, UNIV TEX, AUSTIN, 58- *Mem:* Paleont Soc; Geol Soc Am; Soc Econ Paleont & Mineral; Am Asn Petrol Geol; Am Inst Prof Geologists; Sigma Xi. *Res:* Cretaceous stratigraphy and paleontology of southwestern North America; Cephalopods and rudists. *Mailing Add:* Dept Geol Sci Univ Tex Austin TX 78712

YOUNG, KENNETH CHRISTIE, b Rochester, NY, Nov 9, 41; div. SELF LEARNING SYSTEMS, WEATHER FORECASTING. *Educ:* Ariz State Univ, BS, 65; Univ Chicago, MS, 67, PhD(geophys), 73. *Prof Exp:* Fel, Nat Ctr Atmospheric Res, 73-74; asst prof, 74-80, ASSOC PROF ATMOSPHERIC SCI, UNIV ARIZ, 80- *Res:* Application of self learning systems to weather forecasting problems; development of improved self learning systems; development of expert system weather forecasters; weather modification; numerical simulations of microphysical processes in clouds. *Mailing Add:* Inst Atmospheric Physics Univ Ariz Tucson AZ 85721

YOUNG, KENNETH KONG, b Vancouver, BC, Mar 19, 37; m 67. HIGH ENERGY PHYSICS. *Educ:* Univ Wash, BSc, 59; Univ Pa, PhD(physics), 65. *Prof Exp:* Res assoc physics, Univ Mich, 65-67; from asst prof to assoc prof, 67-77, PROF PHYSICS, UNIV WASH, 77- *Concurrent Pos:* Assoc, Europ Orgn Nuclear Res; sr fel, Univ Col, London. *Mem:* Am Phys Soc. *Res:* Accelerator experiments; test standard model, search for evidence of supersymmetry; neutrino astronomy. *Mailing Add:* Dept Physics Univ Wash Seattle WA 98195

YOUNG, KONRAD KWANG-LEEI, INTEGRATED CIRCUIT TECHNOLOGY, SEMICONDUCTOR DEVICES. *Educ:* Nat Taiwan Univ, BS, 81; Univ Calif, Santa Barbara, MS, 82, Berkeley, PhD(elec eng), 86. *Prof Exp:* Staff mem, Lincoln Lab, Mass Inst Technol, 86-89; MEM TECH STAFF, HEWLETT PACKARD CO, 89- *Mem:* Inst Elec & Electronics Engrs. *Res:* CMOS/BiCMOS integrated circuit technology; silicon on insulator devices. *Mailing Add:* 112 Walter Hays Dr Palo Alto CA 94303

YOUNG, LARRY DALE, b Fountain Head, Tenn, Dec 13, 48; m 70; c 2. PSYCHOPHYSIOLOGY, MEDICAL PSYCHOLOGY. *Educ:* David Lipscomb Col, BA, 70; Univ Ga, MS, 72; Harvard Univ, PhD(psychol), 79. *Prof Exp:* Asst prof psychol, Univ Miss, 78-80; asst prof, 80-88, ASSOC PROF PSYCHOL, BOWMAN GRAY SCH MED, 88- *Concurrent Pos:* Consult, Comn Accreditation Rehab Facil, 84-; consult ed, Annals Behav Med, 91- *Mem:* Soc Psychophysiol Res; Soc Behav Med; Asn Advan Behav Ther; Am Psychol Asn; Int Asn Study Pain; Asn Appl Psychophysiol & Biofeedback. *Res:* Application of behavioral principles to management of stress and pain associated with chronic illness or with stress related medical disorders; disease prevention and health promotion, the latter, specifically with secondary prevention of cervical cancer in minority population. *Mailing Add:* Sect Med Psychol Bowman Gray Sch Med Winston-Salem NC 27103

YOUNG, LAURENCE RETMAN, b New York, NY, Dec 19, 35; m 60; c 3. BIOENGINEERING, INSTRUMENTATION. *Educ:* Amherst Col, AB, 55; Mass Inst Technol, BS, 57, MS, 59, ScD(instrumentation), 62; Univ Paris, cert, 58. *Prof Exp:* Engr, Instrumentation Lab, Mass Inst Technol, 56 & 58-60 & Sperry Gyroscope Co, 57; res asst sch med, Univ PR, 60-61; res asst biol servomech, Electronic Systs Lab, 61-62, from asst prof to assoc prof aeronaut & astronaut, 62-70, PROF AERONAUT & ASTRONAUT, MASS INST TECHNOL, 70- *Concurrent Pos:* Consult to various indust & govt orgns, 62-; mem exec comt, Conf Eng Med & Biol, 67; mem, Eng Biol & Med Training Comt, NIH, 70-73; vis prof, Swiss Fed Inst Technol, 72-73 & Univ Zurich, 72-73; mem staff, Conserv Nat Arts Metiers, France, 72-73; mem eng & clin care subcomt, Nat Acad Eng; mem cardiovasc panel, Nat Acad Sci; mem comt space biol & med, Space Sci Bd, Nat Res Coun, Nat Acad Sci, 74-; chmn vestibular panel, Nat Acad Sci, 77; USAF Scientific Adv Bd, Chmn Airlift Panel, 79-85; CHABA Coun, 82-85; Nat Res Coun, Aeronaut & Space Engr Bd, 82-; Airforce Studies Bd, 82-87; vis prof elec eng, Stanford Univ, 87-88; vis scientist NASA-Ames Res Ctr, 87-88. *Honors & Awards:* Dryden Lectr, Am Inst Aeronaut & Astronaut, 82; Franklin V Taylor Award, Human Factors, Inst Elec & Electronics Engrs, 83. *Mem:* Nat Acad Engr; fel Inst Elec & Electronics Engrs; Biomed Eng Soc (pres, 79); Am Inst Aeronaut & Astronauts; Barany Soc; Aerospace Med Asn; Sigma Xi; Soc Neurosci; Alza Lectr, Biomed Eng Soc, 84. *Res:* Application of control theory to man-vehicle problems, especially orientation; flight simulators; space laboratory experimentation on vestibular function. *Mailing Add:* Rm 37-207 Man-Vehicle Lab Mass Inst Technol Cambridge MA 02139

YOUNG, LAWRENCE DALE, b Lafayette, Ind, Sept 13, 50; m 72; c 3. ANIMAL SCIENCE, ANIMAL BREEDING. *Educ:* Purdue Univ, BS, 72; Okla State Univ, MS, 73, PhD(animal breeding), 75. *Prof Exp:* Res asst, Okla State Univ, 72-75; RES GENETICIST, US MEAT ANIMAL RES CTR, USDA, 76- *Concurrent Pos:* NDEA fel, Okla State Univ, 74-75. *Mem:* Am Soc Animal Sci. *Res:* Animal genetics; physiological genetics. *Mailing Add:* US Meat Animal Res Ctr PO Box 166 Clay Center NE 68933

YOUNG, LAWRENCE DALE, b Hartford, Ky, Apr 18, 51; m 71; c 2. NEMATOLOGY, PLANT BREEDING. *Educ:* Univ Ky, BS, 73; NC State Univ, MS, 75, PhD(plant path), 78. *Prof Exp:* Plant breeder, Pfizer Genetics, 78-79; PLANT PATHOLOGIST, AGR RES SERV, USDA, 79- *Concurrent Pos:* Adj asst prof, Univ Tenn, 80-83, adj assoc prof, 83- *Mem:* Am Phytopath Soc; Crop Sci Am; Soc Nematologist. *Res:* Practical control measures for nematodes that reduce yield of soybeans by developing resistant cultivars and nongenetical methods such as crop rotation. *Mailing Add:* USDA-Agr Res Serv Nematol Res 605 Airways Blvd Jackson TN 38301

YOUNG, LAWRENCE EUGENE, b Waterville, Ohio, Mar 18, 13; m 40; c 4. MEDICINE. *Educ:* Ohio Wesleyan Univ, BA, 35; Univ Rochester, MD, 39; Am Bd Internal Med, dipl. *Hon Degrees:* DSc, Ohio Wesleyan Univ, 67 & Med Col Ohio, Toledo, 77. *Prof Exp:* From intern to asst resident med, Strong Mem Hosp, 39-41, asst bact, Sch Hyg & Pub Health, Johns Hopkins Univ & Hosp, 41-42; chief resident, Strong Mem Hosp, 42-43; instr med, 43-44, 46-47, from asst prof to assoc prof, 48-57, Charles A Dewey prof & chmn dept, 57-74, dir prog internal med, Univ Rochester Assoc Hosps, 74-78, alumni distinguished serv prof med, 74-78, EMER PROF MED, SCH MED & DENT, UNIV ROCHESTER, 78- *Concurrent Pos:* Buswell fel, 46-47; physician-in-chief, Strong Mem Hosp, 57-74; mem comt blood, Nat Res Coun, 51-53; hemat study sect mem, USPHS, 53-57; distinguished vis prof med, Univ S Fla, 78- *Mem:* Am Soc Clin Invest; Asn Am Physicians (vpres, 72-73, pres, 73-74); Am Fedn Clin Res; Asn Profs Med (pres, 66-67); fel Int Soc Hemat. *Res:* Hematology; patient care. *Mailing Add:* Dept Internal Med Box 19 12901 Bruce B Downs Blvd Tampa FL 33612

YOUNG, LEO, b Vienna, Austria, Aug 18, 26; US citizen; m 83; c 3. ELECTRONICS. *Educ:* Cambridge Univ, BA(math), 45, BA(physics), 47, MA, 50; Johns Hopkins Univ, MS, 56, DrEng, 59. *Hon Degrees:* LHD, Johns Hopkins Univ. *Prof Exp:* Lectr physics, Bradford Technol Col, 47-48; engr, A C Cossor, Ltd, 48-51; head, Microwave & Antenna Lab, Decca Radar, Ltd, 51-53; adv engr, Westinghouse Elec Corp, 53-60; head microwave tech prog, Stanford Res Inst, 60-73; staff consult, Naval Res Lab, 73-81; DIR, RES & LAB MGT, DEPT DEFENSE, 81- *Concurrent Pos:* Consult, Westinghouse Elec Corp, Stanford Linear Accelerator Ctr, Varian & TRW Systs, 60-; lectr, Stanford Univ, 63; mem, US comn A, Int Union Radio Sci, 65-, mem, US nat comn, 78-; vis prof, Univ Leeds, 66, Israel Inst Technol, 70-71 & Univ Bologna, 71; Tech Prog Comt chmn, Inst Elec & Electronic Engrs Microwave Soc, 65, pres, 69; mem bd dir, Inst Elec & Electronic Engrs, 71-74 & 79-82, vpres, 79; distinguished lectr summer sch, Eng, 75; chmn eng adv bd, NSF, 79-81; mem US Nat Comt, Int Union Radio Sci, 78-84; assoc mem, Govt Univ Indust Res Roundtable, 82-85; exec secy, Dept Defense, Univ Forum, 82-86; mem adv comt, Johns Hopkins Univ, Mass Inst Technol & Univ Calif, Santa Barbara; Microwave Prize, Inst Elec & Electonics Engrs, 63; AGARD Lectr, Italy, 71. *Mem:* Fel AAAS; fel Inst Elec & Electronics Engrs; Sigma Xi; Am Asn Eng Soc. *Res:* Microwaves; optics; radar; filters; antennas; design and manufacturing. *Mailing Add:* Off Dep Und Sec Res & Adv Tech Pentagon Washington DC 20301-3080

YOUNG, LEONA GRAFF, b New York, NY, Dec 22, 36; m 58; c 3. PHYSIOLOGY. *Educ:* Bryn Mawr Col, BA, 58; Univ SC, MS, 60; Emory Univ, PhD(physiol), 67. *Prof Exp:* Instr physiol, 67-68, USPHS fel biochem, 69-71, instr physiol, 71-72, asst prof, 72-76, ASSOC PROF PHYSIOL, SCH MED, EMORY UNIV, 76- *Concurrent Pos:* USPHS res grant, 72- *Mem:* Biophys Soc; Am Physiol Soc; AAAS; Am Soc Cell Biol. *Res:* Mammalian spermatozoan motility; fertility; immunological infertility; contraception; contractile proteins. *Mailing Add:* Dept Physiol Emory Univ 1364 Clifton Rd NE Atlanta GA 30322

YOUNG, LEONARD M, b Dallas, Tex, Oct 20, 35. GEOLOGY. *Educ:* Rice Univ, BA, 57; Univ Okla, MS, 60; Univ Tex, Austin, PhD(geol), 68. *Prof Exp:* Asst prof, 67-74, ASSOC PROF GEOL, NORTHEAST LA UNIV, 74- *Mem:* Soc Econ Paleontologists & Mineralogists; Nat Asn Geol Teachers. *Res:* Carbonate and terrigenous sedimentary petrology; sedimentary processes; textural parameters; sedimentary structures; fluid inclusion paleotemperatures. *Mailing Add:* Dept Geol Northeast La Univ 700 University Ave Monroe LA 71209

YOUNG, LEWIS BREWSTER, b Los Angeles, Calif, Feb 25, 43; m 66. CATALYSIS, PETROCHEMICALS. *Educ:* Univ Calif, Riverside, BA, 64; Iowa State Univ, PhD(chem), 68. *Prof Exp:* NIH fel chem, Univ Colo, Boulder, 68-70; CHEMIST, MOBIL CHEM CO, MOBIL CORP, 70- *Mem:*

Am Chem Soc. *Res:* Catalysis by zeolites; petrochemicals; synthetic lubricants; additives; dehydrogenation; oxidation of organic compounds by metal ions. *Mailing Add:* Mobil Res & Develop Corp PO Box 1028 Princeton NJ 08540

YOUNG, LI, b Taipei, Taiwan. DESIGN AUTOMATION, AUTOMATED MANUFACTURING. *Educ:* Tunghai Univ, Taiwan, BS, 79; Univ Fla, ME, 85, PhD(mech eng), 86. *Prof Exp:* Res assoc, Univ Fla, 83-86, postdoctoral fel, 86-87, instr robot geom, 87; MECH TECH STAFF, AT&T BELL LABS, 87- *Mem:* Am Soc Mech Engrs. *Res:* Design theory, methodology, and technique of mechanical system; automated manufacturing; robotics; mechanism and machine design and analysis. *Mailing Add:* AT&T Bell Labs Murray Hill NJ 07974

YOUNG, LINDA, b Oakland, Calif, Dec 20, 54. LASER SPECTROSCOPY. *Educ:* Mass Inst Technol, SB, 76; Univ Calif, Berkeley, PhD(chem), 81. *Prof Exp:* Res asst, Univ Calif, Berkeley, 76-81; res assoc chem, James Franck Inst, Univ Chicago, 81-88; STAFF MEM, PHYS DIV, ARGONNE NAT LAB, 88- *Mem:* Am Phys Soc. *Res:* Molecular dynamics using inducer fluorescence as a probe; vibrational energy transfer in the ground electronic state of impurities isolated in inert gas matrices and intramolecular dynamics of aromatic molecules in supersonic free jets. *Mailing Add:* Argonne Nat Lab Physics Div Bldg 203 H-117 Argonne IL 60439

YOUNG, LIONEL WESLEY, b New Orleans, La, Mar 14, 32; m 57; c 3. PEDIATRICS, RADIOLOGY. *Educ:* St Benedict's Col, Kans, BS, 53; Howard Univ, MD, 57. *Prof Exp:* From sr instr to asst prof radiol, 65-69, asst prof pediat, 66-69, assoc prof radiol & pediat, Med Ctr, Univ Rochester, 69-75; PROF RADIOL & PEDIAT, HEALTH CTR, UNIV PITTSBURGH, 75- *Concurrent Pos:* Nat Cancer Inst traineeship grant radiation ther, Med Sch, Univ Rochester, 59-60; Children's Bur, Dept HEW fel pediat radiol, Sch Med, Univ Cincinnati, 63-65; abstractor radiol, Am J Roentgenol & Radium Ther & Nuclear Med, 65-; clin consult, NY State Dept Health, 67-75; mem radiol training comt, Nat Inst Gen Med Sci, NIH, 71-73; pediat radiol consult, Comt Prof Self-Eval & Continuing Educ, Am Col Radiol, 72-82. *Honors & Awards:* Spec Paper Award, Soc Pediat Radiol, 69 & John Caffey Award; Gold Cert, Nat Med Asn, 70. *Mem:* Radiol Soc NAm; Am Roentgen Ray Soc; Soc Pediat Radiol (pres, 84-85); fel Am Col Radiol; Asn Univ Radiologists; fel Am Acad Pediat. *Res:* Magnification radiography and tomagraphy in pediatric radiology; radiology of renal hypoplasias and dysplasias; duodenal, pancreatic and renal injury from blunt trauma; skeletal dysplasias and metabolic bone disease in childhood. *Mailing Add:* Dept Radiol 16-27 Children's Hosp Pittsburgh 3705 Fifth Ave Pittsburgh PA 15213

YOUNG, LLOYD MARTIN, b Merricourt, NDak, Nov 9, 42; m 66; c 2. ACCELERATOR PHYSICS. *Educ:* Univ NDak, BS, 65, MS, 66; Univ Ill, PhD(physics), 72. *Prof Exp:* Res Assoc, Univ Ill, 72-74, asst prof physics, 74-79, sr res physicist, 77-79; PROJ LEADER, LOS ALAMOS NAT LAB, 79- *Mem:* Am Phys Soc. *Res:* Development of an electron accelerator having a 100% duty factor using a superconducting linac through which the beam is recirculated several times. *Mailing Add:* Six Karen Circle Los Alamos NM 87544

YOUNG, LOUIS LEE, b El Paso, Tex, Nov 22, 41. FOOD SCIENCE. *Educ:* Tex A&M Univ, BS, 64, MS, 67, PhD(poultry prod technol), 70. *Prof Exp:* Res asst poultry nutrit, Tex Agr Exp Sta, 64-67; res fel poultry prod technol, Tex A&M Univ, 67-68; res assoc, Tex Agr Exp Sta, 68-71; RES FOOD TECHNOLOGIST, POULTRY PROD TECHNOL, RUSSELL RES CTR, AGR RES SERV, US DEPT AGR, 71- *Mem:* AAAS; Poultry Sci Asn; Inst Food Technol; Sigma Xi. *Res:* Food chemistry and microbiology; poultry processing; recovery and utilization of protein from food processing waste; muscle chemistry. *Mailing Add:* 171 Colonial Dr Athens GA 30606

YOUNG, LOUISE GRAY, b Los Angeles, Calif, Oct 4, 35; m 53, 68; c 2. PLANETARY ATMOSPHERES, MOLECULAR SPECTROSCOPY. *Educ:* Univ Calif, Los Angeles, BS, 58, MS, 59; Calif Inst Technol, PhD(eng), 63. *Prof Exp:* Consult engr, Douglas Aircraft, 62-63; scientist, Jet Propulsion Lab, joint prog, Calif Inst Technol & NASA, 63-65, sr scientist, 66-73; res assoc physics, Tex A&M Univ, 73-79; res assoc atmospheric physics, Geophys Lab, USAF, 79-80; RES ASSOC ASTRON, SAN DIEGO STATE UNIV, 80- *Concurrent Pos:* Asst prof eng, Univ Calif, Los Angeles, 65-66; assoc ed, J Quant Spectros & Radiative Transfer, 66-78; vis astronr, Nat Acad Sci exchange prog, Nicholas Copernicus Univ, Torun, Poland, 72 & Inst Space Res, USSR, 80; res assoc astron, Univ Tex, Austin, 67. *Mem:* Int Astron Union; Optical Soc Am; Am Meterol Soc; Am Astron Soc. *Res:* Calculation of molecular spectra; application of spectroscopic data to the transmission of radiation in planetary atmospheres. *Mailing Add:* 4906 63rd St San Diego CA 92115

YOUNG, LYLE E(UGENE), b Branford, NDak, Oct 16, 19; m 42; c 4. ENGINEERING MECHANICS. *Educ:* Univ Minn, BA, 41, MS, 49. *Prof Exp:* Asst supvr track, Pa, RR, 41-42; instr graphics & surv, Univ Minn, 45-53; prof eng mech, Univ Nebr-Lincoln, 53-84, asst dean, Col Eng, 66-70, assoc dean, Col Eng & Technol, 70-84; RETIRED. *Mem:* Am Soc Eng Educ; Nat Soc Prof Engrs. *Res:* Mechanics of materials; concrete. *Mailing Add:* 3120 N 75 St Ct Lincoln NE 68507

YOUNG, M(ILTON) G(ABRIEL), b Coopersburg, Pa, Nov 29, 11; m 37; c 3. ELECTRONICS ENGINEERING, MATERIALS SCIENCE. *Educ:* Lehigh Univ, BS, 32; Harvard Univ, MS, 33. *Prof Exp:* Elec engr, Saucon Hosiery Mills, 33-34; asst chemist, Devoe & Raynolds Corp, 34-35; res engr, Hamilton Watch Co, 35-36; sr mfg engr, Western Elec Co, 36-40; instr elec eng, 40-42, from asst prof to prof & chmn dept, 42-73, actg dean sch eng, 51-52, dir, Nema-Univ Res Lab, 58-77, exec officer elec eng, 73-74, EMER PROF ELEC ENG, UNIV DEL, 74- *Concurrent Pos:* Consult, Triumph Explosives Co, 40-44, Bio-Chem Res Found, Franklin Inst, 42-45 & Gen Develop Corp, 50-; guest prof, Birla Eng Col, India, 57-58. *Mem:* Sr mem Inst Elec & Electronics Engrs; Sigma Xi. *Res:* Industrial electronics; ultrasonics. *Mailing Add:* Univ Del Newark DE 19711

YOUNG, MARGARET CLAIRE, b Austin, Tex, Sept 23, 43; m 64; c 2. ANATOMY, PHYSIOLOGY. *Educ:* Univ Tex, BA, 64; Univ Tex Med Br Galveston, PhD(physiol), 69. *Prof Exp:* Res technician II physiol, 64-65, res assoc physiol, 69-70, instr anat, 70-71, ASST PROF ANAT, UNIV TEX MED BR, GALVESTON, 71- *Concurrent Pos:* Jeanne B Kempner fel, Univ Tex Med Sch, San Antonio, 69-70. *Mem:* AAAS; Am Asn Anat. *Res:* Autoradiographic evidence of leucocytic participation in nervous system injury; autoradiographic study of pathways of nerve fibers in and out of spinal cord. *Mailing Add:* 917 E 38 1/2 St Austin TX 78751

YOUNG, MARTIN DUNAWAY, b Moreland, Ga, July 4, 09; m 38; c 2. PARASITOLOGY, MALARIOLOGY. *Educ:* Emory Univ, BS, 31, MS, 32; Johns Hopkins Univ, ScD(parasitol), 37; Am Bd Med Microbiol, dipl. *Hon Degrees:* DSc, Emory Univ, 63, Mich State Univ, 75 & Univ Fla, 85. *Prof Exp:* Jr zoologist, NIH, 37-40, dir malaria res lab, 41-50, in charge imported malaria studies, 43-46, dir, Malaria Sur Liberia, 48, sanitarian, 44-48, sr scientist, 48-50, scientist dir, 50-64, head sect epidemiol, Lab Trop Dis, Nat Inst Allergy & Infectious Dis, 50-61, asst chief lab parasite chemother, 61-62, assoc dir extramural prog, 62-64; dir, Gorgas Mem Lab, 64-74, dir res, Gorgas Mem Inst, 72-74; RES PROF PARASITOL, COL VET MED & DEPT IMMUNOL & MED MICROBIOL & DEPT MED, COL MED, UNIV FLA, 74- *Concurrent Pos:* Vis prof, ETenn State Teachers Col, 39, Ala Med Ctr, 65-74 & Mem Univ, St Johns, Newfoundland, 77; coun, AAAS, 47-48 & 52-56; lectr, Meharry Med Sch SC, 60 & Med Sch, Univ Panama, 64-69; clin prof, Sch Med, La State Univ, 67-76; consult, Int Coop Admin, India, 57, WHO, Rumania, 61, CZ Dept, 64-74 & US AID, Africa, 80, PAITO, Jamaica; mem expert adv panel malaria, WHO, 50-84; mem malaria adv panel, Pan Am Health Orgn, 57-68; ed pro team, Am J Trop Med Hyg, 59; mem Columbia Bd Health, SC, 60-61; mem malaria & parasitic dis comns, Armed Forces Epidemiol Bd, Dept Defense, 65-73; hon res assoc, Smithsonian Inst, 66-74; res assoc, Gorgas Mem Lab, 74-77; mem bd dirs, Gorgas Mem Inst, 77-83; prin investr, Effects Insect Pathogens on the Ability of Mosquitoes to transmit Malaria, grant, USAID, 81-85. *Honors & Awards:* Rockefeller Pub Serv Award, 53; Darling Medal & Prize, WHO, 63; Order of Manuel Amador Guerrero, Govt Panama, 74; Gorgas Medal, Asn Mil Surgeons of US, 74; LePrince Award, Am Soc Trop Med & Hyg, 76 & Craig Lectr, 84; Cert Merit, Gargas Mem Inst, 74. *Mem:* AAAS; Am Soc Parasitol (pres, 65); Am Soc Trop Med & Hyg (pres, 52); Royal Soc Trop Med & Hyg; Nat Malaria Soc (secy-treas, 46-50, pres, 52); Soc Protozoologists. *Res:* Malaria parasitology, epidemiology and treatment; parasitic protozoa, helminths; parasitic diseases, especially biology, epidemiology and treatment; parasitic protozoa, mainly human and animal malaria and helminths, with emphasis on life cycles; host parasite relationships; epidemiology; treatment and control. *Mailing Add:* 610 NW 89th St Gainesville FL 32607-1453

YOUNG, MATT, b Brooklyn, NY, Jan 30, 41; m 64; c 2. OPTICS. *Educ:* Univ Rochester, BS, 62, PhD(optics), 67. *Prof Exp:* Res assoc optics, Univ Rochester, 67; asst prof physics, Univ Waterloo, 67-70; asst prof electrophys, Rensselaer Polytech Inst, 70-74; assoc prof natural sci, Verrazzano Col, 74-75; PHYSICIST, ELECTROMAGNETIC TECHNOL DIV, NAT INST STANDARDS & TECHNOL, 76- *Concurrent Pos:* Optics res corresp, Physics Teacher J, 73-76; tech ed, Optical Spectra, 74; consult, Holobeam, Inc, 74, Res & Develop Ctr, Gen Elec Co, 74-75 & NY State Energy Comn, 75; assoc ed, J Optical Soc Am, 74-79; vis scientist, Weizmann Inst Sci, Israel, 84, ed rev bd, Nat Inst Standards & Technol, Boulder, 83-; adj prof elec & comput eng, Univ Colo, 87- *Honors & Awards:* Newton Award, 62; Silver Medal, Dept Com, 83. *Mem:* Optical Soc Am; AAAS; Am Asn Physics Teachers; Fedn Am Scientists; Int Soc Optical Eng. *Res:* Optical fiber measurements; optical processing and holography; lasers and coherent optics; measurement technology; solar energy; laser produced plasmas. *Mailing Add:* Electromagnetic Technol Div Nat Inst Standards & Technol 325 Broadway Boulder CO 80303

YOUNG, MAURICE DURWARD, pediatrics, cardiology; deceased, see previous edition for last biography

YOUNG, MAURICE ISAAC, b Boston, Mass, Feb 10, 27; m 54; c 3. MECHANICAL & AEROSPACE ENGINEERING. *Educ:* Univ Chicago, PhB & SB, 49; Boston Univ, AM, 50; Univ Pa, PhD(eng mech), 60. *Prof Exp:* Sr dynamics engr, Helicopter Div, Bell Aerospace Corp, 51-54; chief dynamics res & develop, Vertol Aircraft Corp, 54-56; prin engr, Piasecki Aircraft Corp, 56-58; sect mgr appl mech, Govt & Indust Div, Philco Corp, 58-61; mgr advan technol, Vertol Div, Boeing Co, 61-68; prof mech & Aerospace Eng, 68-88, PROF EMER MECH & AEROSPACE ENG, UNIV DEL, 88- *Concurrent Pos:* Lectr, Southern Methodist Univ, 53; consult, Vertol Div, Boeing Co, 68-70; E I Du Pont de Nemours & Co, Inc fac fel, Univ Del, 69, Univ Del Res Found grant, 70-71, US Army Res Off NC res grant, 71-; consult, Mech Res, Inc, 69-71, Textile Fibers Dept, E I Du Pont De Nemours & Co, Inc, 72-73, TM Develop, Inc, Pa, 76-, Bigham, Englar, Jones & Houston, NY, 78-81 & Advance Ratio Design Co, Chester, Pa, 79-85; consult, NASA Langley, 87, 88-, NASA ASEE fel, 87- *Mem:* Soc Eng Sci; Int Asn Sci & Technol Develop. *Res:* Dynamics; vibrations; control dynamics; flight mechanics; aeroelasticity; rotary wing and vertical/short takeoff and landing mechanics; alternate energy dynamics; wind and ocean current and thermal. *Mailing Add:* 1454 Todds Lane No B42 Hampton VA 23666

YOUNG, MICHAEL DAVID, Brit citizen; c 1. PHARMACEUTICAL CHEMISTRY. *Educ:* Univ Wales, BSc, 61 & BSc, 62; Univ Col, Cardiff, Wales, PhD(chem), 81; Columbia Univ, MS, 85. *Prof Exp:* House surgeon, Surg Unit, Univ Hosp, Wales, 72, house physician pediat, 73; asst med dir, Delbay Parmaceut, Bloomfield, NJ, 73-75; med dir, Winthrop Labs, NY, 75-76, vpres, 76-77; vpres med affairs, Astra Pharmaceut, Framingham, Mass, 77-79, vpres sci affairs, 79-81; vpres res & develop, Westwood Pharmaceut, Buffalo, 82; vpres med & regulatory affairs, 82-85, VPRES WORLDWIDE REGULATORY OPERS, SMITHKLINE & FRENCH LABS, PHILADELPHIA, 85- *Mem:* Royal Inst Chem; Brit Med Asn; fel Am Inst Chemists; fel Royal Soc Med; Am Soc Regional Anesthesia; Am Acad Dermat. *Res:* Protection of carbohydrate systems from radiation damage and in the free-radical degradation of mucopolysaccharides as a model of disease, for example, arthritis; clinical trial methodology. *Mailing Add:* Dept Biochem Univ Fla Box J-245 JHMHC Gainesville FL 32610

YOUNG, MORRIS NATHAN, b Lawrence, Mass, July 20, 09; m 48; c 2. OPHTHALMOLOGY, SURGERY. *Educ:* Mass Inst Technol, BS, 30; Harvard Univ, MA, 31; Columbia Univ, MD, 35; Am Bd Ophthal, dipl. *Prof Exp:* Intern, Queen's Gen Hosp, NY, 35-37; resident ophthal, Harlem Eye & Ear Hosp, 38-40; asst flight surgeon, Maxwell Field, Ala, 41-42, sr eye, ear, nose & throat officer, Walter Reed Gen Hosp, Washington, DC, 42, chief, eye, ear, nose & throat serv, 69th Ata Hosp, 42-44, chief eye, ear, nose & throat sect, 235th Gen Hosp, 44-45, med officer & mem staff, 301st Logistical Support Brigade, 66, dep comdr, 343rd Gen Hosp, 66-67, dep comr & chief prof serv, 307th Gen Hosp, 67-69, staff med officer, 818th Hosp Ctr, 69; dir ophthal & attend, 69-80, EMER DIR OPHTHAL, BEEKMAN DOWNTOWN HOSP, NEW YORK CITY, 80- *Concurrent Pos:* Ophthalmologist & auth, 45-; attend & prof, Fr & Polyclin Med Sch & Health Ctr, 63-77; med adv, Dir Selective Serv, NY Dist, 65-; consult ophthalmologist, Beth Israel Med Ctr, NY, 72- & St Vincent's Hosp & Med Ctr, 77- *Mem:* AMA; Pan-Am Med Asn; Contact Lens Asn Ophthal; Am Acad Ophthal; Acad Comp Med. *Res:* Medicine; mnemonics and art of memory; science exhibits; illusion practices. *Mailing Add:* Two Fifth Ave Apt 16M New York NY 10011

YOUNG, MYRON H(WAI-HSI), b Shanghai, China, July 3, 29; US citizen; m 59; c 5. NUCLEAR ENGINEERING, MATHEMATICAL MODELING. *Educ:* La State Univ, BS, 53, MS, 57; NC State Col, PhD(nuclear eng), 63. *Prof Exp:* From instr to asst prof eng mech, La State Univ, 56-62; supvry nuclear engr, Air Force Flight Dynamics Lab, Wright-Patterson AFB, 63; from asst prof to assoc prof mech eng, 63-76, prof mech eng, 77-79, PROF MARINE SCI & NUCLEAR ENG, LA STATE UNIV, BATON ROUGE, 80- *Concurrent Pos:* Consult, Eng Physics Lab, Wright Air Develop Ctr, Wright-Patterson AFB, 59-60 & Air Force Flight Dynamics Lab, 64-65; consult, Systs Res Lab, Inc, 66- *Mem:* Am Nuclear Soc; Am Soc Eng Educ; Sigma Xi; Int Asn Math Modeling. *Res:* Estuarine hydrodynamic modeling; cumulative impact study on wet land environment; data management systems. *Mailing Add:* 645 Sunset Blvd Baton Rouge LA 70808

YOUNG, NELSON FORSAITH, b Everett, Wash, Oct 17, 14; m 42; c 4. BIOCHEMISTRY. *Educ:* Univ Wash, BS, 36; NY Univ, PhD(biochem), 45. *Prof Exp:* Res chemist, Mem Hosp, 40-43; asst, Sloan-Kettering Inst, 45-48; from asst prof to prof clin path, Med Col Va, 58-78, lectr biochem, 70-78; RETIRED. *Mem:* Am Soc Clin Path; Am Chem Soc. *Res:* Renal function; protein metabolism; radiation effects. *Mailing Add:* 8505 Rivermont Dr Richmond VA 23229

YOUNG, NORTON BRUCE, b Renton, Wash, Aug 11, 26; m 50; c 3. SPEECH PATHOLOGY, AUDIOLOGY. *Educ:* Univ Wash, BS, 50, MA, 53; Purdue Univ, PhD(speech path, audiol), 57. *Prof Exp:* Sr clinician speech path & audiol, Seattle Speech & Hearing Ctr, 54-55; asst prof, Univ Ore, 55-60, instr audiol, Med Sch, 60-64, from instr to assoc prof audiol & pediat, 64-73, assoc prof speech path & audiol, Crippled Children's Div, 70-73, PROF PEDIAT, MED SCH & PROF SPEECH PATH & AUDIOL, CRIPPLED CHILDREN'S DIV, UNIV ORE, 73- *Mem:* Am Speech & Hearing Asn. *Res:* Pedo-audiology; psychoacoustics; organic disorders of speech; clinical audiology. *Mailing Add:* Dept Pediat Ore Health Sci Univ 3181 SW Sam Jackson Park Rd Portland OR 97201

YOUNG, PAUL ANDREW, b St Louis, Mo, Oct 3, 26; m 49; c 10. NEUROANATOMY. *Educ:* St Louis Univ, BS, 47, MS, 53; Univ Buffalo, PhD(anat), 57. *Prof Exp:* From asst to instr anat, Univ Buffalo, 53-57; from asst prof to assoc prof, 57-72, actg chmn dept, 69-73, PROF ANAT, SCH MED, ST LOUIS UNIV, 72-, CHMN DEPT, 73- *Honors & Awards:* AOA Honor Med Soc, 72, Golden Apple, 74. *Mem:* Am Asn Anat; Soc Neurosci; Sigma Xi. *Res:* Neuroanatomy, human and experimental. *Mailing Add:* St Louis Univ Med Sch 1402 S Grand Blvd St Louis MO 63104

YOUNG, PAUL MCCLURE, b Seaman, Ohio, Feb 13, 16; m 42; c 2. MATHEMATICS. *Educ:* Miami Univ, AB, 37; Ohio State Univ, MA, 39, PhD(math), 41. *Prof Exp:* From instr to asst prof math, Miami Univ, 41-47; from assoc prof to prof, Kans State Univ, 47-62, assoc dean sch arts & sci, 56-62; vpres acad affairs, Univ Ark, 62-66; exec dir, Mid-Am State Univs Asn, 66-78; prof math & vpres univ develop, 70-86, EMER PROF, KANS STATE UNIV, 86- *Mem:* AAAS; Am Math Soc; Math Asn Am. *Res:* Analysis; approximation of functions by integral means; characterization of integral means. *Mailing Add:* Kans State Univ 2023 Arthur Dr Manhattan KS 66502

YOUNG, PAUL RUEL, b St Marys, Ohio, Mar 16, 36; c 2. COMPUTER SCIENCES, MATHEMATICAL LOGIC. *Educ:* Antioch Col, BS, 59; Mass Inst Technol, PhD(math), 63. *Prof Exp:* Asst prof math, Reed Col, 63-66; prof comput info sci & math & chmn dept comput info sci, Univ NMex, 78-79; from asst prof to prof math & comput sci, Purdue Univ, Lafayette, 66-83; chmn dept comput sci, 83-88, VCHMN COMPUT RES BD, UNIV WASH, 87- *Concurrent Pos:* NSF fel, Stanford Univ, 65-66; vis prof, Univ Calif, Berkeley, 72-73 & 82-83; chmn adv bd comput sci, NSF, 79-80; Woodrow Wilson fel; NSF postdoc fel; Brittingham vis prof, Univ Wis, 88-89. *Mem:* Asn Comput Mach; Inst Elec & Electronics Engrs; Asn Symbolic Logic. *Res:* Theory of computational complexity and theory of algorithms; mathematical logic. *Mailing Add:* Dept Comput Sci Univ Wash Seattle WA 98195

YOUNG, PETER CHUN MAN, b Hong Kong, Dec 19, 36; Can citizen; m 67; c 2. BIOCHEMISTRY & ENDOCRINOLOGY. *Educ:* McGill Univ, BS, 61, MS, 63, PhD(biochem), 67. *Prof Exp:* Res assoc endocrinol, St Michael's Hosp, 67-71; asst prof, 71-77, ASSOC PROF OBSTET & GYNEC, SCH MED, IND UNIV, INDIANAPOLIS, 77- *Honors & Awards:* Prize Award, Cent Asn Obstetricians & Gynecologists, 79. *Mem:* AAAS; NY Acad Sci; Tissue Cult Asn. *Res:* Biochemistry of steroid hormones; reproductive endocrinology; biology of cancer. *Mailing Add:* Dept Obstet & Gynec Ind Univ Med Ctr MF 102B Indianapolis IN 46202-5196

YOUNG, PHILLIP GAFFNEY, b Beeville, Tex, July 21, 37; m 60; c 3. NUCLEAR PHYSICS. *Educ:* Univ Tex, Austin, BS, 61, MA, 62; Australian Nat Univ, PhD(nuclear physics), 65. *Prof Exp:* Res fel nuclear physics, Australian Nat Univ, 65-66; res fel, Los Alamos Nat Lab, 66-68, mem staff nuclear cross sect eval, 68-75, group leader nuclear cross sect eval, 75-81, STAFF MEM, LOS ALAMOS NAT LAB, 81- *Mem:* Am Phys Soc; Am Nuclear Soc. *Res:* Low energy nuclear physics; neutron-particle and charged-particle cross sections and polarization. *Mailing Add:* Los Alamos Nat Lab Group T-2 MS B243 PO Box 1663 Los Alamos NM 87545

YOUNG, RALPH ALDEN, b Arickaree, Colo, July 14, 20; m 42; c 2. SOIL FERTILITY. *Educ:* Colo State Univ, BS, 42; Kans State Univ, MS, 47; Cornell Univ, PhD(agron), 53. *Prof Exp:* Instr soils, Kans State Univ, 47-48; asst prof soils, NDak State Univ, Univ & asst soil scientist, Exp Sta, 48-50 & 53-55, from assoc prof & assoc soil scientist to prof soils & soil scientist, 55-62; vis prof, Univ Calif, Davis, 62-63; chmn dept agr biochem & soil sci, Nev Agr Exp Sta, Univ Nev, Reno, 63-65, chmn, Div Plant Soil & Water Sci, 65-75, prof soil sci & univ & soil scientist, 63-75, assoc dir, 75-82, emer prof, 84-; RETIRED. *Mem:* Fel AAAS; Soil Sci Soc Am; Am Soc Agron; Soils Conserv Soc Am; Sigma Xi. *Res:* Fertilizer-water interactions; soil as a waste treatment system. *Mailing Add:* 2229 Stagecoach Rd Grand Junction CO 81503

YOUNG, RALPH HOWARD, b Berkeley, Calif, Mar 22, 42; m 68; c 1. CHEMICAL PHYSICS, SOLID STATE PHYSICS. *Educ:* Calif Inst Technol, BS, 64; Stanford Univ, PhD(chem physics), 68. *Prof Exp:* Lectr chem, Stanford Univ, 68; asst prof, Jackson State Col, 68-70; lectr, Univ Calif, Riverside, 70; RES ASSOC, COPY PRODS RES & DEVELOP, EASTMAN KODAK CO, 71- *Mem:* Sigma Xi. *Res:* Photoconduction in organic solids; quantum chemistry; quantum axiomatics. *Mailing Add:* 270 Cobb Terr Rochester NY 14620

YOUNG, RAYMOND A, b Buffalo, NY, Mar 14, 45; m 62; c 2. WOOD CHEMISTRY. *Educ:* State Univ NY, BS, 66, MS, 68; Univ Wash, PhD(wood chem), 73. *Prof Exp:* Process supvr paper prod, Kimberly-Clark Corp, 68-69; Textile Res Inst fel, Princeton Univ, 73-74, staff scientist textiles,; PROF FORESTRY, UNIV WIS-MADISON, 75- *Concurrent Pos:* Vis scientist, Swedish Forest Prods Lab, Stockholm, Sweden, 72-73; Sr Fulbright res scholar, Aristotelian Univ, Thessalonike, Greece, 89. *Mem:* Tech Asn Pulp & Paper Indust; Fiber Soc; Am Chem Soc. *Res:* Solvent pulping of wood; chemical modification of cellulose and high yield pulp fibers; lignocellulosic based composites. *Mailing Add:* Dept Forestry Univ Wis 120 Russel Lab Madison WI 53706

YOUNG, RAYMOND H(YKES), b Bellefonte, Pa, June 23, 21. ELECTRICAL ENGINEERING. *Educ:* Bucknell Univ, BS, 43; Northwestern Univ, MS, 51, PhD(elec eng), 57. *Prof Exp:* Test engr, Gen Elec Co, 43-46; instr, 47-50, from asst prof to assoc prof, 53-70, PROF ELEC ENG, BUCKNELL UNIV, 70- *Mem:* Inst Elec & Electronics Engrs. *Res:* Switching theory; relation of electrical engineering to medical and biological sciences. *Mailing Add:* 320 N Ninth St Sunbury PA 17801

YOUNG, RAYMOND HINCHCLIFFE, JR, b Pennsauken, NJ, Nov 22, 28; m 53; c 4. CALCINATION OF KAOLIN, BENEFICATION OF MINERALS. *Educ:* Pa Mil Col, BS, 53; Univ Maine, MS, 55, PhD(org chem), 61. *Prof Exp:* Res chemist, Monsanto Co, 60-70, Freeport Kaolin Co, 70-85; RES ASSOC, ENGELHARD CORP, NJ, 85- *Honors & Awards:* A K Doolittle Award, Am Chem Soc, 72. *Mem:* AAAS; Am Chem Soc; Soc Mineral Eng. *Res:* Device and direct basic research projects related to kaolin clay. *Mailing Add:* Engelhard Corp PO Box 337 Gordon GA 31031

YOUNG, REGINALD H F, b Honolulu, Hawaii, May 17, 37; m 59; c 2. SANITARY & CIVIL ENGINEERING. *Educ:* Univ Hawaii, BS, 59, MS, 65; Wash Univ, DSc(environ & sanit eng), 67. *Prof Exp:* Jr engr, Paul Low Eng, 59; jr engr, Sunn, Low, Tom & Hara, Inc, 59, asst proj engr, 62-63; res asst pub health, Univ Hawaii, 63, asst prof environ health & sanit eng, 66-69, assoc prof civil eng, 69-74, asst dir, Water Resources Res Ctr, 73-78, PROF CIVIL ENG, 74-, ASSOC DEAN, COL ENG, 79-, INTERIM DEAN, UNIV HAWAII, 80-81 & 89- *Mem:* Am Acad Environ Engrs; Am Soc Civil Engrs; Water Pollution Control Fedn; Am Water Works Asn; Sigma Xi; Asn Environ Eng Professors. *Res:* Water quality management and pollution control; water and sewage treatment; industrial waste treatment; solid wastes management and control. *Mailing Add:* Col Eng Univ Hawaii Manoa 2540 Dole St Honolulu HI 96822

YOUNG, REUBEN B, b Wilmington, NC, Apr 2, 30; m 53; c 3. MEDICINE, PEDIATRICS. *Educ:* Med Col Va, BS, 53, MD, 57. *Prof Exp:* Instr pediat, Univ Pa, 61-63; from asst prof to assoc prof, 63-71, dir med staff hosp & exec assoc dean, Sch Med, 77-79, PROF PEDIAT, MED COL VA, 71-, PROF GENETICS, 75-, ASSOC DEAN CONTINUING MED EDUC, 79- *Mem:* Am Acad Pediat; Endocrine Soc; Am Diabetes Asn. *Res:* Pediatric endocrinology and diabetology; catecholamine metabolism in children; hypoglycemia in children. *Mailing Add:* Dept Pediat Va Commonwealth Univ Box 163 MCV Scis Richmond VA 23298

YOUNG, RICHARD A, b Providence, RI, Aug 5, 40; m 64; c 2. GEOLOGY. *Educ:* Cornell Univ, BA, 62; Wash Univ, PhD(geol), 66. *Prof Exp:* Asst prof earth sci, 66-72, assoc prof, 72-79, chmn dept, 77-86, PROF GEOL SCI, STATE UNIV NY COL GENESEO, 79- *Concurrent Pos:* Prin investr geol mapping proj using Apollo Mission photog, NASA Contract, 72-75; vis faculty mem, Univ Canterbury, 72; hydrologist, US Geol Surv, 76-; Environ Protection Agency grant, 79-82. *Honors & Awards:* Cole Mem Res Award, Geol Soc Am, 88. *Mem:* AAAS; Geol Soc Am; Sigma Xi. *Res:* Cenozoic geology, including sedimentation, geomorphology, glacial geology and vulcanism; lunar geology. *Mailing Add:* Dept Geol Sci State Univ NY Col Geneseo NY 14454

YOUNG, RICHARD ACCIPITER, solid state physics, for more information see previous edition

YOUNG, RICHARD D, b New York, NY, Mar 16, 24; m 50; c 2. PHYSICS, ELECTRICAL ENGINEERING. *Educ:* Princeton Univ, AB, 45; Calif Inst Technol, MS, 47, PhD(physics), 52. *Prof Exp:* Teaching asst, Calif Inst Technol, 48-50; mem tech staff, Hughes Aircraft Co, 50-54; mem sr staff, Ramo-Wooldridge Corp, TRW Space Technol Labs & Bunker-Ramo Corp, 54-64 & Informatics, Inc, 64-67; asst prog dir, Tracor, Inc, 67-68; dir systs eng, Digilinc Systs Corp, 68-69; mem sr staff, Synergetic Sci, Inc, 69-70; sr sci specialist, Data Systs Div, Litton Industs, Inc, Van Nuys, 70-87; consult, 87-90; RETIRED. *Mem:* Sigma Xi. *Res:* Advanced systems engineering and systems analysis of military command and control systems, electronic countermeasures and radar; theoretical nuclear physics. *Mailing Add:* 54 Saddlebow Rd Bell Canyon CA 91307

YOUNG, RICHARD EDWARD, b Los Angeles, Calif, Aug 20, 38; m 63; c 2. BIOLOGICAL OCEANOGRAPHY. *Educ:* Pomona Col, BA, 60; Univ Southern Calif, MS, 64; Univ Miami, PhD(oceanog), 68. *Prof Exp:* Res asst oceanog, Inst Marine Sci, Univ Miami, 65-68, res scientist, 68; asst prof zool, Ohio Wesleyan Univ, 68-69; from asst prof to assoc prof, 69-82, PROF OCEANOG, UNIV HAWAII, MANOA, 82- *Concurrent Pos:* Mem, Cephalopod Int Adv Coun. *Res:* Cephalopod, deep-sea and invertebrate biology. *Mailing Add:* 182 Kuuhale St Kailua HI 96734

YOUNG, RICHARD L, b Rushville, Ill, Nov 22, 32; m 53; c 3. ORGANIC CHEMISTRY. *Educ:* Univ Ill, BSc, 54; Brown Univ, PhD(org chem), 59; Boston Univ, MBA, 85. *Prof Exp:* Res chemist, Res Inst Med & Chem, 58-61; res assoc biochem, Col Agr, Univ Wis, 61-63; asst prof agr biochem, Univ Hawaii, 64-66; MGR, QUAL SYSTS & NEW TECHNOL, DUPONT, NEN PROD, 66- *Mem:* Am Chem Soc; Sigma Xi. *Res:* Peptide synthesis; tritium and carbon-14 radiochemicals. *Mailing Add:* 880 Chestnut St Newton MA 02168

YOUNG, RICHARD LAWRENCE, analytical biochemistry, neurochemistry, for more information see previous edition

YOUNG, RICHARD WAIN, b Albany, NY, Dec 15, 29; m 55; c 4. ANATOMY. *Educ:* Antioch Col, BA, 56; Columbia Univ, PhD(anat), 59. *Hon Degrees:* Dr Sci, Univ Chicago, 80. *Prof Exp:* From asst prof to assoc prof, 60-68, PROF ANAT, SCH MED, UNIV CALIF, LOS ANGELES, 68- *Concurrent Pos:* NSF fel anat, Univ Bari & Caroline Inst, Sweden, 59-60; Markle scholar med sci, 62-67; guest investr anat, Dept Biol, Saclay Nuclear Res Ctr, France, 66-67; mem, Jules Stein Eye Inst, Univ Calif, Los Angeles. *Honors & Awards:* Fight for Sight Res Citation, 69; Friedenwald Award, Asn Res in Vision & Ophthalmol, 76. *Mem:* Am Asn Anat; Am Soc Cell Biol; Asn Res Vision & Ophthal. *Res:* Cell biology; radioisotope studies of ocular tissues. *Mailing Add:* Dept Anat 73-235 CHS Univ Calif 405 Hilgard Ave Los Angeles CA 90024

YOUNG, ROBERT, JR, b Sept 14, 30; US citizen; m 55; c 3. VETERINARY MEDICINE. *Educ:* Univ Calif, BS, 53, MS, 56, DVM, 61. *Prof Exp:* Vet, Shell Develop Co, 61-67, dept head vet med, 67-69, dept head drug develop, 69-78; RESEARCHER, KEARLEY & YOUNG RES, 78- *Mem:* Am Vet Med Asn; Am Soc Animal Sci; Indust Vet Med Asn. *Res:* Drug development and secondary screening; growth regulators; anthelmintics and ectoparasiticides. *Mailing Add:* Kearley & Young Res PO Box 2089 Turlock CA 95380

YOUNG, ROBERT A, b New York, NY, June 8, 29; m 57, 69; c 1. PHYSICS, PHYSICAL CHEMISTRY. *Educ:* Univ Wash, BS, 51, PhD(physics), 59. *Prof Exp:* Res asst, Univ Wash, 53-59; engr, Boeing Airplane Co, 59-60; from physicist to sr physicist, Stanford Res Inst, 60-67, chmn, Atmospheric Chem Physics Dept, 67-68; prof physics, York Univ, 68-75; dir res, Xonics, Inc, 75-79; chmn & pres, Quantatec Int, Inc, 79-85; chmn & pres, Rocketdyne, Inc, 85-; RETIRED. *Concurrent Pos:* Consult, Dept Physics, Univ Wash, 59-60; vis fel, Joint Inst Lab Astrophys, Univ Colo, 66-67 & Lab Astrophys & Space Physics, 74-75; chmn & pres, Intra-Space Int, 72-75. *Mem:* AAAS; Am Phys Soc; Am Geophys Union. *Res:* Energy transfer; atomic and molecular processes and spectra; rocket experimentation; balloon, aircraft stratospheric measurements; short wavelength chemical lasers. *Mailing Add:* 330 Leisure World Mesa AZ 85206

YOUNG, ROBERT ALAN, b St Cloud, Minn, Jan 24, 21; m 48, 77; c 3. SOLID STATE PHYSICS, CRYSTALLOGRAPHY. *Educ:* Polytech Inst Brooklyn, PhD(physics), 59. *Hon Degrees:* Dr, Univ Toulouse, France, 79. *Prof Exp:* Res asst prof & res physicist, Ga Inst Technol, 51-53; instr physics, Polytech Inst Brooklyn, 53-57; res assoc prof physics & head diffraction lab, 57-63, prof physics & head crystal physics br, 64-82, prof, 82-87, EMER PROF PHYSICS, GA INST TECHNOL, 88- *Concurrent Pos:* Co-ed, J Appl Crystallog, 67-69, ed, 70-78; mem, US Nat Comt Crystallog, 69-74 & 76-78, chmn, 79-81; comn, powder diffraction, Int Union Crystallog, 87- *Mem:* Fel Brit Inst Physics; French Soc Mineral & Crystallog; fel Am Phys Soc; Am Crystallog Asn (treas, 68-71, vpres, 72, pres, 73). *Res:* Crystal physics; x-ray, neutron and electron diffraction theory and applications; structural locations and roles of minor impurities; atomic scale mechanisms; apatites; tooth enamel; extension of methods for greater detail; thermally stimulated current studies. *Mailing Add:* Sch Physics Ga Inst Technol Atlanta GA 30332

YOUNG, ROBERT HAYWARD, b Andover, NB, Mar 18, 40; m 65; c 2. PHOTOCHEMISTRY, ORGANIC POLYMER CHEMISTRY. *Educ:* Mt Allison Univ, BSc, 62, MSc, 63; Mich State Univ, PhD(org chem), 67. *Prof Exp:* Asst prof org chem, Georgetown Univ, 67-73; group leader, Union Carbide Corp, 73-80; MEM STAFF, WEYERHAEUSER CO, 80- *Mem:* Am Chem Soc; Chem Inst Can. *Res:* Organic chemistry and photochemistry of singlet oxygen; polymer chemistry, adhesive bonding fundamentals; phenolic resin chemistry; wood chemistry. *Mailing Add:* Weyerhaeuser Co AUB-17 Tacoma WA 98477

YOUNG, ROBERT JOHN, b Calgary, Alta, Feb 10, 23; m 50; c 2. ANIMAL NUTRITION. *Educ:* Univ BC, BSA, 50; Cornell Univ, PhD(animal nutrit), 53. *Prof Exp:* Asst poultry nutrit, Cornell Univ, 50-53; res assoc, Banting & Best Dept Med Res, Univ Toronto, 53-56; res biochemist, Int Minerals & Chem Corp, Ill, 56-58 & Procter & Gamble Co, Ohio, 58-60; assoc prof animal nutrit & poultry husb, 60-65, head dept poultry sci, 65-76, prof animal nutrit, NY State Col Agr & Life Sci, Cornell Univ, 65-, chmn dept animal sci, 76-; MEM STAFF, INST HEALTH PROM, CHARLOTTE. *Mem:* Poultry Sci Asn; Am Inst Nutrit. *Res:* Mineral metabolism; energy value of fats and fatty acids; nutrition and metabolism of protein and amino acids. *Mailing Add:* 4425 Randolph Rd No 205 Charlotte NC 28211-2348

YOUNG, ROBERT L(YLE), b Neoga, Ill, Apr 3, 25; m 46, 69; c 3. MECHANICAL ENGINEERING, AEROSPACE ENGINEERING. *Educ:* Northwestern Univ, BS, 46, MS, 48, PhD(eng), 53. *Prof Exp:* Instr mech eng, Northwestern Univ, 48-53, asst prof, 53-57; from assoc prof to prof, 57-64, dir, Arnold Eng Develop Ctr Grad Prog, 57-64, assoc dean, Space Inst, 64-79, PROF MECH & AEROSPACE ENG, SPACE INST, UNIV TENN, 64- *Concurrent Pos:* Consult, USAF, 57- *Mem:* Am Soc Mech Engrs; Am Inst Aeronaut & Astronaut; Am Soc Eng Educ; Nat Soc Prof Engrs; Sigma Xi. *Res:* Heat transfer and fluid mechanics with current emphasis on solidification in near zero g environments. *Mailing Add:* 110 Park Circle Tullahoma TN 37388

YOUNG, ROBERT LEE, b Houston, Tex, Apr 10, 34; m 56; c 5. ELECTRICAL ENGINEERING. *Educ:* La State Univ, Baton Rouge, BS, 56, MS, 58; Tex A&M Univ, PhD(elec eng), 66. *Prof Exp:* Instr elec eng, La State Univ, 58-59; from asst prof to assoc prof, Southwestern La Univ, 59-68; prof & chmn dept, Northern Ariz Univ, 68-71, asst dean col eng, 70-71; prof eng, Nicholls State Univ, 71-77; PROF ELEC ENG & ASST TO DEAN COL ENG, UNIV SOUTHWESTERN LA, 77- *Mem:* Inst Elec & Electronics Engrs; Am Soc Eng Educ. *Res:* Multivariable control systems; electronic systems. *Mailing Add:* Dept Elec & Comput Eng Univ Southwestern La PO Box 43980 Lafayette LA 70504

YOUNG, ROBERT M, b Brooklyn, NY, Sept 10, 39. BIOCHEMISTRY, BIOLOGY. *Educ:* Brooklyn Col, BS, 60, MA, 65; Univ Pittsburgh, PhD(biochem), 71. *Prof Exp:* Sr res asst clin enzymol, Beth Israel Med Ctr, 60-65; from asst prof to assoc prof biol, 70-78, chmn dept, 75-79, PROF BIOL, RI COL, 78-, CHMN DEPT, 90- *Concurrent Pos:* Lectr, Brooklyn Col, 62-65; res asst, Creedmoor State Hosp, 63-64. *Mem:* Sigma Xi; AAAS; Am Inst Biol Sci; Am Soc Microbiol; Am Soc Zoologists. *Res:* Ribosomal structure and function; protein and RNA synthesis; biosynthesis of folates and pteridines; use of enzymes for the diagnosis of cancer; lens proteins in anopthalmic mice. *Mailing Add:* Dept Biol RI Col Providence RI 02908

YOUNG, ROBERT RICE, b Washington, Pa, Aug 26, 34; m 59; c 3. NEUROLOGY, NEUROPHYSIOLOGY. *Educ:* Yale Univ, SB, 56; Harvard Med Sch, MD, 61. *Prof Exp:* Intern, Peter Bent Brigham Hosp, Boston, 61-62; resident neurol, Mass Gen Hosp, 62-65; NIH spec fel neurophysiol, Oxford Univ, 65-67; dir, clin neurophysiol lab, Harvard Med Sch & Mass Gen Hosp, Boston, 68-87; CHIEF, SPINAL CORD INJURY SERV, VET ADMIN MED CTR, 87-; PROF NEUROL, HARVARD MED SCH, 86- *Concurrent Pos:* Vis scientist clin neurophys, Swedish Med Res Coun, Uppsala Univ, 79; Josiah Macy, Jr Found fac scholar award, 79. *Mem:* Am Acad Neurol; Am Acad Clin Neurophysiol; Am Neurol Asn; Am Asn Electromyog & Electrodiag; Soc Neurosci. *Res:* Discharge properties of spinal motoneurones, (1) their synchronization to produce tremor, the role played by muscle spindle primary afferent discharge in that synchronization and (2) their abnormalities in spastic paresis including alterations in reflex circuits underlying spasticity; neuro rehabilitation-restorative neurology. *Mailing Add:* Spinal Cord Injury Serv W Roxbury Vet Admin Med Ctr Boston MA 02132

YOUNG, ROBERT WESLEY, glass chemistry; deceased, see previous edition for last biography

YOUNG, ROBERT WILLIAM, b Mansfield, Ohio, May 11, 08; m 47; c 3. ACOUSTICS. *Educ:* Ohio Univ, BS, 30; Univ Wash, PhD(physics), 34. *Prof Exp:* Physicist, C G Conn, Ltd, 34-42; physicist, Div War Res, Univ Calif, 42-46, res assoc, Marine Phys Lab, 46; physicist, USN Electronics Lab, 46-67, consult acoust, Naval Undersea Ctr, 67-74, Naval Ocean Systs Ctr, 74-87; CONSULT ARCHIT ACOUSTICS & COMMUN NOISE, 55- *Mem:* AAAS; fel Acoust Soc Am (vpres, 53-54, pres, 60-61). *Res:* Acoustics of wind musical instruments; piano strings and tuning; underwater ambient and ship noise and sound propagation; acoustical standards; techniques for analyzing sonic booms; architectural acoustics; community noise measurement and description. *Mailing Add:* 1696 Los Altos Rd San Diego CA 92109-1321

YOUNG, ROGER GRIERSON, b Moose Jaw, Sask, Dec 18, 20; nat US; m 54; c 3. COMPARATIVE BIOCHEMISTRY. *Educ:* Univ Alta, BSc, 43, MSc, 48; Univ Ore, PhD(chem), 52. *Prof Exp:* Rockefeller asst, Cornell Univ, 51-52, Geer res fel, 52-53; assoc biochemist, Ethicon, Inc, 53-55; asst prof, 55-60, assoc prof entom, Cornell Univ, 60-; MEM STAFF, HUXLEY COL, WESTERN WASH UNIV. *Concurrent Pos:* Vis prof, Col Agr, Univ Philippines, 68-69. *Mem:* AAAS; Entom Soc Am; Am Chem Soc. *Mailing Add:* 1168 Ellishollow Rd Ithaca NY 14850-2946

YOUNG, ROLAND S, chemical engineering, inorganic chemistry; deceased, see previous edition for last biography

YOUNG, RONALD JEROME, b Hong Kong, Aug 10, 32; m 62; c 2. REPRODUCTIVE BIOCHEMISTRY, REPRODUCTIVE TOXICOLOGY. *Educ:* Univ Sydney, BSc, 54; Univ NSW, PhD, 58. *Prof Exp:* Fel, Univ Wis, 58-62; lectr biochem, Monash Univ, 63-67; fel, Univ Calif, 67-70; assoc prof reproductive biol, Med Ctr, Cornell Univ, 73-80; CONSULT, 80- *Mem:* Am Soc Biochem & Molecular Biol. *Res:* Biochemistry of fertilization; reproductive biology. *Mailing Add:* Toxicol Div Chem Res & Develop Ctr Aberdeen Proving Ground MD 21010-5423

YOUNG, ROSS DARELL, b Osceola, Iowa, July 22, 27; m 72; c 2. MECHANICAL ENGINEERING, BIOENGINEERING. *Educ:* Iowa State Col, BS, 53, MS, 55. *Prof Exp:* Automotive engr, Standard Oil Co Ind, 55-58; asst prof mech eng, Col Eng, Univ Mo-Columbia, 58-65, assoc prof bioeng, 68-72, assoc prof phys med & rehab, Sch Med, 70-78, assoc prof mech & aerospace eng, 65-85; ENG CONSULT, 85- *Res:* Mechanical design; engineering history; blood handling, prosthetic and orthopedic devices; heart assisting devices; medical equipment; product liability consulting; medical engineering consulting. *Mailing Add:* RR 1 Franklin MO 65250

YOUNG, RUSSELL DAWSON, b Huntington, NY, Aug 17, 23; m 54; c 4. OPTICS, SURFACE SCIENCE. *Educ:* Rensselaer Polytech Inst, BS, 53; Pa State Univ, MS, 56, PhD(physics), 59. *Prof Exp:* Res assoc, Pa State Univ, 59-61; mem staff, Nat Bur Standards, 61-73, Chief Mech Processes Div, 73-88; PRES, R D YOUNG CONSULTS INC, 88- *Honors & Awards:* Edward Uhler Condon Award, Nat Bur Standards, 74; Nobel Prize Citation, 86; Presidential Citation, 86. *Mem:* Am Phys Soc; Optical Soc Am. *Res:* Physical characterization of surfaces; micrometrology. *Mailing Add:* R D Young Consults Inc 862 Riverside Dr Pasadena MD 21122

YOUNG, RYLAND F, b Bloomington, Ind, Aug 24, 46; m 75; c 4. BACTERIOPHAGE, EXOTOXINS. *Educ:* Rice Univ, AB, 68; Univ Tex, PhD(molecular biol), 75. *Prof Exp:* From asst prof to assoc prof, Dept Med Biochem & Genetics, 78-86, PROF MOLECULAR GENETICS, DEPT BIOCHEM & BIOPHYS, TEX A&M UNIV, 86- *Mem:* AAAS; Am Soc Metals; Fedn Am Socs Exp Biol. *Res:* Molecular genetics of bacteriophage lysis; genetics of bacterial exotoxins; nucleolar function in fission yeast. *Mailing Add:* Dept Biochem & Biophys Tex A&M Univ College Station TX 77843

YOUNG, SANFORD TYLER, b Chicago, Ill, Apr 14, 36; m 61; c 2. ORGANIC CHEMISTRY. *Educ:* Univ Ill, BS, 58; Univ Rochester, PhD(org chem), 63. *Prof Exp:* Res chemist, 62-68, sr res chemist, 68-74, RES ASSOC, AGR CHEM DIV, FMD CORP, 74- *Mem:* Am Chem Soc. *Res:* Process research; high pressure reactions; laboratory safety. *Mailing Add:* Agr Chem Div FMC Corp PO Box 8 Princeton NJ 08450

YOUNG, SETH YARBROUGH, III, b Victoria, Miss, June 8, 41; m 61; c 3. ENTOMOLOGY, VIROLOGY. *Educ:* Miss State Univ, BS, 63; Auburn Univ, PhD(entom), 67. *Prof Exp:* Res entomologist, Stored Prod Res Lab, Mkt Qual Res Div, USDA, Ga, 66-67; from asst prof to assoc prof, 67-76, PROF ENTOM, UNIV ARK, FAYETTEVILLE, 76- *Mem:* Entom Soc Am; Soc Invert Path. *Res:* Insect virology. *Mailing Add:* Dept Entom Univ Ark Fayetteville AR 72701

YOUNG, SHARON CLAIRENE, b Elk City, Okla, Aug 3, 42. ZOOLOGY, ANATOMY. *Educ:* Bethany Nazarene Col, BS, 64; Okla State Univ, MS, 65, PhD(entom), 69. *Prof Exp:* From asst prof to assoc prof, 68-73, PROF BIOL, SOUTHERN NAZARENE UNIV, 73- *Concurrent Pos:* Rec secy, chmn biol sci sect, Okla Acad Sci. *Mem:* Nat Asn Biol Teachers; AAAS; Sigma Xi. *Res:* Thrips resistance in peanuts; behavior of Mongolian gerbil; very low density lipoproteins. *Mailing Add:* Dept Biol Southern Nazarene Univ Bethany OK 73008

YOUNG, SIMON N, b Godalming, Eng, Feb 6, 45; div; c 1. NEUROSCIENCES. *Educ:* Oxford Univ, BA, 67; London Univ, MS, 68, PhD(biochem), 71. *Prof Exp:* Res asst biochem, Inst Neurol, London Univ, 68-71; fel, 71-75, from asst prof to assoc prof, 75-86, PROF NEUROCHEM, DEPT PSYCHIAT, MCGILL UNIV, 87- *Concurrent Pos:* Fel, Que Med Res Coun, 72-73; J B Collip fel, McGill Univ, 73-75. *Honors & Awards:* Borden Award, Can Soc Nutrit Sci, 89. *Mem:* Soc Neurosci; Can Col Neuropsychophamacol; Can Soc Nutrit Sci. *Res:* Investigation of brain biogenic amine synthesis and function in experimental animals and in man; the effects of diet on brain metabolism and behavior. *Mailing Add:* Dept Psychiat McGill Univ 1033 Pine Ave W Montreal PQ H3A 1A1 Can

YOUNG, STUART, b Haslington, Eng, Dec 30, 25; nat US; m 53. VETERINARY PATHOLOGY. *Educ:* Royal Vet Col, Univ London, MRCVS, 48; Royal Dick Sch Vet Studies, Univ Edinburgh, DVSM, 51; Mich State Univ, MS, 54; Univ Calif, Davis, PhD(path), 63. *Prof Exp:* Asst vet invest officer, Vet Invest Lab, North Scotland Col Agr, 49-55; from asst pathologist to assoc pathologist, Vet Res Lab, Mont State Col, 55-61; assoc prof path, Univ Minn, 63-64; assoc prof path, 64-72, dir, NIH Grad Training Prog, 65-72, dir, NIH Vision Res Training Prog, 79-82, prof, 72-87, EMER PROF PATH, COL VET MED & BIOMED SCI, COLO STATE UNIV, 87- *Concurrent Pos:* NIH spec res fel, Cambridge Univ, 59 & Univ Bern, Switz, 70-71; vis scientist, Univ Bern, Switz, 77. *Mem:* Teratol Soc; Asn Res Vision Ophthal; Am Asn Neuropath; Am Vet Med Asn; Conf Res Workers Animal Dis. *Res:* Infectious, metabolic and developmental disorders of central nervous system; pathology of chronic, progressive pneumonopathies; pathogenesis of nutritional myopathies; retinal developmental disorders; comparative ophthalmic pathology. *Mailing Add:* Dept Path Colo State Univ Ft Collins CO 80523

YOUNG, SUE ELLEN, b Port Arthur, Tex, Nov 28, 39; wid. OPHTHALMOLOGY. *Educ:* Univ Tex, BA, 61; Univ Tex, Galveston, MD, 69. *Prof Exp:* From intern to resident ophthal, Univ Tex, Houston, 69-73; fel neuro ophthal, Johns Hopkins Hosp, 73-74; chief, Ophthal Serv, 75-81, ASST PROF OPHTHAL, UNIV TEX CANCER SYST, 74- *Concurrent Pos:* Fel, Columbia Presby Hosp, 74; asst prof, Univ Tex Med Br, Houston, 74-81; mem Practicing Ophthalmologist Adv Comt, Am Acad Ophthal; pvt pract, 81- *Honors & Awards:* Honor Award, Am Acad Ophthal, 90. *Mem:* Am Acad Ophthal; Am Asn Ophthal; AMA. *Res:* Breast carcinoma metastatic to the choroid its value as a prognosic indicator and its management. *Mailing Add:* 900 Live Oak Ridge Austin TX 78746

YOUNG, SYDNEY SZE YIH, b Fukien, China, Nov 8, 24; m 54; c 1. QUANTITATIVE GENETICS, POPULATION GENETICS. *Educ:* Nantung Univ, BAgrSc, 47; Sydney Tech Col, FSTC, 51; Univ NSW, MSc, 56, DSc(genetics), 66; Univ Sydney, PhD(genetics), 59. *Prof Exp:* From res scientist to prin res scientist, Div Animal Genetics, Commonwealth Sci & Indust Res Orgn, 59-67; PROF GENETICS, OHIO STATE UNIV, 67- *Concurrent Pos:* Commonwealth Sci & Indust Res Orgn overseas fel, 63-64. *Mem:* Biomet Soc; Genetics Soc Am; AAAS. *Res:* Theoretical and experimental quantitative and population genetics; animal breeding; biostatistics. *Mailing Add:* Dept Genetics/391 B&Z Ohio State Univ Columbus OH 43210

YOUNG, THOMAS EDWARD, b Chaves Co, NMex, June 15, 28; m 60; c 2. NUCLEAR PHYSICS. *Educ:* Rice Univ, PhD(physics), 58. *Prof Exp:* Asst prof physics, Pac Col, Calif, 58-60 & Trinity Univ, Tex, 60-61; physicist, Aerojet Nuclear Co, 61-81; PHYSICIST, EG&G, IDAHO, 81- *Mem:* Am Asn Physics Teachers. *Res:* Neutron reactions and low energy charged particle reactions; radiation shielding. *Mailing Add:* 1184 Atlantic Idaho Falls ID 83404

YOUNG, THOMAS EDWIN, b Manheim, Pa, Sept 7, 24; m 45; c 1. ORGANIC CHEMISTRY. *Educ:* Lehigh Univ, BS, 49, MS, 50; Univ Ill, PhD(chem), 52. *Prof Exp:* Asst chem, Lehigh Univ, 49-50; res chemist, E I du Pont de Nemours & Co, 52-55; asst prof chem, Antioch Col, 55-58; from asst prof to assoc prof, 58-66, prof, 66-89, EMER PROF CHEM, LEHIGH UNIV, 89- *Mem:* Am Chem Soc. *Res:* Heterocyclic chemistry; indoles; structure and reactivity of heteroaromatic compounds; organosulfur chemistry. *Mailing Add:* 1952 Pinehurst Rd Bethlehem PA 18018

YOUNG, TZAY Y, b Shanghai, China, Jan 11, 33; m 65; c 2. COMPUTER VISION, IMAGE PROCESSING. *Educ:* Nat Taiwan Univ, BS, 55; Univ Vermont, MS, 59; Johns Hopkins Univ, DrEng, 62. *Prof Exp:* Res assoc, Carlyle Barton Lab, Johns Hopkins Univ, 62-63; mem tech staff, Bell Tel Lab, 63-64; asst prof elec eng, Carnegie-Mellon Univ, 64-68, assoc prof, 68-74; PROF ELEC & COMPUT ENG, UNIV MIAMI, 74-, CHMN DEPT, 88- *Concurrent Pos:* Assoc ed, Inst Elec & Electronics Engrs Trans Comput, 74-76; Adv Bd, Inst Elec & Electronics Engrs Trans, Pattern Anal & Mach Intel, 84- *Mem:* Fel Inst Elec & Electronics Engrs. *Res:* Computer vision and image processing; signal and information theory; pattern recognition; computer processing of biological data. *Mailing Add:* Dept Elec & Comput Eng Univ Miami Coral Gables FL 33124

YOUNG, VERNON ROBERT, b Rhyl, Wales, Nov 15, 37; m 66; c 5. NUTRITION, BIOCHEMISTRY. *Educ:* Univ Reading, BSc, 59; Cambridge Univ, dipl agr, 60; Univ Calif, Davis, PhD(nutrit), 65. *Prof Exp:* Lectr nutrit biochem, 65-66, asst prof physiol chem, 66-72, assoc prof nutrit biochem, 72-76, PROF NUTRIT BIOCHEM, MASS INST TECHNOL, 76-; DIR RES, SHRINERS BURN INST, BOSTON, 88- *Concurrent Pos:* Prog mgt human nutrit, Competitive Res Grants Prog, USDA, 80-81; assoc prog dir, Mass Inst Technol Clin Res Ctr, Cambridge, Mass, 85-87; biochemist, Dept Surg, Mass Gen Hosp & Harvard Med Sch, Boston, 87-; sr vis scientist, USDA Human Nutrit Ctr Aging, Tufts Univ, 88- *Honors & Awards:* Mead Johnson Award, Am Inst Nutrit, 73; Borden Award, Am Inst Nutrit, 82; McCollum Award, Am Soc Clin Nutrit, 87; Rank Prize Nutrit, 89. *Mem:* Nat Acad Sci; Am Inst Nutrit; Nutrit Soc; Am Soc Clin Nutrit; Geront Soc Am; Am Chem Soc. *Res:* Protein and clinical nutrition; muscle protein metabolism; nutrition and aging; study of trace minerals; nutrient bioavailability-use of stable isotopes. *Mailing Add:* Dept Nutrit Biochem Bldg E18 Rm 613 Mass Inst Technol 50 Ames St Cambridge MA 02139

YOUNG, VIOLA MAE (HORVATH), b Allegan, Mich, Oct 9, 15; c 2. MICROBIOLOGY. *Educ:* Mich State Col, BS, 36; Univ Ill, MS, 43; Loyola Univ, Ill, PhD, 53. *Prof Exp:* Technician, Ill Res & Educ Hosp, Chicago, 37-43; instr bact, Univ Chicago Med Sch, 43-45; bacteriologist & parasitologist, Mt Sinai Hosp & Res Found, Chicago, 45-47; res assoc, Sch Trop Med, PR, 47-48; parasitologist, Hektoen Inst, Cook County Hosp, Ill, 48-54; supvry bacteriologist, Dept Bact, Walter Reed Army Inst Res, 54-61; chief microbiol serv, Clin Path Dept, Clin Ctr, NIH, 61-68; head, Res Microbiol Sect, Baltimore Cancer Res Ctr, Clin Br, Nat Cancer Inst, 68-80; RETIRED. *Concurrent Pos:* Asst supv bacteriologist, State Hosp Serv, Ill, 43-44; dir lab, Presby Hosp, San Juan, PR, 47-48; lectr, Loyola Univ, Ill, 49-54; mem fac, Rackham Grad Sch, Univ Mich, 69-; mem, Am Bd Microbiol; consult microbiol & pub health, 81-; microbiologist, Off Drinking Water, Environ Protection Agency, 88- *Mem:* Hon mem Am Soc Microbiol; Am Acad Microbiol; AAAS. *Res:* All infectious agents causing diarrhea; ecology of intestinal tract; Pseudomonas aeruginosa; host-parasite relationships; normal antibodies and immunological response; infection prevention in cancer patients; interrelationships among microorganisms; opportunistic infections. *Mailing Add:* 5203 Bangor Dr Kensington MD 20895

YOUNG, WARREN MELVIN, b Massillon, Ohio, Dec 30, 37; m 60; c 2. ASTRONOMY. *Educ:* Case Inst Technol, BS, 60; Ohio State Univ, MS, 61, PhD(astron), 71. *Prof Exp:* Assoc prof astron & planetarium dir, 62-80, PROF & CHMN DEPT PHYSICS & ASTRON, YOUNGSTOWN STATE UNIV, 80- *Mem:* Am Astron Soc; Int Planetarium Soc. *Res:* Spectrum binary stars; photometry. *Mailing Add:* Dept Physics & Astron Youngstown State Univ Youngstown OH 44555

YOUNG, WILLIAM ALLEN, b St Marys, Ohio, May 21, 30. PETROLEUM GEOCHEMISTRY. *Educ:* Miami Univ, Ohio, BA, 52; Ohio State Univ, MSc, 54, PhD(chem), 57. *Prof Exp:* Res chemist, Carter Oil Co, 57-58, Jersey Prod Res Co, 58-64, Esso Prod Res Co, 64-66, sr res specialist, 66-72, res assoc, 72-75, sr res assoc, 75-78, res adv, 78-81, sr res adv, Exxon Prod Res Co, 81-91. *Mem:* AAAS; Am Asn Petrol Geol; Am Chem Soc. *Res:* Petroleum and organic geochemistry; clay-organic-water interactions; hydrogeology. *Mailing Add:* Exxon Prod Res Co PO Box 2189 Houston TX 77252-2189

YOUNG, WILLIAM ANTHONY, b Cleveland, Ohio, Feb 10, 23; m 54; c 2. PHYSICAL CHEMISTRY, ENGINEERING MANAGEMENT. *Educ:* Univ Wash, Seattle, BS, 49, MS, 53. *Prof Exp:* Res chemist, Am Marietta Co, 53-54; chemist, Thermodyn Sect, Nat Bur Standards, 54-55; test engr, Douglas Aircraft Co, 55-56; res engr, Atomics Int Div, NAm Rockwell Corp, 56-57, sr res chemist, 57-68, mem tech staff, 68-71; mgr anal develop & testing lab, Nuclear Energy Div, 72-74, mgr anal technol develop, 74-81, mgr lab automation, Wilmington Mfg Dept, Gen Elec Co, 81-84; LAB SUPVR, CHEMIST LAB, 84- *Mem:* Am Chem Soc; Am Phys Soc; Am Nuclear Soc; Sigma Xi; fel Am Inst Chem; Am Soc Qual Control; Soc Appl Spectros. *Res:* Metal hydrides; solid state chemistry; high temperature heterogeneous reactions; diffusion in solids; reaction kinetics; thermophysical properties; molecular structure; analytical chemistry; computer applications; environmental sciences; chemical safety; lab automation. *Mailing Add:* 6309 Pintail Ct Wilmington NC 28403

YOUNG, WILLIAM BEN, b Camden, Mich, Oct 29, 34; m 59; c 2. PHYSICAL METALLURGY. *Educ:* Cent Mich Univ, BS, 56; Mich State Univ, BS, 61. *Prof Exp:* Teacher math & sci, Port Austin High Sch, 56-59; staff metallurgist, Linde Div, Union Carbide Corp, 61-67; CHIEF ENGR, MAT SCI DEPT, PERFECT CIRCLE DIV, DANA CORP, 67- *Mem:* Fel Am Soc Metals; Am Ceramic Soc; Soc Automotive Engrs. *Res:* Metallurgical development of wear resistant coatings applied by metallizing processes; plasma, flame spray and detonation gun processes. *Mailing Add:* 599 Baker Rd Hagerstown IN 47346

YOUNG, WILLIAM DONALD, JR, b Glen Ridge, NJ, Nov 2, 38. BACTERIOLOGY, SYSTEMATICS. *Educ:* Fairleigh Dickinson Univ, BS, 60, MS, 74. *Prof Exp:* Res asst parasitol, NY Med Col, 61-62; sci asst clin chem, Walter Reed Army Inst Res, 62-64; scientist bact, Warner Lambert Res Inst, Morris Plains, 64-77; ARTIST-BLACKSMITH, CUSTOM STEEL FORGINGS, W D YOUNG, JR, 77- *Mem:* Am Soc Microbiol; Am Soc Metals; NY Acad Sci. *Res:* Development of rapid biochemical tests for use in diagnostic bacteriology; use of computer technology in the identification of bacterial cultures. *Mailing Add:* RD 1 Box 61 Franklin NY 13775

YOUNG, WILLIAM GLENN, JR, b Washington, DC, Feb 26, 25; m 52; c 4. THORACIC SURGERY. *Educ:* Duke Univ, MD, 48; Am Bd Surg, dipl, 58; Am Bd Thoracic Surg, dipl, 58. *Prof Exp:* Resident surg, Dake Hosp, 56-57; from asst prof to assoc prof, 57-63, PROF SURG, MED CTR, DUKE UNIV, 63- *Concurrent Pos:* Mem sr surg staff, Duke Hosp, 57-; attend surgeon, Vet Admin Hosp, Durham, NC, 57-; consult, Watts Hosp, Durham, 58- & Womack Army Hosp, Ft Bragg, 59- *Mem:* Soc Vascular Surg; Soc Univ Surg; AMA; Am Asn Thoracic Surg; fel Am Col Surg. *Res:* Cardiovascular surgery, application of moderate and profound hypothermia in cardiovascular surgery. *Mailing Add:* Dept Surg Duke Hosp Box 3617 Durham NC 27710

YOUNG, WILLIAM JOHNSON, II, b Lynn, Mass, Dec 10, 25; m 50; c 2. GENETICS, ANATOMY. *Educ:* Amherst Col, BA, 50, MA, 52; Johns Hopkins Univ, PhD(biol), 56. *Prof Exp:* Res assoc biol, Johns Hopkins Univ, 56-57, from asst prof to assoc prof anat, Sch Med, 57-66, assoc prof biophys, 66-68; PROF ANAT & CHMN DEPT, COL MED, UNIV VT, 68- *Mem:* Genetics Soc Am; Am Soc Human Genetics; Am Genetic Asn; Am Asn Anat; Sigma Xi. *Res:* Drosophila biochemical genetics; cytogenetics. *Mailing Add:* 60 Brewer Pkwy S Burlington VT 05401

YOUNG, WILLIAM PAUL, b Spokane, Wash, Oct 11, 13; m 42; c 3. SURGERY. *Educ:* Univ Wis, BS, 37, MS, 39, MD, 41. *Prof Exp:* Intern, Res Hosp, Kansas City, Mo, 41-42; res surg, Univ Wis Hosps, 46-49, instr surg, Univ, 50; chief surg serv, Southeast Fla State Tuberc Hosp, Lantana, 51; chief surg serv, Vet Admin Hosp, Madison, Wis, 52; from asst prof to assoc prof, 53-56, prof surg, Cardiovasc Surg Sect, Med Sch, Univ Wis-Madison, 56-88; RETIRED. *Concurrent Pos:* USPHS trainee, Univ Wis Hosps, 49-50; consult, Vet Admin. *Mem:* AMA; Am Heart Asn; Am Col Surg; Am Asn Thoracic Surg. *Res:* Cardiovascular surgery; pulmonary hypertension associated with congenital heart disease; homografts; myocardial revascularization and studies of anticoagulation techniques. *Mailing Add:* 600 Highland Ave Madison WI 53792

YOUNG, WILLIAM RAE, b Lawton, Mich, Oct 30, 15; m 37; c 3. ELECTRICAL ENGINEERING. *Educ:* Univ Mich, Ann Arbor, BSEE, 37. *Prof Exp:* Mem tech staff, Bell Tel Labs, 37-50, supvr classified mil systs, 50-52, supvr switching syst studies, 52-54, supvr data systs planning, 54-56, head dept, 56-61, head, Dept Switching Systs Studies, 61-70, head, Dept Mobile Systs Eng, 70-79; RETIRED. *Mem:* Fel Inst Elec & Electronics Engrs. *Res:* Mobile radio, propagation studies and system planning for vehicular and portable communications. *Mailing Add:* One Kingfisher Dr Middletown NJ 07748-2935

YOUNG, WILLIAM W, JR, GLYCOSPHINGOLIPID BIOLOGY, IMMUNOLOGY. *Educ:* Washington Univ, PhD(pharmacol), 75. *Prof Exp:* ASSOC PROF PATH, MED CTR, UNIV VA, 80- *Res:* Biochemistry. *Mailing Add:* Dept Path Univ Va Old Med Sch Bldg Charlottesville VA 22908

YOUNGBERG, CHESTER THEODORE, b Seattle, Wash, Mar 26, 17; m 41; c 4. FOREST SOILS. *Educ:* Wheaton Col, BS, 41; Univ Mich, MF, 47; Univ Wis, PhD(soils), 51. *Prof Exp:* Asst soils, Univ Wis, 47-51, forest soils specialist, Weyerhaeuser Timber Co, 51-52; assoc prof soils, Ore State Univ, 52-57; forestry specialist, Monsanto Chem Co, 57-58; prof forest soils, Ore State Univ, 58-82; CONSULT FOREST SOILS, 82- *Concurrent Pos:* Exchange prof, NC State Univ, 69-70. *Mem:* Soc Am Foresters; fel Am Soc Agron; Soil Sci Soc Am; Sigma Xi. *Res:* Soil-vegetation relationships; forest humus; symbiotic nitrogen fixation in non-leguminous plants; tree nutrition; slope-stability. *Mailing Add:* 1963 Manorview Lane Salem OR 97304

YOUNGBLOOD, BETTYE SUE, b Powhatan, Ala, Dec 6, 26. ORGANIC CHEMISTRY. *Educ:* Auburn Univ, BS, 46; Univ Ala, MS, 49, PhD(chem), 57. *Prof Exp:* High sch teacher, Ala, 46-50; instr chem, Univ Miss, 50-52; high sch teacher, Ala, 56-57; asst ed chem nomenclature, Chem Abstr Serv, 57-59, assoc ed, 59-62; asst prof, 62-65, PROF CHEM, JACKSONVILLE STATE UNIV, 65- *Mem:* Int Union Pure & Appl Chem; Am Chem Soc; fel Am Inst Chemists; Sigma Xi; Nat Sci Teacher's Asn. *Res:* Reaction mechanisms of aliphatic sulfonyl chlorides and derivatives; organic nomenclature of steroids and alkaloids. *Mailing Add:* 602 12th St Jacksonville AL 36265

YOUNGBLOOD, DAVE HARPER, b Waco, Tex, Oct 30, 39. PHYSICS. *Educ:* Baylor Univ, BS, 61; Rice Univ, MA, 63, PhD(physics), 65. *Prof Exp:* Fel, Argonne Nat Lab, 65-67; assoc prof, 67-77, PROF PHYSICS, CYCLOTRON INST, TEX A&M UNIV, 77-, DIR, CYCLOTRON INST, 78- *Mem:* Fel Am Phys Soc. *Res:* Nuclear spectroscopy and reaction theory; nuclear physics; grant resonances. *Mailing Add:* Cyclotron Inst Tex A&M Univ College Station TX 77843-3366

YOUNGBLOOD, WILLIAM ALFRED, electronics, communications engineering, for more information see previous edition

YOUNGDAHL, CARL KERR, b Chicago, Ill, Aug 14, 34; m 63; c 3. APPLIED MECHANICS, REACTOR ENGINEERING. *Educ:* Univ Chicago, AB, 53; Ill Inst Technol, BS, 56, MS, 57; Brown Univ, PhD(appl math), 60. *Prof Exp:* Mathematician, 60-74, SR MATHEMATICIAN, ARGONNE NAT LAB, 74- *Concurrent Pos:* Vis lectr, Ill Inst Technol, 63-64. *Mem:* Sigma Xi; Am Soc Mech Engrs; Am Acad Mech. *Res:* Pressure transients in reactor piping systems; dynamic plastic deformation of reactor components; thermoelasticity; fusion reactor structural analysis; approximation methods in dynamic plasticity. *Mailing Add:* Reactor Eng Div Argonne Nat Lab Argonne IL 60439

YOUNGDAHL, PAUL F, b Brockway, Pa, Oct 8, 21; m 43; c 3. MECHANICAL ENGINEERING. *Educ:* Univ Mich, BSE, 42, MSE, 49, PhD(mech eng), 61. *Prof Exp:* Asst sr engr, E I du Pont de Nemours & Co, 42-43 & 46-48; teaching fel & instr, Univ Mich, 48-53; dir res, Mech Handling Systs, Mich, 53-62; from assoc prof to prof mech eng, Univ Mich, Ann Arbor, 62-74; CONSULT MECH ENGR, 74- *Concurrent Pos:* Staff consult, Mech Handling Systs, 62-; design consult, Liquid Drive Corp, 62-; patent & prod liability law suit consult & expert witness; sr lectr mech eng, San Jose State Univ, 77- *Mem:* AAAS; Am Soc Mech Engrs; Am Soc Eng Educ; Nat Soc Prof Engrs. *Res:* System design criteria, specifically directed at automation of mass manufacturing chemical process equipment. *Mailing Add:* 501 Forest Ave Penthouse Four Palo Alto CA 94301

YOUNGDALE, GILBERT ARTHUR, b Detroit, Mich, Jan 15, 29; m 56; c 5. ORGANIC CHEMISTRY. *Educ:* Univ Detroit, BS, 54, MS, 56; Wayne State Univ, PhD(org chem), 59. *Prof Exp:* RES ASSOC MED CHEM, UPJOHN CO, 59- *Mem:* Am Hypoglycemic Agents; Am Chem Soc. *Mailing Add:* 1702 Greenbriar Dr Kalamazoo MI 49008-3514

YOUNGER, DANIEL H, b Flushing, NY, Sept 30, 36; m 65; c 3. MATHEMATICS. *Educ:* Columbia Univ, AB, 57, BS, 58, MS, 59, PhD(elec eng), 63. *Prof Exp:* Sloan fel, Princeton Univ, 63-64; res engr, Res & Develop Ctr, Gen Elec Co, NY, 64-67; assoc prof, 68-75, chmn, dept combinatorics 7 optimization, 78-79, PROF MATH, UNIV WATERLOO, 75- *Concurrent Pos:* Managing ed, J Combinatorial Theory, 68-75; vis scholar, Mass Inst Technol, 81-82; invited prof, Univ Paris-S Orsay, France, 82; dir, Guelph-Waterloo Ctr Grad Work Chem, 82-85. *Honors & Awards:* Rutherford Mem Medal, Royal Soc Can, 84. *Mem:* Sigma Xi; Am Math Soc; Can Math Soc; Am Chem Soc. *Res:* Graph theory, especially minimax theory of directed graphs; algorithms. *Mailing Add:* 130 Dunbar Rd Waterloo ON N2L 2E9 Can

YOUNGER, MARY SUE, b Roanoke, Va, Aug 28, 44. APPLIED STATISTICS, STATISTICAL PROCESS CONTROL. *Educ:* Hollins Col, BA, 66; Va Polytech Inst & State Univ, MS, 69, PhD(statist), 72. *Prof Exp:* Asst prof statist, Va Polytech Inst & State Univ, 72-74; asst prof, 74-76, ASSOC PROF STATIST, UNIV TENN, 76- *Concurrent Pos:* Consult, Elec Consumers Coun, 80-81, St Mary's Hosp, Knoxville, 81-82, Saturn Corp, 90-91; assoc dean, Grad Bus Studies, Univ Tenn, Knoxville, 85-86. *Mem:* Am Statist Asn; Decision Sci Inst. *Res:* Applied statistics; applications of statistical process control; author of books on linear regression and publications in medicine, sociology, psychology. *Mailing Add:* Dept Statist Univ Tenn Knoxville TN 37996-0532

YOUNGGREN, NEWELL A, b River Falls, Wis, Mar 15, 15; m 41; c 2. BIOLOGY. *Educ:* River Falls State Col, BE, 37; Univ Wis, MPh, 40; Univ Colo, PhD, 56. *Prof Exp:* Asst prof biol, Northland Col, 46-48; asst prof, Bradley Univ, 48-55; asst prof, Univ Colo, 55-60, chmn dept, 58-60; head dept biol sci, 68-74, PROF BIOL SCI, UNIV ARIZ, 61- *Concurrent Pos:* Inst dir, NSF, 58-60. *Mem:* Fel AAAS; Am Inst Biol Sci; Nat Asn Biol Teachers. *Res:* Cellular biology; biology education; slime mold physiology. *Mailing Add:* Dept Ecol & Evol Biol Sci Univ Ariz Tucson AZ 85721

YOUNGKEN, HEBER WILKINSON, JR, b Philadelphia, Pa, Aug 13, 13; m 42; c 2. PHARMACOLOGY. *Educ:* Bucknell Univ, AB, 35; Mass Col Pharm, BS, 38; Univ Minn, MS, 40, PhD, 42. *Prof Exp:* Asst biol & pharmacog, Mass Col Pharm, 35-39; asst pharmacog, Col Pharm, Univ Minn, 39-42; from instr to prof, Univ Wash, 42-57; prof pharmacog & dean, Col Pharm, Univ RI, 57-81, provost, Health Sci Affairs, 69-81; RETIRED. *Concurrent Pos:* Chmn, Plant Sci Seminar, 51; Nat Adv Coun Ed Health prof, 64-68; pres, Univ RI Found, 84-87. *Honors & Awards:* E L Newcomb Res Award, 53; Am Soc Pharmacog Award, 70; Egyptian Pharm Award, 77. *Mem:* Fel AAAS; Am Pharmaceut Asn; Soc Exp Biol Med; NY Acad Sci; Sigma Xi; Am Soc Hosp Pharm; Am Soc Pharmacog (pres, 70). *Res:* Plant chemistry and pharmacology of plant constituents; biosynthesis of drug plant glycosides and alkaloids; drugs from the sea. *Mailing Add:* 188 Oakwoods Dr Peace Dale RI 02883

YOUNGLAI, EDWARD VICTOR, b Trinidad, WI, July 15, 40; Can citizen; m 70; c 2. BIOCHEMISTRY, REPRODUCTIVE PHYSIOLOGY. *Educ:* McGill Univ, BSc, 64, PhD(biochem), 67. *Prof Exp:* Res asst biochem, McGill Univ, 64-67; from asst prof to assoc prof, 70-82, PROF OBSTET & GYNEC, MCMASTER UNIV, 82- *Concurrent Pos:* Fel exp med, McGill Univ, 67-68; Med Res Coun Can fel vet clin studies, Cambridge Univ, 68-70; Med Res Coun Can scholar, 70-75 & res grants, McMaster Univ, 70- *Mem:* Endocrine Soc; Brit Soc Study Fertil; Brit Soc Endocrinol; Soc Study Reproduction; Can Biochem Soc; Can Soc Obstet & Gynec. *Res:* Control of gonadal function; gonadal steroid biosynthesis and metabolism; secretion of hormones; puberty. *Mailing Add:* Dept Obstet & Gynec McMaster Univ Health Sci Ctr Hamilton ON L8N 3Z5 Can

YOUNGLOVE, JAMES NEWTON, b Coleman, Tex, Dec 16, 27; m 49; c 3. MATHEMATICS. *Educ:* Univ Tex, BA, 51, PhD(math), 58. *Prof Exp:* Instr math, Univ Tex, 52-58; asst prof, Univ Mo, 58-65; assoc prof, 65-71, PROF MATH, UNIV HOUSTON, 71- *Mem:* Am Math Soc; Math Asn Am. *Res:* Point set topology. *Mailing Add:* Dept Math Univ Houston Univ Pk 4800 Calhoun Rd Houston TX 77204

YOUNGMAN, ARTHUR L, b Chicago, Ill, Oct 24, 37; m 63; c 2. BOTANY. *Educ:* Univ Mont, BA, 59; Western Reserve Univ, MS, 61; Univ Tex, PhD(bot), 65. *Prof Exp:* ASST PROF BIOL, WICHITA STATE UNIV, 65- *Mem:* Ecol Soc Am. *Res:* Physiological ecology of vascular plants; environmental impact of industrial activity on terrestrial ecosystems. *Mailing Add:* Dept Biol Sci Wichita State Univ Wichita KS 67208-1595

YOUNGMAN, EDWARD AUGUST, b Fresno, Calif, May 17, 25; m 47; c 3. ORGANIC CHEMISTRY, POLYMER CHEMISTRY. *Educ:* Univ Wash, BS, 48, PhD(chem), 52. *Prof Exp:* Chemist, Shell Develop Co, Calif, 52-58, res supvr, 58-66, head, Plastics & Resins Res Dept, 66-67, dir, Plastics Technol Ctr, Shell Chem Co, NJ, 67-69, dir phys sci, Biol Sci Res Ctr, 69-80; RETIRED. *Mem:* Am Chem Soc; Soc Chem Indust; Sigma Xi. *Res:* Physical science of biologically active molecules, especially their synthesis and application. *Mailing Add:* 300 Durham Lane Modesto CA 95350

YOUNGMAN, VERN E, b Valley, Nebr, Sept 11, 28; m 54. AGRONOMY. *Educ:* Univ Nebr, BS, 55, MS, 57; Washington State Univ, PhD(agron), 62. *Prof Exp:* Instr agron, Univ Nebr, 56-58; from instr to asst prof, Washington State Univ, 58-67; ASSOC PROF AGRON, COLO STATE UNIV, 67- *Mem:* Am Soc Agron; Soc Econ Bot; Asn Off Seed Analysts; Nat Asn Cols & Teachers of Agr. *Res:* Physiology, ecology and management of crop plants. *Mailing Add:* Dept Agron Colo State Univ Ft Collins CO 80523

YOUNGNER, JULIUS STUART, b New York, NY, Oct 24, 20; m 43, 64; c 2. MEDICAL MICROBIOLOGY. *Educ:* NY Univ, AB, 39; Univ Mich, MS, 41, ScD(bact), 44; Am Bd Med Microbiol, dipl. *Prof Exp:* From asst to instr bact, Univ Mich, 41-44; asst path, Manhattan Proj, Univ Rochester, 45-46; instr, Univ Mich, 46-47; sr asst scientist, Nat Cancer Inst, 47-49; asst res prof virol & bact, Sch Med, Univ Pittsburgh, 49-56, from assoc prof to prof microbiol, 56-89, chmn dept, 66-85, chmn, Dept Microbiol, Biochem & Molecular Biol, 85-89, DISTINGUISHED SERV PROF MICROBIOL, UNIV PITTSBURGH, 90- *Concurrent Pos:* Vis prof, Nat Univ Athens, 63; mem virol & rickettsial study sect, NIH, 66-70; mem comn influenza, Armed Forces Epidemiol Bd, 70-73; mem bd sci counr, Nat Inst Allergy & Infectious Dis, 70-74; nat lectr, Found Microbiol, 72-73; mem study group, Immunol & Infectious Dis, Health Res & Serv Found, 79, Chmn, 78-79; mem clin A fel study sect, NIH, 79-80; mem, microbiol & virol study group, Am Cancer Soc, 81-85, chmn, 84-85; James W McLaughlin vis prof, Med Br, Univ Tex, Galveston, 84. *Honors & Awards:* Lippard Mem Lectr, Col Physicians & Surgeons, Columbia Univ, 80. *Mem:* AAAS; Am Acad Microbiol; Brit Soc Gen Microbiol; Am Soc Microbiol; Infectious Dis Soc Am; Am Soc Virol (pres, 86-); Int Soc Interferon Res; Am Asn Immunol. *Res:* Replication and properties of animal viruses; cellular and host resistance to virus infection; persistent viral infections. *Mailing Add:* Dept Molecular Genetics & Biochem Univ Pittsburgh Sch Med Pittsburgh PA 15261

YOUNGNER, PHILIP GENEVUS, b Nelson, Minn, July 13, 20; m 47; c 4. PHYSICS. *Educ:* St Cloud State Col, BS, 44; Univ Wis, MS, 47, PhD(physics), 58. *Prof Exp:* Pub sch teacher, Minn, 39-41; radium technician, Wis Gen Hosp, 46-47; instr physics, Exten, Univ Wis, 47-49; from asst prof to emer prof physics, St Cloud State Univ, 49-88, chmn dept, 60-80; RETIRED. *Concurrent Pos:* Pres, Ecostill Corp, 80-88. *Mem:* Am Phys Soc; Am Asn Physics Teachers. *Res:* Molecular spectra; solar energy; salt water conversion; atmospheric electricity. *Mailing Add:* 919 SE 18th St St Cloud MN 56304

YOUNGQUIST, R(UDOLPH) WILLIAM, b Minneapolis, Minn, Aug 10, 35; m 59; c 3. FOOD BIOCHEMISTRY. *Educ:* Univ Minn, BChem, 57; Iowa State Univ, MS, 60, PhD(biochem), 62. *Prof Exp:* RES BIOCHEMIST, PROCTER & GAMBLE CO, 62- *Mem:* Am Chem Soc; Am Asn Cereal Chemists. *Res:* Starch and carbohydrate food biochemistry. *Mailing Add:* Procter & Gamble Co Miami Valley Lab Box 398707 Cincinnati OH 45239-8707

YOUNGQUIST, WALTER, b Minneapolis, Minn, May 5, 21; m 43; c 4. GEOLOGY. *Educ:* Gustavus Adolphus Col, BA, 42; Univ Iowa, MS, 43, PhD(geol), 48. *Prof Exp:* Asst geol, Univ Iowa, 42-43; jr geologist, Groundwater Div, US Geol Surv, Iowa, Va & La, 43-44; asst paleont, Univ Iowa, 45-47; asst prof geol, Univ Idaho, 48-51; geologist, Int Petrol Co, Peru, 51-52, sr geologist, 52-53, chief spec studies sect, 53-54; prof geol, Univ Kans, 54-57; from assoc prof to prof, Univ Ore, 57-66; consult, Minerals Dept, Humble Oil & Refining Co, 66-73; GEOTHERMAL RESOURCES CONSULT, EUGENE WATER & ELEC BD, 73- *Mem:* Fel AAAS; Geol Soc Am; Am Asn Petrol Geologists; Geothermal Resources Coun. *Res:* Geology and economics of mineral resources; petroleum geology; geothermal resources. *Mailing Add:* PO Box 5501 Eugene OR 97405

YOUNGS, CLARENCE GEORGE, b Didsbury, Alta, Oct 23, 26; m 51; c 3. FOOD PROCESSING. *Educ:* Univ Alta, BSc, 48; Univ Sask, MSc, 53, PhD(chem), 57. *Prof Exp:* Res officer crop utilization, Prairie Regional Lab, Nat Res Coun, Can, 48-81; RETIRED. *Concurrent Pos:* Consult, Studies Int Develop Res Centre, Food & Agr Orgn, Junta Del Acuerdo de Cartagena, Canola Coun Can, Govt Saskatchewan & Univ Saskatchewan; Can Comt Fats & Oils, Expert Comt Grain Qual; Selection Panel Strategic grants, Food & Agr. *Honors & Awards:* Res Award, Glycerine Producers Asn, 57 & 62. *Mem:* Hon mem Agr Inst Can; emer mem Can Inst Food Sci & Technol. *Res:* Utilization of prairie crops for food and feed, particularly oilseeds and legumes. *Mailing Add:* 821 Wilson Crescent Saskatoon SK S7J 2M3 Can

YOUNGS, ROBERT LELAND, b Pittsfield, Mass, Feb 10, 24; m 49; c 5. FORESTRY, INTERNATIONAL DEVELOPMENT. *Educ:* State Univ NY, BS, 48; Univ Mich, MWT, 50; Yale Univ, PhD(forestry), 57. *Prof Exp:* Forest prod technologist, Forest Prod Lab, US Forest Serv, 51-66, proj leader, fundamental properties, 58-64, chief, div solid wood prod res, 64-66, dir, div forest prod & eng res, 67-70, dir, Southern Forest Exp Sta, 70-72, assoc dep chief res, 72-75, dir, Forest Prod Lab, 75-85; PROF, DEPT FOREST PROD, VA POLYTECH INST & STATE UNIV, 85- *Concurrent Pos:* Foresty consult, 85-; coordr forest prods div, Int Union Forestry Res Orgns, 83-90. *Honors & Awards:* Wood Award, Forest Prod Res Soc, 57; Presidential Rank Award, Sr Exec Serv, 81, 84. *Mem:* Soc Am Foresters; Soc Wood Sci & Technol (secy-treas, 58-59, vpres, 60-61, pres, 62-63); Forest Prod Res Soc; Int Soc Trop Foresters; Sigma Xi; Int Union Forestry Res Orgns. *Res:* Basic physical and mechanical properties of wood and related factors. *Mailing Add:* Dept Forest Prod 210 Cheatham Hall Va Polytech Inst & State Univ Blacksburg VA 24061-0323

YOUNGS, WILEY JAY, b Gouverneur, NY, July 5, 49; m 78; c 2. METALLOCYCLYNES, CONDUCTING POLYMERS. *Educ:* State Univ NY, Albany, BA, 72; State Univ NY, Buffalo, PhD(chem), 80. *Prof Exp:* Postdoctoral fel, Northwestern Univ, Evanston, Ill, 80-83; from asst prof to assoc prof inorg chem, Case Western Res Univ, Cleveland, Ohio, 83-90; ASSOC PROF INORG CHEM, UNIV AKRON, OHIO, 90- *Concurrent Pos:* Collateral fac mem, Ohio Aerospace Inst, Cleveland, Ohio, 90- *Mem:* Am Chem Soc; Am Crystallog Asn; AAAS; Am Asn Univ Professors. *Res:* Synthesis of new Ziegler-Natta polymerization catalysts; synthesis of organometallic molecular conductors and conducting polymers; synthesis of high temperature polymers; synthesis of organometallic compounds with antitumor activity; mechanistic investigation of lithium induced alkyne cyclization reactions; x-ray structural characterization of inorganic, organic and organometallic compounds. *Mailing Add:* Dept Chem Univ Akron Akron OH 44325-3601

YOUNGSTROM, RICHARD EARL, b Durham, NC, Sept 11, 43; m 69, 82; c 2. CHROMATOGRAPHY, RADIOCHEMISTRY. *Educ:* Duke Univ, BS, 65; Wash Univ, MA, 69. *Prof Exp:* Asst scientist steroid chem, Schering Corp, 69-70, assoc scientist, 70-71, scientist radiochem, 71-73, sr scientist, 73-77, res sect leader, 77-85; PRES, CHEM SYSTS, 85- *Mem:* Am Soc Testing & Mat; Am Chem Soc. *Res:* Chromatography solvent theory; optimization methods; approaches to standardized performance testing of sorbents, radiochromatography, liquid scintillation and flow systems; writing training programs in chromatography & radiochemistry. *Mailing Add:* Chem Systs Resolution Inc 11420 Riley St Overland Park KS 66210-2656

YOUNKIN, LARRY MYRLE, b Markleton, Pa, Oct 16, 36; m 69; c 1. CIVIL ENGINEERING. *Educ:* Geneva Col, BS, 56; Univ Pittsburgh, BSCE, 59, MSCE, 62; Va Polytech Inst & State Univ, PhD(civil eng), 71. *Prof Exp:* Instr civil eng, Univ Pittsburgh, 56-59; asst prof, Geneva Col, 59-62; instr, Univ NMex, 62-64; from asst prof to assoc prof, 66-81, PROF CIVIL ENG, BUCKNELL UNIV, 81- *Concurrent Pos:* Consult engr, Pa Dept Transp, 71-77 & USDA Soil Conserv Serv, 82- *Mem:* Am Soc Civil Engrs; Am Geophys Union; Am Water Resources Asn. *Res:* Open channel flow; hydraulic structures; sediment transport and yield from construction; surface water hydrology. *Mailing Add:* Four Hilltop Rd Lewisburg PA 17837

YOUNKIN, STUART G, b US, Jan 16, 12; m 43; c 5. PLANT PATHOLOGY. *Educ:* Iowa State Univ, BS, 36, MS, 39; Cornell Univ, PhD, 43. *Prof Exp:* From asst plant pathologist & geneticist to plant pathologist & geneticist, Campbell Soup Co, 43-52, asst to dir res, 52-53, dir agr res, 53-62, vpres agr res, 62-77, pres, Campbell Inst Agr Res, 66-77; RETIRED. *Concurrent Pos:* Mem agr bd, Nat Acad Sci-Nat Res Coun, 62-68, pres, Agr Res Inst, 64-65; mem panel world food supply, President's Sci Adv Comt, 66-67. *Mem:* Soc Econ Botanists; Am Soc Hort Sci; Am Phytopath Soc. *Res:* Virus diseases of potatoes; vegetable disease control; breeding of tomatoes and peppers. *Mailing Add:* 614 Elm Terr Riverton NJ 08077

YOUNOSZAI, RAFI, b Kabul, Afghanistan, June 13, 30; m 64; c 2. ANATOMY. *Educ:* Univ Calif, Berkeley, BSc, 57; Univ Minn, PhD(anat), 71. *Prof Exp:* From inst to asst prof anat, Dept Anat, Univ Minn, 71-78; assoc prof, 79-85, PROF ANAT, COL OSTEOPATH MED PAC, 86- *Concurrent Pos:* Res coordr, Res Orgn Comt, Am Osteop Asn, Nat Osteop Found, 80-82, dir Off Int Progs. *Mem:* Am Asn Clin Anatomists; Am Diabetes Asn; Nat Coun Int Health. *Res:* Interacellular transport and release of insulin; effect of a low protein and high carbohydrate diet on insulin release; insulin release patterns of induced islet adenomas; electronyographic studies on patients with somatic dysfunctions before and subsequent to manipulative therapy; ossification of cranial sutures. *Mailing Add:* Dept Anat Col Osteopathic Med Pac College Plaza Pomona CA 91766

YOUNSZAI, M KABIR, b Kabul, Aghanistan, Oct 2, 32. GASTROENTEROLOGY. *Educ:* Am Univ Beirut, MD, 62. *Prof Exp:* PROF PEDIAT, UNIV IOWA HOSP, 79- *Mem:* Soc Pediat Res; Am Soc Clin Nutrit; Am Gastroenterol Asn. *Mailing Add:* Dept Pediat JCP Univ Iowa Hosp Iowa City IA 52242

YOUNT, DAVID EUGENE, b Prescott, Ariz, June 5, 35; m 62, 75; c 4. ELEMENTARY PARTICLE PHYSICS, DIVING PHYSIOLOGY. *Educ:* Calif Inst Technol, BS, 57; Stanford Univ, MS, 59, PhD(physics), 63. *Prof Exp:* From instr to asst prof physics, Princeton Univ, 62-64; NSF fel, Linear Accelerator Lab, Orsay, France, 64-65; res assoc, Stanford Linear Accelerator Ctr, 65-69; assoc prof physics, 69-72, chmn dept physics & astron, 79-85, actg asst vpres acad affairs, 85, PROF PHYSICS, UNIV HAWAII, HONOLULU, 72-, VPRES RES & GRAD EDUC, 86- *Concurrent Pos:* 3M Co fel, Princeton Univ, 63; dir, Hawaii Topical Conf Particle Physics, 71. *Honors & Awards:* Stover-Link Award, 87. *Mem:* Am Chem Soc; Am Phys Soc; Undersea & Hyperbaric Med Soc; Sigma Xi. *Res:* Positron scattering, leptonic K-meson decay, hadronic photon absorption, photoproduction of mesons, positron-electron colliding beams, bubble nucleatron, and decompression sickness; instrumentation for particle beams and beam monitors, spark chambers, streamer chambers and multiwire proportional chambers. *Mailing Add:* Dept Physics & Astron Univ Hawaii 2505 Correa Rd Honolulu HI 96822

YOUNT, ERNEST H, b Lincolnton, NC, Feb 23, 19; m 42; c 3. MEDICINE. *Educ:* Univ NC, BA, 40; Vanderbilt Univ, MD, 43. *Prof Exp:* Asst med, Univ Chicago, 45-48; from instr to assoc prof, 48-54, chmn dept, 52-72, PROF MED, BOWMAN GRAY SCH MED, 54- *Concurrent Pos:* Consult, Oak Ridge Inst Nuclear Studies, 50-58; mem dean's comt, Vet Admin Hosp, Salisbury, 54-63; mem, Nat Bd Med Exam, 58-61, chmn, 61. *Mem:* Am Fedn Clin Res; Am Soc Internal Med; Am Col Physicians; Am Diabetes Asn; Asn Prof Med. *Res:* Malaria; adrenal and thyroid function; diabetes. *Mailing Add:* 2800 Greenwich Rd Winston-Salem NC 27104

YOUNT, RALPH GRANVILLE, b Indianapolis, Ind, Mar 25, 32; m 57; c 3. BIOCHEMISTRY. *Educ:* Wabash Col, AB, 54; Iowa State Univ, PhD, 58. *Prof Exp:* Res assoc enzym, Brookhaven Nat Lab, 58-60; from asst prof chem & asst chemist to assoc prof & assoc chemist, 60-72, chmn biochem-biophys prog, 73-78, PROF BIOCHEM & CHEM, WASH STATE UNIV, 72- *Concurrent Pos:* NIH spec fel, Sch Med, Univ Pa & vis prof, Johnson Found, 69-70; mem Comt A & NHLBI, 86-, chmn fel comt, MDA, 86-; mem Comt A & NHLBI, 86-90, chmn fel comt, 86-, vpres, MDA, 87- *Honors & Awards:* MERIT Award, NIH, 86. *Mem:* Fel AAAS; Am Soc Biochem & Molecular Biol; Am Chem Soc; Protein Soc; Biophys Soc. *Res:* Mechanism of enzyme action as it applies to contractile proteins; synthesis of small molecules of biological interest. *Mailing Add:* Dept Biochem & Biophys Wash State Univ Pullman WA 99164-4660

YOUNT, WILLIAM J, b Menominee, Mich, Aug 24, 36; m 63; c 3. ALLERGY, IMMUNOLOGY. *Educ:* Univ Wis-Madison, MD, 60. *Prof Exp:* PROF MED, MICROBIOL & IMMUNOL, SCH MED, UNIV NC, 70- *Concurrent Pos:* Vis prof, Clin Res Centre, London, 78-79; vis Kenan prof, Cambridge Univ, 87-88. *Mem:* Am Soc Clin Invest; Fedn Am Soc Exp Biol. *Res:* Humoral immunodeficiency; IgG subclass immunobiology; systemic lupus erythematosus; rheumatic diseases. *Mailing Add:* Dept Med Div Rheum CB 7280 Univ NC Chapel Hill NC 27514

YOUNTS, SANFORD EUGENE, b Lexington, NC, Aug 29, 30; m 54; c 1. SOIL SCIENCE, AGRONOMY. *Educ:* NC State Univ, BS, 52, MS, 55; Cornell Univ, PhD(agron), 57. *Prof Exp:* Asst prof soils, Univ Md, 57-58; agronomist, Am Potash Inst, 58-60; assoc prof soils, NC State Col, 60-64; regional dir, Am Potash Inst, 64-67, vpres, 67-69; assoc dean col agr & dir rural develop ctr, 69-72, VPRES SERV, UNIV GA, 72-, PROF AGRON, 69- *Mem:* AAAS; fel Am Soc Agron; Int Soc Soil Sci; fel Soil Sci Soc Am. *Res:* Soil fertility and crop physiology; root growth of field crops as influenced by fertilizer and lime placement; chloride nutrition of corn; potash requirements of forage crops; nitrogen sources for turf; copper nutrition of wheat, corn and soybeans. *Mailing Add:* Univ Ga 300 Old College Athens GA 30602

YOURTEE, JOHN BOTELER, b Fredericksburg, Va, Dec 2, 46; m 68; c 2. CHEMICAL ENGINEERING, POLYMER CHEMISTRY. *Educ:* Univ Del, BS, 68; Mass Inst Technol, SM, 69; Univ Wis, PhD(chem eng), 73. *Prof Exp:* Res engr, Union Carbide Corp, 73-75, sr res engr plastics, 75-77; sr group leader, Merck & Co, Pittsburgh, 77-81; staff engr, Exxon Chem Co, 81-85; SECT LEADER, HOECHST CELANESE CORP, NJ, 85- *Mem:* Sigma Xi; Soc Plastics Engrs; Am Ceramic Soc; Am Inst Chem Engrs. *Res:* Product and process development of thermoplastic polymers and water-soluble polymers with current emphasis on engineering plastics. *Mailing Add:* Three Jefferson Dr Flanders NJ 07836

YOURTEE, LAWRENCE KARN, b Brunswick, Md, Mar 6, 17; m 41; c 1. ORGANIC CHEMISTRY. *Educ:* Washington Col, Md, BS, 37; Ga Inst Technol, MS, 39; Univ Tex, PhD(chem), 48. *Prof Exp:* Instr chem, Ga Inst Technol, 40-42 & Univ Tex, 46-47; asst prof, Univ Tenn, 47-48; assoc prof, 48-57, chmn dept, 57-71, Childs prof, 58-80, McEwen prof, 80-82, EMER MCEWEN PROF CHEM, HAMILTON COL, 82- *Mem:* Am Chem Soc. *Res:* Synthesis and properties of heterocyclic nitrogen compounds; Pfitzinger reaction; use of ion-exchange resins in organic synthesis and separations. *Mailing Add:* Clinton House Apts Six 1/2 Kirkland Ave Clinton NY 13323

YOUSE, BEVAN K, b Markle, Ind, Apr 5, 27; m 58; c 1. MATHEMATICAL ANALYSIS. *Educ:* Auburn Univ, BS, 49; Univ Ga, MS, 52. *Prof Exp:* Instr math, Memphis State Univ, 52-53 & Univ Ga, 53-54; asst prof, 54-67, ASSOC PROF MATH, EMORY UNIV, 67- *Concurrent Pos:* NSF fac fel, 60-61. *Mem:* AAAS; Am Math Soc; Math Asn Am; Sigma Xi. *Res:* Mathematical analysis. *Mailing Add:* Dept Math Emory Univ Atlanta GA 30322

YOUSE, HOWARD RAY, b Bryant, Ind, May 22, 15; m 42. BOTANY. *Educ:* DePauw Univ, BA, 37; Ore State Col, MS, 42; Purdue Univ, PhD(bot), 51. *Prof Exp:* From instr to prof bot, DePauw Univ, 40-80, head dept bot & bact, 73-78, chmn dept, 78-80; RETIRED. *Mem:* AAAS; Bot Soc Am. *Res:* Pollen grains; seed germination. *Mailing Add:* PO Box 253 Greencastle IN 46135-0253

YOUSEF, IBRAHIM MOHMOUD, b Egypt, Nov 20, 40; Can citizen; c 1. BIOCHEMISTRY & CLINICAL BIOCHEMISTRY. *Educ:* Univ Ain Shams, Cairo, BSc, 61; Univ Col Wales, MSc, 65; Univ Edinburgh, PhD(biochem), 67. *Prof Exp:* Res assoc, Mich State Univ, 67-69; res assoc, Univ Toronto, 69-70, assoc pathol, 70-72, asst prof, 72-78; assoc prof, 78-83, prof pediat, 83-88, PROF PHARMACOL, UNIV MONTREAL, 88- *Concurrent Pos:* Fel med educ, Univ Toronto, 71; dir labs, Gallstone Study Can, 76-78; consult, Para-Med Educ Cols, Toronto, 74-78; dir, Liver Res Unit, Hosp She-Justine, Montreal, Can, 78-; chmn, dept biochem & clin chem, King Soud Univ Med Sch, Saudi Arabia, 84-85. *Mem:* Am Asn Study Liver Dis; Can Biochem Soc; Soc Exp Biol & Med; Int Asn Study Liver; Int Biochem Soc; Can Soc Clin Chem; Can Acad Clin Biochem; Can Asn Liver Dis; Can Asn Gastroenterol; Soc Toxicol Can. *Res:* Bile formation and secretion; bile acid metabolism; cholestasis in adults and children; experimental cholestasis; clinical chemistry; drug toxicity; experimental pathology. *Mailing Add:* Res Ctr St Justine Hosp 3175 Cote Ste Catherine Montreal PQ H3T 1C5 Can

YOUSEF, MOHAMED KHALIL, b Cairo, Egypt, Aug 19, 35; US citizen; m 63; c 2. ENVIRONMENTAL PHYSIOLOGY. *Educ:* Ain Shams Univ, Cairo, BSc, 59; Univ Mo-Columbia, MS, 63, PhD(environ physiol), 66. *Prof Exp:* Res assoc environ physiol, Univ Mo-Columbia, 66-67; vis asst prof, Inst Arctic Biol, Univ Alaska, 67-68; asst prof, Lab Environ Patho-physiol, Desert Res Inst, Univ Nev, 68-70; from asst prof to assoc prof, 70-74, PROF BIOL & PHYSIOL, UNIV NEV, LAS VEGAS, 74-, COORDR HEALTH PREPROF PROG, 73-, DIR, DESERT BIOL RES CTR, 80- *Honors & Awards:* W F Peterson Found Award, Animal Biometeorol, Neth; Japan Soc, Prom Sci Award. *Mem:* Int Soc Biometeorol; Am Physiol Soc; Soc Exp Biol & Med; Endocrine Soc. *Res:* Physiological adaptations to desert, mountain and arctic environments; role of the respiratory, cardiovascular and endocrine systems in adaptation; comparative thermoregulation during rest and exercise under different environments; comparative adaptations of organisms to various stressful environments; emphasis is on the role of cardiovascular, respiratory and endocrine systems. *Mailing Add:* Dept Biol Sci Univ Nev 4505 Maryland Pkwy Las Vegas NV 89154

YOUSIF, SALAH MOHAMMAD, b Burin, Palestine, Nov 15, 38; m 68; c 3. ELECTRICAL ENGINEERING. *Educ:* Univ Alexandria, BSEE, 62; Mid East Tech Univ, Ankara, MSEE, 64; Ore State Univ, PhD(elec eng), 69. *Prof Exp:* Elec engr, Jordan Broadcasting Serv, 62-63 & Kuwait Broadcasting Serv, 64-65; instr elec eng, Mich State Univ, 66-68 & Ore State Univ, 68-69; asst prof, 69-72, assoc prof, 72-80, PROF ELEC ENG, CALIF STATE UNIV, SACRAMENTO, 80- *Mem:* Inst Elec & Electronics Engrs; Soc Indust & Appl Math; Pattern Recognition Soc. *Res:* Control, information and power systems; pattern recognition. *Mailing Add:* Dept Elec Eng Calif State Univ 6000 J St Sacramento CA 95819

YOUSON, JOHN HAROLD, CELL ULTRASTRUCTURE, ENDOCRINOLOGY. *Educ:* Univ Western Ont, PhD(zool), 69. *Prof Exp:* PROF ZOOL, UNIV TORONTO, 69- *Mailing Add:* Dept Biol Sci Scarborough Col Univ Toronto 1265 Military Trail Scarborough ON M1C 1A4 Can

YOUSSEF, MARY NAGUIB, US citizen; c 1. COMPUTER SCIENCES. *Educ:* Univ Cairo, BS, 58; Columbia Univ, MA, 64; Stanford Univ, MS, 67; Ore State Univ, PhD(statist), 70. *Prof Exp:* Instr celestial mech, Cairo Univ, 58-61; Ford Found fel, 62-63 & 65-67; mem tech staff & researcher, Oper Res Ctr, Inst Nat Planning, Cairo, 63-65; mem tech staff syst anal, Bell Tel Labs, 70-81; assoc prof, City Univ New York, 81-83; adv engr, IBM, 83-85, SR FAC MEM, SRI-IBM, 85-; PROF COMPUT SCI, WESTERN CONN STATE UNIV, 90- *Concurrent Pos:* Assoc prof oper res & comput info systs, Baruch Col, City Univ New York, 81-82. *Honors & Awards:* Centennial Medal, Inst Elec & Electronics Engrs, 84. *Mem:* Sigma Xi; sr mem Inst Elec & Electronics Engrs; Asn Comput Mach. *Res:* Modeling and analyzing queuing systems; data communications and protocols; methods for projecting telecommunication traffic in a special environment; economic and time series forecasting. *Mailing Add:* Three Ivy Lane Monroe CT 06468

YOUSSEF, NABIL NAGUIB, b Cairo, Egypt, Oct 19, 37; US citizen; m 63; c 3. MORPHOLOGY, CELL BIOLOGY. *Educ:* Ain Shams Univ, Cairo, BSc, 58; Utah State Univ, MS, 64, PhD(zool), 66. *Prof Exp:* Asst instr entom & zool, Ain Shams Univ, Cairo, 58-60; from res asst to res assoc, 64-68, asst prof, 68-75, ASSOC PROF ZOOL, UTAH STATE UNIV, 75- *Concurrent Pos:* USDA grant, 66-68. *Mem:* Sigma Xi; AAAS; fel Royal Entom Soc London; Entom Soc Am; Soc Protozoologists. *Res:* Fine structure of Protozoa and Insecta with special emphasis on morphogeneses of normal and abnormal tissues induced by drugs or pathogens. *Mailing Add:* Dept Biol Utah State Univ Logan UT 84322-5305

YOUSTEN, ALLAN A, b Racine, Wis, Nov 9, 36; c 2. MICROBIAL PHYSIOLOGY, MICROBIAL INSECTICIDES. *Educ:* Univ Wis, BS, 58; Cornell Univ, MS, 60, PhD, 63. *Prof Exp:* Microbial biochemist, Int Minerals & Chem Corp, 65-69; NIH spec fel, Univ Wis, 69-71; asst prof, 71-77, assoc prof, 77-86, PROF MICROBIOL, VA POLYTECH INST & STATE UNIV, 86- *Concurrent Pos:* Fulbright-Hays fel, Pasteur Inst, 80; vis prof, Ariz State Univ, 87. *Mem:* AAAS; Am Soc Microbiol; Soc Indust Microbiol; Soc Invert Path. *Res:* Physiology and structure of microorganisms; bacterial spore formation and germination; bacterial insect pathogens; physiology, metabolism, structure, and taxonomy of spore-forming bacteria used as microbial insecticides; fate of these bacteria in the environment, their fermentative production and genetic manipulation. *Mailing Add:* Dept Biol Va Polytech Inst & State Univ Blacksburg VA 24061

YOUTCHEFF, JOHN SHELDON, b Newark, NJ, Apr 16, 25; m 50; c 5. ASTROPHYSICS. *Educ:* Columbia Univ, AB & BS, 50; Univ Calif, Los Angeles, PhD, 54. *Prof Exp:* Dir test staff, US Naval Air Missile Test Ctr, 50-53; opers analyst, Advan Electronics Ctr, Gen Elec Co, 53-56, functional engr, Missile & Space Div, 56-60, consult engr, 60-63, mgr advan reliability concepts oper, 63-72; mgr reliability & maintainability, Litton Industs, 72-73; PROG DIR, US POSTAL SERV HQ, WASHINGTON, DC, 73- *Mem:* Fel AAAS; sr mem Inst Elec & Electronics Engrs; assoc fel Am Inst Aeronaut & Astronaut; Am Soc Mech Engrs; sr mem Am Astron Soc. *Res:* Operations analysis; advanced systems planning; aerospace and environmental systems. *Mailing Add:* 543 Midland Ave Berwyn PA 19312

YOUTSEY, KARL JOHN, b Chicago, Ill, May 6, 39; m 69; c 4. PHYSICS. *Educ:* Loyola Univ, Chicago, BS, 61; Ill Inst Technol, MS, 65, PhD(physics), 68. *Prof Exp:* Physicist, Physics Dept, UOP Inc, 61-64, physicist, Mat Sci Lab, 68-73, dir mat sci, Corp Res, 73-75, dir prod & process develop, Wolverine Div, 75-79, vpres & gen mgr, Automotive Prod Div, 79-85, GEN MGR, ENVIRON CONTROL & MGT SYSTS, UOP INC, 85- *Mem:* Sigma Xi; Am Phys Soc; Am Soc Metals. *Res:* Electronic and physical properties of ceramics; fuel cell technology; solar thermal energy systems; thin film technology; laboratory and industrial automation systems and design; hazardous waste treatment. *Mailing Add:* 1647 Riverside Ct Glenview IL 60025

YOUTZ, BYRON LEROY, b Burbank, Calif, Nov 10, 25; m 51; c 3. NUCLEAR PHYSICS, ENERGY STUDIES. *Educ:* Calif Inst Technol, BS, 48; Univ Calif, Berkeley, PhD(physics), 53. *Prof Exp:* Res physicist, Lawrence Radiation Lab, Univ Calif, 50-53; asst prof physics, Am Univ Beirut, 53-56, actg chmn dept, 55-56; from asst prof to prof, Reed Col, 56-68, actg pres, 67-68; prof physics & acad vpres, State Univ NY Col Old Westbury, 68-70; interim acad dean, 73-74, vpres & provost, 78-83, PROF PHYSICS, DIV SCI, EVERGREEN STATE COL, 70- *Concurrent Pos:* Guest, Japanese Phys Educ Soc & Asia Found Physics Curricula, 61 & 66; mem steering comt, Phys Sci Study Comn, 61-; consult, Educ Serv Inc, Mass, 61-; prin lectr, Sem Sec Sch Physics Curricula, Salisbury, Fedn Rhodesia & Nyasaland, 63 & Sem Advan Topics in Phys Sci Study Comt Physics, Santiago, Chile, 64; mem adv comt, Boston Univ Math Proj, 73-78. *Mem:* AAAS; Am Phys Soc; Am Asn Physics Teachers; Int Solar Energy Soc; Sigma Xi. *Res:* Astrophysics; nuclear structures; energy sources. *Mailing Add:* 6113 Buckthorn Ct NW Olympia WA 98502

YOVANOVITCH, DRASKO D, b Belgrade, Yugoslavia, May 24, 30; US citizen; m 54; c 2. HIGH ENERGY PHYSICS. *Educ:* Belgrade Univ, BSc, 52; Univ Chicago, MSc, 56, PhD(physics), 59. *Prof Exp:* Res assoc physics, Enrico Fermi Inst, Univ Chicago, 59-60; asst physicist, Univ Calif, San Diego, 60-62; assoc physicist, High Energy Physics Div, Argonne Nat Lab, 62-72; PHYSICIST, NAT ACCELERATOR LAB, 72-, CHMN, PHYS DEPT, 79- *Mem:* Fel Am Phys Soc. *Mailing Add:* Nat Accelerator Lab PO Box 500 Batavia IL 60510

YOVICH, JOHN V, b Perth, Australia, Nov 16, 59. EQUINE SURGERY, NEUROPATHOLOGY. *Educ:* Murdoch Univ, Australia, BSc, 80, BVMS, 81; Univ Guelph, Can, dipl surg, 83; Colo State Univ, MS, 86, PhD(neuropath), 88; Am Col Vet Surg, dipl, 87. *Prof Exp:* SR LECTR SURG, MURDOCH UNIV, AUSTRALIA, 88- *Concurrent Pos:* Sr lectr equine surg, Sch Vet Studies, Murdoch Univ, 88- *Mem:* Am Vet Asn; Am Asn Equine Practitioners; Australian Vet Asn; Am Col Vet Surgeons. *Res:* Neuropathology of chronic comprehensive spinal cord injury; osteoarthritis. *Mailing Add:* Murdoch Univ Sch Vet Studies South St Murdoch 6150 Australia

YOVITS, MARSHALL CLINTON, b Brooklyn, NY, May 16, 23; m 52; c 3. COMPUTER SCIENCES. *Educ:* Union Col, BS, 44, MS, 48; Yale Univ, MS, 50, PhD(physics), 51. *Prof Exp:* Physicist, Nat Adv Comt for Aeronaut, Langley Field, Va, 44-46; instr physics, Union Col, 46-48; instr, Yale Univ, 48-50; sr physicist, Appl Physics Lab, Johns Hopkins Univ, 51-56; physicist, Off Naval Res, 56, head info systs br, 56-62, dir naval anal group, 62-66; prof comput & info sci & chmn dept, Ohio State Univ, 66-79; dean, Purdue Sch Sci, 80-88, PROF COMPUTER & INFO SCI, IND UNIV-PURDUE UNIV, INDIANAPOLIS, 89- *Concurrent Pos:* Ed, Adv Comput, 70-; chmn comput sci conf, Columbus, Ohio, 73 & Indianapolis, Ind, 82; mem biomed comt study sect, NIH, 70-74. *Mem:* Fel Inst Elec & Electronics Engrs; Asn Comput Mach; fel AAAS; Sigma Xi; Comput Soc. *Res:* Information systems; management information; self-organizing systems; information science; development of a generalized theory of information flow analysis; relating information to its use in decision-making. *Mailing Add:* 9016 Dewberry Ct Indianapolis IN 46260

YOW, FRANCIS WAGONER, b Asheville, NC, May 1, 31; m 49; c 2. EMBRYOLOGY. *Educ:* Western Carolina Univ, BS, 55; Emory Univ, MS, 56, PhD(protozool), 58. *Prof Exp:* Asst prof biol, Western Carolina Col, 58-60; from asst prof to assoc prof, 60-65, chmn dept, 63-74, PROF BIOL, KENYON COL, 65- *Concurrent Pos:* Consult-examr, NCent Asn Cols & Univs, 71-, comnr, 80-84. *Mem:* AAAS; Soc Protozoologists; Am Soc Zoologists; Soc Develop Biol. *Res:* Morphogenesis in ciliate protozoa; radiation biology and nutrition of invertebrates; nucleic acid synthesis; inhibition of cellular activities by radiation and chemical means; effects of carcinogens on regeneration. *Mailing Add:* Dept Biol Kenyon Col Gambier OH 43022

YOW, MARTHA DUKES, b Talbotton, Ga, Jan 15, 22; m 44; c 3. PEDIATRICS. *Educ:* Univ SC, BS, 40, MD, 43; Am Bd Pediat, dipl. *Prof Exp:* Instr bact, Baylor Col Med, 49-50, instr pediat, 50-52, from instr to asst assoc prof, 55-69, dir, Pediat Infectious Dis Sect, 64-82, PROF PEDIAT, BAYLOR COL MED, 69- *Concurrent Pos:* Res fel pediat, Baylor Col Med, 50-52, Jones fel, 55-; NIH grant; mem bd sci coun, Nat Inst Allergy & Infectious Dis. *Mem:* Am Fedn Clin Res; Am Soc Microbiol; Soc Pediat Res; Infectious Dis Soc Am; Am Acad Pediat; Am Pediat Soc. *Res:* Infectious diseases; applied virology. *Mailing Add:* Dept Pediat Immunol & Microbiol 11948 Adorno Pl San Diego CA 92128

YOZAWITZ, ALLAN, b Brooklyn, NY, Jan 8, 49; m 73; c 3. CLINICAL NEUROPSYCHOLOGY. *Educ:* Polytech Inst Brooklyn, BS, 70; Queens Col, MA, 73; City Univ New York, PhD(neuropsychol), 77; Am Bd Prof Psychol & Am Bd Clin Neuropsychol, dipl, 84. *Prof Exp:* Res asst, Audition Lab Biomet Res, NY State Psychiat Inst, 70-73, res scientist, 73-77; DIR, CLIN NEUROPSYCHOL UNIT, HUTCHINGS PSYCHIAT CTR, SYRACUSE, NY, 77- *Concurrent Pos:* Co-investr, NIMH grant, Biomet Res, NY State Psychiat Inst, 74-77, consult, geront sect, 75-76; co-dir, NIMH training prog in clin neuropsychol, Cornell Univ, Ithaca, 79-83; consult, clin neuropsychol, Syracuse Develop Ctr, 79-83 & Benjamin Rush Ctr, Syracuse, 80-88; asst prof, dept psychiat, Col Med, State Univ NY Health Sci Ctr, Syracuse, 79-89, assoc prof, 89-; adj asst prof, dept psychol, Syracuse Univ, 79-89, adj assoc prof, 89-; consult ed, J Clin & Exp Neuropsychol, 83-; dir, continuing educ, Int Neuropsychol Soc, 85-; consult, NY State Bd Psychol, 87-88; mem, NY State bd psychol, 88-; consult ed, The Clin Neuropsychologist, 91- *Mem:* Int Neuropsychol Soc; Am Psychol Asn; NY Acad Sci; Soc Neurosci; AAAS; NY State Psychol Asn; Int Brain Res Orgn. *Res:* Neuropsychological mediation of psychiatric disorder involving the effects of nondominant hemispheric processes on affective states; clinical practice of neuropsychology with psychiatric patients; neuropsychological rehabilitation of psychiatric vulnerability. *Mailing Add:* Hutchings Psychiat Ctr 620 Madison St Syracuse NY 13210

YOZWIAK, BERNARD JAMES, b Youngstown, Ohio, July 5, 19; m 43; c 4. MATHEMATICS. *Educ:* Marietta Col, AB, 40; Univ Pittsburgh, MS, 51, PhD(math), 61. *Prof Exp:* Clerk, Youngstown Sheet & Tube Co, 40-41; high sch prin & teacher, Ohio, 41-42; civilian instr, US Army Air Forces Tech Training Command, Ill & Wis, 42-44; clerk, Youngstown Sheet & Tube Co, 44-45; high sch prin & teacher, Ohio, 45-47; from asst prof to assoc prof, 47-63, chmn dept, 66-71, PROF MATH, YOUNGSTOWN STATE UNIV, 63-, DEAN, COL ARTS & SCI, 71- *Mem:* Math Asn Am; Sigma Xi. *Res:* Summability methods. *Mailing Add:* 2080 S Schenley Ave Youngstown OH 44511

YPHANTIS, DAVID ANDREW, b Boston, Mass, July 14, 30; m 53; c 5. BIOPHYSICS, BIOCHEMISTRY. *Educ:* Harvard Univ, AB, 52; Mass Inst Technol, PhD(biophys), 55. *Prof Exp:* Am Cancer Soc fel, Mass Inst Technol, 55-56; from asst biophysicist to assoc biophysicist, Argonne Nat Lab, 56-58; from asst prof to assoc prof biochem, Rockefeller Univ, 58-65; prof biol, State Univ NY Buffalo, 65-68, prof biophys & chmn dept biol, 67-68; PROF BIOL, UNIV CONN, 68- *Concurrent Pos:* Consult, Argonne Nat Lab, 58-62 & NIH, 67- *Mem:* AAAS; Am Chem Soc; Biophys Soc; Am Soc Biol Chem; Sigma Xi. *Res:* Physical biochemistry; protein physical chemistry; ultracentrifugation. *Mailing Add:* 99 River Rd Mansfield Center CT 06250

YU, A TOBEY, b Chekiang, China, Jan 6, 21. ENGINEERING. *Educ:* Mass Inst Technol, MA, 46; Lehigh Univ, PhD(civil eng), 49; Columbia Univ, MBA, 72. *Prof Exp:* Design engr, Hewitt Robins Inc, 51-59, vpres opers, Div Litton Indust, Hewitt Robins, 67-71; tech div eng, Compania Minera Santa Fe, 59-67; pres & chmn, Orba Corp, 72-86; RETIRED. *Honors & Awards:* Outstanding Eng Achievement, Am Soc Civil engrs, 77, Am Literal Engrs, 78 & Chinese Inst, 78; Mater Handling Award, Am Soc Mech Eng, 87; Richards Awards, Am Inst Mining Engrs, 87. *Mem:* Nat Acad Eng; Am Inst Mining Engrs; Soc Mining Explor & Metall; Nat Soc Prof Engrs. *Mailing Add:* 3284 Masters Dr Clearwater FL 34621

YU, ANDREW B C, b Nov 24, 45; US citizen. PHARMACEUTICS, BIOPHARMACEUTICS. *Educ:* Albany Col Pharm, BS, 71; Univ Conn, PhD(pharmaceut sci), 76. *Prof Exp:* Asst prof pharm, Northeastern Univ, 76-80; SR RES PHARMACIST, STERLING WINTHROP RES INST, 79- *Concurrent Pos:* Investr, NIH biomed res support grant, 77-78; co-investr res contract, Dooner Labs, 78-79; monograph writer, Am Pharmaceut Asn, Pharmaceut Excipient Codex Comt, 78-79. *Mem:* Am Asn Pharmaceut Scientists; NY Acad Sci. *Res:* Pharmacokinetics of drugs in renal diseased patients; use of antibiotics in the treatment of bacterial endocarditis. *Mailing Add:* 134 Berkshire Blvd Albany NY 12205

YU, BYUNG PAL, b Ham Hung, Korea, June 27, 31; US citizen; m 59; c 1. BIOCHEMISTRY, CELL PHYSIOLOGY. *Educ:* Mo State Univ, BS, 60; Univ Ill, PhD, 65. *Prof Exp:* From res instr to res asst prof, Med Col Pa, 65-68, from asst prof to assoc prof, 68-73; assoc prof, 73-78, PROF PHYSIOL, UNIV TEX HEALTH SCI CTR, SAN ANTONIO, 78- *Honors & Awards:* Henry L Moss Award, Am Diabetes Asn, Res & Career Develop Award. *Mem:* Am Physiol Soc; fel Geront Soc Am; Oxygen Soc. *Res:* Biological membrane structure; biological aspects of aging. *Mailing Add:* Dept Physiol Univ Tex Health Sci Ctr San Antonio TX 78284-7756

YU, CHANG-AN, b Taiwan, Oct 19, 37; US citizen; m 68; c 2. BIOCHEMISTRY. *Educ:* Nat Taiwan Univ, BS, 61, MS, 64; Univ Ill, PhD(biochem), 69. *Prof Exp:* Fel biochem, Univ Ill, 69-70; vis asst prof chem, State Univ NY, 70-75, res assoc prof, 76-80; from assoc prof to prof, 81-85, REGENTS PROF BIOCHEM, OKLA STATE UNIV, 85- *Mem:* Am Soc Biol Chemists; Am Chem Soc; Biophys Soc; AAAS. *Res:* Membrane bioenergetic; biological oxidation; ubiquinone; protein interaction; photosynthesis. *Mailing Add:* Dept Biochem Okla State Univ Stillwater OK 74078

YU, CHIA-NIEN, b Shanghai, China, Aug 5, 31; US citizen; m 66; c 3. ORGANIC CHEMISTRY. *Educ:* Univ Ill, Urbana, BS, 58; Univ Mich, MS, 59. *Prof Exp:* Res chemist, 59-67, RES SCIENTIST, NORWICH PHARMACAL CO, 67- *Mem:* Am Chem Soc. *Res:* Synthesis of organic compounds for biological screenings. *Mailing Add:* 29 Hillview Dr Norwich NY 13815-1007

YU, CHIA-PING, b Kiangsu, China, Aug 21, 34; m 61; c 2. AERONAUTICAL ENGINEERING. *Educ:* Taiwan Norm Technol, BS, 54; Purdue Univ, PhD(aeronaut eng), 64. *Prof Exp:* Instr aeronaut eng, Purdue Univ, 60-64; from asst prof to assoc prof aeronaut & eng sci, 67-70, PROF

ENG SCI, STATE UNIV NY, BUFFALO, 70- *Mem:* Am Inst Aeronaut & Astronaut; Am Phys Soc; AAAS; Am Geophys Union. *Res:* Cosmical and engineering magnetohydrodynamics; waves and instabilities of plasma; atmospheric dynamics; aerosol mechanics; physics of fluids. *Mailing Add:* 46 Candlewood Lane Williamsville NY 14221

YU, CHYANG JOHN, b China, June 20, 48; US citizen; m 74; c 2. SOLID STATE PHYSICS, INORGANIC CHEMISTRY. *Educ:* Taiwan Nat Cheng Kung Univ, BS, 70; Wayne State Univ, MA, 72; Univ Ill, Urbana-Champaign, PhD(ceramic eng), 77. *Prof Exp:* Teaching asst physics, Wayne State Univ, 72-73; res asst, Univ Ill, Urbana-Champaign, 73-77; res engr, Ohio Brass Co, 77-79; res & develop mgr, NAm Philips Co, 79-81; mem fac, Wilkes Univ, 81-82; staff engr, Int Bus Mach Corp, 82-83; mem tech staff, AVX Corp, 84-85; sr res scientist, Corning, Inc, 85-90; MGR, FOXCONN INT, INC, 90- *Concurrent Pos:* Mem fac, Calif Polytech State Univ, 81; vis scientist, Ames Lab, 82; fel, NASA Lewis Res Ctr & Wright Patterson AFB, 84; deleg, People to People Int, 91. *Mem:* Am Ceramic Soc; Nat Inst Ceramic Engrs; Nat Asn Accountants; Inst Cert Mgt Accountants. *Res:* Materials formulation processes and characterizations of oxide semiconductors, dielectrics, ic packaging, high temperature superconductors, other electronic ceramics, fiber or whiskers reinforced composites; electrostatic discharge protections and testing. *Mailing Add:* PO Box 110634 Campbell CA 95011-0634

YU, CLEMENT TAK, b Hong Kong, Aug 31, 48; Can citizen; m 75; c 2. DATA BASE MANAGEMENT, INFORMATION RETRIEVAL. *Educ:* Columbia Univ, BSc, 70; Cornell Univ, MSc, 72, PhD(comput sci), 73. *Prof Exp:* From asst prof to assoc prof comput sci, Univ Alta, Edmonton, 73-78; assoc prof, 78-84, PROF COMPUT SCI, UNIV ILL, CHICAGO, 84- *Concurrent Pos:* Consult, Syst Develop Corp, Santa Monica, Calif, 82-86, Microelectronics Comput Corp, Austin, Tex & various other corp; mem, adv comt, NSF, chmn, ACM Spec Interest Group on Info Retrieval, 85-87. *Mem:* Asn Comput Mach. *Res:* Date base management; information retrieval problems. *Mailing Add:* Dept Elec Eng & Comput Sci Univ Ill Chicago IL 60680

YU, DAVID TAK YAN, b Hong Kong, Feb 20, 43; m 67; c 1. RHEUMATOLOGY. *Educ:* Univ Hong Kong, BS & MB, 66. *Prof Exp:* Intern med, Univ Hong Kong, 66-67 & Long Island Col Hosp, NY, 68-69; pathologist, Hong Kong Govt Inst Path, 67-68; resident, Montefiore Hosp, NY, 69-71; fel rheumatology, 71-74, asst prof med, 74-80, ASSOC PROF RHEUMATOLOGY, UNIV CALIF, LOS ANGELES, 81- *Concurrent Pos:* Guest investr immunol, Rockefeller Univ, New York, 78-79, asst prof, 79-80. *Mem:* Am Asn Immunol; Am Fedn Clin Res; Am Rheumatism Asn; Am Asn Microbiol. *Res:* Cause and mechanisms of the rheumatic diseases; the arthritis conditions Reiter's syndrome and ankylosing spondylitis. *Mailing Add:* Dept Med Rm 35-40 Rehab Ctr Univ Calif 1000 Veteran Ave Los Angeles CA 90024-1670

YU, DAVID U L, b Hong Kong, Aug 27, 40; nat US; m 65; c 2. NUCLEAR PHYSICS, STRUCTURAL MECHANICS. *Educ:* Seattle Pac Col, BSc, 61; Univ Wash, PhD(theoret physics), 64. *Prof Exp:* Res assoc prof theoret physics, Stanford Univ, 64-66; Brit Sci Res Coun fel physics, Univ Surrey, 66-67; from asst prof to assoc prof, Seattle Pac Col, 67-73; mgr, Comput Sci Corp, Richland, Washington & El Segundo, Calif, 73-75; exec vpres & dir, Basic Technol, Inc, 75-83; PRES, DULY CONSULTS, 83- *Concurrent Pos:* Vpres, Int Inst Technol, 77-; fel, Ford Found, 62-63, NASA Jet Propulsion Lab, 69-70 & NSF fel, Ill Inst Tech, 72. *Mem:* Am Phys Soc; Am Asn Physics Teachers; Am Soc Mech Engrs. *Res:* Nuclear structure and reactions; elementary particle physics. *Mailing Add:* 1912 MacArthur St Rancho Palos Verdes CA 90732

YU, FRANCIS T S, b Amoy City, China, Nov 12, 32; US citizen; m 62; c 3. ELECTRICAL ENGINEERING. *Educ:* Mapua Inst Technol, BSEE, 56; Univ Mich, MSE, 58, PhD(elec eng), 64. *Prof Exp:* Res asst, Commun Sci Lab, Univ Mich, 58-64, res assoc engr, 64-66, instr elec eng, 60-64, lectr, 64-65; from asst prof to prof elec eng, Wayne State Univ, 66-80; PROF ELEC ENG, PA STATE UNIV, 80-, EVAN PUGH PROF, 88- *Mem:* Fel Inst Elec & Electronics Engrs; fel Optical Soc Am; fel Soc Photo-Optical Instrumentation Engrs. *Res:* optical communication and filtering; optical information processing; holography; information theory; optical computing; neural network. *Mailing Add:* Dept Elec Eng Pa State Univ University Park PA 16802

YU, FU-LI, b Peking, China, May 2, 34; US citizen; m 80; c 2. CHEMICAL CARCINOGENESIS. *Educ:* Taiwan Chung-Shing Univ, BS, 56; Univ Ala, MS, 62; Univ Calif, San Francisco, PhD(biochem), 65. *Prof Exp:* Instr biochem, Univ NMex, 65-66; fel trainee, Inst Cancer Res, Columbia Univ, 66-69, res assoc, 69-73; asst prof, Jefferson Med Col, 73-79; from asst to assoc prof, 79-85, PROF BIOCHEM, COL MED, UNIV ILL, 85-, HEAD, DEPT BIOMED SCI, 90- *Concurrent Pos:* Res grants, NIH & Am Cancer Soc. *Mem:* Am Soc Biol Chemists; Am Asn Cancer Res; Am Chem Soc; Harvey Soc; NY Acad Sci. *Res:* Chemical carcinogenesis; hormone action; nucleic acid metabolism; RNA polymerase; gene regulation in mammalian cells; mechanisms of action of chemical carcinogens, particularly aflatoxin B1, at the transcriptional level. *Mailing Add:* Dept Biomed Sci Univ Ill Col Med 1601 Parkview Ave Rockford IL 61107

YU, GRACE WEI-CHI HU, b Feb 10, 37; US citizen; m 62; c 2. BIOLOGY, MOLECULAR BIOLOGY. *Educ:* Nat Taiwan Univ, BS, 59; Wash State Univ, MS, 63; Duke Univ, PhD(plant physiol), 67. *Prof Exp:* Res assoc plant physiol, Duke Univ, 66-68; lectr bot, Sch Med, Univ Calif, Los Angeles, 68, res assoc plant physiol, 68-71, asst res biologist, Neuropsychiat Inst, 71-72, ment health training prog fel, Brain Res Inst, 72-78, res assoc hemat & oncol, Dept Pediat, 75-78; ASST PROF PLANT PHYSIOL, DEPT BIOL, CALIF STATE UNIV, DOMINGUEZ HILLS, 78- *Concurrent Pos:* Comput prog cert, Control Data Inst, 82; instr biol, bot, human anat & physiol, Los Angeles Commun Col & Compton Commun Col, 78- *Mem:* AAAS; Am Soc Plant Physiologists; Sigma Xi. *Res:* Plant physiology, especially ion transport; developmental biology; software quality engineering. *Mailing Add:* 30303 Via Borica Rancho Palos Verdes CA 90274

YU, GRETA, b Canton, China, Jan 12, 17; nat US. PHYSICS, OPERATIONS RESEARCH. *Educ:* Sun Yat-Sen Univ, BS, 38; Univ Ill, Univ MS, 40; Univ Cincinnati, PhD(physics), 43. *Prof Exp:* Spectroscopist, Wright Aeronaut Corp, 43-45; physicist, US Naval Ord Plant, 45-49; res physicist, Cornell Aeronaut Lab, Inc, 50-56; sr tech specialist & tech staff, NAm Aviation, Inc, 56-57, sr mem tech staff, NAm Rockwell Corp, 67-71; tech consult econ res, State Ohio, 71-74; STAFF SCIENTIST, OHIO POWER SITTING COMN, 74-, MEM SITTING BD, 80- *Mem:* Am Phys Soc; Opers Res Soc Am; Sigma Xi. *Res:* Energy and environment research; nuclear and fossil power siting and evaluation; applied physics. *Mailing Add:* 601 Fairway Blvd Columbus OH 43213

YU, HWA NIEN, b Shanghai, China, Jan 17, 29; m 55; c 3. ELECTRICAL ENGINEERING. *Educ:* Univ Ill, BS, 53, MS, 54, PhD(elec eng), 58. *Prof Exp:* Assoc engr, Res Lab, IBM Corp, 57-59, staff engr, Adv Systs Develop Div, 59-62, res staff mem semiconductor develop, Res Ctr, 62-80, MGR DEVICE & CIRCUIT TECHNOL, T J WATSON RES CTR, IBM CORP, 80- *Mem:* Sigma Xi. *Res:* Computer design; semiconductor devices; electronic computers; semiconductors. *Mailing Add:* 2849 Hickory Yorktown Heights NY 10598

YU, HYUK, b Kapsan, Korea, Jan 20, 33; m 64; c 3. PHYSICAL CHEMISTRY, BIOPHYSICS. *Educ:* Seoul Nat Univ, BS, 55; Univ Southern Calif, MS, 58; Princeton Univ, PhD(phys chem), 62. *Prof Exp:* Res chemist, Nat Bur Stand, 63-67; from asst prof to assoc prof, 67-77, PROF CHEM, UNIV WIS-MADISON, 78- *Concurrent Pos:* Res assoc, Dartmouth Col, 62-63; consult, Nat Bur Standards, DC, Eastman Kodak Co, NY & Tenn Eastman Co; John Simon Guggenheim Fel, 84-85. *Honors & Awards:* Fulbright-Hays Lectr Korea, 72; fulbright Lectr, 72. *Mem:* Am Chem Soc; Am Phys Soc; NY Acad Sci; Biophys Soc; Mat Res Soc. *Res:* Structure and dynamics of biomembranes; polymer solution characterizations; syntheses of macromolecules; polymer dynamics in bulk and concentrated solutions; interfacial phenomena of polymers. *Mailing Add:* Dept Chem Univ Wis Madison WI 53706

YU, JAMES CHUN-YING, b Hunan, China, Oct 14, 40; US citizen; m 65; c 2. ACOUSTICS, FLUID DYNAMICS. *Educ:* Nat Taiwan Univ, BSc, 62; Syracuse Univ, MSc, 68, PhD(mech eng), 71. *Prof Exp:* Instr mech eng, Syracuse Univ, 70-71; asst res prof, George Washington Univ, 71-75, assoc res prof, 75-76, assoc prof, 76-77; AEROSPACE TECHNOLOGIST ACOUST, NASA LANGLEY RES CTR, 77- *Concurrent Pos:* NASA res grant, Langley Res Ctr, 71- *Mem:* Am Inst Aeronaut & Astronaut; Acoust Soc Am. *Res:* Sound generation from fluid flows; acoustic measurements and instrumentation; turbulent flows. *Mailing Add:* 105 Hoy Ct MS-118 Cary VA 27511

YU, JASON C, b Hupei, China, Feb 5, 36; m 65; c 3. TRANSPORTATION ENGINEERING, CIVIL ENGINEERING. *Educ:* Univ Taiwan, BS, 57; Ga Inst Technol, MS, 63; Univ WVa, PhD(civil eng), 67. *Prof Exp:* Traffic engr, WVa State Rd Comn, 64-65; res assoc civil eng, Univ WVa, 66-67; res specialist transp eng, Univ Pa, 67-68; from asst prof to assoc prof civil eng, Va Polytech Inst & State Univ, 68-74; PROF CIVIL ENG, UNIV UTAH, 74- *Concurrent Pos:* Mem tech comt parking & traffic control devices, Transp Res Bd, Nat Acad Sci-Nat Res Coun, 69-; ed, External Transp, Int Joint Comt Tall Bldg, 73- *Mem:* Am Soc Civil Engrs; Inst Traffic Engrs; Am Soc Eng Educ. *Res:* Transportation energy conservation strategies; urban transit system optimization; transportation system planning, design and operation; effects of transportation on land use development. *Mailing Add:* 4845 Bronbreck St Salt Lake City UT 84117

YU, JEN, b Taipei, Taiwan, Jan 23, 43; US citizen; m 73; c 2. PHYSICAL MEDICINE & REHABILITATION. *Educ:* Nat Taiwan Univ, MD, 68; Univ Pa, PhD(physiol), 72. *Prof Exp:* Asst prof phys med & rehab, Sch Med, Univ Pa, 75-76; from asst prof to assoc prof, Univ Tex, Health Sci Ctr, San Antonio, 76-81; PROF PHYS MED & REHAB, COL MED, UNIV CALIF, IRVINE, 81-, CHAIR DEPT, 82- *Mem:* Am Acad Phys Med & Rehab; Am Cong Rehab Med; Am Asn Anatomists; Soc Neurosci; Asn Acad Physiatrists. *Res:* Neurobiological basis of rehabilitation medicine, including neurobiological studies on learning and memory, control of movement, and recovery after injury. *Mailing Add:* Dept Phys Med & Rehab Univ Calif Irvine Med Ctr 101 The City Dr Orange CA 92668

YU, KAI FUN, b Canton, China, July 19, 50; m 78. STATISTICS, MATHEMATICS. *Educ:* Dartmouth Col, AB, 73; Columbia Univ, PhD(statist), 78. *Prof Exp:* Asst prof statist, Yale Univ, 78-83; ASSOC PROF STATIST, UNIV SC, 84- *Mem:* Am Statist Asn; Inst Math Statist; Sigma Xi. *Mailing Add:* Dept Statist Univ SC Columbia SC 29208

YU, KAI-BOR, b Canton, China, Apr 20, 53; China citizen; m 82; c 2. SIGNAL PROCESSING, COMMUNICATION. *Educ:* Yale Univ, BS, 77; Brown Univ, MS, 79; Purdue Univ, PhD(elec eng), 82. *Prof Exp:* ASST PROF ELEC ENG, VA POLYTECH INST & STATE UNIV, 82- *Mem:* Inst Elec & Electronics Engrs. *Res:* Signal processing; image processing; pattern recognition; communication and radar. *Mailing Add:* Gen Elec Co KWC-606 PO Box 8 Schenectady NY 12301

YU, KARL KA-CHUNG, b Shanghai, China, Aug 31, 36; m 62; c 2. SOLID STATE ELECTRONICS, MATERIALS SCIENCE. *Educ:* Carnegie-Mellon Univ, BS, 57, MS, 59, PhD(elec eng, mat sci), 66. *Prof Exp:* Assoc engr, Westinghouse Elec Corp, 59-61, sr engr, 66-78; prin staff engr, McDonnell Douglas Corp, 78-79; SR STAFF PHYSICIST, HUGHES AIRCRAFT CO, 79- *Mem:* Inst Elec & Electronics Engrs. *Res:* Research and development in the area of solid state semiconductor device physics with emphasis on memory devices. *Mailing Add:* Torrance Res Ctr Hughes Aircraft Co 3100 W Lomita Blvd Torrance CA 90509

YU, LEEPO CHENG, b Shanghai, China, June 25, 39; m 65; c 1. BIOPHYSICS. *Educ:* Brown Univ, BS, 63; Univ Md, PhD(physics), 69. *Prof Exp:* Res assoc muscle physiol, Brown Univ, 69-72; staff fel, 73-75, sr staff fel, 76-77, RES PHYSICIST MUSCLE X-RAY DIFFRACTION, NAT INST ARTHRITIS, MUSCULO-SKELETAL & SKIN, 77- *Concurrent Pos:* Mem ed bd, Biophys J, 86-89. *Mem:* Am Phys Soc; Biophys Soc. *Res:* X-ray diffraction of striated vertebrate muscle; theoretical modelling of force generation. *Mailing Add:* NIH Bldg 6 Rms 114 Bethesda MD 20205

YU, LINDA, b Taiwan, China, Feb 7, 43; m 68; c 2. BIOENERGETICS. *Educ:* Nat Taiwan Norm Univ, Taipei, BS, 66; Univ Ill, Urbana, MS, 68, PhD(microbiol), 70. *Prof Exp:* Res assoc, Dept Chem, State Univ NY, Albany, 70-74, res asst prof, lab bioenergetics, 74-79, dept chem, 79-81; assoc res biochemist, 81-84, res prof, 84-86, ASSOC PROF, DEPT CHEM, OKLA STATE UNIV, STILLWATER, 86- *Concurrent Pos:* Vis instr, Dept Chem, State Univ NY, Albany, 70-74; grants, NIH & Okla Dept Com, 87-88, USDA, 87- *Mem:* Am Soc Biochem & Molecular Biol. *Res:* Membrane bioenergetic, membrane enzymes and phospholipid-protein interaction. *Mailing Add:* Dept Biochem Okla State Univ Stillwater OK 74078-0454

YU, MANG CHUNG, b Hong Kong, Mar 4, 39; US citizen; m 66. ANATOMY, NEUROANATOMY. *Educ:* St Edward's Univ, BS, 63, MS, 66, PhD(anat), 70. *Prof Exp:* Fels, State Univ NY, Buffalo, 70-72; ASST PROF ANAT, COL MED & DENT NJ, 72- *Concurrent Pos:* Consult neuroanat, Vet Admin Hosp, East Orange, NJ, 72- *Mem:* Am Physiol Soc; Am Soc Neurosci; AAAS; Am Asn Anat. *Res:* Neurobiology; neuropathology. *Mailing Add:* Dept Anat Univ Med & Dent NJ 100 Bergen St Newark NJ 07103

YU, MASON K, b Canton, China, Aug 2, 26; US citizen; m 50; c 5. MECHANICAL ENGINEERING. *Educ:* Univ Mich, BSME, 50, MSME, 51. *Prof Exp:* Engr, Borg-Warner Corp, 51-53; res engr, Continental Aviation Eng Corp, 53-55; SR RES ENGR, EMISSIONS RES DEPT, GEN MOTORS RES LABS, 55- *Mem:* Am Soc Mech Engrs; Sigma Xi. *Res:* Computer application to gas turbine thermodynamic turbomachinery research and development. *Mailing Add:* 550 W Brown Birmingham MI 48009

YU, MEI-YING WONG, b Tainan, Taiwan, Repub China; US citizen; c 1. BIOCHEMICAL PHARMACOLOGY, IMMUNOBIOCHEMISTRY. *Educ:* Kachsiung Med Col, Taiwan, Repub China, BS, 64; Univ Ala, Birmingham, MS, 68, PhD(pharmacol), 70. *Prof Exp:* Postdoctoral res fel, Dept Pharmacol, Baylor Sch Med, Houston, Tex, 70-71; res assoc, Dept Zool, Univ Tex, Austin, 71-74; sr staff res fel, Lab Develop Neurobiol, NICHD, NIH, 74-80; RES CHEM, CTR BIOLOGICS EVAL RES, FOOD & DRUG ADMIN, 81- *Mem:* Am Soc Pharmacol & Exp Therapeut; Fedn Am Soc Exp Biol. *Res:* Antithyroid drugs and pyrimidine metabolism; growth factors actions in cultures; human albumin's stability and interaction with hepatitis B virus; recombinant HBV envelope proteins and antibodies; HBV and albumin receptors. *Mailing Add:* Div Hemat CBER/FDA 8800 Rockville Pike Bethesda MD 20892

YU, MING LUN, b Hong Kong, Aug 21, 45; m 78; c 2. SOLID STATE PHYSICS. *Educ:* Hong Kong Univ, BSc, 67; Calif Inst Technol, MSc, 71, PhD(physics), 74. *Prof Exp:* Res assoc, Brookhaven Nat Lab, 73-74, asst physicist, 74-76, assoc physicist, 76-78; RES STAFF MEM, T J WATSON RES CTR, IBM CORP, 78-, MGR ANALYTICAL RES, 84- *Concurrent Pos:* Vis scientist, Physics Dept, State Univ NY, Stony Brook, 76-79. *Mem:* Fel Am Phys Soc; Am Vacuum Soc. *Res:* Superconductivity; Josephson devices; secondary ion mass spectrometry; surface physics; chemistry. *Mailing Add:* IBM Res Div T J Watson Res Ctr PO Box 218 Yorktown Heights NY 10598

YU, MING-HO, b Kaohsiung, Taiwan, May 22, 28; m 56; c 3. ENVIRONMENTAL HEALTH, NUTRITION. *Educ:* Nat Taiwan Univ, BS, 53; Utah State Univ, MS, 64, PhD(plant nutrit & biochem), 67. *Prof Exp:* Res asst agr chem, Taiwan Agr Res Inst, 54-55; asst res fel chem, Inst Chem Acad Sinica, 59-62; fel, Utah State Univ, 67 & Univ Alta, 67-68; vis asst prof plant biochem, 69-70, lectr environ biol, 70 -71, asst prof, to assoc prof, 71-79, PROF ENVIRON BIOL, HUXLEY COL, WESTERN WASH UNIV, 79- *Concurrent Pos:* Fulbright Travel Grant, 62; vpres, Int Soc Fluoride Res, 88-; vis prof, Iwate Med Univ, Japan, 89-90; ed, Environ Sci, 90- *Mem:* AAAS; Int Soc Fluoride Res; NY Acad Sci; Am Chem Soc; Am Inst Nutrit. *Res:* Fluoride effects on the physiology and biochemistry of animals and plants; effects of pollutants on health; vitamin C metabolism. *Mailing Add:* Huxley Col Western Wash Univ Bellingham WA 98225-9079

YU, NAI-TENG, b Pingtung, Formosa, Aug 19, 39; m 66. BIOPHYSICAL CHEMISTRY. *Educ:* Nat Taiwan Univ, BS, 63; NMex Highlands Univ, MS, 66; Mass Inst Technol, PhD(phys chem), 69. *Prof Exp:* Res chemist, Arthur D Little, Mass, 66; res assoc chem, Mass Inst Technol, 69-70; from asst to assoc prof, 70-80, PROF CHEM, GA INST TECHNOL, 80- *Concurrent Pos:* Res Corp res grant, Ga Inst Technol, 71-72, USPHS res grants, 71-; res career develop award, NIH, 76-81; adj prof ophthal, Emory Univ, 82- *Honors & Awards:* Sustained Res Award, Sigma Xi, 86. *Mem:* AAAS; Am Chem Soc; Biophys Soc; Asn Res Vision & Ophthal. *Res:* Laser Raman spectroscopy of biopolymers; metalloporphyrins and hemoproteins; mechanisms of cataract lens formation; development of biomedical instrumentation. *Mailing Add:* Dept Chem Ga Inst Technol Atlanta GA 30332

YU, OLIVER SHUKIANG, b Chentu, China, July 8, 39; m 62; c 1. ELECTRICAL ENGINEERING, PROBABILITY. *Educ:* Nat Taiwan Univ, BSEE, 59; Ga Inst Technol, MSEE, 62; Stanford Univ, MS, 67, PhD(opers res), 72. *Prof Exp:* Res assoc civil defense study, Merrimack Col, 63-64; opers analyst, Opers Anal Dept, Stanford Res Inst, 64-68, res engr, Systs Eval Dept, 68-70, sr res engr, 70-74; proj mgr, Elec Power Res Inst, 74-77; planning specialist, Commonwealth Edison Co, 77; tech asst to mgr, Res & Develop Planning & Assessment Dept, 78, tech mgr, Energy Study Ctr, 78-79, MGR PLANNING ANALYSIS, ELEC POWER RES INST, 79- *Concurrent Pos:* Consult systs anal, SRI Int, 75; adj prof, Univ Calif, Berkeley, 80-85, Santa Clara Univ, 75-81, Calif State Univ, Hayward, 75-79 & San Jose State Univ, 79-81; Fulbright fel, 61-62; consult prof, Stanford Univ, 83-85. *Mem:* Sigma Xi; Inst Elec & Electronics Engrs; Opers Res Soc Am; Int Asn Energy Economists. *Res:* Planning, analysis and evaluation of large-scale systems program development and direction of systems analysis research. *Mailing Add:* SRI Int 333 Ravenswood Ave Menlo Park CA 94025

YU, PAUL N, b Kiangsi, China, Nov 17, 15; nat US; m 44; c 4. INTERNAL MEDICINE, CARDIOLOGY. *Educ:* Nat Med Col Shanghai, China, MD, 39; London Sch Trop Med & Hyg, 46, dipl, 47; Am Bd Internal Med, dipl, 56, cert cardiovasc dis, 57. *Prof Exp:* Instr med, asst resident & chief resident physician, Cen Hosp, Chunking, China, 40; asst resident physician, Hosp, 47, from instr to prof med, 48-69, SARA MCCORT WARD PROF MED, SCH MED, UNIV ROCHESTER, 69-, HEAD CARDIOL UNIT & PHYSICIAN, HOSP, 63- *Concurrent Pos:* Hochstetter fel, Sch Med, Univ Rochester, 48-54; consult, State Depts Health & Social Welfare, NY, 55-, Genesee Hosp, Rochester, 57-, Vet Admin Hosp, Bath, 59-, Highland Hosp, Rochester, Frederick Thompson Mem Hosp, Canandaigua & Newark Community Hosps & St Mary's Hosp; from asst to sr assoc physician, Univ Rochester Hosp, 52-63, founding fel coun clin cardiol, Am Heart Asn; ed, Progress in Cardiol. *Mem:* Fel Am Col Physicians; sr mem Am Fedn Clin Res; Asn Am Physicians; Am Clin & Climatol Asn; Am Heart Asn (pres, 72-73); Sigma Xi. *Res:* Pulmonary circulation, hemodynamics; electrocardiography. *Mailing Add:* Univ Rochester Box 679 Univ Rochester Rochester NY 14642

YU, PETER HAO, b Liaoning, China, May 26, 42; Can citizen; m 70; c 3. NEUROCHEMISTRY, NEUROPSYCHOPHARMACOLOGY. *Educ:* Nat Taiwan Univ, BSc, 64, MSc, 67; Univ Göttingen, DSc Agr(biol), 71. *Prof Exp:* Res scientist, Asn Molecular Biol Res, Ger, 71-73; assoc prof, Nat Taiwan Univ, 73-75; SR RES SCIENTIST, PSYCHIAT RES DIV, CAN DEPT HEALTH, 75- *Concurrent Pos:* Consult, Res Prog Directorate, Health & Welfare, Can, 77-; res assoc, Dept Psychiat, Univ Sask, 78- *Mem:* Int Soc Neurochem. *Res:* Regulation of metabolism of catecholamine and the possible relationship to mental disorder. *Mailing Add:* Neuropsychiat Res Div Rm A102 Med Res Bldg Univ Sask Saskatoon SK S7N 0W8 Can

YU, PETER YOUND, b Shanghai, China, Sept 8, 44; m 71; c 2. EXPERIMENTAL SOLID STATE PHYSICS, SEMICONDUCTORS. *Educ:* Univ Hong Kong, BSc, 66 & 67; Brown Univ, PhD(physics), 72. *Prof Exp:* Fel, Univ Calif, Berkeley, 71-72, lectr physics, 72-73; res staff mem, Thomas J Watson Res Ctr, IBM Corp, 73-79; assoc prof, 79-81, PROF PHYSICS, UNIV CALIF, BERKELEY, 81- *Mem:* Am Phys Soc; AAAS. *Res:* Optical properties and light scattering of semiconductors; picosecond laser spectroscopy; high pressure physics; defects in semiconductors. *Mailing Add:* Phys Dept Univ Calif Berkeley CA 94720

YU, ROBERT KUAN-JEN, b China, Jan 27, 38; m 72; c 2. BIOCHEMISTRY, NEUROCHEMISTRY. *Educ:* Tunghai Univ, Taiwan, BS, 60; Univ Ill, Urbana, PhD(chem), 67. *Hon Degrees:* ScD, Toyko Univ, 80. *Prof Exp:* Res assoc neurol biochem, Albert Einstein Col Med, 68-69; from instr to assoc prof, Yale Univ Med Sch, 69-82, prof neurol biochem, 83-88; PROF & CHMN, MED COL VA, RICHMOND, VA, 88- *Concurrent Pos:* NIH fel, Albert Einstein Col Med, 67-68; Mary Fulton fel, William Randolph Hearst Found; Josiah Macy Found fel. *Honors & Awards:* Kitasoto Medal; Jacob Javits Neuroscience Investr Award; Alexander von Humboldt Stiftung Sr Investr Award. *Mem:* AAAS; Am Chem Soc; Am Soc Neurochem; NY Acad Sci; Int Soc Neurochem; Soc Neurosci; Sigma Xi; Am Soc Biol Chemists. *Res:* Chemistry and metabolism of sphingolipids; sphingolipidosis; properties of lipids in solution and membrane; complex carbohydrates; membrane synthesis and assembly. *Mailing Add:* Dept Biochem & Molecular Biophys Med Col Va Va Commonwealth Univ Richmond VA 23298

YU, RUEY JIIN, b Hsin-chu, Taiwan, Mar 23, 32; m 59; c 3. SKIN PHARMACOLOGY, ACUPUNCTURE. *Educ:* Nat Taiwan Univ, BSc, 56, MSc, 60; Univ Ottawa, PhD(org chem), 65; Sch Int Asn Acupuncture, dipl, 84; Samra Univ, OMD, 86. *Prof Exp:* Lectr chem, Nat Univ Taiwan, 61-62; Nat Res Coun Can fel, 65-67; asst prof, 67-73, ASSOC PROF DERMAT & CONSULT, SKIN & CANCER HOSP, TEMPLE UNIV, 73- *Concurrent Pos:* Clin prac, traditional Chinese acupuncture, dipl, Nat Comn Cert Acupuncture. *Mem:* Am Asn Acupuncture & Oriental Med; Am Acad Dermat. *Res:* Dermatopharmacology for skin disorders, such as age spots and wrinkles; pain control with Chinese acupuncture. *Mailing Add:* Four Lindenwold Ave Ambler PA 19002

YU, SHARON S M, b Taiwan; US citizen. BIOCHEMISTRY. *Educ:* City Univ New York, PhD(biochem), 75. *Prof Exp:* Res assoc, New York Blood Ctr, Lindsley F Kimball Res Inst, 75-85; ADJ PROF, CALIF STATE UNIV, FRESNO, 86- *Mem:* Fedn Am Soc Exp Biol. *Res:* Biosynthesis of plasma proteins. *Mailing Add:* Dept Chem Calif State Univ Fresno CA 93740

YU, SHIH-AN, b Hupei, China, May 10, 27; m 55. BOTANY. *Educ:* Nat Taiwan Univ, BS, 50; Univ NH, MS, 56, PhD(hort), 59. *Prof Exp:* Res assoc forage breeding, Mich State Univ, 59-64; lectr bot, Univ Mich, Ann Arbor, 65-66, res assoc, 66-67; asst prof, 67-70, assoc prof, 70-75, PROF BIOL, EASTERN MICH UNIV, 75- *Mem:* Sigma Xi. *Res:* Fungal genetics and molecular genetics. *Mailing Add:* Dept Biol Eastern Mich Univ Ypsilanti MI 48197

YU, SHIU YEH, b Formosa, China, June 1, 26; nat US; m 60; c 1. BIOCHEMISTRY, ORGANIC CHEMISTRY. *Educ:* Provincial Col Agr, China, BS, 51; Okla State Univ, MS, 56; St Louis Univ, PhD, 63. *Prof Exp:* Chemist & res assoc, Indust Res Inst, Formosa, 51-52; res asst, Okla State Univ, 52-56; res assoc, Inst Exp Path, Jewish Hosp, St Louis, 56-60; fel, Sch Med, St Louis Univ, 61-62; BIOCHEMIST, VET ADMIN HOSP, JEFFERSON BARRACKS, 63- *Concurrent Pos:* Clin biochem consult, St Louis State Hosp & Sch; res asst prof, Sch Med, Washington Univ, 72-78;

instr, Forest Park Community Col, 73-; clin assoc prof, Sch Med, St Louis Univ, 78-83. *Mem:* Geront Soc; Electron Micros Soc Am; Am Soc Exp Path; Am Heart Asn; Brit Biochem Soc; Sigma Xi. *Res:* Immunology; biochemistry of arteriosclerosis; structure of elastin and chemistry of elastase; mechanism of delayed hypersensitivity; mechanism of antibody formation; lung injury; pulmonary emphysema; adult respiratory distress syndrome. *Mailing Add:* Vet Admin St Louis Univ Sch Med Jefferson Barracks St Louis MO 63125

YU, SIMON SHIN-LUN, b Hong Kong, Mar 24, 45; US citizen; m 70. PHYSICS. *Educ:* Seattle Pac Univ, BSc, 67; Univ Wash, Seattle, MSc & PhD(physics), 70. *Prof Exp:* Res asst physics, Seattle Pac Univ, 64-67; fel, Univ Wash, 70-73; res assoc, Univ Pittsburgh, 73-77; PHYSICIST, LAWRENCE LIVERMORE LAB, 77- *Mem:* Am Phys Soc. *Res:* Theoretical physics. *Mailing Add:* Lawrence Livermore Lab L-626 Livermore CA 94550

YU, SIMON SHYI-JIAN, b Lotung, Taiwan, Sept 11, 35; US citizen; m 67; c 2. BIOCHEMICAL TOXICOLOGY, INSECTICIDE TOXICOLOGY. *Educ:* Nat Taiwan Univ, BS, 59; McGill Univ, MS, 65, PhD(entom), 68. *Prof Exp:* Res entomologist, Taiwan Sugar Co, 61-62; res asst toxicol, McGill Univ, 63-68; fel toxicol, Cornell Univ, 68-69; res assoc toxicol, Ore State Univ, 69-74, asst prof, 74-79; asst prof, 80-82, assoc prof, 82-86, PROF, UNIV FLA, 86- *Concurrent Pos:* Prin investr, Univ Fla, 81- *Mem:* Entom Soc Am; AAAS; Am Chem Soc; Sigma Xi. *Res:* Biochemical toxicology of insects and related species, including detoxication mechanisms, enzyme induction, insecticide metabolism, insecticide resistance, insect-host plant interactions and purification of detoxifying enzymes. *Mailing Add:* Dept Entom & Nematol Univ Fla Gainesville FL 32611

YU, THOMAS HUEI-CHUNG, b Taipei, Taiwan, Mar 29, 57; m; c 2. HALOGEN LAMP CHEMISTRY, VACUUM TECHNOLOGY. *Educ:* Nat Taiwan Univ, BS, 80; Univ Rochester, PhD(chem eng), 86. *Prof Exp:* Develop engr, 86-88, advan engr, 88-91, SR ADVAN ENGR, TECHNOL DIV, GEN ELEC LIGHTING, 91- *Concurrent Pos:* Vprin, Chinese Acad Cleveland, 91- *Mem:* Am Inst Chem Engrs. *Res:* High temperature transport phenomenon; kinetics; gas-solid reactions; high temperature lamp chemistry; thermodynamics; Getter chemistry for lamp application; mass spectroscopy. *Mailing Add:* 486 Snavely Rd Richmond Heights OH 44143

YU, TS'AI-FAN, b Shanghai, China, Oct 24, 11; nat US. MEDICINE, METABOLISM. *Educ:* Ginling Col, China, BA, 32; Peking Union Med Col, China, MD, 36. *Prof Exp:* From intern to chief resident med, Peiping Union Med Col, 35-40; instr med, Col Physicians & Surgeons, Columbia Univ, 50-56, assoc, 56-59, asst prof, 60-66; assoc prof, 66-73, res prof, 73-82, EMER PROF RES, MT SINAI SCH MED, 82- *Honors & Awards:* Woman Med Award, 83. *Mem:* Am Physiol Soc; Am Soc Pharmacol & Exp Therapeut; Harvey Soc; Am Soc Nephrol; Am Rheumatism Asn; Int Soc Nephrol. *Res:* Calcium and phosphorous metabolism in osteomalacia; purine metabolism and gout. *Mailing Add:* Mt Sinai Hosp 11 E 100th St Box 1276 New York NY 10029

YU, WEI WEN, b Shantung, China, July 10, 24; US citizen; m 53; c 3. STRUCTURAL ENGINEERING. *Educ:* Taiwan Norm Technol, BS, 50; Okla State Univ, MS, 55; Cornell Univ, PhD(struct eng), 60. *Prof Exp:* Teaching asst civil eng, Taiwan Norm Technol, 50-54; struct designer, T H McKaig & Assocs, NY, 55-56 & 59-60; res asst struct, Cornell Univ, 57-59; engr, Am Iron & Steel Inst, 60-67; staff engr, TRW, Inc, 67-68; from assoc prof to prof civil eng,72-82, CURATORS PROF CIVIL , UNIV MO-ROLLA, 82- *Concurrent Pos:* Lectr, City Col New York, 64-65. *Mem:* Fel Am Soc Civil Engrs; Am Concrete Inst. *Res:* Cold-formed steel structures; semi-rigid connection of steel framing; deflection of reinforced concrete beams; steel structures. *Mailing Add:* Dept Civil Eng Univ Mo Rolla MO 65401

YU, WEN-SHI, b Shanghai, China, Nov 17, 34; US citizen. CHEMICAL ENGINEERING. *Educ:* Nat Taiwan Univ, BS, 56; Pratt Inst, MS, 58; Polytech Inst Brooklyn, PhD(chem eng), 64. *Prof Exp:* ENGR, BROOKHAVEN NAT LAB, 61- *Mem:* Women Engrs; Sigma Xi. *Res:* Heat-transfer of forced-convection with liquid metals in the nuclear and space fields; heat transfer of single-phase and two-phase (boiling) flow; computer coding in applied science field; research development of hydrogen storage systems; thermal storage study and safety studies in the controlled thermal reactor program. *Mailing Add:* 108 81st Ave Kew Gardens NY 11415

YU, YI-YUAN, b China, Jan 29, 23; nat US; m 52; c 2. ENGINEERING MECHANICS, AEROSPACE ENGINEERING. *Educ:* Univ Tientsin, China, BS, 44; Northwestern Univ, MS, 50, PhD(eng mech), 51. *Prof Exp:* Res assoc, Northwestern Univ, 51; asst prof appl mech, Wash Univ, St Louis, 51-54; assoc prof mech eng, Syracuse Univ, 54-57; prof mech eng, Polytech Inst Brooklyn, 57-66; consult engr, Gen Elec Co, 66-71; distinguished prof aeronaut eng, Wichita State Univ, 72-75; mgr eng, Rocketdyne Div, Rockwell Int, 75-79, exec eng, Energy Systs Group, 79-81; dean, Newark Col Eng, 81-85, PROF MECH ENG, NJ INST TECHNOL, 81- *Concurrent Pos:* Guggenheim fel, 59-60; consult, US Naval Appl Sci Lab, 63-69; lectr, Gen Elec Mod Eng Course, 63-73; adv, Mid East Tech Univ, Turkey, 66; mem ad hoc comt dynamic shock anal, USN, 67-68; chmn, Grad Studies Div, Am Soc Eng Educ, 85-86. *Mem:* Assoc fel Am Inst Aeronaut & Astronaut; fel Am Soc Mech Engrs; Am Soc Eng Educ; NY Acad Sci; Inst Elec & Electronics Engrs Comput Soc. *Res:* Stress and vibration analysis; theory of elasticity; theory of plates and shells; dynamics and control of structural and mechanical systems. *Mailing Add:* NJ Inst Technol 323 High St Newark NJ 07102

YU, YUN-SHENG, b I-Hsing, Kiangsu, China, Nov 21, 26; US citizen; m 60; c 4. FLUID MECHANICS, WATER RESOURCES. *Educ:* Nat Taiwan Univ, BS, 53; Univ Iowa, MS, 56; Mass Inst Technol, ScD(civil eng), 60. *Prof Exp:* Res assoc hydraul, Iowa Inst Hydraul Res, 56-57; res assoc hydrodyn lab, Mass Inst Technol, 59-60; assoc prof, 60-64, PROF FLUID MECH, UNIV KANS, 64- *Concurrent Pos:* Consult, US Army Tank-Automotive Ctr, 66; vis prof, Univ Mich, 67; vis prof, Qinghua Univ, Beijing, China, 80-81; tech adv, Inst Water Resources & Environ Res, Bur Environ Protection, Beijing, China, 87- *Honors & Awards:* Li-Found Fel. *Mem:* Math Asn Am; Int Water Resources Asn; Am Geophys Union; Am Water Resources Asn; Am Soc Civil Engrs. *Res:* Theoretical and applied hydrodynamics; water pollution; water resources systems analysis. *Mailing Add:* Dept Civil Eng Univ Kans 208 Learned Hall Lawrence KS 66045

YUAN, EDWARD LUNG, physical chemistry, for more information see previous edition

YUAN, JIAN-MIN, b Chungking, China, Aug 31, 44; m 71; c 2. CHEMICAL & LASER PHYSICS, NONLINEAR DYNAMICS. *Educ:* Nat Taiwan Univ, BS, 66, MS, 68; Univ Chicago, PhD(chem physics), 73. *Prof Exp:* Fel scattering theory, Quantum Theory Proj, Univ Fla, 73-75; instr & res assoc laser interaction, Dept Chem, Univ Rochester, 75-78; asst prof, 78-84, ASSOC PROF PHYSICS, DEPT PHYSICS & ATMOSPHERIC SCI, DREXEL UNIV, 84- *Mem:* Am Phys Soc; Am Chem Soc; Sigma Xi. *Res:* Interaction of atomic and molecular dynamics with an intense laser field; nonlinear dynamics; molecular scattering theories; semiclassical approach; surface physics. *Mailing Add:* Dept Physics & Atmospheric Sci Drexel Univ Philadelphia PA 19104

YUAN, LUKE CHIA LIU, b Changtehfu, China, Apr 5, 12; US citizen; m 42; c 1. PHYSICS. *Educ:* Yenching Univ, BS, 32, MS, 34; Calif Inst Technol, PhD(physics), 40. *Hon Degrees:* DSc, Nanking Univ People's Repub China, 86. *Prof Exp:* Asst physics, Yenching Univ, 32-34; asst, Calif Inst Technol, 37-40, fel, 40-42; res physicist, RCA Labs, 42-46; res assoc, Princeton Univ, 46-49; sr physicist, Brookhaven Nat Lab, Upton, 49-82, consult, 82-87; CHMN BD DIRS, SYNCHROTRON RADIATION RES CTR, TAIWAN, 83- *Concurrent Pos:* Guggenheim fel, 58; vis prof, Europ Orgn Nuclear Res, 72-76, Ctr d'Etude Nucleáre de Saclay, France, 72-76; bd dir, Adelphi Univ Energy Res Ctr, 79-87; vis prof, Inst High Energy Physics, USSR, 79, Univ Paris, France, 82; hon adv, Sci & Technol Workers' Asn, Chinese Acad Sci, People's Repub China. *Honors & Awards:* Hon prof, Chinese Univ Sci & Tech, Hefei, Anhwei Prov, Nankai Univ, Tientsin, Honan Univ, Southeastern Univ, Nanjing, People's Repub China; Sci Achievement Medal, Repub China, 59; Achievement Award, Chinese Inst Elec Eng, 62. *Mem:* Fel Am Phys Soc; Acad Sinica; NY Acad Sci. *Res:* High energy physics; super energy accelerator and particle detection systems; cosmic rays; radio direction finding; frequency modulation radar systems. *Mailing Add:* 15 Claremont Ave New York NY 10027

YUAN, ROBERT L, b Nanking, China, 1937; US citizen; m 68; c 2. ENGINEERING, STRUCTURAL CONCRETE MATERIALS. *Educ:* Cheng Kung Univ, Taiwan, BS, 60; Univ Ill, Urbana, MS, 64, PhD(theoret & appl mech), 68. *Prof Exp:* Res assoc concrete res, Univ Ill, 67-68; from asst prof to assoc prof, 68-81, PROF CIVIL ENG, UNIV TEX, ARLINGTON, 81- *Mem:* Am Concrete Inst; Soc Exp Stress Anal; Am Soc Civil Eng. *Res:* Concrete and new building material research. *Mailing Add:* Dept Civil Eng Univ Tex PO Box 19308 Uta Sta Arlington TX 76019

YUAN, SHAO-YUEN, b Shanghai, China, July 30, 29; US citizen; m 49; c 2. CHEMICAL ENGINEERING. *Educ:* Ill Inst Technol, BS, 50; Univ Louisville, MS, 51. *Prof Exp:* Res engr, E I du Pont de Nemours, 51-56; from res engr to sr res engr, Chevron Res Co, 56-66, supvr prod develop & tech serv, Chevron Chem Co, 66-69, sr proj engr, 69-75, staff engr, 75-78, sr staff engr, 78-81, REGIONAL EXEC, CHEVRON RES CO, STANDARD OIL CALIF, 81- *Mem:* Am Inst Chem Engrs; Am Chem Soc. *Mailing Add:* 70 Heritage Rd San Raphael CA 94901-8308

YUAN, SIDNEY WEI KWUN, b Hong Kong, China, July 30, 57; US citizen; m 81; c 2. CRYOGENICS, THERMODYNAMICS. *Educ:* Univ Calif, Los Angeles, BS, 80, MS, 81, PhD(eng), 85. *Prof Exp:* Res engr cryog, Univ Calif, Los Angeles, 80-85, teaching asst eng, 84; RES SCIENTIST, LOCKHEED MISSILES & SPACE CO, 85- *Concurrent Pos:* Lectr, Univ Calif, Los Angeles, 89; consult, Toyo Sanso Co, Japan, 90- *Honors & Awards:* Nat Excellence Recognition Award, Space Found, 85. *Mem:* Am Inst Aeronaut & Astronaut; Am Inst Chem Engrs; Sigma Xi. *Res:* Heat and mass transport in superfluid; thermal modelling; solid and liquid cryogen dewars; mechanical cryocoolers; computer methodology; author of more than 30 publications in professional journals. *Mailing Add:* 4964 Adagio Ct Fremont CA 94538

YUAN, TZU-LIANG, b Ningpo, China, May 27, 22; nat US; m 60; c 2. SOILS. *Educ:* Nat Univ Chekiang, China, BSc, 45; Ohio State Univ, MSc, 52, PhD(agron), 55. *Prof Exp:* Asst instr soils, Nat Univ Chekiang, China, 45; instr, Nat Univ, Kweichow, 46-48; asst soils chemist, Taiwan Sugar Exp Sta, 48-51; res asst, Ohio Agr Exp Sta, 52-55; from asst prof to prof soil chem, 55-72, PROF SOIL SCI, UNIV FLA, 72- *Mem:* Fel AAAS; Soil Sci Soc Am; Clay Minerals Soc; NY Acad Sci; Int Soil Sci Soc. *Res:* Soil-forming processes; chemical nature of soils; soil properties in relation to plant nutrition. *Mailing Add:* Dept Soil Sci Univ Fla Gainesville FL 32611

YUCEOGLU, YUSUF ZIYA, b Cesme, Turkey, Mar 28, 19; m 47. INTERNAL MEDICINE, CARDIOLOGY. *Educ:* Istanbul Univ, MD, 44. *Prof Exp:* Intern, Istanbul Univ Hosp, 43-44; resident internal med, Ankara Univ Hosp, 47-50, spec asst, 50-52; fel cardiol, Mt Sinai Hosp, New York, 53-55; res fel, Cardiopulmonary Lab, Maimonides Hosp, 55-57; NY Heart Asn fel, 55-57; from res assoc to assoc, Maimonides Hosp, 57-60, from asst attend physician to assoc attend physician, 61-67, dir cardiopulmonary lab, 66-67; ASSOC PROF MED, NY MED COL, 67- *Concurrent Pos:* From instr to asst prof med, State Univ NY Downstate Med Ctr, 55-67; asst attend physician, Kings County Hosp, 66-67; assoc attend physician, Flower & Fifth Ave Hosps, 67- & Metrop Hosps, 67- *Honors & Awards:* Cert Honor, Am Col Angiol & Int Col Angiol, 65. *Mem:* AMA; fel Am Col Angiol. *Res:* Cardiac physiology; vectorcardiography. *Mailing Add:* 30 Essex Dr Northport NY 11768

YUDELSON, JOSEPH SAMUEL, b Philadelphia, Pa, July 20, 25; m 52; c 4. POLYMER CHEMISTRY, PHOTOCHEMISTRY. *Educ:* Univ Pittsburgh, BS, 50; Ill Inst Technol, PhD(chem), 55. *Prof Exp:* Res chemist, Eastman Kodak Res Lab, 54-57, sr res chemist, 57-60, res assoc phys chem, 60-70, sr lab head, 70-80, sr res assoc, Magnetic Media Lab, 80-90; CONTRACTOR, STERLING RES GROUP, 90- *Mem:* Assoc Am Chem Soc; Soc Photog Scientists & Engrs; Sigma Xi. *Res:* Physical chemistry of hydrophilic polymers; photographic chemistry; nonaqueous solvents; inorganic photochemistry; magnetic media; developing ultra fine magnetic particles for chemical separations; investigating new magnetic media for high density information storage; contrast agents for ultrasound and magnetic resonance imaging. *Mailing Add:* 77 Calumet St Rochester NY 14610

YUDKOWSKY, ELIAS B, b Brooklyn, NY, June 3, 32. EDUCATIONAL ADMINISTRATION, DENTISTRY. *Educ:* Northwestern Univ Dent Sch, DDS, 57; Univ Calif, San Francisco, PhD(endocrinol), 68. *Prof Exp:* Dent pract, Chicago, Ill, 58-61; instr oral diag, Northwestern Univ Dent Sch, 59-61; asst secy coun dent res, Am Dent Asn, 67-69; asst dean, Harvard Sch Dent Med, 69-70; asst prof anat, Med Univ SC Col Med, 70-78, assoc prof oral path, 70-78, asst chmn, 74-78; prof & chmn, dept dent hyg, East Tenn State Univ, 78-82; ASSOC PROF ANAT, SCH MED & PROF & DIR, DIV DENT PROGS, UNIV NMEX, 82- *Concurrent Pos:* Consult, Am Dent Asn, 70-75, task force cardiovasc & hypertensive dis, SC Dent Asn, 75-78, Tenn Higher Educ Comt, 79-82, regional dent educ planning, 82-83, col health related profs, Idaho State Univ, 84-85, Prof Sem Consults, Inc, 85- & Col Health, Univ Cent Fla, 86-; assoc ed, Head & Neck Cancer Abstr, 83-84; dir dent progs & mem bd dirs, Prof Seminar Consults, Inc, 82- *Mem:* Am Dent Asn; Am Asn Dent Sch; fel AAAS; Am Asn Univ Prof. *Res:* Nucleic acid levels in rat uterus; normal-abnormal head and neck development; forensic odontology. *Mailing Add:* Dept Allied Health Univ NMex Main Campus Albuquerque NM 87131

YUE, A(LFRED) S(HUI-CHOH), b Canton, China, Nov 12, 20; m 44; c 3. ELECTRONICS ENGINEERING. *Educ:* Chao-Tung Univ, China, BS, 42; Ill Inst Technol, MS, 50; Purdue Univ, PhD(metall eng), 57. *Prof Exp:* Instr metall eng, Purdue Univ, 52-56; res metallurgist, Dow Chem Co, 56-62; res scientist, Lockheed Palo Alto Res Lab, 62-63, staff scientist, 63-66, sr mem, 66-69; PROF MAT SCI, UNIV CALIF, LOS ANGELES, 69- *Concurrent Pos:* Hon prof, Xian Jiao-tong Univ, China; chair prof, Tsing Hwa Univ, Taiwan, ROC. *Honors & Awards:* Apollo-Soyuz Award, NASA; Skylab Award, NASA. *Mem:* Am Inst Mining, Metall & Petrol Eng; AAAS; Am Soc Metals; Mat Res Soc. *Res:* High Tc superconductor films, multilayered structures by MBE and MOCVD techniques for devices applications; preparation of p-n homo- and heterojunctions via the liquid-phase-epitaxial, chemical vapor deposition, close-space-vapor-transport and the directional solidification techniques. *Mailing Add:* 6532 Boelter Hall Univ Calif Los Angeles CA 90024

YUE, ON-CHING, b Macao, Apr 9, 47; US citizen; m 73; c 2. ELECTRICAL ENGINEERING. *Educ:* Cooper Union, BEE, 68; Rochester Inst Technol, MSEE, 71; Univ Calif, San Diego, PhD(info sci), 77. *Prof Exp:* Sr engr, Electronics Div, Gen Dynamics Corp, 68-77; SUPVR, BELL LABS, 77- *Mem:* Inst Elec & Electronics Engrs. *Res:* Communication theory; computer performance analysis. *Mailing Add:* AT&T Bell Lab Rm HO3M-336 Holmdel NJ 07733

YUE, ROBERT HON-SANG, b Canton, China, Sept 9, 37; US citizen; m 70; c 3. BIOCHEMISTRY. *Educ:* ETex Baptist Col, BS, 61; Univ Utah, PhD(biol chem), 68. *Prof Exp:* Res assoc biochem, Lab Study Hereditary & Metab Dis, Univ Utah, 68-70; assoc res scientist, Inst Rehab Med, NY Univ Med Ctr, 70-72, res scientist, 72-73, res asst prof, 73-76, res assoc prof rehab med, 76-81; res biochemist, Revlon Health Care Group, 81-85; dir, 85-90, SR DIR PROD DEVELOP, ENZON, INC, 90- *Mem:* Am Chem Soc; AAAS; NY Acad Sci; Am Heart Asn; Parenteral Drug Asn. *Res:* Physical chemistry of the isolated proteins; mechanism of enzyme action; blood coagulation. *Mailing Add:* 259-15 86th Ave Floral Park New York NY 11001

YUEN, DAVID ALEXANDER, b Hong Kong, June 14, 48; US citizen. GEOPHYSICS, FLUID DYNAMICS & LARGE SCALE COMPUTATIONS. *Educ:* Calif Inst Technol, BS, 69; Scripps Inst Oceanog, Univ Calif, San Diego, MS, 73; Univ Calif, Los Angeles, PhD(geophys), 78. *Prof Exp:* Res geophysicist, Univ Calif, Los Angeles, 74-77; NSF fel geophys, Univ Toronto, 78-79; from asst prof to assoc prof geophys, Ariz State Univ, 80-85; assoc prof, Univ Colo, 85; assoc prof, 85-89, PROF GEOPHYS, UNIV MINN, 89- *Concurrent Pos:* NATO fel, Univ Toronto, 79-80; assoc ed, J Geophys Res, 83-86. *Mem:* Am Geophys Union; Am Chem Soc; Am Phys Soc; Soc Indust & Appl Math. *Res:* Earth's mantle; different time scales of instability mechanisms; rheology of the mantle; seismic attenuation processes in elastic-gravitational free oscillations; rotational dynamics; ice ages; magna chamber processes; mantle convection. *Mailing Add:* 1200 Washington Ave S Minneapolis MN 55415

YUEN, MAN-CHUEN, b Hong Kong, Aug 5, 33; nat US; m 58; c 3. FLUID DYNAMICS, HEAT TRANSFER. *Educ:* Purdue Univ, BS, 56; Mass Inst Technol, MS, 58; Harvard Univ Univ, PhD(eng), 65. *Prof Exp:* From asst prof to assoc prof mech eng & astronaut sci, 64-75, PROF MECH ENG, NORTHWESTERN UNIV, 75- *Concurrent Pos:* Chmn, Mech Eng & Astronaut Sci Dept, Northwestern Univ, 78-79. *Mem:* Am Soc Mech Engrs. *Res:* Multiphase and multicomponent fluid mechanics and heat transfer. *Mailing Add:* McCormick Sch Eng Appl Sci Tech Inst 2145 Sheridan Rd Evanston IL 60208

YUEN, PAUL C(HAN), b Hilo, Hawaii, June 7, 28; m 52; c 2. ELECTRICAL ENGINEERING. *Educ:* Univ Chicago, BS, 52; Ill Inst Technol, MS, 55, PhD(elec eng), 60. *Prof Exp:* Cyclotron technician, Univ Chicago, 50-52; engr, Standard Coil Prod, 53-54; assoc res engr, Armour Res Found, 54-57; res engr, Ill Inst Technol, 57-60, asst prof elec eng, 60-61; assoc prof elec eng, 61-65, actg dean, Col Eng, 69 & 77-78, assoc dean, 70-79, asst to chancellor, 71-72, actg dir, Ctr Eng Res, 76-77, dir, Hawaii Natural Energy Inst, 77-81, PROF ELEC ENG, UNIV HAWAII, MANOA, 65-, DEAN, COL ENG, 81-; ACTG PRES, PAC INT CTR HIGH TECHNOL RES, 83- *Concurrent Pos:* Mem, Geothermal Resources Coun. *Honors & Awards:* Centennial Medal, Inst Elec & Electronics Engrs, 84. *Mem:* Am Geophys Union; Am Soc Eng Educ; Inst Elec & Electronics Engrs; Int Union Radio Sci; Sigma Xi. *Res:* Radio wave propagation; satellite communications; ionospheric physics; renewable energy resources. *Mailing Add:* Col Eng Univ Hawaii Honolulu HI 96822

YUEN, TED GIM HING, b Canora, Sask, Dec 21, 33; m 58; c 4. EXPERIMENTAL PATHOLOGY. *Educ:* Andrews Univ, BA, 56; Univ Southern Calif, PhD(exp path), 69. *Prof Exp:* Fel exp path, Univ Southern Calif, 69-71; RES ASSOC NEUROPATH, HUNTINGTON MED RES INST, 72- *Concurrent Pos:* Res assoc, Dept Path, Univ Southern Calif, 75-76. *Mem:* Am Asn Pathologists; Soc Biomat. *Res:* Ultrastructural study of the effects of electrical stimulation of the brain and peripheral nerves. *Mailing Add:* Dept Neurol Res Huntington Med Res Inst 734 Fairmount Ave Pasadena CA 91105

YUEN, WING, b China, Sept 29, 14; US citizen; c 4. BREAKTHROUGH DEHYDRATION. *Educ:* Lingnan Univ, China, BS, 41; Free Protestant Univ, London, Eng, PhD(food sci & technol), 68; Sussex Col Technol, Eng, DSc(food sci), 69. *Prof Exp:* Lectr agr chem, Lignan Univ, China, 41-42; sr technologist, China Food Industs Corp, 42-45; tech specialist, Agr Mgt & Opers Off, UN, China, 47-49; chemist, Sunkist Growers, Inc, 57-58; res supvr, Ventura Coastal Corp, 58-80; CONSULT, YUEN & ASSOCS, 80- & CHINA FOOD INDUSTS, 84- *Concurrent Pos:* Consult, Inst Food Technologists & Volunteers Int Tech Assistance, 63-; int tech expert, UN Indust Develop Orgn, 68-; vis prof, S China Agr Univ, S China Inst Technol, Che-Jiang Agr Univ, 84, Xinjiang Agr Sci Acad, 86. *Honors & Awards:* IFT Citation for Most Excellent Proposal to Food for Peace Progs in India; VITA Outstanding Award for Supporting Int Develop; Notable Am Award in the Bicentennial Era, Am Biog Inst. *Mem:* Fel AAAS; fel Royal Soc Health; emer mem Inst Food Technologists; emer mem Am Chem Soc. *Res:* Dehydration of natural fruit juices into non-hygroscopic powders with no additive or anti-caking agent, dewaxing, decontaminating, decolorizing and deodorizing essential and edible oils; methods of producing the purest forms of bioflavonoids and of instantly separating emulsions. *Mailing Add:* 3173 Preble Ave Ventura CA 93003

YUH, HUOY-JEN, b Taiwan, China; US citizen; m; c 2. SOLID STATE PHYSICS, PHYSICAL CHEMISTRY. *Educ:* Nat Tsing-Hwa Univ, Taiwan, China, BS, 75; Univ Chicago, PhD(chem), 81. *Prof Exp:* Fel chem, Johns Hopkins Univ, 81-83; MEM RES STAFF, PHOTOCONDUCTOR RES, XEROX CORP, 83- *Res:* Photoconductor material research; foundamental study of charge-transport material properties. *Mailing Add:* 11 Ithaca Dr Pittsford NY 14534

YUHAS, JOSEPH GEORGE, b Cleveland, Ohio, Aug 26, 38; m 60; c 4. ANIMAL BEHAVIOR, ZOOLOGY. *Educ:* Ohio State Univ, BSc Agr, 60, BScEd & MSc, 62, PhD(zool), 70. *Prof Exp:* Res assoc wildlife, Ohio State Univ, 68-69; asst prof biol, Defiance Col, 69-75, chmn dept natural systs, 72-75; assoc prof life sci & dir, Ctr Life Sci, St Francis Col, 75-88; TEACHER SCI, KENNEBUCK HIGH SCH, 88- *Mem:* Animal Behav Soc; Am Inst Biol Sci. *Res:* Effects of environmental contamination on behavior; innate behavior and evolution; endocrine control of behavior; induced ovulation and delayed implantation. *Mailing Add:* Kennebuck High Sch PO Box 222 Bar Mills ME 04004

YUILL, THOMAS MACKAY, b Berkeley, Calif, June 14, 37; m 60; c 2. VIROLOGY, ECOLOGY. *Educ:* Utah State Univ, BS, 59; Univ Wis, MS, 62, PhD(vet sci), 64. *Prof Exp:* Lab officer virol, Walter Reed Army Inst Res, 64-66, med biologist, SEATO Med Res Lab, 66-68; from asst prof to assoc prof, 68-76, chmn, Dept Vet Sci, 78-82, PROF VET SCI & WILDLIFE ECOL, UNIV WIS-MADISON, 76-, PROF PATHOBIOL, ASST DIR AGR EXP STA & ASSOC DEAN RES & GRAD TRAINING, 82- *Concurrent Pos:* Consult, NIH & Environ Protection Agency, 77-; pres, Orgn Trop Studies, 80-85; hon prof vet microbiol, Univ Antioquia Fac Vet Med, Colombia, SAm; consult, US Agency Int Dev, 87- *Honors & Awards:* Distinguished Serv Award, Wildlife Dis Asn. *Mem:* Am Soc Microbiol; Wildlife Soc; Wildlife Dis Asn (treas, 80-85, pres, 85-87); Am Soc Trop Med & Hyg; Royal Soc Trop Med & Hyg; Coun Agr Sci Technol. *Res:* Arbovirus epizootiology; wildlife diseases, especially arthropod-borne; virus ecology; effects of non-lethal infections and intoxications on wildlife. *Mailing Add:* Sch Vet Med Univ Wis Madison WI 53706

YUKAWA, S(UMIO), b Seattle, Wash, Apr 25, 25; m 51; c 1. MECHANICAL METALLURGY. *Educ:* Univ Mich, BSE(chem eng) & BSE(metall eng), 51, MSE, 52, PhD(metall eng), 55. *Prof Exp:* Res assoc metall eng, Univ Mich, 53-54, instr, 54; metallurgist, 54-86, CONSULT, GEN ELEC CO, 86- *Mem:* Fel Am Soc Metals; Am Inst Mining, Metall & Petrol Engrs; fel Am Soc Mech Engrs; Am Soc Testing & Mat; Sigma Xi. *Res:* Mechanical properties and engineering design criteria of metallic materials especially on deformation, fracture and fatigue behavior in power generation applications including steam and gas turbines, electrical generators and nuclear power equipment. *Mailing Add:* 4925 Valkyrie Dr Boulder CO 80301-4360

YUKICH, JOSEPH E, b Cleveland, Ohio. MATHEMATICAL STATISTICS. *Educ:* Oberlin Col, BA, 78; Mass Inst Technol, PhD(math), 83. *Prof Exp:* Vis prof, Univ Strasburg, France, 83-85; ASSOC PROF MATH, LEHIGH UNIV, 85- *Honors & Awards:* Fulbright Jr Lectr Award, France, 83 & 84, Sr Lectr Award, 90. *Mem:* Inst Math Statist. *Res:* Probability theory and mathematical statistics, especially empirical processes, limit theorems, and rates of convergence. *Mailing Add:* Dept Math Lehigh Univ Bethlehem PA 18015

YUKON, STANFORD P, b Kansas City, Mo, Mar 20, 39; m 73. PHYSICS. *Educ:* Mass Inst Technol, BS, 61; Brandeis Univ, PhD(physics), 68. *Prof Exp:* RES PHYSICIST, ROME LAB, 82- *Mem:* Am Phys Soc; AAAS. *Res:* Superconducting electronics. *Mailing Add:* 34 Cliff Rd Wellesley MA 02181

YULE, HERBERT PHILLIP, b Chicago, Ill, Apr 17, 31; m 61; c 2. INFORMATION SCIENCE & SYSTEMS, OTHER COMPUTER SCIENCES. *Educ:* Univ Chicago, PhD(nuclear chem), 60. *Prof Exp:* Res chemist, Calif Res Corp, 57-61; staff mem, Gen Atomic Div, Gen Dynamics Corp, Calif, 62-66; assoc prof activation anal, Tex A&M Univ, 66-69, assoc prof comput sci, 67-69; res chemist, Nat Bur Standards, 69-72; staff consult, NUS Corp, 72-74, mgr comput serv, 74-78, dir info processing, 78-85; mgr spec projs, Chem Div, 85-88, SOFTWARE ENGR, ARGONNE NAT LAB, 89- *Concurrent Pos:* Adj assoc prof biochem, Baylor Col Med, 67-69; lectr, Nat Bur Standards, 68 & NATO, 70; consult, Nat Bur Standards, 74- *Mem:* Sigma Xi. *Res:* Activation analysis, especially computer techniques; numerical analysis of data; computer and system programming. *Mailing Add:* 9 S 450 Parkview Dr Downers Grove IL 60516-4734

YUM, SU IL, b Seoul, Korea, June 25, 39; US citizen; m 68; c 2. CHEMICAL ENGINEERING, BIOMEDICAL ENGINEERING. *Educ:* Yonsei Univ, BS, 62; Univ Minn, MS, 67, PhD(chem eng), 70. *Prof Exp:* Res assoc biomat, Univ Utah, 69-71; develop engr, 71-72, co-proj leader, 72-74, proj leader, 74-75, area dir eng, 75-78, OTS prog dir, 78-81, TTS PROD RES DIR, ALZA RES, ALZA CORP, 81- *Mem:* Am Chem Soc; Am Inst Chem Engrs. *Res:* Hydrodynamics of two-phase flow; diffusion in liquids and polymeric membranes; design of medical devices; stress analysis in plastic products; transdermal mass transfer. *Mailing Add:* 950 Pagemill Rd Los Altos CA 94303

YUN, KWANG-SIK, b Seoul, Korea, July 27, 29; m 60; c 1. PHYSICAL CHEMISTRY. *Educ:* Seoul Nat Univ, BS, 52; Univ Cincinnati, PhD(chem), 61. *Prof Exp:* Res assoc, Inst Molecular Physics, Univ Md, 60-63; Nat Res Coun Can fel, 63-65; scientist, Nat Ctr Atmospheric Res, Colo, 65-67; ASSOC PROF CHEM, UNIV MISS, 67- *Mem:* Am Chem Soc; Sigma Xi. *Res:* Kinetic theory of gases and liquids; gas phase chemical kinetics and chemical reactions of atmospheric gases. *Mailing Add:* Dept Chem Univ Miss University MS 38677

YUN, SEUNG SOO, b Korea, Mar 1, 31; US citizen; m 57; c 3. ACOUSTICS, PHYSICAL ACOUSTICS. *Educ:* Clark Univ, AB, 57; Brown Univ, MSc, 61, PhD(physics), 64. *Prof Exp:* Asst physicist, Ore Regional Primate Res Ctr, 63-65; res physicist, IIT Res Inst, 65-67; from asst prof to assoc prof, 67-86, PROF PHYSICS, OHIO UNIV, 87- *Concurrent Pos:* Vis scientist, Air Force Mats Lab, ADD, Seoul, Korea, 73-74, 77-80; vis prof, Chubu Univ, Kasugai, Japan, 87. *Mem:* Acoust Soc Am; Mats Res Soc; Am Phys Soc; AAAS. *Res:* Physical acoustics and ultrasonics; absorption and dispersion of ultrasound in liquid and solid; critical phenomena; internal frictions, structure and mechanical properties of solids. *Mailing Add:* Dept Physics Ohio Univ Athens OH 45701

YUN, SUK KOO, b Seoul, Korea, Nov 10, 30; US citizen; m 57; c 3. THEORETICAL HIGH ENERGY PHYSICS. *Educ:* Seoul Nat Univ, BS, 55; Univ Chicago, MS, 57; Boston Univ, PhD(physics), 67. *Prof Exp:* Instr, Clarkson Univ, 59-63; res assoc, Boston Univ, 66-67 & Syracuse Univ, 67-69; chmn natural sci div, 72-74, from asst prof to assoc prof, 69-78, PROF PHYSICS, SAGINAW VALLEY STATE UNIV, 78-, CHMN DEPT, 77- *Concurrent Pos:* NSF res grant, 63, 67, 68 & 85-87; res grant, Res Corp New York, 70-71; vis prof, Randall Lab, Univ Mich, 74 & 75, Syracuse Univ, 77 & Fermi Nat Accelerator Lab, 78, 79, 80, 81 & 82; hon res assoc, Harvard Univ, 75 & 76; NSF travel grant to Japan, 78 & East Ger, 84; vis prof, Mass Inst Technol, 80, 84-89. *Honors & Awards:* Warrick Award Excellence Res, 87. *Mem:* Sigma Xi; Am Phys Soc. *Res:* Theoretical high energy (particle) physics. *Mailing Add:* Dept Physics Saginaw Valley State Univ University Center MI 48710

YUN, YOUNG MOK, b Chung Song Co, Korea, Sept 22, 31; m 66; c 2. ENTOMOLOGY, PHYTOPATHOLOGY. *Educ:* Wash State Univ, BS, 61; Ore State Univ, BS, 62; Mich State Univ, MS, 64, PhD(entom), 67. *Prof Exp:* Entomologist, Agr Res Ctr, Great Western Sugar Co, 67-85; sr entomologist, Mono-HY Sugar Beet Seed, Inc, 85-87; sr entomologist, 87-89, QUAL CONTROL MGR, HILLESHOG MONO-HY, INC, 89- *Mem:* Entom Soc Am; Am Soc Sugar Beet Technologists; Am Phytopath Soc. *Res:* Biology, ecology, and control of insects, nematodes, and diseases affecting sugar beets. *Mailing Add:* Hilleshog Mono-Hy Inc 11939 Sugarmill Rd Longmont CO 80501

YUND, E WILLIAM, b Pittsburgh, Pa, Feb 15, 14; m 66. EXPERIMENTAL PSYCHOLOGY & PSYCHOACOUSTICS, VISION. *Educ:* Knox Col, BA, 65; Harvard Univ, MA, 67; Northeastern Univ, PhD(psychol), 70. *Prof Exp:* RES PSYCHOLOGIST, DEPT VET AFFAIRS MED CTR, MARTINEZ, CALIF, 73- *Concurrent Pos:* Res assoc psychol, Univ Calif, Berkeley, 72-80; adj assoc prof neurol, Sch Med, Univ Calif, Davis, 78-82. *Mem:* Acoust Soc Am; Asn Res Vision & Ophthal; Optical Soc Am; AAAS. *Res:* Binaural phenomena; pitch perception; auditory neurophysiology hearing aids; color vision; spatial effects; visual search. *Mailing Add:* 723 Woodhaven Rd Berkeley CA 94708

YUND, MARY ALICE, b Xenia, Ohio, Feb 12, 43; m 66. DEVELOPMENTAL BIOLOGY. *Educ:* Knox Col, BA, 65; Harvard Univ, MA, 67, PhD(biol), 70. *Prof Exp:* NIH fel, Univ Calif, Berkeley, 70-72, trainee, 72-73, res geneticist, 73; asst prof biol, Wayne State Univ, 74-75; asst res geneticist, Univ Calif, Berkeley, 75-88; CONSULT BIOTECHNOL, GRANT PREP, 88- *Concurrent Pos:* Res grants, NSF, 75 & NIH, 77; Adv Panel Develop Biol, NSF, 84-88; lectr, genetics, 86- *Mem:* Sigma Xi; Soc Develop Biol; Genetics Soc Am; Am Soc Zoologists; AAAS; Am Soc Biotechnology. *Res:* Hormonal control of gene activity in differentiation of imaginal discs of Drosophila melanogaster; mechanism of steroid hormone action. *Mailing Add:* 723 Woodhaven Rd Berkeley CA 94708

YUND, RICHARD ALLEN, b Ill, Dec 14, 33; m 57; c 2. MINERALOGY, GEOCHEMISTRY. *Educ:* Univ Ill, PhD(geol), 60. *Prof Exp:* Fel, Geophys Lab, Carnegie Inst, 60-61; from asst prof to assoc prof, 61-68, PROF GEOL, BROWN UNIV, 68- *Concurrent Pos:* Vis prof, Monash Univ, Melbourne, Australia, 73-; Fulbright sr res award, 73-74. *Honors & Awards:* Sr Scientist Award, WGer Gov, 78; Award of Volcanology, Geochem & Petrol Sect, Am Geophys Union, 81. *Mem:* Mineral Soc Am; Geochem Soc; Am Geophys Union. *Res:* Experimental study of diffusion in minerals; kinetics and mechanisms of mineralogical reactions and transformations; ductile deformation of minerals; interaction of these processes. *Mailing Add:* Dept Geol Sci Brown Univ Providence RI 02912

YUNE, HEUN YUNG, b Seoul, Korea, Feb 1, 29; US citizen; m 56; c 3. RADIOLOGY, SURGERY. *Educ:* Severance Union Med Col, MD, 56; Am Bd Radiol, dipl, 64; Korean Bd Radiol, dipl, 65. *Prof Exp:* Resident gen surg, Presby Med Ctr, Korea, 56-60; resident radiol, Vanderbilt Univ Hosp, 60-63, instr, Univ, 62-64; chief radiologist, Presby Med Ctr, Korea, 64-66; from asst prof to assoc prof radiol, Vanderbilt Univ, 66-71; PROF RADIOL, IND UNIV, INDIANAPOLIS, 71- *Concurrent Pos:* Staff radiologist, Indianapolis Vet Admin Hosp, 71 & Wishard Mem Hosp, Indianapolis, 75- *Honors & Awards:* Silver Medal, Am Roentgen Ray Soc, 71, Bronze Medal, 75. *Mem:* Radiol Soc NAm; Am Roentgen Ray Soc; Asn Univ Radiol; Am Soc Head & Neck Radiol; fel Am Col Radiol. *Res:* Vascular radiology; tumor angiography, angiography in trauma, angiography in endocrine disorder and lymphangiography; eye, ear, nose and throat radiology; head and neck tomography and contrast radiography in head and neck. *Mailing Add:* Dept Radiol Ind Univ Med Ctr Indianapolis IN 46202-5253

YUNG, W K ALFRED, b Hong Kong, Apr 8, 48; c 3. NEUROLOGY, NEURO-ONCOLOGY. *Educ:* Univ Minn, BSc, 71; Univ Chicago, MD, 75. *Prof Exp:* Asst prof neuro-oncol, 81-86, asst prof tumor biol, 85-86, ASSOC PROF NEURO-ONCOL & TUMOR BIOL, UNIV TEX M D ANDERSON CANCER CTR & NEUROL, UNIV TEX MED SCH, 86-, FAC MEM BIOMED, GRAD SCH BIOMED SCI, 83- *Concurrent Pos:* Asst neurologist, Univ Tex M D Anderson Cancer Ctr, 81-83, assoc neurologist, 83-, dept chmn, 84- *Honors & Awards:* Franklin McLean Res Award, Univ Chicago, 75. *Mem:* NY Acad Sci; Am Acad Neurol; Am Soc Cell Biol; AAAS; Am Asn Cancer Res; Soc Neurosci. *Res:* Chemotherapy and biologic therapy for primary brain tumor, growth regulation and autocrine growth factor expression in primary brain tumors. *Mailing Add:* Univ Tex M D Anderson Cancer Ctr 1515 Holcombe Blvd Box 118 Houston TX 77030

YUNGBLUTH, THOMAS ALAN, b Warren, Ill, Dec 12, 34. GENETICS, PLANT BREEDING. *Educ:* Univ Ill, BS, 56; Univ Minn, PhD(genetics), 66. *Prof Exp:* Asst prof, 66-72, assoc prof, 72-80, PROF BIOL, WESTERN KY UNIV, 80- *Mem:* Am Soc Agron; Crop Sci Soc Am; AAAS. *Mailing Add:* Dept Biol Western Ky Univ Bowling Green KY 42101

YUNGEN, JOHN A, b Independence, Ore, Dec 30, 21; m 53; c 3. SOILS & SOILS SCIENCE, HORTICULTURE. *Educ:* Ore State Univ, BS, 50, MS, 59. *Prof Exp:* Res asst soil fertil, Ore Agr Exp Sta & USDA, 50-53 & crops-soils, S Ore Exp Sta, Ore State Univ, 54-59; from asst prof to assoc prof, 59-83, prof, 83-88, EMER PROF AGRON, S ORE EXP STA, ORE STATE UNIV, 88- *Concurrent Pos:* Supt, S Ore Exp Sta, Ore State Univ, 81-88. *Mem:* Am Soc Agron. *Res:* Agronomic crops and vegetable crops; crops production; weed control; soil fertility; seed production. *Mailing Add:* 1875 Niedermeyer Dr Central Point OR 97502

YUNGER, LIBBY MARIE, b East Cleveland, Ohio, Feb 20, 44; m 79. PHARMACOLOGY, NEUROCHEMISTRY. *Educ:* Earlham Col, BA, 66; Univ Iowa, MA, 71, PhD(neurosci) 74. *Prof Exp:* Res assoc, Univ Iowa, 73-74; Nat Inst Neurol Dis & Stroke fel, Univ Pittsburgh, 74-75; res biologist, Lederle Labs, Am Cyanamid Co, 75-78; assoc sr investr, Smith Kline & French Labs, 78-83; MGR, BIOANALYSIS RES, PITMAN MOORE INC, 83- *Concurrent Pos:* Adj assoc prof, Dept Animal Sci, Univ Ill, 87-89. *Mem:* Soc Neurosci; Sigma Xi; Am Chem Soc. *Res:* Use of antibodies to characterize proteins; binding and transport of hormones and neurotransmitters; interactions between neurohormones and the immune system. *Mailing Add:* 4100 Laclede Ave No 314 St Louis MO 63108

YUNGHANS, WAYNE N, b Lakewood, Ohio, Dec 10, 45; m 69; c 2. CYTOLOGY. *Educ:* Heidelberg Col, BS, 67; Purdue Univ, MS, 69, PhD(cytol), 74. *Prof Exp:* Res asst microbiol, US Army, 69-71; asst prof, 74-80, ASSOC PROF CYTOL, STATE UNIV NY COL FREDONIA, 80- *Honors & Awards:* State Univ NY Res Found Award, 75; NSF Award, 78. *Mem:* AAAS; Sigma Xi. *Res:* Isolation and purification of cellular membranes including plasma membranes, Golgi apparatus and endoplasmic reticulum; membranes characterized for enzyme activities and protein kinases and phosphoproteins. *Mailing Add:* Dept Biol State Univ NY Fredonia NY 14063

YUNICE, ANDY ANIECE, biochemistry, environmental science, for more information see previous edition

YUNICK, ROBERT P, b Schenectady, NY, Oct 27, 35; m 59; c 3. ORGANIC POLYMER CHEMISTRY. *Educ:* Union Col, NY, BS, 57; Rensselaer Polytech Inst, PhD(org chem), 61. *Prof Exp:* Res chemist, Olefins Div, Union Carbide Chem Corp, 61-63; mgr, 63-72, dir res, 72-76, vpres res, 76-80, VPRES CORP TECH, W HOWARD WRIGHT RES CTR, SCHENECTADY CHEM, INC, 80- *Mem:* Am Chem Soc; Am Ornithologists Union; Sigma Xi. *Res:* Organic synthesis of intermediates for resin synthesis; synthesis of phenolic, hydrocarbon, resorcinolic, polyester and polyesterimide resins; synthesis of high-temperature polymers. *Mailing Add:* Schenectady Chem Inc 2750 Balltown Rd Schenectady NY 12309

YUNIS, ADEL A, b Rahbeh, Lebanon, Mar 17, 30; m 59; c 3. INTERNAL MEDICINE. *Educ:* Am Univ Beirut, BA, 50, MD, 54. *Prof Exp:* Clin fel hemat, Washington Univ, 57-58, res fel, 58-59, res assoc biochem, 59-61, from instr to asst prof med, Med Sch, 61-64; from asst prof to assoc prof & dir hemat, 64-68, PROF MED & BIOCHEM, SCH MED, UNIV MIAMI, 68-, DIR DIV HEMAT, HOWARD HUGHES LABS HEMAT RES, 68- *Concurrent Pos:* Am Leukemia Soc scholar, 61-66; USPHS res career develop award, 66-71. *Mem:* Am Fedn Clin Res; Am Soc Hemat; Am Soc Exp Path; Asn Am Physicians; Am Soc Biol Chemists. *Res:* Colony stimulating factor and modulators of granulopoiesis; mechanism of action of bone marrow toxins and the pathogenesis of chloramphenicol-induced blood dyscrasias. *Mailing Add:* Dept Biochem R-38 Univ Miami Sch Med PO Box 016960 Miami FL 33101

YUNIS, EDMOND J, b Sincelejo, Colombia, Aug 8, 29; US citizen; m 65; c 4. MEDICINE, PATHOLOGY. *Educ:* Nat Univ Colombia, MD, 54. *Prof Exp:* Resident anat path, Univ Kans, 55-57; resident clin path, Univ Hosps, Univ Minn, Minneapolis, 57-59, from instr to prof lab med, 60-71, dir blood bank, 61-68, dir div immunol, 66-76, prof lab med & path, 71-76; CHIEF DIV IMMUNOGENETICS, DANA-FARBER CANCER INST & PROF PATH, HARVARD MED SCH, 76-; DIR HLA LAB, NORTHEAST REGIONAL RED CROSS BLOOD PROG, 76- *Concurrent Pos:* Sr assoc, Ctr Blood Res, 85- *Mem:* Am Soc Exp Path; Am Asn Immunol; Transplantation Soc; Am Asn Pathologists; Am Soc Histocompatibility & Immunogenetics. *Res:* Antigenicity in cells and animals; immunological capacity in animals related to thymus; transplantation immunology and immunogenetics. *Mailing Add:* Dept Path Harvard Med Sch 25 Shattuck St Boston MA 02115

YUNIS, JORGE J, b Sincelejo, Colombia, Oct 5, 33; US citizen. GENETICS, PATHOLOGY. *Educ:* Inst Simon Araujo, Colombia, BS, 51; Cent Univ Madrid, MD, 56, PhD, 57. *Prof Exp:* Intern, Prov Hosp, Barranquilla, Colombia, 57-58, resident internal med, 58-59; resident clin & anat path, 59-62, from instr to assoc prof lab med, 62-69, head div med genetics, 62-77, dir grad studies lab med & path, 69-74, PROF LAB MED & PATH, UNIV MINN, MINNEAPOLIS, 69- *Concurrent Pos:* Fel lab med, Univ Minn, Minneapolis, 62-63, chmn human genetics comt health sci, 72-77; mem bd trustees, Leukemia Soc Am. *Honors & Awards:* Clin Prof Year, Harvard Med Sch, 87; honoree, Columbian Parliament, Bogata, Columbia, 86. *Mem:* Am Soc Human Genetics; Am Soc Cell Biol; Am Asn Path & Bact; Acad Clin Lab Physicians & Sci; Am Soc Hemat; Columbian Acad Med; Leukemia Soc Am. *Res:* Fine structure and molecular organization of human chromosomes; chromosome defects in human cancer. *Mailing Add:* Univ Minn Med Sch Box 198 Minneapolis MN 55455

YUNKER, CONRAD ERHARDT, b Matawan, NJ, Dec 22, 27; m 58; c 4. PARASITOLOGY, MICROBIOLOGY. *Educ:* Univ Md, BS, 52, MS, 54, PhD, 58. *Prof Exp:* Staff mem med zool, US Naval Med Res Unit, Cairo, Egypt, 55-57; res assoc zool, Univ Md, 58, asst prof, 58-59; entomologist, Entom Res Inst, Can Dept Agr, 59-60; scientist, Mid Am Res Unit, NIH, 60-61, scientist dir, Rocky Mountain Lab, Nat Inst Allergy & Infectious Dis, 61-82; prof vet microbiol & pathol, Wash State Univ, 82-85; RES SCIENTIST & CHIEF PARTY, UNIV FLA/USAID ZIMBABWE HEARTWATER PROJ, HARARE, 85- *Concurrent Pos:* Consult, Nat Inst Allergy & Infectious Dis, 59; mem, Bolivian Hemorrhagic Fever Comn, 63; affil prof zool, Univ Mont, 71-75, adj prof Entomol, Univ Idaho, 83-86, adj prof, entomol, Univ Fla, 86-, consult, int prog, Mich State Univ, 78. *Honors & Awards:* Award, Inst Acarology, Ohio State Univ, 72; Sigrid Juselius Found, Helsinki, Finland, 75. *Mem:* Fel AAAS; Entom Soc Am; Tissue Cult Asn; Am Soc Parasitol; Am Soc Trop Med & Hyg. *Res:* Systematic acarology; medical entomology; arthropod tissue culture; arthropod-borne viruses and rickettsias. *Mailing Add:* Dept Infect Dis Col Vet Med Univ Fla Gainesville FL 32610

YUNKER, MARK BERNARD, b Toronto, Ont, Dec, 48; c 1. MARINE CHEMISTRY ORGANICS & METALS. *Educ:* Univ Waterloo, BSc, 71, PhD(chem), 75. *Prof Exp:* Res assoc chem, Univ Hawaii, 75-77; lectr chem, Univ Victoria, 77-79; mem staff, 80-86, RES SCIENTIST, DEPT FISHERIES & OCEANS, 86- *Concurrent Pos:* Nat Res Coun Can fel, 75-77; mem, Comt Marine Anal Chem, Nat Res Coun Can, 84-; vis scientist, dept chem, Univ Victoria, 84-86. *Mem:* Am Chem Soc; Chem Inst Can. *Res:* Synthetic chemistry of bioactive compounds; isolation and structure determination of marine natural products; fundamental studies in marine chemistry, both organics and metals; statistical analysis and interpretation of marine environmental data; determination of the fate and transport of metals and organics in marine and freshwater environments; organic geochemistry. *Mailing Add:* Royal Roads Mil Col Victoria BC V0S 1B0 Can

YUNKER, MARTIN HENRY, b Milton Junction, Wis, Dec 28, 28; m 53; c 2. PHARMACY. *Educ:* Univ Wis, BS, 51, MS, 53, PhD(phys pharm), 57. *Prof Exp:* Tech asst to mgr pharmaceut prod, Merck, Sharp & Dohme Div, Merck, Inc, 57-58, supvr granulation dept, 58-59, supvr qual control, 59-61, res assoc pharmaceut res, 61-63; sr res pharmacist, Abbott Labs, 63-91; RETIRED. *Mem:* Am Pharmaceut Asn; Acad Pharmaceut Sci; Sigma Xi; Am Asn Pharmaceut Scientists. *Res:* Pharmaceutical research, including tablet formulations; biopharmaceutics; in vitro drug dissolution; preformulation characterization of drugs; parenteral formulations. *Mailing Add:* 3035 Burris Ave Waukegan IL 60087

YUNKER, WAYNE HARRY, b Corvallis, Ore, Jan 8, 36; c 3. LIQUID SODIUM COOLANT TECHNOLOGY. *Educ:* Ore State Col, BS, 57; Univ Wash, PhD(chem), 61. *Prof Exp:* Res scientist chem, Gen Elec Co, 63-65; sr res scientist, Battelle Mem Inst, 65-70; sr res scientist chem, 71-77, FEL SCIENTIST CHEM, WESTINGHOUSE HANFORD CO, 78- *Mem:* Am Chem Soc; Sigma Xi. *Res:* Physical chemistry of materials interactions. *Mailing Add:* 1422 Potter Richland WA 99352

YUNUS, MUHAMMAD BASHARAT, b Barisal, Bangladesh, March 1, 42; US citizen; m 81; c 3. INTERNAL MEDICINE, RHEUMATOLOGY. *Educ:* Brojomohun Col, ISC, 59; Chittagong Med Col, MBBS, 64; Am Bd Internal Med, dipl, 79, Am Bd Internal Med, Rheumatology, dipl, 82. *Prof Exp:* Fel gastroenterol, Hull Royal Infirmary, Eng, 74-75; sr med registr, St Marys Hosp, Kettering, Eng, 75-76; resident med, Worcester City Hosp, Mass, 76-77, fel rheumatology, 77-78; sr fel rheumatology, 78-79, asst prof med, 79-85, ASSOC PROF MED, UNIV ILL COL MED, PEORIA, ILL, 85- *Concurrent Pos:* Reviewer, N Eng J Med, J Rheumatology, 81-, Clin Exp Rheumatology, 86- & Pain, 86-; grant reviewer, NIH, 87, Arthritis Soc Can, 87; chmn Nonarticular Rheumatism Study Group, Am Rheumatism Asn, 88. *Mem:* Fel Am Col Physicians; fel Am Rheumatism Asn; Int Asn Study Pain. *Res:* Completed the first detailed controlled clinical study of a rheumatologic conditon called fibromyalgia syndrome with many other original articles in this field subsequently. *Mailing Add:* Dept Med Univ Ill Col Med Bos 1649 Peoria IL 61656

YURA, HAROLD THOMAS, b Buffalo, NY, Nov 20, 37; m 59; c 2. OPTICAL SIGNALS. *Educ:* Calif Inst Technol, BS, 59, PhD(physics), 62. *Prof Exp:* Staff scientist physics, Rand Corp, 62-70; SR SCIENTIST PHYSICS, AEROSPACE CORP, 70- *Concurrent Pos:* Adj prof, Univ Calif, Los Angeles, 75- *Mem:* Fel Optical Soc Am. *Res:* Wave propagation in random media and laser propagation phenomenology; strong optical scintellation effects; extension of the Huygens Fresnel principle to random media. *Mailing Add:* PO Box 92957 Aerospace Corp M2-246 Los Angeles CA 90009

YURCHAK, SERGEI, b Butler Twp, Pa, Feb 11, 43. CHEMICAL ENGINEERING. *Educ:* Pa State Univ, BS, 64; Univ Wis, PhD(chem eng), 68. *Prof Exp:* Res chem engr, 69-72, sr res engr, 72-77, assoc engr, 76-80, res assoc, 80-83, sr res assoc, 83-87, scientist, 87-88, MGR, LIGHT GAS UPGRADING/PETROCHEMICALS GROUP, MOBIL RES & DEVELOP CORP, 88- *Mem:* Am Inst Chem Engrs; Am Chem Soc; Sigma Xi. *Res:* Reaction kinetics and reactor design; process development; synthetic fuels. *Mailing Add:* Mobil Res & Develop Corp Billingsport Rd Paulsboro NJ 08066

YURCHENCO, JOHN ALFONSO, b San Juan, Arg, Feb 22, 15; nat US; m 44; c 4. MEDICAL MICROBIOLOGY. *Educ:* Albion Col, BA, 41; Johns Hopkins Univ, ScD(bact), 49. *Prof Exp:* Chief div microbiol, Eaton Labs, Inc, 48-55; head dept chemother, Squibb Inst Med Res, 55-56; sr res scientist, Wyeth Labs, Inc, 56-82; RETIRED. *Res:* Chemotherapy bacterial infections; bacterial pathogenicity and virulence; low temperature stability of infectious bacterial pools; immunology. *Mailing Add:* 24 Pond Lane Bryn Mawr PA 19010

YURETICH, RICHARD FRANCIS, b Brooklyn, NY, Aug 30, 50; m 74. SEDIMENTOLOGY, GEOCHEMISTRY. *Educ:* New York Univ, BA, 71; Princeton Univ, MA, 76, PhD(geol), 76. *Prof Exp:* Res geologist, Gulf Res & Develop, Gulf Oil Corp, 76-77; asst prof geol, State Univ NY Col Oneonta, 77-80; asst prof, 80-85, ASSOC PROF GEOL, DEPT GEOL-GEOG, UNIV MASS, 85- *Mem:* Geol Soc Am; Soc Econ Paleontologists & Mineralogists; Clay Minerals Soc; Am Asn Petrol Geologists; Am Geophys Union. *Res:* Lacustrine deposits; surface waters; rift valleys; clay minerals. *Mailing Add:* Geol-Geog Dept Univ Mass Amherst MA 01003

YUREWICZ, EDWARD CHARLES, b Philadelphia, Pa, Feb 10, 45; m 67; c 2. GLYCOPROTEINS, ANTIGENS. *Educ:* Univ Del, BA, 66; Sch Med, Johns Hopkins Univ, PhD(physiol chem), 71. *Prof Exp:* Fel biochem, Univ Calif, Davis, 71-74; sr res biochemist, Merck Inst Therapeut Res, 74-75; asst prof, 75-85, ASSOC PROF GYNEC & OBSTET, SCH MED, WAYNE STATE UNIV, 85- *Mem:* AAAS; Am Soc Cell Biol; Am Soc Biochem & Molecular Biol; Soc Complex Carbohydrates; Soc Study Reprod. *Res:* Structural characterization of cervical mucus glycoproteins; biochemical and immunochemical analysis of glycoprotein antigens in mammalian oocyte zona pellucida; biochemistry of sperm-egg interaction; immunocontraception. *Mailing Add:* Dept Gynec & Obstet Sch Med Wayne State Univ CS Mott Ctr Detroit MI 48201

YURKE, BERNARD, b Wittenberg, Ger, Nov 28, 51; US citizen; m 83. NONCLASSICAL STATES OF LIGHT. *Educ:* Univ Tex, Austin, BS, 75, MA, 76; Cornell Univ, PhD(physics), 83. *Prof Exp:* MEM TECH STAFF, AT&T BELL LABS, 83- *Mem:* Am Phys Soc; Optical Soc Am. *Res:* Quantum optics and electronics; the generation and detection of nonclassical states of the electromagnetic field called squeezed states. *Mailing Add:* 1011 Edgewood Ave Plainfield NJ 07060

YURKIEWICZ, WILLIAM J, b Bloomsburg, Pa, Sept 21, 39; m 65; c 2. INSECT PHYSIOLOGY, BIOCHEMISTRY. *Educ:* Bloomsburg State Col, BS, 60; Bucknell Univ, MS, 62; Pa State Univ, PhD(entom), 65. *Prof Exp:* Asst entom, Pa State Univ, 63-65; res entomologist, USDA, Ga, 65-66; PROF BIOL, MILLERSVILLE STATE COL, 66- *Mem:* Entom Soc Am. *Res:* Insect flight physiology; neutral lipid and phospholipid composition and metabolism in insects; neural control of insect flight. *Mailing Add:* Dept Biol Millersville State Col Millersville PA 17551

YURKOWSKI, MICHAEL, b Sask, Can, Sept 1, 28; m 57; c 3. BIOCHEMISTRY, NUTRITION. *Educ:* Univ Sask, BSA, 51, MSc, 59; Univ Guelph, PhD(nutrit), 68. *Prof Exp:* Analytical chemist, Western Potash Corp Ltd, Sask, 51-52; indust chemist, Cereal & Oilseed Processing, Sask Wheat Pool, Can, 52-56; anal chemist, Plant Prod Div, Can Dept Agr, Ont, 59-60; food & drug directorate, Can Dept Nat Health & Welfare, Man, 60-62; res scientist marine lipid biochem, Halifax Lab, Freshwater Inst, 62-65, appl & basic nutrit fish & freshwater organisms, 68-84, res scientist nutrit lipid biochem, Dept Fisheries & Oceans, Can, Artic Mgt Sect, 84-91; RETIRED. *Mem:* Am Oil Chemists' Soc; Can Inst Food Sci & Technol. *Res:* Nutrition and metabolism of arctic marine animals; lipid biochemistry of arctic marine animals; odors in freshwater and freshwater fish. *Mailing Add:* Six Celtic Bay Winnipeg MB R3T 2W9 Can

YUROW, HARVEY WARREN, b New York, NY, Feb 14, 32; m 56; c 3. ANALYTICAL CHEMISTRY. *Educ:* Queens Col (NY), BS, 54; Pa State Univ, PhD(anal chem), 60. *Prof Exp:* Dept Defense fel, Rutgers Univ, 59-60; RES CHEMIST, EDGEWOOD ARSENAL, 60- *Mem:* Am Chem Soc;

Sigma Xi. *Res:* Trace analysis of organic compounds via chromogen formation; structure-activity relationships for physiologically active compounds; organic analysis via chemiluminescence. *Mailing Add:* 3801 Maryland Ave Abingdon MD 21009

YUSHOK, WASLEY DONALD, b Woodbine, NJ, Mar 11, 20; m 50. BIOCHEMISTRY, CANCER. *Educ:* Rutgers Univ, BS, 41, MS, 43; Cornell Univ, PhD(chem embryol), 50. *Prof Exp:* Asst, Rutgers Univ, 41-43 & Cornell Univ, 46-49; asst to ed handbk biol data, Nat Res Coun, 50; res assoc, Univ Tex Med Br, 50-52; res biochemist, Biochem Res Found, 52-59, head div cancer biochem, 59-66; ASSOC MEM DIV BIOCHEM, INST FOR CANCER RES, 66- *Mem:* AAAS; Am Chem Soc; Am Asn Cancer Res; NY Acad Sci. *Res:* Regulation of metabolism and enzymes in cancer cells; nucleotide, protein and carbohydrate metabolism; intracellular energy-dependent protein degradation. *Mailing Add:* Inst Cancer Res 7701 Burholme Ave Philadelphia PA 19111

YUSKA, HENRY, b Brooklyn, NY, Nov 7, 14; m 44; c 2. ORGANIC CHEMISTRY. *Educ:* City Col New York, BS, 35; Polytech Inst Brooklyn, MS, 39; Univ Ill, PhD(org chem), 42. *Prof Exp:* Res chemist, Jewish Hosp, Brooklyn, NY, 35-39; asst chem, Univ Ill, 41; res org chemist, Barrett Div, Allied Chem & Dye Corp, 42-43; resin group leader, Interchem Corp, 43-60, dir dept org chem, Cent Res Labs, 60-63; tech dir, Sun Chem Corp, 63-66; prof 66-85, EMER PROF CHEM, BROOKLYN COL, 85- *Concurrent Pos:* Instr, Eve Div, Hunter Col, 43-46. *Mem:* Am Chem Soc. *Res:* Synthetic resins; organic synthesis of monomers; biochemistry. *Mailing Add:* 113-09 107 Ave Jamaica NY 11419-2501

YUSPA, STUART HOWARD, b Baltimore, Md, July 19, 41; m 65; c 2. CANCER. *Educ:* Johns Hopkins Univ, BS, 62; Univ Md, MD, 66; Am Bd Internal Med, dipl, 72. *Prof Exp:* Intern internal med, Hosp Univ Pa, 66-67; res assoc cancer, Nat Cancer Inst, 67-70; resident internal med, Hosp Univ Pa, 70-72; SR INVESTR CANCER, NAT CANCER INST, 72-, CHIEF, LAB CELLULAR CARCINOGENESIS & TUMOR PROM, DIV CANCER ETIOLOGY, 81- *Concurrent Pos:* Mem biol models segment, Carcinogenesis Prog, Nat Cancer Inst, 72-78, Ed-in-chief, Molecular Carcinogenesis. *Honors & Awards:* Montagna Lectr, 88; Lila Gruber Cancer Res Award, 89; Elizabeth Miller Mem Lectr, 90. *Mem:* Am Asn Cancer Res; AAAS; Amer Soc Cell Biol, Soc Invest Dermat. *Res:* Determine mechanisms whereby chemicals initiate or promote malignant transformation of epithelial cells. *Mailing Add:* Div Cancer Etiology Bldg 37 Rm 3B25 Nat Cancer Inst NIH Bethesda MD 20892

YUST, CHARLES S(IMON), b Newark, NJ, Jan 21, 31; m 55; c 3. PHYSICAL METALLURGY. *Educ:* Newark Col Eng, BS, 52; Univ Tenn, MS, 60. *Prof Exp:* Metall engr, Oak Ridge Gaseous Diffusion Plant, 52-60; METALLURGIST, METALS & CERAMICS DIV, OAK RIDGE NAT LAB, 60- *Mem:* Am Ceramic Soc; Sigma Xi. *Res:* Theory of solid state sintering; deformation mechanisms in ceramics. *Mailing Add:* 106 Newcrest Lane Oak Ridge TN 37830

YUSTER, PHILIP HAROLD, b Fargo, NDak, Nov 7, 17; m 47; c 2. RADIATION PHYSICS. *Educ:* NDak Col, BS, 39; Wash Univ, PhD(phys chem), 49. *Prof Exp:* Jr chemist, Panama Canal, 42-43; asst, Manhattan Dist, Univ Chicago, 43-44; asst, Los Alamos Sci Lab, 44-45; asst, Wash Univ, 45-49; assoc chemist, 49-58, SR CHEMIST, ARGONNE NAT LAB, 58- *Mem:* Am Phys Soc; Sigma Xi. *Res:* Mass spectroscopy; photochemistry; luminescence; color centers. *Mailing Add:* 5831 Washington Downers Grove IL 60516

YUTRZENKA, GERALD J, b Crookston, Minn, Jan 27, 53; m 79; c 2. NEUROPHARMACOLOGY. *Educ:* Moorhead State Univ, BS, 74; Univ N Dak, MS, 77, PhD(physiol), 79. *Prof Exp:* ASSOC PROF PHYSIOL & PHARMACOL, SCH MED, UNIV SDAK, 84- *Concurrent Pos:* Res assoc, dept biochem, Baylor Col Med, 79; fel dept pharmacol, Med Col Va, 80-84. *Mem:* Am Soc Pharmacol & Exp Therapeut; Int Soc Neurochem; Sigma Xi; Am Soc Exp Biol & Med; Indian Acad Neurosci. *Res:* Neuropharmacology and neurochemical mechanisms underlying establishment of physical dependence on central nervous system depressant drugs. *Mailing Add:* Dept Physiol & Pharmacol Sch Med Univ SDak Vermillion SD 57069

YUVARAJAN, SUBBARAYA, b Sowdhapuram, India, Sept 15, 41; m 73; c 2. POWER ELECTRONICS, MOTOR CONTROL. *Educ:* Madras Univ, India, BS, 61, BE, 66; Indian Inst Technol, India, MTech, 69, PhD(elec eng), 81. *Prof Exp:* Lectr elec eng, PSG Col Tech, Coimbatore, 69-74; lectr & asst prof elec eng, Indian Inst Technol, Madras, 74-83; ASSOC PROF ELECTRONICS, NDAK STATE UNIV, FARGO, 83- *Mem:* Inst Elec & Electronics Engrs; Am Soc Eng Educ; Sigma Xi. *Res:* Analyzed a complete variable speed induction motor drive using digital computer simulation; application of new power semiconductor devices for the conversion of electrical power from one form to another. *Mailing Add:* Dept Elec Eng NDak State Univ Fargo ND 58105

YUWILER, ARTHUR, b Mansfield, Ohio, Apr 4, 27; m 50; c 3. NEUROSCIENCES. *Educ:* Univ Calif, Los Angeles, BS, 50, PhD(biochem), 56. *Prof Exp:* Asst chem, Univ Calif, Los Angeles, 50-51, asst physiol chem, 52-54, asst chem, 54-56, res biochemist, 57-59; res neurobiochemist, Vet Admin, Calif, 56-57; res assoc & dir labs & biochem sect, Schizophrenia & Psychopharmacol Res Proj, Ypsilanti State Hosp & Univ Mich, 59-62; from asst prof to assoc prof biochem, 70-76, PROF PSYCHIAT, UNIV CALIF, LOS ANGELES, 76-; CHIEF NEUROBIOCHEM RES, VET ADMIN BRENTWOOD HOSP, 62- *Concurrent Pos:* Res biochemist, Ment Health Res Inst, Univ Mich, 59-62; mem, Brain Res Inst, Univ Calif, Los Angeles, 65-; mem, Basic Sci Res Comt, Vet Admin, 66-69, Career Develop Award Comt, NIMH, 70-76, Calif State Ment Health Adv Comt, 70-71 & Sci Adv Comt, Dystonia Found, 77-83 & 85-88; chair, Career Develop Award Comt, NIMH, 82-83. *Mem:* Am Soc Neurochem; Int Soc Neurochem; Am Col Neuropsychopharmacol; Soc Biol Psychiat; Am Soc Biol Chemists; Sigma Xi; Soc Neurosci. *Res:* Aromatic amino acids; enzymes; biochemistry of mental disease; intermediary metabolism of monoamines; stress; childhood autism; pineal gland. *Mailing Add:* Vet Admin Hosp Neurobiochem Lab T-85 Wilshire/Sawtelle Los Angeles CA 90073

Z

ZAALOUK, MOHAMED GAMAL, b Dessouk, Egypt, Aug 16, 35; m 63; c 2. NUCLEAR & CONTROL ENGINEERING. *Educ:* Cairo Univ, BS, 57; NC State Univ, MS, 62, PhD(nuclear eng), 66. *Prof Exp:* Engr, Egyptian Atomic Energy Estab, 57-60, asst prof reactor anal, 66-68; vis scientist, Inst Atomic Energy, Kjeller, Norway, 68-69; vis asst prof elec eng, NC State Univ, 69-72; prin engr nuclear-mech, Carolina Power & Light Co, 72-81; HEAD, DIV NUCLEAR ENG, HOUSTON LIGHTING & POWER, 81- *Concurrent Pos:* Lectr, Dept Nuclear Eng, Univ Alexandria, 66-68; Norweg Agency Res & Develop fel, Inst Atomic Energy, Kjeller, Norway, 68-69; AEC grants, NC State Univ, 69-71; mem staff proj mgt, Carolina Power & Light Co, 79-81. *Mem:* Am Nuclear Soc; Egyptian Soc Nuclear Sci. *Res:* Nuclear reactor analysis; control theory and systems; boiling dynamics; project management. *Mailing Add:* 4417 Kaplan Dr Raleigh NC 27606

ZABARA, JACOB, b Philadelphia, Pa, May 8, 32; m 70; c 2. PHYSIOLOGY, NEUROPHYSIOLOGY. *Educ:* Johns Hopkins Univ, BS, 53; Univ Pa, MS, 58, PhD(physiol), 59. *Prof Exp:* USPHS fel, Univ Pa, 59-60; instr pharmacol, Dartmouth Col, 60-61; instr pharmacol, Univ Pa, 61-63; USPHS spec fel biomath, 63-64; instr physiol, Univ Pa, 64-65, assoc pharmacol, 65-67, assoc physiol, 65-67; asst prof, 67-72, ASSOC PROF PHYSIOL, GRAD SCH, TEMPLE UNIV, 72- *Concurrent Pos:* Vis prof, Hadassah Med Sch, Hebrew Univ, Stanford Univ. *Mem:* Soc Neurosci; Biophys Soc; Undersea Med Soc; Am Asn Anat; Am Physiol Soc; Am Epilepsy Soc. *Res:* Neurophysiology and cybernetics; epilepsy. *Mailing Add:* Dept Physiol & Biophys Temple Univ Med Sch 3223 N Broad St Philadelphia PA 19140

ZABEL, HARTMUT, b Radolfzell, WGer, Mar 21, 46; m 73; c 3. SOLID STATE PHYSICS. *Educ:* Univ Bonn, Vordipl, 69; Tech Univ Munich, Hauptdipl, 73; Univ Munich, PhD(physics), 78. *Prof Exp:* Fel physics, Univ Houston, 78-79; from asst prof to assoc prof, 79-86, PROF PHYSICS, UNIV ILL, URBANA-CHAMPAIGN, 86- *Concurrent Pos:* Beckman fel, Ctr Advan Studies, Univ Ill, 82; guest scientist, Brookhaven Nat Lab, 85, 88 & Risoe Nat Lab, Denmark, 86; dir, solid state sci prog, Dept Energy, Mat Res Lab, 86- *Mem:* Ger Phys Soc; Am Phys Soc; AAAS; Mat Res Soc. *Res:* X-ray and thermal neutron scattering studies of structural, thermal, magnetic, lattice dynamical properties of condensed matter systems, hydrogen in metals, intercalated graphite, semiconductor and metal superlattices, quasicrystals; thin films and heterostructures. *Mailing Add:* Exp Physics IV Ruhr Univ Bochum Universitatssfrasse 150 D 4630 Bochum Germany

ZABEL, ROBERT ALGER, b Boyceville, Wis, Mar 11, 17; m 44; c 5. FOREST PATHOLOGY. *Educ:* Univ Minn, BS, 38; State Univ NY, MS, 41, PhD(bot), 48. *Prof Exp:* Lab aide, Lake States Forest Exp Sta, US Forest Serv, 40; from asst prof to assoc prof forest path, State Univ NY Col Environ Sci & Forestry, 47-53, head dept bot & forest path, 53-54, assoc dean biol sci & undergrad instr, 64-69, vpres acad affairs, 70-73, prof, 53-85; RETIRED. *Mem:* AAAS; fel Soc Am Foresters; Am Soc Microbiol; Am Phytopath Soc. *Res:* Forest products deterioration; wood decays; lumber stains; evaluation of preservatives; root and heart rots; toxicants and fungicides. *Mailing Add:* 4563 Broad Rd Syracuse NY 13215

ZABIELSKI, CHESTER V, b Schenectady, NY, July 19, 23. METALLURGY. *Educ:* Union Univ, NY, BS, 47; Columbia Univ, BS, 49; Rensselaer Polytech Inst, MS, 54 & 56; Univ Pittsburgh, PhD(metall), 65. *Prof Exp:* Chemist, Gen Aniline & Film Corp, 50-52; Olin Mathieson res asst, Rensselaer Polytech Inst, 52-54; metall engr, Westinghouse Elec Corp, Pa, 57-58; asst prof metall, Univ Pittsburgh, 58-67; SR PHYS METALLURGIST, US ARMY MAT RES AGENCY, 67- *Concurrent Pos:* Ford Found res grant, Univ Pittsburgh. *Mem:* Am Soc Metals; Am Inst Mining, Metall & Petrol Engrs; Am Inst Chem Engrs; Am Chem Soc; Nat Asn Corrosion Engrs; Am Electrochem Soc. *Res:* Physical metallurgy; corrosion; metallurgy of uranium, steel, zirconium, titanium, copper and stainless steels; high temperature alloys; molten salt reactions; kinetics; thermodynamics; optical and x-ray spectroscopy. *Mailing Add:* Box 369 Cambridge MA 02238

ZABIK, JOSEPH, b Chicopee, MA; m 65; c 2. BEHAVIORAL PHARMACOLOGY. *Educ:* Mass Col Pharm, BS, 65, MS, 67; Univ RI, PhD(pharmacol), 72. *Prof Exp:* ASSOC PROF PHARMACOL, SCH PHARM, PURDUE UNIV, 78- *Mem:* Am Soc Pharmacol & Exp Therapeut; Behav Pharmacol Soc; NY Acad Sci; Res Soc Alcoholism. *Res:* The relationships between genetics, stress and biogenic amines, especially serotonin, in the development, maintenance and regulation of alcohol consumption in experimental animals; effects of stress, biochemical mechanisms of drug dependence. *Mailing Add:* Dept Pharmacol & Toxicol Sch Pharm Purdue Univ W Lafayette IN 47907

ZABIK, MARY ELLEN, b Kendallville, Ind, Jan 20, 37; m 58; c 1. FOOD SCIENCE. *Educ:* Purdue Univ, BS, 59; Mich State Univ, MS, 61, PhD(food sci), 70. *Prof Exp:* Instr food res, Mich State Univ, 61-68 & 69-70, from asst prof to prof food chem, 70-90, ASSOC DEAN ACAD AFFAIRS, MICH STATE UNIV, 88-, DISTINGUISHED UNIV, PROF, 90- *Honors & Awards:* Borden Award, Am Home Econ Asn, 83. *Mem:* Fel Inst Food Technologists; Am Oil Chemists Soc; Am Asn Cereal Chemists; Poultry Sci Asn; Am Home Econ Asn; Sigma Xi. *Res:* Food chemistry and rheology; functionality of carbohydrates and proteins in food systems; reduction of environmental contamination from food systems. *Mailing Add:* Five Col Human Ecol Mich State Univ East Lansing MI 48824

ZABIK, MATTHEW JOHN, b South Bend, Ind, Aug 22, 37; m 58; c 1. CHEMISTRY, TOXICOLOGY. *Educ:* Purdue Univ, Lafayette, BS, 59; Mich State Univ, MS, 62, PhD(org chem), 65. *Prof Exp:* From asst prof to assoc prof, 65-72, PROF ENTOM & ASSOC DIR PESTICIDE RES CTR, MICH STATE UNIV, 72- *Mem:* Am Chem Soc; Soc Environ Toxicol & Chem. *Res:* Photochemistry; environmental toxicology of xenobiotics. *Mailing Add:* Dept Entomol Mich State Univ Pesticide Res Ctr Rm 204 East Lansing MI 48823

ZABIN, BURTON ALLEN, b Chicago, Ill, Mar 18, 36; m 72. INORGANIC CHEMISTRY, ANALYTICAL CHEMISTRY. *Educ:* Univ Ill, BS, 57; Purdue Univ, PhD(inorg chem), 62. *Prof Exp:* Res assoc chem, Stanford Univ, 62-63; dir res, 63-72, DIV MGR CHEM, BIO-RAD LABS, 72- *Mem:* Am Chem Soc. *Res:* Separations chemistry, including ion exchange resins, gel filtration materials and other column chromatographic materials. *Mailing Add:* 1414 Harbour Way S Richmond CA 94804-3625

ZABIN, IRVING, b Chicago, Ill, Nov 13, 19; m 42; c 3. BIOCHEMISTRY, MOLECULAR BIOLOGY. *Educ:* Univ Ill, BS, 40; Univ Chicago, PhD(biochem), 49. *Prof Exp:* Res assoc biochem, Univ Chicago, 49-50; res assoc biol chem, 50-51, from lectr to assoc prof, 51-64, PROF BIOL CHEM, SCH MED, UNIV CALIF, LOS ANGELES, 64- *Concurrent Pos:* Nat Multiple Sclerosis Soc scholar, 59-60; Guggenheim fel, 67-68; NATO sr fel sci, 75; vis prof, Pasteur Inst, Paris, 59-60 & 67-68 & Imp Col London, 75. *Mem:* AAAS; Am Soc Biol Chem; Am Soc Microbiol. *Res:* Protein structure, synthesis and control. *Mailing Add:* Dept Biol Chem Univ Calif Sch Med Los Angeles CA 90024-1737

ZABINSKI, MICHAEL PETER, b Haifa, Israel, May 30, 41; US citizen; m 64; c 1. SOLID MECHANICS, BIOMECHANICS. *Educ:* Univ Conn, BS, 62, MS, 63; Yale Univ, MS, 66, MPhil, 68, PhD(eng & appl sci), 69; Univ New Haven, MS, 77. *Prof Exp:* Sr engr, Olin Corp, 63-66; PROF ENG & PHYSICS, FAIRFIELD UNIV, 69- *Concurrent Pos:* Res grant, Fairfield Univ, 69-72; NIH grant internal med, Sch Med, Yale Univ, 72-76. *Mem:* Am Soc Mech Engrs; Am Soc Eng Educ; Am Chem Soc. *Res:* Digital computing; computers in public school education; physical principles of peristaltic phenomena; mechanical properties of tissue. *Mailing Add:* Eight Larkspur Dr Latham NY 12110-4942

ZABLOCKA-ESPLIN, BARBARA, b Warsaw, Poland, Jan 5, 25; m 64; c 4. NEUROPHARMACOLOGY, NEUROPHYSIOLOGY. *Educ:* Med Acad, Warsaw, dipl, 52, MD, 61. *Prof Exp:* Asst prof pharmacol, Med Acad, Warsaw, 52-55, sr asst prof, 55-61; Riker Int fel, Col Med, Univ Utah, 61-62, res assoc, 62-65, asst res prof, 65-68; asst prof, 68-77, ASSOC PROF PHARMACOL & THERAPEUT, MCGILL UNIV, 77- *Mem:* Am Soc Pharmacol & Exp Therapeut; Can Pharmaceut Asn. *Res:* Central excitatory and depressant drugs; central transmitter substances. *Mailing Add:* 428 Claremont Ave Montreal PQ M3Y 2N2 Can

ZABLOW, LEONARD, b New York, NY, Sept 3, 27; m 50; c 2. CARDIAC ELECTROPHYSIOLOGY, NEUROPHYSIOLOGY. *Educ:* Calif Inst Technol, BSc, 48; Columbia Univ, MA, 50. *Prof Exp:* Res worker biochem, Worcester Found Exp Biol, 51-52; res asst neurol, Columbia Univ, 52-60, res assoc, 60-73, sr staff assoc, 73-85, sr staff assoc, Pharmacol Col Physicians & Surgeons, 85-90, SCI PROGRAMMER, HOWARD HUGHES MED INST, COLUMBIA UNIV, 90- *Concurrent Pos:* Lectr, Polytech Inst Brooklyn, 59-66. *Mem:* Fel AAAS; Biophys Soc; Am Phys Soc. *Res:* Spatial electroencephalogram analysis; focal generator size in clinical and experimental epilepsy; electromyography; source distributions in electroencephalography; computer analysis of the electroencephalogram and electromyogram; experimental epilepsy; models in cardiac electrophysiology. *Mailing Add:* 5610 Post Rd Bronx NY 10471

ZABORSKY, OSKAR RUDOLF, b Neuwalddorf, Czech, Oct 6, 41; US citizen; m 68; c 2. BIOTECHNOLOGY, BIOLOGICAL CHEMISTRY. *Educ:* Philadelphia Col Pharm & Sci, BSc, 64; Univ Chicago, PhD(chem), 68. *Prof Exp:* NIH fel, Harvard Univ, 68-69; sr res chemist, Corp Res Labs, Exxon Res & Eng Co, Linden, NJ, 69-74; prog dir, NSF, 74-83; pres & chief exec officer, OMEC Int Inc, 84-88; pres, Ozcom Int Inc, 88-89; DIR BD BIOL, NAT RES COUN, 89- *Concurrent Pos:* Chmn, Comt Biotechnol Nomenclature & Info Orgn, NAS, 85-86, mem, Marine Biotechnol Comt, 83-85, chmn bioexpo, 85 & 86. *Mem:* AAAS; Am Chem Soc; Am Inst Chem Engr; Am Soc Microbiol; Soc Indust Microbiol. *Res:* Biotechnology; enzyme technology; renewable resources; biocatalysis; biomass chemicals and fuels; biochemical engineering; biomaterials; marine biotechnology; information science; environmental technology; bioinformatics; technology transfer. *Mailing Add:* Nat Res Coun Bd Biol 2101 Constitution Ave NW Washington DC 20418

ZABORSZKY, JOHN, b Budapest, Hungary, May 13, 14; nat US; div. ELECTRICAL ENGINEERING. *Educ:* Josef Nador Royal Hungarian Tech Univ, Budapest, dipl, 37, DSc, 42. *Prof Exp:* Chief engr in charge power syst eng, Munic Elec Works Budapest, 44-48; from asst prof to prof eng, Univ Mo, 48-56; chmn, Automatic Control Area, Wash Univ, 59-65, rotating chmn, Dept Systs Mech & Aerospace Eng, 65-68, chmn, Control Systs Sci & Eng, 68-74, chmn, Dept Systs Sci & Math, 74-89, PROF ENG, WASH UNIV, 56- *Concurrent Pos:* Docens, Royal Hungarian Tech Univ, 46-47; consult, Westinghouse Elec Co, E Pittsburg, Pa, 50-51, McDonnell Douglas Corp, St Louis, Mo, 55-75, Emerson Elec Co, St Louis, 62-67, Hi-Voltage Equip Co, Cleveland, Ohio, 57-72; mem, Working Group for Standards on Static Capacitor Switching, Power Eng Soc, 57-58, Power Syst Eng Comt, Subcomt Systs, 77-; mem bd dirs, Inst Elec & Electronic Engrs, 74-75, dir div I, 74-75. *Honors & Awards:* Centennial Medal, Inst Elec & Electronics Engrs, 84; Richard E Bellman Control Heritage Award, Am Automatic Control Coun, 86. *Mem:* Nat Acad Eng; Soc Indust & Appl Math; fel Inst Elec & Electronics Engrs; Am Soc Mech Engrs; hon mem Hungarian Acad Sci; Sigma Xi. *Res:* Control theory; optimal control; functional analysis approaches; identification adaptive control; estimation and filtering bilinear systems; controllability and observability; power systems dynamics, stability and control; switching phenomena; high voltage direct current; author of two books and numerous publications. *Mailing Add:* Sch Eng & Appl Sci Wash Univ PO Box 1040 St Louis MO 63130

ZABRANSKY, RONALD JOSEPH, b Little Ferry, NJ, Mar 18, 35; m 58; c 3. CLINICAL MICROBIOLOGY, CLINICAL BACTERIOLOGY. *Educ:* Rutgers Univ, BS, 56; Ohio State Univ, MS, 61, PhD(microbiol), 63; Am Bd Med Microbiol, dipl, 69. *Prof Exp:* Microbiologist, Battelle Mem Inst, 59-60; teaching asst med microbiol bact, Ohio State Univ, 60-61, res asst, 61-63; assoc consult microbiol, Mayo Clin, 63-64, consult, 64-69; DIR DIV MICROBIOL, MT SINAI MED CTR, MILWAUKEE, 69- *Concurrent Pos:* Asst clin prof, Dept Microbiol, Med Col Wis, 70-74, assoc adj prof microbiol, 75-84, adj prof, 84-; chmn, Nat Registry Microbiologists, 74-79; mem bd gov, Am Acad Microbiol, 74-79; assoc clin prof allied health, Univ Wis-Milwaukee, 78-84, prof, 84-; mem, Nat Comt Clin Lab Standards, 80-; consult, Microbiol Device Panel, Food & Drug Admin, 81-; prof path & lab med, Med Sch, Univ Wis, Milwaukee Clin Campus, 83-; ed, Clin Microbiol Newsletter, 85- *Mem:* Fel Am Acad Microbiol; Am Pub Health Asn; Am Soc Microbiol. *Res:* Isolation and identification of anaerobic bacteria; invitro evaluation of antibiotics; antibiotic testing of anaerobic bacteria. *Mailing Add:* Dept Microbiol 8701 Watertown Pink Rd Milwaukee WI 53226

ZABRISKIE, FRANKLIN ROBERT, b New York, NY, Dec 21, 33; m. ASTRONOMY. *Educ:* Princeton Univ, BSE, 55, MSE, 57, PhD(astron), 60. *Prof Exp:* Asst prof astron, Wesleyan Univ, 60-66; assoc prof astron, Pa State Univ, University Park, 66-79; PRES, ASTRO COMPUT CONTROLS, INC, 80- *Mem:* Am Astron Soc; Int Astron Union; Am Geophys Union. *Res:* Design of instruments for optical telescopes; photoelectric stellar classification; studies of long period variable stars. *Mailing Add:* RD 1 Alexandria PA 16611

ZABRISKIE, JOHN LANSING, JR, b Auburn, NY, June 8, 39; m 63; c 2. ORGANIC CHEMISTRY. *Educ:* Dartmouth Col, BA, 61; Univ Rochester, PhD(chem), 66. *Prof Exp:* Sr chemist process res, Merck & Co, Inc, NJ, 65-68, tech asst to exec dir, animal sci res, 69-70, sr mgr pharmacol qual control, Merck Sharp & Dohme, 70-72, mgr pharmaceut mfg, 72-74, dir pharmaceut mfg, 74-79, secy new prod comt, 79-80, dir mkt planning, 80-81, PRES & DIR MGT ENG, MERCK SHARP & DOHME RES LABS, 81- *Res:* Organic synthesis and reaction mechanisms. *Mailing Add:* PO Box 2000 Rahway NJ 07065

ZABRONSKY, HERMAN, b Brooklyn, NY, April 5, 27; m 57; c 2. RELIABILITY, QUEUEING THEORY. *Educ:* City Univ NY, BS, 48; Univ Pa, MA, 51. *Prof Exp:* Assoc mathematician, Oak Ridge, 51-53; sr engr, Ford Instrument Co, 53-58 & RCA, 58-67; staff scientist, Computer Sci Corp, 67-73, Calculon, 78-79; SR STATISTICIAN, AM SYSTS CORP, 79- *Concurrent Pos:* Lectr math, Stevens Inst Technol, 59-65; consult econ, Int Bus Serv, 80-81. *Mem:* Math Asn Am. *Res:* Applied probability queueing reliability; econometrics; statistical theory of communications; partial differential equations; boundary value problems related to heat and mass transfer; reactors. *Mailing Add:* 10857 Bucknell Dr Wheaton MD 20902

ZABUSKY, NORMAN J, b NY, Jan 4, 29. COMPUTATIONAL FLUID DYNAMICS, VISUALIZATION & MODELING. *Educ:* Col City New York, BEE, 51; Mass Inst Technol, MS, 53; Calif Inst Technol, PhD(physics), 59. *Prof Exp:* Tech staff analog simulation, Raytheon Missile & Radar Div, 53-55; vis assoc plasma physics, Princeton Plasma Physics Lab, 60-61; tech staff reentry physics, Bell Labs, 61-63, supvr plasma physics, 63-68, head, Dept Computational Physics, 68-74, Dept Models & Systs, 74-75; prof math, Univ Pittsburgh, 76-88; PROF FLUID DYNAMICS, RUTGERS UNIV, 88- *Concurrent Pos:* Dir, Int Sch Nonlinear Math & Physics, Max Planen Inst, Munich, 66; consult, Geophys Plasma Dynamics Br, Nat Res Lab, 76-91, Los Alamos Nat Lab, 84-91. *Honors & Awards:* Potts Medal, Franklin Inst, 86; Kiev Nonlinear Conf Medal, 89. *Mem:* Fel Am Phys Soc; Soc Indust & Appl Math; fel AAAS. *Res:* Vortex dynamics and turbulence in two and three dimensions; contour dynamics for inviscid flows; nonlinear dynamical systems and data assimilation in CFD data visualization and scientific modeling. *Mailing Add:* 104 Maple Ct Highland Park NJ 08904

ZACCARIA, ROBERT ANTHONY, b Philadelphia, Pa, May 29, 43; m 64; c 3. DEVELOPMENTAL BIOLOGY, VERTEBRATE ENDOCRINOLOGY. *Educ:* Bridgewater Col, BA, 65; Univ Va, PhD(biol), 73. *Prof Exp:* Asst prof, 73-80, ASSOC PROF BIOL, LYCOMING COL, 80- *Mem:* Am Soc Zool; AAAS; Am Inst Biol Sci; Sigma Xi. *Res:* Interaction among chromatophores in development of pigmentation in amphibians; limb regeneration in urodeles. *Mailing Add:* Dept Biol Lycoming Col Williamsport PA 17701

ZACCHEI, ANTHONY GABRIEL, b Philadelphia, Pa, Mar 31, 40; m 63; c 3. DRUG METABOLISM, BIOCHEMICAL TOXICOLOGY. *Educ:* Villanova Univ, BS, 61, MS, 65; Univ Minn, Minneapolis, PhD(biochem), 68. *Prof Exp:* Res assoc pharmacol, Merck Sharp & Dohme Res Labs, 61-64, sr res pharmacologist, 68-70, res fel, 70-73, sr res fel drug metabolism, 73-78, dir human drug metabolism, 78-85, sr investr, 85-90, DIR, BIOCHEM TOXICOL, MERCK SHARP & DOHME RES LABS, 90- *Concurrent Pos:* Vis asst prof, Inst Lipid Res, Baylor Col Med, 72; adj asst prof med, Jefferson Med Col, 79-85; mem spec study sect, Nat Inst Gen Med Sci, NIH, 78-87. *Mem:* Am Soc Pharmacol & Exp Therapeut; Am Chem Soc; Am Soc Mass Spectrometry. *Res:* Investigations into the physiological disposition of new drug products including absorption, excretion and metabolic studies; investigate mechanism of drug toxicity. *Mailing Add:* Merck Sharp & Dohme Res Labs West Point PA 19486

ZACH, RETO, b Davos-Platz, Switz, Dec 27, 40; Can citizen; m 66; c 1. ZOOLOGY, ECOLOGY. *Educ:* Univ Alta, Edmonton, BSc, 72; Univ Toronto, PhD(zool), 77. *Prof Exp:* Fel zool, Univ BC, Vancouver, 77-78; RESEARCHER ECOL, ATOMIC ENERGY CAN, LTD, 78- *Mem:* Ecol Soc Am; Can Soc Zool; Cooper Ornith Soc. *Res:* Environmental impact of nuclear energy; ecological modelling; animal behavior; birds. *Mailing Add:* Atomic Energy Can Ltd Environ Res Br Pinawa MB R0E 1L0 Can

ZACHARIAH, GERALD L, b McLouth, Kans, June 12, 33; m 53; c 3. AGRICULTURAL ENGINEERING. *Educ:* Kans State Univ, BS, 55, MS, 59; Purdue Univ, PhD(agr eng), 63. *Prof Exp:* Instr agr eng, Kans State Univ, 55-56 & 56-60 & Purdue Univ, 60-63; asst prof eng, Univ Calif, 63-65; from asst prof to prof agr eng, Purdue Univ, Lafayette, 65-75; chmn, Dept Agr Eng, Univ Fla, 75-80, dean resident instr, 80-88, interm vpres, Agr Affairs, 88-89, VPRES, AGR AFFAIRS, UNIV FLA, 89- *Concurrent Pos:* Res grants, 55-; indust consult, 55- *Mem:* Am Soc Agr Engrs; Am Soc Eng Educ; Inst Food Technologists; Nat Soc Prof Engrs. *Res:* Automatic control; food engineering; processing of agricultural products. *Mailing Add:* 1008 McCarty Hall Univ Fla Gainesville FL 32611

ZACHARIAS, DAVID EDWARD, b Philadelphia, Pa, May 16, 26; m 68; c 3. X-RAY CRYSTALLOGRAPHY, ORGANIC CHEMISTRY. *Educ:* Temple Univ, AB, 53, AM, 54; Univ Pittsburgh, PhD(x-ray crystallog), 69. *Prof Exp:* From jr chemist to sr chemist, Smith Kline & French Labs, Pa, 54-71; res assoc, 71-83, SR RES ASSOC, MOLECULAR STRUCT LAB, INST CANCER RES, 83- *Concurrent Pos:* Ed consult, Int Cancer Res Data Bank, 81-89. *Mem:* AAAS; Am Crystallog Asn; Am Chem Soc; Royal Soc Chem. *Res:* Synthesis of heterocyclic compounds; single crystal x-ray structure determination of organic and biologically important compounds; x-ray powder diffraction. *Mailing Add:* Molecular Struct Lab Inst Cancer Res Philadelphia PA 19111-2412

ZACHARIASEN, FREDRIK, b Chicago, Ill, June 14, 31; m 57. THEORETICAL PHYSICS. *Educ:* Univ Chicago, PhB, 50, BS, 51; Calif Inst Technol, PhD(physics), 56. *Prof Exp:* Instr physics, Mass Inst Technol, 55-56, jr res physicist, Univ Calif, 56-57; res assoc physics, Stanford Univ, 57-58, asst prof, 58-60; from asst prof to assoc prof, 60-65, PROF PHYSICS, CALIF INST TECHNOL, 65- *Concurrent Pos:* Consult, Rand Corp, 56-; Sloan Found fel, 60-64; consult, Los Alamos Sci Lab, 61-; Inst Defense Anal, 61-; Guggenheim Found fel, 70-71; Mitre Corp, 78-; assoc dir, Los Alamos Nat Lab, 82-83. *Res:* High energy particle physics. *Mailing Add:* Dept Physics Calif Inst Technol Pasadena CA 91109

ZACHARIASEN, K(ARSTEN) A(NDREAS), b Vardo, Norway, Mar 1, 24; div; c 3. CHEMICAL ENGINEERING. *Educ:* Inst Technol, Norway, Siv Ing, 50. *Prof Exp:* Lab engr, Vestfos Cellulose Works, Norway, 51-52; head, Process & Develop Lab, And H Kiaer & Co, Ltd, 52-55; process engr, Buckeye Cellulose Corp, Proctor & Gamble Co, 56-58, sect head, 59-63, mgr Memphis Plant, 64-68, Foley Plant, 68-72, mgr, Grande Prairie Plant, Proctor & Gamble Cellulose Ltd, 72-75, mgr cell process develop, Buckeye Cellulose Corp, 75-80; RETIRED. *Mem:* Norweg Tech Asn Pulp & Paper Indust. *Res:* Cellulose research; pulping and bleaching. *Mailing Add:* 3967 Grahamdale Circle Memphis TN 38122

ZACHARIUS, ROBERT MARVIN, b New York, NY, Mar 21, 20; m 44; c 4. PLANT CELL TISSUE CULTURE, PLANT-PARASITE INTERACTIONS. *Educ:* NY Univ, BA, 43; Univ Colo, MA, 48; Univ Rochester, PhD(plant physiol), 53. *Prof Exp:* Asst chem, Univ Colo, 47-48; asst bot, Cornell Univ, 51-52; res chemist, Gen Cigar Co, Inc, Pa, 52-54; biochemist, Eastern Utilization Res Br, USDA, 54-57, res chemist, Eastern Utilization Res & Develop Div Pa, 57-71, res chemist, Plant Prod Lab, Eastern Regional Res Ctr, 71-74, res chemist, Eastern Regional Res Ctr, Agr Res Serv, 74-84, res chemist, Beltsville Agr Res Ctr, 84-89; RETIRED. *Concurrent Pos:* Adj biochem lectr, St Joseph's Col, Pa, 67-68. *Mem:* Int Asn Plant Tissue Cult; Phytochem Soc NAm; Biochem Soc; Am Chem Soc; Am Soc Plant Physiol. *Res:* Non-protein nitrogen compounds of plants; plant proteins; Nicotiana alkaloids; ion-exchange and chromatographic methods; electrophoresis; antimetabolites; plant metabolism; proteolytic inhibitors; plant-parasite interactions, stress physiology and phytoalexin induction; plant cell tissue culture. *Mailing Add:* 6567 River Clyde Dr Highland MD 20777

ZACHARUK, R Y, b Yorkton, Sask, May 1, 28; m 52; c 2. INSECT MORPHOLOGY, INSECT PATHOLOGY. *Educ:* Univ Sask, BSA, 50, MSc, 55; Univ Glasgow, PhD(histochem, physiol), 61. *Prof Exp:* Res officer entom, Res Sta, Can Dept Agr, 50-63; assoc prof, 63-65, chmn dept, 65-67, PROF BIOL, UNIV REGINA, 67- *Concurrent Pos:* Agr Inst Can fel, 59-61. *Mem:* AAAS; Soc Invert Path; Asn Chemoreception Sci; Can Soc Entom; Can Soc Zool. *Res:* Sense organ ultrastructure; neurophysiology; entomophagous fungi; histochemistry. *Mailing Add:* Dept Biol Univ Regina Regina SK S4S 0A2 Can

ZACHARY, LOREN WILLIAM, b Colfax, Iowa, Apr 26, 43; m 63; c 2. ENGINEERING. *Educ:* Iowa State Univ, BS, 66, MS, 74, PhD(eng mech), 76. *Prof Exp:* Asst eng, Aerospace Div, Martin Marietta Corp, 66-72; asst prof, 76-80, ASSOC PROF ENG MECH, IOWA STATE UNIV, 80- *Mem:* Soc Exp Mech. *Res:* Experimental stress analysis; non-destructive testing. *Mailing Add:* Dept Eng Sci 210 Lab Mech Iowa State Univ Ames IA 50011

ZACHARY, NORMAN, b New York, NY, Sept 14, 26; m 54; c 3. COMPUTER SCIENCE, SYSTEMS ANALYSIS. *Educ:* NY Univ, AB, 47; Harvard Univ, PhD, 52. *Prof Exp:* Asst prof math, Univ Md, 51-52; mem, Inst Advan Study, 52-54; mem tech staff, Hughes Aircraft Co, 54-55; sect head, Sylvania Elec Prod, Inc, 55-57, lab mgr, Electronic Systs Div, 58-60; staff consult, Otis Elevator Co, 57-58; vpres, Gen Tel Co, Calif, Gen Tel & Electronics Corp, 60-63; dir commun & data systs, NAm Aviation, Inc, 63-64; dir, Harvard Comput Ctr, Mass, 64-71; EXEC VPRES & DIR, DATA ARCHITECTS, INC, 71- *Concurrent Pos:* Consult, banking, ins & financial industs. *Mem:* Am Math Soc; Asn Comput Mach. *Res:* Computers and systems analysis with application to the solution of engineering, administrative and management problems; real time systems; integrated business data processing. *Mailing Add:* 257 Prince St West Newton MA 02165

ZACHMANOGLOU, ELEFTHERIOS CHARALAMBOS, b Thessaloniki, Greece, Mar 19, 34; US citizen; c 4. PARTIAL DIFFERENTIAL EQUATIONS. *Educ:* Rensselaer Polytech Inst, BAeroE, 56, MS, 57; Univ Calif, Berkeley, PhD(appl math), 62. *Prof Exp:* From asst prof to assoc prof, 62-70, PROF MATH, PURDUE UNIV, LAFAYETTE, 70-, ASSOC HEAD DEPT, 81- *Concurrent Pos:* Fulbright res grant, Univ Rome, 65-66. *Mem:* Am Math Soc. *Res:* Partial differential equations; wave propagation. *Mailing Add:* Dept Math Purdue Univ Lafayette IN 47907

ZACHOS, COSMAS K, b Athens, Greece, Sept 8, 51; m 88. PHYSICS, PHYSICAL MATHEMATICS. *Educ:* Princeton Univ, AB, 74; Calif Inst Technol, PhD(physics), 79. *Prof Exp:* Res fel, Calif Inst Technol, 79; res assoc, Univ Wis, Madison, 79-81; res assoc, Fermi Nat Accelerator Lab, 81-83; asst physicist, 83-86, PHYSICIST, ARGONNE NAT LAB, 86- *Mem:* Sigma Xi; Am Phys Soc; Math Asn Am. *Res:* Supergravity, sigma models and superstrings; theoretical high energy physics; infinite dimensional and quantum algebras. *Mailing Add:* HEP 362 Argonne Nat Lab Argonne IL 60439-4815

ZACK, NEIL RICHARD, b Canton, Ohio, Apr 26, 47. INORGANIC CHEMISTRY, FLUORINE CHEMISTRY. *Educ:* Rensselaer Polytech Inst, BS, 69; Marshall Univ, MS, 70; Univ Idaho, PhD(chem), 74. *Prof Exp:* Fel chem, Utah State Univ, 74-75; fel, Univ Idaho, 75-76; sr chemist, Allied Chem Corp, 76-79; group leader, Exxon Nuclear Idaho Co, 79-80; SUPVR, WESTINGHOUSE IDAHO NUCLEAR CO, INC, 81- *Concurrent Pos:* Consult hazardous chem safety. *Mem:* Am Chem Soc; Sigma Xi. *Res:* Inorganic heterocyclics; fluorine containing derivatives of catenated sulfur compounds. *Mailing Add:* Los Alamos Nat Lab PO Box 1663 MS F541 Los Alamos NM 27545

ZACKAY, VICTOR FRANCIS, b San Francisco, Calif, May 2, 20. PHYSICAL METALLURGY. *Educ:* Univ Calif, BS, 47, MS, 48, PhD(metall), 52. *Prof Exp:* Instr phys metall, Univ Calif, 51-52; asst prof, Pa State Univ, 52-54; res scientist, Sci Lab, Ford Motor Co, 54-57, supvr metall, 58-59, asst mgr metall, 60-62; lectr metall, Univ Calif, Berkeley, 62-66, asst dir, Inorg Mat Res Div, Lawrence Berkeley Lab, 64-75, from prof to assoc dean eng, 66-74; pres, Mat & Methods Inc, 80-89; RETIRED. *Honors & Awards:* Howe Award, Am Soc Metals, 60, Albert Sauveur Achievement Award, 71. *Mem:* Fel Am Soc Metals; fel Am Inst Mining, Metall & Petrol Engrs. *Res:* Advanced materials science; engineering, technology and manufacturing for engines; energy conversion and storage; electronics, medical, dental and environmental systems. *Mailing Add:* 1014 West Rd New Canaan CT 06840

ZACKROFF, ROBERT V, b Philadelphia, Pa, Sept 14, 51. INTERMEDIATE FILAMENTS, MICROTUBULES. *Educ:* Temple Univ, AB, 73, MA, 75, PhD(biol), 79. *Prof Exp:* Res asst, Temple Univ, 73-75; fel, Carnegie-Mellon Univ, 79-81; sr res assoc, Med Sch, Northwestern Univ, 81-82, asst prof, 83-84; prin investr, Marine Biol Lab, Woods Hole, Mass, 84-85; ADJ PROF, UNIV RI, 86- *Concurrent Pos:* Lectr biol, Bryant Col, 88- *Mem:* Am Soc Cell Biol. *Res:* Cell motility; biochemistry of the cytoskeleton; self assembly of cytoskeletal proteins; microtubules and intermediate filaments. *Mailing Add:* 120 Central St NARR Narragansett RI 02882

ZACKS, JAMES LEE, b Iron Mountain, Mich, Mar 23, 41; m 66; c 2. VISON RESEARCH, NEUROSCIENCE. *Educ:* Harvard Univ, BA, 63; Univ Calif, Berkeley, MS & PhD(psychol), 67. *Prof Exp:* Asst prof psychol, Univ Pa, 67-71; assoc prof, 72-77, PROF PSYCHOL, MICH STATE UNIV, 77- *Concurrent Pos:* Prog dir, USPHS training grant, 70-76; prin investr, NSF res grant, 70-75; adj prof zool, Mich State Univ, 76-, adj prof radiol, 85-; vis prof, Dept Psychol, Northwestern Univ, 79-80; prin investr, NIH Res Grant, 88- *Mem:* Fel Optical Soc Am; Psychonomic Soc; Asn Res Vision & Ophthal. *Res:* Basic visual capacities and peripheral visual mechanisms which might determine them; visual perceptual changes on aging. *Mailing Add:* Dept Psychol Mich State Univ East Lansing MI 48824

ZACKS, SHELEMYAHU, b Tel Aviv, Israel, Oct 15, 32; m 55; c 2. MATHEMATICAL STATISTICS, STATISTICAL ANALYSIS. *Educ:* Hebrew Univ, Israel, BA, 55; Israel Inst Technol, MSc, 60; Columbia Univ, PhD(indust eng), 62. *Prof Exp:* Sr lectr statist, Israel Inst Technol, 63-65; prof, Kans State Univ, 65-68; prof, Univ NMex, 68-70; prof math & statist, Case Western Reserve Univ, 70-79, chmn dept, 74-79; prof, Va Poly Tech, 79-80; chmn, 80-83, PROF MATH & STATIST, STATE UNIV NY, BINGHAMTON, 80- *Concurrent Pos:* Fel statist, Stanford Univ, 62-63; consult, Inst Mgt Sci & Eng, George Washington Univ, 67-87. *Mem:* Fel Am Statist Asn; fel Inst Math Statist; Int Statist Inst. *Res:* Optimal design of sequential experiments; statistical adaptive processes; analysis of contingency tables; manpower forecasting for large military organizations; stochastic control. *Mailing Add:* Dept Math Sci State Univ NY Binghamton NY 13902-6000

ZACKS, SUMNER IRWIN, b Boston, Mass, June 29, 29; m 53; c 3. PATHOLOGY. *Educ:* Harvard Univ, BA, 51, MD, 55; Am Bd Path, dipl, 61. *Hon Degrees:* MA, Univ Pa & Brown Univ. *Prof Exp:* From intern to asst resident path, Mass Gen Hosp, Boston, 55-58; from asst prof to prof path, Sch Med, Univ Pa, 62-76; prof path & chmn dept, sch med, Brown Univ, 76-82. *Concurrent Pos:* Neuropathologist, Pa Hosp, 61-76, assoc dir, Ayer Lab, 64-76; pathologist-in-chief, Miriam Hosp, 76- *Honors & Awards:* Hektoen Bronze Medal, AMA, 61. *Mem:* Histochem Soc (secy, 65-69); Am Soc Exp Path; Am Soc Cell Biol; Am Asn Neuropath; fel Col Am Path. *Res:* Fine structure of neuromuscular junctions, normal and in disease; fine structure pathology of muscle; molecular pathology of endotoxins; muscle regeneration. *Mailing Add:* Miriam Hosp 164 Summit Ave Providence RI 02906

ZACZEK, NORBERT MARION, b Baltimore, Md, Aug 15, 36; m 73; c 1. ORGANIC CHEMISTRY. *Educ:* Loyola Col, Md, BS, 58; Carnegie-Mellon Univ, PhD(org chem), 62. *Prof Exp:* From instr to assoc prof, 62-71, PROF CHEM, LOYOLA COL MD, 71- *Mem:* AAAS; Am Chem Soc; Royal Soc Chem. *Mailing Add:* Dept Chem Loyola Col 4501 N Charles St Baltimore MD 21210-2600

ZACZEPINSKI, SIOMA, b Grodno, USSR, June 15, 40; US citizen; m 73; c 2. HYDROPROCESSING, RESEARCH MANAGEMENT. *Educ:* Univ Tenn, BS, 63. *Prof Exp:* Process engr, Exxon Res & Eng Co, 63-68, startup adv, 68-69; process adv, Esso Eng Ltd, Europe, 69-71; startup adv, Exxon Res & Eng Co, 71-72, proj leader, 72-74; tech mgr, Esso Thailand Standard, Ltd, 75-76, eng assoc, Europe, 77; sect head, Exxon Res & Eng Co, 77-80, lab dir, 80-84, tech mgr, Esso Italiana, 85-86, lab dir, Synthetic Fuels, 86-87; MGR, BAYTOWN SPECIALTY PROD, 87- *Mem:* Am Inst Chem Engrs; Am Petrol Inst. *Res:* Direct coal liquefaction process; synthetic fuels. *Mailing Add:* PO Box 4255 Baytown TX 77052

ZADEH, L(OTFI) A, b Baku, Russia, Feb 4, 21; nat US; m 46; c 2. ARTIFICIAL INTELLIGENCE EXPERT SYSTEMS. *Educ:* Teheran Univ, BS, 42; Mass Inst Technol, MS, 46; Columbia Univ, PhD(elec eng), 49. *Hon Degrees:* Dr, Paul-Sabatier Univ, Toulouse, France, 86 & State Univ NY, 89. *Prof Exp:* From instr to prof elec eng, Columbia Univ, 46-59; chmn dept, 63-68, PROF ELEC ENG, UNIV CALIF, BERKELEY, 59- *Concurrent Pos:* Chmn US comn, Int Union Radio Sci; ed, Int J Fuzzy Sets & Systs, J Math Anal & Appln & J Computer & Syst Sci; mem, Inst Advan Study, Princeton, NJ, 56, bd trustees, Int Asn Knowledge Engrs; sr postdoctoral fel, NSF, 56-57 & 62-63; vis prof elec eng, Mass Inst Technol, 62 & 68; Guggenheim Found fel, 67-68; vis scientist, Res Lab, IBM, San Jose, Calif, 68, 73 & 77; vis scholar, Artificial Intel Ctr, SRI Int Menlo Park, Calif, 81; vis mem, Ctr Study Lang & Info, Stanford Univ, 88. *Honors & Awards:* Educ Medal, Inst Elec & Electronics Engrs, 73, Centennial Medal, 84; Eringen Medal, Soc Eng Sci Lectr Award, 75; Honda Prize, Honda Found, Japan, 89. *Mem:* Nat Acad Eng; Am Math Soc; Asn Comput Mach; fel Inst Elec & Electronics Engrs; fel AAAS; Int Fuzzy Syst Asn; Soc Eng Sci; Am Asn Artificial Intel; fel World Coun Cybernetics. *Res:* System theory; information processing; theory of fuzziness; artificial intelligence; expert systems; natural language understanding; knowledge representation; theory of evidence; expert systems; author of various publications. *Mailing Add:* Dept Elec Eng & Computer Sci Univ Calif Berkeley CA 94720

ZADNIK, VALENTINE EDWARD, b Cleveland, Ohio, Feb 13, 34; m 58; c 4. GEOLOGY, ENGINEERING GEOLOGY. *Educ:* Western Reserve Univ, BA, 57; Univ Ill, MS, 58, PhD(geol, civil eng), 60. *Prof Exp:* Res geologist, Jersey Prod Res Co, 64-65; sr res geologist, Esso Prod Res Co, 65-66; res geologist, US Army Res Off, 66-74; STAFF GEOLOGIST, OFF ENERGY RESOURCES, US GEOL SURV, 74-, LIAISON GEOLOGIST, NAT PETROL RESERVE ALASKA, 77- *Concurrent Pos:* Fel int affairs, Princeton Univ, 70-71; mem comt rock mech & comt seismol, Nat Acad Sci. *Mem:* Asn Eng Geol; Sigma Xi. *Res:* Petroleum exploration; carbonate rock petrography; solid-earth geophysics, particularly seismology, gravity, geomagnetism and geodesy. *Mailing Add:* 105 S Park Dr Arlington VA 22204

ZADOFF, LEON NATHAN, b Passaic, NJ, Aug 6, 23; m 44; c 2. PHYSICS. *Educ:* Cooper Union, BChE, 48; NY Univ, PhD(physics), 58. *Prof Exp:* Chem engr, Flintkote Co, 48; petrol chemist, Paragon Oil Co, 48-49; protein chemist, Botany Mills, Inc, 49; jr chem engr, City Fire Dept, New York, 49-52; physicist, NY Naval Shipyard, 52-54; proj supvr reactor physics, Ford Instrument Co, 54-58; assoc scientist, Repub Aviation Corp, 58-64, specialist physicist, Repub Aviation Div, Fairchild-Hiller Corp, NY, 64-72; CHIEF SCIENTIST, EMS DEVELOP CORP, FARMINGDALE, 72- *Concurrent Pos:* Lectr, Adelphi Col, 58-; adj assoc prof, NY Inst Technol, 72- *Mem:* AAAS; Am Phys Soc; NY Acad Sci. *Res:* Theoretical, plasma and reactor physics; electromagnetic wave propagation; quantum mechanics. *Mailing Add:* 17 Spruce St Merrick NY 11566

ZADUNAISKY, JOSE ATILIO, b Rosario, Arg, July 15, 32; m 54; c 2. PHYSIOLOGY, BIOPHYSICS. *Educ:* Univ Buenos Aires, MD, 56. *Prof Exp:* Instr, Inst Physiol, Med Sch, Univ Buenos Aires, 52-56; Arg Nat Res Coun estab investr, Dept Biophys, Med Sch, Univ Buenos Aires, 60-63; assoc prof physiol & dir res, Dept Ophthal, Sch Med, Univ Louisville, 64-67; assoc prof ophthal & physiol, Sch Med, Yale Univ, 67-73; PROF PHYSIOL & EXP OPHTHAL, MED SCH, NY UNIV, 73-, DIR SACKLER INST, 81- *Concurrent Pos:* Arg Res Coun fel biochem, Univ Col, Dublin, 58-59 & Inst Biol Chem, Copenhagen, 59-60; USPHS grants, 62-; exec ed, Exp Eye Res, 70-; consult, Visual Sci A Study Sect, USPHS, 76-80; investr, Mount Desert Island Biol Lab, Maine, 75-; ed, Current Topics Eye Res, 80- & Chloride Transport in Biol Membranes, 82- *Honors & Awards:* Alcon Award Outstanding Contrib Vision Res, Alcon Inst, Tex, 84. *Mem:* Fel NY Acad Sci; Biophys Soc; Asn Res Vision & Ophthal; Am Physiol Soc; Int Soc Eye Res (secy, 78-81, pres, 84-88). *Res:* Transport and permeability of biological membranes, especially epithelial tissues. *Mailing Add:* Dept Physiol & Ophthalmol NY Univ Med Sch 550 First Ave New York NY 10016

ZAEHRINGER, MARY VERONICA, b Philadelphia, Pa, May 27, 11. FOODS. *Educ:* Temple Univ, BS, 46; Cornell Univ, MS, 48, PhD(foods), 53. *Prof Exp:* Asst food res, Cornell Univ, 46-48, 51-53; from instr to asst prof home econ res, Mont State Col, 48-50; res prof home econ, Univ Idaho, 53-72, res prof food sci, 72-73, res prof bact & biochem, Agr Exp Sta, 73-88; RETIRED. *Concurrent Pos:* Res fel, Inst Storage & Processing Agr Produce, State Agr Univ, Wageningen, 67-68. *Mem:* Am Asn Cereal Chem; Inst Food Technol; Potato Asn Am. *Res:* Potato texture; quality of food products. *Mailing Add:* 614 Ash St Moscow ID 83843

ZAERR, JOE BENJAMIN, b Los Angeles, Calif, Sept 9, 32; m 54; c 3. FOREST PHYSIOLOGY, GROWTH HORMONES. *Educ:* Univ Calif, Berkeley, BS, 54, PhD(plant physiol), 64. *Prof Exp:* Res assoc, Crops Res Div, Agr Res Serv, USDA, Md, 64-65; vpres, PMS Instrument Co, Corvallis, 67-86; asst prof forestry, 65-71, assoc prof, 71-80, asst dean, Grad Sch, 77-80, PROF FORESTRY, ORE STATE UNIV, 81- *Concurrent Pos:* Vis res prof, Agr Univ, Warsaw, Poland, 73-74, Univ München, WGer, 89-90; Fulbright res scholar, WGer, 81-82. *Mem:* AAAS; Am Soc Plant Physiol; Soc Am Foresters; Scand Soc Plant Physiol. *Res:* Plant growth regulators; forest regeneration; bioelectrical potential in plants; root growth; physiology of forest decline; tissue culture; plant water relations. *Mailing Add:* Forestry Sch Ore State Univ Corvallis OR 97331-5705

ZAFFARANO, DANIEL JOSEPH, b Cleveland, Ohio, Dec 16, 17; m 46; c 6. NUCLEAR PHYSICS. *Educ:* Case Inst Technol, BS, 39; Ind Univ, MS, 48, PhD(physics), 49. *Prof Exp:* Tech liaison, Nat Carbon Co, 40-45; contract liaison, Appl Physics Lab, Johns Hopkins Univ, 45-46; from assoc prof to prof, 49-67, chmn dept physics, 61-71, DISTINGUISHED PROF PHYSICS, IOWA STATE UNIV, 67-, VPRES FOR RES & GRAD DEAN, 71- *Concurrent Pos:* Sci liaison officer, US Off Naval Res, London, 57-58. *Mem:* Fel Am Phys Soc; Sigma Xi. *Res:* Experimental nuclear physics. *Mailing Add:* 3108 Ross Rd Iowa State Univ Ames IA 50010

ZAFFARONI, ALEJANDRO, b Montevideo, Uruguay, Feb 27, 23; m 46; c 2. BIOCHEMISTRY. *Educ:* Univ Montevideo, BS, 41; Univ Rochester, PhD(biochem), 49. *Hon Degrees:* DSc, Univ Rochester, 72; Dr, Univ Repub, Montevideo, Uruguay, 83. *Prof Exp:* Dir biol res, Syntex SA, Mex, 51-54, dir res & develop, 54-56, vpres, 56-61, pres, Syntex Labs, Inc, exec vpres, Syntex Corp & pres, Syntex Res Ctr, Calif, 61-68; pres & dir res, 68-80, FOUNDER, CHMN & CHIEF EXEC OFFICER, ALZA CORP, 80-; FOUNDER & CHMN, DYNAX RES INST MOLECULAR & CELLULAR BIOL, 80- *Concurrent Pos:* Hon prof, Univ Montevideo, 59; pres, Dynapol, 72-83; consult prof pharmacol, Med Sch, Stanford Univ, 78- *Honors & Awards:* President's Award, Weizmann Inst, Israel, 78; Pioneer Award, Am Inst Chemists, 79. *Mem:* Inst Med-Nat Acad Sci; Am Chem Soc; Soc Exp Biol & Med; Endocrine Soc; Am Soc Biol Chemists; Am Soc Microbiol. *Res:* Biochemistry and drug delivery devices. *Mailing Add:* Alza Corp 950 Page Mill Rd Palo Alto CA 94304

ZAFIRATOS, CHRIS DAN, b Portland, Ore, Nov 18, 31; m 57; c 4. NUCLEAR PHYSICS. *Educ:* Lewis & Clark Col, BS, 57; Univ Wash, PhD(physics), 62. *Prof Exp:* Res instr nuclear physics, Univ Wash, 62; staff mem, Los Alamos Sci Lab, 62-64; asst prof physics, Ore State Univ, 64-68; from asst prof to assoc prof, 68-72, chmn dept, 78-82, PROF PHYSICS, UNIV COLO, 72- *Concurrent Pos:* Chmn nuclear physics lab, Univ Colo, 74-76 & 85-86, assoc vchancellor, 86- *Mem:* Fel Am Phys Soc; fel Japan Soc Prom Sci. *Res:* Nuclear reactions; neutron and nuclear structure physics. *Mailing Add:* Dept Physics Univ Colo Boulder CO 80309

ZAFIRIOU, EVANGHELOS, b Athens, Greece, Nov 15, 59. PROCESS CONTROL & OPTIMIZATION, FAULT DIAGNOSIS. *Educ:* Nat Tech Univ Athens, dipl, 83; Calif Inst Technol, PhD(chem eng), 87. *Prof Exp:* Asst prof, 87-91, ASSOC PROF CHEM ENG, UNIV MD, COLLEGE PARK, 91- *Concurrent Pos:* NSF presidential young investr award, 90-95. *Mem:* Am Inst Chem Eng; Inst Elec & Electronics Engrs. *Res:* Control system design, in particular robust control, model predictive control, batch control; optimal operation of chemical processes; process resiliency; fault diagnosis. *Mailing Add:* Systs Res Ctr Univ Md College Park MD 20742

ZAFRAN, MISHA, b Berlin, Ger, Aug 10, 49; US citizen. MATHEMATICAL ANALYSIS. *Educ:* Univ Calif, Riverside, BS, 68, PhD(math), 72. *Prof Exp:* Mem math res, Inst Advan Study, 72-73; asst prof math, Stanford Univ, 73-79; mem staff, Inst Advan Study, 79-80; MEM STAFF, DEPT MATH, UNIV WASH, 80- *Mem:* Am Math Soc. *Res:* Interrelationships between harmonic analysis, spectral theory and interpolation of operators. *Mailing Add:* 1905 NW 100th Seattle WA 98177

ZAGAR, WALTER T, b Brooklyn, NY, Oct 15, 28; m 51; c 4. PHYSICAL CHEMISTRY, POLYMER CHEMISTRY. *Educ:* Manhattan Col, BS, 50; Fordham Univ, MS, 55, PhD(phys chem), 58. *Prof Exp:* Chemist, Dextran Corp, 52-53; instr chem, Notre Dame Col, NY, 55-56; instr, Manhattan Col, 56-57; sr scientist, Polymer Chem Div, W R Grace & Co, 57-66; mgr plastics div, Allied Chem Corp, 66; mgr polymer develop, Chemplex Corp, 66-67; sr res chemist, Plastics Div, NJ, 67-72, La, 72-73, TECH SUPVR, ALLIED CHEM CORP, LA, 73- *Mem:* Soc Plastics Engrs; Am Chem Soc; NY Acad Sci. *Res:* Polymer development in area of polyethylene blends and additive systems for polymers, especially antioxidants, antistats and ultraviolet absorbers. *Mailing Add:* Allied Signal Corp PO Box 53006 Baton Rouge LA 70805

ZAGATA, MICHAEL DEFOREST, b Oneonta, NY, May 28, 42. NATURAL RESOURCES POLICY, ENVIRONMENTAL REGULATION. *Educ:* State Univ NY, Oneonta, BS, 64, MS, 68; Iowa State Univ, PhD(wildlife ecol), 72. *Prof Exp:* Teacher biol, Oneonta Consolidated Sch & Southampton Pub Schs, 64-69; asst prof wildlife ecol, Sch Forest Resources, Univ Maine, 72-75; field dir, Wildlife Soc, 76-77; dir fed rel, Nat Audubon Soc, 77-79; prog develop off, Bd Agr & Nat Resources, Nat Acad Sci, 79-80; mgr ecol sci, Tenneco, Inc, 80-86; DIR ENVIRON & SAFETY, TENNECO OIL CO, 86- *Concurrent Pos:* Prin investr, US Forest Serv, 73-75; co-chmn, Pub Lands Task Force, Nat Asn Mfrs; exec bd, Nat Resources Coun Am; mem, Waterfowl Feeding Adv Comt, US Fish & Wildlife Serv, 77-79; chmn, Conserv Affairs Comt, Wildlife Soc, 79-; mem, Nat Adv Comt Regional Plan Asn, 78 & Nat Comt Fish & Wildlife Res, 78- *Mem:* Wildlife Soc; Soc Petrol Indust Biologists; NY Acad Sci; Am Fisheries Soc; Soc Wetland Scientists. *Res:* Impact of man-induced perturbations on fish and wildlife habitats and on the populations which occupy those habitats, natural resource policy. *Mailing Add:* 14 Hillshire Grove Lane Hillshire Village TX 77055

ZAGER, RONALD, b New York, NY, Dec 27, 34. AROMA CHEMICAL MANUFACTURE, ORGANIC FLUOROCARBONS. *Educ:* Brooklyn Col, BS, 55; Stevens Inst Technol, MS, 69. *Prof Exp:* Res chemist, Charles Pfizer, 56-58; res & develop chemist, Halocarbon Prod Corp, 58-66; develop chemist, Tenneco Chem Corp, 66-71; sr develop chemist, Giuaudan Corp, 71-77; tech dir, Int Flavors & Fragrances, 77-88; CONSULT, RONALD ZAGER ASSOCS, 88- *Mem:* Asn Consult Chemists & Chem Engrs (pres-elect, 90-91, pres, 91-); Am Chem Soc. *Res:* Aroma chemicals; fluorocarbons; medical devices, prostetics; pharmaceutical inhalents; management and general research; manufacturing problem solving; experimental design; project management. *Mailing Add:* One Scenic Dr Highlands NJ 07732

ZAGER, STANLEY E(DWARD), b Sheldon, Iowa, Mar 6, 21; m 49; c 2. CHEMICAL ENGINEERING. *Educ:* Iowa State Univ, BS, 43; Purdue Univ, PhD(chem eng), 50. *Prof Exp:* Jr chemist, Shell Develop Co, 43-46; lab asst, Purdue Univ, 46-47, asst instr, 47-50; res engr, Johns-Manville Corp, 50-57; sr res engr, Res Ctr, B F Goodrich Co, 57-60, sect leader, 60-63, corp task force secy, 64-65, mgr mgt & comput sci, 66-76; mgr process develop & licensing, H K Ferguson Co, 76-77; assoc prof, 77-83, PROF CHEM ENG, YOUNGSTOWN STATE UNIV, 83- *Mem:* Am Chem Soc; Am Inst Chem Engrs; Asn Comput Mach; Sigma Xi. *Res:* Application of computers and mathematics to scientific and business problems; process development, design, economics; improving energy recovery from fossil fuels, process and product development; heterogeneous catalysis; polymer processing. *Mailing Add:* 2874 Cedar Hill Rd Cuyahoga Falls OH 44223

ZAGIER, DON BERNARD, b Heidelberg, West Germany, June 29, 51. MODULAR FORMS, DIOPHANTINE EQUATIONS. *Educ:* Mass Inst Technol, BSc(math) & BSc(physics), 68; Oxford Univ, Eng, PhD(math), 72. *Prof Exp:* Sci staff math res, 71-76, PROF MATH, UNIV BONN, GER, 76-; SCI MEM, MAX-PLANCK INST MATH, BONN, GER, 84- *Concurrent Pos:* Vis researcher, Fed Inst Technol, Zurich, Switz, 72-73 & Inst Higher Sci Studies, Bures-Sur, Yvette, France, 73-74; vis prof, Harvard Univ, 77-78; prof number theory, Dept Math, Univ Md, 79-; prof math, Univ Utrecht, Holland, 90-; prof functional anal, Kyushu Univ, Japan, 90-91. *Honors & Awards:* Carus Medal, Leopoldina Acad, Halle, EGer, 83; Cole Prize, Am Math Soc, 86. *Mem:* Max-Planck Soc; Am Math Soc. *Res:* Number theory; theory of modular forms. *Mailing Add:* Max-Planck Inst Math Gottfried-Claren-Str 26 Bonn D-5300 Germany

ZAGON, IAN STUART, b New York, NY, Mar 28, 43; m 64. ONCOLOGY, DEVELOPMENTAL NEUROBIOLOGY. *Educ:* Univ Wis-Madison, BS, 65; Univ Ill, Urbana, MS, 69; Univ Colo, Denver, PhD(anat), 72. *Prof Exp:* Asst prof biol struct, Med Sch, Univ Miami, 72-74; from asst prof to assoc prof, 74-85, PROF ANAT, MILTON S HERSHEY MED SCH, PA STATE UNIV, 85- *Concurrent Pos:* Biol Stain Comn fel, 68 & 71; grantee, Am Cancer Soc, NIH, Nat Inst Drug Abuse & Pa Res Corp; consult. *Mem:* Soc Neurosci; Am Asn Anat; AAAS; Int Soc Develop Neurosci; Asn Res Vision & Ophthal; Am Soc Cell Biol. *Res:* Developmental neurobiology, focusing on normal and abnormal brain development; relationship of endogenous opioid systems to brain development; biological influences of opioids and opioid receptors in cancer. *Mailing Add:* Dept Neurosci & Anat Hershey Med Ctr Pa State Univ Hershey PA 17033

ZAGZEBSKI, JAMES ANTHONY, b Stevens Point, Wis, Aug 5, 44; m 66; c 2. MEDICAL PHYSICS, MEDICAL IMAGING. *Educ:* St Mary's Col, BS, 66; Univ Wis, Madison, MS, 68, PhD(radiol sci), 72. *Prof Exp:* Res assoc, 72-75, adj asst prof, dept oncol & radiol, 75-77, asst prof, 77-81, PROF, DEPTS MED PHYSICS, RADIOL & HUMAN ONCOL, UNIV WIS-MADISON, 86- *Mem:* Am Asn Physicists Med; Am Inst Ultrasound Med; Inst Elec & Electronics Engrs Ultrasonic & Frequency Control Soc. *Res:* Diagnostic ultrasound imaging and tissue characterization; development of methods for assessing medical ultrasound equipment performance; application of ultrasound in speech research; ultrasound hypothermia. *Mailing Add:* Dept Med Physics Univ Wis Rm 1530 MSC 1300 University Ave Madison WI 53706

ZAHALSKY, ARTHUR C, b New York, NY, Oct 31, 30; div. BIOCHEMISTRY, TOXICOLOGY. *Educ:* McGill Univ, BSc, 52; NY Univ, PhD(microbiol), 63. *Prof Exp:* Res assoc, Haskins Labs, 58-66; res collabr, Brookhaven Nat Labs, 68-74; asst prof microbiol, Queens Col (NY), 66-69; assoc prof biochem, doctoral prog, City Univ New York, 69-71; PROF IMMUNOL, SOUTHERN ILL UNIV, EDWARDSVILLE, 71- *Concurrent Pos:* Consult immunotoxicol, Immunox Res. *Mem:* AAAS; Am Soc Parasitol; Am Soc Microbiol; Am Acad Clin Toxicol; Sigma Xi. *Res:* Immune response to environmental toxicants; immune disorders from halogenated hydrocarbons; host immune response to parasites. *Mailing Add:* Dept Biol Sci Southern Ill Univ Edwardsville IL 62026

ZAHARKO, DANIEL SAMUEL, b New Westminster, BC, Nov 3, 30; US citizen; m 59; c 3. PHARMACOLOGY, PHYSIOLOGY. *Educ:* Univ BC, BPE, 53, dipl educ, 54; Univ Ill, MS, 55, PhD(physiol), 63. *Prof Exp:* Instr, Univ Sask, 55-57 & Univ Ill, 57-59; instr to asst prof pharmacol, Ind Univ, Bloomington, 63-68; USPHS res fel, 68-70, pharmacologist, Lab Chem Pharmacol, 70-84, PHARMACOLOGIST, PHARMAC BR, DTP, NAT CANCER INST, 85 - *Honors & Awards:* Co-winner, Ebert Prize, Am Pharmaceut Asn, 72. *Mem:* Am Asn Cancer Res; AAAS; Am Soc Pharmacol & Exp Therapeut. *Res:* Environmental physiology; biochemical pharmacology; cancer chemotherapy; pharmacokinetics. *Mailing Add:* DCT Nat Cancer Inst NIH Exec Plaza N Rm 841 Rockville MD 20892

ZAHED, HYDER ALI, US citizen; m 81; c 1. SCIENCE COMMUNICATIONS. *Educ:* A P Agr Univ, BS, 69; Univ Delaware, MS, 73; Pratt Inst, MS, 74. *Prof Exp:* Coordr, Kuwait Inst Sci Res, 82-83; chief, User Commun Sec, Biosci Info Ser, 84-85; info specialist, 80-81, HEAD TECH INFO, BURROUGHS WELLCOME CO, 86- *Res:* Computer based training for production personnel; advantages of computer based training over standard lecture methods of training in pharmaceutical industry. *Mailing Add:* 3060 Dartmouth Dr Greenville NC 27858

ZAHED, ISMAIL, b Algeria. MANY-BODY PHYSICS. *Educ:* Mass Inst Technol, MSc, 81, PhD(physics), 83. *Prof Exp:* Res asst physics, Niels Bohr Inst, 83-84; res asst, 84-87, asst prof, 87-90, ASSOC PROF PHYSICS, STATE UNIV NY, STONY BROOK, 90- *Mem:* Am Phys Soc. *Res:* Role of hadronic constituents in nuclear physics; behavior of hadronic matter at high temperature and density. *Mailing Add:* Physics Dept State Univ NY Stony Brook NY 11794-3800

ZAHLER, STANLEY ARNOLD, b New York, NY, May 28, 26; m 52; c 3. MICROBIOL GENETICS. *Educ:* NY Univ, AB, 48; Univ Chicago, MS, 49, PhD, 52. *Prof Exp:* Instr gen bact, Northern Ill Col Optom, 51; USPHS fel bact, Univ Ill, 52-54; instr microbiol, Univ Wash, 54-57, asst prof, 57-58; asst prof, Med Ctr, WVa Univ, 59; from asst prof to assoc prof, 59-79, assoc dir, Div Biol Sci, 75-78, PROF MICROBIOL GENETICS, CORNELL UNIV, 79-, CHAIRPERSON, GENETICS & DEVELOP, 90- *Concurrent Pos:* Consult, Gen Elec Corp, 63-66, Sandoz, Inc, 80-82, Dow Chem, 81-84, Eastman Kodak Co, 85-88; USPHS spec fel, Scripps Clin & Res Found, 66-67. *Mem:* AAAS; Am Soc Microbiol; Genetics Soc Am; Brit Soc Gen Microbiol. *Res:* Bacteriophages; microbial genetics; metabolic controls; genetics of Bacillus subtilis and other bacilli. *Mailing Add:* Sect Genetics 459 Biotech Bldg Cornell Univ Ithaca NY 14853-2703

ZAHLER, WARREN LEIGH, b Springville, NY, June 28, 41; m 68; c 2. BIOLOGICAL CHEMISTRY. *Educ:* Alfred Univ, BA, 63; Univ Wis-Madison, MS, 66, PhD(biochem), 68. *Prof Exp:* NIH fel molecular biol, Vanderbilt Univ, 67-71, res assoc, 71-72; asst prof, 72-77, PROF BIOCHEM, UNIV MO-COLUMBIA, 77- *Mem:* Am Oil Chem Soc; Am Chem Soc; Soc Study Reproduction; Sigma Xi. *Res:* Isolation and characterization of sperm membranes. Study of the function of sperm membranes and membrane bound enzymes in reproduction. *Mailing Add:* Dept Biochem M121 Med Sci Univ Mo Columbia MO 65212

ZAHN, JOHN J, b Beaver Dam, Wis, May 8, 32; div; c 3. ELASTIC STABILITY, STRUCTURAL DESIGN. *Educ:* Univ Wis, BS, 54, MS, 59, PhD(mech), 64. *Prof Exp:* Instr eng mech, Univ Wis, 59-62, lectr, 62-64; ENGR, FOREST PROD LAB, FOREST SERV, USDA, 64- *Honors & Awards:* L J Markwardt Award, Forest Prod Res Soc, 78; Raymond C Reese Res Prize, Am Soc Civil Engrs, 88. *Mem:* Am Soc Civil Engrs; Fedn Am Scientists. *Res:* Theory of elasticity; elastic stability; applied mathematics; probability. *Mailing Add:* Forest Prod Lab One Gifford Pinchot Dr Madison WI 53705

ZAHND, HUGO, b Berne, Switz, May 16, 02; nat US; m 26, 68; c 2. BIOCHEMISTRY. *Educ:* NY Univ, BS, 26; Columbia Univ, AM, 29, PhD(biochem), 33. *Prof Exp:* Tutor chem, 28-34, from instr to prof, 34-72, EMER PROF CHEM, BROOKLYN COL, 72- *Concurrent Pos:* Contrib, Chem & Technol Food & Food Prod, Intersci Publ, Inc, New York, NY, 44, & Acad Am Encycl, Arete Publ Co, Princeton, NJ, 80. *Mem:* Fel AAAS; Am Chem Soc; Hist Sci Soc; fel Am Inst Chem; NY Acad Sci. *Res:* Labile sulfur in proteins; quantitative inorganic and organic analysis; history of chemistry; chromatography as applied to the fields of alkaloids, amino acids and proteins. *Mailing Add:* 42 Herbert Ave Port Washington NY 11050

ZAHNLEY, JAMES CURRY, b Manhattan, Kans, Apr 16, 38; m 65; c 2. BIOCHEMISTRY. *Educ:* Kans State Univ, BS, 58; Purdue Univ, MS, 62, PhD(biochem), 63. *Prof Exp:* Res assoc enzym, Syntex Inst Molecular Biol, Calif, 63-64, assoc biochem, 64-65; asst res biochemist, NIH grant, 65-66; RES CHEMIST, WESTERN REGIONAL RES CTR, AGR RES SERV, USDA, 66- *Mem:* AAAS; Am Soc Biol Chemists; Am Chem Soc; Protein Soc. *Res:* Enzymology; protein chemistry; food, proteins and plant biochemistry. *Mailing Add:* Western Regional Res Ctr USDA 800 Buchanan St Albany CA 94710

ZAHRADNIK, RAYMOND LOUIS, b Ford City, Pa, Sept 18, 36; m 60; c 3. CHEMICAL ENGINEERING, ENERGY SCIENCE. *Educ:* Carnegie Inst Technol, BS, 59, MS, 61, PhD(chem eng), 63. *Prof Exp:* Sr engr res labs, Westinghouse Elec Corp, 61-65, fel engr, 65-66; from assoc prof to prof chem eng, Carnegie-Mellon Univ, 67-74; dir, Div Coal Conversion & Utilization, ERDA, 75-76; pres, Ray Zahradnik Consult, Inc, 76-77; DIR ENERGY RES, OCCIDENTAL OIL & GAS, DIV OCCIDENTAL PETROL CORP, 77- *Concurrent Pos:* Prog mgr energy res & technol, NSF, 72-74; mem res coord panel, Gas Res Inst, 77- *Res:* Energy research. *Mailing Add:* Occidental Petrol Corp PO Box 880408 Steamboat Springs CO 80488

ZAIA, JOHN ANTHONY, b Oneida, NY, Oct 28, 42; m 71; c 3. HEMATOLOGY. *Educ:* Holy Cross Col, AB, 64; Dartmouth Col, BMSc, 66; Harvard Univ, MD, 68. *Prof Exp:* Intern, St Louis Child Hosp & Child Med Ctr, 68-71; sr asst surg, Virol Div, Lab Bur, USPHS, 71-74; res fel infectious dis, Sidney Farber Cancer Inst, 76-77; DIR, DIV PEDIAT, CITY HOPE NAT MED CTR, 80- *Mem:* Am Pediat Soc; Infectious Dis Soc Am. *Res:* Mechanisms of viral pathogenesis; development of anti-viral therapies. *Mailing Add:* Commun Dis Ctr Centers Dis Control Atlanta GA 30322

ZAIDEL, ERAN, b Kibbutz Yagur, Israel, Jan 23, 44; US citizen; m 65; c 2. NEUROPSYCHOLOGY. *Educ:* Columbia Univ, AB, 67; Calif Inst Technol, MSc, 68, PhD(psychobiol), 73. *Prof Exp:* Res fel psychobiol, 73-76, sr res fel, 76-80, VIS ASSOC, DEPT BIOL, CALIF INST TECHNOL, 80- *Mem:* Soc Res Child Develop; Int Neuropsychol Soc; Acad Aphasia. *Res:* Neurolinguistics and psycholinguistics; cognitive and developmental psychology; epistemology and the philosophy of science, of mind and of language. *Mailing Add:* Dept Psychol 1285 Franz Hall Univ Calif 405 Hilgard Ave Los Angeles CA 90024

ZAIDER, MARCO A, b Bacau, Romania, Jan 3, 46; US citizen; m 68; c 3. NUCLEAR PHYSICS, RADIOLOGICAL PHYSICS. *Educ:* Bucharest Univ, MSc, 68; Tel Aviv Univ, PhD(nuclear physics), 76. *Prof Exp:* Teaching & res asst nuclear physics, Tel Aviv Univ, Israel, 71-76; fel, Los Alamos Sci Lab, 76-78, staff mem radiol physics, 78-79; res assoc, Col Physicians & Surgeons, 79-83, res scientist, 83-87, ASSOC PROF RADIATION ONCOL & PUB HEALTH, COLUMBIA UNIV, 87- *Concurrent Pos:* Res grant, Los Alamos Sci Lab, 77-78; Dept Energy grant, 86- *Mem:* Am Asn Physicists Med; Radiation Res Soc. *Res:* Instrumentation; medical physics; biophysics. *Mailing Add:* Col Physicians & Surgeons Columbia Univ 630 W 168th St New York NY 10032

ZAIDI, SYED AMIR ALI, b Lahore, Pakistan, Apr 15, 35; m 62. NUCLEAR PHYSICS. *Educ:* Punjab Univ, BSc, 56; Univ Gottingen, 57-58; Univ Heidelberg, dipl physics, 60, PhD(physics), 64. *Prof Exp:* Vis res scientist, Max Planck Inst Nuclear Physics, 64-66; asst prof, 66-68, ASSOC PROF PHYSICS, UNIV TEX, AUSTIN, 68-, ASSOC DIR, CTR NUCLEAR STUDIES, 67- *Mem:* Fel Am Phys Soc. *Res:* Isobaric analogue resonances; nuclear structure studies using shell model description of reaction theory; heavy ion induced reactions; elementary particle physics. *Mailing Add:* Dept Physics Univ Tex Austin TX 78712

ZAIDINS, CLYDE, ASTROPHYSICS, GAMMA RAY ASTRONOMY. *Educ:* Calif Inst Technol, BS, 61, MS, 63, PhD(physics), 67. *Prof Exp:* PROF PHYSICS, UNIV COLO, DENVER, 71- *Concurrent Pos:* Assoc Ctr Astrophysics & Space Astron, Boulder; vis prof, Instituto de Fisica, Milano, 88-89. *Res:* Astrophysics, nuclear astrophysics; nucleaosynthesis gamma ray astronomy. *Mailing Add:* 1450 Kendall Dr Boulder CO 80309

ZAIDMAN, SAMUEL, b Bucharest, Romania, Sept 4, 33; Can citizen; m 62; c 4. ABSTRACT & PARTIAL DIFFERENTIAL EQUATIONS. *Educ:* Univ Bucharest, Lic, 55; Univ Paris, Dr d'Etat, 70. *Prof Exp:* Asst math, Univ Bucharest, 55-59; vis prof, CNR Italy, 61-64; PROF MATH, UNIV MONTREAL, 64- *Concurrent Pos:* Vis prof, Univ Geneva, 66-68, Univ Padova CNR Italy, 79-80. *Mem:* Am Math Soc. *Res:* Abstract differential equations; pseudo-differential operators; almost-periodic equations. *Mailing Add:* Dept Math Univ Montreal Box 6128 Montreal PQ H3C 3J7 Can

ZAIKA, LAURA LARYSA, b Kharkow, Ukraine, June 23, 38; US citizen. FOOD CHEMISTRY. *Educ:* Drexel Inst Tech, BS, 60; Univ Pa, PhD(org chem), 64. *Prof Exp:* RES CHEMIST FOOD SAFETY, SCI & EDUC ADMIN-AGR RES, USDA, 64- *Mem:* Am Chem Soc; Inst Food Technologists; Asn Off Anal Chemists. *Res:* 2-aryl benzimidazoles; 1, 2, 3-benzotriazines; meat flavor investigations; chromatography; microbial metabolites; fermented meat products; spices. *Mailing Add:* 40 Johns Rd Cheltenham PA 19012

ZAIM, SEMIH, b Bursa, Turkey, Mar 1, 26; US citizen; m 51; c 3. SYNTHETIC RUBBER, RHEOLOGY OF POLYMERS. *Educ:* Univ Istanbul, Turkey, BS, 47, MS, 49. *Prof Exp:* Chem engr aviation fuels & lubricant qual control res develop, Arge Res Labs, Ankara, Turkey, 49-61; NATO res fel aviation fuels res & develop, Carde Res Labs, Que, Can, 61-62; res chemist polymer, res & develop div, Polysar Corp, Sarnia, Ont, Can, 62-69 & Tex US Res Ctr, Parsippany, NJ, 69-76; res supvr adhesives, Essex Chem Corp, Sayreville, NJ, 76-81; VPRES RES & DEVELOP SPECIALTY CHEM, CHESSCO-PROCESS RES PROD, TRENTON, NJ, 81- *Concurrent Pos:* Consult, UN Develop Prog, 79-81. *Res:* Metal working lubricants, specialty chemicals for quartz and silicon wafer and bearing ball manufacturing; metal working lubricants. *Mailing Add:* Chessco Process Res Prod 1013 White Head Rd Ext Trenton NJ 08638-2418

ZAININGER, KARL HEINZ, b Endorf, Ger, Aug 3, 29; US citizen; m 52; c 3. SOLID STATE PHYSICS, PHYSICAL ELECTRONICS. *Educ:* City Col New York, BSEE, 59; Princeton Univ, MSE, 61, MA, 62, PhD(elec eng), 64. *Prof Exp:* Mem tech staff, David Sarnoff Res Ctr, 59-68, group head solid state device technol group RCA Labs, 68-77; dir commercialization, Solar Energy Res Inst, 77-78; dir microelectronics, Electronic Devices & Technol Lab, US Army, 78-80; exec vpres, Siemens Res & Technol Lab, 80-86; PRES, SIEMENS CORP RES & SUPPORT & RES DIR, SIEMENS RES & TECHNOL LABS, PRINCETON, NJ, 86- *Concurrent Pos:* David Sarnoff fel, doctoral studies prog, 62; mem orgn & tech prog comt, Mos Interface Specialist Conf, Las Vegas, 64-68; gen chmn & chmn, Bd Dir Inst Elec & Electronics Engrs Nat Reliability Physics Symp, 71; chmn, Electronics Physics Dept, LaSalle Univ, 73-76; mem conf bd, Adv Panel Radiation Device Technol, 89; ed Inst Elec & Electronics Engrs trans, electron devices; mem, Nat Res Coun Mat Adv Bd; chmn, Pub Serv Elec & Gas Corp, Res & Develop Strategy Comt; mem, Res & Develop Coun NJ; mem, Inst Elec & Electronics Engrs Standards bd; very large scale intergration technol Sci Coordr, UCLA; UN specialist Intergrated Circuit Technol, Jerusalem, Isreal. *Honors & Awards:* RCA Labs Achievement Award, 65. *Mem:* Fel Inst Elec & Electronics Engrs; Am Phys Soc. *Res:* Semiconductor devices; thin film physics; plasmas in solids; physical and electrical properties of SiO2; oxidation and optical properties of Si and GaAs; physics and technology of metal-insulator-semiconductor devices; radiation effects of metal-insulator-semiconductor structures. *Mailing Add:* Siemens Corp Res & Support Princeton Forrestal Ctr 105 College Rd E Princeton NJ 08540

ZAISER, JAMES NORMAN, b Salem, NJ, Jan 23, 34; m 57; c 3. MECHANICAL ENGINEERING. *Educ:* Univ Del, PhD(appl sci), 64. *Prof Exp:* Instr mech eng, Univ Del, 58-63; res engr, Exp Sta, E I du Pont de Nemours & Co, 63; asst prof, 65-76, ASSOC PROF MECH ENG, BUCKNELL UNIV, 76- *Concurrent Pos:* NSF grant, Bucknell Univ, 66-67. *Mem:* Am Soc Mech Engrs; Am Soc Eng Educ. *Res:* Nonlinear analysis and dynamics of particles and rigid bodies; nonlinear analysis of systems described by ordinary and partial differential equations. *Mailing Add:* Dept Mech Eng Bucknell Univ Lewisburg PA 17837

ZAITLIN, MILTON, b Mt Vernon, NY, Apr 2, 27; m 51; c 4. PLANT VIROLOGY. *Educ:* Univ Calif, BS, 49; Univ Calif, Los Angeles, PhD(plant physiol), 54. *Prof Exp:* Res officer, Commonwealth Sci & Indust Res Orgn, Canberra, Australia, 54-58; asst prof hort, Univ Mo, 58-60; asst agr biochem, Univ Ariz, 60-62, from assoc prof to prof agr biochem & plant path, 66-73; assoc dir, biotechnol prog, 83-90, PROF PLANT PATH, CORNELL UNIV, 73-, DIR, BIOTECHNOL PROG, 90- *Concurrent Pos:* Guggenheim & Fulbright fels, 66-67; assoc ed, Virol, 66-71, 82-84, ed, 72-81; bd dir Int Soc Plant Mol Biol, 85-89; sr ed, Molecular Plant-Microbe Interactions, 87-90. *Mem:* Soc Gen Microbiol; fel Am Phytopath Soc; Am Soc Plant Physiologists; Am Soc Virol; Int Soc Plant Molecular Biol. *Res:* Plant viruses; molecular biology of plant virus disease; molecular basis of plant-virus infections. *Mailing Add:* Dept Plant Path Cornell Univ Ithaca NY 14853

ZAJAC, ALFRED, b Vienna, Austria, Feb 18, 17; US citizen; m 50; c 2. PHYSICS. *Educ:* Univ St Andrews, BSc, 46, Hons, 48; NY Univ, MS, 52; Polytech Inst Brooklyn, PhD(physics), 57. *Prof Exp:* From instr to assoc prof, 55-74, PROF PHYSICS & CHMN DEPT, ADELPHI UNIV, 74- *Concurrent Pos:* NSF res grant, 62-64. *Mem:* Am Phys Soc; Am Asn Physics Teachers; Am Crystallog Asn. *Res:* Crystal perfection, thermal motion and anomalous transmission of x-rays. *Mailing Add:* Dept Physics Hofstra Univ 1000 Fulton Ave Hempstead NY 11550

ZAJAC, BARBARA ANN, b Fountain Springs, Pa, Mar 15, 37; m 57; c 1. INTERNAL MEDICINE, INFECTIOUS DISEASES. *Educ:* Univ Pa, BA, 58, PhD(microbiol), 67; Med Col Pa, MD, 79. *Prof Exp:* Assoc pediat/virol, Univ Pa, 69-70; from asst prof to assoc prof microbiol, Med Col Pa, 70-76; resident internal med, Abington Mem Hosp, 79-82; fel infectious dis, Hosp Univ Pa, 82-84; dir clin res, Merck Sharp & Dohme Res Labs, 84-88; GROUP DIR IMMUNOL & INFECTIOUS DIS, DUPONT MERCK PHARMACEUT CO, 88- *Concurrent Pos:* NIH fel, Div Virus Res, Children's Hosp Philadelphia, Pa, 67-69; grants, Res Corp & Anna Fuller Fund, 72-73, Damon Runyon Mem Fund, 73-75 & Nat Cancer Inst, 73-76; affil staff physician, Abington Mem Hosp, 86- *Mem:* AMA; Am Col Physicians; Infectious Dis Soc Am; Am Med Women's Asn; Am Soc Microbiol. *Res:* DNA viruses (EBV, VZV, Hepatitis B) host-parasite relationships and immune responses; clinical investigation, immunology and infectious diseases. *Mailing Add:* PO Box 872 Hockessinn Valley DE 19707-9993

ZAJAC, FELIX EDWARD, III, b Baltimore, Md, Dec 4, 41; m 62; c 2. NEUROPHYSIOLOGY, BIOMEDICAL ENGINEERING. *Educ:* Rensselaer Polytech Inst, BEE, 62; Stanford Univ, MS, 65, PhD(neurosci), 68. *Prof Exp:* Staff assoc, Lab Neural Control, Nat Inst Neurol Dis & Stroke, 68-70; asst prof elec eng, Univ Md, Col Park, 70-73, assoc prof, 73-80, dir, Biomed Res Lab, 71-80; Eng Res & Develop Ctr, Vet Admin Med Ctr, Palo Alto, Calif, 80-; AT DEPT MECH ENG, STANFORD UNIV, CALIF. *Mem:* AAAS; Am Physiol Soc; Soc Neurosci; Inst Elec & Electronics Engrs. *Res:* Neural control and biomechanics of animal movement with emphasis on cat locomotion and jumping. *Mailing Add:* Dept Mech Eng Stanford Univ Stanford CA 94305

ZAJAC, IHOR, b Lwiw, Ukraine, May 26, 31; US citizen; m 57; c 1. MEDICAL MICROBIOLOGY, VIROLOGY. *Educ:* Univ Pa, BA, 58; Hahnemann Med Col, MS, 60, PhD(microbiol), 64. *Prof Exp:* Asst microbiol, Hahnemann Med Col, 64-65, instr, 65; asst prof, Jefferson Med Col, 65-71; assoc sr investr, 71-75, sr investr, Smith Kline & French Labs, 75-83; res assoc, E I Du Pont, 83-90, RES ASSOC, DU PONT MERCK PHARMACOL CO, 90- *Concurrent Pos:* Vis lectr, Med Col Pa, 72-78; vis assoc prof, Hahnemann Med Col, 81-; adj assoc prof, Jefferson Med Col, 81-85. *Mem:* Am Soc Microbiol; Soc Exp Biol & Med; Sigma Xi. *Res:* Enteroviruses; cell-virus interactions; cell membrane; interferon; bacterial and viral chemotherapy; anaerobic bacteria; bacterial receptors; neoplastic chemotherapy; drug metabolism and disposition and interaction; pharmacokinetics. *Mailing Add:* PO Box 872 Hockessin DE 19707

ZAJAC, WALTER WILLIAM, JR, b Central Falls, RI, July 19, 34; m 59; c 6. ORGANIC CHEMISTRY, NATURAL PRODUCTS CHEMISTRY. *Educ:* Providence Col, BS, 55; Va Polytech Inst, MS, 57, PhD(chem), 60. *Prof Exp:* From asst prof to assoc prof, 59-72, PROF CHEM, VILLANOVA UNIV, 72- *Concurrent Pos:* Fel, Univ Alta, 65-66. *Mem:* Am Chem Soc. *Res:* Reduction of organic compounds; synthesis, mechanism of ring closure reactions and conformation of carbocyclic and heterocyclic systems; chemistry of aliphatic nitro compounds; carbohydrates. *Mailing Add:* Dept Chem Villanova Univ Villanova PA 19085

ZAJACEK, JOHN GEORGE, b Allentown, Pa, May 8, 36; m 64; c 3. ORGANIC CHEMISTRY. *Educ:* Lehigh Univ, BA, 58; Cornell Univ, PhD(org chem), 62. *Prof Exp:* MGR OXYGENATED PROF RES, ARCO CHEM CO DIV, ATLANTIC-RICHFIELD CO, 62- *Mem:* Am Chem Soc. *Res:* Oxidation of hydrocarbons; epoxidation of olefins; reactions of carbon monoxide; reduction of aromatic nitro compounds; metal catalyzed reactions; chemistry of selenium reactions; isocyanates; polyols; urethane chemistry; unsaturated polyesters. *Mailing Add:* 669 Clovelly Lane Devon PA 19333-1846

ZAJIC, JAMES EDWARD, microbiology, law; deceased, see previous edition for last biography

ZAJONC, ARTHUR GUY, b Boston, Mass, Oct 11, 49; m 74; c 2. LASER PHYSICS, ATOMIC PHYSICS. *Educ:* Univ Mich, Ann Arbor, BSE, 71, MS, 73, PhD(physics), 76. *Prof Exp:* Fel atomic physics, Joint Inst Lab Astrophys, Nat Bur Standards & Univ Colo, 76-78; asst prof, 78-84, ASSOC PROF PHYSICS, AMHERST COL, 84- *Concurrent Pos:* Vis scientist, Ecole Normale Superieure, 81-82, Max Planck Inst Quantum Optics, 84 & Univ Hanover, Inst Quantum Optics, 86. *Mem:* Am Phys Soc; Am Asn Physics Teachers; Hist Sci Soc; Optic Soc Am; fel Lindisarne Asn. *Res:* Experimental foundations of physics; laser spectroscopy; radiative transfer; electron-atom collisions. *Mailing Add:* Dept Physics Amherst Col Amherst MA 01002

ZAK, BENNIE, b Detroit, Mich, Sept 29, 19; m 46; c 3. SPECTROSCOPY, ELECTROPHORESIS. *Educ:* Wayne State Univ, BS, 48, PhD(chem), 52. *Prof Exp:* From asst prof to prof, 57-90, EMER PROF CLIN CHEM, WAYNE STATE UNIV, 90- *Concurrent Pos:* Res technician, Detroit Receiving Hosp, 50-51, med lab analyst & jr assoc chem, 51-57, head clin chem, 80-; consult, Sinai Hosp, Detroit, Mich, 53-, St John's Hosp, 60-, Vet Admin Hosp, Allen Park, 60-, Holy Cross Hosp, 62- & William Beaumont Hosp, Royal Oak, 65-; fac res award, Sigma Xi, 73. *Honors & Awards:* Ames Award, Am Asn Clin Chem, 74, Gen Diagnostics Lectureship, 81; Benedetti-Pichler Award, Am Microchem Soc, 84. *Mem:* Am Chem Soc; Am Asn Clin Chem. *Res:* Spectrophotometric procedures in clinical chemistry on trace metals; peroxidase-coupled indicator reactions; problems involved with hyperlipidemia-hyperproteinemia measurements; problems in electrophoresis. *Mailing Add:* Wayne State Univ Sch Med 540 E Canfield Detroit MI 48021

ZAK, RADOVAN HYNEK, b C Budejovice, Czech, June 15, 31; US citizen; m 63; c 2. BIOCHEMISTRY, PHYSIOLOGY. *Educ:* Prague Univ, Czech, BS, 52, Dr Nat Sci, 54; Czech Acad Sci, Prague, PhD(biomed sci), 61. *Prof Exp:* Instr org chem, Med Sch, Prague Univ, 51-53; res scientist physiol, Inst Physiol, Czech Acad Sci, 57-61; res fel physiol chem, Dept Med, Northwestern Univ, Chicago, 61-63; fel biochem, 63-65, instr, 65-67, from asst prof to assoc prof, 67-79, PROF, DEPT MED, UNIV CHICAGO, 79- *Honors & Awards:* Nat Award, Czech Acad Sci, 58. *Mem:* AAAS; Am Physiol Soc; Int Soc Heart Res; Am Soc Cell Biol. *Res:* Cardiac hypertrophy; protein synthesis and degradation; proliferation of cardiac myocytes; muscle proteins. *Mailing Add:* Dept Med Pharmacol & Physiol Sci Univ Chicago 950 E 59th St Chicago IL 60637

ZAKAIB, DANIEL D, b Montreal, Que, Apr 1, 25; m 48; c 2. PETROLEUM CHEMISTRY, ANALYTICAL CHEMISTRY. *Educ:* Montreal Tech Inst, dipl chem, 46; Sir George Williams Univ, BSc, 53. *Prof Exp:* Supvr, Montreal Refinery Lab, Brit Am Oil Co, 48-55, asst refinery chemist, 55-58, analysis technologist, Head Off Toronto, 58-63, coordr, Analyticalal Res Labs, Brit Am Res & Develop Co, 63-66; mgr analytical & chem res, Res & Develop Dept, Gulf Oil Can, Ltd, 66-69, supvr petrol chem sect, Phys Sci Div, Gulf Res & Develop Co, 69-71, dir tech opers, Res & Develop Dept, Gulf Can, Ltd, 71-85; CONSULT, D D ZAKAIB & ASSOC, 85- *Mem:* Am Chem Soc; Am Soc Testing & Mat; Can Asn Appl Spectros; fel Chem Inst Can. *Res:* Technical administration of industrial research facility. *Mailing Add:* Two Confederation Way Thornhill ON L3T 5R5 Can

ZAKHARY, RIZKALLA, b Assiut, Egypt, Sept 5, 24; m 66; c 2. HUMAN ANATOMY. *Educ:* Cairo Univ, BS, 49, MS, 54; Tulane Univ, PhD(anat), 64. *Prof Exp:* Technician & res asst biochem, US Naval Med Res Unit 3, Cairo, Egypt, 50-56; instr biol chem & gen sci, Am Univ Cairo, 57-60; asst prof anat & physiol, Sch Dent, Loyola Univ, La, 64-67; asst prof anat, 67-70, ASSOC PROF ANAT, SCH DENT, UNIV SOUTHERN CALIF, 70- *Concurrent Pos:* NIH grant, Tulane Univ, 65-67. *Mem:* AAAS; Am Asn Anatomists; Am Asn Univ Profs. *Res:* Hypothermia, academic and applied aspects; cryobiology; stress and hypothermia; vascular casting. *Mailing Add:* Dept Anat Sch Dent Univ Southern Calif Los Angeles CA 90089-0641

ZAKI, ABD EL-MONEIM EMAM, b Cairo, Egypt, Dec 18, 33; US citizen. HISTOLOGY, ORAL BIOLOGY. *Educ:* Cairo Univ, BChD, 55, DDR, 58; Ind Univ, MSD, 62; Univ Ill, PhD(anat), 69. *Prof Exp:* Dent surgeon, Demonstration & Training Ctr, Qualyub, UAR, 55-59; teaching asst, Ind Univ Sch Dent, 60-61; from instr to asst prof, Fac Dent, Cairo Univ, 62-67; res assoc, Col Dent, 67-70, asst prof histol, Col Dent & lectr, Col Med, 70-72, assoc prof histol, Col Dent & Sch Basic Med Sci, Med Ctr, 72-75, PROF HISTOL, COL DENT, UNIV ILL, 75- *Concurrent Pos:* UAR govt spec mission mem grad study, US, 59-62. *Mem:* AAAS; Am Asn Anat; Electron Micros Soc Am; Int Asn Dent Res; Sigma Xi. *Res:* Cellular control of mineralization using a rodent odontogenic model; cytological and cytochemical aspects of odontogenesis; dental pulp tissue response to therapeutic agents. *Mailing Add:* Dept Histol-MC 690 Univ Ill Box 4348 Chicago IL 60680

ZAKI, MAHFOU H, b Cairo, Egypt, Apr 14, 24; US citizen; m 70. ENVIRONMENTAL MEDICINE, PUBLIC HEALTH. *Educ:* Cairo Univ, MB, ChB, 49; Univ Alexandria, MPH, 58; Columbia Univ, DrPH, 62. *Prof Exp:* Intern med & surg, Kasr-El-Eini Univ Hosp, Cairo, 50-51; med officer, Abu-Sidhom Health Ctr, Minia, 51-56; instr prev med, High Ints Pub Health, 57-59; staff mem edpidemiol, Sch Pub Health, Columbia Univ, 59-62; asst prof prev med, State Univ NY Downstate Med Ctr, 62-68, assoc prof environ med & community health, 68-76; ASSOC PROF ENVIRON MED & COMMUNITY HEALTH, KINGS COUNTY HOSP, 76- *Concurrent Pos:* Grants, Health Res Coun City New York, 64-68 & tuberc br, USPHS & Peace Corps; Peace Corps physician & prog tech adv, US State Dept, Afghanistan, 70-71; dir pub health, Suffolk County, New York; adj prof pub health, City Univ New York, 73-; lectr, Sch Pub Health & Admin Med, Columbia Univ, 73- *Mem:* Fel Am Pub Health Asn; fel Royal Soc Trop Med & Hyg; Am Soc Trop Med & Hyg; fel Am Col Chest Physicians; Am Statist Asn. *Res:* Etiology of sarcoidosis; prevalence of infection with typical and atypical strains of Mycobacterium tuberculosis and of infection with Histoplasma capsulatum in Afghanistan. *Mailing Add:* Dept Comm & Prev Med State Univ NY Health Sci Ctr Stony Brook NY 11794

ZAKIN, JACQUES L(OUIS), b New York, NY, Jan 28, 27; m 50; c 5. CHEMICAL ENGINEERING. *Educ:* Cornell Univ, BChem Eng, 49; Columbia Univ, MSc, 50; NY Univ, DEng Sc, 59. *Prof Exp:* Chem engr res labs, Flintkote Co, 50-51; from res technologist to supvry technologist, Res Dept, Socony Mobil Oil Co, 51-56 & 58-62; from assoc prof to prof chem eng, Univ Mo-Rolla, 62-77; PROF CHEM ENG & CHMN DEPT, OHIO STATE UNIV, 77- *Concurrent Pos:* Adj asst prof, Hofstra Col, 59-60 & 62; Am Chem Soc-Petrol Res Fund Int fel & vis prof, Israel Inst Technol, 68-69; vis scientist, Naval Res Lab, Washington, DC, 75-76; vis prof, Casali Inst, Hebrew Univ, 87. *Honors & Awards:* Tech Person of the Year, Columbus Tech Coun, 87. *Mem:* Am Inst Chem Engrs; Am Chem Soc; Rheology Soc; Am Soc Eng Educ. *Res:* Surfactant drag reduction in turbulent flow; mechanical degradation of high polymers; detailed structure of liquid turbulence; transport of viscous crudes as concentrated oil-in-water emulsions; rheology of polymer solutions. *Mailing Add:* Dept Chem Eng Ohio State Univ Columbus OH 43210

ZAKKAY, VICTOR, b Baghdad, Iraq, Sept 8, 27; US citizen; m 52; c 2. FLUID MECHANICS, COMBUSTION. *Educ:* Polytech Inst Brooklyn, BAeronautEng, 52, MS, 53, PhD(aeronaut), 59. *Prof Exp:* Res asst, Polytech Inst Brooklyn, 52-55, res assoc, 55-58, res group leader, 58-59, from res asst prof to res assoc prof, 59-64; assoc prof aerospace eng, 64-65, prof aeronaut & astronaut, 65-73, asst dir, Aerospace Lab, 70-76, acting chmn, Dept Appl Sci, 76-77, chmn, Dept Appl Sci, 77-84, PROF APPL SCI, NY UNIV, 73-, DIR, ANTONIO FERRI AEROSPACE & ENERGETICS LABS, 76- *Mem:* Am Inst Aeronaut & Astronaut. *Res:* Hypersonic and viscous compressible flow; heat transfer; experimental aerodynamics; combustion; turbulent mixing; fuel combustion fluidized bed coal combustion; hypersonic aerodynamics; wind tunnel technology; solar energy. *Mailing Add:* NY Univ 26-36 Stuyvesant St New York NY 10003

ZAKRAYSEK, LOUIS, b Conemaugh Twp, Pa, Dec 20, 28; m 52; c 3. METALLURGY, MATERIALS SCIENCE. *Educ:* Pa State Univ, BS, 52. *Prof Exp:* Mem staff, Chemet Training Prog, 52-53, metallurgist, 53-60, mgr metall & welding, 60-72, MGR PHYS METALL, GEN ELEC CO, SYRACUSE, 72- *Concurrent Pos:* Chmn joining tech comt, Inst Printed Circuits, 67-68; mem working group 9, Tech Comn 50, Int Electrotech Comn, 68-69; consult eng, EMK Testing Co Inc, Cicero, NY. *Honors & Awards:* President's Award, Inst Printed Circuits, 68. *Mem:* Fel Am Soc Metals; Metall Soc; Am Inst Mining, Metall & Petrol Engrs; Int Metallog Soc. *Res:* Development, selection, specification and application of materials and processes in commerical and military electronic equipment, especially as related to construction analysis, failure analysis, component assembly and electronic packaging. *Mailing Add:* 8432 Brewerton Rd Cicero NY 13039

ZAKRISKI, PAUL MICHAEL, b Amsterdam, NY, July 12, 40; m 64; c 3. ANALYTICAL CHEMISTRY. *Educ:* Univ Rochester, AB, 62; Univ Cincinnati, PhD(org chem), 67. *Prof Exp:* Sr res chemist, 66-74, SECT LEADER, RES DIV, B F GOODRICH CO, BRECKSVILLE, 74- *Mem:* Am Chem Soc; Am Soc Mass Spectrometry. *Res:* Mass spectrometry for structure elucidation; high speed chromatography. *Mailing Add:* 8329 Wyatt Rd Cleveland OH 44147

ZAKRZEWSKI, RICHARD JEROME, b Hamtramck, Mich, Nov 5, 40; m 66; c 2. VERTEBRATE PALEONTOLOGY. *Educ:* Wayne State Univ, BS, 63; Univ Mich, MS, 65, PhD(vert paleont), 68. *Prof Exp:* NSF fel geol, Idaho State Univ-Los Angeles County Mus, 68-69; from asst prof to assoc prof earth sci, 69-78, PROF GEOL, FT HAYS STATE UNIV, 78-, DIR, STERNBERG MEM MUS, 73- *Mem:* Soc Vert Paleont; Paleont Soc; Am Soc Mammal; Soc Syst Zool; Am Quaternary Asn. *Res:* Fossil mammals, particularly rodents; late Cenozoic stratigraphy. *Mailing Add:* Sternberg Mem Mus Ft Hays State Univ Hays KS 67601-4099

ZAKRZEWSKI, SIGMUND FELIX, b Buenos Aires, Arg, Sept 15, 19; m 56; c 1. BIOCHEMISTRY. *Educ:* Univ Hamburg, MS, 52, PhD(biochem), 54. *Prof Exp:* Res asst, Sch Med, Western Reserve Univ, 52-53; res asst, Sch Med, Yale Univ, 53-56; sr cancer res scientist, 56-61, assoc cancer res scientist, 61-71, prin cancer res scientist, Dept Exp Therapeut, 71-76, prin cancer res scientist, Dept Clin Pharm & Therapeut, 78-87, CONSULT, DEPT EXP THER, ROSWELL PARK MEM INST, 87- *Concurrent Pos:* Res prof, Dept Pharmacol, State Univ NY Buffalo, Roswell Park Div, emer prof, Roswell Park Mem Inst. *Mem:* AAAS; Am Asn Cancer Res. *Res:* Cancer chemotherapy; metabolism of folic acid and folic acid antagonist; pharmacokinetics of anticancer drugs in man. *Mailing Add:* 260 Lakewood Pkwy Buffalo NY 14226

ZAKRZEWSKI, THOMAS MICHAEL, b Jackson, Mich, Mar 13, 43; m 68; c 3. AEROSPACE SCIENCES, CHEMISTRY. *Educ:* Univ Mich, BS, 63. *Prof Exp:* Tech staff mem EO & IR sensors, Willow Run Labs, Inst Sci & Technol, Univ Mich, 63-69; tech staff mem IR sensors & systs, Gen Res Corp, 69-73; dir Washington opers IR systs & simulations, Mission Res Corp, 73-74; tech staff mem, Inst Res & Active Optics, 74-75, dir, Space Systs Dept, Gen Res Corp, 75-81, dir, Washington Opers, Flow Gen, Inc, 78-81, dir, Eastern Opers, Technol Appln Group & Space Systs Dept, 81-83; GROUP VPRES, MID-ATLANTIC REGION, NICHOLS RES GROUP, 83- *Concurrent Pos:* Chmn, Comt Optical Measurements for Missile Tests, Air Force Studies Bd, Nat Res Coun, 86-90. *Mem:* AAAS; Sigma Xi; Am Inst Aeronaut & Astronaut; Am Chem Soc; Nat Space Inst; Am Soc Naval Engrs. *Res:* Electro-optical and infrared sensors, target signatures, and phenomenology; advanced offensive and defensive strategic systems; advanced space systems technology. *Mailing Add:* 10917 Howland Dr Reston VA 22091-4903

ZALAY, ANDREW W(ILLIAM), b Budapest, Hungary, May 20, 18; m 46; c 3. ORGANIC CHEMISTRY. *Educ:* Budapest Tech Univ, dipl, 40, EMe, 42. *Prof Exp:* Chem engr, 36-40, Dr Chem Eng, Budapest Tech Univ, 36-42, Pharmacist, Pasmany Peter Univ Sci; Res lab leader org chem, Chinoin Pharmaceut Works, Budapest, 51-56; res chemist, Textile Res Inst, NJ, 57-58; SR RES CHEMIST, STERLING-WINTHROP RES INST, 58- *Mem:* Am Chem Soc. *Res:* Pharmaceuticals. *Mailing Add:* 416 Robbins Ave PO Box 5938 Ewing NJ 08638-3724

ZALAY, ETHEL SUZANNE, b Budapest, Hungary, Sept 1, 19; m 46; c 3. ORGANIC CHEMISTRY. *Educ:* Univ Sci Budapest, Hungary, PhD, 44. *Prof Exp:* Owner, Dr Somody Lab, 45-51; org chemist, Fine Chem Producing Union, 51-54; owner, Dr Somody Lab, 54-56; org chemist, Textile Res Inst, 57; res chemist, Biol Res Lab, Philadelphia Gen Hosp, 57-58; res chemist, 58-69, assoc res chemist, 69-77, RES CHEMIST, STERLING-WINTHROP RES INST, STERLING DRUG, INC, 77- *Mem:* Am Chem Soc. *Res:* Fine organic chemicals; phospholipids; pharmaceuticals; heterocyclic chemistry. *Mailing Add:* 905 Myrtle Ave Albany NY 12208-2219

ZALESAK, JOSEPH FRANCIS, b Fountain Hill, Pa, Jan 2, 42; m 85; c 5. UNDERWATER ACOUSTICS, TRANSDUCTION. *Educ:* LaSalle Col, BA, 63; Lehigh Univ, MS, 67, PhD(solid state physics), 72. *Prof Exp:* Physicist, 63-66, RES PHYSICIST, NAVAL RES LAB, 72- *Concurrent Pos:* Assoc prof, Univ Cent Fla, 80-81. *Mem:* Acoust Soc Am. *Res:* Underwater acoustic calibration techniques and development of calibration systems which implement the above techniques; acoustic radiation from vibrating structures; develop specialized underwater acoustic transducers. *Mailing Add:* Naval Res Lab Underwater Sound Reference PO Box 568337 Orlando FL 32856-8337

ZALESKI, MAREK BOHDAN, b Krzemieniec, Poland, Oct 18, 36; US citizen. IMMUNOGENETICS, TRANSPLANTATION. *Educ:* Sch Med, Warsaw, MD, 60, Dr med sci, 63. *Prof Exp:* From res assoc to assoc prof, Dept Histol, Sch Med, Warsaw, Poland, 55-69; res asst prof, Dept Microbiol, State Univ NY, 69-72; from asst prof to assoc anat & histol, Dept Anat, Mich State Univ, 72-76; assoc prof, 76-79, PROF, DEPT MICROBIOL, STATE UNIV NY, 79- *Concurrent Pos:* Vis scientist, Inst Exp Biol & Genetics, Prague, 65; Brit coun scholar, Res Lab, Queen Victoria Hosp, 66-67; attend physician, Children Hosp, Poland, 60-66; mem, Witebsky Ctr Immunol, 76-; prin invest, NIH res grant, 76-88, NEH grant, 84-86. *Mem:* Transplantation Soc; Int Soc Exp Hemat; Am Asn Immunologists; NY Acad Sci. *Res:* Heterotopic bone tissue induction; cellular aspects of immune response; graft-verus-host reaction; immune response to alloantigens; genetic regulation of the immune response to alloantigens; IL-2-induced proliferation; translation of philosophical works from Polish to English. *Mailing Add:* Dept Microbiol State Univ NY Buffalo NY 14214

ZALESKI, WITOLD ANDREW, b Pyzdry, Poland, Apr 4, 20; Can citizen; m 48; c 4. CHILD & ADOLESCENCE PSYCHIATRY. *Educ:* Univ Edinburgh, MB, ChB, 46; Royal Col Physicians & Surgeons, Ireland, dipl psychol med, 52; Royal Col Physicians & Surgeons Can, cert psychiat, 62; Univ Sask, MD, adeundem, 64; FRCP(C), 72; Royal Col Psychiat, cert, 73. *Prof Exp:* Dep supt & consult psychiatrist, Ment Retardation Insts, Regional Hosp Bd Eng, Birmingham, 54-58; clin dir, Sask Training Sch, Moose Jaw, Can, 58-67; assoc prof, 67-73, prof pediat & assoc prof psychiat, 73-87, dir, Alvin Buckwold Ctr, Univ Hosp, 67-87, EMER PROF, UNIV SASK, 87- *Concurrent Pos:* Vis consult psychiat, Univ Sask Hosp, 62-67, vis consult, St Paul's Hosp & City Hosp, Saskatoon; hon consult, Univ Hosp, Saskatoon, 87-; hon prof psychiatry, Univ BC, 88. *Mem:* Fel Can Pediat Soc; fel Can Acad of Child Psychiat; Can Med Asn; Can Psychiat Asn. *Res:* Etiology and prevention of mental retardation; inborn errors of metabolism; behavioral programs for the retarded; delivery of services in mental retardation. *Mailing Add:* 4195 Rockridge Rd West Vancouver BC V7W 1A3 Can

ZALEWSKI, EDMUND JOSEPH, b Schenectady, NY, July 23, 31; m 58; c 5. ORGANIC POLYMER CHEMISTRY. *Educ:* Union Col, BS, 64. *Prof Exp:* From technician polyester to group leader polymer, 50-67, mgr polymer, 67-69, MGR RES, SCHENECTADY CHEM, INC, 69- *Mem:* Am Chem Soc; Soc Plastics Engrs. *Res:* Development of organic and heterocyclic polymers exhibiting excellent mechanical properties coupled with chemical and thermal resistance for use as electrical insulation. *Mailing Add:* 2761 Maida Lane Schenectady NY 12306

ZALIK, RICHARD ALBERT, b Buenos Aires, Argentina, Nov 20, 43; nat US; m 70; c 2. ROTORDYNAMICS, APPROXIMATION THEORY. *Educ:* Univ Buenos Aires, MS, 68; Technion, Israel Inst Technol, DSc(math), 73. *Prof Exp:* Lectr math, Ben Gurion Univ Negev, Israel, 74-77; vis asst prof, Univ Rhode Island, 77-78; from asst prof to assoc prof math, 78-85, PROF MATH, AUBURN UNIV, 85- *Concurrent Pos:* prin invest, Cray Res Inc, 88, NASA, 89; NASA & Am Soc Eng Educ fac fel, Marshall Space Flight Ctr, 87 & 88. *Mem:* Am Math Soc; Math Assoc Am; Soc Ind & Appl Math; Sigma Xi; Asn Comput Mach. *Mailing Add:* 120 Math Annex Bldg Auburn Univ Auburn AL 36849-5307

ZALIK, SARA E, b Mex, May 23, 39; m 66; c 2. DEVELOPMENTAL BIOLOGY, CELL BIOLOGY. *Educ:* Nat Univ Mex, BS, 59; Univ Ill, PhD(anat), 63. *Prof Exp:* NIH int fel, Biol Div, Oak Ridge Nat Lab, 63-64; asst prof cell biol, Ctr Res & Advan Studies, Nat Polytech Inst, Mex, 64-66; from asst prof to assoc prof, 66-78, PROF ZOOL, UNIV ALTA, 78- *Mem:* AAAS; Can Soc Cell Biol; Am Soc Zool; Soc Develop Biol; Am Soc Cell Biol; Sigma Xi. *Res:* Cell differentiation and metaplasia; cell surface and its role in differentiation and early embryogenesis. *Mailing Add:* Dept Zool Univ Alta Edmonton AB T6G 2E9 Can

ZALIK, SAUL, b Ratcliffe, Sask, May 11, 21; m 66; c 2. PLANT PHYSIOLOGY, BIOCHEMISTRY. *Educ:* Univ Man, BSA, 43, MSc, 48; Purdue Univ, PhD(plant physiol), 52. *Prof Exp:* Lectr plant sci, Univ Man, 48-49; from asst prof to assoc prof, 52-64, PROF PLANT PHYSIOL & BIOCHEM, UNIV ALTA, 64- *Mem:* AAAS; Am Soc Plant Physiol; Can Biochem Soc. *Res:* Metabolism of lipids; nucleic acids and proteins in relation to plant differentiation and development. *Mailing Add:* Dept Plant Sci Univ Alta Edmonton AB T6G 2E2 Can

ZALIPSKY, JEROME JAROSLAW, b Ukraine; US citizen; c 2. ANALYTICAL CHEMISTRY. *Educ:* St Joseph's Col, BS, 58, MS, 62; Univ Pa, PhD(analytical chem), 70. *Prof Exp:* From chemist to group leader analytical chem, Nat Drug Co, Richardson-Merrill Inc, 58-70; from sr scientist to group leader phys chem, 70-74, SECT HEAD PHYS & MICROANALYTICAL CHEM, WILLIAM H RORER INC, 74- *Mem:* Am Chem Soc. *Res:* Chemical structure elucidation of new drug substance; kinetics; characterization of hydrolysis products; analytical and physical profile of drug substance. *Mailing Add:* 7600 Woodlawn Ave Melrose Park Philadelphia PA 19126-1428

ZALISKO, EDWARD JOHN, b Peoria, Ill, Jan 8, 58; m 82; c 2. FUNCTIONAL MORPHOLOGY, HISTOLOGY & HISTOCHEMISTRY. *Educ:* Southern Ill Univ, Carbondale, BA, 80, MA, 82; Wash State Univ, PhD(zool), 87. *Prof Exp:* Asst prof zool, Southeast Mo State Univ, 87-89; PROF BIOL ZOOL, BLACKBURN COL, 89- *Mem:* Sigma Xi; AAAS; Am Soc Ichthyologists & Herpetologists; Am Soc Zoologists; Soc Study Amphibians & Reptiles. *Res:* Functional morphology of amphibian reproductive systems and snake oral glands using light and electron microscopy and histochemistry. *Mailing Add:* Dept Biol Blackburn Col 700 College Ave Carlinville IL 62626

ZALKIN, HOWARD, b New York, NY, Dec 31, 34; m 66; c 3. BIOCHEMISTRY. *Educ:* Univ Calif, Davis, BS, 56, MS, 59, PhD(biochem), 61. *Prof Exp:* Res assoc chem, Harvard Univ, 61-62; res assoc biochem, Pub Health Res Inst New York, 62-64; res assoc, Col Physicians & Surgeons, Columbia Univ, 64-66; from asst prof to assoc prof, 66-72, PROF BIOCHEM, PURDUE UNIV, LAFAYETTE, 72- *Concurrent Pos:* Fels, NSF, 61-63, USPHS, 63-64 & USPHS fel biol sci, Stanford Univ, 72-73; vis scholar, Stanford Univ, 80-81; mem biochem study sect, NIH, 87-91. *Mem:* Am Soc Biochem & Molecular Biol; Am Soc Microbiol. *Res:* Structure, function and regulation of glutamine amidotranferase genes-enzymes; transcriptional and translational regulation of purine nucleotide synthesis. *Mailing Add:* Dept Biochem Purdue Univ Lafayette IN 47907

ZALKOW, LEON HARRY, b Millen, Ga, Nov 27, 29; m 71; c 1. ORGANIC CHEMISTRY. *Educ:* Ga Inst Technol, BCE, 52, PhD(chem), 56. *Prof Exp:* Res fel, Wayne State Univ, 55-56, 57-59; res chemist, E I du Pont de Nemours & Co, 56-57; asst prof chem, Okla State Univ, 59-62, assoc prof, 62-65; assoc prof, 65-69, PROF CHEM, GA INST TECHNOL, 69- *Concurrent Pos:* Prof & head dept chem, Univ of the Negev, 70-72. *Mem:* AAAS; Am Chem Soc; Royal Soc Chem. *Res:* Natural products; conformational analysis; chemistry of bicyclic azides. *Mailing Add:* Dept Chem Ga Inst Technol 225 North Ave NW Atlanta GA 30332-0001

ZALL, LINDA S, b Nov 15, 50; US citizen. GEOLOGY. *Educ:* Cornell Univ, BS, 72, MS, 74, PhD(civil & environ eng), 76. *Prof Exp:* Instr photo-geol, Cornell Univ, 71-75; CONSULT ENVIRON & ENG REMOTE SENSING, EARTH SATELLITE CORP, 75- *Concurrent Pos:* Res engr, Calspan Corp, 71; eng geologist, Trans Alaska Oil Pipeline Proj, 73. *Mem:* Soc Econ Geologists; Am Soc Agron; Am Soc Photogram; Sigma Xi. *Res:* Photo-geology; various remote sensing techniques. *Mailing Add:* 6812 Wilson Lane Bethesda MD 20817

ZALL, ROBERT ROUBEN, b Lowell, Mass, Dec 6, 25; m 49; c 3. FOOD SCIENCE. *Educ:* Univ Mass, BS, 49, MS, 50; Cornell Univ, PhD(food sci), 68. *Prof Exp:* Lab dir dairy prod, Grandview Dairies, Inc, NY, 50-51, mgr, Butter & Cheese Div, 51-53, mgr, Condensed Milks & Powder Div, 53-57, gen mgr corp, 57-66; dir res & prod, Crowley Food Co, 68-71; assoc prof, 71-76, PROF FOOD SCI, CORNELL UNIV, 76- *Concurrent Pos:* Environ Protection Agency pollution abatement demonstration grant, Whey Fractionation Plant, Crowley Foods Co, 69-72, proj dir, 71-; proj dir, farm ultrafiltration, Dairy Res Inc, 84-85. *Honors & Awards:* Howard Marlatt Award Lab Technol, NY State Sanitarians, 79. *Mem:* Inst Food Technologists; Am Soc Agr Engrs; Int Asn Milk, Food & Environ Sanitarians; Int Dairy Fedn. *Res:* Detergents as inhibitors in food; reusing cleaning fluids to reduce consumption of energy and chemicals; reclamation and renovation of food wastes; membrane filtration processing; utilization of whey fractions in foods; on farm ultrafiltration. *Mailing Add:* Dept Food Sci Stocking Hall Cornell Univ Ithaca NY 14853

ZALLEN, EUGENIA MALONE, b Camp Hill, Ala, July 18, 32; m 59. FOOD SCIENCE. *Educ:* Auburn Univ, BS, 53; Purdue Univ, MS, 60; Univ Tenn, PhD(food sci), 74. *Prof Exp:* Dietetic intern, Med Ctr, Duke Univ, 53-54, assoc dietitian, 54-57; asst chief dietary, Univ Hosp, Emory Univ, 57-58; from instr to asst prof food & nutrit, Auburn Univ, 62-66; asst prof food, nutrit & inst admin, Univ Md, 67-72; researcher dairy sci, Okla State Univ, 72-73; researcher, Univ Tenn, 73-74; assoc prof & dir, Sch Home Econ, Univ Okla, 74-80; prof, E Carolina Univ, NC, 80-90; EXEC VPRES, MALONE CEROYS, 90- *Concurrent Pos:* Consult, Head Start Day Care Ctrs, Ala, 65-66; field reader, Bur Res, HEW, DC, 66-; consult, Univ Consult, Inc, Ala, 68 & Optimal Systs, Inc, Ga, 69-; pres, Acad World, Inc, Consults, 75-86. *Mem:* Am Home Econ Asn; Am Dietetic Asn; Inst Food Technologists; Sigma Xi; Soc Nutrit Educ; Am Sch Food Serv Asn. *Res:* Quality factors in quantity food production; analysis of changes in food habits; development of aspertame sweetened jellies; home canning of fish for year round source of Omega 3FA; sensory perceptions of elderly. *Mailing Add:* PO Box 8767 Columbus GA 31908-8767

ZALLEN, RICHARD, b New York, NY, Jan 1, 37; m 64; c 2. CONDENSED MATTER PHYSICS. *Educ:* Rensselaer Polytech Inst, BS, 57; Harvard Univ, AM, 59, PhD(solid state physics), 64. *Prof Exp:* Res asst solid state physics, Harvard Univ, 59-64, res fel, 64-65; staff mem, Xerox Res Labs, 65-83; PROF PHYSICS, VA TECH, 83- *Concurrent Pos:* Vis assoc prof, Technion, 71-72; Sci & Eng Res Coun vis fel, Imperial Col, 90-91. *Mem:* Fel Am Phys Soc. *Res:* Experimental studies of optical properties of solids; vibrational and electronic structure of molecular solids, layer crystals, semiconductors, and amorphous solids; Raman scattering; optical effects in solids at high pressure; percolation theory; phase transitions; solid state theory. *Mailing Add:* Dept Physics Va Tech Blacksburg VA 24061

ZALOSH, ROBERT GEOFFREY, b New York, NY, Oct 10, 44; m 65; c 1. FLUID MECHANICS, HAZARD ANALYSIS. *Educ:* Cooper Union, BE, 65; Univ Rochester, MS, 66; Northeastern Univ, PhD(mech eng), 70. *Prof Exp:* Assoc scientist, Space Systs Div, Avco Corp, 66-67; sr scientist, Mt Auburn Res Assocs, Inc, 70-75; sr res scientist, 75-78, MGR EXPLOSION & ENERGETICS SECT, FACTORY MUTUAL RES CORP, 78-, MGR APPL RES, 88- *Concurrent Pos:* Lectr mech eng, Northeastern Univ, 74-75; adj prof, fire safety, Wooster Polytech. *Mem:* Am Soc Mech Engrs; Combustion Inst; Int Asn Hydrogen Energy. *Res:* Fire and explosion protection; explosion venting; vapor cloud dispersal; blast wave phenomena; advanced battery hazards. *Mailing Add:* 20 Rockland Wellesley MA 02181

ZALUBAS, ROMUALD, b Pandelys, Lithuania, July 20, 11; nat US; m 39; c 1. ASTROPHYSICS. *Educ:* Kaunas State Univ, MA, 36; Georgetown Univ, PhD(astrophys), 55. *Prof Exp:* Asst astron, Vilnius State Univ, 40-44; dir sec sch, Ger, 45-49; instr math, Nazareth Col (NY), 49-51; instr, Georgetown Univ, 52-57; physicist, Nat Bur Standards, 55-87; RETIRED. *Mem:* AAAS; Am Astron Soc; Sigma Xi; Optical Soc Am. *Res:* Description and analysis of atomic spectra; thorium wavelength standards; atomic energy levels of the rare earth elements; critical compilation of energy level data of the first thirty elements (hydrogen through zinc). *Mailing Add:* 908 Roswell Dr Silver Spring MD 20901

ZALUCKY, THEODORE B, b Beleluja, West-Ukraine, Apr 11, 19; US citizen; m 46; c 2. PHARMACY, PHARMACEUTICAL CHEMISTRY. *Educ:* Univ Vienna, MPharm, 42, DSc Nat(pharmaceut chem), 45; Ill, Chicago, BS, 52. *Prof Exp:* Analytical chemist, Control Lab, Chicago Pharm Co, 52-53; res chemist, Foot Prod Lab, Scholl Mfg Co, Inc, 53-55; asst prof pharmaceut chem, 55-63, assoc prof pharm & pharmaceut chem, 63-72, PROF PHARM, COL PHARM, HOWARD UNIV, 72- *Concurrent Pos:* AEC grant, 62-63. *Mem:* Am Chem Soc; Am Pharmaceut Asn; Am Asn Cols Pharm; Acad Pharmaceut Sci; Shevchenko Sci Soc; Sigma Xi. *Res:* Chemistry of morphine alkaloids, epoxy ethers and some benzolypiperidines; structure-chromogenic activity relationship of phenolic compounds with Ehrlich reagent; isolation and structure of some new anhalonium alkaloids; spiro-compounds containing geranium and organo-metallic compounds. *Mailing Add:* Col Pharm Howard Univ 2300 Fourth St NW Washington DC 20059

ZALUSKY, RALPH, b Pawtucket, RI, Oct 11, 31; m 58; c 3. INTERNAL MEDICINE, HEMATOLOGY. *Educ:* Brown Univ, AB, 53; Boston Univ, MD, 57; Am Bd Internal Med, dipl, 64, hemat, 72. *Prof Exp:* From intern to sr resident med, Duke Univ Hosp, 57-62; from asst prof to assoc prof, 66-77, PROF MED, MT SINAI SCH MED, 77-; CHIEF DIV HEMAT-ONCOL, BETH ISRAEL MED CTR, 76- *Concurrent Pos:* USPHS fel, Thorndike Mem Lab, Harvard Univ, 59-61; res assoc med, Mt Sinai Hosp, 64-65, asst attend hematologist, 65-66, actg chief hemat, 72-76, attend hematologist, 77- *Mem:* Am Fedn Clin Res; Am Soc Clin Nutrit; Am Soc Hemat. *Res:* Interrelationships between vitamin B-twelve and folic acid metabolism; sodium and potassium membrane transport; abnormal hemoglobins; erythropoietin physiology. *Mailing Add:* Dept Med Div Hemat-Oncol Beth Israel Med Ctr First Ave & 16th St New York NY 10003

ZAM, STEPHEN G, III, b Toledo, Ohio, Nov 3, 32. PARASITOLOGY. *Educ:* Georgetown Univ, BS, 54; Catholic Univ, MS, 56; Univ Southern Calif, PhD(biol sci), 66. *Prof Exp:* Asst prof biol, Loyola Univ (Calif), 66; from asst prof parasitol to assoc prof zool, 66-77, ASSOC PROF MICROBIOL & CELL SCI, UNIV FLA, 77- *Concurrent Pos:* Consult, Marineland, 67- *Mem:* Am Soc Parasitol; Am Soc Trop Med & Hyg; Int Soc Parasitol. *Res:* Nematode physiology including biochemistry of nematode egg hatching and larval molting; immunology to helminth infections. *Mailing Add:* Dept Microbiol-Cell Sci Univ Fla Gainesville FL 32611

ZAMAN, KHAIRUL B M Q, b Dhaka, Bangladesh, Aug 31, 47; US citizen; m; c 2. AEROSPACE SCIENCES. *Educ:* Univ Bangladesh, BS, 69, MS, 73; Univ Houston, PhD(mech eng), 78. *Prof Exp:* Lectr, Univ Bangladesh, 70-73; res scientist, Univ Houston, 78-82; res assoc, NASA Langley Res Ctr Va, 82-85, AEROSPACE ENGR, NASA LEWIS RES CTR, 86- *Mem:* Am Phys Soc; Am Inst Aeronaut & Astronaut. *Res:* Nurbulent shear flows; separated flows; aeroacoustics. *Mailing Add:* NASA Lewis Res Ctr 21000 Brookpark Rd Cleveland OH 44135

ZAMANZADEH, MEHROOZ, b Tehran, Iran, Mar 12, 50; m 83; c 1. CORROSION ENGINEERING, FAILURE ANALYSIS. *Educ:* San Jose State Univ, BS, 74, MS, 75; Pa State Univ, PhD(metall), 80. *Prof Exp:* Sr metallurgist corrosion, Buehler-NIOC, 80-85; postdoctoral fel corrosion, Carnegie-Mellon Univ, 85-87; MGR METALL DIV, PROF SERV INDUST, PITTSBURGH TESTING LAB DIV, 87- *Concurrent Pos:* Lectr, Carnegie-Mellon Univ, 88-90. *Mem:* Nat Asn Corrosion Engrs; Am Soc Metals; Am Inst Chem Engrs. *Res:* Division management; research and development; technical responsibilities; teaching; conducted and reviewed thousands of metallurgical failure analysis and on site investigations in materials science and engineering; author of 37 publications in materials science and corrosion engineering journals. *Mailing Add:* 1306 Meadowlark Dr Pittsburgh PA 15243

ZAMBERNARD, JOSEPH, b Sept 5, 33; US citizen. CYTOLOGY. *Educ:* Univ Ala, AB, 53, MS, 56; Tulane Univ, PhD(cytol), 64. *Prof Exp:* Fel virol & immunol, Sch Med, Univ Colo, Denver, 64-66, asst prof anat, 66-72; assoc prof, Albany Med Col, 72-75; prof, 75-77, CHMN DEPT ANAT, SCH MED, WRIGHT STATE UNIV, 77- *Mem:* Electron Micros Soc Am; Am Soc Cell Biol; Am Asn Anatomists; Tissue Cult Asn. *Res:* Virology and immunology; ultrastructure. *Mailing Add:* Dept Anat Wright State Univ Colonel Glenn Hwy Dayton OH 45435

ZAMBITO, ARTHUR JOSEPH, b Rochester, NY, Sept 7, 14; m 42; c 7. ORGANIC CHEMISTRY. *Educ:* Univ Mich, BS, 40, MS, 41, PhD(org chem), 47. *Prof Exp:* From res chemist to sr res chemist, Merck & Co Inc, 41-64, sect leader, Res & Develop Lab, 64-75, sr res fel, 75-84; RETIRED. *Mem:* Am Chem Soc. *Res:* Synthesis of pharmaceuticals; synthesis and isolation of amino acids; preparation of parenteral solutions and emulsions; synthesis of anticancer agents; antibiotics. *Mailing Add:* 75 Hillcrest Dr Clark NJ 07066

ZAMBONI, LUCIANO, b San Dona di Piave, Italy, Sept 6, 29; m 57; c 2. PATHOLOGY, ELECTRON MICROSCOPY. *Educ:* Univ Rome, MD, 55, dipl gastroenterol, 58. *Prof Exp:* Instr path, Univ Rome, 55-59; asst resident anat, Univ Calif, Los Angeles, 59-60; instr, McGill Univ, 60-61; asst resident path, Karolinska Inst, Sweden, 61-63; asst prof, 63-65, PROF PATH, UNIV CALIF, LOS ANGELES, 65-; CHIEF DEPT PATH & HEAD ELECTRON MICROS, LOS ANGELES COUNTY HARBOR-UNIV CALIF LOS ANGELES MED CTR, 63- *Mem:* Electron Micros Soc Am; Am Soc Cell Biol; Am Fertil Soc; Soc Study Reproduction; Ital Med Asn. *Res:* Reproductive biology; ultrastructural studies on embryogenesis, early reproduction and fertilization. *Mailing Add:* Dept Path Harbor-Univ Calif Los Angeles Med Ctr 1000 W Carson St Torrance CA 90509

ZAMBRASKI, EDWARD K, b Cold Spring Harbor, NY, May 25, 49. KIDNEY PHYSIOLOGY. *Educ:* Univ Iowa, PhD(physiol & phys educ), 76. *Prof Exp:* ASSOC PROF PHYSIOL, RUTGERS UNIV, 81- *Mem:* Am Physiol Soc; Am Soc Nephrology; Am Col Sports Med. *Mailing Add:* Dept Biol Bartlett Hall Rutgers Univ New Brunswick NJ 08903

ZAMBROW, J(OHN) L(UCIAN), b Milwaukee, Wis, Dec 9, 14; m 42; c 2. METALLURGY. *Educ:* Univ Wis, BS, 40, MS, 46; Ohio State Univ, PhD(metall), 48. *Prof Exp:* Mfg engr, Cutler-Hammer, Inc, 40-44; res engr, Battelle Mem Inst, 44-47; asst res prof metall, Ohio State Univ, 49; res engr, Sylvania Elec Prod, Inc, 49-53, mgr eng, 53-57; mgr eng, Sylvania-Corning Nuclear Corp, 57-59, dir eng, 59-62; mgr metall res, Res Ctr, 62-74, asst dir, 74-77, assoc dir, Res Ctr, 77-80, SR CONSULT, BORG-WARNER CORP, DES PLAINES, 80- *Mem:* Sigma Xi; Am Soc Metals; Am Nuclear Soc; Am Inst Mining, Metall & Petrol Engrs; Am Powder Metall Inst. *Res:* Physical and powder metallurgy; manufacturing process research and development. *Mailing Add:* Res Ctr Borg-Warner Corp Seven Yorkshire Dr Des Plaines IL 60016

ZAME, ALAN, b New York, NY, Aug 16, 41; m 89. MATHEMATICS, COMPUTER SCIENCE. *Educ:* Calif Inst Technol, BS, 62; Univ Calif, Berkeley, PhD(math), 65. *Prof Exp:* Assoc prof, 65-76, PROF MATH, UNIV MIAMI, 76-, CHMN, MATH & COMPUTER SCI, 89- *Mem:* Am Math Soc; Am Math Asn. *Res:* Number theory; functional and combinatorial analysis. *Mailing Add:* Dept Math-Comput Sci Univ Miami University Sta Coral Gables FL 33124

ZAME, WILLIAM ROBIN, b Long Beach, NY, Nov 4, 45; m 90. ANALYSIS & FUNCTIONAL ANALYSIS. *Educ:* Calif Inst Technol, BS, 65; Tulane Univ, MS, 67, PhD(math), 70. *Prof Exp:* Evans instr math, Rice Univ, 70-72; assoc prof, Tulane Univ, 75-78; asst prof, 72-75 assoc prof math, 78-81, PROF MATH, STATE UNIV NY, BUFFALO, 81- *Mem:* Am Math Soc; Am Soc Econometric. *Res:* Several complex variables; Banach algebras, C-algebras Mathmatical Economics. *Mailing Add:* Dept Math Rm 106 Diefendorf Hall State Univ NY Health Sci Ctr 3435 Main St Buffalo NY 14214

ZAMECNIK, PAUL CHARLES, b Cleveland, Ohio, Nov 22, 12; m 36; c 3. MEDICINE. *Educ:* Dartmouth Col, AB, 33; Harvard Univ, MD, 36. *Hon Degrees:* Dr, Univ Utrecht, 66; DSc, Columbia Univ, 71, Harvard Univ, 82, Roger Williams Col, 83 & Dartmouth Col, 88. *Prof Exp:* Resident med, C P Huntington Mem Hosp, Boston, 36-37; intern, Univ Hosps, Cleveland, 38-39; Moseley traveling fel from Harvard Univ, Carlsberg Lab, Copenhagen, 39-40; Finney-Howell fel, Rockefeller Inst, 41-42; from instr to assoc prof, 42-56, Collis P Huntington prof, 56-79, dir, J C Warren Labs, 56-79, EMER PROF ONCOL MED, SCH MED, HARVARD UNIV, 79-; PRIN SCIENTIST, WORCESTER FOUND EXP BIOL, 79- *Concurrent Pos:* Physician, Mass Gen Hosp, 56-79, sr physician, 79-, hon physician, 79- *Honors & Awards:* Nat Medal of Sci, 91; John Collins Warren Triennial Prize, 46 & 50; James Ewing Award, 63; Borden Award, 65; Am Cancer Soc Nat Award, 68; Passano Award, 70. *Mem:* Nat Acad Sci; Asn Am Physicians; Am Acad Arts & Sci; Am Soc Biol Chemists; Am Asn Cancer Res (pres, 64-65). *Res:* Protein synthesis; cancer; nucleic acid metabolism; virology. *Mailing Add:* Worcester Found Exp Biol 222 Maple Ave Shrewsbury MA 01545

ZAMEL, NOE, b Rio Grande, Brazil, Apr 2, 35; m 59; c 3. RESPIRATORY PHYSIOLOGY. *Educ:* Col Med, Fed Univ Rio Grande do Sul, Brazil, MD, 58. *Prof Exp:* From instr to assoc prof med, Col Med, Fed Univ Rio Grande do Sul, Brazil, 62-70; assoc prof med & dir respiratory physiol, Col Med, Univ Nebr, Omaha, 70-72; assoc prof, 72-80, PROF MED, FAC MED & DIR RESPIRATORY PHYSIOL, TRIHOSP RESPIRATORY SERV, UNIV TORONTO, CAN, 80- *Concurrent Pos:* Res fel respiratory physiol, Cardiovasc Res Inst, Univ Calif, San Francisco, 69. *Honors & Awards:* Miguel Couto Award, Col Med, Fed Univ Rio Grande do Sul, 58; Cecile Lehman Mayer Award, Am Col Chest Physicians, 69. *Mem:* Am Col Chest Physicians; Am Thoracic Soc; Am Fedn Clin Res; Can Soc Clin Invest; Am Physiol Soc. *Res:* Lung mechanics. *Mailing Add:* Dept Med Univ Toronto Mount Sinai Hosp 600 University Ave Rm 656 Toronto ON M5G 1X5 Can

ZAMENHOF, PATRICE JOY, b Santa Rosa, Calif, Apr 2, 34; m 61. MOLECULAR GENETICS. *Educ:* Univ Calif, Berkeley, AB, 56, PhD(microbiol), 62. *Prof Exp:* Res assoc biochem, Col Physicians & Surgeons, Columbia Univ, 62-64; asst prof, 64-71, ASSOC PROF BIOL CHEM, SCH MED, UNIV CALIF, LOS ANGELES, 71- *Concurrent Pos:* USPHS res grants, Univ Calif, Los Angeles, 65-68 & 70-, career develop award, Nat Inst Gen Med Sci, 67, mem cancer ctr, 75- *Mem:* Am Soc Microbiol; Genetics Soc Am; Am Soc Biol Chemists; Sigma Xi. *Res:* Mutagenic mechanisms; genetic instability in microorganisms; mutator genes; repair of genetic damage; functional interactions of inactive mutant proteins. *Mailing Add:* Dept Biol Chem Univ Calif Sch Med Los Angeles CA 90024

ZAMENHOF, ROBERT G A, b Kidugala, E Africa, 1946. MEDICAL PHYSICS. *Educ:* Polytech N London, BS, 69; Univ Strathclyde, MS, 71; Mass Inst Technol, PhD(appl radiation physics), 77. *Prof Exp:* HEAD, IMAGING PHYSICS SECT, MED PHYSICS DIV, NEW ENG MED CTR HOSP, 82-, PROF MED PHYSICS, DEPT RADIATION ONCOL, 89- *Mem:* Am Asn Physics Med; Biomed Eng Soc; Am Col Radiol. *Mailing Add:* Dept Radiation Oncol Med Physics Div New Eng Med Ctr Hosp 750 Washington St Boston MA 02111

ZAMENHOF, STEPHEN, b Warsaw, Poland, June 12, 11; US citizen; m 61. NEUROCHEMISTRY. *Educ:* Warsaw Polytech Sch, Dr Tech Sci, 36; Columbia Univ, PhD(biochem), 49. *Prof Exp:* From asst prof to assoc prof biochem, Columbia Univ, 49-64; PROF MICROBIOL GENETICS & BIOL CHEM, SCH MED, UNIV CALIF, LOS ANGELES, 64- *Concurrent Pos:* Guggenheim fel, 58-59. *Mem:* Am Soc Biol Chem; Am Inst Nutrit; Soc Neurosci; Am Asn Anatomists; Int Soc Develop Neurosci; Sigma Xi. *Res:* Microbial genetics; nucleic acids; prenatal brain development. *Mailing Add:* Dept Micro & Immunol Univ Calif Sch Med Los Angeles CA 90024

ZAMES, GEORGE, b Lodz, Poland, Jan 7, 34; m 64; c 2. ELECTRICAL ENGINEERING. *Educ:* McGill Univ, BEng, 54; Mass Inst Technol, ScD(elec eng), 60. *Prof Exp:* Res assoc elec eng, Mass Inst Technol, 54-61, asst prof, 61- 62, 63-65; res fel appl physics, Harvard Univ, 62-63; asst prof elec eng, Mass Inst Technol, 63-65; Nat Acad Sci res fel, Electronic Res Ctr, NASA, 65-68; Guggenheim fel, 67-68; sr scientist, Electronic Res Ctr, NASA, 68-72; vis prof elec eng, Technion, Haifa, 72-74; prof, 74-83, MACDONALD PROF ELEC ENG, MCGILL UNIV, 83- *Concurrent Pos:* Athlone fel, Imperial Col, London Univ, 54-56; RR Associateship, Nat Acad Sci, 66-67; assoc ed, Soc Indust Appl Math J Control, 67-84, Systems & Control Lett, 80-84, Indust Math Asn J Math Control & Info, 83; Guggenheim fel, 67-68; Killiam fel, 84-; sr fel, Can Inst Advan Res, 84-; assoc ed, Math Control, Signals & Systs. *Honors & Awards:* Field Award Sci & Eng, Inst Elec & Electronics Engrs, 85. *Mem:* Fel Inst Elec & Electronics Engrs; Sigma Xi; fel Royal Soc Can. *Res:* Nonlinear systems; feedback organizations; control system theory; communication system theory; functional analysis. *Mailing Add:* Dept Elec Eng 3480 University St Montreal PQ H3A 2A7 Can

ZAMICK, LARRY, b Winnipeg, Man, Mar 15, 35; m 66; c 2. NUCLEAR PHYSICS. *Educ:* Univ Man, BSc, 57; Mass Inst Technol, PhD(physics), 61. *Prof Exp:* Instr physics, Princeton Univ, 62-65, res assoc, 65-66; assoc prof, 66-70, PROF PHYSICS, RUTGERS UNIV, 70- *Mem:* Fel Am Phys Soc. *Res:* High energy deuteron-nucleus scattering; nuclear structure studies with the shell model. *Mailing Add:* Serin Physics Lab Rutgers Univ Piscataway NJ 08854

ZAMIKOFF, IRVING IRA, b Toronto, Ont, Feb 13, 43; m 67; c 2. DENTISTRY, PROSTHODONTICS. *Educ:* Univ Toronto, DDS, 67; Univ Mich, MS, 70; Am Bd Prosthodont, dipl, 72. *Prof Exp:* ASSOC PROF PROSTHODONT, SCH DENT, LA STATE UNIV, NEW ORLEANS, 70- *Concurrent Pos:* Vis dentist, Charity Hosp, New Orleans, 70- *Mem:* Am Dent Asn; Int Asn Dent Res; Can Dent Asn; Am Col Prosthodont. *Mailing Add:* 2103 59th St W Bradenton FL 34209

ZAMIR, LOLITA ORA, b Cairo, Egypt; Israeli & US citizen; m 85; c 1. BIO-ORGANIC CHEMISTRY, ORGANIC CHEMISTRY. *Educ:* Israel Inst Technol, MSc, 66; Yale Univ, PhD(bio-org chem), 73. *Prof Exp:* Teaching asst chem, Yale Univ, 68-73; res fel biochem, Harvard Univ, 73-74; sr res chemist, Merck Inst, 74-75; from asst prof to assoc prof chem & assoc mem ctr biochem res, State Univ NY Binghamton, 75-81; assoc prof, 82-84, PROF, INST ARMAND FRAPPIER, APPL MICROBIOL DEPT, UNIV QUEBEC, 84- *Concurrent Pos:* auxillary prof, Chem Dept, McGill Univ, 82- *Mem:* Am Chem Soc; Am Asn Women Sci; Am Asn Univ Prof. *Res:* Biosynthesis; fungal metabolites; natural products. *Mailing Add:* Univ Quebec Inst Armand Frappier Ctr Bact 531 Blvd Des Prairies Montreal PQ H7N 4Z3 Can

ZAMMUTO, RICHARD MICHAEL, populations, evolution, for more information see previous edition

ZAMORA, ANTONIO, b Nuevo Laredo, Mexico, Dec 6, 42; US citizen; m 67; c 1. COMPUTER SCIENCE, CHEMISTRY. *Educ:* Univ Tex, BS, 62; Ohio State Univ, MS, 69. *Prof Exp:* Med lab technician clin chem, US Army, 62-65; INFO SCIENTIST RES & DEVELOP, CHEM ABSTRACTS SERV, 65- *Mem:* Am Chem Soc; Asn Comput Mach; AAAS. *Res:* Automated language processing; artificial intelligence; information storage and retrieval. *Mailing Add:* 4601 N Park Ave Suite 411 Chevy Chase MD 20815-4521

ZAMORA, CESARIO SIASOCO, b Marikina, Philippines, Nov 1, 38; US citizen; m 66; c 2. VETERINARY HISTOLOGY, VETERINARY GROSS ANATOMY. *Educ:* Univ Philippines, DVM, 59; Univ Minn, MS, 64; Univ Wis, Madison, PhD(vet sci), 72. *Prof Exp:* Instr vet anat, Univ Philippines, 59-62; res asst, Univ Minn, 62-64; instr to asst prof vet anat, Univ Philippines, 64-69; res asst, Univ Wis, 69-72; asst prof, 72-76, assoc prof, 76-81, head anat div, dept vet & comp anat, pharmacol & physiol, Col Vet Med, 81-82, PROF ANAT, WASH STATE UNIV, 81- *Concurrent Pos:* Pvt vet pract (part-time), 59-62, 64-69. *Mem:* Am Vet Med Asn; Am Asn Vet Anatomists; World Asn Vet Anatomists; Asn Am Vet Med Col; Am Asn Anatomists. *Res:* Structure and function of the gastrointestinal tract of pigs; pathophysiology of gastric ulcers and enteric diseases of swine; structure and function of endocrine and reproductive organs of domestic animals. *Mailing Add:* Col Vet Med Wash State Univ Pullman WA 99164-6520

ZAMORA, PAUL O, US citizen. CARCINOGENESIS. *Educ:* Univ NMex, Albuquerque, BA, 72, PhD(cell biol), 80. *Prof Exp:* Lab asst, Dept Biol, Univ NMex, 75-76, teaching asst human anat & physiol, 78, res asst, 74-79; res assoc, Dept Zool, Wash State Univ, 79-80; fel cell toxicol group, Inhalation & Toxicol Res Inst, 80-83; scientist, 83-84, dir cell biol, 84-85, VPRES PRES & DEVELOP, SUMMA MED CORP, 85- *Mem:* Soc Nuclear Med; AAAS. *Res:* Carcinogenesis; colon cancer; tumor-associated antigens; tumor imaging; antibody-based pharmaceuticals. *Mailing Add:* 1514 Vassar Dr NE Albuquerque NM 87106

ZAMRIK, SAM YUSUF, b Damascus, Syria, Dec 11, 32; US citizen; m 54; c 4. ENGINEERING MECHANICS. *Educ:* Univ Tex, BA, 56, BS, 57; Pa State Univ, MS, 61, PhD(eng mech), 65. *Prof Exp:* Design engr, Tex Pipe Line Co, 57-58; proj engr & consult, Gen Petrol Authority, 58-60; from instr to assoc prof eng mech, 60-75, PROF ENG MECH, PA STATE UNIV, UNIVERSITY PARK, 75- *Concurrent Pos:* Consult, Nat Forge Co, 68-72 & Gen Elec Co; fels, NASA & Ford Found; tech reviewer, NSF, NASA, Am Soc Mech Engrs & Soc Exp Stress Anal. *Mem:* Am Soc Testing & Mat; Am Soc Mech Engrs; Soc Exp Stress Anal; Sigma Xi. *Res:* Radiation effects on structural materials; fatigue and fracture mechanics; biaxial creep-fatigue interaction. *Mailing Add:* Dept Eng Sci & Mech Pa State Univ 121 Hammond Bldg University Park PA 16802

ZANAKIS, STELIOS (STEVE) H, b Athens, Greece, Nov 16, 40; m; c 3. MANAGEMENT SCIENCE, APPLIED STATISTICS. *Educ:* Nat Tech Univ, Athens, Dipl, 64; Pa State Univ, MBA, 70, MA, 72, PhD(mgt sci), 73. *Prof Exp:* Engr aerodyn, Greek Res Ctr Nat Defense, 65-66; indust engr mgt consult, Greek Prod Ctr, 67-68; asst prof indust eng & systs analysis, WVa Col Grad Studies, 72-76, prog dir, 73-80, assoc prof, 76-80; from assoc prof to prof & chmn decision sci & info syst, Col Bus Admin, 80-86, FLA INT UNIV, PROF, 86- *Concurrent Pos:* Consult, Ashland Oil, 73, Union Carbide Corp, 75-76, WVa Dept Hwy, 76-77 & Charleston Area Med Ctr, 78-79; mem, WVa State Comprehensive Health Plan Comt, 73-74; prin, Mgt Decision Syst Consults, 76-; guest ed, Mgt Sci & Europ J Oper Res. *Mem:* Decision Sci Inst; Inst Mgt Sci; Hellenic Oper Res Soc. *Res:* Statistics and operations management software development for microcomputers, applications of operations research/management science techniques to solve real management problems; production, inventory, distribution management; hospital management engineering; statistics/optimization interface; simulation and computer applications; prog evaluation and fund allocation under conflicting goals and qualitative or quantitative criteria; forecasting and statistical analysis. *Mailing Add:* Sch Bus Admin Dept Decision Sci & Info Syst Fla Int Univ Miami FL 33199

ZAND, ROBERT, b New York, NY, Jan 7, 30; m 52; c 3. BIOPHYSICAL CHEMISTRY, POLYMER CHEMISTRY. *Educ:* Univ Mo, BS, 51; Polytech Inst Brooklyn, MS, 54; Brandeis Univ, PhD(chem), 61. *Prof Exp:* Res chemist, Irvington Varnish & Insulator, 53; assoc res biophysicist, 63-73, asst prof, 68-73, assoc prof, 73-86, PROF BIOCHEM, UNIV MICH, ANN ARBOR, 86-, RES SCI, 86- *Concurrent Pos:* NIH fel, Harvard Med Sch, 61-63; fel, Brandeis Univ, 61-63; ODOL Found lectr, Univ Buenos Aires, 72; vis prof, Escola Paulista de Med, Sao Paulo, Brazil, 76, Inst Venezolano de Investigaciones Cientificas, Caracas Venezuela, 79; consult Recreational Innovations-Med Prods Div, AISIN Seiki Co, Ltd; proj display prod opers, Gen Elec, 89- *Mem:* Am Chem Soc; Biophys Soc; Am Soc Neurochem; Am Soc Biol Chem; Int Soc Neurochem; Sigma Xi. *Res:* Conformation of proteins, synthetic macromolecules and small molecules by spectroscopic methods; synthesis, physical and biological properties of nucleic acid analogs, polymers; polymer analogs of biomembranes; synthesis and properties of conducting electro-optic polymers. *Mailing Add:* Biophys Res Div Univ Mich Inst Sci & Tech 2200 Bonisteel Blvd Ann Arbor MI 48109-2099

ZANDER, ANDREW THOMAS, b Chicago, Ill, Oct 27, 45; m 77; c 3. SPECTROSCOPY. *Educ:* Univ Ill, Urbana, BS, 68; Univ Md, PhD(analytical chem), 76. *Prof Exp:* Chemist, Chicago Bridge & Iron Co, 66; asst chemist analytical chem, Standard Oil Co, Ind, 68-69; res assoc, Dept Chem, Ind Univ, 76-77; asst prof analytical chem, Cleveland State Univ, 77-79; staff scientist, Spectra Metrics, Inc, 79-84; sr staff engr, Perkins-Elmer Inc, 84-87; DIR MEAS LAB, VARIAN RES CTR, 87- *Mem:* Am Chem Soc; Soc Appl Spectros; Sigma Xi; Optical Soc Am; fel Am Inst Chemists; Inst Elec & Electronics Engrs. *Res:* Design and development of single- and multi-element methods of atomic spectrometric analysis for trace metals; high resolution studies of spectral features in atomic spectroscopy; instrument systems. *Mailing Add:* 10380 Imperial Ave Cupertino CA 95014

ZANDER, ARLEN RAY, b Shiner, Tex, Dec 12, 40; m 64; c 3. NUCLEAR PHYSICS, ATOMIC PHYSICS. *Educ:* Univ Tex, Austin, BS, 64; Fla State Univ, PhD(nuclear physics), 70. *Prof Exp:* Res physicist, Phillips Petrol Co, 64-65; from asst prof to assoc prof, 70-79, external grants coordr, 74-77, PROF PHYSICS, ETEX STATE UNIV, 79-, ASST DEAN ARTS & SCI, 82-, HIGH TECHNOL COORDR, 84- *Mem:* Am Phys Soc; Am Asn Physics Teachers; Europ Physical Soc; Nat Asn Acad Affairs Admin. *Res:* Direct nuclear reaction mechanisms; experimental fast neutron activation studies; x-ray fluorescence studies utilizing charged particle accelerators; atomic collisions. *Mailing Add:* Off Acad Affairs Northeast La Univ Monroe LA 71209

ZANDER, DONALD VICTOR, b Bellingham, Wash, Feb 15, 16; m 45; c 3. AVIAN PATHOLOGY, POULTRY NUTRITION & HUSBANDRY. *Educ:* Univ Calif, Berkeley, BS, 41; Col State Univ, MS, 45, DVM, 50; Univ Calif, Davis, PhD(comp path), 53. *Prof Exp:* Asst specialist & lectr vet med, Univ Calif, Davis, 50-53, asst prof, 53-55; lab instr bact, Colo State Univ, 48; dir poultry health res & serv, H & N Int, 55-89; RETIRED. *Concurrent Pos:* Pres, Western Poultry Dis Conf, 57; dir, Western Dist, Am Asn Avian Path, 84-88. *Honors & Awards:* C A B Bottorff Serv Award, Am Asn Avian Path, 90. *Mem:* Poultry Sci Asn; Am Vet Med Asn; Am Asn Avian Path (pres, 65-66); World Poultry Sci Asn; World Poultry Vet Asn. *Res:* Avian diseases; pathology, diagnosis and epizootiology; poultry husbandry and nutrition; pathogen-free poultry. *Mailing Add:* 18232 160th Ave NE Woodinville WA 98072

ZANDER, VERNON EMIL, b Toledo, Wash, Feb 3, 39; m 66; c 4. MATHEMATICS. *Educ:* Univ Wash, BS, 61; Catholic Univ, MS, 65, PhD(math), 69. *Prof Exp:* Mathematician, NIH, 61-66; from asst prof to assoc prof math, West Ga Col, 68-82; PRES, INTERCOASTAL DATA CORP, 82- *Mem:* Am Math Soc; Math Asn Am. *Mailing Add:* 165 Foggy Bottom Dr Carrollton GA 30117

ZANDI, IRAJ, b Teheran, Iran, June 30, 31; m 58; c 5. CIVIL ENGINEERING. *Educ:* Univ Teheran, BS, 52; Univ Okla, MS, 57; Ga Inst Technol, PhD, 59. *Hon Degrees:* MA, Univ Pa, 71. *Prof Exp:* Dir dept environ sanit, Ministry of Health, Govt Iran, 59-61; assoc prof eng, Abadan Inst Technol, Iran, 61-62; asst prof civil eng, Univ Del, 62-66; from assoc prof to prof, 66-80, actg dir nat ctr energy mgt & power, 72-77, NAT CTR PROF CIVIL ENG, UNIV PA, 80- *Concurrent Pos:* Ed, J Pipeline, Am Soc Civil Engrs, 66-70, Elsevie Sci Publ Co, 79-; ed & publ, J Resource Mgt Technol, 81- *Honors & Awards:* Soc Sigma Xi Ferst Award, Ga Inst Technol, 61. *Mem:* Am Soc Eng Educ; Am Soc Civil Engrs; Am Inst Chem Engrs. *Res:* Pipeline engineering; resource recovery; resources and energy systems. *Mailing Add:* Dept Systs Sci Univ Pa Philadelphia PA 19104-6315

ZANDLER, MELVIN E, b Wichita, Kans, Nov 28, 37; m 59; c 4. THEORETICAL CHEMISTRY. *Educ:* Friends Univ, BA, 60; Univ Wichita, MS, 63; Ariz State Univ, PhD(phys chem), 66. *Prof Exp:* Asst prof, 66-75, ASSOC PROF CHEM, WICHITA STATE UNIV, 75- *Concurrent Pos:* Fel, Univ Utah, 66; Petrol Res Fund res grant, 68-70; vis prof, Univ Calif, Berkeley, 78; NSF grant microcomput, 81-83; res fel, Air Force Acad, 87; Air

Force Off Sci Res Initiation Grant, 88; sabactical, Okla State Univ, 90. *Mem:* Am Chem Soc; Sigma Xi. *Res:* Theory of liquids and liquid mixtures; statistical thermodynamics; semi-empirical and ab-initio molecular orbital calculations; generation and optimization of reactive potential energy surfaces; educational use of microcomputer and workstations. *Mailing Add:* Dept Chem Wichita State Univ Wichita KS 67208

ZANDY, HASSAN F, b Tehran, Iran, Mar 11, 12; US citizen; m 43; c 3. PHYSICS, SPECTROSCOPY. *Educ:* Univ Birmingham, BSc, 35, MSc, 49; Univ Teheran, PhD(physics), 53. *Hon Degrees:* PhD, Univ Teheran, 52. *Prof Exp:* Instr physics, Univ Teheran, 37-46, asst prof, 50-53; lectr, Univ Leicester, 47-50; Fulbright fel, Brooklyn Polytech Inst, 53-54; PROF PHYSICS, UNIV BRIDGEPORT, 54- *Concurrent Pos:* Mem vis scientist prog physics, NSF, 59-; NSF fac fel, 63-64, res grant plasma res, Univ Bridgeport, 72-73. *Mem:* Am Asn Physics Teachers; Sigma Xi. *Res:* High vacuum technique; measurement of temperature of hot plasmas by x-ray spectroscopy. *Mailing Add:* 34 Rosellen Dr Trumbull CT 06611

ZANER, KEN SCOTT, RHEOLOGY, NUCLEAR MAGNETIC RESONANCE. *Educ:* State Univ NY, Downstate Med Ctr, MD & PhD(biophys), 75. *Prof Exp:* Asst prof med-hemat, Mass Gen Hosp, 78-89; ASSOC PROF MED, SCH MED, BOSTON UNIV, 89- *Mailing Add:* Boston Univ Sch Med Bldg S3E 80 E Concord St Boston MA 82118

ZANETTI, NINA CLARE, b Passaic, NJ, May 31, 55. CELL DIFFERENTIATION, CELL BIOLOGY. *Educ:* Muhlenberg Col, BS, 77; Syracuse Univ, PhD(biol), 82. *Prof Exp:* NIH fel develop biol, Univ Iowa, 82-85; ASST PROF DEVELOP BIOL & HISTOL, SIENA COL, 85- *Mem:* AAAS; Am Soc Cell Biol; Soc Develop Biol; Sigma Xi. *Res:* Cell differentiation and histogenesis in embryonic development of the vertebrate limb; the role of extracellular matrix, cell shape, cytoskelton, and epithelial-mesenchymal interactions in regulation of cartilage differentiation. *Mailing Add:* Dept Biol Siena Col Loudonville NY 12211

ZANEVELD, JACQUES RONALD VICTOR, b Leiderdorp, Neth, July 12, 44; US citizen; m 80; c 2. OCEANOGRAPHY. *Educ:* Old Dom Univ, BS, 64; Mass Inst Technol, SM, 66; Ore State Univ, PhD(oceanog), 71. *Prof Exp:* PROF OCEANOG, ORE STATE UNIV, 71- *Concurrent Pos:* Dir res, Sea Tech, Inc, 79- *Mem:* Am Geophys Union; Optical Soc Am; Am Soc Limnol & Oceanog. *Res:* Theoretical and experimental relationships between light attenuation, scattering and absorption properties of the ocean and the properties of suspended and dissolved materials; optical oceanography instrumentation. *Mailing Add:* Sch Oceanog Ore State Univ Corvallis OR 97331

ZANEVELD, LOURENS JAN DIRK, b The Hague, Netherlands, Mar 22, 42; US citizen; div; c 1. REPRODUCTIVE PHYSIOLOGY. *Educ:* Old Dom Col, BSc, 63; Univ Ga, DVM, 67, MS, 68, PhD(biochem), 70. *Prof Exp:* Res assoc biochem, Univ Ga, 69-71; asst prof obstet & gynec, Univ Chicago, 71-74; sci adv & chief pop res ctr, IIT Res Inst, 74-75; from assoc prof to prof physiol, obstet & gynecol, 75-83, actg head, Dept Physiol & Biophys, Univ Ill Med Ctr, 79-83; PROF OBSTET, GYNEC & BIOCHEM, RUSH MED CTR, RUSH UNIV, 83- *Concurrent Pos:* Endowed chair, Boysen professorship, Rush Univ. *Honors & Awards:* Young Andrologist Award, Am Soc Andrology. *Mem:* Am Soc Andrology; Soc Study Reproduction; Am Vet Med Asn. *Res:* Reproduction; biochemistry and physiology of male and female genital tract secretions, spermatozoa and fertilization; infertility; contraception; reproductive toxicology. *Mailing Add:* Obstet/Gynec Res, Rush Med Ctr 1653 W Congress Pkwy Chicago IL 60612-3864

ZANGER, MURRAY, b New York, NY, May 5, 32; m 62; c 2. PHYSICAL ORGANIC CHEMISTRY. *Educ:* City Col New York, BS, 53; Univ Kans, PhD(org chem), 59. *Prof Exp:* Fel chem, Univ Wis, 59-60; res chemist, Marshall Lab, E I du Pont de Nemours & Co, 60-64; from asst prof to assoc prof org chem, 64-72, PROF CHEM, PHILADELPHIA COL PHARM & SCI, 72- *Concurrent Pos:* USPHS res grant phenothiazine chem, 66-69; consult, Sadtler Res Labs, 68- *Mem:* Am Chem Soc. *Res:* Organophosphorus compounds; benzothiazoles; daunomycinone synthesis; sulfa drugs; organic mechanisms; nuclear magnetic resonance spectroscopy. *Mailing Add:* Dept Chem Philadelphia Col Pharm Sci 43rd St & Kingsessing Ave Philadelphia PA 19104

ZANGWILL, ANDREW, b Cleveland, Ohio, Sept 27, 54. THEORETICAL CONDENSED MATTER PHYSICS. *Educ:* Carnegie-Mellon Univ, BS, 76; Univ Pa, PhD(physics), 81. *Prof Exp:* Res scientist, Brookhaven Nat Lab, 81-83; asst prof physics, Polytech Inst NY, 83-85; ASSOC PROF PHYSICS, GA INST TECHNOL, 85- *Mem:* Am Phys Soc. *Res:* Theoretical problems in condensed matter and statistical physics. *Mailing Add:* Sch Physics Ga Inst Technol Atlanta GA 30332

ZANINI-FISHER, MARGHERITA, b Como, Italy, Jan 5, 47; m 76; c 1. SOLID STATE PHYSICS, ELECTROCHEMISTRY. *Educ:* Univ Rome, PhD(physics), 71. *Prof Exp:* Res assoc physics, Italian Nat Res Coun, 71-74, staff scientist, 74-76; res assoc, Moore Sch Elec Eng, Univ Pa, 76-77; PRIN RES SCIENTIST ASSOC, SENSORS & PROCESS DEPT, SCI RES LAB, FORD MOTOR CO, 77- *Concurrent Pos:* Res assoc, Div Eng, Brown Univ, 74-76. *Mem:* Am Phys Soc. *Res:* New materials used for energy storage; electrochemistry; transport properties of solids; soild state sensors. *Mailing Add:* 6999 Castle Dr Birmingham MI 48010

ZANJANI, ESMAIL DABAGHCHIAN, b Resht, Iran, Dec 23, 38; m 63; c 3. HEMATOLOGY, PHYSIOLOGY. *Educ:* NY Univ, BA, 64, MS, 66, PhD(hemat), 69. *Prof Exp:* From asst to res assoc exp hemat, NY Univ, 65-70; asst prof med & physiol, Mt Sinai Sch Med, 70-74, assoc prof physiol, 74-77; PROF MED & PHYSIOL, SCH MED, UNIV MINN, 77- *Mem:* AAAS; Harvey Soc; Am Soc Hemat; Am Soc Zool; NY Acad Sci. *Res:* Experimental hematology; hemopoietic stimulating factor; mechanisms of blood cell production and release; renal involvement in erythropoiesis; erythropoiesis in submammalian species; fetal erythropoiesis. *Mailing Add:* Dept Med Univ Minn Minneapolis MN 55455

ZANKEL, KENNETH L, b New York, NY, Mar 29, 30; m 84; c 1. ATMOSPHERIC SCIENCES. *Educ:* Rutgers Univ, BS, 51; Fla State Univ, MS, 55; Mich State Univ, PhD(physics), 58. *Prof Exp:* Asst res prof physics, Mich State Univ, 58-59; asst prof, Univ Ore, 59-63; Fulbright fel, Univ Heidelberg, 63-64; sr fel, Calif Inst Technol, 64-66; sr res fel, Sect Genetics Develop & Physiol, Cornell Univ, 66-69; scientist, Res Inst For Advan Studies, 69-75, MEM STAFF, MARTIN MARIETTA ENVIRONMENTAL SYSTS, 75- *Mem:* Air Pollution Control Asn; fel Acoust Soc Am. *Res:* Environmental sciences; materials damage; source emissions; receptor modeling; measurements methods. *Mailing Add:* 10714 Mid Summer Lane Columbia MD 21044

ZANNIS, VASSILIS I, b Kourounia Chios, Greece, Nov 18, 40; m 72; c 2. MOLECULAR GENETICS. *Educ:* Univ Athen, Greece, BS, 68; Univ Calif, Berkeley, PhD(biochem), 75. *Prof Exp:* Postdoctorate fel, Univ Calif, San Francisco, 75-77; res assoc, Mass Inst Technol, 77-88; res assoc, Harvard Med Sch, 79-82, asst prof pediat, Harvard Med Sch, 82-84; assoc prof med & biochem, 84-87; PROF MED & BIOCHEM, BOSTON UNIV MED SCH, 87- *Concurrent Pos:* Adj asst mem, Mem Sloan Kittering Cancer Ctr, 82-84; instr, Dept Oral Biol, Harvard Med Sch, 84-; prof biochem, Univ Crete, 86- *Mem:* Biophysical Soc; Am Chem Soc; Coun Arteriosclerosis fel; Am Inst Nutrition; Am Soc Biol Chem. *Res:* Utilization of biochemical cell and molecular biological approaches to investigate the molecular basis of human diseases associated with structural alterations in apolipoprotein genes and the regulation of expression of thes genes; author of 60 scientific publications. *Mailing Add:* Sect Molecular Genetics R420 Boston Univ Med Ctr 80E Concord St Boston MA 02118-2370

ZANNONI, VINCENT G, b New York, NY, June 12, 28. BIOCHEMISTRY, PHARMACOLOGY. *Educ:* City Col New York, BS, 51; George Washington Univ, MS, 56, PhD(biochem), 59. *Prof Exp:* Biochemist, Goldwater Mem Hosp, New York, 51-54 & Nat Heart Inst, 54-56; res chemist, Nat Inst Arthritis & Metab Dis, 57-63; from asst prof to prof biochem pharmacol, Sch Med, NY Univ, 63-74; PROF PHARMACOL, MED SCH, UNIV MICH, ANN ARBOR, 74- *Mem:* AAAS; Am Soc Biol Chemists; Am Soc Pharmacol & Exp Therapeut; NY Acad Sci. *Res:* Inborn errors of metabolism; mechanisms of reactions; amino acid metabolism; enzymology; biochemical pharmacology. *Mailing Add:* Dept Pharmacol Univ Mich Med Sci Bldg I Ann Arbor MI 48109

ZANONI, ALPHONSE E(LIGIUS), b Melrose Park, Ill, July 24, 34; m 60; c 6. ENGINEERING. *Educ:* Marquette Univ, BCE, 56; Univ Minn, MS, 60, PhD(civil eng), 64. *Prof Exp:* Consult engr, Toltz, King, Duvall, Anderson, Inc, Minn, 58-60; instr civil eng, 60-61, from asst prof to assoc prof, 64-76, chmn dept, 70-72, PROF CIVIL ENG, COL ENG, MARQUETTE UNIV, 76- *Concurrent Pos:* Consult water & waste water probs indust & munic. *Mem:* Am Soc Civil Engrs; Am Water Works Asn; Water Pollution Control Fedn; Sigma Xi; Asn Environ Eng Prof. *Res:* Water supply and pollution control. *Mailing Add:* Col Eng Marquette Univ Milwaukee WI 53233

ZANOWIAK, PAUL, b Little Falls, NJ, July 11, 33; m 57; c 4. PHARMACEUTICS, CONTINUING PHARMACEUTICAL EDUCATION. *Educ:* Rutgers Univ, BS, 54, MS, 57; Univ Fla, PhD(pharm), 59. *Prof Exp:* Instr pharm, Col Pharm, Univ Fla, 58-59; res & develop chemist, Noxell Corp, Md, 59-64; from asst prof to assoc prof pharmaceut, Sch Pharm, WVa Univ, 64-71; chmn dept, 71-85, actg dean, 72-74, PROF PHARMACEUT, TEMPLE UNIV, 71-, DIR, DIV CONTINUING PHARMACEUT EDUC, 81- *Concurrent Pos:* Am Found Pharm Educ fel. *Mem:* Am Pharmaceut Asn; Acad Pharmaceut Res & Sci; Am Asn Col Pharm; Am Inst Hist Pharm; Am Asn Pharmaceut Scientists; Sigma Xi. *Res:* Design and evaluation of dosage forms. *Mailing Add:* Dept Pharmaceut Sci Temple Univ Sch Pharm Philadelphia PA 19140

ZANZUCCHI, PETER JOHN, b Syracuse, NY, Apr 21, 41; m 67. ANALYTICAL CHEMISTRY. *Educ:* Le Moyne Col, NY, BS, 63; Univ Ill, Urbana, MS, 65, PhD(chem), 67. *Prof Exp:* SR STAFF CHEMIST, DAVID SARNOFF RES CTR, RCA CORP, 67- *Mem:* Electrochem Soc; Optical Soc Am. *Res:* Measurement of the optical properties of inorganic materials, particularly semiconductor materials in the wavelength range 0.2 to 200 micrometers. *Mailing Add:* 13 Jill Dr W Windsor Township NJ 08561

ZAPATA, PATRICIO, b Santiago, Chile, Oct 26, 37; m 64; c 4. NEUROPHYSIOLOGY, SENSORY PHYSIOLOGY. *Educ:* Cath Univ Chile, Bachellor, 59, Licentiate, 62; Univ Chile, MD, 63. *Prof Exp:* Fel neurophysiol, Nat Comn Med Faculties, Chile, 63-64; postdoctoral fel physiol, Univ Utah, 65-66 & neurol, 67; PROF NEUROBIOL, FAC SCI, CATH UNIV CHILE, 73-, PROF PHYSIOL, FAC MED, 77- *Concurrent Pos:* Vis assoc prof physiol, Univ Utah, 73-74, vis prof, 81-82; assoc ed, Arch Biol Med Exp, 88-; Regional rep, Int Soc Arterial Chemoreception, 88- *Mem:* Am Physiol Soc; Soc Neurosci. *Res:* Mechanisms of chemoreception and reflex control of respiratory and cardiovascular functions. *Mailing Add:* Cath Univ Chile PO Box 114-D Santiago 1 Chile

ZAPFFE, CARL ANDREW, b Brainerd, Minn, July 25, 12; m 37; c 8. METALLURGY, SPACE SCIENCES. *Educ:* Mich Technol Univ, BS, 33; Lehigh Univ, MSc, 34; Harvard Univ, ScD, 39. *Hon Degrees:* DEng, Mich Technol Univ, 60. *Prof Exp:* Metallurgist corrosion, Exp Sta, E I du Pont de Nemours & Co, Inc, 34-36; res assoc metall, Battelle Mem Inst, 38-43; asst tech dir stainless steel, Rustless Iron & Steel Corp, 43-45; CONSULT MAT ENG, 45- *Concurrent Pos:* Civilian scientist, Off Naval Res, 45-54. *Honors & Awards:* Procter Mem Award, Am Electroplaters Soc, 40; Sauveur Mem Award & Lectr, 58. *Mem:* Hon mem Am Soc Metall; Am Chem Soc; Am Phys Soc; Am Geophys Soc; Int Soc Gravitation & Gen Relativity; fel Am Inst Chemists; Am Inst Phys; Nat Asn Corrosion Engrs; Int Geophys Union. *Res:* Hydrogen in metals; stainless steels, superalloys and refractory metals failure analysis; originator of fractography; Pleistocene glaciation and Earth-moon evolution; originator of geohydrothermodynamics; relativistic physics and cosmology; originator of M-space and G-space. *Mailing Add:* 6410 Murray Hill Rd Baltimore MD 21212

ZAPHYR, PETER ANTHONY, b Wheeling, WVa, Sept 4, 26; m 56; c 2. STATISTICS, MANAGEMENT INFORMATION SYSTEMS. *Educ:* Bethany Col, WVa, BS, 48; Univ WVa, MS, 49; Univ Pittsburgh, PhD(math), 57. *Prof Exp:* Instr math, Univ WVa, 49-50; asst, Ill Inst Technol, 50-51 & Univ Pittsburgh, 51-52; analyst, 52-61, mgr digital anal & comput sect, 61-65, asst to dir, Anal Dept, 65-69, mgr eng comput systs, Nuclear Energy Systs, 69-73, mgr, Eng Comput Serv, 73-80, dir, Power Systs Comput Ctr, Westinghouse Elec Corp, 80-87, DIR, WESTINGHOUSE CORP COMPUTER SERV, PITTSBURGH, 87- *Concurrent Pos:* Mem & Officer bd trustees, H C Frick Community Hosp, Mt Pleasant, Pa. *Mem:* Asn Comput Mach. *Res:* Administration of industrial computing services and advanced applications of computers in engineering science, manufacturing and management. *Mailing Add:* 150 Morrison Ave Greensburg PA 15601

ZAPISEK, WILLIAM FRANCIS, b Morris, NY, Mar 29, 35; m 59; c 3. BIOCHEMISTRY, DEVELOPMENTAL BIOLOGY. *Educ:* Syracuse Univ, BA, 60; Univ Conn, MS, 65, PhD(biochem), 67. *Prof Exp:* Fel biochem, Los Alamos Sci Lab, 67-68; from asst prof to assoc prof, 68-77, PROF BIOCHEM, CANISIUS COL, 77- *Mem:* AAAS; Am Chem Soc; Soc Develop Biol. *Res:* Characterization of low molecular weight methylated ribonucleic acid species; dietary induction of cancer, DNA hypomethylation; control of expression of mouse alpha fetoprotein gene. *Mailing Add:* Dept Chem Canisius Col Buffalo NY 14208

ZAPOL, WARREN MYRON, b New York, NY, Mar 16, 42; m 68; c 2. MEDICINE. *Educ:* Mass Inst Technol, BS, 62; Univ Rochester, MD, 66. *Prof Exp:* Res assoc, Nat Heart Inst, Bethesda, Md, 67-70; resident, Harvard Univ, 70-72, asst prof anesthesia, 72-78, assoc prof & dir spec ctr res adult respiratory failure, 78-; AT MASS GEN HOSP; PROF ANESTHESIA, HARVARD MED SCH, 85- *Concurrent Pos:* Investr, US Antarctic Res Prog, 76-78 & 82- *Mem:* Am Physiol Soc; Am Soc Anethesiologists; Am Soc Artificial Internal Organs. *Res:* Circulation and gas exchange in animal models and man with acute lung injury; novel drug therapy in acute respiratory failure; diving seals in antarctica. *Mailing Add:* Dept Anesthesia Mass Gen Hosp Fruit St Boston MA 02114

ZAPOLSKY, HAROLD SAUL, b Chicago, Ill, Dec 24, 35; m 62; c 2. THEORETICAL PHYSICS. *Educ:* Shimer Col, AB, 54; Cornell Univ, PhD(physics), 62. *Prof Exp:* Nat Acad Sci-Nat Res Coun res assoc physics, Goddard Inst Space Studies, New York, 62-63; res assoc, Univ Md, College Park, 63-65, asst prof, 65-70; from assoc prog dir to prog dir theoret physics, NSF, 70-73; chmn, Dept Physics & Astron, 73-79, PROF PHYSICS, RUTGERS UNIV, 73- *Concurrent Pos:* Sr vis scientist, Dept Appl Math & Theoret Physics, Univ Cambridge, 79-80. *Mem:* Am Phys Soc; Am Astron Soc; NY Acad Sci. *Res:* Quantum electrodynamics; astrophysics; general relativity; non-linear systems. *Mailing Add:* Dept Physics Rutgers Univ New Brunswick NJ 08903

ZAPSALIS, CHARLES, b Lowell, Mass, Sept 22, 22; m 48. FOOD SCIENCE, CHEMISTRY. *Educ:* Springfield Col, BS, 52; Univ Mass, PhD(food sci, chem), 63. *Prof Exp:* Teacher, Jr High Sch, Mass, 52-53 & high, NY, 54-55; head sci dept high sch, Mass, 56-60; instrumental chemist, Beechnut Life Savers, Inc, 63, res mgr, 63-65; from asst prof to assoc prof, 65, PROF CHEM, FRAMINGHAM STATE COL, 65-, CHMN DEPT, 66- *Mem:* Am Chem Soc; Inst Food Technol; Sigma Xi. *Res:* Anthocyanins, chemical identification; pesticide methodology and characterization of tea components by gas chromatography; characterization of amino acids and polypeptides. *Mailing Add:* 265 Singletary Lane Framingham Center MA 01701

ZAR, JERROLD HOWARD, b Chicago, Ill, June 28, 41; m 67; c 2. ECOLOGY, PHYSIOLOGY. *Educ:* Northern Ill Univ, BS, 62; Univ Ill, Urbana, MS, 64, PhD(zool), 67. *Prof Exp:* Res assoc physiol ecol, Univ Ill, Urbana, 67-68; from asst prof to assoc prof, 71-78, chmn dept, 78-84, PROF BIOL SCI, NORTHERN ILL UNIV, 78-, ASSOC PROVOST GRAD STUDIES & RES & DEAN GRAD SCH, 84- *Concurrent Pos:* Res assoc, Dept Zool, Univ Ill, Urbana, 67-68; staff consult, Argonne Nat Lab, 73-77, vis scientist, Div Radiol & Environ Res, 74; staff consult, US Environ Protection Agency, 74-84; mem comt to rev methods ecotoxicol, Environ Studies Bd, Nat Res Coun, 79-81; vis ecologist, Biol Sta, Univ Mich, 86; adj prof statist, Northern Ill Univ, 89- *Mem:* Fel AAAS; Ecol Soc Am; Am Physiol Soc; Am Statist Asn; Biometric Soc; Am Soc Zool; Sigma Xi. *Res:* Ecology; ecological animal physiology; statistical data processing environmental assessment; biostatistical analysis. *Mailing Add:* Grad Sch Northern Ill Univ De Kalb IL 60115-2864

ZARAFONETIS, CHRIS JOHN DIMITER, b Hillsboro, Tex, Jan 6, 14; m 43; c 1. INTERNAL MEDICINE. *Educ:* Univ Mich, BA, 36, MS, 37, MD, 41; Am Bd Internal Med, dipl, 50. *Prof Exp:* Externe, Simpson Mem Inst, Univ Mich, 40-41; intern, Boston City Hosp, 41-42; asst prof internal med & res assoc, Univ Mich, 47-50; assoc prof med, Sch Med, Temple Univ, 50-55, clin prof, 55-57, prof clin & res med, 57-60, chief hemat sect, Univ Hosp, 50-60; dir, 60-78, prof, 60-80, EMER PROF INTERNAL MED, SIMPSON MEM INST, MED SCH, UNIV MICH, ANN ARBOR, 80- *Concurrent Pos:* Res fel internal med, Med Sch, Univ Mich, 46-47; consult, Dept Defense, Directorate Res, Develop, Testing & Eval, Dept Army, Sci Adv Panel, Army Sci Bd, Vet Admin Hosp, Mich & hist unit, Dept Surgeon Gen, US Dept Army; mem bd, Med in Pub Interest, Inc & Gorgas Mem Inst, 81-90; asst ed, Am J Med Sci, 51-60. *Honors & Awards:* Henry W Elliot Distinguished Serv Award, Am Soc Clin Pharmacol & Therapeut, 80; Medal Distinguished Pub Serv, Dept Defense, 84; John R Seal Award, Soc Med Consult to Armed Forces, 86. *Mem:* Fel Am Col Physicians; Am Soc Clin Pharmacol & Therapeut (vpres, 65-66, pres, 68-69); fel Int Soc Hemat; hon mem Agr Med Asn; Am Therapeut Soc (pres, 68-69). *Res:* Histoplasmosis; infectious mononucleosis; lymphogranuloma and herpes viruses; rickettsial disease; potassium para-aminobenzoate acid in collagen and bullous disorders and conditions with excess fibrosis; blood dyscrasias; lipid mobilizer hormones. *Mailing Add:* 2721 Bedford Rd Ann Arbor MI 48104

ZARCARO, ROBERT MICHAEL, b Springfield, Mass, Mar 4, 42; m 64; c 3. GENETICS, DEVELOPMENTAL BIOLOGY. *Educ:* Providence Col, BA, 64, MS, 66; Brown Univ, PhD(biol), 71. *Prof Exp:* Asst prof, 66-75, ASSOC PROF BIOL, PROVIDENCE COL, 75- *Mem:* AAAS. *Res:* Role of sulfhydryl compounds in mammalian pigmentation; genetic regulation of the multiple molecular forms of tyrosinase; role of protyrosinase in regulating melanogenesis. *Mailing Add:* Dept Biol Providence Col River Ave & Eaton St Providence RI 02918

ZARCO, ROMEO MORALES, b Caloocan, Philippines, Oct 7, 20; m 48; c 3. IMMUNOCHEMISTRY, PUBLIC HEALTH. *Educ:* Univ Philippines, MD, 43; Johns Hopkins Univ, MPH, 54. *Prof Exp:* Physician internal med, Philippine Gen Hosp, 43-44; from instr to prof microbiol, Univ Philippines, 46-67; assoc dir biochem res, Cordis Corp, 67-73, vpres opers, 77-80, DIR, CORDIS LABS, 73-, PRES OPERS, 80- *Concurrent Pos:* USPHS fel, 60-61; vis investr, Howard Hughes Med Inst, 61-62 & 64-67; asst prof, Univ Miami. *Honors & Awards:* Philippine Med Asn Res Award, 58. *Mem:* AAAS; Am Asn Immunol; NY Acad Sci. *Res:* Immunology of infectious diseases, including typhoid, cholera, leprosy and influenza; complement-anti-complementary factors from snake venom and immunosuppression and the separation and purification of the nine components of complement of human and guinea pig serum. *Mailing Add:* Cordis Labs 2140 N Miami Ave Miami FL 33127

ZARDECKI, ANDREW, b Warsaw, Poland, Aug 26, 42; m 66; c 1. QUANTUM OPTICS, ATMOSPHERIC OPTICS. *Educ:* Univ Warsaw, BSc, 64, MSc, 64; Polish Acad Sci, DSc, 68. *Prof Exp:* From asst to asst prof physics, Warsaw Tech Univ, 64-73; asst prof physics, Laval Univ, 73-79; STAFF MEM, LOS ALAMOS NAT LAB, 81- *Concurrent Pos:* Fel, Laval Univ, 70-72. *Mem:* Can Asn Physicists; Optical Soc Am; Am Phys Soc. *Res:* Functional techniques in the optical coherence theory; radiation theories, transport processes; atmospheric optics; pulse propagation in laser media; light scattering; cosmology; computer modeling and simulation. *Mailing Add:* Los Alamos Nat Lab MS E541 PO Box 1663 Los Alamos NM 87545

ZARE, RICHARD NEIL, b Cleveland, Ohio, Nov 19, 39; m 63; c 3. CHEMICAL PHYSICS, ANALYTICAL CHEMISTRY. *Educ:* Harvard Univ, BA, 61, PhD(chem physics), 64. *Prof Exp:* Fel, Harvard, 64; res assoc & fel, Joint Inst Lab Astrophys, Univ Colo, 64-65; asst prof chem, Mass Inst Technol, 65-66; asst prof physics, Univ Colo, 66-67, from asst prof to assoc prof physics & chem, 67-69; prof chem, Columbia Univ, 69-77; prof chem, Stanford Univ, 77-80, Shell distinguished prof, 80-85, MARGUERITE BLAKE WILBUR PROF CHEM, STANFORD UNIV, 88- *Concurrent Pos:* Mem, Joint Inst Lab Astrophys, Univ Colo, 66-67, fel, 67-69, non-resident fel, 69; Alfred P Sloan res fel, 67-69; consult, Aeronomy Lab, Nat Oceanic & Atmospheric Admin, Radio Standards Physics Div, Nat Bur Standards, 68-77; Higgins prof nat sci, Columbia Univ, 75-77. *Honors & Awards:* Michael Polanyi Medal, 79; Earle K Plyler Prize, Am Phys Soc, 81; Irving Langmuir Prize, Am Phys Soc, 85; Nat Medal Sci, 83; Gibbs Medal, Am Chem Soc, 90; Debye Award, Am Chem Soc, 91. *Mem:* Nat Acad Sci; AAAS; fel Am Phys Soc; Am Chem Soc; Am Acad Arts & Sci; Royal Chem Soc. *Res:* Problems associated with molecular photodissociation, molecular fluorescence and molecular chemiluminescence; application of lasers to chemical problems. *Mailing Add:* Dept Chem Stanford Univ Stanford CA 94305-5080

ZAREM, ABE MORDECAI, b Chicago, Ill, Mar 7, 17; m 41; c 3. ELECTRO-OPTICS, ENGINEERING MANAGEMENT. *Educ:* Ill Inst Technol, BS, 39; Calif Inst Technol, MS, 40, PhD(elec eng), 44. *Hon Degrees:* LLD, Univ Calif, Santa Cruz, 67 & Ill Inst Technol, 68. *Prof Exp:* Group mgr elec eng, US Naval Ord Test Sta, 45-48; assoc dir & mgr, Stanford Res Inst, 48-56; founder, pres & chmn bd, Electro-Optical Systs, Inc, 56-67; vpres, Xerox Corp, 63-67, sr vpres & dir corp develop, 67-69; consult, mgt & eng, 69-75; chmn & chief exec officer, Xerox Develop Corp, 76-81; FOUNDER & MANAGING DIR, FRONTIER ASN, 80- *Concurrent Pos:* Mem adv coun, Sch Eng, Stanford Univ, 66-78; mem, eng adv bd & bd dir, Harvey Mudd Col, 67-69; mem adv comt, Div Eng & Appl Sci, Calif Inst Technol, 69-77, adv comt competitive technol, Dept Com, State Calif, 88-91; mem, Calif Coun Sci & Technol, 89-; trustee, City of Hope, 66-; lectr, Solar & Unconventional Energy Sources, Univ Calif, 56-61. *Honors & Awards:* Sperry Award, Instrument Soc Am, 69. *Mem:* Nat Acad Engrs; fel Inst Elec & Electronics Engrs; Solar Energy Soc; fel Am Inst Elec Eng; fel Am Inst Aeronaut & Astronaut; fel Inst Radio Engrs. *Res:* Determination of the role of socio-biological factors on the development of innovative attitudes, creative thinking and motivational behavioral patterns; author of one book; inventor. *Mailing Add:* Frontier Assoc 9640 Lomitas Ave Beverly Hills CA 90210

ZAREM, HARVEY A, b Savannah, Ga, Feb 13, 32; m 58; c 3. PLASTIC SURGERY. *Educ:* Yale Univ, BA, 53; Columbia Univ, MD, 57. *Prof Exp:* Assoc prof surg, Univ Chicago, 66-73; prof surg & chief div plastic surg, Med Sch, 73-87, EMER PROF, UNIV CALIF, LOS ANGELES, 87- *Concurrent Pos:* Markle scholar, Markle Found, 68. *Mem:* Plastic Surg Res Coun; Soc Univ Surgeons; Am Cleft Palate Asn; Soc Head & Neck Surgeons; Microcirc Soc. *Res:* Microvasculature; microsurgery. *Mailing Add:* 1301 20th St Suite 470 Santa Monica CA 90404

ZAREMSKY, BARUCH, b Cleveland, Ohio, Sept 21, 26; m 51; c 3. ORGANOMETALLIC CHEMISTRY, PLASTICS CHEMISTRY. *Educ:* Western Reserve Univ, BS, 48, MS, 50, PhD(org chem), 54. *Prof Exp:* RES CHEMIST, FERRO CHEM DIV, FERRO CORP, BEDFORD, 53- *Mem:* Am Chem Soc. *Res:* Additives for polyvinyl chloride polypropylene, polycarbonates and polyesters; specialist in synthesis of organotens. *Mailing Add:* 1708 Beaconwood Ave South Euclid OH 44121

ZARET, BARRY L, b New York, NY, Oct 3, 40; m 63; c 3. NUCLEAR CARDIOLOGY, CARDIAC RESEARCH. *Educ:* Queens Col, NY, BS, 62; NY Univ, MD, 66; Yale Univ, MA, 82. *Prof Exp:* Intern & resident internal med, NY Univ-Bellevue Hosp Med Ctr, 66-69; res fel cardiol, Johns Hopkins

Sch Med, 69-71; major, USAF Marine Corps, Travis AFB, 71-73; from asst prof to prof, 73-84, BERLINER PROF INTERNAL MED, YALE UNIV, 84-, CHIEF CARDIOL, 78- *Concurrent Pos:* Assoc ed, Yearbk Nuclear Med, 80-; pres-elect, Asn Prof Cardiol. *Honors & Awards:* Blumgart Award, Soc Nuclear Med, New Eng Chap. *Mem:* Am Physiol Soc; fel Am Col Cardiol; fel Am Heart Asn; Asn Univ Cardiologists; Am Soc Clin Invest; NAm Soc Cardiac Radiol. *Res:* Nuclear cardiology; myocardial metabolism; cardiac imaging; studies of acute myocardial infarction and thrombolysis. *Mailing Add:* Yale Univ Sch Med & Cardiol 333 Cedar St New Haven CT 06510

ZARING, WILSON MILES, b Shelbyville, Ky, Nov 9, 26; m 50; c 2. MATHEMATICS. *Educ:* Ky Wesleyan Col, AB, 50; Univ Ky, MA, 52, PhD(math), 55. *Prof Exp:* From instr to asst prof math, Univ Ill, Urbana, 55-63, grad adv, 79-82, assoc prof, 63-81, dir grad studies, 82-91; RETIRED. *Concurrent Pos:* Consult, CSMP, 74-78. *Honors & Awards:* Max Beberman Award, Ill Coun Teachers Math, 76. *Mem:* Am Math Soc; Math Asn Am; Sigma Xi. *Res:* Analysis; number theory. *Mailing Add:* Dept Math Univ Ill Urbana IL 61801

ZARKOWER, ARIAN, b Tarnopol, Poland, Oct 10, 29; US citizen; m 60; c 2. VETERINARY MEDICINE, IMMUNOLOGY. *Educ:* Ont Vet Col, DVM, 56; Univ Maine, MS, 60; Cornell Univ, PhD(immunochem), 65. *Prof Exp:* Dist vet, NB Prov Vet Serv, 56-57; vet pvt pract, 57-58; res asst animal path, Univ Maine, 58-70; res officer, Animal Dis Res Inst, Can Dept Agr, 60-62; res asst immunochem, Cornell Univ, 62-65; from asst prof to assoc prof vet sci, 65-77, PROF VET SCI, CTR AIR ENVIRON STUDY, PA STATE UNIV, 78-, MEM STAFF CTR, 70- *Mem:* Am Asn Immunol; Am Asn Vet Immunol; Res Workers in Animal Dis; Reticuloendothelial Soc. *Res:* Experimental pathology; immune response and resistance in animals to infections; inflammatory responses. *Mailing Add:* Dept Vet Sci Pa State Univ University Park PA 16802

ZARLING, JOHN P, b Elmhurst, Ill, Mar 15, 42; m 65; c 3. ARCTIC ENGINEERING, HEAT TRANSFER. *Educ:* Mich Tech Univ, BS, 64, MS, 66, PhD(eng mech), 71. *Prof Exp:* Instr eng, Univ Wis-Madison, 66-68, from asst prof to assoc prof, Univ Wis-Parkside, 71-76; assoc prof, 76-80, PROF MECH ENG, UNIV ALASKA, FAIRBANKS, 80- *Concurrent Pos:* Asst vchancellor, Univ Wis-Parkside, 74-76; dir, Inst Northern Eng, Univ Alaska, Fairbanks, 86-91, assoc dean eng, 86-91. *Mem:* Am Soc Eng Educ; Am Soc Mech Engrs; Am Soc Heating, Ventillating & Air Conditioning Engrs. *Res:* Heat transfer problems associated with an arctic climate including perma frost, frost heave, building design etc. *Mailing Add:* 539 Duckering Bldg Univ Alaska Fairbanks AK 99775-0500

ZARNSTORFF, MICHAEL CHARLES, b Denver, Colo, Aug 8, 54. PLASMA PHYSICS. *Educ:* Univ Wis, BS, 76, PhD(physics), 84. *Prof Exp:* Staff physicist, Lawrence Livermore Nat Lab, 76-77; RES STAFF PHYSICIST, PLASMA PHYSICS LAB, PRINCETON UNIV, 84- *Mem:* Am Phys Soc. *Res:* Plasma transport phenomena; experiment on tokamak devices. *Mailing Add:* Plasma Phys Lab Princeton Univ PO Box 451 Princeton NJ 08544

ZAROBILA, CLARENCE JOSEPH, b Cleveland, Ohio, Aug 27, 58; m 82. INTERFEROMETIC FIBER OPTIC SENSORS, FIBER OPTIC INTENSITY SENSORS. *Educ:* John Carroll Univ, Cleveland, Ohio, BS, 80, MS, 82. *Prof Exp:* Engr, Dynamics Systs, Inc, 82-83; PHYSICIST, OPTICAL TECHNOL, INC, 83- *Mem:* Soc Photo-Optical Instrumentation Engrs. *Res:* Fiber optic sensors; amplitude- and phase-modulated types for detection of pressure, acoustics, magnetic fields and other phenomena. *Mailing Add:* 360 Herndon Pkwy Suite 1200 Herndon VA 22070

ZAROMB, SOLOMON, b Belchatow, Poland, Aug 15, 28; US citizen; m 76; c 2. PHYSICAL CHEMISTRY, ELECTROCHEMICAL ENGINEERING. *Educ:* Cooper Union, BChE, 50; Polytech Inst Brooklyn, PhD(chem), 54. *Prof Exp:* Res assoc light scattering, Mass Inst Technol, 53-55; assoc chemist semiconductors, IBM Corp, 56-58; res specialist electrochem devices, Philco Corp, 58-61; scientist res & develop monocrystalline thin films, Martin-Marietta Co, 62-63; pres electrochem & electro-optical systs & devices, Zaromb Res Corp, 63-81; ELECTROCHEM ENGR, TOXIC GAS DETECTION, AGRONNE NAT LAB, 81- *Concurrent Pos:* Pres, Zaromb Res Found, grants from Nat Air Pollution Control Admin & Environ Protection Agency, 69-73. *Mem:* Sigma Xi; Am Chem Soc; Electrochem Soc. *Res:* Electrochemical and electro-optical systems, devices and techniques; aluminum batteries; processing of semiconductor crystals and thin films; solid solutions of ice; remote sensing of air pollutants; lidar spectroscopy; electrochemical and other toxic gas detectors or monitors. *Mailing Add:* 9700 S Cass Ave Argonne Nat Lab ER-203 Argonne IL 60439

ZAROSLINSKI, JOHN F, b Chicago, Ill, Sept 12, 25; m 51; c 2. PHARMACOLOGY, BIOCHEMISTRY. *Educ:* Univ Chicago, PhB, 49; Loyola Univ, Ill, PhD(pharmacol), 65. *Prof Exp:* Chemist, Armour Pharmaceut Co, 51-53; from pharmacologist to sr pharmacologist, Baxter Lab Inc, 53-58; sci dir, Arnar-Stone Labs, Inc, Div Am Hosp Supply Corp, 58-65, vpres res & develop, 65-75, vpres sci planning, 75-78; RETIRED. *Concurrent Pos:* Adj prof, Stritch Sch Med, Loyola Univ, Chicago, 65-88; res consult, US Vet Hosp, Hines, Ill. *Mem:* British Pharmacol Soc; Am Chem Soc; NY Acad Sci; Am Soc Pharmacol & Exp Therapeut. *Res:* Pharmaceutical development; biochemical pharmacology; protein binding of drugs; evaluation of hypnotic drugs; pharmaceutical development and introduction of Intropin (dopamine) into therapy for treatment of shock in humans; supervised scientific and clinical studies leading to approval for use of dopamine in the treatment of shock. *Mailing Add:* 1202 Norman Lane Deerfield IL 60015-3116

ZARTMAN, DAVID LESTER, b Albuquerque, NMex, July 6, 40; m 63; c 2. CYTOGENETICS, REPRODUCTION. *Educ:* NMex State Univ, BS, 62; Ohio State Univ, MS, 66, PhD(cytogenetics), 68. *Prof Exp:* From asst prof to prof animal & range sci, NMex State Univ, 68-84; CHMN & PROF, DEPT DAIRY SCI, OHIO STATE UNIV, COLUMBUS, 84- *Concurrent Pos:* NIH fel, 73; Fulbright-Hays fel, 76; pres, Mary K Zartman Inc, 78-84; consult, Bio-Med Electronics Inc, 84- *Mem:* Fel AAAS; Am Dairy Sci Asn; Sigma Xi; Animal Sci Soc Am; Am Inst Biol Sci. *Res:* Radiation genetics; sex and fertility control; reproduction; animal breeding; radio-telemetry of body temperature. *Mailing Add:* Ohio State Univ Dairy Sci 116 Plumb Hall Columbus OH 43210-1094

ZARTMAN, ROBERT EUGENE, b Lancaster, Pa, May 19, 36; m 75; c 7. GEOCHRONOLOGY, ISOTOPE GEOCHEMISTRY. *Educ:* Pa State Univ, BS, 57; Calif Inst Technol, MS, 59, PhD(geol), 63. *Prof Exp:* Fel geol, Calif Inst Technol, 62; chief, 81-85, GEOLOGIST, ISOTOPE GEOL BR, US GEOL SURV, 62- *Concurrent Pos:* Vis assoc, Calif Inst Technol, 71-72; chmn working group on radiogenic isotopes, Int Asn of Volcanology & Chem of the Earth's Interior, 73-81; mem Lunar Sample Rev Panel, 73-76, working group Precambrian of US & Mex, 76-82, US Geodynamics Comt, 81-84 & Crustal Genesis Rev Panel, 83-84; vis scholar, Univ Chicago, 88. *Honors & Awards:* Meritorious Serv Award, US Dept Interior, 86. *Mem:* Fel Geol Soc Am; fel Mineral Soc Am; Am Geophys Union; Geochem Soc. *Res:* Determination of geologic age by the potassium-argon, rubidium-strontium and uranium-thorium-lead radiometric methods; study of geological processes and crustal/mantle structure by use of natural isotopic tracer systems. *Mailing Add:* Isotope Geol Br US Geol Surv Stop 963 Box 25046 Denver CO 80225-0046

ZARUCKI, TANYA Z, ENDOCRINOLOGY, STEROID RECEPTORS. *Educ:* Columbia Univ, PhD(human genetics & develop), 79. *Prof Exp:* Asst prof cell biol, Baylor Col Med, 83-86; STAFF RESEARCHER, SYNTEX, 86- *Res:* Molecular biology. *Mailing Add:* Sentex Res MS S-3-1 3401 Hillview Ave Palo Alto CA 94304

ZARWYN, B(ERTHOLD), b Austria, Aug 22, 21; nat US. ELECTRONICS ENGINEERING, OPERATIONS RESEARCH. *Educ:* Univ Lwow, ME, 46; Univ Munich, DrEng, 47; NY Univ, PhD(physics), 53; Columbia Univ, EngScD(mech eng), 64. *Prof Exp:* Asst prof mech eng, UNRRA, Univ Munich, 46-48; asst prof, Int Univ, Ger, 48-49; proj engr & dept head, Condenser Serv & Eng Co, NJ, 50-52; proj engr & head, Nuclear Aircraft Study Group, Curtiss-Wright Corp, 52-55; sr res engr, Am Mach & Foundry Co, NY, 55-57; chief physicist, Link Aviation Co, 57-58; dir basic res, Am Bosch-Arma Corp, NY, 58-64, corp consult, Airborne Instrument Lab, 63-65; chief engr, Bell Aerospace Corp, 65-66; sr consult, Mitre Corp, Mass, 66-68; actg chief engr, Hq, US Army Mat Command, 68-69, chief Physics & Electronics Br, Res Div, 69-75, physical scientist, US Army Harry Diamond Labs, 75-77, chief, Syst Analytical Br, 77-79, chief, Technol Div, 79-81, asst tech dir, US Army Electronic Res & Develop Command, 81-85, asst dir, plans & oper, US Army Lab Command, 85-86; PRES, PAN-TECHNOL CORP, 87- *Concurrent Pos:* Mem staff, Columbia Univ, 54-55; vis physicist, Brookhaven Nat Lab & assoc prof, Univ Conn, 56-57; prof, Univ Hartford, 57; consult, Fairchild Engine & Airplane Co, 55-56 & Bendix Aviation Co, 57; dir, Film Micro Electronics Co Inc, 65-67. *Mem:* Am Phys Soc; Inst Elec & Electronics Engrs; NY Acad Sci. *Res:* Nuclear physics and engineering; quantum electronics; solid state physics; propulsion; fluid flow; heat transfer; electronic computers and simulation; navigation and communication. *Mailing Add:* 9727 Mt Pisgah Rd 801 Silver Spring MD 20903-2011

ZARY, KEITH WILFRED, b Sask, Nov 28, 48; m 80. PLANT BREEDING. *Educ:* Univ Sask, BS, 71; Tex A&M Univ, MS, 78, PhD(hort), 80. *Prof Exp:* Res asst, Tex A&M Univ, 77-79; plant breeder, Sun Seeds/Agrigenetics Corp, 80-85; RES DIR, BEAR CREEK GARDENS INC, 85- *Mem:* Am Soc Hort Sci; AAAS; Sigma Xi. *Res:* Horticulturally superior pea (pisum satium) and bean (phascolus vulgaris) varieties for both processing and fresh market consumption. *Mailing Add:* Bear Creek Gardens Inc PO Box 2000 6500 Donlon Rd Somis CA 93066

ZARZECKI, PETER, b Boston, Mass, Aug 29, 45. PHYSIOLOGY, NEUROPHYSIOLOGY. *Educ:* Univ Miami, BS, 68; Duke Univ, PhD(physiol & pharmacol), 74. *Prof Exp:* Res assoc neurophysiol, Rockefeller Univ, 74-77; asst prof, 77-83, ASSOC PROF PHYSIOL, QUEEN'S UNIV, ONT, 83- *Concurrent Pos:* Vis scientist physiol, Gothenburg Univ, 77; Med Res Coun Can res grant, 78-89, res develop award, 78-88. *Mem:* Soc Neurosci; Can Physiol Soc; Can Asn Neurosci. *Res:* Electrophysiological investigations into the control of the mammalian cerebral motor cortex. *Mailing Add:* Dept Physiol Queen's Univ Kingston ON K7L 3N6 Can

ZASADA, ZIGMOND ANTHONY, b Schenectady, NY, May 1, 09; m 37; c 1. FORESTRY. *Educ:* State Univ NY, BS, 31. *Prof Exp:* Forester, Chippewa Nat Forest, US Forest Serv, 33-45, proj leader, Lake States Forest Exp Sta, 45-51, res ctr leader, 51-61, res forester, DC, 61-63, asst dir, NCent Forest Exp Sta, Minn, 63-67; res assoc, Cloquet Forestry Ctr, Col Forestry, Univ Minn, 67-76; FORESTRY CONSULT, 76- *Mem:* Fel Soc Am Foresters. *Res:* Economics of forest management and utilization; silviculture; mechanized timber harvesting. *Mailing Add:* 1015 Third Ave NW Grand Rapids MN 55744

ZASKE, DARWIN ERHARD, b Wadena, Minn, Mar 20, 49. CLINICAL PHARMACOLOGY. *Educ:* Univ Minn, BS, 72, PharmD, 73. *Prof Exp:* clin pharmacologist, St Paul-Ramsey Hosp & Med Ctr, 73-80; asst prof pharmacol, 75-80, MEM FAC, DEPT ADMIN & SOCIAL PHARMACOL, UNIV MINN, 80-, ASST HEAD PHARM PRACT. *Concurrent Pos:* Instr pharmacol, Univ Minn, 73-74. *Mem:* Am Burn Asn; Am Soc Hosp Pharm. *Res:* Clinical application of drug-kinetic principles with the goal being patient individualization of drug therapy to provide more optimal patient therapy. *Mailing Add:* Four Thrush Lane North Oaks MN 55127

ZASLAVSKY, THOMAS, b Brooklyn, NY, Jan 16, 44; m 85; c 2. MATROID THEORY, GRAPH THEORY. *Educ:* City Col New York, BS, 65; Mass Inst Technol, PhD(math), 74. *Prof Exp:* Instr math, Mass Inst Technol, 75-77; asst prof, Ohio State Univ, 77-84; assoc prof, 85-88, PROF MATH, STATE UNIV NY, BINGHAMTON, 88- *Concurrent Pos:* Vis researcher, Univ

Evansville, 84-85. *Mem:* Am Math Soc; Math Asn Am; Soc Indust & Appl Math. *Res:* Matroids, especially invariants; arrangements of hyperplanes; signed and biased graphs, including matroids, coloring and topological signed graph theory. *Mailing Add:* Math Sci Dept State Univ NY Binghamton NY 13902-6000

ZASSENHAUS, HANS J, b Coblenz, Ger, May 28, 12, nat Can; m 42; c 3. PHYSICAL MATHEMATICS. *Educ:* Univ Hamburg, Dr rer nat, 34, Dr habil, 38. *Hon Degrees:* MA, Glasgow Univ, 49; Dr, Univ Ottawa, 66, McGill Univ, 74, Univ Sardrücken, 83. *Prof Exp:* Asst instr, Univ Rostock, 34-36; asst, Univ Hamburg, 36-40; diaeten-docent math, Univ Hamburg, 40-46, assoc prof, 46-50; prof, McGill Univ, 49-59; prof, Univ Notre Dame, 59-64; res prof, 64-82, EMER PROF MATH, OHIO STATE UNIV, 82- *Concurrent Pos:* Vis prof math, Ctr Math Res, Montreal Univ, 77-78 & 83, Univ Calif, Los Angelels, 70 & Warwick Univ, Eng, 72; Fairchild scholarship, Calif Inst Technol, 74-75; hon prof, Univ Linz. *Honors & Awards:* Jeffery-Williams Lect Award, Can Math Cong, 74. *Mem:* Math Asn Am; Am Math Soc; Can Math Cong; fel Royal Soc Can; Ger Math Asn. *Res:* Group theory; Lie algebra; number theory; geometry of numbers; applied mathematics. *Mailing Add:* 942 Spring Grove Lane Columbus OH 43235-3325

ZATKO, DAVID A, b North Tonawanda, NY, Nov 12, 40; m 66; c 1. INORGANIC CHEMISTRY. *Educ:* Colgate Univ, BA, 62; Univ Wis, PhD(chem), 66. *Prof Exp:* Asst prof chem, Univ Ala, 67-74, assoc prof, 74-80; MEM STAFF, UNIVAC DIV, SPERRY CORP, 80- *Mem:* Am Chem Soc. *Res:* Analytic inorganic chemistry; coordination chemistry of silver II and silver III, palladium II, substituted ferrocenes; photoelectron spectroscopy; free radical ligands; oxidation-reductions in nonaqueous solvents. *Mailing Add:* 434 Sailmaker Way Lansdale PA 19446

ZATUCHNI, GERALD IRVING, b Philadelphia, Pa, Oct 5, 35; m 58; c 3. OBSTETRICS & GYNECOLOGY. *Educ:* Temple Univ, AB, 54, MD, 58, MSc, 65. *Prof Exp:* Instr obstet & gynec, Med Sch, Temple Univ, 65-66; dir family planning, Pop Coun, Inc, 66-69; adv, Govt of India, 69-71; Pop Coun, Inc consult family planning & obstet, WHO, 71-73; adv, Govt Iran, 73-75; PROF, DEPT OBSTET & GYNEC, NORTHWESTERN UNIV, 77-, DIR, PROG APPL RES FERTIL REGULATION, 77- *Mem:* AAAS; Am Fedn Clin Res; Am Fertil Soc; Am Col Obstet & Gynec. *Res:* Human reproductive research; contraceptive development; family planning; population study and research; high risk obstetrics. *Mailing Add:* 333 E Superior, Suite 150 Chicago IL 60611

ZATUCHNI, JACOB, b Philadelphia, Pa, Oct 8, 20; m 45; c 4. INTERNAL MEDICINE, CARDIOVASCULAR DISEASE. *Educ:* Temple Univ, AB, 41, MD, 44, MS, 50. *Prof Exp:* Chief, Sect Cardiol, Episcopal Hosp, Philadelphia, 69-82; dir, dept med, 74-82, clin prof, 61-65, prof, 65-87, EMER PROF MED, TEMPLE UNIV SCH MED, 87-; SR DIAGNOSTICIAN & DIR CLIN SERV SECT, PA HOSP, 87-; CLIN PROF MED SCH MED, UNIV PA, 88- *Concurrent Pos:* Teaching chief med, Episcopal Hosp, 58-67; fel coun clin cardiol, Am Heart Asn, 63-; mem, bd dirs, Am Heart Asn, Southeasten Pa, 90-91. *Mem:* AAAS; Am Thoracic Soc; AMA; fel Am Col Physicians; Sigma Xi; fel Am Col Cardiol. *Res:* Cardiovascular diseases. *Mailing Add:* Pa Hosp Eighth & Spruce Sts Philadelphia PA 19107

ZATZ, JOEL L, US citizen. PHARMACEUTICS. *Educ:* Long Island Univ, BS, 56; St John's Univ, MS, 65; Columbia Univ, PhD(pharmaceut Sci), 68. *Prof Exp:* From asst prof to assoc prof, 68-79, PROF PHARMACEUT, RUTGERS UNIV, 79- *Concurrent Pos:* Consult, Merck & Co, 79- *Mem:* Am Pharmaceut Asn; Soc Cosmetic Chemists; Acad Pharmaceut Sci; Am Chem Soc; Am Asn Col Pharm. *Res:* Factors that influence transport of drugs through skin, including the physical chemistry of disperse systems and applications to pharmaceutical and cosmetic products. *Mailing Add:* 77 Woodside Ave Metuchen NJ 08840

ZATZ, LESLIE M, b Schenectady, NY, Nov 2, 28; m 53; c 3. RADIOLOGY. *Educ:* Union Col, NY, BS, 48; Albany Med Col, MD, 52; Univ Pa, MMS, 59. *Prof Exp:* Intern, Univ Chicago Clins, 52-53; resident radiol, Hosp Univ Pa, 55-58; from instr radiol to assoc prof, 59-72, actg dir diag radiol, 66-67, PROF RADIOL, SCH MED, STANFORD UNIV, 72- *Concurrent Pos:* NIH spec res fel, Postgrad Med Sch, Univ London, 65-66; consult, Vet Admin Hosp, Palo Alto, Calif, 60-72, chief radiol, 72- *Mem:* Am Col Radiol; Asn Univ Radiol; AMA; Am Soc Neuroradiol. *Res:* New diagnostic radiologic methods. *Mailing Add:* Radiol Vah Med Ctr 5072 Stanford Univ Stanford CA 94305

ZATZ, MARION M, b New York, NY, Feb 10, 45. IMMUNOLOGY. *Educ:* Barnard Col, BA, 65; Cornell Univ, PhD(immunol & microbiol), 70. *Prof Exp:* Fel immunol, Hosp Spec Surg, 70-71; Damon Runyon fel biochem, Albert Einstein Col Med, 71-72; instr microbiol, Sch Med, Yale Univ, 72-73, asst prof microbiol & path, 73-77, dir, Tissue Typing Lab, 72-74; guest worker, Immunol Br, Nat Cancer Inst, 74-79; assoc res prof biochem, Sch Med, George Washington Univ, 78-84; PROG ADMINR, CELLULAR & MOLECULAR BASIS DIS DIV, NAT INST GEN MED SCI, NIH, 84- *Concurrent Pos:* Asst clin prof med, Sch Med, Georgetown Univ, 78-81. *Mem:* Am Asn Immunologists. *Res:* Investigation of spontaneous leukemogenesis in AKR-J mice; investigation of genetic basis of resistance to growth of lymphoma; T-cell differentiation; thymic hormones. *Mailing Add:* Cell & Molec Basis Dis Prog Westwood Bldg Rm 904 Nat Inst Gen Med Sci Washington DC 20037

ZATZ, MARTIN, b 1944; US citizen. PHARMACOLOGY, PSYCHIATRY. *Educ:* Albert Einstein Col Med, PhD(pharmacol), 70, MD, 72. *Prof Exp:* Resident psychiat, Sch Med, Yale Univ, 72-74; res assoc, 74-76, staff fel, 76-78, MED OFFICER RES, SECT PHARMACOL, LAB CLIN SCI, NIMH, 78- *Res:* Circadian rhythms; cyclic nucleotides; regulation of receptor sensitivity; pineal gland. *Mailing Add:* Dept Biochem Bldg 36 Rm 2A17 NIMH NIH Bethesda MD 20892

ZATZKIS, HENRY, b Holzminden, Ger, Apr 7, 15; nat US; m 51; c 2. THEORETICAL PHYSICS, APPLIED MATHEMATICS. *Educ:* Ohio State Univ, BS, 42; Ind Univ, MS, 44; Syracuse Univ, PhD(physics), 50. *Prof Exp:* Instr physics, Ind Univ, 42-44; instr, Univ NC, 44-46; instr math, Syracuse Univ, 46-51; instr, Univ Conn, 51-53; from asst prof to prof, 53-71, from actg chmn to chmn dept, 59-85, DISTINGUISHED PROF MATH, NEWARK COL, 71-, EMER CHMN DEPT, 85- *Res:* Theory of relativity; heat conduction; acoustics. *Mailing Add:* Five Elliott Pl West Orange NJ 07052

ZATZMAN, MARVIN LEON, b Philadelphia, Pa, Aug 6, 27; m 51; c 2. PHYSIOLOGY. *Educ:* City Col New York, BS, 50; Ohio State Univ, MS, 52, PhD(physiol), 55. *Prof Exp:* Asst prof physiol, Ohio State Univ, 55-56; from asst prof to assoc prof, 56-73, PROF PHYSIOL, MED CTR, UNIV MO-COLUMBIA, 73- *Mem:* AAAS; Biophys Soc; Int Soc Nephrology; Am Physiol Soc. *Res:* Renal, cardiovascular, hibernation. *Mailing Add:* Dept Physiol Univ Mo Med Ctr Columbia MO 65212

ZAUDER, HOWARD L, b New York, NY, Sept 13, 23; m 53; c 2. ANESTHESIOLOGY. *Educ:* Univ Vt, AB, 47, MS, 49; Duke Univ, PhD(physiol, pharmacol), 52; NY Univ, MD, 55. *Prof Exp:* Res assoc pharmacol, Univ Vt, 51-53; from asst prof to assoc prof anesthesiol, Albert Einstein Col Med, Yeshiva Univ, 58-67; prof anesthesiol & chmn dept, Med Sch, Univ Tex, San Antonio & prof pharmacol, Health Sci Ctr, 68-78, assoc dean prof affairs, Univ, 77-78; prof anesthesiol & chmn dept, 78-89, prof pharmacol, 78-89, EMER PROF ANESTHESIOL & PHARMACOL, STATE UNIV NY UPSTATE MED CTR, 89- *Mem:* AAAS; Am Soc Pharmacol & Exp Therapeut; fel Am Col Anesthesiol. *Res:* Pharmacology of anesthetic agents; effects of radiation on response to anesthesia; pulmonary physiology; clinical application of gas chromatography. *Mailing Add:* 1032 Tramway Lane NE Albuquerque NM 87122

ZAUDERER, BERT, b Vienna, Austria, Mar 8, 37; US citizen; m 61; c 5. COAL TECHNOLOGY, ENERGY CONVERSION. *Educ:* City Col New York, BME, 58; Mass Inst Technol, SM, 60, ScD(mech eng), 62. *Prof Exp:* Res engr, Gen Elec Co, 61-67, group leader, 67-70; mgr magnetohydrodyn progs, Space Sci Lab, 70-79 & Energy Dept, 80-81; PRES, COAL TECH CORP, 81- *Mem:* Assoc fel Am Inst Aeronaut & Astronaut; Am Phys Soc; Am Soc Mech Engrs. *Res:* Coal technology; advanced energy conversion; magnetohydrodynamic power. *Mailing Add:* c/o Coal Tech Corp PO Box 154 Merion PA 19066-0154

ZAUDERER, MAURICE, IMMUNE RESPONSES, T LYMPHOCYTE SPECIFICITY. *Educ:* Mass Inst Technol, PhD(cellular biol), 72. *Prof Exp:* ASSOC PROF ONCOL, MICROBIOL & IMMUNOL, UNIV ROCHESTER, 84- *Res:* Regulation of immune responses; cellular immunology. *Mailing Add:* Dept Microbiol Box 704 Univ Rochester Med Ctr Rochester NY 14642

ZAUGG, HAROLD ELMER, b Chicago, Ill, Feb 27, 16; m 40; c 3. ORGANIC CHEMISTRY. *Educ:* Oberlin Col, AB, 37; Univ Minn, PhD(org chem), 41. *Prof Exp:* Asst org chem, Univ Minn, 38-40; res chemist, Abbott Labs, 41-56, res scientist, 56-59, res fel, 59-72, sr res fel, 72-80; RETIRED. *Concurrent Pos:* Fel, Purdue Univ, 58; chmn, Gordon Conf Org Reactions & Processes, 61; mem med study sect, NIH, 64-68; vis prof, Univ Southern Calif, 66; mem med chem study group, Walter Reed Army Inst Res, 77-81. *Mem:* Am Chem Soc. *Res:* Central nervous system drugs; organic syntheses and reaction mechanisms; solvent effects; chemistry of delocalized anions; amidoalkylations; medicinal chemistry of the cannabinoids. *Mailing Add:* 270 E Park Ave Lake Forest IL 60045

ZAUGG, WALDO S, b LaGrande, Ore, Dec 13, 30; m 53; c 4. BIOCHEMISTRY. *Educ:* Brigham Young Univ, BA, 58, PhD(biochem), 61. *Prof Exp:* Fel, Enzyme Inst, Univ Wis, 61-62; fel, Charles F Kettering Res Lab, 62-63, staff scientist, 63-65; biochemist, Western Fish Nutrit Lab, US Fish & Wildlife Serv, 65-76; operator, Mill-A Chem Lab, 76-78; BIOCHEMIST, NAT MARINE FISHERIES SERV, COOK FIELD STA, 78- *Concurrent Pos:* Asst prof, Antioch Col, 63-65; USPHS grant, 64-65. *Mem:* Am Soc Biol Chemists. *Res:* Oxidation-reduction reactions and bioenergetics in photosynthetic bacteria, plants, mammals and poikilotherms; physiology and biochemistry of anadromous fishes. *Mailing Add:* Aquacult Field Sta Nat Marine Fisheries Serv Cook WA 98605

ZAUKELIES, DAVID AARON, b Detroit, Mich, May 22, 25; m 56; c 2. POLYMER, FIBER & FABRIC PHYSICS. *Educ:* Mich State Univ, BS, 46; Northwestern Univ, PhD(phys chem), 50. *Prof Exp:* Asst, Northwestern Univ, 46-49, res assoc phys chem, 49-50; res physicist, Dow Chem Co, 51-54; prof chem, Lee Col (Tenn), 54-55; res physicist, Chemstrand Corp, 55-60, Chemstrand Res Ctr, Inc, NC, 60-61, from assoc scientist to scientist, 61-70; scientist, 70-71, sci fel, Tech Ctr, 71-81, sr Monsanto fel, Monsanto Textiles Co, 81-85; CONSULT, 86- *Mem:* Am Chem Soc; Am Phys Soc. *Res:* Soiling of fibers and fabrics; spinning of nylon yarns; continuous measurement of spun yarn properties and frequency analysis; measurement of fabric color variations and measurement of fabric hand; polyaromatic fibers, cords and composites; carpet, fabrics and fibers; optical properties of yarns and fabrics; statistical analysis of data. *Mailing Add:* PO Box 388 Cantonment FL 32533

ZAUNER, CHRISTIAN WALTER, b July 21, 30; m 57; c 3. EXERCISE PHYSIOLOGY, PULMONARY PHYSIOLOGY. *Educ:* West Chester State Col, BS, 56; Syracuse Univ, MS, 57; Southern Ill Univ, PhD(phys educ), 63. *Prof Exp:* Asst prof exercise physiol & res, Temple Univ, 63-65; assoc prof exercise physiol & res, Univ Fla, 65-71, assoc prof med, 70-71, prof exercise physiol, res & med, 71-84; dir, Sports Med Inst, Mt Sinai Med Ctr, Miami Beach, Fla, 84-87; CHAIR, EXERCISE & SPORT SCI, ORE STATE UNIV, 87- *Concurrent Pos:* Univ Fla fac develop grant & Thordgray Mem Fund, Dept Clin Physiol, Malmo Gen Hosp, Sweden, 71-72 & 78; consult cardiac rehab, Hosp Corp Am; Nat Acad Sci exchange scientist, Czechoslovakia, 86, 88 & 90. *Mem:* Am Physiol Soc; Am Asn Health, Phys Educ & Recreation;

Am Col Sports Med; Sigma Xi. *Res:* Lipid metabolism; exercise and training effects on lipids, work capacity, pulmonary and respiratory function; child athletes; exercise and training effects on circulation. *Mailing Add:* Dept Exercise-Sport Sci Ore State Univ Langton Hall Corvallis OR 97331

ZAUSTINSKY, EUGENE MICHAEL, b Battle Creek, Mich, Oct 19, 26. GEOMETRY. *Educ:* Univ Calif, Los Angeles, AB, 48; Univ Southern Calif, AM, 54, PhD(math), 57. *Prof Exp:* Asst, Univ Southern Calif, 52-54, lectr, 54-57; asst prof math, San Jose State Col, 57; asst prof, Univ Calif, Santa Barbara, 58-61, asst prof, Univ Calif, Berkeley, 61-63; ASSOC PROF MATH, STATE UNIV NY STONY BROOK, 63- *Concurrent Pos:* Vis prof, Kent State Univ, 67 & Rockefeller Univ, 73. *Mem:* Am Math Soc; Math Asn Am; Math Soc France; Danish Math Soc; Swiss Math Soc. *Res:* Differential geometry and topology; synthetic differential geometry. *Mailing Add:* Dept Math State Univ NY Stony Brook NY 11794

ZAVALA, MARIA ELENA, b Pomona, Calif, Jan 9, 50. PLANT DEVELOPMENT. *Educ:* Pomona Col, BA, 72; Univ Calif, Berkeley, PhD(bot), 78. *Prof Exp:* Res assoc, Dept Biol, Ind Univ, 78-80; plant res physiologist, USDA, 80-; AT DEPT BIOL, YALE UNIV. *Concurrent Pos:* Lectr, Dept Bot, Univ Calif, Berkeley, 78. *Mem:* Bot Soc Am; Am Soc Cell Biol; Am Soc Plant Taxonomists; AAAS; Soc Advan Chinese & Native Americans Sci. *Res:* Plant anatomy and cell biology; development of pollen; cryogenic storage of plant cells; high resolution localization of plant compounds in situ. *Mailing Add:* Dept Biol OML Yale Univ New Haven CT 06520

ZAVARIN, EUGENE, b Sombor, Yugoslavia, Feb 21, 24; nat US; m 56; c 5. ORGANIC CHEMISTRY. *Educ:* Univ Gottingen, dipl, 49; Univ Calif, Berkeley, PhD(org chem), 54. *Prof Exp:* Asst, Univ Calif, Berkeley, 52-53, sr lab technician, Forest Prod Lab, Univ Calif, 52-54, asst specialist, 54-56, asst forest prod chemist, 56-62, assoc chemist, 62-68, PROF FORESTRY & FOREST PROD CHEMIST, FOREST PROD LAB, UNIV CALIF, 68- *Concurrent Pos:* NIH fel, Inst Org Chem, Gif-sur-Yvette, France, 63. *Mem:* Am Chem Soc; Int Acad Wood Sci. *Res:* Chemosystematics; chemistry of natural products; polymer chemistry of lignocellulosics; use of terpenoid composition in chemosystematics and genetics of Coniferae; determination of biosynthesis of terpenoids by computer-assisted statistics; thermal analysis of lignocellulosics; reactions of furylic polymers and model compounds; surface chemistry of wood. *Mailing Add:* Forest Prod Lab Univ Calif 1301 S 46th St Richmond CA 94804

ZAVECZ, JAMES HENRY, b Bethlehem, Pa, Dec 15, 46. PHARMACOLOGY. *Educ:* LaSalle Col, BA, 68; Ohio State Univ, PhD(pharmacol), 74. *Prof Exp:* Fel pharmacol, Med Col, Cornell Univ, 74-76, instr, 76-77; res pharmacologist, ICI Americas, Inc, 78-80; ASST PROF, DEPT PHARMACOL, EASTERN VA MED SCH, 80- *Mem:* NY Acad Sci; AAAS; Sigma Xi. *Res:* Effects of histamine on the heart; mechanism of action of cardiac glycosides. *Mailing Add:* Dept Pharmacol Eastern Va Med Sch PO Box 1980 Norfolk VA 23500

ZAVITSAS, ANDREAS ATHANASIOS, b Athens, Greece, July 14, 37; US citizen; m 59; c 1. PHYSICAL ORGANIC CHEMISTRY. *Educ:* City Col New York, BS, 59; Columbia Univ, MA, 61, PhD(chem), 62. *Prof Exp:* Res assoc chem, Brookhaven Nat Lab, 62-64; res chemist, Monsanto Co, Mass, 64-67; from asst prof to assoc prof, 67-73, grad dean, 75-79, PROF CHEM, LONG ISLAND UNIV, BROOKLYN CTR, 73- *Concurrent Pos:* Lectr, New Sch Soc Res, 61-64. *Mem:* Am Chem Soc; NY Acad Sci. *Res:* Organic free-radical chemistry; phenolic resin; electrochemical sensors. *Mailing Add:* Dept Chem Long Island Univ University Plaza Brooklyn NY 11201

ZAVODNEY, LAWRENCE DENNIS, b Akron, Ohio, June 24, 51; m 76; c 4. NONLINEAR STRUCTURAL DYNAMICS, VISCO-ELASTIC DAMPING. *Educ:* Univ Akron, BSME, 74, MSME, 77; Va Polytech Inst & State Univ, PhD(eng mech), 88. *Prof Exp:* Res asst vibrations & acoust, Dept Mech Eng, Univ Akron, 74-77; sr engr, Babcock & Wilcox Res & Develop Div, J Ray McDermott Inc, 77-79; lectr mech, Dept Mech Eng, Yarmouk Univ, 80-82; instr mech, Dept Eng Sci & Mech, Va Polytech Inst & State Univ, 79-80, instr & res asst, 82-87; ASST PROF VIBRATIONS & MECH, DEPT ENG MECH, OHIO STATE UNIV, 88- *Concurrent Pos:* Asst prof, Ohio Aerospace Inst, 90-92. *Mem:* Am Soc Mech Engrs; Am Acad Mech. *Res:* Theoretical and experimental analysis of linear and nonlinear vibrations of multidegree-of-freedom systems; nonlinear system identification; modal analysis; instrumentation; random vibration of nonlinear systems; experimental mechanics. *Mailing Add:* 209 Boyd Lab Ohio State Univ 155 W Woodruff Ave Columbus OH 43210

ZAVODNI, JOHN J, b Gallitzin, Pa, June 17, 43; m 65. PHYSIOLOGY. *Educ:* St Francis Col (Pa), BS, 64; Pa State Univ, PhD(physiol), 68. *Prof Exp:* Instr biol, St Francis Col (Pa), 64-65; asst prof, 68-75, ASSOC PROF ZOOL & CHMN DEPT SCI, PA STATE UNIV, MCKEESPORT CAMPUS, 75- *Mem:* NY Acad Sci; Am Asn Sex Educ & Coun; Sex Educ & Info Coun US. *Res:* Effects of exposure to increased oxygen tensions on the endocrine system; enforcement of water pollution. *Mailing Add:* Dept Zool Pa State Univ University Dr McKeesport PA 15132

ZAVODNI, ZAVIS MARIAN, b Prague, Czech, Aug 17, 41; US citizen; m 81; c 2. ROCK MECHANICS, GEOLOGICAL ENGINEERING. *Educ:* Princeton Univ, BSE, 64; Univ Ariz, MS, 69, PhD(geol eng), 71. *Prof Exp:* Instr math, Univ Sch, 64-66 & Am Sch Paris, 66-67; asst prof geol, Brooklyn Col, 71-74; mining engr, Kennecott Copper Corp, 74-83, chief geotech engr, Kennecott Minerals Co, 83-87; MGR GEOTECH ENG, KENNECOTT, 87- *Concurrent Pos:* Geotech engr, Pincock, Allen & Holt, Inc, Tucson, Ariz, 72-74; appl res award, US Nat Comt Rock Mechanics, 79. *Mem:* Asn Eng Geologists; Can Inst Mining & Metall; Am Inst Mining, Metall & Petrol Engrs; Int Soc Rock Mechanics. *Res:* Slope stability; geomechanics; slope failure kinematics; mine waste dump design. *Mailing Add:* 624 Ninth Ave Salt Lake City UT 84103

ZAVON, MITCHELL RALPH, b Woodhaven, NY, May 9, 23; m 47; c 4. MEDICINE. *Educ:* Boston Univ, MD, 49. *Prof Exp:* From asst prof to clin prof indust med, Kettering Lab, Col Med, Univ Cincinnati, 55-71; assoc dir, Huntingdon Res Ctr, 71-74; med dir, Ethyl Corp, 74-76; health dir, Occidental Chem Corp, 76-86; PRES, AGATHA CORP, 68- *Concurrent Pos:* Dir occup health serv, Cincinnati Health Dept, 55-61, asst health comnr, 61-74; consult, USPHS, 57-59, 66-69, Joint Congressional Comt Atomic Energy, 58-59, Louisville-Jefferson County Health Dept, 59-60 & USDA, 63-69; exec coordr, Miami Valley Proj, Ohio, 68-71; mem staff, St Mary's & Niagara Falls Mem Hosp, Sci Adv Bd, Int Joint Comt, 77-80. *Mem:* AAAS; Am Med Asn; Am Indust Hyg Asn; Am Public Health Asn; Am Col Occup Med. *Res:* Radiation protection; biological effects of agricultural chemicals; occupational health; toxicology. *Mailing Add:* 4497 Lower River Rd Lewiston NY 14092

ZAVORTINK, THOMAS JAMES, b Ravenna, Ohio, May 27, 39; m 89; c 2. ENTOMOLOGY, BIOLOGY. *Educ:* Kent State Univ, BS, 61; Univ Calif, Los Angeles, MA, 63, PhD(zool), 67. *Prof Exp:* Asst res zoologist entom, Univ Calif, Los Angeles, 68-74; asst cur entom, Calif Acad Sci, San Francisco, 74-75; mem fac, dept biol, Univ San Francisco, 75-82; res entom, Walter Reed Army Inst Res, Washington, DC, 82-84; MEM FAC, DEPT BIOL, UNIV SAN FRANCISCO, 84- *Concurrent Pos:* Consult, Southeast Asia Mosquito Proj, Smithsonian Inst, 68-71; lectr, Int Ctr Pub Health Res, Univ SC, 85. *Honors & Awards:* John N Belkin Mem Award. *Mem:* Am Mosquito Control Asn; Pac Coast Entom Soc (pres, 90); Entom Soc Am; Soc Syst Zool. *Res:* Systematics and biology of mosquitoes and bees. *Mailing Add:* Dept Biol Univ San Francisco San Francisco CA 94117-1080

ZAWACKI, BRUCE EDWIN, b Northampton, Mass, Dec 6, 35; m 61; c 3. SURGERY. *Educ:* Col Holy Cross, BS, 57; Harvard Med Sch, MD, 61. *Prof Exp:* Chief trauma study, US Army Inst Surg Res, 67-69; surgeon, Southern Calif Permanente Med Group, 69-71; asst prof, 71-74, ASSOC PROF SURG, SCH MED, UNIV SOUTHERN CALIF, 75-; HEAD PHYSICIAN BURN WARD, LOS ANGELES CO-UNIV SOUTHERN CALIF MED CTR, 71- *Concurrent Pos:* Consult, State Calif Comt Orgn & Delivery Burn Care, 78- *Mem:* Am Burn Asn. *Res:* Inhalation injury; doctor-patient relationship in life-threatening illnesses; burn depth. *Mailing Add:* 1200 N State St Rm 12650 Los Angeles CA 90033

ZAWADA, EDWARD T, JR, b Chicago, Ill, Oct 3, 47; m 77; c 3. NEPHROLOGY, NUTRITION. *Educ:* Loyola Univ, Chicago, BS, 69, Maywood, MD, 73; Am Col Physicians, cert internal med, 76, cert nephrol, 78, cert nutrit, 87, cert critical care, 87, cert geriat, 88. *Prof Exp:* Asst prof med, Sch Med, Univ Calif, Los Angeles, 78-79, Univ Utah, 79-81; assoc prof, Med Col Va, 81-83; assoc prof med, 83-87, PROF MED, PHYSIOL & PHARMACOL, SCH MED, UNIV SDAK, 87-, FREEMAN PROF & CHMN, DEPT INTERNAL MED, 87- *Concurrent Pos:* Prin investr, Va Merit Rev, 82-; chmn, Dept Int Med, Royal C Johnson Va Med Ctr, Sch Med Univ SDak, 87-; reviewer, Nephrology & Geriat Journals, 83- *Mem:* Fel Am Col Physicians; fel Am Col Chest Physicians; fel Am Col Nutrit; Am Soc Pharmacol & Exp Therapeut; Int Soc Nephrology; Am Soc Magnesium Res. *Res:* Role of divalent ions (calcium, magnesium, and phosphate) in blood pressure regulation; geriatric renal-urinary diseases; critical care; geriatrics. *Mailing Add:* Dept Internal Med Univ SDak Sch Med 2501 W 22nd St Sioux Falls SD 57105

ZAWADZKI, JOSEPH FRANCIS, b Withee, Wis, May 30, 35; m 70; c 3. ORGANIC CHEMISTRY. *Educ:* Northland Col, BA, 57; Loyola Univ (Ill), MS, 60, PhD(org chem), 62. *Prof Exp:* Res assoc org chem, Univ Chicago, 62-64; RES INVESTR CHEM PROCESS RES, SEARLE LABS, 64- *Mem:* Am Chem Soc; Sigma Xi. *Res:* Synthetic organic chemistry; reaction mechanisms; molecular rearrangements. *Mailing Add:* Seven Edgewater Dr Rouses Point NY 12979

ZAWADZKI, ZBIGNIEW APOLINARY, b Sosnowiec, Poland, July 23, 21; US citizen; m 47; c 2. MEDICINE. *Educ:* Univ Warsaw, Poland, MD, 47, DrSci, 51; Brown Univ, AM, 75. *Prof Exp:* Intern resident internal med, Dept Med, Univ Warsaw, Poland, 47-51; fel hemat, New Eng Med Ctr, Boston, Mass, 57-59; asst prof hemat, dept hemat, Inst Hemat, Warsaw, 59-60; staff physician hemat & oncol, Vet Admin Hosp, Pittsburgh, Pa, 61-74; assoc prof med immunol & oncol, 74-91, EMER ASSOC PROF MED, BROWN UNIV, PROVIDENCE, RI, 91- *Concurrent Pos:* Staff physician, Hosp Ministry Health, Warsaw, Poland, 47-60; adj asst prof hemat, Dept Hemat, Inst Hemat, Warsaw, Poland, 51-57; asst prof med, Polish Red Cross Hosp, Korea, 54-55; asst prof hemat & assoc prof oncol, Univ Pittsburgh, Pa, 67-74; dir Div Clin Immunol & Oncol, Mem Hosp, Pawtucket, RI, 74- *Mem:* Am Asn Immunologists; Am Soc Clin Oncol; AMA; Am Rheumatism Asn; Int Soc Hemat. *Res:* Paraproteinemias: the long-term studies of serum protein aberrations in myeloma and related disorders (macroglobulinemia, amyloidosis and heavy chain disease), in elderly, asymptomatic patients (idiopathic lanthanic paraproteinemia), and in patients with various chronic diseases, including connective tissue disorders, hepatopathies and myelopathies. *Mailing Add:* 202 Governor St Providence RI 02906

ZAWESKI, EDWARD F, b Jamesport, NY, Nov 2, 33; m 65; c 4. ORGANIC CHEMISTRY. *Educ:* Fordham Univ, BS, 55; Iowa State Univ, PhD(org chem), 59. *Prof Exp:* Chemist, Res Labs, Ethyl Corp, 59-79; mgr lubricant crankcase develop & technol, Edwin Cooper Inc, 79-83; SR RES ASSOC, TECH CTR, ETHYL CORP, BATON ROUGE, 83- *Mem:* Am Chem Soc; Soc Automotive Engrs; Am Oil Chem Soc. *Res:* Synthesis of new components for lubricants including crankcase, industrial, gears, hydraulic oils and fuels; formulation of lubricant blends for crankcase applications and the synthesis of new components for crankcase oils. *Mailing Add:* Ethyl Corp-Tech Ctr 8000 Gsri Ave Baton Rouge LA 70898

ZAWISZA, JULIE ANNE A, b Niles, Mich, Jan 30, 56. TECHNICAL MANAGEMENT, MEDICAL SCIENCES. *Educ:* Ind Univ Northwest, cert, 76; Univ Mich, BS, 84; George Washington Univ, MA, 91. *Prof Exp:* Med technologist, NIH, 84-86, George Washington Med Ctr, 86-90; MGR BIOMED TECH PROGS, HEALTH INDUST MFG ASN, 90- *Mem:* AAAS; Am Soc Med Technologists; Am Soc Clin Path; Am Asn Clin Chem; Nat Comt Clin Lab Standards. *Res:* Biomedical and technology related projects and programs representing the medical device industry; education projects with the Food & Drug Administration; compilation of comments on legislative or regulatory proposals; technical discussions on device related issues. *Mailing Add:* 1030 15th St NW Suite 1100 Washington DC 20005

ZAWORSKI, R(OBERT) J(OSEPH), b Portland, Ore, Jan 24, 26; m 48; c 5. MECHANICAL ENGINEERING. *Educ:* Mass Inst Technol, SB, 47, SM, 58, PhD, 66. *Prof Exp:* From mech engr to head planning develop sect, West Div Gen Eng, Creole Petrol Corp, 47-57; asst prof, 57-58, PROF MECH ENG, ORE STATE UNIV, 58- *Mem:* Sigma Xi. *Res:* Thermodynamics of irreversible processes. *Mailing Add:* 615 NW Witham Dr Corvallis OR 97330

ZAYE, DAVID F, b Toledo, Ohio. INFORMATION SCIENCE, ANALYTICAL CHEMISTRY. *Educ:* Univ Toledo, BS, 62, MS, 64; Univ Hawaii, PhD(anal chem), 68. *Prof Exp:* From assoc ed to sr assoc ed anal chem, 68-74, SR ED, CHEM ABSTR SERV, AM CHEM SOC, 74- *Mem:* Am Chem Soc. *Res:* Scientific vocabulary management processes and information transfer techniques relating to numerical data. *Mailing Add:* Chem Abstr Serv Ohio State Univ PO Box 3012 Columbus OH 43210-0012

ZBAR, BERTON, b Brooklyn, NY, May 22, 38; m; c 1. IMMUNOBIOLOGY. *Educ:* Brooklyn Col, BS, 59; State Univ NY, MD, 63; Am Bd Internal Med, dipl, 83, dipl oncol, 85. *Prof Exp:* Internship & residency, Internal Med, Univ Utah Sch Med, 63-65; clin assoc, Med Br, Nat Cancer Inst, NIH, 65-66, surgeon, Biol Br, 66-70, head, Cellular Immunity Sect, Biol Br, 70-76, sr surgeon, Cellular Immunity Sect, Lab Immunobiol, 76-80, CHIEF, CELLULAR IMMUNITY SECT, LAB IMMUNOBIOL, NAT CANCER INST, NIH, 80- , CHIEF, LAB IMMUNOBIOL, 88- *Concurrent Pos:* USPHS Comn Corps, 65-; Assoc ed J Nat Cancer Inst, 71-80. *Mem:* Am Asn Cancer Res; Am Soc Human Genetics. *Res:* Recessive oncogenes in human renal cell carcinoma and small lung carcinoma; inherited forms of human cancer; Von Hippel-Lindau disease. *Mailing Add:* NIH Nat Cancer Inst Lab Immunobiol Bldg 560 Rm 12-25 Frederick MD 21702

ZBARSKY, SIDNEY HOWARD, b Vonda, Sask, Feb 19, 20; m 44; c 3. BIOCHEMISTRY. *Educ:* Univ Sask, BA, 40; Univ Toronto, MA, 42, PhD(biochem), 46. *Prof Exp:* Res officer, Biol & Med Res Br, Atomic Energy Proj, 46-48; asst prof physiol chem, Univ Minn, 48-49; from assoc prof to prof biochem, Univ BC, 49-85; RETIRED. *Concurrent Pos:* Killam sr fel, Univ BC, 72-73. *Mem:* AAAS; Am Chem Soc; Can Physiol Soc; Can Biochem Soc (vpres, 66-67, pres, 67-68); Am Soc Biol Chemists. *Res:* Detoxication mechanisms; metabolism of British anti-lewisite, purines and pyrimidines; nucleases and nucleic acid enzymes in the intestinal mucosa. *Mailing Add:* 1420 W 49th Ave Vancouver BC V6M 2R5 Can

ZBORALSKE, F FRANK, b Fall Creek, Wis, Aug 2, 32; m 58; c 4. RADIOLOGY. *Educ:* Marquette Univ, MD, 58. *Prof Exp:* Intern, St Joseph's Hosp, Milwaukee, 58-59; resident, Milwaukee County Gen Hosp, 59-62; from instr to asst prof radiol, Sch Med, Marquette Univ, 62-65; actg asst prof, Med Ctr, Univ Calif, San Francisco, 64, from asst prof to assoc prof radiol & dir exp radiol lab, 65-67, chief sect gastrointestinal radiol, 66-67; assoc prof radiol, 67-72, dir div diag radiol, 67-75, PROF RADIOL, SCH MED, STANFORD UNIV, 72- *Concurrent Pos:* Radiologist, Milwaukee County Gen Hosp, 62-65; James Picker Found scholar radiol, 62-65; attend physician & consult, Vet Admin Hosp, Wood, Wis, 65; co-dir NIH res training grants diag radiol, Med Ctr, Univ Calif, San Francisco, 65-67 & dir res training grant, Sch Med, Stanford Univ, 67-76; consult, Vet Admin Hosp, Palo Alto, Calif, 68 & Santa Clara Valley Med Ctr, San Jose, 68- *Mem:* Asn Univ Radiologists; Asn Am Gastroenterol Asn. *Res:* Esophageal motility; esophageal epithelial cell kinetics. *Mailing Add:* Dept Radiol Med Ctr S-068A Stanford Univ Sch Med Stanford CA 94305

ZBOROWSKI, ANDREW, b Broniszewice, Poland, May 12, 36; US citizen; m 61; c 2. OCEAN ENGINEERING, NAVAL ARCHITECTURE. *Educ:* Gdansk Univ Technol, Poland, BSc, 57, MSc, 59, PhD(ship hydrodynamics), 68, DSc, 73. *Prof Exp:* Res asst ship hydrodynamics, Inst Fluid Flow Mach, Polish Acad Sci, 60-64; res officer, Danish Hydro & Aerodynamic Lab, 64-65 & Inst Fluid Flow Mach, 66-69; from asst prof to assoc prof naval architecture, Gdansk Univ Technol, 70-78; prof, Univ Basrah, Iraq, 78-81; eng consult, offshore vehicles design, Broun & Root, Inc, Houston, 82-83; prof naval archit, Fla Inst Technol, 83-85. *Concurrent Pos:* Navy Summer Res Prog, David Taylor Res Ctr, 90. *Mem:* Soc Naval Architects & Marine Engrs. *Res:* Ship hydrodynamics with particular emphasis on ship motions and stability in a seaway; application to offshore vehicles design and enhancement of their efficiency and safety. *Mailing Add:* 3356 Mazur Dr Melbourne FL 32901

ZBUZEK, VLASTA KMENTOVA, b Velka Losenice, Czech, Sept 6, 33; US citizen; m 65. ENDOCRINOLOGY, PHYSIOLOGY. *Educ:* Charles Univ, MS, 63, Cand Sci, 69, Dr rer nat(physiol), 69. *Prof Exp:* Res assoc exp endocrinol, Lab Endocrinol & Metab, Prague, 56-69; res fel reprod physiol, Pop Coun, Rockefeller Univ, 69-71; assoc res scientist physiol & endocrinol, Dept Path, NY Univ, 71-75, res scientist neuroendocrinol, Dept Anesthesiol, 75-78; res physiologist, Vet Admin Med Ctr, New York, 78-80; adj asst prof, 80-85, ASSOC PROF, NEUROENDOCRINOL, DEPT ANESTHESIOL, UNIV MED & DENT NJ, 85- *Mem:* Endocrine Soc; Gerontol Soc; NY Acad Sci; Int Soc Neuroendocrinol. *Res:* Isolation and identification of TRH; gonado-thyroidal relationships; the effects of hormones on cholesterol metabolism; pathophysiology of vasopressin; vasopressin and aging; neuropeptides and aging; the effect of nicotine on vasopressin system; nicotine and pain perception. *Mailing Add:* 100 Manhattan Ave Apt 1314 Union City NJ 07087-5246

ZBUZEK, VRATISLAV, b Prague, Czech, Mar 13, 30; US citizen; m 65. BIOCHEMISTRY, ENDOCRINOLOGY. *Educ:* Charles Univ, Czech, cert, 53, Cand Sci, 65, Dr rer nat, 67. *Prof Exp:* Res assoc biochem & physiol, Phys Cult Res Inst, Czech, 53-64; res assoc biochem & endocrinol, Lab Endocrinol & Metab, Charles Univ, 64-68; res assoc, biochem & endocrinol, 70-75, asst prof, biochem & cell biol, Rockefeller Univ, 75-77; ASST PROF, BIOCHEM & NEUROENDOCRINOL, UNIV NJ, MED & DENT, 80- *Concurrent Pos:* Lectr, Fac Phys Cult & Sports, Charles Univ, 56-62; adj assoc prof, Col Staten Island, City Univ New York, 78-81; res fel, physiology & endocrinology, Population Coun, NY, 68-70; adj fac, St Peters Col, NJ, 87-88. *Mem:* Harvey Soc; Am Chem Soc; NY Acad Sci; Int Soc Neuroendocrinol. *Res:* Biochemistry and physiology of muscular activity; pituitary-thyrotropic function; ultrastructure of C-cells and bone cells; biochemistry of cytotoxic T-lymphocytes; neuropeptides and aging; the effect of nicotine on vasopressin system. *Mailing Add:* 100 Manhattan Ave Apt 1314 Union City NJ 07087-5246

ZDAN, WILLIAM, b New York, NY, June 9, 19; m 49; c 2. SYSTEMS & ELECTRONICS ENGINEERING. *Educ:* Cooper Union, BEE, 42; Polytech Inst Brooklyn, MEE, 55. *Prof Exp:* Proj engr, Western Elec Co, 41-51; proj mgr airborne naval guid control & display systs, Sperry Gyroscope Co, 51-59; sr engr for design of the Atlas ballistic missile guid & control syst, Arma Div, Am Bosch Arma Corp, 59-62; from mgr guid & control space vehicles to head systs anal deep submergence prog, Sperry Gyroscope Co, 62-67, sr res sect head, Sperry Systs Mgt Div, Great Neck, 67-71, prog mgr & tech dir ship res simulators, vessel traffic mgt systs, vehicle maneuvering trainers, comput generated images, display systs & artificial intel, Sperry Div, Sperry Corp, 71-84; mgr res & develop, Expert & Artificial Intel Systs, Systs Mgt Group, Unisys Corp, 84-87; ENG CONSULT, B Z CONSULTS, 87-; BUS COUNR, US SMALL BUS ADMIN, 87- *Mem:* Am Asn Artificial Intel; Inst Elec & Electronics Engrs Comput Soc; Asn Comput Mach. *Res:* Digital computer analysis and evaluation of large real-time military systems; design and development of vehicle simulators and ship handling trainers; system synthesis and design of vessel traffic management systems, real-time computer-generated image display systems and artificial intelligence systems. *Mailing Add:* 30 Appletree Lane East Hills NY 11576

ZDANIS, RICHARD ALBERT, b Baltimore, Md, July 15, 35; m 55; c 2. PHYSICS. *Educ:* Johns Hopkins Univ, AB, 57, PhD(physics), 60. *Prof Exp:* Res assoc physics, Princeton Univ, 60-61, instr, 61-62; from asst prof to assoc prof, 62-69, assoc provost 75-79, vpres admin serv, 77-79, PROF PHYSICS, JOHNS HOPKINS UNIV, 69-, VPROVOST, 79- *Mem:* Am Phys Soc; AAAS. *Res:* Experimental elementary particle research. *Mailing Add:* Provost's Off Case Western Reserve Univ Cleveland OH 44106

ZDERIC, JOHN ANTHONY, b San Jose, Calif, Jan 5, 24; m 49; c 1. PHARMACEUTICAL CHEMISTRY. *Educ:* San Jose State Col, AB, 50; Stanford Univ, MS, 52, PhD(org chem), 55. *Prof Exp:* Squibb fel, Wayne State Univ, 55-56; res chemist, Syntex Corp, 56-59, asst dir chem res, 59-61, dir labs, Syntex Inst Molecular Biol, 61-62, dir corp planning div, 64-66, vpres com develop, Syntex, Int, Mex, 66-70, asst corp vpres, Syntex Corp, 67-70, vpres, Syntex Labs, Inc, Calif, 70-73, VPRES ADMIN & TECH AFFAIRS, SYNTEX RES, 73- *Concurrent Pos:* Mem staff, Swiss Fed Inst Technol, 62-64; mem bd govs, Syva, 74-77. *Mem:* Am Chem Soc. *Res:* Raney nickel catalyzed hydrogenolyses; macrocylic antibiotics; steroidal hormones; nucleosides and nucleotides. *Mailing Add:* Syntex Res 3401 Hillview Ave Palo Alto CA 94304

ZDRAVKOVICH, VERA, b Subotica, Yugoslavia, Dec 19, 39; US citizen; m 62; c 2. HARZARDOUS WASTE TECHNOLOGIES. *Educ:* Univ Belgrade, BS, 62; Univ Novi Sad, MS, 66; George Washington Univ, PhD(organic chem), 79. *Prof Exp:* Teaching asst organic chem, Univ Novi Sad, 63-66; res asst chem, George Washington Univ, 66-67; from asst prof to assoc prof, 67-77, PROF CHEM, PRINCE GEORGE'S COMMUNITY COL, 77- *Concurrent Pos:* Lectr, George Washington Univ, 82; consult, Gattys Chem Co, 86. *Mem:* Am Chem Soc; Nat Sci Teachers Asn; AAAS; Soc Col Sci Teaching. *Res:* Organic and general chemistry; acid rain-receptor mitigation technologies; written laboratory manuals regarding organic, general and biochemistry. *Mailing Add:* 16309 Marlboro Pike Upper Marlboro MD 20772-7785

ZDUNKOWSKI, WILFORD G, b Driesen, Ger, May 4, 29; US citizen; m 51; c 2. PHYSICAL METEOROLOGY, BOUNDARY LAYER METEOROLOGY. *Educ:* Univ Utah, BS, 58, MS, 59; Univ Munich, DSc, 62. *Prof Exp:* Res asst atmospheric fluoride, Univ Utah, 57-58; res meteorologist, Intermountain Weather, Inc, 58-63; from asst prof to prof meteorol, Univ Utah, 63-77; PROF, UNIV MAINZ, GER, 77- *Concurrent Pos:* Res meteorologist, Univ Mainz, 59-60 & Meteorol Inst, Univ Munich, 60-62. *Mem:* Am Meteorol Soc; Meteorol Soc Ger; Royal Meteorol Soc. *Res:* Atmospheric radiation; boundary layer theory. *Mailing Add:* Inst Meteorol Univ Mainz Mainz Germany

ZEALEY, MARION EDWARD, b Augusta, Ga, Mar 26, 13; m 48; c 4. BIOCHEMISTRY, MICROBIOLOGY. *Educ:* Paine Col, AB, 34; Atlanta Univ, MS, 40; Univ Minn, PhD, 60. *Prof Exp:* Instr chem, Miles Col, 36-43; assoc prof, Paine Col, 43-44; assoc prof biochem, 48-59, USPHS fel, 63-65, PROF MICROBIOL, MEHARRY MED COL, 65- *Mem:* AAAS; Am Chem Soc. *Res:* Protein denaturation; x-ray effects; nucleic acids; cell biology. *Mailing Add:* 27500 Franklin Rd No D509 Southfield MI 48034-2359

ZEAMER, RICHARD JERE, b Orange, NJ, May 13, 21; m 44, 69; c 5. SYSTEMS DESIGN & SYSTEMS SCIENCE, HISTORY, PHILOSOPHY & SCIENCE OF HISTORY. *Educ:* Mass Inst Technol, BS, 43, MS, 48; Univ Utah, PhD(mech eng), 75. *Prof Exp:* Prof engr, Morton C Tuttle Co, Boston, Mass, 49-53; process engr, Nekoosa-Edwards Paper Co, Port Edwards, Wis, 53-55; process engr & group leader, WVa Pulp & Paper Co, Luke, Md, 55-60; engr supvr, Allegany Ballistics Lab, Rocket Center, WVa, 60-65; rocket eng supvr, Hercules Powder Co, Magna, Utah, 65-69; sr tech specialist, Hercules

Rocket Plant, Hercules Inc, Magna, Utah, 69-83, proj eng mgr, Hercules Aerospace Div, 83-89; PRES & MGR, APPL SCI ASSOCS, SALT LAKE CITY, UTAH, 89- *Mem:* Assoc fel Am Inst Aeronaut & Astronaut. *Res:* Mechanics; structures; flow; aerodynamics; acoustics; aeroballistics; combustion; detonation; heat transfer; improved material refining, heating, drying; process control; rocket design for greater thrust and reliability; lighter weight; aerodynamics for improved steering; rocket vibration prediction; jet signature prediction; computer calculation of phenomena. *Mailing Add:* 843 13th Ave Salt Lake City UT 84103-3327

ZEBIB, ABDELFATTAH M G, b Cairo, Egypt, Sept 11, 46; US citizen; m 74; c 3. COMPUTATIONAL FLUID MECHANICS, HYDRODYNAMIC STABILITY. *Educ:* Cairo Univ, BS, 67; Univ Colo Boulder, MS, 71 & PhD(mech engr), 75. *Prof Exp:* Instr mech power eng, Cairo Univ, 67-70; fel, Univ Colo Boulder, 75-76; PROF MECH AEROSPACE ENGR, RUTGERS UNIV, 77- *Concurrent Pos:* Res assoc, Univ Calif San Diego, 73; res geophysicist, Univ Calif Los Angeles, 78; vis scholar & res assoc, Stanford Univ, 83-84; consult, IT&T Advan Technol Ctr, 86, IBM Sci Res Ctr, Palo Alto, 84-85 & AT&T Bells, 85- *Mem:* Am Phys Soc; Soc Indust & Appl Math; Am Soc Mech Engrs. *Res:* Computational fluid mechanics which include heat transfer and hydrodynamic instabilities; cooling of microelectronics; material processing; flow control; geophysics. *Mailing Add:* Dept Mech Engr Rutgers Univ New Brunswick NJ 08903

ZEBOLSKY, DONALD MICHAEL, b Chicago, Ill, Aug 20, 33; m 57; c 8. PHYSICAL CHEMISTRY. *Educ:* Northwestern Univ, BA, 56; Kans State Univ, PhD(phys chem), 63. *Prof Exp:* Chemist, Baxter Labs, 56-57; asst prof phys chem, Northern Ill Univ, 63-64; asst prof, 64-68, ASSOC PROF PHYS CHEM, CREIGHTON UNIV, 68- *Mem:* Am Chem Soc; NY Acad Sci. *Res:* Thermodynamics, kinetics and polarography of ionpairs and of transition metal-ion chelate formation; computer modeling of mixing heats in the critical region from equations of state. *Mailing Add:* Dept Chem Creighton Univ Omaha NE 68178-0104

ZEBOUNI, NADIM H, b Beirut, Lebanon, Apr 14, 28; c 2. SOLID STATE PHYSICS. *Educ:* Univ Paris, BS, 53; Nat Sch Advan Telecommun, France, MS, 55; La State Univ, PhD(physics), 61. *Prof Exp:* Eng del to Mid East, Co Gen TSF, 55-57; teaching asst, 57-58, asst prof, 60-65, ASSOC PROF PHYSICS, LA STATE UNIV, BATON ROUGE, 65-, PROF ASTRON, 74- *Mem:* Am Phys Soc. *Mailing Add:* Dept Physics La State Univ 1501 Nicholson Hall Baton Rouge LA 70803

ZEBOVITZ, EUGENE, b Chicago, Ill, Feb 24, 26; m 51; c 7. VIROLOGY. *Educ:* Roosevelt Univ, BS, 49; Univ Chicago, MS, 52, PhD(microbiol), 55. *Prof Exp:* Microbiologist, US Army Biol Labs, Ft Detrick, Md, 55-58; microbiologist, Universal Foods Corp, Wis, 58-62; microbiologist, US Army Biol Labs, 62-70; microbiologist, Naval Med Res Inst, Nat Naval Med Ctr, 70-74; health scientist adminr biol sci, Div Res Grants, 74-84, ASST CHIEF FOR REFERRAL, NIH, 84- *Mem:* AAAS; Am Soc Microbiol; Sigma Xi; Soc Exp Biol & Med. *Res:* Mechanism of virus replication; molecular biology; viral genetics; health science administration. *Mailing Add:* Div Res Grants Rm 248 Westwood Bldg NIH Bethesda MD 20205

ZEBROSKI, EDWIN L, b Chicago, Ill, Apr 1, 21; m 69; c 4. RISK MANAGEMENT, POWER SYSTEMS & MATERIALS. *Educ:* Univ Chicago, BS, 41; Univ Calif, Berkeley, PhD(phys chem), 47. *Prof Exp:* Vis prof nuclear eng, Purdue Univ, 76-77; dir, Nuclear Safety Anal Ctr, 79-81; vpres & dir, Eng Div, Inst Nuclear Opers, 81-83; chief nuclear scientist, Elec Power Res Inst, 83-87; prin engr, 88-90, DIR RISK MGT, APTECH, 90- *Concurrent Pos:* Proj engr, Submarine Advan Reactor, Triton; mgr develop engr, Gen Elec Co; consult, Energy Res Adv Bd, 85-86; dir, syst & mats, Elec Power Res Inst; mem, Nat Res Coun Panels. *Honors & Awards:* Chas A Coffin Award. *Mem:* Nat Acad Eng; fel Am Nuclear Soc; fel AAAS; Am Phys Soc; fel Am Inst Chemists; Sigma Xi. *Res:* Heavy elements; extraction processes; reactor design and analysis; safety analysis; risk management; management information systems; safeguards and accountability of weapons materials; safety and inspections of offshore platforms; research reactors; new production reactors; regulatory reform; radiochemistry; probabilistic risk analysis; physical chemistry; nuclear chemistry; computer applications; decision analysis. *Mailing Add:* 1546 Plateau Ave Los Altos CA 94024

ZEBROWITZ, S(TANLEY), b New York, NY, Nov 28, 27; m 53; c 2. ELECTRICAL ENGINEERING, COMMUNICATIONS. *Educ:* City Col New York, BEE, 49; Univ Pa, MS, 54; Temple Univ, MBA, 81. *Prof Exp:* Engr, Res Div, Philco Corp, 49-61, mgr tech staff, 62-65, mgr systs eng, 65-72, mgr design eng, Command & Eng Div, Ford Aerospace & Commun, 72-81; pres, Stelcom Int Inc, 81-87; RETIRED. *Concurrent Pos:* Consult, Rome Air Develop Ctr, US Air Force, 62-; mem US comt for study group IX, Consult Comt on Int Radio, 66-72; exec comt, Int Solid Circuits Conf, 66-76. *Mem:* Fel Inst Elec & Electronics Engrs. *Res:* Communication systems; microwave and troposcatter propagation and system design; signal processing; integrated communication and computer networks. *Mailing Add:* 1914 Lantern Lane Oreland PA 19075

ZECH, ARTHUR CONRAD, b Julian, Nebr, Aug 24, 27; m 59; c 1. AGRONOMY, BIOCHEMISTRY. *Educ:* Univ Nebr, Lincoln, BS, 58; Kans State Univ, MS, 59, PhD(agron, biochem), 62. *Prof Exp:* Sr agronomist, 61-65, res scientist animal nutrit, 65-69, mgr agr sci res, 69-75, mgr plant sci res, 75-79, SR RES SCIENTIST, FARMLAND INDUSTS, INC, 79- *Mem:* Am Soc Agron; Am Soc Animal Sci; Poultry Sci Asn; Am Genetic Asn; Coun Agr Sci & Technol; Sigma Xi. *Res:* Animal nutrition; forage fertility and quality. *Mailing Add:* Farmland Industs Inc PO Box 7305 Kansas City MO 64116

ZECHIEL, LEON NORRIS, b Wilmington, Del, Sept 23, 23; m 46; c 4. ASTROPHYSICS, OPTICS. *Educ:* DePauw Univ, AB, 48; Ohio State Univ, MA, 51. *Prof Exp:* Res assoc, Res Found, Ohio State Univ, 53-58, assoc supvr & asst to dir, 58-59; sect head, GPL Div, Gen Precision, Inc, NY, 59-63, prin scientist, 63-68; mgr data mgt systs, Sanders Assocs, Inc, NH, 68-72; sr systs engr, NCR-Postal Systs Div, SC, 72-74; prog mgr, Dayton Res Div, Hobart Corp, 74-75, mgr proj planning, 76-86; RETIRED. *Res:* Reentry and space vehicle guidance and navigation; aeronautical charting; optical and infrared instrumentation; stellar photography; air navigation; reconnaissance and surveillance techniques; computerized data management systems; postal systems automation; automated weighing and package labelling systems. *Mailing Add:* 103 Heather Lane Mauldin SC 29662-2015

ZECHMAN, FREDERICK WILLIAM, JR, b Mar 16, 28; m 50; c 2. PHYSIOLOGY. *Educ:* Otterbein Col, BS, 49; Univ Md, MS, 51; Duke Univ, PhD, 56. *Prof Exp:* Asst zool, Univ Md, 49-51; biologist & asst to head biol br, US Off Naval Res, 51-53; instr physiol, Duke Univ, 53-57; from asst prof to assoc prof, Miami Univ, 57-61; from asst prof to assoc prof, med ctr, Univ Ky, 61-68, prof physiol & chmn Dept Physiol & Biophys, 68-80, assoc dean Res & Grad Studies, 82-87, vice chancellor, 87-89, PROF PHYSIOL, MED CTR, UNIV KY, 89- *Concurrent Pos:* Consult biophys br, Aerospace Med Lab, Wright-Patterson AFB, 60-61; vis prof, Univ Hawaii, 71-72. *Mem:* AAAS; Am Physiol Soc; Aerospace Med Asn; Soc Exp Biol & Med. *Res:* Respiratory regulation and mechanics; prolonged and periodic acceleration; effects of lower body negative pressure and posture change; mechanical, reflex and subjective responses to added airflow resistance; bedrest; exercise. *Mailing Add:* Dept Physiol & Biophys Univ Ky Med Ctr Lexington KY 40506

ZECHMANN, ALBERT W, b Sioux City, Iowa, Aug 21, 34; m 65; c 1. MATHEMATICS. *Educ:* Iowa State Univ, BS, 56, MS, 59, PhD(appl math), 61. *Prof Exp:* ASST PROF MATH, UNIV NEBR, LINCOLN, 61- *Mem:* Math Asn Am. *Res:* Solution of visco-elastic problems; unification of the theory of partial differential equations; study of Cauchy problem for elliptic equations. *Mailing Add:* Dept Math-Statist Univ Nebr Lincoln NE 68588-0323

ZEDECK, MORRIS SAMUEL, b Brooklyn, NY, Jan 25, 40; c 3. PHARMACOLOGY, BIOCHEMISTRY. *Educ:* Long Island Univ, BS, 61; Univ Mich, PhD(pharmacol), 65; City Univ NY, MBA, 87. *Prof Exp:* Asst prof pharmacol, Sch Med, Yale Univ, 67-68; assoc pharmacol, Sloan-Kettering Inst Cancer Res, 68-76; from asst prof to assoc prof pharmacol, Grad Sch Med Sci, Sloan-Kettering Div, Cornell Univ, 76-83; assoc mem, Sloan-Kettering Inst Cancer Res, 76-83, vpres opers, Edward Blank Assoc, Inc, 83-88; PRES, ZEDECK ADV GROUP INC, 88-; ADJ ASSOC PROF, JOHN JAY COL CRIMINAL JUSTICE, 90- *Concurrent Pos:* Fel pharmacol, Sch Med, Yale Univ, 65-67. *Mem:* Am Asn Cancer Res; Am Soc Pharmacol & Exp Therapeut; Soc Toxicol; Am Col Toxicol. *Res:* Consultant and expert witness in drug and chemical related matters; mechanism of action studies and preclinical toxicology studies of cancer chemotherapeutic agents; chemical carcinogenesis. *Mailing Add:* 245 E 80th St New York NY 10021

ZEDEK, MISHAEL, b Kaunas, Lithuania, July 16, 26; m 56; c 2. MATHEMATICS. *Educ:* Hebrew Univ, Israel, MSc, 52; Harvard Univ, PhD(math), 56. *Prof Exp:* Asst, Hebrew Univ, Israel, 52-53; asst, Harvard Univ, 53-55; instr math, Univ Calif, Berkeley, 56-58; from asst prof to assoc prof, 58-67, PROF MATH, UNIV MD, COLLEGE PARK, 68- *Mem:* Am Math Soc; Math Asn Am; London Math Soc. *Res:* Mathematical analysis; complex analysis; interpolation and approximation. *Mailing Add:* Dept Math Univ Md College Park MD 20742

ZEDLER, EMPRESS YOUNG, b Abilene, Tex, Nov 9, 08; m 28. SPEECH PATHOLOGY, PSYCHOLOGY. *Educ:* Univ Tex, Austin, BA, 28, MA, 48, PhD(speech path), 52. *Prof Exp:* Prof spec educ, Hearing & Lang Clin, Southwest Tex State Univ, 48-77, chmn dept, 64-77, prof & dir, 77-80, EMER PROF, 80-; PVT PRACT, PSYCHOL & LANG-LEARNING DISABILITIES, 81- *Mem:* Fel Am Speech & Hearing Asn; Am Psychol Asn; Acad Aphasia; fel Am Cong Rehab Med. *Res:* Language; learning disabilities; special research in diagnosis and treatment of dyslexia and related disabilities. *Mailing Add:* PO Box 465 Luling TX 78648

ZEDLER, JOY BUSWELL, b Sioux Falls, SDak, Oct 15, 43; m 65; c 2. ECOLOGY. *Educ:* Augustana Col (SDak), BS, 64; Univ Wis, Madison, MS, 66, PhD(bot), 68. *Prof Exp:* Asst bot, Univ Wis, Madison, 64-66, fel, 66-67, res asst, 67-68; instr, Univ Mo, Columbia, 68-69; lectr, 69-72, from asst prof to assoc prof, 72-80, PROF BIOL, SAN DIEGO STATE UNIV, 80- *Concurrent Pos:* Dir, Pac Estuarine Res Lab, San Diego State Univ, 86- *Mem:* Am Soc Limnol & Oceanog; Ecol Soc Am; Estuarine Res Fedn; Soc Wetland Scientists. *Res:* Coastal wetland structure and functioning, especially salt marsh ecology. *Mailing Add:* Dept Biol San Diego State Univ San Diego CA 92182

ZEDLER, PAUL H(UGO), b Milwaukee, Wis, June 22, 41; m 65; c 2. ECOLOGY, PLANT ECOLOGY. *Educ:* Univ Wis, Milwaukee, BS, 63, Univ Wis, Madison, MS, 66, PhD(bot), 68. *Prof Exp:* Arboretum botanist, Univ Wis, 64-68; fel forestry, Univ Mo, Columbia, 68-69; from asst to assoc prof, 69-78, PROF BIOL, SAN DIEGO STATE UNIV, 78- *Concurrent Pos:* Vis scholar, Univ Col N Wales, Bangor, 80; chmn, Ecol Prog Area, San Diego State Univ, 85-86; vis scientist, Western Australia Wildlife Res Ctr, Woodvale W Australia, 86. *Mem:* AAAS; Brit Ecol Soc; Ecol Soc Am. *Res:* Plant population ecology, fire ecology, successional studies, plant-substrate relationships; temporary wetlands. *Mailing Add:* Dept Biol San Diego State Univ San Diego CA 92182-0057

ZEE, ANTHONY, b China; m 71. THEORETICAL HIGH ENERGY PHYSICS. *Educ:* Princeton Univ, AB, 66; Harvard Univ, AM, 68, PhD(physics), 70. *Prof Exp:* Mem physics, Inst Advan Study, 70-72; asst prof, Rockefeller Univ, 72-73; asst prof physics, Princeton Univ, 73-78; assoc prof physics, Univ Pa, 78-80; prof physics, Univ Wash, 80-85; PROF PHYSICS, UNIV CALIF, SANTA BARBARA, 85- *Concurrent Pos:* A P Sloan Found fel, 73-78; mem, Inst Theoret Physics, Santa Barbara, 85- *Res:* Unification of fundamental interactions; aspects of cosmology and gravity. *Mailing Add:* Inst Theoret Physics Univ Calif Santa Barbara CA 93106

ZEE, DAVID SAMUEL, b Chicago, Ill, Aug 14, 44; c 2. NEUROLOGY, NEUROPHYSIOLOGY. *Educ:* Northwestern Univ, BA, 65; Johns Hopkins Univ, MD, 69. *Prof Exp:* Clin assoc neurol, NIH, 73-75; from asst prof to assoc prof, 75-84, PROF NEUROL, JOHNS HOPKINS UNIV, 85- *Concurrent Pos:* Nat Inst Neurol Dis & Stroke grant, 75-80, Nat Eye Inst res grant, 80-85; NIH res grant, 76-97. *Mem:* Asn Res Vision & Ophthal; Soc Neurosci; Am Acad Neurol; Am Neurol Asn. *Res:* Ocular motor disorders; ocular motor physiology; computer modelling; vestibular disorders. *Mailing Add:* Johns Hopkins Hosp Baltimore MD 21205

ZEE, PAULUS, b Amsterdam, Netherlands, July 2, 28; US citizen; m 57; c 4. PEDIATRICS, BIOCHEMISTRY. *Educ:* Univ Amsterdam, MD, 54; Tulane Univ, PhD(biochem), 65. *Prof Exp:* Resident pediat, Children's Mercy Hosp, Kansas City, Mo, 56-58; asst prof, Univ Tenn, 64-68, assoc prof pediat & physiol, 68-85; PEDIATRICIAN, SWEETWATER HOSP, 85- *Concurrent Pos:* Mem, St Jude Children's Res Hosp, 64-85. *Mem:* Am Oil Chem Soc; AMA; Am Acad Pediat; Am Inst Nutrit; Am Soc Clin Nutrit. *Res:* Lipid metabolism of the newborn; pediatric nutrition. *Mailing Add:* Sweetwater Med Clin 202 Church St Sweetwater TN 37874

ZEE, YUAN CHUNG, b Shanghai, China, Aug 29, 35; m 66. VIROLOGY. *Educ:* Univ Calif, Berkeley, AB, 57, MA, 59, PhD(comp path), 66, Univ Calif, Davis, DVM, 63. *Prof Exp:* Res bacteriologist, Virol Div, Naval Biol Lab, Univ Calif, Berkeley, 63-66, from asst prof to assoc prof, 66-74, PROF VET MICROBIOL & CHMN DEPT, UNIV CALIF, DAVIS, 74- *Mem:* Am Vet Med Asn; Am Soc Microbiol. *Res:* Biological properties of animal viruses; mechanisms of virus replication and electron microscopy. *Mailing Add:* Dept Vet Microbiol Univ Calif Sch Vet Med Davis CA 95616

ZEE-CHENG, ROBERT KWANG-YUEN, b Kashan, Chekiang, China, Sept 2, 25; US citizen; m 49; c 4. ORGANIC CHEMISTRY, MEDICINAL CHEMISTRY. *Educ:* China Tech Inst, BS, 45; NMex Highlands Univ, MS, 57; Univ Tex, Austin, PhD(org chem), 63. *Prof Exp:* Res chemist & chem engr, Taiwan Pulp & Paper Corp, 46-56; asst org chem, NMex Highlands Univ, 56-57; res scientist & teaching asst, Univ Tex, Austin, 57-59; assoc chemist, Midwest Res Inst, 59-61; Welch Found fel, Univ Tex, Austin, 61-62; sr chemist, Celanese Corp Am, Tex, 62-65; sr chemist, Biol Sci Div, Midwest Res Inst, 65-71, prin chemist, 71-80; ASST DIR DRUG DEVELOP LAB, UNIV KANS MED CTR, 80-, RES PROF, DEPT PHARMACOL, TOXICOL & THERAPEUT, 83- *Concurrent Pos:* Contrib ed, Drugs of the Future, 83-, Sci Adv Bd, Drug News & Perspectives, 89-; vis prof, Shanghai Med Univ, 89; consult, Shanghai 12th Pharm Factory, 89- *Honors & Awards:* Award for Outstanding Contrib to Achievement of Sci Knowledge in Cancer Chemother, Coun Prin Scientists Midwest Res Inst, 73. *Mem:* Am Chem Soc; Sigma Xi; AAAS; NY Acad Sci; Am Asn Cancer Res. *Res:* Physical chemistry; chemical engineering; synthesis; identification; reaction mechanism of organic compounds; heterocyclic chemistry; cancer chemotherapy; pharmacology. *Mailing Add:* Univ Kans Med Ctr Rainbow Blvd & 39th Kansas City MO 66103

ZEEMAN, MAURICE GEORGE, b Rockland, Mass, Dec 1, 42; m 78. TOXOCOLOGIST, ENVIRONMENTAL SCIENCES. *Educ:* Calif State Univ, Northridge, BA, 69; Univ Calif, Los Angeles, MA, 72; Utah State Univ, PhD(zool), 80. *Prof Exp:* Teaching asst biol, Dept Zool, Univ Calif, Los Angeles, 69-70 & 71-72; res asst, Dept Biol, Utah State Univ, 75-80; toxicologist, Ctr Vet Med, Food & Drug Admin, 80-88; ADJ PROF, NAT INST HEALTH, GRAD SCH, DEPT PHARMACOL & TOXICOL, 82- *Mem:* AAAS; Am Col Toxicol; Asn Govt Toxicologists; Soc Environ Toxicol Chem; Sigma Xi; NY Acad Sci. *Res:* Environmental toxicology effects of toxic agents on fish immune system, drug uptake and metabolism in fish environmental toxicology; aquatic toxicology; pesticide effects of fish immunology and hematology. *Mailing Add:* 13900 Broomall Lane Silver Spring MD 20906-3052

ZEEVAART, JAN ADRIAAN DINGENIS, b Baarland, Neth, Jan 5, 30; m 56; c 1. PLANT PHYSIOLOGY. *Educ:* State Agr Univ Wageningen, BSc, 53, MSc, 55, PhD(plant physiol), 58. *Prof Exp:* Asst plant physiol, State Agr Univ Wageningen, 55-58; res fel, Calif Inst Technol, 60-63; assoc prof, McMaster Univ, 63-65; assoc prof, 65-70, PROF PLANT PHYSIOL, MICH STATE UNIV, 70- *Concurrent Pos:* Guggenheim fel, Milstead Lab Chem Enzymol, Sittingbourne Res Ctr, 73-74. *Mem:* Am Inst Biol Sci; Am Soc Plant Physiol; corresp mem Royal Dutch Acad Sci; AAAS. *Res:* Physiology of flower formation; plant development as regulated by growth substances; gibberellins and abscisic acid; environmental physiology. *Mailing Add:* Dept Bot S-218 Plant Biol Lab Mich State Univ East Lansing MI 48824

ZEFFREN, EUGENE, b St Louis, Mo, Nov 21, 41; m 64; c 2. BIO-ORGANIC CHEMISTRY. *Educ:* Wash Univ, AB, 63; Univ Chicago, MS, 65, PhD(org chem), 67. *Prof Exp:* Res chemist enzym, Procter & Gamble Co, 67-71, group leader, 71-74, sect head, 74-77, assoc dir, toilet goods div, Winton Hill Tech Ctr, 77-79; VPRES RES & DEVELOP, HELENE CURTIS, INC, 79- *Concurrent Pos:* Vchmn sci adv comt, Cosmetic Toiletry & Fragrance Asn, 84-88, chmn, 88-90. *Mem:* Am Chem Soc; Soc Socmetic Chemists; AAAS. *Res:* Mechanism of enzyme action; model systems for enzymic catalysis; chemistry of hair keratins; protein structure. *Mailing Add:* 4401 W North Ave Helene Curtis Indust Inc Chicago IL 60639

ZEGARELLI, EDWARD VICTOR, b Utica, NY, Sept 9, 12; m 39; c 4. DENTISTRY. *Educ:* Columbia Univ, AB, 34, DDS, 37; Univ Chicago, MS, 42; Am Bd Oral Med, dipl, 56. *Hon Degrees:* DSc, Columbia Univ, 83. *Prof Exp:* Asst dent, 37-38, from instr to asst prof, 38-47, head diag & roentgenol, 47-57, prof, 57-58, dir, Div Stomatol, 58-77, Edwin S Robinson prof dent, Sch Dent & Oral Surg, 58-78, dean, 74-78; dir dent serv, Columbia-Presby Med Ctr, 74-78; EMER PROF DENT & EMER DEAN, SCH DENT & ORAL SURG, COLUMBIA UNIV, 78- *Concurrent Pos:* Dent alumni res award, Columbia Univ, 63; mem univ coun, Columbia Univ, 59-62, cancer coordr & chmn comt dent res, Sch Dent & Oral Surg; mem coun dent therapeut, Am Dent Asn, 63-69, vchmn, 68-69, consult, 69-, consult, coun dent mat & devices, 70-; mem, NY Bd Dent Exam, 63-71, pres, 70-71; attend dent surgeon & dir dent serv, Columbia-Presby Med Ctr & Delafield Inst Cancer Res; cent off consult & dentist in residence, Vet Admin, DC; police surgeon, New York Police Dept; chmn comt exam, NE Regional Bd Dent Examr, 69-91 & joint panel drugs in dent, Nat Acad Sci-Nat Res Coun-Food & Drug Admin; consult, EORange, Kingsbridge & Montrose Vet Admin Hosps, Grasslands, Phelps Mem & Vassar Bros Hosps & USPHS; dir dent serv, Columbia-Presby Med Ctr, 74-78; mem, NY State Health Res Coun, 75-80; Consult dent, Columbia-Presby Med Ctr, 78. *Honors & Awards:* Austin Sniffin Medal Honor, Dent Soc, NY, 61, Jarvie-Burkhardt Medal Honor, 70; Samuel J Miller Medal, Am Acad Oral Med, 76; Henry Spenaded Award, Dent Soc NY, 79; William J Gies Medal, Am Col Dentists, 81. *Mem:* AAAS; Am Cancer Soc; fel Am Col Dentists; hon mem Dent Soc Guatemala; hon fel Acad Gen Dent; fel Int Col Dentists. *Res:* Diseases of the mouth and jaws, especially diagnosis; pharmacotherapeutics of oral diseases. *Mailing Add:* 120 Gory Brook Rd N Tarrytown NY 10591

ZEGEL, WILLIAM CASE, b Port Jefferson, NY, Aug 4, 40; m 62; c 2. ENVIRONMENTAL ENGINEERING, PROJECT MANAGEMENT. *Educ:* Stevens Inst Technol, ME, 61, MS, 62, DSc(chem eng), 65; Environ Eng Intersoc, dipl, 76. *Prof Exp:* Instr chem, Newark Col Eng, 61-64; sr res engr, Allied Chem Corp, 64-68; develop mgr, Scott Res Labs, 68-72; sr res assoc, Ryckman, Edgerly, Tomlinson & Assocs, 72-75; vpres opers, Environ Sci & Eng Inc, 75-79; pres, Strategic Planning & Res Group, 83-86; PRES, WATER & AIR RES INC, 79- *Concurrent Pos:* Vis lectr, Stevens Inst Technol, 65-68; tech consult, Environ Protection Agency Regional Air Pollution Study, 74-75; dir, Air Pollution Control Asn, 83-86; coun chair, Commun & Mkt, Air & Waste Mgt Asn, 88-91. *Mem:* Am Inst Chem Engrs; fel Air & Waste Mgt Asn (vpres, 85-86); AAAS; Nat Soc Prof Engrs. *Res:* Atmospheric chemistry; water chemistry; pollution control technology; environmental impact assessment; process analysis; project management. *Mailing Add:* 11011 NW 12th Pl Gainesville FL 32606

ZEGURA, STEPHEN LUKE, b San Francisco, Calif, July 2, 43; m 83; c 2. HUMAN BIOLOGY, BIOLOGICAL ANTHROPOLOGY. *Educ:* Stanford Univ, BA, 65; Univ Wis-Madison, MS, 69, PhD(human biol), 71. *Prof Exp:* Asst prof anthrop, NY Univ, 71-72; asst prof, 72-77, ASSOC PROF ANTHROP, UNIV ARIZ, 77- *Concurrent Pos:* NY Univ career develop grant, Smithsonian Inst, 72, Irex, 88, Nat Res Coun, 90. *Mem:* AAAS; Am Asn Phys Anthrop; Classification Soc; Am Anthrop Asn; Soc Study Human Biol. *Res:* Multivariate statistics; biological distance; Eskimology; population genetics; evolutionary theory; Adriatic population structure. *Mailing Add:* Dept Anthrop Univ Ariz Tucson AZ 85721

ZEHEB, EZRA, b Haifa, Israel. THEORY & DESIGN OF ROBUSTLY STABLE SYSTEMS. *Educ:* Technion-Israel Inst Technol, BSc, 58, MSc, 62, DSc(elec eng), 66. *Prof Exp:* Mem tech staff, Bell Tel Labs, 68; dir electronics, Appl Res & Develop Ctr, Israel, 70-72; prof elec eng, Tel-Aviv Univ, 85-82; sr lectr, 68-70, PROF ELEC ENG, TECHNION-ISRAEL INST TECHNOL, 88- *Concurrent Pos:* Vis prof, Stevens Inst Technol & Univ Mo, Columbia, 67 & Univ Calif, Davis, 76; consult, Ministry Indust & Com, Israel, 72-80; chmn, Working Group Robust Control, Int Fedn Automatic Control, 87- *Mem:* Fel Inst Elec & Electronics Engrs; Am Math Soc; Sigma Xi. *Res:* Theory and design of robustly stable systems; multidimensional systems theory; signal processing; filter theory. *Mailing Add:* Elec Eng Dept Technion-Israel Inst Technol Haifa Technion City 32000 Israel

ZEHNER, DAVID MURRAY, b Philadelphia, Pa, Aug 7, 43; m 67. SURFACE PHYSICS. *Educ:* Drexel Inst Technol, BS, 66; Brown Univ, PhD(physics), 71. *Prof Exp:* Res asst physics, Brown Univ, 67-71; res physicist surface physics, Oak Ridge Nat Labs, 71-80. *Mem:* Sigma Xi; Am Phys Soc; Am Vacuum Soc. *Res:* Investigation of surface properties of solids using surface sensitive spectroscopic techniques employing electrons, photons and ions as scattering probes; surface damage, chemisorption and catalytic phenomena. *Mailing Add:* Rte 6 Box 396 Brandywine Lenoir City TN 37771

ZEHNER, LEE RANDALL, b Lansdowne, Pa, Mar 15, 47; m 73; c 2. TECHNICAL MANAGEMENT, CHEMICAL ENGINEERING. *Educ:* Univ Pa, BS, 68; Univ Minn, PhD(org chem), 73. *Prof Exp:* Res chemist org chem, Arco Chem Co, 73-75, sr res chemist, 75-78; group leader process develop, Ashland Chem Co, 78-82; mgr org process res, W R Grace & Co, 82-85; DIR BIOTECH PROGS, BIOSPHERICS INC, 85- *Mem:* Am Chem Soc; Am Inst Chem Engrs; Inst Food Technol. *Res:* Specialty chemical and petrochemical processes; carbohydrate chemistry; heterogeneous catalysis; metabolism of carbohydrate derivatives. *Mailing Add:* 131 Brinkwood Rd Brookeville MD 20833

ZEHR, ELDON IRVIN, b Manson, Iowa, June 25, 35; m 57; c 7. BACTERIOLOGY. *Educ:* Goshen Col, BA, 60; Cornell Univ, MS, 65, PhD(plant path), 69. *Prof Exp:* From asst prof to assoc prof, 69-73, PROF PLANT PATH, CLEMSON UNIV, 78- *Mem:* Am Phytopath Soc; AAAS; Soc Nematologists. *Res:* Diseases of apples and peaches; biocontrol of nematodes. *Mailing Add:* Dept Plant Path & Physiol Clemson Univ Clemson SC 29634-0377

ZEHR, FLOYD JOSEPH, b Lowville, NY, June 28, 29; m 57; c 4. EXPERIMENTAL PHYSICS. *Educ:* Eastern Mennonite Col, BS, 54; Goshen Col, BA, 57; Syracuse Univ, MS, 61 & 63, PhD(physics), 67. *Prof Exp:* Teacher jr high sch, PR, 54-56 & sr high sch, NY, 57-59; from asst prof to assoc prof, 65-82, head dept, 69-71, PROF PHYSICS, WESTMINSTER COL, PA, 82- *Concurrent Pos:* Researcher, Argonne Nat Lab, 71-72 & Oak Ridge Nat Lab, 79-80; prof physics, Malaysia, 88-89. *Mem:* Am Asn Physics Teachers; Am Solar Energy Soc. *Res:* Measurement of interatomic potential between lithium ions and helium atoms and lithium ions and hydrogen atoms; nuclear decay studies of LU-174 and Tm-174; neutron capture-gamma ray decay studies; residential solar space heating analyses; energy efficient residential thermal envelope analyses. *Mailing Add:* Dept Physics Westminster Col New Wilmington PA 16142

ZEHR, JOHN E, b Foosland, Ill, Dec 29, 29; m 51; c 4. HYPERTENSION, RENIN-ANGIOTENSIN. *Educ:* Eureka Col, BS, 64; Ind Univ, PhD(physiol), 68. *Prof Exp:* Res assoc physiol, Mayo Grad Sch Med, 68-70; res scientist physiol, Univ Wash, 70-72. *Concurrent Pos:* Mem study sect, NIH, 78-80; vis prof, Jicin Med Sch, Japan, 80-81. *Mem:* Am Physiol Soc; Sigma Xi; Am Heart Asn; Int Soc Hypertension. *Res:* Neural control of the circulation in normal and hypertensive models. *Mailing Add:* Burrill Hall 524 Univ Ill 407 S Goodwin Ave Urbana IL 61801

ZEHR, MARTIN DALE, b Carthage, NY, Sept 30, 50; m 77. NEUROPSYCHOLOGY. *Educ:* State Univ NY, Binghamton, BA, 72; Memphis State Univ, MS, 75, PhD(clin psychol), 79; Univ Mo, Kansas City, JD, 85. *Prof Exp:* Grant coordr victimization, Correctional Res & Eval Ctr, 77-78; psychol intern, Vet Admin Med Ctr, Topeka, Kans, 78-79; neuropsychologist, Vet Admin Med Ctr, Kansas City, Mo, 79-82; ATTY AT LAW, BENSON & MCKAY, KANSAS CITY, MO, 85- *Concurrent Pos:* Adj asst prof, Dept Psychiat, Sch Med, Kans Univ, 80- *Mem:* Am Psychol Asn; Nat Acad Neuropsychologists; NY Acad Sci. *Res:* Relationship between structural or systemic damage to cortical tissues and disruption of behavioral and cognitive-intellectual functions. *Mailing Add:* 663 West 70 Kansas City MO 64113-2070

ZEHRT, WILLIAM H(AROLD), b Racine, Wis, June 9, 22; m 55; c 3. CIVIL ENGINEERING. *Educ:* Univ Wis, BS, 44, MSCE, 58, PhD(civil eng), 62. *Prof Exp:* Gen engr, Gen Eng Co, Wis, 47-48; bridge engr, Wis State Hwy Comn, 48-51; regional bridge engr, US Forest Serv, 51-55; br mgr bldg & related struct, Pub Works Off, Ninth Naval Dist, Ill, 55-57; instr civil eng struct, Univ Wis, 57-61; prof civil eng, Miss State Univ, 61-63; prof, Univ Ala, 63-66; PROF CIVIL ENG & CHMN DEPT, UNIV SALA, 66- *Mem:* Sigma Xi. *Res:* Engineering structures; basic properties of wood as a structural material. *Mailing Add:* 1318 Dauphin St Mobile AL 36604

ZEI, DINO, b Chicago, Ill, Aug 20, 27; m 49; c 2. EXPERIMENTAL PHYSICS, HISTORY OF SCIENCE. *Educ:* Beloit Col, BS, 50; Univ Wis, MS, 52, PhD, 57, MA, 72. *Prof Exp:* Physicist, Nat Bur Standards, 50; instr physics, Beloit Col, 52-53; asst prof, Milton Col, 53-54; assoc prof, St Cloud State Col, 55-57; prof physics, 57-78, chmn dept physics, 57-88, WILLIAM HARLEY BARBER DISTINGUISHED PROF, RIPON COL, 78- *Mem:* Am Asn Univ Profs; Am Asn Physics Teachers; Am Phys Soc; Hist Sci Soc. *Res:* Atomic spectroscopy with lasers. *Mailing Add:* Dept Physics Ripon Col 300 Seward St Ripon WI 54971-0248

ZEICHNER-DAVID, MARGARITA, b Mako, Hungary, July 20, 46; m 79; c 2. MOLECULAR & DEVELOPMENTAL BIOLOGY. *Educ:* Polytech Inst, Mexico City, PhD(cell biol), 74. *Prof Exp:* RES ASSOC PROF BIOL, SCH DENT, UNIV SOUTHERN CALIF, 81- *Mem:* Sigma Xi; AAAS; Am Soc Cell Biol; Am Asn Dent Res; Am Soc Zoologists; NY Acad Sci. *Res:* Molecular genetics of tooth development. *Mailing Add:* 912 12th St No 5 Santa Monica CA 90403

ZEIDE, BORIS, b Moscow, USSR, Mar 30, 37; US citizen; m 81; c 2. GROWTH & YIELD STUDY, FOREST MENSURATION. *Educ:* Moscow Col Forestry, USSR, MSc, 59; All-Union Inst Standards, USSR, PhD(forestry), 70. *Prof Exp:* Forester, Kalmuck Forest Mgt Dist, 59-60; res scientist forestry, All-Union Inst Standards, 66-68, head lab, 68-73; sr lectr ecol, Hebrew Univ, Israel, 74-76; res fel forestry, Harvard Univ, 76-77; asst prof forestry, Rutgers Univ, 77-80; assoc prof, 80-86, PROF FORESTRY, UNIV ARK, 86- *Concurrent Pos:* Vis prof, Univ Joensuu, Finland, 88, Clemson Univ, 89, Inst Global Climate, USSR, 90 & Madrid Univ, Spain, 91. *Mem:* Soc Am Foresters. *Res:* Fractal geometry of tree crowns structure and dynamics of forest stands methodology of ecological modeling; silviculture; biometry. *Mailing Add:* Dept Forestry Univ Ark Monticello AR 71655

ZEIDENBERGS, GIRTS, b Tukums, Latvia, Apr 5, 34; US citizen; m 55; c 3. ELECTRICAL ENGINEERING, SOLID STATE PHYSICS. *Educ:* Univ Conn, BS, 57; Syracuse Univ, MEE, 59, PhD(elec eng), 66. *Prof Exp:* Prog engr, Gen Elec Co, 57-59, engr, 59-63; res engr, Syracuse Univ, 63-67; develop engr, Systs Prod Assurance, 74-77, adv engr, Biomed Systs, 77-80, prod safety prog mgr, Info Records Div, 80-81, PROG MGR PROD SAFETY PROGS, IBM CORP, 81- *Mem:* Inst Elec & Electronics Engrs. *Res:* Network and circuit design; semiconductor devices; heterojunction; biomedical systems. *Mailing Add:* 65 Cedar Rd Katonah NY 10536

ZEIDERS, KENNETH EUGENE, b Sunbury, Pa, Aug 21, 20. MICROBIOLOGY, AGRONOMY. *Educ:* Pa State Univ, BS, 55, MS, 58. *Prof Exp:* Tech asst plant path, 57-59, plant pathologist, 59-75, RES PLANT PATHOLOGIST, US REGIONAL PASTURE RES LAB, AGR RES SERV, USDA, 75- *Concurrent Pos:* Plant pathologist, Dept Plant Path, Pa State Univ, 60- *Mem:* Am Phytopath Soc; Sigma Xi; Am Soc Agron. *Res:* Diseases of forage grasses and legumes; inoculation methods; screening for disease resistance; environmental plant pathology; role of diseases in modeling of forage crop production systems. *Mailing Add:* US Regional Pasture Res Lab Agr Res Serv USDA University Park PA 16802-3701

ZEIDLER, JAMES ROBERT, b Carlinville, Ill, Dec 1, 44; m 68; c 2. SIGNAL PROCESSING, SOLID STATE ELECTRONICS. *Educ:* MacMurray Col, BA, 66; Mich State Univ, MS, 68; Univ Nebr, Lincoln, PhD(physics), 72. *Prof Exp:* Asst physics, Mich State Univ, 66-68; asst, Univ Nebr, Lincoln, 68-73, res assoc & instr optics, 73-74; physicist, Signal Processing Underwater Commun, Naval Undersea Ctr, 74-77, supvry physicist, Naval Ocean Systs Ctr, San Diego, 77-83, tech adv res, eng & systs, asst secy Navy, Wash, DC, 83-84, SCIENTIST, SPACE SYSTS & TECHNOL DIV, NAVAL OCEAN SYSTS CTR, 84-; ADJ PROF, UNIV CALIF, SAN DIEGO, 89- *Concurrent Pos:* Res assoc, Dept Elec & Comp Eng, Univ Calif, San Diego, 88-89. *Mem:* Sr mem Inst Elec & Electronics Engrs. *Res:* Satellite communications; solid state devices; adaptive signal processing techniques; underwater acoustic communications; tracking and localization techniques. *Mailing Add:* Code 7601 Naval Ocean Systs Ctr San Diego CA 92152

ZEIDMAN, BENJAMIN, b New York, NY, Oct 6, 31; m 56, 72; c 3. NUCLEAR PHYSICS. *Educ:* City Col New York, BS, 52; Washington Univ, PhD, 57. *Prof Exp:* SR PHYSICIST, ARGONNE NAT LAB, 57- *Concurrent Pos:* Ford Found fel, Niels Bohr Inst, Copenhagen, Denmark, 63-64; vis prof, State Univ NY, Stonybrook, 72 & Max Plank Inst, Heidelberg, Ger, 75-76. *Honors & Awards:* Sr Scientist Award Alexander Von Humboldt Found, 75. *Mem:* AAAS; fel Am Phys Soc; Sigma Xi. *Res:* Intermediate energy; electron; nuclear reactions, scattering, spectroscopy and structure. *Mailing Add:* Argonne Nat Lab Bldg 203 Argonne IL 60439

ZEIDMAN, IRVING, b Camden, NJ, Mar 17, 18; m 53; c 2. PATHOLOGY. *Educ:* Univ Pa, AB, 37, MD, 41. *Prof Exp:* From instr to assoc prof, 46-66, PROF PATH, SCH MED, UNIV PA, 66- *Res:* Cancer; chemical factors in cell adhesiveness; method of measuring surface area; effect of hyaluronidase on spread of tumors; transpulmonary passage of tumor cells; spread of cancer in lymphatic system. *Mailing Add:* PO Box 3 Barnegat Light NJ 08006

ZEIGEL, ROBERT FRANCIS, b Washington, DC, June 22, 31; m 57, 77; c 3. VIROLOGY, CYTOLOGY. *Educ:* Eastern Ill Univ, BS, 53; Harvard Univ, AM, 55, PhD(biol), 59. *Prof Exp:* Res biologist, Nat Cancer Inst, 59-66; div rep, Roswell Park Mem Inst Div, Dept Microbiol, State Univ NY Buffalo, 70-72; assoc prof microbiol, Roswell Park Mem Inst Div, 68-83,, rep, App & Prom Comt, 76-78, coordr electron micros facil, Cancer Cell Ctr, 80-86, ASSOC CANCER RES SCIENTIST, ROSWELL PARK MEM INST, STATE UNIV NY, BUFFALO, 66-; ASSOC PROF EXPLOR PATH, 83- *Concurrent Pos:* Consult, Nat Cancer Inst, 69-70; mem, Coun Asn Scientists, Roswell Park Mem Inst, 74-76. *Mem:* AAAS; Am Soc Cell Biol; Electron Micros Soc Am; Am Soc Zool. *Res:* Fine structural studies of mode of synthesis of oncogenic viral agents; search for viral agents and their association with human neoplasia; ultrastructure of picean tumors with possible environmental etiology; surface topography of maxillofacial prostheses with relation to their ability to be adapted to various bio-environments; ultrastructure of problems in metastasis; identification of occult viral agents from patients with AIDS or pre-AIDS syndrome. *Mailing Add:* Roswell Park Mem Inst Buffalo NY 14203

ZEIGER, ERROL, b New York, NY, Dec 11, 39; m 63; c 3. ENVIRONMENTAL MUTAGENESIS, MICROBIOLOGY. *Educ:* City Col New York, BS, 60; George Washington Univ, MS, 69, PhD(microbiol), 73. *Prof Exp:* Lab sci asst, Walter Reed Army Inst Res, 63-65; biologist, Lab Parasitic Dis, NIH, 65-66; fel, Dept Microbiol, George Washington Univ, 66-69; res microbiologist, Genetic Toxicol Br, Food & Drug Admin, 69-76; res microbiologist & head microbiol genetics sect, 76-78, RES MICROBIOLOGIST & HEAD ENVIRON MUTAGENESIS GROUP, CELLULAR & GENETIC TOXICOL BR, NAT INST ENVIRON HEALTH SCI, 78- *Mem:* AAAS; Environ Mutagen Soc; Sigma Xi; Genetic Toxicol Asn. *Res:* Microbial systems for the detection of environmental mutagens; metabolism of mutagens to their genetically active forms; use of short-term test systems in evaluating the genetic toxicology of chemicals. *Mailing Add:* Cellular & Genetic Toxicol Br Nat Inst Environ Health Sci PO Box 12233 Research Triangle Park NC 27709

ZEIGER, H PAUL, b Niagara Falls, NY, Nov 12, 36; m 59; c 3. COMPUTER SCIENCE. *Educ:* Mass Inst Technol, SB, 58, SM, 60, PhD(elec eng), 64. *Prof Exp:* Ford Found fel elec eng, Mass Inst Technol, 64-65; asst prof elec eng, Univ BC, 65-66; asst prof aerospace eng, 66-71, ASSOC PROF COMPUT SCI, UNIV COLO, BOULDER, 71-, CHMN DEPT, 78- *Mem:* Sigma Xi. *Res:* Processing and transmission of information; automatic control; applied abstract algebra; art of computer programming; software engineering; formal methods for programmers. *Mailing Add:* 710 Hawthorn Boulder CO 80302

ZEIGER, HERBERT J, b Bronx, NY, Mar 16, 25; m 54; c 3. PHYSICS. *Educ:* City Col New York, BS, 44; Columbia Univ, MA, 48, PhD, 52. *Prof Exp:* Union Carbide & Carbon Corp fel, Columbia Univ, 52-53; res physicist, Lincoln Lab, Mass Inst Technol, 53-90; RETIRED. *Honors & Awards:* Townes Medal, Optical Soc Am, 81. *Mem:* Fel Am Phys Soc. *Res:* Solid state and molecular physics; masers and lasers. *Mailing Add:* Lincoln Lab Mass Inst Technol Lexington MA 02173

ZEIGER, WILLIAM NATHANIEL, b Highland Park, Mich, Sept 7, 46; m 73. NATURAL PRODUCTS CHEMISTRY. *Educ:* Wayne State Univ, BA, 69; Univ Pa, PhD(chem), 72. *Prof Exp:* NIH trainee, 72-74, asst mem, Monell Chem Senses Ctr, 72-; AT MCCORMICK & CO INC. *Mem:* Am Chem Soc. *Res:* Food and flavor chemistry. *Mailing Add:* McCormick & Co Inc 204 Wight Ave Hunt Valley MD 21031-1501

ZEIGHAMI, ELAINE ANN, b Perry, Okla, Nov 13, 44. DATA BASE MANAGEMENT. *Educ:* Okla State Univ, BS, 67, MS, 69; Univ Okla, PhD(epidemiol-biostatist), 74. *Prof Exp:* Asst math, Okla State Univ, 68-70 & dept math, Univ Okla, 71-72; instr, Northwestern State Col, 70-71; asst prof epidemiol, dept community med, Sch Med, Pahlavi Univ, 74-77; res asst prof epidemiol, dept res med, Univ Pa, 77-79; res staff mem, 79-83, group leader, Health Safety Res Div, 83-88, RES STAFF MEM, MEASUREMENTS APPLN & DEVELOP GROUP, OAK RIDGE NAT LAB, 88- *Concurrent Pos:* Vis assoc prof, Dept Community Med, Pahlavi Univ, 78-79. *Mem:* Soc Epidemiol Res. *Res:* Health effects of drinking water contaminants and minerals; applications of statistical techniques and computer techniques to human health effects data; biomonitoring applications to human populations. *Mailing Add:* Oak Ridge Nat Lab X-10 Area Bldg 7503 M/S 6382 PO Box 2008 Oak Ridge TN 37831

ZEIGLER, A(LFRED) G(EYER), b Chambersburg, Pa, Nov 12, 23; m 50; c 2. INORGANIC CHEMISTRY, CHEMICAL ENGINEERING. *Educ:* Bucknell Univ, BSChE, 44; Pa State Univ, cert, 50. *Prof Exp:* Chief chem engr, Cochrane Corp, 46-52; tech dir, Am Water Softener Co, 52-63, gen mgr, 58-63, pres, 60-63, mgr, Water Conditioning div, 63-78, mgr, Indust Group, 78-83, mgr Mkt Develop, Envirex, Inc, 83-86; CONSULT, 86- *Concurrent Pos:* Consult, Elec Boat Div, Gen Dynamics Corp. *Mem:* Am Soc Testing & Mat; Am Chem Soc; Nat Asn Corrosion Engrs. *Res:* Industrial water conditioning; industrial waste treatment; decontamination of radioactive wastes. *Mailing Add:* RD 5 303 Malvern PA 19355

ZEIGLER, BERNARD PHILIP, US citizen; c 3. MODELLING & SIMULATION, KNOWLEDGE-BASED SYSTEMS. *Educ:* McGill Univ, BEng, 62; Mass Inst Tech, MS, 64; Univ Mich, PhD(comput sci), 69. *Prof Exp:* From asst prof to assoc prof comput sci, Univ Mich, 69-75; vis prof, Weizmann Inst Sci, Israel, 76-80; prof, Wayne State Univ, 81-84; PROF ELEC & COMPUT ENG, UNIV ARIZ, 85- *Concurrent Pos:* Vis prof IBM Chair, Univ Ghent, Belgium, 78, Nat Acad Sci, Shanghai Univ, Peoples Republic of China, 85. *Mem:* Inst Elec & Electronics Engrs Comput Soc; Asn Comput Mach; Soc Comput Simulation. *Res:* Methodology and software design for modelling and simulation of discrete event systems; applied artificial intelligence. *Mailing Add:* Dept Elec & Comput Eng Univ Ariz Tucson AZ 85721

ZEIGLER, DAVID WAYNE, b Forest City, Iowa, Jan 30, 49; m 73; c 4. CARDIOVASCULAR PHYSIOLOGY, ENDOCRINOLOGY. *Educ:* Iowa State Univ, BS, 71, MS, 73; Univ Mo, PhD(physiol), 83. *Prof Exp:* Retinal tech opthal, Wills Eye Hosp, Philadelphia, 74-77; res fel physiol, Univ Mo, Columbia, 77-82; postdoctoral fel pharmacol, Univ Minn, Minneapolis, 82-84; asst prof, 84-91, ASSOC PROF PHYSIOL, DEPT PHYSIOL & PHARMACOL, SCH MED UNIV SDAK, 91- *Concurrent Pos:* Chair res comt, Am Heart Asn, Dakota affil, 90-; mem, High Blood Pressure Coun, Am Heart Asn. *Mem:* Am Physiol Soc; Sigma Xi. *Res:* Hormonal factors involved in the development of hypertension; the role of the renin-angiotension system in causing increased vascular reactivity. *Mailing Add:* Dept Physiol & Pharmacol Sch Med Univ SDak Vermillion SD 57069

ZEIGLER, JOHN MARTIN, b Greensburg, Ind, Dec 5, 51. ORGANOSILICON CHEMISTRY, PHYSICAL ORGANIC CHEMISTRY. *Educ:* Wabash Col, BA, 74; Univ Ill, Urbana, PhD(org chem), 79. *Prof Exp:* Res chemist, Am Cyanamid Co, 79-81; mem tech staff, 81-85, SUPVR POLYMERS DIV, SANDIA NAT LABS, 85- *Concurrent Pos:* Consult polysilane. *Mem:* Am Chem Soc; Sigma Xi; AAAS; Soc Photo Optical Instrumentation Engrs. *Res:* Synthesis, characterization and physical and electronic properties of organopolysilanes and other organic electronic materials; organometallic and synthetic organic chemistry; physical chemistry. *Mailing Add:* Silchemy Inc 2208 Lester Dr NE Albuquerque NM 87112

ZEIGLER, JOHN MILTON, geology; deceased, see previous edition for last biography

ZEIGLER, ROYAL KEITH, b Kans, Dec 3, 19; m 43; c 4. MATHEMATICS, APPLIED STATISTICS. *Educ:* Ft Hays Kans State Col, BS, 41; Univ Nev, MS, 46; Univ Iowa, PhD(math statist), 49. *Prof Exp:* Assoc prof math, Bradley Univ, 49-51; statistician, AEC, 51-52; statistician, Theoret Physics Div, Los Alamos Sci Lab, 52-67, group leader statist serv, 68-75, mem staff, 75-79; CONSULT, 80- *Mem:* Fel AAAS; fel Am Statist Asn. *Res:* Sampling theory. *Mailing Add:* Rte 1 Box 97 Chama NM 87520

ZEIKUS, J GREGORY, b Rahway, NJ, Oct 2, 45; m 67; c 2. MICROBIOLOGY. *Educ:* Univ SFla, BA, 67; Ind Univ, Bloomington, MA, 68, PhD(microbiol), 70. *Prof Exp:* NIH fel microbiol, Univ Ill, Champaign, 70-72; from asst prof to assoc prof, 72-80, prof bact, Univ Wis-Madison, 80-; AT MICH BIOTECH INST. *Concurrent Pos:* Vis scientist & fel, Univ Marburg, Ger, 76-77 & Inst Pasteur, France, 81-82. *Mem:* Sigma Xi; Am Soc Microbiol; AAAS. *Res:* Microbial physiology and ecology; anaerobic bacterial metabolism; microbial methance formation; industrial fermentations. *Mailing Add:* Mich Biotech Inst PO Box 27609 Lansing MI 48909-0609

ZEILER, FREDERICK, b Berndorf, Austria, June 28, 21; Can citizen; m 53; c 3. FINITE ELEMENT METHOD, COMPUTER IMPLEMENTATION OF NUMERICAL METHOD. *Educ:* Univ Man, BSc, 45; Univ Alta, MSc, 47; Univ Minn, MA, 55, PhD(math), 67. *Prof Exp:* From lectr to prof math, Univ Man, 67-90; RETIRED. *Concurrent Pos:* Fel, Nat Res Coun, 45 & 46, NSF, 66. *Mem:* Can Math Cong; Can Appl Math Soc; Math Asn Am. *Res:* Numerical analysis; ordinary differential equation; partial differential equation. *Mailing Add:* Group 160 Box 11 RR 1 Vermette MB R0G 2W0 Can

ZEILIK, MICHAEL, b Bridgeport, Conn, Sept 26, 46; m 85. INFRARED ASTRONOMY, BINARY STARS. *Educ:* Princeton Univ, BA, 68; Harvard Univ, MA, 69, PhD(astron), 75. *Prof Exp:* Instr astron, Southern Conn State Col, 69-72; instr astron, Harvard Univ, 74-75; from asst prof to assoc prof, 75-85, PROF ASTRON, UNIV NMEX, 85-, DIR, CTR GRAD STUDIES, 88- *Concurrent Pos:* Vis assoc prof astron, Univ Calif, Berkeley, 80; prin investr, grant Res Corp & educ grant NSF, 80-85, res grant NSF & NASA, 89- *Honors & Awards:* Harlow Shapley Lectr, Am Astron Soc, 76-85. *Mem:* Int Astron Union; Royal Astron Soc; Am Astron Soc; Am Asn Physics Teachers; Am Asn Astron Educ. *Res:* Radio and infrared observations of the planet Mercury; multiwave length observations of active binary star systems; ethnoastronomy and archaeoastronomy of the US southwest. *Mailing Add:* Dept Physics & Astron Univ NMex Albuquerque NM 87131

ZEINER, FREDERICK NEYER, b Finley, NDak, Mar 20, 17; m 42. ZOOLOGY. *Educ:* Univ Denver, BS, 38; Ind Univ, PhD(zool), 42. *Prof Exp:* Asst zool, Ind Univ, 38-40; from asst prof to prof, 46-74, EMER PROF ZOOL, UNIV DENVER, 74- *Concurrent Pos:* Res consult, Martin Co, 59 & 61; mem gov bd, Am Inst Biol Sci, 67-71. *Mem:* AAAS; Am Inst Biol Sci; Sigma Xi. *Res:* Pituitary-ovarian relationships during pregnancy; space physiology. *Mailing Add:* 1417 S Elizabeth St Denver CO 80210

ZEISS, CHESTER RAYMOND, b Evergreen Park, Ill, Jan 4, 41; m 64; c 3. ALLERGY, OCCUPATIONAL MEDICINE. *Educ:* Northwestern Univ, Evanston, BA, 63; Med Sch, Northwestern Univ, Chicago, MD, 67. *Prof Exp:* Asst prof, 74-86, PROF MED & ALLERGY-IMMUNOL, MED SCH, NORTHWESTERN UNIV, CHICAGO, 86-; ASSOC CHIEF STAFF, VET ADMIN LAKESIDE MED CTR, CHICAGO, 81- *Concurrent Pos:* Prof med, Med Sch, Northwestern Univ, 84- *Honors & Awards:* Clin Investr Award, Vet Admin Med Res Serv, 81. *Mem:* Am Acad Allergy & Clin Immunol; Am Asn Immunologists; Am Col Physicians. *Res:* Immunology of occupational lung disease and IgE mediated hypersensitivity. *Mailing Add:* Dept Med Allergy & Immunol Northwestern Univ Med Sch 400 E Ontario Ave Chicago IL 60611

ZEITLIN, JOEL LOEB, b Los Angeles, Calif, July 9, 42; m 72. MATHEMATICS. *Educ:* Univ Calif, Los Angeles, BA, 63, MA, 66, PhD(math), 69. *Prof Exp:* Asst prof math, Wash Univ, 69-72; prof, Cath Univ Valparaiso, 72-73; assoc prof, 73-80, PROF MATH, CALIF STATE UNIV, NORTHRIDGE, 80- *Concurrent Pos:* Fulbright Hays scholar, 72-73. *Mem:* Math Asn Am; Am Math Soc. *Res:* Lie groups; representation theory; special functions; geometry. *Mailing Add:* Dept Math Calif State Univ 18111 Nordhoff St Northridge CA 91330

ZEITZ, LOUIS, b Lakewood, NJ, Jan 22, 22; m 46; c 2. BIOPHYSICS, PHYSICS. *Educ:* Univ Calif, Berkeley, AB, 48; Stanford Univ, PhD(biophys), 62. *Prof Exp:* Res asst x-ray instrumentation, Appl Res Labs, Montrose, Calif, 51-52, res physicist, 52-56; res assoc physics, Univ Redlands, 56-58; res assoc, Biophys Lab, Stanford Univ, 58-59; res assoc, 62-63, ASSOC MEM BIOPHYS DIV, SLOAN-KETTERING INST, 69-; ASSOC PROF, SLOAN-KETTERING DIV, GRAD SCH MED, CORNELL UNIV, 69- *Concurrent Pos:* Nat Inst Child Health & Develop res grant, Sloan-Kettering Inst, 65-67; Dept Energy & NIH-Nat Cancer Inst grants. *Mem:* AAAS; Biophys Soc; Radiation Res Soc; NY Acad Sci. *Res:* X-ray spectrochemical analysis; trace elements in living systems; mechanisms of radiation effects on cell development; in vivo bone mineral content measurement; radiological physics. *Mailing Add:* 131 W 80th St New York NY 10024

ZEIZEL, A(RTHUR) J(OHN), b Waterbury, Conn, Aug 17, 33; m 70; c 3. GEOLOGY, ENVIRONMENTAL SCIENCE. *Educ:* Univ Conn, BA, 56; Univ Ill, MS, 59, PhD(geol), 60. *Prof Exp:* Planning requirements officer, Off Metrop Develop, Dept Housing & Urban Develop, 67-68, dir, Water Resources Res, 68-72, environ scientist, Off Asst Secy for Res & Technol, 72-79; prog mgr water hazard, 79-82, POLICY MGR, EARTHQUAKES & NATURAL HAZARDS, FED EMERGENCY MGT AGENCY, 82- *Concurrent Pos:* Mem comt water resources res, Fed Coun Sci & Technol; sci adv new town construct, US-France Coop, 72-73; liaison, US Decade Natural Disaster Reduction, Nat Acad Sci; proj leader, earthquake loss reduction, NATO, 78-80; tech negotiator US-Mexico agreement for coop in natural disasters, 82-84; vis expert, UN Ctr Regional Develop, Japan, 88. *Mem:* AAAS; fel Geol Soc Am. *Res:* Urban environmental planning; natural hazard reduction; water resources management; hydrogeology. *Mailing Add:* Off Natural Hazard Fed Emergency Mgt Agency 500 C St SW Washington DC 20472

ZEIZEL, EUGENE PAUL, hydrology, water resources, for more information see previous edition

ZELAC, RONALD EDWARD, b Chicago, Ill, Jan 22, 41; m 61; c 2. RADIOLOGICAL HEALTH, RADIOLOGICAL PHYSICS. *Educ:* Univ Ill, Urbana, BS, 62, MS, 64; Univ Mich, Ann Arbor, MS, 65; Univ Fla, PhD(environ sci), 70; Am Bd Health Physics, cert, 71, recert, 81, 85, 89; Am Bd Med Physics, Cert, 90. *Prof Exp:* Res asst solid state physics, Coord Sci Lab, Univ Ill, Urbana, 63-64; chief health physicist, IIT Res Inst, 65-68; asst prof radiation biol, 70-80, DIR RADIOL HEALTH, BIOHAZARDS & CHEM RIGHT TO KNOW CONTROL, 70-, ASSOC PROF RADIOL, TEMPLE UNIV, 80- *Concurrent Pos:* Radiation physicist, Mercy Med Ctr, Chicago, 67-68; consult, Wyeth Labs, Pa, 71-, Presby-Univ Pa Med Ctr, 74-86, Mobil Res & Develop Corp, NJ, 77-, Metropolitan Hosp, Philadelphia, 77-86, Smith Kline & French lab, 79-86, Rorer Group Inc, Pa, 86-, DuPont & Co, 86-87 & Johnson Matthey, Pa, 86-; chmn, Campus Radiation Safety Officer's Conf, 72-74, Am Bd Health Physics Panel Examr, 88-; adj assoc prof radiol, Univ Pa, 80-86; assoc vprovost environ health & safety, Temple Univ, 87- *Mem:* Health Physics Soc; Am Asn Physicists in Med; Sigma Xi. *Res:* Radiation dosimetry and radiological safety in research and health sciences. *Mailing Add:* Temple Univ 3307 N Broad St Philadelphia PA 19140

ZELAZNY, LUCIAN WALTER, b Bristol, Conn, May 8, 42; m 62; c 4. SOIL MINERALOGY. *Educ:* Univ Vt, BS, 64, MS, 66; Va Polytech Inst, PhD(soil chem), 70. *Prof Exp:* Asst prof soils & asst soil chemist, Univ Fla, 70-75; ASSOC PROF & PROF SOIL MINERAL, VA POLYTECH INST & STATE UNIV, 75- *Concurrent Pos:* Nat Acad Sci-Nat Res Coun grant; vis scholar, Fla Inst Phosphate Res, 82; vis scholar, Univ Del, 84; vis scholar, Rutgers Univ, 85; vis scholar, Univ Ky, 85. *Mem:* Fel Am Soc Agron; Clay Mineral Soc; Sigma Xi; fel Soil Sci Soc Am; Clay Minerals Soc; Int Soc Siol Sci. *Res:* The effect of deicing compounds on vegetation and water supplies; chemical, physical and mineralogical analysis of soils. *Mailing Add:* Dept Agron Smyth Hall-0404 Va Polytech Inst & State Univ Blacksburg VA 24061-0404

ZELAZO, NATHANIEL KACHOREK, b Lomza, Poland, Sept, 28, 18; US citizen; m 43; c 2. ROBOTICS, SUPERCOMPUTERS. *Educ:* City Col NY, BS, 40; Univ Wis, MS, 59. *Hon Degrees:* PhD, Milwaukee Sch Eng, 83, Univ Wis, 86. *Prof Exp:* Engr, Dept Defense, 41-52; vpres, Norden Ketay Div, United Technologies, 52-55; dir res & develop, Avionics Div, John Oster Co, 55-59; CHIEF EXEC OFFICER, ASTRONAUT CORP AM, 59-; CHMN BD, KEARFOTT GUID & NAVIG CO, 88- *Concurrent Pos:* Geschaftsf06hrer, Astronaut GmbH; dir, Astronaut C A Ltd, 70-; vpres, Astronaut Foreign Sales Corp, 84- *Honors & Awards:* Albert Einstein Award, Am Technion Soc, 82; Centennial Medal, Inst Elec & Electronics Engrs, 84; Billy Mitchell Award, Air Force Asn. *Mem:* Inst Elec & Electronics Engrs; Nat Soc Prof Engrs; Am Soc Naval Engrs; Am Inst Aeronaut & Astronaut. *Res:* Avionics; cryogenics; navigation. *Mailing Add:* 4115 N Teutonia Ave Milwaukee WI 53209n

ZELBY, LEON W, b Sosnowiec, Poland, Mar 26, 25; US citizen; m 54; c 2. ELECTRICAL ENGINEERING, PHYSICS. *Educ:* Univ Pa, BS, 56, PhD(elec eng & physics), 61; Calif Inst Technol, MS, 57. *Prof Exp:* Res engr physics, RCA, NJ, 59-61; mem fac, Univ Pa, 61-67; PROF ELEC ENG, UNIV OKLA, 67- *Concurrent Pos:* Consol Electrodyn Corp fel, Calif Inst Technol, 56-57; Minn-Honeywell Regulator Co fel, Univ Pa, 57-58, Harrison fel, 58-59; consult, RCA, 61-67 & Moore Sch Elec Eng, Univ Pa, 62-67; chief scientist energy anal, Inst Energy Anal, Oak Ridge, 75-76; NASA fac fel, Lewis Res Ctr, Cleveland, 82, 83, & 85; ed, Inst Elec & Electronics Engrs & Soc Mag, 90- *Mem:* Franklin Inst; Inst Elec & Electronics Engrs; Am Soc Eng Educ. *Res:* Energy analysis and policy; biomedical instrumentation; electromagnetic wave propagation; plasma diagnostics; effects of microwaves on organisms. *Mailing Add:* 1009 Whispering Pines Norman OK 73072

ZELDES, HENRY, b New Britain, Conn, June 11, 21; m 49; c 3. PHYSICAL CHEMISTRY. *Educ:* Yale Univ, BS, 42, MS, 44, PhD(phys chem), 47. *Prof Exp:* Res chemist, Sam Labs, Columbia Univ, Carbide & Carbon Chem Corp, 44-45, Res Chemist, Clinton Labs, Monsanto Chem Corp, Tenn, 47; res chemist, Oak Ridge Nat Lab, Union Carbide Chem Co, 48-84; res chemist, Oak Ridge Nat Lab, Martin Marietta Energy Systs, 84; RETIRED. *Mem:* Am Chem Soc; Sigma Xi. *Res:* Radiochemistry; thermodynamics; electrolyte chemistry; nuclear and electron spin resonance. *Mailing Add:* 173 Surry Lane Hendersonville NC 28739

ZELDIN, ARKADY N, b Moscow, USSR, Aug 8, 39; m 67; c 1. POLYMER SCIENCE. *Educ:* Moscow Chem Technol Inst, MS, 62; Moscow Polytech Inst, PhD(chem eng), 70. *Prof Exp:* Engr polymer chem, All-Union Res Inst Petrol Refining, 62-67; scientist chem eng, All-Union Polytech Inst, 70-73; sr scientist chem, Sci Inst Fertilizers, 73-75; tech mgr plastics, Gibraltar Chem & Plastics Inc, 76; ASSOC CHEM ENGR POLYMER SCI, BROOKHAVEN NAT LAB, 76- *Res:* Polymer-concrete for high temperature and corrosion applications; high temperature polymer and copolymer systems for use in combination with inorganic compounds in geothermal applications. *Mailing Add:* Bldg 526 Brookhaven Nat Lab Upton NY 11973

ZELDIN, MARTEL, b New York, NY, Aug 11, 37; m 58; c 4. POLYMER CHEMISTRY, INORGANIC CHEMISTRY. *Educ:* Queens Col, NY, BS, 59; Brooklyn Col, MA, 62; Pa State Univ, PhD(chem), 66. *Prof Exp:* Chemist, Interchem Corp, 60-62; proj scientist, Union Carbide Corp, 66-68; from asst prof to assoc prof chem, Polytech Inst Brooklyn, 68-80; prof chem & chmn dept, Ind Univ & Purdue Univ, 81-84; CONSULT, 72- *Concurrent Pos:* Ed, J Inorg & Organometallic Polymers; vis scientist, IBM, Yorktown Heights, 74. *Mem:* Am Chem Soc. *Res:* Inorganic and organometallic polymers. *Mailing Add:* Dept Chem PO Box 647 Indianapolis IN 46206

ZELDIN, MICHAEL HERMEN, b Philadelphia, Pa, Mar 25, 38; m 61; c 3. CELL BIOLOGY, VISUAL PHYSIOLOGY. *Educ:* Franklin & Marshall Col, BS, 59; Temple Univ, MA, 61, PhD(biol), 65. *Prof Exp:* Res assoc biol, Brandeis Univ, 65-67; asst prof, Tufts Univ, 67-74; SR RES FEL, HARVARD UNIV, 74- *Concurrent Pos:* NIH fel, 65-67. *Mem:* AAAS; Am Chem Soc; Asn Res Vision & Ophthal. *Res:* Biochemistry and electrophysiology of the vertebrate retina; neurophysiology. structure and function of membranes. *Mailing Add:* Two Clinton St No 37 Cambridge MA 02139

ZELDIS, JEROME B, b Waterbury, Conn, Apr 6, 50; m; c 2. LIVER DISEASES. *Educ:* Yale Univ, MD & PhD(molecular, biophysics & biochem), 78. *Prof Exp:* Res & clin fel med, Mass Gen Hosp, 81-85; instr med, Harvard Med Sch, 85-87; ASSOC PHYSICIAN, BETH ISREAL HOSP, 85-; ASST PROF MED, HARVARD MED SCH, 87- *Mem:* Am Asn Immunologists; Am Gastroenterol Soc; Am Asn Liver; Am Col Physicians. *Res:* viral hepatitis. *Mailing Add:* Dept Gastroenterol Univ Calif-Davis Med Ctr 4301 X St Sacramento CA 95817

ZELEN, MARVIN, b New York, NY, June 21, 27; m 50; c 2. BIOMETRY, MATHEMATICAL STATISTICS. *Educ:* City Col New York, BS, 49; Univ NC, MA, 51; Am Univ, PhD(statist), 57. *Hon Degrees:* AM, Harvard Univ. *Prof Exp:* Mathematician, Nat Bur Standards, 52-61; head math statist & appl math sect, Nat Cancer Inst, 63-67; prof statist, State Univ NY, Buffalo, 67-77, dir statist lab, 71-77; chmn, Dept Biostatist, 81-90, PROF STATIST SCI, SCH PUB HEALTH, HARVARD UNIV, 77-, CHIEF, DIV BIOSTATIST & EPIDEMIOL, DANA-FARBER CANCER INST, 77- *Concurrent Pos:* Vis assoc prof, Univ Calif, Berkeley, 58; assoc prof math, Univ Md, 60-61; permanent mem math res ctr, Univ Wis, 60-63; sr Fulbright scholar, Imp Col & Sch Hyg & Trop Med, Univ London, 65-66; consult, Nat Cancer Inst. *Mem:* Biomet Soc; Am Statist Asn; Inst Math Statist; Royal Statist Soc; Am Soc Clin Oncol; Am Pub Health Asn. *Res:* Probability and mathematical statistics; model building in biomedical sciences; statistical planning of scientific experiments; clinical trials in cancer. *Mailing Add:* Harvard Sch Pub Health 677 Huntington Ave Boston MA 02115

ZELENKA, JERRY STEPHEN, b Cleveland, Ohio, Jan 27, 36; m 58; c 3. ELECTRONICS ENGINEERING, OPTICS. *Educ:* Univ Mich, Ann Arbor, BS, 58, MS, 59, PhD(elec eng), 66. *Prof Exp:* Engr, Res Div, Bendix Corp, 59-61; res engr radar & optics, Univ Mich, Ann Arbor, 61-72; res engr, Environ Res Inst Mich, 72-76; sr scientist, 76-80, asst vpres, Sci Appln Inc, 80-85; vpres, 86-88, CORP VPRES, SCI APPLN INT CORP, 88- *Concurrent Pos:* Lab instr, Univ Mich, Ann Arbor, 58-59, lectr, 72; consult, Westinghouse Elec Corp, 69-72, Gen Motors Corp & IBM Corp, 71-76. *Honors & Awards:* M Barry Carlton Award, Sigma Xi. *Mem:* Inst Elec & Electronics Engrs; Optical Soc Am. *Res:* Systems analysis pertaining to coherent radars and to coherent optical processors. *Mailing Add:* Sci Appln Int Corp 5151 E Broadway Suite 900 Tucson AZ 85711

ZELENKA, PEGGY SUE, b Joplin, Mo, Oct 4, 42; m 66. DEVELOPMENTAL BIOLOGY. *Educ:* Rice Univ, BA, 64; Johns Hopkins Univ, PhD(biophys), 71. *Prof Exp:* Fel pediat, Johns Hopkins Sch Med, 71-72; staff fel develop biol, Nat Inst Child Health & Human Develop, 72-75; sr staff fel develop biol, 75-77, GENETICIST, NAT EYE INST, 77- *Mem:* Asn Res Vision & Ophthal; AAAS; Am Soc Cell Biol. *Res:* Biochemical mechanisms of cellular differentiation during embryonic development; specifically those changes occurring in membranes of developing embryonic chick lenses. *Mailing Add:* Nat Eye Inst Bldg 6 Rm 208 Bethesda MD 20892

ZELENY, LAWRENCE, b Minneapolis, Minn, Apr 30, 04; m 30; c 2. AGRICULTURAL CHEMISTRY, ORNITHOLOGY. *Educ:* Univ Minn, BA, 25, MS, 27, PhD(agr biochem), 30. *Prof Exp:* Asst farm hort, Univ Minn, 28-29, agr biochem, 29, guest fel, 30; agent, US Forest Serv, 31-32; chief chemist, Visual Display, Inc, 33-35; assoc chemist, USDA, 35-41, grain technologist, 41-42, sr grain technologist & chief standardization res sect, grain, feed & seed br, agr mkt serv, 42-43, prin grain technologist & chief standardization & testing br, grain div, 53-66, consult, 67-72; RETIRED. *Honors & Awards:* Patuxent Conserv Award, 77; Paul Bartsch Award, Audubon Naturalist Soc, 88. *Mem:* AAAS; Am Chem Soc; Am Oil Chem Soc; Asn Off Anal Chem; Am Asn Cereal Chem (pres, 56-57); NAm Bluebird Soc; Sigma Xi. *Res:* Chemistry of fats and oils; cereal chemistry; biochemistry; chemical compositon of truck crops; quality evaluation of oilseeds, cereal grains and their products; population restoration of eastern bluebird. *Mailing Add:* 4312 Van Buren St Hyattsville MD 20872

ZELENY, WILLIAM BARDWELL, b Minneapolis, Minn, Mar 14, 34; m 60, 88; c 2. PHYSICS EDUCATION. *Educ:* Univ Md, BS, 56; Syracuse Univ, MS, 58, PhD(physics), 60. *Prof Exp:* Lectr physics, Univ Sydney, 60-62; asst prof, 62-65, ASSOC PROF PHYSICS, NAVAL POSTGRAD SCH, 65- *Concurrent Pos:* Consult, Data Dynamics, Inc, 65-67. *Mem:* Sigma Xi; Am Phys Soc. *Res:* Quantum field theory; electrodynamics. *Mailing Add:* Dept Physics Naval Postgrad Sch Monterey CA 93943

ZELEZNICK, LOWELL D, b Milwaukee, Wis, Feb 1, 35; m 61; c 2. IMMUNOLOGY, ALLERGY. *Educ:* Univ Ill, Chicago, BS, 56, PhD(biochem), 61. *Prof Exp:* Res assoc biochem, Upjohn Co, 64-68; head biochem sect, 68-72, dir allergy dept, Alcon Labs Inc, 72-77; vpres sci & technol, Aerwey Labs, Inc, 77-78; MEM STAFF, CORP DEVELOP, ALLERGAN, 78- *Concurrent Pos:* Ciba fel microbiol, Ciba Pharmaceut Co, 60-63; fel molecular biol, Albert Einstein Col Med, 63-64; USPHS fel, 64-65; adj prof, Tex Christian Univ. *Mem:* AAAS; Am Chem Soc; Am Soc Biol Chem; Am Acad Allergy. *Res:* Drug metabolism; biosynthesis and structure of lipopolysaccharides and bacterial cell walls; immunology-allergy research; quality control and product development. *Mailing Add:* Corp Develop Allergan 2525 Dupont Dr Irvine CA 92713

ZELEZNY, WILLIAM FRANCIS, b Rollins, Mont, Sept 5, 18; m 49. PHYSICAL CHEMISTRY. *Educ:* Mont State Col, BS, 40; Mont Sch Mines, MS, 41; Univ Iowa, PhD(phys chem), 51. *Prof Exp:* Chemist, Anaconda Copper Mining Co, 41-44; instr metall & phys chem, Univ Iowa, 48-49; aeronaut res scientist, Nat Adv Comt Aeronaut, 51-54; asst metallurgist, Div Indust Res, State Col Wash, 54-57; sr scientist, Atomic Energy Div, Phillips Petrol Co, Idaho, 57-66; asst metallurgist, Idaho Nuclear Corp, 66-70; mem staff, Los Alamos Sci Lab, 70-80; RETIRED. *Mem:* Am Chem Soc; Am Soc Metals; Am Inst Mining, Metall & Petrol Eng. *Res:* Kinetics of reaction at high temperatures; x-ray diffraction and spectroscopy; crystal structure; microprobe analysis of irradiated nuclear fuels. *Mailing Add:* PO Box 37 Rollins MT 59931

ZELIGMAN, ISRAEL, b Baltimore, Md, July 24, 13; m 43; c 3. DERMATOLOGY. *Educ:* Johns Hopkins Univ, AB, 33; Univ Md, MD, 37; Columbia Univ, MedScD, 42. *Prof Exp:* From instr to asst prof dermat, Univ Md, 46-56; asst, 46-50, from instr to asst prof, 50-63, ASSOC PROF DERMAT, SCH MED, JOHNS HOPKINS UNIV, 63- *Concurrent Pos:* Pvt pract. *Mem:* Soc Invest Dermat; AMA; Am Acad Dermat; Am Dermat Asn. *Res:* Relationship of porphyrins to dermatoses; dermalogic allergy. *Mailing Add:* 2310 South Rd Baltimore MD 21209

ZELINSKI, ROBERT PAUL, b Chicago, Ill, Jan 13, 20; m 45; c 4. POLYMER CHEMISTRY, RUBBER CHEMISTRY. *Educ:* DePaul Univ, BS, 41; Northwestern Univ, PhD(org chem), 45. *Prof Exp:* Asst, Northwestern Univ, 41-42; from instr to prof chem & chmn dept, DePaul Univ, 43-55; group leader, Rubber Synthesis Br, Phillips Petrol Co, 55-61, from sect mgr to mgr, 61-75, mgr, Eng Plastics Br, Res Div, 75-84; RETIRED. *Concurrent Pos:* Asst, Northwestern Univ, 43-44. *Mem:* Am Chem Soc; Sigma Xi; Soc Plastics Engrs. *Res:* Synthesis of plastics and rubbers. *Mailing Add:* Rte 1 Box 517A Bartlesville OK 74006

ZELINSKY, DANIEL, b Chicago, Ill, Nov 22, 22; m 45; c 3. ALGEBRA. *Educ:* Univ Chicago, SB, 41, SM, 43, PhD(math), 46. *Prof Exp:* Instr math, Univ Chicago, 43-44; asst, Appl Math Group, Columbia Univ, 44-45; Nat Res Coun fel, Univ Chicago, 46, instr math, 46-47; Nat Res Coun fel, Inst Advan Study, 47-49; from asst prof to assoc prof, 49-60, chmn dept, 75-78, PROF MATH, NORTHWESTERN UNIV, EVANSTON, 60- *Concurrent Pos:* Mem exec comt, Nat Res Coun, 66-67; ed jour, Am Math Soc, 61-64, 83-87. *Mem:* AAAS (chmn sect A, 84-87); Am Math Soc; Math Asn Am. *Res:* Rings; homological algebra. *Mailing Add:* Dept Math Northwestern Univ Evanston IL 60208

ZELIS, ROBERT FELIX, b Perth Amboy, NJ, Aug 5, 39; m 60; c 4. CARDIOVASCULAR PHYSIOLOGY, CIRCULATION-CONTROL. *Educ:* Univ Mass, BS, 60; Univ Chicago, MD, 64. *Prof Exp:* Resident, Harvard Univ, Beth Israel Hosp, 66; clin assoc cardiol, Nat Heart Inst, NIH, 66-68; from asst prof to assoc prof med, Univ Calif, Davis, 68-74; PROF MED, COL MED, PA STATE UNIV, 74-, DIR CARDIOL RES, 84- *Concurrent Pos:* Chief cardiol div, M S Hershey Med Ctr, 74-84; vis scientist, Pharmacol Inst, Univ Freiburg, WGer, 81-82; vis prof, Univ Lausanne, Switz, 89-90; Fogarty sr int fel, 89-90. *Mem:* Am Physiol Soc; Am Soc Pharmacol & Exp Therapeut; Asn Am Physicians; Am Soc Clin Invest (vpres, 84-85); Am Fedn Clin Res (pres, 77-78); Am Heart Asn. *Res:* Local and neurohumoral mechanisms controlling regional blood flow, how they alter with congestive heart failure and resulting local and systematic metabolic consequences. *Mailing Add:* M S Hershey Med Ctr Pa State Univ Col Med PO Box 850 Hershey PA 17033

ZELITCH, ISRAEL, b Philadelphia, Pa, June 18, 24; m 45; c 3. BIOCHEMISTRY, PLANT PHYSIOLOGY. *Educ:* Pa State Univ, BS, 47, MS, 48; Univ Wis, PhD(biochem), 51. *Prof Exp:* Nat Res Coun fel, Col Med, NY Univ-Bellevue Med Ctr, 51-52; asst biochemist, 52-54, assoc biochemist, 54-60, biochemist, 60-74, S W JOHNSON DISTINGUISHED SCIENTIST, CONN AGR EXP STA, 74-, HEAD, DEPT BIOCHEM & GENETICS, 63- *Concurrent Pos:* Adj prof, Yale Univ, 58-; Guggenheim fel, Oxford Univ, 60; panel mem, NSF, 62-64 & Physiol Chem Study Sect, NIH, 66-70; Regents lectr, Univ Calif, Riverside, 71; Fulbright distinguished prof, Yugoslavia, 81. *Mem:* Am Chem Soc; Am Soc Biol Chem; Am Soc Plant Physiol (pres, 77-78); fel Am Acad Arts & Sci. *Res:* Plant biochemistry; photosynthesis; respiration; plant productivity. *Mailing Add:* Dept Biochem & Genetics Conn Agr Exp Sta PO Box 1106 New Haven CT 06504

ZELKOWITZ, MARVIN VICTOR, b Brooklyn, NY, Aug 7, 45; m 70; c 2. SOFTWARE ENGINEERING, PROGRAM MEASUREMENT. *Educ:* Rensselaer Polytech Inst, 67; Cornell Univ, MS, 69, PhD (computer sci), 71. *Prof Exp:* Instr computer sci, Ithaca Col, 70-71; from asst prof to assoc prof, 71-90, PROF COMPUTER SCI, UNIV MD, 90-; FAC APPOINTEE, NAT INST STANDARDS & TECHNOL, 76- *Concurrent Pos:* Chair, Spec Interest Group Software Eng, Asn Comput Mach, 79-81, Tech Comt Software Eng, Inst Elec & Electronics Engrs, Computer Soc, 84-86; ser ed, Ablex Ser Software Eng, 85- *Mem:* Asn Comput Mach; Inst Elec & Electronics Engrs, Computer Soc. *Res:* Complexity, understanding and development of computer environments including tools, interfaces and measurement of the underlying development processes. *Mailing Add:* Dept Computer Sci Univ Md College Park MD 20742

ZELL, BLAIR PAUL, b Waterloo, Iowa, Mar 11, 42; m 61; c 2. ADJUSTABLE FREQUENCY DRIVES, DIGITAL & ANALOGY PROCESS CONTROL. *Educ:* Roosevelt Univ, BS & BA, 70. *Prof Exp:* Electronics technician, US Navy, 62-70; indust eng assoc, Western Elec, 70-71; field serv engr, Teledyne Pines, 71-72; electronics technician, Chicago Circuit Drilling, 73-74; systs engr, 74-76, mgr systs eng, 76-84, DIR SYSTS RES & DEVELOP, AURORA PUMP, GEN SIGNALS, 84- *Concurrent Pos:* Chmn tech comt, Am Soc Heating, Refrig & Air Conditioning Engrs, 85-87. *Mem:* Am Soc Heating, Refrig & Air Conditioning Engrs. *Res:* Energy conservation via the efficient application of variable speed drives and controls; solar, water and wind powered electric generators for home use. *Mailing Add:* ON 735 Lea Dr Geneva IL 60134

ZELL, HOWARD CHARLES, b Philadelphia, Pa, Feb 11, 22; m 52. ORGANIC CHEMISTRY. *Educ:* St Joseph's Univ, BS, 43; Univ Del, MS, 51; Univ Pa, PhD(org chem), 64. *Prof Exp:* Chemist, Publicker Industs, Inc, 43-47; res assoc org chem, Merck Sharp & Dohme Res Lab, 48-65; assoc scientist, Ethicon, Inc, 65-67, sr scientist, 67-69, prin scientist, 69-74; chemist, Food & Drug Admin, 74-81, supvy chemist, 81-87; DIR, REGULATORY AFFAIRS, MARSAM PHARMACEUT INC, 88- *Mem:* AAAS; Am Chem Soc; Sigma Xi. *Res:* Medicinal and polymer chemistry. *Mailing Add:* 504 Montgomery Rd Ambler PA 19002

ZELLER, ALBERT FONTENOT, b Oakland, Calif, Jan 10, 47; m 72; c 1. NUCLEAR PHYSICS, APPLIED SUPERCONDUCTIVITY. *Educ:* Univ Wash, BA, 71; Fla State Univ, MS, 73, PhD(nuclear chem), 74. *Prof Exp:* Res assoc nuclear physics, Fla State Univ, 74-75; res fel, dept nuclear physics, Australian Nat Univ, 75-78; res assoc nuclear physics, Cyclotron Inst, Tex A&M Univ, 78-79; STAFF PHYSICIST, NAT SUPERCONDUCTING CYCLOTRON LAB, MICH STATE UNIV, 79- *Concurrent Pos:* Vis asst prof physics, Tex A&M Univ, 79. *Mem:* Am Phys Soc; Inst Elec & Electronics Engrs. *Res:* Constructing superconducting magnets for nuclear physics research. *Mailing Add:* Cyclotron Lab Mich State Univ East Lansing MI 48824-1321

ZELLER, EDWARD JACOB, b Peoria, Ill, Nov 6, 25. GEOCHEMISTRY. *Educ:* Univ Ill, AB, 46; Univ Kans, MA, 48; Univ Wis, PhD(geol), 51. *Prof Exp:* Asst, State Geol Surv, Ill, 45-46; asst instr gen geol, Univ Kans, 45-46; proj assoc, USAEC contract, Wis, 51-56; prof geol & prin investr, USAEC res contract, 56-69, PROF GEOL, PHYSICS & ASTRON, UNIV KANS, 69- *Concurrent Pos:* Mem NSF US Antarctic Res Prog, 59-63; NSF sr fel, Physics Inst, Univ Berne, 60-61; guest scientist, Brookhaven Nat Lab, 65-66; US Air Force res contract, 67-68; prin investr, NASA contract, 72 & Oak Ridge Nat Lab contract, 72; del, 19th & 20th Int Geol Cong; Antarctic Int Radiomet Surv, NSF grant, 75-; co-dir, South Polar Ice Chem Proj, NSF grant, 78-; German Acad Fel, 75. *Honors & Awards:* Distinguished Lectr, Am Asn Petrol Geol, 71. *Mem:* AAAS; Geol Soc Am; Geochem Soc; Am Geophys Union; Int Asn Geochem & Cosmochem; Sigma Xi. *Res:* Thermoluminescence and electron spin resonance in geologic materials; radiation effects from nuclear waste; chemical interactions of fast protons in solid targets; lunar and asteroidal weathering; aerosols and planetary albedo; paleoclimatology. *Mailing Add:* Space Technol Labs Univ Kans 2291 Irving Hill Dr-Campus W Lawrence KS 66044

ZELLER, FRANK JACOB, b Chicago, Ill, Dec 6, 27; m 52; c 3. REPRODUCTIVE ENDOCRINOLOGY. *Educ:* Univ Ill, BS, 51, MS, 52; Ind Univ, PhD(zool), 57. *Prof Exp:* Instr zool, Bryan Col, 52-53; asst, 53-56, res assoc, 56-57, from instr to assoc prof, 57-76, PROF ZOOL, IND UNIV, BLOOMINGTON, 76- *Mem:* Endocrine Soc; Soc Study Reproduction. *Res:* Reproduction in birds; effects of gonadotropins and sex hormones on the anterior pituitary gland and gonads. *Mailing Add:* Dept Biol Ind Univ Bloomington IN 47401

ZELLER, MICHAEL EDWARD, b San Francisco, Calif, Oct 8, 39; m 60; c 2. HIGH ENERGY PHYSICS. *Educ:* Stanford Univ, BS, 61; Univ Calif, Los Angeles, MS, 63, PhD(physics), 68. *Prof Exp:* Res asst physics, Univ Calif, Los Angeles, 63-68, res fel, 68-69; from instr to assoc prof, 69-82, PROF PHYSICS, YALE UNIV, 83- *Mem:* NY Acad Sci; fel Am Phys Soc. *Res:* Polarization phenomena in the kaon-nucleon interaction; high energy, strong and weak interaction polarization phenomena; collider physics; weak interactions-decays. *Mailing Add:* 135 Newton Rd Woodbridge CT 06525

ZELLEY, WALTER GAUNTT, b Camden, NJ, Oct 1, 21; m 46; c 3. ELECTROCHEMISTRY. *Educ:* Univ Pa, BS, 42. *Prof Exp:* Chemist, Lake Ont Ord Works, 42-43; chemist, Burlington Reduction Works, Aluminum Co Am, 43-44, res engr, Res Lab, 44-65, sr scientist, 65-67, eng assoc, 67-71, sect head, Alcoa Tech Ctr, 71-79, Tech Mgr, 80-84; RETIRED. *Mem:* Am Electroplaters Soc; Sigma Xi; Tech Asn Graphic Arts. *Res:* Chemical and electrochemical surface finishing of aluminum; application of aluminum in the graphic arts and packaging; organic coatings for aluminum. *Mailing Add:* 620 Frank St New Kensington PA 15068-4955

ZELLMER, DAVID LOUIS, b Portland, Ore, June 12, 42; m 67; c 1. ANALYTICAL CHEMISTRY. *Educ:* Univ Mich, BSChem, 64; Univ Ill, MS, 66, PhD(anal chem), 69. *Prof Exp:* From asst prof to assoc prof, 69-77, PROF CHEM, CALIF STATE UNIV, FRESNO, 77- *Mem:* Am Chem Soc; Meteoritical Soc. *Res:* Application of radiochemical and electrochemical techniques to the study of semiconducting electrode materials; analysis of extraterrestrial materials; instrumentation automation; computer assisted instruction. *Mailing Add:* Dept Chem Calif State Univ 2555 E San Ramon Ave Fresno CA 93740-0070

ZELLNER, BENJAMIN HOLMES, b Forsyth, Ga, Apr 16, 42; c 3. ASTRONOMY. *Educ:* Ga Inst Technol, BS, 64; Univ Ariz, PhD(astron), 70. *Prof Exp:* Res assoc, Lunar & Planetary Lab, 70-76, res fel, 76-78, ASSOC RES PROF, UNIV ARIZ, 78- *Concurrent Pos:* Res fel, Observ Paris, Meudon, France, 72-73. *Mem:* Int Astron Union; Am Astron Soc. *Res:* Astronomical polarimetry; light scattering by circumstellar and interstellar material; physical properties of minor planets. *Mailing Add:* 5384 Graywing Ct Columbia MD 21045

ZELLNER, CARL NAEHER, organic chemistry; deceased, see previous edition for last biography

ZELLWEGER, HANS ULRICH, b Lugano, Switz, June 19, 09; m 40; c 2. PEDIATRICS. *Educ:* Univ Zurich, MD, 34. *Hon Degrees:* MD, Univ Geneva. *Prof Exp:* Resident, Kanton Hosp, Lucerne, Switz, 34-37; mem staff, Albert Schweitzer Hosp, Lambarene, Gabun, 37-39; asst prof & resident pediat, Children's Hosp, Zurich, 40-51; prof & head dept, Am Univ Beirut, 51-57 & 58-59; res prof, 57-58, prof, 59-78, EMER PROF PEDIAT, UNIV IOWA, 78- *Concurrent Pos:* Gen secy, Int Cong Pediat, Zurich, 50; clin dir, Regional Genetic Consult Serv, Iowa, 77- *Mem:* Am Acad Neurol; Am Acad Cerebral Palsey (pres, 74-75); Am Pediat Soc; hon mem Austrian Pediat Soc; hon mem Swiss Pediat Soc; hon mem, Lebanese Pediat Soc; coord mem, German Pediat Soc. *Res:* Pediatric neurology; genetics; cytogenetics. *Mailing Add:* Dept Pediat Univ Hosps Iowa City IA 52240

ZELMAN, ALLEN, b Los Angeles, Calif, Feb 12, 38; m 72; c 2. BIOPHYSICS, BIOENGINEERING. *Educ:* Univ Calif, Berkeley, BA, 64, PhD(biophys), 71. *Prof Exp:* Asst prof biophys, Meharry Med Col, 72-75; from asst prof to assoc prof biophys & biomed eng, 75-82, PROF BIOMED ENG, RENSSELAER POLYTECH INST, 83- *Concurrent Pos:* Prog dir bioeng & res aid disabled, NSF, 87-89. *Mem:* Am Soc Eng Educ. *Res:* Theoretical and experimental nonequilibrium thermodynamics; micromechanical surgical instrumentation. *Mailing Add:* Dept Biomed Eng Rensselaer Polytech Inst Troy NY 12181

ZELMANOWITZ, JULIUS MARTIN, b New York, NY, Feb 20, 41; m 62; c 1. MATHEMATICS, ALGEBRA. *Educ:* Harvard Univ, AB, 62; Univ Wis-Madison, MS, 63, PhD(math), 66. *Prof Exp:* Instr math, Univ Wis-Madison, 66; from asst prof to assoc prof, 66-77, assoc vchancellor acad affairs, 85-87, PROF MATH, UNIV CALIF, SANTA BARBARA, 77-, ASSOC VCHANCELLOR ACAD PERSONNEL, 88- *Concurrent Pos:* Vis asst prof, Univ Calif, Los Angeles, 69-70, vis assoc prof, 73-74; assoc prof, Carnegie-Mellon Univ, 70-71; vis prof, Univ Rome, 77, The Technion, 79, McGill Univ, 82, Univ Munich, 83, McGill Univ, 87 & Univ Munich, 88. *Mem:* Math Asn Am; Am Math Soc. *Res:* Algebra, rings and modules. *Mailing Add:* Dept Math Univ Calif Santa Barbara CA 93106

ZELNIK, MELVIN, b New York, NY, Sept 22, 28; m 62; c 2. DEMOGRAPHY. *Educ:* Miami Univ, BA, 55; Princeton Univ, PhD(sociol), 59. *Prof Exp:* Instr sociol, Pa State Univ, 58-59; demographic statistician, Pop Div, US Bur Census, 59-61 & 64-66; res assoc, Off Pop Res, Princeton Univ, 61-62; asst prof sociol, Ohio State Univ, 62-64; assoc prof, 66-69, PROF DEMOGRAPHY, JOHNS HOPKINS UNIV, 69- *Concurrent Pos:* Ford Found adv, Nat Inst Public Health, Indonesia, 72-73; Fulbright sr res fel, Univ Philippines, 86-87. *Honors & Awards:* Carl S Schultz Award, Am Public Health Asn, 81. *Mem:* Pop Asn Am; Int Union Sci Study Pop. *Res:* Premarital sexual activity, use or nonuse of contraception, and reproductive behavior of adolescent females and males. *Mailing Add:* 304 Gailridge Rd Timonium MD 21093

ZELSON, PHILIP RICHARD, b Long Beach, Calif, Sept 3, 45; m 67; c 2. BIOCHEMISTRY. *Educ:* Northwestern Univ, DDS, 70; Univ Rochester, PhD(biochem), 75. *Prof Exp:* ASST PROF ORAL MED, SCH DENT MED, UNIV PA, 74- *Concurrent Pos:* Res assoc biochem taste, Monell Chem Senses Ctr, 74-75, asst mem, 75-; res assoc biochem taste, Vet Admin, 75-78, staff dentist/res, 78- *Mem:* AAAS. *Res:* Etiologic and diagnostic factors in periodontal disease. *Mailing Add:* 1630 Farmington Rd Philadelphia PA 19151

ZELTMANN, ALFRED HOWARD, b Brooklyn, NY, Dec 25, 21; m 51; c 3. PHYSICAL CHEMISTRY. *Educ:* State Col Wash, BS, 48; Univ NMex, PhD(chem), 52, MS, 61. *Prof Exp:* Staff mem phys chem, Los Alamos Sci Lab, Univ Calif, 46-84; RETIRED. *Mem:* Am Chem Soc. *Res:* Chemical kinetics; radiation chemistry; high vacuum; preparation and chemistry of gaseous hydrides; nuclear magnetic resonance; complex ions; laser spectroscopy; laser isotope separation. *Mailing Add:* 100 La Cueva Los Alamos NM 87544-2521

ZELTMANN, EUGENE W, b Chicago, Ill, June 26, 40; m 74; c 2. PHYSICAL CHEMISTRY. *Educ:* Beloit Col, BA, 62; Johns Hopkins Univ, MA, 64, PhD(chem), 67. *Prof Exp:* Nuclear chemist, Knolls Atomic Power Lab, Gen Elec Co, NY, 67-70; Alfred E Smith fel in NY State, 70-71; asst to dir power div, NY State Pub Serv Comn, 71-72; mgr environ planning, Gas Turbine Div, 72-80, mgr, oper planning, 80-84, Mkt Support & Prog Develop, 84-88, MGR, MKT DEVELOP COMMUN, GEN ELEC CO, 88- *Mem:* Am Chem Soc; Sigma Xi. *Res:* Chemical kinetics; fission track imaging analyses for purpose of determining presence of minute amounts of fissionable materials. *Mailing Add:* 15 Innisbrook Dr Clifton Park NY 12065

ZEMACH, CHARLES, b Los Angeles, Calif, Sept 15, 30; m 58; c 3. ELEMENTARY PARTICLE PHYSICS. *Educ:* Harvard Univ, BA, 51, PhD(physics), 55. *Prof Exp:* Nat Sci fel, 55-56; instr physics, Univ Pa, 56-57; res assoc, Univ Calif, Berkeley, 57-58, from asst prof to prof, 58-71; officer & spec asst for technol, US Arms Control & Disarmament Agency, State Dept, 70-74, mem policy planning staff, 74-76; dep div leader, 82-86, STAFF MEM, LOS ALAMOS NAT LAB, 76- *Concurrent Pos:* Alfred P Sloan Found fel, 59-63; Guggenheim Found fel, 66-67. *Mem:* Am Phys Soc. *Res:* Thermal neutron diffraction; quantum electrodynamics; strong interactions of elementary particles; fluid dynamics. *Mailing Add:* 740 Canyon Rd Los Alamos NM 87544

ZEMACH, RITA, b Paterson, NJ, Apr 3, 26; m 47; c 2. STATISTICS, PUBLIC HEALTH. *Educ:* Barnard Col, BA, 47; Mich State Univ, MS, 61, PhD(statist, probability), 65. *Prof Exp:* Instr statist & probability, Mich State Univ, 65-66, asst prof systs sci, 66-72; assoc prof biostatist, Univ Mich, Ann Arbor, 72-73; chief prog anal, 73-88, BIOMETRIC SPECIALIST, MICH DEPT PUB HEALTH, 88- *Concurrent Pos:* Mem health care technol study sect, Dept Health, Educ & Welfare, 72-76; adj instr, Col Human Med, Mich State Univ, 77-89. *Mem:* AAAS; fel Am Statist Asn; Am Pub Health Asn; Biomet Soc. *Res:* Analysis of health effects of environmental contamination. *Mailing Add:* Dept Pub Health 3423 N Logan St PO Box 30195 Lansing MI 48909

ZEMAITIS, MICHAEL ALAN, b York, Pa, Aug 21, 46; div; c 2. BIOCHEMICAL PHARMACOLOGY, TOXICOLOGY. *Educ:* Univ Pittsburgh, BS, 69; Pa State Univ, PhD(pharmacol), 75. *Prof Exp:* Instr, 75-76, asst prof, 76-80, ASSOC PROF PHARMACOL, UNIV PITTSBURGH, 80- *Mem:* Am Soc Pharmacol & Exp Therapeut; Soc Toxicol; NY Acad Sci. *Res:* Pathways of drug metabolism in man and laboratory animals; identification of metabolites of drugs and environmental chemicals. *Mailing Add:* 633-1 Salk Hall Univ Pittsburgh Pittsburgh PA 15261

ZEMAN, FRANCES JANE, b Cleveland, Ohio, Mar 5, 25. NUTRITION. *Educ:* Western Reserve Univ, BS, 46, MS, 57; Ohio State Univ, PhD(nutrit), 63. *Prof Exp:* Dietitian, Cleveland City Hosp, Ohio, 49-51, teaching dietitian, Sch Nursing, 51-57; asst prof home econ, Kent State Univ, 57-64; from asst prof to assoc prof, 64-74, PROF NUTRIT, UNIV CALIF, DAVIS, 74- *Mem:* AAAS; Am Physiol Soc; Am Inst Nutrit; Sigma Xi. *Res:* Nutrition in reproduction. *Mailing Add:* Dept Nutrit Univ Calif Davis CA 95616

ZEMANEK, JOSEPH, JR, b Blessing, Tex, Jan 1, 28; m 50; c 2. ACOUSTICS. *Educ:* Univ Tex, BS, 49; Southern Methodist Univ, MS, 57; Univ Calif, Los Angeles, PhD(physics), 62. *Prof Exp:* Test engr, Gen Elec Co, 49-50; jr technologist, Field Res Lab, Magnolia Petrol Co, Mobil Res & Develop Corp, 51-53, from res technologist to sr res technologist, 53-58, sr res technologist, 61-67, res assoc acoustic well logging, 67-81, sr scientist, Petrol Well Log Interpretation, 81-90; RETIRED. *Honors & Awards:* Kauffman Gold Medal, Soc Explor Geophysicists, 71; Gold Medal for Tech Achievement, Soc Prof Well Log Analysts, 90. *Mem:* Sigma Xi; Soc Petrol Engrs; Soc Prof Well Log Analysts. *Res:* Acoustic wave propagation in isotropic media, attenuation and velocity measurements; acoustic well logging, development of instrumentation and methods; wave propagation in boreholes. *Mailing Add:* Zemanek & Assoc 1007 Greenbriar Lane Duncanville TX 75137

ZEMANIAN, ARMEN HUMPARTSOUM, b Bridgewater, Mass, Apr 16, 25; m 58; c 4. APPLIED MATHEMATICS. *Educ:* City Col New York, BEE, 47; NY Univ, MEE, 49, Eng ScD, 53. *Prof Exp:* Tutor elec eng, City Col New York, 47-48; engr, Maintenance Co, 48-52; from instr to assoc prof elec eng, NY Univ, 52-62; chmn dept, 67-68 & 71-74, LEADING PROF ENG, STATE UNIV NY, STONY BROOK, 62- *Concurrent Pos:* Res fel math inst, Univ Edinburgh, 68-69; consult, All-Tronics, Inc & Anaesthesia Assocs, NY; managing ed, Siam Rev, Soc Indust & Appl Math, 69-71, ed-in-chief publs, 70-74, vpres publ, 74-75; vis scholar, Food Res Inst, Stanford Univ, 75-76; NSF fac fel sci, 75-76; co-founder & co-ed jour, Circuits Systs & Signal Processing; co-founder biennial conf, Int Symposia Math Theory Networks & Systs. *Honors & Awards:* Sci Award, Armenian Students Asn, 82. *Mem:* Soc Indust & Appl Math; fel Inst Elec & Electronics Engrs; Am Math Soc; Sigma Xi; for mem Armenian Acad Sci. *Res:* Mathematical systems theory; distribution theory; integral transforms; electrical network theory; periodic marketing systems; computational methods. *Mailing Add:* Dept Elec Eng State Univ NY Stony Brook NY 11794-2350

ZEMBRODT, ANTHONY RAYMOND, b Covington, Ky, Jan 2, 43; m 66; c 3. INDUSTRIAL CHEMISTRY. *Educ:* Thomas More Col, BA, 65; Ohio Univ, PhD(phys chem), 70. *Prof Exp:* scientist anal chem, 69-78, sect mgr, 79-85, MGR, DRACKETT CO, BRISTOL MYERS CO, CINCINNATI, 85- *Concurrent Pos:* Lectr, NKy State Univ, 71-76 & Thomas More Col, 78. *Mem:* Soc Plastic Engrs; Cosmetics, Toiletry & Fragrance Asn; Chem Specialties Mfrs Asn; NAm Thermal Anal Soc; Am Chem Soc. *Res:* Thermal analysis of polymers; instrumental techniques for analyses and x-ray fluorescence of consumer products; industrial chemistry applications. *Mailing Add:* 1004 Park Lane Covington KY 41011

ZEMEL, JAY N(ORMAN), b New York, NY; m 50; c 3. ENERGY & MASS TRANSPORT IN MESOSCALE SYSTEMS, INFORMATION ACQUISITION. *Educ:* Syracuse Univ, AB, 49, MS, 53, PhD, 56. *Hon Degrees:* MA, Univ Pa, 71. *Prof Exp:* Physicist, Naval Ord Lab, 54-58, chief surface & film group, Solid State Div, 58-66; chmn, Elec Eng & Sci Dept, Univ Pa, 72-77, dir, Ctr Chem Electronics, 79-85, dir, Ctr Sensor Technol, 85-90, RCA PROF SOLID STATE ELECTRONICS, MOORE SCH ELEC ENG, UNIV PA, 66- *Concurrent Pos:* Vis scientist, Zenith Radio Res Lab, Ltd, London, 64; coord ed, Thin Solid Films, 70-72, ed-in-chief, 72; vis prof, Ctr Invest & Advan Studies, Nat Polytech Inst, Mex, 71; mem comt predictive testing for mat performance, Nat Mat Adv Bd, 71-72; mem adv comt solidification metals & semiconductors, Univ Space Res Asn, 72-77; chmn, Gordon Conf on MIS Structures, 76; vis prof elec eng, Univ Tokyo, 78; dir, NATO Advan Study Inst on Non-Destructive Eval of Semiconductor Mat & Devices, 78 & NATO Advan Study Inst Chem Sensitive Elec Develop, 80. *Mem:* Am Phys Soc; Inst Elec & Electronics Engrs; Instrument Soc Am. *Res:* Study of and applications of radiative mass transport in micron and submicron structures; flow and pressure sensing. *Mailing Add:* Dept Elec Eng 308 MB D2 Univ Pa Philadelphia PA 19104

ZEMKE, WARREN T, b Fairmont, Minn, Oct 9, 39; m 68; c 3. SPECTROSCOPY, THEORETICAL CHEMISTRY. *Educ:* St Olaf Col, BA, 61; Ill Inst Technol, PhD(chem), 69. *Prof Exp:* From instr to assoc prof, 66-80, PROF CHEM, WARTBURG COL, 80- *Concurrent Pos:* Vis assoc prof, Univ Iowa, 78-79. *Mem:* Sigma Xi; AAAS; Am Chem Soc; Am Phys Soc. *Res:* Molecular electronic wave functions; molecular spectroscopy and structure. *Mailing Add:* Dept Chem Wartburg Col Waverly IA 50677

ZEMLICKA, JIRI, b Prague, Czech, July 31, 33; m 61; c 2. ORGANIC CHEMISTRY, BIOCHEMISTRY. *Educ:* Charles Univ, Prague, MS, 56, RNDr, 66; Czech Acad Sci, PhD(org chem), 59. *Prof Exp:* Res asst anal biochem, Inst Food Technol, Prague, 56; res scientist, Inst Org Chem & Biochem, Czech Acad Sci, 59-68; vis scientist, Mich Cancer Found, 68-69, res scientist, 70-80, assoc mem, 80-83; assoc prof, 71-85, PROF BIOCHEM, SCH MED, WAYNE STATE UNIV, 85-; MEM, MICH CANCER FOUND, 83- *Concurrent Pos:* Ad hoc mem, med chem study sect A, NIH, 79, spec reviewer, bio-org & natural prod chem rev group, 80 & 87. *Mem:* Am Chem Soc; Czech Soc Arts & Sci in Am; NY Acad Sci. *Res:* Chemistry of nucleic acids; protein biosynthesis; cancer and viral chemotherapy. *Mailing Add:* Mich Cancer Found 110 E Warren Ave Detroit MI 48201-9987

ZEMLIN, WILLARD R, b Two Harbors, Minn, July 20, 20; m 54; c 2. SPEECH & HEARING SCIENCES. *Educ:* Univ Minn, BA, 57, MS, 60, PhD(speech), 61. *Prof Exp:* Dir speech & hearing res lab, 61-76, PROF SPEECH & HEARING SCI, UNIV ILL, URBANA-CHAMPAIGN, 76- *Concurrent Pos:* Consult, Lincoln State Sch, 65-67. *Mem:* Am Speech & Hearing Asn; Acoust Soc Am. *Res:* Anatomy and physiology of normal and pathological speech and hearing mechanisms. *Mailing Add:* 1519 W Park Ave Champaign IL 61821

ZEMMER, JOSEPH LAWRENCE, JR, b Biloxi, Miss, Feb 23, 22; m 50; c 3. ALGEBRA. *Educ:* Tulane Univ, BS, 43, MS, 47; Univ Wis, PhD(math), 50. *Prof Exp:* From asst prof to assoc prof, 50-61, chmn dept, 67-70 & 73-76, prof, 61-87, EMER PROF MATH, UNIV MO, COLUMBIA, 87- *Concurrent Pos:* Fulbright lectr, Osmania Univ, India, 63-64. *Mem:* Am Math Soc; Math Asn Am; Can Math Soc; Nat Coun Teachers Math. *Res:* Near-rings, near-fields, and semi-fields. *Mailing Add:* Dept Math Univ Mo Columbia MO 65211

ZEMON, STANLEY ALAN, b Detroit, Mich, June 16, 30; m 67. PHYSICS. *Educ:* Harvard Univ, AB, 52; Columbia Univ, AM, 58, PhD(physics), 64. *Prof Exp:* Res asst physics, Columbia Univ, 58-62, res scientist, 63-64, res assoc, 64-65; MEM TECH STAFF, GTE LABS, INC, 65- *Mem:* Am Phys Soc; Inst Elec & Electronics Engrs; Sigma Xi. *Res:* Superconductivity; low temperature physics; microwaves; acoustoelectric effect; ultrasonics; semiconductors; Brillouin scattering; lasers; acoustic surface waves; nonlinear acoustics; optical guided waves; nonlinear optics. *Mailing Add:* GTE Labs Inc 40 Sylvan Rd Waltham MA 02154

ZEMP, JOHN WORKMAN, b Camden, SC, Sept 28, 31; m 58; c 4. BIOCHEMISTRY. *Educ:* Col Charleston, BS, 53; Med Col SC, MS, 54; Univ NC, PhD(biochem), 66. *Prof Exp:* Asst prof biochem, Ctr Res Pharmacol & Toxicol, Sch Med, Univ NC, Chapel Hill, 67-69; from asst prof to assoc prof biochem, 69-73, actg dean, Col Med, 73-74, actg vpres acad affairs, 74-75, coordr res, 72-73, assoc dean acad affairs, 73-76, exec dir, Charleston Higher Educ Consortium, 77-81, PROF BIOCHEM, MED UNIV SC, 74-, DEAN, COL GRAD STUDIES & UNIV RES, 77- *Concurrent Pos:* NSF res fel, 66-67. *Mem:* AAAS; Am Chem Soc; Soc Neurosci; Soc Exp Biol & Med; NY Acad Sci; Sigma Xi. *Res:* Brain function and biochemistry; developmental neurobiology; nutrition; biochemical pharmacology. *Mailing Add:* Col Grad Studies Med Univ SC Charleston SC 29425

ZEN, E-AN, b Peking, China, May 31, 28; nat US. GEOLOGY, PETROLOGY. *Educ:* Cornell Univ, BA, 51; Harvard Univ, AM, 52, PhD(geol), 55. *Prof Exp:* Res fel, Oceanog Inst, Woods Hole, 55-56, res assoc, 56-58; vis asst prof, Univ NC, 58-59; res geologist, US Geol Surv, 59-89; ADJ PROF, UNIV, MD, 90- *Concurrent Pos:* Vis assoc prof, Calif Inst Technol, 62; Crosby vis prof, Mass Int Technol, 72; Harry Hess sr vis fel, Princeton, 81; pres, Geol Soc Wash, 72 & Mineral Soc Am, 76; counr, Geol Soc Am, 85-88. *Honors & Awards:* Arthur L Day Medal, Geol Soc Am; Roebling Medal, Mineral Soc Am; Distinguished Serv Medal, US Dept Int. *Mem:* Nat Acad Sci; fel AAAS; Geol Soc Am (vpres, 90-91); fel Mineral Soc Am (pres, 76); Mineral Asn Can; fel Am Acad Arts & Sci. *Res:* Phase equilibrium of sedimentary and metamorphic rocks; stratigraphy and structure of NAppalachians and NRockies; igneous petrology. *Mailing Add:* Dept Geol Univ Md College Park MD 20742

ZENCHELSKY, SEYMOUR THEODORE, b New York, NY, July 6, 23. ANALYTICAL CHEMISTRY. *Educ:* NY Univ, BA, 44, MS, 47, PhD(chem), 52. *Prof Exp:* From instr to assoc prof chem, Rutgers Univ, New Brunswick, 51-64, prof, 64-80. *Mem:* AAAS; Am Chem Soc; Soc Appl Spectros; NY Acad Sci. *Res:* Instrumentation; nonaqueous titrations; spectrophotometry; atmospheric aerosols; organic aerosols. *Mailing Add:* 180 C Cedar Lane Highland Park NJ 08904-2628

ZENDER, MICHAEL J, b Austin, Minn, Feb 27, 39. NUCLEAR PHYSICS. *Educ:* St John's Univ, Minn, BA, 61; Vanderbilt Univ, PhD(physics), 66. *Prof Exp:* From asst prof to assoc prof & chmn dept, 66-72, PROF PHYSICS, CALIF STATE UNIV, FRESNO, 73- *Mem:* Am Asn Physics Teachers; AAAS; Health Physics Soc; Am Asn Physicists Med. *Res:* Low energy nuclear, beta and gamma-ray spectroscopy; study of the Auger effect by using a post acceleration Geiger counter in conjunction with very thin counter windows; radiation safety; x-ray fluorescence analysis. *Mailing Add:* 1523 W Robinwood Lane Fresno CA 93711

ZENER, CLARENCE MELVIN, b Indianapolis, Ind, Dec 1, 05; m 31; c 4. PHYSICS. *Educ:* Stanford Univ, AB, 26; Harvard Univ, PhD(physics), 29. *Prof Exp:* Sheldon traveling fel, Ger, 29-30; Nat Res Coun fel, Princeton Univ, 30-32; fel, Bristol Univ, 32-34; instr physics, Wash Univ, 35-37; instr physics, City Col New York, 37-40; assoc prof, State Col Wash, 40-42; from physicist to prin physicist, Watertown Arsenal, 42-45; prof physics, Univ Chicago, 45-51; assoc dir res labs, Westinghouse Elec Corp, 51-56, dir res labs, 56-62, dir sci, 62-65; dean, Col Sci, Tex A&M Univ, 66-68; UNIV PROF PHYSICS, CARNEGIE-MELLON UNIV, 68- *Honors & Awards:* Bingham Award, Soc Rheology, 57; Wetherill Medal, Franklin Inst, 59; Albert Souveur Achievement Award, Am Soc Metals, 65, Gold Medal, 74; von Hippel Award, Mat Res Soc, 82. *Mem:* Nat Acad Sci; fel Am Phys Soc. *Res:* Theoretical physics and engineering, power from the ocean's thermocline. *Mailing Add:* Dept Physics Carnegie-Mellon Univ Pittsburgh PA 15213

ZENGEL, JANET ELAINE, b Baltimore, Md, Feb 21, 48; m 74; c 1. NEUROPHYSIOLOGY. *Educ:* Western Md Col, BA, 70; Univ Miami, PhD(physiol & biophysics), 73. *Prof Exp:* Res instr physiol & biophysics, Sch Med, Univ Miami, 74-78, res asst prof, 78-79; sr assoc anat, Sch Med, Emory Univ, 79-80; RES BIOLOGIST, VET ADMIN MED CTR, GAINESVILLE, 80-; ASSOC PROF NEUROSCI & NEUROSURG, COL MED, UNIV FLA, 80- *Mem:* Biophys Soc; Soc Neurosci; AAAS. *Res:* Neurophysiology; synaptic and neuromuscular transmission; role of calcium 2 in transmitter release; mechanisms of changes in efficacy; spinal cord physiology and anatomy. *Mailing Add:* RR 2 Box 246 Micanopy FL 32667-9606

ZENGEL, JANICE MARIE, b Baltimore, Md, Feb 21, 48; m 78; c 3. MOLECULAR GENETICS, BACTERIAL PHYSIOLOGY. *Educ:* Western Md Col, BA, 70; Univ Wis, PhD(genetics), 76. *Prof Exp:* Fel molecular genetics, Pharmacol Dept, Stanford Univ, 76-78, Neurol Dept, Baylor Col Med, 78-79; RES ASSOC MOLECULAR GENETICS, BIOL DEPT, UNIV ROCHESTER, 79- *Mem:* Am Soc Microbiol. *Res:* E coli molecular basis for the regulation of ribosome synthesis in E coli, using biochemical, genetic and recombinant DNA techniques. *Mailing Add:* Four Grey Fawn Pittsford NY 14622

ZENGER, DONALD HENRY, b Little Falls, NY, July 27, 32; m 57; c 2. GEOLOGY. *Educ:* Union Col, NY, BS, 54; Dartmouth Col, MA, 59; Cornell Univ, PhD(geol), 62. *Prof Exp:* From instr to assoc prof, 62-73, PROF GEOL, POMONA COL, 73-, CHMN, GEOL DEPT, 84-, MINNIE B CAIRNS CHAIR, GEOL, 90- *Concurrent Pos:* Grants, Geol Soc Am, 63 & Am Chem Soc, 66-74; vis geologist, Unocal res, 84-; assoc ed, J Sedimentary Petrol, 82-88, Corbonales & Evaporides, 86- *Mem:* Geol Soc Am; Int Asn Sedimentologists; Am Asn Petrol Geologists; Nat Asn Geol Teachers; Soc Econ Paleont & Mineral. *Res:* Silurian and Devonian stratigraphy, paleontology and carbonate petrology, particularly dolomitization, New York, Wyoming and California. *Mailing Add:* Dept Geol Pomona Col 609 N College Ave Claremont CA 91711-6339

ZENISEK, CYRIL JAMES, b Cleveland, Ohio, Feb 6, 26; m 56; c 3. ZOOLOGY. *Educ:* Ohio State Univ, BSc, 49, MSc, 55, PhD(zool), 63. *Prof Exp:* From asst prof to assoc prof, 60-66, PROF BIOL, INDIANA UNIV PA, 66- *Mem:* Soc Syst Zool; Am Soc Ichthyol & Herpet; Ecol Soc Am; Soc Study Amphibians & Reptiles; Herpetologists League. *Res:* Taxonomy and distribution of amphibians. *Mailing Add:* 670 Diamond Ave Indiana PA 15701

ZENITZ, BERNARD LEON, b Baltimore, Md, Mar 26, 17; m 56; c 2. MEDICINAL CHEMISTRY. *Educ:* Univ Md, BS, 37, PhD(pharmaceut chem), 43. *Prof Exp:* Sr res chemist, Frederic Stearns & Co, Mich, 44-47; sr res chemist & lab head, Sterling-Winthrop Res Inst, 47-81; RETIRED. *Honors & Awards:* Col Medal Gen Excellence & Pharmacog Prize, 37. *Mem:* Am Chem Soc. *Res:* Thymus gland extracts; sympathomimetic amines; quartenary ammonium salts; synthetic detergents; sterols; preparation of B-cyclohexylakylamines; halogen ring substituted propadrines; coronary dilators; local anesthetics; tranquilizers; antioxidants; antiinflammatory agents; antiobesity drugs; bronchodilators; analgesics. *Mailing Add:* N Oaks Apt 620 725 Mt Wilson Lane Baltimore MD 21208

ZENKER, NICOLAS, b Paris, France, Dec 3, 21; nat US; m 52; c 4. PHARMACEUTICAL CHEMISTRY. *Educ:* Cath Univ Louvain, Cand, 48; Univ Calif, MA, 53, PhD(pharmaceut chem), 58. *Prof Exp:* Biochemist, Mt Zion Hosp, San Francisco, 53-60; asst prof pharmaceut chem, 60-63, assoc prof, 63-69, head dept, 69-79, PROF MED CHEM, UNIV MD, BALTIMORE, 69- *Mem:* Am Chem Soc. *Res:* Synthesis and mode of action of metabolic analogues; thyroadrenergic relationships. *Mailing Add:* Dept Med Chem Pharm Univ Md 20 N Pine St Baltimore MD 21201-1041

ZENSER, TERRY VERNON, b Port Clinton, Ohio, Aug 1, 45; m 68; c 2. BIOCHEMISTRY, PHARMACOLOGY. *Educ:* Ohio State Univ, BS, 67; Univ Mo, Columbia, PhD(biochem), 71. *Prof Exp:* Captain, US Army Med Res Inst Infectious Dis, 71-75; lectr, Hood Col, Frederick, Md, 74-75; asst prof biochem & med, 76-80, assoc prof med, 80-85, assoc prof biochem, 80-86, PROF MED, SCH MED, ST LOUIS UNIV, 85-, PROF BIOCHEM, 86- *Concurrent Pos:* Adj asst prof biochem & med, Univ Pittsburgh, 75-76. *Mem:* Geront Soc; Am Asn Cancer Res; Am Soc Biol Chemists; Am Soc Pharmacol & Exp Therapeut; Sigma Xi; Am Fedn Clin Res. *Res:* Drug metabolism; biology of aging; mechanism of initiation of toxic and carcinogenic effects of chemicals. *Mailing Add:* 1200 Dunloe Rd Manchester MO 63011

ZENTILLI, MARCOS, b Santiago, Chile, May 31, 40; m 63; c 3. METALLOGENY, GEOCHRONOLOGY. *Educ:* Univ De Chile, BS, 63; Queen's Univ, PhD(geol), 74. *Prof Exp:* Geologist mineral deposit, Inst Invest Geologicas, 63-68, exploration, Geophys Eng, Ltd, Toronto, 72; from asst prof to assoc prof, 73-86, chmn dept, 84-86, PROF GEOL, DALHOUSIE UNIV, 86- *Concurrent Pos:* Assoc dir, Lester Pearson Inst for Int Develop, 85-86; geol consult, Cuesta Res Ltd, NS, Can. *Mem:* Fel Geol Asn Can; fel Soc Econ Geologists; Can Inst Mining & Metall; Asn Geoscientists Int Develop; Soc Appl Mineral Deposits; Geol Soc Chile. *Res:* Regional metallogenic evolution of the Central Andes; geology of gold deposits of Nova Scotia; metallogenic evolution of Nova Scotia Appalachians; fission track dating applied to tectonics and mineral deposits. *Mailing Add:* Dept Geol Dalhousie Univ Halifax NS B3H 3J5 Can

ZENTMYER, GEORGE AUBREY, JR, b North Platte, Nebr, Aug 9, 13; m 41; c 3. PHYTOPATHOLOGY. *Educ:* Univ Calif, Los Angeles, AB, 35; Univ Calif, MS, 36, PhD(plant path), 38. *Prof Exp:* Asst plant path, Univ Calif, 36-37; asst pathologist, Div Forest Path, Bur Plant Indust, USDA, 37-40 & Conn Exp Sta, 40-44; asst pathologist, Univ Calif, Riverside, 44-48, assoc plant pathologist, 48-55, fac res lectr, 64, chmn, Dept Plant Path, 68-73, prof, 63-81, PLANT PATHOLOGIST, UNIV CALIF, RIVERSIDE, 55-, EMER PROF PLANT PATH, 81- *Concurrent Pos:* Consult, Pineapple Res Inst, 61, Trust Territory, Pac Islands, 64 & 66, Rockefeller Found, Colombia, 67, Australian Govt, 68, US Agency Int Develop, Ghana & Nigeria, 69, NSF Panel, 71-, SAfrican Govt & avacado growers, 80, Israel & avacado growers & Western Australian Govt, 83, Ministry Agr & Univ Cordoba, Spain, 89; Guggenheim fel, 64-65; mem, Nat Res Coun, 68-73; NATO sr sci fel, Eng, 71; mem var comts, Nat Acad Sci-Nat Res Coun; assoc ed, Annual Rev Phytopath, 70-; counr, Int Soc Plant Path, 78-83; Bellagio scholar, Rockefeller Found, 85. *Honors & Awards:* Award Distinction, Am Phytopath Soc, 83. *Mem:* Nat Acad Sci; fel Am Phytopath Soc (secy, 59-62, pres, 65-66); Int Soc Plant Path; Mycol Soc Am; Asn Trop Biol; Sigma Xi; Indian Phytopath Soc; Phillipine Phytopath Soc; Brit Mycol Soc; AAAS. *Res:* Biology and physiology of root pathogens, especially Phytophthora; chemotaxis; chemotherapy; fungicides; diseases of avocado, cacao, other tropicals and subtropicals. *Mailing Add:* Dept Plant Path Univ Calif Riverside CA 92521-0122

ZENTNER, THOMAS GLENN, b Rowena, Tex, Aug 6, 26; m 52; c 5. PULP CHEMISTRY, PAPER CHEMISTRY. *Educ:* Tex A&M Univ, BS, 48; Inst Paper Chem, MS, 50, Lawrence Col, PhD(chem), 52. *Prof Exp:* Mem staff, Gardner Div, Diamond Nat Corp, 52-59, dir res & develop, 59-60; dir forest prod oper, Olin Mathieson Chem Corp, 60-66; vpres res & develop, Olinkraft, Inc, La, 66-68, rep, Forest Prod Div, Olin Corp, West Monroe, 68-75, vpres res, Olinkraft, Inc, 74-79, VPRES RES, MANVILLE FOREST PROD, INC, 80- *Mem:* Am Chem Soc; corp mem Tech Asn Pulp & Paper Indust. *Res:* Paperboard manufacture; coating; converting for packaging; graphic arts. *Mailing Add:* 7116 Whites Ferry Rd West Monroe LA 71291

ZENZ, CARL, b Vienna, Austria, Feb 1, 23; US citizen; m 47; c 3. OCCUPATIONAL MEDICINE. *Educ:* Jefferson Med Col, MD, 49; Univ Cincinnati, ScD, 57. *Prof Exp:* Intern, Ausbury Hosp, Minneapolis, Minn, 49-50; gen pract, Minn & Wis, 50-52; resident occup med, Allis-Chalmers Corp, 56-57, chief clin serv, 57-62, chief physician, 62-63, med dir, 63-65, dir med & hyg serv, 65-70, dir med serv, 70-80; CONSULT, 80- *Mem:* AAAS; Am Occup Med Asn; Am Indust Hyg Asn; fel Am Pub Health Asn; fel Am Acad Occup Med (past pres). *Mailing Add:* 2418 S Root River Pkwy West Allis WI 53227

ZENZ, DAVID R, b Chicago, Ill, Sept 17, 43; m 68; c 1. MUNICIPAL SLUDGE MANAGEMENT, MUNICIPAL SLUDGE PROCESSING. *Educ:* Ill Inst Technol, BS, 65, MS, 66, PhD(environ eng), 68. *Prof Exp:* Coordr res, 68-90, MGR RES & TECH SERV, METROP WATER RECLAMATION, DIST GREATER CHICAGO, 90- *Concurrent Pos:* Adj prof, Ill Inst Technol, 68-72 & 91-; chmn, Sludge Mgt Comt, Water Pollution Control Fedn; pres, Ill Water Pollution Control Asn, 90-91. *Mem:* Water Pollution Control Fedn; Am Acad Environ Engrs. *Res:* Published papers on municipal sludge management, municipal sludge processing, advanced wastewater treatment and wastewater nutrification. *Mailing Add:* 2347 Bayside Ct Lisle IL 60532

ZENZ, FREDERICK A(NTON), b New York, NY, Aug 1, 22; m 49; c 4. CHEMICAL ENGINEERING. *Educ:* Queens Col, BS, 42; NY Univ, MChE, 50; Polytech Inst Brooklyn, DChE, 61. *Prof Exp:* Chem engr, M W Kellogg Co, 42-44, 46, develop engr, 53-56; chem engr, Kellex Corp, 44-45; res engr, Carbide & Carbon Chem Corp, Tenn, 45-46; process engr, Hydrocarbon Res, Inc, 46-53; mgr, Process Dept, Assoc Nucleonics, Inc, 56-62; prof eng, Manhattan Col, 69-87; DIR, PROCESS EQUIP MODELLING & MFG CO, 74- *Concurrent Pos:* Lectr, NY Univ, 51-52; adj prof, Polytech Inst Brooklyn, 59-; consult, 62-; tech dir, Particulate Solid Res, Inc, 70-; vpres, Ducon Co, 75-81; emer prof eng, Manhattan Col, 87- *Honors & Awards:* Tyler Award, Am Inst Chem Engrs, 58. *Mem:* Am Chem Soc; Am Nuclear Soc; Sigma Xi; fel Am Inst Chem Engrs. *Res:* Fluidization; two-phase flow; tower design; distillation; particle filtration; gas-solid reactions; nuclear power reactors; isotope separation; air fractionation; petroleum refining; ammonia. *Mailing Add:* Rte 9D PO Box 241 Manhattan Col Garrison NY 10524

ZEOLI, G(ENE) W(ESLEY), b Los Angeles, Calif, Nov 9, 26; m 56; c 2. ELECTRICAL ENGINEERING. *Educ:* Univ Calif, Berkeley, BS, 48, MS, 49; Univ Calif, Los Angeles, PhD, 71. *Prof Exp:* Lectr elec eng, Univ Calif, 48-49; asst, Mass Inst Technol, 49-50, 51-52; mem tech staff radar systs res, Hughes Aircraft Co, 52-54, group head sensor syst anal, 54-57, sr scientist & head info processing staff, Signal Processing & Display Lab, 57-68, sr scientist, Staff Radar Systs & Signal Processing Lab, 68-76, chief scientist, Advan Progs Div, 76-81; PRES, GENE W ZEOLI, INC, 81- *Concurrent Pos:* Lectr, Univ Calif, Los Angeles, 54-55 & Loyola Univ, 71-72. *Res:* Radar systems design and performance studies. *Mailing Add:* 1405 Granvia Altamira Palos Verdes Estates CA 90274

ZEPF, THOMAS HERMAN, b Cincinnati, Ohio, Feb 13, 35. SURFACE PHYSICS, QUANTUM OPTICS. *Educ:* Xavier Univ, Ohio, BS, 57; St Louis Univ, MS, 60, PhD(physics), 63. *Prof Exp:* From asst prof to assoc prof, 62-75, from actg chmn dept to chmn dept, 63-73, PROF PHYSICS, CREIGHTON UNIV, 75-, CHMN DEPT, 81- *Concurrent Pos:* Vis prof physics, St Louis Univ, 73-74. *Mem:* AAAS; Am Phys Soc; Am Asn Physics Teachers. *Res:* Surface barrier analysis of metals and semiconductors; laser-induced desorption; diagnostics of laser-generated plasmas; laser propagation and particulate interaction; methods of x-ray diffraction; unltra-high vacuum techniques. *Mailing Add:* Dept Physics Creighton Univ Omaha NE 68178

ZEPFEL, WILLIAM F, b Pittsburgh, Pa, Apr 29, 25. IRON & STEEL PRODUCTS. *Educ:* Univ Pittsburgh, BS, 49. *Prof Exp:* SR METALL ENGR, SIDBEC-DOSCO INC, 67- *Concurrent Pos:* Mem bd trustees, Am Soc Metals, 80-83. *Mem:* Fel Am Soc Metals; Am Soc Testing & Mat; Am Inst Mining Metall & Petrol Engrs; Standards Coun Can. *Mailing Add:* Sidbec-Dosco Inc 300 Rue Leo Pariseau Montreal PQ H2W 2S7 Can

ZEPP, EDWIN ANDREW, b Orange, Calif, Sept 15, 45; m 79; c 2. ENDOCRINOLOGY, LIPID METABOLISM. *Educ:* WVa Univ, AB, 67, MS, 72, PhD(pharmacol), 76. *Prof Exp:* Res assoc, 76-78, asst prof, 70-84, ASSOC PROF BIOCHEM, KIRKSVILLE COL OSTEOPATH MED, 84- *Concurrent Pos:* Prin investr, Nat Cancer Inst, NIH, 78-; adj asst prof, Northeast Mo State Univ, 81- *Mem:* Endocrine Soc; Am Diabetes Asn. *Res:* Hormonal control of lipid metabolism and alterations by tumor-bearing and under conditions of dietary manipulation. *Mailing Add:* Dept Biochem Kirksville Col Osteopath Med 204 W Jefferson Kirksville MO 63501

ZEPP, RICHARD GARDNER, b Brooklyn, NY, Nov 20, 41; m 68; c 1. ENVIRONMENTAL CHEMISTRY, BIOGEOCHEMISTRY, PHOTOCHEMISTRY KINETICS. *Educ:* Furman Univ, BS, 63; Fla State Univ, PhD(chem), 68. *Prof Exp:* Res assoc photochem, Mich State Univ, 69-70; RES CHEMIST, US ENVIRON PROTECTION AGENCY, 71- *Concurrent Pos:* Adj prof, Univ Miami; ed adv bd, Environ Sci Technol. *Honors & Awards:* Environ Protection Agency Scientist & Tech Achievement Award, 83, 86. *Mem:* Am Chem Soc; Sigma Xi; Am Soc Limnol & Oceanog; AAAS; Soc Environ Toxicol & Chem; Am Geophys Union. *Res:* Rates and products of chemical reactions and photochemical in water and soil; mathematical simulation of biogeochemical processes in soil and water. *Mailing Add:* Environ Res Lab College Station Rd Athens GA 30613

ZEPPA, ROBERT, b New York, NY, Sept 17, 24; m 52; c 2. THORACIC SURGERY. *Educ:* Columbia Univ, AB, 48; Yale Univ, MD, 52. *Prof Exp:* Intern, Med Ctr, Univ Pittsburgh, 52-53; asst resident surg, Sch Med, Univ NC, 53-56, thoracic resident, 56-57, instr thoracic surg, 58-60, from asst prof surg to assoc prof, 60-65; co-chmn dept, 66-71, CHMN DEPT SURG, SCH MED, UNIV MIAMI, 71-, PROF SURG & PHARMACOL, 65- *Concurrent Pos:* USPHS career trainee, Sch Med, Wash Univ, 56-58; Markle scholar med sci, 59-64; assoc dir clin res unit, NC Mem Hosp, 61-65; chief surg serv, Vet Admin Hosp, Miami, Fla, 65-72. *Mem:* Am Col Surg; Am Surg Asn; Soc Surg Alimentary Tract; Soc Univ Surg. *Res:* Biologically active amines; portal hypertension; gastrointestinal physiology. *Mailing Add:* Dept Surg Univ Miami PO Box 016310 Miami FL 33101

ZERBE, JOHN IRWIN, b Hegins, Pa, June 4, 26; m 51; c 3. WOOD SCIENCE, WOOD TECHNOLOGY. *Educ:* Pa State Univ, BS, 51; State Univ NY, MS, 53, PhD(wood technol), 56. *Prof Exp:* Asst wood technol, State Univ NY, 51-56; res asst prof housing res, Univ Ill, 56-58; mgr, Govt Specifications & Standards Dept, Nat Forest Prod Asn, 58-59, asst vpres, Tech Serv, 59-70; dir forest prod & eng res, 70-76, MGR ENERGY RES, DEVELOP & APPLICATION, FOREST SERV, USDA, 76- *Mem:* Forest Prod Res Soc; Soc Wood Sci & Technol; Soc Am Foresters; Am Soc Testing & Mat. *Res:* Mechanical properties of wood, conversion of wood to solid and liquid fuel, energy from biomass; energy conservation. *Mailing Add:* 3310 Heatherdell Lane Madison WI 53713-3446

ZERBY, CLAYTON DONALD, b Cleveland, Ohio, Jan 27, 24; m 49; c 3. QUALITY IMPROVEMENT. *Educ:* Case Western Reserve Univ, BS, 50; Univ Tenn, MS, 56, PhD(physics), 60. *Prof Exp:* Engr, Oak Ridge Nat Lab, 50-54, group leader physics, 54-63, mgr physics & eng, Defense & Space Systs Dept, 63-66, mgr dept, 66-67, gen mgr, Korad Laser Dept, 67-71, pres, Ocean Systs, Inc, 71-73, Domsea Farms, Inc, 72-74, tech serv mgr, Nuclear Div, 74-76, dir off waste isolation, 76-78; plant mgr, Paducah Gaseous Diffusion Plant, Union Carbide Corp, 78-84; dir mgt syst compliance, Martin Marietta Energy Syst, 84-88; CONSULT, QUALITY IMPROV, 88- *Concurrent Pos:* Lectr, Univ Tenn, 61-63 & Univ Ky, 80-81. *Mem:* Am Phys Soc; Am Nuclear Soc; Sigma Xi. *Res:* Theory of electromagnetic interactions; Monte Carlo methods; nuclear weapons effects; space vehicle radiation shielding; shielding against high energy particles; interaction of high energy particles with complex nuclei; nuclear waste terminal storage; quality improvement techniques. *Mailing Add:* 1102 W Outer Dr Oak Ridge TN 37830

ZEREN, RICHARD WILLIAM, b Baltimore, Md, June 3, 42; m 65; c 3. MECHANICAL ENGINEERING, RESEARCH ADMINISTRATION. *Educ:* Duke Univ, BSME, 64; Stanford Univ, MS, 65, PhD(eng), 70. *Prof Exp:* Actg asst prof mech eng, Stanford Univ, 69; asst prof, Mich State Univ, 70-74; asst to dir, Fossil Fuel & Advan Systs Div, Elec Power Res Inst, 74-77, mgr prog integration & eval, 78-79; sr assoc, Booz-Allen & Hamilton, Inc, 80; dir, res & develop, Elec Power Res Inst, 80-81, Planning & Eval Div, 81-89, mem div, 89-90, SPEC PROJ, ELEC POWER RES INST, 90- *Concurrent Pos:* NSF trainee, 64-68. *Mem:* Am Soc Mech Engrs. *Res:* Research and development. *Mailing Add:* 150 Corona Way Portola Valley CA 94028

ZEREZ, CHARLES RAYMOND, b Aleppo, Syria, July 24, 56; US citizen; m 91. HEMATOLOGY, RED BLOOD CELL METABOLISM. *Educ:* Univ Calif, Los Angeles, BS, 78, CPhil, 82, PhD(biochem), 85, MD, 92; Calif State Univ, MS, 81. *Prof Exp:* Res asst chem, dept chem & biochem, 80-81, teaching assoc biochem, 81-82, res asst, 81-84, res assoc, 84-85, ASST BIOCHEMIST, DEPT MED, UNIV CALIF LOS ANGELES SCH MED, 86-, ASST DIR, HEMAT RES LAB, HARBOR-UNIV CALIF LOS ANGELES, 87- *Concurrent Pos:* Prin investr, Res & Educ Inst, Harbor-Univ Calif Med Ctr, 86- *Mem:* Am Chem Soc; Am Fedn Clin Res; Am Soc Hemat. *Res:* Red blood cell metabolism in hemolytic anemias; sickle cell disease, thalassemia and enzyme deficiency anemias; the mechanisms of red cell hemolysis in these disorders. *Mailing Add:* Harbor-Univ Calif Los Angeles Med Ctr C-1 Rm 12 1124 W Carson St Torrance CA 90502

ZERILLI, FRANK J, b Brooklyn, NY, Dec 25, 42. THEORETICAL PHYSICS. *Educ:* Princeton Univ, PhD(theoret physics), 69. *Prof Exp:* Res assoc, Univ NC, 69-72; asst prof physics, Wash State Univ, 72-75 & Mich State Univ, 75-79; RES PHYSICIST, NAVAL SURFACE WARFARE CTR, 79- *Mem:* Am Phys Soc; Sigma Xi. *Res:* Strength and fracture of materials; equilibrium of reacting energetic materials. *Mailing Add:* Detonation Phys Branch Naval Surface Weapons Ctr Silver Springs MD 20910

ZERLA, FREDRIC JAMES, b Wheeling, WVa, Feb 23, 37; m 66; c 6. HISTORY OF MATHEMATICS. *Educ:* Col Steubenville, BA, 58; Fla State Univ, MS, 60, PhD(math), 67. *Prof Exp:* Asst prof, 63-72, from asst chmn dept to actg chmn dept, 69-74, ASSOC PROF MATH, UNIV SFLA, 72-, UNDERGRAD ADV, DEPT MATH, 74- *Mem:* Math Asn Am. *Res:* Derivations in algebraic fields; field theory; history of mathematics. *Mailing Add:* Dept Math Univ SFla 4202 Fowler Ave Tampa FL 33620

ZERLIN, STANLEY, b New York, NY, Sept 15, 29; m 59; c 2. PSYCHOPHYSIOLOGY, ELECTROPHYSIOLOGY. *Educ:* City Col New York, BS, 51; Columbia Univ, MA, 53; Western Reserve Univ, PhD(audition), 59. *Prof Exp:* Asst proj dir, Cleveland Hearing & Speech Ctr, 57-58; proj dir, Auditory Res Lab, Vet Admin, 59-63, res scientist, 59-62, res dir, 62-63; res assoc auditory processes, Cent Inst for Deaf, 63-65; res assoc auditory evoked responses, Houston Speech & Hearing Ctr, 65-67; ASSOC PROF & RES ASSOC AUDITORY PROCESSES, OTOLARYNGOL LABS, UNIV CHICAGO, 67- *Mem:* Soc Neurosci; Acoust Soc Am; Am Speech & Hearing Asn; Int Soc Audiol. *Res:* Auditory electrophysiology, evoked response; cochlear processes; binaural interaction. *Mailing Add:* 1457 E 55th Pl Chicago IL 60637

ZERNER, MICHAEL CHARLES, b Boston, Mass, Jan 31, 40; m 66; c 2. CHEMISTRY, MOLECULAR ELECTRONIC STRUCTURE. *Educ:* Carnegie-Mellon Univ, BSc, 61; Harvard Univ, MA, 62, PhD, 66. *Prof Exp:* NIH fel, Univ Uppsala, Sweden, 68-70; from asst prof to assoc prof chem, Univ Guelph, Can, 70-80; prof chem, Univ Guelph, 80-82; PROF CHEM, UNIV FLA, GAINESVILLE, 81-, CHAIR CHEM, 88- *Concurrent Pos:* Vis scientist, Univ NC, 75, Stanford Univ, 76; consult, Eastman Kodak, 78-; vis prof quantum chem, Univ Uppsala, 78-79; vis prof, Potificaia Univ, Brazil, 80, Tech Univ, Munich, Ger, 86, Max Planck Inst Physics & Astrophysics, 88 & 89; mem bd dirs, Quantum Chem Prog Exchange, Ind Univ, 80-87; vis prof chem, Univ Fla, Gainesville, 81, assoc dir, Quantum Theory proj, 82-; adj prof, Univ Guelph, 82-86; assoc ed, Int J Quantum Chem, 82-; Fulbright distinguished prof, Ruder Boskovic Inst, Fagreb, 87; ed, Advan Quantum Chem, 88-; vis sr scientist, Brookhaven Nat Lab, 90. *Mem:* Am Inst Physics; Int Soc Quantum Biol; AAAS; Am Chem Soc; NY Acad Sci. *Res:* Molecular electronic structure; molecular bonding, structure, reactivity and spectroscopy predicted from the concepts of theoretical chemistry and physics. *Mailing Add:* Univ Fla 362 Williamson Hall Gainesville FL 32611

ZERNIK, JOSEPH, b Jerusalem, Israel, Oct 24, 55; m 83; c 2. BONE BIOLOGY, EUKANGOTIC GENE EXPRESSION. *Educ:* Univ Tel-Aviv, DMD, 83; Univ Conn, PhD(biomed sci), 88. *Prof Exp:* Asst prof orthod, Univ Conn, 88-91; ASSOC PROF ORTHOD, UNIV SOUTHERN CALIF, 91- *Mem:* Am Soc Bone & Mineral Res; Am Asn Dent Res; AAAS. *Res:* Cloned and characterized rat gene for bone alkaline phosphatase and studied the function of its promoters; developmental regulation of alkaline phosphatase expression. *Mailing Add:* Sch Dent Med Univ Southern Calif Los Angeles CA 90089

ZERNOW, LOUIS, b Brooklyn, NY, Dec 27, 16; m 40; c 4. HIGH STRAIN RATE MATERIAL BEHAVIOR. *Educ:* Cooper Union, BChE, 38; Johns Hopkins Univ, PhD(physics), 53. *Prof Exp:* Mem staff, Ballistic Res Labs, Aberdeen Proving Ground, Ord Dept, US Dept Army, 40-51, chief, Rocket & Ammunition Br, 51-53 & Detonation Physics Br, 53-55; dir res & mgr ord res div, Aerojet-Gen Corp, 55-63; pres, Shock Hydrodynamics, Div Whittaker Corp, 63-81; PRES, ZERNOW TECH SERV INC, 81- *Concurrent Pos:* Consult, US Dept Army, Gould Inc, Gen Dynamics, Bendix, Aerojet & US Air Force, Rockwell Int. *Honors & Awards:* Outstanding Leadership Award, Am Defense Preparedness Asn, 87. *Mem:* Am Inst Aeronaut & Astronaut; Am Inst Mining, Metall & Petrol Eng; Am Soc Metals; Am Phys Soc; Acoust Soc Am; NY Acad Sci; Sigma Xi. *Res:* Detonation and aerosol physics; explosives; high strain-rate behavior of materials; shock waves in solids; effects of super pressure on solids; ordnance systems and explosive metal forming; shaped charge design; undersea defense systems. *Mailing Add:* 1103 E Mountain View Ave Glendora CA 91740

ZERO, DOMENICK THOMAS, b Brooklyn, NY, Sept 24, 49; m. DENTAL CARIES RESEARCH. *Educ:* St Johns Univ, BS, 71; Georgetown Univ, DDS, 75; Univ Rochester, MS, 80. *Prof Exp:* Res asst, Eastman Dent Ctr, 77-78, res assoc, 79-81; asst prof, Va Commonwealth Univ, 81-85; res & clin assoc, 85-90, SR RES & CLIN ASSOC, EASTMAN DENTAL CTR, 90- *Concurrent Pos:* Prin investr, NIH/NIDR grants. *Honors & Awards:* Michael G Buonocere Prize, Am Asn Dent Res; Int Col Dent Award. *Mem:* Int Asn Dent Res; Europ Orgn Caries Res. *Res:* Development of intra-oral models for the study of dental caries; study of microbial virulence related to the development of dental caries; oral retention of topical fluoride products and its relationship to the clinical effectiveness of fluoride products. *Mailing Add:* Eastman Dent Ctr 625 Elmwood Ave Rochester NY 14620

ZEROKA, DANIEL, b Plymouth, Pa, June 22, 41; m 67; c 2. THEORETICAL PHYSICAL CHEMISTRY. *Educ:* Wilkes Col, BS, 63; Univ Pa, PhD(theoret chem), 66. *Prof Exp:* NSF fel statist mech, Yale Univ, 66-67; from asst prof to assoc prof, 67-90, PROF CHEM, LEHIGH UNIV, 90- *Concurrent Pos:* Vis res scientist, DuPont, 84; sabbatical, Cornell, 85. *Mem:* Am Phys Soc; Am Chem Soc. *Res:* Quantum chemistry; statistical mechanics; magnetic resonance; electronic structure of solids. *Mailing Add:* Dept Chem Lehigh Univ Bethlehem PA 18015

ZERONIAN, SARKIS HAIG, b Manchester, Eng, June 30, 32; m 70. TEXTILE CHEMISTRY. *Educ:* Univ Manchester, BScTech, 53, MScTech, 55, PhD(cellulose chem), 62, DSc(polymer & fiber sci), 83. *Prof Exp:* Res officer cellulose chem, Brit Cotton Indust Res Asn, Manchester, Eng, 58-60; res fel, Inst Paper Chem, 62-63; sr res fel nonwoven fabrics, Univ Manchester Inst Sci Technol, 63-66; res assoc cellulose chem, Columbia Cellulose Co, BC, Can, 66-68; from asst prof to assoc prof textile sci, 68-78, chmn, Div Textile & Clothing, 78-86, PROF TEXTILE SCI , UNIV CALIF, DAVIS, 78-, PROF MECH ENG & MAT SCI, 83- *Concurrent Pos:* Chmn, Cellulose, Paper & Textile Div, Am Chem Soc, 83; Fiber Soc lectr, 87-88; counr, Am Chem Soc, 90- *Mem:* Am Chem Soc; Am Asn Textile Chem & Colorists; fel Brit Textile Inst; Fiber Soc; Soc Plastics Engrs. *Res:* Chemical and physical properties of natural and man-made fibers; properties of textile finishes; cellulose chemistry. *Mailing Add:* Div Textiles & Clothing Eng Univ Calif Davis CA 95616-8422

ZERWEKH, CHARLES EZRA, JR, b Galveston, Tex, Aug 24, 22; m 50; c 3. ORGANIC CHEMISTRY. *Educ:* Univ Houston, BS, 44. *Prof Exp:* Chemist, Humble Oil & Refining Co, 44-45, from asst res physicist to res physicist, 45-49, from patent coordr to sr patent coordr, 49-55, supvry patent coordr, 55-57, head tech info sect, Res & Develop Div, 57-63; mgr, Records Mgt Div, Standard Oil Co, NJ, 63-66; tech info specialist, Esso Res & Eng Co, NJ, 66-67; mgr tech info ctr, Polaroid Corp, Cambridge, 67-82; CONSULT COMPUTERIZED LIBR SYSTS, 82- *Mem:* AAAS; Am Chem Soc; NY Acad Sci; Spec Libr Asn; Sigma Xi. *Res:* Technical information, especially patents, machine indexing, literature searching, technical files and information retrieval as applied to petrochemicals, petroleum processing, organic chemistry and photography. *Mailing Add:* W Chenango Rd BC 78 Box 740 Castle Creek NY 13744-9710

ZERWEKH, ROBERT PAUL, b Peoria, Ill, Feb 25, 39; m 74; c 3. MECHANICAL METALLURGY, PHYSICAL METALLURGY. *Educ:* Univ Mo, Rolla, BS, 61; Univ Ill, Urbana, MS, 63; Iowa State Univ, PhD(metall), 70. *Prof Exp:* Sr engr mat, Elec Boat Div, Gen Dynamics Corp, 65-67; from asst prof to assoc prof, 70-82, PROF MECH ENG, UNIV KANS, 82-, ASSOC VCHANCELLOR, 87- *Concurrent Pos:* Fac res fel, NASA-Am Soc Eng Educ, 75; sci grant, NASA/Langley Res Ctr, 76-78 & Gulf & Western Energy Prod Group, 79-83; assoc dean eng, Univ Kans, 80-87. *Mem:* Am Soc Metals; Metall Soc; Am Inst Mining, Metall & Petrol Engrs; Sigma Xi; Am Asn Univ Prof. *Res:* Deformation mechanisms; relationship between microstructure and properties; solid-state phase transformations, composite materials. *Mailing Add:* Dept Mech Eng Univ Kans Lawrence KS 66045

ZETIK, DONALD FRANK, b Brenham, Tex, Nov 28, 38; m 69; c 2. RESERVOIR ENGINEERING, COMPUTER SIMULATOR DEVELOPMENT. *Educ:* Tex A&M Univ, BS, 61; Univ Tex, Austin, PhD(phys chem), 68. *Prof Exp:* Engr petrol prod, Prod Res Lab, Humble Oil Co, 61-63; res assoc chem physics, Wash State Univ, 68-70; sr engr assoc, reservoir mgt, Cities Serv Oil & Gas Corp, 70-85; SR COMPUT SCIENTIST, DOWELL-SCHLUMBERGER, 85- *Concurrent Pos:* Adj prof, Univ Tulsa, 76-77. *Mem:* Soc Petrol Engrs. *Res:* Numerical simulation of fluid flow in porous media; development of new simulation techniques. *Mailing Add:* 9124 E 67th Pl S Tulsa OK 74133-2211

ZETLER, BERNARD DAVID, b New York, NY, Aug 27, 15; m 40; c 2. OCEANOGRAPHY. *Educ:* Brooklyn Col, BA, 36. *Prof Exp:* Jr math scientist, Hydrography Off, US Dept Navy, Washington, DC, 38; computer, US Coast & Geod Surv, 38-39, mathematician, 39-53, oceanogr, 53-56, chief, Currents & Oceanog Br, 56-60 & Oceanog Anal Br, 60-63, res group off oceanog, 63-65, actg dir phys oceanog lab, Inst Oceanog, Environ Sci Serv Admin, Md, 65-68, dir phys oceanog lab, Atlantic Oceanog Labs, Fla, 68-72; RES OCEANOG, INST GEOPHYS & PLANETARY PHYSICS, UNIV CALIF, SAN DIEGO, 72- *Concurrent Pos:* Assoc prof lectr, George Washington Univ, 65-68; mem tsunami comt, Int Union Geod & Geophys; mem tides & mean sea level comt, Int Asn Phys Sci of the Ocean. *Mem:* Am Geophys Union. *Res:* Seismic sea waves; tides; currents; tsunami; earth tides. *Mailing Add:* Inst Geophys & Planetary Phys Univ Calif San Diego A-025 La Jolla CA 92093

ZETLMEISL, MICHAEL JOSEPH, b Baltimore, Md, Feb 26, 42; m 71; c 5. ELECTROCHEMISTRY. *Educ:* Spring Hill Col, BS, 66; Marquette Univ, MS, 67; St Louis Univ, PhD(chem), 71. *Prof Exp:* RES CHEMIST & PROJ LEADER CORROSION & ELECTROCHEM, PETROLITE CORP, 71-, FEL, ELECTROCHEM, 85- *Mem:* Am Chem Soc; Sigma Xi. *Res:* Electrochemistry of high temperature melts as related to corrosion of metals in gas turbines and boilers; refinery corrosion problems. *Mailing Add:* Petrolite Corp 369 Marshall Ave St Louis MO 63119

ZETTEL, LARRY JOSEPH, b Detroit, Mich, Sept 12, 44. COMPUTER SCIENCE EDUCATION. *Educ:* Univ Detroit, BS, 65; Mich State Univ, MS, 66, PhD(math), 70; Univ NMex, MS, 77. *Prof Exp:* Asst prof & chmn dept, Muskingum Col, 69-77, assoc prof math, 77-80; assoc prof, 80-85, PROF COMPUT SCI, LORAS COL, 85- *Mem:* Math Asn Am; Asn Comput Mach; Inst Elec & Electronics Engrs Comput Soc. *Res:* Non-associative algebras; computer usage in undergraduate instruction. *Mailing Add:* Dept Math & Comp Sci Loras Col 1450 Alta Vista Dubuque IA 52001

ZETTER, BRUCE ROBERT, b Providence, RI, Dec 23, 46; m; c 1. CELL & TUMOR BIOLOGY, VASCULARIZATION. *Educ:* Brandeis Univ, BA, 68; Univ RI, PhD(biol), 74. *Prof Exp:* Fel, Mass Inst Technol, 74-76; fel, Salk Inst, 76-77; asst res biochemist, Univ Calif Med Ctr, San Francisco, 77-78; asst prof, 78-85, ASSOC PROF, HARVARD MED SCH, 85- *Concurrent Pos:* Res assoc, Children's Hosp Med Ctr, Boston, Mass, 78- *Honors & Awards:* Fac Res Award, Am Cancer Soc, 83, Merit Award, 88. *Mem:* Am Soc Cell Biol. *Res:* Biology of the cells that comprise the vasculature, the vascular smooth muscle and endothelial cells; interactions of blood vessels with growing tumors with a special interest in the migration of capillary cells toward tumors; mechanisms of tumor metastasis including cell adhesion and cell migration. *Mailing Add:* Dept Surg Harvard Med Sch 25 Shattuck St Boston MA 02115

ZETTL, ANTON, b Gakovo, Yugoslavia, apr 25, 35; US citizen; m 64; c 2. MATHEMATICS. *Educ:* Ill Inst Technol, BS, 59; Univ Tenn, MA, 62, PhD(math), 64. *Prof Exp:* From asst to assoc prof math, La State Univ, Baton Rouge, 64-69; assoc prof, 69-73, PROF MATH, NORTHERN ILL UNIV, 73- *Concurrent Pos:* NASA res grants, 65-67; res mem, Math Res Ctr, Univ Wis, 67-68; vis res fel, Univ Dundee, Scotland, 74-75; Brit Sci Res Coun res grant, 74-75; NSF res grants, 74-75, 75-76 & 76-77; vis res scientist, appl math div, Argonne Nat Lab, 81-82 & 86-87, spec term app consult, 82-86 & 87-88; chmn, Math Northern Ill Univ, 83-86. *Mem:* Am Math Soc; Soc Indust & Appl Math; Math Asn Am; Ger Asn Appl Math & Mech. *Res:* Differential equations; differential operators; norm inequalities for operators. *Mailing Add:* Dept Math Northern Ill Univ De Kalb IL 60115

ZETTLEMOYER, ALBERT CHARLES, colloid chemistry, surface chemistry; deceased, see previous edition for last biography

ZETTLER, FRANCIS WILLIAM, b Easton, Pa, Aug 13, 38; m 61; c 2. PLANT PATHOLOGY, ENTOMOLOGY. *Educ:* Pa State Univ, BS, 61; Cornell Univ, MS, 64, PhD(plant path), 66. *Prof Exp:* From asst prof to assoc prof, 66-75, PROF, UNIV FLA, 75- *Mem:* Am Phytopath Soc. *Res:* Transmission of plant viruses. identification, characterization and control of plant viruses; virus diseases of ornamental plants; research involving edible root crops in the family Araceae. *Mailing Add:* Dept Plant Path Univ Fla Gainesville FL 32601

ZETTNER, ALFRED, b Laibach, Yugoslavia, Nov 21, 28; US citizen; m 59; c 2. CLINICAL PATHOLOGY. *Educ:* Graz Univ, MD, 54. *Prof Exp:* Asst prof path & clin path, Yale Univ, 63-67, assoc prof clin path, 67-68, dir dept clin micros, 63-68; PROF PATH & HEAD DIV CLIN PATH, UNIV CALIF, SAN DIEGO, 68- *Concurrent Pos:* NIH trainee clin path, Yale Univ, 61-63. *Honors & Awards:* Gerald T Evans Award, 79. *Mem:* AAAS; Am Soc Clin Path; Am Fedn Clin Res; Acad Clin Lab Physicians & Sci; Am Asn Clin Chem. *Res:* Competitive binding assays; folates in human serum. *Mailing Add:* 6011 Beaumont Ave La Jolla CA 92037

ZEVNIK, FRANCIS C(LAIR), b Joliet, Ill, Jan 1, 22; m 45; c 5. CHEMICAL ENGINEERING. *Educ:* Univ Wis, BS, 43, MS, 47. *Prof Exp:* Chem engr, Sharples Chem, 43-44; chem supvr, E I du Pont de Nemours & Co, Inc, 47-53; sr engr, C F Braun & Co, 53-56; consult, E I du Pont de Nemours & Co, Inc, 56-65, sr res supvr, 65-69, sr design consult, 70-72; dir res eng, 72-80, Fibre Prod Lab, 80-85, CORP TECH DIR, KENDALL CO, 85- *Mem:* Am Inst Chem Engrs; Am Chem Soc; Tech Asn Pulp & Paper Indust. *Res:* Micro-porous structures and complex multi-component structures employing micro-porous materials; chemical engineering design; engineering evaluations and economics; nonwoven and spunbonded textile sheet structures. *Mailing Add:* 2800 Newport Gap Pike Wilmington DE 19808

ZEVOS, NICHOLAS, b Manchester, NH, June 24, 32; m 66; c 2. PHYSICAL CHEMISTRY. *Educ:* St Anselm's Col, BA, 54; Univ NH, PhD(chem), 63. *Prof Exp:* Instr chem, Univ NH, 63-64; res fel radiation chem, Sloan Kettering Inst, 64-66; fel, Cornell Univ, 66-68; from asst prof to assoc prof, 69-82, CHMN DEPT CHEM, STATE UNIV NY COL POTSDAM, 78- *Concurrent Pos:* Res assoc, Danish Nat Labs, Roskilde, Denmark, Radiation Lab, Univ Notre Dame & Chem Dept, Brookhaven Nat Labs. *Mem:* AAAS; Am Chem Soc; Sigma Xi; InterAm Photochem Soc. *Res:* Gas phase kinetics; radiation and photochemistry. *Mailing Add:* RD 5 Box 390 Potsdam NY 13676

ZEWAIL, AHMED H, b Egypt, Feb 26, 46; m; c 2. LASERS, SOLAR ENERGY. *Educ:* Univ Alexandria, Egypt, BS, 67, MS, 69; Univ Pa, PhD(chem), 74. *Prof Exp:* Instr & researcher chem, Univ Alexandria, 67-69; res fel chem physics, Univ Calif, Berkeley, 74-75, IBM res fel, 75-76; from asst prof to prof, 76-89, LINUS PAULING PROF CHEM PHYSICS, CALIF INST TECHNOL, 90- *Concurrent Pos:* Distinguished vis lectr, Univ Tex, Austin, 77 & Am Chem Soc, Wayne State Univ, 85; chmn, Conf Advan Laser Spectros, San Diego, 77, 29th Ann Conf Mod Spectros, Calif, 82 & Int Conf Photochem & Photobiol, Egypt, 83; Alfred P Sloan Found fel, 78-82; mem, Panel US-France Coop Sci Prog, 79, Comt Infrared & Raman Spetros, Int Union Pure & Appl Chem, 81-, Int Sci Comt Conf Recent Advan Molecular Reaction Dynamics, 85, Phys Chem Workshop, NSF, 87 & Tech Prog Comt, Int Quantum Electronics Conf, 90; John van Geuns vis prof, Univ Amsterdam, Neth, 79; Camille & Henry Dreyfus teacher-scholar award, 79-85; ed, Laser Chem Int J, 80-85, J Phys Chem, 86, 88 & 91, Chem Physics Lett, 90 & 91-; vis prof, Univ Bordeaux, France, 81, Ecole Normale Superieure, France, 83 & Am Univ Cairo, 88; Alexander von Humboldt sr US scientist award, 83; NSF award, 84-86 & 88-90; John S Guggenheim Mem Found fel, 87; Rolf Sammet vis prof, Johann Wolfgang Coethe-Univ, Ger, 90; Christensen prof fel, St Catherine's Col, Oxford, 91. *Honors & Awards:* Buck-Whitney Medal, Am Chem Soc, 85, Harrison Howe Award, 89; W Albert Noyes Jr Mem Lectr, Univ Rochester, 87; Jean Day Lectr, Rutgers Univ, 88; Francis E Blacet Lectr, Univ Calif, Los Angeles, 88; Eyring Lectr, Ariz State Univ, 88 & 89; Harry Emmett Gunning Lectr, Univ Alta, 89; King Faisal Int Prize in Sci, 89; Earnest C Watson Lectr, Calif Inst Technol, 90; Max T Rogers Lectr, Mich State Univ, Am Chem Soc, 90; Reilly Lectr, Univ Notre Dame, 90; Hoechst Prize, 90; Flygare Mem Lectr, Univ Ill, 90; Sir Cyril Hinshelwood Chair Lectr, Oxford, 91; Jacob Bigeleisen Endowed Lectr, Stony Brook, 91; Richard B Bernstein Mem Lectr, Univ Calif, Los Angeles, 91. *Mem:* Nat Acad Sci; Am Inst Physics; InterAm Photochem Soc; Am Chem Soc; Am Phys Soc; Soc Photo-Optical Instrumentation Engrs; Optical Soc Am. *Res:* Nonlinear laser spectroscopy; radiationless processes in molecules; energy transfer in solids; picosecond spectroscopy; solar photovoltaic conversion and laser-induced chemistry; awarded one US patent. *Mailing Add:* Dept Chem Calif Inst Technol Arthur Amos Noyes Lab Chem Physics Mail Code 127-72 Pasadena CA 91125

ZEY, ROBERT L, b California, Mo, Sept 19, 32; m 60; c 2. ORGANIC CHEMISTRY. *Educ:* Cent Methodist Col, AB, 54; Univ Nebr, MS, 59, PhD(chem), 61. *Prof Exp:* Res chemist, Mallinckrodt Chem Works, 61-65; from asst prof to assoc prof, 65-72, PROF CHEM, CENT MO STATE UNIV, 72- *Mem:* Am Chem Soc. *Res:* Cinnolines; aziridinones; x-ray contrast media. *Mailing Add:* RR 1 Warrensburg MO 64093-9801

ZEYEN, RICHARD JOHN, b Mankato, Minn, Jan 17, 43; m 84. PLANT PATHOLOGY, HOST-PARASITE INTERACTIONS. *Educ:* Mankato State Univ, BS, 65, MS, 67; Univ Minn, St Paul, PhD(plant path & physiol), 70. *Prof Exp:* Asst gen biol, Mankato State Univ, 65-67; acad fel, 67-70, res assoc, 70-73, from asst prof to assoc prof, 73-82, PROF, DEPT PLANT PATH, UNIV MINN, ST PAUL, 82- *Concurrent Pos:* Consult, Minn Pollution Control Agency, 77; vis scientist, Imp Col Sci & Technol, London, 80-81 & Tel Aviv Univ, 81. *Mem:* Am Inst Biol Sci; Am Phytopath Soc; AAAS. *Res:* Electron optics; virus-vector relations; histopathology of plants; viral epidemiology; host-parasite relationships, physiology, in situ microanalysis of plant cell responses to fungal parasite attack; electron optical facility director. *Mailing Add:* 495 Borlaug Hall Dept Plant Path Univ Minn St Paul MN 55108

ZFASS, ALVIN MARTIN, b Norfolk, Va, Mar 30, 31; m 63; c 1. GASTROENTEROLOGY, HEPATOLOGY. *Educ:* Univ Va, BA, 53; Med Col Va, MD, 57. *Prof Exp:* Intern med, Bellevue-Cornell Med Ctr, 57-58; resident, Manhattan Vet Hosp, 58-60; mem, Sloan-Kettering Cancer Ctr, 60-61; from instr to assoc prof med, 63-77, DIR ENDOSCOPY, VA COMMONWEALTH UNIV, MED COL, VA, 63-, PROF MED, 77- *Concurrent Pos:* Fel gastroenterol, Manhattan Vet Hosp, 61-63; sabbatical Sloan-Kettering Cancer Clin, 81-82. *Honors & Awards:* Am Cancer Soc Prof Educ Award. *Mem:* Am Gastroenterol Asn; Am Soc Gastroenterol Endoscopy; fel Am Col Physicians; fel Am Col Gastroenterol. *Res:* Smooth muscle physiology of the esophagus; gastrointestinal hormones; biliary endoscopy; laser therapy. *Mailing Add:* Dept Med-Gastroenterol Va Commonwealth Univ Med Col MCV Sta Box 565 Richmond VA 23298

ZGANJAR, EDWARD F, b Virginia, Minn, July 31, 38; m 60; c 4. NUCLEAR PHYSICS. *Educ:* St John's Univ, Minn, BS, 60; Vanderbilt Univ, MS, 63, PhD(physics), 66. *Prof Exp:* AEC fel, Vanderbilt Univ, 60-62 & 64-65, Nat Reactor Testing Sta, 65-66; from asst prof to assoc prof, 70-75, PROF PHYSICS, LA STATE UNIV, BATON ROUGE, 75- *Concurrent Pos:* Sabbatical leave, Oak Ridge Nat Lab, 73-74; chmn, exec comt, UNISOR, 73-75 & 79-80; mem syst nuclear energy comt & radiation safety comt, La State Univ, 75-; mem exec comt, Hollifield Heavy Iron Res Facility Users Groups; invited vis scientist, Brookhaven Nat Lab, 76 & 77 & GSI, Darmstadt, WGer, 81-82. *Mem:* Fel Am Phys Soc; Sigma Xi; Am Chem Soc. *Res:* Low energy nuclear spectroscopy; experimental nuclear structure. *Mailing Add:* Dept Physics La State Univ 243B Nicholson Hall Baton Rouge LA 70803

ZHANG, BING-RONG, b Shanghai, China, Apr 15, 44; m 70; c 2. CHEMICAL VAPOR DEPOSITION, DIELECTRIC THIN FILM. *Educ:* Nanjing Univ, China, MS, 82; Colo State Univ, MS, 87. *Prof Exp:* Engr, Nanjing Commun Works, 78-79; grad res asst & teaching asst elec eng, Colo State Univ, 84-90; DIR & MGR RES & DEVELOP, FAITH INT CORP USA, 90-; INSTR PHYSICS, FRONT RANGE COMMUNITY COL, 91- *Concurrent Pos:* Consult, Faith Int Corp, Saudi Arabia & Kuwait, 91- *Mem:* Am Physics Soc. *Res:* Chemical vapor deposition of dielectric thin films and diagnostics for semiconductors; magnetism and magnetic materials; infrared measurement system. *Mailing Add:* Faith Int 612 S College Ave Suite 1 Ft Collins CO 80524

ZHANG, JOHN ZENG HUI, b Shanghai, China, Feb 15, 61. GAS-PHASE CHEMICAL REACTION, GAS-SURFACE INTERACTIONS. *Educ:* EChina Normal Univ, BS, 82; Univ Houston, PhD(chem physics), 87. *Prof Exp:* Postdoctoral chem physics, Univ Calif, Berkeley, 87-90; ASST PROF CHEM, DEPT CHEM, NY UNIV, 90- *Concurrent Pos:* Camille & Henry Dreyfus Found new fac award, 90. *Mem:* Am Chem Soc; Am Phys Soc. *Res:* Chemical reaction dynamics; molecular collision dynamics; molecule-surface interactions; chemical reaction on metal surfaces; catalysis. *Mailing Add:* Dept Chem NY Univ New York NY 10003

ZHANG, QIMING, b Shaoxin, Zhejiang, China, Jan 14, 57; m 86. SOLID STATE PHYSICS. *Educ:* Nanjing Univ, BS, 81; Pa State Univ, PhD(physics), 86. *Prof Exp:* RES ASSOC MAT SCI, MAT RES LAB, PA STATE UNIV, 86- *Mem:* Am Phys Soc. *Res:* Phase transitions in the reduced dimensional systems; interface phenomena; ferroelectric phase transitions and their applications. *Mailing Add:* Brookhaven Nat Lab Bldg 555 Upton NY 11973

ZHENG, XIAOCI, b Shanghai, China, Apr 21, 49; m 81; c 1. IMPROVEMENT OF WEAR RESISTANCE, IMPROVEMENT OF CORROSION RESISTANCE. *Educ:* Qufu Norm Univ, China, BS, 81; Univ Sci & Technol, China, MS, 84; Univ Wis-Madison, PhD(mat sci), 91. *Prof Exp:* Res asst, Inst Elec Eng, Acad Sinica, 82-84; lectr, Dept Appl Physics, Tsinghua Univ China, 84-86; res asst, Ctr Plasma Aided Mfg, Univ Wis-Madison, 86-91; ASSOC SCIENTIST, UNITED TECHNOLOGIES, 91- *Mem:* Am Phys Soc; Am Soc Metals; Mat Res Soc. *Res:* Surface modification of materials by coating and ion beam techniques; high vacuum; microscopy; surface analysis. *Mailing Add:* 804 Eagle Heights Apt C Madison WI 53705

ZHENG, YUAN FANG, b Shanghai, China, July 2, 46; m 76; c 1. ROBOTICS & AUTOMATION, MULTI-SENSOR INTERACTION. *Educ:* Tsinghua Univ, BS, 70; Ohio State Univ, MS, 80, PhD(elec eng), 84. *Prof Exp:* From asst prof to assoc prof computer & robotics, Clemson Univ, 84-89; ASSOC PROF COMPUTER & ROBOTICS, OHIO STATE UNIV, 89- *Concurrent Pos:* Prin investr, Savannah River Lab, 85-88, NSF, 88- & Off Naval Res, 90-; NSF presidential young investr award, 87. *Mem:* Sr mem Inst Elec & Electronics Engrs. *Res:* Sensor integrated robotics system for advanced manufacturing; computer network for real-time applications; neural computing for intelligent mobile robots; vision for surface reconstruction. *Mailing Add:* Dept Elec Eng Ohio State Univ Columbus OH 43210-1272

ZHOU, SIMON ZHENG, b Shanghai, China, Jan 31, 42; US citizen. GAS LASER, PHOTOCHEMISTRY. *Educ:* Shanghai Univ Sci & Technol, Bachelor, 65, Master, 67. *Prof Exp:* Scientist, Shanghai Inst Laser Technol, 74-87; SR SCIENTIST, FLOROD CORP, 88- *Mem:* Soc Photo-Optical Instrumentation Engrs. *Res:* Developing first compact sealed off excimer laser with long gas lifetime for industry and scientific application; laser induced metal deposition technology and system for semiconductor industry applications. *Mailing Add:* 3220 Merill Dr No 36 Torrance CA 90503

ZIA, PAUL Z, b Changchow, China, May 13, 26; US citizen; m 51; c 2. STRUCTURAL ENGINEERING. *Educ:* Nat Chiao Tung Univ, BSCE, 49; Univ Wash, MSCE, 52; Univ Fla, PhD, 60. *Prof Exp:* Struct designer, Lakeland Eng Assocs, Inc, 51-53, proj engr, 53-54, secy & chief struct engr, 54-55, vpres & chief struct engr, 55; from instr to asst prof civil eng, Univ Fla, 55-61; assoc prof, NC State Univ, 61-64; vis assoc prof, Univ Calif, Berkeley, 64-65; assoc head dept, NC State Univ, 67-79, prof civil eng, 65-88, head dept, 79-88, DISTINGUISHED UNIV PROF CIVIL ENG, NC STATE UNIV, 88- *Concurrent Pos:* Vpres, Am Concrete Inst, 87-89, pres, 89-90; mem, exec comt, Transportation Res Bd, 88-91, vchmn, Concrete & Structures Adv Comt, Strateg Hwy Res Prog, Nat Res Coun, 87-89. *Honors & Awards:* Martin P Korn Award, Prestressed Concrete Inst, 74; T Y Lin Award, Am Soc Civil Engrs, 75, Raymond C Reese Award, 76; Western Elec Fund Award, Am Soc Eng Educ, 76; Joe W Kelly Award, Am Concrete Inst, 84; Lamme Award, Am Soc Eng Educ, 86. *Mem:* Nat Acad Eng; hon mem & fel Am Soc Civil Engrs; Prestressed Concrete Inst; Am Soc Eng Educ; fel Am Concrete Inst. *Res:* Mechanical properties of high performance concrete and prestressed concrete structures. *Mailing Add:* Dept Civil Eng NC State Univ Campus Box 7908 Raleigh NC 27695-7908

ZIA, ROYCE K P, b Shaoyang, China, Dec 1, 43; US citizen; m 71. THEORETICAL PHYSICS. *Educ:* Princeton Univ, AB, 64; Mass Inst Technol, PhD(physics), 68. *Prof Exp:* NATO fel physics, Europ Orgn Nuclear Res, Switz, 68-69; res fels, Univ Birmingham, Eng, 69-72, Rutherford High Energy Lab, 72-73 & Univ Southampton, 73-76; from asst prof to assoc prof, 76-83, PROF, DEPT PHYSICS, VA POLYTECH INST & STATE UNIV, 83- *Concurrent Pos:* Danforth assoc; Alexander von Humboldt fel. *Mem:* Sigma Xi. *Res:* Statistical and high energy physics; phase transitions and critical phenomena; applications of renormalization group analysis to interfacial properties and non-equilibrium steady state systems. *Mailing Add:* Dept Physics Va Polytech Inst Blacksburg VA 24061

ZIAUDDIN, SYED, atmospheric physics, space physics, for more information see previous edition

ZIBOH, VINCENT AZUBIKE, b Warri, Nigeria, Apr 21, 32; m 62; c 3. BIOCHEMISTRY. *Educ:* Doane Col, AB, 58; St Louis Univ, PhD(biochem), 62. *Prof Exp:* Res fel neurochem, Ill State Psychiat Inst, Chicago, 62-64; lectr chem path, Med Sch, Univ Ibadan, 64-67; res assoc dermat, Sch Med, Univ Miami, 67-69, asst prof dermat & biochem, 69-; AT DEPT DERMAT, SCH MED, UNIV CALIF, DAVIS. *Concurrent Pos:* WHO fel clin chem, Bispebjerg Hosp, Copenhagen, Denmark, 66. *Mem:* AAAS; Brit Biochem Soc; Am Chem Soc; Soc Invest Dermat. *Res:* Biochemistry of lipids and steroids; regulation of lipogenesis from glucose in skin; biosynthesis and biochemical basis of prostaglandin action in the skin. *Mailing Add:* Prof Bldg 205 Univ Calif 4301 X St Sacramento CA 95817

ZIC, ERIC A, b New York, NY, Apr 14, 62. INSTRUMENTATION, TEST ENGINEERING. *Educ:* NY Inst Technol, BS, 86. *Prof Exp:* ENGR INSTRUMENT TESTING, GRUMMAN SPACE, 86- *Res:* Infra red technology; control systems-PID or fuzzy logic; vibration static and cryogenic test technology. *Mailing Add:* 42 W Fordham Rd Bronx NY 10468

ZICCARDI, ROBERT JOHN, b New York, NY, Dec 25, 46; m 69. BIOCHEMISTRY, IMMUNOLOGY. *Educ:* Univ Conn, BA, 68; Univ Calif, Los Angeles, PhD(biochem), 73. *Prof Exp:* ASST MEM MOLECULAR IMMUNOL, SCRIPPS CLIN & RES FOUND, 73- *Res:* Complement protein structure, function and interaction with biological membranes. *Mailing Add:* Dept Molecular Immunol Scripps Clin & Res Found 10666 N Torrey Pines Rd La Jolla CA 92037

ZICKEL, JOHN, b Munich, Ger, Feb 9, 19; US citizen; m 42; c 2. MACHINE DESIGN. *Educ:* Lehigh Univ, BS, 48, MS, 49; Brown Univ, PhD(appl math), 53. *Prof Exp:* Specialist, struct anal nuclear reactors, Knolls Atom Power Lab, 52-55, Gen Elec Atomic Power Dept, 55-58; mgr struct res rocket dept, Aerojet Gen Corp, 58-67; prof mech eng, 67-81, CHMN MECH ENG

DEPT, CALIF STATE UNIV, SACRAMENTO, 81- *Concurrent Pos:* Expert witness litigations. *Honors & Awards:* Teeter Award, Soc Automotive Eng, 78. *Mem:* Fel Am Soc Mech Eng; Soc Exp Mech; Sigma Xi; Am Soc Eng Educ; Soc Automotive Eng. *Res:* Automobile collisions; probabilistic design; product liability. *Mailing Add:* 6063 Ranger Way Carmichael CA 95608

ZICKER, ELDON LOUIS, b Milwaukee, Wis, Mar 12, 20; m 54; c 5. SOIL SCIENCE. *Educ:* Univ Wis, BS, 48, PhD(soils), 55. *Prof Exp:* Instr soil surv, Kans State Univ, 49; instr geol & soils, Yakima Valley Col, 59-64; dean, Sch Agr & Home Econ, 70-80, PROF, DEPT PLANT & SOIL SCI, CALIF STATE UNIV, CHICO, 80-, INSTR SOILS, 64-, DIR, SCH AGR, 86- *Mem:* Am Soc Agron; Soil Sci Soc Am; Brit Soc Soil Sci; Sigma Xi; Soil Conserv Soc. *Res:* Soil genesis and development; fertilizers and soil fertility. *Mailing Add:* Plant/Soil Sci Dept Chico State Col First & Normal St Chico CA 95929

ZIDEK, JAMES VICTOR, b Acme, Alta, Sept 26, 39; m 61. MATHEMATICAL STATISTICS. *Educ:* Univ Alta, BSc, 61, MSc, 63; Stanford Univ, PhD(math statist), 67. *Prof Exp:* Lectr probability & statist, Univ Alta, 62-63; from asst prof to assoc prof probability & statist, 67-76, head dept, 84-89, PROF STATIST, UNIV BC, 76- *Concurrent Pos:* Hon res fel, Univ Col London, 71-82; vis sr res scientist, Commonwealth Sci & Indust Orgn, Australia, 76-77; mem, statist grant selection comt, NSERC, 78-81; sr assoc ed, Can J Statist, 80-83; prof statist, Univ Wa, 83-84; vis prof, Imp Col London, 83-84; vis scientist, Coun Sci & Indust Res, 83; mem, adv comt methodology, Statist Can, 85-88; Izaak Walton Killam Mem Fel, 89-90. *Mem:* Fel Inst Math Statist; Stat Soc Can (pres-elect 86, pres, 87); Royal Statist Soc; Am Statist Asn; Int Statist Inst; Int Asn Survey Statist. *Res:* Statistical decision theory; environmetrics. *Mailing Add:* Dept Statist 2021 W Mall Univ BC Vancouver BC V6T 1W5 Can

ZIDULKA, ARNOLD, b Montreal, Que, July 9, 41. RESPIRATORY RESEARCH. *Educ:* McGill Univ, BSc, 62 & MD, 66; Sir George Williams Univ, BA, 64; FRCP(C), 71; Am Bd Internal Med, dipl, 72; dipl respiratory dis, 74. *Hon Degrees:* FCCP, Am Col Chest Physicians, 79. *Prof Exp:* Asst prof, 72-77, ASSOC PROF MED, MCGILL UNIV, 77-; RES DIR RESPIRATORY, MEAKINS-CHRISTIE LABS, 82- *Concurrent Pos:* Internist, pulmonary consult, Queen Mary Vet Hosp, 72-77, Montreal Chest Hosp & Royal Victoria Hosp, 72- & Montreal Gen Hosp, 77-; dir respiratory technol, Montreal Gen Hosp, 82- & Montreal Chest Hosp, 83- *Mem:* Can Soc Clin Invest; Am Physiol Soc; Royal Col Physicians & Surgeons Can; Am Thoracic Soc; Am Col Chest Physicians; Can Lung Asn. *Res:* Pleural pressure measurements; lung chest-wall interdependence; cardiopulmonary resuscitation; attaining ventilation by high-frequency chest wall compression; chest physiotherapy by high frequency chest wall compression; atelectasis induced by chest physiotherapy; negative pressure ventilation; pulmonary wedge pressure; mechanics and gas exchange; obstructive sleep apnea. *Mailing Add:* Montreal Gen Hosp 1650 Cedar Ave Rm 784 Montreal PQ H3G 1A4 Can

ZIEBARTH, TIMOTHY DEAN, b Glendive, Mont, June 10, 46; m 66; c 2. ORGANIC CHEMISTRY, PHYSICAL ORGANIC CHEMISTRY. *Educ:* Mont State Univ, BS, 69; Ore State Univ, PhD(org chem), 73. *Prof Exp:* Fel org chem, Univ Colo, 72-74; CHIEF CHEMIST, HAUSER LABS, 74- *Mem:* Am Chem Soc. *Res:* Photochemistry of allylic amines; novel methods of polymer analysis; forensic chemistry. *Mailing Add:* 4641 Huey Circle Boulder CO 80303

ZIEBUR, ALLEN DOUGLAS, b Shawano, Wis, May 1, 23; m 49; c 2. MATHEMATICS. *Educ:* Univ Wis, PhD, 50. *Prof Exp:* From instr to assoc prof math, Ohio State Univ, 51-61; assoc prof, State Univ NY Binghamton, 61-63, prof math, 63-88; RETIRED. *Mem:* Am Math Soc; Math Asn Am; Soc Indust & Appl Math. *Res:* Differential equations. *Mailing Add:* 48 Kendall Ave Binghamton NY 18903

ZIEG, ROGER GRANT, b McCooke, Nebr, Aug 16, 39; m 70; c 2. PARASITOLOGY, PROTOZOOLOGY. *Educ:* Univ Nebr, BS, 61, MS, 63; Iowa State Univ, PhD(develop biol), 81. *Prof Exp:* In-charge res asst, Tissue Cult Collection, Am Type Cult Collection, 66-68, Protozoan Collection, 68-75; teaching asst biol & phys biochem, Iowa State Univ, 75-80; res fel cell biol, Univ Minn, 80-84; RES SCIENTIST, PETO SEED INC, 84- *Mem:* AAAS; Sigma Xi. *Res:* Corn and soybean tissue culture; morphogenesis in plant cell cultures; isolation of secondary products from plant tissue cultures; protoplast isolation and fusion. *Mailing Add:* Peto Seed Inc Hwy 16 Woodland CA 95695

ZIEGENHAGEN, ALLYN JAMES, b Wautoma, Wis, Sept 5, 35; m 80. CHEMICAL ENGINEERING, ECONOMICS ENGINEERING. *Educ:* Univ Wis, BS, 57, PhD(fluid mech, rheol, chem eng), 62; Mass Inst Technol, SM, 59. *Prof Exp:* Asst prof physics, Univ Wis, Milwaukee, 62-64; vis prof chem eng, Univ Valle, Colombia, 64-65; asst prof eng, Univ Wis, Milwaukee, 65-66; engr, Shell Develop Co, 66-70; sr res engr, Stauffer Chem Co, 70-80, res assoc, process develop dept, 80-84; SR SCIENTIST, CROWN ZELLERBACH, 85- *Mem:* Am Inst Chem Engrs. *Res:* New process research and development in chlorinated hydrocarbons, agricultural chemicals and water pollution control; computer applications; catalyst development; food ingredient extraction from seaweed; exploratory chemical research related to plastic films. *Mailing Add:* Packaging Res & Develop Crown Zellerbach 2199 Williams St San Leandro CA 94577

ZIEGER, HERMAN ERNST, b Philadelphia, Pa, May 17, 35; m 60; c 3. ORGANIC CHEMISTRY. *Educ:* Muhlenberg Col, BS, 56; Pa State Univ, PhD(org chem), 61. *Prof Exp:* Fulbright scholar, Univ Heidelberg, 60-61; from instr to assoc prof, 61-73, PROF CHEM, BROOKLYN COL, 73- *Concurrent Pos:* Vis assoc, Calif Inst Technol, 67-68; Alexander von Humboldt Found fel, 74-75; mem doctoral fac, City Univ New York. *Mem:* Am Chem Soc; Sigma Xi. *Res:* Organolithium chemistry; aryne chemistry; radical anion processes; kinetics and stereochemistry of carbanion coupling processes; fourier transform nuclear magnetic resonance spectroscopy (FT-NMR) in organic structure determination. *Mailing Add:* Dept Chem Brooklyn Col Brooklyn NY 11210

ZIEGLER, ALFRED M, b Boston, Mass, Apr 23, 38; m 64, 86; c 2. PALEONTOLOGY, STRATIGRAPHY. *Educ:* Bates Col, BSc, 59; Oxford Univ, DPhil(paleont), 64. *Prof Exp:* Fel paleont, Calif Inst Technol, 64-66; asst prof, 66-72, ASSOC PROF PALEONT, UNIV CHICAGO, 72-, PROF STRATIG, 76- *Concurrent Pos:* NSF grant, 72-74 & 87. *Mem:* Geol Soc Am; Brit Palaeontol Asn; Sigma Xi. *Res:* Paleontology, paleoecology and stratigraphy of the Silurian age deposits of the British Isles, Norway and eastern North America and world paleogeography. *Mailing Add:* 5492 E Everett Ave Apt 1 Chicago IL 60615-5923

ZIEGLER, CAROLE L, b Chicago, Ill, Sept 12, 46; m 70; c 2. EARTHQUAKE DISASTER PREPAREDNESS. *Educ:* Mundelein Col, BS, 68; George Washington Univ, MS, 79. *Prof Exp:* Qual control chemist, Witco Chem Co, 69-70; chemist, Midwest Res Inst, 71-73; res assoc, Univ Tex Med Br, 80; RES ASSOC, UNIV CALIF, SAN DIEGO, 80-; CONSULT, INST HIGH BLOOD PRESSURE RES, 82- *Concurrent Pos:* Instr environ geol, San Diego State Univ, 87; lectr environ geol, Univ San Diego, 87-; mem & consult, Off Disaster Preparedness, San Diego County-Earthquake Awareness Comt, 90- *Res:* Mineralogy of sediments; trace element analysis using atomic absorption in weathered rock sequences; concentrations of PCBs in the environment using gas- chromatograph techniques. *Mailing Add:* 342 W Lewis St San Diego CA 92103-1924

ZIEGLER, DANIEL, b Quinter, Kans, July 6, 27; m 52; c 4. BIOCHEMISTRY. *Educ:* St Benedict's Col, Kans, BS, 49; Loyola Univ, Ill, PhD(biochem), 55. *Prof Exp:* Fel enzyme chem, Inst Enzyme Res, Univ Wis, 55-59, asst prof, 59-61; assoc prof, 61-68, PROF CHEM, UNIV TEX, AUSTIN, 68-, MEM STAFF, CLAYTON FOUND BIOCHEM INST, 61- *Concurrent Pos:* Estab investr, Am Heart Asn, 60-65; career Develop Award, USPHS, 65-75. *Mem:* AAAS; Am Chem Soc; Am Soc Biol Chem; Sigma Xi; Am Soc Pharmacol & Exp Therapeuts; NY Acad of Sci. *Res:* Synthesis of protein hormones; mammalian mixed-function drug oxidases; flavoproteins of the electron transport system. *Mailing Add:* Clayton Found Biochem Inst Dept Chem Univ Tex Austin TX 78712

ZIEGLER, DEWEY KIPER, b Omaha, Nebr, May 31, 20; m 54; c 3. NEUROLOGY. *Educ:* Harvard Univ, BA, 41, MD, 45. *Prof Exp:* Assoc neurol, Col Physicians & Surgeons, Columbia Univ, 53-55; asst prof, Med Sch, Univ Minn, 55-56; assoc prof, 58-63, chmn dept, 73-85, PROF NEUROL, UNIV KANS MED CTR, KANSAS CITY, 63- *Concurrent Pos:* Consult, US Fed Hosp, Springfield, Mo & Vet Admin Hosp, Kansas City, Mo, 64-; clin prof neurol, Med Sch, Univ Mo-Kansas City; dir, Am Bd Psychiat & Neurol, 74-; mem comt stroke coun, Am Heart Asn, 81-83, mem exec comt. *Mem:* AMA; Soc Neurosci; Am Neurol Asn; fel Am Col Physicians; Am Epilepsy Soc; Am Acad Neurol, (pres, 79-81). *Res:* Epidemiology and natural history of cerebrovascular disease; neurophysiological and biochemical basis of migraine. *Mailing Add:* Dept Neurol Univ Kans Med Ctr Kansas City KS 66103

ZIEGLER, EDWARD N, b Bronx, NY, Aug 15, 38; m 74; c 2. CHEMICAL ENGINEERING, ENVIRONMENTAL ENGINEERING. *Educ:* City Col New York, BChE, 60; Northwestern Univ, MS, 62, PhD(chem eng), 64. *Prof Exp:* Res assoc transport phenomena in fluidized solid systs, Argonne Nat Lab, 62-63; process res engr, Esso Res & Eng Co, 64-65; asst prof chem eng, 65-72, ASSOC PROF CHEM ENG, POLYTECH UNIV, 72- *Concurrent Pos:* Ed, Encycl Environ Sci & Eng, 2nd Ed, 83, & Advan Series Environ Sci & Eng, 79-80; consult, Brookhaven Nat Lab, 74-80; proj engr, Consol Edison Co, 80. *Honors & Awards:* William H White Award. *Mem:* Am Inst Chem Engrs; Sigma Xi; Air & Waste Mgt Asn. *Res:* Applied reaction kinetics; air pollution control; coal conversion; fluidization. *Mailing Add:* Dept Chem Eng Polytech Univ 333 Jay St Brooklyn NY 11201

ZIEGLER, EKHARD E, b Saalfelden, Austria, Apr, 12, 40; m 62; c 3. PEDIATRICS, NEONATOLOGY. *Educ:* Univ Innsbruck, Austria, MD, 64. *Prof Exp:* From asst prof to assoc prof, 73-81, PROF PEDIAT, UNIV IOWA, 81- *Concurrent Pos:* Mem, nutrit study sect, NIH, 88-92. *Honors & Awards:* Nutrit Award, Am Acad Pediat, 88. *Mem:* Am Acad Pediat; Soc Pediat Res; Am Soc Clin Nutrit; Soc Exp Biol & Med; Nutrit Soc; Am Pediat Soc. *Res:* Pediatric nutrition; mineral metabolism; energy metabolism; growth and body composition; nutrition of the premature infant. *Mailing Add:* Dept Pediat Univ Hosp Iowa City IA 52242

ZIEGLER, FREDERICK DIXON, b Waynesboro, Pa, Oct 14, 33. BIOCHEMISTRY, PHYSIOLOGY. *Educ:* Shippensburg State Col, BS, 57; Pa State Univ, MS, 60, PhD(zool), 62. *Prof Exp:* Res trainee virol, Pa State Univ, 62-63; res assoc biochem, State Univ NY Buffalo, 63-65; asst prof, McMaster Univ, 65-69; res scientist, 69-73, SR RES SCIENTIST, DIV LABS & RES, NY STATE DEPT HEALTH, 73- *Mem:* Am Chem Soc; Can Biochem Soc; Sigma Xi. *Res:* Metabolism; coagulation. *Mailing Add:* NY State Dept Health Div Lab Res Albany NY 12209

ZIEGLER, FREDERICK EDWARD, b Teaneck, NJ, Mar 29, 38; m 62; c 2. ORGANIC CHEMISTRY. *Educ:* Fairleigh Dickinson Univ, BS, 60; Columbia Univ, MA, 61, PhD(chem), 64. *Prof Exp:* Eugene Higgins fel, Columbia Univ, 60-61, NIH fel, 61-64; NSF fel, Mass Inst Technol, 64-65; from asst prof to assoc prof, 65-78, PROF CHEM, YALE UNIV, 78- *Concurrent Pos:* Res career develop award, NIH, 73-78. *Mem:* Am Chem Soc; Royal Soc Chem. *Res:* Organic synthetic methods and natural products synthesis. *Mailing Add:* Dept Chem Yale Univ New Haven CT 06520

ZIEGLER, GEORGE WILLIAM, JR, b Cleveland, Ohio, Oct 12, 16; c 2. PHYSICAL CHEMISTRY. *Educ:* Monmouth Col, BS, 39; Ohio State Univ, PhD, 50. *Prof Exp:* Asst physics, Monmouth Col, 37-39; analyst, Grasselli Chem Div, E I du Pont de Nemours & Co, 39-40, tech supvr, 40-42, asst to chief plant technologist, 42-43; leader high temperature calorimetry group, Res Found, Ohio State Univ, 46-50; group leader process develop, Int Resistance Co, 50-51, sect head, 51-53, head dept, 53; res chemist & mem staff, Mat Control Div, Curtis Pub Co, 53-55; dir spec projs lab, AMP, Inc,

55-60, from proj engr to sr proj engr, 60-70, mgr spec prod, 70-76; EMER PROF, PA ACAD SCI, 85- *Concurrent Pos:* Assoc prof, Dickinson Col, 55-59; independent consult, 76- *Mem:* Am Chem Soc; Am Phys Soc; Am Soc Metals; Inst Elec & Electronics Engrs; Inst Mgt Sci. *Res:* High temperature thermodynamics and kinetics; instrumentation; metallurgy; conducting films; heat transfer; color correction theory; photoengraving; radio frequency connectors; aluminum building wire connectors. *Mailing Add:* 17 Circle Dr Carlisle PA 17013

ZIEGLER, HANS K(ONRAD), b Munich, Ger, Mar 1, 11; nat US; m 37; c 3. ELECTRONICS ENGINEERING. *Educ:* Munich Tech Univ, BS, 32, MS, 34, PhD(electronic eng), 36. *Prof Exp:* Asst prof elec eng, Munich Tech Univ, 34-36; chief res & develop dept, Rosenthal Isolatoren, Inc, 36-47; sci consult, US Army Signal Res & Develop Lab, 47-56, asst dir res, 56-58, dir astroelectronics div, 58-59, chief scientist, 59-66, tech dir, 63-66, dep for sci & chief scientist, US Army Electronics Command, 66-71, dir, US Army Electronics Technol & Devices Lab, 71-76; RETIRED. *Concurrent Pos:* Consult, Kohler Co, Wis, 51-55; consult, 76- *Mem:* Fel Am Astronaut Soc; fel Inst Elec & Electronics Engrs; Armed Forces Commun & Electronics Asn. *Res:* Military electronics; communications; space science and technology; energy sources, conversion and transmission; solar energy; geophysics; meteorology; research management. *Mailing Add:* 32 E Larchmont Dr Colts Neck NJ 07722

ZIEGLER, HARRY KIRK, b Greensburg, Pa, Mar 25, 51; m 85. MACROPHAGE FUNCTION, IMMUNE RESPONSE GENE CONTROL. *Educ:* Johns Hopkins Univ, BA, 73, PhD(microbiol), 78. *Prof Exp:* Teaching fel immunol, Med Sch, Harvard Univ, 78-80, instr, 80-81; asst prof, 81-87, ASSOC PROF IMMUNOL, EMORY MED SCH, 87- *Mem:* Am Asn Immunologists. *Res:* Mechanism of lymphocyte and macrophage function related to the regulation of immune responses. *Mailing Add:* Dept Microbiol & Immunol Emory Univ 1364 Clifton Rd NE Atlanta GA 30322

ZIEGLER, JAMES FRANCIS, b Apr 20, 39; US citizen; m 69. NUCLEAR PHYSICS. *Educ:* Yale Univ, BS, 57, MS, 65, PhD(nuclear physics), 67. *Prof Exp:* Mem res staff, 67-69, DIR HIGH ENERGY ACCELERATOR LAB, RES CTR, IBM CORP, 69- *Mem:* Am Phys Soc. *Res:* Solid state analysis using nuclear physics techniques; ion-induced x-rays; nuclear backscattering; nuclear channeling; ion-implantation; optical microcircuits. *Mailing Add:* T J Watson Res Ctr 28-0 PO Box 218 Yorktown Heights NY 10598

ZIEGLER, JOHN BENJAMIN, b Rochester, NY, Jan 2, 17; m 46; c 3. LEPIDOPTERA, ECOLOGY. *Educ:* Univ Rochester, BS, 39; Univ Ill, MS, 40, PhD(org chem), 46. *Prof Exp:* Jr chemist, Merck & Co, Inc, NJ, 40-43; asst chem, Univ Ill, 43-46; res chemist, J T Baker Chem Co, NJ, 46-48; assoc chemist, Develop Div, Ciba Pharmaceut Co, 48-52, supvr develop res labs, 52-62, mgr process res, 62-70, dir chem develop, Pharmaceut Div, Ciba-Geigy Corp, 70-75, dir process res, Pharmaceut Div, 75, sr staff scientist, 76-80; lab & bus admin, Depts Biol & Chem Physics, Seton Hall Univ, 80-82; RETIRED. *Mem:* Am Chem Soc; Sigma Xi; Lepidoptera Soc (treas, 50-53); Lepidoptera Res Found. *Res:* Lepidoptera: distribution, taxonomy, systematics, host plants. *Mailing Add:* 64 Canoe Brook Pkwy Summit NJ 07901-1434

ZIEGLER, JOHN HENRY, JR, b Altoona, Pa, Nov 10, 24; m 49; c 3. MEAT SCIENCE, ANIMAL SCIENCE. *Educ:* Pa State Univ, BS, 50, MS, 52, PhD(animal indust), 65. *Prof Exp:* Soil conservationist, USDA, 49; asst animal husb, Pa State Univ, 50-52; nutritionist, Near's Food Co, NY, 52-54; from instr animal indust to assoc prof meat sci, Pa State Univ, 54-76, prof meat sci, 76-87; RETIRED. *Mem:* Am Soc Meat Sci; Am Soc Animal Sci; Inst Food Technologists; Sigma Xi. *Res:* Meat animal carcass evaluation and utilization; basic adipose tissue anatomy and physiology; meat product development. *Mailing Add:* 200 W Whitehall Rd State College PA 16301

ZIEGLER, MICHAEL GREGORY, b Chicago, Ill, May 12, 46; c 2. PHARMACOLOGY, MEDICINE. *Educ:* Loyola Univ, Chicago, BS, 67; Med Sch, Univ Chicago, MD, 71. *Prof Exp:* Residency med, Med Sch, Univ Kans, 73; pharmacol res assoc, NIMH, 73-75, pharmacologist, 75-76; attend physician med & consult, NIH, 74-76; asst prof med & pharmacol, Univ Tex Med Br Galveston, 76-80; ASSOC PROF MED, UNIV CALIF, SAN DIEGO, 80- *Mem:* Am Fedn Clin Res; Am Soc Nephrology. *Res:* Clinical pharmacology; hypertension; sympathetic nervous system; neurotransmitters; catecholamines. *Mailing Add:* H-781-B Univ Hosp San Diego CA 92103

ZIEGLER, MICHAEL ROBERT, b York, Pa, Oct 31, 42; m 64; c 4. MATHEMATICAL ANALYSIS. *Educ:* Shippensburg State Col, BS, 64; Univ Del, MS, 67, PhD(math), 70. *Prof Exp:* Instr math, Univ Del, Georgetown Exten, 69-70; fel, Univ Ky, 70-71; ASSOC PROF MATH, MARQUETTE UNIV, 71- *Mem:* Am Math Soc; Math Asn Am. *Res:* Complex analysis; geometric function theory; univalent functions. *Mailing Add:* Dept Math & Comp Sci Marquette Univ 1515 W Wisconsin Ave Milwaukee WI 53233

ZIEGLER, MIRIAM MARY, b Gainesville, Fla, Dec 15, 45; m 78; c 2. PROTEIN CHEMISTRY, ENZYMOLOGY. *Educ:* Bucknell Univ, BS, 67; Harvard Univ, MA, 70, PhD(biochem), 72. *Prof Exp:* Res fel biol, Harvard Univ, 72-73, lectr, 73-75, res fel, 75-76; res fel biochem, Univ Ill, Urbana, 76-78, vis asst prof, 78-81; LECTR & RES SCIENTIST, DEPT BIOCHEM & BIOPHYS, TEX A&M UNIV, COLLEGE STATION, 81- *Mem:* Biophys Soc; Am Soc Microbiol. *Res:* Enzymatic mechanism, structure, and subunit function of bacterial luciferase; chemical modification of proteins. *Mailing Add:* Dept Biochem & Biophys Tex A&M Univ College Station TX 77843-2128

ZIEGLER, PAUL FOUT, b Baltimore, Md, Dec 3, 16; m 45; c 2. CHEMISTRY. *Educ:* Otterbein Col, BS, 39; Univ Cincinnati, MS, 47, PhD(chem), 63. *Prof Exp:* Analyst, Am Rolling Mills, Ohio, 39-40, chemist, 40-46; Auburn Univ, 49-82; RETIRED. *Concurrent Pos:* Vis prof, Univ PR, Mayaguez, 82-83. *Mem:* Am Chem Soc; Sigma Xi. *Res:* Analytical and organic chemistry; synthesis and rearrangement. *Mailing Add:* Rte 4 Box 301 Auburn AL 36830

ZIEGLER, PETER, b Vienna, Austria, Mar 26, 22; m 48. ORGANIC CHEMISTRY, BIOCHEMISTRY. *Educ:* Sir Geo Williams Col, BSc, 44; McGill Univ, PhD(biochem), 51. *Prof Exp:* Chemist, Frank W Horner, Ltd, 44-48; asst, McGill Univ, 48-51; group leader, Can Packers Ltd, 51-65, asst dir res, 66-83; RETIRED. *Mem:* Am Chem Soc; fel Chem Inst Can. *Res:* Fine chemicals; pharmaceuticals. *Mailing Add:* 330 Spadina Rd Apt 702 Toronto ON M5R 2V9 Can

ZIEGLER, RICHARD JAMES, b Norristown, Pa, May 30, 43; m 67. VIROLOGY. *Educ:* Muhlenberg Col, BS, 65; Temple Univ, PhD(microbiol), 70. *Prof Exp:* Instr microbiol, Med Sch, Temple Univ, 69; res assoc genetics, Rockefeller Univ, 70-71; asst prof microbiol, 71-77, ASSOC PROF MED MICROBIOL & IMMUNOL, MED SCH, UNIV MINN, DULUTH, 77- *Mem:* AAAS; Tissue Cult Asn; Am Soc Microbiol; Asn Am Med Cols. *Res:* Effects of neurotropic viral multiplication on central nervous system function and effects of herpes simplex virus replication on primary nerve tissue culture cells. *Mailing Add:* Dept Microbiol Med & Immunol Univ Minn Sch Med Duluth MN 55812

ZIEGLER, ROBERT C(HARLES), b Buffalo, NY, Dec 11, 27. PHYSICS, ENVIRONMENTAL SCIENCES. *Educ:* Univ Buffalo, BA, 50, PhD(physics), 57. *Prof Exp:* Lectr physics, Univ Buffalo, 55-56; proj engr, Nucleonics Dept, Bell Aircraft Corp, 56-57; mem staff, Cornell Aeronaut Lab, Inc, 57-66, head remote sensing sect, 66-71, head environ sci sect, 71-73; tech staff environ sci dept, Calspan Corp, 75-81; CONSULT, 81- *Concurrent Pos:* Consult, 74-75. *Mem:* Am Phys Soc; Nat Soc Prof Engrs. *Res:* Geometrical and physical optics; nuclear physics; aerial remote sensing; environmental research. *Mailing Add:* 26 High Park Blvd Buffalo NY 14226

ZIEGLER, ROBERT G, b The Dalles, Ore, Apr 24, 24; m 53; c 3. INORGANIC CHEMISTRY. *Educ:* Ore State Univ, BA, 48, MS, 51; Univ Tenn, PhD, 69. *Prof Exp:* Control chemist, Barium Prod Ltd, 48-49; res chemist, Nitrogen Div, Allied Chem Corp, 52-57; from asst prof to assoc prof, 57-70, head dept, 69-74, chmn, Div Natural Sci, 80-86, PROF CHEM, LINCON MEM UNIV, 70- *Mem:* Am Chem Soc; Am Sci Affil; Sigma Xi; Nat Sci Teachers Asn; Am Asn Physics Teachers. *Res:* Analytical and coordination chemistry; fertilizer technology. *Mailing Add:* Rte 1 Box 19 Harrogate TN 37752

ZIEGLER, THERESA FRANCES, b Budapest, Hungary; nat US; wid; c 1. CHEMISTRY, APPLICATION OF RADIO ISOTOPES IN INDUSTRIAL RESEARCH. *Educ:* Eotvos Lorand Univ, Budapest, BS, 46, PhD(phys chem), 49. *Prof Exp:* Res engr, State Biochem Res Inst, Hungary, 49-50; chemist, Steel Plant Lab, 53-56; radiochemist, Nat Health & Labor Inst, 56-57; radiochemist, radiation safety officer, Stamford Res Labs, Am Cyanamid Co, 57-85; CONSULT, 85- *Mem:* Am Chem Soc; Am Nuclear Soc; NY Acad Sci. *Res:* Radiosyntheses and radio-tracerwork in the following fields-agricultural and industrial chemicals, herbicides, pesticides and insecticides; pharmaceuticals; catalysts; technology of plastics, fibers and papers; surface coatings; biological membranes; industrial hygiene, toxicity studies; autoradiography, experience with sealed sources; radioactive semi-micro and micro syntheses. *Mailing Add:* 1452 Riverbank Rd Stamford CT 06903

ZIEGLER, WILLIAM ARTHUR, b St Louis, Mo, Feb 10, 24; m 51; c 2. INORGANIC CHEMISTRY. *Educ:* Univ Ill, AB, 48, MS, 49, PhD(anal chem), 52. *Prof Exp:* Teaching asst, Univ Ill, 48-52; chemist, Mallinckrodt Chem Works, 52, supvr, 52-55, asst mgr anal dept, Uranium Div, 55-60, mgr anal dept, 60-66, tech asst to dir qual control, 66-67; tech specialist, Nuclear Div, Kerr-McGee Corp, 67-69, mgr qual assurance, 69-70; mgr anal sect, Eastern Res Ctr, Stauffer Chem Co, 71-79; mgr, Anal Lab, 80, MGR SUPPORT GROUP, ADVAN MAT DIV, MAT RES CORP, 81- *Mem:* Am Chem Soc. *Res:* Sampling; uranium chemistry. *Mailing Add:* 61 Pocconock Trail New Canaan CT 06840

ZIEGRA, SUMNER ROOT, b Deep River, Conn, Feb 13, 23; m 45; c 2. PEDIATRICS. *Educ:* Univ Vt, BS, 45; Yale Univ, MD, 47; Am Bd Pediat, dipl, 57. *Prof Exp:* Fel pediat, Sch Med, NY Univ, 49-51; asst prof pediat, State Univ NY Downstate Med Ctr, 56-60; assoc prof, Jefferson Med Col, 60-63; from assoc prof to prof, Hahnemann Med Col, 63-70; prof, Col Med, Thomas Jefferson Univ, 70-72; PROF PEDIAT, MED COL PA, 73-, CHMN DEPT, 75- *Concurrent Pos:* Dir div B, Dept Pediat, Philadelphia Gen Hosp, 63-70, coordr, Hahnemann Div, 66-67; dir dept pediat, Lankenau Hosp, 70-72. *Res:* Infectious diseases. *Mailing Add:* Dept Pediat Hosp Med Col Pa 3300 Henry Ave Philadelphia PA 19129

ZIELEN, ALBIN JOHN, b Chicago, Ill, Dec 22, 25. SYSTEMS ANALYST, PHYSICAL CHEMISTRY. *Educ:* Miami Univ, BA, 50; Univ Calif, PhD(chem), 53. *Prof Exp:* Asst chem, Univ Calif, 51-52; chemist, Radiation Lab, Univ Calif, 52-53; assoc chemist, Argonne Nat Lab, 53-87; RETIRED. *Mem:* Am Chem Soc. *Res:* Computer programming; neptunium solution chemistry; electrochemistry; reaction kinetics; complex ions. *Mailing Add:* 13810 W Springdale Dr Sun City West AZ 85375

ZIELEZNY, MARIA ANNA, b Kaczkowizna, Poland, Sept 20, 39; US citizen; m 69. BIOSTATISTICS. *Educ:* Univ Warsaw, MS, 62; Univ Calif, Los Angeles, PhD(biostatist), 71. *Prof Exp:* Res asst statist, Math Inst, Polish Acad Sci, Warsaw, 62-65; clin asst prof, 71-73, asst prof, 74-80, ASSOC PROF BIOSTATIST, DEPT SOCIAL & PREV MED, SCH MED, STATE UNIV NY, BUFFALO, 80- *Mem:* Am Statist Asn; Am Pub Health Asn; Biomet Soc. *Res:* Statistical applications in particular to medicine, development of statistical methods in evaluation, discriminant analysis, measures of association, techniques for qualitative data. *Mailing Add:* Dept Sociol & Prev Med State Univ NY Health Ctr 2211-3435 Main St Buffalo NY 14214

ZIELEZNY, ZBIGNIEW HENRYK, b Knurow, Poland, Jan 11, 30; m 69. MATHEMATICAL ANALYSIS. *Educ:* Wroclaw Univ, Masters, 54, PhD(math), 59; Polish Acad Sci, docent, 64. *Prof Exp:* Adj math, Wroclaw Tech Univ, 58-61; adj, Polish Acad Sci, 61-64, docent, 64-69; vis prof, Univ Kiel, 69-70; assoc prof, 70-71, PROF MATH, STATE UNIV NY, BUFFALO, 71- *Concurrent Pos:* Vis prof, Univ Kiel, WGer, 79. *Mem:* Am Math Soc. *Res:* Analysis, functional analysis; differential equations; existence and regularity of solutions of convolution equations in various spaces of distributions. *Mailing Add:* Dept Math Rm 106 Diekendorf Hall State Univ NY Health Sci Ctr 3435 Main St Buffalo NY 14214

ZIELINSKI, ADAM, b Jaroslaw, Poland, Oct 2, 43; Can citizen; m 75; c 2. ELECTRICAL ENGINEERING, OCEAN ENGINEERING. *Educ:* Wroclaw Tech Univ, BEE & MEE, 67, PhD(elec eng), 72. *Prof Exp:* Res assoc, Wroclaw Tech Univ, 67-68, asst prof, 68-72; res fel, Tokyo Inst Technol, 72-74 & Univ NB, 74-75; asst prof, 75-80, assoc prof elec eng, Mem Univ NFLD, 80-; ASSOC PROF, UNIV VICTORIA, 85- *Concurrent Pos:* Investr, Nat Res Coun Can operating grants, 76-77 & 78-92, Natural Sci & Eng Res Coun Can strategic grant ocean eng, 78-81 & PRAI grant, 81-82. *Mem:* Sr mem Inst Elec & Electronics Engrs. *Res:* Applied underwater acoustics; acoustic communication; specialized sonar systems; electronic instrumentation and signal processing. *Mailing Add:* Dept Elec Eng Univ Victoria PO Box 1700 Victoria BC V8W 2Y2 Can

ZIELINSKI, THERESA JULIA, b Brooklyn, NY. THEORETICAL CHEMISTRY. *Educ:* Fordham Univ, BS, 63, MS, 68, PhD(chem), 73. *Prof Exp:* ASST PROF CHEM, COL MT ST VINCENT, 72- *Mem:* Am Chem Soc. *Res:* Theoretical chemical study of purines and pyrimidines. *Mailing Add:* 52 Hartford Ave Buffalo NY 14223

ZIELKE, H RONALD, b Pscinno, Poland, June 7, 42; US citizen; m 67; c 2. NEUROCHEMISTRY, CELL BIOLOGY. *Educ:* Univ Ill, BS, 64; Mich State Univ, PhD(biochem), 68. *Prof Exp:* Res assoc, AEC Plant Res Lab, Mich State Univ, 68-71; res fel, Genetics Unit, Mass Gen Hosp, 71-73; asst prof, 73-81, res assoc prof, 81-89, ASSOC PROF PEDIAT, SCH MED, UNIV MD, 89- *Concurrent Pos:* Monsanto fel, 69; dir, Am Type Cult Collection, 81-87. *Mem:* Am Soc Neurochemistry; Am Soc Biochem & Molecular Biol. *Res:* Metabolism and enzymology of brain and cultured mammalian cells; adenosine receptors. *Mailing Add:* Dept Pediat Res Sch Med Univ Md Baltimore MD 21201

ZIEMAN, JOSEPH CROWE, JR, b Mobile, Ala, June 9, 43; m 67. MARINE ECOLOGY, BIOLOGICAL OCEANOGRAPHY. *Educ:* Tulane Univ, BS, 65; Univ Miami, MS, 68, PhD(marine sci), 70. *Prof Exp:* Res asst thermal pollution, Inst Marine Sci, Univ Miami, 68-70; fel syst ecol, Inst Ecol, Univ Ga, 70-71; from asst prof to assoc prof, 71-89, PROF ENVIRON SCI, UNIV VA, 89- *Mem:* Ecol Soc Am; Am Soc Limnol & Oceanog; Sigma Xi; AAAS; Estuarine Res Fedn. *Res:* Comparative studies of tropical and temperate interface zones, seagrasses, coral reefs, mangroves and salt marshes; production, colonization, succession and recovery from disturbances; simulation modeling of growth and succession; marine biogeochemistry. *Mailing Add:* Dept Environ Sci Clark Hall Univ Va Charlottesville VA 22903

ZIEMBA, W T, MATHEMATICS. *Prof Exp:* FAC MEM, DEPT MANAGERIAL SCI, UNIV BC. *Mailing Add:* Dept Managerial Sci Univ BC 2075 Westbrook Pl Vancouver BC V6T 1W5 Can

ZIEMER, PAUL L, b Toledo, Ohio, June 28, 35; m 58; c 4. HEALTH PHYSICS. *Educ:* Wheaton Col, BS, 57; Vanderbilt Univ, MS, 59; Purdue Univ, PhD(bionucleonics), 62; Am Bd Health Physics, dipl, 65. *Prof Exp:* Physicist, US Naval Res Lab, Wash, 57; health physicist, Oak Ridge Nat Lab, 59; radiol control officer, Purdue Univ, 59-82, assoc head bionucleonics, 71-81; from asst prof to prof health physics, Purdue Univ, West Lafayette, 62-90, head, Sch Health Sci, 83-90; ASST SECY ENERGY ENVIRON SAFETY & HEALTH, US DEPT ENERGY, 90- *Concurrent Pos:* Consult, Harrison Steel Castings Co, Ind, 62-66, satellite div, Union Carbide Corp, 66-67, Calif Nuclear, Inc, 66-68, Breed Radium Inst, 68-70, Detroit Diesel Allison Div, Gen Motors Corp, 69 & 77- Mobil Field Res Labs, 70- & Midwest Radiation Protection, Inc, 71-88; mem panel examr, Am Bd Health Physics, 69-71; mem, Sci Comt Oper Health Physics, Nat Coun Radiation Protection, 46-90, Int Standards Orgn, Working Group 6, Sci Comt Radiation Protection, 76-81; ed, Health Physics J, 79-84; vchmn comt N-13 radiation protection standards, Am Nat Standards, 81-82, chmn, 82-89; mem, US Dept Energy, Adv Comt Nuclear Sci, Eng & Health Physics Fels, 81-83; mem, Nat Prog Rev Comt Low-Level Radioactive Waste Mgt, US Dept Energy, 80-87, chmn, 80-83; mem, Bd Dir, N Prk Col, 81-86, chmn, 84-86; mem, Tech Adv Comt, Ill Dept Nuclear Safety, 85-90; mem, Spec Task Force on Bioassay, Oak Ridge Assoc Univs & Nat Cancer Inst, 86-87. *Honors & Awards:* Lederle Pharm Fac Award, 64; Elda E Anderson Award, Health Physics Soc, 71. *Mem:* Health Physics Soc (pres, 75); Int Radiation Protection Asn; AAAS; Am Acad Health Physics (pres, 87). *Res:* Radon emanation rates; radiation dosimetry. *Mailing Add:* US Dept Energy 1000 Independence Ave SW Washington DC 20585

ZIEMER, ROBERT RUHL, b Oklahoma City, Okla, Oct 25, 37. WATERSHED MANAGEMENT, WILDLAND HYDROLOGY. *Educ:* Univ Calif, Berkeley, BS, 59, MS, 63; Colo State Univ, PhD(earth resources), 78. *Prof Exp:* Teaching asst forestry, Univ Calif, Berkeley, 59-60; res forester, 60-77, RES HYDROLOGIST, PAC SOUTHWEST RES STA, USDA FOREST SERV, 77- *Concurrent Pos:* Intel officer, Nev Air Nat Guard, 60-67; adj prof, Humboldt State Univ, 72-; prof forester, State Calif, 75-; chmn, Tech Comt Evaporation & Transpiration, Am Geophys Union, 76-81 & Watershed Mgt, Int Union Forestry Res Orgn, 88-; res fel, East-West Ctr, 85- *Mem:* Am Geophys Union; Int Union Geod & Geophys; Int Union Forestry Res Orgn. *Res:* Effects of forest management on hillslope processes, fishery resources, and stream environments; risk of landslides and erosion; timing and routing of sediment in relation to cumulative effects and anadromous and resident fish habitat. *Mailing Add:* 1700 Bayview Dr Arcata CA 95521

ZIEMER, RODGER EDMUND, b Sargeant, Minn, Aug 22, 37; m 62; c 4. ELECTRICAL ENGINEERING. *Educ:* Univ Minn, BS, 60, MS, 62, PhD(elec eng), 65. *Prof Exp:* From asst prof to prof elec eng, Univ Mo-Rolla, 68-88, grad coordr dept, 78-80,; PROF & CHMN DEPT ELEC ENG, UNIV COLO, COLORADO SPRINGS. *Concurrent Pos:* Consult, Electronics & Space Div, Emerson Elec Co, St Louis, 74-, govt electronics div, Motorola Inc, Scottsdale, Ariz, 80-; on leave, Motorola Inc, Scottsdale, Ariz, 80-81. *Mem:* Fel Inst Elec & Electronics Engrs; Am Soc Eng Educ; Sigma Xi. *Res:* Statistical communication theory; digital signaling in impulsive noise; costas and phase lock loop performance in radio frequency environments; spread spectrum communication techniques; digital signal processing. *Mailing Add:* 8315 Pilot Ct Colorado Springs CO 80920

ZIEMER, WILLIAM P, b Manitowoc, Wis, Mar 26, 34; m 57; c 3. MATHEMATICS. *Educ:* Univ Wis, BS, 56, MS, 57; Brown Univ, PhD(math), 61. *Prof Exp:* Assoc prof, 61-77, PROF MATH, IND UNIV, BLOOMINGTON, 77- *Mem:* Am Math Soc; Math Asn Am. *Res:* Geometric analysis; area theory; surface theory; differential geometry. *Mailing Add:* Dept Math Ind Univ Bloomington IN 47405

ZIEMINSKI, S(TEFAN) A(NTONI), b Zaleszczyki, Poland, Sept 4, 05; nat US; m 50; c 1. CHEMICAL ENGINEERING. *Educ:* Tech Univ, Poland, Chem eng, 27, Dr Tech Sci, 29. *Prof Exp:* Mgr, Chodorow Sugar Refinery, Poland, 34-38; dir, Chodorow Chem Works, 38-39; assoc prof chem eng, Polish Univ Col, Eng, 47-52; res assoc, Eng Res Inst, Univ Mich, 52-54; prof, 54-76, EMER PROF CHEM ENG, UNIV MAINE, ORONO, 76- *Mem:* Am Chem Soc; Am Inst Chem Engrs; Nat Soc Prof Engrs; Brit Inst Chem Engrs. *Res:* Mass transfer operations; beet sugar technology. *Mailing Add:* 16 Main Wood Ave Orono ME 04473

ZIEMNIAK, STEPHEN ERIC, b Rochester, NY, Mar 6, 42; m 70; c 2. AQUEOUS CHEMISTRY, COOLANT TECHNOLOGY. *Educ:* Rensselaer Polytech Inst, BChE, 64, PhD(chem eng), 68. *Prof Exp:* Fel, Los Alamos Sci Lab, 68-69; PRIN ENGR, KNOLLS ATOMIC POWER LAB, GEN ELEC CO, 70- *Concurrent Pos:* Adj fac, Rensselaer Polytech Inst, 77-79. *Res:* Establish principles of corrosion product transport in pressurized water reactor coolants, including thermodynamics of metal oxide solubility behavior; hydrothermal crystallization mechanisms; fluid mechanics and numerical turbulence models. *Mailing Add:* Three Crystal Lane Latham NY 12110

ZIEN, TSE-FOU, b Shanghai, China, Sept 13, 37; US citizen; m 62; c 2. INFORMATION SCIENCE & SYSTEMS. *Educ:* Nat Taiwan Univ, BSc, 58; Brown Univ, MSc, 63; Calif Inst Technol, PhD(aeronaut), 67. *Prof Exp:* Engr, Third Shipyard Chinese Navy, 58-60; res fel aeronaut, Calif Inst Technol, 67; res assoc fluid sci, Case Western Reserve Univ, 67-70; res aeronaut engr, 70-80, BR HEAD, NAVAL SURFACE WARFARE CTR, 80- *Concurrent Pos:* Prof lectr, George Washington Univ, 75-85; mem, Fluid Dynamics Tech Comt, Am Inst Aeronaut & Astronaut, 79-82, Thermophysics Tech Comt, 82-85 & Appl Aerodyn Tech Comt, 90-; chmn, Heat Transfer Panel, Naval Aeroballistics Comn, 81-84; vis prof, Nat Cheng-Kung Univ, 87 & 88; adj prof, Cath Univ Am, 88- *Mem:* Assoc fel Am Inst Aeronaut & Astronaut; Am Phys Soc; Soc Indust & Appl Math. *Res:* Hypersonic flows; holographic interferometry; viscous flow and heat transfer; aerodynamic heating and ablation modelling; bio-fluid dynamics; measurement of viscosity of gases. *Mailing Add:* 3300 Beret Lane Wheaton MD 20906

ZIENIUS, RAYMOND HENRY, b Montreal, Que, Nov 8, 34; m 70; c 2. ANALYTICAL CHEMISTRY. *Educ:* McGill Univ, BSc, 56, PhD, 59. *Prof Exp:* Asst chemist, Dom Tar & Chem Co, Ltd, 55; res chemist, Cent Res Lab, Can Industs Ltd, 59-67; asst prof, 67-72, ASSOC PROF CHEM, SIR GEORGE WILLIAMS CAMPUS, 72-, DIR CHEM & BIOCHEM COOP PROGS, CONCORDIA UNIV, 86- *Mem:* Fel Chem Inst Can; Chemists Asn Que. *Res:* Gas chromatography. *Mailing Add:* Dept Chem Sir George Williams Campus Concordia Univ Montreal PQ H3G 1M8 Can

ZIENTY, FERDINAND B, b Chicago, Ill, Mar 21, 15; m 45; c 3. CHEMISTRY. *Educ:* Univ Ill, BS, 35; Univ Mich, MS, 36, PhD(pharmaceut chem), 38. *Prof Exp:* Res chemist, Monsanto Co, 38-40, res group leader, 40-47, from asst dir res to dir res, 47-60, adv org chem res, Org Chem Div, 60-64, mgr res & develop, food, feed & fine chem, 64-79, dir res & develop, Health Care Develop, 79-83; CONSULT, 83- *Concurrent Pos:* Vpres res, George Lueders & Co, Subsid Monsanto Co, 68-70. *Mem:* Fel AAAS; Am Chem Soc; Am Pharmaceut Asn; Am Inst Chem Eng; fel NY Acad Sci; Inst Food Technol. *Res:* Antispasmodics; sulfa drugs; ethylenediamine derivatives; organic heterocycles, including imidazole and thiophene chemistry; organic acids and anhydrides; nucleophilic substitutions; nucleophilicity of thiols; catalytic oxidations; flavors. *Mailing Add:* 850 Rampart Dr St Louis MO 63122-1644

ZIERDT, CHARLES HENRY, b Pittsburgh, Pa, Apr 24, 22; m 42; c 4. MEDICAL MICROBIOLOGY, EPIDEMIOLOGY. *Educ:* Pa State Univ, BS, 43; Univ Mich, MS, 45; George Washington Univ, PhD(microbiol), 67. *Prof Exp:* Res asst pharmaceut, Parke Davis Pharmaceut, 45-48; bacteriologist clin microbiol, Henry Ford Hosp, 48-53; STAFF MICROBIOLOGIST, HEALTH & HUMAN SERV, NIH, 53- *Concurrent Pos:* Instr gen microbiol, Found Adv Educ Sci, NIH; scientist sponsor, Univ Md. *Mem:* Am Soc Microbiol; Sigma Xi; fel Am Acad Microbiol. *Res:* Taxonomy of anaerobic Corynebacterium; Pseudomonas aeruginosa pathogenicity and typing; Staphylococcus aureus typing and hospital epidemiology; Blastocystis hominis classification ultrastructure, antigenicity, culture and pathogenicity; cultivation of Mycobacterium leprae; microsporidium diagnosis in human infections; author, contributor and patentee. *Mailing Add:* Clin Path Dept Clin Ctr Rm 2C-343 NIH Bethesda MD 20892

ZIERING, LANCE K, b New York, NY, May 17, 39; m 61; c 2. CHEMICAL ENGINEERING. *Educ:* City Col New York, BScChE, 62; Columbia Univ, MScChE, 63; Univ Conn, MBA, 69. *Prof Exp:* Chem engr, Am Cyanamid Co, 63-65, res chem engr, 65-67, group leader supvr, 67-69; asst prod mgr, ICI Am Inc, 69-71, tech sales rep, 71-72, prod mgr films, 72-74, asst to vpres plastics, 74-76, tech serv mgr, 76-79, mkt mgr, 79-80, dir mkt, 80-81, vpres & gen mgr rubicon, 81-89; PRES, SPECIALTY USA, 89- *Res:* Process and product development of fuel cell and battery electrodes and components; cost estimating. *Mailing Add:* 17 Tullamore Dr West Chester PA 19382

ZIERLER, KENNETH, b Baltimore, Md, Sept 5, 17; m 41; c 5. MEDICINE, PHYSIOLOGY. *Educ:* Johns Hopkins Univ, AB, 36; Univ Md, MD, 41. *Prof Exp:* Fel med, Sch Med, Johns Hopkins Univ, 46-48, from instr to assoc prof med, 48-64, prof med, 64-72, prof physiol, 69-72; dir, Inst Muscle Dis, Inc, 72-73; PROF PHYSIOL & MED, SCH MED, JOHNS HOPKINS UNIV & PHYSICIAN, JOHNS HOPKINS HOSP, 73- *Concurrent Pos:* Asst physician, Outpatient Dept, Johns Hopkins Hosp, 46-53, physician, 53-55, physician-in-charge phys ther dept, 50-57, chemist, 57-68, physician, 53-72, assoc prof environ med, Sch Hyg & Pub Health, Johns Hopkins Univ, 54-59; adj prof, Rockefeller Univ, 72-73; adj prof, Med Sch, Cornell Univ, 72-73; chmn, Adv Comt Physiol, Off Naval Res, 64-72; assoc ed, Med, 63-72, Circulation Res, 68-74 & ed, 66 & 68. *Mem:* Am Physiol Soc; Am Soc Clin Invest; Asn Am Physicians; Endocrine Soc. *Res:* Muscle metabolism and function; hormonal action; biomembranes; circulation; tracer kinetics; insulin, water and electrolytes. *Mailing Add:* Rm 918 Taylor Bldg Johns Hopkins Univ Sch Med 720 Rutland Ave Baltimore MD 21205

ZIERLER, NEAL, b Baltimore, Md, Sept 17, 26. MATHEMATICS. *Educ:* Johns Hopkins Univ, AB, 45; Harvard Univ, AM, 49, PhD(math), 59. *Prof Exp:* Mathematician-physicist, Ballistic Res Labs, Aberdeen Proving Ground, Md, 51-52; mem staff, Instrumentation Lab, Mass Inst Technol, 52-54 & Lincoln Lab, 54-60; res group supvr, Jet Propulsion Lab, Calif Inst Technol, 60-61; sr scientist, Arcon Corp, Mass, 61-62; mem staff, Lincoln Lab, Mass Inst Technol, 62; sub-dept head, Mitre Corp, Mass, 62-65; MEM STAFF, CTR COMMUN RES, INST DEFENSE ANALYSIS, 65- *Mem:* Am Math Soc; Math Asn Am; fel Inst Elec & Electronics Engrs; Asn Comput Mach. *Res:* Algebra; mathematical foundations of quantum mechanics; coding and decoding of information; computer applications. *Mailing Add:* Inst Defense Analysis Thanet Rd Princeton NJ 08540-3699

ZIERNICKI, ROBERT S, b 1934; m 57; c 3. GUIDANCE & NAVIGATION SYSTEMS, INFRARED DETECTOR. *Educ:* Cath Univ Am, BA, 57; Univ Ark, MS, 66, Southern Methodist Univ, PhD(physics), 72. *Prof Exp:* Chief, Avionics Div, US Air Force, 74-79; dir guidance navigation, Honeywell Corp, 79-81, dir Space Systs Ctr, 81-83; VPRES BUS DEVELOP, NORTHROP CORP, 83- *Concurrent Pos:* Vis lectr, Univ Calif, Los Angeles, 75-80 & Mass Inst Technol, 85-; adj prof, Tufts Univ, 85- *Mem:* Sigma Xi; Inst Elec & Electronic Engrs; Inst Navig. *Res:* Electronic transport properties of mercury codmum telluride. *Mailing Add:* 824 Harbor Island Dr Clearwater FL 34615

ZIESKE, JAMES DAVID, EPITHELIAL PROTEIN SYNTHESIS. *Educ:* Univ Mich, PhD(biochem), 81. *Prof Exp:* ASST SCIENTIST, DEPT CORNEA RES, EYE RES INST, 84- *Res:* Epithelial protein synthesis during wound repair. *Mailing Add:* Dept Cornea Res Eye Res Inst Retina Found 20 Staniford St Boston MA 02114

ZIETLOW, JAMES PHILIP, b Chicago, Ill, Dec 15, 21; m 52; c 3. PHYSICS. *Educ:* De Paul Univ, BS, 48; Ill Inst Technol, MS, 49, PhD(physics), 55. *Prof Exp:* Sr res physicist, Res & Develop Labs, Pure Oil Co, 51-56; prof physics & math, NMex Highlands Univ, 56-65, head dept physics & math, 56-63, grad dean, 63-64, head dept physics & math, 64-65; assoc dean, Col Arts & Sci, 69-78, PROF PHYSICS, WESTERN MICH UNIV, 65- *Mem:* Am Phys Soc; Coblentz Soc. *Res:* Infrared, Raman, ultraviolet and mass spectroscopies. *Mailing Add:* Dept Physics Western Mich Univ 2218 Evr Kalamazoo MI 49008

ZIETZ, JOSEPH R, JR, b Menominee, Mich, May 12, 25; m 49; c 4. ORGANOMETALLIC CHEMISTRY. *Educ:* Spring Hill Col, BS, 47; Marquette Univ, MS, 49; Tulane Univ, PhD(org chem), 52. *Prof Exp:* Chemist, Warren Petrol Co, 49-50; ASSOC DIR RES & DEVELOP, ETHYL CORP, 52- *Mem:* Am Chem Soc. *Res:* Organic Synthesis; synthesis and chemistry of main group metal-organic compounds. *Mailing Add:* Ethel Tech Ctr PO Box 14799 Baton Rouge LA 70898-4799

ZIEVE, LESLIE, b Minneapolis, Minn, Aug 6, 15; m 41; c 1. MEDICINE. *Educ:* Univ Minn, MA, 39, MD, 43, PhD(med), 52; Am Bd Internal Med, dipl, 51. *Prof Exp:* Resident med, Med Sch, Univ Minn, Minneapolis, 46-49, from instr to prof, 49-62; dir spec cancer lab, Vet Admin Hosp, Minneapolis, 61-72, staff physician, 49-77, assoc chief staff for res, 61-77; dir res, dept med, Hennepin County Med Ctr, 77-85; EMER DISTINGUISHED PHYSICIAN, VET ADMIN HOSP, MINNEAPOLIS, 85- *Concurrent Pos:* Chief Nuclear Med, Vet Admin Hosp, Minneapolis, 50-72; mem exec comt, Grad Sch, Univ Minn, Minneapolis, 65-68; ed, J Lab & Clin Med, 67-70. *Honors & Awards:* Middleton Award, 62. *Mem:* Am Col Physicians. *Res:* Diseases of the liver and pancreas. *Mailing Add:* Dept Med Univ Minn Minneapolis MN 55455

ZIFF, MORRIS, b New York, NY, Nov 19, 13; m 40, 78; c 2. INTERNAL MEDICINE. *Educ:* NY Univ, BS, 34, PhD(chem), 37, MD, 48. *Prof Exp:* Asst chem, NY Univ, 34-39; asst biochem, Col Physicians & Surgeons, Columbia Univ, 39-41, vis scholar, 41-44; instr & lectr, NY Univ, 44, adj asst prof, 48-50, from asst prof med to assoc prof, 54-58; prof internal med, Southwest Med Ctr, Univ Tex Health Sci Ctr, Dallas, 58-87; prof, 58-87, EMER ASHBEL SMITH PROF INTERNAL MED, SOUTHWEST MED CTR, UNIV TEX HEALTH SCI CTR, DALLAS, 88-, EMER MORRIS ZIFF PROF RHEUMATOLOGY, 88- *Concurrent Pos:* Instr, City Col New York, 41-44; from intern to asst resident, Bellevue Hosp, 48-50, chmn clin res sect study group rheumatic dis, 52-58; consult, USPHS, 55-63; founding dir, H C Simmons Arthritis Res Ctr, 83-84. *Honors & Awards:* Heberden Medal, Heberden Soc London, 64; Carol Nachman Prize, 74; Bunim Medal, Am Rheumatism Asn, 82; First Gold Medal, Am Rheumatism Asn, 88. *Mem:* Am Chem Soc; Harvey Soc; Am Soc Clin Invest; Am Rheumatism Asn (pres, 65-66); Am Col Physicians; Asn Am Physicians. *Res:* Chemistry of connective tissue; rheumatic diseases; immunology; chronic inflamation in rheumatoid arthritis, particulary the role of the endothelial cell in the facilitation of emigration of lymphocytes and macrophages into perivascular space. *Mailing Add:* Dept Int Med Univ Tex Southwestern Med Ctr 5323 Harry Hines Blvd Dallas TX 75235-9030

ZIFFER, HERMAN, b New York, NY, Feb 22, 30; m 55; c 3. ORGANIC CHEMISTRY. *Educ:* City Col New York, BS, 51; Ind Univ, MA, 53; Univ Ore, PhD(chem), 55. *Prof Exp:* Res chemist, Nat Aniline Div, Allied Chem Corp, 55-59; RES CHEMIST, NIH, 59- *Concurrent Pos:* Sabbatical, Stanford Univ, 69-70; Sabbatical, Univ Groninger, Netherlands, 85. *Mem:* Am Chem Soc; Royal Soc Chem. *Res:* Photochemistry; asymmetric synthesis and the use of optical rotatory dispersion and other physical measurements for structure determination; biotransformations by enzymes and micro-organisms. *Mailing Add:* Bldg 2 Rm 120 NIH Bethesda MD 20014

ZIFFER, JACK, b New York, NY, Dec 2, 18; m 42; c 2. MICROBIOLOGY, INDUSTRIAL BIOTECHNOLOGY. *Educ:* Brooklyn Col, BS, 40; Univ London, PhD(biochem), 50. *Prof Exp:* Sr res microbiologist, Schenley Res Inst, 42-46; sr res fermentation chemist, A E Staley Co, 46-48; sr res microbiol chemist, Res Labs, Pabst Brewing Co, 50-55, head dept microbiol, Pabst Labs, 55-60, dir microbiol div, P-L Biochem, Inc, 60-67, vpres microbiol div, 67-76, vpres & tech dir, Premier Malt Prod, Inc, 69-76; prof microbiol, Tel-Aviv Univ, 76-77; PROF INDUST BIOTECHNOL, ISRAEL INST TECHNOL, HAIFA, 77- *Concurrent Pos:* Ed, Biotechnol Letters. *Mem:* Am Chem Soc; Am Phytopath Soc; Soc Invert Path; Soc Am Microbiologists. *Res:* Phytoactin; phytostreptin; penicillin; streptomycin; streptothricin; bacitracin; fungal amylase, glucoamylase, protease; bacterial amylase, gumase, protease; vitamin B12; riboflavin; gluconic acid; lactic acid; fungal anthraquinone pigments; cellulose decomposition; plant diseases; microbial insecticides; trans-N-deoxyribosylase; microbial rennet; industrial alcohol. *Mailing Add:* 2548 Sunset Dr NE Athens GA 30604

ZIGHELBOIM, JACOB, b Chernowitz, Rumania, Jan 2, 46; Venezuelan citizen; m 67; c 3. IMMUNOBIOLOGY, HEMATOLOGY. *Educ:* Col Moral & Luces, BS, 63; Univ Cent Venezuela, MD, 69. *Prof Exp:* From intern to resident internal med, Beilinson Hosp, Tel Aviv Sch Med, 70-72; fel immunobiol, Dept Microbiol & Immunol, 72-74, resident, Sch Med, 74-76, clin fel hematol & oncol, 76-81, asst prof microbiol, immunol & med, 81-82, ASSOC PROF, SCH MED, UNIV CALIF, LOS ANGELES, 82- *Mem:* Am Fedn Clin Res; Am Asn Immunologists; Am Soc Contemporary Ophthamol; Am Asn Cancer Res; Royal Entomol Soc. *Res:* Cellular and molecular analysis of regulation of cytotoxic immune responses to histocompatibility antigens; regulation on natural killer cell function; biologic approaches to the control of neoplastic disease. *Mailing Add:* Dept Microbiol 43-239 Chs Univ Calif 405 Hilgard Ave Los Angeles CA 90024

ZIGLER, EDWARD, b Mar 1, 30. PSYCHOLOGY. *Educ:* Univ Mo, BS, 54; Univ Tex, PhD, 58. *Hon Degrees:* MA, Yale Univ, 67; Dr, Boston Col, 85. *Prof Exp:* Staff psychologist, Child Guidance Clin, Tex Univ, Austin, 56-57; intern psychol, Worcester State Univ, Mass, 57-59; from asst prof to prof psychol, 59-76, chmn, Dept Psychol, 73-74, HEAD, PSYCHOL SECT, CHILD STUDY CTR, YALE UNIV, 67-, STERLING PROF PSYCHOL, 76- *Concurrent Pos:* Asst prof, Univ Mo, Columbia, 58-59, dir, Child Diag Ctr, 58-59; dir, Child Develop Prog, Dept Psychol, Yale Univ, 61-76, Inst Social & Policy Studies, 75-, dir, Bush Ctr Child Develop & Social Policy, 77-, Sch Med, Psychol & Child Study Ctr, 82-; dir, Off Child Develop & chief, Children's Bur, US Dept Health, Educ & Welfare, 70-72; consult, Dept Health & Human Serv, Carnegie Corp, Ford Found, Found Child Develop & others; mem numerous comts, adv bds & panels. *Honors & Awards:* Dale Richmond Mem Award, Am Acad Pediat, 76, C Anderson Aldrich Award, 85; G Stanley Hall Award, Am Psychol Asn, 79, Nicholas Hobbs Award, 85, Edgar A Doll Award, 86, Award for Distinguished Prof Contrib to Knowledge, 86, Award for Distinguished Contrib to Community Psychol & Community Ment Health, 89; Nat Achievement Award, Asn Advan Psychol, 85; Blanche F Ittleson Mem Lectr Award, Am Orthopsychiat Asn, 89. *Mem:* Inst Med-Nat Acad Sci; hon mem Am Acad Child & Adolescent Psychiat. *Res:* Cognitive and social-emotional development in children, particularly those who are mentally retarded or from lower-income families. *Mailing Add:* Dept Psychol Yale Univ PO Box 11A Yale Sta New Haven CT 06520

ZIGMAN, SEYMOUR, b Far Rockaway, NY, Nov 21, 32; m 54; c 1. BIOCHEMISTRY. *Educ:* Cornell Univ, BA, 54; Rutgers Univ, MS, 56, PhD(biochem), 59. *Hon Degrees:* DSc, Univ Repub, Montevideo, Uruguay. *Prof Exp:* Fel biochem of the eye, Mass Eye & Ear Infirmary, 59-61, res assoc, 61-62; from instr to asst prof, 62-70, PROF OPHTAL & BIOCHEM SCH MED & DENT, UNIV ROCHESTER, 75- *Concurrent Pos:* Corp & investr, Marine Biol Lab, Woods Hole, Mass; res assoc, Mote Marine Lab, Sarasota, Fla; biol consult, Eastman Kodak Co. *Mem:* Am Chem Geront; Am Soc Biol Chemists; Am Soc Photobiol; Asn Res Vision & Ophthal. *Res:* Photobiology of the lens and retina; role of cataract and retinal degeneration of near-UV radiation; lens pigments to enhance vision. *Mailing Add:* Univ Rochester Dept Opthal & Biochem Univ Rochester Med Ctr Rochester NY 14642

ZIGMOND, MICHAEL JONATHAN, b Waterbury, Conn, Sept 1, 41; m 66; c 2. NEUROCHEMISTRY, NEUROPLASTICITY. *Educ:* Carnegie-Mellon Univ, BS, 63; Univ Chicago, PhD(biopsychol), 68. *Prof Exp:* Teaching asst psychol, Carnegie-Mellon Univ, 62-63; res assoc, Mass Inst Technol, 67-69, instr, 69-70; from asst prof to prof biol & psychol, 78-86, PROF BEHAV NEUROSCI & PSYCHIAT, UNIV PITTSBURGH, 86-, DIR TRAINING CTR, NEUROSCI, 86- *Concurrent Pos:* Nat Inst Ment Health & Nat Inst Neurol & Commun Disorders & Stroke grantee, Univ Pittsburgh, 70-; mem,

Neuropsychol Res Rev Comt, Nat Inst Ment Health, 74-78, res career development awards, 75-; assoc dir basic res, Clin Res Ctr; assoc ed, J Neurosci, 80- & consult ed, Physiol Psychol, 81- *Honors & Awards:* Res Scientist Award, NIMH, 86. *Mem:* AAAS; Am Soc Neurochem; Soc Neurosci; Am Soc Pharmacol & Exp Therapeut; Sigma Xi; NY Acad Sci. *Res:* Interactions between brain neurochemistry, behavior and environment; control of biogenic amine metabolism; neuroplasticity; biological basis of recovery of function following brain damage; animal model of Parkinsonism, biogenic amines and stress. *Mailing Add:* Life Sci Dept Univ Pittsburgh 573 Crawford Hall Pittsburgh PA 15260

ZIGMOND, RICHARD ERIC, b Willimantic, Conn, May 9, 44; m. NEUROPHARMACOLOGY, NEUROBIOLOGY. *Educ:* Harvard Col, BA, 66; Rockefeller Univ, PhD(neurobiol), 71. *Prof Exp:* Fel physiol psychol, Rockefeller Univ, 71-72; fel neurochem, Univ Cambridge, 72-75; from asst prof to assoc prof, Harvard Med Sch, 75-89; PROF, CASE WESTERN RESERVE UNIV, 89- *Concurrent Pos:* Tutor biochem sci, Harvard Col, 75-76; lectr neurobiol course, Marine Biol Lab, 81-84; lectr neurobiol & behav course, Cold Spring Harbor Lab, 79-; lectr, rev & update neurosurgeons, Woods Hole, Mass, 84, 86, 88; chmn, Gordon Conf Neuronal Plasticity, 91. *Mem:* Soc Neurosci; Brit Pharmacol Soc; Am Soc Pharmacol & Exp Therapeut; AAAS; Endocrine Soc. *Res:* Regulation of the levels of enzymes involved in the synthesis of neurotransmitters; recovery of function after neural damage; functional anatomy of the sympathetic nervous system. *Mailing Add:* Dept Neurosci Case Western Reserve Univ 2119 Abington Rd Cleveland OH 44106

ZIGMOND, SALLY H, b Kalamazoo, Mich, Feb 5, 44. CHEMOTOXIS, CELL LOCOMOTION. *Educ:* Wellesley Col, BA, 66; Rockefeller Univ, PhD(biol), 72. *Prof Exp:* From asst prof to assoc prof, 76-87, PROF BIOL, UNIV PA, 87- *Concurrent Pos:* Mem, cell biol study sect, NIH, 82-84; asst ed, J Cell Biol, 86- *Mem:* Am Soc Cell Biol. *Res:* The motile activities of polymorphonuclear leukocytes including chemotoxis, locomotion and pinocytosis; correlation of cell behavior with biochemical events. *Mailing Add:* Dept Biol Leidy Lab Univ Pa Philadelphia PA 19104-6018

ZIGRANG, DENIS JOSEPH, b Livermore, Iowa, May 11, 26; m 48; c 5. SYSTEM DESIGN ENGINEERING, DESIGN ENGINEERING. *Educ:* Iowa State Col, BS, 49 & 50; Univ Tulsa, MS, 70, PhD(chem eng), 76. *Prof Exp:* Supvr qual, Minn Mining & Mfg Co, 50-52; nuclear res engr, NAm Aviation, 52-55, sr res engr aerospace, 55-58; engr specialist, Martin Co, 58-62; chief mech engr, Bendix Corp, 63-66; engr specialist systs, Rockwell Int, 66-78; ASSOC PROF MECH ENG, UNIV TULSA, 78- *Concurrent Pos:* Vis lectr, Univ Colo, 59-61. *Mem:* AAAS; Am Soc Eng Educ; Am Soc Mech Engrs; Am Inst Aeronaut & Astronaut. *Res:* Diffusion of moisture or solvents with composite materials; application of probability in engineering design; corrosion of oil-gas field materials; computational methods, particularly replacement of nomographs with explicit solution approximations. *Mailing Add:* Dept Mech Eng Univ Tulsa 600 S College Ave Tulsa OK 74104

ZIHLMAN, ADRIENNE LOUELLA, b Chicago, Ill, Dec 29, 40. PHYSICAL ANTHROPOLOGY. *Educ:* Univ Colo, Boulder, BA, 62; Univ Calif, Berkeley, PhD(anthrop), 67. *Prof Exp:* From asst prof to assoc prof, 67-79, chmn dept, 75-77, PROF ANTHROP, OAKES COL, UNIV CALIF, SANTA CRUZ, 79- *Concurrent Pos:* Wenner-Gren Found Anthrop Res grants, Transvaal Mus, Pretoria, SAfrica, Univ Witwatersrand, Anthrop Inst, Zurich & Med Sch, Makerere Univ, Uganda, 69, Nat Mus Kenya, Nairobi & Transvaal Mus, 74, Transvaal Mus, Univ Witwatersrand, 79. *Mem:* AAAS; Am Asn Phys Anthrop; Am Anthrop Asn; Am Soc Mammal; Int Primatol soc. *Res:* Locomotor behavior and anatomy of primates; reconstruction of behavior and anatomy of fossil hominoids; ape evolution and human origins; women in evolution. *Mailing Add:* Dept Anthrop Womens Studies Univ Calif Santa Cruz CA 95064

ZIKAKIS, JOHN PHILIP, b Piraeus, Greece, Sept 16, 33; US citizen; m 58; c 1. BIOLOGICAL SCIENCES, BIOCHEMISTRY. *Educ:* Univ Del, BA, 65, MS, 67, PhD(biol, biochem), 70. *Prof Exp:* Res asst nutrit, Stine Lab, E I du Pont de Nemours & Co, Inc, 59-61; res assoc biochem genetics, Univ Del, 68-70, from asst prof to assoc prof biochem genetics, 70-81, prof agr biochem, Col Agr, 81-89, prof marine biol & biochem, Col Marine Sci, 81-89, prof food sci, Col Human Resources, 87-89, EMER PROF, UNIV DEL, 89-; VPRES, UNITED CHITOTECHNOL, INC, 89- *Concurrent Pos:* Sci consult, Fedn Am Socs Exp Biol, 75; trustee, Riverside Hosp, 77-84; Nat Oceanic & Atmospheric Admin grants, 77-84; vis prof & Fulbright scholar, Univ Panama, 84-85, sci adv, 85-87; adv to pres, Univ Panama, 85-89. *Honors & Awards:* Cert & Gold Medal, Univ Patra, Greece, 73. *Mem:* Am Chem Soc; Am Dairy Sci Asn; Am Inst Biol Sci; AAAS; NY Acad Sci; Inst Food Technologists; Sigma Xi; Am Chitosci Soc. *Res:* Various studies with xanthine oxidase as it relates to atherosclerosis; immunological and nutritional studies with xanthine oxidase; biochemical genetic studies on milk and blood protein; polymorphisms; animal product and by-product biochemistry; mammary metabolism and enzymology; marine by-products in nutrition, pharmaceuticals and biotechnology; chitin and chitosan specialists; patentee in field; contributor of over 125 articles in professional journals and author of one book; conducted pioneering research in enzymology which led to the discovery, isolation and purification of human colostral xanthine oxidase; nutritional studies using chitin and whey resulted in the alleviation of lactose intolerance. *Mailing Add:* 6039 Collins Ave No 1605 Miami Beach FL 33140

ZILBER, JOSEPH ABRAHAM, b Boston, Mass, July 27, 23; m 54; c 3. MATHEMATICS. *Educ:* Harvard Univ, AB, 43, MA, 46, PhD, 63. *Prof Exp:* Instr math, Columbia Univ, 48-50 & Johns Hopkins Univ, 50-55; asst prof, Univ Ill, 55-56; lectr, Northwestern Univ, 56-57; res assoc, Brown Univ, 57-62; res assoc, Yale Univ, 62; asst prof, 62-64, ASSOC PROF MATH, OHIO STATE UNIV, 64- *Concurrent Pos:* Assoc ed, Math Rev, Am Math Soc, 57-62. *Mem:* AAAS; Am Math Soc; Math Asn Am. *Res:* Algebraic topology; category theory. *Mailing Add:* Dept Math Ohio State Univ Columbus OH 43210-1174

ZILCH, KARL T, b St Louis, Mo, Nov 14, 21; m 50; c 7. ORGANIC CHEMISTRY. *Educ:* Univ Mo, AB, 43, MA, 47, PhD(chem), 49. *Prof Exp:* Asst chem, Univ Mo, 47-49; res chemist, Northern Utilization Res Br, USDA, 49-55; res chemist & group leader, 55-59, res sect head, 59-61, TECH DIR, EMERY INDUSTS, INC, 61- *Concurrent Pos:* Instr chem, Bradley Univ, 50-51. *Mem:* Am Chem Soc; Am Oil Chem Soc; Swiss Chem Soc. *Res:* Synthesis and processing carboxylic acids; reactions and end use application of carboxylic acids. *Mailing Add:* 7682 Pine Glen Dr Cincinnati OH 45224-1229

ZILCZER, JANET ANN, b New York, NY, Apr 30, 55. CRYSTALLOGRAPHY, EDUCATION. *Educ:* George Wash Univ, BA, 76, MPhil, 79, PhD(geol), 81. *Prof Exp:* Teaching fel geol, George Wash Univ, 76-79; Smithsonian fel, Nat Mus Natural Hist, 79-81; lectr geol, George Mason Univ, 81-82; instrnl asst, 84-87, LECTR, NORTHERN VA COMMUNITY COL, 83-, MGT ANALYST, 88- *Concurrent Pos:* Vis researcher, Nat Mus Natural Hist, Smithsonian Inst, 78; vis asst prof, George Washington Univ, 82-83, asst prof lectr, 83-84; abstractor, Am Mineralogist, 83-85 & Mineral Abstr, 83- *Mem:* Mineral Soc Am; Geol Soc Am; Int Asn Math Geol. *Res:* Feldspar mineralogy; optical mineralogy; crystal physics and chemistry. *Mailing Add:* 2351 N Quantico St Arlington VA 22205

ZILE, MAIJA HELENE, b Latvia, Aug 3, 29; nat US; div; c 3. BIOCHEMISTRY. *Educ:* Univ Md, BS, 54; Univ Wis, MS, 56, PhD(biochem), 59. *Prof Exp:* Res fel biochem, Univ Wis, 59 & Harvard Univ, 59-61; res assoc biochem, Univ Wis, Madison, 61-76, assoc scientist, 76-81; asst prof, 81-85, ASSOC PROF, MICH STATE UNIV, 85- *Mem:* Sigma Xi; Am Inst Nutrit; Soc Exp Biol & Med; NY Acad Sci; Am Asn Cancer Res. *Res:* Metabolism and function of vitamin A; function of vitamin A in cell proliferation and differentiation; anticarcinogenic properties of vitamin A; nutrition and cancer. *Mailing Add:* Dept Food Sci & Human Nutrit Mich State Univ East Lansing MI 48824

ZILINSKAS, BARBARA ANN, b Waltham, Mass, Sept 21, 47. MOLECULAR BIOLOGY. *Educ:* Framingham State Col, BA, 69; Univ Ill, Urbana, MS, 70, PhD(biol), 75. *Prof Exp:* Lab technician biol, Univ Mass Environ Exp Sta, 68-69; NASA fel, Univ Ill, Urbana, 69-72, res & teaching asst, 73-74; fel, Smithsonian Radiation Biol Lab, 75; asst prof, 75-80, assoc prof biochem, 80-86, PROF, COOK COL, RUTGERS UNIV, 86- *Concurrent Pos:* Vis scholar, Harvard Univ, 82-83. *Mem:* Am Soc Plant Physiologists; Plant Molecular Biol Soc; Am Soc Photobiol; AAAS; Sigma Xi. *Res:* Plant response to environmental stress; biochemistry and biophysics of the photosynthetic light reactions; plant molecular biology. *Mailing Add:* Dept Biochem & Microbiol Cook Col Rutgers Univ New Brunswick NJ 08903

ZILKE, SAMUEL, b Chatfield, Man, Nov 4, 14. AGRONOMY, ECOLOGY. *Educ:* Univ Sask, BA, 48, BSEd, 49, BSAgr, 53, MS, 54; SDak State Univ, PhD(agron), 67. *Prof Exp:* Res officer & res asst plant ecol, Univ Sask, 50-55; res asst bot, SDak State Univ, 57-59 & agron & weed sci, 59-61; instr agr, Exten & Col Div, Alta Agr Col, 66-70; agr consult, 70-71; technician, Can Wildlife Serv, 71-72; AGR CONSULT, 72- *Concurrent Pos:* Lectr, Col Agr, Univ Sask, 52-55, consult fertilizers & herbicides, 66-71. *Mem:* Agr Inst Can; Ecol Soc Am; Can Soc Soil Sci; Am Inst Biol Sci; Agron Soc Am. *Res:* Ecological life histories of plants and their physiology under field conditions; effect of variable soil moisture and temperature on seeds; response of field and grass crops to fertilizer and herbicides; effect of aeration of soil on nitrification in late May and June; improvement of habitat for browsers such as Virginia deer, porcupine and rabbits; investigation into the effects of minimum tillage on yields of field crops. *Mailing Add:* Box 147 Springside Saskatchewan SK S0A 3V0 Can

ZILKEY, BRYAN FREDERICK, b Manitou, Man, Apr 14, 41; m 64; c 3. PLANT SCIENCE, BIOCHEMISTRY. *Educ:* Univ Man, BSA, 62, MSc, 63; Purdue Univ, PhD(plant physiol), 69. *Prof Exp:* RES SCIENTIST TOBACCO, RES BR, CAN DEPT AGR, 69- *Mem:* Can Soc Plant Physiol; Am Soc Plant Physiol; Can Fedn Biol Scis; Agr Inst Can; Weed Sci Soc Am. *Res:* Lipid and carbohydrate metabolism and biosynthesis in germinating and developing castor bean endosperm; biochemistry and physiology of tobacco growth; tobacco smoke chemistry and biological properties; weed control in tobacco, ginseng, sweet potatoes; winter cereals; ginseng. *Mailing Add:* Res Sta Agr Can Delhi ON N4B 2W9 Can

ZILL, LEONARD PETER, b Portland, Ore, Oct 9, 20; m 52; c 2. BIOCHEMISTRY. *Educ:* Ore State Col, BS, 42, MS, 44; Ind Univ, PhD(biochem), 50. *Prof Exp:* Asst chem, Ore State Col, 42-44; asst, Forest Prod Lab, US Forest Serv, 42-44; res assoc, Ind Univ, 47-50; biochemist, Oak Ridge Nat Lab, 50-57; sr biochemist, Res Inst Advan Studies, 57-63; chief biol adaptation br, 63-67, mem lunar sample anal planning team, 68-70, chief planetary biol div, 67-74, SR RES SCIENTIST, AMES RES CTR, NASA, 74- *Mem:* AAAS; Am Soc Photobiol; Am Soc Biol Chemists; Sigma Xi; Brit Biochem Soc. *Res:* Growth factors; wood and pulp chemistry; biochemistry of paramecia; chemistry and biochemistry of carbohydrates; photosynthesis; plant lipids. *Mailing Add:* 1525 Samedra Sunnyvale CA 94087

ZILLER, STEPHEN A, JR, b Kansas City, Mo, Nov 2, 38; m 61; c 4. TOXICOLOGY, REGULATORY AFFAIRS. *Educ:* Rockhurst Col, BA, 61; St Louis Univ, PhD(biochem), 67. *Prof Exp:* Res biochemist, Res Div, Procter & Gamble Co, 67-69, res nutritionist, 69-70, nutritionist, Food Prod Develop Div, 71-74, sect head food safety & nutrit, Food Prod Develop Div, 74-77, assoc dir, Indust Food Prod Develop, 77-81, assoc dir prof & regulatory serv, Beverage Prod Develop Div, 81-86, Food Prod Develop, 86-89, ASSOC DIR PROF & REGULATORY SERV, FOOD & BEVERAGE PROD, RES & DEVELOP, PROCTER & GAMBLE CO, 89- *Concurrent Pos:* Chmn, Tech Comt, Inst Shortening & Edible Oils, 89-, Sci Res Comt, Nat Food Processors Asn, 90- *Mem:* Am Chem Soc; Sigma Xi. *Res:* Metabolism of sterols and bile acids; drug metabolism; protein nutrition; food, beverage safety and nutrition. *Mailing Add:* 6071 Center Hill Rd Cincinnati OH 45224-1703

ZILVERSMIT, DONALD BERTHOLD, b Hengelo, Holland, July 11, 19; nat US; m 45; c 3. NUTRITIONAL BIOCHEMISTRY. *Educ:* Univ Calif, BS, 40, PhD(physiol), 48. *Hon Degrees:* Dr, Univ Utrecht, Neth, 80. *Prof Exp:* Clin demonstr, Dent Sch, Univ Calif, 46-48; from instr to prof physiol, Med Units, Univ Tenn, 48-66; prof, 66-90, EMER PROF, DIV NUTRIT SCI & SECT BIOCHEM, MOLECULAR & CELL BIOL, DIV BIOL SCI, CORNELL UNIV, 90- *Concurrent Pos:* Consult, NIH, 55-; ed, J Lipid Res, 59-61; Am Heart Asn career investr, 59-; guest prof, State Univ Leiden, 61-62; vis fel exp path, Australian Nat Univ, 66; ed, Biochimica & Biophysica Acta, 69-80; NIH, Nat Heart, Lung & Blood Inst task forces arteriosclerosis, 70-71, 78-79, 80-82; vis prof biochem, Mass Inst Technol, 72-73; ed, Proceedings Soc Exp Biol & Med, 75- & adv bd, 77-; mem coun arteriosclerosis, Am Heart Asn. *Honors & Awards:* Borden Award, Am Inst Nutrit, 76; George Lyman Duff Mem Lectr, 78. *Mem:* Nat Acad Sci; Am Physiol Soc; Soc Exp Biol & Med; Am Soc Biol Chemists; Am Inst Nutrit; Philos Sci Asn; AAAS. *Res:* Lipid metabolism; lipoproteins; membrane biochemistry; endocytosis; arteriosclerosis; use of isotopes in metabolic work. *Mailing Add:* Div Nutrit Sci Cornell Univ Ithaca NY 14853

ZILZ, MELVIN LEONARD, b Detroit, Mich, Apr 15, 32; m 57; c 3. CELL BIOLOGY, BIOCHEMISTRY. *Educ:* Concordia Teachers Col, Ill, BS, 53; Univ Mich, Ann Arbor, MS & MA, 64; Wayne State Univ, PhD(biol), 70; Concordia Theol Sem, Ft Wayne, colloquy dipl, 78. *Prof Exp:* Teacher pvt sch, Ill, 53-57; instr high sch, Mich, 57-65, chmn dept sci, 58-65; asst prof biol, Concordia Sr Col, 65-72, chmn dept natural sci, 71-72, assoc prof biol & registr admis, 72-77, asst pres & assoc acad dean, 76-78, assoc prof ministry, 76-79, dean admin, 78-83, dir planning & budget admin, 83-85, PROF PASTORAL MINISTRY, CONCORDIA THEOL SEM, 79- *Concurrent Pos:* Instr, Mich Lutheran Col, 62-63; ordained clergyman, 78- *Res:* Cellular research, especially cell division and the anaphase movement of chromosomes; biomedical ethics. *Mailing Add:* Concordia Theol Sem 6600 N Clinton Ft Wayne IN 46825-4996

ZIMAR, FRANK, b Berlin, Wis, Apr 5, 18; m 43, 57; c 7. CHEMISTRY, CERAMICS. *Educ:* Univ Wis, BS, 41; Univ Rochester, PhD(phys chem), 45. *Prof Exp:* Asst, Off Sci Res & Develop, Univ Rochester, 41-45; res chemist, 45-51, res assoc, 51-56, res supvr, 56-70, res assoc, 70-75, SR RES SCIENTIST, CORNING GLASS WORKS, 75- *Mem:* Am Ceramic Soc; Brit Soc Glass Technol; Sigma Xi. *Res:* Kinetics of heterogeneous reactions; surface chemistry of glass; thermal setting solder glass; glass redraw; application of films to glass; glass fiber technology; fiber optic waveguides; catalyst support ceramics. *Mailing Add:* 106 Lake St PO Box 296 Hammondsport NY 14840

ZIMBELMAN, JAMES RAY, Jamestown, NDak, Sept 10, 54; m 76. PLANETARY GEOLOGY, REMOTE SENSING. *Educ:* Northwest Nazarene Col, BA, 76; Univ Calif, Los Angeles, MS, 78; Ariz State Univ, PhD(geol), 84. *Prof Exp:* Postdoctoral fel, Lunar & Planetary Inst, 84-86, staff scientist, 86-88; GEOLOGIST, CEPS/NASM, SMITHSONIAN INST, 88- *Concurrent Pos:* Lectr, Univ Houston, Clear Lake, 86-88; prin investr, Grants Planetary Geol & Geophys, NASA, 86-; dir, Regional Planetary Image Facil, 89-91. *Mem:* Am Geophys Union; Geol Soc Am. *Res:* Analysis of high-resolution imaging and thermal infrared data of Mars, examining opperant geologic processes; sand transport and deposition in the Mojave Desert of California, as a possible indicator of recent climatic change. *Mailing Add:* CEPS/NASM MRC 315 Smithsonian Inst Washington DC 20560

ZIMBELMAN, ROBERT GEORGE, b Keenesburg, Colo, Sept 4, 30; m 52, 87; c 4. REPRODUCTIVE ENDOCRINOLOGY, ANIMAL HEALTH. *Educ:* Colo State Univ, BS, 52; Univ Wis, MS, 57, PhD(endocrinol), 60. *Prof Exp:* Instr genetics, Univ Wis, 57-60; res assoc, Upjohn Co, 60-64, head reproduction & physiol res, Agr Prod Div, 65-71, res mgr reproduction & physiol res, 71-76, exp agr sci, 77-83, dir reproduction & growth physiol res, 83-86, sci affairs, 86-87; EXEC VPRES, AM SOC ANIMAL SCI, 87- *Mem:* Am Soc Animal Sci; Am Inst Biol Sci; Soc Study Reproduction; fel AAAS; NY Acad Sci; Am Dairy Sci Asn. *Res:* Improving fertility and preciseness of breeding time in cattle with synchronization of estrus, improved mammary development and milk production by dairy cattle, endocrinology of growth of farm animals, new approaches to inhibition of estrus in pets, molecular biology in animals. *Mailing Add:* Am Soc Animal Sci 9650 Rockville Pike Bethesda MD 20814

ZIMBRICK, JOHN DAVID, b Dickinson, NDak, Sept 18, 38; m 62; c 2. RADIATION BIOPHYSICS, RADIATION CHEMISTRY. *Educ:* Carleton Col, BA, 60; Univ Kans, MS, 62, PhD(radiation biophys), 67. *Prof Exp:* Asst physicist, IIT Res Inst, 62-64; chief environ studies sect, Health Serv Lab, US AEC, Idaho, 67-68; Nat Inst Gen Med Sci fel lab nuclear med & radiation biol, Univ Calif, Los Angeles, 68, US AEC fel, 68-69; from asst prof to assoc prof, 69-77, prof radiation biophys, Univ Kans, 77-84, chmn, Radiation Biophys, 82-84; exec secy, Radiation Study Sect, NIH, 84-89; MGR, BIOL & CHEM DEPT, BATTELLE PAC NORTHWEST LAB, 89-; PROF RADIOL SCI & MEM GRAD FAC, WASH STATE UNIV, 90- *Concurrent Pos:* Consult mem, Radiation Study Sect, NIH, 78-82, chmn, 80-82; prog mgr radiobiol, US Dept Energy, 81-82; assoc ed, Radiation Res J, 84-88; NSF Fac Sci Fel, Los Alamos Nat Lab, 76-77. *Mem:* Health Physics Soc; Radiation Res Soc (pres, 90-91); Biophys Soc; Sigma Xi; AAAS. *Res:* In vivo studies on DNA base damage induced by gamma radiation; electron spin resonance spectroscopy of radicals produced in biomolecules by radiation; application of electron spin resonance to cancer detection and treatment. *Mailing Add:* Battelle Pac Northwest Lab Battelle Blvd P7-58 Richland WA 99352

ZIMDAHL, ROBERT LAWRENCE, b Buffalo, NY, Feb 28, 35; m 56; c 4. AGRONOMY, WEED SCIENCE. *Educ:* Cornell Univ, BS, 56, MS, 66; Ore State Univ, PhD(agron), 68. *Prof Exp:* Res assoc agron, Cornell Univ, 63-64; from asst prof to assoc prof, 68-77, PROF WEED SCI, COLO STATE UNIV, 77- *Concurrent Pos:* Vis prof, Univ Bologna, Italy, 76; vis scientist, Int Rice Res Inst, Philippines, 84-85; tech adv, USAID, Morocco, 89-90. *Mem:* Fel Weed Sci Soc Am; Am Chem Soc; Am Soc Agron; AAAS. *Res:* Herbicide degradation in soil; environmental pollution by pesticides; weed management in crops. *Mailing Add:* Weed Res Lab Colo State Univ Ft Collins CO 80523

ZIMERING, SHIMSHON, b Kishinev, Romania, July 6, 33; m 66; c 2. MATHEMATICAL ANALYSIS. *Educ:* Univ Geneva, BSc, 56, licence in math, 58; Free Univ Brussels, PhD(math, physics), 65. *Prof Exp:* Res asst, Weizmann Inst Sci, 58-59; res fel, Battelle Inst, Geneva, Switz, 61-66; hon res assoc math & Battelle Inst, Geneva, Switz fel, Harvard Univ, 66-67; mem advan studies ctr, Battelle Inst, Geneva, 67-68; ASSOC PROF MATH, OHIO STATE UNIV, 68- *Concurrent Pos:* Deleg, Int Conf Peaceful Uses Atomic Energy, Geneva, 64; NSF res grant, Ohio State Univ, 73-74. *Honors & Awards:* Award for Res, Battelle Inst, 66. *Mem:* Am Math Soc. *Res:* Real analysis, summability and transform theory; boundary value problems; solid state physics. *Mailing Add:* Dept Math Ohio State Univ 231 W 18th St Ave Columbus OH 43210

ZIMM, BRUNO HASBROUCK, b Kingston, NY, Oct 31, 20; m 44; c 2. BIOPHYSICAL CHEMISTRY, POLYMER CHEMISTRY. *Educ:* Columbia Univ, AB, 41, MS, 43, PhD(chem), 44. *Prof Exp:* Asst chem, Columbia Univ, 41-44; res assoc & instr, Polytech Inst Brooklyn, 44-46; from instr to assoc prof, Univ Calif, 46-52; res assoc, Gen Elec Co, 51-60; PROF CHEM, UNIV CALIF, SAN DIEGO, 60- *Concurrent Pos:* Vis lectr, Harvard Univ, 50-51; vis prof, Yale Univ, 60. *Honors & Awards:* Leo Hendrik Baekland Award, Am Chem Soc, 57; Bingham Medal, Soc Rheol, 60; High Polymer Physics Prize, Am Phys Soc, 63; Chem Sci Award, Nat Acad Sci, 81; Kirkwood Medal, Yale Univ, 82. *Mem:* Nat Acad Sci; Am Chem Soc; Am Phys Soc; Soc Rheol; Am Soc Biol Chemists; Biophys Soc; Am Acad Arts & Sci. *Res:* Theory of macromolecular solutions; properties and structure of high polymers and biological macromolecules. *Mailing Add:* 0317 Dept Chem Univ Calif La Jolla CA 92093-0317

ZIMM, CARL B, b Schenectady, NY, March 31, 54. MAGNETICS MATERIALS, CRYOGENICS. *Educ:* Univ Calif, Santa Cruz, BA, 75; Cornell Univ, PhD (physics), 82. *Prof Exp:* Postdoctoral, Los Alamos Nat Lab, 83; STAFF SCIENTIST, ASTRON CORP, 85- *Mem:* Am Phys Soc; Mat Res Soc. *Res:* Thermal and magnetic properties of materials used in development of magnetic refrigeration. *Mailing Add:* Astronaut Technol Ctr 5800 Cottage Grove Rd Madison WI 53716-1387

ZIMM, GEORGIANNA GREVATT, b Jersey City, NJ, Nov 5, 17; m 44; c 2. DROSOPHILA. *Educ:* Columbia Univ, BA, 40; Univ Pa, MA, 42; Univ Calif, Berkeley, PhD(zool), 50. *Prof Exp:* Teaching asst biol, Univ Del, 40-42; teaching asst zool, Barnard Col, Columbia Univ, 43-45, lectr, 45-46; teaching asst, Univ Calif, Berkeley, 46-50; res asst genetics & bibliogr, 68-75, res assoc biol, 75-85, SPECIALIST GENETICS UNIV CALIF, SAN DIEGO, 85- *Mem:* AAAS; Genetics Soc Am. *Res:* Mutants and cytogenetics of Drosophila. *Mailing Add:* Dept Chem 0317 Univ Calif San Diego La Jolla CA 92093

ZIMMACK, HAROLD LINCOLN, b Chicago, Ill, Feb 12, 25; m 56; c 3. INSECT PATHOLOGY. *Educ:* Eastern Ill Univ, BS, 51; Iowa State Univ, MS, 53, PhD(entom), 56. *Prof Exp:* Asst, Iowa State Univ, 51-56; prof biol, Eastern Ky Univ, 56-63; PROF ZOOL, BALL STATE UNIV, 63- *Concurrent Pos:* Sigma Xi & Ind Acad Sci res grants-in-aid, 67-68. *Mem:* Am Soc Zoologists; Entom Soc Am; AAAS; Soc Invert Path. *Res:* Rapid screening of potential insect pathogen through physiological studies. *Mailing Add:* Dept Biol Ball State Univ Muncie IN 47306

ZIMMER, ARTHUR JAMES, b St Louis, Mo, May 12, 14; m 40; c 1. PHARMACEUTICAL CHEMISTRY. *Educ:* St Louis Col Pharm, BS, 40; Wash Univ, MS, 43, PhD(chem), 46. *Prof Exp:* From instr to prof, St Louis Col Pharm, 41-78, Charles E Caspari prof chem, 78-80; RETIRED. *Concurrent Pos:* Biochemist, Snodgras Lab, City Hosp, 60-; mem, US Pharmacopeial Conv, 70. *Mem:* Am Chem Soc; Am Pharmaceut Asn. *Res:* Instrumental analysis of pharmaceutical compounds. *Mailing Add:* 3401 McAdams Pkwy Godfrey IL 62035-1225

ZIMMER, CARL R(ICHARD), b Syracuse, NY, July 10, 27; m 65. ELECTRICAL ENGINEERING. *Educ:* Cornell Univ, BEE, 49; Syracuse Univ, MEE, 50, PhD(elec eng), 58. *Prof Exp:* Res assoc elec eng, Syracuse Univ, 53-56, instr, 56-59; asst prof, 59-63, ASSOC PROF ENG, ARIZ STATE UNIV, 63- *Concurrent Pos:* Consult, Motorola Aerospace Ctr, 65-67. *Mem:* Inst Elec & Electronics Engrs. *Res:* Solid state devices; active networks. *Mailing Add:* Dept Elec Eng Ariz State Univ Tempe AZ 85281

ZIMMER, DAVID E, b Neoga, Ill, Sept 25, 35; m 56; c 2. PLANT PATHOLOGY, PLANT BREEDING. *Educ:* Eastern Ill Univ, BS, 57; Purdue Univ, MS, 59, PhD(plant path), 61. *Prof Exp:* Res plant pathologist, USDA, NDak State Univ, 61-, adj prof plant path, res leader & tech adv, 77-88; AT RICHARD B RUSSELL AGR RES CTR. *Mem:* AAAS; Am Phytopath Soc; Crop Sci Soc Am. *Res:* Genetics of parasitism with special emphasis on obligate parasites; inheritance and nature of disease resistance in oilseed crops and the improvement of oil-seed crops through disease resistance breeding. *Mailing Add:* 190 Mc Duffie Dr Athens GA 30605

ZIMMER, ELIZABETH ANNE, b Rochester, NY, Dec 19, 51. MOLECULAR EVOLUTION, MOLECULAR GENETICS. *Educ:* Cornell Univ, BS, 73; Univ Calif, Berkeley, PhD(biochem), 81. *Prof Exp:* Postdoctoral fel biol, Stanford Univ, 81-82 & Wash Univ, 83-84; from asst prof to assoc prof, 84-90, ADJ ASSOC PROF BIOCHEM & BOT, LA STATE UNIV, BATON ROUGE, 90-; PRIN INVESTR & BOTANIST, LAB MOLECULAR SYSTEMATICS, SMITHSONIAN, 90- *Concurrent Pos:* Prin investr, NSF grant, Spec Creativity Ext, 87-91; Distinguished lectr, Enhance Visibility Women & Minorities in Sci, Univ Calif, Irvine, 90; USDA panelist, Plant Genome Initiative, 91; lectr, Class 54, Univ NH, 91; ed, Methods Enzymol Vol Molecular Evolution, Molecular Phylogenetics & Evolution, 91-; assoc ed, J Heredity, 90- & Molecular Biol & Evolution, 91-; adj assoc prof bot, Duke Univ, 91-, genetics, George Washington Univ, 91- *Honors & Awards:* Dobzhansky Prize, Soc Study Evolution, 82. *Mem:* Soc Study Evolution; Am Soc Biochem & Molecular Biol; Genetics Soc Am; Int Soc Plant Molecular Biol; AAAS; Asn Women Sci. *Res:* Plant molecular evolution; ribosomal gene evolution; correlations between molecular & organismal characters for tracing evolution of flowering plants; multigene family differentiation. *Mailing Add:* Lab Molecular Systematics MSC NMNH Smithsonian Inst Washington DC 20560

ZIMMER, G(EORGE) A(RTHUR), b Chicago, Ill, Sept 11, 21; m 46; c 1. MECHANICAL ENGINEERING. *Educ:* Purdue Univ, BS, 47; Univ Del, MME, 53. *Prof Exp:* Engr, Fairbanks, Morse & Co, 47-48 & Aberdeen Proving Ground, US Dept Army, 48-54; res engr, Res Ctr, 54-61, res dir, 61-63, chief engr, 63-66, vpres eng, Morse Chain Div, 66-80, group vpres eng, Trans Comp Group, Borg-Warner Corp, 80-83; CONSULT, 88- *Mem:* Am Soc Mech Engrs; Sigma Xi; Am Soc Metals. *Res:* Mechanics and materials; experimental stress analysis; mechanical power transmission design and development. *Mailing Add:* 101 Woolf Lane Ithaca NY 14850

ZIMMER, HANS, b Berlin, Ger, Feb 5, 21; m 46; c 1. ORGANIC CHEMISTRY. *Educ:* Tech Univ, Berlin, Cand, 47, Dipl, 49, DrIng, 50. *Prof Exp:* From asst prof to assoc prof, 54-61, PROF CHEM, UNIV CINCINNATI, 61- *Concurrent Pos:* Vis prof, Univ Mainz, 66-67, Univ Bonn, 67, Univ Bern, 71 & Univ Stuttgart, 83; ed, Methodicum Chemicum; consult, Lithium Corp Am & Matheson, Coleman & Bell, 71-, Morton-Thiocol, 88-; ed, Ann Reports Inorg & Gen Syntheses, 72-78. *Honors & Awards:* Riereschl Award, 90. *Mem:* Fel AAAS; Am Chem Soc; Ger Chem Soc; fel Humboldt Soc. *Res:* Synthetic and metal organic chemistry; organophosphoros chemistry; environmental chemistry. *Mailing Add:* Dept Chem Univ Cincinnati Cincinnati OH 45221-0172

ZIMMER, JAMES GRIFFITH, b Lynbrook, NY, Apr 10, 32; m 71; c 3. PREVENTIVE MEDICINE, COMMUNITY HEALTH. *Educ:* Cornell Univ, BA, 53; Yale Univ, MD, 57; London Sch Hyg & Trop Med, dipl trop pub health, 66; Am Bd Internal Med, dipl, 65. *Prof Exp:* Intern internal med, Grace-New Haven Community Hosp, Conn, 57-58; resident, Strong Mem Hosp, Rochester, NY, 58-60; asst chief dermat, Walter Reed Army Inst Res, 61-63; from sr instr to asst prof prev med & community health, 63-67, actg chmn dept prev med, 68-69, ASSOC PROF PREV MED & COMMUNITY HEALTH, SCH MED & DENT, UNIV ROCHESTER, 68- *Concurrent Pos:* Milbank fac fel, Univ Rochester, 64-71; pres, exec dir, Med Serv Int, Inc, 68-70; pres, Genesee Valley Med Found, 70-79; med dir, Regional Utilization & Med Rev Proj, 71- *Honors & Awards:* Key Award, Am Pub Health Asn, 90. *Mem:* Am Fedn Clin Res; Am Pub Health Asn; Int Epidemiol Asn; fel Am Col Prev Med; Royal Soc Trop Med & Hyg; assoc Am Col Epidemiol. *Res:* Community health; medical care research, especially in areas of utilization and quality of care review and care of chronically ill and aged. *Mailing Add:* Dept Community & Prev Med Univ Rochester Med Ctr Rochester NY 14642

ZIMMER, LOUIS GEORGE, b Marseilles, Ill, Nov 30, 26; m 48; c 2. GEOLOGY. *Educ:* Augustana Col, BA, 50; Univ Iowa, MS, 52. *Prof Exp:* Subsurface geologist, Ohio Oil Co, 52-57; dist geologist, J M Huber Corp, 57-62; partner, 62-84, OWNER, MAGAW & ZIMMER, 84- *Mem:* Am Asn Petrol Geologists; Soc Independent Prof Earth Scientists. *Res:* Petroleum geology. *Mailing Add:* 521 Canyon Rd Edmond OK 73034

ZIMMER, MARTIN F, b Metz, France, Apr 25, 29; US citizen; m; c 3. THERMODYNAMICS, EXPLOSIVES. *Educ:* Univ Munich, BS, 55, MS, 58; Munich Tech Univ, PhD(chem technol), 61. *Prof Exp:* Head fuel res lab, Ger Aeronaut Res Inst, Munich, 60-62; chemist, Naval Ord Sta, 62, head thermodyn br, 62-70; dir high explosive res & develop lab, Eglin AFB, Fla, 70-76; res prog mgr conventional munition, Air Force Off Sci Res, Bolling AFB, Washington, DC, 76-78; tech dir, Munitions Div, Air Force Armament Lab, 80-88, syst prog dir, Int Modular Standoff Weapon (5 nations), MSD, 88-90, TECH DIR, MUNITIONS DIV, WRIGHT LAB ARMAMENT DIRECTORAT, EGLIN AFB, FLA, 90- *Mem:* Combustion Inst. *Res:* Thermodynamics and combustion characteristics of energetic materials; detonation physics and explosive related phenomena; scientific and administrative management. *Mailing Add:* 124 Bayow Dr Niceville FL 32578

ZIMMER, RUSSEL LEONARD, b Springfield, Ill, Nov 7, 31; m; c 2. INVERTEBRATE BIOLOGY. *Educ:* Blackburn Col, AB, 53; Univ Wash, MS, 56, PhD(zool), 64. *Prof Exp:* Instr, Univ Southern Calif, 60-63, vis asst prof, 63-64, asst prof, 64-68, resident dir, Santa Catalina Marine Biol Lab, 68-76, ASSOC PROF BIOL SCI, UNIV SOUTHERN CALIF, 68- *Mem:* AAAS; Am Soc Zool; Soc Syst Zool; Marine Biol Asn UK. *Res:* Reproductive biology; larval development, metamorphosis and systematics of minor invertebrate phyla, especially Phoronida and Bryozoa. *Mailing Add:* Dept Biol Sci 0371 Univ Southern Calif Univ Park Los Angeles CA 90089-0371

ZIMMER, STEPHEN GEORGE, b Trenton, NJ, Oct 26, 42; m 67; c 2. VIROLOGY, MOLECULAR BIOLOGY. *Educ:* Rutgers Univ, AB, 64, MS, 66; Univ Colo, PhD(exp path), 73. *Prof Exp:* Fel molecular biol virol, Washington Univ, 74-76; ASST PROF PATH, UNIV KY, 76- *Mem:* Am Soc Microbiol; AAAS; Am Asn Cancer Res. *Res:* Regulation of viral gene expression; molecular mechanisms of viral transformation. *Mailing Add:* Dept Microbiol Univ Ky Med Sch 800 Rose St Lexington KY 40536

ZIMMER, WILLIAM FREDERICK, JR, b Glouster, Ohio, June 4, 23; m 44; c 6. ORGANIC POLYMER CHEMISTRY. *Educ:* Ohio State Univ, BSc, 48, MSc, 49, PhD(chem), 52. *Prof Exp:* Org chemist, Res Lab, Durez Plastics & Chems, Inc, 52-55; res supvr, Hooker Chem Corp, 55-59, mgr polymer res, 59-62; mgr fiber res, Behr-Manning Div, Norton Co, 63-64, group leader chem appln res & develop, 64-68, asst dir res, Grinding Wheel Div, 68-74, res assoc, 74-86; RETIRED. *Concurrent Pos:* Consult, 86-88. *Mem:* Am Chem Soc. *Res:* Organofluorine chemistry; plastics and polymer chemistry and applications; fiber technology; abrasive systems and materials research; new product research and development. *Mailing Add:* 3629 Carmichael Dr Punta Gorda FL 33950

ZIMMERBERG, HYMAN JOSEPH, b New York, NY, Sept 7, 21; m 43; c 3. MATHEMATICS. *Educ:* Brooklyn Col, BA, 41; Univ Chicago, MS, 42, PhD(math), 45. *Prof Exp:* Instr math, Univ Chicago, 42-45 & NC State Col, 45-46; from instr to assoc prof math, 46-60, PROF MATH, RUTGERS UNIV, NEW BRUNSWICK, 60- *Concurrent Pos:* Vis prof, Univ Ore, 61 & Drew Univ, 62 & 63; dir, Undergrad Res Participation, NSF, Rutgers Univ, 62-69, 71-73 & 77; consult, NSF Student & Coop Prog, 67-70. *Mem:* Am Math Soc; Math Asn Am; Sigma Xi; Am Asn Univ Prof. *Res:* Boundary value problems; linear integro-differential-boundary-parameter problems; self-adjoint systems. *Mailing Add:* Dept Math Hill Ctr Rutgers Univ New Brunswick NJ 08903

ZIMMERER, ROBERT P, b Sheboygan, Wis, Dec 7, 29; m 56; c 3. PLANT PHYSIOLOGY, MICROBIOLOGY. *Educ:* Univ Wis, BS, 54; Cornell Univ, MS, 61; Pa State Univ, PhD(bot), 66. *Prof Exp:* Asst to sales mgr, Stauffer Chem Co, 55-56; asst plant mgr, Hopkins Agr Chem Co, 56-57; chemist, Marathon Div, Am Can Co, 57-59; asst bot, Cornell Univ, 59-61; from instr to assoc prof biol, 61-74, chmn dept, 74-77, DANA PROF BIOL, JUNIATA COL, 74-, CHMN DEPT, 87- *Concurrent Pos:* Res assoc, Hershey Med Ctr, Pa State Univ, 70; vis prof, Univ Maine, 71; consult, USDA, 72-74 & J C Blair Mem Hosp, 74-80. *Mem:* Sigma Xi; Am Soc Microbiol. *Res:* Biogenesis and function of biological membranes; cell cycle; extreme plant stress. *Mailing Add:* Dept Biol Juniata Col Huntingdon PA 16652

ZIMMERER, ROBERT W, b Brooklyn, NY, May 21, 29; m 60; c 2. EXPERIMENTAL PHYSICS, INSTRUMENTATION. *Educ:* Worcester Polytech Inst, BS, 51; NY Univ, MS, 55; Univ Colo, PhD(physics), 60. *Prof Exp:* Design engr, Hazeltine Electronics Corp, 51-55; asst physics, Univ Colo, 55-57, res assoc, 57-60; physicist, Nat Bur Stand, 60-66; chief scientist, Wm Ainsworth & Sons, Inc, Colo, 66-69; pres, Scientech, Inc, 69-72, vpres, 72-81; CONSULT, 81- *Concurrent Pos:* Dept Com sci fel, 65-66; phys sci consult, US Army Fitzsimons Gen Hosp, 69-85. *Mem:* Instrument Soc Am; Am Phys Soc; AAAS. *Res:* Microwave spectroscopy of gases; microwave generation and propagation at very short wavelengths; microwave power measurement; mass measurement by new methods; lung physiology; microcomputer modeling and use in classroom teaching. *Mailing Add:* 131 W 2nd St Port Angeles WA 98362

ZIMMERING, STANLEY, b New York, NY, Apr 14, 24; m 51; c 3. GENETICS. *Educ:* Brooklyn Col, AB, 47; Columbia Univ, AM, 49; Univ Mo, PhD(zool), 53. *Prof Exp:* Lectr biol & res assoc, Univ Rochester, 53-55; asst prof, Trinity Col, Conn, 55-59; res exec zool, Ind Univ, 59-62; assoc prof biol, 62-66, PROF BIOL, BROWN UNIV, 66- *Res:* Segregation mechanisms; radiation genetics; chemical mutagenesis. *Mailing Add:* Dept Biol Brown Univ Brown Sta Providence RI 02912

ZIMMERMAN, ARTHUR MAURICE, b New York, NY, May 24, 29; m 53; c 3. PHYSIOLOGY, CELL BIOLOGY. *Educ:* NY Univ, BA, 50, MS, 54, PhD(cell physiol), 56. *Prof Exp:* Technician, NY Univ, 51-52; res assoc, NY Univ & Marine Biol Lab, Woods Hole, 55-56; Lalor res fel, Marine Biol Lab, Woods Hole, 56; Nat Cancer Inst res fel, Univ Calif, 56-58; from instr to asst prof pharmacol, Col Med, State Univ NY Downstate Med Ctr, 58-64; grad secy, Dept Zool, 70-75, assoc chmn grad affairs, 75-78, assoc dean, Div IV, Sch Grad Studies, 78-81, actg dir, Inst Immunol, 80-81, PROF ZOOL, UNIV TORONTO, 64- *Concurrent Pos:* Mem corp, Marine Biol Lab, Woods Hole; assoc ed, Can J Biochem, 80-85; ed, Cell Biol Series, Acad Press; vis prof anat, Univ Tex Health Sci Ctr, San Antonio; vis scientist, Weizmann Inst Sci, 82; vis prof, Univ Fla, Gainesville, 83; ed, Biochem & Cell Biol, 85- & Exp Cell Res. *Mem:* Am Soc Cell Biol (treas, 74-78); Can Soc Cell Biol (pres, 76); Int Fedn Cell Biol (secy-gen, 85-); Int Cell Cycle Soc (pres, 86-88). *Res:* Cell division; mechanism of cytokinesis and karyokinesis; nuclear-cytoplasmic interrelations; physiological effects of temperature and pressure; amoeboid movement; physicochemical aspects of protoplasmic gels; cell cycle studies and drug action on cells. *Mailing Add:* Dept Zool Univ Toronto Toronto ON M5S 1A1 Can

ZIMMERMAN, BARRY, b New York, NY, Jan 14, 38; m 62; c 3. ORGANIC POLYMER & GENERAL CHEMISTRY. *Educ:* Brooklyn Col, BSc, 59, AM, 61; Fordham Univ, PhD(org chem), 67. *Prof Exp:* Lectr chem, Brooklyn Col, 59-61; instr, Bronx Community Col, 62-66; proj leader chem & plastics, Union Carbide Corp, NY, 62-72,mkt mgr cellular & elastomer mat, 72-76; mkt area sales mgr, Urethane Intermediates, 76-78; bus mgr urethane prods, 78-80, dir res & develop & mkt, 81-83, VPRES PROD MGMT & TECH SERV, GEN FELT INDUST, INC, SADDLEBROOK, NJ, 86-; VPRES & MANAGING DIR, '21' INT HOLDINGS INC, NY, 87- *Concurrent Pos:* Vpres, Foam Prod Group Inc, 87, dir, 88. *Mem:* Am Chem Soc; fel Royal Soc Chem; NY Acad Sci; fel Am Inst Chemists. *Res:* Synthesis and application of new chemical species in rubber processing; cure accelerators; silanes; inorganic bonding in elastomer matricies; insecticide synthesis; microencapsulation; physiochemical properties of polyelectrolytes and other colloids; polyurethane synthesis and catalysis; general environmental sciences. *Mailing Add:* 5 Tara Dr Pomona NY 10970

ZIMMERMAN, BARRY, b Toronto, Ont, Nov 6, 41; m 64; c 2. IMMUNOLOGY. *Educ:* Univ Toronto, MD, 65; FRCP(Can), 80; Am Bd Pediat, dipl, 81. *Prof Exp:* Asst prof pediat & mem staff, Inst Immunol, Univ Toronto, 72-80; assoc prof pediat, Dept Pediat & Mem Staff, Host Resistance Prog, McMaster Univ, 81-83; ASSOC PROF PEDIAT, UNIV TORONTO, 84-, HEAD, DIV ALLERGY, HOSP SICK CHILDREN, 84- *Concurrent Pos:* Mem staff immunol, Hosp Sick Children & scientist, Res Inst, 71-80; grants, Med Res Coun Can, 71- & Nat Cancer Inst Can, 76-78; Med Res Coun Can scholar, 74-79; sr scientist, Res Inst, Hosp Sick Children, 84- *Mem:* Am Asn Immunol; Am Asn Immunologists; Can Soc Immunol; Am Acad Allergy & Clin Immunol. *Res:* Immunochemistry of lymphocyte surface antigens and receptors; transplantation; homograft prolongation by antilymphocyte and enhancing sera; characterization of leukemic lymphocytes; cellular regulation of IgE antibody production; investigation of asthma.

ZIMMERMAN, BEN GEORGE, b Newark, NJ, July 1, 34; m 60; c 2. PHARMACOLOGY. *Educ:* Columbia Univ, BS, 56; Univ Mich, PhD(pharmacol), 60. *Prof Exp:* Pharmacologist, Lederle Labs, Am Cyanamid Co, 60-61; res fel, Cardiovasc Labs, Col Med, Univ Iowa, 61-63; from asst prof to assoc prof pharmacol, 63-72, PROF PHARMACOL, UNIV MINN, MINNEAPOLIS, 72- *Concurrent Pos:* Mem coun high blood pressure res, Coun Circulation & Coun Basic Sci, Am Heart Asn. *Mem:* Am Soc Pharmacol & Exp Therapeut; Soc Exp Biol & Med. *Res:* Vascular effects of angiotensin and other pressor agents; influence of the sympathetic nervous system on various vascular beds; vascular role of prostaglandins. *Mailing Add:* Dept Pharmacol 3-249 Millard Hall Univ Minn 435 Delaware St SE Minneapolis MN 55455

ZIMMERMAN, C DUANE, b Mayville, Wis, Oct 23, 35. COMPUTER SCIENCE. *Educ:* Andrews Univ, BA, 57; Univ Minn, Minneapolis, MS, 60, PhD(comput sci), 69. *Prof Exp:* Instr math, Southern Missionary Col, 61-63; mathematician, Control Data Corp, 64-66; asst prof comput sci, Univ Minn, Minneapolis, 69-70; asst prof biomath, Loma Linda Univ, 70-80; dir syst develop, HBO & Co, 80-83; SR TECH ADV, HEALTH DATA SCIS CORP, 83- *Mem:* Asn Comput Mach; Inst Elec & Electronics Engrs Comput Soc. *Res:* Computer-medical applications. *Mailing Add:* 1261 Kings Way Redlands CA 92373

ZIMMERMAN, CARLE CLARK, JR, b Winchester, Mass, Apr 25, 34; m 60; c 2. CHEMICAL ENGINEERING. *Educ:* Bucknell Univ, BS, 56; Cornell Univ, PhD(chem eng), 63. *Prof Exp:* Res engr process develop, 63-64, advan res engr, 64-70, sr res engr, 70-74, mgr anal dept, 74-77, mgr petrol chem dept, 77-81, mgr oil shale proj group, 81-83, MGR INSTRUMENTATION & ENG DEPT, DENVER RES CTR, MARATHON OIL CO, 83- *Mem:* Am Inst Chem Engrs; Sigma Xi. *Res:* Development and design of new chemical processes. *Mailing Add:* 2539 Ridge Ct Littleton CO 80120

ZIMMERMAN, CAROL JEAN, b Pittsburgh, Pa, Aug 21, 51; m 83; c 2. ENGINEERING GEOPHYSICS. *Educ:* Rensselaer Polytech Inst, BS, 73; Univ Wis, MS, 76, PhD(oceanog), 80. *Prof Exp:* Res asst, Univ Wis-Madison, 74-80; RES SPECIALIST, EXXON PROD RES CO, 80- *Concurrent Pos:* Teaching asst, Univ Wis-Madison, 76-77. *Mem:* Soc Explor Geophysicists; Am Geophys Union; Marine Technol Soc; Sigma Xi. *Res:* Evaluation and design of seismic source and receiver arrays for offshore oil exploration; development of exploration techniques and strategies for offshore placer exploration; design of exploration strategy for vertical seismic profiling; seismic interpretation; processing for direct hydrocarbon indicators. *Mailing Add:* Exxon Prod Res PO Box 2189 ST 584 Houston TX 77001

ZIMMERMAN, CHERYL LEA, b Chippewa Falls, Wis, May 2, 53; m 83. PHARMACOKINETICS. *Educ:* Univ Wis-Madison, BS, 76; Univ Wash, PhD(pharmaceut), 83. *Prof Exp:* Asst prof, 83-90, ASSOC PROF PHARMACEUT, UNIV MINN, MINNEAPOLIS, 90- *Mem:* Am Asn Pharmaceut Scientists; AAAS; Int Soc Study Xenobiotics. *Res:* Development of in vivo and in situ models for the study of first-pass removal of orally administered drugs. *Mailing Add:* 2374 Bourne Ave St Paul MN 55108

ZIMMERMAN, CRAIG ARTHUR, b Painesville, Ohio, Mar 22, 37; m 62; c 2. PHYSIOLOGICAL ECOLOGY, ANIMAL BEHAVIOR. *Educ:* Baldwin-Wallace Col, BS, 60; Univ Mich, MS, 62 & 64, PhD(bot), 69. *Prof Exp:* From instr to asst prof biol, Centre Col Ky, 67-74; environ specialist, Spindletop Res, Inc, 74-75; assoc prof, 75-80, chmn dept, 75-87, PROF BIOL, AURORA UNIV, 80- *Concurrent Pos:* Consult, Deuchler Eng, Kane County Govt & Morton Arboretum; chmn, Acca Bot Prog, Morton Arboretum, 90-; chmn, Biol Div, Assoc Cols Chicago area, 91- *Mem:* Am Inst Biol Sci; AAAS; Ecol Soc Am; Sigma Xi. *Res:* Causes, characteristics, and evolution of weed plants; comparing the biology of the weed with that of related cultivars and narrow endemics; biological indicators of stream pollution; behavior of pheasants and quail. *Mailing Add:* Dept Biol Aurora Univ 347 S Gladstone Ave Aurora IL 60506-4892

ZIMMERMAN, DALE A, b Imlay City, Mich, June 7, 28; m 50; c 1. ORNITHOLOGY, ECOLOGY. *Educ:* Univ Mich, BS, 50, MS, 51, PhD, 56. *Prof Exp:* From asst prof to assoc prof, 57-72, PROF BIOL, WESTERN NMEX UNIV, 72-, CHMN DEPT, 77- *Concurrent Pos:* Mem expeds, Africa, 61, 63, 65 & 66. *Mem:* Am Ornith Union; Wilson Ornith Soc; Cooper Ornith Soc; Brit Ornith Union. *Res:* Taxonomy and ecology of birds and plants. *Mailing Add:* 1011 W Florence Silver City NM 88061

ZIMMERMAN, DANIEL HILL, b Los Angeles, Calif, June 3, 41; m 63; c 3. BIOCHEMISTRY, IMMUNOLOGY. *Educ:* Emory & Henry Col, BS, 63; Univ Fla, MS, 66, PhD(biochem), 69. *Prof Exp:* Jr staff fel biochem, Nat Inst Arthritis, Metab & Digestive Dis, 69-71, sr staff fel, 71-73; cellular immunologist, 73-77, sr res scientist, Res & Develop Dept, 77-80, prog mgr cell immunol, Electro Nucleonics Labs Inc, 81-87; VPRES, RES & DEVELOP, CELL MED, INC, 87- *Mem:* Am Asn Immunol. *Res:* Synthesis and secretion of proteins; differentiation of antibody producing cells. *Mailing Add:* Cell Med Inc PO Box 30115 Bethesda MD 20814

ZIMMERMAN, DEAN R, b Compton, Ill, July 2, 32; m 53; c 2. ANIMAL NUTRITION. *Educ:* Iowa State Univ, BS, 54, PhD(swine nutrit), 60. *Prof Exp:* Res assoc nutrit, Univ Notre Dame, 60-62; asst prof biol, Wartburg Col, 62-65; assoc prof animal sci, Purdue Univ, 65-67; assoc prof, 67-73, PROF ANIMAL SCI, IOWA STATE UNIV, 73- *Mem:* Am Soc Animal Sci; Am Inst Nutrit; Sigma Xi. *Res:* Nutrition research, especially compensatory growth and development; nutrition-disease interrelationships; bioavailability of amino acids; amino acid interrelationships. *Mailing Add:* Dept Animal Sci R337 Kildee Iowa State Univ Ames IA 50011

ZIMMERMAN, DON CHARLES, b Fargo, NDak, Feb 27, 34; m 58; c 1. BIOLOGICAL CHEMISTRY, PLANT PHYSIOLOGY. *Educ:* NDak State Univ, BS, 55, MS, 59, PhD(biochem), 64. *Prof Exp:* RES CHEMIST, SCI & EDUC ADMIN-AGR RES, USDA, 59-, CTR DIR, AGR RES SERV, 88-; ADJ PROF BIOCHEM, NDAK STATE UNIV, 69- *Concurrent Pos:* Asst prof, NDak State Univ, 64-69. *Mem:* AAAS; Am Chem Soc; Am Soc Plant Physiol. *Res:* Metabolism of unsaturated fatty acids; their oxidation by lipoxygenase and other enzymes during early plant growth. *Mailing Add:* US Dept Agr Univ Sta Box 5677 Fargo ND 58105-5677

ZIMMERMAN, DONALD NATHAN, b Somerset, Pa, Dec 30, 32. PHYSICAL INORGANIC CHEMISTRY. *Educ:* Univ Md, College Park, BS, 61; Pa State Univ, MEd, 63; WVa Univ, PhD(chem), 69. *Prof Exp:* Teacher pub sch, Pa, 62-65; assoc prof, Ind Univ Pa, 69-73, prof chem, 73-78; mem staff, Argonne Nat Lab, 78-81; MEM STAFF, DEPT PHYSICS & ASTRON, NORTHWESTERN UNIV, 81- *Mem:* AAAS; Am Chem Soc; Am Inst Physics; Royal Soc Chem. *Res:* Application of magnetochemical techniques to the determination of the structure of coordination compounds. *Mailing Add:* 729 Thomas Ct Libertyville IL 60048-1394

ZIMMERMAN, EARL ABRAM, b Harrisburg, Pa, May 5, 37; m 67, 82. NEUROLOGY, NEUROENDOCRINOLOGY. *Educ:* Franklin & Marshall Univ, BS, 59; Univ Pa, MD, 63. *Prof Exp:* Resident & intern med, Presbyterian Hosp, 63-65; resident neurol, Neurol Inst, NY, 65-68; fel endocrinol, Columbia Univ, 70-72, asst prof neurol, 72-77, assoc prof, 78-81, prof neurol, Col Physicians & Surgeons, 81-85; PROF & CHMN NEUROL, ORE HEALTH SCI UNIV. *Concurrent Pos:* Asst attend physician neurol, Presby Hosp, 72-77, assoc attend physician, 78-81, attend physician, 81-; Lucy Moses Basic Res Prize neurol, Columbia Univ, 74. *Honors & Awards:* Wartenburg Lectr, Am Acad Neurol, 85. *Mem:* Am Neurol Asn; Am Acad Neurol; Endocrine Soc; Soc Neurosci; Histochem Soc. *Res:* Organization and evaluation of neuropeptide pathways in mammalian brain with an emphasis on hypothalmic systems. *Mailing Add:* Dept Neurol Ore Health Sci Univ Sch Med 3181 SW S Jackson Portland OR 97201

ZIMMERMAN, EARL GRAVES, b Detroit, Mich, Feb 15, 43; m 75; c 2. ECOLOGICAL GENETICS, POPULATION GENETICS. *Educ:* Ind State Univ, Terre Haute, BS, 65; Univ Ill, Urbana, MS, 67, PhD(zool), 70. *Prof Exp:* Asst mammal comp anat, Univ Ill, Urbana, 65-69, asst cytogenetics, 69-70; asst prof pop genetics, 70-75, fac res grant, 70-72, assoc prof pop genetics, 75-77, assoc prof biol sci, 77-81, PROF POP GENETICS, N TEX STATE UNIV, 81- *Honors & Awards:* Jackson Award, Am Soc Mammal, 70. *Mem:* Am Soc Mammal; Genetics Soc Am; Soc Study Evolution. *Res:* Population genetics, biochemical variation and evolution of vertebrates. *Mailing Add:* Dept Biol N Tex State Univ Box 5218 Denton TX 76203

ZIMMERMAN, EDWARD JOHN, b Waynetown, Ind, July 12, 24; m 45; c 2. THEORETICAL PHYSICS, PHILOSOPHY OF SCIENCE. *Educ:* Univ Kans, AB, 45; Univ Ill, MS, 47, PhD(physics), 51. *Prof Exp:* Res assoc nuclear physics, Univ Ill, 50-51; from asst prof to assoc prof physics, 51-60, chmn dept, 62-66, PROF PHYSICS, UNIV NEBR-LINCOLN, 60- *Concurrent Pos:* Vis prof, Hamburg Univ, 57-58; NSF sci fac fel philos sci, Cambridge Univ, 66-67. *Mem:* Fel Am Phys Soc; Am Asn Physics Teachers; Philos Sci Asn; Brit Soc Philos Sci. *Res:* Foundations of physics; quantum mechanics. *Mailing Add:* 601 Marshall Ave Univ Nebr Lincoln NE 68510

ZIMMERMAN, ELMER LEROY, b Washington Co, Pa, Feb 5, 21; m 45; c 4. PHYSICS. *Educ:* Washington & Jefferson Col, AB, 42; Syracuse Univ, MA, 44; Ohio State Univ, PhD(physics), 50. *Prof Exp:* Physicist, Tenn Eastman Corp, 44-45, Oak Ridge Nat Lab, 50-55 & Nuclear Develop Corp Am, 55-58; supvr critical exp unit, Atomics Int Div, NAm Aviation, Inc, 58-60; vpres & tech dir, Solid State Radiations, Inc, 60-62; group leader radiation effects & reactor opers, Atomics Int Div, NAm Aviation, Inc, 62-64; asst mgr weapons effects dept, Solid State Physics Lab, TRW Systs, 64-67, mgr electronic & electro-magnetic effects dept, Vulnerability & Hardness Lab, 67-69, mem staff, Opers Res Dept, 69-77; nuclear survivability mgt, Guid & Control Div, Litton Industs, Inc, 77-89; RETIRED. *Mem:* Am Phys Soc; Inst Elec & Electronics Engrs; Am Nuclear Soc. *Res:* Nuclear spectroscopy and resonance; reactor physics and instrumentation; critical experiments; semiconductor radiation detectors; radiation effects and transport; ballistic and cruise missile systems engineering. *Mailing Add:* 22650 MacFarlane Dr Woodland Hills CA 91364

ZIMMERMAN, EMERY GILROY, b Los Angeles, Calif, June 23, 39; m 67; c 3. ANATOMY, MEDICINE. *Educ:* Pomona Col, BA, 61; Baylor Univ, MD, 67, PhD(anat), 71. *Prof Exp:* Intern med, Methodist Hosp, Houston, 68-69; instr anat, Col Med, Baylor Univ, 69-70; surgeon, Addiction Res Ctr, NIMH, Ky, 70-72; ASSOC PROF ANAT, SCH MED, UNIV CALIF, LOS ANGELES, 72- *Mem:* Am Physiol Soc; Endocrine Soc; Soc Exp Biol & Med; AMA; Am Soc Clin Pharmacol & Therapeut. *Res:* Neuroendocrinology, physiology, neuropharmacology. *Mailing Add:* Dept Anat 73-235 Chs Univ Calif 405 Hilgard Ave Los Angeles CA 90024

ZIMMERMAN, ERNEST FREDERICK, b New York, NY, June 2, 33; m 57; c 3. TERATOLOGY, DEVELOPMENTAL BIOLOGY. *Educ:* George Washington Univ, BS, 56, MS, 58, PhD(pharmacol), 61. *Prof Exp:* Res asst pharmacol, George Washington Univ, 56-58; fel biol, Mass Inst Technol, 60-62; from instr to asst prof pharmacol, Sch Med, Stanford Univ, 62-68; assoc prof res pediat, 68-71, dir grad prog develop biol, 71-85, ASSOC PROF RES PHARMACOL, UNIV CINCINNATI, 68-, DIR DIV CELL BIOL, INST DEVELOP RES, CHILDREN'S HOSP RES FOUND, 68-, PROF PEDIAT, COL MED, 71- *Mem:* AAAS; Am Soc Biol Chemists; Teratol Soc (pres, 89-90); Soc Cell Biol; Am Soc Pharmacol & Exp Therapeut. *Res:* role of neurotransmitters in development and differentiation; mechanisms of action of teratogenic drugs. *Mailing Add:* Dept Pediat Children's Hosp Res Found Elland & Bethesda Aves Cincinnati OH 45229

ZIMMERMAN, EUGENE MUNRO, b New Haven, Conn, June 27, 40. HEALTH SCIENCE ADMINISTRATION, VIROLOGY. *Educ:* Yale Univ, BA, 60; Wesleyan Univ, MA, 62; Univ Md, College Park, PhD(microbiol), 68. *Prof Exp:* Microbiologist, Ft Detrick, Md, 69-70; asst proj dir, Microbiol Assocs Inc, Md, 70-73; sr scientist, Litton-Bionetics Inc, 73-76; grants assoc,

76-77, asst prog dir carcinogenesis, Nat Cancer Inst, 77-78, exec secy, Review Br, 78-81, EXEC SECY, ALLERGY & IMMUNOL STUDY SECT, DIV RES GRANTS, NIH, 82- *Concurrent Pos:* Consult, Mt Sinai Sch Med, 75-76. *Mem:* AAAS; Tissue Cult Asn; Am Soc Microbiol; Sigma Xi. *Res:* Grant and contract review; biology of oncogenic herpesviruses and oncarnaviruses; treatment and prophylaxis of leukemia in animal models. *Mailing Add:* 33 Brighton Dr Gaithersburg MD 20877

ZIMMERMAN, GARY ALAN, b Seattle, Wash, Oct 19, 38; m 60, 88; c 2. CLINICAL CHEMISTRY. *Educ:* Calif Inst Technol, BS, 60; Univ Wis, PhD(org chem), 65. *Prof Exp:* From asst prof to prof chem, Seattle Univ, 64-76, dir clin chem, 68-83, dean, Sch Sci & Eng, 73-80, vpres acad affairs, 80-81, exec vpres, 81-87; PROVOST, ANTIOCH UNIV, SEATTLE, WASH, 88- *Concurrent Pos:* Lectr, Univ Wash, 65 & vis sci prog, Pac Sci Ctr, Seattle, 66-68; consult, Gordon Res Conf, 66-70 & Swed Hosp & Med Ctr, Seattle, 68-83; vis prof chem, Univ Idaho, 73. *Mem:* Fel AAAS; Am Chem Soc; Am Asn Clin Chemists (treas, 80-81, pres elect, 82, pres, 83). *Res:* Clinical applications of enzymatic assays; trace metal analyses. *Mailing Add:* Provost Antioch Univ 2607 Second Ave Seattle WA 98121-1211

ZIMMERMAN, GEORGE LANDIS, b Hershey, Pa, Aug 27, 20; m 53. PHYSICAL CHEMISTRY. *Educ:* Swarthmore Col, AB, 41; Univ Chicago, PhD, 49. *Prof Exp:* Res chemist sam labs, Manhattan Dist Proj, Columbia, 42-46; instr, Mass Inst Technol, 49-51; asst prof, 51-55, assoc prof, 55, PROF CHEM, BRYN MAWR COL, 55- *Mem:* Am Chem Soc. *Res:* Molecular spectroscopy. *Mailing Add:* Dept Chem Bryn Mawr Col Bryn Mawr PA 19010

ZIMMERMAN, GEORGE OGUREK, b Katowice, Poland, Oct 20, 35; US citizen; m 64. LOW TEMPERATURE PHYSICS. *Educ:* Yale Univ, BS, 58, MS, 59, PhD(physics), 63. *Prof Exp:* Res asst physics, Yale Univ, 59-62, res assoc, 62-63; from asst prof to assoc prof physics, 63-73, from assoc chmn to chmn dept, 72-83, chmn fac coun, 85-86, PROF PHYSICS, BOSTON UNIV, 73- *Concurrent Pos:* Vis scientist, Nat Magnet Lab, Mass Inst Technol, 65-; assoc physicist, Univ Calif, San Diego, 73; vis physicist, Brookhaven Nat Lab, 80; vis scholar, Harvard Univ, 88, Kamerling-Onnes Lab, Leiden, Neth, 88. *Mem:* AAAS; NY Acad Sci; Am Phys Soc. *Res:* Low temperature phenomena, cryogenics, specifically pertaining to liquid and solid helium three; investigation of paramagnetic phenomena; investigation of phase transitions; two-dimensional magnetism; superconductivity; research grants, patents. *Mailing Add:* Dept Physics Boston Univ 590 Commonwealth Ave Boston MA 02215

ZIMMERMAN, HARRY MARTIN, b Vilna Prov, Russia, Sept 28, 01; nat US; m 30. PATHOLOGY. *Educ:* Yale Univ, BS, 24, MD, 27. *Hon Degrees:* LHD, Yeshiva Univ, 57. *Prof Exp:* Ives fel, Sch Med, Yale Univ, 27-29, from asst prof to assoc prof path, 30-43; from assoc clin prof to prof, Col Physicians & Surgeons, Columbia Univ, 46-64; prof, 64-74, EMER PROF PATH, ALBERT EINSTEIN COL MED, 74- *Concurrent Pos:* Assoc pathologist, New Haven Hosp, Conn, 33-43; consult, Bristol Hosp, Conn, 38-43, US Naval Hosp, St Albans, 46-49, Seton Hosp, 49-55, Beth Israel Hosp, 49-71, Armed Forces Inst Path, 49-71, Long Island Jewish Hosp, 54-82, Vassar Bros Hosp, Poughkeepsie, NY, 61- & Montefiore Med Ctr, 74-; sr consult, Vet Admin Hosp, Bronx, 46-71; chief lab div, Montefiore Hosp, 46-73; prof lectr, Mt Sinai Col Med, 74- *Honors & Awards:* Golden Hope Chest Award, Nat Multiple Sclerosis Soc, 65; Max Weinstein Award, United Cerebral Palsy Found, 72; Gold Headed Cane, Am Asn Path, 82; Middleton Goldsmith Lectr, NY Acad Med, 64. *Mem:* Fel Am Soc Clin Path; Am Asn Pathologists; Am Neurol Asn; Am Asn Neuropath (pres, 44); fel Col Am Path; hon mem Japanese Soc Neurol; hon mem Int Soc Neuropath; hon mem Am Neurol Asn. *Res:* Neuropathology; demyelinating diseases; brain tumors. *Mailing Add:* Montefiore Med Ctr 111 E 210th St Bronx NY 10467-2490

ZIMMERMAN, HOWARD ELLIOT, b New York, NY, July 5, 26; wid; c 3. CHEMISTRY. *Educ:* Yale Univ, BS, 50, PhD(chem), 53. *Prof Exp:* Nat Res Coun postdoctoral fel chem, Harvard Univ, 53-54; from instr to asst prof, Northwestern Univ, 54-60; assoc prof, 60-61, PROF CHEM, UNIV WIS-MADISON, 61-, ARTHUR C COPE CHAIR CHEM, 75-, HILLDALE CHAIR CHEM, 90- *Concurrent Pos:* Mem grants comt, Res Corp, 66-72; chmn, 4th Int Union Pure & Appl Chem, Int Photochem Symp, Baden-Baden, Ger, 72; co-chmn org div, Inter Am Photchem Soc, 76-81, mem exec comt, 81-86; fel, Japan Soc Prom Sci; Sr Humboldt fel, 88; Cope scholar award, Am Chem Soc, 90. *Honors & Awards:* Photochem Award, Am Chem Soc, 75 & James Flack Norris Award Phys-Org Chem, 76; Halpen Award, NY Acad Sci, 79; Chem Pioneering Award, Am Inst Chemists, 86; Hilldale Award in Phys Sci, Univ Wis, 90. *Mem:* Nat Acad Sci; Royal Soc Chem; Ger Chem Soc; Am Chem Soc. *Res:* Organic, physical-organic, synthetic organic chemistry; photochemistry; theoretical organic chemistry; photobiology; reaction mechanisms; stereochemistry; unusual organic phenomena and species. *Mailing Add:* Dept Chem Univ Wis Madison WI 53706

ZIMMERMAN, HYMAN JOSEPH, b Rochester, NY, July 14, 14; m 43; c 4. MEDICINE, HEPATOLOGY. *Educ:* Univ Rochester, AB, 36; Stanford Univ, MA, 38, MD, 43. *Prof Exp:* Intern, Stanford Univ Hosp, 42-43; resident med, Gallinger Munic Hosp & Med Div, George Washington Univ, 46-48, clin instr, Sch Med, 48-51; asst prof, Col Med, Univ Nebr, 51; chief med serv, Vet Admin Hosp, Omaha, 51-53; clin assoc prof med, Col Med, Univ Ill, 53-57; prof & chmn dept, Chicago Med Sch, 57-65; prof, 65-90, EMER PROF MED, GEORGE WASHINGTON UNIV, 90- *Concurrent Pos:* Asst chief med serv, Vet Admin Hosp, DC, 48-49, dir liver & metab res lab, 65-68; chief med serv, Vet Admin West Side Hosp, Chicago, 53-65; chmn dept med, Mt Sinai Hosp, 57-65; prof, Sch Med, Boston Univ, 68-71; lectr, Sch Med, Tufts Univ, 68-71; clin prof med, Sch Med, Sch Med, Howard Univ, Georgetown Univ & Uniformed Serv Univ Health Sci; chief med serv, Vet Admin Hosp, DC, 68-78, distinguished physician, 84-89; distinguished scientist, Armed Forces Inst Pathol, 89-90. *Honors & Awards:* Distinguished Achievement Award, Am Asn Study Liver Dis, 86; Gold Medal, Can Liver Found, 89. *Mem:* Am Soc Clin Invest; Am Asn Study Liver Disease; Am Diabetes Asn; Am Fedn Clin Res; fel Am Col Physicians; Am Gastroenterol Asn. *Res:* Physiology of the liver; toxicology; effect of drugs on the liver; hepatotoxicity. *Mailing Add:* 7913 Charleston Ct Bethesda MD 20817

ZIMMERMAN, IRWIN DAVID, b Philadelphia, Pa, Oct 31, 31; m 57; c 4. NEUROPHYSIOLOGY, BIOPHYSICS. *Educ:* Univ Del, BA, 59; Univ Wash, PhD(physiol), 66. *Prof Exp:* Asst prof, 66-72, ASSOC PROF PHYSIOL & BIOPHYS, MED COL PA, 72- *Concurrent Pos:* NSF grant, 68-70; NIH grant, 72- *Mem:* Am Physiol Soc; Biophys Soc; Soc Neurosci; Soc Math Biol. *Res:* Neural processing and coding of sensory information; membrane phenomena; developmental biology. *Mailing Add:* Dept Physiol & Biophys Womans Med Col 3300 Henry Ave Philadelphia PA 19129

ZIMMERMAN, IVAN HAROLD, b Orland, Calif, Nov 21, 43; m 75; c 1. CHEMICAL PHYSICS. *Educ:* Ore State Univ, BS, 66; Univ Wash, PhD(physics), 72. *Prof Exp:* Fel chem, Univ Rochester, 72-76, res assoc, 77-78; asst prof physics, Clarkson Col, 78-83; CODE PHYSICIST, LAWRENCE LIVERMORE NAT LAB, 84- *Mem:* Am Phys Soc; Am Chem Soc; NY Acad Sci; AAAS. *Res:* Atomic and molecular collisions; effects of intense laser radiation on molecular processes; ion-surface encounters; molecules near solid surfaces; strong-shock hydrodynamics. *Mailing Add:* Lawrence Livermore Nat Lab L-035 PO Box 808 Livermore CA 94550

ZIMMERMAN, JACK MCKAY, b New York, NY, Feb 4, 27; m 53; c 2. SURGERY. *Educ:* Princeton Univ, AB, 49; Johns Hopkins Univ, MD, 53; Univ Kansas City, MA, 63. *Prof Exp:* From intern to resident surg, Johns Hopkins Hosp, 53-59; assoc, Sch Med, Univ Kans, 59-60, asst prof, 60-65; ASSOC PROF SURG, JOHNS HOPKINS UNIV, 65- *Concurrent Pos:* Asst, Johns Hopkins Hosp, 54-59, surgeon, 65-, Halsted fel surg path, Univ, 55-56, instr, 58-59; staff surgeon, Vet Admin Hosp, Kansas City, 59-60, chief surg serv, 60-65; consult, Sch Dent, Univ Kans, 60 & Vet Admin Hosp, Baltimore; chief surg, Church Home & Hosp, 65- *Honors & Awards:* Alumni Achievement Award, Univ Mo-Kansas City. *Mem:* Am Col Surg; Soc Univ Surg. *Res:* Thoracic surgery; wound healing and infections; care of advanced malignancy; medical education; care of terminal illness. *Mailing Add:* 100 N Broadway Baltimore MD 21231

ZIMMERMAN, JAMES KENNETH, b Nelson, Nebr, Aug 23, 43. BIOCHEMISTRY. *Educ:* Univ Nebr, Lincoln, BS, 65; Northwestern Univ, Evanston, PhD(biochem), 69. *Prof Exp:* NIH fel, Univ Va, 69-71; from asst prof to assoc prof, 71-83, PROF BIOCHEM, CLEMSON UNIV, 83-, ASSOC HEAD BIOL SCI, 89- *Mem:* AAAS; Am Chem Soc; Am Soc Biochem & Molecular Biol; Biophys Soc; Biochem Soc Gt Brit. *Res:* Associating protein systems, complement interactions; analytical gel chromatography by direct scanning; analytical ultracentrifugation; computer simulations. *Mailing Add:* Dept Biol Sci Clemson Univ Clemson SC 29634-1903

ZIMMERMAN, JAMES ROSCOE, b Norwood, Ohio, July 12, 28; m 50; c 3. ZOOLOGY, ENTOMOLOGY. *Educ:* Hanover Col, AB, 53; Ind Univ, MA, 55, PhD(zool), 57. *Prof Exp:* Asst prof zool, Univ Wichita, 57-58; assoc prof biol, Ind Cent Col, 58-61, actg chmn dept, 59-61; from asst prof to assoc prof, 61-68, head dept, 74-78, PROF BIOL, N MEX STATE UNIV, 68- *Concurrent Pos:* Vis prof, Escuela Sup De Agric'Herm Escob CD Juarez, 78-81. *Mem:* AAAS; Am Soc Zool; Entom Soc Am; Soc Systs Zool; Am Entom Soc; Sigma Xi. *Res:* Taxonomy of dytiscidae; parasitic hymenoptera. *Mailing Add:* 1960 W Myrtlewood Lane Tucson AZ 85704

ZIMMERMAN, JAY ALAN, b Philadelphia, Pa, Mar 1, 45; m 72; c 2. MAMMALIAN PHYSIOLOGY, GERONTOLOGY. *Educ:* Franklin & Marshall Col, AB, 67; Rutgers Univ, PhD(zool), 75. *Prof Exp:* ASSOC PROF BIOL, ST JOHN'S UNIV, NY, 75- *Concurrent Pos:* Vis scientist, Orentreich Found Advan Sci; pres, Multisciences Assocs. *Mem:* Sigma Xi; Geront Soc; Soc Study Reproduction; AAAS. *Res:* Adaptive mechanisms of organ-system function and biochemistry during aging; chemical carcinogenesis and oncogene expression in senescence; alcohol absorption and physiology. *Mailing Add:* Dept Biol St John's Univ Jamaica NY 11439

ZIMMERMAN, JOHN F, b Monticello, Iowa, June 22, 37; m 59; c 3. ANALYTICAL CHEMISTRY, SCIENCE EDUCATION. *Educ:* Univ Iowa, BS, 59; Univ Kans, PhD(chem), 64. *Prof Exp:* From asst prof to assoc prof, 63-79, PROF CHEM, WABASH COL, 79- *Res:* Use of personal computers in the chemistry lecture and teaching laboratory; development of instrumentation for undergraduate laboratory instruction; multimedia instruction. *Mailing Add:* Dept Chem Wabash Col 301 W Wabash Crawfordsville IN 47933

ZIMMERMAN, JOHN GORDON, b Brooklyn, NY, May 31, 16; m 39; c 4. PHYSICAL CHEMISTRY, INORGANIC CHEMISTRY. *Educ:* Univ Pa, BS, 37, MS, 39; Georgetown Univ, PhD(chem), 71. *Prof Exp:* Instr gen sci, Monmouth Jr Col, 38-42; shift supvr org chem prod, Ala Ord Works, 42-43; instr chem, Drexel Inst, 43-44; develop chemist org chem, Publicker Industs, Inc, 44-47; chmn sci div, St Helena exten, Col William & Mary, 47-48; asst prof, gen & phys chem, Westminster Col, 48-51; from asst prof to assoc prof, 51-72, PROF CHEM, US NAVAL ACAD, 72- *Mem:* Am Chem Soc. *Res:* Kinetics of substitution reactions of transition metal complexes. *Mailing Add:* 1708 Old Generals Hwy Annapolis MD 21401

ZIMMERMAN, JOHN HARVEY, b St Paul, Minn, Feb 11, 45; m 69; c 2. MEDICAL ENTOMOLOGY. *Educ:* Concordia Col, Minn, BA, 67; Univ Del, Newark, MS, 69; Mich State Univ, E Lansing, PhD(entom), 75; Pepperdine Univ, MA, 81. *Prof Exp:* Entomologist, Navy Environ & Prev Med Unit, 75-77; mem staff, Navy Dis Vector Ecol Control Ctr, Naval Air Sta, 77-82; mem staff, US Naval Med Res Unit No 3, Cairo, Egypt, 82-84; command & staff col, 84-85; first med battalion, 85-88, CMNDG & EXEC OFFICER, CAMP PENDLETON, 88- *Mem:* Sigma Xi; Entom Soc Am; Am Mosquito Control Asn. *Res:* The ecology, transmission and control of arthropod-borne diseases of medical and veterinary importance. *Mailing Add:* OIC Navy Dis Vector Ecol Control Ctr Naval Air Sta Jacksonville FL 32212

ZIMMERMAN, JOHN LESTER, b Hamilton, Ont, Feb 17, 33; US citizen; m 55; c 3. ECOLOGY. *Educ:* Mich State Univ, BS, 53, MS, 59; Univ Ill, PhD(zool), 63. *Prof Exp:* Asst prof zool, 63-68, assoc prof biol, 68-76, PROF BIOL, KANS STATE UNIV, 76- *Concurrent Pos:* Sci adv environ protection, Atlantic-Richfield Co, Calif, 74-75. *Mem:* Am Ornith Union; Wilson Ornith Soc; Cooper Ornith Soc. *Res:* Birds, physiological ecology; habitat relection in grassland community ecology. *Mailing Add:* Ackert Hall Kans State Univ Manhattan KS 66506

ZIMMERMAN, JOHN R(ICHARD), b Allentown, Pa, July 25, 25; m 54; c 2. MECHANICAL ENGINEERING. *Educ:* Yale Univ, BE, 46; Boston Univ, STB, 52; Lehigh Univ, MS, 60, PhD, 66. *Prof Exp:* Assoc prof mech eng, Pa State Univ, 66-71; prof, Clarkson Col Technol, 71-74; vis prof, Cornell Univ, 74-76; PROF MECH & AEROSPACE ENG, UNIV DEL, 76- *Mem:* Am Soc Mech Engrs; Am Soc Eng Educ; Am Asn Univ Prof; Sigma Xi; Am Soc Automotive Engrs. *Res:* Mechanical design; computer simulation; random vibrations; land transportation. *Mailing Add:* 922 Quail Ln Newark DE 19711

ZIMMERMAN, JOSEPH, b New York, NY, Aug 9, 21; m 48; c 3. PHYSICAL CHEMISTRY, POLYMER CHEMISTRY. *Educ:* City Col New York, BS, 42; Columbia Univ, AM, 47, PhD(chem), 50. *Prof Exp:* Res chemist, Carothers Res Lab, E I du Pont de Nemours & Co, Inc, 50-53, res assoc, 53-62, res fel, 62-64, res mgr, Carothers Res Lab, 64-71, res mgr, Indust Fibers Div, 71-78, from res mgr to sr res fel, Textile Fibers Dept, Pioneering Res Div, 78-84; CONSULT, 85- *Concurrent Pos:* Consult, 85- *Mem:* Am Chem Soc; Sigma Xi; AAAS; Fiber Soc. *Res:* Polymer and fiber research, especially polyamides, polyesters and aramids. *Mailing Add:* 906 Barley Dr Box 4042 Wilmington DE 19807

ZIMMERMAN, LEONARD NORMAN, b Brooklyn, NY, Sept 13, 23; m 46; c 3. BACTERIOLOGY. *Educ:* Cornell Univ, BS, 48, MS, 49, PhD(bact), 51. *Prof Exp:* Asst, Cornell Univ, 48-51; from asst prof to prof, Pa State Univ, University Park, 51-89, head dept microbiol & cell biol, 75-78, assoc dean res, Col Sci, 78-89, dean, Col Sci, 88-89, EMER DEAN & EMER PROF BACT, PA STATE UNIV, UNIVERSITY PARK, 89- *Mem:* AAAS; Am Soc Microbiol. *Res:* Bacterial genetics and regulatory mechanisms. *Mailing Add:* 203 Frear Lab Pa State Univ University Park PA 16802

ZIMMERMAN, LESTER J, b Conway, Kans, July 16, 18; m 49; c 4. AGRONOMY, MATHEMATICS. *Educ:* Goshen Col, BA, 47; Purdue Univ, MS, 50, PhD(soil fertil), 56; Univ Ill, MA, 61. *Prof Exp:* Instr chem & math, 47-49, from asst prof to prof math, 50-84, EMER PROF MATH, GOSHEN COL, 84- *Concurrent Pos:* Soil survr, Soil Conserv Serv, USDA, 56-58; vis prof, Univ Zambia, 68-69 & 76-78. *Mem:* AAAS; Nat Coun Teachers Math; Soil Sci Soc Am; Soil Conserv Soc Am; Am Sci Affiliation. *Res:* Manganese; plant nutrition. *Mailing Add:* 1512 Greencroft Dr Goshen IN 46526

ZIMMERMAN, LORENZ EUGENE, b Washington, DC, Nov 15, 20; m 45, 59; c 6. PATHOLOGY. *Educ:* George Washington Univ, AB, 43, MD, 45; Am Bd Path, dipl, 52. *Hon Degrees:* DSc, Univ Ill, 81. *Prof Exp:* Chief ophthal path, Armed Forces Inst Path, 53-83; PROF PATH & OPHTHA, GEORGETOWN UNIV SCH MED, 83- *Concurrent Pos:* Assoc prof, Sch Med, George Washington Univ, 54-83, clin prof ophthal path, 63-; lectr, Johns Hopkins Univ, 59-; head, WHO Int Ref Ctr Tumors Eye & Ocular Adnexa, 72-; consult ophthal path, Wash Hosp Ctr, 75-; clin prof path, Uniformed Serv Univ Health Sci, 76- *Honors & Awards:* Ernst Jung Prize, Ernst Jung Found, Hamburg, Ger, 76; F C Donders Medal, Neth Opthal Soc, Groningen, 78; Estelle Doheny mem lectr, Estelle Doheny Eye Found, Los Angeles, 78; Sir William Bowman lectr, Ophthal Soc UK, London, 80; Medalla de Oro, Barraquer Inst, Barcelona, Spain, 69; Leslie Dana Gold Medal, St Louis Soc Blind, 82; Jules Stein Award, Res Prevent Blindness, Inc, 85; Medalla de Oro Jorge Malbran, Fundacion Oftalmologica Arg, Buenos Aires, 86; John Milton McLean Medal, NY Hosp-Cornell Med Ctr, New York, NY, 88; Mericos H Whittier Award, Mericos Eye Ctr-Scripps Mem Hosp, La Jolla, Calif, 89; Lighthouse Pisart Vision Award, New York, NY, 90. *Mem:* Verhoeff Soc; Pan-Am Asn Ophthal; Asn Res Vision & Ophthal; Am Acad Ophthal & Otolaryngol. *Res:* Pathology of diseases of the eye and ocular adnexa. *Mailing Add:* Dept Path 1103 Basic Sci Georgetown Univ 3800 Resevoir Rd NW Washington DC 20007

ZIMMERMAN, MARY PRISLOPSKI, b Bath, NY, Dec 3, 47; m 75. SYNTHETIC ORGANIC CHEMISTRY. *Educ:* State Univ NY Albany, BS, 70; Wesleyan Univ, MA, 73; Univ Rochester, MS, 74, PhD(org chem), 77. *Prof Exp:* Fel, Univ Rochester, 76-78, instr, 76-78; adj asst prof org chem, Clarkson Inst Technol, 78-; RES ASSOC DEPT CHEM, STANFORD UNIV. *Mem:* Am Chem Soc; Sigma Xi. *Res:* Synthesis of natural products; synthetic methods; synthesis of morphinans; sterol biosynthesis. *Mailing Add:* Dept Chem Clarkson Univ Tech Potsdam NY 13676

ZIMMERMAN, MICHAEL RAYMOND, b Newark, NJ, Dec 26, 37; m 60; c 2. CLINICAL PATHOLOGY, PALEOPATHOLOGY. *Educ:* Wash & Jefferson Col, BA, 59; NY Univ, MD, 63; Univ Pa, PhD(anthrop), 76. *Prof Exp:* Major, Walter Reed Army Hosp, US Army, 68-70; pathologist, Lankenau Hosp, 70-72; asst prof path, Univ Pa, 72-77; pathologist, Wayne County Gen Hosp, 77-80; assoc prof path, Hahnemann Univ, 80-82; pathologist, Jeanes Hosp, 82-85; dir clin labs, Coney Island Hosp, 85-87; PROF PATH, HAHNEMANN UNIV, 87- *Concurrent Pos:* Assoc prof, Univ Mich, 77-80; adj assoc prof, Univ Pa, 80-87, adj prof, 86-; clin prof State Univ NY, Brooklyn, 86-87. *Mem:* Col Am Pathologists; Paleopath Asn; US-Can Acad Path; Am Soc Clin Pathologists; Am Asn Phys Anthrobiologists; Asn Clin Scientists. *Res:* Paleopathology, the study of the diseases of ancient peoples; studies of mummies from Alaska and Egypt have provided information on the evolution of various diseases, including atherosclerosis and cancer. *Mailing Add:* Hahnemann Univ MS 113 Broad & Vine Philadelphia PA 19102-1192

ZIMMERMAN, NORMAN, biochemistry, environmental science; deceased, see previous edition for last biography

ZIMMERMAN, NORMAN H(ERBERT), b St Louis, Mo, May 12, 19; m 46; c 2. MECHANICS, MECHANICAL ENGINEERING. *Educ:* Wash Univ, BS, 42, MS, 50, DSc(appl mech), 56. *Prof Exp:* Instr & lectr, Wash Univ, 46-51; mech engr, Sverdrup & Parcel, Inc, 51; staff engr, White-Rodgers Elec Co, 51-53; proj dynamicist, 53-67, scientist, 67-70, br mgr, 70-75, BR CHIEF, MCDONNELL AIRCRAFT CO, MCDONNELL-DOUGLAS CORP, 76- *Concurrent Pos:* Mem, US Navy Aeroballistics Adv Comt, 68-72. *Honors & Awards:* Am Inst Aeronaut & Astronaut Award, 67. *Mem:* Am Inst Aeronaut & Astronaut; Am Chem Soc. *Res:* Dynamics; aeroelasticity; fluid mechanics; vibration and flutter. *Mailing Add:* 2200 NW 37th Ave Pompano Beach FL 33066-2200

ZIMMERMAN, PETER DAVID, b June 15, 41; US citizen; m 67; c 2. NUCLEAR PHYSICS, SCIENCE POLICY. *Educ:* Stanford Univ, BS, 63, PhD(physics), 69; Lund Univ, Sweden, Filosofie Licentiat, 67. *Prof Exp:* Res fel physics, Ger Electron Synchrotron, 69-71; adj asst prof physics & planetary sci, Univ Calif, Los Angeles, 71-73; res assoc physics, Fermi Nat Accelerator Lab, 73-74; from asst prof to assoc prof physics, La State Univ, 74-84; CONSULT, 84- *Concurrent Pos:* Res affil, Mass Inst Technol, 75-86; vis assoc res physicist, Univ Calif, San Diego, 81; consult, var defense related firms; William C Foster fel, US Arms Control & Disarmament Agency, 84-86; mem, US Start Deleg, 85-86; sr assoc, Carnegie Endowment Int Peace, 86-89; distinguished vis prof, George Washington Univ, 90- *Mem:* Fel Am Phys Soc; Sigma Xi; AAAS; Coun Foreign Rels; Int Inst Strategic Studies. *Res:* Arms control, especially non-nuclear defense and new verification technologies; orbital mechanics of large manned satellites; electron scattering experiments from nuclei, principally at large energy loss. *Mailing Add:* 9801 Thunderhill Ct Great Falls VA 22066

ZIMMERMAN, RICHARD HALE, b Bowling Green, Ohio, Apr 11, 34; m 66; c 2. HORTICULTURE, PLANT PHYSIOLOGY. *Educ:* Mich State Univ, BS, 56; Rutgers Univ, MS, 59, PhD(hort), 62. *Prof Exp:* Silviculturist, Tex Forest Serv, 62-64; PLANT PHYSIOLOGIST, AGR RES CTR, USDA, 65- *Honors & Awards:* J H Gourley Award, Am Soc Hort Sci, 72, Stark Award, 78 & 82, Darrow Award, 82; Norman Jay Colman Award, Am Asn Nurserymen, 84. *Mem:* Int Soc Hort Sci; Int Plant Propagators Soc; AAAS; fel Am Soc Hort Sci; Int Asn Plant Tissue Culture; Tissue Culture Asn. *Res:* Tissue culture; juvenility and flower initiation in fruit trees and other woody plants; effects of growth regulators on plant growth and development. *Mailing Add:* Fruit Lab Agr Res Ctr USDA Beltsville MD 20705

ZIMMERMAN, ROBERT ALLAN, b Philadelphia, Pa, June 23, 38; m 60; c 2. RADIOLOGY. *Educ:* Temple Univ, BA, 60; Georgetown Univ Sch Med, MD, 64. *Prof Exp:* Intern, Georgetown Univ Hosp, 64-65; physician resident fel, Hosp Univ Pa, 65-69; radiologist, US Army, Europe, 69-72; asst prof, 72-77, assoc prof, 77-81, PROF RADIOL, HOSP UNIV PA, 81- *Concurrent Pos:* Chief sect neuroradiol, Hosp Univ Pa, 79-; assoc ed, J Comput Tomography, 77-82, Neuroradiol, 81- *Mem:* Am Soc Neuroradiol; Radiol Soc NAm; Asn Univ Radiologists; Am Soc Head & Neck Radiol. *Res:* Medical imaging of central nervous system trauma, using medical imaging in the diagnosis and management of pediatric brain tumors. *Mailing Add:* Dept Radiol/1 Silver/G12 Univ Pa Philadelphia PA 19104

ZIMMERMAN, ROBERT LYMAN, b La Grande, Ore, Dec 30, 35; m 57; c 5. PHYSICS. *Educ:* Univ Ore, BA, 58; Univ Wash, PhD(physics), 63. *Prof Exp:* Physicist, Lawrence Radiation Lab, Univ Calif, Berkeley, 64-66; asst prof, 66-74, PROF PHYSICS, PHYSICS DEPT & INST THEORET SCI, UNIV ORE, 74- *Concurrent Pos:* Res assoc, Univ Ore, 70-74. *Mem:* Am Phys Soc; Sigma Xi. *Res:* Quantum field theory; elementary particle physics; gravitation; astrophysics; general relativity; cosmology; investigation of properties of exact solutions in general relativity, properties of the Big Bang and production of gravitational radiation by energetic astrophysical events. *Mailing Add:* Dept Physics Univ Ore Eugene OR 97403

ZIMMERMAN, ROGER M, b Rehoboth, NMex, May 15, 36; m 56; c 2. CIVIL ENGINEERING, STRUCTURAL MECHANICS. *Educ:* Univ Colo, BS, 59, MS, 61, PhD(civil eng), 65. *Prof Exp:* Instr civil eng, Univ Colo, 59-63, res assoc, 63-64; from asst prof to assoc prof civil eng, NMex State Univ, 64-70, asst dean eng, 67-72, prof, 70-80, assoc dean, 72-74, actg dean, 74-75; sr mem tech staff, 80-90, DISTINGUISHED MEM TECH STAFF, SANDIA NAT LABS, 90- *Concurrent Pos:* Sr engr, Phys Sci Lab, NMex State Univ, 75-79; vis scientist, Rockwell Int Sci Ctr, 79-80. *Mem:* Am Soc Civil Engrs; Nat Soc Prof Engrs; Inst Environ Soc. *Res:* Multiaxial strength properties of plain concrete; deterioration of plain concrete; biomechanical aspects of simulated side and rear automobile impacts; safety aspects of school bus seats; evaluation of structures for seismic response; applications of ultrasonics to predict space shuttle tiles properties; rock mechnics field testing to evaluate tuff for storage medium for high-level radioactive wastes; transient shock testing of electronic components for launchings, payload ejections and stage separations. *Mailing Add:* Sandia Nat Labs Bldg 882 Rm 9 Albuquerque NM 87185

ZIMMERMAN, ROGER PAUL, b Oak Park, Ill, Sept 29, 46; m 74; c 2. NEUROBIOLOGY, VISUAL PHYSIOLOGY. *Educ:* Univ Ill, Chicago, BS, 68; Yale Univ, MPhil, 69, PhD(biol), 77. *Prof Exp:* Res asst paleobot, Dept Biol Sci, Univ Ill, Chicago, 65-68; teaching asst, Dept Biol, Yale Univ, 60 & 73; res fel, Biol Lab, Harvard Univ, 76-78; asst prof neurolsci, 78-84, physiol, 79-89, ASSOC PROF NEUROL SCI, RUSH-PRESBY-ST LUKES MED CTR, RUSH MED COL, 84-, ASSOC PROF PHYSIOL, 89- *Concurrent Pos:* Participant, NSF Summer Res Prog, Univ Ill, Chicago, 68; NSF fel, Yale Univ, 68-73; res asst, Walter Reed Army Inst of Res, US Army, 70-72; NIH fel, Harvard Univ, 76-78, teaching fel, Dept Biol, 78; course dir med neurobiol, Rush Med Col, 81-; STEPS fel, Marine Biol Lab, Woods Hole, MA, 82; lectr, Dept Anat & Cell Biol, Univ Ill, Chicago, 86-; actg dir, Div Cell Biol, Rush Univ, 90-; course dir, molecular cell biol, 90-, vis scientist, cellular, molecular & struct biol, Northwestern Univ, 89-90. *Mem:* Asn Res in Vision & Ophthal; Soc Neurosci; AAAS; Sigma Xi. *Res:* Neurobiology of vision, including the physiology, development and ultrastructure of synaptic interactions in the vertebrate retina; physiology and pharmacology of glial cells; control of gene expression in the adult and developing nervous system. *Mailing Add:* Dept Neurol Sci & Physiol Rush Med Col 1753 W Congress Pkwy Chicago IL 60612

ZIMMERMAN, SARAH E, b Indianapolis, Ind, Oct 29, 37. IMMUNOLOGY, MICROBIOLOGY. *Educ:* Ind Univ, AB, 59, MA, 61; Wayne State Univ, PhD(biochem), 69. *Prof Exp:* Res asst biochem, Sch Med, Wayne State Univ, 68-69; res assoc chem, Ind Univ, Bloomington, 69-71; res assoc microbiol, 71-73, IMMUNOLOGIST, DEPT PATH, SCH MED, IND UNIV, INDIANAPOLIS, 73- *Mem:* Sigma Xi; Am Soc Microbiol; Am Asn Immunologists; Am Chem Soc. *Res:* Structure and specificity of antibody; relation of antibody specificity to genetic markers of antibody molecule; specific antibody class responses to selected pathogens; immunological diagnosis of viral, bacterial and fungal infections. *Mailing Add:* Dept Path Fesler Hall 409 Ind Univ Sch Med 1120 South Dr Indianapolis IN 46202-5113

ZIMMERMAN, SELMA BLAU, b New York, NY, Apr 1, 30; m 53; c 3. EMBRYOLOGY, CELL BIOLOGY. *Educ:* Hunter Col, BA, 50; NY Univ, MS, 54, PhD, 58. *Prof Exp:* Res asst, NY Univ, 53-55; res assoc pharmacol, Col Med, State Univ NY Downstate Med Ctr, 60-61; instr biol, Hunter Col, 61-64 & York Univ, Ont, 65-66; res assoc zool, Univ Toronto, 66-69; asst prof, York Univ, 74-75, coordr, 77-79, 81-82 & 85-86, assoc prof, 74-87, PROF NATURAL SCI, GLENDON COL, YORK UNIV, 87- *Concurrent Pos:* Assoc ed, Biochem & Cell Biol. *Mem:* AAAS; Can Soc Cell & Molecular Biol; Can Asn Women Sci; Sigma Xi. *Res:* Pigment cell physiology; mechanisms of cell division; cell cycle studies and drug action on cells. *Mailing Add:* Natural Sci Div 2275 Bayview Ave Toronto ON M4N 3M6 Can

ZIMMERMAN, SHELDON BERNARD, b New York, NY, Nov 7, 26; m 50; c 1. MICROBIOLOGY. *Educ:* City Col New York, BS, 48; Long Island Univ, MS, 65; NY Univ, PhD(biol), 71. *Prof Exp:* Chemist pharmaceut, Vitamin Corp Am, 49-51; develop microbiologist pharmaceut, Schering Corp, 51-63; sect head pharmaceut, 63-83, ASSOC DIR BASIC MICROBIOL, MERCK INST THERAPEUT RES, 83- *Mem:* Am Soc Microbiol; NY Acad Sci. *Res:* Discovery and mode of action of antibiotics; microbial physiology; microbial ecology; structure-activity relationships of antibiotics; automated microbiological assays; microbial chemotherapeutics. *Mailing Add:* Merck Inst Therapeut Res PO Box 2000 Rahway NJ 07065

ZIMMERMAN, STANLEY WILLIAM, b Detroit, Mich, July 30, 07; m 32; c 4. ELECTRICAL ENGINEERING. *Educ:* Univ Mich, BS & MS, 30. *Prof Exp:* Asst, Res Dept, Detroit Edison Co, 29-30; test man, Gen Elec Co, NY & Mass, 30-32, test man, Pittsfield Works Lab, 32-34, res & develop engr, Lightning Arrester Dept, 35-45; in charge high voltage res lab, 45-58, prof, 45-76, EMER PROF ELEC ENG, SCH ELEC ENG, CORNELL UNIV, 76- *Concurrent Pos:* Mem staff, Eng Dept, Westinghouse Elec Corp, Pa, 52-, Ramo-Wooldridge, Inc, 59-60 & Lawrence Radiation Lab, Univ Calif, 61 & 66-67; mem, Int Conf Large Elec High Tension Systs; consult. *Mem:* Am Soc Eng Educ; sr mem Inst Elec & Electronics Engrs; Brit Inst Elec Engrs. *Res:* Circuit interruption and protection devices; lightning studies; single transient oscillography; wide band transformers; heavy machinery; electrical measurements and insulation; extra-high voltage apparatus design; ionization, pulsed radiation and partial discharge measurements; sulphur hexafluoride power transmission systems; dielectric stress analysis. *Mailing Add:* 102 Valley Rd Ithaca NY 14850

ZIMMERMAN, STEPHEN WILLIAM, b Ironton, Mo, Dec 27, 41. NEPHROLOGY, MEDICINE. *Educ:* Univ of Wis, BS, 63, MD, 66. *Prof Exp:* Chief nephrology Fitzsimmons Army Hosp, 70-72; fel, 69-70, fel nephropath, 72-74, asst prof, 74-80, ASSOC PROF MED, UNIV WIS, 80- *Concurrent Pos:* NIH grant, 77-80. *Mem:* Am Soc Nephrology; Int Soc Nephrology; Am Fedn Clin Res; Nat Kidney Found; Int Soc Peritoneal Dialysis. *Res:* Effects of environmental toxins on the kidney; pathogenesis of renal disease; peritoneal dialysis. *Mailing Add:* Dept Med Nephrology Univ Wis Clin Sci Ctr 600 Highland Ave Madison WI 53792

ZIMMERMAN, STEVEN B, b Chicago, Ill, June 5, 34; m 56; c 2. BIOCHEMISTRY. *Educ:* Univ Ill, BS, 56, MS, 57; Stanford Univ, PhD(biochem), 61. *Prof Exp:* Nat Found res fel, 61-62; RES CHEMIST, NAT INST DIABETES, DIGESTIVE & KIDNEY DIS, 64- *Mem:* Am Soc Biochem & Molecular Biol. *Res:* Nucleic acid synthesis and structure; mechanism of enzyme action; Macromolecular crowding effects. *Mailing Add:* Nat Inst Diabetes Digestive & Kidney Dis Bldg 2 Rm 121 NIH Bethesda MD 20892

ZIMMERMAN, STUART O, b Chicago, Ill, July 27, 35; m 59; c 1. MATHEMATICAL BIOLOGY. *Educ:* Univ Chicago, BA, 54, PhD(math biol), 64. *Prof Exp:* Res assoc, Univ Chicago, 63-65, from instr to asst prof math biol, 65-67; assoc prof biomathematics, M D Anderson Hosp & Tumor Inst, Univ Texas, 68-72, head, div biomed info resources, 81-84, exec dir, info systs, 85-86, PROF BIOMATH & BIOMATHEMATICIAN, UNIV TEX DENT SCI INST, HOUSTON, 72-, KATHRYN O'CONNOR RES PROF, 86- *Concurrent Pos:* Consult, Ill State Dent Soc, 61- & Am Dent Asn, 66-; assoc prof biomath & assoc mem, Univ Tex Dent Sci Inst, Houston, 67-, head dept biomath, Univ Tex M D Anderson Cancer Ctr, 88-; assoc mem, Univ Tex Grad Sch Biomed Sci Houston, 68-70, mem, 70-; actg dir common res comput facil, Univ Tex M D Anderson Hosp & Tumor Inst Houston, 68-73; chmn exec bd, Univ Tex Houston Educ & Res Comput Ctr, 73- *Mem:* AAAS; Am Statist Asn; Asn Comput Mach; Soc Math Biol. *Res:* Biomedical computing; mathematical modeling; image processing; computer karyotyping; information management systems; design and analysis of cancer and dental clinical trials. *Mailing Add:* 9906 Bob White Dr Houston TX 77096

ZIMMERMAN, THEODORE SAMUEL, immunology; deceased, see previous edition for last biography

ZIMMERMAN, THOM J, b Lincoln, Ill, Oct 5, 42; m 70; c 1. OPHTHALMOLOGY, OCULAR PHARMACOLOGY. *Educ:* Univ Ill, Champaign-Urbana, BS, 64, MD, 68; Univ Fla, PhD(pharmacol), 76; Nat Bd Med Examiners, dipl, 69; Am Bd Ophthal, dipl, 78. *Prof Exp:* Intern, Presbyterian St Lukes Hosp, Chicago, 68-69; resident, Dept Ophthal, Col Med, Univ Fla, 71-74, corneal fel, 74-75; glaucoma fel, Dept Ophthal, Washington Univ, St Louis, 76-77; actg chmn, Dept Ophthal, 77, assoc prof, 77-79, PROF OPHTHAL, PHARMACOL & EXP THERAPEUTS & CHMN DEPT OPHTHAL, OSCHNER CLIN, LA STATE UNIV MED CTR, 79- *Concurrent Pos:* Heed Ophthalmic Found fel, 76-77; ophthalmic consult, glaucoma, US Pub Health Hosp, New Orleans, 77-82; Robert E McCormick scholar, Res Prevent Blindness, Inc, 78; consult, Eye Adv Coun, Food & Drug Admin, 79, Nat Eye Adv Coun & Nat Eye Inst, 83 & Handicapped Children's Serv Prog, Off Prev & Pub Health Serv, 85; res career develop award, Nat Eye Inst, 78-; fight for sight dept award, Oschner Clin, 80-81. *Mem:* Fel Am Col Clin Pharmacol; AMA; Asn Res Vision & Ophthal; Am Soc Clin Pharmacol; Am Soc Contemp Ophthal. *Res:* Ocular pharmacology; pharmacology of the glaucoma drugs; clinical care of glaucoma medical and surgical. *Mailing Add:* Dept Ophthal Univ Louisville Sch Med Louisville KY 40292

ZIMMERMAN, THOMAS PAUL, b Plainfield, NJ, Sept 3, 42; m 66; c 2. IMMUNOPHARMACOLOGY, DRUG TRANSPORT. *Educ:* Providence Col, BS, 64; Brown Univ, PhD(biochem), 69. *Prof Exp:* Nat Inst Neurol Dis & Stroke res fel biol & med sci, Brown Univ, 69-71; RES BIOCHEMIST, WELLCOME RES LABS, BURROUGHS WELLCOME & CO, USA, INC, 71-, PRIN SCIENTIST, 84- , ASSOC DIV DIR, 88- *Concurrent Pos:* Mem, Adv Comt on Chemother & Hemat, Am Cancer Soc, 83-87, chair, 87. *Mem:* Am Soc Pharmacol & Exp Therapeut; Am Asn Cancer Res; Int Soc Immunopharmacology. *Res:* Purine metabolism and the mode of action of purine antimetabolities; cyclic nucleotide metabolism; biological methylation reactions; immunosuppression; metabolic studies of purine and pyrimidine antimetabolities; determination of mechanisms by which pharmacological agents modulate immune and inflammatory cell function; nucleoside and drug transport. *Mailing Add:* Div Exp Ther Wellcome Res Labs 3030 Cornwallis Rd Research Triangle Park NC 27709

ZIMMERMAN, TOMMY LYNN, b Lima, Ohio, July 23, 43; m 67; c 2. SOIL & WATER CONSERVATION. *Educ:* Ohio State Univ, BS, 66; Pa State Univ, MS, 69, PhD(agron), 73. *Prof Exp:* From instr to asst prof agron, Delaware Valley Col, 71-75; ASSOC PROF & TECH COORDR SOIL & WATER CONSERV & MGT TECHNOL, AGR TECH INST, OHIO STATE UNIV, 75- *Concurrent Pos:* Consult, soil scientist, 73-75 80; cert prof soil scientists, ARCPACS, 80- *Mem:* Am Soc Agron; Soil & Water Conserv Soc; Coun Agr Sci Technol; Am Soc Agr Engrs. *Mailing Add:* Agr Tech Inst Ohio State Univ 1328 Dover Rd Wooster OH 44691-4099

ZIMMERMAN, WALTER BRUCE, b Evergreen Park, Ill, Nov 27, 33; m 55; c 3. SOLID STATE PHYSICS. *Educ:* Andrews Univ, BA, 55; Mich State Univ, MS, 57, PhD(physics), 60. *Prof Exp:* Asst physics, Mich State Univ, 58-60; sr physicist solar energy conversion, Gen Dynamics Astronaut, 60-62; from asst prof to assoc prof physics, Andrews Univ, 62-69; ASSOC PROF PHYSICS, IND UNIV, SOUTH BEND, 69-, CHMN, DEPT PHYSICS, 74- *Mem:* Am Phys Soc; Am Asn Physics Teachers; Am Vacuum Soc; Sigma Xi. *Res:* Changes that occur in infrared absorption spectrum and lattice constant of lithium hydride as its isotopic composition is varied; magnetic effect in biological processes; agglutination of red blood cells in a magnetic field; solar energy conversion by cadmium sulfide films. *Mailing Add:* PO Box 7111 Ind Univ 1700 Mishawaka Ave South Bend IN 46634

ZIMMERMAN, WILLIAM FREDERICK, b Chicago, Ill, July 7, 38; m 64; c 2. CELL BIOLOGY. *Educ:* Princeton Univ, BA, 60, PhD(biol), 66. *Hon Degrees:* MA, Amherst Col, 80. *Prof Exp:* Instr biol, Princeton Univ, 64-66; asst prof, 66-72, ASSOC PROF BIOL, AMHERST COL, 72- *Concurrent Pos:* NSF res grants, 66-70; NIH spec fel, 69-70; Nat Eye Inst res grants, 70-83; vis res fel, Univ Nijmegen, 73-74 & Cambridge Univ, 79-80. *Mem:* AAAS; Asn Res Vision & Ophthalmol. *Res:* Entrainment of circadian rhythms in insects; action spectra, carotenoid metabolism; cellular biochemistry of visual cycles, photoreceptor cell renewal and retinal pigment epithelium. *Mailing Add:* Dept Biol Amherst Col Amherst MA 01002

ZIMMERMANN, BERNARD, surgery; deceased, see previous edition for last biography

ZIMMERMANN, CHARLES EDWARD, b Juneau, Wis, Nov 15, 30; m 56; c 4. PLANT PHYSIOLOGY. *Educ:* Univ Wis, BS, 53; Ore State Univ, MS, 62. *Prof Exp:* Technician, Wash State Univ, 56-59; res agronomist, USDA, Ore, 59-62, plant physiologist, Agr Res Serv, Ore State Univ, 62-70, res plant physiologist, Sci & Educ Admin-Agr Res, 70-80; pres, Sunny Hops Inc, 80-86; PRES, HOP UNION USA, INC, 86- *Mem:* Am Soc Plant Physiol; Am Soc Agron; Crop Sci Soc Am; Am Soc Brewing Chemists; Master Brewers Asn Am. *Res:* Production and quality of hops; relate gibberellins to morphogenic and physiologic changes in hops. *Mailing Add:* Hop Union USA Inc Box 9697 Yakima WA 98909

ZIMMERMANN, EUGENE ROBERT, oral pathology, microbiology; deceased, see previous edition for last biography

ZIMMERMANN, F(RANCIS) J(OHN), b Jersey City, NJ, Apr 21, 24; m 50; c 2. CRYOGENIC ENGINEERING. *Educ:* Yale Univ, BE, 48; Mass Inst Technol, SM, 50, ME, 51,. *Hon Degrees:* ScD, Mass Inst Technol, 53. *Prof Exp:* Staff engr, Arthur D Little, Inc, 52-55; from asst prof to assoc prof mech eng, Yale Univ, 55-62; dir res & contract progs, 64-68, head dept, 78-82, prof, 62-86, EMER PROF MECH ENG, LAFAYETTE COL, 86- *Concurrent Pos:* Asst prog dir eng prog, NSF, 61-62; consult, Arthur D Little, Inc, 55-61 & Air Prod & Chem, Inc, 63-73. *Mem:* AAAS; Am Soc Mech Engrs; Am Soc Eng Educ; Cryogenic Soc Am. *Res:* Thermodynamics; heat transfer; cryogenic engineering. *Mailing Add:* Dept Mech Eng Lafayette Col Easton PA 18042

ZIMMERMANN, H(ENRY) J(OSEPH), b St Louis, Mo, May 11, 16; m 45; c 4. ELECTRICAL ENGINEERING, ELECTRONICS. *Educ:* Wash Univ, BS, 38; Mass Inst Technol, SM, 42. *Prof Exp:* Instr, Wash Univ, 38-40; from instr to prof, 40-78, assoc dir res lab electronics, 52-61, dir, 61-76, EMER PROF ELEC ENG, MASS INST TECHNOL, 78- *Mem:* Fel Inst Elec & Electronics Engrs. *Res:* Electronic circuits; signal processing and perception. *Mailing Add:* Dept Elec Eng Mass Inst Technol 77 Massachusetts Ave Cambridge MA 02139

ZIMMERMANN, MARK EDWARD, b Weimar, Tex, Sept 29, 52; m 78; c 3. PHYSICS, COMPUTER SCIENCE. *Educ:* Rice Univ, BA, 74; Calif Inst Technol, MS, 76, PhD(physics), 80. *Prof Exp:* Physicist, Inst Defense Analyses, 79-81; PHYSICIST, US GOVT, 81- *Mem:* Am Phys Soc. *Res:* Studies of developments in the physical sciences and information technologies; computer software development. *Mailing Add:* 9511 Gwyndale Dr Silver Spring MD 20910

ZIMMERMANN, MARTIN HULDRYCH, plant anatomy, plant physiology; deceased, see previous edition for last biography

ZIMMERMANN, R ERIK, b Newark, NJ, Oct 29, 41; m 71; c 4. ASTROPHYSICS, SCIENCE EDUCATION. *Educ:* Pomona Col, BA, 63; Univ Calif, Los Angeles, MA, 66, PhD(astron), 70. *Prof Exp:* Asst prof astron, Mich State Univ, 68-71; assoc prof astron, Newark State Col, 71-74; DIR, ROBERT J NOVINS PLANETARIUM, OCEAN COUNTY COL, 74- *Mem:* Int Planetarium Soc; Royal Astron Soc Can; Sigma Xi; Mid Atlantic Planetarium Soc. *Res:* Stellar structure and evolution. *Mailing Add:* Robert J Novins Planetarium Ocean County Col CN 2001 Toms River NJ 08754-2001

ZIMMERMANN, ROBERT ALAN, b Philadelphia, Pa, July 17, 37; m 87; c 1. PROTEIN SYNTHESIS, RNA STRUCTURE. *Educ:* Amherst Col, BA, 59; Mass Inst Technol, PhD(biophys), 64. *Prof Exp:* Vis scientist biochem, Acad Sci USSR, 65-66; res fel microbiol, Med Sch, Harvard Univ, 66-69; res assoc molecular biol, Univ Geneva, 70-73; assoc prof microbiol & biochem, 73-77, head, bioche, dept, 79-86, actg dir, Grad Prog Molecular & Cell Biol, 85-88, PROF BIOCHEM, UNIV MASS, AMHERST, 77- *Concurrent Pos:* Helen Hay Whitney Found fel, 68; sr fel, Europ Molecular Biol Orgn, 71; adv, WHO, 75-78; NIH res career develop award, 75. *Mem:* Am Chem Soc; Am Soc Biochem & Molecular Biol; AAAS; Am Soc Microbiol; Sigma Xi. *Res:* Structure, function and biosynthesis of ribosomes; RNA-protein interaction; primary and secondary structure of ribosomal RNA; photochemical labeling; properties of mutationally-altered ribosomes and their components; gene organization in archaebacteria. *Mailing Add:* Dept Biochem Univ Mass Amherst MA 01003

ZIMMERMANN, WILLIAM, JR, b Philadelphia, Pa, Oct 28, 30; m 62; c 3. PHYSICS. *Educ:* Amherst Col, AB, 52; Calif Inst Technol, PhD(physics), 58. *Prof Exp:* Fulbright fel, Neth, 58-59; lectr, 59-61, asst prof, 61-65, assoc prof, 65-70, PROF PHYSICS, UNIV MINN, MINNEAPOLIS, 70- *Concurrent Pos:* NSF sr fel, Finland, 67-68; vis scientist, Finland, 75-76. *Mem:* Am Phys Soc; Am Asn Physics Teachers; AAAS. *Res:* Low temperature physics; superfluid helium; liquid helium-3/helium-4 mixtures. *Mailing Add:* Sch Physics & Astron Univ Minn 116 Church St SE Minneapolis MN 55455

ZIMMIE, THOMAS FRANK, b Scranton, Pa, Jan 24, 39; m 62; c 3. GEOTECHNICAL ENGINEERING, ENVIRONMENTAL GEOTECHNOLOGY. *Educ:* Worcester Polytech Inst, BS, 60; Univ Conn, MS, 62, PhD(geotech eng), 72. *Prof Exp:* Civil engr, USN Oceanog Off, 61; staff engr, Linde Div, Union Carbide Corp, 64-67; consult, T F Zimmie, PE, 67-72; postdoctoral fel, Norwegian Geotech Inst, 72-73; PROF CIVIL ENG, RENSSELAER POLYTECH INST, 73- *Concurrent Pos:* Eng partner, Wang & Zimmie Consult Engrs, 73-80; geotech engr, NY State Dept Environ Conserv, 83-85; town engr, Town North Greenbush, 85-88; prog mgr geomech, NSF, 88-90. *Honors & Awards:* Spec Serv Award, Am Soc Testing & Mat, 80, Charles B Dudley Award, 84. *Mem:* Am Soc Civil Engrs; Am Soc Testing & Mat; Am Road & Transp Builders Asn; Nat Water Well Asn; Asn Soil & Found Engrs. *Res:* Environmental geotechnology-landfill siting and design, freeze/thaw effects on soil, hydraulic conductivity of soil; geotechnical engineering-laboratory testing, soil dynamics, earthquake engineering. *Mailing Add:* Civil Eng Dept Rensselaer Polytech Inst Troy NY 12180-3590

ZIMMON, DAVID SAMUEL, b Brooklyn, NY, Dec 2, 33; m 62; c 3. GASTROENTEROLOGY, LIVER DISEASE. *Educ:* Harvard Univ, MD, 58; Am Bd Internal Med, dipl, 66 & 68. *Prof Exp:* From intern to sr asst resident med, 2nd Med Div, Bellevue Hosp & Mem Ctr Cancer, 58-61; fel gastroenterol, 2nd Med Div, Bellevue Hosp, 61-62; res asst liver dis, Royal Free Hosp, London, 62-64; instr med, Med Col, 65, asst prof, 66-72, assoc prof, 72-79, PROF CLIN MED, SCH MED, NY, UNIV, 79- *Concurrent Pos:* Chief, Gastroenterol Sect, New York Vet Admin Med Ctr, 65; asst vis physician, Bellevue Hosp, 69. *Mem:* Med Res Soc; Am Fedn Clin Res; Am Asn Study Liver Dis; Am Gastroenterol Asn; Am Soc Gastrointestinal Endoscopy. *Res:* Liver disease; hypertension; biliary and pancreatic disease; endoscopy; endoscopic surgery. *Mailing Add:* 36 Seventh Ave New York NY 10011

ZIMMT, WERNER SIEGFRIED, b Berlin, Ger, Sept 21, 21; nat US; m 47; c 3. POLYMER CHEMISTRY, MUSEUM CONSERVATION. *Educ:* Univ Chicago, PhB & BS, 47, MS, 49, PhD(org chem), 51; Univ Penn, MS, 81. *Prof Exp:* Res chemist, E I du Pont de Nemours & Co, Inc, 51-60, chem assoc, Marshall Lab, 60-62, res assoc, 62-69, res fel, Marshall Lab, 69-85; RETIRED. *Concurrent Pos:* Adj prof Univ Ariz. *Honors & Awards:* George B Heckel Award, Nat Paint & Coatings Asn, 73. *Mem:* Am Chem Soc; Soc Archeol Sci; Soc Am Archeol. *Res:* Free radical chemistry; synthesis and mechanism; paint technology; role of paint solvents in air pollution; use of polymers in museum conservation. *Mailing Add:* 20 Conshohocken State Rd Bala Cynwyd PA 19004

ZIMNY, MARILYN LUCILE, b Chicago, Ill, Dec 12, 27. ANATOMY. *Educ:* Univ Ill, BA, 48; Loyola Univ, Ill, MS, 51, PhD(anat), 54. *Prof Exp:* Asst anat, Med Sch, Loyola Univ, Ill, 51-53; from asst prof to assoc prof, 54-64, actg head dept, 75-76, head dept, 76-90, PROF ANAT, LA STATE UNIV MED CTR, NEW ORLEANS, 64-, VCHANCELLOR ACAD AFFAIRS & DEAN SCH GRAD STUDIES, 90- *Concurrent Pos:* Vis prof, Sch Med, Univ Costa Rica, 61 & 62; mem, Inst Arctic Biol, Univ Alaska, 66. *Mem:* Am Asn Anat (pres, 83); Am Physiol Soc; Electron Micros Soc Am; Am Asn Dent Sch; Am Asn Dent Res. *Res:* Orthopedic research. *Mailing Add:* La State Univ Med Ctr 433 Bolivar St New Orleans LA 70112

ZIMRING, LOIS JACOBS, b Chicago, Ill, Nov 19, 23; div; c 1. SCIENCE FOR THE NON-SCIENCE MAJOR, CHANGING COSMOLOGICAL CONCEPTS. *Educ:* Univ Chicago, BS, 45, MS, 49, PhD(phys chem), 64. *Prof Exp:* Instr chem, Morgan Park Jr Col, 49-51; lectr phys sci, Univ Chicago, 59-61, instr, 61-64; asst prof chem, Univ Minn, 64-66; from asst prof to prof, 66-90, EMER PROF NATURAL SCI, MICH STATE UNIV, 91- *Mem:* Sigma Xi. *Res:* Ultraviolet spectra of conjugated systems; crystal spectra of transition metal halides; solid state mixed alum systems; chirality of prebiotic molecules; changing cosmological views. *Mailing Add:* Ctr Integrative Studies Mich State Univ East Lansing MI 48824-1031

ZINDER, NORTON DAVID, b New York, NY, Nov 7, 28; m 49; c 2. MOLECULAR GENETICS. *Educ:* Columbia Univ, BA, 47; Univ Wis, MS, 49, PhD(med microbiol), 52. *Hon Degrees:* DSc, Univ Wisconsin, 90. *Prof Exp:* Asst, 52-58, from assoc prof to prof, 58-76, JOHN D ROCKEFELLER, JR PROF MICROBIAL GENETICS, ROCKEFELLER UNIV, 76- *Honors & Awards:* Eli Lilly Award Microbiol, 62; US Steel Award Molecular Biol, 66; Sci Freedom & Responsibility Award, AAAS, 82. *Mem:* Nat Acad Sci; AAAS; Genetics Soc Am; Am Soc Microbiol; Am Soc Biol Chem; Sigma Xi. *Res:* Virology; protein biosynthesis; genetics. *Mailing Add:* Rockefeller Univ New York NY 10021-6399

ZINDER, STEPHEN HENRY, b Madison, Wis, Oct 22, 50; m 76; c 3. ANAEROBIC MICROORGANISMS, ENVIRONMENTAL MICROBIOLOGY. *Educ:* Kenyon Col, BA, 72; Colo State Univ, MS, 74; Univ Wis, PhD(bacteriol), 77. *Prof Exp:* Trainee microbiol, Colo State Univ, 72-74; res asst bacteriol, Univ Wis, 74-77; scholar pub health, Univ Calif, Los Angeles, 77-79; asst prof, 80-86, ASSOC PROF MICROBIOL, CORNELL UNIV, 86- *Mem:* Am Soc Microbiol; AAAS; Sigma Xi. *Res:* Physiology and ecology of methanogenic and other anaerobic bacteria. *Mailing Add:* Dept Microbiol Wing Hall Cornell Univ Ithaca NY 14853

ZINDLER, RICHARD EUGENE, b Benton Harbor, Mich, Mar 5, 27; m 58; c 2. MATHEMATICS, OPERATIONS RESEARCH. *Educ:* Mich State Univ, BS & MS, 49, PhD(math), 56. *Prof Exp:* Asst math, Mich State Univ, 49-52; asst prof, 52-57, assoc prof, 57-63, PROF ENG RES, PA STATE UNIV, 63- *Mem:* AAAS; Am Math Soc; Acoust Soc Am; Opers Res Soc Am; Asn Comput Mach. *Res:* Weapon system analysis and synthesis; primate behavior. *Mailing Add:* 639 Stoneledge Rd State College PA 16803

ZINGARO, JOSEPH S, b Mt Morris, NY, Mar 5, 28; m 52; c 4. SCIENCE EDUCATION, CHEMISTRY. *Educ:* State Univ NY Col Geneseo, BS, 51; Syracuse Univ, MS, 55 & 62, PhD, 66. *Prof Exp:* Teacher & chmn dept sci, Vernon-Verona-Sherrill Cent Sch, 51-58; PROF CHEM & CHMN DEPT, STATE UNIV NY, BUFFALO, 58- *Concurrent Pos:* NSF inst grants, 66-81. *Mem:* AAAS; Am Chem Soc. *Res:* Science teaching, especially chemistry teaching; electrical conductance and thermodynamic functions as they relate to solutions. *Mailing Add:* Dept Chem Buffalo State Col Buffalo NY 14222

ZINGARO, RALPH ANTHONY, b Brooklyn, NY, Oct 27, 25; m 50; c 2. INORGANIC CHEMISTRY. *Educ:* City Col New York, BS, 46; Univ Kans, MS, 49, PhD, 50. *Prof Exp:* Sr res chemist, Eastman Kodak Co, 50-52; asst prof, Univ Ark, 52-53; res chemist, Am Cyanamid Co, 53-54; from asst prof to assoc prof, 54-64, PROF CHEM, TEX A&M UNIV, 64- *Concurrent Pos:* NIH spec fel, 68-69; Fulbright lectr, Univ Buenos Aires, 72. *Mem:* Am Chem Soc; NY Acad Sci; Sigma Xi. *Res:* Chemistry and biochemistry of selenium, tellurium and arsenic; trace elements in fossil fuels. *Mailing Add:* Dept Chem Tex A&M Univ College Station TX 77843

ZINGESER, MAURICE ROY, b Birmingham, Ala, Mar 17, 21; m 47; c 3. ANATOMY, ORTHODONTICS. *Educ:* New York Univ, AB, 42; Columbia Univ, DDS, 46; Tufts Univ, MS, 50; Am Bd Orthod, dipl, 63. *Prof Exp:* Intern surg, New York Polyclin Hosp, 46-48; clin assoc orthod, Tufts Univ, 48-50; dent surgeon, USPHS, 50-52; RES PROF ANAT, DENT SCH, ORE HEALTH SCI UNIV, 80- *Concurrent Pos:* Guest lectr, Tufts, Boston & Georgetown Univs, 60-69; vis scientist anthrop & path, Ore Regional Primate Res Ctr, 63-80; contrib, Int Cong Anthrop & Ethnol Sci, 68; chmn crainiofacial biol sect, Int Cong Primatol, 72; guest lectr, Hebrew Univ & Univ London, 72. *Mem:* AAAS; Am Dent Asn; Am Asn Orthod; Am Asn Anat; Am Asn Phys Anthrop. *Res:* Primate odontology, craniofacial embryology and craniology. *Mailing Add:* Dept Anat Dent Sch Ore Health Sci Univ 611 Campus Dr Portland OR 97201

ZINGESSER, LAWRENCE H, b Portchester, NY, Dec 27, 30; div; c 4. RADIOLOGY. *Educ:* Syracuse Univ, AB, 51; Chicago Med Sch, MD, 55. *Prof Exp:* From intern med to resident, Grad Hosp, Univ Pa, 55-57; resident radiol, Grace-New Haven Hosp, Yale Univ, 59-62; from asst prof to prof radiol, Albert Einstein Col Med, 73-77; CHIEF NEURORADIOL & ATTEND PHYSICIAN, ST VINCENT'S HOSP, NY, 77- *Concurrent Pos:* NIH spec fel neuroradiol, Albert Einstein Col Med, 62-64, Nat Inst Neurol Dis & Stroke grant cerebral blood flow, 66-69, clin prof radiol, 77-79; asst attend, Bronx Munic Hosp Ctr, NY, 63-65; clin prof radiol, NY Med Col, 80- *Mem:* Am Soc Neuroradiol; Radiol Soc NAm; Am Col Radiol; French Soc Neuroradiol. *Res:* Regional cerebral blood flow in neurologic disease states; neuroradiology. *Mailing Add:* 108 W Fifth Ave Apt 15 A New York NY 10011

ZINGG, WALTER, b Kloten, Switz, Mar 29, 24; Can citizen; wid; c 4. SURGERY, BIOMEDICAL ENGINEERING. *Educ:* Univ Zurich, MD, 50; Univ Man, MSc, 52; FRCS(C), 58. *Hon Degrees:* DSc, Univ Laval, 86. *Prof Exp:* Lectr physiol, Univ Man, 56-57, lectr surg, 57-61, asst prof, 61-64; from asst prof to assoc prof, Univ Toronto, 64-78, mem inst biomed eng, 72-75, prof & assoc dir, 75-83, prof surg & hon prof dent, 78-89, dir inst biomed eng, 83-89; RETIRED. *Concurrent Pos:* Assoc scientist, Hosp Sick Children, 64-68, sr scientist, 68-80, head, Div Surg Res, 64-88; consult, Ont Crippled Children's Ctr, Toronto, 65-90 & Ont Vet Col, Univ Guelph, 70-79. *Mem:* Am Soc Artificial Internal Organs; Can Physiol Soc; Can Biomat Soc (pres, 78-80); fel Am Col Surg; fel Am Col Cardiol; Can Med Biol Eng Soc (pres, 84-88). *Res:* Surgical research; biomaterials; artificial organs. *Mailing Add:* Inst Biomed Eng Toronto ON M5S 1A4 Can

ZINGMARK, RICHARD G, b San Francisco, Calif, July 4, 41; m 62; c 6. ALGOLOGY. *Educ:* Humboldt State Col, BA, 64, MA, 65; Univ Calif, Santa Barbara, PhD(biol), 69. *Prof Exp:* NSF fel, Marine Lab, Duke Univ, 69-70; ASSOC PROF BIOL & MARINE SCI, UNIV SC, 76-, RES ASSOC MARINE SCI, BELLE W BARUCH COASTAL RES INST, 70- *Concurrent Pos:* Consult, James H Carr & Assoc, 76-; sr Fulbright res award, 89-90. *Mem:* Phycol Soc Am; Int Phycol Soc. *Res:* Physiological ecology of marine algae; biomass, productivity and photosynthesis of phytoplankton, benthic microalgae and macroalgae; red tide research; coastal management. *Mailing Add:* Dept Biol Univ SC Columbia SC 29208

ZINGULA, RICHARD PAUL, b Cedar Rapids, Iowa, May 31, 29; m 53; c 2. PALEONTOLOGY. *Educ:* Iowa State Univ, BS, 51; La State Univ, MS, 53, PhD(geol), 58. *Prof Exp:* Assoc geologist, Humble Oil & Refining Co, 54-60, supvry paleontologist, 60-69, sr prof geologist, 69-72; res paleontologist, Imp Oil Ltd, 72-74; sr prof geologist, Exxon Co USA, 74-76, sr explor geologist, 76-82, geol assoc, 82-86; CONSULT GEOLOGIST, 86- *Mem:* Paleont Soc Am. *Res:* Micropaleontology; stratigraphy. *Mailing Add:* 5134 Lymbar Houston TX 77096-5318

ZINK, FRANK W, b Pullman, Wash, June 17, 23; m 46; c 2. PLANT BREEDING. *Educ:* Univ Calif, Davis, BS, 47, MS, 48. *Prof Exp:* Asst specialist, 48-53, assoc specialist, 54-60, specialist, 61-69, RES SPECIALIST VEG CROPS, CALIF AGR EXP STA, UNIV CALIF, DAVIS, 69-, LECTR, COL AGR, 74- *Concurrent Pos:* Shell Develop grant, 60-61; Veg Growers Asn grant, 60-67; Melon Growers Asn grant, 72-75. *Mem:* Am Soc Hort Sci; Am Phytopath Soc; Am Soc Agron; Int Soc Hort Sci. *Res:* Lettuce breeding, disease resistance; melon breeding for mechanization and disease resistance. *Mailing Add:* Dept Veg Crops Univ Calif Davis CA 95616

ZINK, GILBERT LEROY, b Wheeling, WVa, Aug 14, 42; m 63; c 4. IMMUNOGENETICS. *Educ:* Ohio State Univ, BSc, 65, MSc, 67, PhD(immunogenetics), 71. *Prof Exp:* Res assoc, Ohio State Univ, 67-71, res assoc immunogenetics lab, Dept Dairy Sci, 72; res assoc lectr dept biol sci, 72-75, asst prof, 75-80, ASSOC PROF BIOL SCI, PHILADELPHIA COL PHARM & SCI, 80-, CHAIRPERSON, DEPT BIOL SCI, 83- *Concurrent Pos:* Co-investr, NIH res grant, 77-80. *Res:* Genetics. *Mailing Add:* 1410 Spackman Lane Westchester PA 19380

ZINK, JEFFREY IRVE, b Milwaukee, Wis, Jan 8, 45; m 68. INORGANIC CHEMISTRY, PHOTOCHEMISTRY. *Educ:* Univ of Wis, BS, 66; Univ Ill, PhD(chem), 70. *Prof Exp:* Teaching asst chem, 66-67, res asst chem, Univ Ill, 67-68; from asst prof to assoc prof, 70-82, PROF CHEM, UNIV CALIF, LOS ANGELES, 82- *Concurrent Pos:* Camille & Henry Dreyfus teacher-scholar, 74-79. *Honors & Awards:* Alexander von Humboldt Award, 78; John Simon Guggenheim Fel, 88. *Mem:* Am Chem Soc; Interam Photochem Soc; Nat Audubon Soc. *Res:* Photochemistry, triboluminescence; solar energy conversion and storage; structure and bonding transition metal compounds. *Mailing Add:* Univ Calif Chem/1110 Young Hall 405 Hilgard Ave Los Angeles CA 90024-4199

ZINK, ROBERT EDWIN, b Minneapolis, Minn, Nov 16, 28; m 50; c 3. MATHEMATICS. *Educ:* Univ Minn, BA, 49, MA, 51, PhD(math), 53. *Prof Exp:* Asst math, Univ Minn, 49-53; instr, Purdue Univ, 53-54; lectr, George Washington Univ, 55-56; from asst prof to assoc prof math, 56-66, asst head dept, 65-69, asst dean grad sch, 69-72, PROF MATH, PURDUE UNIV, LAFAYETTE, 66- *Concurrent Pos:* Vis prof, Wabash Col, 61-62 & dept math, Univ Calif, Irvine, 68-69. *Mem:* Am Math Soc; Math Asn Am; Sigma Xi. *Res:* Theory of measure and integration; theory of functions of a real variable; Schauder bases for Banach function spaces; several articles in a variety of journals. *Mailing Add:* Dept Math Purdue Univ Lafayette IN 47907

ZINK, SANDRA, b Dodge City, Kans, Sept 17, 39; c 2. PHYSICS, COMPUTER SCIENCE. *Educ:* Univ NMex, BS, 66, MS, 68, PhD(physics), 73. *Prof Exp:* Assoc Western Univs fel molecular physics, Los Alamos Sci Lab, 71-73, staff mem solar physics, 73-75; guest scientist, Max Planck Inst Extraterrestrial Physics, 75-76; staff mem med physics & comput treatment planning, Los Alamos Sci Lab, 77-85; CANCER EXPERT, NAT CANCER INST, 85- *Mem:* Am Asn Physicists Med. *Res:* Biophysics and medical physics with computer applications. *Mailing Add:* Nat Cancer Inst 6130 Exec Blvd Exec Plaza N Rm 800 Rockville MD 20892

ZINKE, OTTO HENRY, b Webster Groves, Mo, Aug 13, 26; m 55; c 3. PHYSICS. *Educ:* Wash Univ, AB, 50, AM, 53, PhD, 56. *Prof Exp:* Salesman, Nuclear Consults Corp, 52-53; mem res staff, Linde Co Union Carbide Corp, 56-57; asst prof physics, Univ Mo, 57-59; from asst prof to assoc prof physics, Univ Ark, Fayetteville, 59-69, prof, 69-80. *Res:* Transient phenomena in plasmas and metals. *Mailing Add:* 817 N Jackson Fayetteville AR 72701

ZINKE, PAUL JOSEPH, b Los Angeles, Calif, Nov 10, 20; m 47; c 2. FORESTRY. SOIL SCIENCE. *Educ:* Univ Calif, BS, 42, MS, 52, PhD(soil sci), 56. *Prof Exp:* Forester, Tongass Nat Forest, US Forest Serv, 42-43, res forester, Calif Forest & Range Exp Sta, 46-56; PROF FORESTRY & SOIL SCI, UNIV CALIF, BERKELEY, 57- *Concurrent Pos:* In charge, Calif Soil Veg Surv, USDA-US Forest Serv, 59-61; adv, Appl Sci Res Corp Thailand, 67-; adv, Radar Nat Resource Inventory Amazon Basin, Brazil, 71-; mem comt study defoliation effects in SEAsia, Nat Acad Sci, 71- *Mem:* AAAS; Soc Am Foresters; Soil Sci Soc Am; Soil Conserv Soc Am; Soc Range Mgt; Sigma Xi. *Res:* Forest influences and environment; forest soils; soil morphology; soil-vegetation relationships; plant ecology. *Mailing Add:* 805 Hilldale Ave Berkeley CA 94708

ZINKEL, DUANE FORST, b Manitowoc, Wis, Aug 11, 34; m 61; c 5. NATURAL PRODUCTS & NAVAL STORES CHEMISTRY. *Educ:* Univ Wis, BS, 56, PhD(biochem), 61. *Prof Exp:* RES CHEMIST, FOREST PROD LAB, US FOREST SERV, 61- *Mem:* Am Chem Soc; Am Oil Chem Soc. *Res:* Softwood extractives and derived products; structure determination; analytical development and analysis; biosynthesis; specialist in navel stores chemistry (softwood extractives, rosin, turpentine, fatty acids); development of analytical methods, isolation and structure elucidation, chemotoxonomy. *Mailing Add:* 2323 Hollister Ave Madison WI 53705-5315

ZINKHAM, ROBERT EDWARD, b Rochester, Pa, Jan 19, 23; m 56; c 3. MECHANICAL METALLURGY. *Educ:* Geneva Col, BS, 49; Carnegie-Mellon Univ, BSME, 49; Univ Pittsburgh, MS, 56. *Prof Exp:* Develop engr, Jones & Laughlin Steel Corp, 49-57; supvr mech testing, Allegheny Ludlum Steel Corp, 57-60; res engr drilling methods, Gulf Res & Develop Co, 60-63; res scientist mech metall, Reynolds Metals Co, 63-68, dir, 68-86; RETIRED. *Concurrent Pos:* Chmn subcomt fracture toughness, Metal Prop Coun, Inc, 72- *Mem:* Am Soc Mech Engrs; fel Am Soc Metals; Am Soc Testing & Mat. *Res:* Mechanical metallurgy, specializing in fracture mechanics, fatigue, mechanical testing and residual stresses involving aluminum. *Mailing Add:* 9204 Westmoor Dr Richmond VA 23229

ZINKHAM, WILLIAM HOWARD, b Uniontown, Md, May 23, 24; m 52; c 2. PEDIATRICS. *Educ:* Johns Hopkins Univ, AB, 44, MD, 47; Am Bd Pediat, cert pediat, 56, cert hemat-oncol, 74. *Prof Exp:* From instr to prof, 56-77, DISTINGUISHED SERV PROF PEDIAT, SCH MED, JOHNS HOPKINS UNIV, 77- *Mem:* Am Pediat Soc; Soc Clin Invest; Soc Pediat Res. *Res:* Hematology; metabolism of normal and abnormal erythrocytes. *Mailing Add:* Dept Pediat & Oncol Johns Hopkins Univ 720 Rutland Ave Baltimore MD 21205

ZINKL, JOSEPH GRANDJEAN, b Albuquerque, NMex, Aug 30, 39; m 69; c 2. CYTOLOGY, TOXICOLOGY. *Educ:* Univ Calif, Davis, BS, 64, DVM, 66, PhD(comparative path), 71. *Prof Exp:* Sr fel, Nat Inst Environ Health, 71-74; pathologist, US Fish & Wildlife Serv, 74-76; PROF CLIN PATH, SCH VET MED, UNIV CALIF, DAVIS, 76- *Mem:* Am Col Vet Pathologists; Wildlife Dis Asn; Am Soc Vet Clin Path. *Res:* Brain cholinesterase activity in forest birds after aerial application of insecticides; phagocytic, bacteriocidal and chemotactic ability of neutrophils of domestic animals; clinical cytology of diseases of domestic animals; effects of anticholinesterase insecticides on wild mammals, birds and fish. *Mailing Add:* Dept Clin Path Univ Calif Davis CA 95616

ZINMAN, WALTER GEORGE, b New York, NY, Nov 9, 29; m 55; c 2. CHEMISTRY. *Educ:* Rensselaer Polytech Inst, BChE, 51; Harvard Univ, PhD(chem), 55. *Prof Exp:* Res chemist missile & space vehicle div, Gen Elec Co, 56-59; prin res & develop engr, Repub Aviation Corp, 59-65; res group leader, Polytech Inst Brooklyn, 65-66; propulsion engr, Grumman Aircraft Eng Corp, 66-70, consult, 70-72; asst to dir res, Surface Activation Corp, NY, 72-73; CHEM CONSULT, 73- *Mem:* Am Chem Soc; Am Phys Soc; Sigma Xi. *Res:* Reaction of dissociated gases with solids; gaseous detonations; homogeneous kinetics; fluid mechanics and reacting flows; foundations thermodynamics. *Mailing Add:* Eight Conventry Rd Syosset NY 11791

ZINN, BEN T, b Tel-Aviv, Israel, Apr 21, 37; US citizen; c 2. COMBUSTION, FLUID MECHANICS. *Educ:* NY Univ, BS, 61; Stanford Univ, MS, 62; Princeton Univ, MA, 63, PhD(aerospace & mech sci), 66. *Prof Exp:* Asst res combustion instability, Princeton Univ, 64-65; from asst prof to prof, 65-74, REGENTS PROF AEROSPACE ENG, GA INST TECHNOL, 74- *Concurrent Pos:* Consult, Lockheed Ga Res Labs, Naval Weapons Ctr, Calif & adv group aerospace res & develop, NATO; consult, Brazilian Space Res Inst, 77-; ed measurements in combustion systs, Am Inst Aeronaut & Astronaut Progress in Aeronaut & Astronaut J, 77-78; mem bd vis, Nat Acad Fire Prevention & Control, 78-80; assoc ed, Am Inst Aeronaut & Astronaut. *Honors & Awards:* Sustained Res Award, Sigma Xi, 76. *Mem:* Assoc fel Am Inst Aeronaut & Astronaut; Combustion Inst. *Res:* Combustion in energy and propulsion generating devices; fire safety; acoustics. *Mailing Add:* 2006 W Paces Terry Rd NW Atlanta GA 30327

ZINN, DALE WENDEL, b Parkersburg, WVa; m 54; c 2. ANIMAL HUSBANDRY. *Educ:* WVa Univ, MS, 56; Univ Mo, PhD, 53. *Prof Exp:* Asst prof animal husb, NMex State Univ, 57-61; from assoc prof to prof animal sci, Tex Tech Univ, 61-75, chmn dept, 69-74, asst dean, Col Agr Sci & dir div agr serv, 74-75; DEAN COL AGR & FORESTRY & DIR AGR & FORESTRY EXP STA, WVA UNIV, 75- *Mem:* Am Soc Animal Sci; Am Meat Sci Asn; Sigma Xi. *Res:* Production and quality factors affecting quantity and quality of meat and meat products. *Mailing Add:* Harewood Morgantown WV 26505

ZINN, DONALD JOSEPH, b New York, NY, Apr 19, 11; m 41, 87; c 2. INVERTEBRATE ZOOLOGY. *Educ:* Harvard Univ, SB, 33; Univ RI, MS, 37; Yale Univ, PhD(zool), 42. *Prof Exp:* Dir, Bass Biol Lab, Fla, 33-35; tech asst, Ro-Lab, Conn, 38-39; asst, Osborn Zool Lab, Yale Univ, 40-41; naturalist, Marine Biol Lab, Woods Hole, 45-46; from instr to prof zool, 46-74, actg chmn dept, 60-62, chmn dept, 62-65 & 73-74, EMER PROF ZOOL, UNIV RI, 74- *Concurrent Pos:* Chief mosquito control proj, Pine Orchard Asn, Conn, 40; res assoc, Narragansett Marine Lab, 55-74; deleg, Int Cong Zool, London, 58; co-ed, Psammonalia, 66-69; mem, President's Adv Panel Timber & The Environ, 71-74; mem aquatic ecol sect, Int Biol Prog; ecol consult, US Plywood-Champion Papers; Life mem corp, Bermuda Biol Sta &

Marine Biol Lab, Woods Hole Oceanog Inst; pres, Nat Wildlife Fedn, 68-71; trustee & clerk, New Eng Natural Resources Ctr, 72-; mem, Shore Erosion Adv Panel, US Army Corps Engrs, 74-80; corresp, Mus Nat Hist Natural, France, 76; mem, Falmouth, Mass, Conserv Comm, 77-84; pres, Cape Cod Mus Nat Natural Hist, 80-82, sr fel, 82-; trustee, New Eng Environ Mediation Ctr, 83-90; mem, Falmouth Mass Hist Comn, 90- *Mem:* Fel AAAS; Am Inst Biol Sci; Ecol Soc Am; Int Asn Meiobenthologists; Asn Systematics Collections. *Res:* Ecology and taxonomy of marine beaches intertidal interstitial fauna and flora; littoral benthos and fouling organisms; tunicate and entomostracan taxonomy; histological techniques with micrometazoa; conservation education; conservation of natural resources. *Mailing Add:* PO Box 589 Falmouth MA 02541-0589

ZINN, GARY WILLIAM, b Oxford, WVa, Sept 20, 44; m 67; c 2. FOREST ECONOMICS, FOREST MANAGEMENT. *Educ:* WVa Univ, BS, 66; State Univ NY Col Forestry, Syracuse Univ, MS, 68, PhD(forest econ), 72. *Prof Exp:* Instr forestry econ, State Univ NY Col Forestry, Syracuse Univ, 71; asst prof, 72-77, ASSOC PROF FOREST MGT, WVA UNIV, 77- *Mem:* Soc Am Foresters; Southern Econ Asn; Am Foresty Asn. *Res:* Economic contributions of forest-based activity to regions; regional development; forest and natural resource policy; forest land use and management planning. *Mailing Add:* Dept Forestry Mgt WVa Univ Morgantown WV 26506

ZINN, JOHN, b Brooklyn, NY, Feb 28, 28; m 54; c 3. APPLIED PHYSICS. *Educ:* Cornell Univ, AB, 49; Univ Calif, Berkeley, PhD(phys chem), 58. *Prof Exp:* Chemist, Catalin Corp Am, 49-50 & M W Kellogg Co, 52-54; assoc chem, Univ Calif, 55-56, MEM STAFF, LOS ALAMOS SCI LAB, UNIV CALIF, 57- *Concurrent Pos:* Asst prof dept aeronaut eng, Univ Colo, 67-68. *Mem:* Am Phys Soc; AAAS; Am Geophys Union. *Res:* Theoretical research in atmospheric physics and chemistry; atmospheric effects of nuclear explosions. *Mailing Add:* 249 Rio Bravo Los Alamos NM 87544

ZINN, ROBERT JAMES, b Chicago, Ill, Aug 4, 46; m 79; c 3. ASTRONOMY. *Educ:* Case Inst Technol, BS, 68; Yale Univ, PhD(astron), 74. *Prof Exp:* Fel astron, Hale Observs, Carnegie Inst Washington, 74-79; asst prof, 79-82, assoc prof, 82-87, PROF ASTRON, YALE UNIV OBSERV, 87- *Mem:* Am Astron Soc; Int Astron Union. *Res:* Stellar evolution; the chemical compositions of globular cluster stars, variable stars, and the stellar populations of the galaxies of the Local Group. *Mailing Add:* Yale Univ Observ 260 Whitney Ave New Haven CT 06511

ZINN, WALTER HENRY, b Kitchener, Ont, Dec 10, 06; nat US; m 33, 66; c 2. PHYSICS. *Educ:* Queen's Univ, Ont, BA, 27, MA, 29; Columbia Univ, PhD(physics), 34. *Hon Degrees:* DSc, Queen's Univ, Ont, 57. *Prof Exp:* Asst physics, Queen's Univ, Ont, 27-28; asst, Columbia Univ, 31-32; from instr to asst prof, City Col New York, 32-41; physicist, Metall Lab, Manhattan Dist, Univ Chicago, 42-46; dir, Argonne Nat Lab, 45-56; vpres, Combustion Eng, Inc, 59-71; RETIRED. *Concurrent Pos:* Spec consult, Joint Cong Comt Atomic Energy, 56; spec mem, President's Sci Adv Comt; pres, Gen Nuclear Eng Corp, 56-64. *Honors & Awards:* Enrico Fermi Award, 69. *Mem:* Nat Acad Sci; Nat Acad Eng; AAAS; fel Am Phys Soc; Am Nuclear Soc (pres, 55). *Res:* Nuclear physics and reactor development. *Mailing Add:* 2940 Bay Meadow Ct Clearwater FL 34619

ZINNER, STEPHEN HARVEY, b New York, NY, Apr 29, 39; m 66; c 2. INFECTIOUS DISEASES, EPIDEMIOLOGY. *Educ:* Northwestern Univ, BA, 61; Sch Med, Univ Pa, MD, 65; Am Bd Internal Med, cert internal med, 72, cert infectious dis, 74. *Prof Exp:* Res fel bacterio & immunol, Channing Lab, Med Sch, Harvard Univ, 67-68, res assoc med, 68-69, res fel, Thorndike Lab, 61-71, instr med, Med Sch, 71-72; asst prof biol & med sci, 72-76, assoc prof, 76-81, PROF MED, MED SECT, BROWN UNIV, 81- *Concurrent Pos:* Field officer, Nat Heart Dis Control, USPHS, 67-69; clin instr med, Med Sch, Harvard Univ, 72-; consult infectious dis, RI Hosp, Women & Infants Hosp & Vet Admin Hosp, 72-, Miriam Hosp, Providence, 76-; head, Div Infectious Dis, Roger Williams Med Ctr & Brown Univ, 72-, Rhode Island Hosp, 89- *Mem:* Infectious Dis Soc Am; Am Soc Microbiol; Am Fedn Clin Res; Soc Epidemiol Res; Sigma Xi; Am Soc Clinc Investr. *Res:* Epidemiology of blood pressure in infants and children; infections in the immunosuppressed patient; antibiotic combinations; new methods for in vitro antibiotic activity determinations; reactive antibodies to Gram-negative infecting organisms. *Mailing Add:* Dept Med Roger Williams Med Ctr 825 Chalkstone Ave Providence RI 02908

ZINNES, HAROLD, b New York, NY, Apr 7, 29; m 53; c 3. ORGANIC CHEMISTRY. *Educ:* Rutgers Univ, BS, 51; Univ Mich, MS, 52, PhD, 55. *Prof Exp:* Res assoc, E R Squibb & Co, 56-58; scientist, 58-63, sr scientist, 63-68, sr res assoc, Warner-Lambert Res Inst, 68-77, DIR PHARMACEUT TECH DEVELOP, WARNER-LAMBERT INT, 77- *Mem:* Am Chem Soc; Am Pharmaceut Asn. *Res:* Medicinals; natural products; antibiotics; heterocycles; indoles; benzothiazines; anti-inflammatory agents; international pharmaceutical development. *Mailing Add:* 16 Calument Ave Rockaway NJ 07866-1822

ZINS, GERALD RAYMOND, b New Ulm, Minn, Oct 23, 32; m 81; c 3. PHARMACOLOGY. *Educ:* SDak State Univ, BS, 54; Univ Chicago, PhD(pharmacol), 58. *Prof Exp:* Res assoc & instr, Univ Chicago, 58-59; res scientist, Upjohn Co, 59-66; vis assoc prof, State Univ NY Upstate Med Ctr, 66-67; sr scientist, 67-68, head, 68-81, MGR CARDIOVASC RES, UPJOHN CO, 81- *Mem:* AAAS; Am Soc Pharmacol & Exp Therapeut; Am Soc Nephrol. *Res:* Renal and cardiovascular pharmacology; renal prostaglandins; drug metabolism; hypertension; intermediary carbohydrate metabolism. *Mailing Add:* Upjohn Co 301 Henrietta St Kalamazoo MI 49001

ZINSER, EDWARD JOHN, b Toronto, Ont, Mar 13, 41; m 66; c 3. ANALYTICAL CHEMISTRY. *Educ:* Univ Toronto, BSc, 65, PhD(anal chem), 69. *Prof Exp:* Asst, Univ Toronto, 65-68; fel, Queen's Univ, Ont, 69-70; res chemist, Marshall Lab, E I du Pont de Nemours & Co Inc, 70-71 & Exp Sta, Wilmington, 72, sr prod specialist, 72-73, tech objectives mgr, Finishes Div, 73-74, res supvr, Marshall Res & Develop Lab, 74-75, prod mgr, Finishes Div, 76-77, nat mkt mgr, Maintenance Finishes, 78-79, worldwide mkt & prod planning mgr, 79-80, worldwide bus mgr, Packaging Finishes, 80-82, mgr, Health Prod, Cent Res Dept, 82-84, planning mgr, Latin Am Int Dept, 84-88, BUS MGR, ENG SERV, E I DU PONT DE NEMOURS & CO, INC, 88- *Concurrent Pos:* Nat Res Coun Can scholar, Univ Toronto, 68-70. *Mem:* Am Chem Soc; Chem Inst Can; Fedn Soc Coatings Technol; Nat Asn Corrosion Engrs; Steel Struct Painting Coun. *Res:* Environmental analytical chemistry; electrochemistry; atomic absorption; spectroscopy; organic coatings; fluorocarbon coatings. *Mailing Add:* Automotive Prod DuPont Co Wilmington DE 19898

ZINSMEISTER, GEORGE EMIL, b Huntington, NY, Dec 27, 39; m 65. MECHANICAL ENGINEERING, HEAT TRANSFER. *Educ:* Rensselaer Polytech Inst, BME, 61; Purdue Univ, MSME, 63, PhD(mech eng), 65. *Prof Exp:* Res engr, E I du Pont de Nemours & Co, Inc, Del, 65-66; asst prof, 66-69, ASSOC PROF MECH ENG, UNIV MASS, AMHERST, 69- *Concurrent Pos:* NSF res grants, 67-70 & 72-73. *Mem:* Am Soc Mech Engrs. *Res:* Heat transfer in composite materials; prediction of thermal processing conditions in foods; engineering education. *Mailing Add:* Dept Mech Eng Univ Mass Amherst MA 01003

ZINSMEISTER, PHILIP PRICE, b Columbus, Ohio, May 15, 40; m 68; c 1. DEVELOPMENTAL BIOLOGY. *Educ:* Wittenberg Univ, BS, 62; Univ Ill, MS, 66, PhD(zool), 69. *Prof Exp:* Master biol, US Peace Corps, Ghana, 62-64; asst prof biol, Northeast Mo State Col, 70; lectr, Univ Sci & Technol, Kumasi, Ghana, 70-72; vis asst prof, Univ Ill, 72-73; assoc prof & chmn div sci & math, 73-80, PROF BIOL, OGLETHORPE UNIV, 80- *Concurrent Pos:* Res assoc, Emory Univ, 80-81 & 87-88. *Mem:* Am Inst Biol Sci; Soc Develop Biol; AAAS. *Res:* Insect oogenesis and development; nerve cell development. *Mailing Add:* Oglethorpe Univ 4484 Peachtree Rd NE Atlanta GA 30319

ZINSMEISTER, WILLIAM JOHN, b Nogales, Ariz, May 6, 43; m 66; c 2. BIOGEOGRAPHY, EVOLUTION. *Educ:* Calif State Col, Long Beach, BS, 69; Univ Calif, Riverside, MS, 73, PhD(geol), 74. *Prof Exp:* Res assoc paleontol, 75-80, sr res assoc, Inst Polar Studies, 80-85; RES ASSOC GEOL, PURDUE UNIV, 84- *Concurrent Pos:* NSF grant, 74-76 & 77-; Nat Geog Soc grant, 76-77; adv, Earth Sci Adv Comt, Brit Antarctic Survey, Nat Environ Res Coun, UK, 81, Int Comn Cretaceous Climates, 83- *Honors & Awards:* Antarctic Serv Medal, NSF, 83. *Mem:* Paleont Soc; Int Paleont Union; Sigma Xi; AAAS. *Res:* Changes in the distribution of shallow-water marine faunas in the southern hemisphere in response to the final fragmentation of Gondwanaland during the late Cretaceous and early Tertiary; role polar regions played in the evolution of modern marine fauna. *Mailing Add:* 2100 Edgewood Dr West Lafayette IN 47906

ZINSSER, HARRY FREDERICK, b Pittsburgh, Pa, May 1, 18; m 43; c 4. CARDIOLOGY. *Educ:* Univ Pittsburgh, BS, 37, MD, 39. *Hon Degrees:* LLD, Univ Pa, 71. *Prof Exp:* Teaching fel internal med, Sch Med, Univ Pittsburgh, 40-42, asst instr, 46-47; fel, 47-48, from asst instr to assoc, 48-51, asst prof clin med, 51-53, asst prof med, 53-55, assoc prof clin med, 55-58, assoc prof med, 58-68, prof cardiol, Div Grad Med, 63-68, dir cardiol, Grad Hosp, 63-78, chmn, Dept Med, 70-79, prof, 68-85, EMER PROF MED, SCH MED, UNIV PA, 85- *Concurrent Pos:* Secy, Subspecialty Bd Cardiovasc Dis, 79-81; vpres, Am Heart Asn, 79-80. *Mem:* Am Fedn Clin Res; Am Soc Clin Invest; Am Clin & Climat Asn; Asn Univ Cardiol; AAAS. *Res:* Cardiovascular diseases. *Mailing Add:* 1112 Woodmont Rd Gladwyne PA 19035

ZINTEL, HAROLD ALBERT, b Akron, Ohio, Dec 27, 12; m 38; c 5. SURGERY. *Educ:* Univ Akron, BS, 34; Univ Pa, MD, 38, DSc(med), 46. *Hon Degrees:* DSc, Univ Akron, 56. *Prof Exp:* Asst biol, Univ Akron, 33-34; asst path, Med Sch, Univ Pa, 35-36, asst chief med officer, Hosp, 39-40, from asst instr surg to instr, Med Sch, 40-47, asst prof, Med Sch & Grad Sch Med, 47, lectr, Grad Sch Nursing, 47-48, asst prof, Grad Sch Med, 50-54, assoc prof, Med Sch, 51-52, prof clin surg, 52-54; clin prof, Col Physicians & Surgeons, Columbia Univ, 54-69; dir, Dept Spec Educ Proj, Am Col Surgeons, 81-84; prof, 70-81, EMER PROF SURG, MED SCH, NORTHWESTERN UNIV, CHICAGO, 81- *Concurrent Pos:* Consult, Camden Munic Hosp, 48-54, Children's Hosp of Philadelphia, 51-54 & Vet Admin Hosp, Philadelphia, 53-54; attend surgeon & dir surg, St Luke's Hosp Ctr, 54-69; consult, Off Surgeon Gen, US Army, 58-67; asst dir, Am Col Surgeons, 69-80; trustee, Comn Prof Hosp Activities, 69- *Mem:* Am Col Surg; Soc Vascular Surg; Soc Univ Surgeons (pres, 55-56); Am Surg Asn. *Res:* Antibiotics; wound healing; peripheral vascular hypertension; antiseptics; carcinoma of head of pancreas; nutrition; portal hypertension; trauma; computers in medicine; allied health manpower; continuing medical education. *Mailing Add:* 891 Vernon Ave Winnetka IL 60093

ZIOCK, KLAUS OTTO H, b Herchen, Germany, Feb 4, 25; nat US; m 52; c 4. EXPERIMENTAL PHYSICS. *Educ:* Univ Bonn, Dipl, 49, Dr rer nat, 56. *Prof Exp:* Physicist, E Leybold's Nachfolger, Germany, 50-55; res assoc, Univ Bonn, 56-58; res assoc physics, Yale Univ, 58-60, asst prof, 60-62; assoc prof, 60-72, actg dir, Va Assoc Res Ctr, 62-64, PROF PHYSICS, UNIV VA, 72- *Concurrent Pos:* Vis scientist, Europ Orgn Nuclear Res, 69-70. *Honors & Awards:* Alexander von Humboldt Award, 77. *Mem:* Am Phys Soc. *Res:* Nuclear physics; physics of elementary particles; atomic physics. *Mailing Add:* Dept Physics Univ Va Charlottesville VA 22901

ZIOLKOWSKI, RICHARD WALTER, b Warsaw, NY, Nov 22, 52; m 81; c 2. ELECTROMAGNETICS, NUMERICAL MODELING. *Educ:* Brown Univ, ScB, 74; Univ Ill, MS, 75, PhD(physics), 80. *Prof Exp:* Engr, Lawrence Livermore Nat Lab, 81-84, computational electronics & electromagnetics thrust area leader, 84-90; ASSOC PROF ELEC ENG, DEPT ELEC & COMPUTER ENG, UNIV ARIZ, 90- *Concurrent Pos:* Vchmn, Int Inst Elec & Electronics Engrs AP-S, Symp & Nat Radio Sci, 89-; mem, Union Radio Sci Int, Comn B, Tech Activ Comt, 89- *Mem:* Inst Elec & Electronics Engrs;

Int Union Radio Sci; Am Phys Soc; Acoust Soc Am. *Res:* Application of new mathematical methods to linear and nonlinear problems dealing with the interaction of acoustic and electromagnetic waves with scattering objects, plasmas and dielectric materials. *Mailing Add:* Dept Elec & Computer Eng Univ Ariz Tucson AZ 85721

ZIOLO, RONALD F, b Philadelphia, Pa, Aug 16, 44; m 67; c 3. CHEMISTRY. *Educ:* Univ Calif, Los Angeles, BS, 66; Temple Univ, PhD(chem), 70. *Prof Exp:* Res fel chem, Calif Inst Technol, 71-72; assoc scientist, 73, scientist, 74-78, SR SCIENTIST, XEROX CORP, 78- *Mem:* AAAS; Am Chem Soc; Sigma Xi; Am Crystallog Asn; Mat Res Soc. *Res:* Inorganic and organometallic chemistry; x-ray crystallography; structure and bonding; magnetic, optical and electrical properties; chemistry of high surface area materials; nanometer structures and cluster chemistry. *Mailing Add:* Webster Res Ctr Xerox Corp 0114-39D 800 Phillips Rd Webster NY 14580

ZIOMEK, CAROL A, b Wilkes-Barre, Pa, Sept 26, 50. TRANSGENIC EXPRESSION, EMBRYO CULTURE. *Educ:* Wilkes Col, Wilkes-Barre, Pa, BS, 72; Johns Hopkins Univ, PhD(biol), 78. *Prof Exp:* Fel embryol, Dept Anat, Univ Cambridge, Eng, 78-82; staff scientist cell biol, Worcester Found Exp Biol, 82-90; PRIN SCIENTIST, GENZYME CORP, 90- *Mem:* Am Soc Cell Biol; AAAS; Soc Develop Biol; Soc Study Reprod. *Mailing Add:* Genzyme Corp One Mountain Rd Framingham MA 01701

ZIONY, JOSEPH ISRAEL, b Los Angeles, Calif, Apr 6, 35; m 61; c 3. GEOLOGY. *Educ:* Univ Calif, Los Angeles, AB, 56, MA, 59, PhD(geol), 66. *Prof Exp:* Geologist, Mil Geol Br, US Geol Surv, DC, 57-59, Fuels Br, Calif, 59-60, Southwestern Br, 65-69 & Eng Geol Br, 69-73, dep chief, Off Earthquake Studies, Reston, Va, 73-76, Earthquake Hazards Br, Menlo Park, 76-77, asst chief geologist, Western Region, 77-81, Eng Seismol & Geol Br, Menlo Park, Calif, 81-86; ASST DIR MINING & GEOL, CALIF DEPT CONSERV, SACRAMENTO, CALIF, 88- *Honors & Awards:* E B Burwell Award Eng Geol, Geol Soc Am, 87. *Mem:* Geol Soc Am; Seismol Soc Am; Asn Eng Geol. *Res:* Earthquake hazards assessment; delineation of active faults in southern California; evaluation of their relative activity using late Quaternary slip histories; estimation of their earthquake potential. *Mailing Add:* 1640 Escobita Ave Palo Alto CA 94306

ZIPES, DOUGLAS PETER, b White Plains, NY, Feb 27, 39; m 61; c 3. MEDICINE. *Educ:* Dartmouth Col, BA, 61; Dartmouth Med Sch, BMed Sci, 62; Harvard Med Sch, MD, 64. *Prof Exp:* Intern & resident med, Duke Univ Med Ctr, 64-66, fel cardiol, 66-68; vis prof electrophysiol, Masonic Res Lab, 70-71; from asst prof to assoc prof, 70-76, PROF MED, IND UNIV SCH MED, 76-; SR RES ASSOC MED & CARDIOL, KRANNERT INST CARDIOL, 74- *Concurrent Pos:* Fel, Coun Clin Cardiol, 72 & 83; steering comt res, Med & Community Prog, Am Heart Asn, 88-89; chmn, subcomt assess clin intracardiac electrophysiol studies, Am Heart Asn & Am Cl Cardiol; sci pub comt, Am Heart Asn, 86-89, res study comt, 87-; chmn, Young Investr Awards, Am Col Cardiol; mem subspecialty bd cardiovasc dis, Am Bd Internal Med; chmn, Test Comt Clin Cardiac Electrophysiol; Stockholm lectr cardiol, 90. *Honors & Awards:* Distinguished Achievement Award, Am Heart Asn, 89; Balfour Lectr, Mayo Clinic, 89; Dan May Lectr, Vanderbilt Univ, 90; Carl J Wiggers Lectr, Case Western, 91. *Mem:* Fel Am Col Cardiol; Am Heart Asn; Am Soc Clin Invest; Asn Univ Cardiologists; Asn Am Physicians; NAm Asn Pacing & Electrophysiol (pres, 89). *Res:* Clinical and animal investigations into mechanisms responsible for cardiac arrhythmias with special emphasis on the roll of autonomic nervous system. *Mailing Add:* Krannert Inst Cardiol 1001 W Tenth St Indianapolis IN 46202

ZIPF, ELIZABETH M(ARGARET), b Barrington, NJ, Nov 17, 27. BIOLOGY, INFORMATION SCIENCE. *Educ:* Univ Va, BA, 50, PhD(biol), 59; Univ Pa, MA, 52. *Prof Exp:* Res biologist cancer res, Med Sch, Univ Va, 55; res asst biol, Princeton Univ, 56-57; assoc ed, 57-62, sr assoc ed, 62-63, actg supvry ed, biol & bio-med subj, Biol Abstracts, 63-64, supvry ed, 64-71, head, Ed Dept, 71-80, actg dir, Sci Div, 80-81, TECH CONSULT TO PRES, BIOSCI INFO SERV, 81- *Concurrent Pos:* Fel, Univ Pa, 62-63; consult in prep Water Resources Thesaurus, Off Water Resources Res, US Dept Interior, 66; mem Z-39 comt, Am Standards Asn, 66-67 & Nat Fed Sci Abstracting & Indexing Serv. *Mem:* Fel AAAS; Am Inst Biol Sci; Coun Biol Ed; Am Soc Zool. *Res:* Biological research in invertebrate and vertebrate embryology; teratoma formation in salamanders; information science in biology and biomedical fields. *Mailing Add:* PO Box 127 Barrington NJ 08007

ZIPFEL, CHRISTIE LEWIS, b Detroit, Mich, Oct 2, 41; m 64; c 1. LIGHTWAVE DEVICES RELIABILITY. *Educ:* Vassar Col, AB, 63; Univ Mich, MS, 65, PhD(physics), 69. *Prof Exp:* Instr physics, State Univ NY Stony Brook, 69-72; asst prof, Towson State Col, 72-74; res assoc physics, 74-76, MEM TECH STAFF, BELL LABS, MURRAY HILL, 76- *Res:* LED reliability. *Mailing Add:* Bell Labs 7D408 Murray Hill NJ 07974-2070

ZIPFEL, GEORGE G, JR, b Richmond, Va, Dec 23, 38; m 64; c 1. ACOUSTICS. *Educ:* Mass Inst Technol, BS, 60, BEE, 61; Univ Fla, MSE, 62; Univ Mich, PhD(physics), 68. *Prof Exp:* Res assoc, Inst Theoret Physics, State Univ NY, Stony Brook, 68-71; Nat Res Coun res assoc, US Naval Res Lab, 71-73; mem tech staff, 73-83, SUPVR, AT&T BELL LABS, 83- *Concurrent Pos:* Distinguished mem tech staff, Bell Labs, 82. *Mem:* AAAS; Am Phys Soc. *Res:* Statistical field theory; scattering theory; particle physics; applied classical field theory; statistical field theory; electro-mechanical transducers; flow noise; structural acoustics. *Mailing Add:* AT&T Bell Labs Whippany Rd Whippany NJ 07981

ZIPP, ARDEN PETER, b Dolgeville, NY, July 14, 38; m 89; c 2. PHYSICAL INORGANIC CHEMISTRY. *Educ:* Colgate Univ, AB, 60; Univ Pa, PhD, 64. *Prof Exp:* Asst prof chem, Drew Univ, 64-66; from asst prof to assoc prof, 66-73, DISTINGUISHED TEACHING PROF CHEM, STATE UNIV NY COL CORTLAND, 85- *Mem:* Am Chem Soc; Sigma Xi; Nat Sci Teachers Asn. *Res:* Photochemistry of transition metal complexes; oxidation reduction reactions of transition metal ions. *Mailing Add:* Dept Chem State Univ NY Col Cortland NY 13045

ZIPPIN, CALVIN, b Albany, NY, July 17, 26; m 64; c 2. BIOSTATISTICS, EPIDEMIOLOGY. *Educ:* State Univ NY, AB, 47; Johns Hopkins Univ,. *Hon Degrees:* DSC (biostatist), John Hopkins univ, 53. *Prof Exp:* Res asst statist, Sterling-Winthrop Res Inst, 47-50; res asst biostatist, Johns Hopkins Univ, 50-53; instr, Sch Pub Health, Univ Calif, Berkeley, 53-55, from asst res biostatistician to res biostatistician, Cancer Res Inst, 55-67, asst prof prev med, 58-60, lectr, 60-67, lectr path, 61-67, PROF EPIDEMIOL, CANCER RES INST, DEPT EPIDEMIOL & INT HEALTH & DEPT PATH, SCH MED, UNIV CALIF, SAN FRANCISCO, 67- *Concurrent Pos:* Consult, US Naval Biol Lab, 55-66, Letterman Gen Hosp, 58-75, WHO, 69- & Am Joint Comt on Cancer, 69-89; vis assoc prof, Stanford Univ, 62; NIH spec fel, London Sch Hyg & Trop Med, 64-65; temporary adv, WHO, 69, 72 & 74; Eleanor Roosevelt Int Cancer fel, Univ London, 75; vis res worker, Middlesex Hosp Med Sch, London, 75; fac advisor, Regional Cancer Ctr, Trivandrum, India, 83- *Mem:* AAAS; fel Am Statist Asn; Biomet Soc; fel Royal Statist Soc; fel Am Col Epidemiol. *Res:* Identification of environmental and other factors associated with the risk of cancer as well as study of patient and disease characteristics which influence pattern of survival following diagnosis of a malignancy; biometry and epidemiology in cancer research. *Mailing Add:* Dept Epidemiol & Biostatist Univ Calif San Francisco CA 94143-0746

ZIPSER, DAVID, b New York, NY, May 31, 37; m 65; c 2. MOLECULAR GENETICS. *Educ:* Cornell Univ, BS, 58; Harvard Univ, PhD(biochem), 63. *Prof Exp:* Prof biol, Columbia Univ, 65-69; investr genetics, Cold Spring Harbor Lab, 70-82; PROF, DEPT COGNITIVE SCI, UNIV CALIF, SAN DIEGO, 82- *Res:* Lactose operon function in bacteria. *Mailing Add:* 2558 Torrey Pines Rd La Jolla CA 92037

ZIRAKZADEH, ABOULGHASSEM, b Isfahan, Iran, Feb 7, 22; m 51; c 1. MATHEMATICS. *Educ:* Univ Teheran, BS, 44; Univ Mich, MS, 49; Okla State Univ, PhD(math), 53. *Prof Exp:* Instr math, Okla State Univ, 52-53, Univ Colo, 53-54 & Wash State Univ, 54-55; asst prof, Univ Teheran, 55-56; asst prof, 57-64, ASSOC PROF MATH, UNIV COLO, BOULDER, 64- *Mem:* Am Math Soc; Math Asn Am. *Res:* Geometry; convexity. *Mailing Add:* 868 Sixth Boulder CO 80302

ZIRIN, HAROLD, b Boston, Mass, Oct 7, 29; m 57; c 2. ASTRONOMY. *Educ:* Harvard Univ, AB, 50, MA, 51, PhD(astrophys), 53. *Prof Exp:* Physicist, Rand Corp, 52-53; instr astron, Harvard Univ, 53-55; mem sr res staff, High Altitude Observ, Univ Colo, 55-64; PROF ASTROPHYS, CALIF INST TECHNOL, 64- *Concurrent Pos:* Sloane fel, 58-60; Guggenheim fel, 61; dir, Big Bear Solar Observ, 69-; mem staff, Hale Observ, 64-80; dir, Aura, 78-83. *Mem:* Am Astron Soc. *Res:* Solar physics; stellar spectroscopy; interstellar matter; geophysics. *Mailing Add:* Dept Astrophys 105-24 Calif Inst Technol Pasadena CA 91125

ZIRKER, JACK BERNARD, b New York, NY, July 19, 27; m 51; c 3. SOLAR PHYSICS. *Educ:* City Col New York, BME, 49; Harvard Univ, PhD(astron), 56. *Prof Exp:* Mech eng labs, Radio Corp Am, 49-53, astrophysicist, Sacramento Peak Observ, 56-64; astrophysicist & prof physics, Univ Hawaii, 64-78; actg dir, Sacramento Peak Observ, 76-78, dir, 78-82; actg dir, 82-84, ASTRONR, NAT SOLAR OBSERV, 84- *Concurrent Pos:* Consult, NASA, 68-; mem, Astron Adv Panel, NSF, 73-76 & mem comt, Solar-Terrestrial Res, Nat Res Coun, 81-82. *Mem:* Am Astron Soc; Int Astron Union. *Res:* Physics of the outer atmosphere of the sun; physics of corona, solar wind, prominences; analysis of spectroscopic, polarimetric solar observations. *Mailing Add:* Sacramento Peak Observ Sunspot NM 88349

ZIRKIN, BARRY RONALD, b Bronx, NY, May 17, 42; m 65; c 2. CELL BIOLOGY, REPRODUCTIVE BIOLOGY. *Educ:* State Univ NY, Binghamton, BA, 63; Univ Rochester, MS, 65, PhD(cell biol), 69. *Prof Exp:* Asst cell biol, Univ Calif, Davis, 69-71; asst prof biol, Ill Inst Technol, 71-74; from asst prof to assoc prof, 74-81, PROF REPRODUCTIVE BIOL, JOHNS HOPKINS UNIV, 81-, HEAD DEPT, 84- *Concurrent Pos:* Asst ed, Biol Reproduction J, 81-85; mem, Clin 3 Study Sect, NIH, 80-84, Reproductive Biol Study Sect, 84-88. *Mem:* Soc Study Reproduction (treas, 85-88); Am Soc Cell Biol; Am Asn Anatomists; Soc Study Reproduction; Am Soc Andrology. *Res:* Hormonal regulation of spermatogenesis; quantitative relationship between cell structure and function. *Mailing Add:* Dept Pop Dynamics Sch Pub Health Johns Hopkins Univ 615 N Wolfe St Baltimore MD 21205

ZIRKIND, RALPH, b New York, NY, Oct 20, 18; m 40; c 3. PHYSICS. *Educ:* City Col New York, BS, 40; Ill Inst Technol, MS, 46; Univ Md, College Park, PhD(physics), 59. *Hon Degrees:* DSc, Univ RI, 68. *Prof Exp:* Tech asst metall, Naval Inspector Ord, US Dept Navy, 41-42, physicist, 42-45, physicist, Bur Aeronaut, 45-52, chief physicist, 52-60; physicist, Advan Res Projs Agency, US Dept Defense, 60-64; prof aerospace eng, Polytech Inst Brooklyn, 64-70; prof elec eng, Univ RI, 70-72; physicist, Advan Projs Res Agency, US Dept Defense, 72-74; prin scientist, Gen Res Corp, McLean, Va, 74-81; PROF PHYSICS, UNIV RI, 73- *Concurrent Pos:* Consult, Jet Propulsion Lab, 64-76; consult, 81- *Honors & Awards:* Sigma Xi. *Mem:* Am Phys Soc. *Res:* Optical and radiation physics; atmospheric sciences; optical physics; lasers, atmospheric physics. *Mailing Add:* 820 Hillsboro Dr Silver Spring MD 20902-3202

ZIRKLE, LARRY DON, b Wheeler, Tex, Nov 11, 36; m 57; c 3. MECHANICAL ENGINEERING. *Educ:* Okla State Univ, BS, 59, MS, 60; Univ Tex, Austin, PhD(eng mech), 69. *Prof Exp:* Assoc engr, Tex Instruments Inc, 60-61; asst prof eng mech, Univ Tex, Austin, 69-70; asst prof, 70-74, assoc prof, 74-87, DIR STUDENT SERV, ENG, TECHNOL & ARCHIT, OKLA STATE UNIV, 77-, PROF MECH ENG, 87- *Mem:* Am Soc Mech Engrs; Am Soc Eng Educ; Nat Soc Prof Engrs. *Res:* Random vibrations with particular interest in nonlinear systems; application of engineering to biomedical problems; control theory; nonlinear analysis; dynamics; student counseling and advisement, academic student affairs; co-op education; accident reconstruction. *Mailing Add:* EN101 Okla State Univ Stillwater OK 74074

ZISCHKE, JAMES ALBERT, b Sioux Falls, SDak, Sept 18, 34; m 61; c 1. INVERTEBRATE ZOOLOGY, AQUATIC ECOLOGY. *Educ:* Univ Wis, BS, 57; Univ SDak, MA, 60; Tulane Univ, PhD(parasitol), 66. *Prof Exp:* Instr, 63-65, from asst prof to assoc prof, 66-78, PROF BIOL, ST OLAF COL, 78- *Concurrent Pos:* Partic, AEC Res Prog, PR Nuclear Ctr, 67; Duke Univ res fel, Cent Univ Venezuela, 67-68; NSF fac fel, Univ Miami, 71-72 & Argonne Nat Lab, 76; aquatic biologist, US Environ Protection Agency, 76-80 & Oak Ridge Nat Lab, 78-79; prin investr, US Environ Protection Agency, 82-85 & US Fish & Wildlife Serv, 85-87, Minnesota Pollution Control Agency, 89- *Mem:* Am Soc Zool; Ecol Soc Am. *Res:* Marine littoral ecology; stream ecology; parasitology; aquatic toxicology. *Mailing Add:* Dept Biol St Olaf Col Northfield MN 55057

ZISK, STANLEY HARRIS, b Boston, Mass; c 3. RADAR ASTRONOMY, PLANETARY SCIENCE. *Educ:* Mass Inst Technol, SB & SM, 53; Stanford Univ, PhD(elec eng & radio astron), 65. *Prof Exp:* Res assoc elec eng, 65-68, mem sci staff, Haystack Observ, Mass Inst Technol, 68-88; res prof, 88-90, PROF DEPT GEOL & GEOPHYS, INST GEOPHYS, UNIV HAWAII, 90- *Concurrent Pos:* Prin investr, Haystack Observ, 69-; vis scientist, Ltapetinga Radio Observ, Brazil, 74-84, Brown Univ, 80, Univ Hawaii, 79-88. *Mem:* Am Astron Soc; Am Geophys Union; Union Radio Int Sci; Inst Elec & Electronics Engrs. *Res:* Planetary surfaces research, geological history and current physical/geochemical state; analysis of planetary radar data: Earth, moon, Mars and Venus; analysis of sea floor acoustic imaging data. *Mailing Add:* Sch Ocean & Earth Sci Univ Hawaii 2525 Correa Rd Honolulu HI 96822

ZISKIN, MARVIN CARL, b Philadelphia, Pa, Oct 1, 36; m 60; c 3. BIOMEDICAL ENGINEERING. *Educ:* Temple Univ, AB, 58, MD, 62; Drexel Inst, MSBmE, 65. *Prof Exp:* Intern, West Jersey Hosp, Camden, 62-63; NIH fel, Drexel Inst, 63-65; NASA fel theoret biophys, 65; instr radiol & res assoc diag ultrasonics, Hahnemann Med Col, 65-66; from asst prof to assoc prof radiol & med physics, 68-76, PROF RADIOL & MED PHYSICS, MED SCH, TEMPLE UNIV, 76-, CHMN COMT BIOPHYS & BIOENG, 74- *Concurrent Pos:* Lectr biomed eng, Drexel Univ, 65-71, adj assoc prof, 71-; NSF fel analog & digital electronics, 72; mem comt on sci & arts, Franklin Inst, 72-; mem bd dirs, Inst Ultrasonics in Med. *Mem:* Am Inst Ultrasound in Med (pres, 82-84); Am Heart Asn; Inst Elec & Electronics Engrs; Soc Photo-Optical Instrument Eng; NY Acad Sci. *Res:* Biomathematics; diagnostic ultrasonics; thermography; image processing; vision; hearing; information processing in the nervous system. *Mailing Add:* 900 Abington Rd Cherry Hill NJ 08034

ZISMAN, WILLIAM ALBERT, physical chemistry; deceased, see previous edition for last biography

ZISON, STANLEY WARREN, environmental engineering, statistics, for more information see previous edition

ZISSIS, GEORGE JOHN, b Lebanon, Ind, Dec 31, 22; m 54; c 4. OPTICS. *Educ:* Purdue Univ, BS, 46, MS, 50, PhD(physics), 54. *Prof Exp:* Instr eng physics, Purdue Univ, 46-50, 52-54; assoc scientist, Atomic Power Div, Westinghouse Elec Corp, 54-55; mem spec air defense study, Off Naval Res, 57; alt head infrared lab, Willow Run Labs, Inst Sci & Technol, 55-64, head lab, 64-69, chief scientist, Infrared & Optics Div, 69-73; chief scientist, Infrared & Optics Div, 73-89, EMER SR RES PHYSICIST, ENVIRON RES INST MICH, 89- *Concurrent Pos:* Vis lectr, Univ Mich, 61-62, lectr, Dept Elec Eng, 70-72, adj prof elec & comput eng, 73-; mem staff, Res Eng, Support Div, Inst Defense Anal, DC, 62-64, consult, 64-72; consult, Army Res Off, 64-72; mem comt space prog earth observ, Nat Res Coun-Nat Acad Sci, chmn, 69-72; adv, Div Earth Sci, US Geol Surv; ed-in-chief, J Remote Sensing of the Environ, 71-78; consult, Infrared Technol, 89- *Honors & Awards:* Pub Serv Award, US Dept Interior, 72. *Mem:* Fel AAAS; fel Optical Soc Am; Sigma Xi; fel Int Soc Optical Eng. *Res:* High resolution spectroscopy; infrared; radiometry; optical radiation physics; precision measurements. *Mailing Add:* 1549 Stonehaven Rd Ann Arbor MI 48104

ZITARELLI, DAVID EARL, b Chester, Pa, Aug 12, 41; m 66; c 2. MATHEMATICS. *Educ:* Temple Univ, BA, 63, MA, 65; Pa State Univ, PhD(math), 70. *Prof Exp:* Asst prof, 70-77, ASSOC PROF MATH, TEMPLE UNIV, 77- *Concurrent Pos:* Vis prof, Vanderbilt Univ, 76. *Mem:* Am Math Soc; Math Asn Am. *Res:* History of mathematics; algebraic theory of semigroups. *Mailing Add:* Dept Math Temple Univ Philadelphia PA 19122

ZITNAK, AMBROSE, b Bratislava, Czech, Dec 30, 22; nat Can; m 50; c 3. PLANT BIOCHEMISTRY, FOOD TECHNOLOGY. *Educ:* Slovak Inst Tech, Czech, BSA, 46; Univ Alta, MSc, 53, PhD, 55. *Prof Exp:* Res officer, Agr Res Inst, Czech, 46-47; asst forage chem, Agr Exp Sta, Swiss Fed Inst Technol, 47-48; asst plant biochem, Univ Alta, 51-55, chief analyst feed & soil chem, 55-57; assoc prof hort biochem, Univ Guelph, 57-88; RETIRED. *Mem:* Can Inst Food Technol; Can Soc Plant Physiol; Can Soc Hort Sci; Can Geog Soc; Agr Inst Can. *Res:* Solanaceous glycoalkaloids; plant growth substances; technology of fruit and vegetable preservation; heat-browning of potato products; physiology of the potato tuber; cyanogenic glucosides. *Mailing Add:* 47 Walnut Dr Guelph ON N1G 2W1 Can

ZITNEY, STEPHEN EDWARD, b Homestead, Pa, Aug 2, 61; m 88; c 2. SUPERCOMPUTING, PARALLEL PROCESSING. *Educ:* Carnegie-Mellon Univ, BS, 83; Univ Ill, Urbana-Champaign, MS, 86, PhD(chem eng), 89. *Prof Exp:* SR CHEM ENGR, CRAY RES, INC, 89- *Concurrent Pos:* Consult, Univ Ill, 87. *Mem:* Am Inst Chem Engrs; Soc Indust & Appl Math; Instrument Soc Am. *Res:* Chemical process synthesis, design, optimization, and control; supercomputing strategies for chemical process engineering; applied mathematics, especially sparse matrix methods. *Mailing Add:* Cray Res Inc 655E Lone Oak Dr Eagan MN 55121

ZITOMER, RICHARD STEPHEN, b New York, NY, Sept 29, 46; m 69; c 4. MOLECULAR BIOLOGY. *Educ:* Univ Pa, BA, 68, PhD(biol), 72. *Prof Exp:* Fel, Dept Biochem, Univ Pa, 72-73 & Dept Genetics, Univ Wash, 73-75; from asst prof to assoc prof, 76-89, PROF GENETICS, DEPT BIOL SCI, STATE UNIV NY, 89- *Concurrent Pos:* Res career develop award, NIH, 81-86; Alexander von Humbolt fel, 81-82; mem, Genetics Study Sect, NIH, 89-92. *Mem:* Am Soc Microbiol. *Res:* Regulation of the expression of oxygen regulated genes of yeast. *Mailing Add:* Dept Biol Sci State Univ NY 1400 Washington Ave Albany NY 12222

ZITRIN, ARTHUR, b NY, Apr 10, 18; m 42; c 2. PSYCHIATRY. *Educ:* City Col New York, BS, 38; NY Univ, MS, 41, MD, 45. *Prof Exp:* Instr physiol, Hunter Col, 48-49; from clin asst psychiat to clin instr, 49-54, from asst clin prof to assoc prof, 54-67, PROF PSYCHIAT, SCH MED, NY UNIV, 67- *Concurrent Pos:* Pvt pract; sr psychiatrist, Bellevue Hosp, 50-, asst dir psychiat div, 54-55, dir, 55-69. *Mem:* Am Psychiat Asn; Am Psychoanal Asn; Asn Res Nerv & Ment Dis. *Res:* Physiological psychology; psychoanalysis; clinical psychiatry. *Mailing Add:* 56 Ruxton Rd Great Neck NY 11020

ZITRIN, CHARLOTTE MARKER, b New York, NY, Sept 30, 18; m 43; c 2. PHOBIC DISORDER, PANIC DISORDER. *Educ:* New York Univ, BA, 39, MD, 43. *Prof Exp:* Intern, Kings County Hosp, 43-44; resident, pediat, Bellevue Hosp, 44-45; asst pediat, New York Univ Bellevue Med Ctr, 45-46, res fel, 48-49, instr pediat, 49-54, asst prof, 54-58, asst clin prof pediat, New York Univ Med Ctr, 58-60; resident psychiat, Hillside Hosp, 60-63, clin asst, 64-65; SUPV PSYCHIATRIST, HILLSIDE HOSP, DIV LONG ISLAND JEWISH MED CTR, 65-, DIR BEHAV THER CLIN, 70- *Concurrent Pos:* Attend physician pediat, Long Island Jewish Hosp, 58-60; dir, Phobia Clinic, Hillside Hosp, Div Long Island Jewis Med Ctr, 72-; asst prof clin psychiat, State Univ NY, Sony Brook, 73-81, clin assoc prof psychiat, 82-89; assoc prof psychiat, Albert Einstein Col Med, 89- *Mem:* Am Psychopath Asn; NY Acad Sci; fel Am Psychiat Asn; AAAS; Asn Advan Behav Ther. *Mailing Add:* 56 Ruxton Rd Great Neck NY 11023

ZITRON, NORMAN RALPH, b New York, NY, May 22, 30. APPLIED MATHEMATICS. *Educ:* Cornell Univ, AB, 52; NY Univ, MS, 56, PhD(math), 59. *Prof Exp:* Lab asst physics, NY Univ, 52-53, res asst, Courant Inst Math Sci, 54-58; tech assoc appl physics, Harvard Univ, 58-59; res assoc eng, Brown Univ, 59-60; asst prof eng sci, Fla State Univ, 60-61; asst prof math, Res Ctr, Univ Wis, 61-62; ASSOC PROF MATH, PURDUE UNIV, 62- *Concurrent Pos:* Fulbright sr res scholar, Tech Univ Denmark, 64-65; assoc res mathematician, Radiation Lab, Univ Mich, 65-66; sr res fel, Univ Surrey, 68; res assoc, Univ Calif, Berkeley, 71; vis scholar, Stanford Univ, 72 & 78-79. *Mem:* Am Math Soc; Sigma Xi. *Res:* Propagation, diffraction and scattering of electromagnetic, acoustic and elastic waves; asymptotoic solutions of partial differential equations; optimization. *Mailing Add:* 268 Monmouth Ave New Milford NJ 07646

ZITTER, ROBERT NATHAN, b New York, NY, Oct 3, 28; m 64. SOLID STATE PHYSICS, QUANTUM ELECTRONICS. *Educ:* Univ Chicago, BA, 50, MS, 52 & 60, PhD(physics), 62. *Prof Exp:* Mathematician, US Naval Proving Grounds, 52; res physicist, Chicago Midway Labs, 52-60; mem tech staff, Bell Tel Labs, 62-67; PROF PHYSICS, SOUTHERN ILL UNIV, CARBONDALE, 67- *Mem:* Am Phys Soc. *Res:* Semiconductors and semimetals; photoconductivity; gaseous lasers; infrared detection. *Mailing Add:* Dept Physics Southern Ill Univ Carbondale IL 62901

ZITTER, THOMAS ANDREW, b Saginaw, Mich, Dec 30, 41; m 66; c 3. PLANT VIROLOGY. *Educ:* Mich State Univ, BS, 63, PhD(plant path), 68. *Prof Exp:* Asst prof & asst plant pathologist, Agr Res & Educ Ctr, Univ Fla, 68-74, assoc prof, 74-79; MEM STAFF, DEPT PLANT PATH, CORNELL UNIV, 79- *Mem:* Am Phytopath Soc; Entom Soc Am; Asn Appl Biologists. *Res:* Isolation and identification of vegetable viruses; establishment of plant-vector-virus relationships; determination of epidemiology and control of virus diseases; vegetable diseases. *Mailing Add:* Dept Plant Path Cornell Univ Ithaca NY 14853

ZITZEWITZ, PAUL WILLIAM, b Chicago, Ill, June 5, 42; m 66; c 2. EXPERIMENTAL ATOMIC PHYSICS. *Educ:* Carleton Col, BA, 64; Harvard Univ, AM, 65, PhD(physics), 70. *Prof Exp:* Scholar physics, Univ Western Ont, 70-72; res fel & sr physicist, Corning Glass Works, 72-73; from asst prof to assoc prof, 73-83, PROF PHYSICS, UNIV MICH, DEARBORN, 83-, ASSOC DEAN, COL ARTS, SCI & LETTERS, 88- *Concurrent Pos:* Alexander von Humboldt fel, 79-80. *Mem:* Am Phys Soc; Am Asn Physics Teachers; Sigma Xi; Nat Asn Sci Teachers. *Res:* Positrons; positron interactions in solids; positronium; atom-surface interactions; atomic spectroscopy; fundamental constants. *Mailing Add:* Col Arts Sci & Letters Univ Mich Dearborn Dearborn MI 48128-1491

ZIVI, SAMUEL M(EISNER), b St Louis, Mo, June 4, 25; m 49; c 3. OPTICAL ENGINEERING, NUCLEAR ENGINEERING. *Educ:* Iowa State Univ, BSME, 46; Wash Univ, MSME, 48. *Prof Exp:* Mech engr, Kennard Corp, 48-52; sr mech engr, Midwest Res Inst, 52-55, sect head, 55-56; staff engr, Atomic Energy Div, Am-Standard Corp, 56-57; develop engr, Kennard Div, Am Air Filter Co, Inc, 57-58; Singer Librascope, 81-83; mem tech staff, TRW Inc, 58-71 & 84-89; RETIRED. *Concurrent Pos:* Lectr, Kjeller, Norway, 62. *Mem:* Am Soc Mech Engrs; Am Nuclear Soc; Sigma Xi. *Res:* Thermodynamics; heat transfer; optical engineering; fluid mechanics. *Mailing Add:* 2016 Euclid-4 Santa Monica CA 90405

ZLATKIS, ALBERT, b Pomorzany, Poland, Mar 27, 24; nat US; m 47; c 3. ANALYTICAL CHEMISTRY, ORGANIC CHEMISTRY. *Educ:* Univ Toronto, BASc, 47, MASc, 48; Wayne State Univ, PhD(chem), 52. *Prof Exp:* Demonstr chem eng, Univ Toronto, 47-48; instr, Wayne State Univ, 49-52, res assoc, 52-53; res chemist, Shell Oil Co, Tex, 53-55; from instr to assoc prof chem, 54-62, chmn dept, 58-62, PROF CHEM, UNIV HOUSTON, 62- *Concurrent Pos:* Chmn, Int Symp Advances in Chromatography, 63-; adj prof, Baylor Col Med, 75- *Honors & Awards:* Chromatog Award, Am Chem

Soc, 73; Chromatog Commemorative Medal, 80. *Mem:* Am Chem Soc. *Res:* Gas chromatography and mass spectrometry of biological metabolites; clinical chemistry; environmental studies; chemistry of flavors and natural products; reactions of thermal electrons with organic compounds. *Mailing Add:* Dept Chem Univ Houston Houston TX 77004

ZLETZ, ALEX, b Detroit, Mich, Mar 28, 19; m 48; c 4. PHYSICAL CHEMISTRY, ORGANIC CHEMISTRY. *Educ:* Wayne Univ, BS, 40, MS, 48; Purdue Univ, PhD(chem), 50. *Prof Exp:* Electroplating chemist, Auto City Plating Co, 41; res chemist, Stand Oil Co, 50-54, group leader, 54-61; group leader, Am Oil Co, 61-79; mem staff, Amoco Chem Corp, 79-84. *Concurrent Pos:* Guest scientist, Free Radicals Proj, Nat Bur Stand, 57. *Mem:* Am Chem Soc; Royal Soc Chem; AAAS. *Res:* High vacuum; extreme pressure; aryl borons; polyolefins; liquid rocket fuels; free radical and hydrocarbon chemistry; catalysis; fluids and lubricants; greases; railway diesel lubricating oil; hydrocarbon oxidation; homogeneous and heterogeneous oxidation of hydrocarbons to chemical intermediates. *Mailing Add:* 1004 Mill St Apt 106 Naperville IL 60563-2529

ZLOBEC, SANJO, b Brezicani, Yugoslavia, Nov 16, 40; m 65; c 2. APPLIED MATHEMATICS. *Educ:* Univ Zagreb, BEng, 63, MSc, 67; Northwestern Univ, PhD(appl math), 70. *Prof Exp:* Res engr, Northwestern Univ, 68-69, lectr math, 69-70; Nat Res Coun Can grants, 70-91, from asst prof to assoc prof, 70-84, PROF MATH, MCGILL UNIV, 84- *Concurrent Pos:* Vis asst prof, Univ Zagreb, 71-72; vis prof, Univ Del, 78 & Univ Witwatersrand, Johannesburg, 85-86; vis scientist, Coun Sci Indust Res, Pretoria, 78-79 & 85-86. *Mem:* Am Math Soc. *Res:* Optimization theory and applications; applied functional analysis; numerical analysis; mathematics of operations research. *Mailing Add:* Dept Math & Statist McGill Univ 805 Sherbrooke St W Montreal PQ H3A 2K6 Can

ZLOT, WILLIAM LEONARD, b New York, NY, June 20, 29; m 56; c 2. MATHEMATICS. *Educ:* City Col New York, BS, 50; Columbia Univ, MA, 52, MBA, 53, PhD(math educ), 57, MA, 59. *Prof Exp:* Lectr math, City Col New York, 53-57, instr, 57-59; assoc prof, Paterson State Col, 59-61; from asst prof to assoc prof math educ, Yeshiva Univ, 61-67; assoc prof math, City Col New York, 67-69; PROF MATH EDUC, DIV SCI & MATH EDUC, SCH EDUC, NY UNIV, 69- *Mem:* Am Math Soc; Math Asn Am. *Mailing Add:* 110 Bleecker St New York NY 10012

ZLOTNICK, MARTIN, b New York, NY, Feb 16, 28; m 82; c 4. AERONAUTICAL ENGINEERING. *Educ:* NY Univ, BAeroE, 48; Univ Va, MAeroE, 51. *Prof Exp:* Aeronaut res scientist, Nat Adv Comt Aeronaut, Va, 48-53; prin aerodynamicist, Repub Aviation Corp, NY, 53-55; 53-55; prin staff scientist, Avco Res & Adv Develop, Mass, 55-62; mem prof staff, Hudson Inst, NY, 62-65; prin res scientist, Heliodyne Corp, Calif, 65-66 & Avco Everett Res Lab, Mass, 66-68; phys scientist, US Army Advan Ballistic Missile Defense Agency, 68-75; proj officer, Dept Energy, 75-80, br chief Advan Concepts, 81-82; TECH STAFF, NICHOLS RES CORP, 82- *Concurrent Pos:* Consult, Hudson Inst, 66-68, 82-85. *Mem:* Am Inst Aeronaut & Astronaut; Sigma Xi. *Res:* Interaction of technology with political-strategic problems; aerospace weapon-system analysis; applied research in energy technology. *Mailing Add:* 2500 Que St NW Apt 413 Washington DC 20007

ZLOTNIK, ALBERT, b Mex, Nov 29, 54; US citizen; m 82; c 3. T CELL ONTOGENY, CYTOKINE BIOLOGY. *Educ:* Univ Colo, PhD(immunol), 81. *Prof Exp:* Postdoctoral fel immunol, Nat Jewish Hosp, Denver, 82-83; staff scientist, 84-88, SR STAFF SCIENTIST, DNAX RES INST, 89- *Concurrent Pos:* Assoc ed, J Immunol, 87- *Mem:* Soc Leukocyte Biol; Am Asn Immunologists; AAAS. *Res:* Identification of novel cytokines; role of cytokines in T cell ontogeny. *Mailing Add:* Dept Immunol DNAX Res Inst 901 California Ave Palo Alto CA 94304

ZMESKAL, OTTO, physical metallurgy, for more information see previous edition

ZMIJEWSKI, CHESTER MICHAEL, b Buffalo, NY, June 3, 32; m 54; c 4. IMMUNOLOGY. *Educ:* Univ Buffalo, BA, 55, MA, 57, PhD(immunol), 60; Millard Fillmore Hosp, cert med tech, 55. *Hon Degrees:* MA, Univ Pa, 78. *Prof Exp:* Asst bact & immunol, Sch Med, Univ Buffalo, 55-58, instr & res fel, 60-61; asst prof clin path & dir blood bank, Med Col Va, 61-63; from asst prof immunol to assoc prof, Sch Med, Duke Univ, 63-70; dir transplantation immunol, Ortho Res Found, 70-73; assoc prof path, 75-84, PROF PATH, SCH MED, UNIV PA, 84-; ASSOC DIR, WILLIAM PEPPER LAB, UNIV PA HOSP, 83- *Concurrent Pos:* Lectr, Approved Sch Med Technol, 59-61; consult blood bank & serol labs, Millard Fillmore Hosp, Buffalo, 60-61; mem ad hoc subcomt stand adv panel collab res in transplantation & immunol, NIH, 64; immunologist, Yerkes Regional Primate Res Ctr, 65; assoc res prof, Sch Med, State Univ NY Buffalo; dir clin immunol, 73- *Mem:* Affil Royal Soc Med; Am Asn Immunol; NY Acad Sci; Am Asn Blood Banks. *Res:* Immunohematology; immunogenetics; histocompatibility testing for human allo-transplantation; cancer research and tissue culture. *Mailing Add:* William Pepper Lab Hosp Univ Pa Philadelphia PA 19104-4283

ZMOLA, PAUL C(ARL), b Chicago, Ill, Nov 21, 23; wid; c 1. ENGINEERING. *Educ:* Purdue Univ, BS, 44, MS, 47, PhD(eng), 50. *Prof Exp:* Mfg engr, Western Elec Co, 44-45; instr mech eng, Purdue Univ, 45-48; sr develop engr, Oak Ridge Nat Lab, 50-55; mgr reactor eng, Small Submarine Reactor Proj, Combustion Eng, Inc, 56-59, mgr adv develop, Combustion Div, 59-66, mgr thermal design, 67-71, mgr res & develop prod sales, Power Systs Group, 72-76, dir, Tech Liaison Corp, Group, 77-80; CONSULT, 81- *Mem:* AAAS; Am Soc Mech Engrs; Am Nuclear Soc. *Res:* Nuclear reactor design and development; heat transfer; thermodynamics; fluid mechanics. *Mailing Add:* 5409 Newington Rd Bethesda MD 20816

ZMOLEK, WILLIAM G, b Toledo, Iowa, July 3, 21; m 45; c 5. ANIMAL SCIENCE. *Educ:* Iowa State Col, BS, 44, MS, 51. *Prof Exp:* County exten dir, Iowa State Univ, 44-47, Extent Livestock Specialist, 48-84, prof animal sci, agr exten serv, 63-84; RETIRED. *Concurrent Pos:* Agr rep, Newton Nat Bank, 48. *Honors & Awards:* Outstanding Educ Accomplishments Extension Award, Am Animal Sci Asn, 78. *Mem:* Am Soc Animal Sci. *Mailing Add:* 2022 McCarthy Rd Ames IA 50010

ZOBACK, MARK D, b Brooklyn, New York, 48. GEOPHYSICS. *Educ:* Univ Ariz, BS, 69; Stanford Univ, MS, 73, PhD(geophys), 75. *Prof Exp:* Res assoc, NRC, 75-76; geophysicist, US Geol Surv, 76-84; PROF GEOPHYS, STANFORD UNIV, 84- *Concurrent Pos:* Assoc ed, J Geophys Res, 81-84. *Mem:* Fel Geol Soc Am. *Res:* Author of numerous technical publications. *Mailing Add:* Dept Geophys Stanford Univ Stanford CA 94305

ZOBACK, MARY LOU CHETLAIN, b Sanford, Fla, July 5, 52; m 73; c 2. TECTONICS. *Educ:* Stanford Univ, BS, 74, MS, 75, PhD(geophys), 78. *Prof Exp:* Nat Res Coun fel, 78-79, GEOPHYSICIST, US GEOL SURV, 79- *Concurrent Pos:* Mem, US Geodynamics Comt, Nat Res Coun, 85-89; chmn, World Stress Map Proj, Inter-Union Comn Lithosphere, 86-90. *Honors & Awards:* Macelwane Award, Am Geophys Union, 87. *Mem:* Fel Am Geophys Union; fel Geol Soc Am. *Res:* Pattern and sources of the in-situ stress field; structural style and tectonism in extensional regimes, particularly the Basin and Range province. *Mailing Add:* US Geol Surv MS 977 345 Middlefield Rd Menlo Park CA 94025

ZOBEL, BRUCE JOHN, b Calif, Feb 11, 20; m 41; c 4. FOREST GENETICS. *Educ:* Univ Calif, BS, 43, MF, 49, PhD(forest genetics), 51. *Hon Degrees:* DSc, NY State Univ, Syracuse, 86. *Prof Exp:* Asst to logging engr, Pac Lumber Co, 43-44; sr lab asst, Univ Calif, 46-49; silviculturist, Tex Forest Serv, 51-56; from assoc prof to prof forest genetics, 56-62, Edwin F Conger distinguished prof forestry, 62-80, EMER PROF & CONSULT, NC STATE UNIV, 80-; PRES, ZOBEL FORESTRY ASSOC. *Honors & Awards:* Biol Res Award, Am Soc Foresters, 68; Res Award, Tech Asn Pulp & Paper Asn, 73, Gold Medal, 75; Oliver Max Gardner Award, 72; Outstanding Exten Serv Award, NC State Univ, 73 & Forest Farmer Award, 78. *Mem:* Fel Am Soc Foresters; fel Tech Asn Pulp & Paper Asn; fel Int Acad Wood Sci. *Res:* Silviculture. *Mailing Add:* Col Forest Resources NC State Univ Box 8002 Raleigh NC 27650

ZOBEL, C(ARL) RICHARD, b Pittsburgh, Pa, Aug 29, 28; m 55; c 2. MOLECULAR BIOPHYSICS, BIOPHYSICAL CHEMISTRY. *Educ:* Purdue Univ, BS, 51; Univ Rochester, PhD(phys chem), 54. *Prof Exp:* Res assoc physics, Univ Mich, 54-56; asst prof chem, Am Univ Beirut, 56-59; res assoc biophys, Johns Hopkins Univ, 59-62; asst prof, 62-68, ASSOC PROF BIOPHYS SCI, SCH MED, STATE UNIV NY, BUFFALO, 68- *Concurrent Pos:* Du Pont fel infrared spectros, 55-56. *Mem:* AAAS; Electron Micros Soc Am; Biophys Soc; Am Soc Cell Biol. *Res:* Electron microscopy, particularly of macromolecules and ordered complexes of macromolecules; development of staining methods for electron microscopy; ultra structure and function of motile systems, particularly platelets; cell adhesion. *Mailing Add:* Dept Biophys Sci Sch Med State Univ NY Health Sci Ctr 3435 Main St Buffalo NY 14214

ZOBEL, DONALD BRUCE, b Salinas, Calif, July 17, 42; m 66; c 2. PLANT ECOLOGY. *Educ:* NC State Univ, BS, 64; Duke Univ, MA, 66, PhD(bot), 68. *Prof Exp:* From asst prof to assoc prof, 68-82, PROF BOT, ORE STATE UNIV, 82- *Concurrent Pos:* Lectr, Fulbright Prog, Nepal, 84-85. *Mem:* Ecol Soc Am; AAAS; Brit Ecol Soc. *Res:* Forest ecology; autecology of conifers; ecology of forest understory plants. *Mailing Add:* Dept Bot Ore State Univ Corvallis OR 97331-2902

ZOBEL, EDWARD C(HARLES), engineering mechanics, for more information see previous edition

ZOBEL, HENRY FREEMAN, b Ft Scott, Kans, Mar 13, 22; m 44; c 3. PHYSICAL CHEMISTRY, ANALYTICAL CHEMISTRY. *Educ:* Univ Ill, Champaign, BS, 50, MS, 51. *Prof Exp:* Chemist, Northern Regional Lab, Ill, 51-67; sr res scientist, Moffett Tech Ctr, CPC Int Inc, 67-86; RETIRED. *Concurrent Pos:* Corn Industs res fel, Northern Regional Lab, Ill, 65-67; mem sci adv comt, Am Inst Baking; consult, US Feed Grains Coun & consult Starch Struct, properties, utilization & bread staling. *Honors & Awards:* Plenary Lectr, Starch Conf, Detmold, Ger, 86; E A Day Mem Lectr, Penn State Univ, University Park, Pa. *Mem:* Am Asn Cereal Chem. *Res:* Physical properties and structure of natural polymers such as granular starches, starch derived products and synthetic polymers; writer, Major Subject-Starch. *Mailing Add:* ABCV Starch Assoc 1105 Bel Air Dr Darien IL 60559

ZOBELL, CLAUDE E, microbiology; deceased, see previous edition for last biography

ZOBRIST, GEORGE W(INSTON), b Highland, Ill, Feb 13, 34; m 55; c 3. ELECTRICAL ENGINEERING, COMPUTER SIMULATION. *Educ:* Univ Mo, BSEE, 58, PhD(elec eng), 65; Wichita State Univ, MSEE, 61. *Prof Exp:* Electronic scientist, US Naval Ord Test Sta, Calif, 58-59; assoc engr, Boeing Co, Kans, 59-60; res assoc radar, Res Dept, Wichita State Univ, 60-61; from instr to assoc prof elec eng, Univ Mo-Columbia, 61-69, undergrad prog dir, 68-69; assoc prof elec & electronic systs, Univ SFla, 69-70; prof elec eng & chmn dept, Univ Miami, 70-71; prof elec & electronic systs, Univ SFla, 71-76; prof elec eng & chmn dept, Univ Toledo, 76-79; dir computer sci & eng, Sanborn, Steketse, Otis & Evans, 79- 82; prof computer sci, Grad Eng Ctr, St Louis, 82-85, PROF COMPUTER SCI, UNIV MO-ROLLA, 85- *Concurrent Pos:* Lectr, Stevens Inst Technol, 67; NASA res grant, 67-68; sr vis res fel, Univ Edinburgh, 72-73; elec engr & consult, US Naval Ord Test Sta, Calif, 65; consult, Wilcox Elec, Mo, 66-69, M Jones Assoc, Calif, 67-68, Med Serv Bur, Fla, 69-73, Defense Commun Agency, 71-72, US Naval Res Lab, 71-72, NASA Kennedy Space Ctr, Fla, 73-76, 88 & 89, Prestolite Co, Toledo, 77-79, Patrick AFB, 87, Wright Patterson AFB, Ohio, 86, IBM,

Lexington, KY, 84-86. *Mem:* Inst Elec & Electronics Engrs; Nat Soc Prof Engrs. *Res:* Radar circuitry design; electronic countermeasure research; topological analysis of networks; electrical engineering education; computer-aided circuit design; software engineering 661 computer science education; computer science education. *Mailing Add:* Dept Comp Sci Univ Mo-Rolla MO 65401

ZOCHOLL, STANLEY E, b Philadelphia, Pa, July 23, 29; m; c 3. ELECTRICAL POWER SYSTEMS, MICRO-PROCESSOR TECHNOLOGY. *Educ:* Drexel Univ, BS, 58, MS, 73. *Prof Exp:* DIR PROTECTION TECHNOL, ABB ASEA-BROWN BOVERI INC, 47- *Mem:* Fel Inst Elec & Electronics Engrs. *Mailing Add:* Asea-Brown Boveri Inc Protective Relay Div 35 N Snowdrift Rd Allentown PA 18106

ZODROW, ERWIN LORENZ, b Deutsch Krone, Ger, Jan 5, 34; Can citizen; m 63; c 1. MATHEMATICAL GEOLOGY, SULFATE MINERALOGY. *Educ:* St Francis Xavier Univ, BSc, 62; Pa State Univ, MSc, 67; Univ Western Ont, PhD(geol), 73. *Prof Exp:* Asst party chief geol, Que Cartier Mining Co, 61; field engr, Algoma Ore Properties, 62-63; mem staff mine & develop, Iron Ore Co Can, 64-68; lectr & asst prof, Xavier Col, 70-78; assoc prof geol, 78-87, PROF GEOL, UNIV COL CAPE BRETON, 87- *Concurrent Pos:* Res assoc, NS Mus, 77-; curator fossil plants, Univ Col Cape Breton, 83. *Mem:* Systematics Asn; Paleont Soc Am; Can Inst Mining & Metall; Mineral Soc Am; fel Geol Asn Can. *Res:* Upper Carboniferous phytostratigraphy and paleobiology of eastern Canada and European correlation; trace-elemental stratigraphy of Sydney Coalfield (Upper Carboniferous), N S Canada; mathematical geology, distributions and factor analysis (applied), secondary sulfate mineralogy and pyrite oxidation products in coals; trace elements in coal. *Mailing Add:* Dept Geol Univ Col Cape Breton Sydney NS B1P 6L2 Can

ZOELLER, GILBERT NORBERT, b St Louis, Mo, Sept 30, 31; m 53; c 3. DENTISTRY. *Educ:* St Louis Univ, BS, 53, DDS, 57. *Prof Exp:* Instr pedodontics, St Louis Univ, Sch Dent, 57-62, assoc prof prosthodontics, 66-70; assoc prof & dir clin, Dent Fac, 74-76, PROF & DIR GEN PROG DENT, SCH DENT MED, SOUTHERN ILL UNIV, 78-; HOSP STAFF DENT, ST ANTHONY'S MED CTR, 75-; PVT PRACT GEN DENT, ST LOUIS, 57- *Concurrent Pos:* Several USPHS grants, 68-70. *Mem:* Int Asn Dent Res; Sigma Xi. *Res:* Dental materials; in vitro penetration of acid etchant into dentin; biomedical bases for dental therapy; effect of x-rays upon fetal rat calvarial bone cells. *Mailing Add:* 12010 Theiss Rd Affton MO 63128

ZOELLNER, JOHN ARTHUR, b Ames, Iowa, July 27, 27; m 51; c 3. STATISTICS. *Educ:* Iowa State Col, BS, 51, MS, 53. *Prof Exp:* Mgr statist methods, Gen Elec Co, 56-61; supvr anal & prog, IIT Res Inst, 61-63, mgr opers dept, 63-66, asst dir opers, 66-69, dep dir opers, 69-75; DEP & TECH DIR, ELECTROMAGNETIC COMPATIBILITY ANALYSIS CTR, DEPT DEFENSE, 76- *Mem:* Inst Elec & Electronics Engrs; Asn Comput Mach; Opers Res Soc Am. *Res:* Analytical engineering; operational and statistical analysis. *Mailing Add:* Five Chase Rd Annapolis MD 21401

ZOGG, CARL A, b Belleville, Ill, Feb 9, 27; m 52; c 6. MEDICAL PHYSIOLOGY, NUTRITION. *Educ:* Univ Ill, BS, 49, MS, 60, PhD(nutrit in dairy sci), 62. *Prof Exp:* Milk sanitarian, Dressel Young Dairy, 49-58; asst dairy sci, Univ Ill. 58-62, res assoc physiol reprod, 62-63; asst prof physiol, 63-71, assoc prof physiol & pharmacol, 71-78, PROF PHYSIOL, SCH MED, UNIV NDAK, 78- *Mem:* Am Physiol Soc; Am Mil Soc. *Res:* Ruminant nutrition and physiology; male reproductive physiology; physiology of lactation of simple stomached animals; hyperbaric physiology, nutrition and microbiology; nutritional studies conducted on subjects exposed to hyperbaric helium-molecular oxygen environmental conditions; cecal function in monogastric animals and gastrointestinal motility. *Mailing Add:* Dept Physiol Univ NDak Box 8155 Univ Sta Grand Forks ND 58202

ZOGLIO, MICHAEL ANTHONY, b Providence, RI, June 27, 36; m 59; c 3. PHARMACEUTICAL CHEMISTRY. *Educ:* Univ RI, BSc, 58; Univ Minn, PhD(med chem), 66. *Prof Exp:* Mgr pharm res, Sandoz Wander Inc, 64-70; dir pharmaceut develop, Hoechst Pharmaceut Inc, 70-72; dir pharmaceut res & develop, Merrell Nat Labs, 72-81, res assoc, 81-84, ASSOC SCIENTIST, MERRELL-DOW, PHARMACEUT, 84- *Concurrent Pos:* Adj asst prof biopharmaceut, Col Pharm, Univ Cincinnati, 74-84, adj assoc prof, 84-; vis prof pharm, Univ Wis, 75- *Mem:* Fel Am Pharmaceut Asn; Am Asn Col Pharm. *Res:* Drug stability; drug absorption; physicochemical characterization of pharmaceuticals; physical aspects of dosage form processing. *Mailing Add:* 6835 Stonington Rd Cincinnati OH 45230

ZOGORSKI, JOHN STEWARD, b Trenton, NJ, Mar 5, 46; c 4. ENVIRONMENTAL SCIENCE, WATER RESOURCES. *Educ:* Drexel Univ, BS, 69; Rutgers Univ, MS, 72, PhD(environ sci), 75. *Prof Exp:* Hydrologist water resources div, US Geol Surv, 65-79; proj mgr, Rice Univ, 73-74; asst prof civil & environ eng, Univ Louisville, 74-77; ASSOC PROF ENVIRON SCI, IND UNIV, 77-, DIR, ENVIRON SYSTS APPLN CTR, 79- *Concurrent Pos:* Co-prin investr, NJ Water Resources Res Inst, Rutgers Univ, 72-73 & Clark Maritime Proj, Sverdrup & Parcel & Assocs, Univ Louisville, 74-75; proj dir & prin investr, Colgate-Palmolive, Inc, 75; proj dir & prin investr, Louisville Water Co, 74-77 & Ky Water Resources Res Inst, 77-78; co-prin investr, Ohio State Univ & Ind Univ, 77-78, Indianapolis Water Co, 78-80, Ind State Bd Health, 78-80, Ind Dept Natural Resources, 78-79 & 80-81, US Geol Surv, 80-81 & City Columbus, Ind, 80-81; prin investr, US Environ Protection Agency, 81-82; consult, Union Carbide Corp, 80 & US Environ Protection Agency, 81. *Mem:* Am Water Works Asn; Water Pollution Control Fedn; Int Asn Water Pollution Res; Am Water Resources Asn; Int Water Resources Asn. *Res:* Environmental sciences and environmental engineering with emphasis on water resources and water quality assessment; applied hydrology; management of river basins; water and wastewater treatment; aquatic chemistry; impacts of surface mining; removal of trace organics from drinking water. *Mailing Add:* SDak Sch Mines & Technol Dept Civil Eng Rapid City SD 57701

ZOGRAFI, GEORGE, b New York, NY, Mar 13, 36; m 57; c 4. PHARMACEUTICAL CHEMISTRY, SURFACE CHEMISTRY. *Educ:* Columbia Univ, BS, 56; Univ Mich, MS, 58, PhD(pharmaceut chem), 61. *Hon Degrees:* DS, Columbia Univ, 76. *Prof Exp:* Asst prof pharm, Columbia Univ, 61-64; from asst prof to assoc prof, Univ Mich, 64-72; dean, 75-80, PROF PHARM, UNIV WIS-MADISON, 72- *Concurrent Pos:* Am Found Pharmaceut Educ Pfeiffer Mem res fel, Utrecht Univ, 70-71. *Honors & Awards:* Ebert Prize, Am Pharmaceut Asn, 84. *Mem:* AAAS; Am Pharmaceut Asn; fel Am Acad Pharmaceut Sci; Am Chem Soc; Sigma Xi; Am Asn Pharmaceut Scientists; Int Pharmaceut Fedn; Int Asn Colloid & Interface Scientists; Am Inst Hist Pharm. *Res:* Physical chemical basis for therapeutic activity of drugs; interfacial activity of drugs; lipids and proteins emphasizing structure and function of biological membranes. *Mailing Add:* Sch Pharm Univ Wis-Madison 425 N Charter St Madison WI 53706

ZOLBER, KATHLEEN KEEN, b Walla Walla, Wash, Dec 9, 16; m; m 37. NUTRITION. *Educ:* Walla Walla Col, BS, 41; Wash State Univ, MA, 61; Univ Wis, PhD(food systs admin), 68. *Prof Exp:* Food serv dir, Walla Walla Col, 41-50, mgr col store, 51-59, from asst prof to assoc prof foods & nutrit, 59-64; dir dietetic internship, 67-71, dir dietetic educ, 71-84, dir nutrit serv, 72-84, PROF NUTRIT, MED CTR, LOMA LINDA UNIV, 64-, PROG DIR NUTRIT, SCH PUB HEALTH, 84- *Concurrent Pos:* Mead Johnson fel, Am Dietetic Asn. *Mem:* Am Dietetic Asn (pres-elect, 81-82, pres, 82-83); Am Mgt Asn; Am Pub Health Asn; Am Home Econ Asn; AAAS. *Res:* Productivity in food service systems; role of technicians in health care professions; health and nutrition. *Mailing Add:* Box 981 Loma Linda CA 92354

ZOLLA-PAZNER, SUSAN BETH, b Chicago, Ill, Feb 25, 42. IMMUNOLOGY. *Educ:* Stanford Univ, BA, 63; Univ Calif, San Francisco, PhD(microbiol), 67. *Prof Exp:* NIH fel, 67-69, asst prof, 69-78, ASSOC PROF PATH, MED SCH, NY UNIV, 78-; RES MICROBIOLOGIST, MANHATTAN VET ADMIN HOSP, 69-, CHIEF CLIN IMMUNOL, 79-, CO-DIR, ACQUIRED IMMUNE DEFICIENCY SYNDROME CTR, 83- *Mem:* AAAS; Am Asn Immunol. *Res:* Regulation of the immune response; effects of malignancies on immune function. *Mailing Add:* Dept Path NY Univ Sch Med 550 First Ave New York NY 10016

ZOLLARS, RICHARD LEE, b Minneapolis, Minn, Nov 9, 46; m 68; c 2. INTERFACIAL PHENOMENA, HETEROPHASE REACTOR DESIGN. *Educ:* Univ Minn, BChe, 68; Univ Colo, MS, 72, PhD(chem eng), 74. *Prof Exp:* Sr engr, Union Carbide Corp, 74-77; from asst prof to assoc prof chem eng, Univ Colo, 77-83; prog dir, NSF, 83-84; assoc prof, 84-89, CHEM ENG, WASH STATE UNIV, 89- *Mem:* Am Inst Chem Engrs; Am Chem Soc; Am Soc Eng Educ; AAAS; Sigma Xi. *Res:* Fundamental aspects of hetero phase reactions; interfacial phenomenon; stability analysis; particulate systems; adsorption from the liquid phase. *Mailing Add:* Dept Chem Eng Wash State Univ Pullman WA 99164-2710

ZOLLER, PAUL, b Lucerne, Switz, Nov 23, 39; nat US; m 67; c 2. POLYMER PHYSICS. *Educ:* Swiss Fed Inst Technol, dipl phys, 65; Univ Wis-Madison, MSc, 67, PhD(physics), 68. *Prof Exp:* Res assoc physics, Univ Wis, 68-69; res physicist polymers, Film Dept, E I du Pont de Nemours & Co, 69-72; prof polymer sci, Neu-Technikum Buchs, Switz, 72-77; mem staff polymer sci, Cent Res & Develop Dept, E I Du Pont de Nemours & Co, Inc, 77-86; PROF, DEPT MECH ENG, UNIV COLO, 86- *Mem:* Fel Am Phys Soc. *Res:* Pressure, volume and temperature relationships of polymers; polymer rheology; characterization and low-temperature properties of polymers; polymer processing. *Mailing Add:* Dept Mech Eng Univ Colo Campus Box 427 Boulder CO 80309

ZOLLER, WILLIAM H, b Cedar Rapids, Iowa, Mar 3, 43; m 69; c 4. NUCLEAR CHEMISTRY. *Educ:* Univ Alaska, BS, 65; Mass Inst Technol, PhD(nuclear chem), 69. *Prof Exp:* Tech asst, Inst Geophys, Univ Hawaii, 65; res asst, Arthur A Noyes Nuclear Chem Ctr, Mass Inst Technol, 65-69, res assoc, 69; res assoc, Inst Geophys, Univ Hawaii, 69-70; from asst prof to assoc prof nuclear & environ chem, Univ Md, College Park, 70-79, prof nuclear & anal chem, 79-; AT DEPT CHEM, UNIV WASH, SEATTLE. *Mem:* AAAS; Am Phys Soc; Am Meteorol Soc; Am Chem Soc; Am Geophys Union. *Res:* Nuclear phenomenon and environmental chemical problems, especially with respect to air and water pollution; instrumental neutron and photon activation analysis of air pollutants; atmospheric chemical studies in Antarctica the Arctic and Hawaii. *Mailing Add:* Dept Chem Univ Wash Seattle WA 98105

ZOLLINGER, JOSEPH LAMAR, b Salt Lake City, Utah, July 9, 27; m 52; c 7. ORGANIC CHEMISTRY. *Educ:* Univ Utah, BS, 51, PhD(chem), 54. *Prof Exp:* Res org chemist, M W Kellogg Co, 54-57; res specialist, 3M CO, 57-77, sr res specialist, 77-79, sr patent liaison, 79-85, DIV SCIENTIST, INTELLECTUAL PROPERTY, 3M CO, 85- *Mem:* Am Chem Soc. *Res:* Organic fluorine compounds; elastomers; fluoronitrogen compounds; polymers; silicon compounds; abrasion resistant coatings. *Mailing Add:* 3M Co 3M Ctr Bldg 236-1B-133 St Paul MN 55144

ZOLLMAN, DEAN ALVIN, b Kendallville, Ind, Oct 13, 41; m 74; c 2. PHYSICS EDUCATION. *Educ:* Ind Univ, BS, 64, MS, 65; Univ Md, PhD(physics), 70. *Prof Exp:* Asst prof physics, Kans State Univ, 70-75; staff physicist, Am Asn Physics Teachers, 75-77; assoc prof physics, Kans State Univ, 77-81, PROF PHYSICS, KANS STATE UNIV, 81- *Concurrent Pos:* Film repository ed, Am Asn Physics Teachers, 76-; vis assoc prof physics, Univ Utah, 81-82; guest prof, Univ Munich, 89. *Honors & Awards:* Distinguished Serv Award, Am Asn Physics Teachers. *Mem:* Am Asn Physics Teachers; Nat Sci Teachers Asn; Am Phys Soc; Int Asn Comput Educ; Soc Appl Learning Technol. *Res:* Applications of multi-medic, interactive video and computer technologies to physics instruction. *Mailing Add:* Dept Physics Kans State Univ Manhattan KS 66506

ZOLLO, RONALD FRANCIS, b Brooklyn, NY, Sept 13, 41. CIVIL ENGINEERING, MATERIALS SCIENCE. *Educ:* Carnegie Inst Technol, BS, 63, MS, 66; Carnegie-Mellon Univ, PhD(civil eng), 71. *Prof Exp:* Lectr civil eng, City Col New York, 69, asst prof, 70-72; ASSOC PROF CIVIL ENG, UNIV MIAMI, 72- *Concurrent Pos:* Consult, Roman Stone Construct Co, Inc, 71-76; res fel, Battelle Mem Inst, 72-76; NSF res grant, City Col New York & Battelle Mem Inst, 72-76; consult prof engr. *Mem:* Am Soc Civil Engrs; Am Soc Eng Educ; Am Concrete Inst. *Res:* Hybrid computer techniques applied to civil engineering problems; structural and solid mechanics; fiber reinforced concrete; ferro cement. *Mailing Add:* Dept Civil Eng Univ Miami Univ Sta Coral Gables FL 33124

ZOLLWEG, JOHN ALLMAN, b Rochester, NY, July 3, 42; m 67; c 2. PHYSICAL CHEMISTRY. *Educ:* Oberlin Col, AB, 64; Cornell Univ, PhD(phys chem), 69. *Prof Exp:* NSF fel, Mass Inst Technol, 68-70; from asst prof to assoc prof chem, Univ Maine, Orono, 70-82; sr res assoc, 82-86, ASSOC PROF CHEM ENG, CORNELL UNIV, 86- *Concurrent Pos:* Vis res fel, Cornell Univ, 80-81. *Res:* Experimental and theoretical study of fluids, including excess enthalpies of mixtures at cryogenic temperatures, vapor-liquid equilibrium, supercritical extraction, sound speed and surface tension; simulation of two dimensional systems on a digital computer. *Mailing Add:* Sch Chem Eng Olin Hall Cornell Univ Ithaca NY 14853-5201

ZOLLWEG, ROBERT JOHN, b Medina, NY, Aug 1, 24; m 46; c 2. EXPERIMENTAL PHYSICS. *Educ:* Northwestern Univ, BS, 49, MS, 50; Cornell Univ, PhD(physics), 55. *Prof Exp:* Asst, Cornell Univ, 50-54; sr physicist, Res Labs, Westinghouse Elec Corp, 54-68, adv physicist, 68-80, consult physicist, 80-89; RETIRED. *Mem:* Am Phys Soc. *Res:* Solid state optics; photoemission and optical absorption; electron reflection from metal surfaces; optical properties of plasmas; thermionic energy conversions; arc discharges; thermodynamics and material properties. *Mailing Add:* 4560 Bulltown Rd Murrysville PA 15668

ZOLMAN, JAMES F, b Dayton, Ohio, Aug 1, 36; m 56; c 2. NEUROPSYCHOLOGY, PSYCHOPHARMACOLOGY. *Educ:* Denison Univ, BS, 58; Univ Calif, Berkeley, PhD(psychol), 63. *Prof Exp:* NIMH fel, Dept Psychiat & Brain Res Inst, Univ Calif, Los Angeles, 62-64; asst prof, 64-69, assoc prof, 69-79, PROF PHYSIOL & BIOPHYS, MED CTR, UNIV KY, 79- *Concurrent Pos:* Found Fund fel res psychiat, Inst Psychiat, London, Eng, 71-72; Fogarty sr int fel, Dept Animal Behaviour, Cambridge, Eng, 79-80; vis prof, Dept Psychol, Latrobe Univ, Bundoora, Victoria, Australia, 87. *Mem:* AAAS; Am Psychol Asn; Psychonomic Soc; Soc Neurosci; Int Soc Develop Psychobiol. *Res:* Developmental psychopharmacology. *Mailing Add:* 1800 Bon Air Dr Lexington KY 40502

ZOLNOWSKI, DENNIS RONALD, b Buffalo, NY, Mar 9, 43; m 71. NUCLEAR PHYSICS. *Educ:* Canisius Col, NY, BS, 64; Univ Notre Dame, PhD(physics), 71. *Prof Exp:* Res assoc, 71-73, res scientist physics, Cyclotron Inst, Tex A&M Univ, 74-; AT BELL LABS. *Concurrent Pos:* Welch Found fel, 71-72. *Mem:* Am Phys Soc. *Res:* Heavy-ion reactions; structure; in-beam gamma ray and conversion electron experiments; isomeric states; radioactivity. *Mailing Add:* Bell Labs MS 6F415 600 Mountain Ave Murray Hill NJ 07974

ZOLOTOROFE, DONALD LEE, b Bronx, NY, Sept 5, 46. CHEMICAL ENGINEERING, POLYMER CHEMISTRY. *Educ:* Cornell Univ, BS, 67, MS, 69, PhD(chem eng), 71. *Prof Exp:* CHEM ENGR, ROHM & HAAS CO, 71- *Res:* Physical properties of polymeric systems; bulk solution and emulsion polymerization processes (product and process development); high pressure process development; monomer process engineering. *Mailing Add:* Rohm & Haas Co 727 Norris Town Rd Spring Mills PA 19477

ZOLTAI, TIBOR, b Gyor, Hungary, Oct 17, 25; m 50; c 3. MINERALOGY, CRYSTALLOGRAPHY. *Educ:* Univ Toronto, BASc, 55; Mass Inst Technol, PhD(mineral), 59. *Hon Degrees:* Hon Dr. Tech Univ Heavy Indust, Miskolc, Hungary, 89. *Prof Exp:* From asst prof to assoc prof, 59-62, chmn dept geol & geophys, 63-71, PROF MINERAL, UNIV MINN, MINNEAPOLIS, 64- *Concurrent Pos:* Consult, US Bur Mines, Ft Snelling, Minn, 59-90; vis prof, Univ Saarlandes, Saarborücken, Ger, 68; vis prof, Univ Cent Venezuela, Caracas, 78-79; vis prof, Univ Chem Indust, Veszprém, Hungary, 89; consult & mem, NAS/NRC Geochem Div, Comt Fibrous Mat as Health Hazards, 80-83; mem, NAS/NRC Life Sci Div, Comt Non-Occup Health Risks Asbestiform Fibers, 82-84; mem, rev panel NIH res grant, Mt Sinai Sch Med, NY, 83; mem, range studies adv panel, Minn Dept Health, 85. *Mem:* Geol Soc Am; Mineral Soc Am; Am Crystallog Asn; Mineral Asn Can; Hungarian Acad Sci. *Res:* Crystal structures and crystal chemistry of minerals; mineralogy and physical properties of asbestos. *Mailing Add:* Dept Geol & Geophys Univ Minn Minneapolis MN 55455

ZOLTAN, BART JOSEPH, b Dec 26, 46; US citizen; m 80; c 3. INSTRUMENTATION, ELECTRO-OPTICS. *Educ:* Fairleigh Dickinson Univ, BS, 69, MS, 73. *Prof Exp:* Sr engr, Kearfott Div, Singer Co, 69-77; PRIN RES SCIENTIST, MED RES DIV, AM CYANAMID, 77- *Mem:* Inst Elec & Electronics Engrs; Instrument Soc Am; AAAS. *Res:* Development of instrumentation in support of biomedical and pharmacological research; use of electro-optical techniques for measurements or analysis; design of biomedical devices. *Mailing Add:* 152 De Wolf Rd Old Tappan NJ 07675

ZOLTEWICZ, JOHN A, b Nanticoke, Pa, Dec 5, 35; m 65; c 3. ORGANIC CHEMISTRY. *Educ:* Princeton Unib, AB, 57, PhD(org chem), 60. *Prof Exp:* NATO fel, Univ Munich, 60-61; Shell Corp Fund fel, Brown Univ, 61-62, NIH fel, 62-63; from asst prof to assoc prof chem, 63-73, PROF CHEM, UNIV FLA, 73- *Mem:* Am Chem Soc. *Res:* Heterocyclic chemistry; kinetics; rates and mechanism of hydrogen-deuterium exchange in heterocycles; radical-anion heteroaromatic nucleophilic substitution; thiamine model studies; covalent amination. *Mailing Add:* Dept Chem Univ Fla Gainesville FL 32611-2046

ZOLTON, RAYMOND PETER, b Jersey City, NJ, May 21, 39; m 60; c 3. HEMATOLOGY. *Educ:* Newark Col Eng, BS, 63; Univ Del, MS, 67; Purdue Univ, PhD(biochem), 72. *Prof Exp:* Engr dyes, E I du Pont de Nemours & Co, Inc, 63-68; res assoc, Wayne State Univ, 72-74; SR SCIENTIST COAGULATION, ORTHO DIAG INC, 74- *Concurrent Pos:* Mem thrombosis coun, Am Heart Asn. *Mem:* Am Soc Coagulation; Am Chem Soc. *Res:* Development of diagnostic test for coagulation and fibrinolysis parameters. *Mailing Add:* PO Box 62 Bethlehem NH 03574-0062

ZOMBECK, MARTIN VINCENT, b Peekskill, NY, Aug 14, 36; m 63; c 2. PHYSICS. *Educ:* Mass Inst Technol, BS, 57, PhD(physics), 69. *Prof Exp:* Observer, Smithsonian Astrophys Observ, 61-64; res asst physics, Mass Inst Technol, 64-69; proj scientist, Am Sci & Eng, Inc, 69-73; head physics & instrumentation, Damon Corp, 73-74; res assoc, Harvard Col Observ, 74-78; PHYSICIST, SMITHSONIAN ASTROPHYS OBSERV, 78- *Concurrent Pos:* Fulbright scholar, WGer. *Mem:* Am Phys Soc; Am Astron Soc; Int Astron Union. *Res:* Solar and stellar x-ray astronomy from rockets and space observatories; x-ray imaging instrumentation; x-ray optics; x-ray spectroscopy. *Mailing Add:* Ctr Astrophys 60 Garden St Cambridge MA 02138

ZOMPA, LEVERETT JOSEPH, b Lawrence, Mass, May 31, 38; m 66; c 6. INORGANIC CHEMISTRY. *Educ:* Merrimack Col, BS, 59; Col Holy Cross, MS, 60; Boston Col, PhD(inorg chem), 64. *Prof Exp:* Asst prof chem, Boston Col, 64-65; fel, Mich State Univ, 65-66; from asst prof to assoc prof, Univ Mass, Boston, 66-71, chmn dept, 73-83, dir grad prog, 84-88, PROF CHEM, UNIV MASS, BOSTON, 77-, VCHANCELLOR ACAD AFFAIRS & PROVOST, 88- *Mem:* Am Chem Soc; Royal Soc Chem; Sigma Xi. *Res:* Transition metal complexes of stereorestrictive amines and amino acids; complexes of anions-structive, bonding and selectivity. *Mailing Add:* Six Kathleen Dr Andover MA 01810

ZOMZELY-NEURATH, CLAIRE ELEANORE, b Newark, NJ, July 4, 24; m 63. NEUROBIOLOGY. *Educ:* Columbia Univ, BS, 50; Harvard Univ, MS, 56, PhD(biochem, nutrit), 58. *Prof Exp:* Res technician, Columbia Univ, 50-53; res technician obstet & gynec, Sch Med, NY Univ, 53-54; Fulbright fel, Lab Comp Biochem, Col France, 58-59; res assoc med, Sch Med, NY Univ, 59-60; res biochemist, Monsanto Chem Co, Mo, 60-62; asst res biochemist, Sch Med, Univ Calif, Los Angeles, 62-70; mem biochem, Roche Inst Molecular Biol, 70-80; res dir, Queen's Med Ctr, Honolulu, Hawaii, 80-84; liason scientist, Dept Navy London Br, Off Naval Res, 84-89; RETIRED. *Concurrent Pos:* Adj prof, Col Med & Dent NJ; adj prof biochem, Dept Physiol, John A Burns Sch Med, Univ Hawaii, 80-84. *Mem:* AAAS; Am Physiol Soc; Am Soc Biol Chem; Am Soc Neurochem; Int Soc Neurochem. *Res:* Regulation of cerebral protein synthesis; effects of environmental and behavioral factors on cerebral protein and nuclei acid synthesis during development and aging. *Mailing Add:* 14809 Priscilla St San Diego CA 92129

ZON, GERALD, b Buffalo, NY, Apr 3, 45. ORGANIC CHEMISTRY. *Educ:* Canisius Col, BS, 67; Princeton Univ, PhD(org chem), 71. *Prof Exp:* Res assoc chem, Ohio State Univ, 71-72, NIH fel, 72-73; assoc prof chem, Cath Univ Am, 73-; AT DIV BIOCHEM & BIOPHYS, FOOD & DRUG ADMIN. *Concurrent Pos:* Sr res assoc, Andrulis Res Corp, 75- *Mem:* Am Chem Soc; Fedn Am Supporting Sci & Technol. *Res:* Synthesis, stereochemistry and mechanism of anticancer drugs; novel reactions of organometalloid systems. *Mailing Add:* Appl Bio Systs 850 Lincoln Ctr Dr Foster City CA 94404-1155

ZONIS, IRWIN S(AMUEL), b Boston, Mass, June 22, 30; m 55; c 3. CHEMICAL ENGINEERING. *Educ:* Rensselaer Polytech Inst, BChE, 50; Mass Inst Technol, MS, 52. *Prof Exp:* Proj mgr, Nat Res Corp, 51-57; chief process engr, Columbia-Nat Corp, 57-59; supvr chem eng, Dixon Chem & Res Inc, 59-62; plant mgr, 62-63, opers mgr, 63-64, vpres opers, chem div, 64-69, GEN MGR CHEM DIV & VPRES, ESSEX CHEM CORP, 69- *Concurrent Pos:* Mem, NJ State Air Pollution Control Comn, 66-67 & NJ State Clean Air Coun, 68, vchmn, 70, 71, chmn, 72- *Mem:* AAAS; Am Chem Soc; Am Inst Chem Engrs; Air Pollution Control Asn; Sigma Xi. *Res:* Heavy inorganic chemicals; production management; air quality control; extractive metallurgy. *Mailing Add:* 71 Crestmont Rd West Orange NJ 07052

ZOOK, ALMA CLAIRE, b Los Angeles, Calif, June 27, 50. SPECTROSCOPY OF DIATOMIC MOLECULES, NONLINEAR OPTICS. *Educ:* Pomona Col, Calif, BA, 72; Univ Calif, Berkeley, PhD(astron), 79. *Prof Exp:* Asst prof physics, Hamilton Col, 78-82; asst prof, 82-87, ASSOC PROF PHYSICS, POMONA COL, 87- *Mem:* Am Astron Soc; Astron Soc Pac; Optical Soc Am; Am Women Sci; AAAS; Sigma Xi. *Res:* Laboratory spectroscopy of astronomically interesting molecules; optical phase conjugation in barium titanate. *Mailing Add:* Dept Physics Pomona Col Claremont CA 91711

ZOOK, BERNARD CHARLES, b Beach, NDak, Nov 1, 35; m 55; c 3. RADIATION PATHOLOGY, CONGENITAL DISEASES. *Educ:* Colo State Univ, BS, 62, DVM, 63; Am Col Vet Pathologists, cert, 68. *Prof Exp:* Res fel path, Med Sch, Harvard Univ, 63-67, asst, 67-69, res fel electron micros, Sch Pub Health, 67-68; from asst prof to assoc prof, 69-82, DIR, ANIMAL RES FACIL, MED CTR, GEORGE WASHINGTON UNIV, 72-, PROF PATH, 82- *Concurrent Pos:* Res fel path, Angell Mem Hosp, 63-67, assoc pathologist, 67-69; res assoc, Smithsonian Inst, 69-; mem bd dirs, nat Soc Med Res, 82-85; Consult population coun. *Mem:* Am Vet Med Asn; Am Col Vet Pathologists; Radiation Res Soc; Am Asn Pathologists; Soc Tox Pathologists; Am Asn Lab Animal Sci. *Mailing Add:* Med Ctr George Washington Univ 2300 Eye St NW Washington DC 20037

ZOOK, HARRY DAVID, b Milroy, Pa, Feb 8, 16; m 37; c 2. ORGANIC CHEMISTRY, ACADEMIC ADMINISTRATION. *Educ:* Pa State Col, BS, 38, PhD(org chem), 42; Northwestern Univ, MS, 39. *Prof Exp:* Asst org chem, Northwestern Univ, 38-39; asst chem, 39-41, from instr to assoc prof, 41-60, asst to vpres res, 65-70, prof chem, 60-81, asst vpres res & assoc dean grad sch, 70-81, EMER PROF, PA STATE UNIV, UNIVERSITY PARK, 81- *Concurrent Pos:* Vis lectr, Stanford Univ, 62. *Mem:* Am Chem Soc. *Res:* Kinetics and mechanism of organic reactions. *Mailing Add:* PO Box 10091 State College PA 16805-0091

ZOOK, HERBERT ALLEN, b Laurel, Mont, Apr 21, 32; m 69; c 1. OPTICAL DETECTION OF SMALL BODIES IN SPACE, PLANETARY RINGS. *Educ:* Mont State Col, BS, 58, MS, 60. *Prof Exp:* SPACE SCIENTIST, SOLAR SYST EXPLOR DIV, NASA-JOHNSON SPACE CTR, 74- *Mem:* Meteoritical Soc; Planetary Soc; Am Geophys Union. *Res:* Origin and evolution of meteoroids and meteorites and their relationships to comets and asteroids; orbital and collisional evolution of meteoroids are theorectically studied and experimental spacecraft impact data are evaluated. *Mailing Add:* NASA Johnson Space Ctr SN3 Houston TX 77058

ZOON, KATHRYN C, CYTOKINE BIOLOGY. *Educ:* Rensselaer Polytech Inst, BS, 70; Johns Hopkins Univ, PhD(biochem), 75. *Prof Exp:* Postdoctoral fel, NIH, 75-77, staff fel & sr staff fel, 77-80; DIR, DIV CYTOKINE BIOL, CTR BIOLOGICS EVAL & RES, FOOD & DRUG ADMIN, 80- *Concurrent Pos:* Ed, J Inteferon Res. *Res:* Regulatory issues related to cytokines and growth factors; interferon purification, characterization and interferon receptors; author numerous scientific papers. *Mailing Add:* Div Cytokine Biol Food & Drug Admin Ctr Biologics Eval & Res Bldg 29A Rm 2D-16 Bethesda MD 20892

ZOPF, DAVID ARNOLD, b St Louis, Mo; m; c 4. IMMUNOCHEMISTRY, CARBOHYDRATE ANTIGENS. *Educ:* Wash Univ, St Louis, MD, 69. *Prof Exp:* Chief, Sect Biochem Path, Path Lab, Nat Cancer Inst, 79-88; VPRES & COO, BIOCARB INC, 88- *Mem:* Asn Am Physician; Am Soc Biochem & Molecular Biol. *Res:* Biochemistry and immunochemistry of complex carbohydrates. *Mailing Add:* Biocarb Inc 300 Professional Dr Gaithersburg MD 20879

ZORDAN, THOMAS ANTHONY, b Rockford, Ill, Nov 21, 43. PHYSICAL CHEMISTRY. *Educ:* Northern Ill Univ, BS, 65; Univ Louisville, PhD(phys chem), 69. *Prof Exp:* Res chemist, Gulf Res & Develop Co, 69-72; sr engr, Nuclear Ctr, Westinghouse Elec Corp, 72-74, mgr safeguards eval, 74-77, mgr reliability & safety, 77-78, mgr standard plant eng, 78-80; proj mgr, D'Appolonia Consult Engrs, 80-; AT ZORDAN ASSOCS. *Mem:* Am Chem Soc; AAAS; Am Inst Chem Engrs. *Res:* Thermodynamics of solutions and pure substances; theory of the liquid state; calorimetry of chemical reactions; nuclear power reactor safety; nuclear power plant design. *Mailing Add:* SAIC 4055 Monroeville Blvd Monroeville PA 15146

ZORN, GUS TOM, b Ada, Okla, June 18, 24; wid. HIGH ENERGY PHYSICS. *Educ:* Okla State Univ, BS, 48; Univ NMex, MS, 52; Univ Padua, PhD(physics), 54. *Prof Exp:* Res assoc physics, Brookhaven Nat Lab, 54-56, assoc physicist, 56-62; assoc prof physics, 62-72, PROF PHYSICS, UNIV MD, COLLEGE PARK, 72- *Concurrent Pos:* Vis scientist, Max Planck Inst Physics, 58; vis physicist, Frascat Nat Lab, Italy, 67-68, 76-77 & DESY Lab, Hamburg, WGermany, 84. *Mem:* Fel Am Phys Soc; Ital Phys Soc; NY Acad Sci. *Res:* Experimental high energy and elementary particle physics, using nuclear emulsions, bubble chambers, counters, spark chambers, and streamer and proportional wire chambers. *Mailing Add:* Dept Physics & Astron Univ Md College Park MD 20742

ZORN, JENS CHRISTIAN, b Halle, Ger, June 19, 31; US citizen; m 54; c 2. ATOMIC PHYSICS. *Educ:* Miami Univ, AB, 55; Yale Univ, MS, 57, PhD(physics), 61. *Prof Exp:* Engr, Sarkes Tarzian, Inc, Ind, 53-54; res asst physics, Univ Tubingen, 55-56; consult, Sch Med, Yale Univ, 59-61, instr, 61-62; from asst prof to assoc prof, 62-69, dir Residential Col, 72-74, assoc dean, Col Arts & Sci, 80-82, PROF PHYSICS, UNIV MICH, ANN ARBOR, 70- *Concurrent Pos:* Vis prof, Univ Puebla, 64-65; Phoenix fac fel, Univ Mich, 65-66, Ombudsman, 82-90. *Mem:* Fel Am Phys Soc; Am Asn Physics Teachers. *Res:* Atomic and molecular structure; atomic beams; microwave spectroscopy; laboratory astrophysics; space physics; scientific manpower administration. *Mailing Add:* Randall Lab Physics Univ Mich Ann Arbor MI 48109

ZORNETZER, STEVEN F, b New York, NY, Jan 21, 45; m 69; c 1. NEUROBIOLOGY, SCIENCE MANAGEMENT. *Educ:* State Univ NY, Stony Brook, BA, 66; Univ Wis-Madison, MA, 67; Univ Calif, Irvine, PhD(biol), 71. *Prof Exp:* From asst prof to assoc prof neurosci, Col Med Univ, Fla, 71-80; assoc prof pharmacol, Univ Calif, Irvine, 80-82; DIR, LIFE SCI, OFF NAVAL RES, 82- *Concurrent Pos:* Sloan res fel, 75; adj prof psychiat, Uniformed Serv Univ Health Sci, 85- *Honors & Awards:* William P Beck Sci Res Award, Interstate Postgrad Med Asn, 74. *Mem:* AAAS; Soc Neurosci; NY Acad Sci; Int Brain Res Orgn. *Res:* Neural information processing; memory; synaptic plasticity; neural changes during aging. *Mailing Add:* 7735 Rocton Ct Chevy Chase MD 20815

ZORNIG, JOHN GRANT, b Davenport, Iowa, Apr 22, 44. COMMUNICATIONS ENGINEERING, DIGITAL SYSTEMS. *Educ:* Yale Univ, BS, 66, MS, 72, PhD(eng), 74. *Prof Exp:* Lectr, 74-75, assoc prof, 75-79, assoc prof elec eng, dept eng & appl sci, 79-; AT BBN COMMUN CORP, CAMBRIDGE. *Concurrent Pos:* Consult, Xybion Corp, 76-77, Periphonics Inc & Allied Corp, 81. *Mem:* Inst Elec & Electronics Engrs; Acoust Soc Am. *Res:* Sound propagation in the ocean for application to communication and detection systems by physical scale modelling; vehicular robotics. *Mailing Add:* Digital Equip Corp Lkgi-2/E19 550 King St Littleton MA 01460

ZOROWSKI, CARL F(RANK), b Pittsburgh, Pa, July 14, 30; m 85; c 3. MECHANICAL ENGINEERING. *Educ:* Carnegie Inst Technol, BS, 52, MS, 53, PhD(mech eng), 56. *Prof Exp:* From instr to assoc prof mech eng, Carnegie Inst Technol, 54-62; from assoc prof to prof, 62-69, assoc head dept, 66-72, head dept, 72-79, assoc dean, Sch Eng, 79-85, R J REYNOLDS PROF MECH ENG & AEROSPACE ENG, NC STATE UNIV, 69-; DIR, INT SOC MUSIC EDUC, 86- *Concurrent Pos:* Orgn Europ Econ Coop sr vis fel, Brit Iron & Steel Res Asn, Sheffield, Eng, 62; consult, Army Res Off, Durham, 71 & Monsanto Co, 63-80, Int Bus Mach, engr, 85. *Honors & Awards:* Western Elec Award, Am Soc Eng Educ, 68; Fiber Soc Res Award, 70; Charles Russ Richards Award, Am Soc Mech Engrs, 75. *Mem:* Am Soc Mech Engrs (vpres, 80-83); Am Soc Eng Educ; Fiber Soc; Soc Rheology; Soc Mfg Engr. *Res:* Design for manufacturability; computer aided design for automated manufacturing; applied mechanics and mechanical design with emphasis on fibers, textile and composite property characterization and deformation mechanics; metal forming mechanics; mechanical system and component design and dynamic response anaylsis. *Mailing Add:* Int Mfg Systs Eng Inst NC State Univ Raleigh NC 27695-7915

ZORY, PETER S, JR, b Syracuse, NY, Oct 9, 36; m 61; c 2. LASERS. *Educ:* Syracuse Univ, BS, 58; Carnegie-Mellon Univ, PhD(physics), 64. *Prof Exp:* Physicist, Gyroscope Div, Sperry Rand Corp, 64-66, sr physicist, 66-68; mem prof staff, T J Watson Res Ctr, IBM Corp, 68-78; sr scientist, Optical Info Systs Inc, 78-; AT MCDONNELL-DOUGLAS ASTRONAUT, OPTO-ELECTRONICS CTR, ELMSFORD, NY. *Mem:* Inst Elec & Electronics Engrs; Optical Soc Am. *Res:* Laser research. *Mailing Add:* Dept Elec Eng Univ Fla 129 Larsen Hall Gainesville FL 32611

ZORZOLI, ANITA, b New York, NY, Dec 27, 17. BIOCHEMISTRY, PHYSIOLOGY. *Educ:* Hunter Col, AB, 38; Columbia Univ, AM, 40; NY Univ, PhD(biol), 45. *Prof Exp:* Asst zool, Columbia Univ, 40-42; asst instr biol, NY Univ, 44-45; res asst path, Sch Med, Wash Univ, 45-46, instr biochem, Sch Dent, 46-48, res assoc pharmacol, Sch Med, 48-49, asst prof biochem, Sch Dent, 48-52; asst prof, Southern Ill Univ, 52-55; from assoc prof to prof biol, 55-73, PROF BIOL, JOHN GUY VASSAR CHAIR, VASSAR COL, 73- *Concurrent Pos:* Mem corp, Marine Biol Lab, Woods Hole. *Mem:* Fel AAAS; fel Geront Soc (vpres, 65-66); Am Physiol Soc; Int Asn Geront; Am Aging Asn; Sigma Xi. *Res:* Biochemistry of aging in mice. *Mailing Add:* 18 Wilbur Blvd Poughkeepsie NY 12603

ZOSS, ABRAHAM OSCAR, b South Bend, Ind, Feb 17, 17; m 39; c 3. INDUSTRIAL CHEMISTRY, TECHNOLOGY TRANSFER. *Educ:* Univ Notre Dame, BSChE, 38, MS, 39, PhD(org chem), 41. *Prof Exp:* Asst, Univ Notre Dame, 39-41; res chemist, Gen Aniline & Film Corp, NJ, 41-43, Pa, 43-47, dept chemist, NJ, 47-49, area supt, 49-51, prod mgr, 51-54, tech mgr, 54-55, plant mgr, 55-57; mgr mfg admin, Chem Div, Minn Mining & Mfg Co, 57-58, div prod mgr, 58-60; vpres, Photek, Inc, 60-62; asst corp tech dir, Celanese Corp, 62-65, corp tech dir, 65-66, corp dir com develop, 66-69; vpres, Tenneco Chem, Inc, NY, 69-71; vpres corp dev, Universal Oil Prod Co, 71-72; group vpres, Engelhard Minerals & Chem Corp, 72-75, vpres bus develop, 75-77; vpres corp develop, CPS Chem Co, Inc, 77-84; PRES, BUS DEVELOP INT, 84- *Concurrent Pos:* Mem field info agency, Off Tech Serv, US Dept Com, Ger, 46; consult, 84- *Honors & Awards:* Centennial of Sci Award, Univ Notre Dame, 65. *Mem:* AAAS; Am Chem Soc; Am Inst Chem Engrs; Com Develop Asn; NY Acad Sci; Am Sect Societe de Chimie Industrielle (pres, 91-). *Res:* Acetylene and high polymer chemistry; petrochemicals; synthetic fibers; plastics; technology transfer; coatings; commercial development; catalytic chemistry; specialty monomers; polymers. *Mailing Add:* 1530 Palisade Ave Suite 22M Ft Lee NJ 07024

ZOSS, LESLIE M(ILTON), b Lockport, NY, Nov 23, 26; m 49; c 4. MECHANICAL ENGINEERING. *Educ:* Purdue Univ, BS, 49, MS, 50, PhD(mech eng), 52. *Prof Exp:* Instr mech eng lab, Purdue Univ, 49-50; res engr, Taylor Instrument Co, 52-55, tech training dir, 55-58; from asst prof to prof, Valparaiso Univ, 58-66, head dept, 65-76, res prof mech eng, 66-86; CONSULT, 58- *Concurrent Pos:* Consult, Esso Res & Eng Co, 59-65, Am Oil Co, 66-68, Shell Chem Co, 69-70 & Eli Lilly & Co, 71- *Honors & Awards:* Donald P Eckman Educ Award, Instrument Soc Am, 68. *Mem:* Am Soc Mech Engrs; fel Inst Measurement & Control; fel Instrument Soc Am. *Res:* Instrumentation for automatic control. *Mailing Add:* 2990 Horse Hill Dr E Indianapolis IN 46214

ZOTOS, JOHN, b Brockton, Mass, Jan 12, 32; m 58; c 3. MATERIALS SCIENCE, METALLURGY. *Educ:* Northeastern Univ, BSChE, 54; Mass Inst Technol, MSMet, 56, Metall Engr, 67. *Prof Exp:* Eng asst metall, Raytheon Mfg Co, 54; proj metallurgist, Watertown Arsenal's Rodman Lab, 55-56, asst chief exp foundry br, 56-60; from asst prof to assoc prof, 60-84, PROF MECH ENG, NORTHEASTERN UNIV, 84- *Concurrent Pos:* Mat & metall consult & lectr to indust, 60-; consult & mem bd dir, Indust Magnetics, Inc, Mass, 63-70; NSF sci fac fel, 63-70; Welding Res Coun study grant, 69; consult & mem tech adv bd, Thermo Magnetics, Inc, Mass, 70-73. *Honors & Awards:* Adams Mem Award, Am Welding Soc, 72. *Mem:* Am Foundrymen's Soc; Am Defense Preparedness Asn; Am Inst Mining, Metall & Petrol Engrs; Am Inst Chem Engrs; Am Soc Metals; Am Powder Metall Inst. *Res:* Development of mathematical models which describe the chemical, mechanical and physical properties of both ferrous and non-ferrous casting alloys; use of electromagnetic solid state joining for bonding similar and dissimilar metals and alloys; the role and responsibilities of the professional person in his civic and religious communities. *Mailing Add:* 28 Old Coach Rd Cohasset MA 02025

ZOTTOLA, EDMUND ANTHONY, b Gilroy, Calif, June 25, 32; m 60; c 4. FOOD SCIENCE. *Educ:* Ore State Univ, BS, 54, MS, 58; Univ Minn, St Paul, PhD(dairy tech), 64. *Prof Exp:* Res asst food sci, Ore State Univ, 56-58; res fel food sci & industs, Univ Minn, St Paul, 58-64; bacteriologist, Nat Dairy Prod Corp, Ill, 64-65; microbiologist, Nodaway Valley Foods, Iowa, 65-66; assoc prof food sci & industs, 66-72, exten food microbiologist, 66-84, PROF FOOD SCI & NUTRIT, UNIV MINN, ST PAUL, 72- *Honors & Awards:* Educr Award, Int Asn Milk, Food & Environ Sanitarians, 88, Sherman Award, 89. *Mem:* Fel Inst Food Technologists; Am Dairy Sci Asn; Int Asn Milk, Food & Environ Sanitarians. *Res:* Spoilage and pathogenic microorganisms in food; thermal destruction of microorganisms; airborne microorganisms; food plant and equipment sanitation; detection of microorganisms in raw and processed foods; attachment of microorganisms to food and food contact surfaces; microorganisms involved in cheese manufacturing and the technology of cheese manufacture; membrane processing liquid foods, UF/RO/MF. *Mailing Add:* Dept Food Sci & Nutrit Univ Minn St Paul MN 55108

ZOTTOLI, ROBERT, b Boston, Mass, Apr 17, 39; m; c 3. INVERTEBRATE ZOOLOGY. *Educ:* Bowdoin Col, BA, 60; Univ NH, MS, 63, PhD(zool), 66. *Prof Exp:* Asst biol, Univ NH, 61-63; from asst prof to assoc prof, 65-75, PROF BIOL, FITCHBURG STATE COL, 75- *Mem:* AAAS; Sigma Xi; Am Soc Limnol Oceanogr; Am Soc Zool. *Res:* Natural history of polychaetous annelid worms; polychaete family ampharetidae. *Mailing Add:* Fitchburg State Col Fitchburg MA 01420

ZOTTOLI, STEVEN JAYNES, b Boston, Mass, Aug 28, 47; m; c 4. NEUROBIOLOGY. *Educ:* Bowdoin Col, BA, 69; Univ Mass, Amherst, MS, 72, PhD(zool), 76. *Prof Exp:* Lectr cell physiol, Col Our Lady of the Elms, 73-74; Nat Inst Neurol, Commun Dis & Stroke fel, Res Inst Alcoholism, Buffalo, NY, 75-77, res scientist II, 77-78; asst prof physiol, State Univ NY, Buffalo, 78-80; asst prof, 80-87, ASSOC PROF BIOL, WILLIAMS COL, WILLIAMSTOWN, MASS, 87- *Concurrent Pos:* Grass fel, 78; trustee, Grass Found, 87-90. *Mem:* Soc Neurosci; Am Soc Zoologists; Sigma Xi; AAAS. *Res:* Neurophysiological, morphological and behavioral studies of Mauthner cell function in teleosts. *Mailing Add:* Dept Biol Williams Col Williamstown MA 01267

ZOUROS, ELEFTHERIOS, b Lesbos, Greece, Aug 31, 39; m 68; c 2. POPULATION GENETICS. *Educ:* Agr Col Athens, BSc, 63, PhD(biol), 68; Univ Chicago, PhD(biol), 72. *Prof Exp:* Res assoc biol, Agr Col Athens, 65-68; fel pop biol, Univ Chicago, 69-73; PROF BIOL, DALHOUSIE UNIV, 73- & UNIV CRETE, 83- *Concurrent Pos:* Scholar, Greek Nat Found Scholars, 66; Ford Found fel, 69. *Mem:* Genetics Soc Am; Am Soc Naturalists; Genetics Soc Can; Soc Study Evolution. *Res:* Genetic basis of the evolutionary process. *Mailing Add:* Dept Biol Dalhousie Univ Halifax NS B3H 4J1 Can

ZRAKET, CHARLES ANTHONY, b Lawrence, Mass, Jan 9, 24; m 61; c 4. ELECTRICAL & SYSTEMS ENGINEERING. *Educ:* Northeastern Univ, BS, 51; Mass Inst Technol, SM, 53. *Hon Degrees:* DEng, Northeastern Univ, 88. *Prof Exp:* Tech staff mem, Digital Comput Lab, Mass Inst Technol, 51-53 & Lincoln Labs, 53-54, sect leader, 54-56, group leader, 56-58; dept head systs design, Mitre Corp, 58-61, tech dir systs planning & res div, 61-67, vpres & tech dir, 67-69, vpres, 69-74, sr vpres, 75-78, exec vpres, 78-86, pres & chief exec officer, 86-90; RETIRED. *Concurrent Pos:* Consult, Dept Defense. *Mem:* Nat Acad Eng; fel Inst Elec & Electronics Engrs; Am Mgt Asn; Am Inst Aeronaut & Astronaut; NY Acad Sci; fel Am Acad Arts & Sci; fel AAAS. *Res:* Digital computers; digital computer programming; electronic control systems for large-scale, real-time use; information systems; communications systems; transportation systems; environmental control; energy; educational technology; law enforcement. *Mailing Add:* 71 Sylvan Lane Weston MA 02193

ZRUDSKY, DONALD RICHARD, b Cedar Rapids, Iowa, Apr 4, 34; div; c 2. SOLID STATE PHYSICS, ELECTRICAL ENGINEERING. *Educ:* Iowa State Univ, BS, 56, MS, 59; Univ Iowa, PhD(physics), 68. *Prof Exp:* Asst engr, Rockwell Corp, 56-57; res helper physics, Ames Lab, 57-59; res staff, Rockwell Corp, 59-62, assoc engr, 62-65; NASA trainee physics, Univ Iowa, 65-68; asst prof elec eng, Univ Toledo, 68-71, assoc prof, 71-80; AT COLUMBIA SCI INDUSTS, AUSTIN, TEX. *Concurrent Pos:* Dir NSF grant, 70-71; consult, Keithley Instruments Corp, 73 & Los Alamos Sci Lab, 78. *Mem:* Am Phys Soc; Inst Elec & Electronics Engrs. *Res:* Solid state materials, specifically, magnetic alloys, magnetic compounds and semiconductors; solid state devices; integrated circuits; field effect devices; stirling cycle magnetic heat engines; electronic circuits; instrumentation. *Mailing Add:* PO Box 1358 Cedar Park TX 78613

ZSCHEILE, FREDERICK PAUL, JR, b Burlington, Kans, May 11, 07; m 33; c 3. PLANT PHYSIOLOGY, PHYTOPATHOLOGY. *Educ:* Univ Calif, BS, 28, PhD(plant physiol), 31. *Prof Exp:* Nat Res Coun fel biol, Univ Chicago, 31-33, asst pediat, 33-34, res assoc chem, 34-37, res assoc bot, 44-46; from asst prof agr chem & asst chemist to assoc prof & assoc chemist, Purdue Univ, 37-44; from assoc prof to prof, 46-74, from assoc biochemist to biochemist, Exp Sta, 46-74, EMER PROF AGRON, COL AGR, UNIV CALIF, DAVIS, 74- *Concurrent Pos:* Guggenheim fel, 58-59. *Mem:* AAAS; Am Chem Soc; Am Soc Plant Physiol; fel Am Inst Chem; Am Phytopath Soc. *Res:* Spectrophotometry of plant pigments; preservation of carotene and vitamin A; biochemical nature of disease resistance in plants; phytotron design; gas chromatography of amino acids; lysine in food plants. *Mailing Add:* Apt 22 Stollwood Convalescent Hosp 135 Woodland Ave Woodland CA 95695-2759

ZSIGMOND, ELEMER K, b Budapest, Hungary, May 16, 30; US citizen; m 63; c 1. PHARMACOLOGY, BIOCHEMISTRY. *Educ:* Univ Budapest, MD, 55. *Prof Exp:* Intern med, Clins, Med Sch, Univ Budapest, 54-55; resident internal med, Sztalinvarosi Korhaz, Hungary, 55-56; cardiol res, Balatonfured, Hungary, 56-57; intern med, Allegheny Gen Hosp, Pittsburgh, Pa, 60-61, resident anesthesiol, 61-63, dir anesthesiol res lab, 66-68; prof anesthesiol, Med Ctr, Univ Mich, Ann Arbor, 68-79; PROF ANESTHESIOL, MED CTR, UNIV ILL, 79- *Concurrent Pos:* Res asst/assoc anesthesiol lab, Mercy Hosp, Pittsburgh, Pa, 57-59. *Mem:* Am Soc Anesthesiol; Int Anesthesia Res Soc; NY Acad Sci; Am Soc Clin Pharmacol; AMA; Soc Acad Anesthesiol. *Res:* Plasmacholinesterase studies to determine susceptibility to drugs used in anesthesia; malignant hyperpyrexia determinations; intravenous anesthetics on cardio-respi system; muscle relaxants and reversing drugs. *Mailing Add:* Dept Anesthesiol M-C 515 Univ Ill Col Med PO Box 6998 Chicago IL 60680

ZSIGRAY, ROBERT MICHAEL, b Glen Rogers, WVa, Mar 22, 39; m 70; c 4. MICROBIOLOGY. *Educ:* Miami Univ, AB, 61; Georgetown Univ, MS, 67, PhD(biol), 69. *Prof Exp:* Microbiologist, Wis State Lab Hyg, 61; res technologist genetics, US Army Biol Lab, Ft Detrick, 62-64, Nat Res Coun res assoc, 68-70; asst prof, 70-76, ASSOC PROF MICROBIOL, UNIV NH, 76- *Mem:* AAAS; Am Soc Microbiol. *Res:* Entry of exogenous DNA into cells. *Mailing Add:* Star Route Canaan Rd Barrington NH 03825

ZSOTER, THOMAS, b Budapest, Hungary, Dec 27, 22; m 53; c 2. INTERNAL MEDICINE. *Educ:* Univ Budapest, MD, 47; FRCP(C). *Prof Exp:* Resident med, Univ Budapest, 47-49, instr internal med, 49-51; from asst prof to assoc prof med, Univ Szeged, 50-57; head circulation lab, Ayerst, McKenna & Harrison, Ltd, 57-60; ASSOC PROF PHARMACOL, UNIV TORONTO, 66-, ASSOC PROF MED, 68- *Concurrent Pos:* French Govt fel, Paris, 48-49; staff mem, Div Internal Med, Toronto Western Hosp, Ont, 66-, dir, 73-75. *Mem:* Fel Am Col Physicians; Can Pharmacol Soc; Can Fedn Biol Soc; fel Royal Col Can; Can Cardiovasc Soc; Can Soc Clin Invest. *Res:* Circulation; pathophysiology; congestive heart failure; hemodynamics in experimental valvular defects; microscopic circulation; cardiovascular pharmacology; hypertension; calcium antagonistic drugs; catecholemines. *Mailing Add:* 446 Otter Rye Cresent Toronto ON M5S 1A8 Can

ZUBAL, I GEORGE, b Lorain, Ohio, July 9, 50. MEDICAL IMAGING, THERAPY PLANNING. *Educ:* Ohio State Univ, BS, 72, MS, 74; Univ des Saarlandes, WGer, PhD(biophysics), 81. *Prof Exp:* Fel, Brookhaven Nat Lab, 81-; ASST PROF, STATE UNIV NY, 81-; AT YALE UNIV, NEW HAVEN, CONN. *Mem:* Inst Elec & Electronics Engrs. *Res:* Development of software programs for analyzing medical patient pictures and application of these results to the radio therapy planning with nonconventional radiation. *Mailing Add:* Yale Univ 333 Cedar Sr PO Box 3333 New Haven CT 06510

ZUBALY, ROBERT B(ARNES), b Philadelphia, Pa, Apr 20, 33; m 56; c 3. NAVAL ARCHITECTURE, MECHANICAL ENGINEERING. *Educ:* Webb Inst Naval Archit, BS, 55; Columbia Univ, MS, 59. *Prof Exp:* From instr to assoc prof, 55-66, PROF ENG, STATE NY MARITIME COL, 66- *Concurrent Pos:* Res engr, Davidson Lab, Stevens Inst Technol, 57-61; res assoc, Webb Inst Naval Archit, 61- *Mem:* Soc Naval Archit & Marine Engrs; Am Soc Eng Educ. *Res:* Ship design; ship response to sea; ocean transportation; naval architecture, hydrodynamics. *Mailing Add:* Dept Eng State Univ NY Maritime Col Ft Schuyler Bronx NY 10465

ZUBAY, GEOFFREY, b Chicago, Ill, Nov 15, 31. BIOLOGY. *Educ:* Univ Chicago, PhB, 49, MS, 52; Harvard Univ, PhD(phys chem), 57. *Prof Exp:* NSF res fel molecular biol, King's Col, Univ London, 57-59, NIH fel, 59-60; res assoc, Rockefeller Inst, 60-61; asst biochemist, Brookhaven Nat Lab, 61-63; assoc prof, 63-71, PROF MOLECULAR BIOL, COLUMBIA UNIV, 72- *Res:* Molecular biology of gene regulation; reactions leading to the origin of life. *Mailing Add:* Dept Biol Sci Fairchild Ctr Columbia Univ New York NY 10027

ZUBECK, ROBERT BRUCE, b Minneapolis, Minn, Oct 12, 44; m 69. LOW TEMPERATURE PHYSICS, APPLIED PHYSICS. *Educ:* Univ Minn, BPhys, 66; Stanford Univ, MS, 68, PhD(appl physics), 73. *Prof Exp:* Consult, Stanford Res Inst, 69-70, res assoc appl physics, Stanford Univ, 73-77; STAFF ENGR, IBM CORP, 77- *Mem:* Am Phys Soc. *Res:* High resolution low temperature calorimetry; magnetic flux pinning in high field superconductors; electronbeam co-deposition techniques; growth morphology and crystallographic ordering of A-15 superconductors; synthesis of new superconductors; thin film magnetics; process reliability. *Mailing Add:* 1102 Lisa Lane Los Altos CA 94022

ZUBECKIS, EDGAR, food technology, for more information see previous edition

ZUBER, B(ERT) L, b Houston, Tex, Oct 27, 38; div; c 2. BIOENGINEERING. *Educ:* Univ Pa, BA, 60, BS, 61; Mass Inst Technol, MS, 63, PhD(bioeng), 65. *Prof Exp:* From asst prof to assoc prof bioeng & physiol, Univ Ill, Chicago, 65-73, actg head bioeng prog, 70-71, prof physiol, Med Ctr, 73-78, PROF BIOENG, UNIV ILL, CHICAGO, 73- *Concurrent Pos:* Consult, Biosysts, Inc, Mass, 65-66, Aerospace Med Res Lab, Wright-Patterson AFB, Ohio, 71, Pollution Monitors, Inc, Chicago, 71, Dept Psychiat, Univ Chicago, 76-77, Manteno Mental Health Ctr, Ill, 77-79 & Eng Design Assocs, San Francisco, 76-79; asst attend, Presby-St Luke's Hosp, 65-69, assoc attend, 69-76; dir, Biomed Eng Consult Serv, 66-67; USPHS res grant, 68-73; NSF res grant, 74-76; mem bd dir, Nat Asn Bioengrs, 72-75; vis scientist, Bell Labs, NJ, 75; consult, Chicago Habits Clin, Ltd, Evanston, Ill, 79 & Rehab Eng Res & Design Ctr, Hines Vet Admin Hosp, Hines, Ill, 80-85; lectr, Dept Orthopedics, Stritch Sch Med, Loyola Univ, Maywood, Ill, 80-; mem comt, Biomed Eng Soc, 82-84; organizer, OMS Int Res Conf on the Oculomotor Syst, 82; scientific prog co-chair, 38th Annual Conf on Engr in Med & Biol, 85; mem publ bd, Biomed Eng Soc, 86-90, chmn publ bd, 87-90, mem, Bd of dirs, 87-90; vis scientist, Univ de Provence, Marseille, France, 90. *Mem:* Am Physiol Soc; Inst Elec & Electronics Engrs; Biomed Eng Soc; Int Brain Res Orgn. *Res:* Physiological control systems; neurophysiological and control systems aspects of visual and oculo-motor function; bioinstrumentation; bioengineering education; information processing in reading; reading aids for the blind. *Mailing Add:* Dept Bioeng MC 063 Univ Ill Chicago Chicago IL 60680

ZUBER, MARCUS STANLEY, b Gettysburg, SDak, Jan 10, 12; m 41; c 1. AGRONOMY. *Educ:* SDak State Univ, BS, 37; Iowa State Univ, MS, 40, PhD(plant breeding), 50. *Hon Degrees:* DSc, SDak State Univ, 83. *Prof Exp:* Agent corn invests, Div Cereal Crops & Dis, Mo Agr Exp Sta, USDA, 37-42, assoc agronomist & in charge corn breeding, Div Cereal Crops, 46-50, res agronomist & in charge, 50-73, res leader, Corn Breeding, Div Cereal Crops, 73-76; prof, 56-82, EMER PROF AGRON, UNIV MO-COLUMBIA, 82- *Mem:* Fel AAAS; fel Am Soc Agron. *Res:* Corn breeding and genetics; physiology; insect and disease resistance; cereal chemistry. *Mailing Add:* 1408 Bus Loop 70 W Columbia MO 65202

ZUBER, NOVAK, b Belgrade, Yugoslavia, Dec 4, 22; nat US; m 58. ENGINEERING. *Educ:* Univ Calif, Los Angeles, BS, 51, MSc, 54, PhD(eng), 59. *Prof Exp:* From asst res engr to assoc res engr, Univ Calif, Los Angeles, 51-58; mem tech staff, Res Lab, Ramo-Wooldridge Corp, 58-60; thermal engr, Gen Eng Lab, Gen Elec Co, 60-62, sr thermal engr, Adv Tech Labs, 62-65, consult engr, Res & Develop Ctr, 65-67; prof mech eng, NY

Univ, 67-69; Fuller E Callaway Prof, Ga Inst Technol, 69-77; SR REACTOR ANALYST REACTOR SAFETY RES, NUCLEAR REGULATORY COMN, 77- *Concurrent Pos:* Consult, Mech Div, Off Sci Res, US Air Force, 63 & Los Alamos Sci Lab, 67. *Honors & Awards:* Mem Award, Am Soc Mech Engrs, 62. *Mem:* Am Inst Aeronaut & Astronaut; Am Inst Chem Engrs; Am Soc Mech Engrs; Sigma Xi. *Res:* Heat transfer; fluid dynamics; thermodynamics; transport phenomena in multi-phase systems. *Mailing Add:* 703 New Mark Esplanade Rockville MD 20850

ZUBER, WILLIAM HENRY, JR, b Memphis, Tenn, Sept 26, 37; m 59; c 3. PHYSICAL CHEMISTRY. *Educ:* Memphis State Univ, BS, 60; Univ Ky, PhD(phys chem), 64. *Prof Exp:* Asst prof chem, Murray State Univ, 64-66; asst prof, 66-70, ASSOC PROF CHEM, MEMPHIS STATE UNIV, 70- *Mem:* Am Chem Soc. *Res:* Nonaqueous solution chemistry. *Mailing Add:* Dept Chem Memphis State Univ Memphis TN 38152

ZUBERER, DAVID ALAN, b Paterson, NJ, Feb 9, 47; m 68; c 1. SOIL MICROBIOLOGY, MICROBIAL ECOLOGY. *Educ:* WVa Univ, AB, 69, MS, 71; Univ SFla, PhD(biol), 76. *Prof Exp:* Res scientist microbiol, Univ Fla, 76-78; from asst prof to assoc prof, 78-90, PROF MICROBIOL, TEX A&M UNIV, 90- *Mem:* Am Soc Microbiol; Am Soc Agron; Soil Sci Soc Am; AAAS; Soc Indust Microbiol. *Res:* Biological nitrogen fixation, including associative N2-fixation in grasses and N2-fixation by legumes; use of beneficial microorganisms to enhance plant nutrient uptake; mycorrhizae and reclamation microbiology. *Mailing Add:* Dept Soil & Crop Sci Tex A&M Univ College Station TX 77843

ZUBIETA, JON ANDONI, b New York, June 16, 45; m 69; c 2. INORGANIC CHEMISTRY. *Educ:* Fordham Univ, BS, 66; Columbia Univ, PhD(chem), 71. *Prof Exp:* Res assoc chem, Univ Sussex, 71-73; asst prof, 73-80, ASSOC PROF CHEM, STATE UNIV NY, ALBANY, 80- *Concurrent Pos:* NIH fel, 72-73; NATO travelling fel, 81-82. *Mem:* Am Chem Soc. *Res:* Inorganic models for molybdoenzymes; electroanalytical chemistry; structure and reactivity of cluster compounds. *Mailing Add:* Dept Chem Syracuse Univ Syracuse NY 13244

ZUBIN, JOSEPH, experimental psychopathology, experimental epidemiology; deceased, see previous edition for last biography

ZUBKO, L(EONARD) M(ARTIN), b Bridgeport, Conn, July 8, 20; m 41; c 1. MECHANICAL ENGINEERING. *Educ:* Rutgers Univ, BSME, 42; Rensselaer Polytech Inst, MME, 49. *Prof Exp:* Instr mech eng, Rensselaer Polytech Inst, 46-49; assoc prof, Univ Ill, 49-51 & 52-55; proj engr, Sverdrup Parcel Consult Engrs, 51-52; mgr propulsion appl res, Flight Propulsion Lab Dept, 55-60, mgr adv eng, 60-73, MGR USSR PROGS, GEN ELEC CO, 73- *Concurrent Pos:* Mem subcomt combustion, NASA, 56-58; adj assoc prof, Univ Vt, 64-71. *Mem:* Am Soc Mech Engrs; Am Inst Aeronaut & Astronaut; Am Ord Asn; Nat Soc Prof Engrs; Sigma Xi. *Res:* Gas dynamics; fluid mechanics; heat transfer; combustion; mechanisms. *Mailing Add:* 51 Laurel Dr Mt Dora FL 32757

ZUBKOFF, MICHAEL, b June 2, 44; c 3. MEDICINE. *Educ:* Am Int Col, BA, 65; Columbia Univ, MA, 66, PhD(econ), 69. *Prof Exp:* Instr econ, Columbia Univ, 67-69; Woodrow Wilson teaching fel, Meharry Med Col & Fisk Univ, 69-70; from asst prof to assoc prof health econ, Meharry Med Col & Vanderbilt Univ, 70-75; PROF & CHMN, DEPT COMMUN & FAMILY MED, DARTMOUTH MED SCH, 75- *Concurrent Pos:* Deleg & health spokesman, White House Summit Inflation, Washington, DC, 74; prof health econ & mgt, Amos Tuck Sch Bus Admin & adj prof econ & policy studies, Dartmouth Col, 75-; consult, Domestic Coun, White House, 76-80, Nat Ctr Health Serv Res, Health Care Financing Admin, House Subcomt Health & Environ, Senate Subcomt Health, Off Secy Health & Human Serv & Geriat Assessment & Planning Prog, South Shore Hosp & Med Ctr, Miami, 80-; mem, var med & educ comts, Inst Med-Nat Acad Sci, 77- *Mem:* Inst Med-Nat Acad Sci. *Res:* Health economics and management; author of numerous technical publications. *Mailing Add:* Community Family Med Prog Dartmouth Med Sch Hanover NH 03755

ZUBKOFF, PAUL L(EON), b Niagara Falls, NY, Nov 24, 34; m 60; c 2. BIOCHEMISTRY, ENVIRONMENTAL SCIENCES. *Educ:* Univ Buffalo, BA, 56; George Washington Univ, MS, 58; Cornell Univ, PhD(biochem), 62. *Prof Exp:* Res asst biochem, George Washington Univ, 56-58; res asst, Cornell Univ, 58-61; res biol chemist, Univ Calif, Los Angeles, 61-63; NIH trainee & res assoc biophys, Mass Inst Technol, 63-66; asst prof biochem, Ohio State Univ, 66-70; sr marine scientist & head environ physiol, Va Inst Marine Sci, 70- 82; assoc prof marine sci, Col William & Mary & Univ Va, 70-83; PRES & SR CONSULT, MICRO SCI CONSULT, 83-; CHEMIST, SURFACE WATER HYDROL, US ENVIRON PROTECTION AGENCY, 89- *Concurrent Pos:* Adj prof marine sci, Va Inst Marine Sci-Col William & Mary, 83-; vis fel, Dept Agron, Cornell Univ, Ithaca, NY, 88-89. *Mem:* Atlantic Estuarine Res Soc; Am Chem Soc; Am Nuclear Soc; Am Soc Limnol & Oceanog; Sigma Xi; Estuarine Res Fedn; Marine Biol Asn UK. *Res:* Dynamics of aquatic ecosystems; comparative biochemistry of macromolecules; metabolism of marine invertebrates; influences of terrestrial environments on surface and ground water hydrodynamics. *Mailing Add:* 113 W Queens Dr Williamsburg VA 23185

ZUBLENA, JOSEPH PETER, b Englewood, NJ, Nov 26, 51; m 78; c 2. PRODUCTION COUNSELING, AGENT & GROWER TRAINING. *Educ:* Rutgers Univ, BS, 73, MS, 76, PhD(agron), 79. *Prof Exp:* Teaching asst crops & soils, Rutgers Univ, 73-74, res assoc, 75-77, res intern, 77-79; assoc prof agron & soils, 79-87, prof agron & soils, 87-88, PROF & EXTEN SPECIALIST IN-CHARGE SOIL SCI, CLEMSON UNIV, 88- *Concurrent Pos:* Lectr, Clemson Univ, 83. *Mem:* Agron Soc Am; Soil Sci Soc Am; Crop Sci Soc Am. *Res:* Improved corn fertilizer efficiencies through placement, timing and source and rate selections; herbicide accelerated degradation phenomena; cultivar testing; aflatoxin as affected by production practices; land application of waste productions. *Mailing Add:* NC State Univ PO Box 7619 Raleigh NC 27695-7619

ZUBLER, EDWARD GEORGE, b Lackawanna, NY, Mar 12, 25; m 50; c 3. INORGANIC CHEMISTRY. *Educ:* Canisius Col, BS, 49; Univ Notre Dame, PhD(phys chem), 53. *Prof Exp:* Res phys chemist, Gen Elec Co, 53-65, tech leader, 65-72, res adv, 72-87; CONSULT, LAMP CHEM, 87- *Concurrent Pos:* Lectr, Fenn Col, 60-65 & Cleveland State Univ, 65-66; exec comt, High & Temperature Mat Div, Electrochem Soc, 84- *Honors & Awards:* Elenbaas Award, Philip's Gloeilampenfabrieken-Netherlands, 81; Steinmetz Award, Gen Elec, 73. *Mem:* Am Chem Soc; Electrochem Soc. *Res:* High temperature gas-metal reactions; chemical transport processes; mass spectrometry; high vacuum and high purity gas techniques; microbalance techniques; computer themodynamic calculations. *Mailing Add:* 28430 Hidden Valley Dr Chagrin Falls OH 44022

ZUBRISKI, JOSEPH CAZIMER, b Goodman, Wis, July 7, 19; m 46; c 3. SOIL SCIENCE. *Educ:* Univ Wis, BS, 47, MS, 48, PhD(soils), 51. *Prof Exp:* From asst prof to assoc prof, 51-63, PROF SOILS, NDAK STATE UNIV, 63- *Mem:* Soil Sci Soc Am; Am Soc Agron. *Res:* Soil fertility; effect of plant population and fertilizers on yield and quality of sunflower seeds; soil phosphorus; fertilizer and water management of irrigated crops. *Mailing Add:* Dept Soils ND State Univ Fargo ND 58105

ZUBROD, CHARLES GORDON, b New York, NY, Jan 22, 14; m 40; c 5. CANCER. *Educ:* Col of the Holy Cross, AB, 36; Columbia Univ, MD, 40. *Hon Degrees:* DSc, Col of the Holy Cross, 69. *Prof Exp:* Intern, Cent Islip State Hosp, NJ, 40-41 & Jersey City Hosp, NY, 41-42; intern & asst resident med, Presby Hosp, New York, 42-43; instr med, Sch Med, Johns Hopkins Univ, 46-49, asst prof med & pharmacol, 49-53; assoc prof med & dir res, Dept Med, St Louis Univ, 53-54; chief gen med br, Nat Cancer Inst, 54-55, clin dir, 55-61, chmn med bd, 57-58, dir int res, 61-65, sci dir chemother, 65-72, dir div cancer treatment, 72-74; PROF ONCOL & CHMN DEPT, PROF MED & DIR COMPREHENSIVE CANCER CTR, SCH MED, UNIV MIAMI, 74- *Concurrent Pos:* Roche res fel chemother bact dis, Johns Hopkins Hosp, 46-49; mem, Mt Desert Island Biol Lab; mem & mem exec comt, Lerner Marine Lab, Bimini. *Honors & Awards:* Lasker Award, 72. *Mem:* Am Soc Clin Invest; Am Soc Pharmacol & Exp Therapeut; Am Asn Cancer Res (pres, 77); Am Asn Cancer Insts (pres, 78); Asn Am Physicians. *Res:* Pharmacology, especially of cancer chemotherapeutic agents; marine biology. *Mailing Add:* Univ Miami Comp Cancer Ctr PO Box 016960 Miami FL 33101

ZUBRZYCKI, LEONARD JOSEPH, b Camden, NJ, Feb 25, 32; m 54; c 1. MEDICAL MICROBIOLOGY. *Educ:* Temple Univ, AB, 53, PhD(med microbiol), 58. *Prof Exp:* Sr scientist, Wyeth, Inc, 58-61; PROF MICROBIOL, SCH MED, TEMPLE UNIV, 61- *Mem:* AAAS; Am Soc Microbiol; Am Venereal Dis Soc. *Res:* Bacterial genetics; diagnostic bacteriology. *Mailing Add:* Dept Microbiol Temple Univ Philadelphia PA 19122

ZUCCA, RICARDO, b Trieste, Italy, Feb 7, 36; US citizen; m 58; c 4. SEMICONDUCTOR PHYSICS, SEMICONDUCTOR DEVICES. *Educ:* Univ Rosario, Arg, MA, 60; Univ Calif, Berkeley, PhD(physics), 71. *Prof Exp:* Prof physics, Univ Rosario, Arg, 69-72; MEM TECH STAFF, SCI CTR, ROCKWELL INT, 72- *Mem:* Am Phys Soc; sr mem Inst Elec & Electronics Engrs. *Res:* Semiconductor physics and semiconductor devices; development of the gallium arsenic digital high-speed technology; infrared semiconductor mercury cadmium tellurium and its applications to infrared devices. *Mailing Add:* Rockwell Int 1049 Camino Dos Rios Thousand Oaks CA 91360

ZUCCARELLI, ANTHONY JOSEPH, b New York, NY, Aug 11, 44; m 68; c 2. MOLECULAR BIOLOGY, MOLECULAR GENETICS. *Educ:* Cornell Univ, BS, 66; Loma Linda Univ, MS, 68; Calif Inst Technol, PhD(biophys), 74. *Prof Exp:* Fel molecular biol, Am Cancer Soc, 74-76; asst prof biol, 76-80, ASSOC PROF MICROBIOL, LOMA LINDA UNIV, 80- *Concurrent Pos:* Roberts' scholar, Cornell Univ, 64-65; assoc fac, biol dept, Loma Linda Univ, 80-, biochem dept, 85- *Mem:* Am Soc Microbiol; AAAS; Sigma Xi. *Res:* Molecular biology of single-stranded DNA bacterial viruses, particularly the enzymology of DNA replication and mechanisms of gentic recombination; investigation of functions of prokaryotic genes using recombinant DNA technology; determinants of bacterial virulence; DNA sequencing instrumentation. *Mailing Add:* Dept Microbiol AH115 Loma Linda Univ Loma Linda CA 92350

ZUCCHETTO, JAMES JOHN, b Brooklyn, NY, Mar 11, 46; m 78; c 2. SYSTEMS ECOLOGY, ENVIRONMENTAL SCIENCE. *Educ:* Polytech Inst Brooklyn, BS, 66; NY Univ, MS, 69; Univ Fla, PhD(syst ecol), 75. *Prof Exp:* Mem tech staff, Bell Tel Labs, Inc, 68-71; asst syst ecol, Univ Fla, 72-75, assoc eng energy in transp, 75-76; guest researcher environ, Univ Stockholm, 76-78; asst prof regional sci, Univ Pa, 78-85; NAT ACAD SCI, 85- *Concurrent Pos:* Researcher, Rockefeller Found, 77 & Univ Pa, 78-79; consult, US Nat Res Coun, 76 & Fed Energy Admin, 75. *Mem:* Regional Sci Asn; Sigma Xi. *Res:* Ecological models; regional energy-ecological-economic interactions; energy analysis; general systems; environmental impact evaluation; systems analysis. *Mailing Add:* Nat Acad Sci GF 286 2101 Const Ave NW Washington DC 20418

ZUCHELLI, ARTLEY JOSEPH, b Alexandria, Va, Nov 3, 34. PHYSICS. *Educ:* Univ Va, BA, 55, PhD(physics), 58. *Prof Exp:* NSF fel, Univ Birmingham, 58-59; assoc prof physics, Univ Miss, 59-63; assoc prof, 63-67, PROF PHYSICS, GEORGE WASHINGTON UNIV, 67- *Res:* Theoretical, classical and quantum fields. *Mailing Add:* Dept Physics George Washington Univ 2121 Eye St NW Washington DC 20052

ZUCK, DONALD ANTON, b Hafford, Sask, Dec 27, 18; m 44, 58; c 3. PHARMACEUTICAL CHEMISTRY. *Educ:* Univ Alta, BSc, 48; Univ Wis, MS, 50, PhD(pharm), 52. *Prof Exp:* Lectr, Sch Pharm, Univ BC, 48-50; from pharmaceut chemist to sr pharmaceut chemist, Eli Lilly & Co, 52-65; prof pharm, Col Pharm, Univ Sask, 65-86; RETIRED. *Concurrent Pos:* Examr, Pharm Exam Bd Can, 72-77; sci ed, Can J Pharmaceut Sci, 72-82; mem, Drug Qual Assessment Comt, 74- *Mem:* Am Pharmaceut Asn; fel Am Found Pharm Educ; Can Pharm Asn; Am Asn Cols Pharm; Asn Fac Pharm Can. *Res:* Physical chemistry as applied to pharmaceutical problems. *Mailing Add:* Dept Pharm Univ Sask Saskatoon SK S7N 0W0 Can

ZUCK, ROBERT KARL, b Rochester, NY, Oct 8, 14; m 38; c 4. BOTANY. *Educ:* Oberlin Col, AB, 37; Univ Tenn, MS, 39; Univ Chicago, PhD(bot), 43. *Prof Exp:* Asst plant pathologist, Exp Sta, Univ Tenn, 38-41; asst bot, Univ Chicago, 41-43; instr biol, Evansville Col, 43-44; plant pathologist, Bur Plant Indust, Soils & Agr Eng, USDA, Md, 44-46; from asst prof to assoc prof bot, 46-55, chmn dept, 47-77, prof, 55-80, EMER PROF BOT, DREW UNIV, 80- *Concurrent Pos:* Adj prof dept bot & cur, Florence & Robert Zurk Arbortum, Drew Univ, 80- *Mem:* Fel AAAS; Am Phytopath Soc; Bot Soc Am; Mycol Soc Am; World Acad Arts & Sci; Int Dendrol Soc. *Res:* Mycology; plant evolution. *Mailing Add:* Dept Bot Drew Univ Madison NJ 07940

ZUCKEKANDL, EMILE, b Vienna, Austria, July 4, 22; French citizen; m 50. REPETITIVE SEQUENCE FAMILIES, REGULATORY DIFFERENCES BETWEEN STATES OF CANCER CELLS. *Educ:* Univ Ill-Urbana, MS, 47; Sorbonne, PhD(biochem), 59. *Prof Exp:* Postdoctoral fel, Calif Inst Technol, 59-64; res dir, Nat Sci Res Ctr, Montpellier, France, 67-80; dir, Ctr Macromolecular Biochem, 65-75; PRES & RES PROF, LINUS PAULING INST SCI & MED, 80- *Concurrent Pos:* Consult genetics, Stanford Univ, 63, vis prof, 64; vis prof biol, Univ Del, 76; ed-in-chief, J Molecular Evolution, 71-; appointee, Molecular Biol Comn, DGRST, Paris, 67-70, Medical Biochem INSERM, Paris, 68-74, comt consult, Univ Paris. *Honors & Awards:* Order of Merit, French Govt. *Mem:* Fel AAAS; Soc Study Origins Life. *Res:* Molecular evolution: relation to organismal evolution; evolution of gene-gene interaction; issue of the molecular clock; questions of functions in non-coding DNA; interaction of the environment with genetic systems. *Mailing Add:* Linus Pauling Inst Sci & Med 440 Page Mill Rd Linus Pauling Inst CA 94306

ZUCKER, ALEXANDER, b Zagreb, Yugoslavia, Aug 1, 24; nat US; m 53; c 3. NUCLEAR PHYSICS. *Educ:* Univ Vt, BA, 47; Yale Univ, MS, 48, PhD(physics), 50. *Prof Exp:* Res asst physics, Yale Univ, 48-50; physicist, Oak Ridge Nat Lab, 50-53, sr physicist, 53-70, assoc dir, Electronuclear Div, 60-70; Ford prof physics, Univ Tenn, 68-72; exec dir, Environ Studies Bd, Nat Acad Sci-Nat Acad Eng, 70-72; dir heavy ion proj, Oak Ridge Nat Lab, 72-74, assoc dir phys sci, 73-88, actg dir, 88, ASSOC DIR NUCLEAR TECHNOLOGIES, OAK RIDGE NAT LAB, 89- *Concurrent Pos:* Guggenheim fel & Fulbright res scholar, 66-67; del, Pugwash Conf, 71; mem comt nuclear sci & nuclear physics panel, Physics Surv, Nat Res Coun; US del peaceful uses of atomic energy, USSR; mem-at-large, US Nat Comt, Int Union Pure & Appl Physics, 76-78; mem, res coordr coun, Gas Res Inst, 78-85; mem, Nuclear Physics Deleg to People's Repub China, Nat Acad Sci, 79; ed, Nuclear Sci Applications, 80; mem, adv panel technologies to reduce US mat import vulnerability, Off Technol Assessment, 82-85; mem, comt manpower, Nat Res Coun, 82-83; mem, coun energy eng res, Dept Energy, 83-; mem, White House Steel Indust/Nat Labs Initiative, 84-86; mem, Am Phys Soc, Panel Pub Affairs, 86-, Subpanel Int Sci Affairs, 88-; chmn, Am Soc Mech Engrs Nat Lab Technol Transfer Comt, 87- *Mem:* Fel AAAS; fel Am Phys Soc; fel Sigma Xi; Am Soc Metals. *Res:* Nuclear reactions with heavy ions; medium energy nuclear physics, including scattering and polarization of protons; few-nucleon interactions; high-current electronuclear machines; AVF and heavy ion cyclotrons; environment and public policy; managing physical research programs related to energy. *Mailing Add:* Oak Ridge Nat Lab PO Box 2008 Oak Ridge TN 37831-6248

ZUCKER, GORDON L(ESLIE), b Providence, RI, Nov 24, 29; m 51; c 3. MINERAL PROCESSING ENGINEERING. *Educ:* Mass Inst Technol, SB, 51; Univ Wis, MS, 54; Columbia Univ, DEngS, 59. *Prof Exp:* Engr, Semiconductor & Mat Div, Radio Corp, Am, 58-59; mem tech staff, Cent Res Labs, Tex Instruments Inc, 59-62; sr scientist, Sperry Rand Res Ctr, 62-64; group mgr, Res Div, Cadillac Gage Co, 64-66; dept head, Electronics Div, Union Carbide Corp, 66-70; tech dir, Tansitor Electronics Inc, 70-72; mgr engr, Seimens-USA, 73-75; tech dir, Mineral Res Ctr, Mont Tech Alumni Found, 76-79, PROF MINERAL PROCESSING ENG, MONT COL MINERAL SCI & TECHNOL, 75- *Mem:* Am Chem Soc; Am Inst Mining Metall & Petrol Engrs; Sigma Xi. *Res:* Digital process control; gold processing; industrial minerals. *Mailing Add:* Mont Col Mineral Sci & Technol Butte MT 59701

ZUCKER, IRVING, b Montreal, Que, Oct 2, 40; m 63; c 2. BIOLOGICAL RHYTHMS, NEUROENDOCRINOLOGY. *Educ:* McGill Univ, BSc, 61; Univ Chicago, PhD(biopsychol), 64. *Prof Exp:* Res assoc reprod physiol, Ore Regional Primate Res Ctr, 64-65; vis scientist, Sch Med, Univ Wis, 65; res assoc reprod physiol, Ore Regional Primate Res Ctr, 66; from asst prof to assoc prof psychol, 66-74, PROF PSYCHOL, UNIV CALIF, BERKELEY, 74- *Mem:* AAAS; Animal Behav Soc; Am Soc Mammalogists; Neurosci Soc; Soc Study Biol Rhythms. *Res:* Seasonal reproductive cycles; Circadian clocks; behavioral endocrinology; hibernation; circannual rhythms. *Mailing Add:* Dept Psychol Univ Calif Berkeley CA 94720

ZUCKER, IRVING H, b Bronx, NY, July 13, 42; m 70; c 3. PHYSIOLOGY. *Educ:* City Col New York, BS, 65; Univ Mo-Kansas City, MS, 67; New York Med Col, PhD(physiol), 72. *Prof Exp:* USPHS fel, 72-73, asst prof, 73-76, assoc prof, 76-83, PROF PHYSIOL, UNIV NEBR MED CTR OMAHA, 83- *Concurrent Pos:* Estab investr, Am Heart Asn, 77-82; mem, Great Planes regional rev group, Am Heart Asn; Masua Honor lectr, 81. *Mem:* Am Physiol Soc; Am Heart Asn; Soc Exp Biol & Med; Sigma Xi; Animal Welfare Comn. *Res:* Cardiovascular receptors and the neural control of blood volume. *Mailing Add:* Dept Physiol & Biophys Univ Nebr Med Ctr 42nd & Dewey Omaha NE 68105

ZUCKER, JOSEPH, b New York, NY, Apr 11, 28; m 53; c 3. SOLID STATE PHYSICS. *Educ:* Univ Miami, BS, 51; NY Univ, MS, 55, PhD(physics), 61. *Prof Exp:* Res assoc elec eng, Res Div, Col Eng, NY Univ, 51-55; engr, Sylvania Elec Prod, Inc, 55-57, sr engr, 57-59, res engr, 59-63, adv res engr, 63-65, eng specialist. 65-66, adv eng specialist, 66-68, MEM TECH STAFF, GEN TEL & ELECTRONICS LABS, 68- *Concurrent Pos:* Lectr, Polytech Inst Brooklyn, 60-65, adj prof, 65-72; adj prof, Hofstra Univ, 65-68. *Honors & Awards:* IR 100 Award, Indust Res Mag, 65. *Mem:* AAAS; NY Acad Sci; Am Phys Soc; Inst Elec & Electronics Engrs; Sigma Xi. *Res:* Transport properties of bulk semiconductors; optical probing of acoustoelectric interactions in piezoelectric semiconductors; microwave detectors, modulators and harmonic generators; photoelastic and electrooptic effect; acoustic surface wave devices; light emitting diodes; integrated optical devices; optical communication systems. *Mailing Add:* 4010 Calle Sonor Deste Apt 1B Laguna Hills CA 92653

ZUCKER, MARJORIE BASS, b New York, NY, June 10, 19; m 38; c 4. PHYSIOLOGY, HEMATOLOGY. *Educ:* Vassar Col, AB, 39; Columbia Univ, PhD(physiol), 44. *Prof Exp:* Instr physiol, Col Physicians & Surgeons, Columbia Univ, 42-44, res asst, 45-49; from asst prof to assoc prof, Col Dent, NY Univ, 49-55; assoc mem, Sloan-Kettering Inst Cancer Res, 55-63; assoc prof path, 63-71, PROF PATH, SCH MED, NY UNIV, 71- *Concurrent Pos:* Asst res dir, Eastern Div Res Lab, Am Nat Red Cross, 63-70; mem hemat study sect, NIH, 70-74, Int Comt Haemostasis & Thrombosis, 70-76 & review comt B, Nat Heart, Lung & Blood Inst, 76-80. *Mem:* Am Physiol Soc; Soc Exp Biol & Med; Am Soc Hemat; Int Soc Hemat; Int Soc Thrombosis & Haemostasis. *Res:* Platelets and blood coagulation. *Mailing Add:* Dept Path NY Univ Med Ctr New York NY 10016

ZUCKER, MARTIN SAMUEL, b New York, NY, Mar 15, 30; m 58; c 3. PLASMA PHYSICS, RADIATION PHYSICS. *Educ:* Cornell Univ, BEngPhys, 52; Univ Wis, MSc, 53, PhD(nuclear physics), 61. *Prof Exp:* Asst physicist, 58-62, consult, Radiation Div, Nuclear Eng Dept, 61-63, ASSOC PHYSICIST, BROOKHAVEN NAT LAB, 63- *Mem:* AAAS; Am Phys Soc; Am Asn Physics Teachers; Am Nuclear Soc; Sigma Xi. *Res:* Fast neutron polarization; electrical effects of nuclear radiations on matter; direct conversion of energy to electricity; statistical mechanics; chemical physics; accelerator development; scientific applications of computers. *Mailing Add:* Brookhaven Nat Lab 830 Upton NY 11973

ZUCKER, MELVIN JOSEPH, b Charleston, SC, May 6, 29; div; c 2. SOLID STATE PHYSICS. *Educ:* Brooklyn Col, BS, 51; Rutgers Univ, PhD(physics), 57. *Prof Exp:* Physicist, Airborne Instruments Lab, Cutler-Hammer, Inc, 57-59; mem tech staff, Semiconductor Div, Hughes Aircraft Co, 58-62; mem tech staff, Am-Standard Corp, NJ, 62-68; chmn dept math & physics, 69-77, ASSOC PROF, MERCER COUNTY COMMUNITY COL, NJ, 69- *Concurrent Pos:* Consult, Off Promoting Tech Innovation, State NJ. *Mem:* Am Phys Soc. *Res:* Impurities in superconductors; semiconductor devices; paramagnetic resonance; peizoresitivity studies in semiconductors. *Mailing Add:* Dept Math & Physics Mercer Community Col 1200 Old Trenton Rd PO Box B Trenton NJ 08690

ZUCKER, OVED SHLOMO FRANK, b Jerusalem, Isreal, June 24, 39; US citizen; m 60; c 3. ELECTRICAL ENGINEERING, PLASMA PHYSICS. *Educ:* City Univ NY, BEE, 65. *Prof Exp:* Mem tech staff elec eng, Lawrence Livermore Lab, 65-78; STAFF PHYSICIST, PHYSICS INT CO, 78-; PRES, ENERGY COMPRESSION RES CORP. *Concurrent Pos:* Consult physics, Lawrence Livermore Lab, 78-79 & La Jolla Inst, 79; ed, Energy Storage Compression & Switching Conf & Proceedings, 74. *Mem:* Am Phys Soc; Inst Elec & Electronics Engrs. *Res:* Relativistic electron beam; space-time energy compression; solid state physics. *Mailing Add:* Energy Compression Res Corp 990 Highland Dr Suite 101 Solana Beach CA 92075

ZUCKER, PAUL ALAN, b New York, NY, Nov 20, 44; m 73; c 1. PHYSICS, THEORETICAL PHYSICS. *Educ:* Univ Chicago, BS, 66; Stanford Univ, MS, 67, PhD(physics), 71. *Prof Exp:* Res assoc physics, Univ Ore, 70-72 & Univ Minn, 72-75; SR PHYSICIST, APPL PHYSICS LAB, JOHNS HOPKINS UNIV, 75- *Mem:* Am Phys Soc; AAAS; Inst Elec & Electronics Engrs; Sigma Xi. *Res:* Electroproduction and weak production of nucleon resonances; signal processing; Kalman filtering; system identification; estimation; model validation. *Mailing Add:* 112 Finale Terr Silver Spring MD 20901-5036

ZUCKER, ROBERT ALPERT, b New York, NY, Dec 9, 35; m 79; c 4. CLINICAL PSYCHOLOGY, PSYCHOPATHOLOGY. *Educ:* City Col New York, BCE, 56; Harvard Univ, PhD(clin psychol), 66. *Prof Exp:* Asst prof psychol, Rutgers Univ, 63-68; asst prof, 68-70, assoc prof, 70-75, PROF CLIN PSYCHOL, MICH STATE UNIV, 75-, CO-DIR CLIN TRAINING, 82- *Concurrent Pos:* Consult ed, J Studies Alcohol, 64-84; prin investr grants adolescent drinking, Nat Inst Mental Health, 66-71, study alcoholism & antisocial behav, Nat Inst Alcohol Abuse & Alcoholism, 87-; consult, Vet Admin Hosp, Battle Creek, Mich, 74-76; vis prof, Univ Tex, Austin, 75; bd dir, Nat Coun Alcoholism, 78-81; vis scholar alcohol problems, Nat Inst Alcohol Abuse & Alcoholism, 80-81; lectr, Nebr Symp Motivation, 86, Vanderbilt Alcohol Symp, 88; mem psychol rev comt, Nat Inst Alcohol Abuse & Acoholism, 88-; vis prof psychol, Univ Mich, 90-91. *Mem:* Fel Am Psychol Asn; fel Am Orthopsychiat Asn; Soc Life Hist Res Psychopath. *Res:* Longitudinal development of psychopathology with special interest in alcohol and drug abuse; personality influences on behavior; personality theory; psychotherapy. *Mailing Add:* Dept Psychol Mich State Univ East Lansing MI 48824

ZUCKER, ROBERT MARTIN, b New York, NY, May 13, 43; m. BIOPHYSICS, BIOLOGY. *Educ:* Univ Calif, Los Angeles, BS, 65, MS, 66, PhD(biophys), 70. *Prof Exp:* Res scientist hematol, Univ Calif, Los Angeles, 66-70; res, Max Planck Inst Protein & Leather Res, 70-72; sr scientist biophys & biol, Papanicolaou Cancer Res Inst, 72-84; SR SCIENTIST, MANTECH ENVIRON SCI, 85- *Concurrent Pos:* Assoc prof med, Univ Miami, 74-84. *Mem:* Am Chem Soc; Am Soc Cell Biol; NY Acad Sci; Am Asn Cancer Res. *Res:* Cellular biophysics; flow cytometry; image analysis. *Mailing Add:* USEPA Mail Drop 67 Research Triangle Park SC 27711

ZUCKER, ROBERT STEPHEN, b Philadelphia, Pa, Apr 18, 45; m 83; c 3. NEUROPHYSIOLOGY. *Educ:* Mass Inst Technol, SB, 66; Stanford Univ, PhD(neurol sci), 71. *Prof Exp:* From asst prof to assoc prof, 73-85, PROF NEUROBIOL, UNIV CALIF, BERKELEY, 85- *Concurrent Pos:* Hon asst res assoc, dept biophys, Univ Col, Univ London, Eng, 71-73; temp investr, Cellular Neurobiol Lab, Nat Ctr Sci Res, France, 73-74; res fel, Alfred P Sloan Found, 76-80; prin investr, NSF & NIH res grants, 77-; sr investr, Marine Biol Lab, Woods Hole, Mass, 80 & 81; Javit neurosci investr award, NIH, 87-94. *Mem:* Soc Neurosci; AAAS; Biophys Soc; Sigma Xi; Fed Am Scientists; Union Concerned Scientists. *Res:* Synaptic transmission; synaptic plasticity; excitable membrane biophysics; neurophysiological basis of behavior; neuronal calcium metabolism; egg fertilization and activation. *Mailing Add:* Univ Calif Dept Molecular & Cell Biol Box 111 Life Sci Addn Berkeley CA 94720

ZUCKER, STEVEN MARK, b New York, NY, Sept 12, 49. MATHEMATICS. *Educ:* Brown Univ, ScB, 70; Princeton Univ, PhD(math), 74. *Prof Exp:* Asst prof math, Rutgers Univ, 74-80; assoc prof math, Ind Univ, 81-84; PROF DEPT MATH, JOHNS HOPKINS UNIV. *Concurrent Pos:* NSF grant, 76- *Mem:* Am Math Soc. *Res:* Analytic methods in algebraic geometry; Hodge theory and the cohomology of projective varieties; differential geometry. *Mailing Add:* Dept Math Johns Hopkins Univ Baltimore MD 21218

ZUCKER, STEVEN WARREN, b Philadelphia, Pa, Apr 20, 48; m 74; c 2. COMPUTER VISION, ROBOTICS. *Educ:* Carnegie-Mellow Univ, BS, 69; Drexel Univ, MEGN, 72 & PhD(biomed eng), 75. *Prof Exp:* Fel Univ Md, 74-76, asst prof elect eng, 76-80, assoc prof, 80-84, PROF, MCGILL UNIV, 85- *Concurrent Pos:* Prin, Lognicom Corp. *Mem:* Fel Can Inst Advan Res; fel Inst Elec & Electronics Engrs. *Res:* Computational vision and biological perception. *Mailing Add:* Dept Elec Eng McGill Univ Montreal PQ H3A 2A7 Can

ZUCKERBERG, HYAM L, b New York, NY, Dec 5, 37; m 64; c 1. MATHEMATICS. *Educ:* Yeshiva Univ, BA & BHL, 59, MA, 61, PhD(math), 63. *Prof Exp:* Res mathematician, Davidson Lab, Stevens Inst Technol, 63-80; PROF MATH, LONG ISLAND UNIV, 80- *Mem:* Am Math Soc. *Res:* Conformal mapping; potential theory; Hilbert spaces; topology; Bergman kernel function. *Mailing Add:* Dept Math Long Island Univ Brooklyn NY 11201

ZUCKERBROD, DAVID, b Brooklyn, NY, Aug 27, 54; m 79. ELECTROCHEMISTRY. *Educ:* Rensselaer Polytech Inst, BS, 75, PhD(inorg chem), 82. *Prof Exp:* Sr engr, Westinghouse Elec Corp, 80-88; SR RES CHEMIST, W R GRACE & CO, COLUMBIA, MD, 88- *Mem:* Am Chem Soc. *Mailing Add:* 2705 Copperfield Ct Baltimore MD 21209

ZUCKER-FRANKLIN, DOROTHEA, b Berlin, Ger, Aug 9, 29; US citizen; m 56; c 1. CELL BIOLOGY. *Educ:* Hunter Col, BA, 52; New York Med Col, MD, 56. *Prof Exp:* Intern med, Philadelphia Gen Hosp, Pa, 56-57; resident, Montefiore Hosp, New York, 57-59, USPHS res fel hemat, 59-61; USPHS res fel electron micros, 61-63, from asst prof to assoc prof, 63-74, PROF MED, SCH MED, NY UNIV, 74- *Concurrent Pos:* Attend physician, Bellevue Hosp, New York, 63-; attend physician, Univ Hosp, New York, 63-; spec consult path training comt, Nat Inst Gen Med Sci; USPHS res career scientist award, 66-76; Nat Inst Arthritis & Metab Dis res grant; assoc ed, Blood, 64-75 & 81, J Reticuloendothelial Soc, 65-73, Am J Path, 78-, Am J Med & J Hemat Oncol; mem blood prod comt, Food & Drug Admin, 81. *Mem:* Am Fedn Clin Res; Am Soc Hemat; Am Soc Exp Path; Am Soc Clin Invest; Am Asn Physicians. *Res:* Hematology, including white blood cells, coagulation of blood, and platelets; immunology; electron microscopy. *Mailing Add:* Dept Med NY Univ Sch Med New York NY 10016

ZUCKERMAN, BENJAMIN MICHAEL, b New York, NY, Aug 16, 43. ASTRONOMY. *Educ:* Mass Inst Technol, SB & SM, 63; Harvard Univ, PhD(astron), 68. *Prof Exp:* From asst prof to prof physics & astron, Univ Md, College Park, 68-83; PROF ASTRON, UNIV CALIF, LOS ANGELES, 82- *Concurrent Pos:* Alfred P Sloan res fel, 72-74; consult, Jet Propulsion Lab; Guggenheim Found fel, 77. *Honors & Awards:* Helen B Warner Prize, Am Astron Soc, 75; Muhlmann Prize, Astron Soc Pac, 86. *Mem:* Int Astron Union; Int Union Radio Sci; Am Astron Soc. *Res:* Infrared and radio astronomy. *Mailing Add:* Dept Astron Univ Calif 405 Hilgard Ave Los Angeles CA 90024

ZUCKERMAN, ISRAEL, b St Louis, Mo, Oct 10, 24; m 54; c 2. MATHEMATICS. *Educ:* City Col New York, BBA, 46; Brooklyn Col, MA, 58; Rutgers Univ, PhD(math), 63. *Prof Exp:* Instr math, Brooklyn Col, 56-58 & Rutgers Univ, 59-63; asst prof, Queens Col, 63-65 & Vassar Col, 65-66; assoc prof, 66-77, PROF MATH, LONG ISLAND UNIV, BROOKLYN CTR, 77- *Mem:* Math Asn Am; Am Math Soc. *Res:* Differential algebra; ring theory. *Mailing Add:* Dept Math Long Island Univ Brooklyn Ctr Brooklyn NY 11201

ZUCKERMAN, JOAN ELLEN, IMMUNO-PHARMACOLOGY, MONOCLONAL ANTIBODIES. *Educ:* Univ Calif, Davis, PhD(pharmacol & toxicol), 80. *Prof Exp:* RES ASSOC, M D ANDERSON & TUMOR INST, 84- *Mailing Add:* 326 Cobble Creek Curve Newark DE 19702

ZUCKERMAN, LEO, b Brooklyn, NY, July 3, 17; m 42; c 3. LABORATORY MEDICINE, TECHNICAL MANAGEMENT. *Educ:* Brooklyn Col, BA, 42. *Prof Exp:* Foreman blood fractionation, E R Squibb & Sons, 42-44, res assoc group leader, 44-52; res assoc, Ortho Res Found, 52-58, supvr blood prod & biol mfg, Ortho Pharmaceut Corp, 58-62, mgr fractionation dept, Ortho Diag, 62-72, mfg dir biochem prod, 72-73, mfg dir serol prod, 73-76, dir mfg serv int, 76-78, dir tech serv int, 78-82, dir regulatory compliance, Ortho Diag Systs, 82-84; RETIRED. *Concurrent Pos:* Consult, 84- *Mem:* Am Chem Soc; Am Inst Chem; Am Asn Clin Chem; Soc Cryobiol; NY Acad Sci; Am Asn Blood Banks. *Res:* Coagulation; immunology; serology; chromatography; electrophoresis; lyophilization; protein isolation and characterization; clinical chemisty; cryobiology; blood fractionation. *Mailing Add:* 1983 Holland Brk Rd W Somerville NJ 08876-3845

ZUCKERMAN, MARTIN MICHAEL, b Brooklyn, NY, June 27, 34; m 60. MATHEMATICAL LOGIC. *Educ:* Brandeis Univ, BA, 55; Brown Univ, MA, 60; Yeshiva Univ, PhD(math), 67. *Prof Exp:* Mathematician, Int Elec Co, 61-63; instr math, NY Univ, 63-66; asst prof, Hunter Col, 67-68; asst prof, 68-72, ASSOC PROF MATH, CITY COL NEW YORK, 72- *Mem:* Am Math Soc; Asn Symbolic Logic; Math Asn Am. *Res:* Set theory. *Mailing Add:* Dept Math City Col New York Convent & 138th St New York NY 10031

ZUCKERMAN, SAMUEL, b New York, NY, Oct 22, 15; m 38; c 2. ORGANIC CHEMISTRY. *Educ:* City Col, BS, 37; Polytech Inst Brooklyn, MS, 42, PhD, 50. *Prof Exp:* Chemist, H Kohnstamm & Co, 36-50, tech dir & plant mgr, Brooklyn Div, 50-60, vpres & dir, 59-84; RETIRED. *Honors & Awards:* Medal Award, Soc Cosmetic Chem, 70. *Mem:* Am Chem Soc; Soc Cosmetic Chem. *Res:* Organic synthesis of dyestuffs; chemical microscopy; cosmetic colors for camouflage; certified food, drug and cosmetic colors. *Mailing Add:* 18 Dubonnet Rd Valley Stream NY 11581

ZUCKERMAN, STEVEN H, SOMATIC CELL GENETICS, MICROPHAGE BIOLOGY. *Educ:* Univ Minn, PhD(microbiol & immunol), 77. *Prof Exp:* SR SCIENTIST, ELI LILLY & CO, 83- *Mailing Add:* Dept Immunol Lilly Res Labs Bldg 98 Merrill St Indianapolis IN 46285

ZUCKERMANN, MARTIN JULIUS, b Berlin, Ger, July 7, 36; m 60; c 4. PHYSICS. *Educ:* Oxford Univ, BA, 60, PhD(phyiscs), 64. *Prof Exp:* Fel, Univ Chicago, 64-65; asst prof physics, Univ Va, 65-67; lectr, Imp Col, Univ London, 67-69; assoc prof, 69-80, PROF PHYSICS, MCGILL UNIV, 80- *Mem:* Am Inst Physics. *Res:* Theoretical solid state physics; investigation into superconductivity and magnetism in disordered systems; weather physics. *Mailing Add:* Dept Physics McGill Univ 853 Sherbrooke St W Montreal PQ H3A 2M5 Can

ZUDKEVITCH, DAVID, b Hadera, Israel, Jan 15, 30; US citizen; c 1. CHEMICAL ENGINEERING. *Educ:* Israel Inst Technol, BSc, 53, Dipl Ing, 54; Polytech Inst Brooklyn, MChE, 58, PhD(chem eng), 59. *Prof Exp:* Plant engr, Fertilizers & Chem Co, Israel, 53; lab dir, Alliance Tire & Rubber Co, 53-55; engr, Exxon Res & Eng Co, 59-62, eng assoc & consult thermodyn, 62-75; sr eng assoc, Allied Chem Corp, 75-85; SR SCIENTIST, DEPT CHEM ENG, COLUMBIA UNIV, 79- *Concurrent Pos:* Adj prof, NJ Inst Technol; mem comt tech data, Am Petrol Inst; mem, Nat Acad Rev Panel for Nat Bur Standards Heat Div, 76- 77. *Mem:* Am Chem Soc; Am Inst Chem Engrs; Am Petrol Inst; Gas Processors Asn. *Res:* Development of correlations for physical and thermodynamic properties, especially phase equilibria; evaluation and development of separation and transport processes; evaluation and design of pollution abatement processes and energy utilization. *Mailing Add:* 24 Magnolia Ave Denville NJ 07834-9328

ZUECH, ERNEST A, b Frontenac, Kans, Nov 17, 34; m 56; c 3. ORGANIC CHEMISTRY. *Educ:* Kans State Col, Pittsburg, BS, 55, MS, 56; Iowa State Univ, PhD(org chem), 60. *Prof Exp:* Asst org chem, Ohio State Univ, 60-61; chemist, 61-68, sect mgr, 68-78, br mgr, 78-80, DIR PETROL RES, PHILLIPS PETROL CO, 80- *Mem:* Am Chem Soc. *Res:* Organometallic chemistry. *Mailing Add:* 1317 Harned Dr Bartlesville OK 74006

ZUEHLKE, CARL WILLIAM, b Bonduel, Wis, Oct 28, 16; wid; c 2. ANALYTICAL CHEMISTRY. *Educ:* Univ Wis, BS, 38; Univ Mich, MS, 40, PhD(anal chem), 42. *Prof Exp:* Chief chemist, Methods Lab, Gen Chem Div, Allied Chem & Dye Corp, 42-45, E St Louis Works, 45-46, asst mgr, Chem Control Div, 46-48; res assoc, Eastman Kodak Co, 48-61, asst head, Chem Div, Res Labs, 61-68, head, Anal Sci Div, Res Labs, 68-79; RETIRED. *Concurrent Pos:* Chmn, Gordon Res Conf Anal Chem, 67. *Mem:* Am Chem Soc. *Res:* Analytical chemistry of germanium; analysis for micro amounts of mercury. *Mailing Add:* 92 Skyview Lane Rochester NY 14625

ZUEHLKE, RICHARD WILLIAM, b Milwaukee, Wis, June 17, 33; m 55; c 3. INFORMATION MANAGEMENT. *Educ:* Lawrence Col, BS, 55; Univ Minn, PhD(chem), 60. *Prof Exp:* From instr to asst prof chem, Lawrence Univ, 58-68; chmn dept chem, Univ Bridgeport, 68-73, acad liaison officer, 73-74, Remington prof, 68-80; assoc marine scientist, Grad Sch Oceanog, 80-85, ASST TO DIR, GORDON RES CONFERENCES, GORDON RES CTR, UNIV RI, 90- *Concurrent Pos:* Consult, Kimberly-Clark Corp, 60-62; NSF fac fel, Univ Pittsburgh, 66-67; consult, United Illum Co, 69-71 & Wooster Davis & Cifelli Chem Specialties Corp, 70-73; consult, Oxford Univ Press, 74-76 & Sperry Remington Co, 76-77; vis prof oceanog, Univ RI, 76-77; pres, TRC Consults, 85-90. *Mem:* AAAS; Am Chem Soc; fel Am Inst Chem. *Res:* Computer simulation and modeling; instrument design; laboratory automation; information management software. *Mailing Add:* Gordon Res Conferences Gordon Res Ctr Univ RI Kingston RI 02881-0801

ZUG, GEORGE R, b Carlisle, Pa, Nov 16, 38; m 60; c 2. HERPETOLOGY, MORPHOLOGY. *Educ:* Albright Col, BS, 60; Univ Fla, MS, 63; Univ Mich, PhD(zool), 68. *Prof Exp:* Instr zool, Univ Mich, 68; asst cur herpet, 69-73, chmn, Dept Vert Zool, 77-83, CUR HERPET, MUS NATURAL HIST, SMITHSONIAN INST, 73- *Mem:* Soc Study Evolution; Soc Study Amphibians & Reptiles; Soc Syst Zool; Am Soc Ichthyol & Herpet; Herpetologists League. *Res:* Systematics and evolution of reptiles and amphibians, particularly chelonians; morphology of reptiles and amphibians and functional relationships; locomotion of vertebrates. *Mailing Add:* Div Amphibians & Reptiles MRC-NH162 Nat Mus Natural Hist Smithsonian Inst Washington DC 20560

ZUGIBE, FREDERICK T, b Garnerville, NY, May 28, 28; m 51; c 7. FORENSIC MEDICINE, CARDIOVASCULAR DISEASES. *Educ:* St Francis Col, BS, 52; Univ Chicago, MS, 59, PhD(anat), 60; WVa Univ, MD, 68; Am Bd Path & Family Practice, dipl, 78, cert anat path, 80 & forensic path, 82. *Prof Exp:* Res histologist, Lederle Labs, Am Cyanamid Co, 50-52, chemist, 53-55; asst ophthal, Col Physicians & Surgeons, Columbia Univ, 55; res histochemist in chg atherosclerosis sect, Vet Admin Hosp, Downey, Ill, 56-60; dir cardiovasc res, Vet Admin Hosp, Pittsburgh, 61-69; CHIEF MED

EXAMR, ROCKLAND COUNTY, NY, 69- *Concurrent Pos:* Adj prof, Duquesne Univ; asst res prof, Sch Med, Univ Pittsburgh, 61-69; adj assoc prof path, Columbia Univ, 69-, Col Physicians & Surgeons, 75-; consult path, ABC Labs, 72-79; consult, comt connective tissue, skeletal & muscle, Vet Admin & sci res technol, 74-77, dir, Angelus Path Lab, 72- & Acad Labs, 76-78; consult path & physician, Police Surgeon Fire, Surgeon & Ambulance Corp Physician; mem, coun arteriosclerosis, Am Heart Asn; continuing educ award path, Am Soc Clin Path & Am Col Path, 75-77, 77-80, 81-84 & 85-88; vis prof, Dept Path, Univ Wis Med Sch, 86-; res grants, NIH, Am Heart Asn & Am Cancer Soc. *Honors & Awards:* Physicians Recognition Award, AMA, 71-74, 74-77, 77-80 & 81-84, 85-88; Shields Law Enforcement Award, 73; Distinguished Serv Award, Rockland County & Am Legion Award for Serv in Nicaragua During Earthquake, 73; Am Heart Asn Serv Recognition Award, 74; Law Enforcement Award, Police Chiefs Asn, 80. *Mem:* Histochem Soc; fel Am Heart Asn; NY Acad Sci; fel Am Acad Forensic Sci; fel Am Col Cardiol; fel Col Am Pathologists; Nat Asn Med Examr; Asn Scientist & Scholars Int (pres); AAAS; Sigma Xi. *Res:* Atherosclerosis and aging research; carbohydrate, lipid and enzyme histochemistry; ultramicrohistochemistry; cardiovascular research; histochemistry and forensic pathology research; Shroud of Turn research; crucifixion research; numerous publications in medical and scientific journals and several book chapters in medical and scientific books and textbooks. *Mailing Add:* One Angelus Dr Garnerville NY 10923

ZUHR, RAYMOND ARTHUR, b New York, NY, May 20, 40; m 82; c 3. SURFACE PHYSICS, THIN FILM DEPOSITION. *Educ:* Rensselaer Polytech Inst, BME, 62, MS, 72, PhD(physics), 74. *Prof Exp:* Assoc, Rensselaer Polytech Inst, 74-76; res assoc, State Univ NY, Albany, 76-77; MEM STAFF, OAK RIDGE NAT LAB, 77- *Mem:* Am Phys Soc; Am Vacuum Soc; Mat Res Soc. *Res:* Solid state physics and formation of thin films; study of single crystal surfaces using Rutherford backscattering and electron spectroscopy; plasma-wall interactions in magnetic confinement fusion devices using Rutherford backscattering and nuclear reaction analysis. *Mailing Add:* Oak Ridge Nat Lab Bldg 3137 PO Box 2008 Oak Ridge TN 37831-6057

ZUICHES, JAMES J, b Eau Claire, Wis, Mar 24, 43; m 67; c 2. DEMOGRAPHY. *Educ:* Univ Portland, BA, 67; Univ Wisc, Madison, MS, 69, PhD (sociol), 73. *Prof Exp:* Asst prof sociol, Mich State Univ, 71-77, assoc prof sociol, 71-82, prof sociol, 71-82; prog dir, Nat Sci Found, 79-82; assoc dir & prof rural sociol, Cornell Univ Ag Exp Station & Office for Res Rural Sociol, 82-86; DIR, AGR RES CTR, WASH STATE UNIV, 86- *Concurrent Pos:* Vis prof sociol, Univ Surrey, Guildford, 78-79; assoc prog dir, Nat Sci Found, 79-82; consult, Univ Ga, 81-87, Univ Minn, 81-87, Cornell Univ, 81-87. *Mailing Add:* 815 Meadow Vale Dr Pullman WA 99163

ZUIDEMA, GEORGE DALE, b Holland, Mich, Mar 8, 28; m 53; c 4. GENERAL SURGERY. *Educ:* Hope Col, AB, 49; Johns Hopkins Univ, MD, 53; Am Bd Surg, dipl, 60. *Hon Degrees:* DSc, Hope Col, 69. *Prof Exp:* Intern surg, Mass Gen Hosp, 53-54, asst resident, 54 & 57-58, chief resident, 59; surgeon-in-chief, Johns Hopkins Hosp, 64-84, prof surg & dir dept, Sch Med, Johns Hopkins Univ, 64-84; from asst prof to prof, 60-64, PROF SURG, MED SCH & VPROVOST MED AFFAIRS, UNIV MICH, 84- *Concurrent Pos:* Fel, Harvard Med Sch, 59; attend surgeon, Ann Arbor Vet Admin Hosp, 60-64; USPHS sr res fel, 61, career develop award, 63; Markle scholar acad med, 61-66; consult, Walter Reed Army Med Ctr, Sinai & Baltimore City Hosps & clin ctr, NIH; asst ed, J Surg Res, 60, ed, 66-; co-ed-in-chief, Surg, 75- *Honors & Awards:* Russel Award, Univ Mich, 63; hon fel, Royal Col Surg, Ireland, 72. *Mem:* Inst Med-Nat Acad Sci; fel Am Col Surgeons; Asn Am Med Cols; Soc Univ Surgeons; Am Surg Asn; Am Soc Clin Surg. *Res:* Cardiovascular and acceleration physiology; space medicine; gastrointestinal and hepatic physiology. *Mailing Add:* Med Sci 1 Bldg M3246 Univ Mich Ann Arbor MI 48109

ZUK, MARLENE, b Philadelphia, Pa, May 20, 56; m 86. BEHAVIORAL ECOLOGY, EVOLUTIONARY BIOLOGY. *Educ:* Univ Calif, Santa Barbara, BA, 77; Univ Mich, MS, 83, PhD(zool), 86. *Prof Exp:* Postdoctoral, Univ NMex, 87-89; ASST PROF BIOL, UNIV CALIF, RIVERSIDE, 89- *Honors & Awards:* Outstanding Young Investr, Am Soc Naturalists, 88. *Mem:* Soc Study Evolution; Animal Behav Soc. *Res:* Evolution of sexual behavior, especially the effect of parasites on host sexual selection; mate choice; insect song. *Mailing Add:* Dept Biol Univ Calif Riverside CA 92521

ZUK, WILLIAM, b New York, NY, July 6, 24; m 48; c 4. STRUCTURAL ENGINEERING. *Educ:* Cornell Univ, BSc, 44, PhD(struct mech), 55; Johns Hopkins Univ, MSc, 47. *Prof Exp:* Asst prof civil eng, Univ Denver, 50-52; from assoc prof to prof civil eng, 55-64, PROF ARCHIT & DIR ARCHIT TECHNOL, UNIV VA, 64- *Concurrent Pos:* Consult, Martin Aircraft Co, 55 & Va Hwy Res Coun, 56- *Mem:* Fel Am Soc Civil Engrs. *Res:* Structural dynamics; kinetic structures; construction robotics; knowledge engineering. *Mailing Add:* Campbell Hall Univ Va Charlottesville VA 22903

ZUKAS, EUGENE G, b Armstrong, Pa, Dec 19, 21. ALLOYS. *Educ:* Univ Pittsburgh, BS, 43, MS, 50; Lehigh Univ, PhD(metall), 52. *Prof Exp:* Staff mem, Los Alamos Sci Lab, 52-86; RETIRED. *Mem:* Fel Am Soc Metals Int. *Mailing Add:* 1847 W Camino Urbano Green Valley AZ 85614

ZUKEL, WILLIAM JOHN, b Northampton, Mass, June 8, 22. CARDIOVASCULAR DISEASES. *Educ:* Univ Mass, BS, 43; Hahnemann Med Col, MD, 47; London Sch Hyg & Trop Med, dipl pub health, 61. *Hon Degrees:* DSc, Univ PR, 85. *Prof Exp:* Intern & resident med, Newton-Wellesley Hosp, Newton, Mass, 47-49; from asst med officer in chg to med officer in chg, Newton Heart Prog, USPHS, Mass, 49-51, from asst chief to actg chief, Heart Dis Control Prog, DC, 51-52, asst med, Mass Gen Hosp, 52-53, asst med, Albany Med Col, 53-55, chief oper res, Heart Dis Control Prog, DC, 55-57, asst dir, Nat Heart Inst, 57-58, prog asst, Off Surgeon Gen, 58-60, assoc dir collab studies, Nat Heart Inst, 62-67, assoc dir epidemiol & biomet, 67-69, assoc dir clin appln & prev, 69-79, assoc dir prog coord & plan, 79-81, assoc dir sci progs, Div Heart & Vascular Dis, Nat Heart & Lung Inst, 81-84, dep dir, 84-88; RETIRED. *Concurrent Pos:* USPHS fel cardiol, NY State Dept Health, 53-55; assoc clin prof prev med & community health, Sch Med, George Washington Univ, 63-72, assoc clin prof epidemiol & environ health, 72-80. *Honors & Awards:* White Award , Asn Mil Surgeons,85. *Mem:* AMA; fel Am Heart Asn; fel Am Pub Health Asn; fel Am Col Cardiol; fel Int Soc Cardiol; Asn Mil Surgeons US. *Res:* Clinical trials, prevention and epidemiological studies. *Mailing Add:* 6600 Pyle Rd Bethesda MD 20817

ZUKER, MICHAEL, b Montreal, Que, Apr 1, 49; m 78; c 2. BIOMATHEMATICS. *Educ:* McGill Univ, BSc, 70; Mass Inst Technol, PhD(probability theory), 74. *Prof Exp:* Asst res off biomath, 74-80, assoc res off biomath, 87, SR RES OFF BIOMATH, NAT RES COUN CAN, 87- *Concurrent Pos:* Sessional lectr, Carleton Univ, 75-76; assoc ed, Bull Math Biol, 84-86 & Computer Applications Biosci, 87- *Mem:* Soc Math Biol; Can Appl Math Soc; fel Can Inst Advan Res. *Res:* Algorithms for molecular genetics; biomolecular sequence analysis. *Mailing Add:* Nat Res Coun M-54 Ottawa ON K1A 0R6 Can

ZUKIN, STEPHEN R, b Philadelphia, Pa, Aug 15, 48; c 2. PSYCHIATRY, NEUROSCIENCE. *Educ:* Haverford Col, BA, 70; Johns Hopkins Univ, Baltimore, MD, 74. *Prof Exp:* Asst prof psychiat, State Univ NY, 77-79 & Mt Sinai Sch Med, 79-82; assoc prof psychiat, 82-87, PROF PSYCHIAT & NEUROSCI, ALBERT EINSTEIN COL MED, YESHIVA UNIV, 87- *Concurrent Pos:* Dir res, Bronx Psychiat Ctr, 83-; assoc prof neurosci, Albert Einstein Col Med, Yeshiva Univ, 84-87. *Honors & Awards:* Citation Classic, Inst Sci Info, 91. *Mem:* Am Col Neuropsychopharmacol; Am Soc Pharmacol & Exp Therapeut; Soc Neurosci; Am Psychiat Asn. *Res:* Function and regulation of brain NMDA receptors; molecular mechanisms of psychotic illness; mechanisms of action of psychotomimetic drugs; development of NMDA-related treatments of neuropsychiatric diseases. *Mailing Add:* Dept Psychiat & Neurosci Albert Einstein Col Med 1300 Morris Park Ave F111 Bronx NY 10461

ZUKOSKI, CHARLES FREDERICK, b St Louis, Mo, Jan 26, 26; m 53; c 3. SURGERY. *Educ:* Univ NC, AB, 47; Harvard Med Sch, MD, 51. *Prof Exp:* From intern surg to resident, Roosevelt Hosp, New York, 51-52; resident, Univ Ala Hosp, 55-58, instr, Univ, 58-59; res fel, Med Col Va, 59-61; from asst prof to assoc prof, Sch Med, Vanderbilt Univ, 61-68; assoc prof, Univ NC, Chapel Hill, 68-69; PROF SURG, COL MED, UNIV ARIZ, 69- *Concurrent Pos:* Nat Inst Neurol Dis & Blindness spec trainee, 59-61; Nat Inst Allergy & Infectious Dis spec fel, 66-67; Josiah Macy Jr sr fac fel, 76-77. *Mem:* Am Col Surg; Soc Univ Surg; Am Surg Asn; Am Soc Exp Path; Transplantation Soc. *Res:* Homotransplantation; renal allografts; experimental and clinical research. *Mailing Add:* Dept Surg Univ Ariz Col Med Tuscon AZ 85724

ZUKOSKI, EDWARD EDOM, b Birmingham, Ala, June 29, 27; m 60; c 3. FLUID MECHANICS, COMBUSTION. *Educ:* Harvard Univ, BS, 50; Calif Inst Technol, MS, 51, PhD(aeronaut), 54. *Prof Exp:* Sr res engr, Jet Propulsion Lab, Calif Inst Technol, 54-57, from asst prof to assoc prof eng, 57-66, PROF JET PROPULSION & MECH ENG, CALIF INST TECHNOL, 66- *Concurrent Pos:* Mem fire res comt, Nat Acad Sci, 65-71. *Mem:* Fel Am Inst Aeronaut & Astronaut; Int Combustion Inst. *Res:* Combustion of air-fuel mixtures; interaction of transverse jets with supersonic flows; separation of turbulent boundary layers; electrical phenomena in high density seeded plasma; uncontrolled fires in buildings. *Mailing Add:* Dept Jet Propulsion Calif Inst Technol 1201 E California Blvd Pasadena CA 91125

ZUKOTYNSKI, STEFAN, b Warsaw, Poland, Feb 26, 39; m 68; c 1. SOLID STATE PHYSICS, ELECTRICAL ENGINEERING. *Educ:* Univ Warsaw, Mag, 61, PhD(physics), 66. *Prof Exp:* Nat Res Coun Can fel, Univ Alta, 66-68; from asst prof to assoc prof, 68-81, PROF ELEC ENG, UNIV TORONTO, 81- *Concurrent Pos:* Consult, Elec Eng Consociates Ltd, 71-; resident vis, Bell Labs, 77-78; pres, Torion Plasma Corp. *Mem:* Can Asn Physicists; Am Phys Soc; Inst Elec & Electronics Engrs. *Res:* Electrical and optical properties of semiconductors and semiconductor devices. *Mailing Add:* Dept Elec Eng Univ Toronto Toronto ON M5S 1A4 Can

ZUKOWSKI, CHARLES ALBERT, b Buffalo, NY, Aug 17, 59; m 83; c 2. INTEGRATED CIRCUIT DESIGN, CIRCUIT SIMULATION. *Educ:* Mass Inst Technol, BS & MS, 82, PhD(elec eng), 85. *Prof Exp:* Res assoc elec eng, Mass Inst Technol, 85; asst prof, 85-90, ASSOC PROF ELEC ENG, COLUMBIA UNIV, 90- *Concurrent Pos:* NSF presidential young investr, 87; mem, Tech Comt, Int Conf Computer Design, 88- & Custom Integrated Circuits Conf, 90-; consult, Int Bus Mach Corp, 89- *Mem:* Inst Elec & Electronics Engrs. *Res:* Design of high-performance digital integrated circuits, especially for the application of telecommunications; computer-aided design challenges such as circuit simulation, optimization and timing analysis. *Mailing Add:* Dept Elec Eng Columbia Univ New York NY 10027-6699

ZULALIAN, JACK, b New York, NY, Apr 21, 36. METABOLISM. *Educ:* Queens Col, NY, BS, 57; Purdue Univ, Lafayette, PhD(chem), 62. *Prof Exp:* Res assoc natural prod biosynthesis, Korman Res Labs, Albert Einstein Med Ctr, Philadelphia, 62-66; SR RES CHEMIST, METAB, AGR DIV, AM CYANAMID CO, 66- *Mem:* Am Chem Soc. *Res:* Organic synthesis; metabolism of organic compounds designed for use in agriculture as herbicides; pesticides and animal health; radiotracter synthesis. *Mailing Add:* Am Cyanamid Co PO Box 400 Princeton NJ 08540

ZULEEG, RAINER, b Erlangen, Ger, Sept 23, 27; US citizen; m 58; c 2. RADIATION EFFECTS IN SEMICONDUCTORS, THIN-FILM FERROELECTRIC MATERIALS. *Educ:* Tohoku Univ, Japan, PhD(solid state physics), 72. *Prof Exp:* Staff dir, McDonnell Douglas, 67-90; CONSULT SEMICONDUCTORS, 90- *Concurrent Pos:* Lectr, Univ Calif, Irvine, 72-89; guest prof, Royal Melbourne Inst Technol, 75 & 82 & Univ Davisburg, Ger, 87; sr fel, McDonnell Douglas, 86; adj prof, Univ Colo, 90- *Mem:* Am Phys Soc; Inst Elec & Electronics Engrs; Electrochem Soc. *Res:* Semiconductor devices and integrated circuits; radiation effects of microelectronic components and thin-film ferroelectric materials. *Mailing Add:* 33571 Avenida Calita San Juan Capistrano CA 92675

ZULL, JAMES E, b North Branch, Mich, Sept 29, 39; m 61, 68; c 3. BIOCHEMISTRY, CELLULAR BIOLOGY. *Educ:* Houghton Col, BA, 61; Univ Wis, MS, 63, PhD(biochem), 66. *Prof Exp:* Fel biochem, Univ Wis, 65-66; from asst prof to assoc prof, 66-72, PROF BIOL, CASE WESTERN RESERVE UNIV, 77- *Concurrent Pos:* NIH career develop award, 71-76; vis prof biochem, Inst Pathophysiol, Bern, Switz, 74-75. *Mem:* Am Soc Bone Mineral Res; Am Soc Biol Chemists; AAAS; Protein Soc. *Res:* Membrane-hormone interactions; hormone mechanisms; peptide structure; lysosome function; calcium wet abolisms; molecular biology of peptide hormone receptors. *Mailing Add:* Dept Biol & Biochem Case Western Reserve Univ 2040 Adelbert Rd Cleveland OH 44106

ZULLO, VICTOR AUGUST, b San Francisco, Calif, July 24, 36; m 87. INVERTEBRATE PALEONTOLOGY, ZOOLOGY. *Educ:* Univ Calif, AB, 58, MA, 60, PhD(paleont), 63. *Prof Exp:* Fel systs-ecol prog, Woods Hole Marine Biol Lab, 62-63, resident systematist, 63-67, asst dir prog, 64-66; assoc cur, dept geol, Calif Acad Sci, 67-70, chmn, 68-70; dir, Prog Environ Sci, 71, chmn dept geol, 83-89, PROF GEOL, UNIV NC, WILMINGTON, 71- *Concurrent Pos:* Res assoc, Los Angeles County Mus Natural Hist; fel, Calif Acad Sci. *Mem:* Geol Soc Am; Sigma Xi; Paleont Soc; Paleont Res Inst; Soc Econ Paleontologists & Mineralogists; Crustacean Soc. *Res:* Systematics, evolution and biogeography of Cirripedia; paleontology, sequence stratigraphy and biostratigraphy of Cenozoic marine deposits. *Mailing Add:* Dept Earth Sci Univ NC Wilmington NC 28403

ZUMAN, PETR, b Prague, Czech, Jan 13, 26; m 51; c 2. ELECTROCHEMISTRY, PHYSICAL ORGANIC CHEMISTRY. *Educ:* Charles Univ, Prague, RNDr(chem), 50; Czech Acad Sci, DrSc, 62; Univ Birmingham, DSc, 68. *Prof Exp:* Head org polarography div, J Heyrovsky Inst Polarography, Czech Acad Sci, Prague, 50-68; PROF CHEM, CLARKSON UNIV, 70- *Concurrent Pos:* Sr vis fel, Univ Birmingham, 66-70; distinguished vis prof, Brooklyn Polytech Inst, 67; consult, Xerox Corp, 72-76, Technicon, 74-, IBM Corp, 75-76 & 81 & Texaco, 85-; vis prof, Univ Amsterdam, 77, Free Univ Brussels, 77, Univ Utrecht, 79 & 81, Tech Univ Lyngby, 83, Univ Linz, 83, Deakin Univ, Australia, 87, Univ Bologna, 87, 88, 89. *Honors & Awards:* Heyrovsky Medal, Czech Acad Sci, 60 & 90; Coover Lectr, Am Chem Soc, 73; Theophilus Redwood Lectr, The Chem Soc, 75; Benedetti-Pichler Award, Am Microchem Soc, 75; Gold Medal Award, Am Electroplaters Soc, 76. *Mem:* Am Chem Soc; Sigma Xi; fel Royal Soc Chem; Electrochem Soc; Int Soc Electrochem; fel Chem Inst Can; Int Union Pure & Appl Chem. *Res:* Use of polarography and other electrochemical and optical methods for study of reactivity, equilibria, kinetics and mechanisms of reactions of organic compounds. *Mailing Add:* Dept Chem Clarkson Univ Potsdam NY 13699

ZUMBERGE, JAMES FREDERICK, b Ann Arbor, Mich, Jan 26, 53; m 84; c 2. GEODESY. *Educ:* Univ Mich, BS, 74; Calif Inst Technol, PhD(physics), 81. *Prof Exp:* mem tech staff, MDH Industs, Inc, 81-82, appl sci mgr, 82-86, dir, 86-85, vpres, 88-90; MEM TECH STAFF, JET PROPULSION LAB, 90- *Mem:* Am Phys Soc. *Res:* Applications of the global positioning system to geodesy. *Mailing Add:* Jet Propulsion Lab 238-625 Pasadena CA 91109

ZUMBERGE, JAMES HERBERT, b Minneapolis, Minn, Dec 27, 23; m 47; c 4. US ARCTIC RESEARCH POLICY, ANTARCTIC GEOPOLITICS. *Educ:* Univ Minn, BA, 46, PhD(geol), 50. *Hon Degrees:* LLD, Grand Valley Col, Mich, 70 & Kwansei Gakuin Univ, Japan, 79; LHD, Nebr Wesleyan Univ, 72; DSc, Chapman Col, Calif, 82. *Prof Exp:* Instr geol, Duke Univ, 46-47; prof, Univ Mich, 50-62; pres & prof, Grand Valley Col, Mich, 62-68; dean & prof, Col Earth Sci, Univ Ariz, 68-72; pres & prof, Univ Nebr, Lincoln, 72-75 & Southern Methodist Univ, 75-80; PRES & PROF GEOL, UNIV SOUTHERN CALIF, 80- *Concurrent Pos:* Chief glaciologist, Ross Ice Shelf Proj, Antarctica, 57-58; chmn, Polar Res Bd, Nat Acad Sci, 70-80; US deleg, Sci Comt Antarctic Res, 70-, pres, 82-86; mem, Nat Sci Bd, NSF, 72-80; chmn, Arctic Res Comn, 85- *Honors & Awards:* Antarctic Serv Medal, NSF, 66. *Mem:* Geol Soc Am; Sigma Xi; Soc Econ Geologists; Am Geophys Union; Int Glaciological Soc; AAAS. *Res:* Glacial geology of Great Lakes region; origin of lakes of Minnesota; mechanical properties and dynamics of lake ice; mass balance and deformation of Ross Ice Shelf, Antarctica; potential mineral resources of Antarctica; geopolitics of Antarctica. *Mailing Add:* 3201 La Encina Way Pasadena CA 91107

ZUMBRUNNEN, CHARLES EDWARD, b Grafton, WVa, Oct 29, 21; m 85; c 2. DENTISTRY. *Educ:* WVa Wesleyan Col, BS, 43; Northwestern Univ, Chicago, DDS, 45; Univ NC, Chapel Hill, MPH, 64. *Prof Exp:* Pvt dent pract, Huntington, WVa, 48-51 & 54-63; DIR, BUR DENT PUB HEALTH, NH DEPT HEALTH & WELFARE, 64- *Concurrent Pos:* Instr, Sch Dent Med, Tufts Univ, 69-; prof, NH Tech Inst, Concord, 70-71 & 80-; exec secy, NH Bd Dent Examrs, 78- *Mem:* Asn State & Territorial Dent Dirs; Am Col Dent; Am Dent Asn; Am Asn Pub Health Dentists. *Res:* Dental health education methodology in elementary schools. *Mailing Add:* Riverhill Ave Penacook NH 03303

ZUMINO, BRUNO, b Rome, Italy, Apr 28, 23; div. GENERAL RELATIVITY, PARTICLE PHYSICS. *Educ:* Res assoc physics, New York Univ, 51-53, from asst prof to prof, 53-68. *Prof Exp:* sr researcher, Europ Orgn Nuclear Res, 68-81; PROF PHYSICS, UNIV CALIF, BERKELEY, 81- *Honors & Awards:* Dirac Medal, Int Centre Theoret Physics, Trieste, Italy, 87; Heineman Prize Math Physics, Am Phys Soc, 88; Max-Planck Medal, Ger Phys Soc, 89. *Mem:* Nat Acad Sci; fel Am Acad Arts & Sci; fel Am Phys Soc; Italian Phys Soc. *Res:* Relativity and gravitation; particle physics. *Mailing Add:* Dept Physics Univ Calif Berkeley CA 94720

ZUMOFF, BARNETT, b Brooklyn, NY, June 1, 26; m 51; c 3. MEDICINE, MEDICAL RESEARCH. *Educ:* Columbia Univ, AB, 45; Long Island Col Med, MD, 49. *Prof Exp:* Res fel, Sloan-Kettering Inst, 55-57, from asst to assoc, 57-61; from asst prof to prof med, Albert Einstein Col Med, 65-82; ATTEND PHYSICIAN & CHIEF DIV ENDOCRINOL & METAB, BETH ISRAEL MED CTR, 81-; PROF MED, MT SINAI SCH MED, 82- *Concurrent Pos:* Asst dir, Clin Res Ctr, Montefiore Hosp, 61-76, dir, 76-81 & attend med & oncol, 61-82. *Mem:* Am Soc Clin Invest; Am Diabetes Asn; Aerospace Med Asn; Asn Mil Surg US; Am Fedn Clin Res; Endocrine Soc. *Res:* Human steroid metabolism; cholesterol metabolism and atherosclerosis; radioisotope tracer studies in man; hormones in breast and prostate cancer; obesity; reproductive biology; hormonal chronobiology; diabetes mellitus; psycho endocrinology; hormones in coronary artery disease. *Mailing Add:* Div Endocrinol & Metab Dept Med Beth Israel Med Ctr New York NY 10003

ZUMSTEG, FREDRICK C, JR, b Mansfield, Ohio, Apr 24, 43; m 68. EXPERIMENTAL SOLID STATE PHYSICS. *Educ:* Univ Ill, BS, 64; Univ Rochester, PhD(physics), 72. *Prof Exp:* Res assoc, Cornell Univ, 71-73; RES PHYSICIST, E I DU PONT DE NEMOURS & CO, INC, 73- *Mem:* Am Phys Soc. *Res:* Development and characterization of new electrooptic and nonlinear optic materials. *Mailing Add:* Dupont PO Box 80356 Wilmington DE 19880-0356

ZUMWALT, GLEN W(ALLACE), b Vinita, Okla, Apr 21, 26; m 52; c 5. AERONAUTICAL ENGINEERING. *Educ:* Univ Tex, BS, 48 & 49, MS, 53; Univ Ill, PhD(mech & aeronaut eng), 59. *Prof Exp:* Instr eng mech, Univ Tex, 53-55; res assoc mech eng, Univ Ill, 55-59; from asst prof to prof aeronaut eng, Okla State Univ, 59-68; DISTINGUISHED PROF AEROSPACE ENG, WICHITA STATE UNIV, 68- *Concurrent Pos:* Fel, Inst Aerophys, Univ Toronto, 62; consult various industs. *Mem:* Assoc fel Am Inst Aeronaut & Astronaut; Am Soc Eng Educ. *Res:* Gas dynamics; aerodynamics; wind tunnel test and design; aircraft icing protection. *Mailing Add:* Dept Aerospace Eng Wichita State Univ Wichita KS 67208

ZUMWALT, LLOYD ROBERT, b Richmond, Calif, Sept 4, 14; m 60, 78; c 2. PHYSICAL CHEMISTRY. *Educ:* Univ Calif, BS, 36; Calif Inst Technol, PhD(phys chem), 39. *Prof Exp:* Asst, Calif Inst Technol, 36-39, Noyes fel, 39-41; res chemist, Shell Develop Co, 41-42; sr chemist, Oak Ridge Nat Lab, 46-48; dir, Western Div, Tracerlab, Inc, 48-56; vpres, Nuclear Sci & Eng Corp, 56-57; res staff mem, Gen Atomic Div, Gen Dynamics Corp, 57-60; sr res adv, 60-67; prof, 67-80, EMER PROF NUCLEAR ENG, NC STATE UNIV, 80- *Concurrent Pos:* Consult, Gen Atomic Co, 67-79, Los Alamos Sci Lab & Brookhaven Nat Lab, 73-78, Oak Ridge Nat Lab, 81, US Nuclear Regulatory Comn, 82-84, EG&G, Idaho Nat Eng Lab, 90. *Mem:* Am Chem Soc; fel Am Nuclear Soc; Sigma Xi. *Res:* Fission product and tritium diffusion and sorption in materials; high temperature and nuclear reactor chemistry; effect of radiaiton on materials; nuclear fuel recycle; HTGR nuclear fuel. *Mailing Add:* Ten Dixie Trail Raleigh NC 27607

ZUND, JOSEPH DAVID, b Ft Worth, Tex, Apr 27, 39. GEOMETRY, GEODESY. *Educ:* Agr & Mech Col Tex, BA & MS, 61; Univ Tex Austin, PhD(math), 64. *Prof Exp:* Res assoc, Southwest Ctr, Advan Studies, Tex, 64 & 65; from asst prof to assoc prof math, NC State Univ, 65-69; assoc prof, Va Polytech Inst, 69-70; assoc prof math sci, 70-71, PROF MATH & MATH SCI, NMEX STATE UNIV, 72- *Concurrent Pos:* Res assoc, Inst Field Physics, Univ NC, Chapel Hill, 64-65, vis lectr, Dept Math, 65; vis prof, Cambridge Univ, 68-70. *Mem:* Tensor Soc; Am Meteorol Soc; Geol Soc Am; Am Geophys Union. *Res:* Differential and projective geometry; general relativity; electromagnetic theory; differential geodesy; geophysics. *Mailing Add:* Dept Math Sci NMex State Univ Las Cruces NM 88003

ZUNDE, PRANAS, b Kaunas, Lithuania, Nov 26, 23; US citizen; wid; c 5. INFORMATION MEASURES, SYSTEMS. *Educ:* Hannover Tech Univ, MS, 47; George Washington Univ, MS, 65; Ga Inst Technol, PhD(indust eng), 68. *Prof Exp:* Consult info systs, Europe, 47-61; syst analyst, Document Inc, Washington, DC, 61-63; proj mgr info systs, 63-64, dep head mgt systs, 64-65; sr res scientist, 65-68, assoc prof info sci & indust eng, 68-72, PROF INFO SCI, GA INST TECHNOL, 72- *Concurrent Pos:* Consult, Document, Inc, 65-66, Lockheed-Ga Co, 66-67, Ga State Govt, 71 & Nat Inst Technol, Quito, Ecuador, 72; prin investr, HEW, 69-71 & NSF, 72-; Fulbright prof, Nat Acad Sci, 76; vis prof, Univ Simon Bolivar, Caracas, Venezuela, 78, J Kepler Univ, Linz, Austria, 81 & Riso Nat Lab, Roskilde, Denmark, 83. *Honors & Awards:* Lead Award, Soc Man Eng, 86. *Mem:* Sigma Xi; Am Soc Info Sci; Semiotic Soc Am. *Res:* Design of information and communication systems; human factors in systems design; control and socioeconomic systems; operations research; systems theory; design of educational systems; foundations of information science. *Mailing Add:* Col Comput Ga Inst Tech 225 North Ave Atlanta GA 30332

ZUNG, WILLIAM WEN-KWAI, b Shanghai, China; US citizen. PSYCHIATRY, PSYCHOPHARMACOLOGY. *Educ:* Univ Wis, BS, 49; Union Theol Sem, NY, MDiv, 52; Trinity Univ, MS, 56; Univ Tex Med Br Galveston, MD, 61. *Prof Exp:* Instr, 65-66, assoc, 66-67, from asst prof to assoc prof, 67-72, PROF PSYCHIAT, MED CTR, DUKE UNIV, 73- *Concurrent Pos:* Clin investr, Vet Admin Hosp, Durham, NC, 65-67; NIMH res career develop award, 67-72; consult, Ctr Studies Suicide Prev, NIMH, 70-71, Career Develop Rev Div, Vet Admin Cent Off, 72- & Res Task Force, NIMH, 72-; consult pub health serv, Bur Drugs, Food & Drug Admin, 73-75 & mem geriat adv panel, 73-75. *Honors & Awards:* Award for Excellence, Am Acad Gen Pract, 69. *Mem:* Asn Psychophysiol Study Sleep; Am Asn Suicidol; Am Psychiat Asn; Acad Psychosom Med; fel Int Col Psychosom Med; Am Col Neuropsychopharmacol. *Res:* Affective disorders, especially depressive disorders, anxiety disorders and suicide; psychopharmacology of depression, neurophysiological aspects, including sleep disturbances in psychiatric disorders; biometric approach to psychopathology; mathematical models of psychiatric illness. *Mailing Add:* Duke Univ Med Ctr Box 2914 Durham NC 27710

ZUNKER, HEINZ OTTO HERMANN, b Berlin, Ger, May 27, 24; m 63; c 2. PATHOLOGY. *Educ:* Free Univ Berlin, MD, 54; Am Bd Path, cert, 66. *Prof Exp:* From intern to resident internal med, Free Univ Berlin, 54-56; head pharmacol res, Pharmaceut Co, 57-59; res assoc pharmacol, Columbia Univ, 60-61, assoc path, Col Physicians & Surgeons, 65-68, asst prof, 68; chief path & dir labs, Deaconess Hosp, Evansville, Ind, 68-73; assoc prof path & dir clin path, Col Med, Univ S Fla, 74-75; chief pathologist & dir labs, Beaumont Med Surg Hosp, Beaumont, 75-82; ASSOC PATHOLOGIST, ST ELIZABETH HOSP, BEAUMONT, TEX, 86- *Concurrent Pos:* Vis fel path, Columbia-Presby Med Ctr, 62-65; asst attend pathologist, Presby Hosp & consult pathologist, Harlem Hosp, New York, 66-68; assoc prof allied health sci, Univ Evansville & Ind State Univ, Evansville, 72-73; consult, surg path. *Mem:* Col Am Path; Am Soc Clin Path; NY Acad Sci. *Res:* Pharmacology; experimental pathology; electron microscopy; clinical chemistry. *Mailing Add:* 6840 Hialeah St Beaumont TX 77706

ZUPERKU, EDWARD JOHN, b Sept 14, 42; m 75; c 4. BREATHING CONTROL. *Educ:* Marquette Univ, BEE, 65, MS, 67, PhD(biomed eng), 70. *Prof Exp:* From asst prof to assoc prof, 74-89, PROF BIOMED ENG, MED COL WIS, 89-; BIOMED ENGR, VET ADMIN MED CTR, MILWAUKEE, 75- *Concurrent Pos:* Prin investr, Vet Admin Med Ctr, 78- *Mem:* Am Physiol Asn; Am Thoracic Soc; Am Heart Asn; Neurosci Soc. *Res:* Mathematical modeling of neural circuits; processing of afferent input patterns by the central nervous system. *Mailing Add:* Res Serv 151 Zablocki Vet Admin Med Ctr Milwaukee WI 53295

ZUPPERO, ANTHONY CHARLES, b Lakewood, Ohio; m; c 2. PHYSICS, SYSTEMS ENGINEERING. *Educ:* Case Western Reserve Univ, BS, 65, PhD(physics), 70. *Prof Exp:* Staff mem systs eng, Sandia Nat Labs, Albuquerque, NMex, 70-74, staff mem physics, Livermore, Calif, 74-77, mem tech staff systs eng, 77-79, real-time comput algorithm design, Albuquerque, 79-88; PROG MGR, SATELLITE-SUBMARINE LASER COMMUN & SPACE SYST ENG & MGT, 88- *Mem:* AAAS. *Res:* Laser photochemistry; information systems; pulse power systems; applied engineering; laser strain seismometry; energy research and development; astrophys; space resources transp systs (exofuel). *Mailing Add:* 17225 Cuvee Ct Poway CA 92064-1214

ZURAWSKI, VINCENT RICHARD, JR, b Irvington, NJ, June 10, 46; m 68; c 2. BIOCHEMISTRY, IMMUNOLOGY. *Educ:* Montclair State Col, BA, 68; Purdue Univ, PhD(chem), 73. *Prof Exp:* Res assoc biochem, Purdue Univ, 74; res fel med virol, immunochem & cell biol, Harvard Med Sch, 75-78, instr path, 78-79; co-founder & vpres, 79-82, sr vpres, 82-83, exec vpres, 83-87, TECH DIR, CENTOCOR, 79-, CORP SECY, 81-, SR VPRES & CHIEF SCI OFFICER, 87- *Concurrent Pos:* Res fel med virol, immunochem & cell biol, Cardiac Biochem Lab, Mass Gen Hosp, Boston, 75-78, res fel, Cell & Molecular Res Lab, 78-79, NIH, 76-78 & Med Found, 78-79; lectr, dept obstet & gynec, Harvard Med Sch, 85-; prin investr, Nat Cancer Inst res grant. *Mem:* AAAS; Am Chem Soc; Am Soc Microbiol; Am Asn Immunologists; Tissue Culture Asnt; Soc Nuclear Med; Am Asn Cancer Res; Am Soc Biochem & Molecular Biol. *Res:* Immunochemical and immunobiological research aimed at producing monoclonal antibodies of predetermined specificity both in vivo and in vitro for structural studies; diagnostic and therapeutic applications and applications to cellular immunology; cell biological and molecular genetics studies with particular emphasis on problems associated with ovarian cancer. *Mailing Add:* Centocor 244 Great Valley Pkwy Malvern PA 19355

ZURIER, ROBERT B, b Passaic, NJ, Feb 19, 34; m 62; c 1. CELL BIOLOGY. *Educ:* Rutgers Univ, BS, 55; Southwestern Med Sch, Univ Tex, MD, 62; Univ Pa, MA, 82. *Prof Exp:* Asst, assoc & prof med, Sch Med, Univ Conn, 73-80; PROF MED & CHIEF RHEUMATOLOGY SECT, SCH MED, UNIV PA, 80- *Concurrent Pos:* Guggenheim Found fel, 86-87. *Mem:* Am Soc Clin Invest; Am Rheumatism Asn; Am Asn Immunologists; Am Asn Advan Sci. *Res:* Role of prostaglandins and fatty acids in immune responses and inflammatory reactions. *Mailing Add:* Rheumatology Sect 570 Maloney Hosp Univ Pa 3600 Spruce St Philadelphia PA 19104-4283

ZURMUHLE, ROBERT W, b Lucerne, Switz, Nov 27, 33; US citizen. NUCLEAR PHYSICS. *Educ:* Univ Zurich, PhD(physics), 60. *Prof Exp:* Res assoc, 61-63, from asst prof to assoc prof, 63-76, PROF PHYSICS, UNIV PA, 76- *Mem:* Fel Am Phys Soc. *Res:* Nuclear structure and nuclear reactions. *Mailing Add:* Dept Physics Univ Pa Philadelphia PA 19104

ZUSMAN, FRED SELWYN, b Boston, Mass, July 24, 31; m 54; c 2. MATHEMATICS. *Educ:* Harvard Univ, AB, 52, MA, 55. *Prof Exp:* Analyst, Nat Security Agency, 52-54; sr mathematician appl physics lab, Johns Hopkins Univ, 55-61; sr scientist, Opers Res Inc, 62; dir comput lab, Nat Biomed Res Found, 63; dir comput ctr, Opers Res Inc, 63-67, sr scientist, 67-69; vpres, Sci Mgt Systs, Inc, 69-72; EXEC SCIENTIST & DIR COMPUT SCI, ARC PSG, INC, 72- *Concurrent Pos:* Lectr, Sch Hyg & Pub Health, Johns Hopkins Univ, 60-63. *Mem:* Asn Comput Mach; Opers Res Soc Am. *Res:* Operations research; computer technology; system simulation; computer sciences. *Mailing Add:* ARC PSG Inc 2440 Research Blvd Rockville MD 20850

ZUSMAN, JACK, b Brooklyn, NY, Jan 6, 34; m 55; c 4. PSYCHIATRY, PUBLIC HEALTH. *Educ:* Columbia Univ, AB, 55, MPH, 66; Ind Univ, MA, 56; Albert Einstein Col Med, MD, 60. *Prof Exp:* Intern, USPHS Hosp, New Orleans, La, 60-61, epidemic intel serv officer, Communicable Dis Ctr, 61-62, ment health career develop officer, NIMH, 62-66, staff psychiatrist, Epidemiol Studies Br, 66-67, chief, Ctr Epidemiol Studies, 67-68; assoc prof psychiat, State Univ NY Buffalo, 68-71, dir div community psychiat, 69-74, prof, 71-75; PROF PSYCHIAT, UNIV SOUTHERN CALIF, 75- *Concurrent Pos:* Trainee community psychiat, Columbia Univ, 64-66; adj prof law & psychiat, 74-75 & Univ Southern Calif, 79- *Mem:* AMA; Am Acad Psychiat & Law; Am Psychiat Asn. *Res:* Social factors which influence the course of mental illness and methods of their control; interaction of law and psychiatry; methods of organizing and providing medical care. *Mailing Add:* Dept Psychiat Univ SFla Col Med 12901 N 30th Tampa FL 33612

ZUSPAN, FREDERICK PAUL, b Richwood, Ohio, Jan 20, 22; m 43; c 3. OBSTETRICS & GYNECOLOGY. *Educ:* Ohio State Univ, BA, 47, MD, 51. *Prof Exp:* Chief dept obstet & gynec, McDowell Mem Hosp, Ky, 56-58, chief clin serv, 57-58; asst prof obstet & gynec, Sch Med, Western Reserve Univ, 59-60; prof & chmn dept, Med Col Ga, 60-66; Joseph Bolivar DeLee prof & chmn dept, Univ Chicago, 66-74; chmn dept obstet & gynec, 74-87, PROF, R L MEILING CHAIR, OBSTET & GYNEC, OHIO STATE UNIV, 88- *Concurrent Pos:* Oglebay fel obstet & gynec, Sch Med, Western Reserve Univ, 58-60; assoc examr & dir, Am Bd Obstet & Gynec, 65-80; gynecologist-in-chief, Chicago Lying-In Hosp, 66-74; pres, Barren Found, 74-76; founding ed, J Reprod Med; ed, Am J Obstet & Gynec & Current Concepts in Obstet & Gynec; consult ed, Acta Cytologica, Exerpta Medica, Obstet & Gynec Surv, Hypertension, J Obstet & Gynec (Mex) & J Reprod Med. *Res:* Human reproductive physiology; epinephrine and norephinephrine in the obstetric patient; maternal-fetal medicine. *Mailing Add:* Dept Obstet & Gynec Ohio State Univ Columbus OH 43210

ZUSPAN, G WILLIAM, b Richwood, Ohio, Mar 24, 26; m 48; c 5. METALLURGICAL ENGINEERING. *Educ:* Ohio State Univ, BMetE & MS, 51. *Prof Exp:* Plant metallurgist, E I du Pont de Nemours & Co, 51-53; res metallurgist, Battelle Mem Inst, 53-54; asst prof metall eng, 54-67, dean freshmen, 67-80, ASSOC PROF ENG, DREXEL UNIV, 67-, ASST VPRES STUDENT AFFAIRS, 72- *Mem:* Am Soc Metals; Am Soc Eng Educ; Nat Asn Corrosion Engrs. *Res:* Chemical metallurgy and corrosion. *Mailing Add:* Dept Eng Drexel Univ 32nd & Chestnuts Sts Philadelphia PA 19104

ZUSSMAN, MELVIN PAUL, b Boston, Mass Aug 6, 56; m 80; c 2. POLYMER CHEMISTRY. *Educ:* Haverford Col, BA, 78; Carnegie-Mellon Univ, PhD(chem), 82. *Prof Exp:* Res chemist, US Steel Res Ctr, 82-83; sr engr, Res & Develop Labs, Westinghouse Elec Corp, 84-87; RES CHEMIST, EXP STA, E I DU PONT DE NEMOURS & CO, 87- *Concurrent Pos:* Sr lectr, chem dept, Carnegie-Mellon Univ, 84. *Mem:* Am Chem Soc; AAAS; Sigma Xi. *Res:* Development of new or modified polymers for electronics applications; effects of high energy particles on polymeric materials; preparation and characterization of ordered organic films. *Mailing Add:* 2627 Epping Rd Wilmington DE 19810

ZUSY, DENNIS, b Milwaukee, Wis, Dec 21, 28. ECOLOGY, PHILOSOPHY OF SCIENCE. *Educ:* Aquinas Inst, MA, 52 & 56; Northwestern Univ, MS, 64, PhD(biol), 67. *Prof Exp:* Asst prof philos, St Xavier Col, Ill, 56-62; from asst prof to prof biol, Clarke Col, 71-81; PROF PHILOS & RELIGION, AQUINAS INST THEOL, 67- *Concurrent Pos:* Vis asst prof biol, Concordia Teachers Col, Ill, 67-71. *Mem:* AAAS; Sigma Xi. *Res:* Biological rhythms; freshwater ecology; philosophical implications and history of scientific concepts. *Mailing Add:* St Pius Priory 1909 S Ashland Ave Chicago IL 60608-2994

ZUZACK, JOHN W, b St Louis, Mo, Sept 24, 38; m 62; c 3. ORGANIC CHEMISTRY. *Educ:* St Louis Univ, BS, 61, MS, 64, PhD(org chem), 67. *Prof Exp:* Lab instr freshman & org chem, St Louis Univ, 61-65, res asst med chem, 65-66; from asst prof to assoc prof, 66-89, PROF MED CHEM, ST LOUIS COL PHARM, 90- *Mem:* Am Chem Soc. *Res:* Leukemia chemotherapy; structure activity relationships in rickettsiostatic pyrrolidine analgesic agents and the diels alder reaction. *Mailing Add:* Dept Med Chem 4588 Park View Pl St Louis MO 63110

ZUZOLO, RALPH C, b Italy, Sept 5, 29; US citizen. CELL PHYSIOLOGY, CELLULAR MICROSURGERY. *Educ:* NY Univ, BA, 56, MS, 60, PhD(biol), 65. *Prof Exp:* Res fel, Dept Path Lab, Univ Tex, 66-67; res scientist, Guggenheim Inst Dent Res, NY Univ, 78-80; SUPVR DEPT BIOL, SCH GEN STUDIES, CITY COL NEW YORK, 68-, INSTR, ROBERT CHAMBERS LAB CELLULAR MICROSURG, 74-, ASSOC PROF BIOL, 75-, LAB CO-DIR, 80- *Concurrent Pos:* Grants, NASA, NY Univ, 66, Biomed Support, City Col New York, 71, Ellis Phillips Found, 81 & Olympus Corp Am, 83 & 85; consult cellular microsurgery & microinjection, 72-; from adj asst prof to adj assoc prof, NY Univ, 74-80; vis scientist, Boyce Thompson Inst Plant Res, 77-78; dir Course & Standings, SGS, City Col New York, 84- *Mem:* AAAS; Am Inst Physics; Soc Appl Spectros; Sigma Xi; NY Acad Sci; Tissue Cult Soc Am. *Res:* Microsurgery, the application of the laser as a microsurgical tool and the effect of lasers irradiation on cells; effect of chemical carcinogens and bisulfite on nucleic acid in living cells; the effect and detection of insect nuclear polyhedrosis virus after microinjection into early vertebrate embryos; design and construction of instruments for cellular microsurgery; development of techniques and methods for single cell manipulation and analysis; image analysis as applied to protozoa. *Mailing Add:* 131 Valentine St Marshak Bldg City Col New York Mt Vernon NY 10550

ZVAIFLER, NATHAN J, b Newark, NJ, Nov 26, 27; m 52, 83; c 4. MEDICINE, IMMUNOLOGY. *Educ:* Haverford Col, BS, 48; Jefferson Med Col, MD, 52. *Prof Exp:* Resident med, Univ Mich, 55-58, instr, Med Sch, 58-59; from instr to prof, Georgetown Univ, 60-70; PROF RHEUMATOL, UNIV CALIF, SAN DIEGO, 70- *Concurrent Pos:* NIH fel arthritis, Univ Mich, 58-60; Macy Found Scholar, 76-77; vis prof, Rockefeller Univ, 83- *Res:* Arthritis. *Mailing Add:* H-811-G Univ Calif San Diego Box 109 San Diego CA 92103

ZVEJNIEKS, ANDREJS, b Rauna, Latvia, Jan 6, 22; m 51. ORGANIC CHEMISTRY. *Educ:* Latvia Univ, BS, 43, MS, 44; Royal Inst Technol, Sweden, PhD, 55. *Prof Exp:* Res engr, Liljeholmens Stearinfabriks, AB, Sweden, 45-48, sect leader, 48-56; qual control supvr, Conn Adamant Plaster Co, 56; tech adv to plant mgr, Chem Div, Gen Mills, Inc, 57-58, tech dir, Petrol Chem Dept, 58-60, dir appln res, 60-62, sr scientist, 62; pres, AZ Prod, Inc, Fla, 62-72; PRES & CHIEF EXEC OFFICER, AZS CORP, 72- *Concurrent Pos:* Mem, Hwy Res Bd, Nat Acad Sci-Nat Res Coun, 59- *Mem:* Am Chem Soc. *Res:* Organic ammonium compounds; ore flotation reagents; petroleum chemicals. *Mailing Add:* 2337 Christopher Walk Atlanta GA 30327

ZVENGROWSKI, PETER DANIEL, b New York, NY, Sept 8, 39; m 84; c 2. MATHEMATICS. *Educ:* Rensselaer Polytech Inst, BS, 59; Univ Chicago, MS, 60, PhD(math), 65. *Prof Exp:* Asst prof math, Univ Ill, Urbana, 64-70; grom asst prof to assoc prof 70-85, chmn div pure math, 72-74, PROF MATH, UNIV CALGARY, 85- *Res:* Algebraic topology and homotopy theory; G-invariant homotopy theory; span of differentiable manifolds, in particular flag manifolds, Grassmann manifolds, projective Stiefel manifolds; applications in Quantum Field Theory. *Mailing Add:* Dept Math Univ Calgary Calgary AB T2N 1N4 Can

ZWAAN, JOHAN THOMAS, b Gorinchem, Neth, Sept 28, 34; m 60; c 2. PEDIATRIC OPHTHALMOLOGY, HUMAN EMBRYOLOGY. *Educ:* Univ Amsterdam, MedDrs, 60, Dr(embryol), 63. *Prof Exp:* From asst anat to head asst, Lab Anat & Embryol, Univ Amsterdam, 58-63; fel pediat, Sch Med, Johns Hopkins Univ, 63-64; from asst prof to assoc prof, Sch Med, Univ Va, 64-71; assoc prof anat, 71-78, asst prof ophthal, 78-82, ASSOC PROF OPHTHAL, HARVARD MED SCH, 82- *Concurrent Pos:* Lectr, Acad Phys Educ, Amsterdam, 62-63; res assoc ophthal, Children's Hosp Med Ctr, Boston, 71-75; assoc prof anat & resident ophthal, Albany Med Col, 75-78; asst ophthal, Mass Eye & Ear Infirmary, 78-, dir pediat & ocular motility serv, 81-, asst surg, 83-; res assoc, Mass Inst Technol, 81- *Mem:* AAAS; Soc Develop Biol; Asn Res Vision & Ophthal; Am Acad Ophthal. *Res:* Chemical and morphological changes in differentiation of vertebrate cells (model system; eye lens); developmental genetics; ophthalmic genetics; teratology; normal and abnormal development of the visual system. *Mailing Add:* Univ Tex Health Sci Ctr 7703 Floyd Curl Dr San Antonio TX 78234-7779

ZWADYK, PETER, JR, b Kansas City, Kans, Apr 3, 41; m 63; c 4. MICROBIOLOGY. *Educ:* Univ Kans, BS, 62; Univ Iowa, MS, 66, PhD(microbiol), 71. *Prof Exp:* Microbiologist, Sci Assocs, 62-63; scientist microbiol, Mead Johnson & Co, 66-69; asst prof path, 71-75, asst prof microbiol, 71-76, ASSOC PROF PATH, DUKE UNIV, 75-, ASSOC PROF MICROBIOL, 76-; CHIEF MICROBIOL, VET ADMIN HOSP, DURHAM, 71- *Mem:* Am Soc Microbiol; Sigma Xi; Southeastern Asn Clin Microbiol; Am Acad Microbiol. *Res:* DNA probes; antibiotics; general clinical microbiology. *Mailing Add:* Vet Admin Hosp Fulton St & Erwin Rd Durham NC 27705

ZWANZIG, FRANCES RYDER, b South Amboy, NJ, Oct 22, 29; m 53; c 2. CHEMISTRY. *Educ:* Columbia Univ, BA, 51; Yale Univ, MS, 53, PhD(chem), 56. *Prof Exp:* Res asst physiol chem, Sch Med, Johns Hopkins Univ, 55-58; asst ed, Rev Mod Physics, Am Phys Soc, Washington, DC, 69-73; asst ed, Transp Res Bd, 77-80, staff officer, Off Chem & Chem Technol, 79-81, assoc ed proc, 80-83, MANAGING ED PROC, NAT ACAD SCI, WASHINGTON, DC, 83- *Mem:* Am Chem Soc; Coun Biol Ed. *Mailing Add:* 5314 Sangamore Rd Bethesda MD 20816

ZWANZIG, ROBERT WALTER, b Brooklyn, NY, Apr 9, 28; m 53; c 2. CHEMICAL PHYSICS. *Educ:* Polytech Inst Brooklyn, BS, 48; Univ Southern Calif, MS, 50; Calif Inst Technol, PhD(chem), 52. *Prof Exp:* Res fel theoret chem, Yale Univ, 51-54; asst prof chem, Johns Hopkins Univ, 54-58; phys chemist, Nat Bur Stand, 58-66; res prof, Inst Phys Sci & Technol, Univ Md, Col Park, 66-79, distinguished prof, 79-88; RESEARCHER, BIOPHYS, NIH, 88- *Concurrent Pos:* Sherman Fairchild scholar, Calif Inst Technol, 74-75. *Honors & Awards:* Peter Debye Award Phys Chem, Am Chem Soc, 76, Irving Langmuir Award, Chem Physics, 85. *Mem:* Nat Acad Sci; Am Chem Soc; Am Phys Soc; Am Acad Arts & Sci. *Res:* Theoretical chemical physics; statistical mechanics; theory of liquids and gases. *Mailing Add:* NIH Bldg 2 Rm 112 Lab Chem Physics Bethesda MD 20892

ZWANZIGER, DANIEL, b New York, NY, May 20, 35. THEORETICAL PHYSICS. *Educ:* Columbia Univ, BA, 55, PhD(physics), 60. *Prof Exp:* NSF fel physics, Univ Calif, Berkeley, 60-61, lectr, 61-62; scientist, Univ Rome, 62-63; vis, Saclay Ctr Nuclear Studies, France, 63-65; vis scientist, 65-67, assoc prof physics, 67-74, PROF PHYSICS, NY UNIV, 74- *Concurrent Pos:* Assoc, Europ Orgn Nuclear Res, Switz, 80; vis prof, Ecole Normale Superieure, France, 81; vis mem, Inst Avan Study, 84. *Mem:* Am Phys Soc; Fedn Am Scientists; Am Geophys Union. *Res:* Quantum field theory; mathematical physics. *Mailing Add:* Dept Physics NY Univ Four Washington Sq New York NY 10003

ZWARICH, RONALD JAMES, b Kamloops, BC, Apr 26, 36. PHYSICAL CHEMISTRY. *Educ:* Univ BC, BSc, 63, PhD(chem), 69. *Prof Exp:* Asst prof chem, State Univ NY Albany, 71-75; asst prof, 75-80, ASSOC PROF CHEM, UNIV PETROL & MINERALS, SAUDI ARABIA, 80- *Mem:* Am Chem Soc. *Res:* Molecular spectroscopy; vibrational structure and electronic properties of aromatic and heterocyclic compounds. *Mailing Add:* 3259 Bank Rd Kamloops BC V2B 6Z9 Can

ZWARUN, ANDREW ALEXANDER, b Pidvolochyska, Ukraine, Feb 9, 43; US citizen; m 67; c 2. MEDICAL DEVICE DEVELOPMENT, STERILIZATION. *Educ:* Ohio State Univ, BSc, 65, MSc, 67; Univ Ky, PhD(soil microbiol), 70. *Prof Exp:* Asst prof agron, Univ Md, Eastern Shore, 70-71; chief microbiologist, Johnston Labs, Inc, 71-74; res microbiologist, Betz Labs, Inc, 74-76; DIR RES, PROPPER MFG CO, 77- *Mem:* Am Soc Microbiol; Health Indust Mfg Asn; Sigma Xi. *Res:* Automation of microbial and biochemical procedures; trace metal toxicities; microbial activity in soils and water; industrial water treatment. *Mailing Add:* Ten Schoolhouse Lane Roslyn Heights NY 11577

ZWASS, VLADIMIR, b Lvov, USSR, Feb 3, 45; US citizen; m 77; c 1. MANAGEMENT INFORMATION SYSTEMS. *Educ:* Moscow Inst Energetics, MS, 69; Columbia Univ, MPh & PhD(computer sci), 75. *Prof Exp:* Mem prof staff, Int Atomic Energy Agency, 70; from asst prof to assoc prof, 75-84, PROF COMPUTER SCI, FAIRLEIGH DICKINSON UNIV, 84- *Concurrent Pos:* Consult, Metrop Life Ins Co, Diebold Group & Citicorp; NSF grant, 82-83; co-prin investr, USN, 85-86; ed-in-chief, J Mgt Info Systs, 84- *Mem:* Inst Elec & Electronic Engrs; Asn Comput Mach; Sigma Xi. *Res:* Management information systems; software engineering; operating systems; management information systems and computer science as scientific disciplines. *Mailing Add:* 538 Churchill Rd Teaneck NJ 07666

ZWEBEN, CARL HENRY, b Albany, NY. COMPOSITE MATERIALS & STRUCTURES. *Educ:* Cooper Union, BCE, 60; Columbia Univ, MS, 61; Polytech Inst, Brooklyn, PhD(appl mech), 66. *Prof Exp:* Res engr, Space Sci Lab, Gen Elec Co, 66-69; sr res engr, Jet Propulsion Lab, 69-70; prog mgr, Mat Sci Corp, 70-72; res assoc, E I du Pont de Nemours & Co, Inc, 72-78; ADVAN TECHNOL MGR, SPACE DIV, GEN ELEC CO, 80- *Concurrent Pos:* Consult, Nat Mat Adv Bd, Nat Acad Sci, 69 & 81-82 & Ctr Composite Mat, Univ Del, 78-; lectr, Univ Calif, Los Angeles, 73- & Univ Cambridge, Eng, 79-; consult, Metal Matrix Composites Info Anal Ctr, US Dept Defense, Congress Off Technol Assessment. *Honors & Awards:* Tech Brief Award, NASA, 72. *Mem:* Am Soc Test & Mat; Am Soc Mech Eng; Am Soc Civil Eng; Soc Advan Mat & Process Eng; Am Inst Aeronaut & Astronaut. *Res:* Properties and application of polymer, metal and ceramic matrix composite materials, including, design analysis, development, test methods, micromechanics, fatigue, fracture, impact, failure mechanics, material development structural test electronic packaging. *Mailing Add:* Gen Elec Co PO Box 8555 Philadelphia PA 19101

ZWEBEN, STUART HARVEY, b New York, NY, Apr 21, 48; m 71; c 1. COMPUTER SCIENCE. *Educ:* City Col NY, BS, 68; Purdue Univ, MS, 71, PhD(comput sci), 74. *Prof Exp:* Syst analyst comput sci, IBM Corp, 69-70; instr, Purdue Univ, 74; asst prof, 74-80, actg chmn dept, 83-84, ASSOC PROF COMPUT & INFO SCI, OHIO STATE UNIV, 80- *Concurrent Pos:* Prin investr res grant, Ohio State Univ, 75, Dow Chem Co, 78-79, US Army Res Off, 80-83, NSF, 81-83 & 88-90, Dept Educ, 82-85, AT&T, 84, 86-88 & Appl Info Tech Res Ctr, 89-90; dept coordr grants, Dept Health, Educ & Welfare, 77-79; secy-treas, Comput Sci Accred Bd, 86-87, vpres, 87-89, pres, 89-91; assoc ed, Inst Elec & Electronic Engrs Trans Software Engr, 90- *Mem:* Asn Comput Mach; Inst Elec & Electronics Engrs; Am Asn Univ Prof. *Res:* Software engineering; programming methodology; software testing; software reuse. *Mailing Add:* Dept Comput & Info Sci 2036 Neil Ave Mall Columbus OH 43210

ZWEIBEL, ELLEN GOULD, b New York, NY, Dec 20, 52. THEORETICAL ASTROPHYSICS. *Educ:* Univ Chicago, BA, 73; Princeton Univ, PhD(astrophys sci), 77. *Prof Exp:* Vis mem, Inst Advan Study, 77-78; STAFF MEM SOLAR PHYSICS, HIGH ALTITUDE OBSERV, NAT CTR ATMOSPHERIC RES, 78- *Mem:* Am Astron Soc. *Res:* Plasma astrophysics, especially cosmic rays and solar and interplanetary physics. *Mailing Add:* Apas Dept Campus Box 391 Univ Colo Boulder CO 80309

ZWEIDLER, ALFRED, b Switz; m; c 3. CHROMOSOMES, NUCLEAR PROTEINS. *Educ:* Univ Zurich, PhD(cell biol), 66. *Prof Exp:* MEM, INST CANCER RES, 71- *Concurrent Pos:* Adj assoc prof biochem, Univ PA, 85- *Mem:* AAAS; ASCB; Protein Soc; Swiss Genetic Soc; Sigma Xi. *Res:* chromosome structure and function. *Mailing Add:* Inst Cancer Res 7701 Burholme Ave Philadelphia PA 19111

ZWEIFACH, BENJAMIN WILLIAM, b New York, NY, Nov 27, 10; m 37; c 3. PHYSIOLOGY, BIOENGINEERING. *Educ:* City Col New York, BS, 31; NY Univ, MS, 33, PhD(anat), 36. *Prof Exp:* Asst biol, NY Univ, 31-33, asst anat, Col Med, 35-36, res assoc, 38-45; res assoc med, Med Col, Cornell Univ, 45-47, asst prof physiol, 47-52; assoc prof biol, NY Univ, 52-55, from assoc prof path to prof, Sch Med, 55-66; prof, 66-82, EMER PROF BIOENG, UNIV CALIF, SAN DIEGO, 82- *Concurrent Pos:* Charlton res fel, Med Sch, Tufts Col, 37-38; ed, Josiah Macy, Jr Found Conf Factors Regulating Blood Pressure, 48-53; mem cardiovasc study sect, USPHS & subcomt shock, Nat Res Coun; estab investr, Am Heart Asn, 55-60; career investr, Health Res Coun, New York, 60-66; Prof, Peking Union Med Col, Beijing, China. *Honors & Awards:* Landis Award, 71; Malpighi Gold Medal, 80. *Mem:* Am Physiol Soc; Soc Exp Biol & Med; Histochem Soc; Microcirc Soc (pres, 79-80); fel NY Acad Sci. *Res:* Microcirculation, shock, tissue injury. *Mailing Add:* Dept Bioeng 5028 Bsb M-005 Univ Calif San Diego La Jolla CA 92093

ZWEIFEL, GEORGE, b Rapperswil, Switz, Oct 2, 26; m 53; c 2. ORGANIC CHEMISTRY. *Educ:* Swiss Fed Inst Technol, Dr sc tech, 55. *Prof Exp:* Res asst carbohydrate chem, Univ Edinburgh, 55-56; res fel, Univ Birmingham, 56-58; res assoc boron chem, Purdue Univ, 58-63; assoc prof chem, 63-72, PROF CHEM, UNIV CALIF, DAVIS, 72- *Mem:* Am Chem Soc; Sigma Xi. *Res:* Chemistry of natural products; utilization of organoboranes and organoalanes in organic syntheses. *Mailing Add:* Dept Chem Univ Calif Davis CA 95616

ZWEIFEL, PAUL FREDERICK, b New York, NY, June 21, 29; m 60, 67; c 4. MATHEMATICAL PHYSICS, NUCLEAR SCIENCE. *Educ:* Carnegie Inst Technol, BS, 48; Duke Univ, PhD, 54. *Prof Exp:* Asst physicist, Chem Lab, Am Brake Shoe Co, 48; teacher, Malcom Gordon Sch, NY, 48-49; asst, Duke Univ, 50-52; res assoc, Knolls Atomic Power Lab, Gen Elec Co, 53-56, mgr theoret physics, 56-57; consult physicist, 57-58; assoc prof nuclear eng, Univ Mich, Ann Arbor, 58-60, prof, 60-68; from prof to univ prof, 68-75, DISTINGUISHED UNIV PROF PHYSICS & NUCLEAR ENG, VA POLYTECH INST & STATE UNIV, 75- *Concurrent Pos:* Fel Duke Univ, 52-53; mem adv comt reactor physics, Atomic Energy Comn, 57-64; vis prof, Middle East Tech Univ, Ankara, 64-65; consult, indust orgns & govt labs; vis prof, Rockefeller Univ, 73, 74-75; fel, J S Guggenheim Mem Found, 74-75; vis prof, Univ Florence, 75, 81-83, 85, vis prof, Univ Ulm, WGer, 78, Univ Milan, 78; consult, Los Alamos Sci Lab, 75- & Nuclear Regulatory Comn, 76. *Honors & Awards:* E O Lawrence Award, 72. *Mem:* Am Phys Soc; Am Nuclear Soc; Fedn Am Sci (secy, 57-58); Am Math Soc; Int Asn Math Physics. *Res:* Mathematical physics; neutron transport theory and nuclear energy; foundations of quantum mechanics. *Mailing Add:* Dept Physics Va Polytech Inst Blacksburg VA 24060

ZWEIFEL, RICHARD GEORGE, b Los Angeles, Calif, Nov 5, 26; m 56; c 3. HERPETOLOGY. *Educ:* Univ Calif, Los Angeles, BA, 50; Univ Calif, Berkeley, PhD(zool), 54. *Prof Exp:* From asst cur to assoc cur, Am Mus Natural Hist, 54-65, chmn dept, 68-80, cur herpet, 65-89, EMER CUR, AM MUS NATURAL HIST, 89- *Mem:* Am Soc Ichthyol & Herpet; Soc Study Evolution; Ecol Soc Am; Soc Study Amphibians & Reptiles; Herpet League. *Res:* Ecology and systematics of amphibians and reptiles. *Mailing Add:* PO Box 354 Portal AZ 85632

ZWEIFLER, ANDREW J, b Newark, NJ, Feb 2, 30; m 54; c 5. INTERNAL MEDICINE. *Educ:* Haverford Col, AB, 50; Jefferson Med Col, MD, 54. *Prof Exp:* Intern, Mt Sinai Hosp, New York, 54-55; resident internal med, 57-60, Nat Heart & Lung Inst fel, 60-63, from instr to assoc prof, 60-72, PROF INTERNAL MED, MED CTR, UNIV MICH, ANN ARBOR, 72- *Concurrent Pos:* Vis prof, Meharry Med Col, 67-68; fel coun arteriosclerosis & coun thrombosis, Am Heart Asn. *Mem:* Am Fedn Clin Res. *Res:* Thrombosis; vascular disease; hypertension. *Mailing Add:* Dept Internal Med Univ Mich Box 0356 Ann Arbor MI 48104

ZWEIG, FELIX, b Ft Wayne, Ind, Sept 25, 16; m 45; c 3. ELECTRICAL ENGINEERING. *Educ:* Yale Univ, BE, 38, PhD(elec eng), 41. *Prof Exp:* From asst prof to assoc prof, 43-55, chmn dept, 61-66, PROF ENG & APPL SCI, YALE UNIV, 66- *Concurrent Pos:* Res assoc, Mass Inst Technol, 46; consult, Gen Elec Co, 46-48, Gen Precision, Inc, 51-59, Burroughs Corp, 53, Bristol Co, 60-61 & Autonetics Div, NAm Aviation, Inc, 61. *Mem:* Inst Elec & Electronics Engrs; Sigma Xi. *Res:* Feedback control systems and inertial navigation. *Mailing Add:* Dept Elec Eng Yale Univ Becton Ctr PO Box 2157 Yale Sta New Haven CT 06520

ZWEIG, GILBERT, b New York, NY, Apr 5, 38; c 3. ENGINEERING PHYSICS, PHOTO-OPTICS. *Educ:* NY Univ, BA & BME, 60, MS, 65. *Prof Exp:* Res staff mem, IBM, T J Watson Res Ctr, 61-63; supvr advan res, Pitney-Bowes, Inc, 63-73; vpres & cofounder, Imtex, Inc, 73-74; dep tech dir, TWT Labs Inc, Div Arkwright Inc, 74-78, dir tech bus develop, 78-81; vpres, MCI Optonix Inc, Subsid Mitsubishi Chem Industs Am, 81-83; PRES, GLENBROOK TECHNOLOGIES, 83- *Mem:* Sigma Xi; Soc Photog Sci & Eng; Am Asn Physicists Med; Optical Soc Am; Soc Photo-Optical Instrument Eng. *Res:* Imaging systems of non-silver classification; including electrophotographic diazo and radioluminescent image technologies. *Mailing Add:* 24 Stiles Ave Morris Plains NJ 07950

ZWEIG, HANS JACOB, mathematical statistics, operations research, for more information see previous edition

ZWEIG, JOHN E, b Poestenkill, NY, June 24, 36; m 66; c 3. MECHANICAL ENGINEERING. *Educ:* Rensselaer Polytech Inst, BME, 57, MSE, 60, PhD(mech eng), 70. *Prof Exp:* Instr mech eng, Rensselaer Polytech Inst, 57-59; mech engr, Watervliet Arsenal, 60-64, chief, Exp Mech & Thermodyn Lab, 64-71, chief, Appl Math & Mech Div, 71-77, CHIEF DEVELOP ENGR, WATERVLIET ARSENAL, 77- *Mem:* Am Soc Mech Engrs; Soc Exp Stress Anal; Sigma Xi. *Res:* Fluid dynamics; heat transfer; experimental mechanics. *Mailing Add:* RR 1 Box 1597 Poestenkill NY 12140

ZWEIG, RONALD DAVID, biology, limnology, for more information see previous edition

ZWEIMAN, BURTON, b New York, NY, June 7, 31; m 62; c 2. MEDICINE. *Educ:* Univ Pa, AB, 52, MD, 56. *Prof Exp:* Assoc med, 63-67, from asst prof to assoc prof, 67-75, co-chief allergy & immunol, 69-74, CHIEF ALLERGY & IMMUNOL SECT, UNIV PA SCH MED, 74-, PROF MED, 75-, PROF NEUROL, 80- *Concurrent Pos:* mem exec comt, Am Acad Allergy, 75-80; mem bd dirs, Am Bd Allergy & Immunol, co-chmn, 79-80; ed, J Allergy & Clin Immunol, 88- *Honors & Awards:* Lindback Award, 67. *Mem:* Am Acad Allergy; Am Asn Immunol; Am Col Physicians; Am Fedn Clin Res. *Res:* Cellular inflammatory reactions in allergic diseases; immunologic mechanisms in demyelinating disease and systemic lupus erythematosus. *Mailing Add:* Univ Pa Sch Med 36th & Hamilton Walk Philadelphia PA 19104

ZWEMER, THOMAS J, b Mishawaka, Ind, Mar 23, 25; m 49; c 3. ORTHODONTICS. *Educ:* Univ Ill, DDS, 50; Northwestern Univ, MSD, 54. *Prof Exp:* Instr pedodont, Marquette Univ, 50-52, from asst prof to assoc prof oral rehab, 52-58; from asst prof to assoc prof orthod, Sch Dent, Loma Linda Univ, 58-66, chmn dept, 60-66; prof dent & assoc dean clin sci, Sch Dent, 66-84, VPRES ACAD AFFAIRS, MED COL GA, 84- *Concurrent Pos:* Mem attend staff, Wood's Vet Admin Hosp, 54-56; consult, Cerebral Palsy Clin, 54-58; chief dent serv, Milwaukee Children's Hosp, 55-58. *Mem:* Am Asn Orthod; Am Dent Asn; Am Col Dent; Int Asn Dent Res; Sigma Xi. *Res:* Health care delivery systems; physical anthropology. *Mailing Add:* Med Col Ga Augusta GA 30912

ZWERDLING, SOLOMON, b New York, NY, Jan 31, 22; m 44; c 3. SOLID STATE PHYSICS, PHYSICAL CHEMISTRY. *Educ:* Drew Univ, BA, 43; Johns Hopkins Univ, MA, 44; Columbia Univ, MA, 47, PhD(chem), 52. *Prof Exp:* Instr chem, Johns Hopkins Univ, 43-44; asst prof naval sci & tactics, Columbia Univ, 46, asst chem, 46-49; sr res chemist, Lever Bros Co, 51-52; asst group leader & div staff physicist, Lincoln lab, Mass Inst Technol, 52-63, staff physicist, Div Sponsored Res, Ctr Mat Sci & Eng, 63-68; res scientist, Douglas Advan Res Labs, Calif, 68-70, mgr res, Solid State Sci Dept, McDonnell Douglas Res Labs, 70-74; dir solar energy prog, Argonne Nat Lab, 75-77; mgr res & develop, Northeast Solar Energy Ctr, 77-79; SUPVR ADVAN PHOTOVOLTAIC DEVELOP GROUP, JET PROPULSION LAB, 79- *Concurrent Pos:* Mem bd dir, A D Jones Optical Co Inc, Burlington, Mass. *Mem:* Fel Optical Soc Am; Int Solar Energy Soc; sr mem Inst Elec & Electronics Engrs; Am Phys Soc. *Res:* Infrared absorption of solids; infrared magneto-optical effects in semiconductors at liquid helium temperature; physics of semiconductor electronic energy band structure; excitons in semiconductors; far infrared spectroscopy and detectors; metal physics, strength and corrosion; general solar energy conversion; photovoltaic solar cells; chemical vapor deposition technology for thin film electronic devices. *Mailing Add:* 1339 Riviera Dr Pasadena CA 91107

ZWERLING, ISRAEL, b New York, NY, June 12, 17; m 40; c 2. PSYCHIATRY. *Educ:* City Col New York, BS, 37, MS, 38; Columbia Univ, PhD(psychol), 47; State Univ NY, MD, 50. *Prof Exp:* High sch instr, NY, 39-42; instr psychol, City Col New York, 47-49 & Hunter Col, 49-50; from asst prof to prof psychiat, Albert Einstein Col Med, 54-73, exec chmn psychiat, 68-71; dir, Bronx State Hosp, 66-73; prof ment health sci & chmn dept, Med Col, 73-85, DEAN, SCH MED, HAHNEMANN UNIV, 85- *Concurrent Pos:* Lectr & psychiatrist-in-chg alcohol clin, State Univ NY, 54-55. *Mem:* Am Psychosom Soc; Am Psychol Asn; Am Psychoanal Asn. *Res:* Social and community psychiatry. *Mailing Add:* 1326 Spruce St Apt 804 Philadelphia PA 19107

ZWICK, DAAN MARSH, b New York, NY, July 28, 22; m 48; c 3. PHOTOGRAPHIC CHEMISTRY, PHYSICS. *Educ:* Univ Vt, BSChem, 43. *Prof Exp:* Instr physics, Univ Vt, 43-44; res chemist photog chem, Eastman Kodak Co Res Labs, 44-56, res assoc photog sci, 56-73, sr lab head color physics, 73-78, sr res assoc, 78-86; RETIRED. *Honors & Awards:* Kalmus Gold Medal, Soc Motion Picture & TV Engrs, 72, Agfa-Gevaert Gold Medal, 75, Progress Medal, 81. *Mem:* Fel Soc Motion Picture & TV Engrs; fel Soc Photog Scientists & Engrs; hon fel Brit Kinematograph Sound & TV Soc. *Res:* Physics of color photography, with emphasis on image structure and the psychophysics of imaging. *Mailing Add:* 15 Nunda Blvd Rochester NY 14612

ZWICK, EARL J, b Canton, Ohio, May 20, 31. MATHEMATICS. *Educ:* Kent State Univ, BS, 53, MS, 57; Ohio State Univ, PhD(math ed), 64. *Prof Exp:* Teacher pub sch, Ohio, 53-61; instr math, Ohio State Univ, 62-63; assoc prof, 63-72, PROF MATH, IND STATE UNIV, TERRE HAUTE, 72- *Mem:* Nat Coun Teachers Math. *Res:* Teaching methods. *Mailing Add:* Dept Math Ind State Univ 217 N Sixth St Terre Haute IN 47809

ZWICKEL, FRED CHARLES, b Seattle, Wash, Dec 18, 26; m 51; c 3. WILDLIFE ECOLOGY, ZOOLOGY. *Educ:* Wash State Univ, BSc, 50, MSc, 58; Univ BC, PhD(zool), 65. *Prof Exp:* Biologist, State Dept Game, Wash, 50-61; asst prof wildlife ecol, Ore State Univ, 66-67; from asst prof to prof, 67-85, EMER PROF ZOOL, UNIV ALTA, 85- *Concurrent Pos:* Nat Res Coun Can-NATO overseas res fel natural hist, Aberdeen Univ, 65-66; sabbatical study, Inst Appl Zool, Univ Hokkaido, Sapporo, Japan, 73-74, Wau Ecol Inst, Papua, New Guinea, 80-81 & Colo Div Wildlife, Ft Collins, 84. *Honors & Awards:* Roberts Award, Cooper Ornith Soc, 65. *Mem:* Am Ornith Union; Cooper Ornith Soc; Wildlife Soc; Am Soc Mammal; Can Soc Zool. *Res:* Population ecology; general biology of gallinaceous birds and land mammals. *Mailing Add:* Dept Zool Univ Alta Box 81 Manson's Landing BC V0P 1K0 Can

ZWICKER, BENJAMIN M G, b Pendleton, Ore, July 11, 15; m 42; c 2. CHEMISTRY. *Educ:* Whitman Col, AB, 35; Univ Wash, MS, 38, PhD(phys chem), 40. *Prof Exp:* Res chemist, B F Goodrich Chem Co, 40-43, mgr, Akron Exp Sta, 43-50, dir new prod planning, 50-60, dir planning, 60-78, in-charge environ sci, 58-78; res assoc, Scripps Inst Oceanog, Univ Calif, San Diego, 79-85; RETIRED. *Concurrent Pos:* Mem tech & res comts, Off Rubber Reserve, 43-50; consult indust chem & toxicol, 78- *Mem:* Am Chem Soc; Soc Chem Indust; Am Inst Chem Eng; Asn Indust Hygienists Am. *Res:* Analytical, physical and organic chemistry; academic oceanography; geology; microscopy; polymer chemistry; industrial chemistry and economics; research, development and business management toxicology, biodegradation; industrial hygiene. *Mailing Add:* 8581 Tulane St San Diego CA 92122-3221

ZWICKER, GARY M, VETERINARY PATHOLOGY, CHEMICAL TOXICOLOGY. *Educ:* Wash State Univ, DMV, 72; Purdue Univ, PhD(vet path), 73. *Prof Exp:* PRIN PATHOLOGIST, STAUFFER CHEM GROUP CHESEBROUGH-POND'S, INC, 78- *Mailing Add:* Dept Toxicol Dow Inc 9550 Zionsville Rd PO Box 68470 Indianapolis IN 46268

ZWICKER, WALTER KARL, b Vienna, Austria, Sept 5, 23; US citizen; m 56. SOLID STATE CHEMISTRY. *Educ:* Univ Vienna, PhD(mineral chem), 54. *Prof Exp:* Res fel chem, Harvard Univ, 54-55; res chemist, Metals Res Labs, Union Carbide Corp, 55-59, & Res Ctr, Union Carbide Nuclear Co, 59-64; sr res chemist, Thiokol Chem Corp, 64-65; sr proj leader, Philips Labs, NAm Philips Corp, 65-90; PVT CONSULT, 91- *Mem:* Emer mem Electro Chem Soc; fel Am Mineral Soc. *Res:* Solid state and materials science; crystal growth; mineral synthesis and phase equilibria; thin films; semiconductors and dielectrics. *Mailing Add:* 76 River Rd Scarborough Briarcliff Manor NY 10510-2412

ZWIEP, DONALD N, b Hull, Iowa, Mar 18, 24; m 48; c 4. MECHANICAL ENGINEERING. *Educ:* Iowa State Univ, BS, 48, MS, 51. *Hon Degrees:* DE, Worcester Polytech Inst, 65. *Prof Exp:* From asst prof to assoc prof mech eng, Colo State Univ, 51-57; prof mech eng & head dept, 57-88, actg provost & vpres acad affairs, 88-90, EMER PROF & DEPT HEAD, WORCESTER POLYTECH INST, 90- *Concurrent Pos:* Chief consult, Aviation Div, Forney Mfg Co, 56-57; chmn trustees, James F Lincoln Arc Welding Found, 76- *Honors & Awards:* Dedicated Serv Medal, Am Soc Mech Engrs, Centennial Medal. *Mem:* Am Soc Mech Engrs (pres, 79-80); Am Soc Eng Educ; Am Welding Soc; Sigma Xi; Soc Mfg Engrs. *Res:* Design and mechanics. *Mailing Add:* Dept Mech Eng Worcester Polytech Inst 100 Inst Rd Worcester MA 01609

ZWIER, PAUL J, b Denver, Colo, Oct 24, 27; m 51; c 6. MATHEMATICS. *Educ:* Calvin Col, AB, 50; Univ Mich, MA, 51; Purdue Univ, PhD(math), 60. *Prof Exp:* Instr math, Purdue Univ, 58-60; from asst prof to assoc prof, 60-74, chmn dept, 77-80, PROF MATH, CALVIN COL, 74- *Mem:* Math Asn Am; Am Math Soc. *Res:* Non-associative rings. *Mailing Add:* Dept Math Calvin Col 3201 Burton St Grand Rapids MI 49506

ZWILLING, BRUCE STEPHEN, b Brooklyn, NY, Jan 16, 43; m 67; c 2. IMMUNOBIOLOGY. *Educ:* Fairleigh Dickinson Univ, BS, 65; NY Univ, MS, 68; Univ Mo, PhD(microbiol), 71. *Prof Exp:* Guest worker immunol, Biol Br, Nat Cancer Inst, 72-73; from asst prof to assoc prof, 74-88, PROF MICROBIOL, COL BIOL SCI, OHIO STATE UNIV, 88- *Concurrent Pos:* Chmn, Div E, Am Soc Microbiol; gov coun, Soc Leukemia Biol; dir, Flow cytometer, Monoclonal Antibody Facil. *Mem:* AAAS; Am Soc Microbiol; Am Asn Cancer Res; Am Asn Immunol; Sigma Xi; Soc Leukemia Biol. *Res:* Regulation of MHC Class II expression; macrophage activating interferongamma mediated second signal generation. *Mailing Add:* Dept Microbiol Ohio State Univ 1484 N 12th Ave Columbus OH 43210

ZWILSKY, KLAUS M(AX), b Berlin, Ger, Aug 16, 32; US citizen; m 56; c 2. METALLURGY, MATERIALS SCIENCE. *Educ:* Mass Inst Technol, BS, 54, MS, 55, ScD(metall), 59. *Prof Exp:* Dir metall res, New Eng Mat Lab, Inc, 59-62; sr res assoc metall, Pratt & Whitney Aircraft Div, United Aircraft Corp, 62-63; sr scientist solid state, Melpar, Inc, Westinghouse Air Brake Co, 63-64; head alloy develop br, David Taylor Res Ctr, Annapolis, Md, 64-67; phys metallurgist, AEC, Dept of Energy, 67-73, chief mat & radiation effects br, Off of Fusion Energy, 73-81; EXEC DIR, NAT MAT ADV BD, NAT ACAD SCI, 81- *Concurrent Pos:* Trustee, Am Soc Metals Int, 84-87, 88-91, pres, 90. *Honors & Awards:* Nat Mat Advan Award, Fdn Mat Socs, 90. *Mem:* Fel Am Soc Metals Int; Minerals, Metals & Mat Soc; Sigma Xi; Mat Res Soc. *Res:* Physical and nuclear metallurgy including high temperature, composite, and powder metallurgy; nuclear materials for fission reactors and materials for fusion reactors; strategic materials; materials policy and materials science and engineering. *Mailing Add:* Nat Mat Adv Bd 2101 Constitution Ave Washington DC 20418

ZWISLOCKI, JOZEF JOHN, b Lwow, Poland, Mar 19, 22; nat US; m 54. AUDITORY BIOPHYSICS, PSYCHOPHYSICS. *Educ:* Swiss Fed Tech Inst, EE, 44, ScD, 48. *Prof Exp:* Res asst & head electroacoustical lab dept otolaryngol, Univ Basel, 45-51; res fel, Psychoacoustics Lab, Harvard Univ, 51-57; res assoc prof audiol, Inst Sensory Res, 57-62, from assoc prof to prof elec eng, 60-74, founding dir, Lab Sensory Commun, 63-74, prof sensory sci & dir, 73-84, prof neurosci 84-88, AFFIL PROF BIOENG, INST SENSORY RES, 86-, DISTINGUISHED PROF NEUROSCI, 88- *Concurrent Pos:* Assoc res prof, State Univ NY Upstate Med Ctr, 61-67, res prof, 67-; mem comt hearing & bio-acoustics, Nat Res Coun, 53-, mem exec coun, 64-68; mem rev panel commun sci, NIH, 66-70, chmn, 69-70, mem communicative disorders rev comt, 71-75; mem, Comt on Hearing & Bio-Acoustics, Nat Res Coun-Nat Acad Sci, mem exec coun, 65-68, chmn, 67-68; mem Tech coun, Tech Comt Psychol & Physiol Acoustics, chmn, 62 & 63, mem exec coun, 82-85, mem, long Range Planning Comt, 82-86, chmn, 83-86; chmn bd sci adv, Univ Wis Health Sci Ctr, 75-78; mem adv bd, Univ Fla Inst Advan Study Communiocative Processes, 76; mem, Comn Auditory Physiol, Int Union Physiol Sci, 82-; assoc mem, Int Comn Acoustics, Int Union Pure & Applied Physics, 82. *Honors & Awards:* Sigma Xi Fac Res Award, 73; Amplifon Prize, Int Res & Study, Milan, Italy, 76; Jarvits Neurosci Investigators Award, 84; First Berkesy Medal, Acoust Soc Am, 85; Award of Merit, Asn Res Otolaryngol, 88. *Mem:* Nat Acad Sci; fel Acoust Soc Am; Int Soc Audiol; NY Acad Sci; Asn Res Otolaryngol; fel Am Speech & Hearing Asn; assoc mem Am Otol Soc; AAAS. *Res:* Sound transmission in middle and inner ear; psychoacoustics; mathematical analysis of auditory system; audiological diagnostic methods; acoustic instruments. *Mailing Add:* Inst Sensory Res Syracuse Univ Syracuse NY 13244-5290

ZWOLENIK, JAMES J, b Cleveland, Ohio, Dec 31, 33. CHEMISTRY, ADMINISTRATION. *Educ:* Western Reserve Univ, AB, 56; Yale Univ, PhD(phys chem), 61; Univ Cambridge, PhD(phys chem), 64. *Prof Exp:* NSF fel, Queens Col, Univ Cambridge, 60-62; res chemist, Fundamental Res Group, Chevron Res Co, 63-67; prog officer, chem dynamics & chem thermodyn progs, Chem Sect, 67-70, policy analyst, Off Policy Studies, 70-71, Spec Anal Sect, Div Sci Resources Studies, 71-74, off div dir, Div Sci Resources Studies, 74-75, staff dir & exec secy, Comt Eighth Nat Sci Bd, 75-76, spec asst, Nat Sci Bd, 76-79, head, oversight sect, Off Audit & Oversight, 79-88, ASST INSPECTOR GEN OVERSIGHT, OFF INSPECTOR GEN, NSF, 89- *Concurrent Pos:* Instr, Exten Div, Univ Calif, Berkeley, 65-67; collabr, Smithsonian Radiation Biol Lab, 68-70; vpres, Higher Educ Group, Wash, DC, 76-77, pres, 77-78. *Mem:* Am Chem Soc; Am Phys Soc; AAAS; Cosmos Club; Sigma Xi. *Res:* Photochemistry; kinetic spectroscopy; science policy; higher education. *Mailing Add:* NSF 1800 G St NW Washington DC 20006

ZWOLINSKI, BRUNO JOHN, b Buffalo, NY, Nov 4, 19; m 52; c 3. PHYSICAL CHEMISTRY. *Educ:* Canisius Col, BS, 41; Purdue Univ, MS, 43; Princeton Univ, AM, 44, PhD(phys chem), 47. *Prof Exp:* Instr chem eng, sci & mgt war training, Purdue Univ, 42; asst, Princeton Univ, 43-44; res scientist, Manhattan Proj, Columbia Univ, 44-45; Am Chem Soc fel, Univ Utah, 47-48, asst prof chem, 48-53; sr physicist, Stanford Res Inst, 53-57; prin res chemist & dir res projs chem & petrol res lab & lectr chem, Carnegie Inst Technol, 57-61; dir, Thermodyn Res Ctr, 61-78, PROF CHEM, TEX A&M UNIV, 61- *Concurrent Pos:* Asst dir chem prog, NSF, 54-57; mem adv bd off critical tables, Nat Res Coun. *Honors & Awards:* US Calorimetry Award; Crane-Patterson Award. *Mem:* Fel NY Acad Sci; fel Am Inst Chemists; fel AAAS. *Res:* Compilation of selected values of physical, thermodynamic and spectral data of chemical compounds; dynamic properties of liquids; charge or electron transfer phenomena in chemical kinetics; statistical thermodynamics. *Mailing Add:* Chem Dept Tex A&M Univ College Station TX 77840

ZWOLINSKI, MALCOLM JOHN, b Winchester, NH, Oct 23, 37; m 59; c 2. FIRE ECOLOGY, NATURAL RESOURCE MANAGEMENT. *Educ:* Univ NH, BS, 59; Yale Univ, MF, 61; Univ Ariz, PhD(watershed mgt), 66. *Prof Exp:* Res assoc watershed mgt, 64-65, from asst prof to assoc prof, 66-72, PROF WATERSHED MGT, UNIV ARIZ, 72-, ASSOC DIR SCH RENEWABLE NATURAL RESOURCES, 75- *Mem:* Fel AAAS; Soc Am Foresters; Soil Conserv Soc Am; Soc Range Mgt. *Res:* Fire ecology; effects of fire on natural ecosystems, including vegetation, soils and water; prescribed burning and fire management; watershed hydrology; watershed management including vegetation and soil influences. *Mailing Add:* Sch Renewable Natural Resources Col Agr Univ Ariz Tucson AZ 85721

ZWOYER, EUGENE, b Plainfield, NJ, Sept 8, 26; m 46; c 3. STRUCTURAL ENGINEERING. *Educ:* Univ NMex, BS, 47; Ill Inst Technol, MS, 49; Univ Ill, PhD(eng), 53. *Prof Exp:* Assoc prof civil eng, Univ NMex, 48-61, res prof civil eng & dir Eric H Wang Civil Eng Facil, 61-71; exec dir & secy, Am Soc Civil Engrs, 72-82; pres, Am Asn Eng Soc, 82-84; pres, T Y Lin Int, 84-89; chief operating officer, 90, MGT CONSULT TO PRES, POLAR MOLECULAR CORP, 91- *Concurrent Pos:* Res assoc, Univ Ill, 51-53, consult engr, Eugene Zwoyer & Assocs, 53-72, 90-; mem, Interprof Coun Environ Design, 72-82, secy, 72 & 78; trustee, People-to-People Int, 74-86 & Small Bus Res Corp, 76-79; mem, Engr Joint Coun Int Comn, 75-79, Cert Comn, 75-79 & Finance Comt, 75-79, dir, Coun, 77-79, chmn, Sub-Comt for 1979 UN Conf Sci & Technol for Develop, 78-80; dir, Eng Socs Comn Energy, 77-82, secy, treas & mem exec comt, 78; vpres, World Fedn Eng Orgn, 82-88. *Mem:* AAAS; Am Soc Civil Engrs; Nat Soc Prof Engrs; Am Concrete Inst; Am Soc Eng Educ. *Res:* Ultimate strength of structures, particularly on structures to resist effects of nuclear weapons. *Mailing Add:* 6363 Christie Ave No 1362 Emeryville CA 94608

ZYCH, ALLEN DALE, b Cleveland, Ohio, Apr 8, 38; m 75; c 2. ASTROPHYSICS. *Educ:* Case Inst Technol, BS, 61, MS, 65; Case Western Reserve Univ, PhD(physics), 68. *Prof Exp:* Fel physics, Case Western Reserve Univ, 68-70; from asst prof to assoc prof, 70-82, PROF PHYSICS, UNIV CALIF, RIVERSIDE, 82- *Mem:* Am Phys Soc; Am Geophys Union; Am Astron Soc; Inst Elec & Electronics Engrs. *Res:* Experimental high energy astrophysics. *Mailing Add:* Inst Geophys-Planetary Physics Univ Calif Riverside CA 92502

ZYGMONT, ANTHONY J, b Philadelphia, Pa, Sept 16, 37. SYSTEMS ENGINEERING. *Educ:* Villanova Univ, BEE, 59; Drexel Univ, MSEE, 63; Univ Pa, PhD(elec eng), 71. *Prof Exp:* Jr engr commun systs, Philco Corp, 59-62; engr, Radio Corp Am, 62-63; sr engr radar systs, Philco Corp, 63; ASSOC PROF ELEC ENG SYSTS, VILLANOVA UNIV, 63- *Concurrent Pos:* Educ consult, Philadelphia Elec Co, 70- *Mem:* Inst Elec & Electronics Engrs; Instrument Soc Am; Am Soc Eng Educ. *Res:* Application of systems engineering to large scale problems, especially the problems of identification and optimization. *Mailing Add:* 56 Wellfleet Lane Wayne PA 19087

ZYGMUND, ANTONI, b Warsaw, Poland, Dec 26, 00; nat US; m 25; c 1. MATHEMATICS. *Educ:* Univ Warsaw, PhD(math), 23. *Hon Degrees:* DSc, Wash Univ, 72, Univ Torun, Poland, 73, Univ Paris, 74, Univ Upsala, 77. *Prof Exp:* Instr math, Warsaw Polytech Sch, 22-30; privat docent, Univ Warsaw, 26-30; prof, Wilno Univ, 30-39; vis lectr, Mass Inst Technol, 39-40; from asst prof to assoc prof, Mt Holyoke Col, 40-45; prof, Univ Pa, 45-47; prof, 47-67, Swift distinguished serv prof, 67-82, EMER GUSTAVUS & M M SWIFT PROF MATH, UNIV CHICAGO, 82- *Concurrent Pos:* Rockefeller fel, Oxford Univ & Cambridge Univ, 29-30; Guggenheim fel, 53-54. *Honors & Awards:* Prize, Polish Acad Sci, 39. *Mem:* Nat Acad Sci; Am Acad Arts & Sci; fel Am Math Soc (vpres, 54-); hon mem London Math Soc; Polish Acad Sci. Spanish Acad Sci. *Res:* Fourier series; real variables. *Mailing Add:* Dept Math Univ Chicago 5734 University Ave Chicago IL 60637

ZYGMUNT, WALTER A, b Calumet City, Ill, Mar 24, 24; m 52; c 2. MICROBIOLOGY, BIOCHEMISTRY. *Educ:* Univ Ill, BS, 47, MS, 48, PhD(bact), 50. *Prof Exp:* Res microbiologist res labs, Merck & Co, Inc, 50-53; res microbiologist & biochemist, Mead Johnson Res Ctr, 53-75, assoc dir biol res, 75-78, assoc dir drug regulatory affairs, 78-87; RETIRED. *Mem:* Fel AAAS; Am Chem Soc; Am Soc Microbiol; fel Am Acad Microbiol; Soc Indust Microbiol. *Res:* Microbial chemistry; chemotherapy; amino acid antagonists; immunology; pulmonary biochemistry; metabolic diseases. *Mailing Add:* 7729 Meadowview Dr Evansville IN 47710